CONSOLIDATED INDEX OF TRANSLATIONS INTO ENGLISH

compiled by the
National Translations Center
(formerly the SLA Translations Center)
The John Crerar Library

SPECIAL LIBRARIES ASSOCIATION
New York
1969

Standard Book Number 87111-197-7
Library of Congress Catalog Card Number 76-101337

235 Park Avenue South, New York 10003

Printed in the United States of America

Foreword

Persons who need a translation for the first time, are usually astonished when they learn how expensive custom translations are or how difficult it is to locate a copy of an existing translation. In an attempt to alleviate part of this problem Special Libraries Association established the SLA Translations Center in 1952, through contract with The John Crerar Library. Financial support for the Center was primarily from grants and contracts. Such support was negotiated over the years by Headquarters staff of Special Libraries Association with various agencies. (National Science Foundation, U. S. Public Health Service, Clearinghouse for Federal Scientific and Technical Information, and American Iron and Steel Institute).

By pooling translations which the Association's members had commissioned and by soliciting deposits of translations from industry, government agencies, universities and other sources, the Center has amassed a collection that now exceeds 140,000 items. Approximately 60% of the translations are from Russian and 40% are from other languages.

The translations may be obtained by loan or photocopy from the National Translations Center, which is the successor to the original SLA Translations Center. Since 1968 the National Translations Center has been sponsored by The John Crerar Library with National Science Foundation support.

Many lists of translations available from numerous sources have been issued by various organizations through the years. The existence of so many lists caused serious complications in determining not only whether a given item exists in translation but also its actual availability, cost, etc. In 1963 the SLA Translations Activities Committee proposed and made initial estimates for a consolidation of such listings into a single "availability index." By 1965 more detailed studies by the Committee led to specific proposals for compilation and publication. These efforts were rewarded in 1966 with a grant to SLA from the National Science Foundation; work began at the SLA Translations Center under the continuing general guidance of the SLA Translations Activities Committee. This publication—the final product of such efforts—includes information on an estimated 142,000 translations. The support of the National Science Foundation is particularly acknowledged for making this compilation possible, for warm encouragement, and for continuing advice and assistance.

To the Translations Center staff, who have resolved so many problems during compilation of this volume, goes praise for a task well done—in particular, to Mrs. Ildiko D. Nowak, chief of the Center, to Mrs. Helen Byrnes and Donald Hinton, bibliographers, and to Mr. Mustafa Akalin, systems analyst.

Three chairmen of SLA's Translations Activities Committee must be cited for their continuing advice, interest and efforts on behalf of the Center during the planning and compilation of this volume. They are Elizabeth M. Walkey (1962-64), Mrs. Irma Johnson (1964-66), and Roger M. Martin (1966-68).

The *Introduction* to this volume explains its use, its basis, and its scope. No compilation of this type, whether manually prepared or, as in this case, by computer processing can be complete and perfect. However, it is the conviction of all associated with it that a useful tool has been produced, and that their past labors will minimize the future labors of many others.

"Have you a catalogue
Of all the voices that we have procured?"

Sicinus, in *Coriolanus* iii:3,9

Introduction

Purpose and Coverage

The object of this volume is to bring together in one compilation information on the availability of translations which have appeared in a number of different lists issued by different agencies at different times. Publications used as sources for this compilation are listed at the end of this *Introduction.* Included are translation bibliographies, selective translation journals, and collections of translations. Cover-to-cover translations of journals are not listed article by article—on the assumption that such journals-in-translation are either completely self-indexed or that they are adequately identified in the conventional secondary indexing services. A STAR ★ is used in this volume as the symbol to identify those journals for which cover-to-cover translations are available. Information regarding the availability and time period covered by these translated journals can be found in the compilation by Carl J. Himmelsbach and Grace E. Boyd: *A Guide to Scientific and Technical Journals in Translation* (N.Y., Special Libraries Association, 1968).

Items included in this volume are restricted to translations (into English) of serially published originals such as journals, patents, standards, etc. Due to limitations of space and cost, non-serial publications (monographs, conference proceedings, etc.) are not included. However, information about translations of such publications may be obtainable from the National Translations Center. Many bibliographic references in source lists were found to be incomplete or inaccurate or both; and only a limited effort could be devoted to correction or completion. The use of computer processing required certain minimum input data. Basic criteria for inclusion were:

1) an authentic verifiable serial title;
2) volume number or year of publication;
3) beginning (or inclusive) pagination of the translated article.

The World List of Scientific Periodicals (4th ed., 1963) was used as the basic authority for identification and entry forms of serial titles; other standard lists were also employed in expanding the final authority file.

How to Use This Volume

This volume is divided into two sections: the *Serial Citation Index* and the *Patent Citation Index.* The *Serial Citation Index* lists journals, technical reports, standards, newspapers and other such publications arranged alphabetically by title. Under each serial title, translations are arranged chronologically. When applicable and when available, the following data are given in this order: year of publication; series (preceded by S); volume (preceded by V); issue or number (preceded by N); pages (preceded by P); a unique reference number assigned to each translation; a symbol or code indicating the source from which copies may be obtained; and additional symbols giving specific details about the nature of the translations. Serial titles preceded by a STAR ★ are available wholly or in part as cover-to-cover translation journals.

The *Patent Citation Index* lists translations of patents under the name of the country issuing the patent. The citations include translations of both original patents and patent applications, as well as of patent additions (ADD) and supplements (SUP).

In using this volume to determine if a particular article, etc., has been translated, the serial title is first located in the alphabetic sequence. Except for Oriental serials, titles contemporary to the original article are used for all journals. That is, if a journal has changed its name, the citation will be found under the title used at the time of publication of the original foreign language article. Oriental titles have been cross-referenced from the English or Latin title to the current romanized title of the vernacular (whenever it is known). It is advisable to examine all entries under the title to be sure that a specific citation is, or is not, listed. Due to occasional references with incomplete bibliographic information, computer processing sometimes produced a list that is not in strictly chronological order.

The example below demonstrates the elements in the listings in this volume:

Also available as cover-to-cover translation — Supplementary information

★ ZAVODSKAYA LABORATORIYA [Journal]						BELGIUM [Patent]		
1937	v3	p300-302	67-3687	*	O	563,247	67-2113	*
1953	v9	p133-135	67-4112	*	P	640,076	67-5011	*

Translation reference number — Source symbol

ABBREVIATIONS USED

ADD	Addition
M	Machine translation
O	Figures, refs. omitted
P	Partial translation
SUP	Supplement

Precautions to Keep in Mind

1. Serial publications of organizations and institutions are listed under publication or series title according to the *World List* system. Thus, the name of the issuing organization often follows the name of the publication (e.g., *Doklady Akademii Nauk SSSR, Berichte der Deutschen Chemischen Gesellschaft, Annales de l'Institut Pasteur, Paris*), in contrast with the system of *Union List of Serials*, or with the main entry form in most card catalogs.

2. Some journals have no volume numbering; some have a double numbering system and either may have been used. Publication numbers may have been handled as a volume number or as an issue number. However, the date is the key factor; and all citations for a given year should be examined.

3. Title variations are often found, particularly with Soviet organizations and institutes; for example, *Sbornik* may also be identified as *Trudy*, or *Nauchnye Raboty* as *Raboty*.

Official entries of titles in the Baltic Republics of the USSR and in the Ukrainian SSR may appear under either or both the Russian title and the corresponding title in Ukrainian, Estonian, Latvian, or Lithuanian. Complete cross-references or unification was not possible due to lack of satisfactory verification.

How to Obtain Translations

All of the listed translations were available at some time in the past from one source or another. Anyone interested in acquiring retention or loan copies may find that a few of them are no longer available. Each translation entry is coded so as to designate the organization from which a copy may be ordered. The *Directory of Sources*, found elsewhere in this volume, gives the complete names and addresses of such organizations. When copies are requested, it is important to identify the item ordered as completely as possible. The citation should include, whenever possible, author and title of an article, as well as the specific journal or patent reference, plus the unique reference number assigned to each translation. Users of this volume should also feel free to query the National Translations Center, where more complete information is usually available.

Translations Register-Index

Items listed in this volume include citations from listings published through 1966. In 1967 Special Libraries Association initiated publication of a new periodical,

Translations Register-Index (*TRI*), which is effectively a continuing supplement to this volume. The "Register Section" of *TRI* lists in COSATI subject categories, new translations deposited at the Translations Center. The "Index Section" (journal citation, patent citation, monograph listing) includes the above deposits, plus citations derived from *Technical Translations* (until publication ceased in December 1967), *U. S. Government Research and Development Reports* (beginning January 1968), commercial translators, and other listings in the *Directory of Sources.*

The computer index arrangement in *TRI* is identical with the listing in this volume. It is hoped that further overall cumulations can be published in the future. Subscriptions to *Translations Register-Index* should be addressed to:

Special Libraries Association
235 Park Avenue South
New York 10003.

National Translations Center

Special Libraries Association contributed its collection of translations in the SLA Translations Center to The John Crerar Library in 1968. The Center has been renamed as the National Translations Center and is now directly sponsored by The John Crerar Library. An Advisory Board has as its members representatives of professional associations in science, engineering and librarianship—including Special Libraries Association. The continuing contribution of translations is actively solicited from all sources—from societies, industrial libraries and other special libraries, government agencies, universities and other institutions in the United States and abroad. Effective service by the Center depends on the continuing cooperation and support of an ever increasing number of users of the translations on deposit. Complete translations (preferably incorporating copies of original figures or illustrations) may be submitted in typescript or other legible form. If neither full-size copies nor microfilm copies are available for permanent deposit, the Center can arrange to film a copy and then return the original to the contributor. Such material is returned by insured mail within two weeks after it is received at the Center. Permission to lend or reproduce is a condition of deposit; in return, the Center agrees to obliterate all identification of the donor of a translation. Thus the anonymous input of many organizations is made available to all.

Services provided by the Center include the following:

1. Upon receipt of a complete bibliographic citation (by mail, telephone, or teletype), the Center will search its files to determine if an English language translation exists and where it is located.
2. The Center provides either paper or microfilm copies of translations in its collection. The charges for this service include a small fee which helps to support the operation of the Center.
3. The Center conducts literature searches in its files for all translated works of a given author or all translations from a given journal (excluding those in cover-to-cover translation). There is a nominal charge for this search service.

For information about contributions to the Center, translation searches, services, or prices, write, telephone, or teletype:

National Translations Center
The John Crerar Library
35 West 33rd Street
Chicago, Illinois 60616
Telephone: 312/225-2526
TWX: 910/221-5131

Publications Indexed In This Volume

Basic Sources

Bibliography of Translations from Russian Scientific and Technical Literature. 1953–56 (nos. 1–36).

SLA Author List of Translations and Supplement. 1953–54.

Technical Translations. 1959–66 (1–16).

Translation Monthly. 1955–58 (v. 1–4).

Selected Translation Sources

AIAA Journal: Russian Supplement. 1963–64.

ARS Journal. 1959–62.

American Mathematical Society Translations. 1949–66.

Federal Proceedings Translation Supplement. 1963–65.

Geochemistry International. 1964–65.

International Chemical Engineering. 1961–66.

International Geology Review. 1959–66.

Journal of the Astronautical Sciences. 1964–65.

NLL Translations Bulletin. 1960–66.

Rubber Chemistry and Technology. 1946–66.

Selected Translations in Mathematical Statistics and Probability. 1961–66.

Soviet Geography: Review and Translation. 1960–66.

Soviet Hydrology: Selected Papers. 1962 and 1966.

DIRECTORY OF SOURCES

*
National Translations Center
The John Crerar Library
35 West 33rd Street
Chicago, Illinois 60616
(formerly SLA Translations Center)

=
U.S. Department of Commerce
Clearinghouse for Federal Scientific
and Technical Information
Port Royal and Braddock Roads
Springfield, Virginia 22151
(formerly Office of Technical Services)

$
European Translations Centre
Doelenstraat 101
Delft, The Netherlands

+
Photoduplication Service
Publication Board Project
Library of Congress
Washington, D.C. 20450

AAS
American Astronautical Society, Inc.
428 East Preston Street
Baltimore, Maryland 21202

AC
A. Capelo
P.O. Box 46354
Hollywood, California 90046

ACI
Australian Consolidated Industries
Technical Library
813 Dowling Street
Waterloo, New South Wales
Australia

ACS
American Chemical Society
Prince and Lemon Streets
Lancaster, Pennsylvania 17603
Source for:
RCT-Rubber Chemistry and Technology

ACSI see NLL

AGI
American Geological Institute
1444 N Street, N.W.
Washington, D.C. 20005
Source for:
GEI-Geochemistry International
IGR-International Geology Review

AGS
American Geographical Society
Broadway at 156th Street
New York, New York 10032
Source for:
SGRT-Soviet Geography:
Review and Translation

AGU
American Geophysical Union
Suite 506
1145 19th Street, N.W.
Washington, D.C. 20036
Source for:
SHSP-Soviet Hydrology: Selected Papers
SO-Soviet Oceanography

AIAA
American Institute of Aeronautics
and Astronautics
1290 Avenue of the Americas
New York, New York 10019
Source for:
AIAAJ-AIAA Journal: Russian Supplement
ARSJ-ARS Journal

AIBS
American Institute of Biological Sciences
2000 P Street, N.W.
Washington, D.C. 20006

AINA
Arctic Institute of North America
1619 New Hampshire, N.W.
Washington, D.C. 20009
Source for:
AON-Anthropology of the North

AIP
American Institute of Physics
335 East 45th Street
New York, New York 10017

AIS
Argus Information Service
624 N. Eutaw Street
Baltimore, Maryland 21201

AM
Acta Metallurgica
122 East 55th Street
New York, New York 10022

AMS
American Mathematical Society
190 Hope Street
P.O. Box 6248
Providence, Rhode Island 02904
Source for:
AMST-American Mathematical
Society Translations
STMSP-Selected Translations in
Mathematical Statistics and Probability

ANSP
Academy of Natural Sciences of Philadelphia
19th Street and the Parkway
Philadelphia, Pennsylvania 19103

APC
Astex Publishing Company
Guildford
Surry, England

API
Academic Press, Inc.
111 Fifth Avenue
New York, New York 10003

ARS see AIAA

ATS
Associated Technical Services, Inc.
855 Bloomfield Avenue
Glen Ridge, New Jersey 07028

BCIR
British Cast Iron Research Association
Bordesley Hall, Alvechurch
Birmingham, England

BCRA
British Ceramic Research Association
Queens Road Penkhull
Stoke-On-Trent, England

BGIR
British Glass Industry Research Association
Northumberland Road
Sheffield 10, England

BI
Boreal Institute
University of Alberta
Edmonton, Alberta, Canada

BISI
British Iron and Steel Industry
Translation Service
The Iron and Steel Institute
4, Grosvenor Gardens
London, S.W.1, England

BR
Technical Library
U.S. Bureau of Reclamation
Commissioner's Office
Building 53, Denver Federal Center
Denver, Colorado 80202

BRS
Building Research Station
Garston, England

BSRA
British Ship Research Association
Prince Consort House
27/29 Albert Embankment
London, S.E.1, England

BWRA
British Welding Research Association
Abington Hall, Abington
Cambridge, England

CAOA
Catherine A.O. Aalders
83 Elm Avenue
Kentville, Nova Scotia, Canada

CB see PLPC

CBPB
Commonwealth Bureau of Plant Breeding and Genetics
School of Agriculture
Cambridge, England

CCA
Cement and Concrete Association
52, Grosvenor Gardens
London, S.W.1, England

CCT see PLPC

CEGB
Central Electricity Generating Board
Grindall House
25 Newgate Street
London, S.E.1, England

CEMB
The European Cement Association
2, Rue Saint Charles
Paris 15, France

CHP
Cleaver-Hume Press Ltd.
31 Wright's Lane
London, W.8, England

CLAI
C.L.A.I.R.A. Laboratories
Church Street
Welwyn, Herts, England

CP
College of Physicians of Philadelphia
Library
19 South 22nd Street
Philadelphia, Pennsylvania 19103

CRIC
Centre National de Recherches Scientifiques et Techniques pour l'Industrie Cimentiere
127 Avenue Adolphe Buyl
Brussels 5, Belgium

CRL
Central Research Laboratories
Broken Hill Proprietary Company, Ltd.
500 Bourke Street
Melbourne C.1, Australia

CSIR
Commonwealth Scientific and Industrial Research Organization Information Service
314 Albert Street
East Melbourne C.2, Victoria
Australia

CSSA
Council for Scientific and Industrial Research
The Head
Information Division
P.O. Box 395
Pretoria, Republic of South Africa

CTRA
The Librarian
The Coal Tar Research Association
Oxford Road
Gomersal
Leeds, England

CTS
Chemical Translating Service
2243 Greer Road
Palo Alto, California 94300

CTT
Columbia Technical Translations
5 Vermont Avenue
White Plains, New York 10606

DAP
Division of Air Pollution, PHS
Department of Health, Education and Welfare
Washington, D.C. 20201

DI see DIL

DIL
Circulation Unit
Department of the Interior
Central Library
Room 1149
Washington, D.C. 20025

DTC
Danish Translations Centre
% Library of the Danish Atomic Energy Commission
Riso. Roskilde
Denmark

EEUP
Department of Electrical Engineering
University of Pittsburgh
Pittsburgh, Pennsylvania 15213

EIS
Engineering Information Services
Kirkham, Preston, Lancashire
England

EP
Euromed Publications
97, Moore Park Road
London, S.W.6, England

ES
Erica Stux
584 Moreley Avenue
Akron, Ohio 44320

ETHB
Eidgenossische Technische Hochschulbibliothek
Leonhardstrasse 33
Zurich, Switzerland

ETS
Express Translation Service
P.O. Box 428
56 Wimbeldon Hill Road
London, S.W.19, England

EURO
Euro-Translation Service
20 Rue Europe
Grivegnee (Liege), Belgium

EZ
Electrolytic Zinc Company of Australia Ltd.
G.P.O. Box 377D
Hobart, Tasmania
Australia

FASE
Federation of American Societies for Experimental Biology
9650 Rockville Pike
Bethesda, Maryland 20014
Source for:
FPTS-Federation Proceedings Translations Supplement

FP
Faraday Press, Inc.
84 Fifth Avenue
New York, New York 10011

FRI
Fulmer Research Institute, Ltd.
Library
Stoke Poges
Buckinghamshire, England

FT see FP

GB
Gordon and Breach Publishers, Inc.
150 Fifth Avenue
New York, New York 10011

GL
Geigy (U.K.) Ltd.
Hawkshead Road
Paisley, Scotland

GPO
Superintendent of Documents
Government Printing Office
Washington, D.C. 20402

GPRC
Geophysical & Polar Research Center
6021 South Highland Road
Madison, Wisconsin 53706

GSL
U.S. Geological Survey Library
Room 1033, GSA Building
Washington, D.C. 20025
(Translations available on loan only)

HB
Henry Brutcher
P.O. Box 157
Altadena, California 91001

HMSO
Her Majesty's Stationery Office
P.O.B. 569
London, S.E.1, England

IAI
Institute for Asbestos Information
298 Rokeby Road
Subiaco, Western Australia
Australia

IASP
International Arts and Sciences Press
108 Grand Street
White Plains, New York 10601
Source for:
SAA-Soviet Anthropology and Archeology
SP-Soviet Psychiatry
SPSY-Soviet Psychology
SPP-Soviet Psychology and Psychiatry
SR-Soviet Review

ICE
International Chemical Engineering
345 East 47th Street
New York, New York 10017
Source for:
ICE-International Chemical Engineering

ICS
Intercontinental Services
Washington Court House, Ohio 43160

IEEE
Institute of Electrical and Electronics Engineers
275 Madison Avenue
New York, New York 10016

IICH
International Information Clearing House
615 Linwood Drive
Midland, Michigan 48642

INFO
Infoserv
P. O. Box 217
West Hartford, Connecticut 06097

INSD
Indian National Scientific Documentation Center
New Dehli 12, India

INT
International Patent Service N. V.
3 Pondsmeade
Redbourne, Hertfordshire
England

IP
Interscience Publishers, Inc.
250 Fifth Avenue
New York, New York 10001

IPIX
International Physical Index, Inc.
1909 Park Avenue
New York, New York 10035

IPST
Israel Program for Scientific Translation
14 Shammai Street
P. O. Box 7145
Jerusalem, Israel

ISA
Instrument Society of America
530 William Penn Place
Pittsburgh, Pennsylvania 15219

JBS
John B. Southard
100 East Palisade Avenue
Apt. B31
Englewood, New Jersey 07605

JCS
Joseph Crosfield & Sons, Ltd.
Bank Quay
Warrington, England

JLRD
John Laing Research and Development Ltd.
Hertfordshire, England

JTBC
J. T. Baker Chemical Co.
Phillipsburg, New Jersey 08865

JW
J. Woroncow
5230 Clairemont Mesa Boulevard
San Diego, California 92117

K-H or KH
Kresge-Hooker Science Library Associates
Wayne State University
Detroit, Michigan 48202

LAL
Laboratoire d'Astronomie de Lille
1, Impasse d'Observatoire
Lille, France
Source for:
ANL-Astronomical News Letter

LAND
Land-Air, Inc.
Pacific Missile Range
Point Mugu, California 93041

LS
Lingua Scientiae
Box 21086 Campus Station
Cincinnati, Ohio 45221

LSA
Literature Service Associates
Route 1
Bound Brook, New Jersey 08805

LSB
Language Service Bureau
3093 East Derbyshire Road
Cleveland, Ohio 44118

MACL
Maclaren and Sons, Ltd.
Maclaren House
131 Great Suffolk Street
London, S.E.1, England

MACM
MacMillan Co.
60 Fifth Avenue
New York, New York 10011

MDF see

MFA
Metal Finishing Abstracts Translation Service
85 Udney Park Road
Teddington, Middlesex, England

MPBW
Ministry of Public Building and Works
Library
London, England

MT
M. de O. Tollemache
Pennybrook Cottage
Burrington
Umberleigh, Devon, England

MUL
Mulholland Engineering Translations
26 Burton Woods Lane
Cincinnati, Ohio 45229

NASA
National Aeronautics and Space Administration
1520 H Street, N.W.
Washington, D.C. 20025

NIAE
National Institute of Agricultural Engineering
Wrest Park, Silsoe
Bedford, England

NLL
National Lending Library for Science and Technology
Boston Spa, Yorkshire
England
Source for:
NLLTB-NLL Translations Bulletin

NRCC see NRC

NRC
National Research Council
Library
Ottawa 2, Canada

NS
Neuroscience Translations
9650 Rockville Pike
Bethesda, Maryland 20014
Source for:
NS-Neuroscience Translations

OB
Oliver & Boyd, Ltd.
Tweeddale Court
14 High Street
Edinburgh 1, Scotland

OSA
Optical Society of America, Inc.
33 University Road
Cambridge, Massachusetts

PAD see PANS

PANS
Pakistan National Scientific and Technical Documentation Centre
2/141-S
P.E.C.H.S.
Karachi 29, Pakistan

PAR
Pargas Kalkbergs Aktiebolag
Fredrikinkatu 47
Helsinki, Finland

PERA
The Information Manager
Production Engineering Research Association
Melton Mowbray
Leicestershire, England

PLP see PLPC

PLPC
Plenum Publishing Corporation
227 West 17th Street
New York, New York 10011

PP
Pergamon Press
44-01 21st Street
Long Island City, New York 11101
Source for:
PS-Polymer Science USSR

PS
Primary Sources
11 Bleecker Street
New York, New York 10012
Source for:
SMT-Soviet Metal Technology

RAP
Rubber and Plastics Research
Association of Great Britain
Shawbury
Shrewsbury, Shropshire, England

RCT
Rubber Chemistry and Technology
Prince and Lemon Streets
Lancaster, Pennsylvania 17603
Source for:
RCT-Rubber Chemistry and Technology

RHC
R.H. Chandler, Ltd.
42 Grays Inn Road
London, W.C.1, England

RIA
Research International Associates
1522 Connecticut Avenue, N.W.
Washington, D.C. 20006

RIS
Research Information Service
44-01 21st Street
Long Island City, New York 11101

RRIM
Rubber Research Institute of Malaya
P.O. Box 150
Kuala Lumpar, Malaysia

RT
Reilly Translations
3690 Jasmine Avenue
Los Angeles, California 90034

SA
Seizaburo Aoki
Japanese Language Translation Service
15/2027 Chogo
Fujisawa, Japan

SAO
Smithsonian Institution
Astrophysical Observatory
60 Garden Street
Cambridge, Massachusetts 02138

SI see SAO

SIAM
Business Manager
SIAM Publications
P.O. Box 7541
Philadelphia, Pennsylvania 19100
Source for:
JSAC-Journal, Series A: Control

SIC
Scientific Information Consultants Ltd.
661 Finchley Road
London, N.W.2, England

SM
S. Muller
1614-27th Street
Orlando, Florida 32805

SNS
Swedish Natural Science Research Council
Stockholm, Sweden

SSSA
Soil Science Society of America, Inc.
Program Director, Soviet Soil Science
677 South Segoe Road
Madison, Wisconsin 53711

ST
Scripta Technica, Inc.
1000 Vermont Avenue, N.W.
Washington, D.C. 20005

STA
Science and Technology Agency
3, Dyers Building
Holborn, London, E.C.1
England

STC
Swedish Translation Centre
Kunglige Tehniske Hogskolans
Bibliotek
Stockholm, Sweden

STS
Scientific Translation Service
1938 Verdugo Knolls Drive
Glendale, California 91200

STT
SCIENTECH
Scientific and Technical Translations
P.O. Box 3530
Cleveland, Ohio 44118

TA
Translavic Associates
6363 Wilshire Boulevard
Room 228
Los Angeles, California 90048

TC
Chemical Translating Service
Trans-Chem Incorporated
820 Tulip Street
Knoxville, Tennessee 37921

TCT
Technical and Commercial Translations
2135 Spruce Street
Philadelphia, Pennsylvania 19103

TD
Mr. John Wiley
Technical Development Laboratories
P.O. Box 769
Savannah, Georgia 31402

TF
Taylor and Francis Ltd.
Red Lion Court, Fleet Street
London, E.C.4, England

TH
Therapia Hungarica
P.O. Box 64
Budapest, Hungary
Source for:
TH-Therapia Hungarica

TIC
Technical Information Company
Patents Department, Chancery House
Chancery Lane
London, W.C.2, England

TIDC
Technical Information-Documentation
Consultants Ltd.
1549 Burnside Place
Montreal 25, Quebec, Canada

TLS
Tobacco Literature Service
North Carolina State University
205 D.H. Hill Library
P.O. Box 5007
Raleigh, North Carolina 27607

TP
Taurus Press, Inc.
207 East 37th Street
New York, New York 10016

TTIS see RHC

UBO
Universitets Biblioteket. Oslo
Drammensveien 42B
Oslo, Norway

UKSM see NLL

WBI
Warren Birkenhead, Inc.
901 Northern Life Tower
Seattle, Washington 98101

WFK
William F. Kelly
915 Naudain Avenue, Northridge
Claymont, Delaware 19703

YLM
Yale L. Meltzer
84-23 Manton Street
Jamaica, New York 11435

A.B.M.
SEE BOLETIM. ASSOCIACAO BRASILEIRA DE METAIS

ACEC REVIEW
1955 N2/3 P3-27 T1818 <INSD>

A.E.G. MESS-UND FERNMELDETECHNIK. BERLIN
1938 N1 P36-39 58-1118 <*>

A.E.G. MITTEILUNGEN
1938 N9 P453-459 58-482 <*>
1939 N7 P347-353 2174 <*>
1951 V41 N11/2 P302-309 72G6G <ATS>
1955 V45 P346-357 T1824 <INSD>
1958 V48 P101-103 60-00595 <*>
1958 V48 N8/9 P432-441 18K28G <ATS>
1960 V50 P331-335 2438 <BISI>
1960 V50 P335-341 2439 <BISI>
1960 V50 P342-345 2440 <BISI>
1960 V50 P388-391 2441 <BISI>
1960 V50 P419-426 2442 <BISI>
1960 V50 P427-432 2443 <BISI>

A.M.A. ARCHIVES OF OPHTHALMOLOGY
1959 V62 P795-803 65-00128 <*> O

ATA. ASSOCIAZIONE TECNICA DELL'AUTOMOBILE. TORINO
1963 P332-339 65-17204 <*>

ATB METALLURGIE
SEE METALLURGIE. MONS

ATM
SEE ARCHIV FUER TECHNISCHES MESSEN

AWF-MITTEILUNGEN
1939 N1/2 P1-6 66-12397 <*>

ABHANDLUNGEN DER AKADEMIE DER WISSENSCHAFTEN ZU GOETTINGEN. MATHEMATISCH-PHYSIKALICHE KLASSE
1958 N2 P5-16 64-00019 <*>

ABHANDLUNGEN DER AKADEMIE DER WISSENSCHAFTEN ZU MAINZ
SEE ABHANDLUNGEN. MATHEMATISCH-NATURWISSEN-SCHAFTLICHE KLASSE. AKADEMIE DER WISSENSCHAFTEN UND DER LITERATUR, MAINZ

ABHANDLUNGEN DER BRAUNSCHWEIGISCHEN WISSENSCHAFT-LICHEN GESELLSCHAFT
1952 V4 P117-126 59-17549 <*>
1953 V5 P164-186 NP-TR-1146 <*>

ABHANDLUNGEN DER DEUTSCHEN AKADEMIE DER WISSEN-SCHAFTEN ZU BERLIN. KLASSE FUER CHEMIE, GEOLOGIE UND BIOLOGIE
1957 N2 P1-17 ORNL-TR-755 <*>
1963 N1 P91-98 66-11760 <*> O
1963 N1 P137-151 66-10756 <*>
1963 N1 P195-196 65-11303 <*>
1963 N1 P317-323 65-11304 <*>
1963 N1 P343-346 65-13982 <*>

ABHANDLUNGEN DER DEUTSCHEN AKADEMIE DER WISSEN-SCHAFTEN ZU BERLIN. KLASSE FUER MATHEMATIK, PHYSIK UND TECHNIK
1963 N1 P187-192 65-12063 <*>

ABHANDLUNGEN DER DEUTSCHEN AKADEMIE DER WISSEN-SCHAFTEN ZU BERLIN. KLASSE FUER MEDIZIN
1960 N1 P86-91 AEC-TR-4981 <*>
1960 N3 P187-197 64-10672 <*>
1960 N3 P268-271 61-00627 <*>
1960 N3 P272-278 61-00654 <*>
1960 N3 P323-325 61-00619 <*>

ABHANDLUNGEN DES DEUTSCHEN KAELTETECHNISCHEN VEREINES
1952 N6 P1-57 3419 <HB>
1952 V3 N6 P1-57 64-30836 <*> O

ABHANDLUNGEN DES GEOLOGISCHEN LANDESAMTES IN BADEN-WUERTTEMBERG
1953 V1 P1-60 I-948 <*>

ABHANDLUNGEN. MATHEMATISCH-NATURWISSENSCHAFTLICHE KLASSE. AKADEMIE DER WISSENSCHAFTEN UND DER LITERATUR, MAINZ
1956 N1 P3-26 03M47G <ATS>
1958 N3 P117-194 65-12124 <*> P
1959 N8 ENTIRE ISSUE E-541 <RIS>
1959 N8 P537-766 65-12508 <*>
1960 N3 P187-197 11499-E <K-H>

ABHANDLUNGEN. MATHEMATISCH-NATURWISSENSCHAFTLICHE KLASSE. AKADEMIE DER WISSENSCHAFTEN. MUNICH
1920 V36 N4 P433-510 61-14725 <*>
1956 N79B ENTIRE ISSUE 62-25593 <=*>

ABHANDLUNGEN DER NATURFORSCHENDEN GESELLSCHAFT ZU HALLE
1886 V16 P256- 57-2597 <*>

ABHANDLUNGEN. OESTERREICHISCHE GEOGRAPHISCHE GESELLSCHAFT
1951 V17 N2 P5-85 59-12004 <=*> O

ABHANDLUNGEN DES PREUSSISCHEN METEOROLOGISCHEN INSTITUTS. BERLIN
1932 V10 N2 ENTIRE ISSUE 60-10694 <*>

ABHANDLUNGEN HRSG. VON DER SENCKENBERGISCHEN NATURFORSCHENDEN GESELLSCHAFT
1912 V31 N2 P109-150 II-716 <*>
1952 N486 P6-18 84J16G <ATS>
1952 N486 P26-38 84J16G <ATS>

ACAROLOGIA
1959 V1 P56-85 63-01130 <*>

ACIERS SPECIAUX, METAUX ET ALLIAGES
1929 V4 P3-13 64-16722 <*> PO
1934 V9 N110 P408-414 57-631 <*>

ACQUE, BONIFICHE, COSTRUZIONE
1961 V29 N2 P27-33 59P61I <ATS>

ACTA ACADEMIAE ABOENSIS. MATHEMATICA ET PHYSICA
1964 V24 N6 P3-15 66-14603 <*>

ACTA ADRIATICA
1957 V6 N6 ENTIRE ISSUE C-4471 <NRCC>
1957 V6 N6 P1- 63-01084 <*>
1957 V8 N7 P1-26 60-21713 <=>

ACTA AEROPHYSIOLOGICA
1934 V1 N2 P65-78 61-14571 <*>

ACTA AGRICULTURA SUECANA
1946 V2 P1-157 66-13018 <*> P

ACTA AGRICULTURAE SINICA
SEE NUNG YEH HSUEH PAO

ACTA AGROBOTANICA
1957 V6 P117-143 60-21239 <=>

ACTA ALLERGOLOGICA
1953 V6 P295-303 64-14604 <*>

ACTA ANAESTHESIOLOGICA
1955 V6 P151-156 57-2971 <*>

ACTA ANAESTHESIOLOGICA BELGICA
1955 V2 P94-101 58-1183 <*>
1960 V11 N2 P163-176 62-16114 <*>

ACTA ANATOMICA
1950 V11 P329-347 65-00404 <*> O
1950 V11 N1 P300-328 57-1945 <*>

1956 V26 P94-109 61-20002 <*> 0

ACTA ANATOMICA NIPPONICA
SEE KAIBOGAKU ZASSHI

ACTA ANATOMICA SINICA
SEE CHIEH-POU HSUEH-PAO

ACTA ASTRONOMICA. WARSZAWA, KRAKOW
1961 V11 N3 P165-169 65-00226 <*>

ACTA ASTRONOMICA SINICA
SEE T'IEN WEN HSUEH PAO

ACTA AUTOMATICA SINICA
SEE TZU TUNG HUA HSUEH PAO

ACTA BIOCHIMICA ET BIOPHYSICA SINICA
SEE SHENG WU HUA HSUEH YU SHENG WU WU LI HSUEH PAO

★ACTA BIOCHIMICA POLONICA
1954 V1 P307-312 T-1755 <INSD>
1954 V1 P313-327 T-1756 <INSD>
1954 V1 N1 P47-58 II-905 <*>
1955 V2 P39-46 1202 <*>
1955 V2 P259-278 T-1757 <INSD>
1955 V2 P321-327 T1758 <INSA>
1955 V2 P329-341 T1754 <INSD>
1955 V2 N1 P59-71 60-21570 <=>
1958 V5 N4 P431-436 61-00642 <*>
1959 V6 N2 P227-233 61-11325 <=>
1960 V7 N1 P85-91 61-00622 <*>
1960 V7 N2/3 P235-238 61-00368 <*>

ACTA BIOLOGIAE EXPERIMENTALIS
1937 V11 P229-271 3043 <*>

ACTA BIOLOGIAE EXPERIMENTALIS SINICA
SEE SHIH YEH SHENG WU HSUEH PAO

ACTA BIOLOGICA ET MEDICA GERMANICA
1959 V3 P276-283 61-00369 <*>
1959 V3 P515-517 65-00087 <*>
1960 V4 P39-54 61-00197 <*>
1960 V4 P55-70 61-00191 <*>
1960 V4 P216-221 33P60G <ATS>
1960 V5 N3 P299-303 61-00623 <*>
1960 V5 N3 P304-306 61-00630 <*>
1960 V5 N3 P311-313 61-00621 <*>
1960 V5 N4 P409-412 61-00618 <*>
1960 V5 N4 P413-416 61-00652 <*>
1961 V6 P491-497 63-00080 <*>
1961 V6 N1 P86-96 62-24563 <=*>
1961 V6 N5 P395-405 62-24568 <=*>
1961 V7 N2 P139-144 62-24541 <=*>
1961 V7 N2 P145-160 62-15958 <=*>
1962 V9 P161-166 594 <TC>
1962 V9 P386-410 AEC-TR-6140 <*>
1962 V9 N1 P79-82 63-00550 <*>
1963 V10 N3 P357-374 64-31388 <=>
1963 V10 N1/2 P63-78 63-01100 <*>
1964 V12 P502-506 65-10971 <*> 0
1964 V13 N4 P504-512 65-30503 <=$>

ACTA BOTANICA SINICA
SEE CHIH WU HSUEH PAO

ACTA CARDIOLOGICA
1948 V3 N1 P1-47 65-00113 <*> 0
1957 V12 N3 P269-286 59-15973 <*>
1958 V13 N2 P153-172 61-00267 <*>
1962 V17 N4 P335-378 63-01136 <*>

ACTA CHEMICA SCANDINAVICA
1947 V1 P529-538 64-14605 <*>
1947 V1 P547-565 59H12G <ATS>
1949 V3 P487-492 64-15596 <=*$>
1950 V4 P1375-1385 26R76G <ATS>
1950 V4 N3 P422-434 59-15177 <*> 0
1953 V6 P805- AL-230 <*>
1955 V9 N3 P547- 1067 <*>
1957 V11 N10 P1633 61-18711 <*>
1958 V12 P1456-1475 63-01425 <*>
1961 V15 P1667-1675 C-4003 <NRC>
1962 V16 P529-540 62-01398 <*>
1962 V16 N3 P678-682 64-16724 <*> 0
1963 V17 N1 P73-78 63-01230 <*>
1965 V19 P2139-2150 9000 <IICH>

ACTA CHIMICA ACADEMIAE SCIENTIARUM HUNGARICAE
1954 V4 P245-258 58-570 <*>
1955 V6 P443-451 <DIL>
1955 V7 N1/2 P93-115 64-16726 <*>
1955 V8 P295-408 59-10652 <*>
1955 V8 P383-394 59-10656 <*>
1955 V8 P409-422 59-10653 <*>
1956 V10 P99-110 59-10655 <*>
1957 V13 P71-81 63-16376 <*>
1958 V14 N1 P173-196 59-10747 <*> 0
1958 V15 N3 P315-323 66-10286 <*>
1958 V16 P83-89 382 <TC>
1959 V18 P35-44 61-20844 <*>
1959 V21 N4 P351-361 65-12910 <*> 0
1960 V22 N1 P99-105 65-13201 <*>
1961 V26 P511-518 62-20192 <*>
1961 V26 N1/4 P451-458 64-21414 <=>
1961 V27 N1/4 P451-461 62-16296 <*>
1961 V28 N1/3 P17-27 2741 <BISI>
1962 V33 P381-385 63-14446 <*>
1962 V33 N1 P77-85 63-20392 <*> 0
1963 V39 N2 P253-270 749 <TC>
1964 V40 N3 P289-294 1810 <TC>

ACTA CHIMICA SINICA
SEE HUA HSUEH HSUEH PAO

ACTA CHIRURGICA. BUDAPEST
1963 V4 N3 P181-188 AD-635 854 <=$>

ACTA CHIRURGICA BELGICA
1960 V59 N8 P762-780 65-14220 <*> 0

ACTA CHIRURGICA SCANDINAVICA
1943 V88 P49-72 60-15639 <=*> 0

ACTA CLINICA BELGICA
1954 V9 N[illegible] P347-352 62-00287 <*>
1957 V12 N2 P195-204 59-18137 <+*> 0

ACTA CRYSTALLOGRAPHICA
1948 V1 P315-323 AD-621 937 <=$>
1948 V1 N27 P27-34 57-3307 <*>
1949 V2 P298-304 AI-TR-20 <*>
1949 V2 N3 P133-138 59-17909 <*> 0
1950 V3 P14-18 87Q73F <ATS>
1951 V4 P70-71 59-10743 <*>
1951 V4 P453-457 AD-621 938 <=$>
1952 V5 P117-121 I-35 <*>
2112 <*>
1952 V5 P351-356 58-1593 <*>
1953 V6 N4 P352-356 I-983 <*>
1955 V8 P412-419 58-2427 <*>
1956 V9 P95-108 59-17764 <*> P
1957 V10 P287-290 66-12145 <*>
1958 V11 N3 P145-148 61-18433 <*>
1961 V14 N8 P866-872 66-12997 <*>
1963 V16 P301-306 65-13073 <*> 0
1963 V16 P681-689 8803 <IICH>
1963 V16 P989-993 UCRL TRANS-1055(L) <*>
1963 V16 N8 P836 65-13525 <*>
1965 V19 N4 P504-513 66-12853 <*>
1965 V19 N4 P513-524 66-12967 <*>

ACTA DERMATOLOGICA
SEE HIFUKA KIYO

ACTA ELETRONICA
1957 V2 N1/2 P364-370 66-12990 <*> 0
1961 V5 P39-51 UCRL TRANS-1013(L) <*>

1961 V5 P53-60 UCRL TRANS-1006(L) <*>
1961 V5 N1 P7-30 UCRL TRANS-879(L) <*>
1961 V5 N4 P409-425 65-10822 <*>
1962 V6 N1 P7-66 64-11809 <=>
1963 V7 N1 P16-29 4167 <BISI>

ACTA ELECTRONICA SINICA
SEE TIEN TZU HSUEH PAO

ACTA ENDOCRINOLOGICA
1951 V7 P54-59 64-10315 <*>
1953 V12 N1 P23-27 64-10299 <*>
1953 V12 N1 P41-46 64-10300 <*>
64-14606 <*>
1953 V14 N1 P12-26 65-12546 <*>
1954 V17 P54-57 II-32 <*>
1954 V17 P375-384 I-527 <*>
1955 V19 N1 P11-20 2093 <*>
1955 V20 P371-378 64-14607 <*> O
1957 V26 P204-208 58-15 <*>
1958 V27 P1-35 58-2438 <*>
1958 V28 P219-226 62-01002 <*>
1959 V30 P1-21 63-20975 <*> O
1959 V31 N3 P341-348 63-10357 <*> O
1959 V31 N3 P433-441 62-00323 <*>
1959 V32 N2 P243-254 64-18413 <*>
1960 V34 P163-168 66-12179 <*>
1961 V36 P375-392 61-00982 <*>
1961 V37 P103-109 63-00926 <*>
1961 V37 P253-262 61-01061 <*>
1961 V37 N3 P394-404 62-16121 <*>
1962 V40 P217-231 62-16905 <*>
1962 V41 P259-264 63-00740 <*>

ACTA ENTOMOLOGICA SINICA
SEE K'UN CH'UNG HSUEH PAO

ACTA FOCALICA SINICA
SEE JAN LIAO HSUEH PAO

ACTA FORESTALIA FENNICA
1934 V40 P461-483 57-1242 <*>

ACTA GASTRO-ENTEROLOGICA BELGICA
1952 V15 P93-98 57-204 <*>
1957 V20 P738-742 33N51F <ATS>

ACTA GENETICA ET STATISTICA MEDICA. BASEL NEW YORK
1960 V10 N4 P267-294 62-00282 <*>
1960 V10 N1/3 P183-190 61-00863 <*>

ACTA GENETICAE MEDICAE ET GEMELLOLOGIAE
1959 V8 P47-68 SUP 2 66-12603 <*>

ACTA GEODETICA ET CARTOGRAPHICA SINICA
SEE T'SE LIANG CHI SHU HSUEH PAO

ACTA GEOGRAPHICA SINICA
SEE TI-LI-HSUEH-PAO

ACTA GEOLOGICA ACADEMIAE SCIENTIARUM HUNGARICAE
1956 V4 N2 P143-156 IGR V1 N6 P74 <AGI>
1958 V5 P187-195 62-16593 <*>

ACTA GEOLOGICA SINICA
SEE TI CHIH HSUEH PAO

ACTA GEOPHYSICA SINICA
SEE TI CH'IU WU LI HSUEH PAO

ACTA GERONTOLOGICA
1957 V7 P162-163 58-1810 <*>

ACTA GERONTOLOGICA JAPONICA
SEE YOKUFUEN CHOSA KENKYU KIYO

ACTA HAEMATOLOGICA. BASEL, NEW YORK
1950 V3 P135-150 62-00233 <*>
1950 V3 N3/4 P170-173 59-17331 <*> P
1950 V3 N3/4 P174-178 59-17332 <*> P
1950 V4 N2 P97-109 I-996 <*>
1952 V7 P69-105 AEC-TR-5114 <*>
1952 V8 P63-73 1337 <*>
1956 V15 P145-174 1113 <*>
1956 V15 N5 P323-334 58-277 <*>
1956 V16 P1-10 57-744 <*>
1957 V17 P16-24 AEC-TR-3636 <*>
1957 V18 N2 P126-136 60-10511 <*>
1960 V24 N6 P334-340 66-11758 <*> O
1963 V29 P187-190 63-26240 <=$>

ACTA HAEMATOLOGICA JAPONICA
SEE NIPPON KETSUEKI GAKKAI ZASSHI

ACTA HEPATOLOGICA JAPONICA
SEE KANZO

ACTA HEPATO-SPLENOLOGICA
1960 V7 P24-31 64-20011 <*>

ACTA HIPNOLOGICA LATINOAMERICANA
1960 V1 N2 P9-19 64-00187 <*>

ACTA HISTOCHEMICA
1955 V2 N1/2 P47-67 II-489 <*>
1956 V2 P196-207 58-913 <*>
1957 V4 P5-8 59-12693 <+*> O
1957 V4 P102-116 59-12694 <+*> O
1957 V4 P314-324 59-12693 <+*> O
1958 V5 P351-368 59-12246 <+*> O
1960 V9 N1/4 P66-68 61-10423 <*> O
1960 V10 P44-82 65-60319 <=$>

ACTA HYDROBIOLOGICA
1959 V1 N1 P5-36 63-11403 <=>
1960 V2 N2 P69-124 63-11353 <=>
1962 V4 N2 P151-204 65-50362 <=>
1962 V4 N3/4 P345-391,413-420
65-50362 <=>
1964 V6 N3 P187-205 65-50362 <=>

ACTA OF THE INTERNATIONAL UNION AGAINST CANCER
SEE ACTA UNIONIS INTERNATIONALIS CONTRA CANCRUM

ACTA ISOTOPICA
1963 V2 P391-400 HW-TR-79 <*>

ACTA MATHEMATICA. STOCKHOLM, UPPSALA
1892 V16 N1/3 P153-215 65-10279 <*>
1912 V35 P105-179 63-14670 <*>
1932 V58 P57-85 T-1501 <INSD>
1936 V66 P147-251 I-559 <*>
1945 V76 P157-184 1391 <*>

ACTA MATHEMATICA ACADEMIAE SCIENTIARIUM HUNGARICAE
1956 V7 P99-106 STMSP V2 P229-235 <AMS>

ACTA MECHANICA SINICA
SEE LI HSUEH HSUEH PAO

ACTA MEDICA. FUKUOKA
SEE IGAKU KENKYU

ACTA MEDICA ACADEMIAE SCIENTIARUM HUNGARICAE
1960 V16 N3 P233-236 11095-I <K-H>
1964 V20 N3 P357-363 66-10931 <*>

ACTA MEDICA ITALICA DI MALATTIE INFETTIVE E PARASSITARIE
1955 V10 N7 P171-175 62-00324 <*>

★ ACTA MEDICA IUGOSLAVICA
1958 V12 P1-43 SUP.1 60-21665 <=>
1961 V15 P470-476 12539 G <K-H>

ACTA MEDICA NAGASAKIENSIA
SEE NAGASAKI IGAKKAI ZASSHI

ACTA MEDICA PATAVINA

1954 V14 P69-109 II-232 <*>

ACTA MEDICA SCANDINAVICA
1930 V35 P331-338 57-1602 <*>
1931 V75 N3/4 P198-204 I-584 <*>
1931 V76 P4-6 66-11623 <*> 0
1931 V76 P491-520 66-11623 <*> 0
1941 V107 P547-578 57-1534 <*>
1941 V107 N6 P579-583 57-1624 <*>
1941 V109 N1/2 P177-210 59-19782 <=*> 0
1942 V112 P478-514 1400-A <K-H>
1961 V169 N2 P201-204 62-14913 <*>

ACTA MEDICA SCANDINAVICA. SUPPLEMENT
1931 N38 P1-65 1145G <KH>
1956 N312 P129-139 63-20869 <*>

ACTA MEDICA U.R.S.S.
1940 V3 N3 P187-199 62-13927 <=*> 0
62-13930 <=*>

ACTA MEDICINAE LEGALIS ET SOCIALIS
1957 V10 N4 P577-593 63-16941 <*>
8650-A <KH>
1959 V12 N3/4 P147-155 11053-A <KH>
63-16946 <*>

ACTA METALLURGICA
1953 V1 P79-92 3509 <BISI>
64-14586 <*> 0P
1953 V1 P519-530 61-20851 <*>
1953 V1 N11 P711-719 58-1513 <*>
1954 V2 P752-769 C-2582 <NRC>
1954 V2 P770-796 C-2583 <NRC>
1955 V3 P199-200 AEC-TR-6065 <*>
1955 V3 P380-391 60-21002 <=> 0
1955 V3 P542-548 C-2548 <NRC>
UCRL TRANS-711 <*>
1955 V3 N4 P347-353 64-16727 <*> 0
1955 V3 N4 P354-360 3013 <*>
1956 V4 P201-205 81H10F <ATS>
1956 V4 N2 P201-205 57-551 <*>
1957 V5 P175-176 C2358 <NRC>
UCRL TRANS-712 <*>
1957 V5 P695-702 58-1360 <*>
1957 V5 N1 P24-28 58-32 <*>
1957 V5 N3 P175-176 62-14562 <*> 0
1958 V6 P439-445 1080 <BISI>
1958 V6 N3 P184-194 61-18898 <*> 0
1958 V6 N3 P219-221 61-18405 <*>
1958 V6 N3 P221-223 61-18395 <*>
1958 V6 N6 P439-445 61-20592 <*>
1958 V6 N12 P778-780 62-10801 <*>
1959 V7 P43-50 59-17757 <*>
1959 V7 P93-100 98L32G <ATS>
1959 V7 P227-228 2655 <BISI>
1959 V7 P305-318 AEC-TR-3918 <*>
1959 V7 P523 1549 <BISI>
1959 V7 P589-598 1543 <BISI>
1959 V7 N7 P523-525 62-14547 <*>
1959 V7 N9 P589-598 62-14167 <*>
1960 V8 P384-387 AI-TRANS-87 <*>
1961 V9 N2 P155-161 62-00203 <*>
1961 V9 N11 P1001-1003 63-18180 <*>
1962 V10 P348-357 AEC-TR-5107 <*>
1962 V10 N3 P247-251 65-14592 <*> 0
1962 V10 N9 P901-913 65-14601 <*>
1963 V11 P617-622 AEC-TR-5970 <*>
1963 V11 N5 P475-484 AEC-TR-5879 <*>
1964 V12 P270-272 AI-TRANS-68 <*>
1964 V12 P649-664 AI-TR-47 <*>
1964 V12 P964-966 WAPD-TRANS-5 <*>
1964 V12 P1449-1453 66-11986 <*>
1964 V12 N4 P431-441 AI-TRANS-53 <*>
1964 V12 N5 P649-663 253-R <TC>
1964 V12 N6 P761-762 66-11858 <*>
1965 V13 P419-431 65-18101 <*>
1966 V14 P193-200 66-11868 <*>
1966 V14 P783-790 66-13612 <*>

ACTA METEOROLOGICA SINICA
SEE CHI HSIANG HSUEH PAO

ACTA MICROBIOLOGICA ACADEMIAE SCIENTARUM HUNGARICAE
1954 V2 N1/2 P191-194 61-00582 <*>

ACTA MICROBIOLOGICA POLONICA
1951 V1 N1 P42-51 65-23232 <$>
1952 N1 P93-106 AEC-TR-4527 <*>
1953 V2 N2/3 P129-132 RJ-290 <ATS>
1954 V3 N4 P395-397 60-21245 <=>
1955 V4 N3 P205-217 60-21570 <=>
1955 V4 N4 P271-279 64-30666 <*>
1956 V5 P79-80 58-1340 <*>
1956 V5 N1/2 P141-145 64-30667 <*>
1956 V5 N1/2 P217-224 60-21570 <=>
1957 V6 N4 P367-376 60-21570 <=>
1958 V7 N1 P35-44 64-30662 <*>
1958 V7 N2 P131-138 60-21570 <=>
1959 V8 P283-287 63-11362 <=>
1962 V11 N3 P277-282 63-18938 <*>
1962 V11 N1/2 P21-26 65-11059 <*> 0

ACTA MICROBIOLOGICA SINICA
SEE WEI SHEN WU HSUEH PAO

ACTA MORPHOLOGICA ACADEMIAE SCIENTIARUM HUNGARICAE
1954 V14 N1 P1-7 1893 <*>

ACTA NEUROCHIRURGICA
1955 P90-97 SUP 3 UCRL TRANS-881 <*>
1961 V9 P398-428 63-16383 <*>

ACTA NEUROLOGICA
1946 V1 P102-110 60-10681 <*>
1946 V1 P318-330 58-403 <*>
1947 V2 P296-305 60-10680 <*> 0
1951 V6 P618-633 I-570 <*>
1957 V12 N4 P475-493 64-10428 <*>
9054-C <K-H>

ACTA NEUROLOGICA ET PSYCHIATRICA BELGICA
1949 V49 N5 P327-330 60-10509 <*>
1955 V55 P399-408 II-857 <*>
1957 V57 P667-672 58-17 <*>
1959 V59 N6 P693-701 60-17570 <=*>
1960 V60 N5 P520-524 62-10266 <*>
1960 V60 N7 P631-637 61-00986 <*>

ACTA NEUROPSIQUIATRICA ARGENTINA
1960 V6 P68-71 65-17246 <*>

ACTA NEUROVEGETATIVA
1953 V7 P189-196 I-901 <*>
1957 V16 N1/4 P82-85 64-16938 <*> 0
1958 V18 N1/4 P320-334 65-00156 <*> 0

ACTA ONCOLOGICA
1963 V2 N1 P43-50 65-17249 <*>

ACTA OPHTHALMOLOGICA
1928 V6 P237 61-10950 <*>
1935 V13 P192-224 61-14199 <*> 0

ACTA ORTHOPAEDICA BELGICA
1950 V16 P332-352 58-1771 <*>

ACTA ORTHOPAEDICA SCANDINAVICA
1962 V32 P499-508 63-26238 <=$>

ACTA OTO-LARYNGOLOGICA
1921 V3 P183-189 57-1541 <*>
1929 N8 P1-48 61-14599 <*>
1935 V22 P557-583 61-14899 <*>
1939 V27 P107-122 61-14648 <*>
1952 N100 SUP. P119-133 60-18549 <*>
1952 V42 P334-344 64-10311 <*> 0
1954 V44 P456-461 2705 <*>
1954 V44 N5/6 P525-531 3546-A <K-H>

1955 V45 P402-415 57-3571 <*>
1955 V45 N5 P402-415 60-18550 <*>
1955 V45 N5 P440-454 65-12551 <*>
1958 V49 N6 P483-494 60-14082 <*>
1960 V51 N3/4 P319-331 64-30304 <*>
1960 V52 P473-476 63-18036 <*>
1962 V54 N1 P38-48 65-11670 <*>

ACTA OTO-RHINO-LARYNGOLOGICA BELGICA
1964 V18 P79-408 66-13014 <*> P

ACTA OTO-RINO-LARINGOLOGICA IBERO-AMERICANA
1957 V8 N2 P270-273 63-10706 <*>

ACTA PAEDIATRICA. STOCKHOLM, UPPSALA
1956 V45 P618-632 57-3160 <*>

ACTA PAEDIATRICA BELGICA
1959 V13 N5 P280-286 65-14423 <*>
1960 V14 N2 P80-86 64-18659 <*>

ACTA PAEDIATRICA JAPONICA
SEE NIPPON SHONIKA GAKKAI ZASSHI

ACTA PAEDIATRICA LATINA
1950 V3 P275-319 60-16614 <*>
1953 V6 P627-640 UCRL TRANS-789 <*>

ACTA PAEDIATRICA SINICA
SEE CHUNG HUA MIN KUO HSIAO ERH K'O I HSUEH HUI

ACTA PALAEONTOLOGICA POLONICA
1960 V5 N4 P367-432 61-11312 <=>

ACTA PALAEONTCLOGICA SINICA
SEE KU SHENG WU HSUEH PAO

ACTA PARASITOLOGICA POLONICA
1955 V2 N19 P361-404 61-31289 <=>
1956 V4 N19 P627-661 60-21226 <=>
1956 V4 N20 P663-750 60-21225 <=>
1957 V5 N28 P613-644 60-21514 <=>
1958 V6 N4 P143-178 60-21231 <=>
1959 V7 N15 P315-342 61-31290 <=$>

ACTA PATHOLOGICA ET MICROBIOLOGICA SCANDINAVICA
1930 V7 P250-257 57-1876 <*>
1936 P1-295 SUP30 2808 <*>
1936 V13 P512-531 2945 <*>
1962 V54 P473-478 64-00359 <*>
1962 V56 P341-342 64-00191 <*>

ACTA PEDIATRICA ESPANOLA
1962 V20 P26-36 66-11599 <*>

ACTA PEDOLOGICA SINICA
SEE T'U JUNG HSUEH PAO

ACTA PHARMACEUTICA HUNGARICA
1961 V31 N2 P75-82 62-19677 <=*> P
1961 V31 N3 P128-132 62-19677 <=*> P
1964 V34 N4 P175-179 65-14124 <*>

ACTA PHARMACEUTICA SINICA
SEE YAO HSUEH HSUEH PAO

ACTA PHYSICA ACADEMIAE SCIENTIARUM HUNGARICAE
1952 V2 N2 P129-139 08Q72G <ATS>
1953 V3 N1 P11-14 II-1042 <*>
1955 V4 N3 P197-208 57-1359 <*>
1955 V5 P123-129 UCRL-TRANS-720(L) <*>
1955 V5 P339-345 UCRL-TRANS-721(L) <*>
1957 V7 N1 P141-149 57-3205 <*>
1957 V8 N1/2 P75-81 62-18118 <*>
1958 V8 P321-358 UCRL TRANS-722(L) <*>
1960 V11 N4 P323-331 62-11512 <=>
62-11512 <=>
1961 V13 P233-238 UCRL TRANS-751(L) <*>
1962 V14 N4 P345-380 63-23321 <=*>

ACTA PHYSICA AUSTRIACA
1949 P245-254 2839 <*>
1949 V2 P379-400 AL-753 <*>
58-662 <*>
64-71612 <=>
1951 V5 P12-29 3185 <*>
1953 V6 P241-245 I-944 <*>
1953 V7 N1 P52-90 60-16780 <*>
1954 V8 N3 P285-288 59-17318 <*>
1957 V10 N4 P448-452 61-20533 <*>
1957 V11 P70-74 C-3592 <NRC>
1962 V15 N1/2 P57-63 63-10161 <*>
1963 V16 P234-244 AEC-TR-6283 <*>
1963 V16 N2 P115-124 64-16280 <*>
1963 V16 N3/4 P296-303 65-14934 <*>
1964 V19 N3 P268-286 66-10789 <*>

ACTA PHYSICA POLONICA
1936 V5 P229-253 I-717 <*>
1936 V5 P335-347 NP-TR-771 <*>
1961 V20 N7 P553-561 62-19566 <=*>
1961 V20 N7 P563-565 62-19567 <=*>
1962 V22 N2 P199-208 64-16491 <*> 0
1962 V22 N3 P265-283 64-16492 <*> 0
1964 V26 N3 P331-343 AD-646 452 <=>
1964 V26 N4 P683-688 AD-646 452 <=>

ACTA PHYSICA SINICA
SEE WU LI HSUEH PAO

ACTA PHYSICO-CHIMICA URSS
1934 V1 N3/4 P429-448 64-19359 <=$>
1935 V2 P621-632 AEC-TR-4251 <*>
1935 V2 N2 P239-271 59-17448 <*>
1935 V3 N6 P767-778 R-216 <*>
1935 V3 N6 P791-818 63-23291 <*=>
1936 V4 N6 P911-928 63-23290 <*=>
1936 V5 N1 P23-44 C-4334 <NRC>
1937 V6 N3 P401-418 63-19207 <=*>
1938 V8 P811-847 R-348 <*>
1938 V9 N3/4 P381-394 59-17685 <*> P
1938 V9 N3/4 P421-452 R-819 <*>
1938 V9 N3/4 P581-592 59-17686 <*>
1938 V9 N3/4 P593-619 58-496 <*>
1938 V9 N3/4 P593-620 59-17687 <*> P
1939 V11 P865-882 2047 <*>
1939 V11 N6 P883-898 53M47G <ATS>
1940 V12 P67-72 II-199 <*>
1940 V12 P485-488 II-193 <*>
1940 V12 N3 P327-356 RJ-308 <ATS>
1940 V12 N6 P921-930 59-17477 <*>
1940 V13 P111- R-497 <*>
1940 V13 P443-453 R-2624 <*>
1940 V13 N3 P405-522 07K21G <ATS>
1940 V13 N4 P565-568 64-16729 <=*$>
1941 V14 N1 P105-118 61-16873 <*=>
1941 V14 N2 P271-278 61-16871 <*=>
1941 V14 N3 P365-370 61-16872 <*=>
1942 V16 P363-375 61-16642 <*=>
1942 V16 N1/2 P1-11 61-16667 <*=>
1942 V16 N1/2 P88-96 64-16730 <=*$>
1942 V17 P83-92 61-16643 <*=>
1942 V17 P212-217 61-16644 <*=>
1942 V17 P308-313 61-16653 <*=>
1943 V18 N1 P58-68 62-14615 <=*> P
1944 V19 P225-247 61-18172 <*=>
1944 V19 P266-285 61-18171 <*=>
1944 V19 P329-359 61-18191 <*=>
1945 V20 P3-30 61-18187 <*=>
1945 V20 P117-126 61-18334 <*=>
1945 V20 P259-267 61-18353 <*=>
1946 V21 P283-288 61-20157 <=*>
1946 V21 P335-344 61-20158 <=*>

ACTA PHYSIOLOGICA ACADEMIAE SCIENTIARUM HUNGARICAE
1952 V3 P513-524 61-00392 <*>
1953 V4 P31-44 61-00657 <*>
1953 V4 P235-240 61-00384 <*>
1954 V5 P393-399 65-00159 <*> 0
1954 V6 N2/3 P191-198 59-10899 <*>

1955 V7 N1/2 P119-125 C-2397 <NRC>
1957 V12 P1-8 61-00375 <*>
1957 V12 P9-12 61-00394 <*>
1959 V15 N2 P161-177 60-17782 <=*>

ACTA PHYSIOLOGICA LATINOAMERICANA
1962 V12 N3 P302-311 65-00249 <*>

ACTA PHYSIOLOGICA ET PHARMACOLOGICA NEERLANDICA
1957 V6 P43-52 60-17330 <+*> O

★ACTA PHYSIOLOGICA POLONICA
1954 V5 N2 P131-145 57-547 <*>
1954 V5 N2 P147-160 57-546 <*>
1954 V5 N2 P229-230 57-545 <*>
1957 V8 P312-313 9002 <K-H>
1957 V8 N3/3A P308-309 63-16960 <*> O
1957 V8 N3/3A P312-313 63-16959 <*> O
1960 V11 P840-841 C-4183 <NRC>
1960 V11 P909-911 93Q66P <ATS>
1961 V12 P267-273 64-00351 <*>
1961 V12 P275-279 64-00370 <*>
1961 V12 P441-449 64-00363 <*>
1961 V12 N1 P129-132 62-19381 <=*>
1961 V12 N1 P173-180 62-19576 <=*>

ACTA PHYSIOLOGICA SCANDINAVICA
1940 V1 N3 P220-232 63-16236 <*>
1956 V35 N3/4 P203-214 57-3459 <*>
58-685 <*>
1956 V36 N3 P18 58-614 <*>
1956 V36 N3 P219-228 58-614 <*>
1956 V37 N2/3 P204-214 57-3458 <*>
58-616 <*>
1960 V49 N2/3 P139-147 63-20284 <*> O

ACTA PHYSIOLOGICA SINICA
SEE SHENG LI HSUEH PAO

ACTA PHYTOCHIMICA JAPONICA
SEE SHOKUBUTSU KAGAKU ZASSHI

★ACTA POLONIAE PHARMACEUTICA
1961 V18 N3 P177-181 51P63P <ATS>
1961 V18 N4 P339-342 11627-A <KH>
63-16649 <*>
1963 V20 N2 P199-204 38Q72P <ATS>

ACTA PSIQUIATRICA Y PSICOLOGICA DE AMERICA LATINA
1964 V10 P177-185 47S86SP <ATS>

ACTA PSIQUIATRICA Y PSICOLOGICA ARGENTINA
1963 V9 P136-143 65-17290 <*>
1963 V9 N3 P226-232 65-17289 <*>

ACTA PSYCHIATRICA ET NEUROLOGICA SCANDINAVICA
1939 V14 P223-231 61-00856 <*>
1956 V31 N2 P160-162 57-3457 <*>
58-615 <*>

ACTA PSYCHIATRICA SCANDINAVICA
1962 V38 N2 P108-111 63-10512 <*>
1962 V38 N2 P112-116 63-10707 <*>

ACTA PSYCHOLOGICA
1950 V7 P164-189 64-10049 <*>
1951 V8 P129-146 64-10041 <*>
1951 V8 N1 P25-34 64-10028 <*>

ACTA PSYCHOLOGICA SINICA
SEE HSIN LI HSUEH PAO

ACTA PSYCHOSOMATICA
1948 N1 P3-42 63-00922 <*>
63-27745 <$>

ACTA RADIOLOGICA
1941 V22 P252-259 AEC-TR-4017 <*>
1944 V25 P13-32 64-14572 <*> O
9523 <K-H>
1955 V44 P57-82 61-01002 <*>

ACTA RADIOLOGICA: DIAGNOSIS
1963 V1 P468-480 64-16931 <*> O
1963 V1 N5 P1111-1122 NP-TR-1135 <*>

ACTA SCIENTIARIUM MATHEMATICARUM
1964 V25 P90-123 AMST S2 V51 P155-188 <AMS>

ACTA SCIENTIARUM NATURALIUM. CHI-LIN UNIVERSITY
SEE CHI-LIN TA HSUEH TZU JAN K'O HSUEH HSUEH PAO

ACTA SCIENTIARIUM NATURALIUM UNIVERSITATIS PEKINENSIS
SEE PEI-CHING TA HSUEH HSUEH PAO-TZU JAN K'O HSUEH

ACTA SOCIETATIS BOTANICORUM POLONIAE
1926 V4 N1 P1-54 65-11084 <*> O
1927 V5 N1 P79-98 65-11050 <*> O
1928 V5 N6 P46-90 65-11083 <*> O
1930 V7 P507-519 65-11044 <*> O
1930 V7 N2 P250-273 65-11049 <*> O
1953 V22 N1 P97-125 60-21390 <=>
1953 V22 N1 P133-168 61-11332 <=>
1953 V22 N3 P561-576 61-11333 <=>
1953 V22 N4 P737-752 61-11334 <=>
1954 V23 N3 P487-504 61-11330 <=>
1954 V23 N3 P589-616 60-21527 <=>
1955 V24 N1 P65-93 60-21522 <=>
1956 V25 N1 P3-26 60-21384 <=>
1956 V25 N2 P245-274 60-21221 <=>
1957 V26 N1 P79-103 60-21519 <=>
1957 V26 N2 P413-466 60-21388 <=>
1958 V27 N1 P75-82 60-21389 <=>
1958 V27 N2 P313-342 61-11322 <=>
1958 V27 N4 P523-540 61-11324 <=>

ACTA SOCIETATIS OPHTHALMOLOGICAE JAPONICAE
SEE NIPPON GANKA GAKKAI ZASSHI

ACTA SOCIETATIS PATHOLOGICAE JAPONICAE
SEE NIPPON BYORIGAKKAI KAISHI

ACTA TECHNICA. PRAGUE
1963 V8 N6 P509-523 66-12868 <*>
1964 V9 P1-16 SC-T-65-714 <*>

ACTA TECHNICA ACADEMIAE SCIENTIARUM HUNGARICAE
1951 N2 P449-459 LC-1268 <BRS>
1951 V2 N2 P449-459 LC-1268 <NLL>
1953 V7 N3/4 P413-423 T-2261 <INSD>
T-2261 <INSD>
1954 V8 N3/4 P319-360 58-1056 <*>
1956 V15 N1/2 P179-189 62-20142 <*>
1956 V15 N1/2 P205-218 61-16300 <*>
1957 V17 P67-112 58-205 <*>
1957 V17 N3/4 P361-380 2393 <BISI>
1957 V18 N1/2 P103-115 63-41074 <=> O
1957 V19 P115-125 58-2304 <*>
1957 V19 P199-241 58-2314 <*>
1958 V21 N1/2 P123-139 AEC-TR-3922 <*>
1958 V21 N3/4 P275-289 89L36F <ATS>
1958 V22 N3/4 P395-412 1564 <BISI>
19M39G <ATS>
1959 V25 N1/2 P87-117 63-11730 <=>
1960 V31 N3/4 P359-389 STT-181-468 <STT>
1961 V32 N1/2 P163-183 2476 <BISI>
1961 V35/6 P219-236 32P61G <ATS>
1961 V35/6 P555-574 3100 <BISI>
1962 V38 N1/2 P235-257 63-18340 <*>
1962 V39 N1/2 P135-162 66-32154 <=$>

ACTA TROPICA
1947 V4 N1 P1-9 63-00775 <*>
1947 V4 N1 P10-20 63-00734 <*>
1948 V5 N3 P249-251 1605 <*>
1948 V5 N3 P257-260 1607 <*>
1952 V9 N3 P272-276 63-01120 <*>
1957 V14 N3 P208-217 63-00576 <*>

ACTA TUBERCULOSEA SCANDINAVICA
1951 V25 N2/3 P263-304 2768 <*>

ACTA UNIONIS INTERNATIONALIS CONTRA CANCRUM. BRUXELLES, PARIS
1956 V12 N3 P326-328 5796-I <K-H>
1958 V14 N4 P408-412 61-00186 <*>
1959 V15 N3/4 P737-747 62-00897 <*>
1960 V16 P935-936 61-00417 <*>
1961 V17 P181-197 AEC-TR-6160 <*>

ACTA UNIVERSITATIS CAROLINAE. MATHEMATICA ET PHYSICA. (TITLE VARIES)
1960 N2 P62-67 64-14782 <*>
1964 V5 P1-48 SUP 1 AMST S2 V57 P85-112 <AMS>

ACTA UNIVERSITATIS CAROLINAE. MEDICA. SUPPLEMENTUM
1961 V14 P207-218 AEC-TR-5816 <*>
1961 V14 P219-222 AEC-TR-5815 <*>

ACTA UNIVERSITATIS LUNDENSIS
1940 V36 N4 P31- NACA TM-1432 <NASA>

ACTA UNIVERSITATIS PALACKIANAE OLOMUCENSIS. FACULTAS RERUM NATURALIUM
SEE SBORNIK PRACI PRIRODOVEDECKE FACULTY PALACKEHO UNIVERSITY V OLOMOUCI

ACTA UNIVERSITATIS SZEGEDIENSIS. ACTA CHEMICA, MINERALOGICA ET PHYSICA
1939 V7 P7-25 58-723 <*>

ACTA UNIVERSITATIS VORONEGIENSIS
SEE TRUDY VORONEZHSKOGO UNIVERSITETA

ACTA UROLOGICA JAPONICA
SEE HINYOKIKA KIYO

ACTA VETERINARIA
1951 V1 N1 P128-131 60-21623 <=>
1956 V6 N2 P3-13 60-21624 <=>
1960 V10 N3 P123-128 62-19465 <=*>
1960 V10 N4 P73-79 62-19465 <=*>
1960 V10 N4 P103-109 62-19465 <=*>
1961 V11 N1 P49-54 62-19598 <=*>

ACTA VETERINARIA ACADEMIAE SCIENTIARUM HUNGARICAE
1958 V8 N4 P375-388 62-00671 <*>

ACTA VIROLOGICA
1958 V2 P198-200 60-14296 <*> ,

ACTA VITAMINOLOGICA
1947 V1 P130-135 60-10834 <*>
1948 V2 P49-52 II-564 <*>
1948 V2 N6 P127-130 1546 <K-H>
1954 V8 N2 P85-87 60-10508 <*> 0
1954 V8 N5 P209-213 II-300 <*>
1957 V11 N6 P257-260 58-1355 <*>
1959 V13 N8 P149-154 00P61I <ATS>
1963 V17 N6 P239-254 64-18067 <*>

ACTA ZOOLOGICA. STOCKHOLM
1920 V1 P321-336 <DIL>
1926 V7 P101-115 65-12123 <*>
1926 V7 N1 P101-115 57-2285 <*>

ACTAS DERMO-SIFILOGRAFICAS
1942 P510-512 57-1608 <*>

ACTAS ESPANOLAS DE NEUROLOGIA Y PSIQUIATRIA
1959 V18 P108-113 65-17296 <*>

ACTAS LUSO-ESPANOLAS DE NEUROLOGIA Y PSIQUIATRIA
1957 V16 N4 P305-307 59-15885 <*>

ACTIVITAS NERVOSA SUPERIOR
1962 V4 N1 P36-40 66-10976 <*>
1963 V5 N1 P4-12 64-23604 <=$>

ACTUALIDAD PEDIATRICA
1961 V20 N7 P1-6 63-16528 <*>

ACTUALITES NEPHROLOGIQUES DE L'HOPITAL NECKER
1963 P127-141 63-20471 <*> 0

ACTUALITES NEUROPHYSIOLOGIQUES
1959 V1 P39-61 65-00098 <*> 0
1959 V1 P71-88 65-00282 <*> 0
1959 V1 P166-181 63-01766 <*>

ACTUALITES ODONTOSTOMATOLOGIQUES
1954 N25 P7-26 64-16111 <*> 0

ACTUALITES SCIENTIFIQUES ET INDUSTRIELLES
1938 N737 P43-66 I-560 <*>
1946 N1009 P37-41 II-68 <*>
1949 N1071 P68-79 II-910 <*>

ACTUALITES SPATIALE. L'HOMME ET L'ESPACE
1963 N16 P7-44 63-41144 <=>

ACUSTICA
1952 V2 N1 P1-17 21Q74G <ATS>
1952 V2 N1 P18-19 22Q74G <ATS>
1952 V2 N6 P263-281 61-18912 <*>
1953 P434-440 59-10200 <*>
1954 V4 P396-402 58-1233 <*>
1954 V4 P433-444 NS-68 <TTIS>
61-19983 <=*>
1954 V4 P445-455 59-20942 <*> 0
1954 V4 P671-676 58-1224 <*>
1954 V4 N3 P365-379 NACA TM 1404 <NASA>
1955 V5 P163-166 1217 <*>
1955 V5 N3 P149-163 NACA TM 1409 <NASA>
1955 V5 N6 P323-330 C-2722 <NRC>
58-1671 <*>
1956 V6 P1-4 62-18334 <*>
1956 V6 N4 P365-381 57-3048 <*>
1957 V7 P75-83 63-18883 <*>
1957 V7 P101- 60-21421 <=>
1958 V8 P102- 60-21184 <=>
1958 V8 P273-280 61-20535 <*>
1958 V8 N1 P31- 60-21422 <=>
1958 V8 N2 P86-90 58-1788 <*>
1959 V9 P275-288 62-20333 <*>
1959 V9 P359-364 AEC-TR-4025 <*>
1959 V9 P359- 60-21423 <=>
1959 V9 P387-398 62-20331 <*>
1959 V9 N5 P345-352 60-16915 <*>
1960 V10 N1 P30-34 62-10078 <*>
1961 V11 P161-177 48N56G <ATS>
1961 V11 N1 P8-17 62-14437 <*>
1961 V11 N1 P50-64 65-13911 <*>
1961 V11 N3 P121-126 62-14436 <*>
64-16330 <*>
1961 V11 N3 P127-136 62-14435 <*>
1961 V11 N5 P366-367 65-14210 <*>
1962 V12 N5 P322-334 63-20618 <*>
1962 V12 N6 P386-397 64-16584 <*>
1962 V12 N6 P404-410 65-00199 <*>
1963 V13 P75-85 65-11454 <*> 0
1963 V13 N2 P92-108 63-20653 <*> 0
1964 V14 N5 P254-266 66-10893 <*>
1964 V14 N6 P337-346 55S83G <ATS>
1964 V14 N6 P360-364 65-11744 <*>
1965 V15 P332-338 02S87G <ATS>
1965 V15 N4 P199-206 POED-TRANS-2217 <NLL>
1965 V15 N5 P271-284 65-14612 <*>
1965 V16 N3 P159-165 66-13609 <*>

ADHAESION
1960 V4 N10 P514-517 62-00080 <*>
1961 V5 N5 P225-230 64-14776 <*>
1962 V6 N5 P219-222 65-11014 <*>
1962 V6 N7 P332-338 64-10411 <*>
1963 V7 N11 P503-512 163 <TC>
272 <TC>
64-20947 <*> 0
1964 V8 N6 P253-254 NS-300 <TTIS>

1964 V8 N6 P256 NS-300 <TTIS>
1964 V8 N6 P258 NS-300 <TTIS>
1964 V8 N6 P276-277 3882 <BISI> P
1964 V8 N7/8 P326-331 65-11643 <*>
1965 V9 N7/8 P289-307 4507 <BISI>

ADKOARAS
1956 V60 P384-389 C-2654 <NRC>

ADVANCED ENERGY CONVERSION; AN INTERNATIONAL JOURNAL. OXFORD
1963 V3 P569-590 64-16193 <*>
64-18127 <*> O

ADVANCES IN CANCER RESEARCH
1955 V1 P3-6 R-137 <*>
1955 V2 P3-8 R-137 <*>

ADVANCES IN ELECTRONICS AND ELECTRON PHYSICS
1960 V12 P5-16 65-10828 <*>

ADVANCES IN ENZYMOLOGY
1954 V15 P49-77 59-15479 <*> OP

ADVANCES IN PHYSICS
1962 V11 N43 P233-279 AEC-TR-6378 <*>
63-18030 <*>

AEROLOGIYA
1951 P341-350 R-38 <*>

AEROMEDICA ACTA
1956 V5 P347-353 64-10589 <*>

AERONAUTICAL KNOWLEDGE. CHINESE PEOPLES REPUBLIC
SEE HANG KUNG CHIH SHIH

AERO-SPORT
1964 N3 P77-83 64-71540 <=>
1965 N9 P290-297 AD-638 881 <=$>

AEROTECNICA. ROMA
1935 V15 N3 P237-275 63-20557 <*>
1935 V15 N7/8 P735-778 63-20557 <*>
1937 V17 P507-519 63-20491 <*>
1937 V17 N5 P415-442 I-871 <*>
1937 V17 N6 P519-534 63-20556 <*>
1937 V17 N10 P850-861 AL-546 <*>
1938 V18 N1 P18-21 63-18829 <*> O
1940 V20 P815-833 65-12962 <*>
1941 V21 N9/10 P19 AL-145 <*>
1952 V32 N3 P135-142 2988 <*>
1952 V32 N4 P206-219 2823 <*>
1955 V35 P230-234 1535 <*>
1956 V36 P101-111 57-3050 <*>
1956 V36 P127-131 57-3051 <*>
1956 V36 N2 P68-94 61-31611 <=>
1956 V36 N3 P160-176 59-10336 <*>
1956 V36 N3 P177-217 59-10337 <*>
1956 V36 N5 P315-332 59-16777 <+*>
1957 V37 P311-320 60-17365 <*+>
1960 V40 N3 P172-177 61-14910 <*>

AERZTLICHE FORSCHUNG
1953 V7 P129-133 AL-672 <*>
1954 V8 P268-272 1396 <*>
1954 V8 N2 P85-88 64-14608 <*>
1955 V9 N5 P211-228 62-00802 <*>
1957 V11 N11 P537-551 63-01526 <*>
1958 V12 N2 P53-60 64-14609 <*> P
1958 V12 N3 P148-151 59-10624 <*> O
1958 V12 N10 P471-475 62-00686 <*>
1960 V14 N9 P454-460 61-10760 <*>
1961 V15 N1 P6-9 62-00477 <*>
1963 V17 N1 P1-16 63-01119 <*>
1963 V17 N7 P383-391 66-10101 <*> O
1964 V18 N8 P427-429 66-10102 <*>
1965 V19 N1 P3-7 66-12039 <*>
1965 V19 N4 P209-219 68S88G <ATS>

AERZTLICHE KORRESPONDENZ
1934 N5 1-6 SPEC. NO. 61-10438 <*>
1935 N11 1-8 SPEC. NO. 61-10439 <*>

AERZTLICHE KOSMETIK
1956 V4 P13-38 62-00837 <*>

AERZTLICHE LABORATORIUM
1942 V8 N5 P1-3 63-00437 <*>

AERZTLICHE MITTEILUNGEN. STRASSBURG
1955 V40 N14 P417-427 63-16005 <*>
6801-A <K-H>

AERZTLICHE PRAXIS
1953 V5 P3 AL-493 <*>
1953 V5 P3- I-785 <*>
1955 V7 P20-21 57-1829 <*>
1963 V15 P1338 66-10105 <*>
1963 V15 P1753-1754 66-10104 <*>
1963 V15 P2942-2943 66-10103 <*>
1964 V16 P339-400 66-10107 <*>
1964 V16 N49 P2153-2156 66-10106 <*>

AERZTLICHE WOCHENSCHRIFT
1948 V3 N41 P644-646 1724 <*>
1950 V5 N32 P599-601 C2353 <NRC>
3441 <*>
1951 V6 N27 P644-645 59-17124 <*>
1951 V6 N38 P905-908 2577 <*>
1953 V8 P110-114 62-00833 <*>
1953 V8 P345-350 AL-866 <*>
1953 V8 N17 P416-420 3138-E <K-H>
1953 V8 N35 P841-843 71F3G <ATS>
1953 V8 N35 P843-845 70F3G <ATS>
1954 V9 P344-345 2005 <*>
1954 V9 N43/4 P1013-1016 3390-C <K-H>
1954 V9 N43/4 P1039-1041 3390-C <K-H>
1955 V10 P394-397 II-872 <*>
1955 V10 P647-648 57-3332 <*>
1955 V10 P947-949 58-1825 <*>
1955 V10 N18 P410- 1461 <*>
1955 V10 N31 P708-713 II-571 <*>
1956 V11 P886-888 57-270 <*>
1956 V11 N42 P941-943 59-15756 <*>
1956 V11 N25/6 P550-551 64-14610 <*> O
1957 V12 N11 P225-230 62-00454 <*>
1957 V12 N37 P822-823 64-20029 <*>
1957 V12 N42 P939-944 61-00854 <*>
1958 V13 P110-113 R-5123 <*>
1958 V13 N5 P97-105 62-14917 <*> O
1960 N7 P118-123 10324-D <K-H>

AESCULAPE
1912 V2 P183-185 57-1668 <*>
1928 V18 P116-119 57-1610 <*>

AESTHETISCHE MEDIZIN
1962 V11 N1 P5-14 65-14607 <*> O

AFINIDAD
1949 V26 P200-201 63-16769 <*>
1951 V28 P264-269 63-14675 <*>

AFRIQUE FRANCAISE CHIRURGICALE
1955 V13 P255-257 58-234 <*>
1958 V16 P55-57 59-18144 <+*>

AGE NUCLEAIRE
1958 N11 P187-193 AEC-TR-3739 <=*>
1958 N11 P210-215 AEC-TR-3642 <=*>

AGGIORNAMENTI SULLE MALATTIE DA INFEZIONE
1961 V7 N3 P141-148 63-10502 <*>
1961 V7 N4 P213-220 63-10503 <*>

AGGIORNAMENTO PEDIATRICO
1957 V8 P577-580 64-18706 <*>
1961 V12 N10 P477-484 63-10501 <*>

AGITATOR
1960 N15 P13-15 62-23483 <*=>
1961 N5 P22-25 61-27130 <=>
1961 N19 P39-42 62-15485 <*=>
1962 10/19 P44-47 63-15100 <=>
1962 N15 P35-37 62-33025 <=*>
1963 N4 P41-44 63-31115 <=>
1963 N5 P50-51 63-21789 <=>
1963 N8 P25 63-31424 <=>
1963 N11 P44-46 63-31406 <=>
1963 N12 P52-54 63-31601 <=>
1963 N12 P64 63-31625 <=>
1963 N21 P44-47 64-21516 <=>

AGRARIO LEVANTINO
1945 V11 N128 P34-36 3268 <*>

AGRARTUDOMANYI EGYETEM AGRONOMIAI KAR KIADVANYAI
1957 V4 N4 P3-8 63-16279 <*>
1957 V4 N4 P9- 63-16280 <*>

AGRARWIRTSCHAFT
1952 V1 P11-14 64-10046 <*>
1953 V2 P1C1-102 64-10037 <*>
1953 V2 N8 P237-256 65-12537 <*>
1953 V2 N8 P241-248 64-10360 <*>
1954 V3 P289-294 65-12374 <*>
1954 V3 N2 P49-52 64-10359 <*>
1955 V4 P65-68 66-12526 <*>
1956 V5 N2 P34-38 59-20980 <*>
1957 V6 N3 P70-79 59-20865 <*>
1958 V7 N3 P80-86 60-18023 <*>

AGRESSOLOGIE
1961 V2 P26-29 65-10989 <*> O
1961 V2 P575-580 65-10963 <*>
1961 V2 N6 P605-607 62-16906 <*>
1962 V3 N4 P653-657 65-10962 <*>
1964 V5 P279-288 66-13024 <*>
1965 V6 P505-515 66-10988 <*>

AGRICOLTURA ITALIANA
1960 V15 N9 P287-313 55Q731 <ATS>

AGRICOLTURA DELLE VENEZIE
1957 V11 P594-608 63-16163 <*> O

AGRICULTOR VENEZOLANO
1945 V10 N111 P20-22 3406 <*>
1945 V10 N111 P43 3406 <*>

AGRICULTURA. COMERCIO E INDUSTRIAS
1942 V3 N6 P17-22 3352 <*>

AGRICULTURA SOCIALISTA
1963 V1 N36 P8 63-41094 <=>
1965 V3 N9 P8-9 65-30784 <=$>
1966 V4 N8 P9 66-31048 <=$>

AGRICULTURA TECNICA EN MEXICO
1955 V1 P7-15 59-17250 <*>

AGRICULTURAL MAGAZINE OF JAPAN
SEE NOGYO SEKAI

AGRICULTURAL TECHNOLOGY. CHINESE PEOPLES REPUBLIC
SEE NUNG YEH CHI SHU

AGRICULTURAL TECHNOLOGY. JAPAN
SEE NOGYO GIJUTSU

AGRICULTURE. PARIS
1963 V26 N257 P191-195 65-14145 <*> O

AGRICULTURE IN HOKKAIDO
SEE HOKUNO

AGRICULTURE AND HORTICULTURE. JAPAN
SEE NOGYO OYOBI ENGEI

AGROBIOLOGIYA
1948 N4 P48-56 RT-2728 <*> O
1949 N1 P131-147 60-21871 <=>
1949 N1 P148-160 60-21874 <=>
1949 N5 P113-126 RT-2434 <*>
1949 N6 P141-146 RT-2764 <*>
1950 N4 P133-137 60-21873 <=>
1951 N1 P140-145 RT-4026 <*>
1952 N2 P88-108 RT-2796 <*>
1952 N4 P12-25 AEC-TR-5237 <=*>
1952 N4 P170-172 53/0822 <NLL>
1952 N5 P80-90 RT-2765 <*>
1952 N6 P123-130 60-21886 <=>
1953 N4 P89-99 RT-1669 <*>
1954 N1 P60-67 <INSD>
1954 N4 P45-64 R-5279 <*>
1954 N4 P81-85 R-4387 <*>
1954 V6 P36-45 CSIRO-3381 <CSIR>
R-4587 <*>
1955 N1 P100-106 RT-4634 <*>
1955 N2 P130-134 RT-3915 <*>
1955 N3 P93-105 RT-4633 <*>
1955 N6 P146-147 RT-4233 <*>
1955 V4 P215-219 C-2323 <NRC>
1955 V4 P241-247 R-114 <*>
1956 N1 P18-27 RT-4141 <*>
1956 N1 P38-39 RT-4156 <*>
1956 N3 P114-116 C-2648 <NRC>
1956 N4 P132-135 60-21902 <=>
1957 N3 P118-122 60-16665 <+*>
1958 N3 P3-18 59-16794 <+*>
1958 N5 P141-156 66-61885 <=$>
1959 P115-141 60-16523 <+*>
1959 N1 P79-85 60-16685 <+*>
1959 N2 P171-176 61-19375 <+*>
1959 N2 P252-257 66-51086 <=>
1960 N3 P402-405 63-11038 <=>
1960 N4 P624-634 61-15855 <+*>
1960 N5 P713-734 65-50045 <=>
1960 N5 P766-768 65P62R <ATS>
1960 N5 P778-779 63-27821 <$>
1960 N6 P903-910 63-23125 <=*$> O
1961 N2 P259-269 61-31605 <=>
1961 N3 P323-356 61-27368 <*=>
1961 N4 P493-499 63-15386 <=*>
1961 N6 P854-869 64-14432 <=*$>
1962 N1 P148-151 64-31305 <=>
1962 N1 P154-155 63-19973 <=*>
1962 N2 P258-267 66-51080 <=>
1962 N4 P610-616 63-26958 <=$>
1962 N5 P756-768 64-15414 <=*$>
1962 N6 P861-865 64-19127 <=*$>
1963 N6 P876-879 RTS-2796 <NLL>
1963 N6 P903-919 64-31479 <=>
1963 N6 P943-949 RTS-2797 <NLL>
1965 N4 P633-636 65-32324 <=*$>

AGROCHIMICA
1959 V3 N2 P148-164 61-20432 <*>

AGROKEMIA ES TALAJTAN
1955 V4 N3 P251-264 C-2711 <NRC>
1955 V4 N3 P251-263 CSIRO-3027 <CSIR>
1957 V6 N1 P65-68 66-25097 <*$> O
1961 V10 N2 P285-290 66-10252 <*>

AGROKHIMIYA
1964 N1 P3-4 64-31684 <=>
1964 N1 P145-148 64-31684 <=>
1964 N2 P22-31 64-41117 <=>
1964 N2 P42-52 64-41117 <=>
1964 N2 P56-63 64-41117 <=>
1964 N3 P96-101 65-30947 <=$>
1964 N5 P3-16 64-51988 <=$>
1964 N5 P83-90 65-30432 <=$>
1964 N5 P130-136 65-30093 <=$>
1964 N5 P165-169 64-51852 <=$>
1964 N8 P3-17 65-30261 <*=$>
1964 N8 P28-34 65-30261 <*=$>
1964 N8 P49-56 65-30261 <*=$>

1964 N8 P60-80 65-30261 <*=$>
1964 N8 P114-120 65-30261 <*=$>
1964 N8 P145-154 65-30261 <*=$>
1964 N9 P63-74 65-30590 <=*$>
1964 N9 P98-105 65-30590 <=*$>
1964 N10 P3-16 65-30381 <=*$>
1964 N10 P73-81 RTS-3249 <NLL>
1965 N1 P102-104 65-31011 <=$>

AGRONOMIA. LA MOLINA
1943 V8 N30 P5-10 3365 <*>
1955 V21 P74-87 60-18521 <*>

AGRONOMIE TROPICALE. NOGENT-SUR-MARNE
1957 V12 N1 P67-113 CSIRO-3532 <CSIR>

AGRONOMSKI GLASNIK. ZAGREB
1964 V14 N6 P417-429 64-51350 <=>
1966 N10 P20-29 66-35301 <=>

AGROTECNIA. HAVANA
1955 V10 N1 P46-50 63-00163 <*>

AGROTEKHNIKA
1956 V3 P139-146 60-21246 <=>

AIR. PARIS
1958 V39 N735 P24-25 59-11906 <=*> 0
1959 V40 N744 P17-20 59-21079 <=*> 0

AIR CLEANING. JAPAN
SEE KUKI SEIJO

AIR ET COSMOS
1965 N107 P1 65-32768 <=$>
1965 N107 P16-33,45 65-32768 <=$>
1965 N114 P14-15,46 65-32293 <=>

AIR REVUE
1957 P546-547 59-21024 <=> 0
1958 P383 59-21030 <=> 0

AIR AND WATER POLLUTION; AN INTERNATIONAL JOURNAL
1961 N1/2 P1-23 63-01725 <*>
1962 V6 P27-48 64-00040 <*>
1962 V6 P277-282 3004 <BISI>

AKTUALNYYE VOPROSY PERELIVANIYA KROVI
1959 N7 P105-109 61-19897 <=*>
1959 N7 P109-114 61-19899 <=*>
1959 N7 P115-118 61-23061 <=*>
1959 N7 P118-121 61-19901 <=*>
1959 N7 P122-128 61-19900 <=*>
1959 N7 P128-134 61-19898 <=*>

AKUSHERSTVO I GINEKOLOGIYA
1937 V2 N7 P29-33 63-19702 <=*$>
1956 V32 N1 P75-76 63-19686 <=*>
1956 V32 N2 P3-5 62-15342 <=*> 0
1958 V34 N4 P49-53 61-15811 <+*>
1958 V34 N4 P98-99 64-23536 <=$>
1958 V34 N6 P16-20 61-15122 <=>
1959 V35 N1 P24-26 61-19173 <+*>
1959 V35 N1 P93-94 61-15121 <*+>
1959 V35 N3 P59-63 61-15812 <+*>
1959 V35 N4 P48-53 64-13180 <=*$>
1960 V36 N2 P124-125 63-26232 <=*>
1961 V37 N2 P78-80 C-4162 <NRC>
1962 N1 P51- FPTS V22 N2 P.T376 <FASE>
1962 V38 N1 P101-102 63-15985 <=*>
1962 V38 N1 P106-107 63-15984 <=*>
1962 V38 N6 P110 65-13827 <*>
1963 V39 N4 P132 65-60993 <=$> 0
1964 V40 N1 P86-91 65-61082 <=$>
1964 V40 N1 P91-95 65-61083 <=$>
1964 V40 N5 P22-28 65-62833 <=$>
1964 V40 N5 P28-35 65-62832 <=$>
1964 V40 N5 P122-126 65-63582 <=$>

★AKUSTICHESKII ZHURNAL
1955 V1 P40-47 R-3787 <*>
1955 V1 N1 P23-30 R-3817 <*>
RT-4125 <*>
62-23013 <=*>
1955 V1 N1 P40-47 R-1010 <*>
1955 V1 N2 P138-143 62-11298 <=>
1955 V1 N2 P165-170 62-11634 <=>
1955 V1 N2 P171-180 62-15895 <=*>
1955 V1 N2 P182-190 R-4075 <*>
3777 <HB>
1955 V1 N4 P326-338 R-2071 <*>
62-15860 <=*>
1955 V1 N4 P326-337 62-25344 <=*>
1955 V1 N4 P348-351 62-11616 <=>
1956 V2 N3 P255-266 CSIRO-3608 <CSIR>
1957 V3 N3 P282-285 R-2378 <*>
63-15205 <=*$>
1958 V4 N2 P125-127 59-15796 <+*>
1958 V4 N3 P239-243 59-15779 <+*>
1958 V4 N3 P292-294 59-15780 <+*>
1959 V5 N1 P14-20 60-13892 <+*>
1959 V5 N1 P51-57 60-13814 <+*>
1959 V5 N1 P64-69 60-13662 <+*>
1959 V5 N2 P146-150 60-17190 <+*>
1959 V5 N2 P196-201 60-17191 <+*>
1959 V5 N2 P206-211 60-17192 <+*>
1959 V5 N2 P255-256 60-11184 <=>
1959 V5 N3 P376-378 61-13551 <*+>
1959 V5 N3 P378-379 61-13550 <+*>
1959 V5 N4 P403-407 61-13333 <+*>
1959 V5 N4 P445-449 61-13333 <+*>
1959 V5 N4 P498-501 61-13512 <*+>
1959 V5 N4 P504-505 60-11624 <=>
60-31026 <=>
1960 V6 N1 P139-141 60-11766 <=>
1960 V6 N2 P205-212 61-28152 <*=>
1960 V6 N2 P263-264 60-31786 <=>
1960 V6 N3 P292-298 61-23512 <=*>
1960 V6 N3 P311-320 61-27616 <*=>
1960 V6 N3 P326-334 61-27614 <=*>
1960 V6 N3 P370-373 61-27615 <*=>
1960 V6 N4 P462-467 61-23863 <*=> 0
1960 V6 N4 P507-508 5430 <HB>
1961 V7 N1 P18-20 62-15690 <*=>
1961 V7 N1 P87-89 62-15698 <*=>
1961 V7 N1 P106-108 62-15690 <*=>
1961 V7 N1 P109-114 61-21603 <=>
1961 V7 N2 P165-173 62-25639 <=*>
1961 V7 N2 P236-241 62-24876 <=*>
1961 V7 N4 P415-420 62-25951 <=*>
1961 V7 N4 P421-427 63-15171 <=*>
1961 V7 N4 P428-435 62-25914 <=*>
1962 V8 N1 P34-41 63-15224 <=*>
1962 V8 N1 P42-48 63-23356 <=*>
1962 V8 N1 P49-55 64-16308 <=*$>
1962 V8 N1 P124-128 63-15223 <=*>
1962 V8 N3 P367-369 63-23359 <=*$>
1962 V8 N4 P447-453 63-21564 <=>
1962 V8 N4 P492-493 63-21564 <=>
1965 V11 N2 P251-252 66-12509 <*>
1965 V11 N4 P417-426 66-30548 <=*$>

AKUSTISCHE BEIHEFTE
1952 N3 P149-170 C-2417 <NRC>
1952 N4 P208-218 57-2408 <*>
1953 N2 P274-278 57-3374 <*>
1955 N1 P67-74 58-539 <*>
1956 V6 N2 P489-493 94J19G <ATS>
1956 V6 N2 P494-511 95J19G <ATS>
1958 N1 P273-280 90N52G <ATS>

AKUSTISCHE ZEITSCHRIFT
1937 P153-168 58-980 <*>
60-00791 <*>
1937 V2 P135-147 57-2151 <*>
1937 V2 P273-295 59-20070 <*> 0
1937 V2 N5 P217-224 57-361 <*>
1938 V3 N1 P21-31 57-360 <*>
1938 V3 N4 P169-175 66-10781 <*>
98S88G <ATS>

1938 V3 N5 P250-258 57-725 <*>
1939 V4 N4 P238-252 57-322 <*>
1940 V5 N1 P7-10 57-3 <*>
1940 V5 N3 P145-147 57-302 <*>
1941 V6 N1 P1-16 59-17243 <*>
1941 V6 N5 P265-278 57-324 <*>
1942 V7 N3 P81-104 C-2395 <NRC>
1942 V7 N5/6 P173-186 57-363 <*>
1943 V8 P49-63 58-985 <*>
1943 V8 P91-119 64-16573 <*>
1943 V8 P145-168 AEC-TR-3638 <+*> 0
1943 V8 N2 P66-76 57-362 <*>

ALBRECHT V. GRAEFES ARCHIV FUER OPHTHALMOLOGIE
1860 V7 N2 P58-71 II-871 <*>
1868 V14 N2 P228-246 61-14348 <*>
1873 V19 P156-162 61-14741 <*>
1876 V22 P1046-1067 61-14355 <*>
1889 V35 P157-171 61-14226 <*> 0
1889 V35 N1 P137-146 61-16028 <*>
1890 V36 P193-216 61-10979 <*>
1891 V37 P55-85 61-14082 <*>
1891 V37 P97-136 61-14183 <*>
1892 V38 P110-117 61-10937 <*>
1892 V38 P169-183 61-14230 <*>
1892 V38 P184-190 61-14232 <*> 0
1892 V38 N2 P204-220 61-14172 <*>
1895 V41 P135-157 61-14193 <*>
1895 V41 P283-292 61-14225 <*> 0
1896 V42 P134-178 61-14738 <*>
1896 V42 P249-256 61-14228 <*>
1896 V42 P316 61-14175 <*>
1897 V44 P87-104 PT.1 61-14856 <*>
1898 V45 P90-96 61-14262 <*>
1898 V45 P357-373 61-14242 <*>
1898 V46 P621-629 61-10967 <*>
1900 V51 P146-173 61-14212 <*>
1900 V51 P453-460 61-14115 <*>
1901 V52 P294-301 61-16023 <*>
1901 V52 P387-401 61-14990 <*>
1902 V53 P401-422 61-10933 <*>
1902 V54 P201-210 61-14439 <*>
1902 V54 P411-429 61-14159 <*>
1902 V54 P430-435 61-14434 <*>
1904 V57 P1-23 61-14404 <*>
61-14640 <*>
1904 V59 P459-471 61-14438 <*>
1904 V59 P581-586 61-14440 <*>
1906 V62 P400-463 61-14123 <*> 0
1907 V66 P477-496 61-14244 <*>
1910 V75 P561-585 61-14387 <*>
1910 V76 P1-97 61-14785 <*>
1912 V80 P500-513 61-14294 <*>
1919 V99 P174-194 61-14609 <*>
1919 V101 N2/3 P145-164 61-14461 <*>
1920 V103 P262-269 61-14251 <*>
1921 V105 P799-843 61-14618 <*>
1921 V105 P844-850 61-10946 <*>
1921 V105 P964-972 61-14417 <*>
1921 V105 P1091-1108 61-14078 <*>
1921 V106 P152-156 61-10934 <*>
1922 V108 P398-400 61-10957 <*>
1922 V110 P128-152 61-14622 <*>
1924 V114 N3/4 P441-464 61-14176 <*>
1926 V117 P511-537 61-14277 <*>
1927 V118 P292-310 61-14268 <*>
1927 V118 P633-680 61-14593 <*>
1928 V119 P711-718 61-14797 <*>
1928 V121 P163-165 61-14999 <*>
1928 V121 P756-780 61-14085 <*>
1929 V121 P740-755 58-406 <*>
1929 V122 P562-571 61-14354 <*>
1929 V122 N1 P59-74 61-14222 <*>
1930 V125 P493-553 61-14117 <*> 0
1932 V128 N3 P460-471 60-10507 <*> 0
1933 V129 N3 P413-425 60-10506 <*> 0
1934 V131 P452-457 61-14266 <*>
1934 V131 N1 P25-31 60-10505 <*>
1934 V133 P121-130 61-14202 <*>
1934 V133 P231-253 61-14206 <*>
1936 V136 P377-386 61-14139 <*>
1937 V137 P506-509 61-14620 <*>
1939 V140 P86-115 61-14654 <*>
1939 V140 P553-560 61-14436 <*>
1940 V142 P367-388 58-984 <*>
1944 V146 P110-127 1324-A <K-H>
1949 V149 P407-412 57-1934 <*>
1953 V153 P459-470 II-332 <*>
1954 V155 P457-484 II-1002 <*>
1961 V163 P397-402 UCRL TRANS-880 <*>

ALGEBRA I LOGIKA, SEMINAR. AKADEMIYA NAUK SSSR. SIBIRSKOE OTDELENIE. SEMINAR ALGEBRY I LOGIKI
1963 V2 N4 P47-66 AMST S2 V59 P56-72 <AMS>
1964 V3 N1 P5-39 AMST S2 V46 P165-192 <AMS>

ALGERIE MEDICALE
1950 V54 N1 P42 2203D <K-H>
1952 V56 N9 P569-574 II-1012 <*>
1956 V60 P651-659 57-2712 <*>
1957 V61 N12 P1297-1298 63-16938 <*>
8619-B <KH>

ALLAM- ES JOGTUDOMANYI INTEZET TUDOMANYOS KONYVTARA. (TITLE VARIES)
1965 V8 N3 P389-407 66-32342 <=$>

ALLATORVOSI KOZLONY
1933 V30 P7-10 61-20444 <*>

ALLERGIE
1956 V5 N3 P17-22 58-116 <*>

ALLERGIE UND ASTHMA
1955 V1 P31-40 57-1960 <*>
1960 V6 N3 P145-149 10573-B <KH>
63-16070 <*>
1961 V7 P271-278 62-18177 <*>
1961 V7 N6 P345-352 62-01511 <*>
1961 V7 N4/5 P264-271 62-14348 <*>

ALLGEMEINE BAUZEITUNG
1954 V9 N420 P10-14 T-1738 <INSD>
1960 V15 N623 P1-6 62-18087 <*>

ALLGEMEINE FISCHEREIZEITUNG
1949 V74 N4 P49-50 3105 <*>

ALLGEMEINE FORSTZEITSCHRIFT
1956 V11 P297-304 C-2674 <NRC>
1961 V16 N31 P448-449 63-10527 <*>
1964 V19 N29/0 P433-439 C-5950 <NRC> 0
1964 V19 N29/0 P442-443,446
C-5950 <NRC> 0
1964 V19 N29/0 P448-449 C-5950 <NRC> 0
1964 V19 N29/0 P452 C-5950 <NRC> 0
1964 V19 N29/0 P456 C-5950 <NRC> 0
1965 V20 N36 P561-565 C-6004 <NRC>

ALLGEMEINE FORST-U. JAGDZEITUNG
1954 V126 N6 P189-190 C-2705 <NRC>

ALLGEMEINE OEL-UND FETTZEITUNG
1914 V11 P590-603 58-956 <*>
1931 V28 N1 P3-7 60-10521 <*> P
1932 V29 N11 P597-598 60-10523 <*>
1935 V32 P57-81 60-10522 <*>
1941 V38 P360-361 60-10524 <*>

ALLGEMEINE PAPIERRUNDSCHAU
1951 P864 57-3510 <*>
1954 N7 P280-290 57-832 <*>
1954 N12 P625-627 2446 <*>
1954 N13 P670-674 2446 <*>
1954 N14 P714-715 2446 <*>
1954 N21 P1040-1042 2468 <*>
1956 P198-206 57-3285 <*>
1956 N12 P588- 57-1715 <*>
1956 N17 P922 58-762 <*>
1957 N4 P154-156 63-14496 <*>

1957 N21 P1077-1081 58-2167 <*>
1957 N23 P1188-1189 62-18803 <*>
1958 N4 P152-153 63-18348 <*>
1958 V6 N12 P617-620 59-10799 <*> 0
1958 V6 N13 P674-678 59-10799 <*> 0
1958 V6 N14 P722-726 59-10799 <*> 0
1958 V6 N19 P978-979 59-10295 <*>
1958 V6 N22 P1126-1129 59-15970 <*> 0
1959 N12 P613-614 61-20885 <*>
9808 <K-H>
1959 V7 N12 P613-614 60-18924 <*>
1960 N8 P443-447 61-20193 <*> 0
1960 N10 P550-551 62-16589 <*>
1960 V8 N10 P550-551 61-20012 <*>
1962 N3 P134 62-16499 <*> 0
1962 N6 P283-285 62-18885 <*>

ALLGEMEINE WAERMETECHNIK
1951 V2 N6/7 P129-142 NACA TM 1408 <NASA>
1952 V3 N8/9 P167-174 BGIRA-554 <BGIR>
1956 P104-108 1381 <BISI>
1956 V7 N10 P211-219 1898 <BISI>
1956 V7 N5/6 P104-108 62-10471 <*>
1958 V9 N9 P185-189 1899 <BISI>
1961 V10 N11 P205-212 63-20707 <*>

ALLUMINIO (NUOVA METALLURGIA; AND NUOVA METALLURGIA. TITLE VARIES)
1936 V5 P37-45 59-17039 <*> OP
1946 V15 P4-8 I-214 <*>
1947 V16 N7/8 P293-297 I-980 <*>
1949 V18 P147-152 60-14480 <*>
1950 V19 P133-145 63-20458 <*>
1950 V19 P215-224 58-2002 <*>
63-20451 <*>
1950 V19 P541-547 63-20456 <*>
1950 V19 N1 P19-26 6133 <HB>
1950 V19 N2 P133-145 I-781 <*>
1951 V20 N1 P29-38 I-780 <*>
1952 V21 P252-256 I-590 <*>
1952 V21 N2 P135-140 I-1013 <*>
1952 V21 N6 P573-593 64-20297 <*>
1953 V22 P635-643 I-765 <*>
1953 V22 P645-652 I-513 <*>
1953 V22 P655-662 1422 <*>
1953 V22 P663-671 1500 <*>
1953 V22 P672-677 I-487 <*>
1953 V22 P672-678 II-377 <*>
1953 V22 P679-689 I-1007 <*>
1953 V22 P699-707 I-513 <*>
1953 V22 P709-720 C-2109 <NRC>
1953 V22 P718-720 I-465 <*>
1953 V22 P721-729 C-2110 <NRC>
II-382 <*>
1953 V22 N5 P495-506 64-20830 <*> 0
1954 V23 P35-39 1292 <*>
1954 V23 P255-260 2304 <*>
1954 V23 P515-532 62-16108 <*>
1954 V23 P533-545 I-192 <*>
57-1439 <*>
66-12080 <*> 0
1954 V23 N1 P23-34 I-505 <*>
1954 V23 N4 P399-411 63-18714 <*>
1954 V23 N4 P473-479 57-713 <*>
1954 V23 N5 P503-513 57-1809 <*>
1955 V24 P121-127 66-12081 <*> 0
1955 V24 N4 P335-343 11H91 <ATS>
1955 V24 N5 P459-463 58-52 <*>
1955 V24 N6 P543-554 1348 <*>
1955 V24 N6 P555-560 61-16286 <*>
1956 V25 P79-84 II-1027 <*>
1956 V25 P127-134 C-2262 <NRC>
1514 <*>
1956 V25 P275-278 63-10282 <*> 0
1956 V25 P333-336 62-16091 <*>
1956 V25 N5 P225-230 58-1330 <*>
1956 V25 N9 P373-384 57-1173 <*>
1957 V26 N10 P422-425 02L35I <ATS>
1958 V27 P7-12 GAT-Z-5033 <*>
1958 V27 P55-61 66-12082 <*> 0
1958 V27 P321-329 66-12083 <*> 0
1958 V27 N7/8 P321-329 60-10109 <*>
1959 V28 N3 P115-120 61-10380 <*>
1960 V29 N1 P5-20 1867 <BISI>
1963 V32 N8 P397-407 3758 <BISI>
1963 V32 N12 P579-588 11S86I <ATS>
1964 V33 N2 P63-70 1008 <TC>
1964 V33 N4 P173-184 4031 <BISI>
1964 V33 N12 P613-625 1009 <TC>
1965 V34 P331-349 1417 <TC>

ALTA FREQUENZA
1932 V1 N4 P485-499 62-23072 <*>
1933 V2 N4 P500-515 62-20415 <*>
1939 V8 N8/9 P560-564 I-162 <*>
1941 V10 P70-86 57-1001 <*>
1941 V10 P97-98 57-1001 <*>
1941 V10 N8/9 P470-515 57-757 <*>
1942 V11 N11/2 P452-556 57-759 <*>
1949 V18 N2 P73-82 65-12381 <*>
1952 V21 N2 P102-104 57-2765 <*>
1954 V23 N3/4 P115-138 57-2831 <*>
1954 V23 N3/4 P157-177 24N48I <ATS>
61-14918 <*>
1955 V24 N3 P238-245 57-2743 <*>
1955 V24 N4/5 P309-338 57-3118 <*>
1957 V26 N1 P41-89 18K22I <ATS>
1957 V26 N6 P580-602 58-2633 <*>

ALUMINIUM. BUDAPEST
1951 V3 N2 P25-28 2238 <K-H>
63-20432 <*>

ALUMINIUM. ALUMINIUM ZENTRALE. BERLIN
1934 V17 P37-39 II-624 <*>
1934 V17 P191-195 66-12084 <*> 0
1935 V17 P3-6 58-2103 <*>
1935 V17 P559 58-2025 <*>
1936 V17 P562 58-2023 <*>
1936 V18 P305-306 66-12085 <*> 0
1937 V19 P302-305 66-12086 <*> 0
1937 V19 P381-384 66-12087 <*> 0
1937 V19 P699- I-60 <*>
1937 V19 N8 P502-503 60-18745 <*>
1938 P69 58-2007 <*>
1938 P71 58-2024 <*>
1938 P73 58-2065 <*>
1938 P75 58-2058 <*>
1938 P85-94 58-2038 <*>
1938 N3 P168-181 58-1999 <*>
1938 N6 P389-394 58-2048 <*>
1938 V20 N2 P109-117 59-20829 <*> 0
1938 V20 N5 P315-320 60-14356 <*>
1938 V20 N6 P365-378 58-1967 <*>
1938 V20 N12 P839-843 I-753 <*>
1939 V21 P183- I-506 <*>
1939 V21 P192- I-873 <*>
1939 V21 P210-213 I-565 <*>
1939 V21 N2 P135-136 I-882 <*>
1939 V21 N3 P192-201 58-1892 <*>
1939 V21 N7 P521-528 62-18435 <*> 0
1939 V21 N9 P637-639 58-1065 <*>
1940 V22 P9-12 I-890 <*>
1940 V22 P343- I-19 <*>
1940 V22 P344- I-18 <*>
1940 V22 N5 P266-270 60-14413 <*> 0
1940 V22 N7 P341-343 61-20771 <*>
1940 V22 N9 P421-435 II-115 <*>
1940 V22 N10 P575-578 60-14353 <*>
1940 V22 N10 P578-579 60-14354 <*>
1941 P205-208 I-971 <*>
1941 V23 P121-130 66-12088 <*> 0
1941 V23 P239-246 58-894 <*>
1941 V23 P281-284 I-556 <*>
1941 V23 N3 P121-130 58-2006 <*>
1942 V24 P176-178 I-687 <*>
1942 V24 P178- I-195 <*>
1942 V24 P267- I-1004 <*>
1942 V24 P428-432 60-18074 <*>
1942 V24 N2 P69-71 62-18488 <*> 0

Citation	Translation
1942 V24 N4 P131-140	I-852 <*>
1942 V24 N6/7 P209-215	2298 <*>
1943 P97-100	I-969 <*>
1943 V25 P62-69	2323 <*>
1943 V25 P106-110	I-396 <*>
1943 V25 P110-112	I-608 <*>
1943 V25 P130-133	I-176 <*>
1943 V25 P253-256	I-175 <*>
1943 V25 P283-291	60-18825 <*>
1943 V25 P358-359	58-855 <*>
	63-16611 <*>
1943 V25 P359-360	58-848 <*>
1943 V25 P413-417	I-399 <*>
1943 V25 N3 P106-110	60-14570 <*>
1943 V25 N3 P116-126	58-487 <*>
1943 V25 N5 P194-202	64-10756 <*> 0
1943 V25 N6 P240-246	58-1965 <*>
1944 V26 P10-13	I-374 <*>
1944 V26 P76-82	58-2033 <*>
1944 V26 P190-194	58-2055 <*>
1944 V26 P230-240	58-2034 <*>
1944 V26 P240-241	58-2008 <*>
1944 V26 N1 P2-10	58-2049 <*>
1944 V26 N1 P13-17	58-1943 <*>
1944 V26 N1 P17-22	58-1942 <*>
1952 V28 P391-400	57-3252 <*>
1952 V28 N4 P146-150	2732 <*>
1952 V28 N10 P341-346	I-333 <*>
1952 V28 N7/8 P238-245	I-458 <*>
1953 V29 P51-58	I-655 <*>
1953 V29 P315-317	II-224 <*>
1953 V29 P497-508	II-237 <*>
1953 V29 N4 P164-165	62-18421 <*> 0
1953 V29 N5 P194-197	AL-884 <*>
1953 V29 N5 P198-200	AL-115 <*>
1953 V29 N5 P203-206	AL-585 <*>
1953 V29 N9 P361-370	1891 <*>
1953 V29 N11 P451-461	65-11396 <*>
1953 V29 N12 P497-508	I-384 <*>
1953 V29 N7/8 P305-307	II-223 <*>
1954 P163-164	2440 <*>
1954 V30 P95-98	2318 <*>
1954 V30 P279-283	II-296 <*>
1954 V30 N3 P98-100	II-346 <*>
1954 V30 N5 P195-199	1969 <*>
1954 V30 N7 P279-283	II-700 <*>
1954 V30 N12 P534-535	1961 <*>
1955 V31 P8-9	2726 <*>
1955 V31 P224-226	1955 <*>
1955 V31 N1 P10-14	59-17346 <*>
1955 V31 N2 P62-69	59-17346 <*>
1955 V31 N4 P151-156	2971 <*>
	62-16956 <*> 0
	64-16947 <*> 0
1955 V31 N6 P260-266	62-16957 <*> 0
1955 V31 N6 P275-281	62-18339 <*> 0
1955 V31 N10 P484-489	C-2111 <NRC>
	II-773 <*>
1955 V31 N11 P544-553	7K26G <ATS>
1955 V31 N7/8 P321-327	57-77 <*>
1955 V31 N7/8 P328-334	57-2546 <*>
1955 V31 N7/8 P338-341	62-16994 <*> 0
1955 V31 N7/8 P341-342	62-18228 <*> 0
1955 V31 N7/8 P344-348	62-16955 <*> 0
1956 V32 P350-356	58-1483 <*>
1956 V32 P545-549	66-12090 <*> 0
1956 V32 P610-616	T-1995 <INSD>
1956 V32 P617-621	T-1996 <INSD>
1956 V32 P633-636	1997 <INSD>
1956 V32 N1 P6-12	C-2829 <NRC>
	57-1421 <*>
	57-220 <*>
	66-10164 <*> 0
1956 V32 N2 P64-70	C-2829 <NRC>
	57-1421 <*>
1956 V32 N2 P64-71	65-17324 <*> 0
1956 V32 N3 P126-135	57-52 <*>
1956 V32 N3 P145-146	64-20465 <*>
1956 V32 N4 P208-209	66-10166 <*> 0
1956 V32 N6 P333-339	C-2255 <NRC>
	1811 <*>
1956 V32 N6 P333-336	57-211 <*>
1956 V32 N6 P333-339	66-12089 <*> 0
1956 V32 N6 P340-343	C2360 <NRC>
	57-842 <*>
1956 V32 N8 P476-479	58-47 <*>
1956 V32 N8 P496-497	57-135 <*>
1956 V32 N10 P610-616	57-3439 <*>
1957 N8 P514-520	57-3560 <*>
1957 V33 P234-240	57-3545 <*>
1957 V33 P514-520	62-19916 <=>
1957 V33 N1 P16-32	58-555 <*>
1957 V33 N2 P78-91	58-43 <*>
1957 V33 N2 P101-103	58-1504 <*>
1957 V33 N4 P263	64-16917 <*> 0
1957 V33 N5 P306-317	58-1329 <*>
1957 V33 N6 P372-376	57-3302 <*>
1957 V33 N9 P589-591	58-2081 <*>
1957 V33 N9 P606-612	58-252 <*>
1957 V33 N11 P721-723	58-2068 <*>
1957 V33 N11 P730-739	SCL-T-312 <*>
1958 V34 P128-130	HW-TR-20 <*>
1958 V34 P510-518	66-12091 <*> 0
1958 V34 N3 P120-130	59-15074 <*> 0
1958 V34 N3 P128-130	58-1614 <*>
1958 V34 N3 P130-134	58-1764 <*>
1958 V34 N5 P276-279	59-10366 <*>
1958 V34 N7 P410-416	59-10851 <*>
	59-17629 <*>
1958 V34 N9 P518-521	59-00883 <*>
1958 V34 N9 P530-537	65-17327 <*> 0
1958 V34 N11 P634-642	59-17408 <*>
1959 V35 N10 P560-563	60-16867 <*>
1960 V36 P457-461	AD-635 648 <=$>
1960 V36 P568-579	61-20494 <*> 0
1960 V36 N5 P265-267	63-10857 <*>
1960 V36 N5 P267-271	AD-635 649 <=$>
1960 V36 N6 P312-317	34M45G <ATS>
1960 V36 N6 P337-338	AD-635 646 <=$>
1960 V36 N6 P339	61-10547 <*> 0
1961 V37 P204-205	63-10166 <*> 0
1961 V37 P401-411	TRANS-172 <FRI>
1961 V37 P633-644	66-12093 <*> 0
1961 V37 N1 P19-28	AD-631 432 <=*$>
1961 V37 N3 P143-149	66-12092 <*> 0
1961 V37 N4 P215-221	65-10060 <*> 0
1961 V37 N5 P267-273	66-10626 <*> 0
1961 V37 N6 P351-352	62-16226 <*> 0
1961 V37 N7 P401-411	62-14836 <*> 0
	62-18307 <*>
1961 V37 N9 P569-575	62-18018 <*> 0
1961 V37 N10 P633-644	62-14833 <*> 0P
1962 N4 P207-221	AD-636 191 <=$>
1962 V38 P718-721	63-16472 <*> 0
1962 V38 N1 P13-19	62-18103 <*> 0
1962 V38 N3 P150-154	63-10145 <*> 0
1962 V38 N3 P161	63-10146 <*>
1962 V38 N5 P295-299	66-12094 <*> 0
1962 V38 N9 P590-593	66-12095 <*> 0
1963 V39 P33-41	AD-635 687 <=$>
1963 V39 N7 P424-428	AD-641 932 <=$>
1963 V39 N11 P675-682	66-12231 <*> 0
1963 V39 N12 P753-759	AD-440 100 <=>
1964 V40 N5 P290-294	90R77G <ATS>
1964 V40 N8 P504-506	65-11189 <*>
1965 V41 N4 P242-244	65-14643 <*>
1965 V41 N10 P629-633	66-12665 <*> 0
1966 V42 N2 P101-104	MFA/TS-825 <MFA>

ALUMINIUM SUISSE

Citation	Translation
1951 V1 P75-81	63-18624 <*> 0
1955 V5 N4 P134-138	1241 <*>
1956 V6 N3 P82-89	57-3312 <*>
1957 V7 N1 P23-28	58-2082 <*>
1960 V10 N3 P184-190	TRANS-124 <MT>
1962 V12 N1 P19-23	65-13973 <*>
1962 V12 N2 P75-83	62-18822 <*>
1963 V13 N2 P49-65	64-14595 <*>

ALUMINIUM-ARCHIV
1938 V12 ENTIRE ISSUE I-844 <*>
1938 V12 P24 58-1879 <*>

ALYUMINIEVYE SPLAVY
1963 V1 P139-149 AD-620 803 <=$>

AMERICAN JOURNAL OF OPHTHALMOLOGY
1930 S3 V13 N2 P113-117 61-14196 <*>

AMINO ACIDES, PEPTIDES, PROTEINES
1960 N4 P111-119 62-00455 <*>

AMTLICHE VEROEFFENTLICHUNG DES METEOROLOGISCHEN AMTS FUER NORDWEST-DEUTSCHLAND
1949 P1-70 II-1515 <*>
1782 <*>

ANAESTHESIST
1952 V1 N4 P107-110 59-17140 <*> 0
1953 V2 N5 P180-183 62-18952 <*>
1954 V3 P25-31 I-895 <*>
1954 V3 N2 P82-84 59-17851 <*> 0
1960 V9 N10 P315-318 62-10917 <*>
1961 V10 N8 P242-245 62-14327 <*>

ANAIS DA ACADEMIA BRASILEIRA DE CIENCIAS
1954 V26 N2 P17- 58-795 <*>
1954 V26 N3/4 P22-23 58-796 <*>
63-16570 <*>

ANAIS DA ASSOCIACAO BRASILEIRA DE QUIMICA
1961 V20 P49-53 65-12904 <*> 0

ANAIS DA ASSOCIACAO QUIMICA DO BRASIL
1944 V3 P152-158 99P64PT <ATS>
1948 V7 P196-199 RCT V23 N2 P457 <RCT>
1948 V7 N4 P204-207 RCT V23 N4 P972 <RCT>
1950 V9 P65-74 II-190 <*>

ANAIS DO INSTITUTO DE MEDICINA TROPICA
1956 V13 N1/2 P199-208 57-3505 <*>

ANAIS PAULISTAS DE MEDICINA E CIRURGIA
1942 V43 P409-413 57-1665 <*>
1942 V43 P416-423 57-1611 <*>
1960 V80 P294-295 61-20239 <*>
1960 V80 N6 P342 62-16080 <*>
1960 V80 N6 P344 62-16080 <*>
1961 V81 P146-148 61-20775 <*>

ANAIS DA SOCIEDADE DE BIOLOGIA DE PERNAMBUCO
1957 V15 N1 P93-125 59-12699 <+*> 0
1957 V15 N2 P269-281 60-13060 <+*> 0

ANALELE ROMINO SOVIETICE. SERIA MATEMATICA-FIZICA. ACADEMIA REPUBLICI POPULARE ROMINE. BUCHAREST
1959 S3 V13 N1 P5-33 AMST S2 V33 P291-321 <AMS>

ANALELE STIINTIFICE DE UNIVERSITATII AL. I. CUZA DIN IASI. SECT I. MATEMATICA, FIZICA, CHIMIE
1955 SNS V1 P231-242 AEC-TR-4108 <*>
1957 V3 N1/2 P281-286 81M38RU <ATS>
1958 V4 N1 P39-44 62-11507 <=>
1958 S1 V4 P199-202 66-14198 <*$>
1960 V6 P183-186 66-14197 <*$>
1960 V6 P585-598 64-15680 <=*$>

ANALELE UNIVERSITATII C. I. PARHON. SERIA STIINTELOR NATURII. BUCHAREST
1958 V7 N20 P37-47 STMSP V4 P87-97 <AMS>

ANALES DE LA ASOCIACION QUIMICA ARGENTINA
1929 V17 P95-141 63-10392 <*>
1936 V24 P141 63-10343 <*>
1936 V24 P142 63-10348 <*>
1936 V24 P147 63-10345 <*>
1938 V26 P249-253 63-10339 <*>
1939 V27 P64-73 59-17347 <*>
63-10347 <*>
1939 V27 P127-129 63-10119 <*>
1939 V27 P130-139 63-10349 <*>
1942 V30 P146-157 60-18438 <*>
1947 V35 P137-148 61-18810 <*>
1949 V37 P246-262 57-3532 <*>
59-15941 <*>
1955 V43 P215-225 58-742 <*>
63-16378 <*>

ANALES DE BROMATOLOGIA
1951 V3 P53-65 60-10515 <*> P
1954 V6 P473-557 T1825 <INSD>
1957 V9 P114-116 58-227 <*>
1958 V10 P183-193 58-1824 <*>

ANALES DE EDAFOLOGIA Y FISIOLOGIA VEGETAL
1951 V10 P595-602 I-495 <*>
1954 V13 P1-18 T1671 <INSD>
1959 V18 P389-405 62-01238 <*>

ANALES DE LA ESCUELA NACIONAL DE CIENCIAS BIOLOGICAS. MEXICO CITY
1946 V4 P149-164 914-D <K-H>
1947 V4 P391-403 1350-A <K-H>

ANALES ESPANOLES DE ODONTOESTOMATOLOGIA
1953 V12 P597-600 62-20212 <*>

ANALES DE LA ESTACION EXPERIMENTAL DE AULA DEI
1952 V2 N3/4 P270-276 59-17234 <*>

ANALES DE LA FACULTAD DE MEDICINA. UNIVERSIDAD DE MONTEVIDEO
1952 V37 P467-469 59-12130 <=*>
1956 V41 P53-58 62-10712 <*> 0

ANALES DE LA FACULTAD DE ODONTOLOGIA. UNIVERSIDAD DE LA REPUBLICA ORIENTAL DEL URUGUAY. MONTEVIDEO
1955 V1 139-157,PT.2 58-637 <*>
1957 V3 N5 P81-89 PT.1 65-00362 <*> OP

ANALES DE FARMACIA Y BIOQUIMICA. BUENOS AIRES
1930 V1 P8-19 66-13113 <*>
1941 V12 P126-141 3677-A <K-H>

ANALES DE FISICA Y QUIMICA
1945 V41 P101-110 I-787 <*>
1947 V43 P655-670 2240 <*>

ANALES HIDROGRAFICOS
1952 V11 P223-232 I-456 <*>

ANALES DEL INSTITUTO DE INVESTIGACIONES CIENTIFICAS Y TECNOLOGICAS. UNIVERSIDAD NACIONAL DEL LITORAL
1932 V1 P15-29 63-10361 <*>
1932 V1 P30-37 63-10366 <*>
1932 V1 P49-53 63-10134 <*>
1932 V1 P54-57 63-10138 <*>
1932 V1 P58-63 63-10137 <*>
1933 V2 P33-49 63-10135 <*>
1934 V3/4 P23-31 63-10136 <*>
1937 V7 P10-16 63-10133 <*>
1937 V7 P34- 63-10368 <*>
1937 V7 P37-38 63-10369 <*>
1943 P119-146 63-14404 <*>
1943 V12/3 P119-146 58-1267 <*>

ANALES DEL INSTITUTO DE INVESTIGACIONES VETERINARIAS. MADRID
1954 V6 P115-126 9016 <IICH>

ANALES DEL INSTITUTO DE NEUROLOGIA
1953 V10 P7-14 2215 <*>

ANALES DEL INSTITUTO DE ONCOLOGIA 'ANGEL H. ROFFO'
1956 V7 P27 65-00252 <*>
1956 V7 P29-38 65-00291 <*>
1956 V7 P49-80 65-00331 <*> 0

ANALES DE MECANICA Y ELECTRICIDAD
1959 S2 V36 N6 P452-463 2542 <BISI>

ANALES R. ACADEMIA DE FARMACIA. MADRID
1951 V17 P139-142 AEC-TR-4340 <*>
57-3046 <*>
1954 V20 N5 P429-438 3546-C <K-H>
1963 V29 P171-180 65-14543 <*>

ANALES DE LA R. SOCIEDAD ESPANOLA DE FISICA Y QUIMICA
1920 V18 P184-206 60-10526 <*> 0
1928 V26 P365-371 63-10381 <*>
1931 V29 P235-246 58-1532 <*>
1936 V34 P501-506 I-534 <*>
1943 V39 P743-750 61-18649 <*>
1945 V41 P138-161 442 <TC>
1945 V41 P414-423 60-10741 <*> 0

ANALES DE LA R. SOCIEDAD ESPANOLA DE FISICA Y QUIMICA. SERIE A. FISICA
1952 V48A P133-134 II-187 <*>
1953 V49A P77-86 II-176 <*>
1953 V49A P311-314 II-177 <*>
1954 V50A N11/2 P215-220 90J16SP <ATS>
1959 V55A N11/2 P283-294 66-13589 <*>
1961 V57 P29-38 AEC-TR-6334 <*>

ANALES DE LA R. SOCIEDAD ESPANOLA DE FISICA Y QUIMICA. SERIE B. QUIMICA
1948 V44B N5 P571-576 59-17327 <*>
1949 V45B P381-391 62-10461 <*>
1949 V45B N9/10 P1151-1170
65-13293 <*>
1950 V46B P225-232 58-1227 <*>
1951 V47B P229-234 I-391 <*>
1952 V48B P89-98 61-10894 <*> 0
1952 V48B P131-139 63-18980 <*>
1952 V48B P825-840 II-239 <*>
1952 V48B P841-850 II-240 <*>
1952 V48B N2 P179-182 63-18979 <*> 0
1953 V49B P539-546 60-10525 <*>
1953 V49B N9/10 P557-564 3710-C <K-H>
1954 V50B P413-420 3236-A <K-H>
1955 V51B P101-116 CSIRO-3123 <CSIR>
1955 V51B N2 P173-186 2749 <*>
1955 V51B N3 P247-260 2750 <*>
1956 V52B P237-250 NP-TR-775 <*>
1956 V52B P601-616 58-516 <*>
1956 V52B N1 P51-62 5258-B <K-H>
1956 V52B N5 P291-294 59-10498 <*>
1958 V54B P677-688 44Q68SP <ATS>
1960 V56B P231-252 358 <TC>
1960 V56B P253-262 359 <TC>
1962 V58B P295-314 360 <TC>
1962 V58B P431-436 63-18429 <*>
1963 V59B N6 P401-414 65-11319 <*> 0

ANALES DE LA SOCIEDAD DI BIOLOGIA DE BOGOTA
1954 V6 N3 P109-126 64-00277 <*>

ANALES DE LA SOCIEDAD CIENTIFICA ARGENTINA. BUENOS AIRES
1934 V118 P227-241 63-10342 <*>
1935 V120 P98-103 63-10344 <*>
1936 V122 P339-348 63-10338 <*>
1937 V124 P229-234 63-10351 <*>
1937 V124 P234-240 63-10388 <*>
1937 V124 P240-247 63-10367 <*> 0

ANALES DE LA SOCIEDAD CIENTIFICA ARGENTINA. SECCION SANTA FE
1934 V6 P17-19 63-10107 <*>
1934 V6 P20-22 63-10106 <*>

ANALES DE LA SOCIEDAD MEXICANA DE OFTALMOLOGIA Y OTO-RINO-LARINGOLOGIA
1949 V23 P25-31 65-17295 <*>

ANALES DE LA SOCIEDAD QUIMICA ARGENTINA
1919 V7 P121-123 5448-C <K-H>
1920 V8 P105-117 63-10486 <*> 0

ANALES DE LA UNIVERSIDAD DE SANTO DOMINGO
1947 V12 P171-180 61-00639 <*>

ANALYTICA CHIMICA ACTA. AMSTERDAM, NEW YORK
1947 V1 P254-259 61-14000 <*>
1947 V1 P408-416 63-18576 <*>
1949 V3 P422-427 II-66 <*>
1950 V4 P247-255 63-16908 <*>
1950 V4 P309-315 63-18251 <*>
1950 V4 P328-331 II-67 <*>
1950 V4 N3 P242- I-992 <*>
1950 V4 N3 P242-244 62-16987 <*> 0
1951 V5 P317-320 63-14199 <*> 0
1951 V5 P368-374 64-30063 <*>
1951 V5 N4 P375-379 RCT V25 N2 P371 <RCT>
1951 V5 N5 P463-471 60-18116 <*>
1953 V8 P57-65 II-575 <*>
1953 V8 P65-87 II-594 <*>
1953 V8 P229-234 AL-678 <*>
64-10127 <*>
1953 V8 N1 P22-28 64-18347 <*>
1953 V8 N6 P510-525 59-10137 <*>
1954 V10 N3 P265-272 59-10551 <*>
1955 V13 P396-403 62-16715 <*>
1956 V14 P430-438 1085 <*>
57-3451 <*>
1956 V14 N5 P430-438 CSIRO-3385 <CSIR>
64-16470 <*>
1957 V16 P592-596 62-16068 <*>
1957 V16 N6 P530-540 AEC-TR-3483 <*>
1957 V17 P234-246 AEC-TR-6346 <*>
500 <TC>
1957 V17 N6 P530-534 61-16437 <*>
1958 V18 P404-411 63-10254 <*>
1958 V18 N4 P282-290 64-2110 <*>
1958 V19 P542-547 86N48F <ATS>
1958 V19 N5 P375-382 95N48R <ATS>
1960 V23 P523-530 4037 <BISI>
1962 V27 P36-39 HW-TR-45 <*>
1962 V27 N4 P402-404 66-13252 <*>
1963 V29 N4 P350-357 65-13450 <*>
65-18126 <*>
1964 V30 N3 P273-279 65-13251 <*>
1964 V31 P24-32 487 <TC>
1965 V32 P355-360 66-11374 <*>

ANATOMISCHER ANZEIGER
1891 V6 P219-220 1892 <*>
1908 V8 P818-822 T-2522 <INSD>
1909 V35 N11/2 P257-280 57-2348 <*>
1928 V64 P383-395 57-2349 <*>
1951 V98 N18/0 P313-316 57-666 <*>
1954 V100 P18-20 65-00153 <*> 0
1954 V100 P258-264 65-00153 <*> 0
1956 V102 N22/4 P434-442 1288 <*>
1957 V103 P162-171 SUP 59-19051 <+*> 0

ANESTHESIE ET ANALGESIE
1951 V8 N4 P663-672 64-14730 <*>
1952 V9 N2 P232-239 64-14731 <*> P
1952 V9 N2 P240-249 64-14732 <*>

ANESTHESIE, ANALGESIE, REANIMATION
1957 V14 N1 P164-166 58-622 <*>
1958 V15 N3 P480-494 62-14589 <*> 0
62-16219 <*> 0
1959 V16 P797-807 G-270 <RIS>
1959 V16 N2 P401-407 64-10192 <*> 0
1959 V16 N5 P942-951 G-237 <RIS>
1959 V16 N5 P966-978 G-238 <RIS>
1962 V19 N4 P813-825 64-18680 <*>

ANGEWANDTE BOTANIK
1934 V16 N4 P363-372 59-10970 <*>
1937 V19 P172-181 I-979 <*>
1952 V26 N1 P1-13 57-558 <*>
1954 V28 P192-203 64-20846 <*> 0

Citation	Translation
1957 V31 P93-105	59-20875 <*>
1960 V34 P134-157	63-10778 <*>
1962 V36 P203-257	66-10228 <*>

★ANGEWANDTE CHEMIE

Citation	Translation
1932 V45 P108-109	57-591 <*>
1932 V45 P245-249	66-12098 <*> 0
1932 V45 P429-431	I-845 <*>
1932 V45 N33 P537-538	63-10468 <*>
1933 V46 P106-109	66-12099 <*> 0
1933 V46 P473-477	66-12100 <*> 0
1933 V46 N5 P90-91	63-10456 <*>
1933 V46 N38 P596-610	63-16225 <*>
1934 V47 P11-13	66-13127 <*>
1934 V47 P64-	II-399 <*>
1934 V47 P149-151	II-165 <*>
1934 V47 N5 P71-74	27L31G <ATS>
1934 V47 N15 P227-228	64-20873 <*>
1934 V47 N20 P315-318	64-14733 <*> 0
1935 V48 P261-263	63-18459 <*> 0
1935 V48 P394-396	63-20798 <*>
1935 V48 P557-561	66-13049 <*>
1935 V48 P642-644	66-13050 <*>
1935 V48 N26 P394-396	66-13128 <*>
1935 V48 N37 P593-595	64-20872 <*> P
1935 V48 N44 P689-691	59-20451 <*> 0
1936 V49 N8 P145-153	64-20869 <*> 0
1936 V49 N28 P455-460	64-14708 <*>
1936 V49 N28 P463-464	65-14500 <*>
1936 V49 N30 P499-502	64-14901 <*>
1936 V49 N31 P553	63-10495 <*>
1936 V49 N46 P822-826	59-20301 <*>
1936 V49 N52 P926-928	AL-777 <*>
1937 P891-895	I-278 <*>
1937 V50 P127-132	AL-820 <*>
1937 V50 P193-204	ACE-TR-3709 <+*>
1937 V50 P593-595	66-13051 <*>
1937 V50 P831-836	57-375 <*>
1937 V50 N4 P101-104	1629 <*>
1937 V50 N15 P278-279	AL-653 <*>
1937 V50 N37 P354-355	59-20401 <*> 0
1938 V51 P243-251	63-20049 <*>
1938 V51 P432	66-13052 <*>
1938 V51 P433	66-13053 <*>
1938 V51 P570-574	57-1362 <*>
1938 V51 P603-621	AL-527 <*>
1938 V51 P703-707	66-13054 <*> 0
1938 V51 P719-721	9067 <IICH>
1938 V51 N32 P531-537	58-2116 <*>
	59-12937 <+*> 0
1939 V52 P89-96	66-13056 <*>
1939 V52 P407	66-13057 <*>
1939 V52 P637-638	AEC-TR-3744 <+*>
1939 V52 N11 P215-219	64-20868 <*> 0
1939 V52 N37 P591-598	11592 <K-H>
	65-13582 <*> 0
1939 V52 N48 P687-693	59-15148 <*>
1940 V53 P35-39	63-16748 <*> 0
1940 V53 P60-65	64-10064 <*>
1940 V53 P425-428	66-13058 <*> 0
1940 V53 N7/8 P69-74	AL-405 <*>
1940 V53 N7/8 P98-99	60-10549 <*>
1940 V53 N29/0 P313-319	60-14407 <*> 0
1940 V53 N45/6 P520-522	64-14735 <*> 0
1940 V53 N47/8 P537-549	66-13327 <*> 0
1940 V53 N47/8 P550-552	RCT V14 N1 P914 <RCT>
1940 V53 N47/8 P550-551	60-16846 <*>
1941 V54 P307-308	759 <TC>
1941 V54 N1/2 P33-35	64-20867 <*> 0
1941 V54 N1/2 P36	64-14736 <*>
1941 V54 N7/8 P99-100	65-11624 <*>
1947 N5/6 P150-	2835 <*>
1947 V59 P42-48	I-591 <*>
1947 V59 P83-87	AL-648 <*>
1947 V59 N9 P257-272	59-15888 <*> 0
	61-20990 <*>
1947 V59 N5/6 P158-159	64-10841 <*> 0
1947 V59 N5/6 P163-	63-20861 <*>
1947 V59A P37-41	I-395 <*>
1947 V59A N3 P61-63	63-10020 <*> 0
1948 P45-46	C-2624 <NRC>
1948 N9 P225-231	AL-817 <*>
1948 V60 P98-99	60-23245 <*=>
1948 V60 P159-161	64-10055 <*>
1948 V60 P200-204	64-10062 <*>
1948 V60 P316-320	2085 <BISI>
1948 V60 N3 P78	64-18099 <*>
1948 V60 N5 P113-115	64-14737 <*>
1948 V60 N9 P245	65-13200 <*>
1948 V60 N7/8 P175-179	63-14977 <*>
1948 V60A P88-96	57-2344 <*>
1948 V60A N4 P88-96	87J16G <ATS>
1948 V60A N10 P263-267	60-10550 <*>
1948 V60A N10 P267-271	48K22G <ATS>
1949 V61 P84-89	AL-742 <*>
1949 V61 P229-241	II-558 <*>
1949 V61 P352-357	II-606 <*>
1949 V61 P357-365	II-630 <*>
1949 V61 P410-411	60-23245 <*=>
1949 V61 N5 P168-179	61-10062 <*>
1949 V61 N6 P229-241	64-10056 <*>
1949 V61 N6 P241-245	64-10054 <*>
1949 V61 N10 P411-412	77J16G <ATS>
1949 V61 N11 P458-	76J16G <ATS>
1950 V62 P121-125	I-960 <*>
1950 V62 P166-167	60-18055 <*> 0
	62-18934 <*>
1950 V62 P231-236	T1847 <INSD>
	8679 <IICH>
1950 V62 P264-268	63-20452 <*>
1950 V62 P302-305	63-18224 <*>
1950 V62 P312-315	AL-857 <*>
1950 V62 P342	62-20463 <*> 0
1950 V62 P413-414	UCRL TRANS-122 <*>
1950 V62 P519-526	63-18236 <*> 0
1950 V62 N3 P57-66	RCT V23 N4 P812 <RCT>
	66-13062 <*>
1950 V62 N5 P105-118	59-17052 <*>
	63-18968 <*> P
	64-10060 <*> 0
1950 V62 N5 P112-114	57-1897 <*>
1950 V62 N11 P264-268	2770-G <K-H>
1950 V62 N12 P291-292	63-16770 <*>
1950 V62 N16 P382-385	57-358 <*>
	64-18097 <*>
1950 V62 N17 P405-409	I-840 <*>
1950 V62 N17 P406	66-13063 <*>
1950 V62 N18 P423-426	66-13064 <*> 0
1950 V62 N13/4 P305-311	57-3065 <*>
1950 V62 N23/4 P556-558	C3L36G <ATS>
1950 V62 N23/4 P567-572	64-16591 <*>
1950 V62 N9/10 P231-236	AEC-TR-4392 <*>
1951 V63 P193-194	60-23245 <*=>
1951 V63 P201-206	62-25811 <=*>
1951 V63 P467-469	62-25810 <=*>
1951 V63 P508-511	62-25809 <=*>
1951 V63 P527-530	62-25808 <=*>
1951 V63 P531-	62-10068 <*>
1951 V63 N1 P1-7	58-2630 <*> 0
1951 V63 N2 P41-43	3280-C <K-H>
	61-10781 <*>
1951 V63 N2 P47-48	60-10555 <*>
1951 V63 N11 P258-259	60-10554 <*> 0
1951 V63 N11 P260-263	57-1901 <*>
1951 V63 N11 P263-267	64-18924 <*>
1951 V63 N15 P345-349	I-350 <*>
1951 V63 N17/8 P385-395	II-249 <*>
	64-14378 <*> 0
	65-10016 <*> 0
1952 V64 P10-11	62-25807 <=*>
1952 V64 P133-135	8197 <K-H>
1952 V64 P136-	1855 <*>
	59-19197 <+*>
1952 V64 P164-165	60-23245 <*=>
1952 V64 P213-220	63-18691 <*> 0
1952 V64 P323-329	AL-841 <*>
1952 V64 P448-	<ES>
1952 V64 P523-531	RCT V26 N3 P493 <RCT>
1952 V64 P533-536	1637 <*>
	62-25806 <=*>

Citation	Translation
1952 V64 N1 P10-11	I-987 <*>
1952 V64 N1 P28-29	I-816 <*>
1952 V64 N2 P54-58	1871 <*>
1952 V64 N3 P65-76	64-20295 <*> PO
1952 V64 N5 P133-135	58-2417 <*>
	64-30942 <*> O
1952 V64 N8 P213-220	190 <TC>
1952 V64 N12 P323-329	58-1373 <*>
	60-18391 <*>
	62-16697 <*>
	63-18666 <*> O
	64-18850 <*> P
	64-20863 <*>
1952 V64 N16 P437-447	64-18912 <*> O
1952 V64 N16 P448-452	58-791 <*>
1952 V64 N19 P539-553	66-13065 <*> P
1952 V64 N24 P679-685	3029 <*>
1952 V64 N19/2 P523-531	59-17910 <*> O
1952 V64 N9/10 P233-247	57-2369 <*>
1952 V64 N9/10 P265-269	62-10040 <*>
1953 V65 P16-33	I-833 <*>
	II-408 <*>
	63-10571 <*>
1953 V65 P59-61	60-23245 <*=>
1953 V65 P78-81	62-25805 <=*>
1953 V65 P159-161	62-25804 <=*>
1953 V65 P173-178	62-25803 <=*>
1953 V65 P183-186	3280-D <K-H>
1953 V65 P293-299	1137 <*>
1953 V65 P453-457	62-25800 <=*>
1953 V65 N1 P16-33	64-14370 <*>
1953 V65 N4 P107-109	I-237 <*>
1953 V65 N11 P299-303	37G6G <ATS>
1953 V65 N11 P303-304	62-18702 <*>
1953 V65 N23 P578-581	39L30G <ATS>
	62-25801 <=*>
1954 V66 P65-75	57-1430 <*>
1954 V66 P285-292	AEC-TR-2151 <*>
1954 V66 P386-390	II-138 <*>
1954 V66 P407-417	C-2572 <NRC>
	58-175 <*>
	58-562 <*>
1954 V66 P693-701	1860 <*>
	3382-C <K-H>
	64-14685 <*> O
1954 V66 P711-712	II-273 <*>
1954 V66 P765-768	I-596 <*>
1954 V66 N1 P17-27	64-20493 <*>
1954 V66 N4 P102-106	63-14615 <*>
1954 V66 N7 P189-192	57-245 <*>
1954 V66 N7 P198-210	57-1747 <*>
1954 V66 N9 P241-249	66-13066 <*>
1954 V66 N12 P339-340	60-16819 <*>
1954 V66 N15 P435-436	57-2558 <*>
	65-63403 <=$>
1954 V66 N19 P586-589	59-17330 <*>
1954 V66 N19 P611	62-10416 <*>
1954 V66 N20 P636-638	62-10904 <*>
1954 V66 N21 P658-677	58-1449 <*>
1954 V66 N21 P678-679	58-1594 <*>
1954 V66 N22 P701-710	AEC-TR-4756 <*>
1955 N19/0 P548-556	87H10G <ATS>
1955 V67 P1-8	60-23430 <=*>
1955 V67 P53-56	3307 <*>
1955 V67 P89-107	60-23434 <=*>
1955 V67 P141-145	60-23428 <=*>
1955 V67 P257-260	AEC-TR-5422 <*>
1955 V67 P289-300	62-20060 <*>
1955 V67 P425-426	II-818 <*>
1955 V67 P426-	II-819 <*>
1955 V67 P426	64-16435 <*>
1955 V67 P483-493	18G8G <ATS>
1955 V67 P548-	<ES>
1955 V67 P561-572	1188 <*>
1955 V67 P582-592	8986 <IICH>
1955 V67 P603-606	57-810 <*>
1955 V67 P613-615	57-2911 <*>
1955 V67 P748-749	57-1867 <*>
1955 V67 N1 P13-16	59-17351 <*>
1955 V67 N1 P16-25	58-385 <*>
1955 V67 N2 P45-52	64-20864 <*>
1955 V67 N2 P53-56	66-13067 <*>
1955 V67 N2 P61-68	60-18921 <*>
1955 V67 N7 P189-207	49G6G <ATS>
	57-2819 <*>
1955 V67 N7 P213-214	24L34G <ATS>
1955 V67 N8 P217-231	60-10537 <*> P
1955 V67 N11 P305	78M37G <ATS>
1955 V67 N12 P326	62-00679 <*>
1955 V67 N12 P327	62-00680 <*>
1955 V67 N13 P337-344	50K21G <ATS>
1955 V67 N16 P424	64-14931 <*>
1955 V67 N16 P424-425	64-14941 <*>
1955 V67 N16 P424	64-14942 <*>
	64-14944 <*>
1955 V67 N16 P425-426	60-10538 <*>
1955 V67 N16 P425	64-14938 <*>
1955 V67 N16 P425-426	64-16449 <*>
1955 V67 N17 P483-492	62-16761 <*> O
1955 V67 N17 P524	63-10620 <*>
	63-10621 <*>
1955 V67 N17 P541-547	57-2805 <*>
1955 V67 N19 P548-556	3805 <K-H>
	62-20064 <*>
1955 V67 N21 P637-642	CSIRO-3063 <CSIR>
1955 V67 N22 P685-694	II-963 <*>
1955 V67 N22 P694-698	CSIRO-3536 <CSIR>
1955 V67 N22 P705	64-18545 <*>
1955 V67 N23 P737-739	II-607 <*>
1955 V67 N23 P740-741	II-650 <*>
	60-14052 <*>
1955 V67 N17/8 P475-482	64-16114 <*> O
1955 V67 N17/8 P483-493	63-14201 <*>
1955 V67 N19/0 P541-547	57-768 <*>
	61-16727 <*>
	64-16126 <*> O
	64-20865 <*> O
	65-10014 <*>
	88H10G <ATS>
1955 V67 N19/0 P548-556	57-2818 <*>
	62-10013 <*>
	64-16151 <*> O
1955 V67 N19/0 P561-572	189 <TC>
1955 V67 N9/10 P245-251	57-1168 <*>
1955 V67 N9/10 P266-270	59-17547 <*> O
1955 V67 N9/10 P270-273	04L36G <ATS>
1956 N12 P393-403	64-30814 <*>
1956 V68 P133-150	II-956 <*>
1956 V68 P212-213	5299-B <K-H>
1956 V68 P238	AEC-TR-3943 <*>
1956 V68 P272-284	57-241 <*>
1956 V68 P291-295	57-1167 <*>
1956 V68 P349-352	1373 <*>
1956 V68 P352-354	62-18946 <*>
1956 V68 P393-403	<ES>
1956 V68 P438	C-2200 <NRC>
1956 V68 P473-485	3449 <*>
	655 <TC>
1956 V68 P581	58-2088 <*>
1956 V68 P615-616	59-10683 <*>
1956 V68 P633-660	8771 <IICH>
1956 V68 P641-643	64-16773 <*> O
1956 V68 P721-729	57-1997 <*>
1956 V68 N1 P6-13	57-213 <*>
1956 V68 N1 P13-20	60-14043 <*> O
1956 V68 N1 P18-20	58-155 <*>
1956 V68 N2 P70-71	1204 <*>
1956 V68 N7 P243-	57-1729 <*>
1956 V68 N10 P352-354	65-11247 <*> O
	9852 <K-H>
1956 V68 N11 P371-373	C-2520 <NRC>
	58-48 <*>
	60-23474 <=*>
1956 V68 N12 P393-424	57-2048 <*>
1956 V68 N12 P393-403	61-16700 <*>
	64-16056 <*>
	64-20866 <*>
1956 V68 N12 P406-411	64-20881 <*> P
	64-20921 <*> PO
	65-12416 <*>

Citation	Translation
1956 V68 N16 P505-508	62-16696 <*>
1956 V68 N19 P615-616	64-16315 <*>
1956 V68 N20 P633-640	3227 <*>
	42H13G <ATS>
	57-1796 <*>
	61-16739 <*>
	61053 <RIS>
	62-16455 <*> 0
	62-20083 <*>
	63-14202 <*> 0
1956 V68 N20 P641-643	43H13G <ATS>
1956 V68 N22 P689-692	64-16298 <*>
1956 V68 N23 P721-729	<ES>
	59-15975 <*>
	63-10587 <*>
1956 V68 N24 P761-776	38J16G <ATS>
1957 N13/4 P478-	58-1190 <*>
1957 V69 P94-95	<ES>
1957 V69 P105	62-25802 <=*>
1957 V69 P213-219	3225 <*>
1957 V69 P226-236	58-1540 <*>
1957 V69 P252-257	57-2967 <*>
1957 V69 P313-321	T-2587 <INSD>
1957 V69 P341-359	6360-A <K-H>
1957 V69 P397-398	64-16894 <*>
1957 V69 P480	AEC-TR-4368 <*>
1957 V69 P482-	57-2698 <*>
1957 V69 P600-608	8621 <IICH>
1957 V69 P634-638	64-18150 <*>
1957 V69 P689-	58-828 <*>
1957 V69 P689	63-16434 <*>
1957 V69 P699-706	64-18509 <*>
1957 V69 P717	60-19174 <=*>
1957 V69 P780	63-10548 <*>
1957 V69 P781	63-10596 <*>
1957 V69 P783-784	582 <TC>
1957 V69 P783	63-10597 <*>
1957 V69 N3 P77-82	62-10383 <*>
1957 V69 N3 P82-86	64-16819 <*>
1957 V69 N3 P86-90	3421 <*>
1957 V69 N3 P91-93	64-16824 <*>
1957 V69 N3 P94	58-2267 <*>
1957 V69 N3 P94-95	61-20932 <*>
	63-10608 <*>
1957 V69 N3 P94	64-30098 <*>
1957 V69 N3 P95	64-16825 <*> 0
1957 V69 N4 P124-131	64-18382 <*>
1957 V69 N5 P153-171	63-14200 <*>
1957 V69 N5 P172-174	6456-C <K-H>
1957 V69 N5 P179	60-10717 <*>
1957 V69 N6 P205	59-10490 <*>
1957 V69 N7 P213-244	59-15929 <*>
1957 V69 N7 P213-219	61-20651 <*>
	62-14036 <*> 0
	63-14203 <*> 0
	64-16552 <*>
	93J15G <ATS>
1957 V69 N7 P219-226	RCT V30 N4 P1103 <RCT>
1957 V69 N7 P226-236	57-2972 <*>
1957 V69 N8 P245-251	6095 A <K-H>
	64-20216 <*>
1957 V69 N8 P262-266	52K21G <ATS>
1957 V69 N10 P313-321	59-12936 <+*>
1957 V69 N11 P341-359	58-2406 <*>
	64-20244 <*> 0
1957 V69 N11 P359-361	59-19886 <+*> 0
1957 V69 N11 P362-371	58-2066 <*>
1957 V69 N11 P397-398	58-2509 <*>
1957 V69 N12 P419-422	59-10141 <*>
1957 V69 N14 P480	59-15886 <*>
1957 V69 N15 P497-505	61-10562 <*>
1957 V69 N16 P523-529	64-20242 <*>
1957 V69 N16 P530-533	64-10178 <*>
	64-18100 <*> 0
1957 V69 N17 P551-557	59-22282 <+*> 0
1957 V69 N20 P634-638	60-10724 <*>
1957 V69 N20 P640	60-15722 <*=> 0
1957 V69 N22 P699-706	60-16808 <*>
1957 V69 N1/2 P16-23	64-14984 <*> 0
1957 V69 N1/2 P50-58	64-14739 <*>
1957 V69 N13/4 P482	60-15909 <*=>
	65-00397 <*>
1957 V69 N18/9 P579-599	65-12442 <*>
1957 V69 N18/9 P608-614	AEC-TR-3634 <+*>
1958 V70 P307-309	63-18866 <*>
1958 V70 P311	58-2485 <*>
1958 V70 P327-339	59-10235 <*>
	61-10454 <*>
1958 V70 P496-502	R-5124 <*>
1958 V70 P552-570	59-14716 <*=>
1958 V70 P625-627	8529 A <K-H>
1958 V70 P738-742	59-12913 <+*> 0
1958 V70 N1 P1-5	60-14261 <*>
1958 V70 N1 P14-20	62-10595 <*> 0
1958 V70 N2 P52	64-30097 <*>
1958 V70 N3 P65-71	3K23G <ATS>
	63-20300 <*>
	64-30112 <*> P
1958 V70 N3 P71-72	64-30125 <*> 0
1958 V70 N4 P97-98	63-10572 <*>
1958 V70 N4 P98-104	58-2348 <*>
	65-10428 <*>
1958 V70 N4 P106-107	64-18814 <*>
1958 V70 N5 P117-123	61-10446 <*>
1958 V70 N5 P123-128	61-10447 <*>
1958 V70 N5 P128-133	61-10445 <*>
1958 V70 N7 P179-180	64-30119 <*>
1958 V70 N9 P251-266	C-3178 <NRC>
1958 V70 N9 P266-268	1739 <TC>
	61-20648 <*>
	97N53G <ATS>
1958 V70 N10 P285-298	77P60G <ATS>
1958 V70 N12 P351-367	415 <TC>
1958 V70 N12 P367-371	66-10910 <*> 0
1958 V70 N12 P371-372	66-13071 <*>
1958 V70 N13 P408-409	65-11444 <*>
1958 V70 N13 P413-414	65-11849 <*>
1958 V70 N14 P434-437	60-17572 <*=> 0
1958 V70 N16 P496-500	2K26G <ATS>
	58-2352 <*>
	59-15914 <*>
1958 V70 N16 P500-502	16K26G <*>
1958 V70 N20 P625-627	64-30973 <*>
1958 V70 N24 P738-742	59-20134 <*>
	60M39G <ATS>
1958 V70 N17/8 P532-539	59-10720 <*> 0
	61-20649 <*>
	64-14550 <*>
	9298-E <K-H>
1958 V70 N22/3 P667-683	66-13922 <*> 0
1958 V70 N22/3 P697-699	65-10177 <*>
1959 V71 P31-32	63-10564 <*>
1959 V71 P70-71	65-10309 <*>
1959 V71 P176-182	59-15333 <*>
1959 V71 P521	63-10562 <*>
1959 V71 P691-697	64-10933 <*>
1959 V71 N3 P110-119	59-15873 <*>
1959 V71 N3 P125-126	65-10176 <*>
1959 V71 N3 P127	59-17936 <*>
1959 V71 N4 P153-160	63-14204 <*> 0
1959 V71 N4 P162	65-14807 <*>
1959 V71 N5 P176-182	56L32G <ATS>
	60-14238 <*>
	61-18807 <*>
	64-30682 <*>
	65-10320 <*> 0
1959 V71 N6 P205-210	59-15761 <*> 0
	59-17204 <*>
	62-14035 <*>
	63-14208 <*> 0
1959 V71 N6 P211-215	61-16186 <*>
	62-20452 <*>
1959 V71 N7 P229-236	59-17266 <*>
	65-10175 <*>
1959 V71 N8 P274-276	66-13072 <*>
1959 V71 N8 P276-283	62-01102 <*>
1959 V71 N11 P366-371	59-17497 <*>
1959 V71 N17 P537-541	60-14071 <*> 0
1959 V71 N17 P541-545	12385-B <K-H>
	65-10174 <*>

	65-14498 <*>
1959 V71 N17 P545-549	74M45G <ATS>
1959 V71 N18 P574	65-10940 <*>
1959 V71 N19 P608-612	43L36C <ATS>
	65-10173 <*>
1959 V71 N19 P612-618	64-16535 <*>
	77M37G <ATS>
1959 V71 N19 P618-623	61-16711 <*>
	65-10172 <*>
	96M45G <ATS>
1959 V71 N21 P657-662	60-15711 <*+>
1959 V71 N21 P663-667	65-12143 <*> 0
1959 V71 N22 P685-691	62-00529 <*>
1959 V71 N15/6 P481-486	60-13414 <*+>
1959 V71 N15/6 P508	60-19908 <=*> 0
1959 V71 N15/6 P512-514	61-10855 <*>
1959 V71 N15/6 P520	60-10994 <*>
	65-11222 <*>
1960 V72 P40	63-10619 <*>
1960 V72 P225-231	62-16901 <*>
1960 V72 P236-249	AEC-TR-4565 <*>
1960 V72 P564	N.S.1 <TTIS>
1960 V72 P719-725	66-12056 <*>
1960 V72 P750-751	62-10379 <*>
1960 V72 P896-902	888 <TC>
1960 V72 N2 P74-76	65-11236 <*>
1960 V72 N2 P78	61-20937 <*>
	65-11253 <*>
1960 V72 N4 P127-131	65-12260 <*>
1960 V72 N4 P135-137	65-12259 <*>
1960 V72 N4 P138-139	60-16904 <*>
	63-10607 <*>
1960 V72 N4 P138	65-11249 <*>
1960 V72 N5 P168-169	63-18274 <*>
1960 V72 N6 P183-193	59M43G <ATS>
1960 V72 N6 P193-197	65-14724 <*>
1960 V72 N9 P287-294	NP-TR-772 <*>
1960 V72 N11 P379-384	19M43G <ATS>
1960 V72 N12 P401-413	UCRL TRANS-654 <*>
	08M47G <ATS>
1960 V72 N15 P563	63-10563 <*>
1960 V72 N16 P535-554	82N48G <ATS>
1960 V72 N16 P559-562	89N48G <ATS>
1960 V72 N16 P581-582	61-10789 <*>
1960 V72 N17 P618-626	65-13990 <*>
1960 V72 N20 P738-740	65-13188 <*> 0
1960 V72 N21 P775	AEC-TR-5287 <*>
1960 V72 N21 P775-777	3727 <BISI>
1960 V72 N22 P802-805	65-13186 <*> 0
1960 V72 N22 P829-835	10610 B <K-H>
	21N49G <ATS>
	61-20688 <*>
	65-12932 <*>
1960 V72 N22 P836-845	62-10408 <*>
1960 V72 N22 P856-864	61-20687 <*>
1960 V72 N24 P934-939	61-20991 <*> 0
	62-10053 <*>
	62-20026 <*>
1960 V72 N24 P994-1000	63-10581 <*>
1960 V72 N24 P1000	65-13226 <*> 0
	65-13227 <*> 0
1960 V72 N7/8 P249-258	62-10111 <*>
1960 V72 N19/0 P719-725	65-12510 <*>
1960 V72 N19/0 P730-738	62-10418 <*> 0
1960 V72 N19/0 P738-740	10546 <K-H>
	61-10788 <*>
1961 V73 P49-56	AEC-TR-4816 <*>
1961 V73 P59-63	583 <TC>
1961 V73 P186-197	62-10018 <*>
1961 V73 P309-322	HW-TR-33 <*>
1961 V73 P371-383	63-10612 <*>
1961 V73 P525-530	62-10464 <*>
1961 V73 P589-597	63-18761 <*>
1961 V73 N1 P11-14	AEC-TR-5258 <*>
1961 V73 N1 P23-25	63-20726 <*>
1961 V73 N1 P27	NP-TR-811 <*>
1961 V73 N2 P66	63-10681 <*>
1961 V73 N3 P81-107	10895 B <K-H>
	62-16363 <*>
	63-10545 <*>
	65-13543 <*> 0
1961 V73 N3 P107-113	61-18687 <*>
1961 V73 N4 P136-142	61-20034 <*> 0
1961 V73 N5 P161-166	62-18718 <*>
	63-20725 <*>
1961 V73 N5 P171	63-10565 <*>
1961 V73 N6 P177-185	11308-F <K-H>
	62-10671 <*>
1961 V73 N6 P177-186	65-12928 <*>
	65-13573 <*> 0
	93N51G <ATS>
1961 V73 N6 P186-197	61-18648 <*>
	65-12925 <*>
1961 V73 N6 P197-208	61-18688 <*>
1961 V73 N6 P215-218	65-14505 <*>
1961 V73 N9 P277-290	64-10717 <*>
1961 V73 N9 P290-298	63-10125 <*>
	65-13013 <*>
	95N53G <ATS>
1961 V73 N10 P305-309	63-16222 <*> 0
1961 V73 N10 P309-322	63-16221 <*> 0
1961 V73 N11 P353-364	90N55G <ATS>
1961 V73 N11 P371-383	65-13547 <*>
1961 V73 N11 P388-393	63-10603 <*>
1961 V73 N12 P423-432	61-20579 <*> 0
1961 V73 N15 P513-519	61-20583 <*> 0
	62-00604 <*>
1961 V73 N15 P525-530	65-12417 <*>
	65-13349 <*>
1961 V73 N15 P540-541	63-10578 <*>
1961 V73 N17 P619	62-14241 <*>
1961 V73 N19 P641-646	17N58G <ATS>
1961 V73 N21 P695-706	63-01677 <*>
1961 V73 N23 P745-751	64-23592 <=$>
1961 V73 N23 P755-756	62-18076 <*>
1961 V73 N23 P756	62-18077 <*>
1961 V73 N17/8 P612-615	62-10958 <*>
	65-12415 <*>
1961 V73 N17/8 P620	63-10556 <*>
1962 V74 P633-637	12346-B <K-H>
1962 V74 N6 P216-217	63-18797 <*>
1962 V74 N13 P469	62-20107 <*>
1962 V74 N14 P484-488	63-00066 <*>
1962 V74 N14 P508-509	59P65G <ATS>
1962 V74 N15 P551-562	63-16557 <*>
1962 V74 N16 P599-605	84Q69G <ATS>
1962 V74 N16 P606-612	95Q70G <ATS>
1962 V74 N16 P633-637	65-14485 <*>
1963 V75 N9 P417	87Q70G <ATS>
1963 V75 N12 P508-515	64-30651 <*> 0
1963 V75 N19 P894-900	65-12125 <*>
1963 V75 N22 P1068-1079	32R76G <ATS>
1964 V76 N1 P9-19	64-30065 <*>

ANGEWANDTE CHEMIE. AUSGABE B., TECHNISCH-WIRTSCHAFTLICHER TEIL

1947 V19 N5/6 P135-142	62-16983 <*> 0
1948 V20 P331-332	64-20870 <*>
1948 V20 N9 P225-230	64-20871 <*> P
1948 SB V20 N10 P257-288	58-1406 <*>

ANGEWANDTE MESS UND REGELTECHNIK

1962 V2 N8 P63-66	BGIRA-545 <BGIR>

ANGEWANDTE METEOROLOGIE. BERLIN

1958 V3 P161-163	SCL-T-274 <*>
1958 V3 N6 P170-173	SCL-T-279 <*>

ANGEWANDTE PFLANZENSOZIOLOGIE

1954 V2 P1207-1222	CSIRO-3285 <CSIR>

ANILINOKRASOCHNAYA PROMYSHLENNOST

1932 V2 N7 P12-15	63-20326 <=*>
1933 V3 P405-412	RJ-132 <ATS>
1933 V3 N9 P405-412	RT-132 <*>
1934 V4 P195-	R-99 <*>
1934 V4 P461-472	64-18558 <*>
1934 V4 P523-531	64-18558 <*>
1934 V4 N1 P70-76	RT-1437 <*>

ANNALEN DER CHEMIE
SEE JUSTUS LIEBIG'S ANNALEN DER CHEMIE

ANNALEN DER CHEMIE UND PHARMACIE
1841 V39 P160-168 60-10835 <*>
1845 V55 P249-297 2193 <*>
1846 V58 N3 P273-300 SCL-T-252 <+*>
1855 V94 P57-77 1062 <*>
1857 V104 P251-287 60-10837 <*>
1859 V111 P242-252 64-20879 <*>
1860 V114 P213-217 58-111 <*>
1861 V120 P288-295 60-10836 <*>
1863 V126 P195-217 66-13077 <*> P
1864 V129 P81-87 65-14843 <*>
1866 V138 P129-164 MF-6 <*>
1867 V144 P37-42 66-13078 <*> P
1867 V144 P268-271 63-14206 <*>
1868 V148 P24-30 66-13079 <*> P
1868 V148 P376-380 1660 <*>
1870 V154 P59-62 64-10053 <*>
1873 V165 P328-349 57-94 <*>

ANNALEN DER HYDROGRAPHIE U. MARITIME METEOROLOGIE
1915 P169-178 II-934 <*>
1915 V43 P111-124 II-934 <*>
1933 P233-236 I-414 <*>
1934 V63 P195-204 1483 <*>
1936 P23- 58-973 <*>
1936 V64 N10 P469-473 C-2687 <NRC>
1938 V66 N9 P455-459 57-626 <*>
1944 P193-205 38K20G <ATS>
1944 V72 N10 P277-279 II-1019 <*>

ANNALEN DER METEOROLOGIE
1951 N1/6 P78-86 57-2382 <*>
1951 V4 N10 P487-492 AEC-TR-3082 <*>
1951 V4 N7/9 P358-374 59-18846 <=*>
1955 V7 N3/4 P213-218 1538 <*>

ANNALEN DES NATURHISTORISCHEN MUSEUMS IN WIEN
1902 V17 P112-116 05K23G <ATS>

ANNALEN DER PHARMACIE
1832 V3 P288-310 66-13073 <*> P

ANNALEN DER PHYSIK
1829 V92 P403- 57-746 <*>
1854 V6 N14 P412-425 C-2589 <NRC>
1855 S2 V84 P59-86 AEC-TR-4009 <*>
1861 S2 V113 P337-339 59-20874 <*>
1863 S2 V120 P118-158 61-14276 <*>
1866 V127 P430-431 63-14112 <*>
1869 S2 V136 P144-151 63-10715 <*> O
1883 V18 P79-94 58-2141 <*>
58-2322 <*>
1885 V25 P337-357 58-2461 <*>
1887 V31 N5 P109-126 57-2473 <*>
1889 V37 P144-172 63-14009 <*>
1890 S3 V39 P481-554 64-00037 <*>
1890 S3 V39 N2 P161-186 65-00301 <*>
1891 V43 P43-60 T1891 <INSD>
1892 V46 P584-636 63-18864 <*>
1894 V53 P959-960 57-706 <*>
1894 S3 V53 P499-504 62-18358 <*> O
1898 V65 P873-876 57-1416 <*>
1899 V67 N2 P233-236 58-65 <*>
1899 V67 N2 P287-290 58-65 <*>
1899 S3 V68 P569-573 66-13100 <*> P
1900 S4 V3 P733-743 63-16498 <*>
1903 S4 V11 P114-126 59-20395 <*> PO
1904 S4 V14 P539-546 58-336 <*>
1905 S4 V16 P172-177 57-3235 <*>
1905 S4 V17 P132-148 57-2236 <*>
1905 S4 V17 P359-363 58-2148 <*>
59-10044 <*>
1905 S4 V17 P549-560 65-13501 <*>
1906 V19 P371-381 57-1051 <*>
1908 V25 P377-445 I-407 <*>
1908 V25 N5 P921-982 66-12895 <*>
1908 S4 V25 N3 P377-445 57-2920 <*>
1908 S4 V27 P233-300 63-01048 <*>
1909 S4 V28 P75-130 58-1659 <*>
1909 S4 V28 P665-736 57-1021 <*>
1909 S4 V28 P959-1016 AEC-TR-3715 <=*> O
1909 S4 V29 P566-584 62-18484 <*> PO
1909 S4 V29 N6 P664-678 AEC-TR-5938 <*>
1910 V32 P739-748 65-12877 <*>
1910 S4 V33 P1275-1298 AL-220 <*>
1911 S4 V35 P617-678 63-01519 <*>
1911 S4 V35 P1009-1033 62-18485 <*> O
1912 V39 N16 P1133-1138 57-579 <*>
1912 S4 V39 P307-345 65-00327 <*>
1913 S4 V41 P289-336 II-725 <*>
1913 S4 V42 P303-344 NP-TR-277 PT.1 <*>
1913 S4 V42 P796-806 UCRL TRANS-873(L) <*>
1913 S4 V42 P1061-1063 UCRL TRANS-874(L) <*>
1914 V43 P1186-1194 63-14904 <*>
1914 S4 V43 P1079-1100 65-10779 <*>
1914 S4 V44 P177-202 1573 <*>
1914 S4 V44 P203-240 AD-90(I) <*>
1914 S4 V45 P1253-1256 57-2087 <*>
1915 S4 V48 P273-306 57-3347 <*>
1916 S4 V49 P1-38 2453 <*>
1916 S4 V49 P117-143 2453 <*>
1916 S4 V50 P769-795 63-10088 <*>
1918 V57 P541-567 66-12391 <*> O
1918 S4 V56 N11 P161-207 63-10165 <*>
1919 V59 N16 P689-742 58-2464 <*>
1919 S4 V58 P55-104 50T94G <ATS>
1919 S4 V60 P454-480 58-130 <*>
1919 S4 V60 P734-762 63-20349 <*>
1919 S4 V60 N20 P297-323 63-14686 <*>
1920 S4 V61 P471-500 63-00263 <*>
1920 S4 V62 P623-648 63-00264 <*>
1920 S4 V63 P57-84 58-145 <*>
1921 S4 V64 P253-287 AI-TRANS-43 <*>
1922 V69 N19 P205-230 66-12388 <*> O
1922 S4 V68 N15/6 P553-573
59-20245 <*>
1923 V71 N13 P317-376 66-13012 <*> O
1924 S4 V74 N15 P577-627 AL-874 <*>
1925 V76 P39- 57-3023 <*>
1925 S4 V78 P157-166 AL-342 <*>
1926 S4 V80 P672-706 66-13101 <*> OP
1926 S4 V81 P1135-1153 UCRL TRANS-950(L) <*>
1927 V84 P457 I-56 <*>
1927 S4 V82 P138-142 65-17398 <*>
1927 S4 V82 P836-840 65-17397 <*>
1927 S4 V84 N20 P457-484 57-1355 <*>
1928 V86 N10 P291-318 57-346 <*>
1928 S4 V87 P850-876 66-61846 <=*$>
1929 V3 N2 P133-208 57-218 <*>
57-2315 <*>
1929 S5 V1 P977-1077 NP-TR-277 PT.2 <*>
1929 S5 V2 N3 P249-263 63-20354 <*>
1929 S5 V2 N6 P617-630 55L29G <ATS>
1929 S5 V3 P1055-1101 I-472 <*>
1930 V6 P852-874 C-2317 <NRC>
57-2356 <*>
1930 S5 V4 P33-48 1567 <*>
1930 S5 V4 P149-184 UCRL TRANS-634(L) <*>
1930 S5 V4 P863-897 1933 <*>
1930 S5 V4 N2 P252-272 59-20832 <*> O
1930 S5 V5 P325-400 65-00086 <*>
1930 S5 V5 N1 P1-35 II-152 <*>
1930 S5 V5 N6 P762-792 59-20252 <*> O
1930 S5 V7 P489-568 AEC-TR-36 <+*> O
1931 V8 N6 P737-776 57-305 <*>
1931 V10 P888-904 57-1774 <*>
1931 V11 N1 P1-39 58-2177 <*>
1931 V11 N1/8 P579-605 57-2401 <*>
1931 S5 V9 P273-294 63-20480 <*> O
1931 S5 V10 P789-824 58-346 <*>
1931 S5 V10 P905-926 58-358 <*>
1931 S5 V11 P579-605 63-14008 <*> O
1932 V13 N2 P191-211 57-2136 <*>
1932 S5 V12 P1-51 60-14642 <*> O
1932 S5 V12 P961-989 1557 <*>
63-20121 <*> O

Citation	Translation
1932 S5 V13 P681-702	AL-297 <*>
1932 S5 V13 P967-984	66-10713 <*>
1932 S5 V13 N7 P802-810	63-14688 <*>
1932 S5 V14 P712-728	63-01023 <*>
1932 S5 V14 P729-754	63-01228 <*>
1932 S5 V14 N6 P617-643	63-14687 <*>
1932 S5 V15 P903-925	57-1452 <*>
1933 V16 P174-189	58-720 <*>
1933 V16 N7 P844-860	57-327 <*>
1933 V18 N5 P457-485	I-199 <*>
1933 S5 V18 P625-655	UCRL TRANS-749(L) <*>
1933 S5 V18 N4 P417-456	58-153 <*>
1933 S5 V18 N5 P457-485	58-101 <*>
1934 V19 P585-622	57-915 <*>
1934 S5 V19 P99-117	63-01229 <*>
1934 S5 V19 P434-464	3198 <*>
1934 S5 V20 P828-842	63-01021 <*>
1934 S5 V21 N2 P113-138	57-317 <$>
1935 V22 P375-401	57-3058 <*>
1935 S5 V22 P281-310	UCRL TRANS-125 <*>
1935 S5 V22 P375-401	66-10775 <*> 0
1935 S5 V24 P545-590	AEC-TR-5124 <*>
1935 S5 V24 P636-664	44L31G <ATS>
1935 S5 V24 P719	AEC-TR-5683 <*>
1935 S5 V24 P719-752	II-160 <*>
	24J14G <ATS>
1936 S5 V25 P359-372	I-28 <*>
1936 S5 V25 P517-526	UCRL TRANS-926(L) <*>
1936 S5 V25 N4 P373-384	II-159 <*>
1936 S5 V26 P69-96	II-143 <*>
1936 S5 V26 P233-257	NP-TR-1032 <*>
1936 S5 V26 P449-464	63-10739 <*>
1936 S5 V26 N6 P567-584	65-14433 <*>
1936 S5 V27 P243-255	63-10740 <*>
1936 S5 V27 P299-311	UCRL TRANS-1050(L) <*>
1937 V29 N8 P665-697	AL-593 <*>
1937 S5 V29 P239-245	I-500 <*>
1937 S5 V29 P394-406	AEC-TR-5316 <*>
1937 S5 V29 P605-635	SCL-T-436 <*>
1937 S5 V29 N6 P514-526	58-2191 <*>
1938 S5 V31 P365-376	I-492 <*>
	63-10676 <*>
1938 S5 V31 N6 P531-539	58-2192 <*>
1938 S5 V32 P128-140	AEC-TR-6158 <*>
1938 S5 V32 N4 P347-352	58-2193 <*>
1938 S5 V33 N5 P389-403	NP-TR-361 <*>
1939 V36 P625-650	07H9G <ATS>
	57-1372 <*>
1939 S5 V34 N1 P96-112	59-20244 <*>
1939 S5 V35 P73-83	I-491 <*>
1939 S5 V36 P284-294	59-20164 <*>
1940 S5 V37 P509-533	63-14691 <*>
1940 S5 V38 N6 P469-485	66-10140 <*> 0
1941 V39 N2 P81-128	57-11 <*>
1942 V42 P534-560	57-3056 <*>
1942 S5 V41 N1 P1-36	57-3387 <*>
1942 S5 V41 N6 P43-46	AL-512 <*>
1942 S5 V42 P241-253	E-656 <RIS>
1942 S5 V42 P254-272	58-299 <*>
1943 V43 N6/7 P393-403	23T94G <ATS>
1943 S5 V43 P187-192	AL-200 <*>
1943 S5 V43 N5 P369-382	59-17466 <*>
1943 S5 V43 N6/7 P393-403	AL-513 <*>
1947 S6 V1 P41-58	61-20037 <*>
1947 S6 V1 P95-118	61-18765 <*>
1947 S6 V1 P95-106	62-10070 <*> 0
1947 S6 V1 P153-168	61-10127 <*> 0
1947 S6 V1 P190-206	63-14465 <*>
1947 S6 V1 P275-	I-501 <*>
1948 V3 N12 P393-457	I-378 <*>
1948 S6 V2 P55-75	2996 <*>
1948 S6 V2 P60-75	I-180 <*>
1948 S6 V2 P103-112	UCRL TRANS-760(L) <*>
1948 S6 V3 P223-229	57N49G <ATS>
1948 S6 V3 P230-239	58N49C <ATS>
1948 S6 V3 P240-254	59N49G <ATS>
1948 S6 V4 P95-163	UCRL TRANS-399(L) <*>
1948 S6 V4 P432-440	I-958 <*>
1948 S6 V4 N3/4 P95-126	65-10773 <*>
1948 S6 V4 N3/4 P127-135	65-10772 <*>
1948 S6 V4 N3/4 P136-142	65-10774 <*>
1948 S6 V4 N3/4 P143-149	65-10771 <*>
1948 S6 V4 N3/4 P150-160	65-10769 <*>
1949 S6 V5 P33-50	28P66G <ATS>
1949 S6 V5 P71-88	AEC-TR-3552 <+*>
1949 S6 V5 N3/5 P251-252	SC-T-64-2000 <*>
1949 S6 V6 P63-70	AEC-TR-3596 <+*>
1950 S6 V7 N1/2 P1-17	63-10741 <*>
1950 S6 V8 P3-4	AL-253 <*>
1950 S6 V8 P129-133	AL-253 <*>
1950 S6 V7/8 P88-92	61-20209 <*>
1951 V9 N5 P235-244	65-12337 <*>
1952 V10 N6 P336-348	66-13803 <*>
1952 S6 V10 N4/5 P201-210	62-14908 <*>
1952 S6 V11 N2/3 P99-112	2139 <BISI>
1953 V12 P401-409	57-549 <*>
1953 V13 P305-321	22H10G <ATS>
1953 V13 P322-326	23H10G <ATS>
1953 S6 V11 N4/7 P362-367	59-17931 <*>
1953 S6 V11 N4/7 P368-376	59-17930 <*>
1953 S6 V11 N4/7 P377-382	59-17870 <+*>
1953 S6 V12 P57-83	AEC-TR-4262 <*>
1953 S6 V12 P227-236	63-00269 <*>
1953 S6 V12 P361-378	57-2499 <*>
1953 S6 V12 N7/8 P361-378	59-17905 <*>
1953 S6 V13 P166-177	UCRL TRANS-587(L) <*>
1953 S6 V13 P229-252	63-00902 <*>
1953 S6 V13 P404-420	I-625 <*>
1954 S6 V14 P33-53	60-18050 <*>
1954 S6 V14 P129-134	65-12875 <*>
1954 S6 V14 N6 P233-252	63-10912 <*>
1954 S6 V14 N1/2 P17-32	65-11354 <*> 0
1955 V15 P311-324	70H9G <ATS>
1955 S6 V15 P279-287	65-13100 <*>
1955 S6 V15 N6 P268-278	AEC-TR-3554 <+*>
1955 S6 V15 N5/6 P288-301	65-13102 <*>
1955 S6 V16 N5/8 P193-209	63-14689 <*>
1956 S6 V17 P126-140	57-886 <*>
1956 S6 V17 N2/3 P57-77	66-10346 <*>
1956 S6 V19 P19-30	65J16G <ATS>
1957 S6 V19 P283-297	TT-828 <NRC>
1957 S6 V19 P322-328	6738 <K-H>
1957 S6 V19 N6/7 P322-328	58-1467 <*>
1957 S6 V20 P349-354	60-17368 <=*>
1957 S6 V20 P368-380	60-17369 <=*>
1957 S6 V20 P386-389	UCRL TRANS-698(L) <*>
1958 V1 N6/8 P305-318	28T95G <ATS>
1958 S7 V1 N1/3 P173-182	66-14493 <*>
1958 S7 V1 N4/5 P281-295	SCL-T-381 <*>
1958 S7 V2 N3/4 P130-145	NP-TR-840 <*>
1958 S7 V2 N5/6 P236-262	66-11878 <*>
1959 S7 V2 N7/8 P380-386	62-11451 <=>
1959 S7 V3 N1/2 P37-47	65-10838 <*>
1959 S7 V4 N1/5 P140-144	73Q67G <ATS>
1959 S7 V4 N1/5 P145-153	NP-TR-471 <*>
1959 S7 V4 N6/8 P373-387	65-10877 <*>
1959 S7 V5 N1/2 P57-69	61-20357 <*>
1960 V5 P414-428	65-12494 <*>
1960 S7 V5 N3/4 P113-128	43N48G <ATS>
1960 S7 V5 N5/6 P252-267	66-11241 <*>
1960 S7 V5 N5/6 P308-326	61-10790 <*>
1960 S7 V5 N7/8 P373-387	63-14927 <*>
1960 S7 V6 N5/6 P236-240	62-24538 <=*>
1960 S7 V6 N5/6 P267-287	62-24539 <=*>
1960 S7 V6 N5/6 P298-306	62-24540 <=*>
1960 S7 V6 N5/6 P307-310	62-24536 <=*>
1961 S7 V7 P1-7	UCRL TRANS-716(L) <*>
1961 S7 V7 P302-311	64-14745 <*>
1961 S7 V7 N5/6 P326-332	62-00663 <*>
1961 S7 V7 N5/6 P333-341	62-00665 <*>
1961 S7 V8 N1/2 P28-30	62-24537 <=*>
1961 S7 V8 N3/4 P146-157	62-00809 <*>
1961 S7 V8 N3/4 P196-203	63-10863 <*>
1962 S7 V9 P168-173	UCRL TRANS-860(L) <*>
1962 S7 V9 N7/8 P382-387	63-10918 <*>
1962 S7 V10 N1/2 P1-30	64-71541 <=>
1962 S7 V10 N1/2 P64-73	64-10938 <*>
1963 S7 V11 P83-100	66-11750 <*>
1963 S7 V11 N1/6 P83-100	64-15967 <*>

1963 S7 V12 N3/4 P133-155 AEC-TR-6370 <*>
1964 S7 V13 N1/2 P25-40 65-11498 <*>
1965 S7 V15 P30-48 ORNL-TR-679 <*>
1965 S7 V15 N3/4 P113-143 33S88G <ATS>
65-18111 <*>

ANNALEN DER SCHWEIZERISCHEN METEOROLOGISCHEN ZENTRALANSTALT
1939 V75 N9 P1-13 T-2287 <INSD>

ANNALES ACADEMIAE SCIENTARUM FENNICAE
SEE SUOMALAISEN TIEDEAKATEMIAN TOIMITUKSIA. SARJA A.

ANNALES AGRONOMIQUES
1943 V13 P8-11 63-18471 <*>
1949 V19 N6 P925-937 59-17224 <*>
1950 V1 N3 P368-381 64-10306 <*>
1953 V4 N1 P27-43 19Q72F <ATS>
1955 SA V6 N3 P363-371 65-12622 <*>
1957 V8 N2 P5-15 58-2229 <*>

ANNALES DE L'AMELIORATION DES PLANTES
1954 V4 N1 P5-20 64-30305 <*>

ANNALES D'ANATOMIE PATHOLOGIQUE
1956 V1 N3 P299-328 63-16948 <*> 0
9667 <KH>

ANNALES D'ASTROPHYSIQUE
1943 V6 N3/4 P81-89 58-2324 <*>
1944 V7 N1/2 P31-79 65-00254 <*>
1951 N3 P249-265 58-2462 <*>
1951 V14 N1 P115-151 63-11695 <=>
1956 V19 N1 P1-8 1543 <*>
1960 V23 N5 P797-801 62-10152 <*>
1965 V28 N2 P399-411 65-14569 <*>

ANNALES DE BIOLOGIE CLINIQUE
1958 V16 N3/4 P181-189 65-00118 <*>
1958 V16 N7/9 P471-480 64-16879 <*>
1958 V16 N7/9 P481-485 64-16880 <*>
1959 V17 P486-497 64-18645 <*>
1960 V18 P504-507 61-16481 <*>
1963 V21 N5/6 P453-456 64-18958 <*>
1963 V21 N5/6 P475-479 64-18957 <*> 0
1965 V23 N7/9 P987-991 66-13207 <*>

ANNALES DE CHIMIE
1914 S9 V1 P469-519 63-10474 <*> P
1915 S9 V4 P5-27 5513-A <KH>
1915 S9 V4 P58-136 I-107 <*>
1915 S9 V4 P168-224 I-107 <*>
1916 V5 P52-108 I-107 <*>
1916 V5 P194-224 I-107 <*>
1920 S9 V13 P229- 2048 <*>
1920 S9 V14 P1-58 AEC-TR-3841 <*>
1920 S9 V14 P215-321 66-13110 <*>
1921 S9 V15 P221-252 3279 <*>
1921 S9 V15 P291-332 66-13112 <*>
1922 V17 P38- 58-1344 <*>
1922 V17 P41-43 2195 <*>
1922 S9 V17 P294-303 I-1050 <*>
1929 S10 V12 P480-484 58-2284 <*>
1930 S10 V14 P227-282 63-18128 <*> PO
1931 V16 P25-26 58-213 <*>
1932 S10 V18 P5-31 63-14549 <*> OP
1932 S10 V18 P410-418 63-10582 <*>
1936 S11 V5 P83-146 64-20875 <*> PO
1937 S11 V7 P225-297 10005 <HB>
1943 S11 V18 P97-116 AEC-TR-4652 <*>
1944 V19 P345-361 8983 <IICH>
1944 S11 V19 P345-361 I-751 <*>
1945 S11 V20 P5-72 SCL-T-483 <*>
1945 S11 V20 P367-420 63-16092 <*> 0
1946 S12 V1 P446-473 65-10145 <*> 0
1947 S12 V2 P527-598 II-419 <*>
1947 S12 V2 P790-843 1271H <K-H>
1950 S12 V5 P819-881 5083-B <K-H>
1951 S12 V6 P165-219 00217 <*>
1951 S12 V6 N9/10 P705-716
61-10158 <*>
1951 S12 V6 N9/10 P723-737
61-10171 <*> 0
1952 S12 V7 P265-310 AEC-TR-5515 <*>
1952 S12 V7 P808-848 60-23457 <+*>
1953 S12 V8 P450-497 63-16872 <*> P
1953 S12 V8 P826-840 60-23426 <=*>
1955 V10 P903-967 1334 <*>
1955 V12 P225-254 II-278 <*>
1955 S12 V10 P1018-1060 65-12667 <*> P
1956 S13 V1 P381-398 AEC-TR-4838 <*>
1957 S13 V2 P145-188 61-00655 <*>
1957 S13 V2 P616-625 58-1232 <*>
1958 S13 V3 N9/10 P645-655
60-16903 <*>
1960 V5 P1459-1502 65-11740 <*> 0
1960 S13 V5 P5-64 AEC-TR-4564 <=*>
1960 S13 V5 P469-527 63-18585 <*>
1960 S13 V5 P707-764 AEC-TR-5687 <*>
1960 S13 V5 P1459-1502 UCRL TRANS-791(L) <*>
63-10012 <*>
1961 S13 V6 N3/4 P245-283 65-00085 <*>
1961 S13 V6 N5/6 P575-622 63-20662 <*> 0
1963 S13 V8 P117-149 ORNL-TR-201 <*>

ANNALES DE CHIMIE ET DE PHYSIQUE
1809 V69 P204-220 66-13102 <*>
1816 S2 V1 P334-335 64-14296 <*>
1830 V45 N2 P203-208 58-2260 <*>
1831 S2 V48 P337-353 66-13103 <*>
1854 S3 V40 P156-221 64-71482 <=>
1860 S3 V58 P254-256 1572 <*>
1860 S3 V59 P124-128 1575 <*>
1863 S3 V67 P340-349 66-13105 <*>
1866 S4 V7 P73 64-14170 <*>
1876 S5 V7 P452-479 58-1978 <*>
1883 S5 V28 P508-569 63-10425 <*>
1885 S6 V6 P502-504 66-13106 <*>
1886 S6 V9 P156-162 64-20046 <*>
1887 S6 V12 P358-384 AL-341 <*>
1892 S6 V25 P433-519 64-18166 <*> P
1898 S7 V13 P480-555 299 <TC>
1899 S7 V17 P527-532 66-13107 <*>
1905 S8 V5 P245-288 60-23980 <*=> 0
1905 S8 V6 P205-209 66-13108 <*>
1907 S8 V10 P145-198 64-20389 <*> P
1908 S8 V13 P276-277 2075 <*>
1910 S8 V20 P289-352 66-13109 <*> P

ANNALES CHIRURGIAE ET GYNAECOLOGIAE FENNIAE. HELSINKI
1962 V51 P466-474 65-00107 <*>

ANNALES DE CHIRURGIE
1962 V16 N25/6 PC1827-C1834
64-18695 <*>
1963 V17 N17/8 P1185-1214 66-12164 <*>

ANNALES DE DERMATOLOGIE ET DE SYPHILIGRAPHIE
1910 S5 V1 P188-190 66-10862 <*>
1948 S8 V8 N2 P146-148 64-14740 <*>
1950 V10 P241-262 57-72 <*>
1956 V83 N4 P369-404 63-20265 <*> 0

ANNALES DE L'ECOLE NATIONALE DES EAUX ET FORETS ET DE LA STATION DE RECHERCHES ET EXPERIENCES FORESTIERES
1928 V2 P139-154 57-1237 <*>

ANNALES D'ENDOCRINOLOGIE
1945 V6 N1 P51-54 63-01375 <*>
1948 V9 N2 P136-148 59-15481 <*> 0
1949 V10 N1 P47-49 63-01392 <*>
1951 V12 N3 P317-332 60-10498 <*> 0
1952 V13 N1 P1-8 60-10497 <*> 0
1952 V13 N3 P485-486 62-00197 <*>
1952 V13 N4 P687-690 61-10930 <*> 0
1953 V14 N6 P939-947 64-14741 <*> 0
1955 V16 P77-85 61-01009 <*>

1955 V16 P576-581 57-271 <*>
1955 V16 N3 P334-347 61-00404 <*>
1957 V18 P171-173 0A-139 <RIS>
1958 V19 P128-133 60-16598 <*>
1958 V19 P134-136 60-16599 <*>
1958 V19 P792-803 63-00207 <*>
1958 V19 N3 P572-577 63-01760 <*>
1958 V19 N6 P1143-1156 60-13219 <=*> 0
1959 V20 P377-381 61-16100 <*>
1959 V20 N5 P754-760 60-18754 <*>
1960 V21 P556-574 62-01270 <*>
1960 V21 N1 P75-78 61-00179 <*>
1961 V22 P230-238 62-20104 <*>
1961 V22 P239-247 62-16078 <*>
1961 V22 N2 P118-125 62-16079 <*>
1961 V22 N4 P549-555 62-01203 <*>
1962 V23 P65-77 62-18050 <*>
1962 V23 N3 P344-348 63-00082 <*>
1965 V26 N2 P161-167 66-10739 <*>

ANNALES DES FALSIFICATIONS ET DES FRAUDES
1923 V21 P532-534 I-266 <*>
1951 V44 P103-110 57-1131 <*>
1953 V46 P350-358 58-2440 <*>

ANNALES DES FERMENTATIONS
1937 V3 P257-275 63-14882 <*> 0

ANNALES FRANCAISES DE CHRONOMETRIE
1934 N1 P45-55 66-13324 <*> 0
1958 V27 N3 P183-188 60-13235 <=*>

ANNALES DE GEOPHYSIQUE
1952 V8 N1 P33-54 62-00167 <*>
1952 V8 N3 P328-332 SCL-T-348 <*>
1956 V12 N2 P113-143 62-10139 <*>
1957 V13 N3 P222-230 62-10153 <*>
1957 V13 N4 P286-306 64-71475 <=>
1957 V13 N4 P317-320 62-10142 <*>
1959 V15 N2 P179-216 60-17772 <=*>
1959 V15 N2 P266-271 60-17774 <*+>
1961 V17 N3 P319-337 SCL-T-400 <*>
1962 V18 N4 P351-359 65-12402 <*>
1963 V19 N1 P21-30 65-11609 <*> 0
1964 V20 N3 P319-324 65-14339 <*>
1964 V20 N4 P473-479 N65-19513 <=$>
66-14978 <*> 0

ANNALES D'HISTOCHIMIE
1963 V8 P405-410 65-14632 <*> 0

ANNALES D'HYGIENE PUBLIQUE, INDUSTRIELLE ET SOCIALE
1929 V7 P733-740 66-13114 <*>
1932 V10 P209-212 63-10455 <*>
1941 V18 P133-137 63-14559 <*>

ANNALES DE L'INSTITUT EXPERIMENTAL DU TABAC DE BERGERAC
1951 V1 P97-125 2318-D <K-H>
1951 V1 N2 P36-54 2318-A <K-H>
1952 V1 N3 P61-69 3180-B <K-H>
1956 V2 N3 P121-124 60-18476 <*>
1956 V2 N3 P125-129 60-18475 <*>
1956 V2 N3 P149-164 59-10211 <*>
1958 V3 N1 P67-76 9092-B <K-H>
1958 V3 N1 P77-83 9039-B <K-H>

ANNALES DE L'INSTITUT FOURIER
1951 V3 P301-319 57-1190 <*>

ANNALES DE L'INSTITUT HENRI POINCARE
1952 V13 P1-42 57-798 <*>

ANNALES DE L'INSTITUT PASTEUR. PARIS
1921 V35 P487-495 65-11060 <*> 0
1921 V35 P893-901 65-12559 <*>
1924 V38 P1-10 66-11531 <*>
1929 V43 P549-633 65-11081 <*> 0
1931 V46 P202-221 I-610 <*>
1931 V47 P339-357 66-10933 <*>
1932 V48 N2 P208-267 63-14800 <*>
1938 V61 N2 P121-171 61-00299 <*>
1939 V63 P159-189 64-16391 <*> 0
1940 V64 P301- 2605 <*>
1945 V71 N9/10 P431-439 59-20299 <*>
1946 V72 N5/6 P441-444 2510 <*>
1947 V73 P695-696 I-713 <*>
1947 V73 N2 P188-191 2509 <*>
1947 V73 N8 P713-724 2625 <*>
1948 V74 N6 P507 60-17665 <*+>
1948 V75 P33-51 3169 <*>
1949 V77 P389-394 58-809 <*>
63-16584 <*> 0
1949 V77 N3 P310-313 2603 <*>
1949 V77 N6 P653-656 2782 <*>
1949 V77 N6 P764-767 2783 <*>
1950 V78 N2 P283-285 2602 <*>
1950 V79 N4 P390-410 59-17834 <*> 0
59-18840 <+*> 0
1950 V80 P644-659 58-2225 <*>
1951 V81 P126- 2518 <*>
1951 V81 P665-667 65-00307 <*>
1951 V81 N126 P23- 2545 <*>
1952 V82 P145-158 2020 <*>
1952 V83 N4 P449-454 MF-2 <*>
1952 V83 N4 P455-463 MF-2 <*>
1953 V84 P370-375 58-163 <*>
1953 V84 P937-945 66-12242 <*>
1953 V84 N5 P890-894 MF-2 <*>
1953 V85 P336-347 AEC-TR-3536 <*>
1953 V85 P492-496 60-17576 <*+>
1953 V85 P690-705 2211 <*>
1953 V85 N5 P544-575 84F2F <ATS>
1953 V85 N5 P671-674 06F3F <ATS>
1954 V86 N1 P38-52 55F3F <ATS>
1954 V86 N2 P197-203 80F3F <ATS>
1954 V86 N2 P236-237 II-655 <*>
1954 V86 N3 P356-369 27F3F <ATS>
1954 V86 N4 P418-424 57-1138 <*>
1954 V86 N5 P618-635 59-14529 <+*> 0
1954 V87 N1 P46-49 57-860 <*>
1954 V87 N5 P522-535 70G5F <ATS>
1955 V88 N1 P44-59 71G5F <ATS>
1955 V89 P206-215 36H11F <ATS>
58-310 <*>
1955 V89 P531-552 57-1812 <*>
1955 V89 P554-555 57-1812 <*>
1955 V89 P654-665 58-2339 <*>
1955 V89 N5 P505-513 58-1559 <*>
1956 V90 N5 P660-665 59-20322 <*>
1956 V91 P693-720 59-16858 <+*> 0
1956 V91 N2 P269-273 57-1985 <*>
1956 V91 N2 P274-277 57-2011 <*>
1956 V91 N5 P614-622 58-1603 <*>
1957 V92 P62-73 58-2338 <*>
1958 V94 P636-642 58-2199 <*>
1958 V94 N5 P670-673 59-15911 <*>
1958 V95 P793-795 62-00152 <*>
1958 V95 N1 P73-87 62-00379 <*>
1959 V96 P79-92 63-18508 <*>
1959 V96 N1 P72-78 59-12912 <+*> 0
1959 V96 N3 P289-294 60-18660 <*>
1959 V96 N3 P295-302 60-18659 <*>
1959 V96 N3 P303-332 59-18467 <+*> 0
1959 V96 N3 P365-368 NP-TR-882 <*>
1959 V97 P718-732 63-00212 <*>
1959 V97 N1 P41-72 63-16514 <*>
1960 V98 P229-235 63-00155 <*>
1960 V98 N1 P112-118 60-17662 <+*> 0
1960 V98 N2 P173-203 62-00160 <*>
1961 V100 SUP. P71-76 63-16508 <*>
1961 V100 N4 P32-43 SUP 64-18699 <*>
1961 V100 N4 P7-13 SUP 64-18696 <*>
1961 V100 N4 P14-31 SUP 64-18625 <*>
1961 V100 N4 P53-58 SUP 64-18704 <*>
1961 V100 N4 P59-66 SUP 64-18673 <*>
1961 V100 N4 P67-70 SUP 64-18703 <*>
1961 V100 N4 P77-83 SUP 63-16513 <*>
1961 V100 N4 P84-88 SUP 64-18678 <*>

1961 V100 N6 P805-811 62-16234 <*> 0
1961 V100 N6 P100-113 SUP 63-16511 <*>
1961 V101 P104-119 63-20865 <*>
1961 V101 P703-721 63-00617 <*>
1961 V101 N2 P203-210 63-16509 <*>
1961 V101 N3 P313-317 62-01001 <*>
1962 V102 N2 P137-152 63-00311 <*>
1962 V103 P24-31 63-18799 <*>
1962 V103 P841-865 63-01685 <*>
1963 V104 P26-42 63-01133 <*>
1963 V104 P137-141 66-12628 <*>
1963 V104 P313-334 63-01678 <*>
1963 V104 P734-745 65-00117 <*>
1963 V105 N5 P849-865 64-00273 <*>
1965 V108 N2 P196-214 66-12168 <*>

ANNALES DE L'INSTITUT PHYTOPATHOLOGIQUE BENAKI
1935 V1 P51-56 58-1025 <*>
1957 SNS V1 P44-79 60-18545 <*>

ANNALES DE L'INSTITUT POLYTECHNIQUE DE GRENOBLE
1952 P7-21 I-95 <*>
1952 N4 P187-209 57-2746 <*>

ANNALES DE L'INSTITUT TECHNIQUE DU BATIMENT ET DES TRAVAUX PUBLICS
1948 N8 P1-3 65-12411 <*> P
1948 N9 P1-8 57-2791 <*>
1950 N140 P1-22 58-2584 <*>
1951 N204 P3-18 65-10414 <*>
1952 N58 P972-990 3387 <*>
1952 N59 P1131-1146 3386 <*>
1952 V5 N51 P385-390 64-10007 <*>
1952 V5 N59 P1131-1146 1850 <*>
1952 V5 N235 P9-12 59-17925 <*> 0
1955 P320-336 SUP 1GRL-T/R-72 <*>
1955 N86 P171-194 58-2605 <*>
1955 V8 N86 P171-194 SUP 61-18959 <*>
1956 N103 P693-698 C-2546 <NRC>
58-610 <*>
1956 V9 N105 P785-822 61-18962 <*>
1957 N109 P34-62 58-2586 <*>
1957 V10 N109 P34-62 SUP 61-18964 <*>
1958 V11 N121 P78-98 63-20617 <*>
1959 P1369-1390 61-20561 <*> 0
1960 V12 N145 P15-28 SUP 61-18983 <*>
1960 V12 N154 P1003-1016 C-3979 <NRC>
1961 N159 P377-388 C-3857 <NRC>
1963 V16 N181 P1259-1276 65-00310 <*>
1963 V16 N185 P496-509 <CCA>
65-13523 <*>
1963 V16 N186 P590-620 65-00311 <*>
1963 V16 N187 P775-811 65-13521 <*>
1963 V16 N190 P1054-1062 C-5112 <NRC>
1963 V16 N192 P1303-1329 65-13141 <*>
1964 V17 N201 P957-958 66-14559 <*>
1964 V17 N201 P959-972 66-14558 <*> P

ANNALES DES MALADIES DE L'OREILLE ET DU/LARYNX
1928 V47 P652-659 61-14643 <*>

ANNALES DE MEDECINE
1954 V55 P660-673 57-2966 <*>
1954 V55 N7 P660-673 1707 <*>

ANNALES DE MEDECINE LEGALE, DE CRIMINOLOGIE ET DE POLICE SCIENTIFIQUE
1930 V27 N2 P124-131 66-12154 <*>
1958 V38 P150-154 64-10612 <*>
8500-C <K-H>
1959 V39 N6 P535-537 62-16010 <*>
1960 V40 N1 P40-43 10401-E <KH>
64-14525 <*>
1962 V42 P359-364 64-00263 <*>

ANNALES DE MEDECINE VETERINAIRE
1954 V98 N7 P432-440 3173 <*>

ANNALES MEDICALES DE NANCY
1964 V3 P546-556 66-13036 <*>

ANNALES MEDICINAE EXPERIMENTALIS ET BIOLOGIAE FENNIAE
1947 V24 N1 P115-116 64-10193 <*>

ANNALES MEDICINAE INTERNAE FENNIAE
1947 V36 P516-522 1305 <K-H>

ANNALES MEDICO-PSYCHOLOGIQUES
1932 S14 V90 P565-572 62-00591 <*>
1956 V114 P830-834 5525-E <K-H>
1958 V116 P965-979 PT.2 64-10363 <*>
1958 V116 N3 P560-561 PT.2
60-13038 <+*>
1958 V116 N3 P562-566 PT.2
60-13136 <+*>
1960 V118 P145-152 64-00069 <*>
1960 V118 N1 PT1 P119-133 65-10964 <*>
1961 V119 P140-146 PT.1 64-10362 <*>
1961 V119 P378 64-00051 <*>
1961 V119 N1 PT2 P145-149 65-10968 <*>
1961 V119 N2 PT2 P352-360 66-10975 <*>
1961 V119 N5 PT2 P968-976 65-10981 <*>
1961 V119 N5 P968-976 PT.2
62-16223 <*>
1963 V121 N2 PT1 P193-204 65-10955 <*> 0
1963 V121 N2 P191-200 65-60301 <=$>
1964 V122 N4 P585- PT2 66-12016 <*>
1965 V123 N2 P237-245 PT2 66-11833 <*>
1965 V123 N2 P223-236 66-12612 <*>

ANNALES DES MINES DE BELGIQUE
1953 V3 P387-389 1848 <*>
1956 P31-37 SPEC.NO. 58-2699 <*>
1956 P31-37 SPEC NO 59-18538 <+*> 0
1956 SPEC NO P102- 59-18568 <+*>
1956 SPEC.NO. P102- 58-2500 <*>
1957 N5 P420-423 63-16868 <*>
1959 N5 P478-534 62-11114 <=>

ANNALES DES MINES ET DES CARBURANTS
1856 S5 V9 P505- 64-14240 <*>
1950 V139 N1 P35-63 63-14594 <*>
1957 P95-101 CSIRO-3475 <CSIR>
1958 V147 P93-111 TS-1757 <BISI>
1962 V167 P121-132 2980 <BISI>

ANNALES DE LA NUTRITION ET DE L'ALIMENTATION
1955 V9 N1 P93-130 57-855 <*>
1956 V10 N5/6 P239-252 CSIRO-3415 <CSIR>

ANNALES DE L'OBSERVATOIRE DE STRASBOURG
1933 V2 N3 P205-339 64-10816 <*>

ANNALES D'OCULISTIQUE
1881 V85 P182-185 61-14107 <*> 0
1923 V160 P652-657 61-14442 <*> 0
1926 V163 N9 P673-683 61-14157 <*> 0
1927 V164 N10 P770-780 61-14280 <*>
1961 V194 N3 P236-258 63-01525 <*>
64-16904 <*> 0

ANNALES D'OTO-LARYNGOLOGIE
1948 V65 N3 P137-142 9092-A <K-H>
1948 V65 N3/4 P137-142 63-16605 <*>
1948 V65 N3/4 P155-158 63-10713 <*>
1954 V71 P257-274 5795-B <KH>
1955 V72 N11/2 P928-930 63-10714 <*>
1959 V76 N4/5 P366-390 60-15724 <=*>
1960 V77 N6 P492-496 64-18676 <*>
1961 V78 N12 P773-778 64-16940 <*>
1963 V80 N9 P817-819 65-14675 <*>
1964 V81 P449-460 66-12605 <*>
1964 V81 N4/5 P259-263 66-13219 <*>

ANNALES PAEDIATRICI
1946 V167 P188-198 61-01053 <*>
1948 V170 N2 P57-95 2547 <*>
1948 V171 N5/6 P286-290 2516 <*>
1950 V174 P87-96 63-00081 <*>

1953 V180 N4/5 P279-281 2221 <*>
1955 V184 N2 P64-88 I-1006 <*>

ANNALES DE PARASITOLOGIE HUMAINE ET COMPAREE
1949 V24 N5/6 P424-435 59-20887 <*>
1960 V35 N5/6 P687-703 62-01505 <*>

ANNALES DE PEDIATRIE
1960 V7 P429-433 64-18705 <*>
1961 V8 P19-25 62-00161 <*>
1963 V10 P252-254 65-14591 <*>
1963 V10 P326-329 65-14451 <*>
1965 V12 P324-329 66-13185 <*>

ANNALES PHARMACEUTIQUES FRANCAISE
1944 P1-43 2925 <*>
1944 V2 SUP 4, P3-43 62-16040 <*>
1944 V2 P29-33 1922 <K-H>
1944 V2 P78-81 1922 <K-H>
1948 V6 P338-347 63-14049 <*>
1949 V7 P77-89 64-10005 <*>
1950 V8 P261-273 09F3F <ATS>
1951 V9 P338-343 64-14763 <*> 0
1952 V10 P435-440 63-14589 <*>
1953 V11 P588-594 57-1870 <*>
1953 V11 P615-620 59-22098 <=*>
1953 V11 N9/10 P569-578 64-18978 <*>
1954 V12 P799-814 64-14762 <*> 0
1954 V12 N3 P193-203 64-20007 <*> 0
1954 V12 N3 P203-210 64-18999 <*> 0
64-20005 <*> 0
1954 V12 N7/8 P509-518 5604-B <K-H>
1955 V13 P192-199 3821-B <K-H>
1956 V14 P470-475 1595 <TC>
1957 V15 P39-54 75S85F <ATS>
1958 V16 N2 P86-90 58-2655 <*>
1958 V16 N2 P125-133 59-15604 <*>
1959 V17 N10/2 P585-592 61-20438 <*>
1960 V18 N6 P454-456 62-00852 <*>
1960 V18 N1C P759-762 65-14639 <*>
1961 V19 N1 P24-36 64-18619 <*>
1961 V19 N9/10 P604-609 62-14533 <*>
1962 V20 N5 P409-419 63-00521 <*>
1963 V21 N11 P723-726 65-14884 <*>
1965 V23 P117-124 66-13198 <*>

ANNALES DE PHYSIOLOGIE ET DE PHYSICOCHIMIE BIOLOGIQUE
1928 V4 P739-743 57-3157 <*>
1937 V13 P971-973 59-10635 <*>
1940 V16 P298-312 60-10841 <*>

ANNALES DE PHYSIOLOGIE VEGETALE. PARIS
1959 V1 N2 P179-187 64-30288 <*>

ANNALES DE PHYSIQUE
1932 S10 V17 P283-314 62-00373 <*>
1935 S11 V3 P321-369 SCL-T-365 <*>
1938 V9 P645-722 58-600 <*>
1939 S11 V12 P161-237 57-3073 <*>
1948 S12 V3 P137-198 57-3472 <*>
1948 S12 V3 P322-375 65-10807 <*> P
1948 S12 V3 P521-549 AL-157 <*>
1949 S12 V4 P269-309 AEC-TR-3646 <*>
1951 S12 V6 P561-609 <AC>
1952 V7 P185-237 PT12 II-301 <*>
1952 V7 P710-747 66-12251 <*>
1952 S12 V7 P185-237 MF-1 <*>
1955 V10 P830-873 57-887 <*>
1956 S13 V1 P344-394 UCRL TRANS-592(L) <*>
1957 S13 V2 P214-266 E-655 <RIS>
1957 S13 V2 P309-317 62-18537 <*>
1958 S13 V3 P424-462 62-25260 <=*>
1960 S13 V5 N3/4 P469-508 65-17095 <*>
1961 S13 V6 N1/2 P5-19 64-15683 <=*$>
1962 V7 N9/10 P469-504 65-12338 <*>
1962 S13 V7 N7/8 P423-461 65-10426 <*> 0
1962 S13 V7 N9/10 P535-544
64-14001 <*>
1963 S13 V8 P531-576 335 <TC>
1963 S13 V8 P733-746 66-12197 <*>
1965 S13 V10 N1/2 P13-21 65-17148 <*>

ANNALES DES POIDS ET MESURES
1907 V13 P1-48 63-14704 <*>

ANNALES DES PONTS ET CHAUSSEES
1887 S7 P338-388 65-12283 <*> 0
1901 S8 V4 N50 P129-190 AD-607 734 <=>
1932 V102 P1434-1441 64-14849 <*>
1932 V102 N3 P408-416 64-14351 <*>
1934 V104 N6 P474-498 64-18284 <*> 0
1949 V114 N1 P25-78 AL-488 <*>
1952 V122 N5 P543-562 64-71381 <=>
1952 V122 N52 P525-542 62-16474 <*>
1952 S8 V122 N4 P463-494 65-12369 <*>
1953 V123 N3 P281-286 AL-605 <*>
1953 V123 N3 P543-562 64-18272 <*> 0

ANNALES DES POSTES, TELEGRAPHES ET TELEPHONES
1913 V4 N2 P198-209 57-2240 <*>
1914 V4 N4 P566-603 57-1575 <*>
1918 V7 P281- 57-2199 <*>
1920 P366-412 57-2210 <*>
1920 V9 P37- 57-1546 <*>
1921 V10 P193-207 57-556 <*>
1923 V12 P1016-1061 57-2183 <*>
1924 V13 P1235-1249 58-349 <*>
60-10898 <*>
1924 V13 N11 P1215-1217 66-14137 <*>
1925 V14 P709-728 57-2122 <*>
1925 V14 N1 P67-73 66-12690 <*>
1926 P13-24 57-1048 <*>
1926 V15 P768-825 57-2469 <*>
1927 V16 P1-16 57-875 <*>
1927 V16 P932-951 66-12250 <*> 0
1927 V16 N7 P592-594 57-908 <*>
1930 V19 P1009- 57-1582 <*>
1931 V20 N11 P882-902 57-2083 <*>
1932 P329-341 57-2374 <*>
1934 V23 N3 P201-221 57-2237 <*>
1935 V24 N6 P573-599 57-321 <*>
1936 V25 N6 P595-603 57-2046 <*>
1937 V26 N11 P965-982 57-945 <*>
1939 P319-361 57-2062 <*>
1939 V28 N3 P175-210 57-1037 <*>

ANNALES DE RADIOELECTRICITE
1948 V3 P339-343 58-1450 <*>
1948 V3 N12 P71-106 60-14339 <*>
1948 V3 N14 P281-289 63-00563 <*>
1948 V3 N14 P339-343 62-18566 <*> 0
6688 <K-H>
1949 V4 P76-84 58-1272 <*>
1949 V4 P117-130 58-1272 <*>
1949 V4 N15 P54-67 60-16850 <*>
1950 V5 P155-167 I-723 <*>
1950 V5 N19 P8-11 57-1482 <*>
1951 V6 P99-105 T-1562 <INSD>
1951 V6 N23 P51-83 58-1269 <*>
1951 V6 N24 P114-155 58-1269 <*>
1952 V7 P75-77 T-1561 <INSD>
1952 V7 N29 P225-234 58-1290 <*>
1953 V8 P183-193 57-2490 <*>
1953 V8 N31 P3-19 66-12706 <*>
1953 V8 N32 P103-110 R-991 <*>
1954 V9 P83-97 II-1048 <*>
1954 V9 P219-226 II-947 <*>
1954 V9 N36 P180-192 57-3324 <*>
1954 V9 N36 P193-216 59-10209 <*> 0P
1954 V9 N38 P311-328 57-3323 <*>
1955 V10 N39 P3-19 62-23076 <=*>
1955 V10 N39 P42-63 65-10806 <*>
1955 V10 N39 P64-73 57-602 <*>
1955 V10 N41 P227-255 57-3265 <*>
1955 V10 N42 P387-398 58-219 <*>
1956 V11 N43 P3-28 1536 <*>
1956 V11 N43 P29-56 30N55F <ATS>
61-00099 <*>
1956 V11 N44 P165-183 61-10417 <*>

1958 V13 N52 P107-129	AEC-TR-3704 <+*>
	65-11364 <*>
1958 V13 N52 P130-139	61-10455 <*>
1958 V13 N53 P212-223	59-20696 <*>
1959 V14 N55 P3-16	62-25709 <=*>
1959 V14 N58 P1-	63-00185 <*>
1960 V15 N60 P1-	63-00185 <*>
1962 V17 N67 P3-12	63-10247 <*>
1962 V17 N69 P223-226	64-14628 <*>
1965 V20 N81 P233-245	66-11525 <*>

ANNALES DES SCIENCES NATURELLES. BOTANIQUE

1933 V15 P100-136	65-17280 <*>

ANNALES SCIENTIFIQUES DE L'ECOLE NORMALE SUPERIEURE

1883 V12 P47-88	94H14F <ATS>
1916 V33 P17-69	AL-90-II <*>

ANNALES SCIENTIFIQUES TEXTILES BELGES

1956 V4 N1 P176-194	61-16796 <*>
1956 V4 N4 P7-30	65-14746 <*>
1960 V8 N3 P7-29	61-16795 <*>
1961 V9 N3 P63-68	63-10794 <*>

ANNALES DE LA SOCIETE BELGE DE MEDECINE TROPICALE

1959 V39 P653-662	60-16957 <*>
1962 V42 N1 P133-136	66-11600 <*>

ANNALES DE LA SOCIETE GEOLOGIQUE DU NORD

1904 V33 P50-63	3K22F <ATS>
1904 V33 P63-75	1064-FJ <ATS>
	45K23F <ATS>

ANNALES. SOCIETE R. DES SCIENCES MEDICALES ET NATURELLES DE BRUXELLES

1954 V7 P129-138	II-897 <*>
1963 V16 N1 P5-36	66-12966 <*>

ANNALES DE LA SOCIETE SCIENTIFIQUE DE BRUXELLES

1910 V35 P1-16	2243 <*>
1950 S1 V64 P188	AEC-TR-5322 <*>
1958 S1 V72 P159-169	64-10609 <*>

ANNALES DE TECHNOLOGIE AGRICOLE

1954 V3 P241-258	65-12936 <*>

ANNALES DES TELECOMMUNICATIONS

1946 V1 N12 P270-282	66-13344 <*> 0
1947 V2 N4 P118-124	66-12263 <*>
1947 V2 N4 P129-136	62-18068 <*> 0
	66-13339 <*> 0
1947 V2 N7 P222-230	57-306 <*>
1951 V6 P325-330	64-14236 <*> 0
1951 V6 N5 P122-126	57-980 <*>
1952 V7 P150-204	58-1305 <*>
1952 V7 N2 P75-84	57-3523 <*>
1952 V7 N2 P99-108	65-12355 <*>
1952 V7 N5 P206-216	58-1307 <*>
1952 V7 N6 P262-275	58-1309 <*>
1953 V8 N5 P171-183	57-2501 <*>
	57-650 <*>
1953 V8 N11 P373-376	57-578 <*>
1953 V8 N8/9 P271-	1177 <*>
1954 V9 N2 P44-45	59-17035 <*> 0
1954 V9 N7/8 P201-203	C-2316 <NRC>
	57-2354 <*>
1955 V10 N10 P206-216	65-14209 <*> 0
1955 V10 N11 P237-242	57-1975 <*>
1957 V12 N6 P234-244	39K22F <ATS>
1958 V13 P114	R-2708 <*>
1959 V14 N1/2 P21-32	71M40F <ATS>
1963 V18 N11/2 P294-300	65-00276 <*>
1964 V19 N1/2 P49-59	65-10431 <*> 0
1964 V19 N7/8 P169-172	66-13831 <*>
1964 V19 N7/8 P173-187	66-13869 <*>
1965 V20 N1/2 P2-4	66-10487 <*>

ANNALES TEXTILES

1955 N4 P12-	60-10792 <*>
1956 N3/9 P62-77	T-2098 <INSD>

ANNALES DES TRAVAUX PUBLICS DE BELGIQUE

1949 V102 P649	61-18950 <*>
1949 V102 P687-716	61-18950 <*>
1949 V102 N5 P481-512	61-18919 <*>
1949 V102 N5 P513-533	58-2570 <*>
	61-18870 <*>
1949 V102 N6 P687-716	58-2555 <*>
1950 V103 N2 P201-256	58-2594 <*>
	61-18946 <*>
1950 V103 N2 P261-266	61-18947 <*>
1959 V112 N2 P28-29	62-18015 <*>

ANNALES UNIVERSITATIS FENNICAE ABOENSIS. ABO, HELSINKI

1932 SB V17 P1-114	65-00216 <*> P

ANNALES UNIVERSITATIS MARIAE CURIE-SKLODOWSKA

1947 V2 N10 P167-183	59-15831 <*>
1951 V6 N9 P193-211	60-21516 <=>
1952 V7 N15 P473-486	60-21406 <=>
1953 V8 N3 P41-59	60-21524 <=>
1954 V9 N4 P59-68	57-1184 <*>
1955 V10 N9 P239-266	60-21223 <=>
1956 V11 N6 P124-186	63-13855 <=*>
1956 V11 N8 P223-225	60-21407 <=>
1958 V13 N4 P35-44	10693-G <K-H>
	64-10671 <*>
1958 V13 N7 P127-130	60-21403 <=>
1958 V13 N14 P231-259	61-11343 <=>
1959 V14 N20 P181-206	10938-A <K-H>
	63-16607 <*>

ANNALES DE L'UNIVERSITE DE GRENOBLE

1946 V22 P299-343	57-31 <*>
	57H12F <ATS>

ANNALES ZOOLOGICI

1958 V4 N9 P135-230	60-21404 <=>

ANNALES DE ZOOTECHNIE

1953 V2 P275-284	65-12679 <*>
1954 V3 P1-7	59-20883 <*>
1955 V4 P133-145	65-12673 <*>
1955 V4 P146-152	59-20883 <*>
1955 V4 P153-163	59-20171 <*>
1956 V5 P237-253	61-20442 <*>
1956 V5 N1 P12-19	59-20882 <*>
1957 V6 N1 P29-40	65-12681 <*>

ANNALI DI CHIMICA

1950 V40 P195-204	71M47I <ATS>
1950 V40 P222-233	45P61I <ATS>
1950 V40 N3/5 P135-142	65-12483 <ATS>
1950 V40 N3/5 P143-148	65-12484 <*> 0
1951 V41 P51-60	63-16843 <*>
1951 V41 P293-308	6255-B <KH>
1951 V41 P309-319	6196-B <K-H>
1951 V41 P455-464	5876-A <KH>
	63-16350 <*>
1952 V42 P227-238	60-23421 <*+>
1952 V42 P349-355	64-14233 <*>
1952 V42 P598-606	NP-TR-555 <*>
1952 V42 P632-644	62-14043 <*> 0
1953 V43 P853-864	66-11039 <*>
1953 V43 P865-872	66-11038 <*>
1953 V43 N2 P113-118	64-20072 <*> 0
1954 P787-796	T-2357 <INSD>
1954 V44 P11-19	2439 <*>
1954 V44 P59-62	6256 <K-H> 0
	64-20205 <*> 0
1954 V44 P299-304	3737-C <K-H>
	64-16118 <*>
1954 V44 P443-453	75H10I <ATS>
1954 V44 N3/4 P288-298	63-00757 <*>
1955 V45 P614-617	66-11033 <*>
1956 V46 P194-198	63-14706 <*>
1956 V46 P793-805	C-2426 <NRC>
	57-3029 <*>

1957 V47 P410-414 60-18056 <*>
1957 V47 P639-645 58-835 <*>
1957 V47 P797-805 11R801 <ATS>
1957 V47 N2 P150-154 10216 <KH>
1957 V47 N2 P167-177 C-3196 <NRC>
1958 V48 N2 P193-197 59-15006 <*>
1958 V48 N6/7 P501-509 60-14148 <*>
64-30964 <*> 0
65-12934 <*>
8549 <K-H>
1959 V49 P902-910 61-18710 <*>
1959 V49 P1143-1154 62-16445 <*> 0
1960 V49 P1639-1648 60-18525 <*>
1960 V50 P83-98 42N55I <ATS>
1960 V50 P1540-1546 9023 <IICH>
1960 V50 P1573-1580 46Q68I <ATS>
1960 V50 N12 P1713-1720 66-13116 <*> 0
1960 V50 N1/2 P147-155 66-13115 <*>
1961 V51 P1399-1403 67S811 <ATS>
1961 V51 N12 P1399-1403 65-13059 <*>
1962 V52 P620-628 63-20351 <*>
1962 V52 N7 P620-628 C-4545 <NRCC>
1963 V53 P96-116 41S87I <ATS>
1963 V53 N7 P894-900 65-12075 <*>
1963 V53 N1/2 P70-77 (CHEM)-351 <CTS>
1964 V54 P1081-1087 66-14064 <*>
1964 V54 N5 P384-411 (CHEM)-416 <CTS>

ANNALI DI CHIMICA APPLICATA
1929 V19 N6 P273-282 66-13117 <*>
1933 V23 P235- I-386 <*>
1935 V25 P163-173 1996 <*>
1935 V25 P620-628 63-16082 <*>
1937 V27 P505-518 64-18193 <*> 0
1938 V28 P405-408 63-16138 <*>
1941 V31 P20-24 63-16080 <*>
1941 V31 P68-73 1904A <K-H>
1941 V31 P129-130 64-20394 <*>
1941 V31 P235-254 63-20248 <*> P0
1941 V31 N6 P221-227 1906 <K-H>
1941 V31 N10 P453-456 60-10536 <*>
1944 V34 P99-126 CSIRO-3472 <CSIR>
1944 V34 P99-136 59-10906 <*>
1946 V36 P273 64-10397 <*>
1946 V36 N3/4 P69-78 61-10258 <*>
1947 V37 P24-34 65-10130 <*> 0
1949 V39 P142-148 I-457 <*>
1949 V39 N6/7 P311-320 65-12482 <*>

ANNALI DELLA FACOLTA DI AGRARIA DI PORTICI DELLA R. UNIVERSITA DI NAPOLI
1950 S3 V19 P79-88 3359-D <K-H>

ANNALI DELLA FACOLTA DI MEDICINAE CHIRURGIA, UNIVERSITA DI PERUGIA
1956 V47 N4 P343-362 57-2343 <*>

ANNALI DELLA FACOLTA DI MEDICINA VETERINARIA DI TORINO
1958 V8 P181-192 62-10287 <*> 0P

ANNALI DI FARMACOTERAPIA E CHIMICA BIOLOGICA
1887 S4 V28 P297-301 58-1243 <*>

ANNALI DI FITOPATOLOGIYA
1952 V1 P26-29 MF-5 <*>

ANNALI DI GEOFISICA
1949 V2 P92-102 57-1271 <*>
1949 V2 P137-141 57-889 <*>
1954 V7 N1 P451-462 59-17608 <*>
1956 V9 N4 P469-524 64-30660 <*>
1957 V10 P61-70 63-01142 <*>
1960 V13 P297-368 62-00738 <*>
1960 V13 P389-437 62-00738 <*>
1960 V13 N1 P85-128 62-00738 <*>

ANNALI DELL'ISTITUTO CARLO FORLANINI
1956 V16 N2 P233-243 5998 A <K-H>
1959 V19 N1 P1-17 61-00202 <*>

ANNALI ITALIANI DI CHIRURGIA
1936 V15 P104-117 57-1671 <*>
1956 V33 N2 P172-193 60-15151 <=*> 0
1957 V34 N2 P117-143 59-10346 <*>
1960 V37 P363-383 63-20201 <*> 0

ANNALI DI MATEMATICA PURA ED APPLICATA
1953 S4 V35 P327-341 57-212 <*>
1955 S4 V39 P97-119 65-18124 <*>

ANNALI DI MEDICINA NAVALE E COLONIALE. ROME
1932 V38 P641-663 64-10832 <*> 0

ANNALI DI MEDICINA NAVALE E TROPICALE
1960 V65 P535- 10860-A <K-H>

ANNALI DI MICROBIOLOGIA ED ENZIMOLOGIA
1954 V6 P18-40 63G61 <ATS>
1954 V6 P51-60 59-15252 <*>
1955 V6 P183-199 59-15253 <*>
1957 V7 N5/6 P164-182 58-2667 <*>
1960 V10 P77-81 62-10332 <*>

ANNALI DI OSTETRICIA E GINECOLOGIA
1939 V61 P371-386 57-1635 <*>
1951 V73 P644-659 58-1768 <*>
1961 V83 P1083-1100 64-14742 <*>

ANNALI DI OTTALMOLOGIA E CLINICA OCULISTA
1910 V39 P767-787 61-14359 <*>
1925 V53 N7 P130-138 61-14131 <*>
1925 V53 N7 P235-254 61-14112 <*>
1927 V55 N11/2 P923-925 61-14142 <*>
1949 V75 P45-54 57-2969 <*>
1959 V85 N8 P372-386 10N52I <ATS>
1960 V86 P87-97 63-16614 <*>
1963 V89 P566-571 65-14633 <*> 0

ANNALI DELLA SANITA PUBBLICA
1951 V12 N1 P67-73 1828 <*>
1959 V20 N4 P689-709 9915-C <K-H>
1960 V21 P1395-1444 11234 <KH>
1964 V25 N3 P439-516 66-10936 <*>

ANNALI SCLAVO
1961 V3 N1 P37-48 62-00274 <*>
1961 V3 N1 P63-95 62-00975 <*>
1961 V3 N6 P692-702 63-01749 <*>
1962 V4 N1 P103-106 62-00803 <*>
1963 V5 N1 P33-44 64-18683 <*>

ANNALI DELLA SPERIMENTAZIONE AGRARIA
1948 V2 N1 P69-78 1303B <K-H>

ANNALI DELLA STAZIONE CHIMICO-AGRARIA SPERIMENTALE DI ROMA
1955 N116 P1-9 5547-D <K-H>
1955 N117 P1-8 5547-D <K-H>
1955 N119 P1- 5547-D <K-H>

ANNALI DELL'UNIVERSITA DI FERRARA. SEZIONE 9. SCIENZE GEOLOGICHE E PALEONTOLOGICHE
1957 V3 N1 P1-36 87N50I <ATS>

ANNALS OF THE ENTOMOLOGICAL SOCIETY OF QUEBEC
1958 V4 P25-32 65-11680 <*>

ANNALS OF MATHEMATICS
1956 V64 N1 P20-82 65-11690 <*> P

ANNALS OF THE PHYTOPATHOLOGICAL SOCIETY OF JAPAN
SEE NIHON SHOKUBUTSU BYORIGAKU KAIHO

ANNEE PSYCHOLOGIQUE
1925 V26 P72-78 61-14574 <*>
1926 V27 P203-206 61-14575 <*>
1940 V41/2 P46-64 61-14710 <*>
1948 V49 P341-358 65-12621 <*>
1951 V50 P507-519 58-2243 <*>

1951 V50 P705-711 64-10022 <*>
1951 V51 P35-52 64-10018 <*>
1951 V51 P237-248 64-10337 <*>
1952 V52 N1 P39-46 65-12667 <*>
1952 V52 N1 P127-135 59-20167 <*>
1953 V53 N2 P539-551 65-12371 <*>
1953 V53 N2 P553-562 59-20168 <*>
1954 V54 N2 P345-356 65-12567 <*>
1955 V55 N2 P329-348 65-12586 <*>
1956 V56 N2 P453-460 59-20889 <*>
1956 V56 N2 P461-474 62-00496 <*>
1957 V57 P23-32 60-18636 <*>
1959 V59 P381-394 61-20439 <*>

ANNUAL REPORT
SEE REPORT

ANTIBIOTICA ET CHEMOTHERAPIA
1960 V8 P17-31 62-10337 <*>
1963 V11 P136-147 65-14422 <*>
1963 V11 P194-228 64-18698 <*>
1964 V12 P253-259 66-13556 <*>

★ANTIBIOTIKI
1956 V1 N1 P15-18 R-457 <*>
1956 V1 N1 P18-20 59-12758 <+*>
1956 V1 N2 P47-50 R-1570 <*>
1956 V1 N4 P9-13 59-16795 <+*>
1956 V1 N4 P44-48 60-10985 <+*>
1956 V1 N5 P26-30 59-12759 <+*> 0
1957 N3 P49-53 58-1836 <*>
1957 V2 N1 P5-10 59-12760 <+*>
1957 V2 N3 P14-17 40M43R <ATS>
60-15485 <+*>
1957 V2 N3 P24-27 60-15486 <+*>
1957 V2 N3 P35-37 60-15487 <+*>
1957 V2 N3 P37-40 60-15488 <+*>
1957 V2 N3 P40-44 60-15489 <+*>
1957 V2 N3 P44-48 60-15490 <+*>
1957 V2 N3 P49-53 64-19397 <=$>
1957 V2 N4 P32-35 60-15491 <+*>
1957 V2 N4 P35-37 60-15492 <+*>
1957 V2 N5 P3-7 T-2629 <INSD>
1957 V2 N5 P8-11 T-2630 <INSD>
1957 V2 N5 P35-39 T-2633 <INSD>
1957 V2 N5 P40-44 PB 141 167T <=>
1957 V2 N5 P44-49 PB 141 237T <=>
1957 V2 N5 P49-53 59-13065 <=>
1957 V2 N6 P4-8 86M44R <ATS>
1958 V3 N1 P10-13 59-11655 <=>
1958 V3 N1 P18-22 59-11655 <=>
1958 V3 N1 P31-40 59-11655 <=>
1958 V3 N2 P29-34 41M43R <ATS>
59-13266 <=> 0
1958 V3 N2 P63-67 59-13267 <=>
1958 V3 N2 P94-99 55Q71R <ATS>
1958 V3 N2 P114-122 59-13268 <=>
1958 V3 N3 P8-12 4K26R <ATS>
1958 V3 N3 P18-22 5K26R <ATS>
61-10431 <*+>
1958 V3 N4 P3-6 59-15710 <+*> 0
1958 V3 N6 P24-26 60-10983 <+*>
1958 V3 N6 P36-41 32L30R <ATS>
59-20806 <+*>
1959 V4 N1 P10-15 60-10953 <+*>
1959 V4 N1 P21-25 60-10955 <+*>
1959 V4 N1 P21-26 60-15401 <+*> 0
1959 V4 N1 P26-31 60-10954 <+*>
1959 V4 N1 P37-39 60-10956 <*>
1959 V4 N2 P9-14 60-10957 <+*>
1959 V4 N2 P34-39 60-10959 <+*>
1959 V4 N2 P40-45 60-10958 <+*>
1959 V4 N3 P8-12 60-10961 <*>
1959 V4 N3 P20-23 60-10960 <+*>
1959 V4 N3 P44-48 60-15715 <+*> 0
1959 V4 N3 P90-94 60-00801 <*>
60-10962 <+*>
93L35R <ATS>
1959 V4 N4 P29-33 60-10965 <+*>
1959 V4 N4 P37-43 60-10943 <+*>
1959 V4 N4 P46-50 60-10964 <+*>
1960 V5 N1 P3-9 96M40R <ATS>
1960 V5 N1 P9-14 97M40R <ATS>
1960 V5 N2 P83-85 59Q69R <ATS>
1960 V5 N4 P3-6 62-32143 <=*>
1960 V5 N4 P6-10 62-32144 <=*>
1960 V5 N5 P41-44 58Q69R <ATS>
1960 V5 N5 P52-53 61-10430 <*+>
1960 V5 N6 P114-116 61-21692 <=>
1961 V6 P552-561 64-18631 <*>
1961 V6 P1013-1916 AD-638 191 <=$>
1961 V6 N2 P111-115 66-11632 <*> 0
1961 V6 N4 P374-377 61-28748 <*=>
1961 V6 N5 P402-406 62-24748 <=*> 0
1961 V6 N7 P585-594 51P59R <ATS>
1961 V6 N7 P623-629 49P59R <ATS>
1961 V6 N7 P629-635 50P59R <ATS>
1961 V6 N7 P649-659 62-14590 <=*> 0
1961 V6 N8 P738-751 21Q68R <ATS>
1961 V6 N10 P904-908 10N58R <ATS>
1961 V6 N11 P1006-1009 62-23409 <=*>
1961 V6 N12 P1070-1073 62-32140 <=*>
1962 V7 N1 P27-31 62-23583 <=*>
1962 V7 N1 P32- FPTS V22 N3 P.T569 <FASE>
1962 V7 N3 P34-38 63-13016 <=*>
1962 V7 N3 P39-44 63-13017 <=*>
1962 V7 N3 P57-59 63-13705 <=*>
1962 V7 N3 P60- FPTS V22 N6 P.T1238 <FASE>
1962 V7 N3 P64- FPTS V22 N6 P.T1236 <FASE>
1962 V7 N3 P67- FPTS V22 N6 P.T1253 <FASE>
1962 V7 N6 P483-491 62-18749 <=*> 0
1962 V7 N7 P571-597 62-33036 <=>
1962 V7 N9 P771-776 61-13070 <=>
1962 V7 N9 P786-789 61-13070 <=>
1962 V7 N9 P841-852 63-21153 <=>
1962 V7 N10 P912-916 63-13270 <=*>
1962 V7 N12 P1085-1090 63-21128 <=>
1963 V8 N1 P29-33 63-24475 <=*$>
1963 V8 N1 P39-45 FPTS,V23,P.T75-T78 <FASE>
63-21496 <=>
1963 V8 N1 P49-53 64-13504 <=*$>
1963 V8 N1 P53-57 64-13503 <=*$> 0
1963 V8 N1 P58- FPTS V22 N5 P.T953 <FASE>
1963 V8 N1 P69- FPTS V22 N5 P.T795 <FASE>
1963 V8 N1 P104-110 69C69R <ATS>
1963 V8 N2 P111- FPTS V23 N2 P.T349 <FASE>
1963 V8 N2 P119- FPTS V23 N2 P.T384 <FASE>
1963 V8 N2 P144-147 63-24501 <=*$>
1963 V8 N2 P152-153 64-13535 <=*$>
1963 V8 N2 P154- FPTS V23 N1 P.T43 <FASE>
FPTS,V23,P.T43-T45 <FASE>
1963 V8 N4 P304- FPTS V23 N2 P.T380 <FASE>
1963 V8 N4 P314-319 37Q72R <ATS>
1963 V8 N4 P332- FPTS V23 N3 P.T484 <FASE>
FPTS,V23,N3,P.T484 <FASE>
1963 V8 N4 P336-343 64-15440 <=*$> 0
1963 V8 N4 P373-376 63-20948 <=*$>
1963 V8 N4 P449-456 64-13433 <=*$>
1963 V8 N5 P444-449 64-11758 <=> 0
1963 V8 N6 P503-507 64-19247 <=*$>
1963 V8 N6 P516-520 64-15475 <=*$>
1963 V8 N6 P525-527 64-15476 <=*$>
1963 V8 N6 P533-535 64-19291 <=*$>
1963 V8 N6 P540- FPTS V23 N4 P.T841 <FASE>
FPTS,V23,N4,P.T841 <FASE>
1963 V8 N7 P593- FPTS V23 N5 P.T989 <FASE>
FPTS,V23,N5,P.T989 <FASE>
1963 V8 N7 P604-609 64-19292 <=*$>
1963 V8 N7 P611-613 64-19293 <=*$>
1963 V8 N7 P614-618 64-15467 <=*$>
1963 V8 N8 P723-728 63-31813 <=>
64-19216 <=*$>
1963 V8 N8 P728-732 56Q72R <ATS>
1963 V8 N9 P796- FPTS,V23,N4,P.T817 <FASE>
1963 V8 N12 P1059-1064 64-16109 <*> 0
1963 V8 N12 P1071-1074 65-60174 <=$>
1963 V8 N12 P1091- FPTS V23 N6 P.T1358 <FASE>
1963 V8 N12 P1091-1095 FPTS V23 P.T1358 <FASE>
1963 V8 N12 P1096-1099 FPTS V23 P.T1267 <FASE>
1963 V8 N12 P1100-1105 64-21302 <=>

65-60175 <=$>
1963 V8 N12 P1113-1116 65-60167 <=$>
1964 V9 N1 P7-13 65-62167 <=$>
1964 V9 N1 P13- FPTS V24 N1 P.T77 <FASE>
1964 V9 N1 P13-16 FPTS V24 P.T77-T79 <FASE>
1964 V9 N1 P33- FPTS V23 N6 P.T1214 <FASE>
1964 V9 N1 P33 FPTS V23 P.T1214 <FASE>
1964 V9 N1 P53-60 65-62179 <=$>
1964 V9 N1 P69- FPTS V24 N2 P.T281 <FASE>
1964 V9 N1 P69-72 FPTS V24 P.T281-T282 <FASE>
1964 V9 N2 P99- FPTS V23 N6 P.T1333 <FASE>
1964 V9 N2 P99-104 FPTS V23 P.T1333 <FASE>
1964 V9 N2 P110-115 65-62177 <=$>
1964 V9 N3 P205- FPTS V24 N1 P.T119 <FASE>
1964 V9 N3 P205-207 FPTS V24 P.T119-T120 <FASE>
1964 V9 N3 P220-225 64-16539 <=*$>
1964 V9 N3 P228-232 64-16538 <=*$>
1964 V9 N3 P244-249 65-60170 <=$>
1964 V9 N5 P403-408 64-31958 <=>
1964 V9 N5 P472-475 64-31958 <=>
1964 V9 N6 P506- FPTS V24 N4 P.T725 <FASE>
1964 V9 N6 P528- FPTS V24 N4 P.T696 <FASE>
1964 V9 N6 P528-537 64-41565 <=>
1964 V9 N6 P545-551 66-11646 <0>
1964 V9 N6 P565-572 66-10885 <*>
1964 V9 N7 P596-602 65-63583 <*>
1964 V9 N7 P641- FPTS V24 N4 P.T622 <FASE>
1964 V9 N8 P690- FPTS V24 N4 P.T731 <FASE>
1964 V9 N8 P692- FPTS V24 N4 P.T732 <FASE>
1964 V9 N8 P695-701 65-11746 <*>
1964 V9 N8 P711-716 65-11404 <*>
1964 V9 N8 P719-722 64-51475 <=$>
1964 V9 N8 P722-726 65-11405 <*>
1964 V9 N8 P744-748 65-11938 <*>
1964 V9 N8 P748-750 65-11939 <*>
1964 V9 N9 P801-806 65-63584 <=$>
1964 V9 N9 P821-824 65-63585 <*>
1964 V9 N9 P828-832 65-63586 <*>
1964 V9 N9 P846-851 65-63587 <=$>
1964 V9 N10 P919- FPTS V24 N5 P.T919 <FASE>
1964 V9 N10 P926-932 64-51806 <*>
1964 V9 N10 P946-950 64-51806 <*>
1964 V9 N11 P984-987 65-63588 <*>
1964 V9 N11 P987-993 65-63589 <*>
1964 V9 N11 P1007-1012 65-64548 <=*$>
1964 V9 N12 P1066-1070 65-64547 <=*$>
1964 V9 N12 P1070-1072 65-11975 <*>
1964 V9 N12 P1073-1076 65-11976 <*>
1964 V9 N12 P1077-1080 65-11974 <*>
1964 V9 N12 P1094-1096 65-64546 <=*$>
1965 V10 N1 P3- FPTS V24 N6 P.T1063 <FASE>
1965 V10 N1 P87-94 65-30755 <=$>
1965 V10 N2 P99-104 66-11725 <*> 0
1965 V10 N2 P112- FPTS V24 N6 P.T1049 <FASE>
1965 V10 N2 P117-121 65-13802 <*>
1965 V10 N2 P117-122 66-60320 <=*$>
1965 V10 N2 P137- FPTS V25 N1 P.T83 <FASE>
1965 V10 N3 P207-210 66-60873 <=*$>
1965 V10 N3 P250- FPTS V25 N2 P.T321 <FASE>
1965 V10 N3 P255-259 66-60875 <=*$>
1965 V10 N4 P307- FPTS V25 N2 P.T329 <FASE>
1965 V10 N4 P314-317 66-60861 <=*$>
1965 V10 N4 P338-340 66-60860 <=*$>
1965 V10 N5 P420- FPTS V25 N2 P.T309 <FASE>
1965 V10 N5 P425- FPTS V25 N2 P.T312 <FASE>
1965 V10 N6 P507- FPTS V25 N4 P.T707 <FASE>
1965 V10 N6 P511-517 65-14605 <*> 0
1965 V10 N6 P547- FPTS V25 N3 P.T518 <FASE>
1965 V10 N7 P613- FPTS V25 N4 P.T695 <FASE>
1965 V10 N8 P675-678 65-33039 <=*$>
1965 V10 N8 P713- FPTS V25 N5 P.T836 <FASE>
1965 V10 N9 P816-819 66-11768 <*>
1965 V10 N9 P845-847 66-12735 <*>
1965 V10 N9 P848- FPTS V25 N4 P.T654 <FASE>
1965 V10 N10 P880- FPTS V25 N6 P.T1109 <FASE>
1965 V10 N10 P919-924 66-61996 <=*$>
1965 V10 N11 P1011- FPTS V25 N6 P.T1083 <FASE>
1965 V10 N12 P1118-1121 66-30631 <=*$>
1966 V11 N1 P60-61 66-31369 <=$>
1966 V11 N2 P99-107 66-12880 <*>
1966 V11 N3 P245-247 66-32113 <=$>
1966 V11 N3 P253-257 AD-637 626 <=$>

ANTIBIOTIKI; SBORNIK PEREVODOV, OBZOROV I REFERATOV INOSTRANNOI PERIODICHESKOI LITERATURY
1955 N6 P39-50 R-459 <$>
1955 V8 N4 P3-11 62-15154 <=*>

ANUAR INSTITUTULUI DE PATOLOGIE SI IGIENA ANIMALA
1957 V7 P5-27 60-00826 <*>

ANUARIO BRASILEIRO DE ECONOMIA FLORESTAL
1957 V9 P3-15 58-2073 <*>

ANUARIO. FACULTAD DE MEDICINA VETERINARIA. UNIVERSIDAD NACIONAL DE LA PLATA
1941 V4 P25-28 57-2052 <*>

ANZEIGER FUER DEN MASCHINEN-U. WERKZEUG-HANDEL VON UHLAND
1935 V57 P38-40 II-297 <*>

ANZEIGER DER OESTERREICHISCHEN AKADEMIE DER WISSENSCHAFTEN. MATEMATISCH-NATURWISSENSCHAFTLICHE KLASSE. WIEN
1952 P217-222 I-947 <*>
1964 N570A P1- NP-TR-1198 <*>

ANZEIGER FUER SCHAEDLINGSKUNDE
1937 V13 P26-28 62-11558 <=>
1949 V22 P116-119 AL-946 <*>
2860B <K-H>
63-20759 <*>
1955 V28 N4 P51-57 C-2677 <NRC>
1958 V31 N12 P9-10 66-10126 <*>

APARATO RESPIRATORIO Y TUBERCULOSIS
1947 V12 N2 P149-161 2500 <*>

APLIKACE MATEMATIKY
1956 V1 P431-444 58-1387 <*>
1956 V1 N3 P200-214 66-11161 <*>
1957 V2 N5 P370-386 58-1635 <*>
1958 V3 P124-140 AEC-TR-4543 <*>
1959 V4 P35-52 STMSP V4 P71-86 <AMS>
1959 V4 P290-302 STMSP V3 P277-289 <AMS>
1963 V8 N3 P180-196 65-12776 <*>
1963 V8 N6 P399-410 AD-609 780 <=>

APPLIED OPTICS
1962 V1 P67-74 64-14648 <*>
1962 V1 N1 P17-24 62-16224 <*>
1962 V1 N3 P267-268 63-15905 <=*>
1962 V1 N5 P575-585 66-13902 <*>
1963 V2 N1 P23-30 66-10714 <*> 0
1963 V2 N1 P23-29 66-11265 <*>
1963 V2 N1 P79-87 63S83F <ATS>

APPLIED SCIENTIFIC RESEARCH. MECHANICS, HEAT, CHEMICAL ENGINEERING, MATHEMATICAL METHODS
1955 V5 P349 62-24057 <=*>

APPLIED SCIENTIFIC RESEARCH. ELECTOPHYSICS, ACOUSTICS, OPTICS
1953 V4 P13-24 59-15078 <*>
1955 V5 P210-218 24L30G <ATS>

APPLIED STATISTICS IN METEOROLOGY. JAPAN
SEE KISHO TO TOKEI, OYO SUIKEIGAKU ZASSHI

★APTECHNOE DELO
1952 N2 P67-68 RT-4428 <*>
1954 V3 N5 P7-8 <INSD>
1955 V4 N5 P3-6 RT-4136 <*>
1955 V4 N6 P27-28 RT-3557 <*>
1956 V5 N1 P25-30 R-2022 <*>
1957 V6 N4 P48-51 60-15501 <+*>
1958 V7 N2 P19-22 62-18053 <=*>
1958 V7 N2 P31-33 60-13466 <+*>
1959 N2 P8-13 NLLTB V1 N10 P12 <NLL>

1959 V8 N1 P29-34 59-11669 <=>
1959 V8 N3 P3-10 59-13890 <=>
1959 V8 N3 P31-38 59-13943 <=>
1959 V8 N5 P30-33 61-19105 <+*>
1959 V8 N6 P44-47 60-11886 <=>
1959 V8 N6 P83-87 60-11758 <=>
1960 V9 N2 P3-5 60-31722 <=>
1960 V9 N2 P6-11 60-31723 <=>
1960 V9 N2 P75-76 60-41707 <=>
1960 V9 N5 P6-8 61-11701 <=>
1961 V10 N1 P4-7 61-23041 <=*>
1961 V10 N3 P3-13 62-23335 <=*>
62-25527 <=*>
1961 V10 N5 P3-8 62-23646 <=*>
1961 V10 N5 P59-66 62-23598 <=*>
1962 V11 N1 P3-11 62-25055 <=*>
1962 V11 N1 P3-18 62-25621 <=>
1962 V11 N1 P46-49 62-25053 <=*>
1963 V12 N3 P3-7 NLLTB V5 N11 P1014 <NLL>
1964 N4 P3-11 NLLTB V7 N4 P303 <NLL>
1964 V13 N1 P48-53 64-31155 <=>
1964 V13 N4 P3-11 64-51772 <=$>
1964 V13 N6 P5-8 65-30837 <=$>
1965 V14 N2 P3-8 65-31219 <=$>
65-31356 <=$>
1965 V14 N2 P83-87 65-31357 <=$>
1965 V14 N3 P3-10 65-32659 <=$*>
1965 V14 N3 P46-49 65-32659 <=$*>
1965 V14 N4 P3-8 65-33057 <=*$>
1965 V14 N5 P3-15 66-30124 <=*$>
1965 V14 N6 P7-10 66-31066 <=$>
1966 N2 P60-64 NLLTB V8 N9 P704 <NLL>

ARBEITEN AUS DER BIOLOGISCHEN BUNDESANSTALT FUER LAND- U. FORSTWIRTSCHAFT
1925 V14 N2 P163-169 ORNL-TR-285 <*>

ARBEITEN AUF DEM GEBIET DER PATHOLOGISCHEN ANATOMIE UND BAKTERIOLOGIE AN DEM PATHOLOGISCH-ANATOMISCHEN INSTITUT ZU TUEBINGEN
1914 V8 P114-128 62-16300 <*> 0

ARBEITEN AUS DEM K. GESUNDHEITSAMTE
1915 V68 P1-79 63-00736 <*>

ARBEITEN. LANDWIRTSCHAFTLICHEN HOCHSCHULE, HOHENHEIM
1962 V12 P109-121 65-12119 <*>

ARBEITEN AUS DEM NEUROLOGISCHEN INSTITUT AN DER WIENER UNIVERSITAET
1929 V31 P129-152 1091 <*>

ARBEITEN AUS DER PHYSIOLOGISCHEN ANSTALT ZU LEIPZIG. PHYSIOLOGISCHES INSTITUT. LEIPZIG UNIVERSITY
1871 V6 P139-176 61-00187 <*>

ARBEITSPHYSIOLOGIE
1930 V2 P461-473 N66-22280 <=$>
1943 V12 P85-91 1263 <*>
1950 V14 P137-146 C-2423 <NRC>
1950 V14 N2 P137-146 57-2542 <*>
1950 V14 N2 P166-235 2492 <*>
1952 V14 P469-476 3178-E <K-H>
1953 V15 P111-126 1733 <*>
1953 V15 P201-206 1474 <*>
1953 V15 P223-230 1266 <*>

ARCHIEF VOOR DE RUBBERCULTUUR IN NEDERLANDSCH-INDIE
1939 V23 N2 P130-139 RCT V13 N1 P422 <RCT>
1939 V23 N3 P176-188 RCT V13 N1 P400 <RCT>
1939 V23 N4 P239-255 RCT V13 N1 P505 <RCT>
1939 V23 N4 P256-296 RCT V13 N1 P761 <RCT>
1940 V24 P598-632 RCT V16 N1 P318 <RCT>
1940 V24 N2 P58-73 RCT V13 N1 P750 <RCT>
1940 V24 N2 P74-97 RCT V13 N1 P728 <RCT>
1940 V24 N5 P403-412 RCT V14 N1 P137 <RCT>
1940 V24 N5 P413-430 RCT V14 N1 P144 <RCT>
1940 V24 N6 P431-441 RCT V14 N1 P300 <RCT>
1940 V24 N8 P647-654 RCT V14 N1 P657 <RCT>
1941 V25 N1 P1-13 RCT V14 N1 P664 <RCT>
1941 V25 N5 P483-502 RCT V16 N1 P388 <RCT>

ARCHIEF VOOR DE SUIKERINDUSTRIE IN NEDERLANDSCH-INDIE
1934 V42 N6 P167-209 63-16370 <*>

ARCHITECTURE. CHINESE PEOPLES REPUBLIC
SEE CHIEN CHU

ARCHITEKTURA C S R
1961 V20 N6 P433-438 64-15888 <=*$>
1963 N6 P373-376 LC-1285 <BRS>
1963 V22 N2 P128-132 LC-1270 <NLL>
1964 V23 N2 P140-142 LC-1267 <NLL>

ARCHIV FUER ANATOMIE UND PHYSIOLOGIE
1843 P314-317 64-18954 <*> 0
1852 V19 P85-92 63-10126 <*>
1859 V26 P784-816 58-1343 <*>
1864 N1 P27-51 61-14216 <*>
1870 V37 P300-332 63-00038 <*>
1883 P328-360 SUP II-628 <*>
1891 P350-351 1261 <*>
1893 P201-248 1731 <*>
1902 V90 P33-72 N66-13484 <=$>
1909 P102-105 57-2284 <*>
1909 P108-109 57-2284 <*>
1909 V129 P69-81 65-00260 <*>

ARCHIV FUER AUGENHEILKUNDE
1889 V19 P123-158 61-14173 <*> 0
1903 V49 P338-350 61-14263 <*>
1921 V88 P149-154 61-14292 <*>
1928 V98 P401-447 61-14831 <*>
1928 V99 P587-610 61-14418 <*>
1936 V109 P462-467 59-15711 <*> 0
1937 V110 P330-356 61-14377 <*> 0
1937 V110 N4 P397-404 61-14153 <*> 0

ARCHIV FUER DERMATOLOGIE UND SYPHILIS
1910 V102 P191-206 62-00854 <*>
1932 V167 P141-153 2946 <*>
1937 V176 P42-62 2203 <*>
1950 V192 P2-6C 57-1949 <*>
1951 V192 N6 P516-529 59-17032 <*> 0
1951 V193 P518-526 57-223 <*>
1952 V193 P610-627 57-2555 <*>
1952 V195 P99-104 57-770 <*>
1953 V195 N5 P549-578 II-338 <*>
1953 V196 P375-402 II-368 <*>
1955 V200 P370-377 62-00280 <*>
1955 V201 P73-82 1203 <*>
1955 V201 P361-377 60-13220 <=*> 0

ARCHIV DER DEUTSCHEN SEEWARTE UND DES MARINE-OBSERVATORIUMS
1941 V61 N1 P74-87 AL-269 <*>

ARCHIV FUER DAS EISENHUETTENWESEN
1929 V2 P625-674 AD-640 397 <=$>
1930 V3 N11 P717-720 59-20420 <*> 0
1930 V4 N3 P145-150 59-20228 <*> 0
1930 V4 N4 P185-192 60-18269 <*>
1931 V4 N7 P333-342 57-3228 <*>
1931 V4 N8 P367-374 60-20252 <*> 0
1931 V4 N8 P375-382 63-16738 <*>
1931 V5 P173-176 58-2097 <*>
1932 V5 P355-366 63-16774 <*> 0
1932 V5 P427-430 58-1928 <*>
1932 V5 N8 P413-418 59-20595 <*>
1932 V5 N8 P431-440 59-20229 <*> 0
1932 V6 N1 P17-23 60-14629 <*> 0
1932 V6 N6 P247-251 59-20597 <*> 0
1933 V6 N7 P277-281 AL-467 <*>
1933 V6 N7 P283-288 59-20822 <*>
1933 V6 N8 P335-340 59-20590 <*>

Citation	Translation
1933 V7 N3 P165-174	63-16772 <*>
1933 V7 N4 P223-227	63-16773 <*> 0
1933 V7 N5 P315-317	59-20591 <*>
1933 V7 N6 P359-363	II-179 <*>
	59-20596 <*> 0
1933 V7 N6 P365-366	60-10067 <*> 0
1934 V7 P627-632	II-548 <*>
1934 V7 N7 P427-430	59-20593 <*>
1934 V7 N11 P637-640	59-20425 <*> 0
1934 V7 N12 P665-672	II-105 <*>
1934 V8 N4 P169-171	59-20055 <*> 0
1934 V8 N5 P181-186	64-14297 <*>
1934 V8 N5 P219-222	59-20589 <*>
1935 V8 P379-389	66-13123 <*> 0
1935 V8 N7 P319-324	57-2443 <*>
1935 V8 N10 P433-444	66-13340 <*> 0
1935 V8 N11 P507-509	57-2114 <*>
1935 V9 P31-35	58-938 <*>
1935 V9 N2 P73-90	57-1389 <*>
1935 V9 N5 P231-239	62-18933 <*> P
1935 V9 N5 P267-272	59-20410 <*>
1936 V10 P45-52	GB70 4085 <NLL>
1936 V10 P101-107	57-921 <*>
1936 V10 P155-160	2566 <BISI>
1936 V10 N2 P47-52	I-236 <*>
1936 V10 N2 P53-58	59-20409 <*> 0
1936 V10 N3 P109-110	59-20408 <*> 0
1936 V10 N4 P165-169	II-164 <*>
1936 V10 N6 P257-261	59-20570 <*>
1937 V10 N7 P297-306	60-10182 <*> 0
1937 V10 N7 P321-325	I-822 <*>
1937 V10 N12 P555-562	57-2033 <*>
1937 V11 P131-138	57-2616 <*>
1937 V11 N1 P10-16	63-16869 <*>
1937 V11 N4 P203-213	59-20424 <*> 0
1938 V11 P443-448	AL-219 <*>
1938 V11 P563-568	58-939 <*>
1938 V11 P569-582	63-16864 <*>
1938 V11 N8 P375-383	AL-35 <*>
1938 V11 N8 P393-400	59-20574 <*> 0
1938 V11 N9 P417-419	62-18133 <*>
1938 V11 N9 P449-454	59-20423 <*> 0
1938 V12 P195-198	57-2110 <*>
1938 V12 N4 P185-193	59-17669 <*>
1939 V12 N9 P459-464	57-168 <*>
1939 V12 N11 P553-564	60-10173 <*> 0
1939 V12 N11 P565-569	58-435 <*>
	59-17670 <*>
1940 V13 P345-354	65-12346 <*>
1940 V13 N10 P437-444	II-756 <*>
1940 V13 N12 P543-547	SC-T-64-47 <*>
1940 V14 P27-34	4122 <BISI>
1940 V14 P59-66	57-2106 <*>
1940 V14 P223-231	61-18556 <*>
1940 V14 P271-278	57-3019 <*>
1940 V14 N2 P59-66	59-15108 <*> 0
1941 V14 P429-438	AEC-TR-4125 <*>
1941 V14 N10 P478-488	60-00192 <*>
1941 V14 N10 P521-526	63-20833 <*>
1941 V14 N11 P551-553	59-15109 <*>
1941 V15 P201-202	57-2999 <*>
1941 V15 N6 P271-283	61-20753 <*> 0
1942 V15 P357-361	57-2996 <*>
1942 V15 P379-387	61-18652 <*>
1942 V15 N9 P407-412	59-20573 <*>
1943 V16 N7 P253-260	57-2809 <*>
1943 V16 N7 P261-268	60-10174 <*> 0
1943 V16 N9 P333-340	61-18654 <*>
1944 N7/8 P185-191	58-904 <*>
1944 V17 N3 P221-226	4332 <BISI>
1944 V17 N3 P227-234	3087 <BISI>
1944 V18 P125-130	AEC-TR-4022 <*>
1944 V18 N1/2 P1-6	60-17301 <=*> 0
1944 V18 N1/2 P13-22	59-15187 <*>
1945 V18 P161-167	AEC-TR-4020 <*>
1948 V19 P119-124	63-14375 <*>
1949 V20 N1/2 P59-68	60-18285 <*>
1949 V20 N3/4 P125-134	57-2617 <*>
1949 V20 N3/4 P135-138	62-16958 <*> 0
1949 V20 N3/4 P139-150	60-16761 <*> 0
1949 V20 N11/2 P337-344	60-18286 <*>
1949 V20 N11/2 P359-363	63-16837 <*>
1949 V20 N11/2 P359-362	65-12368 <*>
1949 V20 N11/2 P375-383	01L34G <ATS>
1949 V20 N9/10 P275-286	62-16948 <*> 0
1950 V21 P105-118	57-2620 <*>
1950 V21 P273-282	I-45 <*>
1950 V21 N3/4 P77-88	60-18628 <*>
1950 V21 N3/4 P123-127	62-10453 <*>
1950 V21 N3/4 P137-142	63-20490 <*>
1950 V21 N5/6 P143-152	60-18629 <*> 0
1950 V21 N5/6 P175-179	AL-272-II <*>
1950 V21 N5/6 P181-185	AL-204 <*>
1950 V21 N9/10 P327-336	65-17155 <*>
1951 V22 N1/2 P9-14	59-10339 <*>
1951 V22 N1/2 P65-71	I-295 <*>
1951 V22 N3/4 P87-98	60-18630 <*> 0
1951 V22 N7/8 P205-213	62-16959 <*> 0
1951 V22 N7/8 P215-224	1934 <*>
	64-18893 <*> 0
1952 V23 P163-172	63-16833 <*>
1952 V23 N1/2 P1-15	57-1764 <*>
1952 V23 N1/2 P17-20	60-18631 <*> 0
1952 V23 N3/4 P83-93	60-18632 <*>
1952 V23 N3/4 P107-111	60-18287 <*>
1952 V23 N5/6 P173-181	66-13124 <*>
1952 V23 N7/8 P272-276	57-2636 <*>
1952 V23 N7/8 P293-297	AL-108 <*>
1952 V23 N11/2 P417-425	60-18288 <*> 0
1953 V24 N1/2 P1-10	60-18633 <*>
1953 V24 N1/2 P43-45	60-18393 <*>
1953 V24 N3/4 P161-171	1121 <*>
	66G7G <ATS>
1953 V24 N11/2 P483-495	59-15819 <*> 0
1953 V24 N11/2 P519-522	59-10099 <*>
1954 V25 P89-92	62-18950 <*>
1954 V25 P421-433	61-16301 <*>
1954 V25 P569-578	2444 <BISI>
1954 V25 N1 P1-10	61-18408 <*>
1954 V25 N1/2 P33-38	60-18634 <*>
1954 V25 N3/4 P159-164	II-943 <*>
	57-1761 <*>
1954 V25 N5/6 P225-230	60-18635 <*>
1954 V25 N7/8 P307-314	2712 <BISI>
1954 V25 N7/8 P373-375	5342 <KH>
1954 V25 N7/8 P377-381	62-16993 <*> 0
1955 P455-461	TS-647 <BISI>
1955 P535-540	GB66 <NLL>
1955 V26 P455-461	61-16256 <*>
1955 V26 P475-481	57-51 <*>
1955 V26 P497-506	TS-1812 <BISI>
1955 V26 P583-587	61-16302 <*>
1955 V26 P599-602	61-18444 <*>
1955 V26 N2 P71-98	61-20481 <*> 0
1955 V26 N2 P99-104	65-12388 <*>
1955 V26 N2 P117-123	62M40G <ATS>
1955 V26 N3 P127-130	4197 <HB>
1955 V26 N4 P213-227	61-16329 <*>
1955 V26 N5 P243-251	60-18289 <*>
1955 V26 N5 P253-266	1503 <*>
1955 V26 N5 P279-285	66-10435 <*> 0
1955 V26 N5 P287-289	66-10436 <*> 0
1955 V26 N7 P393-404	61-16433 <*>
	80M41G <ATS>
1955 V26 N9 P535-540	61-16292 <*>
1955 V26 N9 P561-562	3924 <HB>
1955 V26 N12 P777-782	63-16856 <*>
1956 P449-452	61-16416 <*>
1956 V27 P127-133	AEC-TR-4209 <*>
1956 V27 P149-152	GB70 4192 <NLL>
1956 V27 P343-346	UCRL TRANS-913(L) <*>
1956 V27 P413-420	UCRL TRANS-883(L) <*>
1956 V27 P469-474	T-2131 <INSD>
1956 V27 P521-529	TS 1333 <BISI>
	62-14360 <*>
1956 V27 N1 P45-54	64-18521 <*>
1956 V27 N2 P119-126	C-2579 <NRC>
1956 V27 N3 P153-160	TS-997 <BISI>
	61-18417 <*>
	61-18903 <*>

Citation	Translation
1956 V27 N3 P161-163	63-16845 <*>
1956 V27 N5 P289-295	3803 <HB>
1956 V27 N5 P303-309	3761 <HB>
1956 V27 N7 P423-428	T2253 <INSD>
1956 V27 N7 P433-440	3876 <HB>
1956 V27 N7 P440-448	3928 <HB>
1956 V27 N7 P475-486	T-2239 <INSD>
1956 V27 N8 P513-520	60-18290 <*>
1956 V27 N11 P673-679	4832 <HB>
1956 V27 N11 P695-699	61-16320 <*>
1956 V27 N11 P701-706	61-18900 <*>
1956 V27 N11 P707-713	4377 <BISI>
1956 V27 N11 P715-724	61-16319 <*>
1956 V27 N12 P753-760	61-16322 <*>
1956 V27 N12 P793-799	2534 <*>
1957 V28 P123-125	TS-630 <BISI>
1957 V28 P401-416	61-18720 <*>
1957 V28 P531-541	58-2330 <*>
1957 V28 P679-685	61-16307 <*>
1957 V28 P687-694	61-16203 <*>
1957 V28 N1 P7-12	61-16328 <*>
1957 V28 N1 P49-51	4162 <HB>
	57-2594 <*>
1957 V28 N2 P57-64	61-16420 <*>
1957 V28 N2 P65-66	3934 <HB>
1957 V28 N3 P123-125	61-16318 <*>
1957 V28 N3 P153-156	61-16316 <*>
1957 V28 N3 P167-177	63-16877 <*>
1957 V28 N4 P201-205	2840 <BISI>
1957 V28 N4 P223-226	61-16283 <*>
1957 V28 N4 P229-235	61-18434 <*>
1957 V28 N8 P433-444	61-16326 <*>
1957 V28 N8 P445-459	61-16327 <*>
	714 <BISI>
1957 V28 N8 P461-468	61-16324 <*>
1957 V28 N8 P505-516	4207 <BISI>
1957 V28 N9 P523-530	61-16325 <*>
1957 V28 N9 P557-566	61-18439 <*>
1957 V28 N9 P567-574	4329 <HB>
1957 V28 N10 P615-621	61-18787 <*>
1957 V28 N10 P623-624	61-18367 <*>
1957 V28 N10 P633-639	61-16312 <*>
1957 V28 N1C P641-644	4231 <HB>
1957 V28 N10 P655-662	64-18520 <*>
	66-10434 <*> 0
1957 V28 N11 P673-677	61-16467 <*>
1957 V28 N11 P695-702	61-16500 <*>
1957 V28 N11 P7C2-706	61-16501 <*>
1957 V28 N11 P707-710	61-16502 <*>
1957 V28 N11 P710-712	61-16533 <*>
1957 V28 N11 P712-713	61-16503 <*>
1957 V28 N11 P713-715	61-16504 <*>
1957 V28 N11 P717-719	61-16505 <*>
1957 V28 N11 P719-726	61-16506 <*>
1957 V28 N11 P726-729	61-16507 <*>
1957 V28 N11 P729-730	61-16508 <*>
1957 V28 N11 P731-738	61-16311 <*>
1957 V28 N12 P777-783	61-16313 <*>
1957 V28 N12 P785-794	6C-21039 <=*>
	61-16314 <*>
1957 V28 N12 P807-823	TS 1197 <BISI>
	61-20595 <*>
1957 V28 N5/6 P245-252	61-16305 <*>
1957 V28 N5/6 P253-258	61-16317 <*>
1957 V28 N5/6 P259-267	61-16304 <*>
1957 V28 N5/6 P269-285	61-16306 <*>
1957 V28 N5/6 P287-304	61-16310 <*>
1957 V28 N5/6 P305-310	61-16308 <*>
1957 V28 N5/6 P311-316	61-16315 <*>
1957 V28 N5/6 P317-323	61-16309 <*>
1957 V28 N5/6 P325-337	61-16321 <*>
1957 V28 N5/6 P339-344	61-16323 <*>
1957 V28 N5/6 P345-353	4136 <HB>
1958 P471-472	TS-1893 <BISI>
1958 V29 P25-34	2271 <BISI>
1958 V29 P107-113	TS-1332 <BISI>
1958 V29 P115-118	TS-1196 <BISI>
1958 V29 P153-158	TS-1902 <BISI>
1958 V29 P4C1-410	61-18750 <*>
1958 V29 P423-432	61-18752 <*>
1958 V29 P543-546	61-18744 <*>
1958 V29 N1 P1-9	4416 <HB>
1958 V29 N1 P41-46	4239 <BISI>
1958 V29 N1 P47-56	66-10586 <*> 0
1958 V29 N1 P57-63	61-18409 <*>
1958 V29 N1 P65-72	61-18380 <*>
1958 V29 N2 P83-88	4238 <BISI>
1958 V29 N2 P89-94	TS-1315 <BISI>
	61-20591 <*>
1958 V29 N2 P107-113	62-14362 <*>
1958 V29 N2 P115-118	62-10514 <*>
1958 V29 N3 P147-152	61-18440 <*>
1958 V29 N3 P159-164	TS-1226 <BISI>
	61-20603 <*>
1958 V29 N3 P179-187	61-18897 <*> 0
1958 V29 N3 P193-203	61-18427 <*>
1958 V29 N4 P219-224	58-2687 <*>
	61-18438 <*>
1958 V29 N4 P225-229	60-21072 <=*>
1958 V29 N4 P231-234	60-21176 <=*>
1958 V29 N4 P235-240	60-21071 <=*>
1958 V29 N5 P275-281	TS-1459 <BISI>
	62-14359 <*>
1958 V29 N6 P329-337	4473 <HB>
1958 V29 N6 P339-349	4440 <HB>
1958 V29 N6 P353-357	3189 <BISI>
1958 V29 N6 P377-390	61-18431 <*>
1958 V29 N7 P433-447	61-18749 <*>
1958 V29 N8 P479-484	61-18386 <*>
1958 V29 N8 P485-488	60-16763 <*> 0
1958 V29 N8 P495-504	66-10433 <*> 0
1958 V29 N9 P547-552	4501 <HB>
1958 V29 N9 P575-584	4870 <HB>
1958 V29 N10 P603-609	5191 <HB>
1958 V29 N10 P619-626	TS 1212 <BISI>
1958 V29 N10 P627-629	61-20596 <*>
1958 V29 N10 P631-642	TS-1227 <BISI>
1958 V29 N10 P643-651	60-14214 <*>
1958 V29 N11 P715-721	60-16764 <*> 0
	61-18733 <*>
1958 V29 N11 P723-729	61-18794 <*>
1958 V29 N12 P737-744	TS-1214 <BISI>
	62-10488 <*>
1959 V30 P1-6	AEC-TR-3962 <*>
1959 V30 P71-78	2419 <BISI>
1959 V30 P205-209	TS-1356 <BISI>
1959 V30 P233-238	TS-1373 <BISI>
1959 V30 P275-282	61-00224 <*>
1959 V30 P283-292	TS-1390 <BISI>
1959 V30 P293-297	1904 <BISI>
1959 V30 P299-309	TS-1627 <BISI>
1959 V30 P311-314	TS-1389 <BISI>
1959 V30 P321-327	TS-1453 <BISI>
1959 V30 P329-335	TS-1428 <BISI>
1959 V30 P337-344	TS-1414 <BISI>
1959 V30 P351-360	TS-1465 <BISI>
1959 V30 P361-370	TS-1413 <BISI>
1959 V30 P391-396	TS 1423 <BISI>
1959 V30 P397-405	TS 1429 <BISI>
1959 V30 P429-434	TS 1438 <BISI>
1959 V30 P451-460	1461 <BISI>
1959 V30 P461-472	TS 1462 <BISI>
1959 V30 P473-476	TS-1463 <BISI>
1959 V30 P513-518	TS 1464 <BISI>
1959 V30 P549-552	TS 1523 <BISI>
1959 V30 P553-564	TS-1551 <BISI>
1959 V30 P585-587	TS 1554 <BISI>
1959 V30 P589-593	TS 1604 <BISI>
1959 V30 P659-660	2254 <BISI>
1959 V30 P693-703	TS 1566 <BISI>
1959 V30 P731-735	TS 1636 <BISI>
1959 V30 P743-746	2211 <BISI>
1959 V30 P747-750	TS 1637 <BISI>
1959 V30 N1 P15-20	2432 <BISI>
1959 V30 N3 P137-144	4558 <HB>
1959 V30 N4 P187-195	4697 <HB>
1959 V30 N4 P195-198	4725 <HB>
1959 V30 N4 P2C5-209	62-10495 <*>
1959 V30 N5 P299-309	62-14191 <*>
1959 V30 N5 P311-314	62-10497 <*>

Citation	Translation
1959 V30 N6 P321-327	TRANS-20 <MT>
	62-14332 <*>
1959 V30 N6 P329-335	62-14333 <*>
1959 V30 N6 P337-344	62-10469 <*>
1959 V30 N6 P351-360	62-14334 <*>
1959 V30 N6 P361-370	62-14383 <*>
1959 V30 N7 P397-405	62-14385 <*>
1959 V30 N7 P429-434	62-14317 <*>
1959 V30 N7 P435-439	66-10428 <*>
1959 V30 N8 P461-472	62-14564 <*>
1959 V30 N8 P473-476	62-14386 <*>
1959 V30 N9 P539-540	62-14336 <*>
1959 V30 N9 P549-552	62-14183 <*>
1959 V30 N10 P581-584	4802 <HB>
1959 V30 N10 P585-587	62-14659 <*>
1959 V30 N10 P601-603	3987 <BISI>
1959 V30 N10 P613-619	1971 <BISI>
1959 V30 N10 P637-640	3559 <BISI>
1959 V30 N11 P641-648	62-14198 <*>
1959 V30 N11 P693-703	62-16847 <*>
1959 V30 N12 P713-714	4793 <HB>
1959 V30 N12 P723-730	62-14660 <*>
1959 V30 N12 P731-735	62-14199 <*>
1959 V30 N12 P737-742	3240 <BISI>
1959 V30 N12 P747-750	62-14197 <*>
1960 V31 P11-17	TS-1641 <BISI>
1960 V31 P19-24	TS 1658 <BISI>
1960 V31 P59-65	TS 1659 <BISI>
1960 V31 P97-102	TS 1706 <BISI>
1960 V31 P121-128	2675 <BISI>
1960 V31 P177-182	1836 <BISI>
1960 V31 P201-213	TS-1817 <BISI>
1960 V31 P227-235	TS-1779 <BISI>
1960 V31 P299-308	2335 <BISI>
1960 V31 P337-342	1932 <BISI>
1960 V31 P351	AEC-TR-5284 <*>
1960 V31 P385-391	2181 <BISI>
1960 V31 P427-432	2366 <BISI>
1960 V31 P459-470	2171 <BISI>
1960 V31 N1 P1-9	4792 <HB>
1960 V31 N1 P11-17	62-16899 <*>
1960 V31 N1 P19-24	62-14519 <*>
1960 V31 N1 P25-32	60-18906 <*>
1960 V31 N1 P39-46	2053 <BISI>
1960 V31 N1 P59-65	62-16866 <*>
1960 V31 N2 P67-81	2916 <BISI>
1960 V31 N2 P87-90	2068 <K-H>
	61-10546 <*>
	63-18426 <*>
1960 V31 N2 P91-96	4801 <HB>
1960 V31 N2 P97-102	62-16865 <*>
1960 V31 N2 P103-111	2054 <BISI>
1960 V31 N2 P121-128	63-14886 <*> O
1960 V31 N3 P135-143	1806 <BISI>
1960 V31 N3 P153-160	1990 <BISI>
1960 V31 N4 P237-242	1768 <BISI>
1960 V31 N5 P279-286	1780 <BISI>
1960 V31 N6 P351-354	4887 <HB>
1960 V31 N6 P355-358	4888 <HB>
1960 V31 N7 P405-410	1905 <BISI>
1960 V31 N7 P411-417	1906 <BISI>
	69M46G <ATS>
1960 V31 N7 P423-426	1907 <BISI>
1960 V31 N8 P451-458	1959 <BISI>
1960 V31 N8 P491-496	3491 <BISI>
1960 V31 N9 P509-519	4921 <HB>
1960 V31 N9 P521-529	1950 <BISI>
1960 V31 N10 P577-582	5033 <HB>
1960 V31 N10 P587-594	5034 <HB>
1960 V31 N10 P595-601	2471 <BISI>
1960 V31 N10 P603-606	2471 <BISI>
1960 V31 N11 P639-643	5076 <HB>
1960 V31 N11 P645-648	5077 <HB>
1960 V31 N12 P691-702	2635 <BISI>
1960 V31 N12 P723-730	61-10864 <*> O
1960 V31 N12 P743-745	AEC-TR-4637 <*>
1961 V32 P19-29	2133 <BISI>
1961 V32 P77-78	2309 <BISI>
1961 V32 P89-94	2325 <BISI>
1961 V32 P95-101	2305 <BISI>
1961 V32 P173-185	2190 <BISI>
1961 V32 P251-260	2505 <BISI>
1961 V32 P297-302	2411 <BISI>
1961 V32 P361-368	2447 <BISI>
1961 V32 P369-373	2469 <BISI>
1961 V32 P376-377	2469 <BISI>
1961 V32 P437-450	2445 <BISI>
1961 V32 P475-482	2427 <BISI>
1961 V32 P493-500	2460 PT.1 <BISI>
1961 V32 P513-520	2499 <BISI>
1961 V32 P541-556	2501 <BISI>
1961 V32 P681-687	4314 <BISI>
1961 V32 P823-841	2677 <BISI>
1961 V32 P879-882	4403 <BISI>
1961 V32 N1 P1-10	4627 <BISI>
1961 V32 N1 P11-18	5118 <HB>
	63-00955 <*>
1961 V32 N1 P19-28	5362 <HB>
1961 V32 N1 P31-38	5409 <HB>
1961 V32 N3 P133-144	2743 <BISI>
1961 V32 N3 P153-164	2197 <BISI>
1961 V32 N4 P213-219	2782 <BISI>
1961 V32 N4 P269-273	3111 <BISI>
1961 V32 N5 P275-280	5333 <HB>
1961 V32 N5 P275-286	78N56G <ATS>
1961 V32 N5 P280-286	5585 <HB>
1961 V32 N5 P287-296	06N56G <ATS>
1961 V32 N5 P287-292	5334 <HB> P
1961 V32 N5 P292-296	5748 <HB>
1961 V32 N5 P331-336	2633 <BISI>
1961 V32 N6 P421-429	5246 <HB>
	62-14442 <*>
1961 V32 N7 P489-491	2704 <BISI>
1961 V32 N8 P521-523	3006 <BISI>
1961 V32 N8 P573-579	2460 <BISI>
1961 V32 N9 P597-606	C-4057 <NRC>
1961 V32 N9 P637-643	2999 <BISI>
1961 V32 N9 P651-653	38T95G <ATS>
1961 V32 N10 P667-673	2598 <BISI>
1961 V32 N10 P689-699	4313 <BISI>
1961 V32 N10 P729-731	5412 <HB>
1961 V32 N11 P753-760	2597 <BISI>
	55P59G <ATS>
1961 V32 N11 P779-790	2668 <BISI>
1961 V32 N12 P799-808	2660 <BISI>
1961 V32 N12 P809-822	2685 <BISI>
1961 V32 N12 P843-850	2807 <BISI>
1961 V32 N12 P857-861	2762 <BISI>
1961 V32 N12 P883-892	2809 PT-2 <BISI>
1962 V33 N1 P5-15	2912 <BISI>
1962 V33 N1 P17-22	5558 <HB>
1962 V33 N1 P23-26	2805 <BISI>
1962 V33 N1 P27-34	2873 <BISI>
1962 V33 N1 P35-47	2874 <BISI>
1962 V33 N1 P49-60	2875 <BISI>
1962 V33 N1 P67-76	3324 <BISI>
1962 V33 N2 P85-89	5562 <HB>
1962 V33 N2 P91-100	2850 <BISI>
1962 V33 N2 P101-106	5563 <HB>
1962 V33 N2 P107-114	2839 <BISI>
1962 V33 N2 P131-140	2806 <BISI>
1962 V33 N2 P141-143	5627 <HB>
1962 V33 N3 P141-143	BGIRA-568 <BGIR>
1962 V33 N3 P145-146	2935 <BISI>
1962 V33 N3 P157-166	2857 <BISI>
1962 V33 N3 P167-175	2919 <BISI>
1962 V33 N3 P195-210	2931 <BISI>
1962 V33 N4 P211-216	5636 <HB>
1962 V33 N4 P223-226	5637 <HB>
1962 V33 N4 P229-232	5584 <HB>
1962 V33 N4 P251-259	2897 <BISI>
1962 V33 N4 P261-267	2898 <BISI>
1962 V33 N4 P269-275	2914 <BISI>
1962 V33 N4 P271-276	3760 <BISI>
1962 V33 N5 P281-284	2930 <BISI>
1962 V33 N5 P311-316	5655 <HB>
1962 V33 N5 P317-325	5656 <HB>
1962 V33 N6 P347-353	2968 <BISI>
1962 V33 N6 P355-358	5668 <HB>
1962 V33 N6 P359-367	3065 <BISI>

Citation	Translation
	65-12863 <*>
1962 V33 N6 P371-383	3070 <BISI>
1962 V33 N6 P393-403	2945 <BISI>
1962 V33 N7 P417-420	5708 <HB>
1962 V33 N7 P441-451	2968-II <BISI>
1962 V33 N7 P453-455	5583 <HB>
1962 V33 N8 P527-531	5716 <HB>
1962 V33 N8 P559-566	3095 <BISI>
1962 V33 N9 P589-591	3145 <BISI>
1962 V33 N9 P593-600	5751 <HB>
1962 V33 N9 P611-616	5750 <HB>
1962 V33 N10 P643-648	65-10768 <*>
1962 V33 N10 P649-659	3102 <BISI>
1962 V33 N10 P661-668	5763 <HB>
1962 V33 N10 P669-678	3089 <BISI>
1962 V33 N10 P699-704	5764 <HB>
1962 V33 N10 P704-710	5872 <HB>
1962 V33 N10 P711-713	3090 <BISI>
1962 V33 N10 P715-722	3091 <BISI>
1962 V33 N10 P723-727	3223 <BISI>
1962 V33 N11 P729-743	3166 <BISI>
1962 V33 N11 P745-755	62Q69G <ATS>
1962 V33 N11 P757-760	5810 <HB>
1962 V33 N11 P771-774	61Q69G <ATS>
1962 V33 N11 P775-781	3232 <BISI>
1962 V33 N12 P805-808	3361 <BISI>
1962 V33 N12 P809-816	5840 <HB>
1962 V33 N12 P817-826	3175 <BISI>
1962 V33 N12 P831-839	6576 <HB>
1962 V33 N12 P841-851	3152 <BISI>
1962 V33 N12 P853-862	3177 <BISI>
1962 V33 N12 P863-872	3178 <BISI>
1962 V33 N12 P873-876	3179 <BISI>
	64-14006 <*>
1962 V33 N12 P877-881	3180 <BISI>
1963 V34 P555-560	AEC-TR-6165 <*>
1963 V34 N1 P9-16	3537 <BISI>
1963 V34 N1 P17-26	3161 <BISI>
1963 V34 N1 P55-59	3162 <BISI>
1963 V34 N1 P61-68	5860 <HB>
1963 V34 N2 P93-100	3248 <BISI>
1963 V34 N2 P101-108	3221 <BISI>
1963 V34 N2 P135-146	3287 <BISI>
1963 V34 N3 P159-166	3261 <BISI>
1963 V34 N3 P173-179	3547-I/II <BISI>
1963 V34 N3 P181-185	3262 <BISI>
1963 V34 N3 P187-194	3295 <BISI>
1963 V34 N3 P195-204	3263 <BISI>
1963 V34 N3 P205-210	3264 <BISI>
1963 V34 N3 P217-221	3695 <BISI>
1963 V34 N3 P223-226	3615 <BISI>
1963 V34 N4 P235-241	3297 <BISI>
1963 V34 N4 P243-246	5959 <HB>
1963 V34 N4 P259-267	3298 <BISI>
1963 V34 N4 P269-277	5960 <HB>
1963 V34 N4 P307-316	3615 <BISI>
1963 V34 N5 P325-340	3334 <BISI>
1963 V34 N5 P361-366	6003 <HB>
1963 V34 N5 P381-390	3320 <BISI>
1963 V34 N6 P425-434	3391 <BISI>
1963 V34 N6 P435-438	6269 <HB>
1963 V34 N6 P453-463	4113 <BISI>
1963 V34 N7 P519-525	6067 <HB>
1963 V34 N7 P531-546	3439 <BISI>
1963 V34 N7 P547-554	3437 <BISI>
1963 V34 N8 P565-570	3464 <BISI>
1963 V34 N8 P571-581	3505 <BISI>
1963 V34 N8 P583-590	3455 <BISI>
1963 V34 N8 P591-594	3754 <BISI>
1963 V34 N8 P601-604	3433 <BISI>
1963 V34 N8 P617-624	3435 <BISI>
1963 V34 N9 P647-652	3519 <BISI>
1963 V34 N9 P659-672	3520 <BISI>
1963 V34 N9 P679-683	3544 <BISI>
1963 V34 N9 P701-712	3588 <BISI>
1963 V34 N10 P713-726	3575 <BISI>
1963 V34 N10 P733-744	3702 <BISI>
1963 V34 N10 P755-759	3705 <BISI>
1963 V34 N10 P795-811	3600 <BISI>
1963 V34 N11 P831-837	3565 <BISI>
1963 V34 N11 P839-843	3566 <BISI>
1963 V34 N11 P853-858	64-16369 <*>
1963 V34 N11 P867-870	5733 <HB>
1963 V34 N11 P871-878	6200 <HB>
1963 V34 N12 P935-950	3621 <BISI>
1964 V35 P21-26	4462 <BISI>
1964 V35 P247-253	6273 <HB>
1964 V35 P839-846	4197 <BISI>
1964 V35 P1011-1018	4196 <BISI>
1964 V35 N1 P9-14	3769 <BISI>
1964 V35 N1 P15-20	3675 <BISI>
1964 V35 N1 P37-38	64-16651 <*>
1964 V35 N1 P39-44	3677 <BISI>
1964 V35 N1 P45-55	3989 <BISI>
1964 V35 N1 P57-64	3676 <BISI>
1964 V35 N1 P65-73	3990 <BISI>
1964 V35 N1 P75-83	3601 <BISI>
1964 V35 N2 P91-108	3851 <BISI>
1964 V35 N2 P109-114	3678 <BISI>
1964 V35 N2 P133-140	4175 <BISI>
1964 V35 N2 P141-151	3729 <BISI>
1964 V35 N2 P159-167	3675 <BISI>
1964 V35 N3 P193-201	6275 <HB>
1964 V35 N3 P203-208	6274 <HB>
1964 V35 N3 P221-240	4024 <BISI>
1964 V35 N4 P269-277	C-5579 <NRC>
	4019 <BISI>
1964 V35 N4 P279-285	6276 <HB>
1964 V35 N4 P307-316	3840 <BISI>
1964 V35 N4 P329-337	3867 <BISI>
1964 V35 N4 P339-351	3865 <BISI>
1964 V35 N4 P353-358	6281 <HB>
1964 V35 N4 P359-365	6278 <HB>
1964 V35 N4 P367-369	6280 <HB>
1964 V35 N4 P371-372	3756 <BISI>
1964 V35 N4 P373-380	3755 <BISI>
1964 V35 N5 P381-388	3799 <BISI>
1964 V35 N5 P389-390	3932 <BISI>
1964 V35 N5 P391-399	3831 <BISI>
1964 V35 N5 P401-405	6302 <HB>
1964 V35 N5 P407-411	6285 <HB>
1964 V35 N5 P411-	6339 <HB>
1964 V35 N5 P447-453	6492 <HB>
1964 V35 N5 P454-458	6512 <HB>
1964 V35 N6 P475-483	3949 <BISI>
1964 V35 N6 P487-494	4401 <BISI>
1964 V35 N6 P507-515	4071 <BISI>
1964 V35 N6 P517-526	3977 <BISI>
1964 V35 N6 P541-549	3841 <BISI>
1964 V35 N6 P551-559	6325 <HB>
1964 V35 N6 P561-566	4173 <BISI>
1964 V35 N7 P577-584	3943 <BISI>
1964 V35 N7 P585-601	3905,I-II <BISI>
1964 V35 N7 P603-612	3873 <BISI>
1964 V35 N7 P613-617	3870 <BISI>
1964 V35 N7 P633-636	3871 <BISI>
1964 V35 N7 P637-647	3909 <BISI>
1964 V35 N7 P647-657	3911 <BISI>
1964 V35 N8 P677-698	4030 <BISI>
1964 V35 N8 P699-711	4400 <BISI>
1964 V35 N8 P713-724	3954 <BISI>
1964 V35 N8 P725-737	3942 <BISI>
1964 V35 N8 P739-747	4416 <BISI>
1964 V35 N8 P749-751	3956 <BISI>
1964 V35 N8 P781-800	4587 <BISI>
1964 V35 N9 P803-819	4038 <BISI>
1964 V35 N9 P821-825	4039 <BISI>
1964 V35 N9 P855-860	4010 <BISI>
1964 V35 N9 P861-866	4062 <BISI>
1964 V35 N9 P867-869	4059 <BISI>
1964 V35 N9 P871-875	4060 <BISI>
1964 V35 N9 P903-908	6206 <HB>
1964 V35 N9 P909-918	4240 <BISI>
1964 V35 N10 P941-954	4099 <BISI>
1964 V35 N10 P963-976	4418 <BISI>
1964 V35 N10 P983-986	6421 <HB>
1964 V35 N10 P987-991	4029 <BISI>
1964 V35 N10 P1009-1010	6420 <HB>
1964 V35 N11 P1089-1095	6452 <HB>
1964 V35 N11 P1111-1113	6453 <HB>

1964 V35 N12 P1123-1131 4131 <BISI>
1964 V35 N12 P1133-1144 4132 <BISI>
1964 V35 N12 P1151-1160 4117 <BISI>
1964 V35 N12 P1173-1180 6489 <HB>
1965 V36 P1-8 4268 <BISI>
1965 V36 P99-108 4456 <BISI>
1965 V36 P221-236 4399 <BISI>
1965 V36 P257-267 4664 <BISI>
1965 V36 P285-291 66-11909 <*>
1965 V36 P297-406 4464 <BISI>
1965 V36 P301-309 4665 <BISI>
1965 V36 P415-422 4674 <BISI>
1965 V36 P457-462 4391 <BISI>
1965 V36 P633-642 4894 <BISI>
1965 V36 P649-653 4581 <BISI>
1965 V36 P655-666 4558 <BISI>
1965 V36 P677-682 4583 <BISI>
1965 V36 P683-693 4663 <BISI>
1965 V36 P737-743 4618 <BISI>
1965 V36 P799-807 4635 <BISI>
1965 V36 N1 P47-51 N66-11608 <=$>
1965 V36 N3 P167-173 6547 <HB>
1965 V36 N3 P183-190 4249 <BISI>
1965 V36 N3 P191-200 4461 <BISI>
1965 V36 N4 P277-284 66-13003 <*>
1965 V36 N4 P311-316 4304 <BISI>
1965 V36 N5 P333-342 66-11763 <*>
1965 V36 N5 P343-349 6586 <HB>
1965 V36 N6 P449-455 6630 <HB>
1965 V36 N7 P485-487 6602 <HB>
1965 V36 N7 P505-508 6609 <HB>
1965 V36 N7 P509-511 4424 <BISI>
1965 V36 N7 P513-516 6610 <HB>
1965 V36 N7 P529-531 6611 <HB>
1965 V36 N7 P533-536 6612 <HB>
1965 V36 N8 P591-596 4496 <BISI>
1965 V36 N8 P603-608 66-12341 <*> 0
1965 V36 N9 P609-610 4585 <BISI>
1965 V36 N9 P643-648 6370 <HB>
1965 V36 N9 P667-676 4582 <BISI>
1965 V36 N10 P695-699 6518 <HB>
1965 V36 N10 P725-736 6824 <HB>
1965 V36 N12 P873-876 6795 <HB>
1965 V36 N12 P877-885 6869 <HB>
1965 V36 N12 P897-900 6794 <HB>
1966 V37 P27-41 4777 <BISI>
1966 V37 N1 P43-47 6830 <HB>
1966 V37 N2 P127-131 6847 <HB>
1966 V37 N3 P245-252 4964 <BISI>
1966 V37 N3 P259-262 6862 <HB>

ARCHIV DER ELEKTRISCHEN UEBERTRAGUNG
1947 V1 N1/2 P2-13 57-2275 <*>
1948 V2 P136-139 57-130 <*>
1949 V3 P155-159 C-2433 <NRC>
1950 V4 P207-212 II-711 <*>
1950 V4 P493-508 63-14429 <*>
1950 V4 N12 P494-508 C-2429 <NRC>
1952 V6 P187-194 58-1312 <*>
1952 V6 P414-418 2089 <*>
1952 V6 P507-513 C-2246 <NRC>
57-259 <*>
1952 V6 N12 P532-534 66-11163 <*>
1953 V7 P63-70 II-140 <*>
1953 V7 P236-248 58-740 <*>
1953 V7 P415-420 II-786 <*>
1953 V7 P429-440 64-00247 <*>
1953 V7 P592-596 57-3534 <*>
58-1304 <*>
1953 V7 N8 P375-378 C-2390 <NRC>
1953 V7 N10 P501-504 57-958 <*>
1953 V7 N11 P531-536 57-958 <*>
1954 V8 P111-122 58-1311 <*>
1954 V8 P132-136 63-14491 <*>
1954 V8 P143-154 C-2378 <*>
1954 V8 P173-178 57-2648 <*>
1954 V8 P217-222 58-1587 <*>
1954 V8 P223-228 57-2808 <*>
1954 V8 P269-278 59-17482 <*>
1954 V8 P363-369 57-2630 <*>
1954 V8 P367-403 57-2800 <*>
1954 V8 P404-410 58-1622 <*>
1954 V8 P499-504 57-2631 <*>
1954 V8 P523-529 58-34 <*>
1954 V9 P207-220 58-2649 <*>
1955 P20-28 II-759 <*>
1955 P39-46 II-745 <*>
1955 N1 P20-28 57-2276 <*>
1955 V9 P241-245 57-2784 <*>
1955 V9 P259-269 58-1377 <*>
1955 V9 P285-290 57-3053 <*>
1955 V9 N1 P13-19 57-1379 <*>
1955 V9 N2 P93-98 C-2332 <NRC>
57-2389 <*>
1955 V9 N4 P157-161 63-10248 <*>
1955 V9 N7 P299-306 57J16G-A <ATS>
57J16G-B <ATS>
1955 V9 N8 P369-374 57-1455 <*>
1955 V9 N8 P381-387 57J16G-A <ATS>
1955 V9 N11 P528-532 62-16968 <*> 0
1956 V10 P261-265 ACSIL-925 <ACSI>
57-3354 <*>
1956 V10 N3 P107-116 57-2931 <*>
1956 V10 N4 P145-150 57-2689 <*>
1956 V10 N8 P343-352 C-2453 <NRC>
1956 V10 N9 P389-391 58-785 <*>
1956 V10 N12 P505-511 7819G <ATS>
1956 V10 N12 P535-540 55J15G <ATS>
57-3002 <*>
1957 V11 N1 P1-7 59-10874 <*>
1957 V11 N3 P97-100 62-14203 <*> 0
1957 V11 N5 P183-194 57-3352 <*>
1957 V11 N10 P387-396 59-10873 <*>
1957 V11 N12 P485-494 25S85G <ATS>
1958 V12 P193-202 62-00114 <*>
1958 V12 N1 P15-25 59-10864 <*>
1958 V12 N3 P138-148 SCL-T-453 <*>
1958 V12 N4 P176-182 SCL-T-452 <*>
1958 V12 N9 P419-427 17K28G <ATS>
1958 V12 N11 P510-514 62-11532 <=>
1959 V13 N2 P58-62 59-17758 <*>
1959 V13 N3 P121-131 80L33G <ATS>
1959 V13 N5 P211-218 60-14483 <*>
1960 V14 P373-379 62-24058 <=*>
1960 V14 N5 P204-216 66-12760 <*>
1960 V14 N10 P451-467 02N54G <ATS>
61-20847 <*>
1960 V14 N10 P468-476 62-24059 <=*>
1960 V14 N12 P531-538 21N53G <ATS>
61-20525 <*>
1961 V15 P273-284 62-10450 <*>
1961 V15 N2 P94-100 65-13647 <*>
1961 V15 N7 P303-307 62-16683 <*>
1961 V15 N10 P437-443 65-13646 <*>
1961 V15 N10 P444-454 66-13874 <*>
1962 V16 N3 P117-128 65-13117 <*>
1962 V16 N4 P173-188 65-13649 <*>
1962 V16 N8 P365-380 65-12222 <*>
1963 V17 P163-176 64-14634 <*>
1963 V17 N5 P254-260 64-16585 <*>
1964 V18 N1 P1-4 65-12311 <*>
1964 V18 N1 P43-50 SC-T-65-700 <*>
1964 V18 N1 P60-66 65-12221 <*>
1964 V18 N3 P181-188 66-13875 <*>
1964 V18 N3 P197-203 65-11499 <*>
1964 V18 N5 P293-308 65-10434 <*>
1964 V18 N9 P555-564 16S82G <ATS>
65-13110 <*>
1965 V19 P453-458 01T92G <ATS>
1965 V19 N2 P112-118 POED-TRANS-2008 <NLL>
1965 V19 N7 P384-394 66-12737 <*> 0
1965 V19 N9 P453-458 66-12696 <*>
1965 V19 N9 P483-491 66-12695 <*>

ARCHIV FUER ELEKTROTECHNIK
1912 V1 N4 P141-150 NP-TR-503 <*>
1914 V2 N9 P371-387 64-10995 <*> 0
1914 V2 N11 P475-489 57-1396 <*>
1915 V3 N10/1 P332-344 59-20447 <*> 0
1918 V7 P1-17 AL-56 <*>

1941 V11 P106-116 2290 <*>
1941 V11 P117-130 2351 <*>
1956 V14 N16 P626-630 64-20237 <*> 0
1960 V18 P317-326 62-01424 <*>
1960 V18 P327-336 AEC-TR-4759 <*>
1961 V18 P538-555 66-10323 <*>
66-14249 <*>

ARCHIV FUER GYNAEKOLOGIE
1923 V118 P59-100 63-00737 <*>
1931 V145 P2-69 58-1337 <*>
1933 V153 P584-592 63-18446 <*> 0
1940 V170 P413-421 1590 <*>
1955 V186 P384-388 57-2961 <*>
1955 V186 P392-393 57-2961 <*>
1956 V187 P556-594 57-391 <*>
1960 V192 P293-303 61-00991 <*>
1960 V192 P423-427 62-10041 <*>
1964 V199 P671-678 65-17384 <*>

ARCHIV FUER HYDROBIOLOGIE
1906 V1 N3 P385-391 I-986 <*>

ARCHIV FUER HYGIENE UND BAKTERIOLOGIE
1899 V34 P210-243 63-00745 <*>
1902 V37 P48-72 63-00744 <*>
1924 V93 P126-150 62-00537 <*>
1931 V106 P271-298 63-18447 <*>
1943 V129 P286-292 UCRL TRANS-757(L) <*>
1952 V136 P129-138 5513-B <KH>
1952 V136 P265-274 1300 <TC>
1952 V136 N2 P139-158 62-18727 <*> 0
1955 V139 P551 <CSIR>
1955 V139 N5 P349-404 5009-B <KH>
1956 V140 P253-263 64-14851 <*>
1956 V140 P335-344 59-19375 <*=>
1956 V140 N4 P288-291 58-811 <*>
1956 V140 N8 P605-609 58-814 <*>
1963 V147 P289-297 65-17291 <*>
1963 V147 N1 P72-80 64-16359 <*>

ARCHIV FUER JAPANISCHE CHIRURGIE
SEE NIPPON GEKA HOKAN

ARCHIV DER JULIUS KLAUS-STIFTUNG FUER VEREBUNGS-FORSCHUNG, SOZIALANTHROPOLOGIE UND RASSENHYGIENE
1941 V16 N3/4 P469-486 II-983 <*>
1941 V16 N3/4 P492-494 II-983 <*>

ARCHIV FUER KINDERHEILKUNDE
1897 V22 P61-74 62-01227 <*>
1955 V151 N3 P238-245 59-10847 <*>
1958 V158 N3 P248-252 60-17648 <+*> 0
1962 V166 P164-174 63-00199 <*>
63-26699 <=$>
1962 V166 N3 P209-215 63-01773 <*>

ARCHIV FUER KLINISCHE CHIRURGIE
1873 V15 P279-281 57-1606 <*>
1885 V32 P446- 57-1626 <*>
1900 V62 P147-156 59-10698 <*> 0
1910 V92 P536-595 57-1646 <*>
1911 V95 N1 P89-111 60-10489 <*>
1924 V133 P226-231 57-1694 <*>
1926 V142 P374-418 57-1522 <*>
1926 V143 P125-146 1335-A <KH>
1929 V156 P66- 2443A <K-H>
1932 V172 P240-276 57-1683 <*>
1934 V178 P607-628 66-11629 <*> 0

ARCHIV FUER KLINISCHE CHIRURGIE
SEE LANGENBECK'S ARCHIV FUER KLINISCHE CHIRURGIE VEREINIGT MIT DEUTSCHE ZEITSCHRIFT FUER CHIRURGIE

ARCHIV FUER KLINISCHE UND EXPERIMENTELLE DERMATOLOGIE
1956 V203 P203-216 58-1610 <*>
1957 V205 P312-320 58-1177 <*>
1957 V205 N3 P312-320 63-18470 <*>
1957 V206 P589-596 66-10329 <*>
1957 V206 P730-738 63-00295 <*>
1958 V207 N3 P261-267 59-10699 <*> P
1958 V208 P44-52 61-10648 <*> 0
1960 V211 P18-35 63-00061 <*>
1962 V214 P274-288 75R74G <ATS>

ARCHIV FUER KREISLAUFFORSCHUNG
1953 V20 N1/6 P25-44 61-00211 <*>
1959 V29 P242- W-60 <RIS>
1961 V34 P201-244 63-00771 <*>
1961 V34 N4 P287-303 64-00267 <*>

ARCHIV FUER KRIMINOLOGIE
1959 V123 P65-87 60-31066 <=>

ARCHIV FUER LEBENSMITTELHYGIENE
1960 V11 N3 P60-64 NP-TR-494 <*>

ARCHIV FOR MATHEMATIK OG NATURVIDENSKAB
1938 V42 N1/2 P1-71 57-2439 <*>

ARCHIV FUER MATHEMATISCHE LOGIK UND GRUNDLAGENFORSCHUNG
1958 V9 P82-93 UCRL TRANS-557(L) <*>

ARCHIV FUER METALLKUNDE
1946 V1 P16-25 59-20585 <*> PO
65-11925 <*> 0
1947 N3 P124-125 II-116 <*>
1947 V1 N3 P137-138 62-18417 <*> 0
1947 V1 N10 P456-460 62-18416 <*> 0
1947 V1 N7/8 P335-345 63-16759 <*>
1947 V1 N7/8 P359-361 63-16747 <*>
1948 V2 N2 P137-139 57-2637 <*>
1948 V2 N7 P223-230 62-20337 <*> 0
1949 V3 P253-255 57-1704 <*>
1949 V3 P409-413 62-18163 <*> 0
1949 V3 N8 P287-288 62-18375 <*> 0
1949 V3 N12 P432-436 63-10019 <*>

ARCHIV FUER METEOROLOGIE, GEOPHYSIK UND BIOKLIMATOLOGIE
1949 SA V1 P39-61 61-01018 <*>
1949 SB V2 P66-85 AL-823 <*>
1949 SB V2 P427-440 66-12125 <*> 0
1950 SA V2 N2/3 P308-314 C-3464 <NRC>
1950 SA V2 N2/3 P314-324 C-3465 <NRC>
1952 SB V4 N1 P65-84 3097-B <K-H>
1953 SB V5 N1 P66-102 59-15297 <*>
1955 SB V6 P443-451 CSIRO-3338 <CSIR>
1957 SB V8 P143-148 63-01028 <*>
1958 SB V9 N1 P73-79 63-10678 <*>
1959 SA V11 P93-108 89L30G <ATS>
1961 SA V12 N2 P262-270 63-31556 <=>
1963 SA V13 N3/4 P450-460 63-31890 <=> 0

ARCHIV FUER MIKROBIOLOGIE
1933 V4 P394-408 65-11073 <*> 0
1933 V4 N2 P247- 2882 <*>
1933 V4 N2 P248-256 2881 <*>
1935 V6 P350-358 57-3063 <*>
1936 V7 P584-589 59-10289 <*>
1938 V9 N5 P477-485 65-12658 <*>
1950 V14 N4 P602-634 64-18876 <*>
1951 V16 N2 P163-176 65-11067 <*> 0
1952 V17 N4 P403-407 65-11064 <*> C
1953 V18 N3 P245-272 65-11072 <*> 0
1953 V19 P166-172 T-1899 <INSD>
1953 V19 N2 P206-246 65-11070 <*> 0
1953 V19 N4 P365-371 65-11062 <*> 0
1954 V20 N2 P176-182 59-15277 <*>
1954 V20 N4 P362-390 64-16882 <*>
1956 V24 P1-7 3042 <*>
1956 V25 P39-57 C-2281 <NRC>
58-1750 <*>
1957 V26 P373-414 63-20682 <*> 0
1957 V27 N1 P63-81 64-10704 <*> 0
1958 V30 N1 P30-44 C-3580 <NRCC>
1959 V34 N1 P204-210 65-13272 <*> 0

1960 V37 P7-17 63-10506 <*> 0
1960 V37 P57-94 63-10947 <*>
1961 V39 P374-422 62-18780 <*> 0
62-25600 <=*>
1962 V43 N3 P262-279 65-11069 <*>

ARCHIV FUER MIKROSKOPISCHE ANATOMIE UND ENTWICKLUNGSMECHANIK
1869 V5 P1-24 61-00979 <*>
1901 V57 P676-702 62-00533 <*>
1905 V67 P1-17 57-2333 <*>
1912 V80 P306-395 C-2638 <NRC>

ARCHIV FUER NATURGESCHICHTE
1913 V79 P1-10 T-1425 <INSD>

ARCHIV FUER OHRENHEILKUNDE
1873 V1 P56- 57-1643 <*>
1897 V31 P234-294 61-14884 <*>
1906 V68 P1-30 61-14770 <*>

ARCHIV FUER OHREN-, NASEN- U. KEHLKOPFHEILKUNDE
1922 V110 P6-14 61-14645 <*>
1932 V131 P134-166 61-14653 <*>
1937 V143 P52-74 61-14328 <*>
1937 V143 P257-270 61-14332 <*>
1939 V146 P16 61-14315 <*>
1939 V146 P302-333 61-14316 <*>
1940 V147 P1-4 61-14204 <*> 0
1941 V149 P210-218 61-14306 <*>
1941 V150 P25-30 61-14723 <*>
1951 V157 P474-484 57-231 <*>
1958 V173 P529-533 61-10587 <*> 0
1958 V173 N2 P222-225 62-00169 <*>
1961 V177 N5 P427-431 11363D <K-H>
64-14549 <*>
1961 V178 P1-104 63-20017 <=*>

ARCHIV FUER ORTHOPAEDISCHE UND UNFALLCHIRURGIE
1936 V37 P223-231 65-10708 <*>
1936 V37 P232-245 65-10003 <*> 0
1956 V48 P115 61-25663 <*>

ARCHIV FOR PHARMACI OG CHEMI
1940 V47 P287-300 64-16349 <*> P
1940 V47 P753-755 3151 <*>
1951 V58 N9 P268-281 II-8 <*>

ARCHIV DER PHARMAZIE
1892 V230 P561-589 81Q70G <ATS>
1897 V235 P435-441 62-14920 <*>
1907 V245 P25-28 66-13119 <*>
1924 V262 P3-7 66-13120 <*>
1924 V262 P126-127 63-18552 <*>
1925 V263 P181-186 1180D <K-H>
1928 V266 N7 P492-501 63-18442 <*>
1932 V270 P109-114 66-12787 <*>
1932 V270 P476-493 11455-E <K-H>
1933 V271 P1-35 66-11686 <*>
1936 V274 P10-16 59-19209 <+*> 0
1940 V278 P350-360 64-10027 <*>
1942 V280 P76-85 64-16343 <*>
1943 V281 P8-22 63-20498 <*> 0
1943 V281 P171-185 64-16344 <*>
1944 V282 P1-9 63-20264 <*>
1950 V283 N1 P127-136 60-10499 <*>
1951 V284 P6-7 3902-C <K-H>
1953 V286 P218-219 64-16345 <*>
1953 V286 N7 P330-337 64-10346 <*>
1954 V287 P96-98 64-16346 <*>
1954 V287 P142-147 63-20287 <*>
65-12145 <*>
1954 V287 N9/10 P582-590 64-16347 <*>
1955 V288 N3 P145-148 63-16278 <*> 0
1955 V288 N4 P178-179 61-18641 <*>
1955 V288 N4 P186-194 59-15429 <*>
1956 V289 N11 P651-663 223 <TC>
1957 V290 N1 P1-8 57-2823 <*>
1957 V290 N3 P109-117 9817-A <K-H>
1957 V290 N4 P204-206 12159-B <K-H>
1959 V292 P411-416 SC-T-64-611 <*>
1959 V292 N1 P9-14 10265-F <K-H>
61-10904 <*>
1960 V293 P117-120 61-10244 <*> 0
1960 V293 P598-609 63-20218 <*> 0
1960 V293 N2 P187-194 66N55G <ATS>
1960 V293 N5 P531-537 88Q70G <ATS>
1961 V294 P153-163 63-10930 <*>
1961 V294 N6 P366-370 64-16348 <*>
1961 V294 N12 P783-794 65-12110 <*>

ARCHIV FUER PHYSIKALISCHE THERAPIE, BALNEOLOGIE UND KLIMATOLOGIE
1954 V6 N2 P115-119 63-01624 <*>
1956 V8 P262-272 5892 <KH>
1956 V8 N6 P382-387 65-11320 <*> C
1958 V10 N4 P250-252 9681-C <K-H>

ARCHIV FUER POST UND TELEGRAPHIE
1936 V64 N7 P190- 57-2321 <*>

ARCHIV FUER PROTISTENKUNDE
1913 V28 P141-242 64-10305 <*>
1924 V47 N2 P253-307 66-11952 <*>
1924 V50 P67-88 65-11058 <*> PO
1933 V79 N1 P50-71 CSIRO-3183 <CSIR>
1940 V94 N1 P1-50 CSIRO-3182 <CSIR>

ARCHIV FUER PSYCHIATRIE UND NERVENKRANKHEITEN
1885 V16 P698-710 2039 <*>
1932 V97 P644-659 61-14305 <*>
1934 V101 P195-230 63-00618 <*>
1936 V105 P475-483 63-00166 <*>

ARCHIV FUER PSYCHIATRIE UND NERVENKRANKHEITEN VEREINIGT MIT ZEITSCHRIFT FUER DIE GESAMTE NEUROLOGIE UND PSYCHIATRIE
1951 V186 P327-340 57-2713 <*>
1952 V187 P521-536 57-3038 <*>
1952 V188 P345-378 62-20438 <*> 0
1954 V192 P115-127 1369 <*>
1956 V194 N3 P263-288 1719 <*>
1959 V199 P172-185 65-00266 <*> 0
1959 V199 P553-572 62-01229 <*>
1960 V200 N5 P520-530 61-00185 <*>
1960 V201 P332-354 62-00449 <*>
1962 V203 P648-666 63-01017 <*>
1964 V206 N2 P260-271 66-11696 <*>
1965 V206 N4 P454-473 66-11701 <*>

ARCHIV FUER RASSEN- U. GESELLSCHAFTSBIOLOGIE EINSCHLIESSEND RASSEN- U. GESELLSCHAFTSHYGIENE
1930 V24 P91-92 3367-H <KH>

ARCHIV FUER SCHIFFS- U. TROPENHYGIENE
1933 V37 P125-211 SUP 3 63-01240 <*>
1933 V37 P256-271 64-00357 <*>
1935 V39 N6 P257-260 57-1663 <*>
1938 V42 N1 P21-22 61-00462 <*>

ARCHIV FUER TECHNISCHES MESSEN
1932 V2 N15 PT131-T132 II-478 <*>
1933 V2 N21 PT34-T35 66-13338 <*> 0
1934 V3 N34 PT56 59-20365 <*> 0
1934 V3 N35 PF14- 57-2452 <*>
1935 V4 N43 PT8-T9 57-304 <*>
1937 N73 PT89-T90 65-12967 <*>
1938 P124-126 57-4 <*>
1939 N94 PT49-T51 59-20915 <*> 0
1939 N95 PT61-T63 59-20915 <*> C
1940 V112 PT117-T118 66-13425 <*> 0
1941 PT19-T20 57-641 <*>
1941 N115 PF1 57-2091 <*>
1941 N118 PT47 59-20916 <*> 0
1950 V150 N175 PT95-T96 62-20270 <*> 0
1951 N184 PT53-T54 57-1105 <*>
1952 V199 P175-178 II-744 <*>
1953 P91-94 62-18603 <*> C
1953 N210 P165-168 1232 <*>
1953 V213 P219-222 I-119 <*>

1955 V233 P125-128 1129 <*>
1955 V239 P281-284 62-14454 <*>
1956 N247 P173-174 CSIRO-3337 <CSIR>
1956 N248 P203-204 58-2169 <*>
1956 N249 P217-220 59-17156 <*>
1956 V243 P89-92 CSIRO-3337 <CSIR>
1956 V247 P175-178 62-10055 <*>
1957 P221-224 58-1666 <*>
1958 V274 P229-232 60-00439 <*>
1959 V283 P161-164 TS 1590 <BISI>
62-14661 <*>
1960 V293 P109-112 10797-A <K-H>
61-20445 <*>
1960 V293 P125-128 84N48G-B <ATS>
1960 V295 P153-156 10797-B <KH>
61-20445 <*>
1960 V299 NV116 P245-248 63-00471 <*>
1963 N326 PR25-R36 ANL-TRANS-84 <*>
1964 V336 P17-22 65-12839 <*>
1964 V337 P43-46 65-12840 <*>

ARCHIV FUER TIERERNAEHRUNG
1960 V10 N6 P401-413 63-10848 <*> 0

ARCHIV FUER TOXIKOLOGIE
1956 V16 P208-214 10325-C <KH>
64-10611 <*>
1961 V19 P196-198 62-16009 <*>
1961 V19 P199-204 64-14314 <*>
1963 V20 P127-140 8662 <IICH>
1964 V20 N4 P235-241 66-11553 <*>

ARCHIV FUER VIRUSFORSCHUNG
1961 V10 N5 P551-559 62-00977 <*>
1961 V10 N5 P560-577 62-00899 <*>

ARCHIV FUER WAERMWIRTSCHAFT UND DAMPFKESSELWESEN
1937 V18 P75-77 AEC-TR-3246 <*>
58-2389 <*>
1937 V18 N7 P191-194 AEC-TR-5326 <*>
1938 V19 P129-131 60-10572 <*> 0
1939 V20 P135-138 I-317 <*>

ARCHIV FUER WISSENSCHAFTLICHE U. PRAKTISCHE TIERHEILKUNDE
1931 V63 P275-282 65E1G <ATS>

ARCHIVES D'ANATOMIE, D'HISTOLOGIE ET D'EMBRYOLOGIE
1955 V34 N1/8 P145-148 60-18015 <*>

ARCHIVES D'ANATOMIE MICROSCOPIQUE
1899 V3 P3-10 59-15545 <*> 0

ARCHIVES D'ANATOMIE MICROSCOPIQUE ET DE MORPHOLOGIE EXPERIMENTALE
1954 V43 P236-271 59-18114 <+*> 0
1955 V44 N1 P20-26 1330 <*>
1960 V49 P307-332 11095-B <KH>
63-16606 <*>

ARCHIVES BELGES DE MEDECINE SOCIALE, HYGIENE, MEDECINE DU TRAVAIL ET MEDECINE LEGALE
1956 V14 N5 P275-278 5968-B <K-H>

ARCHIVES OF BIOCHEMISTRY AND BIOPHYSICS
1958 V78 P573-586 61-00210 <*>
1959 V83 P275-282 60-13068 <=*> 0P
1960 SUP 1 P260-263 AEC-TR-5671 <*>

ARCHIVES DE BIOLOGIE
1921 V31 P173-298 2734 <*>
1931 V41 P1-35 2200 <*>
1935 N46 P764-775 58-372 <*>
1935 N46 P812-814 58-372 <*>

ARCHIVES FRANCAISES DE PEDIATRIE
1952 V9 P978-983 2042 <*>
1956 V13 N8 P880-888 57-2347 <*>
1958 V15 N9 P1204-1212 64-18740 <*> 0
1958 V15 N9 P1227-1230 64-18616 <*> 0
1959 V16 N2 P145-157 60-15716 <=*> 0
1960 V17 P199-204 65-17288 <*>
1960 V17 P413-416 51N54F <ATS>
1964 V21 P753-761 65-14720 <*>
1965 V22 N10 P1183-1200 66-12618 <*>

ARCHIVES FRANCO-BELGES DE CHIRURGIE
1923 V26 P1-12 65-14733 <*> P

ARCHIVES GENERALES DE MEDECINE
1826 V11 P292-294 N66-14383 <=*$>

ARCHIVES DE L'INSTITUT PASTEUR D'ALGERIE
1945 V23 N3 P169-172 2503 <*>
1950 V28 N2 P130-144 60-10678 <*> 0
1951 V29 N4 P280-286 20E2F <ATS>
1952 V30 P51-54 T-1925 <INSD>
1954 V32 P71-86 2017 <*>

ARCHIVES DE L'INSTITUT PASTEUR DE TUNIS
1931 V20 P426-429 64-10827 <*>

ARCHIVES INTERNATIONALES DE PHARMACODYNAMIE ET DE THERAPIE
1896 V2 P127-194 64-10835 <*> 0
1919 V25 P391-401 5389-A <KH>
1930 V37 P34-60 59-10345 <*> 0
1931 V40 P54-91 1145-C <K-H>
1932 V43 N1 P86-110 1241-D <K-H>
1933 V44 P156-163 2928 <*>
1935 V51 P151-170 58-1865 <*>
1937 V55 P15-31 64-16097 <*> 0
1937 V55 N2 P235-256 58-549 <*>
1938 V59 P149-194 64-16098 <*> 0
1939 V62 P234-260 64-16099 <*>
1939 V62 P330-341 64-16204 <*>
1939 V62 P381-389 1291-M <K-H>
1941 V65 N2 P125-195 64-16100 <*> 0
1943 V69 N2 P181-185 64-16101 <*>
1946 V72 P405-429 1291N <KH>
1948 V75 N3/4 P307-324 5028 <K-H>
1948 V75 N3/4 P413-414 5028 <K-H>
1948 V76 N1 P102-108 63-20976 <*>
1948 V76 N4 P351-366 64-16102 <*> 0P
1950 V81 P53-60 65-10975 <*> 0
1950 V82 P247-296 58-2656 <*>
1950 V82 N8 P360-363 64-16103 <*>
1950 V84 P257-268 AL-637 <*>
1950 V84 P276-282 64-16104 <*>
1951 V87 N3 P290-300 58-159 <*>
1951 V88 P159-202 64-16105 <*> 0
1952 V89 N1 P55-64 64-16106 <*> 0
1952 V89 N4 P365-379 63-18867 <*> 0
1952 V89 N4 P380-398 63-18871 <*> 0
1952 V91 N3/4 P404-411 59-17141 <*> 0
1952 V92 N1 P13-38 64-16107 <*> 0
1952 V92 N1 P44-70 64-16108 <*> 0
1952 V92 N2 P129-162 64-16084 <*> 0
1953 V92 P305-361 I-108 <*>
1953 V92 N3/4 P368-407 64-16085 <*>
1953 V92 N3/4 P471-481 1253-B <K-H>
1953 V93 N2 P163-166 60-10491 <*>
1953 V93 N2 P167-176 60-10490 <*>
1953 V95 P30-48 61-10337 <*>
1954 V96 P426-446 II-204 <*>
1954 V96 N3/4 P327-355 86F3F <ATS>
1954 V97 P141-148 II-428 <*>
1954 V97 P251-266 II-683 <*>
1954 V97 N2 P206-213 64-16086 <*> 0
1954 V98 P452-458 59-22099 <+*> 0
1954 V98 N4 P411-420 3390-B <K-H>
1954 V99 P481-486 II-379 <*>
1954 V100 P127-160 57-2560 <*>
1955 V101 P1-16 57-1917 <*>
1955 V101 P158-170 II-1050 <*>
1955 V102 P187- II-1049 <*>
1955 V102 N3 P314-334 64-16089 <*> 0
1955 V102 N1/2 P1-16 64-16087 <*> 0P
1955 V102 N1/2 P33-54 64-16088 <*> 0
1955 V104 P121-136 2117 <*>

1955 V104 P137-145 1146 <*>
1955 V105 P73-80 2631 <*>
1956 V107 N1 P33-44 59-15412 <*>
1956 V107 N2 P138-149 64-16090 <*> 0
1956 V107 N3/4 P271-274 57-261 <*>
1957 V109 P55-77 57-686 <*>
1957 V109 N1/2 P55-77 64-16091 <*> 0
1957 V109 N3/4 P329-331 58-230 <*>
63-16386 <*>
1957 V112 P319-341 58-75 <*>
1958 V115 P1-31 61-20520 <*>
1958 V115 N1/2 P164-168 64-16092 <*>
1958 V115 N1/2 P169-174 64-16093 <*>
1958 V117 P185-196 58-2613 <*>
1959 V119 P189-193 62-18048 <*>
1959 V119 P497-501 59-15890 <*>
1959 V122 N3/4 P463-473 65-14307 <*> 0
1959 V123 P78-114 61-20899 <*> 0
1960 V124 N1/2 P53-65 64-16094 <*> 0
1960 V125 N1/2 P30-47 61-10221 <*>
1960 V125 N1/2 P172-193 64-18996 <*> 0
1960 V126 N1/2 P126-139 60-16945 <*>
1960 V126 N1/2 P140-153 64-16095 <*> 0
1960 V127 P180-202 6? 16107 <*>
1960 V127 N112 P180-202 64 16096 <*> 0
1960 V129 P343-363 11071-A <KH>
1960 V129 N3/4 P343-363 64-10486 <*>
1961 V131 N3/4 P257-261 62-00503 <*>
1961 V132 P287-294 62-18047 <*>
1961 V132 N1/2 P126-138 65-10100 <*> 0
1961 V132 N3/4 P372-377 64-16354 <*>
1961 V133 P89-100 62-18779 <*> 0
1961 V134 P255-291 64-16070 <*>
1962 V138 N3/4 P505-533 63-01248 <*>
1963 V142 N1/2 P117-124 63-20629 <*>
65-10958 <*>
1963 V142 N1/2 P125-140 63-18014 <*>
1963 V146 N3/4 P561-578 65-11741 <*>
1964 V152 N3/4 P298-306 764 <TC>
1965 V153 N2 P323-333 66-10277 <*> 0
1965 V154 N2 P297-312 66-11707 <*> 0

ARCHIVES INTERNATIONALES DE PHYSIOLOGIE
1946 V54 N1 P30-48 1291-D <KH>
1946 V54 N1 P107-116 64-10019 <*>
1954 V62 N4 P556 65-10133 <*>

ARCHIVES INTERNATIONALES DE PHYSIOLOGIE ET DE BIOCHIMIE
1956 V64 N2 P234-250 57-1084 <*>
1956 V64 N4 P564-570 60-17331 <*+>
1958 V66 N4 P604-609 61-01017 <*>
1958 V66 N4 P633-652 61-01007 <*>
1961 V69 N5 P645-667 65-00234 <*>
1962 V70 N3 P431-435 65-11785 <*>

ARCHIVES ITALIENNES DE BIOLOGIE
1910 V54 P425-428 T2108 <INSD>
1912 V58 P333-336 2752 <*>
1914 V61 N1 P79- 2817 <*>
1958 V96 N1 P78-110 66-11828 <*>

ARCHIVES DES MALADIES DE L'APPAREIL DIGESTIF ET DE LA NUTRITICN
1912 N6 P278-299 65-13843 <*>
1953 N42 P1222-1226 57-2716 <*>
1953 N42 P1283-1285 57-2609 <*>
1954 V43 P1095-1100 58-238 <*>
63-16084 <*>
1955 V44 N6 P99-115 60-14098 <*> 0
1955 V44 N6 P117-131 60-14095 <*>
1955 V44 N6 P132-159 60-14099 <*>
1955 V44 N9/10 P1081-1083 59-15752 <*>
1957 V46 N5 P414-418 66-12537 <*> 0
1960 V49 P43-54 63-16625 <*>
1960 V49 N12 P1585-1600 66-12538 <*> 0
1963 V52 P669-678 64-20918 <*>
1963 V52 N4 P309-315 63-01687 <*>
1963 V52 N6 P690-695 66-12534 <*>
1963 V52 N6 P695-699 66-12535 <*> 0
1963 V52 N11 P1149-1152 66-12536 <*>

ARCHIVES DES MALADIES DU COEUR, DES VAISSEAUX ET DU SANG
1934 V27 P725-741 61-14490 <*>
1938 V31 P824-827 64-16340 <*>
1951 V44 P550-554 1896 <*>
1951 V44 P555-557 1895 <*>
1951 V44 N8 P673-686 59-15542 <*>
1952 V45 P252-258 I-982 <*>
1953 V46 N10 P898-904 2911 <K-H>
1955 V4. P1156-1166 66-12001 <*>
1959 V52 N2 P121-132 62-00495 <*>
1959 V52 N9 P10C -1019 62-00481 <*>
1961 V54 P460-462 66-11795 <*>
1961 V54 N1 P20-36 62-01392 <*>
1961 V54 N5 P481-500 62-00451 <*>
1962 V55 P1290-1303 65-17284 <*>
1962 V55 N6 P610-618 63-00175 <*>

ARCHIVES DES MALADIES PROFESSIONELLES DE MEDECINE DU TRAVAIL ET DE SECURITE SOCIALE
1946 V7 P19-23 64-16200 <*>
1946 V7 P23-27 64-16221 <*>
1946 V7 P27-28 64-16232 <*>
1953 V14 N3 P288-290 2923-A <KH>
1954 V15 P229-233 3318-A <KH>
1955 V16 N1 P5-19 59-20960 <*> P
1958 V19 P602-606 60-18559 <*>
1958 V19 N1 P5-20 SC-T-64-43 <*>
1961 V22 P63-66 62-10288 <*> 0
1964 V25 P248-253 65-14332 <*>

ARCHIVES DE MEDECINE CES ENFANTS
1933 V36 P713-719 63-20202 <*> 0

ARCHIVES DE MEDECINE EXPERIMENTALE ET D'ANATOMIE PATHOLOGIQUE
1912 V24 P585-608 65-00144 <*> 0

ARCHIVES DE MEDECINE SOCIALE
1946 V2 N9 P577-594 2804 <*>

ARCHIVES MENSUELLES D'OBSTETRIQUE ET DE GYNECOLOGIE
1912 V1 N10 P177- 2810 <*>

ARCHIVES NEERLANDAISES DE PHONETIQUE EXPERIMENTALE
1931 V6 P105-114 64-10295 <*>
1940 V16 P122-142 65-12539 <*>

ARCHIVES NEERLANDAISES DE PHYSIOLOGIE DE L'HOMME ET DES ANIMAUX
1928 V12 P259-264 60-10488 <*> PO
1947 V28 N3/4 P475-480 60-10471 <*>
1947 V28 N3/4 P510-520 60-10455 <*>

ARCHIVES NEERLANDAISES DES EXACTES ET NATURELLES. SERIES 3A
1933 V14 P84-117 UCRL TRANS-728(L) <*>

ARCHIVES D'OPHTHALMOLOGIE
1883 V4 P245-251 61-14287 <*>
1897 V17 P465-474 61-14713 <*>
1921 V38 P98-112 61-14368 <*>
1921 V38 P547-561 61-14369 <*>
1922 V39 P83-1C7 61-14366 <*> 0
1926 V43 P415-423 61-14370 <*>
1926 V43 P551-557 61-14143 <*>
1927 V44 P205-218 61-14367 <*> 0
1928 V45 P515-521 61-14132 <*>
1930 V47 N6 P363-377 61-14362 <*>
1930 V47 N7 P428-433 61-14145 <*>
1930 V47 N8 P560-562 61-16036 <*>
1930 V47 N9 P6C1-605 61-14353 <*>

ARCHIVES D'OPHTHALMOLOGIE ET REVUE GENERALE D'OPHTHALMOLOGIE
1964 V24 N1 P1C-22 66-12120 <*>

ARCHIVES DE PHYSIOLOGIE NORMALE ET PATHOLOGIQUE
1868 V1 P110-127 61-16135 <*>
1868 V1 P643-665 61-16135 <*>
1868 V1 P725-734 61-16135 <*>
1883 V1 P200-212 II-899 <*>
1884 V3 N3 P125-141 60-18943 <*>
1896 V5 P572-579 66-10232 <*>

ARCHIVES PORTUGAISES DES SCIENCES BIOLOGIQUES
1936 V5 P44-47 2226 <*>

ARCHIVES OF PRACTICAL PHARMACY. JAPAN
SEE YAKUZAIGAKU

ARCHIVES DE PSYCHOLOGIE
1943 V29 P267-276 59-18482 <+*>
1946 V31 P300-306 59-18484 <+*>
1950 V33 N129 P49-70 64-10023 <*>
1951 V33 P81-130 59-14244 <+*> 0

ARCHIVES FOR RATIONAL MECHANICS AND ANALYSIS
1963 V13 P330-347 66-12135 <*>

ARCHIVES ROUMAINES DE PATHOLOGIE EXPERIMENTALE ET DE MICROBIOLOGIE
1938 V11 N2 P247-272 60-10492 <*>
1963 V23 P515-522 65-17025 <*>
1964 V23 N3 P749-762 66-10973 <*>

ARCHIVES DES SCIENCES
1948 V1 P195-199 85J15G <ATS>
1957 V10 P247-249 SP.NO. AEC-TR-5328 <*>
1957 V10 P62 59-20685 <*>
1957 V10 P197-199 58-2522 <*>
1958 V11 P143-149 60-16231 <*>
1962 V15 P137-146 64-18681 <*>

ARCHIVES DES SCIENCES PHYSIOLOGIQUES
1948 V2 N1/4 P243-255 59-17216 <*>
1950 V4 N1 P13-29 61-20731 <*> 0
1952 V6 N1 P53-61 65-12532 <*>
1956 V10 N3 P249-275 62-01221 <*>

ARCHIVES DES SCIENCES PHYSIQUES ET NATURELLES
1929 S5 V11 P239-259 59-14724 <*=>

ARCHIVES DE ZOOLOGIE EXPERIMENTALE ET GENERALE
1886 V4 P535-624 61-14573 <*>
1933 V75 P353-358 65-00132 <*> 0

ARCHIVIO DI ANTROPOLOGIA CRIMINALE, PSICHIATRIA E MEDICINA LEGALE
1940 V60 P762-779 62-16149 <*>

ARCHIVIO PER L'ANTROPOLOGIA E LA ETNOLOGIA
1959 V89 P148-167 62-00304 <*>

ARCHIVIO 'DE VECCHI' PER L'ANATOMIA PATOLOGICA E LA MEDICINA CLINICA
1960 V32 P681-706 63-26231 <=$>
1962 V37 N1 P1-11 64-00272 <*>
1963 V39 N1 P259-301 63-01388 <*>

ARCHIVIO E. MARAGLIANO DI PATOLOGIA E CLINICA
1950 V5 P771-777 I-675 <*>
1958 V14 P69-74 64-16341 <*>
1960 V16 P491-502 98N52I <ATS>

ARCHIVIO DI FARMACOLOGIA SPERIMENTALE E SCIENZE AFFINI
1909 V8 P63-80 65-14334 <*> 0
1912 V14 N2 P81-84 58-630 <*>
1917 V24 P280-288 57-181 <*>
1945 V75 P129-141 64-20015 <*> 0

ARCHIVIO DI FISIOLOGIA
1929 V27 N4 P519-542 61-00628 <*>
1929 V27 N4 P543-557 61-00629 <*>
1929 V27 N4 P558-580 61-00634 <*>
1938 V38 P26-35 64-20014 <*> 0
1947 V47 N4 P254-261 61-00646 <*>
1955 V55 N3 P349-374 63-00925 <*>

ARCHIVIO DELL'ISTITUTO BIOCHIMICO ITALIANO
1937 V9 P3-18 64-16339 <*>

ARCHIVIO ITALIANO DI ANATOMIA E DI EMBRIOLOGIA
1924 V21 P55-86 61-01056 <*>
1939 V42 P506-564 59-14523 <=*> 0
1941 V45 P360- 57-2608 <*>
1954 V59 N2 P57-82 59-18954 <=*> 0
1959 V64 N2 P131-148 62-01037 <*>

ARCHIVIO ITALIANO DI ANATOMIA E ISTOLOGIA PATOLOGICA
1932 V3 P707-717 60-13022 <*+> 0
1936 V7 P419-427 57-1687 <*>
1942 V15 P214-232 58-329 <*>
1962 V36 P99-113 63-20226 <*> 0

ARCHIVIO ITALIANO DI CHIRURGIA
1933 V34 N6 P810-816 3062 <*>
1938 V53 P135-157 65-14281 <*> 0
1953 V76 P300-320 58-1246 <*>

ARCHIVIO ITALIANO DI DERMATOLOGIA, SIFILOGRAFIA E VENEREOLOGIA
1955 V27 N3 P271-281 59-15854 <*> 0

ARCHIVIO ITALIANO DEI MALATTIE DELL'APPARATO DIGERENTE
1960 V27 P403-432 66-11656 <*> 0
1960 V27 P469-481 66-11655 <*> 0
1963 V30 N5 P417-421 66-11718 <*> 0

ARCHIVIO ITALIANO DI OTOLOGIA, RINOLOGIA E LARINGOLOGIA
1934 V46 P43-62 58-553 <*>
1954 V62 P235-262 64-10668 <*>
6871-C <K-H>
1959 V70 N1 P84-95 64-14531 <*>
9440-A <K-H>
1960 V71 P164-189 10809-D <K-H>
64-10667 <*>
1960 V71 P376-408 10938-B <K-H>
64-14513 <*>
1961 V72 N2 P1-8 62-01211 <*>
1963 V74 P378-386 66-13013 <*>

ARCHIVIO ITALIANO DI PSICOLOGIA
1920 V1 P107-182 61-16081 <=>

ARCHIVIO ITALIANO DI SCIENZE FARMACOLOGICHE
1950 S2 V3 N6 P229-231 63-16409 <*>
1952 V2 P117-133 AL-1 <*>
1952 V2 P345-350 T-2405 <INSD>
1955 S3 V5 P209-217 5285 <K-H>
1956 S3 V6 P254-256 5968-A <K-H>
1959 V9 P56-60 61-01055 <*>
1959 S3 V9 P479-501 10281-E <K-H>
64-14771 <*>
1959 S3 V9 N2 P176-177 64-10590 <*>

ARCHIVIO DI ORTOPEDIA
1935 V51 P413-423 57-1615 <*>
1951 V64 P16-33 57-1862 <*>
1956 V69 N6 P527-564 60-14191 <*> 0

ARCHIVIO DI OSTETRICIA E GINECOLOGIA
1933 V20 P93-100 1361E <K-H>

ARCHIVIO DI OTTALMOLOGIA
1895 V2 P205-226 61-14358 <*> 0
1896 V4 P397-405 61-14130 <*>
1928 V35 N2 P49-71 61-14299 <*>
1936 V43 P146-148 <K-H>

ARCHIVIO DI PATOLOGIA E CLINICA MEDICA

1940 N21 P228-262 58-323 <*>

ARCHIVIO DI PSICOLOGIA, NEUROLOGIA E PSICHIATRIA
1958 V20 P325-355 60-14541 <*>
1959 V20 N5/6 P423-433 61-00203 <*>

ARCHIVIO "PUTTI" DI CHIRURGIA DEGLI ORGANI DI MOVIMENTO
1952 V2 P173-183 57-577 <*>

ARCHIVIO DI SCIENZE BIOLOGICHE
1930 V14 P379-390 AL-769 <*>
1933 V19 P76-80 65-14777 <*>
1935 V21 P510-521 64-16222 <*>
1941 V27 P291-326 60-10843 <*>
1941 V27 P327-332 60-10842 <*>
1942 V28 N4 P317-340 63-14552 <*>
1950 V34 P294-301 2248 <*>
1955 V39 P474-486 58-1730 <*>
1955 V39 P525-532 65-00305 <*> O
1957 V42 P39-45 58-636 <*>
1958 V42 N5 P490-494 62-00887 <*>

ARCHIVIO PER LE SCIENZE MEDICHE
1909 V33 P1-37 T-8082 <INSD>
1957 V104 N5 P656-676 66-11621 <*>
1957 V104 N5 P677-692 66-11634 <*>
1957 V105 N4 P361-392 59-18964 <+*> O
1958 V106 N6 P877-885 64-18622 <*>
1960 V110 N2 P228-253 64-18632 <*>
1961 V112 P501-509 UCRL TRANS-1027(L) <*>

ARCHIVIO PER LO STUDIO DELLA FISIOPATOLOGIA E CLINICA DEL RICAMBIO
1960 V24 N3/4 P321-325 61-20912 <*>
1960 V24 N3/4 P396-402 61-20913 <*>
1960 V24 N3/4 P403-409 61-20914 <*>

ARCHIVIO DI TISIOLOGIA E DELLE MALATTIE DELL'APARATO RESPIRATORIO
1947 V2 N3 P252-262 2573 <*>
1948 V3 P211-225 2914 <*>
1949 V4 P12- 2546 <*>
1952 V7 N7 P237-246 1876 <*>
1955 V10 P126-132 59-18489 <+*> O

ARCHIVO DE ANATOMIA E ANTHROPOLOGIA
1921 V7 P293-319 57-1620 <*>

ARCHIVO DE BIOLOGIA VEGETAL TEORICA Y APLICADA
1943 V1 P64-84 65-12385 <*>

ARCHIVOS ARGENTINOS DE PEDIATRIA
1942 V17 N3 P1-20 2515 <*>
1961 V55 P269-272 66-11592 <*>

ARCHIVOS DE BIOLOGIA. SAO PAULO
1945 V29 P119-124 66-14257 <*>
1948 V32 N285 P49-52 60-19556 <*+>

ARCHIVOS DE BIOQUIMICA QUIMICA Y FARMACIA, TUCUMAN
1957 V8 P143-1446 63-20998 <*>

ARCHIVOS BRASILEIROS DE CARDIOLOGIA
1956 V9 P186-187 57-1851 <*>

ARCHIVOS BRASILEIROS DE ENDOCRINOLOGIA E METABOLOGIA
1959 V8 N1 P1-28 60-17328 <*=> O

ARCHIVOS BRASILEIROS DE MEDICINA
1942 V32 P1-20 1070-D <K-H>

ARCHIVOS CUBANOS DE CANCEROLOGIA
1954 V13 P188-216 3289-A <K-H>

ARCHIVOS DEL INSTITUTO DE CARDIOLOGIA DE MEXICO
1950 V20 P147-181 I-1035 <*>
1953 V23 P131-159 1827 <*>
1962 V32 P430-451 63-01683 <*>

ARCHIVOS DE OFTALMOLOGIA DE BUENOS AIRES
1965 V38 P365-370 65-00295 <*>

ARCHIVOS DE PEDIATRIA. RIO DE JANEIRO
1954 V25 N1 P5-21 1916 <*>

ARCHIVOS DE PEDIATRIA DEL URUGUAY
1960 V31 N9 P496-498 09N48SP <ATS>

ARCHIVOS DE LA SOCIEDAD AMERICANA DE OFTALMOLOGIA Y OPTOMETRIA
1963 V4 P229-262 65-12009 <*>

ARCHIVOS DE LA SOCIEDAD DE BIOLOGIA DE MONTEVIDEO
1937 V1 P1-7 II-120 <*>
1949 V16 P1-4 1910 <K-H>

ARCHIVOS DE LA SOCIEDAD OFTALMOLOGICA HISPANO-AMERICANA
1962 V22 N5 P441-455 63-00213 <*>

ARCHIVOS URUGUAYOS DE MEDICINA, CIRUGIA Y ESPECIALIDADES
1942 V21 P90-100 66-11637 <*> O
1950 V36 P483-494 I-967 <*>

ARCHIVUM SOCIETATIS ZOOLOGICAE-BOTANICAE FENNICAE VANAMO'
SEE SUOMALAISEN ELAIN- JA KASVITIETEELLISEN SEURAN VANAMON TIEDONANNOT

ARCHIWUM AUTOMATYKI I TELEMECHANIKI
1961 V6 N1 P3-21 62-19481 <=*>
1961 V6 N1 P23-32 62-19549 <=*>
1961 V6 N1 P33-36 62-19488 <=*>
1961 V6 N1 P39-45 62-19487 <=*>
1961 V6 N1 P71-77 62-19491 <=*>
1961 V6 N1 P93-117 62-19485 <=*>
1961 V6 N1 P121-125 62-19531 <=*>
1961 V6 N2/3 P143-356 62-19555 <=*>
1963 V8 N3 P335-346 AD-635 835 <=$>

ARCHIWUM BUDOWY MASZYN
1959 V6 N4 P493-520 62-20428 <*>
1960 V7 N3 P283-294 63-16630 <*>
1961 V8 N1 P73-79 62-20053 <*> OP
1961 V8 N2 P139-192 AD-627 092 <=$>
1963 V10 N2 P103-140 64-71209 <=>

ARCHIWUM CHEMJI I FARMACJI
1936 V3 P33-49 66-13118 <*>

ARCHIWUM ELEKTROTECHNIKI
1961 V10 N1 P129-146 65-64227 <=*$>
1961 V10 N2 P441-467 63-01032 <*>
1961 V10 N3 P723-727 64-00375 <*>
1964 V13 N3 P697-711 66-12309 <*> O

ARCHIWUM GORNICTWA
1957 V2 N4 P249-265 60-21305 <=>
1964 V9 N1 P85-93 85S82P <ATS>

ARCHIWUM GORNICTWA I HUTNICTWA
1954 V2 P243-352 58-2246 <*>
1954 V2 N1 P71-122 60-21561 <=>
1955 V3 N1 P3-9 60-21289 <=>
1955 V3 N1 P11-42 2005 <BISI>
1955 V3 N1 P69-87 60-21561 <=>
1955 V3 N3 P379-400 63-11388 <=>
1955 V3 N3 P401-435 63-14209 <*> O

ARCHIWUM HUTNICTWA
1958 V3 N1 P59-78 65-64225 <=$*>
1958 V3 N2 P113-123 AD-621 802 <=$>
1958 V3 N2 P125-146 4860 <HB>
1959 V4 N2 P122-159 4820 <HB>
1960 V5 N4 P363-387 5043 <HB>
1961 V6 N1 P61-82 4092 <BISI>
1962 V7 N3 P251-264 3790 <BISI>

1962 V7 N4 P333-344 3660 <BISI>
1963 V8 N2 P119-140 3617 <BISI>

ARCHIWUM HYDROTECHNIKI
1958 V5 N1 P3-19 61-31298 <=>
1958 V5 N2 P251-266 60-21521 <=>
1959 V6 N2 P149-151 60-21566 <=>
1959 V6 N4 P483-487 60-21566 <=>
1960 V7 N1 P129-136 60-21566 <=>
1960 V7 N2 P143-209 60-21566 <=>

★ARCHIWUM IMMUNOLOGII I TERAPII DOSWIADCZALNEJ
1962 V10 N1 ENTIRE ISSUE 62-11069/1 <=>

ARCHIWUM MECHANIKI STOSOWANEJ
1955 V7 N1 P151-168 66-13835 <*>
1960 V12 N1 P13-27 66-11061 <*>
1964 V16 N2 P517-519 AD-640 289 <=$>
1964 V16 N4 P1009-1021 N66-14055 <=$>

ARCHIWUM MINERALOGICZNE
1955 V19 P1-8 39H12P <ATS>

ARCISPEDALE S. ANNA DI FERRARA; RIVISTA TRIMESTRALE DI SCIENZE MEDICHE
1963 V16 N2 P363-366 64-16941 <*>

ARCOS
1938 N89 P1951-1967 96L34F <ATS>
1951 V28 N121 P3027-3043 57-1318 <*>

ARERUGI
1954 V3 N1 P73-81 2386 <*>

ARHIV BIOLOSKIH NAUKA
1949 V1 N3 P253-257 61-11252/1-5 <=>
1950 V2 N1 P33-50 60-21625 <=>
1950 V2 N2 P109-113 61-11252/1-5 <=>
1950 V2 N3/4 P211-221 61-11252/1-5 <=>
1951 V3 N1/2 P27-32 61-11252/1-5 <=>
1951 V3 N3/4 P117-120 61-11252/1-5 <=>

ARHIV ZU FARMACIJU
1960 V10 N6 P411-413 62-19697 <*=> P

ARHIV ZA HIGIJENU RADA
1954 V5 N1 P49-56 3199-F <K-H>

ARHIV ZA KEMIJU
1946 V18 P12-36 63-18555 <*> P
1949 V21 P218-225 I-942 <*>
1950 V22 P85-100 57-1099 <*>
1951 V23 P22-29 57-351 <*>
1951 V23 P104-118 CR-640 <ATS>
1951 V23 N112 P104-118 64-16762 <*>
1953 V25 P141-150 3183-B <KH>

ARHIV MINISTARSTVA POLJOPRIVREDE I SUMARSTVA
1937 V4 N7 P88-108 60-21692 <=>

ARHIV ZA POLJOPRIVREDNE NAUKE I TEHNIKU
1951 V4 N6 P3-53 60-21617 <=>
1951 V4 N6 P76-86 61-11209 <=>
1951 V4 N6 P176-179 60-21695 <=>
1952 V5 N7 P104-117 60-21676 <=>
1954 V7 N18 P1-20 60-21621 <=>
1963 V16 N51 P3-27 63-11453/1 <=>
1963 V16 N51 P29-35 63-11453/1 <=>
1963 V16 N51 P37-49 63-11453/1 <=>
1963 V16 N51 P52-57 63-11453/1 <=>
1963 V16 N51 P60-76 63-11453/1 <=>
1963 V16 N51 P79-90 63-11453/1 <=>
1963 V16 N51 P92-100 63-11453/1 <=>
1963 V16 N51 P102-111 63-11453/1 <=>
1963 V16 N51 P114-122 63-11453/1 <=>
1963 V16 N51 P124-130 63-11453/1 <=>
1963 V16 N53 P3-140 63-11453/3 <=>
1963 V16 N54 P3-158 63-11453/4 <=>
1964 V17 N56 P3-154 64-11453/2 <=$>
1964 V17 N58 P3-77 64-11453/4 <=>
1964 V17 N58 P79-162 64-11453/4 <=>

ARKHIMEDES. HELSINKI
1964 N1 P27-36 66-14447 <*>

ARKHITEKTURA SSSR
1955 N9 P29-35 R-5289 <*>
1958 N5 P11-12 65-63004 <=$>
1959 N1 P10-16 C-5158 <NRC>
1959 N2 P4-6 61-13995 <*+>
1959 N2 P6-11 61-15924 <+*>
1959 N5 P23-25 60-17682 <+*>
1959 N7 P14-15 59-13939 <=>
1959 N10 P26-30 62-24617 <=*>
1960 N5 P16-19 62-24614 <=*$>
1960 N12 P10-15 63-24214 <=*$>
1961 N1 P45-48 62-25917 <=*>
1961 N2 P1-4 62-25918 <=*>
1961 N2 P28-30 62-25916 <=*>
1962 N1 P51-53 63-24220 <=*$>
1962 N2 P29-30 63-24222 <=*$>
1963 N5 P16-18 LC-1280 <NLL> O
1963 N11 P23-31 <MPBW>
1963 V11 N5 P10-12 M.5196 <NLL>
1964 N2 P22-29 RTS-2848 <NLL>
1964 N3 P56-58 RTS-2847 <NLL>
1964 N5 P33-37 65-63809 <=$>
1965 N3 P40-49 RTS-3080 <NLL>

ARKHITEKTURA I STROITELSTVO. MOSCOW
1949 V4 N7 P23-24 59-19133 <+*>

ARKHITEKTURA I STROITELSTVO MOSKVY
SEE STROITELSTVO I ARKHITEKTURA MOSKVY

ARKHIV ANATOMII, GISTOLOGII I EMBRIOLOGII
1954 V31 N4 P25-32 RT-3796 <*>
1954 V31 N4 P63-65 RT-3806 <*> O
1955 V32 N2 P66-71 R-67 <*>
1955 V32 N4 P6-18 62-15153 <=*> C
1955 V32 N4 P19-22 62-15131 <*=>
1955 V32 N4 P54-62 62-15149 <*=> O
1956 V33 N3 P3-14 62-15150 <=*>
1957 V34 N1 P22-28 62-15735 <*=>
1957 V34 N1 P90-99 61-28685 <*=>
1957 V34 N2 P42-46 61-15407 <+*>
1957 V34 N5 P3-7 62-15093 <*=>
1957 V34 N5 P8-18 62-15092 <*=>
1958 V35 N2 P3-16 59-13825 <=>
1958 V35 N2 P30-38 59-13826 <=> C
1958 V35 N3 P8-18 59-13838 <=>
1958 V35 N3 P69-72 59-13883 <=> O
59-17367 <+*>
1958 V35 N5 P99-101 RTS 2840 <NLL>
1958 V35 N6 P3-21 59-13827 <=> O
1958 V35 N6 P52-57 59-13828 <=> O
1959 V36 N1 P3-6 59-11760 <=>
1959 V36 N1 P55-62 61-15311 <+*> C
1959 V36 N1 P90-98 59-31046 <=>
1959 V36 N1 P115-120 59-11761 <=>
1959 V36 N4 P3-15 60-18362 <+*>
60-41174 <=>
1959 V36 N5 P3-6 59-13857 <=>
1959 V36 N5 P19-31 59-13875 <=> O
1959 V36 N6 P3-10 59-13969 <=>
1959 V36 N8 P117-132 60-11534 <=>
1959 V36 N12 P104-118 60-11474 <=>
1959 V37 N7 P19-28 62-13920 <*=>
1959 V37 N10 P67-73 C-182 <RIS>
1959 V37 N12 P3-26 60-31398 <=>
1960 V38 N1 P3-18 60-11692 <=>
1960 V38 N1 P123-127 60-31425 <=>
1960 V38 N2 P3-23 60-11894 <=>
1960 V38 N2 P104-120 60-11993 <=>
1960 V38 N3 P3-15 60-41721 <=>
1960 V38 N3 P53-62 60-41720 <=>
1960 V38 N3 P63-71 61-11095 <=>
1960 V38 N3 P72-79 60-41722 <=>
1960 V38 N3 P107-112 60-41719 <=>
1960 V38 N4 P117-125 61-27237 <*=>

Citation	Translation
1960 V38 N5 P56-59	60-31715 <=>
1960 V38 N5 P94-103	60-41454 <=>
1960 V38 N5 P124-126	60-41455 <=>
1960 V38 N6 P117-122	61-11021 <=>
1960 V39 N10 P24-36	61-21381 <=>
1960 V39 N10 P37-42	61-21400 <=>
1960 V39 N1C P66-77	61-21378 <=>
1960 V39 N10 P79-93	61-21304 <=>
1960 V39 N11 P3-10	61-21430 <=>
1960 V39 N11 P11-22	61-21429 <=>
1960 V39 N11 P23-32	61-21484 <=>
1960 V39 N11 P33-42	61-21437 <=>
1960 V39 N11 P44-50	61-21362 <=>
1960 V39 N11 P51-59	61-21361 <=>
1960 V39 N11 P60-67	61-21405 <=>
1960 V39 N11 P68-73	61-21671 <=>
1960 V39 N11 P74-82	61-21654 <=>
1960 V39 N11 P121-124	61-21673 <=>
1961 V40 N2 P17-23	61-23125 <*=>
1961 V40 N4 P72-79	61-27068 <=*>
1961 V40 N5 P3-17	61-27869 <*=>
1961 V40 N5 P18-26	61-27753 <*=>
1961 V40 N5 P27-33	61-27756 <*=>
1961 V40 N5 P34-38	61-27757 <*=>
1961 V40 N6 P23-30	61-28396 <*=>
1961 V40 N6 P31-40	61-28412 <*=>
1961 V40 N6 P41-45	61-31609 <=>
1961 V40 N6 P46-53	61-28401 <=>
1961 V40 N6 P54-59	61-28415 <*=>
1961 V41 N7 P28-37	62-13308 <=*>
1961 V41 N7 P54-57	64-13051 <=*$>
1961 V41 N7 P72-84	62-13268 <=*>
1961 V41 N8 P125-126	62-15467 <=*>
1961 V41 N10 P114-118	62-25495 <=*>
1961 V41 N1C P122-127	62-25186 <=*>
1961 V41 N11 P106-111	62-15788 <=*>
1962 V42 N1 P3-21	64-23601 <=$>
1962 V42 N2 P12-	FPTS V22 N3 P.T389 <FASE>
1962 V42 N2 P61-	FPTS V22 N1 P.T41 <FASE>
1962 V42 N2 P61-65	62-24973 <=*>
1962 V42 N2 P66-	FPTS V22 N1 P.T48 <FASE>
1962 V42 N2 P69-82	63-15431 <=*> 0
1962 V42 N2 P83-93	63-19023 <=*>
1962 V42 N2 P112-121	62-33022 <=*>
1962 V42 N3 P3-29	62-32204 <=*>
1962 V42 N3 P30-	FPTS V22 N4 P.T597 <FASE>
1962 V42 N3 P55-60	63-23908 <=*> 0
1962 V42 N3 P61-69	63-19112 <=*>
1962 V42 N3 P103-	FPTS V22 N4 P.T684 <FASE>
1962 V42 N3 P108-114	63-19372 <=*>
1962 V42 N4 P3-92	62-32673 <=>
1962 V42 N5 P3-13	62-32444 <=*>
1962 V42 N5 P14-	FPTS V22 N4 P.T773 <FASE>
1962 V42 N5 P29-34	63-23048 <=*>
1962 V42 N5 P83-88	63-23047 <=*>
1962 V42 N5 P90-	FPTS V22 N5 P.T886 <FASE>
1962 V42 N6 P3-23	62-33518 <=*>
1962 V42 N6 P24-	FPTS V23 N1 P.T155 <FASE>
1962 V42 N6 P78-	FPTS V22 N2 P.T379 <FASE>
1962 V42 N6 P84-87	63-24470 <=*$>
1962 V42 N6 P88-	FPTS V23 N1 P.T129 <FASE>
1962 V42 N6 P97-102	64-13506 <=*$>
1962 V43 N8 P11-	FPTS V22 N4 P.T712 <FASE>
1962 V43 N8 P29-	FPTS V22 N4 P.T724 <FASE>
1962 V43 N8 P35-42	63-23046 <=*>
1962 V43 N8 P43-	FPTS V22 N4 P.T746 <FASE>
1962 V43 N9 P23-	FPTS V23 N1 P.T117 <FASE>
1962 V43 N9 P46-	FPTS V23 N1 P.T183 <FASE>
1962 V43 N9 P66-75	63-24520 <=*$> 0
1962 V43 N9 P83-87	63-24521 <=*$>
1962 V43 N1C P3-	FPTS V22 N5 P.T872 <FASE>
1962 V43 N10 P13-	FPTS V22 N5 P.T879 <FASE>
1962 V43 N10 P19-35	63-23883 <=*$> 0
1962 V43 N1C P36-54	63-23043 <=*>
1962 V43 N10 P61-70	63-23848 <=*>
1962 V43 N11 P3-11	63-23053 <=*>
1962 V43 N11 P12-18	63-23860 <=>
1962 V43 N11 P18-	FPTS V22 N5 P.T931 <FASE>
1962 V43 N11 P29-45	63-24452 <=*$> 0
1962 V43 N11 P45-52	63-23861 <=*$> 0
1962 V43 N11 P52-58	64-13487 <=*$> 0
1962 V43 N11 P58-67	63-23862 <=*>
1962 V43 N11 P68-76	63-23863 <=*>
1962 V43 N12 P3-	FPTS V22 N5 P.T994 <FASE>
1962 V43 N12 P29-35	63-21470 <=>
1962 V43 N12 P50-	FPTS V22 N4 P.T740 <FASE>
1962 V43 N12 P77-83	63-24456 <=*$> 0
1963 V44 N1 P69-77	63-23829 <=*>
1963 V44 N1 P77-	FPTS V22 N6 P.T1120 <FASE>
1963 V44 N2 P66-72	64-13826 <=*$> 0
1963 V44 N2 P72-	FPTS V23 N1 P.T134 <FASE>
1963 V44 N2 P80-	FPTS V23 N2 P.T308 <FASE>
1963 V44 N2 P87-93	64-13412 <=*$> 0
1963 V44 N2 P93-102	63-31921 <=>
1963 V44 N3 P3-18	64-13401 <=*$> 0
1963 V44 N3 P19-28	64-13400 <=*$> 0
1963 V44 N3 P28-48	64-13399 <=*$>
1963 V44 N3 P48-62	64-13398 <=*$> 0
1963 V44 N3 P68-	FPTS V23 N1 P.T159 <FASE>
1963 V44 N3 P91-	FPTS V23 N1 P.T168 <FASE>
1963 V44 N3 P115-	FPTS V23 N1 P.T150 <FASE>
1963 V44 N3 P121-127	64-13479 <=*$> 0
1963 V44 N4 P26-	FPTS V23 N2 P.T404 <FASE>
1963 V44 N4 P1C0-105	64-13829 <=*$> C
1963 V44 N5 P26-36	63-31802 <=>
1963 V44 N5 P57-62	64-15504 <=*$> 0
1963 V44 N5 P96-105	63-31802 <=>
1963 V44 N6 P21-29	63-31706 <=>
1963 V44 N6 P3C-	FPTS V23 N4 P.T695 <FASE>
1963 V44 N6 P40-47	63-31716 <=>
1963 V44 N6 P54-61	64-15505 <=*$> 0
1963 V44 N6 P80-86	64-15455 <=*$>
1963 V45 N7 P3-	FPTS V23 N3 P.T647 <FASE>
1963 V45 N7 P3-17	FPTS,V23,N3,P.T647 <FASE>
	63-31738 <=>
1963 V45 N7 P18-26	64-15506 <=*$> 0
1963 V45 N7 P34-41	64-15507 <=*$> 0
1963 V45 N7 P64-	FPTS V23 N3 P.T617 <FASE>
	FPTS,V23,N3,P.T617 <FASE>
1963 V45 N7 P77-	FPTS V23 N3 P.T499 <FASE>
	FPTS,V23,N3,P.T499 <FASE>
1963 V45 N7 P89-95	64-15508 <=*$> 0
1963 V45 N8 P3-18	64-15510 <=*$> 0
1963 V45 N8 P44-54	64-15511 <=*$> 0
1963 V45 N8 P55-	FPTS V23 N3 P.T596 <FASE>
	FPTS,V23,N3,P.T596 <FASE>
1963 V45 N8 P84-89	64-15485 <=*$> C
1963 V45 N9 P51-	FPTS V23 N4 P.T706 <FASE>
1963 V45 N9 P59-	FPTS V23 N4 P.T711 <FASE>
1963 V45 N9 P72-83	64-15509 <=*$> 0
1963 V45 N11 P3-13	64-21420 <=>
1963 V45 N11 P53-61	64-21419 <=>
1963 V45 N11 P62-65	64-19648 <=$>
1963 V45 N11 P81-	FPTS V23 N5 P.T1099 <FASE>
1963 V45 N12 P41-	FPTS V23 N6 P.T1231 <FASE>
1963 V45 N12 P41-50	FPTS V23 P.T1231 <FASE>
1964 V46 N1 P3-49	64-21905 <*>
1964 V46 N1 P62-70	64-19728 <=$>
	64-21905 <*>
1964 V46 N1 P8C-87	64-21905 <*>
1964 V46 N2 P14-	FPTS V24 N2 P.T300 <FASE>
1964 V46 N2 P14-22	FPTS V24 P.T300-T304 <FASE>
1964 V46 N2 P23-	FPTS V24 N1 P.T166 <FASE>
1964 V46 N2 P23-31	FPTS V24 P.T166-T170 <FASE>
1964 V46 N2 P36-42	64-31148 <=>
1964 V46 N2 P48-	FPTS V24 N1 P.T137 <FASE>
1964 V46 N2 P48-59	FPTS V24 P.T137-T141 <FASE>
1964 V46 N2 P7C-	FPTS V24 N1 P.T171 <FASE>
1964 V46 N2 P70-75	FPTS V24 P.T171-T174 <FASE>
1964 V46 N3 P3-27	64-31209 <=>
1964 V46 N3 P43-53	65-61089 <=$>
1964 V46 N3 P80-	FPTS V24 N2 P.T2225 <FASE>
1964 V46 N3 P8C-90	FPTS V24 P.T225-T230 <FASE>
1964 V46 N4 P27-	FPTS V24 N3 P.T397 <FASE>
1964 V46 N4 P63-	FPTS V24 N3 P.T459 <FASE>
1964 V46 N4 P87-92	64-31570 <=>
1964 V46 N5 P3-11	64-41312 <=>
	65-62163 <=$>
1964 V46 N5 P16-24	66-61764 <=$>
1964 V46 N5 P5C-57	64-41320 <=>

1964 V46 N5 P58- FPTS V24 N4 P.T584 <FASE>
1964 V46 N5 P83-87 64-41321 <=>
1964 V46 N6 P2-8 65-62839 <=$>
1964 V46 N6 P56- FPTS V24 N5 P.T804 <FASE>
1964 V47 N7 P77- FPTS V24 N4 P.T629 <FASE>
1964 V47 N8 P59- FPTS V24 N5 P.T899 <FASE>
1964 V47 N8 P81- FPTS V24 N4 P.T571 <FASE>
1964 V47 N8 P87- FPTS V24 N4 P.T605 <FASE>
1964 V47 N9 P26-29 65-62323 <=$>
1964 V47 N9 P84-92 65-62322 <=$>
1964 V47 N9 P92-99 65-62316 <=$>
1964 V47 N10 P37-43 65-63590 <=$>
1964 V47 N10 P62- FPTS V24 N6 P.T941 <FASE>
1964 V47 N10 P88- FPTS V24 N6 P.T948 <FASE>
1964 V47 N11 P41- FPTS V24 N5 P.T812 <FASE>
1964 V47 N11 P41-44 65-30204 <=$>
1964 V47 N11 P95-98 65-30203 <=$>
1965 V48 N1 P46-57 65-64523 <=*$>
1965 V48 N1 P57-70 65-64522 <=*$>
1965 V48 N1 P96-102 65-30817 <=$>
1965 V48 N2 P3-16 65-30894 <=$>
1965 V48 N2 P51-60 66-60321 <=*$>
1965 V48 N2 P60-67 66-60322 <=*$>
1965 V48 N2 P67-74 65-30864 <=$>
1965 V48 N2 P99-104 65-30895 <=$>
1965 V48 N3 P25-31 66-60843 <=*$>
1965 V48 N3 P31- FPTS V25 N2 P.T204 <FASE>
1965 V48 N4 P24- FPTS V25 N2 P.T196 <FASE>
1965 V48 N4 P67- FPTS V25 N2 P.T258 <FASE>
1965 V48 N4 P80- FPTS V25 N2 P.T295 <FASE>
1965 V48 N4 P117-125 65-31458 <=$>
1965 V48 N5 P3- FPTS V25 N2 P.T211 <FASE>
1965 V48 N5 P34- FPTS V25 N2 P.T201 <FASE>
1965 V48 N6 P3-11 66-61069 <=$>
1965 V48 N6 P11-21 66-61743 <=*$>
1965 V48 N6 P29-37 66-61072 <=*$>
1965 V48 N6 P80-87 66-61078 <=*>
1965 V48 N6 P1112-1115 65-32327 <=$>
1965 V49 N7 P54- FPTS V25 N4 P.T595 <FASE>
1965 V49 N8 P13- FPTS V25 N4 P.T555 <FASE>
1965 V49 N8 P21- FPTS V25 N4 P.T559 <FASE>
1965 V49 N8 P60- FPTS V25 N3 P.T499 <FASE>
1965 V49 N9 P3- FPTS V25 N5 P.T889 <FASE>
1965 V49 N9 P27- FPTS V25 N4 P.T585 <FASE>
1965 V49 N10 P33-42 66-30138 <=*$>
1965 V49 N11 P23- FPTS V25 N4 P.T661 <FASE>
1965 V49 N12 P29- FPTS V25 N5 P.T771 <FASE>
1966 V50 N1 P37- FPTS V25 N6 P.T989 <FASE>

★ARKHIV BIOLCGICHESKIKH NAUK
1934 V36 N1 P25-52 60-21786 <=>
1936 V41 N2 P113-136 3289-B <KH>
1938 V49 N2 P138 65-60072 <=$>
1938 V52 N1 P126-131 62-25137 <=*>
1938 V53 N1 P138-144 64-15701 <=*$>
1940 V57 N2/3 P94-105 RT-1299 <*>
1940 V60 N2 P42-55 61-27447 <*=>
1940 V60 N3 P3-13 RT-1831 <*>
1940 V60 N3 P15-24 RT-1823 <*>
1940 V60 N3 P25-28 RT-1822 <*>
1940 V60 N3 P86-92 65-60059 <=$> 0
1940 V60 N3 P93-99 65-60060 <=$> 0
1940 V60 N3 P100-101 65-60080 <=$>
65-62914 <=>
1941 V61 N2 P3-12 65-63245 <=>
1941 V62 N2 P21-29 65-62912 <=$>
1941 V62 N2 P30-34 65-62910 <=$>

ARKHIV PATOLOGII
1947 V9 N2 P59-64 61-28091 <*=> 0
1949 V11 N1 P3-17 60-13740 <+*>
1949 V11 N1 P40-46 RT-3803 <*>
1951 V13 P79 60-17180 <+*> 0
1953 V15 N2 P3-14 RT-1271 <*>
1953 V15 N2 P15-24 RT-1272 <*>
1953 V15 N2 P55-60 RT-1273 <*>
1953 V15 N2 P88-89 RT-1291 <*>
1953 V15 N3 P3-14 61-00978 <*>
1953 V15 N4 P94-95 RT-4231 <*>
1953 V15 N5 P58-66 64-15563 <=*$> 0
1953 V15 N5 P79-81 RT-2515 <*>
1953 V15 N6 P38-43 R-3441 <*>
1955 V17 N4 P3-14 R-576 <*>
62-23517 <=*> 0
1955 V17 N4 P20-27 60-16929 <*+>
1956 N6 P58-68 59-12172 <+*>
1956 V18 N1 P54-56 59-18147 <+*>
1956 V18 N2 P52-57 R-431 <*>
64-19503 <=$>
1956 V18 N3 P95-97 62-15345 <=*> 0
1956 V18 N4 P3-15 62-15138 <*=>
1956 V18 N5 P69- 62-15299 <*=> P
1956 V18 N6 P18-28 65-63994 <=*$>
1956 V18 N6 P33-45 R-4704 <*>
1956 V18 N6 P35- 62-15300 <*=> P
1956 V18 N7 P8-20 62-15143 <*=>
1956 V18 N7 P84-92 61-19538 <=*> 0
1957 V19 N2 P3-19 60-23232 <+*> 0
1957 V19 N3 P3-15 62-15144 <*=>
1957 V19 N3 P15-25 62-15151 <*=>
1957 V19 N5 P3-13 62-15140 <=*>
1957 V19 N7 P18-23 64-10613 <=*$>
8043-B <K-H>
1957 V19 N7 P44-52 61-15119 <+*>
1957 V19 N9 P27-35 AEC-TR-3550 <+*> 0
1957 V19 N9 P50-57 59-11397 <=> 0
1957 V19 N9 P84-85 R-4260 <*>
65-61610 <=$> 0
1957 V19 N10 P38-60 62-15083 <*=>
1958 V20 N1 P12-22 59-11102 <=> 0
1958 V20 N1 P32-38 59-11101 <=> 0
1958 V20 N1 P39-43 59-11145 <=> 0
1958 V20 N1 P44-48 59-11144 <=> 0
1958 V20 N2 P17-21 PB 141 165T <=>
1958 V20 N3 P75-79 59-11499 <=> 0
1958 V20 N4 P89-95 59-13334 <=>
1958 V20 N5 P21-27 59-13185 <=> 0
1958 V20 N5 P39-45 PB 141 324T <=>
1958 V20 N6 P20-25 59-11053 <=>
1958 V20 N6 P25-31 62-14717 <=*> C
1958 V20 N11 P53-58 59-13193 <=> 0
1959 V21 N2 P28-31 59-11688 <=> 0
1959 V21 N5 P3-11 60-11154 <=>
1959 V21 N5 P12-18 59-13786 <=> 0
1959 V21 N5 P19-24 59-13787 <=> 0
1959 V21 N5 P25-30 59-13788 <=> 0
1959 V21 N5 P30-34 65-61137 <=$>
1959 V21 N6 P39-43 59-13798 <=> 0
1959 V21 N7 P37-43 59-13899 <=> 0
1959 V21 N10 P25-30 60-11151 <=>
1959 V21 N10 P72-74 AD-608 614 <=>
1959 V21 N12 P74-75 60-15966 <+*>
1960 V22 N1 P60-63 60-11666 <=>
1960 V22 N2 P73-75 60-11585 <=>
1960 V22 N4 P3-14 60-41071 <=>
1960 V22 N4 P15-21 60-41515 <=>
1960 V22 N4 P21-28 60-41089 <=>
1960 V22 N4 P29-33 60-41287 <=>
1960 V22 N4 P34-42 60-41449 <=>
1960 V22 N5 P26-34 60-41258 <=>
1960 V22 N5 P58-63 60-41259 <=>
1960 V22 N5 P83-84 60-41260 <=>
1960 V22 N6 P48-54 63-19110 <=*>
1960 V22 N8 P3-17 61-11744 <=>
1960 V22 N8 P18-33 61-11662 <=>
1960 V22 N9 P44-49 61-11695 <=>
1960 V22 N10 P3-16 61-28038 <*=>
1960 V22 N10 P17-23 61-27086 <=*>
1960 V22 N10 P47-49 61-23607 <=*>
1960 V22 N10 P53-59 61-21693 <=>
64-00115 <*>
64-15201 <=*$>
1960 V22 N11 P5-12 61-11986 <=>
1960 V22 N11 P13-18 61-11984 <=>
1960 V22 N11 P18-24 61-11993 <=>
1960 V22 N11 P24-33 61-11987 <=>
1960 V22 N11 P34-44 61-23001 <=*>
1960 V22 N11 P44-50 61-21462 <=>
1960 V22 N11 P50-58 61-21709 <=>
1960 V22 N11 P58-63 61-21609 <=>

1960 V22 N11 P64-67	61-21242 <=>
1960 V22 N11 P68-72	61-21243 <=>
1960 V22 N11 P72-77	61-21253 <=>
1960 V22 N11 P78-84	61-21731 <=>
1960 V22 N11 P85-91	61-31119 <=>
1960 V22 N12 P3-19	61-21499 <=>
	62-23791 <=*>
1960 V22 N12 P23-29	61-21498 <=>
1960 V22 N12 P29-35	61-21501 <=>
1961 V23 N1 P3-23	61-31127 <=>
1961 V23 N1 P51-57	61-31121 <=>
1961 V23 N2 P51-59	61-23551 <=*>
1961 V23 N3 P3-33	61-31521 <=>
1961 V23 N3 P34-37	62-23867 <=*> 0
1961 V23 N4 P15-23	61-27194 <*=>
1961 V23 N4 P76-79	61-27194 <*=>
1961 V23 N4 P92-94	61-28720 <=>
1961 V23 N5 P3-18	61-31551 <=>
1961 V23 N5 P19-26	61-27345 <*=>
1961 V23 N6 P10-16	61-28056 <*=>
1961 V23 N6 P38-47	1141-F <K-H>
	64-10622 <=*$>
1961 V23 N7 P3-19	62-13325 <=*>
1961 V23 N7 P51-55	62-13306 <=*>
1961 V23 N8 P24-49	62-15377 <*=>
1961 V23 N12 P57-68	62-23390 <=*>
1962 V24 N1 P3-	FPTS V22 N6 P.T1160 <FASE>
1962 V24 N1 P35-41	63-23904 <=*$> 0
1962 V24 N1 P55-	FPTS V22 N5 P.T906 <FASE>
1962 V24 N1 P63-	FPTS V22 N3 P.T470 <FASE>
1962 V24 N1 P63-69	62-23772 <=*>
1962 V24 N1 P70-78	63-23027 <=*>
1962 V24 N2 P19-	FPTS V22 N3 P.T497 <FASE>
1962 V24 N2 P33-	FPTS V22 N3 P.T463 <FASE>
1962 V24 N2 P37-42	62-24384 <=*>
1962 V24 N3 P13-	FPTS V22 N3 P.T502 <FASE>
1962 V24 N3 P27-	FPTS V22 N3 P.T518 <FASE>
1962 V24 N3 P39-44	63-15997 <=*>
1962 V24 N3 P44-	FPTS V22 N3 P.T531 <FASE>
1962 V24 N3 P53-60	63-15998 <=*>
1962 V24 N3 P60-68	63-15999 <=*>
1962 V24 N3 P77-79	63-19000 <=*>
1962 V24 N4 P13-	FPTS V22 N1 P.T37 <FASE>
1962 V24 N4 P19-24	63-19008 <=*>
1962 V24 N4 P32-37	63-15429 <=*>
1962 V24 N4 P38-	FPTS V22 N1 P.T59 <FASE>
1962 V24 N4 P43-	FPTS V22 N1 P.T160 <FASE>
1962 V24 N4 P50-	FPTS V22 N2 P.T317 <FASE>
1962 V24 N4 P55-61	63-15430 <=*> 0
1962 V24 N4 P73-77	64-51287 <=>
1962 V24 N4 P78-79	63-19025 <=*>
1962 V24 N5 P3-	FPTS V22 N5 P.T822 <FASE>
1962 V24 N5 P3-7	62-32375 <=*>
1962 V24 N5 P7-	FPTS V22 N3 P.T565 <FASE>
1962 V24 N5 P13-18	63-19378 <=*>
1962 V24 N5 P57-63	57P66R <ATS>
	63-19379 <=*> 0
1962 V24 N5 P63-	FPTS V22 N3 P.T459 <FASE>
1962 V24 N6 P3-	FPTS V22 N2 P.T331 <FASE>
1962 V24 N6 P3-12	62-32670 <=*>
1962 V24 N6 P12-17	63-19373 <=*>
1962 V24 N6 P21-28	63-19001 <=*>
1962 V24 N6 P29-	FPTS V22 N4 P.T657 <FASE>
1962 V24 N6 P35-41	65-61139 <=$>
1962 V24 N6 P41-47	63-19374 <=*> 0
1962 V24 N6 P47-	FPTS V22 N2 P.T262 <FASE>
1962 V24 N6 P53-	FPTS V22 N2 P.T341 <FASE>
1962 V24 N6 P53-60	62-32670 <=*>
1962 V24 N6 P60-	FPTS V22 N2 P.T236 <FASE>
1962 V24 N7 P3-18	62-33726 <=>
1962 V24 N7 P19-26	63-19365 <=*> 0
1962 V24 N7 P26-	FPTS V22 N6 P.T1147 <FASE>
1962 V24 N7 P34-42	63-23026 <=*>
1962 V24 N7 P42-54	62-33726 <=>
1962 V24 N7 P42-49	63-19364 <=*> 0
1962 V24 N7 P60-	FPTS V22 N3 P.T440 <FASE>
1962 V24 N8 P7-18	62-33298 <=*>
1962 V24 N8 P18-26	63-19380 <=*> 0
1962 V24 N8 P56-	FPTS V22 N3 P.T488 <FASE>
1962 V24 N9 P24-	FPTS V22 N5 P.T918 <FASE>
1962 V24 N9 P33-39	63-23029 <=*>
1962 V24 N9 P50-	FPTS V22 N6 P.T1150 <FASE>
1962 V24 N10 P38-44	63-23881 <=*>
1962 V24 N10 P45-	FPTS V22 N5 P.T923 <FASE>
1962 V24 N10 P52-	FPTS V22 N5 P.T928 <FASE>
	FPTS V22 N5 PT2 <FASE>
1962 V24 N10 P64-69	63-23880 <=*$> 0
1962 V24 N10 P80-82	63-23058 <=*>
1962 V24 N11 P23-	FPTS V22 N5 P.T829 <FASE>
1962 V24 N11 P23-28	63-15083 <=*>
1962 V24 N11 P29-34	64-13502 <=*$> 0
1962 V24 N11 P34-	FPTS V22 N5 P.T850 <FASE>
1962 V24 N11 P42-46	63-23844 <=*>
1962 V24 N11 P57-63	63-23843 <=*>
1962 V24 N12 P15-21	63-21277 <=>
	64-13495 <=*$> 0
1962 V24 N12 P21-26	63-24467 <=*$>
1962 V24 N12 P32-38	63-24469 <=*$>
1962 V24 N12 P61-	FPTS V22 N6 P.T1139 <FASE>
1962 V24 N12 P69-	FPTS V22 N6 P.T1157 <FASE>
1963 V25 N1 P25-29	64-13507 <=*$> 0
1963 V25 N1 P41-	FPTS V22 N6 P.T1083 <FASE>
1963 V25 N1 P46-	FPTS V23 N1 P.T71 <FASE>
1963 V25 N1 P60-65	63-24472 <=*$> 0
1963 V25 N2 P17-24	64-13475 <=>
1963 V25 N2 P24-	FPTS V23 N2 P.T312 <FASE>
1963 V25 N2 P47-53	64-13474 <=*$>
1963 V25 N2 P53-	FPTS V23 N3 P.T519 <FASE>
1963 V25 N2 P53-58	FPTS,V23,N3,P.T519 <FASE>
1963 V25 N2 P59-	FPTS V23 N1 P.T189 <FASE>
1963 V25 N2 P66-70	64-13892 <=*$>
1963 V25 N3 P3-9	63-31571 <=>
1963 V25 N3 P34-	FPTS V23 N2 P.T343 <FASE>
1963 V25 N3 P38-44	64-13827 <=*$>
1963 V25 N3 P56-	FPTS V23 N2 P.T287 <FASE>
1963 V25 N3 P62-65	63-31570 <=>
	64-13909 <=*$>
1963 V25 N3 P65-70	64-13828 <=*$> 0
1963 V25 N4 P38-44	64-19651 <=$> 0
1963 V25 N4 P62-69	64-19652 <=$> 0
1963 V25 N5 P35-40	64-13888 <=*$> 0
1963 V25 N5 P40-	FPTS V23 N3 P.T526 <FASE>
1963 V25 N5 P40-44	FPTS,V23,N3,P.T526 <FASE>
	63-31307 <=>
1963 V25 N5 P72-78	64-13831 <=*$> 0
1963 V25 N6 P3-	FPTS V23 N4 P.T679 <FASE>
1963 V25 N6 P16-	FPTS V23 N3 P.T572 <FASE>
1963 V25 N6 P16-26	FPTS,V23,N3,P.T572 <FASE>
1963 V25 N6 P27-	FPTS V23 N3 P.T565 <FASE>
1963 V25 N6 P27-32	FPTS,V23,N3,P.T565 <FASE>
1963 V25 N6 P33-38	64-13854 <=*$> 0
1963 V25 N6 P38-	FPTS V23 N3 P.T475 <FASE>
1963 V25 N6 P38-44	FPTS,V23,N3,P.T475 <FASE>
1963 V25 N6 P45-52	FPTS,V23,N3,P.T450 <=*$>
1963 V25 N6 P53-57	64-15446 <=*$>
1963 V25 N6 P67-71	63-31709 <=>
1963 V25 N7 P48-53	64-19277 <=*$>
1963 V25 N7 P49-53	63-31750 <=>
1963 V25 N7 P72-80	63-31739 <=>
1963 V25 N8 P14-23	64-19294 <=*$>
1963 V25 N8 P23-30	64-19261 <=*$>
1963 V25 N8 P38-	FPTS V23 N6 P.T1351 <FASE>
	FPTS V23 P.T1351 <FASE>
1963 V25 N9 P69-	FPTS,V23,N5,P.T941 <FASE>
1963 V25 N10 P16-24	64-19636 <=$> 0
1963 V25 N10 P24-28	64-19637 <=$>
1963 V25 N10 P28-	FPTS V23 N5 P.T1047 <FASE>
	FPTS,V23,N5,P.T1047 <FASE>
1963 V25 N11 P21-	FPTS V23 N6 P.T1243 <FASE>
	FPTS V23 P.T1243 <FASE>
1963 V25 N11 P29-37	64-19655 <=$> 0
1963 V25 N11 P37-43	64-19656 <=$> 0
1963 V25 N11 P50-58	64-19657 <=$>
1963 V25 N12 P3-22	64-21729 <=>
1963 V25 N12 P29-	FPTS V23 N6 P.T1355 <FASE>
	FPTS V23 P.T1355 <FASE>
1963 V25 N12 P35-	FPTS V23 N6 P.T1250 <FASE>
	FPTS V23 P.T1250 <FASE>
1963 V25 N12 P44-50	64-21729 <=>
1963 V25 N12 P51-56	64-21729 <=>

1964 V26 N1 P29-39 65-61076 <=$>
1964 V26 N1 P39- FPTS V24 N2 P.T213 <FASE>
1964 V26 N1 P39-46 FPTS V24 P.T213-T217 <FASE>
1964 V26 N1 P71-76 65-60227 <=$>
1964 V26 N2 P10-16 65-60161 <=$>
1964 V26 N2 P44-50 65-60171 <=$>
1964 V26 N2 P56-62 65-61077 <=$>
1964 V26 N3 P3-15 64-31623 <=>
1964 V26 N3 P16- FPTS V24 N3 P.T493 <FASE>
1964 V26 N3 P26- FPTS V24 N3 P.T515 <FASE>
1964 V26 N3 P51- FPTS V24 N2 P.T231 <FASE>
FPTS V24 P.T231-T234 <FASE>
1964 V26 N3 P63-68 65-61088 <=$>
1964 V26 N4 P3-17 64-31557 <=>
1964 V26 N4 P32-40 64-31556 <=>
1964 V26 N4 P32-41 65-61112 <=$>
1964 V26 N4 P71-75 65-61085 <=$>
1964 V26 N5 P25-31 65-62182 <=$>
1964 V26 N5 P63- FPTS V24 N3 P.T382 <FASE>
1964 V26 N6 P18-25 65-62838 <=$>
1964 V26 N6 P34-39 65-62180 <=>
1964 V26 N6 P65-73 65-62164 <=$>
1964 V26 N7 P18-23 RTS-2880 <NLL>
1964 V26 N8 P3-13 64-51462 <=>
1964 V26 N8 P42- FPTS V24 N4 P.T627 <FASE>
1964 V26 N9 P15- FPTS V24 N5 P.T829 <FASE>
1964 V26 N9 P21- FPTS V24 N5 P.T833 <FASE>
1964 V26 N9 P42- FPTS V24 N5 P.T826 <FASE>
1964 V26 N10 P37-43 65-63591 <=$>
1964 V26 N11 P3-11 65-30341 <=$>
1964 V26 N11 P23-31 65-63592 <=$>
1964 V26 N11 P59-65 65-63593 <=$>
1964 V26 N12 P18-26 66-60302 <=*$> 0
1964 V26 N12 P38-51 65-30783 <=$>
1964 V26 N12 P52-58 66-60323 <=*$>
1965 V27 N1 P74-81 65-64524 <=*$>
1965 V27 N1 P89-91 65-30908 <=$> 0
1965 V27 N2 P6-13 65-31032 <=$>
1965 V27 N2 P19- FPTS V25 N1 P.T107 <FASE>
1965 V27 N2 P52-60 65-31033 <=$>
1965 V27 N2 P60-67 66-60864 <=*$>
1965 V27 N2 P86-87 65-31034 <=$>
1965 V27 N3 P18- FPTS V25 N1 P.T157 <FASE>
1965 V27 N3 P43-48 66-60842 <=*$>
1965 V27 N3 P88-90 65-30960 <=$>
1965 V27 N3 P93-94 65-30960 <=$>
1965 V27 N4 P57-66 66-60883 <=*$>
1965 V27 N4 P66-74 66-60884 <=*$>
1965 V27 N5 P8-15 66-61071 <=*$>
1965 V27 N5 P24-30 66-61066 <=$*>
1965 V27 N5 P30-36 66-61073 <=*$>
1965 V27 N6 P58- FPTS V25 N3 P.T463 <FASE>
1965 V27 N7 P19- FPTS V25 N4 P.T623 <FASE>
1965 V27 N9 P61-63 66-30268 <=*$>
1965 V27 N9 P78-79 66-30268 <=*$>
1965 V27 N10 P38-44 66-61997 <=*$>
1966 V28 N1 P3-8 66-31198 <=$>
1966 V28 N11 P89-90 66-33310 <*>

ARKIV FOR FYSIK
1951 V3 N37 P577- II-401 <*>

ARKIV FOR GEOFYSIK
1954 V2 N6 P109-138 2109 <*>

ARKIV FOR KEMI
1952 V3 N62 P571-580 59-15746 <*> 0
1952 V4 N26 P381-384 57-836 <*>
1953 V5 N29 P313-316 59-17022 <*> 0
1954 V7 N46 P417-426 C-132 <RIS>
1957 V10 N5 P473-482 59-10011 <*>
1957 V10 N6 P497-505 59-10018 <*>
1957 V10 N6 P507-511 59-10019 <*>
1957 V10 N6 P513-521 59-10020 <*>
1957 V10 N6 P541-547 59-10012 <*>
1959 V15 N40 P433-438 62-00288 <*>

ARKIV FOR KEMI, MINERALOGI OCH GEOLOGI
1918 V7 N14 P1-9 MF-6 <*>
1920 V7 N31 P1-15 II-47 <*>
1921 V8 N14 P1-23 59-17317 <*>
1940 V13A N27 P38-50 58-971 <*>
1941 P282-287 58-964 <*>
1942 V15A N19 P1-10 65-13214 <*>
1943 V16A N19 P1-17 I-978 <*>
1944 V18B N3 P1-8 60-10566 <*>
1945 V20A P1-26 AL-845 <*>
1945 V20A N18 P1-26 14J16G <ATS>
1946 V23B N6 P1-5 RCT V20 N4 P978 <RCT>

ARKIV FOR MATEMATIK
1949 V1 N13 P141-160 63-10779 <*>

ARKIV FOR MATEMATIK, ASTRONOMI OCH FYSIK
1924 V18 N30 P1-11 I-945 <*>
1928 V21A N4 P291- 57-2328 <*>
1934 V25A N1 P1-13 57-3489 <*>
1944 V31A N18 P1-17 I-6 <*>
NP-2234 <*>

ARKTIKA
1934 N2 P25-37 C-2213 <NRC>
RT-20 <*>
RT-4599 <*>

ARMEE RUNDSCHAU
1964 N7 P36-38 64-41516 <=$>

ARMEYSKI PREGLAD
1958 N4 P83-89 59-13182 <=> 0
1966 N9 P14-17 66-35623 <=>
1966 N9 P53-56 66-35623 <=>
1966 N9 P67-70 66-35723 <=>

ARMIANSKII KHIMICHESKII ZHURNAL
1966 V19 N3 P167-173 AD-638 671 <=$>
1966 V19 N3 P184-187 66-14176 <*>

ARS MEDICI BRUXELLES
1947 V2 P140-142 64-16350 <*>

ARS MEDICI. LIESTHAL
1952 V42 P550-554 AL-725 <*>
1955 V45 P338-339 2146 <*>

ARTILLERI-TIDSKRIFT
1958 V87 N2 P49-52 59-11913 <=> 0

ARTS ET METIERS
1936 N191 P164-166 60-10249 <*>

ARZNEIMITTEL-FORSCHUNG
1952 V2 P250-251 03K21G <ATS>
1952 V2 N7 P304-307 64-16778 <*> 0
1953 V3 N1 P19-23 64-16779 <*> 0
1953 V3 N6 P267-276 AL-359 <*>
1954 P607-614 58-360 <*>
1954 V4 N1 P75-79 3614-A <K-H>
1954 V4 N2 P249-257 07N48G <ATS>
1954 V4 N4 P232-242 64-16780 <*> OP
1954 V4 N4 P262-268 C-2391 <NRC>
1954 V4 N5 P341-344 59-15246 <*> 0
1954 V4 N6 P367-368 3225-B <KH>
1954 V4 N6 P388-391 2657 <*>
1954 V4 N7 P442-443 3145 <*>
1955 V5 P176- II-632 <*>
1955 V5 P204-208 2143 <*>
1955 V5 P295-296 2676 <*>
1955 V5 P488-489 61-20573 <*>
1955 V5 P502-505 3467 <*>
1955 V5 N3 P105-109 3155 <*>
1955 V5 N5 P295-296 64-16781 <*> 0
1955 V5 N6 P331-335 58-170 <*>
1955 V5 N6 P342-348 57-2565 <*>
1955 V5 N6 P348-350 3808-C <K-H>
1955 V5 N7 P367-370 64-16293 <*> 0
1955 V5 N7 P380-390 64-16783 <*> 0
1955 V5 N8 P432-436 64-16784 <*> 0
1955 V5 N12 P715-719 64-16785 <*> 0
1956 V6 P61-66 1133 <*>

1956 V6 P322-330	5525-H <K-H>
1956 V6 P423-426	58-2432 <*>
1956 V6 P426-430	62-18887 <*> 0
1956 V6 P430-432	62-18889 <*>
1956 V6 P752-756	60-18640 <*>
1956 V6 N1 P22-25	64-16786 <*>
1956 V6 N5 P265-269	5525-G <K-H>
1956 V6 N8 P454-457	64-16787 <*> P
1956 V6 N9 P565-568	58-1338 <*>
1957 V7 P663-664	58-829 <*>
	63-16591 <*>
1957 V7 N2 P123-128	65-12688 <*>
1957 V7 N4 P233-237	59-10344 <*>
1957 V7 N6 P344-349	64-16788 <*> 0
1957 V7 N7 P402-404	51J19G <ATS>
1957 V7 N7 P436-439	63L36G <ATS>
1957 V7 N11 P651-653	58-1405 <*>
1957 V7 N11 P653-658	58-1403 <*>
1957 V7 N11 P658-662	58-1324 <*>
1957 V7 N12 P756	64-16789 <*>
1958 V8 P197-199	58-2210 <*>
1958 V8 P507-511	59-22176 <+*> 0
1958 V8 P675-676	59-15391 <*>
1958 V8 N1 P42-44	64-16790 <*> 0
1958 V8 N4 P238-240	58-2657 <*>
	59-15247 <*>
1958 V8 N7 P376-379	65-11970 <*>
1958 V8 N8 P481-484	59-19271 <+*> 0
1958 V8 N8 P553-554	59-20701 <*> 0
	64-16791 <*> P
1958 V8 N9 P622	64-16792 <*>
1958 V8 N12 P750-753	64-30284 <*>
1959 V9 P81-89	59-15595 <*> 0
1959 V9 P341-346	AEC-TR-4027 <*>
1959 V9 P672-678	UCRL TRANS-937 <*>
1959 V9 N1 P39-45	64-16793 <*> 0
1959 V9 N1 P45-47	64-16794 <*>
1959 V9 N1 P47-49	64-16795 <*> 0
1959 V9 N1 P49-59	64-16796 <*> 0
1959 V9 N6 P363-365	63-10397 <*> 0
1959 V9 N6 P368-375	59-20966 <*> 0
1959 V9 N7 P411-419	60-10744 <*>
1959 V9 N9 P539-544	64-16868 <*>
1959 V9 N9 P581-585	60-10847 <*> 0
1959 V9 N10 P647-653	64-16869 <*>
1959 V9 N12 P728-734	66-11724 <*> 0
1959 V9 N12 P767-769	64-16797 <*>
1959 V9 N12 P778-785	64-16870 <*>
1960 V10 P921-923	63-10329 <*> 0
1960 V10 N1 P54-58	64-16871 <*>
1960 V10 N2 P139-142	64-16872 <*>
1960 V10 N3 P188-192	64-16873 <*>
1960 V10 N4 P229-231	60-18557 <*>
1960 V10 N6 P468-474	64-16874 <*>
1960 V10 N7 P547-554	64-16875 <*>
1960 V10 N8 P653-656	65-14200 <*> 0
1960 V10 N9 P728-731	63-10325 <*> 0
1960 V10 N10 P831-834	63-10356 <*> 0
1960 V10 N10 P834-836	61-20565 <*> 0
1961 V11 P110-113	63-14768 <*> 0
1961 V11 P682-684	62-14098 <*>
1961 V11 P684-694	62-14097 <*>
1961 V11 P695-700	62-14096 <*>
1961 V11 P726-736	62-14343 <*>
1961 V11 P1011-1016	12057-C <K-H>
1961 V11 P1149-1157	12057-D <K-H>
1961 V11 N2 P131-132	66-11436 <*>
1961 V11 N3 P247-255	66-12964 <*>
1961 V11 N4 P395-400	63-14769 <*>
1961 V11 N7 P602-606	63-10327 <*> 0
1961 V11 N7 P606-612	63-10326 <*> 0
1961 V11 N8 P695-700	62-16218 <*> 0
1961 V11 N8 P701-707	62-16256 <*> 0
1961 V11 N8 P707-720	62-16257 <*> 0
1961 V11 N9 P867-874	66-12957 <*>
1961 V11 N9 P874-877	65-11676 <*>
1961 V11 N10 P909-911	65-14201 <*>
1962 V12 P1065-1070	64-16069 <*>
1962 V12 N1 P82-84	62-18180 <*>
1962 V12 N2 P202-206	12057E <K-H>
1962 V12 N3 P228-230	63-10513 <*> 0
1962 V12 N3 P286-289	63-20952 <*> 0
1962 V12 N3 P289-294	63-10187 <*>
1962 V12 N9 P899-902	64-18960 <*> 0
1962 V12 N11 P1086-1087	63-18802 <*>
1963 V13 P3-7	64-00276 <*>
1963 V13 N2 P160	64-16798 <*>
1963 V13 N3 P166-168	64-16799 <*>
1963 V13 N3 P177-178	15Q70G <ATS>
1963 V13 N3 P185-191	63-20956 <*>
1963 V13 N4 P288-294	65-17027 <*>
1963 V13 N5 P370-371	64-00198 <*>
1963 V13 N6 P436-438	64-20913 <*>
1963 V13 N8 P728-734	66-10083 <*>
1963 V13 N10 P908-913	50S86G <ATS>
1964 V14 P461-468	95S89G <ATS>
1964 V14 N2 P57-77	65-00406 <*>
1964 V14 N3 P199-203	66-10082 <*>
1964 V14 N3 P226-230	64-20150 <*>
1964 V14 N3 P238	53S86G <ATS>
1964 V14 N4 P252-258	66-12118 <*>
1964 V14 N5 P415-424	51S86G <ATS>
1964 V14 N5 P424-428	52S86G <ATS>
1964 V14 N8 P898-902	65-12796 <*>
1964 V14 N12 P1332-1334	66-13558 <*>
1964 V14 N12 P1334-1343	66-13557 <*>
1965 V15 P278-281	1015 <TC>
1965 V15 P595-604	66-13025 <*>
1965 V15 P852-856	66-12035 <*>
1965 V15 N1 P69-75	66-11532 <*>
1965 V15 N1 P83-88	66-10536 <*>
1965 V15 N4 P382-387	65-14165 <*>
1965 V15 N4 P443-445	65-14943 <*>
1965 V15 N6 P659-666	66-11437 <*>
1965 V15 N7 P777-782	66-10087 <*>
1965 V15 N7 P785-786	66-11533 <*> 0
1965 V15 N9 P1088-1091	66-11624 <*> 0
1965 V15 N9 P1091-1096	66-13211 <*>
1965 V15 N10 P1155-1157	66-10533 <*>
1965 V15 N10 P1157-1158	66-10534 <*>
1965 V15 N10 P1159-1162	66-10535 <*>

ASAHI GARASU KENKYU HOKOKU

1953 V3 N1 P34-43	62-18204 <*>
1959 V9 N1 P1-29	62-14500 <*>
1959 V9 N6 P49-65	66-11359 <*>

ASIATIC MAINLAND AFFAIRS. JAPAN
SEE TAIRIKU MONDAI

ASKERI VETERINER DERGISI

1955 V131 P341-352	58-89 <*>

ASKLEPIOS. AUSTRIA

1958 V5 P162-164	64-16800 <*>
1959 V6 P217-221	64-16801 <*> P
1961 V8 P350-352, 357-360	64-16802 <*> P
1962 V9 P365-372	64-16803 <*> P
1962 V9 P375-378	64-16804 <*>
1962 V9 P422-427	64-16805 <*> P

ASTRONAUTICA ACTA. WIEN, LONDON

1956 V2 N1 P30-47	61-15075 <=*>
1957 V3 N4 P241-280	59-21108 <=>
1960 V6 N1 P1	61-10910 <*>
1960 V6 N1 P4-15	61-10910 <*>
1965 V11 N1 P36-42	66-10831 <*>

ASTRONAUTYKA

1961 V4 N3 P18-19	62-23915 <*=>

ASTRONOMICHESKII TSIRKULYAR. ASTRONOMICHESKAYA OBSERVATORIYA, KIEVSKII GOSUDARSTVENNYII UNIVERSITET

1952 N123 P10-12	R-4873 <*>
1952 N130 ENTIRE ISSUE	59-10123 <+*>
1953 N141 P5-8	61-23418 <=*>
1953 N144 P11-12	64-71485 <=>
1954 N147 P11-13	63-11670 <=>
1955 N156 P19-21	63-11639 <=>

Citation	Translation
1955 N157 P8-9	61-19443 <+*>
1956 N172 P10-12	61-15270 <*+> 0
1957 N176 P18-20	60-31711 <=>
1957 N184 P24-26	63-14912 <=*>
1957 N187 P5-7	61-15322 <=>
1958 N180 P1-2	R-4719 <*>
1958 N190 P3-5	59-11743 <=> 0
1958 N190 P11-17	59-11743 <=> 0
1958 N190 P16-19	59-11518 <=>
1958 N190 P25-26	59-11518 <=>
1958 N191 P3-9	59-11743 <=> 0
1958 N192 P20-21	64-71487 <=>
1958 N195 P8-9	61-13961 <*+>
1958 N198 P1-2	AD-608 538 <=>
1959 N199 P9-10	61-13962 <*+>
1959 N200 P22-27	60-41413 <=>
1959 N204 P13-14	61-15231 <*+>
1959 N205 P1	61-28243 <*=>
1959 N205 P2-3	61-27813 <*=>
1959 N205 P7-8	AD-609 159 <=>
1960 N208 P25-26	62-25821 <=*>
1960 N211 P14-16	62-15837 <=*>
1960 N215 P11-15	62-23720 <=*>
1963 N240 P1	65-60673 <=$>
1963 N255 P1-5	64-19192 <=*$>
1963 N273 P1-3	64-18757 <*>
1964 N289 P1-4	65-62985 <=*$>
1964 N293 P1-4	65-63021 <=$>
1964 N298 P3-4	65-63764 <=$>
1964 N299 P1-4	65-63728 <=*$>
1964 N303 P3-6	65-63020 <=$>
1964 N305 P2-4	65-63751 <=*$>
1964 N310 P1-4	65-17387 <*>
1964 N310 P5-8	65-63729 <=*$>
1964 V291 P3-4	65-63736 <=$>
1965 N327 P2-4	66-61026 <=*$>
1965 N327 P5-7	66-61016 <=*$>
1965 N329 P2-4	66-61018 <=*$>
1965 N333 P1-6	TRANS-F.135. <NLL>

ASTRONOMICHESKII ZHURNAL

Citation	Translation
1933 V10 N4 P465-486	63-14485 <=*>
1936 V13 N5 P435-454	63-20285 <=*>
1938 V15 N3 P226-231	N65-17114 <=$>
1940 V17 N6 P21-25	RT-3883 <*>
1941 V18 N1 P10-25	61-23459 <=*>
1943 V20 P5-7	63-14911 <*>
1944 V21 N2 P131-136	RT-4453 <*>
1946 V23 N2 P69-81	63-14915 <=*> 0
1946 V23 N6 P333-347	RT-663 <*>
	TT.105 <NRC>
1947 V24 N3 P167-177	RT-3978 <*>
	63-14916 <=*> 0
1947 V24 N6 P329-343	59-19527 <+*>
1948 V25 N2 P101-108	RT-1243 <*>
1948 V25 N3 P172-179	62-23175 <=>
1948 V25 N4 P209-215	59-19526 <+*>
1948 V25 N5 P316-326	60-23858 <+*>
1949 V26 N1 P10-14	RT-1411 <*>
1949 V26 N1 P28-37	59-19471 <+*>
1949 V26 N2 P84-96	RT-109 <*>
1949 V26 N4 P207-218	59-19528 <+*>
1949 V26 N6 P346-354	R-4873 <*>
1949 V26 N6 P355-362	RT-2200 <*>
	61-13177 <*+>
1950 V27 N2 P89-96	RT-2201 <*>
	61-13177 <*+>
1950 V27 N2 P97-104	R-4873 <*>
1950 V27 N6 P351-354	61-13177 <*+>
1951 V28 N1 P64-67	RT-1054 <*>
1951 V28 N4 P219-220	61-23859 <*=>
1951 V28 N4 P234-243	59-19474 <+*>
1951 V28 N6 P443-449	59-19531 <+*>
1952 V29 N1 P49-56	R-4873 <*>
1952 V29 N1 P57-61	TT.325 <NRC>
1952 V29 N1 P86-90	52/3430 <NLL>
1952 V29 N2 P144-153	RT-1412 <*>
1952 V29 N2 P162-170	62-10653 <=*>
1952 V29 N3 P350-362	63-20240 <=*>
1952 V29 N4 P406-417	59-17527 <*>
1952 V29 N4 P418-449	RT-2408 <*>
1952 V29 N4 P450-462	62-25135 <=*>
1952 V29 N4 P463-471	59-19530 <+*>
1952 V29 N5 P538-555	64-00471 <*>
	64-11835 <=>
1952 V29 N6 P633-637	61-13177 <*+>
1952 V29 N6 P638-648	59-19763 <+*>
1952 V29 N6 P738-741	TT.429 <NRC>
1953 V30 N1 P3-14	RT-4138 <*>
1953 V30 N1 P15-36	RT-3440 <*>
1953 V30 N1 P37-49	RT-2766 <*>
1953 V30 N1 P68-75	RT-3182 <*>
1953 V30 N2 P161-179	R-2096 <*>
1953 V30 N3 P279-285	59-19529 <+*>
1953 V30 N3 P286-294	RT-951 <*>
1953 V30 N3 P295-301	62-10149 <=*>
	65-17089 <*>
1953 V30 N4 P414-425	59-19519 <+*>
1953 V30 N5 P495-507	RT-2767 <*>
1954 V31 P112-	R-3731 <*>
	64-71362 <=>
1954 V31 P141-153	R-856 <*>
1954 V31 N1 P51-59	60-13113 <+*>
	61-28769 <*=>
1954 V31 N1 P85-89	59-19461 <+*>
1954 V31 N2 P131-136	R-4624 <*>
1954 V31 N2 P132-140	R-4370 <*>
1954 V31 N2 P137-140	CSIRO-3421 <CSIR>
1954 V31 N2 P141-153	RT-3412 <*>
1954 V31 N2 P191-196	60-27014 <=*>
1954 V31 N3 P217-223	RT-2247 <*>
1954 V31 N4 P327-334	59-19463 <+*>
1954 V31 N4 P335-357	R-2362 <*>
1954 V31 N5 P433-435	62-10652 <=*>
1954 V31 N5 P442-452	AD-609 693 <=>
1954 V31 N6 P511-528	59-20683 <+*>
	65-10849 <*>
1954 V31 N6 P529-532	RT-3733 <*>
1954 V31 N6 P533-536	RT-3734 <*>
1955 V32 P201-208	59-19464 <+*>
1955 V32 N1 P29-32	RT-2873 <*>
1955 V32 N1 P33-34	RT-4204 <*>
1955 V32 N1 P61-71	61-15304 <+*>
1955 V32 N1 P79-89	R-4873 <*>
1955 V32 N1 P90-92	62-19836 <=>
1955 V32 N2 P139-149	59-19473 <+*>
1955 V32 N2 P150-164	R-3622 <*>
	59-19473 <+*>
1955 V32 N2 P165-176	59-19473 <+*>
1955 V32 N2 P177-191	59-19473 <+*>
1955 V32 N3 P226-238	3033 <*>
	59-15040 <+*>
1955 V32 N3 P255-264	AD-624 914 <=*$> M
1955 V32 N3 P265-281	AD-610 965 <=$>
	C-2682 <NRC>
	R-4270 <*>
1955 V32 N4 P327-337	59-19462 <+*>
1955 V32 N4 P354-358	59-19462 <+*>
1955 V32 N4 P359-372	AD-610 968 <=$>
	C-2656 <NRC>
	R-4023 <*>
1955 V32 N6 P477-488	RT-4361 <*>
1955 V32 N6 P503-513	61-15174 <+*>
1955 V32 N6 P545-549	R-3952 <*>
1955 V32 N6 P550-554	CSIRO-3278 <CSIR>
	R-3902 <*>
	59-20684 <+*>
1956 V33 N1 P14-19	59-19606 <+*>
1956 V33 N1 P62-73	R-3642 <*>
1956 V33 N1 P62-83	59-19532 <+*>
1956 V33 N1 P84-86	59-19520 <+*>
1956 V33 N1 P87-92	59-19532 <+*>
1956 V33 N1 P93-100	62-10168 <=*>
	63-11639 <=>
1956 V33 N2 P129-136	59-19540 <+*>
1956 V33 N2 P222-235	RTS-2719 <NLL>
1956 V33 N3 P315-339	RTS-2720 <NLL>
1956 V33 N4 P549-555	AD-616 537 <=$>
1956 V33 N4 P588-598	59-19533 <+*>
1956 V33 N5 P641-645	59-19496 <+*>

1956 V33 N6 P785-799 63-11527 <=>
1956 V33 N6 P893-903 59-19534 <+*>
1956 V33 N6 P953-956 59-19534 <+*>
1957 V34 P247-249 R-1843 <*>
1957 V34 N2 P241-246 R-1774 <*>
59-20693 <+*>
1957 V34 N2 P247-249 C-2401 <NRC>
1957 V34 N2 P267-275 64-13339 <=*$>
1957 V34 N3 P313 R-2679 <*>
1957 V34 N3 P328-335 R-4699 <*>
1957 V34 N3 P419-423 59-20691 <+*>
1957 V34 N3 P435-439 59-19495 <+*>
1957 V34 N3 P440-441 C-2507 <NRC>
R-2312 <*>
1957 V34 N3 P469-473 59-11387 <=> 0
1957 V34 N4 P505-515 59-19539 <+*>
1957 V34 N4 P539-556 59-19539 <+*>
1957 V34 N4 P557-567 59-10981 <+*>
1957 V34 N4 P621-624 62-32903 <=*>
1957 V34 N4 P671-674 59-19539 <+*>
1957 V34 N5 P684-693 R-4875 <*>
59-19537 <+*>
1958 V35 N1 P129-136 R-5051 <*>
59-20692 <+*>
1958 V35 N1 P157-159 59-14572 <+*>
1958 V35 N1 P160-164 R-5050 <*>
1958 V35 N1 P166-168 C-2681 <NRC>
R-4265 <*>
1958 V35 N2 P297-300 62-11305 <=>
1958 V35 N2 P301-304 62-23218 <=>
1958 V35 N3 P323-326 59-13258 <=>
1958 V35 N3 P323-334 59-13258 <=>
1958 V35 N3 P327-334 59-13258 <=>
1958 V35 N4 P657-659 59-10198 <+*>
1959 V36 N1 P33-40 59-21089 <=>
1959 V36 N1 P190-192 59-18585 <=>
1959 V36 N2 P337-339 59-13963 <=>
1959 V36 N3 P427-433 62-32653 <=*>
1959 V36 N3 P481-486 62-32625 <=*>
1959 V36 N3 P544-546 62-32629 <=*>
1959 V36 N4 P564-572 60-11344 <=>
1959 V36 N4 P579-584 60-21042 <=>
1959 V36 N4 P623-625 62-32628 <=*>
1959 V36 N4 P626-628 60-11344 <=>
1959 V36 N4 P723-733 61-10637 <*+>
61-15288 <+*>
1959 V36 N4 P734-740 60-11344 <=>
1959 V36 N5 P883-889 64-16379 <=*$>
1959 V36 N5 P890-901 63-10149 <=*>
1959 V36 N6 P1073-1077 61-19015 <=*>
1959 V36 N6 P1144-1146 60-11404 <=>
1960 V37 N1 P185-186 60-11990 <=>
1960 V37 N2 P281-283 60-41553 <=>
1960 V37 N2 P297-300 62-13117 <*=>
1960 V37 N2 P301-305 60-23841 <+*>
1960 V37 N2 P354-356 60-41554 <=>
1960 V37 N2 P362-368 60-41139 <=>
1960 V37 N3 P517-535 61-27641 <*=>
1960 V37 N3 P550-554 60-41610 <=>
1960 V37 N5 P785-793 61-19203 <+*>
1960 V37 N5 P908-917 61-10749 <+*>
1960 V37 N5 P918-926 61-16545 <*=>
1960 V37 N5 P931-934 62-16100 <=*>
1961 V38 N1 P125-130 61-27181 <*=>
1961 V38 N2 P325-335 62-10452 <*=>
1961 V38 N4 P577-592 65-17162 <*>
1961 V38 N6 P1099-1113 63-10331 <=*>
1962 V39 N1 P158-159 N65-22588 <=$>
1962 V39 N2 P355-361 64-10159 <=*$>
1962 V39 N3 P418-427 N65-23903 <=$>
1962 V39 N3 P428-438 N65-23902 <=$>
1962 V39 N3 P527-531 63-10598 <=*>
1962 V39 N4 P569-582 63-14732 <=*>
1962 V39 N4 P653-659 63-14428 <=*>
1962 V39 N4 P724-735 63-10330 <=*>
1962 V39 N4 P736-745 63-14427 <=*>
1962 V39 N5 P798-812 63-14854 <=*> 0
1962 V39 N5 P931-937 63-14730 <=*>
1962 V39 N5 P938-950 63-14736 <=*>
1962 V39 N6 P961-964 63-14889 <=*>
1962 V39 N6 P1112-1123 63-14729 <=*>
1963 V40 N1 P158-160 63-16549 <=*>
1963 V40 N2 P363-372 63-18710 <=*>
63-21832 <=>
1963 V40 N3 P466-476 <IPIX>
1963 V40 N4 P757-766 64-21554 <=>
1963 V40 N5 P812-818 64-16020 <=*$>
1963 V40 N5 P842-846 64-15431 <=*$>
1964 V41 N2 P282-287 64-31588 <=>
1964 V41 N5 P937-941 N65-11449 <=$>
1964 V41 N5 P942-950 N65-11445 <=$>
1964 V41 N5 P992-995 N65-14610 <=$*>
1964 V41 N5 P995-997 N65-14604 <=$>
1964 V41 N6 P1001-1006 N65-15744 <=$*>
1964 V41 N6 P1084-1089 N65-15891 <=$*>
1965 V42 P1070-1074 AD-634 941 <=$>
1965 V42 N1 P136-144 AD-623 439 <*=$>
1965 V42 N2 P217-232 65-31277 <=$*>
1965 V42 N6 P1287-1295 N66-23606 <=$>

ASTRONOMIE UND RAUMFAHRT
1965 N2 P46-55 66-12526 <*>

ASTRONOMISCHE MITTEILUNGEN
1939 P131-136 R-4936 <*>
1939 N13/4 P556-567 58-2612 <*> 0
1939 V13 P131-136 57-1269 <*>
1939 V13/4 P470-481 57-1440 <*>
1939 V13/4 P556-567 57-1270 <*>
1942 N141 P31-38 59-10107 <*>

ASTRONOMISCHE NACHRICHTEN
1825 N86 P241- 3212 <*>
1937 V263 P121-134 AL-40 <*>

ATENEO PARMENSE
1943 V14 N5 P337-349 57-1220 <*>
1959 V30 P150-168 SUP 1 10332-B <K-H>
1959 V30 SUP 1 P150-168 64-10673 <*>

ATHENA: RASSEGNA MENSILE DI BIOLOGIA, CLINICA E TERAPIA
1954 V20 P147-150 3225-H <K-H>

ATOM UND STROM
1960 V6 N12 P103-107 NP-TR-632 <*>

ATOM WIRTSCHAFT
1958 V3 P341-342 AEC-TR-4779 <*> P
61-18815 <*>
1958 V3 P496-501 62-01484 <*>
1959 V4 P345-348 AEC-TR-4094 <*>
1960 V5 P204-208 AEC-TR-4317 <*>
1960 V5 P511-518 AEC-TR-4639 <*>
1961 V6 P66-72 2534 <BISI>
1961 V6 N1 P73-79 2535 <BISI>
1962 V7 N5 P256-260 SCL-T-491 <*>
1962 V7 N7 P338-340 AEC-TR-6245 <*>
1963 V8 P418-423 ORNL-TR-196 <*>
1963 V8 N3 P165-170 AEC-TR-6233 <*>
1963 V8 N6 P370 AEC-TR-6080 <*>
1964 V9 P108-113 66-11121 <*>
1964 V9 N2 P46-53 AEC-TR-6385 <*>
1965 V10 P376-379 66-11978 <*>
1965 V10 P386-387 66-11977 <*>
1965 V10 N3 P140-141 66-11883 <*>

ATOMES
1956 V11 N128 P363-364 NP-TR-891 <*>
1957 V12 N132 P131-135 NP-TR-892 <*>

ATOMKERNENERGIE
1956 V1 N7/8 P237-244 AEC-TR-2893 <*>
58-1520 <*>
59-10215 <*>
1957 V2 P181-186 58-206 <*>
1957 V2 N6 P207-213 58-2307 <*>
1958 V3 P21 58-1680 <*>
1958 V3 P328- AEC-TR-3641 <+*>

Citation	Translation
1958 V3 N2 P72-74	59-10259 <*> 0
1958 V3 N7 P249-255	62-14502 <*>
1958 V3 N10 P382-388	1900 <BISI>
1958 V3 N8/9 P321-328	61-20477 <*>
1958 V3 N8/9 P328-331	1882 <BISI>
1959 V4 N2 P75-80	AEC-TR-3802 <*>
1959 V4 N3 P112-115	NP-TR-421 <*>
1960 V5 P18-22	AEC-TR-4278 <*>
1960 V5 P217-222	UCRL TRANS-629(L) <*>
1960 V5 P397-400	AEC-TR-4995 <*>
1960 V5 N1 P15-18	63-01194 <*>
1960 V5 N3 P100-107	63-01194 <*>
1960 V5 N6 P209-217	UCRL TRANS-730(L) <*>
1960 V5 N12 P453-455	62-00610 <*>
1961 V6 P98-100	AEC-TR-5062 <*>
1961 V6 P100-103	63-10243 <*>
1961 V6 P152-161	AEC-TR-6264 <*>
1961 V6 P165-170	AEC-TR-4440 <=>
1961 V6 P434-440	AEC-TR-5480 <*>
1961 V6 N3 P117-122	NP-TR-818 <*>
1961 V6 N6 P250-260	NP-TR-863 <*>
1961 V6 N9 P350-353	AEC-TR-5204 <*>
	62-18172 <*>
1962 V7 N1 P1-14	AEC-TR-5485 <*>
1963 V8 P350-353	ORNL-TR-5 <*>
1963 V8 N2 P41-51	ORNL-TR-70 <*>
1963 V8 N2 P74-77	64-10947 <*>
1964 V9 N3/4 P81-92	66-11859 <*>

ATOMKI KOZLEMENYEK

Citation	Translation
1962 V4 P45-49	NP-TR-968 <*>

★ ATOMNAYA ENERGIYA

Citation	Translation
1956 N3 P5-	41H11R <ATS>
1956 N3 P11-	42H11R <ATS>
1956 N3 P13-	43H11R <ATS>
1956 N3 P21-	44H11R <ATS>
1956 N3 P27-	45H11R <ATS>
1956 N3 P31	46H11R <ATS>
1956 N3 P33-	47H11R <ATS>
1956 N3 P40-	48H11R <ATS>
1956 N3 P45-	49H11R <ATS>
1956 N3 P56-	51H11R <ATS>
1956 N3 P61-	52H11R <ATS>
1956 N3 P65-	53H11R <ATS>
1956 N3 P76-	54H11R <ATS>
1956 N3 P81-	55H11R <ATS>
1956 N3 P84-	56H11R <ATS>
1956 N3 P88-	57H11R <ATS>
1956 N3 P97-	58H11R <ATS>
1956 N3 P107-	59H11R <ATS>
1956 N3 P117-	60H11R <ATS>
1956 N3 P122-132	61H11R <ATS>
1956 N3 P132-	62H11R <ATS>
1956 N3 P135-	63H11R <ATS>
1956 N3 P137-	64H11R <ATS>
1956 N4 P5-	82H12R <ATS>
1956 N4 P13-	83H12R <ATS>
1956 N4 P22-	84H12R <ATS>
1956 N4 P31-	85H12R <ATS>
1956 N4 P34-	86H12R <ATS>
1956 N4 P38-	87H12R <ATS>
1956 N4 P46-	88H12R <ATS>
1956 N4 P51-	89H12R <ATS>
1956 N4 P57-	90H12R <ATS>
1956 N4 P67-	92H12R <ATS>
1956 N4 P71-	93H12R <ATS>
1956 N4 P80-	94H12R <ATS>
1956 N4 P92-	95H12R <ATS>
1956 N4 P107-	96H12R <ATS>
1956 N4 P113-	97H12R <ATS>
1956 N4 P118-	98H12R <ATS>
1956 N4 P131-	99H12R <ATS>
1956 N4 P139-	H13R <ATS>
1956 N4 P147-	1H13R <ATS>
1956 N4 P149-	2H13R <ATS>
1956 N4 P150-	3H13R <ATS>
1956 N4 P155-	4H13R <ATS>
1956 N4 P158-	5H13R <ATS>
1956 N4 P166-	8H13R <ATS>
1956 N5 P113-114	R-1659 <*>
1957 V2 P66-68	UCRL TRANS-1198(L) <=$>
1957 V2 P231-239	R-4890 <*>
1957 V2 P272-274	R-4526 <*>
1957 V2 N1 P54-59	62-16530 <=*> 0
1957 V2 N2 P146-151	R-3241 <*>
1957 V2 N2 P152-156	R-3238 <*>
1957 V2 N5 P417-419	R-3239 <*>
1957 V3 P135-140	R-4686 <*>
1957 V3 P341-344	R-4046 <*>
1957 V3 P398-408	R-4811 <*>
1957 V3 N7 P11-14	R-2309 <*>
1957 V3 N7 P23-27	63-18888 <=*>
1957 V3 N9 P187-203	AEC-TR-3514 <+*>
1957 V3 N9 P215-221	R-4213 <*>
1957 V3 N9 P238-244	R-4861 <*>
1957 V3 N11 P413-443	AEC-TR-3592 <*=> 0
1957 V3 N12 P483-491	R-3614 <*>
1958 V4 P57-62	R-5038 <*>
1958 V4 P154-160	R-5313 <*>
1958 V4 N1 P5-21	AEC-TR-3515 <*>
1958 V4 N1 P52-56	R-5088 <*>
1958 V4 N1 P71-81	AEC-TR-4243 <*+>
1958 V4 N1 P88-90	PB 141 200T <=>
1958 V4 N1 P92-96	PB 141 210T <=>
1958 V4 N4 P376-377	AEC-TR-3497 <+*>
1958 V4 N4 P389-391	59-10817 <+*>
	64-71267 <=>
1958 V4 N5 P422-436	60-13100 <+*>
1958 V4 N5 P437-442	AEC-TR-3469 <+*>
1958 V4 N6 P571-575	59-21000 <=>
1958 V4 N6 P576-580	59-22558 <+*>
1958 V5 N1 P71	59-22543 <+*>
1958 V5 N2 P105-	59-11932 <=>
1958 V5 N2 P130-134	59-22548 <+*>
1958 V5 N2 P175-176	59-22472 <+*>
1958 V5 N3 P257-276	59-21082 <=> 0
1958 V5 N4 P446-451	63-19593 <=*$>
1958 V5 N6 P643-644	59-21005 <=>
1958 V5 N6 P646-647	59-21048 <=>
1959 V6 N1 P70-71	60-31121 <=>
1959 V6 N2 P135	AEC-TR-3956 <+*>
1959 V6 N2 P140-144	AEC-TR-3957 <+*>
1959 V6 N4 P453-457	AEC-TR-3913 <+*>
1959 V6 N4 P472-474	59-21205 <=>
1959 V6 N5 P528-532	61-23152 <=*>
1959 V7 P479-	AEC-TR-4526 <=*>
1959 V7 N2 P163-175	AEC-TR-3961 <+*>
1959 V7 N2 P172-173	60-23737 <+*>
1959 V7 N3 P268-272	AEC-TR-4091 <+*>
1959 V7 N4 P399	AEC-TR-4227 <*+>
1959 V7 N5 P429-444	AEC-TR-3982 <+*>
1959 V7 N6 P531-536	61-20698 <*=> 0
1960 V8 P573-575	65-28858 <$>
1960 V8 N1 P49-51	60-18762 <+*>
1960 V8 N1 P64-65	61-11131 <=>
1960 V8 N2 P121-126	60-31455 <=>
1960 V8 N2 P148-150	60-41048 <=>
1960 V8 N3 P239-247	AEC-TR-4443 <=*>
	61-27580 <*=>
1960 V8 N4 P340-347	60-31309 <=>
1960 V8 N4 P348-353	60-31308 <=>
1960 V8 N4 P367-370	61-23435 <*=> 0
1960 V8 N4 P378-380	60-41052 <=>
1960 V8 N5 P420-424	63-23151 <=*>
1960 V8 N6 P514-518	AEC-TR-4328 <*=>
1960 V8 N6 P555-556	64-30084 <*>
1960 V8 N6 P559-561	AEC-TR-4568 <*>
1960 V9 N1 P27-32	AEC-TR-4407 <*+>
1960 V9 N1 P49-51	AEC-TR-4621 <*>
1960 V9 N2 P98-103	AEC-TR-4323 <=*>
1960 V9 N2 P104	AEC-TR-4446 <=*>
1960 V9 N3 P194-200	AEC-TR-4425 <=>
1960 V9 N3 P221-242	61-21246 <=>
1960 V9 N4 P262-269	61-27846 <*=>
1960 V9 N4 P286-296	61-21266 <=>
1960 V9 N4 P313-315	63-16990 <*=>
1960 V9 N4 P347-348	61-28380 <*=>
1960 V9 N5 P392-398	61-21122 <=>
1960 V9 N6 P477-482	AEC-TR-4615 <*>

Citation	Translation
	AEC-TR-4615 <*=>
1961 V10 P43-49	AEC-TR-4620 <*>
1961 V10 N1 P50-57	AEC-TR-4619 <=*>
1961 V10 N1 P75-76	61-28228 <=*>
1961 V10 N2 P138-142	61-18573 <=*>
1961 V10 N2 P158-160	62-11178 <=>
1961 V10 N2 P170-172	61-27833 <*=>
1961 V10 N3 P211-221	AEC-TR-4611 <=*>
1961 V10 N3 P262-264	62-11178 <=>
1961 V10 N3 P264-265	62-11178 <=>
1961 V10 N3 P282-285	61-28302 <*=>
1961 V10 N4 P317-342	61-27937 <*=>
1961 V10 N4 P343-346	62-13570 <=*>
1961 V10 N4 P362-367	62-15175 <*=>
1961 V10 N4 P372-373	62-13570 <=*>
1961 V10 N4 P377-392	62-13570 <=*>
1961 V10 N4 P388-389	63-23397 <=*>
1961 V10 N4 P392-393	62-32146 <=*>
1961 V10 N4 P396	62-13570 <=*>
1961 V10 N4 P403-404	62-13570 <=*>
1961 V10 N4 P406-407	62-13570 <=*>
1961 V10 N4 P412-415	62-13570 <=*>
1961 V10 N4 P420-422	62-13570 <=*>
1961 V10 N5 P508-513	62-11178 <=>
1961 V10 N5 P513-515	63-19824 <=*>
1961 V10 N5 P549-551	62-13258 <=>
	62-24787 <=*>
1961 V10 N6 P565-633	62-13576 <=*>
1961 V10 N6 P587-591	62-14792 <=*>
1961 V10 N6 P597-605	61-28786 <*=>
1961 V11 N1 P12-18	62-13584 <*=>
1961 V11 N2 P109-121	63-13448 <=*>
1961 V11 N2 P122-125	62-24933 <=*>
1961 V11 N2 P133-139	63-13448 <=*>
1961 V11 N2 P184-185	62-11178 <=>
1961 V11 N2 P184-189	63-13448 <=*>
1961 V11 N2 P190-194	62-13913 <*=>
1961 V11 N2 P199	62-13913 <*=>
1961 V11 N3 P251-255	62-11178 <=>
1961 V11 N4 P301-344	62-19771 <=>
1961 V11 N4 P356-369	62-32528 <=*>
1961 V11 N4 P370-394	62-19771 <=>
1961 V11 N4 P379-394	AEC-TR-5238 <=*>
1961 V11 N4 P395-399	62-11178 <=>
1961 V11 N4 P404-415	62-19771 <=>
1961 V11 N5 P426-430	62-24270 <=*>
1961 V11 N5 P435-439	AEC-TR-5133 <=*>
1961 V11 N5 P442-447	62-11178 <=>
1961 V11 N6 P493-497	62-24285 <=*>
1961 V11 N6 P498-505	62-24915 <=*>
1961 V11 N6 P515-527	62-24915 <=*>
1961 V11 N6 P533-538	62-23833 <=*>
1962 V12 N1 P56-57	62-24921 <=*>
1962 V12 N1 P58-62	62-25841 <=*>
1962 V12 N1 P62-64	62-32995 <=*>
1962 V12 N1 P64-66	62-25841 <=*>
1962 V12 N2 P123-128	AEC-TR-5420 <=*>
1962 V12 N2 P129-139	62-32419 <=*>
1962 V12 N2 P156-159	62-32419 <=*>
1962 V12 N3 P193-197	63-13204 <=*>
1962 V12 N3 P216-229	62-32998 <=*>
1962 V12 N5 P378-391	62-32376 <=>
1962 V12 N5 P404-407	62-32376 <=>
1962 V12 N5 P415-421	62-32376 <=>
1962 V12 N6 P503-513	AEC-TR-5327 <=*>
1962 V13 N1 P68-72	63-19083 <=*>
1962 V13 N2 P155-161	63-15916 <=*>
1962 V13 N4 P388-390	63-3158 <=>
1962 V13 N4 P393-394	63-31581 <=>
1962 V13 N5 P458-466	AEC-TR-5835 <=$>
1962 V13 N5 P497-500	63-31581 <=>
1962 V13 N6 P521-529	63-23446 <=*>
1962 V13 N6 P576-580	63-21147 <=>
1962 V13 N6 P597-599	AEC-TR-5910 <*>
1962 V13 N6 P603-606	63-15363 <=*>
1962 V13 N6 P608-609	64-71456 <=> M
1963 V14 N1 P18-26	63-21256 <=>
1963 V14 N1 P23-29	AEC-TR-6014 <*>
1963 V14 N2 P221-224	63-31581 <=>
1963 V14 N3 P319-320	RTS-3006 <NLL>
1963 V14 N3 P330-332	63-31581 <=>
1963 V14 N4 P353	63-31581 <=>
1963 V14 N4 P383-394	AEC-TR-5870 <=$>
1963 V14 N4 P405-407	64-16012 <=*$>
1963 V14 N5 P502-505	64-21815 <=>
1963 V14 N5 P505	64-21815 <=>
1963 V14 N5 P506	64-21815 <=>
1963 V14 N5 P506-508	64-21815 <=>
1963 V14 N6 P591-593	63-31581 <=>
1963 V14 N6 P593-594	64-71440 <=> M
1963 V14 N6 P595-600	63-31581 <=>
1963 V15 P409-410	64-19569 <=$>
1963 V15 P422-423	ORNL-TR-527 <=*$>
1963 V15 N1 P23-29	AEC-TR-6014 <=$>
1963 V15 N1 P48-52	64-15841 <=*$>
1963 V15 N1 P79-80	AEC-TR-6237 <=$>
1963 V15 N2 P115-120	64-11958 <=>
1963 V15 N2 P120-130	64-11886 <=>
1963 V15 N3 P262-264	KFK-TR-146 <*>
1963 V15 N4 P318-319	64-18013 <=*$>
	64-19309 <=$>
1963 V15 N4 P383-392	64-16384 <=*$>
1963 V15 N5 P409-410	64-19569 <=$>
1963 V15 N5 P411-413	66-12214 <*>
1963 V15 N6 P499-504	64-21606 <=>
1964 V16 N3 P195-206	64-19306 <=$>
1964 V16 N4 P279-282	ORNL-TR-387 <*>
1964 V16 N4 P291-295	AD-611 010 <=$>
1964 V16 N5 P484-488	AEC-TR-6463 <=$>
1964 V16 N6 P426-432	64-41560 <=>
1964 V16 N6 P489-496	AEC-TR-6462 <=$>
1964 V17 N1 P17-22	UCRL-TRANS-1181 <=*$>
1964 V17 N1 P53-56	65-14598 <*> 0
1964 V17 N1 P70-71	AD-621 010 <=*$>
1964 V17 N4 P269-278	AD-620 864 <=$>
1964 V17 N5 P406-408	648 <TC>
	65-12914 <*>
1964 V17 N6 P497-500	65-30247 <=$>
1964 V17 N6 P508-509	66-11002 <*>
1965 V18 P636-640	65-23422 <$>
1965 V18 N1 P5-14	AD-629 349 <=*$>
1965 V18 N1 P14-18	N65-18342 <=$>
1965 V18 N1 P46-48	65-14260 <*>
1965 V18 N2 P175-719	AD-623 818 <=*$>
1965 V18 N3 P203-209	AD-621 002 <=*$>
1965 V18 N3 P239-242	AD-633 536 <=$>
1965 V18 N3 P242-245	65-31185 <*=$>
1965 V18 N4 P373-378	IGR V8 N8 P912 <AGI>
1965 V18 N5 P447-451	65-31546 <=$>
1965 V18 N5 P519-520	65-64112 <=*$>
1965 V18 N6 P636-638	65-17106 <*>
1965 V19 N1 P36-38	65-32858 <=*$>
1965 V19 N2 P144-153	66-30616 <=$> 0
1965 V19 N3 P269-272	66-30616 <=$> 0
1965 V19 N3 P297-298	66-14351 <*$> 0
1965 V19 N6 P507-510	66-23806 <$>
1966 V20 N2 P123-126	66-33649 <=$>
1966 V21 N6 P450-465	67-31327 <=>
1966 V21 N6 P476-478	67-31327 <=>

ATOMPRAXIS

Citation	Translation
1957 V3 P362-368	AEC-TR-3986 <*>
1957 V3 P369-372	AEC-TR-3981 <*>
1957 V3 P377-382	60-00184 <*>
1957 V3 P382-388	AEC-TR-3969 <*>
1957 V3 P389-398	AEC-TR-3980 <*>
1957 V3 N1 P1-6	58-106 <*>
1957 V3 N7 P258-259	NP-TR-440 <*>
1957 V3 N10 P382-388	60-31446 <=*> 0
1958 V4 P177-178	59-10389 <*>
1958 V4 P375-381	AEC-TR-4861 <*>
1958 V4 N6 P212-227	59-11633 <=> 0
1959 V5 P91-92	NP-TR-469 <*>
1959 V5 P182-187	NP-TR-469 <*>
1959 V5 P280-284	AEC-TR-4003 <*>
1959 V5 P310-313	62-32408 <=*>
1959 V5 P421-425	AEC-TR-4859 <*>
1959 V5 N9 P339-342	62-32409 <=*>
1959 V5 N12 P475-481	NP-TR-443 <*>
1960 V6 P1-15	AEC-TR-5573 <*>

1960 V6 P221-225 AEC-TR-4353 <*>
1960 V6 P301-308 NP-TR-776 <*>
1960 V6 P308-316 AEC-TR-5052 <*>
1960 V6 P320-322 NP-TR-553 <*>
1960 V6 P357-361 AEC-TR-4860 <*>
1960 V6 N4/5 P129-132 3074 <BISI>
1960 V6 N10/1 P385-391 NP-TR-831 <*>
1961 V7 P279-284 64-00043 <*>
1961 V7 P332-336 AEC-TR-5904 <*>
1961 V7 P401-407 66-14264 <*>
1961 V7 P465-467 8915 <IICH>
1961 V7 N6 P213-216 NP-TR-846 <*>
1961 V7 N8 P284-288 AEC-TR-4955 <*>
1961 V7 N11 P401-407 AEC-TR-5041 <*>
1961 V7 N11 P412-419 AEC-TR-5102 <*>
1961 V7 N12 P467-470 62-16555 <*> 0
1962 V8 P1-8 AEC-TR-5216 <*>
1962 V8 P87-92 AEC-TR-6013 <*>
1962 V8 N1 P6 AEC-TR-5016 <*>
1962 V8 N3 P93-94 AEC-TR-6047 <*>
1962 V8 N5 P175-182 AEC-TR-5557 <*>
1963 V9 P389-395 AEC-TR-6381 <*>
1963 V9 N5 P179-188 66-12940 <*> 0
1964 V10 N6 P268-276 65-17102 <*> 0
1964 V10 N9/10 P434-436 66-11874 <*>
1965 V11 N2 P80-84 66-11123 <*>

ATOOMENERGIE EN HAAR TOEPASSINGEN
1959 P1-87 62-19290 <=*>
1962 V4 N3 P47-59 NP-TR-934 <*>
1963 V5 P187-194 AEC-TR-6358 <*>

ATTI DELL'ACCADEMIA DEI FISIOCRITICI DI SIENA
1911 S5 V3 P155-161 63-26957 <=*>

ATTI DELL'ACCADEMIA DEI FISIOCRITICI DI SIENA. SEZIONE AGRARIA
1948 V11 N3/4 P49-58 64-16710 <*> 0
64-20298 <*>

ATTI DELL'ACCADEMIA DEI FISIOCRITICI DI SIENA SEZIONE MEDICO-FISICA
1957 S13 V4 P221-230 <LSA>

ATTI DELL'ACCADEMIA LIGURE DI SCIENZE E LETTERE
1954 V10 P196-203 58-806 <*>

ATTI DELL'ACCADEMIA NAZIONALE DEI LINCEI. MEMORIE
1902 S5 V11 P16-17 60-10071 <*>
1905 S5 V14 P699-703 60-10070 <*>
1955 S8 V4 N4 P61-71 SEZ II
61-17641 <=$>
1955 S8 V4 N5 P73-80 SEZ II
61-17647 <=$>
1955 S8 V4 N5 P73-80 C-2198 <NRC>
1958 S8 V5 P117-148 65-12342 <*>

ATTI DELL'ACCADEMIA NAZIONALE DEI LINCEI. RENDICONTI CLASSE DI SCIENZE FISICHE, MATEMATICHE E NATURALI
1887 V3 P97-105 65-12339 <*>
1889 V5 P804-810 2217 <*>
1893 S5 V2 P415-423 63-16088 <*> P
1907 V16 P806-810 57-728 <*>
1910 S5 V19 N1 P228-231 63-18949 <*>
1911 S5 V20 P433-440 64-16712 <*> 0
1912 V21 N5 P695-701 57-2946 <*>
1914 S5 V23 P959-965 63-14308 <*> 0
1914 S5 V23 N2 P464-471 62-16297 <*>
1915 V24 P625-631 57-2930 <*>
1916 V25 P105-111 57-3045 <*>
1916 V25 P221-227 57-3133 <*>
1916 V25 P326-331 57-2956 <*>
1920 S5 V29 P62-66 3923-A <KH>
1923 V32 P569-572 58-1315 <*>
1924 S5 V33 P258-261 58-146 <*>
1927 S6 V5 P614-618 65-13249 <*>
1928 S6 V7 P885-891 63-10427 <*> 0P
1929 S6 V10 P193-196 60-18884 <*>
64-16718 <*>
64-20370 <*>
1930 S6 V11 P1108- AL-982 <*>
1930 S6 V12 P216-222 AEC-TR-1950 <*>
1931 S6 V13 P128-133 58-126 <*>
1931 S6 V13 P779-784 64-16713 <*>
1931 S6 V13 P809-813 64-16714 <*>
1933 S6 V18 P156-161 64-16715 <*> 0
1934 S6 V19 P415-420 64-16716 <*> 0
1934 S6 V20 P384-390 63-16022 <*> 0
63-18210 <*>
1935 S6 V22 P146-149 61-18259 <*>
1936 V24 P459-464 57-40 <*>
1936 S6 V24 P381-388 59-10813 <*> 0
61-10453 <*> 0
63-14122 <*> 0
64-16719 <*>
1937 V25 P129-132 57-188 <*>
1937 V26 P233-238 57-42 <*>
1938 S6 V27 P292-297 59-12023 <+*>
1939 V29 P664-671 57-2741 <*>
1941 S7 V2 N12 P1107-1110 AL-535 <*>
1949 V7 P103-106 II-183 <*>
1949 V7 P106-108 I-88 <*>
1950 V8 P417-422 58-404 <*>
1950 S8 V8 N2 P108-112 64-18590 <*>
1951 V10 N4 P275-280 AL-644 <*>
1951 S8 V11 N5 P265-267 59-17961 <*> 0
1952 S8 V12 N4 P444-448 64-16806 <*> 0
1952 S8 V13 P265-268 AEC-TR-6387 <*>
1953 S8 V14 N5 P680-687 62-16018 <*>
1954 S8 V16 N2 P249-257 65-13829 <*> 0
1954 S8 V17 P315-322 C-4459 <NRC>
1955 V18 P19 C-2199 <NRC>
1955 V18 P19-27 18G8I <ATS>
1955 V19 P229-230 C-2318 <NRC>
1955 V21 N1 P216-221 2111 <*>
1955 S8 V18 P19-27 57-1387 <*>
57-3524 <*>
61-17359 <=$>
1955 S8 V18 N1 P19-27 57-999 <*>
1955 S8 V19 P229-237 61-17644 <=$>
1955 S8 V19 P397-403 61-17350 <=$>
1955 S8 V19 P404-411 5408-B <K-H> 0
60-18496 <*>
61-17348 <=$>
1955 S8 V19 P453-459 61-17346 <=$>
1955 S8 V19 P497-498 60-17497 <*=>
1955 S8 V19 N6 P397-403 60-18499 <*>
64-16768 <*>
1955 S8 V19 N6 P404-411 64-16769 <*> 0
1955 S8 V19 N6 P453-459 64-16757 <*> 0
1956 V21 N6 P365-372 17K22I <ATS>
1956 S8 V20 P408-413 61-17646 <=$>
1956 S8 V20 P728-734 61-17636 <=$>
1956 S8 V21 N5 P365-372 59-15931 <*> 0
1957 V22 N1 P11-17 80J18I <ATS>
1957 S8 V22 P11-17 59-15002 <*>
1957 S8 V22 P726-730 61-20542 <*>
1957 S8 V22 N1 P11-17 65-11601 <*>
1957 S8 V23 P263-274 AEC-TR-5121 <*>
1957 S8 V23 N6 P363-370 03P60I <ATS>
1958 V24 P254-260 30K25I <ATS>
1958 V24 P479-487 58-2326 <*>
1958 V24 N2 P121-129 63K23I <ATS>
1958 S8 V24 P254-260 59-15001 <*>
1958 S8 V24 P488-493 C-4413 <NRCC>
1958 S8 V24 N1 P43-49 59-10834 <*>
1958 S8 V24 N3 P246-255 59-10809 <*>
1958 S8 V24 N3 P246-253 60-10044 <*> 0
1958 S8 V25 P3-12 65-12500 <*> 0
1958 S8 V25 P424-430 (CHEM)-348 <CTS>
1958 S8 V25 N6 P417-423 61-16791 <*>
1958 S8 V25 N6 P424-430 65-11600 <*>
1958 S8 V25 N6 P498-508 59-20891 <*>
61-16797 <*>
1958 S8 V25 N6 P509-516 61-16790 <*>
1958 S8 V25 N6 P517-519 59-17072 <=>
1958 S8 V25 N6 P520-527 59-20890 <*>
1958 S8 V25 N1/2 P3-12 59-15708 <*>
1958 S8 V25 N1/2 P70-74 59-17065 <*>

1959 V26 N1 P14-17 65-11595 <*>
1959 S8 V26 P155-163 63-18378 <*>
1959 S8 V26 P774-776 (CHEM)-113 <CTS>
1959 S8 V26 N1 P14-17 REPT. 97069 <RIS>
64-20920 <*> 0
1959 S8 V26 N2 P150-154 (CHEM)-260 <CTS>
1959 S8 V26 N2 P155-163 64-14957 <*>
1959 S8 V26 N4 P431-434 62-10352 <*>
1959 S8 V27 P274-280 (CHEM)-264 <CTS>
1959 S8 V27 P392-396 (CHEM)-265 <CTS>
1959 S8 V27 N5 P162-167 62-10351 <*>
1959 S8 V27 N3/4 P107-112 (CHEM)-262 <CTS>
65-11242 <*>
1960 V28 N1 P8-17 30M45I <ATS>
1960 S8 V28 P284-292 63-10821 <*>
1960 S8 V28 P452-460 (CHEM)-263 <CTS>
10630 <K-H> 0
1960 S8 V28 P539-544 (CHEM)-258 <CTS>
65-12467 <*>
1960 S8 V28 P632-638 (CHEM)-257 <CTS>
1960 S8 V28 N1 P8-17 64-16720 <*>
1960 S8 V28 N1 P18-26 (CHEM)-259 <CTS>
1960 S8 V28 N4 P452-460 65-13190 <*> 0
1960 S8 V28 N5 P539-544 72P59I <ATS>
1960 S8 V29 P155-161 62-14844 <*>
1960 S8 V29 P225-231 62-14853 <*>
1960 S8 V29 P257-264 (CHEM)-222 <CTS>
1960 S8 V29 P465-471 62-14854 <*>
1960 S8 V29 N5 P491-496 (CHEM)-221 <CTS>
61-18845 <*>
1960 S8 V29 N5 P562-565 (CHEM)-220 <CTS>
1961 S8 V30 P53-54 73N54I <ATS>
1961 S8 V31 N6 P350-356 65-13830 <*> 0
1962 S8 V32 N5 P666-673 (CHEM)-345 <CTS>
1963 V34 P524-529 66-11085 <*>
1963 V34 N6 P659-664 65-18061 <*>
1963 V35 P565- RCT V38 N3 P620 <RCT>
1963 V35 N6 P565-574 65-12157 <*>
1963 S8 V35 N6 P555-557 76R80I <ATS>
1963 S8 V35 N6 P565-574 80S81I <ATS>

ATTI DELL'ACCADEMIA DELLE SCIENZE. CLASSE DI SCIENZE FISICHE, MATEMATICHE E NATURALI. TORINO
1934 V69 P162-165 64-16721 <*>
1934 V69 P358-363 59-17987 <*>
1940 V75 P445-453 I-409 <*>
1950 V84 N1 P3-18 63-20659 <*>
1954 V89 P66-67 2096 <BISI>
1954 V89 P345-349 2097 <BISI>
1955 V89 P350-358 60-21000 <=>
1956 V90 P533-551 59-21194 <=>
1959 V94 P67-76 66-14106 <*>
1959 V94 P424-431 AI-TR-8 <*>
1962 V97 P87-102 UCRL TRANS-971(L) <*>

ATTI DELL,ACCADEMIA DELLE SCIENZE DELL,ISTITUTO DI BOLOGNA. RENDICONTI. CLASSE DI SCIENZE FISICHE
1958 S11 V5 P88-94 63-01030 <*>
1961 V9 N1 P194-214 66-11956 <*>

ATTI. ASSOCIAZIONE GENETICA ITALIANA. ROMA
1960 V5 P275-280 62-00305 <*>
1961 V6 P99-100 63-00290 <*>

ATTI. CONVEGNO DI CORROSIONE
1953 P119-120 57-761 <*>

ATTI DELLA FONDAZIONE E CONTRIBUTI DELL'ISTITUTO NAZIONALE DI OTTICA
1953 V8 N6 P385-392 65-11430 <*>
1955 V10 N5 P356-370 65-11432 <*>
1955 V10 N6 P452-467 65-11431 <*>
1961 V16 N4 P289-291 65-11546 <*>
1963 V18 N1 P1-6 65-11547 <*>
1963 V18 N6 P589-596 65-11548 <*>

ATTI DEL ISTITUTO VENETO DI SCIENZE, LETTERE ED ARTI
1915 V75 N2 P465-479 5448-D <K-H>
1937 V97 P623-674 PT2 I-966 <*>
1942 V102 P104-117 2902 <*>
1958 V116 P113-122 66-10285 <*>
1962 V121 P83-107 66-10983 <*>

ATTI E MEMORIE DELL'ACCADEMIA PATAVINA DI SCIENZE, LETTERE ED ARTI
1949 V62 P35-38 60-18200 <*> P

ATTI E NOTIZIE. ASSOCIAZIONE ITALIANA DI METALLURGIA
1959 N1 P17-19 TS-1475 <BISI>
62-14394 <*>

ATTI DELLA R. ACCADEMIA D'ITALIA. RENDICONTI. CLASSE DI SCIENZE FISICHE, MATEMATICHE E NATURALI
SEE ATTI DELL'ACCADEMIA NAZIONALE DEI LINCEI. RENDICONTI. CLASSE DI SCIENZE FISICHE, MATEMATICHE E NATURALI

ATTI DEL SEMINARIO MATEMATICO E FISICO DELL'UNIVERSITA DI MODENA
1959 V9 P13-23 <LSA>

ATTI DELLA SOCIETA ITALIANA DI ANATOMIA
1953 P1-4 58-1837 <*>
1954 P27-30 58-1797 <*>

ATTI DELLA SOCIETA ITALIANA DI CARDIOLOGIA
1949 V11 P301-302 60-10504 <*>
1954 V16 P105-107 63-16927 <*>
8619-A <KH>
1962 V22 N2 P133-134 63-01377 <*>

ATTI DELLA SOCIETA ITALIANA DI PATOLOGIA. PAVIA
1957 V5 P487-501 65-00338 <*> 0

ATTI DELLA SOCIETA ITALIANA PER IL PROGRESSO DELLE SCIENZE
1931 V20 N2 P164-166 62-20169 <*>

ATTI DELLA SOCIETA ITALIANA DELLE SCIENZE VETERINARIE
1947 V1 P190-196 60-10503 <*>
1947 V1 P197-203 60-10502 <*>
1952 V6 P289-296 57-2014 <*>
1953 V7 P764-768 C-2483 <NRC>
1955 V9 P432-434 57-2051 <*>
1964 V17 P765-768 66-10925 <*>

ATTI DELLA SOCIETA LOMBARDA DI SCIENZE MEDICHE E BIOLOGICHE
1954 V9 P422-429 2965 <*>
1955 V10 N1 P102-106 58-90 <*>
1956 V11 P417-420 58-2700 <*>
1957 V12 N3 P452-461 59-10708 <*> 0
1959 V14 P840-851 60-16613 <*>
1960 V15 P582-601 SUP 62-16210 <*>
1961 V16 P257-260 64-14621 <*>
1962 V17 P378-384 64-16358 <*>

ATTI DELLA SOCIETA MEDICO-CHIRURGICA DI PADOVA E BOLLETINO DELLA FACOLTA DI MEDICINA E CHIRURGIA DELLA UNIVERSITA DI PADOVA
1939 V17 P214-218 60-10676 <*>

ATTI DELLA SOCIETA DEI NATURALISTI E MATEMATICI
1961 V92 P3-14 C-4963 <NRC>

ATTI. SOCIETA PELORITANA DI SCIENZE FISICHE, MATEMATICHE E NATURALI
1960 V6 N1 P75-81 61-00981 <*>
1961 V7 N3/4 P399-407 63-01615 <*>

ATTUALITA MEDICA. ROME
1952 V17 P15-19 AL-48 <*>

AUFBEREITUNGS-TECHNIK
1960 V1 P429-435 2412 <BISI>
1960 V1 P518-527 2412 <BISI>

1960 V1 N12 P509-517 63-16210 <*> 0
1961 V2 P241-249 2412 <BISI>
1961 V2 N5 P198-201 63-20314 <*>
1963 N6 P243-249 66-10755 <*>
1963 V4 P363-380 3543 <BISI>
1963 V4 N5 P219-223 4337 <BISI>
1963 V4 N11 P462-466 3715 <BISI>
1963 V4 N11 P507-510 3635 <BISI>
1963 V4 N11 P511-513 3636 <BISI>
1964 V5 N9 P471-491 4127 <BISI>
1964 V5 N11 P580-605 4081 <BISI>
1965 V6 P18-22 4316 <BISI>
1965 V6 P90-95 4405 <BISI>
1965 V6 N2 P45-49 4448 <BISI>

AUSSENHANDEL
SEE AUSSENHANDEL UND INNERDENTSCHE HANDEL

AUSSENHANDEL UND INNERDENTSCHE HANDEL
1964 V14 N6 P35-42 64-41029 <=>

AUSTRIA STANDARDS
B-3302 58-2710 <*>
1952 M-1365 62-20423 <*>

AUTOGENE METALLBEARBEITUNG
1930 V23 N14 P218-226 60-10203 <*> 0
1932 V25 N9 P135-138 66-14266 <*$> 0
1934 V27 N13 P209-215 62-20213 <*> 0
1934 V27 N13 P225-232 62-20213 <*> 0
1936 V29 P225-232 62-18330 <*> 0
1936 V29 P241-246 62-18330 <*> 0
1938 V31 N5 P67-71 60-10761 <*> 0
1938 V31 N6 P89-97 60-10761 <*> 0

AUTOMATICA SI ELECTRONICA
1958 V2 N4 P139-144 62-10138 <*>
1960 V4 N4 P150-155 62-19432 <=*>
1960 V4 N6 P255-259 62-19462 <=*>
1960 V4 N6 P262-269 62-19462 <=*>
1960 V4 N6 P281 62-19462 <=*>
1961 V5 N3 P105-109 65-12286 <*>
1963 V7 N1 P45-46 63-31066 <=>
1963 V7 N3 P100-106 63-41009 <=>
1964 V8 N4 P146-149 64-51973 <=$>

AUTOMATION. CHINESE PEOPLES REPUBLIC
SEE TZU TUNG HUA

AUTOMATISME
1960 V5 N3 P102-108 66-13241 <*>
1960 V5 N10 P381-388 66-13241 <*>
1960 V5 N7/8 P273-278 66-13241 <*>

AUTOMATIZACE
1960 V3 N4 P117-119 61-18776 <*> 0
1960 V3 N4 P120-122 61-18775 <*> 0
1960 V3 N4 P122 61-18774 <*> 0
1960 V3 N4 P123 61-18773 <*> 0
1960 V3 N4 P124 61-18772 <*> 0
1962 V5 N3 P58-60 64-71542 <=>

AUTOMOBIL-INDUSTRIE
1960 V8 P53-83 61-20040 <*>
1963 V11 N19F P129-142 64-16277 <*>

AUTOMOBILE. CHINESE PEOPLES REPUBLIC
SEE CH'I-CH'E

AUTOMOBILE. PARIS
1962 N198 P75-81 63-18006 <*> 0

AUTOMOBILES AND TRACTORS. CHINESE PEOPLES REPUBLIC
SEE CH'I-CH'E YU T'O-LA-CHI

AUTOMOBIL-REVUE
1962 V57 N32 P19 62-20302 <*>
1962 V57 N32 P23 62-20302 <*>

AUTOMOBILTECHNISCHE ZEITSCHRIFT
1930 V33 N29 P705-707 59-20417 <*>
1934 V37 N9 P250-256 66-14267 <*> 0
1935 V38 N10 P243-252 65-13001 <*>
1935 V38 N20 P499-504 66-13356 <*>
1937 V40 N19 P473-475 63-20078 <*>
1938 V41 N14 P372-374 63-10452 <*> 0
1938 V41 N23 P616-620 64-18181 <*>
1939 V42 N2 P41-44 63-10479 <*>
1939 V42 N5 P151-153 59-20418 <*> 0
1939 V42 N8 P213-221 63-14278 <*> 0
1939 V42 N13 P364-370 63-10461 <*> 0
1939 V42 N14 P397-406 65-13733 <*>
1939 V42 N19 P515-522 57-01941 <*>
1939 V42 N20 P539-541 63-14277 <*> 0
1940 N16 P403-406 63-20542 <*>
1940 V43 N17 P426-428 63-20543 <*>
1941 N23 P581-586 57-1938 <*>
1942 V45 P331-334 58-686 <*>
1942 V45 P454-457 63-14286 <*> 0
1942 V45 N12 P326-327 58-686 <*>
1943 V46 P36-39 63-16162 <*> 0
1943 V46 N2 P31-36 58-459 <*>
63-16464 <*>
1947 V49 N2 P17-18 62-10316 <*>
1947 V49 N2 P26-28 62-10317 <*>
1948 V50 N1 P4-9 62-10324 <*>
6877-C <K-H>
1949 V51 N4 P95-97 59-17620 <*>
1951 V53 N2 P25-32 II-771 <*>
1951 V53 N4 P74-78 3047 <*>
1952 V54 P123-128 AD-636 945 <=$>
1952 V54 N6 P123-128 59-17148 <*> 0
1953 V55 N1 P3-10 59-17136 <*> 0
1954 P173-179 58-896 <*>
1954 V56 P141-150 58-899 <*>
1954 V56 N5 P134-136 65-11952 <*>
1954 V56 N7 P169-173 64-14897 <*> 0
1954 V56 N11 P293-299 65-00238 <*>
1954 V56 N11 P309-312 26G5G <ATS>
1955 V57 P127-132 58-868 <*>
1955 V57 N2 P31-34 58-1465 <*>
1955 V57 N3 P78-83 62-18789 <*>
1955 V57 N4 P107-113 58-1156 <*>
1955 V57 N5 P135-142 61-10421 <*>
1955 V57 N6 P164-170 59-17122 <*> 0
1955 V57 N7 P191-196 59-17070 <*>
1955 V57 N8 P213-227 55H9G <ATS>
1956 V58 P219-223 57-136 <*>
1956 V58 P251-258 58-873 <*>
1956 V58 N1 P20-22 58-2385 <*>
1956 V58 N1 P22-24 58-2386 <*>
1956 V58 N5 P134-138 59-10139 <*>
1956 V58 N6 P153-161 59-10139 <*>
1956 V58 N9 P251-258 63-16623 <*> P
1956 V58 N9 P258-260 58-872 <*>
1957 P119-126 57-2811 <*>
1957 N3 P68-75 57-3300 <*>
1957 N9 P239-246 58-168 <*>
1957 V58 P319-320 57-3568 <*>
1957 V59 N1 P1-7 59-10138 <*>
1957 V59 N9 P239-246 60-18113 <*>
1957 V59 N9 P250-255 58-167 <*>
1957 V59 N10 P289-292 59-17369 <*> 0
1957 V59 N10 P293-297 59-17398 <*> 0
1957 V59 N10 P297-302 59-17397 <*>
1957 V59 N10 P302-303 59-17368 <*>
1957 V59 N10 P304-307 64-30967 <*> 0
1957 V59 N10 P307-313 59-17396 <*> 0
1957 V59 N12 P359-365 58-1187 <*>
60-18113 <*>
1958 V60 N1 P1-6 61-10420 <*>
1958 V60 N6 P149-152 59-10169 <*>
1958 V60 N6 P168-174 59-17385 <*>
59-17736 <*>
1958 V60 N8 P225-228 60-10747 <*>
1958 V60 N9 P251-255 59-17759 <*>
1959 V61 P103-107 61-16152 <*> 0
1959 V61 N5 P131-134 62-18755 <*>
1959 V61 N5 P134-136 60-16643 <*>
1959 V61 N9 P237-246 60-14498 <*>

9751 <K-H>
1959 V61 N10 P296-301 60-14067 <*>
9525 <K-H>
1960 V62 N1 P6-9 10M40G <ATS>
1960 V62 N8 P201-210 61-10289 <*>
1960 V62 N12 P311-320 61-20079 <*>
1960 V62 N12 P320-325 61-20203 <*>
1960 V62 N12 P325-326 61-20084 <*>
1960 V62 N12 P326-328 61-20078 <*>
1960 V62 N12 P328-329 61-20082 <*>
1961 V63 N2 P33-41 16N53G <ATS>
1961 V63 N3 P78-84 65-10041 <*> 0
1961 V63 N3 P88 61-20081 <*>
1961 V63 N9 P279-293 62-10741 <*>
1962 V64 N3 P70-73 66-13511 <*>
1962 V64 N5 P149-152 63-10275 <*>
1962 V64 N5 P152-158 62-18674 <*>
1962 V64 N6 P175-179 65-11350 <*>
1962 V64 N6 P179-184 62-18916 <*>
1962 V64 N11 P317-325 65-17124 <*> 0
1962 V64 N11 P317-324 65Q73G <ATS>
1963 V65 N1 P11-16 63-16287 <*>
1964 V66 N3 P76-80 65-11343 <*>
1964 V66 N12 P353-359 65-11982 <*>
1964 V66 N12 P359-364 65-12161 <*>
1965 V67 N7 P215-218 66-10715 <*>
1966 V68 N2 P38-40 66-13391 <*>

AUTO-TECHNIK
1907 V10 P289-292 58-248 <*>

AVIA VLIEGWERELD
1958 V7 N12 P330 59-11896 <=> 0
1959 V8 N19 P486-487 60-19085 <=*>

AVIAPROMYSHLENNOST. MOSKVA
1937 V6 N1 P27-33 66-14268 <*> 0
1937 V6 N4 P27-32 66-14269 <*$>
1937 V6 N12 P24-32 66-14270 <*$>

AVIATION MAGAZINE
1957 03/15 P31-41 59-16951 <=*> 0
1957 N2381 P1-14 60-21767 <=*>
1958 05/01 P19-26 59-11209 <=*>
1958 09/15 P41-42 59-11944 <=*> 0
1958 10/01 P32 59-21023 <=*>
1958 N264 P11 59-21021 <=*> 0
1959 05/01 P20-21 59-21156 <=*> 0
1959 05/01 P24-25 59-21156 <=*> 0
1961 N1 P19-23 62-32342 <=*>
1965 N423 P38-39 65-32379 <=$>

AVIATSIONNAYA PROMYSHLENNOST
1958 N2 P41 61-13548 <*=>
1958 N3 P19-26 59-19697 <+*>
1958 N3 P26-27 59-19693 <+*>
1958 N4 P25-26 59-19692 <+*>

★ AVIATSIYA I KOSMONAVTIKA
1962 V44 N2 P24-30 62-24739 <=*>
1962 V44 N4 P20-27 AD-611 110 <=$>
1962 V44 N5 P7-13 64-19048 <=*$>
1962 V44 N5 P22-27 62-33695 <=*>
1962 V44 N6 P8-12 63-13076 <=*>
1962 V44 N7 P29-34 62-33443 <=*>
1962 V44 N7 P90-91 62-33443 <=*>
1963 V45 N4 P26-32 AD-611 109 <=$>
1963 V46 N1 P16-19 64-11939 <=> M
1963 V46 N7 P26-32 63-31959 <=>
1963 V46 N8 P27-29 64-14128 <=*$>
1963 V46 N10 P8-19 64-21478 <=>
1963 V46 N12 P36-42 JAS V11,N4,P108-112 <AAS>
65-14539 <*>
65-63689 <=$>
1964 V47 SPEC NO P59-64 65-30405 <=$>
1964 V47 N2 P58-63 64-31286 <=>
1964 V47 N5 P71-79 66-11233 <*>
1964 V47 N7 P48-53 N66-22284 <=$>
1964 V47 N9 P77-82 64-51827 <=$>
1964 V47 N11 P24-36 65-30486 <=$>
1964 V47 N11 P50-59 66-62402 <=*$>
1964 V47 N12 P34-38 65-30700 <=$>
1965 N12 P65-71 AD-636 407 <=$>
1965 V48 N1 P44-57 65-31006 <=$>
1965 V48 N1 P64-66 65-31006 <=$>
1965 V48 N1 P68-70 65-31006 <=$>
1965 V48 N3 P34-40 65-30951 <=$>
1965 V48 N5 P84-87 65-31368 <=$>
1965 V48 N7 P33-35 65-32574 <=$>
1965 V48 N7 P75-77 N66-13297 <=$>
1965 V48 N8 P64-69 N66-13485 <=$>
1965 V48 N9 P68-73 65-32829 <=$*>
1965 V48 N10 P78-79 66-30113 <=*$>
1966 N5 P71-76 66-32715 <=$>
1966 N7 P17-24 66-33736 <=$>
1966 N7 P28-30 66-33736 <$=>

AVTOGENNOE DELO
1941 V12 N5 P25-26 63-18125 <=*> 0
1947 N9 P12-17 TT.103 <NRC>
1947 V18 N7 P1-10 4331 <HB>
1948 V19 N6 P29-30 R-1878 <*>
1948 V19 N9 P14-19 4822 <HB>
1948 V19 N10 P30-31 4861 <HB>
1949 V20 N4 P1-7 R-1864 <*>
1949 V20 N5 P16-17 R-1882 <*>
2999 <HB>
1950 V21 N10 P13-16 3983 <HB>
1951 V22 N3 P21-23 R-1881 <*>
3075 <HB>
1951 V22 N4 P26-27 R-1863 <*>
1951 V22 N10 P13-16 62-20161 <=*>
1952 N10 P1-5 <ATS>
1952 V23 N1 P3-6 2919 <HB>
1952 V23 N1 P14-16 61-15738 <+*>
1952 V23 N8 P19- 60-17194 <+*>
1952 V23 N8 P22- 60-17195 <+*>
1952 V23 N8 P22-23 63-16840 <=*>
1952 V23 N9 P5-8 62-20159 <=*>
1952 V23 N10 P5-7 3007 <HB>
1952 V23 N11 P10-13 3225 <HB>
62-20160 <=*>
1952 V23 N12 P20-22 RT-1212 <*>
1953 V24 N1 P4-8 3140 <*>
1953 V24 N2 P25-26 3016 <HB>
1953 V24 N4 P7-9 3154 <HB>
1953 V24 N6 P25-26 3153 <HB>

★ AVTOMATICHESKAYA SVARKA
1951 V4 N5 P3-17 61-15008 <*+>
1952 V5 N4 P5-10 60-23591 <+*>
1953 V6 N1 P10-18 3072 <HB>
1953 V6 N1 P19-26 3270 <HB>
1953 V6 N1 P27-33 3272 <HB>
1953 V6 N2 P3-14 3233 <HB>
1953 V6 N3 P46-49 3188 <HB>
1953 V6 N4 P3-9 R-1920 <*>
1953 V6 N4 P3-23 3370 <HB>
3371 <HB>
1953 V6 N4 P18-23 R-1926 <*>
1953 V6 N6 P3-10 4607 <HB>
1953 V6 N6 P52-55 4059 <HB>
1954 V7 N3 P26-40 R-1949 <*>
3516 <HB>
1954 V7 N4 P12-28 3400 <HB>
1954 V7 N4 P46-52 65-12021 <*> 0
1954 V7 N5 P38-43 4437 <HB>
1954 V7 N6 P27-32 3550 <HB>
1954 V7 N6 P44-51 3591 <HB>
1954 V7 N6 P52-58 60-17204 <+*>
1954 V7 N6 P59-72 3572 <HB>
1954 V7 N6 P73-76 3536 <HB>
1955 N2 P91-93 3552 <HB>
1955 N4 P58-62 TS-1336 <BISI>
1955 V8 N2 P79-90 3554 <HB>
1955 V8 N3 P13-25 59-22671 <+*> 0
1955 V8 N3 P51-59 3606 <HB>
1955 V8 N4 P31-41 TS 1335 <BISI>
62-14395 <=*>
62-25330 <=*>

Citation	Translation
1955 V8 N4 P58-62	62-24230 <=*>
1955 V8 N4 P90-94	3992 <HB>
1955 V8 N5 P14-24	63-00473 <*>
1955 V8 N5 P47-49	4438 <HB>
1955 V8 N5 P74-77	3774 <HB>
1955 V8 N5 P82	3690 <HB>
1955 V8 N6 P3-18	61-19394 <+*>
1956 V9 N2 P12-17	62-13189 <=*>
1956 V9 N2 P58-66	3802 <HB>
1956 V9 N3 P26-35	65-12022 <*> 0
1956 V9 N3 P65-71	3978 <HB>
1956 V9 N4 P30-49	64-31336 <=>
1956 V9 N5 P19-21	4890 <HB>
	62-20140 <=*>
1956 V9 N5 P22-30	60-15885 <+*>
1956 V9 N5 P80-83	TS-1284 <BISI>
	62-24217 <=*>
1956 V9 N5 P84-89	4489 <HB>
1956 V9 N6 P1-30	61-19691 <=*>
1956 V9 N6 P55-63	4168 <HB>
1956 V9 N6 P65-76	3991 <HB>
1957 V10 N1 P31-36	60-21032 <=>
1957 V10 N1 P73-76	62-24095 <=*>
1957 V10 N1 P88-102	22L36R <ATS>
	61-23550 <=*> 0
1957 V10 N2 P1-10	4412 <HB>
1957 V10 N2 P32-45	4464 <HB>
	60-00135 <*>
1957 V10 N3 P22-27	60-13703 <+*>
1957 V10 N3 P58-63	60-13469 <+*>
1957 V10 N3 P85-91	06L33R <ATS>
	62-20145 <=*>
1957 V10 N3 P97-104	60-13432 <+*>
1957 V10 N4 P29-32	59-16068 <+*> 0
1957 V10 N4 P33-47	4866 <HB>
1957 V10 N4 P107-112	59-11323 <=>
1957 V10 N5 P38-48	63-19425 <=*>
1957 V10 N5 P86-94	AEC-TR-4203 <*+>
1957 V10 N6 P99-102	60-13703 <+*>
1958 N8 P1-3	NLLTB V1 N2 P16 <NLL>
1958 N11 P5-15	M-5748 <NLL>
1958 V11 N1 P44-47	6043 <HB>
1958 V11 N2 P20-29	59-22657 <+*>
1958 V11 N2 P37-41	59-22656 <+*>
1958 V11 N3 P24-34	59-22655 <+*>
1958 V11 N3 P69-78	3780 <BISI>
1958 V11 N4 P24-31	64-31411 <=>
1958 V11 N4 P67-71	62-20141 <=*>
1958 V11 N4 P72-83	62-20144 <=*>
1958 V11 N5 P72-82	61-15915 <+*>
1958 V11 N6 P3-12	62-24213 <=*>
1958 V11 N6 P13-31	4804 <HB>
1958 V11 N6 P32-41	TRANS-21 <MT>
1958 V11 N6 P51-55	3779 <BISI>
1958 V11 N6 P76-83	64-31165 <=>
1958 V11 N6 P84-87	4413 <HB>
1958 V11 N6 P88-91	4462 <HB>
1958 V11 N7 P52-59	<LSA>
	61-19109 <+*>
1958 V11 N7 P83-84	4421 <HB>
1958 V11 N8 P6-18	4516 <HB>
1958 V11 N8 P84-88	59-22269 <+*>
1958 V11 N9 P3-12	<MT>
	62-24189 <=*>
1958 V11 N9 P20-23	<MT>
	62-24190 <=*>
1958 V11 N9 P98-110	TRANS-22 <MT>
	60-13629 <+*>
1958 V11 N10 P67-74	4636 <HB>
1958 V11 N10 P75-80	TS 1747 <BISI>
	5289 <HB>
	60-15654 <+*>
1958 V11 N10 P86-89	TS-1347 <BISI>
	62-24233 <=*>
1958 V11 N11 P16-31	61-19695 <=*>
1958 V11 N11 P32-36	60-19571 <+*>
1958 V11 N11 P57-60	4685 <HB>
1958 V11 N11 P71-80	4786 <HB>
1958 V11 N11 P81-84	60-17318 <+*>
1958 V11 N12 P17-27	63-41198 <=>
1958 V11 N12 P50-56	60-19179 <+*>
1958 V11 N12 P57-62	64-31411 <=>
1959 V12 N1 P49-52	59-31027 <=> 0
1959 V12 N1 P87-91	59-31028 <=> 0
1959 V12 N2 P3-19	61-23163 <*=>
1959 V12 N4 P36-46	61-23872 <*=> 0
1959 V12 N4 P92-93	61-23872 <*=> 0
1959 V12 N8 P57-66	61-23868 <*=>
1959 V12 N10 P38-39	62-20143 <=*> 0
1960 V13 N9 P93-96	61-21422 <=>
1960 V13 N10 P36-41	61-27166 <*=> 0
1960 V13 N11 P31-38	61-27167 <*=> 0
1961 V14 N1 P26-42	2613 <BISI>
1961 V14 N3 P3-11	5463 <HB>
1961 V14 N4 P47-53	62-15801 <=*> 0
1961 V14 N5 P95-96	61-31579 <=>
1961 V14 N10 P91-93	62-25752 <=*>
1961 V14 N11 P86-94	62-25851 <=*> 0
1962 V15 N3 P54-57	5574 <HB>
1962 V15 N5 P1-24	63-19700 <=*>
1962 V15 N5 P57-63	63-19700 <=*>
1962 V15 N5 P78-88	63-19700 <=*>
1962 V15 N9 P92-94	63-13675 <=*>
1962 V15 N11 P1-7	63-21474 <=> 0
1962 V15 N11 P71-76	63-21465 <=>
1962 V15 N12 P56-59	65-30848 <=$>
1963 V16 N2 P41-42	5917 <HB>
1963 V16 N3 P7-12	63-41080 <=>
1963 V16 N4 P27-33	65-64600 <=*$>
1963 V16 N4 P34-40	74S83R <ATS>
1963 V16 N4 P95	63-41026 <=>
1963 V16 N4 P95-96	63-41026 <=>
1963 V16 N8 P83-86	64-21099 <=>
1963 V16 N11 P66-71	6151 <HB>
1964 V17 N1 P71-74	64-31231 <=>
1964 V17 N2 P54-58	64-41028 <=>
1964 V17 N2 P67-74	64-41028 <=>
1964 V17 N2 P77-80	64-41028 <=>
1964 V17 N7 P50-53	64-51148 <=>
1964 V17 N7 P91-92	AD-620 121 <=*$>
	64-51148 <=>
1965 V18 N2 P23-33	AD-615 536 <=$>
1966 V19 N1 P78-80	66-34725 <=$>
1966 V19 N5 P41-42	66-34725 <=$>
1966 V19 N5 P79-80	66-34725 <=$>
1966 V19 N6 P46	66-34725 <=$>

AVTOMATICHESKII KONTROL I IZMERITELNAYA TEKHNIKA

Citation	Translation
1960 N4 P102-108	61-28696 <*=>

AVTOMATICHESKOE REGULIROVANIE AVIADVIGATELEI: SBORNIK STATEI

Citation	Translation
1960 V2 P66-106	61-27260 <*=>
1961 V3 P33-50	63-19175 <=*>
1962 V4 P3-4	63-13200 <=*>
1962 V4 P70-94	63-13200 <=*>

AVTOMATICHESKOE UPRAVLENIE I VYCHISLITELNAYA TEKHNIKA

Citation	Translation
1958 P136-145	59-13704 <=> 0
1958 V1 P166-181	62-15403 <=*>
1958 V1 P204-222	60-41691 <=>
1958 V1 P223-242	60-41693 <=>
1958 V1 P375-418	61-28461 <*=>
1960 V3 P5-35	61-19995 <=*>
1960 V3 P419-444	63-13206 <=*>
1961 V4 P1-383	62-24261 <=*>
1962 V5 P365-442	63-23666 <=*$>

AVTOMATIKA

Citation	Translation
1957 V2 N3 P32-36	59-19909 <=>
1958 N2 P30-47	61-15736 <+*>
1958 N4 P1-17	60-11594 <=>
1958 V3 N1 P20-36	59-11694 <=>
1959 N1 P39-50	60-31091 <=>
1959 N1 P94-	61-19351 <+*>
1959 N2 P1-17	61-19100 <+*>
1959 N4 P1-24	63-19124 <=*>
1960 N1 P63-68	61-28374 <*=>
1960 N1 P78-86	60-31779 <=>

1960 N2 P20-35	62-25912 <=*>
1960 N3 P70-75	61-28702 <*=>
1960 N3 P76-92	61-21439 <=>
1960 N4 P14-31	61-18803 <*=>
1961 N1 P3-24	62-23776 <=*>
	63-13681 <=>
1961 N1 P26-31	62-23776 <=*>
	63-13681 <=>
1961 N1 P33-45	62-23776 <=*>
	63-13681 <=>
1961 N1 P47-54	62-23347 <=*>
1961 N2 P3-9	62-15678 <*=>
1961 N2 P24-28	61-27943 <=*>
1961 N2 P44-52	61-27927 <*=>
1961 N2 P83-89	62-15678 <*=>
1961 N2 P94-97	62-15678 <*=>
1961 N3 P3-27	61-28903 <*=>
1961 N3 P3-29	63-15233 <=*>
1961 N3 P30-34	62-13444 <*=>
1961 N3 P36-43	62-13444 <*=>
1961 N3 P45-54	62-13682 <=>
1961 N3 P56-78	62-15938 <=*>
1961 N3 P64-73	2834 <BISI>
	62-25909 <=*>
1961 N3 P79-80	62-13682 <=>
1961 N3 P82-87	62-15938 <=*>
1961 N3 P88-95	NLLTB V4 N1 P1 <NLL>
1961 N3 P88-92	62-13682 <=>
1961 N5 P3-12	62-24273 <=*>
1961 N5 P13-96	62-25718 <=>
1961 N5 P42-48	63-15234 <=*>
1961 N6 P3-8	62-24392 <=>
1961 N6 P10-13	62-24392 <=>
1961 N6 P15-24	62-24392 <=>
1961 N6 P26-50	62-24392 <=>
1961 N6 P52-59	62-24392 <=>
1961 N6 P61-68	62-24392 <=>
1961 N6 P70-75	62-24392 <=>
1962 N1 P10-23	62-32837 <=*>
1962 N1 P26-45	62-32837 <=*>
1962 N1 P55-64	62-33016 <=*>
1962 N4 P34-59	63-13639 <=*>
1962 V7 N1 P82-84	63-15168 <=*>
1962 V7 N2 P26-41	64-71469 <=>
1962 V7 N2 P63-67	63-15241 <=*>
1962 V7 N2 P69-72	63-15235 <=*>
1962 V7 N4 P80-85	63-21043 <=>
1962 V7 N5 P27-33	63-21236 <=>
1962 V7 N5 P59-73	63-21236 <=>
1962 V7 N6 P3-9	63-31445 <=>
1962 V7 N6 P10-19	63-21441 <=>
1962 V7 N6 P20-39	63-31445 <=>
1962 V7 N6 P60-64	63-21441 <=>
1962 V7 N6 P69-78	63-21398 <=>
1963 V8 N1 P3-9	63-21907 <=>
1963 V8 N1 P10-23	63-21968 <=>
1963 V8 N1 P39-53	63-21968 <=>
1963 V8 N1 P63-66	63-21968 <=>
1963 V8 N1 P70-72	63-21968 <=>
1963 V8 N1 P90-93	63-21907 <=>
1963 V8 N2 P31-40	63-41027 <=>
1963 V8 N2 P41-52	63-31902 <=>
1963 V8 N2 P82-88	63-31903 <=>
1963 V8 N3 P47-58	64-71237 <=>
1963 V8 N3 P84-90	63-41106 <=>
1963 V8 N4 P85-88	64-31577 <=>
1963 V8 N6 P3-10	AD-615 729 <=>
1963 V8 N6 P11-16	64-21867 <=>
1963 V8 N6 P17-83	AD-615 729 <=>
1964 N5 P31-35	AD-633 815 <=$>
1964 V9 N1 P19-41	64-21928 <*>
1964 V9 N1 P43-63	64-21928 <*>
1964 V9 N2 P42-57	64-31490 <=>
1964 V9 N2 P59-70	AD-635 849 <=$>
1964 V9 N2 P59-69	64-31490 <=>
1964 V9 N2 P71-75	AD-621 425 <=*$>
1964 V9 N2 P79-88	64-31490 <=>
1964 V9 N4 P3-9	AD-622 458 <=*$>
1964 V9 N4 P3-48	64-51732 <=$>
1964 V9 N4 P72-74	64-51732 <=$>
1964 V9 N4 P80-82	64-51732 <=$>
1964 V9 N5 P3-14	AD-636 654 <=$>
1965 V10 N2 P8-16	65-31847 <=$>
1965 V10 N2 P21-28	65-31847 <=$>
1965 V10 N2 P39-60	65-31847 <=$>
1965 V10 N2 P76-78	65-31847 <=$>
1965 V10 N3 P40-84	65-33828 <=*$>
1966 V11 N2 P34-64	66-32959 <=$>
1966 V11 N2 P65-71	66-32260 <=$>
1966 V11 N3 P24-48	66-34055 <=$>
1966 V11 N5 P3-4	67-30525 <=>
1966 V11 N5 P15-26	67-30525 <=>
1966 V11 N5 P28-36	67-30525 <=>
1966 V11 N5 P45-55	67-30525 <=>
1966 V11 N5 P90-94	67-30525 <=>
1966 V11 N6 P46-52	67-30731 <=$>

AVTOMATIKA I PRIBOROSTROENIE

1962 V1 N2 P3-6	62-11733 <=>
1962 V1 N2 P28-45	62-11733 <=>
1962 V1 N3 P51-53	63-13475 <=>
1962 V1 N3 P60-65	63-13475 <=>
1962 V1 N3 P85-86	63-13475 <=>
1963 V2 N2 P90-92	63-31482 <=>
1964 V3 N1 P45-47	AD-610 324 <=$>
1964 V3 N2 P62-65	64-51843 <=$>
1964 V3 N2 P68-70	64-51843 <=$>
1964 V3 N2 P80-81	64-51843 <=$>
1964 V3 N3 P24-25	AD-623 209 <=*$>

★AVTOMATIKA I TELEMEKHANIKA

1947 V8 N4 P225-242	RT-1051 <*>
1947 V8 N5 P349-383	RT-4651 <*>
1948 V9 N2 P144-151	RT-1195 <*>
1948 V9 N3 P190-203	61-23648 <*=>
1949 V10 N1 P32-50	51/0100 <NLL>
1949 V10 N6 P437-451	59-18649 <+*> 0
1950 V11 N3 P161-163	RT-3040 <*>
1950 V11 N4 P286-288	59-14171 <+*>
1950 V11 N5 P289-299	RT-3582 <*>
1950 V11 N5 P300-319	RT-3624 <*>
1951 V12 N1 P15-27	RT-4652 <*>
1951 V12 N5 P389-397	RT-3581 <*>
1952 V13 N2 P109-120	RT-3587 <*>
1952 V13 N2 P121-133	RT-3515 <*>
1952 V13 N2 P134-144	RT-3330 <*>
	65-12811 <*>
1952 V13 N2 P145-151	RT-3517 <*>
1952 V13 N2 P176-192	61-31603 <=> 0
1952 V13 N2 P212-216	RT-3518 <*>
1952 V13 N2 P217-224	RT-3374 <*>
1952 V13 N3 P227-281	61-13467 <*=>
1952 V13 N4 P405-416	RT-3585 <*>
1952 V13 N4 P425-428	RT-3329 <*>
1952 V13 N4 P429-444	RT-3513 <*>
1952 V13 N4 P465-468	RT-3516 <*>
1952 V13 N4 P488-490	RT-3136 <*>
1952 V13 N5 P585-591	59-12065 <=>
1952 V13 N6 P650-663	RT-4169 <*>
1952 V13 N6 P744-746	RT-3584 <*>
1952 V13 N6 P747-749	RT-3578 <*>
1953 V14 N6 P712-728	65-13811 <*>
1954 V15 N4 P332-335	62-23192 <=*>
1954 V15 N4 P367-374	RT-3562 <*>
1955 N3 P300-305	R-320 <*>
1955 N4 P364-365	62-33207 <=*>
1955 V16 N1 P47-63	RT-3459 <*>
1955 V16 N1 P64-70	62-11585 <=>
1955 V16 N1 P87-95	RT-3458 <*>
	62-23178 <=*>
1955 V16 N2 P172-183	RJ-411 <ATS>
	62-15848 <=*>
1955 V16 N2 P203-205	R-627 <*>
1955 V16 N2 P219-224	R-1893 <*>
1955 V16 N3 P293-299	62-15922 <=*>
1955 V16 N4 P344-355	62-23075 <=*>
	63-18708 <=*>
1955 V16 N4 P372-381	RT-4263 <*>
1955 V16 N4 P372-382	62-19867 <=*>
1955 V16 N4 P411-412	R-4082 <*>

1955 V16 N5 P421-501 59-14153 <+*>
1955 V16 N5 P421-430 62-23158 <=*>
1955 V16 N5 P497-500 62-15855 <=*>
1955 V16 N6 P530-535 RJ-461 <ATS>
1955 V16 N6 P548-553 RJ-462 <ATS>
1956 V17 P97-106 R-4477 <*>
1956 V17 N2 P97-106 RT-3952 <*>
1956 V17 N2 P180-190 R-3732 <*>
1956 V17 N6 P500-512 63-20079 <=*>
1956 V17 N7 P590-600 5J18R <ATS>
1956 V17 N10 P890-896 59-18707 <+*> O
1957 V18 N2 P126-136 R-4631 <*>
1957 V18 N2 P145-162 60-13586 <+*>
1957 V18 N3 P201-222 R-4630 <*>
1957 V18 N3 P240-255 60-13586 <+*>
1957 V18 N4 P324-335 R-4632 <*>
1957 V18 N5 P437-443 60-13586 <+*>
1957 V18 N6 P514-528 R-4633 <*>
1957 V18 N9 P841- 59-11236 <=>
1957 V18 N10 P899-910 R-3702 <*>
1957 V18 N10 P950-952 62-18271 <=*>
1957 V18 N11 P985-998 65-60877 <=$>
1957 V18 N12 P1136-1138 R-4968 <*>
1958 V19 N1 P75-84 60-13598 <+*> O
1958 V19 N1 P99-100 60-11195 <=>
1958 V19 N2 P135-147 60-13560 <+*>
1958 V19 N3 P217-220 59-11958 <=> O
1958 V19 N3 P257-267 59-11959 <=>
1958 V19 N4 P325-333 59-11273 <=> O
1958 V19 N4 P360-365 59-11274 <=>
1958 V19 N6 P614-620 60-13585 <+*>
1958 V19 N9 P836-848 62-23944 <=*>
1958 V19 N11 P997-1009 60-13793 <+*>
1958 V19 N12 P1118-1125 59-21148 <=>
1959 V20 N1 P3-15 61-14929 <=*>
1959 V20 N1 P100-106 59-13411 <=>
1959 V20 N3 P361-375 61-23671 <*=> O
1959 V20 N4 P447-467 59-21162 <=> O
1959 V20 N7 P848-855 61-10720 <*=>
1959 V20 N7 P988-991 61-13078 <*+> O
1959 V20 N10 P1435-1438 60-11260 <=>
1959 V20 N11 P1483 60-17623 <+*> O
1960 V21 N1 P42-47 60-31179 <=>
61-10620 <+*> O
1960 V21 N1 P139-142 62-24793 <=*>
1960 V21 N1 P145-148 60-11969 <=>
1960 V21 N2 P161-166 60-41723 <=>
1960 V21 N2 P180-190 61-28196 <*=>
1960 V21 N3 P359-368 60-21916 <=>
1960 V21 N3 P369-373 62-11401 <=>
1960 V21 N3 P429-430 60-41247 <=>
1960 V21 N4 P465-473 61-16447 <*=>
1960 V21 N4 P474-480 61-16766 <*=>
1960 V21 N6 P729-739 62-10179 <*=>
1960 V21 N7 P997-1006 61-11125 <=>
1960 V21 N7 P1084-1087 61-27434 <*=> O
1960 V21 N9 P1320-1322 61-21052 <=>
1960 V21 N9 P1326-1331 61-21052 <=>
1960 V21 N10 P1375-1386 61-14962 <=*>
61-21648 <=>
1961 V22 N1 P111-118 61-31640 <=> O
1961 V22 N1 P119-120 61-31677 <=>
1961 V22 N2 P129-142 61-28534 <*=>
1961 V22 N4 P536-538 62-15406 <*=>
1961 V22 N4 P539-542 62-23379 <=*>
1961 V22 N4 P543-551 63-24339 <=*$>
1961 V22 N8 P986-1001 62-11117 <=>
62-19162 <=*>
1961 V22 N8 P1027-1037 62-13560 <=*>
1961 V22 N8 P1080-1087 62-13560 <=*>
1961 V22 N8 P1108-1116 AD-267 715 <=>
1961 V22 N9 P1235- ICE V2 N2 P169-173 <ICE>
1962 V23 N1 P49-59 63-23106 <=*>
1962 V23 N1 P127 62-23623 <=*>
1962 V23 N2 P214-221 62-32302 <=*>
1962 V23 N2 P222-241 62-32517 <=*>
1962 V23 N3 P349-364 62-24731 <=*>
1962 V23 N5 P661-684 62-20262 <=*>
1962 V23 N6 P713-720 62-32072 <=*>
1962 V23 N8 P993-1007 62-33342 <=*>
1962 V23 N9 P1117-1140 63-21184 <*>
1962 V23 N9 P1144-1153 63-21184 <=>
1962 V23 N9 P1165-1178 63-21184 <=>
1962 V23 N9 P1237-1242 63-21184 <=>
1962 V23 N9 P1268-1269 63-21184 <=>
1962 V23 N12 P1571-1583 63-14731 <*=>
1962 V23 N12 P1643-1653 63-18064 <=*>
1962 V23 N12 P1720-1723 63-21051 <=>
1963 V24 N1 P3-6 63-31242 <=>
1963 V24 N5 P581-598 65-11462 <*>
1963 V24 N6 P785-798 64-71338 <=> M
65-11004 <*>
1963 V24 N7 P962-974 64-15302 <=*$>
1963 V24 N8 P1021-1036 AD-607 910 <=> M
1963 V24 N8 P1155-1162 63-31761 <=>
1963 V24 N9 P1292-1294 64-71608 <*>
1963 V24 N11 P1474-1486 64-11873 <=>
1963 V24 N11 P1514-1518 64-21520 <=>
1964 V25 N3 P416-423 RTS-2898 <NLL>
1964 V25 N4 P539-546 64-51653 <=$>
1964 V25 N6 P747-831 64-51615 <=*$>
1965 V26 N1 P188-190 65-30521 <=$>
1965 V26 N10 P1746-1756 65-34014 <=*$>
1965 V26 N12 P2214-2218 66-31047 <=$>
1966 V27 N2 P152-155 66-12593 <*>
1966 V27 N5 P204-206 66-33098 <=$>

AVTOMATIKA, TELEMEKHANIKA I SVYAZ
1958 N7 P12-13 PB 141 273T <=>
1958 N11 P8-11 61-13481 <*+>
1959 N3 P11-15 60-31051 <=>
1959 N4 P13-19 60-11318 <=>
1959 N10 P4-6 61-11594 <=>
61-15398 <*+>
1960 N2 P21-24 60-31240 <=> O
1960 N4 P40-42 60-41247 <=>
1961 V5 N1 P40-41 62-23476 <=>
1961 V5 N4 P13-16 62-15445 <*=>
1961 V5 N11 P42-47 62-23337 <=*>
1961 V5 N12 P37-43 62-23337 <=*>
1962 N2 P1-3 NLLTB V4 N7 P643 <NLL>
1962 N10 P1-2 NLLTB V5 N3 P190 <NLL>
1962 V6 N7 P44 62-33657 <=*>
1962 V6 N10 P46 63-21381 <=>
1964 V8 N5 P4-8 64-41421 <=$>
1964 V8 N5 P37-38 64-41421 <=$>
1964 V8 N9 P9-12 65-31205 <=$>
1966 N4 P8-11 66-32638 <=$>
1966 N6 P1-2 66-33517 <=$>
1966 N7 P18-20 66-34022 <=$>
1967 N3 P43 67-31402 <=$>

★AVTOMATIKA I VYCHISLITELNAYA TEKHNIKA
1962 N2 P29-38 63-21839 <=>

AVTOMATIZATSIYA I PRIBOROSTROENIE
1960 N1 P34-38 64-31191 <=>
1961 N1 P3-5 65-60961 <=$>
1961 N1 P5-10 65-60773 <=$>
1961 N4 P12-16 65-61232 <=$>
1961 V2 P31-41 64-15108 <=*$>
1961 V2 P153-164 63-23675 <=*$>
1964 V3 N3 P8-10 65-30124 <=$>
1964 V3 N3 P67-68 65-30124 <=$>

AVTOMATIZATSIYA PROIZVODSTVENNYKH PROTSESSOV
1958 N2 P83-93 NLLTB V1 N8 P17 <NLL>
59-22641 <+*> C
1960 N3 P3-43 62-23578 <=>
1960 N3 P90-101 62-23578 <=>
1960 N3 P97-101 NLLTB V3 N1 P12 <NLL>
1962 V2 P152-157 3700 <BISI>
1964 N4 P5-21 ICE V6 N4 P608-618 <ICE>

AVTOMATIZATSIYA PROTSESSOV MASHINOSTROENIYA. MOSCOW.
1963 V3 P85-90 AD-631 578 <=$>

★AVTOMETRIYA
1965 N4 P1-33 65-33418 <=*$>

1965 N6 P3-51	66-31395 <=$>

AVTOMOBIL

1941 V19 N2 P15-17	61-18062 <*=>
1946 V24 N1 P4-7	61-20415 <*=> P
1948 V26 N7 P12-13	RT-1424 <*>
1951 V29 N2 P31-34	93P60R <ATS>

AVTOMOBILNAYA PROMYSHLENNOST

1946 N7/8 P18-19	R-1884 <*>
1955 N1 P7-9	62-26261 <=$>
1956 N3 P38-41	GKN-TRANS-2012 <NLL>
1956 N5 P13-15	61-28432 <*=>
1956 N12 P22-25	60-13322 <+*>
1957 N3 P46	60-13228 <+*> P
1957 N12 P1-7	59-11292 <=>
1958 N1 P1-2	59-11283 <=>
1958 N1 P25-27	8598 <K-H>
1958 N1 P39-40	59-11284 <=> 0
1958 N2 P1-6	59-18857 <+*>
1958 N3 P5-8	60-15811 <+*> 0
1958 N3 P16-19	59-11280 <=> 0
1958 N3 P22-27	59-22256 <+*>
1958 N4 P16-18	59-22403 <+*>
1958 N5 P5-10	59-14891 <+*> 0
1958 N5 P26-28	59-14892 <+*> 0
1958 N6 P7-9	59-11910 <=> 0
1958 N10 P12-15	59-22589 <+*>
1958 N11 P17-21	60-19873 <+*>
1958 N11 P46	59-21070 <=> 0
1958 N12 P13-15	59-21093 <=> 0
1958 N12 P29-33	61-13136 <*=>
1959 N1 P20-23	61-23027 <=*>
1959 N2 P3-6	60-21733 <=>
1959 N2 P36-39	60-13830 <+*>
1959 N3 P1-4	59-22267 <+*>
1959 N4 P3-4	60-21217 <=>
1959 N4 P42-45	63-19816 <=*$>
1959 N6 P29-31	TRANS-84 <MT>
1959 N7 P27-30	61-13486 <*+>
1959 N8 P3-5	60-17621 <+*>
1959 N9 P19-20	10397 <K-H>
	61-10310 <*+> 0
1959 N9 P24-26	60-21583 <=> 0
1959 N10 P10-13	61-13570 <*=>
1959 N11 P25-28	62-13091 <*=>
1959 N12 P29	E-743 <RIS>
1959 N12 P29-30	E-744 <RIS>
1959 S4 N11 P38-39	60-17742 <+*> 0
1960 N1 P4-8	60-19865 <+*> 0
1960 N2 P32-33	61-13754 <*=>
1960 N4 P18-23	61-11100 <=> 0
1960 N4 P23-24	63-19822 <=*$>
1960 N7 P7-9	62-15265 <*=> 0
1960 N9 P1-3	61-11863 <=>
1960 N9 P5-8	61-11863 <=>
1960 N10 P13-14	62-19132 <=*>
1960 N11 P6-15	61-21969 <=> 0
1960 N12 P4-7	63-15651 <=*>
1961 N4 P34-35	61-27232 <=>
1961 V27 N4 P8-12	62-11668 <=>
1961 V27 N4 P12-14	62-25323 <=*>
1961 V27 N4 P14-16	63-24028 <=*>
1961 V27 N7 P8-9	63-19781 <=*>
1961 V27 N10 P29-33	62-32535 <=*>
1961 V27 N12 P33-36	64-19387 <=$>
1962 N10 P1-2	NLLTB V5 N3 P197 <NLL>
1962 V28 N1 P1-3	62-33023 <=*>
1962 V28 N1 P4-7	64-13290 <=*$>
1962 V28 N1 P17-20	64-13057 <=*$>
1962 V28 N2 P6-8	62-33023 <=*>
1962 V28 N2 P9-12	64-13031 <=*$>
1962 V28 N2 P44	62-32057 <=>
1962 V28 N3 P1-3	64-13032 <=*$>
1962 V28 N6 P13-15	64-13030 <=*$>
1962 V28 N7 P8-12	65-61403 <=>
1962 V28 N8 P10-11	2005 <NLL>
1962 V28 N12 P21-23	63-24560 <=*$>
1963 V29 N1 P3-5	64-21683 <=>
1963 V29 N1 P6-7	63-21707 <=>
	63-21707 <=>
1963 V29 N1 P12-14	63-21707 <=>
1963 V29 N2 P1-7	63-21434 <=>
1963 V29 N7 P13-15	65-60756 <=>
1963 V29 N8 P5-7	65-60355 <=$>
1963 V29 N10 P18-20	65-60955 <=$>
1963 V29 N12 P30-31	64-31461 <=>
1964 V30 N1 P15-18	RTS-2781 <NLL>
1964 V30 N5 P42-45	AD-640 288 <=$>
1964 V30 N6 P16-19	TRANS-2103 <NLL>
1964 V30 N6 P38	64-41406 <=$>
1964 V30 N6 P47-48	64-41406 <=$>
1964 V30 N7 P42-45	64-51009 <=>
1964 V30 N9 P5-7	AD-636 677 <=$>
1964 V30 N10 P1-4	65-30291 <=$>
1964 V30 N10 P5-10	AD-639 154 <=$>
1964 V30 N12 P28-31	65-30584 <=$>
1965 N9 P37-39	NLLTB V8 N2 P102 <NLL>
1965 V31 N1 P1-3	65-30753 <=$>
1965 V31 N1 P43-44	65-30753 <=$>
1965 V31 N5 P4-7	66-11477 <*>
1966 V32 N3 P1-3	66-32027 <=$>
1966 V32 N9 P1-2	66-34937 <=$>
1966 V32 N9 P9-15	66-34937 <=$>

AVTOMOBILNAYA I TRAKTORNAYA PROMYSHLENNOST

1951 N2 P12-15	RJ-59 <ATS>
	RJ59 <ATS>
1954 N1 P6-8	RT-2310 <*>
1954 N1 P20-21	RT-2647 <*>
1954 N1 P22-24	RT-2839 <*>
1955 N3 P21-24	3588 <HB>
1955 N7 P4-6	R-3686 <*>
1955 N8 P19-21	R-3679 <*>
1955 N8 P21-24	R-3679 <*>
1955 N10 P123-125	R-3814 <*>
1955 N12 P20-26	R-3680 <*>

AVTOMOBILNYE DOROGI. MOSCOW

1957 V20 N2 P20-21	59-19628 <+*>
1957 V20 N4 P8-9	59-18524 <+*> 0
1958 V21 N3 P18-20	CSIR-TR.132 <CSSA>
	CSIR-TR-132 <CSSA>
1958 V21 N4 P15-16	61-23390 <=*>
1958 V21 N5 P23-24	59-10794 <+*>
1958 V21 N7 P14-15	60-15660 <+*>
1958 V21 N7 P22-24	59-22722 <+*>
1958 V21 N9 P4-6	59-11883 <=> 0
1958 V21 N9 P18-19	59-22564 <+*>
1958 V21 N12 P25-26	61-13907 <*+>
1959 V22 N1 P10	59-22559 <+*>
1959 V22 N3 P10-12	59-13865 <=>
	60-21212 <=> 0
1959 V22 N3 P23	CSIR-TR.115 <CSSA>
1959 V22 N4 P7-9	60-14800 <+*> 0
1959 V22 N4 P17-19	59-13865 <=>
1959 V22 N4 P22-23	60-14801 <+*> 0
1959 V22 N4 P25-27	60-14798 <+*> 0
	61-13537 <*=> 0
1959 V22 N8 P16-17	61-23663 <=*>
1960 V23 N6 P33	61-23666 <*=>
1960 V23 N10 P16-17	61-31118 <=>
1960 V23 N11 P27	62-25281 <=*>
1961 V24 N4 P22-23	AEC-TR-5512 <=*>
	CSIR-TR.303 <CSSA>
1961 V24 N8 P24-25	81P62R <ATS>
1961 V24 N8 P25-26	82P62R <ATS>
1961 V24 N9 P32	65-61166 <=$>
1962 N10 P4-6	NLLTB V5 N6 P534 <NLL>
1962 V25 N1 P27-28	63-15341 <=*>
1962 V25 N2 P4-7	63-13093 <=>
1962 V25 N4 P23-25	63-23236 <=*>
1962 V25 N7 P13-14	65-60360 <=$>
1962 V25 N9 P9-11	64-11842 <=> 0
1962 V25 N12 P22-23	64-13392 <=*$>
1962 V25 N12 P27-28	64-00140 <*>
1963 V26 N1 P15-17	64-19125 <=*$>
1963 V26 N11 P9-10	RTS-2745 <NLL>

AVTOMOBILNYI TRANSPORT

1955 V33 N11 P6-8 59-11587 <=>
1958 V36 N3 P8 59-11915 <=> O
1958 V36 N6 P21 59-11288 <=>
1958 V36 N6 P36-39 61-21963 <=>
1958 V36 N8 P38-43 59-21016 <=> O
1958 V36 N9 P28-29 59-22253 <+*>
1958 V36 N12 P4-5 59-21210 <=>
1958 V36 N12 P29-32 59-21211 <=> O
1958 V36 N12 P33 59-21212 <=> O
1958 V36 N12 P35 59-21213 <=> O
1959 N5 P36-41 NLLTB V1 N11 P37 <NLL>
1959 V37 N2 P47 60-13745 <+*>
1959 V37 N2 P48 60-13746 <+*>
1959 V37 N3 P17- 60-21057 <=> O
1959 V37 N4 P42 60-13747 <+*>
1959 V37 N4 P46 60-13748 <+*>
61-13543 <*=>
1959 V37 N5 P16-18 60-21431 <=>
1959 V37 N5 P36-41 60-21431 <=>
1959 V37 N9 P13-15 61-23343 <=*> P
1960 V38 N6 P47-48 61-21293 <=>
1960 V38 N11 P41-42 61-31118 <=>
1961 V39 N1 P36-41 62-11698 <=>
1961 V39 N3 P30-32 62-25283 <=*>
1961 V39 N6 P29-31 62-11675 <=>
1962 V40 N3 P14 62-33697 <=>
1962 V40 N3 P19-23 62-33697 <=>
1962 V40 N3 P29-30 62-33697 <=>
1962 V40 N5 P24-26 63-19736 <=*> O
1962 V40 N5 P44-45 62-33121 <=>
1962 V40 N6 P26 62-33697 <=>
1962 V40 N6 P32-33 62-33697 <=>
1962 V40 N6 P56 62-33697 <=>
1962 V40 N7 P13-14 63-13850 <=>
1962 V40 N7 P37 63-13850 <=>
1962 V40 N7 P40 63-13850 <=>
1962 V40 N7 P55 63-13850 <=>
1962 V40 N9 P59 63-13850 <=>
1962 V40 N10 P58 63-15395 <=*>
1963 V41 N3 P41-43 63-21942 <=>
1963 V41 N3 P47 63-21942 <=>
1963 V41 N5 P19-21 64-23518 <=$>
1963 V41 N5 P41-42 64-11517 <=> O
1963 V41 N6 P25 63-31927 <=>
1964 V42 N2 P15-17 64-31762 <=>
1964 V42 N2 P22-24 64-31762 <=>
1964 V42 N3 P3-5 64-41634 <=>
1964 V42 N3 P38-40 64-41103 <=>
1964 V42 N4 P1-4 64-41187 <=>
1964 V42 N4 P4-5 64-41255 <=>
1964 V42 N10 P32-33 65-30931 <=$>
1965 V43 N3 P23-29 65-30889 <=*$>

AZERBAIDZHANSKII KHIMICHESKII ZHURNAL
1961 N1 P31-38 RCT V36 N3 P747 <RCT>
1961 N4 P23-29 63-15247 <*=>
1962 N5 P29-40 65-10235 <*>
1962 N6 P3-7 65-11957 <*>
1963 N1 P3-9 65-11958 <*>
1963 N1 P11-16 65-11956 <*>
1963 N5 P39-43 65-12197 <*>
1964 N1 P125-131 42T91R <ATS>
1964 N1 P133-138 44T91R <ATS>
1964 N4 P336 1735 <TC>
1964 N5 P55-64 15S88R <ATS>
1964 N5 P81-86 72T91R <ATS>
1964 N5 P103-107 43T91R <ATS>
1965 N1 P31-34 68S86R <ATS>
1965 N1 P53-56 69S86R <ATS>
1965 N2 P91-94 ICE V6 N4 P674-676 <ICE>
16T90R <ATS>

AZERBAIDZHANSKII MEDITSINSKII ZHURNAL
1959 N3 P47-53 59-13629 <=>
1959 N5 P51-56 59-13862 <=>
1959 N11 P3-7 60-11420 <=>
1959 N12 P10-13 60-11422 <=>
1959 N12 P64-66 60-11422 <=>
1960 N1 P3-5 60-31707 <=>
1960 N1 P30-34 60-31706 <=>
1960 N1 P56-59 60-31708 <=>
1960 N2 P28-32 60-31280 <=>
1960 N4 P66-71 60-31871 <=>
1960 N4 P118-120 60-31872 <=>
1960 N4 P125-128 60-31873 <=>
1960 N5 P88-91 61-27373 <*=>
1960 N6 P16-21 60-31790 <=>
1960 N7 P63-67 61-11578 <=>
1960 N9 P3-9 61-11548 <=>
1960 N10 P73-82 61-21241 <=>
1960 N11 P9-17 61-21235 <=>
1960 N11 P56-59 61-21252 <=>
1960 N12 P27-33 61-21534 <=>
1961 N1 P40-43 62-25514 <=*>
1961 N1 P52-55 61-21586 <=>
1961 N5 P3-9 61-27359 <*=>
1961 N7 P3-8 62-19798 <=*>
1961 N10 P78-83 62-24256 <=*>
1962 N1 P17-20 62-24407 <=*>
1962 N2 P67-71 62-32443 <=*>
1962 N3 P16-23 62-25608 <=*>
1962 N4 P18-22 62-33331 <=*>
1962 N4 P98-105 62-33331 <=*>
1962 N10 P16-25 63-15403 <=*>
63-15403 <=*>
1962 N10 P31-42 63-15403 <=*>
1962 N10 P81-83 63-15403 <=*>
1962 N11 P15-20 63-15393 <=*>
1963 N9 P3-11 63-41246 <=>
1963 N12 P72-75 64-31204 <=>

AZERBAIDZHANSKOE NEFTYANOE KHOZYAISTVO
1929 V9 N8/9 P83- 63-10704 <=*> O
1930 V10 N2 P72-77 63-10340 <*>
1930 V10 N9 P102-103 63-10693 <=*>
1930 V10 N11 P4-15 63-10448 <*=>
1932 N11 P15-18 62-18532 <=*>
1932 V12 N12 P86-87 61-17988 <=$>
1934 V14 N1 P60-62 63-10695 <=*>
1934 V14 N5 P62-69 63-10409 <*=> O
1934 V14 N11/2 P113-114 61-17989 <=$>
1935 N5 P91-94 61-18146 <=*>
1935 V15 N1 P93-99 63-10408 <*=>
1935 V15 N7/8 P119-125 61-16690 <=*$>
1936 V16 N2/3 P90-100 61-18055 <=*$>
1938 V18 N4 P49-53 61-17987 <=$>
1939 N12 P10-15 R-2107 <*>
R-829 <*>
59-10397 <+*>
1939 V19 N10/1 P55-60 61-16856 <=*$>
1940 N9 P29-32 66-14271 <*$>
1940 V20 N2/3 P62-66 61-16868 <*=>
1947 N10 P1-2 RT-1236 <*>
1947 V26 N5 P23-26 59-15057 <+*>
1947 V26 N8 P17-21 RJ-20 <ATS>
1953 N10 P94- 63-10692 <=*>
1953 N8/9 P92- 63-10692 <=*>
1956 N2 P18-22 59-17246 <+*>
1956 N10 P27-31 50J19R <ATS>
1958 V37 N3 P19-21 60-10990 <+*>
1959 V38 N7 P36-38 62-10437 <*=>
1959 V38 N8 P37-40 62-10437 <*=>
1961 V40 N1 P47-48 61-27941 <*=>
1961 V40 N4 P35-36 52Q72R <ATS>
1961 V40 N5 P1-2 62-23345 <=*>
1964 V43 N1 P36-38 AD-622 942 <=$>
1964 V43 N2 P30-32 AD-622 942 <=$>

AZIYA I AFRIKA SEGODNYA
1961 N9 P20-21 62-15776 <=*>
1964 V7 N5 P26-29 64-41488 <=$>

B.B.C. NACHRICHTEN
1960 V42 P228-231 2089 <BISI>
1965 P3-12 4370 <BISI>

BELGICATOM
SEE BULLETIN D'INFORMATION DE L'ASSOCIATION BELGE POUR LE DEVELOPMENT PACIFIQUE DE L'ENERGIE ATOMIQUE

BTA BUEROTECHNIK UND AUTOMATION
1965 N3 P114-120 65-13259 <*>

BAENDER, BLECHE, ROHRE
1960 P105-111 4330 <BISI>
1960 P126-127 4346 <BISI>
1961 P1-5 2267 <BISI>
1961 P8-15 2409 <BISI>
1961 P56-62 2409 <BISI>
1961 P153-162 2536 <BISI>
1961 P492-494 2621 <BISI>
1961 N1 P5-7 3426 <BISI>
1961 N1 P21-24 2790 <BISI>
1961 N2 P64-67 3206 <BISI>
1961 N4 P162-169 2972 <BISI>
1961 N6 P273-278 2972 <BISI>
1961 N10 P485-491 2768 <BISI>
1961 N10 P494-497 2769 <BISI>
1961 N10 P500-507 3387 <BISI>
1961 N11 P558-564 2790 <BISI>
1961 N12 P600-605 2790 <BISI>
1962 P49-61 2749 <BISI>
1962 N1 P17-20 2978 <BISI>
1962 N4 P181-186 2871 <BISI>
1962 N4 P187-189 2971 <BISI>
1962 N5 P228-237 3109 <BISI>
1963 N1 P13-16 3243 <BISI>
1963 N1 P16-21 3225 <BISI>
1963 N2 P58-61 3845 <BISI>
1963 N2 P73-80 3916 <BISI>
1963 N4 P165-173 3616 <BISI>
1963 N4 P174-178 3359 <BISI>
1963 N5 P226-232 3373 <BISI>
1963 N6 P282-287 6341 <HB>
1963 N6 P305-314 3373 <BISI>
1963 N10 P529-534 3832 <BISI>
1964 P566-576 4074 <BISI>
1964 N1 P5-8 66-13949 <*> 0
1964 N3 P136-139 3732 <BISI>
1964 N4 P191-195 6272 <HB>
1964 N5 P245-252 4098 <BISI>
1964 N8 P432-439 4004 <BISI>
1964 N9 P493-500 6400 <HB>
1964 N9 P500-502 4080 <BISI>
1964 N10 P551-553 6440 <HB>
1965 N6 P148-152 4451 <BISI>
1965 N6 P184-189 4478 <BISI>

BALNEOLOGIA ET BALNEOTHERAPIA
1959 V19 P371-387 62-00418 <*> 0

BANSKY OBZOR
1949 V3 P97-99 60-18827 <*>
1949 V3 P121-123 60-18827 <*>

BANYASZATI ES KOHASZATI LAPOK
1950 V5 P700-704 60-18826 <*>
1950 V5 N1 P68-70 60-18822 <*>

BANYASZATI LAPOK
1958 V13 N5 P295-309 59-11362 <=> 0
1958 V13 N6 P373-374 59-11362 <=> 0
1958 V13 N6 P378 59-11362 <=> 0
1958 V13 N6 P380 59-11362 <=> 0
1962 V17 P54-59 62-25606 <=*>
1964 V97 N5 P353-359 07T94H <ATS>
1964 V97 N11 P757-759 65-13078 <*>
1965 V98 N2 P136-142 44S84H <ATS>
1965 V98 N4 P267-272 01S86H <ATS>
1965 V98 N10 P703-706 43T92H <ATS>

BATTERIEN; GALVANISCHE ELEMENTE, AKKUMULATOREN
1939 P1029-1031 59-20371 <*> 0
1939 P1039-1041 59-20371 <*> 0
1939 P1055-1057 59-20371 <*> 0
1940 P1071-1073 59-20371 <*> 0

BAUGILDE
1939 V21 P466-469 AL-131 <*>

BAUINGENIEUR
1920 V1 N22 P631-636 65-12361 <*>
1924 P417 64-18261 <*>
1932 V13 N19/2 P261-267 59-20886 <*>
1933 V14 N3/4 P50-53 59-10283 <*>
1935 V16 N35/6 P374-381 59-15084 <*>
1938 V19 P424-426 II-669 <*>
1938 V19 N5/6 P69-75 59-17186 <*> 0
1939 V20 N17/8 P231-232 64-71234 <=>
1951 N21 P467-470 60-10786 <*>
1954 V29 N8 P278-294 61-18953 <*>
1954 V29 N9 P355-361 57-3179 <*>
1954 V29 N11 P405-409 58-2603 <*>
61-20033 <*>
1956 V31 P47-48 T-1725 <INSD>
1956 V31 N4 P134-135 61-10798 <*>
1957 V32 N7 P262-275 58-2401 <*>
1958 V33 N7 P256-265 61-18976 <*>
1958 V33 N8 P287-294 42M41G <ATS>
1958 V33 N9 P344-351 1120-GJ <ATS>
43M41G <ATS>
1960 V35 N1 P6-11 60-14554 <*>
1960 V35 N5 P169-178 61-18970 <*>
1961 V36 N2 P66-69 62-18016 <*>
1961 V36 N8 P293-300 C-4342 <NRC>
1964 V39 N2 P50-64 65-11733 <*> 0

BAUMASCHINE UND BAUTECHNIK
1955 V2 P247-253 T1772 <INSD>
1956 V3 N3 P257-268 58-2608 <*>

BAUMATERIALIENKUNDE
1899 V4 P161-189 64-18216 <*> 0
1899 V4 P205-221 64-18216 <*> 0
1899 V4 P237-253 64-18216 <*> C

BAUPLANUNG UND BAUTECHNIK
1955 V9 N8 P343-349 II-393 <*>
1956 V10 N11 P441-447 58-2582 <*>
61-18961 <*>
1967 V21 N2 P105-106 67-31447 <=$>
1967 V21 N2 P111 67-31447 <=$>

BAUSTOFFINDUSTRIE
1966 V9 N11 P339-342 67-30236 <=>

BAUTECHNIK
1933 V11 P579-582 AL-748 <*>
1935 V13 N18 P226-229 I-153 <*>
1937 V15 N1 P14-16 1394 <TC>
1941 V19 N50/1 P537-540 CSIRO-3299 <CSIR>
57-1711 <*>
58-1627 <*>
1949 V26 N10 P300-306 58-2587 <*>
61-18993 <*>
1950 V27 N1 P16-18 57-195 <*>
1951 V28 N12 P309-310 I-842 <*>
1954 V31 N6 P287-292 3000 <*>
1954 V31 N10 P324-330 65-11677 <*>
1956 V33 P246-250 T-2500 <INSD>
1956 V33 N2 P47-54 59-10746 <=>
1956 V33 N8 P268-273 59-15063 <*>
1957 V34 N1 P16-20 C-3589 <NRCC>
1957 V34 N3 P89-94 C-3589 <NRCC>
1957 V34 N12 P458-465 63-10825 <*>
1958 V35 N7 P261-267 59-20046 <*>
1960 V37 N11 P419-428 62-00079 <*>
1961 V38 N1 P2-14 65-13524 <*>
1961 V38 N2 P60-65 65-13524 <*>
1962 V39 N10 P325-328 C-5792 <NRC>
1963 V40 N12 P401-408 C-5086 <NRC>

BAUZEITUNG
1957 V11 N13 P372-373 PB-141-196T <=*>
1957 V11 N13 P391-392 PB-141-185T <=*>
1957 V11 N14 P406-409 PB-141-187T <=*>
1957 V11 N14 P420-422 PB-141-189T <=*>
1957 V11 N14 P425-426 PB-141-188T <=*>

1957 V11 N15 P430-431 PB-141-190T <=*>
1957 V11 N15 P454-455 PB-141-191T <=*>
1957 V11 N16 P90-91 PB-141-194T <=*>
1957 V11 N16 P475-476 PB-141-192T <=*>
1957 V11 N16 P477-478 PB-141-193T <=*>
1957 V11 N18 P109-110 PB-141-197T <=*>
1957 V11 N18 P520-522 PB-141-186T <=*>
1957 V11 N18 P527 PB-141-187T <=*>
1957 V11 N18 P536-538 PB-141-198T <=*>

BEHRINGWERK-MITTEILUNGEN
1957 V33 P11-38 61-00597 <*>

BEIHEFTE ZUM BOTANISCHEN ZENTRALBLATT
1935 SA V53 P591-594 I-817 <*>

BEIHEFTE ZUM GESUNDHEITS-INGENIEUR
1927 V20 ENTIRE ISSUE 62-16103 <*>

BEIHEFTE ZU DEN ZEITSCHRIFTEN DES SCHWEIZERISCHEN FORSTVEREINS
1960 N30 P13-20 63-18525 <*> 0
1960 N30 P21-29 63-16782 <*> 0

BEITRAEGE ZUR AGRARWISSENSCHAFT
1948 N3 P24-31 61-10555 <*>

BEITRAEGE ZUR BIOLOGIE DER PFLANZEN
1957 V33 N3 P437-458 NP-TR-860 <*>

BEITRAEGE ZUR EXPERIMENTELLEN PSYCHOLOGIE
1899 N2 P182-234 61-14863 <*>

BEITRAEGE ZUR GEOLOGIE DER SCHWEIZ. GEOTECHNISCHE SERIE
1939 V3 ENTIRE ISSUE II-153 <*>
1948 N6 P1-113 1578 <*>

BEITRAEGE ZUR GEOLOGISCHEN GARTE DER SCHWEIZ
1952 N8 P1-36 59-18580 <=*>

BEITRAEGE ZUR GEOPHYSIK
1900 V5 P98-104 58-1931 <*>
1930 V27 P217-225 II-203 <*>
1931 V31 P173-216 58-2479 <*>
1932 V35 P46-50 II-202 <*>
1932 V35 P295-340 65-00146 <*>
1937 V51 P146-166 59-00763 <*>
1958 V68 N1 P21-30 17L34G <ATS>
1960 V69 N4 P206-239 62-20278 <*>

BEITRAEGE ZUR KLINIK DER TUBERKULOSE UND ZUR SPEZIFISCHEN TUBERKULOSEFORSCHUNG
1926 V63 N6 P895-899 2627 <*>
1934 V84 P508-558 2608 <*>
1938 V92 P275-370 2556 <*>
1940 V95 P112-119 64-18016 <*>
1948 V101 P365-394 64-18017 <*> OP
1952 V107 P325-337 AL-199 <*>
1952 V107 N4 P325-337 2615 <*>
1957 V117 N2 P259-264 59-15431 <*>

BEITRAEGE ZUR KLINISCHEN CHIRURGIE
1901 V31 P99-116 1698 <*>
1953 V187 N2 P191-194 64-18015 <*> 0

BEITRAEGE ZUR MEDIZINISCHEN STATISTIK UND STAATSARZNEIKUNDE
1835 V2 P3-20 65-00111 <*>
1835 V2 P43-55 65-00111 <*>
1835 V2 P65-70 65-00111 <*>
1835 V2 P156-170 65-00111 <*>

BEITRAEGE ZUR MINERALOGIE UND PETROGRAPHIE
1962 V8 N5 P323-338 64-16404 <*>
1963 V9 N2 P95-110 64-16424 <*>
1965 V11 P586-613 66-14161 <*>
1965 V11 N6 P614-620 66-14127 <*>

BEITRAEGE ZUR PATHOLOGISCHEN ANATOMIE UND ZUR ALLGEMEINEN PATHOLOGIE
1903 V33 P480-584 57-1544 <*>
1908 V44 P88-149 62-00170 <*>
1912 V52 P406-439 63-01758 <*>
1921 V68 P161-169 65-13867 <*> 0
1921 V68 P213-233 57-2606 <*>
1933 V91 P515-553 1186 <*>
1938 V101 P253-267 66-10929 <*>
1939 V102 P36-68 61-00352 <*>
1954 V114 N1 P48-64 1094 <*>
1955 V115 P1-6 59-14626 <=*> 0
1955 V115 P13-31 59-14626 <=*> 0
1956 V116 N2 P177-199 3041 <*>
1957 V117 N1 P17-31 59-19185 <=*> 0
1957 V118 N1 P1-23 59-12132 <=*> 0
1957 V118 N2 P228-240 62-01228 <*>
1959 V121 N3 P442-469 62-01052 <*>
1960 V122 N2 P313-344 62-01215 <*>
1961 V125 P148-172 62-01232 <*>
1962 V127 N3 P450-473 64-00270 <*>

BEITRAEGE ZUR PHYSIK DER ATMOSPHAERE
1956 V29 N4 P90-96 59-12147 <=*>
1956 V29 N4 P97-99 59-12147 <=*>
1957 V29 N3 P219-233 62-01384 <*>
1958 V30 P177-188 58-1959 <*>
58-2679 <*>
1958 V30 P189-199 58-2678 <*>
1958 V30 P200-206 58-2665 <*>
1958 V30 P207-210 58-2677 <*>
1958 V30 P211-214 58-2675 <*>
1959 V32 P43-52 63-01541 <*>
1959 V32 N1/2 P65-77 61-00306 <*>
1959 V32 N1/2 P78-83 61-00308 <*>
1961 V33 N3/4 P225-231 64-00039 <*>
1961 V34 P259-273 62-01406 <*>

BEITRAEGE ZUR PHYSIK DER FREIEN ATMOSPHAERE
1925 V12 P171-181 C-2377 <NRC>
C2377 <NRC>
1934 V21 N2 P129-142 CSIRO-3353 <CSIR>
1941 V27 P49-61 57-624 <*>

BEITRAEGE AUS DER PLASMAPHYSIK
1960 V1 P179-201 4553 <BISI>
1964 V4 N1 P33-39 AD-636 656 <=$>

BEITRAEGE ZUR SILIKOSE-FORSCHUNG
1958 P619-643 SP.NC. 62-00871 <*>
1960 V4 P231-247 63-00208 <*>
1960 V4 P249-263 63-01757 <*>
1960 V4 P265-287 63-00279 <*>

BEITRAEGE ZUR TABAKFORSCHUNG
1961 N1 P15-18 63-16663 <*>
1961 N1 P19-29 63-16621 <*>
1961 N2 P35-38 11033-C <K-H>
1961 N3 P75-81 11401D <K-H>
63-16910 <*>
1961 N3 P83-91 11411(5) <K-H>
1961 N3 P93-96 11411-1 <K-H>
63-16620 <*>
1961 N3 P97-100 11411-3 <K-H>
63-16660 <*>
63-18425 <*>
1961 N3 P101-106 11411-4 <K-H>
63-16911 <*>
1961 N3 P107-116 11411-2 <K-H>
63-16658 <*>
1961 N4 P125-131 64-10628 <*>
1961 N4 P144 63-16665 <*>
1962 N4 P171-175 11816D <K-H>
1962 N5 P155-163 11816-B <K-H>
1962 N5 P155-164 64-14534 <*> 0
1962 N5 P165-169 11816-C <K-H>
64-14533 <*>
1962 N5 P171-175 63-16955 <*> 0
1962 N5 P177-180 11816A <K-H>
63-16957 <*> 0
1962 N5 P180-185 63-16655 <*> 0

1962 N5 P187-191 11816-G <K-H>
1962 N5 P193-194 11816-H <K-H>
1962 N5 P196-198 11816-I <K-H>
1962 N7 P275-284 12359 A <K-H>
1962 N7 P285-290 12539B <K-H>
63-14394 <*> O
1962 N7 P291-297 12539C <K-H>
63-14393 <*>
1962 N7 P298-304 12539 D <K-H>
1962 N7 P305-306 12539 E <K-H>
1962 N8 P307-314 12539 F <K-H>
1965 V3 N2 P109-127 65-13976 <*>
860 <TC>
1965 V3 N2 P151-156 1445 <TC>
1965 V3 N4 P243-250 1907 <TC>

BEITRAEGE ZUR WELTRAUMFORSCHUNG UND ASTRONAUTIK
1952 N1 P1 3128 <*>

BEIZTECHNIK
1955 P36-39 2701 <BISI>
1957 V6 N10 P109-113 58-1417 <*>

BELGISCH TIJDSCHRIFT VOOR GENEESKUNDE
1953 V9 P9-16 2854B <K-H>
1953 V9 P145-152 2840D <K-H>
1957 V13 N26 P1335-1339 65-60305 <=$>
1961 V17 P881-898 63-20225 <*> O
1961 V17 N10 P524-530 AEC-TR-4956 <*>
1962 V18 P434-437 66-11826 <*>
1962 V18 P438-442 66-11825 <*>
1963 V19 P638-643 64-00049 <*>
64-26244 <$>
1964 V20 P804-811 66-11628 <*> O

BERGAKADEMIE
1953 N4 P121-130 63-16263 <*> O
1953 V5 P271-274 T1814 <INSD>
1958 V10 P19-25 58-1320 <*>
59-22339 <=*> O
1958 V10 N1 P40-42 59-11117 <=>
1959 V11 N5 P297-309 63-16266 <*>
1959 V11 N12 P699-703 15P63G <ATS>
1960 V12 N7 P391-392 07N49G <ATS>
1960 V12 N8 P440-453 63-16272 <*> O
1960 V12 N10 P559-567 2165 <BISI>

BERGBAU. GELSENKIRCHEN
1937 P362-367 57-2310 <*>

BERGBAU UND ENERGIEWIRTSCHAFT
1949 V2 N7/8 P243-250 1881 <*>
1950 V3 N1 P70-73 CSIRO-3528 <CSIR>

BERGBAU-ARCHIV
1946 V2 P5-57 66-10231 <*> O

BERGBAUTECHNIK
1954 V4 N3 P132-139 63-16264 <*> O
1954 V4 N4 P230-232 61-20961 <*> O
1955 V5 N4 P201-209 4404 <HB>
1957 V7 P90-92 628 <TC>
1961 V11 N5 P260-265 62-20163 <*>
1965 V15 N10 P506-510 65-33876 <=$>
1965 V15 N11 P571-576 66-30184 <=$>
1966 V16 N12 P617-621 67-30429 <=>
1966 V16 N12 P635-637 67-30429 <=>

BERGBAUWISSENSCHAFTEN
1958 V5 N3 P70-77 63-16146 <*> O
1961 V8 N19 P461-468 63-16262 <*> O
1962 V9 N15/6 P341-351 63-16269 <*> O
1962 V9 N15/6 P356-361 63-16270 <*> O

BERGCULTURES
1930 N9 P223-228 3346 <*>
1930 V4 N9 P218- 57-2837 <*>
1941 V15 N32 P1091-1094 3266 <*>

BERGENS MUSEUMS ARBOG
1944 N5 P1-20 63-14210 <*> O

BERG- UND HUETTENMAENNISCHE MONATSHEFTE
1939 V87 P142-148 57-2211 <*>
1947 V92 N6 P106-109 57-3057 <*>
1947 V92 N10 P166-178 62-18405 <*> O
1948 V93 N6 P87-88 63-16744 <*>
1948 V93 N12 P248-254 61-20205 <*> O
1949 V94 N1 P10-17 61-14762 <*> O
1949 V94 N2 P34-37 63-10039 <*>
1950 V95 N1 P14-16 64-20065 <*> O
1950 V95 N8 P150-156 57-2910 <*>
1950 V95 N11 P217-222 57-2910 <*>
1950 V95 N12 P368-370 57-2645 <*> P
1952 V97 P81-91 60-18095 <*>
1953 V98 P168-174 57-1894 <*>
1955 V100 N5 P166-170 3701 <HB>
1955 V100 N7/8 P230-238 61-16207 <*>
1955 V100 N7/8 P238-244 61-16395 <*>
1956 V101 N12 P277-285 58-1385 <*>
1956 V101 N12 P292-300 61-16294 <*>
1956 V101 N12 P353-359 57-3454 <*>
1957 V102 P115-125 TS-918 <BISI>
1958 V103 P114-118 TRANS-138 <FRI>
61-10149 <*>
1958 V103 P163-175 TS 1470 <BISI>
1958 V103 N9 P163-175 62-14368 <*>
1959 V104 N2 P18-21 3736 <BISI>
1959 V104 N2 P21-26 4644 <HB>
1959 V104 N2 P26-31 4718 <HB>
1959 V104 N2 P31-40 4645 <HB>
1959 V104 N2 P41-50 4717 <HB>
1959 V104 N2 P50-55 4646 <HB>
61-20949 <*> O
1959 V104 N10/1 P222-224 4772 <HB>
1959 V104 N10/1 P225-227 4729 <HB>
1960 V105 N6 P133-135 4892 <HB>
1960 V105 N6 P135-140 4863 <HB>
1960 V105 N6 P140-142 4894 <HB>
1960 V105 N6 P144-149 4895 <HB>
1960 V105 N6 P149-150 5026 <HB>
1960 V105 N6 P151-157 4896 <HB>
1960 V105 N6 P157-161 4897 <HB>
1960 V105 N11 P261-268 5050 <HB>
1960 V105 N11 P268-278 5185 <HB>
1960 V105 N11 P302-313 5051 <HB>
1961 V106 P273-280 2572 <BISI>
1961 V106 P286-299 2624 <BISI>
1961 V106 P307-310 2625 <BISI>
1961 V106 P311-315 2626 <BISI>
1961 V106 P315-317 2627 <BISI>
1961 V106 P318-322 2628 <BISI>
1961 V106 P322-327 2629 <BISI>
1961 V106 N8 P248-251 65-17368 <*> O
1962 V107 P73-76 2936 <BISI>
1962 V107 P96-106 2938 <BISI>
1962 V107 P127-133 2937 <BISI>
1962 V107 P134-144 2939 <BISI>
1962 V107 N1 P1-12 2899 <BISI>
1962 V107 N4 P76-82 5640 <HB>
1962 V107 N4 P82-87 5697 <HB>
1962 V107 N4 P87-96 5641 <HB>
1962 V107 N4 P107-118 5642 <HB>
1962 V107 N4 P118-127 2921 <BISI>
1962 V107 N4 P152-157 5643 <HB>
1962 V107 N4 P157-162 5644 <HB>
1962 V107 N4 P162-168 5706 <HB>
1962 V107 N4 P168-173 5645 <HB>
1962 V107 N4 P173-182 2923 <BISI>
1962 V107 N9 P313-323 3817 <BISI>
1962 V107 N10 P335-341 3329 <BISI>
1962 V107 N12 P377-382 5809 <HB>
1963 V108 N1 P1-8 5895 <HB>
1963 V108 N1 P8-14 5894 <HB>
1963 V108 N5 P215-222 3899 <BISI>
1963 V108 N5 P222-228 3900 <BISI>
1963 V108 N6 P237-240 634 <TC>
1963 V108 N11 P380-391 3651 <BISI>
1963 V108 N11 P391-401 6235 <HB>

1964 V109 N1 P1-13 3663 <BISI>
1964 V109 N1 P13-25 3664 <BISI>
1964 V109 N3 P73-76 3892 <BISI>
1964 V109 N3 P95-97 4035 <BISI>
1964 V109 N3 P105-107 3893 <BISI>
1964 V109 N3 P114-119 3907 <BISI>
1964 V109 N7 P237-242 6352 <HB>
1965 V110 N3 P66-71 6526 <HB>
1965 V110 N9 P291-304 4867 <BISI>

BERG- UND HUETTENMAENNISCHE ZEITUNG
1897 P204-207 58-1971 <*>

BERG- UND HUETTENMAENNISCHES JAHRBUCH DER K.K. MONTANISTISCHEN HOCHSCHULEN ZU LEOBEN UND PRIBRAM
1883 P250-252 60-10205 <*>
1927 V75 N2 P69-81 66-10181 <*>
1933 V81 N4 P135-139 57-598 <*>

BERICHTE. AKADEMIE DER WISSENSCHAFTEN
1920 V19 P380-385 60-10796 <*>

BERICHTE AUS DER BAUFORSCHUNG
1962 V26 P9-15 64-00252 <*>
1962 V26 P31-39 64-00254 <*>
1962 V26 P55-65 64-00326 <*>

BERICHT DES DEUTSCHEN AUSSCHUSSES FUER STAHLBAU
1954 V18 ENTIRE ISSUE 61-16206 <*>

BERICHT DER DEUTSCHEN BOTANISCHEN GESELLSCHAFT
1929 V47 N5 P313-320 63-10128 <*>
1937 V55 N9 P514-529 64-16015 <*>
1940 V57 P506-515 1135B <K-H>
1950 V63 P35-36 3166B <K-H>
1952 V65 P239-245 AEC-TR-5571 <*>
1956 V69 N4 P177-188 58-64 <*>
1956 V69 N9 P429-434 65-12106 <*>
1957 V70 P11-20 65-12652 <*>
1957 V70 P227-232 65-11063 <*> 0
1957 V70 N2 P51-56 59-18390 <+*>
1960 V73 N8 P349-357 61-01038 <*>
1961 V74 N9 P436-440 65-12127 <*>
1962 V75 N11 P90-95 65-11075 <*> 0
1963 V76 N8 P329 AEC-TR-6362 <*>

BERICHT DER DEUTSCHEN BUNSENGESELLSCHAFT FUER PHYSIKALISCHE CHEMIE
1963 V67 P164-167 63-18937 <*>
1963 V67 P486-493 63-20722 <*>
1963 V67 N1 P62-67 445 <TC>
1963 V67 N2 P142-151 63-18759 <*>
65-12310 <*>
1963 V67 N2 P151-156 63-18760 <*>
65-12309 <*>
1963 V67 N2 P156-164 63-18758 <*>
65-12308 <*>
1963 V67 N2 P164-167 66-14877 <*>
1963 V67 N2 P229-235 64-16691 <*>
1963 V67 N3 P267-280 64-10366 <*> 0
1963 V67 N5 P486-493 65-12276 <*>
1963 V67 N6 P593-601 64-30699 <*>
1963 V67 N7 P639-645 64-16190 <*>
1963 V67 N7 P657-671 64-16191 <*>
1963 V67 N7 P698-703 66-12993 <*>
1963 V67 N8 P791-795 64-18541 <*>
1963 V67 N9/10 P958-964 3735 <BISI>
1964 V68 P601-608 535 <TC>
1964 V68 P964-972 739 <TC>
1964 V68 N1 P64-70 164 <TC>
65-17080 <*> 0
1964 V68 N2 P163-169 38R76G <ATS>
1964 V68 N7 P646-652 1060 <TC>
1965 V69 N2 P160-167 66-14601 <*>
814 <TC>
1965 V69 N6 P503-513 66-13396 <*>
1965 V69 N7 P566-576 66-10059 <*>

BERICHTE DER DEUTSCHEN CHEMISCHEN GESELLSCHAFT
1870 V3 P261-262 66-14278 <*>
1870 V3 P612-613 66-14279 <*>
1871 V4 P162-164 66-14280 <*>
1871 V4 P267-268 64-20898 <*>
1872 V5 P409-410 63-14211 <*> 0
1874 V7 P1180-1181 64-18018 <*>
1874 V7 P1223-1228 SCL-T-237 <+*>
1875 V8 P143-147 63-14083 <*> 0
1875 V8 P1426-1427 66-14283 <*>
1875 V8 P1428-1429 66-14284 <*>
1876 V9 P824-828 66-14285 <*>
1876 V9 P1749-1752 AL-304 <*>
1877 V10 P542-544 62-20464 <*>
1877 V10 P1365-1375 T-1769 <INSD>
1878 V11 P374-381 66-14286 <*>
1879 V12 P417-423 66-14287 <*>
1879 V12 P426-428 58-115 <*>
1879 V12 P503-506 66-14288 <*>
1880 V13 P58-61 58-2263 <*>
63-18176 <*>
1881 V14 P420-428 64-14666 <*> P
1881 V14 P1545-1552 60-10561 <*>
1882 V15 P180 58-2264 <*>
1883 V16 P558-560 60-10543 <*>
1883 V16 P1536-1544 60-10544 <*> P
1883 V16 P1655-1659 8944 <IICH>
1883 V16 P2284 63-10425 <*>
1884 V17 P412-415 66-14289 <*>
1884 V17 P936-938 98M41G <ATS>
1884 V17 P2681-2699 T2151 <INSD>
1885 V17 N2 P3316-3319 NP-TR-942 <*>
1885 V18 P335- 1182 <*>
1885 V18 P335-337 63-18241 <*> P
1885 V18 P519-520 66-14290 <*>
1885 V18 P591-601 64-18019 <*>
1885 V18 P1760-1762 63-14212 <*> 0
1885 V18 P2755-2781 63-10301 <*>
1885 V18 P3436-3441 64-18021 <*>
1886 V19 P2482-2493 57-2601 <*>
1887 V20 P660-664 64-20508 <*>
1887 V20 P2027-2031 8630 <IICH>
1887 V20 P2251-2259 57-30 <*>
1888 V21 P1717-1726 64-18022 <*>
1888 V21 P1772-1777 64-18023 <*>
1888 V21 P3347-3355 65-10157 <*>
1888 V21 P3499-3501 63-10442 <*> P
1889 V22 P467-470 64-18511 <*>
1889 V22 P3086-3096 60-10563 <*>
1890 V23 P1265-1312 61-16452 <*> P
1890 V23 P2084-2110 63-14085 <*>
1890 V23 P2318-2321 63-14213 <*>
1890 V23 P2657-2664 I-843 <*>
1890 V23 P3023-3033 64-15685 <=*$>
1890 V23 P3169-3174 66-14291 <*>
1890 V23 P3269-3276 66-14292 <*>
1890 V23 P3734-3735 64-18922 <*>
1891 V24 P768-772 66-14293 <*>
1891 V24 P3889-3894 66-14294 <*> 0
1892 V25 P3602-3604 66-14295 <*>
1893 V26 P1545-1558 PT 2 64-18898 <*> P
1893 V26 P2197-2209 PT2 64-18852 <*> OP
1893 V26 P2756-2759 64-10063 <*>
1893 V26 P2940-2945 AEC-TR-6105 <*>
1894 V27 P55-58 64-20895 <*> 0
1894 V27 P244-249 58-2265 <*>
1894 V27 P244-262 63-10573 <*>
1894 V27 P2507-2520 64-18913 <*> P
1894 V27 P3235-3238 66-14297 <*>
1895 V28 P1135-1140 66-14298 <*>
1895 V28 P1323-1327 58-2253 <*>
1895 V28 N1 P890-895 60-13282 <=*>
1895 V28 N2 P1633-1637 60-13283 <+*>
1896 V29 P1136-1139 60-10562 <*>
1896 V29 P1474-1477 60-10546 <*>
1897 V30 P2369-2381 PT2 8829 <IICH>
1897 V30 P586-593 64-15681 <=*$>
1897 V30 P1860-1862 60-10547 <*>
1898 V31 P18-20 66-14299 <*>
1898 V31 P347-348 66-14300 <*>
1898 V31 P502 1975 <*>
1898 V31 P2103-2105 64-16124 <*> 0

Citation	Number
1899 V32 P1597-1598	AEC-TR-4006 <*>
1900 V33 P2226-2232 PT2	1412 <*>
1900 V33 P815-816	8951 <IICH>
1901 V34 P68-71	AL-750 <*>
1901 V34 P2604-2607	63-10590 <*> O
1902 V35 P151-157	64-18911 <*>
1902 V35 P4C0-401,404	66-14301 <*> O
1902 V35 P534-535	62-18252 <*>
1902 V35 P535-539	2348 <*>
	62-18266 <*>
1902 V35 P817-830	2854 <*>
1902 V35 P3463-3470	58-599 <*>
1902 V35 P3470-3476	58-268 <*>
1902 V35 P3480-3485	58-870 <*>
1902 V35 P4073-4079	64-18854 <*> P
1902 V35 P4153-4157	66-14302 <*>
1903 V36 P2219-2225	66-14303 <*>
1903 V36 P3902-3911	64-14749 <*>
1903 V36 N3 P3213-3221	59-20982 <*>
1904 V37 P100-123	60-14017 <*>
1904 V37 P4021-4022	66-14305 <*>
1905 V38 P266-270	64-20897 <*>
1905 V38 P1130-1137	51K21G <ATS>
1906 V39 P161-162	66-14306 <*>
1906 V39 P794-802	66-14307 <*>
1906 V39 P2333-2334	64-10061 <*>
1906 V39 P4428-4436	66-14308 <*>
1907 V40 P641-648	63-14380 <*> P
1907 V40 P997-999	58-262 <*>
1907 V40 P1176-1193	64-15684 <=*$>
1907 V40 P1829-1830	66-14309 <*>
1908 V41 P2C95-2099	2338 <*>
1908 V41 P2250-2264	64-18842 <*>
1908 V41 P2634-2645	II-166 <*>
19C8 V41 P3095-3099	58-2258 <*>
1909 V42 P495	63-10439 <*>
1909 V42 P900	64-18844 <*> O
1909 V42 P1198-1203	62-01206 <*>
1909 V42 P1299-1302	59-10690 <*>
1909 V42 P1701-1707	58-244 <*>
1909 V42 P1723-1725	58-76 <*>
1909 V42 P3090-3096	58-2270 <*>
1909 V42 P3873-3878	59-10687 <*>
1909 V42 P3912-3925	64-18921 <*> P
1909 V42 P4728-4747	66-14310 <*>
1909 V42 N4 P4728-4747	AL-929 <*>
1909 V42 N8 P4581-4586	64-14249 <*> O
1910 V43 P297-306	II-167 <*>
1910 V43 P1528-1532	66-14311 <*>
1910 V43 P1957-1962	64-18896 <*> P
1910 V43 P2574-2581	58-276 <*>
1910 V43 P3384	63-10689 <*>
1911 V43 P763-772	I-739 <*>
1911 V44 P1164	66-14312 <*>
1911 V44 P1608-1619	62-10554 <*> O
1911 V44 P2504-2522	71J15G <ATS>
1911 V44 P3141-3145	I-739 <*>
1911 V44 P3333-3336	I-134 <*>
1911 V44 P3473-3480	64-20894 <*>
1912 V45 P231-247	63-14388 <*> O
	63-18230 <*>
1912 V45 P411-	1180 <*>
1912 V45 P5C1-509	63-10604 <*>
1912 V45 P921-928	57-2792 <*>
1913 V46 P103-110	66-11371 <*>
1913 V46 P342-347	1657 <*>
1913 V46 P2117-2131	62-16301 <*>
1913 V46 P2588-2590	57-2641 <*>
1913 V46 P3937-3946	58-241 <*>
1914 V47 P646-656	66-14313 <*>
1914 V47 P656-660	66-14314 <*>
1914 V47 P961-965	63-10600 <*>
1914 V47 P3156-3159	66-14315 <*> O
1915 V48 P1183-1195	58-242 <*>
1915 V48 P1847-1865	64-20293 <*>
1916 V49 P1C41-1C44	65-13436 <*>
1916 V49 P1415-1428	64-14933 <*>
1917 V50 P586-596	AD-479 637 <=$>
1917 V50 P1808-1813	57-2689 <*>
	64-14715 <*>
1918 V51 P669-672	66-14316 <*>
1919 V52 P656-665	AD-623 635 <=*$>
1919 V52 P753-761	66-14317 <*> O
1919 V52 P1272-1284	66-14318 <*> O
1919 V52 P1641-1652	63-14215 <*> P
1920 V53 P2096-2113	60-10544 <*>
1921 V54 P406-413	66-14319 <*>
1921 V54 P550-553	66-14320 <*> O
1921 V54 P560-566	2347 <*>
	63-18174 <*>
1921 V54 P1393-1396	66-14321 <*>
1921 V54 P1626-1644	63-10631 <*>
1921 V54 P1979-1987	66-14322 <*>
1921 V54 N2 P179-186	AD-636 934 <=$>
1921 V54B P560-566	58-2278 <*>
1922 V55 P457-463	63-14723 <*>
1922 V55 P1261-1265	58-2266 <*>
1922 V55 P1529-1534	66-14323 <*> O
1922 V55 P3710-3726	66-14324 <*> O
1922 V55B N4 P934-935	1223 <*>
1923 V56 P937-947	63-14217 <*> O
1923 V56 P2076-2082	58-2482 <*>
1923 V56B P1262-1269	2232 <*>
1924 V57 P32-42	58-2536 <*>
1924 V57 P95-99	12190-B <KH>
	65-14467 <*>
1924 V57 P851-853	1155 <*>
1924 V57 P1023-1038	63-20670 <*>
1925 V58 P1-3	63-10686 <*>
1925 V58 P459-463	64-16876 <*>
1925 V58 P464-467	64-16877 <*> O
1925 V58 P479-481	78J16G <ATS>
1925 V58 P566-571	63-20077 <*>
1925 V58 P864-869	1674 <*>
1925 V58 P941-961	63-10586 <*>
1925 V58 P953-955	I-854 <*>
1925 V58 P1320-1323	63-18233 <*>
1925 V58 N4 P732-736	59-15480 <*>
1925 V59 P103-104	1654 <*>
1925 V58B P1320-1323	3243 <*>
1925 V58B N11 P1601-	1271 <*>
1926 V59 P458-461	64-20893 <*>
1926 V59 P565-589	64-15686 <=*$>
1926 V59 P625-630	64-20504 <*>
1926 V59 P777-783	63-16027 <*> O
1926 V59 P777-785	64-18153 <*>
1926 V59 P859-865	64-18437 <*>
1926 V59 P883-886	65-14471 <*>
1926 V59 P1166-1171	64-00050 <*> P
	66-14325 <*> P
	66-15363 <*>
1926 V59 P1412-1426	3284 <*>
	3308 <*>
1926 V59 P2025-2028	64-18440 <*>
1926 V59 P2433-2444	66-14326 <*> O
1926 V59 P2778-2784	63-14381 <*> OP
1926 V59B P1043-1048	59-15430 <*>
1927 V60 P50-57	T-1485 <INSD>
1927 V60 P1186-1190	63-18985 <*>
1927 V60 P1658-1663	59-10691 <*>
1927 V60 P1963-1971	63-10690 <*>
1927 V60 P2005-2018	99K22G <ATS>
1927 V60 P2335-2341	AL-762 <*>
1927 V60 N1 P225-	AL-235 <*>
1928 V61 P2521-2525 PT2	I-380 <*>
1928 V61 P253-263	64-14902 <*>
1928 V61 P558-565	66-14327 <*>
1928 V61 P799-801	63-16836 <*>
1928 V61 P1057-1060	64-18917 <*> P
1928 V61 P1328-1334	63-10591 <*>
1928 V61 P2097-2101	66-14328 <*>
1928 V61 P2142-2148	63-14216 <*>
1928 V61 P2451-2459	63-14425 <*> O
1929 V62 P289-316	58-2233 <*>
1929 V62 P592	66-14329 <*>
1929 V62 P628-634	II-45 <*>
1929 V62 P658-677	63-10417 <*>
1929 V62 P1554-1561	66-14330 <*> O
1929 V62 P1954-1959	64-10131 <*> O

Citation	Translation
1929 V62 P2102-2106	63-14218 <*> P
1929 V62 P2395-2405	64-10059 <*>
1929 V62 P2612-2620	1851 <*>
	3411-C <KH>
	64-14686 <*>
1929 V62 P2742-2758	AD-633 399 <=$>
1929 V62 P2844-2850	66-14611 <*> O
1929 V62B P90-99	58-2256 <*>
1929 V62B P640-641	64-20896 <*>
1929 V62B P1059-1065	65-13816 <*>
1929 V62B P1478-1482	61-20984 <*> O
1930 V63 P335-342	64-14927 <*> O
1930 V63 P1019-1024	I-24 <*>
1930 V63 P1869-1871	2083 <*>
1930 V63 P1941-1944	64-10070 <*>
1930 V63 N6 P1446-1455	64-18839 <*> P
1930 V63B P99-102	AEC-TR-5021 <*>
1931 V64 P1049-1056	62-01494 <*>
1931 V64 P2584-2590	66-14331 <*>
1931 V64 P2739-2748	80K20G <ATS>
1931 V64B P667-678	I-233 <*>
1932 V65 P187-190	64-18024 <*>
1932 V65 P315-320	1663 <*>
1932 V65 P463-467	65-13817 <*> O
1932 V65 P1253-1257	66-14332 <*>
1932 V65B P432-445	8669 <IICH>
1933 V66 P19-27	63-14379 <*>
1933 V66 P484-485	63-10464 <*>
1933 V66 P1048-1061	2865 <*>
1933 V66 P1545-1556	64-18025 <*>
1933 V66 P1892-1900	T-1288 <INSD>
	63-10454 <*>
1933 V66 N2 P1274-1280	63-00126 <*>
1934 V67 P563-573	AL-10 <*>
1934 V67 P1164-1172	66-14333 <*>
1934 V67 P1942-1946	AL-12 <*>
1934 V67B P1696-1712	60-10567 <*> P
1934 V67B P2027-2031	60-10822 <*>
1935 V68 P343-349	63-10610 <*>
1935 V68 P1210-1216	66-14335 <*>
1935 V68 P1435-1438	AL-11 <*>
1935 V68 P1667-1670	3366-B <KH>
1935 V68 P2070-2083	63-14219 <*>
1935 V68 N3 P455-471	64-18897 <*> P
1935 V68 N5 P769-775	59-15794 <*>
1936 V69 P32-40	1312 <TC>
1936 V69 P549-552	63-14220 <*>
1936 V69 P553-559	63-14221 <*>
1936 V69 P1066-1074	65-12250 <*>
1936 V69 P1456-1469	57-1712 <*>
1936 V69 P2207-2210	64-10052 <*> O
1936 V69 P2244-2251	79K20G <ATS>
1936 V69 N1 P579-585	59-15079 <*>
1936 V69 N7/8 P1713-1721	66-14336 <*> O
1936 V69B P2100-2102	1306 <*>
1936 V69B P2112-2123	64-16558 <*>
1937 V70 P979-993	65-10493 <*>
1937 V70 P1393-1402	63-20301 <*> O
1937 V70 N2 P218-223	58-2453 <*>
1937 V70 N2 P223-227	58-2295 <*>
1937 V70 N11 P2318-2330	59-10061 <*> O
1938 V71 P1267-1272	65-12251 <*>
1938 V71 P2015-2021	AL-53 <*>
1938 V71 P2451-2461	64-18020 <*> P
1938 V71 P2627-2636	64-10051 <*>
1938 V71 N7 P1489-1492	1898 <*>
1938 V71 N7 P1492-1497	64-18026 <*> P
1939 V72 P327-330	64-30657 <*>
1939 V72B P1346-1353	60-10534 <*>
1939 V72B N6 P1753-1762	59-15363 <*> O
1940 V73 P167-171	63-14198 <*> O
1940 V73 P1080-1091	64-18027 <*> OP
1940 V73 P1092-1094	64-18028 <*> P
1940 V73 P1095-1100	64-18029 <*> OP
1940 V73 P1100-1105	64-18030 <*> P
1940 V73 P1105-1109	64-18031 <*> P
1940 V73 P1109-1113	64-18032 <*> P
1940 V73 P1278-1283	AEC-TR-6262 <*>
1940 V73 N6 P701-708	61-00573 <*>
1940 V73 N7 P754-757	57-3543 <*>
1940 V73 N9 P941-949	64-18846 <*>
1940 V73B P391-404	8807 <IICH>
1941 V74 P941-948	64-18034 <*> OP
1941 V74 P949-952	64-18035 <*>
1941 V74 P1285-1296	58-603 <*>
1941 V74 P1456-1459	63-14222 <*>
1941 V74 P1700-1701	64-18919 <*>
1941 V74 P1818-1824	64-18036 <*>
1941 V74 P1825-1829	1358 <*>
1941 V74 N1 P73-78	65-17258 <*>
1941 V74 N2 P239-241	1345 <*>
1941 V74 N3 P387-397	6234-A <KH>
1941 V74 N4 P636-647	64-18033 <*> P
1941 V74 N5 P756-758	63-00608 <*>
1941 V74B P163-170	1128 <*>
1941 V74B P1700-1701	6192 <*>
	66-15239 <*>
1941 V74B N6 P1232-1236	59-15364 <*>
1942 V75 P369-378	64-18037 <*>
1942 V75 P547-554	63-14223 <*>
1942 V75 P660-663	67N49G <ATS>
1942 V75 P1055-1061	61-10578 <*>
1942 V75 P1489-1491	57-53 <*>
1942 V75 P1610-1622	AL-475 <*>
1942 V75 P1636-1643	738 <TC>
1942 V75 N3 P215-226	30F3G <ATS>
1942 V75 N5 P505-509	57-2510 <*>
1942 V75 N6 P656-660	62-00528 <*>
	66N49G <ATS>
1942 V75 N7 P909-920	11999-B <KH>
1942 V75 N12 P1517-1522	64-10050 <*>
1943 V76 P80-89	64-10069 <*>
1943 V76 P373-386	63-10626 <*> O
1943 V76 P466-479	66-13633 <*>
1943 V76 P479-483	59-12185 <+*>
1943 V76 P1196-1208	62-18701 <*>
1943 V76 N1/2 P121-127	31F3G <ATS>
1944 V77 P257-264	59-12935 <+*>
1944 V77B P484	AEC-TR-3964 <*>

BERICHTE DER DEUTSCHEN GESELLSCHAFT FUER HOLZFORSCHUNG

Citation	Translation
1958 N3 P41-50	60-18478 <*>

BERICHT DER DEUTSCHEN KERAMISCHEN GESELLSCHAFT

Citation	Translation
1930 V11 P333-363	63-20829 <*>
1933 V14 N7 P259-279	57-3068 <*>
1934 V15 N3 P101-110	AL-401 <*>
1935 V16 P111-117	AEC-TR-5019 <*>
1936 V17 N5 P237-264	60-10287 <*> O
1937 V18 N10 P433-443	64-10489 <*>
1938 V19 N2 P38-54	60-10202 <*> O
1939 V20 N12 P508-522	NP-TR-1004 <*>
1942 V23 P243-260	T-1678 <INSD>
1943 V24 P335-339 PT1	I-78 <*>
1944 V25 P17-27 PT2	I-79 <*>
1944 V25 P95-112	T-1677 <INSD>
1951 V28 P299-302	I-137 <*>
1952 V29 N3 P73-77	59-10974 <*> O
	62-10530 <*>
1953 V30 P31-34	T-1744 <INSD>
1953 V30 P168-169	T-1618 <INSD>
1953 V30 N3 P47-61	22L31G <ATS>
1953 V30 N4 P71-82	23L31G <ATS>
1954 V31 N2 P45-48	2339 <*>
1954 V31 N12 P396-401	57-3003 <*>
1954 V31 N12 P404-409	62-10556 <*> O
1955 V32 P229-250	57-1906 <*>
1955 V32 P257-261	T1680 <INSD>
1955 V32 N4 P114-119	CSIRO-2883 <CSIR>
1955 V32 N9 P257-261	59-15014 <*>
1955 V32 N10 P281-285	CSIRO-2939 <CSIR>
1956 V33 P321-	58-1284 <*>
1956 V33 N10 P321-329	62-14452 <*>
1957 V34 P397-402	58-1282 <*>
1957 V34 N6 P183-189	AEC-TR-3751 <+*>
1957 V34 N11 P353-362	61-14466 <*>
1958 V35 N3 P69-77	59-10629 <*> O
	65-13610 <*>
1958 V35 N4 P108-116	46K28G <ATS>

1958 V35 N6 P187-193 62-10531 <*>
1958 V35 N6 P193-204 NP-TR-837 <*>
1958 V35 N8 P249-258 61-16176 <*> 0
1960 V37 P11-22 TS 1735 <BISI>
1960 V37 P368-371 2029 <BISI>
1960 V37 N1 P11-22 62-16876 <*>
1961 V38 P451-457 2599 <BISI>
1961 V38 N1 P1-8 2110 <BISI>
1961 V38 N6 P258-267 66-11292 <*>
1961 V38 N6 P258-268 97S88G <ATS>
1961 V38 N11 P495-511 64-10960 <*> 0
1962 V39 P131-135 64-10808 <*>
1962 V39 P294-296 64-00246 <*>
1962 V39 P385-386 AEC-TR-6397 <*>
1962 V39 N3 P162-167 65-13517 <*>
1962 V39 N5 P280-285 64-00250 <*>
1962 V39 N7 P376-381 BGIRA-571 <BGIR>
1962 V39 N10 P489-494 65-12878 <*>
1963 V40 N2 P159-172 63-20486 <*> 0
1963 V40 N5 P294-299 BGIRA-583 <BGIR>
1963 V40 N5 P300-326 BGIRA-580 <BGIR>
1963 V40 N5 P304-315 BGIRA-586 <BGIR>
1963 V40 N5 P327-329 BGIRA-582 <BGIR>
1963 V40 N5 P332-336 BGIRA-581 <BGIR>
1963 V40 N5 P337-343 BGIRA-579 <BGIR>
1963 V40 N7 P399-408 64-16036 <*> 0
1963 V40 N8 P451-459 BGIRA-601 <BGIR>
1964 V41 P112-119 66-12343 <*>
1964 V41 P301-309 4532 <BISI>
1964 V41 N2 P98-107 66-12362 <*>
1964 V41 N9 P484-486 BGIRA-639 <BGIR>
1964 V41 N11 P632-638 4771 <BISI>
1965 V42 P357-362 4708 <BISI>
1965 V42 P462-467 66-12193 <*>
1965 V42 N5 P222-232 6583 <HB>
1966 V43 N4 P271-279 4935 <BISI>

BERICHT DER DEUTSCHEN OPHTHALMOLOGISCHEN GESELLSCHAFT
SEE BERICHT UBER DIE VERSAMMLUNG DER DEUTSCHEN OPHTHALMOLOGISCHEN GESELLSCHAFT

BERICHT DER DEUTSCHEN PHYSIKALISCHEN GESELLSCHAFT
1910 V11 P430-433 58-83 <*>
1919 V20 P401-403 57-3521 <*>
1919 V21 P85-99 AEC-TR-3725 <+*>

BERICHT DER DEUTSCHEN VERSUCHSANSTALT FUER LUFT -UND RAUMFAHRT. KOELN
1956 N7 P104 59-10286 <*>
1960 N108 P5-115 62-10479 <*> 0P
1960 N130 P1-60 C-3997 <NRCC>
1962 N190 ENTIRE ISSUE AD-608 096 <=>
1964 N268 P1-21 C-5134 <NRC>

BERICHT DES DEUTSCHEN WETTERDIENSTES. BAD KISSINGEN
1954 V2 N13 P1-47 59-11979 <=>
1955 V3 P2-12 64-00021 <*>
1955 V3 N18 P1-45 CSIRO-2912 <CSIR>
1955 V3 N19 P2-12 64-00021 <*>
1956 V4 N22 P43-54 57-2870 <*>
1956 V4 N22 P134-139 59-00657 <*>
59-18999 <=*>
1956 V5 P4-53 CSIRO-3760 <CSIR>
1957 V5 P1-13 SCL-T-281 <*>
1965 V14 N100 P3-23 66-12694 <*>

BERICHT DES DEUTSCHEN WETTERDIENSTES IN DER U.S. ZONE
1950 V2 P200-201 2465 <*>
1952 N38 P127-136 57-3314 <*>
58-2449 <*>
1952 V35 P277-284 CSIRO-2910 <CSIR>
1952 V38 P127-136 59-12026 <=*>
1952 V38 P298-302 1484 <*>
1952 V42 P195-199 57-1177 <*>

BERICHT DER EIDGENOESSISCHEN MATERIALPRUEFUNGS- UND VERSUCHSANSTALT FUER INDUSTRIE BAUWESEN UND GEWERBE
1938 N114 P1-22 64-14205 <*> 0

BERICHT DES FORSCHUNGSINSTITUTES FUER KRAFTFAHR- WESEN UND FAHRZEUGMOTOREN, TECHNISCHE HOCHSCHULE. STUTTGART
1942 V9 P109-121 63-14293 <*>

BERICHT DER GESELLSCHAFT KUER KOHLENTECHNIK
1926 V2 P30-53 64-16325 <*>
1930 V3 P133-134 MF-6 <*>
1930 V3 P135-144 MF-6 <*>
1930 V3 P145-148 MF-6 <*>
1930 V3 N2 P101-132 MF-6 <*>
1930 V4 N1 P73-81 MF-6 <*>
1950 V5 P314-346 II-597 <*>

BERICHT DES INSTITUTS FUER TABAKFORSCHUNG
1955 V2 N1 P23-29 58-1033 <*>
6445-B <K-H>
1955 V2 N1 P94-111 6146-D <K-H>
1955 V2 N1 P112-119 6146-F <K-H>
1955 V2 N2 P127-141 6146-E <K-H>
1956 V3 N1 P60-76 6146-B <K-H>
1956 V3 N2 P77-87 6146-C <K-H>
1956 V3 N2 P248-281 6171-B <K-H>
1956 V3 N2 P310-328 6171-A <K-H>
1958 V5 N1 P145-154 64-14553 <*> 0
9019-C <K-H>
1959 V6 N2 P231-240 10281-F <K-H>
64-14536 <*>
1959 V6 N2 P241-264 10281-G <K-H>
62-20208 <*>
1959 V6 N2 P265-277 64-14554 <*>
1960 V7 N1 P81-102 10573-E <K-H>
64-14571 <*>
1960 V7 N2 P187-197 10969-B <K-H>
64-10619 <*>
1960 V7 N2 P198-215 10969-D <K-H>
64-14574 <*>
1960 V7 N2 P216-240 10969-E <K-H>
64-14583 <*>
1961 V8 N1 P73-86 63-14369 <*>
1961 V8 N1 P87-101 11711 <KH>
63-16131 <*>
1963 V10 N1 P69-106 64-18060 <*>
1963 V10 N2 P203-237 64-18525 <*> 0
1963 V10 N2 P238-263 64-18526 <*>

BERICHT DES OHARA INSTITUTS FUER LANDWIRTSCHAFT- LICHE BIOLOGIE
SEE NOGAKU KENKYU

BERICHT. SAECHSISCHE AKADEMIE DER WISSENSCHAFTEN
1899 V51 P16-24 61-14218 <*>

BERICHT DER SCHWEIZERISCHEN BOTANISCHEN GESELL- SCHAFT
1947 V57 P132-148 59-15455 <*> 0
1947 V57 P227-241 57-3239 <*>
65-12348 <*>
1950 V60 P199-230 59-10053 <*> 0
1952 V62 P123-163 58-844 <*>
1952 V62 P509-526 3097-A <K-H>
1953 V63 P90-102 T1921 <INSD>
1954 V64 P487-494 10M42G <ATS>
1958 V68 P239-248 61-20441 <*>

BERICHT DES SCHWEIZERISCHEN VERBANDES FUER DIE MATERIALPRUEFUNGEN DER TECHNIK
1928 V13 P5 59-20088 <*>

BERICHT DES VEREINS DEUTSCHER INGENIEURE
SEE VDI-BERICHTE

BERICHT UBER DIE VERSAMMLUNG DER DEUTSCHEN OPHTHALMOLOGISCHEN GESELLSCHAFT
1922 N43 P101-106 61-14441 <*>
1927 V46 P328-335 61-14402 <*>
1928 V47 P374-379 61-14401 <*> 0

1929 V47 P391-394 61-14210 <*>
1932 V49 P36-44 61-14180 <*>
1961 V63 P167-178 63-00214 <*>

BERICHT UEBER DIE VERSAMMLUNG DER OPHTHALMOLOGISCHEN GESELLSCHAFT
1884 V16 P60-63 2764 <*>
1911 N37 P11-15 61-14160 <*>

BERITA PERIKANAN. DJAKARTA
1955 N6 P92-93 58-2123 <*>
1955 N6 P96-98 58-2123 <*>

BERLINER KLINISCHE WOCHENSCHRIFT
1866 V3 P269-270 62-01041 <*>
1891 V38 P926-933 63-00773 <*>
1905 V42 N15 P427-428 63-00043 <*>
1912 V49 N6 P261-262 65-18031 <*>

BERLINER MEDIZIN
1958 P88-90 62-00688 <*>
1959 V10 P103-106 59-17665 <*>
1959 V10 P152- 60-18159 <*>
1959 V10 P265-271 62-01214 <*>
1960 V11 P72-74 60-18160 <*> O
1961 V12 P555-558 63-01761 <*>
1962 V13 P166-171 64-18962 <*>
1963 V14 P300 66-11613 <*>

BERLINER UND MUENCHENER TIERAERZTLICHE WOCHENSCHRIFT
1947 P141-143 II-586 <*>
1953 V66 P1-12 MF-2 <*>
1954 V67 P335-338 65-12615 <*>
1955 V69 N5 P94-96 CSIRO-3573 <CSIR>

1962 V75 N22 P423-425 64-00080 <*>
1963 V76 N6 P101-105 64-00075 <*>
1964 V77 N19 P373-379 66-12626 <*>
1964 V77 N20 P401-408 66-12111 <*>

BERLINER TIERAERZTLICHE WOCHENSCHRIFT
1929 V80 N31 P433-436 66-12160 <*>

BERUFSDERMATOSEN
1958 V6 P171-192 62-00868 <*>
1963 V11 N3 P125-140 65-18011 <*>

BETON UND EISEN
1925 V24 N4 P51-56 64-14451 <*> O
1938 V37 N11 P188-191 II-874 <*>
1939 V38 N13 P221-223 II-874 <*>
1939 V38 N15 P248-255 II-874 <*>
1939 V38 N22 P342-344 II-874 <*>
1940 V39 N9 P17-20 II-874 <*>
1942 V41 P86-91 C-4204 <NRC>
1942 V41 P103-108 C-4204 <NRC>
1942 V41 P127-130 C-4204 <NRC>

BETON HERSTELLUNG-VERWENDUNG
1960 V10 N1 P1-8 63-10208 <*> O
1960 V10 N6 P256-259 61-20560 <*>
1962 V12 N8 P363-364 63-20311 <*>
1962 V12 N11 P496-500 63-14708 <*>
1962 V12 N12 P545-552 63-18500 <*> O
1963 V13 N1 P11-18 64-14279 <*>
1963 V13 N2 P81-84 64-14279 <*>

BETON UND STAHLBETONBAU
1944 V43 N1/2 P3-8 II-874 <*>
1944 V43 N9/10 P52-55 57-621 <*>
1950 V45 P276-279 T-1408 <INSD>
1950 V45 N1 P3-6 58-2550 <*>
1950 V45 N1 P3-10 61-18945 <*>
1950 V45 N1 P52-56 61-18945 <*>
1950 V45 N3 P52-56 58-2550 <*>
1950 V45 N3 P66-68 58-2551 <*>
1950 V45 N4 P73-77 58-2610 <*>
61-18944 <*>
1950 V45 N5 P108-116 58-2588 <*>
61-18988 <*>
1950 V45 N5 P116-118 58-2595 <*>
61-18986 <*>
1950 V45 N7 P149-157 58-2590 <*>
61-18987 <*>
1952 V47 N2 P36-42 58-2568 <*>
1952 V47 N3 P36-42 61-18929 <*>
1952 V47 N10 P225-231 61-18930 <*>
1952 V47 N10 P245-247 61-18985 <*>
1953 V48 P260-266 T-1457 <INSD>
1953 V48 N1 P1-5 58-2566 <*>
61-18860 <*>
1953 V48 N2 P11-14 13J17G <ATS>
1953 V48 N4 P78-83 14J17G <ATS>
1953 V48 N5 P120-123 14J17G <ATS>
1953 V48 N6 P140-144 14J17G <ATS>
1953 V48 N7 P162-164 15J17G <ATS>
1953 V48 N8 P177-182 61-18984 <*>
1954 V49 P165-171 T-1741 <INSD>
1954 V49 P188-193 T-1742 <INSD>
1955 V50 N2 P51-54 C-2336 <NRC>
57-2390 <*>
1955 V50 N2 P61-64 16J17G <ATS>
1955 V50 N2 P64-71 58-2573 <*>
61-20023 <*>
1955 V50 N3 P89-92 58-2573 <*>
1955 V50 N3 P89-93 61-20023 <*>
1955 V50 N6 P158-163 II-367 <*>
1956 V51 P49-55 T-1765 <INSD>
1956 V51 N2 P29-36 65-13530 <*>
1956 V51 N3 P59-62 65-13530 <*>
1957 V52 N3 P67-71 58-2558 <*>
61-18965 <*>
1957 V52 N12 P292-294 58-789 <*>
63-16562 <*>
1958 V53 N3 P49-56
1958 V53 N4 P73-85 61-18975 <*>
1959 V54 N1 P7-9 63-01818 <*>
1959 V54 N3 P49-63 61-18971 <*>
1959 V54 N4 P89-92 61-18971 <*> P
1959 V54 N5 P123-128 60-16826 <=> O
1959 V54 N9 P223-228 60-14256 <*>
1959 V54 N9 P247-248 60-14256 <*>
1959 V54 N10 P240-247 C-3652 <NRCC>
1960 V55 N3 P49-58 61-18978 <*>
1960 V55 N6 P121-132 65-13280 <*>
1960 V55 N7 P151-162 65-13507 <*>
1960 V55 N9 P193-205 103 <CCA>
65-13531 <*>
1961 V56 N1 P6-10 2170 <BISI>
1961 V56 N2 P32-37 62-18783 <*>
1961 V56 N12 ENTIRE ISSUE TRANS-111 <CCA>
1961 V56 N12 P277-290 C-5459 <NRC>
65-13538 <*>
1962 V57 N2 ENTIRE ISSUE TRANS-111 <CCA>
1962 V57 N2 P32-44 65-13538 <*>
1962 V57 N3 ENTIRE ISSUE TRANS-111 <CCA>
1962 V57 N3 P54-64 C-5459 <NRC>
65-13538 <*>
1962 V57 N5 P97-104 65-13532 <*>
1962 V57 N6 ENTIRE ISSUE TRANS-111 <CCA>
1962 V57 N6 P97-104 CACA-LIB-TRANS-108 <CCA>
1962 V57 N6 P141-149 C-5459 <NRC>
65-13538 <*>
1962 V57 N7 ENTIRE ISSUE TRANS-111 <CCA>
1962 V57 N7 P161-173 C-5459 <NRC>
65-13538 <*>
1962 V57 N8 ENTIRE ISSUE TRANS-111 <CCA>
1962 V57 N8 P184-188 C-5459 <NRC>
65-13538 <*>
1962 V57 N9 P218-220 66-11077 <*>
1962 V57 N10 P233-239 C-5016 <NRC>
1962 V57 N11 P261-271 CACA-LIB-TRANS-110 <CCA>
65-13533 <*>
1963 V58 N1 P6-11 63-18318 <*>
1963 V58 N2 P40-45 63-18788 <*> O
1963 V58 N5 P112-120 TRANS-114 <CCA>
65-13607 <*>
1963 V58 N11 P268-270 65-14891 <*> O
1964 V59 N3 P49-56 N66-10550 <=$>

Citation	Translation
1964 V59 N4 P91-94	N66-10550 <=$>
1964 V59 N5 P111-119	N66-10550 <=$>
1964 V59 N10 P224-230	65-14394 <*> 0

BETON I ZHELEZOBETON

Citation	Translation
1956 N2 P59-64	61-23394 <*=>
1956 N4 P130-132	R-4153 <*>
1956 N7 P264-266	61-20032 <*=>
1956 N8 P277-280	R-4155 <*>
1956 N12 P435-440	R-4389 <*>
1957 N1 P28-29	64-10177 <=*$>
1957 N3 P107-108	R-5321 <*>
1957 N5 P202-205	63-24211 <=*$>
1957 N8 P322-325	59-19278 <+*>
1957 N10 P413-415	59-19285 <+*>
1957 N12 P486-491	63-14711 <=*>
1958 N1 P4-13	60-15378 <+*>
1958 N1 P19-22	60-15377 <+*>
1958 N3 P100-103	61-13891 <*+>
1958 N3 P107-111	61-13890 <*+>
1958 N5 P183-186	61-13276 <*+>
1958 N5 P187-190	61-19378 <+*>
1958 N6 P221-222	59-19277 <+*>
1958 N6 P226-228	61-19382 <+*>
1958 N7 P259-263	61-13953 <+*>
1958 N7 P278-280	60-23667 <+*>
	61-28138 <*=>
1958 N10 P372-378	60-13744 <+*>
1958 N10 P388	60-15527 <+*>
1958 N10 P389	60-15530 <+*>
1959 N1 P25-29	61-19381 <+*>
1959 N2 P62-67	61-15178 <*+>
1959 N2 P71-75	61-13889 <*+>
1959 N2 P82-85	61-13278 <*+>
	61-15070 <+*>
1959 N3 P100-103	61-15292 <+*>
1959 N3 P109-113	62-14253 <=*>
1959 N3 P127-130	61-27053 <=*>
1959 N4 P169-174	61-15710 <+*>
1959 N5 P198-202	60-17687 <+*>
1959 N6 P262-266	64-19890 <=$>
1959 N7 P299-303	61-15177 <*+>
1959 N7 P326-329	61-13950 <=*>
1959 N7 P329-330	61-28642 <*=>
1959 N8 P342-344	64-19131 <=*$>
1959 N9 P413-416	61-15504 <+*>
1959 N10 P442-446	61-13954 <=*>
1959 N12 P531	60-11470 <=>
1960 N1 P1-2	60-11569 <=>
1960 N3 P105-111	62-13278 <=*> 0
	62-24611 <=*>
1960 N5 P193-194	60-41063 <=>
1960 N8 P341-343	61-11722 <=>
1960 N9 P424-425	63-24209 <=*$>
1960 N10 P450-454	63-24207 <=*$>
1960 N10 P480-481	63-23401 <=*$>
1960 N11 P501-503	61-21388 <=> P
1960 N12 P533-537	64-13761 <=*$>
1961 N1 P1-3	61-21686 <=>
1961 N1 P14-18	62-32806 <=*>
1961 N2 P51-63	62-32808 <=*>
1961 N2 P68-83	62-32807 <=*>
1961 N4 P146	61-23945 <=>
1961 N4 P162-164	61-23945 <=>
1961 N4 P170-174	63-24217 <=*$>
1961 N5 P234-235	62-25371 <=*>
1961 N8 P361-364	62-33552 <=*>
1961 N8 P364-369	63-13597 <=*>
1961 N9 P414-417	3269 <BISI>
1961 N9 P417-418	62-33617 <=*>
1961 N11 P501-507	63-24216 <=*$>
1961 N11 P518-520	63-24216 <=*$>
1961 N12 P531-536	62-33620 <=*>
1961 V7 N2 P58-63	<BRS>
1962 N6 P265-268	63-10195 <=*>
1962 V8 N1 P17-21	64-13352 <=*$>
1962 V8 N1 P21-25	64-19989 <=$>
1962 V8 N1 P29-32	64-13351 <=*$>
1962 V8 N2 P79-81	64-15867 <=*$>
1962 V8 N2 P82-85	64-19103 <=*$>
1962 V8 N3 P93-97	64-13350 <=*$>
1962 V8 N5 P219-222	63-24224 <=*$>
	64-19104 <=*$>
1962 V8 N5 P233-237	64-19105 <=*$>
1962 V8 N6 P271-274	63-14714 <=*>
1962 V8 N7 P306-311	63-23255 <=*>
1962 V8 N7 P312-314	63-23253 <=*>
1962 V8 N7 P331-333	65-14926 <*> 0
1962 V8 N8 P339-342	63-14712 <=*>
1962 V8 N9 P389	<BRS>
	65-63885 <=$>
1962 V8 N9 P390-393	65-63886 <=$>
1962 V8 N9 P397-402	<BRS>
1962 V8 N9 P401-402	65-63885 <=$>
1962 V8 N9 P402-408	<BRS>
1962 V8 N9 P408	65-63885 <=$>
1962 V8 N9 P414-418	65-29502 <$>
1962 V8 N9 P414	65-63885 <=$>
1962 V8 N9 P422-424	65-29503 <$>
1962 V8 N11 P491-498	64-19101 <=*$>
1962 V8 N11 P501-503	64-19100 <=*$>
1962 V8 N11 P503-507	64-13307 <=*$>
1962 V8 N11 P511-531	TR.10 <JLRD>
1962 V8 N11 P513-517	<JLRD>
1963 V9 N1 P27-30	TR.6 <JLRD>
1963 V9 N2 P53-60	64-13738 <=*$>
1963 V9 N2 P78-80	64-15869 <=*$>
1963 V9 N2 P89-92	64-15887 <=*$>
1963 V9 N4 P145-151	<BRS>
	65-63794 <=$>
1963 V9 N5 P196-199	64-15868 <=*$>
1963 V9 N5 P215-219	65-11707 <*>
1963 V9 N6 P265-267	65-63896 <=$>
1963 V9 N7 P292-298	TR.11 <JLRD>
1963 V9 N7 P319-322	TR.9 <JLRD>
1963 V9 N8 P369-371	64-15732 <=$>
1963 V9 N10 P449-450	TR.16 <JLRD>
1963 V9 N10 P459-462	RTS-3376 <NLL>
1964 V10 N1 P8-12	<CEMB>
1964 V10 N1 P12-14	65-60892 <=$>
1964 V10 N1 P21-25	65-60894 <=$>
1964 V10 N1 P30-35	64-18009 <=*$> 0
1964 V10 N2 P64-65	65-11703 <*>
1964 V10 N2 P75-78	65-11712 <*>
	65-60886 <=$>
1964 V10 N2 P92-94	RTS-2971 <NLL>
1964 V10 N3 P116-117	65-60512 <=$>
1964 V10 N3 P117-120	65-60513 <=$>
1964 V10 N3 P120-122	65-60517 <=$>
1964 V10 N7 P295-297	65-60879 <=$>
1964 V10 N7 P330-333	65-64381 <=*$>
1964 V10 N7 P334	65-29614 <$>
1964 V10 N8 P368-375	C-5567 <NRC>
1964 V10 N10 P465-466	<JLRD>
1965 V11 N2 P1-5	65-64388 <=*$>
1965 V11 N2 P12-16	4578 <BISI>
1965 V11 N3 P1-4	LC-1305 <BRC>
	LC-1305 <NLL>
1965 V11 N10 P29-33	66-12818 <*>

BETONG

Citation	Translation
1948 N3 P125-138	58-2589 <*>
1948 V33 N3 P125-128	61-18949 <*>
1955 N1 P15-43	58-2571 <*>
1955 V40 N1 P15-43	61-18957 <*>
	64-10191 <*>
1955 V40 N2 P135-149	58-2609 <*>
	61-18958 <*>
	61-18958 <*>
1955 V40 N3 P241-275	58-2607 <*>
	61-18992 <*>

BETONGEN IDAG

Citation	Translation
1956 V21 N6 P140-148	57-613 <*>
1956 V21 N6 P148-158	2092 <*>
1959 V24 N1 P11-16	61-18966 <*> 0

BETONSTEINZEITUNG

Citation	Translation
1952 V18 N1 P17-18	61-20235 <*>
1952 V18 N5 P180-182	61-20262 <*>

1952 V18 N8 P289-294 58-2567 <*>
61-20234 <*>
1953 V19 N8 P305-307 AL-684 <*>
1957 V23 N2 P83-89 58-2556 <*>
61-20214 <*>
1959 V25 N9 P376-382 60-14259 <*>
1963 V29 P409-418 1670 <TC>

BETCNSTRASSE
1936 V11 N9 P193-203 65-17422 <*>

BETON-TEKNIK
1950 V16 N1 P1-4 58-2577 <*>
1950 V16 N4 P110-118 T-1507 <INSD>
1958 V24 N1 P3-22 C-3962 <NRCC>
1959 V25 N4 P119-141 61-18778 <*> O
1965 N1 P10-16 15T93D <ATS>

BEZOPASNCST TRUDA V PROMYSHLENNOSTI
1958 V2 N12 P22-23 64-13275 <=*$>
1960 N4 P21-23 60-41304 <=>
1960 N6 P20-21 60-41676 <=>
1960 V4 N3 P7-9 61-13359 <=*>
1960 V4 N6 P8-10 61-28178 <*=>
1960 V4 N9 P7-10 61-23905 <*=>
1960 V4 N9 P7-9 62-24082 <=*>
1960 V4 N9 P38-39 61-23908 <*=>
1960 V4 N11 P20-21 61-28179 <*=>
1960 V4 N12 P8-9 62-240-81 <=*>
1961 V5 N2 P26-27 64-13034 <=*$>
1961 V5 N4 P37-38 61-27732 <*=>
1961 V5 N7 P5-6 61-28627 <*=>
1961 V5 N9 P21-22 63-19860 <=*$>
1961 V5 N10 P21-23 UCRL TRANS-935(L) <=*$>
1961 V5 N12 P1-2 62-23342 <=>
1961 V5 N12 P8-11 62-25274 <=*>
1961 V5 N12 P18-19 62-23342 <=>
1961 V5 N12 P36 62-23342 <=>
1961 V5 N12 P38-39 62-23342 <=>
1962 V6 N1 P9-10 62-32060 <=*>
1962 V6 N5 P10-12 64-13209 <=*$>
1962 V6 N6 P16-17 M-5250 <NLL>
1963 V7 N5 P13 64-13210 <=*$>
1963 V7 N9 P10-14 64-13211 <=*$>
1963 V7 N10 P25-26 64-13208 <=*$>
1963 V7 N11 P17-18 64-19161 <=*$>
1963 V7 N11 P36-37 64-19153 <=*$>
1963 V7 N11 P64-19152 64-19152 <=*$>
1964 V8 N4 P13-15 SMRE-TRANS-5072 <NLL>
1964 V8 N5 P23-25 SMRE-TRANS-5025 <NLL>
1964 V8 N6 P9-11 SMRE-TRANS-5039 <NLL>
1964 V8 N6 P32 SMRE-TRANS-5110 <NLL>
1964 V8 N7 P50-51 SMRE-TRANS-5041 <NLL>
1964 V8 N11 P2-3 SMRE-TRANS-5176 <NLL>
1964 V8 N11 P51-52 SMRE-TRANS-5109 <NLL>
1965 V8 N5 P46-49 65-33106 <=*$>
1965 V9 N1 P7-9 SMRE-TRANS-5136 <NLL>
1965 V9 N9 P40-42 SMRE-TRANS-5212 <NLL>

BIBLIOGRAFIA POLAROGRAFICA
1957 V3 P184-202 62-10382 <*>

BIBLIOTECHKA PO OVCSHEVODSTVY
1957 N15 P1-156 60-21165 <=>

BIBLIOTECHKA ZHURNALA SOVETSKII VOIN
1965 V47 N6 P10-12 AD-619 315 <=$>
65-31354 <=$>
1965 V47 N6 P37-40 AD-619 315 <=$>
65-31354 <=$>
1966 N13 P30-31 66-34804 <=$> O

BIBLIOTEKA PO AVTOMATIKE
1960 V17 P1-120 61-28511 <*=>
1960 V20 ENTIRE ISSUE 63-13256 <*=>
1961 V40 P1-95 62-25716 <=*>

BIBLIOTEKAR USSR
1956 N9 P5-9 61-15645 <+*>
1958 N10 P43-46 NLLTB V1 N2 P44 <NLL>
59-22636 <=*>
1958 N10 P60-61 NLLTB V1 N2 P41 <NLL>
59-22642 <*=>
1959 N9 P47-51 NLLTB V2 N3 P176 <NLL>
1960 N10 P5-7 NLLTB V3 N4 P320 <NLL>

BIBLIOTEKOVEDENIE I BIBLIOGRAFIA ZA RUBEZHOM
1962 N9 P52-78 64-30715 <*>
1962 N9 P79-97 64-30718 <*>

BIBLIOTHECA HAEMATOLOGICA
1956 N4 P174-182 57-2781 <*> P
1959 V9 P92-102 62-00441 <*>
1961 V12 P245-259 62-00888 <*>

BIBLIOTHECA MICROBIOLOGICA
1964 N4 P95-105 64-20144 <*>

BILD UND TON
1955 V8 N2 P34-35 62-14210 <*>
1955 V8 N5 P128-131 62-14209 <*>
1955 V8 N10 P271-273 94N48G <ATS>
1958 V11 N3 P62-63 62-14211 <*>
1958 V11 N12 P310-312 62-10409 <*> O
1959 V12 N6 P176-177 60-23950 <=*>
1961 V14 N9 P266-271 10R80G <ATS>
1962 V15 N2 P34-38 C8Q71G <ATS>
1962 V15 N2 P34-39 63-18069 <*>
1962 V15 N5 P133-135 65-13058 <*>
1963 V16 N1 P7-10 16Q69G <ATS>
64-10599 <*> O
1963 V16 N2 P39-44 79Q70G <ATS>
1963 V16 N3 P66-69 64-16045 <*>
1965 V18 N8 P233-239 07S88G <ATS>

BILDMESSUNG UND LUFTBILCWESEN
1961 N1 P8-15 62-19736 <=*>
1961 N3 P97-100 62-19718 <+*>

*BILTEN INSTITUTA ZA NUKLEARNE NAUKE BORIS KIDRIC. VINCA, YUGOSLAVIA
1962 V13 N1 P1-88 AEC-TR-5031/1 <=>
1962 V13 N2 P1-83 AEC-TR-5031/2 <=>
1962 V13 N3 P1-55 AEC-TR-5031/3 <=>
1962 V13 N4 P1-53 AEC-TR-5031/4 <=>

BIOCHEMICAL PHARMACOLOGY. LONDON
1959 V2 N4 P243-254 61-00395 <*>
1962 V11 P813-822 63-18800 <*>

BIOCHEMISCHE ZEITSCHRIFT
1908 V13 P305-320 57-293 <*>
1910 V26 P312-324 62-01030 <*>
1910 V29 P279-293 57-295 <*>
1913 V49 P333-369 63-00140 <*>
1914 V55 P88- 57-539 <*>
1914 V59 P77-99 62-01518 <*>
1915 V69 P461-466 3964 <K-H>
1915 V70 P426- 57-421 <*>
1923 V136 P128- 57-300 <*>
1923 V136 P403-410 62S85G <ATS>
1923 V138 P156-160 60-10456 <*>
1924 V146 P226-238 1693 <*>
1924 V146 P239-244 1411 <*>
1924 V152 P309-317 63-01759 <*> P
1925 V158 P193-196 64-18789 <*>
1925 V163 P412-421 60-10457 <*> P
1925 V163 N1/3 P41-50 61-00181 <*>
1926 V174 P333-340 64-16215 <*>
1927 V182 P424-433 8697 <IICH>
1927 V184 P453- 57-411 <*>
57-430 <*>
1927 V185 P344-348 66-13596 <*>
1927 V186 N1 P181-193 57-284 <*>
1927 V187 P307- 57-429 <*>
1927 V188 P241-258 AL-404 <*>
1927 V188 P259-269 1592 <*>
1927 V189 P233-241 66-14519 <*>
1927 V190 P28-41 57-288 <*>
1928 V192 P428-430 60-10466 <*>

1928 V197 P136- 57-428 <*>
1928 V197 P197-209 57-13 <*>
1928 V203 P22-49 58-1558 <*>
1929 V204 P14-27 57-420 <*>
1929 V205 P144-153 57-388 <*>
1929 V205 P360-368 60-10467 <*>
1929 V208 P32-44 57-285 <*>
1929 V208 P45-49 57-287 <*>
1929 V211 P323-325 61-00985 <*>
1929 V212 P53-60 R-4947 <*>
1929 V214 P175-186 57-286 <*>
1930 V221 P403-417 63-16535 <*> O
1930 V228 P272- 57-290 <*>
1930 V228 P286-303 57-390 <*>
1931 V231 P347- 57-409 <*>
1931 V232 P338-345 63-16536 <*> O
1931 V233 P58-61 61-01005 <*>
1931 V233 P460-469 60-10469 <*> P
1931 V234 P139-141 64-16217 <*>
1931 V237 P247-275 64-18788 <*>
1931 V239 P250-256 57-410 <*>
1931 V243 P292-309 AL-352 <*>
1932 V244 P4-8 65-12611 <*>
1932 V244 P258-267 65-12610 <*>
1932 V250 P281-304 60-10470 <*> P
1932 V252 P1-7 60-10461 <*> O
1932 V252 P185-200 57-401 <*>
1932 V254 P381-397 60-10462 <*> P
1932 V255 P247-277 57-405 <*>
1933 V257 P337-343 57-406 <*>
1933 V258 P154-157 I-264 <*>
1933 V258 P340-346 I-417 <*>
1933 V259 P1- 57-427 <*>
1933 V263 N4/5 P400-409 I-823 <*>
1933 V266 N4/6 P329- I-824 <*>
1934 V270 N1 P52-62 58-229 <*>
1934 V271 P357-369 60-10464 <*>
1934 V272 P180- 57-301 <*>
1935 V275 P367- 57-426 <*>
1935 V279 P417-423 65-14589 <*>
1935 V280 P448-457 63-20170 <*>
66-15254 <*>
1935 V282 N1/2 P104-108 60-10479 <*> P
1935 V282 N1/2 P109-119 60-10478 <*>
1936 V285 N1/2 P67-71 60-10480 <*>
1936 V286 N3/4 P160-181 59-17043 <*> O
1936 V287 N5/6 P291-328 60-10481 <*> P
1936 V288 N1/2 P39-40 60-10482 <*>
1937 V289 P217-233 62-00598 <*>
1937 V289 P243- 57-291 <*>
1937 V290 P419-427 66-10142 <*> O
1937 V292 P221-229 2304A <K-H>
1938 V298 N3/4 P150-168 60-10473 <*>
1938 V299 P32-57 2304B <K-H>
1939 V301 P429-436 62-00595 <*>
1940 V304 N4 P259-265 57-432 <*>
1940 V305 P150-161 I-737 <*>
1940 V305 N1 P136-144 60-14265 <*>
1940 V305 N5/6 P418-442 PT1
57-431 <*>
1940 V306 P316-336 I-738 <*>
1942 V310 P384-421 64-16244 <*>
1942 V312 P277-288 3923-C <K-H>
1942 V313 P377-387 1593 <*>
1943 V315 N1/2 P83-96 60-10477 <*>
1943 V316 P255-263 I-924 <*>
1943 V316 N1/2 P96-107 60-18329 <*>
1951 V321 P451-461 58-813 <*>
1951 V321 N4 P354-356 57-1160 <*>
1952 V322 P174- 57-3243 <*>
1952 V322 P467-470 58-1572 <*>
1952 V322 N6 P467-470 64-10335 <*>
1952 V323 P181-191 64-10287 <*>
1952 V323 N4 P245-250 64-10291 <*> O
1952 V323 N4 P265 64-10292 <*>
1953 V324 P125-127 57-3242 <*>
1953 V324 N1 P19-31 64-10347 <*>
1953 V324 N2 P138-143 65-12540 <*>
1953 V324 N3 P186-194 65-12543 <*>
1953 V324 N7 P536-543 65-12541 <*>
1953 V325 N1 P1-11 60-10485 <*>
1954 V326 P150-160 58-2235 <*>
1955 N326 P433-435 II-838 <*>
1955 V326 P311-316 58-2222 <*>
1955 V326 P436-441 57-3411 <*>
1955 V326 P484-492 65-12548 <*>
1955 V327 P93-108 57-2549 <*>
1955 V327 P136-148 62-00119 <*>
1955 V327 N1 P39-61 65-12609 <*>
1955 V327 N3 P189-194 65-12552 <*>
1956 V327 N7 P473-483 CSIRO-3240 <CSIR>
1956 V328 P117-125 61-00292 <*>
1956 V328 P267-284 58-1347 <*>
1956 V328 P309-322 CSIRO-3462 <CSIR>
1957 V328 N6 P458-564 59-17231 <*> O
1957 V329 P320-331 62-01523 <*>
1957 V329 N1 P59-74 65-14448 <*> P
1958 V330 P521-537 542 <TC>
1959 V332 N1 P47-66 61-00169 <*>
1959 V332 N2 P195-212 61-00188 <*>
1960 V333 P1-9 61-20523 <*>
1960 V333 P10-32 61-00200 <*>
1961 V334 P203-217 62-01034 <*>
1961 V334 P227-244 11325 <KH>
63-16952 <*>
1961 V334 N5 P431-440 62-00884 <*>
1962 V335 P400-407 AEC-TR-5600 <*>
1962 V335 P519-539 63-00033 <*>
1962 V335 P540-547 63-00034 <*>
1962 V336 N4 P351-370 65-13023 <*>
1963 V336 P455-459 63-00844 <*>
1963 V336 P510-525 63-00777 <*>
1963 V336 P545-556 AEC-TR-5874 <*>
1965 V341 P129-138 66-11093 <*>

BIOCHEMISCHES ZENTRALBLATT

1903 V2 N1 P1- 2807 <*>

BIOCHIMICA ET BIOPHYSICA ACTA

1950 V4 P144-155 65-12345 <*>
1951 V7 P439-445 64-00352 <*>
1952 V9 P161-169 I-549 <*>
1953 V11 P138-146 59-22222 <+*>
1953 V11 N2 P190-198 II-631 <*>
1953 V12 N3 P424-431 60-10501 <*> O
1953 V12 N3 P487-488 60-10500 <*>
1954 V14 P401-406 62-25795 <=*$>
1954 V15 P237-245 2059 <*>
1955 V16 P183-197 2616 <*>
1955 V16 P410-417 5998-B <KH>
1955 V17 P67-74 57-781 <*>
1955 V18 N1 P71-82 65-12666 <*>
1956 V21 P500-506 61-00153 <*>
1956 V21 N2 P349-359 3259 <*>
1956 V21 N2 P378-380 59-12364 <+*> O
1957 V25 N1 P100-109 57-3443 <*>
1957 V25 N1 P110-117 57-3444 <*>
1958 V27 P247-255 59-16575 <+*> O
1958 V27 N3 P598-608 59-20881 <*>
1960 V39 N1 P122-139 60-23054 <=*> O
61-10424 <*>
1960 V41 P345-348 61-01004 <*>
1960 V41 N2 P192-203 61-00172 <*>
1960 V41 N2 P204-216 61-00171 <*>
1961 V47 P561-568 61-00987 <*>
1961 V47 N2 P307-316 65-12277 <*>
1961 V48 P400-402 62-00101 <*>
1961 V48 N3 P562-572 61-00588 <*>
1961 V50 P186-188 62-00421 <*>
1961 V52 P254-265 AEC-TR-5743 <*>
1961 V52 P552-565 63-00941 <*>
1961 V53 P11-18 AEC-TR-5047 <*>
1961 V53 P221-223 63-10792 <*>
1961 V54 P455-468 63-10660 <*>
1962 V59 P227-228 AEC-TR-5715 <*>
1962 V59 P261-272 63-00074 <*>
1962 V61 P852-854 63-18372 <*>
1962 V61 P857-864 64-00432 <*>
1962 V63 P489-495 63-00950 <*>
1963 V71 P737-738 AEC-TR-6363 <*>

Citation	Translation
1963 V75 N1 P1-11	66-12123 <*>
1964 V91 N4 P549-558	65-17254 <*>
1964 V91 N4 P559-572	65-17255 <*>
1964 V91 N4 P573-583	65-17261 <*>
1965 V108 N2 P194-201	66-11575 <*>

BIOCHIMICA E TERAPIA SPERIMENTALE

Citation	Translation
1939 V26 P168-170	60-10677 <*>

*BIOFIZIKA

Citation	Translation
1956 V1 N5 P435-437	C-3660 <NRCC>
1956 V1 N5 P448-451	R-2727 <*>
	R-3524 <*>
1956 V1 N6 P513-524	R-2728 <*>
	R-3528 <*>
1956 V1 N6 P585-592	R-1785 <*>
	64-19450 <=$> O
1956 V1 N7 P628-632	59-10054 <*>
1956 V1 N7 P633-636	ORNL-TR-84 <=$>
1956 V1 N8 P721-728	62-32198 <=*>
1957 V2 P313-317	R-2126 <*>
1957 V2 P649-659	R-4300 <*>
1957 V2 P756-763	R-4299 <*>
1957 V2 N1 P67-78	60-18560 <+*>
1958 V3 P377-380	R-5258 <*>
1958 V3 P521-523	R-5268 <*>
1958 V3 P547-557	R-5263 <*>
1958 V3 N1 P108-110	62-13797 <=*> O
1958 V3 N3 P358-363	R-5231 <*>
1958 V3 N4 P385-390	62-13798 <=*>
1958 V3 N5 P558-561	62-15733 <*=>
1958 V3 N5 P606-609	62-15797 <=*>
1959 V4 P3-18	59-17413 <+*>
1959 V4 P19-26	59-17414 <+*>
1959 V4 P124-128	59-17415 <+*>
1959 V4 P232-237	59-17781 <+*>
1959 V4 P289-299	59-20556 <+*>
1959 V4 P521-532	60-16000 <+*> O
1959 V4 P601-605	60-16001 <+*>
1959 V4 P641-649	60-16002 <+*>
1959 V4 N1 P119-120	60-10941 <+*>
1959 V4 N1 P122-123	60-10942 <+*>
1959 V4 N2 P153-157	60-18562 <+*>
1959 V4 N2 P224-227	UCRL TRANS-521 <+*>
1959 V4 N5 P601-605	62-15796 <*=>
1959 V4 N6 P641-649	UCRL TRANS-527 <*=>
1960 V5 P308-317	61-10352 <*+>
1960 V5 P599-608	61-10345 <*+>
1960 V5 N2 P121-126	UCRL TRANS-560 <+*>
	61-10343 <*+>
1960 V5 N2 P235-238	UCRL TRANS-561 <+*>
1960 V5 N2 P244-248	60-41311 <=>
1960 V5 N3 P339-345	65-60254 <=$> O
1960 V5 N6 P740-744	61-28225 <=*>
1961 V6 P424-435	65-12951 <*>
1961 V6 N1 P30-39	62-18633 <=*>
1961 V6 N2 P231-233	62-19931 <*=>
1961 V6 N2 P242-243	62-18636 <=*>
1961 V6 N3 P294-299	62-18635 <=*>
1961 V6 N3 P300-316	62-19754 <=*>
1961 V6 N4 P392-402	62-18634 <=*>
1961 V6 N5 P563-571	62-18632 <=*>
1961 V6 N6 P645-649	63-16733 <=*>
1962 V7 P561-567	63-16722 <=*>
1962 V7 P571-577	63-16728 <=*>
1962 V7 N1 P80-85	62-25552 <=*>
1962 V7 N2 P225-226	63-23411 <=*$> O
1962 V7 N3 P270-280	64-13150 <=*$> O
1962 V7 N3 P281-291	64-18833 <*>
1962 V7 N3 P306-310	62-33403 <=*>
1962 V7 N4 P426-432	65-61134 <=$>
1962 V7 N4 P480-483	63-21866 <=>
1962 V7 N5 P523-528	M-5650 <NLL>
1962 V7 N5 P561-567	AEC-TR-5662 <=*>
1962 V7 N5 P578-591	63-16727 <=*>
1963 V8 P561-568	64-14794 <=*$>
1963 V8 P664-676	64-14792 <=*$>
1963 V8 N1 P3-8	63-11762 <=>
	64-19724 <=$>
1963 V8 N1 P19-27	64-31008 <=>
1963 V8 N1 P28-33	64-19725 <=$>
1963 V8 N1 P40-44	65-60185 <=$>
1963 V8 N1 P90-100	63-21454 <=>
1963 V8 N1 P124-125	64-19674 <=$>
1963 V8 N1 P124	64-31008 <=>
1963 V8 N1 P132-	FPTS V23 N6 P.T1177 <FASE>
	FPTS V23 P.T1177 <FASE>
1963 V8 N1 P14C-	FPTS V23 N5 P.T1009 <FASE>
1963 V8 N1 P14C-141	FPTS,V23,N5,P.T1009 <FASE>
	64-31008 <=>
1963 V8 N2 P145-146	NLLTB V5 N11 P1008 <NLL>
	63-21897 <=>
1963 V8 N2 P147-153	64-31015 <=>
1963 V8 N2 P172-18C	AEC-TR-5966 <=*$>
	63-18602 <=*$>
1963 V8 N2 P181-190	63-18603 <=*>
1963 V8 N2 P201-211	63-18605 <=*>
1963 V8 N3 P288-300	64-31105 <=>
1963 V8 N3 P335-343	63-18609 <=*>
	64-13785 <=*$>
1963 V8 N3 P361-366	M.5772 <NLL>
1963 V8 N3 P387-393	64-31105 <=>
1963 V8 N3 P394-395	63-23124 <=*$>
1963 V8 N4 P433-440	64-31259 <=>
1963 V8 N4 P446-456	64-13115 <=*$>
	64-14798 <=*$>
1963 V8 N4 P475-486	64-31260 <=>
1963 V8 N4 P526	64-31261 <=>
1963 V8 N5 P536-542	64-21990 <=>
1963 V8 N5 P556-560	64-21990 <=>
1963 V8 N6 P715-721	64-21892 <=>
1964 V9 P293-298	64-18836 <*>
1964 V9 P414-422	65-12949 <*>
1964 V9 P537-544	65-63207 <=$>
1964 V9 P625-627	65-12945 <*>
1964 V9 P739-741	65-12943 <*>
1964 V9 N1 P33-39	64-21960 <=>
1964 V9 N1 P40-47	64-31159 <=>
964 V9 N1 P118-121	64-31161 <=>
1964 V9 N1 P124-127	64-21962 <=>
1964 V9 N1 P131-134	AD-625 857 <=$>
1964 V9 N2 P148-159	RTS-2892 <NLL>
	64-31924 <=>
1964 V9 N2 P172-183	64-31924 <=>
1964 V9 N2 P204-225	64-31924 <=>
1964 V9 N2 P233-236	64-31924 <=>
1964 V9 N2 P242-254	64-31924 <=>
1964 V9 N3 P293-305	64-41004 <=>
1964 V9 N3 P372-375	64-51137 <=$>
1964 V9 N3 P376-381	65-31505 <=$>
1964 V9 N4 P423-427	64-51172 <=>
1964 V9 N4 P500-501	65-63080 <=$>
1964 V9 N5 P589-596	65-33482 <=*$>
1964 V9 N5 P633-634	65-63075 <=$>
1964 V9 N6 P678-689	65-30206 <=$>
1964 V9 N6 P695-700	AD-629 865 <=$>
1965 V10 P242-245	65-12944 <*>
1965 V10 N1 P11-	FPTS V24 N6 P.T1041 <FASE>
1965 V10 N1 P32-36	65-63435 <=$>
1965 V10 N1 P148-	FPTS V24 N6 P.T1015 <FASE>
1965 V10 N2 P217-221	66-60870 <=*$>
1965 V10 N2 P232-236	66-60872 <=*$>
1965 V10 N2 P236-245	65-31779 <=$>
1965 V10 N2 P261-267	AD-638 597 <=$>
1965 V10 N2 P268-	FPTS V25 N2 P.T359 <FASE>
1965 V10 N2 P272-296	65-31779 <=$>
1965 V1C N2 P288	65-32158 <=$>
1965 V10 N2 P347-349	AD-638 589 <=$>
1965 V10 N2 P365-366	65-31271 <=$>
1965 V10 N2 P386-387	65-31270 <=$>
1965 V10 N3 P221-226	66-60871 <=*$>
1965 V10 N3 P404-408	66-61081 <=*$>
1965 V10 N3 P413-420	66-61076 <=*>
1965 V10 N3 P470-	FPTS V25 N2 P.T227 <FASE>
1965 V10 N3 P476-	FPTS V25 N2 P.T285 <FASE>
1965 V10 N4 P567-	FPTS V25 N3 P.T476 <FASE>
1965 V10 N4 P625-	FPTS V25 N3 P.T447 <FASE>
1965 V10 N4 P634-	FPTS V25 N3 P.T443 <FASE>
1965 V10 N4 P641-	FPTS V25 N3 P.T535 <FASE>
1965 V10 N4 P681-688	66-61309 <=*$>

Citation	Translation
1965 V10 N5 P723-	FPTS V25 N5 P.T847 <FASE>
1965 V10 N5 P805-	FPTS V25 N4 P.T716 <FASE>
1965 V10 N5 P826-	FPTS V25 N5 P.T901 <FASE>
1965 V10 N6 P918-	FPTS V25 N6 P.T1045 <FASE>
1965 V10 N6 P961-	FPTS V25 N5 P.T868 <FASE>
1965 V10 N6 P1003-	FPTS V25 N6 P.T961 <FASE>
1965 V10 N6 P1030-1036	66-61994 <=*$>
1966 V11 N1 P96-	FPTS V25 N6 P.T953 <FASE>

*BIOKHIMIYA

Citation	Translation
1938 V3 N4 P500-521	RT-1512 <*>
	64-10842 <=*$> 0
1939 V4 N5 P516-535	65-63431 <=$>
1940 V5 N5 P547-556	61-19739 <=*>
1940 V5 N5 P557-566	TT.141 <NRC>
1941 V6 N1 P29-36	5179-C <K-H>
1941 V6 N3 P302-311	65-62949 <=> 0
1943 V8 N1 P9-36	RT-490 <*>
	65-63374 <=*$>
1943 V8 N2/3 P97-107	65-63441 <=*$>
1943 V8 N5/6 P234-283	65-63001 <=>
1944 V9 P312-321	60-10829 <+*>
1945 V10 N1 P45-53	C-4487 <NRC>
1945 V10 N1 P79-81	RT-1302 <*>
	64-18569 <*>
	65-62964 <=$>
1945 V10 N2 P117-124	65-62952 <=$> 0
1945 V10 N2 P130-134	65-63805 <=>
1945 V10 N4 P360-362	61-28092 <*=> 0
1947 V12 P59-67	T-2647 <INSD>
1947 V12 N2 P141-152	63-13584 <=*>
1947 V12 N2 P153-162	RT-1687 <*>
1947 V12 N3 P201-208	65-63240 <=$>
1947 V12 N4 P285-290	RT-2160 <*>
	64-15369 <=*$>
1947 V12 N5 P389-405	60-10860 <+*>
1947 V12 N5 P421-436	2144 <K-H>
1947 V12 N5 P452-464	1730B <K-H>
1948 V13 P173-178	T-2648 <INSD>
1948 V13 P370-	R-4545 <*>
1948 V13 N1 P7-16	52/3064 <*>
1948 V13 N1 P39-41	RT-1710 <*>
1948 V13 N1 P55-60	RT-2400 <*>
	64-15062 <=*$> 0
1948 V13 N1 P88-94	R-667 <*>
1948 V13 N2 P124-126	RT-1686 <*>
1948 V13 N3 P219-224	RT-1025 <*>
1948 V13 N3 P236-243	R-2676 <*>
	64-19486 <=$>
1948 V13 N4 P370-377	RT-131 <*>
	1771 <K-H>
1948 V13 N5 P409-416	RT-1509 <*>
1948 V13 N6 P508-515	2602B <K-H>
	60-13733 <+*>
	60-18913 <+*>
	63-14577 <=*>
1948 V13 N6 P530-537	2150C <K-H>
1949 V14 P124-129	RT-173 <*>
1949 V14 P163-179	RT-174 <*>
1949 V14 P223-229	RT-172 <*>
1949 V14 P467-477	2207 <K-H>
1949 V14 N1 P14-19	RT-163 <*>
1949 V14 N1 P20-25	65-62953 <=$> 0
1949 V14 N2 P163-178	60-13862 <+*>
1949 V14 N3 P201-210	TT.163 <NRC>
1949 V14 N3 P223-229	63-14621 <=*>
1949 V14 N3 P256-258	RT-2895 <*>
1949 V14 N5 P413-418	RT-1458 <*>
1949 V14 N5 P424-431	RT-1711 <*>
1949 V14 N5 P449-451	RT-1688 <*>
1949 V14 N5 P466-477	R-567 <*>
1949 V14 N6 P487-498	61-19554 <=*>
1949 V14 N6 P499-502	RT-2080 <*>
1949 V14 N6 P524-537	59-20355 <+*>
	60-10828 <+*> 0
1950 V15 N1 P52-57	RT-2081 <*>
1950 V15 N3 P243-248	RT-1459 <*>
1950 V15 N3 P287-290	61-28112 <*=> 0
1950 V15 N4 P297-298	RT-1024 <*>
1950 V15 N4 P337-345	T-2280 <INSD>

Citation	Translation
1950 V15 N6 P528-533	2509 <K-H>
1951 V16 N1 P62-73	61-19553 <=*>
1951 V16 N2 P171-175	60-10824 <+*> 0
1951 V16 N2 P176-185	60-13713 <+*>
1951 V16 N4 P350-351	60-15934 <+*>
1951 V16 N5 P399-409	C-2495 <NRC>
	R-3110 <*>
	R-4138 <*>
	64-19435 <=$> 0
1951 V16 N5 P449-452	RT-3926 <*>
1951 V16 N5 P461-470	R-4386 <*>
1951 V16 N6 P522-	59-19103 <+*>
1951 V16 N6 P579-583	63-11043 <=>
1951 V16 N6 P604-610	61-21472 <=>
1951 V16 N6 P621-626	RT-2147 <*>
1952 P230-245	T1935 <INSD>
1952 V17 P97-107	1934 <INSD>
1952 V17 P195-197	RT-130 <*>
1952 V17 P230-245	T-2283 <INSD>
1952 V17 P456-461	RT-128 <*>
1952 V17 P469-475	RT-129 <*>
1952 V17 P664-675	RT-127 <*>
1952 V17 N1 P44-55	R-2694 <*>
1952 V17 N1 P82-90	RT-1595 <*>
1952 V17 N2 P195-197	61-28090 <*=>
1952 V17 N4 P456-461	61-28106 <*=> 0
1952 V17 N4 P469-475	61-28097 <*=> 0
1952 V17 N5 P551-556	64-10289 <=*$>
1952 V17 N5 P593-597	RT-1117 <*>
1952 V17 N5 P598-610	RT-1728 <*>
1952 V17 N6 P664-675	61-28102 <*=> 0
1952 V17 N6 P729-733	T-2282 <INSD>
1953 V18 P305-	60-18018 <+*>
1953 V18 P340-347	R-5271 <*>
1953 V18 N1 P7-11	RT-2740 <*>
1953 V18 N1 P51-55	RT-1950 <*>
1953 V18 N3 P311-314	RT-1175 <*>
1953 V18 N3 P351-353	RT-2013 <*>
	64-15557 <=*$>
1953 V18 N4 P393-411	RT-1532 <*>
	64-15577 <=*$>
1953 V18 N4 P448-451	RT-438 <*>
1953 V18 N4 P462-474	RT-4299 <*>
	1485 <*>
	64-19056 <=*$>
1953 V18 N4 P475-479	RT-439 <*>
1953 V18 N4 P480-483	1705 <K-H>
1953 V18 N5 P522-530	RT-1639 <*>
	64-15659 <=*$> 0
1953 V18 N5 P571-575	RT-2014 <*>
	64-15581 <=*$>
1953 V18 N5 P594-602	RT-2170 <*>
	64-15373 <=*$> C
1953 V18 N5 P618-625	65-12352 <*>
1953 V18 N6 P688-695	59-20356 <+*>
1953 V18 N6 P696-700	RT-2136 <*>
	64-15370 <=*$> 0
1953 V18 N6 P732-738	RT-2137 <*>
	64-15559 <=*$> 0
1953 V18 N6 P743-747	R-268 <*>
	RT-2423 <*>
	64-15067 <=*$>
1953 V18 N6 P748-752	R-95 <*>
	RT-2417 <*>
	64-15068 <=*$> 0
1954 V19 P578-585	R-5368 <*>
1954 V19 P641-644	57-1263 <*>
1954 V19 N1 P1-2	R-1130 <*>
	RT-3369 <*>
	64-19065 <=*$>
1954 V19 N1 P16-18	RT-2480 <*>
1954 V19 N1 P45-49	R-72 <*>
	RT-2664 <*>
	64-15056 <=*$> C
1954 V19 N1 P68-79	R-157 <*>
	RT-2568 <*>
	62-15297 <=*> C
1954 V19 N1 P126-128	RT-2481 <*>
1954 V19 N2 P208-215	63-11044 <=>
1954 V19 N3 P319-331	63-11045 <=>

Citation	Translation
1954 V19 N3 P332-335	C-3843 <NRCC>
1954 V19 N4 P440-448	59-19740 <+*>
1954 V19 N4 P478-484	R-619 <*>
	64-19505 <=$>
1954 V19 N5 P610-615	R-4143 <*>
	64-19434 <=$>
1955 V20 P57-65	T2002 <INSD>
1955 V20 P126-	R-3862 <*>
1955 V20 P496-506	R-399 <*>
1955 V20 P657-664	60-17232 <+*>
1955 V20 N2 P146-151	RJ-366 <ATS>
1955 V20 N2 P152-157	R-228 <*>
	1522 <*>
	64-19079 <=*$>
1955 V20 N2 P173-178	RT-3728 <*>
	1463 <*>
	64-19067 <=*$> 0
1955 V20 N3 P336-338	CSIRO-3393 <CSIR>
	R-1502 <*>
1955 V20 N3 P460-464	R-903 <*>
1955 V20 N4 P425-430	R-2703 <*>
	R-304 <*>
1955 V20 N4 P444-449	R-4140 <*>
	64-19436 <=$>
1955 V20 N4 P490-494	RT-3834 <*>
	64-19096 <=*$>
1955 V20 N5 P571-575	65-62128 <=$>
1955 V20 N6 P701-704	61-27647 <*=>
1955 V20 N6 P740-748	R-556 <*>
1956 V21 P126-136	R-1361 <*>
1956 V21 P380-384	T-2643 <INSD>
1956 V21 N1 P108-110	R-3778 <*>
	5484-A <K-H>
1956 V21 N6 P709-714	62-14709 <=*>
1956 V21 N6 P715-721	R-4615 <*>
1956 V21 N6 P777-783	R-4620 <*>
1957 V22 P33-40	R-4848 <*>
1957 V22 P351-367	R-4619 <*>
1957 V22 P546-553	R-4817 <*>
1957 V22 P830-837	R-4621 <*>
	62-15042 <=*>
1957 V22 P918-922	R-5331 <*>
1957 V22 N3 P537-	R-5358 <*>
1957 V22 N4 P707-714	R-4863 <*>
1957 V22 N4 P725-729	C-2564 <NRC>
1957 V22 N5 P776-788	59-10465 <+*>
1957 V22 N5 P807-812	R-4611 <*>
1957 V22 N1/2 P51-58	59-17380 <+*> 0
	62-15009 <*=>
1957 V22 N1/2 P359-368	59-17382 <+*> 0
1958 V23 N3 P347-350	60-18377 <+*> 0
1958 V23 N3 P388-389	108 <LSB>
1958 V23 N4 P635-638	59-20323 <+*>
1958 V23 N5 P674-682	59-17381 <+*>
1958 V23 N5 P751-754	59-17379 <+*> 0
1958 V23 N6 P840-844	60-18378 <+*>
1959 V24 N1 P63-66	60-10949 <+*>
1959 V24 N2 P291-300	59-13647 <=> 0
1959 V24 N3 P414-420	60-10948 <+*>
1959 V24 N3 P496-502	60-18369 <+*>
1959 V24 N3 P528-534	62-14691 <=*>
1959 V24 N3 P563-565	60-10950 <+*>
1959 V24 N4 P729-737	60-18370 <+*>
1959 V24 N5 P826-832	60-18371 <+*>
1959 V24 N5 P866-871	60-18368 <+*>
1959 V24 N5 P891-898	60-18367 <+*>
1960 V25 N2 P288-295	UCRL TRANS-562 <+*>
1960 V25 N2 P310-317	62-14690 <=*>
1960 V25 N6 P1099-1104	61-23058 <*=>
	62-14727 <=*> 0
1961 V26 N1 P82-85	11071-B <KH>
	64-10637 <=*$>
1961 V26 N1 P126-131	61-23830 <*=>
1961 V26 N2 P244-248	62-24648 <=*>
1961 V26 N2 P276-280	62-20435 <=*>
1961 V26 N6 P1027-1033	62-20367 <=*>
1961 V26 N6 P1090-1094	UCRL TRANS-815 <=*>
1962 V27 N2 P293-303	62-20355 <=*>
1962 V27 N5 P788-793	63-15111 <=*>
1963 V28 N2 P193-203	64-13793 <=*$>
1963 V28 N2 P340-344	64-15573 <=*$>
1963 V28 N2 P353-360	63-23416 <=*$>
1963 V28 N4 P616-621	64-16650 <=*$> 0
1963 V28 N4 P625-634	65-60620 <=$>
1963 V28 N5 P920-930	65-63213 <=*$>
1963 V28 N6 P1047-1052	65-61130 <=$>
1964 V29 N1 P116-125	64-15872 <=*$>
1964 V29 N2 P246-254	65-61138 <=$>
1964 V29 N2 P273-282	64-31547 <=>
1964 V29 N2 P300-311	65-60316 <=$>
1964 V29 N2 P353-374	64-31547 <=>
1964 V29 N3 P393-398	65-63760 <=>
1964 V29 N3 P477-486	64-31949 <=>
1964 V29 N3 P502-507	65-60313 <=$>
1964 V29 N4 P596-601	65-63396 <=$> 0
1964 V29 N5 P905-909	65-63077 <=$>
1965 V30 N2 P235-240	N66-13483 <=*>
1965 V30 N2 P344-349	65-33452 <=*$>
1965 V30 N2 P415-422	65-33452 <=*$>
1965 V30 N5 P980-984	65-34107 <=*$>
1965 V30 N6 P1251	66-12243 <*>
1966 V31 N2 P424-430	66-32164 <=$>

BIOKHIMIYA CHAINOGO PROIZVODSTVA

Citation	Translation
1962 V9 P177-181	64-13742 <=*$>
1962 V9 P185-188	RTS-2491 <NLL>
1962 V9 P189-191	65-60927 <=$>
1962 V9 P192-195	64-19117 <=*$>
1962 V9 P196-200	64-19952 <=$>
1962 V9 P201-203	64-19935 <=$>

BIOKHIMIYA PLODOV I OVOSHCHEI

Citation	Translation
1958 P59-68	60-17469 <+*>
1958 V4 P5-23	60-19875 <+*>
1958 V4 P42-50	59-22481 <+*> 0
	63-11085 <=>
1959 V5 P5-101	63-11011 <=>
1959 V5 P113-132	63-13574 <=*>
	63-19457 <=*>
1959 V5 P259-276	63-11025 <=>
1961 V6 P5-57	62-23420 <=*>
1961 V6 P77-95	63-11084 <=>
1961 V6 P175-207	63-11084 <=>
1961 V6 P219-227	63-11084 <=>
1961 V6 P252-261	63-11084 <=>

BIOKHIMIYA VINODELIYA

Citation	Translation
1959 V5 P27-37	9874-A <K-H>
1960 V6 P223-234	62-32238 <=*>

BIOKHIMIYA ZERNA

Citation	Translation
1958 N4 P229-240	60-19928 <+*>
1960 V5 P47-64	61-16092 <=*>

BIOKHIMIYA ZERNA I KHLEBOPECHENIYA

Citation	Translation
1960 V6 P66-74	65-63237 <=*$> 0
1964 V7 P117-138	RTS-2902 <NLL>
1964 V7 P151-158	RTS-2907 <NLL>
1964 V7 P159-166	65-61569 <=$> 0
1964 V7 P195-201	RTS-2906 <NLL>
1964 V7 P271-274	RTS-2905 <NLL>
	65-61570 <=$> 0

BIOLOGIA. BRATISLAVA

Citation	Translation
1955 V10 P773-777	59J18C <ATS>
1961 P31-39	AEC-TR-5847 <*>
1962 V17 N1 P71-73	62-25489 <=*>

BIOLOGIA NEONATORUM

Citation	Translation
1959 V1 N1 P37-60	62-00605 <*>

BIOLOGICA LATINA

Citation	Translation
1959 V12 N1 P119-123	60-10862 <*> 0

BIOLOGICHESKAYA NAUKA: SELSKOMUI LESNOMU KHOZYAYSTVU

Citation	Translation
1957 V3 P141-142	59-16804 <+*>

BIOLOGICHESKIE VOPROSY OVTESEVODSTVA I BOLEZNI OVETS

1955 N4 P79-87 65-63982 <=$>

BIOLOGICHESKII ZHURNAL. GOSUDARSTVENNOE BIOLOGICHESKOE I MEDITSINSKOE IZDATELSTVO
1934 V3 N2 P294-306 62-23844 <=*>
1935 V4 N5 P923-928 RT-2761 <*>
1935 V4 N6 P997-1004 61-31009 <=>
1938 V7 N2 P335-358 61-11485 <=>
1938 V7 N4 P837-865 61-11486 <=>

BIOLOGICHESKII ZHURNAL ARMENII
1966 V19 N2 P2-28 66-32508 <=*>

BIOLOGICHESKCE DEISTVIE RADIATSII
1962 N1 P81-83 64-21531 <=>

BIOLOGICKE PRACE SLOVENSKEJ AKADEMIE VIED
1956 V2 N11 P5 61-16440 <*> 0
6456B <K-H>
1963 V9 N8 P87-126 64-00428 <*>

BIOLOGICO
1950 V16 N2 P25-34 57-2589 <*>
1960 V26 P52-55 63-10096 <*>

BIOLOGIE MEDICALE
1956 V45 P231-246 58-1818 <*>
1957 V46 N1 P40-95 58-45 <*>
1957 V46 N2 P97-140 58-1075 <*>

BIOLCGISCHES ZENTRALBLATT
1893 V13 N21/2 P641-656 N66-13481 <=$>
1937 V57 P355-363 58-2155 <*>
1937 V57 P522-550 57-296 <*>
1938 V58 N5/6 P273-301 TT-800 <NRCC>
1949 V68 P232-243 60-18645 <*>
1956 V75 P129-149 TT-802 <NRCC>
1956 V75 P476-499 TT-803 <NRCC>
1956 V75 P576-596 TT-804 <NRCC>
1956 V75 N9/10 P597-611 57-3233 <*>

BIOLOGISKE SKRIFTER
1940 V1 N2 P5-19 60-10856 <*>

BIOLOGIYA BELCGC MORYA
1962 V1 P248-255 NASA-TT-F-189 <=>

BIOLOGIYA I KHIMIYA. SOFIA
1963 V6 N5 P1-8 64-21424 <=>

BIOLOGIYA V SHKOLE
1961 N3 P76-80 62-23635 <=*>
1961 N4 P3-6 61-28257 <*=>
1961 N4 P73-79 61-28261 <*=>
1961 N4 P90-91 61-28251 <*=>
1961 N4 P91-92 61-28298 <*=>
1965 N4 P86-90 65-33498 <=*$>

*BIULOSKI GLASNIK
1962 V15 N2 P55-132 62-11762/2 <*>

BIOMETRICS
1951 V7 P180-184 64-10012 <=>
1951 V7 P275-282 64-10025 <*>

BIOMETRISCHE ZEITSCHRIFT
1961 V3 P135-142 65-11877 <*>
1962 V4 N4 P263-273 AEC-TR-6236 <*>

BIOPHYSIK. EAST GERMANY
1963 V1 P20-32 UCRL TRANS-1016 <*>
1963 V1 P33-50 UCRL TRANS-1017 <*>
1964 V1 P218-219 ANL-TRANS-115 <*>
1964 V1 P225-259 ANL-TRANS-127 <*>
1964 V1 P289-296 ANL-TRANS-97 <*>

BI-SEDEH HA-BENIYAH
1955 N23 P1-14 PANSDOC-TR.260 <PANS>

BITAMIN
1953 V6 N1 P7-12 57-843 <*>
1953 V6 N1 P17-21 57-845 <*>
1953 V6 N2 P233-241 57-844 <*>
1953 V6 N5 P783-786 2708 <*>
1953 V6 N6 P951-955 2710 <*>
1953 V6 N6 P956-960 2707 <*>
1955 V9 P58-60 61-16087 <*>
1956 V10 N4 P217-223 61-00637 <*>

BITUMEN. BERLIN, HAMBURG
1939 V9 N5 P88-92 60-10293 <*> 0
1939 V9 N10 P31-35 59-17272 <*> 0
1953 V15 N5 P1C5-109 62-18689 <*>
1954 N1 P3-15 59-1C924 <*>

BITUMEN, TEERE, ASPHALTE, PECHE UND VERWANDTE STOFFE
1954 P123-124 T-1614 <INSD>
1954 V5 P129-131 T1669 <INSD>
1954 V5 P165-166 T-1615 <INSD>
1954 V5 P252-253 T2007 <INSD>
1955 V6 P118-119 T-1612 <INSD>
1955 V6 P185-187 T-1613 <INSD>
1955 V6 N1 P12-20 61-10208 <*> 0
1955 V6 N2 P56-57 CSIRO-2774 <CSIR>
1956 V7 P141-145 57-723 <*>
1956 V7 N5 P198-200 CSIRO-3384 <CSIR>
1959 V10 P12-21 57Q72G <ATS>
1959 V10 N12 P475-476 65-13C44 <*>
1960 V11 N1 P18-21 51M40G <ATS>
1960 V11 N7 P275-280 79M46G <ATS>
1961 V12 N1 P6-8 64-14014 <*>
1961 V12 N8 P329-335 89N57G <ATS>
1963 V14 N12 P628-633 2033 <TC>
1964 V15 P220-222 65-11946 <*>
1964 V15 P224-225 65-11946 <*>
1965 V16 P483-490 66-12062 <*>
1966 V17 N3 P112-114 66-13004 <*>

BIULETYN. GLOWNY INSTYTUT GORNICTWA
1961 N2 P11-14 CSIR-TR.323 <CSSA>

BIULETYN INFCRMACYJNY CENTRALNOGO LABORATORIA TECHNOLOGII PRZSTWORSTWA I PRZECHOWALNICTWA ZBOZ. WARSAW
1961 V5 N1 P4-9 AEC-TR-5795 <*>
1961 V5 N2 P29-34 AEC-TR-5797 <*>

BIULETYN INSTYTUT HODOWLI I AKLIMATYZACJI ROSLIN
1959 N4/5 P37-38 60-21579 <=>

BIULLETEN ASTRONOMICHESKIKH INSTITUTOV CHEKHOSLOVAKII. CESKOSLOVENSKA AKADEMIA
1952 V3 P6-12 65-10770 <*>

BLAD FOR BERGHANDTERINGENS VANNER
1960 N1 P17-47 1967 <BISI>

BLAETTER. DEUTSCHE GESELLSCHAFT FUER VERISCHERUNGSMATHEMATIK
1960 V5 N1 P55-59 66-11798 <*>

BLAETTER FUER PARADENTOSEFORSCHUNG UND DENTALE ANATOMIE UND PATHOLOGIE
1936 N32 P733-741 61-00648 <*>

BLAETTER FUER ZAHNHEILKUNDE
1956 V17 N1 P19-20 1738 <*>

BLECH
1957 P81-85 TS 971 <BISI>
1957 V4 N11 P81-85 61-18419 <*>
1957 V4 N11 P88-92 62-18402 <*>
1957 V4 N11 P92-98 61-18418 <*>
1958 V5 P17-23 2546 <BISI>
1959 V6 N10 P62-14685 62-14685 <=*>
1960 V7 P299-309 1982 <BISI>
1960 V7 P654-657 2229 <BISI>
1960 V7 N10 P650-653 2204 <BISI>
1961 V8 P6-8 2544 <BISI>

1961 V8 P179-183 2545 <BISI>
1961 V8 P494-497 2879 <BISI>
1961 V8 N3 P173-178 2981 <BISI>
1961 V8 N4 P243-247 2884 <BISI>
1961 V8 N5 P4- 65-17369 <*>
1961 V8 N8 P628-632 2976 <BISI>
1961 V8 N10 P740-744 2977 <BISI>
1961 V8 N11 P804-810 3138 <BISI>
1961 V8 N11 P810-820 3139 <BISI>
1962 V9 N5 P241-248 3082 <BISI>
1962 V9 N5 P258-260 3083 <BISI>
1962 V9 N11 P595-597 3691 <BISI>
1963 V10 N2 P58-67 3234 <BISI>
1963 V10 N2 P75-80 3235 <BISI>
1963 V10 N4 P170-180 3314 <BISI>
1963 V10 N7 P433-442 3695 <BISI>
1963 V10 N10 P739-741 3648 <BISI>
1963 V10 N11 P795-801 3917 <BISI>
1963 V10 N11 P815-820 4109 <BISI>
1963 V10 N11 P853-860 3844 <BISI>
1964 V11 N11 P5-12 4287 <BISI>
1964 V11 N11 P563-650 4319 <BISI>
1965 V12 RP1/2-1/7 4490 <BISI>
1965 V12 P6-11 4555 <BISI>
1965 V12 P55-56 4557 <BISI>
1965 V12 P118 4382 <BISI>
1965 V12 N2 P77-80 4301 <BISI>

BLUT
1956 P277-287 57-1814 <*>
1957 V3 P77-85 57-3116 <*>
1957 V3 P270-275 59-12107 <+*>
1958 V4 N1 P1-7 59-14576 <+*>
1960 V6 P399-413 62-00315 <*>
1960 V6 P414-423 62-00632 <*>

BOCHU KAGAKU
1949 N13 P11-13 57-772 <*>
1954 V19 P61-69 96K22J <ATS>

BODEN WAND UND DECKE
1964 V10 N10 P766-770 65-14923 <*>

BODENKULTUR
1952 V6 N3 P225-234 59-10359 <*>

BODENKUNDE UND PFLANZENERNAEHRUNG
1937 V4 P294-327 MF-6 <*>
1942 V28 N3 P129-156 CSIRC-3909 <CSIR>
1943 V31 P295-306 57-3428 <*>

BOEI EISEI
1962 V9 N5 P251-254 63-01124 <*>
1962 V9 N9 P399-402 63-01126 <*>

BOERENKRANT
1945 V1 N2 P2- 3267 <*>

BOIS ET FORETS DES TROPIQUES
1954 V33 P41-50 I-984 <*>
1958 N62 P41-45 60-16480 <*> 0
1960 N74 P35-49 62-00218 <*>
1962 N82 P53-58 63-00128 <*>
1963 N91 P41-52 65-00342 <*>
1965 V100 P5-19 66-13393 <*> 0

BOLETIM. ASSOCIACAO BRASILEIRA DE METAIS
1950 V6 P151-178 2804-C <K-H>
63-20443 <*> 0
1951 V7 P20-35 I-64 <*>
1961 V17 P71-80 4022 <BISI>
1962 V18 P793-808 3666 <BISI>
1964 V20 P109-164 4164 <BISI>
1964 V20 P175-212 4165 <BISI>
1964 V20 N85 P681-715 81S88PT <ATS>

BOLETIM CULTURAL DA GUINE PORTUGUESA
1952 V26 P351-357 T1875 <INSD>

BOLETIM. DEPARTMENTO DE ESTATISTICA. SAO PAULO
1953 N1 P25-36 59-11916 <=*>

BOLETIM. ESCOLA DE FARMACIA, COIMBRA
1961 V21 P153-163 64-18642 <*>

BOLETIM DO INSTITUTO JOAQUIM NABUCO DE PESQUISAS SOCIAIS
1963 V12 P5-22 66-31773 <=$>

BOLETIM DO MINISTERIO DA AGRICULTURA. DIVISAO DE GEOLOGIA E MINERALOGIA. BRAZIL
1955 N157 P1-54 63-16183 <*> 0

BOLETIM DO MINISTERIO DA AGRICULTURA, INDUSTRIA E COMERCIO. BRAZIL
1945 V1 N7/8 P303-315 3264 <*>

BOLETIM DA PESCA
1954 V43 P14-25 II-689 <*>

BOLETIM DO SERVICO NACIONAL DE LEPRA
1958 V17 P5-11 59-10967 <*>

BOLETIM DOS SERVICOS DE SAUDE PUBLICA
1955 V2 N1 P4-6 II-856 <*>
1955 V2 N1 P8-10 II-856 <*>
1955 V2 N1 P23-25 II-856 <*>

BOLETIN. ASOCIACION GENERAL ANTITUBERCULOSA DE PUERTO RICO
1950 V2 N9 P1-2 2523 <*>
1951 V3 N9 P1-2 2512 <*>

BOLETIN DE LA ASOCIACION MEDICA DE PUERTO RICO
1963 V55 N12 P481-488 65-14282 <*> 0

BOLETIN. ASOCIACION MEXICANA DE GEOLOGOS PETROLERAS
1952 V4 P153-262 62-20276 <*>
1960 V12 N7/8 P221-242 65-12757 <*>

BOLETIN. CENTRO ANTIRREUMATICO. UNIVERSIDAD NACIONAL BUENOS AIRES
1937 V1 P185-187 64-18786 <*>

BOLETIN CHILENO DE PARASITOLOGIA
1954 V9 P94-99 I-991 <*>

BOLETIN. COMISION HONORARIA PARA LA LUCHA ANTITUBERCULOSA
1948 V3 P4-5 2904 <*>

BOLETIN DEL DEPARTAMENTO FORESTAL Y DE CAZA Y PESCA DE MEXICO
1937 V2 N6 P165-202 57-1233 <*>

BOLETIN EPIDEMIOLOGICO
1963 V27 N1 P76-80 65-17294 <*>

BOLETIN. ESTACION EXPERIMENTAL AGRICOLA DE LA MOLINA. LIMA
1954 V90 P32-45 58-1195 <*>

BOLETIN DE GEOLOGIA. DIRECCION DE GEOLOGIA, VENEZUELA
1951 V1 N3 P289-293 2344 <*>

BOLETIN DE INFORMACIONES PETROLERAS. BUENOS AIRES (TITLE VARIES)
1935 V12 N125 P39-50 63-18115 <*>
1941 N3 P49-56 64N50SP <ATS>

BOLETIN INFORMATIVO. INSTITUTO NACIONAL DEL CARBON OVIEDO
1955 N19 P3-16 63-14061 <*>
1956 V5 P49-70 T1786 <INSD>
1956 V5 P71-87 T1787 <INSD>
1956 V5 N6 P71-86 CSIRO-3457 <CSIR>

BOLETIN DEL INSTITUTO DE ESTUDIOS MEDICOS Y

BIOLOGICOS, UNIVERSIDAD NACIONAL DE MEXICO
1961 V19 N2 P125-153 45P60SP <ATS>

BOLETIN. INSTITUTC FORESTAL DE INVESTIGACIONES Y EXPERIENCIAS. MADRID
1939 N17 P13-21 I-955 <*>
1939 N17 P29-33 I-955 <*>

BOLETIN DEL INSTITUTO DE MEDICINA EXPERIMENTAL PARA EL ESTUDIO Y TRATAMIENTO DEL CANCER
1938 V15 N47 P5-22 64-14515 <*>
9353-D <K-H>
1941 V18 P1003-1011 60-10831 <*>

BOLETIN DEL INSTITUTO NACIONAL DE FOMENTO TABACALERO. CENSO TABACELERO DE COLOMBIA. BOGOTA
1955 P16 58-1023 <*>

BOLETIN DEL INSTITUTO NACIONAL DE INVESTIGACIONES AGRONOMICAS
1945 V5 N12 P285-328 1997 <K-H>
1945 V5 N13 P17-67 1582A <K-H>
1946 V6 N15 P97-128 1691 <K-H>
1949 V9 P165-193 1986 <K-H>
1951 V11 N25 P1-27 2762-B <K-H>

BOLETIN INTERAMERICANO DE HOSPITALES
1956 V1 N1 P1-11 1736 <*>

BOLETIN DE LA LIGA CONTRA EL CANCER
1956 V31 P17-18 5339-D <K-H>
1956 V31 P85-86 5795-A <K-H>
1957 V32 N1 P1-4 64-14560 <*>
1957 V32 N3 P78 64-14561 <*>
1957 V32 N6 P175-187 8650-F <K-H>

BOLETIN MEDICC DEL HOSPITAL INFANTIL
1960 V17 P913-936 57N52SP <ATS>

BOLETIN DE MINAS Y PETROLEO. MEXICO
1945 V16 N1 P8-14 63-16179 <*> 0

BOLETIN MINERC. MEXICO
1922 V13 N3 P312-318 63-16180 <*>

BOLETIN OFICIAL DE LA ASCCIACION DE TECNICOS AZUCAREROS DE CUBA
1956 V15 P215-227 63-16347 <*>

BOLETIN DE LA OFICINA SANITARIA PAN-AMERICANA
1935 V14 N12 P1143-1146 62-00828 <*>
1950 V29 N6 P647-652 2624 <*>
1956 V40 P187-191 II-1120 <*>
1958 V44 P10-18 58-1803 <*>

BOLETIN DE LA SOCIEDAD ESPANOLA DE HISTORIA NATURAL. SECCION BIOLOGICA. MADRID
1958 V56 N1 P5-20 63-01135 <*>

BOLETIN DE RADIACTIVIDAD
1951 V24 P18-33 57-1222 <*>

BOLETIN SANITARIO. BUENOS AIRES
1942 V6 P731-737 SUP.1 64-14773 <*>
1942 V6 731-737 SUP.1 10265-E <KH>

BOLETIN DE LA SCCIEDAD DE GEOGRAFIA Y ESTADISTICA DE LA REPUBLICA MEXICANA
1949 V68 P19-44 63-16181 <*>
1949 V68 P45-62 63-16178 <*>

BOLETIN DE LA SCCIEDAD GECGRAFICA DE COLOMBIA
1965 V23 N85 P12-45 66-31798 <=>
1965 V23 N86 P67-77 66-31798 <=>

BOLETIN DE LA SOCIEDAD GEOLOGICA DEL PERU
1957 V32 N1 P5-19 63-16243 <*>

BOLETIN DE LA SOCIEDAD MEXICANA DE GEOGRAFIA Y ESTADISTICA
SEE BOLETIN DE LA SOCIEDAD DE GEOGRAFIA Y ESTADISTICA DE LA REPUBLICA MEXICANA

BOLETIN DE LA SOCIEDAD NACIONAL DE MINERIA. SANTIAGO DE CHILI
1947 V59 P34-41 II-427 <*>

BOLETIN DE LA SOCIEDAD QUIMICA DEL PERU
1952 V4 P43-59 MF-6 <*>

BOLETIN DE LA SOCIEDAD VENEZOLANA DE CIRUGIA
1955 V9 N42 P471-481 6012-A <K-H>

BOLETIN Y TRABAJOS DE LA SOCIEDAD DE CIRUGIA DE BUENOS AIRES
1936 V20 P749-760 57-1514 <*>

BOLLETTINO ED ATTI DELLA ACCADEMIA MEDICA
1938 V64 P165-171 57-1511 <*>

BOLLETINO CHIMICC-FARMACEUTICO
1932 V71 N2 P45-47 60-10826 <*>
1953 V92 P49-50 59-15740 <*>
1955 V94 P485-493 57-48 <*>
60-19000 <*+>
1958 V97 P560-566 60-10001 <*> 0
1961 V100 P200-201 63-10568 <*>
1962 V101 P948-954 63-20731 <*>
1963 V102 N11 P784-788 65-14279 <*> 0
1964 V103 P37-40 65-13746 <*>
1964 V103 N4 P164-167 65-13289 <*>
1964 V103 N9 P679-680 65-14751 <*>

BOLLETTINO DI GECFISICA TEORICA ED APPLICATA
1961 V3 N11 P2C9-214 61P62I <ATS>

BOLLETTINO DI INFORMAZICNI DELL'ASSOCIAZIONE NAZIONALE FRA INDUSTRIE AUTOMOBILISTICHE E AFFINI. TURIN
1958 V25 N2 ENTIRE ISSUE 59-11214 <=>

BOLLETTINO DELL'ISTITUTC CENTRALE DEL RESTAURO
1957 N31/2 P109-113 61-18270 <*>
1958 N33 P19-26 61-18620 <*>

BOLLETTINO DELL'ISTITUTO SIEROTERAPICO MILANESE
1949 V28 P244-251 58-393 <*>
1949 V28 P366-371 58-396 <*>
1950 V29 P194-201 58-413 <*>
1950 V29 P225-227 58-429 <*>
1950 V29 P252-257 58-419 <*>
1951 V30 P67-75 58-405 <*>
1953 V32 P482-488 64-18785 <*>
1954 V33 P643-647 3143 <*>
1954 V33 N11/2 P590-596 2225 <*>
1954 V33 N11/2 P612-623 61-00580 <*>
1954 V33 N11/2 P643-647 2266 <*>
1955 V34 N1/2 P1-4 2377 <*>
1955 V34 N3/4 P186-191 2264 <*>
1956 V35 P77-85 62K201 <ATS>
1956 V35 N11/2 P1-7 63-01629 <*>
1957 V36 P565-573 58-1830 <*>
1959 V38 N11/2 P441-445 62-00122 <*>
1960 V39 N9 P479-482 63-00162 <*>
1964 V43 P29-32 65-14713 <*>

BOLLETINO DELLE MALATTIE DELL'ORECCHIO, DELLA GOLA E DEL NASO
1951 V69 N11/2 P506-509 61-10585 <*>
1957 V75 N2 P175-179 63-16673 <*>
6821-A <K-H>
1959 V77 N2 P3-14 61-10586 <*>
1963 V81 P234-244 66-12751 <*>

BOLLETTINO E MEMORIE DELLA SOCIETA PIEMONTESE DI CHIRURGIA
1954 V24 P634-638 2460 <*>

BOLLETTINO E MEMCRIE DELLA SCCIETA TOSCOUMBRA DI CHIRURGIA

1955 V16 P47-49 II-830 <*>

BOLLETTINO MENSILE. OSSERVATORIO METEORICO-AEROLOGICO-GEODINAMICO DI MONTECASSINO
1938 V19 P53-72 UCRL TRANS-506(L) <*>

BOLLETTINO DI OCULISTICA
1936 V15 P1215-1231 58-420 <*>
1947 V26 P699-709 58-412 <*>
1948 V27 P41-53 58-417 <*>
1951 V30 N3 P129-134 64-18784 <*>
1954 V33 N1 P31-39 3252-A <K-H>
1962 V41 P3-12 66-12172 <*>

BOLLETTINO DI ONCOLOGIA
1957 V31 N5 P649-674 63-16893 <*> 0
8664 <K-H>

BOLLETTINO. R. ISTITUTO SUPERIORE AGRARIA. PISA
1930 V6 P263-299 3965-B <K-H>

BOLLETTINO SCHERMOGRAFICO
1951 V4 N1/2 P11-24 2538 <*>
1951 V4 N3/4 P61-70 2506 <*>

BOLLETINO SCIENTIFICO DELLA FACOLTA DI CHIMICA INDUSTRIALE. UNIVERSITA DI BOLOGNA
1940 N1/2 P1-11 3285 <*>
1950 V8 N4 P115-118 II-676 <*>
1954 V12 N2 P21-29 22G8I <ATS>

BOLLETTINO DELLE SCIENZE MEDICHE
1948 V120 P406-415 406-415 <*>

BOLLETTINO. SERVIZIO GEOLOGICO D'ITALIA
1955 V77 N1 P131-166 03H12I <ATS>

BOLLETINO DELLA SOCIETA ITALIANA DI BIOLOGIA SPERIMENTALE
1931 V6 P903-907 AL-937 <*>
1932 V7 P509-512 64-16223 <*>
1932 V7 P632-638 1378D <K-H>
1933 V8 P18-21 64-16224 <*>
1933 V8 P21-22 64-16225 <*>
1936 V11 P31-32 64-16226 <*>
1936 V11 P33-34 64-16227 <*>
1938 V13 P973-976 1147-A <K-H>
1938 V13 P1004-1006 1139-A <K-H>
1938 V13 N10 P976-978 61-00576 <*>
1938 V13 N10 P979-981 61-00575 <*>
1938 V13 N10 P981-983 61-00577 <*>
1938 V13 N10 P983-985 61-00578 <*>
1939 V14 P338-340 64-16205 <*>
1939 V14 P623-625 2751 <*>
1939 V14 N11 P740-741 59-20069 <*>
1940 V15 P491-492 T-1648 <INSD>
1940 V15 P1054-1056 60-10663 <*>
1940 V15 P1193-1194 60-10879 <*>
1941 V16 N2 P157 60-10655 <*>
1941 V16 N7 P417-419 60-10654 <*>
1941 V16 N12 P768-770 60-10653 <*> 0
1942 V17 P136-138 922A <K-H>
1942 V17 N2 P138-140 64-16242 <*>
1942 V17 N3 P206- I-566 <*>
1942 V17 N3 P206-208 60-10659 <*>
1943 V18 P77-79 69S88I <ATS>
1945 V20 N11 P678-680 59-20064 <*> 0
1946 V21 P182-184 AL-560 <*>
1946 V22 P57-59 58-399 <*>
1946 V22 P1272-1273 58-422 <*>
63-16385 <*>
1946 V22 N5 P485-486 3074 <*>
1947 V23 P907-908 60-10670 <*>
1947 V23 P1149-1153 58-409 <*>
1947 V23 P1176-1179 58-400 <*>
1948 V24 P449-455 58-407 <*>
1948 V24 P692-693 58-428 <*>
1949 V25 P337- AL-73 <*>
1949 V25 P1082-1084 I-510 <*>
1950 V26 P269-271 63-20442 <*>
1950 V26 N7 P1051-1054 II-1013 <*>
1951 N27 P651-653 CSIRO-3917 <CSIR>
1951 V27 P94-96 59-12101 <=*>
1951 V27 P430-433 60-10877 <*>
1951 V27 P1096-1098 9850 <K-H>
1951 V27 P1233 58-557 <*>
63-16428 <*>
1951 V27 P1234-1236 58-558 <*>
1951 V27 P1236-1238 58-556 <*>
63-16495 <*>
1951 V27 N3 P491-493 2191 <*>
1951 V27 N3 P494-496 2192 <*>
1951 V27 N6 P1096-1098 65-11243 <*>
1951 V27 N9/11 P1456-1459 1456 <*>
1952 V28 P612-615 3324-B <K-H>
1952 V28 P1224-1225 1694 <*>
1952 V28 N2 P147-149 2736 <*>
1952 V28 N5 P1030-1032 II-302 <*>
1953 V29 P9-11 1382 <*>
1953 V29 P782-785 09L29I <ATS>
1953 V29 P792-795 1939 <*>
1953 V29 N1 P34-41 1829 <*>
1953 V29 N1 P41-43 2030 <*>
1953 V29 N2 P152-155 3169-C <K-H>
1953 V29 N4 P504-505 II-546 <*>
1953 V29 N4 P523-525 II-432 <*>
1953 V29 N4 P657-659 1604 <*>
1953 V29 N5 P1096-1098 58-1840 <*>
1953 V29 N7 P1458-1460 1830 <*>
1953 V29 N7 P1511-1514 59-17854 <*>
1953 V29 N8 P1613-1614 58-1799 <*>
1953 V29 N12 P1876-1877 58-1798 <*>
1953 V29 N12 P1991-1993 64-18783 <*>
1953 V29 N9/11 P1739-1741 II-939 <*>
1954 V30 P4-5 II-97 <*>
1954 V30 P261-263 63-18765 <=*>
1954 V30 P308-311 II-97 <*>
1954 V30 N1/2 P34-35 1169 <*>
1954 V30 N1/2 P35-37 1713 <*>
1954 V30 N1/2 P37-39 1721 <*>
1954 V30 N1/2 P824-825 CSIRO-3787 <CSIR>
1954 V30 N1/2 P825-827 CSIRO-3756 <CSIR>
1954 V30 N4/5 P308-311 58-383 <*>
1955 V31 P157-159 3156 <*>
1955 V31 P422-425 1403 <*>
1955 V31 P426-428 1599 <*>
1955 V31 P883-885 5352-B <K-H>
1955 V31 N1/2 P101-102 57-236 <*>
1955 V31 N3/4 P293-295 39M43I <ATS>
1955 V31 N3/4 P357-358 II-634 <*>
1955 V31 N7/8 P906-909 57-1218 <*>
1955 V31 N9/10 P1159-1163 64-18782 <*> 0
1955 V31 N9/10 P1261-1263 62-18291 <*>
1955 V31 N9/10 P1263-1264 62-18290 <*>
1955 V31 N9/10 P1265-1266 62-18286 <*>
1956 V32 N7/8 P654-655 64-18779 <*>
1956 V32 N7/8 P727-729 58-376 <*>
1957 V33 P327-330 59-10343 <*> 0
1957 V33 P1258-1260 95K22I <ATS>
1957 V33 N7 P1142-1145 61-00162 <*>
1958 V34 P1556 03N48I <ATS>
1958 V34 N9 P424-426 8500-E <K-H>
1958 V34 N14 P713-715 60-19491 <*+>
1958 V34 N19 P1214-1216 59-10622 <*>
1958 V34 N20 P1326-1328 64-14601 <*>
9331-A <K-H>
1959 V35 P469-470 60-15085 <*+>
1959 V35 P471-472 60-15086 <*+>
1959 V35 P473-474 60-15084 <*+>
1959 V35 P1791-1793 78N48I <ATS>
1960 V36 P794-796 62-00753 <*>
1960 V36 P955-956 61-16108 <*>
1960 V36 P1160-1162 C-4629 <NRCC>
64-10804 <*>
1960 V36 P1163-1165 C-4630 <NRCC>
64-10803 <*>
1960 V36 P1357-1361 65-00145 <*>
1960 V36 N6 P262-266 60-16916 <*> 0
1960 V36 N8 P379-383 62-14095 <*>
1960 V36 N8 P790-794 61-10814 <*>

1960 V36 N1C P452-454 10481-A <KH>
63-18464 <*>
1960 V36 N10 P454-458 60-18756 <*>
1960 V36 N10 P458-461 60-18755 <*>
1960 V36 N12 P598-602 10607-A <KH>
63-16895 <*>
1960 V36 N14 P727-729 10764-A <K-H>
63-16930 <*>
1961 V37 P189-192 62-10731 <*>
1961 V37 P192-195 61-20778 <*>
1961 V37 P675-678 11548-A <KH>
63-16587 <*>
1961 V37 P766-769 11517-C <KH>
1961 V37 P769-771 11537-C <KH>
1961 V37 P826-827 63-20876 <*>
1961 V37 P990-992 63-01391 <*>
1961 V37 P1310-1312 62-20035 <*>
1961 V37 N2 P41-43 66-12759 <*> 0
1961 V37 N6 P260-262 62-10196 <*>
1961 V37 N7 P325-329 62-01233 <*>
1961 V37 N8 P359-363 61-20735 <*> 0
1961 V37 N14 P633-635 62-18178 <*>
62-20034 <*>
1961 V37 N15 P752-754 62-14978 <*> 0
1961 V37 N16 P761-762 62-18060 <*>
1962 V38 P37-40 63-20892 <*>
1962 V38 P610-611 63-10093 <*>
1962 V38 N21 P1076-1078 65-11332 <*> 0
1962 V38 N21 P1104-1107 65-11331 <*> 0
1963 V39 P435-437 65-00141 <*>
1963 V39 N14 P810-812 65-10988 <*>
1963 V39 N2C P1207-1209 65-10979 <*>
1963 V39 N20 P1209-1210 65-10980 <*>
1963 V39 N20 P1211-1213 66-10963 <*>
1963 V39 N24 P1937-1940 65-13447 <*>
1963 V39 N24 P1940-1943 65-13446 <*>
1963 V39 N24 P1974-1977 65-17253 <*>
1964 V40 N18 P1065-1067 65-13461 <*>
1964 V40 N18 P1067-1069 65-17252 <*>
1964 V40 N2C P1260-1262 65-14952 <*>
66-10961 <*>
1964 V40 N2C P1262-1264 65-17028 <*>
1964 V40 N21 P1321-1324 65-13462 <*>
1964 V40 N24 P2232-2235 66-11787 <*>
1965 V41 N1 P2-4 66-11434 <*>
1965 V41 N6 P335-336 66-10738 <*>
1965 V41 N8 P417-418 66-11280 <*>
1965 V41 N11 P581-583 65-14818 <*>
1965 V41 N12 P682-684 56S85I <ATS>

BOLLETTINO DELLA SOCIETA ITALIANA DI ENDOCRINOLOGIA. ROME
1952 V2 P97-98 2382 <*>

BOLLETINC DELLA SOCIETA ITALIANA DI GEOFISICA E METEOROLOGIA
1956 V4 N1/2 P8-10 R-2869 <*>

BOLLETTINO DELLA SOCIETA ITALIANA DI PATOLOGIA
1957 V5 N3 P145-148 63-00305 <*> P

BOLLETTINO DELLA SOCIETA MEDICO-CHIRURGICA DI MODENA
1932 V46 P699-722 58-397 <*>
1947 V47 P277-282 AL-791 <*>
1948 V48 P3-8 I-915 <*>

BOLLETTINO DELLA SOCIETA MEDICO-CHIRURGICA DI PAVIA
1957 V71 P921-936 62-14956 <*>

BOLLETTINO. SOCIETA MEDICO-CHIRURGICA DI PISA
1946 V12 P1-3 60-10821 <*>
1959 V27 P122-135 62-16608 <*>
1961 V29 N6 P1-4 AEC-TR-5550 <=*$>

BOLLETTINO DELLA SOCIETA DI NATURALISTI I NAPOLI
1890 V4 P189-208 T-1390 <INSD>

BOLLETTINO TECNICO FINSIDER
1965 N217 P281-294 4534 <BISI>
1965 N217 P295-303 4533 <BISI>

BOLLETTINO DELL'UNIONE MATEMATICA ITALIANA
1956 S3 V11 P158-167 62-20228 <*>

BOLLETTINC DI ZOOLOGIA, PUBBLICATO DALL'UNIONE ZOOLOGICA ITALIANA
1964 V31 N2 P573-581 66-12672 <*>

BOL'NICHNAYA GAZETA BOTKINA
1897 V8 P406 60-19750 <=*$>
1897 V8 N17 P634-639 60-19789 <=*$>

BOLOGNA MEDICA
1961 V7 P87-99 64-16932 <*>

BOR ES CIPOTECHNIKA
1962 V12 N5/6 P133-135 63Q73H <ATS>

BORBA
1966 P8 66-32721 <=$>
1966 P12-13 66-33804 <=$>

BORBA S SILIKOZOM
1953 P141-150 R-43 <*>
1953 P151-161 R-59 <*>
1953 P157-175 R-60 <*>
1953 P162-166 R-94 <*>
1953 P176-179 R-33 <*>
1953 P180-185 R-260 <*>

BORBA S TUBERKULEZOM
1932 N10 P633-642 65-60086 <=$>

BORDEAUX CHIRURGICAL
1960 N2 P87-91 64-18617 <*>

BORGYOGYASZATI ES VENEROLCGIAI SZEMLE
1959 V13 N2 P88-90 62-14982 <*> 0

BOSHOKU GIJUTSU
1958 V7 P223-228 <BRS>
1959 V8 N4 P147-149 62-18775 <*>
1964 V13 N2 P65-69 65-11784 <*>
1964 V13 N7 P303-308 1386 <TC>

BOTANICAL MAGAZINE. TOKYO
SEE SHOKUBUTSUGAKU ZASSHI

BOTANICHESKIE MATERIALY GERBARIYA BOTANICHESKOGO INSTITUTA V. A. KOMAROVA. TASHKENT
1954 V16 P29-38 66-51078 <=>

BOTANICHESKII ZHURNAL
1935 V20 N4 P414-417 60-51009 <=>
1935 V20 N6 P600-616 RT-2705 <*>
1936 V21 N2 P189-195 63-11042 <=>
1943 V28 N4 P154-170 63-23409 <=*$>
1946 V31 N3 P13-21 66-10246 <*> 0
1946 V31 N8 P1205-1208 62-25169 <=*>
1947 V32 N2 P61-64 66-61883 <=*$>
1948 V33 N5 P487-495 60-21901 <=>
63-13599 <=*>
1950 V35 N5 P445-460 RT-1666 <*>
1950 V35 N6 P647-655 61-15958 <*=> 0
1951 V36 N1 P5-20 RT-1932 <*>
RT-2212 <*>
1951 V36 N1 P21-28 62-10746 <=*>
1951 V36 N5 P517-522 RT-1933 <*>
1951 V36 N6 P597-606 AEC-TR-5012 <=*>
1952 V37 N5 P585-593 RT-2266 <*> P
1954 V39 P48-57 GB146 <NLL>
1954 V39 N1 P122-125 RT-4001 <*>
1954 V39 N2 P180-186 61-15003 <*+> P
1954 V39 N2 P2C2-203 RT-3668 <*>
1954 V39 N2 P202-223 RT-3900 <*>
1954 V39 N3 P317-325 61-15383 <+*>
1954 V39 N4 P482-487 63-11014 <=>
1954 V39 N6 P797-808 RT-4617 <*>

1955 V40 N1 P64-90	R-115 <*>
1955 V40 N1 P155-156	R-65 <*>
1955 V40 N1 P156-157	R-64 <*>
1955 V40 N3 P408-410	R-3658 <*>
1955 V40 N4 P481-507	RT-3870 <*>
1955 V40 N4 P542-547	R-4584 <*>
1955 V40 N4 P587-592	R-3648 <*>
1955 V40 N4 P592-596	R-340 <*>
1955 V40 N4 P603-604	RT-3892 <*>
1955 V40 N5 P696-702	CSIRO-3432 <CSIR>
	R-4567 <*>
	R-5098 <*>
	62-10744 <=*>
1955 V40 N5 P727-728	61-15927 <+*>
1956 V41 N1 P64-80	60-51182 <=>
1956 V41 N1 P81-84	60-51185 <=>
1956 V41 N1 P85-89	60-51186 <=>
1956 V41 N2 P193-205	RT-4264 <*>
1956 V41 N2 P254-257	R-4381 <*>
1956 V41 N6 P797-809	59-10244 <+*>
	59-10867 <+*>
1956 V41 N6 P836-854	R-3914 <*>
1956 V41 N11 P1647-1652	R-4703 <*>
1957 V42 P1035-1043	65-64099 <=*$>
1957 V42 P1097-1099	61-10299 <*+>
1957 V42 N1 P41-56	60-51001 <=>
1957 V42 N1 P68-72	63-11183 <=>
1957 V42 N1 P72-78	61-15928 <+*>
1957 V42 N1 P95-97	R-3161 <*>
1957 V42 N2 P276-280	60-51000 <=>
1957 V42 N4 P517-534	60-21149 <=>
1957 V42 N4 P639-641	61-15925 <+*>
1957 V42 N5 P691-702	61-19567 <*=> 0
1957 V42 N5 P741-745	60-51043 <=>
1957 V42 N7 P1011-1034	62-13467 <=*>
1957 V42 N8 P1172-1181	60-21109 <=>
1957 V42 N8 P1182-1195	65-63559 <=*$>
1957 V42 N11 P1573-1595	R-4186 <*>
1958 V43 P1630-1633	60-51005 <=>
1958 V43 N1 P50-60	59-12761 <+*>
1958 V43 N1 P146-153	59-12764 <+*>
1958 V43 N1 P153-155	59-12765 <+*>
1958 V43 N1 P156-157	59-12762 <+*>
1958 V43 N1 P157-158	59-12763 <+*>
1958 V43 N2 P178-193	62-13466 <=*>
1958 V43 N2 P242-246	61-15004 <*+>
1958 V43 N3 P317-336	59-12767 <+*>
1958 V43 N4 P612-617	59-13263 <=>
1958 V43 N5 P679-683	60-51071 <=>
1958 V43 N6 P831-836	60-15564 <+*>
1958 V43 N6 P869-876	63-11006 <=>
1958 V43 N7 P989-997	59-12756 <+*>
1958 V43 N7 P1054-1056	59-18851 <+*>
1958 V43 N8 P1093-1107	59-22319 <+*>
1958 V43 N8 P1135-1145	59-22317 <+*>
1958 V43 N10 P1445-1459	60-21882 <=>
1958 V43 N11 P1654-1657	59-16806 <+*>
1958 V43 N12 P1708-1712	60-21911 <=>
1958 V43 N12 P1763-1765	NLLTB V1 N5 P31 <NLL>
1959 V44 N1 P101-104	59-16826 <+*>
1959 V44 N2 P1-6	59-22340 <*>
1959 V44 N2 P202-209	60-51022 <=>
1959 V44 N2 P215-219	60-21893 <=>
1959 V44 N4 P536-543	61-13100 <+*>
1959 V44 N5 P645-647	60-15357 <+*> 0
1959 V44 N6 P854-860	63-11009 <=>
1959 V44 N6 P868-872	AD-639 491 <=$>
1959 V44 N9 P1260-1270	60-41378 <=>
1959 V44 N9 P1291-1298	61-27252 <*=>
1959 V44 N9 P1311-1314	61-28144 <*=>
1959 V44 N9 P1364-1371	60-11838 <=>
1959 V44 N10 P1425-1436	60-19896 <+*>
1959 V44 N10 P1437-1444	60-19897 <+*>
1959 V44 N10 P1515-1518	60-31087 <=>
1959 V44 N12 P1730-1734	63-19605 <=*>
1959 V44 N12 P1783-1785	60-31157 <=>
1960 V45 P1781-1786	65-61247 <=$>
1960 V45 N1 P34-47	61-28640 <*=>
1960 V45 N1 P48-63	62-32236 <=*>
1960 V45 N3 P382-393	62-32971 <=*>
1960 V45 N3 P1169-1175	61-28643 <*=>
1960 V45 N4 P472-479	60-41303 <=>
1960 V45 N4 P480-491	60-11988 <=>
1960 V45 N4 P564-566	61-15308 <+*>
1960 V45 N5 P637-648	63-13604 <=*>
1960 V45 N5 P778-782	61-11070 <=>
1960 V45 N7 P1063-1066	64-19134 <=*$> 0
1960 V45 N9 P1350-1356	63-13589 <=*>
	63-19548 <=*>
1960 V45 N9 P1395-1399	63-23183 <=*>
1960 V45 N10 P1476-1487	64-18592 <*>
1960 V45 N10 P1502-1503	61-28215 <*=>
1960 V45 N10 P1506-1511	63-15258 <=*>
1960 V45 N10 P1538-1540	61-28218 <*=>
1960 V45 N12 P1819-1823	61-21580 <=>
1960 V45 N12 P1828-1833	61-21580 <=>
1961 V46 P673-676	UCRL TRANS-737 <=*>
1961 V46 N1 P3-20	62-13606 <*=>
1961 V46 N1 P112-115	62-13458 <=*>
1961 V46 N1 P115-119	62-23713 <=*>
1961 V46 N1 P135-139	61-28641 <*=>
1961 V46 N2 P161-173	AEC-TR-4808 <=*>
1961 V46 N3 P377-386	64-13602 <=*$>
1961 V46 N3 P447-454	63-19547 <=*>
1961 V46 N3 P463-465	62-24401 <=>
1961 V46 N7 P1001-1006	98P59R <ATS>
1961 V46 N7 P1006-1008	62-23569 <=*> 0
1961 V46 N7 P1055-1057	62-23570 <=*>
1961 V46 N10 P1385-1401	NLLTB V4 N4 P305 <NLL>
1961 V46 N10 P1533-1537	65-60345 <=$>
1961 V46 N11 P1574-1583	62-24529 <=*>
1961 V46 N12 P1774-1780	62-33328 <=*>
1962 V47 P96-99	CCL-65/4 <STC>
1962 V47 N1 P92-95	65-60401 <=$>
1962 V47 N5 P626-	AEC-TR-5492 <=*>
1962 V47 N6 P802-807	64-15729 <=$>
1962 V47 N7 P982-986	63-15970 <=*>
1962 V47 N7 P1029-1032	65-23229 <$>
1962 V47 N7 P1046-1047	<CAOA>
1962 V47 N9 P1318-1323	64-15296 <=*$>
1962 V47 N12 P1858-1864	63-31646 <=>
1963 V48 N1 P16-34	64-11757 <=>
1963 V48 N1 P113-118	AD-616 725 <=$> 0
1963 V48 N1 P126	AD-616 974 <=$>
1963 V48 N4 P486-501	65-60891 <=$>
1963 V48 N4 P557-563	65-63523 <=*$> 0
1963 V48 N5 P724-726	63-24034 <=*$>
1963 V48 N6 P867-869	65-63749 <=$>
1963 V48 N8 P1151-1160	64-13132 <=*$>
1963 V48 N9 P1271-1281	AD-617 077 <=$>
1963 V48 N9 P1404-1406	AD-625 641 <=$>
1963 V48 N10 P1484-1489	RTS-2704 <NLL>
1963 V48 N10 P1500-1511	64-21092 <=>
1963 V48 N11 P1653-1659	65-60887 <=$>
1963 V48 N12 P1839-1840	64-21491 <=>
1963 V48 N12 P1841-1844	64-21492 <=>
1963 V48 N12 P1857-1860	64-21493 <=>
1964 V49 N1 P156-159	64-31456 <=>
1964 V49 N2 P161-176	64-15319 <=*$>
1964 V49 N2 P280-286	64-15318 <=*$>
1964 V49 N3 P403-404	C-5707 <NRC>
1964 V49 N7 P980-987	RTS-2916 <NLL>
1964 V49 N9 P1237-1247	65-20427 <*>
1964 V49 N9 P1272-1278	65-23234 <$>
1964 V49 N9 P1381-1382	NLLTB V7 N5 P454 <NLL>
	TRANS.BULL.V7N5 <NLL>
1964 V49 N12 P1826-1838	65-31298 <=$> 0
1965 V50 N3 P368-370	65-25352 <$>
1965 V50 N5 P730-733	65-31610 <=$>
1965 V50 N9 P1205-1247	66-30117 <=*$>

BOTANISCHES ARCHIV

1940 V41 P159-167	5227 <K-H>

BOTANY AND ZOOLOGY. TOKYO
SEE SHOKUBUTSU OYOBI DOBUTSU

BOTYU-KAGAKU. SCIENTIFIC INSECT CONTROL

1949 V14 P10-19	63-16794 <*> 0
1950 V15 P97-103	63-14838 <*>

1950 V15 P149-155 <ATS>
1950 V15 P201-206 2546 <K-H>
63-18951 <*>
1952 V17 P64-74 63-18671 <*>
1953 V18 P122- 2828 <K-H>
63-20736 <*>

BOUW
1950 V5 P398-405 T1570 <INSD>
1955 V10 P961-963 T1855 <INSD>
1958 V13 N34 P862-868 C-3513 <NRCC>
1959 V14 N39 P1103-1105 C-3359 <NRCC>
1960 V15 N6 P154-160 C-3613 <NRCC>
1961 V16 N41 P1290-1292 C-4053 <NRCC>

BOUWBEDRIJF
1932 V9 N22 P275-283 2068 <*>

BRAGANTIA
1944 V4 P523-540 1240D <K-H>

BRAIN AND NERVE. TOKYO
SEE NO TO SHINKEI

BRANDSCHUTZ
1956 V10 N1 P23- II-682 <*>
1957 P107-108 58-818 <*>
1957 V11 N4 P77-79 58-2410 <*>
1957 V11 N4 P156-160 58-2409 <*>
1957 V11 N10 P224-226 58-2408 <*>

BRANNTWEINWIRTSCHAFT
1948 V2 N8 P113-114 59-15118 <*>
1962 V102 N14 P379-384 37S83G <ATS>

BRASIL ACUCAREIRO
1948 V31 P633-639 T1585 <INSD>
1954 V44 P60-62 T-1915 <INSD>

BRASIL-MEDICO
1936 V50 N37 P793-798 64-10833 <*>
1942 V56 P355-370 57-1528 <*>
1944 V58 P72-74 57-1518 <*>
1953 V67 N2 P43-46 62-16182 <*>
1954 V68 N14/7 P11-18 5069-A <K-H>
1955 V69 P33 T1646 <INSD>
1955 V69 N14/8 P173-177 3799-B <K-H>

BRATISLAVSKE LEKARSKE LISTY
1960 V40 N3 P156-160 62-19459 <=*>
1962 V42 N6 P321-322 62-25047 <=*>

BRAUNKOHLE
1937 V36 P832-839 57-1579 <*>
1940 V39 N8 P71-75 64-16199 <*>
1941 V40 P49- 61-10638 <*> 0

BRAUNKOHLE, WAERME UND ENERGIE
1953 V5 N1/2 P1-12 63-16126 <*>
1953 V5 N3/4 P49-60 63-16126 <*>
1956 V8 N11/2 P221-227 63-16204 <*> 0
1956 V8 N9/10 P165-174 63-16204 <*> 0
1957 V9 N23/4 P486-493 59-17463 <*> 0
1958 V10 P163-165 <DIL>
1958 V10 P193-205 <DIL>
1959 V11 N1 P1-14 63-16203 <*> 0
1960 V12 P53-59 2230 <BISI>
1960 V12 N10 P472-478 61-10361 <*>
1961 V13 N7 P276-282 62-10210 <*>

BRAUNKOHLENARCHIV
1930 N31 P1-10 63-10432 <*>

BRAUWELT. ED. B
1951 N48 P57-59 78H12G <ATS>

BRAUWISSENSCHAFT
1960 V13 P91-92 61-10152 <*>

BRENNSTOFF-CHEMIE
1921 V2 N1 P9 1959 <*>
1923 V4 P353-357 II-581 <*>
1923 V4 N23 P353-357 3411-B <K-H>
64-14684 <*>
1927 V8 N15 P241-244 1942 <*>
1927 V8 N20 P321-323 64-18815 <*>
1928 V9 N7 P105-113 1843 <*>
1928 V9 N8 P121-122 1844 <*>
1929 V10 P406-408 63-10472 <*>
1929 V10 N10 P203-205 64-10773 <*>
1929 V10 N16 P324-329 63-14226 <*> 0
64-20059 <*>
1930 V11 P1-9 60-00240 <*>
1930 V11 P489-500 63-10694 <*> 0
1930 V11 N14 P277-281 63-10059 <*>
1930 V11 N22 P449-452 63-10445 <*>
1931 V12 P69-71 65-13130 <*>
1931 V12 N7 P122-127 64-20310 <*> 0
1931 V12 N18 P345-348 64-20860 <*> 0
1932 V13 N2 P21-27 1985 <*>
1933 V14 P121- 3244 <*>
1933 V14 P463-468 63-14227 <*>
1934 V15 N18 P341-347 1869 <*>
1934 V15 N21 P404-405 1946 <*>
1935 V16 P368-369 63-20532 <*>
1935 V16 P466-469 63-20533 <*>
1936 V17 P1-11 63-10463 <*> 0
1936 V17 P203-206 63-20538 <*>
1936 V17 P221-228 63-20634 <*>
1936 V17 P301- 2423 <*>
1936 V17 P466-470 65-13129 <*>
1936 V17 N18 P341-351 3064 <*>
1936 V17 N24 P461-465 58-2216 <*>
1937 V17 N2 P61-67 63-10460 <*> 0
1938 V19 N12 P226-230 64-20134 <*>
1939 V20 P317-319 62-20127 <*>
1939 V20 N3 P41-48 64-30655 <*> 0
1939 V20 N13 P247-250 64-18168 <*> 0
1940 V21 P85-87 62-18949 <*>
1940 V21 N4 P37-42 60-00460 <*>
1940 V21 N12 P133-141 64-20861 <*> 0
1940 V21 N14 P157-167 60-18405 <*>
1940 V21 N15 P169-177 60-18404 <*>
1940 V21 N22 P257-264 64-20128 <*>
1941 V22 N20 P229-236 59-17460 <*>
1942 V23 N6 P67-73 65-12581 <*>
1943 V24 N4 P39-40 60-18058 <*>
1949 V30 P60-68 2424 <*>
1949 V30 P81-84 2425 <*>
1949 V30 N1/2 P13-22 2426 <*>
1949 V30 N17/8 P285-299 64-10282 <*> 0
1950 V31 P10- 2421 <*>
1950 V31 P143-145 63-14823 <*> P
1950 V31 P148 63-14823 <*> P
1950 V31 N1/2 P14-22 61-20856 <*>
1950 V31 N11/2 P173-180 1787 <*>
1950 V31 N21/2 P337-350 AL-259 <*>
1950 V31 N23/4 P361-374 64-20862 <*> 0
1951 V32 P65-69 AL-744 <*>
1951 V32 N1/2 P1-32 58-684 <*>
1951 V32 N3/4 P5-16 58-684 <*>
1951 V32 N5/6 P69-74 AL-468 <*>
1951 V32 N11/2 P161-174 AL-382 <*>
58-2354 <*>
1951 V32 N13/4 P193-198 AL-198 <*>
1951 V32 N9/10 P33 AL-957 <*>
1951 V32 N9/10 P129-133 AL-526 <*>
1952 V33 P1-12 AL-746 <*>
1952 V33 P13-21 AL-386 <*>
1952 V33 P21-30 1648 <*>
2428 <*>
1952 V33 P193-200 63-16805 <*>
1952 V33 P296-307 3292 <*>
1952 V33 N11/2 P193-200 61-00389 <*>
1952 V33 N21/2 P370-375 42N56G <ATS>
1953 V34 N1/2 P6-11 1949 <*>
1953 V34 N3/4 P37-45 57-1902 <*>
1953 V34 N5/6 P83-87 63-18973 <*>
1953 V34 N7/8 P118-122 64-18161 <*>

Citation	Translation
1953 V34 N21/2 P330-333	3K26G <ATS>
1954 V35 P308-310	T1490 <INSD>
1954 V35 P321-325	64-20825 <*>
1954 V35 P325-334	86G4G <ATS>
1954 V35 N21 P321-325	62-20123 <*>
1954 V35 N21 P334-337	CSIRO-3112 <CSIR>
1954 V35 N23 P363-368	63-10013 <*> 0
1954 V35 N3/4 P41-44	47F3G <ATS>
1954 V35 N7/8 P105-112	65L35G <ATS>
1954 V35 N13/4 P202-211	1952 <*>
1954 V35 N15/6 P225-231	I-718 <*>
1954 V35 N15/6 P236-246	1952 <*>
1954 V35 N17/8 P269-275	1952 <*>
1954 V35 N19/0 P298-304	1952 <*>
1954 V35 N21/2 P321-325	I-878 <*>
	6G5G <ATS>
1954 V35 N23/4 P353-362	59-20056 <*> P
1954 V35 N23/4 P368-371	66L35G <ATS>
1955 V36 P33-37	TS 1247 <BISI>
1955 V36 P225-228	T1516 <INSD>
1955 V36 P353-358	T-1683 <INSD>
1955 V36 N19 P304-314	IGR V2 N1 P68 <AGI>
1955 V36 N1/2 P1-11	53T91G <ATS>
1955 V36 N3/4 P33-37	61-20668 <*> 0
1955 V36 N11/2 P176-181	37N49G <ATS>
1955 V36 N13/4 P193-199	CSIRO-3201 <CSIR>
	CSIRO-3201 <CSIR>
1955 V36 N19/2 P304-314	60-12604 <=>
1955 V36 N21/2 P321-328	64-16515 <*>
	88H8G <ATS>
1955 V36 N9/10 P151-155	58-1097 <*>
1956 V37 P47-53	59-10643 <*> 0
1956 V37 N3/4 P33-41	75K23G <ATS>
1956 V37 N3/4 P42-46	18H10G <ATS>
1956 V37 N11/2 P161-171	CSIRO-3409 <CSIR>
1956 V37 N11/2 P182-186	CSIRO-3408 <CSIR>
	IGR V1 N11 P89 <AGI>
1956 V37 N11/2 P186-189	CSIRO-3407 <CSIR>
1956 V37 N17/8 P263-267	64-16549 <*>
1956 V37 N19/C P301-310	CSIRO-3537 <CSIR>
1956 V37 N19/0 P336-341	60-16661 <*>
1956 V37 N19/2 P310-317	60-12606 <=>
1956 V37 N23/4 P391-395	CSIRO-3647 <CSIR>
1956 V37 N23/4 P404-408	93J14G <ATS>
1957 V38 P173-174	6699 <K-H> 0
1957 V38 N11 P173-174	64-30105 <*> 0
1957 V38 N1/2 P2-9	07J16G <ATS>
1957 V38 N1/2 P9-14	64-16500 <*>
1957 V38 N3/4 P51-54	CSIRO-3755 <CSIR>
1957 V38 N3/4 P58-60	C-2431 <NRC>
1957 V38 N7/8 P107-116	61-16205 <*>
1957 V38 N19/0 P289-297	61-18443 <*>
1957 V38 N21/2 P321-329	61-10497 <*>
1957 V38 N23/4 P372-373	59-17288 <*> 0
1958 V39 P1-6	<DIL>
1958 V39 P17-19	62-16878 <*>
1958 V39 P65-74	9027 <IICH>
1958 V39 P84-8'	<DIL>
1958 V39 P141-145	<DIL>
1958 V39 P146-148	<DIL>
1958 V39 P161-163	<DIL>
1958 V39 P353-359	<DIL>
1958 V39 PS1-S6	<DIL>
1958 V39 PS17-S19	TS 1694 <BISI>
1958 V39 N3/4 P33-43	59-10173 <*>
1958 V39 N13/4 P213-217	60-14161 <*>
1958 V39 N17/8 P257-268	02L30G <ATS>
	61-10499 <*>
	64-30977 <*> 0
1958 V39 N17/8 P271-273	60-00600 <*>
1958 V39 N19/C P302-306	61-10498 <*>
	65-11481 <*>
1958 V39 N21/2 P329-331	59-10426 <*> 0
1958 V39 N23/4 P368-369	76L30G <ATS>
1959 V40 P261-263 PT.2	60-00458 <*>
1959 V40 P76-85	TS 1484 <BISI>
1959 V40 P198-201	<DIL>
1959 V40 P346-354	60-19849 <*=>
1959 V40 N2 P33-41	38L31G <ATS>
1959 V40 N2 P52-55	60-10993 <*> 0
1959 V40 N3 P76-85	62-14365 <*>
1959 V40 N4 P97-104	59-17649 <*>
	60-16911 <*> 0
1959 V40 N4 P115-122	60-21716 <=>
1959 V40 N5 P160-161	60-23541 <*=>
1959 V40 N5 P164-166	64-30658 <*>
1959 V40 N6 P177-186	65-10932 <*> 0
1960 V41 N1 P14-18	61-00917 <*>
1960 V41 N4 P113-119	1868 <BISI>
1960 V41 N4 P263-272	61-20052 <*>
1960 V41 N9 P257-263	61-20053 <*>
1960 V41 N10 P304-308	70N50G <ATS>
1960 V41 N11 P321-325	10613 <K-H>
1960 V41 N11 P321-352	22N49G <ATS>
1960 V41 N11 P321-325	65-12521 <*>
	65-13043 <*>
1961 V42 P295-296	ORNL-TR-186 <*>
1961 V42 P312-319	62-01156 <*>
1961 V42 P385-394	62-01157 <*>
1961 V42 N1 P11-17	65-13042 <*>
1961 V42 N2 P50-54	65-12954 <*>
1961 V42 N4 P136-140	61-20077 <*>
1961 V42 N5 P167	65-13016 <*>
1961 V42 N7 P220-223	92N55G <ATS>
1961 V42 N8 P261-267	65-12419 <*>
1961 V42 N9 P273-277	49N58G <ATS>
1961 V42 N9 P278-283	90P58G <ATS>
1961 V42 N10 P305-336	63-00100 <*>
1961 V42 N12 P369-375	65-12430 <*>
1962 V43 P161-168	9041 <IICH>
1962 V43 N3 P71-78	62-18847 <*>
1962 V43 N4 P97-105	63-20195 <*>
1962 V43 N4 P106-108	62-01153 <*>
1962 V43 N5 P129-134	65-12294 <*>
1962 V43 N6 P31-34	63-10011 <*>
1962 V43 N6 P173-176	63-10131 <*>
1962 V43 N8 P251-252	4544 <BISI>
1962 V43 N9 P260-263	53P66G <ATS>
1962 V43 N9 P266-268	64-30041 <*>
1962 V43 N10 P302-308	54P66G <ATS>
1963 V44 N1 P1-5	65-12470 <*>
1963 V44 N2 P33-37	RAPRA-1257 <RAP>
1963 V44 N2 PW22-W25	65-13230 <*>
1963 V44 N4 P97-104	3694 <BISI>
1963 V44 N4 P110-118	87Q71G <ATS>
1963 V44 N5 P144-149	3422 <BISI>
1963 V44 N5 P149-154	3423 <BISI>
1963 V44 N6 P184-186	65-14800 <*>
1963 V44 N10 P309-310	64-16499 <*>
1963 V44 N10 P310-	3709 <BISI>
1963 V44 N11 P339-343	31R78G <ATS>
1963 V44 N12 P383-387	3710 <BISI>
	64-16908 <*>
1964 V45 N3 P89-92	65-10479 <*> 0
1964 V45 N5 P9-12	65-00280 <*>
1964 V45 N5 P132-138	66-11400 <*>
1964 V45 N5 P144-150	65-00429 <*> 0
1964 V45 N5 P151-155	4257 <BISI>
1964 V45 N6 P161-165	CHEM-391 <CTS>
	65-13399 <*>
1964 V45 N6 P178-181	3976 <BISI>
1964 V45 N7 P194-200	560 <TC>
	66-11401 <*>
1964 V45 N7 P200-206	66-11402 <*>
1964 V45 N9 P258-261	CHEM-392 <CTS>
	65-13400 <*>
1964 V45 N9 P275-280	4088 <BISI>
1964 V45 N10 P295-299	CHEM-393 <CTS>
	65-13401 <*>
1964 V45 N11 P321-330	CHEM-394 <CTS>
	65-17450 <*>
1964 V45 N12 P360-371	65-00273 <*>
1965 V46 N1 P17-20	4272 <BISI>
1965 V46 N2 P48-51	4368 <BISI>
1965 V46 N3 P65-66	4425 <BISI>
1965 V46 N5 P129-133	99S86G <ATS>
1965 V46 N8 P251-252	4544 <BISI>
1965 V46 N9 P275-276	66-12932 <*>
1965 V46 N11 P377	66-14358 <*>
1965 V46 N12 P387-392	94T92G <ATS>

1965 V46 N12 P397-399 66-13847 <*>
1966 V47 N6 P161-169 80T95G <ATS>

BRENNSTOFF-WAERME-KRAFT
1949 V1 N8 P203-208 57-2640 <*>
1951 V3 P144-148 AEC-TR-4965 <*>
1951 V3 P422-426 TS 1488 <BISI>
1951 V3 N5 P141-143 60-18804 <*>
1951 V3 N12 P422-425 62-14330 <*>
1952 V4 P2-6 1857 <*>
1952 V4 P62-66 60-18006 <*>
1953 V5 P116 61-20918 <*> O
1953 V5 N2 P45-49 64-16246 <*> O
1953 V5 N7 P237-239 48F4G <ATS>
1953 V5 N10 P333-337 59-17853 <*> O
1954 V6 P244-249 1839 <*>
1955 V7 N1 P1-10 59-17521 <*>
1956 V8 P1-9 58-1829 <*>
1956 V8 P264-269 T2255 <INSD>
1956 V8 N1 P1-9 64K27G <ATS>
1956 V8 N11 P521-524 61-16208 <*>
1958 V10 N6 P298-299 61-10126 <*>
1959 V11 P223-233 60-17111 <=*>
1959 V11 N9 P407-413 AEC-TR-3977 <*>
1959 V11 N10 P455-462 61-00077 <*>
1960 V12 N1 P9-13 AEC-TR-5172 <*>
84P60G <ATS>
1960 V12 N2 P62-64 60-00788 <*>
1960 V12 N4 P142-144 1908 <BISI>
1960 V12 N6 P238-243 61-00916 <*>
1960 V12 N8 P347-350 61-20051 <*>
1960 V12 N8 P356-358 61-00915 <*>
66-15269 <*>
1961 V13 P105-115 AEC-TR-5922 <*>
1961 V13 N4 P168-170 88P58G <ATS>
1962 P105-115 62-01573 <*>
1962 V14 N8 P361-367 65-13611 <NLL>
1963 V15 N7 P331-337 3518 <BISI>
1964 V16 N3 P132-137 90T92G <ATS>
1965 V17 N6 P292-295 65-14208 <*>
1965 V17 N6 P296-298 65-14205 <*>
1965 V17 N6 P298-304 4430 <BISI>
1966 V18 P76-79 4837 <BISI>

BRENNSTOFF- U. WAERMEWIRTSCHAFT
1929 V12 P427-435 63-10438 <*>

BRODOGRADNJA. ZAGREB
1966 V17 N5/6 P185-189 67-30335 <=>

BROT UND GEBAECK
1954 V8 P114-115 63-16478 <*>
1955 V9 P103-105 63-14878 <*>
1955 V9 P139-141 63-14877 <*>
1957 V11 N9 P187-189 63-14966 <*> P
1959 V13 N11 P205-209 NP-TR-926 <*>
1960 V14 N11 P215-223 63-14979 <*>
1962 V16 P64-70 63-16567 <*>
1962 V16 N6 P108-116 63-10721 <*>

BROWN BOVERI MITTEILUNGEN
1960 V47 N12 P860-867 137-593 <STT>
1966 V53 N1/2 P142-148 91T95G <ATS>

BRUNS BEITRAEGE ZUR KLINISCHEN CHIRURGIE
1932 V155 P109-124 66-10867 <*>
1957 V194 N4 P395-407 65-12010 <*>

BRUXELLES MEDICAL
1954 V34 P1745-1750 II-435 <*>
1954 V34 N10 P422-428 64-18781 <*> P
1957 V37 N35 P1285-1304 58-1357 <*>
1957 V37 N46 P1752-1756 64-18780 <*>
1958 V38 N18 P719-733 W-59 <RIS> O
1958 V38 N19 P765-777 W-62 <RIS>
1961 V41 N4 P101-115 62-00375 <*>
1961 V41 N43 P1530-1535 63-00312 <*>

BUDAPESTI MUSZAKI EGYETEM MEZOGAYDASAGI KEMAI TECHNOLOGIAI TANSZGAKENEK EVKONYVE
1952 V3/8 P92-96 5857-A <K-H>

BUDIVELNI MATERIALY I KONSTRUKTSII
1962 V4 N4 P13-15 63-21125 <=> O

BUDOWNICTWO OKRETOWE
1957 V3 P117-122 60-13229 <=*>
1961 N3 P90-92 63-11394 <=>
1964 V9 N7 P230 2043 <BSRA>
1964 V9 N8 P268 2043 <BSRA>
1964 V9 N9 P303 2043 <BSRA>
1964 V9 N10 P340 2053 <BSRA>
1965 V10 P20 2431 <BSRA>
1965 V10 P79 2270 <BSRA>
1965 V10 P112 2270 <BSRA>

BUDOWNICTWO WIEJSKIE
1958 V10 N11 P24-26 60-21224 <=>
1958 V10 N12 P16-17 60-21242 <=>

BUILDING ENGINEERING. TOKYO
SEE KENCHIKU GIJUTSU

BUKHGALTERSKII UCHET
1960 N9 P5-10 61-31565 <=>
1960 N10 P5-11 61-31565 <=>
1962 N9 P16-20 63-13663 <=>
1963 V20 N7 P1-4 63-31680 <=>
1963 V20 N7 P44-48 63-31599 <=>
1965 V22 N5 P3-9 65-31586 <=$>
1965 V22 N10 P19-21 65-33604 <=*$>
1966 V23 P77-79 66-31686 <=$>
1966 V23 N7 P3-6 66-34606 <=$>

BULETIN STIINTIFIC. ACADEMIA REPUBLICII POPULARE ROMANE. SERIA: GEOLOGIE, GEOGRAFIE, BIOLOGIE, STIINTE TEHNICE SI AGRICOLE. SECTIUNEA DE STIINTE BIOLOGICE, AGRONOMICE, GEOLOGICE SI GEOGRAFICE. BUCHAREST
1948 V30 N9 P1-7 I-66 <*>
58-1517 <*>

BULETIN STIINTIFIC. ACADEMIA REPUBLICII POPULARE ROMANE. SECTIA DE GEOLOGIE SI GEOGRAFIE
1957 V2 N3/4 P605-612 59-11438 <=*>

BULETIN STIINTIFIC. ACADEMIA REPUBLICII POPULARE ROMANE. SECTIA DE STIINTE MATEMATICE SI FIZICE
1952 V4 N3 P617-627 59-15993 <*>
63-14494 <*>

BULETIN STIINTIFIC. ACADEMIA REPUBLICI POPULARE ROMANE. SECTIA DE STIINTE TEHNICE SI CHIMICE
1954 V6 P251-275 UCRL TRANS-836(L) <*>
1954 V6 N4 P1043-1083 86P62RU <ATS>

BULETIN STIINTIFIC. ACADEMIA REPUBLICII POPULARE ROMANE. SERIA: STIINTE MEDICALE
1949 V1 N10 P847-849 64-19524 <=$>

BULETINUL CULTIVAREI SI FERMENTAREI TUTUNULUI
1953 V40 P3-10 3084 D <K-H>

BULETINUL INSTITUTULUI POLITEHNIC BUCURESTI
1957 V19 P2-3 AEC-TR-3899 <*>
1959 V21 N2 P97-105 CSIR-TR.283 <CSSA>
1963 V25 N2 P35-46 ICE V4 N3 P511-516 <ICE>

BULETINUL INSTITUTULUI POLITEHNIC DIN IASI
1958 V4 P23-32 STMSP V3 P341-349 <AMS>
1958 V4 N8 P227-236 61-18269 <*> O
1958 V4 N3/4 P19-68 AMST S2 V33 P59-121 <AMS>
1958 SNS V4 N1/2 P227-236 61-10397 <*>
1959 V5 P3-4 11477 <K-H>
62-20256 <*> O
1959 V5 N1/2 P51-100 AMST S2 V33 P123-187 <AMS>
1959 V5 N3/4 P177-193 63-15708 <=*>
1962 V8 N1/2 P17-19 63-31867 <=>
1962 V8 N3/4 P121-128 19R76RU <ATS>

BULETINUL OFICIAL AL REPUBLICII SOCIALISTE ROMANIA
1965 N5 P37-38 PT.1 65-33990 <=*$>
1965 N13 P106-112 66-30035 <=*$>
1966 P393-394 PT1 66-35213 <=>

BULETINUL SOCIETATII DE STIINTE DIN BUCURESTI
1905 V14 P49- AL-984 <*>
1905 V14 P288 AL-985 <*>

BULETINUL SOCIETATII DE STIINTE DIN CLUJ
1929 V4 N1 P521- AL-983 <*>

BULETINUL STIINTIFIC. INSTITUTUL POLITEHNIC, CLUJ
1961 V4 P45-54 64-71303 <=>
1962 V5 P39-49 AD-609 692 <=>

BULGARSKI KNIGOPIS
1960 V64 P44-48 62-19458 <*=>

BULGARSKI TITUTIUM
1958 V3 N2 P70-73 59-17799 <*>
1959 V4 P454-457 11499-C <K-H>
63-16899 <*>
1959 V4 P500-503 11499-B <K-H>
1959 V4 P500-502 63-16898 <*>
1960 V5 P150-154 11411-C <KH>
63-16648 <*>
1960 V5 N9/10 P429-439 10825 <K-H>
1961 V6 N1 P29-32 63-16646 <*>
1961 V6 N1 P33-43 63-16887 <*>
1961 V6 N5 P26-33 11547-A <K-H>
63-16674 <*>
1961 V6 N6 P39-44 11499-A <KH>
63-16612 <*>
1961 V6 N7 P42-44 11537-D <KH>
1961 V6 N9 P30-46 2629 <K-H>
1962 V7 N1 P9-12 63-41073 <=>
1962 V7 N5 P23-27 63-41073 <=>
1962 V7 N9 P17-20 13074-B <K-H>
1962 V7 N9 P21-29 12900-B <K-H>
1962 V7 N7/8 P30-33 63-16493 <*>
1964 V9 N8/9 P4-19 64-51156 <=>
1964 V9 N8/9 P38-41 64-51156 <=>

BULLETIN DE L'ACADEMIE MALGACHE
1954 V31 P71-77 63-00514 <*>

BULLETIN DE L'ACADEMIE DE MEDECINE
1918 S3 V80 P655- 57-1525 <*>
1920 S3 V84 P93-102 59-15539 <*>
1930 V103 P256-266 3072 <*>
1932 S3 V107 N4 P126-128 59-15541 <*>
1932 S3 V108 P1678-1681 1109E <KH>
1934 S3 V111 P58-63 60-10853 <*>
1935 S3 V113 P60-65 61-14491 <*>
1936 V115 P385-392 63-00599 <*>
1937 S3 V117 P299-310 I-521 <*>
1945 S3 V129 P564-575 58-415 <*>

BULLETIN DE L'ACADEMIE NATIONALE DE MEDECINE
1947 S3 V131 N15/6 P265-267
59-15912 <*>
1951 V135 P95-97 I-133 <*>
1951 V135 P169-173 I-127 <*>
1951 V135 N1/2 P7-8 2564 <*>
1951 V135 N1/2 P29-32 2557 <*>
1951 S3 V135 P314-317 64-18950 <*>
1953 S3 V137 P441-442 64-18943 <*>
1958 V142 N9/10 P229-235 61-00295 <*>
1958 S3 V142 P573-575 64-14599 <*>
8739 <K-H>
1959 S3 V143 N32 P731-734 63-16693 <*>
1959 S3 V143 N5/6 P106-110
60-15640 <=*> 0
1959 S3 V143 N15/6 P367-370
61-15905 <=*>
1961 S3 V145 P197-201 11170-A <K-H>
1961 S3 V145 N20/1 P416-423
11433-A <KH>
63-20125 <*>
1963 S3 V147 N29 P610-620 64-16929 <*> 0

BULLETIN DE L'ACADEMIE POLONAISE DE SCIENCES. SERIE DES SCIENCES MATHEMATIQUES, ASTRONOMIQUES ET PHYSIQUES
1961 V9 N2 P69-74 62-23267 <=*>

BULLETIN DE L'ACADEMIE POLONAISE DE SCIENCES. SERIE DES SCIENCES TECHNIQUES
1960 V8 N11/2 P655-660 63-14448 <*>
64-16953 <*>

BULLETIN DE L'ACADEMIE R. DE BELGIQUE. CLASSE DES SCIENCES
1902 N6 P445-494 64-16708 <*>
1909 P728-743 66-10328 <*>
1950 S5 V37 N1 P56-78 66-11141 <*>
1951 P784-791 I-951 <*>
1951 S5 V37 N12 P1019-1036
65-11661 <*>
1952 V38 N5 P770-779 I-426 <*>
1952 V38 N1/2 P197-218 65-14095 <*>
1952 S5 P770-779 I-426 <*>
1952 S5 V38 P106-112 AL-945 <*>
1952 S5 V38 P169-175 T-2127 <INSD>
1952 S5 V38 P197-218 63-20057 <*>
1952 S5 V38 P711-717 63-20652 <*> 0
1952 S5 V38 P851-860 63-20651 <*>
1955 S5 V41 P759 65-10054 <*>
1960 S5 V46 N5 P390-395 62-16459 <*> 0

BULLETIN DE L'ACADEMIE R. DE MEDECINE DE BELGIQUE
1932 S6 V12 P252-265 <K-H>
1962 S7 V2 P561-572 63-20900 <*> 0
1964 S7 V4 N4 P267-311 65-11628 <*>

BULLETIN DE L'ACADEMIE VETERINAIRE DE FRANCE
1949 V22 P145-150 61-20580 <*>
1953 N5 P267-278 I-1009 <*>
1953 V25 P351-360 60-18541 <*> 0
1953 V26 P417-420 20F3F <ATS>
1953 V26 N1 P73-78 MF-2 <*>
1953 V26 N2 P135-140 MF-2 <*>
1953 V26 N8 P391-399 01F3F <ATS>
1953 V26 N8 P421-422 00F3F <ATS>
1954 V27 P441-446 3168 <*>
1956 V29 P367-371 58-169 <*>

BULLETIN AGRICOLE DU CONGO BELGE
1952 V43 P999-1009 3084-E <K-H>
1956 V47 N1 P93-111 CSIRO-3696 <CSIR>

BULLETIN AGRONOMIQUE. MINISTERE DE LA FRANCE D'OUTRE MER
1951 V6 P19-32 II-174 <*>

BULLETIN ALGERIEN DE CARCINOLOGIE
1953 V6 P19-24 3247-B <K-H>
1953 V6 P401-403 3183-C <K-H>

BULLETIN DES ANCIENS ELEVES DE L'ECOLE FRANCAISE DE MEUNERIE
1956 N155 P223- 63-14972 <*>
1957 N158 P65-69 63-16477 <*>

BULLETIN ANNUEL DE LA SOCIETE SUISSE DE CHRONOMETRIE ET DU LABORATOIRE SUISSE DE RECHERCHES HORLOGERES
1948 V2 P507-546 59-20096 <*> 0

BULLETIN DE L'ASSOCIATION DES CHIMISTES DE SUCRERIE, DE DISTILLERIE ET DES INDUSTRIES AGRICOLES DE FRANCE ET DES COLONIES
1902 V20 P1475-1476 63-14969 <*>
1932 V49 P153-163 66-14258 <*>
1932 V49 P189-195 64-20905 <*>
1933 V50 P224-231 66-14259 <*>
1938 V55 P275-285 58-621 <*>

BULLETIN DE L'ASSOCIATION DES DIPLOMES DE MICRO-

BIOLOGIE DE LA FACULTE DE PHARMACIE DE NANCY
1960 N80 P16-19 63-10531 <*>

BULLETIN DE L'ASSOCIATION FRANCAISE POUR L'ETUDE DU CANCER
1932 V21 P385-396 66-10947 <*> O
1932 V21 P570-573 I-16 <*>
1950 V37 P15-19 58-361 <*>
1950 V37 P52-56 60-10861 <*> O
1952 V39 N4 P450-460 2196 <*>
1954 V41 N1 P46-64 57-1158 <*>
1954 V41 N4 P423-444 1194 <*>
1955 V42 P247-278 63-16934 <*>
6871-B <KH>
1956 V43 N2 P180-198 5709-A <K-H>
1957 V44 P409-421 58-1822 <*>
1957 V44 N1 P92-105 62-01032 <*>
1957 V44 N1 P197-202 63-01764 <*>
1957 V44 N2 P336-361 60-17483 <=*>
6871E <K-H>
1957 V44 N3 P387-408 63-16695 <*> O
8342-A <K-H>
1957 V44 N3 P426-439 8346B <K-H>
1957 V44 N3 P440-443 58-1695 <*>
1957 V44 N4 P483-492 58-1688 <*>
1958 V45 P1-37 63-16935 <*>
8619-D <KH>
1958 V45 P400-416 63-16697 <*> O
9061-A <K-H>
1958 V45 N3 P289-300 SCL-T-277 <*>
1958 V45 N4 P434-444 62-01005 <*>
1958 V45 N4 P454-459 61-00218 <*>
1959 V46 P295-309 63-16698 <*> O
9728 <K-H>
1959 V46 N1 P75-91 60-13095 <=*> O
1959 V46 N2 P212-252 62-01062 <*>
1959 V46 N2 P336-346 62-01040 <*>
1959 V46 N2 P356-381 62-01033 <*>
1960 V47 N2 P291-307 62-01076 <*>
1961 V48 P112-121 11279-G <KH>
63-16696 <*>
1962 V49 N4 P416-420 63-01104 <*>

BULLETIN DE L'ASSOCIATION FRANCAISE POUR L'ETUDE DU SOL
1959 V4 P193-198 62-20244 <*> O

BULLETIN DE L'ASSOCIATION FRANCAISE DES TECHNICIENS DU PETROLE
1951 N87 P53-64 <ATS>
1956 N118 P291-310 62-18521 <*>
1959 N133 P43-61 91L35F <ATS>
1960 N140 P3-18 65-14730 <*>
1961 N145 P59-72 35N53F <ATS>
1961 N146 P259-272 70N53F <ATS>
1961 N148 P471-484 65-12806 <*>
1963 N157 P127-138 78Q69F <ATS>
1964 N163 P3-14 66-10652 <*>
1964 N167 P591-604 04S81F <ATS>

BULLETIN DE L'ASSOCIATION DES GEOGRAPHES FRANCAIS
1965 N334 P2-12 66-32613 <=$>

BULLETIN DE L'ASSOCIATION DES MEDECINS DE LANGUE FRANCAISE
1954 V83 P1087- 1835 <*>

BULLETIN DE L'ASSOCIATION PERMANENTE DES CONGRES BELGES DE LA ROUTE
1935 V23 P17-35 T-2437 <INSD>

BULLETIN DE L'ASSOCIATION SUISSE DES ELECTRICIENS
SEE BULLETIN DES SCHWEIZERISCHEN ELEKTROTECHNISCHEN VEREINS

BULLETIN DE L'ASSOCIATION TECHNIQUE DE LA FONDERIE
1932 P519-524 AL-485 <*>
1933 V7 P383-390 66-14262 <*> O
1937 P112-116 57-2311 <*>
1937 P208-215 57-1117 <*>

BULLETIN DE L'ASSOCIATION TECHNIQUE DE L'INDUSTRIE PAPETIERE
1953 N4 P93-106 61-10714 <*>
1954 N7 P280-290 57-1726 <*>
1955 N1 P22-26 61-10293 <*> P
1955 N6 P179-186 57-2307 <*>
60-14025 <*> O
1955 N4/5 P95-104 2369 <*>
1956 N1 P9-12 C-2287 <NRC>
1956 N2 P33- 57-1723 <*>
1956 N2 P50- 57-1741 <*>
1956 N2 P50 57-837 <*>
1956 N2 P56 58-287 <*>
1956 N3 P67-77 59-10006 <*> O
1956 N4 P106-112 60-18885 <*>
1956 N6 P177-184 61-10748 <*> O
1957 N3 P90-97 59-10442 <*> O
1957 N5 P192 58-1805 <*>
1958 N6 P195-203 60-14137 <*> O
61-10407 <*>
66-14263 <*>
1959 N1 P32-45 80L36F <ATS>
1959 N4 P201-214 63-10913 <*> O
1959 N5 P265-275 60-18886 <*> O
1960 N2 P66-83 61-20185 <*> O
63-10941 <*> O
1960 N4 P133-149 62-16509 <*> O
1961 N1 P37-53 62-10610 <*> O
1961 N2 P132-139 62-10609 <*>
1961 N4 P287-304 62-16496 <*> O
1961 N5 P331-338 63-10757 <*>
1961 N5 P353-371 63-10530 <*> O
1961 N5 P404-424 62-16211 <*>
1961 N6 P480-496 62-20189 <*> O
1962 N16 P370-386 63-18789 <*>

BULLETIN DE L'ASSOCIATION TECHNIQUE MARITIME ET AERONAUTIQUE
1963 N63 P687-704 65-17046 <*>

BULLETIN OF THE ASTRONOMICAL INSTITUTES OF CZECHOSLOVAKIA
1950 V2 P22-24 58-2313 <*>
1950 V2 P138-140 58-2313 <*>
1954 V5 P79-82 58-2313 <*>

BULLETIN BIMENSUEL. SOCIETE DE MEDECINE MILITAIRE FRANCAISE
1955 V49 P119-126 62-14574 <*> O

BULLETIN OF THE BIOGEOGRAPHICAL SOCIETY OF JAPAN
SEE NIPPON SEIBUTSU CHIRIGAKKAI KAIHO

BULLETIN BIOLOGIQUE DE LA FRANCE ET DE LA BELGIQUE
1947 V81 N3/4 P274-284 64-00368 <*>
1947 V81 N3/4 P297-304 64-00368 <*>
1947 V81 N3/4 P412-422 64-00368 <*>
1950 V84 N3 P217-224 1937 <*>
1953 P1-90 SUP38 59-15477 <*> O

BULLETIN OF BIOLOGY. CHINESE PEOPLES REPUBLIC
SEE SHENG WU HSUEH T'UNG PAO

BULLETIN. CENTRE BELGE D'ETUDE ET DE DOCUMENTATION DES EAUX
1950 V8 P494-500 I-782 <*>
1952 N15 P49- C-3244 <NRCC>
1952 N16 P115-119 57-3273 <*>
1954 V25 P146-155 II-283 <*>
1956 N65 P204-209 58-271 <*>
1960 N3 P76-81 2088 <BISI>
1960 N111 P95-100 65-10094 <*>

BULLETIN DE CENTRE D'ETUDES DE RECHERCHES ET D'ESSAIS SCIENTIFIQUES DU GENIE CIVIL ET D'HYDRAULIQUE FLUVIALE
1953 V6 P99-232 60-18331 <*> P
1956 V8 P155-170 61-10803 <*>

BULLETIN. CENTRE D'ETUDES ET RECHERCHES PSYCHOTECHNIQUES
1953 V2 N1 P2-12 65-12534 <*>
1954 V3 N2 P2-9 59-20868 <*>
1955 V4 P125-139 65-12692 <*>
1955 V4 P154-173 65-12250 <*>
1955 V4 N1 P33-34 58-2231 <*>
1955 V4 N3 P307-312 59-20166 <*>
1955 V4 N4 P141-151 65-11697 <*>
1955 V4 N4 P371-377 64-30301 <*>
1955 V4 N4 P379-391 61-20436 <*>
1956 V5 N1 P25-32 61-20434 <*>
1956 V5 N4 P407-422 61-16209 <*>
1957 V6 N3 P305-311 60-18020 <*>
1959 V8 P111-120 61-20430 <*>
1961 V10 N2 P117-129 65-12111 <*>
1961 V10 N4 P433-444 64-00279 <*>
1962 V11 N1 P1-11 65-11683 <*>

BULLETIN. CENTRE POLONAIS DE RECHERCHES SCIENTIFIQUES DE PARIS
1961 N19 P3-12 63-21305 <=>

BULLETIN. CENTRE DE RECHERCHES ET D'ESSAIS DE CHATOU
1963 N5 P3-27 65-14689 <*>

BULLETIN. CERCLE D'ETUDES DES METAUX
1962 V8 P325-332 3270 <BISI>
1962 V8 P351-356 3196 <BISI>
1964 V9 N8 P285-303 4129 <BISI>
1965 V9 N11 P439-482 4743 <BISI>

BULLETIN OF THE CHEMICAL RESEARCH INSTITUTE OF NON-AQUEOUS SOLUTIONS, TOHOKU UNIVERSITY
SEE TOHOKU DAIGAKU HISUI YOEKI KAGAKU KENKYU-SHO HOKOKU

BULLETIN OF THE CHEMICAL SOCIETY OF JAPAN
SEE NIHON KAGAKU-KAI

BULLETIN OF THE CIVIL ENGINEERING EXPERIMENT STATION. TOKYO
SEE DOBOKU SHIKENJO IHO

BULLETIN CLIMATOLOGIQUE. OBSERVATOIRE DE PUY-DE-DOME
1956 N4 P113-125 59-22175 <=*> 0
1959 N1 P7-10 63-00583 <*>
63-27133 <$>
1961 N2 P93-99 64-10814 <*>

BULLETIN OF THE COLLEGE OF AGRICULTURE, UTSUNOMIYA UNIVERSITY. JAPAN
SEE UTSUNOMIYA DAIGAKU NOGAKUBU GAKUJUTSU HOKOKU

BULLETIN. COMITE CENTRAL HOUILLERES DE FRANCE
1938 N410 P61-80 AL-17 <*>

BULLETIN OF THE DIVISION OF PLANT BREEDING AND CULTIVATION, TOKAI-KINKI NATIONAL AGRICULTURAL EXPERIMENT STATION
SEE TOKAI-KINKI NOGYO SHIKENJO KENKYU HOKOKU SAIBAI DAI-I-BU

BULLETIN OF THE EARTHQUAKE RESEARCH INSTITUTE, TOKYO UNIVERSITY
SEE TOKYO DAIGAKU JISHIN KENKYUSHO IHO

BULLETIN OF THE ELECTROTECHNICAL LABORATORY. JAPAN
SEE DENKI SHIKEN-SHO IHO

BULLETIN OF THE ENGINEERING RESEARCH INSTITUTE, KYOTO UNIVERSITY
SEE KYOTO DAIGAKU KOGAKU KENKYOJU IHO

BULLETIN OF EPIZOOTIC DISEASES OF AFRICA
1957 V5 P211-221 62-01263 <*>

BULLETIN OF EXPERIMENTAL ANIMALS. JAPAN
SEE JIKKEN DOBUTSU

BULLETIN OF THE FACULTY OF AGRICULTURE, KAGOSHIMA UNIVERSITY
SEE KAGOSHIMA DAIGAKU NOGAKUBU GAKUJUTSU HOKOKU

BULLETIN OF THE FACULTY OF ENGINEERING, HIROSHIMA UNIVERSITY
SEE HIROSHIMA DAIGAKU KOGAKUBU KENKYU HOKOKU

BULLETIN OF THE FACULTY OF ENGINEERING, HOKKAIDO UNIVERSITY
SEE HOKKAIDO DAIGAKU KOGAKUBU KENKYU HOKOKU

BULLETIN OF THE FACULTY OF ENGINEERING, TOYAMA UNIVERSITY
SEE TOYAMA DAIGAKU KOGAKUBU KIYO

BULLETIN OF THE FACULTY OF FISHERIES, HOKKAIDO UNIVERSITY
SEE HOKKAIDO DAIGAKU SUISANGAKUBU KENKYU IHO

BULLETIN DE LA FEDERATION DES SOCIETES DE GYNECO-GYNECOLOGIE ET D'OBSTETRIQUE DE LANGUE FRANCAISE
1952 V4 N3 P568-577 64-18949 <*>
1958 V10 P384-388 63-10311 <*> 0
1965 V17 N2 P147-150 65T91F <ATS>

BULLETIN OF FORMOSAN GEOGRAPHY
SEE TAIWAN TIGAKU KIZI

BULLETIN OF THE FRESHWATER FISHERIES RESEARCH LABORATORY. JAPAN
SEE TANSUI KU SUISAN KENKYUSHO KENKYU HOKOKU

BULLETIN OF THE FUKUI PREFECTURE AGRICULTURAL EXPERIMENT STATION
SEE FUKUI-KEN NOGYO SHIKENJO HOKOKU

BULLETIN GALENICA. BERN
1954 V17 P50-61 II-536 <*>

BULLETIN GENERAL DE THERAPEUTIQUE MEDICALE, CHIRURGICALE, OBSTETRICALE ET PHARMACEUTIQUE
1869 V77 P303-306 63-18042 <*>
1917 V169 P728-744 64-18948 <*>

BULLETIN OF THE GEOLOGICAL SURVEY OF JAPAN
SEE CHISHITSU CHOSOJO GEPPO

BULLETIN OF THE GOVERNMENT FOREST EXPERIMENT STATION. TOKYO
SEE RINGYO SHIKENJO KENKYU HOKOKU

BULLETIN OF THE GOVERNMENT INDUSTRIAL RESEARCH INSTITUTE. NAGOYA
SEE NAGOYA KOGYO GIJUTSU SHIKEN-SHO HOKOKU

BULLETIN OF THE GOVERNMENT RESEARCH INSTITUTE. OSAKA
SEE OSAKA KOGYO GIJUTSU SHIKEN-SHO KIHO

BULLETIN DU GROUPE FRANCAIS DES ARGILES
1955 V6 N1 P13-18 22J17F <ATS>
1955 V6 N1 P23-29 23J17F <ATS>
1955 V8 N1 P41-42 24J17F <ATS>

BULLETIN DU GROUPEMENT INTERNATIONAL POUR LA RECHERCHE SCIENTIFIQUE EN STOMATOLOGIE
1962 V5 N4 P523-542 63-01105 <*>

BULLETIN OF THE HATANO TOBACCO EXPERIMENT STATION
SEE HATANO TABAKO SHIKENJO HOKOKU, HOBUN

BULLETIN D'HISTOLOGIE APPLIQUEE A LA PHYSIOLOGIE ET A LA PATHOLOGIE ET DE TECHNIQUE MICROSCOPIQUE
1928 V5 P253-259 4072-B <KH>

BULLETIN OF THE HOKKAIDO PREFECTURAL AGRICULTURAL EXPERIMENT STATION

SEE HOKKAIDORITSU NOGYO SHIKENJO SHUHO

BULLETIN OF THE HOKKAIDO REGIONAL FISHERIES RESEARCH LABCRATORY
SEE HOKKAIDC-KU SUISAN KENKYU-SHO KENKYU HOKOKU

BULLETIN OF THE HOKURIKU AGRICULTURAL EXPERIMENT STATION
SEE HOKURIKU NOGYO SHIKENJO HOKOKU

BULLETIN D'INFORMATION DE L'ASSOCIATION BELGE POUR LE DEVELOPMENT PACIFIQUE DE L'ENERGIE ATOMIQUE
1962 N32 P5-11 AEC-TR-4954 <*> P

BULLETIN D'INFORMATION. COMITE CENTRAL D'OCEANOGRAPHIE ET D'ETUDE DES COTES
1955 V7 N5 P196-200 2995 <*>

BULLETIN D'INFORMATION COMITE EUROPEEN DE BETON
1957 N15 ENTIRE ISSUE 61-10141 <*>
1958 N8 ENTIRE ISSUE 62-18774 <*>
1959 N18 ENTIRE ISSUE 62-20456 <*>
1960 N21 ENTIRE ISSUE 61-10791 <*>
61-10792 <*>
1961 N12 ENTIRE ISSUE 61-10140 <*>
1961 N14 ENTIRE ISSUE 61-10142 <*>

BULLETIN D'INFORMATION CCCPERATION CENTRE FOR SCIENTIFIC RESEARCH RELATIVE TO TOBACCO
1961 N1 P89-91 63-14370 <*>

BULLETIN OF INFORMATION AND DOCUMENTATION
1952 V1 N4 P21-26 60-21698 <=>

BULLETIN D'INFORMATION. LABORATOIRE CENTRAL DES INDUSTRIES ELECTRIQUES
1962 N33 ENTIRE ISSUE 63-15906 <=*$>

BULLETIN D'INFORMATIONS SCIENTIFIQUES ET TECHNIQUES. COMMISSARIAT A L'ENERGIE ATOMIQUE. FRANCE
1958 N20 P56-68 AEC-TR-5828 <*>
1958 N23 P11-12 62-32406 <=$>
1958 V21 P13-25 62-32405 <=*$>
1960 N41 P14-27 NP-TR-712 <*>
1960 N43 P32-43 AEC-TR-4629 <*>
1960 N43 P52-57 AEC-TR-4551 <*>
1960 N45 P2-12 NP-TR-841 <*>
1961 N42 P2-6C AEC-TR-4771 <*>
1961 N47 P2-22 HW-TR-32 <*>
1961 N49 P67-79 AEC-TR-5780 <*>
1961 N53 P2-6 AEC-TR-5452 <*>
1962 N61 P6-15 AEC-TR-5893 <*>
1962 N61 P16-23 AEC-TR-5896 <*>
1962 N61 P24-36 AEC-TR-5895 <*>
1962 N61 P46-50 AEC-TR-5894 <*>
1962 N61 P123-138 AEC-TR-5897 <*>
1962 N63 P9-24 AEC-TR-5823 <*>
1962 N63 P25-30 AEC-TR-5817 <*>
1962 N63 P31-42 AEC-TR-5818 <*>
1962 N63 P43-48 AEC-TR-5859 <*>
1962 N63 P83-95 AEC-TR-5860 <*>
1962 N64 P29-36 66-11852 <*>
1963 N69 P3-6 AEC-TR-6294 <*>
1963 N69 P7-9 AEC-TR-6295 <*>
1963 N69 P11-17 AEC-TR-6296 <*>
1963 N69 P19-25 AEC-TR-6297 <*>
1963 N69 P27-35 AEC-TR-6298 <*>
1963 N69 P37-41 AEC-TR-6299 <*>
1963 N69 P43-48 AEC-TR-6300 <*>
1963 N75 P101-104 NP-TR-1140 <*>
1963 N72/3 P83-89 HW-TR-72 <*>
1965 N92 P1-12 66-10425 <*> 0
1965 N95 P79-115 66-14065 <*>

BULLETIN DE L'INSTITUT AGRONOMIQUE ET DES STATIONS DE RECHERCHES DE GEMBLOUX
1936 V5 N3/4 P215-296 I-776 <*>
1948 V17 P26-42 62-01427 <*>

BULLETIN DE L'INSTITUT D'EGYPTE
1932 V14 P77-87 I-617 <*>
1932 V14 P141-151 I-619 <*>
1932 V14 P153-176 I-620 <*>

BULLETIN DE L'INSTITUT D'HYGIENE DU MAROC
1941 SNS V1 P35-44 64-10826 <*> 0

BULLETIN DE L'INSTITUT INTERNATIONAL DE STATISTIQUE. ROME
1958 V36 N3 P87-101 64-00441 <*>

BULLETIN DE L'INSTITUT NATIONAL D'HYGIENE
1957 V12 N1 P137-150 59-16070 <+*>
1958 V13 N2 P373-380 59-19006 <+*>

BULLETIN. INSTITUT ZA NUKLEARNE NAUKE, BORIS KIDRIC. BELGRAD
1952 P95-101 I-728 <*>
1952 P111-113 I-727 <*>
1954 V4 P37-44 I-138 <*>
1954 V4 P71-73 63-10521 <*>

BULLETIN DE L'INSTITUT PASTEUR
1932 V30 P369-379 64-10330 <*>

BULLETIN DE L'INSTITUT DU PIN
1927 N41 P241-246 AL-205 <*>
1928 N52 P205-207 AL-208 <*>
1930 N7 P155-157 AL-619 <*>
1931 N2 P159-160 AL-206 <*>
1931 N2 P173-176 AL-206 <*>
1931 N2 P199-200 AL-206 <*>
1931 N2 P219-221 AL-206 <*>
1936 V3 N23/4 P256-258 AL-865 <*>

BULLETIN DE L'INSTITUT R. DES SCIENCES NATURELLES DE BELGIQUE
1960 V36 N12 P10-15 62-00819 <*>

BULLETIN. INSTITUT DE RECHERCHES ECONOMIQUES. LAUVIAN UNIVERSITY
1946 P7-11 65-12699 <*>
1946 P212-214 65-12699 <*>

BULLETIN DE L'INSTITUT TEXTILE DE FRANCE
1948 N7 P9-19 61-14042 <*>
1950 N17 P9-32 61-14043 <*>
1951 N26 P23-64 T2148 <INSD>
1952 P7 T-2137 <INSD>
1952 N30 P37-53 61-14044 <*> 0
1952 N30 P405-414 61-14045 <*> P
1952 N31 P19-2C T-2147 <INSD>
1952 V35 P51-73 64-20299 <*> 0
1953 P23 T-2137 <INSD>
1953 N40 P63-74 61-14046 <*> 0
1954 N46 P69-75 61-14516 <*> 0
1955 N46 P7-19 CSIRO-3426 <CSIR>
1955 N55 P7-22 CSIRO-3426 <CSIR>
1955 V45 P45-53 CSIRO-2958 <CSIR>
1956 N58 P49-56 CSIRO-3249 <CSIR>
1957 N65 P15-40 61-16733 <*>
1957 N66 P7-14 58-1496 <*>
1958 N75 P7-25 61-14974 <*>
1958 N77 P55-62 54M38F <ATS>
1958 N78 P27-36 61-20713 <*>
1958 N78 P27-28 65-14191 <*>
1959 N82 P113-125 D-661 <RIS>
1959 N83 P99-126 66-13020 <*>
1959 N84 P49-68 AEC-TR-5091 <*>
1960 N87 P69-81 61-14517 <*>
1962 N102 P887-944 63-18346 <*>
1962 N102 P963-975 63-18361 <*>
1962 N102 P977-980 66-10874 <*>
1962 N102 P945-962 63-18296 <*>
1963 V17 N106 P541-556 12Q74F <ATS>
1965 V19 N119 P535-578 2816 <TC>
1965 V19 N120 P725-739 2816 <TC>

BULLETIN DE L'INSTITUT DU VERRE

1946 N3 P1-10 62-18142 <*>
1946 N5 P15-21 60-18772 <*>
62-10545 <*> O

BULLETIN OF THE INSTITUTE OF AGRICULTURAL RESEARCH, TOHOKU UNIVERSITY
SEE TOHOKU DAIGAKU NOGAKU KENKYUSHO HOKOKU

BULLETIN OF THE INSTITUTE FOR CHEMICAL RESEARCH, KYOTO UNIVERSITY
1947 V16 P6-8 60-18127 <*>
1947 V16 P29-31 MF-6 <*>
1950 V20 P22-27 57-1846 <*>
1956 N34 P316-320 61-10157 <*> O

BULLETIN OF THE INSTITUTE OF NUCLEAR SCIENCES 'BORIS KIDRICH'
SEE BILTEN INSTITUTA ZA NUKLEARNE NAUKE BORIS KIDRIC, VINCA, YUGOSLAVIA

BULLETIN OF THE INSTITUTE OF PHYSICAL AND CHEMICAL RESEARCH. JAPAN
SEE RIKAGAKU KENKYUSHO HOKOKU

BULLETIN OF THE INSTITUTE OF PUBLIC HEALTH. TOKYO
SEE KOSHU EISEIIN KENKYU HOKOKU

BULLETIN OF THE INSTITUTE OF SPACE AND AERONAUTICS UNIVERSITY OF TOKYO
SEE TOKYO DAIGAKU UCHU KOKU KENKYUJO HOKOKU

BULLETIN INTERNATIONAL DE L'ACADEMIE POLONAISE DES SCIENCES ET DES LETTRES. CLASSE DES SCIENCES MATHEMATIQUES ET NATURELLES. SERIE A. SCIENCES MATHEMATIQUES
1925 SA P81-92 I-402 <*>
1934 P315-328 63-10708 <*> O

BULLETIN INTERNATIONAL DE L'ACADEMIE POLONAISE DES SCIENCES ET DE LETTRES. CLASSE DE MEDICINE
1932 P239-258 2463 <*>
1936 P555-562 64-18947 <*>

BULLETIN INTERNATIONAL. ACADEMIE TCHEQUE DES SCIENCES. PRAGUE
1950 V51 N13 ENTIRE ISSUE AL-393 <*>

BULLETIN OF THE INTERNATIONAL INSTITUTE OF REFRIGERATION
1956 P41-49 58-2373 <*>
1956 V2 P51-62 60-00134 <*>

BULLETIN INTERNATIONAL POLSKA AKADEMII NAUK
SEE BULLETIN INTERNATIONAL DE L'ACADEMIE POLONAISE DES SCIENCES ET DE LETTRES

BULLETIN INTERNATIONAL. SERIES B. SCIENCES NATURELLES. POLSKA AKADEMIA UMIEJETNOSCI. WYDZIAL MATEMATYCZNO-PRZYRODNICZY
1937 SB N1 P11-31 65-11051 <*> O
1937 SB N1 P33-59 65-11052 <*> O

BULLETIN INTERNATIONAL DES SERVICES DE SANTE DES ARMEES DE TERRE, DE MER ET DE L'AIR
1939 V12 P140-151 AL-82 <*>

BULLETIN OF J.S.M.E. TOKYO
1960 V3 N12 P444-448 13R76G <ATS>

BULLETIN OF THE JAPAN INSTITUTE OF METALS
SEE NIHON KINZOKU GAKKAI KAIHO

BULLETIN OF THE JAPAN SEA REGIONAL FISHERIES RESEARCH LABORATORY
SEE NIPPON KAIKU SUISAN KENKYUSHO KENKYU HOKOKU

BULLETIN OF THE JAPANESE SOCIETY OF SCIENTIFIC FISHERIES
SEE NIPPON SUISAN GEKKAISHI

BULLETIN DU JARDIN BOTANIQUE DE BUITENZORG
1912 S2 N4 P1-48 3417 <*>
1932 V30 P573-588 R-954 <*>

BULLETIN OF THE KOBAYASI INSTITUTE OF PHYSICAL RESEARCH
SEE KOBAYASHI RIGAKU KENKYUSHO HOKOKU

BULLETIN OF THE KOBE MEDICAL COLLEGE
SEE KOBE IKA DAIGAKU KIYO

BULLETIN OF THE KYUSHU AGRICULTURAL EXPERIMENT STATION
SEE KYUSHU NOGYO SHIKENJO IHO

BULLETIN OF THE KYUSHU INSTITUTE OF TECHNOLOGY. SCIENCE AND TECHNOLOGY
SEE KYUSHU KOGYO DAIGAKU KENKYU HOKOKU, KOGAKU

BULLETIN DU LABORATOIRE DE RECHERCHES ET DU CONTROLE DU CAOUTCHOUC
1957 N52 P16- 59-18718 <+*>

BULLETIN ET MEMOIRES DE LA FACULTE NATIONALE DE MEDECINE ET DE PHARMACIE DE DAKAR
1963 V11 P238-257 66-12008 <*> O

BULLETIN ET MEMOIRES DE LA SOCIETE ANATOMIQUE DE PARIS
1861 V36 P398-407 63-00298 <*>

BULLETIN ET MEMOIRES DE LA SOCIETE D'ANTHROPOLOGIE DE PARIS
1861 V2 P235-238 63-00303 <*>

BULLETIN ET MEMOIRES DE LA SOCIETE FRANCAISE D'OPHTALMOLOGIE
1898 V16 P199-207 61-10927 <*>
1926 V39 P51-60 61-14365 <*>
1926 V39 P60-65 61-14158 <*> O

BULLETIN ET MEMOIRES DE LA SOCIETE MEDICALE DES HOPITAUX DE BUCAREST
1931 V13 P1-4 66-12114 <*>

BULLETIN ET MEMOIRES DE LA SOCIETE MEDICALE DES HOPITAUX DE PARIS
1907 V3 N24 P1203-1211 II-988 <*>
1912 V33 P545-555 57-3162 <*>
1921 S3 V45 P484-488 57-1537 <*>
1922 V46 P463-469 62-01390 <*>
1934 S3 V50 P1248-1255 61-16498 <*> O
1936 S3 V52 P936-941 61-14492 <*>
1936 S3 V52 P941-944 61-14493 <*>
1943 S3 V59 N13/4 P167-168 61-16844 <*>
1944 S3 V60 N1 P6-8 61-16567 <*>
1949 V65 P826-832 64-18945 <*>
1949 V65 P1008-1013 64-18944 <*> O
1950 S4 V66 N1/2 P76-79 61-00208 <*>
1950 S4 V66 N3/4 P85-96 59-17383 <*> P
1950 S4 V66 N17/8 P779-782 60-10850 <*> O
1950 S4 V66 N33/4 P1701-1716 61-16461 <*> O
1950 S4 V66 N9/10 P405-407 60-10851 <*> O
1952 S4 V68 N9/10 P368-375 61-16470 <*>
1953 V68 P866-868 1859 <*>
1953 V69 P1030-1048 64-18942 <*> O
1953 V69 N17/8 P530 AL-925 <*>
1955 N19 P726-732 R-1822 <*>
1956 S4 V72 P227-237 II-1258 <*>
1956 S4 V72 P237-246 II-1205 <*>
1958 S4 V74 N13/4 P308-319 59-16642 <+*>
1958 S4 V74 N28/9 P689-692 60-16473 <*> O
1959 S4 V75 P9-11 59-18133 <+*>

1959 S4 V75 P319-322 59-18133 <+*>
1960 S4 V76 N21/2 P812-819
62-10748 <*> O
1962 V113 P767-784 65-11152 <*>
1962 V113 P1199-1205 64-18612 <=>
1962 V113 N4 P317-323 66-12167 <*>
1962 V113 N10 P753-766 66-10971 <*>
1963 V114 P249-254 64-18615 <*>
1964 V115 N8 P693-708 66-13188 <*>
1965 V116 P25-37 66-13179 <*>
1965 V116 P39-44 66-13178 <*>
1965 V116 P45-50 66-13214 <*>

BULLETIN ET MEMOIRES DE LA SOCIETE NATIONALE DE CHIRURGIE. PARIS
1933 V59 P4-7 61-14488 <*>

BULLETIN MENSUEL D'INFORMATION. INSTITUT TECHNIQUE D'ETUDES ET DE RECHERCHES DES CORPS GRAS
1953 V7 P490-494 60-10696 <*> O

BULLETIN MENSUEL DE L'OFFICE INTERNATIONAL D'HYGIENE PUBLIQUE
1935 V27 P313-314 64-18946 <*>

BULLETIN MENSUEL DE LA SOCIETE BELGE D'ELECTRICIENS
1944 V60 N3 P88-93 62-16949 <*> O
1957 V73 N2 P85-105 61-18379 <*>

BULLETIN MENSUEL. SOCIETE DE MEDECINE MILITAIRE FRANCAISE
1938 V32 P132-151 AL-71 <*>
1958 V52 P139-149 62-14586 <*> O

BULLETIN MENSUEL DE LA SOCIETE VETERINAIRE PRATIQUE DE FRANCE
1964 V48 N8 P295-301 65-13771 <*>

BULLETIN DE MICROSCOPIE APPLIQUEE
1954 V4 N1/2 P6-8 II-637 <*>
1954 V4 N11/2 P139-143 2205 <*>
1956 V6 P151-154 58-1234 <*>
1957 V7 N4 P73-75 62-00693 <*>
1958 V8 N6 P135-147 66-11666 <*> OP
1958 S2 V8 N3 P64-70 59-18111 <=*> O
8817-B <K-H>
1958 S2 V8 N6 P129-148 66-11112 <*> O

BULLETIN OF THE MISAKI MARINE BIOLOGICAL INSTITUTE JAPAN
SEE KYOTO DAIGAKU. MISAKI RINKAI KENKYUJO.

BULLETIN DU MUSEUM NATIONAL D'HISTOIRE NATURELLE. PARIS
1948 V20 P232-234 I-762 <*>

BULLETIN OF THE NAGANO PREFECTURAL AGRICULTURAL EXPERIMENT STATION
SEE NAGANO-KEN NOGYO SHIKENJO HOKOKU

BULLETIN OF THE NAGOYA INSTITUTE OF TECHNOLOGY
SEE NAGOYA KOGYA DAIGAKU GAKUHO

BULLETIN OF THE NAIKAI REGIONAL FISHERIES RESEARCH LABORATORY. JAPAN
SEE NAIKAI-KU SUISAN KENKYUSHO KENKYU HOKOKU

BULLETIN OF THE NATIONAL INSTITUTE OF AGRICULTURAL SCIENCES. SERIES C. PLANT PATHOLOGY AND ENTOMOLOGY
SEE NOGYO GIJUTSU KENKYUJO HOKOKU C. BYORI, KONCHU

BULLETIN OF THE NATIONAL INSTITUTE OF HYGIENIC SCIENCES
SEE EISEI SHIKEN-SHO HOKOKU

BULLETIN. OFFICE INTERNATIONAL DES EPIZOOTIES
1949 V32 P22-35 92F2F <ATS>
1949 V32 P222-231 95F2F <ATS>
1950 V33 P189-200 94F2F <ATS>
1952 V37 N9/10 P514-529 I-522 <*>
MF-2 <*>
1955 V43 P159-165 3163 <*>
1955 V43 P354-355 3162 <*>
1955 V43 P745- 3177 <*>

BULLETIN OF PLANT PHYSIOLOGY. CHINESE PEOPLES REPUBLIC
SEE CHIH WU SHEN LI HSUEH T'UNG HSUN

BULLETIN. POLSKA AKADEMIIA NAUK
SEE BULLETIN DE L'ACADEMIE POLONAISE DES SCIENCES

BULLETIN POZNOINSKIEGO TOWARZYSTWA PRZYJACIOL NAUK. SERIES B. SCIENCES MATHEMATIQUES ET NATURELLES
1957 V4 N2 P1-66 60-21229 <=>

BULLETIN OF THE RESEARCH INSTITUTE OF APPLIED ELECTRICITY, HOKKAIDO UNIVERSITY
SEE OYO DENKI KENKYUJO IHO

BULLETIN OF THE RESEARCH INSTITUTE OF MINERAL DRESSING AND METALLURGY, TOHOKU UNIVERSITY
SEE TOHOKU DAIGAKU SENKC SEIREN KENKYUSHU IHO

BULLETIN OF THE RESEARCH INSTITUTE FOR SCIENTIFIC MEASUREMENTS, TOHOKU UNIVERSITY
SEE TOHOKU DAIGAKU KAGAKU KEISOKU KENKYUSHO HOKOKU

BULLETIN DER SCHWEIZERISCHEN AKADEMIE DER MEDIZINISCHEN WISSENSCHAFTEN
1951 V7 N5/6 P418-429 60-17484 <*=>
1954 V10 N3/4 P249-259 II-605 <*>
1955 V11 N4/5 P346-351 60-15717 <+*> O
65-00398 <*> O
1958 V14 P230-233 58-2198 <*>
1958 V14 P571-579 HW-TR-17 <*>
1959 V15 N4 P346-359 60-14192 <*>
1964 V20 N4/6 P336-359 66-11814 <*>

BULLETIN DES SCHWEIZERISCHEN ELEKTROTECHNISCHEN VEREINS
1924 V15 P106-116 66-13642 <*>
1926 V17 N5 P155-159 TS-1698 <BISI>
1927 V18 P113-122 58-1964 <*>
1930 V21 N23 P756-788 57-331 <*>
1937 N3 P54-57 58-245 <*>
1945 V36 N15 P459-460 63-10053 <*> O
1945 V36 N15 P533-534 63-10053 <*> O
1946 V37 N7 P175-185 2459 <*>
1946 V37 N11 P291-298 62-18395 <*> O
1946 V37 N11 P298-302 66-12253 <*> O
1947 V38 P363-371 58-671 <*>
1947 V38 P374-377 63-10052 <*> O
1952 V43 N18 P721-727 37S88G <ATS>
1952 V43 N26 P1069-1074 57-2518 <*>
1955 V46 P721-725 T1820 <INSD>
1955 V46 P873-878 T1819 <INSD>
1957 V48 P273-282 60-10055 <*>
1958 V49 N2 P37-45 58-1670 <*>
1958 V49 N2 P45-50 SC-T-65-712 <*>
1958 V49 N7 P278-289 65-10239 <*>
1958 V49 N9 P408-412 59-15618 <*>
1959 V50 N1 P2-9 59-00795 <*>
1959 V50 N13 P601-613 60-21723 <=>
1959 V50 N18 P890-897 62-00516 <*>
1961 V52 P23- 62-10966 <*> O
1961 V52 N19 P757-764 63-14990 <*>
1961 V52 N23 P915-923 63-14931 <*>
1964 V55 N21 P1059-1065 66-11382 <*>

BULLETIN DER SCHWEIZERISCHEN VEREINIGUNG FUER ATOMENERGIE
1966 N14 P1-6 SUP 67-30355 <=>
1966 N17 P1-3 SUP 67-30355 <=>

BULLETIN DES SCIENCES PHARMACOLOGIQUES
1921 V28 P545-549 64-18941 <*>
1927 V34 P417- 66-12300 <*> 0
1936 V43 P289-292 64-18940 <*>
1936 V43 P696-708 64-18939 <*>
1938 V45 P241-252 64-18938 <*>

BULLETIN SCIENTIFIQUE A.I.M.
1952 V65 P1104-1116 1987 <*>
1958 V71 N11/2 P1161-1178 61-10656 <*>
1959 V72 N4/5 P323-342 1869 <BISI>

BULLETIN SCIENTIFIQUE DE L'ASSOCIATION DES INGENIEURS ELECTRICIENS SORTIS DE L'INSTITUT ELECTROTECHNIQUE. MONTEFIORE
SEE BULLETIN SCIENTIFIQUE A.I.M.

BULLETIN SCIENTIFIQUE DE L'ECOLE POLYTECHNIQUE. TIMISOARA
1934 V5 P225-232 60-10250 <*> 0

BULLETIN DE LA SECTION SCIENTIFIQUE DE L'ACADEMIE ROUMAINE
1939 N22 P1-19 T-1375 <INSD>

BULLETIN OF THE SERICULTURAL EXPERIMENT STATION. JAPAN
SEE SANSHI SHIKENJO HOKOKU

BULLETIN DU SERVICE DE LA CARTE GEOLOGIQUE DE L'ALGERIE
1941 S3 P6-18 3379 <*>
1941 S3 P29-32 3379 <*>

BULLETIN DES SERVICES ZOOTECHNIQUES ET DES EPIZOOTIES DE L'AFRIQUE OCCIDENTAL FRANCAISE
1938 V1 N3 P57-58 3269 <*>

BULLETIN OF THE SHIMANE AGRICULTURAL COLLEGE. JAPAN
SEE SHIMANE NOKA DAIGAKU KENKYU HOKOKU

BULLETIN DE LA SILVA MEDITERRANEA
1930 P1-46 57-1225 <*>

BULLETIN. SOCIETE DES AMIS DES SCIENCES ET DES LETTRES DE POZNAN. SERIE B. SCIENCES MATHEMATIQUE ET NATURELLES
1956 SB N13 P125-130 58-2296 <*>
1956 SB N13 P135-140 58-2455 <*>

BULLETIN DE LA SOCIETE BELGE D'ELECTRICIENS
SEE BULLETIN MENSUEL DE LA SOCIETE BELGE D'ELECTRICIENS

BULLETIN DE LA SOCIETE BELGE D'OTOLOGIE, DE LARYNGOLOGIE ET DE RHINOLOGIE
1937 P673-680 61-14579 <*>
1939 P21-24 61-14591 <*>

BULLETIN. SOCIETE BOTANIQUE DE FRANCE
1955 V102 N9 P519-527 66-10221 <*>
1959 V106 N9 P414-417 66-10219 <*>

BULLETIN DE LA SOCIETE DE CHIMIE BIOLOGIQUE
1933 V15 P63-74 58-605 <*>
1934 V16 P1366-1371 64-16209 <*>
1936 V18 N5 P841-867 I-1030 <*>
1936 V18 N5 P868-876 I-1031 <*>
1937 V19 P1697-1710 58-1597 <*>
1937 V19 N12 P1676-1682 64-20287 <*>
1944 V26 P99-105 63-00076 <*>
1945 V27 P411-415 63-16766 <*>
1947 V29 P427-444 58-2463 <*>
1947 V29 N4/6 P460-461 II-9 <*>
1948 V30 N5/6 P336-346 60-10640 <*> 0
1949 V30 P497-500 63-20526 <*>
1949 V31 P1297-1300 63-14550 <*> 0
1949 V31 N2 P256-264 60-10639 <*>
1950 V32 P30-39 I-552 <*>
1950 V32 P375-381 II-825 <*>
1951 V33 P846-856 2378B <K-H>
1951 V33 P1041-1058 57-82 <*>
1951 V33 P1061-1074 57-773 <*>
1951 V33 P1075-1112 57-807 <*>
1951 V33 P1113-1146 57-786 <*>
1951 V33 N7 P771-778 64-20286 <*>
1952 V34 P380-387 57-2782 <*>
1952 V34 P872-885 I-1016 <*>
1952 V34 P897-899 I-1010 <*>
1952 V34 N9 P886-896 1928 <*>
1952 V34 N1/2 P65-76 59-16141 <+*>
1952 V34 N1/2 P204-214 2021 <*>
1952 V34 N3/4 P366-379 I-906 <*>
1953 V35 P1157-1165 3160-F <K-H>
1954 V36 P51-63 I-551 <*>
1954 V36 P143-157 60-10685 <*> PO
1954 V36 P1163-1172 57-1932 <*>
1954 V36 N1 P65-77 66-12716 <*> 0
1954 V36 N8 P1093-1100 62-20151 <*>
1954 V36 N8 P1163-1172 61-10929 <*> 0
1954 V36 N9 P1311-1317 2214 <*>
1954 V36 N10 P1391-1406 2384 <*>
1954 V36 N10 P1461-1471 3046 <*>
1954 V36 N11/2 P1641-1651 3045 <*>
1955 V37 P819-829 II-917 <*>
1955 V37 N7/8 P783-796 61-00647 <*>
1957 V39 P593-605 58-1576 <*>
62-18100 <*> 0
1957 V39 N2/3 P337-341 58-1596 <*>
1958 V40 N1 P59-65 62-18079 <*> 0
1958 V40 N11 P1387-1413 59-18465 <+*> 0
1958 V40 N2/3 P423-429 C-3189 <NRCC>
1959 V41 N1 P79-87 63-00942 <*>
1959 V41 N11 P1391-1425 61-01010 <*>
1959 V41 N12 P1693-1705 10281-J <K-H>
1959 V41 N5/6 P805-812 63-00036 <*>
1960 V42 P99-114 63-18764 <*> 0
1960 V42 P505-518 61-16154 <*> 0
1960 V42 P611-622 62-01059 <*>
1960 V42 P913-933 61-20522 <*>
1960 V42 N12 P1399-1427 63-00205 <*>
1960 V42 N5/6 P505-518 61-00650 <*>
1960 V42 N5/6 P633-641 62-01061 <*>
1960 V42 N5/6 P643-654 62-01060 <*>
1961 V43 P467-470 62-10336 <*>
1961 V43 N4 P495-504 64-18648 <*>
1961 V43 N2/3 P447-451 62-00998 <*>
1961 V43 N5/6 P827-840 63-00065 <*>
1962 V44 N1 P83-90 64-16935 <*>
1962 V44 N11 P997-1007 63-00610 <*>
1964 V46 N9/10 P1035-1045 65-17259 <*>

BULLETIN. SOCIETE CHIMIQUE DE BELGIQUE
1910 V24 P328-337 63-16484 <*>
1919 V28 P381-392 64-20904 <*>
1923 V32 P179-194 65-10150 <*>
1925 V34 P363-398 8769 <IICH>
1930 V39 P395-401 2002 <*>
1931 V40 P295-304 61-16453 <*> 0
1933 V42 N1 P114-118 62-16479 <*> 0
1935 V44 P387-394 50G6F <ATS>
1935 V44 P473-503 1329 <*>
1940 V49 P103-122 58-910 <*>
1940 V49 N6/7 P167-180 1138 <*>

BULLETIN DE LA SOCIETE CHIMIQUE DE FRANCE
1907 V1 P733-740 CSIRO-3509 <CSIR>
58-2371 <*>
1910 V7 P90-99 I-39 <*>
1910 S4 V7 P638-645 64-20468 <*> P
1912 S4 N11 P646-648 62-18675 <*> 0
1912 S4 V11 P382-388 6196-C <K-H>
1915 S4 V17 P10-14 II-949 <*>
1919 S4 V25 N1 P4-9 63-20161 <*>
1920 V27 P290-292 1331 <*>
1920 S4 V27 P737-739 64-20900 <*>
1921 S4 V29 P21-29 64-20899 <*> 0
1922 S4 V31 P102-108 64-14393 <*>
1922 S4 V31 P169-181 64-14393 <*>

Citation	Translation
1922 S4 V31 P394-412	64-20892 <*> 0
1923 V33 P1808-1823	58-2281 <*>
1924 S4 V35 P29-31	64-18246 <*>
1924 S4 V35 P164-168	SC-T-64-906 <*>
1924 S4 V35 P550-584	63-16826 <*>
1924 S4 V35 P1141-1144	64-20207 <*>
1925 S4 V37 P276-277	64-18974 <*>
1925 S4 V37 P1576-1577	64-20902 <*>
1926 S4 V39 P1263-1265	59-10672 <*>
1926 S4 V39 P1584-1589	59-10670 <*>
1926 S4 V40 P106-107	63-14233 <*>
1927 V4 N41 P454-457	57-1987 <*>
1927 S4 V41 P939-940	59-10681 <+*>
1927 S4 V41 P1217-1224	60-10321 <*> 0
1928 V43 P683-696	I-1044 <*>
1928 S4 V43 P881-883	60-10320 <*>
1928 S4 V43 P942-957	65-11438 <*> 0
1930 S4 V47 P704-730	64-20070 <*> P
1930 S4 V47 P894-900	64-20834 <*>
1930 S4 V47 P1128-1131	AEC-TR-5254 <*>
1930 S4 V47 P1300-1314	60-10319 <*> P
1931 S4 V49 P54-70	63-16166 <*> 0
1932 S4 V51 P597-615	64-30620 <*>
1933 S4 V53 P222-234	63-14234 <*>
1934 V1 P833-852	I-381 <*>
1935 S5 V2 P2128-2134	AT-1528 <*>
1936 V3 P1530-1539	57-1143 <*>
1936 S5 V3 P169-176	60-10329 <*>
1936 S5 V3 P476-488	8660 <IICH>
1936 S5 V3 P498-500	64-20288 <*>
1936 S5 V3 P1338-1343	63-18101 <*>
1937 V4 P132-	1639 <*>
1937 S5 V4 P944-950	63-14232 <*> 0
1938 V5 P187-191	57-1142 <*>
1938 V5 N5 P105-113	66-11766 <*>
1938 S5 V5 P1085-1091	63-10522 <*>
1938 S5 V5 P1092-1106	63-10389 <*>
1938 S5 V5 P1106-1120	63-10390 <*>
1938 S5 V5 P1153-1158	AEC-TR-4895 <*>
1939 V6 P55-70	I-889 <*>
1939 S5 V6 P191-200	60-10366 <*>
1939 S5 V6 P302-316	58-2383 <*>
1939 S5 V6 P664-672	62-16093 <*>
1939 S5 V6 P672-676	64-10579 <*>
1939 S5 V6 P1255-1257	59-18564 <*=>
1939 S5 V6 P1434-1435	AEC-TR-4896 <*>
1940 V7 P743-750	2481 <*>
1940 S5 V7 P296-345	II-513 <*>
	63-10424 <*>
1941 V8 P660-664	66-12911 <*>
1941 S5 V8 P695-699	64-20891 <*>
1942 S5 V9 P146-152	65-10887 <*>
1943 S5 V10 P98-102	57-2574 <*>
1943 S5 V10 P315-322	65-10890 <*>
1944 S5 V11 P2-6	63-14746 <*>
1944 S5 V11 P6-17	2285 <*>
1944 S5 V11 P38-40	MF-6 <*>
1944 S5 V11 P561-564	AL-167-I <*>
1944 S5 V11 N10/1 P553-561	RCT V20 N4 P938 <RCT>
1945 V12 P89-91	58-233 <*>
1945 V12 P120-124	65-11955 <*>
1945 V12 N5 P1004-1110	58-2382 <*>
1945 S5 V12 P412-430	63-18884 <*>
1945 S5 V12 P568-581	54G67F <ATS>
1946 P261-265	63-14743 <*>
1946 P428-436	AL-421 <*>
1946 P674-676	62-10253 <*>
1947 P122-123	AEC-TR-5209 <*>
1947 N1/2 P122-123	64-20432 <*> 0
1947 N1/2 P286-289	63-20496 <*>
1948 P305-324	I-543 <*>
1948 P354-357	63-18264 <*>
1948 P678-679	60-17566 <*+>
1948 P680-681	60-17567 <*+>
1948 P681-682	60-17568 <*+>
1948 P682-683	60-17569 <*+>
1948 N3/4 P305-324	63-10458 <*>
1948 N5/6 P528-529	AL-155 <*>
1948 N7/8 P764-769	I-357 <*>
	1679 <*>
1948 N9/10 P963-968	59-15731 <*>
1949 P117-119	1441 <*>
1949 P437-439	I-698 <*>
1949 PD110-D116	63-16858 <*>
1949 PD117-D119	63-16857 <*>
1949 N5/6 P437-439	59-15316 <*> 0
1949 S5 V16 P231-237	I-182 <*>
1950 P322-326	58-2515 <*>
1950 P849-851	II-197 <*>
1950 P873-876	63-16030 <*>
1950 PD51-D65	63-20325 <*>
1950 PD83-D92	64-20814 <*> PO
1950 N12 P1165-1167	59-21117 <=>
1950 N3/4 P278-282	60-10337 <*> 0
1950 N3/4 P358-361	60-10336 <*>
1950 N8/9 P476-489	64-20341 <*> 0
1950 V17 P1165-1167	I-3 <*>
1950 V17 P1253-1260	58-2381 <*>
1951 P140-142	I-147 <*>
1951 P465-467	54L31F <ATS>
1951 P854-857	65-10002 <*>
1951 P895-899	63-14042 <*>
1951 N5 P565-567	AEC-TR-4899 <*>
1951 N1/2 P73-76	59-15658 <*>
1951 N5/6 P347-348	65-10916 <*>
1951 N7/8 P565-567	62-10235 <*>
1951 N9/10 P773-778	RCT V25 N2 P275 <RCT>
1951 V18 P50-59	1K23F <ATS>
1951 V18 P45-50	K23F <ATS>
1952 P55-60	AL-29 <*>
1952 P169-171	AEC-TR-4972 <*>
1952 P369-372	58-2512 <*>
1952 P490-501	2096 <*>
1952 P966-967	61-20957 <*>
1952 N5 P569-573	60-13247 <=*>
1952 N11/2 P928-937	60-10328 <*> 0
1952 V19 P418-421	T-1395 <INSD>
1953 P569-610	64-16489 <*> 0
1953 P754-756	63-10622 <*> 0
1953 P781-788	I-888 <*>
1953 P840-841	I-476 <*>
1953 N9 P821-826	5199 <K-H>
1953 N7/8 P693-695	62-14204 <*>
1953 N10 P1010-1012	I-749 <*>
1953 V20 N5 P433-437	3291 <*>
1953 S5 V20 P256-263	04F4F <ATS>
1954 P1298-1304	59-15092 <*>
1954 P1458-1463	59-10606 <*>
1954 N5 P646-647	65-10110 <*>
1954 N6 P811-812	64-20441 <*> 0
1954 N9 P1142-1148	64-20555 <*> P
1954 N7/8 P936-940	3261 <*>
1954 N11/2 P1347-1349	63-16133 <*> 0
1954 N11/2 P1477-1480	64-20549 <*> P
1954 V21 N5 P142-143	21J18F <ATS>
1955 P339-345	60-10327 <*>
1955 P448-452	57-2686 <*>
	8G6F <ATS>
1955 P946-947	61-20956 <*>
1955 P983-991	86H8F <ATS>
1955 N3 P367-369	AEC-TR-3997 <*>
1955 N3 P455-456	60-10326 <*> 0
1955 N7/8 P999-1012	64-14934 <*> OP
1955 N7/8 P1013-1017	62-10011 <*>
1955 N10 P1302-1303	57-1697 <*>
1955 V7/8 P983-991	63-10453 <*>
1956 P967-970	AEC-TR-3551 <+*>
1956 P1679-1682	AD-609 162 <=>

1956 N1 P1-6	62-20057 <*>
1956 N1 P103-106	1112 <*>
1956 N4 P542-546	64-20926 <*>
1956 N6 P878-881	64-20463 <*> 0
1956 N6 P967-970	64-20480 <*>
1956 N7 P1020-1028	1680 <*>
1956 N7 P1040-1049	65-10084 <*> 0
1956 N10 P1392-1398	65-10047 <*> 0
1956 N8/9 P1200-1202	60-18148 <*>
1956 N8/9 P1234-1238	62-16262 <*> 0
	64-20470 <*> 0
1956 N3 P497-527	3430 <*>
1957 P486-487	60-19173 <*+>
1957 P529-531	64R74F <ATS>
1957 P1394-1403	65-10028 <*>
1957 N2 P153-158	62-14568 <*>
1957 N3 P436-440	61-10526 <*>
1957 N4 P529-531	64-20890 <*> 0
1957 N5 P728-733	61-16743 <*>
1957 N10 P1108-1110	60-18987 <*>
1957 N10 P1280-1288	65-12212 <*>
1957 N1/2 P5-25	59-10585 <*>
1957 N11/2 P1388-1393	58-1723 <*>
1958 P566-569	5K25F <ATS>
1958 P853-854	78Q70F <ATS>
1958 P855-856	99Q71F <ATS>
1958 P958-964	91K25F <ATS>
1958 P1415-1417	AEC-TR-5210 <*>
1958 N1 P69-75	60K26F <ATS>
1958 N3 P319-323	65-12429 <*>
1958 N6 P766-772	62-00141 <*>
1958 N6 P772-780	AEC-TR-3504 <*>
1958 N6 P853-854	65-14077 <*>
1958 N6 P855-856	65-14078 <*>
1958 N7 P973-980	59-15762 <*> 0
	62-00142 <*>
1958 N2/3 P297-314	63-00071 <*>
1958 N8/9 P1088-1093	AEC-TR-3584 <+*>
1958 N8/9 P1167-1174	59-10566 <*>
1959 P97-102	63M38F <ATS>
1959 P1570-1577	AEC-TR-5676 <*>
1959 N4 P594-597	62-00140 <*>
1959 N4 P637-642	59-17468 <*> 0
1959 N6 P810-815	61-10169 <*>
1959 N10 P1543-1544	AEC-TR-4815 <*>
1960 P1827-1830	61-20855 <*> 0
1960 P1831-1834	61-20859 <*> 0
1960 P1835-1838	62-10246 <*> 0
1960 P1839-1843	62-10245 <*> 0
1960 P1909-1914	AD-609 163 <=>
1960 N2 P217-221	64-18444 <*>
1960 N2 P242-245	86M46F <ATS>
1960 N2 P250-259	61-10861 <*> 0
1960 N2 P313-321	61-10862 <*>
1960 N3 P473-481	64-18443 <*>
1960 N3 P525-533	63-14236 <*> 0
1960 N3 P551-562	61-10386 <*> 0
1960 N5 P906-910	63-20232 <*>
	65-17439 <*>
1960 N5 P982	63-16619 <*>
1960 N10 P1740	61-16543 <*>
1960 N10 P1849-1867	65-10065 <*> 0P
1960 N11 P2139-2147	10969-C <K-H>
1960 N8/9 P1654-1659	63-20233 <*>
1960 N11/2 P11-16	AEC-TR-5233 <=*>
1960 N11/2 P2139-2147	64-14598 <*>
1961 P177-180	AEC-TR-4844 <*>
1961 N1 P70-74	5260 <HB>
1961 N1 P78-79	5259 <HB>
1961 N3 P449-460	62-16013 <*>
1961 N3 P553-558	65-12315 <*>
1961 N4 P810-812	62-18722 <*> 0
1961 N4 P813-818	62-18721 <*> 0
1961 N4 P818-822	62-18720 <*> 0
1961 N5 P973-975	11N58F <ATS>
1961 N7 P1264-1265	66-14155 <*>
1961 N11 P2295-2296	63-10244 <*>
1961 N12 P2447-2452	62-18065 <*>
1961 N12 P2453-2475	18P60F <ATS>
1961 N8/9 P1550-1559	62-16584 <*>
1962 P411-418	NP-TR-954 <*>
1962 P1774-1782	65-12238 <*>
1962 P1837-1842	12806 B <K-H>
1962 P2103-2107	66-12079 <*>
1962 N1 P53-55	63-20289 <*>
1962 N4 P816-822	64-18480 <*>
1962 N4 P844-850	62-01264 <*>
1962 N6 P1224-1236	66-10196 <*> 0
1962 N6 P1237-1243	66-10197 <*> 0
1962 N6 P1243-1246	66-10198 <*> 0
1962 N7 P1404-1412	63-20053 <*>
1962 N10 P1837-1842	65-14849 <*>
1962 N10 P1887-1892	TRANS-185 <FRI>
	63-20189 <*>
1962 N10 P1918-1923	65-12734 <*>
1962 N10 P1962-1970	63-01076 <*>
1962 N8/9 P1688-1694	65-10201 <*>
1962 N8/9 P1727-1735	65-12485 <*>
1962 N11/2 P2048-2054	65-10791 <*>
1963 P1854	66-13566 <*>
1963 N1 P93-99	65-12791 <*>
1963 N2 P210-211	64-14753 <*>
1963 N2 P244-247	AI-TR-27 <*>
1963 N2 P248-251	65-12790 <*>
1963 N2 P304-309	65-12638 <*>
1963 N2 P362-365	63-01316 <*>
1963 N2 P388-391	63-01317 <*>
1963 N3 P464-470	HW-TR-56 <*>
1963 N7 P1456-1461	65-18160 <*>
1963 N10 P2025-2034	64-18777 <*>
1963 N10 P2376-2392	65-10200 <*>
1963 N11 P2546-2550	64-16170 <*> 0
1963 N11 P2643-2651	64-16686 <*>
1963 N12 P2712-2721	65-00264 <*>
1963 N12 P2735-2737	AEC-TR-6438 <*>
1963 N12 P2762-2768	64-30693 <*>
1963 N12 P2787-2793	AD-616 337 <=$>
1963 N12 P2903-2909	286 <TC>
1963 N8/9 P1855-1860	66-10032 <*>
1964 P31-32	66-11094 <*>
1964 P196-203	30R78F <ATS>
1964 N1 P19-22	66-10817 <*> 0
1964 N2 P236-240	NS-340 <TTIS>
1964 N2 P334-348	65-10167 <*>
1964 N5 P912-913	65-11298 <*> 0
1964 N6 P1240-1244	65-12317 <*>
1964 N6 P1302-1305	65-12488 <*>
1964 N6 P1393-1405	65-13309 <*>
1964 N7 P1641-1646	621 <TC>
1964 N8 P1845-1852	8692 <IICH>
1964 N9 P2093-2095	65-12201 <*>
1964 N9 P2095-2097	65-12200 <*>
1964 N9 P2221-2224	65-17216 <*>
1964 N11 P2974-2986	66-10654 <*> 0
	667 <TC>
1964 N11 P2987-2998	670 <TC>
1965 P3077-3083	2044 <TC>
1965 P3590-3596	8746 <IICH>
1965 N4 P27-29	65-14956 <*>
1965 N4 P1001-1007	66-13522 <*>
1965 N4 P1021-1039	65-17039 <*>
1965 N4 P1138-1141	66-11231 <*>
1965 N7 P1979-1985	66-10653 <*>
1965 V2 P386-388	66-11996 <*>
1966 N1 P173-177	66-12973 <*>
1966 N4 P1456-1468	66-13811 <*>

BULLETIN DE LA SOCIETE CHIMIQUE DU NORD DE LA FRANCE

1892 V2 P58-84	63-14231 <*>

BULLETIN DE LA SOCIETE CHIMIQUE DE PARIS

1877 V27 N2 P558-559	65-14465 <*>
1885 N2 P26-32	64-20901 <*>
1887 S2 P95-96	65-13579 <*>
1889 V1 N3 P280-283	I-89 <*>
1889 V2 P388-391	58-810 <*>
1892 V7 N3 P791-793	59-15091 <*>
1897 S3 V17 P230-234	63-18573 <*>
1897 S3 V17 P599-609	9036 <IICH>

1898 S3 V19 P488-494 63-16789 <*>
1898 S3 V19 P598-603 63-16790 <*>
1899 S3 V21 P532 63-16791 <*>
1902 S3 V27 P797-803 82H11F <ATS>
1903 S3 V29 P35-47 83H11F <ATS>
1903 S3 V29 P169 64-14293 <*>
1903 S3 V35 P781 64-14424 <*>
1906 S3 V35 P778 64-14243 <*>
1906 S3 V35 P781 64-14424 <*>

BULLETIN DE LA SOCIETE D'ENCOURAGEMENT POUR L'INDUSTRIE NATIONALE
1900 V98 P49 64-14292 <*>
1911 V116 P280-321 64-18219 <*> PO

BULLETIN DE LA SOCIETE FRANCAISE DE CERAMIQUE
1950 N9 P1-17 AL-36 <*>
1950 N9 P24-36 AL-142 <*>
1955 N29 P41-52 CSIRO-3168 <CSIR>
1959 N43 P97-112 65-13045 <*>
1960 N46 P79-83 65-14369 <*>
1960 V46 P3-11 UCRL TRANS-1042(L) <*>
1961 N50 P25-33 2863 <BISI>
1961 N51 P53-68 66-13933 <*>
1963 N59 P25-45 66-13559 <*>
1964 N65 P51-66 66-13495 <*>
1964 V63 P15-25 65-18110 <*>

BULLETIN DE LA SOCIETE FRANCAISE DE DERMATOLOGIE ET DE SYPHILIGRAPHIE
1934 V41 P1674-1678 65-13468 <*>
1939 V46 P344-348 64-20285 <*>
1949 V56 P431-432 MF-2 <*>
1950 V57 P570- MF-2 <*>
1951 V61 N58 P292-294 T1817 <INSD>
1960 V67 N2 P234-238 10809-C <KH>
63-16904 <*>
1962 V69 N3 P493-494 66-10865 <*>
1964 V71 N4 P541-543 65-13481 <*>
1964 V71 N5 P645-646 66-11699 <*>

BULLETIN DE LA SOCIETE FRANCAISE D'ECONOMIE RURALE
1952 V4 N4 P121-187 64-10290 <*> P

BULLETIN DE LA SOCIETE FRANCAISE DES ELECTRICIENS
1919 V9 P85- 57-2415 <*>
1920 V10 P375-386 57-3330 <*>
1924 V4 P185-202 57-2167 <*>
1924 V4 P463-513 57-3329 <*>
1925 V5 P27-40 57-2182 <*>
1927 V5 N75 P1354-1357 57-926 <*>
1929 V9 N90 P195-206 57-2148 <*>
57-3511 <*>
1929 V9 N98 P1146-1158 57-1398 <*>
1930 V10 P973-1001 57-3109 <*>
1930 V10 N111 P1249-1255 57-334 <*>
1930 S4 V10 N109 P1002-1025
59-17277 <*> 0
1931 S5 V1 N3 P288-305 57-2227 <*>
1933 V3 P882-884 57-366 <*>
1934 V4 N45 P847-883 57-2486 <*>
1934 S5 V4 N39 P229-248 57-2412 <*>
1936 S5 V6 P775-786 AL-520 <*>
1936 S5 V6 N65 P547-554 59-20251 <*> 0
1937 P1045-1056 2179 <*>
1938 V8 N96 P1033-1037 57-1568 <*>
1939 V5 N9 P911-928 57-132 <*>
1939 V9 P345-353 57-1003 <*>
1941 V1 N3 P132-146 57-698 <*>
1944 V4 N35 P1-23 57-928 <*>
1945 V5 N6 P19-27 58-2031 <*>
1946 S6 V6 N62 P498-509 62-18157 <*> 0
1947 S6 V7 P540-544 AL-266 <*>
63-20782 <*>
1949 V9 P53-57 C-2357 <NRC>
1949 V9 P53-71 3443 <*>
1955 P197-204 T-2052 <INSD>
1956 V5 N49 P19-26 12H13F <ATS>
1956 V6 P73-82 65-11931 <*>
1957 S7 V7 P576-583 SCL-T-313 <*>
1957 S7 V7 N80 P499-510 65-10255 <*>
1957 S7 V9 P564-573 60-21030 <=> 0
1959 S7 V9 N97 P35-42 64-14082 <*>
1959 S7 V9 N97 P43-47 64-14081 <*> 0
1959 S7 V9 N106 P575-596 62-16490 <*>
1959 S7 V9 N107 P649-665 61-10381 <*>
1959 S7 V11 N102 P1-12 60-10352 <*>
1960 S8 V1 P584-591 HW-TR-35 <*>
1960 S8 V1 N5 P269-277 63-16702 <*>
1960 S8 V1 N12 P840-847 2821 <BISI>
1963 V4 N45 P523-540 3748 <BISI>
1964 S8 V5 P640-643 65-17042 <*>
1964 S8 V5 N59 P685-692 65-14372 <*>

BULLETIN DE LA SOCIETE FRANCAISE DE MINERALOGIE
SEE BULLETIN DE LA SOCIETE FRANCAISE DE MINERALOGIE ET DE CRISTALLOGRAPHIE

BULLETIN DE LA SOCIETE FRANCAISE DE MINERALOGIE ET DE CRISTALLOGRAPHIE
1887 V10 P63-69 65-17423 <*>
1936 V59 P286-308 II-429 <*>
1950 V73 N10/2 P511-600 65-10275 <*> P
1953 V76 P391-414 AI-TR-2 <*>
1953 V76 N7/8 P237-293 SCL-T-426 <*>
1954 V77 P797-814 98L36F <ATS>
1954 V77 P1084-1101 63-00904 <*>
63-27389 <$>
1954 V77 N1/3 P611-630 59-17441 <*> 0
1954 V77 N4/6 P262-266 61-18827 <*>
1955 V78 P461-474 AI-TR-4 <*>
1955 V78 P497-520 AI-TR-7 <*>
1958 V81 P79-102 61-10176 <*> 0
1958 V81 P266-273 AI-TR-9 <*>
1959 V82 P335-340 61-14761 <*>
1959 V82 P367-373 63-16244 <*> 0
1960 V83 P254-256 63-16241 <*> 0
1961 V84 P292-311 AI-TR-13 <*>
1964 N87 P149-156 GI N5 1964 P1025 <AGI>
1964 N87 P163-165 GI N5 1964 P1035 <AGI>
1964 N87 P385-392 GI N6 1964 P1196 <AGI>
1964 V87 N1 P28-30 GI N3 1964 P570 <AGI>
1964 V87 N1 P105-108 GI N3 1964 P575 <AGI>
1964 V87 N2 P241-272 66-10176 <*>

BULLETIN DE LA SOCIETE FRANCAISE DE PHOTOGRAPHIE ET DE CINEMATOGRAPHIE
1927 S3 V14 N1 P12-30 63-10891 <*> 0

BULLETIN DE LA SOCIETE GEOLOGIQUE DE FRANCE
1930 V30 N4 P173-190 25J18F <ATS>
1935 S5 V5 P621-628 58K22F <ATS>
1936 V6 N5 P487-493 32K22F <ATS>
1937 V7 N5 P632-635 31K22F <ATS>
1941 S5 V11 P183-193 II-514 <*>
1941 S5 V11 N1 P183-193 3391 <*>
1947 V16 P133-145 83J16F <ATS>
1950 V20 P399-402 T-1421 <INSD>
1954 S6 V4 N7/9 P467-473 3390 <*>
1954 S6 V4 N7/9 P609-619 3389 <*>
1959 S7 V1 N5 P500-510 1872 <BISI>
1959 S7 V1 N7 P649-650 25M44F <ATS>

BULLETIN DE LA SOCIETE D'HISTOIRE NATURELLE DE TOULOUSE
1950 V85 P383-386 1899 <*> P

BULLETIN DE LA SOCIETE INDUSTRIELLE DE MULHOUSE
1854 V26 N128 P188-237 58-1441 <*>
1929 V95 P325-343 63-16218 <*>
1939 V115 N4 P169-174 RCT V13 N1 P133 <RCT>

BULLETIN DE LA SOCIETE MATHEMATIQUE DE FRANCE
1954 V82 P22-23 58-2228 <*>

BULLETIN DE LA SOCIETE MINERALOGIQUE DE FRANCE
SEE BULLETIN DE LA SOCIETE FRANCAISE DE MINERALOGIE ET DE CRISTALLOGRAPHIE

BULLETIN DE LA SOCIETE NEUCHATELOISE DE GEOGRAPHIE

1946 N4 P1-7 3338 <*>

BULLETIN DE LA SOCIETE NEUCHATELOISE DES SCIENCES NATURELLES
1946 V52 N4 P1-7 PT1 AL-606 <*>

BULLETIN DE LA SOCIETE DE PATHOLOGIE EXOTIQUE
1922 V15 P36-39 65-00326 <*>
1934 V27 P666-668 63-01682 <*>
1937 V30 P193-203 63-00856 <*>
1948 V41 N1/2 P47-59 1700 <*>
1949 V42 P466-470 63-01625 <*>
1949 V42 N1/2 P11-12 1704 <*>
1949 V42 N1/2 P52-62 60-17657 <*+>
1949 V42 N5/6 P197-209 60-17658 <*+> O
1949 V42 N5/6 P210-215 1705 <*>
1950 V43 N7/8 P398-403 60-17656 <*+>
1951 V44 N7/8 P413-415 59-15416 <*>
1952 V45 N1 P19-23 59-15419 <*>
1952 V45 N3 P304-305 60-17574 <*+>
1952 V45 N5 P592-594 60-17666 <*+>
1953 V46 N1 P24-26 60-17664 <*+>
1953 V46 N3 P353-359 59-15415 <*>
1953 V46 N6 P866-870 59-15418 <*>
1955 V48 N5 P6C2-607 61-00372 <*>
1955 V48 N6 P810-814 58-1711 <*>
1958 V51 N1 P35-38 61-00297 <*>
1958 V51 N4 P5C0-501 59-2C963 <*>
1958 V51 N4 P582-589 61-00584 <*>
1958 V51 N6 P897-901 61-00161 <*>
1959 V52 N3 P276-281 61-00160 <*>
1959 V52 N6 P759-764 62-00534 <*>
1960 N3 P398-401 61-18766 <*> O
1960 V53 P352-357 64-10392 <*>
1960 V53 P836-841 62-00672 <*>
1960 V53 P950-960 62-10015 <*>
1960 V53 N1 P13-16 62-10912 <*>
1960 V53 N1 P72-84 61-00275 <*>
1960 V53 N3 P563-581 66-11997 <*>
1961 V54 N5 P1164-1183 65-17297 <*>
1963 V56 P880-886 65-14544 <*>
1964 V57 P588-611 66-10940 <*>
1964 V57 P611-618 66-10941 <*>

BULLETIN DE LA SOCIETE DE PATHOLOGIE VEGETALE DE FRANCE
1919 V6 P68-71 62-01201 <*>

BULLETIN DE LA SOCIETE DE PHARMACIE DF BORDEAUX
1952 V90 P123-130 I-440 <*>
1955 V94 P193-195 5709-C <K-H>
1962 V101 P51-55 36Q69F <ATS>
1965 V104 N1 P7-12 66-13667 <*>

BULLETIN DE LA SOCIETE DE PHARMACIE DE MARSEILLE
1953 N7 P184-186 64-20284 <*>

BULLETIN DE LA SOCIETE R. BELGE DES INGENIEURS ET DES INDUSTRIELS
1953 N6 ENTIRE ISSUE I-363 <*>

BULLETIN DE LA SOCIETE R. DES SCIENCES DE LIEGE
1942 V11 P598-605 60-10564 <*>
1949 V18 P1C8-112 UCRL TRANS-793(L) <*>
1950 V19 P343-348 58-2446 <*>
1953 P437-443 I-215 <*>
1954 V23 N11 P377-394 59-1C594 <*>
1956 V25 N3 P183-195 59-10596 <*>
1959 V28 N9/1C P207-221 62-10627 <*>
1962 V31 N9/10 P637-649 64-10931 <*>
1963 V32 N1/2 P54-61 64-71231 <=>
1965 V34 P284- 1881 <TC>
1965 V34 N5/6 P266-283 1880 <TC>

BULLETIN DE LA SOCIETE SCIENTIFIQUE ET MEDICALE DE L'OUEST
1903 V12 P41C-411 T-2548 <*>

BULLETIN DE LA SOCIETE ZCCLOGIQUE DE FRANCE
1944 V69 P94-97 T-1634 <INSD>
1955 V80 P275-287 65-12659 <*>

BULLETIN DES SOCIETES CHIMIQUES BELGES
1947 V56 P349-368 64-20359 <*>
1948 V57 P381-399 57-3533 <*>
58-1268 <*>
1949 V58 P103-111 65-10112 <*>
1949 V58 P210-246 03M43F <ATS>
1949 V58 P285-300 63-20931 <*> O
1950 V59 P40-71 63-14228 <*> O
1950 V59 P476-489 65-14096 <*> O
1950 V59 P573-580 61-20208 <*>
1950 V59 N8/9 P476-489 59-17134 <*> O
64-20903 <*> O
1951 V60 P137-155 64-14964 <*> O
1951 V60 P357-384 1159 <*>
1952 V61 P167-180 61-01060 <*>
1952 V61 P651-682 57-3366 <*>
1954 V63 P70-81 58-1390 <*>
1954 V63 P261-284 64-14965 <*> O
1954 V63 N11/2 P500-524 59-10550 <*>
1955 V64 P203-209 58-92 <*>
1955 V64 P333-351 1305 <*>
1955 V64 P470-483 57-141 <*>
1956 V65 P769-793 63-14230 <*> O
1956 V65 P957-959 T2269 <INSD>
1957 V66 N1/2 P26-32 59-1C588 <*>
1957 V66 N1/2 P33-54 59-10587 <*>
1957 V66 N7/8 P512-524 59-10586 <*>
1958 V67 P676-684 65-10053 <*>
1959 V68 P336-343 62-19449 <=*>
1961 V70 P415-422 63-20928 <*> O
1963 V72 P264-275 85Q70F <ATS>

BULLETIN DES SOCIETES D'OPHTALMOLOGIE DE FRANCE
1955 P569- 5131-C <K-H>

BULLETIN OF THE SOCIETY OF NAVAL ARCHITECTS OF JAPAN
SEE ZOSEN KYOKAI-SHI

BULLETIN OF THE SOCIETY CF SALT SCIENCE. TOKYC
SEE NIHON SHIO GAKKAISHI

BULLETIN TECHNIQUE DU BUREAU VERITAS
1964 V46 N9 P145 2110 <BSRA>
1964 V46 N9 P163 2110 <BSRA>

BULLETIN TECHNIQUE DE LA SUISSE ROMANDE
1940 V66 N14 P149-154 58-495 <*>
1940 V66 N16 P169-174 58-491 <*>
1945 N1 P13-22 3382 <*>
1945 N4 P41-50 3382 <*>
1945 V71 N1 P13-22 03382 <*>
3382 <*>
1945 V71 N4 P41-50 03382 <*>
1946 N22 P3-10 58-784 <*>
1950 V76 P225-237 T-1411 <INSD>

BULLETIN OF TELECOMMUNICATICNS SCIENCE. CHINESE PEOPLES REPUBLIC
SEE TIEN HSIN CHI SHU T'UNG HSUN

BULLETIN OF THE TOHOKU NATIONAL AGRICULTURAL EXPERIMENTAL STATION
SEE TOHOKU NOGYO SHIKENJO KENKYU HOKOKU

BULLETIN OF THE TOHOKU REGIONAL FISHERIES RESEARCH LABORATORY
SEE TOHOKU KAIKU SUISAN KENKYUSHO KENKYU HOKOKU

BULLETIN OF THE TOKAI REGIONAL FISHERIES RESEARCH LABORATORY
SEE TOKAI-KU SUISAN KENKYUSHO KENKYU HOKOKU

BULLETIN OF THE TOKYO INSTITUTE OF TECHNOLOGY. SERIES A
SEE TOKYO KOGYC DAIGAKU GAKUHO

BULLETIN OF THE TOKYO UNIVERSITY FORESTS
SEE TOKYO DAIGAKU NOGAKUBU ENSHURIN HOKOKU

BULLETIN TRIMESTRIEL. CENTRE REGIONAL DES ETUDES ECONOMIQUES
1953 N2 P45-52 65-12680 <*>

BULLETIN DE L'UNION DES AGRICULTEURS D'EGYPTE
1945 V43 N358 P131-139 3413 <*>

BULLETIN DE L'UNIVERSITE DE L'ASIE CENTRALE. TASHKENT
1923 P152-154 T1916 <INSD>

BULLETIN. VEREINIGUNG SCHWEIZER PETROLEUM-GEOLOGEN UND -INGENIEURE
1958 V25 N68 P29-35 IGR V1 N6 P1 <AGI>

BULLETIN OF THE VOLCANOLOGICAL SOCIETY OF JAPAN
SEE KAZAN

BULLETIN VOLCANOLOGIQUE
1951 S2 V12 P33-47 63-20784 <*> O

BULLETIN. VYSKUMNY USTAV PAPIERA A CELULOZY
1962 V5 N4 ENTIRE ISSUE 64-14713 <*>
1962 V5 N4 P24- 64-14713 <*>
1962 V7 P269-284 64-10161 <*>

BULLETIN OF THE WOOD RESEARCH INSTITUTE. KYOTO UNIVERSITY
SEE MOKUZAI KENKYU

BULLETIN OF THE WORLD HEALTH ORGANIZATION
1951 V4 N1 P131-139 2499 <*>
1965 V32 N1 P73-82 66-10989 <*>

BUMAZHNAYA PROMYSHLENNOST
1934 V13 N10/1 P57-65 RT-1471 <*>
1935 V14 N1 P70-73 R-2883 <*>
1938 V16 N2 P52-55 60-14602 <+*>
1946 V21 N5/6 P24-31 RT-2925 <*>
1947 V22 N7 P16-22 61-10683 <+*> O
1949 V24 N5 P6-10 28S89R <ATS>
1949 V24 N6 P8-13 29S89R <ATS>
1950 V25 N3 P20-25 TT.597 <NRC>
1951 V26 N4 P6-13 TT.598 <NRC>
1953 V28 N1 P6-13 63-11012 <=>
1953 V28 N4 P23-26 RT-1671 <*>
1953 V28 N7 P6-11 61-20893 <*=> O
1953 V28 N9 P15-19 64-20453 <*>
1953 V28 N12 P6-9 R-1506 <*>
RT-3701 <*>
1955 V30 N5 P10-13 60-18347 <+*> O
1955 V30 N9 P9-13 M-5576 <NLL>
1955 V30 N10 P5-7 M-5577 <NLL>
1955 V30 N10 P7-10 60-18443 <*> O
1955 V30 N1C P23 62-16510 <=*>
1956 N3 P6-10 59-10227 <+*>
1956 N11 P2-6 R-2349 <*>
1956 V31 N1 P18-19 CSIRO-3430 <CSIR>
R-1899 <*>
1956 V31 N4 P13-14 60-14667 <*> O
1956 V31 N4 P19-21 R-4106 <*>
1956 V31 N4 P21-22 R-3446 <*>
R-4575 <*>
1957 N9 P8- R-4254 <*>
1957 V32 N1 P15-16 R-3449 <*>
R-4161 <*>
R-5280 <*>
1957 V32 N3 P2-5 59-10509 <*>
1957 V32 N9 P2-5 57M42R <ATS>
61-10309 <*+> O
1957 V32 N9 P5-7 60-10138 <+*> O
1957 V32 N12 P5-10 59-10298 <+*>
1958 V33 N1 P9-11 59-17163 <+*> O
1958 V33 N4 P6-9 59-17751 <*>
1958 V33 N4 P13-15 59-15834 <+*> O
1958 V33 N5 P10-13 61-13999 <*+>
1958 V33 N6 P2-5 60-16517 <+*>
1958 V33 N7 P2-5 59-10269 <*>
1958 V33 N8 P2-3 61-18640 <*=>
1958 V33 N8 P6-9 60-14013 <+*> O
1958 V33 N8 P12-14 59-22258 <+*>
1958 V33 N11 P7-9 59-15835 <+*> O
1958 V33 N12 P8-10 60-14062 <*> O
1959 V34 N1 P21-22 59-15812 <+*> O
1959 V34 N3 P2-5 62-16139 <=*>
1959 V34 N5 P4-7 59-17481 <+*>
1959 V34 N6 P12-15 63-11088 <=>
1959 V34 N7 P5-7 61-14919 <=*>
1959 V34 N7 P9 63-11090 <=>
1959 V34 N8 P16-17 60-14291 <+*> O
1959 V34 N10 P20-21 61-20013 <*=> O
1959 V34 N12 P7-8 61-10680 <=*>
1959 V34 N12 P12 62-16514 <=*>
1960 V35 N1 P4-8 61-13360 <+*>
1960 V35 N2 P2- 61-16535 <*=>
1960 V35 N4 P22-24 62-11163 <=>
1960 V35 N7 P10 62-16513 <=*>
1960 V35 N8 P22-23 61-18569 <*> O
1960 V35 N11 P15-17 61-20011 <*=>
1960 V35 N12 P5-7 61-20866 <*=>
66-12779 <*>
1961 V36 N4 P10-11 62-10655 <=*>
1961 V36 N5 P7-8 62-16501 <=*>
1961 V36 N5 P9-10 62-14358 <=*>
1961 V36 N9 P4-7 62-16074 <*=>
1961 V36 N10 P4-5 62-16075 <=*>
1961 V36 N12 P12-14 62-18875 <=*> O
1962 V37 N1 P17-19 63-13110 <=*>
1962 V37 N3 P14-16 62-20470 <=*> O
1962 V37 N5 P14- 63-20824 <=*$>
1962 V37 N5 P19 62-32448 <=>
1962 V37 N5 P20 62-32448 <=>
1962 V37 N10 P4-6 29Q72R <ATS>
1962 V37 N10 P6-10 63-13680 <=>
1962 V37 N10 P11-13 28Q72R <ATS>
1962 V37 N10 P23-26 65-12089 <*>
1962 V37 N11 P18-20 63-18696 <=*$> O
1963 V38 N1 P7-8 63-21601 <=>
1963 V38 N4 P2C 63-18574 <=*$>
1963 V38 N6 P3-5 65-12079 <*>
1963 V38 N7 P12-13 63-31920 <=>
1964 N1 P22-24 65-14604 <*>
1964 N2 P1-2 NLLTB V6 N6 P557 <NLL>
1964 V39 N1 P13-15 64-16906 <=*$>
1964 V39 N2 P1-2 TRANS.BULL.1964 P557 <NLL>
1964 V39 N3 P7-9 66-11424 <*>
1964 V39 N5 P3-5 65-13133 <*>
1964 V39 N5 P8-9 RTS-2888 <NLL>
1964 V39 N5 P30-31 RTS-2889 <NLL>
1964 V39 N6 P17-19 RTS-2890 <NLL>
1964 V39 N7 P14-16 23S84R <ATS>
1965 N9 P12-14 ICE V6 N4 P677-680 <ICE>
1965 V40 N6 P16-18 M.5807 <NLL>
1965 V40 N9 P12-14 85T92R <ATS>
1965 V40 N10 P3-5 66-13994 <*>

BUNDESBAHN
1955 N21 P900-905 T1960 <INSD>

BUNDESGESUNDHEITSBLATT
1960 V4 N1 P11-12 83N50G <ATS>
1961 N10 P149-156 61-01003 <*>

BUNGEI SHUNJU
1960 N3 P158-166 60-31142 <=*>

BUNSEKI KAGAKU
1952 V1 N2 P126-130 AEC-TR-6249 <*>
1953 V2 P428-432 92K23J <ATS>
1953 V2 P436-439 32H12J <ATS>
1953 V2 P456-460 57-1202 <*>
1954 V3 P107-114 58-312 <*>
77H11J <ATS>

1954 V3 P195-196 58-313 <*>
76H11J <ATS>
1954 V3 P196-199 58-314 <*>
76H11J <ATS>
1954 V3 P224-228 57-714 <*>
1954 V3 P299-304 4145 <BISI>
1954 V3 P361-363 1250 <*>
1954 V3 P364-367 1289 <*>
1954 V3 N3 P213-215 08H14J <ATS>
1954 V3 N4 P333-334 AEC-TR-2104 <*>
1251 <*>
1954 V3 N4 P335-348 1303 <*>
1954 V3 N4 P349-356 1290 <*>
1954 V3 N4 P356-361 1273 <*>
1954 V3 N4 P368- 1270 <*>
1955 V4 P14-16 48H9J <ATS>
1955 V4 P148-152 T-2015 <INSD>
1955 V4 N1 P158-162 61-16767 <*>
1955 V4 N5 P290-293 64-18128 <*> 0
1955 V4 N6 P349-353 64-18550 <*> 0
1955 V4 N2/3 P163-166 61-16768 <*>
1955 V4 N2/3 P167-171 61-16769 <*>
1956 N5 P264-267 61-16240 <*>
1956 V5 P404-407 58-839 <*>
1956 V5 P449-452 61-10502 <*> 0
1956 V5 N1 P7-11 37Q71J <ATS>
1956 V5 N4 P203-205 59-19346 <+*>
1956 V5 N4 P220-221 AEC-TR-3947 <*>
1956 V5 N7 P395-398 63-14237 <*> 0
1957 V6 P787-790 75N56J <ATS>
1958 V7 P33-37 UCRL TRANS-621(L) <*>
1958 V7 N4 P205-210 AEC-TR-3848 <*>
1958 V7 N7 P445-449 66-14684 <*>
1959 V8 P180-185 37N51J <ATS>
1959 V8 P456-457 AEC-TR-3897 <*>
1959 V8 P471-484 AEC-TR-3898 <*>
1959 V8 N4 P257-259 59-00867 <*>
1959 V8 N8 P491-495 62-14858 <*>
1960 V9 P33-37 AEC-TR-4038 <*>
1960 V9 P257-264 <JTB>
1960 V9 N7 P628-629 63-26206 <=*>
1961 V10 P1327-1331 AEC-TR-5548 <*>
1961 V10 P1331-1335 AEC-TR-5580 <*>
1961 V10 N6 P580-585 65-18004 <*>
1962 V11 P533-543 INT-9 <INT>
1962 V11 P587-595 AEC-TR-5419 <*>
1962 V11 P943-951 AEC-TR-5768 <=*$>
1962 V11 P21R-36R AEC-TR-5880 <*>
1962 V11 P37R-46R AEC-TR-6011 <*>
1962 V11 P47R-50R AEC-TR-5929 <*>
1962 V11 P71R-75R AEC-TR-6072 <*>
1962 V11 P75R-85R AEC-TR-5975 <*>
1962 V11 N1 P140-147 AEC-TR-5194 <*>
1963 V12 P15-19 N66-13875 <=$>
1963 V12 P26-32 AEC-TR-5837 <*>
1963 V12 P150-155 68R76J <ATS>
1963 V12 P313-325 65-10832 <*>
1963 V12 P475-483 64-15674 <=*$>
1963 V12 N2 P125-130 65-13433 <*> 0
1963 V12 N2 P137-143 65-13432 <*> 0
65-18073 <*>
1963 V12 N3 P247-252 67Q73J <ATS>
1963 V12 N10 P961-963 64-14643 <*>
1964 V13 P1262-1264 1061 <TC>
1964 V13 N1 P68-70 65-11326 <*>
1964 V13 N3 P282-288 64-18544 <*> 0
1964 V13 N5 P445-449 36S86J <ATS>
1964 V13 N5 P469-471 AD-612 691 <=$>
1964 V13 N12 P1259-1261 1081 <TC>
1965 V14 P519-526 AD-623 633 <=*$>
1965 V14 P732-735 66-11102 <*>

BURGENLANDISCHE HEIMABLATTER
1955 V17 N2 P49-55 06K26G <ATS>

BUSSEIRON KENKYU
1950 N30 P15-26 1298 <*>
1951 N34 P44-51 64-15593 <=*$>
1952 P10-15 I-393 <*>
1952 N51 P15-24 63-00677 <*>
1952 N51 P73-79 57-14 <*>
1953 N58 P31-39 66-10223 <*> 0
1953 N69 P20-25 AEC-TR-5660 <*>
1953 N69 P145-180 57-603 <*>
1954 N79 P101-128 UCRL TRANS-1026(L) <=$>
1956 N102 P95-105 61-20708 <*>
1957 S2 V1 N103 P165-169 UCRL TRANS-771(L) <*>
1959 S2 V5 N132 P560-574 UCRL TRANS-772(L) <*>
1959 S2 V6 N138 P648-658 UCRL TRANS-773(L) <*>
1960 S2 V7 N2 P123-141 10N56J <ATS>

BUTSURI KAGAKU NO SHINPO
1943 V17 P1-11 AEC-TR-3640 <+*>
1945 V19 P17-19 AEC-TR-3657 <=*>
1945 V19 P21-24 AEC-TR-3657 <=*>

BYGGMASTAREN. UPPLAGA A.
1951 V17 P277-284 C-2191 <NRC>
1364 <*>
1954 V33 N8 P129-134 CSIRO-2651 <CSIR>
1956 P3-5 T1981 <INSD>

BYGGMASTAREN. UPPLAGA B.
1946 V17 P311-314 C-2689 <NRC>
1956 V18 P19-24 T-1982 <INSD>

BYULLETEN. ABASTUMANSKAYA ASTROFIZICHESKAYA OBSERVATORIYA
1953 N15 P169-260 59-16048 <+*>
1954 N17 P3-24 61-19114 <+*> 0
1954 N17 P149-272 59-16049 <+*>
1959 N24 P161-173 63-13228 <=*>
1964 N30 P137-138 AD-622 345 <=$*>

BYULLETEN ASTRONOMICHESKOGO TEORETICHESKOGO INSTITUTA. LENINGRAD
V3 N45 P115-152 63-14920 <=*>
1929 N23 P21-28 62-25440 <=*>
1941 N52 P407-435 AD-623 442 <=*$>
1949 V4 N3 P103-141 62-15965 <=*>
1949 V4 N6 P270-286 62-25441 <=*>
1950 V4 N8 P375-407 62-15963 <=*>
1950 V4 N8 P408-413 60-15510 <+*>
1953 V5 N8 P461-559 62-15988 <=*>
1957 V6 N8 P505-523 64-13355 <=*$>
1957 V6 N8 P550-565 60-23063 <+*>
1957 V6 N8 P566-576 64-13356 <=*$>
1958 V6 N9 P592-629 61-16439 <*=>
1958 V7 N1 P72-75 63-20613 <=*$>
1959 V7 N4 P257-280 61-16449 <*=>
1959 V7 N4 P257-286 61-28369 <*=>
1959 V7 N4 P257-280 61-31561 <=>
1959 V7 N4 P281-286 61-16448 <*=>
1959 V7 N5 P321-326 61-23996 <*=>
1959 V7 N6 P422-432 62-15832 <=*>
1960 V7 N7 P511-520 ARSJ V31 N9 P1345 <AIAA>
62-33635 <=*>
1960 V7 N7 P521-536 65-63646 <=$*>
1960 V7 N7 P537-548 ARSJ V31 N1 P117 <AIAA>
64-13799 <=*$>
1960 V7 N7 P554-569 ARSJ V30 N9 P865 <AIAA>
61-28244 <*=>
63-24312 <=*$>
1960 V7 N7 P570-580 ARSJ V30 N9 P859 <AIAA>
63-24313 <=*$>
1960 V7 N8 P599-638 61-28378 <*=>
1960 V7 N10 P757-765 62-33650 <=*>
1961 V8 N2 P134-152 ARSJ V32 N9 P1479 <AIAA>
1962 V8 N5 P324-335 RAE-LIB-TRANS-1108 <NLL>
1962 V8 N5 P335-342 AD-612 377 <=$> M
1962 V8 N6 P381-395 63-15406 <=*>
1962 V8 N6 P402-404 65-60672 <=$>
1962 V8 N6 P405-420 64-11922 <=> M
1963 V9 N1 P1-10 AIAAJ V2 N1 P203 <AIAA>
65-10808 <*>
1963 V9 N1 P11-14 AIAAJ V1 N12 P2909 <AIAA>
1963 V9 N1 P15-45 AIAAJ V1 N12 P2893 <AIAA>
1963 V9 N2 P144-153 65-10842 <*>
1963 V9 N2 P154-167 65-13677 <*>
1963 V9 N3 P185-203 65-60395 <=$>

1963 V9 N4 P274-282 AD-631 566 <=$>
1963 V9 N5 P295-309 AD-616 282 <=$>
1963 V9 N5 P323-329 AD-617 666 <=$>
1965 V10 P109-117 AD-635 922 <=$>
1965 V10 N3 P173-180 66-11242 <*>

BYULLETEN ASTRONOMICHESKOI OBSERVATORII IM. V.P. ENGELGARDTA. KAZAN
1960 N34 P1-23 62-13056 <=*>

BYULLETEN ASTRONOMICHESKOI OBSERVATORII. EREVAN
1945 N6 P3-9 AEC-TR-1409 <*>

BYULLETEN ASTRONOMICHESKOI OBSERVATORII KHARKOVSKOGO GOSUDARSTVENNOGO UNIVERSITETA IMENI A. M. GORKOGO
1946 N6 P9-11 RT-2706 <*>

*BYULLETEN EKSPERIMENTALNOI BIOLOGII I MEDITSINY
1939 V8 N2 P119-122 R-415 <*>
RT-3843 <*>
64-15094 <=*$>
1939 V8 N3/4 P226-227 61-19725 <=*>
1939 V8 N3/4 P282-284 65-63449 <=>
1940 V10 N3 P107-109 65-62911 <=$>
1940 V10 N3 P110-111 61-28020 <*=>
1941 V11 N5 P432-434 1184A <K-H>
1942 V14 N4 P11-13 65-63448 <=>
1942 V14 N11/2 P50-53 69F2R <ATS>
1943 V15 N3 P17-19 RT-583 <*>
1943 V15 N3 P61-63 60-19829 <+*>
1943 V15 N1/2 P29-32 RT-581 <*>
1943 V15 N1/2 P32-35 RT-582 <*>
1943 V15 N1/2 P60-62 RT-1832 <*>
1944 V17 N3 P58-60 65-60088 <=$>
1944 V18 N4 P15-18 60-19709 <=*$> 0
1944 V18 N4/5 P64-67 63-15955 <=*>
1945 V19 N3 P12-15 64-23635 <=$>
1946 V21 N5 P40-42 65-63252 <=>
1946 V21 N1/2 P15-17 65-60049 <=$>
1946 V21 N1/2 P28-31 65-62969 <=$>
1946 V21 N1/2 P31-33 65-62954 <=$>
1946 V21 N1/2 P33-36 RT-102 <*>
65-60058 <=$>
1946 V21 N1/2 P37-40 65-62997 <=>
1946 V21 N1/2 P42-44 64-23629 <=$>
1946 V22 N1 P6-8 65-60052 <=$> P
1946 V22 N2 P9-12 65-62963 <=>
1946 V22 N5 P20-23 65-63375 <=$>
1947 V23 N1 P32-35 RT-2111 <*>
1947 V23 N2 P83-86 AD-616 973 <=$>
1947 V23 N3 P197-198 R-69 <*>
RT-4658 <*>
64-19075 <=*$> 0
1947 V23 N3 P198-201 61-11477 <=>
1947 V23 N6 P429-433 RT-236 <*>
1952 V33 N3 P16-20 60-10104 <*>
1954 V37 N1 P32-37 RT-2545 <*>
1954 V37 N1 P44-47 RT-2544 <*>
1954 V37 N2 P7-12 62-10460 <=*>
1954 V37 N2 P74-75 RT-2280 <*>
1954 V37 N2 P78-79 RT-2530 <*>
1954 V37 N3 P37-41 RT-2531 <*>
1954 V37 N6 P7-11 3374-A <K-H>
1954 V37 N7 P23-26 61-10872 <+*> 0
1954 V38 N8 P33-36 R-2753 <*>
R-3512 <*>
R-84 <*>
1954 V38 N8 P48-51 R-2752 <*>
R-3513 <*>
RT-2560 <*>
1954 V38 N8 P63-65 63-16515 <=*>
1954 V38 N9 P32-36 RT-2875 <*> 0
1954 V38 N11 P14-18 RT-4644 <*>
1954 V38 N11 P43-46 R-2695 <*>
1955 V39 N1 P19-21 62-20177 <=*> 0
1955 V39 N1 P42-45 RT-4319 <*>
1955 V39 N2 P7-11 RT-3125 <*>
1955 V39 N2 P21-23 62-10713 <=*> 0
1955 V39 N3 P27-29 61-28686 <*=>
1955 V39 N3 P59-62 27H13R <ATS>
1955 V39 N4 P32-35 RT-3896 <*>
1955 V39 N4 P44-46 R-276 <*>
1955 V39 N4 P46-49 R-321 <*>
1955 V39 N4 P76-79 RT-3768 <*>
1955 V39 N5 P32-35 5225-B <K-H>
1955 V39 N5 P39 R-3395 <*>
1955 V39 N5 P45-50 R-3395 <*>
1955 V39 N5 P53-56 59-15394 <+*>
1955 V40 N8 P42-46 RT-4206 <*>
1714 <*>
64-19050 <=*$> 0
1955 V40 N8 P49-52 RT-4384 <*>
1955 V40 N9 P3-6 RT-4432 <*>
1955 V40 N9 P51-54 R-2750 <*>
R-82 <*>
1955 V40 N9 P58-62 R-2748 <*>
R-3447 <*>
R-85 <*>
1955 V40 N10 P47-48 RJ-426 <*>
RJ-426 <ATS>
1956 V40 N1 P20-23 6012-B <K-H>
1956 V40 N8 P47-52 R-2747 <*>
R-3443 <*>
1956 V41 N1 P78-79 62-16050 <=*>
1956 V41 N2 P52-57 RT-4430 <*>
1956 V41 N2 P57-60 RT-4431 <*>
1957 V44 N9 P10-14 63-11594 <=>
1958 V45 N11 P55-57 81L30R <ATS>
1958 V46 N7 P30-34 63-16417 <=*>
1958 V46 N7 P63-67 59-17377 <+*>
1958 V46 N9 P96-100 63-13015 <=*>
1958 V46 N9 P105-107 62-32650 <=*>
1959 V47 N2 P60-62 62-14722 <=*>
1959 V47 N2 P62-69 60-18418 <*>
1959 V47 N3 P34-38 60-18564 <*>
1959 V47 N8 P56-59 60-18566 <+*>
1959 V48 N9 P34-37 62-14721 <=*>
1959 V48 N9 P120-130 60-19895 <+*>
1959 V48 N11 P59-61 62-14740 <=*>
1959 V48 N11 P65-69 62-14741 <=*>
1959 V48 N11 P69-73 62-14703 <=*>
1960 V49 N2 P77-80 62-14713 <=*>
1960 V49 N3 P41-46 63-19708 <=*>
1960 V49 N3 P46-51 60-11762 <=>
1960 V49 N3 P73-76 60-23721 <*+>
1960 V49 N3 P81-84 60-11763 <=>
1960 V49 N3 P113-117 60-23173 <+*>
1960 V50 N10 P87-91 62-14777 <=*>
1960 V50 N10 P97-100 61-27363 <*=>
1960 V50 N10 P101-105 61-27362 <*=>
1960 V50 N10 P105-110 61-27361 <*=>
1960 V50 N10 P111-113 61-27346 <*=>
1960 V50 N11 P3-7 61-27670 <*=>
1960 V50 N11 P8-13 61-27667 <*=>
1960 V50 N11 P51-56 61-27675 <*=>
1960 V50 N11 P66-70 61-27725 <*=>
1960 V50 N11 P71-76 61-27672 <*=>
1960 V50 N11 P76-79 61-27685 <*=>
1960 V50 N11 P130-132 61-27679 <*=>
1960 V50 N12 P44-47 62-14720 <=*>
1961 V51 N2 P32-35 62-20312 <=*>
1961 V51 N2 P42-45 62-14747 <=*> 0
1961 V51 N2 P49-54 62-20313 <=*>
1961 V51 N3 P18-23 11279-C <KH>
64-10604 <=*$>
1961 V51 N4 P57-61 61-23574 <*=>
1961 V52 N8 P59-62 62-14716 <=*>
1961 V52 N8 P110-111 62-14738 <=*>
1961 V52 N9 P56-59 62-20362 <=*>
1961 V52 N12 P43-46 62-23392 <=*> 0
1961 V52 N12 P47-50 62-25829 <=*>
1962 N7 P88-90 65-64511 <=*$>
1962 V53 N1 P25-31 62-25555 <=*>
1962 V53 N1 P31-36 63-15301 <=*>
1962 V53 N1 P41-45 62-25830 <=*>
1962 V53 N1 P74-76 63-24512 <=*$>
1962 V53 N1 P112-117 62-24922 <=*>
63-23006 <=*$>
1962 V53 N2 P9-13 63-19685 <=*$>

1962 V53 N2 P43-47 62-25859 <=*>
1962 V53 N2 P89-92 63-15991 <=*>
1962 V53 N2 P93- FPTS V22 N2 P.T314 <FASE>
1962 V53 N2 P97- FPTS V22 N3 P.T554 <FASE>
1962 V53 N2 P123-125 63-15667 <=*>
1962 V53 N3 P3-7 62-25604 <=*>
1962 V53 N3 P14-19 63-23415 <=*$>
1962 V53 N3 P34-38 63-13210 <=*>
1962 V53 N3 P96-101 62-33120 <=*>
1962 V53 N3 P109-116 63-13210 <=*>
1962 V53 N4 P38-46 63-13898 <=*>
1962 V53 N4 P42-46 62-33107 <=*>
1962 V53 N5 P124-127 62-33120 <=*>
1962 V54 N2 P39-42 62-32934 <=*>
1962 V54 N2 P79-84 62-32934 <=*>
1962 V54 N2 P94-96 62-32934 <=*>
1962 V54 N7 P11-13 63-11615 <=> 0
1962 V54 N12 P61-63 64-18749 <*> 0
1963 V55 N1 P21- FPTS V23 N2 P.T397 <FASE>
1963 V55 N1 P26-30 63-31582 <=>
1963 V55 N1 P31- FPTS V23 N1 P.T147 <FASE>
1963 V55 N1 P35- FPTS V22 N6 P.T1127 <FASE>
1963 V55 N1 P43- FPTS V22 N6 P.T1112 <FASE>
1963 V55 N1 P43-47 63-24237 <=*$> 0
1963 V55 N1 P56-60 64-13516 <=*$>
1963 V55 N1 P77-80 63-31586 <=>
1963 V55 N1 P81- FPTS V22 N6 P.T1245 <FASE>
1963 V55 N1 P88-92 63-24493 <=*$> 0
1963 V55 N1 P93-95 64-21258 <=>
1963 V55 N1 P96-100 63-24494 <=*$>
1963 V55 N1 P114-117 63-31585 <=>
1963 V55 N1 P122-123 64-13515 <=*$> 0
1963 V55 N2 P13-17 63-24504 <=*$>
1963 V55 N2 P45- FPTS V23 N2 O.T401 <FASE>
1963 V55 N2 P60-63 64-13426 <=*$>
1963 V55 N2 P68-73 63-24505 <=*$>
1963 V55 N2 P73-78 64-13425 <=*$>
1963 V55 N2 P78-83 64-13902 <=*$> 0
1963 V55 N2 P100- FPTS V23 N1 P.T93 <FASE>
1963 V55 N2 P116-121 63-24506 <=*$>
1963 V55 N2 P121-123 64-13424 <=*$>
1963 V55 N2 P123-126 64-13423 <=*$> 0
1963 V55 N4 P9-14 64-13862 <=*$> 0
1963 V55 N4 P14- FPTS V23 N2 P.T238 <FASE>
1963 V55 N4 P29- FPTS V23 N2 P.T296 <FASE>
1963 V55 N4 P47-49 64-13863 <=*$>
1963 V55 N4 P49-53 64-13883 <=*$>
1963 V55 N4 P66-70 AD-617 374 <=$>
64-13896 <=*$>
1963 V55 N4 P82-91 64-13815 <=*$> 0
1963 V55 N4 P91-94 64-13624 <=*$> 0
1963 V55 N4 P94-100 64-13884 <=*$> 0
1963 V55 N4 P100-104 64-13819 <=*$>
1963 V55 N4 P108-110 64-13820 <=*$> 0
1963 V55 N4 P110-114 64-13821 <=*$> 0
1963 V55 N4 P114- FPTS V23 N4 P.T747 <FASE>
1963 V55 N5 P3-4 63-23545 <=*$>
1963 V55 N5 P5- FPTS V23 N3 P.T624 <FASE>
1963 V55 N5 P5-8 FPTS,V23,N3,P.T624 <FASE>
1963 V55 N5 P9- FPTS V23 N3 P.T627 <FASE>
FPTS,V23,N3,P.T627 <FASE>
1963 V55 N5 P29-33 64-13823 <=*$> 0
1963 V55 N5 P37-40 64-13864 <=*$> 0
1963 V55 N5 P48-54 64-13899 <=*$>
1963 V55 N5 P54-57 64-13906 <=*$>
1963 V55 N5 P57-60 64-13907 <=*$>
1963 V55 N5 P73-78 64-15516 <=*$>
1963 V55 N5 P82-84 64-13927 <=*$>
1963 V55 N5 P88- FPTS V23 N3 P.T555 <FASE>
FPTS,V23,N3,P.T555 <FASE>
1963 V55 N5 P91- FPTS V23 N3 P.T539 <FASE>
1963 V55 N5 P91-92 FPTS,V23,N3,P.T539 <FASE>
1963 V55 N5 P93- FPTS V23 N3 P.T465 <FASE>
1963 V55 N5 P93-95 FPTS,V23,N3,P.T465 <FASE>
1963 V55 N5 P96- FPTS V23 N3 P.T523 <FASE>
FPTS V23,N3,P.T523 <FASE>
1963 V55 N5 P104-108 64-13830 <=*$>
1963 V55 N5 P108-111 64-13908 <=*$> 0
1963 V55 N6 P8-13 64-19219 <=*$> 0
1963 V55 N6 P14-24 63-31801 <=>
1963 V55 N6 P29-34 64-19259 <=*$>
1963 V55 N6 P56-59 64-19241 <=*$>
1963 V55 N6 P76- FPTS,V23,N5,P.T925 <FASE>
1963 V55 N6 P76-79 FPTS,V23,N5,P.T925 <FASE>
1963 V55 N6 P102- FPTS V23 N4 P.T725 <FASE>
1963 V55 N7 P79-82 AD-617 388 <=$>
1963 V56 N7 P8-12 64-19220 <=*$>
1963 V56 N7 P24-29 64-13316 <=*$>
64-19278 <=*$>
1963 V56 N7 P67- FPTS V23 N5 P.T995 <FASE>
FPTS,V23,N5,P.T995 <FASE>
1963 V56 N7 P83- FPTS V23 N5 P.T1051 <FASE>
FPTS,V23,N5,P.T1051 <FASE>
1963 V56 N8 P17- FPTS V23 N5 P.T1142 <FASE>
1963 V56 N8 P28-32 63-41158 <=>
1963 V56 N8 P33-37 63-41158 <=>
64-19279 <=*$> 0
1963 V56 N8 P56-57 64-19619 <=$>
1963 V56 N8 P64- FPTS V23 N4 P.T750 <FASE>
1963 V56 N8 P75- FPTS,V23,N5,P.T931 <FASE>
1963 V56 N8 P75-76 FPTS,V23,N5,P.T931 <FASE>
1963 V56 N8 P77- 64-19271 <=*$>
1963 V56 N8 P116-120 64-31037 <=>
1963 V56 N9 P3- FPTS V23 N5 P.T1153 <FASE>
1963 V56 N9 P3-7 FPTS,V23,N5,P.T1153 <FASE>
1963 V56 N9 P8- FPTS V23 N5 P.T1136 <FASE>
1963 V56 N9 P8-12 FPTS,V23,N5,P.T1136 <FASE>
1963 V56 N9 P42-46 64-19716 <=$> 0
1963 V56 N9 P56-60 64-13638 <=*$>
1963 V56 N9 P56-61 64-19244 <=*$>
1963 V56 N9 P73- FPTS V23 N5 P.T987 <FASE>
1963 V56 N9 P73-75 FPTS,V23,N5,P.T987 <FASE>
1963 V56 N9 P76-80 AD-618 860 <=$>
1963 V56 N9 P76-81 64-19297 <=*$>
1963 V56 N9 P81-84 AD-617 387 <=$>
1963 V56 N9 P84-88 AD-617 376 <=$>
1963 V56 N9 P112-113 64-14644 <=*$>
1963 V56 N10 P18- 64-19260 <=*$>
1963 V56 N10 P29- FPTS V23 N6 P.T1291 <FASE>
FPTS V23 P.T1291 <FASE>
1963 V56 N10 P55-57 64-19274 <=*$>
1963 V56 N10 P65-67 64-19298 <=*$>
1963 V56 N10 P100-105 64-19717 <=$>
1963 V56 N10 P110-113 64-19224 <=*$> 0
1963 V56 N11 P3- FPTS V23 N5 P.T1129 <FASE>
1963 V56 N11 P11- FPTS V23 N5 P.T1127 <FASE>
1963 V56 N11 P18- FPTS V23 N6 P.T1215 <FASE>
FPTS V23 P.T1215 <FASE>
1963 V56 N11 P39- FPTS,V23,N5,P.T932 <FASE>
1963 V56 N11 P39-42 FPTS,V23,N5,P.T932 <FASE>
64-15906 <=*$>
1963 V56 N11 P44- FPTS V23 N6 P.T1240 <FASE>
FPTS V23 P.T1240 <FASE>
1963 V56 N11 P56-61 64-19658 <=$>
1963 V56 N11 P61-64 64-19720 <=$>
1963 V56 N11 P82-85 64-19659 <=$>
1963 V56 N11 P89-93 64-19705 <=$>
1963 V56 N11 P97-101 64-19620 <=$>
1963 V56 N11 P104-107 64-19621 <=$>
1963 V56 N11 P107-110 AD-618 869 <=$>
1963 V56 N12 P14- FPTS V23 N5 P.T1085 <FASE>
1963 V56 N12 P36-38 64-31304 <=>
1963 V56 N12 P39-42 64-19672 <=$>
1963 V56 N12 P44- FPTS V23 N5 P.T972 <FASE>
FPTS,V23,N5,P.T972 <FASE>
1963 V56 N12 P49-52 64-19690 <=$>
64-31304 <=>
1963 V56 N12 P52-55 65-60218 <=$>
1963 V56 N12 P66-69 64-19624 <=$>
1963 V56 N12 P97-100 64-19708 <=$>
64-31347 <=>
1963 V56 N12 P100-103 64-31304 <=>
1964 V57 N1 P24-28 64-19673 <=$> 0
1964 V57 N1 P36-41 64-19634 <=$>
1964 V57 N1 P49-54 64-19664 <=$>
1964 V57 N1 P58-63 64-19665 <=$>
1964 V57 N1 P88-91 64-19695 <=$> 0
1964 V57 N2 P29-40 64-31131 <=>
1964 V57 N2 P59- FPTS V24 N1 P.T43 <FASE>
1964 V57 N2 P59-60 FPTS V24 P.T43-T44 <FASE>

1964 V57 N2 P68-73 65-61053 <=$>
1964 V57 N2 P93-101 64-31131 <=>
1964 V57 N2 P93-97 65-61054 <=$>
1964 V57 N2 P104-107 64-31131 <=>
1964 V57 N2 P117-121 64-31131 <=>
1964 V57 N3 P3-5 64-31736 <=>
1964 V57 N3 P11-15 65-61056 <=$>
1964 V57 N3 P30-34 65-61057 <=$>
1964 V57 N3 P59-61 64-31736 <=>
1964 V57 N3 P67-70 64-31736 <=>
1964 V57 N3 P71-75 65-61058 <=$>
1964 V57 N3 P82-85 64-31736 <=>
1964 V57 N3 P94-97 64-31736 <=>
65-60165 <=$>
1964 V57 N3 P121-125 64-31736 <=>
1964 V57 N4 P3-8 64-41094 P.1-8 <=>
66-60624 <=*$>
1964 V57 N4 P16-19 64-41094 P.9-14 <=>
1964 V57 N4 P84-87 65-61064 <=$>
1964 V57 N5 P55-58 65-62168 <=>
1964 V57 N5 P112-116 65-62169 <=>
1964 V57 N6 P5-10 65-62175 <=>
1964 V57 N6 P26-30 65-61094 <=$>
1964 V57 N6 P34-37 65-67172 <=>
1964 V57 N6 P54- FPTS V24 N3 P.T387 <FASE>
1964 V57 N6 P64-69 65-62171 <=>
1964 V57 N8 P6-9 64-41982 <*>
1964 V57 N8 P13- FPTS V24 N4 P.T659 <FASE>
1964 V57 N8 P13-20 64-41982 <*>
1964 V57 N8 P16- FPTS V24 N4 P.T661 <FASE>
1964 V57 N8 P34-38 64-41982 <*>
1964 V57 N8 P57- FPTS V24 N4 P.T625 <FASE>
1964 V57 N8 P60-68 64-41982 <*>
1964 V58 N7 P19- FPTS V24 N4 P.T665 <FASE>
1964 V58 N7 P120-121 RTS-3100 <NLL>
1964 V58 N8 P69-71 65-62170 <=>
1964 V58 N9 P3- FPTS V24 N4 P.T676 <FASE>
1964 V58 N9 P12-16 65-62349 <=$>
1964 V58 N9 P34-38 65-30400 <=$>
1964 V58 N9 P116- FPTS V24 N5 P.T858 <FASE>
1964 V58 N10 P41-44 65-63547 <=$>
1964 V58 N1C P53- FPTS V24 N5 P.T863 <FASE>
1964 V58 N10 P98-100 66-13153 <*>
1964 V58 N11 P3-8 65-63637 <=$>
1964 V58 N11 P29-33 65-64555 <=*$>
1964 V58 N11 P82-86 65-63611 <=$>
1964 V58 N11 P116-119 65-12328 <*>
1964 V58 N11 P119-122 65-12329 <*>
1964 V58 N12 P17-22 65-64534 <=*$>
1964 V58 N12 P27-35 65-30938 <=$>
1964 V58 N12 P80-87 65-30938 <=$>
1965 V59 N1 P29-33 66-60851 <=*$>
1965 V59 N1 P36- FPTS V24 N6 P.T957 <FASE>
1965 V59 N1 P78-81 65-13899 <*>
1965 V59 N2 P3- FPTS V25 N1 P.T43 <FASE>
1965 V59 N2 P50-55 66-60404 <=*$>
1965 V59 N2 P55- FPTS V25 N1 P.T67 <FASE>
1965 V59 N3 P6-9 66-60892 <=*$>
1965 V59 N3 P23-25 66-60891 <=*$>
1965 V59 N3 P52-54 66-60885 <=*$>
1965 V59 N3 P85-88 66-61751 <=*>
1965 V59 N3 P103- FPTS V25 N2 P.T267 <FASE>
1965 V59 N3 P110-114 66-60886 <=*$>
1965 V59 N4 P3- FPTS V25 N3 P.T381 <FASE>
1965 V59 N4 P17-19 65-31496 <=$>
1965 V59 N4 P77-81 66-61077 <=*>
1965 V59 N4 P117-118 65-31495 <=$>
1965 V59 N5 P3- FPTS V25 N4 P.T574 <FASE>
1965 V59 N5 P67-71 65-32383 <=*$>
1965 V59 N5 P85- FPTS V25 N3 P.T504 <FASE>
1965 V59 N5 P93- FPTS V25 N5 P.T857 <FASE>
1965 V59 N6 P16- FPTS V25 N3 P.T404 <FASE>
1965 V59 N6 P62-64 66-61312 <=*$>
1965 V59 N6 P98- FPTS V25 N4 P.T612 <FASE>
1965 V60 N7 P7-12 AD-623 272 <=*$>
1965 V60 N7 P7- FPTS V25 N4 P.T601 <FASE>
1965 V60 N7 P54-56 66-61313 <=*$>
1965 V60 N8 P85- FPTS V25 N6 P.T1091 <FASE>
1965 V60 N8 P89- FPTS V25 N5 P.T905 <FASE>
1965 V60 N9 P38- FPTS V25 N4 P.T648 <FASE>
1965 V60 N9 P51- FPTS V25 N5 P.T860 <FASE>
1965 V60 N9 P75-78 65-17200 <*>
66-10382 <*>
1965 V60 N10 P39- FPTS V25 N6 P.T969 <FASE>
1965 V60 N10 P50- FPTS V25 N5 P.T778 <FASE>
1965 V60 N11 P3- FPTS V25 N6 P.T966 <FASE>
1965 V60 N11 P50-53 66-30632 <=$>
1965 V60 N11 P118-121 66-30632 <=$>
1966 N1 P95-98 66-32713 <=$>
1966 N2 P75-78 66-32750 <=$>
1966 N2 P97-101 66-32750 <=$>

BYULLETEN GLAVNOGO BOTANICHESKOGO SADA. MOSCOW, PUSHKINSKOE
1950 N7 P27-36 R-4571 <*>
1950 V7 P27-36 CSIRO-2782 <CSIR>
1952 N14 P3-12 RT-4608 <*> 0
1952 N14 P12-23 RT-4607 <*>
1954 N17 P91-94 60-51016 <=>
1956 N24 P58-63 61-11427 <=>
1956 N26 P38-44 60-23026 <+*>
1957 N28 P56-62 63-15947 <=*>
1960 N36 P10-18 64-21008 <=>
1960 V38 P104-108 NLLTB V5 N6 P483 <NLL>
1961 N41 P101-106 66-51085 <=>
1963 N49 P3-6 NLLTB V6 N5 P450 <NLL>
TRANS.BULL.1964 P450 <NLL>
1964 N55 P3-16 TRANS.BULL. V7 N8 <NLL>
1964 V55 P3-16 NLLTB V7 N8 P667 <NLL>

BYULLETEN INSTITUTA ASTROFIZIKI
1958 N26 P3-11 61-28503 <=*>
1959 N27 P37-39 61-28977 <*=>

BYULLETEN INSTITUTA BIOLOGII. BELARUSKAYA AKADEMIYA NAVUK
1959 V4 N5 P58-62 66-51091 <=>

BYULLETEN INSTITUTA BIOLOGII VODOKHRANILISHCH
1960 N8/9 P46-49 64-19881 <=$>
1960 V3 P231-237 63-11125 <=>
1961 N10 P28-30 64-15989 <=*$>
1961 N10 P31-34 64-15988 <=*$>
1962 N12 P17-20 63-21761 <=>
63-23345 <=*$>
1962 N12 P32-33 63-21761 <=>
1962 N12 P53-56 63-15157 <=*>

BYULLETEN INSTITUTA METALOKERAMIKY I SPETSIALNYKH SPLAVIV AKADEMIYA NAUK URSR, KIEV
1959 N4 P5-37 63-24235 <=*$>
1959 N4 P65-71 62-18956 <=*>
1960 N5 P1-65 63-21045 <=>
1963 P141-151 M-5624 <NLL>
1963 N8 P199-204 AD-615 223 <=$>

BYULLETEN INSTITUTA TEORETICHESKOI ASTRONOMII
SEE BYULLETEN ASTRONOMICHESKOGO TEORETICHESKOGO INSTITUTA

BYULLETEN ISPOLNITELNOGO KOMITETA MOSKOVSKOGO GORODSKOGO SOVETA DEPUTATOV TRUDIASHCHIKHSIA
1963 04/07 P20-21 63-31424 <=>
1963 N5 P71 63-31594 <=>

BYULLETEN IZOBRETENII
1958 N6 P83-84 59-19480 <+*>
1958 N7 P22-24 59-14145 <+*>
1959 N16 P59- 61-21968 <=>
1960 N8 P16 62-13078 <*=>
1960 N9 P56 61-28981 <*=>
1960 N9 P57 61-28980 <*=>
1960 N11 P24 61-23837 <*=>
1960 N11 P25 61-23836 <*=>
1960 N15 P59 62-13067 <*=>
1960 N15 P60 61-28978 <*=>
1960 N16 P23-24 62-24037 <=*>
1960 N18 P25-26 62-15186 <*=>
1960 N19 P13 62-24752 <=*>
1960 N21 P4 62-13066 <*=>

1961 N2 P29 61-19865 <*=>
1961 N3 P1-2 62-13005 <*=>
1961 N17 P5-7 62-24737 <=*>
1962 N13 P2C 63-13042 <=*>
1962 N20 P1-3 64-11550 <=>

*BYULLETEN IZOBRETENII I TOVARNYKH ZNAKOV
1964 N13 P75-85 64-51925 <=$>
1964 N13 P106-107 64-51925 <=$>
1964 N13 P117-118 64-51925 <=$>
1965 N5 P16-17 66-13878 <*>

BYULLETEN KAZAKHSKOGO FILIALA AKADEMII NAUK SSSR
1943 N2 P48-49 RT-2564 <*>

BYULLETEN KCMISSII PO KOMETAM I METEORAM
1958 N2 P44-45 63-19208 <=*>
1959 N4 P35-41 63-15543 <=*>
1961 N5 P37-44 65-10834 <*>

BYULLETEN KOMISSII PO OKHRANE PRIRODY
1960 N4 P1-136 61-31054 <=>

BYULLETEN. KOMISSIYA PC OPREDELENIYU ABSOLYUTNOGO VOZRASTA GEOLOGICHESKIKH FORMATSII. AKADEMII NAUK SSSR
1957 N2 P5-7 44L29R <ATS>
1957 N2 P8-27 52L29R <ATS>
1961 N4 P144-147 AEC-TR-5858 <=*$>
1962 N5 P12-25 IGR V6 N12 P2100 <AGI>
1962 N5 P35-42 63-19598 <=*$>
1962 N5 P43-47 63-19599 <=*$>
1962 N5 P97-135 65-61567 <=$>

BYULLETEN. MINISTERSTVO VYSSHEGO I SREDNOGO SPETSIALNOGO OBRAZOVANIYA. USSSR
1962 N7 P1-5 62-33427 <=*>
1962 N8 P15-17 63-31210 <=>
1962 N11 P6-11 63-21429 <=>
1963 N1 P3-5 63-21210 <=>
1963 N2 P1-19 63-21521 <=>
1963 N3 P1-2 63-21210 <=>
1963 N3 P12-13 63-21210 <=>
1965 N11 P2 66-30619 <=$>
1965 N11 P5-9 66-30619 <=$>
1965 N11 P11-13 66-30619 <=$>
1966 N4 P10-15 66-32229 <=$>

BYULLETEN MCSKOVSKOGO GOSUDARSTVENNOGO UNIVERSITETA
1941 V2 P1-40 RT-1876 <*>

BYULLETEN MOSKOVSKOGO OBSHCHESTVA ISPYTATELEI PRIRODY. OTDEL BIOLOGICHESKII
1936 V45 N5 P331-337 RT-3522 <*>
1937 V46 P36-42 R-5323 <*>
1939 V48 N4 P70-78 64-11099 <=>
1940 N3/4 P173-179 RTS-2750 <NLL>
1948 V53 N2 P67-79 63-13605 <=*>
1950 V55 N2 P1-10 M.5133 <NLL>
1950 V55 N6 P15-20 M.5134 <NLL>
1953 V58 N1 P35-54 63-15260 <=*>
1954 V59 N6 P3-9 R-4878 <*>
1954 V59 N6 P11-25 R-4697 <*>
1955 V60 N2 P91-98 RT-3910 <*>
1955 V60 N2 P109-116 RT-3911 <*>
1955 V60 N5 P113-119 R-5322 <*>
1955 V60 N6 P87-98 C-2445 <NRC>
1956 V31 N3 P43-50 61-27448 <*=> 0
1956 V31 N4 P23-35 61-27251 <*=>
1957 V32 N6 P31-34 62-00821 <*>
1957 V62 N5 P77-91 RTS-2957 <NLL>
1957 V62 N6 P25-29 59-15902 <+*>
1957 V62 N6 P31-34 64-15620 <=*$>
1958 V33 N1 P35-50 62-13480 <*=>
1958 V33 N2 P5-36 65-50017 <=$>
1958 V33 N4 P5-12 61-28220 <*=>
1959 V34 N6 P5-21 63-24128 <=*$>
1959 V64 N3 P35-45 60-11452 <=>
1960 V35 N3 P101-113 62-15382 <*=>
1960 V35 N3 P132-135 62-24797 <=*>
1960 V35 N3 P150- 62-15376 <*=>
1960 V35 N3 P167 63-10218 <=*>
1960 V35 N5 P101-102 62-33454 <=*>
1960 V35 N5 P125-127 61-23644 <=*>
1960 V35 N5 P137-140 62-33454 <=*>
1960 V35 N6 P109-128 61-28066 <*=>
1960 V65 N3 P46-52 62-23950 <=*>
1960 V65 N3 P93-102 63-19606 <=*>
1960 V65 N4 P27-33 65-60344 <=$>
1961 N1 P80-88 66-51084 <=>
1961 V36 N2 P135-152 61-27349 <*=>
1961 V36 N4 P42-52 65-63748 <=$>
1961 V36 N4 P66-71 62-32959 <=*>
1961 V36 N4 P119-133 63-21414 <=>
1961 V66 N2 P33-38 64-51878 <=$>
1961 V66 N3 P32-39 63-21432 <=>
1961 V66 N6 P110-115 63-21443 <=>
1962 V37 N1 P15-22 63-21488 <=>
1962 V37 N1 P34-58 63-21488 <=>
1962 V37 N1 P128-131 63-21488 <=>
1962 V37 N2 P86-107 63-21355 <=>
1962 V37 N3 P82-93 63-01660 <*>
1962 V37 N3 P138-139 63-31213 <=>
1962 V37 N3 P153-154 63-24458 <=*$>
1962 V37 N3 P156-157 63-24461 <=*$>
1962 V37 N5 P1C-18 63-21354 <=>
1962 V37 N5 P120-127 63-21354 <=>
1962 V67 N1 P96-114 63-13233 <=*>
1962 V67 N1 P158- FPTS V22 N2 P.T260 <FASE>
1962 V67 N3 P148-149 TR-410 <CSIR>
1962 V67 N4 P109- FPTS V22 N3 P.T547 <FASE>
1963 V38 N2 P126-129 63-23182 <=*$>
1963 V38 N3 P98-109 64-13391 <=*$>
1963 V38 N4 P37-43 64-21341 <=>
1963 V38 N4 P152-154 64-21341 <=>
1963 V68 N1 P52-62 63-31160 <=>
1963 V68 N1 P133- FPTS V22 N6 P.T1218 <FASE>
1963 V68 N1 P133-137 63-31160 <=>
1963 V68 N2 P18-42 63-31509 <=>
1964 V39 N1 P51-58 64-31073 <=>
1964 V39 N5 P153-159 65-30422 <=$>
1964 V39 N6 P137-140 65-30965 <=$>
1964 V69 N1 P41-50 64-31502 <=>
1964 V69 N2 P5-21 64-31376 <=>
1964 V69 N2 P22-38 NLLTB V7 N2 P99 <NLL>
64-31369 <=>
1964 V69 N3 P20-24 64-41326 <=>
1964 V69 N3 P110-126 64-41326 <=>
1964 V69 N3 P110-127 65-62337 <=$>
1964 V69 N4 P15- FPTS V24 N5 P.T851 <FASE>
1964 V69 N5 P103- FPTS V24 N5 P.T903 <FASE>
1964 V69 N6 P110-120 65-64552 <=*$>
1965 V40 N1 P166-176 65-31080 <=$>

*BYULLETEN MOSKCVSKOGO OBSHCHESTVA ISPYTATELEI PRIRODY. OTDEL GEOLOGICHESKII
1934 V12 N2 P263-277 R-1653 <*>
60-12742 <DIL>
1948 V23 N2 P77-81 52K23R <ATS>
1949 V24 N2 P89-100 11N53R <ATS>
1952 V27 N4 P28-45 37L33R <ATS>
1954 V29 N1 P49 65-61577 <=$>
1954 V29 N1 P55 65-61577 <=$>
1954 V29 N1 P67-70 30N51R <ATS>
1954 V29 N2 P3-20 RJ-373 <ATS>
1956 V31 N1 ENTIRE ISSUE 61-11402/1 <=>
1956 V31 N1 P95-97 DSIS-T-365-R <NRC>
1956 V31 N2 ENTIRE ISSUE 61-11402/2 <=>
1956 V31 N3 ENTIRE ISSUE 61-11402/3 <=>
1956 V31 N4 ENTIRE ISSUE 61-11402/4 <=>
1956 V31 N5 ENTIRE ISSUE 61-11402/5 <=>
1956 V31 N6 ENTIRE ISSUE 61-11402/6 <=>
1957 V32 N1 P21-37 C-4275 <NRC>
1957 V32 N4/6 ENTIRE ISSUE
63-11134/4-6 <=>
1958 V33 N1 P97-105 62-10209 <*=>
1958 V33 N2 P93-102 61-15881 <=*> 0
1958 V63 N4 P15-23 66-30729 <=$>
1959 N1 P3-38 IGR V2 N9 P781 <AGI>

1959 V34 N1 P61-71 IGR V2 N2 P107 <AGI>
1959 V34 N1 P93-107 IGR V2 N2 P144 <AGI>
1959 V34 N1 P109-116 IGR V2 N2 P167 <AGI>
1960 V35 N4 P22-28 IGR V4 N6 P649 <AGI>
1961 V36 N1 P76-87 62-23281 <=*>
1962 V37 N1 P8-24 IGR V6 N3 P439 <AGI>
1962 V37 N1 P25-42 IGR V5 N8 P957 <AGI>
1963 V38 N1 P17-30 <JBS>
1963 V38 N3 P3-31 <JBS>
1963 V38 N5 P65-74 <JBS>
1963 V68 N3 P33-44 64-26428 <$>
1964 V34 N4 P3-20 <JBS>
1965 V70 N2 P121-139 94S86R <ATS>
1965 V70 N3 P5-16 66-62217 <=*$>

BYULLETEN NAUCHNO INFORMATSII. TRUD I ZARABOTNAYA PLATA
1959 N3 P34- 60-11127 <=>
1960 N1 P10-16 60-41195 <=>
1960 N8 P9-15 61-11538 <=>
1960 N12 P52-56 62-13531 <*=>
1961 N6 P3-8 62-13679 <*=>
1961 N6 P14-15 62-13679 <*=>

BYULLETEN NAUCHNO-ISSLEDOVATELSKII INSTITUTA PO BEZOPASNOSTI RABOT V GORNOI PROMYSHLENNOSTI MAKEYEVKA
1959 N10 P1-5 63-24153 <=$>
1959 N10 P7-13 63-24152 <=$>

BYULLETEN NAUCHNO-TEKHNICHESKOI INFORMATSII PO AGRONOMICHESKOI FIZIKE
1956 N1 P5-14 59-13257 <=>
1956 N1 P15-17 R-5232 <*>
1956 N1 P18-20 59-13195 <=>
1957 N3 P11-17 59-13246 <=>

BYULLETEN NAUCHNO-TEKHNICHESKOI INFORMATSII. LENINGRADSKII NAUCHNO ISSLEDOVATELSKII VETERNARNYI INSTITUT
1959 N7 P5-6 61-28145 <=*>
1959 N7 P18-20 61-28146 <=*>

BYULLETEN NAUCHNO-TEKHNICHESKOI INFORMATSII UZBEKSKOGO NAUCHNO ISSLEDOVATELSKOGO VETERINARNOGO INSTITUTA. TASHKENT
1956 V1 P3-8 R-3504 <*>

BYULLETEN. NAUCHNO-TEKHNICHESKOI INFORMATSII. VSESOUINOGO NAUCHNO-ISSLEDOVATELSKOGO INSTITUTA TABAKAI MAKHORKI
1958 N4 P51-56 59-17001 <+*> 0

BYULLETEN NAUCHNO-TEKHNICHESKOI INFORMATSII VSESOYUZNOGO NAUCHNO-ISSLEDOVATELSKOGO INSTITUTA VETERINARNOI SANITARII I EKTOPARAZITOLOGII. MOSCOW
1956 N1 P48-49 R-4433 <*>
1956 N1 P50 R-4422 <*>
1956 N1 P51 R-4439 <*>

BYULLETEN OBMENA OPYTOM LAKOKRASOCHNOI PROMYSHLENNOSTI. LENINGRAD
1940 N2 P19-21 R-2923 <*>

BYULLETEN. ODESSA VSESOYUZNOGO SELEKTSIONNO GENETICHESKOGO INSTITUTA IM LYSENKO
1956 N1 P7-10 CSIR-3489 <CSSA>
R-4595 <*>
1956 N1 P11-13 CSIRO-3490 <CSIR>
R-4377 <*>
1956 N1 P15-22 CSIRO-3491 <CSIR>
R-4580 <*>
1956 N1 P23-27 CSIR-3492 <CSSA>
R-3919 <*>
1956 N1 P29-33 CSIRO-3493 <CSIR>
R-4371 <*>
1956 N1 P35-37 CSIRO-3494 <CSIR>
R-4594 <*>

BYULLETEN OKEANOGRAFICHESKOI KOMISSII
1958 N1 P60-63 60-23768 <+*>
1958 N2 P57-60 60-11174 <=>
1958 N2 P65-74 60-11174 <=>
1961 V8 P75-80 AD-635 461 <=$>

BYULLETEN SOVETA PO SEISMOLOGII
1956 N2 P3-7 62-16170 <=*>
1956 N2 P8-11 59-13259 <=>
1957 N6 P71-75 64-11093 <=>
1958 N4 P1-60 62-16173 <=*> 0
1960 N8 P5-27 IGR V4 N5 P505 <AGI>
1960 N8 P217-225 01N54R <ATS>
1961 N9 P23-55 62-16172 <=*>
1963 N14 P14-27 66-62216 <=$*>
1963 N14 P122-127 AD-617 154 <=$>
1963 N15 P81-94 65-14347 <*>
1963 V15 P72-80 65-14819 <*>

BYULLETEN STALINABADSKOI ASTRONOMICHESKOI OBSERVATORII
1952 N1 P15-26 R-3567 <*>
RT-4000 <*>
64-71280 <=>
1954 N9 P22-24 61-15280 <+*>

BYULLETEN STANTSI OPTICHESKOGO NABLYUDENIYA ISKUSSTVENNYKH SPUTNIKOV ZEMLI. AKADEMIYA NAUK SSSR. ASTRONOMICHESKII SOVET
1959 N7 ENTIRE ISSUE 60-21608 <=>
1959 N8 P1-10 60-41044 <=>
1959 N8 P12-16 60-41044 <=>
1959 N10 ENTIRE ISSUE 61-11195 <=>
1959 N10 P4-7 63-24299 <=*$>
1960 N2 P1-24 63-24309 <=*$>
1960 N4 P1-6 63-24323 <=*$>
1960 N5 P1-9 63-21190 <=>
1960 N7 ENTIRE ISSUE 62-11607 <=>
1960 N7 P3-6 63-24329 <=*$>
1960 N8 P14-20 63-24330 <=*$>
1960 N10 P8-10 63-19654 <=*>
1960 N10 P12-23 63-23644 <=*$>
1962 N29 P33-37 AD-618 640 <=$>
1962 N31 P16-34 65-33229 <=*$>
1962 V25 P5-6 63-31260 <=>
1962 V25 P13-15 63-31260 <=>
1963 N32 P3-7 AD-640 248 <=$>
1963 N35 P34 AD-633 667 <=$> M
1963 N36 P21-22 AD-637 419 <=$>
1963 N36 P23-29 N66-16146 <=$>

BYULLETEN STROITELNOI TEKHNIKI
1957 N5 P10-13 60-15563 <+*>
1959 V16 N9 P28-32 61-27054 <*=>
1960 N7 P34 61-21307 <=>
1960 V17 N12 P3-4 61-23953 <=>
1961 N5 P3-5 NLLTB V3 N12 P973 <NLL>
1962 N4 P14 62-32057 <=>
1962 V19 N11 P43-45 63-15367 <=*>
1963 V20 N11 P29-31 64-21206 <=>
1964 N5 P71-78 AC/66/II-148 <CEMB>
1964 V21 N6 P29-30 65-30302 <=$>
1964 V21 N6 P36-38 65-30302 <=$>
1965 N2 P3-6 NLLTB V7 N8 P696 <NLL>
TRANS.BULL.V7 N8 <NLL>
1966 N6 P19-21 66-34364 <=$>
1966 N8 P1-3 66-35305 <=>
1966 N8 P5-8 66-35305 <=>
1966 N11 P1-5 66-35662 <=>
1966 N11 P7-9 66-35662 <=>
1966 N11 P21-23 66-35662 <=>
1966 N11 P30 66-35662 <=>
1967 N1 P1-8 67-31427 <=$>
1967 N1 P27-29 67-31427 <=$>
1967 N2 P14 67-31450 <=$>

BYULLETEN TEKHNICHESKOI INFORMATSII
1959 N7 P7-11 59-13939 <=>
1961 N2 P78-80 61-23610 <=*>

BYULLETEN TEKHNIKO-EKONOMICHESKOI INFORMATSII

Citation	Source
1956 N2 P11-16	61-15974 <=*>
1957 N1 P8	59-11302 <=> 0
1957 N2 P25-26	59-11302 <=>
1957 N2 P29-30	PB 141 483T <=> 0
1957 N2 P36-37	59-11302 <=>
1957 N2 P68-69	59-11026 <=>
1957 N2 P70-72	59-11026 <=>
1957 N3 P4-6	59-11302 <=> 0
1957 N3 P33-34	59-11302 <=> 0
1957 N3 P75-76	59-11026 <=>
1957 N4 P22-25	59-11302 <=> 0
1957 N4 P71-72	59-11026 <=>
1957 N4 P74-76	59-11026 <=>
1957 N5 P36-37	59-11302 <=> 0
1957 N5 P37-39	59-11302 <=> 0
1957 N5 P73-75	59-11026 <=>
1957 N6 P61-62	59-11026 <=>
1957 N7 P14-15	59-11302 <=> 0
1957 N7 P23-29	PB 141 483T <=> 0
1957 N8 P26-28	PB 141 483T <=> 0
1957 N8 P69-70	59-11026 <=>
1957 N9 P3-4	59-11302 <=> 0
1957 N9 P37-39	59-11302 <=> 0
1957 N9 P69-71	59-11026 <=>
1957 N9 P76-77	59-11026 <=>
1957 N10 P13-14	59-11302 <=> 0
1957 N10 P30-31	59-11302 <=> 0
1957 N11 P4-5	59-11302 <=> 0
1957 N11 P16-18	59-11302 <=> 0
1957 N11 P18-20	59-11302 <=> 0
1957 N11 P69-70	59-11026 <=>
1957 N11 P75-77	59-11026 <=>
1957 N12 P57-58	59-11026 <=>
1957 N12 P63-64	59-11026 <=>
1958 N2 P51-52	60-15422 <+*>
1958 N9 P77-79	NLLTB V1 N3 P5 <NLL>
1958 N10 P80-82	NLLTB V1 N3 P9 <NLL>
1958 N12 P70-72	NLLTB V1 N5 P1 <NLL>
1959 N1 P75-78	NLLTB V1 N7 P1 <NLL>
1959 N2 P63-66	NLLTB V1 N7 P32 <NLL>
1959 N2 P84-85	NLLTB V1 N7 P6 <NLL>
1959 N3 P37-38	59-13865 <=>
1959 N3 P40-42	59-13865 <=>
1959 N3 P78-80	NLLTB V1 N8 P5 <NLL>
1959 N5 P78-81	NLLTB V1 N12 P12 <NLL>
1959 N6 P9-10	66-13898 <*$>
1959 N6 P19-20	62-11702 <=> 0
1959 N6 P73-76	NLLTB V1 N11 P7 <NLL>
1959 N8 P81-83	NLLTB V1 N12 P7 <NLL>
1959 N10 P17-27	60-11860 <=>
1959 N10 P78-81	NLLTB V2 N5 P343 <NLL>
1959 N11 P21-23	60-41002 <=>
1959 N11 P26-37	60-41002 <=>
1959 N11 P40-45	60-41173 <=>
1959 N12 P13-26	60-41312 <=>
1959 N12 P26-32	60-41312 <=>
1959 N12 P35-36	60-41212 <=>
1959 N12 P39-40	60-41212 <=>
1960 N1 P59-61	61-11709 <=>
1960 N2 P6-8	60-11933 <=>
1960 N2 P12-13	60-11933 <=>
1960 N2 P15-25	60-41002 <=>
1960 N2 P28-31	60-41002 <=>
1960 N2 P33-39	60-41173 <=>
1960 N3 P5-7	60-41313 <=>
1960 N3 P9-28	60-41313 <=>
1960 N4 P5-8	60-41513 <=>
1960 N4 P13-28	60-41513 <=>
1960 N4 P32-33	60-41513 <=>
1960 N4 P35-41	60-41513 <=>
1960 N4 P43-44	60-41513 <=>
1960 N4 P53-56	61-11734 <=> P
1960 N4 P56-58	61-11574 <=> P
1960 N5 P15-16	61-11605 <=>
1960 N5 P22-29	61-11605 <=>
1960 N5 P31-33	61-11605 <=>
1960 N5 P37-38	61-11605 <=>
1960 N6 P4-6	61-21174 <=>
1960 N6 P12-30	61-21174 <=>
1960 N6 P33-34	62-32413 <=*>
1960 N6 P36-37	61-21174 <=>
1960 N6 P40-43	61-21174 <=>
1960 N7 P7-10	61-21735 <=>
1960 N7 P17-25	61-21735 <=>
1960 N7 P27-29	61-21735 <=>
1960 N7 P37-40	61-27001 <*=> 0
1960 N8 P19-27	61-21735 <=>
1960 N8 P29-31	61-21687 <=>
1960 N8 P40-44	61-21735 <=>
1960 N8 P40-42	61-23414 <*=>
1960 N8 P52-53	61-11518 <=>
1960 N8 P72	NLLTB V3 N5 P392 <NLL>
1960 N8 P73-76	NLLTB V3 N1 P1 <NLL>
1960 N8 P78-82	61-23414 <*=>
1960 N9 P4-6	61-21324 <=>
	61-23955 <*=>
	61-28531 <=>
1960 N9 P8-9	61-21324 <=>
1960 N9 P13-17	61-21324 <=>
1960 N9 P20-34	61-21324 <=>
1960 N9 P34-37	61-21687 <=>
1960 N9 P40-41	61-21687 <=>
1960 N9 P43-45	61-28531 <=>
1960 N9 P56-60	61-28531 <=>
1960 N9 P63-66	62-24782 <=>
1960 N9 P68	61-21324 <=>
1960 N9 P69-70	NLLTB V3 N3 P219 <NLL>
1960 N9 P81-82	NLLTB V3 N3 P173 <NLL>
1960 N10 P4-6	61-21324 <=>
1960 N10 P10-12	61-11791 <=>
1960 N10 P12-14	61-21324 <=>
1960 N10 P16-35	61-21324 <=>
1960 N10 P35-37	61-11791 <=>
1960 N10 P39-40	61-21324 <=>
1960 N10 P45-46	61-11791 <=>
1960 N10 P68	61-11791 <=>
1960 N11 P13-17	61-21685 <=>
1960 N11 P19-25	61-21685 <=>
1960 N11 P34-36	61-21685 <=>
1960 N11 P42-44	61-21685 <=>
1960 N12 P3-86	61-31683 <=>
1960 N12 P7	61-23019 <=>
1960 N12 P12	61-23019 <=>
1960 N12 P15	61-23019 <=>
1960 N12 P24	61-23019 <=>
1960 N12 P36	61-23019 <=>
1960 N12 P49	61-23019 <=>
1961 N1 P7-9	61-31157 <=>
1961 N1 P28-29	61-31157 <=>
1961 N1 P32-33	61-31157 <=>
1961 N1 P35-36	61-31157 <=>
1961 N1 P38-40	61-31157 <=>
1961 N1 P62-63	61-31157 <=>
1961 N2 P26-35	61-27083 <=>
1961 N2 P38-40	61-27083 <=>
1961 N2 P75-77	61-27083 <=>
1961 N3 P17-19	61-27232 <=>
1961 N3 P21-23	61-27232 <=>
1961 N3 P27-31	61-27232 <=>
1961 N3 P73-76	61-27232 <=>
1961 N7 P40-41	62-15835 <=*>
1961 N8 P18-20	AD-638 870 <=$>
1961 N10 P23-29	62-32415 <=*> 0
1961 N11 P42-44	63-19486 <=*>
1961 N11 P49-51	62-33324 <=>
1961 N12 P21-22	62-32868 <=>
1961 N12 P43-49	63-13091 <=*>
1961 N12 P75-77	62-24284 <=*>
1961 N12 P80-83	NLLTB V4 N8 P714 <NLL>
1962 N2 P55-56	62-32009 <=>
1962 N3 P23-24	63-15537 <=*>
1962 N3 P24	62-25986 <=>
1962 N3 P26	62-25986 <=>
1962 N3 P30-32	62-25986 <=>
1962 N3 P34-35	62-33115 <=>
1962 N3 P52-55	62-32194 <=*>
1962 N3 P57-60	62-32194 <=*>
1962 N3 P63-80	62-32194 <=*>
1962 N3 P72-76	62-32076 <=*>

1962 N4 P3-7 63-13113 <=*>
1962 N4 P9-11 62-32461 <=>
1962 N4 P11-13 62-32209 <=>
1962 N4 P19-21 62-25986 <=>
1962 N4 P28-29 62-25986 <=>
1962 N4 P31-36 62-25986 <=>
1962 N4 P40-43 62-25986 <=>
1962 N4 P45-54 62-33048 <=$>
1962 N4 P46-54 62-32201 <=*>
1962 N4 P82-83 62-25985 <=>
1962 N4 P84-89 62-25986 <=>
1962 N5 P3-13 62-32450 <=>
1962 N5 P65-70 62-33333 <=>
1962 N5 P73-76 62-33697 <=>
1962 N5 P77-80 62-32844 <=*>
1962 N6 P3-8 62-32894 <=*>
1962 N6 P17-19 62-32894 <=*>
1962 N6 P24-39 62-33024 <=$>
1962 N6 P31-35 65-61523 <=$>
1962 N6 P79-80 62-32894 <=*>
1962 N6 P84-86 62-33019 <=>
1962 N7 P3-6 63-13084 <=>
1962 N7 P32-35 62-33519 <=*>
1962 N7 P40-46 62-33519 <=*>
1962 N7 P57-66 62-33602 <=>
1962 N8 P5-8 63-13084 <=>
1962 N8 P19-22 63-13320 <=> P
1962 N8 P45-48 63-13475 <=>
1962 N8 P53-54 63-13459 <=>
1962 N8 P80-82 62-33601 <=>
1962 N9 P18-19 63-21353 <=>
1962 N9 P29-30 63-21353 <=>
1962 N9 P32-34 63-13291 <=>
1962 N9 P38-40 63-13291 <=>
1962 N9 P44-46 63-21057 <=>
1962 N9 P46-47 63-21281 <=>
1962 N9 P47-48 63-21353 <=>
1962 N10 P3-16 63-21058 <=>
1962 N10 P37-38 63-21058 <=>
1962 N10 P69-71 63-21393 <=>
1962 N10 P76 63-15103 <=>
1962 N11 P3-19 63-21525 <=>
1962 N11 P61-64 63-21311 <=>
1962 N11 P65-69 63-13843 <=*>
1962 N11 P77-83 63-15105 <=*>
1962 N11 P86-87 63-13843 <=*>
1962 N11 P90-92 63-21053 <=>
1962 N12 P3-19 63-21624 <=>
1962 N12 P27-29 63-21624 <=>
1962 N12 P71-72 63-21792 <=>
1963 N1 P3-11 63-21610 <=>
1963 N1 P26-28 63-21769 <=>
1963 N1 P37-41 63-21649 <=>
1963 N1 P64-69 63-21814 <=>
1963 N1 P78-79 63-21610 <=>
1963 N2 P3-8 63-21901 <=>
1963 N2 P16-18 63-31012 <=>
1963 N2 P26 63-21694 <=>
1963 N2 P35-40 62-21621 <=> 0
63-21621 <=>
1963 N2 P44-46 62-21621 <=> 0
63-21621 <=>
1963 N2 P45-50 63-31452 <=>
1963 N2 P58-60 63-21783 <=>
1963 N2 P65-72 63-21674 <=>
1963 N2 P77-83 63-21792 <=>
1963 N3 P3-9 63-31014 <=>
1963 N3 P22-23 63-21922 <=>
1963 N3 P26-34 63-21922 <=>
1963 N3 P41-45 63-31059 <=>
1963 N3 P68-71 63-21943 <=>
1963 N3 P77-79 63-31059 <=>
1963 N4 P2 63-31144 <=>
1963 N4 P3-8 63-31435 <=>
1963 N4 P8-18 63-31102 <=>
1963 N4 P27-39 63-31130 <=>
1963 N4 P42-51 63-31234 <=>
1963 N4 P56-65 63-31503 <=>
1963 N4 P70-71 63-31144 <=>
1963 N5 P10-11 63-31579 <=>
1963 N5 P19-24 63-31524 <=>
1963 N5 P24-28 63-31607 <=>
1963 N5 P52-58 63-31499 <=> 0
1963 N5 P67-72 63-31643 <=>
1963 N6 P22-23 63-31900 <=>
1963 N6 P27-38 63-31636 <=>
1963 N6 P38-39 63-31687 <=>
1963 N6 P76-85 63-31733 <=>
1963 N6 P82-85 NLLTB V5 N11 P1026 <NLL>
1963 N7 P19-25 63-31843 <=>
1963 N7 P42-48 63-31923 <=>
1963 N7 P49-50 63-31920 <=>
1963 N7 P64-68 63-31807 <=>
1963 N7 P72-73 63-31920 <=>
1963 N8 P3-6 64-19147 <=*$>
1963 N9 P7-11 AD-624 905 <=*$>
1963 N11 P32-34 64-21136 <=>
1963 N12 P54-58 64-31028 <=>
1963 N12 P89-90 64-31028 <=>
1963 V5 P14-18 63-31678 <=>
1963 V5 P36-43 63-31678 <=>
1964 N2 P5-6 <SIC>
1964 N2 P76-77 AD-615 288 <=$>
1964 N3 P35-45 64-31729 <=>
1964 N3 P62-64 64-41104 <=>
1964 N3 P65-67 64-41104 <=>
1964 N3 P67-69 64-41104 <=>
1964 N5 P30-32 64-41226 <=>
1964 N5 P32-34 64-41226 <=>
1964 N5 P35-36 64-41226 <=>
1964 N5 P42-43 64-41447 <=$>
1964 N6 P1-6 NLLTB V6 N10 P913 <NLL>
TRANS.BULL.1964 P913 <NLL>
1964 N6 P56-57 64-41447 <=$>
1964 N7 P3 64-51660 <=$>
1964 N7 P5 64-51660 <=$>
1964 N7 P81-82 64-51660 <=$>
1964 N9 P25-28 AD-640 285 <=$>
1964 N12 P20 65-30270 <=$>
1964 N12 P53-56 65-30583 <=$>
1965 N1 P10-17 65-31058 <=$>
1965 N1 P67-68 65-31058 <=$>
1965 N2 P10-12 AD-620 098 <=*$>
1965 N2 P55-56 AD-630 856 <=$>
1965 N8 P3-4 65-33197 <=$>
1965 N8 P57 65-33197 <=$>
1965 N9 P49-51 4703 <BISI>
1966 N5 P15-17 66-34606 <=$>
1966 N5 P47-49 66-34610 <=$>

BYULLETEN TEKHNICHESKOI INFORMATSII. MOLDAVSKII NAUCHNO-ISSLEDOVATELSKOGO INSTITUTA SADOVODSTVA, VINOGRADARSTVA I VINODELIIA, KISHINEV
1961 V2 N11 P46-49 64-23649 <=$>

BYULLETEN TSENTRALNOGO INSTITUTA INFORMATSII TSVETNOI METALLURGII. MOSCOW
1957 N18 P18-23 60-11463 <=>
1958 N5 P32 60-11311 <=>
1958 N5 P35-36 60-11311 <=>
1958 N6 P19-22 CSIR-TR-429 <CSSA>
1958 N7 P21-25 60-11463 <=>
1958 N7 P32 60-11311 <=>
1958 N7 P34-35 60-11311 <=>
1958 N10 P10-13 60-41001 <=>
1958 N10 P20-23 60-41001 <=>

BYULLETEN TSENTRALNOI GENETICHESKOI LABORATORII IMENI I. V. MICHURINA
1957 N3 P35-37 65-60917 <=$>

BYULLETEN UCHENOGO MEDITSINSKOGO SOVETA MINISTERSTVA ZDRAVOOKHRANENIYA RSFSR
1961 V2 N2 P3-47 61-28285 <=>
1961 V2 N3 P3-6 62-23799 <=>
1961 V2 N3 P10-16 62-23799 <=>
1961 V2 N3 P20-48 62-23799 <=>
1961 V2 N4 P3-22 62-23338 <=>
1961 V2 N4 P27-44 62-23338 <=>
1961 V2 N5 P16-24 62-25526 <=*>

1962 V3 N1 P7-10 62-24727 <=*>
1962 V3 N2 P22-27 62-33310 <=*>
1962 V3 N2 P34-40 62-33135 <=*>
1962 V3 N4 P45-47 63-21827 <=>
1962 V13 N3 P3-45 62-32891 <=>
1962 V13 N3 P50-51 62-32891 <=>

BYULLETEN VSESOYUZNOGO ASTRONOMO-GEODEZICHESKOGO OBSHCHESTVA. AKADEMIYA NAUK SSSR
1954 N15 P1C-24 61-23442 <*=>
1957 N20 P3-71 60-11456 <=>
1960 N26 P3-14 64-11801 <=>
1960 N27 P32-36 AD-612 373 <=$>
1960 N28 P51-55 62-25174 <=*>
1962 N31 P3-14 64-13362 <=*$>
1963 N34 P20-28 64-19372 <=$>
1965 N36 P66-73 66-31511 <=$>

BYULLETEN VSESOYUZNOGO INSTITUTA RASTENIEVODSTVA
1961 N9 P3-9 64-13650 <=*$>
1961 N9 P34-39 63-13028 <=*>
1961 N9 P44-47 64-15079 <=*$>

BYULLETEN VSESOYUZNOGO KHIMICHESKOGO OBSHCHESTVA IMENI D.I. MENDELEEVA
1939 N3/4 P40-41 5K21R <ATS>

BYULLETEN VULKANOLOGICHESKOISTANTSII NA KAMCHATKE
1947 N11 P26-35 65-11118 <*>
1959 N28 P79-91 61-28000 <*=>

BYULLETEN. ZAKAVKAZSKII OPYTNO-ISSLEDOVATELSKII INSTITUT VODNOGO KHOZYAISTVA, TIFLIS
1931 N7 P119-129 60-16750 <*> O

C. C. SCHMIDTS JAHRBUECHER DER IN- UND AUSLAENDISCHEN GESAMTEN MEDIZIN
1861 N110 P204-209 2330 <*>

CERN REPORT
SEE REPORT. CENTRE EUROPEAN DE RECHERCHE NUCLEARE. GENEVA

CNRM PUBLICATIONS
SEE PUBLICATIONS DU CENTRE NATIONAL DE RECHERCHES METALLURGIQUES. BRUXELLES

CNRN NOTIZIARIO
SEE NOTIZIARIO. COMITATO NAZIONALE PER L'ENERGIA NUCLEARE. ITALY

CABLES ET TRANSMISSION
1947 V1 P261-27C 62-16940 <*> O
1947 V1 N1 P39-60 65-13121 <*>
1948 V2 N1 P61-74 65-12215 <*>
1948 V2 N4 P285-304 58-343 <*>
1950 V4 N1 P9-23 62-20206 <*>
1950 V4 N2 P89-125 64-18105 <*>
1950 V4 N4 P336-351 57-2262 <*>
1951 V5 N1 P25-30 57-982 <*>
1951 V5 N1 P31-39 57-952 <*>
1952 V6 N1 P96-102 57-993 <*>
1953 V7 N1 P54-77 57-1031 <*>
1953 V7 N1 P78-80 57-2528 <*>
1953 V7 N3 P175-184 26T95F <ATS>
1953 V7 N3 P185-217 57-1029 <*>
1953 V7 N4 P325-358 57-2530 <*>
1955 V9 N4 P287-292 66-10905 <*> O
1956 V10 N2 P145-151 88J15F <ATS>
1957 V11 N1 P22-31 61-10641 <*>
1958 V12 N1 P3-9 59-10965 <*>
1958 V12 N2 P136-143 59-17774 <*>
66-11784 <*> O
1958 V12 N2 P162-169 72L30F <ATS>
1959 V13 N3 P145-156 60-16806 <*>
1960 V14 N1 P12-29 49M41F <ATS>
1961 V15 N1 P3-26 61-20536 <*>
91N52F <ATS>
1961 V15 N2 P148-159 62-16583 <*>
1961 V15 N4 P395-402 62-20173 <*>
1962 V16 N1 P3-10 64-14121 <*>
1962 V16 N3 P200-205 63-18877 <*>
1962 V16 N3 P249-253 63-18879 <*> O
1962 V16A N3 P143-147 63-18029 <*>
1963 V17 N1 P3-13 64-16565 <*>
1963 V17 N3 P179-206 64-14632 <*>
1963 V17 N4 P227-238 66-13236 <*>
1963 V17 N4 P250-263 64-16616 <*>
1964 V18 N3 P269-281 65-14311 <*>
1964 V18 N4 P315-323 35S84F <ATS>
1964 V18 N4 P337-340 36S84F <ATS>
1964 V18 N4 P341-345 33S84F <ATS>
1965 V19 N1 P9-19 37S84F <ATS>
1965 V19 N1 P27-29 39S84F <ATS>

CACAO EN COLOMBIA
1953 V2 P155-161 65-18143 <*>

CAHIERS D'ART
1955 P153-166 57-99 <*>

CAHIERS DU CENTRE SCIENTIFIQUE ET TECHNIQUE DU BATIMENT
1950 V52 N5 ENTIRE NO. 61-20029 <*>
1950 V73 P1-15 T-1652 <INSD>
1956 N217 ENTIRE ISSUE 62-01112 <*>
1960 N334 P1-2C C-5075 <NRC>
1960 N348 ENTIRE ISSUE C-3837 <NRC>

CAHIERS. CENTRE TECHNIQUE DU BOIS
1956 N17 P1-25 58-2071 <*>
1959 N37 P2-36 62-20191 <*>

CAHIERS DU CESSID
1957 N1 ENTIRE ISSUE 61-18363 <*>
1957 N1 P5-40 61-18362 <*>
1957 N2 P1-80 PT.2 61-18364 <*> O
1957 N2 P1-80 PT.4 945 <BISI>
1958 P1-52 1436 <BISI>

CAHIERS FRANCO-CHINOIS
1961 V10 P73-83 61-28873 <*=>
1962 N15/6 P23-45 63-21625 <=>

CAHIERS INTERNATIONAUX DE SOCIOLOGIE
1959 V26 P165-17C 64-30289 <*>
1961 V57 P81-94 65-11679 <*>

CAHIERS D'OUTRE-MER
1964 V17 N68 P370-420 66-31609 <=$>

CAHIERS DE PHYSIQUE
1942 V9 P1-15 63-14238 <*> O
1942 V10 P30-42 58-592 <*>
1943 V13 P5-17 58-579 <*>
1943 V18 P18-55 T-2710 <INSD>
1943 V18 P18-63 T-2710 <INSD>
1944 N23 P43-56 62-18373 <*> O
1953 V46 P1-18 58-1419 <*>
1954 N49 P1-22 62-23126 <=>
1954 N51 P43-51 3007 <*>

CALORE
1959 N2 P73-83 AEC-TR-4045 <*>
1959 V30 N2 P73-83 64-30659 <*> O

CAMINOS
1955 N143 P11-13 I-701 <*>
1962 V44 N4 P371-378 65-14481 <*> O

CAMPO Y MECANICA
1965 N31 P17-19 66-11474 <*> O

CANADIAN JOURNAL OF BIOCHEMISTRY AND PHYSIOLOGY
1955 V33 N6 P1018-1032 65-12626 <*>

CANADIAN JOURNAL OF MICRCBICLCGY
1958 V4 N5 P543-550 59-15393 <*>

60-13137 <*+>

CANADIAN JOURNAL OF PHYSICS
1951 V29 P193-202 UCRL TRANS-103 <*>
1961 V39 P409-418 62-10622 <*>
1964 V42 N11 P2102-2120 65-11425 <*>

CANADIAN JOURNAL OF RESEARCH. SERIES B. CHEMICAL SCIENCES
1949 V27B P716-720 66-11908 <*>

CANADIAN JOURNAL OF RESEARCH. SERIES C.
1941 V19C N2 P27-39 65-12350 <*>

CANADIAN MEDICAL ASSOCIATION JOURNAL
1962 V87 P1367-1374 63-00730 <*>
1963 V89 P933-936 64-00364 <*>
1964 V91 P335-342 66-11807 <*>

CANADIAN METALLURGICAL QUARTERLY
1963 V2 N3 P281-289 AEC-TR-6366 <*>

CANADIAN MINING AND METALLURGICAL BULLETIN
1946 V39 N414 P516-523 63-16804 <*>

CANCEROLOGIE
1953 V1 N3 P128-139 60-16469 <*>
1957 V3 N2 P63-75 63-16616 <*>
1957 V3 N3 P63-75 9561-B <KH>

CANCRO. TORINO
1956 V9 N4 P3-36 65-00166 <*> 0

CAOUTCHOUC ET LA GUTTA-PERCHA
1925 P12666-12669 2860 <*>
1925 P12672-12673 2934 <*>
1935 P17097-17099 57-1111 <*>

CAOUTCHOUCS ET PLASTIQUES
V2 P37 RCT V13 N1 P604 <RCT>

CAPITA ZOOLOGICA
1934 V4 N5 P271 62-01399 <*>

CARBON. AN INTERNATIONAL JOURNAL. OXFORD
1964 V2 P149-161 ORNL-TR-481 <*>

CARBONS. JAPAN
SEE TANSO

CARDIOLOGIA
1946 V10 N3 P130-138 61-20241 <*>
1953 V23 P349-358 I-607 <*>
1954 V24 P285-310 3225-5 <K-H>
1959 V34 N2 P120-130 59-15493 <*>
1959 V34 N2 P131-138 59-15492 <*>
1959 V35 N3 P139-154 60-10982 <*> 0
1959 V35 N5 P316-323 60-14179 <*> 0
1960 V36 P49-54 10281-H <KH>
1961 V38 P112-114 63-01019 <*>
1961 V38 P197-204 61-20777 <*>

CASOPIS CESKOSLOVENSKEHO LEKARNICTVA
1949 V62 N10/2 P103-107 65-11017 <*>
1950 V63 P36-49 59-15853 <*> 0
1950 V63 P264-265 60-18570 <*>
1958 V97 P341-343 62-01226 <*>

CASOPIS CESKOSLOVENSKYCH VETERINARU
1947 V2 P160-163 61-00902 <*>

CASOPIS LEKARU CESKYCH
1903 V41 N24 P651-656 23J15C <ATS>
1903 V42 P656-661 24J15C <ATS>
1903 V42 N24 P661-663 25J15C <ATS>
1937 N11 P325-330 64-10844 <*> 0
1946 V85 P785-851 63-00139 <*>
63-26696 <=$>
1951 V90 N42 P1232-1235 II-720 <*>
1951 V90 N51/2 P1516-1518 65-00180 <*>
1953 V92 P867-870 PANSDOC-TR.615 <PANS>
1954 V93 N10 P242-248 57-1886 <*>
1954 V93 N36/7 P1007-1012 3457 <K-H>
1957 V96 P1618-1624 8535-D <K-H>
1958 V97 P1553-1555 9298-A <K-H>
1958 V97 N5 P1553-1555 63-16925 <*>
1958 V97 N34 P1077-1078 60-10848 <*> 0
1959 N16 P502-506 61-00294 <*>
1959 V98 N24 P754-758 60-15721 <*+> 0
1959 V98 N49/0 P1549-1552 64-20038 <*> 0
1959 V99 N1 P1-4 60-31060 <*=>
1960 V99 P1146-1150 66-11631 <*>
1960 V99 N9 P44-47 60-31283 <=*>
1960 V99 N3/4 P101-104 62-14779 <*>
1961 V100 N21 P662-664 62-19626 <=*>
1961 V100 N42 P1332-1335 09P61C <ATS>
1961 V100 N27/8 P888-890 62-19627 <=*>
1962 V101 N21 P642-643 62-25189 <=*>
1962 V101 N24/5 P738-739 62-32067 <=*>
1963 V102 P51-54 09R75C <ATS>
1963 V102 N10 P277 64-71149 <=>
64-71150 <=>

CASOPIS PRO MINERALOGII A GEOLOGII
1958 V3 N1 P73-91 65-11132 <*>

CASOPIS MORAVSKEHO MUSEA V BRNE
1955 V40 P89-92 66-14472 <*>
1960 V45 P25-30 66-14097 <*$>

CASOPIS NARODNIHC MUSEA
1964 V133 N1 P17-20 26S82C <ATS>

CASOPIS PRO PESTCVANI MATEMATIKY
1955 V80 P17-31 STMSP V2 P27-40 <AMS>
1955 V80 N3 P261-273 59-11017 <=*>
1957 V82 P47-75 STMSP V4 P43-69 <AMS>
60-17178 <=*>
1957 V82 P182-194 STMSP V2 P63-74 <AMS>
1958 V83 P327-329 STMSP V2 P75-77 <AMS>
1958 V83 P425-439 52M40C <ATS>
1959 V84 P140-149 STMSP V4 P175-182 <AMS>

CASTING WORKS. CHINESE PECPLES REPUBLIC
SEE CHU KUNG. PEKING

CATALYSEURS
1964 P334-338 65-14583 <*>
1964 P349-352 65-14584 <*>

CATALYST. TOKYO
SEE SHOKUBAI. TOKYO

CELLULE
1932 V41 P205-216 66-11813 <*>

CELLULOSA E CARTA
1959 V9 N11 P9-22 60-14109 <*>
1959 V9 N12 P15-27 60-14109 <*>

CELLULOSE INDUSTRY. TOKYO
SEE SEN-I-SO KCGYO

CELLULOSE-CHEMIE
1924 V5 P6-7 AL-726 <*>
1932 V13 P58-64 8958 <IICH>
1932 V13 P71-74 8864 <IICH>
1933 V14 N6 P81-83 65-12873 <*>
1942 V20 N3 P61-72 60-16580 <*>
1943 V21 N4 P73-84 61-16800 <*> 0

CELULOZA SI HIRTIE
1959 V8 N10 P318-320 61-18572 <*>
1960 V9 N3 P73-76 61-10406 <*>
1960 V9 N9 P296-299 88N57RU <ATS>
1961 V10 N7/8 P265-267 37R77RU <ATS>
1962 V11 N5 P186-191 64-16601 <*> 0

CEMENT. AMSTERDAM
1949 V1 N5/6 P94-97 61-20216 <*>
1950 V2 N13/4 P251-255 61-20217 <*>
1955 V7 N1/2 P39- I-595 <*>
1956 V8 N13/4 P327-330 T-2413 <INSD>
1956 V8 N17/8 P405-407 1534 <*>
1958 V10 N13/4 P559-560 64-14756 <*>
1959 V11 N3 P293-294 60-14260 <*>
1959 V11 N6 P503-506 60-18431 <*> 0
1961 V13 N1 P13-18 TRANS-107 <CCA> 65-13534 <*>
1961 V13 N2 P66-73 TRANS-107 <CCA> 65-13534 <*>
1961 V13 N3 P148-155 TRANS-107 <CCA> 65-13534 <*>
1961 V13 N4 P186-190 TRANS-107 <CCA> 65-13534 <*>

CEMENT OCH BETONG
1948 V23 N1 P17-27 61-20479 <*> 0
1948 V23 N2 P64-78 61-18877 <*>
1948 V23 N3 P114-119 61-20473 <*>
1951 V26 N2 P130-133 61-18936 <*>
1951 V26 N2 P134-147 58-2544 <*>
1954 V29 P135-143 T1804 <INSD>
1956 V31 N1 P22-32 CSIRO-3243 <CSIR>

CEMENT AND CONCRETE. TOKYO
SEE SEMENTO KONKURITO

CEMENT, WAPNO, GIPS
1953 V9 N5/6 P97-98 87Q68P <ATS>
1954 V10 N4 P66-76 60-21236 <=>
1955 V11 N3 P62-64 60-21233 <=>
1955 V11 N4 P75-80 58-1681 <*>
1955 V11/2 N3 P57-62 60-21237 <=>
1957 N1 P1-7 PB 141 31CT <=>
1960 N3 P70-78 CSIR-TR-151 <CSSA>
1960 N6 P166-168 CSIR-TR.158 <CSSA>
1960 V15 N3 P61-64 CSIR-TR.149 <CSSA>
1960 V15 N3 P65-69 CSIR-TR.150 <CSSA>
1960 V15 N3 P70-78 CSIR-TR.151 <CSSA>
1960 V15 N4 P93-107 CSIR-TR.157 <CSSA>
1960 V15 N4 P93-106 CSIR-TR157 <CSSA>
1960 V15 N6 P166-168 CSIR-TR.158 <CSSA>
1960 V16 N5 P121-123 63-11389 <=>
1961 V16 N4 P89-97 CSIR-TR.193 <CSSA>
1962 V17 N6 P157-171 65-64234 <=*$>
1962 V17 N9 P252-255 64-16843 <*> 0
1962 V17 N7/8 P187-195 CSIR-TR.151 <CSSA>
1963 N9 P186-190 65-23225 <$>
1963 V18 N5 P99-104 65-23224 <$>
1963 V18 N7/8 P159-167 CSIR-TR.350 <CSSA> CSIR-TR-350 <CSSA>
1964 V19 N6 P137-142 64-30617 <*>
1964 V19 N10 P278-281 CSIR-504 <CSSA>

CENICAFE
1960 N11 P251-258 52Q67SP <ATS>

CENTRAL SOCIETY FOR VETERINARY MEDICINE. JAPAN
SEE CHUO JUIKWAI ZASSHI

CERAMICA. ROMA
1957 SNS V12 N12 P175-178 60-16839 <*>
1959 SNS V14 N4 P71-73 60-10031 <*>
1960 V15 N9 P61-68 66-10686 <*>
1963 V18 N1 P58-60 52Q73I <ATS>

CERAMIQUE
SEE CERAMIQUE ET LES MATERIAUX DE CONSTRUCTION

CERAMIQUE ET LES MATERIAUX DE CONSTRUCTION
1923 V26 P305 64-18251 <*>
1924 V27 P1 64-18257 <*>

CERVELLO
1955 V31 P271-295 3048 <*>
1955 V31 P271-292 3176 <*>

CESKOSLOVENSKA BIOLOGIE
1953 V2 P265-277 II-205 <*>
1953 V2 N1 P41-49 62-25258 <=*>
1953 V2 N2 P68-75 62-25259 <=*>
1953 V2 N5 P267-282 RJ-424 <ATS>
1953 V2 N5 P290-297 PANSDOC-TR.207 <PANS>
1953 V2 N6 P323-334 62-25257 <=*>
1954 V3 N2 P82-91 59-15388 <*>
1954 V3 N3 P173-181 RJ-288 <ATS>
1955 V4 N1 P1-6 II-196 <*>

CESKOSLOVENSKA DERMATOLOGIE
1954 V29 P80-90 1938 <*>
1962 V37 N3 P203-206 34P63C <ATS>

CESKOSLOVENSKA EPIDEMIOLOGIE, MIKROBIOLOGIE, IMUNOLOGIE
1958 V7 N6 P361-364 63-00772 <*>
1958 V7 N6 P387-395 61-00590 <*>
1959 V9 N5 P289-298 32N54C <ATS>
1960 V9 N2 P111-121 98N54C <ATS>
1961 V10 N3 P148-157 61-01059 <*>
1962 V11 N1 P53-57 63-01379 <*>
1962 V11 N2 P73-77 63-00460 <*>
1962 V11 N2 P135-139 65-00151 <*>
1964 V13 N1 P20-27 65-13492 <*> 0
1964 V13 N6 P343-350 65-13478 <*>

CESKOSLOVENSKA FARMACIE
1953 V3 P38-43 63-14239 <*> 0
1955 V4 P65-68 88P65C <ATS>
1956 V5 P516-519 T-2575 <INSD>
1956 V5 P591-593 64-14318 <*> 0
1956 V5 N6 P321-323 61-16480 <*> 0
1957 V6 P77-82 T-2577 <INSD>
1960 V9 P396-397 34N50C <ATS>
1961 V10 N3 P126-130 70N57C <ATS>
1963 V12 N1 P4-6 26R79SK <ATS>
1963 V12 N6 P309-310 66-11700 <*>
1963 V12 N6 P312-316 66-12007 <*>

CESKOSLOVENSKA FOTOGRAFIE
1958 N6 P66- 59-15353 <*>

CESKOSLOVENSKA FYSIOLOGIE
1953 V2 P303-306 62-14743 <*>
1953 V2 N4 P363-366 62-14742 <*>
1954 V3 P306-311 R-264 <*>
1954 V3 N3 P306-311 58-617 <*>
1958 V7 N2 P216-220 CSIR-TR.253 <CSSA>
1959 V8 N2 P81-91 62-01205 <*>
1959 V8 N6 P574-582 60-31088 <*=>
1960 V9 P59-60 AEC-TR-4736 <*>
1961 N10 P267-268 62-00589 <*>
1963 V12 P1-25 13134B <K-H>
1964 V13 N1 P33- FPTS V24 N5 P.T786 <FASE>
1964 V13 N2 P126-146 65-63610 <=$>
1964 V13 N2 P146-165 65-62497 <=$>
1964 V13 N2 P170-175 65-62495 <=$>
1964 V13 N4 P386- FPTS V24 N5 P.T821 <FASE>
1964 V13 N4 P389- FPTS V24 N4 P.T581 <FASE>
1965 V14 N2 P98- FPTS V25 N3 P.T524 <FASE>

CESKOSLOVENSKA GASTROENTEROLOGIE A VYZIVA
1961 V15 P572-575 592 <TC>
1962 V16 P273-277 66-11630 <*>
1963 V17 N8 P477-480 66-13388 <*> 0

CESKOSLOVENSKA GYNEKOLOGIE
1958 V23 P198-202 59-18490 <=*> 0
1958 V44 P476-485 63-00209 <*>

CESKOSLOVENSKA HYGIENA
1961 V6 N2/3 P129-134 62-19522 <=>
1961 V6 N2/3 P144-148 62-19522 <=>
1961 V6 N2/3 P162-166 62-19522 <=>
1962 V7 N8 P475-481 AEC-TR-6188 <*> 14Q73C <ATS>
1964 N8 P481-490 LC-1264 <NLL>

CESKOSLOVENSKA MIKROBIOLOGIE
1956 V1 P129-134 59-10850 <*>
1956 V1 P223-226 59-10180 <*>
1956 V1 N6 P263-267 59-15386 <*>
1957 V2 P175-182 59-10925 <*>
1957 V2 P183-184 59-15411 <*>
1957 V2 N1 P43-47 8J17C <ATS>
1957 V2 N5 P300-305 2K23C <ATS>
1958 V3 N2 P82-91 59-21180 <=*> O
1958 V3 N3 P197-201 PANSDOC-TR.453 <PANS>
65-11759 <*>

CESKOSLOVENSKA MORFOLOGIE
1955 V3 N2 P157-167 59-14769 <=*> O
1956 V4 N4 P341-347 57-1893 <*>
1961 V9 N1 P79-81 63-00172 <*>
1963 V11 N1 P57-67 AD-632 027 <=$>
1963 V11 N1 P81-82 RTS-3045 <NLL>
1964 V12 N1 P50-63 65-62192 <=$>
1964 V12 N1 P63-74 65-62189 <=$>
1964 V12 N1 P85-92 65-61045 <=$>
1964 V12 N2 P141-151 65-62191 <=$>
1964 V12 N3 P268- FPTS V24 N5 P.T815 <FASE>
1964 V12 N4 P418-429 65-63612 <=$>

CESKOSLOVENSKA NEUROLOGIE
1960 V23 N6 P406-411 63-00602 <*>
1962 V25 N3 P183-186 63-00616 <*>

CESKOSLOVENSKA OFTALMOLOGIE
1950 V6 P5-8 65-00272 <*>
1961 V17 N4/5 P292-298 63-00742 <*>
1963 V19 N5 P338-342 66-12116 <*>

CESKOSLOVENSKA ONKOLOGIA
1954 N3/4 P249-253 1529 <*>
1954 V1 P275-282 PANSDOC-TR.616 <PANS>
1955 V2 N1 P48-53 58-1693 <*>

CESKOSLOVENSKA PARASITOLOGIE
1954 V1 P15-22 60-14297 <*>
1954 V1 P23-44 60-14298 <*>
1954 V1 P45-76 60-14299 <*>
1955 V2 P191-199 C-2652 <NRC>
1957 V4 P359-367 C-2651 <NRC>
1958 V5 N2 P121-124 62-00151 <*>

CESKOSLOVENSKA PEDIATRIE
1955 V10 N4 P274-277 62-00297 <*>
1957 V12 N1 P68-74 59-13037 <=*> O
1957 V12 N4 P308-311 62-00322 <*>
1961 V16 N10 P935-938 64-00186 <*>
1962 V17 N12 P1065-1070 65-00325 <*>

CESKOSLOVENSKA PSYCHOLOGIE
1957 V1 N1 P25-38 65-00241 <*> O
1957 V1 N1 P39-44 65-00257 <*> O

CESKOSLOVENSKA STOMATOLOGIE
1965 V65 N2 P137-140 65-14944 <*>

CESKOSLOVENSKE SPOJE
1961 V6 N4 P3-7 62-19689 <=*>

CESKOSLOVENSKE ZDRAVOTNICTVI
1960 V8 N1 P1-10 60-31409 <*=>
1960 V8 N2/3 P124-138 62-19457 <*=>
1961 V9 N6 P366-375 62-19621 <*=>

CESKOSLOVENSKY CASOPIS PRO FYSIKU
1956 V6 N3 P264-276 60-21610 <=*>
1958 V8 P566-574 80M38C <ATS>
1958 V8 N1 P19-22 60-14659 <*>
1960 N1 P35-40 64-16019 <*>
1960 V10A P312-315 AEC-TR-5120 <*>
1961 V11 N4 P314-322 65-10757 <*>
1962 V12 P583-589 SC-T-532 <*>
1962 V12 N5/6 P653-665 AD-639 423 <=$>
1963 V13 N5 P345-366 65-11210 <*>
1964 V14 P428-434 65-17138 <*> O
1964 V14 P467-479 65-18119 <*>

CHALEUR ET INDUSTRIES
1926 V7 P673-678 3281 <*>
1937 V18 P91-94 63-14503 <*> O
1937 V18 N201 P36-43 TT-801 <NRCC>
1948 V29 N272 P74-76 60-10166 <*> O
1949 V30 P159-168 I-126 <*>
1949 V30 P304-308 II-665 <*>
1951 V32 P85-94 UCRL TRANS-775(L) <*>
1952 V33 P425-430 1799 <*>
1954 V35 N343 P41-58 61-20926 <*> O
1955 V36 N354 P3-13 57-1702 <*>
1955 V36 N356 P95-100 74Q70F <ATS>

CHALMERS TEKNISKA HOGSKOL. HANDLINGAR
1942 V5 N11 P3-47 57-796 <*>
1944 N36 P1-58 57-2370 <*>
1946 N51 P3-9 8749 <IICH>
1946 V54 P1-59 57-811 <*>
1956 N175 P1-7 58-1643 <*>
1956 N176 P1-16 AEC-TR-3289 <+*>
1961 V248 P4-49 63-00129 <*>

CHARITE-ANNALEN
1887 V12 P599 59-10364 <*>
1889 V14 P367-369 1308 <*>

CHEKHOSLOVATSKAYA BIOLOGIYA
1953 V2 N1 P18- AEC-TR-5423 <=*>
1954 V3 N5 P298-307 R-1537 <*>
RT-4561 <*>

CHEKHOSLOVATSKII FIZICHESKII ZHURNAL
1955 V5 N2 P121-132 59-17509 <*>
1955 V5 N3 P369-383 AEC-TR-3510 <+*>
AEC-TR-3512 <*>
1955 V5 N4 P480-501 62-23010 <=*>
1956 V6 N1 P84-90 60-12613 <=>
1956 V6 N5 P487-495 59-10590 <*>
1957 V7 N2 P191-201 59-10955 <*>
1957 V7 N6 P744-747 60-14074 <*>
1958 V8 P612 60-14022 <*>
1958 V8 P685-688 TS 1815 <BISI>
1959 V9 N1 P37-46 66-11030 <*>
1959 V9 N3 P377-387 62-19917 <=>
1959 V9 N4 P505-511 62-19919 <=*>
1960 V10 N8 P612-613 66-13765 <*>
1963 V13 N7 P509-517 64-11869 <=>
1964 V14 N2 P137-141 65-10297 <*>
1965 V15 N8 P602-605 66-13752 <*>

CHEMIA ANALITYCZNA
1956 V1 P91-100 58-739 <*>
63-16534 <*>
1956 V1 P246-254 61-31280 <=>
1956 V1 P285-293 62-14021 <*> O
1957 V2 P176-182 61-31259 <=>
1958 V3 P515-530 61-31281 <=>
1958 V3 P729-736 60-21457 <=> O
1958 V3 P745-751 60-21457 <=> O
1958 V3 P817-820 10L34P <ATS>
1958 V3 P865-869 60-21457 <=> O
1958 V3 P893-895 09L34P <ATS>
1958 V3 N3/4 P475-482 61-11377 <=>
1959 V4 P123-134 AEC-TR-4076 <*>
1959 V4 P803-807 61-31293 <=>
1959 V4 P809-818 61-31293 <=>
1959 V4 P849-854 61-11386 <=>
1959 V4 N1/2 P25-28 61-11370 <=>
1960 V5 P407-411 65-12809 <*>
1960 V5 N4 P687-689 62-19456 <=*>
1961 V6 P411-417 62-16008 <*>
1961 V6 P443-448 CSIR-TR.298 <CSSA>
1961 V6 P551- 11634 <KH>
1961 V6 N4 P551-558 64-14759 <*>
1961 V6 N6 P903-913 CSIR-TR.304 <CSSA>
1961 V6 N6 P915-927 CSIR-TR.305 <CSSA>
1962 V7 P579-582 65-12495 <*>
1962 V7 N1 P7-46 CSIR-TR.306 <CSSA>

1962 V7 N1 P134-147 3823 <BISI>
1962 V7 N2 P333-341 3822 <BISI>
1962 V7 N2 P429-433 13Q71G <ATS>
1963 V8 P607-612 66-14226 <*$>
1963 V8 P961-964 M.5804 <=$>
1963 V8 N5 P685-690 66-11188 <*>
1964 V9 P1063-1070 763 <TC>
1964 V9 P1129-1131 762 <TC>
1964 V9 N4 P753-760 66-13782 <*>
1964 V9 N5 P953-958 65-11310 <*>
1965 V10 P129 1660 <TC>
1965 V10 N1 P113-116 66-13406 <*>

CHEMIA STOSOWANA
1958 V2 P387-399 10L35P <ATS>
66-15291 <*>
1959 V3 P339-351 65P66P <ATS>
1959 V3 N3 P365-375 75S88P <ATS>
1961 V5 N2 P299-310 76S88P <ATS>
1961 V5 N3 P427-437 62-19672 <=*>
1962 N2 P279- ICE V2 N4 P597-603 <ICE>
1962 V4 N2 P153-165 ICE V2 N4 P560-566 <ICE>
1962 V6 P133-146 AEC-TR-5562 <*>
1962 V6 N1 P191-200 77S88P <ATS>
1963 N3 P329-368 ICE V3 N4 P533-556 <ICE>

CHEMIA STOSOWANA. SERIA A: KWARTALNIK POSWIECONY ZAGADNIENIOM TECHNOLOGII
1964 V8 P439-452 65-14954 <*>
1964 V8 N1 P45-58 66-11518 <*>
1964 V8 N1 P93-103 ICE V4 N4 P688-694 <ICE>
1964 V8 N4 P419-430 1428 <TC>
1966 N1A P105-112 ICE V6 N4 P604-608 <ICE>

CHEMIA STOSOWANA. SERIA B. KWARTALNIK POSWIECONY ZAGADNIENION INZYNIERII I APARATURY CHEMICZNEJ
1964 N2 P161-179 ICE V5 N3 P45-54 <ICE>
1964 N4 P467-488 ICE V5 N3 P524-532 <ICE>
1965 N1 P3-43 ICE V5 N4 P672-694 <ICE>
1965 N1 P101-125 ICE V5 N4 P695-710 <ICE>
1965 N2B P129-153 ICE V6 N1 P85-98 <ICE>
1965 N3B P309-330 ICE V6 N2 P204-217 <ICE>

CHEMICAL ENGINEERING, JAPAN
SEE KAGAKU KOGAKU

CHEMICAL ENGINEERING SCIENCE. LONDON
1953 V2 N6 P233-239 57-2685 <*>
1956 V5 P127-139 62-16649 <*$
1957 V6 P160-169 AEC-TR-6318 <*>
1961 V14 P183-189 <LSA>
1961 V15 P1-38 AEC-TR-4973 <*>
1961 V15 P39-74 AEC-TR-4976 <*>
1961 V15 P75-99 AEC-TR-4974 <*>
1961 V15 N1/2 P1-38 66-13793 <*>

CHEMICAL INDUSTRY. CHINESE PEOPLES REPUBLIC
SEE HUA HSUEH KUNG YEH

CHEMICAL INDUSTRY. JAPAN
SEE KAGAKU KOGYO

CHEMICAL INDUSTRY AND ENGINEERING. CHINESE PEOPLES REPUBLIC
SEE HUA KUNG HSUEH PAO

CHEMICAL REVIEW, KYOTO UNIVERSITY. JAPAN
SEE KAGAKU HYORON

CHEMICAL WORLD . CHINESE PEOPLES REPUBLIC
SEE HUA HSUEH SHIH CHIEH

CHEMICKE LISTY
1930 V24 P181-185 60-10707 <*>
1934 V27 P230-233 63-18143 <*>
1936 V30 N13 P169-173 88L36C <ATS>
1937 V31 P15-20 63-10441 <*>
1942 V35 P253-255 64-18932 <*>
1942 V36 N23/4 P313-317 58-1698 <*>
1943 V37 P289-290 3980A <K-H>
64-16433 <*>
1944 V38 P17-21 C-2241 <NRC>
1950 V44 P259-262 62-10046 <*> O
1950 V44 P283-291 1531 <*>
1950 V44 P576-592 73R75C <ATS>
1951 V45 P257-259 18Q69C <ATS>
1951 V45 P272-274 57-1698 <*>
1951 V45 P300-303 3322 <*>
1951 V45 P379-380 19Q69C <ATS>
1952 V46 P331-337 57-3275 <*>
58-2212 <*>
60-19910 <=*>
1952 V46 P337-340 58-1792 <*>
1952 V46 P341-344 58-79 <*>
63-14815 <*>
1952 V46 P382-383 2671 <K-H>
63-20446 <*>
1952 V46 P400-403 T1803 <INSD>
1952 V46 P688-698 58-290 <*>
1952 V46 N2 P92-94 65-13404 <*>
1952 V46 N3 P375-379 58-1514 <*>
1952 V46 N6 P341-344 65-13508 <*> C
1952 V46 N11 P688-708 60-18539 <*> P
63-14507 <*> OP
1953 V47 P464-467 3963-A <K-H>
1953 V47 P503-511 39Q72C <ATS>
1953 V47 P718- 2880 <*>
1953 V47 P817-827 CSIRO-3516 <CSIR>
3963-B <K-H>
1953 V47 P828-836 3963-C <K-H>
1953 V47 P837-841 3963-D <K-H>
1953 V47 P1113-1119 20P63C <ATS>
1953 V47 P1173-1183 59-10980 <*> O
61-20252 <*>
1953 V47 P1333-1337 6209 <K-H>
1953 V47 P1580-1590 724 <TC>
1953 V47 P1591-1597 63P64C <ATS>
1953 V47 P1745-1748 725 <TC>
1953 V47 N9 P1333-1337 64-20228 <*> O
1953 V47 N11 P1621-1632 63-10583 <*>
1954 V48 P383-385 61-20251 <*>
1954 V48 P397-400 57-1765 <*>
1954 V48 P414-416 2246 <*>
57-1967 <*>
1954 V48 P469-470 2245 <*>
1954 V48 P825-827 63-18159 <*>
1954 V48 P843-846 CJ-408 <ATS>
1954 V48 P1263- II-532 <*>
1954 V48 P1354-1359 CJ-409 <ATS>
1954 V48 P1585-1586 57-3450 <*>
1954 V48 N1 P41-44 6233 <K-H>
64-20223 <*>
1954 V48 N4 P492-497 65-13050 <*>
1954 V48 N6 P825-827 65-10894 <*>
1954 V48 N6 P828-838 60-19914 <=*>
1954 V48 N6 P917-919 II-281 <*>
1954 V48 N7 P1066-1070 1277 <*>
1954 V48 N8 P1189-1196 1136 <*>
1954 V48 N8 P1261-1262 2259 <*>
1955 V49 P1433-1441 PT5 RCT V29 N4 P1245 <RCT>
1955 V49 P1587-1597 PT4 RCT V29 N4 P1245 <RCT>
1955 V49 P60-62 43L31C <ATS>
1955 V49 P134- 57-1086 <*>
1955 V49 P149-157 II-1028 <*>
2822 <*>
1955 V49 P317-324 UCRL TRANS-569(L) <*>
1955 V49 P325-327 UCRL TRANS-800(L) <*>
1955 V49 P328-332 UCRL TRANS-769(L) <*>
1955 V49 P552-554 58-594 <*>
1955 V49 P862-868 5577-B <K-H>
1955 V49 P1185-1187 CJ-403 <ATS>
1955 V49 P1403-1405 CJ-406 <ATS>
1955 V49 P1442-1447 59-15921 <*> O
1955 V49 P1508-1516 64-18551 <*>
1955 V49 P1612-1616 15K28C <ATS>
1955 V49 P1848-1850 1440 <*>
1955 V49 N6 P858-861 64-18373 <*>
1955 V49 N6 P862-868 64-18371 <*> O
1955 V49 N11 P1656-1660 57-3202 <*>
1955 V49 N12 P1891-1894 65-12043 <*>

1956 V50 P3-10 SCL-T-407 <*>
1956 V50 P88-93 2908 <*>
1956 V50 P162-163 1644 <*>
1956 V50 P221-226 20K26C <ATS>
1956 V50 P308- 9019 B <K-H>
1956 V50 P395-397 58-1808 <*>
1956 V50 P716-720 61-10228 <*> 0
1956 V50 P726- AEC-TR-3633 <+*>
1956 V50 P748-751 63-14244 <*> 0
1956 V50 P752-755 63-14243 <*> 0
1956 V50 P1103- 59-15439 <*>
1956 V50 P1120-1125 5577-A <K-H>
1956 V50 P1215-1218 57-3153 <*>
60-19911 <=*>
1956 V50 P1306-1307 58-2376 <*>
1956 V50 P1331-1334 NP-TR-1053 <*>
1956 V50 P1406-1409 AEC-TR-4847 <*>
1956 V50 P2C25 129TT <CCT>
58-199 <*>
1956 V50 N1 P23-28 4543 <BISI>
1956 V50 N2 P314-316 131TM <CTT>
1956 V50 N5 P817-820 1413 <*>
1956 V50 N7 P1120-1125 64-18372 <*> 0
65-13893 <*>
1956 V50 N7 P1141-1146 57-240 <*>
1956 V50 N8 P1236-1240 1437 <*>
1956 V50 N8 P1329-1330 1527 <*>
1956 V50 N9 P1406-1409 62-16201 <*> 0
1956 V50 N10 P1624-1626 57-1882 <*>
1956 V50 N11 P1702-1711 59-15013 <*>
1956 V50 N11 P1818-1821 57-224 <*>
1956 V50 N11 P1834-1839 57-737 <*>
1956 V50 N12 P1988-1994 59-11570 <=*>
1956 V50 N12 P2040-2041 57-3493 <*>
1957 V51 P13-20 2TM <CTT>
57-3386 <*>
1957 V51 P287-291 64-30109 <*> 0
6818 <K-H>
1957 V51 P378-380 128TM <CTT>
67J16C <ATS>
1957 V51 P470-473 468TM <CTT>
57-3365 <*>
1957 V51 P672-675 115TM <CTT>
1957 V51 P672 58-511 <*>
1957 V51 P768-770 62-16710 <*> 0
1957 V51 P1159-1164 77K26C <ATS>
1957 V51 P1202-1203 23K21C <ATS>
1957 V51 P1467-1470 UCRL TRANS-536 <*+>
1957 V51 P1885-1893 1TM <CTT>
1957 V51 P1894-1905 60TM <CTT>
1957 V51 P1971-1985 58-2625 <*>
1957 V51 P1998-2001 63-16186 <*> 0
1957 V51 P2304-2308 58-2094 <*>
1957 V51 N2 P376-378 518 <MT>
1957 V51 N3 P459-462 65-11453 <*>
1957 V51 N4 P716-721 C-3216 <NRCC>
1957 V51 N5 P8C3-817 65-10166 <*>
1957 V51 N5 P818-822 60-11678 <=*> 0
1957 V51 N7 P1379-1381 61-10914 <*>
6609-A <K-H>
1958 V52 P105-108 61-20089 <*>
1958 V52 P262-268 AEC-TR-3662 <+*>
1958 V52 P631-635 77L31C <ATS>
1958 V52 P896- 60-10133 <*> 0
1958 V52 P1108-1112 64-14362 <*> 0
1958 V52 P1201-1205 59-15533 <*>
1958 V52 P1422-1427 AEC-TR-3810 <+*>
1958 V52 P1726-1734 53L31C <ATS>
1958 V52 P2018-2021 57P64C <ATS>
1958 V52 P2296-2310 R-5125 <*>
1958 V52 N1 P7-15 AEC-TR-3827 <*>
1958 V52 N4 P618-622 AEC-TR-3666 <+*>
1958 V52 N4 P623-630 AEC-TR-3665 <+*>
1958 V52 N7 P1289-1298 61-10466 <*>
1958 V52 N7 P1370-1372 60-23056 <=*> 0
1958 V52 N8 P1431-1434 65-10210 <*>
1958 V52 N8 P1613-1621 50K28C <ATS>
1959 V53 P311-364 NP-TR-1046 <*>
1959 V53 P465-480 UCRL TRANS-599(L) <*>
62-00406 <*>
1959 V53 P537-539 C-3367 <NRCC>
1959 V53 P757-761 AEC-TR-4174 <*>
1959 V53 P821-828 61-10383 <*>
62-10197 <*> 0
1959 V53 P847-848 SCL-T-490 <*>
1959 V53 P928-940 24L37C <ATS>
1959 V53 P941-944 63-16548 <*>
1959 V53 P950-951 65-10117 <*>
1959 V53 P997-1028 AEC-TR-4477 <*>
1959 V53 N1 P1-5 62-10577 <*>
1959 V53 N5 P512-526 60-19863 <*+>
1959 V53 N8 P829-843 65-12461 <*>
1959 V53 N10 P1054-1057 61-10627 <*>
1960 V54 P764-790 AEC-TR-4761 <*>
1960 V54 P1173-1183 16N50C <ATS>
1960 V54 P1279-1298 ORNL-TR-565 <*>
1961 V55 P139-153 62-10623 <*>
1961 V55 P919-929 66-12192 <*>
1961 V55 P1439-1443 AEC-TR-5494 <*>
1961 V55 N4 P389-399 63-18156 <*>
1961 V55 N5 P536-541 62-25223 <=*>
1961 V55 N7 P765-776 66-12189 <*>
1961 V55 N7 P777-788 29N57C <ATS>
1961 V55 N8 P974-982 62-00626 <*>
1961 V55 N9 P1C62-1067 3007 <BISI>
1962 N10 P1222 ICE V3 N3 P388-424 <ICE>
1962 V56 P495-504 63-00045 <*>
1962 V56 P728-762 AEC-TR-5813 <*>
1962 V56 N3 P334-354 C-4111 <NRCC>
1962 V56 N4 P476 62-32205 <=*>
1962 V56 N5 P574-576 62-32212 <=*>
1962 V56 N5 P619 62-32205 <=*>
1962 V56 N8 P987-1028 63-00539 <*>
1962 V56 N12 P1445-1447 CSIR-TR.338 <CSSA>
1963 V57 P419-439 65-10856 <*>
1963 V57 N8 P8C3-811 492 <MT>
1964 V58 P163-178 UCRL TRANS-1243(L) <*>
1964 V58 P349-375 66-14093 <*$>
1964 V58 P911-945 AD-612 879 <=$>
1964 V58 P1471-1497 11S89C <ATS>
1964 V58 N5 P573-576 4100 <BISI>
1964 V58 N6 P651-656 66-10341 <*>
1964 V58 N11 P1325-1328 1423 <TC>
1965 V59 N7 P800-820 68S87C <ATS>

CHEMICKE ZVESTI
1947 V1 P72-76 63-10980 <*>
1950 V4 N2 P60-63 65-11353 <*>
1951 V5 P322-330 65-12861 <*>
1953 V7 N7 P385-408 RCT V28 N4 P968 <RCT>
1953 V7 N3/4 P179-187 65-13844 <*>
1953 V7 N5/6 P257-288 RCT V28 N2 P383 <RCT>
1955 V8 N10 P867-877 32J19C <ATS>
1956 V10 N3 P183-187 17L32C <ATS>
1956 V10 N7 P460-467 126TT <CCT>
1956 V10 N10 P604-611 62-14571 <*>
1957 V11 P358-368 62-26052 <=$> P
1957 V11 N12 P724-730 59-10583 <*>
1958 V12 N4 P213-220 61-00586 <*>
1958 V12 N4 P244-251 61-16780 <*>
1958 V12 N6 P353-365 AEC-TR-3940 <*=>
1960 V14 P91-94 AEC-TR-4222 <*>
1960 V14 N11/2 P757-761 62-19692 <=*>
1960 V14 N11/2 P818-821 62-19692 <=*>
1961 V15 P181-190 ICE V1 N1 P117-121 <ICE>
14N52C <ATS>
1961 V15 P309-314 AEC-TR-4957 <*>
1961 V15 N8 P547-553 32T94SK <ATS>
1962 V16 N7 P505-515 33Q72SK <ATS>
1962 V16 N7 P562-573 65-17341 <*>
1962 V16 N4/5 P242-244 63-10737 <*>
1963 V17 P569-574 65-23237 <$>
1963 V17 P592-596 1101 <TC>
1963 V17 P656-665 RAPRA-1271 <RAP>
1963 V17 P912-915 C-5559 <NRC>
1963 V17 N10/1 P709-716 65-14336 <*>
1964 V18 P294-298 65-64102 <*=$> 0
1964 V18 P614-619 12S89SL <ATS>
1964 V18 P620-628 13S89SL <ATS>
1965 V19 N11 P865-879 66-30085 <=*$>

Citation	Translation
1966 V20 N3 P169-179	66-13376 <*>

CHEMICKY OBZOR

Citation	Translation
1928 V3 P330-336	AL-64 <*>
1941 V16 P150-154	64-10467 <*> 0
1948 V23 P217-221	59-15243 <*> P

CHEMICKY PRUMYSL

Citation	Translation
1953 V3 P100-102	61-18901 <*>
1953 V3 N9 P313-316	61-18835 <*>
1953 V3 N10 P356-363	61-18835 <*>
1953 V3 N7/8 P255-258	61-18835 <*>
1954 V4 N5 P182-183	63-18256 <*>
1954 V4 N10 P381-383	62-10672 <*>
1955 V5 P330-332	66-11567 <*>
1955 V5 N1 P24-27	3882 C <K-H>
	64-16157 <*> 0
1955 V5 N2 P72-74	35P62C <ATS>
1955 V5 N5 P212-213	RCT V29 N2 P647 <RCT>
1955 V5 N6 P239-243	57-1976 <*>
1955 V5 N30 P219-	25J16C <ATS>
1956 V6 P10-14	62-32395 <=*>
1956 V6 P426-427	62-32396 <=*>
1956 V6 N7 P293	21Q72C <ATS>
1956 V6 N10 P410-413	85M47C <ATS>
1956 V6 N12 P490-496	86M47C <ATS>
1956 V6 N31 P45-49	50H13C <ATS>
1957 V7 P78-79	1384 <TC>
1957 V7 P651-652	62-10691 <*>
1957 V7 N1 P4-8	60-19977 <=*>
1957 V7 N5 P265-268	04P66C <ATS>
1957 V7 N6 P321-326	(CHEM)-377 <CTS>
1957 V7 N8 P393-396	58-2448 <*>
1957 V7 N11 P579-581	70K23C <ATS>
1958 V8 P545-551	63-14023 <*> 0
1958 V8 N1 P50-51	07L31C <ATS>
1958 V8 N3 P157-163	61-16801 <*>
1958 V8 N5 P229-233	18L32C <ATS>
1958 V8 N6 P321-324	91K26C <ATS>
1958 V8 N9 P455-457	72P61C <ATS>
1958 V8 N10 P516-523	NP-TR-896 <*>
1959 V9 P160-164	60-16963 <*>
1959 V9 N2 P101-104	59-20802 <*>
1959 V9 N8 P398-402	73R74C <ATS>
1960 V10 P100-102	CHEM-155 <CTS>
1960 V10 N2 P84-85	82N55C <ATS>
1960 V10 N2 P100-102	64-20885 <*>
1960 V10 N3 P123-126	62-25224 <=*> 0
1960 V10 N3 P155-161	62-14265 <*>
1960 V10 N6 P331-334	42R79C <ATS>
1960 V10 N9 P492-495	62-10124 <*> 0
1960 V10 N10 P521-525	AEC-TR-5650 <=*>
	64P66C <ATS>
1961 N9 P457	ICE V2 N2 P216-220 <ICE>
1961 N10 P509-	ICE V2 N3 P306-314 <ICE>
1961 N11 P638-	ICE V2 N2 P279-282 <ICE>
1961 V11 N3 P113-117	ICE V2 N1 P12-17 <ICE>
1961 V11 N5 P225-229	65-10118 <*> 0
1961 V11 N6 P281-	ICF V2 N1 P57-60 <ICE>
1961 V11 N6 P281-284	62-10302 <*> 0
1961 V11 N6 P328-332	48P60C <ATS>
1961 V11 N7 P345-348	12Q73SK <ATS>
1961 V11 N8 P433-438	62-20422 <*>
1961 V11 N8 P439-444	CSIR-TR.269 <CSSA>
1961 V11 N9 P472-473	CTS-417 <CTS>
1961 V11 N10 P521-523	63-14396 <*>
1961 V11 N11 P604-607	57Q70SK <ATS>
1962 N1 P12-	ICE V2 N3 P365-371 <ICE>
1962 N2 P73-	ICE V2 N3 P394-399 <ICE>
1962 N4 P209-	ICE V2 N4 P482-486 <ICE>
1962 N5 P232-	ICE V2 N4 P531-536 <ICE>
1962 N6 P321-	ICE V3 N1 P12-19 <ICE>
1962 N7 P355-	ICF V3 N1 P76-79 <ICE>
1962 N9 P473-	ICE V3 N1 P143-149 <ICE>
1962 N10 P575-578	ICE V4 N1 P36-40 <ICF>
1962 V12 P237-239	62-20241 <*>
1962 V12 P649-652	1390 <TC>
1962 V12 P689-691	7001 <TTIS>
1962 V12 N2 P97-102	69P63G <ATS>
1962 V12 N2 P1C2-106	(CHEM)-284 <CTS>
	64-20884 <*>
1962 V12 N4 P170-174	65-13031 <*>
1962 V12 N4 P213-219	65-12907 <*> 0
1962 V12 N9 P518-520	58Q68C <ATS>
1962 V12 N10 P529-534	63-18854 <*> 0
1962 V12 N11 P637-638	<TTIS>
1962 V12 N12 P655-658	NS-105 <TTIS>
1962 V12 N12 P681-685	37Q69C <ATS>
1963 N1 P49-53	ICE V3 N4 P586-590 <ICE>
1963 N3 P160-165	ICE V3 N4 P591-596 <ICE>
1963 N8 P436-441	ICE V4 N1 P147-152 <ICE>
1963 N11 P578-581	ICE V4 N3 P456-460 <ICE>
1963 V13 N1 P49-53	65-11139 <*>
1963 V13 N2 P63-67	64-16423 <*>
1963 V13 N2 P77-78	20S85C <ATS>
1963 V13 N3 P165-167	NS-78 <TTIS>
1963 V13 N4 P169-173	65-10462 <*> 0
1963 V13 N6 P295-299	65-10461 <*> 0
1963 V13 N6 P321-325	22R76C <ATS>
	65-14652 <*>
1963 V13 N6 P325-328	1100 <TC>
1963 V13 N7 P351-356	ICE V4 N1 P93-99 <ICE>
1963 V13 N8 P393-397	ICE V4 N1 P104-109 <ICE>
1963 V13 N8 P433-436	ICE V4 N1 P100-103 <ICE>
1963 V13 N9 P449-453	26Q74C <ATS>
1963 V13 N9 P460-464	64-18010 <*>
1963 V13 N9 P492-497	RAPRA-1267 <RAP>
1963 V13 N9 P498-501	RAPRA-1273 <RAP>
1963 V13 N11 P575-577	86R75C <ATS>
1963 V13 N12 P658-662	66-10601 <*>
1963 V14 N1 P30-33	NS-201 <TTIS>
1964 N1 P12-16	ICE V5 N2 P245-251 <ICE>
1964 N1 P17-19	ICE V4 N4 P565-567 <ICE>
1964 N2 P75-78	ICE V4 N4 P568-572 <ICE>
1964 N7 P366-369	ICE V5 N1 P152-156 <ICE>
1964 V14 P19-25	1617 <TC>
1964 V14 P184-188	NS-413 <TTIS>
1964 V14 N1 P5-8	NEL-TRANS-1697 <NLL>
1964 V14 N1 P17-19	ICE V4 N4 P565-567 <ICE>
1964 V14 N2 P75-78	ICE V4 N4 P568-572 <ICE>
1964 V14 N2 P81-86	RAPRA-1245 <RAP>
1964 V14 N3 P113-119	64-18548 <*>
1964 V14 N5 P238-241	65-11408 <*>
1964 V14 N5 P263-265	65-14890 <*>
1964 V14 N6 P296-299	65-11325 <*>
1964 V14 N7 P362-365	NS-447 <TTIS>
1964 V14 N7 P366-369	ICE V5 N1 P.152-156 <ICE>
1964 V14 N7 P376-378	66-13369 <*>
1964 V14 N8 P395-398	1420 <TC>
1964 V14 N10 P511-515	576 <TC>
1964 V14 N11 P576-581	NS-352 <TTIS>
1964 V14 N12 P604-605	NS-446 <TTIS>
1965 N1 P9-15	ICE V5 N3 P453-460 <ICE>
1965 N1 P25-28	ICE V5 N3 P481-484 <ICE>
1965 N3 P165-170	ICE V5 N4 P647-653 <ICE>
1965 N7 P385-387	ICE V6 N1 P21-24 <ICE>
1965 N8 P450-456	ICE V6 N1 P130-137 <ICE>
1965 N11 P641-644	ICE V6 N3 P402-406 <ICE>
1965 V3 N14 P142-147	844 <TC>
1965 V15 N2 P80-85	66-11285 <*>
1965 V15 N2 P1C6-107	65-14064 <*>
1965 V15 N3 P142-147	66-10818 <*>
1965 V15 N4 P223-226	66-10776 <*> 0
1965 V15 N5 P283-287	66-11289 <*> 0
1965 V15 N7 P421-424	89T90SL <ATS>
1965 V15 N8 P498-500	82S89SL <ATS>
1965 V15 N12 P719-922	24T92C <ATS>
1966 N1 P4-9	ICE V6 N3 P488-494 <ICE>
1966 N1 P16-19	ICE V6 N3 P481-485 <ICE>
1966 N2 P73-78	ICE V6 N3 P522-529 <ICE>
1966 N2 P85-88	ICE V6 N3 P518-521 <ICE>
1966 N3 P141-146	ICE V6 N4 P657-663 <ICE>
1966 N5 P262-267	ICE V6 N4 P687-694 <ICE>
1966 N6 P344-347	ICE V6 N4 P695-699 <ICE>

CHEMIE. BERLIN

Citation	Translation
1942 V48 N11 P640-646	MF-4 <*>
1942 V55 P7-11	2668 <*>
1942 V55 P356-359	II-812 <*>
1942 V55 N1/2 P7-11	63-10984 <*>

1942 V55 N3/4 P24-28	63-10984 <*>
	63-14000 <*>
1942 V55 N5/6 P37-38	60-14524 <*>
1942 V55 N15/6 P115-130	1624 <*>
1943 V56 P225-230	2429 <*>
1943 V56 N11/2 P67-71	60-18305 <*>
1944 V57 P95-100	57-960 <*>
1944 V57 N13/6 P90-94	62-10446 <*>
1944 V57 N21/4 P159	AEC-TR-869 <*>
1945 V58 P25-30	60-10258 <*> 0

CHEMIE. PRAGUE

1948 N4 P200-201	2828 <*>
1958 N10 P635-646	AEC-TR-3958 <*>
1958 V10 P499-502	66-14153 <*$>
1958 V10 N8 P631-634	65-11607 <*>

CHEMIE DER ERDE

1932 V7 N1 P113-120	AL-347 <*>
1935 V8 P252-315	58-1875 <*>
1942 V14 P239-252	57-3414 <*>
1958 V19 P275-285	AEC-TR-5119 <=*>
1958 V19 N3 P207-229	IGR V2 N1 P8 <AGI>
1964 N23 P279-311	GI N6 1964 P1205 <AGI>
1965 V24 N2 P115-146	66-14415 <*>

CHEMIE EN TECHNIEK

1963 V18 N1 P12-19	862 <TC>

CHEMIEFASERN

1961 V11 P184-186	64-10871 <*>
1961 V11 N1 P38-42	64-10809 <*> 0
1961 V11 N2 P121-124	64-10809 <*> 0
1961 V11 N3 P166-173	63-16312 <*>
1961 V11 N3 P176-179	64-10809 <*> 0
1961 V11 N4 P248-252	64-10809 <*> 0
1962 V12 N3 P184-186	65-17308 <*> 0
1962 V12 N6 P398-401	63-18021 <*> 0
1963 V13 N9 P676-685	65-12094 <*>
1963 V13 N10 P768-776	64-20945 <*> 0
1964 V14 N1 P25-39	65-12176 <*>
1964 V14 N2 P121-126	64-18346 <*>
1964 V14 N4 P260-264	65-13866 <*> 0
1964 V14 N8 P558-562	579 <TC>
	65-11826 <*>
	66-10568 <*>
1964 V14 N10 P728-729	439 <TC>
1964 V14 N10 P728-731	65-12172 <*>
1965 N2 P646-656	1714 <TC>
1965 V15 N3 P168-177	65-10077 <*>
1965 V15 N7 P502-509	66-10584 <*>
1965 V15 N11 P840-845	66-14604 <*>
1965 V15 N12 P923-928	66-14352 <*>

CHEMIE-INGENIEUR-TECHNIK

1949 N19/0 P382-383	57-3069 <*>
1949 V21 P191-193	62-18424 <*> 0
1949 V21 P331-334	I-899 <*>
1949 V21 N15 P298-304	62-20354 <*> 0
1949 V21 N11/2 P227-229	I-343 <*>
1949 V21 N9/10 P191-193	I-398 <*>
1950 V22 P39-40	II-800 <*>
1950 V22 P396-397	AL-300-II <*>
1950 V22 N3 P54-56	1163 <*>
1950 V22 N4 P77-80	AEC-TR-3837 <*>
1950 V22 N12 P253-258	63-16227 <*> 0
1950 V22 N17 P361-373	64-18856 <*> 0
	64-20498 <*>
1950 V22 N20 P437-442	64-20499 <*>
1950 V22 N13/4 P273-283	64-20497 <*>
1950 V22 N23/4 P527-539	64-20500 <*> 0
1951 V23 P39-41	63-18966 <*>
1951 V23 P59-64	58-1242 <*>
1951 V23 P161-165	62-25796 <=*>
1951 V23 P289-293	62-25797 <=*>
1951 V23 P321-324	I-537 <*>
	1164 <*>
1951 V23 N3 P59-64	1162 <*>
1951 V23 N13 P321-324	64-14408 <*> 0
1952 P237-247	1649 <*>
1952 N6 P339-347	AL-628 <*>
1952 V24 P82-91	2230 <*>
1952 V24 P609-610	1189 <*>
1952 V24 N1 P1-2	65-10015 <*>
1952 V24 N1 P22-25	<ATS>
1952 V24 N2 P92-	1953 <*>
1952 V24 N3 P153-154	65-14118 <*>
1952 V24 N6 P333-338	AL-482 <*>
1952 V24 N9 P494-500	39H13G <ATS>
1953 V25 P61-65	8781 <IICH>
1953 V25 P144-148	II-613 <*>
1953 V25 P201-203	57-2056 <*>
1953 V25 P249-252	I-518 <*>
1953 V25 P277-285	57-1780 <*>
1953 V25 P477-480	I-589 <*>
1953 V25 N2 P61-65	64-20496 <*> 0
1953 V25 N3 P114-124	14M40G <ATS>
1953 V25 N3 P137-141	60-10370 <*> P
1953 V25 N5 P229-237	15M40G <ATS>
1953 V25 N5 P245-248	64-18847 <*> 0
1953 V25 N6 P285-291	64-18855 <*>
1953 V25 N6 P313-316	65-12655 <*>
1953 V25 N6 P317-327	1558 <*>
1953 V25 N6 P328-330	1416 <*>
	65-12919 <*>
1953 V25 N6 P331-341	59-17522 <*>
	65-10018 <*> 0
1953 V25 N10 P620-622	3245 <*>
	65-10023 <*>
1953 V25 N12 P697-701	62-20044 <*>
	64-20886 <*> 0
1953 V25 N12 P735-737	57-2008 <*>
	62-20062 <*>
1953 V25 N8/9 P474-476	AEC-TR-5841 <*>
1953 V25 N8/9 P481-484	57-2750 <*>
1953 V25 N8/9 P534-536	62-20185 <*>
1954 N2 P90-94	57-1742 <*>
1954 V26 P29-34	9028 <IICH>
1954 V26 P189-201	I-741 <*>
1954 V26 P245-253	57-1317 <*>
1954 V26 P253-258	3258 <*>
	57-1316 <*>
1954 V26 P301-309	3034 <*>
1954 V26 P421-	57-1783 <*>
1954 V26 P661-667	62-14037 <*> 0
1954 V26 N2 P83-89	2034 <*>
	64-14672 <*> 0
	65-12660 <*>
1954 V26 N2 P90-94	63-14976 <*>
1954 V26 N2 P97-100	II-1023 <*>
	64-14410 <*> 0
1954 V26 N3 P150-155	T1885 <INSD>
	59-17316 <*>
	65-10224 <*>
1954 V26 N4 P189-201	64-10432 <*>
1954 V26 N5 P245-253	2826 <*>
1954 V26 N5 P253-258	2825 <*>
1954 V26 N5 P259-264	I-536 <*>
	64-14663 <*> 0
1954 V26 N6 P331-337	59-20132 <*>
	65-10223 <*>
1954 V26 N7 P421-422	60-10774 <*>
1954 V26 N10 P555-561	64-20492 <*> 0
1954 V26 N10 P572-	57-789 <*>
1954 V26 N11 P614-630	62-10230 <*>
1954 V26 N12 P684-686	AEC-TR-3582 <+*>
1955 V27 N1 P1-4	3424-B <K-H>
	63-14626 <*> 0
1955 V27 N1 P13-17	59-10640 <*>
1955 V27 N2 P71-75	3485-A <K-H>
	64-14908 <*>
1955 V27 N2 P79-83	62-20072 <*>
1955 V27 N3 P1942-1948	83H10G <ATS>
1955 V27 N4 P190-192	59-17532 <*>
1955 V27 N4 P193-194	59-17534 <*>
1955 V27 N4 P209-213	3663 <K-H>
	64-16119 <*> 0
	65-11850 <*> 0
1955 V27 N10 P599-601	64-18155 <*> 0
1955 V27 N10 P602-604	62-18249 <*>

Citation	Translation
1955 V27 N11 P661-668	62-20130 <*>
1955 V27 N8/9 P507-512	62-20129 <*>
1956 V28 P81-87	57-219 <*>
1956 V28 P141-152	58-577 <*>
1956 V28 P155-161	AEC-TR-4663 <*>
1956 V28 P221-225	3255 <*>
1956 V28 P405-410	58-1386 <*>
1956 V28 P473-475	66-12308 <*> 0
1956 V28 P763-768	<ES>
1956 V28 N1 P25-30	3141 <NLL>
1956 V28 N1 P49-53	93H9G <ATS>
1956 V28 N1 P703-706	57-1970 <*>
1956 V28 N2 P88-93	59-20281 <*> 0
1956 V28 N3 P155-161	61-20964 <*>
1956 V28 N3 P165-174	AEC-TR-5844 <*>
1956 V28 N3 P181-189	57-754 <*>
1956 V28 N3 P190-195	57-354 <*>
1956 V28 N3 P203-213	61-20886 <*>
1956 V28 N5 P337-342	57-2925 <*>
1956 V28 N5 P350-365	59-17211 <*> 0
1956 V28 N7 P475-480	5486-A <K-H> 0
	64-18362 <*> 0
1956 V28 N7 P489-495	59-10641 <*>
1956 V28 N10 P625-688	57-2975 <*>
1956 V28 N10 P646-654	57-2580 <*>
1956 V28 N11 P689-697	34J14G <ATS>
1956 V28 N12 P786-793	10J17G <ATS>
	64-16563 <*>
1957 N4 P250-254	57-3408 <*>
1957 V29 P219	C2327 <NRC>
1957 V29 P267-275	57-3117 <*>
1957 V29 P505	61-18737 <*>
1957 V29 N1 P32-38	64-20115 <*> 0
1957 V29 N4 P277-279	10J18G <ATS>
1957 V29 N5 P345-347	3246 <*>
1957 V29 N5 P351-352	83J18G <ATS>
1957 V29 N6 P393-397	61-10458 <*>
1957 V29 N9 P595-599	61-10465 <*>
1957 V29 N9 P603-614	58-2380 <*>
1957 V29 N11 P709-721	66-10222 <*> 0
1957 V29 N11 P721-726	63-16117 <*>
	90K26G <ATS>
1957 V29 N11 P727-732	66-11751 <*>
1958 V30 P329-336	60-19155 <*+>
1958 V30 N2 P85	58-2681 <*>
1958 V30 N2 P85-95	64-18816 <*>
1958 V30 N3 P144-158	64-10179 <*>
1958 V30 N3 P159-165	65-13416 <*>
1958 V30 N3 P171-180	35N56G <ATS>
1958 V30 N3 P171-	61-10896 <*>
1958 V30 N4 P226-228	64-18796 <*> 0
1958 V30 N5 P288-292	59-10046 <*>
1958 V30 N5 P305-310	12K24G <ATS>
1958 V30 N7 P433-440	79Q67G <ATS>
1958 V30 N7 P440-446	59-15309 <*>
	62-20084 <*>
1958 V30 N8 P529-532	48K27G <ATS>
1958 V30 N9 P553-559	06L29G <ATS>
	59-10567 <*> 0
	59-10736 <*> 0
1958 V30 N9 P567-572	1594 <BISI>
	59-15836 <*> 0
1959 V31 P73-79	61-18289 <*> 0
1959 V31 P553-560	62-25798 <*=> 0
1959 V31 N1 P22-30 PT.2	NP-TR-400 <*>
1959 V31 N1 P50-54	65-11971 <*>
1959 V31 N3 P148-153	68L33G <ATS>
1959 V31 N3 P166-173	61-20483 <*> 0
1959 V31 N4 P260-261	15L34G <ATS>
1959 V31 N5 P310-318	GAT-Z-5016 <*>
1959 V31 N5 P323-337	71M46G <ATS>
1959 V31 N5 P345-351	49L34G <ATS>
1959 V31 N8 P525-526	61-20267 <*>
1959 V31 N9 P553-560	65-10196 <*> 0
1959 V31 N9 P583-587	51L36G <ATS>
1959 V31 N9 P588-598	52L36G <ATS>
1959 V31 N10 P677-679	60-13259 <=*>
1959 V31 N12 P761-765	S-2120 <RIS>
	61-20895 <*> 0
1960 V32 P285-288	AEC-TR-4952 <*>
1960 V32 P448-454	61-00904 <*>
1960 V32 P591-594	AEC-TR-5487 <*>
1960 V32 P747-749	AEC-TR-5707 <*>
1960 V32 P765-773	AEC-TR-4774 <*>
1960 V32 N1 P8-16	63M43G <ATS>
	66-14376 <*> 0
1960 V32 N2 P164-170	60-00704 <*>
1960 V32 N3 P129-135	70M46G <ATS>
1960 V32 N3 P164-171	2152 <BISI>
1960 V32 N4 P253-257	2927 <BISI>
1960 V32 N4 P291-297	SC-T-64-1647 <*>
1960 V32 N4 P297-298	64-10914 <*>
1960 V32 N5 P335-342	87M42G <ATS>
1960 V32 N5 P349-354	61-10326 <*> 0
1960 V32 N6 P413-417	64-18006 <*>
	65-11654 <*>
1960 V32 N9 P582-584	64M47G <ATS>
1960 V32 N12 P773-781	NP-TR-647 <*>
1960 V32 N12 P782-788	65-13025 <*>
	81N51G <ATS>
1960 V32 N12 P789-795	38N51G <ATS>
1960 V32 N12 P806-811	61-18301 <*>
1961 V33 N1 P27-31	61-20478 <*>
1961 V33 N3 P139-145	NP-TR-796 <*>
1961 V33 N5 P301-311	65-12929 <*>
1961 V33 N5 P327-335	4225 <BISI>
1961 V33 N6 P431-448	4225 <BISI>
1961 V33 N7 P469-478	63-16220 <*> 0
1961 V33 N8 P556-558	94N56G <ATS>
1961 V33 N11 P715-718	NP-TR-935 <*>
1962 V34 P567-571	63-16407 <*>
1962 V34 P580	AEC-TR-5986 <*>
1962 V34 P603-609	63-18434 <*>
1962 V34 P841-842	63-18948 <*>
1962 V34 N1 P45-61	11777-B <K-H>
	62-18671 <*> 0
	65-14010 <*> 0
1962 V34 N4 P269-282	65-12299 <*>
1962 V34 N4 P309-311	5890 <HB>
1962 V34 N5 P353-357	64-10483 <*>
1962 V34 N5 P376-378	63-20467 <*>
1962 V34 N6 P432-436	12054 <K-H>
	65-14035 <*> 0
1962 V34 N8 P546-551	4527 <BISI>
1962 V34 N10 P692-696	AEC-TR-5648 <*>
	63-18732 <*>
1962 V34 N10 P697-701	63-10904 <*> 0
1962 V34 N11 P782-786	69R75G <ATS>
1962 V34 N12 P841-842	65-14121 <*>
1963 V35 P343-352	63-20602 <*>
1963 V35 P372-376	508 <TC>
1963 V35 P405-410	880 <TC>
1963 V35 P856-860	222 <TC>
1963 V35 N1 P7-10	65-17449 <*>
1963 V35 N1 P29-36	3582 <BISI>
1963 V35 N1 P37-44	66-10327 <*> 0
1963 V35 N2 P78-80	66-13160 <*>
1963 V35 N4 P262-266	63-20315 <*> 0
1963 V35 N5 P325-331	03R75G <ATS>
	64-16331 <*>
	65-18156 <*>
1963 V35 N5 P353-361	66-12970 <*>
1963 V35 N7 P542	01R75G <ATS>
1963 V35 N8 P580-582	64-16862 <*>
1963 V35 N8 P586-589	08Q73G <ATS>
	64-18081 <*>
1963 V35 N9 P637-640	64-16978 <*>
1963 V35 N11 P785-788	65-11959 <*>
1963 V35 N12 P844-850	75R76G <ATS>
1964 V36 P546-552	3862 <BISI>
1964 V36 N1 P9-14	64-18125 <*> 0
1964 V36 N2 P131-147	3935 <BISI>
1964 V36 N3 P169-174	65-13407 <*>
1964 V36 N3 P175-185	65-13325 <*> 0
	66-14896 <*>
1964 V36 N3 P277-282	27R79G <ATS>
1964 V36 N4 P325-330	3874 <BISI>
1964 V36 N5 P417-422	66-11906 <*>
1964 V36 N5 P422-429	66-11907 <*>

1964 V36 N5 P523-537 ORNL-TR-272 <*>
73S82G <ATS>
1964 V36 N7 P717-729 65-13408 <*>
1964 V36 N8 P858-865 66-11404 <*>
1964 V36 N9 P947-956 84T94G <ATS>
1964 V36 N10 P1011-1019 4097 <BISI>
66-14842 <*> 0
1964 V36 N1C P1028-1033 65-10466 <*>
1964 V36 N10 P1046-1050 66-10326 <*> 0
1964 V36 N11 P1085-1089 66-11914 <*>
1965 V37 P1055-1062 2086 <TC>
1965 V37 N1 P57-61 66-10340 <*>
1965 V37 N2 P99-101 757 <TC>
1965 V37 N2 P101-107 65-13405 <*>
1965 V37 N3 P187-202 66-13859 <*>
1965 V37 N3 P218-226 804 <TC>
1965 V37 N3 P252-255 66-10342 <*>
804 <TC>
1965 V37 N4 P361-367 8616 <IICH>
1965 V37 N4 P400-402 65-17428 <*>
1965 V37 N4 P402-405 69S84G <ATS>
1965 V37 N5 P498-500 1092 <TC>
1965 V37 N6 P573-581 1096 <TC>
66-14839 <*>
1965 V37 N6 P581-586 1098 <TC>
66-14841 <*>
1965 V37 N6 P587-590 1095 <TC>
66-14840 <*>
1965 V37 N6 P601-606 1097 <TC>
1965 V37 N6 P616-621 1099 <TC>
60S87G <ATS>
1965 V37 N6 P626-631 1094 <TC>
1965 V37 N6 P639-643 1093 <TC>
65-17466 <*>
66-14839 <*>
1965 V37 N7 P709-714 66-12923 <*>
1965 V37 N9 P944-951 10T94G <ATS>
1965 V37 N11 P1136-1139 66-12236 <*>
1965 V37 N11 P1139-1143 66-10819 <*>
1966 V38 N2 P134-136 66-14529 <*>
1966 V38 N3 P350-355 66-13455 <*>
1966 V38 N4 P417-422 95T95G <ATS>
1966 V38 N4 P439-443 66-13472 <*>

CHEMIK
1954 V7 N10 P279-281 1645 <*>
1957 V10 P38-41 63-16236 <*> 0
1958 V11 P60-61 64-30968 <*>
8597 <K-H>
1959 V12 P238-240 65-10088 <*>
1962 V15 N5 P173-176 62-33047 <=*>
1963 V16 N1 P10-13 1242 <TC>
1964 V17 P164-165 66-11519 <*>
1965 V18 N6 P201-209 65-32907 <=*$>
1965 V18 N6 P226-227 65-32907 <=$*>
1965 V18 N10 P343-346 66-30070 <=$>

CHEMIKA CHRONIKA
SEE HEMIKA HRONIKA

CHEMIKERZEITUNG
1898 P243-244 63-16174 <*> P
1901 V25 P292- 58-1388 <*>
1905 N15 P195-198 AL-16 <*>
1907 V31 P67-69 63-14245 <*>
1908 V32 P633 63-14590 <*>
1911 V35 P1125-1126 63-14246 <*>
1915 V39 P397-398 3923F <K-H>
1915 V39 P406-407 3923F <K-H>
1915 V39 N3 P14-15 58-2641 <*>
1916 V40 P1013 3964-1 <K-H>
1918 V42 P221- 84H11G <ATS>
1921 V45 P309-313 64-18563 <*>
1921 V45 P335-337 64-18563 <*>
1921 V45 P360-364 64-18563 <*>
1921 V45 P411-412 64-18563 <*>
1923 V47 P528-529 58-1982 <*>
1923 V47 P616-617 57-1115 <*>
60-14576 <*> 0
1924 V48 P157-158 64-14294 <*>
1925 P467- 2935 <*>
1925 V49 P821 64-18308 <*>
1925 V49 N31 P229-230 60-10358 <*>
1926 N95 P761- AL-832 <*>
1926 N97 P781- AL-832 <*>
1926 V50 P165-167 64-14427 <*> 0
1926 V50 P202-204 64-14427 <*> 0
1926 V50 P239-241 64-14427 <*> 0
1926 V50 P246-248 64-14427 <*> 0
1928 V52 P5-6 II-118 <*>
1929 V53 P1-32 AL-718 <*>
1929 V53 P737-739 59-20222 <*>
1929 V53 N73 P705-706 60-10355 <*>
1929 V53 N74 P717-718 3998-C <K-H>
1930 V54 P871-872 II-601 <*>
1930 V54 N27 P259-260 59-20246 <*>
1930 V54 N95 P914-915 59-20400 <*>
1931 V55 N21 P201-203 59-20399 <*>
1931 V55 N48 P462-463 4069-B <KH>
1931 V55 N87 P837- 57-2169 <*>
1932 V56 P889-890 II-673 <*>
1622 <*>
1932 V56 N47 P462- 5497-C <K-H>
1932 V56 N63 P623-624 5497-D <K-H>
1932 V56 N98 P991 2261-F <K-H>
1935 V59 P113-114 63-14247 <*>
1935 V59 P316-317 63-18250 <*>
1935 V59 N48 P485-488 52113 <*>
1937 V61 N39 P408-409 63-16750 <*>
1939 V63 P70- 57-1302 <*>
1939 V63 P117-121 II-577 <*>
1939 V63 P737-740 63-14248 <*> 0
1939 V63 P752-754 63-14248 <*> 0
1940 N59/0 P282-285 58-2044 <*>
1940 V64 P420-423 59-10648 <*>
1940 V64 P498 63-10991 <*>
1941 V65 N59 P276 63-16139 <*>
1941 V65 N57/8 P264-267 63-16139 <*>
1941 V65 N87/8 P414 64-18889 <*>
1942 V66 N19 P196-199 63-14557 <*>
1943 N5/6 P49-52 57-142 <*>
1944 V68 N6 P106-107 63-16776 <*> 0
1950 V74 P649-651 AL-274 <*>
1950 V74 N20 P256-257 59-15709 <*>
1950 V74 N27 P353-355 64-20817 <*> 0
1950 V74 N40 P606-607 33J19G <ATS>
1950 V74 N50 P745-749 59-15718 <*> 0
1952 N11/2 P252-256 I-696 <*>
1952 V76 P471-475 57-1301 <*>
1952 V76 N28 P811-814 64-20364 <*> 0
1952 V76 N29 P841-844 64-20364 <*> 0
1952 V76 N11/2 P248-251 66-10622 <*>
1954 V78 P285-287 I-754 <*>
1954 V78 P834-838 62-20070 <*>
1954 V78 N2 P41-42 59-15552 <*>
1955 V79 P335-337 3763-A <K-H>
1955 V79 N10 P335-337 64-16136 <*>
1956 V80 N18 P624-627 57-1206 <*>
1958 V82 N14 P490-493 60-18386 <*>
1958 V82 N18 P651-656 59-10396 <*>

CHEMIKERZEITUNG-CHEMISCHE APPARATUR
1959 V83 P645-650 712 <TC>
1959 V83 N3 P76-81 63-16148 <*> 0
1959 V83 N12 P399-406 59-20697 <*>
1959 V83 N14 P478-481 62-10010 <*>
1959 V83 N24 P819-823 NP-TR-470 <*>
1960 V84 N8 P259-261 60-16638 <*> 0
1960 V84 N11 P365-366 12351-B <K-H>
1960 V84 N16 P539-560 62-10675 <*> 0
1960 V84 N16 P539-540 63-10290 <*> 0
1961 V85 P535-538 SC-T-64-615 <*>
1961 V85 N10 P327-333 62-10242 <*>
1961 V85 N17 P635-639 62-14835 <*>
1961 V85 N18 P672-682 15P62G <ATS>
1962 V86 P76-82 63-20925 <*> 0
1962 V86 N8 P263-270 63-16245 <*>
1963 V87 P315-318 9026 <IICH>
1964 V88 P37-43 4605 <BISI>
1964 V88 N5 P155-159 64-18761 <*>

1964 V89 P553-554 1406 <TC>
1965 V89 P265-274 1674 <TC>
1965 V89 N4 P107-112 918 <TC>
1965 V89 N7 P212-217 66-11726 <*> O

CHEMINS DE FER
1957 N206 P129-135 59-11222 <=> O

CHEMISCH WEEKBLAD
1903 V1 P7-12 62-10400 <*>
1908 V5 N7 P93-95 63-20171 <*>
1908 V5 N7 P96-101 63-20499 <*>
1917 P34-44 2249 <*>
1917 V14 P544-547 59-18131 <+*>
1919 V16 P510-526 58-2502 <*>
1927 V24 N9 P1C2-105 59-15224 <*>
1931 V28 P82-86 64-20439 <*>
1931 V28 P238-259 86L31DU <ATS>
1931 V28 P288-289 64-20438 <*>
1931 V28 N5 P82-86 25L31DU <ATS>
1934 V31 P490-492 65-12423 <*>
1934 V31 P497-504 I-516 <*>
1934 V31 P558-561 87L31DU <ATS>
1936 V33 P230-233 58-2158 <*>
1936 V33 P358-362 NP-TR-516 <*>
1937 V34 P221-222 57-3188 <*>
1938 V35 P868-872 II-559 <*>
1941 V38 P42-43 62-10089 <*>
1941 V38 P1C6-113 58-1043 <*>
1946 V42 P134-135 I-788 <*>
1946 V43 P713-714 I-666 <*>
1948 V44 P437-445 AL-402 <*>
1949 V45 P57-1208 57-1208 <*>
1950 V46 P301-302 II-622 <*>
1950 V46 P358-362 AL-523 <*>
1950 V46 P373-378 3303 <*>
1950 V46 P902-906 64-18307 <*>
1950 V46 N7 P282-287 65-13686 <*>
1950 V46 N5C P902-905 57-2681 <*>
1951 V47 P169-178 62-25793 <=*>
1951 V47 P281-287 I-733 <*>
1951 V47 P623-626 62-25794 <=*>
1951 V47 P845-848 T-1396 <INSD>
1951 V47 N27 P427-435 62-14907 <*>
1952 V48 P2C2-208 64-18927 <*>
1952 V48 P247-259 59-17135 <*> O
64-10792 <*> O
1952 V48 P356-361 64-20395 <*>
1952 V48 P821- I-757 <*>
1952 V48 N16 P237-242 64-18857 <*>
1952 V48 N16 P259-264 64-18885 <*> O
1952 V48 N16 P264-273 64-18886 <*> O
1952 V48 N38 P699-703 59-00791 <*>
1953 V49 P164-169 761 <TC>
1953 V49 P341-348 64-20369 <*>
1953 V49 P733-735 I-261 <*>
2980-A <K-H>
64-10438 <*>
1953 V49 N38 P717-721 85N51D4 <ATS>
1954 V50 P21-23 I-260 <*>
2980-B <K-H>
64-10436 <*>
1954 V50 P581-589 2638 <*>
1954 V50 P593-600 2638 <*>
1954 V50 N2 P21-23 T-2291 <INSD>
1955 P2-6 2742 <*>
1955 V51 P387-391 SC-T-64-1628 <*>
1955 V51 P519-527 57-1268 <*>
1955 V51 P547-548 T1558 <INSD>
1955 V51 P6C7-6C8 3969 <HB>
1955 V51 N10 P186-188 T2162 <INSD>
1955 V51 N12 P211-219 39P66DU <ATS>
1955 V51 N18 P319-320 CSIRO-3314 <CSIR>
1956 V52 P193-197 57-187 <*>
1956 V52 P265-267 760 <TC>
1956 V52 P481-490 64-16322 <*>
1956 V52 P526-532 58-1525 <*>
1957 V53 N23 P307-310 61-10468 <*>
1958 V54 P1-7 76N51D4 <ATS>
1958 V54 P277 AEC-TR-5261 <*>
1958 V54 P687-691 63-10733 <*> O
1958 V54 P721-726 63-10499 <*>
1958 V54 N21 P277 62-18700 <*>
1959 V55 P109-114 60-31312 <*=>
1959 V55 P571-579 62-18139 <*>
1959 V55 P661-665 17M44DU <ATS>
1959 V55 N10 P93-101 60-14562 <*>
65-10323 <*>
8939 <K-H>
1959 V55 N45 P632-633 60-18150 <*>
1960 V56 N11 P161-172 65-12248 <*>
1960 V56 N23 P339-349 C-4160 <NRCC>
1960 V56 N40 P547-548 22M47DU <ATS>
1961 V57 P377-382 63-20567 <*>
1961 V57 P501-505 UCRL TRANS-1000(L) <*>
1961 V57 P577-580 63-00125 <*>
1962 V58 P177-183 63-00198 <*>
1962 V58 P189-195 63-00198 <*>
1962 V58 P397-404 824 <TC>
1962 V58 P460-462 94Q70DU <ATS>
1964 V60 N26 P345-358 66-11142 <*>

CHEMISCHE APPARATUR
1931 V4 N33 P343- 63-10426 <*> O
1940 V27 P145-150 II-596 <*>
1940 V27 P161-165 II-596 <*>

CHEMISCHE BERICHTE
1947 V80 P413-417 AEC-TR-5854 <*>
1947 V80 P417-423 AEC-TR-5853 <*>
1947 V80 N5 P417-423 61-10008 <*>
1948 V81 P327-340 64-10498 <*> O
1948 V81 P380-381 8952 <IICH>
1948 V81 P527-531 3282 <*>
1949 V82 N6 P495-514 58-2690 <*>
1950 V83 P354-358 3456 I <K-H>
64-14709 <*>
1951 V84 P571-576 61-20512 <*>
1951 V84 N1 P4-12 64-18851 <*> P
1951 V84 N4 P376-381 63-18231 <*> O
1951 V84 N5/6 P427-433 63-18232 <*>
1952 V85 P267-278 I-2 <*>
57-275 <*>
1952 V85 P508-514 62-25790 <=*>
1952 V85 P593-605 652-0001 <*> O
1952 V85 P1989-1992 II-44 <*>
1952 V85 N1 P9-19 60-10371 <*> P
1952 V85 N4 P344-345 64-18923 <*>
1952 V85 N6 P641-647 AL-257 <*>
1952 V85 N9/10 P924-932 60-16872 <*>
1953 V86 P88-96 II-295 <*>
1953 V86 P96-109 II-402 <*>
1953 V86 P563-572 62-25791 <=*>
1953 V86 P700-710 75N51G <ATS>
1953 V86 P722-724 63-01408 <*>
1953 V86 P827-830 I-342 <*>
1953 V86 N5 P563-572 22H9G <ATS>
1953 V86 N12 P1514-1523 59-17349 <*> P
1954 V87 P35-37 63-01407 <*>
1954 V87 P38-45 63-01662 <*>
1954 V87 P640-645 II-162 <*>
1954 V87 P645-651 3C52 <*>
1954 V87 P690-691 90S81G <ATS>
1954 V87 P755-758 II-648 <*>
1954 V87 P1916-1922 63-14027 <*> O
1954 V87 N3 P289-300 65-12362 <*>
1954 V87 N5 P668-676 62-16760 <*> O
1954 V87 N5 P747-754 60-14670 <*>
1954 V87 N10 P1605-1616 64-14918 <*> O
1955 V88 P875-878 57-3353 <*>
1955 V88 P962-976 58-2254 <*>
1955 V88 P1C43-1048 2848 <*>
1955 V88 P1048-1053 57-1826 <*>
1955 V88 P1839-1846 58-2252 <*>
1955 V88 P1997-2002 66-14231 <*>
1955 V88 P2003-2011 3056 <*>
1955 V88 N1 P156-163 65-12244 <*>
1955 V88 N2 P294-301 65-14024 <*>
1955 V88 N3 P445-451 65-11823 <*>
1955 V88 N5 P6C1-914 C-3299 <NRCC>

Citation	Translation
1955 V88 N5 P601-614	64-14919 <*> O
1955 V88 N6 P742-763	3224 <*>
	64-18505 <*>
1955 V88 N7 P1043-1048	64-16128 <*>
1955 V88 N7 P1043-1061	65-10017 <*>
1955 V88 N7 P1048-1053	64-16148 <*>
1955 V88 N10 P1507-1510	64-16127 <*> O
1955 V88 N10 P1551-1555	60-10186 <*>
1955 V88 N11 P1771-1777	57-3415 <*>
1956 V89 P140-146	T2118 <INSD>
1956 V89 P434-	61-20291 <*>
1956 V89 P731-736	3445 <*>
1956 V89 P1496-1502	3260 <*>
1956 V89 P1768-1775	62-25792 <=*>
1956 V89 P2887-2896	191 <TC>
1956 V89 N2 P193-201	65-13578 <*>
1956 V89 N2 P263-270	60-16882 <*>
1956 V89 N2 P434-443	65-10908 <*> O
1956 V89 N2 P526-534	CSIRO-3340 <CSIR>
	57-1709 <*>
1956 V89 N3 P616-619	65-13577 <*> O
1956 V89 N5 P1263-1270	1656 <*>
1956 V89 N7 P1768-1775	61-10178 <*> O
1956 V89 N8 P1978-1988	59-10526 <*>
1956 V89 N9 P2013-2025	63-00613 <*>
1956 V89 N9 P2174-2185	CSIRO-3356 <CSIR>
1956 V89 N10 P2397-2400	59-15935 <*>
1957 V90 P151-153	62-25792 <=*>
1957 V90 P438-443	57-3391 <*>
1957 V90 P1100-1107	60-13454 <=*> O
1957 V90 P1188-1201	78P58G <ATS>
1957 V90 P1673-1683	60-16905 <*>
1957 V90 P1790-1797	58-2384 <*>
1957 V90 P2819-2832	1628 <TC>
1957 V90 N1 P151-153	61-10147 <*> O
1957 V90 N1 P711-723	10693-E <K-H>
1957 V90 N2 P171-181	62-10387 <*> O
1957 V90 N3 P382-386	65-14037 <*> P
1957 V90 N3 P425-437	60-16907 <*>
1957 V90 N3 P446-448	61-20658 <*>
1957 V90 N4 P481-485	59-14708 <+*>
1957 V90 N7 P1337-1342	64-30123 <*>
1957 V90 N8 P1411-1418	58-1497 <*>
1957 V90 N9 P2058-2071	64-18149 <*>
1957 V90 N10 P2325-2338	64-20235 <*> O
1957 V90 N12 P2713-2719	64-20222 <*> O
1958 N91 P873-882	8881 <IICH>
1958 V91 P938-943	8917 <IICH>
1958 V91 P1146-1155	73N51G <ATS>
1958 V91 P1156-1161	74N51G <ATS>
1958 V91 P1165-1169	58-1758 <*>
1958 V91 P1266-1273	63-00248 <*>
1958 V91 P1681-1687	04L30G <ATS>
1958 V91 P1725-1731	03L30G <ATS>
1958 V91 P1801-1805	65-11301 <*>
1958 V91 P1841-1846	61-20656 <*> O
1958 V91 P1851-1854	61-20657 <*> O
1958 V91 P1950-1955	60-18051 <*>
1958 V91 P1955-1960	65-11302 <*>
1958 V91 P1981-1982	65-17203 <*>
1958 V91 P2143-2150	65-17378 <*>
1958 V91 P2157-2167	C-5367 <NRC>
1958 V91 P2528-2531	60-18078 <*>
1958 V91 P2670-2681	C-3251 <NRCC>
1958 V91 P2682-2692	C-3253 <NRCC>
1958 V91 N1 P122-129	59-17511 <*> P
1958 V91 N6 P1357-1358	64-30983 <*>
1958 V91 N10 P2117-2121	65-11445 <*>
1959 N11 P2716-2723	65-14490 <*>
1959 V92 P837-849	66L33G <ATS>
1959 V92 P1667-1671	63-00545 <*>
1959 V92 P2320-2332	9575 <K-H>
1959 V92 P2716-2723	12385-A <K-H>
1959 V92 P3050-3052	63-14225 <*>
1959 V92 N1 P10-17	60-18537 <*>
1959 V92 N1 P192-202	59-17008 <*>
1959 V92 N2 P370-378	59-17007 <*>
1959 V92 N2 P424-429	59-17006 <*>
1959 V92 N4 P780-791	65-10337 <*> O
1959 V92 N4 P837-843	59-17498 <*>
1959 V92 N4 P837-849	60-14070 <*> O
1959 V92 N4 P958-970	51L34G <ATS>
	60-14534 <*>
1959 V92 N4 P988-998	60-14069 <*> OP
1959 V92 N5 P1055-1062	64-30688 <*>
1959 V92 N5 P1088-1089	65-10920 <*>
1959 V92 N5 P1115-1117	96M46G <ATS>
1959 V92 N5 P1188-1195	62-10380 <*> O
1959 V92 N6 P1423-1427	65-10335 <*>
1959 V92 N7 P1587-1593	61-10302 <*> O
1959 V92 N7 P1594-1599	61-10303 <*> C
1959 V92 N7 P1624-1628	65-10923 <*>
1959 V92 N8 P1810-1813	61-20514 <*>
1959 V92 N9 P2099-2106	60-15728 <=*> O
	61-00903 <*>
1959 V92 N9 P2163-2171	61-10409 <*>
	64-30677 <*>
	65-11228 <*> O
1959 V92 N9 P2278-2293	65-10946 <*> O
1959 V92 N9 P2320-2332	65-11220 <*> O
1959 V92 N9 P2364-2371	63-14224 <*>
1959 V92 N12 P3064-3075	65-13877 <*>
1959 V92 N12 P3122-3150	61-10304 <*> O
1960 V93 P181-186	62Q67G <ATS>
1960 V93 P706-719	64-10160 <*>
1960 V93 P1078-1084	63-10570 <*>
1960 V93 P1220-1230	60-16869 <*>
1960 V93 P1632-1642	62-10423 <*>
1960 V93 P1956-1960	C-5335 <NRC>
1960 V93 N1 P88-94	62-16294 <*> P
1960 V93 N1 P116-127	62-16266 <*> O
1960 V93 N3 P689-693	61-20975 <*>
1960 V93 N3 P689-700	65-14052 <*>
1960 V93 N4 P765-774	60-16774 <*>
1960 V93 N4 P803-808	60-18173 <*>
1960 V93 N9 P2085-2086	65-13587 <*>
1960 V93 N9 P2087-2088	65-13550 <*>
1960 V93 N12 P2819-2823	62-16258 <*>
1960 V93 N12 P2891-2897	61-16429 <*>
1960 V93 N12 P2938-2950	11313 <K-H>
	65-13575 <*>
1960 V93 N12 P2951-2965	64-10482 <*>
1961 V94 P290-297	86P59G <ATS>
1961 V94 P1891-1898	66-13569 <*>
1961 V94 P2122-2125	62-16191 <*>
1961 V94 P2314-2327	C-3939 <NRCC>
1961 V94 P2937-2942	62-16317 <*> O
1961 V94 P2942-2950	62-16316 <*> O
1961 V94 P3309-3317	192 <TC>
1961 V94 P3317-3327	1054 <TC>
	66-13625 <*>
1961 V94 N1 P164-168	65-13199 <*>
1961 V94 N2 P398-406	65-13211 <*> O
1961 V94 N3 P642-650	64-16685 <*> O
1961 V94 N3 P761-765	65-13548 <*> O
1961 V94 N4 P1116-1121	65-13221 <*> O
1961 V94 N4 P1157-1158	65-13220 <*>
1961 V94 N6 P1403-1409	64-16983 <*>
1961 V94 N6 P1417-1425	65-13558 <*> O
1961 V94 N8 P2166-2173	62-10203 <*> O
	62-16556 <*>
1961 V94 N8 P2187-2193	65-12409 <*>
1961 V94 N8 P2314-2327	63-10387 <*> O
	65-11526 <*>
1961 V94 N9 P2416-2429	28P60G <ATS>
1961 V94 N9 P2416-2419	62-14298 <*>
1961 V94 N9 P2416-2429	65-12414 <*>
1962 V95 P389-393	798 <TC>
1962 V95 P394-402	799 <TC>
1962 V95 P881-888	AEC-TR-5302 <*>
1962 V95 P2764-2768	64-16684 <*>
1962 V95 N3 P676-691	64-18063 <*>
1962 V95 N3 P777-782	65-14006 <*> O
1962 V95 N3 P795-802	64-18094 <*>
1962 V95 N4 P1049-1051	63-18086 <*>
1962 V95 N5 P1155-1169	65-14029 <*>
1962 V95 N5 P1170-1178	65-14030 <*>
1962 V95 N5 P1179-1185	65-14031 <*>
1962 V95 N5 P1186-1196	65-14032 <*>
1962 V95 N11 P2755-2763	64-16984 <*>

1963 V96 P866-880 63-00918 <*>
1963 V96 P1036-1045 8685 <IICH>
1963 V96 N1 P179-183 63-16531 <*>
1963 V96 N2 P550-555 63-18853 <*>
1964 V97 P1286-1293 29R78G <ATS>
1964 V97 P1857-1862 8908 <IICH>
1964 V97 P1863-1869 336 <TC>
1964 V97 P3162-3172 754 <TC>
1964 V97 N10 P2903-2916 66-13358 <*>
1964 V97 N1C P2917-2925 66-12827 <*>
1964 V97 N10 P2926-2933 66-12824 <*> P
1964 V97 N12 P3517-3523 66-13590 <*>
1966 V99 P421-430 64T92G <ATS>
1966 V99 N4 P1149-1152 66-14367 <*>

CHEMISCHE FABRIK
1928 N9 P102-103 57-2411 <*>
1929 V2 P407-408 63-18551 <*>
1929 V2 P417-419 63-18551 <*>
1930 P342-345 59-20534 <*> 0
1930 P353-355 59-20534 <*> 0
1930 P445-446 60-10108 <*> 0
1930 P458-460 60-10108 <*> 0
1930 V3 P61-63 T-1389 <INSD>
1933 V6 P127-142 58-1920 <*>
1935 V8 N21/2 P193-196 57-2744 <*>
1936 V9 P217-220 57-1136 <*>
1937 V10 P13-17 62-14214 <*> 0
1937 V10 N35/6 P371-372 64-20887 <*> 0
1938 P10-20 1073 <*>
1938 P385-390 II-522 <*>
II-670 <*>
1938 V11 P293-299 II-767 <*>
1938 V11 P449-464 57-2618 <*>
58-1902 <*>
1939 P49- 57-2462 <*>
1939 V12 P358-359 II-913 <*>
1940 P384-387 II-11 <*>
1940 V13 P238-241 1908 <*>
1940 V13 N1 P3-9 64-18928 <*>
1940 V13 N6 P101-104 1074 <*>
1940 V13 N19 P344-349 63-16228 <*>

CHEMISCHE INDUSTRIE, BERLIN
1950 V2 P405-408 63-16114 <*>
1952 V4 N12 P1010-1012 63-16061 <*>
1953 V5 N2 P66-68 63-14241 <*>
1953 V5 N3 P125-131 63-14240 <*>
1953 V5 N5 P385-392 63-14242 <*>
1957 V9 P658-666 63-10436 <*>
1960 V12 P545-546 75N52G <ATS>
1961 V13 N2 P9-17 SUP 65-10111 <*>
1961 V13 N3 P97-100 61-18814 <*>
1961 V13 N3 P133-134 61-18816 <*>
1965 V17 P219-221 NS-412 <TTIS>

CHEMISCHE INDUSTRIE, DUSSELDORF
1956 V8 N3 P101-102 58-2274 <*>
1960 V12 N2 P47-52 65-10696 <*>

CHEMISCHE INDUSTRIE INTERNATIONAL
1958 V10 P507-508 58-2692 <*>

CHEMISCHE RUNDSCHAU
1959 V2 P185-189 TS 1662 <BISI>
1959 V12 N2 P185-189 SUP2 62-14520 <*>
1959 V12 N18 P491-493 64-10406 <*>
1964 V17 N3 P45-46 64-30133 <*> 0

CHEMISCHE RUNDSCHAU FUER MITTELEUROPA UND DEN BALKAN
1929 V6 N14 P83-84 63-20460 <*>

CHEMISCHE TECHNIK. 1942-1945
1942 V14 N5 P91-94 60-14658 <*> 0
1942 V15 N9 P95-99 64-20883 <*> P
1943 V16 P153-157 57-2924 <*>
1943 V16 P249-260 I-455 <*>

CHEMISCHE TECHNIK
1949 V1 P66-68 II-208 <*>
1949 V1 N4 P107-111 60-14172 <*>
1950 V2 N1 P27-29 63-20151 <*>
1950 V2 N2 P35-39 2326 <*>
1950 V2 N4 P116-118 66-10034 <*>
1950 V2 N5 P157-161 63-14822 <*>
1950 V2 N5 P181-183 60-16816 <*>
1950 V2 N6 P173-178 63-18549 <*> 0
1950 V2 N6 P179-181 1838 <*>
1951 V3 P13-22 61-18258 <*> 0
1951 V3 P163-168 57-1956 <*>
1951 V3 N1 P5-10 59-15249 <*>
1951 V3 N5 P151-153 2325 <*>
1952 V4 P491-495 I-349 <*>
1952 V4 N3 P124 66-10050 <*>
1952 V4 N4 P165-169 66-10051 <*>
1952 V4 N5 P193-199 66-10036 <*> 0
1952 V4 N6 P247-251 63-18679 <*>
1952 V4 N6 P264 66-10052 <*>
1952 V4 N6 P265 66-10053 <*>
1952 V4 N6 P265-266 66-10054 <*>
1952 V4 N6 P266 66-10055 <*>
1952 V4 N9 P391-393 17N54G <ATS>
66-10037 <*>
1952 V4 N10 P471-476 64-16758 <*>
1952 V4 N12 P582-583 66-10019 <*>
1953 P503-507 <INSD>
1953 V5 P359-363 II-57 <*>
1953 V5 P716-717 3140-C <KH>
1953 V5 N4 P187-192 59-17296 <*> 0P
1953 V5 N7 P359-363 64-10448 <*>
1953 V5 N7 P363-364 63-20429 <*>
1953 V5 N8 P442-444 3236-C <K-H>
1954 V6 P489-494 3277 <*>
1954 V6 P644-647 3278 <*>
1954 V6 N1 PS5 66-10028 <*>
1954 V6 N2 P70-77 65-10044 <*>
1954 V6 N8 P462 66-10020 <*>
1954 V6 N9 P489-494 80M45G <ATS>
1954 V6 N12 P639-644 60-10920 <*>
64-20478 <*>
1954 V6 N12 P649-658 3471-D <K-H>
1955 V7 P471-472 T-1731 <INSD>
1955 V7 P682-683 58-573 <*>
1955 V7 N9 P526-529 65-12440 <*>
1956 V8 N1 P36-42 59-10571 <*> 0
1956 V8 N2 P80 57-1700 <*>
1956 V8 N4 P214-217 62-15864 <=*>
1956 V8 N5 P274-279 62-00224 <*>
1956 V8 N6 P333-337 77J19G <ATS>
1956 V8 N6 P341-346 61-20516 <*>
1956 V8 N7 P395-401 59-15651 <*> 0P
1956 V8 N7 P430 66-10021 <*>
1956 V8 N12 P702-704 65-10031 <*>
1957 V9 P198-202 6485 <K-H>
1957 V9 P283-286 63-00635 <*>
1957 V9 N3 P139-150 62-14065 <*>
65-10121 <*>
77K20G <ATS>
1957 V9 N3 P151-155 59-20396 <*> C
1957 V9 N4 P198-202 64-20245 <*> 0
1957 V9 N4 P209-213 59-17464 <*>
1957 V9 N7 P419-422 64-30113 <*>
1957 V9 N7 P419-420 6912-A <K-H>
1957 V9 N9 P542-543 66-10023 <*>
1957 V9 N10 P584-594 58-2673 <*>
1958 V10 N2 P75-78 59-15822 <*> 0
1958 V10 N6 P351-353 59-17060 <*>
1959 V11 P260 AEC-TR-5386 <=*>
1959 V11 N1 P24-28 65-10114 <*>
1959 V11 N5 P246-250 61-10405 <*>
65-12441 <*>
1959 V11 N6 P304-308 66-10038 <*>
1959 V11 N8 P431-440 63-00491 <*>
1959 V11 N10 P546-553 63-20389 <*> 0
1960 V12 N4 P187-188 61-20745 <*>
1960 V12 N6 P327-331 62-18839 <*>
1960 V12 N7 P414-418 63-00867 <*>
1960 V12 N8 P478-482 71N50G <ATS>
1960 V12 N10 P575-580 23N50G <ATS>

1960 V12 N12 P706-714 18N54G <ATS>
1961 V13 N2 P99-104 60N53G <ATS>
1961 V13 N3 P132-139 65-13542 <*> 0
87S87G <ATS>
1961 V13 N4 P233-236 62-24550 <=*>
1961 V13 N5 P306-307 5576 <HB>
1961 V13 N6 P317-319 62-24549 <=*> P
1961 V13 N6 P364-365 62-24548 <=*>
1961 V13 N6 P365-366 5577 <HB>
1961 V13 N6 P376-377 62-24557 <=*>
1961 V13 N6 P377-379 62-24534 <=*>
1961 V13 N9 P522-529 62-24533 <=*>
1961 V13 N9 P560-563 62-24535 <=*>
1961 V13 N10 P587-590 65-13872 <*>
1961 V13 N10 P594-595 56P59G <ATS>
1961 V13 N10 P598-603 63-14383 <*>
1961 V13 N11 P662-665 1038 <TC>
65-14222 <*>
1961 V13 N7/8 P394-399 64-30040 <*>
1962 V14 N4 P194-196 65-13805 <*>
1962 V14 N4 P214-221 64P64G <ATS>
87Q67G <ATS>
1962 V14 N8 P469-471 63-10670 <*> 0
1962 V14 N9 P541-545 65-12077 <*> 0
1962 V14 N10 P596-599 65-12076 <*>
1963 V15 N1 P34-35 65-18020 <*>
1963 V15 N5 P294-296 64-71299 <=>
1963 V15 N8 P477-479 91Q73G <ATS>
1963 V15 N10 P589-591 65-10395 <*>
65-10465 <*>
1963 V15 N10 P595-600 65-12039 <*> 0
1963 V15 N12 P708-712 <ES>
65-12192 <*>
1964 V16 P98-104 182 <TC>
1964 V16 P560-561 513 <TC>
1964 V16 P745 4519 <BISI>
1964 V16 N4 P203-208 64R77G <ATS>
1964 V16 N5 P257-262 66-11329 <*>
1964 V16 N6 P346-353 TRANS-547 <MT>
1964 V16 N7 P392-394 64-41774 <=>
1964 V16 N9 P529-533 64-51761 <=$>
1964 V16 N10 P605-607 66-12928 <*>
1964 V16 N12 P757 65-12957 <*>
1965 N10 P631-634 65-33656 <=$>
1965 V17 P129-139 8654 <IICH>
1965 V17 P148-154 1675 <TC>
1965 V17 N3 P139-146 1249 <TC>
1965 V17 N3 P147-149 8655 <IICH>
1965 V17 N4 P205-210 1298 <TC>
1965 V17 N4 P213-217 47T92G <ATS>
1965 V17 N5 P296-297 66-13405 <*>
1965 V17 N8 P489-491 66-10076 <*>
1965 V17 N9 P518-524 66-12578 <*>
1965 V17 N10 P624-625 65-33669 <=*$>
1965 V17 N12 P49-52 66-14516 <*>
1965 V17 N12 P738-742 66-14691 <*>
1966 V17 N10 P588-594 66-35552 <=>
1966 V18 N3 P138-142 66-14339 <*>

CHEMISCHE UMSCHAU AUF DEM GEBIETE DER FETTE, OELE, WACHSE U. HARZE
1926 V33 N24 P285-291 60-10451 <*>

CHEMISCHE ZEITSCHRIFT
1908 V32 P1029-1031 64-18280 <*> 0

CHEMISCHES ZENTRALBLATT
1899 P42- 1962 <*>
1907 V78 N2 P971- 2811 <*>
1921 V92 N3 P1498 59-17990 <*>
1950 V121 P1843 60-10006 <*>

CHEMISCH-TECHNISCHE FABRIKANT
1935 V22 P456-457 62-16627 <*>

CHEMISTRY. FORMOSA
SEE HUA HSUEH. FORMOSA

CHEMISTRY. KYOTO
SEE KAGAKU. KYOTO

CHEMISTRY AND BIOLOGY. JAPAN
SEE KAGAKU TO SEIBUTSU

CHEMISTRY BULLETIN. CHINESE PEOPLES REPUBLIC
SEE HUA HSUEH TUNG PAO

CHEMISTRY AND CHEMICAL INDUSTRY. NORTH KOREA
SEE HWAHAK KWA HWAHAK KONGOP

CHEMISTRY AND CHEMICAL INDUSTRY. TOKYO
SEE KAGAKU TO KOGYO. TOKYO

CHEMISTRY OF HIGH POLYMERS. JAPAN
SEE KOBUNSHI KAGAKU

CHEMOTHERAPIA. BASEL, NEW YORK
1960 V1 N2 P66-80 07P60F <ATS>
1961 V3 N1 P25-34 62-14094 <*>
1962 V5 N2 P126-135 64-18723 <*>
1962 V5 N2 P136-143 64-18724 <*>
1963 V6 N1/2 P75-81 64-18674 <*>
1963 V7 N2 P77-84 64-18679 <*>
1963 V7 N6 P603-605 65-14421 <*>

CHEMOTHERAPIA. JAPAN
SEE KAGAKU RYOHO

CHEMOTHERAPY. JAPAN
SEE NIPPON KAGAKU RYOHOGAKKAI ZASSHI

CHEN SHIH CHIEN SHE
1959 N9 P1-2 60-41210 <=>
1959 N9 P4 60-41210 <=>
1959 N9 P12-13 60-41210 <=>
1959 N9 P24 60-41210 <=> P
1959 N9 P28 60-41210 <=>
1959 N9 P32 60-41210 <=>
1959 N9 P34-36 60-41210 <=> P
1959 N10 P1-2 60-41144 <=>
1959 N10 P5-21 60-41144 <=>
1959 N10 P25 60-41144 <=>
1959 N10 P28-31 60-41144 <=> P
1959 N10 P32-43 60-41144 <=>

CH'I-CH'E
1958 P18 59-19234 <+*>

CH'I-CH'E YU T'O-LA-CHI
1958 N12 P5-11 59-13398 <=>

CHI HSIANG HSUEH PAO
1955 V26 N3 P167-180 61-15264 <*+>
1955 V26 N3 P183-193 61-15265 <*+>
1955 V26 N3 P195-209 61-15266 <*+> 0
1955 V26 N1/2 P1-23 61-11041 <=>
1955 V26 N1/2 P35-64 59-22198 <+*> 0
1957 V28 N3 P234-246 61-23293 <*=> 0
1957 V28 N4 P264-278 61-23294 <=*>
1958 V29 N1 P57-62 AD-636 925 <=$>
1958 V29 N2 P73-82 AD-631 060 <=$>
1959 V30 N1 P1-3 61-21255 <=>
1959 V30 N1 P5-9 61-21255 <=>
1959 V30 N1 P11-27 61-21255 <=>
1959 V30 N1 P39-43 61-21255 <=>
1959 V30 N1 P99-113 61-21255 <=>
1959 V30 N2 P186-190 66-32735 <=$>
1959 V30 N3 P197-201 60-11451 <=>
1959 V30 N3 P202-205 60-11490 <=>
1959 V30 N3 P206-211 60-11524 <=>
1959 V30 N3 P212-217 60-11544 <=>
1959 V30 N3 P223-235 60-11449 <=>
1959 V30 N3 P231-235 60-11565 <=>
1959 V30 N3 P236-242 60-11469 <=>
1959 V30 N3 P243-250 60-11471 <=>
1959 V30 N3 P263-276 60-11630 <=>
1959 V30 N3 P286-290 60-11490 <=>
1959 V30 N4 P350-360 61-21127 <=>
1959 V30 N4 P405-413 63-01730 <*>
64-13570 <=*$>

1962 V32 N4 P308-321 AD-637 060 <=$>
1963 V33 N1 P78-96 66-32735 <=$>
1963 V33 N1 P97-108 66-30900 <=$>
1963 V33 N2 P131-144 AD-623 117 <=*>
1963 V33 N2 P145-151 63-31933 <=> 0
1963 V33 N2 P257-270 AD-623 110 <=*>
1963 V33 N2 P271-280 AD-611 032 <=$>
1963 V33 N3 P281-288 65-14340 <*> 0
1963 V33 N3 P290-296 AD-611 279 <=$>
1963 V33 N3 P375-381 AD-631 059 <=$>
1963 V33 N4 P426-434 AD-636 620 <=$>
1963 V33 N1/2 P51-63 AD-623 116 <=*>
1963 V33 N1/2 P153-162 AD-623 116 <=*>
1964 V34 N1 P87-93 AD-623 112 <=*>
1964 V34 N1 P94-102 AD-628 223 <=*$>
1964 V34 N2 P146-164 AD-636 918 <=$>
1964 V34 N2 P211-224 AD-637 414 <=$>
1964 V34 N2 P225-232 AD-631 118 <=$>
1964 V34 N2 P233-241 AD-631 117 <=$>
1964 V34 N3 P316-328 AD-632 407 <=$>
1964 V34 N3 P378-382 AD-631 116 <=$>
1964 V34 N4 P397-408 AD-632 406 <=$>
1965 V35 N1 P18-32 65-31293 <=$>
1965 V35 N3 P343-351 AD-640 029 <=$>
1965 V35 N4 P383-397 66-31363 <=$>
1965 V35 N4 P519-526 66-31292 <=$>

CHI HSIEH KUNG CH'ENG HSUEH PAO
1963 V11 N2 P36-56 AD-475 424 <=$>
1964 V12 N3 P94-96 AD-625 788 <=*$>

CHI-HSIEH KUNG YEH
1954 N33 P10-14 57-179 <*>
1963 N3 P3-10 64-51581 <=$*>

CHI HSIEH KUNG YEH CHOU-PAO
1960 N60 P8 60-41465 <=>

CHIANGSU HUA PAO
1959 N9 P15-16 60-21433 <=>

CHIAO-T'UNG UNIVERSITY: JOURNAL. CHINESE PEOPLES REPUBLIC
SEE CHIAO T'UNG TA HSUEH HSUEH PAO

CHIBA DAIGAKU KOGAKUBU KENKYU HOKOKU
1959 V10 N18 P22-25 60-16474 <*>

CHIEH-POU HSUEH-PAO
1965 V8 N3 P314-325 67-30552 <=$>

CHIEN CHU
1958 N18 P1-3 59-12978 <+*>
1958 N18 P5-6 59-12978 <+*>

CHIEN CHU HSUEH PAO
1962 N1 P35-36 62-33274 <=>
1962 N7 P10-18 63-21595 <=>
1962 N7 P25-27 63-21595 <=>

CHIEN CHU SHE CHI
1965 N4 P24-32 65-34108 <=>
1965 N8 P17-18,29 65-34108 <=>
1965 N9 P2-6 65-34108 <=>

CHIEN CHU TS'AI LIAO KUNG YEH
1959 N3 P13-14 60-00196 <*>
1959 N19 P3-4 60-11562 <=>
1959 N20 P3-8 60-11562 <=>
1959 N20 P15-20 60-11562 <=>
1959 N21 P5-16 60-11562 <=>
1959 N22 P3-7 60-11562 <=>
1959 N22 P14-17 60-11562 <=>
1959 N22 P19-21 60-11562 <=>
1959 N22 P24-25 60-11562 <=>
1963 N12 P35 AD-619 465 <=$>

CHIEN SHE YUEH K'AN
1957 N11 P6-9 60-31044 <=>

CHIFFRES
SEE REVUE FRANCAISE DE TRAITEMENT DE L'INFORMATION CHIFFRES

CHIGAKU ZASSHI
1951 V60 N1 P9-19 58-1651 <*>
1956 V65 N3 P15-20 58-1656 <*>

CHIH SHA YEN CHIU
1962 N4 ENTIRE ISSUE 63-31681 <=>

CHIH WU HSUEH PAO
1964 V12 N1 P100-104 65-63670 <=>
1964 V12 N4 P384 65-30944 <=$>

CHIH WU SHEN LI HSUEH T'UNG HSUN
1959 N4 P60-62 60-11403 <=>

CHIJIKI KANSOKUJO YOHO
1954 V7 N1 P49-54 2838 <*>
1957 V8 N1 P87-92 C-2662 <NRC>
58-1409 <*>

CHIKYU BUTSURI
1940 V4 N3 P161-169 64-00461 <*>

CHI-LIN TA HSUEH TZU JAN K'O HSUEH HSUEH PAO
1957 N2 P69-80 UCRL TRANS-487 <*>

CHIMIA
1948 V2 P56-59 57-1947 <*>
63-20180 <*> 0
1949 V3 N9 P209-213 1220 <*>
1950 V4 P233-235 57-1420 <*>
1951 V5 P49-60 AEC-TR-5022 <*>
1952 V6 P177-180 T2349 <INSD>
62-20081 <*>
1952 V6 N1 P3-13 62-16990 <*> 0
1954 V8 P109-122 3014 <*>
1954 V8 P145-156 3014 <*>
1954 V8 N5 P109-122 59-17323 <*>
1954 V8 N6 P145-156 59-17323 <*>
1955 V9 P49-55 57-2577 <*>
1955 V9 P119- 2120 <*>
1955 V9 P250-255 57-3364 <*>
1957 V11 N6 P141-172 63-14783 <*> 0
1961 V15 N1 P156-163 65-17458 <*> 0
1961 V15 N1 P186-192 C-5069 <NRC>
1962 V16 P226-231 168 <TC>
1963 V17 P145-157 64-14722 <*>
1964 V18 N12 P406-407 65-14676 <*>
1965 V19 P416-424 1569 <TC>
1965 V19 N3 P128-131 65-13250 <*>
1965 V19 N5 P322-324 73T91G <ATS>

CHIMICA. MILANO
1953 V8 P119-121 12595 <K-H>
1955 V11 P55-57 64-20550 <*>
1956 V12 P325-328 3436 <*>
1956 V12 P511-513 59-10332 <*>
6124-B <K-H>
1957 V33 P10-15 6124-C <K-H>
1957 V33 P326-335 58-1039 <*>
1959 V35 N5 P333-340 9929-B <K-H>
1959 V35 N10 P589-594 9929-C <K-H>
1960 V36 N3 P124-133 61-16088 <*>

CHIMICA E L'INDUSTRIA
1935 V17 P592-594 II-521 <*>
1936 V18 P468 1647 <*>
1936 V18 P511-513 II-646 <*>
1936 V18 N5 P232-235 AL-299 <*>
1936 V18 N8 P396-397 II-641 <*>
1937 V19 P574-578 1174 <*>
1937 V19 N3 P123-125 63-20163 <*> 0
1937 V19 N5 P252-254 I-365 <*>
1938 P737- II-578 <*>
1938 V20 P8-12 II-433 <*>
1938 V20 P8- 1628 <*>
1938 V20 N2 P81- II-508 <*>

1938 V20 N3 P133-136	63-14249 <*>
1939 P478-	II-652 <*>
1939 P572-	1145 <*>
1939 V21 N2 P65-	1156 <*>
1940 P743-746	58-2691 <*>
1940 V22 N10 P457-463	64-20882 <*> 0
1941 V23 P117-123	T1748 <INSD>
1944 V26 P97-101	<ATS>
1944 V26 P167-170	AL-146 <*>
1944 V26 N7/8 P97-101	64-20525 <*>
1945 V27 N1/2 P6-10	<ATS>
1947 V29 N1 P4-5	61-00173 <*>
1948 V30 N3 P73-78	64-20547 <*> 0
1949 V31 P147	AL-80 <*>
1949 V31 P233-241	AL-81 <*>
1951 V33 P71-76	AL-153 <*>
	57-3396 <*>
1951 V33 P129-132	2149E <K-H>
1951 V33 P193-	1104 <*>
1951 V33 P619-623	T-1423 <INSD>
1951 V33 P708-711	57-3395 <*>
1951 V33 N1 P17-24	64-18937 <*> 0
1951 V33 N2 P71-76	9196-A <K-H>
1951 V33 N5 P272-282	(CHEM)-271 <CTS>
1951 V33 N10 P613-619	1393 <TC>
	60-18059 <*>
1951 V33 N11 P685-694	I-75 <*>
1951 V33 N12 P782-789	64-10440 <*> 0
1952 V34 P4-6	63-18955 <*> 0
1952 V34 P266-268	AL-599 <*>
1952 V34 P391-403	CSIRO-3213 <CSIR>
1952 V34 P449-453	62-20049 <*>
1952 V34 N5 P265-266	I-171 <*>
1952 V34 N8 P449-453	64-20344 <*>
1952 V34 N9 P520-570	58-2087 <*>
1953 V35 P13-15	58-747 <*>
1953 V35 N1 P13-15	62-14941 <*>
1953 V35 N6 P397-402	I-86 <*>
1954 P560-	57-1699 <*>
1954 V36 P611-617	CSIRO-2809 <CSIR>
1954 V36 N2 P108-110	57-75 <*>
1954 V36 N9 P693-698	65-10179 <*>
1954 V36 N12 P883-889	66-11572 <*>
1955 V37 P183-189	57-1324 <*>
1955 V37 P273-275	T-1541 <INSD>
1955 V37 P881-882	T-1993 <INSD>
1955 V37 P888-900	61-17645 <=$>
1955 V37 P927-932	61-17351 <=$>
1955 V37 N2 P109-112	3596-A <K-H>
1955 V37 N2 P113-114	86M37I <ATS>
1955 V37 N3 P169-175	65-14677 <*>
1955 V37 N7 P541-545	64-18512 <*>
1955 V37 N11 P888-900	5388 <K-H>
	64-16774 <*>
	64-20459 <*> 0
1955 V37 N12 P927-932	59-10767 <*>
	86L28I <ATS>
1955 V37 N12 P960-965	6178 <KH>
1956 V38 P124-127	REPT. 96959 <RIS>
1956 V38 N1 P1-5	RCT V30 N1 P274 <RCT>
1956 V38 N2 P102-108	57T91I <ATS>
1956 V38 N2 P124-127	1502 <*>
	5335 B <K-H>
	64-16749 <*> 0
	64-20925 <*> 0
1956 V38 N4 P298-303	3425 <*>
	61-16720 <*>
	64-16750 <*> 0
1956 V38 N11 P932-937	65-10180 <*>
1957 V39 P19-20	REPT. 96964 <RIS>
1957 V39 P275-283	96958 <RIS> 0
1957 V39 P832-841	63-16473 <*> 0
1957 V39 N1 P17-18	12J18I <ATS>
1957 V39 N1 P19-20	62-14202 <*>
	64-18165 <*>
	64-20922 <*> 0
1957 V39 N2 P81-83	57-1995 <*>
1957 V39 N4 P275-283	34K20I <ATS>
	3447 <*>
	59-17763 <*>
	61-16713 <*>
	64-20923 <*> 0
	65-11594 <*>
1957 V39 N9 P733-743	C-2699 <NRC>
	43J19I <ATS>
	58-732 <*>
	61-10476 <*>
1957 V39 N9 P743-748	C-2697 <NRC>
	44J19I <ATS>
	58-735 <*>
	61-10473 <*>
	64-18519 <*>
1957 V39 N10 P821-824	61-10474 <*>
1957 V39 N10 P825-831	REPT. 96963 <RIS> 0
	35K20I <ATS>
	58-733 <*>
	61-10475 <*>
	64-20924 <*> 0
1957 V39 N11 P902-904	64-30120 <*>
	7064 <K-H>
1957 V39 N12 P993-1001	59-15512 <*>
	61-10477 <*>
	64-16596 <*>
1957 V39 N12 P1002-1012	59-15513 <*>
	61-10478 <*>
1957 V39 N12 P1032-1033	58-751 <*>
	63-16555 <*>
1958 V40 P445-450	58-2392 <*>
1958 V40 P470-475	59-15986 <*> P
1958 V40 P717-724	59-15642 <*>
1958 V40 N2 P97-103	59-20711 <*>
	61-10469 <*>
	61-16771 <*>
1958 V40 N2 P103-111	59-20710 <*> 0
	61-10472 <*>
1958 V40 N3 P183-188	59-15515 <+*>
	61-10470 <*>
1958 V40 N4 P267-273	61-10471 <*>
	61-16701 <*>
1958 V40 N4 P287-289	61-10700 <*>
	64-30943 <*>
	65-14053 <*>
	8254 <K-H>
1958 V40 N5 P362-371	58-1759 <*>
	58-2292 <*>
1958 V40 N6 P445-450	59-17209 <*>
	63-14693 <*>
	67K24I <ATS>
1958 V40 N6 P450-457	28K25I <ATS>
	59-15514 <*>
1958 V40 N7 P552-556	1K26I <ATS>
1958 V40 N7 P552-561	61-18715 <*> 0
1958 V40 N7 P556-560	29K25I <ATS>
1958 V40 N8 P628-633	59-15229 <*>
1958 V40 N8 P638-640	59-20219 <*>
1958 V40 N9 P717-724	14K27I <ATS>
	59-10989 <*>
	61-10581 <*>
1958 V40 N10 P813-815	59-10172 <*>
	59-15227 <*>
	65-11144 <*>
1958 V40 N10 P816-819	59-10170 <*>
	59-15226 <*>
1958 V40 N10 P822-826	66-12244 <*>
1958 V40 N11 P896-905	59-15120 <*>
	59-17265 <*>
	63K28I <ATS>
1958 V40 N11 P909-914	59-15978 <*>
1958 V40 N12 P1003-1007	59-10425 <*>
	8891 <K-H>
1959 V41 N1 P13-18	59-15033 <*>
	59-17387 <*>
1959 V41 N2 P116-122	59-15340 <*>
	59-20931 <*>
1959 V41 N2 P123-128	59-15341 <*>
	59-15628 <*>
	61-10459 <*>
1959 V41 N3 P189-194	59-15342 <*>
	65-11848 <*>
1959 V41 N4 P277-280	60-14493 <*> 0

Citation	Translation
1959 V41 N4 P281-286	59-17205 <*>
	59-20221 <*>
	59-20933 <*>
1959 V41 N4 P287-292	59-17206 <*>
	60-10597 <*>
	65-10187 <*>
1959 V41 N4 P310-318	60-16659 <*>
	65-14040 <*>
1959 V41 N5 P387-397	59-20081 <*>
	59-20935 <*>
1959 V41 N5 P398-404	59-17384 <*>
	59-20934 <*>
1959 V41 N5 P404-407	65-10938 <*>
1959 V41 N5 P408-413	62-01253 <*>
1959 V41 N5 P414-420	62-01363 <*>
1959 V41 N6 P515-518	60-10599 <*>
1959 V41 N6 P519-526	60-10598 <*>
	60-16908 <*>
	61-10464 <*>
1959 V41 N6 P526-533	60-10487 <*>
	64-30656 <*>
1959 V41 N6 P534-539	59-20850 <*>
	59-20932 <*>
1959 V41 N6 P539-547	62-01364 <*>
1959 V41 N7 P598-602	62-1362 <*>
1959 V41 N7 P647-652	59-20861 <*>
	60-14118 <*>
1959 V41 N8 P731-737	63-14191 <*> 0
1959 V41 N8 P737-740	60-14217 <*>
	61-16702 <*>
	65-10186 <*>
1959 V41 N8 P741-748	60-14209 <*>
	60-14216 <*>
	65-10185 <*>
	77L36I <ATS>
1959 V41 N8 P748-757	26L37I <ATS>
	60-14218 <*>
	61-16738 <*>
	65-10184 <*>
1959 V41 N8 P758-762	27L37I <ATS>
	60-14213 <*>
	64-30668 <*>
1959 V41 N8 P764-775	60-10901 <*>
	60-14212 <*>
	65-10182 <*>
	76L36I <ATS>
1959 V41 N9 P875-879	60-14239 <*>
1959 V41 N9 P879-881	CHEM-101 <CTS>
	16M38I <ATS>
	60-14215 <*>
1959 V41 N9 P882-886	61-10898 <*>
1959 V41 N10 P964-967	60-14125 <*>
	60-18097 <*>
1959 V41 N10 P968-974	60-14211 <*>
1959 V41 N10 P975-980	60-14560 <*>
	62-26056 <=$>
1959 V41 N10 P984-987	60-14210 <*>
1959 V41 N12 P1163-1169	CHEM-103 <CTS>
	65-10181 <*>
1959 V41 N12 P1170-1175	CHEM-102 <CTS>
1959 V41 N12 P1176-1180	CHEM-104 <CTS>
	60-18429 <*>
1959 V41 N12 P1180-1184	10995 <K-H>
	65-13225 <*> 0
1959 V41 N12 P1185-1188	2023 <BISI>
1960 V42 P854-858	63-10819 <*>
1960 V42 N1 P52	CHEM-109 <CTS>
1960 V42 N2 P125-132	CHEM-114 <CTS>
	60-16775 <*>
	65-11596 <*>
1960 V42 N2 P133-137	CHEM-115 <CTS>
	62-26057 <=$>
1960 V42 N2 P137-142	CHEM-118 <CTS>
	62-26058 <=$>
	65-13224 <*>
1960 V42 N3 P225-259	CHEM-120 <CTS>
1960 V42 N3 P243-248	46M41I <ATS>
1960 V42 N3 P248-251	62-16061 <*>
	79P61I <ATS>
1960 V42 N4 P337-347	59M42I <ATS>
1960 V42 N4 P348-352	CHEM-129 <CTS>
	61-16703 <*>
1960 V42 N5 P457-462	CHEM-133 <CTS>
	62-26055 <=$>
1960 V42 N5 P463-467	CHEM-134 <CTS>
	62-26054 <=$>
1960 V42 N5 P468-474	135 <CTS>
1960 V42 N5 P487	136 <CTS>
1960 V42 N6 P587-598	65-14054 <*>
1960 V42 N6 P588-598	63-14402 <*>
1960 V42 N7 P712-721	CHEM-148 <CTS>
1960 V42 N7 P724-727	11310 <K-H>
	62-10318 <*>
1960 V42 N7 P735-744	C-5095 <NRC>
1960 V42 N8 P843-848	CHEM-157 <CTS>
1960 V42 N9 P959-964	65-12888 <*>
1960 V42 N9 P967-974	(CHEM)-315 <CTS>
1960 V42 N9 P974-977	12005 <K-H>
	65-14028 <*>
1960 V42 N9 P978-981	(CHEM)-174 <CTS>
1960 V42 N10 P1091-1099	(CHEM)-178 <CTS>
	G-259 <RIS>
1960 V42 N11 P1207-1225	CHEM-179 <CTS>
	61-16729 <*>
	65-13020 <*>
1960 V42 N11 P1226-1230	CHEM-180 <CTS>
1960 V42 N11 P1231-1234	CHEM-181 <CTS>
	65-13208 <*>
1960 V42 N11 P1234-1237	CHEM-182 <CTS>
1960 V42 N12 P1349-1361	CHEM-190 <CTS>
1960 V42 N12 P1361-1363	CHEM-188 <CTS>
	84N49I <ATS>
1960 V42 N12 P1363-1364	85N49I <ATS>
1961 V43 P259-262	61-18689 <*>
1961 V43 P412-414	C-4199 <NRCC>
1961 V43 N1 P8-9	45P59I <ATS>
1961 V43 N2 P137-142	CHEM-204 <CTS>
	61-16534 <*>
	63-20056 <*>
1961 V43 N2 P142-146	65-12887 <*>
1961 V43 N2 P161-162	CHEM-205 <CTS>
1961 V43 N4 P365-368	61-18690 <*>
1961 V43 N4 P368-375	CHEM-206 <CTS>
1961 V43 N4 P376-381	CHEM-207 <CTS>
1961 V43 N4 P414	CHEM-218 <CTS>
1961 V43 N5 P509-512	CHEM-215 <CTS>
1961 V43 N5 P529	CHEM-216 <CTS>
1961 V43 N6 P625-630	CHEM-230 <CTS>
	65-10447 <*>
1961 V43 N6 P638-640	CHEM-233 <CTS>
	65-13030 <*>
	66-10015 <*>
1961 V43 N6 P641-643	(CHEM)-256 <CTS>
1961 V43 N6 P644-648	CHEM-234 <CTS>
1961 V43 N7 P735-740	CHEM-235 <CTS>
	04P60I <ATS>
	62-14311 <*>
1961 V43 N7 P741-743	CHEM-240 <CTS>
1961 V43 N7 P744-749	65-12448 <*>
1961 V43 N8 P861-866	65-12447 <*>
1961 V43 N8 P871-874	CHEM-236 <CTS>
	63-10800 <*>
1961 V43 N8 P875-880	CHEM-248 <CTS>
1961 V43 N8 P880-884	90N56I <ATS>
1961 V43 N8 P885-891	64-16572 <*> C
1961 V43 N9 P993-998	CHEM-249 <CTS>
1961 V43 N9 P999-1003	CHEM-252 <CTS>
1961 V43 N12 P1394-1397	(CHEM)-266 <CTS>
1962 V44 P262	(CHEM)-281 <CTS>
1962 V44 P851-855	AEC-TR-5716 <*>
1962 V44 P1350-1353	RCT V36 N3 P660 <RCT>
1962 V44 N1 P1-9	(CHEM)-272 <CTS>
1962 V44 N1 P10-17	(CHEM)-273 <CTS>
	62-18918 <*> 0
1962 V44 N1 P18-23	(CHEM)-274 <CTS>
1962 V44 N1 P32-33	(CHEM)-275 <CTS>
1962 V44 N2 P121-126	(CHEM)-270 <CTS>
1962 V44 N2 P131-135	65-12293 <*>
1962 V44 N3 P229-234	(CHEM)-278 <CTS>
1962 V44 N3 P235-240	(CHEM)-279 <CTS>

Citation	Translation
1962 V44 N3 P241-246	(CHEM)-280 <CTS>
1962 V44 N4 P352-359	(CHEM)-282 <CTS>
1962 V44 N4 P371-378	(CHEM)-288 <CTS>
	12245-B <K-H>
	62-18673 <*> 0
	62-18868 <*>
1962 V44 N4 P383-384	(CHEM)-283 <CTS>
	63-27099 <=$>
1962 V44 N5 P463-	(CHEM)-292 <CTS>
1962 V44 N5 P463-473	62-18672 <*>
	65-12290 <*>
1962 V44 N5 P474-482	(CHEM)-293 <CTS>
	63-26634 <=$>
1962 V44 N5 P489-493	(CHEM)-294 <CTS>
	65-12291 <*>
1962 V44 N5 P529	(CHEM)-295 <CTS>
1962 V44 N5 P532-533	(CHEM)-296 <CTS>
1962 V44 N6 P611-620	(CHEM)-297 <CTS>
	63-18281 <*>
	63-26635 <=$>
1962 V44 N6 P621-626	(CHEM)-298 <CTS>
	63-26633 <=$>
	65-11597 <*>
1962 V44 N6 P627-630	(CHEM)-299 <CTS>
	63-26632 <=$>
1962 V44 N6 P631-635	62-18817 <*>
1962 V44 N6 P636-641	65-11592 <*>
1962 V44 N7 P725-729	(CHEM)-302 <CTS>
1962 V44 N9 P990-996	(CHEM)-309 <CTS>
1962 V44 N9 P990-995	12427 <K-H>
1962 V44 N9 P990-996	65-14493 <*> 0
1962 V44 N10 P1091-1094	(CHEM)-311 <CTS>
1962 V44 N10 P1095-1100	(CHEM)-312 <CTS>
1962 V44 N10 P1114-1120	(CHEM)-313 <CTS>
	RCT V36 N2 P459 <RCT>
	RUB.CH.TECH. V36 N2 <ACS>
1962 V44 N11 P1203-1211	(CHEM)-317 <CTS>
1962 V44 N11 P1212-1216	(CHEM)-318 <CTS>
1962 V44 N11 P1217-1220	(CHEM)-319 <CTS>
1962 V44 N11 P1217-1219	12589 <K-H>
	65-14841 <*>
1962 V44 N11 P1220-1227	(CHEM)-320 <CTS>
1962 V44 N11 P1228-1236	(CHEM)-321 <CTS>
1962 V44 N12 P1344-1349	CHEM-325 <CTS>
1962 V44 N12 P1350-1353	CHEM-322 <CTS>
1962 V44 N12 P1362-1366	CHEM-323 <CTS>
1962 V44 N12 P1383-1389	CHEM-324 <CTS>
1963 V45 N2 P203-208	12952 <K-H>
	63-27097 <=$>
	65-14865 <*>
1963 V45 N3 P299-314	CHEM-334 <CTS>
1963 V45 N4 P401-404	CHEM-336 <CTS>
1963 V45 N4 P405-416	17Q74I <ATS>
1963 V45 N4 P448-450	CHEM-335 <CTS>
1963 V45 N5 P522-528	339 <CTS>
1963 V45 N5 P528-532	340 <CTS>
1963 V45 N6 P651-656	341 <CTS>
1963 V45 N6 P657-	RCT V38 N2 P343 <RCT> 0
1963 V45 N6 P657-659	342 <CTS>
1963 V45 N6 P665-673	64-18774 <*>
1963 V45 N6 P690-695	343 <CTS>
1963 V45 N7 P806-812	(CHEM)-347 <CTS>
1963 V45 N8 P923-926	(CHEM)-349 <CTS>
1963 V45 N8 P927-936	(CHEM)-350 <CTS>
1963 V45 N8 P944-948	64-16971 <*>
1963 V45 N9 P1069-1075	65-11764 <*>
1963 V45 N10 P1212-1215	64-16866 <*>
1963 V45 N12 P1478-1482	(CHEM)-370 <CTS>
	53S83I <ATS>
1964 V46 P151-	RCT V38 N3 P590 <RCT>
1964 V46 N1 P9-15	65-11435 <*> 0
	66-10046 <*>
1964 V46 N1 P16-20	66-10026 <*>
	66-14973 <*>
1964 V46 N1 P57	(CHEM)-361 <CTS>
1964 V46 N2 P151-155	(CHEM)-369 <CTS>
1964 V46 N2 P166-171	66-10039 <*> 0
1964 V46 N2 P173	(CHEM)-362 <CTS>
1964 V46 N3 P245-251	(CHEM)-365 <CTS>
	65-10178 <*>
1964 V46 N3 P252-257	65-12836 <*>
1964 V46 N3 P258-262	(CHEM)-364 <CTS>
1964 V46 N4 P363-370	711 <TC>
1964 V46 N4 P371-375	66-10043 <*>
1964 V46 N4 P376-381	(CHEM)-365 <CTS>
1964 V46 N4 P428-429	(CHEM)-366 <CTS>
1964 V46 N5 P546-548	(CHEM)-368 <CTS>
1964 V46 N7 P746-751	(CHEM)-372 <CTS>
1964 V46 N7 P761-767	CHEM-380 <CTS>
1964 V46 N8 P883-894	N65-16305 <=$>
1964 V46 N8 P894-901	CHEM-374 <CTS>
1964 V46 N8 P910-914	65-14674 <*>
1964 V46 N9 P1042-1048	1411 <TC>
1964 V46 N9 P1054-1063	65-13406 <*>
1964 V46 N10 P1131-1142	CHEM-378 <CTS>
1964 V46 N10 P1152-1157	704 <TC>
1964 V46 N10 P1158-1164	CHEM-379 <CTS>
1964 V46 N11 P1275-1279	49S84I <ATS>
1964 V46 N11 P1287-1296	CHEM-381 <CTS>
1964 V46 N11 P1306-1311	CHEM-382 <CTS>
	65-12064 <*>
1964 V46 N11 P1311-1316	65-17125 <*>
1964 V46 N11 P1338-1339	CHEM-383 <CTS>
1964 V46 N11 P1339-1342	CHEM-384 <CTS>
1964 V46 N12 P1429-1435	CHEM-386 <CTS>
	66-11326 <*>
	66-11923 <*>
1964 V46 N12 P1455-1457	CHEM-387 <CTS>
1964 V46 N12 P1464-1469	CHEM-388 <CTS>
1964 V46 N12 P1474-1483	CHEM-389 <CTS>
1964 V46 N12 P1525-1526	CHEM-390 <CTS>
1965 V47 P384-	RCT V39 N5 P1667 <RCT>
1965 V47 P1060-1063	66-11684 <*>
1965 V47 N1 P1-9	CHEM-398 <CTS>
1965 V47 N1 P14-19	399 <CTS>
	65-17016 <*> 0
1965 V47 N1 P30-48	(CHEM)-400 <CTS>
1965 V47 N1 P45-51	NS-338 <TTIS>
	65-13800 <*>
1965 V47 N1 P52-54	CHEM-401 <CTS>
1965 V47 N2 P136-147	50S84I <ATS>
1965 V47 N2 P148-155	65-13799 <*>
1965 V47 N3 P282-290	51S84I <ATS>
1965 V47 N4 P378-383	408 <CTS>
	890 <TC>
1965 V47 N4 P384-390	407 <CTS>
	51S88I <ATS>
1965 V47 N4 P468-472	406 <CTS>
1965 V47 N5 P500-505	03S88I <ATS>
	66-13261 <*>
1965 V47 N5 P524-525	409 <NLL>
	66-11559 <*>
1965 V47 N6 P581-585	410 <CTS>
1965 V47 N6 P585-587	411 <CTS>
1965 V47 N7 P716-721	412 <CTS>
1965 V47 N7 P722-731	413 <CTS>
1965 V47 N7 P736-743	66-12204 <*>
1965 V47 N8 P839-841	414 <CTS>
	66-11560 <*>
1965 V47 N8 P862-864	415 <CTS>
1965 V47 N9 P955-959	66-13279 <*>
1965 V47 N9 P960-965	(CHEM)-418 <CTS>
1965 V47 N10 P1064-1067	1913 <TC>
1965 V47 N11 P1176-1183	(CHEM)-420 <CTS>
1965 V47 N11 P1196-1200	76T91I <ATS>
1965 V47 N11 P1204-1207	(CHEM)-421 <CTS>
1965 V47 N12 P1298-1302	(CHEM)-422 <CTS>
1965 V47 N12 P1313-1321	(CHEM)-423 <CTS>
	66-13996 <*>
1965 V47 N12 P1322-1324	(CHEM)-424 <CTS>
1966 V48 P458-465	66-14546 <*>
1966 V48 N1 P1-8	31T93I <ATS>
1966 V48 N1 P9-17	66-13884 <*>
1966 V48 N2 P144	66-12937 <*>

CHIMICA NELL'INDUSTRIA NELL'AGRICOLTURA, NELLA BIOLOGIA E ALTRE SUE REALIZZAZONI APPLICAZIONI

Citation	Translation
1955 V31 P234-240	321 <TC>
1959 V35 N5 P333-340	64-14462 <*>
1959 V35 N10 P589-594	64-14463 <*>

Citation	Translation
1960 V36 N10 P506-509	64-14460 <*>

CHIMIE ANALYTIQUE

Citation	Translation
1942 V24 N4 P118-121	AEC-TR-2010 <*> 0
	AL-881 <*>
1942 V24 N4 P137-139	AEC-TR-2010 <*> 0
	AL-881 <*>
1946 V28 P32-	II-29 <*>
1948 V30 P1C9-113	1271G <K-H>
1948 V30 N3 P60	64-18871 <*>
1949 V31 P201-203	52E1F <ATS>
1950 V32 P7-13	II-804 <*>
1950 V32 P35-41	II-804 <*>
1950 V32 P109-114	II-804 <*>
1950 V32 P156-161	II-804 <*>
1950 V32 P185-189	II-804 <*>
1950 V33 P48-5C	63-16823 <*>
1952 V34 P189-192	63-18969 <*> 0
1954 V36 P182-186	57-1325 <*>
1954 V36 N6 P145-152	61-20769 <*>
1954 V36 N7 P182-186	65-10368 <*>
1954 V36 N11 P294-301	59-15290 <*>
	63-14879 <*> 0
1955 V37 P31-38	57-1701 <*>
1957 V39 P63-66	11J18F <ATS>
1957 V39 N11 P413-417	64-30094 <*>
1958 V40 P72-76	2458 <BISI>
1958 V40 P120-126	45K26F <ATS>
1958 V40 N3 P80-86	64-30626 <*> 0
1958 V40 N4 P120-126	64-16326 <*>
1959 V41 P56-62	4134 <BISI>
1959 V41 N6 P225-239	1794 <BISI>
1959 V41 N9 P351-358	41N50F <ATS>
1960 V42 P61-68	2435 <BISI>
1960 V42 P236-244	2073 <BISI>
1960 V42 P329-335	4135 <BISI>
1960 V42 N6 P287-299	65-12450 <*>
1960 V42 N7 P336-345	65-14181 <*> 0
1960 V42 N7/8 P346-354	64-16864 <*> 0
1960 V42 N7/8 P381-387	64-16864 <*> 0
1961 V43 P391-397	4136 <BISI>
1962 V44 P208-213	3510 <BISI>
1962 V44 P3C2-308	3227 <BISI>
1963 V45 N2 P53-59	66-10344 <*>
1964 V46 N9 P457-460	65-13985 <*>
1965 V47 N1 P10-16	66-10276 <*>
1965 V47 N3 P120-124	4373 <BISI>
1965 V47 N1C P502-511	66-11953 <*>

CHIMIE ET INDUSTRIE

Citation	Translation
1921 V5 N3 P239-256	59-10668 <*> 0
1921 V5 N4 P398-408	59-10668 <*> 0
1921 V5 N5 P518-528	59-10668 <*> 0
1922 V8 P296-304	64-14246 <*> 0
1924 P406-4C7 SP NO	64-18302 <*>
1928 V20 P414-428	II-439 <*>
1929 P453-457	<ES>
1929 V21 P7C1-707	II-616 <*>
1929 V21 N2 P243-251	43K25F <ATS>
1930 245-246 SPEC NO	60-10290 <*> 0
1930 293-294 SPEC NO	60-10291 <*>
1930 445-451 SPEC NO	60-10292 <*>
1930 V24 N1 P3-19	AL-209 <*>
1930 V24 N2 P271-279	63-10466 <*>
1930 V24 N3 P526-545	63-10461 <*>
1931 P317-324	59-10748 <*>
1931 V25 P26-32	T-2308 <INSD>
1931 V25 P556-569	1630 <*>
1931 V25 P1C47-1057	58-2248 <*>
1931 V25 N3 P543-555	65-10596 <*> 0
1932 P329-	57-196 <*>
1932 V27 N1 P3T-	1072 <*>
1934 V31 N5 P1C31-1039	60-10315 <*>
	63-01024 <*>
1934 V32 N3 P517-527	II-643 <*>
1935 V34 P255-	II-452 <*>
1935 V34 P373-	2430 <*>
1935 V34 N3 P526-529	58-2217 <*>
1936 V36 N5 P879-887	3397 <*>
1940 V43 P283-286	63-18218 <*>
1941 V45 P417-419	64-18160 <*> 0
1943 V50 N1 P16-20	64-16044 <*> 0
1945 V53 P27-32	1396A <K-H>
1946 V55 P180-191	62-18891 <*> 0
1946 V56 N5 P373-381	CSIRO-2882 <CSIR>
1946 V56 N6 P449-455	II-914 <*>
1947 V57 N2 P117-125	65-10400 <*> 0
1947 V57 N6 P540-544	57-873 <*>
1947 V58 P545-547	II-567 <*>
1947 V58 N5 P433-	1641 <*>
1947 V58 N6 P545-547	57-1300 <*>
1948 V59 P548-551	1786 <*>
1948 V59 N3 P240-246	64-18201 <*> 0
1948 V60 P22-35	II-588 <*>
1948 V60 N1 P22-35	66-10241 <*> 0
1949 V61 N2 P144-145	03360 <*>
	3360 <*>
1949 V62 N2 P135-142	64-20437 <*> 0
1949 V62 N3 P243-248	60-18384 <*> 0
1949 V62 N4 P362-370	60-18384 <*> 0
1950 V63 N3 P246-251	1627 <*>
1950 V63 N3 P396-397	62-14864 <*>
1951 V66 P3-11	<ATS>
1952 V67 N2 P335-338	64-18222 <*> 0
1952 V67 N6 P909-919	12083-A <K-H>
	63-16880 <*>
1952 V68 P55-6C	12083-B <K-H>
	63-16879 <*>
1952 V68 N2 P191-195	66-10201 <*>
1952 V68 N3 P333-350	AL-591 <*>
1953 V69 N4 P653-657	60-16852 <*>
1953 V70 P1081-1085	T-1400 <INSD>
1953 V70 N1 P64	12083-C <K-H>
	63-16878 <*>
1953 V70 N2 P213-215	1224 <*>
1953 V70 N3 P383-396	1956 <*>
1954 V72 P458	57-2572 <*>
1955 V73 N3 P531-540	33Q69F <ATS>
	60-23449 <=*>
1955 V73 N3 P541-	57-1703 <*>
1955 V73 N6 P1149-1158	65-10022 <*> 0
1956 V75 N6 SUP P29-40	61-16184 <*>
1956 V76 P754-757	57-822 <*>
1957 V77 N5 P1C09-1031	64-18085 <*> 0
1957 V77 N6 P1288-1291	60-19172 <=*>
1957 V78 N3 P206-213	61-18404 <*>
1958 V79 N1 P3-10	61-10467 <*>
1958 V79 N1 P11-14	64-18800 <*>
1958 V79 N5 P563-578	65-11474 <*> P
1958 V80 N3 P220-233	62-16017 <*> 0
1958 V80 N3 P248-260	65-11300 <*>
1958 V80 N4 P429-436	62-16732 <*> P
1959 V81 P895-901	65-11232 <*>
	9672 <K-H>
1959 V82 N3 P309-328	62-14861 <*>
1959 V82 N4 P513-516	65-10171 <*>
1959 V82 N5 P663-675	10 731 <K-H>
	61-20448 <*>
1959 V82 N6 P878-880	62-25256 <=*>
1960 V83 P541-548	61-20087 <*> 0
1960 V83 N2 P223-231	62-14859 <*>
1960 V83 N4 P557-563	12693-C <K-H>
	65-14842 <*>
1961 V85 N5 P733-739	61-20674 <*>
1962 V87 P43-62	65-13605 <*> 0
1962 V87 N1 P43-62	62-18059 <*> 0
1962 V87 N2 P231-239	3112 <BISI>
1962 V87 N3 P371-387	63-27410 <$>
1962 V88 N3 P233-238	64-16011 <*>
1963 V90 N3 P145-146	64-16386 <*>
	66-10017 <*>
1963 V90 N4 P358-369	64-30695 <*>
1963 V90 N5 P430-433	64-16338 <*>
1963 V90 N6 P636-638	65-11339 <*>
1964 V91 N1 P47-56	64-18431 <*>
1964 V91 N5 P519-528	65-11734 <*>
1965 V93 N2 P125	66-12809 <*>

CHIMIE ET INDUSTRIE. SPECIAL ISSUE

Citation	Translation
1928 P537-540 SP NO	66-12582 <*>

1936 P702-710 1625 <*>
1939 V41 N4 P260-270 57-157 <*>

CHIMIE MODERNE
1959 V4 N27 P91-93 64-18482 <*>

CHIMIE DES PEINTURES
1949 V12 N6 P184-190 1441 <TC>
1951 V14 P230-234 64-20517 <*>
1953 V16 P69-74 63-14192 <*>
1957 V20 P105-111 65-10107 <*> O

CHIN JIH HSIN WEN
1963 07/01 P7-8 63-31766 <=>

CHIN SHU HSUEH PAO
1958 N2 P98-109 TS 1657 <BISI>
1958 V3 N2 P85-98 1657 <BISI> P
62-14674 <=*> P
62-25879 <=*>
1958 V3 N2 P138-180 TS 1624 <BISI>
62-25699 <=*>
1959 N1 P69-74 TS-1397 <BISI>
1959 V4 P195-205 2015 <BISI>
1959 V4 P206-216 2016 <BISI>
1959 V4 N1 P16-21 TS-1445 <BISI>
62-25334 <=*>
1959 V4 N1 P69-74 62-24234 <=*>
1966 V9 N2 P176-180 67-30405 <=>

CHINA'S FORESTRY
SEE CHUNG KUO LIN YEH

CHINA SCIENCE
SEE CHUNG-KUC K'O HSUEH

CHINA YOUTH MAGAZINE
SEE CHUNG KUO CHING NIEN PAO

CHINESE AGRICULTURAL BULLETIN
SEE CHUNG KUO NUNG PAO

CHINESE AGRICULTURAL SCIENCE
SEE CHUNG KUO NUNG YEH K'O HSUEH

CHINESE AUTOMCTIVE INDUSTRY
SEE CHIANGSU HUA PAO

CHINESE FORESTRY SCIENCE
SEE LIN YEH KOI HSUEH

CHINESE JOURNAL OF CIVIL ENGINEERING
SEE TU MU KUNG CHENG HSUEH PAO

CHINESE JOURNAL OF INTERNAL MEDICINE
SEE CHUNG HUA NEI K'O TSA CHIH

CHINESE JOURNAL OF MECHANICAL ENGINEERING
SEE CHI HSIEH KUNG CH'ENG HSUEH PAO

CHINESE JOURNAL OF MECHANICS
SEE LI HSUEH HSUEH PAO

CHINESE JOURNAL OF NEURCLOGY AND PSYCHIATRY
SEE CHUNG HUA SHEN-CHING CHING-SHEN K'O TSA CHIH

CHINESE JOURNAL OF OBSTETRICS AND GYNECOLOGY
SEE CHUNG HUA FU CH'AN K'O TSA CHIH

CHINESE JOURNAL OF OPTICS
SEE CHUNG HUA YEN K'O TSA CHIH

CHINESE JOURNAL OF PEDIATRICS
SEE CHUNG HUA ERH K'O TSA CHIH

CHINESE JOURNAL OF PATHOLOGY
SEE CHUNG HUA PING LI HSUEH TSA CHIH

CHINESE JOURNAL OF PHYSIOLOGY
SEE SHENG LI HSUEH PAO

CHINESE JOURNAL OF RADIOLOGY
SEE CHUNG HUA FANG SHE HSUEH TSA CHIH

CHINESE JOURNAL CF SURGERY
SEE CHUNG HUA WAI K'C TSA CHIH

CHINESE JOURNAL OF TUBERCULOSIS
SEE CHUNG HUA CHIEH HO PING K'O TSA CHIH

CHINESE LINGUISTICS
SEE CHUNG KUO YU WEN

CHINESE SCIENCE BULLETIN
SEE K'O HSUEH T'UNG PAO

CHINESE SCIENTIFIC NEWS
SEE CHUNG KUO K'O HSUEH HSIN WEN

CHING CHI YEN CHIU
1958 N3 P29-38 59-16431 <+*>
1966 N3 P29-36 66-32334 <=$>
1966 N3 P59-64 66-32334 <=$>
1966 N3 P74-77 66-32334 <=$>

CHIRURG. BERLIN
1953 V24 P218-223 II-970 <*>

CHIRURGIA ITALIANA
1961 V13 N5 P405-420 63-00538 <*>

CHIRURGIA DEGLI ORGANI DI MOVIMENTO
1937 V22 P479-486 58-236 <*>
1938 V23 P499-508 66-10868 <*>
1938 V23 N6 P499-524 66-12687 <*> OP
1951 V35 P651-661 57-1820 <*>
1960 V49 N3 P230-237 62-00865 <*>

CHIRURGIEN-DENTISTE FRANCAIS
1963 V23 N14 P695-701 63-20197 <*>

CHIRURGISCHE PRAXIS
1963 V7 P307-309 64-18961 <*>

CHIRYO
1954 V36 P1301-1304 64-18741 <*>
1962 V44 P2305-2308 64-00073 <*>
1963 V45 P1835-1842 66-11577 <*>
1963 V45 P1849-1856 66-13044 <*>
1963 V45 N2 P656-665 66-13192 <*>
1964 V46 P967-971 65-14701 <*>
1964 V46 N4 P745-752 66-11679 <*> C

CHISHITSU CHOSCJC GEPPO
1960 V11 N6 P1-20 IGR V5 N8 P999 <AGI>
1960 V11 N6 P33-48 IGR V6 N3 P459 <AGI>
IGR,V6,V3,P459-477 <AGI>
1960 V11 N11 P1-20 IGR V5 N10 P1243 <AGI>
1960 V11 N12 P26-29 IGR V5 N10 P1264 <AGI>

CHISTITSUGAKU ZASSHI
1950 V56 N660 P423-432 57-1843 <*>
1951 V56 N670 P331 58-1649 <*>
1951 V62 N670 P331 58-1650 <*>
1956 V62 N731 P415-43C 3465 <*>
1959 V65 N761 P71-79 IGR V4 N4 P468 <AGI>
1959 V65 N762 P165-17C IGR V4 N5 P590 <AGI>
1959 V65 N763 P222-226 IGR V4 N6 P719 <AGI>
1959 V65 N765 P343-348 IGR V4 N9 P1000 <AGI>
1959 V65 N766 P406-411 IGR V3 N11 P975 <AGI>
1959 V65 N770 P688-700 IGR V4 N9 P987 <AGI>
1962 V68 N801 P301-312 IGR V6 N12 P2150 <AGI>

CHOLLYOK
1959 N7 P47-51 60-11350 <*=>

CHOSON KWAHAKWAN TONGBO
1959 N5 P4-9 62-19453 <=*>
1959 N5 P59-61 62-19453 <=*>

CHRIST UND WELT
1966 P28 66-33922 <=$>

CHROME DUR
1959 P33-39 3332 <BISI>

CHROMOSOMA
1942 V2 P407-456 1611 <*>

CHRONIQUE DES MINES COLONIALES
1954 V22 N215 P114-117 57-663 <*>

CHRONIQUE DES MINES ET DE LA RECHERCHE MINIERE
1961 V23 N297 P9-14 63-16215 <*> O

CHU KUNG. PEKING
1959 N9 P1 66-11543 <=>
1959 N9 P10 66-11543 <=>
1959 N9 P27 66-11543 <=>
1959 N9 P31 66-11543 <=>
1959 N9 P36 66-11543 <=>
1959 N9 P42 66-11543 <=>
1959 N10 P1-4 66-11543 <=>
1959 N10 P7-8 66-11543 <=>
1959 N10 P11 66-11543 <=>
1959 N10 P20 66-11543 <=>
1959 N10 P28 66-11543 <=>
1959 N10 P50 65-11543 <=>
1959 N10 P53 65-11543 <=>
1959 N11 P2 65-11543 <=>
1959 N11 P18 65-11543 <=>
1959 N11 P21 65-11543 <=>
1959 N11 P24 65-11543 <=>
1959 N11 P30 65-11543 <=>
1959 N11 P33 65-11543 <=>
1959 N12 P6-8 65-11543 <=>
1959 N12 P23-24 65-11543 <=>
1959 N12 P29 65-11543 <=>
1959 N12 P48 65-11543 <=>

CHUGOKU AGRICULTURAL RESEARCH REPORT
SEE CHUGOKU NOGYO KENKYU

CHUGOKU NOGYO KENKYU
1964 N29 P47-50 66-13572 <*>
1964 N29 P51-52 66-12670 <*>

CHUNG CHI I KAN
1959 N10 P16-18 60-31322 <=>

CHUNG HUA CHIEH HO PING K'O TSA CHIH
1959 N2 P92-98 63-16608 <=*>
9396-B <KH>

CHUNG HUA ERH K'O TSA CHIH
1959 V10 N5 P408-411 60-41290 <=>
1959 V10 N5 P431-432 60-41291 <=>
1960 V11 N2 P140-141 60-41147 <=>

CHUNG HUA FANG SHE HSUEH TSA CHIH
1958 N6 P401-402 59-11532 <=> P
1958 N6 P444-445 59-11533 <=> P
1959 V7 N5 P317-326 60-11322 <=>
1959 V7 N5 P327-333 60-11321 <=>
1959 V7 N5 P379-383 60-11312 <=>

CHUNG HUA I HSUEH TSA CHIH
1956 V42 N7 P680-698 59-11456 <=>
1958 V44 N4 P343-348 59-11552 <=> O
1960 V46 N1 P8-14 61-31654 <=>
1960 V46 N1 P15-25 62-13548 <=*>
1960 V46 N1 P25-32 62-11079 <=>
1960 V46 N1 P32-40 62-13548 <=*>
1960 V46 N1 P40-48 62-13616 <=>
1960 V46 N1 P48-55 62-11079 <=>
1960 V46 N1 P55-59 62-13369 <*=>
1960 V46 N1 P60-64 62-13616 <=>
1960 V46 N1 P69-72 62-13616 <=>
1960 V46 N1 P72-74 62-13015 <*=>
1960 V46 N1 P75-79 62-13220 <*=>
1960 V46 N1 P84 62-13015 <*=>

CHUNG HUA MIN KUO HSIAO ERH K'O I HSUEH HUI
1962 V3 P108-113 64-18713 <*>

CHUNG HUA NEI K'O TSA CHIH
1958 V6 N8 P790-795 64-10635 <=*$>
9681-A <K-H>
1959 V7 N9 P813-833 60-31761 <=> O
1959 V7 N11 P1013 60-11464 <=>
1959 V7 N11 P1028 60-11464 <=>
1959 V7 N11 P1044-1047 60-11464 <=>
1960 V8 N1 P7-10 62-13849 <*=>
1960 V8 N3 P203-209 61-23943 <=>
1960 V8 N3 P211-237 61-23943 <=>
1960 V8 N3 P237-279 61-27758 <=>
1960 V8 N3 P294 61-27758 <=>
1960 V8 N3 P296 61-27758 <=>
1960 V8 N5 P414-418 61-11946 <=>
1960 V8 N5 P434-440 61-11946 <=>
1960 V8 N5 P454-456 61-11946 <=>
1960 V8 N5 P486-488 61-11946 <=>

CHUNG HUA PING LI HSUEH TSA CHIH
1959 V5 N3 P127-134 61-11061 <=>
1959 V5 N3 P136-138 61-11061 <=>
1959 V5 N3 P169-174 61-11061 <=>

CHUNG HUA SHEN-CHING CHING-SHEN K'O TSA CHIH
1959 V5 N1 P19-62 61-27229 <=>
1959 V5 N1 P70-73 61-27229 <=>
1959 V5 N1 P80-84 61-27229 <=>

CHUNG HUA WAI K'O TSA CHIH
1959 V7 N9 P848-855 60-11800 <=>
1959 V7 N9 P894-899 60-11800 <=>
1959 V7 N9 P928-931 60-11800 <=>
1962 V10 N2 P79-85 62-33017 <=$>
1962 V10 N2 P114-117 62-33017 <=$>
1962 V10 N2 P126-132 62-33017 <=$>
1962 V10 N3 P188-190 62-32059 <=>
1962 V10 N10 P612-641 63-21592 <=>
1962 V10 N10 P644-650 63-21592 <=>
1962 V10 N10 P657-674 63-21592 <=>
1962 V10 N11 P634 63-21592 <=>
1962 V10 N11 P681-684 63-21592 <=>
1962 V10 N11 P687-706 63-21592 <=>
1962 V10 N11 P711-731 63-21592 <=>
1962 V10 N11 P733-738 63-21592 <=>
1962 V10 N12 P767 63-21847 <=>
1962 V10 N12 P769 63-21847 <=>
1962 V10 N12 P773-775 63-21847 <=>
1962 V10 N12 P777 63-21847 <=>
1962 V10 N12 P781 63-21847 <=>
1962 V10 N12 P790-792 63-21847 <=>
1962 V10 N12 P798-799 63-21847 <=>
1963 V11 N1 P6-9 63-31516 <=>
1963 V11 N1 P35-40 63-31516 <=>
1963 V11 N1 P45-49 63-31516 <=>
1963 V11 N1 P78-79 63-31516 <=>
1963 V11 N2 P87-88 63-31516 <=>
1963 V11 N2 P97 63-31516 <=>
1963 V11 N2 P102-116 63-31516 <=>
1963 V11 N2 P120-127 63-31516 <=>
1963 V11 N2 P132 63-31516 <=>
1963 V11 N2 P158 63-31516 <=>
1963 V11 N2 P166-168 63-31516 <=>
1963 V11 N5 P339-418 63-31806 <=>
1963 V11 N7 P514-515 63-31998 <=>
1963 V11 N7 P530-531 63-31998 <=>
1963 V11 N7 P536 63-31998 <=>
1963 V11 N7 P570 63-31998 <=>

CHUNG-KUO CH'ING KUNG-YEH
1958 N1 P13-14 59-13727 <=>
1958 N2 P2-5 59-13727 <=>
1958 N4 P4 59-13727 <=>
1958 N8 P3-8 59-13727 <=>
1958 N14 P2-8 60-11349 <*>
1958 N15 P3-7 60-11349 <*>

1958 N16 P2 60-41185 <=>
1958 N16 P4 60-41185 <=>
1958 N17 P4-13 60-11349 <*>
1958 N17 P4 60-41185 <=>
1958 N17 P25 60-41185 <=>
1958 N17 P28 60-41185 <=>
1958 N19 P3-5 59-13262 <=> 0
1958 N19 P10-16 59-13262 <=> 0
1958 N19 P22-26 59-13262 <=> 0
1958 N20 P2-5 59-13262 <=> 0
1958 N20 P16-17 59-13262 <=> 0
1958 N20 P20-21 59-13262 <=> 0
1958 N22 P7-11 59-11762 <=>
1958 N22 P13-14 59-11764 <=>
1958 N23 P12-14 59-11763 <=>
1958 N24 P23-24 59-11764 <=>
1958 N24 P27-29 59-11764 <=>
1959 V7 N10 P929-931 60-31129 <=>

CHUNG KUO CHING NIEN PAO
1962 P4 62-33318 <=>

CHUNG KUO HSIN WEN
1937 V50 P305-340 59-15896 <*> 0

CHUNG-KUO K'O HSUEH
1951 V2 N2 P191-192 57-1674 <*>
57-821 <*>

CHUNG-KUO K'O HSUEH YUAN YING YUNG HUA HSUEH YEN CHIU SO CHI K'AN
1964 N11 P9-22 AD-636 653 <=$>

CHUNG KUO LIN YEH
1958 N12 P1 59-11455 <=>
1958 N12 P9-10 59-11455 <=>
1958 N12 P21-22 59-11455 <=>
1958 N12 P27 59-11455 <=>

CHUNG KUO NUNG PAO
1958 N2 P1-3 PB 141 316T <=>
1958 N2 P6 PB 141 316T <=>
1958 N2 P33 PB 141 316T <=>
1964 N1 P9-12 64-51762 <=$>
1964 N1 P17-30 64-51762 <=$>
1964 N3 P35-37 64-51762 <=$>
1965 N3 P1-3 65-32402 <=$>
1965 N3 P28-30 65-32402 <=$>

CHUNG KUO NUNG YEH CHI
1965 P26 65-34031 <=$>
1965 P30-32 65-34031 <=$>

CHUNG KUO NUNG YEH K'O HSUEH
1962 N2 P15-20 64-51466 <=$>
1962 N11 P7-14 64-51463 <=$>
1963 N2 P26-30 63-31846 <=>
1963 N4 P45-48 63-31795 <=>
1963 N5 P9-15 65-30142 <=$>
1963 N5 P53 63-31800 <=>
1963 N5 P54 63-31800 <=>
1964 N4 P22-41 65-30648 <*>
1964 N5 P31-33 65-30506 <=$>
1964 N5 P48-50 65-30506 <=$>
1965 N6 P13-16 66-30171 <=$>
1965 N6 P29-30 66-30171 <=$>
1965 N7 P31-32 66-30171 <=$>

CHUNG KUO SHUI LI
1958 N4 P24-33 59-11742 <=>
1958 N5 P6-10 59-11497 <=>
1958 N6 P39-45 59-11497 <=>
1958 N7 P12-20 59-11742 <=>

CHUNG KUO TI CHIH
1962 N7 P1-4 63-31142 <=>

CHUNG KUO TI SSU CHI YEN CHIU
1958 V1 N2 P46-60 62-25969 <=*>

CHUNG KUO TSAO CHUAN
1959 N2 P17-38 66-30513 <=*$>
1959 N4 P1-17 60-11667 <=>
1959 N4 P18-60 60-11773 <=>
1964 N1 P75- TRANS-1913 <NLL>
1964 N3 P62- 2111 <BSRA>
1964 N4 P7-20 2186 <BSRA>

CHUNG KUO YU WEN
1962 N10 P439-458 63-21413 <=>

CIEL ET TERRE
1956 V72 N3/4 P101-110 T-1917 <INSD>

CIENCIA E INVESTIGACION
1950 V6 P540-549 65-11181 <*>
1965 V21 N3 P124-131 66-13944 <*>
1965 V21 N3 P132-133 66-12608 <*>

CIENCIAS. ANALES DE LA ASOCIACION ESPANOLA PARA EL PROGRESO DE LAS CIENCIAS. MADRID
1956 V21 N3 P405-411 SCL-T-423 <*>

CIENCIAS SOCIALES
1954 V3 N29 P220-227 57-233 <*>

CIMENT ET CERAMIQUE
1921 V26 P291 64-14419 <*>
1923 V28 P355-360 64-14327 <*>
1924 N8 P335 64-14420 <*>
1925 P174 64-18208 <*>
1925 P405 64-14417 <*>
1927 P82 64-18211 <*>
1928 V33 P429-430 64-18326 <*>
1931 V36 P149-151 64-18341 <*>

CIRCULAIRE D'INFORMATIONS TECHNIQUES. CENTRE DE DOCUMENTATION SIDERURGIQUE
1948 V5 N1 P45-48 5126 <HB>
1952 V9 N4 P534-542 60-18828 <*> 0
1953 V10 N3 P465-492 3463 <BISI>
1953 V10 N9 P1421-1429 2859 <BISI>
1955 N5 P1031-1045 3119 <*>
1955 V12 N1 P169-192 61-16211 <*>
1955 V12 N6 P1209-1213 61-16215 <*>
1955 V12 N12 P2413-2419 62-14545 <*> 0
1956 V13 N3 P549-554 3399 <BISI>
1956 V13 N7 P1415-1444 61-16210 <*>
1956 V13 N10 P1973-1982 61-16213 <*>
1956 V13 N12 P2425-2442 61-16216 <*>
1957 V14 N2 P289-295 59-10897 <*>
1957 V14 N8 P1641-1645 61-16212 <*>
1957 V14 N11 P2259-2266 61-16214 <*>
1958 V15 N4 P819-833 61-18724 <*>
1958 V15 N7 P1479-1486 2092 <BISI>
1959 V16 P2583-2587 2215 <BISI>
1959 V16 N1 P135-155 1288 <BISI>
1959 V16 N3 P651-678 2118 <BISI>
1959 V16 N4 P899-902 3060 <BISI>
1960 V17 P429-436 2328 <BISI>
1960 V17 P2327-2429 2560 <BISI>
1960 V17 N3 P673-683 2740 <BISI>
1960 V17 N3 P703-714 2880 <BISI>
1960 V17 N11 P2415-2425 2559 <BISI>
1961 V18 N3 P723-735 2507 <BISI>
1961 V18 N3 P757-778 2707 <BISI>
1961 V18 N9 P2033-2038 3218 <BISI>
1961 V18 N10 P2247-2262 3013 <BISI>
1961 V18 N11 P2415-2420 3040 <BISI>
1961 V18 N7/8 P1715-1724 2706 <BISI>
1961 V18 N7/8 P1777-1789 2645 <BISI>
1962 V19 P401-404 2847 <BISI>
1962 V19 P405-411 2848 <BISI>
1962 V19 N2 P361-366 3057 <BISI>
1962 V19 N2 P393-399 2846 <BISI>
1962 V19 N2 P413-429 2849 <BISI>
1962 V19 N2 P441-446 3058 <BISI>
1962 V19 N4 P917-946 3421 <BISI>
1962 V19 N5 P1259-1276 3276 <BISI>
1962 V19 N6 P1459-1476 64-16037 <*>

1962 V19 N12 P2677-2718 3187 <BISI>
1963 N5 P1203-1212 4475 <BISI>
1963 V5 P1165-1187 3826 <BISI>
1963 V20 P2223-2224 3561 <BISI>
1963 V20 P2231-2236 3562 <BISI>
1963 V20 N3 P743-748 3291 <BISI>
1963 V20 N3 P749-766 3292 <BISI>
1963 V20 N5 P1243-1246 3825 <BISI>
1963 V20 N6 P1509-1523 3394 <BISI>
1963 V20 N9 P1967-2004 3655 <BISI>
1964 V21 N1 P163-174 3759 <BISI>
1964 V21 N2 P497-499 3895 <BISI>
1964 V21 N2 P501-503 3896 <BISI>
1964 V21 N2 P505-515 3897 <BISI>
1964 V21 N2 P517-521 3898 <BISI>
1964 V21 N3 P797-800 6345 <HB>
1964 V21 N6 P1501-1508 3901-I <BISI>
1964 V21 N6 P1509-1526 3901-II <BISI>
1965 V22 N1 P175-204 4372 <BISI>
1965 V22 N4 P975-987 4408 <BISI>
1966 N3 P621-648 5006 <BISI>

CIRCULAIRE. INSTITUT TECHNIQUE DU BATIMENT ET DES TRAVAUX PUBLICS
1945 N8 ENTIRE NO. 61-20092 <*>
1946 N18 P7- 61-20475 <*>
1946 SD V16 ENTIRE ISSUE I-63 <*> P
1947 N38 ENTIRE ISSUE 61-20474 <*>

CITY CONSTRUCTION. CHINESE PEOPLES REPUBLIC
SEE CHEN SHIH CHIEN SHE

CIVIL ENGINEERING. JAPAN
SEE DOBOKU GOJUTSU

CIVILNA ODBRANA
1966 N3 P54-60 66-33789 <=$>

CIVILTA DELLE MACCHINE
1955 P25-31 57-682 <*>

CLINICAL ALL-ROUND. JAPAN
SEE SOGO RINSHO

CLINICA. BOLOGNA
1956 N16 P315-327 58-322 <*>

CLINICA CHIMICA ACTA
1956 V1 N2 P167-177 57-894 <*>
1956 V1 N3 P210-224 57-895 <*>
1959 V4 P127-133 60-13036 <=*>
1959 V4 N1 P62-67 61-00589 <*>
1960 V5 P423-430 62-00232 <*>
1962 V7 P322-333 63-00293 <*>

CLINICA CHIRURGICA
1930 V6 P318-321 57-1614 <*>

CLINICA Y LABORATORIO
1960 V69 P347-352 62-16038 <*>

CLINICA MEDICA ITALIANA
1940 V71 P355-369 64-16235 <*> O

CLINICA NUOVA
1947 V4 N3 P142- 2813 <*>
1948 V6 N6 P33-35 1609A <K-H>

CLINICA ODONTOIATRICA E GINECOLOGIA
1957 V12 N1 P10-18 58-632 <*>

CLINICA OTORINO-LARINGO-IATRICA
1957 V9 N4 P295-312 61-16130 <*>

CLINICA PEDIATRICA
1955 V37 N6 P447-456 62-00494 <*>
1956 V38 N6 P462-471 62-00432 <*>
1959 V41 P462-470 64-18700 <*>
1960 V42 N6 P405-426 62-00279 <*>
1962 V44 P291-295 66-12651 <*> O

CLINICA TERAPEUTICA
1955 V9 N2 P129-153 58-179 <*>
1961 V21 P636-644 63-10723 <*>
1962 V23 P355-374 64-18643 <*>
1963 V25 N4 P301-336 65-13959 <*>
1963 V27 P641-660 66-11625 <*>
1964 V29 P442-445 65-14950 <*>

CLINICAL ENDOCRINOLOGY. JAPAN
SEE HORUMON TO RINSHO

CLINICAL GYNECOLOGY AND OBSTETRICS. JAPAN
SEE RINSHO FUJINKA SANKA

CLINICAL NEUROLOGY. JAPAN
SEE RINSHO SHINKEIGAKU

CLINICAL PSYCHIATRY. JAPAN
SEE SEISHIN IGAKU

CLINICAL SURGERY. JAPAN
SEE RINSHO GEKA

COAL INDUSTRY. CHINESE PEOPLES REPUBLIC
SEE MEI-T'AN KUNG-YEH

COAL MINING TECHNIQUE. CHINESE PEOPLES REPUBLIC
SEE MEI K'UNG CHI SHU

COAL TAR. JAPAN
SEE KORU TARU

COEUR ET MEDECINE INTERNE
1962 V1 N4 P447-452 63-01618 <*>

COLLANA DI ATTUALITA SCIENTIFICHE DEI QUADERNI DI ANATOMIA PRATICA
1957 S12 N1/4 P336-349 65-00152 <*> O

COLLECTED PAPERS OF THE AERONAUTICAL RESEARCH INSTITUTE. UNIVERSITY OF TOKYO
SEE TOKYO DAIGAKU KOKU KENKYUSHO SHUHO

COLLECTED PAPERS ON GEOGRAPHY. CHINESE PEOPLES REPUBLIC
SEE TI-LI CHI K'AN

COLLECTION OF ARTICLES OF THE JAPANESE PRINTING RESEARCH ASSOCIATION
SEE NIHON INSASTSU GAKKAI RONBUNSHU

COLLECTION OF CZECHOSLOVAK CHEMICAL COMMUNICATIONS
1934 V6 P211-223 21F3F <ATS>
1936 V8 P377-389 II-654 <*>
1936 V10 P182-189 SCL-T-369 <*>
1939 V11 P592-613 64-30046 <*>
1953 V18 P798-803 2442 <*>
1953 V18 N5 P597-610 57-672 <*>
1953 V18 N6 P783-797 1510 <*>
1954 V19 P98-106 I-512 <*>
1954 V19 P234-237 36Q72R <ATS>
1954 V19 P444-456 RJ497 <ATS>
1954 V19 P679-683 64-14401 <*>
1954 V19 P684-699 II-467 <*>
57-2254 <*>
1954 V19 P700-711 II-468 <*>
1954 V19 P917-924 II-468 <*>
1954 V19 P1111-1122 RJ-498 <ATS>
1954 V19 P1171-1174 63-18159 <*>
1954 V19 P1258-1263 58-1809 <*>
1954 V19 N1 P16-23 65-12956 <*>
1954 V19 N2 P349-356 2618 <*>
1954 V19 N2 P393-400 R-905 <*>
RT-3962 <*> C
1954 V19 N3 P417-427 65-10495 <*> O
1954 V19 N6 P1111-1121 R-68 <*>
1955 V20 P116-123 64-30045 <*>
1955 V20 P162-169 T1706 <INSD>
1955 V20 P636-639 31M45R <ATS>

COMMISSARIAT A L'ENERGIE ATOMIQUE
SEE RAPPORT CEA. COMMISSARIAT A L'ENERGIE ATOMIQUE. FRANCE

COMMUNICATIONS DE LA FACULTE DES SCIENCES DE L'UNIVERSITE D'ANKARA

Citation	Translation
1949 SC V2 P89-109	57-88 <*>

COMMUNICATICNS FROM THE PHYSICAL LABORATORY AT THE UNIVERSITY OF LEIDEN

Citation	Translation
V15 N157 P3-10	3194 <*>

COMMUTATION ET ELECTRONIQUE

Citation	Translation
1962 N3 P20-35	64-14000 <*>
1962 N3 P36-43	63-20640 <*>
1962 N3 P53-62	63-18482 <*>
1962 V3 P5-19	63-01426 <*>
1965 V10 N2 P80-89	66-13430 <*>

COMPTE RENDU DE L'ACADEMIE BULGARE DES SCIENCES
SEE DOKLADY. BULGARSKA AKADEMIYA NA NAUKITE

COMPTE RENDU DE L'ASSOCIATION DES ANATOMISTES

Citation	Translation
1933 V28 P277-279	57-862 <*>
1954 P12-14	58-1801 <*>

COMPTE RENDU DU CONGRES DE CHIMIE INDUSTRIELLE
SEE CHIMIE ET INDUSTRIE. SPECIAL ISSUE

COMPTE RENDU HEBDOMADAIRE DES SEANCES DE L'ACADEMIE D'AGRICULTURE DE FRANCE

Citation	Translation
1933 V19 P34-44	II-666 <*>
1938 V24 P669-672	63-18118 <*>
	64-16705 <*>
1951 V37 P48-52	65-12571 <*>
1952 V39 P548-552	63-16054 <*>
1953 V39 P292-295	66-10202 <*>
1954 V40 P40-48	62-16637 <*>
1954 V40 N5 P181-194	65-12536 <*>
1955 V41 P515-520	65-12588 <*>
1962 V48 P40-44	AEC-TR-5455 <*>

COMPTE RENDU HEBDOMADAIRE DES SEANCES DE L'ACADEMIE DES SCIENCES. PARIS

Citation	Translation
1843 V17 P942-954	N65-23684 <=*$>
1850 V32 P289-292	64-71235 <=>
1851 V33 P576-579	64-11807 <=>
1852 V36 P276-278	64-11807 <=>
1857 V45 N21 P857-861	63-16149 <*>
1859 V48 P627	64-18306 <*>
1861 V53 P161-164	AL-193 <*>
1865 V60 P993	64-14170 <*>
1869 V73 P382-385	<DI>
1869 V73 P464-467	<DI>
1871 V73 P443-447	58-1973 <*>
1871 V73 P563-571	58-1976 <*>
1877 V84 P946-949	58-1977 <*>
1879 V88 P1361-1364	59-15525 <*>
1879 V89 P667-669	59-15560 <*>
1880 V90 P210-211	61-20764 <*>
1880 V91 P294-	57-2137 <*>
1880 V91 P576-	II-858 <*>
1881 V93 P133-136	<DI>
1881 V93 P340-341	320 <TC>
1881 V93 P613-619	66-14664 <*> 0
1884 V98 P920	64-14169 <*>
1885 V100 P1219-1220	63-14250 <*>
1885 V100 P1497-1499	59-20449 <*>
1887 V105 P702-707	63-20887 <*>
1887 V105 P1621-1624	63-20319 <*>
1888 V106 P1441-1443	63-20877 <*>
1888 V106 P1800-1803	<LSA>
	59-10950 <*>
	64-14629 <*>
1888 V107 P681-684	57-3268 <*>
1888 V107 P836-837	57-3267 <*>
1888 V107 P1143-1145	57-3269 <*>
1891 V113 P624-627	60-10869 <*>
1891 V113 P726-729	60-10868 <*>
1891 V113 P787-788	60-10867 <*>
1892 V114 P319-324	58-2262 <*>
1893 V115 P122-124	59-14131 <=*>
1893 V116 P641	65-10155 <*>
1893 V117 P732-734	65-10160 <*>
1895 V121 P206-208	58-2271 <*>
1896 V122 P362-363	62-18669 <*>
1896 V123 P135-137	63-14135 <*>
1896 V123 P523-530	66-14661 <*> 0
1897 V124 P1000-1024	66-14663 <*> 0
1898 V126 P646-648	65-14480 <*>
1898 V126 P1510-1513	66-14666 <*>
1898 V126 P1719-1722	58-2273 <*>
1898 V126 P1753-1758	AL-495 <*>
1898 V126 P1756-1757	63-20820 <*>
1899 V128 P114-115	57-1299 <*>
1899 V128 P333-339	66-14665 <*>
1899 V128 P1234	63-16792 <*>
1899 V128 P1460-1463	8710 <IICH>
1899 V129 P417-420	66-11046 <*>
1900 V131 P187-190	64-20855 <*>
1900 V131 P541-544	64-14225 <*>
1901 V133 P786-789	57-2127 <*>
1902 V134 P175-177	66-14668 <*>
1902 V135 P791-794	23M38F <ATS>
1904 V138 P521-526	AEC-TR-5947 <*>
1904 V138 P1095-1097	57-1298 <*>
1904 V138 P1225-1227	64-16291 <*> 0
1904 V139 P721-724	65-13008 <*>
1905 V140 P936-937	63-20633 <*>
1905 V141 P238-241	63-10109 <*>
1905 V141 P317-319	UCRL TRANS-841 <*>
1905 V141 P349-351	UCRL TRANS-841 <*>
1906 V142 P142	UCRL TRANS-841 <*>
1906 V142 P201-203	UCRL TRANS-841 <*>
1906 V142 P203-205	UCRL TRANS-841 <*>
1907 V144 P853-856	62-16488 <*>
1908 V146 P697-699	64-10399 <*>
1910 V150 P175-177	I-87 <*>
1910 V150 N26 P1344-1347	63-00900 <*>
1910 V151 N20 P859-861	63-00900 <*>
1911 V152 P1735-1738	57-89 <*>
1912 V154 P427-428	57-1406 <*>
1912 V154 P1612-1614	66-14094 <*>
1912 V155 P353-354	63-14283 <*>
1912 V155 P1227-1229	2341 <*>
1912 V155 N5 P353-355	65-10717 <*> 0
1913 V156 P258-261	62-00171 <*>
1913 V156 P548-550	2341 <*>
1913 V156 P777-779	58-190 <*>
1913 V156 P1044-1048	66-14095 <*>
1913 V156 P1179-1181	61-00217 <*>
1913 V156 P1954-1958	AEC-TR-6146 <*>
1913 V156 N5 P409-411	62-00874 <*>
1913 V157 P77-78	65-11061 <*>
1913 V157 P1433-1436	I-36 <*>
	2619 <*>
1913 V157 P1533-1536	T1763 <INSD>
1914 V158 P201-204	I-41 <*>
1914 V158 P473-474	57-1586 <*>
1914 V158 P1763-1766	64-16125 <*> 0
1914 V158 P1885-1887	57-2065 <*>
1917 V165 P557-559	60-10364 <*>
1918 V166 P36-38	60-10363 <*>
1918 V166 P121-123	60-10362 <*>
1918 V166 P215-217	60-10361 <*>
1918 V167 P77-	57-2171 <*>
1918 V167 P711-718	59-20624 <*> 0
1919 V168 P945-947	60-10360 <*>
1919 V168 P1321-1323	57-1116 <*>
1919 V169 P24-	57-2465 <*>
1919 V169 P1391-1393	II-754 <*>
1920 V171 P566-569	AL-643 <*>
1920 V171 P1396-1397	63-14632 <*>
1921 V172 P1133-1134	58-225 <*>
1922 V174 P388-391	AEC-TR-5866 <*>
1922 V174 P1317-1318	57-2187 <*>
1922 V174 P1424-1426	64-20283 <*>
1922 V175 P615-617	57-3328 <*>
	57-3536 <*>
1923 V176 P355-358	63-10112 <*>

Citation	Translation
1923 V176 P1715-1716	64-14196 <*>
1923 V177 P113-116	64-14226 <*>
1923 V177 P313-316	59-20239 <*>
1923 V177 P818-821	AEC-TR-5944 <*>
1923 V177 P1110-1112	AEC-TR-5943 <*>
1923 V177 P1116-1118	66-14444 <*>
1923 V177 P1300-1302	64-18287 <*>
1924 V178 P481-483	64-18253 <*>
1924 V178 P493-495	65-13826 <*>
1924 V178 P564-566	64-14229 <*>
1924 V178 P569-571	64-14228 <*> 0
1924 V178 P1082-1084	64-14195 <*>
1924 V178 N26 P2173-2176	63-16736 <*> P
	63-16860 <*>
	64-14218 <*>
1924 V179 P44-46	64-14227 <*>
1924 V179 P237-243	64-20131 <*>
1924 V179 P718-721	64-18212 <*>
1924 V179 P1153-1156	57-2204 <*>
1924 V179 N7 P394-397	57-2022 <*>
1925 V180 P68-70	63-16167 <*>
1925 V180 P370-373	59-20442 <*> 0
1925 V180 P1029-1031	64-14230 <*>
1925 V180 P1264-1266	64-14217 <*>
1925 V180 P1328	57-1466 <*>
1925 V180 P1843-1845	64-14231 <*> 0
1925 V180 P1853-1855	64-14232 <*>
1925 V180 P1855-1858	64-18223 <*>
1925 V180 P2074-2077	317-D <K-H>
1925 V181 P15-17	63-10111 <*>
1925 V181 P243-244	II-950 <*>
1925 V181 P463-465	64-20424 <*>
1925 V181 P509-510	64-14194 <*>
1926 V182 P128-129	64-14366 <*>
1926 V182 P236-238	61-14700 <*>
1926 V182 P516-517	64-20425 <*>
1926 V182 P777-779	64-14193 <*>
1926 V182 P1221-1223	II-421 <*>
1926 V182 P1270-1272	66-13825 <*>
1926 V182 P1340-1342	64-20461 <*>
	73S81F <ATS>
1926 V183 P507-508	59-20061 <*>
1926 V183 P975-978	63-10688 <*>
1926 V183 N5 P352-354	57-946 <*>
1926 V183 N25 P1289-1291	63-20546 <*>
1927 V184 P493-497	64-10328 <*>
1927 V184 P1376-1378	I-976 <*>
1927 V184 P1454-1456	58-2269 <*>
1927 V184 N21 P1250-1252	63-20545 <*>
1927 V185 P773-774	57-2198 <*>
1928 V186 P135-137	66-14446 <*>
1928 V186 P1116-1118	98G4F <ATS>
1928 V186 P1537-1539	57-1558 <*>
1928 V186 P1846-1848	I-712 <*>
1928 V186 N13 P867-869	57-903 <*>
1928 V187 P115-117	I-977 <*>
1928 V187 P326-329	64-10329 <*>
1928 V187 P381-382	64-18188 <*>
1928 V187 P661-663	65-13818 <*>
1928 V187 P1044-1046	57-2071 <*>
1929 V188 P167-169	57-1464 <*>
1929 V188 P790-792	63-10382 <*>
1929 V188 P860-861	59-15225 <*> 0
1929 V188 P991-992	63-10379 <*>
1929 V188 P1250-1253	57-1872 <*>
1929 V189 P248-250	59 20443 <*>
1929 V189 P266-268	60-10375 <*>
1929 V189 P551-553	II-324 <*>
1929 V189 P1279-1281	58-2283 <*>
	65-10158 <*>
	8696 <IICH>
1930 V190 P457-460	64-71441 <=>
1930 V190 P798-800	64-20045 <*>
1930 V190 P876-	1659 <*>
1930 V191 P332-333	63-18106 <*>
1930 V191 P851-854	65-11475 <*>
1930 V191 P924-925	59-20445 <*>
1930 V191 P1328-1330	60-10170 <*>
1930 V191 N23 P1128-1130	59-20444 <*>
1930 V192 P1448-1451	II-455 <*>

Citation	Translation
1931 V192 P837-	57-1145 <*>
1931 V192 P1235-1237	57-3094 <*>
1931 V192 P1378-1381	66-11379 <*>
1931 V192 P1453-1454	I-114 <*>
1931 V192 P1645-1647	I-228 <*>
1931 V192 N7 P1035-1037	57-1122 <*>
1931 V192 N13 P802-804	57-1419 <*>
1931 V193 P587-589	57-1144 <*>
1931 V193 N15 P593-594	57-2214 <*>
1932 V194 P975-977	57-1297 <*>
1932 V194 P1573-1574	58-2282 <*>
	63-18175 <*>
1932 V195 P41-43	I-648 <*>
1932 V195 P75-77	58-214 <*>
1932 V195 P131-133	I-224 <*>
1932 V195 P778-779	AEC-TR-4261 <=*>
1933 V196 P170-172	I-481 <*>
1933 V196 P264-266	I-646 <*>
1933 V196 P346-348	1090 <K-H>
1933 V196 P418-420	1681 <*>
1933 V196 P536-538	AEC-TR-5522 <*>
1933 V196 P935-936	II-667 <*>
1933 V196 P1020-1022	AEC-TR-5941 <*>
1933 V196 P1577-1579	57-2474 <*>
1933 V196 P1788-	I-230 <*>
1933 V196 P1898-1899	T-1649 <INSD>
1933 V196 N21 P1577-1579	66-14613 <*>
1933 V197 P49-51	I-544 <*>
	64-14662 <*> 0
1933 V197 P447-448	50E2F <ATS>
1933 V197 P915-917	50E2F-B <ATS>
1933 V197 P1604-1606	51E2F <ATS>
1934 V198 P185-187	65-10151 <*>
1934 V198 P254	AEC-TR-5976 <*>
1934 V198 P254-256	I-297 <*>
1934 V198 P731-733	I-647 <*>
1934 V198 P1097-1100	62-00498 <*>
1934 V198 P1239	63-10447 <*>
1934 V198 P1911-1913	I-217 <*>
1934 V198 P1958-1960	I-216 <*>
1934 V199 P789-	II-765 <*>
1934 V199 P1109-	2286 <*>
1934 V199 P1397-1399	I-226 <*>
1935 V200 P123-125	II-657 <*>
1935 V200 P317-319	2105 <*>
1935 V200 P662-665	NP-TR-289 <*>
1935 V200 P1749-1751	II-376 <*>
1935 V200 P1765-1767	1197 <*>
	64-16731 <*> 0
1935 V200 P2173-2175	2107 <*>
1935 V200 N3 P212-215	57-934 <*>
1935 V201 P74-75	66-11016 <*>
1935 V201 P394-	II-173 <*>
1935 V201 P451-453	59-10716 <*>
1935 V201 P903-905	61-18784 <*>
1935 V201 N1 P45-47	57-3470 <*>
1935 V201 N1 P769-771	AD-608 369 <=>
1936 V202 P138-141	2427 <*>
1936 V202 P474-476	57-902 <*>
1936 V202 P759-760	63-10113 <*>
1936 V202 P1072-1074	1509 <*>
1936 V202 P1496-1498	T-1569 <INSD>
1936 V202 P2159-2161	63-20503 <*>
1936 V203 P1514-	57-158 <*>
1936 V203 N1 P78-80	64-14197 <*>
1937 V204 P759-761	II-69 <*>
1937 V204 P870-871	66-11955 <*>
1937 V204 P1004-1005	II-734 <*>
1937 V204 P1142-1143	57-1148 <*>
1937 V204 P1339-1340	63-14214 <*> 0
1937 V204 P1820-1822	58-135 <*>
1937 V204 P1822-1823	63-14841 <*>
1937 V204 N6 P1740-1741	61-10456 <*>
1937 V205 P141-143	63-10355 <*>
1937 V205 P181-182	64-20282 <*>
1937 V205 P251-254	64-20281 <*>
1937 V205 P299-300	64-20280 <*>
1937 V205 P315-316	I-923 <*>
1937 V205 P321-322	63-10354 <*>
1937 V205 P395-396	64-20279 <*>

Citation	Translation
1937 V205 P1018-1020	64-20278 <*>
1937 V205 P1108-1110	64-20277 <*>
1938 V206 P1644-1647	II-31 <*>
	58-1519 <*>
1938 V206 P1972-1974	57-904 <*>
1938 V207 P71-73	I-97 <*>
1938 V207 P193-195	57-1149 <*>
1938 V207 P257-259	II-574 <*>
1938 V207 P345-347	64-20276 <*>
1938 V207 P356-357	40Q68F <ATS>
1938 V207 P543-545	64-20275 <*>
1938 V207 P1042-1044	AL-735 <*>
1938 V207 P1220-1221	65-12249 <*>
1939 V208 P90-92	61-20998 <*>
1939 V208 P267-269	62-20093 <*>
1939 V208 P499-501	AL-472 <*>
1939 V208 P898-900	AEC-TR-6371 <*>
1939 V208 P1396-1398	3428 <*>
1939 V208 P1492-1494	AL-940 <*>
1939 V209 P105-106	II-444 <*>
1939 V209 P682-684	60-10373 <*>
1941 V212 P251-253	64-20274 <*>
1941 V212 P612-614	I-98 <*>
1941 V212 N19 P797-800	RCT V18 N1 P607 <RCT>
1941 V213 P619-620	63-18103 <*>
1941 V213 N23 P833-836	59-10636 <*>
1942 V214 P282-	I-777 <*>
1942 V214 P309-	I-299 <*>
1942 V214 P665-666	2283 <*>
	63-20172 <*>
1942 V214 P797-799	I-475 <*>
1942 V215 P114-115	I-218 <*>
1942 V215 P181-182	RCT V17 N1 P941 <RCT>
1942 V215 P187-188	63-18104 <*>
1942 V215 P386-387	64-16231 <*>
1942 V215 P496-498	64-20273 <*>
1942 V215 P506-508	62-01407 <*>
1942 V215 P534-536	2282 <*>
1942 V215 N15 P299-301	RCT V18 N1 P8 <RCT>
1943 V216 P348-350	03J20F <ATS>
1943 V216 P433-445	2284 <*>
1943 V216 P449-451	24K20F <ATS>
1943 V216 P505-507	64-20272 <*>
1943 V216 P649-652	62-16962 <*>
1943 V216 N7 P249-250	60-14452 <*>
1943 V217 P347-349	1682 <*>
1943 V217 P459-461	64-20271 <*> 0
1943 V217 P530-532	57-1411 <*>
1943 V217 P580-582	64-20270 <*>
1943 V217 P603-605	AL-764 <*>
1943 V217 N13 P297-299	RCT V18 N1 P22 <RCT>
1943 V217 N18 P424-426	57-1410 <*>
1944 V218 P49-56	3330 <*>
1944 V218 P458-461	2289 <*>
1944 V218 P610-612	63-14598 <*>
1944 V219 P209-210	2275 <*>
1944 V219 P349-451	T2270 <INSD>
1944 V219 N18 P447-448	RCT V19 N1 P1088 <RCT>
1945 V220 P242-243	62-18152 <*>
1945 V220 N6 P822-823	61-10159 <*>
1945 V220 N13 P456-458	RCT V19 N1 P392 <RCT>
1945 V221 P296-298	ORNL-TR-673 <*>
1945 V221 P361-363	64-20269 <*> 0
1945 V221 P498-500	1447 <*>
1946 V222 P605-607	AL-183 <*>
1946 V222 P733-734	59-15763 <*> 0
1946 V222 P791-793	63-14588 <*>
1946 V222 P1117-1118	64-20405 <*>
1946 V222 N2 P151-153	64-20268 <*>
1946 V222 N4 P261-263	64-20267 <*>
1946 V222 N5 P297-299	64-20266 <*>
1946 V222 N7 P403-405	3341 <*>
1946 V222 N12 P621-622	64-20265 <*>
1946 V222 N13 P762-763	64-20264 <*>
1946 V223 P8-11	62-00434 <*>
1946 V223 P82-84	AL-140 <*>
1946 V223 P141-142	60-14322 <*>
1946 V223 P198-199	60-14323 <*>
1946 V223 P205-206	63-14137 <*>
1946 V223 P242-243	35J18F <ATS>
1946 V223 P624-625	AEC-TR-4977 <*>
1946 V223 P860-862	63-14138 <*>
1946 V223 N19 P724-726	RCT V20 N2 P427 <RCT>
1946 V223 N24 P1029-1031	64-20263 <*>
1947 V224 P45-47	57-1412 <*>
1947 V224 P104-105	57-1404 <*>
1947 V224 P404-406	63-14139 <*>
1947 V224 P483-484	57-1953 <*>
1947 V224 P821-822	63-14140 <*>
1947 V224 P1345-1346	63-14563 <*>
1947 V224 P1395-1396	03F4F <ATS>
1947 V224 P1492-1494	64-10379 <*>
1947 V224 P1494-1496	AL-21 <*>
1947 V224 P1550	62-14061 <*>
1947 V224 P1551-1553	64-10378 <*>
1947 V224 P1558-1560	57-963 <*>
1947 V224 N1 P45-47	62-10169 <*> 0
1947 V224 N9 P656-657	64-20262 <*>
1947 V224 N16 P1164-1165	64-20261 <*>
1947 V224 N16 P1174-1175	AL-127 <*>
	3326 <*>
1947 V224 N21 P1511-1514	RCT V21 N2 P344 <RCT>
1947 V225 P249-250	AL-880 <*>
1947 V225 P394-396	2611 <*>
1947 V225 P462-464	63-14586 <*>
1947 V225 P537-539	I-1045 <*>
1947 V225 P569-571	MF-1 <*>
1947 V225 P669-	60-18139 <*> C
1947 V225 P734-	II-299 <*>
1947 V225 P954-956	3189-C <K-H>
1947 V225 P1156-1158	I-750 <*>
	II-684 <*>
1947 V225 P1358-1360	AL-175 <*>
1947 V225 N11 P541-542	64-16945 <*>
1947 V225 N15 P629-631	RCT V21 N3 P682 <RCT>
1947 V225 N16 P690-692	65-00131 <*>
1948 V226 P191-193	62-18431 <*> 0
1948 V226 P341-342	AL-518 <*>
	57-1283 <*>
1948 V226 P656-659	60-10767 <*> 0
1948 V226 P805-807	57-1141 <*>
1948 V226 P922-923	59-15061 <*>
1948 V226 P1636-1637	63-14591 <*>
1948 V226 P1720-1721	63-18983 <*>
1948 V226 P1808-1810	MF-1 <*>
1948 V226 P1904-1905	II-356 <*>
1948 V226 N4 P324-326	62-18430 <*> 0
1948 V226 N16 P295-296	3335 <*>
1948 V226 N24 P1964-1965	AL-828 <*>
1948 V226 N24 P1983-1984	62-18432 <*> 0
1948 V227 P51-52	T-2304 <INSD>
1948 V227 P299-300	MF-1 <*>
1948 V227 P1080-1082	63-14824 <*>
1948 V227 N14 P726-728	62-18489 <*>
1949 V228 P318-320	57-3471 <*>
1949 V228 P383-385	I-1002 <*>
1949 V228 P473-475	I-533 <*>
1949 V228 P487-489	62-14868 <*>
1949 V228 P492-494	I-401 <*>
	2114 <*>
1949 V228 P545-547	AL-590 <*>
1949 V228 P553-555	MF-1 <*>
1949 V228 P651-653	62-18799 <*>
1949 V228 P686-688	I-247 <*>
1949 V228 P1118-1120	57-3535 <*>
	58-1285 <*>
1949 V228 P1224-1226	II-736 <*>
1949 V228 P1437-1439	60-18043 <*>
1949 V228 P1781-1783	57-1296 <*>
1949 V228 P1992-1994	57-1296 <*>
1949 V228 N5 P1699-1701	61-10461 <*>
1949 V228 N6 P467-469	AL-177 <*>
1949 V228 N9 P738-739	AL-652 <*>
1949 V228 N9 P751-753	AL-174 <*>
1949 V228 N14 P1220-1222	59-20560 <*>
1949 V228 N16 P1337-1339	57-2653 <*>
1949 V228 N19 P1490-1492	57-901 <*>
1949 V228 N23 P1773-1775	59-20288 <*>
1949 V229 P71-73	64-10316 <*>
1949 V229 P199-201	87Q66F <ATS>

Citation	Translation
1949 V229 P354-356	3385 <*>
1949 V229 P389-391	60-10376 <*>
1949 V229 P547	AEC-TR-4656 <*>
1949 V229 P640-643	59-10334 <*>
1949 V229 P827-829	58-2186 <*>
1949 V229 P833-834	AL-517 <*>
1949 V229 P1009-1011	59-10335 <*> 0
1949 V229 P1232-1234	T-1988 <INSD>
1949 V229 N5 P353-354	57-2215 <*>
1949 V229 N8 P453-455	I-38 <*>
1949 V229 N11 P533-535	I-40 <*>
1949 V229 N11 P549-551	57-2749 <*>
1949 V229 N13 P609-610	I-37 <*>
1949 V229 N16 P749-751	63-14052 <*>
1949 V229 N20 P1007-1009	59-10333 <*>
1950 V230 P89-91	57-2643 <*>
	62-16972 <*> 0
1950 V230 P196-198	I-467 <*>
1950 V230 P534-536	62-32245 <=*>
1950 V230 P545-547	63-14142 <*>
1950 V230 P595-598	62-32246 <=*>
1950 V230 P735	AEC-TR-4655 <*>
1950 V230 P947-949	64-14303 <*>
1950 V230 P1000-1002	64-10317 <*>
1950 V230 P1050-1051	MF-1 <*>
1950 V230 P1156-1158	I-225 <*>
1950 V230 P1256-	57-2410 <*>
1950 V230 P1270-1272	I-956 <*>
1950 V230 P1367-1369	63-01320 <*>
1950 V230 P1411-1413	59-20503 <*>
1950 V230 P1671-1673	63-20254 <*>
1950 V230 P1858-1860	62-32247 <=*>
1950 V230 P1864-1865	62-32248 <=*>
1950 V230 P2025-2027	62-32249 <=*>
1950 V230 P2190-2192	I-1008 <*>
1950 V230 P2213-2214	SCL-T-411 <*>
1950 V230 N3 P258-260	65-12684 <*>
1950 V230 N3 P298-299	RCT V24 N1 P197 <RCT>
1950 V230 N12 P1181-1183	57-2624 <*>
1950 V230 N13 P1256-1258	62-18337 <*> 0
1950 V230 N16 P1438-1441	65-12651 <*>
1950 V230 N16 P1466-1467	60-10378 <*>
1950 V230 N19 P1677-1679	65-10387 <*>
1950 V230 N21 P1844-1845	<LSA>
	59-20899 <*>
1950 V231 P73-75	63-01528 <*>
1950 V231 P145-147	RCT V24 N3 P638 <RCT>
1950 V231 P220-222	59-20235 <*>
1950 V231 P230-232	RCT V24 N2 P229 <RCT>
1950 V231 P371-373	64-10318 <*>
1950 V231 P522-524	HW-61492 <*>
1950 V231 P618-619	63-20530 <*>
1950 V231 P759-761	57-1903 <*>
1950 V231 P1218-1220	I-779 <*>
1950 V231 P1220-1221	57-26 <*>
1950 V231 P1220	58-2688 <*>
1950 V231 P1232-1233	521 <TC>
1950 V231 P1440-1442	57-1904 <*>
1950 V231 N2 P126-217	AL-615 <*>
1950 V231 N4 P261-263	66-13382 <*> 0
1950 V231 N7 P441-443	64-18210 <*> 0
1950 V231 N8 P461-464	59-10118 <*> 0
1950 V231 N15 P728-729	64-20248 <*>
1950 V231 N16 P764-765	AL-790 <*>
1950 V231 N16 P766-767	AL-55 <*>
1950 V231 N19 P1004-1006	64-20249 <*>
1950 V231 N20 P1065-1066	RCT V24 N3 P662 <RCT>
1950 V231 N22 P1237-1238	60-10367 <*>
1950 V231 N25 P1392-1394	66-13383 <*>
1950 V231 N25 P1578-1580	64-20250 <*>
1951 V232 P100-102	8972 <IICH>
1951 V232 P231-232	86Q72F <ATS>
1951 V232 P334-336	8968 <IICH>
1951 V232 P498-499	II-14 <*>
1951 V232 P501-503	II-134 <*>
1951 V232 P800-802	II-305 <*>
1951 V232 P855-857	I-489 <*>
1951 V232 P949-950	T2242 <INSD>
1951 V232 P1100-1101	57-1562 <*>
1951 V232 P1272-1274	59-10764 <*>
1951 V232 P1378-1380	62-00125 <*>
1951 V232 P1929-1930	62-32250 <=*>
1951 V232 P2214-2216	86Q66F <ATS>
1951 V232 P2456-2458	63-18249 <*>
1951 V232 N1 P82-84	RCT V24 N4 P914 <RCT>
1951 V232 N11 P1088-1089	II-306 <*>
	MF-1 <*>
1951 V232 N20 P1826-1828	II-288 <*>
	MF-1 <*>
1951 V232 N20 P1837-1839	65-13448 <*>
1951 V233 P372-374	60-16896 <*>
1951 V233 P657-659	AEC-TR-4898 <*>
1951 V233 P1281	I-62 <*>
1951 V233 P1676-1677	66-10220 <*>
1951 V233 N3 P277-279	60-10720 <*>
1951 V233 N14 P745-747	59-17943 <*>
1951 V233 N17 P919-921	II-41 <*>
1951 V233 N18 P1092-1094	60-18091 <*>
1952 V234 P255-257	57-1295 <*>
1952 V234 P260-262	MF-6 <*>
1952 V234 P276-278	58-2185 <*>
1952 V234 P946-948	AL-966 <*>
1952 V234 P1049-	I-112 <*>
1952 V234 P1165-1166	1853 <*>
1952 V234 P1777-1779	62-32251 <=*>
1952 V234 P1920-1921	57-1294 <*>
1952 V234 P2055-2057	59-11953 <=*>
1952 V234 P2064-2066	UCRL TRANS-1019(L) <*>
1952 V234 P2131-2133	57-1298 <*>
1952 V234 P2285-2287	62-32252 <=*>
1952 V234 P2316-2318	I-1039 <*>
1952 V234 P2363-2365	SCL-T-236 <+*>
1952 V234 P2589-2591	I-700 <*>
	64-10133 <*>
1952 V234 P2636-2638	1910 <$>
1952 V234 N1 P113-115	59-20036 <*>
1952 V234 N2 P173-174	65-12685 <*>
1952 V234 N3 P345-347	59-20039 <*>
1952 V234 N3 P347-349	59-20035 <*>
1952 V234 N3 P372-374	64-20251 <*>
1952 V234 N8 P896-897	64-20253 <*> 0
1952 V234 N13 P1351-1354	59-20600 <*>
1952 V234 N15 P1549-1551	57-876 <*>
1952 V234 N18 P1773-1775	60-10159 <*> 0
1952 V234 N19 P1877-1879	RCT V26 N1 P78 <RCT>
1952 V234 N19 P1883-1885	59-17946 <*>
1952 V234 N23 P2283-2285	59-17921 <*>
1952 V234 N23 P2292-2294	65-13891 <*> 0
1952 V235 P287-289	8739 <IICH>
1952 V235 P470-472	58-1270 <*>
1952 V235 P609-611	AI-TR-11 <*>
1952 V235 P619-621	57-1969 <*>
1952 V235 P652-655	AI-TR-11 <*>
1952 V235 P658-659	62-18667 <*>
1952 V235 P1621-1623	AI-TR-11 <*>
1952 V235 N2 P139-140	59-20926 <*> 0
1952 V235 N2 P154-156	65-10480 <*>
1952 V235 N3 P236-238	<LSA>
	59-15517 <*>
1952 V235 N5 P376-377	64-20535 <*>
1952 V235 N7 P472-473	59-17919 <*>
1952 V235 N21 P1286-1288	II-34 <*>
1953 V236 P85-87	AL-938 <*>
1953 V236 P170-172	58-2183 <*>
1953 V236 P469-471	62-18296 <*> 0
	62-32253 <=*>
1953 V236 P484-486	UCRL TRANS-1020(L) <*>
1953 V236 P486-488	62-32254 <=*>
1953 V236 P593-594	62-18295 <*> 0
1953 V236 P736-738	I-242 <*>
1953 V236 P809-810	65-10008 <*> 0
1953 V236 P811-813	AL-333 <*>
1953 V236 P1018-1021	63-20430 <*>
1953 V236 P1025-1028	67M39F <ATS>
1953 V236 P1028-1030	62-32255 <=*>
1953 V236 P1038-1041	AL-332 <*>
1953 V236 P1145-1146	I-640 <*>
1953 V236 P1275-1278	AL-30 <*>
1953 V236 P1361-1363	57-1292 <*>

1953 V236 P1417-1419 UCRL TRANS-583(L) <*>
1953 V236 P1547-1549 AEC-TR-4199 <*>
1953 V236 P1571 62-10004 <*>
1953 V236 N1 P85-87 60-10158 <*>
1953 V236 N2 P170-172 65-12677 <*>
1953 V236 N7 P704-706 I-493 <*>
II-256 <*>
1953 V236 N9 P909-911 65-10481 <*>
1953 V236 N9 P931-933 64-20402 <*>
1953 V236 N10 P1021-1023 60-10157 <*> 0
1953 V236 N11 P1169-1171 57-987 <*>
1953 V236 N12 P1275-1278 59-20257 <*> 0
1953 V236 N14 P1412-1413 59-17945 <*>
1953 V236 N16 P1538-1540 59-17915 <*>
1953 V236 N16 P1565-1567 60-10171 <*>
1953 V236 N18 P1741-1743 59-20927 <*> 0
1953 V236 N18 P1743-1745 64-16051 <*>
1953 V236 N21 P2076-2078 64-20536 <*> 0
1953 V236 N23 P2234-2236 59-20571 <*>
1953 V236 N26 P2486-2488 59-17914 <*>
1953 V236 N26 P2553-2555 64-20006 <*>
1953 V237 P258-260 62-32257 <=*>
1953 V237 P262-264 II-309 <*>
1953 V237 P308- 57-2958 <*>
1953 V237 P564-566 57-2733 <*>
1953 V237 P714-715 II-353 <*>
1953 V237 P808-810 57-216 <*>
1953 V237 P1002-1003 II-406 <*>
1953 V237 P1409-1411 AEC-TR-3579 <=*>
1953 V237 P1415-1417 T1990 <INSD>
1953 V237 P1800-1802 63-14033 <*>
1953 V237 N1 P59-62 60-10719 <*> 0
1953 V237 N1 P62-64 62-18790 <*> 0
1953 V237 N2 P194-196 I-345 <*>
1953 V237 N3 P238-240 62-32256 <=*>
1953 V237 N4 P310-313 59-17903 <*>
1953 V237 N14 P1468-1470 <LSA>
59-15521 <*>
1953 V237 N15 P812-813 II-25 <*>
1953 V237 N16 P869-871 60-10189 <*>
1953 V237 N17 P1641-1643 62-14077 <*>
1953 V237 N23 P1316-1318 I-50 <*>
1953 V237 N25 P1686-1688 64-20803 <*> 0
1954 V238 P188-191 57-91 <*>
58-2477 <*>
1954 V238 P224-226 62-32258 <=*>
1954 V238 P251 62-10001 <*>
1954 V238 P253 62-14047 <*>
1954 V238 P282-284 64-71437 <=>
1954 V238 P358-360 AEC-TR-4897 <*>
1954 V238 P413-414 II-103 <*>
1954 V238 P527-528 II-104 <*>
1954 V238 P573-575 62-32259 <=*>
1954 V238 P576-578 62-32260 <=*>
1954 V238 P660-662 62-32261 <=*>
1954 V238 P690-692 8741 <IICH>
1954 V238 P784-786 62-32262 <=*>
1954 V238 P888-890 62-32263 <=*>
1954 V238 P1005-1007 62-32264 <=*>
1954 V238 P1198-1200 68G5F <ATS>
1954 V238 P1414-1416 61-00636 <*>
1954 V238 P1532-1534 2237 <*>
1954 V238 P1578-1580 UCRL TRANS-576(L) <*>
1954 V238 P1661-1663 I-420 <*>
1954 V238 P1710-1711 62-32265 <=*>
1954 V238 P1775- 58-1639 <*>
1954 V238 P1993-1995 I-519 <*>
1954 V238 P2053 62-10112 <*> 0
1954 V238 P2061-2063 57-2881 <*>
1954 V238 P2255-2257 62-32266 <=*>
1954 V238 P2402-2403 57-2880 <*>
1954 V238 N1 P98-99 60-10180 <*> 0
1954 V238 N4 P460-462 57-989 <*>
1954 V238 N4 P471-473 76F3F <ATS>
1954 V238 N6 P686-688 2529 <*>
1954 V238 N7 P814-815 62-18791 <*>
1954 V238 N7 P815-817 57-2769 <*>
1954 V238 N8 P894-896 RCT V28 N2 P606 <RCT>
1954 V238 N12 P1324-1325 60-10179 <*>
1954 V238 N14 P2413-2414 AEC-TR-3195 <*>
1954 V238 N17 P1733-1735 60-10434 <*>
1954 V238 N21 P2238-2239 60-10586 <*>
1954 V238 N22 P2162-2165 60-10178 <*> 0
1954 V238 N24 P2318-2320 62-18787 <*>
1954 V238 N26 P2514-2516 30P66F <ATS>
1954 V239 P31-33 II-912 <*>
1954 V239 P33-35 2531 <*>
1954 V239 P51-52 62-32267 <=*>
1954 V239 P89-91 57-92 <*>
1954 V239 P155-157 57-2409 <*>
1954 V239 P162-164 63-10550 <*>
1954 V239 P168-170 8974 <IICH>
1954 V239 P249-251 62-32268 <=*>
1954 V239 P282- II-380 <*>
1954 V239 P822-824 57-1308 <*>
1954 V239 P1220-1222 63-18178 <*>
1954 V239 P1287-1289 I-389 <*>
1954 V239 P1303- II-316 <*>
1954 V239 P1303 64-14398 <*>
1954 V239 P1350-1352 36G5F <ATS>
1954 V239 P1408-1410 57-1307 <*>
1954 V239 P1474-1476 62-32253 <=*>
1954 V239 P1478-1480 62-32269 <=*>
1954 V239 N3 P274-275 60-10187 <*>
1954 V239 N5 P402-404 62-16235 <*>
1954 V239 N12 P732-734 57-1156 <*>
1954 V239 N14 P794-796 64-10400 <*>
1954 V239 N15 P879-880 60-10160 <*>
1954 V239 N17 P1042-1043 59-20930 <*>
1954 V239 N17 P1089-1091 61-16789 <*>
1954 V239 N22 P1474-1476 62-18325 <*> 0
1954 V239 N22 P1495-1497 62-18788 <*> 0
1955 V240 P308-310 62-32270 <=*>
1955 V240 P312-314 2851 <*>
1955 V240 P412-415 59-11956 <=*>
1955 V240 P610-612 57-2848 <*>
1955 V240 P615-617 AEC-TR-4201 <*>
1955 V240 P756-758 57-1326 <*>
1955 V240 P768-770 II-236 <*>
1955 V240 P808-810 3157 <*>
1955 V240 P862-864 57-1315 <*>
1955 V240 P975-977 T-2293 <INSD>
1955 V240 P1085-1087 1394 <*>
1955 V240 P1195-1197 CSIRO-3016 <CSIR>
1955 V240 P1210- 57-1868 <*>
1955 V240 P1634-1636 II-1021 <*>
1955 V240 P2235-2237 AEC-TR-6260 <*>
1955 V240 P2324 AEC-TR-4849 <*>
1955 V240 P2436-2438 63-01284 <*>
1955 V240 P2522-2544 58-1644 <*>
1955 V240 N5 P524-526 59-20929 <*>
1955 V240 N7 P776-778 59-20928 <*> 0
1955 V240 N8 P864-866 64-16992 <*>
1955 V240 N11 P1221-1223 5756-B <KH>
1955 V240 N13 P1470-1472 64-20254 <*>
1955 V240 N16 P1630-1632 59-10117 <*> 0
1955 V240 N17 P1719-1720 64-20255 <*>
1955 V240 N18 P1800-1801 64-20256 <*> 0
1955 V240 N25 P2456-2457 64-20257 <*>
1955 V240 N26 P2571-2573 3254 <*>
1955 V241 P40-42 II-219 <*>
II-291 <*>
1955 V241 P42-44 62-32271 <=*>
1955 V241 P732-734 C-3809 <NRCC>
1955 V241 P865-867 63-18094 <*>
1955 V241 P905-907 58-183 <*>
1955 V241 P1298-1299 57-1800 <*>
1955 V241 P1772-1775 64L29F <ATS>
1955 V241 P1947-1949 08L29F <ATS>
1955 V241 N2 P180-182 3038 <*>
1955 V241 N6 P533-536 57-1358 <*>
1955 V241 N16 P1055-1057 65-14479 <*>
1955 V241 N17 P1129-1130 60-10716 <*>
1955 V241 N22 P1568-1571 64-16990 <*>
1955 V241 N22 P1589-1592 IGR V1 N6 P86 <AGI>
1955 V241 N24 P1755-1758 64-16991 <*>
1955 V241 N24 P1760-1762 60-10715 <*>
1956 V242 P124-126 4844 <BISI>
1956 V242 P632-635 4748 <BISI>
1956 V242 P1154-1156 62-32272 <=*>

Citation	Translation
1956 V242 P1172-1175	4747 <BISI>
1956 V242 P1222-1225	58-1380 <*>
1956 V242 P1233-1235	II-1054 <*>
1956 V242 P1433-1436	AEC-TR-4200 <*>
1956 V242 P1651-1653	57-1854 <*>
1956 V242 P1824-1828	61-20088 <*>
1956 V242 P1996-1998	58-598 <*>
1956 V242 P2308-2309	57-2849 <*>
1956 V242 P2517-2519	UCRL TRANS-575(L) <*>
1956 V242 P2528-2531	57-1066 <*>
1956 V242 P2531-2533	98M38I <ATS>
1956 V242 P3081-3083	AEC-TR-5931 <*>
1956 V242 P3089-3092	57-1282 <*>
1956 V242 N2 P254-256	1542 <*>
1956 V242 N3 P382-384	<LSA>
	59-15223 <*>
	60-18126 <*> 0
1956 V242 N4 P486-489	UCRL TRANS-283(L) <*>
1956 V242 N4 P508-510	57-2944 <*>
1956 V242 N10 P1309-1311	57-2973 <*>
1956 V242 N13 P1720-1722	57-1069 <*>
1956 V242 N15 P1849-1852	59-20312 <*>
1956 V242 N16 P1989-1990	66-14382 <*>
1956 V242 N19 P2312-2315	59-20602 <*>
1956 V242 N19 P2355-2357	61-16277 <*>
1956 V242 N20 P2451-2453	57-2360 <*>
1956 V242 N21 P2525-2528	64-16952 <*>
1956 V242 N23 P2712-2715	59-20601 <*>
1956 V243 P41-44	62-32273 <=*>
1956 V243 P50-53	63-14143 <*> 0
1956 V243 P410-412	57-1855 <*>
1956 V243 P898-901	99M38F <ATS>
1956 V243 P982-985	62-32274 <=*>
1956 V243 P1115	57-2734 <*>
1956 V243 P1217-1219	62-32275 <=*>
1956 V243 P1315-	57-2927 <*>
1956 V243 P1499-1502	<LSA>
	60-14481 <*>
1956 V243 P1737-1740	00M39F <ATS>
1956 V243 P2063-2065	57-2933 <*>
1956 V243 N5 P493-495	62-20147 <*>
1956 V243 N6 P576-578	57-1068 <*>
1956 V243 N6 P585-588	64-16748 <*> 0
1956 V243 N8 P695-698	IGR V1 N6 P88 <AGI>
1956 V243 N13 P893-895	58-2398 <*>
1956 V243 N17 P1208-1209	59-15933 <*>
1956 V243 N22 P1769-1772	65-10896 <*>
1957 V244 P459-461	RCT V30 N4 P1166 <RCT>
1957 V244 P680-683	59-16017 <=*> 0
1957 V244 P1029-1031	58-1198 <*>
1957 V244 P1478-1481	66-14109 <*>
1957 V244 P2077-2080	62-10141 <*>
1957 V244 P2788-2791	66-14110 <*>
1957 V244 P2827-2830	3444 <*>
1957 V244 P2925-2928	63-00469 <*>
1957 V244 P2928-2929	63-00480 <*>
1957 V244 P2944-2946	58-369 <*>
1957 V244 N1 P52-54	64-16314 <*>
1957 V244 N1 P77-80	58-1488 <*>
1957 V244 N1 P80-82	59-15926 <*>
1957 V244 N3 P364-367	62-10137 <*>
1957 V244 N5 P577-580	64-18502 <*>
1957 V244 N8 P956-970	57-3030 <*>
1957 V244 N8 P965-970	<LSA>
1957 V244 N8 P1027-1029	63-18095 <*>
1957 V244 N8 P1039-1040	61-10919 <*>
	9353-B <K-H>
1957 V244 N9 P1192-1193	<LSA>
	57-3171 <*>
1957 V244 N9 P1193-1195	57-3169 <*>
1957 V244 N9 P1212-1214	<LSA>
	57-3170 <*>
1957 V244 N11 P1505-1507	57-3327 <*>
1957 V244 N12 P1577-1579	<LSA>
	59-15522 <*>
1957 V244 N16 P2152-2155	64-18501 <*>
1957 V244 N18 P2326-2329	64-14551 <*> 0
1957 V244 N19 P2363-2366	59-10949 <*>
1957 V244 N20 P2469-2471	SCL-T-388 <*>
1957 V245 P62-64	66-11104 <*>

Citation	Translation
1957 V245 P950-952	58-70 <*>
1957 V245 P1135-1138	4480 <BISI>
1957 V245 P1424-1427	8963 <K-H>
1957 V245 P1708-1710	59-17225 <*>
1957 V245 P1758-1760	65-17251 <*>
1957 V245 P2235-2236	UCRL TRANS-574(L) <*>
1957 V245 P2467-2470	61-00954 <*>
1957 V245 P2493-2496	59-17225 <*>
1957 V245 P2507-2510	AEC-TR-5930 <*>
1957 V245 N8 P837-839	58-1261 <*>
1957 V245 N14 P1128-1129	AEC-TR-3534 <=*>
1957 V245 N17 P1424-1427	59-17737 <*>
1957 V245 N22 P1931-1933	60-00691 <*>
1958 V245 P2313-2320	59-20406 <*>
1958 V246 P102-104	64-10195 <*>
1958 V246 P110-113	AEC-TR-5249 <*>
1958 V246 P113-116	AEC-TR-5932 <*>
1958 V246 P332-334	63-00746 <*>
1958 V246 P499-501	58-1696 <*>
1958 V246 P611-614	61-01057 <*>
1958 V246 P851-852	63-14873 <*>
1958 V246 P853-855	58-1694 <*>
1958 V246 P916-918	66-10795 <*> 0
1958 V246 P1109-1111	61-00176 <*>
1958 V246 P1214-1217	RCT V31 N3 P664 <RCT>
1958 V246 P1809-1811	62-10154 <*=>
1958 V246 P2023-2026	62-00526 <*>
1958 V246 P2720-2722	9123 <K-H>
1958 V246 P2753-2756	AEC-TR-3522 <*>
1958 V246 P2793-2795	<LSA>
1958 V246 P2845-2848	UCRL TRANS-905(L) <*>
1958 V246 P2963-2968	59-20405 <*>
1958 V246 P3073-3076	62-01014 <*>
1958 V246 P3347-3350	AEC-TR-5086 <*>
1958 V246 P3422-3425	SCL-T-455 <*>
1958 V246 N1 P82-84	<LSA>
	59-10960 <*>
1958 V246 N1 P190-192	59-14243 <=*>
1958 V246 N5 P741-744	ORNL-TR-413 <*>
1958 V246 N5 P766-769	65-10029 <*> 0
1958 V246 N6 P913-916	AEC-TR-3556 <=*>
1958 V246 N6 P3341-3344	61-10460 <*>
1958 V246 N7 P1040-1042	8342-B <KH>
1958 V246 N9 P1440-1441	63-16884 <*>
	8467-B <K-H>
1958 V246 N12 P1838-1840	59-10961 <*>
1958 V246 N15 P2281-2283	62-10134 <*>
1958 V246 N16 P2374-2377	64-18804 <*>
1958 V246 N17 P2481-2484	AEC-TR-3677 <+*>
1958 V246 N18 P2571-2574	61-18263 <*>
1958 V246 N18 P2649-2695	60-14059 <*>
1958 V246 N19 P2720-2722	65-10913 <*>
1958 V246 N19 P2737-2740	59-10440 <*>
1958 V246 N22 P3161-3164	65-00140 <*>
1958 V246 N23 P3230-3232	59-10950 <*>
	64-20125 <*>
1958 V246 N24 P3341-3344	64-20126 <*>
1958 V246 N25 P3454-3457	1427 <BISI>
	62-10537 <*>
1958 V246 N26 P3641-3644	60-10999 <*>
1958 V247 P80-83	AEC-TR-6288 <*>
1958 V247 P152-154	63-18016 <*>
1958 V247 P795-796	62-00900 <*>
1958 V247 P1802-1805	63-00507 <*>
1958 V247 P1836	AEC-TR-5313 <*>
1958 V247 P1869-1872	AEC-TR-3711 <+*>
1958 V247 N1 P83-85	AEC-TR-3846 <*>
1958 V247 N3 P369-371	59-14577 <=*>
1958 V247 N3 P371-372	59-14578 <=*>
1958 V247 N9 P738-741	60-21462 <=>
1958 V247 N10 P778-780	60-21459 <=>
1958 V247 N14 P1014-1016	64-30989 <*>
1958 V247 N16 P1199-1201	AEC-TR-3844 <*>
1958 V247 N17 P1418-1420	59-15410 <*>
1958 V247 N19 P1608-1611	<LSA>
	59-17227 <*>
1958 V247 N19 P1612-1613	<LSA>
	59-15720 <*>
1958 V247 N25 P351-2354	AEC-TR-3727 <*>
1959 V248 P154-156	59-18997 <=*>

Citation	Reference
1959 V248 P160-163	63-00944 <*>
1959 V248 P410-413	41L30F <ATS>
1959 V248 P555-558	11M38F <ATS>
1959 V248 P672-673	AEC-TR-4221 <*>
1959 V248 P999-1001	59-21053 <=*>
1959 V248 P1248-1249	59-18576 <=*>
1959 V248 P1819-1822	94L31F <ATS>
1959 V248 P2003-2005	AEC-TR-3845 <*>
1959 V248 P2006-2008	59-17127 <*>
1959 V248 P2012-2014	63-18158 <*>
1959 V248 P2421-2423	60-16660 <*>
1959 V248 P2472-2474	AEC-TR-4268 <*>
1959 V248 P2807-2808	62-00275 <*>
1959 V248 N1 P54-57	63-14493 <*>
1959 V248 N2 P205-207	60-14040 <*>
	60-18063 <*> 0
1959 V248 N4 P602-603	66-11302 <*>
1959 V248 N9 P1333-1335	<LSA>
1959 V248 N10 P1528-1530	26P60F <ATS>
1959 V248 N11 P1655-1658	<LSA>
	60-10126 <*>
1959 V248 N11 P1672-1675	65-10812 <*>
1959 V248 N11 P1721-1722	66-10420 <*>
1959 V248 N13 P1988-1990	<LSA>
	60-10125 <*>
1959 V248 N15 P2168-2170	NP-TR-318 <*>
1959 V248 N15 P2179-2181	60-16999 <*>
1959 V248 N15 P2182-2184	60-18008 <*>
1959 V248 N16 P2345-2347	60-10124 <*>
1959 V248 N16 P2348-2350	65-11639 <*>
1959 V248 N17 P2528-2530	62-10922 <*>
1959 V248 N18 P2564-2566	60-18320 <*>
1959 V248 N21 P2976-2978	64-16921 <*>
1959 V248 N22 P3151-3153	60-16996 <*>
1959 V248 N23 P3301-3303	64-16965 <*>
1959 V248 N23 P3317-3319	60-18041 <*>
1959 V248 N24 P3418-3420	60-16997 <*>
1959 V248 N24 P3424-3426	60-18001 <*>
1959 V248 N25 P3543-3545	60-16998 <*>
1959 V248 N26 P3696-3698	60-18136 <*>
1959 V248 N26 P3702-3704	64-30654 <*>
1959 V249 P407	AEC-TR-4664 <*>
1959 V249 P982-984	AEC-TR-3886 <*>
1959 V249 P1872-1874	62-16699 <*> 0
1959 V249 P2237-2238	61-00166 <*>
1959 V249 P2249-2250	61-00174 <*>
1959 V249 P2769-2771	AEC-TR-6109 <*>
1959 V249 N1 P53-55	60-18323 <*>
1959 V249 N1 P61-63	61-20289 <*>
1959 V249 N1 P97-98	64-30689 <*>
1959 V249 N2 P245-247	60-18322 <*>
1959 V249 N2 P274-276	59-17563 <*> 0
1959 V249 N3 P435-437	<LSA>
	60-14489 <*>
1959 V249 N5 P680-681	61-10218 <*>
1959 V249 N10 P956-958	<LSA>
	60-16251 <*>
1959 V249 N13 P1133-1135	62-16101 <*>
1959 V249 N14 P1196-1198	66-11348 <*> 0
1959 V249 N14 P1282-1284	65-00208 <*>
1959 V249 N22 P2329-2331	AEC-TR-4049 <*>
1959 V249 N23 P2549-2551	<LSA>
	60-18098 <*>
1960 V250 P115-117	HW-TR-38 <*>
1960 V250 P126-127	42N50F <ATS>
1960 V250 P943-945	10033 <K-H>
	64-10614 <*>
1960 V250 P979-981	62-01078 <*>
1960 V250 P1134-1136	AEC-TR-4044 <*>
1960 V250 P1468-1470	SCL-T-339 <*>
1960 V250 P1552-1554	AEC-TR-4601 <*>
1960 V250 P1824-1826	63-10246 <*>
1960 V250 P2853-2855	63-10777 <*>
1960 V250 P2948-2950	62-00909 <*>
1960 V250 N1 P109-111	61-10220 <*>
	64-14359 <*> 0
1960 V250 N2 P3476-3478	66-13384 <*>
1960 V250 N4 P683-685	60-18088 <*>
1960 V250 N5 P819-821	60-18087 <*>
1960 V250 N7 P1258-1260	65-12239 <*>

Citation	Reference
1960 V250 N8 P1569-1571	11411-G <K-H>
	64-10615 <*>
1960 V250 N9 P1615-1617	61-10273 <*>
1960 V250 N9 P1651-1652	61-16185 <*>
1960 V250 N9 P1727-1729	65-00207 <*>
1960 V250 N11 P2070-2072	60-18086 <*>
1960 V250 N12 P2267-2268	60-18089 <*>
1960 V250 N13 P2471-2473	65-00096 <*>
1960 V250 N14 P2538-2540	61-10192 <*>
1960 V250 N17 P2853-2855	65-12240 <*>
1960 V250 N17 P2892-2894	61-10191 <*>
1960 V250 N17 P2956-2958	64-10169 <*>
1960 V250 N18 P3025	AEC-TR-5167 <=*>
1960 V250 N22 P3608-3610	66-11659 <*>
1960 V250 N22 P3650-3652	61-10332 <*>
1960 V250 N25 P4157-4159	TRANS-144 <FRI>
	61-10657 <*>
1960 V251 P725-726	61-18700 <*> 0
1960 V251 P747-749	64-14057 <*>
1960 V251 P959-961	62-10757 <*>
1960 V251 P1001-1003	61-14754 <*>
1960 V251 P2199-2201	66-13001 <*>
1960 V251 P2248-2249	61-10892 <*>
1960 V251 P2341-2343	AEC-TR-4534 <*>
1960 V251 P2347-2349	2392 <BISI>
1960 V251 N1 P155-157	65-00298 <*>
1960 V251 N3 P324-326	61-14757 <*>
1960 V251 N3 P343-345	64-14130 <*>
1960 V251 N4 P495-497	61-14758 <*>
1960 V251 N4 P529-531	T61-6 <MUL>
1960 V251 N5 P684-685	61-14749 <*>
1960 V251 N8 P1010-1012	66-12989 <*> 0
1960 V251 N10 P1127-1129	62-10458 <*>
1960 V251 N13 P1277-1279	61-18280 <*>
1960 V251 N14 P1322-1324	61-20019 <*>
1960 V251 N19 P2035-2037	288 <TC>
1960 V251 N22 P2353-2355	62-14315 <*>
1960 V251 N23 P2653-2655	65-12285 <*>
1960 V251 N25 P2930-2932	3016 <BISI>
1960 V251 N25 P3079-3081	65-18077 <*>
1960 V251 N119 P2035-2037	61-18614 <*>
1961 V252 P273-275	AEC-TR-5024 <*>
1961 V252 P614-615	62-18057 <*>
1961 V252 P683-685	62-10239 <*>
1961 V252 P698-700	AEC-TR-4999 <*>
1961 V252 P1008-1010	UCRL TRANS-781(L) <
1961 V252 P1127-1128	UCRL TRANS-877(L) <
1961 V252 P1305-1307	AEC-TR-6395 <*>
1961 V252 P1320-1322	AEC-TR-5059 <*>
1961 V252 P1616-1618	64-10165 <*>
1961 V252 P1904-1906	63-10297 <*>
1961 V252 P1922	AEC-TR-5312 <*>
1961 V252 P2119-2121	UCRL TRANS-739 <*>
1961 V252 P2221-2222	AEC-TR-4658 <*>
1961 V252 P3011-3014	63-01214 <*>
1961 V252 P3154-3156	63-00510 <*>
1961 V252 P4139-4141	63-00590 <*>
	63-27141 <$>
1961 V252 N8 P1155-1157	55N52F <ATS>
1961 V252 N9 P1288-1290	62-00104 <*>
1961 V252 N9 P1347-1349	62-14201 <*> 0
1961 V252 N12 P1800-1802	62-14450 <*> C
1961 V252 N13 P1945-1947	65-10042 <*>
1961 V252 N22 P3462-3464	66-13427 <*>
1961 V252 N25 P3989-3991	25P60F <ATS>
1961 V253 P590-592	62-00445 <*>
1961 V253 P1100-1102	66-11090 <*>
1961 V253 P1229-1231	62-00447 <*>
1961 V253 P1281-1282	62-00505 <*>
1961 V253 P1331-1333	AEC-TR-5095 <*>
1961 V253 P1852-1853	SCL-T-404 <*>
1961 V253 P2427-2429	63-00511 <*>
1961 V253 P2648-2650	62-24774 <=*$>
1961 V253 P2967-2969	AEC-TR-5933 <*>
1961 V253 N1 P9-12	62-10345 <*>
1961 V253 N1 P67-69	SCL-T-395 <*>
1961 V253 N2 P203-208	62-10346 <*>
1961 V253 N4 P641-643	65-14244 <*>
1961 V253 N15 P1524-1526	62-18782 <*>
1961 V253 N16 P1699-1701	62-14535 <*>

Citation	Translation
1961 V253 N17 P1880-1882	66-12621 <=*>
1961 V253 N22 P2568-2570	79P60F <ATS>
1961 V253 N23 P2685-2687	TRANS-178 <FRI>
1962 V254 P272-275	4172 <BISI>
1962 V254 P299-301	62-18781 <*>
1962 V254 P462-464	AEC-TR-5456 <*>
1962 V254 P585-587	63-00573 <*>
1962 V254 P1052-1054	AEC-TR-5738 <*>
1962 V254 P1765-1767	66-10675 <*>
1962 V254 P1774-1776	SC-T-526 <*>
1962 V254 P1794-1796	62-18812 <*>
1962 V254 P1946-1947	12510 <K-H>
	63-14376 <*>
1962 V254 P1991-1993	4171 <BISI>
1962 V254 P2328-2330	AEC-TR-5740 <*>
1962 V254 P2978-2979	63-18910 <*>
1962 V254 P3173-3175	62-20042 <*>
1962 V254 P3357-3359	AI-TR-50 <*>
1962 V254 P3360-3362	63-18092 <*>
1962 V254 P3603-3605	63-00560 <*>
1962 V254 P3606-3608	63-00512 <*>
1962 V254 P3834-3835	36P65F <ATS>
1962 V254 P4114-4116	NP-TR-970 <*>
1962 V254 P4179-4181	AEC-TR-5353 <*>
1962 V254 P4214-4216	63-00304 <*>
1962 V254 P4225-4227	63-00601 <*>
1962 V254 P4467-4469	AEC-TR-5646 <*>
1962 V254 N2 P240-242	65-00285 <*>
1962 V254 N2 P249-251	64-30646 <*> 0
1962 V254 N4 P674-676	62-20243 <*>
1962 V254 N17 P3090-3092	62-20441 <*> 0
1962 V254 N18 P3216-3218	65-12284 <*>
1962 V254 N21 P3671-3673	62-20440 <*> 0
1962 V254 N26 P4404-4406	64-00048 <*>
1962 V255 P524-526	63-00505 <*>
1962 V255 P665-667	66Q69F <ATS>
1962 V255 P887-889	AEC-TR-6064 <*>
1962 V255 P912-913	73Q70F <ATS>
1962 V255 P1511-1513	NP-TR-958 <*>
1962 V255 P1741-1743	63-18395 <*> 0
1962 V255 P2098-2099	SC-T-528 <*>
1962 V255 P2247-2249	63-18738 <*>
1962 V255 P2539	AEC-TR-6380 <*>
1962 V255 P2757-2759	63-20344 <*>
1962 V255 N2 P252-254	63-14744 <*>
1962 V255 N3 P524-526	64-16609 <*>
1962 V255 N4 P698-700	NP-TR-990 <*>
1962 V255 N5 P901-902	SCL-T-476 <*>
1962 V255 N8 P1269-1271	3105 <BISI>
1962 V255 N14 P1592-1594	63-20921 <*>
1962 V255 N16 P1893-1895	66-12991 <*> 0
1962 V255 N18 P2247-2249	65-10388 <*>
1962 V255 N21 P2760-2762	AEC-TR-5940 <*>
1963 V256 P90-92	66-12947 <*>
1963 V256 P298-300	AEC-TR-5705 <*>
1963 V256 P634-637	10Q70F <ATS>
1963 V256 P681-683	AEC-TR-5718 <*>
1963 V256 P2258-2260	63-01102 <*>
1963 V256 P2597-2600	UCRL TRANS-1246(L) <*>
1963 V256 P2834-2836	SC-T-516 <*>
1963 V256 P2846-2849	64-16317 <*>
1963 V256 P3029-3030	SC-T-510 <*>
1963 V256 P3074-3077	1240 <TC>
1963 V256 P3118	65-18132 <*>
1963 V256 P3632-3635	65-18088 <*>
1963 V256 P3892-3894	AEC-TR-5936 <*>
1963 V256 P4189-4191	64-10950 <*> 0
1963 V256 P4227-	AEC-TR-5939 <*>
1963 V256 P4419-4421	66-14468 <*>
1963 V256 P5366-5369	AEC-TR-6256 <*>
1963 V256 N1 P307-308	65-13840 <*> 0
1963 V256 N2 P427-429	63-18163 <*>
1963 V256 N4 P953-955	63-18164 <*>
1963 V256 N7 P1507-1510	65-10992 <*>
1963 V256 N12 P2588-2590	SCL-T-504 <*>
1963 V256 N12 P2616-2619	HW-TR-55 <*>
1963 V256 N13 P2841-2843	64-30705 <*>
1963 V256 N14 P3210-3212	65-13963 <*> 0
1963 V256 N15 P3359-3362	64-20258 <*> PO
1963 V256 N17 P3618-3621	64-16406 <*>
1963 V256 N23 P4489-4490	66-12357 <*>
1963 V256 N24 P5239-5241	64-20259 <*>
1963 V256 N25 P5302-5304	65-13962 <*>
1963 V257 P1267-1270	66-11748 <*>
1963 V257 P1276-1279	AEC-TR-6307 <*>
	NP-TR-1157 <*>
1963 V257 P1285-1287	HW-TR-61 <*>
1963 V257 P1566-1569	66-11796 <*>
1963 V257 N1 P48-51	64-10687 <*>
1963 V257 N2 P394-397	65-17347 <*>
1963 V257 N3 P635-638	65-14360 <*>
1963 V257 N3 P671-673	64-16285 <*>
1963 V257 N16 P2258-2259	65-17144 <*>
1963 V257 N17 P2431-2433	65-17441 <*>
1963 V257 N23 P3609-3611	64-30694 <*>
1964 V258 P101-102	66-12942 <*>
1964 V258 P155-157	66-11836 <*>
1964 V258 P570-573	66-11835 <*>
1964 V258 P1069-1071	64-00435 <*>
1964 V258 P1096-1098	66-10981 <*>
1964 V258 P2376-2378	N65-14429 <=$>
1964 V258 P2525-2528	66-14103 <*>
1964 V258 P3461-3464	66-11089 <*>
1964 V258 P3672-3675	65-12322 <*> 0
1964 V258 P3690-3693	65-12326 <*>
1964 V258 P4171-4173	66-12611 <=*>
1964 V258 P4528-4530	3868 <BISI>
1964 V258 P4531-4534	66-13605 <*>
1964 V258 P5217-5219	4159 <BISI>
1964 V258 N3 P841-844	65-11497 <*>
1964 V258 N5 P1423-1425	65-11504 <*>
1964 V258 N6 P1752-1755	65-11502 <*>
1964 V258 N8 P2268-2270	65-13421 <*>
1964 V258 N9 P2482-2485	65-11503 <*>
1964 V258 N11 P2995-2998	64-30645 <*> 0
1964 V258 N12 P3228-3231	65-14649 <*>
1964 V258 N13 P3595-3601	66-11820 <*>
1964 V258 N14 P3672-3675	65-12255 <*>
1964 V258 N14 P3686-3689	65-11426 <*>
1964 V258 N18 P4539-4541	65-11501 <*>
1964 V258 N18 P4570-4572	65-10202 <*>
1964 V258 N19 P4706-4709	66-14145 <*>
1964 V258 N20 P4932-4934	65-13107 <*>
1964 V258 N23 P5573-5576	66-13871 <*>
1964 V258 N23 P5763-5766	66-12004 <*>
1964 V258 N26 P6399-6402	66-60573 <=*$>
1964 V258 N26 P6500-6502	66-11215 <*>
1964 V259 P1159-1162	GI N4 1964 P828 <AGI>
1964 V259 P1627-1630	4072 <BISI>
1964 V259 P1756-1759	GI N2 1964 P336 <AGI>
	N65-13542 <=$>
1964 V259 P2203-2206	65-14718 <*>
1964 V259 P2401-2403	66-11146 <*>
1964 V259 P2442-2444	66-11091 <*>
1964 V259 P4191-4194	66-12960 <*>
1964 V259 P4285-4288	66-11096 <*>
1964 V259 P4323-4326	GI V2 N1 1965 P171 <AGI>
1964 V259 P4773-4775	N65-18341 <=$>
1964 V259 N1 P77-79	N64-30854 <=$>
1964 V259 N1 P162-165	65-18133 <*>
1964 V259 N6 P3227-3228	N65-16310 <=$>
1964 V259 N10 P3584-3587	N65-19707 <=$>
1964 V259 N10 P4776-4778	N65-17268 <=$*>
1964 V259 N11 P1889-1890	GI N3 1964 P573 <AGI>
1964 V259 N11 P1891-1893	GI N5 1964 P1022 <AGI>
1964 V259 N13 P2105-2107	65-17131 <*>
1964 V259 N14 P2185-2186	66-11223 <*>
1964 V259 N14 P4415-4427	66-12152 <*>
1964 V259 N15 P2359-2360	65-13862 <*>
1964 V259 N18 P2977-2979	66-10788 <*>
1964 V259 N20 P3470-3473	65-17014 <*>
1964 V259 N23 P4233-4236	65-14090 <*>
1964 V259 N25 P4671-4674	66-10796 <*>
1965 V260 P1174-1177	66-10364 <*>
1965 V260 P1897-1900	54T90F <ATS>
1965 V260 P1957-1959	66-12811 <*>
1965 V260 P2405-2408	66-13904 <*>
1965 V260 P2443-2446	65-18131 <*>
1965 V260 P3359-3362	65-18122 <*>
1965 V260 P3390-3392	LA-TR-65-15 <*>

1965 V260 P3694-3695 66-10381 <*>
1965 V260 P3961 ANL-TRANS-197 <*>
1965 V260 P4917-4920 N66-13299 <=$>
1965 V260 P5047-5048 66-12130 <*>
1965 V260 P5307-5309 66-13942 <*>
1965 V260 P5809-5811 66-10380 <*>
1965 V260 N1 P152-154 65-14375 <*>
1965 V260 N2 P465-467 65-14430 <*>
1965 V260 N3 P1867-1869 N65-21636 <=$*>
1965 V260 N6 P1642-1645 65-14906 <*>
1965 V260 N6 P1681-1685 GI N3 1964 P567 <AGI>
1965 V260 N8 P2342-2343 65-13841 <*> 0
1965 V260 N9 P2617-2618 65-13839 <*>
1965 V260 N10 P1465-1467 N65-19708 <=$>
1965 V260 N10 P3219-3231 N65-31637 <=*$>
1965 V260 N12 P3393-3395 66-10365 <*>
1965 V260 N14 P3949-3952 66-14491 <*>
1965 V260 N21 P5538-5541 NLCO-TR-6 <*>
1965 V260 N22 P5795-5798 66-11279 <*>
1965 V260 N23 P6051-6053 66-10787 <*>
1965 V260 N25 P6562-6564 66-10889 <*>
1965 V261 P939 AD-632 676 <=$>
1965 V261 P1859-1861 66-11105 <*>
1965 V261 P2891-2894 4828 <BISI>
1965 V261 P4559-4562 66-11829 <*>
1965 V261 N5 P5007-5010 66-13917 <*>
1965 V261 N10 P2102-2105 66-12704 <*>
1965 V261 N18 P3650-3652 66-14166 <*>
1965 V261 N20 P4003-4006 66-14616 <*>
1965 V261 N21 P4373-4376 66-14680 <*>
1965 V261 N21 P4377-4380 66-11144 <*>
1965 V261 N22 P4729-4731 66-14006 <*>
1965 V261 N23 P5239-5242 66-12241 <*>
1965 V261 N25 P5501-5504 66-14681 <*>
1966 V262 P519-522 66-13863 <*>
1966 V262 P1024-1027 66-13864 <*>

COMPTE RENDU DES RECHERCHES EFFECTUEES. LABORATOIRES DU BATIMENT ET DES TRAVAUX PUBLICS
1943 P40-62 03372 <*>
1944 P49-56 AL-414 <*>
1947 P153-170 I-173 <*>
I473 <*>
3383 <*>

COMPTE-RENDU DES REUNIONS. COMITE INTERNATIONAL DE THERMODYNAMIQUE ET DE CINETIQUE ELECTROCHIMIQUES
1950 P185-197 1965 <*>
1951 P379-391 64-16453 <*> 0
1952 P344-347 II-424 <*>

COMPTE RENDU DE LA REUNION ANNUELLE. SOCIETE DE CHIMIE PHYSIQUE. PARIS
1952 P344-347 64-10457 <*> 0
1952 P354-356 64-10450 <*>

COMPTE RENDU DES SEANCES DE LA SOCIETE DE BIOLOGIE
1893 V45 N16 P63-96 59-10750 <*> 0
1902 V54 P170-172 64-30806 <*>
1902 V54 P788-790 64-30805 <*>
1904 V56 P852-855 63-20878 <*>
1916 V79 P352-354 64-30804 <*>
1922 V86 P1140-1142 57-3163 <*>
1924 V90 P1380-1382 1834 <*>
1924 V91 P351-354 II-81 <*>
1924 V91 P592-593 2376 <*>
1924 V91 P1067-1069 1833 <*>
1924 V91 P1069-1071 1831 <*>
1925 V92 P477-479 1832 <*>
1927 V97 P863-864 58-266 <*>
1928 V97 P1731-1733 65-00213 <*>
1928 V99 P805-810 59-15844 <*> 0
1929 V100 P501-504 59-18996 <=*>
1929 V101 P756-757 60-10628 <*>
1929 V102 P371-373 60-10630 <*>
1929 V102 P528-531 60-10629 <*>
1930 V103 P60-62 UCRL TRANS-1025 <*>
1930 V103 P776-778 2741 <*>
1930 V104 P831-832 64-10104 <*>
1931 V106 N1 P1080-1081 63-14851 <*>
1931 V106 N1 P1081-1084 63-14850 <*>
1931 V108 P27-30 64-30739 <*>
1931 V108 P1046-1048 64-30803 <*>
1932 V109 P923-925 64-30802 <*>
1932 V111 P545-547 I-355 <*>
1932 V111 P779-781 I-856 <*>
1932 V111 P996-998 I-618 <*>
1933 V112 P714-716 61-16842 <*>
1933 V114 P643-645 63-20875 <*>
1933 V114 P817-821 60-10627 <*> 0
1933 V114 P1162-1164 64-30801 <*>
1933 V114 P1164-1166 64-30800 <*>
1933 V114 P1297-1299 64-30799 <*>
1934 V115 P1624-1626 63-01774 <*>
1935 V118 P38-42 60-10631 <*>
1935 V118 P248-250 2827-D <K-H>
1935 V118 P577-580 64-30798 <*>
1935 V118 P967-969 63-14542 <*>
1935 V118 P1235-1241 61-00983 <*>
1935 V118 P1241-1244 61-00989 <*>
1935 V120 P577-580 64-30797 <*>
1936 V121 P1060-1062 64-30796 <*> 0
1936 V121 N1 P41-42 2496 <*>
1936 V122 P150-152 64-30795 <*> 0
1936 V122 P373-374 64-30794 <*> 0
1936 V122 P652-654 64-30793 <*>
1937 V124 P133-134 58-2289 <*>
1937 V124 N2 P164-165 2498 <*>
1937 V125 P1027-1028 64-30792 <*>
1937 V126 P31-32 3158 <*>
1937 V126 P635-637 64-30791 <*> 0
1937 V126 P1264-1266 64-30790 <*>
1937 V126 P1279-1280 60-10647 <*>
1937 V127 P1381-1382 1284G <K-H>
1937 V127 P1389-1392 1284E <K-H>
1938 V127 P39-42 64-30789 <*>
1938 V127 P113-116 64-30788 <*>
1938 V127 P116-118 64-30787 <*>
1938 V127 P345-346 64-30784 <*> 0
1938 V127 P1131-1133 2827-E <K-H>
1938 V127 P1195-1197 63-01622 <*>
1938 V127 P1464-1467 2776I <K-H>
1938 V127 P1493-1495 2827-C <K-H>
1938 V128 P995-997 64-10312 <*>
1938 V129 P32-37 2261 <*>
1938 V129 P838-840 64-10314 <*>
1938 V129 P1224-1225 64-30785 <*>
1939 V130 P27-29 64-30734 <*>
1939 V130 P29-31 64-30735 <*>
1939 V130 P207-209 64-30736 <*>
1939 V130 P209-211 64-30737 <*>
1939 V130 P211-214 64-30738 <*>
1939 V130 P429-430 64-30786 <*> 0
1939 V130 P1132-1134 2835C <K-H>
1939 V130 P1292-1294 2751B <K-H>
1939 V130 P1294-1295 2751A <K-H>
1939 V131 P222-224 59-12096 <=*>
1939 V131 P380-382 2766F <K-H>
1939 V131 P1186-1187 2766-D <K-H>
1939 V132 P87-90 2421A <K-H>
1939 V132 P375-377 62-00607 <*>
1941 V135 P113-115 II-38 <*>
1941 V135 P508-510 60-10644 <*>
1941 V135 P1133-1135 64-30783 <*>
1941 V135 P1309-1312 65-18099 <*>
1942 V136 P322 60-10646 <*>
1942 V136 P349-350 64-30782 <*>
1942 V136 P715-716 64-30781 <*>
1942 V136 P752-753 64-30780 <*>
1943 V137 P111 64-30731 <*> 0
1943 V137 P134-135 64-30730 <*> 0
1943 V137 P143 64-30779 <*>
1943 V137 P171-172 64-30778 <*>
1943 V137 P178-179 64-30729 <*> 0
1943 V137 P214-215 64-30777 <*>
1943 V137 P292-293 64-30776 <*> 0
1943 V137 P305-307 65-30775 <*>
1943 V137 P329-330 64-30774 <*>
1943 V137 P332 64-30773 <*>
1943 V137 P347-348 64-30772 <*>

1943 V137 P396 64-30771 <*>
1943 V137 P400-402 64-30770 <*> 0
1943 V137 P500-501 64-30728 <*>
1943 V137 P515-516 64-30769 <*>
1943 V137 P630-631 64-30768 <*>
1943 V137 P638-639 64-30767 <*>
1943 V137 P657-659 64-30727 <*> 0
1943 V137 P675-676 64-30726 <*> 0
1943 V137 P688-689 64-30766 <*> 0
1943 V137 P716-717 64-30765 <*>
1943 V137 P744-746 64-30764 <*> 0
1943 V137 N1/2 P74-75 64-30722 <*> 0
1943 V137 N1/2 P101-102 64-30810 <*>
1944 V138 P82-84 64-30762 <*> 0
1944 V138 P109-111 64-30761 <*>
1944 V138 P119-120 63-01527 <*>
1944 V138 P275-277 64-10820 <*>
1944 V138 P330-332 64-30760 <*>
1944 V138 P480-482 64-30759 <*>
1944 V138 P517-518 64-16228 <*>
1944 V138 P751-753 1395-A <K-H>
1944 V138 N1/2 P58-59 64-30763 <*>
1944 V139 03/15 P748-750 1284D <K-H>
1945 V139 P942-949 750 <K-H>
1945 V139 P946-947 64-30758 <*>
1946 V140 P644-646 64-16230 <*>
1946 V140 P646-647 64-16229 <*>
1946 V140 P974-976 1253-B <K-H>
1946 V140 N23/4 P955-956 64-16203 <*>
1947 V141 P20-22 3165 <*>
1947 V141 P111-114 AEC-TR-5829 <*>
1947 V141 P152-153 1166 <K-H>
1947 V141 P747-749 63-01376 <*>
1947 V141 N3/4 P159-161 64-30757 <*>
1947 V141 N11/2 P573-574 64-30756 <*>
1948 V142 P933-935 65-00143 <*>
1948 V142 P1074-1076 60-10642 <*> 0
78J17F <ATS>
1948 V142 P1089-1092 II-486 <*>
1948 V142 P1144-1146 II-518 <*>
1948 V142 N5/6 P325-327 64-30755 <*> 0
1948 V142 N5/6 P406-408 I-994 <*>
1949 V143 P500-501 1962-B <K-H>
1949 V143 P605-606 2360-B <K-H>
1949 V143 N13/4 P1032-1033
64-30724 <*>
1949 V143 N17/8 P1217-1218
64-30723 <*>
1949 V143 N17/8 P1259-1261
64-30754 <*>
1949 V143 N17/8 P1264-1266
64-30732 <*>
1949 V143 N17/8 P1366-1369
60-10689 <*> 0
1950 V144 P861-863 1984 <K-H>
1950 V144 P1340-1341 2536-A <K-H>
1950 V144 P1341-1343 2536-B <K-H>
1950 V144 P1579-1581 I-132 <*>
1950 V144 N1/2 P95-96 64-30753 <*>
1950 V144 N3/4 P230-231 64-30752 <*>
1951 V145 P417-421 64-10296 <*>
1951 V145 P1014-1017 60-10641 <*>
1951 V145 P1480-1483 64-20001 <*> 0
1951 V145 N1 P41-43 64-10297 <*>
1951 V145 N13/4 P1088-1091
61-00907 <*>
1951 V145 N15/6 P1151-1154
64-30740 <*> 0
1951 V145 N21/2 P1733-1735
60-15714 <=*>
1952 V146 P378-380 63-16254 <*>
1952 V146 P979-980 64-10313 <*>
1952 V146 N5/6 P470-471 64-30751 <*>
1952 V146 N13/4 P1095-1098
65-12624 <*>
1952 V146 N13/4 P1099-1102
63-01322 <*>
1952 V146 N13/4 P1121-1124
59-17141 <*> 0
1952 V146 N19/0 P1472-1473
2223 <*>
1953 V147 P204-206 57-1288 <*>
1953 V147 P207-210 57-1289 <*>
1953 V147 P295-299 57-1291 <*>
1953 V147 P603-605 AL-611 <*>
1953 V147 P647-649 AL-348 <*>
1953 V147 P1223-1225 AL-850 <*>
1953 V147 P1358-1360 I-104 <*>
1953 V147 P1361-1363 I-548 <*>
1953 V147 P1372-1375 I-106 <*>
1953 V147 P1410-1412 I-985 <*>
1953 V147 P1434-1437 I-463 <*>
1953 V147 P1465-1468 I-680 <*>
1953 V147 P1568-1571 AL-549 <*>
1953 V147 P1577-1579 AL-849 <*>
1953 V147 P1753-1756 63-01381 <*>
1953 V147 P1889-1891 II-142 <*>
1953 V147 P1891-1893 II-21 <*>
1953 V147 P1930-1933 II-82 <*>
1953 V147 P1950-1954 II-430 <*>
1953 V147 P2012-2016 II-92 <*>
1953 V147 N2 P1675-1677 63-01531 <*>
1953 V147 N3/4 P186-191 59-20170 <*>
1953 V147 N19/0 P1610-1616
64-30750 <*>
1954 V148 P66-68 60-17765 <*+>
1954 V148 P227-229 II-98 <*>
1954 V148 P251-254 II-131 <*>
1954 V148 P268-270 II-185 <*>
1954 V148 P279-280 II-184 <*>
1954 V148 P315-317 II-241 <*>
1954 V148 P355-358 58-388 <*>
1954 V148 P419-422 64-30749 <*>
1954 V148 P611- 57-823 <*>
1954 V148 P743- 57-1082 <*>
1954 V148 P907-909 II-76 <*>
1954 V148 P1177-1180 II-453 <*>
1954 V148 P1182-1184 II-363 <*>
1954 V148 P1196-1199 57-1135 <*>
1954 V148 P1213-1216 II-703 <*>
1954 V148 P1241-1243 57-1130 <*>
1954 V148 P1243-1246 1863 <*>
1954 V148 P1293- 57-1183 <*>
1954 V148 P1448-1450 2134 <*>
1954 V148 P1521-1523 II-436 <*>
1954 V148 P1523-1524 II-253 <*>
1954 V148 P1867-1870 II-955 <*>
1954 V148 P2005-2007 2630 <*>
1954 V148 P2024-2026 2628 <*>
1954 V148 P2037-2038 57-710 <*>
1954 V148 P2039-2040 57-709 <*>
1954 V148 P2123-2125 57-711 <*>
1954 V148 N5/6 P431-435 64-30748 <*>
1954 V148 N9/10 P981-984 59-17036 <*> 0
1955 V149 P377-378 57-1333 <*>
1955 V149 P398-400 57-1754 <*>
1955 V149 P462- 2992 <*>
1955 V149 P465-467 57-1840 <*>
1955 V149 P475-480 3026 <*>
1955 V149 P513-515 57-1830 <*>
1955 V149 P568-570 57-1836 <*>
1955 V149 P653-655 1165 <*>
1955 V149 P730-732 2993 <*>
1955 V149 P937-939 57-1321 <*>
1955 V149 P1251-1254 1209 <*>
1955 V149 P1319-1322 II-886 <*>
1955 V149 P1398-1402 1086 <*>
1955 V149 P1539-1541 II-849 <*>
1955 V149 P1634-1636 1142 <*>
1955 V149 P1642-1646 II-633 <*>
1955 V149 P1664-1665 II-999 <*>
1955 V149 P1722-1725 II-860 <*>
1955 V149 P1746-1749 II-719 <*>
1955 V149 P1759-1762 II-704 <*>
1955 V149 P1775-1778 II-726 <*>
1955 V149 P2199-2200 57-395 <*>
1955 V149 N12 P2269 5276 <K-H>
1955 V149 N5/6 P540-543 59-15550 <*>
1955 V149 N9/0 P1075-1077 64-30747 <*> 0
1955 V194 P495-498 57-1841 <*>

1956 P160-161 1356 <*>
1956 V150 P129-131 1448 <*>
1956 V150 P158-160 1712 <*>
1956 V150 P162-164 1322 <*>
1956 V150 P173-175 1314 <*>
1956 V150 P243-246 1193 <*>
1956 V150 P272-274 5525-B <K-H>
1956 V150 P290-292 1711 <*>
1956 V150 P429-431 1586 <*>
1956 V150 P664-666 58-953 <*>
1956 V150 P959-963 57-687 <*>
1956 V150 P963-967 57-689 <*>
1956 V150 P1462-1464 57-688 <*>
1956 V150 N1 P32-35 63-01385 <*>
1956 V150 N2 P444-447 58-2337 <*>
1956 V150 N4 P675-677 64-30746 <*>
1956 V150 N4 P790-792 64-30733 <*>
1956 V150 N5 P959-963 64-30745 <*> 0
1956 V150 N8/9 P1541-1544 64-30744 <*>
1957 V151 P82-85 57-2954 <*>
1957 V151 P402-404 58-73 <*>
1957 V151 P539-540 63-01679 <*>
1957 V151 P683-689 63-01681 <*>
1957 V151 P1773-1776 58-1725 <*>
1957 V151 N1 P166-169 60-18016 <*>
1957 V151 N6 P1149-1150 59-10355 <*>
1957 V151 N7 P1383-1386 62-14605 <*> 0
1957 V151 N11 P1888-1889 64-18621 <*>
1957 V151 N11 P1954-1956 66-10946 <*>
1958 V151 N12 P2135-2137 61-00156 <*>
1958 V152 P1086-1088 59-15409 <*>
1958 V152 N1 P29-31 60-18752 <*>
1958 V152 N1 P227 62-00644 <*>
1958 V152 N2 P238-240 59-10357 <*> 0
1958 V152 N2 P337-339 61-00155 <*>
1958 V152 N2 P367-369 60-18753 <*>
1958 V152 N3 P441-443 60-16680 <*>
1958 V152 N3 P533-534 61-10880 <*>
1958 V152 N4 P588-589 61-00157 <*>
1958 V152 N6 P1030-1034 59-15491 <*>
1958 V152 N6 P1034-1038 59-15490 <*>
1958 V152 N11 P1627-1629 61-00204 <*>
1958 V152 N12 P1828 64-30743 <*>
1959 V153 P314-318 63-16342 <*>
1959 V153 P744-747 60-14874 <*>
1959 V153 P817-818 60-18175 <*>
1959 V153 P1859-1863 62-10194 <*> 0
1959 V153 N2 P370-372 65-14638 <*>
1959 V153 N3 P489-490 60-14186 <*>
1959 V153 N3 P496 60-14185 <*>
1959 V153 N4 P704-706 60-14187 <*>
1959 V153 N4 P706-708 60-14188 <*>
1959 V153 N10 P1653- 61-00177 <*>
1959 V153 N10 P1669 65-00201 <*>
1959 V153 N11 P1728-1730 64-30742 <*> 0
1959 V153 N12 P1962-1965 63-01616 <*>
1959 V153 N8/9 P1346-1350 64-18647 <*>
1960 V153 N10 P1621-1624 60-16681 <*>
1960 V154 P202-206 63-00842 <*>
1960 V154 P1216-1219 61-18661 <*>
1960 V154 N3 P564-566 64-18998 <*> 0
1960 V154 N3 P567-570 64-18997 <*> 0
1960 V154 N7 P1359-1361 62-14989 <*>
1960 V154 N7 P1536-1539 61-10925 <*>
1960 V154 N12 P2371-2372 61-20578 <*>
1961 V154 P1694-1697 62-14675 <*>
1961 V154 P1878-1880 61-20464 <*>
1961 V154 P2399-2401 62-10736 <*>
1961 V155 P425-426 62-16600 <*>
1961 V155 P1099-1102 62-18000 <*>
1961 V155 P1417-1419 62-16609 <*>
1961 V155 P2356-2360 63-10184 <*>
1961 V155 N1 P18-21 62-16012 <*>
1961 V155 N1 P194-196 61-20974 <*>
1961 V155 N2 P335-338 62-14914 <*>
1961 V155 N2 P349-350 62-14979 <*>
1961 V155 N4 P790-792 62-14322 <*>
1961 V155 N4 P803-806 62-16904 <*>
1961 V155 N4 P807-810 62-14997 <*>
1961 V155 N4 P897-904 63-00732 <*>
1961 V155 N4 P934-938 62-14999 <*>
1961 V155 N5 P1034-1036 62-16261 <*>
1961 V155 N5 P1178-1181 62-01047 <*>
1961 V155 N5 P1181-1183 62-01029 <*>
1961 V155 N5 P1187-1189 62-01048 <*>
1961 V155 N5 P1192-1195 62-01046 <*>
1961 V155 N12 P2235-2240 62-20036 <*>
1962 V155 P1450-1452 63-10074 <*>
1962 V155 P1516-1519 62-16614 <*>
1962 V155 P1628-1630 62-16601 <*>
1962 V156 P183-186 63-20880 <*>
1962 V156 P400-402 64-30741 <*>
1962 V156 P593-597 63-10073 <*>
1962 V156 P670-672 63-10072 <*>
1962 V156 P765-768 63-18015 <*>
1962 V156 P1238-1245 63-18746 <*>
1962 V156 P1316-1319 65-00110 <*>
1962 V156 P1876-1880 63-20627 <*>
1962 V156 N2 P232-234 62-20106 <*>
1962 V156 N8/9 P1407-1410 63-01103 <*>
1962 V156 N8/9 P1517-1522 63-00606 <*>
1962 V156 N8/9 P1522-1525 63-00607 <*>
1963 V156 P1797-1802 63-18803 <*>
1963 V156 P1836-1839 63-18801 <*>
1963 V156 P1935-1936 63-18791 <*>
1963 V157 P631-634 C-1351 <=*$>
1963 V157 P780-782 64-14617 <*>
1963 V157 P866-868 64-14616 <*>
1963 V157 N1 P5-8 64-10382 <*>
1963 V157 N2 P300-303 64-10383 <*>
1963 V157 N2 P447-451 64-16072 <*>
1963 V157 N3 P548-552 64-18650 <*> 0
1963 V157 N4 P826-830 64-16859 <*>
1963 V157 N4 P931-933 64-16860 <*>
1963 V157 N5 P1039-1041 64-16853 <*>
1963 V157 N6 P1236-1240 64-16351 <*>
1963 V157 N10 P1830-1832 65-12795 <*>
1963 V157 N11 P1879-1882 65-11287 <*> 0
1963 V157 N11 P2005-2007 65-14458 <*>
1963 V157 N8/9 P1596-1599 65-18076 <*>
1964 V158 N2 P414-418 65-14395 <*>
1964 V158 N3 P483-486 65-11869 <*>
1964 V158 N3 P610-612 65-11291 <*>
1964 V158 N7 P1493-1495 65-14048 <*>
1964 V158 N8/9 P1754-1756 65-14049 <*>
1965 V159 N2 P316-319 65-18139 <*>
1965 V159 N2 P431-435 66-10532 <*>

COMPTE RENDU DES SEANCES DE LA SOCIETE DE PHYSIQUE ET D'HISTOIRE NATURELLE DE GENEVE
1937 V54 P122-123 60-10830 <*>

COMPTE RENDU. SOCIETE GEOLOGIQUE DE FINLANDE
1943 V15 P40-75 64-18279 <*> 0

COMPTE RENDU SOMMAIRE DES SEANCES DE LA SOCIETE GEOLOGIQUE DE FRANCE
1957 N14 P322-325 34K28F <ATS>
1961 V55 P1170-1172 62-16076 <*>

COMUNICARILE ACADEMIEI REPUBLICII POPULARE ROMANE
1952 V2 N3/4 P243-248 62-14870 <*>
1955 V5 N5 P821-825 5796-A <K-H>
1956 N1 P115-122 08I31RU <ATS>
1956 V6 P385-386 SCL-T-477 <*>
1956 V6 N2 P225-229 SCL-T-479 <*>
1956 V6 N2 P231-234 SCL-T-488 <*>
1956 V6 N3 P505-508 SCL-T-478 <*>
1956 V6 N6 P971-973 SCL-T-485 <*>
1957 V7 N1 P65-72 66-11469 <*>
1958 V8 P1127-1128 SCL-T-487 <*>
1960 V10 N2 P159-164 66-26094 <$*>
1961 V11 N2 P155-159 62-19452 <=*>
1961 V11 N2 P161-165 62-19452 <=*>
1961 V11 N2 P180-185 62-19429 <=*>
1962 V12 N9 P985-993 64-71614 <=>

CONCOURS MEDICAL
1953 V75 N36 P2001- II-853 <*>
1955 V77 N10 P1218-1222 4089-A <KH>

1955 V77 N10 P1225 4089-A <KH>
1955 V77 N11 P1107 3746 <KH>
1955 V77 N11 P1109-1111 3746 <KH>
1955 V77 N11 P1113-1114 3746 <KH>
1958 V80 N1 P19-32 59-14550 <+*> 0
1958 V80 N20 P2455- 59-10837 <*>
1960 V82 P189-196 62-13387 <*=>
1960 V82 N21 P2645-2650 62-32052 <=*> 0
1961 V83 P4067-4069 63-16539 <*>
1961 V83 N25 P3601-3608 63-00520 <*>

CONFINIA NEUROLOGICA
1944 V6 P317-340 66-10863 <*>
1954 V14 N1 P1-7 65-13955 <*> 0
1955 V15 N2 P73-83 57-829 <*>
1956 V16 N4/5 P238-242 5920-B <K-H>
1958 V18 P144-149 65-00392 <*> 0
1958 V18 P167-171 62-00285 <*>

CONGRES DE CHIMIE INDUSTRIELLE
1936 V16 P45-49 63-14144 <*>

CONSERVA. DEN HAAG
1961 V10 N6 P115-117 AEC-TR-5773 <*>

CONSERVE E DERIVATI AGRUMARI
1958 V7 N27 P154-162 59-21134 <=*> 0

CONSTRUCTION. BRUSSELS
1960 V15 N6 P270-276 B-64 <RIS>

CONSTRUCTION AND INDUSTRY
1938 V16 N2 P43-44 1792 <*>

CONSTRUCTION MATERIAL INDUSTRY. CHINESE PEOPLES REPUBLIC
SEE CHIEN CHU TS'AI LIAO KUNG YEH

CONSTRUCTION MONTHLY. CHINESE PEOPLES REPUBLIC
SEE CHIEN SHE YUEH K'AN

CONSTRUCTION ET TRAVAUX PUBLICS. PARIS
1933 V17 N1 P3-17 64-14203 <*>

CONSTRUCTORUL
1964 V16 P2 64-51224 <=>
1964 V16 N252 P1-2 64-31770 <=>
1964 V16 N751 P1-2 64-31770 <=>

CONTEMPORANUL
1962 N5 P7 63-21482 <=>
1962 N35 P1 62-33521 <=*>
1962 N35 P7 62-33521 <=*>
1963 N34 P7 63-41128 <=>
1965 V6 P7 65-30614 <=$>

CONTI ELEKTRO BERICHTE
1961 V7 P23-33 2450 <BISI>

CONTINENTALER EISENHANDEL
1963 V13 P41-44 3633 <BISI> P

CONTRIBUTI TEORICI E SPERIMENTALI DI POLAROGRAFIA
1955 V2 P1-15 57-2691 <*>
79H13I <ATS>

CONTRIBUTION FROM THE INSTITUTE OF GEOLOGY AND PALEONTOLOGY, TOHOKU UNIVERSITY
SEE TOHOKU DAIGAKU RIGAKUBU CHISHITSUGAKU KOSEIBUTSUGAKU KYOSHITSU KENKYU HOBUN HOKOKU

COOPERADOR DENTAL
1959 V26 N154 P6-10 62-00290 <*>
1959 V27 N155 P1-3 62-00290 <*>

CORRIERE SANITARIO
1903 V14 P78-80 10324-B <K-H>
64-14578 <*>

CORROSION ET ANTI-CORROSION
1954 V2 P240-244 II-347 <*>
1956 V4 N7 P233-240 T-2248 <INSD>
1956 V4 N7 P252-258 T2249 <INSD>
1957 N5 P112-118 38L32F <ATS>
1957 V5 N5 P153-155 59-10954 <*>
1958 V6 N1 P9-14 58-2374 <*>
1959 P56-62 TS-1407 <BISI>
1959 V7 P134-145 3782 <BISI>
1959 V7 N2 P56-62 62-10472 <*>
1959 V7 N3 P110-113 66-10830 <*> 0
1960 V8 N2 P49-58 4310 <BISI>
1961 V9 P341-356 HW-TR-36 <*>
1961 V9 P385-395 HW-TR-36 <*>
1961 V9 N5 P145-159 62-14434 <*>
1962 V10 N12 P401-407 AD-612 652 <=$>
1963 V11 N9 P313-318 4033 <BISI>
1964 V12 N1 P9-14 3918 <BISI>
1964 V12 N5 P213-226 65-17230 <*>

CORROSION ENGINEERING. JAPAN
SEE BOSHOKU GIJUTSU

CORROSION SCIENCE
1962 V2 P1-20 4148 <BISI>
1962 V2 P59-69 4187 <BISI>
1962 V2 P71-84 67P63G <ATS>
1962 V2 P119-131 AEC-TR-6082 <*>

COSMETOLOGIE
SEE ARCHIVES DE BIOCHIMIE ET COSMETOLOGIE

COULEURS
1955 N12 P1-9 57-2771 <*>

CRIMINOLOGIA. MILAN
1959 V12 N2 P119-149 60-31783 <=> C

CUIVRE ET LAITON
1939 P233-234 57-1104 <*>

CUIVRE, LAITONS, ALLIAGES
1956 N34 P9-13 66-10911 <*> 0
1961 N62 P9-17 66-10845 <*>

CULTURE
1958 V39 P7-12 58-2091 <*>
58-2465 <*>

CULTUUR GIDS
1905 V8 P626-630 57-2838 <*>
1911 S2 V13 P1-13 57-3359 <*>

CUORE E CIRCULAZIONE
1942 V26 N9 P257-269 64-10629 <*>
1942 V26 N9 P257-259 6999-B <K-H>
1954 V38 P164-170 3027 <*>
1957 V41 P103-115 63-01108 <*>
1962 V45 N1 P1-17 63-01239 <*>

CURRENT PROBLEMS IN DERMATOLOGY
1959 V1 P382-411 60-16947 <*>

CYTOLOGIA. TOKYO
1954 V19 N1 P1-10 60-18027 <*>

CZASOPISMO GEOGRAFICZNE
1961 V32 N1 P81-87 62-19480 <*=>

CZASOPISMO STOMATOLOGICZNE
1965 V18 N3 P214-220 66-12961 <*>

CZECHOSLOVAK JOURNAL OF PHYSICS
1953 V3 P170-171 AEC-TR-4660 <*>
1954 V4 N1 P53-66 57-2587 <*>
57-2909 <*>
58-2450 <*>
61-27272 <=*>
1954 V4 N4 P486-494 CSIRO-3402 <CSIR>
1954 V5 N4 P480-499 10K25R <ATS>
1955 V5 P521-530 AEC-TR-4400 <*>

1955 V5 N1 P11-17 62-11610 <=>
1955 V5 N1 P80-86 57-2587 <*>
58-2526 <*>
61-27272 <=*>
1955 V5 N1 P96-98 57-601 <*>
1955 V5 N2 P224-238 AEC-TR-5875 <*>
1956 V6 N3 P246-255 57-2960 <*>
1956 V6 N4 P310-321 3373 <*>
1956 V6 N4 P359-363 57-1198 <*>
62-01103 <*>
1956 V6 N5 P468-471 57-2388 <*>
1957 V7 P20-25 AEC-TR-5132 <*>
1957 V7 P110-111 AEC-TR-3755 <=*>
1957 V7 P592-598 AEC-TR-5128 <*>
1957 V7 N1 P120-122 62-14421 <*> 0
1957 V7 N2 P127-151 58-86 <*>
1957 V7 N4 P468-475 62-15997 <=*> 0
1958 V8 P592-599 UCRL TRANS-715(L) <*>
1958 V8 N1 P113-118 59-14902 <=*>
1958 V8 N2 P256-257 R-5121 <*>
1958 V8 N3 P322-329 61-14988 <*>
1958 V8 N3 P330-331 61-14987 <*> 0
1958 V8 N3 P530-543 AEC-TR-4433 PT.1 <*>
1958 V8 N4 P490 R-5118 <*>
1958 V8 N6 P665-684 AEC-TR-4433 PT.2 <*>
1959 V9 P258-259 65-12739 <*>
1959 V9 P652-665 AEC-TR-5461 <*>
1959 V9 P717-720 61-20882 <*>
1959 V9 N5 P590-596 SCL-T-373 <*>
1960 V10 P931-948 61-20549 <*>
1960 V10 P949-953 61-20547 <*>
1960 V10 P969-970 AEC-TR-4858 <*>
1960 V10 N9 P659-662 66-12131 <*>
1961 V11 N9 P664-667 13Q68R <ATS>
1962 V12 N5 P382-391 65-10884 <*>
1963 V13B P219-221 65-18090 <*>
1964 V14 P509-521 AD-618 724 <=$>

CZECHOSLOVAK MATHEMATICAL JOURNAL
1952 V2 P221-232 STMSP V1 P145-155 <AMS>
1953 V3 P154-157 AMST S2 V14 P55-57 <AMS>
1953 V3 N3 P328 STMSP V1 P145-155 <AMS>
1954 V4 P372-380 STMSP V2 P79-86 <AMS>
1955 V5 P451-461 AMST S2 V24 P1-8 <AMS>
1956 V6 P94-117 STMSP V2 P41-61 <AMS>
1956 V6 P217-259 AMST S2 V24 P19-77 <AMS>
1956 V6 P455-484 AMST S2 V24 P19-77 <AMS>
1956 V6 N1 P94-117 57-654 <*>
1956 V6 N81 P26-29 58-1259 <*>
1957 V7 P130-153 STMSP V2 P87-107 <AMS>
1957 V7 P254-272 AMST S2 V29 P271-288 <AMS>
1958 V8 P448-459 STMSP V2 P241-251 <AMS>
1958 V8 P610-617 STMSP V1 P245-252 <AMS>

DEW TECHNISCHE BERICHTE
1961 N1 P17-19 2178 <BISI>
1961 V1 N1 P2-5 6117 <HB>
1961 V1 N1 P15-16 5595 <HB>
1961 V1 N1 P17-19 5591 <HB>
1961 V1 N4 P150-155 5561 <HB>
1962 V2 P166-173 3260 <BISI>
1962 V2 N1 P16-24 3086 <BISI>
1963 V3 N2 P58-62 3438 <BISI>
1963 V3 N4 P136-149 3713 <BISI>
1964 V4 N1 P1-8 3686 <BISI>
1964 V4 N1 P9-12 6439 <HB>
1964 V4 N1 P13-23 6491 <HB>
1964 V4 N2 P80-84 6475 <HB>
1964 V4 N3 P107-110 6395 <HB>
1965 V5 N1 P13-20 65-17231 <*>
1965 V5 N2 P58-64 6631 <HB>
1965 V5 N3 P106-110 6751 <HB>

DIN NORMENHEFTE
1952 V1 P4114 57-3469 <*>

DAGENS NYHETER
1966 P1,14 66-33793 <=$>

DAI-4-KI KENKYU

1958 V1 N2 P45-48 59-15759 <*>

DAINICHI NEWS
SEE DAINICHI DENSEN JIHO

DAINICHI NIPPON CABLES REVIEW. JAPAN
SEE DAINICHI NIPPON DENSEN JIHO

DANSK LANDBRUG. AARHUS
1952 V71 N12 P144-145 C2381 <NRC>

DANSK TIDSSKRIFT
1950 V24 P1-8 II-612 <*>
1950 V24 P195-202 64-20000 <*>
1951 V25 P153-163 57-2663 <*>
1951 V25 P164-173 57-2660 <*>
1956 V30 P52-55 59-18483 <+*> P
1958 V32 P62-74 59-15584 <*>
1964 V38 P95-103 77S83D <ATS>

DANSKE VIDENSKABERNES SELSKAB. MATHEMATISK-FYSISKE MEDDELELSER. COPENHAGEN
1933 V12 N8 P1-65 UCRL TRANS-957(L) <*>

DANSKE VIDENSKABERNES SELSKAB. SELSKABS SKRIFTER. SERIES 2. NATURVIDENSKABELIG OG MATHEMATISK AFDELING
1924 S8 V5 N4 P237-343 65-17301 <*> P

DECHEMA-MONOGRAPHIEN
1939 V10 P78-86 II-154 <*>
1952 V21 P340-359 64-20308 <*> 0
1955 N24 P146-169 AEC-TR-3648 <+*>
1955 V24 N286 P105-145 60-16302 <*>
1956 V26 P238-259 63-16216 <=*>
1956 V26 P279-300 93J17G <ATS>
1956 V27 P32-44 09Q68G <ATS>
1956 V27 P275-285 61-00672 <*>
1956 V27 P328-336 3420 <*>
1957 V32 P39-45 NP-TR-362 <*>
1959 V32 P46-60 65-14238 <*> 0
1959 V32 P74-82 61-10224 <*>
1959 V33 N477 P269-277 63-16492 <*>
1962 V40 P41-76 3085 <BISI>
1962 V44 P59- NP-TR-1064 <*>
1962 V47 N805 P823-828 65-12759 <*>

DEKORATIVNOYE ISKUSSTVO
1960 N2 P17-21 61-15745 <+*>
1961 N1 P43-45 61-27292 <*=>
1962 N6 P22-25 63-15951 <=*> C

DEMPUNTO GIJUTSU KENKYU KAIHO
1962 N26 P44-57 64-18984 <*>
1963 N27 P13-17 720 <TC>
1963 N27 P26-30 64-51375 <=$>
1965 N31 P20-29 1882 <TC>

DENKI GAKKAI
1956 V24 P557-561 58-2261 <*>

★DENKI GAKKAI ZASSHI
1932 V52 N6 P498-499 66-14133 <*$>
1932 V52 N8 P633 66-13918 <=$>
1932 V52 N9 P736 66-14620 <*>
1932 V52 N11 P929 66-13919 <*$>
1933 V53 P408-412 59-20924 <*>
1933 V53 N10 P917-918 66-14246 <*> 0
1933 V53 N12 P1141-1142 66-14245 <*$> C
1934 V54 N1 P82-83 66-14212 <*$> 0
1936 V56 P394-395 65-10241 <*>
1936 V56 P1036-1041 SCL-T-420 <*>
1952 V72 N4 P41-46 59-12020 <+*>
1953 V73 P717-721 62-14085 <*> 0
1954 V74 N3 P271-277 SCL-T-294 <*>
1955 V75 P465-470 SCL-T-292 <*>
1960 V80 N860 P589-597 65-10831 <*>
1961 V81 N874 P1100-1108 47S83J <ATS>

★DENKI KAGAKU

1934 V2 P15-18 C-3388 <NRCC>
1935 V3 P127-141 I-306 <*>
2928-G <K-H>
64-10156 <*> 0
1942 V10 P409-413 T-1643 <INSD>
1945 V13 P2-3 64-20518 <*>
1945 V13 P3-6 64-20519 <*>
1947 V15 P39-42 I-446 <*>
I-448 <*>
64-14388 <*> 0
1947 V15 P61-63 T-1644 <INSD>
1947 V15 P72-74 I-447 <*>
I-449 <*>
64-14386 <*> P
1948 V16 P9-13 T-1645 <INSD>
1948 V16 P15-16 63-20466 <*>
1948 V16 P31-36 63-20465 <*>
1948 V16 P47-54 64-14683 <*> 0
1948 V16 N1/4 P1-5 64-14692 <*> 0
1948 V16 N1/4 P23-27 59-17183 <*> 0
1949 V17 P31-32 64-20520 <*>
1949 V17 P127-132 64-14151 <*> 0
1949 V17 P254-258 63-20464 <*>
1949 V17 N5 P76-79 64-16434 <*> 0
1950 V18 P12-14 64-14665 <*> 0
1950 V18 P158-160 I-433 <*>
1950 V18 P198-200 59-10604 <*>
1950 V18 P215-217 57-3011 <*>
1950 V18 P239-246 AL-451 <*>
2867G <K-H>
63-20774 <*>
1950 V18 P268-271 66-14553 <*>
1950 V18 P329-331 I-307 <*>
2953-H <K-H>
64-10435 <*>
1950 V18 P382-385 MF-6 <*>
1950 V18 N4 P110-112 64-20816 <*> 0
1950 V18 N6 P187-189 2385 <*>
1950 V18 N8 P266-268 64-20815 <*> 0
1951 V19 P107-110 AL-632 <*>
63-20762 <*>
1951 V19 P374-376 II-331 <*>
3000-D <K-H>
64-14141 <*> 0
1951 V19 N10 P328-331 64-20808 <*> 0
1952 V20 P108-111 T1784 <INSD>
1952 V20 P164-170 T-1770 <INSD>
1952 V20 P495-502 UCRL TRANS-515 <*>
1954 V22 P121-126 65-10019 <*> 0
1954 V22 P608-611 57-255 <*>
1954 V22 P612-615 57-256 <*>
1954 V22 P615-618 57-257 <*>
1954 V22 P662-667 II-887 <*>
1955 V23 P632-636 T-1905 <INSD>
1956 V24 P300-306 T-1906 <INSD>
1957 V25 P126-131 95N49J <ATS>
1957 V25 P238-248 16TM <CTT>
1959 V27 P391-394 64-14360 <*>
1961 V29 P8-14 62-10621 <*>
1961 V29 P697-701 62-18322 <*>
1962 N30 P82-88 88S85J <ATS>
1962 V30 P817-821 1693 <TC>
1962 V30 N8 P582-586 88Q69J <ATS>
1964 V32 P167-170 IS-TRANS-7 <*>
1964 V32 P376-378 1053 <TC>
1964 V32 P378-381 1052 <TC>
1964 V32 P664-667 65-14728 <*>
1964 V32 N4 P378-381 66-10491 <*>
1964 V32 N8 P586-589 1050 <TC>
66-14779 <*$>
1964 V32 N10 P747-752 1572 <TC>
1964 V32 N10 P757-762 1573 <TC>
1965 V33 P269-272 35T92J <ATS>
1965 V33 P346-350 66-10347 <*>
1965 V33 P518-522 66-12187 <*>

DENKI KORON
1962 V33 P117-128 3507 <BISI>
1962 V33 P280-285 3462 <BISI>
1963 V34 P384-392 3812 <BISI>

DENKI SEIKO
1962 V33 P117-128 3507 <BISI>
1962 V33 N4 P280-285 3462 <BISI>
1963 V34 N4 P384-392 3812 <BISI>

DENKI SHIKEN-SHO IHO
1955 V10 P825-832 57-2798 <*>
1960 V24 P925-939 AEC-TR-4793 <*>
1963 V27 N5 P350-360 65-10295 <*> 0
1963 V27 N7 P491-495 AD-609 439 <=>
1963 V27 N12 P928-934 66-11267 <*>

*DENKI TSUSHIN GAKKAI ZASSHI
1943 V26 P598-610 59-17476 <*>
1950 V33 N6 P305-312 59-17795 <*>
1952 V35 N5 P211-218 57-554 <*>
1952 V35 N11 P501-506 57-555 <*>
1953 V36 N2 P59-67 57-1008 <*>
1953 V36 N11 P620-624 66-14151 <*$> 0
1954 V37 N3 P177-182 66-14150 <*$> 0
1954 V37 N5 P359-365 57-962 <*>
1955 V38 P614-615 T-2068 <INSD>
1955 V38 N10 P770-775 57-562 <*>
58-29 <*>
1956 V39 N2 P112-117 63-18398 <*>
1956 V39 N6 P586-590 58-140 <*>
1957 V40 N2 P162-169 58-2391 <*>
59-15691 <*>
1957 V40 N11 P1151-1157 62-20162 <*>
1957 V40 N11 P1171-1177 AEC-TR-3654 <+*>
1958 V41 N11 P1132-1141 59-17793 <*>
1958 V41 N11 P1141-1149 59-17792 <*>
1958 V41 N11 P1149-1155 59-17794 <*>
1959 V42 N2 P151-155 66-11225 <*>
1959 V42 N8 P731-737 C-4860 <NRC>
1960 V43 N2 P146-153 56N49J <ATS>
1960 V43 N3 P271-280 02N53J <ATS>
61-20548 <*>
62-13702 <*=>
1960 V43 N3 P298-305 62-20180 <*>
1960 V43 N6 P711-718 64-14758 <*> 0
1961 V44 N4 P500-507 62-18841 <*>
1961 V44 N5 P811-814 62-18096 <*>
1961 V44 N5 P838-843 65-13290 <*>
1961 V44 N5 P934-941 62-16605 <*>
1961 V44 N7 P1033-1036 62-14790 <*>
1961 V44 N7 P1036-1040 62-14789 <*>
1961 V44 N8 P1207-1216 62-16603 <*>
1961 V44 N9 P1322-1328 62-18842 <*>
1961 V44 N9 P1328-1336 62-18093 <*>
1961 V44 N10 P1450-1456 63-14740 <*>
1962 V45 N2 P144-150 SC-T-64-1616 <*>
1962 V45 N5 P637-641 64-16954 <*>
1962 V45 N5 P641-647 64-18603 <*>
1962 V45 N8 P1071-1075 AD-640 519 <=$>
1963 V46 N5 P675-684 65-10293 <*> 0
1963 V46 N5 P685-693 65-10249 <*>
1964 V47 N1 P55-64 65-11530 <*>

DENKI TSUSHIN KENKYUSHO KENKYU JITSUYOKA HOKOKU
1965 V14 N8 P1479-1487 86S88J <ATS>
1965 V14 N8 P1507-1513 84S88J <ATS>
1965 V14 N8 P1571-1580 85S88J <ATS>

DENKSCHRIFTEN DER AKADEMIE DER WISSENSCHAFTEN
1915 V91 P362-377 2418 <*>

DENSHI-KEMBIKYO GAKKAISHI
1956 V4 N3 P166-173 59-15436 <*>
1961 V10 N2 P65-76 SCL-T-430 <*>

DENSHI SHASHIN
1960 V2 N1 P24-31 64-16498 <*> 0
1960 V2 N2 P13-16 08R75J <ATS>
1960 V2 N2 P26-29 25S88J <ATS>
1961 V3 N1 P12-15 53N57J <ATS>

1961 V3 N1 P15-18 54N57J <ATS>
1961 V3 N1 P18-24 55N57J <ATS>
1961 V3 N1 P25-26 56N57J <ATS>
1961 V3 N1 P27-31 57N57J <ATS>
1961 V3 N1 P32-38 58N57J <ATS>
1961 V3 N2 P16-21 52P6CJ <ATS>
1961 V3 N2 P22-26 78Q72J <ATS>
1961 V3 N2 P27-32 80Q72 <ATS>
1961 V3 N2 P32-37 96P61J <ATS>
1961 V3 N3 P9-17 46P62J <ATS>
1961 V3 N3 P18-19 47P62J <ATS>
1961 V3 N3 P26-28 48P62J <ATS>
1961 V3 N3 P29-32 49P62J <ATS>
1961 V3 N3 P33-38 50P62J <ATS>
1961 V3 N3 P49-54 51P62J <ATS>
63-18322 <*>
1961 V3 N3 P55-58 52P62J <ATS>
63-18321 <*>
1962 V4 N1 P3-6 66Q67J <ATS>
1962 V4 N1 P7-9 67Q67J <ATS>
1962 V4 N1 P13-17 69Q67J <ATS>
1962 V4 N1 P18-22 63-18805 <*>
70Q67J <ATS>
1962 V4 N1 P23-26 71Q67J <ATS>
1962 V4 N1 P27-30 63-18806 <*>
1962 V4 N2 P1-4 11Q71J <ATS>
1962 V4 N2 P5-12 91Q69J <ATS>
1962 V4 N2 P13-19 92Q69J <ATS>
1962 V4 N2 P20-24 93Q69J <ATS>
1962 V4 N2 P32-36 94Q69J <ATS>
1962 V4 N2 P37-39 10Q71J <ATS>
1962 V4 N3 P3-11 42Q73J <ATS>
1962 V4 N3 P12-17 40Q73J <ATS>
1962 V4 N3 P18-20 43Q73J <ATS>
1962 V4 N3 P21-27 41Q73J <ATS>
1962 V4 N3 P28-32 39Q73J <ATS>
1963 V5 N1 P5-10 64-16655 <*>
1963 V5 N1 P11-13 64-16656 <*>
1963 V5 N1 P14-20 64-16657 <*>
1963 V5 N1 P21-24 64-16658 <*>
64-20949 <*>
1963 V5 N2 P26-32 34S82J <ATS>
1963 V5 N2 P33-37 61R76J <ATS>
1963 V5 N2 P38-41 62R76J <ATS>
1963 V5 N2 P42-45 63R76J <ATS>
1963 V5 N2 P46-51 64R76J <ATS>
1963 V5 N2 P52-58 65R76J <ATS>
1963 V5 N3 P89-97 75S82J <ATS>
1964 V5 N3 P7-13 65-11951 <*>
1964 V5 N3 P26-30 65-11950 <*>
1964 V5 N3 P31-38 65-11949 <*>
1964 V6 N1 P2-6 87S80J <ATS>
1964 V6 N1 P7-10 88S80J <ATS>
1964 V6 N1 P11-17 89S80J <ATS>
1964 V6 N1 P18-22 33S81J <ATS>
1964 V6 N1 P23-25 74S81J <ATS>
1965 V6 N2 P8-15 78S88J <ATS>
1965 V6 N2 P16-21 79S88J <ATS>
1965 V6 N2 P22-28 80S88J <ATS>
1965 V6 N2 P26-31 82R84J <ATS>
1965 V6 N2 P32-39 66-10377 <*> 0
1966 V6 N3 P58-64,71 59T95J <ATS>
1966 V6 N3 P65-71 60T95J <ATS>
1966 V6 N3 P72-78 61T95J <ATS>
1966 V6 N3 P79-84 62T95J <ATS>
1966 V6 N3 P85-92 63T95J <ATS>

DENTAL JOURNAL OF JAPAN
SEE NIPPON NO SHIKAI

★DEREVOOBRABATYVAYUSHCHAYA PROMYSHLENNOST
1955 V4 N3 P24-25 59-15039 <+*>
1955 V4 N9 P23-25 CSIRO-3224 <CSIR>
R-1900 <*>
65-62680 <=*$>
1955 V4 N12 P9-12 R-4384 <*>
R-4410 <*>
1956 V5 N2 P17 CSIRO-3293 <CSIR>
1956 V5 N2 P17- R-1501 <*>
1956 V5 N9 P6-8 R-4151 <*>
1956 V5 N10 P3-5 R-4409 <*>
1956 V5 N10 P6-8 CSIRO-3802 <CSIR>
R-4986 <*>
1956 V5 N10 P9-19 CSIRO-3759 <CSIR>
1956 V5 N10 P9-10 R-4971 <*>
1957 V6 N2 P11-13 60-21896 <=>
1957 V6 N3 P17-18 60-51036 <=>
1957 V6 N4 P8-11 60-51019 <=>
1957 V6 N4 P11-13 60-51034 <=>
1957 V6 N5 P9-10 R-5156 <*>
1957 V6 N5 P10-12 60-51019 <=>
1957 V6 N5 P19-20 59-10184 <+*>
1957 V6 N7 P13-16 64-00325 <*>
1957 V6 N7 P17-18 60-51033 <=>
1957 V6 N8 P11-14 59-15402 <+*>
1957 V6 N9 P14-15 59-15402 <+*>
1957 V6 N9 P17-18 59-15401 <+*>
1957 V6 N9 P21 59-15400 <+*>
1958 N8 P11-13 60-10699 <+*> 0
1958 V7 N1 P4-7 59-15322 <+*>
1958 V7 N2 P9-11 59-15323 <+*>
1958 V7 N2 P13 60-00811 <*>
1958 V7 N2 P18 75L33R <ATS>
1958 V7 N3 P7-8 60-00810 <*>
1958 V7 N3 P16-17 60-51035 <=>
1958 V7 N5 P21-23 60-15763 <+*>
1958 V7 N6 P22 60-16515 <+*>
1958 V7 N9 P13-15 60-10934 <*>
1958 V7 N12 P4-5 60-10933 <*>
1959 V8 N2 P13-15 61-11422 <=>
1959 V8 N6 P25 60-16516 <+*>
1960 V9 N10 P9-10 66-51095 <=>
1961 V10 N8 P11-12 66-51097 <=>
1962 V11 N12 P16-17 65-63668 <=*$>
1962 V11 N12 P17-18 65-63691 <=*$>
1963 V12 N2 P17-19 65-63018 <=*$>
1963 V12 N4 P6-8 64-19542 <=$>
1963 V12 N9 P19 66-11552 <*>
1965 V14 N3 P13-15 66-12198 <*>

DEREVOPERERABATYVAYUSHCHAYA I LESOKHIMICHESKAYA PROMYSHLENNOST
1953 V2 N3 P6-8 R-1505 <*>
RT-3738 <*>
1954 V3 N2 P3-6 CSIRO-3316 <CSIR>
R-1442 <*>
R-3945 <*>
1954 V3 N2 P29 CSIRO-3317 <CSIR>
1954 V3 N2 P29- R-3936 <*>
1954 V3 N2 P29 R-832 <*>
1954 V3 N4 P3-8 RT-3737 <*>
1954 V3 N10 P5-8 R-777 <*>

DERMATOLOGIA. NAPOLI
1955 V6 P33-34 II-852 <*>

DERMATOLOGICA. BASEL
1948 V97 N1/2 P25-34 64-16934 <*> 0
1950 V100 N4/6 P304-309 59-17862 <*> 0
1950 V101 N2 P90-93 65-13240 <*> 0
1951 V102 N4/6 P302-306 59-17863 <*> 0
1952 V104 N4/5 P267-272 59-17864 <*> 0
1953 V106 P300- AL-486 <*>
1953 V106 P357-378 AL-646 <*>
1953 V106 N6 P357-378 63-18479 <*> 0
1954 V108 N4/6 P235-256 61-16478 <*> OP
1954 V109 N5 P295-305 II-1140 <*>
1955 V110 P315-322 58-1608 <*>
1955 V110 P323-331 58-1609 <*>
1955 V110 P426-438 57-1834 <*>
1957 P540-546 1224 <TC>
1957 V115 N4 P540-546 66-11771 <*>
1961 V122 N3/4 P274-278 65-13836 <*> 0
1961 V123 P277-287 64-14309 <*>
1962 V124 N2 P81-98 62-01200 <*>
1964 V128 N4 P297-317 64-20911 <*>
1965 V130 N2 P107-112 65-17031 <*>
1965 V131 N3 P213-223 66-10737 <*>

DERMATOLOGISCHE WOCHENSCHRIFT

1926 V83 P1C20-1025 64-00201 <*>
1952 V125 P565-571 2942 <*>
1952 V125 N1 P1-6 61-00365 <*>
1952 V126 N28 P657-662 64-10503 <*>
1953 V128 N34 P829-837 64-20036 <*>
1956 V133 N5 P105-108 65-17236 <*>
1956 V134 N45 P1206-1208 65-10730 <*>
1957 V136 P1289-1303 65-60302 <=$>
1957 V136 N41 P1085-1092 61-00147 <*>
1963 V148 P585-588 65-13820 <*>
1964 V150 N52 P673-675 65-13910 <*>

DERMATOLOGISCHE ZEITSCHRIFT
1928 V52 P30-39 881-V <KH>

DERMATOLOGY AND UROLOGY. JAPAN
SEE HIFU TO HINYO

DEUTSCHE AGRARTECHNIK
1964 V14 N6 P257-261 66-11710 <*>

DEUTSCHE APOTHEKERZEITUNG
1938 V53 P719 65-10728 <*>
1938 V53 P1245 5564 <K-H>
1942 V57 P155-156 60-10672 <*>
1942 V57 P163-164 60-10672 <*>
1951 V91 P319-321 65-10729 <*> 0
1953 V93 N5 P81-82 64-10307 <*>
1954 V94 P721-723 57-1432 <*>
1955 V95 N7 P153-157 64-16943 <*>
1961 V101 P1011-1015 65-14599 <*>
1961 V101 N46 P1481-1482 62-20170 <*>
1962 V102 N27 P827-832 65-11783 <*>
1963 V103 N16 P467-474 65-17208 <*>

DEUTSCHE BAUZEITSCHRIFT
1954 V4 P231-235 T-1536 <INSD>

DEUTSCHE BUNDESBAHN
1955 N15 P629-648 T-2088 <INSD>

DEUTSCHE DROGISTENZEITUNG
1950 V5 P565-566 07L32G <ATS>

DEUTSCHE EISENBAHNTECHNIK
1955 N3 P235-236 T-2023 <INSD>
1955 V3 P237-240 T-2024 <INSD>
1955 V3 P249-254 T2025 <INSD>
1956 V4 P41-45 T2091 <INSD>
1958 V6 N6 P267-275 59-11009 <=*>
1958 V6 N6 P276-282 59-11024 <=*>
1958 V6 N6 P298-304 59-11044 <=*>
1960 N2 P84-86 1825 <BISI>
1960 V8 N9 P429-432 146-113 <STT>
1961 V9 N3 P125-129 155-623 <STT>
1962 V10 N3 P113-118 62-32776 <=*>
1962 V10 N6 P268-269 62-32776 <=*>
1963 V11 N1 P7-12 63-21530 <=>
1963 V11 N1 P40 63-21530 <=>
1964 V12 N5 P193-196 64-41195 <=>
1964 V12 N8 P343-346 65-14169 <*> 0
1964 V12 N8 P351-356 64-51109 <=>
1965 V13 N7 P300-303 65-32238 <=$>

DEUTSCHE ELEKTROTECHNIK
1952 V6 N5 P199-202 62-16932 <*> 0
1953 V7 N8 P381-383 57-2830 <*>
1955 P129-134 T-2058 <INSD>
1956 V10 N8 P286-290 58-1717 <*>

DEUTSCHE FARBEN-ZEITSCHRIFT
1954 V8 N3 P88-91 59-15650 <*> 0
1955 V10 P377-387 61-16180 <*> 0
1958 V12 N1 P15-24 59-15659 <*> 0
1958 V12 N2 P56-61 59-15660 <*> 0
1958 V12 N2 P63-64 59-15674 <*>
1958 V12 N3 P103-108 59-15670 <*> 0
1958 V12 N4 P139-143 59-15669 <*> 0
1958 V12 N10 P397-401 59-15977 <*> 0
1958 V12 N11 P437-442 60-16573 <*> 0

1960 V14 P268-271 2217 <TTIS>
1960 V14 N6 P229-233 <TTIS>
1960 V14 N10 P393-394 2239 <TTIS>
1960 V14 N10 P394-395 2240 <TTIS>
1960 V14 N10 P397-398 NP-TR-847 <*>
1962 V16 N9 P379-387 64-10391 <*> 0
1963 V17 P201-203 NS 131 <TTIS>
1963 V17 N1 P38-43 71Q73G <ATS>
1963 V17 N9 P417-424 65-13528 <*>
1963 V17 N11 P497-502 64-14261 <*>
1964 V18 N1 P8-16 224 <TC>
1964 V18 N3 P117-121 225 <TC>
1964 V18 N9 P395-401 NS-350 <TTIS>
1964 V18 N9 P401-405 65-11621 <*> 0
1965 V19 N2 P57-58 NS-351 <TTIS>
65-13359 <*>
1965 V19 N5 P183-187 4643 <BISI>

DEUTSCHE FLUGTECHNIK
1961 N9 P345-346 62-13288 <*=>
1961 V5 N11 P418-419 63-15921 <=*>

DEUTSCHE GEFLUEGELZEITUNG. BERLIN
1957 V6 N1 P1-2 58-2441 <*>

DEUTSCHE GEODASTISCHE KOMMISSION
1955 N10B P7-16 27L35G <ATS>
1955 N10B P33-35 26L35G <ATS>
1955 N10B P37-48 62-25585-2 <=*>
1956 N28 P1- 62-25589 <=*> P

DEUTSCHE GEODAETISCHE KOMMISSION. REIHE A.
1957 N19 ENTIRE ISSUE 62-25586 <=*>
1957 N22 P1- 62-25587 <=*> C
1957 N22 P35- 62-25587 <=*> 0
1957 N26 ENTIRE ISSUE 62-25588 <*$>

DEUTSCHE GEODAETISCHE KOMMISSION. REIHE B.
1959 N54 ENTIRE ISSUE 62-25591 <=*>

DEUTSCHE GESUNDHEITSWESEN
1950 V5 P776-780 65-10731 <*>
1953 V8 N22 P654-657 3138-B <K-H>
1955 V10 P808-812 3808-B <K-H>
1955 V10 P1250-1251 57-3400 <*>
1955 V10 N47 P1536-1540 64-20020 <*>
1956 V11 N40 P1363-1365 59-12324 <=*>
1958 V13 N27 P855-859 8835-C <K-H>
1959 V14 P1638-1642 60-21029 <=*>
1959 V14 N4 P152-157 60-10846 <*> 0
1959 V14 N17 P778-788 60-21060 <=*>
1959 V14 N24 P1111-1126 61-00909 <*>
1959 V14 N42 P1944-1951 64-10636 <*> 0
9874-C <K-H>
1960 V15 P59-69 10281-C <K-H>
1960 V15 P165-167 62-00157 <*>
1960 V15 P1724-1726 62-18776 <*>
1960 V15 N1 P59-69 63-16670 <*>
1960 V15 N12 P630-631 60-16687 <*> 0
1960 V15 N51 P2477-2483 62-19716 <=*>
1960 V15 N51 P2494-2495 63-16956 <*>
1960 V15 N49/0 P2387-2403 62-19714 <=*>
1960 V15 N49/0 P2387 62-19715 <=>
1960 V15 N49/0 P2419 62-19715 <=>
1961 V16 P701-708 63-26233 <=$>
1961 V16 N1 P9-11 11095-F <KH>
64-10417 <*>
1961 V16 N14 P615-618 62-19746 <=*>
1961 V16 N15 P686-690 11335 <KH>
1961 V16 N15 P690-694 62-19713 <*=>
1961 V16 N16 P717-719 62-19726 <=*>
1961 V16 N16 P733-739 62-19717 <=*> P
1961 V16 N19 P873-878 62-19747 <=*>
1961 V16 N21 P964-972 62-19727 <=*>
1961 V16 N21 P972-977 62-19728 <=*>
1961 V16 N36 P1671-1673 62-11088 <=>
1961 V16 N44 P2079-2084 62-19729 <=*>
1962 V17 N3C P1266-1270 62-33683 <=*>
1962 V17 N48 P2090-2093 63-21118 <=>
1963 V18 N42 P1813-1819 63-41251 <=>

1964 V19 N14 P635-637 66-11821 <*>
1964 V19 N28 P1293-1297 64-41291 <=>
1964 V19 N34 P1576-1580 65-17293 <*>
1965 V2C N2 P53-62 65-30430 <=$>
1965 V20 N3 P101-107 65-30430 <=$>
1965 V20 N9 P409-412 65-30632 <=$>
1965 V20 N32 P1479-1485 65-32830 <=*$>

DEUTSCHE GEWAESSERKUNDLICHE MITTEILUNGEN
1960 V4 N1 P17-27 NP-TR-943 <*>

DEUTSCHE GOLDSCHMIEDEZEITUNG
1958 P193-197 UCRL TRANS-914(L) <*>
1958 P249-251 UCRL TRANS-914(L) <*>
1958 P327-329 UCRL TRANS-914(L) <*>
1958 P381-385 UCRL TRANS-914(L) <*>

DEUTSCHE HYDROGRAPHISCHE ZEITSCHRIFT
1949 V2 N1/3 P44-51 II-877 <*>
1950 V3 N1/2 P69-77 62-00576 <*>
1950 V3 N3/4 P169-183 57-1322 <*>
1951 V4 N1/2 P13-17 I-303 <*>
1951 V4 N4/6 P129-149 II-307 <*>
1952 V5 N5 P277-285 I-730 <*>
1952 V5 N2/3 P114-131 AL-757 <*>
1952 V5 N5/6 P231-245 I-120 <*>
1953 V6 N1 P18-32 AL-429 <*>
1953 V6 N3 P107-123 2233 <*>
1953 V6 N4/6 P145-170 62-00574 <*>
1954 N3/4 P129-140 57-1163 <*>
1954 V7 N3/4 P139-140 1585 <*>
1955 V8 P186-194 58-741 <*>
1955 V8 N1 P29-30 II-474 <*>
1957 V10 N5 P169-175 59-10052 <*>
1964 V17 N5 P232-235 AD-623 257 <=$*>

DEUTSCHE KRAFTFAHRTFORSCHUNG
1938 N7 P1-14 63-14263 <*> 0
1938 N14 P1-35 59-20599 <*> P
1939 N21 P1-35 63-18205 <*> PO
1939 N27 P15- 63-10975 <*> OP
1941 N53 P31-44 63-10996 <*>
1941 N63 P1-2C 63-10993 <=*> 0
63-14279 <*> 0

DEUTSCHE KRAFTFAHRTFORSCHUNG UND STRASSENVERKEHRS-TECHNIK
1960 N142 P3-12 61-20083 <*>

DEUTSCHE LEBENSMITTEL-RUNDSCHAU
V47 P221-228 57-1929 <*>
1948 V44 P197 60-10810 <*>
1949 V45 P29-40 T1837 <INSD>
1953 V49 N2 P33-34 60-10809 <*>
1955 V51 P23-25 60-10808 <*>
1955 V51 P1C6-111 19G8G <ATS>
1955 V51 N4 P106-111 CSIRO-3204 <CSIR>
1955 V51 N5 P121-124 61-16719 <*>
1955 V51 N5 P130-132 67G8G <ATS>
1958 V54 N2 P25-28 63-16255 <*>
1958 V54 N7 P151-155 60-14171 <*> 0
1960 V56 N1 P1-6 NP-TR-524 <*>
1961 V57 N3 P57-60 11128-C <KH>
1962 V58 N10 P293-296 63-20734 <*>
1963 V59 N11 P317-320 64-20380 <*>
1964 V60 N2 P38-43 94R75G <ATS>

DEUTSCHE LUFTWACHT
1939 V6 N2 P47-53 I-836 <*>
1940 V7 N4 P101-103 63-20187 <*> 0
1940 V7 N4 P113-114 65-12964 <*>
1940 V7 N8 P282-283 65-12969 <*>
1940 V7 N11 P388-392 65-12963 <*>
1941 V8 N10 P305-309 61-18558 <*>
1942 V9 N9 P263-267 61-18465 <*>
1942 V9 N11 P311-312 61-18546 <*>
1942 V9 N12 P356-360 58-488 <*>

DEUTSCHE MEDIZINISCHE WOCHENSCHRIFT
1893 V9 N15 P341-344 63-00037 <*>
1899 V25 N45 P733-735 62-16298 <*>
1901 V27 N14 P219- 1596 <*>
1904 V30 P1497-1499 MF-2 <*>
1905 V31 P1667-1671 58-1072 <*>
1907 V33 N50 P2082-2083 64-26247 <$>
1907 V34 N47 P2012-2014 64-10878 <*>
1912 V38 P1495-1497 66-12634 <*>
1912 V38 P1651 66-12632 <*>
1912 V38 P2035 66-12631 <*>
1913 V39 P1366 2670 <*>
1913 V39 P1586-1588 63-00747 <*>
1916 V42 N1 P1-6 57-574 <*>
1916 V42 N14 P434- 57-573 <*>
1922 V48 P1100-1102 65-12013 <*>
1924 V50 P944-945 61-14167 <*> 0
1924 V50 P980-983 61-14167 <*> 0
1925 V51 N13 P520-521 59-17217 <*>
1932 V58 N39 P1525-1527 59-14575 <+*>
1932 V58 N49 P1929-1932 59-18148 <+*>
1932 V58 N49 P1932-1934 59-18139 <+*>
1932 V58 N49 P1934-1935 59-18143 <+*>
1934 V60 P561-564 65-10735 <*> 0
1934 V60 N31 P1167-1173 61-14184 <*>
1935 V61 P829-832 65-10736 <*>
1935 V61 P2079-2081 65-10733 <*>
1936 V62 N16 P636-641 60-10616 <*> 0
1936 V62 N17 P679-683 6C-10615 <*>
1937 V63 P893-894 64-16214 <*>
1937 V63 N27 P1029-1033 63-14576 <*>
1938 V64 P671 65-10732 <*>
1938 V64 P1174-1178 65-10737 <*>
1938 V64 N17 P6C4-605 60-10613 <*>
1939 V65 P967-968 65-10734 <*>
1940 V66 N35 P959-962 60-10611 <*>
1941 P315-319 57-1653 <*>
1941 V67 P393-399 57-1679 <*>
1942 V68 P888-890 PT2 57-1676 <*>
1943 V69 P245-246 3561-A <K-H>
1944 N8 P126-136 59-19272 <+*> 0
1949 V74 N6 P161-167 65-10724 <*> P
1949 V74 N15 P465-468 60-10610 <*>
1949 V74 N31/2 P969-970 2607 <*>
1950 V75 N35 P1150-1153 60-10607 <*>
1950 V75 N27/8 P950 60-10609 <*>
1950 V75 N29/3 P987-990 60-10608 <*>
1951 V76 N9 P277 65-10725 <*> 0
1951 V76 N18 P605-608 2578 <*>
1951 V76 N33/4 P1019-1021 59-17191 <*> 0

1952 V77 N1 P13-15 65-10726 <*> 0
1952 V77 N1 P15-17 65-10718 <*>
1952 V77 N42 P1284-1287 65-10720 <*> 0
1952 V77 N43 P1317-1320 66-11715 <*> 0
1952 V78 P322-327 AL-816 <*>
1953 V78 P565-568 AL-13 <*>
1953 V78 P879-883 I-14 <*>
1953 V78 P1129-1134 57-1290 <*>
1953 V78 N10 P330-332 65-10721 <*>
1953 V78 N36 P1209-1210 2219 <*>
1953 V79 N45 P1566-1569 59-16299 <+*>
1953 V79 N46 P1600-1604 59-16299 <+*>
1954 V79 P39-41 AL-166 <*>
1954 V79 P239-241 62-00915 <*>
1954 V79 P356-359 2037 <*>
1954 V79 P1257-1261 I-514 <*>
1954 V79 P1664-1666 2480 <*>
1954 V79 P1738 5550-A <K-H>
1954 V79 N16 P6C1-604 65-10723 <*> 0
1954 V79 N16 P630-636 65-13239 <*> 0
1954 V79 N22 P879-880 65-10722 <*> 0
1954 V79 N41 P1522- II-787 <*>
1954 V79 N41 P1535-1536 II-787 <*>
1955 V80 P47-54 57-1153 <*>
1955 V80 P929-930 2116 <*>
1955 V80 P1380-1382 1111 <*>
1955 V80 P1396-1397 1111 <*>
1955 V80 P1449-1452 II-864 <*>
1955 V80 P1452-1455 II-821 <*>
1955 V80 P1455-1460 II-833 <*>
1955 V80 P1489-1494 57-1959 <*>

1955 V80 P1654-1655 II-561 <*>
1955 V80 N14 P465-467 1488 <*>
1955 V80 N18 P718-724 3991 <K-H>
1955 V80 N40 P1452-1455 65-12980 <*>
1955 V80 N40 P1455-1460 65-12982 <*>
1955 V80 N47 P1717-1721 1465 <*>
1956 V81 P495-501 5299-A <K-H>
1956 V81 P801- <RIS>
1956 V81 P8C6- <RIS>
1956 V81 P1084-1086 3057 <*>
1956 V81 P1130-1133 57-1252 <*>
1956 V81 P1613-1614 57-735 <*>
1956 V81 P2058-2061 57-2963 <*>
1956 V81 N1 P4-8 65-00230 <*>
1956 V81 N5 P173-174 6405-A <K-H>
1956 V81 N16 P617-621 59-14874 <+*> 0
1956 V81 N21 P823-835 62-14949 <*>
1956 V81 N24 P964-965 57-2934 <*>
1956 V81 N21/2 P835-846 62-14948 <*>
1956 V81 N21/2 P887-905 62-14948 <*>
1957 N28 P1172-1174 59-13038 <=>
1957 V82 P217-218 3393 <*>
1957 V82 P1377-1382 59-19056 <+*> 0
1957 V82 P1465-1468 58-2319 <*>
1957 V82 N9 P311-315 0A-78 <RIS>
1957 V82 N17 P644-649 60-17340 <*+> 0
1957 V82 N36 P1514-1515 62-14946 <*>
1957 V82 N36 P1515-1518 62-14944 <*>
1957 V82 N36 P1518-1524 62-14945 <*>
1957 V82 N36 P1524-1525 62-14947 <*>
1957 V82 N36 P1525-1526 62-14932 <*>
1957 V82 N36 P1526-1528 62-14933 <*>
1957 V82 N36 P1528-1531 62-14931 <*>
1957 V82 N36 P1531-1533 62-14930 <*>
1957 V82 N36 P1533-1537 62-14929 <*>
1957 V82 N36 P1537-1539 62-14928 <*>
1957 V82 N36 P1539-1541 62-14927 <*>
1957 V82 N36 P1541-1542 62-14926 <*>
1957 V82 N36 P1544-1551 62-14925 <*>
1957 V82 N36 P1554-1556 62-14924 <*>
1957 V82 N36 P1568-1574 62-14922 <*>
1957 V82 N43 P1826-1828 64-20026 <*>
1958 V83 P500-501 58-2616 <*>
1958 V83 P1399-1400 96L30G <ATS>
1958 V83 P1488-1491 58-2696 <*>
1958 V83 P1497-1500 58-2706 <*>
1958 V83 P2063-2064 62-16611 <*>
1958 V83 N1 P12-17 63-00859 <*>
1958 V83 N1 P25-26 63-00859 <*>
1958 V83 N5 P174-180 59-14240 <+*>
1958 V83 N15 P617-619 60-17327 <*+> 0
1958 V83 N19 P834-838 59-14624 <+*> 0
1958 V83 N21 P905-911 61-20001 <*>
1958 V83 N27 P1167 65-12992 <*>
1958 V83 N29 P1227-1230 61-01048 <*>
1958 V83 N34 P1426-1428 59-16414 <+*> 0
1958 V83 N43 P1908-1911 59-16155 <+*>
1958 V83 N45 P2CC2-2005 61-20634 <*>
1958 V83 N50 P2261-2265 59-21015 <=> 0
1959 V84 P428-433 65-00323 <*> 0
1959 V84 P1934-1939 60-19903 <*+> 0
1959 V84 N22 P1022-1028 62-01395 <*>
1959 V84 N27 P1328-1330 10732 <K-H>
1959 V84 N37 P1687-1689 01L37G <ATS>
1959 V84 N48 P2158-2160 60-19486 <*=> 0
1960 V85 P2242-2245 63-16500 <*>
1960 V85 N12 P463-467 61-00192 <*>
1960 V85 N16 P586-593 61-00214 <*>
1960 V85 N39 P1714-1717 63-16501 <*>
1960 V85 N39 P1717-1719 63-16499 <*>
1961 V86 P1517-1521 63-14767 <*>
1961 V86 P1596-1599 63-14766 <*>
1961 V86 P1606-1608 62-16613 <*>
1961 V86 P1621-1625 62-10084 <*> P
1961 V86 N6 P264-266 63-00557 <*>
1961 V86 N11 P499 62-00863 <*>
1961 V86 N12 P531-532 62-14916 <*>
1961 V86 N12 P535 62-14916 <*>
1961 V86 N13 P607-610 61-00842 <*>
1961 V86 N18 P897-904 62-00905 <*>
1961 V86 N32 P1497-1503 61-01000 <*>
1961 V86 N40 P1893-1899 63-00438 <*>
1961 V86 N45 P2170-2176 62-20030 <*>
1961 V86 N48 P2316-2322 62-00907 <*>
1961 V86 N48 P2327-2332 63-10543 <*>
1962 V87 P884-887 63-01751 <*>
1962 V87 P1388-1394 63-20903 <*>
1962 V87 P2288-2290 63-10924 <*>
1962 V87 P2661-2667 63-16343 <*>
1962 V87 N1 P23-27 65-11738 <*>
1962 V87 N3 P1613-1614 63-10042 <*>
1962 V87 N5 P246-249 63-14773 <*>
1962 V87 N8 P4C6-408 62-16122 <*>
1962 V87 N19 P992-1002 63-10328 <*> 0
1962 V87 N33 P1597-1607 62-20105 <*>
1962 V87 N46 P2378-2381 66-12495 <*>
1962 V87 N47 P2408-2419 66-12749 <*> 0
1963 V88 P313-320 63-20908 <*> 0
1963 V88 P836-844,847 65-12802 <*>
1963 V88 N10 P485-490 64-16858 <*> 0
1963 V88 N11 P505-514 63-00923 <*>
1963 V88 N33 P1598-1603 65-10976 <*>
1963 V88 N34 P1627-1633 64-00262 <*>
1963 V88 N34 P1638-1643 64-18635 <*>
1963 V88 N39 P1878-1886 65-13473 <*>
1964 V89 N13 P636-637 65-14404 <*>
1964 V89 N27 P1303-1306 65-14704 <*>
1964 V89 N34 P1596-1597 66-10113 <*>
1964 V89 N43 P2044-2047 65-14705 <*>
1964 V89 N45 P2117-2121 96S89G <ATS>
1964 V89 N50 P2374-2379 65-13477 <*>
1965 V90 P258-261 66-13203 <*>
1965 V90 P1290-1294 66-15119 <*>
1965 V90 N5 P2C0-204 65-14428 <*>
1965 V90 N5 P214-219 65-14427 <*>
1965 V90 N24 P1110 65-17420 <*>
1965 V90 N29 P1290-1294 66-11668 <*>
1965 V90 N43 P1922-1923 66-13210 <*>

DEUTSCHE MILITAERARZT
1941 V6 P272-274 II-993 <*>
1941 V6 N1 P1-9 57-1659 <*>

DEUTSCHE MOLKEREI-ZEITUNG
1955 V76 P1273-1275 57-1213 <*>

DEUTSCHE MOTOR-ZEITSCHRIFT
1939 V16 P6 II-642 <*>
1939 V16 P8- II-642 <*>
1941 V18 N1 P3-4 65-13774 <*>

DEUTSCHE OPTISCHE WOCHENSCHRIFT
1921 V7 N38 P702- 61-14261 <*> 0

DEUTSCHE PAPIERWIRTSCHAFT
1964 N4 P46-59 66-11747 <*>

DEUTSCHE POST
1962 V7 N2 P33-37 62-24714 <=*>
1962 V7 N2 P47-48 62-24703 <=*>
1964 V9 N9 P259-261 64-51746 <=$>

DEUTSCHE TABAKBAU
1953 P11-12 3030 <*>
58-937 <*>
1953 P30-31 II-225 <*>
58-1219 <*>
1953 N8 P14-15 58-1319 <*>
1953 N8 P73-74 57-1937 <*>
1953 V14 P127-132 1684 <*>
1958 N14 P127-132 58-991 <*>

DEUTSCHE TEXTILTECHNIK
1957 V7 N5 P300-303 61-16735 <*>
1957 V7 N10 P543-547 61-14503 <*> 0
63-18493 <*>
1957 V7 N10 P573-574 61-14504 <*>
1958 V8 N1 P7-8 61-14505 <*> 0
1958 V8 N1 P23-25 SCL-T-262 <=*>
1958 V8 N3 P149-151 59-15258 <*>

1958 V8 N4 P161-163 PB-141-300T <=>
1958 V8 N4 P173-178 61-16718 <*>
1958 V8 N4 P201-204 59-15792 <*>
1958 V8 N6 P273-277 PB 141 306T <=>
1959 V9 N1 P43-44 59-15269 <*>
1959 V9 N5 P258-263 05M39G <ATS>
1959 V9 N11 P588-590 61-16792 <*> 0
1960 V10 N7 P373-379 61-16734 <*>
1960 V10 N11 P594-597 C-3867 <NRCC>
1962 V12 N10 P533-534 64-14118 <*>
1963 V13 N11 P603-604 65-10370 <*>
1965 V15 N4 P216-218 1389 <TC>
1965 V15 N10 P548-551 2953 <TC>
1965 V15 N1C P614-618 2953 <TC>
1965 V15 N10 P672-677 2953 <TC>
1966 V16 N1 P1-5 66-31210 <=$>
1966 V16 N12 P705-709 67-30322 <=>

DEUTSCHE TIERAERZTLICHE WOCHENSCHRIFT
1907 V15 P417-421 66-11681 <*>
1955 V62 P489-493 II-713 <*>
1958 V65 N9 P229-235 60-10668 <*> 0
1960 V67 P558-560 8817 <IICH>

DEUTSCHE ZAHNAERZTLICHE WOCHENSCHRIFT
1935 P461-465 II-878 <*>
1936 V39 N23 P526-529 II-907 <*>

DEUTSCHE ZAHNAERZTLICHE ZEITSCHRIFT
1952 V7 P467-476 65-00142 <*> 0
1954 V9 N16 P899-905 1883 <*>
1956 V11 P614-621 65-00177 <*> 0
1956 V11 P1387-1391 58-1828 <*>
1956 V11 N5 P259-271 II-1101 <*>
1957 P477-489 65-00154 <*> 0
1964 V19 N7 P577-585 66-10949 <*>

DEUTSCHE ZEITSCHRIFT FUER CHIRURGIE
1899 V51 P361-369 57-1931 <*>
1909 V98 P233-257 C-2449 <NRC>
1932 V235 N8 P450-467 65-14766 <*> 0

DEUTSCHE ZEITSCHRIFT FUER DIE GESAMTE GERICHTLICHE MEDIZIN
1928 V12 P259-269 57-1642 <*>
1935 V24 P401-405 3510-F <K-H>
1939 V31 P55-59 2736G <K-H>

DEUTSCHE ZEITSCHRIFT FUER NERVENHEILKUNDE
1898 V12 P199-214 66-10966 <*>
1904 V27 P414-423 61-14211 <*>
1911 V41 P146-172 66-10967 <*>
1925 V84 P179-233 61-16062 <*>
1930 V112 P1-19 1706 <*>
1937 V144 P141-159 61-14138 <*>
1939 V149 P93-106 57-1651 <*>
1950 V164 P381-394 63-18869 <*>
1952 V168 P499-517 3054 <*>
1953 N121 P1-19 58-328 <*>
1953 V170 P179-208 57-2706 <*>
1954 V171 P169-172 II-359 <*>
1955 V172 P495-525 2401 <*>
1955 V173 P426-447 57-2707 <*>
1960 V181 P159-173 64-10831 <*> 0
1962 V183 P363-376 64-30050 <*> 0

DEUTSCHE ZEITSCHRIFT FUER PHILOSOPHIE
1960 V8 N10 P1266-1277 62-24562 <=*>
1965 V13 N1 P5-31 65-30961 <=$>
1965 V13 N3 P290-314 65-31221 <=*$>

DEUTSCHE ZEITSCHRIFT FUER VERDAUUNGS-UND STOFFWECHSELKRANKHEITEN
1952 V12 N1 P26-30 3166 <*>
1960 V20 N6 P257-264 62-01431 <*>

DEUTSCHER VERBAND FUER DIE MATERIALPRUEFUNGEN DER TECHNIK
1927 V78 P3-13 AL-691 <*>

DEUTSCHER ZIMMERMEISTER
1951 V53 N3 P1-9 CSIRO-3459 <CSIR>
57-3405 <*>
1954 N18 P1-8 II-829 <*>

DEUTSCHES ARCHIV FUER KLINISCHE MEDIZIN
1917 V123 P403-434 61-00385 <*>
1932 V173 P339-358 57-2700 <*>
1933 V175 P453-483 66-12643 <*>
1937 V180 P274-287 57-1081 <*>
1937 V180 P318-326 65-14330 <*> 0
1943 V191 P378-398 58-2437 <*>
1952 V199 P431-442 2697-G <K-H>
1955 V202 P320-346 5322-B <KH>
1957 V204 P253-264 59-12325 <=*> 0
1962 V208 P340-360 63-00313 <*>

DEUTSCHES MEDIZINISCHES JOURNAL
1956 V7 P133-137 5470-A <K-H>
1956 V7 N5/6 P138-140 6049-B <KH>
1962 V13 N16 P475-480 63-18722 <*>
1963 V14 N11 P368-369 66-10091 <*>
1964 V15 N13 P470-471 66-10092 <*>
1964 V15 N14 P493-494 66-10093 <*>
1965 V16 P411-420 66-13216 <*>
1965 V16 N1 P22-26 66-10741 <*> 0

DEVOTEES ASTRONOMY. CHINESE PEOPLES REPUBLIC
SEE TIEN WEN AI HAO CHE

DIA MEDICO. ARGENTINA
1945 V17 N3 P41-43 66-10948 <*>
1958 V30 N24 P800-806 58-1613 <*>
1960 V32 P542-544 61-20045 <*>
1960 V32 P546 61-20045 <*>
1961 V33 P1466-1468 63-18005 <*>
1961 V33 P2829-2835 63-10539 <*>
1961 V33 P2904-2908 63-10541 <*>
1961 V33 N26 P1324 63-16520 <*>
1962 V34 P288-290 63-10098 <*>
1962 V34 P409-414 63-10671 <*>
1962 V34 P1046 63-10919 <*>
1962 V34 P1614-1616 65-13951 <*> 0

DIA MEDICO. URUGUAY
1949 V16 P317-325 60-10833 <*>

DIABETE
1965 N2 P77-79 66-11714 <*> 0

DIAGNOSIS AND THERAPY. JAPAN
SEE SHINRYO

DIAGNOSIS AND TREATMENT. TOKYO
SEE SHINDAN TO CHIRYO

DIAGNOSTICA E TECNICA DI LABORATORIO
1934 V5 P928-932 57-1691 <*>

DIALECTICA
1958 V12 N3/4 P451-464 61-20530 <*>
77N52G <ATS>

DIAMANT. GLASINDUSTRIE-ZEITUNG. LEIPZIG
1915 V37 N20 P405-406 57-560 <*>

DIFESA SOCIALE
1951 V30 N3/4 P198-200 2773 <*>

★ DIFFERENTSIALNYE URAVNENIYA
1965 V1 N2 P219-226 65-17350 <*>
1965 V1 N2 P1493-1508 66-12733 <*>

DINGLER'S POLYTECHNISCHES JOURNAL
1865 V177 P58-76 63-10444 <*>
1870 V197 P343-346 65-10154 <*>

DISSERTATIONES PHARMACEUTICAE
1958 V10 N3 P183-189 64-18993 <*> 0
1963 V15 N2 P179-187 64-14807 <*>

DNEVNIK LETCHIKA KOSMONAVTA
1962 N2 P22-26 63-31700 <=>

DOBOKU GAKKAISHI
1955 V40 N5 P236-241 58-1279 <*>
1955 V40 N6 P15-20 TR.221 <MPBW>
1959 V44 N1 P9-13 61-20099 <*>

DOBOKU GIJUTSU
1953 V8 P12-16 T-1930 <INSD>

DOBOKU SHIKENJO IHO
1959 N20 P1-88 C-3915 <NRC>

DOBUTSUGAKU ZASSHI
1915 V27 P157-159 C-2719 <NRC>
1916 V28 P125-136 C-2688 <NRC>
1924 V36 N432 P397-408 AL-737 <*>
1931 V43 P276-280 C-2683 <NRC>
1947 V56 P57-61 T1929 <INSD>
1951 V60 P74-78 T-1851 <INSD>
1957 N5 P65-66 1119 <*>

DOCAERO
1954 N25 P59-63 3125 <*>
1954 N29 P17-22 64-10506 <*> 0
1954 N29 P23-29 64-10505 <*> 0
1954 N29 P30-34 64-10504 <*> 0
1956 N40 P45-64 62-10389 <*> P
1956 V40 P45-64 57-2899 <*>
1957 N42 P13-28 62-10252 <*>
1957 N45 P35-52 58-1331 <*>

DOCUMENT. SUD-AVIATION. FRANCE
1964 N68 EI-515/64 N65-16452 <=$>

DOCUMENTA OPHTHALMOLOGICA
1959 V13 P359-388 63-01315 <*>

DOCUMENTATION METALLURGIQUE
1956 P256-277 62-14711 <*> P

DOHANYIPAR
1959 V1 N2 P1-7 10969-A <K-H>
61-20774 <*>
1959 V1 N7 P134-137 64-14564 <*>

DOITAI TO HOSHASEN
1960 V3 N1 P53-61 NP-TR-991 <*>

DOKLADY AKADEMII NAUK ARMYANSKOI SSR
1947 V6 N1 P23-26 66-61479 <=*>
1954 V18 N1 P7-10 R-311 <*>
1954 V18 N4 P125-128 RT-3430 <*>
1954 V19 N1 P23-27 RT-3556 <*>
1955 V20 N5 P161-164 61-19340 <+*> 0
1956 V23 P81-85 60-21076 <=>
1957 V25 N3 P107-116 R-3417 <*>
1957 V25 N4 P185-192 63-14900 <=*>
1958 V26 N1 P11-20 59-10494 <*>
1958 V26 N3 P135-140 STMSP V4 P289-294 <AMS>
1958 V27 P37-39 59-15626 <+*>
1958 V27 N1 P13-22 61-15970 <+*>
1958 V27 N2 P81-85 93L30R <ATS>
1958 V27 N3 P145-147 61-15968 <+*>
1958 V27 N4 P251-256 62-00917 <*>
1959 V28 P183-186 63-01661 <*>
1959 V28 N5 P217-221 62-25335 <=*>
1959 V29 N3 P133-136 62-13479 <*=>
1960 V30 N3 P139-149 63-15911 <=*>
1960 V30 N4 P211-218 61-11067 <=>
1960 V30 N5 P257-263 64-19941 <=$>
1960 V31 N5 P261-265 62-25102 <=*>
1961 V32 N2 P117-122 65-60898 <=$>
1961 V32 N2 P123-127 62-15787 <=*>
1961 V32 N2 P202-204 66-26092 <=*>
1961 V33 N2 P49-52 63-15456 <=*>
1961 V33 N2 P79-82 63-24068 <=*$>
1962 V35 P177-180 63-23703 <=*$>
1962 V35 N1 P21-31 64-13600 <=*$>
1962 V35 N4 P181-183 CSIR-TR-439 <CSSA>
1963 V36 N2 P105-109 65-11127 <*>
1963 V36 N3 P147-151 AD-608 046 <=>
1963 V37 N3 P151-156 64-71313 <=>
1963 V37 N4 P221-225 65-14316 <*>
1963 V37 N4 P227-229 65-12631 <*>
1964 V38 N2 P71-76 AD-615 954 <=$>
1964 V38 N2 P87-92 AD-615 219 <=$>
1964 V38 N4 P225-230 M-5708 <NLL>
1964 V39 N4 P217-219 AD-638 901 <=$>
1965 V40 N4 P193-196 AD-635 733 <=$>
1965 V40 N4 P205-207 AD-627 072 <=$>
1965 V41 P27-32 AD-631 434 <=*$>
1965 V41 N2 P65-72 AD-635 574 <=$>
1965 V41 N3 P129-134 AD-637 533 <=$>

DOKLADY AKADEMII NAUK AZERBAIDZHANSKOI SSR
1946 V2 N8 P332-335 61-15302 <+*>
1947 V3 N1 P3-7 AEC-TR-3796 <+*> 0
1951 V7 N12 P565-570 61-28213 <*=>
1954 V10 N1 P3-10 RJ-219 <ATS>
1954 V10 N4 P297-302 R-160 <*>
1955 V11 N1 P13-19 RJ-575 <ATS>
1955 V11 N5 P313-318 RTS-2430 <NLL>
1955 V11 N12 P811-817 10010-A <K-H>
65-11269 <*> 0
1955 V11 N12 P819-823 59-22673 <+*>
1955 V11 N12 P857-859 R-3445 <*>
1956 V12 N8 P547-552 10010B <K-H>
65-11259 <*>
1956 V12 N8 P563-564 61-14730 <=*>
1956 V12 N12 P923-933 64-26425 <$>
1957 V13 N3 P253-257 64-26424 <$>
1957 V13 N8 P843-845 62-16668 <=*>
1958 V14 N3 P245-247 62-15630 <=*> 0
1958 V14 N4 P287-290 62-16429 <=*>
1958 V14 N11 P831-834 65-10739 <*>
1959 N2 P131-134 NLLTB V1 N8 P43 <NLL>
1959 V15 P1019-1023 66-12891 <*>
1959 V15 N2 P119-123 61-19341 <+*>
1959 V15 N2 P131-135 27L34R <ATS>
1959 V15 N8 P681-683 96M39R <ATS>
1959 V15 N9 P775-779 33T91AZ <ATS>
1959 V15 N9 P781-786 61-14957 <=*>
1959 V15 N10 P891-895 61-27332 <*=>
1959 V15 N10 P897-899 95S84AZ <ATS>
1959 V15 N12 P1091-1095 49M43R <ATS>
1960 V16 N1 P19-22 12M45R <ATS>
1960 V16 N1 P49-52 55M44R <ATS>
1960 V16 N2 P121-125 62-11547 <=>
1960 V16 N2 P127-131 07M43R <ATS>
1960 V16 N5 P443-445 59M46R <ATS>
1960 V16 N5 P447-451 60M46R <ATS>
1960 V16 N6 P535-539 ARSJ V32 N9 P1452 <AIAA>
1960 V16 N7 P643-646 65N49R <ATS>
1960 V16 N7 P655-658 23N49R <ATS>
1960 V16 N11 P1067-1070 65-13429 <*>
1961 V17 N1 P25-29 63-19049 <=*>
1961 V17 N4 P331-334 62-11647 <=>
1961 V17 N6 P451-456 65-13665 <*>
1961 V17 N6 P471-477 66N56R <ATS>
1961 V17 N8 P671-672 09P60R <ATS>
1961 V17 N8 P745-748 20P59R <ATS>
1961 V17 N9 P773-777 98P64R <ATS>
1961 V17 N10 P901-906 65-13395 <*>
1962 V18 N1 P11-15 66-11212 <*>
1963 N1 P13-18 ICE V4 N3 P382-385 <ICE>
1963 V19 N1 P13-18 ICE V4 N3 P382-385 <ICE>
86Q71R <ATS>
1963 V19 N1 P31-32 64-21504 <=>
1963 V19 N4 P37-41 64-21514 <=>
1963 V19 N11 P59-63 42S84AZ <ATS>
73R76R <ATS> P
1964 N1 P31-33 ICE V4 N4 P586-587 <ICE>
1964 N7 P21-24 ICE V5 N3 P489-490 <ICE>
1964 V20 N1 P21-26 93R77R <ATS>
1964 V20 N1 P31-33 AD-615 989 <=$>
23R77R <ATS>
1964 V20 N1 P69-73 32R77R <ATS>
1964 V20 N5 P11-14 54R80R <ATS>

1964 V20 N6 P73-75 30S81R <ATS>
1964 V20 N7 P21-24 18S82R <ATS>
1964 V20 N9 P13-15 N65-29733 <=*$>
1964 V20 N9 P37-40 65-14833 <*>
1964 V20 N10 P13-16 27S88R <ATS>
1964 V20 N10 P27-31 93T90R <ATS>
1965 V21 N4 P17-21 66-12209 <*>
1965 V21 N9 P16-19 66-14361 <*$>
1965 V21 N9 P32-34 66-13672 <*>
1965 V21 N10 P14-18 16T93R <ATS>

DOKLADY AKADEMII NAUK BELORUSSKOI SSR
1957 V1 P57-60 60-17718 <+*>
1957 V1 N1 P13-16 59-15627 <+*>
1957 V1 N3 P92-95 61-16799 <*=>
1958 V2 N1 P37-39 62-20431 <=*>
1959 V3 N5 P197-201 62-14076 <=*>
1959 V3 N5 P202-204 65-29625 <$>
1959 V3 N5 P208-210 63-18022 <=*> 0
1959 V3 N7 P318-321 60-21776 <=>
1959 V3 N11 P445-448 66-13002 <*>
1959 V3 N12 P479-483 ARSJ V31 N11 P1632 <AIAA>
1959 V3 N12 P488-491 62-18882 <=*> 0
1960 V4 N6 P241-243 62-19857 <*=>
1960 V4 N9 P369-371 AIAAJ V1 N7 P1745 <AIAA>
1960 V4 N9 P372-375 62-14263 <=*>
1960 V4 N11 P447-449 62-11565 <=>
1961 V5 N3 P118-121 C-4276 <NRC>
1961 V5 N6 P253-255 2827 <BISI>
1961 V5 N6 P263-265 5665 <HB>
1961 V5 N7 P291-294 1354 <TC>
1961 V5 N7 P311-314 65-20003 <$>
1961 V5 N8 P324-326 63-19531 <=*>
1961 V5 N8 P336-338 49Q73R <ATS>
1961 V5 N9 P375-379 65-60515 <=$>
1961 V5 N9 P387-388 63-19815 <=*>
1961 V5 N10 P442-447 65-64403 <=*$>
1961 V5 N11 P489-491 63-11685 <=>
1961 V5 N12 P541-544 63-11764 <=>
1962 V6 P223-225 63-11714 <=>
1962 V6 N1 P7-8 65-60514 <=$>
1962 V6 N1 P26-30 63-19405 <=*>
1962 V6 N3 P147-150 63-11716 <=>
1962 V6 N3 P151-154 63-11722 <=>
1962 V6 N4 P209-211 AIAAJ V1 N7 P1746 <AIAA>
1962 V6 N4 P237-239 63-15957 <=*>
1962 V6 N4 P247-250 66-10290 <*> 0
1962 V6 N5 P288-292 63-11672 <=>
1962 V6 N5 P297-300 63-11674 <=>
1962 V6 N5 P301-304 63-11673 <=>
1962 V6 N6 P355-359 64-11509 <=>
1962 V6 N6 P366-369 63-19671 <=*>
1962 V6 N7 P418-422 64-11538 <=>
1962 V6 N8 P494-496 63-19423 <=*>
65-13397 <*>
1962 V6 N9 P372-375 63-11721 <=>
1962 V6 N10 P629-632 63-11724 <=>
1962 V6 N10 P633-637 64-11547 <=>
1963 V7 P453-455 AD-618 318 <=$>
1963 V7 P721-723 RTS-2976 <NLL>
1963 V7 N1 P30-32 63-18152 <=*>
1963 V7 N2 P95-98 63-24137 <=*$>
1963 V7 N4 P230-232 AD-639 340 <=$>
1963 V7 N5 P305-308 AD-634 231 <=$>
1963 V7 N5 P309-312 66-11762 <*>
1963 V7 N5 P347-349 64-15149 <=*$>
1963 V7 N6 P376-377 N65-29739 <=*$>
1963 V7 N6 P387-390 68Q73R <ATS>
1963 V7 N6 P422-424 64-41566 <=>
1963 V7 N7 P446-448 64-11846 <=>
1964 V8 N1 P5-9 N65-32728 <=$>
65-14741 <*>
65-63180 <=$>
1964 V8 N2 P104-107 65-29598 <$>
1964 V8 N6 P394-397 65-14717 <*>
1964 V8 N7 P438-440 AD-635 879 <=$>
1964 V8 N8 P501-504 64-51908 <=$>
1964 V8 N8 P505-508 65-13352 <*>
1964 V8 N9 P568-571 64-51922 <=$>
1964 V8 N9 P575-578 65-14059 <*>
65-62957 <=$>
1964 V8 N9 P604-606 65-30023 <=$>
1964 V8 N10 P627-631 65-13157 <*>
1964 V8 N10 P632-633 AD-640 303 <=$>
1964 V8 N10 P675-676 65-30622 <=$>
1964 V8 N10 P680-681 65-30622 <=$>
1964 V8 N11 P689-692 N65-19511 <=$>
1964 V8 N11 P713-716 AD-625 311 <=*$>
1964 V8 N12 P784-787 AD-635 878 <=$>
1964 V8 N12 P788-792 66-11252 <*>
1964 V8 N12 P792-794 65-14057 <*>
65-30519 <=$>
65-63190 <=$>
1965 V9 N1 P11-14 65-30892 <=*$>
1965 V9 N1 P22-26 AD-635 890 <=$>
1965 V9 N1 P157 89S85R <ATS>
1965 V9 N3 P164-166 65-63895 <=$>
1965 V9 N4 P224-227 AD-634 248 <=$>
1965 V9 N6 P360-363 AD-634 246 <=$>
1965 V9 N6 P364-366 AD-634 245 <=$>
1965 V9 N7 P432-434 AD-634 247 <=$>
1965 V9 N7 P435-437 07S89R <ATS>
1965 V9 N9 P578-580 AD-635 889 <=$>
1965 V9 N9 P581-584 66-11214 <*>
1965 V9 N12 P822-824 64T93R <ATS>

★ DOKLADY AKADEMII NAUK SSSR
1934 V1 N3 P97-99 62-32138 <=*>
1934 V1 N6 P312-314 64-16728 <*>
1934 V1 N2/3 P389-394 59-18324 <+*> 0
1934 V2 N8 P451-454 RJ-539 <ATS>
1934 V3 P601-602 R-2495 <*>
1937 V16 N8 P133-137 AD-634 928 <=$>
1938 V18 P337 61-28430 <=*> C
1938 V18 P559-563 63-19560 <=*>
1938 V19 N4 P309-312 RT-3204 <*>
1939 V22 N5 P263-267 62-13103 <=*>
1939 V24 N6 P562-564 63-10110 <=*>
1939 V24 N6 P565-567 64-20113 <*>
1939 V25 P743-746 R-4494 <*>
1940 V27 N5 P458-459 4522 <BISI>
1940 V27 N7 P658-663 61-20818 <*=>
1940 V27 N7 P664-669 61-20823 <*=>
1940 V27 N7 P670-672 59-17968 <*> 0
1940 V27 N9 P960-963 61-20179 <*=>
1940 V29 N8/9 P575-576 62-14636 <=*>
1941 V30 N1 P21-22 61-20820 <*=>
1941 V30 N1 P23-25 61-20817 <*=>
1941 V30 N1 P26-28 61-18328 <*=>
1941 V30 N1 P29-31 61-20819 <*=>
1941 V30 N1 P32-36 61-18327 <*=>
1941 V30 N1 P37-39 61-18326 <*=>
1941 V30 N2 P144-147 61-18325 <*=>
1941 V30 N2 P148-151 61-18331 <*=>
1941 V30 N8 P717-720 61-18330 <*=>
1941 V30 N8 P726-727 61-18333 <=*>
1941 V30 N8 P728-731 61-18332 <*=>
1941 V30 N9 P854-856 PANSDOC-TR.507 <PANS>
1941 V31 N3 P264-265 PANSDOC-TR.59 <PANS>
1941 V31 N5 P448-452 61-18329 <*=>
1941 V31 N8 P765-766 61-16883 <*=>
1941 V31 N9 P855-897 R-1149 <*>
R-4447 <*>
1941 V32 P50-52 61-18014 <*=>
1941 V32 P135-138 61-18013 <*=>
1941 V32 P551-554 61-16641 <*=>
1941 V33 P28-33 61-16940 <*=>
1941 V33 P217-222 1338E <K-H>
61-16941 <*=>
1941 V33 P223-226 61-16942 <*=>
1941 V33 N1 P34-36 62-16779 <=*>
1941 V33 N1 P41-44 60-41558 <=>
1941 V33 N7/8 P445-449 91F2R <ATS>
1942 V34 N3 P88-92 61-16668 <*=>
1942 V34 N7 P196-198 61-16692 <*=>
1942 V37 P223-225 61-16656 <*=>
1943 V39 P26-30 61-16983 <*=>
1943 V39 P261-264 61-16979 <*=>
1943 V39 N5 P209-212 PANSDOC-TR.380 <PANS>
61-19219 <=*>

Citation	Translation
1943 V40 P75-78	R-3372 <*>
1943 V40 P254-256	61-16928 <*=>
1943 V40 P402-404	61-16914 <*=>
1943 V40 N2 P70-72	61-16975 <*=>
1943 V40 N4 P171-174	62-16771 <=*>
1943 V41 P71-73	61-16920 <*=>
1943 V41 P275-277	AMST S1 V11 P322-324 <AMS>
1943 V41 N7 P307-309	65-60246 <=$>
1943 V41 N7 P310-311	62-16765 <=*>
1944 V42 N6 P276-278	61-11773 <=>
1944 V43 N8 P360-366	61-11781 <=>
1944 V44 P152-153	61-18175 <*=>
1944 V44 P375-377	61-18166 <*=>
1944 V44 N1 P21-26	61-21078 <=>
1944 V44 N2 P68-73	61-11782 <=>
1944 V44 N6 P265-269	61-11768 <=>
1944 V44 N9 P372-374	61-18167 <*=>
1944 V45 P19-21	61-18165 <*=>
1944 V45 P24-26	61-18164 <*=>
1944 V45 P244-247	61-18197 <*=>
1944 V45 P331-333	61-18196 <*=>
1944 V45 N4 P164-166	61-11770 <=>
1945 V46 P150-153	61-18198 <*=>
1945 V46 P368-369	61-18241 <*=>
1945 V46 N3 P113-116	61-11780 <=>
1945 V46 N9 P399-402	64-15197 <=*$>
1945 V47 P103-105	61-18247 <*=>
1945 V47 P106-109	61-18245 <*=>
1945 V47 P348-350	61-18244 <*=>
1945 V47 P410-411	61-18243 <*=>
1945 V47 N5 P344-347	62-10787 <*=> P
1945 V47 N7 P472-474	63-14136 <=*>
1945 V47 N7 P486-489	N65-21627 <=$>
1945 V47 N7 P501-503	60-41584 <=>
1945 V47 N8 P610-612	PANSDOC-TR.383 <PANS>
1945 V48 P259-262	61-18352 <*=>
1945 V48 P339-342	61-18351 <*=>
1945 V48 P482-483	61-18349 <*=>
1945 V48 N1 P32-35	27N58R <ATS>
1945 V48 N3 P198-201	61-11071 <=>
1945 V48 N6 P471-473	PANSDOC-TR.381 <PANS>
1945 V49 P191-193	61-18342 <=*>
1945 V49 P418-420	61-20134 <*=>
1945 V49 P652-654	61-20135 <*=>
1945 V49 P655-658	61-20136 <*=>
1945 V49 N1 P38-40	61-11713 <=>
1945 V49 N2 P116-118	61-11714 <=>
1945 V49 N4 P519-522	R-4 <*>
1945 V49 N5 P364-368	R-1621 <*>
1945 V49 N8 P568-571	62-10798 <*=>
1945 V50 P89-93	R-4349 <*>
1945 V50 P261-264	3182 <HB>
1945 V50 P285-288	61-11939 <=>
1945 V50 P289-291	RT-974 <*>
1945 V50 P295-297	61-11716 <=>
1945 V50 P299-301	61-11760 <=>
1945 V50 P303-305	61-11719 <=>
1945 V50 P429-432	63-14606 <=*>
1945 V50 N5 P261-264	59-19090 <+*>
1946 V51 P213-216	61-20165 <*=>
1946 V51 N4 P247-249	AD-619 563 <=$>
1946 V52 P227-229	61-20173 <*=>
1946 V52 P239-240	61-20174 <*=>
1946 V52 P313-316	61-20175 <*=>
1946 V52 P421-423	61-20176 <*=>
1946 V52 N1 P43-46	61-20170 <=*>
1946 V52 N2 P147-150	60-41505 <=>
1946 V52 N3 P235-238	62-10809 <*=> P
1946 V52 N4 P317-319	62-10810 <*=> P
1946 V52 N6 P507-509	60-41512 <=>
1946 V52 N9 P783-785	60-41517 <=>
1946 V53 N1 P17-19	59-11866 <=>
1946 V53 N2 P107-110	SCL-T-456 <=*>
1946 V53 N4 P373-376	61-28013 <*=> O
1946 V53 N6 P515-518	62-24685 <=*>
1946 V53 N7 P595-596	59-11859 <=>
1946 V54 P205-208	833TM <CTT>
1946 V54 N1 P65-68	60-15536 <+*>
1946 V54 N2 P109-112	64-71489 <=> M
1946 V54 N2 P121-124	832TM <CTT>
1946 V54 N5 P457-460	PANSDOC-TR.342 <PANS>
	61-15212 <*+>
1946 V54 N6 P507-509	60-41358 <=>
1946 V54 N6 P515-518	65-60380 <=$>
1946 V54 N9 P769-772	RT-205 <*>
1946 V54 N9 P787-789	62-11192 <=>
1947 V55 N2 P115-118	RT-3167 <*>
1947 V55 N2 P141-143	59-20368 <+*>
1947 V55 N3 P207-210	63-19966 <=*>
1947 V55 N6 P513-515	RT-231 <*>
1947 V55 N8 P745-748	62-23003 <=>
1947 V55 N9 P813-816	60-14316 <+*>
1947 V55 N9 P825-828	61-10281 <*+>
1947 V56 P49-52	R-5058 <*>
1947 V56 N2 P145-148	AD-610 899 <=$>
	RT-1915 <*>
1947 V56 N3 P245-247	RT-1551 <*>
1947 V56 N3 P253-254	AD-610 899 <=$>
	RT-1916 <*>
1947 V56 N3 P255-258	RT-3111 <*>
1947 V56 N3 P259-260	AD-610 882 <=$>
	RT-1651 <*>
1947 V56 N4 P355-358	RT-3112 <*>
1947 V56 N4 P391-392	R-1219 <*>
1947 V56 N5 P485-486	154TM <CTT>
1947 V56 N5 P491-494	63-14390 <=*>
	63-20657 <=*$>
1947 V56 N5 P545-547	50/2059 <NLL>
1947 V56 N6 P571-574	RT-266 <*>
1947 V56 N6 P583-586	AD-610 899 <=$>
	RT-1917 <*>
	64-11967 <=> M
1947 V56 N6 P587-588	RT-243 <*>
1947 V56 N7 P699-702	AD-610 899 <=$>
	RT-1918 <*>
1947 V56 N7 P703-705	RT-1240 <*>
1947 V56 N7 P727-729	RT-2674 <*>
1947 V56 N7 P755-757	RJ-45 <ATS>
1947 V57 P833-836	59-15894 <+*>
1947 V57 P1013-1016	NP-2075 <*>
1947 V57 N1 P53-56	4846 <HB>
1947 V57 N2 P129-132	61-19462 <+*>
	62-16674 <=*>
1947 V57 N3 P271-274	RT-4053 <*>
1947 V57 N4 P361-363	RT-2383 <*>
1947 V57 N4 P365-368	60-41347 <=>
1947 V57 N5 P475-477	R-71 <*>
	RT-4657 <*>
	64-19074 <=*$> O
1947 V57 N7 P697-700	17N49R <ATS>
1947 V57 N8 P789-792	60-41346 <=>
1947 V57 N8 P833-836	R-3915 <*>
	3117 <*>
1947 V57 N9 P905-906	RT-1511 <*>
	64-10154 <=*$>
1947 V57 N9 P915-918	TT.166 <NRC>
1947 V57 N9 P955-958	63-11039 <=>
1947 V58 P261-263	59-19241 <+*> O
1947 V58 P1077-1079	AEC-TR-3785 <+*>
1947 V58 P1357-1359	R-3271 <*>
1947 V58 N1 P61-64	RT-1072 <*>
	61-18905 <*=>
1947 V58 N1 P83-84	R-1197 <*>
1947 V58 N1 P119-122	60-17781 <+*>
1947 V58 N2 P229-231	61-18926 <*=>
1947 V58 N2 P241-244	60-41345 <=>
1947 V58 N2 P245-248	60-41432 <=>
1947 V58 N3 P365-368	65-60795 <=>
1947 V58 N3 P381-384	RT-1071 <*>
1947 V58 N3 P389-392	60-10568 <+*>
1947 V58 N4 P543-546	RT-925 <*>
1947 V58 N4 P607-610	RT-1237 <*>
1947 V58 N4 P647-650	R-1376 <*>
1947 V58 N5 P757-760	RT-2894 <*>
1947 V58 N5 P761-762	N65-27720 <=*$>
1947 V58 N5 P775-778	RT-688 <*>
1947 V58 N6 P1073-1075	RT-1489 <*>
	64-10155 <=*$>
1947 V58 N7 P1313-1316	60-13127 <+*>
1947 V58 N7 P1369-1372	RT-2708 <*> O

	64-10454 <=*$> 0
1947 V58 N7 P1377-1378	R.582 <RIS>
	R.582 <RIS>
1947 V58 N7 P1379-1381	R.583 <RIS>
1947 V58 N7 P1407-1410	R.584 <RIS>
1947 V58 N8 P1693-1695	TT.181 <NRC>
	65-13591 <*>
1947 V58 N9 P1935-1938	RT-856 <*>
1947 V58 N9 P1969-1972	TT.180 <NRC>
1948 V59 P99-102	RJ664 <ATS>
1948 V59 P265-268	R-3808 <*>
1948 V59 P279-280	73F2R <ATS>
1948 V59 P945-947	59-19080 <+*>
1948 V59 P1313-	R-5107 <*>
1948 V59 P1579-1582	<ATS>
1948 V59 N1 P29-31	RT-191 <*>
1948 V59 N1 P41-43	RT-211 <*>
1948 V59 N1 P57-60	RT-644 <*>
1948 V59 N1 P61-64	RT-645 <*>
1948 V59 N1 P83-85	RT-139 <*>
1948 V59 N1 P99-102	72F2R <ATS>
1948 V59 N2 P227-231	RT-1095 <*>
1948 V59 N2 P231-234	RT-204 <*>
	RT-2419 <*>
1948 V59 N2 P239-242	<INSD>
1948 V59 N2 P279-280	73F2R <ATS>
1948 V59 N2 P293-294	61-13105 <*+>
1948 V59 N3 P423-425	64-15204 <=*$>
1948 V59 N3 P467-469	RT-209 <*>
1948 V59 N3 P471-474	62-20461 <=*>
1948 V59 N3 P475-478	RT-643 <*>
1948 V59 N3 P479-482	RT-71 <*>
1948 V59 N3 P483-486	RT-1165 <*>
1948 V59 N3 P487-488	RT-206 <*>
1948 V59 N3 P527-530	RT-382 <*>
1948 V59 N3 P535-536	63-23213 <=*>
1948 V59 N4 P659-662	60-41331 <=>
	60-41442 <=>
1948 V59 N4 P675-678	RT-1622 <*>
1948 V59 N4 P683-686	RT-1450 <*>
1948 V59 N4 P697-700	RT-193 <*>
1948 V59 N5 P879-882	RT-899 <*>
	63-24372 <*>
	63-24372 <=*>
1948 V59 N5 P889-892	RT-2126 <*>
1948 V59 N5 P899	66-11209 <*>
1948 V59 N6 P1073-1076	RT-210 <*>
1948 V59 N6 P1125-1128	27K22R <ATS>
1948 V59 N6 P1129-1132	RT-1893 <*>
1948 V59 N6 P1137-1140	62-23889 <=*>
1948 V59 N6 P1183-1186	61-11478 <=>
1948 V59 N6 P1221-1224	35L37R <ATS>
1948 V59 N7 P1261-1264	R-5009 <*>
1948 V59 N7 P1277-1279	RT-808 <*>
1948 V59 N7 P1281-1282	RT-97 <*>
1948 V59 N7 P1283-1286	RT-691 <*>
1948 V59 N7 P1291-1292	R.585 <RIS>
1948 V59 N7 P1293-1296	R.586 <RIS>
1948 V59 N7 P1297-1300	R.587 <*>
1948 V59 N7 P1301-1304	R.554 <RIS>
1948 V59 N7 P1305-1308	R.545 <RIS>
1948 V59 N7 P1309-1312	R.546 <RIS>
1948 V59 N7 P1321-1324	59-19092 <+*>
1948 V59 N9 P1521-1524	RT-202 <*>
1948 V59 N9 P1541-1544	R-5008 <*>
1948 V59 N9 P1555-1557	63-20793 <=*$>
1948 V59 N9 P1563-1566	62-16214 <=*>
1948 V59 N9 P1579-1582	R-3214 <*>
1948 V59 N9 P1635-1638	65-61578 <=$>
1948 V59 N9 P1665-1668	RT-1782 <*>
1948 V60 P421-427	R-108 <*>
1948 V60 P623-626	R-108 <*>
1948 V60 N1 P5-8	RT-252 <*>
1948 V60 N1 P49-52	RT-846 <*>
1948 V60 N1 P53-56	RT-623 <*>
1948 V60 N1 P57-58	R.550 <RIS>
1948 V60 N1 P59-61	R.551 <RIS>
	61-17303 <=$>
1948 V60 N1 P63-65	R.552 <RIS>
1948 V60 N1 P63-66	59-10731 <+*> 0
1948 V60 N1 P67-72	R.553 <RIS>
1948 V60 N1 P81-84	R-654 <*>
	64-19493 <=$>
1948 V60 N2 P203-206	R-220 <*>
1948 V60 N2 P207-209	RT-865 <*>
1948 V60 N2 P211-212	RT-866 <*>
1948 V60 N2 P213-215	RT.879 <*>
1948 V60 N2 P235-237	RT-192 <*>
1948 V60 N2 P239-241	R.555 <RIS>
1948 V60 N2 P247-250	28P62R <ATS>
	3124 <HB>
1948 V60 N3 P361-364	TT.118 <NRC>
1948 V60 N3 P371-374	R-4489 <*>
	RT-1858 <*>
1948 V60 N3 P387-390	TT.178 <NRC>
1948 V60 N3 P399-400	R.556 <RIS>
1948 V60 N3 P403-408	59-19121 <+*>
1948 V60 N3 P421-427	RT-3708 <*>
1948 V60 N3 P469-472	50/2217 <NLL>
	61-11479 <=>
1948 V60 N4 P567-568	RT-809 <*>
1948 V60 N4 P569-570	66-12287 <*>
1948 V60 N4 P591-594	R.557 <RIS>
1948 V60 N4 P599-601	TT.155 <NRC>
1948 V60 N4 P607-610	R-559 <RIS>
1948 V60 N4 P611-613	59-19122 <+*>
1948 V60 N4 P615-618	R-560 <RIS>
	TT-490 <NRC>
1948 V60 N4 P623-626	RT-3705 <*>
1948 V60 N4 P635-636	50/3264 <NLL>
	61-13375 <+*>
1948 V60 N4 P637-639	R.561 <RIS>
1948 V60 N4 P649-652	R-1205 <*>
1948 V60 N5 P783-784	RT-848 <*>
1948 V60 N5 P811-812	RT-58 <*>
1948 V60 N5 P821-824	63-20792 <=*$>
1948 V60 N5 P825-828	RT-2774 <*>
1948 V60 N5 P833-835	65-29154 <$>
1948 V60 N5 P837-839	R-233 <*>
	R-70 <*>
	64-19073 <=*$>
1948 V60 N5 P909-912	60-18002 <+*> 0
1948 V60 N6 P957-960	RT-908 <*>
1948 V60 N6 P989-992	RT-3119 <*>
	60-13714 <+*>
1948 V60 N6 P993-996	R.537 <RIS>
1948 V60 N6 P999-1002	R.538 <RIS>
	61-18636 <*=>
1948 V60 N6 P1003-1004	R.539 <RIS>
1948 V60 N6 P1005-1008	R.540 <RIS>
1948 V60 N6 P1015-1017	R.542 <RIS>
1948 V60 N6 P1033-1036	62-16426 <=*>
1948 V60 N7 P1161-1164	RT-1623 <*>
1948 V60 N7 P1177-1180	R.607 <RIS>
1948 V60 N7 P1185-1188	AEC-TR-2022 <+*>
1948 V60 N7 P1205-1207	TT.197 <NRC>
	1722B <K-H>
1948 V60 N7 P1213-1215	RT-3982 <*>
1948 V60 N7 P1219-1221	R.612 <RIS>
1948 V60 N8 P1343-1345	TT.179 <NRC>
1948 V60 N8 P1347-1350	50/3022 <NLL>
	61-13620 <+*>
1948 V60 N8 P1353-1355	TT.196 <NRC>
	1722A <KH>
1948 V60 N8 P1357-1359	1594 <K-H>
1948 V60 N9 P1519	60-13886 <+*>
1948 V60 N9 P1523-1526	59-19079 <+*>
1948 V60 N9 P1527-1529	59-19071 <+*>
1948 V60 N9 P1531-1534	59-19073 <+*>
1948 V60 N9 P1535-1538	RT-1959 <*>
1948 V61 P91-94	R-108 <*>
1948 V61 P197-199	66-10890 <*>
1948 V61 P293-296	59-22703 <+*>
1948 V61 P561-564	R-5261 <*>
1948 V61 P933-934	60-15812 <+*>
1948 V61 N1 P31-34	59-16040 <+*>
1948 V61 N1 P35-38	61-17476 <=$>
1948 V61 N1 P47	60-13887 <+*>
1948 V61 N1 P79-82	AD-610 897 <=$>
	RT-1516 <*>

Citation	Translation
1948 V61 N1 P83-86	RT-1473 <*>
1948 V61 N1 P83	61-23302 <*>
1948 V61 N1 P83-86	61-23302 <*=>
1948 V61 N1 P91-94	RT-3709 <*>
1948 V61 N1 P713-715	60-15532 <+*>
1948 V61 N2 P243	RT-689 <*>
	65-60077 <=$>
1948 V61 N2 P297-300	RT-1894 <*>
1948 V61 N2 P301-304	1729 <K-H>
1948 V61 N2 P329-331	25E2R <ATS>
	61-17985 <=$>
1948 V61 N3 P467-470	61-15743 <+*>
1948 V61 N4 P617-620	R-7 <*>
1948 V61 N4 P625-628	E-640 <RIS>
1948 V61 N4 P641-644	RT-1217 <*>
1948 V61 N4 P649-652	59-19127 <+*> 0
1948 V61 N4 P653-656	RJ-209 <ATS>
	64-19508 <=$>
1948 V61 N4 P657-660	RT-2021 <*>
1948 V61 N4 P661-664	RT-1753 <*>
	59-15979 <+*> 0
1948 V61 N4 P713-715	61-11480 <=>
1948 V61 N4 P741-744	65-12364 <*>
1948 V61 N4 P757-770	<CTT>
1948 V61 N5 P799-802	RT-2386 <*>
1948 V61 N5 P817-820	RT-855 <*>
1948 V61 N5 P845-848	RJ.120 <ATS>
	RJ-120 <ATS>
1948 V61 N5 P849-852	RT-1396 <*>
1948 V61 N5 P869-872	RT-618 <*>
1948 V61 N6 P993-996	65-60782 <=>
1948 V61 N6 P1009-1012	62-10584 <*=>
1948 V61 N6 P1013-1015	RT-2615 <*>
1948 V61 N6 P1017-1018	R-4490 <*>
	RT-1892 <*>
1948 V61 N6 P1023-1026	R-4499 <*>
	RT-100 <*>
1948 V61 N6 P1125-1127	RT-519 <*>
1948 V61 N6 P1125-1128	307TM <CTT>
1948 V62 P255-258	R-4542 <*>
1948 V62 P607-609	UCRL-TRANS-489 <+*>
1948 V62 P761-764	59-15112 <+*>
1948 V62 N1 P51-54	61-17448 <=$>
1948 V62 N1 P93-96	RT-2777 <*>
	59-10787 <+*> 0
	59-10787 <*>
1948 V62 N1 P121-124	RT-3865 <*>
1948 V62 N2 P223-225	RTS-3077 <NLL>
1948 V62 N2 P247-250	R-3777 <*>
1948 V62 N2 P255-257	RT-132 <*>
1948 V62 N2 P271-274	65-12365 <*>
1948 V62 N3 P301-304	RT-1692 <*>
	59-15446 <+*> 0
1948 V62 N3 P323-324	RT-1998 <*>
1948 V62 N3 P345-348	RT-3524 <*>
1948 V62 N3 P353-356	RT-1895 <*>
1948 V62 N3 P357-360	61-17914 <=$>
1948 V62 N3 P369-372	RT-1833 <*> 0
1948 V62 N3 P409-412	62-10754 <*=> 0
1948 V62 N3 P417-420	R-4133 <*>
1948 V62 N4 P469-471	RT-619 <*>
1948 V62 N4 P537-540	62-10906 <=*>
1948 V62 N5 P595-598	RT-2028 <*>
1948 V62 N5 P615-617	C-2186 <NRC>
1948 V62 N5 P661-663	<DIL>
	R-4681 <*>
1948 V62 N5 P665-667	TT.247 <NRC>
1948 V62 N5 P689-692	R-2655 <*>
1948 V62 N5 P705-712	62-10756 <*=> 0
1948 V62 N5 P713-716	62-10755 <*=> 0
1948 V62 N6 P795-798	62-10198 <*=>
1948 V63 P163-165	R-108 <*>
1948 V63 P499-502	T-2210 <INSD>
1948 V63 P693-696	<ATS>
1948 V63 N1 P23-26	RT-621 <*>
1948 V63 N1 P33-36	RT-1606 <*>
1948 V63 N1 P53-56	59-17402 <+*>
1948 V63 N1 P57-60	59-17401 <+*>
1948 V63 N1 P73-75	RT-3762 <*>
1948 V63 N2 P119-122	AD-610 893 <=$>
	R-1763 <*>
	RT-1613 <*>
1948 V63 N2 P143-145	63-20521 <*>
	64-20192 <*>
1948 V63 N2 P151-154	RT-612 <*>
	61-17454 <=$>
	84N51R <ATS>
1948 V63 N2 P155-158	RT-1910 <*>
	59-17171 <+*> 0
1948 V63 N2 P163-165	RT-3719 <*>
1948 V63 N2 P187-190	61-19565 <=*>
1948 V63 N3 P239-242	RT-620 <*>
	61-17455 <=$>
1948 V63 N3 P251-254	TT.104 <NRC>
1948 V63 N3 P255-257	63-23943 <=*$>
1948 V63 N3 P297-299	RT-4091 <*>
1948 V63 N3 P301-304	64-20190 <*> 0
1948 V63 N3 P305-306	RT-3511 <*>
	TT.261 <NRC>
1948 V63 N4 P411-413	62-15884 <=>
1948 V63 N4 P455-458	RT-1220 <*>
1948 V63 N4 P469-472	RT-524 <*>
1948 V63 N5 P507-510	3221 <HB>
1948 V63 N5 P511-513	RT-101 <*>
	61-17456 <=$>
1948 V63 N5 P553-556	R-1175 <*>
1948 V63 N5 P557-560	RT-1515 <*>
1948 V63 N6 P629-630	59-16039 <+*>
1948 V63 N6 P649-651	RT-1773 <*>
1948 V63 N6 P693-696	52/2940 <NLL>
1948 V63 N6 P705-708	R-1036 <*>
	RT-3491 <*>
1948 V63 N6 P709-712	59-16743 <+*>
1948 V63 N6 P713-715	59-16376 <+*>
1949 V64 P81-84	T-2415 <INSD>
1949 V64 P99-102	R-4543 <*>
1949 V64 P145-148	1584B <K-H>
1949 V64 P849-852	C-2631 <NRC>
1949 V64 N1 P69-72	20L29R <ATS>
1949 V64 N1 P73-76	RJ-13 <ATS>
1949 V64 N1 P85-86	2316 <HB>
	61-13369 <+*>
1949 V64 N1 P99-102	RT-175 <*>
1949 V64 N3 P301-304	RT-318 <*>
1949 V64 N3 P309-312	RT-3030 <*>
1949 V64 N3 P321-323	RT-613 <*>
	61-17457 <=$>
1949 V64 N3 P333-335	63-14846 <=*> 0
1949 V64 N3 P345-347	RT-494 <*>
1949 V64 N3 P357-360	63-20794 <=*$>
1949 V64 N3 P361-364	39P62R <ATS>
1949 V64 N3 P405-408	1584A <K-H>
1949 V64 N3 P425-428	RT-1615 <*>
1949 V64 N4 P483-486	59-17715 <+*>
1949 V64 N4 P495-498	64-20189 <*> 0
1949 V64 N4 P499-502	R-685 <*>
	61-17479 <=$>
1949 V64 N4 P507-509	61-28019 <*=>
1949 V64 N4 P525-528	RT-1084 <*>
	85K26R <ATS>
1949 V64 N4 P533-536	RT-2956 <*>
1949 V64 N4 P583-586	61-13307 <+*> 0
1949 V64 N5 P803-805	61-17480 <=$>
1949 V64 N6 P689-692	RT-1834 <*>
1949 V64 N6 P779-782	62-10162 <*=>
1949 V64 N6 P783-786	63-24428 <=*$>
1949 V64 N6 P795-797	59-19538 <+*>
1949 V64 N6 P803-805	RT-1108 <*>
1949 V64 N6 P835-838	62-11643 <=>
1949 V65 P97-100	C-2633 <NRC>
1949 V65 P229-232	C-2621 <NRC>
1949 V65 P793-796	AMST S1 V11 171-177 <AMS>
1949 V65 N1 P33-36	61-17481 <=$>
1949 V65 N1 P41-44	59-11653 <=>
1949 V65 N1 P81-84	59-19125 <+*>
1949 V65 N2 P149-150	RT-1652 <*>
	61-17473 <=$>
1949 V65 N2 P151-154	RT-1427 <*>
1949 V65 N2 P155-158	RT-806 <*>

1949 V65 N2 P163-165 TT.143 <NRC>
1949 V65 N3 P291-294 62-23099 <=*>
1949 V65 N3 P307-310 R-3483 <*>
RT-2781 <*>
1949 V65 N3 P385-388 61-13358 <+*> 0
1949 V65 N3 P397-399 R-261 <*>
1949 V65 N3 P405-408 R-267 <*>
1949 V65 N4 P527-530 RT-134 <*>
1949 V65 N4 P577-580 R-2373 <*>
1949 V65 N5 P621-624 59-20194 <+*>
1949 V65 N5 P629-631 50/2304 <NLL>
1949 V65 N5 P645-648 59-10312 <+*>
1949 V65 N5 P711-714 R-4150 <*>
1949 V65 N5 P719- R-4493 <*>
1949 V65 N5 P719-722 RT-135 <*>
1949 V65 N5 P749-752 27J19R <ATS>
1949 V65 N6 P793-796 RT-1440 <*>
1949 V65 N6 P815-818 50/1337 <NLL>
1949 V65 N6 P831-834 RT-807 <*>
1949 V65 N6 P861-864 R-4511 <*>
RT-2006 <*>
1949 V65 N6 P871-874 60-41533 <=>
1949 V65 N6 P883-886 TT.258 <NRC>
1949 V66 P645-646 57-50 <*>
1949 V66 P663-666 R-108 <*>
1949 V66 P821-824 R-1133 <*>
1949 V66 P825-828 R-1184 <*>
1949 V66 P1133-1136 R-108 <*>
1949 V66 N1 P25-28 RT-4066 <*>
1949 V66 N1 P41-44 TT.430 <NRC>
1949 V66 N1 P49-51 64-15287 <=*$>
65-60076 <=$> 0
1949 V66 N1 P53-54 RT-693 <*>
306TM <CTT>
1949 V66 N1 P55-57 RT-4159 <*>
1949 V66 N1 P95-97 62-24170 <=*>
1949 V66 N2 P181-184 RT-3971 <*>
1949 V66 N2 P187-189 RT-901 <*>
1949 V66 N2 P195-198 RT-1752 <*>
1949 V66 N2 P277-280 62-10745 <=*>
1949 V66 N3 P343-345 62-20325 <=*>
1949 V66 N3 P359-360 RT-1533 <*>
1949 V66 N3 P373-376 AD-610 902 <=$>
RT-1610 <*>
50/1806 <NLL>
1949 V66 N3 P413-416 TT.109 <NRC>
1949 V66 N4 P575-576 62-24309 <=*>
1949 V66 N4 P609-612 RT-610 <*>
61-17474 <=$>
1949 V66 N4 P617-620 8921-C <KH>
1949 V66 N4 P641-644 R-699 <*>
RT-1718 <*>
59-20210 <+*>
59-20210 <*>
1949 V66 N4 P647-650 1618C <K-H>
1949 V66 N4 P657-660 2638 <HB>
1949 V66 N4 P663-666 RT-2087 <*>
1949 V66 N4 P673-676 60-13712 <+*>
1949 V66 N4 P729-732 1618D <K-H>
1949 V66 N4 P733-736 61-13615 <+*>
1949 V66 N5 P825-828 C-2216 <NLL>
R-235 <*>
60-17228 <+*>
60-23010 <+*>
1949 V66 N5 P829-832 RT-2025 <*>
61-17475 <=$>
1949 V66 N5 P835-838 07L37P <ATS>
1949 V66 N5 P863-866 1435TM <CTT>
59-16712 <+*>
1949 V66 N5 P867-870 R-4041 <*>
1949 V66 N5 P885-888 TT.113 <NRC>
1949 V66 N5 P893-894 RT-611 <*>
61-17452 <=$>
1949 V66 N5 P901-904 RT-92 <*>
1949 V66 N5 P905-908 RT-3493 <*>
1949 V66 N5 P1013-1016 RT-521 <*>
1949 V66 N6 P1111-1112 TT.121 <NRC>
1949 V66 N6 P1121-1124 TT.122 <NRC>
1949 V66 N6 P1133-1136 RT-3706 <*>
1949 V66 N6 P1211-1214 K/ 1594 <NLL>
50/1594 <NLL>
1949 V67 P325-328 R-108 <*>
1949 V67 P511-512 308TM <CTT>
1949 V67 P597-599 AMST S1 V8 P1-5 <AMS>
1949 V67 P773-776 AMST S1 V8 P6-10 <AMS>
1949 V67 P791-794 AMST S1 V6 P424-429 <AMS>
1949 V67 P961-964 AMST S1 V5 P389-395 <AMS>
1949 V67 N1 P45-48 59-12453 <*>
1949 V67 N1 P53-55 RT-609 <*>
61-17453 <=$>
1949 V67 N1 P65-67 R-8 <*>
1949 V67 N1 P85-88 TT.364 <NRC>
1949 V67 N1 P93-95 63-18208 <=*> 0
64-18859 <*> 0
64-20188 <*> 0
1949 V67 N1 P101-104 60-10334 <*> 0
1949 V67 N1 P153-155 60-51057 <=>
1949 V67 N1 P193-196 RT-520 <*>
1949 V67 N2 P255-258 59-17457 <*>
1949 V67 N2 P271-273 C-2643 <NRC>
R-4257 <*>
1949 V67 N2 P293-295 RJ-231 <ATS>
RT-1089 <*>
59-10558 <+*>
1949 V67 N2 P305-308 TT.462 <*>
1949 V67 N2 P309-312 52/2941 <NLL>
61-13614 <+*>
1949 V67 N2 P313-315 60-10333 <+*> 0
1949 V67 N2 P325- R-4500 <*>
1949 V67 N2 P325-328 RT-133 <*>
1949 V67 N2 P345-347 R-101 <*>
RT-4052 <*>
64-19078 <=*$>
1949 V67 N2 P393-396 60-51139 <=>
1949 V67 N3 P447-450 <ATS>
1949 V67 N3 P459-462 R-2395 <*>
309TM <CTT>
1949 V67 N3 P471-474 RT-304 <*>
1949 V67 N3 P479-481 RJ-382 <ATS>
1949 V67 N3 P491-494 2732 <HB>
1949 V67 N4 P597-599 RT-315 <*>
1949 V67 N4 P639-642 R-21 <*>
1949 V67 N4 P659-661 RT-1774 <*>
1949 V67 N4 P675-677 2722 <HB>
1949 V67 N4 P683-685 RCT V25 N1 P12 <RCT>
1949 V67 N5 P773-776 RT-316 <*>
1949 V67 N5 P791-794 RT-345 <*>
1949 V67 N5 P831-834 52M47R <ATS>
1949 V67 N5 P835-838 RT-1835 <*>
1949 V67 N5 P839-842 TT-165 <NRC>
1949 V67 N5 P855-858 86M41R <ATS>
1949 V67 N5 P859-862 52/2942 <NLL>
61-13504 <+*>
1949 V67 N5 P867-869 RT-117 <*>
1949 V67 N5 P875-878 60-41511 <=>
1949 V67 N5 P909-912 RT-767 <*>
1949 V67 N6 P813 60-15764 <+*>
1949 V67 N6 P961-964 RT-1181 <*>
1949 V67 N6 P1009-1012 RT-2150 <*>
1949 V67 N6 P1013-1016 65-10863 <*>
1949 V67 N6 P1021-1023 R-219 <*>
1949 V67 N6 P1029-1031 RJ-21 <ATS>
1949 V67 N6 P1053-1056 RJ-22 <ATS>
61-13595 <+*>
1949 V67 N6 P1073-1076 2703 <HB>
1949 V68 P31-32 R-4517 <*>
1949 V68 P537-539 UCRL TRANS-996(L) <=*$>
1949 V68 N1 P31-32 RT-2263 <*>
1949 V68 N1 P37-40 RT-4237 <*>
3239 <HB>
1949 V68 N1 P49-52 RT-3390 <*>
T-525 <INSD>
1949 V68 N1 P53-56 62-11280 <=>
1949 V68 N1 P177-180 RT-934 <*>
1949 V68 N2 P257-260 RT-2005 <*>
1949 V68 N2 P297-300 59-19141 <+*>
1949 V68 N2 P305-308 RT-3551 <*>
61-10268 <*+>
1949 V68 N2 P333-336 RT-3249 <*>
1949 V68 N2 P421-424 R-275 <*>

Citation	Translation
1949 V68 N3 P457-460	RT-683 <*>
1949 V68 N3 P461-463	RT-684 <*>
1949 V68 N3 P469-472	R-1630 <*>
1949 V68 N3 P477-479	RT-1764 <*>
1949 V68 N3 P483-485	RT-608 <*>
1949 V68 N3 P505-508	RT-4070 <*>
1949 V68 N3 P519-	61-13594 <+*>
1949 V68 N3 P561-564	R-5084 <*>
	59-10067 <+*>
1949 V68 N4 P665-668	TT.129 <NRC>
	61-10128 <+*>
1949 V68 N4 P729-732	64-18406 <*>
	78H13R <ATS>
1949 V68 N5 P829-831	60-23860 <+*>
1949 V68 N5 P865-867	63-11012 <=>
1949 V68 N5 P881-884	61-17880 <=$>
1949 V68 N6 P1009-1011	60-27016 <*+>
1949 V68 N6 P1017-1020	4 <FRI>
1949 V68 N6 P1029-1032	RT-1804 <*>
1949 V68 N6 P1037-1039	2907 <HB>
1949 V68 N6 P1079-1080	62-32151 <=*>
1949 V69 P161-	59-19126 <+*>
1949 V69 P257-260	C-2632 <NRC>
1949 V69 P261-263	C-2628 <NRC>
1949 V69 P301-304	STMSP V1 P1-5 <AMS>
1949 V69 P393-396	R-108 <*>
1949 V69 P821-824	RJ-76 <ATS>
1949 V69 N1 P45-48	50/3391 <NLL>
1949 V69 N1 P49-52	50 2970 <NLL>
1949 V69 N1 P81-83	R-904 <*>
	RT-3205 <*>
	64-19089 <=*$>
1949 V69 N1 P85-88	TT-359 <NRC>
1949 V69 N2 P157-160	RT-607 <*>
1949 V69 N2 P161-164	RT-1997 <*>
	50/1340 <NLL>
1949 V69 N2 P169-171	R-4504 <*>
	RT-1871 <*>
1949 V69 N2 P175-179	RT-606 <*>
1949 V69 N2 P205-207	RT-156 <*>
1949 V69 N3 P329-332	TT.139 <NRC>
1949 V69 N3 P337-340	2 <FRI>
1949 V69 N3 P345-347	RT-2165 <*>
1949 V69 N3 P357-359	64-15202 <=*$>
1949 V69 N3 P373-376	RCT V27 N1 P12 <RCT>
1949 V69 N3 P377-380	TT.192 <NRC>
1949 V69 N3 P393-396	RT-2086 <*>
1949 V69 N4 P519-520	62-11328 <=>
1949 V69 N4 P521-522	R-656 <*>
	RT-2151 <*>
	53/0199 <NLL>
1949 V69 N4 P539-541	RT-4057 <*>
1949 V69 N4 P555-556	AEC-TR-2730 <+*>
	R-1736 <*>
	R-1912 <*>
1949 V69 N4 P569-572	RT-1070 <*>
1949 V69 N5 P627-628	R-387 <*>
	RT-1719 <*>
1949 V69 N5 P695-697	C-3646 <NRCC>
1949 V69 N6 P743-746	RT-772 <*>
1949 V69 N6 P767-769	C-2169 <NRC>
1949 V69 N6 P813-815	RJ-75 <ATS>
1949 V69 N6 P821-824	RJ-76 <ATS>
1949 V69 N6 P849-851	04S83R <ATS>
1950 V70 P5-8	AMST S2 V37 P39-43 <AMS>
1950 V70 P261-264	R-108 <*>
1950 V70 P425-428	R-827 <*>
1950 V70 P569-572	AMST S2 V23 P1-5 <AMS>
1950 V70 P605-608	61-17464 <=$>
1950 V70 P621-624	R-823 <*>
1950 V70 P625-627	R-2785 <*>
	R-4039 <*>
1950 V70 P839-841	R-3243 <*>
1950 V70 N1 P43-45	61-17459 <=$>
1950 V70 N2 P205-206	R-1128 <*>
1950 V70 N2 P215-218	RJ-590 <ATS>
	60-14160 <+*>
	61-17312 <=$>
	66-12877 <*>
1950 V70 N2 P225-228	RT-690 <*>
1950 V70 N2 P245-248	2811 <HB>
1950 V70 N2 P253-256	R-1743 <*>
	RT-1614 <*>
1950 V70 N2 P257-259	R-1188 <*>
	R-4663 <*>
1950 V70 N2 P261-264	RT-3707 <*>
1950 V70 N2 P285-286	61-13373 <+*>
1950 V70 N3 P401-404	62-14050 <=*> O
1950 V70 N3 P417-419	65-60398 <=$>
1950 V70 N3 P421-423	R-1003 <*>
	RT-4286 <*>
	60-13616 <+*>
	62-23217 <=*>
1950 V70 N3 P437-440	61-13010 <+*>
1950 V70 N3 P445-448	62-24878 <=*>
1950 V70 N4 P593-596	R-1548 <*>
1950 V70 N4 P601-603	RT-369 <*>
	RT-687 <*>
1950 V70 N4 P617-620	51/0093 <NLL>
1950 V70 N4 P621-624	RJ-356 <ATS>
1950 V70 N4 P655-657	RT-2157 <*>
1950 V70 N5 P773-775	60-13711 <+*>
1950 V70 N5 P821-824	RT-152 <*>
	59-19089 <+*>
	60-13694 <+*>
1950 V70 N5 P825-827	RT-1846 <*>
1950 V70 N5 P851-853	RJ-23 <ATS>
	2614 <HB>
1950 V70 N5 P859-862	61-16707 <*=>
1950 V70 N5 P867-870	61-17226 <=$>
1950 V70 N6 P989-990	RT-487 <*>
1950 V70 N6 P999-1000	RT-4145 <*>
	63-00761 <*>
1950 V70 N6 P1001-1003	RT-920 <*>
1950 V70 N6 P1013-1016	64-20365 <*>
1950 V70 N6 P1017-1019	RCT V24 N1 P140 <RCT>
1950 V70 N6 P1021-1024	RCT V24 N1 P95 <RCT>
1950 V70 N6 P1025-1028	RT-2290 <*>
	63-20656 <=*$> O
1950 V70 N6 P1033-1036	60-17202 <+*>
1950 V70 N6 P1037-1040	UCRL TRANS-116 <+*>
1950 V70 N6 P1045-1048	R-1460 <*>
	62-25372 <=*>
	64-18333 <*> O
1950 V71 P721-723	R-2560 <*>
1950 V71 P825-828	AEC-TR-6132 <=*$>
1950 V71 P875-878	AEC-TR-3702 <+*>
1950 V71 P907-910	C-2543 <NRC>
1950 V71 P1017-1020	AEC-TR-6134 <=*$>
1950 V71 N1 P39-40	59-15447 <+*>
1950 V71 N1 P41-44	RT-1721 <*>
1950 V71 N1 P53-55	RT-3919 <*>
1950 V71 N1 P69-71	R-1024 <*>
1950 V71 N1 P85-88	59-19087 <+*>
1950 V71 N1 P105-107	R-241 <*>
	R-74 <*>
	64-19076 <=*$>
1950 V71 N1 P205-208	R-5329 <*>
1950 V71 N2 P273-274	59-11706 <=>
1950 V71 N2 P275-276	R-4373 <*>
	3040 <NLL>
1950 V71 N2 P303-305	RT-1100 <*>
1950 V71 N2 P311-313	60-19918 <+*>
1950 V71 N2 P319-322	RCT V24 N2 P250 <RCT>
1950 V71 N2 P343-346	65-13809 <*>
1950 V71 N3 P447-450	T2104 <INSD>
1950 V71 N3 P451-452	3066 <HB>
1950 V71 N3 P453-456	RT-464 <*>
1950 V71 N3 P481-484	3028 <*>
1950 V71 N3 P485-487	RT-1034 <*>
1950 V71 N3 P493-496	65-60926 <=$>
1950 V71 N3 P501-504	62-01495 <*>
	62-11746 <=>
1950 V71 N3 P521-522	R-239 <*>
	R-73 <*>
	64-19077 <=*$>
1950 V71 N3 P533-536	R-1193 <*>
1950 V71 N3 P549	61-13692 <+*> P
1950 V71 N4 P641-642	RT-736 <*>
1950 V71 N4 P655-658	60-21884 <=>

Citation	Translation
1950 V71 N4 P685-688	64-19948 <=$>
1950 V71 N4 P701-704	63-20441 <=*$>
1950 V71 N4 P785-788	C-2999 <NRCC>
1950 V71 N5 P879-882	R-12 <*>
1950 V71 N5 P887-890	50/0099 <NLL>
1950 V71 N5 P905-906	R-1211 <*>
1950 V71 N5 P911-914	RT-2253 <*>
1950 V71 N5 P937-940	RT-2088 <*>
1950 V71 N5 P985-987	R-5375 <*>
1950 V71 N6 P1057-1060	R-1481 <*>
1950 V71 N6 P1061-1064	R-11 <*>
1950 V71 N6 P1073-1075	58J19R <ATS>
1950 V72 P145-148	R-1037 <*>
1950 V72 P351-353	R-4515 <*>
1950 V72 P663-666	R-4514 <*>
1950 V72 P699-701	AEC-TR-1024 <*>
1950 V72 P915-918	64-20184 <*>
1950 V72 N1 P23-26	RT-924 <*>
1950 V72 N1 P35-38	TT.472 <NRC>
1950 V72 N1 P41-44	RT-2092 <*>
1950 V72 N1 P53-56	TT.390 <NRC>
1950 V72 N1 P73-76	R-1199 <*>
	RJ-384 <ATS>
1950 V72 N1 P77-80	RT-2089 <*>
1950 V72 N1 P93-95	63-16755 <=*>
1950 V72 N1 P125-128	59-22352 <+*>
1950 V72 N2 P161-164	62-15876 <=*>
1950 V72 N2 P265-267	R-4604 <*>
	61-13593 <*+>
1950 V72 N2 P273-275	60-13876 <+*>
1950 V72 N2 P301-304	RT-3614 <*>
1950 V72 N2 P311-313	3055 <HB>
1950 V72 N2 P319-322	52/2935 <*>
1950 V72 N2 P331-334	60-13704 <+*>
1950 V72 N2 P335-338	AD-610 893 <=$>
	R-1745 <*>
	RT-1920 <*>
1950 V72 N2 P343-345	RT-1461 <*>
1950 V72 N2 P351-353	R-1198 <*>
1950 V72 N2 P365-368	R.703 <RIS>
1950 V72 N2 P413-415	RT-2091 <*>
1950 V72 N2 P433-435	62-15371 <*=> 0
1950 V72 N3 P477-480	9 <FRI>
1950 V72 N3 P485-487	25 <FBJ>
1950 V72 N3 P515-517	59-15347 <+*>
	64-10877 <=*$>
1950 V72 N3 P543-546	TT.593 <NRC>
1950 V72 N3 P555-558	RT-466 <*>
1950 V72 N3 P561-564	59-19088 <+*>
1950 V72 N3 P587-590	RT-2095 <*>
1950 V72 N4 P659-660	59-16038 <+*>
1950 V72 N4 P661-662	RT-1722 <*>
1950 V72 N4 P663-666	RT-1723 <*>
1950 V72 N4 P671-674	R-1030 <*>
1950 V72 N4 P687-690	61-13142 <+*>
1950 V72 N4 P707-709	2709 <HB>
1950 V72 N4 P711-712	R-4854 <*>
	R-901 <*>
	RT-1033 <*>
1950 V72 N4 P713-715	21L29R <ATS>
1950 V72 N4 P725-728	63-20518 <=*$>
	64-20185 <*>
1950 V72 N4 P779-780	62-10639 <=*>
1950 V72 N5 P903-906	RT-72 <*>
1950 V72 N5 P915-918	63-20517 <=*$>
1950 V72 N5 P923-926	60-23192 <+*> 0
1950 V72 N5 P937-939	R-1204 <*>
1950 V72 N5 P957-960	61-16556 <=*> 0
1950 V72 N6 P1037-1039	TT.202 <NRC>
1950 V72 N6 P1059-1061	TT.273 <NRC>
1950 V72 N6 P1071-1074	R.709 <*>
1950 V72 N6 P1075-1078	RT-3613 <*>
1950 V72 N6 P1175-1177	RT-3250 <*>
1950 V73 P499-502	19K22R <ATS>
1950 V73 P1239-1242	R-108 <*>
1950 V73 P1263-1266	C-2544 <NRC>
1950 V73 N1 P53-54	R-4873 <*>
1950 V73 N1 P55-58	R-4522 <*>
	RT-178 <*>
1950 V73 N1 P111-112	61-13592 <*+>
1950 V73 N1 P113-115	60-13709 <+*>
1950 V73 N1 P129-132	RCT V24 N3 P569 <RCT>
1950 V73 N2 P291-294	RT-1616 <*>
1950 V73 N2 P295-297	RT-1693 <*>
1950 V73 N2 P327-330	RT-3766 <*>
1950 V73 N2 P333-336	RT-3767 <*>
1950 V73 N2 P359-362	RT-3848 <*>
1950 V73 N2 P381-384	52/2939 <NLL>
	61-13591 <+*>
1950 V73 N3 P475-478	RT-69 <*>
1950 V73 N3 P479-481	RT-67 <*>
1950 V73 N3 P487-489	R-824 <*>
	RT-3836 <*>
1950 V73 N3 P499-502	RT-3859 <*>
1950 V73 N3 P511-513	3195 <HB>
1950 V73 N3 P515-518	R-1880 <*>
1950 V73 N3 P535-538	TT.214 <NRC>
1950 V73 N3 P601-604	RT-1783 <*>
1950 V73 N3 P621-624	RT-522 <*>
1950 V73 N4 P679-682	RT-4144 <*>
1950 V73 N4 P683-684	60-19920 <+*>
1950 V73 N4 P689-692	3043 <*>
1950 V73 N4 P697-700	2876 <HB>
1950 V73 N4 P701-704	RCT V24 N3 P616 <RCT>
1950 V73 N4 P771-774	TT.216 <NRC>
1950 V73 N4 P861-863	RT-523 <*>
1950 V73 N5 P905-908	R-776 <*>
	RT-3261 <*>
1950 V73 N5 P921-924	60-15915 <+*>
1950 V73 N5 P925-927	RT-488 <*>
1950 V73 N5 P937-940	RT-1736 <*>
1950 V73 N5 P945-947	RT-605 <*>
1950 V73 N5 P957-958	61-28027 <*=>
1950 V73 N5 P959-961	64-18933 <*>
	64-20187 <*>
1950 V73 N5 P971-973	51/3311 <NLL>
1950 V73 N5 P975-978	62-13077 <*=>
1950 V73 N5 P987-990	52/2585 <NLL>
1950 V73 N5 P1093-1095	RT-3258 <*>
1950 V73 N6 P1143-1144	RT-3212 <*>
1950 V73 N6 P1149-1151	RT-2462 <*>
1950 V73 N6 P1157-1160	TT.369 <NRC>
1950 V73 N6 P1217-1220	RT-2406 <*>
	64-14390 <=*$> 0
1950 V73 N6 P1221-1224	RT-465 <*>
	TT.245 <NRC>
1950 V73 N6 P1225-1228	RT-1231 <*>
1950 V73 N6 P1239-1242	RT-2034 <*>
1950 V73 N6 P1305-1308	RT-116 <*>
	61-28103 <*=> 0
1950 V74 P103-106	R-108 <*>
1950 V74 P315-318	R-108 <*>
1950 V74 P781-783	R-108 <*>
1950 V74 P1077-1080	C-2644 <NRC>
1950 V74 N1 P29-32	TT.359 <NRC>
1950 V74 N1 P49-52	2075 <BISI>
1950 V74 N1 P57-59	RJ-357 <ATS>
	61-20927 <*=>
	65-13212 <*>
1950 V74 N1 P91-94	RT-1031 <*>
	63-18995 <=*>
1950 V74 N1 P103-106	RT-3716 <*>
1950 V74 N1 P153-156	RJ-62 <ATS>
1950 V74 N2 P197-200	62-10800 <=*>
1950 V74 N2 P213-216	RT-2022 <*>
	60-14159 <+*>
	61-17289 <=$>
1950 V74 N2 P237-240	R-658 <*>
1950 V74 N2 P251-254	RT-1937 <*>
1950 V74 N2 P315-318	RT-1938 <*>
1950 V74 N2 P343-344	C-2979 <NRCC>
	59-15312 <+*>
1950 V74 N3 P461-463	C-3221 <NRCC>
1950 V74 N3 P477-480	TT.370 <NRC>
1950 V74 N3 P485-488	RT-1484 <*>
1950 V74 N3 P489-492	R-4042 <*>
1950 V74 N3 P501-504	R-97 <*>
	RT-2468 <*>
	64-15073 <=*$>
1950 V74 N3 P521-524	61-13691 <+*>

1950 V74 N3 P591-594 60-13702.<+*>
1950 V74 N4 P677-680 RT-1607 <*>
1950 V74 N4 P687-690 RJ-82 <ATS>
RT-2381 <*>
1950 V74 N4 P703-705 RT-717 <*>
1950 V74 N4 P723-724 62-11431 <=>
1950 V74 N4 P725-727 RCT V25 N1 P33 <RCT>
1950 V74 N4 P743-746 07M40R <ATS>
1950 V74 N4 P755-757 59-18330 <+*>
1950 V74 N4 P767-769 RT-2164 <*>
1950 V74 N4 P771-775 R-1668 <*>
1950 V74 N4 P779-780 60-15936 <+*>
1950 V74 N4 P781-783 RT-3710 <*>
1950 V74 N4 P833-835 RT-3942 <*>
1950 V74 N5 P909-912 RT-1485 <*>
1950 V74 N5 P917-919 59-17348 <+*>
1950 V74 N5 P921-923 RT-1708 <*>
59-17352 <+*>
1950 V74 N5 P955-958 RJ-394 <ATS>
1950 V74 N5 P979-981 RT-2090 <*>
1950 V74 N6 P1073-1076 51/3312 <NLL>
1950 V74 N6 P1077-1080 R-4258 <*>
1950 V74 N6 P1101-1104 64-18154 <*>
1950 V75 P189-192 R-107 <*>
1950 V75 N1 P79-82 64-15059 <=*$>
1950 V75 N1 P91-94 RT-346 <*>
1950 V75 N2 P189-192 13 <FRI>
1950 V75 N2 P201-203 62-23042 <=*>
1950 V75 N2 P213-214 R-1879 <*>
RT-486 <*>
1950 V75 N2 P215-217 64-20183 <*>
1950 V75 N2 P219-222 R-5054 <*>
155TM <CTT>
7J19R <ATS>
1950 V75 N2 P223-226 RCT V24 N4 P853 <RCT>
1950 V75 N2 P243-246 RT-3856 <*>
1950 V75 N3 P367-370 R-481 <*>
1950 V75 N3 P371-374 RT-4042 <*>
1950 V75 N3 P387-390 14 <FRI>
62-11364 <=>
1950 V75 N3 P445-447 59-19093 <+*>
1950 V75 N4 P535-538 15 <FRI>
62-11360 <=>
1950 V75 N5 P617-620 RT-948 <*>
1950 V75 N5 P629-630 59-19460 <+*>
1950 V75 N5 P639-641 R-871 <*>
1950 V75 N5 P647-650 R-3231 <*>
1950 V75 N5 P655-658 R-4553 <*>
RT-177 <*>
1950 V75 N5 P665-667 TT.513 <NRC>
1950 V75 N5 P681-683 64-18379 <*>
1950 V75 N5 P685-687 62-15862 <=*>
1950 V75 N6 P769-772 RT-332 <*>
1950 V75 N6 P789-792 RT-4167 <*>
1950 V75 N6 P793-795 RT-1921 <*> 0
1950 V75 N6 P815-818 RT-1032 <*>
63-20444 <=*$>
1950 V75 N6 P819-822 06M40R <ATS>
1950 V75 N6 P833-835 RT-2097 <*>
1950 V75 N6 P855-858 RTS-3067 <NLL>
1950 V75 N6 P863-864 60-51139 <=>
1951 V76 P69-72 R-2778 <*>
1951 V76 P69-73 R-4036 <*>
1951 V76 P221-222 R-2862 <*>
1951 V76 P223-225 M-6223 <NLL>
1951 V76 P673-676 T-2112 <INSD>
1951 V76 P685-687 R-2780 <*>
R-2784 <*>
R-3978 <*>
1951 V76 N1 P49-52 R-1465 <*>
61-13176 <+*>
1951 V76 N1 P57-60 R-1465 <*>
61-13176 <+*>
1951 V76 N1 P61-64 RT-2036 <*>
1951 V76 N1 P85-88 R-1025 <*>
R-3235 <*>
1951 V76 N1 P111-113 15Q72R <ATS>
1951 V76 N1 P133-135 RJ-56 <ATS>
62-32197 <=*>
1951 V76 N1 P141-144 62-25634 <=*>
1951 V76 N1 P157-160 59-15901 <+*>
1951 V76 N2 P177-180 RT-1093 <*>
60-12602 <=>
1951 V76 N2 P197-200 RT-1922 <*>
70M38R <ATS>
1951 V76 N2 P201-204 RJ-604 <ATS>
1951 V76 N2 P215-217 R-3098 <*>
R-4038 <*>
RT-4506 <*>
1951 V76 N2 P273-276 RT-2127 <*>
1951 V76 N2 P303-304 RT-3845 <*>
1951 V76 N2 P317-320 62-32197 <=*>
1951 V76 N3 P355-358 RT-333 <*>
1951 V76 N3 P367-370 RT-926 <*>
60-12498 <=>
1951 V76 N3 P375-376 59-16037 <+*>
1951 V76 N3 P377-380 RT-1546 <*>
1951 V76 N3 P395-397 2708 <HB>
1951 V76 N3 P427-429 RJ-52 <ATS>
1951 V76 N3 P455-457 R-1200 <*>
1951 V76 N3 P475-478 62-11219 <=>
1951 V76 N4 P513-514 R-3089 <*>
1951 V76 N4 P519-522 RT-302 <*>
1951 V76 N4 P523-526 C-4746 <NRC>
1951 V76 N4 P539-542 65-17358 <*> 0
1951 V76 N4 P543-546 RJ-49 <ATS>
1951 V76 N4 P547-550 RJ-50 <ATS>
1951 V76 N4 P551-554 RJ-391 <ATS>
156TM <CTT>
1951 V76 N4 P593-596 RT-2064 <*>
1951 V76 N5 P673-676 R-1639 <*>
TT.570 <NRC>
1951 V76 N5 P743-745 RT-3252 <*>
1951 V76 N5 P755-758 RT-614 <*>
1951 V76 N6 P821-823 R-3177 <*>
1951 V76 N6 P825-826 RT-604 <*>
1951 V76 N6 P839-842 RT-3839 <*>
1951 V76 N6 P855-858 T.708 <INSD>
1951 V77 P93-95 C-2542 <NRC>
1951 V77 P285-288 R-2240 <*>
1951 V77 P565-568 AMST S1 V3 P281-286 <AMS>
1951 V77 P855-858 R-3714 <*>
1951 V77 N1 P45-48 19 <FRI>
1951 V77 N1 P49-51 61-19445 <+*> C
1951 V77 N1 P57-60 RT-4143 <*>
1951 V77 N1 P57-63 64-19059 <=*$>
1951 V77 N1 P69-72 59-15504 <+*>
1951 V77 N1 P75-76 RT-3741 <*>
1951 V77 N1 P81-84 RCT V24 N4 P763 <RCT>
1951 V77 N1 P85-86 RT-3063 <*>
1951 V77 N1 P137-140 R-4689 <*>
59-18555 <+*> 0
1951 V77 N1 P157-160 RJ-174 <ATS>
1951 V77 N1 P565-568 RT-1112 <*>
1951 V77 N2 P197-200 NASA-TT-F-192 <=>
POED-TRANS-2223 <NLL>
1951 V77 N2 P217-220 59-16064 <+*>
1951 V77 N2 P221-224 R-354 <*>
1951 V77 N2 P225-227 RT.602 <*>
61-17468 <=$>
1951 V77 N2 P233-236 C-2187 <NRC>
1951 V77 N2 P241-244 61-18798 <*=>
1951 V77 N2 P249-252 RT-281 <*>
1951 V77 N2 P261-264 18 <FRI>
1951 V77 N2 P273-275 RT-2062 <*>
1951 V77 N2 P281-284 RT-3891 <*>
1951 V77 N2 P285-288 RT-2318 <*>
RTS-2659 <NLL>
1951 V77 N2 P301-303 65-29158 <=>
96J18R <ATS>
1951 V77 N2 P365-368 R-4125 <*>
1951 V77 N3 P395-398 RT-74 <*>
1951 V77 N3 P399-402 RT-685 <*>
1951 V77 N3 P415-418 61-20293 <*=> 0
1951 V77 N3 P427-428 26 <FRI>
61-13590 <+*>
1951 V77 N3 P429-432 RT-3745 <*>
1951 V77 N3 P433-434 62-10523 <*=>
1951 V77 N3 P435-438 59-10383 <+*> 0
61-13690 <+*> C

Citation	Translation
1951 V77 N3 P461-464	R-1217 <*>
1951 V77 N4 P557-560	UCRL-TRANS-643 <=*>
1951 V77 N4 P561-564	RT-2118 <*>
1951 V77 N4 P585-588	RT-898 <*>
1951 V77 N4 P597-598	RT-601 <*>
	61-17472 <=$>
1951 V77 N4 P599-602	RT-603 <*>
	61-17470 <=$>
1951 V77 N4 P621-624	94J18R <ATS>
1951 V77 N4 P645-647	95J18R <ATS>
1951 V77 N4 P681-684	60-13582 <+*>
1951 V77 N4 P741-744	RT-3523 <*>
1951 V77 N5 P787-789	64-71335 <=> M
1951 V77 N5 P815-818	RT-662 <*>
1951 V77 N5 P827-830	3159 <HB>
1951 V77 N5 P831-834	59-13514 <=>
1951 V77 N5 P835-837	RT-912 <*>
	60-13738 <+*>
1951 V77 N5 P843-846	TT.259 <NRC>
1951 V77 N5 P847-850	RT-65 <*>
1951 V77 N5 P855-858	RT-847 <*>
1951 V77 N5 P863-865	TT.299 <NRC>
1951 V77 N5 P871-874	59-18369 <+*>
1951 V77 N6 P985-988	RT-2027 <*>
1951 V77 N6 P997-999	RT-70 <*>
1951 V77 N6 P1023-1026	28L30R <ATS>
1951 V77 N6 P1031-1034	R-3979 <*>
1951 V77 N6 P1055-1058	61-13083 <+*>
1951 V77 N6 P1115-1118	UCRL TRANS-106 <+*>
1951 V78 P137-139	R-1247 <*>
1951 V78 P259-262	R-1248 <*>
1951 V78 P327-329	R-1249 <*>
1951 V78 P405-408	AMST S1 V3 P287-293 <AMS>
1951 V78 P435-438	R-1414 <*>
	1372TM <CTT>
1951 V78 P531-534	C-2541 <NRC>
1951 V78 P949-951	R-1250 <*>
1951 V78 N1 P29-32	RT-2177 <*>
1951 V78 N1 P33-36	R-3830 <*>
	R-3848 <*>
	RT-207 <*>
1951 V78 N1 P59-62	RT-3990 <*>
1951 V78 N1 P67-69	05M40R <ATS>
1951 V78 N1 P83-86	59-17365 <+*> 0
1951 V78 N1 P91-94	RT-1057 <*>
1951 V78 N2 P225-228	RT-1849 <*>
1951 V78 N2 P245-248	R-4838 <*>
1951 V78 N2 P263-266	RJ-469 <ATS>
1951 V78 N2 P267-270	61-13141 <+*>
1951 V78 N2 P271-274	61-13687 <+*>
1951 V78 N2 P299-302	3143 <HB>
	61-13689 <+*>
1951 V78 N2 P303-306	TT.265 <NRC>
1951 V78 N2 P327-329	RT-2269 <*>
1951 V78 N3 P405-408	RT-600 <*>
1951 V78 N3 P411-414	66-60625 <=*$>
1951 V78 N3 P427-429	RT-75 <*>
1951 V78 N3 P431-433	RT-68 <*>
1951 V78 N3 P447-450	RT.599 <*>
	60-13708 <+*>
	61-17471 <=$>
1951 V78 N3 P457-459	RT-45 <*>
	RT-66 <*>
1951 V78 N3 P469-472	RT-686 <*>
1951 V78 N3 P473-476	R-817 <*>
	RJ-83 <ATS>
1951 V78 N3 P481-483	63-11012 <=>
1951 V78 N3 P489-492	62-18668 <=*> 0
1951 V78 N3 P497-500	RJ-107 <ATS>
1951 V78 N3 P505-507	TT.268 <NRC>
1951 V78 N3 P523-526	TT.321 <NRC>
	59-15732 <+*>
	60-13870 <+*>
1951 V78 N3 P547-550	R-402 <*>
	64-19506 <=$>
1951 V78 N3 P573-575	RT-2178 <*>
1951 V78 N4 P657-660	61-23262 <*=>
1951 V78 N4 P665-668	RT-733 <*>
1951 V78 N4 P669-672	RT-73 <*>
1951 V78 N4 P709-712	RT-2382 <*>
	64-15376 <=*$>
1951 V78 N4 P713-716	RT-2085 <*>
	64-20182 <*> 0
1951 V78 N4 P749-752	R-713 <*>
	R-974 <*>
1951 V78 N4 P825-827	T-2020 <INSD>
1951 V78 N5 P837-840	65-61237 <=$>
1951 V78 N5 P875-877	R-741 <*>
	RT-485 <*>
1951 V78 N5 P883-885	UCRL TRANS-666(L) <*=>
1951 V78 N5 P889-891	UCRL TRANS-825 <=*>
1951 V78 N5 P909-911	30 <FRI>
1951 V78 N5 P949-951	RT-2158 <*>
1951 V78 N5 P955-957	RT-2179 <*>
1951 V78 N6 P1105-1108	TT.411 <NRC>
1951 V78 N6 P1131-1134	52/2518 <NLL>
1951 V78 N6 P1145-1148	R-813 <*>
1951 V78 N6 P1169-1172	60-13865 <+*>
1951 V78 N6 P1181-1184	R-4282 <*>
1951 V79 P273-276	64-18061 <=*$>
1951 V79 P775-	R-1127 <*>
1951 V79 P783	R-5079 <*>
1951 V79 P811-813	63-14091 <=*> 0
1951 V79 P993-996	R-4003 <*>
1951 V79 N1 P29-32	65-61363 <=$>
1951 V79 N1 P53-56	RT-303 <*>
1951 V79 N1 P69-72	R-657 <*>
1951 V79 N1 P109-112	T.829 <INSD>
	202TM <CTT>
1951 V79 N2 P225-228	63-20524 <=*$> 0
	64-20181 <*>
1951 V79 N2 P237-240	61-13589 <*+>
1951 V79 N2 P241-244	RT-4061 <*>
	63-00768 <*>
1951 V79 N2 P257-260	62-10474 <=*>
1951 V79 N2 P273-276	TT.313 <NRC>
1951 V79 N2 P287-288	3160 <HB>
1951 V79 N2 P295-298	RT-2214 <*>
1951 V79 N2 P303-306	52/2600 <NLL>
	63-13582 <=*>
1951 V79 N2 P315-318	T-2265 <INSD>
1951 V79 N3 P431-434	RT-742 <*>
1951 V79 N3 P439-442	R-721 <*>
1951 V79 N3 P455-458	TT.263 <NRC>
1951 V79 N3 P467-470	RCT V25 N2 P230 <RCT>
1951 V79 N3 P467	60-13866 <+*>
1951 V79 N3 P487-489	R-3779 <*>
	2233A <K-H>
1951 V79 N3 P507-508	RT-254 <*>
1951 V79 N3 P521-524	60-18029 <+*>
1951 V79 N4 P581-584	RT-253 <*>
1951 V79 N4 P585-588	RT-2632 <*>
1951 V79 N4 P589-590	59-20217 <+*>
1951 V79 N4 P605-608	RT-3866 <*>
1951 V79 N4 P629-632	RT-2193 <*>
1951 V79 N4 P637-638	63-13583 <=*>
1951 V79 N5 P747-750	RT-1097 <*>
1951 V79 N5 P755-758	RT-1957 <*>
1951 V79 N5 P759-762	RT-1958 <*>
1951 V79 N5 P763-766	59-19465 <+*>
1951 V79 N5 P771-774	R-1996 <*>
1951 V79 N5 P775-777	RT-1720 <*>
1951 V79 N5 P779-781	RT-4507 <*>
1951 V79 N5 P783-786	RT-1153 <*>
	60-16497 <+*>
1951 V79 N5 P819-821	R-789 <*>
1951 V79 N5 P823-826	R-3990 <*>
1951 V79 N5 P827-830	61-17132 <=$>
1951 V79 N6 P945-948	M.5109 <NLL>
1951 V79 N6 P957-960	59-19466 <+*>
1951 V79 N6 P961-964	RT-4160 <*>
1951 V79 N6 P993-996	TT.260 <NRC>
1951 V79 N6 P1001-1004	R-4375 <*>
1951 V79 N6 P1025-1027	RT-4446 <*>
1951 V80 P261-263	R-1251 <*>
1951 V80 P369-372	R-2773 <*>
	R-4867 <*>
1951 V80 P385-388	R-1253 <*>
1951 V80 P449-451	R-1252 <*>
1951 V80 P525-528	STMSP V1 P13-16 <AMS>

Citation	Translation
1951 V80 P739-742	RJ-605 <ATS>
1951 V80 P907-	R-3751 <*>
1951 V80 N1 P33-35	1457TM <CTT>
	59-16686 <+*>
1951 V80 N1 P53-56	RT-2691 <*>
1951 V80 N1 P57-60	RT-3920 <*>
1951 V80 N1 P69-72	61-15459 <+*>
1951 V80 N2 P153-156	RTS-2581 <NLL>
1951 V80 N2 P189-192	R-10 <*>
1951 V80 N2 P193-195	RT-2031 <*>
1951 V80 N2 P197-200	61-13688 <*+>
	61-17113 <=$>
1951 V80 N3 P325-327	64-11888 <=> M
1951 V80 N3 P345-347	R-763 <*>
1951 V80 N3 P361-364	R-9 <*>
1951 V80 N3 P381-384	61-13507 <+*>
1951 V80 N3 P401-403	65-10188 <*>
1951 V80 N3 P405-408	RT-1667 <*>
	64-15575 <=*$> P
1951 V80 N3 P409-412	2318E <K-H>
1951 V80 N3 P421-424	3098 <HB>
1951 V80 N3 P425-428	R-2563 <*>
1951 V80 N4 P525-528	RT-434 <*>
1951 V80 N4 P577-578	C-2573 <NRC>
	R-3440 <*>
1951 V80 N4 P587-590	RT-2185 <*>
1951 V80 N4 P599-601	RT-1976 <*>
1951 V80 N4 P611-613	RT-1058 <*>
1951 V80 N4 P693-695	64-13740 <=*$>
1951 V80 N5 P735-738	RT-1725 <*>
1951 V80 N5 P791-792	63-19145 <=*>
1951 V80 N6 P857-860	RT-692 <*>
1951 V80 N6 P867-870	59-16063 <+*>
1951 V80 N6 P875-878	RT-3448 <*>
1951 V80 N6 P879-880	R-1465 <*>
	61-13176 <+*>
1951 V80 N6 P884-892	59-19106 <+*>
1951 V80 N6 P897-898	R-3983 <*>
1951 V80 N6 P899-902	R-812 <*>
1951 V80 N6 P903-905	TT.298 <NRC>
1951 V80 N6 P907-910	RT-2380 <*>
1951 V80 N6 P937-939	RJ-377 <ATS>
1951 V81 P223-226	R-1254 <*>
1951 V81 P879-882	R-108 <*>
	R-1255 <*>
1951 V81 P1105-1108	R-108 <*>
	R-1256 <*>
1951 V81 N1 P27-30	RT-1751 <*>
	1482TM <CTT>
1951 V81 N1 P39-42	36 <FRI>
1951 V81 N1 P43-45	RT-2101 <*>
1951 V81 N1 P63-66	RCT V30 N2 P544 <RCT>
1951 V81 N1 P63	59-19142 <+*>
1951 V81 N2 P167-170	59-16062 <+*>
1951 V81 N2 P175-177	RT-2163 <*>
1951 V81 N2 P183-186	RT-1924 <*>
1951 V81 N2 P215-218	TT.358 <NRC>
1951 V81 N2 P223-226	TT.331 <NRC>
	45Q68R <ATS>
1951 V81 N2 P235-237	64-14813 <*>
1951 V81 N2 P243-245	63-23408 <=*>
1951 V81 N2 P255-258	R-3657 <*>
1951 V81 N3 P363-366	59-19467 <+*>
1951 V81 N3 P367-370	RT-4447 <*>
1951 V81 N3 P371-374	RT-2884 <*>
1951 V81 N3 P375-377	R-4094 <*>
1951 V81 N3 P411-413	61-10897 <+*>
1951 V81 N3 P427-430	61-17300 <=$>
1951 V81 N3 P435-438	RT-2345 <*>
1951 V81 N3 P461-464	RT-3264 <*>
1951 V81 N4 P525-528	RT-1620 <*>
1951 V81 N4 P537-539	RT-2106 <*>
	60-13759 <+*>
1951 V81 N4 P549-551	R-2076 <*>
	RJ-249 <ATS>
	RT-2063 <*>
	62-19897 <=*>
1951 V81 N4 P565-568	RT-1925 <*>
	2844 <HB>
1951 V81 N4 P597-600	R-2344 <*>
	64-71418 <=> M
1951 V81 N4 P605-607	62-24116 <=*>
1951 V81 N4 P613	62-24116 <=*>
1951 V81 N4 P637-640	91K20R <ATS>
1951 V81 N4 P645-648	60-41379 <=>
1951 V81 N4 P651-654	RJ-665 <ATS>
1951 V81 N5 P765-766	RT-2633 <*>
	61-31475 <=>
	62-11231 <=>
1951 V81 N5 P767-769	RT-2125 <*>
1951 V81 N5 P791-794	RT-2206 <*>
1951 V81 N5 P807-810	RT-2811 <*>
1951 V81 N5 P811-814	RT-1621 <*>
1951 V81 N5 P849-852	R-4095 <*>
1951 V81 N5 P855-858	2850 <HB>
	62-10170 <*=>
1951 V81 N5 P863-866	64L35R <ATS>
1951 V81 N5 P879-882	RT-3711 <*>
1951 V81 N5 P887-890	3190 <HB>
1951 V81 N6 P1023-1026	AD-619 431 <=$>
1951 V81 N6 P1055-1057	RT-3976 <*>
1951 V81 N6 P1069-1072	TT.342 <NRC>
1951 V81 N6 P1081-1083	RT-2346 <*>
1951 V81 N6 P1085-1088	76L32R <ATS>
1951 V81 N6 P1093-1096	65-60900 <=$>
1951 V81 N6 P1101-1104	RT-3269 <*>
	59-17517 <+*>
1951 V81 N6 P1105-1108	RT-3712 <*>
1952 V82 P289-291	R-1257 <*>
1952 V82 P513-516	STMSP V1 P69-72 <AMS>
1952 V82 P661-663	STMSP V1 P73-75 <AMS>
1952 V82 P739-742	R-1258 <*>
1952 V82 P761-764	R-108 <*>
	R-1259 <*>
1952 V82 P837-840	STMSP V1 P77-81 <AMS>
1952 V82 P841-843	STMSP V1 P83-85 <AMS>
1952 V82 P947-950	R-108 <*>
	R-1260 <*>
1952 V82 N1 P21-28	59-19469 <+*>
1952 V82 N1 P29-32	62-11418 <=>
1952 V82 N1 P33-36	RT-2146 <*>
	62-00408 <*>
1952 V82 N1 P37-40	TT.451 <NRC>
1952 V82 N1 P45-47	RT-4130 <*>
1952 V82 N1 P57-60	62-11198 <=>
1952 V82 N1 P61-63	RT-3769 <*>
1952 V82 N1 P93-96	RT-1268 <*>
1952 V82 N1 P97-100	RT-3268 <*>
1952 V82 N1 P131-133	RJ-133 <ATS>
1952 V82 N1 P131-	60-23036 <+*>
1952 V82 N1 P135-138	RT-2356 <*>
1952 V82 N2 P229-231	61-18927 <*=>
	62-20284 <=*> O
	62-23186 <=>
1952 V82 N2 P249-251	62-11197 <=>
1952 V82 N2 P273-276	R-3994 <*>
1952 V82 N2 P281-284	RT-2722 <*>
1952 V82 N3 P377-378	32K20R <ATS>
1952 V82 N3 P379-380	RT-928 <*>
1952 V82 N3 P415-417	61-17239 <=$>
1952 V82 N3 P423-426	60-15822 <+*> O
1952 V82 N3 P427-430	RT-2725 <*>
1952 V82 N4 P513-516	R-3587 <*>
	RT-340 <*>
	64-71259 <=>
1952 V82 N4 P545-548	65-60968 <=$> O
1952 V82 N4 P571-574	R-882 <*>
1952 V82 N4 P579-580	AD-610 889 <=$>
	RT-1727 <*>
1952 V82 N4 P589-591	63-18667 <=*> C
1952 V82 N4 P597-600	TT.333 <NRC>
1952 V82 N4 P601-602	64-23517 <=$>
1952 V82 N4 P603-605	R-2319 <*>
1952 V82 N4 P623-624	R-1212 <*>
1952 V82 N5 P661-663	RT-339 <*>
1952 V82 N5 P697-700	792TM <CTT>
1952 V82 N5 P705-708	RT-4050 <*>
1952 V82 N5 P717-718	RT-4043 <*>
1952 V82 N5 P747-750	RCT V26 N3 P589 <RCT>
1952 V82 N5 P751-753	R-3691 <*>

Citation	Translation
1952 V82 N5 P755-756	RT-3740 <*>
1952 V82 N5 P761-764	RT-2093 <*>
1952 V82 N5 P765-768	IGR V2 N2 P120 <AGI>
	R-1203 <*>
1952 V82 N6 P837-840	RT-338 <*>
1952 V82 N6 P841-843	RT-337 <*>
1952 V82 N6 P861-864	RT-1759 <*>
1952 V82 N6 P885-887	62-11752 <=>
1952 V82 N6 P909-912	R-4050 <*>
1952 V82 N6 P927-929	R-4051 <*>
1952 V82 N6 P935-938	RCT V26 N1 P98 <RCT>
1952 V82 N6 P939-942	RT-2582 <*>
1952 V82 N6 P947-950	RT-3717 <*=>
1952 V82 N6 P961-964	157TM <CTT>
	28H13R <ATS>
1952 V82 N6 P1001	R-4935 <*>
1952 V83 P59-61	R-4362 <*>
1952 V83 P157-159	R-1261 <*>
1952 V83 P353-355	STMSP V1 P59-61 <AMS>
1952 V83 P477-480	R-1263 <*>
1952 V83 P481-484	R-1264 <*>
1952 V83 P485-487	R-1265 <*>
1952 V83 P561	R-5063 <*>
1952 V83 P929-930	R-1266 <*>
1952 V83 N1 P79-80	R-4000 <*>
1952 V83 N1 P81-83	R-1952 <*>
	RJ-157 <ATS>
1952 V83 N1 P101-104	01L36R <ATS>
1952 V83 N1 P109-110	RT-3764 <*>
	99K24R <ATS>
1952 V83 N1 P111-114	RCT V26 N2 P352 <RCT>
1952 V83 N1 P115-116	RT-61 <*>
1952 V83 N2 P173-174	RT-837 <*>
1952 V83 N2 P227-230	RT-4071 <*>
1952 V83 N2 P243-246	RT-1040 <*>
1952 V83 N2 P247-250	63-23307 <=*>
1952 V83 N2 P253-255	3116 <HB>
1952 V83 N2 P265-267	3042 <HB>
1952 V83 N3 P353-355	RT-787 <*>
1952 V83 N3 P385-388	RT-473 <*>
	63-10773 <=*>
1952 V83 N3 P397-398	3520 <HB>
1952 V83 N3 P443-446	59-15157 <+*>
	60-41443 <=>
	61-13588 <=*>
1952 V83 N3 P447-450	RT-3333 <*>
1952 V83 N3 P477-480	60-15935 <+*>
1952 V83 N4 P541-544	59-22065 <+*> 0
1952 V83 N4 P549-553	R-2360 <*>
1952 V83 N4 P557-560	RT-3052 <*>
1952 V83 N4 P565-568	R-4182 <*>
1952 V83 N4 P585-588	RT-2454 <*>
1952 V83 N5 P675-676	R-476 <*>
	2979 <HB>
1952 V83 N5 P689-692	RT-3763 <*>
	73P64R <ATS>
1952 V83 N5 P713-716	RJ-92 <ATS>
1952 V83 N5 P717-720	RJ-94 <ATS>
	RT-1961 <*>
1952 V83 N6 P841-842	3169 <HB>
1952 V83 N6 P843-845	3435 <HB>
1952 V83 N6 P863-864	R-4097 <*>
1952 V83 N6 P865-868	RT-3765 <*>
1952 V83 N6 P873-876	RT-3051 <*>
1952 V83 N6 P901-902	RJ-386 <ATS>
1952 V84 P37-40	R-4091 <*>
1952 V84 P205-208	R-800 <*>
1952 V84 P321-324	R-2553 <*>
1952 V84 P555-558	R-5357 <*>
1952 V84 P607-610	R-1267 <*>
1952 V84 P883-886	AMST S2 V2 P141-145 <AMS>
1952 V84 P895-898	AMST S2 V38 P1-4 <AMS>
1952 V84 P919-921	65-11182 <*>
1952 V84 P1021-1024	T-1632 <INSD>
1952 V84 N1 P31-32	61-17235 <=$>
1952 V84 N1 P37-40	C-2325 <*>
	R-1485 <*>
	R-1913 <*>
	R-851 <*>
1952 V84 N1 P41-44	RT-4508 <*>

Citation	Translation
1952 V84 N1 P55-58	62-11372 <=>
1952 V84 N1 P63-66	TS-932 <BISI>
	3045 <HB>
	62-13211 <=*>
1952 V84 N1 P67-70	3542 <HB>
1952 V84 N1 P89-92	59-11657 <=>
1952 V84 N2 P201-204	64-11960 <=> M
1952 V84 N2 P245-248	RT-1618 <*>
1952 V84 N2 P283-286	59-13537 <=>
1952 V84 N2 P317-319	TT.332 <*>
1952 V84 N3 P491-493	R-196 <*>
1952 V84 N3 P519-522	RT-3770 <*>
1952 V84 N3 P539-542	59-11656 <=>
1952 V84 N3 P607-610	RT-2724 <*>
1952 V84 N4 P689-692	RCT V28 N1 P19 <RCT>
1952 V84 N4 P709-712	193 <TC>
1952 V84 N4 P717-720	RT-3633 <*>
	64-19099 <=*$>
1952 V84 N4 P725-727	<ATS>
1952 V84 N4 P737-740	61-15208 <*+>
1952 V84 N4 P745-748	RT-63 <*>
1952 V84 N4 P765-768	63-11085 <=>
1952 V84 N4 P777-779	RT-112 <*>
	61-28098 <*=> 0
1952 V84 N4 P821-824	62-33095 <=*>
1952 V84 N5 P887-890	R-1170 <*>
1952 V84 N5 P913-915	59-16061 <+*>
1952 V84 N5 P923-925	RT-732 <*>
1952 V84 N5 P951-953	RT-3772 <*>
1952 V84 N5 P959-961	RJ-96 <ATS>
1952 V84 N5 P979-980	RJ-98 <ATS>
1952 V84 N5 P981-984	R-2037 <*>
	RJ-95 <ATS>
1952 V84 N5 P981-985	RJ-95 <ATS>
1952 V84 N5 P989-992	59-13750 <=>
1952 V84 N5 P993-996	TT.365 <NRC>
1952 V84 N5 P997-1000	RJ-91 <ATS>
1952 V84 N5 P1017-1020	RJ-362 <ATS>
1952 V84 N5 P1089-1092	61-28022 <*=> 0
1952 V84 N6 P1139-1142	R-679 <*>
1952 V84 N6 P1159-1162	60-11610 <=>
1952 V84 N6 P1175-1178	RT-2736 <*>
1952 V84 N6 P1183-1186	RT-3771 <*>
	65-12196 <*>
1952 V84 N6 P1187-1190	TT.355 <NRC>
1952 V84 N6 P1195-1196	61-17225 <=$>
1952 V84 N6 P1235-1238	C.T.S.NO.12 <NLL>
1952 V84 N6 P1235-1237	RT-2355 <*>
1952 V84 N6 P1281-1284	59-13750 <=>
1952 V85 P25-27	STMSP V1 P55-57 <AMS>
1952 V85 P137-139	R-2851 <*>
1952 V85 P177-180	R-108 <*>
	R-1268 <*>
1952 V85 P373-376	R-3763 <*>
1952 V85 P389-392	R-108 <*>
	R-1269 <*>
1952 V85 P485-488	STMSP V1 P63-67 <AMS>
1952 V85 P517-520	R-1271 <*>
1952 V85 P607-610	R-1270 <*>
1952 V85 P753-755	R-938 <*>
	836TM <CTT>
1952 V85 P1073-1076	R-1272 <*>
1952 V85 P1357-1360	R-1273 <*>
1952 V85 P1395-1398	R-1274 <*>
1952 V85 N1 P5-8	R-4821 <*>
1952 V85 N1 P25-27	RT-335 <*>
1952 V85 N1 P49-52	61-17238 <=$>
1952 V85 N1 P55-58	65-60974 <=$> 0
1952 V85 N1 P85-86	RT-3402 <*>
1952 V85 N1 P91-93	24R77R <ATS>
1952 V85 N1 P131-133	RCT V26 N4 P858 <RCT>
1952 V85 N1 P153-156	RT-4108 <*>
1952 V85 N1 P177-180	RT-3720 <*=>
1952 V85 N1 P181-183	RT-1923 <*> 0
	64-15660 <=*$>
1952 V85 N2 P321-324	AEC-TR-5784 <=*$>
	47 <FBI>
1952 V85 N2 P331-333	RJ-255 <ATS>
1952 V85 N2 P345-347	RJ-217 <ATS>
1952 V85 N2 P349-352	RT-114 <*>

Citation	Translation
1952 V85 N2 P357-360	RJ-380 <ATS>
	61-13686 <+*>
1952 V85 N2 P373-376	RT-62 <*>
1952 V85 N2 P377-379	RJ-159 <ATS>
	63-10380 <*=>
1952 V85 N2 P389-392	RT-3721 <*>
1952 V85 N2 P393-395	RJ-196 <ATS>
	RJ-96 <ATS>
	2589D <K-H>
1952 V85 N2 P397-400	RT-2735 <*>
1952 V85 N3 P345-347	RJ-217 <ATS>
1952 V85 N3 P485-488	RT-336 <*>
1952 V85 N3 P509-512	RT-64 <*>
1952 V85 N3 P517-520	RT-481 <*>
1952 V85 N3 P525-528	RT-1536 <*>
1952 V85 N3 P539-542	RT-380 <*>
1952 V85 N3 P547-550	TT.529 <NRC>
	64-23573 <=$>
1952 V85 N3 P555-558	<JBS>
	61-17240 <=$>
1952 V85 N3 P583-586	<INSD>
	R-1552 <*>
1952 V85 N3 P599-602	RT-212 <*>
1952 V85 N3 P607-610	RT-2032 <*>
1952 V85 N3 P641-644	59-17860 <+*> 0
1952 V85 N3 P645-647	RT-2033 <*>
1952 V85 N3 P649-652	RT-2162 <*>
1952 V85 N3 P665-668	RT-110 <*>
	61-28104 <*=> 0
1952 V85 N4 P749-752	RT-916 <*>
1952 V85 N4 P753-755	R-1430 <*>
1952 V85 N4 P757-760	RT-379 <*>
1952 V85 N4 P773-775	60-16989 <*> 0
1952 V85 N4 P797-800	61-20938 <*=> 0
1952 V85 N4 P815-818	65-13434 <*>
	65-13434 <*> 0
1952 V85 N4 P823-826	65-18186 <*>
1952 V85 N4 P827-830	RT-164 <*>
1952 V85 N4 P839-841	R-4685 <*>
1952 V85 N4 P843-846	RT-1951 <*>
1952 V85 N4 P859-862	63-11085 <=>
1952 V85 N4 P889-891	R-1220 <*>
	RT-3554 <*>
1952 V85 N4 P897-900	44P62R <ATS>
1952 V85 N4 P913-916	63-15308 <=*>
1952 V85 N5 P797-800	R-3996 <*>
1952 V85 N5 P977-980	RT-735 <*>
1952 V85 N5 P981-984	62-16703 <=*> 0
1952 V85 N5 P997-999	3080 <HB>
1952 V85 N5 P1013-1016	RT-1391 <*>
1952 V85 N5 P1025-1028	AEC-TR-6232 <=*$>
1952 V85 N5 P1029-1031	RT-2219 <*>
1952 V85 N5 P1037-1040	10179 <KH>
	65-11833 <*>
1952 V85 N5 P1C49-1051	65-18176 <*>
1952 V85 N5 P1057-1060	TT.424 <NRC>
1952 V85 N5 P1065-1068	RJ-208 <ATS>
1952 V85 N5 P1069-1072	RCT V26 N4 P759 <RCT>
1952 V85 N5 P1089-1092	TT.423 <NRC>
1952 V85 N5 P1C93-1095	TT.380 <NRC>
1952 V85 N5 P1123-1126	63-11085 <=>
1952 V85 N5 P1181-1184	RT-257 <*>
	61-28023 <*=>
1952 V85 N6 P1227-1230	R-200 <*>
	R-5022 <*>
1952 V85 N6 P1231-1234	RT-543 <*>
1952 V85 N6 P1235-1238	R-1035 <*>
	60-14340 <+*>
1952 V85 N6 P1247-1250	AD-618 320 <=$>
1952 V85 N6 P1251-1255	TT.452 <NRC>
1952 V85 N6 P1277-1280	RT-111 <*>
1952 V85 N6 P1301-1304	C-2649 <NRC>
1952 V85 N6 P1301-1404	R-4295 <*>
1952 V85 N6 P1313-1316	TT-448 <NRC>
1952 V85 N6 P1329-1332	TT.453 <*>
1952 V85 N6 P1341-1343	R-1218 <*>
1952 V86 P43-46	R-1013 <*>
1952 V86 P369-372	R-1275 <*>
1952 V86 P429-432	R-1276 <*>
1952 V86 P493-495	R-1282 <*>
1952 V86 P745-747	62-01596 <*>
1952 V86 P853-856	R-1277 <*>
1952 V86 P1223-1226	R-1278 <*>
1952 V86 N1 P39-42	RT-1619 <*>
1952 V86 N1 P43-46	RT-475 <*>
1952 V86 N1 P47-50	63-21303 <=>
1952 V86 N1 P51-54	TT.428 <NRC>
1952 V86 N1 P71-73	RT-378 <*>
1952 V86 N1 P75-78	RT-3407 <ATS>
1952 V86 N1 P95-98	RT-2171 <*>
1952 V86 N1 P121-124	60-41434 <=>
1952 V86 N1 P125-128	<INSD>
1952 V86 N1 P129-131	RCT V27 N2 P468 <RCT>
1952 V86 N2 P231-234	R-326 <*>
1952 V86 N2 P251-253	RT-2141 <*>
1952 V86 N2 P263-265	62-24497 <=*>
1952 V86 N2 P293-296	RT-1059 <*>
1952 V86 N2 P341-344	R-4869 <*>
1952 V86 N2 P345-347	2966 <HB>
1952 V86 N2 P349-352	TT.391 <NRC>
1952 V86 N2 P393-396	62-14629 <=*>
1952 V86 N3 P489-492	R-921 <*>
	RT-834 <*>
	61-15697 <=*> 0
	62-14455 <=*>
1952 V86 N3 P5C1-504	RT-734 <*>
	61-10132 <*+>
1952 V86 N3 P501-	61-15700 <+*> 0
1952 V86 N3 P509-512	RT-3861 <*>
1952 V86 N3 P517-520	R-1034 <*>
	RT-381 <*>
1952 V86 N3 P521-523	63-19908 <=*$>
1952 V86 N3 P525-528	62-15893 <=*>
1952 V86 N3 P561-564	TT.409 <NRC>
	26J16R <ATS>
1952 V86 N3 P569-571	RCT V29 N4 P1369 <RCT>
1952 V86 N3 P577-580	3048 <HB>
1952 V86 N3 P581-583	RJ-177 <ATS>
1952 V86 N3 P589-592	TT.396 <NRC>
1952 V86 N3 P593-595	45H12R <ATS>
1952 V86 N3 P6C1-604	60-17244 <+*>
1952 V86 N3 P633-636	RT-3422 <*>
1952 V86 N4 P669	61-13684 <+*>
1952 V86 N4 P687-690	R-920 <*>
1952 V86 N4 P717-720	63-10580 <=*> 0
1952 V86 N4 P745-747	RT-3255 <*>
	T.674 <INSD>
	63-15729 <=*>
1952 V86 N4 P749-752	60-13869 <+*>
1952 V86 N4 P753-756	64-20180 <*>
	64-20401 <*>
1952 V86 N4 P759-762	RT-1016 <*>
1952 V86 N4 P775-777	RT-4112 <*>
1952 V86 N4 P787-788	63-11085 <=>
1952 V86 N4 P805-808	R-1194 <*>
1952 V86 N4 P873-875	60-51139 <=>
1952 V86 N5 P9C9-912	RJ-440 <ATS>
1952 V86 N5 P913-916	RT-1740 <*>
1952 V86 N5 P917-920	RT-1739 <*>
1952 V86 N5 P925-928	RT-511 <*>
1952 V86 N5 P945-947	RT-2264 <*>
1952 V86 N5 P989-992	RJ-162 <ATS>
1952 V86 N5 P993-995	59-22382 <+*>
	61-17223 <=$>
1952 V86 N5 P1C01-1004	RT-115 <*>
1952 V86 N6 P1097-1099	59-20485 <+*>
1952 V86 N6 P1101-1104	RT-631 <*>
1952 V86 N6 P1105-1108	RT-2148 <*>
1952 V86 N6 P1137-1140	3127 <HB>
	61-23259 <*=>
1952 V86 N6 P1147-1150	RCT V29 N1 P126 <RCT>
1952 V86 N6 P1151-1153	3015 <HB>
1952 V86 N6 P1159-1161	RJ-139 <*>
	TT.456 <NRC>
	61-17224 <=$>
1952 V87 P9-10	T1894 <INSD>
1952 V87 P109-112	R-108 <*>
	R-1279 <*>
1952 V87 P245-247	R-1280 <*>
1952 V87 P301-304	R-1281 <*>

Citation	Translation
	3160-B <K-H>
1952 V87 P453-455	R-1283 <*>
1952 V87 N1 P5-8	RTS-2580 <NLL>
1952 V87 N1 P33-35	61-10130 <*+>
1952 V87 N1 P41-43	3014 <HB>
1952 V87 N1 P45-47	RT-2566 <*>
1952 V87 N1 P49-52	3027 <*>
1952 V87 N1 P65-68	RJ-126 <ATS>
1952 V87 N1 P73-76	64-15383 <=*$>
1952 V87 N1 P93-95	3078 <HB>
1952 V87 N1 P109-112	RT-1672 <*>
1952 V87 N1 P113-116	RT-4075 <*>
1952 V87 N2 P187-190	RT-1537 <*>
1952 V87 N2 P195-196	TT.447 <NRC>
1952 V87 N2 P197-199	RT-3567 <*>
1952 V87 N2 P215-218	62-19823 <=>
	65-12263 <*>
1952 V87 N2 P237-240	RT-3423 <*>
	97L28R <ATS>
1952 V87 N2 P241-244	TT.384 <NRC>
1952 V87 N2 P273-276	C-2314 <NRC>
1952 V87 N3 P361-364	RT-3568 <*>
1952 V87 N3 P369-372	R-6 <*>
	RT-1451 <*>
1952 V87 N3 P409-412	3101 <HB>
1952 V87 N3 P417-420	64-15378 <=*$>
1952 V87 N3 P421-422	RT-922 <*>
1952 V87 N3 P423-426	97J19R <ATS>
1952 V87 N3 P437-440	TT.386 <NRC>
1952 V87 N3 P441-443	60-13011 <+*>
1952 V87 N3 P457-460	RT-1547 <*>
1952 V87 N4 P523-526	TT-450 <NRC>
1952 V87 N4 P535-538	TT.497 <NRC>
1952 V87 N4 P547-550	RT-2957 <*>
1952 V87 N4 P551-554	RT-1540 <*>
1952 V87 N4 P555-558	RT-1233 <*>
1952 V87 N4 P567-570	62-24828 <=*>
1952 V87 N4 P571-574	RT-1452 <*>
1952 V87 N4 P593-595	RT-3863 <*>
1952 V87 N4 P605-608	RT-2143 <*>
1952 V87 N4 P617-620	3161 <HB>
1952 V87 N4 P627-630	RT-3857 <*>
	61-17230 <=$>
1952 V87 N4 P631-634	RT-441 <*>
1952 V87 N4 P635-637	AD-610 883 <=$>
	RT-1538 <*>
1952 V87 N4 P665-668	RT-309 <*>
	61-28024 <*=> 0
1952 V87 N5 P715-718	RT-474 <*>
	59-15106 <+*>
1952 V87 N5 P719-722	AD-619 327 <=$>
	R-1942 <*>
	RT-3050 <*>
	60-12603 <=>
	62-11304 <=>
1952 V87 N5 P735-737	R-1767 <*>
	3062 <HB>
1952 V87 N5 P743-746	60-41421 <=>
	61-13685 <+*>
1952 V87 N5 P771-773	62-20286 <=*> 0
1952 V87 N5 P783-785	TT.446 <NRC>
1952 V87 N5 P791-793	RT-2156 <*>
1952 V87 N5 P795-796	3036 <HB>
	64-18384 <*>
1952 V87 N5 P797-800	3202 <HB>
1952 V87 N5 P801-803	TT.385 <NRC>
1952 V87 N5 P805-808	RT-3855 <*>
	59-14190 <+*>
1952 V87 N5 P821-824	RT-4115 <*>
	246TM <CTT>
1952 V87 N5 P825-827	AD-610 883 <=$>
1952 V87 N5 P833-835	RT-1069 <*>
1952 V87 N6 P931-934	64-23616 <=>
1952 V87 N6 P935-938	TT.467 <NRC>
1952 V87 N6 P943-946	62-33489 <=*>
1952 V87 N6 P957-960	IGR V2 N2 P125 <AGI>
	R-1651 <*>
1952 V87 N6 P991-994	RT-758 <*>
	64-15382 <=*$>
1952 V87 N6 P1009-1012	TT.565 <NRC>
	64-23623 <=>
1952 V87 N6 P1029-1031	3222 <HB>
	63-18409 <=*>
1952 V87 N58 P25-827	RT-1539 <*>
1953 V88 P257-259	RT-11 <*>
1953 V88 P349-351	R-1284 <*>
1953 V88 P567-570	R-1285 <*>
1953 V88 P611-614	AD-638 625 <=$>
1953 V88 P725-727	R-1286 <*>
1953 V88 P781-784	R-649 <*>
1953 V88 P785-786	64-00024 <*>
1953 V88 P891-893	R-770 <*>
1953 V88 P937-940	R-1287 <*>
1953 V88 P1031-1034	T1889 <INSD>
1953 V88 N1 P33-36	RT-1513 <*>
1953 V88 N1 P37-40	RT-26 <*>
1953 V88 N1 P41-44	RT-3 <*>
1953 V88 N1 P45-48	RT-22 <*>
1953 V88 N1 P49-52	RT-46 <*>
1953 V88 N1 P57-60	RT-47 <*>
1953 V88 N1 P73-76	13077-A <K-H>
	64-18438 <*>
	65-14872 <*>
1953 V88 N1 P79-82	RT-1015 <*>
	T.867 <INSD>
	59-19143 <+*>
	64-10150 <=*$>
1953 V88 N1 P83-85	10M41R <ATS>
1953 V88 N1 P87-89	RJ-110 <ATS>
1953 V88 N1 P91-94	T-2266 <INSD>
1953 V88 N1 P99-102	R-1385 <*>
1953 V88 N1 P103-104	RJ-113 <ATS>
1953 V88 N1 P113-116	RT-113 <*>
1953 V88 N2 P197-200	RT-771 <*>
1953 V88 N2 P229-232	RT-1738 <*>
1953 V88 N2 P233-236	RT-9 <*>
1953 V88 N2 P237-240	RT-48 <*>
1953 V88 N2 P241-243	RT-19 <*>
1953 V88 N2 P245-247	RT-23 <*>
1953 V88 N2 P249-252	RT-44 <*>
1953 V88 N2 P253-256	RT-21 <*>
	RT-27 <*>
1953 V88 N2 P265-268	3088 <HB>
1953 V88 N2 P273-276	59-13750 <=>
1953 V88 N2 P277-279	R-848 <*>
1953 V88 N2 P285-287	65-18179 <*>
1953 V88 N2 P297-300	RT-2296 <*>
1953 V88 N2 P301-304	60-15827 <+*> 0
1953 V88 N2 P317-320	RT-472 <*>
	61-28096 <*=>
1953 V88 N3 P397-400	RT-770 <*>
1953 V88 N3 P431-434	RJ-150 <ATS>
	RT-49 <*>
1953 V88 N3 P435-438	R-4390 <*>
	R-872 <*>
1953 V88 N3 P439-440	RT-6 <*>
1953 V88 N3 P441-443	RT-34 <*>
1953 V88 N3 P445-448	RJ-123 <ATS>
	RT-28 <*>
1953 V88 N3 P449-452	RT-29 <*>
1953 V88 N3 P457-459	RJ-118 <ATS>
1953 V88 N3 P471-473	RJ-115 <ATS>
1953 V88 N3 P475-478	RJ-109 <ATS>
	61-17313 <=$>
	63-14145 <=*> 0
	64-18439 <*>
1953 V88 N3 P487-489	RJ-108 <ATS>
1953 V88 N3 P507-510	RT-2692 <*>
1953 V88 N3 P515-518	TT.387 <NRC>
1953 V88 N3 P527-529	RJ-121 <ATS>
1953 V88 N3 P559-560	RJ-122 <ATS>
1953 V88 N4 P635-637	RT-7 <*>
1953 V88 N4 P639-642	RT-179 <*>
1953 V88 N4 P643-646	RT-184 <*>
1953 V88 N4 P647-650	RT-227 <*>
1953 V88 N4 P651-652	RT-31 <*>
1953 V88 N4 P653-656	R-846 <*>
1953 V88 N4 P661-663	RT-1584 <*>
1953 V88 N4 P665-668	RT-10 <*>
1953 V88 N4 P669-672	RT-222 <*>

Citation	Translation
1953 V88 N4 P673-675	RT-225 <*>
1953 V88 N4 P677-678	RT-2298 <*>
1953 V88 N4 P705-708	RT-60 <*>
1953 V88 N4 P713-716	TT.578 <NRC>
1953 V88 N5 P769-772	RT-180 <*>
1953 V88 N5 P773-776	RT-20 <*>
1953 V88 N5 P777-780	RT-24 <*>
	64-23574 <=>
1953 V88 N5 P781-784	RT-188 <*>
1953 V88 N5 P785-786	RT-183 <*>
1953 V88 N5 P799-802	RT-214 <*>
1953 V88 N5 P803-806	RT-223 <*>
1953 V88 N5 P829-832	65-61171 <=>
1953 V88 N5 P849-952	200TM <CTT>
1953 V88 N5 P859-862	59-10644 <+*>
	61-13638 <+*>
1953 V88 N5 P871-874	RT-4101 <*>
1953 V88 N5 P891-893	201TM <CTT>
1953 V88 N6 P965-966	RT-30 <*>
1953 V88 N6 P967-970	RT-502 <*>
1953 V88 N6 P971-973	RT-39 <*>
1953 V88 N6 P983-986	63-14147 <=*>
	64-20179 <*>
	64-20416 <*>
	65-18157 <*>
1953 V88 N6 P991-993	62-10568 <*=>
	65-61306 <=$>
1953 V88 N6 P999-1002	62-18970 <=*>
1953 V88 N6 P1015-1018	RCT V28 N3 P891 <RCT>
1953 V88 N6 P1019-1022	61-23258 <*=>
1953 V88 N6 P1023-1026	RT-59 <*>
1953 V88 N6 P1031-1034	61-13682 <+*>
1953 V89 P101-104	R-3286 <*>
1953 V89 P197-200	R-1288 <*>
1953 V89 P495-498	T-2414 <INSD>
1953 V89 P523-526	R-108 <*>
	R-1289 <*>
1953 V89 P527-530	R-108 <*>
	R-1290 <*>
1953 V89 P651-653	R-648 <*>
1953 V89 P673-675	R-646 <*>
1953 V89 P677-679	R-647 <*>
1953 V89 N1 P9-11	R-3669 <*>
1953 V89 N1 P33-36	RT-54 <*>
1953 V89 N1 P37-40	RT-15 <*>
1953 V89 N1 P41-44	RT-52 <*>
1953 V89 N1 P45-48	RT-18 <*>
1953 V89 N1 P49-52	RT-12 <*>
1953 V89 N1 P53-56	RT-53 <*>
1953 V89 N1 P57-60	RT-36 <*>
1953 V89 N1 P61-64	RT-40 <*>
1953 V89 N1 P77-80	24P66R <ATS>
1953 V89 N1 P89-92	RT-3627 <*>
1953 V89 N1 P109-112	60-10058 <+*>
1953 V89 N1 P133-136	<INSD>
1953 V89 N1 P141-142	3170 <HB>
1953 V89 N2 P229-232	RT-2297 <*>
1953 V89 N2 P241-244	61-17276 <=$>
1953 V89 N2 P245-247	RJ-149 <ATS>
	RT-2 <*>
1953 V89 N2 P249-252	RT-13 <*>
1953 V89 N2 P257-260	RT-25 <*>
1953 V89 N2 P261-264	RT-14 <*>
1953 V89 N2 P265-267	RJ-146 <ATS>
1953 V89 N2 P269-270	RT-5 <*>
1953 V89 N2 P271-274	RJ-148 <ATS>
	RT-32 <*>
1953 V89 N2 P275-278	RT-35 <*>
1953 V89 N2 P287-289	R-2767 <*>
	R-4868 <*>
1953 V89 N2 P291-292	RT-1848 <*>
	64-15580 <=*$>
1953 V89 N2 P309-312	RT-1457 <*>
1953 V89 N2 P317-320	RJ-125 <ATS>
	RJ-147 <*>
1953 V89 N2 P325-328	RCT V29 N2 P568 <RCT>
1953 V89 N2 P325-	61-13681 <+*>
1953 V89 N3 P415-418	61-23261 <*=>
1953 V89 N3 P419-422	RT-17 <*>
1953 V89 N3 P423-426	RT-33 <*>
1953 V89 N3 P427-430	RT-41 <*>
1953 V89 N3 P431-434	RT-42 <*>
1953 V89 N3 P435-438	RT-38 <*>
1953 V89 N3 P439-442	RT-16 <*>
1953 V89 N3 P447-449	RT-8 <*>
	62-10290 <=*>
1953 V89 N3 P451-453	RT-50 <*>
1953 V89 N3 P455-458	RT-37 <*>
1953 V89 N3 P479-482	61-13680 <+*>
	65-63513 <=$>
1953 V89 N3 P483-486	61-17196 <=$>
1953 V89 N3 P487-490	R-5194 <*>
1953 V89 N3 P495-498	RT-1729 <*>
1953 V89 N3 P499-500	RT-3824 <*>
1953 V89 N3 P505-508	RCT V28 N3 P833 <RCT>
	59-15922 <+*> 0
1953 V89 N3 P515-517	RT-4110 <*>
	534 <TC>
1953 V89 N3 P523-526	RT-3718 <*=>
1953 V89 N3 P527-530	RT-3713 <*>
1953 V89 N4 P601-604	66-11230 <*>
1953 V89 N4 P647-650	60-19021 <+*>
1953 V89 N4 P651-653	RT-216 <*>
1953 V89 N4 P659-662	RT-517 <*>
1953 V89 N4 P663-664	RT-185 <*>
1953 V89 N4 P665-668	RT-43 <*>
1953 V89 N4 P669-672	RT-219 <*>
1953 V89 N4 P673-675	RT-229 <*>
1953 V89 N4 P677-679	RT-215 <*>
1953 V89 N4 P685-687	R-1875 <*>
	RJ.138 <ATS>
	RJ-138 <ATS>
1953 V89 N4 P693-695	AD-610 900 <=$>
	RT-1737 <*>
1953 V89 N4 P705-707	RT-1679 <*>
1953 V89 N4 P709-712	RJ-321 <ATS>
1953 V89 N4 P709-711	60-15750 <+*>
1953 V89 N4 P737-740	R-4178 <*>
1953 V89 N5 P771-774	RT-1668 <*>
1953 V89 N5 P809-812	RT-396 <*>
1953 V89 N5 P813-816	RT-395 <*>
1953 V89 N5 P817-819	RT-500 <*>
1953 V89 N5 P821-824	RT-501 <*>
1953 V89 N5 P825-828	NSF-TR-119 <+*>
	RT-498 <*>
1953 V89 N5 P829-832	RT-505 <*>
1953 V89 N5 P833-835	RT-536 <*>
	RT-832 <*>
1953 V89 N5 P837-840	RT-542 <*>
1953 V89 N5 P841-844	RT-459 <*>
1953 V89 N5 P845-848	RT-499 <*>
1953 V89 N5 P849-852	RT-394 <*>
1953 V89 N5 P865-866	RT-3996 <*>
1953 V89 N6 P991-993	RT-1026 <*>
	RT-181 <*>
1953 V89 N6 P995-997	RT-220 <*>
1953 V89 N6 P999-1002	RT-221 <*>
	61-20999 <*=>
1953 V89 N6 P1003-1006	RT-4 <*>
1953 V89 N6 P1015-1016	R-2766 <*>
	R-4853 <*>
1953 V89 N6 P1037-1040	TT.573 <NRC>
1953 V89 N6 P1111-1114	63-11009 <=>
1953 V90 P425-428	R-1196 <*>
1953 V90 P525-528	62-26017 <=$>
1953 V90 P767-769	R-1182 <*>
1953 V90 P1023-1026	R-1511 <*>
1953 V90 N1 P49-50	RT-51 <*>
	60-13760 <+*>
1953 V90 N1 P51-54	RT-182 <*>
	60-13761 <+*>
1953 V90 N1 P55-57	RT-1 <*>
	60-13762 <+*>
1953 V90 N2 P149-151	RT-224 <*>
	60-13763 <+*>
1953 V90 N2 P153-156	RT-393 <*>
	60-13764 <+*>
1953 V90 N2 P157-158	RT-228 <*>
	60-13765 <+*>
1953 V90 N2 P159-162	RT-484 <*>

Citation	Translation
	60-13766 <+*>
1953 V90 N2 P163-166	RT-198 <*>
	60-13767 <+*>
1953 V90 N2 P171-174	68K20R <ATS>
1953 V90 N2 P201-204	RCT V29 N1 P63 <RCT>
1953 V90 N2 P209-212	RCT V29 N2 P598 <RCT>
1953 V90 N2 P213-216	61-17206 <=$>
1953 V90 N2 P225-226	RJ-388 <ATS>
	59-19629 <+*> 0
1953 V90 N3 P355-358	60-13768 <+*>
1953 V90 N3 P359-362	RT-824 <*>
	60-13769 <+*>
1953 V90 N3 P363-366	RT-392 <*>
	15K27R <ATS>
	60-13770 <+*>
1953 V90 N3 P367-370	59-18523 <+*>
1953 V90 N3 P371-374	61-15224 <+*>
	63-16460 <=*>
1953 V90 N3 P375-377	RT-570 <*>
	60-13771 <+*>
1953 V90 N3 P379-382	RT-226 <*>
	60-13772 <+*>
1953 V90 N3 P387-390	60-17203 <+*>
1953 V90 N3 P405-407	61-17207 <=$>
1953 V90 N3 P409-412	RCT V29 N2 P530 <RCT>
1953 V90 N3 P437-440	62-24171 <=*>
1953 V90 N4 P503-506	60-15847 <+*>
1953 V90 N4 P521-523	60-13773 <+*>
	61-20980 <*>
1953 V90 N4 P525-528	62-25869 <=*>
1953 V90 N4 P529-531	17K27R <ATS>
	60-13774 <+*>
1953 V90 N4 P537-540	61-13194 <+*>
1953 V90 N4 P565-567	R-894 <*>
	RT-3211 <*>
	64-19091 <=*$>
1953 V90 N4 P573-576	61-23263 <*=>
1953 V90 N4 P595-598	RT-3583 <*>
1953 V90 N4 P599-602	20178 <*>
1953 V90 N4 P681-683	66-10240 <*> 0
1953 V90 N5 P723-725	64-71163 <=>
1953 V90 N5 P745-748	RT-504 <*>
1953 V90 N5 P749-751	RT-391 <*>
1953 V90 N5 P753-756	RT-756 <*>
1953 V90 N5 P757-760	RT-755 <*>
1953 V90 N5 P761-764	RT-749 <*>
1953 V90 N5 P767-769	RT-584 <*>
1953 V90 N5 P775-776	R-4363 <*>
1953 V90 N5 P795-797	RT-4107 <*>
1953 V90 N5 P799-802	R-103 <*>
	RT-4033 <*>
	64-19058 <=*$>
1953 V90 N5 P819-882	RT-3669 <*>
1953 V90 N5 P823-826	RT-4090 <*>
1953 V90 N5 P843-845	RT-1105 <*>
1953 V90 N5 P847-849	M-6267 <NLL>
1953 V90 N5 P871-874	R-3634 <*>
1953 V90 N6 P983-986	RT-883 <*>
1953 V90 N6 P987-990	RT-539 <*>
1953 V90 N6 P991-994	RT-751 <*>
1953 V90 N6 P995-998	RT-754 <*>
1953 V90 N6 P999-1001	RT-586 <*>
1953 V90 N6 P1003-1004	RT-831 <*>
1953 V90 N6 P1005-1008	RT-594 <*>
1953 V90 N6 P1009-1010	RT-545 <*>
1953 V90 N6 P1011-1014	RT-593 <*>
1953 V90 N6 P1015-1018	RT-753 <*>
1953 V90 N6 P1019-1022	RT-757 <*>
1953 V90 N6 P1023-1026	RT-830 <*>
1953 V90 N6 P1027-1029	RT-748 <*>
1953 V90 N6 P1051-1054	199TM <CTT>
	63-10406 <=*>
	80K25R <ATS>
1953 V90 N6 P1055-1058	62-10107 <*=>
1953 V90 N6 P1063-1066	RCT V29 N2 P593 <RCT>
1953 V90 N6 P1067-1070	63-26688 <=$>
1953 V90 N6 P1083-1085	C-2547 <NRC>
1953 V91 P47-50	R-650 <*>
1953 V91 P51-54	R-653 <*>
1953 V91 P55-58	R-456 <*>
1953 V91 P63-66	62-26016 <=$>
1953 V91 P67-70	R-651 <*>
1953 V91 P79-82	R-193 <*>
	R-652 <*>
	62-26015 <=$>
1953 V91 P103-106	R-2765 <*>
	R-4856 <*>
1953 V91 P339-341	R-1291 <*>
1953 V91 P343-346	R-108 <*>
	R-1292 <*>
1953 V91 P535-538	R-682 <*>
1953 V91 P581-587	R-5354 <*>
1953 V91 P599-600	R-1293 <*>
1953 V91 P737-740	AMST S2 V23 P103-107 <AMS>
1953 V91 P899-902	R-108 <*>
	R-1294 <*>
1953 V91 P1279-1280	AMST S2 V8 P19-20 <AMS>
1953 V91 N1 P21-33	61-23327 <*=>
1953 V91 N1 P31-33	RT-2294 <*>
1953 V91 N1 P39-42	RT-850 <*>
1953 V91 N1 P43-45	RT-186 <*>
1953 V91 N1 P47-50	R-1415 <*>
	RT-187 <*>
	1373TM <CTT>
1953 V91 N1 P51-54	RT-200 <*>
1953 V91 N1 P55-58	RT-230 <*>
	62-16577 <=*>
1953 V91 N1 P59-62	RT-390 <*>
1953 V91 N1 P63-66	RT-201 <*>
1953 V91 N1 P67-70	AEC-TR-5782 <=*$>
	RT-199 <*>
1953 V91 N1 P75-77	RT-218 <*>
1953 V91 N1 P79-82	RT-213 <*>
1953 V91 N1 P83-85	RT-217 <*>
1953 V91 N1 P89-91	RT-3703 <*>
1953 V91 N1 P93-94	RT-1847 <*>
1953 V91 N1 P99-102	RT-2837 <*>
1953 V91 N1 P107-109	RT-2295 <*>
1953 V91 N1 P125-127	63-18460 <=*>
1953 V91 N1 P129-131	RT-3332 <*>
1953 V91 N2 P221-223	RT-460 <*>
1953 V91 N2 P225-228	RT-851 <*>
1953 V91 N2 P229-232	RT-454 <*>
1953 V91 N2 P241-244	RT-2812 <*>
1953 V91 N2 P245-248	RT-389 <*>
1953 V91 N2 P249-251	RT-456 <*>
1953 V91 N2 P253-255	RT-455 <*>
1953 V91 N2 P269-270	65-11638 <*> 0
1953 V91 N2 P281-283	RT-4087 <*>
1953 V91 N2 P285-286	3167 <HB>
1953 V91 N2 P291-293	61-13396 <+*>
1953 V91 N2 P295-298	60-17704 <+*>
1953 V91 N2 P315-317	RT-3580 <*>
	16P66R <ATS>
1953 V91 N2 P327-329	R-1512 <*>
1953 V91 N2 P335-337	45T91R <ATS>
1953 V91 N2 P343-346	RTS-3722 <*>
1953 V91 N2 P347-349	64-16219 <=*$>
1953 V91 N2 P355-358	65-13621 <*>
1953 V91 N2 P359-362	R-1039 <*>
	RT-4482 <*>
1953 V91 N3 P475-478	RT-1495 <*>
1953 V91 N3 P479-482	RT-458 <*>
1953 V91 N3 P483-485	61-13172 <*+>
1953 V91 N3 P487-490	RT-436 <*>
1953 V91 N3 P491-494	RT-503 <*>
1953 V91 N3 P495-498	RT-437 <*>
1953 V91 N3 P499-502	RT-457 <*>
1953 V91 N3 P503-505	RT-377 <*>
1953 V91 N3 P507-510	RT-789 <*>
1953 V91 N3 P511-513	RT-398 <*>
1953 V91 N3 P515-518	61-13173 <+*>
1953 V91 N3 P519-522	61-13174 <+*>
1953 V91 N3 P523-526	RT-829 <*>
1953 V91 N3 P549-552	3201 <HB>
1953 V91 N3 P553-556	61-17197 <=$>
1953 V91 N3 P577-580	61-17208 <=$>
1953 V91 N3 P597-598	893TM <CTT>
1953 V91 N3 P609-611	64-14044 <=*$>
1953 V91 N3 P625-628	RT-2052 <*>

Citation	Translation
1953 V91 N4 P757-758	RT-917 <*>
1953 V91 N4 P763-765	IGR V2 N6 P527 <AGI>
	RT-788 <*>
1953 V91 N4 P767-769	RT-540 <*>
1953 V91 N4 P771-774	RT-585 <*>
1953 V91 N4 P775-778	RT-587 <*>
1953 V91 N4 P779-782	RT-752 <*>
1953 V91 N4 P783-786	RT-815 <*>
1953 V91 N4 P787-790	RT-817 <*>
1953 V91 N4 P791-794	RT-550 <*>
1953 V91 N4 P795-798	RT-747 <*>
1953 V91 N4 P803-806	RT-588 <*>
1953 V91 N4 P807-810	NSF-TR-167 <+*>
	RT-816 <*>
1953 V91 N4 P811-812	RT-589 <*>
1953 V91 N4 P817-820	RT-3838 <*>
1953 V91 N4 P825-827	195TM <CTT>
1953 V91 N4 P829-832	63-14146 <=*>
	64-20177 <*>
	65-18158 <*>
1953 V91 N4 P837-839	RT-3847 <*>
1953 V91 N4 P841-843	UCRL TRANS-969 <=*$>
1953 V91 N4 P861-864	65-13435 <*>
1953 V91 N4 P877-880	62-24861 <=*>
1953 V91 N4 P891-893	RT-3537 <*>
1953 V91 N4 P899-902	RT-3714 <*>
	RT-3714 <*=>
1953 V91 N4 P903-905	R-2701 <*>
1953 V91 N4 P953-955	RT-2415 <*>
1953 V91 N5 P1039-1042	61-17284 <=$>
1953 V91 N5 P1043-1045	R-4120 <*>
	RT-2495 <*>
1953 V91 N5 P1047-1050	RT-4074 <*>
1953 V91 N5 P1051-1054	RT-881 <*>
1953 V91 N5 P1055-1058	RT-828 <*>
1953 V91 N5 P1059-1062	RT-827 <*>
1953 V91 N5 P1063-1065	RT-867 <*>
1953 V91 N5 P1067-1070	RT-819 <*>
1953 V91 N5 P1071-1074	RT-595 <*>
1953 V91 N5 P1075-1078	RT-820 <*>
1953 V91 N5 P1079-1082	RT-2669 <*>
1953 V91 N5 P1083-1084	RT-590 <*>
1953 V91 N5 P1085-1088	RT-818 <*>
1953 V91 N5 P1089-1090	RT-591 <*>
1953 V91 N5 P1091-1094	RT-592 <*>
1953 V91 N5 P1099-1102	01L33R <ATS>
1953 V91 N5 P1107-1110	RCT V27 N4 P974 <RCT>
1953 V91 N5 P1137-1140	SCL-T-251 <=*$>
1953 V91 N5 P1141-1144	196TM <CTT>
	61-13679 <+*>
1953 V91 N5 P1155-1158	197TM <CTT>
1953 V91 N5 P1187-1190	TT.579 <NRC>
1953 V91 N6 P1289-1292	RT-921 <*>
1953 V91 N6 P1293-1296	61-17268 <=$>
1953 V91 N6 P1301-1303	AD-622 385 <=$*>
	R-3624 <*>
	62-11611 <=>
	63-10191 <=*>
1953 V91 N6 P1313-1316	RT-548 <*>
1953 V91 N6 P1317-1320	RT-506 <*>
1953 V91 N6 P1321-1324	RT-826 <*>
	65-17385 <*>
1953 V91 N6 P1329-1332	92K20R <ATS>
1953 V91 N6 P1341-1344	RT-2258 <*>
	64-15379 <=*$>
1953 V91 N6 P1357-1360	RT-1985 <*>
1953 V92 P165-168	R-1295 <*>
1953 V92 P381-384	R-108 <*>
	R-1296 <*>
1953 V92 P1201-1204	R-108 <*>
	R-1297 <*>
1953 V92 N1 P17-20	61-17311 <=$>
1953 V92 N1 P25-28	RT-4073 <*>
1953 V92 N1 P29-32	RT-814 <*>
	61-17193 <=$>
1953 V92 N1 P33-35	RT-744 <*>
1953 V92 N1 P45-48	RT-813 <*>
1953 V92 N1 P49-52	RT-812 <*>
1953 V92 N1 P53-56	RT-890 <*>
1953 V92 N1 P57-60	RT-2270 <*>
1953 V92 N1 P61-64	RCT V27 N4 P920 <RCT>
1953 V92 N1 P69-71	63-19133 <=*>
1953 V92 N1 P109-110	61-13678 <+*>
1953 V92 N1 P115-118	222TT <CCT>
	3330 <HB>
	61-13349 <+*>
1953 V92 N1 P127-130	TT.438 <NRC>
1953 V92 N1 P151-152	53S87R <ATS>
	61-17199 <=$>
	66-10648 <*>
1953 V92 N1 P153-155	R-1210 <*>
1953 V92 N1 P165-168	61-17194 <=$>
1953 V92 N1 P389-391	<INSD>
1953 V92 N2 P255-257	RT-823 <*>
1953 V92 N2 P259-262	RT-822 <*>
1953 V92 N2 P263-264	RT-825 <*>
1953 V92 N2 P265-267	RT-821 <*>
1953 V92 N2 P269-271	RT-868 <*>
1953 V92 N2 P273-275	RT-869 <*>
1953 V92 N2 P277-279	RT-887 <*>
1953 V92 N2 P281-284	RT-889 <*>
1953 V92 N2 P285-288	RT-842 <*>
1953 V92 N2 P289-291	RT-884 <*>
1953 V92 N2 P301-302	RT-895 <*>
1953 V92 N2 P303-306	RT-858 <*>
1953 V92 N2 P311-314	09T90R <ATS>
1953 V92 N2 P329-331	R-1924 <*>
	RJ-156 <ATS>
1953 V92 N2 P345-348	R-1923 <*>
	RJ-155 <ATS>
1953 V92 N2 P349-352	R-4989 <*>
	61-31100 <=>
1953 V92 N2 P353-355	RJ.262 <ATS>
	RJ-262 <ATS>
1953 V92 N2 P365-368	RT-3183 <*>
1953 V92 N2 P369-372	61-13676 <+*>
1953 V92 N2 P381-384	RT-3715 <*>
1953 V92 N2 P385-388	RT-1196 <*>
1953 V92 N3 P507-510	RT-918 <*>
1953 V92 N3 P511-513	RT.596 <*>
1953 V92 N3 P515-517	RT-839 <*>
1953 V92 N3 P519-521	RT.513 <*>
1953 V92 N3 P523-524	RT-745 <*>
1953 V92 N3 P525-528	NSF-TR-200 <+*>
	RT-897 <*>
1953 V92 N3 P529-530	RT-597 <*>
1953 V92 N3 P531-534	RT-746 <*>
1953 V92 N3 P535-536	RT-530 <*>
1953 V92 N3 P537-540	RT-544 <*>
	65-10860 <*>
1953 V92 N3 P541-543	RT-888 <*>
1953 V92 N3 P545-548	RT-844 <*>
1953 V92 N3 P549-552	RT-882 <*>
1953 V92 N3 P553-555	RT-885 <*>
1953 V92 N3 P557-560	60-41093 <=>
1953 V92 N3 P561-564	RT-1627 <*>
1953 V92 N3 P565-568	RT-598 <*>
	64-20176 <*>
1953 V92 N3 P569-572	61-13677 <+*>
1953 V92 N3 P593-595	3248 <HB>
1953 V92 N3 P603-605	3252 <HB>
1953 V92 N3 P607-610	60-10163 <+*>
1953 V92 N3 P621-624	65-60346 <=$>
1953 V92 N3 P625-627	60-15064 <+*>
1953 V92 N3 P637-640	RT-3352 <*>
1953 V92 N4 P719-721	RT-785 <*>
1953 V92 N4 P723-725	RT-893 <*>
1953 V92 N4 P727-730	RT-706 <*>
1953 V92 N4 P731-734	61-13175 <+*>
1953 V92 N4 P735-738	RT-864 <*>
1953 V92 N4 P739-742	RT-995 <*>
1953 V92 N4 P743-746	RT-845 <*>
1953 V92 N4 P755-758	RT-838 <*>
1953 V92 N4 P759-762	RT-841 <*>
1953 V92 N4 P763-766	RT-992 <*>
1953 V92 N4 P773-776	66-14682 <*>
1953 V92 N4 P777-779	65-60270 <=$>
1953 V92 N4 P781-784	92S86R <ATS>
1953 V92 N4 P793-796	T2113 <INSD>
1953 V92 N4 P797-798	RJ-472 <ATS>

	3403 <HB>
	64-16752 <=*$>
1953 V92 N4 P807-810	RT-1996 <*>
1953 V92 N4 P851-853	RT-1681 <*>
1953 V92 N5 P907-910	RT-863 <*>
1953 V92 N5 P911-914	RT-952 <*>
1953 V92 N5 P915-917	RT-835 <*>
1953 V92 N5 P919-922	RT-988 <*>
1953 V92 N5 P923-926	RT-966 <*>
1953 V92 N5 P927-930	RT-993 <*>
1953 V92 N5 P931-933	RT.994 <*>
1953 V92 N5 P935-937	RT-741 <*>
1953 V92 N5 P939-942	RT-996 <*>
	62-16581 <=*>
1953 V92 N5 P947-950	RT-984 <*>
1953 V92 N5 P951-953	RT-990 <*>
1953 V92 N5 P955-957	RT-991 <*>
1953 V92 N5 P963-966	62-15256 <*=>
1953 V92 N5 P979-982	RT-1419 <*>
1953 V92 N5 P987-990	RJ-165 <ATS>
	61-13675 <+*>
	61-17198 <=$>
	894TM <CTT>
1953 V92 N5 P999-1002	RT-1995 <*>
	64-15375 <=*$>
1953 V92 N5 P1015-1018	61-13674 <+*>
1953 V92 N5 P1023-1025	R-1213 <*>
1953 V92 N6 P1113-1116	RT-1388 <*>
1953 V92 N6 P1125-1128	RT-983 <*>
1953 V92 N6 P1129-1132	RT-985 <*>
1953 V92 N6 P1133-1136	RT-1389 <*>
1953 V92 N6 P1137-1140	RT-986 <*>
1953 V92 N6 P1141-1144	RT-987 <*>
1953 V92 N6 P1153-1156	59-15992 <+*>
1953 V92 N6 P1163-1165	RT-2937 <*>
1953 V92 N6 P1171-1173	RT-2326 <*>
	59-17319 <+*>
1953 V92 N6 P1193-1195	RJ-405 <ATS>
1953 V92 N6 P1197-1199	RT-4100 <*>
1953 V92 N6 P1201-1204	RT-3723 <*=>
1953 V92 N6 P1225-1228	60-51139 <=>
1953 V93 P115-117	R-1298 <*>
1953 V93 P511-514	R-1299 <*>
1953 V93 P519-522	RCT V29 N3 P770 <RCT>
1953 V93 P911-914	R-1300 <*>
1953 V93 P915-917	R-1301 <*>
1953 V93 P999-1002	UCRL TRANS-668(L) <*>
1953 V93 N1 P47-50	60-41420 <=>
1953 V93 N1 P71-74	81J19R <ATS>
1953 V93 N1 P89-92	RJ-656 <ATS>
	61-17267 <=$>
1953 V93 N1 P93-95	R-2100 <*>
	RJ-271 <ATS>
	3287 <HB>
1953 V93 N1 P97-99	RT-2283 <*>
1953 V93 N1 P105-107	59-17321 <+*>
1953 V93 N1 P115-117	RJ-363 <ATS>
1953 V93 N1 P131-134	3217 <HB>
1953 V93 N1 P197-200	RT-4069 <*>
1953 V93 N2 P249-252	RT-1312 <*>
1953 V93 N2 P253-256	R-4558 <*>
	3200 <NLL>
	69K20R <ATS>
1953 V93 N2 P257-260	RT-1598 <*>
1953 V93 N2 P261-264	RT-2813 <*>
1953 V93 N2 P281-283	3219 <HB>
	60-23037 <+*>
1953 V93 N2 P297-300	98J19R <ATS>
1953 V93 N2 P329-332	3218 <HB>
1953 V93 N2 P353-356	RT-4356 <*>
1953 V93 N3 P429-430	R-599 <*>
1953 V93 N3 P431-434	R-849 <*>
1953 V93 N3 P467-470	R-1378 <*>
1953 V93 N3 P471-473	RT-3822 <*>
	T.849 <INSD>
1953 V93 N3 P581-584	RT-2327 <*>
	1980 <*>
	64-15071 <=*$>
1953 V93 N4 P595-597	RT-1199 <*>
1953 V93 N4 P637-639	R-598 <*>
1953 V93 N4 P671-674	R-3077 <*>
1953 V93 N4 P689-692	3237 <HB>
1953 V93 N5 P795-798	64-13383 <=*$>
1953 V93 N5 P799-802	59-17339 <*>
1953 V93 N5 P845-846	RT-2658 <*>
1953 V93 N5 P847-850	60-31122 <=>
1953 V93 N5 P859-861	3285 <HB>
1953 V93 N6 P965-967	RT-1148 <*>
1953 V93 N6 P993-996	RT-1315 <*>
1953 V93 N6 P1003-1006	AD-619 430 <=*$>
1953 V93 N6 P1011-1014	RT-2284 <*>
1953 V93 N6 P1025-1027	RT-3225 <*>
1953 V93 N6 P1033-1035	R-4248 <*>
1953 V93 N6 P1069-1072	RT-1597 <*>
	64-15574 <=*$> 0
1953 V93 N6 P1077-1079	R-4383 <*>
	3256 <HB>
1954 V94 P13-16	STMSP V4 P127-131 <AMS>
1954 V94 P101-104	R-3750 <*>
1954 V94 P113-116	60-15820 <+*>
1954 V94 P229	AEC-TR-6139 <*>
1954 V94 P301-304	R-1302 <*>
1954 V94 P451-454	AEC-TR-2405 <*>
1954 V94 P1137-1139	R-1303 <*>
1954 V94 N1 P37-39	63-10068 <=*>
1954 V94 N1 P61-64	RT-4097 <*>
1954 V94 N1 P101-104	RT-2053 <*>
1954 V94 N1 P109-112	RT-2391 <*>
	20L35R <ATS>
	60-21117 <=>
1954 V94 N2 P203-206	AD-619 429 <=$>
1954 V94 N2 P229-231	AEC-TR-6139 <=*$>
1954 V94 N2 P241-244	RJ-215 <ATS>
	RT-1735 <*>
1954 V94 N2 P253-256	RCT V27 N4 P925 <RCT>
1954 V94 N2 P269-272	RT-4352 <*>
1954 V94 N2 P273-276	59-13758 <=>
1954 V94 N2 P277-279	R-1033 <*>
	837TM <CTT>
1954 V94 N2 P285-287	RT-4109 <*>
1954 V94 N2 P315-318	R-1039 <*>
	RT-4483 <*>
1954 V94 N3 P397-400	RT-1387 <*>
1954 V94 N3 P401-404	RT-1386 <*>
1954 V94 N3 P451-454	RT-3827 <*>
1954 V94 N3 P467-470	AD-610 961 <=$>
1954 V94 N4 P635-638	62-11766 <=>
1954 V94 N4 P639-641	59-19468 <+*> 0
1954 V94 N4 P655-658	60-23678 <+*>
1954 V94 N4 P659-662	62-11196 <=*>
1954 V94 N4 P663-665	3333 <HB>
1954 V94 N4 P667-669	RT-3958 <*>
1954 V94 N4 P673-676	61-13819 <=*>
1954 V94 N4 P681-684	3331 <HB>
1954 V94 N4 P689-691	3319 <HB>
1954 V94 N4 P693-696	3318 <HB>
1954 V94 N4 P697-699	RJ-280 <ATS>
1954 V94 N4 P729-732	R-1817 <*>
1954 V94 N4 P733-735	RJ639 <ATS>
	RT-2727 <*>
	63-16107 <=*> 0
1954 V94 N5 P801-803	CHEM-127 <CTS>
	62-11517 <=>
	62-16089 <=*>
1954 V94 N5 P807-810	RT-2720 <*>
1954 V94 N5 P811-812	RT-1542 <*>
1954 V94 N5 P813-816	R-3949 <*>
	3111 <NLL>
	59-10021 <+*>
1954 V94 N5 P829-832	64-10408 <=*$>
1954 V94 N5 P833-834	3312 <HB>
1954 V94 N5 P839-842	RT-2485 <*>
1954 V94 N5 P843-844	RJ-242 <ATS>
1954 V94 N5 P845-848	3416 <HB>
	65-12001 <*>
1954 V94 N5 P891-893	RJ-228 <ATS>
1954 V94 N5 P909-912	RJ-222 <ATS>
1954 V94 N5 P919-921	59-17320 <*>
1954 V94 N5 P923-926	61-17273 <=$>
1954 V94 N5 P969-971	C-3623 <NRCC>

Citation	Translation
	61-00659 <*>
1954 V94 N6 P1017-1022	RT-2393 <*>
1954 V94 N6 P1021-1024	RT-1860 <*>
1954 V94 N6 P1029-1032	R-4397 <*>
1954 V94 N6 P1045-1048	AD-610 961 <=$>
1954 V94 N6 P1045-	R-3738 <*>
1954 V94 N6 P1045-1048	70K20R <ATS>
1954 V94 N6 P1061-1063	3352 <HB>
1954 V94 N6 P1065-1067	3332 <HB>
1954 V94 N6 P1075-1078	62-33209 <=*>
1954 V94 N6 P1093-1096	158TM <CTT>
1954 V94 N6 P1101-1104	60-15819 <+*>
1954 V94 N6 P1120-1124	RT-3681 <*>
1954 V94 N6 P1121-1124	3J16R <ATS>
1954 V94 N6 P1145-1147	63-11096 <=>
1954 V94 N6 P1177-1180	RT-4337 <*>
1954 V94 N6 P1181-1184	NIOT-84 <NLL>
1954 V94 N6 P1189-1192	RT-2138 <*>
1954 V95 P451-454	AMST S2 V42 P7-12 <AMS>
	R-3595 <*>
1954 V95 P493	R-5229 <*>
1954 V95 P611-614	R-1304 <*>
1954 V95 P669-671	R-1305 <*>
1954 V95 P841-844	R-1306 <*>
1954 V95 P849-851	R-2263 <*>
1954 V95 P969-970	R-842 <*>
1954 V95 P1025-1027	R-1307 <*>
1954 V95 N1 P37-39	RT-1859 <*>
1954 V95 N1 P41-44	R-919 <*>
	RJ-599 <ATS>
	59-20215 <+*>
	62-11424 <=>
1954 V95 N1 P61-64	R-877 <*>
	RT-2378 <*>
1954 V95 N1 P77-80	RJ-227 <ATS>
1954 V95 N1 P85-87	RJ-200 <ATS>
1954 V95 N1 P93-96	RT-2484 <*>
	218TT <CCT>
1954 V95 N1 P101-103	RJ-204 <*>
	RJ-204 <ATS>
1954 V95 N1 P105-108	3315 <HB>
1954 V95 N1 P109-110	RJ-216 <ATS>
1954 V95 N1 P111-113	RJ-212 <ATS>
1954 V95 N1 P163-166	RT-2139 <*>
1954 V95 N1 P187-190	R-3900 <*>
	62-10747 <=*>
1954 V95 N2 P229-232	RT-2394 <*>
1954 V95 N2 P253-255	60-13043 <+*> 0
	64-15315 <=*$> 0
1954 V95 N2 P257-259	60-10772 <*>
1954 V95 N2 P297-299	RT-2906 <*>
	61-17274 <=$>
	795TM <CTT>
1954 V95 N2 P305-308	3075 <BISI>
	62-11208 <=>
1954 V95 N2 P329-331	RT-4469 <*>
1954 V95 N2 P333-336	RT-2558 <*>
1954 V95 N2 P337-340	63-11006 <=>
1954 V95 N3 P451-454	AMS-TRANS V42 P7-12 <AMS>
	64-71243 <=>
1954 V95 N3 P467-470	31K20R <ATS>
1954 V95 N3 P481-484	60-23600 <+*>
1954 V95 N3 P489-492	RT-2275 <*>
1954 V95 N3 P501-503	RJ-202 <ATS>
	RT-2273 <*>
1954 V95 N3 P521-523	R-879 <*>
1954 V95 N3 P529-530	R-994 <*>
	96J19R <ATS>
1954 V95 N3 P543-545	61-19369 <+*> 0
1954 V95 N3 P567-570	RJ-407 <ATS>
1954 V95 N3 P571-574	RCT V29 N3 P789 <RCT>
1954 V95 N3 P575-577	RT-2208 <*>
1954 V95 N3 P583-586	63-11007 <=>
1954 V95 N3 P591-594	M-6266 <NLL>
1954 V95 N3 P595-598	RT-3424 <*>
	60-13758 <+*> 0
1954 V95 N3 P603-606	97S86R <TTIS>
1954 V95 N3 P649-652	RT-2140 <*>
1954 V95 N3 P693-696	R-2734 <*>
	RT-2548 <*>

Citation	Translation
1954 V95 N4 P717-720	RJ-602 <ATS>
1954 V95 N4 P729-731	66-12913 <*> 0
1954 V95 N4 P749-751	RT-2491 <*>
1954 V95 N4 P753-756	RT-2541 <*>
	59-20214 <+*>
1954 V95 N4 P757-760	62-11395 <=>
1954 V95 N4 P769-771	CSIRO-3287 <CSIR>
	R-3182 <*>
	R-3954 <*>
1954 V95 N4 P793-796	RJ-250 <ATS>
1954 V95 N4 P813-815	RJ-345 <ATS>
1954 V95 N4 P829-832	65-61313 <=$>
	896TM <CTT>
1954 V95 N4 P837-840	C-3586 <NRCC>
1954 V95 N4 P841-844	62-10340 <*=> 0
1954 V95 N4 P849-851	RT-3184 <*>
1954 V95 N5 P961-963	R-2776 <*>
1954 V95 N5 P983-986	RT-2486 <*>
	59-16080 <+*>
1954 V95 N5 P1043-1045	3415 <HB>
1954 V95 N6 P1177-1180	GB43 <NLL>
1954 V95 N6 P1207-1210	AEC-TR-5911 <=*$>
1954 V95 N6 P1235-1238	RJ-272 <ATS>
	RT-2487 <*>
	60-21424 <=>
	64-18094 <*>
	65-60711 <=>
1954 V95 N6 P1239-1241	RJ-210 <ATS>
1954 V95 N6 P1271-1274	81K27R <ATS>
1954 V95 N6 P1317-1320	62-13924 <*=>
1954 V95 N6 P1329-1331	RT-2986 <*>
1954 V95 N6 P1359-1362	RJ-234 <ATS>
1954 V96 P9-12	AMST S2 V23 P83-87 <AMS>
1954 V96 P81-83	R-2852 <*>
1954 V96 P141-142	T.1483 <NLL>
1954 V96 P287-288	R-2836 <*>
1954 V96 P449-451	R-1308 <*>
1954 V96 P613-616	R-1598 <*>
1954 V96 P653-656	R-1309 <*>
1954 V96 P785-788	R-1310 <*>
1954 V96 P1025-1028	R-1311 <*>
1954 V96 P1187-1189	899TM <CTT>
1954 V96 P1201-1204	R-1312 <*>
1954 V96 P1209-1212	R-1313 <*>
1954 V96 N1 P5-7	<INSD>
1954 V96 N1 P25-28	GB82 3917 <NLL>
1954 V96 N1 P67-68	59-20924 <+*> 0
1954 V96 N1 P69-72	RJ-304 <ATS>
1954 V96 N1 P81-83	RJ-303 <ATS>
	RT-2260 <*>
	64-1555 <=*$>
1954 V96 N1 P91-93	AEC-TR-3879 <+*>
1954 V96 N1 P107-110	R-2315 <*>
	898TM <CTT>
1954 V96 N1 P143-144	3423 <HB>
1954 V96 N1 P197-200	60-51010 <=>
1954 V96 N2 P213-216	R-3418 <*>
1954 V96 N2 P261-264	GB43 <NLL>
	59-10381 <+*>
1954 V96 N2 P269-272	TT.546 <NRC>
1954 V96 N2 P277	60-15749 <+*>
1954 V96 N2 P287-288	RJ-257 <ATS>
	RT-2259 <*>
	64-14693 <=*$>
	64-15558 <=*$>
1954 V96 N2 P311-314	RT-4078 <*>
1954 V96 N2 P315-318	RT-2745 <*>
1954 V96 N2 P323-326	13L35R <ATS>
1954 V96 N2 P331-334	RJ-289 <ATS>
	61-17272 <=$>
1954 V96 N2 P343-346	RT-2246 <*>
	64-15372 <=*$> 0
1954 V96 N3 P433-436	59-16716 <+*>
1954 V96 N3 P441-444	CSIRO-3288 <CSIR>
	R-4368 <*>
1954 V96 N3 P453-456	GB125 IB 1435 <NLL>
1954 V96 N3 P463-464	RT-2332 <*>
1954 V96 N3 P469-472	RT-2614 <*>
1954 V96 N3 P495-498	RT-4099 <*>
1954 V96 N3 P507-510	3375 <HB>

Citation	Translation
1954 V96 N3 P557-560	3376 <HB>
1954 V96 N3 P565-567	RT-2231 <*>
	64-15576 <=*$>
1954 V96 N3 P577-580	3382 <HB>
1954 V96 N3 P609-612	61-17270 <=$>
1954 V96 N4 P717-719	RT-4226 <*>
	59-17553 <*>
1954 V96 N4 P725-728	RT-2928 <*>
1954 V96 N4 P729-731	RT-2612 <*>
1954 V96 N4 P733-735	RT-2068 <*>
1954 V96 N4 P741-743	3385 <HB>
1954 V96 N4 P777-779	RJ-383 <ATS>
1954 V96 N4 P781-784	59-15113 <+*>
1954 V96 N4 P841-844	63-11009 <=>
1954 V96 N4 P857-859	RT-2985 <*>
1954 V96 N4 P861-864	RT-2611 <*>
1954 V96 N5 P933-936	61-23309 <*=>
	62-23033 <=*>
1954 V96 N5 P941-943	RT-2020 <*>
1954 V96 N5 P945-948	RT-1880 <*>
1954 V96 N5 P959-961	RT-2602 <*>
1954 V96 N5 P963-966	61-13673 <=*>
	62-11299 <=>
1954 V96 N5 P975-978	R-3998 <*>
1954 V96 N5 P991-994	RT-2488 <*>
1954 V96 N5 P999-1002	RT-3267 <*>
	61-13059 <+*>
1954 V96 N5 P1015-1016	61-13672 <*+>
1954 V96 N5 P1025-1028	RT-3909 <*>
1954 V96 N5 P1057-1060	64-19055 <=*$>
1954 V96 N6 P1139-1142	R-868 <*>
1954 V96 N6 P1143-1145	R-866 <*>
1954 V96 N6 P1143	61-13671 <+*>
1954 V96 N6 P1147-1150	3421 <HB>
1954 V96 N6 P1159-1160	RT-2204 <*>
1954 V96 N6 P1161-1164	59-19613 <+*>
1954 V96 N6 P1179-1181	RJ-516 <ATS>
	61-13670 <*+>
1954 V96 N6 P1187-1189	RCT V29 N1 P131 <RCT>
1954 V96 N6 P1187-	60-23041 <*+>
1954 V96 N6 P1197-1199	RT-2572 <*>
	64-13079 <=*$>
1954 V96 N6 P1201-1204	62-10341 <*=> 0
1954 V96 N6 P1213-1216	3571 <HB>
1954 V96 N6 P1225-	61-15876 <*=> 0
1954 V97 P5-7	58-816 <*>
1954 V97 P77-80	64-00373 <*>
1954 V97 P487-489	R-1314 <*>
1954 V97 P577-579	T-1728 <NLL>
1954 V97 P667-670	T-2340 <INSD>
1954 V97 P973-976	AMST S2 V32 P159-162 <AMS>
1954 V97 N1 P5-7	64-71242 <=>
1954 V97 N1 P45-47	59-16074 <+*> 0
1954 V97 N1 P49-52	62-10651 <=*>
1954 V97 N1 P57-60	C2292 <NRC>
	R-982 <*>
1954 V97 N1 P77-80	64-19011 <=*$>
	92K21R <ATS>
1954 V97 N1 P85-87	GB4 T/1033 <NLL>
1954 V97 N1 P95-98	61-17263 <=$>
1954 V97 N1 P111-114	RCT V29 N1 P121 <RCT>
1954 V97 N1 P121-124	R-4139 <*>
	59-17465 <*>
	64-19477 <=$> 0
1954 V97 N2 P193-196	R-3458 <*>
	1455TM <CTT>
	59-16695 <+*>
1954 V97 N2 P217-219	TT.523 <NRC>
1954 V97 N2 P237-238	R-880 <*>
1954 V97 N2 P239-242	RT-2874 <*>
1954 V97 N2 P261-264	65-61432 <=>
1954 V97 N2 P265-266	50L31R <ATS>
1954 V97 N2 P269-272	CSIRO-3144 <CSIR>
	R-3938 <*>
1954 V97 N2 P273-275	00S82R <ATS>
1954 V97 N2 P277-279	RJ-416 <ATS>
1954 V97 N2 P293-295	R-4179 <*>
1954 V97 N2 P329-332	CSIRO-3239 <CSIR>
	R-3955 <*>
1954 V97 N2 P365-368	C-3239 <NRCC>

Citation	Translation
1954 V97 N2 P393-395	65-63167 <=$>
1954 V97 N3 P411-414	60-12504 <=>
	62-10161 <=*>
	62-11308 <=>
1954 V97 N3 P445-448	RT-2655 <*>
1954 V97 N3 P459-461	RJ-252 <ATS>
	59-10645 <+*>
	62-14279 <=*>
	64-14679 <=*$>
	64-20175 <*>
	65-12579 <*>
1954 V97 N3 P467-470	192TM <CTT>
1954 V97 N3 P495-498	61-28912 <*=>
1954 V97 N3 P543-545	64-15980 <=*$> 0
1954 V97 N4 P613-617	62-10366 <*=>
1954 V97 N4 P617-620	59-12458 <=>
1954 V97 N4 P621-628	59-19542 <+*>
1954 V97 N4 P629-632	63-24051 <=*>
1954 V97 N4 P639-642	62-11217 <=>
	62-15204 <*=>
1954 V97 N4 P663-665	3469 <HB>
1954 V97 N4 P667-670	193TM <CTT>
	61-17261 <=$>
1954 V97 N4 P671-673	RT-3277 <*>
1954 V97 N4 P703-706	60-11644 <=>
1954 V97 N4 P711-714	RJ-261 <ATS>
1954 V97 N4 P715-718	RT-3621 <*>
1954 V97 N5 P785-788	RT-2395 <*>
1954 V97 N5 P789-792	60-13119 <+*>
1954 V97 N5 P813-816	40J17R <ATS>
1954 V97 N5 P821-822	RJ-241 <ATS>
1954 V97 N5 P843-846	<ES>
	RJ-230 <ATS>
	64-18441 <*>
1954 V97 N5 P859-862	63-11094 <=>
	63-15789 <=*>
1954 V97 N5 P871-874	60-21125 <=>
1954 V97 N5 P915-918	RT-4383 <*>
1954 V97 N5 P919-922	62-32359 <=*>
	64-14273 <=*$>
1954 V97 N5 P927-930	63-11013 <=>
1954 V97 N6 P995-997	RT-2489 <*>
1954 V97 N6 P1013-1014	RJ-330 <ATS>
	59-20213 <+*>
1954 V97 N6 P1023-1026	RT-3413 <*>
1954 V97 N6 P1037-1038	RT-2834 <*>
1954 V97 N6 P1061-1064	63-11028 <=>
1954 V97 N6 P1069-1073	CSIRO-3099 <CSIR>
	R-3637 <*>
1954 V98 P111-114	R-1315 <*>
1954 V98 P427-430	R-774 <*>
1954 V98 P569-571	R-4509 <*>
1954 V98 P731-734	STMSP V1 P157-161 <AMS>
1954 V98 P735-738	STMSP V1 P163-167 <AMS>
1954 V98 P747	60-13879 <+*>
1954 V98 P885-888	AMST S2 V26 P1-4 <AMS>
1954 V98 P893-895	AMST S2 V32 P155-158 <AMS>
1954 V98 P1017-1020	R-1316 <*>
1954 V98 N1 P31-33	RT-2961 <*>
1954 V98 N1 P35-38	RT-2695 <*>
1954 V98 N1 P39-42	RT-4509 <*>
1954 V98 N1 P43-45	RT-4510 <*>
	1291TM <CTT>
1954 V98 N1 P47-50	62-10645 <=*>
1954 V98 N1 P55-57	R-19 <*>
1954 V98 N1 P59-62	RT-4511 <*>
	1290TM <CTT>
1954 V98 N1 P89-91	C-3733 <NRCC>
1954 V98 N1 P93-96	RT-4278 <*>
	T2226 <INSD>
1954 V98 N1 P99-102	AEC-TR-6265 <=*$>
1954 V98 N1 P107-110	R-3147 <*>
1954 V98 N1 P115-118	60-13757 <+*>
1954 V98 N1 P141-144	CSIR-TR 144 <CSSA>
1954 V98 N1 P153-154	62-24046 <=*>
1954 V98 N1 P159-162	RT-4640 <*>
1954 V98 N2 P189-192	62-19906 <=*>
1954 V98 N2 P205-206	RT-2392 <*>
1954 V98 N2 P211-213	RT-2719 <*>
1954 V98 N2 P223-226	RJ-220 <ATS>

Citation	Translation
1954 V98 N2 P227-228	RT-2637 <*>
1954 V98 N2 P237-240	R-3641 <*>
1954 V98 N2 P251-252	62-13916 <*=>
1954 V98 N2 P317-319	RT-3736 <*>
	60-10998 <+*>
1954 V98 N3 P333-336	62-25864 <=*>
1954 V98 N3 P353-356	RT-2377 <*>
1954 V98 N3 P357-360	UCRL TRANS-1002(L) <=*$>
1954 V98 N3 P373-376	59-17438 <*>
1954 V98 N3 P389-390	RT-3224 <*>
1954 V98 N3 P395-397	59-11711 <=> 0
1954 V98 N3 P399-402	RT-3342 <*>
1954 V98 N3 P431-434	RT-2922 <*>
1954 V98 N3 P435-438	59-21042 <=>
1954 V98 N3 P491-494	3390-D <K-H>
1954 V98 N3 P495-496	5604-A <K-H>
1954 V98 N3 P497-499	62-25644 <=*>
1954 V98 N4 P557-560	RT-3375 <*>
	65-10804 <*>
1954 V98 N4 P575-578	RT-4512 <*>
1954 V98 N4 P583-584	3446 <HB>
1954 V98 N4 P593-595	189TM <CTT>
1954 V98 N4 P601-604	61-15384 <*+>
1954 V98 N4 P605-607	3441 <HB>
1954 V98 N4 P611-612	RCT V29 N1 P145 <RCT>
1954 V98 N4 P617-618	UCRL TRANS-573(L) <+*>
1954 V98 N4 P633-636	CSIRO-314C <CSIR>
	R-3925 <*>
1954 V98 N4 P641-644	55H12R <*>
1954 V98 N5 P749-752	RT-3418 <*>
1954 V98 N5 P757-759	799TM <CTT>
1954 V98 N5 P765-768	59-15488 <+*> 0
1954 V98 N5 P769-771	C-2488 <NRC>
1954 V98 N5 P781-782	61-18910 <*=> 0
1954 V98 N5 P791-794	<ATS>
1954 V98 N5 P795-798	RJ-350 <ATS>
1954 V98 N5 P819-820	RT.3106 <*>
	3022 <NLL>
	59-15046 <+*>
1954 V98 N5 P849-852	RT-4161 <*>
1954 V98 N5 P869-872	RT-4174 <*>
1954 V98 N5 P873-875	RT-4163 <*>
1954 V98 N6 P941-944	62-11617 <=>
1954 V98 N6 P953-956	62-11297 <=>
1954 V98 N6 P957-960	C-2491 <NRC>
	R-2086 <*>
1954 V98 N6 P961-964	R-2736 <*>
1954 V98 N6 P965-967	RT-2687 <*>
1954 V98 N6 P989-992	R-1007 <*>
1954 V98 N6 P993-995	60-15829 <+*> 0
1954 V98 N6 P1005-1006	43J15R <ATS>
1954 V98 N6 P1011-1012	T-2279 <INSD>
1954 V98 N6 P1017-1020	62-10339 <*=> 0
1954 V99 P27-30	AMST S2 V42 P1-6 <AMS>
1954 V99 P69-72	R-258 <*>
1954 V99 P81-84	R-773 <*>
1954 V99 P125-128	R-5086 <*>
1954 V99 P137-140	CSIR-TR.256 <CSSA>
1954 V99 P141-144	R-1317 <*>
1954 V99 P213-216	61-00010 <*>
1954 V99 P451-454	R-1612 <*>
1954 V99 P471-473	R-1318 <*>
1954 V99 P1041-1044	AD-629 045 <=*$>
1954 V99 N1 P27-30	AMS-TRANS V42 P1-6 <AMS>
	UCRL TRANS-963(L) <=*$>
1954 V99 N1 P45-46	RT-4285 <*>
1954 V99 N1 P51	RT-2603 <*>
1954 V99 N1 P53-55	R-775 <*>
1954 V99 N1 P65-68	RT-2954 <*>
1954 V99 N1 P73-75	RJ-470 <ATS>
1954 V99 N1 P77-80	60-16980 <*> 0
1954 V99 N1 P93-95	AEC-TR-5131 <=*>
	62-16141 <=*>
1954 V99 N1 P101-104	R-3675 <*>
1954 V99 N1 P105-108	63-11007 <=>
1954 V99 N1 P125-128	61-10129 <+*>
1954 V99 N1 P129-132	RT-4105 <*>
1954 V99 N1 P137-140	15P66R <ATS>
1954 V99 N2 P209-212	R-3285 <*>
	838TM <CTT>
1954 V99 N2 P225-226	<INSD>
1954 V99 N2 P227-230	65-63655 <=$>
1954 V99 N2 P239-242	59-14184 <+*>
1954 V99 N2 P243-246	RT-3213 <*>
1954 V99 N2 P301-304	65-11138 <*>
1954 V99 N2 P305-306	RJ-316 <ATS>
1954 V99 N2 P315-316	RT-4084 <*>
1954 V99 N2 P317-320	RT-3729 <*>
1954 V99 N2 P325-328	RT-3730 <*>
1954 V99 N2 P333-336	59-22705 <+*> 0
1954 V99 N3 P349-352	61-23310 <*=>
1954 V99 N3 P361-364	RT-3179 <*>
1954 V99 N3 P385-388	RT-3376 <*>
	T-1617C <INSD>
1954 V99 N3 P389-390	RT-3799 <*>
1954 V99 N3 P391-394	RT-3805 <*>
1954 V99 N3 P395-398	R-580 <*>
	RT-3810 <*>
1954 V99 N3 P407-410	17H13R <ATS>
1954 V99 N3 P411-414	RJ-471 <ATS>
	3448 <HB>
1954 V99 N3 P427-430	R-412 <*>
1954 V99 N3 P431-434	RCT V29 N3 P829 <RCT>
	60-41433 <=>
1954 V99 N4 P519-522	R-914 <*>
	1288TM <CTT>
	62-11426 <=>
1954 V99 N4 P529-531	RT-2878 <*>
1954 V99 N4 P537-538	R-1438 <*>
	59-12169 <+*>
1954 V99 N4 P539-542	RJ-251 <ATS>
	64-14680 <=*$>
1954 V99 N4 P547-549	RJ-402 <ATS>
1954 V99 N4 P573-579	60-16820 <+*>
1954 V99 N4 P585-588	RJ-524 <ATS>
1954 V99 N4 P589-592	RT-3731 <*>
1954 V99 N4 P593-596	65-63164 <=$>
1954 V99 N4 P617-619	R-3923 <*>
	R-4210 <*>
1954 V99 N4 P637-640	CSIRO-3145 <CSIR>
	R-3909 <*>
1954 V99 N4 P649-652	RT-4085 <*>
1954 V99 N4 P653-656	RT-3775 <*>
1954 V99 N5 P685-687	61-19437 <+*> 0
1954 V99 N5 P699-702	UCRL TRANS-554 <+*>
	64-13556 <=*$>
1954 V99 N5 P715-718	60-10059 <+*> 0
1954 V99 N5 P735-736	R-1006 <*>
	62-20283 <=*> 0
1954 V99 N5 P741-743	RT-3873 <*>
1954 V99 N5 P745-748	RT-3417 <*>
	800TM <CTT>
1954 V99 N5 P749-751	60-15751 <+*>
1954 V99 N5 P761-763	R-2340 <*>
1954 V99 N5 P769-771	RJ-301 <ATS>
	61-17262 <=$>
1954 V99 N5 P793-796	61-13669 <+*>
1954 V99 N5 P797-800	RT-3793 <*>
	19J18R <ATS>
1954 V99 N5 P801-804	3603 <HB>
	64-18360 <=*$>
1954 V99 N5 P805-808	RJ-313 <ATS>
	RT-2872 <*>
1954 V99 N5 P809-812	R-2056 <*>
	65-10483 <*>
1954 V99 N5 P819-822	RCT V29 N2 P511 <RCT>
1954 V99 N5 P831-834	R-132 <*>
1954 V99 N5 P853	R-4934 <*>
1954 V99 N5 P865-868	RT-2953 <*>
1954 V99 N5 P869-872	R-75 <*>
1954 V99 N6 P921-923	RT-4513 <*>
1954 V99 N6 P929-930	<CB>
1954 V99 N6 P929	59-17358 <+*>
1954 V99 N6 P929-930	61-17482 <=$>
	62-24768 <=*>
1954 V99 N6 P931-933	<CB>
	1242TT <CCT>
	61-17536 <=$>
1954 V99 N6 P939-942	<CB>
	1243TT <CCT>

Citation	Translation
	61-17483 <=$>
1954 V99 N6 P943-946	<CB>
	1244TT <CCT>
	61-17535 <=$>
	62-24767 <=*>
1954 V99 N6 P947-950	62-24800 <=*>
1954 V99 N6 P955-958	<CB>
	1245TT <CCT>
	61-17534 <=$>
	62-24766 <=*>
1954 V99 N6 P959-961	<CB>
	1246TT <CCT>
	59-17359 <+*>
	61-17533 <=$>
	62-24769 <=*>
1954 V99 N6 P967-969	RJ-238 <ATS>
	1247TT <CCT>
	20G5R <ATS>
	61-17532 <=$>
	62-24760 <=*>
1954 V99 N6 P987-990	62-23209 <=*>
1954 V99 N6 P1003-1006	61-17265 <=$>
1954 V99 N6 P1019-1022	64-20217 <*> 0
1954 V99 N6 P1025-1027	RJ-318 <ATS>
	3519 <HB>
	64-16753 <=*$>
1954 V99 N6 P1029-1032	R-3474 <*>
	61-23335 <*=>
1954 V99 N6 P1037-1039	RCT V29 N2 P451 <RCT>
1954 V99 N6 P1053-1056	RCT V29 N2 P602 <RCT>
1954 V99 N6 P1061-1063	RJ-260 <ATS>
1954 V99 N6 P1069-1071	R-2347 <*>
1954 V99 N6 P1091-1093	R-361 <*>
1954 V99 N6 P1103-1106	R-362 <*>
1954 V99 N6 P1111-1114	R-363 <*>
1954 V99 N6 P1115-1117	R-352 <*>
1954 V99 N6 P1165-1168	3414 <HB>
1955 V100 P131-134	R-1319 <*>
1955 V100 P525-528	R-1320 <*>
1955 V100 P563-566	GB146 <NLL>
1955 V100 P1009-1011	C-2620 <NRC>
1955 V100 P1159-1160	R-1611 <*>
1955 V100 N1 P29-32	R-918 <*>
1955 V100 N1 P33-36	60-16529 <+*>
1955 V100 N1 P43-44	R-3065 <*>
1955 V100 N1 P53-56	RT-3466 <*>
1955 V100 N1 P57-60	R-339 <*>
1955 V100 N1 P69-72	3511 <HB>
1955 V100 N1 P93-96	62-25897 <=$>
1955 V100 N1 P97-100	59-22610 <+*>
1955 V100 N1 P112-114	GB4 T/1032 <NLL>
1955 V100 N1 P115-118	64-14898 <=*$>
1955 V100 N1 P119-122	900TM <CTT>
1955 V100 N1 P127-130	UCRL TRANS-225(L) <=*$>
1955 V100 N1 P155-157	R-2662 <*>
1955 V100 N1 P179-182	R-334 <*>
1955 V100 N1 P183-186	RT-4646 <*>
1955 V100 N2 P225-228	RT-4497 <*>
	1356TM <CTT>
1955 V100 N2 P229-232	RJ-235 <ATS>
	59-17544 <*>
1955 V100 N2 P233-235	<CB>
	R-1043 <*>
1955 V100 N2 P233-236	1248TT <CCT>
1955 V100 N2 P233-235	61-17543 <=$>
1955 V100 N2 P237-240	60-13856 <+*>
1955 V100 N2 P241-242	RT-2856 <*>
	1249TT <CCT>
	61-17542 <=$>
	62-24763 <=*>
1955 V100 N2 P243-246	R-1238 <*>
	R-1476 <*>
	R-1692 <*>
	1371TM <CTT>
1955 V100 N2 P251-253	TT.539 <NRC>
1955 V100 N2 P255	AEC-TR-2158 <*>
1955 V100 N2 P255-257	RT-2923 <*>
	59-17531 <*>
1955 V100 N2 P275-278	RCT V29 N2 P423 <RCT>
1955 V100 N2 P315-318	TRC-TRANS-1103 <NLL>
1955 V100 N2 P327-330	R-1042 <*>
	3502 <HB>
	62-11386 <=>
	96P60R <ATS>
1955 V100 N2 P331-334	64-19054 <=*$>
1955 V100 N2 P373-376	CSIRO-3029 <CSIR>
	R-3930 <*>
1955 V100 N2 P389-391	RT-2833 <*>
1955 V100 N3 P381-384	3205 <NLL>
1955 V100 N3 P413-416	62-15925 <=*>
1955 V100 N3 P425-428	62-15851 <=*>
1955 V100 N3 P437-440	62-11425 <=>
1955 V100 N3 P441-444	<CB>
	1255TT <CCT>
	61-17516 <=$>
1955 V100 N3 P445-448	C-2208 <NLL>
	R-884 <*>
	62-15244 <*=>
1955 V100 N3 P459-600	CSIR-TR.143 <CSSA>
1955 V100 N3 P477-480	RCT V29 N2 P504 <RCT>
1955 V100 N3 P481-484	R-4378 <*>
	62-11216 <=>
1955 V100 N3 P485-486	R-2678 <*>
1955 V100 N3 P521-524	R-4570 <*>
1955 V100 N3 P583-586	R-3629 <*>
	3082 <*>
1955 V100 N3 P597-600	R-2691 <*>
1955 V100 N4 P631-633	62-23154 <=*>
1955 V100 N4 P647-650	R-1919 <*>
	62-15858 <=*>
1955 V100 N4 P651-654	RJ-442 <ATS>
1955 V100 N4 P655-658	RT-2775 <*>
	1250TT <CCT>
	61-17541 <=$>
1955 V100 N4 P661-663	<CB>
	RT-4451 <*>
1955 V100 N4 P661-664	1251TT <CCT>
1955 V100 N4 P661-663	61-17540 <=$>
1955 V100 N4 P665-667	R-1413 <*>
	1370TM <CTT>
1955 V100 N4 P669-672	<CB>
	61-17517 <=$>
1955 V100 N4 P673-676	RT-2776 <*>
	1252TT <CCT>
	61-17539 <=$>
1955 V100 N4 P677-679	RT-2914 <*>
	1253TT <CCT>
	61-17538 <*=>
	62-24762 <=*>
1955 V100 N4 P681-683	R-1412 <*>
	1369TM <CTT>
1955 V100 N4 P689-692	RT-2913 <*>
	1254TT <CCT>
	59-17462 <*>
	61-17537 <=$>
	62-24761 <=*>
1955 V100 N4 P701-703	<ES>
	63-14148 <=*> 0
	65-18180 <*>
1955 V100 N4 P707-709	R-772 <*>
	64-15058 <=*$>
1955 V100 N4 P711-714	RJ-281 <ATS>
1955 V100 N4 P731-733	R-2350 <*>
1955 V100 N4 P745-748	29K24R <ATS>
1955 V100 N4 P761-764	62-11237 <=>
1955 V100 N4 P829-832	75H12R <ATS>
1955 V100 N5 P853-856	R-3450 <*>
1955 V100 N5 P867-870	62-23045 <=*>
1955 V100 N5 P871-874	62-23161 <=*>
1955 V100 N5 P875-878	RT-3529 <*>
1955 V100 N5 P879-881	60-21964 <=>
1955 V100 N5 P893-896	8921-B <K-H>
1955 V100 N5 P897-900	62-10646 <=*>
1955 V100 N5 P901-903	R-3265 <*>
	RT-3256 <*>
1955 V100 N5 P909-912	62-11409 <=>
1955 V100 N5 P921-923	RT-3463 <*>
1955 V100 N5 P933-936	RT-4037 <*>
1955 V100 N5 P937-938	RJ-530 <ATS>
1955 V100 N5 P939-942	60-11487 <=>

Citation	Translation
1955 V100 N5 P1021-1023	74H12R <ATS>
1955 V100 N6 P1067-1069	60-13128 <+*> 0
1955 V100 N6 P1069-1072	RT-3278 <*>
1955 V100 N6 P1073-1075	3518 <HB>
1955 V100 N6 P1077-1078	3541 <HB>
1955 V100 N6 P1099-1101	62-14278 <=*>
	64-20174 <*> 0
	64-20231 <*> 0
	8062 <K-H>
1955 V100 N6 P1103-1106	60-17148 <+*>
1955 V100 N6 P1107-1110	64-14914 <=*$> 0
1955 V100 N6 P1123-1126	R-4605 <*>
1955 V100 N6 P1175-1178	C-2276 <NRC>
	R-4814 <*>
1955 V100 N6 P1191-1193	59-18625 <+*>
1955 V101 P35-38	AMST S2 V10 P341-344 <AMS>
1955 V101 P181-183	R-1321 <*>
1955 V101 P351-354	59-22698 <+*> 0
1955 V101 P531-534	R-1322 <*>
1955 V101 P937-940	R-1605 <*>
1955 V101 P955-957	R-1323 <*>
1955 V101 P1113-1116	R-1608 <*>
	R-4654 <*>
1955 V101 N1 P43-46	62-23160 <=*>
1955 V101 N1 P47-49	61-17518 <=$>
	62-23054 <=*>
1955 V101 N1 P55-57	R-3135 <*>
	62-25068 <=*>
1955 V101 N1 P63-64	RT-3387 <*>
1955 V101 N1 P65-67	3623 <*>
1955 V101 N1 P97-98	4091 <HB>
1955 V101 N1 P107-109	RT-3271 <*>
1955 V101 N1 P115-118	61-13339 <+*>
1955 V101 N1 P119-122	62-23101 <=*>
1955 V101 N1 P125-128	RT-2857 <*>
1955 V101 N1 P151-153	33J18R <ATS>
1955 V101 N2 P209-212	64-11961 <=> M
1955 V101 N2 P233-236	<CB>
	61-17519 <=$>
1955 V101 N2 P237-240	63-20241 <=*>
1955 V101 N2 P245-248	<CB>
1955 V101 N2 P245	R-4862 <*>
1955 V101 N2 P245-248	RT-4501 <*>
	61-17546 <=$>
1955 V101 N2 P259-261	RT-3038 <*>
	3626-C <K-H>
	64-14932 <=*$>
1955 V101 N2 P289-292	62-11230 <=>
1955 V101 N2 P297-300	R-3918 <*>
	3206 <NLL>
	62-11215 <=>
1955 V101 N2 P325-326	3545 <HB>
1955 V101 N2 P327-329	60-23752 <+*> 0
1955 V101 N2 P391-392	R-217 <*>
	64-23602 <=$>
1955 V101 N3 P429-432	R-3656 <*>
1955 V101 N3 P433-435	RJ-387 <ATS>
	62-15861 <=*>
1955 V101 N3 P441-444	62-15867 <=*>
1955 V101 N3 P445-447	61-28011 <*=>
	85N53R <ATS>
1955 V101 N3 P449-452	RT-2911 <*>
1955 V101 N3 P453-455	R-1402 <*>
1955 V101 N3 P457-459	62-11416 <=>
1955 V101 N3 P465-468	59-16708 <+*>
1955 V101 N3 P477-478	TS 1266 <BISI>
	62-24221 <=*>
1955 V101 N3 P517-520	R-2099 <*>
	RJ.266 <ATS>
1955 V101 N3 P517	RJ-266 <ATS>
1955 V101 N4 P619-622	62-15850 <=*>
1955 V101 N4 P627	59-19470 <+*>
1955 V101 N4 P629-632	62-11213 <=>
1955 V101 N4 P637-639	RT-2912 <*>
1955 V101 N4 P641-644	R-1190 <*>
1955 V101 N4 P649-652	62-11430 <=>
1955 V101 N4 P657-660	62-15852 <=*>
1955 V101 N4 P661-664	62-11212 <=>
1955 V101 N4 P687-688	59-10092 <+*>
1955 V101 N4 P703-706	RT-3968 <*>
1955 V101 N4 P711-714	RT-3388 <*>
1955 V101 N4 P715-717	R-3910 <*>
1955 V101 N4 P727-730	61-28671 <*=>
1955 V101 N4 P783-785	R-2118 <*>
1955 V101 N5 P813-816	62-15877 <=*>
1955 V101 N5 P821-823	RT-3675 <*>
1955 V101 N5 P825-828	RT-4619 <*>
	134TM <CTT>
1955 V101 N5 P833-836	R-1189 <*>
	62-11303 <=>
1955 V101 N5 P841-844	62-11202 <=>
	62-14581 <=*>
1955 V101 N5 P845-847	R-1192 <*>
1955 V101 N5 P853-856	59-17487 <*> 0
1955 V101 N5 P857-859	3556 <HB>
1955 V101 N5 P861-864	RT-3478 <*>
1955 V101 N5 P889-892	RT-4457 <*>
	64-19053 <=*$>
1955 V101 N5 P899-900	3557 <HB>
1955 V101 N5 P911-912	05S83R <ATS>
1955 V101 N5 P917-920	GB21/LC 728 <NLL>
	RT-3370 <*>
1955 V101 N5 P961-964	R-238 <*>
1955 V101 N6 P1009-1012	AD-619 325 <=$>
1955 V101 N6 P1017-1018	59-10122 <+*> 0
	59-16060 <+*>
1955 V101 N6 P1019-1021	62-15880 <=*>
1955 V101 N6 P1023-1025	RJ-331 <ATS>
	62-11238 <=>
1955 V101 N6 P1027-1030	RT-3101 <*>
1955 V101 N6 P1031-1034	61-28226 <*=>
	86N53R <ATS>
1955 V101 N6 P1065-1067	RT-3429 <*>
	61-13108 <+*>
1955 V101 N6 P1083-1084	RT-3416 <*>
1955 V101 N6 P1085-1088	R-5360 <*>
1955 V101 N6 P1109-1112	60-21965 <=>
1955 V101 N8 P947-949	R-240 <*>
1955 V102 P53-56	1442TM <CTT>
1955 V102 P73-75	AEC-TR-2233 <*>
1955 V102 P81-83	R-5056 <*>
1955 V102 P165-167	R-1324 <*>
1955 V102 P378-381	R-1325 <*>
1955 V102 P521-524	GB2 <NLL>
1955 V102 P539-542	R-1326 <*>
1955 V102 P579-582	R-1327 <*>
1955 V102 P767-770	R-1328 <*>
1955 V102 P985-988	R-1329 <*>
1955 V102 P1131-1134	GB39 <NLL>
1955 V102 P1143-1145	RCT V29 N2 P419 <RCT>
1955 V102 P1215-1218	R-1330 <*>
1955 V102 N1 P49-51	62-11589 <=>
1955 V102 N1 P53-56	59-16719 <+*>
1955 V102 N1 P57-60	RT-2963 <*>
1955 V102 N1 P61-64	R-1401 <*>
	R-1733 <*>
1955 V102 N1 P69-71	1432TM <CTT>
	59-16709 <+*>
1955 V102 N1 P73-75	RT-3526 <*>
	59-17543 <*>
1955 V102 N1 P85-88	R-2051 <*>
	159TM <CTT>
1955 V102 N1 P97-99	59-18112 <+*>
1955 V102 N1 P105-108	59-15311 <+*>
	62-16632 <=*> 0
1955 V102 N1 P117-119	RT-3735 <*>
1955 V102 N1 P121-124	AEC-TR-2234 <*>
	RT-3475 <*>
1955 V102 N1 P133-136	RTS-2731 <NLL>
1955 V102 N1 P145-147	RJ481 <ATS>
1955 V102 N1 P165-167	5613-A <K-H>
1955 V102 N1 P185-187	RT-3547 <*>
1955 V102 N2 P245-248	61-17507 <=$>
1955 V102 N2 P253-256	<CB>
	RT-3036 <*>
	59-17443 <*>
	61-17544 <*=>
1955 V102 N2 P265-266	60-15509 <+*>
1955 V102 N2 P301-304	RT-3494 <*>
	35Q68R <ATS>

Citation	Report
1955 V102 N2 P305-306	RT-3409 <*>
1955 V102 N2 P1103-1106	61-17489 <=$>
1955 V102 N3 P393-394	61-17506 <=$>
1955 V102 N3 P431-434	62-23011 <=*>
1955 V102 N3 P469-472	AD-619 328 <=$>
	R-999 <*>
1955 V102 N3 P477-480	46H12R <ATS>
	59-20207 <+*>
1955 V102 N3 P481-484	R-1509 <*>
	69P66R <ATS>
1955 V102 N3 P495-497	<CB>
	RT-4601 <*>
	59-17459 <*>
	599 <NLL>
	61-17545 <=$>
1955 V102 N3 P503-506	59-16689 <+*>
1955 V102 N3 P519-520	RT-3164 <*>
1955 V102 N3 P531-534	59-10015 <+*>
1955 V102 N3 P535-538	61-20228 <*=>
	64-30121 <*> 0
	65-12208 <*>
	7082 B <K-H>
1955 V102 N3 P547-549	R-993 <*>
1955 V102 N3 P551-554	TT.577 <NRC>
1955 V102 N3 P567-570	62-14706 <=*>
1955 V102 N3 P571-573	63-11085 <=>
1955 V102 N3 P583-586	GB70 4196 <NLL>
1955 V102 N3 P595-597	48H12R <ATS>
1955 V102 N4 P707-710	R-4400 <*>
	59-12192 <+*>
1955 V102 N4 P711-714	1434TM <CTT>
	59-16711 <+*>
1955 V102 N4 P715-718	RT-3104 <*>
1955 V102 N4 P719-721	RT-3102 <*>
	61-17492 <=$>
	810TM <CTT>
1955 V102 N4 P723-725	RT-3103 <*>
	59-17533 <*>
1955 V102 N4 P727-728	60-13416 <+*> 0
1955 V102 N4 P737-740	59-16137 <+*>
1955 V102 N4 P813-814	62-16427 <=*>
1955 V102 N5 P903-906	R-2348 <*>
	18H13R <ATS>
1955 V102 N5 P911-914	RT-4239 <*>
	44H12R <ATS>
	805TM <CTT>
1955 V102 N5 P915-917	RT-3105 <*>
1955 V102 N5 P919-920	RT-4456 <*>
	1343TM <CTT>
1955 V102 N5 P921-923	RT-3174 <*>
1955 V102 N5 P933-934	61-17491 <=$>
	62-11629 <=>
	811TM <CTT>
1955 V102 N5 P935-938	RT-4146 <*>
1955 V102 N5 P947-948	RJ-343 <ATS>
1955 V102 N5 P949-952	RT-4203 <*>
1955 V102 N5 P953-956	R-153 <*>
	RT-4200 <*>
	64-19057 <=*$> 0
1955 V102 N5 P969-972	RT-3530 <*>
1955 V102 N6 P1083-1084	AEC-TR-2310 <*>
1955 V102 N6 P1099-1100	RT-4118 <*>
	1310TM <CTT>
	61-17490 <=$>
1955 V102 N6 P1101-1102	RT-3100 <*>
1955 V102 N6 P1103-1106	R-718 <*>
	1333TM <CTT>
1955 V102 N6 P1107-1110	RT-3123 <*>
1955 V102 N6 P1111-1114	RT-3549 <*>
1955 V102 N6 P1119-1122	84L30R <ATS>
1955 V102 N6 P1159-1162	RT-3600 <*>
1955 V102 N6 P1163-1164	4050-B <K-H>
1955 V102 N6 P1163-1165	64-16447 <*> 0
1955 V103 P97-100	R-1331 <*>
1955 V103 P207-208	1338TM <CTT>
1955 V103 P219	R-5015 <*>
1955 V103 P247-250	R-907 <*>
1955 V103 P283-286	R-1332 <*>
1955 V103 P333-335	R-1333 <*>
1955 V103 P391-394	R-548 <*>
1955 V103 P507-510	R-1334 <*>
1955 V103 P573-576	1268TM <CTT>
1955 V103 P717-720	C-2476 <NRC>
1955 V103 P717-719	59-22699 <+*> 0
1955 V103 P875-877	R-1335 <*>
1955 V103 P1061-1063	R-1657 <*>
1955 V103 P1073-1076	T-1776 <INSD>
1955 V103 N1 P35-36	59-22665 <+*>
1955 V103 N1 P41-43	69J15R <ATS>
1955 V103 N1 P45-47	RT-3432 <*>
	59-17536 <*> 0
1955 V103 N1 P49-51	59-17548 <*>
1955 V103 N1 P53-56	62-23100 <=*>
1955 V103 N1 P61-64	65-61425 <=>
1955 V103 N1 P73-76	1399TM <CTT>
1955 V103 N1 P77-80	GB4 T/1040 <NLL>
	R-3255 <*>
1955 V103 N1 P81-82	AD-623 781 <=*$>
1955 V103 N1 P95-96	GB39 <NLL>
1955 V103 N1 P101-104	RT-3198 <*>
1955 V103 N1 P173-176	RT-3546 <*>
1955 V103 N2 P203-206	RT-4460 <*>
1955 V103 N2 P207-208	RJ-437 <ATS>
	61-17488 <=$>
1955 V103 N2 P209-212	RT-3352 <*>
1955 V103 N2 P227-228	RT-3149 <*>
	61-17515 <=$>
1955 V103 N2 P229-232	RT-3312 <*>
1955 V103 N2 P233	61-17514 <=$>
1955 V103 N2 P235-238	RJ-444 <ATS>
1955 V103 N2 P251-252	R-3933 <*>
	3148 <*>
1955 V103 N2 P261-263	RT-3591 <*>
1955 V103 N2 P277-280	RJ-415 <ATS>
1955 V103 N2 P299-302	42J15R <ATS>
1955 V103 N3 P287	GB43 <NLL>
1955 V103 N3 P387-390	R-3111 <*>
	62-23224 <=*>
	840TM <CTT>
1955 V103 N3 P395-397	RT-3411 <*>
	59-17550 <*>
1955 V103 N3 P399-402	NP-TR-272 <*>
	RJ-439 <ATS>
1955 V103 N3 P399	1273TM <CTT>
1955 V103 N3 P399-402	61-17505 <=$>
1955 V103 N3 P403-405	1368TM <CTT>
1955 V103 N3 P407	RT-3470 <*>
	61-17513 <=$>
1955 V103 N3 P409-411	RT-4304 <*>
	61-17512 <=$>
1955 V103 N3 P413-416	61-17511 <=$>
1955 V103 N3 P417-419	RT-3310 <*>
	61-17510 <=$>
1955 V103 N3 P421-424	RT-3311 <*>
1955 V103 N3 P425-426	RT-3257 <*>
1955 V103 N3 P427-429	62-13484 <=*>
1955 V103 N3 P431-432	62-11627 <=> 0
1955 V103 N3 P439-442	C-5381 <NRC>
1955 V103 N3 P461-464	RCT V30 N1 P334 <RCT>
1955 V103 N3 P461-4644	59-17770 <+*>
1955 V103 N3 P465-468	61-21130 <=>
1955 V103 N3 P469-472	R-3148 <*>
1955 V103 N3 P479-482	62-16430 <=*>
1955 V103 N4 P573-575	RT-4008 <*>
	59-17556 <*>
	61-17509 <=$>
1955 V103 N4 P581-584	RT-3465 <*>
1955 V103 N4 P589-592	RT-3309 <*>
1955 V103 N4 P593-595	RT-4111 <*>
1955 V103 N4 P593	1269TM <CTT>
1955 V103 N4 P593-595	61-17508 <=$>
1955 V103 N4 P601-604	62-23201 <=*>
1955 V103 N4 P605-608	62-11381 <=>
1955 V103 N4 P609-610	3672 <HB>
	62-11379 <=>
1955 V103 N4 P623-626	65-61539 <=$>
1955 V103 N4 P647-650	R-4837 <*>
	59-14516 <+*>
1955 V103 N4 P667-668	61-17344 <=$>
1955 V103 N5 P781-782	R-602 <*>

Citation	Reference
1955 V103 N5 P783-786	RT-3800 <*>
1955 V103 N5 P803-806	RJ-443 <ATS>
1955 V103 N5 P803-808	RJ-443 <ATS>
1955 V103 N5 P807-809	127TM <CTT>
	61-17487 <=$>
1955 V103 N5 P815-818	60-41110 <=>
1955 V103 N5 P839-841	RT-3443 <*>
1955 V103 N5 P917-920	C-2479 <NRC>
1955 V103 N6 P989-992	IGR V2 N6 P530 <AGI>
	R-838 <*>
	59-16705 <+*>
1955 V103 N6 P997-999	RT-3460 <*>
1955 V103 N6 P1005-1008	RT-3427 <*>
1955 V103 N6 P1009-1011	C-2165 <NRC>
	RT-4302 <*>
1955 V103 N6 P1017-1020	R-3257 <*>
1955 V103 N6 P1021-1024	R-3256 <*>
1955 V103 N6 P1033-1034	60-23749 <*+>
1955 V103 N6 P1057-1059	RCT V29 N4 P1300 <RCT>
1955 V103 N6 P1073-1076	R-4975 <*>
1955 V103 N6 P1077-1080	AD-610 906 <=>
	RT-3468 <*>
1955 V104 P88-90	GB4 T/1000 <NLL>
1955 V104 P108-111	R-1654 <*>
1955 V104 P272-275	R-1336 <*>
1955 V104 P315-318	59-22704 <+*> 0
1955 V104 P380-383	R-668 <*>
1955 V104 P387-388	R-1411 <*>
1955 V104 P413	R-5007 <*>
1955 V104 P440-443	R-1337 <*>
1955 V104 P517-519	3120 <NLL>
1955 V104 P546-548	R-1011 <*>
1955 V104 P575-578	R-1338 <*>
1955 V104 P622-625	R-2682 <*>
1955 V104 P746-749	R-1339 <*>
1955 V104 P882-885	R-1340 <*>
1955 V104 N1 P38-39	R-4638 <*>
1955 V104 N1 P40-43	RT-3464 <*>
1955 V104 N1 P44-46	62-11583 <=>
1955 V104 N1 P51-53	RT-3433 <*>
	61-17503 <=$>
1955 V104 N1 P56-59	61-17504 <=$>
1955 V104 N1 P68-71	64-18361 <*> 0
1955 V104 N1 P76-77	R-1206 <*>
1955 V104 N1 P82-84	GB4 T/999 <NLL>
1955 V104 N1 P91-92	160TM <CTT>
1955 V104 N1 P98-100	R-1129 <*>
1955 V104 N1 P101-103	RJ-344 <ATS>
1955 V104 N1 P112-117	60-13798 <+*>
1955 V104 N1 P114-	R-1663 <*>
1955 V104 N1 P141-143	C-2478 <NRC>
1955 V104 N2 P177-179	61-18302 <*=>
1955 V104 N2 P183-185	62-33208 <=*>
1955 V104 N2 P201-204	R-3471 <*>
	RT-4647 <*>
1955 V104 N2 P209-210	62-23079 <=*>
1955 V104 N2 P227-228	GB4 T/1056 <NLL>
	R-3248 <*>
	64-18389 <*>
1955 V104 N2 P239-241	62-19908 <=*>
1955 V104 N2 P249-252	60-17713 <+*>
1955 V104 N2 P260-263	59-10017 <+*>
1955 V104 N3 P376-379	61-21406 <=> 0
1955 V104 N3 P384-386	62-11584 <=>
	62-20285 <*> 0
1955 V104 N3 P389-390	62-11639 <=>
1955 V104 N3 P391-392	61-17500 <=$>
1955 V104 N3 P393-396	RT-4147 <*>
1955 V104 N3 P405-408	RT-4300 <*>
	3661 <HB>
	62-11385 <=>
1955 V104 N3 P415-417	RJ-354 <ATS>
1955 V104 N3 P422-426	88H11R <ATS>
1955 V104 N3 P440-443	C-2566 <NRC>
1955 V104 N3 P462-465	RT-3467 <*>
	59-15699 <+*>
1955 V104 N3 P843-845	61-17502 <=$>
1955 V104 N4 P209-210	R-2073 <*>
1955 V104 N4 P505-508	R-1243 <*>
1955 V104 N4 P509-512	62-23115 <=*>
1955 V104 N4 P513-516	R-4334 <*>
1955 V104 N4 P517-519	R-3928 <*>
	RT-3545 <*>
1955 V104 N4 P527-529	R-787 <*>
1955 V104 N4 P530-532	R-2072 <*>
	62-15854 <=*>
1955 V104 N4 P537-539	3236 <NLL>
1955 V104 N4 P543-545	R-4974 <*>
	3674 <HB>
1955 V104 N4 P555-558	187TM <CTT>
1955 V104 N4 P567-570	R-191 <*>
1955 V104 N4 P581-583	63-19330 <=*>
1955 V104 N4 P587-588	C-2409 <NRC>
	R-2054 <*>
1955 V104 N4 P593-596	R-4651 <*>
1955 V104 N4 P634-637	C-2106 <NLL>
1955 V104 N5 P695-698	UCRL TRANS-638 <=*>
1955 V104 N5 P710-712	RT-3906 <*>
1955 V104 N5 P717-720	RT-3479 <*>
1955 V104 N5 P721-724	63-15728 <=*>
1955 V104 N5 P746-749	RT-4305 <*>
1955 V104 N5 P779-782	C-2107 <NLL>
1955 V104 N5 P854-856	RJ-355 <ATS>
1955 V104 N6 P805-809	R-1242 <*>
1955 V104 N6 P809-812	61-18303 <*=>
1955 V104 N6 P840-842	RT-4152 <*>
1955 V104 N6 P843-845	RT-3492 <*>
	62-11406 <=>
1955 V104 N6 P850-853	C2365 <NRC>
	R-787 <*>
	62-20391 <=*>
	62-25919 <=*>
	63-26592 <=$>
1955 V104 N6 P854-856	RJ-355 <ATS>
1955 V104 N6 P857-860	R-3813 <*>
	RJ-518 <ATS>
1955 V104 N6 P872-875	59-10028 <+*>
1955 V104 N6 P876-879	R-3551 <*>
	37N56R <ATS>
1955 V104 N6 P882-885	C-2565 <NRC>
1955 V105 P46-49	R-4256 <*>
1955 V105 P115-118	T-1572 <INSD>
1955 V105 P248-249	1689 <NLL>
1955 V105 P586-587	R-1341 <*>
1955 V105 P693-695	R-1041 <*>
1955 V105 P741-743	R-242 <*>
1955 V105 P812-813	R-1658 <*>
1955 V105 P976-977	AEC-TR-4000 <+*>
1955 V105 P1252-1255	R-501 <$>
1955 V105 P1328-1331	R-2681 <*>
	R-4666 <*>
1955 V105 N1 P46-49	R-979 <*>
1955 V105 N1 P57	1278TM <CTT>
1955 V105 N1 P57-60	61-17501 <=$>
1955 V105 N1 P69-72	RT-3542 <*>
1955 V105 N1 P73-76	RT-3541 <*>
1955 V105 N1 P80	1279TM <CTT>
1955 V105 N1 P80-82	51-17499 <=$>
1955 V105 N1 P83-86	62-11199 <=> 0
1955 V105 N1 P94-95	59-10088 <+*>
1955 V105 N1 P108-111	C-3345 <NRCC>
1955 V105 N1 P119-122	RT-3599 <*>
1955 V105 N1 P123-125	01M43R <ATS>
1955 V105 N1 P126-128	R-4799 <*>
	37H12R <ATS>
1955 V105 N2 P225-228	R-1627 <*>
1955 V105 N2 P229-232	62-23119 <=*>
1955 V105 N2 P244	1280TM <CTT>
1955 V105 N2 P244-247	61-17498 <=$>
1955 V105 N2 P256-259	CSIRO-3143 <CSIR>
	R-3944 <*>
1955 V105 N2 P264-267	RT-4311 <*>
1955 V105 N2 P275-278	RCT V29 N4 P1363 <RCT>
	RT-4229 <*>
1955 V105 N2 P282-285	RJ-485 <ATS>
1955 V105 N2 P298-300	161TM <CTT>
1955 V105 N2 P301-304	C-2549 <NRC>
1955 V105 N2 P339-342	RT-4301 <*>
	59-15700 <+*>
1955 V105 N2 P367-369	NIOT-85 <NLL>

Citation	Report
1955 V105 N2 P390-392	63-24630 <=*$>
1955 V105 N3 P442-444	RT-4127 <*>
1955 V105 N3 P454-457	61-17497 <=$>
1955 V105 N3 P496-498	RJ-489 <ATS>
1955 V105 N3 P499	4797 <HB>
	59-10016 <+*>
	62-11652 <=>
1955 V105 N3 P514-516	RCT V31 N1 P27 <RCT>
1955 V105 N3 P606-609	C-2275 <NRC>
	R-4569 <*>
	R-4813 <*>
	3116 <*>
1955 V105 N4 P664-667	62-23118 <=*>
1955 V105 N4 P676-679	C-2702 <NRC>
	R-4262 <*>
1955 V105 N4 P680-682	62-24798 <=*>
1955 V105 N4 P693-695	RT-3544 <*>
1955 V105 N4 P706-708	RJ-589 <ATS>
1955 V105 N4 P706	1282TM <CTT>
1955 V105 N4 P706-708	61-17496 <=$>
1955 V105 N4 P713-715	R-4972 <*>
1955 V105 N4 P741-743	R-1174 <*>
1955 V105 N4 P744-756	RJ-527 <ATS>
1955 V105 N4 P779-	61-23315 <*=>
1955 V105 N4 P828-831	C-2481 <NRC>
1955 V105 N4 P848-851	23K24R <ATS>
1955 V105 N4 P873-876	C-3240 <NRCC>
1955 V105 N5 P931-934	R-1719 <*>
	1417TM <CTT>
1955 V105 N5 P935-938	62-11420 <=>
1955 V105 N5 P943-946	64-21710 <=>
1955 V105 N5 P947-949	R-806 <*>
1955 V105 N5 P951	1283TM <CTT>
1955 V105 N5 P951-954	61-17495 <=$>
1955 V105 N5 P955-957	807TM <CTT>
1955 V105 N5 P965-967	R-2066 <*>
	62-23032 <=*> 0
1955 V105 N5 P993-996	59-10027 <+*>
1955 V105 N5 P1000-1002	R-932 <*>
	59-16550 <+*>
	61-15313 <+*>
1955 V105 N5 P1007-1009	60-16656 <+*> 0
1955 V105 N5 P1021-1023	RT-3965 <*>
1955 V105 N5 P1024-1027	60-11606 <=>
1955 V105 N5 P1028-1030	62-14567 <=*>
1955 V105 N5 P1031-1033	162TM <CTT>
	62-16667 <=*>
1955 V105 N5 P1036-1039	R-377 <*>
1955 V105 N5 P1036-1038	65-63065 <=$>
1955 V105 N5 P1057-	61-23320 <*=>
1955 V105 N5 P1104-1105	51K25R <ATS>
1955 V105 N5 P1118-1120	R-2702 <*>
	R-407 <*>
	64-19478 <=$>
1955 V105 N6 P1170-	61-23324 <*=>
1955 V105 N6 P1184-1187	61-23076 <=*>
1955 V105 N6 P1192-1195	RT-3872 <*>
	61-17494 <=$>
1955 V105 N6 P1196-1199	64-21710 <=>
1955 V105 N6 P1200-1203	R-197 <*>
1955 V105 N6 P1204-1207	RT-3887 <*>
1955 V105 N6 P1218-1220	C-2709 <NRC>
	R-4277 <*>
1955 V105 N6 P1221-1224	RT-3814 <*>
1955 V105 N6 P1225-1228	R-384 <*>
	4103 <HB>
1955 V105 N6 P1229-1232	CSIRO-3439 <CSIR>
	R-4379 <*>
1955 V105 N6 P1238-1240	R-381 <*>
1955 V105 N6 P1252-1255	188TM <CTT>
1955 V105 N6 P1256-1257	RT-4005 <*>
1955 V105 N6 P1258-1261	RT-4010 <*>
1955 V105 N6 P1285-1288	RJ-480 <ATS>
	61-17356 <=$>
	62-25892 <=*>
1956 V106 P355-357	R-1342 <*>
1956 V106 P453-456	C2309 <NRC>
1956 V106 P565-568	R-1343 <*>
1956 V106 P805-806	R-4809 <*>
1956 V106 P822-825	T-1690 <NLL>
1956 V106 P955-958	AMST S2 V26 P5-10 <AMS>
1956 V106 N1 P23-26	62-23184 <=*>
1956 V106 N1 P27-30	62-23162 <=*>
1956 V106 N2 P211-213	R-1638 <*>
1956 V106 N2 P237-238	60-15508 <+*>
1956 V106 N2 P239-241	16H13R <ATS>
1956 V106 N2 P246-249	62-25297 <=*> 0
1956 V106 N2 P250-253	RJ-435 <ATS>
1956 V106 N2 P267-	60-15765 <+*>
1956 V106 N2 P271-274	RT-3967 <*>
1956 V106 N2 P283-285	RJ-410 <ATS>
1956 V106 N2 P295-298	R-4680 <*>
1956 V106 N2 P299-302	C-2550 <NRC>
	R-3438 <*>
1956 V106 N2 P303-306	RJ-521 <ATS>
1956 V106 N2 P311-312	36H12R <ATS>
1956 V106 N2 P328-330	RJ-376 <ATS>
1956 V106 N2 P343-344	61-15880 <*=> 0
1956 V106 N3 P385-388	63-14480 <=*>
1956 V106 N3 P401-404	CSIRO-3270 <CSIR>
	R-3715 <*>
1956 V106 N3 P422-424	64-18605 <*>
1956 V106 N3 P441-444	62-19907 <=*>
1956 V106 N3 P445-448	RJ499 <ATS>
1956 V106 N3 P453-456	R-1165 <*>
1956 V106 N3 P460-461	43J17R <ATS>
1956 V106 N3 P473-475	59-22664 <+*>
1956 V106 N3 P479-481	61L35R <ATS>
1956 V106 N3 P487-490	<HB>
	R-229 <*>
	61-17352 <=$>
1956 V106 N3 P494-496	RJ-465 <ATS>
	64-30122 <*> 0
	7082-A <K-H>
1956 V106 N3 P497-500	59-15178 <+*>
1956 V106 N3 P573-576	R-4887 <*>
	62-15045 <=*>
1956 V106 N4 P595-597	64-11871 <=> M
1956 V106 N4 P619-622	CSIRO-3368 <CSIR>
	R-3946 <*>
1956 V106 N4 P655-667	GB4 T/1009 <NLL>
1956 V106 N4 P655-657	N66-14381 <=*$>
	5055 <HB>
	64-20173 <*> C
1956 V106 N4 P679-682	R-1921 <*>
	RJ-508 <ATS>
	RS-508 <ATS>
1956 V106 N4 P687-690	GB4 T/1010 <NLL>
	63-14007 <=*>
1956 V106 N4 P691-692	R-1666 <*>
1956 V106 N4 P706-707	C-2123 <NLL>
	RT-4094 <*>
1956 V106 N5 P770-772	64-11870 <=> M
1956 V106 N5 P839-840	RT-4077 <*>
1956 V106 N5 P845-847	60-14515 <+*>
1956 V106 N5 P848-850	RJ-483 <ATS>
1956 V106 N5 P851-854	CSIR-TR.147 <CSSA>
	61-15731 <+*>
	64-10198 <=*$>
1956 V106 N5 P859-861	RJ-507 <ATS>
	61-17354 <=$>
1956 V106 N5 P904-906	C-2344 <NRC>
	R-1167 <*>
1956 V106 N5 P937-940	R-4834 <*>
1956 V106 N6 P981	GB4 T/1044 <NLL>
	R-452 <*>
	RT-4006 <*>
	4075 <HB>
1956 V106 N6 P982-985	R-1640 <*>
	R-1990 <*>
1956 V106 N6 P1011-1014	63-14665 <=*>
1956 V106 N6 P1031-1034	98H13R <ATS>
1956 V106 N6 P1046-1049	58-80 <*>
1956 V107 P105-107	62-16677 <=*>
1956 V107 P245-	R-3753 <*>
1956 V107 P249-251	R-3473 <*>
1956 V107 P289-290	R-3803 <*>
1956 V107 P329-332	R-1344 <*>
1956 V107 P449-451	R-1345 <*>
1956 V107 P463-466	R-2090 <*>

Citation	Translation
1956 V107 P516-519	R-3758 <*>
1956 V107 P737-739	R-1604 <*>
1956 V107 P757-760	R-1346 <*>
1956 V107 N1 P43-46	R-929 <*>
	RT-4137 <*>
1956 V107 N1 P47-50	59-22597 <+*>
	62-32639 <=*>
1956 V107 N1 P51-53	RT-4140 <*>
1956 V107 N1 P59-62	C-2511 <NRC>
	R-2359 <*>
1956 V107 N1 P63-66	R-928 <*>
1956 V107 N1 P85-88	R-927 <*>
	61-17622 <=$>
1956 V107 N2 P217-220	62-11373 <=>
1956 V107 N2 P221-224	62-23048 <=*>
1956 V107 N2 P262-264	64-30947 <*>
1956 V107 N2 P265-268	RJ-482 <ATS>
1956 V107 N2 P280-283	AEC-TR-5802 <=*$>
	64-20172 <*> 0
1956 V107 N2 P297-298	61-19324 <+*>
1956 V107 N2 P325-327	60-51030 <=>
1956 V107 N3 P405-407	C-2460 <NRC>
	R-2088 <*>
1956 V107 N3 P425-427	C-2354 <*>
	R-1620 <*>
1956 V107 N3 P428-431	RCT V32 N2 P562 <RCT>
	86H13R <ATS>
1956 V107 N3 P436-439	C-2221 <NLL>
1956 V107 N3 P463-466	C-2463 <NRC>
	60-13227 <+*>
1956 V107 N3 P467-470	CSIRO-3461 <CSIR>
	163-TT <CCT>
1956 V107 N3 P471-472	RJ-562 <ATS>
1956 V107 N4 P516-519	64-71371 <=>
1956 V107 N4 P533-536	IGR V2 N7 P577 <AGI>
1956 V107 N4 P551-553	60-13781 <+*>
	64-16472 <=*$>
1956 V107 N4 P555-561	R-4785 <*>
1956 V107 N5 P657-660	62-23223 <=*>
	64-16863 <=*$> 0
1956 V107 N5 P679-682	R-1465 <*>
	61-13176 <+*>
1956 V107 N5 P685-688	59-18329 <+*>
1956 V107 N5 P700-701	93H13R <ATS>
1956 V107 N5 P702-705	59L35R <ATS>
1956 V107 N5 P719-722	61-16708 <*=>
1956 V107 N5 P757-760	CSIRO-3397 <CSIR>
	R-4597 <*>
1956 V107 N6 P839-842	60L35R <ATS>
1956 V107 N6 P843-846	R-961 <*>
	164-TT <CCT>
1956 V107 N6 P863-866	3940 <HB>
1956 V107 N7 P579-580	60-23740 <+*>
1956 V108 P115-117	58-611 <*>
1956 V108 P179-182	AMST S2 V17 P369-373 <AMS>
1956 V108 P183-186	R-3757 <*>
1956 V108 P507-510	R-1347 <*>
1956 V108 P655-658	R-1348 <*>
1956 V108 P795-798	R-1349 <*>
1956 V108 P823	R-3463 <*>
1956 V108 P1004-1006	STMSP V4 P39-42 <AMS>
1956 V108 P1083-1085	3961 <HB>
1956 V108 N1 P95-98	90M45R <ATS>
1956 V108 N1 P112-115	R-4411 <*>
1956 V108 N1 P126-129	60-15823 <+*>
1956 V108 N1 P131-134	C2355 <NRC>
1956 V108 N2 P183-186	64-71239 <=>
1956 V108 N2 P205-207	59-14122 <+*> 0
1956 V108 N2 P232-235	3939 <HB>
1956 V108 N2 P251-252	64-16319 <=*$>
1956 V108 N2 P270-273	59-10549 <+*>
1956 V108 N2 P294-297	60-31131 <=>
1956 V108 N3 P385-388	R-845 <*>
	66-12667 <*>
1956 V108 N3 P423-424	CSIRO-3343 <CSIR>
1956 V108 N3 P425-426	63-20717 <=*$>
1956 V108 N3 P428-431	C-2217 <NLL>
	R-192 <*>
	RT-4613 <*>
1956 V108 N3 P451-454	60-14512 <+*>
1956 V108 N3 P465-468	54K24R <ATS>
1956 V108 N3 P481-483	94H13R <ATS>
1956 V108 N3 P511-514	R-3795 <*>
	5575-B <K-H>
	59-10145 <+*>
1956 V108 N3 P518-521	C-2529 <NRC>
	59-18678 <+*>
1956 V108 N4 P623-625	R-1737 <*>
1956 V108 N4 P627-628	<ATS>
1956 V108 N4 P629-632	62-11354 <=>
1956 V108 N4 P633-635	AEC-TR-3830 <*>
1956 V108 N4 P648-650	95H13R <ATS>
1956 V108 N4 P665-667	C-2528 <NRC>
1956 V108 N4 P711-714	16L35R <ATS>
1956 V108 N5 P799-801	AD-610 962 <=$>
	C-2506 <NRC>
1956 V108 N5 P825-828	85J14R <ATS>
1956 V108 N5 P829-832	R-4659 <*>
1956 V108 N5 P843-845	245TT <CCT> 0
1956 V108 N5 P861-863	64-16556 <=*$>
1956 V108 N5 P864-867	60-10308 <+*>
1956 V108 N5 P885-888	30H13R <ATS>
1956 V108 N5 P892-894	31H13R <ATS>
1956 V108 N6 P997-1000	R-972 <*>
1956 V108 N6 P1069-1071	R-869 <*>
1956 V108 N6 P1093-1095	24H12R <ATS>
1956 V108 N6 P1098-1101	54H12R <ATS>
1956 V108 N6 P1106-1108	65-11872 <*>
1956 V108 N6 P1117-1119	32H13R <ATS>
1956 V108 N6 P1137-1139	59-18519 <+*>
1956 V108 N6 P1152-1155	63-19592 <=*$>
1956 V108 N6 P1179-1181	59-12749 <+*>
1956 V109 P152-155	R-4219 <*>
1956 V109 P260-263	62-11591 <=>
1956 V109 P351-353	R-1131 <*>
1956 V109 P477-480	R-1169 <*>
	R-1811 <*>
1956 V109 P525-527	R-1186 <*>
1956 V109 P582-	R-1744 <*>
1956 V109 P586-588	R-3747 <*>
1956 V109 P701-703	R-4145 <*>
1956 V109 P777-780	R-1185 <*>
1956 V109 P807-810	R-1655 <*>
1956 V109 P811-812	R-1656 <*>
1956 V109 P849-850	R-1603 <*>
1956 V109 P929-930	R-1350 <*>
1956 V109 N1 P49-52	UCRL TRANS-1003(L) <=*$>
	65-61522 <=$>
1956 V109 N1 P53-56	R-429 <*>
1956 V109 N1 P77-79	30K20R <ATS>
1956 V109 N1 P113-116	165-TT <CCT>
1956 V109 N1 P124-126	4549 <HB>
1956 V109 N1 P135-138	AEC-TR-5042 <=*>
1956 V109 N1 P144-	61-15875 <*=>
1956 V109 N1 P213-216	33H13R <ATS>
1956 V109 N2 P271-273	66-11200 <*>
1956 V109 N2 P275-278	28H12R <ATS>
1956 V109 N2 P285-288	R-1245 <*>
1956 V109 N2 P289-291	CSIRO-3369 <CSIR>
	R-3901 <*>
1956 V109 N2 P312-314	65-13319 <*>
1956 V109 N2 P315-318	59-18531 <+*>
1956 V109 N2 P322-324	R-3647 <*>
1956 V109 N2 P325-328	20H12R <ATS>
1956 V109 N2 P347-350	R-1246 <*>
1956 V109 N2 P351-353	19H12R <ATS>
1956 V109 N2 P358-360	61-00440 <*>
1956 V109 N2 P387-388	R-4973 <*>
1956 V109 N2 P397-399	66-60917 <=*$>
1956 V109 N2 P400-402	60-18031 <+*>
1956 V109 N3 P469-471	61-19327 <+*> 0
1956 V109 N3 P473-476	R-731 <*>
1956 V109 N3 P519-	R-479 <*>
1956 V109 N3 P519-520	4916 <HB>
1956 V109 N3 P543-545	64-30948 <*>
1956 V109 N3 P582-585	60-17903 <+*>
1956 V109 N3 P586-588	R-246 <*>
	65H12R <ATS>
1956 V109 N3 P589-592	R-1195 <*>
	61-19917 <=*>

Citation	Translation
1956 V109 N3 P614-616	28J14R <ATS>
1956 V109 N3 P617-620	R-1607 <*>
1956 V109 N4 P687-689	64-71389 <=> M
1956 V109 N4 P774-776	62-25379 <=*>
1956 V109 N4 P777-780	PANSDOC-TR.54 <PANS>
1956 V109 N4 P799-801	R-1484 <*>
	65-61326 <=$>
1956 V109 N4 P803-806	R-4636 <*>
1956 V109 N4 P813-815	R-4784 <*>
	62-11195 <=>
1956 V109 N4 P821-823	62-11587 <=> 0
1956 V109 N4 P828-831	C-2723 <NRC>
	59-18628 <+*> 0
1956 V109 N4 P842-845	63-19589 <=*$>
1956 V109 N4 P859-861	66-61559 <=*$>
1956 V109 N5 P939-942	2301 <NRC>
	57-786 <*>
1956 V109 N5 P962-965	R-4621 <*>
1956 V109 N5 P971-974	R-666 <*>
1956 V109 N5 P979-981	R-3387 <*>
1956 V109 N5 P982-985	46H13R <ATS>
1956 V109 N5 P990-992	65-64404 <=*$>
1956 V109 N5 P1012-1014	R-4290 <*>
	59-16546 <+*>
1956 V109 N6 P1098-1101	61-14982 <=*>
1956 V109 N6 P1115-1118	63-10666 <=*>
1956 V109 N6 P1123-1125	66K20R <ATS>
1956 V109 N6 P1129-1132	94K20R <ATS>
1956 V109 N6 P1133-1135	AD-625 653 <=*$>
	59-10094 <+*>
1956 V109 N6 P1163-1166	62-16578 <=*> 0
1956 V110 P64-	R-1756 <*>
1956 V110 P79-82	61-20577 <*=> 0
1956 V110 P408-410	R-1351 <*>
1956 V110 P411-413	504TM <CTT>
1956 V110 N1 P7-10	61-15814 <+*>
	61-23140 <=*>
	62-10633 <*=>
	64-30293 <*>
1956 V110 N1 P53-56	62-16671 <=*>
1956 V110 N1 P57-60	63-27823 <$>
	64-15313 <=*$>
1956 V110 N1 P61-	R-1741 <*>
1956 V110 N1 P61-63	62-11575 <=>
1956 V110 N1 P79-82	73H13R <ATS>
1956 V110 N1 P87-88	R-791 <*>
1956 V110 N1 P89-92	64-16507 <=*$>
	70K25R <ATS>
1956 V110 N1 P101-104	RCT V32 N2 P557 <RCT>
1956 V110 N1 P112-115	47H13R <ATS>
	65-14744 <*>
1956 V110 N1 P119-121	C2345 <NRC>
	R-1166 <*>
1956 V110 N1 P125-128	R-1981 <*>
1956 V110 N2 P201-202	R-329 <*>
1956 V110 N2 P207-208	82S80R <ATS>
1956 V110 N2 P209-211	3893 <HB>
1956 V110 N3 P348-350	62-11191 <=>
1956 V110 N3 P358-361	R-678 <*>
1956 V110 N3 P362-365	R-423 <*>
1956 V110 N3 P390-392	41J15R <ATS>
1956 V110 N3 P401-403	RCT V30 N4 P1162 <RCT>
	60-16657 <+*>
1956 V110 N3 P453-456	47J15R <ATS>
1956 V110 N3 P457-460	61-19673 <=>
1956 V110 N3 P476-479	65-64366 <=*$>
1956 V110 N3 P483-486	60-13315 <+*>
1956 V110 N4 P527-530	63-26596 <=$>
1956 V110 N4 P559-561	61-16425 <*=>
	9K26R <ATS>
1956 V110 N4 P562-565	3872 <HB>
1956 V110 N4 P581-584	508TT <CCT>
	61-17385 <=$>
	62-14346 <=*> 0
1956 V110 N4 P589-592	62-14578 <=*>
1956 V110 N4 P593-596	62-19000 <=*>
1956 V110 N4 P600-602	59-22609 <+*>
1956 V110 N4 P646-650	59-15114 <+*>
1956 V110 N4 P688-691	59-14183 <+*>
1956 V110 N5 P723-726	62-11638 <=>
1956 V110 N5 P742-745	62-16440 <=*>
1956 V110 N5 P746-749	R-3621 <*>
	62-14577 <=*>
1956 V110 N5 P750-753	R-4639 <*>
1956 V110 N5 P755-757	R-436 <*>
1956 V110 N5 P765-768	60-16985 <*> 0
1956 V110 N5 P780-782	61-15275 <*+> 0
1956 V110 N5 P858-861	89K25R <ATS>
1956 V110 N5 P874-876	62-32585 <=*>
1956 V110 N6 P970-971	R-5248 <*>
1956 V110 N6 P985-988	3943 <*>
1956 V110 N6 P1015-1017	46J16R <ATS>
1956 V110 N6 P1018-1021	59-22557 <+*> 0
	60-16986 <*> 0
1956 V110 N6 P1026-1029	R-3434 <*>
1956 V110 N6 P1030-1033	R-3744 <*>
1956 V110 N6 P1133-1136	60-10997 <+*> 0
1956 V111 P486-488	R-1352 <*>
1956 V111 P561-563	R-1353 <*>
1956 V111 P745-748	R-5080 <*>
1956 V111 P1087-1090	1613 <*>
1956 V111 P1375-1377	T-2271 <INSD>
1956 V111 N1 P85-88	93K20R <ATS>
1956 V111 N1 P98-101	3953 <HB>
1956 V111 N1 P102-104	62-33210 <=*>
1956 V111 N1 P105-106	62-23274 <=*>
1956 V111 N1 P107-109	67J14R <ATS>
1956 V111 N1 P121-124	03J15R <ATS>
	59-14182 <+*>
1956 V111 N1 P136-139	59-21089 <=>
1956 V111 N1 P148-150	59-10146 <+*> 0
1956 V111 N1 P206-208	R-1566 <*>
	R-1895 <*>
1956 V111 N2 P7	R-2772 <*>
1956 V111 N2 P193-196	60-13129 <+*>
1956 V111 N2 P291-294	UCRL TRANS-965(L) <=*$>
1956 V111 N2 P312-315	59-15506 <+*>
	62-23229 <*=>
1956 V111 N2 P328-330	R-418 <*>
1956 V111 N2 P353-354	61-18789 <*=> 0
1956 V111 N2 P362-364	64-30091 <*>
1956 V111 N2 P365-368	59-16541 <+*>
1956 V111 N2 P376-379	R-4483 <*>
	22J16R <ATS>
1956 V111 N2 P384-387	60-31132 <=>
1956 V111 N2 P395-397	R-4738 <*>
	12J19R <ATS>
1956 V111 N2 P404-406	60-23991 <+*>
1956 V111 N2 P410-412	R-4668 <*>
1956 V111 N2 P449-451	97M38R <ATS>
1956 V111 N2 P470-472	R-4189 <*>
1956 V111 N3 P517-520	UCRL TRANS-550(L) <*+>
	61-14980 <=*>
1956 V111 N3 P554-556	99J14R <ATS>
1956 V111 N3 P557-559	05J15R <ATS>
1956 V111 N3 P561-563	62-20384 <=*>
1956 V111 N3 P605-608	59-10557 <+*>
1956 V111 N3 P613-616	61-15719 <+*>
1956 V111 N3 P621-622	SCL-T-208 <+*>
	85J16R <ATS>
1956 V111 N3 P637-639	60-18778 <+*>
1956 V111 N3 P640-643	PANSDOC-TR.31 <PANS>
1956 V111 N3 P644-646	R-4657 <*>
1956 V111 N3 P659-662	64-18229 <=*$>
1956 V111 N3 P687-689	R-1596 <*>
1956 V111 N3 P706-708	R-1546 <*>
	64-19431 <$=> 0
1956 V111 N4 P745-748	R-810 <*>
1956 V111 N4 P753-756	R-844 <*>
	61-10782 <=*>
1956 V111 N4 P763-786	98J14R <ATS>
1956 V111 N4 P777-779	R-3448 <*>
1956 V111 N4 P780-782	R-1433 <*>
1956 V111 N4 P821-823	6J15R <ATS>
1956 V111 N4 P824-826	60-14513 <+*>
1956 V111 N4 P851-854	61-13795 <*+>
1956 V111 N4 P855-858	59-2228 <+*>
1956 V111 N4 P884-886	59-10142 <+*>
1956 V111 N5 P969-971	R-1240 <*>
	97J14R <ATS>

Citation	Translation
1956 V111 N5 P981-984	60-13800 <+*>
1956 V111 N5 P1000-10003	R-1239 <*>
1956 V111 N5 P1014-1016	51K20R <ATS>
1956 V111 N5 P1039-1041	<ATS>
1956 V111 N5 P1042-1044	64-16220 <=*$>
1956 V111 N5 P1098-1100	50K25R <ATS>
1956 V111 N6 P1171-1174	62-23220 <=*>
1956 V111 N6 P1193-1196	62-20390 <=*>
1956 V111 N6 P1197-1200	60-13801 <+*>
1956 V111 N6 P1219-1222	R-3457 <*>
1956 V111 N6 P1234-1237	62-11268 <=>
1956 V111 N6 P1252-1254	59-19624 <+*>
1956 V111 N6 P1257-1259	R-1498 <*>
1956 V111 N6 P1278-1281	62-16665 <=*>
	73N50R <ATS>
1956 V111 N6 P1286-1289	43K28R <ATS>
1956 V111 N6 P1403-1405	AD-625 625 <=*$>
1957 V112 P124-126	R-1610 <*>
	R-4664 <*>
1957 V112 P206-	R-3755 <*>
1957 V112 P249-252	R-4070 <*>
1957 V112 P422-424	R-4451 <*>
1957 V112 P449-452	RCT V31 N2 P348 <RCT>
1957 V112 P599-602	AMST S2 V33 P41-46 <AMS>
1957 V112 P623-625	R-5006 <*>
1957 V112 P645-648	R-2672 <*>
1957 V112 P655-658	R-1648 <*>
1957 V112 P772-773	R-1852 <*>
1957 V112 P880-881	R-1650 <*>
1957 V112 P911-914	R-1354 <*>
1957 V112 P1079-1081	R-1853 <*>
1957 V112 N1 P16-19	R-948 <*>
1957 V112 N1 P29	R-3456 <*>
1957 V112 N1 P55-57	62-23044 <=*>
1957 V112 N1 P62-65	62-23207 <=*>
1957 V112 N1 P70-72	60-16987 <*> 0
1957 V112 N1 P104-106	4002 <HB>
1957 V112 N2 P206-	R-1164 <*>
1957 V112 N2 P206	64-71438 <=>
1957 V112 N2 P207-210	R-3085 <*>
1957 V112 N2 P211-212	61-13079 <*+>
1957 V112 N2 P213-216	61-13090 <*+>
1957 V112 N2 P217-220	R-4922 <*>
1957 V112 N2 P228-231	R-5180 <*>
	61-15616 <+*>
1957 V112 N2 P253-256	62-11636 <=>
1957 V112 N2 P283-286	61-13166 <+*>
1957 V112 N3 P405-406	R-1985 <*>
1957 V112 N3 P407-410	29K20R <ATS>
1957 V112 N3 P449-452	RCT V31 N4 P751 <RCT>
1957 V112 N3 P465-466	59-17787 <+*>
1957 V112 N3 P467-469	59-19279 <+*>
1957 V112 N3 P519-521	R-4667 <*>
1957 V112 N4 P587-590	60-21963 <=>
1957 V112 N4 P607-610	58K28R <ATS>
1957 V112 N4 P626-627	R-2535 <*>
	67K20R <ATS>
1957 V112 N4 P628-631	AD-610 958 <=$>
	C-2337 <NRC>
	R-1179 <*>
1957 V112 N4 P636-639	62-24126 <=*>
1957 V112 N4 P662-665	63-14090 <=*> 0
1957 V112 N4 P669-672	59-15747 <+*>
	88J16R <ATS>
1957 V112 N4 P692-695	64-20194 <*>
1957 V112 N4 P724-727	4465 <HB>
1957 V112 N4 P763-765	04J17R <ATS>
1957 V112 N5 P815-818	R-1488 <*>
1957 V112 N5 P846-848	AD-610 958 <=$>
	C-2337 <NRC>
	R-3470 <*>
1957 V112 N5 P864-867	62-25380 <=*>
1957 V112 N5 P886-889	R-3774 <*>
	41K28R <ATS>
	61-23513 <*=> 0
1957 V112 N5 P957-960	M.5139 <NLL>
	R-5281 <*>
1957 V112 N6 P1005-1007	62-15881 <=*>
1957 V112 N6 P1012-1015	R-1373 <*>
1957 V112 N6 P1033-1036	AD-611 033 <=$>
	95K20R <ATS>
1957 V112 N6 P1043-1046	59-15719 <+*>
1957 V112 N6 P1050-1052	95J16R <ATS>
1957 V112 N6 P1071-1074	16K23R <ATS>
1957 V112 N6 P1098-1100	62-16663 <=*>
1957 V113 P112-115	R-1645 <*>
1957 V113 P116-119	R-1646 <*>
1957 V113 P138-141	60-23581 <+*>
1957 V113 P436-439	62-15010 <=*>
1957 V113 P454-457	60-21128 <=>
1957 V113 P646-649	R-1854 <*>
1957 V113 P695-698	R-1855 <*>
1957 V113 P742-745	61-10649 <*=> P
1957 V113 P752-755	STMSP V4 P183-187 <AMS>
1957 V113 P832-835	R-2673 <*>
1957 V113 P1203-1205	AMST S2 V32 P311-314 <AMS>
1957 V113 P1324-1327	R-2120 <*>
1957 V113 N1 P28-31	64-16162 <=*$>
1957 V113 N1 P39-42	R-1635 <*>
	R-3544 <*>
	60-18779 <+*>
1957 V113 N1 P62-64	R-4650 <*>
1957 V113 N1 P94-96	R-1375 <*>
1957 V113 N1 P123-126	R-4119 <*>
1957 V113 N1 P134-137	59-14305 <+*>
1957 V113 N1 P152-155	59-19138 <+*>
1957 V113 N2 P258-260	R-1783 <*>
1957 V113 N2 P283-286	R-1841 <*>
	R-3542 <*>
	61-10173 <*+>
1957 V113 N2 P291-293	62-19889 <=*>
1957 V113 N2 P297-300	R-1782 <*>
1957 V113 N2 P307-310	AD-610 962 <=$>
	C-2499 <NRC>
	R-2659 <*>
1957 V113 N2 P391-394	15J19R <ATS>
1957 V113 N2 P436-439	R-4828 <*>
1957 V113 N3 P339-342	61-17521 <=$>
1957 V113 N3 P509-512	61-15314 <=*>
	62-23057 <=*>
1957 V113 N3 P544-547	R-1549 <*>
1957 V113 N3 P548-549	62-24751 <=*>
1957 V113 N3 P556-559	59-20209 <+*>
	63-16636 <=*>
1957 V113 N3 P564-566	4410 <HB>
1957 V113 N3 P571-572	R-2322 <*>
	9K20R <ATS>
1957 V113 N3 P581-584	94J16R <ATS>
1957 V113 N3 P588-589	26J17R <ATS>
1957 V113 N3 P624-626	45J17R <ATS>
1957 V113 N3 P627-630	R-3791 <*>
1957 V113 N3 P635-638	60-17450 <+*>
1957 V113 N3 P688-691	R-4829 <*>
1957 V113 N3 P695-698	R-3726 <*>
	64-19442 <=$> C
1957 V113 N4 P731-733	59-17803 <+*>
1957 V113 N4 P738-741	R-3545 <*>
1957 V113 N4 P738-	59-11934 <=>
1957 V113 N4 P746-747	R-1969 <*>
	R-3541 <*>
	59-16342 <+*>
1957 V113 N4 P760-761	60-18780 <+*>
1957 V113 N4 P762-765	R-2015 <*>
1957 V113 N4 P777-779	R-1955 <*>
1957 V113 N4 P784-786	61-16706 <*=>
1957 V113 N4 P787-790	PB 141 218T <=>
1957 V113 N4 P795-796	61-19362 <+*>
1957 V113 N4 P803-805	65-11469 <*>
1957 V113 N4 P824-827	59R74R <ATS>
1957 V113 N4 P842-845	65-64368 <=*$>
1957 V113 N4 P869-872	60-13174 <+*>
1957 V113 N5 P951-954	59-20208 <+*>
1957 V113 N5 P959-961	R-1998 <*>
1957 V113 N5 P970-973	61-10650 <*+>
1957 V113 N5 P974-976	R-2018 <*>
1957 V113 N5 P987-990	59-18263 <+*>
1957 V113 N5 P1006-1009	60-15073 <+*> 0
	61-13340 <*+>
1957 V113 N5 P1010-1012	59-12222 <+*>
	61-13585 <+*>

1957 V113 N5 P1C23-1024 R-4239 <*>
1957 V113 N5 P1050-1052 59-10273 <+*>
1957 V113 N5 P1057-1060 61-19436 <+*> 0
1957 V113 N5 P1070-1072 4086 <HB>
1957 V113 N5 P1130-1132 59-14719 <+*>
1957 V113 N5 P1163-1164 R-4705 <*>
1957 V113 N6 P1293-1294 R-4342 <*>
64-18793 <*> 0
1957 V114 P11-13 AMST S2 V24 P9-12 <AMS>
1957 V114 P29-32 AMST S2 V29 P289-293 <AMS>
1957 V114 P199-202 59-15905 <+*>
1957 V114 P375-378 R-2121 <*>
1957 V114 P634-636 R-4661 <*>
1957 V114 P662-665 R-2122 <*>
1957 V114 P679-681 AMST S2 V28 P51-54 <AMS>
1957 V114 P751-753 R-2123 <*>
1957 V114 P953-956 AMST S2 V28 P55-59 <AMS>
1957 V114 P1005-1007 64-13762 <=*$>
1957 V114 P1066-1069 R-2124 <*>
1957 V114 P1080-1083 R-2125 <*>
1957 V114 P1195- R-4067 <*>
1957 V114 N1 P21-24 R-2110 <*>
1957 V114 N1 P67-69 R-4018 <*>
1957 V114 N1 P86-97 64-21710 <=>
1957 V114 N1 P106-109 60-14514 <+*>
60-17146 <+*>
1957 V114 N1 P113-115 RJ-871 <ATS>
61-10047 <*+>
61-17360 <=$>
1957 V114 N1 P139-142 60-13826 <+*>
1957 V114 N1 P143-145 R-3213 <*>
1957 V114 N1 P150-153 R-3727 <*>
64-19420 <=$>
1957 V114 N2 P271-471 C-5150 <NRC>
1957 V114 N2 P271-274 62-23198 <=*>
1957 V114 N2 P293-296 R-3080 <*>
4268 <HB>
1957 V114 N2 P311-313 60-13047 <+*>
64-18510 <*>
1957 V114 N2 P335-338 64-30951 <*>
1957 V114 N2 P343-346 61-13347 <+*>
1957 V114 N2 P365-368 59-10274 <+*>
1957 V114 N2 P417-418 71J18R <ATS>
1957 V114 N3 P494-497 R-3210 <*>
1957 V114 N3 P513-516 R-1780 <*>
1957 V114 N3 P545-548 R-4242 <*>
59-18370 <+*>
1957 V114 N3 P568-570 60-15695 <+*>
1957 V114 N3 P575-578 6K23R <ATS>
61-17353 <=$>
1957 V114 N3 P586-589 RCT V31 N4 P747 <RCT>
1957 V114 N3 P594-597 <NRC>
1957 V114 N3 P620-622 72J18R <ATS>
1957 V114 N3 P645-647 73J18R <ATS>
1957 V114 N3 P648-651 74J18R <ATS>
1957 V114 N3 P662-665 59-1C392 <+*>
1957 V114 N4 P721-724 62-23229 <*=>
1957 V114 N4 P737-740 62-11640 <=>
1957 V114 N4 P745-747 61-28201 <*=>
1957 V114 N4 P751-753 UCRL TRANS-375 <+*>
1957 V114 N4 P757-759 R-4358 <*>
1957 V114 N4 P800-802 64-30946 <*>
1957 V114 N4 P844-847 63-16981 <=*>
1957 V114 N5 P930-933 R-3133 <*>
1957 V114 N5 P949-952 R-2357 <*>
1957 V114 N5 P953-956 R-3134 <*>
65-11006 <*>
1957 V114 N5 P968-971 R-4780 <*>
1957 V114 N5 P976-979 R-4173 <*>
1957 V114 N5 P997-1000 59-10276 <+*>
1957 V114 N5 P1C01-1003 4338 <HB>
1957 V114 N5 P1008-1011 59-10385 <+*>
1957 V114 N5 P1C21-1024 65-10764 <*>
1957 V114 N5 P1046-1048 R-3130 <*>
1957 V114 N5 P1053-1057 60-15567 <+*>
1957 V114 N5 P1C66-1069 59-1C387 <+*>
1957 V114 N5 P1113-1115 59-10143 <+*>
59-15748 <+*>
1957 V114 N6 P594-597 R-4292 <*>
1957 V114 N6 P1185-1188 59-20216 <+*>
1957 V114 N6 P1199-1202 R-5162 <*>
1957 V114 N6 P1203-1205 R-5163 <*>
61-15614 <+*>
1957 V114 N6 P1220-1223 4305 <HB>
1957 V114 N6 P1224-1227 4293 <HB>
1957 V114 N6 P1228-1230 59-22468 <+*>
1957 V114 N6 P1231-1234 60-13313 <+*>
1957 V114 N6 P1254-1265 R-3127 <*>
1957 V114 N6 P1269-1271 R-3674 <*>
1957 V115 P100-102 R-4216 <*>
1957 V115 P190-192 R-4305 <*>
1957 V115 P462-465 AMST S2 V47 P31-35 <AMS>
1957 V115 P516-517 RCT V33 N3 P636 <RCT>
1957 V115 P530-533 R-4304 <*>
1957 V115 P623-625 R-4303 <*>
1957 V115 P674-676 AMST S2 V37 P283-285 <AMS>
1957 V115 P747-750 RCT V33 N1 P240 <RCT>
1957 V115 P1111-1114 R-4008 <*>
1957 V115 P1142-1145 R-5133 <*>
1957 V115 N1 P78-79 63-19767 <=*>
1957 V115 N1 P84-87 R-2311 <*>
59-11569 <=> 0
1957 V115 N1 P88-90 62-16436 <=*>
1957 V115 N1 P91-93 R-2669 <*>
1957 V115 N1 P97-99 50K20R <ATS>
64-20171 <*> 0
1957 V115 N1 P100-102 SCL-T-464 <=*>
1957 V115 N1 P103-106 40K20R <ATS>
1957 V115 N1 P110-113 6C-16940 <*> 0
1957 V115 N1 P126-129 64-10197 <=*$>
1957 V115 N2 P267-270 59-11871 <=>
61K23R <ATS>
1957 V115 N2 P308-311 59-22712 <+*>
60-15573 <+*>
64-18794 <*>
1957 V115 N3 P431-433 AD-614 917 <=$>
R-3121 <*>
61-13050 <+*>
65-13269 <*>
1957 V115 N3 P441-444 R-4726 <*>
64-71365 <=>
1957 V115 N3 P445-446 61-13341 <+*>
1957 V115 N3 P454-457
AMS-TRANS V42 P13-17 <AMS>
1957 V115 N3 P473-474 R-4170 <*>
1957 V115 N3 P483-485 37K23R <ATS>
1957 V115 N3 P486-487 64-20170 <*> 0
1957 V115 N3 P497-500 65-14342 <*>
1957 V115 N3 P501-503 64-21710 <=>
1957 V115 N3 P516-517 R-4245 <*>
6K20R <ATS>
60-13317 <+*>
61-20629 <*=>
64-18807 <*>
1957 V115 N3 P537-540 63-14334 <=*>
1957 V115 N3 P545-547 R-4247 <*>
1957 V115 N3 P548-551 AEC-TR-5799 <=*$>
1957 V115 N3 P552-553 59-22234 <+*>
62-18384 <=*>
1957 V115 N3 P580-582 60-17790 <+*> 0
1957 V115 N3 P583-585 R-2668 <*>
60-10066 <+*> 0
1957 V115 N3 P586-587 49L29R <ATS>
1957 V115 N4 P639-642 R-4246 <*>
1957 V115 N4 P681-683 C-2637 <NRC>
62-23230 <*>
1957 V115 N4 P702-705 R-4830 <*>
1957 V115 N4 P710-713 60-15677 <+*>
1957 V115 N4 P731-733 59-22235 <+*>
63P60R <ATS>
1957 V115 N4 P734-736 65-61282 <=$>
1957 V115 N4 P747-750 60-13051 <+*>
1957 V115 N4 P768-770 R-4176 <*>
1957 V115 N4 P813-815 39K24R <ATS>
1957 V115 N4 P816-818 40K24R <ATS>
1957 V115 N4 P833-836 62-14705 <=*>
1957 V115 N5 P855-857 R-3118 <*>
1957 V115 N5 P887-890 59-20212 <+*>
1957 V115 N5 P901-903 R-4431 <*>
59-12045 <+*>

Citation	Translation
1957 V115 N5 P934-937	60-17451 <+*>
1957 V115 N5 P942-945	R-4692 <*>
	41J19R <ATS>
	60-15698 <+*>
1957 V115 N5 P999-1001	62K21R <ATS>
1957 V115 N6 P1080-1083	R-5104 <*>
1957 V115 N6 P1104-1106	59-20211 <+*>
1957 V115 N6 P1107-1110	61-19522 <=*>
1957 V115 N6 P1126-1128	62-25286 <=*>
1957 V115 N6 P1129-1130	R-4231 <*>
	62-25292 <=*>
1957 V115 N6 P1165-1168	03K28R <ATS>
	60-17234 <+*>
1957 V115 N6 P1193-1196	60-15430 <+*>
1957 V115 N6 P1197-1199	61-14965 <=*>
	62-16425 <=*>
1957 V116 P78-80	R-4742 <*>
1957 V116 P105-108	RCT V32 N1 P118 <RCT>
1957 V116 P157-160	R-4301 <*>
1957 V116 P197-199	UCRL TRANS-1176(L) <*>
1957 V116 P813-816	RCT V32 N1 P278 <RCT>
1957 V116 N1 P9-11	61-23141 <=*>
	64-30292 <*>
1957 V116 N1 P15-17	64-11867 <=>
1957 V116 N1 P56-59	R-3278 <*>
1957 V116 N1 P71-73	R-5155 <*>
1957 V116 N1 P74-77	R-5311 <*>
1957 V116 N1 P85-88	52K20R <ATS>
1957 V116 N1 P101-104	59-22345 <+*>
	64-18795 <*>
1957 V116 N1 P105-108	59-10272 <+*>
1957 V116 N1 P153-156	60-17449 <+*>
1957 V116 N2 P193-196	62-11625 <=>
1957 V116 N2 P203-206	61-10131 <*+>
1957 V116 N2 P217-220	C-2578 <NRC>
	R-3451 <*>
1957 V116 N2 P236-238	39K21R <ATS>
	60-17147 <+*>
1957 V116 N2 P244-247	64-18806 <*>
1957 V116 N2 P251-254	59-10277 <+*>
1957 V116 N2 P277-279	63-13581 <=*>
1957 V116 N2 P301-303	02K28R <ATS>
1957 V116 N3 P365-368	62-11641 <=>
1957 V116 N3 P373-376	59-18755 <+*>
1957 V116 N3 P415-418	4557 <HB>
1957 V116 N3 P422-424	39K23R <ATS>
1957 V116 N3 P443-446	63K21R <ATS>
1957 V116 N3 P447-450	64K21R <ATS>
1957 V116 N3 P474-476	60-17447 <+*>
1957 V116 N4 P538-541	PB 141 459T <=>
1957 V116 N4 P609-612	R-4228 <*>
	R-5192 <*>
1957 V116 N4 P637-640	62-00195 <*>
1957 V116 N4 P671-672	65K21R <ATS>
1957 V116 N4 P684-687	61-28214 <*=>
1957 V116 N4 P710-712	CSIR-TR 277 <CSSA>
1957 V116 N5 P766-768	C-2698 <INSD>
1957 V116 N6 P913-916	59-11247 <=>
1957 V116 N6 P917-919	R-3436 <*>
	61-10179 <*+>
1957 V116 N6 P920-922	R-3857 <*>
1957 V116 N6 P933-936	59-11995 <=>
	60-12492 <=>
1957 V116 N6 P949-951	C-2696 <NRC>
	R-4263 <*>
1957 V116 N6 P956-959	R-3435 <*>
1957 V116 N6 P961-964	R-4902 <*>
	60-19566 <+*>
1957 V116 N6 P965-968	68K23R <ATS>
1957 V116 N6 P976-978	4273 <HB>
	60-15597 <+*>
1957 V116 N6 P979-982	43K23R <ATS>
1957 V116 N6 P983-985	R-4988 <*>
1957 V116 N6 P986-989	61-17357 <=$>
1957 V116 N6 P990-993	R-4226 <*>
1957 V116 N6 P990	R-4832 <*>
1957 V116 N6 P990-993	59-15115 <+*>
1957 V116 N6 P1009-1011	05K28R <ATS>
1957 V116 N34 P22-424	R-4227 <*>
1957 V117 P44-46	AMST S2 V25 P1-4 <AMS>
1957 V117 P106-109	R-4298 <*>
1957 V117 P117-120	R-4642 <*>
1957 V117 P174-176	R-3493 <*>
1957 V117 P184-187	AMST S2 V25 P5-9 <AMS>
1957 V117 P359-362	R-3562 <*>
1957 V117 P605-608	R-4297 <*>
1957 V117 P745-747	AMST S2 V17 P365-367 <AMS>
1957 V117 P987-989	R-5197 <*>
1957 V117 N1 P9-12	R-3854 <*>
1957 V117 N1 P44-46	62-10171 <=*>
1957 V117 N1 P47-49	59-16057 <+*> 0
1957 V117 N1 P57-60	R-4928 <*>
	61-15618 <*+>
1957 V117 N1 P78-80	59-16057 <+*> 0
1957 V117 N1 P92-94	R-4175 <*>
1957 V117 N1 P102-105	R-3459 <*>
1957 V117 N1 P111-114	06K28R <ATS>
1957 V117 N1 P138-141	R-3799 <*>
	64-19444 <=$>
1957 V117 N2 P167-170	R-4024 <*>
1957 V117 N2 P184-187	62-10172 <=*>
1957 V117 N2 P191-194	R-3414 <*>
1957 V117 N2 P199-202	60-12495 <=>
1957 V117 N2 P225-226	R-5041 <*>
1957 V117 N3 P380-383	R-5106 <*>
1957 V117 N3 P433-436	60-19567 <+*>
1957 V117 N4 P576-577	59-17804 <+*>
1957 V117 N4 P578-581	CSIR-TR.156 <CSSA>
1957 V117 N4 P582-585	59-19472 <+*>
1957 V117 N4 P609-615	59-19472 <+*>
1957 V117 N4 P619-622	59-10314 <+*>
1957 V117 N4 P635-637	61-20628 <*=>
	65-61303 <=$>
1957 V117 N4 P638-640	R-5283 <*>
1957 V117 N4 P641-644	R-4548 <*>
	61-15380 <+*>
1957 V117 N4 P645-647	59-18740 <+*>
1957 V117 N4 P655-657	59-18751 <+*> 0
1957 V117 N4 P675-677	59-10078 <+*>
1957 V117 N5 P735-738	64-16380 <=*$>
1957 V117 N5 P739-741	R-3492 <*>
1957 V117 N5 P745-747	R-3707 <*>
1957 V117 N5 P777-780	60-12494 <=>
	62-11312 <=>
1957 V117 N5 P788-791	R-3721 <*>
1957 V117 N5 P802-803	R-4733 <*>
1957 V117 N5 P804-807	R-4792 <*>
1957 V117 N5 P823-825	60-21018 <=>
1957 V117 N5 P826-828	R-5037 <*>
1957 V117 N5 P829-832	R-4914 <*>
1957 V117 N5 P881-884	61-10768 <+*>
1957 V117 N5 P896-899	41K24R <ATS>
1957 V117 N5 P910-913	60-23770 <+*>
1957 V117 N6 P931-934	65-60924 <=$>
1957 V117 N6 P938-986	62-23053 <=>
1957 V117 N6 P947-948	62-19828 <=>
1957 V117 N6 P953-955	59-18729 <+*>
1957 V117 N6 P959-962	CSIR-TR.148 <CSSA>
1957 V117 N6 P963-964	59-12180 <+*>
1957 V117 N6 P971-974	M.5107 <NLL>
1957 V117 N6 P979-982	59-12180 <+*>
1957 V117 N6 P1003-1006	R-5082 <*>
1957 V117 N6 P1007-1009	15K23R <ATS>
1957 V117 N6 P1034-1036	63-18498 <=*>
1957 V117 N6 P1077-1080	59-10529 <+*> 0
1957 V117 N98 P101-	R-4472 <*>
1958 V118 P472-475	R-5314 <*>
1958 V118 P520-522	R-5142 <*>
1958 V118 P778-781	R-5254 <*>
1958 V118 P935-937	R-5315 <*>
1958 V118 P1146-1149	R-5253 <*>
1958 V118 N1 P17-19	64-30707 <*>
1958 V118 N1 P92-95	36K22R <ATS>
1958 V118 N1 P125-127	35K22R <ATS>
1958 V118 N1 P128-131	60-17794 <+*>
1958 V118 N1 P132-134	97K22R <ATS>
1958 V118 N1 P135-138	62-14686 <=*> 0
1958 V118 N1 P149-152	85K23R <ATS>
1958 V118 N1 P171-173	98K27R <ATS>
1958 V118 N1 P189-191	R-4452 <*>

64-19473 <=$> 0
1958 V118 N2 P269-272 R-4031 <*>
1958 V118 N2 P273-276 71K28R <ATS>
1958 V118 N2 P280-283 R-5306 <*>
1958 V118 N2 P309-311 59-18159 <+*>
60-16655 <+*>
7K25R <ATS>
1958 V118 N2 P328-330 41L29R <ATS>
1958 V118 N3 P427-430 R-4244 <*>
1958 V118 N3 P483-484 R-3887 <*>
1958 V118 N3 P488-491 60-13107 <+*>
1958 V118 N3 P493-496 AEC-TR-3496 <+*>
1958 V118 N3 P505-508 59-10313 <+*>
1958 V118 N3 P509-511 59-10315 <+*>
1958 V118 N3 P515-516 AEC-TR-3523 <+*>
60-17160 <+*>
1958 V118 N3 P523-525 R-4028 <*>
R-5181 <*>
61-20684 <*=>
1958 V118 N3 P534-536 61-19545 <=*>
1958 V118 N3 P546-548 62-13599 <*=>
1958 V118 N3 P573-576 65-13594 <*>
84K24R <ATS>
1958 V118 N4 P646-649 R-4779 <*>
1958 V118 N4 P683-686 59-22445 <+*>
1958 V118 N4 P713-715 R-5183 <*>
22K23R <ATS>
1958 V118 N4 P720-722 4409 <HB>
1958 V118 N4 P738-739 59-22410 <+*>
1958 V118 N4 P744-746 R-5178 <*>
61-15615 <*+>
1958 V118 N4 P782-784 61-13106 <*+>
1958 V118 N4 P803-806 61-15874 <*=>
1958 V118 N4 P815-817 59-19387 <+*> 0
1958 V118 N4 P819-822 00K28R <ATS>
1958 V118 N5 P884-887 R-4924 <*>
59-10917 <+*>
59-20193 <+*>
1958 V118 N5 P913-916 R-5203 <*>
64-71627 <=>
1958 V118 N5 P950-953 K25R <ATS>
1958 V118 N5 P977-979 59-22439 <+*>
1958 V118 N5 P980-982 62-16711 <=*>
94L28R <ATS>
1958 V118 N5 P1031-1033 59-19604 <+*>
1958 V118 N5 P1034-1035 97K27R <ATS>
1958 V118 N6 P1053-1055 R-4015 <*>
1958 V118 N6 P1094-1097 59-20192 <+*>
63-14668 <=*>
1958 V118 N6 P1124-1127 R-5171 <*>
1958 V118 N6 P1138-1141 15K24R <ATS>
1958 V119 P23-26 STMSP V4 P147-151 <AMS>
1958 V119 P125-128 R-5256 <*>
1958 V119 P219-222 AMST S2 V48 P151-155 <AMS>
1958 V119 P333-335 R-5255 <*>
1958 V119 P515-517 R-5139 <*>
1958 V119 P652-654 STMSP V2 P171-174 <AMS>
1958 V119 P982-985 RCT V33 N3 P623 <RCT>
1958 V119 P1088-1091 AMST S2 V47 P37-41 <AMS>
1958 V119 N1 P9- 59-11253 <=>
1958 V119 N1 P23-26 62-23219 <=>
1958 V119 N1 P79-82 5569 <HB>
1958 V119 N1 P101-103 97P59R <ATS>
1958 V119 N1 P117-120 AEC-TR-3416 <+*>
1958 V119 N2 P215-218 61-10444 <+*>
1958 V119 N2 P257-260 62-11474 <=>
1958 V119 N2 P267 59-19170 <+*>
1958 V119 N2 P278-281 62-11445 <=>
1958 V119 N2 P282-284 R-5172 <*>
65-61198 <=>
1958 V119 N2 P302-304 61-17638 <=$>
1958 V119 N2 P307-310 AEC-TR-3417 <+*>
1958 V119 N2 P322-325 32K27R <ATS>
1958 V119 N2 P336-338 AEC-TR-3498 <+*>
1958 V119 N2 P339-341 62-10260 <*> 0
1958 V119 N3 P478-480 59-22623 <+*>
1958 V119 N3 P488-489 62-18925 <=*>
1958 V119 N3 P498-500 4261 <HB>
1958 V119 N3 P501-503 4269 <HB>
1958 V119 N3 P511-514 60-16575 <+*>
1958 V119 N3 P523-525 AEC-TR-3734 <+*> 0
1958 V119 N3 P533-535 60-10910 <+*>
1958 V119 N3 P575-578 59-19068 <+*> 0
1958 V119 N3 P617-620 59-15158 <+*>
1958 V119 N4 P636-639 62-14831 <=*>
1958 V119 N4 P644-647 61-16466 <*=>
1958 V119 N4 P694-697 M5135 <NLL> 0
1958 V119 N4 P705-707 74M44R <ATS>
1958 V119 N4 P716-719 59-15116 <+*>
59-20144 <+*>
68K26R <ATS>
1958 V119 N4 P720-723 60-16555 <+*>
1958 V119 N4 P727-730 71K26R <ATS>
1958 V119 N4 P753-755 59-15160 <+*>
1958 V119 N4 P776-778 04K28R <ATS>
1958 V119 N5 P847-850 R-4710 <*>
1958 V119 N5 P851-853 R-4627 <*>
1958 V119 N5 P861-864 R-4628 <*>
1958 V119 N5 P922-925 58K24R <ATS>
1958 V119 N5 P926-928 59-18852 <+*>
1958 V119 N5 P936-937 59-20140 <+*>
1958 V119 N5 P957-960 60-16574 <+*>
1958 V119 N5 P971-974 66-60458 <=*$>
1958 V119 N5 P982-985 60-10065 <+*>
1958 V119 N5 P1006-1008 60-17241 <+*>
61-15964 <*=>
1958 V119 N6 P1070-1073 61-23560 <*=>
1958 V119 N6 P1106- 59-11938 <=>
1958 V119 N6 P1110-1112 62-11206 <=>
1958 V119 N6 P1149-1151 63R79R <ATS>
1958 V119 N6 P1159-1161 59-10229 <+*>
1958 V119 N6 P1167-1169 59-18859 <+*>
1958 V119 N6 P1170-1173 59-22168 <+*>
1958 V119 N6 P1174-1176 26K26R <ATS>
1958 V119 N6 P1177-1179 59-13117 <=> 0
1958 V119 N6 P1187-1190 80K27R <ATS>
1958 V119 N6 P1225-1258 59-18231 <+*> 0
1958 V120 P25-28 AMST S2 V42 P19-23 <AMS>
1958 V120 P581-584 R-5269 <*>
1958 V120 P764-767 R-5270 <*>
1958 V120 P1021-1023 R-5300 <*>
1958 V120 P1062-1064 RCT V32 N2 P527 <RCT>
1958 V120 P1141-1143 R-5272 <*>
1958 V120 P1263-1266 R-5173 <*>
1958 V120 P1267-1270 R-5157 <*>
1958 V120 P1307-1310 R-5267 <*>
1958 V120 N1 P9-12 R-4823 <*>
1958 V120 N1 P21-24 61-20006 <*=>
1958 V120 N1 P25-28 AMS-TRANS V42 P19-23 <AMS>
1958 V120 N1 P43-46 60-15104 <+*>
1958 V120 N1 P103-106 60-13180 <+*>
1958 V120 N1 P107-110 R-5160 <*>
1958 V120 N1 P148-150 24K25R <ATS>
1958 V120 N1 P187-190 AEC-TR-3468 <+*>
1958 V120 N1 P215-218 59-15366 <+*>
1958 V120 N1 P219-222 59-18232 <+*> 0
1958 V120 N2 P231-234 R-4714 <*>
R-5028 <*>
1958 V120 N2 P239-241 59-10394 <+*>
1958 V120 N2 P297-300 41K27R <ATS>
59-20792 <+*>
1958 V120 N2 P314-315 59-18510 <+*>
1958 V120 N2 P323-325 25K25R <ATS>
1958 V120 N2 P336-338 61-10355 <*+>
1958 V120 N2 P346-348 59-11441 <=> 0
59-11862 <=>
1958 V120 N2 P372-375 59-15707 <+*>
60-17686 <+*>
1958 V120 N2 P405-408 66M40R <ATS>
1958 V120 N3 P485-486 62-11309 <=>
1958 V120 N3 P544-547 59-18406 <+*>
1958 V120 N3 P552-553 64M41R <ATS>
1958 V120 N3 P633-636 01K28R <ATS>
1958 V120 N3 P641-643 27K27R <ATS>
1958 V120 N3 P677-680 63-20943 <=*$>
1958 V120 N4 P693-696 59-21115 <=>
1958 V120 N4 P734-737 60-27022 <*+> 0
1958 V120 N4 P743-746 62-18809 <=*>
1958 V120 N4 P761-763 CSIR-TR.171 <CSSA>
59-13248 <=>

Citation	Translation
1958 V120 N4 P789-792	59-10893 <+*>
	59-17403 <+*>
1958 V120 N4 P819-822	TS-1398 <BISI>
	62-24231 <=*>
1958 V120 N4 P830-833	62K26R <ATS>
1958 V120 N4 P834-837	84M44R <ATS>
1958 V120 N4 P853-856	60-17685 <+*>
1958 V120 N4 P875-878	62-13281 <*=>
1958 V120 N4 P879-881	71L30R <ATS>
1958 V120 N5 P945-948	SC-T-534 <=*$>
1958 V120 N5 P980-983	62-18850 <=*>
1958 V120 N5 P984-986	62-19838 <=>
1958 V120 N5 P1007-1010	59-10395 <+*>
1958 V120 N5 P1C18-1020	R-5305 <*>
1958 V120 N5 P1021-1023	60-10912 <+*>
1958 V120 N5 P1031-1034	60-10911 <+*>
1958 V120 N5 P1049-1051	59-14727 <+*>
1958 V120 N6 P1178-1179	R-5350 <*>
1958 V120 N6 P1187-1190	R-5335 <*>
1958 V120 N6 P1194-1195	R-5333 <*>
1958 V120 N6 P1196-1199	R-5334 <*>
1958 V120 N6 P1231-1233	DSIS-T-305-R <NRC>
	R-4954 <*>
	59-11294 <=>
1958 V120 N6 P1234-1237	R-5136 <*>
	59-11375 <=> 0
	59-11940 <=>
	59-19137 <+*>
1958 V120 N6 P1246-1248	M.5106 <NLL> 0
1958 V120 N6 P1287-1290	61-13208 <*+>
1958 V120 N6 P1298-1301	59-14627 <+*>
	59-17405 <+*>
	59-20139 <+*>
	60-16556 <*+>
	61-19204 <+*>
1958 V121 P13-15	STMSP V2 P1-4 <AMS>
1958 V121 P52-55	STMSP V2 P207-210 <AMS>
1958 V121 P117-118	R-5164 <*>
1958 V121 P753-754	R-5265 <*>
1958 V121 P859-861	RCT V33 N3 P696 <RCT>
1958 V121 N1 P67-69	59-15750 <+*>
	65-14400 <*>
1958 V121 N1 P107-110	60-10721 <+*>
1958 V121 N1 P115-116	59-20145 <+*>
1958 V121 N1 P117-118	59-22447 <+*>
1958 V121 N1 P129-132	59-10326 <+*>
1958 V121 N1 P133-135	59-10894 <+*>
1958 V121 N1 P136-137	59-18014 <+*>
1958 V121 N1 P145-148	60-14397 <+*>
1958 V121 N2 P299-302	59-20522 <+*>
1958 V121 N2 P307-310	59-22535 <+*>
1958 V121 N2 P311-314	R-5161 <*>
	61-15611 <+*>
1958 V121 N2 P315-318	15K26R <ATS>
1958 V121 N2 P322-325	62-14553 <=*>
1958 V121 N2 P330-333	60-23990 <*+>
	64-14521 <=*$>
	8945-B <K-H>
1958 V121 N2 P339-342	59-19818 <+*>
1958 V121 N2 P343-345	59-22591 <+*>
1958 V121 N3 P474-476	20K28R <ATS>
	59-15507 <+*>
1958 V121 N3 P485-487	45K28R <ATS>
1958 V121 N3 P488-491	AEC-TR-3516 <+*>
1958 V121 N3 P495-498	AEC-TR-4520 <*=>
	61-17629 <=$>
1958 V121 N3 P499-502	23K28R <ATS>
1958 V121 N3 P511-514	61-15378 <+*>
	64-14520 <=*$> 0
	8995-A <K-H>
1958 V121 N3 P523-526	60-13565 <+*> 0
	60-18928 <+*>
1958 V121 N4 P6C6-609	59-12415 <=>
1958 V121 N4 P610-612	59-12067 <=>
1958 V121 N4 P627-630	59-15180 <+*>
1958 V121 N4 P637-639	AEC-TR-3517 <+*>
1958 V121 N4 P641-643	59-15370 <+*>
1958 V121 N4 P671-673	43L37R <ATS>
1958 V121 N5 P775-777	59-17769 <+*>
1958 V121 N5 P845-847	60-15740 <+*>
1958 V121 N5 P848-849	59-20138 <+*>
	59-22679 <+*>
1958 V121 N5 P852-854	59-20137 <+*>
1958 V121 N5 P855-857	AEC-TR-3464 <+*>
	59-20136 <+*>
1958 V121 N5 P859-861	59-15016 <+*>
1958 V121 N5 P862-864	59-15906 <+*>
1958 V121 N5 P865-868	59-10517 <+*>
1958 V121 N5 P877-880	61-15376 <+*>
1958 V121 N5 P881-884	61-18625 <*=>
1958 V121 N5 P889-891	59-17766 <+*>
1958 V121 N5 P932-935	61-15303 <+*>
	66-10225 <*> 0
1958 V121 N6 P987-990	62-19837 <=>
1958 V121 N6 P1009-1011	M5108 <NLL>
1958 V121 N6 P1012-1014	65-60688 <=$> 0
1958 V121 N6 P1015-1018	AEC-TR-3518 <+*>
1958 V121 N6 P1019-1020	AEC-TR-3519 <+*>
	59-20135 <+*>
1958 V121 N6 P1038-1040	59-17219 <+*>
1958 V121 N6 P1043-1044	AEC-TR-3520 <+*>
1958 V121 N6 P1060-1062	62-13156 <*=>
	64-10926 <=*$>
	87R79R <ATS>
1958 V121 N6 P1097-1100	59-22071 <+*>
1958 V121 N6 P1105-1108	C-3245 <NRCC>
1958 V122 P111-113	R-5262 <*>
1958 V122 P179-182	59-22025 <+*>
1958 V122 P762-765	AMST S2 V42 P31-36 <AMS>
1958 V122 N1 P48-50	59-15805 <+*>
1958 V122 N1 P51-53	59-15804 <+*>
1958 V122 N1 P94-96	60-17172 <+*>
1958 V122 N1 P103-105	59-10427 <*>
	59-10632 <+*>
1958 V122 N2 P204-207	59-12736 <+*>
1958 V122 N2 P211-214	65-10879 <*>
1958 V122 N2 P231-234	59-10633 <+*>
1958 V122 N2 P250-253	1003 <TC>
1958 V122 N2 P297-299	48K28R <ATS>
1958 V122 N3 P385-388	59-20143 <+*>
1958 V122 N3 P389-392	59-20142 <+*> 0
1958 V122 N3 P420-423	59-10175 <+*>
1958 V122 N3 P424-427	59-20141 <+*>
1958 V122 N3 P434-436	59-20091 <+*>
1958 V122 N3 P481-484	81M42R <ATS>
1958 V122 N4 P551-554	AMS-TRANS V42 P25-30 <AMS>
1958 V122 N4 P575-577	59-22687 <+*>
1958 V122 N4 P578-581	M5125 <NLL>
1958 V122 N4 P600-602	60-13576 <+*>
1958 V122 N4 P609-611	59-17628 <+*>
1958 V122 N4 P629-631	60-10907 <+*>
	65-10189 <*>
1958 V122 N4 P668-670	60-18401 <*+>
1958 V122 N5 P762-765	AMS-TRANS V42 P31-36 <AMS>
1958 V122 N5 P785-787	59-17406 <+*>
1958 V122 N5 P806-809	AEC-TR-4535 <=*>
1958 V122 N5 P821-824	61-10640 <+*>
1958 V122 N5 P863-866	59-22167 <+*>
1958 V122 N6 P1007-1010	62-16424 <=*>
1958 V122 N6 P1014-1017	85L29R <ATS>
1958 V122 N6 P1018-1020	M.5105 <NLL>
	84L29R <ATS>
1958 V122 N6 P1029-1031	61-20645 <*=>
1958 V122 N6 P1035-1038	59-17404 <+*>
1958 V122 N6 P1042-1045	59-20486 <+*>
1958 V122 N6 P1049-1052	59-20133 <+*>
1958 V122 N6 P1061-1064	59-17410 <+*>
1958 V122 N6 P1068-1070	64-14522 <=*$> 0
	8945-C <K-H>
1958 V122 N6 P1076-1078	59-10539 <+*>
	59-15757 <+*> 0
1958 V123 P685-687	RCT V23 N4 P1036 <RCT>
1958 V123 N1 P83-86	59-17437 <+*>
	60-10064 <+*>
1958 V123 N1 P113-116	60L31R <ATS>
1958 V123 N1 P156-158	59-10363 <+*>
1958 V123 N1 P185-188	59-17411 <+*>
1958 V123 N2 P227-230	59-00667 <*>

1958 V123 N2 P246-248 62-11787 <=>
1958 V123 N2 P256-257 N66-16575 <=$>
1958 V123 N2 P279-281 59-14571 <+*>
1958 V123 N2 P282-284 59-15501 <+*>
60-10063 <+*>
1958 V123 N2 P289-291 60-10590 <+*>
1958 V123 N2 P331-334 59-20504 <+*>
1958 V123 N2 P343-345 63-23344 <=*$>
1958 V123 N2 P353-356 16L30R <ATS>
1958 V123 N2 P371-374 C-3241 <NRCC> 0
1958 V123 N3 P453-456 1797 <BISI>
1958 V123 N3 P468-470 60-10908 <+*>
1958 V123 N3 P475-478 59-15503 <+*>
1958 V123 N3 P479-482 59-15502 <+*>
1958 V123 N3 P492-494 62-11441 <=>
1958 V123 N4 P655-658 C-3272 <NRCC>
1958 V123 N4 P696-699 59-15508 <+*>
1958 V123 N5 P849-852 59-15509 <+*>
1958 V123 N5 P874-877 60-10061 <+*>
1958 V123 N5 P891-894 59-15505 <+*>
80L30R <ATS>
1958 V123 N6 P978-980 64-11851 <=>
1958 V123 N6 P1003-1005 60-13657 <+*>
1958 V123 N6 P1041-1043 62L31R <ATS>
1958 V123 N6 P1068-1070 61-31643 <=> 0

1959 V124 P1211-1214 AMST S2 V19 P167-171 <AMS>
1959 V124 N1 P19-21 59-14685 <+*>
1959 V124 N1 P26-28 62-23043 <=>
1959 V124 N1 P51-52 62-11276 <=>
1959 V124 N1 P94-97 59-14570 <+*>
1959 V124 N1 P133-134 60-10277 <+*>
1959 V124 N2 P285-287 66-12912 <*> 0
1959 V124 N2 P295-297 60-10915 <+*>
1959 V124 N2 P311-313 64-13338 <=*$>
1959 V124 N2 P314-317 SCL-T-240 <+*>
1959 V124 N2 P321-323 60-10914 <+*>
61-18874 <*=> 0
1959 V124 N2 P350-353 59-16139 <+*>
60-10913 <+*>
1959 V124 N2 P370-372 65-10126 <*>
1959 V124 N2 P373-376 62-14356 <=*> 0
1959 V124 N2 P398-401 35L36R <ATS>
1959 V124 N2 P432-435 65-60690 <=$>
1959 V124 N2 P462-465 65-29643 <$>
1959 V124 N3 P529-532 UCRL-TRANS-878(L) <=*>
1959 V124 N3 P571-574 60-13831 <+*>
1959 V124 N3 P583-585 36L34R <ATS>
62-11485 <=>
1959 V124 N3 P595-597 60-14333 <+*>
1959 V124 N3 P609-612 59-14684 <+*>
1959 V124 N3 P621-624 62-11306 <=>
1959 V124 N3 P625-627 83L32R <ATS>
1959 V124 N3 P632-634 16L36R <ATS>
1959 V124 N3 P674-677 59-14959 <+*> 0
1959 V124 N3 P691-694 59-17416 <+*>
1959 V124 N4 P796-799 59-17776 <+*>
1959 V124 N4 P803-805 59-21118 <=>
1959 V124 N4 P842-845 33L32R <ATS>
1959 V124 N4 P865-868 00L36R <ATS>
1959 V124 N4 P873-875 77L33R <ATS>
1959 V124 N4 P915-918 61-19241 <+*> 0
1959 V124 N5 P1001-1004 62-11633 <=>
1959 V124 N5 P1061-1064 28L35R <ATS>
1959 V124 N5 P1076-1079 60-14251 <+*>
1959 V124 N5 P1080-1082 45L35R <ATS>
1959 V124 N5 P1083-1084 44L37R <ATS>
1959 V124 N5 P1144-1146 59-18363 <+*> 0
59-21201 <=>
1959 V124 N6 P1226-1228 62-11612 <=>
62-11618 <=>
1959 V124 N6 P1301-1304 M.5118 <NLL>
1959 V124 N6 P1305-1308 M.5119 <NLL>
1959 V124 N6 P1309-1312 M.5124 <NLL>
1959 V125 P485-487 AMST S2 V47 P43-46 <AMS>
1959 V125 P711-714 STMSP V4 P99-103 <AMS>
1959 V125 P974-977 AMST S2 V19 P173-178 <AMS>
1959 V125 N1 P59-61 61-16544 <*=> 0
1959 V125 N1 P118-121 60-10278 <+*>
1959 V125 N1 P126-128 61-13388 <*+>
1959 V125 N1 P208-209 60-15026 <+*> 0
1959 V125 N1 P236-239 62-14688 <=*>
1959 V125 N2 P304- 59-21045 <=>
1959 V125 N2 P341-344 60-16895 <+*>
1959 V125 N2 P348-350 60-10909 <+*>
1959 V125 N2 P351-353 01L32R <ATS>
1959 V125 N3 P475-478 61-23038 <=*>
1959 V125 N3 P485-487 64-16576 <=*$>
1959 V125 N3 P523-525 21L32R <ATS>
1959 V125 N3 P570-572 60-13847 <+*>
1959 V125 N3 P584-587 62-16569 <=*>
1959 V125 N3 P639-642 59-13959 <=> 0
1959 V125 N4 P775-778 62-32045 <=*>
1959 V125 N4 P829-830 42L32R <ATS>
1959 V125 N4 P895-897 09M39R <ATS>
1959 V125 N4 P925-927 44L32R <ATS>
1959 V125 N4 P928-930 45L32R <ATS>
1959 V125 N4 P931-934 98L33R <ATS>
1959 V125 N5 P1027-1029 56L33R <ATS>
1959 V125 N5 P1051-1052 60-10906 <+*>
1959 V125 N5 P1073-1076 66L34R <ATS>
1959 V125 N5 P1097-1099 65M37R <ATS>
1959 V125 N5 P1119-1122 AD-635 372 <=$>
1959 V125 N5 P1134-1136 60-10905 <+*>
1959 V125 N5 P1151-1153 UCRL TRANS-496 <+*>
59-20552 <+*>
1959 V125 N5 P1158-1161 59-20557 <+*>
1959 V125 N6 P1242-1245 62-32642 <=*>
1959 V125 N6 P1292-1293 60-11527 <=>
1959 V126 P22-25 STMSP V4 P277-281 <AMS>
1959 V126 P271-273 STMSP V4 P171-174 <AMS>
1959 V126 P719-722 AMST S2 V48 P156-160 <AMS>
1959 V126 P784-786 66-14163 <*$>
1959 V126 N1 P22-25 62-16442 <=*>
1959 V126 N1 P111-114 60-12607 <=>
62-11506 <=>
1959 V126 N1 P151-154 61-13149 <*+>
1959 V126 N2 P244-247 61-10681 <+*>
1959 V126 N2 P307-309 60-18917 <+*>
1959 V126 N2 P318-321 59-13997 <=>
1959 V126 N2 P337 60-15807 <+*>
1959 V126 N2 P337-340 60-17689 <+*>
1959 V126 N2 P410-413 59-20551 <+*>
1959 V126 N2 P442-445 62-14730 <=*> 0
1959 V126 N2 P450-453 59-13970 <=>
1959 V126 N3 P524-527 63-16260 <=*> 0
1959 V126 N3 P532-533 60-18914 <+*>
1959 V126 N3 P571-574 56L36R <ATS>
60-16894 <*+>
1959 V126 N3 P575-578 57L33R <ATS>
1959 V126 N3 P582-585 60-18039 <+*>
1959 V126 N3 P594-597 60-18038 <+*>
83L34R <ATS>
1959 V126 N3 P608-611 60-18037 <+*>
1959 V126 N3 P612-615 60-18072 <+*>
61-23425 <*=> 0
1959 V126 N3 P641-644 60-13859 <+*> 0
1959 V126 N3 P675-677 60-18375 <+*>
1959 V126 N3 P696-698 59-19055 <+*> 0
62-14749 <=*>
1959 V126 N3 P699-702 62-14687 <=*>
1959 V126 N4 P759-762 60-18036 <+*>
1959 V126 N4 P774-776 60-21064 <=>
1959 V126 N4 P787-790 60-16893 <+*>
1959 V126 N4 P794-797 61-28289 <*=>
1959 V126 N4 P806-808 60-14016 <+*>
1959 V126 N4 P809-812 60-16892 <=*>
1959 V126 N4 P813-816 60-11475 <=>
60-15809 <+*>
1959 V126 N4 P817-820 60-16891 <+*>
1959 V126 N4 P821-822 60-10388 <+*>
1959 V126 N4 P853-854 60-13860 <+*>
65-12493 <*>
1959 V126 N4 P867-869 60-18374 <+*>
1959 V126 N5 P927-930 63-14774 <*=>
1959 V126 N5 P1021-1024 60-10353 <+*>
1959 V126 N5 P1025-1028 63-14335 <=*>
80P58R <ATS>
1959 V126 N5 P1062-1065 47M39R <ATS>
1959 V126 N6 P1227-1228 59-13963 <=>

61-11134 <=>
1959 V126 N6 P1239-1241 60-16890 <*+>
1959 V126 N6 P1268-1269 60-19999 <+*>
61-10457 <*+>
1959 V126 N6 P1296-1299 62-11405 <=>
1959 V126 N6 P1371-1374 68L34R <ATS>
1959 V127 P17-19 STMSP V3 P327-330 <AMS>
1959 V127 P749-752 AMST S2 V45 P19-22 <AMS>
1959 V127 P977-979 AMST S2 V37 P287-290 <AMS>
1959 V127 P1128-1131 59-20555 <+*>
1959 V127 N1 P17-19 63-18097 <=*>
1959 V127 N1 P74 65-60692 <=$>
1959 V127 N1 P78-81 CHEM-105 <CTS>
60-21182 <=>
60-21758 <=>
62-16025 <=*>
1959 V127 N1 P97-99 50M37R <ATS>
1959 V127 N1 P100-103 51M37R <ATS>
1959 V127 N1 P141-144 16M40R <ATS>
1959 V127 N1 P145-148 AEC-TR-3929 <*>
60-16554 <+*> 0
1959 V127 N1 P198-201 59-20550 <+*>
1959 V127 N2 P254-257 64-30709 <*>
1959 V127 N2 P348-351 60-18436 <+*>
1959 V127 N2 P352-355 66M38R <ATS>
1959 V127 N2 P359-361 41S85R <ATS>
1959 V127 N2 P377-379 62-33082 <=>
1959 V127 N2 P384-385 60-14224 <+*>
1959 V127 N2 P392-395 62-16144 <=*>
1959 V127 N3 P509-512 63-18755 <=*>
1959 V127 N3 P520-523 61-23142 <=*>
1959 V127 N3 P529-530 66-11165 <*>
1959 V127 N3 P595-598 63-10229 <=*>
1959 V127 N3 P677-680 MAFF-TRANS-NS-37 <NLL>
1959 V127 N3 P699-701 59-17719 <+*>
1959 V127 N3 P7C2-7C5 60-18373 <+*>
1959 V127 N4 P772-773 62-23887 <=*$> 0
1959 V127 N4 P822-824 NP-TR-363 <*>
61-13534 <+*>
1959 V127 N4 P831-834 61-13752 <+*>
1959 V127 N4 P881-883 65-60686 <=$>
1959 V127 N4 P907-910 59-20553 <+*>
1959 V127 N4 P913-916 59-20554 <+*>
1959 V127 N5 P945-948 61-10710 <=*>
1959 V127 N5 P961-964 61-16465 <*=>
1959 V127 N5 P965-968 64-16383 <=*$>
1959 V127 N5 P1C05-1008 60-11169 <=>
1959 V127 N5 P1027-1028 60-14250 <+*>
95M37R <ATS>
1959 V127 N5 P1C33-1036 61-17551 <=$>
1959 V127 N5 P1099-1102 66-20964 <$>
1959 V127 N6 P1207-1209 RCT V34 N1 P119 <RCT>
1959 V127 N6 P1217-1220 37L35R <ATS>
60-10389 <+*>
1959 V127 N6 P1228-1230 17M38R <=>
1959 V128 P243-245 STMSP V4 P283-287 <AMS>
1959 V128 P647-65C AMST S2 V49 P162-166 <AMS>
1959 V128 P657-660 AMST S2 V46 P149-152 <AMS>
1959 V128 P873-875 AMST S2 V49 P167-17C <AMS>
1959 V128 N1 P21-24 60-11642 <=>
1959 V128 N1 P71-72 60-14163 <+*>
1959 V128 N1 P85-88 AEC-TR-4035 <+*>
1959 V128 N1 P92-94 60-12547 <=>
62-11396 <=>
1959 V128 N1 P95-98 61-18595 <*=>
1959 V128 N1 P113-116 51L35R <ATS>
1959 V128 N1 P117-120 75T94R <ATS>
1959 V128 N1 P136-139 03M42R <ATS>
1959 V128 N1 P165-168 89M39R <ATS>
1959 V128 N1 P172-175 60-11489 <=> 0
1959 V128 N2 P231-234 65-60923 <=$>
1959 V128 N2 P265-268 60-11440 <=>
1959 V128 N2 P3C2-304 60-10391 <+*>
1959 V128 N2 P309-311 15L36R <ATS>
1959 V128 N2 P312-315 60-10390 <+*>
60-14273 <+*>
60-21773 <=>
61-10370 <*+>
1959 V128 N2 P323-325 CSIR-TR.141 <CSSA>
1959 V128 N2 P348-351 60-17690 <+*>
1959 V128 N2 P355-358 90M39R <ATS>
1959 V128 N3 P485-487 61-10374 <*+>
1959 V128 N3 P488-490 60-15452 <+*>
61-10379 <*+>
1959 V128 N3 P521-524 36N52R <ATS>
1959 V128 N3 P575-577 63-15132 <=*>
1959 V128 N3 P590-593 95M39R <ATS>
1959 V128 N3 P628-631 62L36R <ATS>
1959 V128 N4 P657-660 AMS TRANS V.46 P149 <AMS>
66-11159 <*>
1959 V128 N4 P732-735 61-16710 <*=>
1959 V128 N4 P736-739 RCT V33 N2 P357 <RCT>
1959 V128 N4 P740-743 21L36R <ATS>
1959 V128 N4 P744-747 36L36R <ATS>
1959 V128 N4 P752-754 60-14254 <+*>
1959 V128 N4 P769-772 75M39R <ATS>
1959 V128 N4 P823-826 80M42R <ATS>
1959 V128 N5 P876-879 61-11168 <=>
1959 V128 N5 P890-892 61-10451 <=*>
62-10156 <*=>
1959 V128 N5 P917-920 60-11290 <=>
1959 V128 N5 P921-923 63-24301 <=*$>
1959 V128 N5 P956-959 35M39R <ATS>
1959 V128 N5 P966-969 61-10384 <+*>
1959 V128 N5 P973-976 63-24303 <=*$>
1959 V128 N5 P999-1002 52M39R <ATS>
1959 V128 N5 P1012-1015 63-24302 <=*$>
1959 V128 N5 P1063-1065 6C-10991 <+*>
1959 V128 N6 P1153-1156 98N53R <ATS>
1959 V128 N6 P1167-1170 M5130 <NLL>
1959 V128 N6 P1175-1178 16M39R <ATS>
64-10098 <=*$>
1959 V128 N6 P1188-1191 34M40R <ATS>
1959 V128 N6 P1198-1200 71M38R <ATS>
1959 V128 N6 P1234-1237 65-18049 <*>
1959 V128 N6 P1271-1273 C-3406 <NRCC> 0
1959 V128 N6 P1298-1301 61-23974 <*=>
1959 V129 P488-491 STMSP V3 P331-335 <AMS>
1959 V129 P687-690 62-13586 <=*>
1959 V129 P1430-1433 60-16003 <+*>
1959 V129 N1 P52-55 60-41156 <=>
1959 V129 N1 P77-80 60-31013 <=>
61-23282 <=>
1959 V129 N1 P88-90 60-41157 <=>
1959 V129 N1 P102-104 61-10360 <*+>
1959 V129 N1 P109-112 64-20168 <*>
1959 V129 N1 P128-130 09M44R <ATS>
61-10378 <*+>
1959 V129 N1 P141-144 44M39R <ATS>
1959 V129 N1 P145-148 61-20060 <*=>
1959 V129 N1 P165-167 60-14240 <+*>
1959 V129 N1 P187-190 61-15891 <*=>
1959 V129 N2 P280-283 62-11119 <=>
1959 V129 N2 P3C3-306 60-16303 <+*>
1959 V129 N2 P310-313 61-23630 <*=>
1959 V129 N2 P334-336 61-19368 <=*>
1959 V129 N2 P361-364 RCT V33 N4 P1005 <RCT>
61-10356 <+*>
1959 V129 N2 P428-430 02M42R <ATS>
1959 V129 N3 P525-528 63-13201 <=*>
1959 V129 N3 P533-535 AEC-TR-4050 <+*> 0
60-14241 <+*> 0
1959 V129 N3 P565-568 55M40R <ATS>
1959 V129 N3 P623-626 62-19891 <=>
1959 V129 N4 P781-784 63-19916 <=*>
1959 V129 N4 P785-788 <GPRC>
1959 V129 N4 P799-801 60-31238 <=>
61-10369 <*+>
1959 V129 N4 P824-826 U-138 <RIS>
61-13602 <*+>
1959 V129 N4 P827-830 60-12506 <=>
1959 V129 N4 P940-943 63-16461 <*=> 0
1959 V129 N5 P1008-1011 60-14660 <+*>
1959 V129 N5 P1C28-1030 61-13605 <*+> C
1959 V129 N5 P1035-1037 62-25396 <=*>
1959 V129 N5 P1C46-1048 14M38R <ATS>
61-10197 <*+>
1959 V129 N5 P1068-1069 61-10368 <*+>
1959 V129 N5 P1079-1081 61-19343 <*+>
1959 V129 N5 P1089-1092 39M38R <ATS>

Citation	Translation
1959 V129 N5 P1096-1099	61-19343 <*+>
1959 V129 N5 P1104-1106	61-13282 <+*>
1959 V129 N5 P1126-1129	73N48R <ATS>
1959 V129 N5 P1254-1256	60-11797 <=>
1959 V129 N6 P1199-1202	61-10267 <*+>
1959 V129 N6 P1257-1260	60-41276 <=>
1959 V129 N6 P1283-1286	SCL-T-295 <+*>
	62-32910 <=*>
1959 V129 N6 P1293-1296	61-10367 <*+>
1959 V129 N6 P1303-1305	60-31638 <=> 0
1959 V129 N6 P1306-1308	61-10743 <+*>
1959 V129 N6 P1309-1312	78T94R <ATS>
1959 V129 N6 P1362-1365	61-27617 <*=>
	66-13387 <*>
1960 V130 P1228-1231	62-15699 <*=>
1960 V130 N1 P47-56	60-11488 <=>
1960 V130 N1 P90-93	61-10193 <+*> 0
1960 V130 N1 P102-104	61-10639 <+*>
1960 V130 N1 P126-128	61-10198 <+*>
1960 V130 N1 P183-186	60-11649 <=>
	63-15150 <=*>
1960 V130 N2 P269-271	61-15497 <+*>
1960 V130 N2 P287-289	60-31210 <=>
1960 V130 N2 P303-306	60-31211 <=>
1960 V130 N2 P338-340	60-16773 <+*>
1960 V130 N2 P353-355	60-14533 <+*>
1960 V130 N2 P356-358	60-16549 <+*>
1960 V130 N2 P362-366	62-10098 <*=>
1960 V130 N2 P362-365	62-13697 <*=>
	62-13697 <=>
1960 V130 N2 P408	82M45R <ATS>
1960 V130 N2 P450-452	61-27255 <*=>
1960 V130 N3 P556-558	36S88R <ATS>
1960 V130 N3 P649-652	62-33566 <*=> 0
1960 V130 N3 P663-666	60-16004 <+*>
1960 V130 N4 P804-806	60-41489 <=>
1960 V130 N4 P820-823	60M42R <ATS>
	61-15471 <+*>
	65-17084 <*>
1960 V130 N5 P1001-1003	60-31457 <=>
1960 V130 N5 P1055-1058	61-15470 <=*>
1960 V130 N5 P1081-1084	03M40R <ATS>
1960 V130 N5 P1089-1090	61-31104 <=>
1960 V130 N5 P1148-1152	60-16545 <+*>
1960 V130 N6 P1232-1235	63-14509 <=*>
1960 V130 N6 P1248-1251	60-11677 <=>
1960 V130 N6 P1273-1276	62-25920 <=*>
1960 V130 N6 P1284-1287	36M41R <ATS>
1960 V130 N6 P1294-1297	63-14149 <=*> 0
1960 V130 N6 P1317-1318	08M43R <ATS>
	66-10650 <*>
1960 V130 N6 P1345-1348	60-41694 <=>
1960 V130 N6 P1366-1369	60-16802 <+*> 0
1960 V130 N6 P1382-1384	63-20942 <=*$>
1960 V131 N1 P64-67	61-10190 <*+>
1960 V131 N1 P78-81	60-31683 <=>
1960 V131 N1 P113-116	66-10647 <*>
1960 V131 N1 P125-128	63-15130 <=*>
	63-19818 <=*>
1960 V131 N1 P191-194	61-11521 <=>
1960 V131 N2 P279-282	61-10195 <+*>
	77M40R <ATS>
1960 V131 N2 P293-296	62-24892 <=*>
1960 V131 N2 P300-302	60-31851 <=>
	61-23375 <=*>
1960 V131 N2 P312-315	61-10344 <*+>
1960 V131 N2 P321-324	61M41R <ATS>
1960 V131 N2 P342-345	61-10194 <*+>
1960 V131 N2 P346-347	60-21929 <=>
	63-13579 <=*>
1960 V131 N2 P445-448	63-13580 <=*>
1960 V131 N3 P561-562	84M42R <ATS>
1960 V131 N3 P563-565	RCT V33 N4 P1188 <RCT>
1960 V131 N3 P566-567	AEC-TR-4310 <*+>
1960 V131 N3 P591-592	60-41561 <=> 0
1960 V131 N3 P597-600	81M46R <ATS>
1960 V131 N4 P789-792	TRANS-173 <FRI>
	62-25931 <=*>
1960 V131 N4 P793-796	62-25351 <=*>
1960 V131 N4 P827-829	86M43R <ATS>

Citation	Translation
1960 V131 N4 P830-832	RCT V33 N4 P985 <RCT>
1960 V131 N4 P936-939	49M45R <ATS>
1960 V131 N5 P1038-1041	61-10711 <+*>
1960 V131 N5 P1060-1063	60-41628 <=>
1960 V131 N5 P1069-1071	75M43R <ATS>
1960 V131 N5 P1092-1095	90M43R <ATS>
1960 V131 N5 P1109-1112	87Q72R <ATS>
1960 V131 N5 P1140-1142	62-25349 <=*>
1960 V131 N5 P1156-1158	61-15093 <*+>
1960 V131 N6 P1283-1286	07M42R <ATS>
	60-11951 <=>
1960 V131 N6 P1287-1290	60-11950 <=>
1960 V131 N6 P1359-1360	66-10876 <*>
1960 V131 N6 P1444-1446	61-10348 <*+>
1960 V132 N1 P78-81	60-31824 <=>
1960 V132 N1 P85-88	ARSJ V30 N11 P1060 <AIAA>
	61-23437 <=*> 0
1960 V132 N1 P157-159	41M46R <ATS>
1960 V132 N1 P168-171	61-28141 <*=>
1960 V132 N2 P311-314	13M47R <ATS>
1960 V132 N2 P319-322	61-13980 <+*>
1960 V132 N2 P353-356	61-27537 <*=>
	82M42R <ATS>
1960 V132 N2 P360-363	05M44R <ATS>
	63-15181 <=*>
1960 V132 N2 P367-370	61-14983 <=*>
	95M43R <ATS>
1960 V132 N2 P395-398	62-18298 <=*>
1960 V132 N2 P406-408	31M43R <ATS>
1960 V132 N3 P553-556	14M47R <ATS>
1960 V132 N3 P557-560	60-23178 <+*>
1960 V132 N3 P577-580	09M43R <ATS>
1960 V132 N3 P677-680	62-23890 <=*>
1960 V132 N3 P723-725	62-19004 <=*>
1960 V132 N4 P803-805	60-21933 <=>
1960 V132 N4 P846-849	60-31632 <=>
1960 V132 N4 P918-920	62-24894 <=*> 0
1960 V132 N4 P921-924	62-15292 <*=>
1960 V132 N4 P929-931	61-15872 <=*> 0
1960 V132 N5 P1049-1050	61-11745 <=>
1960 V132 N5 P1062-1065	60-41212 <=>
1960 V132 N5 P1078-1081	54M45R <ATS>
	61-10610 <*+>
1960 V132 N5 P1082-1085	61-16705 <*=>
	62-10433 <*=>
	73M45R <ATS>
1960 V132 N5 P1103-1106	63-15714 <=*>
1960 V132 N5 P1129-1131	43M44R <ATS>
1960 V132 N5 P1135-1138	36N48R <ATS>
1960 V132 N5 P1136-1139	63-15315 <=*>
1960 V132 N5 P1140-1143	61-27637 <*=>
1960 V132 N5 P1179-1182	22M43R <ATS>
1960 V132 N5 P1187-1190	C-3625 <NRCC> 0
1960 V132 N6 P1291-1294	61-11182 <=>
1960 V132 N6 P1299-1302	UCRL TRANS-624 <*+>
	61-20059 <*=>
1960 V132 N6 P1319-1321	61-10013 <+*>
1960 V132 N6 P1339-1340	60-41737 <=>
	63-19935 <*=>
1960 V132 N6 P1356-1359	61-10765 <+*>
1960 V132 N6 P1360-1363	61-10120 <+*>
1960 V132 N6 P1376-1377	60-41736 <=>
1960 V133 P230-232	62-25675 <=*>
1960 V133 P1358-1360	61-27503 <*=>
1960 V133 P1472-1475	61-10373 <*+>
1960 V133 N1 P20-23	61-31645 <=>
1960 V133 N1 P71-73	61-11139 <=>
1960 V133 N1 P106-107	61-10884 <=*>
1960 V133 N1 P138-140	61-10014 <+*>
1960 V133 N1 P141-143	66-10649 <*>
1960 V133 N1 P144-147	85P59R <ATS>
1960 V133 N1 P178-181	61-28382 <*=>
1960 V133 N2 P269-272	61-21727 <=>
1960 V133 N2 P341-344	RCT V35 N1 P178 <RCT>
1960 V133 N2 P370-373	64-20167 <=*$>
1960 V133 N2 P377-380	62-24697 <=*>
	83N51R <ATS>
1960 V133 N2 P399-400	61-21725 <=>
1960 V133 N2 P413-416	61-21726 <=>
1960 V133 N2 P431-434	35N40R <ATS>

Citation	Translation
1960 V133 N3 P531-534	63-18063 <=*>
1960 V133 N3 P581-584	63-13722 <=*>
1960 V133 N3 P630-632	63-20313 <=*$>
1960 V133 N3 P645-648	61-14984 <*=>
1960 V133 N4 P804-806	61-11115 <=>
1960 V133 N4 P838-840	61-13897 <+*>
1960 V133 N4 P878-881	63-13719 <=*>
1960 V133 N5 P1041-1044	61-21054 <=>
1960 V133 N5 P1060-1063	61-18306 <*=>
1960 V133 N5 P1C73-1076	62-16575 <=*>
1960 V133 N5 P1C84-1085	61-10800 <+*>
1960 V133 N5 P1098-1101	61-10723 <+*>
1960 V133 N5 P1125-1127	61-14966 <=*>
1960 V133 N5 P1213-1215	61-10346 <+*>
1960 V133 N5 P1227-1230	09N51R <ATS>
1960 V133 N6 P1327-1330	62-24623 <=*$>
1960 V133 N6 P1344-1346	61-28204 <*=>
1960 V133 N6 P1350-1353	62-10430 <*=>
1960 V133 N6 P1388-1390	63-13720 <=*>
1960 V133 N6 P1451-1454	61-28142 <*=>
1960 V133 N6 P1465-1467	36N50R <ATS>
1960 V134 P1486-1489	61-10353 <*+>
1960 V134 N1 P59-61	61-14979 <=*>
	89N50R <ATS>
1960 V134 N1 P106-109	61-16430 <*=>
1960 V134 N1 P110-113	38N48R <ATS>
1960 V134 N1 P125-127	61-11746 <=>
1960 V134 N1 P134-136	63-13587 <=*>
1960 V134 N1 P145-148	65-13593 <*>
1960 V134 N1 P168-170	61-19240 <+*>
1960 V134 N2 P296-299	61-21314 <=>
1960 V134 N2 P300-303	61-21320 <=>
1960 V134 N2 P3C4-3C7	61-21321 <=>
	61M46R <ATS>
1960 V134 N2 P308-310	62-10173 <=*>
1960 V134 N2 P341-344	61-21720 <=>
1960 V134 N2 P349-350	62-24612 <=*>
1960 V134 N2 P374-375	00N50R <ATS>
1960 V134 N2 P435-438	18N50R <ATS>
1960 V134 N3 P511-513	62-32626 <=*>
1960 V134 N3 P591-594	62-24613 <=*>
1960 V134 N3 P607-608	61-10634 <+*>
1960 V134 N3 P612-614	64M46R <ATS>
1960 V134 N3 P618-620	63-13721 <=*>
1960 V134 N3 P654-657	20N51R <ATS>
1960 V134 N3 P684-687	21N51R <ATS>
1960 V134 N3 P710-712	61-10347 <*+>
1960 V134 N4 P778-781	62-10174 <=*>
1960 V134 N4 P812-815	61-28241 <*=>
1960 V134 N4 P836-839	61-18599 <*=>
	88N48R <ATS>
1960 V134 N4 P864-866	62-11323 <=>
1960 V134 N4 P955-958	61-21371 <=>
1960 V134 N4 P963-964	61-10349 <*+>
1960 V134 N5 P1041-1043	61-21249 <=>
1960 V134 N5 P1044-1047	61-21229 <=>
1960 V134 N5 P1065-1068	61-21225 <=>
1960 V134 N5 P1073-1075	61-27498 <*=>
1960 V134 N5 P1C85-1086	82N49R <ATS>
1960 V134 N5 P1098-1099	ICE V1 N1 P17-18 <ICE>
	52N53R <ATS>
	61-27645 <*=>
1960 V134 N5 P1123-1126	37M47R <ATS>
	61-11879 <=>
	62-13147 <*=>
1960 V134 N5 P1153-1154	61-18904 <=*>
1960 V134 N5 P1166-1168	21N50R <ATS>
1960 V134 N5 P1232-1235	61-10351 <*+>
1960 V134 N5 P1236-1239	62-20436 <=*>
1960 V134 N6 P1334-1336	TRANS-147 <FRI>
	62-25385 <=*>
1960 V134 N6 P1374-1377	AD-612 961 <=$>
1960 V134 N6 P1381-1383	56M47R <ATS>
1960 V134 N6 P1387-1389	61-10992 <+*>
1960 V134 N6 P1390-1393	38M47R <ATS>
1960 V134 N6 P1443-1446	64-19775 <=$>
1960 V134 N6 P1501-1503	63-15138 <=*>
1960 V135 P133-136	RCT V34 N3 P879 <RCT>
1960 V135 N1 P45-47	62-11249 <=>
1960 V135 N1 P48-51	61-21595 <=>
1960 V135 N1 P65-68	61-18598 <*=>
1960 V135 N1 P84-86	01N50R <ATS>
1960 V135 N1 P98-100	64-10173 <=*$> 0
1960 V135 N1 P109-112	62N48R <ATS>
1960 V135 N1 P113-116	63-13576 <=*>
1960 V135 N1 P121-124	63-14320 <=*>
1960 V135 N2 P277-279	61-21568 <=>
1960 V135 N2 P283-286	61-21569 <=>
1960 V135 N2 P294-297	62-32677 <+*>
1960 V135 N2 P301-304	61-21407 <=>
1960 V135 N2 P309-311	AEC-TR-4566 <=*>
1960 V135 N2 P354-356	83N49R <ATS>
1960 V135 N2 P357-360	64-20166 <*> 0
1960 V135 N2 P361-364	62-10700 <*=>
1960 V135 N3 P581-583	62-13153 <*=>
1960 V135 N3 P591-594	65-60486 <=$>
1960 V135 N3 P603-605	61-10866 <+*>
1960 V135 N3 P713-716	61-18604 <*=>
1960 V135 N3 P743-746	UCRL TRANS-646 <*=>
1960 V135 N4 P847-848	61-18605 <*=>
1960 V135 N4 P849-852	61-10865 <+*>
1960 V135 N4 P853-856	61-18603 <*=>
1960 V135 N4 P893-895	62-23981 <=*>
1960 V135 N4 P899-901	63-23378 <=*$>
1960 V135 N4 P1009-1011	RTS-2893 <NLL>
1960 V135 N5 P1064-1067	61-23480 <=*>
1960 V135 N5 P1123-1126	62-33490 <=*> 0
1960 V135 N5 P1147-1149	63-24325 <=*$>
1960 V135 N5 P1160-1163	63-15316 <=*>
1960 V135 N5 P1164-1167	61-19204 <+*>
1960 V135 N5 P1168-1171	22N51R <ATS>
1960 V135 N5 P1172-1175	ICE V1 N1 P58-60 <ICE>
	61-31547 <=>
1960 V135 N5 P1226-1228	39N51R <ATS>
1960 V135 N5 P1258-1261	62-18625 <=*>
1960 V135 N5 P1266-1269	61-10991 <+*>
1960 V135 N6 P1395-1398	56N52R <ATS>
1960 V135 N6 P1399-1401	61-27441 <*=> 0
1960 V135 N6 P1442-1445	63-13865 <=*>
1960 V135 N6 P1528-1531	62-18626 <=*>
1960 V135 N6 P1532-1535	62-18627 <=*>
1961 V136 N1 P26-28	61-31453 <=>
1961 V136 N1 P39-42	61-21513 <=>
1961 V136 N1 P72-78	61-28553 <*=>
1961 V136 N1 P84-87	61-23499 <*=>
	65-13505 <*>
1961 V136 N1 P125-128	62-10438 <*=>
1961 V136 N1 P136-139	62-16733 <=*>
1961 V136 N1 P140-142	61-20676 <*=> 0
1961 V136 N2 P308-310	61-21657 <=>
1961 V136 N2 P322-324	61-21657 <=>
1961 V136 N2 P332-335	63-21442 <=>
1961 V136 N2 P336-338	61-31542 <=>
1961 V136 N2 P342-345	47P64R <ATS>
1961 V136 N2 P360-363	62-15247 <*=>
1961 V136 N2 P377-380	62-15245 <*=>
1961 V136 N2 P381-383	61-21657 <=>
1961 V136 N2 P394-397	64-10109 <=*$>
1961 V136 N2 P398-400	62-25377 <=*>
1961 V136 N2 P401-404	62-10434 <*=>
1961 V136 N2 P4C5-407	62-18638 <=*>
1961 V136 N2 P432-433	94N52R <ATS>
1961 V136 N2 P445-448	95N52R <ATS>
1961 V136 N2 P468-471	48N50R <ATS>
1961 V136 N3 P587-590	61-21562 <=>
1961 V136 N3 P591-594	42S85R <ATS>
1961 V136 N3 P599-602	62-10440 <*=>
	62-24651 <=*>
1961 V136 N3 P663-666	63-13862 <=*>
1961 V136 N3 P714-717	47N50R <ATS>
1961 V136 N4 P765-767	61-23050 <=*>
1961 V136 N4 P783-786	61-23049 <=*>
1961 V136 N4 P791-794	61-23005 <*=>
1961 V136 N4 P8C3-806	61-23007 <*=>
1961 V136 N4 P810-812	61-23111 <*=>
1961 V136 N4 P849-851	62-20392 <=*>
1961 V136 N4 P852-855	24N54R <ATS>
1961 V136 N4 P856-859	54N52R <ATS>
1961 V136 N4 P921-923	63-21526 <=>
1961 V136 N5 P1041-1042	61-27684 <*=>

1961 V136 N5 P1043-1046 61-27683 <*=>
1961 V136 N5 P1047-1051 62-13333 <*=>
1961 V136 N5 P1051-1054 61-27843 <*=>
1961 V136 N5 P1066-1068 61-27160 <*=>
61-31613 <=>
1961 V136 N5 P1078-1081 61-28559 <*=>
1961 V136 N5 P1127-1129
ICE V1 N1 P122-124 <ICE>
1961 V136 N5 P1137-1138 62-25384 <=*>
1961 V136 N5 P1142-1145 62-19281 <=*>
1961 V136 N5 P1150-1153 62-23673 <=*>
1961 V136 N5 P1216-1218 61-28406 <*=>
1961 V136 N5 P1223-1226 62-18628 <=*>
1961 V136 N5 P1227-1230 62-18629 <=*>
1961 V136 N5 P1231-1234 61-28719 <*=>
1961 V136 N6 P1306-1309 62-13550 <*=>
1961 V136 N6 P1321- 61-27168 <*=>
1961 V136 N6 P1325-1327 61-28899 <*=>
1961 V136 N6 P1339-1341 63-13423 <=*>
1961 V136 N6 P1342-1344 61-28928 <*=>
1961 V136 N6 P1376-1379 63-13864 <=*>
1961 V136 N6 P1396-1398 61-18634 <*=> 0
61-20986 <*=> 0
1961 V137 N1 P109-110 63-13859 <=*>
1961 V137 N1 P113-115 62-16117 <=*>
62-24846 <=*>
1961 V137 N1 P120-122 63-13863 <=*>
1961 V137 N1 P123-125 61-31536 <=>
1961 V137 N1 P130-133 62-25361 <=*>
1961 V137 N1 P154-157 34N54R <ATS>
1961 V137 N1 P158-161 62-15738 <=*>
1961 V137 N1 P162-165 35N55R <ATS>
1961 V137 N1 P178-181 63-19858 <*=>
1961 V137 N1 P196-198 99N51R <ATS>
1961 V137 N1 P199-202 UCRL TRANS-693 <*=>
1961 V137 N2 P295-298 61-23619 <=*>
1961 V137 N2 P303-306 61-27173 <*=>
1961 V137 N2 P327-330 61-27141 <*=>
1961 V137 N2 P363-365 62-25364 <=*>
1961 V137 N2 P400-402 62-15737 <=*>
1961 V137 N3 P660-662 61-23208 <*=>
1961 V137 N3 P706-709 UCRL TRANS-695 <=*>
1961 V137 N3 P715-718 UCRL-TRANS-696 <=*>
1961 V137 N4 P804-806 61-28752 <*=>
1961 V137 N4 P807-809 61-28614 <*=>
1961 V137 N4 P810-813 61-27844 <*=>
1961 V137 N4 P822-825 61-27842 <*=>
1961 V137 N4 P833-835 TRC-TRANS-1099 <NLL>
65-28879 <=*>
1961 V137 N4 P893-895 95N54R <ATS>
1961 V137 N4 P900-903 62-11522 <=>
1961 V137 N4 P968-969 53P64R <ATS>
1961 V137 N4 P972-975 62-25544 <=*>
1961 V137 N5 P1102-1105 61-27821 <*=>
1961 V137 N5 P1174-1176 36N55R <ATS>
1961 V137 N6 P1334-1335 62-14443 <=*>
1961 V137 N6 P1354-1355 97R76R <ATS>
1961 V137 N6 P1370-1373 63-24334 <=*$>
64-10922 <=*$>
1961 V137 N6 P1481-1484 C-3728 <NRCC>
1961 V138 N1 P119-122 45N55R <ATS>
1961 V138 N1 P125-126 61-18685 <*=>
1961 V138 N1 P173-176 35N56R <ATS>
1961 V138 N2 P320-321 62-15352 <*=>
1961 V138 N2 P344-347 62-19904 <=>
1961 V138 N2 P351-354 63-19952 <=*> 0
1961 V138 N2 P581-583 62-13019 <*=>
1961 V138 N3 P529-532 62-32293 <=*>
1961 V138 N3 P584-586 62-11521 <=>
62-13632 <*=>
62-15428 <*=>
1961 V138 N3 P591-594 62-25674 <=*>
63-24110 <=*$>
1961 V138 N3 P619-620 62-33368 <=*>
1961 V138 N3 P621-624 63-13861 <=*>
1961 V138 N3 P651-654 61-28441 <*=>
1961 V138 N4 P813-816 62-33203 <=*>
1961 V138 N4 P839-842 02N57R <ATS>
1961 V138 N4 P843-845 63-15131 <=*>
1961 V138 N4 P852-855 63-15129 <=*>
1961 V138 N4 P856-858 62-23004 <*=>
1961 V138 N4 P866-869 71N56R <ATS>
1961 V138 N4 P870- ICE V2 N1 P126-128 <ICE>
1961 V138 N4 P880- ICE V2 N1 P129-131 <ICE>
1961 V138 N4 P880-883 62-32510 <=*>
1961 V138 N4 P886-889 61-31629 <=>
1961 V138 N4 P958-961 62-18630 <=*>
1961 V138 N5 P1043-1046 61-28969 <*=>
1961 V138 N5 P1050-1053 63-11750 <=>
1961 V138 N5 P1095-1098 72N54R <ATS>
1961 V138 N5 P1126-1129 62-14628 <=*>
1961 V138 N5 P1245-1247 62-15458 <=*>
1961 V138 N6 P1317-1320 62-23423 <*=>
1961 V138 N6 P1349-1352 93N56R <ATS>
1961 V138 N6 P1399-1401 96N54R <ATS>
1961 V138 N6 P1453-1455 62-18631 <=*>
1961 V139 P1396-1399 62-10682 <*=>
1961 V139 N1 P60- ICE V2 N1 P123-125 <ICE>
1961 V139 N1 P67-70 62-11118 <=>
1961 V139 N1 P91-93 62-13262 <*=>
64-13363 <=*$>
1961 V139 N1 P99-101 64-10095 <=*$>
1961 V139 N1 P124-127 82N56R <ATS>
1961 V139 N1 P214-216 08P59R <ATS>
1961 V139 N1 P217-218 09P59R <ATS>
1961 V139 N2 P263-266 62-16567 <=*>
1961 V139 N2 P275-278 62-13018 <=*>
62-32294 <=*>
1961 V139 N2 P320-323 61-28740 <*=>
1961 V139 N2 P324-326 61-28886 <*=>
1961 V139 N2 P327-330 62-13046 <*=>
1961 V139 N2 P355-358 62-10650 <=*>
1961 V139 N2 P398-401 63-19796 <=*$>
1961 V139 N2 P510-512 63-23232 <=*$>
1961 V139 N3 P556-559 61-28897 <*=>
1961 V139 N3 P560-565 61-31690 <*=>
1961 V139 N3 P566-569 61-28862 <*=>
1961 V139 N3 P570-573 61-28893 <*=>
62-25911 <=*>
1961 V139 N3 P585-586 62-13043 <*=>
1961 V139 N3 P605-607 25N55R <ATS>
63-23354 <=*$>
1961 V139 N3 P626-629 62-16568 <=*>
1961 V139 N3 P630-633 62-24348 <=*>
1961 V139 N3 P723-725 62-15363 <*=>
1961 V139 N3 P726-728 62-15354 <*=>
1961 V139 N4 P859-862 AEC-TR-4961 <=*>
91N57R <ATS>
1961 V139 N4 P874-876 54P58R <ATS>
1961 V139 N4 P895-898 2816 <BISI>
1961 V139 N4 P899-902 65-11953 <*>
1961 V139 N4 P916-918 62-16570 <=*>
1961 V139 N4 P938-941 62-19272 <=*>
1961 V139 N4 P967-969 65-25354 <$> 0
1961 V139 N5 P1063-1066 62-16596 <=*>
1961 V139 N5 P1071-1074 63-18068 <=*>
1961 V139 N5 P1098-1100 62-25943 <=*>
1961 V139 N5 P1132-1135 RTS-2787 <NLL>
1961 V139 N5 P1145-1148 62-18383 <=*>
1961 V139 N5 P1153-1156 62-10648 <=*>
1961 V139 N5 P1157- ICE V2 N2 P198-199 <ICE>
1961 V139 N6 P1325-1328 63-18066 <=*>
1961 V139 N6 P1359-1362 62-10244 <=*>
1961 V139 N6 P1371 30N58R <ATS>
1961 V139 N6 P1389-1391 34P60R <ATS>
1961 V139 N6 P1396-1399 63-18433 <*=>
1961 V139 N6 P1405-1408 62-14349 <=*>
1961 V139 N6 P1409-1412 55P58R <ATS>
1961 V139 N6 P1445-1448 64-23568 <=$>
1961 V140 N1 P23-26 62-13652 <*=>
1961 V140 N1 P77-80 62-23416 <=*>
1961 V140 N1 P81-83 62-11084 <=>
1961 V140 N1 P122-124 62-32818 <=*>
1961 V140 N1 P141-144 62-15423 <*=>
1961 V140 N1 P145-148 63P58R <ATS>
1961 V140 N1 P159-161 63-13716 <=*>
1961 V140 N1 P168- ICE V2 N2 P153-155 <ICE>
1961 V140 N1 P172-175 62-10635 <=*>
1961 V140 N1 P179-180 64-13722 <=*$>
1961 V140 N1 P219-222 63-15142 <*=>

Citation	Translation
1961 V140 N2 P371-373	63-19870 <=*>
1961 V140 N2 P381-383	65-13998 <*>
1961 V140 N2 P409-411	62-15836 <=*>
1961 V140 N2 P412-	ICE V2 N2 P297-299 <ICE>
1961 V140 N3 P575-578	49N57R <ATS>
	62-23597 <=*>
1961 V140 N3 P575-582	62-25950 <=*>
1961 V140 N3 P579-582	50N57R <ATS>
1961 V140 N3 P583-586	63-11710 <=>
1961 V140 N3 P591-594	51N57R <ATS>
	62-23597 <=*>
	62-25950 <=*>
1961 V140 N3 P598-600	63-15196 <*=>
1961 V140 N3 P637-640	88P60R <ATS>
1961 V140 N4 P787-790	61-00948 <*>
1961 V140 N4 P787-7900	62-15228 <*=>
1961 V140 N4 P818-821	62-33367 <=*>
1961 V140 N4 P822-824	63-13717 <=*>
1961 V140 N4 P859-	ICE V2 N3 P333-336 <ICE>
1961 V140 N4 P867-869	62-15831 <=*>
1961 V140 N4 P870-873	62-10636 <=*>
1961 V140 N4 P935-937	62-15667 <*=>
1961 V140 N4 P942-945	62-15946 <=*>
1961 V140 N4 P950-951	62-15941 <=*>
1961 V140 N5 P1041-1044	61-00949 <*>
	62-15228 <*=>
1961 V140 N5 P1070-1072	75P59R <ATS>
1961 V140 N5 P1087-1089	65-17334 <*>
1961 V140 N5 P1100-1101	62-24352 <=*>
1961 V140 N5 P1110-	ICE V2 N3 P345-348 <ICE>
1961 V140 N5 P1121-	ICE V2 N3 P303-305 <ICE>
1961 V140 N5 P1121-1124	62-18710 <=*>
	73P59R <ATS>
1961 V140 N5 P1125-1127	62-20388 <=*>
1961 V140 N5 P1195-1198	62-16295 <=*> 0
	62-24944 <=*>
1961 V140 N6 P1274-1277	62-13664 <*=>
1961 V140 N6 P1307-1309	62-20389 <=*>
1961 V140 N6 P1317-1320	SCL-T-431 <=*>
1961 V140 N6 P1321-1323	63-16723 <=*>
1961 V140 N6 P1327-1329	62-14991 <=*>
1961 V140 N6 P1348-1351	12P59R <ATS>
1961 V140 N6 P1376-1379	64-16320 <=*$>
1961 V140 N6 P1445-1447	63-16720 <=*>
1961 V141 P461-463	62-23576 <=*>
1961 V141 P505-508	62-23576 <=*>
1961 V141 N1 P74-76	AD-633 829 <=$>
	63-15172 <=*>
1961 V141 N1 P90-93	96P58R <ATS>
1961 V141 N1 P117-	ICE V2 N3 P412-415 <ICE>
1961 V141 N1 P117-120	62-20393 <=*>
1961 V141 N1 P125-128	62-14990 <=*>
1961 V141 N1 P139-142	62-24925 <=*>
1961 V141 N1 P165-167	63-16980 <=*>
1961 V141 N2 P285-287	62-11663 <=>
1961 V141 N2 P330-333	62-24924 <=*>
1961 V141 N2 P334-337	RCT V36 N1 P64 <RCT>
1961 V141 N2 P357-360	62-14985 <=*>
1961 V141 N2 P378-380	62-10230 <=*>
1961 V141 N2 P384-386	62-23584 <=*>
1961 V141 N2 P477-480	63-16719 <=*>
1961 V141 N3 P613-615	64-20165 <*> 0
1961 V141 N3 P632-635	62-26051 <=$>
	64-20164 <*> 0
1961 V141 N3 P652-654	63-13714 <=*>
1961 V141 N3 P659-	ICE V2 N3 P450-452 <ICE>
1961 V141 N3 P659-661	62-24028 <=*>
1961 V141 N3 P662-664	10P60R <ATS>
1961 V141 N3 P679-	ICE V2 N3 P419-421 <ICE>
1961 V141 N4 P869-871	62-14995 <=*>
1961 V141 N4 P887-890	SCL-T-429 <=*>
1961 V141 N4 P987-990	63-16726 <=*>
1961 V141 N5 P1078-1081	62-32120 <=*>
1961 V141 N5 P1115-1157	98P60R <ATS>
1961 V141 N5 P1117-1119	79T94R <ATS>
1961 V141 N5 P1120-1123	62-25545 <=*>
1961 V141 N5 P1131-1134	62-25545 <=*>
1961 V141 N5 P1135-1338	62-16183 <=*>
1961 V141 N5 P1253-1256	63-16459 <=*>
1961 V141 N6 P1518-1521	63-16458 <*=>
1961 V141 N6 P1522-1524	62-20304 <=*> 0
1962 V142 P633-636	RCT V35 N4 P877 <RCT>
1962 V142 N1 P59-62	63-10852 <=*>
1962 V142 N1 P67-68	62-18696 <=*>
1962 V142 N1 P88-91	62-11686 <=>
	62-14992 <=*>
1962 V142 N1 P113-116	63-13424 <=*>
1962 V142 N1 P149-151	63-14329 <=*>
	99P60R <ATS>
1962 V142 N1 P208-218	62-25504 <=>
1962 V142 N1 P233-236	62-25504 <=>
1962 V142 N1 P249-252	62-18056 <=*>
1962 V142 N2 P313-316	62-11776 <=>
	62-18054 <=*>
1962 V142 N2 P322-325	66P60R <ATS>
1962 V142 N2 P337-339	62P62R <ATS>
1962 V142 N2 P347-350	64-20163 <*> 0
1962 V142 N2 P354-357	62-14994 <=*>
	63-13181 <=*>
1962 V142 N2 P377-379	62-18055 <=*>
1962 V142 N2 P407-410	RCT V36 N1 P143 <RCT>
1962 V142 N2 P484-	FPTS V22 N3 P.T401 <FASE>
1962 V142 N3 P569-599	62-16287 <=*> 0
1962 V142 N3 P572-575	63-10158 <=*>
1962 V142 N3 P587-588	65-61275 <=$> C
1962 V142 N3 P589-592	64-20162 <*>
1962 V142 N3 P612-614	62-14993 <=*>
1962 V142 N3 P615-618	62-18033 <=*>
1962 V142 N3 P639-641	63-10319 <=*>
1962 V142 N3 P699-705	62-25722 <=>
1962 V142 N4 P785-787	62-24705 <=*>
1962 V142 N4 P838-840	62-18061 <=*>
1962 V142 N4 P866-869	62-18031 <=*>
1962 V142 N5 P1050-1053	62-32527 <=*>
1962 V142 N5 P1054-1057	62-33063 <=*>
1962 V142 N5 P1077-1080	62-18029 <=*>
	62-32988 <=*>
1962 V142 N5 P1130-1133	63-10775 <=*>
1962 V142 N5 P1212-1215	64-19374 <=$>
1962 V142 N6 P1294-1297	RTS-2873 <NLL>
1962 V142 N6 P1327-1330	CSIR-TR.271 <CSSA>
1962 V142 N6 P1342-1345	63-24212 <=*$>
1962 V142 N6 P1405-1408	63-10180 <=*> C
1962 V143 N1 P69-71	302 <LS>
1962 V143 N1 P162-165	63-10181 <=*>
1962 V143 N2 P301-304	N65-22585 <=$>
1962 V143 N2 P312-315	63-23508 <=*$>
1962 V143 N2 P345-347	63-23509 <=*$>
1962 V143 N2 P362-365	88P61R <ATS>
1962 V143 N2 P456-459	63-16725 <=*>
1962 V143 N2 P460-463	63-16729 <=*>
1962 V143 N2 P475-478	62-33641 <=*>
1962 V143 N3 P559-562	63-10898 <=*>
1962 V143 N3 P610-612	62-18820 <=*> 0
1962 V143 N3 P636-639	AEC-TR-5184 <=*>
1962 V143 N4 P818-821	62-32898 <=*>
1962 V143 N4 P822-824	14Q68R <ATS>
	62-20019 <=*>
	63-20018 <=*>
1962 V143 N4 P825-828	15Q68R <ATS>
1962 V143 N4 P951-954	63-19302 <=*>
1962 V143 N4 P976-979	62-32511 <=*>
1962 V143 N5 P1064-1066	62-25485 <=*>
1962 V143 N5 P1087-1089	94D62R <ATS>
1962 V143 N5 P1112-1115	16P63R <ATS>
1962 V143 N5 P1123-1126	17P62R <ATS>
1962 V143 N5 P1156-1158	64-13291 <=*$>
1962 V143 N5 P1211-1218	62-33034 <=*>
1962 V143 N6 P1243-1245	62-32446 <=>
1962 V143 N6 P1304-1307	62-20259 <=*>
1962 V143 N6 P1304-1308	62-32446 <=>
1962 V143 N6 P1317-1320	62-32446 <=>
1962 V143 N6 P1325-1327	64-23569 <=>
1962 V143 N6 P1336-1339	62-32446 <=>
1962 V143 N6 P1345-1347	63-16718 <*=> 0
1962 V143 N6 P1355-1357	62-32446 <=>
	63-23507 <=*$>
1962 V143 N6 P1380-1383	63-19557 <=*>
1962 V143 N6 P1399-1402	62-18227 <=*>
1962 V143 N6 P1409-1412	63-23358 <=*>

Citation	Report
	64-23557 <=>
1962 V144 N1 P129-131	63-14454 <=*>
1962 V144 N1 P180-181	63-24219 <=*$>
1962 V144 N1 P226-229	63-16730 <=*>
1962 V144 N2 P334-337	62-32458 <=*>
1962 V144 N2 P338-340	N65-24109 <=*$>
1962 V144 N2 P347-348	63-18852 <=*> 0
1962 V144 N2 P435-437	03P65R <ATS>
1962 V144 N3 P516-519	62-33028 <=*>
1962 V144 N3 P581-584	63-14312 <=*>
	83P64R <ATS>
1962 V144 N3 P592-595	62P63R <ATS>
1962 V144 N3 P602-605	64-13224 <=*$>
1962 V144 N3 P662	AEC-TR-5489 <=*>
1962 V144 N3 P665-668	63-23405 <=*>
	64-00290 <*>
	64-15713 <=$> 0
1962 V144 N3 P675-677	62-32564 <=*>
1962 V144 N4 P747-750	N65-22855 <=*$>
1962 V144 N4 P778-780	75P62R <ATS>
1962 V144 N4 P795-797	62-18641 <=*>
1962 V144 N4 P817-820	63-18423 <*=>
1962 V144 N5 P1003-1010	62-33459 <=*>
1962 V144 N5 P1022-1025	63-19963 <=*>
1962 V144 N5 P1053-1055	63-10095 <=*>
	63-14317 <=*>
	77P64R <ATS>
1962 V144 N5 P1069-1072	63-19963 <=*>
1962 V144 N5 P1115-1118	63-19587 <=*>
1962 V144 N5 P1160-1167	62-32874 <=*>
1962 V144 N6 P1237-1244	63-21063 <=>
1962 V144 N6 P1245-1250	63-13350 <=*>
1962 V144 N6 P1262-1265	62-20429 <=*>
	63-15570 <=*>
1962 V144 N6 P1366-1368	04P65R <ATS>
1962 V144 N6 P1380-1383	63-21867 <=>
1962 V144 N6 P1414-1417	SNSRC-144 <SNSR>
1962 V145 N1 P48-51	63-10333 <=*>
1962 V145 N1 P48-55	63-13267 <=*>
1962 V145 N1 P78-81	63-20021 <=*>
	93Q72R <ATS>
1962 V145 N1 P85-88	576 <BGIR>
1962 V145 N1 P89-92	63-18440 <*=>
1962 V145 N1 P112-114	65-61424 <=>
1962 V145 N1 P122-124	AEC-TR-5673 <=*$> 0
1962 V145 N1 P125-128	63-18289 <=*>
1962 V145 N1 P129-132	63-16724 <=*>
1962 V145 N1 P136-139	63-18422 <*=>
1962 V145 N1 P140-143	86P65R <ATS>
1962 V145 N1 P195-205	63-13293 <=>
1962 V145 N1 P202-205	63-16734 <=*>
1962 V145 N1 P225-228	SNSRC-145 <SNSR>
1962 V145 N2 P330-331	64-16545 <=*$>
1962 V145 N3 P487-490	63-14414 <=*>
1962 V145 N3 P565-566	63-18419 <=*>
1962 V145 N3 P598-601	63-10705 <=*>
1962 V145 N4 P748-751	62-33678 <=*>
1962 V145 N4 P782-785	63-13481 <=*>
1962 V145 N4 P787-788	63-18416 <*=>
1962 V145 N4 P793-796	64-20161 <*> 0
1962 V145 N4 P822-824	63-20237 <=*$>
1962 V145 N5 P993-995	63-13078 <=*>
1962 V145 N5 P1035-1038	62-20397 <=*>
	63-13554 <=*>
	63-23352 <=*>
1962 V145 N5 P1064-1067	AEC-TR-5766 <*=>
1962 V145 N5 P1081-1084	AEC-TR-5730 <*=>
1962 V145 N5 P1089-1091	63-15455 <=*>
1962 V145 N5 P1092-1094	63-16731 <=*>
1962 V145 N5 P1123-1126	63-20719 <=*$>
1962 V145 N6 P1239-1242	66-20966 <$>
1962 V145 N6 P1255-1258	63-11553 <=>
1962 V145 N6 P1269-1270	64-10940 <=*$>
1962 V145 N6 P1285-1287	63-18424 <*=>
1962 V145 N6 P1328-1330	63-23194 <=*>
1962 V146 N1 P58-61	N65-24661 <=*$>
	63-13476 <=*>
1962 V146 N1 P82-85	63-20704 <=*$>
1962 V146 N1 P86-88	63-13371 <=*>
1962 V146 N1 P115-117	63-18310 <=*> 0

Citation	Report
	63-19195 <=*>
1962 V146 N1 P125-128	69Q72R <ATS>
1962 V146 N1 P217-220	62-33721 <=*>
1962 V146 N1 P242-245	63-13480 <=*>
1962 V146 N2 P263-266	63-15102 <=*>
1962 V146 N2 P344-346	63-21109 <=>
1962 V146 N2 P413-414	63-19651 <=*>
1962 V146 N3 P581-584	N65-24658 <=$>
1962 V146 N3 P596-599	N65-24666 <=*$>
	63-13737 <=*>
1962 V146 N3 P625-627	63-11763 <=>
1962 V146 N3 P638-641	63-23208 <=*>
1962 V146 N3 P666-668	21Q67R <ATS>
1962 V146 N3 P728-730	63-13318 <=*>
1962 V146 N3 P734-736	63-13319 <=*>
1962 V146 N4 P813-815	63-23158 <=*>
1962 V146 N4 P837-839	64-15265 <=*$>
1962 V146 N4 P877-880	63-23157 <=*>
1962 V146 N4 P925-932	63-15361 <=*>
1962 V146 N4 P963-966	64-19554 <=$>
1962 V146 N5 P983-986	63-13484 <=*>
1962 V146 N5 P1141-1142	63-18836 <=*>
1962 V146 N5 P1189-1192	63-21097 <=>
1962 V146 N5 P1193-1196	63-13830 <=*>
1962 V146 N5 P1197-1200	63-21097 <=>
1962 V146 N6 P1305-1308	63-23469 <=*$>
1962 V146 N6 P1327-1330	22Q68R <ATS>
1962 V146 N6 P1335-1336	63-20212 <=*$>
1962 V146 N6 P1337-1339	63-23367 <=*$>
1962 V146 N6 P1347-1349	63-23366 <=*$> 0
1962 V147 P150-152	RCT V36 N4 P1003 <RCT>
1962 V147 P951-957	63-21359 <=>
1962 V147 N1 P92-95	66-11063 <*>
1962 V147 N1 P123-126	63-21765 <=>
1962 V147 N1 P217-220	63-21290 <=>
1962 V147 N1 P252-254	64-15907 <=*$>
1962 V147 N1 P255-258	63-21290 <=>
1962 V147 N2 P350-352	AEC-TR-5969 <=*$>
1962 V147 N2 P399-402	24Q68R <ATS>
1962 V147 N2 P414-417	AEC-TR-5719 <*=>
1962 V147 N2 P474-476	63-21868 <=>
1962 V147 N2 P488-489	63-16732 <=*>
1962 V147 N3 P609-611	AEC-TR-5616 <=*>
1962 V147 N3 P622-624	63-14456 <=*>
	63-16624 <=*>
1962 V147 N3 P636-638	64-10924 <=*$>
1962 V147 N3 P735-737	63-16721 <=*>
1962 V147 N4 P751-754	63-18065 <=*>
1962 V147 N4 P776-778	63-14413 <=*>
1962 V147 N4 P817-818	63-21343 <=>
1962 V147 N4 P826-828	64-15365 <=*$>
1962 V147 N4 P833-834	63-20568 <=*$>
1962 V147 N4 P860-862	64-10707 <=*$>
1962 V147 N4 P870-873	64-14272 <=*$>
1962 V147 N4 P985-988	63-21167 <=>
1962 V147 N5 P1075-1087	66-11062 <*>
1962 V147 N5 P1108-1111	52Q69R <ATS>
1962 V147 N5 P1122-1125	AEC-TR-5714 <=*$>
	63-18309 <=*> 0
	63-24174 <=*$>
1962 V147 N5 P1200-1203	63-11580 <=>
1962 V147 N5 P1247-1249	63-21361 <=>
1962 V147 N6 P1300-1305	63-21323 <=>
1962 V147 N6 P1320-1323	63-18149 <=*>
	63-21357 <=>
	64-13009 <=*$>
1962 V147 N6 P1365-1368	64-15280 <=*$>
1962 V147 N6 P1386-1388	63-16975 <=*> 0
1962 V147 N6 P1409-1412	64-71470 <=>
1962 V147 N6 P1476-1479	63-21476 <=>
1962 V147 N6 P1484-1486	UCRL TRANS-943 <=*$>
1963 V148 P338-341	64-18775 <*>
1963 V148 N1 P47-53	63-21370 <=>
1963 V148 N1 P54-56	63-14415 <*=>
1963 V148 N1 P91-94	65-63557 <=$>
1963 V148 N1 P122-125	64-16679 <=*$>
1963 V148 N1 P132-135	82Q70R <ATS>
1963 V148 N1 P140-143	63-11772 <=>
1963 V148 N2 P296-303	63-21461 <=>
1963 V148 N2 P433-436	64-16414 <=*$>

Citation	Translation
1963 V148 N2 P477-480	64-15503 <=*$> 0
1963 V148 N3 P538-540	63-31016 <=>
1963 V148 N3 P577-580	63-21583 <=>
1963 V148 N3 P585-588	63-21791 <=>
1963 V148 N3 P641-643	64-14948 <=*$>
	65-61329 <=$>
1963 V148 N4 P803-805	64-11929 <=> M
1963 V148 N4 P810-813	63-31269 <=>
1963 V148 N4 P825-828	64-18769 <*>
1963 V148 N4 P829-831	64-10097 <=*$>
1963 V148 N4 P853-855	63-20205 <=*>
1963 V148 N4 P935-938	63-11745 <=>
	63-18600 <=*>
1963 V148 N5 P1102-1105	64-30692 <*>
1963 V148 N5 P1106-1109	64-15993 <=*$>
1963 V148 N5 P1141-1144	64-00087 <*>
	64-15147 <=*$>
1963 V148 N5 P1199-1201	63-31393 <=>
1963 V148 N5 P1207-1209	63-31393 <=>
1963 V148 N6 P1309-1311	64-10706 <=*$>
1963 V148 N6 P1316-1319	63-16416 <*=>
1963 V148 N6 P1389-1391	63-21953 <=>
1963 V148 N6 P1392-1393	C-4725 <NRC>
	64-26254 <$>
1963 V148 N6 P1397-1399	63-31256 <=>
1963 V148 N6 P1425-1427	63-31256 <=>
1963 V149 N1 P52-53	63-21840 <=>
1963 V149 N1 P65-67	63-11647 <=>
1963 V149 N1 P114-116	13070 <K-H>
	65-14871 <*>
1963 V149 N1 P117-119	63-16408 <=*>
1963 V149 N1 P142-145	63-31273 <=>
1963 V149 N1 P185-188	63-31273 <=>
1963 V149 N1 P194-201	63-31273 <=>
1963 V149 N1 P205-206	63-31273 <=>
1963 V149 N1 P213-216	63-31273 <=>
1963 V149 N2 P268-271	63-31320 <=>
1963 V149 N2 P276-279	63-20939 <=*$>
1963 V149 N2 P284-287	63-19624 <=*$>
1963 V149 N2 P284-291	63-31119 <=>
1963 V149 N2 P292-294	64-11922 <=> M
1963 V149 N2 P321-323	63-31320 <=>
1963 V149 N2 P363-366	63-31320 <=>
1963 V149 N2 P431-441	63-31320 <=>
1963 V149 N2 P438-441	63-16410 <=*>
1963 V149 N2 P456-459	63-18601 <=*>
1963 V149 N2 P469-472	63-31659 <=>
1963 V149 N3 P580-582	63-20993 <=*$>
1963 V149 N3 P624-643	64-71355 <=> M
1963 V149 N3 P644-647	64-14435 <=*$>
	97Q71R <ATS>
1963 V149 N3 P692-695	30Q73R <ATS>
1963 V149 N4 P759-762	63-18707 <=*>
1963 V149 N4 P960-962	63-31467 <=>
1963 V149 N4 P973-975	63-31466 <=>
1963 V149 N5 P1067-1070	64-14886 <=*$>
1963 V149 N5 P1084-1087	63-11765 <=>
1963 V149 N5 P1134-1136	63-24569 <=*$>
1963 V149 N5 P1157	AEC-TR-6125 <=*$>
1963 V149 N5 P1194-1196	63-31864 <=>
1963 V149 N5 P1213-1216	63-31865 <=>
1963 V149 N6 P1311-1314	63-31292 <=>
1963 V149 N6 P1384-1386	RCT V37 N1 P99 <RCT>
1963 V149 N6 P1414-1415	85Q71R <ATS>
1963 V149 N6 P1428-1431	63-18606 <=*$>
1963 V150 N1 P21-22	63-31456 <=>
	64-13937 <=*$>
1963 V150 N1 P48-51	64-13937 <=*$>
1963 V150 N1 P99-101	63-18890 <=*>
1963 V150 N1 P140-142	C-4925 <NRC>
1963 V150 N1 P174-175	63-18607 <=*$>
1963 V150 N1 P195-198	65-60651 <=$>
1963 V150 N2 P275-278	63-18622 <=*> 0
	65-10819 <*> 0
1963 V150 N2 P317-320	64-14757 <=*$>
1963 V150 N2 P411-413	63-18608 <=*$>
1963 V150 N3 P455-458	64-11936 <=> M
1963 V150 N3 P499-502	<IPIX>
1963 V150 N3 P547-550	53Q71R <ATS>
	66-11003 <*>
1963 V150 N3 P574-577	66-10542 <*>
1963 V150 N3 P584-587	29Q73R <ATS>
1963 V150 N3 P635-638	51Q72R <ATS>
1963 V150 N3 P665-679	63-41228 <=>
1963 V150 N4 P788-790	63-31694 <=>
	64-15435 <=*$>
1963 V150 N4 P823-825	64-18483 <*>
1963 V150 N4 P829-832	65-13402 <*>
1963 V150 N4 P856-858	63-20607 <=*$>
1963 V150 N4 P920-923	64-14797 <=*$>
1963 V150 N4 P924-927	64-14796 <=*$>
1963 V150 N5 P967-970	64-71495 <=> M
1963 V150 N5 P1066-1068	64-10705 <=*$>
1963 V150 N5 P1077-1080	AEC-TR-6178 <=*$>
1963 V150 N5 P1146-1148	63-31819 <=>
1963 V150 N6 P1281-1284	63-20470 <=*$> 0
1963 V150 N6 P1311-1314	63-31971 <=>
1963 V150 N6 P1366-1369	63-41017 <=>
1963 V150 N6 P1370-1372	63-31973 <=>
1963 V150 N6 P1378-1381	64-14795 <=*$>
1963 V150 N6 P1397-1400	63-31974 <=>
1963 V151 N1 P9-79	<AMS>
1963 V151 N1 P80-83	64-19362 <=$>
1963 V151 N1 P110-113	65-18154 <*>
1963 V151 N1 P203-205	63-41258 <=>
1963 V151 N1 P209-212	63-41258 <=>
1963 V151 N1 P213-216	63-41258 <=>
1963 V151 N1 P217-219	63-41258 <=>
1963 V151 N2 P247-294	<AMS>
1963 V151 N2 P323-325	63-31692 <=>
1963 V151 N2 P365-368	64-10948 <=*$> 0
1963 V151 N2 P369-372	N65-15158 <=$*>
1963 V151 N2 P392-395	64-16402 <=*$>
1963 V151 N3 P479-514	<AMS>
1963 V151 N3 P493-496	63-31938 <=>
1963 V151 N3 P515-518	63-31938 <=>
1963 V151 N3 P519-521	65-10814 <*>
1963 V151 N3 P532-535	65-10750 <*>
1963 V151 N3 P560-563	63-20853 <=*$>
	64-71154 <=> M
	65-10776 <*>
1963 V151 N3 P620-623	64-16412 <=*$>
1963 V151 N3 P634-637	64-15994 <=*$>
1963 V151 N3 P687-690	63-41024 <=>
1963 V151 N3 P700-703	63-41022 <=>
1963 V151 N3 P712-713	63-41283 <=>
1963 V151 N3 P714-717	64-31198 <=>
1963 V151 N3 P718-721	63-41023 <=>
1963 V151 N3 P722-724	RTS-2732 <NLL>
1963 V151 N3 P732-736	64-15175 <=*$> 0
1963 V151 N4 P751-758	<AMS>
1963 V151 N4 P811-814	65-10749 <*>
1963 V151 N4 P822-825	64-14177 <=*$> C
1963 V151 N4 P841-845	AD-611 119 <=$>
1963 V151 N4 P841-844	M.5431 <NLL> 0
	63-31961 <=>
1963 V151 N4 P886-889	AD-608 052 <=>
1963 V151 N4 P982-985	N64-26782 <=>
1963 V151 N4 P986-988	64-00046 <*>
	64-13795 <=*$>
1963 V151 N5 P999-1037	<AMS>
1963 V151 N5 P1042-1045	64-11843 <=> M
1963 V151 N5 P1046-1049	64-11952 <=> M
1963 V151 N5 P1053-1055	63-20570 <=*$>
1963 V151 N5 P1074-1075	64-15226 <=*$>
1963 V151 N5 P1089-1092	63-41058 <=>
	65-61360 <=$> C
1963 V151 N5 P1093-1096	<SIC>
1963 V151 N5 P1104-1107	63-41020 <=>
1963 V151 N5 P1110-1113	64-10715 <=*$>
1963 V151 N5 P1118-1119	64-16680 <=*$>
1963 V151 N5 P1127-1130	64-18436 <*>
	85R74R <ATS>
1963 V151 N5 P1139-1142	64-16677 <=*$>
1963 V151 N5 P1195-1197	63-41065 <=> 0
1963 V151 N5 P1222-1224	63-41064 <=> 0
	65-60024 <=$> 0
	65-60649 <=$> 0
1963 V151 N5 P1236-1237	64-21009 <=>
1963 V151 N6 P1247-1294	<AMS>

Citation	Translation
1963 V151 N6 P1295-1298	63-41260 <=>
1963 V151 N6 P1315-1318	63-41050 <=>
1963 V151 N6 P1322-1325	64-16979 <*>
1963 V151 N6 P1399-1401	63-41055 <=>
1963 V151 N6 P1437-1440	64-21051 <=>
1963 V151 N6 P1450-1452	64-21014 <=> 0
1963 V151 N6 P1456-1457	64-21051 <=>
1963 V151 N6 P1475-1478	64-21051 <=>
1963 V152 P1383-1386	66-12233 <*>
1963 V152 N1 P67-70	63-41154 <=>
1963 V152 N1 P75-77	64-21212 <=>
1963 V152 N1 P114-116	64-16980 <=*$>
	65-18168 <*>
1963 V152 N1 P127-130	63-41153 <=>
1963 V152 N1 P131-133	AD-611 549 <=$>
1963 V152 N1 P143-146	TRC-TRANS-1107 <NLL>
1963 V152 N1 P147-150	96R75R <ATS>
1963 V152 N1 P168-170	65-11441 <*>
1963 V152 N1 P202-204	63-41257 <=>
1963 V152 N1 P215-217	63-41257 <=>
1963 V152 N1 P218-220	64-13649 <=*$> 0
1963 V152 N1 P241-243	63-41202 <=>
1963 V152 N1 P246-248	63-41201 <=>
1963 V152 N1 P845-848	64-21164 <=>
1963 V152 N2 P299-301	64-21053 <=>
1963 V152 N2 P324-326	65-10858 <*>
1963 V152 N2 P359-362	65-10482 <*>
1963 V152 N2 P441-442	64-21238 <=>
1963 V152 N2 P450-453	64-21245 <=>
1963 V152 N2 P457-460	64-21242 <=>
1963 V152 N2 P471-475	64-19540 <=$>
1963 V152 N2 P492-493	64-21244 <=> 0
1963 V152 N2 P501-504	64-21243 <=>
1963 V152 N3 P515-518	64-21861 <=>
1963 V152 N3 P598-601	64-21515 <=>
1963 V152 N3 P629-632	86R74R <ATS>
1963 V152 N3 P671-673	65-10874 <*>
1963 V152 N3 P724-726	64-21600 <=>
1963 V152 N3 P737-743	64-21994 <=>
1963 V152 N4 P911-914	36Q74R <ATS>
1963 V152 N4 P992-994	64-21167 <=>
1963 V152 N4 P1001-1004	64-21166 <=>
	65-63063 <=$>
1963 V152 N4 P1005-1008	65-61215 <=$>
1963 V152 N5 P1086-1088	64-21508 <=>
	65-64599 <=*$>
1963 V152 N5 P1089-1091	64-71568 <=>
1963 V152 N5 P1108-1110	64-11670 <=>
1963 V152 N5 P1143-1146	63R74R <ATS>
1963 V152 N5 P1181-1184	64-18442 <*>
1963 V152 N5 P1192-1195	65-11092 <*>
1963 V152 N5 P1222-1224	64-21275 <=>
1963 V152 N5 P1225-1226	64-21280 <=>
1963 V152 N5 P1227-1230	64-11660 <=>
	64-21194 <=>
1963 V152 N5 P1231-1234	64-14793 <=*$>
	64-21471 <=>
1963 V152 N6 P1461-1464	64-21260 <=> 0
1963 V152 N6 P1465-1466	64-21259 <=>
1963 V153 N1 P61-63	64-21639 <=>
1963 V153 N1 P111-113	66-10001 <*>
1963 V153 N1 P140-143	71R74R <ATS>
1963 V153 N1 P162-165	65-27473 <$>
1963 V153 N1 P212-215	64-14791 <=*$>
1963 V153 N1 P243-245	64-21777 <=>
1963 V153 N2 P303-305	M-5664 <NLL>
1963 V153 N2 P323-325	64-21381 <=>
1963 V153 N2 P363-365	65-60571 <=$>
1963 V153 N2 P383-385	65-11756 <*>
1963 V153 N2 P408-411	64-19852 <=$>
1963 V153 N2 P412-415	65-27472 <$>
1963 V153 N2 P427-428	64-21383 <=>
1963 V153 N2 P444-446	04R75R <ATS>
1963 V153 N2 P450-453	64-21877 <=>
1963 V153 N2 P485-488	64-21878 <=>
1963 V153 N3 P578-580	AD-610 295 <=$>
1963 V153 N3 P585-587	64-31078 <=>
1963 V153 N3 P631-633	64-16827 <*>
1963 V153 N3 P631-634	65-63788 <=> 0
1963 V153 N3 P661-663	46R76R <ATS>
1963 V153 N3 P664-667	64-21496 <=>
1963 V153 N3 P668-671	64-21495 <=>
1963 V153 N3 P718-720	64-14790 <=*$>
1963 V153 N3 P721-724	64-14789 <=*$>
1963 V153 N4 P747-750	64-21761 <*>
1963 V153 N4 P768-782	64-21761 <*>
1963 V153 N4 P863-864	65-11954 <*>
1963 V153 N4 P926-929	64-31075 <=>
1963 V153 N4 P930-932	64-21992 <=>
1963 V153 N4 P933-935	64-21819 <=>
1963 V153 N4 P933-938	64-27226 <$>
1963 V153 N4 P936-938	64-21822 <=>
1963 V153 N4 P947-949	64-21820 <=>
1963 V153 N4 P950-953	64-31000 <=>
1963 V153 N4 P970-973	64-21821 <=>
1963 V153 N4 P981-983	64-21818 <=>
1963 V153 N5 P1020-1023	64-31334 <=>
1963 V153 N5 P1027-1029	64-71550 <=>
1963 V153 N5 P1067-1070	64-21581 <=>
	65-61262 <=$> 0
1963 V153 N5 P1071-1072	65-13019 <*>
1963 V153 N5 P1097-1100	36R80R <ATS>
1963 V153 N5 P1119-1121	3806 <BISI>
1963 V153 N5 P1125-1128	92R78R <ATS>
1963 V153 N5 P1129-1131	65-12397 <*>
1963 V153 N5 P1132-1135	64-14788 <=*$>
1963 V153 N5 P1193-1194	64-21849 <=>
1963 V153 N5 P1199-1201	64-21942 <=>
1963 V153 N5 P1202-1203	64-21943 <=>
1963 V153 N5 P1204-1206	64-21941 <=>
1963 V153 N5 P1213-1215	64-21944 <=>
1963 V153 N6 P1284-1287	AD-610 298 <=$>
1963 V153 N6 P1299-1302	AD-610 294 <=$> 0
	64-21578 <=>
1963 V153 N6 P1303-1306	64-21500 <=>
	65-61364 <=$> 0
1963 V153 N6 P1307-1309	64-21501 <=>
1963 V153 N6 P1313-1314	65-10852 <*>
1963 V153 N6 P1327-1329	53R75R <ATS>
1963 V153 N6 P1338-1341	64-31052 <=>
1963 V153 N6 P1395-1397	76R75R <ATS>
1963 V153 N6 P1398-1399	64-21499 <=>
1963 V153 N6 P1450-1453	64-31052 <=>
1964 V154 N1 P38-40	RTS-2831 <NLL>
1964 V154 N1 P80-82	65-12398 <*>
1964 V154 N1 P121-124	64-26426 <$>
1964 V154 N2 P287-289	64-16652 <=*$>
1964 V154 N2 P300-301	65-10895 <*>
1964 V154 N2 P325-328	65-12399 <*>
1964 V154 N2 P407-410	34R76R <ATS>
1964 V154 N2 P460-462	64-18829 <*>
1964 V154 N2 P463-466	65-60257 <=$> 0
1964 V154 N2 P471-472	64-19898 <=$>
1964 V154 N3 P495-496	ORNL-TR-42 <=$>
1964 V154 N3 P695-698	56R76R <ATS>
1964 V154 N3 P714-717	64-18830 <*>
1964 V154 N4 P854-856	65-12647 <*>
1964 V154 N4 P877-880	RTS 2743 <NLL>
1964 V154 N4 P890-893	65-13014 <*>
1964 V154 N4 P933-935	64-26395 <$>
1964 V154 N4 P946-949	64-18826 <*>
1964 V154 N4 P950-952	64-18827 <*>
1964 V154 N4 P974-977	64-18828 <*>
	64-31395 <=>
1964 V154 N5 P1082-1083	64-16681 <=*$>
1964 V154 N5 P1113-1115	64-23582 <=$>
1964 V154 N5 P1125-1127	65-63214 <=$> C
1964 V154 N5 P1132-1134	65-61180 <=>
1964 V154 N5 P1145-1148	65-11591 <*>
1964 V154 N6 P1303-1305	65-10778 <*>
1964 V154 N6 P1325-1327	64-31421 <=>
1964 V154 N6 P1414-1416	50R77R <ATS>
1964 V154 N6 P1417-1420	11R77R <ATS>
1964 V154 N6 P1425-1428	64-31493 <=>
1964 V154 N6 P1448-1451	64-31613 <=>
1964 V155 P640-643	SC-T-65-706 <=$>
1964 V155 N1 P35-37	AD-619 432 <=$>
1964 V155 N1 P70-71	64-11875 <=>
1964 V155 N1 P160-163	65-10418 <*>

Citation	Translation
1964 V155 N2 P295-301	64-31582 <=>
1964 V155 N2 P316-319	64-23642 <=$>
1964 V155 N2 P320-322	TRC-TRANS-1095 <NLL>
	65-28107 <$>
1964 V155 N2 P385-388	64-31574 <=>
1964 V155 N2 P454-456	64-31572 <=>
1964 V155 N2 P457-460	64-31575 <=>
1964 V155 N3 P626-628	64-31497 <=>
1964 V155 N3 P632-636	64-31497 <=>
1964 V155 N3 P640-643	65-14257 <*>
1964 V155 N3 P662-665	64-31497 <=>
1964 V155 N3 P688-690	64-41095 <=>
1964 V155 N3 P698-701	64-41095 <=>
	65-63673 <=$>
1964 V155 N3 P711-714	64-41095 <=>
1964 V155 N4 P783	65-10817 <*>
1964 V155 N4 P784-787	AD-609 155 <=>
1964 V155 N4 P788-791	64-31345 <=>
1964 V155 N4 P839-842	66-11640 <*>
1964 V155 N4 P900-903	64-51654 <=$>
1964 V155 N4 P937-939	ANL-TRANS-146 <=$>
1964 V155 N5 P1039-1041	64-31672 <=>
1964 V155 N5 P1054-1057	M-5580 <NLL>
1964 V155 N5 P1062-1065	64-31671 <=>
1964 V155 N5 P1066-1069	AD-621 055 <=*$>
1964 V155 N5 P1108-1110	64-18546 <*>
1964 V155 N5 P1212-1215	64-18832 <*>
1964 V155 N6 P1306-1309	M-5585 <NLL>
1964 V155 N6 P1310-1313	AD-618 903 <=$>
1964 V155 N6 P1314-1316	64-31504 <=>
1964 V155 N6 P1341-1344	65-12400 <*>
1964 V155 N6 P1345-1347	33R77R <ATS>
1964 V155 N6 P1354-1356	65-14716 <*>
1964 V155 N6 P1429-1431	64-51990 <=$>
1964 V155 N6 P1457-1459	65-60572 <=$>
1964 V156 N1 P47-49	N65-13541 <=$>
1964 V156 N1 P125-127	64-18834 <*>
1964 V156 N2 P308-311	TRANS-318(M) <NLL>
1964 V156 N2 P372-374	65-13079 <*> 0
1964 V156 N2 P430-433	65-10847 <*>
1964 V156 N2 P471-473	64-18835 <*>
1964 V156 N3 P604-607	64-30038 <*> 0
1964 V156 N3 P637-640	64-41934 <=$>
1964 V156 N3 P647-649	AD-620 076 <=$*>
1964 V156 N3 P673-676	76R79R <ATS>
1964 V156 N4 P763-765	83R78R <ATS>
1964 V156 N4 P792-794	84R78R <ATS>
1964 V156 N4 P803-805	ORNL-TR-456 <=$>
1964 V156 N4 P810-813	64-51964 <=$>
	65-11355 <*>
1964 V156 N4 P920-923	AD-619 321 <$=>
1964 V156 N4 P972-975	AD-620949 <=*$>
1964 V156 N5 P1057-1060	65-20550 <$>
1964 V156 N5 P1065-1067	66-11269 <*>
1964 V156 N5 P1102-1104	65-10417 <*>
	65-13488 <*>
1964 V156 N5 P1121-1123	65-60878 <=$>
1964 V156 N5 P1156-1158	M-5722 <NLL>
	65-17445 <*>
	66-13465 <*>
1964 V156 N5 P1244-1247	65-61140 <=$>
1964 V156 N6 P1277-1280	65-23804 <$>
1964 V156 N6 P1336-1338	64-41356 <=>
	65-14445 <*>
	65-60615 <=$> 0
	65-63428 <=$>
1964 V156 N6 P1341-1342	C-5040 <NRC>
1964 V156 N6 P1375-1378	65-60880 <=$>
1964 V156 N6 P1402-1405	65-10416 <*>
1964 V156 N6 P1472-1475	N65-34507 <=$>
1964 V157 N2 P309-312	N64-30841 <=$>
1964 V157 N2 P384-387	ICE V5 N1 P.79-82 <ICE>
	ICE V5 N1 P79-82 <ICE>
1964 V157 N2 P422-425	65-64407 <=*$>
1964 V157 N2 P426-429	38R80R <ATS>
1964 V157 N3 P554-556	N64-30842 <=$>
1964 V157 N3 P580-582	65-12207 <*>
1964 V157 N4 P822-825	N65-13572 <=$>
1964 V157 N4 P826-829	64-41779 <=>
1964 V157 N4 P913-916	65-11486 <*> 0
	65-62126 <=$>
1964 V157 N4 P975-981	64-51494 <=>
1964 V157 N5 P1153-1155	65-10444 <*>
1964 V157 N6 P1459-1462	64-51445 <=$>
1964 V158 N1 P74-77	AD-625 065 <=*$>
1964 V158 N1 P126-129	65-31323 <*>
1964 V158 N1 P141-142	AD-624 912 <=*$>
1964 V158 N2 P302-304	AD-613 981 <=$>
	64-51812 <=$>
1964 V158 N2 P427-428	65-13339 <*>
1964 V158 N2 P452-455	AD-625 150 <=*$>
	AD-629 864 <=*$>
1964 V158 N2 P460-463	65-12948 <*>
1964 V158 N3 P562-565	N65-11448 <=$>
1964 V158 N3 P598-601	17S81R <ATS>
1964 V158 N3 P682-684	AD-619 322 <=$>
1964 V158 N3 P706-709	828 <TC>
1964 V158 N3 P722-725	N65-32973 <=$>
1964 V158 N3 P730-733	65-12946 <*>
1964 V158 N4 P798-801	AD-619 324 <=$>
1964 V158 N4 P808-810	N65-14605 <=$*>
1964 V158 N4 P811-814	N65-14298 <=$>
1964 V158 N4 P827-830	N65-14611 <=*$>
1964 V158 N4 P872-875	65-63983 <=>
1964 V158 N4 P880-883	65-30125 <=$>
1964 V158 N5 P1108-1111	66-61666 <=*$>
1964 V158 N6 P1278-1280	AD-619 323 <=$>
1964 V158 N6 P1291-1294	N65-14612 <=*>
1964 V158 N6 P1310-1313	65-30566 <=$>
1964 V158 N6 P1320-1322	N65-14568 <=$>
1964 V158 N6 P1320-1323	65-12065 <*>
1964 V159 N1 P39-42	AD-619 484 <=$>
1964 V159 N1 P57-59	N65-14614 <=$*>
1964 V159 N1 P106-108	50S82R <ATS>
1964 V159 N1 P182-185	65-22213 <$>
1964 V159 N1 P196-197	AD-623 972 <=*$>
	65-12950 <*>
1964 V159 N2 P294-297	N65-14701 <=$*>
1964 V159 N2 P306-309	65-23421 <$>
1964 V159 N2 P327-329	18S83R <ATS>
1964 V159 N2 P439-441	N65-15163 <=*$>
1964 V159 N3 P636-639	65-14312 <*>
1964 V159 N3 P690-692	65-30712 <=$>
1964 V159 N4 P761-764	AD-625 156 <=*$>
1964 V159 N4 P779-781	N65-15736 <=$>
1964 V159 N4 P786-788	M-5669 <NLL>
	65-12401 <*>
	65-30262 <=$>
1964 V159 N4 P814-816	49S82R <ATS>
1964 V159 N4 P843-846	AD-619 433 <=$>
1964 V159 N4 P885-886	RAPRA-1226 <RAP>
1964 V159 N4 P923-925	65-30586 <=$>
1964 V159 N5 P1003-1006	SMRE-TRANS-5161 <NLL>
1964 V159 N5 P1013-1016	N65-16311 <=$>
1964 V159 N5 P1055-1058	17S83R <ATS>
1964 V159 N5 P1117-1119	656 <TC>
1964 V159 N5 P1154-1157	65-63758 <=> 0
	65-64302 <=*$> 0
	66-60563 <=*$>
1964 V159 N6 P1247-1248	AD-619 330 <=$>
1964 V159 N6 P1272-1275	65-30304 <=$>
1964 V159 N6 P1367-1370	65-14314 <*>
1964 V159 N6 P1381-1384	65-14748 <*>
1964 V159 N6 P1385-1388	N65-29737 <=*$>
1964 V159 N6 P1402-1404	65-30699 <=$>
1965 V159 N5 P1087-1090	45S85R <ATS>
1965 V160 P1417-1420	65-12941 <*>
1965 V160 P1427-1429	65-12947 <*>
1965 V160 N1 P54-56	N65-21232 <=$>
1965 V160 N1 P61-63	N65-18184 <=$>
1965 V160 N1 P115-118	25S84R <ATS>
1965 V160 N1 P139-142	617 <TC>
	65-13015 <*>
	65-17127 <*>
1965 V160 N1 P154-157	1126 <TC>
1965 V160 N2 P259-262	619 485 <=$>
1965 V160 N2 P355-358	AD-615 870 <=$>
1965 V160 N2 P394-397	AD-636 611 <=$>
	62S84R <ATS>
1965 V160 N2 P405-408	AD-625 153 <=*$>

Citation	Translation
1965 V160 N3 P582-585	65-13935 <*>
1965 V160 N3 P591-593	65-14715 <*>
1965 V160 N3 P629-632	65-13680 <*>
1965 V160 N3 P633-634	16S87R <ATS>
1965 V160 N3 P713-716	AD-619 329 <=$>
1965 V160 N3 P720	65-12952 <*>
1965 V160 N3 P724-727	65-23223 <$>
1965 V160 N4 P815-818	TRANS-351(M) <NLL>
1965 V160 N4 P853-856	AD-625 158 <=*$>
1965 V160 N4 P861-863	65-17465 <*>
1965 V160 N4 P871-874	65-13896 <*>
1965 V160 N4 P960-963	65-30669 <=$>
1965 V160 N4 P972-975	65-31963 <=*$>
1965 V160 N5 P1042-1045	AD-625 058 <=*$>
1965 V160 N5 P1057-1060	N65-29738 <=*$>
1965 V160 N5 P1C93-1096	66-12220 <*>
1965 V160 N6 P1283-1284	N65-22625 <=$>
1965 V16C N1/3 P218-226	65-31196 <=$>
1965 V160 N1/3 P731-733	65-31196 <=$>
1965 V161 N1 P63-65	N66-11143 <=*$>
1965 V161 N1 P66-69	65-31210 <=*$>
1965 V161 N1 P81-83	N66-11134 <=*$>
1965 V161 N1 P107-110	65-13690 <*>
1965 V161 N1 P132-135	59S89R <ATS>
1965 V161 N1 P241-243	66-61544 <=*$>
1965 V161 N2 P315-317	N66-11147 <=*$>
1965 V161 N2 P324-327	27S87R <ATS>
1965 V161 N2 P332-335	AEC-TR-6559 <=$>
1965 V161 N2 P343-346	65-31211 <=*$>
1965 V161 N2 P377-379	66-61862 <=*$>
1965 V161 N2 P392-394	66-13378 <*>
1965 V161 N2 P483-486	65-12942 <*>
1965 V161 N3 P540-546	65-31193 <=$>
1965 V161 N3 P551-553	N66-11138 <=*$>
1965 V161 N3 P556-559	AD-618 389 <=$>
1965 V161 N3 P617-619	65-14161 <*> 0
1965 V161 N3 P637-640	N66-21689 <=$>
1965 V161 N3 P673-675	51S87R <ATS>
1965 V161 N3 P721-723	65-31119 <=$>
1965 V161 N4 P789-790	65-31118 <=$>
1965 V161 N4 P791-794	RAE-LIB-TRANS-1112 <NLL>
1965 V161 N4 P795-798	65-14414 <*>
	65-63713 <=$>
1965 V161 N4 P799-801	AD-627 995 <=*$>
1965 V161 N4 P828-831	65-31373 <=$>
1965 V161 N4 P851-852	60S89R <ATS>
	65-17448 <*>
1965 V161 N4 P986-988	N65-27686 <=*$>
1965 V161 N4 P989-991	65-63963 <=*$>
	66-11284 <*>
1965 V161 N5 P1060-1062	65-17057 <*> 0
1965 V161 N5 P1067-1068	AD-627 12C <=$>
1965 V161 N5 P1114-1117	65-14909 <*>
1965 V161 N5 P1118-	TRC-TRANS-1113 <NLL>
1965 V161 N5 P1121-1123	66-13783 <*>
1965 V161 N5 P1149-1151	66-13379 <*>
1965 V161 N6 P1301-1302	N66-10592 <=$>
1965 V161 N6 P1362-1364	65-17462 <*>
1965 V161 N6 P1365-1367	66-12222 <*>
1965 V161 N1/3 P328-331	N66-11171 <=$>
1965 V162 P546-548	65-22512 <=$>
1965 V162 N1 P33-35	65-32057 <=*$>
1965 V162 N1 P64-69	65-31722 <=$>
1965 V162 N1 P133-135	65-64615 <=$>
1965 V162 N1 P140-143	RAPRA-1251 <RAP>
1965 V162 N1 P201-204	65-32057 <=*$>
1965 V162 N2 P441-445	66-61756 <*=$>
1965 V162 N2 P447-450	65-31960 <=*$>
1965 V162 N3 P516-522	65-31990 <=*$>
1965 V162 N3 P559-562	AD-623 106 <=*$>
1965 V162 N3 P629-631	66-21307 <$>
1965 V162 N3 P688-693	65-32043 <=*$>
1965 V162 N3 P713-718	65-32043 <=*$>
1965 V162 N4 P847-850	66-14171 <*$>
1965 V162 N5 P1015-1018	65-32161 <=$>
1965 V162 N5 P1101-1104	65-17447 <*>
1965 V162 N5 P1105-1108	66-14683 <*$>
1965 V163 N1 P40-42	65-32257 <=$>
1965 V163 N2 P365-368	65-32748 <=*$>
1965 V163 N2 P398-401	66-13377 <*>
1965 V163 N3 P667-670	66-13381 <*>
1965 V163 N4 P841-844	65-33137 <=$>
1965 V163 N4 P873-876	AD-640247 <=$>
1965 V163 N5 P1100-1103	65-64505 <=*$>
	66-10483 <*>
1965 V163 N5 P1121-1123	66-31280 <=$>
1965 V163 N5 P1131-1133	66-13158 <*>
1965 V163 N5 P1134-1137	65-33206 <=*$>
	65-33979 <=*$>
1965 V163 N6 P1331-1333	65-32751 <=*$>
1965 V163 N6 P1377-1380	65-33207 <=*$>
1965 V163 N6 P1481-1483	65-33919 <=*$>
1965 V164 P1069-1072	66-12510 <*>
1965 V164 N1 P64-67	57T90R <ATS>
1965 V164 N1 P75-77	66-11001 <*>
1965 V164 N1 P144-146	66-14568 <*>
1965 V164 N2 P289-291	65-33517 <=*$>
1965 V164 N2 P331-333	66-13287 <*>
1965 V164 N2 P344-346	66-30640 <=*$>
1965 V164 N2 P374-377	32T90R <ATS>
1965 V164 N3 P515-518	65-33727 <=*$>
1965 V164 N3 P545-548	65-33724 <=*$>
1965 V164 N3 P606-6C9	N66-16147 <=$>
1965 V164 N3 P614-617	66-13830 <*>
1965 V164 N3 P629-632	66-12823 <*>
1965 V164 N4 P757-760	65-33516 <=*$>
1965 V164 N4 P796-799	65-33786 <=*$>
1965 V164 N4 P845-848	65-33138 <=$>
1965 V164 N4 P910-912	65-33787 <=*$>
1965 V164 N5 P993-996	65-34072 <=*$>
1965 V164 N5 P1115-1118	66-12822 <*>
1965 V164 N6 P1256-1259	66-30058 <=*$>
1965 V164 N6 P1366-1369	66-30339 <=*$>
1965 V165 P130-132	66-14887 <*>
1965 V165 N1 P47-54	66-30260 <=*$>
1965 V165 N1 P55-57	66-30340 <=*$>
1965 V165 N1 P91-94	66-30657 <=$>
1965 V165 N1 P130-132	66-13972 <*>
1965 V165 N3 P575-	AD-636 566 <=$>
1965 V165 N3 P629-632	66-30221 <=*$>
1965 V165 N3 P692-695	66-30274 <=*$>
1965 V165 N4 P780-782	66-30858 <=$>
1965 V165 N5 P1059-1061	66-31280 <=$>
1965 V165 N5 P1069-1070	66-12988 <*> 0
1965 V165 N6 P1275-1277	66-31280 <=$>
1965 V165 N6 P1290-1293	66-12470 <*>
1965 V165 N4/6 P1287-1289	66-30671 <=$>
1965 V165 N4/6 P1290-1293	66-30671 <=$>
1966 V166 N1 P49-52	66-61979 <=*$>
1966 V166 N1 P63-66	CSIR-TR.571 <CSSA>
1966 V166 N2 P353-355	66-12303 <*>
1966 V166 N2 P370-373	52T92R <ATS>
1966 V166 N3 P570-573	66-31746 <=$>
1966 V166 N4 P961-964	66-32787 <=$>
1966 V166 N5 P1062-1065	66-32090 <=$>
1966 V166 N5 P1088-1090	AD-636 717 <=$>
1966 V166 N5 P1129-1131	66-14196 <*$>
1966 V166 N6 P1312-1314	66-32261 <=$>
1966 V166 N6 P1338-1341	66-34226 <=$>
1966 V167 N1 P59-62	AD-634 380 <=$>
1966 V167 N1 P63-64	AD-634 379 <=$>
1966 V167 N1 P99-101	00T95R <ATS>
1966 V167 N2 P339-341	37T94R <ATS>
1966 V167 N3 P579-582	66-12304 <*>
1966 V167 N3 P639-641	66-14120 <*$>
1966 V167 N4 P772-774	66-32966 <=$>
1966 V167 N4 P807-810	66-32957 <=$>
1966 V167 N4 P863-866	93T94R <ATS>
1966 V167 N5 P1008-1011	66-33687 <=$>
1966 V167 N6 P1245-1250	66-33650 <=$>
1966 V167 N6 P1280-1282	66-14187 <*$> 0
1966 V168 N4 P844-845	AD-641 866 <=$>
1966 V168 N6 P1283-1286	66-62308 <=*$>
1966 V168 N1/3 P569-572	66-33582 <$=>
1966 V169 N1 P52-54	66-34602 <=$>
1966 V169 N2 P295-298	66-34137 <=$>
1966 V169 N3 P565-568	66-34284 <=$>
1966 V169 N6 P1289-1295	67-30731 <=$>
1967 V172 N2 P353-356	67-30916 <=$>
1967 V172 N2 P383-385	67-30916 <=$>

DOKLADY AKADEMII NAUK TADZHIKSKOI SSR. DUSHANBE
1958 V1 N2 P11-15 59-18010 <+*>
1958 V1 N2 P23-25 E-740 <RIS>

1959 V2 N1 P45-51 65-64101 <=*$>
1961 V4 N2 P57-60 64-19107 <=*$>
1962 V5 N3 P3- ICE V3 N4 P464-466 <ICE>
1962 V5 N3 P8-11 65-11180 <*>
1962 V5 N4 P39-42 65-60676 <=$>
1962 V5 N4 P43-47 65-63508 <=*$>
1962 V5 N6 P18-19 RTS-2901 <NLL>
1963 V6 N3 P22-24 4482 <BISI>
1963 V6 N5 P24-27 CSIR-507 <CSSA>

DOKLADY AKADEMII NAUK UKRAINSKOI SSR.
SEE DOPOVIDI AKADEMIYI NAUK UKRAYINSKOYI RSR

DOKLADY AKADEMII NAUK UZBEKSKOI SSR
1951 N9 P32-34 63-19153 <=*>
1951 N12 P33-35 63-15259 <=*>
1954 N9 P33-36 20L33R <ATS>
1958 N2 P5-10 SCL-T-393 <=*>
1959 N2 P7-10 59-00656 <*>
59-16559 <+*>
1959 N3 P28-31 62-24452 <=*>
1959 N5 P43-45 61-00439 <*>
1959 N11 P9-12 60-25580 <=>
1959 N47 P77-94 61-00599 <*>
1960 N10 P11-14 64-15053 <=*$>
1961 V18 N1 P35-36 65-63395 <=$> 0
1961 V18 N6 P38-40 65-63397 <=$>
1961 V18 N7 P39-40 65-63394 <=$> 0
1962 V19 N1 P56-58 64-19836 <=$>
1962 V19 N3 P39-41 66-12077 <*>
1962 V19 N11 P23-25 65-63993 <=>
1963 V20 N2 P19-21 65-60882 <=$>
1964 N2 P61-63 66-61836 <=*$>
1965 V22 N7 P23-25 66-12890 <*>
1965 V22 N7 P29-32 66-14409 <*>

DOKLADY AKADEMIIA PEDAGOGICHESKIKH NAUK RSFSR. MOSCOW
1960 N3 P79-85 62-13266 <=>
1960 N3 P91-102 62-13266 <=>
1960 N3 P121-124 62-13266 <=>
1960 N4 P73-76 61-28560 <=>
1960 N4 P87-91 61-28560 <=>
1960 N4 P93-100 61-28560 <=>
1960 N4 P105-108 61-28560 <=>
1960 N4 P109-112 62-13039 <*=>
1961 N1 P117-122 63-19693 <=*>
1961 N2 P109-113 63-19692 <=*>
1961 N3 P101-104 63-19453 <=*$>
1961 N3 P105-110 63-19456 <=*$>
1962 N2 P75-80 64-13553 <=*$>
1963 N1 P5-12 64-21560 <=>
1963 N1 P25-26 64-21560 <=>
1963 N1 P55-56 64-21560 <=>
1963 N1 P73-80 64-21560 <=>
1963 N1 P97-100 64-21560 <=>
1963 N1 P113-116 64-21560 <=>

DOKLADY. BULGARSKA AKADEMIYA NA NAUKITE
1953 V5 N2/3 P9-12 63-18946 <*> 0
1957 V10 N2 P157-160 59-12010 <*>
1958 V11 N6 P493-496 9856-A <KH>
1959 V12 N6 P525-528 62-18834 <*>
1960 V12 N6 P525-528 10833-A <KH>
1960 V13 N1 P67-70 62-19288 <*=>
1960 V13 N3 P363-364 64-10699 <*>
1961 N103 P363-364 11095-A <KH>
1961 V14 N1 P39-41 62-19492 <=*>
1961 V14 N1 P103-106 62-19493 <=*>
1961 V14 N2 P143-145 62-19773 <=*>
1961 V14 N2 P167-170 62-19784 <=*>
1961 V14 N2 P179-182 62-19156 <=*>
1961 V14 N4 P381-384 66-51092 <=>
1961 V14 N6 P591-594 63-20927 <*> 0
1962 V15 P747-750 8871 <IICH>
1962 V15 N2 P155-158 56S84G <ATS>
1962 V15 N7 P715-718 63-18473 <*>
1963 V16 N3 P321-324 N66-11615 <=*$>
1963 V16 N5 P485-488 RTS-2979 <NLL>
1964 V17 N1 P53-56 GI N2 1964 P314 <AGI>
1965 V18 N2 P117-120 44S87G <ATS>
1965 V18 N7 P631-634 AD-624 462 <=$>

DOKLADY. INSTITUTA GEOGRAFII SIBIRI I DAL'NEGO VOSTOKA. SIBIRSKOGO OTDELENIA. AKADEMIIA NAUK SSR
1963 N4 P33-41 SGRT V5 N5 P56 <AGS>
SOV.GEOGR. V5 N5 P56 <AGS>
1963 N4 P49-55 SGRT V5 N5 P64 <AGS>
SOV.GEOGR. V5 N5 P64 <AGS>
1963 N4 P70-77 SGRT V5 N5 P3 <AGS>
SOV.GEOGR. V5 N5 P3 <AGS>

DOKLADY MOSKOVSKOI SELSKO-KHOZYAISTVENNOI AKADEMII IM. K. A. TIMIRYAZEVA
1949 V11 P180-187 RT-1029 <*>
1950 V12 P131-133 RT-1755 <*>
1956 N22 P393-397 60-17721 <+*>
1957 N28 P71-77 61-28636 <*=>
1958 N32 P512-517 61-20697 <*=>
1958 N35 P168-175 62-18895 <*>
1958 N36 P161-164 65-60677 <=$>
1958 N37 P140-145 60-23150 <+*>
1958 N38 P41-46 60-23730 <+*>
1958 N38 P169-179 61-13775 <*+>
1959 N45 P193-198 RTS-3145 <NLL>
1959 N47 P77-94 61-15850 <+*> 0
1960 N52 P65-70 65-50063 <=$>
1961 N61 P235-241 70P63R <ATS>
1961 N62 P303-313 63-23231 <=*$>
1961 N64 P15-18 64-15418 <=*$>
1961 N70 P107-110 63-15243 <=*>
1962 N77 P79-84 64-13682 <=*$>
1963 N84 P253-260 64-19843 <=$>
1963 N87 P17-28 66-60981 <=*$>
1963 N87 P119-126 65-64387 <=*$>

DOKLADY NA EZHEGODNOM CHTENII PAMIATI N. A. KHOLODKOVSKOGO
1951 V4 P29-52 RT-2829 <*>

DOKLADY VSESOYUZNOI AKADEMII SELSKO-KHOZYAISTVENNYKH NAUK IM. V. I. LENINA
1939 V4 N23/4 P56-59 64-13759 <=*$>
1940 N13 P3-6 RT-430 <*>
1940 N22 P29-32 RT-431 <*>
1941 N2 P15-17 RT-432 <*>
1941 N9 P3-6 RT-429 <*>
1941 V6 N11 P15-17 61-15176 <*+>
1942 N9/10 P7-10 RT-4638 <*>
1947 N6 P44-48 RT-1284 <*>
1947 V12 N11 P3-12 RT-1779 <*>
1948 N1 P34-38 RT-1204 <*>
1948 N1 P39-42 RT-1205 <*>
1948 V13 N5 P8-13 RT-2653 <*>
1949 V14 N3 P39-43 RT-3153 <*>
1949 V14 N10 P31-33 RT-1788 <*>
1950 N3 P30-32 61-17301 <=$>
1950 V15 N5 P36-40 61-13376 <=*>
1950 V15 N7 P33-35 RT-3815 <*>
1950 V15 N10 P35-38 RT-3290 <*>
1950 V15 N11 P42-47 60-21143 <=>
1951 V16 N1 P34-38 RT-3324 <*>
1951 V16 N2 P25-31 C-2268 <NRC>
1951 V16 N10 P43-48 RT-3586 <*>
1952 V17 N4 P41-48 RT-2583 <*>
1954 V19 N2 P16-20 RT-3000 <*>
1955 V20 N5 P35-41 R-3935 <*>
1956 N3 P23-26 RT-4632 <*>
1956 V21 N4 P42-44 R-2408 <*>
1956 V21 N11 P22-29 R-5364 <*>
1957 V22 N1 P11-14 60-18032 <+*>
1958 V23 N2 P43-48 60-13433 <+*>
1958 V23 N3 P11-16 65-50077 <=>
1958 V23 N4 P3-6 61-15437 <+*>
1959 V24 N9 P14-17 63-11011 <=>
1959 V24 N9 P27-28 64-15866 <=*$>

DOKUMENTACIJA ZA GRADJEVINARSTVO I ARHITEKTURU
1961 V32 P216- 65-14927 <*>

DOKUMENTATION
1961 V8 N5 P154-155 64-30716 <*>
1962 V9 N5 P131-136 63-31530 <=>
1962 V9 N5 P139-141 63-31598 <=>
1962 V9 N5 P165-169 63-31530 <=>
1962 V9 N6 P178-181 63-31383 <=>
1963 V10 N1 P9-16 63-31598 <=>

DOCUMENTATION STUDY. TOKYO
SEE DOKUMENTESHON KENKYU

DOKUMENTESHON KENKYU
1965 V15 N1 P1-5 65-31344 <=$>

DON
1960 N4 P151-160 60-41419 <=>
1966 N8 P151-165 66-34855 <=$>

DOPOVIDI AKADEMIYI NAUK UKRAYINSKOI RSR
1951 P135-137 66-14541 <*>
1953 N3 P215-219 TT.532 <NRC>
1953 N4 P289-293 AD-625 305 <=*$>
1953 N5 P323-326 17P61U <ATS>
1954 P331-334 R-5057 <*>
1955 N1 P63-66 61-23694 <*=>
1955 N1 P77-81 86J14U <ATS>
1955 N2 P123-125 N65-12088 <=$>
1955 N6 P540-544 64-16958 <=*$>
1956 N5 P433-438 60-10800 <+*>
1956 N5 P454-456 20M43U <ATS>
1957 P227-230 AMST S2 V34 P165-167 <AMS>
1957 N2 P183-186 R-4818 <*>
1957 N3 P247-249 60-19152 <+*>
1957 N4 P362- 61-19441 <+*>
1957 N5 P448-452 R-3064 <*>
1957 N5 P453-456 R-5340 <*>
61-15609 <*+>
1957 N5 P461-464 R-3136 <*>
1957 N5 P466-469 59-16309 <+*>
1957 N5 P478-479 AEC-TR-3653 <+*>
59-16564 <+*>
1957 N6 P602-604 60-17154 <+*>
1957 N6 P605-608 60-17155 <+*>
1958 P239-242 STMSP V4 P303-306 <AMS>
1958 P810-812 STMSP V4 P123-126 <AMS>
1958 N1 P33-36 R-4450 <*>
1958 N1 P37-40 60-23989 <+*>
1958 N1 P45-48 R-4520 <*>
1958 N2 P175-177 AEC-TR-3423 <+*>
63R77U <ATS>
1958 N2 P213-215 59-18511 <+*>
1958 N3 P243-245 62-23107 <=>
1958 N3 P343-346 66-51093 <=>
1958 N4 P372-380 60-13005 <+*>
1958 N5 P477-478 R-4626 <*>
1958 N5 P505-507 59-11943 <=>
1958 N5 P512-514 59-11971 <=>
1958 N5 P531-534 61-23632 <*=>
1958 N7 P7 61-23517 <=*>
1958 N7 P712-715 61-23517 <=*>
1958 N7 P740-742 60-23877 <+*>
62-24357 <=*>
1958 N8 P838-839 60-13092 <+*>
1958 N9 P974-976 63-24300 <=*$>
1958 N10 P1075-1077 60-23935 <+*>
1958 N10 P1079-1082 60-23876 <+*>
1958 N11 P1157-1160 59-22733 <+*>
60-12187 <=>
1958 N11 P1216-1220 61-23445 <*=>
1958 V1 P41-44 R-4486 <*>
1959 P120-124 STMSP V3 P315-319 <AMS>
1959 P347-350 STMSP V3 P337-340 <AMS>
1959 P355-358 STMSP V3 P321-325 <AMS>
1959 P571-573 STMSP V4 P13-16 <AMS>
1959 N1 P32-36 63-23729 <=*$>
1959 N1 P46-48 60-19037 <+*>
1959 N3 P281-285 61-23415 <=*>
1959 N4 P392-394 61-19677 <=*>
1959 N5 P502-504 61-27878 <*=>
1959 N7 P756-759 5774 <HB>
62-16734 <=*>
1959 N8 P866-868 AEC-TR-4077 <+*>
61-13511 <=$>
65-63639 <=$>
1959 N9 P985-990 61-23436 <=*>
1959 N10 P1119-1125 61C101 <CTT>
1960 N1 P122-123 60-11721 <=>
1960 N2 P173-176 62-23687 <=*>
1960 N2 P232-234 60-41143 <=>
1960 N2 P235-237 60-41142 <=>
1960 N3 P380-391 61-11550 <=>
1960 N4 P547-548 61-11001 <=>
1960 N5 P630-633 62-33067 <=*>
1960 N5 P642-645 62-33067 <=*>
1960 N6 P840-857 61-11652 <=>
1960 N7 P984-995 61-27075 <*=>
1960 N8 P1124-1129 61-31141 <=>
1960 N9 P1185-1189 61-21988 <=> 0
1960 N9 P1236-1240 AD-621 051 <=$>
1960 N10 P1224-1226 AD-628 417 <=*$>
1960 N10 P1393-1395 AD-628 416 <=*$>
1960 N10 P1408-1412 62-18645 <=*>
1960 N11 P1485-1491 64-13123 <=*$>
1961 N1 P10-14 AD-611 525 <=$>
1961 N1 P25-28 5123 <HB>
1961 N1 P40-43 5196 <HB>
1961 N1 P54-58 AD-611 525 <=$>
1961 N1 P120-122 61-28592 <=*>
1961 N2 P173-174 5059 <HB>
1961 N2 P183-187 5197 <HB>
1961 N2 P192-196 5411 <HB>
1961 N2 P251-255 61-27858 <*=>
1961 N2 P256-258 61-27961 <*=>
1961 N3 P316-320 AEC-TR-4998 <=*>
1961 N3 P321-324 5198 <HB>
1961 N3 P325-327 5199 <HB>
1961 N3 P335-339 5128 <HB>
1961 N3 P396-400 62-13017 <*=>
1961 N4 P415-419 62-23785 <=*>
1961 N4 P429-436 62-23785 <=*>
1961 N4 P469-472 61-27677 <*=>
1961 N4 P486-489 4555 <HB>
1961 N4 P490-494 4335 <HB>
1961 N4 P538-541 61-27868 <*=>
1961 N5 P553 62-15209 <*=>
1961 N5 P554-557 62-15212 <*=>
1961 N5 P558-577 62-15211 <*=>
1961 N5 P577-581 62-15224 <*=>
1961 N5 P582-589 62-15235 <*=>
62-15493 <*=>
1961 N5 P608-610 61-28274 <*=>
1961 N6 P699-707 62-13542 <*=>
1961 N6 P708-712 62-19165 <=*>
1961 N6 P745-748 5773 <HB>
1961 N6 P766-768 62-10617 <=*>
1961 N6 P812-815 62-13541 <*=>
1961 N6 P825-829 62-13528 <*=>
1961 N6 P830-835 62-13544 <*=>
1961 N7 P839-844 62-15674 <*=>
1961 N7 P850-853 62-15681 <*=>
1961 N7 P911-914 5464 <HB>
1961 N7 P970-975 62-13663 <*=>
1961 N8 P1049-1051 63-13272 <=*>
1961 N9 P1297-1306 61-28062 <*=>
1961 N10 P1290-1295 64-13285 <=*$> 0
1961 N10 P1317-1322 63-13919 <=*>
1961 N11 P1495-1497 63-13214 <=*>
1961 N12 P1556-1560 62-25546 <=*> 0
1961 N12 P1578-1582 63-13885 <=*>
1962 N1 P50-53 3493 <BISI>
1962 N1 P69-73 AEC-TR-6244 <=*$>
1962 N1 P131-133 62-25058 <=*>
1962 N2 P196-200 64-13727 <=*$>
1962 N3 P283-322 62-33350 <=>
1962 N3 P350-352 63-11574 <=>
1962 N3 P370-373 5687 <HB>
1962 N3 P374-377 5814 <HB>

1962 N3 P414-422	62-33350 <=>
1962 N4 P540-542	63-13536 <=*>
1962 N5 P575-577	62-11759 <=>
1962 N5 P624-626	62-33326 <=*>
1962 N5 P681-684	63-13348 <=>
1962 N6 P743-744	62-32838 <=*>
1962 N6 P789-791	62-33344 <=*>
1962 N6 P837-839	63-19306 <=*>
1962 N7 P863-873	63-13316 <=>
1962 N7 P901-905	63-23633 <=*$>
1962 N7 P955-957	63-13311 <=*>
1962 N7 P970-971	63-13316 <=>
1962 N7 P970-975	63-21466 <=>
1962 N7 P974-975	63-13316 <=>
1962 N8 P1044-1047	63-23423 <=*>
1962 N8 P1053-1056	63-19319 <=*>
1962 N8 P1066-1067	5813 <HB>
1962 N8 P1066-1068	63-24433 <=*$>
1962 N8 P1117-1121	63-21660 <=>
1962 N10 P1305-1307	63-21239 <=>
1962 N10 P1392-14402	63-21660 <=>
1962 N12 P1553-1554	63-21388 <=>
1963 N2 P59-61	5152 <HB>
1963 N2 P156-161	66-13834 <*>
1963 N2 P178-183	63-23726 <=*$>
1963 N2 P188-192	AD-630 558 <=$>
1963 N2 P202-205	65-32600 <=$>
1963 N2 P271-272	63-21880 <=>
1963 N3 P295-298	64-30057 <*>
1963 N4 P447-453	63-21983 <=>
1963 N4 P466-472	64-21211 <=>
1963 N5 P592-595	TIL/T.5576 <NLL>
1963 N5 P618-623	AD-625 781 <=*$>
1963 N5 P665-678	63-31753 <=>
1963 N6 P782-785	6487 <HB>
1963 N6 P798-801	AD-615 232 <=$>
1963 N7 P839-851	63-41312 <=>
1963 N7 P875-878	63-20630 <=*$>
1963 N7 P896-899	6120 <HB>
1963 N7 P924-927	63-41090 <=>
1963 N7 P927-929	63-41089 <=>
1963 N7 P973-979	63-41312 <=>
1963 N8 P1018-1021	64-71196 <=>
1963 N9 P1190-1193	6164 <HB>
	64-15196 <=*$>
1963 N9 P1253-1259	63-31753 <=>
1963 N9 P1262	64-21117 <=>
1963 N9 P1263	64-21117 <=>
1963 N9 P1265-1267	64-21117 <=>
1963 N10 P1326-1330	64-14181 <=*$>
1963 N11 P1470-1473	64-71167 <=>
1963 N11 P1486-1489	6248 <HB>
1963 N11 P1489-1491	AD-614 953 <=$>
1963 N12 P1600-1602	6295 <HB>
1964 N1 P10-14	65-13173 <*>
1964 N1 P67-70	AD-610 792 <=$>
	6423 <HB>
1964 N1 P133-134	64-31054 <=>
1964 N1 P134-135	64-31056 <=>
1964 N1 P136-137	64-31055 <=>
1964 N1 P138-140	64-31066 <=>
1964 N1 P141-143	64-31067 <=>
1964 N2 P155-158	64-31049 <=>
1964 N2 P181-184	64-31049 <=>
1964 N2 P185-187	65-63199 <=$>
1964 N2 P188-193	AD-613 467 <=$>
1964 N3 P412-416	64-31217 <=>
1964 N4 P427-441	64-51323 <=$>
1964 N4 P499-501	AD-624 910 <=*$>
1964 N4 P553-555	64-31445 <=>
1964 N4 P553-556	64-51323 <=$>
1964 N5 P580-585	64-41023 <=>
1964 N5 P600-602	66-14098 <*$>
1964 N5 P603-606	AD-617 322 <=$>
1964 N5 P692-693	64-51033 <=$>
1964 N5 P695-699	64-51033 <=$>
1964 N6 P745-748	AD-615 874 <=$>
1964 N7 P873-877	65-13171 <*>
1964 N7 P893-896	AD-637 439 <=$>
1964 N7 P919-922	6563 <HB>
1964 N7 P925-929	AD-639 334 <=$>
1964 N8 P1035-1037	AD-631-851 <=$>
1964 N8 P1038-1042	AD-631-851 <=$>
1964 N8 P1043-1046	AD-631-851 <=$>
1964 N8 P1063-1066	5911 <HB>
1964 N8 P1075-1076	6451 <HB>
1964 N9 P1145-1149	65-12650 <*>
1964 N9 P1168-1172	6422 <HB>
1964 N9 P1179-1182	6424 <HB>
1964 N9 P1200-1205	GI N6 1964 P1190 <AGI>
1964 N9 P1250-1255	65-30179 <=$>
1964 N9 P1264	65-30179 <=$>
1964 N10 P1301-1305	AD-619 332 <=$>
1964 N10 P1326-1330	AD-637 418 <=$>
1964 N11 P1456-1459	65-13156 <*>
1964 N11 P1460-1462	65-14420 <*>
	65-63686 <=$>
1964 N11 P1478-1482	6434 <HB>
1964 N11 P1494-1497	6614 <HB>
1964 N12 P1590-1593	6535 <HB>
1964 N12 P1595-1599	6536 <HB>
1964 V6 P745-748	65-13172 <*>
1965 N2 P204-206	AD-625 265 <=*$>
1965 N3 P296-299	65-31351 <=$>
1965 N3 P317-322	65-14560 <*>
	65-63695 <=$>
1965 N3 P336-338	6562 <HB>
1965 N3 P339-341	6561 <HB>
1965 N3 P389-390	65-31352 <=$>
1965 N4 P426-429	AD-627 859 <=*$>
1965 N4 P460-463	65-14457 <*>
	65-63222 <=*$>
1965 N4 P474-477	6613 <HB>
1965 N5 P547-550	AD-627 860 <=*$>
1965 N5 P570-572	65-31447 <=$>
1965 N6 P715-716	66-32408 <=$>
1965 N7 P864-867	65-64111 <=*$>
	66-10478 <*>
1965 N7 P873-875	6693 <HB>
1965 N7 P897-900	6691 <HB>
1965 N8 P1016-1019	66-61740 <=*$>
1965 N8 P1021-1023	66-10476 <*>
1965 N9 P1139-1145	64-10477 <*>
1965 N9 P1145-1148	66-10467 <*>
1965 N9 P1150-1152	66-10468 <*>
1965 N9 P1179-1182	6801 <HB>
1965 N9 P1248-1250	65-33541 <=*$>
1965 N10 P1295-1297	66-61702 <=*$>
1965 N10 P1321-1326	6799 <HB>
1965 N10 P1336-1339	6797 <HB>
1965 N10 P1342-1344	6798 <HB>
1965 N10 P1396-1399	66-31283 <=$>
1965 N10 P1403-1406	4467 <HB>
1965 N11 P1410-1413	66-61364 <=*$>
1965 N11 P1430-1433	66-12318 <*>
	66-60487 <=$>
1965 N11 P1438-1442	66-12319 <*>
	66-60588 <=*>
1966 N1 P76-79	AD-639 110 <=$>
1966 N1 P129-132	AD-635 576 <=$>
1966 N4 P437-440	66-61695 <=$>
1966 N4 P441-447	66-13473 <*>
	66-61735 <=*$>
1966 N6 P733-737	66-62303 <=*$>
1966 N6 P737-740	66-62307 <=*$>
1966 N6 P762-766	66-34138 <=$>
1966 N7 P875-878	66-62301 <=*$>

DOPOVIDI TA POVIDOMLENNYA. LVIVSKYI DERZHAVNYI UNIVERSYTET IM. I. FRANKA

1957 V7 PT.3 P183-	59-10891 <+*>
1957 V7 P180-183 PT.3	DC-59-3-176 <=*>
1957 V7 P180-188	59-10890 <+*>

DOSHISHA DAIGAKU RIKOGAKU KENKYU HOKOKU

1964 V5 N2 P82-97	66-12132 <*>

DOSHISHA ENGINEERING REVIEW. KYOTO
SEE DOSHISHA KOGAKU KAISHI

DOSHISHA KOGAKU KAISHI
1957 V8 N2 P76-79 47P60J <ATS>

DOSHKOLNCE VOSPITANIE
1960 N9 P59-65 61-28063 <*=>
1963 V36 N7 P29-49 64-31394 <=>
1963 V36 N7 P51-54 64-31394 <=>
1963 V36 N7 P69-74 64-31394 <=>

DOSTIZHENIYA NAUKI I PEREDOVOGO OPYTA V SELSKOM KHOZYAISTVE
1951 V2 P6-8 RT-1926 <*>
1953 V5 P26-31 RT-4182 <*>
1954 N1 P78-81 RT-3341 <*>
1954 N7 P27-34 R-116 <*>
1955 N5 P92-94 RT-4162 <*>

DOSTIZHENIYA NAUKI I TEKHNIKI I PEREDOVOGO OPYT V PROMYSLENNOSTI I STROITELSTVI
1959 N4 P3-71 61-27006 <=*>
1959 N4 P77-79 61-27006 <=*>
1959 N4 P97 61-27006 <=*>
1959 N4 P125-129 61-27006 <=*>
1959 N4 P148-156 61-27006 <=*>
1960 N3 P95-146 62-24747 <=*> P

DRAEGER HEFTE
1940 N206 P4396-4398 59-15126 <*> O

DRAHT
1952 V3 N2 P42-44 62-18149 <*>
1952 V3 N10 P330-335 62-16919 <*> O
1954 V5 N2 P50-53 57-1998 <*>
62-16950 <*> O
1954 V5 N5 P167-172 62-16951 <*> O
1954 V5 N6 P212-216 62-16952 <*> O
1954 V5 N8 P288-289 62-18229 <*> O
1954 V5 N10 P375-378 62-16953 <*> O
1954 V5 N11 P418-420 4863 <HB>
62-16954 <*> O
1955 V6 N1 P12-13 62-16997 <*> O
1956 V7 N3 P85-87 62-16792 <*> O
1956 V7 N6 P216-219 61-10808 <*> O
1956 V7 N11 P419-427 2886 <BISI>
1956 V7 N12 P473-480 2886 <BISI>
1957 V8 N11 P465-470 1059 <BISI>
1958 V9 N1 P13-15 61-18381 <*>
1958 V9 N5 P163-167 66-11347 <*> O
1958 V9 N6 P218-219 66-10141 <*> O
1958 V9 N9 P335-342 TS-1277 <BISI>
61-20606 <*>
1958 V9 N10 P390-391 62-18536 <*>
1959 V10 P225-229 TS 1450 <BISI>
1959 V10 N8 P370-374 1663 <BISI>
66-10158 <*> O
1959 V10 N9 P465-471 60-18444 <*>
1959 V10 N10 P511-516 1744 <BISI>
1959 V10 N11 P571-577 1744 <BISI>
1960 V11 N1 P1-4 61-10567 <*> O
1960 V11 N2 P55-56 61-16162 <*>
1960 V11 N2 P59-60 1979 <BISI>
1960 V11 N7 P335-338 66-11345 <*> O
1960 V11 N12 P739-743 146-167 <STT>
1961 V12 N6 P263-269 2421 <BISI>
1961 V12 N8 P352-356 62-16189 <*> O
1962 V13 N1 P13-15 2765 <BISI>
1962 V13 N9 P534-541 63-10999 <*>
1962 V13 N10 P571-577 63-18351 <*>
1962 V13 N11 P634-636 63-18046 <*>
1963 V14 N8 P447-456 3602 <BISI>
1963 V14 N8 P470-477 3603 <BISI>
1963 V14 N8 P499-504 3646 PT1 <BISI>
1963 V14 N9 P572-577 64-14854 <*>
1963 V14 N10 P632-637 3718 <BISI>
1963 V14 N10 P650-654 3646 PT2 <BISI>
1964 V15 P701-711 4228 <BISI>
1964 V15 P731-736 4229 <BISI>
1964 V15 N2 P51-56 3791 <BISI>
1964 V15 N8 P496-499 4043 <BISI>
1964 V15 N9 P632-635 65-12155 <*>

DRAHTWELT
1962 V48 P123-135 3150 <BISI>
1964 V50 N1 P2-4 65-17367 <*> O
1965 P215-221 4415 <BISI>
1965 V51 P669-674 4790 <BISI>
1966 V52 P39-44 4790 <BISI>

DREVARSKY VYSKUM
1956 V1 N1/2 P171-191 59-17419 <*>
1957 V2 N2 P235-249 59-10183 <*>
1958 V3 N2 P263-282 60-16518 <*>
1959 V4 N6 P107-120 C-3582 <NRCC> C

DREVO
1959 V14 P99-102 61-28939 <=*> O

DRUCK UND REPRODUKTION
1958 V7 N10 P146-148 61-14927 <*> O
1958 V7 N11 P159-160 61-14926 <*> O

DRUCKFARBE
1959 N2 P249-250 62-14074 <*>
1959 N2 P252 62-14074 <*>

DRUCKSPIEGEL
1958 V13 N5 P279-289 PB 141 176T <=>
1961 V16 N10 P636-641 62-20250 <*> O

DRVNA INDUSTRIJA
1959 V10 N5/6 P86-93 60-00613 <*>
1960 V11 N1/2 P24-25 62-00081 <*>

DUKLIA
1966 V14 N1 P47-50 66-31985 <=$>

DUODECIM
1938 V54 P433-449 63-26248 <=$>
1951 V67 P84-87 66-12017 <*>
1953 V69 P631-643 62-10589 <*>
1955 V71 N11 P1115-1134 5223-A <K-H>
1959 V75 N1 P1-9 62-00283 <*>
64-10593 <*>
9372-C <K-H>
1960 V76 P487-492 10607-B <KH>
63-16685 <*>
1961 V77 P657-658 63-20906 <*>
1963 V79 P804-807 64-00365 <*>
1964 V80 N16 P693-700 65-63594 <=$>

DURFERRIT-HAUSMITTEILUNGEN
1951 N24 P7-17 4307 <HB>

DURZHAVEN VESTNIK
1963 N3 P1-3 63-21032 <=>
1963 N4 P6-7 63-21312 <=>
1965 P2 65-31575 <=$>
1965 N46 P1-2 65-31469 <=$>
1965 N63 P2-6 C8/10 65-32242 <=$>
1965 N95 P1-7 65-34076 <=*$>

DUVAN
1957 V7 N5 P159-167 63-16686 <*> O
6769-B <K-H>
1960 V10 N12 P337-351 11092-D <K-H>
1961 V11 N6/7 P211-216 11627-B <KH>
63-16617 <*>

DYNA
1951 V26 P243-253 II-664 <*>

DZIENNIK USTAW
1960 N52 P500-504 62-19431 <*=>

ECAFE INFORMATION. JAPAN
SEE EKAFE TSUSHIN

EKAFE TSUSHIN
1960 N248 P24-40 62-19428 <=*>

EARTHQUAKE
SEE ZISIN

EAU
1950 V37 N4 P57-61 57-2718 <*>
1954 V41 N2 P27-32 64-16129 <*> 0
1954 V41 N4 P71-72 64-16129 <*> 0
1954 V41 N7 P127-133 64-16129 <*> 0
1954 V41 N9 P163-166 64-16129 <*> 0
1954 V41 N12 P233-237 64-16129 <*> 0

ECHO MEDICAL DU NORD
1933 V37 P229-238 57-833 <*>

ECHO DES MINES ET DE LA METALLURGIE
1954 V20 P3464 T-1440 <NLL>

ECHO DES RECHERCHES. CENTRE NATIONAL D'ETUDES DES TELECOMMUNICATIONS
1951 N2 P2-8 57-595 <*>
1952 N7 P7-12 57-2398 <*>
1952 N8 P31-38 57-2829 <*>
1953 N10 P14-29 57-2829 <*>
1954 N17 P19-22 57-1009 <*>
1955 N19 P28-50 49H9F <ATS>
57-2770 <*>
1955 N21 P39-43 57-24 <*>
1959 N34 P4-15 60-16937 <*>
1961 N40 P1-9 62-14803 <*>

ECLAIRAGE ELECTRIQUE
1907 V51 P37-49 57-01477 <*>

ECLOGAE GEOLOGICAE HELVETIAE
1944 V37 N2 P385-400 AL-498 <*>
3333 <*>
1961 V54 N2 P283-334 65-14211 <*>

ECOLE DES PARENTS
1955 N10 P14-22 65-12653 <*>

ECONOMIC RESEARCH
SEE CHING CHI YEN CHIU

ECONOMIE APPLIQUEE
1955 V8 N1/2 P67-84 65-12556 <*>

ECONOMIE RURALE
1953 N16 P29-33 65-12535 <*>
1954 P3-7 65-12387 <*>
1954 N20 P17-23 65-12683 <*>
1954 N21 P5-10 59-20151 <*>
1954 N22 P7-12 65-12597 <*>
1955 N23 P21-35 59-20162 <*>
1955 V23 P15-20 65-12674 <*>
1956 N27 P87-92 59-20160 <*>
1956 N27 P95-99 65-12657 <*>
1956 N27 P159-164 59-20871 <*>
1956 N29 P45-52 65-12668 <*>
1956 N30 P35-37 59-20157 <*>
1957 N33 P37-41 65-12698 <*>
1958 N35 P2-8 64-30282 <*>
1958 N35 P19-22 59-20888 <*>
1958 V37 P3-8 61-20418 <*>

EESTI NSV TEADUSTE AKADEEMIA FUUSIKA JA ASTRONOOMIA INSTITUUDI UURIMUSED
1957 N6 P63-80 AEC-TR-4354 <+*>
1958 V7 P85-111 59-21105 <=>
1959 N10 P97-121 AEC-TR-5002 <=*>
1959 V8 N10 P220-232 60-16755 <=*>
1960 N12 P249-261 62-10975 <=*>
1960 N12 P271-274 61-18307 <*=>
1961 N17 P3-26 AEC-TR-5311 <=*>
1962 N21 P173-194 64-20099 <*>
1962 V21 P12-19 AD-631 579 <=$>
1962 V21 P139-172 64-20092 <*>
1963 N23 P38-59 65-60323 <=$>
1963 V23 P3-17 64-15840 <=*$>
1963 V23 P221-224 64-15385 <=*$>
1963 V23 P224-225 64-15384 <=*$>
64-18540 <*>
1964 N27 P23-56 66-12646 <*>
1964 N27 P57-68 66-10411 <*>
1964 N27 P69-84 66-10410 <*>
1964 N27 P99-107 66-10408 <*>
1964 N28 P93-110 65-14534 <*>
1964 N28 P111-120 65-14535 <*>

EESTI NSV TEADUSTE AKADEEMIA TOIMETISED
1953 V2 N1 P103-107 61-14958 <=*>
1955 V4 N1 P113-115 R-1561 <*>
1958 V7 N2 P83-93 60-21905 <=>
1959 N3 P211-215 60-11149 <=>
1959 N3 P216-219 60-11150 <=>
1960 V9 N2 P167-176 61-15934 <+*>
1961 V10 N1 P28-32 62-32283 <=*>

EESTI NSV TEADUSTE AKADEEMIA TOIMETISED. BIOLOOGILINE SEER
1963 V12 N2 P148-158 64-19696 <=$>
1964 V13 N1 P3-9 64-31777 <=>

EESTI NSV TEADUSTE AKADEEMIA TOIMETISED. TEHNILISTE JA FUUSIKALIS-MATEMAATILISTE TEADUSTE SEER
1960 V9 N1 P26-31 62-10981 <=*>
1960 V9 N1 P69-74 64-71515 <=> M
1960 V9 N3 P250-255 62-25084 <=*>
1961 N4 P329-339 63-00266 <*>
1961 V10 N4 P329-339 62-24704 <=*>
63-15664 <=*$>
1962 V11 N1 P16-23 63-18700 <=*>
1962 V11 N4 P253-262 63-23191 <*=>
1963 V12 N1 P57-74 AD-609 150 <=>
1963 V12 N3 P227-237 AD-634 811 <=$>
1965 N3 P444-454 ICE V6 N4 P681-687 <ICE>
1965 V14 N3 P444-454 61S89R <ATS>

EGESZSEGTUDOMANY
1960 V4 P317-324 11181 <K-H>
64-14576 <*>

EINHEIT
1962 V17 N12 P146-150 63-21047 <=>
1965 V20 N2 P93-104 65-30640 <=$>
1966 P593-604 66-33249 <=$>
1966 P719-724 66-33249 <=$>

EINZELVEROEFFENTLICHUNGEN. SEEWETTERAMT, DEUTSCHER WETTERDIENST
1953 N1 P1-48 I-558 <*>
1954 N5 P3-20 57-2559 <*>
1954 N6 P1-18 II-772 <*>

EISEI DOBUTSU
1957 V8 N3 P147-159 62-00814 <*>
1959 V10 N1 P46-48 62-00685 <*>
1960 V11 N2 P94 62-00812 <*>

EISEI SHIKEN-SHO HOKOKU
1963 V81 P83-85 66-11586 <*>
1963 V81 P85-87 66-11585 <*>
1964 V82 P90-92 66-11604 <*>

EISENBAHNER. DUESSELDORF
1955 V8 P41-43 <INSD>
1955 V8 P107-110 <INSD>

EISENBAHNFACHMANN
1955 N21 P491-492 T2092 <INSD>
1956 P417-418 T1958 <INSD>

EISENBAHNINGENIEUR
1955 P328-329 T-1959 <INSD>
1955 V6 N12 P1-7 T-2054 <INSD>
1960 V11 N7 P203-207 1829 <BISI>

EISENBAHNTECHNISCHE RUNDSCHAU
1955 V4 P561-566 T-2056 <INSD>

Citation	Translation
1956 N6 P239-248	T-2212 <INSD>
1956 V5 P1-11	T1861 <INSD>
1956 V5 P38-40	T1860 <INSD>
1956 V5 P42-43	T1858 <INSD>
1956 V5 P81-85	T1859 <INSD>
1956 V5 P113-120	T1842 <INSD>
1956 V5 N7 P281-288	3859 <HB>
1957 N7 P266-273	61-16217 <*>
1957 N10 P384-388	TRANS-23 <MT>
1958 N5 P200-209	61-20027 <*>
1958 N10 P402-408	TRANS-24 <MT>
1959 V8 P57-67 SPEC 11	1944 <BISI>
1959 V8 P73-78 SPEC 11	1945 <BISI>
1959 V8 N4 P163-169	62-14387 <*>

EISZEITALTER UND GEGENWART

Citation	Translation
1951 N1 P16-26	IGR V1 N9 P72 <AGI>
1951 V1 P16-26	57-3212 <*>
1954 V4/5 P189-208	57-3211 <*>
1955 V6 P110-115	57-3210 <*>
1956 V7 P78-86	57-3224 <*>
1956 V7 P87-104	57-3223 <*>

EIYO TO SHOKURYO

Citation	Translation
1953 V6 N3 P89-96	60-10804 <*>
1955 V8 P170-173	63-16359 <*>

EIYOGAKU ZASSHI

Citation	Translation
1963 V21 N2 P29-31	64-10598 <*> O

EKOLOGIA POLSKA. SER. A

Citation	Translation
1956 SA V4 N8 P225-292	60-21248 <=>
1957 SA V5 N7 P213-256	61-11329 <=>
1958 V6 N7 P205-291	63-11396 <=>
1960 V8 N7 P155-168	63-11400 <=>

EKOLOGIA POLSKA. SER. B

Citation	Translation
1955 SB V1 N3/4 P107-111	63-11351 <=>
1959 SB V5 N1 P23-34	63-11351 <=>
1959 SB V5 N1 P47-54	63-11351 <=>
1959 SB V5 N2 P139-145	63-11351 <=>
1960 SB V6 N1 P1-9	63-11351 <=>

EKONOMICHESKAYA GAZETA

Citation	Translation
1960 09/14 P4	NLLTB V2 N12 P1046 <NLL>
1960 09/20 P3	NLLTB V2 N12 P1053 <NLL>
1960 10/25 P2-3	NLLTB V3 N2 P115 <NLL>
1961 01/07 P3	NLLTB V3 N6 P518 <NLL>
1961 01/13 P3	NLLTB V3 N6 P497 <NLL>
1961 01/29 P2	NLLTB V3 N7 P578 <NLL>
1961 02/01 P2	NLLTB V3 N7 P589 <NLL>
1961 03/16 P4	NLLTB V3 N7 P594 <NLL>
1961 04/12 P1	NLLTB V3 N7 P569 <NLL>
1961 06/04 P1	NLLTB V3 N9 P767 <NLL>
1961 06/04 P3	NLLTB V3 N9 P767 <NLL>
1961 06/13 P2-3	NLLTB V3 N10 P855 <NLL>
1961 06/17 P1	NLLTB V3 N9 P761 <NLL>
1961 09/18 P20	NLLTB V4 N2 P132 <NLL>
1961 10/09 P14	NLLTB V4 N3 P243 <NLL>
1961 N7 P12	62-13762 <*=>
1961 N14 P32	62-32571 <=>
1961 N18 P21-22	NLLTB V4 N6 P510 <NLL>
1961 N18 P26-27	NLLTB V4 N6 P542 <NLL>
1961 N20 P4-5	NLLTB V4 N5 P446 <NLL>
1962 01/29 P6	NLLTB V4 N5 P421 <NLL>
1962 01/29 P7	NLLTB V4 N5 P421 <NLL>
1962 07/21 P3	NLLTB V4 N12 P1038 <NLL>
1962 N4 P4-5	NLLTB V4 N6 P528 <NLL>
1962 N13 P10	NLLTB V4 N8 P727 <NLL>
1962 N19 P25	62-33018 <=>
1962 N22 P11	62-33332 <=>
1962 N27 P5	NLLTB V4 N11 P1029 <NLL>
1962 N28 P36	62-32845 <=>
1962 N31 P3-4	NLLTB V4 N11 P999 <NLL>
1962 N33 P15	62-33601 <=>
1963 01/26 P12-13	NLLTB V5 N5 P365 <NLL>
1963 02/03 P20	4550 <BISI>
1963 04/27 P10	NLLTB V5 N8 P726 <NLL>
1963 10/19 P12-13	NLLTB V6 N2 P150 <NLL>
1963 11/09 P7	NLLTB V6 N3 P213 <NLL>
1963 11/09 P4-5	NLLTB V6 N3 P223 <NLL>
1963 11/16 P4	NLLTB V6 N3 P241 <NLL>
1963 N3 01/19 P14	NLLTB V5 N5 P386 <NLL>
1963 N23 07/08 P11	NLLTB V5 N10 P904 <NLL>
1964 P42	AD-621 059 <=$>
1964 N4 P4-42	AD-618 901 <=$>
1964 N5 P3	NLLTB V6 N8 P738 <NLL>
1964 N25 P14	NLLTB V6 N10 P887 <NLL>
1964 N30 P5	64-51016 <=$>
1964 N33 P14-15	64-51035 <=>
1964 N37 P13	AD-621 060 <=$>
1964 N38 P9	NLLTB V7 N4 P352 <NLL>
	TRANS.BULL. V7 N4 <NLL>
1964 N39 P8	64-51521 <=>
1964 N39 P14	64-51521 <=>
1964 N39 P15	64-51521 <=>
1964 N41 P14	64-51757 <=$>
1964 N46 P11	65-30419 <=$>
1964 N46 P16	65-30419 <=$>
1964 N51 P13	65-30130 <=$>
1964 N51 P29	65-30248 <=$>
1964 N51 P43	RTS-3064 <NLL>
1965 P7	AD-630 264 <=*$>
1965 P39	NLLTB V7 N6 P530 <NLL>
1965 N2 P6	65-30942 <=$>
1965 N2 P41	AD-622 467 <=*$>
1965 N4 P17	65-31254 <=$>
1965 N5 P10	65-30997 <=$>
1965 N8 P11	65-30990 <=*$>
1965 N17 P5-6	NLLTB V7 N9 P763 <NLL>
1965 N18 P14	65-32019 <=$>
1965 N22 P5-6	65-31896 <=$*>
1965 N28 P6	NLLTB V7 N12 P1052 <NLL>
1965 N33 P5-6	65-32631 <=$>
1965 N33 P38	AD638 898 <=$>
1965 N34 P5-7	65-32692 <=$*>
1965 N34 P7	NLLTB V7 N12 P1045 <NLL>
1965 N38 P17	65-32999 <=*$>
1965 N39 P18	65-33433 <=*$>
1965 N41 P26-28	65-33747 <=*$>
1965 N42 P4-5	NLLTB V8 N2 P87 <NLL>
1965 N43 P22	65-33864 <=*$>
1965 N43 P25-29	65-33164 <=$>
1965 N45 P6-13	65-33740 <=*$>
1965 N45 P25-28	65-33681 <=*$>
1965 N45 P31-32	66-30189 <=*$>
1965 N46 P4-5	NLLTB V8 N3 P177 <NLL>
1965 N47 P15	66-30188 <=*$>
1965 N47 P40-41	AD-633 210 <=*>
1965 N48 P5-6	66-30108 <=*$>
1965 N49 P21	66-30577 <=$>
1965 N49 P23	66-30306 <=*$>
1965 N50 P11-12	NLLTB V8 N5 P399 <NLL>
1965 N51 P6-7	66-30468 <=*$>
1965 V19 P7	65-31252 <=$>
1966 N2 P33-34	66-30541 <=*$>
1966 N6 P31-35	66-30673 <=$>
1966 N13 P25	NLLTB V8 N9 P694 <NLL>
1966 N14 P24-25	NLLTB V8 N7 P595 <NLL>
1966 N17 P11	66-32343 <=$>
1966 N24 P16	66-32754 <=$>
1966 N26 P20	66-33219 <=$>
1966 N28 P6	66-33654 <=$>
1966 N35 P35-36	66-34253 <=$>

EKONOMICHESKIE NAUKI

Citation	Translation
1965 N6 P9-16	66-30727 <=$>

EKONOMIKA I MATEMATICHESKIE METODY

Citation	Translation
1965 V1 N3 P391-409	66-30057 <=*$>
1965 V1 N4 P481-501	65-33396 <=*$>
1965 V1 N5 P641-650	65-33630 <=*$>
1965 V1 N5 P699-717	65-33647 <=*$>
1965 V1 N5 P718-729	65-33915 <=*$>
1965 V1 N5 P739-745	65-33894 <=*$>
1965 V1 N5 P805-807	65-33790 <=*$>
1965 V1 N6 P911-915	66-30596 <=$>
1965 V1 N6 P940-947	66-30596 <=$>
1965 V1 N6 P948-958	66-30596 <=$>
1966 V2 N2 P164-170	66-32253 <=$>

1966 V2 N3 P321-326 66-33652 <=$>
1967 N1 P47-60 67-31057 <=$>

EKONOMIKA RADYANSKOI UKRAINY
1963 N1 P11-17 63-21815 <=>
1963 N1 P32-35 63-21815 <=>
1963 N1 P36-39 63-21610 <=>
1963 N2 P10-14 63-21973 <=>
1963 N5 P13-21 64-21482 <=>
1963 N5 P28-33 64-21006 <=>
1964 N3 P3-8 64-41058 <=>
1964 N5 P30-33 65-30120 <=$>
1964 N5 P71-74 65-30120 <=$>
1965 N4 P33-37 66-30220 <=$>
1965 N10 P32-33 66-32407 <=$>

EKONOMIKA SOVETSKOI UKRAINY
1965 N6 P45-48 65-32847 <=$>
1966 N5 P30-34 66-34752 <=$>
1966 N5 P40-44 66-34752 <=$>
1966 N5 P62-64 66-34752 <=$>
1966 N9 P58-62 67-30082 <=>

EKONOMIKA ZEMEDELSTVI
1966 V6 N8 P205-208 66-34801 <=$>
1967 N1 P7-11 67-30645 <=>

EKONOMIKA I ZHIZN
1965 N10 P8-12 65-33929 <=*$>
1966 N3 P35-38 66-34808 <=$>

EKONOMIKA SELSKOGO KHOZYAISTVA
1958 V29 N5 P42-48 59-11774 <=>
1959 V30 N3 P38-43 59-31004 <=>
1959 V30 N4 P10-20 60-31001 <=>
1959 V30 N6 P10-22 60-31018 <=>
1959 V30 N6 P46-51 60-31018 <=>
1960 V31 N3 P1-10 60-31408 <=>
1960 V31 N3 P37-44 60-31408 <=>
1960 V31 N5 P10-15 60-31693 <=>
1960 V31 N5 P37-43 60-31693 <=>
1960 V31 N7 P83-89 61-21146 <=>
1960 V31 N9 P67-70 61-21339 <=>
1960 V31 N12 P39-45 61-21538 <=>
1961 V32 N1 P5-7 61-23585 <=>
1961 V32 N1 P25-30 61-23585 <=>
1961 V32 N1 P53-64 61-23585 <=>
1961 V32 N1 P79-89 61-23585 <=>
1962 V33 N5 P70-73 62-33018 <=> P
62-33018 <=*>
1962 V33 N8 P85-88 63-13230 <=*>
1964 V35 N6 P16-25 64-51393 <=>
1964 V35 N8 P31-67 64-51845 <=$>
1964 V35 N8 P110-113 64-51417 <=>
64-51845 <=$>
1964 V35 N11 P36-49 65-30776 <=$>
1965 V36 N2 P114-117 65-30809 <=$>
1965 V36 N4 P1-9 65-31393 <=$>
1965 V36 N4 P24-27 66-30380 <=$>
1965 V36 N12 P19-29 66-30950 <=$>
1965 V36 N12 P47-50 66-30950 <=$>
1965 V36 N12 P119-120 66-30950 <=$>
1967 N2 P3-18 67-31361 <=$>

EKONOMIKA STROITELSTVA
1959 N6 P24-28 59-13909 <=>
1959 N6 P37-40 59-13909 <=>
1959 N6 P43-52 59-13909 <=>
1959 N10 P41-47 60-11359 <=>
1959 V11 N11 P7-11 60-11359 <=>
1960 N8 P13-17 61-11722 <=>
1960 N8 P36-42 61-21094 <=>
1960 N12 P3-7 61-21536 <=>
1960 N12 P17-22 61-21536 <=>
1960 N12 P27-29 61-21538 <=>
1963 N3 P35-42 63-31082 <=>
1964 N5 P34-54 66-30479 <=*$>
1965 N2 P1-4 65-33093 <=$>
1965 N8 P30-32 65-33093 <=$>
1965 N9 P38-41 65-33093 <=$>
1965 N9 P53-58 65-33093 <=$>
1965 N9 P62 65-33093 <=$>
1966 N7 P16-24 66-34704 <=$>
1966 N7 P56-57 66-34704 <=$>
1966 N12 P5-12 67-30977 <=$>
1966 N12 P25-28 67-30977 <=$>
1966 N12 P33-37 67-30977 <=$>
1966 N12 P46-47 67-30977 <=$>
1967 N1 P23-27 67-31161 <=$>
1967 N2 P3-10 67-31251 <=$>
1967 N2 P14-21 67-31251 <=$>
1967 N2 P65-66 67-31251 <=$>

EKONOMIKO-MATEMATICHESKIE METODY
1963 N1 P63-106 64-41526 <=>

EKONOMISTA
1964 N3 P587-597 64-41449 <=>

EKONOMSKA POLITIKA
1963 V12 N580 P565 63-31620 <=>

EKONOMSKI PREGLED. ZAGREB
1965 V16 N7/8 P55-56 65-33005 <=$>
1966 N11/2 P701-726 67-31096 <=$>

EKSPERIMENTALNA MEDITSINA I MORFOLOGIYA
1964 V3 N2 P81-95 64-51792 <=$>
1965 V4 N1 P60-65 65-32335 <=*$>

EKSPERIMENTALNAYA KHIRURGIYA I ANESTEZIOLOGIYA
1956 V1 N1 P53-59 R-218 <*>
1957 V2 N4 P3-7 62-11155 <=>
62-15071 <=*>
1959 V4 N1 P6-11 59-11672 <=>
1959 V4 N2 P39-42 60-18364 <+*>
1959 V4 N4 P12-19 63-19458 <=*>
1959 V4 N4 P53 64-19552 <=$>
1960 V5 N1 P3-7 60-31427 <=>
1960 V5 N1 P58-64 60-31437 <=>
1960 V5 N2 P8-18 60-41176 <=>
1960 V5 N2 P26-36 60-41176 <=>
1960 V5 N2 P39-45 60-41176 <=>
1960 V5 N2 P52-54 60-41146 <=>
1960 V5 N3 P57-58 60-41501 <=>
1960 V5 N3 P59-60 60-41502 <=>
1960 V5 N3 P60-61 60-41503 <=>
1960 V5 N6 ENTIRE ISSUE 61-21470 <=>
1961 V6 N1 P12-18 61-23131 <=*>
1961 V6 N1 P19-21 61-23134 <*=>
1961 V6 N1 P22-26 61-23129 <*=>
1961 V6 N1 P37-42 61-23127 <=*>
1961 V6 N1 P43-47 61-31158 <=>
1961 V6 N1 P48-51 61-23133 <=*>
1961 V6 N2 P3-14 61-27084 <*=>
1961 V6 N2 P41-46 61-31513 <=>
1961 V6 N2 P50 61-23481 <=*>
1961 V6 N3 P3-12 62-13225 <*=>
1961 V6 N3 P23-26 62-13251 <=*>
1961 V6 N3 P26-30 62-13215 <=*>
1961 V6 N3 P30-32 62-13227 <*=>
1961 V6 N3 P48-51 62-13244 <*=>
1961 V6 N4 P3-15 62-15362 <*=>
1961 V6 N4 P15-22 62-10093 <=>
1961 V6 N4 P22-24 62-13246 <*=>
1961 V6 N4 P52-53 62-13245 <*=>
1961 V6 N4 P53-57 62-13353 <*=>
1961 V6 N5 P3-10 62-19756 <=>
1961 V6 N5 P44-48 62-19756 <=>
1961 V6 N5 P55-56 62-19756 <=>
1962 V7 N1 P3-11 62-24391 <=*>
1962 V7 N1 P41-42 62-24709 <=>
1962 V7 N1 P50-53 62-24709 <=>
1962 V7 N1 P55-61 62-24709 <=>
1962 V7 N2 P3-8 62-25046 <=>
1962 V7 N2 P16-19 62-25046 <=>
1962 V7 N2 P66-70 62-25046 <=>
1962 V7 N2 P86-91 62-32769 <=*> O
1962 V7 N2 P94-95 62-25046 <=>
1962 V7 N3 P3-15 62-33462 <=*>

1962 V7 N3 P58-61 62-33462 <=*>
1962 V7 N4 P62-67 63-13086 <=*>
1962 V7 N5 P42-46 63-13546 <=$>
1962 V7 N5 P65-68 63-13546 <=$>
1963 V8 N2 P30-31 63-31245 <=>
1963 V8 N2 P46-47 63-31245 <=>
1963 V8 N2 P57-64 AD-615 873 <=$>
AD-617 391 <=$> 0
1963 V8 N3 P16-19 63-31742 <=>
1963 V8 N3 P20-22 64-15512 <=*$> 0
1963 V8 N3 P41- FPTS V23 N4 P.T686 <FASE>
1963 V8 N3 P41-44 63-31758 <=>
1963 V8 N3 P52-56 64-15448 <=*$> 0
1963 V8 N3 P58-62 64-19203 <=*$>
1963 V8 N3 P71-73 63-31740 <=>
1963 V8 N3 P74-76 63-31741 <=>
1963 V8 N3 P77-82 63-31757 <=>
1963 V8 N4 P3-6 63-41041 <=>
63-41081 <=>
1963 V8 N4 P6-9 63-41041 <=>
1963 V8 N4 P10-12 63-41041 <=>
1963 V8 N4 P12-13 63-41041 <=>
1963 V8 N4 P14-15 63-41041 <=>
1963 V8 N4 P15-19 63-41059 <=>
1963 V8 N4 P42-46 63-41059 <=>
1963 V8 N4 P54-55 64-14615 <=*$>
1963 V8 N4 P55-58 63-41081 <=>
1963 V8 N4 P83-86 63-41081 <=>
1963 V8 N5 P21-28 63-41162 <=>
1963 V8 N5 P45-47 63-41162 <=>
1963 V8 N5 P47-48 63-41162 <=>
1963 V8 N5 P57-60 63-41162 <=>
1963 V8 N5 P75-78 63-41162 <=>
1963 V8 N6 P3-5 64-21438 <=>
1963 V8 N6 P66-67 64-21439 <=>
1964 V9 P26-33 AD-645 567 <=>
1964 V9 N2 P3-13 64-31354 <=>
1964 V9 N2 P13-18 64-31355 <=>
1964 V9 N2 P61-64 65-31251 <=$>
1964 V9 N2 P77-80 64-31348 <=>
1964 V9 N3 P18-19 65-31355 <=*$>
1964 V9 N4 P32-36 N66-12263 <=$>

EKSPERIMENTINES MEDICINOS INSTITUTO DARBAI
1955 V3 P61-67 64-31531 <=>
1955 V3 P67-72 64-31531 <=>
1958 V3/4 P185-192 65-50109 <=>
1958 V4/5 P177-181 64-31531 <=>
1958 V4/5 P183-191 64-31630 <=>
1958 V4/5 P193-205 64-31531 <=>
1962 N27 P53-62 RTS-3056 <NLL>

EKSPERYMENTALNA MEDYTSYNA
1937 N2 P69-74 62-11722 <=>

ELECTRIC FURNACE STEEL, JAPAN
SEE DENKI SEIKO

ELECTRIC MACHINERY INDUSTRY. CHINESE PEOPLES REPUBLIC
SEE TIEN CHI KUNG YEH

ELECTRIC POWER. NORTH KOREA
SEE CHOLLYOK

ELECTRICAL COMMUNICATION LABORATORY, TECHNICAL JOURNAL. JAPAN
SEE DENKI TSUSHIN KENKYUSHO KENKYU JITSUYOKA HOKOKU

ELECTRICAL REVIEW
SEE DENKI KORON

ELECTRICAL WORLD. CHINESE PEOPLES REPUBLIC
SEE TIEN SHIH CHIEH

ELECTRICIEN
1930 V61 N1504 P514-518 57-3131 <*>

ELECTRICITE. PARIS
1936 V20 N23 P289-292 60-10208 <*> 0
1937 P351-352 2152 <*>
1937 P353-356 2166 <*>
1946 V30 N120 P199-205 62-20288 <*> 0
1947 V31 N5 P105-110 62-16977 <*> 0
1950 V34 N165 P254-256 59-17033 <*> 0
1954 P311-315 57-2820 <*>
1955 V39 P59-65 62-20267 <*> 0

ELECTRO MAGAZINE
1956 V7 N54 P26-31 59-18099 <=*>

ELECTRO-CHEMISTRY AND INDUSTRIAL PHYSICAL CHEMISTRY
SEE DENKI KAGAKU OYOBI KOGYO BUTSURI KAGAKU

ELECTROCHIMICA ACTA
1959 P70-81 TS 1388 <BISI>
1959 V1 P70-82 62-14370 <*>
1959 V1 P177-189 AEC-TR-6023 <*>
1959 V1 N2/3 P283-290 62-16739 <*> 0
1960 V2 P50-96 SC-T-64-918 <*>
1960 V2 N4 P287-310 41N49G <ATS>
1960 V2 N1/3 P1-21 11M43G <ATS>
1960 V3 P106-114 AEC-TR-5679 <*>
1961 V4 P274-287 AEC-TR-5675 <*>
1961 V5 P105-111 AEC-TR-5657 <*>
1963 V8 N10 P795-803 64-16915 <*> 0
1963 V8 N12 P949-959 64-18488 <*>
1964 V9 P1373-1390 1811 <TC>
66-12147 <*>
1964 V9 P1391-1404 65-114 <*>
1965 V10 P773-782 66-14670 <*> 0
1965 V10 P1067-1075 66-10725 <*>
1965 V10 N6 P605-615 65-14139 <*>
1966 V11 P421-434 8993 <IICH>

ELECTROENCEPHALOGRAPHY AND CLINICAL NEUROPHYSIOLOGY. MONTREAL
1954 V6 P119-144 58-1356 <*>
1955 V7 P179-192 59-10212 <*>
1956 V8 N3 P353-369 61-10238 <*> 0
1957 V9 N1 P1-34 58-113 <*>

ELECTRONIC TECHNOLOGY. CHINESE PEOPLES REPUBLIC
SEE TIEN TZU CHI SHI

ELECTRONIQUE ET AUTOMATISME (1960-1964, CONTINUED AS ELECTRONIQUE 1965-)
1962 P270 63-10324 <*> 0
1962 P272 63-10324 <*> 0

ELECTROPHOTOGRAPHY. JAPAN
SEE DENSHI SHASHIN

ELECTROTECHNICAL JOURNAL OF JAPAN
SEE DENKI GAKKAI

ELECTROTECHNIEK
1951 V29 N15 P281-284 57-888 <*>
1955 V33 N15 P278-282 1418 <*>
1956 V34 N22 P455-467 NP-TR-56 <*>
1962 V40 N5 P99-110 64-16233 <*> 0
1963 V41 P502-508 3934 <BISI>

ELECTROTEHNICA
1959 V7 P249-259 64-14979 <*> 0
1960 V8 P57-62 E-746 <RIS>
1960 V8 N3 P75-89 E-745 <RIS>

ELEKTRICHESKAYA I TEPLOVOZNAYA TYAGA
1957 N12 P15-16 59-11037 <=>
1958 N1 P7-10 59-18746 <+*> 0
1958 N3 P9-10 59-22426 <+*>
1958 N3 P35-36 59-22425 <+*>
60-13429 <+*>
1958 N4 P20-22 59-22427 <+*> 0
1958 N5 P4-9 59-18747 <+*> 0
1958 N7 P10-12 59-18749 <+*> 0
1958 N8 P19-21 60-15756 <+*>

Citation	Translation
1958 N8 P25-26	59-22257 <+*> O
1958 N8 P32-33	59-18748 <+*> O
1958 N9 P20-21	60-15424 <+*>
1958 V2 N6 P23-26	62-18784 <=*> O
1959 N3 P29-31	61-23289 <*=>
1959 N4 P27-29	61-15387 <+*>
1959 N4 P44-45	61-28459 <*=>
1959 N8 P37-39	60-19940 <*+>
1959 N10 P7-10	61-23647 <*=> O
1959 N11 P25-27	62-15253 <*=>
1959 N12 P26-28	62-15252 <*=>
1959 N12 P35-37	61-28458 <*=>
1960 N1 P23-25	62-13733 <*=>
1960 N1 P33-37	60-19864 <+*>
1960 N12 P15-17	62-24819 <=*>
1960 N12 P19-22	62-24818 <=*>
1960 N12 P33-35	62-24835 <=*>
1960 N12 P35-37	62-24836 <=*>
1961 N4 P22-28	62-24903 <=*>
1961 N4 P31-34	62-24904 <=*>
1961 V5 N7 P28-35	63-19760 <=*>
1961 V5 N8 P36-39	65-60719 <=>
1961 V5 N9 P35-41	65-61446 <=>
1961 V5 N11 P39-42	64-23545 <=$>
1962 V6 N1 P1-4	63-23413 <=*>
1962 V6 N1 P32-37	65-60707 <=>
1962 V6 N4 P27-32	65-60726 <=>
1962 V6 N7 P33-35	64-13015 <=*$>
1963 V7 N2 P20-25	65-61308 <=$>
1963 V7 N2 P30-34	65-61462 <=>
1963 V7 N6 P38-41	65-61460 <=>
1964 V8 N1 P22-25	TP/T-3560 <NLL>
1964 V8 N2 P26-30	TP/T 3568 <NLL>
1964 V8 N8 P4-7	TP/T-3658 <NLL>
1964 V8 N9 P3-5	TP/T-3665 <NLL>
1964 V8 N9 P9-11	TP/T-3666 <NLL>
1966 V12 N8 P32-33	66-35297 <=>

ELEKTRICHESKIE STANTSII

Citation	Translation
1949 V20 N9 P61-62	61-13668 <+*>
1952 V23 N2 P12-14	RT-3456 <*>
1952 V23 N8 P34-37	59-18664 <+*>
1953 V24 N11 P3-5	RT-1329 <*>
1953 V24 N11 P13-16	64-13022 <=*$>
1954 N9 P18-21	RT-2698 <*>
1954 V25 N4 P36-41	63-23598 <=*>
1954 V25 N10 P21-24	62-10116 <*=>
1955 V26 N6 P6-9	63-24580 <=*$>
1955 V26 N12 P25-30	59-22490 <+*>
1956 N2 P31-36	R-970 <*>
1956 V27 N3 P21-24	61-28162 <*=>
1956 V27 N3 P44-47	59-18669 <+*> O
1956 V27 N7 P14-17	63-15139 <=*> O
1957 V28 N3 P19-21	63-23973 <=*$>
1957 V28 N4 P27-30	59-22713 <+*>
1957 V28 N6 P19-22	59-22500 <+*>
1957 V28 N6 P42-44	59-22540 <+*>
1957 V28 N8 P4-7	63-19768 <=*>
1957 V28 N10 P60-63	60-19974 <+*>
1958 P55-59	46K25R <ATS>
1958 N3 P2-6	PB 141 243T <=>
1958 N3 P32-38	PB 141 244T <=>
1958 N3 P81-82	PB 141 245T <=>
1958 V29 N2 P71-74	59-22512 <+*>
1958 V29 N3 P23-27	RTS-2761 <NLL>
1958 V29 N3 P62-64	59-22510 <+*>
1958 V29 N3 P87	59-22621 <+*>
1958 V29 N4 P22-25	60-17917 <+*>
1958 V29 N5 P2-10	59-22620 <+*>
1958 V29 N6 P15-20	64-15936 <=*$>
1958 V29 N6 P42-43	59-22551 <+*>
1958 V29 N6 P54-57	59-18121 <+*>
1958 V29 N6 P67-70	59-19149 <+*>
1958 V29 N6 P70-72	59-19150 <+*>
1958 V29 N6 P72-75	59-19162 <+*>
1958 V29 N7 P60-67	59-22531 <+*>
1958 V29 N7 P70-73	59-22530 <+*>
1958 V29 N8 P22-25	59-18947 <+*>
1958 V29 N8 P61-63	59-18738 <+*> O
1958 V29 N8 P63-66	60-15787 <+*>
1958 V29 N8 P73-76	59-22454 <+*>
1958 V29 N9 P7-11	61-13572 <+*>
1958 V29 N9 P40-43	60-13434 <+*>
1958 V29 N9 P47-48	59-19163 <+*> C
1958 V29 N9 P50-52	59-19152 <+*> O
1958 V29 N10 P55-57	60-13636 <+*>
	60-19938 <*+>
1958 V29 N10 P58-64	60-15813 <+*>
1958 V29 N12 P35-38	60-15788 <+*>
1958 V29 N12 P42-48	60-13566 <+*>
1958 V29 N12 P68-69	TRANS-25 <MT>
1959 V30 N1 P65-70	60-15007 <+*>
1959 V30 N2 P2-8	60-13845 <+*>
1959 V30 N2 P55-58	60-13645 <+*>
1959 V30 N3 P28-30	63-15147 <=*> O
1959 V30 N3 P51-53	61-13019 <+*>
1959 V30 N5 P52-53	60-15839 <+*>
1959 V30 N5 P54-55	60-17225 <+*>
1959 V30 N6 P5-7	61-13076 <+*>
1959 V30 N6 P38-41	60-17219 <+*>
1959 V30 N6 P79-82	60-13856 <+*>
1959 V30 N7 P5-12	61-27494 <*=>
1959 V30 N7 P42-44	60-15775 <+*>
1959 V30 N7 P49-52	62-24661 <=*>
1959 V30 N7 P8C-82	61-13325 <+*>
1959 V30 N8 P37-40	61-13121 <+*>
1959 V30 N9 P2C-26	61-13324 <+*>
1959 V30 N9 P42-45	60-15834 <+*>
1959 V30 N9 P45-47	60-15833 <+*>
1959 V30 N9 P48-54	61-13420 <+*>
1959 V30 N9 P54-59	64-13111 <=*$>
1959 V30 N10 P10-15	62-15251 <*=>
1959 V30 N11 P33-41	63-24318 <=*$>
1959 V30 N11 P57-60	61-13468 <+*>
1959 V30 N11 P60-62	61-15463 <+*>
1959 V30 N12 P15-16	E-749 <RIS>
1959 V30 N12 P76-78	62-23925 <=*>
1960 V31 N1 P38-41	62-15249 <*=>
1960 V31 N1 P81-82	61-13130 <+*>
1960 V31 N2 P78-81	62-19248 <=*>
1960 V31 N3 P10-17	62-33221 <=*>
1960 V31 N4 P32-36	63-19889 <=*>
1960 V31 N4 P6C-64	61-23958 <*=>
	85N48R <ATS>
1960 V31 N4 P81-82	62-15250 <*=>
1960 V31 N4 P82-85	62-23927 <=*>
1960 V31 N4 P85-87	62-23929 <=*>
1960 V31 N4 P87-89	62-23928 <=*>
1960 V31 N6 P8-24	61-23957 <*=>
1960 V31 N6 P6C-64	61-27623 <*=>
1960 V31 N6 P78-79	62-23930 <=*>
1960 V31 N6 P90-91	62-32813 <=*>
1960 V31 N7 P17-20	61-28194 <*=>
1960 V31 N7 P20-25	63-19891 <=*>
1960 V31 N7 P42-46	RTS-2857 <NLL>
1960 V31 N8 P48-54	61-21094 <=>
1960 V31 N8 P65-72	61-10669 <=*>
1960 V31 N8 P81-83	61-28192 <*=>
1960 V31 N9 P6-10	45R76R <ATS>
	62-11749 <=>
1960 V31 N9 P3C-35	61-21094 <=>
1960 V31 N9 P54-59	62-13126 <*=>
1960 V31 N9 P81	61-21094 <=>
1960 V31 N9 P91-93	63-19887 <=*>
1960 V31 N10 P17-22	62-24864 <=*>
1960 V31 N10 P89-91	61-27585 <*=>
1960 V31 N11 P2-4	61-21538 <=> P
1960 V31 N11 P36-40	62-24675 <=*>
1960 V31 N11 P56-59	62-23924 <=*>
1960 V31 N12 P5-9	61-31155 <=>
1960 V31 N12 P37-41	62-19144 <=*>
1961 V32 N1 P24-27	62-24671 <=*>
1961 V32 N1 P44-49	62-25960 <=*>
1961 V32 N1 P63-69	65-60377 <=$>
1961 V32 N1 P78-85	62-24860 <=*>
1961 V32 N2 P72-74	62-24657 <=*>
1961 V32 N3 P2-6	61-31582 <=>
1961 V32 N3 P16-21	62-25963 <=*>
1961 V32 N3 P26-33	62-25387 <=*>
1961 V32 N3 P49-55	62-23980 <=*>

Citation	Translation
1961 V32 N4 P28-33	63-23395 <=*$>
1961 V32 N4 P33-34	62-15716 <*=>
1961 V32 N5 P12-15	64-15938 <=*$>
1961 V32 N5 P27-34	63-19756 <=*>
1961 V32 N5 P35-41	63-19758 <=*>
1961 V32 N5 P62-65	62-25887 <=*>
1961 V32 N5 P71-75	62-24299 <=*>
1961 V32 N6 P84-85	63-19759 <=*>
1961 V32 N7 P40-45	63-24688 <=*>
1961 V32 N7 P57-61	62-25934 <=*>
1961 V32 N8 P58-62	63-23393 <=*$> 0
1961 V32 N8 P85-86	64-19889 <=$>
1961 V32 N8 P86	63-23952 <=*>
1961 V32 N9 P21-25	63-19762 <=*>
1961 V32 N9 P55-59	65-60718 <=>
1961 V32 N10 P18-22	62-32804 <=*>
1961 V32 N10 P43-47	63-19886 <=*> 0
1961 V32 N10 P61-65	62-25932 <=*>
1961 V32 N1C P84-86	63-19763 <=*>
1961 V32 N11 P33-36	63-19764 <=*$>
1961 V32 N11 P50-53	63-19765 <=*>
1961 V32 N11 P53-56	65-60721 <=>
1961 V32 N11 P73-76	63-23392 <=*$>
1961 V32 N11 P76-79	63-23394 <=*$>
1961 V32 N12 P47-50	63-23946 <=*$>
1962 V33 P35-37	65-64222 <=*$>
1962 V33 N1 P51-54	65-60722 <=>
1962 V33 N1 P60-64	63-15191 <=*$>
1962 V33 N1 P81-82	63-19825 <*=>
1962 V33 N1 P86-87	65-60381 <=$>
1962 V33 N1 P92	63-13091 <=*>
1962 V33 N2 P13-16	63-15699 <=*>
1962 V33 N2 P59-62	65-60723 <=>
1962 V33 N2 P75-83	63-24079 <=*>
1962 V33 N3 P11-13	63-19769 <=*>
1962 V33 N3 P14-17	63-15697 <=*>
1962 V33 N3 P53-57	64-13026 <=*$>
1962 V33 N5 P2-6	64-15933 <=*$>
1962 V33 N5 P19-21	65-61408 <=>
1962 V33 N5 P45-48	63-23979 <=*$>
1962 V33 N5 P48-50	63-23977 <=*>
1962 V33 N5 P51-52	63-23970 <=*$>
1962 V33 N6 P2-8	65-61478 <=>
1962 V33 N6 P27-31	65-61466 <=>
1962 V33 N6 P73-74	65-60733 <=>
1962 V33 N7 P38-42	65-60731 <=>
1962 V33 N7 P59-64	64-15952 <=*$>
1962 V33 N7 P84-85	63-23967 <=*>
1962 V33 N8 P9-12	63-24582 <=*$>
1962 V33 N8 P47-49	65-61296 <=$>
1962 V33 N8 P60-68	65-61294 <=$>
1962 V33 N9 P69-73	64-23565 <=$>
1962 V33 N9 P77-85	63-24559 <=*$>
1962 V33 N10 P29-33	63-24596 <=*$>
1962 V33 N1C P33-35	63-24589 <=*$>
1962 V33 N10 P35-37	63-24099 <=*$>
1962 V33 N10 P60-62	64-23619 <=>
1962 V33 N10 P72-75	63-24593 <=*$>
1962 V33 N10 P77-78	64-15937 <=*$>
1962 V33 N11 P20-27	64-15935 <=*$>
1962 V33 N12 P10-13	64-15278 <=*$>
1962 V33 N12 P46-51	63-23941 <=$>
1962 V33 N12 P51-54	63-23971 <=*>
1962 V33 N12 P58-62	63-23976 <=*$>
1962 V33 N12 P85	63-21467 <=>
1963 N8 P73-76	65-60818 <=>
1963 V34 P40-45	65-64221 <=*$>
1963 V34 N1 P43-45	63-24594 <=*$>
1963 V34 N1 P45-48	63-24093 <=*$>
	65-61483 <=>
1963 V34 N1 P81-83	63-24595 <=*$>
1963 V34 N1 P83-84	64-15949 <=*$>
1963 V34 N2 P20-26	RTS-2820 <NLL>
1963 V34 N3 P69-76	64-23613 <=>
1963 V34 N4 P2-5	64-15934 <=*$>
1963 V34 N4 P26-28	64-15275 <=*$>
1963 V34 N4 P48-53	C-5140 <NRC>
	64-19357 <=$>
1963 V34 N5 P26-28	TP/T-3538 <NLL>
1963 V34 N5 P46-50	64-19358 <=$> 0
1963 V34 N5 P51-54	65-60653 <=$>
1963 V34 N6 P13-17	64-15950 <=*$>
1963 V34 N6 P35-38	64-15951 <=*$>
1963 V34 N6 P38-42	65-60392 <=$>
1963 V34 N7 P24-27	65-60384 <=$>
1963 V34 N7 P47-54	65-60626 <=$>
1963 V34 N8 P13-15	66-60460 <=*$>
1963 V34 N8 P15-19	TP/T-3539 <NLL>
1963 V34 N8 P76-78	64-15953 <=*$>
1964 V35 N1 P43-51	65-64240 <=*$>
1964 V35 N2 P66-68	66-61877 <=*$>
1964 V35 N3 P25-27	CE-TRANS-3733 <NLL>
1964 V35 N3 P41-43	65-64219 <=*$>
1964 V35 N3 P74-76	65-64220 <=*$>
1964 V35 N3 P80-81	65-64241 <=*$>
1964 V35 N4 P41-44	TP/T-3613 <NLL>
1964 V35 N4 P70-73	65-64233 <=*$>
1964 V35 N4 P89-90	TP/T-3614 <NLL>
1964 V35 N6 P15-20	TP/T-3640 <NLL>
1964 V35 N6 P58-63	TP/T-3769 <NLL>
1964 V35 N7 P57-60	65-64211 <=*$>
1964 V35 N7 P74-77	65-64208 <=*$>
1964 V35 N9 P34-38	TP/T-3676 <NLL>
1964 V35 N9 P38-43	TP/T-3677 <NLL>
1964 V35 N9 P77-82	TP/T-3710 <NLL>
1964 V35 N10 P28-31	65-64215 <=$*>
1964 V35 N10 P54-59	66-60461 <=*$>
1964 V35 N11 P35-37	TP/T-3711 <NLL>
1965 N10 P27-30	65-33646 <=*$>
1965 N10 P72-73	65-34034 <=*$>
1965 N11 P7-10	65-34069 <=*$>
1965 V36 N1 P58-61	66-60465 <=*$>
1965 V36 N3 P8-12	66-61523 <=*$>
1965 V36 N3 P13-16	66-61521 <=*$>
1965 V36 N6 P53-55	66-61525 <=*$>
1965 V36 N8 P51-53	65-32557 <=$>
1965 V36 N10 P2-8	66-303139 <=*$>
1966 N6 P7-8	66-33072 <=$>
1967 N3 P6-8	67-31552 <=$>

*ELEKTRICHESTVO

Citation	Translation
1937 N6 P11-22	64-71453 <=> 0
1938 N4 P67-68	63-15190 <=*>
1938 N6 P34-40	65-60727 <=>
1939 N3 P55-59	C-2328 <NRC>
1940 N8 P62-64	RT-2660 <*>
1945 N3 P9-11	63-24577 <=*$>
1945 N5 P50-52	62-24298 <=*>
1946 N8 P28-32	RT-167 <*>
1946 N8 P46-50	59-18668 <+*>
1946 N10 P43-50	59-22549 <+*>
1947 N2 P14-22	R-4002 <*>
1947 N2 P54-59	60-13628 <+*>
1947 N4 P5-13	60-10899 <*> 0
	60-16583 <+*>
1947 N5 P36-41	60-17200 <+*>
1947 N6 P28-34	RT-90 <*>
1947 N7 P86-87	RT-2659 <*>
1947 N9 P37-44	66-61675 <=*$>
1947 N10 P67	66-61524 <=*$>
1948 N1 P60-62	63-10027 <=*> 0
1948 N6 P30-35	C-2347 <NRC>
1948 N8 P34-40	RT-2011 <*>
1948 N11 P74-79	RT-3844 <*>
1948 V6 P30-35	R-1491 <*>
1949 N1 P33-48	C-4978 <NRC>
1949 N2 P43-46	59-20682 <+*> 0
1949 N4 P29-38	59-18642 <+*> 0
1949 N7 P30-36	61-27544 <*=>
1950 N2 P53-59	63-23607 <=*$>
1950 N4 P56-65	RT-3917 <*>
1950 N4 P88-89	RT-1270 <*>
1950 N4 P89-90	RT-1269 <*>
1950 N8 P53-56	RT-84 <*>
1950 N9 P39-45	TT.231 <NRC>
1950 N11 P33-37	62-32947 <=*>
1950 N12 P19-20	R-180 <*>
1950 N12 P87	RT-1661 <*>
1950 V6 P26-30	R-5045 <*>
1950 V12 P85-86	RT-1662 <*>

Citation	Translation
1951 N2 P16-20	R-1427 <*>
	RT-1543 <*>
1951 N3 P85-86	RT-1050 <*>
1951 N3 P86-88	RT-2135 <*>
1951 N3 P89-90	RT-1049 <*>
1951 N5 P57-63	RT-85 <*>
1951 N6 P25-29	59-18648 <+*> 0
1951 N7 P18-24	61-13906 <*+>
1951 N7 P33-38	62-32814 <=*>
1951 N9 P54-57	61-15198 <+*>
1952 N4 P71-75	R-4044 <*>
1952 N5 P5-10	53/0722 <NLL>
1953 N3 P19-29	62-19995 <=*>
1953 N5 P35-40	63-15787 <=*>
1953 N7 P12-19	RTS 2709 <NLL>
1953 N7 P87-92	RT-1042 <*>
1953 N12 P40-43	RT-2640 <*>
1954 N2 P21-25	63-23699 <=*$> 0
1954 N6 P46-52	62-13754 <*=>
1954 N7 P60-62	63-10776 <=*>
1954 N8 P3-9	GB125 BE 772 <NLL>
1954 N8 P52-57	3508 <HB>
1954 N9 P26-28	63-23700 <=*$> 0
1954 N10 P3-7	59-22709 <+*>
1954 N10 P29-32	59-18662 <+*>
1954 N10 P43-49	59-12174 <+*>
1954 N10 P55-62	59-18661 <+*>
1954 N10 P68-72	59-18657 <+*>
1954 N11 P63-68	RT-3065 <*>
1954 N11 P73-76	<INSD>
1955 N1 P42-47	GB125 IB 14443 <NLL>
1955 N3 P65-68	C-2193 <NRC>
1955 N4 P62-68	59-18655 <+*> 0
1955 N5 P18-23	RT-3890 <*>
1955 N5 P24-27	RT-3895 <*>
1955 N5 P49-51	62-24817 <=*>
1955 N5 P59-63	59-18647 <+*>
1955 N5 P87-88	RT-3869 <*>
1955 N6 P25-32	59-18654 <+*>
1955 N6 P37-43	R-3689 <*>
1955 N7 P93-99	C-2340 <NRC>
1955 N7 P107	RT-3343 <*>
1955 N7 P108-113	RT-3382 <*>
1955 N7 P114-122	RT-3381 <*>
1955 N7 P123-129	RT-4367 <*>
1955 N8 P11-19	C-3633 <NRCC>
1955 N8 P63-68	R-3116 <*>
1955 N8 P68-74	R-4985 <*>
	T1816 <INSD>
1955 N9 P1-8	59-18629 <+*> 0
1955 N9 P18-22	59-18627 <+*> 0
1955 N9 P44-54	59-18709 <+*>
1955 N9 P54-59	59-18652 <+*> 0
1955 N9 P60-62	59-18710 <+*>
1955 N9 P63-64	59-18711 <+*> 0
1955 N9 P73-76	RJ-417 <*>
	RJ-417 <ATS>
1955 N10 P7-10	59-18712 <+*> 0
1955 N10 P18-23	59-18651 <+*> 0
1955 N10 P24-28	59-18650 <+*> 0
1955 N10 P35-39	RT-4429 <*>
1955 N10 P40-44	59-18653 <+*> 0
1955 N10 P79-85	RT-3667 <*>
1955 N11 P15-26	59-18643 <+*> 0
1955 N12 P9-18	59-16994 <+*> 0
1955 N12 P19-24	59-14177 <+*>
1955 N12 P76-	R-4076 <*>
1955 V2 P32-36	R-1889 <*>
1955 V4 P53-55	R-1885 <*>
1955 V7 P93-99	R-1486 <*>
1955 V8 P75-78	T1823 <INSD>
1955 V10 P63-66	R-4080 <*>
1956 N3 P13-22	59-10191 <*>
1956 N4 P65-67	59-14180 <+*>
1956 N6 P31-34	6118 <K-H>
1956 N6 P65-70	60-17187 <+*>
1956 N8 P22-26	59-18630 <+*> 0
1956 N8 P55-57	60-23023 <+*>
1956 N9 P14-23	R-732 <*>
1956 N9 P23-32	63-11138 <=>
1956 N10 P1-6	59-22525 <+*>
1956 N11 P8-14	59-22504 <+*>
1956 N11 P30-33	59-22625 <+*> 0
1956 N11 P34-37	59-22573 <+*> 0
1956 N11 P55-62	59-18660 <+*> 0
1956 N11 P68-73	38M44R <ATS>
1956 V6 P31-34	R-1809 <*>
1957 N1 P30-34	96K21R <ATS>
1957 N1 P38-41	RTS-2975 <NLL>
1957 N2 P25-32	59-18714 <+*> 0
1957 N3 P24-27	59-17049 <+*>
	59-18677 <+*>
1957 N3 P62-67	60-13173 <+*>
1957 N4 P57-60	59-18745 <+*> 0
1957 N4 P75-79	59-18744 <+*> 0
1957 N5 P72-74	TRANS-26 <MT>
1957 N6 P32-35	59-18624 <+*> 0
1957 N6 P35-38	59-22520 <+*>
1957 N7 P20-25	61-31117 <=>
1957 N7 P32-34	59-22484 <+*>
1957 N7 P39-45	59-22541 <+*>
1957 N7 P67	59-19066 <+*> P
1957 N8 P44-46	59-18659 <+*> 0
1957 N8 P47-48	59-18658 <+*> 0
1957 N9 P10-13	59-22488 <+*>
1957 N9 P59-60	60-17457 <+*>
1957 N10 P12-19	59-22491 <+*>
1957 N10 P54-56	59 22482 <+*> 0
1957 N10 P57-61	59-22496 <+*>
1957 N11 P34-40	59-18724 <+*> 0
1957 N11 P58-64	59-18413 <+*>
1957 N11 P64-72	59-22497 <+*>
1957 N12 P31-34	63-24547 <=*$>
1957 N12 P50-54	3097 <NRCC>
1958 N1 P39-44	59-22483 <+*> 0
1958 N1 P64-67	59-18756 <+*>
1958 N3 P27-32	59-18731 <+*> 0
1958 N3 P32-35	59-18732 <+*> 0
1958 N3 P35-36	59-22574 <+*> 0
1958 N3 P37-39	61-10382 <=*> 0
1958 N3 P40-44	NP-TR-215 <+*>
	59-22708 <+*>
1958 N3 P52-56	63-19433 <=*>
1958 N4 P1-9	59-20694 <+*> P
1958 N5 P15-20	<MT>
	62-24191 <=*>
1958 N5 P20-24	TRANS-10 <MT>
1958 N5 P31-35	59-22514 <+*>
1958 N5 P51-54	AEC-TR-3561 <+*>
1958 N5 P67-70	65-12014 <*> 0
1958 N6 P82-83	59-22515 <+*>
1958 N7 P1-6	AEC-TR-4972 <=*>
	62-23897 <=*>
1958 N7 P35-41	60-17185 <+*>
1958 N7 P41-45	59-22556 <+*> 0
1958 N7 P51-55	59-22555 <+*>
1958 N7 P56-58	59-18017 <+*>
1958 N8 P5-11	59-19165 <+*> 0
1958 N8 P21-28	59-22622 <+*> 0
1958 N8 P37-41	59-19166 <+*> 0
1958 N8 P46-49	59-19167 <+*> 0
1958 N8 P64-68	59-22441 <+*>
1958 N9 P30-34	59-22424 <+*> 0
1958 N9 P60-62	60-13567 <+*> 0
1958 N9 P63-66	60-15731 <+*>
1958 N10 P14-20	61-13575 <+*>
1958 N10 P43-47	59-19456 <+*>
1958 N11 P1-7	60-23003 <*+>
1958 N11 P18-24	60-15757 <+*>
	60-23004 <*+> 0
1958 N11 P55-58	59-19153 <+*> 0
1958 N11 P69-71	60-17911 <+*>
1958 N12 P9-13	61-13074 <+*>
1958 N12 P39-43	60-17205 <+*>
1958 N12 P52-55	60-13810 <+*>
1959 N1 P9-16	61-27563 <*=>
1959 N1 P13-17	62-24897 <=*>
1959 N1 P25-29	60-13640 <+*>
1959 N1 P41-45	61-13225 <*+>
1959 N1 P64-68	59-19811 <+*>

Citation	Translation
1959 N1 P72-75	59-22721 <+*>
1959 N2 P6-10	61-28184 <*=>
1959 N2 P30-35	61-13018 <+*>
1959 N2 P64-66	60-13804 <+*>
1959 N2 P66-69	60-17186 <+*>
1959 N2 P78-84	61-19357 <+*> 0
1959 N3 P24	60-13854 <+*>
1959 N3 P37-41	61-15933 <+*>
1959 N4 P31-37	61-13021 <+*>
1959 N5 P5-9	61-13127 <+*>
1959 N5 P14-17	60-13785 <+*>
1959 N5 P54-56	60-15789 <+*>
	61-13128 <+*>
1959 N5 P60-65	61-15730 <+*>
1959 N5 P72-77	61-13126 <+*>
1959 N6 P40-47	61-13129 <+*>
1959 N6 P48-54	61-13391 <+*>
1959 N6 P71-76	TRANS-27 <MT>
	61-15392 <*+>
1959 N6 P76-77	60-15838 <+*>
1959 N6 P78-80	60-23002 <+*>
1959 N7 P13-18	62-23940 <=*>
1959 N7 P35-41	62-24810 <=*>
1959 N7 P65-70	61-13073 <+*>
1959 N7 P70-72	61-13227 <*+>
1959 N8 P26-29	61-15464 <+*>
1959 N8 P30-35	63-24321 <=*$>
1959 N8 P57-62	0-7 <RIS>
	SCL-T-352 <=*>
1959 N8 P69-72	61-13119 <+*>
1959 N9 P1-5	61-15460 <+*>
1959 N9 P5-8	61-13421 <*+>
1959 N9 P46-50	61-15642 <+*>
1959 N9 P84-88	61-15502 <+*>
1959 N10 P43-47	61-15916 <+*>
1959 N10 P47-53	62-23926 <=*>
1959 N10 P50-56	62-24641 <=*>
1959 N11 P52-56	63-24310 <=*$>
1959 N12 P10-13	61-23267 <*=>
1959 N12 P61-63	61-19136 <+*>
1959 N12 P64-69	61-23265 <*=>
1960 N1 P1-5	60-11860 <=>
1960 N1 P43-47	61-23529 <=*>
1960 N1 P63-68	61-23250 <*=>
1960 N1 P73-78	62-24656 <=*>
1960 N2 P1-8	61-23266 <*=>
1960 N2 P13-18	62-23425 <=*>
1960 N2 P35-40	61-23527 <=*>
1960 N2 P64-68	62-24636 <=$>
1960 N2 P73-77	61-28190 <*=>
1960 N2 P89-90	60-41173 <=>
1960 N3 P1-13	61-27583 <*=>
1960 N3 P14-20	62-23945 <=*>
1960 N3 P48-54	ARSJ V32 N9 P1442 <AIAA>
1960 N3 P73-74	61-23525 <=*>
1960 N3 P77-81	64-16567 <=*$>
1960 N4 P7-12	61-27584 <*=>
1960 N4 P30-37	63-24327 <=*$>
1960 N4 P37-42	61-27581 <*=>
	63-24331 <=*$>
1960 N5 P6-13	61-27609 <*=>
1960 N5 P13-15	61-23526 <=*>
1960 N5 P22-27	62-25953 <=*>
1960 N5 P39-45	61-23526 <=*>
1960 N5 P50-53	62-24660 <=*>
1960 N5 P55-59	62-24639 <=*>
1960 N5 P60-66	NP-TR-521 <*>
	61-23295 <*=>
	61-23526 <=*>
1960 N6 P6-10	64-13811 <=*$>
1960 N6 P14-16	61-27608 <*=>
1960 N6 P25-29	62-25924 <=*>
1960 N6 P91-93	62-13565 <=*>
	65-17400 <*>
1960 N7 P47-50	62-23920 <=*>
1960 N7 P56-61	60-18764 <+*> 0
1960 N7 P61-65	60-41444 <=*>
1960 N7 P77-82	63-24064 <=*>
1960 N7 P89-93	61-21307 <=>
1960 N8 P16-22	61-11537 <=> P
1960 N8 P22-28	60-31834 <=>
1960 N8 P68-75	65-13062 <*> 0
1960 N8 P76-78	62-23922 <=*>
1960 N9 P4-11	62-15717 <*=>
1960 N9 P11-14	62-32771 <=*>
1960 N9 P29-34	62-23360 <=*>
1960 N9 P63-67	AEC-TR-6400 <=$>
1960 N9 P73-80	62-23931 <=*>
1960 N10 P63-64	62-24816 <=*>
1960 N11 P1-7	61-28197 <*=>
1960 N12 P47-58	63-19826 <*=>
1960 N12 P48-52	63-23375 <=*$>
1961 N1 P56-61	63-19766 <=*>
1961 N1 P61-66	65-61423 <=>
1961 N1 P77-88	62-24851 <=*>
1961 N2 P1-9	61-31550 <=>
1961 N2 P73-76	63-16639 <=*>
1961 N3 P14-17	61-31550 <=>
	62-13149 <*=>
1961 N3 P28-35	65-61297 <=$>
1961 N3 P51-56	63-24340 <=*$>
1961 N3 P73-76	63-23062 <=*$>
1961 N3 P83-86	62-23943 <=*>
1961 N4 P13-20	62-19129 <=*>
	63-14930 <=*>
1961 N4 P43-48	63-15952 <=*>
1961 N4 P48-50	62-32278 <=*>
1961 N4 P68-73	TRANS-195 <MT>
1961 N4 P76-81	63-19470 <=*>
1961 N5 P38-44	M.5638 <NLL>
1961 N5 P64-68	63-14929 <=*>
1961 N5 P70	62-25926 <=*>
1961 N6 P5-10	63-23596 <=*>
1961 N6 P58-60	63-15945 <=*$>
1961 N7 P7-13	65-60724 <=>
1961 N7 P24-30	62-25959 <=*>
1961 N7 P31-34	62-25937 <=*>
1961 N7 P68-72	63-19892 <=*$>
1961 N8 P76-78	66-62399 <=*$>
1961 N9 P10-16	63-19300 <=*>
1961 N9 P57-63	62-24679 <=*>
	62-24680 <=*>
1961 N11 P7-14	63-19482 <=*>
1961 N11 P48-55	65-60720 <=>
1961 N11 P70-73	63-15184 <=*>
1961 N12 P16-19	63-19490 <=*>
1961 N12 P40-44	63-23951 <=*>
1961 N12 P59-63	63-23959 <=*>
1961 V8 N6 P25-28	09N54R <ATS>
1962 N1 P10-11	63-23960 <=*$>
1962 N1 P12-15	63-23961 <=*$>
1962 N1 P49-51	62-11777 <=>
1962 N1 P73-75	65-60693 <=$>
1962 N2 P48-51	65-60330 <=$>
	65-60725 <=>
1962 N2 P77-83	64-13270 <=*$>
1962 N3 P6-9	66-20967 <$>
1962 N3 P24-28	65-23702 <$>
1962 N3 P76-81	65-23695 <$>
1962 N4 P7-19	62-25998 <=*>
1962 N4 P33-36	62-32000 <=*>
1962 N4 P36-40	63-24084 <=*>
	64-23567 <=$>
1962 N5 P1-7	65-23696 <$>
1962 N5 P20-27	62-33332 <=>
1962 N5 P64-70	64-23564 <=>
1962 N5 P78-83	65-61401 <=>
1962 N6 P52-57	63-24592 <=*$>
1962 N6 P73-75	65-61409 <=>
1962 N7 P21-23	63-24588 <=*$>
1962 N7 P28-29	AD-621 795 <=*$>
1962 N7 P30-35	64-13029 <=*$>
1962 N7 P59-66	65-60735 <=>
1962 N8 P18-20	63-23983 <=*>
1962 N8 P20-25	63-24086 <=*$>
1962 N8 P65-68	63-23937 <=$>
	63-24359 <=*$>
1962 N8 P69-72	63-23936 <=$>
1962 N9 P60-62	63-23934 <=$>
1962 N9 P232	63-23935 <=$>

1962 N10 P32-36 M.5639 <NLL>
1962 N10 P52-55 65-61479 <=>
1962 N11 P7-13 63-19408 <=*>
1962 N11 P13-20 65-60815 <=>
1962 N11 P20-25 63-24100 <=*>
1962 N11 P50-53 65-61454 <=>
1962 N11 P78-82 65-61482 <=>
1962 N12 P46-51 63-24540 <=*$>
1963 N1 P17-21 AD-615 255 <=$>
1963 N3 P9-14 63-24541 <=*$>
64-15923 <=*$>
1963 N3 P39-45 64-23546 <=>
1963 N3 P48-51 RTS-3176 <NLL>
1963 N3 P66-71 65-61461 <=>
1963 N4 P61-62 64-23614 <=$>
1963 N4 P86-88 64-13112 <=*$>
1963 N7 P49-55 64-15924 <=*$>
1963 N7 P94-95 64-71179 <=>
1963 N8 P66-70 AD-610 368 <=$>
1963 N8 P70-72 65-60819 <=>
1963 N9 P27-33 65-61254 <=$>
1963 N9 P44-48 65-60741 <=>
1963 N10 P22-31 65-60743 <=>
1963 N10 P57-61 66-60459 <=*$>
1963 N12 P63-66 3613 <BISI>
1963 V4 P1-5 63-31092 <=>
1963 V11 P9-12 RTS-2730 <NLL>
1964 N2 P18-23 CE-TRANS-3770 <NLL>
1964 N2 P32-37 TP/T-3558 <NLL>
1964 N2 P58-62 TP/T-3559 <NLL>
1964 N4 P1-4 AD-610 331 <=$>
1964 N4 P90-92 64-41677 <=>
1964 N5 P17-22 TP/T-3630 <NLL>
1964 N6 P1-6 64-51337 <=>
1964 N9 P53-59 TP/T-3659 <NLL>
1964 N10 P52-54 65-30856 <=$>
1964 N10 P89-96 65-30964 <=$>
1964 N11 P19-21 66-60455 <=*$>
1964 N12 P58-60 TP/T-3799 <NLL>
1964 N12 P68-70 66-61520 <=*$>
1964 V8 P67-70 65-11732 <*>
1965 N1 P33-42 4454 <BISI>
1965 N1 P48-53 AD-632 069 <=$>
1965 N2 P70-73 RTS-3078 <NLL>
1965 N2 P74-75 TP/T-3756 <NLL>
1965 N3 P52-58 TP/T-3772 <NLL>
1965 N4 P8 TP/T-3790 <NLL>
1965 N5 P38-44 TP/T-3809 <NLL>
1965 N5 P67-70 TP/T-3810 <NLL>
1965 N6 P1-5 TP/T-3836 <NLL>
1965 N6 P81-84 EECL-LIB-TRANS-1337 <NLL>
1965 N7 P49-52 TP/T-3869 <NLL>
1966 N1 P13-16 66-31351 <=$>
1966 N1 P22-27 66-31493 <=$>
1966 N4 P1-11 66-32727 <=$>

ELEKTRIE
1960 V14 N8 P278-281 61-20493 <*>
1961 V15 P166-168 61-20739 <*> 0
1961 V15 N2 P12-15 62-14665 <*>
62-16709 <*>
1961 V15 N3 P78-81 62-24546 <=>
1961 V15 N3 P92-96 62-24546 <=>
1961 V15 N6 P176-181 3037 <BISI>
1962 V16 N3 P82-88 3282 <BISI>
1963 V17 N4 P120-124 64-14988 <*> 0
1965 V19 P328-333 NS-543 <TTIS>
1966 N6 P247-248 66-33280 <=$>
1966 N4/5 P148-149 66-32612 <=$>
1966 N4/5 P181-185 66-32612 <=$>

ELEKTRISCHE BAHNEN
1926 P144-151 57-1408 <*>
1926 P227-233 57-696 <*>
1932 V8 P69-73 AEC-TR-5173 <*>
85P58G <ATS>
1954 P219-222 T1623 <INSD>
1955 V26 P121-133 T1980 <INSD>
1955 V50 P50-58 T2060 <INSD>

ELEKTRISCHE-NACHRICHTEN-TECHNIK
1925 V2 N4 P96-103 57-2413 <*>
1925 V2 N5 P132-145 66-13646 <*>
1925 V2 N10 P330-334 58-134 <*>
1925 V2 N12 P454-456 66-12379 <*>
1926 V3 P161-171 57-7 <*>
1926 V3 N3 P97 58-339 <*>
1926 V3 N6 P220-229 66-13329 <*> 0
1926 V3 N6 P229-235 58-341 <*>
1927 V4 P106-115 58-192 <*>
1927 V4 N3 P125- 57-2380 <*>
1927 V4 N9 P385-387 57-938 <*>
1927 V4 N10 P405-426 66-12399 <*> 0
1928 V5 N2 P121- 60-10817 <*>
1928 V5 N4 P163- 57-2066 <*>
1928 V5 N4 P171-176 58-123 <*>
1928 V5 N5 P214-217 57-1487 <*>
1928 V5 N8 P312-333 57-2113 <*>
1928 V5 N11 P413-421 57-2099 <*>
1928 V5 N12 P522-529 58-191 <*>
1929 V6 P80-86 57-2203 <*>
1929 V6 P165-181 1554 <*>
165-181 <*>
1929 V6 N1 P9-17 57-649 <*>
1929 V6 N5 P165-181 II-691 <*>
1553 <*>
1929 V6 N9 P358-365 57-2428 <*>
1929 V6 N12 P467-479 66-13330 <*> 0
1930 V7 N2 P49-64 57-2231 <*>
1930 V7 N2 P72-78 57-697 <*>
1930 V7 N3 P108-119 66-12580 <*> 0
1930 V7 N4 P147- 57-1469 <*>
1930 V7 N6 P226-231 57-3253 <*>
1930 V7 N8 P307-317 60-10271 <*> 0
1930 V7 N9 P362-368 57-1363 <*>
1930 V7 N11 P443-448 60-10270 <*> 0
1931 V8 N1 P39-42 57-3209 <*>
1931 V8 N2 P49-62 60-10268 <*> 0
1931 V8 N2 P77-88 60-10267 <*> 0
1931 V8 N11 P480-488 57-1485 <*>
1931 V8 N12 P516-527 2180 <*>
1932 V9 N11 P412-420 57-1463 <*>
1933 V10 N6 P258-276 57-1348 <*>
1933 V10 N8 P317-332 57-39 <*>
1933 V10 N10 P416-422 57-879 <*>
1934 V11 N7 P238-245 60-10819 <*>
1934 V11 N7 P257-261 57-2168 <*>
1934 V11 N8 P281-288 60-10818 <*>
1934 V11 N9 P319-329 57-2519 <*>
1934 V11 N10 P338-341 29S85G <ATS>
1935 V12 P83-86 57-594 <*>
1935 V12 P87-91 57-594 <*>
1935 V12 N1 P2-16 66-12398 <*>
1935 V12 N2 P55-60 2185 <*>
1935 V12 N9 P278-288 57-2397 <*>
1935 V12 N11 P355-362 2184 <*>
1935 V12 N11 P368-379 57-2156 <*>
1936 V13 P162-163 61-10087 <*> 0
1936 V13 P419-425 59-17276 <*> 0
1936 V13 N1 P1-12 57-2483 <*>
1936 V13 N2 P47-73 57-2399 <*>
1936 V13 N4 P111- 57-2234 <*>
1936 V13 N5 P149-161 60-10269 <*> 0
1936 V13 N6 P205-216 57-1057 <*>
1936 V13 N12 P414-419 AL-84 <*>
1937 V14 N1 P13-23 57-1109 <*>
1938 V15 N3 P78-101 57-2489 <*>
1939 V16 N2 P48-52 60-14612 <*> 0
1939 V16 N3 P73-85 64-14647 <*>
1939 V16 N4 P96-120 57-847 <*>
1939 V16 N10 P258-273 57-961 <*>
1940 V17 P57-69 <LSA>
1940 V17 P93-107 63-14015 <*> 0
1940 V17 N1 P1-5 E-635 <RIS>
1941 V18 N11 P239-246 66-12396 <*>
1942 V19 N10 P199-218 57-953 <*>
1942 V19 N3/4 P45-62 52-03500 <*>
1943 V20 N4 P102-111 58-479 <*>
1943 V20 N11/2 P270-276 59-15147 <*> 0

ELEKTRIZITAET IM BERGBAU
1930 V5 N11 P216-219 60-10191 <*> 0

ELEKTRIZITAETSVERWERTUNG
1962 V37 N9 P279-285 49R75G <ATS>
64-16914 <*>

ELEKTRIZITAETSWIRTSCHAFT
1938 V37 N15 P389-392 60-10288 <*> 0
1940 V39 N4/5 P53-57 57-1080 <*>
1949 V48 N11 P264-267 62-16585 <*>
1951 V50 P66-68 AL-680 <*>
1954 V53 N24 P791-798 CSIRO-3238 <CSIR>
57-2385 <*>
1954 V53 N24 P820-821 CSIRO-3077 <CSIR>
1466 <*>
1955 V54 P761-768 T-2363 <INSD>
1955 V54 P779-784 2548 <BISI>
1957 V56 N7 P207-213 59-10957 <*>
1960 V59 N18 P642-646 2161 <BISI>
1960 V59 N24 P854-864 62-14827 <*>
1961 V60 N3 P78-82 65-13952 <*>

ELEKTRO-ANZEIGER
1959 N1 P5-7 60-31327 <=$>
1959 N41 P411-413 60-31284 <=$>
1959 N42/3 P426-427 60-31261 <=$> 0
1959 N42/3 P431-433 60-31262 <=$>
1960 N5/6 P45-49 60-31256 <=$> 0

ELEKTROENERGETIKA
1960 V2 P94-104 63-15349 <=*>
1960 V2 P191-214 62-23611 <=*> 0
1960 V2 N2 P105-114 62-15242 <=*>

ELEKTROENERGIYA
1964 V15 N6 P1-9 64-41965 <=>
1964 V15 N6 P25-26 64-41965 <=>
1964 V15 N7/8 P2-9 64-51487 <=>
1964 V15 N7/8 P23-27 64-51487 <=>

ELEKTROFIZICHESKAYA APPARATURA
1963 N1 ENTIRE ISSUE AEC-TR-6636 <=>
1963 V1 P119-133 AD-617 662 <=$>

ELEKTROIZMERITELNYE PRIBORY
1964 N7 P1-96 AD-619 478 <=$> M

*ELEKTROKHIMIYA
1965 N1 P84-89 ICE V6 N1 P24-28 <ICE>
1965 V1 N1 P84-89 21S86R <ATS>
1965 V1 N7 P800-805 66-12731 <*>
1965 V1 N7 P818-821 65-17172 <*>
1965 V1 N9 P1118-1123 66-10423 <*>
1965 V1 N10 P1167-1173 66-10424 <*>
1965 V1 N11 P1311-1318 66-10903 <*>
1965 V1 N12 P1485-1487 66-11049 <*>
1966 V2 N1 P87-88 66-14523 <*>

ELEKTRON. AMSTERDAM
1955 V10 P224-226 59-21170 <=*> 0
1955 V10 P232 59-21172 <=*>
1956 V11 P233 59-19638 <=*> 0
1956 V11 P238 59-19637 <=*> 0

ELEKTRON. LINZ
1947 V1 N3 P104-107 AL-847 <*>
1948 V2 N2 P220-224 57-1385 <*>
1951 N13/4 P429-439 57-2793 <*>
1957 V10 N8 P208 60-31215 <=*>
1957 V10 N9 P240 60-31216 <=*>
1957 V10 N12 P342-343 60-31217 <=*>
1957 V10 N12 P364 60-31217 <=*>
1958 V11 N6 P120-121 60-31219 <=*>
1958 V11 N6 P139 60-31219 <=*>
1958 V11 N6 P141 60-31219 <=*>
1958 V11 N1/2 P4 60-31218 <=*>
1958 V11 N1/2 P9 60-31218 <=*>
1959 V12 N6 P132-134 60-31379 <=*>

ELEKTRON IN WISSENSCHAFT UND TECHNIK
1948 V2 N10 P213-214 66-11195 <*>

ELEKTRONIK
1958 V7 N12 P375-378 59-21152 <=> 0
1960 V9 N8 P232-234 63-18875 <*> C
1962 V11 N2 P45-46 62-16169 <*>

ELEKTRONIKA
1958 N1/2 P3-41 60-31876 <=>
1958 V4 N4/5 P141-147 59-00645 <*>

ELEKTRONIKA BOLSHIKH MOSHCHNOSTEI. MOSCOW
1963 N2 P148-156 AD-631 855 <=$>

ELEKTRONISCHE RECHENANLAGEN
1959 V1 N3 P127-133 60-12267 <=>
1961 V3 N4 P167-175 66-11166 <*>
66-11657 <*>
1961 V3 N5 P197-205 62-20175 <*>
1963 V5 N6 P257-261 66-10922 <*>
1964 V6 N1 P20-26 66-10921 <*>

ELEKTRONISCHE RUNDSCHAU. BERLIN
1955 V9 P238-241 T-2470 <*>
1955 V9 N10 P365-368 63-16602 <*> 0
1956 V10 N2 P43-46 58-2674 <*>
1956 V10 N5 P133-135 57-3129 <*>
1957 V11 N1 P23-24 57-2470 <*>
1957 V11 N3 P65-67 58-510 <*>
1957 V11 N4 P102-105 59-10978 <*>
1957 V11 N10 P302-305 58-2394 <*>
1958 V12 P414-416 NP-TR-456 <*>
1959 V13 N4 P122-123 60-16856 <*>
1959 V13 N5 P179-180 60-13254 <=*>
1959 V13 N11 P407-408 60-14501 <*>
1960 V14 N4 P121-125 63-20698 <*> C
1960 V14 N5 P181-183 63-20699 <*> 0
1960 V14 N7 P273-275 61-16095 <*> 0
1960 V14 N7 P273 66-11781 <*> 0
1960 V14 N9 P367-369 61-20039 <*>
1960 V14 N9 P371-373 61-18619 <*>
1961 V15 N3 P91-95 62-10072 <*>
1961 V15 N4 P149-152 62-10072 <*>
1961 V15 N12 P567-573 62-16237 <*> 0
1962 V16 N1 P18-20 64-10407 <*> 0
1962 V16 N3 P111-114 63-18713 <*>
1963 V17 N2 P63-64 64-10909 <*>
1963 V17 N8 P401-403 65-11032 <*>

ELEKTRONNAYA OBRABOTKA MATERIALOV
1965 N1 P3-11 NLLTB V8 N4 P272 <NLL>
1965 N3 ENTIRE ISSUE 66-33842 <=$>

ELEKTRONNYE VYCHISLITEKNYE MASHINY I IKH PRIMENENIE
1959 ENTIRE ISSUE 61-27035 <=*> 0
1965 N4 P1-87 66-32406 <=$>

ELEKTRO-POST
1955 N23 P422-425 57-16 <*>
62-16996 <*> 0

ELEKTROPRIVREDA
1960 V13 N2 P101-104 65-60627 <=$>

ELEKTROSCHWEISSUNG
1930 V1 N10 P185-189 60-10078 <*> C
1931 V2 N2 P28-31 59-20663 <*> C
1933 V4 N10 P181-187 57-3187 <*>
1934 V5 N2 P21-25 57-2128 <*>
1937 V8 N6 P101-106 I-802 <*>
1937 V8 N7 P125-128 I-802 <*>

ELEKTROSILA
1956 N14 P5-11 59-18670 <+*>
1956 N14 P12-18 59-18641 <+*>
1956 N14 P19-24 59-18640 <+*> 0
1956 N14 P24-27 59-18634 <+*>
1956 N14 P27-32 59-18639 <+*> 0

1956 N14 P33-35 59-18633 <+*>
1956 N14 P35-40 59-18674 <+*>
1956 N14 P40-44 59-18632 <+*>
1956 N14 P44-51 59-18638 <+*> 0
1956 N14 P52-64 59-18637 <+*>
1956 N14 P71-82 59-18636 <+*> 0
1956 N14 P83-92 59-18635 <+*>
1956 N14 P93-94 59-18671 <+*>
1956 N14 P98-102 59-18631 <+*> 0
1956 N14 P103-104 59-18672 <+*>
1956 N14 P1C4-1C8 59-18673 <+*>
1961 N2C P1C-14 65-61160 <=>
1961 N20 P44-47 65-61162 <=>
1962 N21 P6-12 64-61159 <=>
1962 N21 P22-26 65-61161 <=>
1962 N21 P32-37 65-61430 <=>
1962 N21 P49-54 65-61158 <=>
1962 N21 P65-68 65-61400 <=>

★ELEKTROSVYAZ
1956 P95-99 R-1637 <*>
1956 N4 ENTIRE ISSUE 59-12389 <+*>
1956 V10 N1 P1-81 59-14112 <+*>
1956 V10 N1 P10-20 R-714 <*>
1956 V10 N1 P26-34 R-1447 <*>
1956 V10 N1 P62-71 R-942 <*>
1956 V10 N2 P41-49 28J18R <ATS>
58-142 <*>
1956 V10 N3 P1-80 59-12383 <+*>
1956 V10 N3 P13-20 R-475 <*>
1956 V10 N4 P1-79 59-12389 <+*>
1956 V10 N4 P28-34 R-1956 <*>
62-23166 <=*>
1956 V10 N4 P62-67 R-1379 <*>
1956 V10 N5 P1-122 59-12388 <+*>
1956 V10 N5 P28-31 R-1383 <*>
1956 V10 N6 P3-13 R-1812 <*>
1956 V10 N7 P1-80 59-14095 <+*>
1956 V10 N8 P25-35 R-1925 <*>
37J15R <ATS>
1956 V10 N8 P39-51 R-669 <*>
1956 V10 N9 P1-103 59-14107 <+*>
1956 V10 N9 P26-45 27J18R <ATS>
58-143 <*>
1956 V10 N10 P3-80 59-14151 <+*>
1956 V10 N11 P3-80 59-14150 <+*>
1956 V10 N11 P65-75 64-14004 <=*$>
1956 V10 N12 P27-37 R-1153 <*>
1957 N2 P5-9 R-1840 <*>
1957 N6 P3-9 R-1495 <*>
1957 N12 P45-49 R-3735 <*>
1957 V11 N1 P3-10 62-23029 <=*>
1957 V11 N1 P21-23 R-1551 <*>
1957 V11 N2 P5-9 773TM <CTT>
1957 V11 N2 P47-56 59-18951 <+*>
1957 V11 N2 P57-66 59-18016 <+*>
1957 V11 N5 P7-14 R-2019 <*>
1957 V11 N8 P3-13 R-2385 <*>
1957 V11 N9 P33-41 59-10871 <+*>
1957 V11 N1C P1C-12 64-14980 <=*$>
1957 V11 N10 P11-13 R-3856 <*>
1957 V11 N11 P42-46 R-3128 <*>
1958 V12 N1 P3-8 R-3888 <*>
59-19064 <+*>
1958 V12 N5 P5-15 R-5137 <*>
1958 V12 N5 P23-27 59-10503 <+*>
1958 V12 N6 P30-39 C-3398 <NRCC>
1958 V12 N7 P6-10 62-23199 <=>
1958 V12 N9 P71-73 59-22720 <+*>
1958 V12 N9 P74-75 59-19171 <+*>
1958 V12 N1C P38-46 61-27497 <*=>
1958 V12 N11 P3-8 59-17461 <+*>
1959 N12 P35-42 62-14208 <=*> 0
1959 V13 N2 P43-54 61-16150 <*=>
66M39R <ATS>
1959 V13 N2 P64-71 21L35R <ATS>
1959 V13 N3 P55-62 61-19458 <+*>
1959 V13 N6 P65-7C 60-17188 <+*>
1959 V13 N7 P3-9 61-27497 <*=>
1959 V13 N7 P10-16 60-31207 <+*>
1959 V13 N8 P14-23 60-16952 <+*>
1959 V13 N10 P3-12 60-12597 <=>
6C-15854 <+*>
62-11515 <=>
1959 V13 N11 P12-16 62-11184 <=>
1959 V13 N11 P17-23 60-31071 <=>
62-11541 <=>
1959 V13 N11 P40-49 6C-25087 <=>
1959 V13 N11 P74-77 60-31070 <=>
1960 V14 N1 P4C-45 62-32657 <=*>
1960 V14 N1 P45-55 62-32658 <=*>
1960 V14 N2 P3-13 62-32659 <=*>
1960 V14 N2 P14-19 62-32661 <=*>
1960 V14 N2 P20-27 61-28126 <*=>
1960 V14 N2 P66-70 62-32660 <=*>
1960 V14 N4 P3-6 62-11556 <=>
62-32634 <=*>
1960 V14 N4 P49-61 62-32635 <=*>
1960 V14 N4 P62-71 62-32636 <=*>
1960 V14 N5 P1C-16 61-23299 <*=>
1960 V14 N5 P45-50 62-32663 <=*>
1960 V14 N6 P3-9 61-23244 <*=>
1960 V14 N6 P66-68 61-11674 <=>
1960 V14 N7 P3-12 62-24472 <=*>
1960 V14 N8 P19-25 62-32656 <=*>
1960 V14 N8 P65-73 62-32655 <=*>
1960 V14 N9 P42-51 62-32651 <=*>
1960 V14 N1C P21-26 62-32598 <=*>
1960 V14 N1C P47-52 62-32597 <=*>
1960 V14 N10 P53-61 61-21469 <=>
62-32599 <=*>
1960 V14 N10 P62-69 62-32600 <=*>
1960 V14 N11 P3-14 61-28125 <*=>
1960 V14 N11 P15-2C 62-32601 <=*>
1960 V14 N11 P26-33 62-32603 <=*>
1960 V14 N11 P48-53 62-32604 <=*>
1960 V14 N11 P54-61 62-32605 <=*>
1960 V14 N11 P62-70 62-32606 <=*>
1960 V14 N11 P74-75 62-32607 <=*>
1960 V14 N12 P3-10 62-32630 <=*>
1960 V14 N12 P11-18 62-32608 <=*>
1960 V14 N12 P19-28 62-32609 <=*>
1960 V14 N12 P29-37 62-32610 <=*>
1960 V14 N12 P38-44 62-32611 <=*>
1960 V14 N12 P56-60 62-32612 <=*>
1960 V14 N12 P61-68 62-32613 <=*>
1960 V14 N12 P69-71 62-32614 <=*>
1960 V14 N12 P72-75 61-21410 <=>
62-32615 <=*>
1961 V15 N2 P23-30 61-23591 <*=>
1961 V15 N3 P3-7 62-32624 <=*>
1961 V15 N3 P8-17 62-11210 <=>
62-13840 <*=>
1961 V15 N5 P18-25 62-32633 <=*>
1961 V15 N7 P30-36 63-10853 <=*>
1961 V15 N9 P1-2 62-10637 <=*>
1962 V16 N11 P3-10 63-21089 <=>
1962 V16 N12 P26-32 63-23514 <=*>
1963 V17 N4 P28-32 65-60362 <=$>
1963 V17 N6 P10-15 65-11464 <*>
1963 V17 N8 P32-48 63-31976 <=>
1964 N11 P27-32 AD-626 067 <=*$>
1964 V18 N2 P8-14 65-30225 <=$>
1964 V18 N3 P5-16 64-31232 <=>
1964 V18 N7 P26-32 64-41528 <=>
1964 V18 N7 P60-68 64-41528 <=>
1964 V18 N8 P8 64-51342 <=>
1964 V18 N8 P20 64-51342 <=>
1964 V18 N8 P51 64-51342 <=>
1964 V18 N8 P57 64-51342 <=>
1964 V18 N8 P70 64-51342 <=>
1964 V18 N8 P77-78 64-41787 <=>
1964 V18 N10 P41-46 65-30075 <=$>
1965 V19 N8 P54-61 65-33344 <=*$>
1965 V19 N11 P17-32 66-30324 <=$>
1965 V19 N11 P48-54 66-30324 <=$>
1965 V19 N12 P19-26 66-31129 <=$>
1966 N1 P60-66 66-31352 <=$>
1966 N3 P33-41 66-32462 <=$>

ELEKTROTECHNICKY CASOPIS
1965 V16 N5 P487-495 NEL-TRANS-1782 <NLL>

ELEKTROTECHNICKY OBZOR
1938 V27 P554-558 60-10201 <*> 0
1947 V36 N18 P332-337 62-20353 <*>
1953 V42 N5 P267-275 2444 <*>
1960 V49 N4 P169-173 AEC-TR-6306 <=*$>
    75Q73C <ATS>
1960 V49 N7 P350-355 63-16703 <*>
1961 V50 N1 P85-90 63-16705 <*>
1961 V50 N4 P192-196 63-14928 <*>
1961 V50 N6 P332-335 34P59C <ATS>
1962 V51 N4 P168-172 65-23227 <$>
1962 V51 N8 P401-406 65-12203 <*>
1963 V52 N6 P284-287 66-12352 <*>
1963 V52 N12 P641-647 CE-TRANS-3673 <NLL>
1964 V53 N1 P8-13 65-64231 <=*$>
1964 V53 N4 P202-208 65-64237 <=*$>
1964 V53 N4 P209-216 65-64228 <=*$>
1964 V53 N7 P370-372 65-11760 <*>
1965 V54 N4 P155-159 EECL-LIB-TRANS-1319 <NLL>
1965 V54 N8 P371-374 EECL-LIB-TRANS-1352 <NLL>

ELEKTROTECHNIK. BERLIN
1947 V1 N4 P97-105 57-2795 <*>
1950 V4 N4 P104-108 62-20475 <*> 0
1950 V4 N5 P183-188 62-20476 <*> 0
1952 V6 N1 P11-17 61-14001 <*> 0
1952 V6 N2 P81-82 66-10072 <*> 0
1952 V6 N3 P115-118 59-15175 <*>

ELEKTROTECHNIK. PRAHA
1954 V8 N11 P383-387 58-515 <*>

ELEKTROTECHNIK UND MASCHINENBAU
1910 V28 P837-840 T1811 <INSD>
1913 P919- 57-943 <*>
1913 V31 P920 58-212 <*>
1913 V31 N2 P54- 57-2152 <*>
1914 V32 P10-13 57-2377 <*>
1914 V32 P30-37 57-2377 <*>
1914 V32 N32 P686 57-1394 <*>
1915 V33 N12 P149 57-2477 <*>
1923 V41 P181-186 57-2144 <*>
1923 V41 P193-200 57-2144 <*>
1923 V41 P417-424 66-12390 <*>
1924 V42 P529-534 57-2308 <*>
1925 V43 N11 P62- 57-8 <*>
1927 P557-588 57-948 <*>
1929 V47 N48 P1048-1051 66-12275 <*>
1929 V47 N49 P1090-1091 57-2024 <*>
1931 V49 N6 P108-110 66-14617 <*> 0
1938 V56 N8 P98-100 59-20372 <*> 0
1943 V61 N39/0 P479-486 66-13333 <*>
1946 V63 N7/8 P179-185 62-16790 <*> 0
1946 V63 N11/2 P254-256 62-18687 <*> 0
1948 V65 N1/2 P14-17 62-16982 <*> 0
1949 V66 N4 P88-92 62-18371 <*> 0
1952 V69 P484-489 66-13234 <*>
1953 V70 N10 P224-230 57-2523 <*>
1953 V70 N15/6 P348-353 66-10127 <*>
1954 V71 N24 P569-573 II-902 <*>
1958 V75 N4 P69-73 AEC-TR-3557 <+*>
1958 V75 N5 P89-92 AEC-TR-3555 <*>
1958 V75 N1/2 P17-24 62-18398 <*> 0
1959 V76 N1 P1-5 61-10653 <*>
1959 V76 N2 P37-42 61-10653 <*>
1959 V76 N5 P103-108 66-10137 <*> 0
1959 V76 N15/6 P373-378 64-16612 <*>
1960 V77 N11 P253-261 C-3737 <NRCC>
1961 V78 N1/2 P75-89 14S87G <ATS>
    66-11521 <*>
1963 V80 N19/0 P475-479 66-13873 <*>

ELEKTROTECHNIKA
1954 V47 P267-274 T1812 <INSD>
1955 V48 N10 P305-309 57-1201 <*>
1957 V50 N4 P121-130 63-14461 <*>
1958 V51 N1/2 P73-74 59-11453 <=>
1962 V55 N7 P291-302 65-17087 <*>
1962 V55 N12 P567-573 63-21216 <=>
1963 V56 N7/8 P325-330 65-31535 <$=>
1964 N10 P15-16 65-11731 <*>
1964 V57 P37-38 EECL-LIB-TRANS-1266 <NLL>
1964 V57 N1/2 P15-26 64-41357E <=$>
1965 V58 N5 P214-216 65-31603 <=$>
1965 V58 N5 P216-217 65-31571 <=$>

ELEKTROTECHNISCHE ZEITSCHRIFT
1881 P61-63 57-3538 <*>
1890 V11 N24 P333-335 66-13597 <*>
1902 V23 N9 P165- 57-2495 <*>
1904 V25 N12 P241- 57-2446 <*>
1907 N21 P527-531 57-3237 <*>
1908 N42 P1019-1023 62-25234 <=*>
1912 N39 P1006-1009 57-2181 <*>
1913 P116- 57-2247 <*>
1913 P142- 57-2247 <*>
1913 P175- 57-2247 <*>
1913 V34 N49 P1395-1396 57-1483 <*>
1914 V35 P442-447 66-12900 <*> 0
1914 V35 N23 P646-649 57-1002 <*>
1915 V36 N8 P85-88 65-10242 <*>
1915 V36 N9 P99-102 65-10242 <*>
1915 V36 N20 P241-244 57-1459 <*>
1915 V36 N16/7 P189-191 57-2070 <*>
1915 V36 N16/7 P200-201 57-2070 <*>
1917 V38 P553-566 66-13138 <*>
1919 P330- 57-2225 <*>
1920 P806-809 57-3232 <*>
1920 V41 P604-605 57-1572 <*>
1920 V41 P785-788 57-2157 <*>
1920 V41 N2 P91 57-3240 <*>
1920 V41 N2 P670-672 57-3344 <*>
1920 V41 N7 P125-128 66-12389 <*>
1920 V41 N37 P726-727 59-17952 <*> 0
1921 V42 P588-591 57-2249 <*>
1921 V42 P616-622 57-2249 <*>
1921 V42 P673-878 57-2023 <*>
1921 V42 P695-697 57-1587 <*>
1921 V42 P714-716 57-2238 <*>
1921 V42 P1025-1029 57-2143 <*>
1921 V42 N1 P7-8 57-918 <*>
1921 V42 N14 P333-337 66-12742 <*> 0
1921 V42 N15 P370-374 66-12742 <*> 0
1922 V43 P1305-1307 66-133000 <*>
1923 V44 P237-241 66-13322 <*> 0
1923 V44 P257-260 57-2468 <*>
1923 V44 P289-291 57-3254 <*>
1923 V44 P1027-1030 57-606 <*>
1923 V44 N21 P481-484 59-20378 <*> 0
1923 V44 N31 P732-733 57-2038 <*>
1923 V44 N34 P809-810 57-970 <*>
1923 V44 N35 P830-833 59-15646 <*> 0
1924 P21- 57-1488 <*>
1924 P266- 57-1472 <*>
1924 V45 P417-428 57-1554 <*>
1924 V45 P532- 57-1390 <*>
1924 V45 P817-819 57-2186 <*>
1924 V45 N6 P89-91 57-1552 <*>
1924 V45 N11 P210-213 59-17954 <*> 0
1924 V45 N13 P261-266 66-13313 <*>
1925 V46 P1342-1346 AEC-TR-4193 <*>
1925 V46 P1577-1580 57-02201 <*>
    57-2202 <*>
1925 V46 P1617-1626 57-02201 <*>
    57-2202 <*>
1925 V46 N11 P368-374 57-1462 <*>
1926 P133-134 57-2229 <*>
1926 P500-505 2130 <*>
1926 P1514-1519 57-818 <*>
1926 P1539-1544 57-818 <*>
1926 V47 P380-385 58-105 <*>
1926 V47 P717-719 66-12903 <*>
1926 V47 P1065-1067 66-14240 <*> 0
1926 V47 P1453-1458 59-20384 <*> 0
1926 V47 P1479-1482 66-13321 <*> 0
1926 V47 N34 P985-989 57-693 <*>

Citation	Translation
1926 V47 N36 P1050-1051	59-20385 <*> 0
1927 V48 P905-916	57-2438 <*>
1927 V48 N16 P535-537	62-18282 <*> 0
1927 V48 N26 P950	66-14612 <*>
1927 V48 N31 P1114-	57-2079 <*>
1927 V48 N33 P1173-1176	57-644 <*>
1928 V49 P455-460	66-13819 <*> 0
1928 V49 P1478-1480	57-2029 <*>
1928 V49 P1780-1784	57-1336 <*>
1929 V50 P959-963	2153 <*>
1929 V50 P1016-1018	II-789 <*>
1929 V50 N28 P1019-1024	58-194 <*>
1929 V50 N32 P1156-1159	57-3255 <*>
1929 V50 N52 P1873-1875	57-210 <*>
1930 P503-505	57-33 <*>
1930 N45 P1543-1545	57-3465 <*>
1930 V51 P797-800	59-20374 <*> 0
1930 V51 N13 P454-457	59-20375 <*> 0
1930 V51 N16 P571-573	58-125 <*>
1930 V51 N27 P983-984	59-15644 <*>
1930 V51 N33 P1158-1160	59-20377 <*> 0
1930 V51 N36 P1257-1262	57-3348 <*>
1930 V51 N38 P1322-1323	59-20376 <*> 0
1930 V51 N45 P1543-1545	57-923 <*>
1930 V51 N47 P1610-1613	57-1561 <*>
1931 N2 P43-45	SCL-T-506 <*>
1931 V52 P1432	58-137 <*>
1931 V52 N7 P205-209	58-308 <*>
1931 V52 N11 P347-348	57-2364 <*>
1931 V52 N14 P439-440	57-3092 <*>
1931 V52 N32 P1026-1029	57-2414 <*>
1931 V52 N37 P1157-1161	66-13136 <*> 0
1932 V53 N1 P1-5	66-14622 <*>
1932 V53 N9 P204-205	57-2077 <*>
1932 V53 N21 P497-499	66-13312 <*>
1932 V53 N22 P532-534	66-13312 <*>
1933 V54 P1259-1261	AL-356 <*>
1933 V54 N4 P73-76	66-13343 <*>
1933 V54 N32 P774-777	57-2043 <*>
1933 V54 N34 P815-818	57-2043 <*>
1933 V54 N42 P1017-1019	57-2096 <*>
1933 V54 N50 P1219	66-14233 <*>
1934 V55 N30 P742-744	65-14298 <*>
1934 V55 N41 P997-999	57-3345 <*>
1934 V55 N45 P1097-1100	57-3247 <*>
1935 N20 P570-	57-2193 <*>
1935 V56 N3 P49-52	57-914 <*>
1935 V56 N6 P121-122	57-3249 <*>
1935 V56 N50 P1355-1356	57-3250 <*>
1936 V57 N12 P329-332	66-13146 <*> 0
1936 V57 N14 P385-387	66-13146 <*> 0
1936 V57 N18 P489-491	57-968 <*>
1937 V58 P767	59-10582 <*> 0
1937 V58 N19 P499-503	57-627 <*>
1937 V58 N20 P535-538	57-627 <*>
1937 V58 N42 P1129-1133	57-2270 <*>
1937 V58 N43 P1158-1160	57-2270 <*>
1937 V58 N49 P1309-1313	59-10943 <*>
1937 V58 N41/2 P1111-1115	62-18366 <*> 0
1937 V58 N41/2 P1138-1142	62-18366 <*> 0
1938 V59 P11-	57-1418 <*>
1938 V59 N28 P1285-1289	57-2147 <*>
1938 V59 N48 P1285-1289	66-12896 <*>
1938 V59 N48 P1295-1298	59-10996 <*>
1938 V59 N48 P1311-1312	57-1564 <*>
1939 P892-896	57-3516 <*>
1939 N3 P55-59	R-1424 <*>
1939 V60 P825-831	57-1075 <*>
1939 V60 P1113-1115	64-16206 <*> 0
1939 V60 N4 P89-92	SCL-T-153 <*>
1939 V60 N17 P498-503	59-20381 <*> 0
1939 V60 N18 P532-538	59-20381 <*> 0
1939 V60 N27 P793-798	1492 <*>
1940 V61 N5 P97-100	57-2244 <*>
1940 V61 N8 P163-165	57-2376 <*>
1940 V61 N10 P237-240	59-20382 <*> 0
1940 V61 N21 P461-463	57-345 <*>
1940 V61 N42 P945-948	59-20379 <*> 0
1940 V61 N43 P969-973	59-20379 <*> 0
1940 V61 N49 P1126-1131	57-900 <*>
1941 V62 N1 P3-16	63-14575 <*> C
1941 V62 N4 P73-75	62-18231 <*> 0
1941 V62 N5 P85-92	57-949 <*>
1941 V62 N9 P214-215	2173 <*>
	57-3460 <*>
1941 V62 N12 P305-308	58-982 <*>
	63-16672 <*>
1941 V62 N15 P372	66-14141 <*>
1941 V62 N22 P493-497	57-1113 <*>
1941 V62 N26 P589-591	62-18365 <*> 0
1941 V62 N33 P706-709	66-11364 <*>
1941 V62 N48/9 P953-955	62-18160 <*> 0
1942 V63 P349-351	AEC-TR-3948 <*>
1942 V63 N35 P405-409	62-20292 <*> 0
1942 V63 N41 P503-504	62-18168 <*>
1942 V63 N49 P587-591	62-18394 <*> C
1942 V63 N29/0 P341-348	57-691 <*>
1942 V63 N31/2 P367-372	I-105 <*>
1943 V64 N39 P529-532	62-18367 <*> C
1943 V64 N25/6 P347-349	59-15163 <*> 0
1943 V64 N37/8 P507-510	58-461 <*>
1943 V64 N43/4 P579-584	58-203 <*>
1943 V64 N9/10 P124-125	57-881 <*>
1944 V65 P76-77	59-15292 <*> C
1944 V65 N23 P233-237	62-18368 <*> 0
1944 V65 N11/2 P93-95	62-16678 <*>
1944 V65 N15/6 P139-142	58-856 <*>
1944 V65 N27/4 P277-284	59-20383 <*> 0
1950 V71 P171-172	AL-875 <*>
1950 V71 P191-192	AL-280 <*>
1950 V71 N8/9 P196-197	62-18369 <*> 0
1951 V72 N15 P465-469	62-18602 <*> 0
1952 V73 P649-653	57-2807 <*>
1952 V73 N9 P285-287	AL-124 <*>
1952 V73 N10 P338-339	57-2531 <*>
1952 V73 N22 P708-711	57-2422 <*>

ELEKTROTECHNISCHE ZEITSCHRIFT. AUSGABE A

Citation	Translation
1953 V74 N1 P4-10	62-18246 <*> 0
1953 V74 N4 P98-101	66-11007 <*>
1953 V74 N10 P281-289	64-10350 <*>
1954 V75 N2 P45-48	59-15795 <*>
1954 V75 N5 P194-197	96Q67G <ATS>
1955 P382-386	GB125 CE947 <NLL>
1955 V67 P13-16	C-2304 <NRC>
1955 V76 N15 P525-532	57-2799 <*>
1955 V76 N17 P508-604	57-2728 <*>
1955 V76 N22 P802-806	1135 <*>
1955 V76 N24 P857-859	57-1425 <*>
1955 V76A P13-16	57-1278 <*>
1955 V76A P17-25	C-2225 <NRC>
	1766 <*>
1956 P41-47	TS-1433 <BISI>
1956 V77 P326-329	58-1621 <*>
1956 V77 N4 P101-105	57-3183 <*>
1956 V77 N4 P105-107	58-512 <*>
1956 V77 N14 P487-490	58-1093 <*>
	59-16034 <+*>
1956 V77 N17 P578-581	58-1669 <*>
1956 V77 NA14 P487-490	59-12039 <+*>
1956 V77 NA17 P578-581	59-12040 <+*>
1957 V78 P182-187	59-12037 <+*>
1957 V78 P869-873	TS-1699 <BISI>
1957 V78 N20 P734-736	59-16032 <+*>
1958 V79 N7 P235-240	69L29G <ATS>
1958 V79 N7 P245-248	11K28G <ATS>
1958 V79 N7 P249-254	68L29G <ATS>
1958 V79 N11 P385-388	C-3098 <NRCC>
1958 V79 N16 P553-560	TT-811 <NRCC>
1958 V79 N16 P572-576	62-18261 <*> 0
1958 V79 N22 P885-893	59-11977 <=> 0
1959 V80 P576-582	TS 1740 <BISI>
1959 V80 P588-593	2155 <BISI>
1959 V80 P593-599	2156 <BISI>
1959 V80 N3 P78-82	66-10790 <*> 0
1959 V80 N6 P175-180	UCRL TRANS-491 <*>
1959 V80 N16 P536-541	77M43G <ATS>
1959 V80 N17 P576-582	62-16864 <*>
1959 V80 N17 P582-588	25N49G <ATS>
	61-14753 <*>

1959 V80 N20 P705-710 62-10548 <*> 0
1960 V81 P317-323 2403 <BISI>
1960 V81 N3 P97-101 63-16700 <*>
1960 V81 N3 P102-108 63-16701 <*>
1960 V81 N9 P332-338 NP-TR-927 <*>
1960 V81 N23 P801-807 61-18629 <*> 0
1960 V81 N26 P937-944 154-161 <STT>
1960 V81 N20/1 P722-729 2361 <BISI>
1960 V81 N20/1 P735-740 171-03 <STT>
1960 V81 N20/1 P744-749 96N56G <ATS>
1961 V82 N1 P1-9 61-16555 <*> 0
1961 V82 N15 P475-480 65-18091 <*>
1961 V82 N19 P605-609 62-20385 <*>
1961 V82 N26 P833-838 62-14736 <*> 0
1962 V83 N1 P16-23 2744 <BISI>
1963 V84 N15 P485-493 65-13645 <*>
1963 V84A N2 P39-44 65-13648 <*>
1963 V84A N8 P252-256 65-13652 <*>
1963 V84A N13 P428-435 65-13651 <*>
1963 V84A N13 P436-441 65-13650 <*>
1963 V84A N26 P883-886 65-10281 <*>
1964 V85 N12 P381-382 65-11539 <*> 0
1964 V85 N20 P658-666 POED-TRANS-2226 <NLL>
1965 V86 N5 P129-133 50S85G <ATS>
1965 V86 N12 P407-409 65-14271 <*> 0
1965 V86A N5 P129-133 65-17145 <*>

ELEKTROTECHNISCHE ZEITSCHRIFT. AUSGABE B.
1952 V4 N1 P1-8 62-16976 <*> 0
1956 V8 N2 P41-47 <MT>
62-14419 <*>
1957 V9 N6 P245-247 65-12018 <*>
1957 V9 N10 P385-388 C-2713 <NRC>
58-1425 <*>
1959 V11 N10 P412-415 62-10955 <*>
1961 N3 P55-56 AEC-TR-5472 <*>
1961 V13 N1 P1-7 188-313 <STT>
1961 V13 N3 P55-56 130-441 <STT>
1961 V13 N6 P135-141 139-595 <STT>
1961 V13 N12 P334-338 63-14421 <*> 0
1961 V13 N13 P353-357 62-10990 <*>
1962 V14 P700-704 2442 <TTIS>
1962 V14 N3 P49-55 62-18173 <*>
1962 V14 N5 P124-128 64-14117 <*>
1963 V15 P603-607 NS-170 <TTIS>
1963 V15 N21 P615-617 65-10280 <*>
1965 V17 N5 P103-108 66-14694 <*>
1965 V18 N11 P435-463 36T91G <ATS>

ELEKTROTECHNISCHER ANZEIGER
1939 V56 P557-559 63-18304 <*>
1940 V57 N4 P88-92 60-18589 <*> 0

ELEKTROTEHNISKI VESTNIK
1957 V25 N9/10 P341-345 59-15343 <*>
1964 N35 P110-111 65-30596 <=$>
1965 N3/5 P32-38 66-34252 <=$>

ELEKTROTEKHNICHESKII VESTNIK
1955 N5/6 P170-176 R-2074 <*>

*ELEKTROTEKHNIKA
V35 N7 P40-42 AD-627 013 <=$>
V35 N7 P44-46 AD-627 013 <=$>
1963 V34 N9 P20-25 RTS-2739 <NLL>
1963 V34 N9 P71-72 RTS-2738 <NLL>
1963 V34 N11 P14-19 RTS-2751 <NLL>
1963 V34 N12 P41-45 RTS-2753 <NLL>
1963 V34 N12 P54-55 SMRE-TRANS-4955 <NLL>
1964 N9 P8-14 AD-639 488 <=$>
1964 V35 N7 P1-4 RTS-2977 <NLL>
1964 V35 N7 P1-10 64-51962 <=$>
1964 V35 N7 P19-22 RTS-2828 <NLL>
1964 V35 N7 P36-38 RTS-2829 <NLL>
1964 V35 N7 P62-64 AD-625 295 <=*$>
1964 V35 N8 P11-13 TP/T-3775 <NLL>
1964 V35 N8 P40-43 RTS-2874 <NLL>
1964 V35 N9 P1-6 64-51842 <=$>
1964 V35 N11 P8-11 AD-614 765 <=$>
1964 V35 N11 P41-44 AD-620 812 <=*$>
1965 N10 P8-11 AD-638 666 <=$>
1965 V36 N1 P5-7 AD-613 464 <=$>
1965 V36 N3 P3-5 AD-640 297 <=$>
1966 N1 P58-59 AD-634 516 <=$>

ELEKTROTEKNIKEREN
1930 V26 N13 P253-257 59-15712 <*>
1954 V50 N24 P583-590 57-2527 <*>

ELEKTROTEKNISK TIDSSKRIFT
1925 V38 P51-54 57-2450 <*>
1925 V38 P57- 57-2164 <*>
1925 V38 N8 P59-62 57-1046 <*>
1935 V48 N7 P85-87 57-596 <*>
1935 V48 N8 P98-99 57-596 <*>
1957 V70 N31 P413-423 59-22492 <*+>
1963 V73 N10 P165-168 62-10091 <*> 0

ELEKTROWAERME. DUESSELDORF, ESSEN
1938 V8 N10 P261-266 63-18569 <*>
1941 V11 N6 P106-109 65-10136 <*> 0
1958 V16 N5 P163-169 62-20380 <*> 0
1958 V16 N6 P189-194 62-20379 <*> 0P

ELEKTROWAERME-TECHNIK
1953 N1 P7-11 2309 <*>
1953 V4 P50-53 2310 <*>
1955 V6 N1 P1-6 58-537 <*>

ELELMEZESI IPAR
1954 V8 N4 P114-117 26J19H <ATS>
1962 V16 N8 P240-243 63-20846 <*>

ELET ES TUDOMANY
1961 V16 N8 P232-235 62-19430 <=*>

ELETTRONICA
1956 V5 N3 P116-122 C-3278 <NRC>

ELETTROTECNICA
1914 V1 N30 P66-13328 66-13328 <*> 0
1920 V7 N19 P348-353 66-12255 <*> 0
1921 V8 P682- 57-3316 <*>
1924 V11 N33 P914-922 57-2767 <*>
1925 P589-594 57-3377 <*>
1925 V12 N20 P484-487 66-12383 <*>
1925 V12 N8/9 P189-194 57-2433 <*>
1925 V12 N8/9 P213-216 57-2433 <*>
1927 V14 N19 P413-419 66-14149 <*> 0
1931 V18 N1 P2-8 57-333 <*>
1932 V19 N13 P346-347 57-3317 <*>
1932 V19 N30 P755-757 57-1013 <*>
1933 V20 P9- 57-3196 <*>
1933 V20 N6 P118-121 63-14696 <*>
1945 V32 P182-193 62-14684 <*>
1948 V35 N5 P239-243 57-3008 <*>
1949 V36 P543-546 58-1291 <*>
1950 V37 N12 P551-555 57-1842 <*>
1954 V41 N3 P150-163 C-3514 <NRCC>
1960 V47 N2 P1-12 61-18924 <*>
1961 V48 P217-226 64-14719 <*>

ELIN-ZEITSCHRIFT
1958 N3 P12-29 1701 <BISI>

ELLENIKE
SEE HELLENIKE

ELOQUENCE FRANCAISE
1940 V2 P361-371 65-12656 <*>
1940 V2 P379-390 65-12591 <*>
1947 V1 P327-335 65-12359 <*>

ELTEKNIK
1958 N7 P3-10 58-2302 <*>
1961 V4 N5 P73-77 63-01432 <*>
1961 V4 N7 P127-131 62-18086 <*> 0
1963 V6 N7 P123-129 65-11459 <*>

EMBALLAGES

1953 V23 P15-17 MF-2 <*>
1957 V27 P263-265 66-13348 <*>
1960 N200 P163-169 57P6CF <ATS>
1961 V31 P195 65-12864 <*>
1962 V32 N205 P127-129 05Q70F <ATS>
1962 V32 N205 P131-133 05Q70F <ATS>
1964 V34 N216 P94-99 66-10746 <*>

ENCEPHALE
1931 V26 N2 P97-109 63-01777 <*>
1931 V26 N9 P659-670 63-00611 <*>
1939 V34 N2 P339-350 62-01429 <*>
1952 V41 P243-288 59-14130 <+*>
1957 V46 N3 P281-298 62-10715 <*> P
1960 V49 P428-434 66-11808 <*>
1962 V51 P181-197 63-20868 <*>
1962 V51 N3 P205-231 63-00553 <*>
1962 V51 N4 P301-344 64-00371 <*>
1964 V53 P543-552 66-12956 <*>

ENDOKRINOLOGIE
1952 V29 N5/6 P288-293 64-10348 <*>
1952 V29 N5/6 P305-311 64-10349 <*> 0
1953 V30 P176-184 65-12568 <*>
1954 V31 P70- 1904 <*>
1954 V32 P38-45 65-12598 <*>
1955 V33 N1/2 P1-8 65-12570 <*>
1961 V40 N5/6 P257-266 63-20966 <*> 0
66-15516 <*>
1962 V43 N3/4 P117-123 63-01245 <*>

ENERGETICA
1957 V5 N115 P22-524 PB 141 280T <=>
1964 V12 N7 P301-310 EECL-LIB-TRANS-1307 <NLL>
1965 N10 P461-464 66-30184 <=$>
1966 V14 N5 P221-224 66-33831 <=$>

ENERGETICHESKII BYULLETEN MINISTERSTVA NEFTYANOI PROMYSHLENNOSTI
1946 N1 P26-30 61-20414 <*=>
1947 N1 P19-22 2580 <HB>
1955 N4 P10-14 R-1230 <*>
1957 N7 P26-31 59-18682 <+*> 0
1958 N1 P10-16 59-22508 <+*>

ENERGETICHESKOE STROITELSTVO
1961 N22 P95-98 63-21213 <=>

ENERGETIK
1955 V3 N11 P1-4 RTS-3185 <NLL>
1956 N12 P1-3 59-10992 <+*>
1957 V5 N2 P32-34 62-24776 <=*>
1957 V5 N4 P31-33 59-22485 <+*>
1957 V5 N12 P7-8 59-22498 <+*>
1958 N1 P13-15 59-18157 <+*>
1958 V6 N1 P13-15 4K25R <ATS>
1959 N8 P40 60-11305 <=>
1959 V7 N2 P10-13 60-15008 <+*>
1959 V7 N7 P4-7 61-13352 <+*>
1959 V7 N7 P8-9 62-13112 <*=>
1959 V7 N9 P39-40 60-11860 <=>
1959 V7 N10 P40 60-11860 <=>
1960 V7 N12 P33-34 60-11860 <=>
1960 V8 N1 P39-40 60-11860 <=>
1960 V8 N2 P27-33 60-11759 <=>
1960 V8 N4 P38 61-11033 <=>
1960 V8 N8 P1-3 61-11537 <=> P
1961 V9 P17 61-31584 <=>
1961 V9 N4 P1-3 61-27119 <=>
1961 V9 N4 P36-39 61-31582 <=>
1961 V9 N5 P18-20 62-32014 <=*>
1961 V9 N9 P6-9 62-32815 <=*>
1962 V10 N1 P1-3 63-23948 <=*$>
1962 V10 N3 P14-16 64-23519 <=$>
1962 V10 N3 P37-38 63-13116 <=*>
1962 V10 N6 P38-40 62-33688 <=*>
1962 V10 N8 P23-26 63-23962 <=*$>
1962 V10 N10 P38 63-13365 <=>
1962 V10 N11 P24-25 63-23969 <=*$>
1962 V10 N82 P3-26 37R80R <ATS>
1962 V10 N1/2 P39 63-13116 <=*>
1963 V11 N2 P39-40 63-21687 <=>
1963 V11 N5 P9-11 65-60821 <=>
1963 V11 N5 P45 63-31538 <=>
1963 V11 N7 P31-32 63-31687 <=>
1964 V12 N5 P44-47 64-41123 <=>
1964 V12 N5 P47-48 64-41123 <=>
1964 V12 N7 P32-33 65-64206 <=*$>
1964 V12 N11 P4-7 TP/T-3694 <NLL>
1964 V12 N11 P35-38 TP/T-3695 <NLL>
1965 N10 P1-3 65-33843 <=*$>
1965 V13 N3 P8-10 66-61669 <=*$>
1965 V13 N5 P3-5 TP/T-3811 <NLL>
1965 V13 N8 P3-5 66-61672 <=*$>
1965 V13 N11 P4-5 66-30101 <=*$>
1966 N5 P1-2 NLLTB V8 N11 P927 <NLL>
1966 V14 N3 P36-37 67-30113 <=>
1966 V14 N8 P1-3 66-34956 <=$>
1966 V14 N8 P6-8 66-34956 <=$>
1966 V14 N8 P13 66-34956 <=$>

ENERGETIKA. KHARKOV
1964 V18 N9 P274-276 65-64212 <=$*>

ENERGETIKA I ELEKTRIFIKATSIYA
1966 N5 P3-4 67-30346 <=>
1966 N5 P46-48 67-30346 <=>
1967 N1 P6-7 67-31552 <=$>
1967 N1 P18-19 67-31552 <=$>

ENERGETIKA. SOFIA
1962 N1 P13-17 62-32050 <=>
1962 N2 P22-23 62-32465 <=*>
1964 N1 P1-2 NLLTB V6 N9 P803 <NLL>
TRANS.BULL.1964 P803 <NLL>
1964 N4 P3-6 65-30484 <=$>
1965 N3 P49-50 65-33986 <=*$>
1965 N3 P50-52 65-34077 <=*$>

ENERGETYKA. WARSZAWA
1964 V18 N7 P193-197 64-51223 <=>
1965 N2 P36-39 CE-TRANS-4182 <NLL>
1965 N10 P289-290 65-33964 <=$>

ENERGIA ES ATOMTECHNIKA
1957 N11/2 P597-603 60-17366 <*=>
1959 V12 P245-254 K-50/1 <RIS>
1959 V12 P482-492 K-50/2 <RIS>
1963 V16 N3 P142-143 63-31261 <=>
1963 V16 N9 P407-414 64-21075 <=> 0
1964 N2 P62-66 AD-637 442 <=$>
1964 N8 P385-390 ORNL-TR-431 <*>
1965 N9 P401-411 65-33750 <=$>

ENERGIA ELETTRICA
1949 V26 P461-468 21G81 <ATS>
1955 P433-440 1095 <*>
1955 V32 P185-197 T-1498 <INSD>
1955 V32 N7 P554-567 60-13256 <*=> 0
1959 V36 N3 P243-249 AEC-TR-3931 <*>
1959 V36 N4 P367-368 AEC-TR-3930 <*> P
1959 V36 N7 P641-651 60-13258 <=*> 0
1963 V40 P559-565 AEC-TR-6355 <*>

ENERGIA NUCLEAR. MADRID
1957 V1 N3 P53-66 AEC-TR-3719 <+*>
1958 V2 N1 P2-20 58-2367 <*>
1961 V5 N19 P24-47 HW-TR-44 <*>

ENERGIA NUCLEARE
1955 V2 N16 P426-433 59-17523 <*>
1956 V3 P23-31 65-10273 <*>
1956 V3 P74-81 58-298 <*>
1956 V3 N2 P113-118 65-11211 <*>
1957 V4 N4 P267-282 62-32402 <=*>
1957 V4 N4 P293-300 62-32401 <=*>
1958 V5 P815-823 62-32403 <=*>
1959 V6 P307-321 AEC-TR-4545 <*>
1959 V6 P376-392 AEC-TR-4559 <*>
1959 V6 N2 P141-146 AEC-TR-3843 <*>

1959 V6 N8 P521-531 AEC-TR-4753 <*>
1960 V7 N2 P111-120 33M43I <ATS>
1961 N3 P251-260 65-10499 <*> 0
1961 V8 N3 P209-212 C-3812 <NRCC>
1961 V8 N7 P467-473 ORNL-TR-135 <*>
1962 V9 N4 P212-225 AEC-TR-5200 <*>
1962 V9 N10 P581-594 UCRL TRANS-1049(L) <*>

ENERGIA TERMICA
1939 V7 P114-117 28J17I <ATS>

ENERGIE. MUENCHEN
1956 V8 N7 P260-264 61-16218 <*> 0
1958 V10 P267-277 AEC-TR-4103 <*>
1959 V11 N6 P241-248 GA-TR-2704 <*>
1959 V11 N8 P344-350 74P60G <ATS>
1960 V12 N4 P156-161 61-00031 <*>
1961 V13 N5 P231-235 62-00810 <*>
1962 V14 N4 P130-136 63P66G <ATS>
1964 V16 N2 P1-8 3937 <BISI>
1964 V16 N9 P365-370 4206 <BISI>

ENERGIE NUCLEAIRE. 1957-58
1958 V2 N3 P195-201 64-20114 <*>

ENERGIE NUCLEAIRE
1959 N1 P77-84 62-32404 <=*>
1959 V1 P217-221 AEC-TR-4029 <*>
1960 V2 N3 P160-165 AEC-TR-5643 <*>
1961 V3 N2 P128-138 HW-TR-29 <*>
1962 V4 P16-23 AEC-TR-5546 <*>
1962 V4 P600-606 AEC-TR-6020 <*>
1962 V4 N1 P40-42 AEC-TR-5464 <*>
1962 V4 N2 P109-119 NP-TR-945 <*>
1963 V5 P26-30 AEC-TR-6019 <*>
1963 V5 P257-262 AEC-TR-6156 <*>
1963 V5 P263-270 AEC-TR-6152 <*>
1963 V5 P271-281 AEC-TR-6154 <*>
1963 V5 P282-290 AEC-TR-6164 <*>

ENERGIETECHNIK
1957 V7 P287-293 61-18414 <*>
1959 V9 N3 P113-116 60-18462 <*>
1965 V15 N12 P529-534 66-30806 <=$>
1966 N11 P505-511 67-30236 <=>

ENERGOKHOZYAYSTVO ZZ RUBESHOM
1958 N3 P1-10 PB 141 236T <=>
1965 N2 P36-48 65-31478 <=$>

ENERGOMASHINOSTROENIE
1955 V1 N3 P14-18 60-13817 <+*>
1956 V2 N1 P14-18 61-28324 <*=>
1956 V2 N6 P11-15 64-15926 <=*$>
1956 V2 N9 P23-25 61-27179 <*=>
1956 V2 N10 P13-17 NP-TR-65 <+*> 0
1957 V3 P22-25 R-4085 <*>
1957 V3 N4 P10-12 61-13463 <+*>
1957 V3 N5 P22-25 PB 131 79CT <=>
R-3859 <*>
59-22527 <+*>
1957 V3 N7 P31-32 59-15766 <+*>
1957 V3 N8 P1-4 59-19618 <+*>
1957 V3 N9 P18-21 63-19948 <=*$>
1957 V3 N9 P30-34 61-13465 <+*>
1957 V3 N12 P42 62-24242 <=>
1958 V4 N1 P21-26 59-15827 <+*>
1958 V4 N1 P27-30 59-15826 <+*>
1958 V4 N1 P47 59-15823 <+*>
1958 V4 N1 P48 59-15825 <+*>
1958 V4 N3 P15-19 RTS-2740 <NLL>
1958 V4 N3 P24-27 60-23034 <*+>
1958 V4 N5 P6-9 63-19949 <=*$>
1958 V4 N7 P5-9 59-19442 <+*>
1958 V4 N7 P16-18 60-23561 <+*>
1958 V4 N7 P32-35 61-23657 <*=>
1958 V4 N8 P17-19 75L36R <ATS>
1958 V4 N9 P8-13 62-32696 <=*>
1958 V4 N9 P23-27 59-18009 <+*>
1958 V4 N11 P20-24 59-22402 <+*>
1958 V4 N11 P40-42 59-13249 <=>
1958 V4 N12 P15-20 60-15648 <+*>
1959 V5 N1 P3-8 62-15838 <*=>
1959 V5 N1 P29-32 59-19812 <+*>
1959 V5 N2 P35-39 12M38R <ATS>
1959 V5 N3 P13-17 61-15702 <+*>
1959 V5 N6 P1-7 65-61467 <=>
1959 V5 N6 P23-26 AD-625 242 <=*$>
64-71435 <=>
1959 V5 N6 P30 60-11198 <=>
1959 V5 N7 P1-7 61-13071 <*+>
1959 V5 N7 P7-12 61-13207 <*+>
1959 V5 N7 P12-14 61-13154 <*+>
1959 V5 N7 P14-18 61-13123 <+*>
1959 V5 N7 P18-22 60-17229 <+*> 0
1959 V5 N7 P26-27 61-13072 <*+>
1959 V5 N7 P28-29 61-13070 <+*>
1959 V5 N8 P20-23 61-13503 <*+>
1959 V5 N8 P31-34 61-13473 <+*>
1959 V5 N9 P6-8 61-23089 <=*>
1959 V5 N9 P14-16 61-23665 <*=>
1959 V5 N12 P33-37 6C-51134 <=>
1960 N3 P1-4 60-11933 <=>
1960 V6 N1 P17-21 62-25961 <=*>
1960 V6 N1 P26-32 61-13378 <+*>
1960 V6 N1 P33-35 61-13210 <+*>
1960 V6 N2 P29-31 62-32519 <=*>
1960 V6 N3 P23-27 61-23274 <=*>
1960 V6 N3 P32-35 61-15465 <+*>
1960 V6 N4 P8-12 61-27177 <*=>
1960 V6 N4 P12-15 61-28199 <*=>
1960 V6 N4 P26-30 62-11660 <=>
1960 V6 N4 P31 AD-611 108 <=$>
1960 V6 N5 P1-6 61-23534 <=*>
1960 V6 N5 P21-23 61-23535 <=*>
1960 V6 N6 P17-21 62-15176 <*=>
1960 V6 N6 P33-36 61-19945 <*>
1960 V6 N7 P1-5 61-23533 <=*>
63-24548 <=*$>
1960 V6 N7 P9-12 61-28182 <*=>
1960 V6 N7 P18 62-23483 <*=>
1960 V6 N7 P19-21 61-28188 <*=>
62-23678 <=*>
1960 V6 N7 P27-28 62-23483 <*=>
1960 V6 N7 P29-31 61-19222 <+*>
61-27551 <*=>
1960 V6 N7 P34 61-11641 <=>
1960 V6 N7 P35-38 61-28186 <*=>
1960 V6 N8 P22-26 61-27300 <*=>
62-24937 <=*>
1960 V6 N9 P8-11 61-28499 <*=>
1960 V6 N10 P29-32 62-15714 <*=>
1960 V6 N11 P16-19 62-25925 <=*>
1960 V6 N11 P24-25 62-15715 <*=>
1960 V6 N11 P28-29 62-24635 <=$>
64-23611 <=>
1960 V6 N11 P36-39 2066 <BISI>
1960 V6 N11 P43 61-31115 <=>
1960 V6 N12 P22-25 62-24886 <=*>
1960 V6 N12 P43-45 61-27193 <*=>
1961 V7 N1 P8-12 62-24938 <=*>
1961 V7 N1 P15-18 62-24643 <=*>
1961 V7 N1 P22-25 64-19985 <=$>
1961 V7 N1 P38-40 62-24938 <=*>
1961 V7 N2 P1-5 62-24637 <=$>
1961 V7 N2 P12-16 62-24841 <=*>
1961 V7 N2 P17-21 62-25412 <=*>
1961 V7 N2 P36-39 62-24839 <=*>
1961 V7 N3 P24-27 62-24633 <=$>
1961 V7 N3 P32-34 61-31527 <=>
1961 V7 N4 P32-34 2316 <BISI>
62-24645 <=*>
1961 V7 N4 P40 62-24645 <=*>
1961 V7 N4 P42-44 62-24684 <=*>
1961 V7 N5 P1-6 62-24642 <=*>
1961 V7 N6 P1-7 63-23933 <=$>
1961 V7 N6 P7-11 63-24549 <=*$>
1961 V7 N7 P16-19 65-61277 <=$>
1961 V7 N8 P5-8 62-24632 <=$>
1961 V7 N8 P42-45 62-33061 <=*>

1961 V7 N10 P23-26 63-13389 <=*>
1961 V7 N10 P27-30 64-13805 <=*$>
1961 V7 N1C P33-36 63-23974 <=*>
1961 V7 N10 P48 63-13389 <=*>
1961 V7 N11 P1-5 64-13016 <=*$>
1961 V7 N11 P5-9 63-24078 <=*>
1961 V7 N11 P26 64-13014 <=*$>
1961 V7 N11 P27-30 63-15185 <=*>
1961 V7 N11 P34-38 63-24080 <=*>
1961 V7 N11 P39-40 63-23396 <=*$>
1961 V7 N12 P1-5 63-24544 <=*$>
1961 V7 N12 P22-24 65-61458 <=>
1961 V7 N12 P25-26 63-15183 <=*>
64-13269 <=*$>
1961 V7 N12 P46-47 63-24561 <=*$>
1961 V7 N12 P48 63-24550 <=*$>
1961 V8 N12 P22-24 63-19182 <=*>
1962 V8 N1 P6-9 63-24558 <=*$>
1962 V8 N1 P24-29 63-19947 <*=>
1962 V8 N1 P33-36 3005 <BISI>
1962 V8 N1 P45-46 63-13198 <=*>
63-24546 <=*$>
1962 V8 N2 P34-37 63-11628 <=>
1962 V8 N3 P18-21 65-61412 <=>
1962 V8 N3 P26-29 63-24545 <=*$>
1962 V8 N4 P5-10 63-23984 <=*$>
1962 V8 N4 P33-37 63-15189 <=*>
1962 V8 N5 P1-4 65-61481 <=>
1962 V8 N5 P24-27 3002 <BISI>
1962 V8 N6 P4-10 64-23621 <=$>
1962 V8 N6 P22-25 65-61480 <=>
1962 V8 N6 P26-28 63-11622 <=>
1962 V8 N6 P29-32 63-15979 <=*>
1962 V8 N6 P32-35 63-13114 <=*>
1962 V8 N6 P40-41 63-13114 <=*>
1962 V8 N7 P16-18 65-30448 <=$>
1962 V8 N7 P20-32 63-23964 <=*$>
1962 V8 N7 P38-40 62-33441 <=*> P
1962 V8 N7 P39 63-24087 <=*>
1962 V8 N8 P16-18 64-23610 <=>
1962 V8 N8 P28-29 64-23612 <=$>
1962 V8 N8 P31-32 65-61261 <=>
1962 V8 N9 P1-5 63-13519 <=>
1962 V8 N9 P35-37 63-13519 <=>
1962 V8 N9 P41 63-13519 <=>
1962 V8 N10 P6-10 64-23620 <=$>
1962 V8 N11 P6-10 65-60391 <=$>
1962 V8 N11 P24-28 64-19945 <=$>
1963 V9 N3 P1-2 63-21816 <=>
1963 V9 N3 P43 63-21816 <=>
1963 V9 N3 P45 63-21816 <=>
1963 V9 N4 P9-13 63-24576 <=*$>
1963 V9 N4 P18-22 TP/T-3532 <NLL>
1963 V9 N5 P1-6 64-15276 <=*$>
1963 V9 N6 P33-35 63-24575 <=*$>
1963 V9 N7 P14-16 65-60822 <=>
1963 V9 N7 P21 64-71131 <=>
1963 V9 N8 P14-17 AD-617 668 <=$>
AD-625 182 <=*$>
1963 V9 N8 P20-21 TP/T-3570 <NLL>
1963 V9 N8 P36-37 63-41123 <=>
1963 V9 N8 P38-39 63-41123 <=>
1963 V9 N9 P14-18 TP/T-3533 <NLL>
1963 V9 N9 P43-45 65-60652 <=$>
1963 V9 N11 P11-14 AD-615 430 <=$>
1963 V9 N11 P18-22 64-19377 <=$>
1963 V9 N12 P8-11 TP/T-3534 <NLL>
1963 V9 N12 P18-21 TP/T-3535 <NLL>
64-71560 <=>
1963 V9 N12 P29-31 TP/T-3536 <NLL>
65-60353 <=>
1963 V9 N12 P39-40 TP/T-3537 <NLL>
1964 N10 P43-44 AD-635 832 <=$>
1964 N11 P14-17 AD-625 291 <=*$>
1964 V10 N1 P35-37 TP/T-3748 <NLL>
1964 V10 N1 P38-40 RTS-2666 <NLL>
1964 V10 N1 P41-43 AD-614 954 <=$>
1964 V10 N1 P46-47 65-60656 <=$>
1964 V10 N3 P4-7 65-64226 <=*$>
1964 V10 N3 P8-12 65-30749 <=$>
1964 V10 N3 P12-16 AD-610 786 <=$>
1964 V10 N3 P25-30 64-31492 <=>
1964 V10 N4 P1-5 CE-TRANS-3661 <NLL>
65-64230 <=*$>
1964 V10 N4 P14-16 AD-622 374 <=*$>
1964 V10 N4 P17-24 TP/T-3618 <NLL>
1964 V10 N4 P25-30 65-30799 <=*$>
1964 V10 N4 P28-30 CE TRANS-3776 <NLL>
TP/T-3619 <NLL>
1964 V10 N4 P30-33 RTS-2711 <NLL>
1964 V10 N5 P6-9 TP/T-3667 <NLL>
65-63829 <=>
1964 V10 N6 P8-11 AD-625 056 <=*$>
1964 V10 N6 P34-35 TP/T-3634 <NLL>
1964 V10 N6 P39-40 TP/T-3635 <NLL>
1964 V10 N7 P11-15 AD-624 852 <=*$>
1964 V10 N7 P19-21 65-30723 <=*$>
1964 V10 N7 P41-45 AD-620 807 <=$>
1964 V10 N8 P15-18 TP/T-3747 <NLL>
1964 V10 N8 P26-29 65-30664 <=$>
65-64210 <=*$>
1964 V10 N8 P41-43 AD-620 769 <=$*>
1964 V10 N9 P18-22 TP/T-3671 <NLL>
1964 V10 N9 P36-38 TP/T-3664 <NLL>
1964 V10 N10 P8-11 TP/T-3681 <NLL>
1964 V10 N10 P12-14 AD-622 460 <=$*>
1964 V10 N10 P17-20 AD-624 836 <=*$>
1964 V10 N10 P26-32 65-30520 <=*$>
1964 V10 N10 P29-30 65-64210 <=$*>
1964 V10 N10 P30-32 66-61670 <=*$>
1964 V10 N10 P35-37 TP/T-3682 <=$>
1964 V10 N12 P27-31 EECL-LIB-TRANS-1317 <NLL>
1964 V10 N12 P31-34 TP/T-3705 <NLL>
1964 V10 N12 P34-37 TP T-3706 <NLL>
1964 V10 N12 P45-46 65-30484 <=$>
1965 V10 N2 P7-11 66-61526 <=*$>
1965 V11 N1 P7-10 TP/T-3725 <NLL>
1965 V11 N1 P24-28 66-61665 <=*$>
1965 V11 N1 P32-36 TP/T-3817 <NLL>
66-60466 <=*$>
1965 V11 N2 P7-11 AD-620 951 <=$>
TP/T-3744 <NLL>
1965 V11 N2 P14-17 AD-617 679 <=$>
AD-625 194 <=*$>
1965 V11 N2 P23-26 66-60457 <=*$>
1965 V11 N2 P30-32 AD-625 290 <=*$>
1965 V11 N2 P36-37 TP/T-3745 <NLL>
1965 V11 N2 P44-45 NEL-TRANS-1679 <NLL>
1965 V11 N4 P19-22 TP/T-3792 <NLL>
1965 V11 N4 P33-34 NEL-TRANS-1737 <NLL>
1965 V11 N5 P14-17 AD-637 386 <=$>
1965 V11 N5 P29-32 AD-640 293 <=$>
1965 V11 N5 P35-37 AD-631 172 <=*$>
1965 V11 N5 P46 NEL-TRANS-1740 <NLL>
1965 V11 N7 P17-21 EECL-LIB-TRANS-1354 <NLL>
1965 V11 N9 P40-42 66-30342 <=*$>

ENFANCE. PSYCHOLOGIE, PEDAGOGIE, NEURO-PSYCHIATRIE SOCIOLOGIE
1952 V5 P250-261 64-10358 <*>
1955 V8 N5 P509-514 59-20154 <*>
1956 V9 N4 P21-27 58-2239 <*>
65-12691 <*>
1956 V9 N4 P49-59 60-18648 <*>
1957 V10 P71-77 65-12678 <*>
1957 V10 P377-382 65-12682 <*>
1957 V10 N2 P173-178 65-12695 <*>
1959 V12 N2 P143-152 65-11686 <*>

ENGENHARIA, MINERACAO E METALURGIA
1963 V38 P13-18 4193 <BISI>
1963 V38 P85-89 4193 <BISI>

ENGINEERING CONSTRUCTION. CHINESE PEOPLES REPUBLIC
SEE KUNG CHENG CHIEN SHE

ENGRAIS
1935 V50 P245-247 61-00201 <*>

ENKA AND BREDA RAYON REVIEW

1954 V8 N4 P125-134 1900 <*>

*ENTCMCLOGICHESKCE OBOZRENIE
1935 V2C N3/4 P206-220 63-11032 <=>
1949 V30 N3/4 P208-215 63-24357 <=*$>
1950 V31 N1/2 P121-122 R-3651 <*>
64-15612 <=*$>
1951 V31 N3/4 P323-335 C-2663 <NRC>
1951 V31 N3/4 P336-346 C-2382 <NRC>
1951 V31 N3/4 P467-473 61-19198 <+*> 0
1952 V32 P15-26 CSIRO-3345 <CSIR>
1952 V32 P15-25 59-18527 <+*>
1953 V33 P17-31 C2371 <NRC>
1955 V34 P131-136 R-513 <*>
1955 V34 P222-226 R-511 <*>
1955 V34 P227-230 R-509 <*>
1955 V34 P231-239 R-510 <*>
1956 V35 N1 P28-42 R-4695 <*>
1956 V35 N1 P132-138 62-15625 <=*>
1956 V35 N2 P201-209 62-15641 <=*> OP
1956 V35 N2 P262-284 62-15644 <=*> 0
1956 V35 N2 P324-333 61-19196 <+*>
1956 V35 N2 P371-376 61-19197 <+*> 0
1956 V35 N2 P377-396 61-19190 <+*>
1956 V35 N2 P397-420 R-1862 <*>
1956 V35 N2 P462-472 61-19188 <+*> 0
1956 V35 N3 P560-569 R-1977 <*>
1956 V35 N3 P716-723 62-15629 <=*> 0
1956 V35 N4 P856-882 62-15645 <=*> 0
1956 V35 N4 P944-955 64-15705 <=*$>
1957 V36 N1 P116-124 R-4126 <*>
1957 V36 N1 P125-133 R-4882 <*>
1957 V36 N1 P231-232 R-4122 <*>
1957 V36 N2 P418-435 61-19195 <+*>
1957 V36 N2 P436-450 61-19692 <=*>
1957 V36 N2 P501-537 61-19186 <+*> 0
1957 V36 N3 P625-631 63-23597 <=*$>
1957 V36 N3 P671-694 63-23591 <=*> 0
1957 V36 N4 P829-844 59-11780 <=>
1957 V36 N4 P860-868 65-61127 <=$>
1957 V36 N4 P877-894 63-23592 <=*$>
1958 V37 N2 P369-373 60-51078 <=>
1958 V37 N3 P616-629 64-15123 <=*$>
1958 V37 N3 P705-707 CSIR-TR.218 <CSSA>
1959 V38 N1 P8-17 60-17776 <+*>
1959 V38 N2 P424-434 65-61126 <=$>
1959 V38 N3 P652-654 62-15640 <=*>
1959 V38 N3 P675-681 62-15623 <=*> 0
1959 V38 N4 P774-789 63-23593 <=*> 0
1960 V39 N1 P77-85 61-15972 <+*>
1960 V39 N2 P275-283 60-41279 <=>
1960 V39 N2 P296-299 61-28212 <*=>
1960 V39 N2 P477-485 60-41279 <=>
1962 V41 N1 P99-108 63-13474 <=*>
1962 V41 N1 P109-124 63-21923 <=>
1962 V41 N2 P306-309 64-13086 <=*$>
1963 V42 N1 P39-48 63-31708 <=>
1963 V42 N1 P49-55 65-61452 <=>
1963 V42 N1 P219-225 63-31558 <=>
1963 V42 N2 P329-336 63-31763 <=>
1963 V42 N2 P351-363 65-64610 <=*$>
1963 V42 N3 P516-519 63-41234 <=>
1964 V43 P568-576 65-64606 <=*$>
1964 V43 N1 P118-130 64-31735 <=>
1964 V43 N2 P484-487 64-41689 <=>

ENTOMOLOGISCHE BERICHTEN
1956 V16 P173-175 61-00364 <*>

ENTOMOPHAGA
1959 V4 P193-199 64-14165 <*>
64-16840 <*> 0

ENTREPRISE
1958 N161 P14-21 59-11907 <=*> 0
1958 N161 P38-39 59-11908 <=*> 0

ENZYMOLOGIA: ACTA BIOCATALYTICA
1936 V1 P191-198 64-10013 <*>
1937 V3 P16-20 60-10854 <*>
1937 V4 P198-204 65-13977 <*>
1939 V7 P25- 57-380 <*>

ENZYMOLOGIA BIOLOGICA ET CLINICA
1962 V2 P175-187 64-00079 <*>
1962 V2 N2 P127-135 65-11316 <*> 0
1965 V5 N2 P65-76 66-10884 <*>

EPATOLOGIA
1962 V8 P48-50 66-11638 <*>

EPITOANYAG
1952 V4 N3/4 P58-63 63-18614 <*>
1958 V10 N3 P86-89 63-18594 <*>
1961 V13 P188-194 66-11327 <*>

ERDKUNDE
1947 V1 P162-175 57-3288 <*>
1948 V2 P1-21 57-3289 <*>
1950 V4 P81-88 57-3315 <*>
1953 V7 P249-266 IGR V1 N3 P1 <AGI>

ERDOE. BUDAPEST
1957 V6 N1 P3-8 62-00074 <*>

ERDOEL, ERDGAS; ZEITSCHRIFT FUER BOHR UND FOERDER TECHNIK
1965 V81 P480-496 62T94G <ATS>

ERDOEL UND KOHLE
1948 V1 N7/8 P232-240 57-3037 <*>
1949 N2 P52-59 AL-385 <*>
1949 V2 N9 P389-396 64-18369 <*> 0
1950 N5 P231-233 AL-86-II <*>
1950 V3 N1 P16-21 66-12785 <*>
1950 V3 N7 P315-317 58-1044 <*>
1950 V3 N7 P321-327 57-2651 <*>
1950 V3 N7 P330-333 44S86G <ATS>
1951 V4 N1 P9-11 61-18821 <*>
1951 V4 N4 P180-186 58-1081 <*>
1952 V5 P47 AL-934 <*>
1952 V5 P80-83 AL-262 <*>
1952 V5 P293-296 62-16650 <*>
1952 V5 P407-412 I-720 <*>
1952 V5 P552-560 AL-679 <*>
1952 V5 N3 P177-180 63-18663 <*>
1952 V5 N4 P177-180 63-18663 <*>
1952 V5 N6 P337-341 91J17G <ATS>
1953 V6 P613-616 62-18948 <*>
1953 V6 P693 <DIL>
1953 V6 N2 P65-66 29F3G <ATS>
1953 V6 N5 P266-270 <ATS>
1953 V6 N7 P375-380 3248 <K-H>
1953 V6 N8 P462-467 87M37G <ATS>
1953 V6 N10 P613-616 1950 <*>
1954 V7 P156-160 T-1746 <INSD>
1954 V7 P335-336 14H9G <ATS>
1954 V7 P496-501 8764 <IICH>
1954 V7 N3 P183-184 2821 <*>
1954 V7 N10 P640-642 40F4G <ATS>
1955 N12 P895-898 58-1612 <*>
1955 V8 P294-298 T1665 <INSD>
1955 V8 P864-873 T-1605 <INSD>
1955 V8 N2 P86-88 64-14687 <*>
1955 V8 N6 P393-401 99H8G <ATS>
1955 V8 N6 P407-411 62-10066 <*> 0
1955 V8 N6 P414-419 57-3442 <*>
64-14939 <*> 0
1955 V8 N9 P651-655 64-16733 <*> 0
1955 V8 N10 P700-702 33G8G <ATS>
1955 V8 N10 P706-711 CSIRO-3015 <CSIR>
59-20058 <*> 0
1956 N7 P441-447 57-2298 <*>
1956 N9 P690-693 T2102 <INSD>
1956 V9 P616-620 <DIL>
1956 V9 P686-690 T2101 <INSD>
1956 V9 N5 P280-283 65-10195 <*>
1956 V9 N6 P380-382 59-20846 <*>
1956 V9 N7 P447-450 59-20057 <*> PO
1956 V9 N8 P511-515 62-16477 <*>

Citation	Translation
1956 V9 N8 P516-520	57-2299 <*>
1956 V9 N9 P580-583	62-16524 <*>
1956 V9 N12 P839-843	57-2297 <*>
1956 V9 N12 P853-857	64-18383 <*> C
1957 P238-243	60-00701 <*>
1957 V10 P151-157	<DIL>
1957 V10 P228-231	<DIL>
1957 V10 P430-433	58-737 <*>
1957 V10 N2 P61-65	62-16646 <*>
1957 V10 N2 P75-79	59-10593 <*>
1957 V10 N3 P141-146	62-16647 <*>
1957 V10 N4 P218-221	62-16648 <*>
1957 V10 N4 P231	61-10495 <*>
1957 V10 N5 P303-305	64-20238 <*> 0
1957 V10 N6 P372-375	64-16899 <*>
1957 V10 N7 P428-429	61-10481 <*>
1957 V10 N7 P433-441	61-10480 <*>
1957 V10 N7 P442-444	61-10479 <*>
1957 V10 N7 P444-449	C-2692 <NRC>
1957 V10 N9 P558-592	60-00673 <*>
1957 V10 N9 P584-587	16J19G <ATS>
1957 V10 N9 P588-592	61-18713 <*>
1957 V10 N10 P675-677	58-267 <*>
	63-16399 <*>
1957 V10 N11 P747-752	62-14556 <*>
	87L35G <ATS>
1957 V10 N11 P754-757	64-18518 <*>
1957 V10 N11 P757-758	64-18503 <*>
1957 V10 N12 P825-826	62-16472 <*>
1957 V10 N12 P826-830	43K27G <ATS>
	62-14555 <*>
1957 V10 N12 P853-856	58-1203 <*>
1958 V11 P636-639	1903 <BISI>
1958 V11 P702-705	65-18127 <*>
1958 V11 N1 P6-11	90K24G <ATS>
1958 V11 N1 P13-18	61-10494 <*> 0
1958 V11 N2 P69-72	64-18818 <*> 0
1958 V11 N2 P72-75	62-18522 <*>
	64-18817 <*>
	89K24G <ATS>
1958 V11 N5 P312-317	64-18803 <*>
1958 V11 N7 P460-461	59-10178 <*>
1958 V11 N8 P511-515	61-10493 <*>
1958 V11 N8 P515-521	38K28G <ATS>
1958 V11 N9 P618-621	1419 <TC>
	59-10568 <*>
	60-16776 <*>
	61-10492 <*>
1958 V11 N10 P700-702	59-10722 <*>
1958 V11 N10 P705-708	65-11983 <*>
1958 V11 N11 P766-772	65-10305 <*>
1958 V11 N11 P778-781	59-15121 <*>
1958 V11 N12 P849-852	83Q71G <ATS>
1958 V11 N12 P868-872	60-18445 <*>
1959 V12 P65-71	62-20279 <*>
1959 V12 N1 P17-20	59-17061 <*>
1959 V12 N2 P71-73	38L33G <ATS>
1959 V12 N2 P71-77	60-00714 <*>
1959 V12 N2 P92-100	70N52G <ATS>
1959 V12 N3 P140-144	59-20047 <*>
	65-11962 <*>
	97L33G <ATS>
1959 V12 N3 P144-150	65-11636 <*>
1959 V12 N4 P216-218	44L34G <ATS>
	65-10198 <*>
1959 V12 N5 P423-431	74L34G <ATS>
1959 V12 N6 P467-471	64-30672 <*> P
	92L36G <ATS>
1959 V12 N6 P472-480	55L35G <ATS>
1959 V12 N7 P542-547	64-30671 <*>
1959 V12 N7 P552-558	64-30676 <*>
1959 V12 N8 P614-619	61-10496 <*>
1959 V12 N8 P635-640	65-11227 <*> 0
1959 V12 N9 P706-712	62-16422 <*>
1959 V12 N10 P805-815	62-16434 <*>
1959 V12 N12 P957-958	41S88G <ATS>
1959 V12 N12 P987-989	65-11245 <*>
1960 V13 N1 P5-11	62-14563 <*>
1960 V13 N1 P11-18	83M41G <ATS>
1960 V13 N1 P27-32	62-14433 <*>
1960 V13 N2 P79-83	84M41G <ATS>
1960 V13 N5 P305-312	45P64G <ATS>
1960 V13 N7 P470-472	72N50G <ATS>
1960 V13 N8 P541-544	76N48G <ATS>
1960 V13 N10 P787-788	38N49G <ATS>
ERDOEL UND KOHLE, ERDGAS, PETROCHEMIE	
1960 V13 N2 P79-83	65-12253 <*>
1960 V13 N3 P160-163	10217 <K-H>
	65-11839 <*> 0
1960 V13 N4 P263-266	62-10118 <*>
1960 V13 N6 P382-387	65-12443 <*>
1960 V13 N9 P658-664	63-01274 <*>
1960 V13 N10 P740-742	65-13041 <*> P
1960 V13 N10 P742-748	65-12931 <*>
1960 V13 N10 P806-	62-10133 <*>
1960 V13 N11 P836-845	65-12476 <*>
1960 V13 N11 P850-852	65-12922 <*>
1960 V13 N12 P955-960	65-12920 <*>
1961 V14 N1 P11-17	63-10221 <*> 0
1961 V14 N1 P24-26	11096 <K-H>
	65-13557 <*> C
1961 V14 N4 P268-271	65-14011 <*> 0
1961 V14 N5 P346-354	65-13040 <*>
1961 V14 N7 P527-537	13N58G <ATS>
1961 V14 N7 P542-544	2866 <BISI>
	59N57G <ATS>
1961 V14 N9 P686-692	00P63G <ATS>
1961 V14 N9 P714-716	59R76G <ATS>
1961 V14 N9 P725-731	50N58G <ATS>
1961 V14 N10 P804-809	65-10449 <*>
1961 V14 N10 P809-812	84P58G <ATS>
1961 V14 N11 P910-914	65-10446 <*>
1961 V14 N11 P915-917	63-16974 <*> 0
	65-11570 <*>
1961 V14 N12 P1011-1018	65-12214 <*>
	65-13993 <*> 0
1962 V15 P790-799	65-12156 <*>
1962 V15 N1 P27-31	65-12312 <*>
1962 V15 N3 P176-179	62-20472 <*> 0
1962 V15 N4 P262-268	62-18870 <*>
1962 V15 N4 P270-274	65-12292 <*>
1962 V15 N4 P274-282	CHEM-395 <CTS>
1962 V15 N5 P339-346	66-14637 <*>
1962 V15 N5 P348-352	CHEM-396 <CTS>
1962 V15 N6 P426-429	63-20729 <*>
	63-21310 <=>
1962 V15 N6 P441-451	3358 <BISI>
1962 V15 N6 P451-455	40P65G <ATS>
1962 V15 N6 P535-538	60P65G <ATS>
1962 V15 N7 P515-519	64-16966 <*>
1962 V15 N7 P523-529	64-18446 <*>
1962 V15 N7 P538-541	63-10284 <*>
1962 V15 N8 P619-622	65-14477 <*> 0
1962 V15 N8 P640-643	63-10285 <*> 0
1962 V15 N9 P699-702	65-11932 <*>
1962 V15 N9 P702-707	63-20075 <*>
1962 V15 N11 P880-887	20Q73G <ATS>
	63-20715 <*>
1962 V15 N11 P898-906	65-00215 <*>
1963 V16 P944-946	64-00018 <*>
1963 V16 N4 P284-287	66-11905 <*>
1963 V16 N5 P371-378	3734 <BISI>
	66Q71G <ATS>
1963 V16 N5 P379-382	169 <TC>
	67Q71G <ATS>
1963 V16 N7 P754-759	CHEM-397 <CTS>
1963 V16 N8 P843-846	64-18770 <*>
1963 V16 N9 P937-940	81R75G <ATS>
1963 V16 N10 P1000-1004	65-11093 <*>
1963 V16 N11 P1116-1119	64-18767 <*>
1963 V16 N11 P1121-1123	70R75G <ATS>
1963 V16 N6/1 P504-511	65S85G <ATS>
1963 V16 N6/1 P619-629	65-12405 <*>
1964 V17 N1 P2-9	43R79G <ATS>
1964 V17 N1 P22-28	03R76G <ATS>
1964 V17 N2 P74-83	65-14673 <*>
1964 V17 N4 P305-306	44R78G <ATS>
1964 V17 N5 P346-348	65-12406 <*>
1964 V17 N5 P352-356	66-10986 <*>

1964 V17 N5 P367-369 65-14725 <*>
1964 V17 N7 P550-553 73R79G <ATS>
1964 V17 N8 P605-609 65-12404 <*>
83S81G <ATS>
1964 V17 N8 P609-613 509 <TC>
65-14258 <*>
1964 V17 N1C P802-811 65-00232 <*>
1965 V18 P525-527 1439 <TC>
1965 V18 N2 P77-80 66S84G <ATS>
1965 V18 N2 P127-136 65-14315 <*>
1965 V18 N4 P273-281 1023 <TC>
1965 V18 N7 P523-524 66-14002 <*>
1965 V18 N7 P545-548 66-10459 <*> 0
1965 V18 N7 P548-553 66-10821 <*>
1965 V18 N9 P713-715 41T90G <ATS>
1965 V18 N9 P716-719 66-10822 <*>
1965 V18 N1C P770-775 66-12680 <*>
1965 V18 N10 P776-779 66-12934 <*>
1965 V18 N10 P780-787 66-12679 <*>
1965 V18 N1C P787-791 66-12933 <*> 0
1965 V18 N11 P865-890 66-10815 <*>
1965 V18 N11 P885-893 66-14003 <*> 0
1965 V18 N12 P950-957 66-14165 <*>
1965 V18 N12 P964-972 66-13985 <*> 0
1965 V18 N12 P972-976 66-13431 <*>
1965 V18 N12 P973-976 66-13998 <*> 0
1966 V19 N3 P171-172 74T93G <ATS>
1966 V19 N3 P182-185 66-14525 <*>
1966 V19 N4 P251-258 36T94G <ATS>
1966 V19 N4 P281-287 66-14535 <*>

ERDOEL-ZEITSCHRIFT FUER BOHR- UND FOERDERTECHNIK
1957 V73 P271-278 14K23G <ATS>
1957 V73 N4 P79-83 62-14151 <*>
1959 N11 P436-438 1878 <BISI>
1959 N12 P498-504 1875 <BISI>
1960 V76 N10 P333-340 10N50G <ATS>
1961 V77 P315-325 65-12425 <*>
1961 V77 N12 P581-602 48Q73G <ATS>
1961 V77 N12 P614-623 65-10096 <*> 0
1962 V78 P435-440 65-11401 <*>
1962 V78 N1 P17-44 56P61G <ATS>
1962 V78 N6 P339-368 53P63G <ATS>
1962 V78 N8 P481-486 84P64G <ATS>
1963 V79 P343-349 64-30696 <*>
1963 V79 P381-387 8901 <IICH>
1963 V79 N8 P343-349 35Q74G <ATS>
1963 V79 N12 P563-569 18R76G <ATS>
1964 V80 N8 P297-305 786 <TC>

ERGEBNISSE DER ALLGEMEINEN PATHOLOGIE UND PATHOLOGISCHEN ANATOMIE DES MENSCHEN UND DER TIERE
1931 V24 P451-553 63-00479 <*>

ERGEBNISSE DER ANGEWANDTEN MATHEMATIK
1958 N5 P148-156 65-14401 <*>

ERGEBNISSE DER BLUTTRANSFUSIONSFORSCHUNG
1957 V3 P108-120 65-00262 <*> 0

ERGEBNISSE DER ENZYMFORSCHUNG
1934 V3 P303-308 MF-2 <*>
1935 V4 P274-296 58-2117 <*>

ERGEBNISSE DER EXAKTEN NATURWISSENSCHAFTEN
1930 V9 P275-341 66-12674 <*> 0
66-12708 <*>
66-13133 <*>
1932 V11 P31- 63-14903 <*> 0
1932 V11 P134-175 NP-TR-916 <*>
1932 V11 P176-218 NP-TR-885 <*>
1936 V15 P160-188 NP-TR-1184 <*>
1939 V18 P155-156 AEC-TR-4394 <*>
1939 V18 P155-228 I-810 <*>
1939 V18 P175-191 AEC-TR-4394 <*>
1939 V18 P206-213 AEC-TR-4394 <*>
1939 V18 P219-223 AEC-TR-4394 <*>
1940 V19 P170-236 I-394 <*>
1950 V23 P127-134 23Q74G <ATS>

ERGEBNISSE DER GESAMTEN TUBERKULOSE-FORSCHUNG
1953 V11 P340-372 3697-B <K-H>

ERGEBNISSE DER INNEREN MEDIZIN UND KINDERHEILKUNDE
1922 V21 P204-250 UCRL TRANS-802 <*>
1958 V9 P591-621 65-00396 <*> 0
1958 V9 N1 P1-37 60-14180 <*>
1958 SNS V9 P1-37 62-00229 <*>

ERGEBNISSE DER PHYSIOLOGIE (BIOLOGISCHEN CHEMIE UND EXPERIMENTELLEN PHARMAKOLOGIE)
1905 V4 P517-564 PT2 61-14426 <*>
1921 V19 P211 PT2 57-299 <*>
1928 V26 P577-775 57-2538 <*>
1928 V27 P832-863 57-123 <*>
1950 V46 P71-125 61-00980 <*>
1950 V46 P126-260 57-576 <*>
58-44 <*>
1963 V52 P157-204 N65-11451 <=$>

ERGEBNISSE DER VITAMIN UND HORMONFORSCHUNG
1938 V1 P344-370 60-10686 <*>

ERICSSON TECHNICS
1943 N44 P56-66 62-10321 <*>

ERKENNTNIS
1938 V7 P211-225 1539 <*>

ERNAEHRUNGSFORSCHUNG
1959 V4 N2 P127-166 60-21048 <=>
1960 V5 P32-39 61-18256 <*>

ERNAEHRUNGSWIRTSCHAFT
1958 N3 P45-47 LSA-ER-1-58 <LSA>

ERZBERGBAU UND METALLHUETTENWESEN
1955 V8 P14-18 T-1654 <INSD>

ESCHER WYSS NEWS
1940 V13 P83-90 59-17716 <*>

ESSENZE, DERIVATI AGRUMARI
1960 V30 N1 P3-25 12N51I <ATS>

EST ET QUEST
1965 N347 P9-23 65-32882 <=*$>

ESTAMPAGE ET FORGE
1964 V2 P31-36 3995 <BISI>

ESTUDIO. INSTITUTO MEDICO-NACIONAL. MEXICO CITY
1890 P11-49 66-12663 <*>

ETUDE. CENTRE NATIONAL D'ETUDES DES TELE-COMMUNICATIONS. ISSY-LES-MOULINEAUX
1960 N578 P13 62-00164 <*>

ETUDES TECHNIQUES. CENTRE NATIONAL DU BOIS
1953 DOCUMENT S1 60-10925 <*>
1953 DOCUMENT S2 60-10926 <*>

EUROPA-ARCHIV
1961 V16 N18 P518-524 62-23579 <=*>

EUROPAEISCHER FERNSPRECHDIENST
1927 P46-50 66-13634 <*>
1927 P51-62 66-12894 <*>
1929 P261-262 57-1107 <*>
1929 N11 P50-55 57-2451 <*>
1929 V11 P61-63 58-354 <*>
1930 N16 P83-91 57-2445 <*>
1930 N19 P287-288 60-10299 <*>
1930 N19 P317-322 57-2441 <*>
1930 N20 P376-378 57-871 <*>
1930 V20 P379-385 66-12596 <*>
66-13294 <*>
1931 N21 P27-47 57-2435 <*>
1932 N28 P112-117 57-2455 <*>

1932 N29 P188-193	66-12745 <*> 0
1933 N31 P14-18	66-13660 <*>
1933 V33 P209-210	57-2178 <*>
1935 V41 P235-239	2169 <*>
1937 N45 P15-25	57-2323 <*>
1937 N45 P40-48	57-1061 <*>
1938 P13-20	57-2324 <*>
1938 N48 P33-40	57-313 <*>
1938 V50 P308-312	2168 <*>
1939 N51 P43-49	57-2040 <*>
1939 N52 P191-202	57-3341 <*>
1939 V52 P155-162	57-927 <*>

EUROPEAN ATOMIC ENERGY COMMUNITY. EURATOM
SEE REPORT. EUROPEAN ATOMIC ENERGY COMMUNITY. EURATOM

EUROPEAN POLYMER JOURNAL

1966 V2 P151-162	66-14159 <*>
1966 V2 N2 P163-171	66-14533 <*>

EVIDENTA CONTABILA

1964 V9 N8 P11-20	64-51331 <=$>

EXCERPTA MEDICA. SECTION 8. NEUROLOGY AND PSYCHIATRY

1958 V3 N4 P361-373	65-00333 <*> 0

EXPERIENTIA

1945 V1 N7 P207-212	II-262 <*>
1946 V2 P70-71	65-12994 <*>
1946 V2 P153-159	65-00281 <*>
1946 V2 P451-453	AEC-TR-2274 <*>
	II-260 <*>
1946 V2 N11 P438-448	58-2478 <*>
1947 V3 P198-199	65-12995 <*>
1947 V3 N1 P26-27	RCT V21 N1 P112 <RCT>
1947 V3 N11 P462-464	C-5097 <NRC>
1947 V3 N12 P490-492	RCT V22 N2 P316 <RCT>
1948 V4 P305-307	2090 <*>
	63-20495 <*>
1948 V4 N1 P6-22	60-10383 <*>
1948 V4 N2 P56-59	60-10382 <*>
1948 V4 N2 P64-66	AL-799 <*>
1949 V5 N5 P200	59-15887 <*>
1949 V5 N7 P271-277	C-2634 <NRC>
1949 V5 N11 P455-460	CSIRO-3217 <CSIR>
1950 V6 P302-304	58-2458 <*>
1950 V6 P432-433	I-131 <*>
1950 V6 N1 P19-21	65-12987 <*>
1950 V6 N5 P186-187	65-12981 <*>
1950 V6 N10 P388-389	65-00134 <*>
1950 V6 N10 P392-393	65-12985 <*>
1950 V6 N12 P469-471	60-10381 <*> 0
	63-18484 <*>
	65-12997 <*> 0
1951 V7 P121-127	I-912 <*>
1951 V7 P134-135	II-345 <*>
1951 V7 P258	2901 <*>
1952 V8 P220-221	61-00579 <*>
1952 V8 P221-223	61-00574 <*>
1952 V8 P432-434	AL-879 <*>
1952 V8 N4 P153-154	65-12990 <*>
1953 V9 P333-	II-533 <*>
1953 V9 P405-412	AL-394 <*>
1953 V9 N10 P379-380	65-12991 <*>
1954 V10 P74-76	AL-478 <*>
1954 V10 P76-77	AL-776 <*>
1954 V10 P132-	II-276 <*>
1954 V10 P210-212	58-1811 <*>
1954 V10 P259-261	I-174 <*>
1954 V10 P261-262	I-748 <*>
1954 V10 P315-316	II-161 <*>
1954 V10 P374-376	3144 <*>
1954 V10 P388-389	6234-B <K-H>
1954 V10 N6 P261-262	65-12986 <*>
1954 V10 N11 P472-473	65-12993 <*>
1955 N11 P417-	1737 <*>
1955 V11 P442-	1697 <*>
1955 V11 P446-	II-1008 <*>
1955 V11 P485-489	1376 <*>
1955 V11 N3 P98-99	65-12984 <*>
1955 V11 N5 P205-208	II-900 <*>
1955 V11 N7 P185-186	63-20965 <*> 0
1955 V11 N7 P272-	57-1918 <*>
1955 V11 N7 P281-282	62-18262 <*>
1955 V11 N9 P365-368	59-15377 <*>
1956 N5 P107-133	59-17123 <*> 0
1956 V12 P7-14	65-00363 <*> 0
1956 V12 P21-22	II-348 <*>
1956 V12 P31-33	II-294 <*>
1956 V12 P57-58	58-237 <*>
1956 V12 P103-104	1597 <*>
1956 V12 P235-236	57-3322 <*>
1956 V12 P365-368	65-10702 <*>
1956 V12 P411-418	60-18025 <*>
1956 V12 P479-480	57-1092 <*>
1956 V12 N1 P34-	1728 <*>
1956 V12 N6 P235-236	65-12996 <*>
1957 V13 P249-250	3069 <*>
1957 V13 P312-313	58-2108 <*>
1957 V13 P399-402	58-1034 <*>
1957 V13 P400-401	58-22 <*>
1957 V13 P401-403	58-27 <*>
1957 V13 P469-471	66-12335 <*>
1957 V13 P492-493	58-602 <*>
	63-16506 <*>
1957 V13 N1 P20-21	65-12983 <*>
1957 V13 N2 P74-	57-2983 <*>
1957 V13 N5 P187-	57-2984 <*>
1957 V13 N5 P197-	57-2991 <*>
1957 V13 N7 P291-305	64-18801 <*>
1957 V13 N10 P400-401	65-12989 <*>
1957 V13 N10 P401-403	59-15435 <*>
	65-12988 <*>
1958 V14 P70-71	58-832 <*>
	63-16603 <*>
1958 V14 P288-289	58-2442 <*>
1959 V15 P181-182	59-17128 <*>
1959 V15 N7 P258-260	60-10710 <*>
1959 V15 N11 P436-438	62-00524 <*>
1959 V15 N12 P468-470	60-18176 <*>
1960 V16 P141-142	60-18178 <*>
1960 V16 P546-547	61-20192 <*>
1960 V16 N1 P36-38	61-10893 <*>
1960 V16 N2 P75-76	60-18180 <*>
1960 V16 N3 P112-113	60-16639 <*>
1960 V16 N4 P129-133	61-10909 <*>
	62-10087 <*>
1960 V16 N4 P147	60-18177 <*>
1960 V16 N5 P204-205	60-18179 <*>
1960 V16 N8 P378-383	63-01746 <*>
1961 V17 P299-302	62-00422 <*>
1961 V17 P371-372	63-18733 <*>
1961 V17 P430-431	62-10732 <*>
1961 V17 P495-496	63-18796 <*>
1961 V17 N1 P22-23	NP-TR-825 <*>
1961 V17 N1 P27-28	61-10755 <*>
1961 V17 N6 P241-251	62-00412 <*>
1961 V17 N8 P329-343	63-10090 <*>
1962 V18 P233-239	62-20136 <*>
1962 V18 P504-506	64-14310 <*>
1963 V19 P421-423	63-20636 <*>
1963 V19 N5 P234-236	63-20626 <*>
1963 V19 N7 P374-376	64-16073 <*> 0
1963 V19 N12 P609-618	64-16376 <*>
1964 V20 N8 P451-452	65-11935 <*>
1964 V20 N8 P452	66-10854 <*>
1965 V21 N7 P417-418	65-14885 <*>
1965 V21 N12 P726-728	66-12045 <*> 0

EXPERIMENTAL CELL RESEARCH

1952 V3 N2 P485-488	57-771 <*>
1954 V6 N2 P560-562	3225-4 <K-H>
1962 V26 P62-69	65-13637 <*>

EXPERIMENTAL MEDICINE AND SURGERY

1954 V12 P209-214	63-14642 <*>

EXPERIMENTELLE TECHNIK DER PHYSIK

1956 V4 P160-162 UCRL TRANS-786(L) <*>
1956 V4 N2 P50-62 59-10055 <*>
1956 V4 N6 P253-262 SCL-T-361 <*>
1958 V6 P49-62 AEC-TR-3619 <+*>
1958 V6 N2 P49-62 59-17241 <*> 0
1959 V7 N4 P168-181 64-14783 <*> 0
1961 V9 P227-235 64-10840 <*>
1961 V9 N1 P1-12 62-24544 <=*>
1961 V9 N1 P13-22 62-15952 <=*>
1961 V9 N1 P33-36 62-24545 <=*>
1961 V9 N1 P37-43 62-15955 <=*>
1961 V9 N1 P43-45 62-11132 <=>
1961 V9 N3 P106-109 AD-610 118 <=$>
1962 V10 N6 P380-396 C-6005 <NRC>
1965 V13 N1 P1-9 AD-621 936 <=$>

EXPERIMENTELLE VETERINAERMEDIZIN
1950 V3 P8-17 62-01421 <*>

EXPLOSIFS
1961 V14 N2 P36-47 62-16160 <*>

EXPLOSIVSTOFFE
1953 N1/2 P15-18 59-12440 <+*>
1955 V3 N8 P109-113 II-547 <*>
59-12438 <+*>
1955 V3 N9 P109-113 1117 <*>
59-12437 <+*>
1955 V3 N10 P153-156 1295 <*>
1955 V3 N9/10 P129-134 59-12435 <+*>
1955 V3 N9/10 P153-156 59-12435 <+*>
1956 V4 N1 P1-10 57-65 <*>
59-12434 <+*>
1956 V4 N4 P68-78 C-220 <NRC>
1956 V4 N4 P69-78 57-273 <*>
59-12432 <+*>
1956 V4 N10 P221-226 58-519 <*>
59-12427 <+*>
1956 V4 N6/7 P119-125 59-10682 <*>
1956 V4 N6/7 P143-148 59-10682 <*>
1957 V5 N2 P29-32 59-10591 <*>
1957 V5 N8 P161-176 59-18471 <+*>
1959 V7 N4 P71-76 62-25778 <=*>
1959 V7 N5 P91-95 62-25778 <=*>
1960 N12 P275-284 66-13563 <*> 0
1960 V8 N8 P177-179 62-32333 <=*>
1961 P121-133 62-16161 <*> P
1964 V12 N9 P197-199 65-10485 <*>

EXPOSES ANNUELS DE BIOCHIMIE MEDICALE
1939 V2 P138-160 I-857 <*>
1950 V17 P172-190 65-00211 <*> 0
1959 V21 P81-95 63-00073 <*>

EXPOSES ANNUELS D'OTO-RHINO-LARYNGOLOGIE
1956 P107-126 58-722 <*>

EZHEGODNIK. INSTITUTA EKSPERIMENTALNOGO MEDITSINSKOGO AKADEMIIA MEDITSINSKIKH NAUK SSSR
1958 V4 P25-31 64-19818 <=$> 0
1958 V4 P38-42 64-19817 <=$> 0
1958 V4 P42-49 64-13633 <=*$>
1958 V4 P197-205 61-28835 <*=>
1958 V4 P410-415 61-28426 <*=>

EZHEGODNIK ZOOLOGICHESKAGO MUZEYA
1909 V14 N3/4 P29-32 RT-508 <*>

F. AND G. RUNDSCHAU
1950 N28 P34-45 57-2513 <*>
1959 N44 P149-155 66-14619 <*> 0

F.I.A.T. INFORMATION BULLETIN-REPORT. SOCIETA PER AZIONI. TURIN
1946 N865-I P6-11 66-12725 <*> 0
1946 N865-II P13-19 66-12726 <*>
1946 N865-III P22-29 66-12727 <*>
1946 N865-IV P31-39 66-12728 <*>
1946 N865-V P46-58 66-12729 <*>
1946 N865-VI P62-72 66-12730 <*>

F.I.A.T. REVIEW OF GERMAN SCIENCE 1939-1946
1948 P12-25 PT1 I-927 <*>
1948 P192-202 PT.2 62-00555 <*>
1948 P215-238 PT1 64-20247 <*>
1948 P80-94 PT2 I-925 <*>
1948 P19-24 I-699 <*>
1948 P295 1930 <*>

FACHAUSSCHUSSBERICHT. DEUTSCHE GLASTECHNISCHE GESELLSCHAFT
1936 N38 P1-25 64-16518 <*>
1936 N38 P203-227 64-16518 <*>
1958 N56 P519-531 62-14862 <*>
1961 N60 P599-607 65-14410 <*>

FACHBERATER FUER DAS DEUTSCHE KLEINGARTENWESEN
1961 V11 N42 P21-22 63-10500 <*>

FACHHEFTE FUER DIE CHEMIEGRAPHIE, LITHOGRAPHIE UND DEN TIEFDRUCK
1954 P143-149 62-16750 <*>

FACHLICHE MITTEILUNGEN DER OESTERREICHISCHEN TABAKREGIE
1915 P116-120 3928-A <K-H>
1916 P1-3 5461-A <K-H>
1931 N3 P1-10 58-1238 <*>
1935 N3 P2-7 58-1313 <*>
1951 N2 P1-5 3055-A <K-H>
1951 N2 P21-25 3236-B <K-H>
1952 N1 P1-6 3084-B <K-H>
1954 N1 P1-9 3449-D <K-H>
1954 N1 P10-13 3449-B <K-H>
1954 N1 P14-16 3449-C <K-H>
1954 N2 P1-7 3504-B <K-H>
1955 P1 58-935 <*>
1955 P1-11 58-935 <*>
1955 N1 P12-18 3923-E <K-H>
1955 N2 P1-13 5448-A <K-H>
58-1021 <*>
1956 N1 P6-13 6508 <K-H>
1956 N1 P13-16 5691-B <K-H>
1956 N1 P18-19 5691-A <K-H>
1957 P25-29 PT1 58-1016 <*>
1957 N2 P8-21 6690-B <K-H>
1957 N2/8 P21- 58-1019 <*>
1958 N2 P3-18 59-10258 <*> 0
1960 N1 P1-11 10684-B <KH>
64-10429 <*>
1965 N4 P1-11 SPEC IS 65-14352 <*>

FAERBER-ZEITUNG. CHEMISCHREINIGER UND FAERBER
1965 V18 N9 P412-418 1593 <TC>

FAIPAR
1958 V8 N11/2 P358-364 61-00667 <*>

FAMILLES DANS LE MONDE
1954 V7 N3 P185-199 65-12379 <*>
1954 V7 N3 P256-260 65-12380 <*>

FARBE
1955 V3 N5/6 P165-174 1239 <*>
1955 V4 P285-288 58-2457 <*>
1955 V4 N1 P3-14 1240 <*>
1956 V5 N1/2 P41-68 C-2558 <NRC>
58-564 <*>
62-18306 <*> 0
1958 V7 N1/3 P65-91 59-10443 <*>
1958 V7 N1/3 P93-119 63-20967 <*> 0
1959 V8 N2/3 P65-82 94M46G <ATS>
1960 V9 P53-62 65-14570 <*>
1960 V9 N4/6 P259-266 63-16338 <*> 0
1963 V12 P6-58 496 <TC>
1963 V12 P95-104 397 <TC>

FARBE UND LACK
1926 V24 P451-452 59-20363 <*>
1931 N13/5 P149-150 66-13314 <*>

1931 N13/5 P158-159 66-13314 <*>
1931 N13/5 P175 66-13314 <*>
1940 V46 N38 P333-334 59-20231 <*> O
1949 V55 N9 P326 59-20067 <*>
1950 V56 P447-449 AL-859 <*>
1952 V58 N1 P5-10 40J19G <ATS>
61-14002 <*>
1952 V58 N2 P51-54 5268-H <K-H>
1952 V58 N4 P143-150 40J19G <ATS>
61-14074 <*>
1952 V58 N5 P221-222 61-10000 <*>
1952 V58 N8 P356 60-18998 <*>
1952 V58 N12 P522- 40J19G <ATS>
1953 V59 P181-186 2091 <TC>
1953 V59 N1 P7-16 59-15974 <*> O
1954 V60 P315-352 2436 <TTIS>
1954 V60 N4 P144-145 64-20328 <*>
1955 V61 N7 P315-323 C-2452 <NRC>
1955 V61 N12 P552-560 60-10429 <*>
1956 V62 P323-331 130 <TC>
1956 V62 N2 P51-58 58-1480 <*>
1956 V62 N6 P260-267 129 <TC>
1956 V62 N8 P361-366 64-16323 <*>
1956 V62 N12 P581-583 59-15663 <*> O
64-16505 <*>
1957 V63 N4 P154-161 59-15673 <*> OP
1957 V63 N8 P384-387 59-15661 <*> O
1957 V63 N9 P435-443 59-15675 <*> OP
1957 V63 N11 P535-541 59-15671 <*> PO
1958 V64 N3 P107-108 59-15664 <*> O
1958 V64 N7 P368-372 59-15672 <*> O
1958 V64 N7 P377-381 61-16741 <*>
1958 V64 N8 P421-430 61-10125 <*> O
1958 V64 N12 P642-649 59-15656 <*> O
1959 V65 N1 P17-24 62-14053 <*> O
1959 V65 N1 P25-36 59-15662 <*> O
1959 V65 N2 P64-71 60-10110 <*> O
1959 V65 N2 P72-75 59-15875 <*> O
1959 V65 N2 P79-80 59-17802 <*>
1959 V65 N4 P179-184 60-14494 <*> O
1959 V65 N4 P184-185 60-10514 <*>
1959 V65 N5 P241-243 60-10528 <*> O
1959 V65 N7 P370-373 59-20851 <*> O
1959 V65 N8 P440-449 60-16500 <*> O
1959 V65 N8 P450-458 60-16499 <*> O
1959 V65 N11 P634-646 60-14531 <*> O
1959 V65 N12 P713-714 2167 <TTIS>
1960 V66 P454-457 2225 <TTIS>
1960 V66 N1 P10-13 62-20454 <*> O
1960 V66 N3 P142-145 60-16479 <*> O
1960 V66 N1C P560-569 62-10032 <*> O
1961 V67 N2 P71-80 62-10031 <*> O
1961 V67 N3 P148-153 62-10033 <*> O
64-30619 <*> O
1961 V67 N5 P294-298 NS-336 <TTIS>
1961 V67 N5 P299-301 31P65G <ATS>
1961 V67 N5 P302-306 62-10030 <*> O
1961 V67 N7 P434-437 66-14167 <*>
1961 V67 N9 P560-565 63-20268 <*>
1961 V67 N11 P693-699 63-20236 <*>
1961 V67 N11 P703-704 65-12906 <*>
1961 V67 N12 P748-758 2428 <TTIS>
1962 V68 P847-848 2437 <TTIS>
1962 V68 N1 P23-31 <ES>
2421 <TTIS>
1962 V68 N3 P155-162 63-20933 <*> O
1962 V68 N5 P315-321 3123 <BISI>
1962 V68 N9 P595-598 2443 <TTIS>
1962 V68 N9 P613-615 2419 <TTIS>
1962 V68 N10 P682-688 64-30623 <*> O
1962 V68 N11 P765-773 64-30624 <*> O
1962 V68 N11 P774-777 64-14951 <*> O
1962 V68 N12 P849-852 66-14168 <*>
1962 V68 N12 P853-859 2440 <TTIS>
1962 V68 N12 P859-866 64-30625 <*> O
1963 V69 P820-827 8927 <IICH>
1963 V69 N1 P15-21 64-30622 <*> O
1963 V69 N1 P21-26 NS 111 <TTIS>
1963 V69 N2 P80-87 NS-80 <TTIS>
1963 V69 N2 P87-92 NS 116 <TTIS>
1963 V69 N3 P171-174 NS-102 <TTIS>
1963 V69 N4 P373-378 NS 132 <TTIS>
1964 V70 P341-344 309 <TC>
1964 V70 P522-532 1232 <TC>
1964 V70 P532-538 322 <TC>
1964 V70 P595-599 8903 <IICH>
1964 V70 P779-787 NS-297 <TTIS>
1964 V70 N1 P34-40 65-12758 <*>
1964 V70 N1 P40-43 NS-203 <TTIS>
1964 V70 N4 P258-263 65-14644 <*> O
1964 V70 N4 P271-279 NS 236 <TTIS>
1964 V70 N10 P791-797 NS-335 <TTIS>
1965 V71 P39-44 536 <TC>
1965 V71 P185-187 NS-415 <TTIS>
951 <TC>
1965 V71 P353-365 883 <TC>
1965 V71 P977-984 NS-544 <TTIS>
1965 V71 N1 P39-44 65-12913 <*>
1965 V71 N2 P113-118 1666 <TC>
1965 V71 N4 P282-291 65-17470 <*>
1965 V71 N6 P445-448 65-14924 <*>
1965 V71 N8 P625-632 65-17473 <*>
1965 V71 N8 P632-643 65-17104 <*>
1965 V71 N8 P649-655 28T90G <ATS>
1966 V72 P218-224 66-12333 <*> O
1966 V72 N1 P36-50 66-12194 <*>
66-13582 <*>
1966 V72 N2 P111-118 66-12195 <*>

FARBEN, LACKE, ANSTRICHSTOFFE
1948 V2 N8 P123-129 59-15725 <*> O
1949 V3 P39-46 63-18182 <*> O
1950 V4 P160-162 63-20512 <*>

FARBEN-CHEMIKER
1932 V3 P144-145 60-10212 <*>
1932 V3 P180-182 60-10212 <*>

FARBENZEITUNG
1927 V33 P801-803 AL-567 <*>
1930 V35 P1661-1662 T-1998 <INSD>
1930 V35 N20 P1019-1022 60-10592 <*>
1930 V35 N36 P1824-1825 59-20517 <*> O
1930 V35 N41 P2099-2101 59-20516 <*> O
1930 V35 N49 P2493-2495 59-20515 <*> O
1930 V35 N50 P2543-2545 59-20515 <*> O
1930 V36 N10 P453 59-20519 <*>
1930 V36 N10 P454-456 59-20518 <*>
1930 V36 N11 P504-505 59-20519 <*>
1931 V36 P1090 60-10082 <*>
1931 V36 N26 P1176-1177 57-3089 <*>
1933 N40 P1011-1012 57-3189 <*>
1936 V41 N38 P939-940 59-20530 <*> O
1938 V43 P670 63-18690 <*>
1939 V44 P1230-1232 66-12908 <*> O
1941 V46 P280-281 57-1073 <*>

FARBIGE MEDIZIN
1963 V9 P1-8 65-00123 <*> O

FARMACEUTICKY OBZOR
1961 V30 N2 P58-64 62-19387 <=*>
1961 V30 N4 P97-103 62-19694 <=*>
1961 V30 N6 P161-167 62-19693 <=*> P
1961 V30 N6 P185-188 62-19693 <=*> P
1963 V32 P97-1C5 547 <TC>

FARMACEUTSKI GLASNIK
1960 V16 P119-126 AEC-TR-5126 <*>
1961 V17 N4/5 P141-149 62-19751 <=*> P
1961 V17 N7/8 P289 62-19154 <=*>
1962 V18 N2/3 P95-97 62-25472 <=*>

FARMACEVTSKI VESTNIK. LJUBLJANA
1960 V11 N1/4 P9-12 62-19224 <=>
1960 V11 N1/4 P17-25 62-19224 <=>
1960 V11 N1/4 P33 62-19224 <=>

FARMACIA. BRATISLAVA
1960 V29 N5 P154-156 62-19388 <=*>

FARMACIA. BUCURESTI

Citation	Translation
1961 V9 N1 P19-23	62-19389 <=*>
1961 V9 N2 P89-92	62-19390 <=*>
1961 V9 N4 P195-203	62-23266 <=*> P
1964 V12 N1 P1-6	64-31176 <=>

FARMACIA NOVA

Citation	Translation
1951 V16 P103-108	T-1609 <INSD>
1951 V16 P161-166	T-1609 <INSD>

FARMACJA POLSKA

Citation	Translation
1955 N11 P73-79	T-1522 <INSD>
1955 V11 P80-84	T-1523 <INSD>
1955 V11 P84-89	T-1524 <INSD>
1960 V16 N12 P246-247	60-31862 <*=>
1960 V16 N23/4 P479-481	62-19391 <=*>
1961 V17 N8 P159-163	62-19532 <=*>
1961 V17 N9 P173-174	62-19523 <=*> P
1961 V17 N12 P245-247	62-19613 <*=>
1961 V17 N12 P250-257	62-23341 <=*>
1961 V17 N13 P262-264	62-23341 <=> P
1961 V17 N13 P266-271	62-23341 <=> P
1961 V17 N13 P277-278	62-23341 <=> P
1961 V17 N14 P298-300	62-23341 <=> P
1961 V17 N18 P371-376	62-19608 <=>
1961 V17 N18 P390-391	62-23341 <=> P
1961 V17 N15/6 P314-317	62-19608 <=> P
1961 V17 N15/6 P330-331	62-19608 <=> P
1961 V17 N15/6 P332-335	62-23341 <=> P
1961 V17 N15/6 P341	62-23341 <=> P
1962 V18 N15 P372-376	63-15084 <=*>
1964 V20 N21/2 P827-828	65-13752 <*>

FARMACO. SCIENZA E TECNICA

Citation	Translation
1947 V2 N2 P89	2815 <*>
1947 V2 N2 P98	2812 <*>
1948 V3 N4 P389-396	62-10689 <*>
1949 V4 P515-525	1727 <KH>
1951 V6 P291-299	65-14959 <*>
1952 V7 P418-429	58-1761 <*>

FARMACO. EDIZIONE SCIENTIFICA

Citation	Translation
1954 V9 N11 P629-641	II-1060 <*>
1955 V10 P179-186	57-854 <*>
1955 V10 P337-345	58-174 <*>
1955 V10 N1 P37-46	25G7I <ATS>
1956 V11 P244-252	65-14960 <*>
1956 V11 N8 P695-698	57-2780 <*>
1957 V12 N2 P57-72	59-15426 <*> 0
1957 V12 N2 P103-119	59-10031 <*>
1957 V12 N11 P899-929	59-10023 <*>
1958 V13 N4 P286-293	96K25I <ATS>
1958 V13 N10 P726-731	59-10829 <*>
1959 V14 N1 P70-74	59-17590 <*>
1960 V15 N7 P488-490	64-10386 <*>
1961 V16 P23-31	65-14961 <*>
1963 V18 N10 P763-772	56R74I <ATS>

FARMACO. EDIZIONE PRATICA

Citation	Translation
1954 V9 N12 P623-628	3779-G <K-H>
1957 V12 N2 P57-72	57-2872 <*>
1959 V14 N1 P3-31	59-15852 <*>
1959 V14 N2 P73-108	59-17485 <*>
1959 V14 N11 P690-696	60-14873 <*>
1960 V15 N5 P311-316	65-14242 <*> 0
1960 V15 N8 P453-470	61-10923 <*>
1960 V15 N10 P606-616	63-00171 <*>
1961 V16 P487-498	63-16433 <*>
1961 V16 N2 P65-79	63-20870 <*> 0
1961 V16 N5 P232-239	65-14241 <*> 0
1961 V16 N7 P338-343	62-20031 <*>
1961 V16 N8 P349-370	64-14652 <*>
1961 V16 N8 P583-590	62-14957 <*> 0
1962 V17 P404-415	65-10455 <*>
1964 V19 N1 P52-59	65-12095 <*>
1964 V19 N10 P507-511	65-13063 <*>

FARMACOTERAPIA ACTUAL

Citation	Translation
1945 V2 N2 P313-318	59-10833 <*> 0
1946 V3 P748-751	1171E <K-H>

*FARMAKOLOGIYA I TOKSIKOLOGIYA

Citation	Translation
1939 V2 N5 P38-50	1271-E <K-H>
1943 V6 N2 P61-65	65-63274 <=>
1944 V7 N2 P41-46	63-18531 <=*> 0
1947 V10 N1 P3-17	RT-2621 <*>
1947 V10 N6 P12-16	RT-4579 <*>
1953 V16 N1 P5-10	R.717 <RIS>
1953 V16 N1 P42-44	RJ-144 <ATS>
1953 V16 N2 P43-47	RJ-145 <ATS>
1954 V17 N2 P55-57	GB39 <NLL>
1954 V17 N3 P11-14	R-2083 <*>
	64-19451 <=$> 0
1954 V17 N4 P10-14	RT-3016 <*>
1954 V17 N6 P3-5	R-2687 <*>
	RT-3505 <*>
1954 V17 N6 P6-12	R-2688 <*>
	RT-3507 <*>
1954 V17 N6 P12-18	RT-3506 <*>
1954 V17 N6 P39	62-15305 <*=>
1955 N3 P37-40	RT-3195 <*>
1955 V18 N1 P3-7	RT-3073 <*>
1955 V18 N1 P23-26	RJ-297 <ATS>
	5367-B <K-H>
1955 V18 N1 P37-38	RJ-298 <ATS>
1955 V18 N2 P21-37	RT-4079 <*>
1955 V18 N2 P48-50	81TM <CTT>
1955 V18 N2 P61	RT-3058 <*>
1955 V18 N3 P3-9	RT-4316 <*>
1955 V18 N4 P21-27	R-382 <*>
	59-18838 <+*> 0
1955 V18 N5 P30-34	66-11717 <*> 0
1956 V19 P29-30 SUP	64-19460 <=$>
1956 V19 N2 P46-49	RT-4433 <*>
1956 V19 N4 P17-19	61-19555 <=*>
1956 V19 N4 P24-26	65-14228 <*> 0
1956 V19 N4 P36-41	59-18837 <+*> 0
1956 V19 N4 P57-62	R-1827 <*>
	R-935 <*>
	60-19496 <+*>
	60-19497 <+*>
	62-23869 <=*>
	64-19483 <=$>
1956 V19 N6 P10-17	64-21643 <=>
1957 V20 P49-53 SUP	62-33096 <=*>
1957 V20 N3 P23-29	60-15493 <+*>
1957 V20 N3 P59-63	60-15497 <+*>
1957 V20 N3 P69-74	60-15494 <+*>
1957 V20 N4 P7-13	60-15500 <+*>
1957 V20 N4 P20-26	60-15498 <+*>
1957 V20 N4 P27-31	60-15496 <+*>
1957 V20 N4 P42-48	60-15499 <+*>
1957 V20 N4 P48-53	60-15495 <+*>
1958 N1 P53-57	R-5359 <*>
1958 V21 N6 P28-30	63-16654 <*=>
	9298-C <K-H>
1959 V22 P555-557	UCRL TRANS-567 <+*>
1959 V22 N1 P20-27	50L32R <ATS>
1959 V22 N2 P113-116	60-18358 <+*>
1959 V22 N2 P154-158	60-18357 <+*>
1959 V22 N3 P273-274	60-10968 <+*>
1959 V22 N3 P280-281	60-10967 <+*>
1959 V22 N6 P504-505	60-18356 <+*>
1960 V23 P99-105	62-24320 <=*>
1960 V23 P549-557	62-23881 <=*> 0
1960 V23 N1 P64-66	62-14726 <=*>
1960 V23 N1 P67-71	63-31382 <=>
1960 V23 N2 P125-127	60-41186 <=>
1960 V23 N2 P128-129	60-41187 <=>
1960 V23 N2 P130-132	60-41188 <=>
1960 V23 N2 P136-139	60-41189 <=>
1960 V23 N2 P140-142	60-41190 <=>
1960 V23 N2 P169-173	60-41191 <=> 0
1960 V23 N2 P173-174	60-41192 <=>
1960 V23 N2 P178-182	60-31700 <=>
1960 V23 N3 P206-215	64-21643 <=>
1960 V23 N5 P379-384	61-21647 <=>
1960 V23 N5 P417-421	61-21646 <=>
1960 V23 N5 P421-426	61-21622 <=>
1960 V23 N5 P450-454	61-21541 <=>

Citation	Translation
1960 V23 N5 P459-464	61-21724 <=>
1960 V23 N5 P498-506	64-21643 <=>
1960 V23 N6 P493-499	64-21575 <=>
1961 V24 P515-518	11679-A <KH>
	64-10666 <=*$>
1961 V24 N1 P3-13	62-24016 <=*>
1961 V24 N1 P18-22	61-23897 <=>
1961 V24 N1 P30-32	61-23897 <=>
1961 V24 N1 P40-44	61-20545 <*=> 0
1961 V24 N1 P125	61-23897 <=>
1961 V24 N2 P191-196	61-27146 <*=>
1961 V24 N2 P247-251	61-27154 <=*>
1961 V24 N3 P275-279	64-21575 <=>
1961 V24 N3 P304-309	62-24016 <=*>
1961 V24 N3 P318-324	62-13527 <=*>
1961 V24 N3 P357-371	62-13037 <*=>
1961 V24 N4 P432-436	64-21643 <=>
1961 V24 N4 P499-507	64-21623 <=>
1961 V24 N5 P534-540	11679-B <KH>
1961 V24 N5 P534-548	62-24016 <=*>
1961 V24 N5 P534-540	63-16919 <*=>
1961 V24 N5 P557-561	62-24016 <=*>
1961 V24 N5 P635-638	62-24008 <=*>
1961 V24 N6 P687-690	62-24016 <=*>
1961 V24 N6 P700-713	62-24016 <=*>
1961 V24 N6 P754-761	62-23809 <=>
1962 N3 P291-297	12447 <KH>
1962 V25 N1 P32-37	64-21643 <=>
1962 V25 N3 P301-303	62-32085 <=*>
1962 V25 N3 P320-335	63-21987 <=>
1962 V25 N3 P339-345	63-21987 <=>
1962 V25 N4 P395-401	64-21643 <=>
1962 V25 N4 P401-410	63-13301 <=*>
1962 V25 N4 P428-433	64-21643 <=>
1962 V25 N4 P462-466	63-13301 <=*>
1962 V25 N4 P478-482	63-13260 <=*>
1962 V25 N5 P519-530	63-21987 <=>
1962 V25 N5 P533-538	63-21056 <=>
1962 V25 N5 P564-569	63-21056 <=>
1962 V25 N5 P579-584	63-21056 <=>
1962 V25 N6 P679-698	63-21987 <=>
1963 V26 N1 P10-	FPTS V23 N1 P.T125 <FASE>
1963 V26 N1 P10-17	63-21987 <=>
1963 V26 N1 P28-35	63-18768 <=*>
	63-24502 <=*$>
1963 V26 N1 P47-51	63-24503 <=*$>
1963 V26 N1 P66-72	64-13857 <=*$>
1963 V26 N1 P75-	FPTS V23 N3 P.T493 <FASE>
	FPTS,V23,N3,P.T493 <FASE>
1963 V26 N1 P102-	FPTS V23 N1 P.T39 <FASE>
1963 V26 N2 P131-138	63-31250 <=>
	64-13461 <=*$> 0
1963 V26 N2 P145-150	64-13460 <=*$>
1963 V26 N2 P150-	FPTS V23 N1 P.T113 <FASE>
1963 V26 N2 P157-163	63-31250 <=>
1963 V26 N2 P164-169	64-13459 <=*$>
1963 V26 N2 P172-	FPTS V23 N2 P.T272 <FASE>
1963 V26 N2 P172-179	63-31250 <=>
1963 V26 N2 P179-	FPTS V23 N1 P.T122 <FASE>
1963 V26 N2 P184-188	63-31250 <=>
1963 V26 N2 P197-201	64-13458 <=*$>
1963 V26 N2 P219-225	64-13457 <=*$> 0
1963 V26 N2 P233-	FPTS V23 N1 P.T55 <FASE>
1963 V26 N3 P289-292	64-15535 <=*$>
1963 V26 N3 P301-	FPTS V23 N4 P.T870 <FASE>
	FPTS,V23,N4,P.T870 <FASE>
1963 V26 N3 P322-323	64-15456 <=*$>
1963 V26 N3 P327-333	64-15490 <=*$>
1963 V26 N3 P338-349	64-15457 <=*$> 0
1963 V26 N3 P355-	FPTS V23 N4 P.T863 <FASE>
	FPTS,V23,N4,P.T863 <FASE>
1963 V26 N3 P370-	FPTS,V23,N4,P.T905 <FASE>
1963 V26 N3 P381-382	64-15536 <=*$>
1963 V26 N4 P403-405	63-31968 <=>
1963 V26 N4 P406-410	64-15458 <=*$>
1963 V26 N4 P435-439	64-15532 <=*$>
1963 V26 N4 P439-445	FPTS,V23,N5,P.T927 <FASE>
1963 V26 N4 P446-	64-19269 <=*$>
1963 V26 N4 P452-454	64-15468 <=*$>
1963 V26 N4 P455-	FPTS V23 N3 P.T487 <FASE>

Citation	Translation
	FPTS,V23,N3,P.T487 <FASE>
1963 V26 N4 P467-472	79R76R <ATS>
1963 V26 N4 P472-476	64-19286 <=*$>
1963 V26 N4 P476-	64-19270 <=*$>
1963 V26 N4 P494-498	64-15459 <=*$>
1963 V26 N5 P515-518	64-19275 <=*$>
1963 V26 N5 P518-525	64-19276 <=*$>
1963 V26 N5 P531-537	64-19660 <=$>
1963 V26 N5 P556-	FPTS V23 N6 P.T1325 <FASE>
	FPTS V23 P.T1325 <FASE>
1963 V26 N5 P578-583	FPTS,V23,N5,P.T921 <FASE>
1963 V26 N5 P597-602	64-19225 <=>
1963 V26 N6 P650-	FPTS V23 N6 P.T1330 <FASE>
1963 V26 N6 P656-661	64-21702 <=>
1963 V26 N6 P661-666	FPTS,V23,N5,P.T917 <FASE>
1963 V26 N6 P674-676	65-60215 <=$>
1963 V26 N6 P677-678	64-19661 <=$>
1963 V26 N6 P678-684	65-60216 <=$>
1963 V26 N6 P684-687	64-21721 <=>
1963 V26 N6 P702-707	FPTS V23 P.T1327 <FASE>
	64-21721 <=>
1963 V26 N6 P715-718	64-19649 <=$>
1963 V26 N6 P729-732	64-19662 <=$>
1963 V26 N6 P732-737	64-21721 <=>
1963 V26 N6 P737-742	64-21721 <=>
1963 V26 N6 P753-757	64-19663 <=$>
1963 V26 N6 P753-756	64-21721 <=>
1964 V27 P32-35	64-31339 <=>
1964 V27 N1 P15-16	65-60210 <=$>
1964 V27 N1 P22-25	65-62181 <=$>
1964 V27 N1 P36-42	65-60238 <=$>
1964 V27 N1 P48-53	64-31306 <=>
1964 V27 N1 P63-68	65-61047 <=$>
1964 V27 N1 P68-72	65-61048 <=$>
1964 V27 N1 P73-76	FPTS V24 P.T192-T194 <FASE>
1964 V27 N1 P73-82	64-31356 <=>
1964 V27 N1 P100-102	65-60232 <=$>
1964 V27 N1 P107-121	64-31205 <=>
1964 V27 N2 P138-143	64-41974 <=$>
1964 V27 N2 P165-167	64-41974 <=$>
1964 V27 N2 P174-179	65-62166 <=$>
1964 V27 N2 P184-186	64-41974 <=$>
1964 V27 N2 P189-193	64-41974 <=$>
1964 V27 N3 P275-	FPTS V24 N3 P.T554 <FASE>
1964 V27 N3 P282-292	65-30300 <=$>
1964 V27 N3 P293-295	65-62309 <=$>
1964 V27 N3 P295-300	65-30300 <=$>
1964 V27 N3 P331-	FPTS V24 N4 P.T591 <FASE>
1964 V27 N3 P343-345	65-31105 <=$>
1964 V27 N3 P362-363	65-62178 <=$>
1964 V27 N3 P603-	FPTS V24 N6 P.T1116 <FASE>
1964 V27 N4 P387-391	65-62836 <=$>
1964 V27 N4 P424-426	65-63595 <=$>
1964 V27 N4 P429-432	65-63238 <=$>
1964 V27 N4 P451-	FPTS V24 N4 P.T579 <FASE>
1964 V27 N4 P472-475	65-63596 <=$>
1964 V27 N4 P498-501	65-63597 <=$>
1964 V27 N5 P519-	FPTS V24 N4 P.T673 <FASE>
1964 V27 N5 P533-537	65-64557 <=*$>
1964 V27 N6 P675-692	65-30300 <=$>
1964 V27 N6 P681-	FPTS V24 N6 P.T1108 <FASE>
1964 V27 N6 P690-692	65-33415 <=*$>
	66-60304 <=*$>
1964 V27 N6 P692-	FPTS V24 N6 P.T1119 <FASE>
1964 V27 N6 P729-732	66-60303 <=*$>
1964 V27 N6 P732-735	65-30300 <=$>
	65-63598 <=$>
1964 V27 N6 P746-756	65-30300 <=$>
1965 V28 P20-23	65-31045 <=$>
1965 V28 P37-39	65-31045 <=$>
1965 V28 N1 P23-27	66-60401 <=*$>
1965 V28 N1 P37-40	65-64518 <=*$>
1965 V28 N1 P83-87	66-60405 <=*$>
1965 V28 N1 P92-	FPTS V25 N1 P.T93 <FASE>
1965 V28 N1 P108-111	66-60406 <=*$>
1965 V28 N2 P176-180	66-60876 <=*$>
1965 V28 N2 P224-228	66-60877 <=*$>
1965 V28 N2 P228-230	66-60894 <=*$>
1965 V28 N2 P234-238	66-61314 <=*$>
1965 V28 N2 P238-241	65-31474 <=$>

1965 V28 N2 P245-250 65-31474 <=$>
1965 V28 N3 P282-284 66-61315 <=*$>
1965 V28 N3 P298- FPTS V25 N3 P.T512 <FASE>
1965 V28 N3 P305- FPTS V25 N3 P.T521 <FASE>
1965 V28 N3 P316-320 66-61316 <=*$>
1965 V28 N3 P324- FPTS V25 N3 P.T458 <FASE>
1965 V28 N3 P347-349 65-32427 <*$>
1965 V28 N3 P351-355 65-32427 <*$>
1965 V28 N3 P368-371 65-32428 <=*$>
1965 V28 N4 P389-393 65-33388 <=*$>
1965 V28 N4 P466 FPTS V25 N4 P.T686 <FASE>
1965 V28 N4 P495-497 65-33400 <=*$>
1965 V28 N5 P542- FPTS V25 N5 P.T747 <FASE>
1965 V28 N5 P587- FPTS V25 N4 P.T666 <FASE>
1965 V28 N5 P603-608 66-61317 <=*$>
1965 V28 N6 P744- FPTS V25 N5 P.T910 <FASE>
1966 V29 N1 P25- FPTS V25 N6 P.T1051 <FASE>
1966 V29 N1 P31- FPTS V25 N6 P.T1029 <FASE>
1966 V29 N1 P44- FPTS V25 N6 P.T997 <FASE>
1966 V29 N1 P118-124 66-31413 <=$>

FARMATSEVTYCHNII ZHURNAL
1959 V13 N2 P27-28 64-13733 <=*$>
1960 V14 N6 P15-20 61-11063 <=>
1960 V15 N1 P85-87 61-27243 <*=>
1960 V15 N3 P8-13 60-41418 <=>
1960 V15 N5 P48-49 64-19124 <=*$>
1961 V16 N4 P15-16 64-19966 <=$>
1961 V16 N4 P16-20 64-18354 <*>
1963 V18 N1 P10-12 66-12533 <*>
1963 V18 N1 P31-37 65-14331 <*> 0
1965 V20 N3 P76-82 65-32247 <=*$>

FARMATSIYA. MOSKVA
1939 V2 N6 P1-8 PANSDOC-TR.592 <PANS>
1940 N11 P16-19 61-20732 <*=>
1943 V6 N6 P16-22 64-23631 <=$>
1944 V7 N6 P35-36 PANSDOC-TR.268 <PANS>
1946 V9 N5 P20 59-15218 <+*>

FARMATSIYA. SOFIYA
1960 V10 N1 P15-17 62-19392 <*=> 0
1960 V10 N1 P23-27 62-19392 <*=> 0
1960 V10 N2 P26-30 62-19393 <=*>
1961 V11 N1 P3-4 62-19168 <=*>
1961 V11 N5 P3-5 62-23614 <=*>
1966 V17 N1 P1-4 66-31984 <=$>

FARMING MECHANIZATION. JAPAN
SEE KIKAIKA NOGYO

FASERFORSCHUNG
1924 V4 P124-129 2878 <*>

FASERFORSCHUNG UND TEXTILTECHNIK
1951 V2 P299-303 I-259 <*>
1951 V2 P492-497 64-20363 <*> 0
1951 V2 N8 P299-303 60-10773 <*>
1951 V2 N8 P308-321 64-18879 <*>
1951 V2 N10 P383-390 64-18891 <*> 0
1951 V2 N18 P409-418 1986 <*>
1952 N10 P381-390 SCL-T-314 <*>
1952 V3 N2 P58-66 59-15930 <*> 0
1952 V3 N4 P127-141 59-10151 <*>
61-16595 <*>
1952 V3 N9 P341-344 59-15318 <*>
1952 V3 N9 P354-356 62-18433 <*> 0
1952 V3 N10 P412-417 59-15317 <*>
1953 V4 P499-507 II-869 <*>
1953 V4 N12 P499-507 64-20436 <*>
1953 V4 N12 P510-511 I-957 <*>
1954 V5 P1-8 2661 <*>
59-15510 <*>
1954 V5 N1 P14-21 62-14270 <*>
1954 V5 N2 P59-64 59-15319 <*>
1954 V5 N4 P171-172 1971 <*>
1954 V5 N6 P269-270 61-14003 <*> 0
1954 V5 N7 P277-284 59-15783 <*> 0
1954 V5 N8 P337-347 59-17023 <*> 0
1954 V5 N11 P493-497 59-20825 <*>
1955 P398-401 T-2246 <INSD>
1955 V6 P105-113 62-10371 <*> 0
1955 V6 N2 P45-53 59-20824 <*>
1955 V6 N6 P277-286 57-1794 <*>
1955 V6 N9 P398-401 57-1792 <*>
60-10765 <*>
1956 V7 P420-422 61-16788 <*>
1956 V7 N1 P1-13 C-3517 <NRCC>
1956 V7 N2 P53-63 61-16586 <*>
1956 V7 N4 P165-170 57-1955 <*>
60-18395 <*>
1956 V7 N8 P339-345 62-14585 <*>
1956 V7 N9 P408-412 60-18394 <*>
1956 V7 N10 P468-476 65-14175 <*>
69M37G <ATS>
1956 V7 N12 P546-556 57-1993 <*>
1956 V7 N12 P561-564 59-17400 <*>
1957 V8 N1 P32-34 60-10766 <*>
1957 V8 N3 P91-98 59-10168 <*>
1957 V8 N3 P99-108 57-3390 <*>
1957 V8 N4 P143-150 88J18G <ATS>
1957 V8 N5 P179-184 15J18G <ATS>
1957 V8 N6 P230-239 C-3204 <NRCC>
1957 V8 N7 P262-267 59-10165 <*> 0
1957 V8 N9 P348-354 61-16587 <*>
1957 V8 N11 P444-447 60-18425 <*>
61-14507 <*> 0
1957 V8 N11 P467-469 63-18089 <*>
1957 V8 N12 P487-494 59-10166 <*>
1957 V8 N12 P521-523 62-10022 <*> 0
1958 V9 N1 P1-10 87M40G <ATS>
1958 V9 N2 P67-75 61-16594 <*>
1958 V9 N5 P163-167 81K25G <ATS>
1958 V9 N5 P189-193 59-17386 <*>
1958 V9 N6 P226-231 60-18400 <*>
61-20633 <*>
1958 V9 N7 P262-272 59-17444 <*> 0
1958 V9 N8 P307-321 60-10777 <*>
61-16588 <*>
1958 V9 N9 P351-361 62-10656 <*>
1958 V9 N10 P405-416 59-17264 <*>
59-17767 <*>
63-15589 <*>
1958 V9 N11 P476-484 59-20826 <*>
1958 V9 N11 P488-498 62-16712 <*> 0
1958 V9 N12 P513-519 C-3322 <NRCC>
61-16585 <*> 0
1959 V10 N2 P53-62 61-16596 <*>
1959 V10 N2 P62-67 59-15815 <*>
61-16591 <*> 0
1959 V10 N3 P104-114 59-17263 <*>
61-16584 <*>
1959 V10 N3 P129-137 C-3572 <NRCC>
1959 V10 N4 P155-158 35L34G <ATS>
61-10168 <*>
1959 V10 N5 P214-224 63-18345 <*>
69L33G <ATS>
1959 V10 N5 P224-231 59-20892 <*>
1959 V10 N6 P245-248 97L35G <ATS>
1959 V10 N6 P275-282 61-16593 <*>
1959 V10 N7 P297-308 61-10216 <*>
1959 V10 N8 P369-371 18L37G <ATS>
1959 V10 N8 P387-392 62-16460 <*> 0
1959 V10 N10 P464-472 62-16622 <*> 0
1959 V10 N12 P578-587 61-16590 <*>
1960 V11 N1 P7-15 62-14079 <*> 0
1960 V11 N2 P53-62 23M40G <ATS>
1960 V11 N2 P90-91 61-10161 <*>
88M46G <ATS>
1960 V11 N3 P107-113 61-16542 <*>
1960 V11 N3 P113-117 61-16592 <*>
1960 V11 N3 P118-124 61-16589 <*>
1960 V11 N5 P219-220 63-10809 <*>
1960 V11 N7 P312-319 62-16740 <*> 0
1960 V11 N8 P353-359 62-20378 <*> 0
1960 V11 N8 P365-373 04M46G <ATS>
1960 V11 N9 P401-408 C-147 <RIS>
1960 V11 N9 P423-427 C-147 <RIS>
1960 V11 N11 P513-523 <LSA>
1960 V11 N12 P583-590 61-20632 <*>

1961 V12 N1 P9-18 63-15585 <*> 0
1961 V12 N1 P29-30 63-10804 <*>
1961 V12 N2 P49-55 63-14717 <*>
1961 V12 N4 P133-140 62-10243 <*>
1961 V12 N5 P196-237 65-12332 <*>
1961 V12 N5 P208-213 62-10241 <*>
1961 V12 N5 P395-401 <LSA>
1961 V12 N8 P361-369 62-16006 <*>
1962 V13 P437-442 520 <TC>
1962 V13 P442-449 63-20862 <*>
1962 V13 P481-490 63-20983 <*> 0
1962 V13 P564-570 63-18901 <*>
1962 V13 N2 P70-79 63-10818 <*>
1962 V13 N3 P112-124 32P63G <ATS>
1962 V13 N6 P256-261 63-18275 <*>
1962 V13 N6 P264-269 63-18336 <*>
1962 V13 N7 P293-299 63-10752 <*>
1962 V13 N7 P304-309 63-18355 <*>
64-14423 <*>
1962 V13 N9 P385-392 63-18341 <*>
1962 V13 N9 P393-396 63-18338 <*>
1962 V13 N10 P437-442 65-12182 <*>
1962 V13 N11 P502-511 63-18384 <*>
1963 V14 P117-120 324 <TC>
1963 V14 N4 P131-140 65-12087 <*>
1963 V14 N6 P219-227 66-10539 <*>
1963 V14 N7 P274-280 64-16174 <*> 0
64-16364 <*>
65-12086 <*>
1963 V14 N8 P307-313 66-10541 <*>
1963 V14 N8 P313-319 64-10710 <*>
65-12171 <*>
1963 V14 N9 P386-388 65-12173 <*>
1963 V14 N11 P485-489 C-5503 <NRC>
1964 V15 P215-224 433 <TC>
1964 V15 P386-390 748 <TC>
1964 V15 P527-532 692 <TC>
1964 V15 P533-537 693 <TC>
1964 V15 P545-554 809 <TC>
1964 V15 N1 P21-29 65-12175 <*>
1964 V15 N1 P30-39 66-10554 <*>
1964 V15 N1 P39-43 21R78G <ATS>
1964 V15 N3 P101-109 65-13503 <*>
1964 V15 N4 P153-157 66-10555 <*>
1964 V15 N5 P215-224 65-12037 <*>
65-12140 <*>
1964 V15 N7 P289-297 66-10547 <*>
1964 V15 N7 P304-315 65-12073 <*>
1964 V15 N7 P321-325 807 <TC>
1964 V15 N11 P537-542 808 <TC>
1964 V15 N12 P590-597 694 <TC>
1964 V15 N12 P626-631 65-17438 <*>
1965 V16 P400-408 1584 <TC>
1965 V16 N1 P37-41 65-14055 <*$=>
1965 V16 N2 P68-72 65-13892 <*>
897 <TC>
1965 V16 N2 P83-88 1855 <TC>
1965 V16 N3 P121-128 1856 <TC>
1965 V16 N3 P160-161 1857 <TC>
1965 V16 N6 P290-297 30S89G <ATS>
1965 V16 N6 P312-318 66-10374 <*>
1965 V16 N7 P330-348 1585 <TC>
1965 V16 N8 P395-400 90T91G <ATS>
1965 V16 N8 P400-408 65-17021 <*>
1965 V16 N9 P433-438 66-11048 <*>
1965 V16 N9 P438-443 66-11306 <*>
1965 V16 N9 P443-449 66-10538 <*>
1965 V16 N10 P495-501 66-13010 <*>
1966 V17 N4 P142-147 66-13129 <*>
66-13476 <*>
66-14501 <*>

FAUNA VON DEUTSCHLAND
1953 V25 N5 P206-219 C-2675 <NRC>

FAUNA SSSR
1948 V34 ENTIRE ISSUE 63-11071 <=>
1948 V34 P198-204 C-2306 <NRC>
1948 V35 ENTIRE ISSUE 63-11163 <=>
1951 V47 P1-143 64-11025 <=>
1952 N55 P247 60-21156 <=>
1952 V50 P425 64-11026 <=>
1952 V54 ENTIRE ISSUE 63-11060 <=$>
1956 V6 N6 P505 62-15643 <=*> 0
1957 V24 N1 P31-33 61-19189 <=*> 0
1957 V24 N1 P143-144 61-19189 <=*> 0
1960 P263 65-20006 <$>
1960 N75 P1-271 <API>

FEINGERAETE-TECHNIK
1955 V4 N12 P558-562 59-17187 <*> 0
1961 V10 N5 P199-209 62-32345 <=*>
1963 V12 N2 P59-64 63-21653 <=>
1963 V12 N3 P97-98 63-21637 <=>
1963 V12 N12 P529-531 64-21671 <=>
1964 V13 N9 P386-388 64-51984 <=$>
1965 V14 N7 P289-296 65-32527 <=$>
1965 V14 N7 P304-307 65-32527 <=$>
1965 V14 N9 P401-404 65-33652 <=*$>
1966 V15 N2 P45-53 66-31659 <=$>

FEINMECHANIK UND PRAEZISION
1938 V46 N21 P289-291 59-17018 <*> 0
1942 V50 N17/8 P255-260 59-17013 <*> 0P

FEINWERK-TECHNIK
1949 V53 N6 P167-172 63-14753 <*>
1950 V54 N4 P88-91 59-17493 <*>
1951 V55 N6 P138-142 58-714 <*>
1952 V56 N2 P29-40 57-2748 <*>
1958 V62 P81-86 62-10126 <*> 0
1958 V62 N1 P1-8 59-17761 <*> 0
1961 V65 N2 P63-66 62-10901 <*>

FELDSHER. MOSKVA
1952 N11 P56-57 RT-1348 <*>
1953 N8 P31-34 R.713 <RIS>
1954 N2 P45-46 3225-A <KH>
1954 N10 P60-61 RT-3802 <*>
1956 V21 N12 P32-33 63-41249 <=>
1957 V22 N3 P46-48 8258 <KH>
1960 N1 P25-30 61-21661 <=>
1960 N3 P54-55 60-31676 <=>
1961 N2 P56-57 61-28250 <*=>
1961 N10 P3-6 62-25012 <=*>
1962 V27 N8 P41-48 62-33425 <=*>
1963 V28 N5 P36-39 65-60953 <=$>
1965 N11 P20-22 66-30838 <=$>

FELDSHER I AKUSHERKA
1948 N6 P45-49 RT-4338 <*>

FELDWIRTSCHAFT
1966 N8 P401-403 66-34708 <=$>
1966 V7 N12 P662-664 67-30270 <=>
1967 N2 P111-112 67-30775 <=$>
1967 N3 P167-168 67-31096 <=$>

FELSOOKTATOSI SZEMLE
1959 V8 N1 P58-61 60-31879 <*=>
1959 V8 N4 P271-275 60-31879 <*=>
1959 V8 N6 P390-396 60-31879 <*=>
1960 V9 N9 P554-555 62-19417 <*=>
1960 V9 N10 P635-638 62-19418 <*=>
1960 V9 N10 P646-649 62-19415 <*=>
1961 V10 N1/2 P9-12 62-19385 <*=>
1961 V10 N1/2 P15-19 62-19386 <*=>
1962 V11 N7/8 P423-426 63-13281 <=*>
1963 V12 N4 P197-207 63-31342 <=>
1964 V13 N3 P166-181 64-31675 <=>
1964 V13 N12 P727-734 65-30424 <=$>

FELTEN UND GUILLEAUME CARLSWERK-RUNDSCHAU
1933 N11/2 P59- 57-2467 <*>

FERMENTATIO
1955 N1 P30-38 63-14971 <*>

FERNKABEL
1925 N9 P23-32 66-13298 <*>

FERNMELDE-PRAXIS
1953 V30 N8 P257-259 57-699 <*>

FERNMELDETECHNISCHE ZEITSCHRIFT
1949 V2 N6 P179-188 60-16586 <*>
1949 V2 N11 P359-368 57-2509 <*>
1950 V3 P94-100 58-256 <*>
1951 V4 P125-132 58-255 <*>
1951 V4 N11 P481-485 57-2491 <*>
1952 N7 P318- 3137 <*>
1952 V5 N2 P67-78 64-14637 <*>
1952 V5 N8 P349-356 62-20347 <*> 0
1952 V5 N8 P372-377 62-18491 <*> 0
1952 V5 N10 P447-455 57-1000 <*>
1952 V5 N10 P456-467 63-14406 <*>
1952 V5 N11 P502-511 63-14431 <*>
1952 V5 N12 P545-550 62-16970 <*> 0
1953 V6 P254-261 I-251 <*>
1953 V6 N2 P60-65 57-1448 <*>
1953 V6 N3 P101-103 57-600 <*>
1953 V6 N4 P165-168 63-10046 <*> 0
1953 V6 N5 P204-207 57-1007 <*>
1953 V6 N5 P214-217 62-20348 <*> 0
1953 V6 N6 P279-281 62-18921 <*> 0
1953 V6 N7 P297-301 57-1014 <*>
1953 V6 N8 P389-395 57-2257 <*>
1953 V6 N9 P413-416 62-20349 <*>
1953 V6 N12 P571-577 64-14649 <*>
1954 N3 P122-128 I-572 <*>
1954 V7 N2 P57-64 62-23008 <=>
1954 V7 N4 P193-198 57-2272 <*>
1954 V7 N5 P221-226 57-2502 <*>
1954 V7 N7 P327-333 62-20350 <*> 0
1954 V7 N10 P522-528 46J19G <ATS>
1954 V7 N11 P577-581 57-1450 <*>
1954 V7 N12 P670-677 57-1384 <*>
1954 V7 N12 P678-682 57-1346 <*>
57-586 <*>
1955 V8 N3 P165-167 58-2106 <*>
1955 V8 N11 P578-586 58-2126 <*>

FERNSEHEN
1930 N5 P193-197 57-2032 <*>
1930 V1 N10 P448-452 66-13837 <*>
1931 V2 N1 P1-5 60-10195 <*>

FERNSEHEN UND TONFILM
1938 V1 N2 P51-55 57-2205 <*>
1939 N3 P17-22 57-1034 <*>
1939 V1 N3 P82-88 57-2223 <*>
1939 V1 N6 P220-226 57-1585 <*>
1940 V2 N1 P1-6 57-367 <*>
1940 V2 N1 P12-16 57-563 <*>

FERRUM
1913 V10 N4 P97-112 86L34G <ATS>

FERTIGUNGSTECHNIK. BERLIN
1954 V4 N4 P182-183 16N51G <ATS>
1956 V6 P321-322 TS-1279 <BISI>
1956 V6 N7 P321-323 61-20608 <*>
1957 V7 P413-417 62-14369 <*>
1957 V7 N2 P63-64 61-16219 <*>
1957 V7 N9 P413-417 TS-1400 <BISI>

FERTIGUNGSTECHNIK UND BETRIEB
1962 V12 N8 P520-524 48R80G <ATS>
1964 V14 N8 P469-472 64-51051 <=>
1966 V16 N5 P278-282 66-32753 <=$>
1967 V17 N2 P125-126 67-30894 <=$>

FESTSCHRIFT AUS ANLASS DES 100 JAEHRIGEN JUBILAEUMS DER FIRMA W. C. HERAEUS, GMBH. HANAU
1951 P124-146 AEC-TR-4418 <*>

FESTSCHRIFT KARL WURSTER ZUM 60. GEBURTSTAG 1960
1960 P231-247 9043 <IICH>
1960 P279-292 9065 <IICH>

FETTCHEMISCHE UMSCHAU
1935 N3 P46-48 58-1962 <*>
1935 N3 P75-76 58-1962 <*>

FETTE UND SEIFEN
1937 V44 P228-229 II-155 <*>
1938 V45 N11 P626-629 58-1064 <*>
1940 V47 N11 P510-514 60-10421 <*>
1940 V47 N11 P542-543 63-10524 <*>
1940 V47 N12 P595-598 60-10420 <*> 0
1941 V48 N2 P51-53 60-10419 <*>
1942 V49 P81-88 66-14543 <*>
1942 V49 N10 P733-735 62-20335 <*> 0
1943 V50 P279-288 62-20334 <*> 0
1943 V50 N8 P396-398 60-10414 <*>
1950 V52 P306-308 58-970 <*>
1950 V52 N7 P415-419 60-10396 <*>
1950 V52 N8 P474-476 58-1215 <*>
1950 V52 N9 P517-528 64-20398 <*> 0
1950 V52 N10 P581-587 PT1 58-1140 <*>
1950 V52 N11 P675-680 PT2 58-1140 <*>
1950 V52 N12 P725-728 PT3 58-1140 <*>
1951 V53 P548-551 58-968 <*>
1951 V53 N4 P191-193 60-10395 <*>
1951 V53 N4 P207-209 60-10394 <*>
1951 V53 N7 P390-399 I-1018 <*>
1951 V53 N9 P525-531 61-16090 <*>
1952 V54 P321-324 61-10296 <*>
1952 V54 P673-674 57-1287 <*>
1952 V54 N1 P10-12 63-10822 <*> 0
1952 V54 N3 P147-149 63-14961 <*>
1953 V55 P178-179 58-967 <*>
1953 V55 P596-600 57-200 <*>
1953 V55 N1 P10-16 64-20833 <*>
1953 V55 N3 P170-173 59-15527 <*> 0
1953 V55 N3 P178-180 60-10393 <*>
1953 V55 N7 P435-440 63-10787 <*>
1953 V55 N8 P529-532 58-966 <*>
1953 V55 N12 P847-851 60-10392 <*>

FETTE, SEIFEN, ANSTRICHMITTEL
1954 V56 P9- <ES>
1954 V56 P9-13 58-736 <*>
1954 V56 P218-220 I-998 <*>
1954 V56 N3 P145-149 50F3G <ATS>
1954 V56 N4 P218-220 58-965 <*>
1954 V56 N4 P242-245 63-20930 <*> 0
1954 V56 N5 P286-291 61-16089 <*> 0
1954 V56 N10 P775-784 5105-B <K-H>
1955 V57 P96-99 3320 <*>
1955 V57 N1 P24-32 60-18452 <*>
1955 V57 N1 P36-42 60-10413 <*>
1955 V57 N3 P168-172 60-18451 <*>
1955 V57 N4 P231-235 58-957 <*>
1955 V57 N4 P236-240 61-16187 <*> 0
1955 V57 N6 P405-407 60-10711 <*>
1955 V57 N6 P413-415 63-14892 <*>
1955 V57 N6 P423-425 63-14784 <*>
1955 V57 N7 P474-478 59-17067 <*>
1955 V57 N9 P686-691 59-15657 <*> 0
1955 V57 N12 P1010-1017 80H13G <ATS>
1956 V58 P553-556 T-2366 <INSD>
1956 V58 P736-739 8786 <IICH>
1956 V58 P849-852 3253 <*>
1956 V58 P977-984 57-1205 <*>
1956 V58 N2 P91-94 63-16295 <*>
1956 V58 N7 P528-534 59-15668 <*> P
1956 V58 N8 P585-592 60-10412 <*>
1956 V58 N10 P879-885 63-16392 <*>
1956 V58 N10 P886-890 63-16394 <*>
1956 V58 N12 P1073-1076 59-15653 <*> 0
1957 V59 P589-594 57-3549 <*>
1957 V59 N3 P140-142 63-10296 <*> 0
1957 V59 N5 P321-328 61-16091 <*>
1957 V59 N7 P493-498 17N55G <ATS>
1957 V59 N7 P509-514 58-522 <*>
63-16480 <*>
1957 V59 N8 P646-651 60-14262 <*>
1957 V59 N10 P811-814 63-16494 <*>
1957 V59 N10 P852-856 61M39G <ATS>
1957 V59 N11 P961-966 61-20891 <*>

1957 V59 N12 P1078-1084 08R78G <ATS>
1958 V60 N1 P12-16 59-20437 <*> 0
64-30685 <*> 0
1959 V61 N2 P93-95 AEC-TR-3765 <*>
1959 V61 N2 P134-138 61-16110 <*> 0
1959 V61 N3 P177-181 61-16112 <*> 0
1959 V61 N4 P257-264 61-16111 <*>
1959 V61 N9 P782-784 61-10196 <*>
1959 V61 N11 P1127-1129 2166 <TTIS>
1959 V61 N11 P1131-1138 60-18501 <*>
1959 V61 N11 P1142-1149 66-14866 <*>
1960 V62 P607-610 2231 <TTIS>
1960 V62 N1 P31-36 60-16767 <*>
60-18502 <*> 0
1960 V62 N3 P175-182 61-10269 <*>
1960 V62 N3 P197-204 65M44G <ATS>
1960 V62 N4 P326-332 63-16786 <*>
1960 V62 N11 P1024-1030 61-16122 <*>
1960 V62 N11 P1078-1082 61-18809 <*>
1961 V63 P630-632 62-16192 <*> 0
1961 V63 N1 P49-55 62-10738 <*> 0
1961 V63 N5 P445-451 65-13039 <*>
1961 V63 N5 P451-455 65-14903 <*>
1961 V63 N9 P843-844 NS-298 <TTIS>
1961 V63 N10 P950-960 58P58G <ATS>
1962 V64 N1 P27-40 65-11625 <*>
1962 V64 N2 P107-110 63-14416 <*> 0
1962 V64 N2 P110-113 62-20375 <*>
1962 V64 N3 P218-231 65-11626 <*>
1962 V64 N3 P270-279 63-14401 <*>
1962 V64 N6 P521-524 62-18844 <*> 0
1962 V64 N9 P807-813 C-5619 <NRC>
1962 V64 N12 P1109-1114 63-18098 <*>
1963 V65 P11-13 63-14395 <*>
1963 V65 N2 P117-121 65-13526 <*>
1963 V65 N6 P479-482 64-10181 <*>
1963 V65 N9 P717-721 65-13527 <*>
1963 V65 N9 P755-759 64-16678 <*>
1963 V65 N10 P834-838 64-16842 <*> 0
1964 V66 P123-132 619 <TC>
1964 V66 P214-215 295 <TC>
1964 V66 N1 P59-61 64-18136 <*>
1964 V66 N2 P97-100 64-30059 <*>
1964 V66 N2 P112-122 65-12407 <*>
1964 V66 N3 P222-225 65-11519 <*> 0
1964 V66 N10 P763-773 65-11620 <*>
65-11620 <*> 0
1964 V66 N11 P951-952 800 <TC>
1964 V66 N11 P952-954 658 <TC>
1964 V66 N12 P1032-1040 07S86G <ATS>
1964 V66 N12 P1059-1062 65-13807 <*>
1965 V67 N2 P89-94 1932 <TC>
1965 V67 N2 P115-120 80S89G <ATS>
1965 V67 N2 P130-132 66-10753 <*>
1965 V67 N12 P983-990 66-11490 <*>
1965 V67 N12 P1000-1003 66-13956 <*>
1966 V68 N4 P293-301 66-13788 <*>

FEUERFEST
1925 V1 P25-27 64-14336 <*>
1925 V1 P48-49 64-14336 <*>
1925 V1 P67-68 64-14336 <*>

FEUERUNGSTECHNIK
1934 V22 N6 P65-70 I-136 <*>
1939 V27 P99-102 61-18505 <*>

FIBRE WORLD. JAPAN
SEE SEN-I KAI

FICHES DE PHYTOPATHOLOGIE TROPICALE
1956 V149 P1-7 65-12686 <*>

FILMTECHNIK
1930 P16-17 05/03 60-10215 <*>
1930 V6 N7 P13- 57-1079 <*>

FILOSOFSKA MISUL
1959 V15 N5 P81-93 60-41051 <=*>
1963 V19 N1 P144-147 63-21892 <=>

FILOSOFSKIE NAUKI
1965 N5 P79-87 65-33789 <=*$>

FINANSI SSSR
1960 N11 P20-26 61-21697 <=>
1961 N2 P59-67 62-23476 <=>
1962 V23 N3 P47-50 62-32579 <=>
1962 V23 N5 P30-36 62-33307 <=>
1962 V23 N5 P43-47 62-33307 <=>
1962 V23 N8 P69-71 63-13266 <=>
1962 V23 N10 P51-58 63-21186 <=>
1963 V24 N3 P70-72 63-31122 <=>
1963 V24 N7 P69 63-31752 <=>
1963 V24 N12 P65-66 64-31218 <=>
1965 N9 P3-13 66-30564 <=*$>
1965 N9 P75-78 66-30564 <=*$>
1965 N9 P95-96 66-30564 <=*$>
1965 N10 P25-39 66-30408 <=*$>
1965 N10 P54-64 66-30186 <=*$>
1965 V25 N2 P35-42 65-31093 <*>
1966 N1 P21-27 67-30890 <=$>
1966 N4 P15-19 66-32330 <=$>
1966 N4 P74-78 66-32330 <=$>
1966 N4 P95-96 66-32330 <=$>
1966 N7 P80-83 67-30890 <=$>
1967 N2 P3-7 67-31079 <=$>
1967 N2 P17-51 67-31079 <=$>

FINANSI I KREDIT
1966 N4 P43-52 66-33221 <=$>

FINANSIJA
1960 N5 P255-256 60-31763 <*=>
1960 N5 P256-259 60-31764 <*=>
1966 N7/8 P340-356 66-34938 <=$>
1966 N7/8 P390-403 63-34938 <=$>

FINNISH PAPER AND TIMBER JOURNAL
SEE PAPERI JA PUU

FINSK VETERINARTIDSKRIFT
1960 V66 N9 P499-510 62-19346 <=*> 0

FINSKA LAEKARESAELLSKAPETS HANDLINGAR
1926 V68 N2 P87-112 66-12645 <*>
1947 V90 N8 P1795-1800 57-3576 <*>
1965 V109 P43-47 66-12020 <*>

FISCHWAREN-UND FEINKOSTINDUSTRIE
1953 V5 N12 P290-292 I-553 <*>

FISHERIES SCIENCE
SEE SUISAN KAGAKU

FISHERIES SCIENCE SERIES. RESEARCH DIVISION, FISHERIES AGENCY. JAPAN
SEE GYOGYO KAGAKU SOSHO

FISHERY
SEE SUISAN KOZA

FISIOLOGIA E MEDICINA
1954 V18 N22 P179-193 II-839 <*>

FISKERIDIREKTORATETS SKRIFTER. SERIE HAVUNDERSOKELSER
1938 V5 P6-19 II-96 <*>

FISKETS GANG
1959 V45 N38 P522-525 60-17113 <=*>
1959 V45 N43 P593-596 60-17104 <=*>

FIZICHESKIE PROBLEMY SPEKTROSKOPII
1960 V1 P39-41 66-61859 <=$>
1960 V1 P51-54 66-61858 <=*$>
1960 V1 P114-116 65-63651 <=*$>

1963 N2 P94-96 AD-647 714·<=>
1963 N2 P109-113 AD-647 714 <=>

FIZICHESKII SBORNIK
1957 N3 P161-167 97N57R <ATS>
1957 N3 P272-274 87M38R <ATS>
1957 N3 P428-430 RCT V33 N1 P208 <RCT>
1957 N3 P437-439 65-11212 <*>
1958 N4 P395-402 2369 <BISI>
1958 N4 P421-422 62-13915 <*=>
1958 N4 P451-452 2370 <BISI>

FIZICHESKII ZHURNAL
1941 V4 N5 P473-478 59-15210 <*>

*FIZIKA GORENIYA I VZRYVA
1965 N3 P3-9 66-31966 <=$>
1965 N3 P20-40 66-31966 <=$>
1965 N3 P45-53 66-31966 <=$>
1965 N3 P83-97 66-31966 <=$>
1965 N4 P20-23 66-33105 <=$>
1965 N4 P31-51 66-33105 <=$>
1965 N4 P63-82 66-33105 <=$>
1965 V1 N4 P3-9 66-33988 <=$>
1965 V1 N4 P20-70 66-33988 <=$>
1965 V1 N4 P78-87 66-33988 <=$>

*FIZIKA METALLOV I METALLOVEDENIE
1955 V1 N1 P185-192 3628 <HB>
1955 V1 N2 P251-260 NACA TM 1411 <NASA>
1955 V1 N2 P269-272 R-383 <*>
1955 V1 N2 P298-302 R-289 <*>
1955 V1 N2 P366-367 4946 <HB>
1955 V1 N3 P467-478 R-4996 <*>
1955 V1 N3 P529-537 GB4 <*>
1956 V2 N1 P3-9 R-1770 <*>
1956 V2 N1 P10-15 R-1496 <*>
61-15325 <+*>
1956 V2 N1 P33-42 R-1141 <*>
1956 V2 N1 P88-92 63-19127 <=*>
1956 V2 N1 P93-99 R-1804 <*>
1956 V2 N1 P100-104 R-1803 <*>
1956 V2 N1 P105-119 R-1805 <*>
1956 V2 N1 P168-171 R-965 <*>
865TM <CTT>
1956 V2 N1 P172-175 R-964 <*>
866TM <CTT>
1956 V2 N1 P176-180 60-16995 <*> 0
1956 V2 N2 P193-205 R-1136 <*>
1956 V2 N2 P215-221 R-3743 <*>
1956 V2 N2 P225-236 64-16918 <=*$>
1956 V2 N2 P277-284 R-1776 <*>
1956 V2 N2 P303-308 61-19669 <=*>
1956 V2 N2 P309-319 R-4810 <*>
3902 <HB>
62-24144 <=*>
1956 V2 N2 P320- AEC-TR-4120 <+*>
1956 V2 N2 P339-350 66L29R <ATS>
1956 V2 N2 P361-369 4104 <HB>
1956 V2 N3 P385-391 R-1493 <*>
1956 V2 N3 P397-402 R-195 <*>
1956 V2 N3 P403-405 R-1492 <*>
1956 V2 N3 P447-453 R-873 <*>
1956 V2 N3 P494-503 61-19434 <+*> 0
1956 V2 N3 P509-513 4406 <HB>
1956 V2 N3 P531-537 3981 <HB>
1956 V2 N3 P538-545 63-24620 <=*$>
1956 V2 N3 P546-551 R-1777 <*>
4011 <HB>
61-15326 <+*>
1956 V2 N3 P567-568 64-23554 <=>
1956 V3 P26-30 UCRL TRANS-997(L) <=*$>
1956 V3 P70-75 R-1988 <*>
1956 V3 P238- R-1761 <*>
1956 V3 P513-515 3952 <HB>
1956 V3 P564-565 4154 <HB>
1956 V3 N1 P3-10 R-662 <*>
1956 V3 N1 P11-15 R-1550 <*>
1956 V3 N1 P15-17 R-867 <*>
1956 V3 N1 P18-21 R-428 <*>
1956 V3 N1 P55-61 R-1553 <*>
1956 V3 N1 P66-69 3976 <HB>
1956 V3 N1 P87-96 R-1771 <*>
1956 V3 N1 P183-184 59-22667 <+*>
1956 V3 N1 P184-185 3942 <HB>
1956 V3 N1 P185-186 3923 <HB>
1956 V3 N1 P186-188 4258 <HB>
1956 V3 N1 P190-191 R-850 <*>
1956 V3 N2 P193-199 63-10650 <=*>
1956 V3 N2 P208-215 63-14447 <=*>
1956 V3 N2 P247-253 61-10665 <+*> 0
1956 V3 N2 P254-268 3913 <HB>
1956 V3 N2 P269-277 R-3532 <*>
59-22487 <+*>
1956 V3 N2 P309-319 R-3190 <*>
717TM <CTT>
1956 V3 N2 P314-320 4009 <HB>
60-16993 <*> 0
1956 V3 N2 P360-362 R-663 <*>
1956 V3 N3 P385-394 R-1973 <*>
1956 V3 N3 P395-405 59-10308 <+*>
1956 V3 N3 P406-410 R-1137 <*>
1956 V3 N3 P411-421 R-1592 <*>
1956 V3 N3 P422-432 R-1443 <*>
1956 V3 N3 P444-448 R-4250 <*>
1956 V3 N3 P460-467 59-22506 <+*>
1956 V3 N3 P468-470 59-22507 <+*>
1956 V3 N3 P471-476 60-16990 <*> 0
1956 V3 N3 P477-485 R-4249 <*>
1956 V3 N3 P483-485 62-13198 <=*>
1956 V3 N3 P503-507 60-16991 <*> 0
1956 V3 N3 P530-536 R-2358 <*>
1956 V3 N3 P540-546 4234 <HB>
1956 V3 N3 P547-548 R-2114 <*>
1956 V3 N3 P549-550 R-2043 <*>
1956 V3 N3 P550-551 R-2115 <*>
1956 V3 N3 P553-555 3994 <HB>
1956 V3 N3 P555-557 60-16992 <*> 0
1956 V3 N3 P557-560 60-16988 <*> 0
1956 V3 N3 P560-561 R-2112 <*>
1957 V4 P94-102 R-4269 <*>
1957 V4 N1 P9-13 R-1781 <*>
1957 V4 N1 P14-16 R-1633 <*>
1957 V4 N1 P28-35 60-16984 <*> 0
1957 V4 N1 P36-40 R-4030 <*>
1957 V4 N1 P54-59 61-13458 <=*> 0
1957 V4 N1 P76-83 R-3897 <*>
1957 V4 N1 P84-88 3996 <HB>
1957 V4 N1 P171-176 60-13314 <+*>
1957 V4 N1 P183-184 R-2045 <*>
1957 V4 N1 P185-186 R-2044 <*>
1957 V4 N1 P187-189 60-16983 <*> 0
1957 V4 N1 P189-190 4257 <HB>
1957 V4 N2 P232-238 59-18742 <+*> 0
1957 V4 N2 P267-277 R-4065 <*>
59-18348 <+*> 0
1957 V4 N3 P566-567 R-2363 <*>
1957 V5 N1 P137-141 64-14054 <=*$> 0
1957 V5 N1 P168-169 R-3704 <*>
1957 V5 N2 P193-202 59-18741 <+*> 0
1957 V5 N2 P234-240 61-13459 <+*> 0
1957 V5 N3 P412-420 61-13460 <=*> 0
1957 V5 N3 P421-427 61-13462 <=*> 0
1957 V5 N3 P552-553 R-4711 <*>
1958 V6 N1 P3-14 22M39R <ATS>
1958 V6 N1 P181-183 62-23317 <=*>
1958 V6 N3 P420-425 61-13461 <=*> 0
1958 V6 N3 P569 R-5332 <*>
1958 V6 N5 P832-837 61-19430 <+*>
1958 V6 N5 P929-930 59-14737 <+*>
1958 V6 N5 P932-934 59-14738 <+*>
1958 V6 N5 P951-952 62-11388 <=>
1958 V6 N6 P919-923 60-12504 <=>
62-11410 <=>
1958 V6 N6 P1100-1104 61-20950 <*=>
1959 V7 N1 P91-94 4682 <HB>
1959 V7 N1 P128-132 62-19820 <=>
1959 V7 N2 P288-289 59-16505 <+*>
1959 V7 N3 P475-476 61-27582 <*=>
1959 V7 N3 P476-477 TRANS-139 <FRI>

Citation	Translation
1959 V7 N4 P559-564	61-23448 <=*> 0
1959 V7 N6 P825-831	AEC-TR-3863 <+*>
1959 V7 N6 P880-884	61-19401 <+*> 0
1959 V8 P904-907	AEC-TR-4012 <**>
1959 V8 N2 P165-169	60-12570 <=>
	62-11408 <=>
	62-13694 <*=>
1959 V8 N2 P170-175	62-11446 <=>
1959 V8 N2 P235-239	62-11534 <=>
1959 V8 N2 P240-248	62-19888 <=>
1959 V8 N2 P309-310	60-12539 <=>
	62-11331 <=>
1959 V8 N2 P316-318	60 12540 <=>
1959 V8 N2 P318-320	62-11514 <=>
1959 V8 N3 P330-336	62-11254 <=>
1959 V8 N3 P342-345	62-11334 <=>
1959 V8 N3 P412-416	62-11182 <=>
1959 V8 N3 P463-465	62-23174 <=>
1959 V8 N4 P557-561	62-11345 <=>
1959 V8 N4 P562-568	62-23153 <=>
1959 V8 N4 P595-598	61-27003 <*=> 0
1959 V8 N4 P622-630	61-27003 <*=> 0
1959 V8 N4 P639-640	62-11432 <=>
1959 V8 N5 P678-684	62-11287 <=>
1959 V8 N5 P685-688	62-11288 <=>
1959 V8 N5 P694-699	62-19839 <=>
1959 V8 N5 P766-776	61-23896 <*=> 0
1959 V8 N6 P820-828	60-11429 <=>
1959 V8 N6 P885-891	62-11413 <=>
1960 V9 N1 P48-52	61-31562 <=>
1960 V9 N2 P212-215	61-13608 <*+>
1960 V9 N3 P374-378	62-11511 <=>
1960 V9 N6 P810-814	61-31106 <=>
1960 V9 N6 P823-827	62-23169 <=>
1960 V9 N6 P940-942	61-31105 <=>
1960 V10 N1 P3-8	62-14771 <=*> 0
1960 V10 N1 P122-130	61-27567 <*=>
1960 V10 N2 P216-222	63-10086 <=*>
1960 V10 N3 P495-496	62-11433 <=>
1960 V10 N4 P633-634	5039 <HB>
1960 V10 N4 P635-636	63-13575 <=*>
1960 V10 N6 P835-837	61-27403 <=*>
1960 V10 N6 P862-865	61-27439 <*=> 0
1960 V10 N6 P873-878	61-31476 <=>
1961 V11 N1 P29-33	61-28237 <*=> 0
1961 V11 N1 P115-122	61-28817 <*=>
1961 V11 N2 P186-202	62-13180 <=*> 0
1961 V11 N2 P252-260	62-13180 <=*> 0
1961 V11 N2 P272-280	62-13180 <=*> 0
1961 V11 N3 P400-403	62-24942 <=*>
1961 V11 N3 P461-464	61-27831 <*=>
1961 V11 N3 P476-477	61-27828 <*=>
	62-16128 <=*>
1961 V11 N4 P564-567	5236 <HB>
1961 V11 N4 P568-574	62-25849 <=*>
1961 V11 N4 P628-629	61-28505 <*=>
1961 V11 N4 P630-633	62-25543 <=*>
1961 V11 N5 P650-663	62-23189 <=*>
1961 V11 N6 P856-863	62-32692 <=*>
1961 V11 N6 P899-909	62-25301 <=*>
1961 V11 N6 P945-947	62-15448 <*=>
1961 V11 N6 P955-957	62-25838 <=*>
1961 V12 N1 P78-96	62-32992 <=*>
1961 V12 N1 P84-90	62-16181 <=*>
1961 V12 N2 P176-182	62-25844 <=*>
1961 V12 N2 P217-222	62-24347 <=*>
1961 V12 N3 P409-416	62-32977 <=*>
1961 V12 N3 P455-457	62-25315 <=*> 0
1962 V13 N2 P306-308	62-33215 <=*>
1962 V13 N3 P415-426	63-23954 <=*$>
1962 V13 N3 P470-474	63-15539 <=*>
1962 V13 N4 P550-554	63-13922 <=*>
1962 V13 N4 P626-631	64-13284 <=*$> 0
1962 V13 N4 P636-639	63-13379 <=*>
1962 V13 N5 P658-662	63-20728 <=*$>
1962 V13 N5 P693-700	63-15284 <=*>
1962 V13 N5 P701-709	64-13279 <=*$>
1962 V13 N5 P799-800	63-19871 <=*>
1962 V13 N6 P832-841	M5200 <NLL>
	63-19186 <=*>
1962 V13 N6 P942-944	5713 <HB>
1962 V14 N1 P10-16	63-21122 <=>
1962 V14 N1 P26-29	5735 <HB>
1962 V14 N1 P55-60	M-5019 <NLL>
1962 V14 N1 P85-91	M-5018 <NLL>
1962 V14 N1 P137-139	63-21122 <=>
1962 V14 N1 P148-150	63-21122 <=>
1962 V14 N1 P157-160	5734 <HB>
1962 V14 N2 P283-286	64-13763 <=*$>
1962 V14 N3 P348-357	64-23553 <=>
1962 V14 N3 P374-377	64-23555 <=>
1962 V14 N4 P535-541	M-5332 <NLL>
1962 V14 N4 P542-547	M-5196 <NLL>
1962 V14 N4 P574-577	63-41079 <=>
1962 V14 N4 P582-588	63-21658 <=>
1962 V14 N4 P608-612	64-23551 <=>
1962 V14 N4 P613-617	64-71458 <=> M
1962 V14 N5 P687-692	63-21333 <=>
1962 V14 N5 P745-749	63-21333 <=>
1962 V14 N5 P783-784	63-21333 <=>
1962 V14 N6 P852-856	63-21148 <=>
1962 V14 N6 P899-903	M-5333 <NLL>
1962 V14 N6 P935-936	5902 <HB>
1963 V15 N1 P29-32	63-41045 <=>
1963 V15 N1 P100-104	63-23929 <=>
1963 V15 N1 P148-150	M-5331 <NLL>
1963 V15 N2 P210-214	63-21879 <=>
1963 V15 N2 P280-284	63-21658 <=>
1963 V15 N4 P544-553	64-13038 <=*$>
1963 V15 N4 P554-564	63-23918 <=*$>
1963 V15 N5 P658-663	64-13213 <=*$>
1963 V15 N5 P673-677	6004 <HB>
1963 V15 N5 P710-716	6001 <HB>
1963 V15 N5 P791-793	64-18953 <*>
1963 V15 N5 P793-795	64-11669 <=>
1963 V15 N6 P860-866	64-15255 <=*$>
1963 V15 N6 P873-879	64-13220 <=*$>
1963 V15 N6 P940-941	64-13219 <=*$>
1963 V15 N6 P943-944	63-41026 <=>
1963 V16 N1 P134-135	AD-615 238 <=$>
1963 V16 N3 P493-494	AEC-TR-6388 <=*$>
1963 V16 N3 P495-496	AEC-TR-6374 <=*$>
1963 V16 N6 P872-876	6210 <HB>
1963 V16 N6 P877-885	65-61436 <=>
1963 V16 N6 P895-903	64-19144 <=$>
1964 V17 N1 P146-148	65-10880 <*>
1964 V17 N2 P296-298	65-61176 <=>
1964 V17 N3 P390-399	M-5589 <NLL>
1964 V17 N3 P400-407	65-61193 <=>
1964 V17 N3 P435-439	65-61222 <=$>
1964 V17 N3 P471-474	65-61177 <=>
1964 V17 N3 P474-476	65-61417 <=>
1964 V17 N4 P512-518	58S81R <ATS>
1964 V17 N4 P519-526	65-61168 <=>
1964 V17 N4 P527-535	65-61175 <=>
1964 V17 N4 P613-614	M-5590 <NLL>
1964 V17 N6 P862-865	TIL/T.5569 <=$>
1964 V17 N6 P892-897	N65-23685 <=$>
1964 V17 N6 P903-908	65-13046 <*>
1964 V18 N1 P148-149	M-5585 <NLL>
1964 V18 N2 P300-303	65-13047 <*>
1964 V18 N3 P396-400	AD-636 640 <=$>
1964 V18 N3 P454-456	M-5618 <NLL>
1964 V18 N3 P473-475	M.5617 <NLL> 0
1964 V18 N4 P502-505	65-14528 <*>
1964 V18 N5 P740-745	AD-622 477 <=$*>
	N65-27681 <=*$>
1964 V18 N5 P796-798	M-5681 <NLL>
1964 V18 N6 P858-861	AD-625 161 <=*$>
1965 V19 P311-313	N65-29735 <=*$>
1965 V19 N1 P141-144	65-17206 <*>
1965 V19 N2 P226-240	65-30858 <=$>
1965 V19 N2 P299-301	M-5695 <NLL>
1965 V19 N3 P360-366	65-17207 <*>
1965 V19 N3 P411-417	65-31212 <=$>
1965 V19 N4 P521-529	M.5809 <NLL>
1965 V19 N4 P592-595	M.5810 <NLL>
1965 V19 N5 P735-740	6482 <HB>
1965 V19 N5 P757-761	M.5811 <NLL>
1965 V19 N5 P793-796	M-5749 <NLL>

1965 V20 N1 P103-110 66-11171 <*>
1965 V20 N1 P114-119 66-11180 <*>
1965 V20 N1 P155-157 66-11182 <*>
1965 V20 N2 P243-250 66-11162 <*>
1965 V20 N4 P630-632 66-13373 <*>
1966 V21 N5 P785-786 AD-641 284 <=$>

FIZIKA V SHKOLE
1957 N5 P23-36 59-21198 <=>
1957 N6 P16-17 59-19478 <+*>
1957 V17 N1 P32-33 63-23718 <=*$>
1963 V23 N5 P6-18 64-21722 <=>

*FIZIKA TVERDOGO TELA
1959 V1 P114-121 62-23094 <*=>
1959 V1 N4 P572-573 AEC-TR-4938 <=*>
13N56R <ATS>
1959 V1 N6 P990-992 60-13904 <+*>
1959 V1 N7 P1027-1034 SCL-T-344 <=> P
1959 V1 N8 P1213-1220 62-19825 <=>
1959 V1 N10 P1562-1582 61-23449 <*=>
1959 V1 N12 P1871-1873 60-16512 <+*>
1960 V2 P88-90 62-10665 <=*> O
1960 V2 N2 P377-379 62-24875 <=*>
1960 V2 N5 P781-792 61-21198 <=>
1960 V2 N8 P1739-1740 61-27545 <*=>
1960 V2 N8 P1945-1948 61-27546 <*=>
1960 V2 N12 P3048-3049 62-15692 <*=>
1961 V3 P1815-1820 62-25104 <=*>
1961 V3 N1 P61-72 62-32652 <=*>
1961 V3 N2 P456-458 62-15691 <*=>
1961 V3 N2 P632-641 61-27835 <*=>
1961 V3 N2 P660-662 61-27142 <*=>
1961 V3 N3 P677-686 61-20495 <*=>
62-24870 <=*>
1961 V3 N3 P786 62-32621 <=*>
1961 V3 N3 P832-840 62-32617 <=*>
1961 V3 N3 P991-994 62-11477 <=>
1961 V3 N4 P1019-1030 61-27837 <*=>
1961 V3 N4 P1061-1065 62-32618 <=*>
1961 V3 N4 P1144-1151 62-19922 <*=>
1961 V3 N6 P1662-1667 61-28657 <=*>
1961 V3 N6 P1683-1687 62-15723 <=*> O
1961 V3 N7 P2031-2040 62-10618 <=*>
1961 V3 N9 P2567-2572 AEC-TR-5037 <=*>
64P58R <ATS>
1961 V3 N10 P3181-3186 62-19271 <=*>
1962 V4 P2458-2460 RTS-2612 <NLL>
1962 V4 N2 P449-453 62-25648 <=*>
1962 V4 N2 P524-529 63-13629 <=*>
1962 V4 N5 P1196-1205 63-10157 <=*>
1962 V4 N6 P1449-1454 63-21006 <=>
1962 V4 N6 P1627-1631 AEC-TR-5532 <=*>
1962 V4 N7 P1735-1742 63-21152 <=>
1962 V4 N7 P1846-1852 63-21152 <=>
1962 V4 N7 P1878-1881 63-21152 <=>
1962 V4 N7 P1959-1960 63-21152 <=>
1962 V4 N8 P2109-2115 63-21173 <=>
1962 V4 N8 P2151-2159 66-13007 <*>
1962 V4 N8 P2258-2261 63-21170 <=>
1962 V4 N9 P2447-2449 63-21272 <=>
1962 V4 N9 P2585-2596 63-21272 <=>
1962 V4 N10 P2733-2737 63-21298 <=>
1962 V4 N10 P2901-2907 63-21298 <=>
1962 V4 N10 P2917-2920 63-21298 <=>
1962 V4 N11 P3350-3351 63-14453 <=*>
1962 V4 N12 P3471-3481 63-14450 <=*>
1963 V5 N1 P211-219 63-31566 <=>
1963 V5 N1 P308-314 63-31566 <=>
1963 V5 N1 P361-362 64-71416 <=> M
1963 V5 N2 P373-380 63-19028 <=*>
1963 V5 N3 P921-927 64-13612 <=*$>
1963 V5 N4 P1077-1081 64-16833 <=*$>
1963 V5 N4 P1222-1225 64-14003 <=*$>
1963 V5 N5 P1368-1372 65-10269 <*>
1963 V5 N6 P1496-1510 65-10848 <*>
1963 V5 N6 P1537-1547 64-10784 <=*$> O
1963 V5 N6 P1728-1730 63-31915 <=>
1963 V5 N6 P1737-1740 65-10752 <*>
1963 V5 N6 P1756-1759 64-10785 <=*$> O
1963 V5 N7 P1940-1945 65-10785 <*>
1963 V5 N8 P2332-2338 63-20575 <=*$>
1963 V5 N9 P2409-2419 72R74R <ATS>
1963 V5 N9 P2580-2586 64-71415 <=> M
1963 V5 N9 P2653-2655 65-10854 <*> O
1963 V5 N10 P2871-2876 65-10754 <*>
1964 V6 P2376-2388 65-13138 <*>
1964 V6 N1 P316-317 64-16910 <=*$>
1964 V6 N1 P328-329 65-11033 <*>
1964 V6 N3 P962-964 65-10748 <*>
1964 V6 N4 P1104-1114 65-10760 <*>
1964 V6 N4 P1234-1235 65-10853 <*>
1964 V6 N5 P1406-1412 64-31617 <=>
1964 V6 N5 P1449-1452 65-10265 <*> O
1964 V6 N5 P1511-1519 65-12230 <*>
1964 V6 N5 P1563-1565 65-10251 <*>
1964 V6 N7 P1939-1945 60R78R <ATS>
1964 V6 N7 P2003-2009 AD-607 855 <=>
1965 V7 P24-27 65-17149 <*>
1965 V7 N5 P1517-1518 65-18070 <*>
1965 V7 N5 P1567-1568 66-10416 <*>
1965 V7 N6 P1876-1877 66-10417 <*>
1965 V7 N9 P2673-2677 68T90R <ATS>
1965 V7 N9 P2740- 66-10870 <*>
1965 V7 N10 P3054-3062 06S89R <ATS>
1965 V7 N11 P3255-3259 35R90R <ATS>
1965 V7 N11 P3421-3422 66-11204 <*>
1965 V7 N11 P3451-3452 66-14894 <*$>
1966 V8 P111-114 66-21294 <$>
1966 V8 N1 P286-287 66-11226 <*>
1966 V8 N4 P1312-1314 66-14390 <*$>

FIZIKA TVERDOGO TELA. SBORNIK STATEI
1959 V1 P170 AEC-TR-5541 <=*>
1959 V1 P211-227 64-13329 <=*$>
1959 V2 P96-98 61-18319 <=$>
1959 V2 P158-161 61-21986 <=>
1959 V2 P235-241 64-10780 <=*$>
1959 V2 P306-316 61-21983 <=>

FIZIKAI SZEMLE
1960 V10 P291-298 62-19422 <*=>
1960 V10 N5 P131-140 14N48H <ATS>
1960 V10 N6 P184-189 62-19416 <*=> O
1960 V10 N6 P190-191 62-19423 <*=>
1960 V10 N11 P348-352 62-19421 <=*>
1960 V10 N12 P376-379 62-19420 <*=>
1961 V11 N1 P3-12 62-19383 <*=>
1961 V11 N2 P60-61 62-19384 <=*>
1961 V11 N5 P135-141 62-19517 <*=> P
1961 V11 N7 P210-214 62-19673 <=*>
1961 V11 N12 P380-384 62-23617 <=*>
1962 V12 N1 P32-33 62-23588 <=*>
1962 V12 N5 P138-143 62-33325 <=*>
1962 V12 N5 P159-160 62-33337 <=*>
1962 V12 N12 P357-360 63-21513 <=>
1963 V13 N1 P13-20 63-21534 <=>
1963 V13 N3 P71-78 63-21976 <=>
1963 V13 N3 P95-96 63-21976 <=>
1963 V13 N5 P131-143 63-31910 <=>
1965 V15 N4 P97-100 65-32066 <=$>

FIZIKAS INSTITUTA RAKSTI
1956 V8 ENTIRE ISSUE 60-19148 <+*>
1956 V9 P105-135 62-25382 <=*>
1960 V9 N3 P374-378 62-11511 <=>

*FIZIKO-KHIMICHESKAYA MEKHANIKA MATERIALOV
1965 N1 P54 2199 <BSRA>
1965 N1 P67-72 65-31002 <=*$>
1965 V1 N2 P151-161 M.5812 <NLL>

FIZIKO-KHIMICHESKIE PROBLEMY FORMIROVANIYA GORNYKH POROD I RUD
1961 V1 P78-80 01Q73R <ATS>
1961 V1 P622-640 00R78R <ATS>
1961 V1 P641-646 AD-637 467 <=$>

*FIZIOLOGICHESKII ZHURNAL SSSR.
1936 V21 N5/6 P809 62-20166 <=*>

Citation	Translation
1938 V24 P624-629	63-20481 <=*$>
1938 V25 N4 P418-425	63-16811 <=*>
1939 V27 P559-563	65-61608 <=$>
1939 V27 N5 P559-563	59-15398 <+*>
1940 V28 P679-685	R-1046 <*>
1940 V28 N1 P29-33	RT-4341 <*>
1940 V28 N1 P104-112	R-273 <*>
1940 V28 N4 P402-403	R-272 <*>
1940 V28 N4 P404-405	R-270 <*>
1940 V28 N6 P571-595	RT-4256 <*>
1940 V28 N2/3 P147-156	RT-4441 <*>
1940 V28 N2/3 P271-272	R-274 <*>
1940 V29 N5 P401-411	RT-4260 <*>
1940 V29 N6 P526-535	60-18895 <*+> 0
1940 V29 N1/2 P3-14	60-18893 <+*> 0
	60-18893 <*> 0
1940 V29 N1/2 P15-25	60-18894 <*+> 0
1941 V30 N6 P772-783	<ATS>
1946 V32 N1 P76-88	63-15966 <=*>
1947 V33 N3 P267-	60-13888 <+*> 0
1947 V33 N6 P689-698	52/2934 <NLL>
1948 V34 N1 P131-134	RT-2059 <*>
	60-10902 <*>
1949 N2 P236-241	RT-4576 <*>
1949 V35 N1 P16-26	RT-1563 <*>
1949 V35 N2 P236-241	64-19082 <=*$>
1951 V37 N4 P431-438	63-11607 <=>
1952 V38 N3 P395-403	RT-2535 <*>
1952 V38 N4 P423-433	RTS-2321 <NLL>
1952 V38 N6 P751-755	R-2723 <*>
	RT-4571 <*>
1953 P282-293	RT-1946 <*>
1953 P294-298	RT-1947 <*>
1953 V39 P3-16	R-2698 <*>
1953 V39 N2 P210-217	RT-1990 <*>
1953 V39 N4 P423-431	RT-1394 <*>
1953 V39 N5 P533-539	R-416 <*>
	RT-2662 <*>
	60-19493 <+*> 0
1953 V39 N5 P540-548	RT-1988 <*>
1953 V39 N5 P618-622	RT-1888 <*>
1953 V39 N5 P622-626	RT-1989 <*>
1953 V39 N6 P677-684	RT-2550 <*>
1954 V40 N1 P86-89	RT-2549 <*>
1954 V40 N1 P115-127	RT-2479 <*>
1954 V40 N2 P148-161	61-27257 <=*>
1954 V40 N2 P221-223	RT-3750 <*>
1954 V40 N2 P226-230	61-27257 <=*>
1954 V40 N2 P235-236	RTS-2839 <NLL>
1954 V40 N3 P280-288	63-11606 <=>
1954 V40 N3 P332-337	RT-2967 <*>
1954 V40 N4 P424-430	R-266 <*>
1954 V40 N4 P453-457	R-269 <*>
1954 V40 N5 P555-565	R-2697 <*>
1955 V41 N1 P19-24	62-15057 <*=>
1955 V41 N2 P178-186	PB 141 248T <=>
1955 V41 N4 P568-574	CSIRO-3268 <CSIR>
	R-3906 <*>
1955 V41 N4 P575-581	R-2692 <*>
	RT-4313 <*>
1955 V41 N5 P660-665	R-386 <*>
1955 V41 N5 P671-675	R-2705 <*>
	R-376 <*>
1955 V41 N6 P718-724	62-10165 <*=>
1955 V41 N6 P742-747	PB 141 249T <=>
1956 V42 N2 P142-148	59-11051 <=>
	62-15053 <*=>
1956 V42 N2 P242-244	59-11119 <=>
1956 V42 N3 P312-316	63-11603 <=>
1956 V42 N6 P477-486	63-18530 <=*>
1956 V42 N7 P541-545	61-23860 <*=>
1956 V42 N8 P660-667	63-11608 <=>
1956 V42 N10 P854-860	62-15054 <*=> 0
1956 V42 N11 P981-988	61-28672 <*=>
	62-15066 <=*> 0
1956 V42 N11 P988-992	PB 141 268T <=>
1957 V43 N2 P107-116	R-4870 <*>
1957 V43 N7 P611-618	60-18361 <+*>
1957 V43 N7 P651-656	60-23624 <+*>
1957 V43 N7 P664-671	62-14734 <=*> 0
1957 V43 N10 P983-994	62-15047 <=*>
1957 V43 N11 P1008-1020	PB 141 256T <=>
1957 V43 N11 P1021-1036	PB 141 257T <=>
1957 V43 N11 P1086-1097	62-15043 <=*> 0
1958 V44 N1 P23-28	62-13790 <=*>
1958 V44 N1 P77-82	62-13775 <=*>
1958 V44 N3 P243-248	R-5330 <*>
	62-13781 <=*>
1958 V44 N6 P595-598	62-13774 <=*>
1958 V44 N9 P829-838	62-13780 <=*> 0
1958 V44 N9 P873-881	62-13786 <=*>
1958 V44 N10 P922-927	62-13772 <=*> 0
1958 V44 N11 P1017-1025	59-11473 <=>
1958 V44 N11 P1040-1048	62-13779 <=*> 0
1958 V44 N12 P1131-1136	62-13789 <=*>
1959 V45 N9 P1045-1052	RTS-2320 <NLL>
1960 V46 N1 P135-138	60-31714 <=>
1960 V46 N2 P261-264	60-11963 <=>
1960 V46 N4 P429-433	10402-E <KH>
	64-10419 <=*$>
1960 V46 N5 P647-650	60-31682 <=>
1960 V46 N11 P1373-1379	62-14693 <=*>
1960 V46 N11 P1423-1425	61-11854 <=>
1961 V47 N1 P19-29	62-20307 <=*>
1961 V47 N3 P301-309	62-14724 <=*>
1961 V47 N7 P934-937	62-15380 <*=>
1961 V47 N11 P1352-1359	65-61214 <=$>
1961 V47 N11 P1432-1444	62-24389 <=*>
1962 V48 P150-158	62-20306 <=*>
1962 V48 N1 P11-15	62-32313 <=*>
	63-23910 <=*$> 0
1962 V48 N1 P55-	FPTS V22 N1 P.T80 <FASE>
1962 V48 N1 P64-	FPTS V22 N1 P.T8 <FASE>
1962 V48 N1 P95-97	62-32522 <=*>
1962 V48 N1 P97-103	62-25530 <=*>
1962 V48 N1 P108-111	62-25530 <=*>
1962 V48 N2 P121-125	62-25161 <=*>
1962 V48 N2 P145-	FPTS V22 N1 P.T127 <FASE>
1962 V48 N2 P150-158	63-15428 <=*>
1962 V48 N2 P207-	FPTS V22 N1 P.T114 <FASE>
1962 V48 N2 P214-218	NASA-TT-F-203 <=>
1962 V48 N3 P279-	FPTS V22 N1 P.T13 <FASE>
1962 V48 N3 P279-289	62-25860 <=*>
1962 V48 N3 P290-	FPTS V22 N1 P.T19 <FASE>
1962 V48 N3 P314-	FPTS V22 N1 P.T118 <FASE>
1962 V48 N3 P369-370	62-32987 <=*>
1962 V48 N4 P413-421	63-15427 <=*>
1962 V48 N4 P436-443	63-15759 <=*>
1962 V48 N4 P444-454	62-33075 <=*>
1962 V48 N4 P449-454	63-24054 <=*$>
1962 V48 N4 P455-	FPTS V22 N1 P.T86 <FASE>
1962 V48 N4 P455-463	63-15759 <=*>
1962 V48 N4 P464-	FPTS V22 N1 P.T44 <FASE>
1962 V48 N4 P470-	FPTS V22 N1 P.T99 <FASE>
1962 V48 N4 P499-501	62-33075 <=*>
1962 V48 N5 P510-	FPTS V22 N3 P.T421 <FASE>
1962 V48 N5 P534-	FPTS V22 N4 P.T728 <FASE>
1962 V48 N5 P540-	FPTS V22 N3 P.T443 <FASE>
1962 V48 N5 P545-	FPTS V22 N3 P.T405 <FASE>
1962 V48 N5 P554-562	63-19359 <=*>
1962 V48 N5 P563-	FPTS V22 N4 P.T732 <FASE>
1962 V48 N5 P579-	FPTS V22 N3 P.T416 <FASE>
1962 V48 N5 P587-592	63-19019 <=*>
1962 V48 N5 P598-	FPTS V22 N3 P.T428 <FASE>
1962 V48 N6 P629-637	63-19010 <=*>
	63-19091 <=*>
	63-23911 <=*$> 0
1962 V48 N6 P646-653	63-19356 <=*>
1962 V48 N6 P677-	FPTS V22 N2 P.T266 <FASE>
1962 V48 N6 P684-	FPTS V22 N2 P.T197 <FASE>
1962 V48 N6 P692-	FPTS V22 N2 P.T202 <FASE>
1962 V48 N6 P722-	FPTS V22 N2 P.T255 <FASE>
1962 V48 N6 P728-	FPTS V22 N2 P.T306 <FASE>
1962 V48 N6 P735-	FPTS V22 N2 P.T301 <FASE>
1962 V48 N6 P742-747	63-19011 <=*>
1962 V48 N6 P754-759	63-19012 <=*>
1962 V48 N7 P796-	FPTS V22 N2 P.T271 <FASE>
1962 V48 N7 P813-822	63-19020 <=*>
1962 V48 N7 P823-	FPTS V22 N5 P.T982 <FASE>
	1963,V22,N5,PT2 <FASE>

1962 V48 N7 P842- FPTS V22 N2 P.T219 <FASE>
1962 V48 N7 P850- FPTS V22 N5 P.T969 <FASE>
1962 V48 N8 P889-892 63-19013 <=*>
1962 V48 N8 P907- FPTS V22 N2 P.T284 <FASE>
1962 V48 N8 P922- FPTS V22 N2 P.T290 <FASE>
1962 V48 N8 P942-952 63-19017 <=*>
1962 V48 N8 P967- FPTS V22 N2 P.T207 <FASE>
1962 V48 N8 P976-982 63-19014 <=*>
1962 V48 N8 P989- FPTS V22 N2 P.T298 <FASE>
1962 V48 N8 P994- FPTS V22 N2 P.T282 <FASE>
1962 V48 N8 P997-999 RTS-3153 <NLL>
1962 V48 N9 P1010-1016 63-19391 <=*>
1962 V48 N9 P1017- FPTS V22 N3 P.T447 <FASE>
1962 V48 N9 P1026- FPTS V22 N3 P.T411 <FASE>
1962 V48 N9 P1042- FPTS V22 N3 P.T453 <FASE>
1962 V48 N9 P1051-1063 63-19392 <=*>
1962 V48 N9 P1064- FPTS V22 N3 P.T436 <FASE>
1962 V48 N9 P1085- FPTS V22 N3 P.T537 <FASE>
1962 V48 N9 P1099-1104 63-19393 <=*>
63-23548 <=*$>
1962 V48 N9 P1105-1112 63-19394 <=*>
1962 V48 N9 P1113- FPTS V22 N4 P.T623 <FASE>
1962 V48 N10 P1178- FPTS V23 N2 P.T275 <FASE>
1962 V48 N10 P1203-1217 63-15299 <=*>
1962 V48 N10 P1240- FPTS V23 N2 P.T388 <FASE>
1962 V48 N10 P1265-1269 RTS-2885 <NLL>
1962 V48 N11 P1301- FPTS V22 N4 P.T706 <FASE>
1962 V48 N11 P1311- FPTS V22 N5 P.T987 <FASE>
1962 V48 N11 P1316- FPTS V22 N6 P.T1101 <FASE>
1962 V48 N11 P1350- FPTS V22 N4 P.T749 <FASE>
1962 V48 N11 P1359- FPTS V22 N4 P.T755 <FASE>
1962 V48 N11 P1377- FPTS V22 N5 P.T973 <FASE>
1962 V48 N11 P1382-1391 63-23847 <=*$>
1962 V48 N11 P1392- FPTS V22 N5 P.T941 <FASE>
1962 V48 N12 P1444- FPTS V22 N4 P.T651 <FASE>
1962 V48 N12 P1454- FPTS V22 N4 P.T761 <FASE>
1962 V48 N12 P1466-1471 63-23057 <=*>
1962 V48 N12 P1479- FPTS V22 N4 P.T661 <FASE>
1962 V48 N12 P1488- FPTS V22 N4 P.T668 <FASE>
1962 V48 N12 P1498- FPTS V22 N4 P.T737 <FASE>
1962 V48 N12 P1504-1507 63-23878 <=*$>
1962 V48 N12 P1517-1521 63-23879 <=*$>
1962 V48 N12 P1523-1525 63-21022 <=>
1963 V49 N1 P42- FPTS V23 N1 P.T88 <FASE>
1963 V49 N1 P60- FPTS V23 N2 P.T218 <FASE>
1963 V49 N1 P85- FPTS V23 N2 P.T317 <FASE>
1963 V49 N1 P115- FPTS V23 N2 P.T365 <FASE>
1963 V49 N3 P269- FPTS V23 N2 P.T247 <FASE>
FPTS,V23,N1,T247-251 <FASE>
1963 V49 N3 P277- FPTS V23 N2 P.T235 <FASE>
FPTS,V23,P.T235-T237 <FASE>
1963 V49 N3 P281- FPTS V23 N2 P.T252 <FASE>
FPTS,V23,P.T252-T259 <FASE>
1963 V49 N3 P299- FPTS V23 N1 P.T101 <FASE>
FPTS,V23,P.T101-T104 <FASE>
1963 V49 N3 P322- FPTS V23 N2 P.T280 <FASE>
1963 V49 N3 P322 FPTS,V23,P.T280-T283 <FASE>
1963 V49 N3 P346-352 RTS-2912 <NLL>
1963 V49 N3 P359-365 65-60247 <=$> 0
1963 V49 N3 P366-369 64-13404 <=*$>
1963 V49 N3 P379-381 NASA TT-F-199 <=>
1963 V49 N4 P398- FPTS V23 N3 P.T613 <FASE>
FPTS,V23,N3,P.T613 <FASE>
1963 V49 N4 P440- FPTS V23 N2 P.T230 <FASE>
1963 V49 N4 P470- FPTS V23 N3 P.T468 <FASE>
FPTS,V23,N3,P.T468 <FASE>
1963 V49 N4 P482-489 64-13440 <=*$> 0
1963 V49 N4 P494- FPTS V23 N2 P.T222 <FASE>
1963 V49 N4 P503- FPTS V23 N2 P.T420 <FASE>
1963 V49 N5 P542-547 63-31589 <=>
1963 V49 N5 P575-583 64-19221 <=*$> 0
1963 V49 N5 P603- FPTS V23 N5 P.T1103 <FASE>
1963 V49 N5 P621- FPTS V23 N4 P.T661 <FASE>
1963 V49 N5 P639-642 63-31590 <=>
1963 V49 N5 P643-647 63-31611 <=>
1963 V49 N6 P649- FPTS V23 N5 P.T1111 <FASE>
1963 V49 N6 P666- FPTS V23 N5 P.T1075 <FASE>
1963 V49 N6 P767-770 63-31789 <=>
1963 V49 N6 P770-773 63-31790 <=>
1963 V49 N6 P774-777 63-31713 <=>
1963 V49 N6 P778-779 63-31723 <=>
1963 V49 N7 P790- FPTS V23 N5 P.T1148 <FASE>
FPTS,V23,N5,P.T1148 <FASE>
1963 V49 N7 P805- FPTS V23 N5 P.T1107 <FASE>
1963 V49 N7 P812- FPTS V23 N5 P.T1139 <FASE>
1963 V49 N7 P817- FPTS V23 N5 P.T1072 <FASE>
1963 V49 N7 P830- FPTS V23 N4 P.T689 <FASE>
FPTS,V23,N4,P.T689 <FASE>
1963 V49 N7 P834- FPTS V23 N5 P.T975 <FASE>
1963 V49 N7 P863-869 64-19227 <=*$>
1963 V49 N7 P870-872 63-24236 <=*$>
1963 V49 N7 P886-888 63-24236 <=*$>
1963 V49 N8 P919- FPTS V23 N3 P.T641 <FASE>
FPTS,V23,N3,P.T641 <FASE>
1963 V49 N9 P1026-1029 64-19222 <=*$> 0
1963 V49 N9 P1036-1043 64-13545 <=*$>
64-21292 <=>
1963 V49 N9 P1044-1049 64-21292 <=>
1963 V49 N9 P1059- FPTS V23 N5 P.T1156 <FASE>
1963 V49 N9 P1071 FPTS V23 N4 P.T692 <FASE>
1963 V49 N9 P1089-1091 64-19230 <=*$>
1963 V49 N9 P1099-1104 64-19223 <=*$> 0
1963 V49 N9 P1109-1114 RTS-2918 <NLL>
1963 V49 N9 P1115-1116 64-15223 <=*$>
1963 V49 N9 P1116-1120 64-15247 <=*$>
1963 V49 N9 P1122-1125 64-15237 <=*$>
1963 V49 N10 P1145-1149 64-13717 <=*$>
1963 V49 N10 P1163- FPTS V23 N6 P.T1201 <FASE>
1963 V49 N10 P1221-1229 64-15566 <=*$>
1963 V49 N10 P1230-1234 65-60212 <=$>
1963 V49 N10 P1234- FPTS V24 N1 P.T45 <FASE>
FPTS V24 P.T45-T47 <FASE>
1963 V49 N11 P1310- FPTS V24 N1 P.T155 <FASE>
1963 V49 N11 P1310-1317
FPTS V24 P.T155-T158 <FASE>
1963 V49 N11 P1318- FPTS V24 N1 P.T159 <FASE>
1963 V49 N11 P1318-1329
FPTS V24 P.T159-T165 <FASE>
1963 V49 N11 P1330-1337 65-60229 <=$>
1963 V49 N11 P1370- FPTS V24 N3 P.T421 <FASE>
1963 V49 N12 P1391-1399 64-19709 <=$>
1963 V49 N12 P1400- FPTS V23 N6 P.T1195 <FASE>
FPTS V23 P.T1195 <FASE>
1963 V49 N12 P1410- FPTS V23 N6 P.T1206 <FASE>
FPTS V23 P.T1206 <FASE>
1963 V49 N12 P1447- FPTS V23 N5 P.T1089 <FASE>
1963 V49 N12 P1468- FPTS V23 N6 P.T1173 <FASE>
FPTS V23 P.T1173 <FASE>
1964 V50 N1 P10- FPTS V24 N1 P.T145 <FASE>
FPTS V24 P.T145-T149 <FASE>
1964 V50 N1 P81- FPTS V24 N1 P.T142 <FASE>
FPTS V24 P.T142-T144 <FASE>
1964 V50 N1 P87- FPTS V23 N6 P.T1181 <FASE>
FPTS V23 P.T1181 <FASE>
1964 V50 N1 P106-112 64-19712 <=$>
1964 V50 N2 P129- FPTS V24 N3 P.T426 <FASE>
1964 V50 N2 P145-160 64-31094 <=>
1964 V50 N2 P153-161 65-60146 <=$>
1964 V50 N2 P177-182 AD-609 135 <=>
1964 V50 N2 P187- FPTS V24 N2 P.T325 <FASE>
1964 V50 N2 P187-192
FPTS V24 P.T325-T328 <FASE>
1964 V50 N2 P193-198 AD-611 052 <=$>
1964 V50 N2 P193- FPTS V24 N2 P.T209 <FASE>
1964 V50 N2 P193-198
FPTS V24 P.T209-T212 <FASE>
1964 V50 N2 P211-216 65-60209 <=$>
1964 V50 N3 P252- FPTS V24 N2 P.T347 <FASE>
FPTS V24 P.T347-T350 <FASE>
1964 V50 N3 P268- FPTS V24 N2 P.T368 <FASE>
1964 V50 N3 P268-271
FPTS V24 P.T368-T370 <FASE>
1964 V50 N3 P272- FPTS V24 N2 P.T343 <FASE>
FPTS V24 P.T343-T346 <FASE>
1964 V50 N3 P280-287 65-31273 <=$>
1964 V50 N3 P301-305
FPTS V24 P.T189-T191 <FASE>
1964 V50 N3 P306-313 65-61059 <=$>
1964 V50 N3 P328- FPTS V24 N2 P.T340 <FASE>
FPTS V24 P.T340-T342 <FASE>

1964 V50 N3 P373- FPTS V24 N2 P.T329 <FASE>
1964 V50 N3 P373-380 FPTS V24 P.T329-T332 <FASE>
1964 V50 N3 P381-383 64-71537 <=>
1964 V50 N4 P393- FPTS V24 N2 P.T321 <FASE>
FPTS V24 P.T321-T324 <FASE>
1964 V50 N4 P418-425 65-61060 <=$>
1964 V50 N4 P426- FPTS V24 N2 P.T357 <FASE>
FPTS V24 P.T357-T362 <FASE>
1964 V50 N4 P444- FPTS V24 N2 P.T333 <FASE>
1964 V50 N4 P444-456 FPTS V24 P.T333-T339 <FASE>
1964 V50 N4 P457- FPTS V24 N2 P.T371 <FASE>
FPTS V24 P.T371-T374 <FASE>
1964 V50 N4 P479-487 65-62183 <=>
1964 V50 N4 P514-519 65-31510 <=$>
1964 V50 N5 P531- FPTS V24 N3 P.T417 <FASE>
1964 V50 N5 P557-563 65-61063 <=$>
1964 V50 N5 P603- FPTS V24 N3 P.T408 <FASE>
1964 V50 N5 P618- FPTS V24 N3 P.T403 <FASE>
1964 V50 N6 P649-654 AD-616 311 <=$>
1964 V50 N6 P762-764 65-31367 <=*$>
1964 V50 N7 P773-778 65-62176 <=>
1964 V50 N7 P828-834 65-62184 <=>
1964 V50 N7 P855-861 65-62185 <=>
1964 V50 N7 P894- FPTS V24 N4 P.T563 <FASE>
1964 V50 N8 P913- FPTS V24 N4 P.T668 <FASE>
1964 V50 N8 P924-933 AD-612 376 <=$>
65-62356 <=$>
1964 V50 N8 P941-951 65-62357 <=$>
1964 V50 N8 P1017- FPTS V24 N4 P.T679 <FASE>
1964 V50 N9 P1079- FPTS V24 N5 P.T751 <FASE>
1964 V50 N9 P1089- FPTS V24 N4 P.T567 <FASE>
1964 V50 N9 P1104- FPTS V24 N5 P.T768 <FASE>
1964 V50 N9 P1150-1158 65-62358 <=$>
1964 V50 N9 P1191-1193 65-31226 <=$>
1964 V50 N10 P1264- FPTS V24 N5 P.T777 <FASE>
1964 V50 N10 P1270- FPTS V24 N4 P.T600 <FASE>
1964 V50 N10 P1276- FPTS V24 N5 P.T865 <FASE>
1964 V50 N11 P1321- FPTS V24 N5 P.T763 <FASE>
1964 V50 N11 P1342- FPTS V24 N6 P.T1091 <FASE>
1964 V50 N11 P1350- FPTS V24 N6 P.T1096 <FASE>
1964 V50 N11 P1377- FPTS V24 N6 P.T1101 <FASE>
1964 V50 N12 P1424- FPTS V24 N6 P.T1033 <FASE>
1964 V50 N12 P1424-1433 65-30728 <=$>
1965 V51 N1 P47- FPTS V24 N6 P.T999 <FASE>
1965 V51 N1 P99- FPTS V24 N6 P.T1008 <FASE>
1965 V51 N2 P159- FPTS V25 N1 P.T1 <FASE>
1965 V51 N2 P164- FPTS V25 N1 P.T18 <FASE>
1965 V51 N2 P251- FPTS V25 N1 P.T5 <FASE>
1965 V51 N2 P275-277 65-30693 <=$>
1965 V51 N2 P278-280 65-30956 <=$>
1965 V51 N3 P285-292 66-60424 <=*$>
1965 V51 N3 P309- FPTS V25 N1 P.T23 <FASE>
1965 V51 N3 P405-412 65-30875 <=$>
1965 V51 N4 P413-420 66-60887 <=*$>
1965 V51 N4 P441- FPTS V25 N2 P.T231 <FASE>
1965 V51 N4 P449-457 66-60893 <=*$>
1965 V51 N4 P457- FPTS V25 N2 P.T187 <FASE>
1965 V51 N4 P479- FPTS V25 N3 P.T465 <FASE>
1965 V51 N4 P501- FPTS V25 N2 P.T236 <FASE>
1965 V51 N6 P670- FPTS V25 N3 P.T407 <FASE>
1965 V51 N6 P693- FPTS V25 N3 P.T413 <FASE>
1965 V51 N6 P703- FPTS V25 N3 P.T438 <FASE>
1965 V51 N6 P717- FPTS V25 N3 P.T419 <FASE>
1965 V51 N7 P793-798 65-33634 <=*$>
1965 V51 N7 P799-805 65-33633 <=*$>
1965 V51 N7 P832-837 66-61760 <=$>
1965 V51 N7 P893-894 65-33632 <=*$>
1965 V51 N8 P909- FPTS V25 N3 P.T423 <FASE>
1965 V51 N8 P918- FPTS V25 N3 P.T428 <FASE>
1965 V51 N8 P926-935 66-61322 <=$*>
1965 V51 N8 P936- FPTS V25 N4 P.T551 <FASE>
1965 V51 N8 P943-951 66-61302 <=*$>
1965 V51 N8 P952- FPTS V25 N3 P.T433 <FASE>
1965 V51 N8 P966-973 65-33643 <=*$>
1965 V51 N9 P1037- FPTS V25 N4 P.T543 <FASE>
1965 V51 N9 P1057-1065 65-33425 <=$>
1965 V51 N9 P1080- FPTS V25 N4 P.T589 <FASE>
1965 V51 N9 P1094-1099 66-60568 <=*$>
1965 V51 N9 P1116-1122 66-61323 <*=$>
1965 V51 N9 P1133-1136 65-33425 <=$>
1965 V51 N9 P1141-1142 65-33425 <=$>
1965 V51 N10 P1169- FPTS V25 N5 P.T734 <FASE>
1965 V51 N10 P1188- FPTS V25 N5 P.T763 <FASE>
1965 V51 N10 P1192- FPTS V25 N5 P.T743 <FASE>
1965 V51 N10 P1231- FPTS V25 N5 P.T739 <FASE>
1965 V51 N12 P1397-1402 66-61998 <=*$>
1965 V51 N12 P1420- FPTS V25 N6 P.T937 <FASE>
1965 V51 N12 P1434- FPTS V25 N5 P.T729 <FASE>
1965 V51 N12 P1442- FPTS V25 N6 P.T931 <FASE>
1965 V51 N12 P1474-1477 66-31137 <=$>
1965 V51 N12 P1501-1506 66-31137 <=$>
1966 V52 P172-178 66-31425 <=$>
1966 V52 P201-204 66-31425 <=$>
1966 V52 N1 P40- FPTS V25 N6 P.T927 <FASE>
1966 V52 N1 P99- FPTS V25 N6 P.T948 <FASE>
1966 V52 N2 P131- FPTS V25 N6 P.T957 <FASE>

FIZIOLOGICHNII ZHURNAL
SEE FYZYOLOHICHNII ZHURNAL

*FIZIOLOGIYA RASTENII
1954 V1 N2 P156-163 R-4146 <*>
1955 V2 N4 P320-333 63-11029 <=>
1955 V2 N4 P346-353 60-51025 <=>
1955 V2 N4 P354-357 60-23207 <*+>
1955 V2 N4 P373-377 63-11009 <=>
1955 V2 N5 P505-508 65-10030 <*>
1955 V2 N6 P539-548 CSIRO-3186 <CSIR>
R-4376 <*>
1955 V2 N6 P549-557 CSIRO-3187 <CSIR>
R-3939 <*>
1955 V2 N6 P558-564 R-3170 <*>
1955 V2 N6 P565-572 59-19601 <+*>
1955 V2 N6 P573-577 5044-A <K-H>
1956 V3 N1 P79-86 61-14952 <=*>
1956 V3 N1 P90-93 60-21880 <=>
1956 V3 N2 P121-124 59-12757 <+*> 0
1956 V3 N4 P306-312 63-19600 <=*>
1956 V3 N4 P343-351 R-3152 <*>
1956 V3 N6 P534-538 R-3175 <*>
85R75R <ATS>
1957 V4 P393-396 R-5264 <*>
1957 V4 N2 P124-133 C-2721 <NRC>
R-4302 <*>
1957 V4 N4 P320-325 64-11006 <=>
1959 V6 P474-475 61-10924 <+*>
1959 V6 N2 P183-189 62-23721 <=*> 0
1959 V6 N4 P474-475 9793-A <K-H>
1960 V7 N2 P247-255 60-41332 <=>
1960 V7 N3 P340-357 60-41618 <=>
1960 V7 N5 P610-614 62-25645 <=*>
1960 V7 N6 P724-726 66-10239 <*>
1960 V7 N6 P734-735 65-61245 <=$>
1961 V8 N1 P141-142 62-24401 <=>
1961 V8 N3 P358-360 63-31040 <=>
1961 V8 N4 P518-520 62-23774 <=*>
1962 V9 N2 P129-132 62-33046 <=*>
1962 V9 N3 P270- AEC-TR-5491 <=*>

FIZKULTURA I SPORT. MOSCOW
1965 N9 P26-27 65-33211 <=$>
1966 N1 P29-31 66-30836 <=$>

FIZYKA NAUKOVI POVIDOMLENNIA KIEV UNIVERSITET
1956 N1 P15-17 SCL-T-438 <=*>

FLACARA
1964 V13 P4-7 64-41480 <=>
1966 V15 N19 P10-12 66-32978 <=$>

FLAMME ET THERMIQUE
1949 N12 P7-14 I-936 <*>
1964 V3 N25 P57-69 65-10413 <*> 0

FLEISCHWIRTSCHAFT
1950 V2 N12 P273-274 60-10801 <*>
1955 V7 N10 P589-592 CSIRO-3246 <CSIR>
1956 V8 N5 P266-271 CSIRO-3216 <CSIR>

1956 V8 N10 P600-602 CSIRO-3400 <CSIR>
1956 V8 N12 P733-736 CSIRO-3464 <CSIR>
1957 V9 P552-554 58-2511 <*>
1957 V9 N11 P687-691 10K23G <ATS>
1957 V9 N11 P696-701 CSIRO-3855 <CSIR>
1958 V10 N3 P162-164 59-15344 <*>
1958 V10 N6 P405-407 59-10223 <*>
1959 V11 N5 P361-366 60-17499 <+*>
1959 V11 N6 P453-454 60-17500 <+*>
1960 V12 P443-446 2209 <BISI>

FLIEGER
1958 N10 P415-417 59-21055 <=> O

FLORA, ODER ALLGEMEINE BOTANISCHE ZEITUNG
1864 V47 P65-75 59-14877 <=*>
1923 V116 P73-84 64-10968 <*> O
1955 V142 P295-306 65-12689 <*>
1956 V144 P105-120 60-18026 <*>
1961 V150 N1 P73-86 11095-E <K-H>
63-16682 <*>
1961 V151 P368-397 63-10192 <*>

FLORA SSSR
1935 V3 P629 61-31221 <=>
1961 V150 N2/3 P175-178 NP-TR-995 <*>

FLORES DO BRASIL
1957 V2 N3 P129-131 63-00308 <*>

FLUGKOERPER
1960 V2 N6 P185-193 61-18823 <*>

FLUG-REVUE
1965 P29-33 65-17374 <=*$>

FLUGWEHR UND TECHNIK
1940 V2 N9 P206-210 58-988 <*>
1943 V5 N12 P328-330 59-17728 <*>
1946 N12 P307-311 63-20559 <*>
1950 N7 P155- 58-777 <*>

FLUGWELT
1953 P349-353 57-702 <*>
58-680 <*>
1953 V2 P349-353 64-71465 <=>
1956 V8 N8 P487-490 59-12386 <+*>
1956 V8 N8 P516 59-12386 <+*>
1956 V8 N8 P519-521 59-12386 <+*>
1957 V9 N2 P96-98 C-3658 <NRCC>
1963 N6 P494-497 66-11789 <*>

FOCUS. BLOEMENDAAL
1955 V40 N9 P194-198 63-10313 <*> O

FOERDERN UND HEBEN
1956 V6 N10 P1050-1054 59-10432 <*> O
1958 V8 P58-61 TS 1269 <BISI>
61-20619 <*>
1960 N2 P91-93 61-00220 <*>
1964 V14 P151-154 4189 <BISI>
1964 V14 P243-248 3904 <BISI>
1964 V14 P404-406 4189 <BISI>
1964 V14 P630-633 4189 <BISI>
1965 V15 P89-99 4551 <BISI>

FOERDERTECHNIK UND FRACHTVERKEHR
1930 P191-193 60-10282 <*> O
1935 V28 N3/4 P31-34 59-20812 <*>

FOERDERUNGSDIENST
1957 V5 P65-70 59-20161 <*>

FOGORVOSI SZEMLE
1965 V58 P42-48 66-12954 <*>
1965 V58 P100-106 66-12839 <*>

FOGRA MITTEILUNGEN
1961 N29 P14-22 63-10169 <*> O

FOLDRAJZI ERTESITO
1960 V9 N2 P129-138 62-19426 <=*>
1961 V10 N1 P117-121 62-19425 <=*>
1961 V10 N2 P153-169 62-19499 <=*>
1962 V11 N1 P1-13 62-25060 <=*> P
1962 V11 N2 P203-215 63-31664 <=>
1962 V11 N4 P487-497 63-31623 <=>

FOLDRAJZI KOZLEMENYEK
1958 V4 N3 P237-261 59-13487 <=> O
1960 V8 N4 P372-374 62-19424 <=*>
1961 V9 N2 P109-121 62-19494 <=*> P
1961 V9 N2 P153-168 62-15365 <=*>
1961 V9 N2 P176-178 62-19503 <=*>

FOLDTANI KOZLONY
1955 V85 N3 P386-389 AEC-TR-4084 <*>
1956 V86 P284-286 62-14625 <*>

FOLHA MEDICA
1937 V18 P46- 57-1690 <*>
1953 V34 P153-154 AL-851 <*>
1954 V35 P63-65 I-789 <*>
1955 V36 P129- 59-10830 <*> O
1959 V40 P44- 60-14540 <*>
1960 V41 P155-157 62-20039 <*>
1962 V44 P321-325 63-10659 <*>

FOLIA ALLERGOLOGICA
1954 V1 N3/4 P249-258 1587 <*>
1954 V1 N3/4 P259-265 II-980 <*>

FOLIA BIOLOGICA. BUENOS AIRES
1938 N79 P347-351 60-10692 <*>

FOLIA BIOLOGICA. PRAHA
1955 V1 N1 P29-40 65-29146 <$>
1956 V2 N2 P100-111 60-21112 <=>

FOLIA CARDIOLOGICA
1951 V10 N4 P261-272 2821C <K-H>
1951 V10 N6 P419-422 63-01612 <*>
1952 V10 P272-282 SUP 3178-B <K-H>
1960 V19 N3 P193-207 66-11803 <*>

FOLIA CLINICA INTERNACIONAL
1952 V2 N4 P160-164 2504 <*>
1956 V6 P61-72 5448-B <KH>

FOLIA ENDOCRINOLOGICA
1952 V5 P593-624 58-1773 <*>
1954 V7 P135-163 II-862 <*>
1955 V8 N4 P605-609 62-20443 <*> O
1961 V14 P266-271 66-12540 <*>
1962 V15 N3 P363-369 66-12633 <*>
1963 V16 N2 P151-167 66-10935 <*>
1964 V17 N4 P505-514 65-14774 <*>

FOLIA ENDOCRINOLOGICA JAPONICA
SEE NIPPON NAIBUNPI GAKKAI ZASSHI

FOLIA FORESTALIA POLONICA. SERIA B. DRZEWNICTWO
1959 SB V1 P103-154 62-00071 <*>

FOLIA HAEMATOLOGICA. LEIPZIG
1906 N3 P332-339 58-335 <*>
1906 N3 P339-343 58-332 <*>
1913 V15 N3 P383-390 65-00100 <*>
1960 V77 P495-501 66-11587 <*>

FOLIA HEREDITARIA ET PATHOLOGICA
1964 V13 N4 P395-410 66-10930 <*>
1964 V14 N1 P35-47 66-11603 <*> O

FOLIA MEDICA. NAPOLI
1932 V18 P594-613 810A <K-H>
1937 V23 P729-748 3065 <*>
65-14962 <*> O
1942 N11 P393-397 58-402 <*>
1947 V30 P257-263 60-10878 <*>

1952 V35 P471-756 57-1286 <*>
1952 V35 N11 P741-756 57-2296 <*>
1954 V37 P1095-1099 <CP>
1955 V38 P1014-1024 4036-A <KH>
1959 V42 N9 P1025-1060 80P61I <ATS>
1964 V47 P238-248 65-18115 <*>

FOLIA MEDICA NEERLANDICA
1959 V2 N2 P79-90 61-00391 <*>

FOLIA OPHTHALMOLOGICA JAPONICA
SEE NIPPON GANKA KIYO

FOLIA PHARMACOLOGICA JAPONICA
SEE NIPPON YAKURIGAKU ZASSHI

FOLIA PHONIATRICA
1951 V3 N3 P170-177 65-12554 <*>
1958 V10 N1 P44-50 61-10339 <*>
1958 V10 N3 P154-166 61-10450 <*> O
1964 V16 N3 P228-238 65-12129 <*>

FOLIA PSYCHIATRICA ET NEUROLOGICA JAPONICA
SEE NIHON SEISHIN SHINKEI GAKKAI

FOLLETO. SERVICIO NACIONAL DE INVESTIGACION AGRICOLA. PANAMA
1956 V25 P1-10 57-3198 <*>
58-1026 <*>

FONDERIA: RASSEGNA TECNICA DELL' INDUSTRIA FUSORIA. MILANO
1957 V6 P19-21 62-20076 <*>
1959 P47-51 TS-1325 <BISI>
1959 N162 P323-325 57M47F <ATS>
1964 V13 P137-140 3955 <BISI>

FONDERIA ITALIANA
1959 V8 N2 P47-51 62-10522 <*>
1961 V10 N8 P317-323 39N58I <ATS>
1965 V14 P385-386 4662 <BISI>

FONDERIE
1948 N7 P1217-1235 61-10012 <*>
1948 N32 P1271-1282 62-18606 <*> O
1949 N48 P1873-1875 59-15861 <*> O
1951 N67 P2565-2577 60-18803 <*> O
1953 V85 P3307-3312 I-189 <*>
1954 V107 P4281-4284 64-20533 <*>
1955 N109 P4353-4372 62-14906 <*>
1956 N125 P232-237 61-10580 <*> O
1956 N131 P496-508 66-10907 <*> O
1958 N148 P125- 61-10642 <*> PO
1958 N149 P249-262 61-16157 <*>
1958 N150 P319-337 61-16156 <*>
1958 N151 P374-384 63-20361 <*>
1958 N155 P581-585 4128 <BISI>
1959 N167 P531-536 60-18774 <*> O
1960 N169 P74-76 62-10283 <*>
1960 N172 P206-209 61-20831 <*>
1960 N179 P512-519 61-20832 <*>
1961 N181 P70-84 62-10218 <*>
1961 N184 P171-177 2837 <BISI>
1961 N185 P209-219 2810 <BISI>
1961 N187 P317-327 80P62F <ATS>
1961 N189 P385-396 2666 <BISI>
1962 N191 P1-14 62-18898 <*>
1963 N204 P54-62 4162 <BISI>
1964 P431-448 4479 <BISI>
1964 N215 P1-16 3859 <BISI>
1964 N222 P255-268 4248 <BISI>
1965 N233 P240-245 6770 <HB>

FONDERIE BELGE
1957 N5 P92-105 2465 <BISI>
1960 V30 N12 P313-325 62-18904 <*>

FORCES AERIENNES FRANCAISE
1949 N37 P1- AL-149 <*>
1960 P45-66 UCRL TRANS-663(L) <*>
1963 N193 P883-896 AD-632 862 <=$>

FORHANDLINGER I DET NORSKE VIDENSKABERS SELSKAB. TRONDHJEM
1929 V1 N45 P131-134 64-18858 <*>
1935 N19 P66-69 1459 <*>
1939 V12 P153-155 AL-248 <*>

FORMOSAN FISHERIES MAGAZINE
SEE TAIWAN SUISAN ZASSHI

FORSCHUNG AUF DEM GEBIETE DES INGENIEURWESENS
1931 V2 P109 58-2114 <*>
1931 V2 P213-217 820 <TC>
1931 V2 N5 P165-178 59-10282 <*>
1931 V2 N6 P213-217 59-10282 <*>
1931 V2 N11 P395-407 64-16996 <*> P
1932 V3 N4 P177-180 65-12016 <*>
1932 V3 N5 P277-286 64-10331 <*>
1933 V4 P128-137 58-2105 <*>
63-14510 <*> OP
1933 V4 N1 P27-30 60-16760 <*> O
1933 V4 N2 P53-63 AD-608 384 <=>
1933 V4 N4 ENTIRE ISSUE 1970 <*>
1933 V4 N4 P193-198 58-1037 <*>
58-919 <*>
1934 V5 P186-191 63-18154 <*>
1934 V5A P105-117 58-2404 <*>
1935 V6 P9-22 PT.A II-458 <*>
1935 V6 N4 P175-183 60-13123 <=*> O
1935 V6 N5 P240-244 58-471 <*>
63-16034 <*>
1935 V6 N6 P273-280 R-500 <*>
58-475 <*>
1936 V7 N2 P66-74 58-644 <*>
1936 V7 N3 P140-157 60-14596 <*> O
1937 V8 N3 P118-130 63-20387 <*>
1938 V9 N2 P57-67 II-549 <*>
1938 V9 N6 P290-306 58-871 <*>
1939 V10 N1 P2-14 58-643 <*>
1939 V10 N1 P15-27 58-699 <*>
1939 V10 N5 P201-211 60-10048 <*>
1940 V11 P67-75 65-12958 <*>
1940 V11 P334-339 58-866 <*>
1940 V11 N2 P58-66 65-12965 <*>
1940 V11 N4 P149-158 AL-471 <*>
1940 V11 N5 P237-245 65-12959 <*>
1941 V12 P16-20 66-14429 <*>
1942 V13 N5 P177-185 1286 <*>
1943 V14 P77-81 66-11092 <*> O
1943 V14 P124-131 58-932 <*>
1943 V14 N1 P24-29 58-692 <*>
1943 V14 N1 P30-34 59-17335 <*>
1943 V14 N3 P82-84 58-1152 <*>
58-651 <*>
1943 V14 N3 P85-87 58-1153 <*>
1943 V14 N5 P113-123 58-928 <*>
1943 V14 N6 P148-158 58-864 <*>
1944 V15 P1-11 65-13337 <*>
1944 V15 N1 P12-17 64-11817 <=>
1949 V16 N2 P33-42 57-1319 <*>
64-10308 <*>
1951 V17 N1 P1-8 1750 <*>
1951 V17 N3 P65-76 59-20075 <*> O
1952 V18 P17-28 58-1087 <*>
1952 V18 P515- 58-1139 <*>
1952 V18 N1 P25- 3215 <*>
1952 V18 N4 P97-105 58-1103 <*>
1953 V19 N3 P65-80 II-1287 <*>
65-13357 <*>
1953 V19B N4 P101-104 81G5G <ATS>
1954 V20 P155-157 2994 <*>
1954 V20 N3 P81-93 57-2697 <*>
1955 V21 N1 P1-8 64-16122 <*> O
1955 V21 N2 P50-62 57-1442 <*>
1956 V22 N1 P1-8 66-13022 <*>
1956 V22 N2 P56-62 57-2725 <*>
1957 V23 N3 P81-90 60M37G <ATS>
1957 V23 N1/2 P49-54 59-10331 <*> O
1958 V24 N2 P50-58 2246 <BISI>

1958 V24 N2 P59-69 66-11676 <*> 0
1958 V24 N5 P137-148 59-17262 <*>
1958 V24 N6 P169-177 86N52G <ATS>
1958 V24 N6 P187-192 59-17262 <*>
1959 V25 P37-43 2558 <BISI>
1959 V25 N1 P17-25 87N52G <ATS>
1959 V25 N1 P26-32 64-30290 <*> 0
1959 V25 N2 P44-54 TS 1639 <BISI>
62-16890 <*>
1959 V25 N2 P55-63 64-30290 <*> 0
1965 V31 N4 P115-118 NEL-TRANS-1789 <NLL>

FORSCHUNGEN UND FORTSCHRITTE
1937 V13 N35/6 P425-426 65-12618 <*>
1949 V25 N11/2 P126-128 61-16038 <*>
1950 V26 P21-22 SPEC NO 64-14387 <*>
1950 V26 P21-22 II-228 <*>
1955 V29 N7 P193-198 38M46G <ATS>
1961 V35 N9 P262-264 71P59G <ATS>

FORSCHUNGEN AUS DEM GEBIET DER PFLANZENKRANKHEITEN
1955 V5 P83-84 C-2590 <NRC>

FORSCHUNGEN ZUR GESCHICHTE DER OPTIK
1936 V2 P251-252 61-14390 <*>
1939 V3 N1 P21-22 61-14395 <*>

FORSCHUNGSARBEITEN AUF DEM GEBIET DES INGENIEURWESENS
1923 N263 P5-14 58-650 <*>
1923 N263 P43-46 58-650 <*>
1926 N300 P15-32 58-862 <*>
1930 N330 P1-31 59-11904 <=>

FORSCHUNGSARBEITEN AUS DEM STRASSENWESEN
1954 N5 P58-61 1781 <*>
1955 N6 P55-60 57-620 <*>
1958 N37 P30-41 62-18752 <*>

FORSCHUNGSBERICHTE DES LANDES NORDRHEIN-WESTFALEN
1958 V499 P5-22 61-16736 <*>
1960 N822 P13-32 65-11657 <*> 0
1960 N897 P72-75 63-10899 <*>
1961 N973 P1-27 65-14507 <*>
1964 N1347 P27- 4470 <BISI>
1964 N1355 P9-14 4347 <BISI>
1964 N1413 P22- 4420 <BISI>

FORSCHUNGSBERICHTE DES WIRTSCHAFTS- UND VERKEHRS-MINISTERIUMS, NORDRHEIN-WESTFALEN. KOELN
1958 N461 P1-33 63-00760 <*>

FORSCHUNGSGESELLSCHAFT FUER DAS STRASSENREIHE DER ARBEITSGRUPPE BETONSTRASSEN
1955 N6 P22-25 II-1310 <*>

FORSCHUNGSHEFTE. VEREIN DEUTSCHER INGENIEURE
1933 N361 ENTIRE ISSUE 1972 <*>
1934 N367 ENTIRE ISSUE 57-599 <*>
63-14926 <*>
1934 V5 N364 P20 58-1909 <*>
1935 N375 P9-20 65-10159 <*> 0
1936 N379 ENTIRE ISSUE 63-10867 <*=> 0
63-10867 <=*> 0
1939 N397 P1-20 59-17725 <*>
1951 N431 P5-31 59-20074 <*>
1952 N434 P1-15 60-10234 <*> 0
1954 N441 P1-19 58-1086 <*>
1954 N442 P1-19 66-14427 <*>
1954 N442 P19-48 66-14428 <*>
1954 V20 N441 ENTIRE ISSUE
57-2012 <*>
1954 V20 N444 ENTIRE ISSUE
58-872 <*>
1954 V20 N445 P1-44 17G6G <ATS>
1954 V442 P1-24 1751 <*>
1954 V442 P25-48 1753 <*>
1954 V445 P1-44 II-398 <*>
1955 N447 ENTIRE ISSUE 59-16784 <*+>
1955 V21 N451 ENTIRE ISSUE
59-16776 <+*>
1956 N455 ENTIRE ISSUE 58-2142 <*>
1956 N455 P5-39 58-2142 <*>
1957 N459 ENTIRE ISSUE 59-10063 <*>
1957 N459 P1- 61-10643 <*>
1957 N462 P7-13 60-10047 <*> 0
1958 N469 P1-44 AEC-TR-3875 <=>
1958 V24 N467 P32 65-13419 <*>
1961 V27 N483 P1-7 ANL-TRANS-126 <*>

FORSTARCHIV
1929 V5 P321-326 57-1240 <*>
1955 V7 P149-157 C-2712 <NRC>
1955 V12 N15 P274-283 C-2700 <NRC>
1957 V28 N11 P223-227 59-17233 <*>

FORSTLIGE FORSOKSVAESEN I DANMARK
1932 V12 P343-380 57-1224 <*>

FORST- UND HOLZWIRT
1956 V11 P427-429 C-2678 <NRC>
1965 V20 N16 P9 C-6002 <NRC>
1965 V20 N17 P382-386 C-6003 <NRC>

FORSTWISSENSCHAFTLICHES ZENTRALBLATT
1939 V61 P329-342 CSIRO-3054 <CSIR>
1950 V69 N8 P441-465 59-10854 <*> 0
1950 V69 N8 P441-466 64-20837 <*> 0
1952 V71 N11/2 P322-349 CSIRO-2905 <CSIR>
59-10921 <*>
1956 V75 N9/10 P257-512 C-2704 <NRC>

FORTSCHRITTE DER BOTANIK
1952 V15 P313-347 3902-E <K-H>

FORTSCHRITTE DER CHEMIE ORGANISCHER NATURSTOFFE
1952 V9 P1-40 65-12545 <*>
1956 V13 P444-459 CSIRO-3513 <CSIR>

FORTSCHRITTE DER CHEMISCHEN FORSCHUNG
1951 V2 P311-374 63-18689 <*>
63-20637 <*> P
1952 V2 N3 P484-527 I-438 <*>
1953 V2 N4 P670-757 59-17538 <*> 0
1955 V3 P430-502 58-572 <*>
1955 V3 P603-656 57-659 <*>

FORTSCHRITTE UND FORSCHUNGEN IM BAUWESSEN
1942 SB N2 P74-79 AL-129 <*>
1956 N24 P5-14 C-3412 <NRCC>
1956 N24 P15-31 C-3413 <NRCC>

FORTSCHRITTE AUF DEM GEBIETE DER ROENTGENSTRAHLEN
1906 V10 N3 P155-160 58-1518 <*>
1936 V54 N6 P590-595 59-17478 <*>
1949 V71 N1 P134-142 II-37 <*>
1950 V73 P492-500 58-753 <*>
1951 V74 P354-358 63-20562 <*> C
1951 V75 N6 P700-712 65-00321 <*> 0
1954 V80 N6 P754-762 II-535 <*>
57-780 <*>
1954 V81 P55-58 57-373 <*>
1955 V82 N4 P504-505 II-892 <*>
1955 V82 N5 P618-625 63-10712 <*> 0
1955 V83 N3 P396-402 II-889 <*>
1955 V83 N4 P476-489 II-936 <*>
1955 V83 N4 P580-584 1229 <*>

FORTSCHRITTE AUF DEM GEBIETE DER ROENTGENSTRAHLEN UND DER NUKLEARMEDIZIN
1956 V84 N4 P447-451 1506 <*>
1956 V85 P433-447 1401 <FT>
1956 V85 N2 P207-211 57-1740 <*>
1958 V88 N4 P465-474 59-17229 <*>
1958 V89 N6 P746-752 C-3137 <NRCC>
1959 V90 N1 P101-109 TT-823 <NRC>
1959 V91 N5 P643-648 65-10106 <*>
1961 V94 N2 P243-260 NP-TR-684 <*>
1961 V94 N3 P409-420 NP-TR-813 <*>
1962 V97 N1 P63-81 AEC-TR-5608 <*>

FORTSCHRITTE IN DER GEOLOGIE VON RHEINLAND UND WESTFALEN
1963 V10 P239-253 GI N1 1964 P105 <AGI>

FORTSCHRITTE DER HOCHPOLYMEREN-FORSCHUNG
1960 V2 P363-400 61-18371 <*>
1961 V3 N1 P59-105 65-11500 <*>
1963 V3 N3 P394-507 N66-18127 <=$>

FORTSCHRITTE DER KIEFER-UND GESICHTS-CHIRURGIE
1957 V3 P314-318 58-1562 <*>
1960 N6 P194-200 62-00599 <*>
1960 V6 P1-15 62-00872 <*>

FORTSCHRITTE DER KIEFERORTHOPAEDIE
1954 V15 P205-212 57-1157 <*>

FORTSCHRITTE DER LANDWIRTSCHAFT
1928 V3 N10 P441-456 CSIRO-2914 <CSIR>
1930 V23 N5 P797-800 57-1228 <*>

FORTSCHRITTE DER MEDIZIN
1895 V13 N18 P725-738 3344-D <K-H>
1895 V13 N18 P765-789 3344-D <K-H>
1953 V71 N2 P31-32 65-11334 <*> O
1954 V72 N7/8 P167-168 II-476 <*>
1957 V75 N4 P91-92 64-20040 <*>
1959 V77 P181-182 63-16389 <*>
1961 V79 N19 P499-504 62-00482 <*>
1963 V81 N20 P799-800 65-10978 <*>
1964 V82 P913-916 65-13942 <*>

FORTSCHRITTE DER MINERALOGIE
1956 V34 P40-43 61-20719 <*> O
1956 V34 N2 P85-150 60-16501 <*>
1957 V35 P30-33 03S83G <ATS>
1961 V39 N1 P148-149 62-20296 <*>
1965 V42 N1 P87-112 66-12681 <*>

FORTSCHRITE DER MINERALOGIE, KRISTALLOGRAPHIE UND PETROGRAPHIE
1936 V20 P168-195 58-1620 <*>

FORTSCHRITTE DER NEUROLOGIE UND PSYCHIATRIE UND IHRER GRENZGEBIETE
1951 V19 P386-452 57-739 <*>
1956 V24 P7-24 59-14315 <=*>
65-00196 <*> O
1957 V25 N5 P265-297 60-15903 <=*> PO
1962 V30 P155-165 63-20227 <*>

FORTSCHRITTE DER PHYSIK
1958 V6 N6 P271-334 62-10632 <*> O
1959 V7 N12 P641-674 AEC-TR-5465 <*>
1961 V9 P393-454 63-10771 <*>
63-18716 <*>
1963 V11 P325-356 65-00088 <*>
66-11119 <*>
1964 V12 P271-346 N66-11606 <=*$>
1964 V12 N11 P597-642 65-12788 <*> P

FORTSCHRITTE DER PSYCHOLOGIE UND IHRER ANWENDUNGEN
1917 V4 P257-326 61-14374 <*>

FORTSCHRITTE DER THERAPIE
1939 V15 P540-543 60-10780 <*>

FORTSCHRITTE DER VERFAHRENSTECHNIK
1962 V6 P509-514 66-13512 <*> O

FORTSCHRITTSBERICHTE DER CHEMIKER-ZEITUNG
1928 V3 P102-104 3879-B <K-H>

FOTOGRAFIE. HALLE
1960 V14 N7 P259-260 61-20057 <*>
1960 V14 N7 P279-281 61-20057 <*>

FRAGMENTA FAUNISTICA
1931 V1 N13 ENTIRE ISSUE 61-31264 <=>
1947 V5 N2 P25-27 61-31262 <=>

FRANCE MEDICALE. REVUE GENERALE DE MEDECINE ET CHIRURGIE
1954 V17 N11 P23-26 3730 <K-H>

FRANCE. MEDICAL STANDARDS
1963 NFC 74-300 66-11608 <*> O

FRANCE, STANDARDS
1953 NFT 12-005 66-13805 <*>
1958 RTF SN 401 C 62-16727 <*>
1961 NF Q 03-020 62-16512 <=*>

FRANKFURTER ALLGEMEINE
1965 P29 65-33926 <=$>

FRANKFURTER ZEITSCHRIFT FUER PATHOLOGIE
1911 V7 P83-111 66-10968 <*>
1913 V13 P45-76 9500-B <K-H>
1913 V14 P359-366 57-1498 <*>
1921 V26 P356-381 65-00415 <*>
1924 V30 P377-394 65-13744 <*>
1934 V46 P513-522 2864A <K-H>
1935 V47 P52-62 2864B <K-H>
1938 V52 P114-124 66-10993 <*>
1947 V59 P143-150 64-16213 <*> O
1950 V61 N3 P417-429 I-858 <*>
1953 P522-531 57-1660 <*>
1953 V64 P252-260 2383 <*>
1953 V64 P405-437 9614 <K-H>
1955 V66 P201-225 57-3431 <*>
1956 V67 P599-620 58-261 <*>
1957 V68 P49-54 58-1796 <*>
1957 V68 P404-409 63-20910 <*>
1957 V68 N4 P361-382 63-18467 <*> O
6821-B <K-H>
1958 V68 P710-727 61-00195 <*>
1959 V69 P685-694 62-01031 <*>

FREIBERGER FORSCHUNGSHEFTE. REIHE A. BERGBAU
1952 N14 P12-18 66-14277 <*> O
1952 V14 P12-18 64-20049 <*>
1953 V17 P29-37 <DIL>
1953 V17 P38-43 <DIL>
1954 V23 P5-12 <DIL>
1954 V23 P90-101 CSIRO-3095 <CSIR>
1955 V33 P24-31 T1863 <INSD>
1955 V36 P31-41 <DIL>
1956 V51 P35-65 59-22341 <+*> O
1956 V51 P89-108 60-10788 <*> OP
1956 V51 P109-126 61-10422 <*>
1956 V51A P35-65 58-2501 <*>
1958 V100 ENTIRE ISSUE 63-16268 <*> O
1960 V140 P48-73 TRANS-553 <MT>
1961 V222 P51-60 63-18025 <*> O
1961 V267 P195-205 9042 <IICH>
1963 V296 P431-440 66-10826 <*>
1963 V246A P109-117 65-00354 <*>
1963 V270A P19-31 65-00279 <*>

FREIBERGER FORSCHUNGSHEFTE. REIHE B. HUETTENWESEN, METALLURGIE
1956 V7 P80-108 61-16220 <*>
1956 V11 P218-267 62-20077 <*>
1956 V17 P34-49 62-20078 <*>
1956 V17B P18-33 NP-TR-1130 <*>
1960 V48 P5-25 1941 <BISI>
1960 V48 P26-56 1942 <BISI>
1960 V48 P59-88 1943 <BISI>
1960 V48 P130-141 TRANS-117 <MT>
1961 V44 P79-91 2841 <BISI>
1961 V44 P92-103 2842 <BISI>
1961 V44 P104-115 2843 <BISI>
1961 V61 P5-17 4392 <BISI>
1961 V62 P1-36 3761 <BISI>
1962 V68 P5-18 3192 <BISI>
1962 V68 P19-26 3193 <BISI>
1962 V68 P47-70 3050 <BISI>
1962 V68 P71-79 3051 <BISI>
1962 V68 P81-103 3716 <BISI>

1962 V69 P5-17 3690 <BISI>
1963 V76 P51-71 3765 <BISI>
1963 V76 P73-85 6161 <HB>
1963 V76 P147-148 6160 <HB>
1963 V95 P139-148 6198 <HB>
1964 V88 P49-80 4233 <BISI>
1964 V96 P43-55 4078 <BISI>

FREIBERGER FORSCHUNGSHEFTE. REIHE C. ANGEWANDTE NATURWISSENSCHAFTEN: GEOPHYSIK
1957 V32 P5-19 62-16423 <*>
1961 V114 P1-98 63-20724 <*>
1963 V181 P109-116 226 <TC>

FREQUENZ
1947 V1 N1 P2-15 59-10786 <*>
1947 V1 N1 P16-22 58-668 <*>
64-71625 <=>
1949 V3 N8 P244-249 63-10028 <*> 0
1950 V4 P193-195 62-19617 <*> 0
1950 V4 N10 P271-278 57-3309 <*>
1950 V4 N11 P289-294 57-3309 <*>
1951 V5 P27-29 C-2194 <NRC>
1375 <*>
1951 V5 N6 P145-155 59-10995 <*>
63-10001 <*> 0
1951 V5 N7 P186-190 63-10001 <*> 0
1952 V6 N9 P257-266 62-18357 <*> 0
1952 V6 N5/6 P168-182 57-2261 <*>
1954 V8 N6 P180-182 57-2763 <*>
1954 V8 N10 P306-313 57-2752 <*>
96H8G <ATS>
1955 V9 N1 P15-25 57-1065 <*>
1955 V9 N1 P25-28 57-2747 <*>
1955 V9 N6 P199-205 57-359 <*>
1955 V9 N8 P264-273 57-1169 <*>
1955 V9 N9 P319-324 57-1769 <*>
1956 V10 N5 P137-142 89J19G <ATS>
1957 N1 P27- 3427 <*>
1958 V12 N5 P159-163 62-19918 <=>
1958 V12 N6 P191-198 62-19918 <=>
1958 V12 N9 P273-284 59-10630 <*>
1958 V12 N11 P348-353 66-11689 <*> 0
1958 V12 N11 P353-355 63-10025 <*> 0
1959 V13 N2 P52-56 63-18167 <*>
1959 V13 N4 P97-102 60-18920 <*>
1960 V14 N5 P157-162 C-3845 <NRCC>
1960 V14 N5 P167-181 65-12223 <*>
78R80G <ATS>
1961 V15 N2 P39-47 65-13140 <*>
1961 V15 N4 P111-121 63-01020 <*>
1961 V15 N4 P111-122 63-18084 <*> 0
1961 V15 N5 P141-155 63-20616 <*>
1961 V15 N10 P324-332 62-18861 <*>
1962 V16 N3 P69-76 63-18085 <*>
1962 V16 N3 P91-96 62-18840 <*>
1962 V16 N4 P121-125 SCL-T-460 <*>
1962 V16 N9 P347-354 65-10365 <*>
1962 V16 N10 P391-401 65-12302 <*>
1962 V16 N11 P423-428 65-00324 <*>
1962 V16 N11 P442-453 65-12302 <*>
1963 V17 N1 P23-32 64-14640 <*>
1963 V17 N3 P88-94 66-14394 <*>
1963 V17 N3 P103-107 64-16281 <*>
1963 V17 N3 P113-122 39R74G <ATS>
64-16926 <*>
1963 V17 N4 P133-139 64-13747 <=*$>
1963 V17 N4 P158-164 40R74G <ATS>
64-16949 <*>
1963 V17 N5 P165-172 64-16588 <*>
1963 V17 N5 P189-200 64-14626 <*>
1963 V17 N9 P319-328 64-15951 <*>
1963 V17 N10 P364-370 65-10284 <*>
1963 V17 N11 P417-423 65-11456 <*>
1963 V17 N11 P462-465 65-12224 <*>
1964 V18 N1 P14-20 65-10286 <*> 0
1964 V18 N6 P183-186 65-12229 <*>
1964 V18 N11 P358-367 34S88G <ATS>
1964 V18 N11 P367-374 65-17140 <*>
1964 V18 N12 P395-409 35S88G <ATS>
1965 V19 N5 P156-162 POED-TRANS-2214 <NLL>
1965 V19 N8 P265-282 27T90G <ATS>

FRUITS
1958 V13 P269- 59-21134 <=*> 0
1963 V18 N2 P69-70 64-18187 <*>

FRUKT I AR
1955 P184-185 II-471 <*>

FUEHLINGS LANDWIRTSCHAFTLICHE ZEITUNG
1920 V69 N9/10 P161-166 63-16128 <*>

FUJI SEITETSU GIHO
1962 V11 P326-328 3470 <BISI>
1963 V12 N2 P162-174 CRNL-TR-350 <*>
1963 V12 N4 P445-453 3877 <BISI>
1963 V12 N4 P519 64-30064 <*>
1964 V13 N2 P269-277 4589 <BISI>

FUJIKURA TECHNICAL REVIEW
SEE FUJIKURA DENSEN GIHO

FUKUI-KEN NOJI SHIKENJO HOKOKU
SEE FUKUI-KEN NOGYO SHIKENJO HOKOKU

FUKUOKA ACTA MEDICA
SEE FUKUOKA IGAKU ZASSHI

FUKUOKA IGAKU ZASSHI
1929 V22 N7 P1519-1537 1427 <K-H>

FUKUSHIMA MEDICAL JOURNAL. JAPAN
SEE FUKUSHIMA IGAKU ZASSHI

FUNK UND TON
1948 N9 P435-442 58-06 <*>
1948 N11 P564-578 1748 <*>
1948 V2 P607-621 57-2162 <*>
1949 V3 N8 P423-428 57-912 <*>
1951 V5 N2 P65-75 62-18243 <*> 0
1951 V5 N2 P76-78 62-16971 <*> 0
1951 V5 N7 P374-376 63-10034 <*>
1952 V6 N1 P25-28 II-741 <*>
1953 N7 P333-341 57-2737 <*>
1954 V8 N2 P79-86 57-2392 <*>
1954 V8 N4 P169-173 57-552 <*>
1954 V8 N4 P202-214 57-992 <*>
1954 V8 N5 P253-265 57-992 <*>
1954 V8 N10 P537-548 58-538 <*>

FUNKSCHAU
1948 V20 N1 P1-2 57-1451 <*>
1962 V34 N9 P222-224 63-10850 <*>

FUNKSTII ORGANIZMA V USLOVIIAKH IZMENENNOI GAZOVOI SREDY
1955 V1 P118-160 AD-620 970 <=*$>

FUNKTECHNIK
1951 V6 N24 P668-670 62-18613 <*>

FUNKTECHNISCHE MONATSHEFTE
1934 N8 P315- 2177 <*>
1934 N10 P395-396 57-1584 <*>
1937 N10 P302-306 57-595 <*>
1941 N4 P55-61 57-905 <*>

FUNTAI OYOBI FUNMATSU YAKIN
1962 V9 N5 P163-168 AEC-TR-6168 <*>
1962 V9 N6 P217-221 AEC-TR-6167 <*>
1962 V9 N6 P222-227 24T90J <ATS>
66-11522 <*>
1963 V10 N1 P7-12 25T90J <ATS>
65-14392 <*>
66-11523 <*>
1963 V10 N14 P172-180 66-14238 <*$>
1964 V11 N2 P66-72 26T90J <ATS>

FUSEES ET RECHERCHE AERONAUTIQUE

Citation	Translation
1956 V1 N2 P171-178	59-14110 <+*>
1957 V2 N3 P219-228	58-1668 <*>

FYSISK TIDSSKRIFT

Citation	Translation
1941 V39 P2-32	I-52 <*>

FYZYCHNY ZAPYSKY

Citation	Translation
1941 V8 N2 P161-165	RT-1441 <*>

FYZYCLOHICHNII ZHURNAL

Citation	Translation
1955 V1 N3 P123-129	AD-630 483 <=*$>
1955 V1 N4 P70-74	R-708 <*>
1956 V2 N3 P155-156	59-13707 <=>
1957 V3 N3 P112-115	5347 <*>
1958 V4 N2 P249-255	59-13194 <=>
1958 V4 N3 P408-420	PB 141 215T <+*> 0
1958 V4 N5 P704-705	59-13204 <=>
1958 V4 N6 P775-781	59-13463 <=> 0
1958 V4 N6 P814-820	59-13467 <=> 0
1959 V5 N1 P3-6	59-11750 <=>
1959 V5 N1 P16-21	59-11751 <=> 0
1959 V5 N1 P99-100	59-11752 <=>
1959 V5 N1 P110-118	63-23218 <=*$>
1959 V5 N1 P132-146	59-11754 <=>
1959 V5 N2 P261-269	59-13853 <=>
1959 V5 N5 P634-638	60-18360 <+*>
1959 V5 N5 P670-678	60-11275 <=>
1959 V5 N6 P769-773	60-11833 <=>
1959 V5 N6 P781-786	60-11834 <=>
1959 V5 N6 P848-854	60-31143 <=>
1960 V6 N1 P3-20	60-11940 <=>
1960 V6 N1 P44-49	60-11941 <=>
1960 V6 N1 P73-77	60-11942 <=>
1960 V6 N1 P114-116	60-11943 <=>
1960 V6 N1 P139-141	60-11944 <=>
1960 V6 N2 P280-281	60-31885 <=>
1960 V6 N3 P293-302	61-21502 <=>
1960 V6 N3 P372-376	61-21502 <=>
1960 V6 N3 P378-384	61-21502 <=>
1960 V6 N3 P391-397	61-21502 <=>
1960 V6 N3 P405-413	61-21502 <=>
1960 V6 N3 P427-428	60-31822 <=>
1960 V6 N4 P450-458	61-11841 <=>
1960 V6 N4 P498-506	61-11978 <=>
1960 V6 N4 P509-517	61-11840 <=>
1960 V6 N5 P563-570	61-28782 <*=>
1960 V6 N5 P577-583	61-21315 <=>
1960 V6 N5 P622-629	61-21312 <=>
1960 V6 N5 P694-697	61-11849 <=>
1960 V6 N6 P801-807	61-21664 <=>
1961 V7 N1 P3-8	61-21758 <=>
1961 V7 N1 P9-17	61-31491 <=>
1961 V7 N1 P76-81	61-23615 <*=>
1961 V7 N1 P83-91	61-27067 <*=>
1961 V7 N1 P93-99	61-27065 <*=>
1961 V7 N1 P113-119	61-23561 <*=>
1961 V7 N1 P151-153	61-23562 <*=>
1961 V7 N2 P159-164	61-28037 <*=>
1961 V7 N2 P165-176	61-28060 <*=>
1961 V7 N2 P187-195	61-28068 <*=>
1961 V7 N2 P197-205	61-28058 <*=>
1961 V7 N2 P214-219	61-28042 <*=>
1961 V7 N2 P266-270	61-28253 <*=>
1961 V7 N2 P277-283	61-28070 <*=>
1961 V7 N2 P287-289	61-28069 <*=>
1961 V7 N3 P327-332	62-13265 <*=>
1961 V7 N3 P333-342	62-13272 <*=>
1961 V7 N3 P343-351	61-28891 <*=>
1961 V7 N3 P352-361	62-13028 <*=>
1961 V7 N3 P362-368	62-13022 <*=>
1961 V7 N3 P369-370	62-13012 <*=>
1961 V7 N3 P371-375	62-13213 <*=>
1961 V7 N3 P385-394	62-13307 <*=>
1961 V7 N3 P417-423	62-13013 <*=>
1961 V7 N4 P443-449	62-23440 <=*>
1961 V7 N4 P563-566	61-28405 <*=>
1961 V7 N4 P569-570	62-23440 <=*>
1961 V7 N4 P571-572	61-28404 <*=>
1961 V7 N5 P602-606	62-15488 <=>
1961 V7 N5 P617-629	62-15488 <=>
1961 V7 N5 P667-670	62-15488 <=>
1961 V7 N5 P672-674	62-15488 <=>
1961 V7 N5 P709-710	62-15488 <=>
1961 V7 N5 P711-713	62-25185 <=*>
1961 V7 N6 P846-848	62-19802 <=*>
1962 V8 N1 P3-20	62-32445 <=>
1962 V8 N1 P38-43	62-32445 <=>
1962 V8 N1 P71-78	63-24509 <=*$> 0
1962 V8 N1 P86-	FPTS V22 N5 P.T990 <FASE>
1962 V8 N1 P93-	FPTS V22 N6 P.T1090 <FASE>
1962 V8 N1 P100-	FPTS V22 N5 P.T978 <FASE>
1962 V8 N1 P107-111	64-13413 <=*$>
1962 V8 N1 P120-	FPTS V22 N6 P.T1094 <FASE>
1962 V8 N1 P139-142	62-32463 <=*>
1962 V8 N2 P147-158	64-13499 <=*$> 0
1962 V8 N2 P159-167	63-23903 <=*$>
1962 V8 N2 P193-	FPTS V22 N6 P.T1098 <FASE>
1962 V8 N3 P283-291	63-24459 <=*$>
1962 V8 N3 P375-381	64-13410 <=*$>
1962 V8 N3 P382-396	63-13878 <=*>
1962 V8 N4 P433-441	62-33276 <=*>
1962 V8 N4 P442-	FPTS V23 N1 P.T28 <FASE>
1962 V8 N4 P471-480	63-23902 <=*$> 0
1962 V8 N4 P481-487	63-23901 <=*$> 0
1962 V8 N4 P532-542	64-13408 <=*$> 0
1962 V8 N5 P581-	FPTS 22 N6 P.T1107 <FASE>
1962 V8 N5 P615-621	63-24523 <=*$> 0
1962 V8 N5 P622-629	63-24525 <=*$>
1962 V8 N5 P630-637	63-24526 <=*$>
1962 V8 N5 P638-643	63-24532 <=*$>
1962 V8 N5 P664-	FPTS V22 N6 P.T1241 <FASE>
1962 V8 N6 P701-708	63-21417 <=>
1962 V8 N6 P790-795	63-41315 <=>
1962 V8 N6 P815-817	63-21417 <=>
1962 V8 N6 P827-829	63-21417 <=>
1963 V9 N1 P48-55	64-13909 <=*$>
1963 V9 N1 P77-88	63-21652 <=>
1963 V9 N1 P90-95	64-13531 <=*$>
1963 V9 N2 P151-157	AD-447 341 <=$>
1963 V9 N2 P215-220	64-13905 <=*$>
1963 V9 N2 P240-244	FPTS,V23,N3,P.T531 <FASE>
	63-21852 <=>
1963 V9 N2 P273-274	AD-447 341 <=$>
1963 V9 N2 P274-276	AD-447 341 <=$>
1963 V9 N3 P306-310	64-19713 <=$>
1963 V9 N3 P325-329	64-13943 <=*$>
1963 V9 N3 P357-362	64-15445 <=*$>
1963 V9 N3 P394-395	63-31815 <=>
1963 V9 N5 P596-600	64-00282 <*>
	64-15569 <=*$>
1963 V9 N6 P707-715	64-21477 <=>
1963 V9 N6 P824-826	64-21476 <=>
1964 V10 N2 P147-	FPTS V24 N2 P.T363 <FASE>
1964 V10 N2 P156-162	65-31511 <=$>
1964 V10 N2 P163-170	64-31865 <=>
1964 V10 N2 P177-182	64-31871 <=>
1964 V10 N2 P206-	FPTS V24 N2 P.316 <FASE>
1964 V10 N2 P227-234	
	FPTS V24 P.T195-T198 <FASE>
	64-31872 <=>
1964 V10 N2 P274-277	64-31873 <=>
1964 V10 N3 P301-306	64-41702 <=>
1964 V10 N3 P308-312	64-41702 <=>
1964 V10 N3 P360-365	64-41702 <=>
1964 V10 N3 P379-382	64-41702 <=>
1964 V10 N3 P390-396	64-41702 <=>
1964 V10 N3 P403-411	64-41702 <=>
1964 V10 N3 P416-418	64-41702 <=>
1964 V10 N4 P450-	FPTS V24 N4 P.T650 <FASE>
1964 V10 N4 P450-458	64-51246 <=$>
1964 V10 N4 P460-467	64-51246 <=$>
1964 V10 N4 P469-474	64-51246 <=$>
1964 V10 N4 P494-499	64-51246 <=$>
1964 V10 N4 P508-514	64-51246 <=$>
1964 V10 N4 P518-521	64-51246 <=$>
1964 V10 N4 P521-523	65-31765 <=$>
1964 V10 N4 P554-557	64-51246 <=$>
1964 V10 N4 P558-560	65-31766 <=$>
1964 V10 N5 P609-614	65-31106 <=$>

1964 V10 N5 P636-670 64-51997 <=$>
1964 V10 N5 P647-654 65-64516 <=*$>
1964 V10 N5 P689-694 65-31626 <=$>
1964 V10 N6 P709-733 65-30339 <=$>
1964 V10 N6 P813-814 65-30339 <=$>
1965 V11 N1 P3-9 65-31001 <=$>
1965 V11 N1 P10- FPTS V25 N1 P.T10 <FASE>
1965 V11 N1 P19- FPTS V25 N1 P.T28 <FASE>
1965 V11 N1 P32-36 65-31001 <=$>
1965 V11 N1 P45-51 66-60425 <=*$>
1965 V11 N1 P104-110 65-31001 <=$>
1965 V11 N1 P137-138 65-31001 <=$>
1965 V11 N2 P151- FPTS V25 N1 P.T15 <FASE>
1965 V11 N2 P265-268 65-31239 <=$>
1965 V11 N3 P283-288 65-32256 <=*$>
1965 V11 N4 P526- FPTS V25 N4 P.T580 <FASE>
1966 V12 N1 P56-68 66-31624 <=$>
1966 V12 N2 P138-164 66-32599 <=$>
1966 V12 N2 P269-276 66-32599 <=$>

GEA: REVISTA VENEZOLANA DE GEOGRAFIA
1963 V3 N5 P9-45 66-32910 <=$>

G.E.N. SOCIEDAD VENEZOLANA DE GASTROENTEROLOGIA, ENDOCRINOLOGIA Y NUTRICION. CARACAS
1961 V15 N3 P185-198 61-00591 <*>

G-I-T
1963 V7 P281-282 64-14727 <*>
1963 V7 P285-286 64-14727 <*>
1963 V7 P289 64-14727 <*>

GACETA MEDICA. GUAYAQUIL
1959 V14 N4 P502-504 32M46SP <ATS>

GACETA MEDICA DE CARACAS
1934 V41 P376-389 57-1532 <*>
1935 V42 P12-16 57-1532 <*>
1935 V42 P46-48 57-1532 <*>
1954 N11/2 P691-708 II-528 <*>

GACETA MEDICA DE MEXICO
1957 V87 N10 P751-753 61-00583 <*>
1959 V89 P753-765 62-10959 <*> 0
1959 V89 N10 P841-850 64-18981 <*> 0
1965 V95 N2 P163-177 65-14945 <*>

GACETA VETERINARIA
1941 V3 P67-70 59-10819 <*> 0

GAKUJUTSU GEPPO
1959 V12 N5 P34-46 59-31053 <+*>
1959 V12 N5 P46-51 59-31052 <=> 0

GALENIKA
1960 V7 N1 P21-22 62-19427 <*=>
1960 V7 N2 P43-50 62-19707 <=*>

GALVANO
1957 V26 N241 P37-39 63-16866 <*>
1958 V27 N256 P35-38 58-2129 <*>
1961 V30 P481-484 FA 79 <TTIS>

GALVANOTECHNIK. SAULGAU
1962 V53 N2 P62-74 SCL-T-495 <*>
1962 V53 N2 P87-92 SCL-T-505 <*>
1962 V53 N2 P95-96 62-18003 <*>
1962 V53 N3 P122-123 64-16608 <*>
1962 V53 N5 P228-235 64-16610 <*>
1963 V54 N2 P66-80 TRANS-313 <MT>
1964 V55 P668-672 8798 <IICH>
1964 V55 N10 P612-620 4590 <BISI>
1965 V56 N1 P14-19 MFA/TS 713 <MFA>
1965 V56 N1 P20-22 TS-714 <MFA>
1965 V56 N1 P86-91 MFA/TS 715 <MFA>
1965 V56 N2 P92-95 MFA/TS 716 <MFA>
1965 V56 N2 P105-108 MFA/TS 712 <MFA>
1965 V56 N6 P372 MFA/TS-743 <MFA>
1965 V56 N7 P390-408 MFA/TS-742 <MFA>
1965 V56 N7 P421-428 MFA/TS-741 <MFA>
1965 V56 N7 P428-432 MFA/TS-745 <MFA>
1965 V56 N8 P481-482 TRANS-575 <MT>
1965 V56 N11 P650-654 MFA/TS-761 <MFA>

GALVANOTECHNIK UND OBERFLAECHENSCHUTZ
1962 V3 P69-72 64-16580 <*>
1962 V3 P199-202 PN-223 <TTIS>
1965 V6 N6 P154-157 MFA/TS-744 <MFA>
1965 V6 N8 P187-190 2107 <TC>
1966 V7 N2 P39-49 MFA/TS-824 <MFA>
1966 V7 N2 P50-59 MFA/TS-823 <RAP>

GALVANOTECNICA
1951 V2 N9 P199-202 97P60I <ATS>
1958 V9 P301-308 29P61I <ATS>

GAN NO RINSHO
1962 V8 P97-101 66-11705 <*>
1964 V10 P550-553 65-14770 <*>

GANN, ERGEBNISSE DER KREBSFORSCHUNG IN JAPAN
1937 V31 N3 P220-223 1393 <*>
1938 V32 N2 P100-106 66-13154 <*>
1939 V33 N3 P203-205 57-149 <*>
1952 V43 N2/3 P178-180 58-273 <*>
1953 V44 P361-362 62-32361 <=*>
1953 V44 N2/3 P216-218 2250 <*>
1953 V44 N2/3 P357-359 58-803 <*>
63-16577 <*>
1954 V45 P107-108 57-1910 <*>
1954 V45 P138-139 57-1911 <*>
1954 V45 N2/3 P451-452 II-551 <*>
1954 V45 N2/3 P457-459 II-543 <*>
1955 V46 N2/3 P153-156 II-817 <*>
1955 V46 N2/3 P180-183 II-798 <*>
1955 V46 N2/3 P205-206 II-799 <*>
1956 V47 P397-398 62-00684 <*>
1956 V47 N3/4 P292-294 62-25262 <=*>
1956 V47 N3/4 P359-360 58-629 <*>
63-16527 <*>
1956 V47 N3/4 P360-363 58-1200 <*>
1956 V47 N3/4 P465-467 58-1201 <*>
1957 V48 P494-496 62-00643 <*>
1957 V48 N4 P331-332 59-16656 <+*>
1957 V48 N4 P397-399 59-12911 <+*>
1957 V48 N4 P399-401 59-12910 <+*>
1957 V48 N4 P425-427 59-12103 <+*>
1957 V48 N4 P443-445 59-12909 <+*>
1957 V48 N4 P465-466 59-12102 <+*>
1957 V48 N4 P496-498 59-14316 <+*>
1957 V48 N4 P565-566 58-1691 <*>
1957 V48 N4 P581-582 58-1687 <*>
1958 V49 P27-28 59-17927 <*>
1958 V49 P28-30 59-17926 <*>
1958 V49 P27-30 SUP 62-16049 <*>
1959 V49 P81-82 59-22095 <+*>
1959 V49 P100-101 59-22096 <+*>
1959 V49 P101-102 59-22089 <+*>
1959 V49 P108-109 59-22394 <+*>
1959 V49 P112-113 61-10426 <*>
1959 V49 P130-131 59-22090 <+*>
1959 V49 P132-133 59-22102 <+*>
1959 V49 P293-294 59-22579 <+*>
1959 V49 P293-294 SUP 65-00190 <*>

GANSEKI KOBUTSU KOSHOGAKU ZASSHI
1944 V32 P60-71 57-2579 <*>
1956 V40 N3 P104-115 57-2576 <*>

GARTENWELT. HAMBURG
1961 V61 N13 P258-260 63-10507 <*>

GAS WAERME
1956 V5 P8-13 3563 <BISI>
1957 P348-351 TS-1216 <BISI>
61-20672 <*>
1961 V10 P347-359 3860 <BISI>
1964 V13 P387-396 4252 <BISI>

GAS, WASSER, WAERME

1948 V2 N7 P143-149 62-18241 <*> O
1949 V3 N6 P101-107 AL-416 <*>
1952 V6 N11 P251- 1887 <*>

GAS-UND ELEKTROWAERME
1944 P74-75 59-17700 <*>
1944 N4 P74-75 58-493 <*>

GAS UND WASSERFACH
1926 V70 P40-41 61-10027 <*>
1927 V70 P873-877 AI-TR-40 <*>
1928 V71 N10 P222-224 57-1310 <*>
1930 V73 P440-444 63-10485 <*> O
1936 V79 N38 P694-697 59-20531 <*>
1937 V80 P619-625 58-68 <*>
1937 V80 P925-928 65-13128 <*>
1940 V83 P6-9 65-13874 <*>
1941 V84 N1 P9-13 62-16313 <*>
1941 V84 N40 P549-552 60-18303 <*>
1942 V85 P321-327 58-1171 <*>
58-1323 <*>
59-18540 <=*> O
1944 V87 P61-67 58-67 <*>
1944 V87 N2 P31-37 AI-TR-42 <*>
1947 V88 N3 P86-89 66-11368 <*>
1948 V89 N6 P183-187 66-11369 <*>
1948 V89 N12 P367-371 62-16453 <*> O
1949 V90 N8 P169-174 62-16762 <*> O
1949 V90 N24 P634-641 64-71481 <=>
1951 N19 P273-277 AL-629 <*>
1951 V92 N5 P53-54 57-1921 <*>
1951 V92 N17 P238-242 64-18872 <*> O
1952 V93 N9 P255-257 63-16226 <*> O
1952 V93 N19 P537-547 1788 <*>
1953 V94 N7 P178-180 64-20368 <*>
1953 V94 N21 P645-647 00N57G <ATS>
1955 V96 N23 P761-766 74M43G <ATS>
1955 V97 N17 P559-562 57-20 <*>
1957 V98 N21 P517-524 59-15064 <*> O
1957 V98 N26 P661-663 48M45G <ATS>
62-10097 <*>
1958 V99 N41 P1045-1054 59-10852 <*> O
59-20043 <*>
1959 V100 P189-193 61-18812 <*> O
1959 V100 P681-687 TS 1449 <BISI>
1959 V100 N13 P309-317 60-17595 <*=>
61-20931 <*>
1959 V100 N27 P681-687 62-14323 <*>
1960 V101 N36 P909-914 NP-TR-557 <*>
1960 V101 N38 P969-972 NP-TR-557 <*>
1961 V102 P1288-1290 2950 <BISI>
1961 V102 N8 P181-186 65-11664 <*>
1963 V104 N44 P1261-1271 65-10231 <*> O

GASMASKE
1938 V10 P120-127 60-10756 <*> O

GASSCHUTZ UND LUFTSCHUTZ
1932 V2 P10-14 64-16218 <*>

GASTROENTEROLOGIA. BERLIN, BASEL
1953 V79 P257-281 58-2709 <*>
1955 V84 P225-251 65-00149 <*> O
1957 V88 P133-171 62-10192 <*> O
1958 V90 P29-38 58-2202 <*>
1960 V94 N6 P366-379 65-14966 <*> P
1961 V95 P270-276 62-00864 <*>
1964 V101 N4 P239-246 65-11289 <*> O

GAZ, WODA I TECHNIKA SANITARNA
1949 V23 N9 P280-285 PJ-479 <ATS>
1957 V31 N8 P287-293 63-20924 <*>
1959 V33 N9 P360-363 NP-TR-1049 <=*$>
1961 V35 N9 P340-347 62-19595 <=*>
1966 V40 N10 P325-327 66-35541 <=>

GAZETA CUKROWNICZA
1955 V57 P157-159 63-14883 <*> O

GAZETA MATEMATICA
1964 V69 N7 P241-246 65-30022 <=$>
1964 V69 N7 P246-250 64-51993 <=$>
1964 V69 N8 P281-285 64-51816 <=$>
1964 V69 N8 P285-291 64-51870 <=$>
1964 V69 N8 P291-297 64-51928 <=$>
1964 V69 N9 P321-324 65-30342 <=$>

GAZETA MEDICA PORTUGUESA
1955 V8 P33-44 58-797 <*>
1955 V8 P225-240 1398 <*>

GAZETA OBSERWATORA. PANSTWOWY INSTYTUT HYDROLOGICZNO-METEOROLOGICZNY
1958 V11 N10 P12-15 61-31294 <=>

GAZETTE HEBDOMADAIRE DE MEDECINE ET DE CHIRURGIE
1887 V24 N31 P501-502 65-14608 <*>

GAZETTE MEDICALE DE FRANCE
1936 V43 P93-96 63-00299 <*>
1940 V47 P48-52 59-19210 <=*>
1952 V59 N20 P1252-1253 65-14967 <*>
1957 V64 N1 P91-92 65-14968 <*>
1957 V64 N7 P519-524 OA 10 <RIS>
1958 V65 P2029-2030 64-18626 <*>
1960 V67 P2699-2700 61-20571 <*>
1960 V67 P2703 61-20571 <*>
1961 V68 P1169-1180 63-18491 <*>
1962 V69 N8 P1395-1399 66-11438 <*>
1965 V72 N5 P1035-1040 66-13212 <*>

GAZOVAYA PROMYSHLENNOST
1956 N5 P10-13 R-5328 <*>
1956 N5 P29 R-3194 <*>
1956 N5 P35-37 61-14928 <*=>
1956 N7 P8-14 60-15781 <+*>
62-25288 <=*>
1956 N8 P19-25 AD-615 246 <=$>
1956 N12 P7-9 59-22478 <+*>
1957 N2 P21-25 60-17456 <+*>
1957 N4 P22-27 61-15375 <+*>
1957 N6 P32-40 49K20R <ATS>
1957 N9 P31-34 59-18717 <+*>
1957 N9 P35-37 82K22R <ATS>
1957 N10 P32-39 53K23R <ATS>
1957 V2 N6 P11-13 62-11574 <=>
1958 N5 P36-40 59-18334 <+*>
1958 N8 P8-12 59-22608 <+*>
1958 N8 P43-47 59-22193 <+*>
1958 V3 N11 P12-18 64-13229 <=*$>
1959 N5 P36-41 61-10729 <+*>
1959 N7 P24-27 NLLTB V1 N12 P55 <NLL>
1959 N7 P29-33 61-10730 <+*>
1959 V4 N1 P38-41 58L31R <ATS>
1959 V4 N1 P51-56 59L31R <ATS>
1959 V4 N2 P43-49 05L33R <ATS>
1959 V4 N3 P34-39 60-23614 <+*>
1959 V4 N3 P36-41 96L35R <ATS>
1959 V4 N5 P4-7 95L35R <ATS>
1959 V4 N7 P1-5 59-19322 <+*>
1959 V4 N8 P8-12 61-15931 <+*>
95M38R <ATS>
1959 V4 N9 P36-41 43M39R <ATS>
1959 V4 N10 P17-20 62-15267 <*=> P
1959 V4 N11 P10-13 55M41R <ATS>
1959 V4 N12 P7-9 56M41R <ATS>
1959 V4 N12 P17-24 61-14954 <=*>
70N56R <ATS>
1960 V5 N2 P45-52 07P66R <ATS>
65-10236 <*>
1960 V5 N4 P30-37 60-31695 <=>
1960 V5 N7 P1-5 99N49R <ATS>
1960 V5 N7 P5-13 58N48R <ATS>
1960 V5 N8 P15-19 62-23953 <=*>
1960 V5 N9 P6-12 81M47R <ATS>
1960 V5 N11 P50-52 61-21694 <=>
1961 V6 N3 P22-25 62-13431 <*=>
1961 V6 N3 P26-29 62-13427 <*=>
1961 V6 N3 P54-55 61-28889 <*=>
1961 V6 N8 P46-50 62-23454 <=*>

91P58R <ATS>
1961 V6 N11 P18-22 32P59R <ATS>
1961 V6 N11 P35-38 65P59R <ATS>
1961 V6 N11 P39-46 30P63R <ATS>
1962 N8 P1-3 NLLTB V5 N1 P11 <NLL>
1962 V7 N6 P9-13 08P65R <ATS>
1963 N5 P32-56 ICE V4 N1 P5-9 <ICE>
1963 N9 P28-29 NLLTB V6 N2 P122 <NLL>
1963 V8 N2 P49-50 63-21517 <=>
1963 V8 N4 P35-39 66-11892 <*>
1963 V8 N4 P40-42 63-27411 <=$>
1963 V8 N4 P43-48 64-30134 <*> 0
1963 V8 N5 P1-2 63-31321 <=>
1963 V8 N5 P32-36 18Q72R <ATS>
1963 V8 N8 P52-53 63-41037 <=>
1963 V8 N9 P44-46 66-11916 <*>
1964 N1 P1-4 NLLTB V6 N7 P664 <NLL>
1964 V9 N3 P38-41 64-26431 <$>
1964 V9 N3 P41-42 64-26430 <$>
1964 V9 N4 P48-51 01R78R <ATS>
1964 V9 N5 P8-11 58R78R <ATS>
1964 V9 N6 P10-14 57R79R <ATS>
1964 V9 N7 P38-43 65-27637 <$>
1964 V9 N8 P41-43 LA-TR-65-1 <=*$>
1964 V9 N9 P40-43 66-10339 <*>
1964 V9 N12 P7-9 12S83R <ATS>
1965 N11 P24-26 ICE V6 N2 P250-252 <ICE>
1965 V10 N1 P20-21 39T95R <ATS>
1965 V10 N6 P5-12 23S88R <ATS>
1965 V10 N9 P1-6 92T90R <ATS>
1965 V10 N9 P7-11 46T91R <ATS>
1965 V10 N9 P11-14 47R91R <ATS>
1965 V10 N9 P53-56 48T91R <ATS>
1965 V10 N11 P1-3 66-30594 <=$>

GAZOVOE DELO
1963 N11 P50-53 07S81R <ATS>

GAZZETTA CHIMICA ITALIANA
1883 V13 P358-363 3002 <*>
1883 V13 P378-381 12178-B <K-H>
65-14473 <*>
1911 V41 P35-38 61-18296 <*>
1923 V53 P64-74 63-18333 <*>
1923 V53 P795-800 60-10057 <*> P
1925 P468-501 AL-426 <*>
1925 V55 P294-300 63-14845 <*>
1925 V55 P301-306 63-14844 <*>
1925 V55 P306-310 63-14843 <*>
1927 V57 P560-561 59-17768 <*>
1927 V57 P584-592 1858 <TC>
1928 V58 P6-25 58-1100 <*>
1928 V58 P209-216 61-10175 <*>
1928 V58 P597-618 58-1786 <*>
1929 V59 P189-198 AL-505 <*>
1929 V59 P198-200 AL-505 <*>
1929 V59 P578-590 II-1025 <*>
12806 <K-H>
63-14197 <*>
65-14848 <*>
1929 V59 P591-600 1565 <*>
57-352 <*>
1930 V60 P648-664 AL-506 <*>
1930 V60 P762-767 AEC-TR-3743 <*> 0
1930 V60 P833-838 I-1051 <*>
1930 V60 P839-842 I-1052 <*>
1931 V61 P191-215 I-439 <*>
1931 V61 P286-293 64-18864 <*>
1931 V61 P305-311 AL-181 <*>
63-20745 <*>
1931 V61 P578-580 <LSA>
1931 V61 P828-832 58-2535 <*>
1931 V61 P915-921 62-00298 <*>
1932 V62 P317-332 AL-507 <*>
1932 V62 P577-582 64-20346 <*>
1932 V62 P873-877 <LSA>
1932 V62 P937-944 59-12309 <=*>
1932 V62 P1041-1048 59-17967 <*>
1933 V63 P761-763 59-17837 <*>
1934 V64 P774-778 AL-616 <*>
1934 V64 P792-800 AL-508 <*>
1934 V64 P824-832 AL-510 <*>
1934 V65 P556- 3297 <*>
1935 V65 P229-241 AL-976 <*>
1935 V65 P349-356 AL-977 <*>
1935 V65 P357-366 AL-509 <*>
1935 V65 P766-772 61-10913 <*> 0
1936 V66 P48-57 AL-978 <*>
1936 V66 P808-812 65-12300 <*>
1937 V67 P501-510 68H91 <ATS>
1937 V67 P628-633 59-17893 <*>
1937 V67 P664-668 AL-986 <*>
1938 V68 P343-352 C-2273 <NRC>
1668 <*>
1939 V69 P97-104 1977 <*>
1939 V69 P275-284 57-3070 <*>
1939 V69 P315-322 59-17765 <*>
1939 V69 P408-416 3286 <*>
1939 V69 P453-459 60-10060 <*>
1940 V70 P483-490 59-10545 <*> 0
1941 V71 P73-81 63-18970 <*> 0
1941 V71 P117-128 T-1747 <INSD>
1941 V71 P311-319 AL-134 <*>
1941 V71 P696-713 63-18971 <*> 0
1942 V72 P83-89 63-18861 <*>
1942 V72 P109-124 1935 <*>
1942 V72 P370-377 53J15I <ATS>
58-05 <*>
1942 V72 N7 P305-312 61-20756 <*>
1944 V74 P23-25 63-18670 <*>
1946 V76 P272-282 63-18992 <*>
1946 V76 P410-418 2237B <K-H>
1946 V76 N7/8 P272-282 64-20419 <*>
1947 V77 P238-247 5391 <K-H>
1947 V77 P238-240 64-18184 <*>
1947 V77 P240-247 64-18185 <*>
1947 V77 P312-318 57-2821 <*>
1948 V78 P301- 3311 <*>
1949 V79 P164-175 AL-609 <*>
1949 V79 N1 P3-13 64-16512 <*>
1949 V79 N5 P364-368 RCT V23 N1 P151 <RCT>
1950 V80 P347-351 T-1415 <INSD>
1950 V80 N7 P527-532 60-14248 <*>
1950 V80 N11/2 P855-863 59-17038 <*> P
1950 V80 N8/10 P658-662 60-14249 <*>
1951 V81 P106-116 64-11644 <=>
1951 V81 P236-244 II-971 <*>
1951 V81 P245-250 II-972 <*>
1951 V81 P272-275 AL-610 <*>
1951 V81 P596-608 3256 <*>
1951 V81 P625-634 <ATS>
1951 V81 P635-645 <ATS>
1951 V81 P664-667 11001 <K-H>
1951 V81 N6 P507-510 63-10938 <*>
1951 V81 N7 P664-667 65-13229 <*>
1952 V82 P79-85 2731 <K-H>
63-18986 <*> 0
1952 V82 P227-242 64-20062 <*>
1952 V82 P394-405 86K231 <ATS>
1952 V82 P558-563 65-11219 <*>
1954 V84 P729-734 3241 <*>
1954 V84 P1171-1176 NP-TR-1251 <*>
1954 V84 N9 P874-878 60-16885 <*>
1955 V85 P61-68 64-16456 <*>
1955 V85 P137-144 63-16426 <*>
1955 V85 P349-363 60-15407 <=*> 0
1955 V85 P1111-1117 64-20240 <*>
1955 V85 P1118-1140 57-3575 <*>
1955 V85 P1224-1231 II-901 <*>
03H9I <ATS>
1955 V85 N1/2 P103-110 65-12210 <*>
1955 V85 N5/6 P614-619 65-12211 <*>
1955 V85 N5/6 P620-627 65-12213 <*>
1956 V86 P350-357 57-3389 <*>
1956 V86 P406-414 58-1389 <*>
1956 V86 P705-709 76J14I <ATS>
1956 V86 N5/7 P350-357 59-12439 <+*>
1957 V87 P90-99 72N56I <ATS>
1957 V87 P159-170 63-18010 <*> 0
1957 V87 P446-453 59-14707 <=*>

1957 V87 P503-509 06M38I <ATS>
1957 V87 P805-829 58-551 <*>
1957 V87 N3 P293-309 62-14355 <*> O
1957 V87 N4 P438-445 59-14706 <+*>
1957 V87 N4 P519-527 80J17I <ATS>
1957 V87 N5 P528-548 49L32I <ATS>
59-15049 <*>
61-17648 <=$>
65-11598 <*>
1957 V87 N5 P549-569 54J17I <ATS>
1957 V87 N5 P570-585 55J17I <ATS>
1957 V87 N6 P673-681 57-3477 <*>
59-10522 <*>
1957 V87 N6 P682-687 57-3397 <*>
1957 V87 N6 P830-836 59-10555 <*>
1957 V87 N10 P1199-1202 58-1365 <*>
1957 V87 N7/8 P876-884 31K24I <ATS>
1957 V87 N7/8 P931-934 61-20210 <*>
1957 V87 N7/8 P936 61-20210 <*>
1957 V87 N7/8 P942 61-20210 <*>
1958 V88 P125-148 C-4945 <NRC>
1958 V88 P1170-1182 10L33I <ATS>
1958 V88 N2 P219-228 59-17561 <*>
61-16730 <*>
64-20116 <*>
1958 V88 N3 P279-295 32K24I <ATS>
1958 V88 N11 P1063-1080 59-15059 <*>
1958 V88 N8/9 P746-754 59-10721 <*>
1959 V89 P502-504 63-16406 <*>
1959 V89 P798-808 REPT. 97070 <RIS>
54L34I <ATS>
1959 V89 N2 P465-494 10508 <K-H>
61-10713 <*> O
1959 V89 N3 P713-731 NP-TR-466 <*>
1959 V89 N3 P750-755 62-25788 <=*$>
1959 V89 N3 P756-760 62-10054 <*>
1959 V89 N3 P761-774 59-20016 <*>
60-10134 <*>
1959 V89 N3 P775-783 60-10120 <*>
1959 V89 N3 P784-797 60-10701 <*>
1959 V89 N3 P798-808 65-11599 <*>
1959 V89 N3 P809-817 65-10934 <*>
1959 V89 N3 P854-865 02M38I <ATS>
1959 V89 N3 P866-877 04M38I <ATS>
1959 V89 N3 P878-887 03M38I <ATS>
1959 V89 N5/6 P1313-1323 65-10068 <*> O
1959 V89 N5/6 P1332-1337 09M38I <ATS>
62-20183 <*>
1959 V89 N7/8 P1543-1547 08M38I <ATS>
1959 V89 N7/8 P1681-1686 43M42I <ATS>
1960 V90 P831-840 61-00862 <*>
1960 V90 P1394-1398 63-01533 <*>
1960 V90 P1505-1515 AEC-TR-4786 <*>
1960 V90 P1516-1521 AEC-TR-4784 <*>
1960 V90 P1682-1699 UCRL TRANS-861(L) <*>
1960 V90 P1800-1806 AEC-TR-4798 <*>
1960 V90 N9/10 P1253-1265 37N48I <ATS>
1961 V91 P121-144 UCRL TRANS-861(L) <*>
1961 V91 P270-279 AD-626 821 <=*$>
1961 V91 P412-427 UCRL TRANS-861(L) <*>
1961 V91 P493-503 CHEM-239 <CTS>
1961 V91 P512-528 UCRL TRANS-741 <*>
1961 V91 P529-536 UCRL TRANS-944(L) <*>
1961 V91 P787-803 UCRL TRANS-902(L) <*>
1961 V91 P825-940 65-10076 <*>
1961 V91 P1063-1084 UCRL TRANS-903(L) <*>
1961 V91 N5 P545-561 62-14911 <*>
1961 V91 N5 P567-570 62-14959 <*> P
1961 V91 N12 P1357-1364 62-18911 <*>
1961 V91 N6/7 P655-664 62-14910 <*>
1961 V91 N6/7 P665-671 62-14909 <*> P
1961 V91 N6/7 P686-705 62-14958 <*> PO
1961 V91 N8/9 P964-972 65-12069 <*>
1962 V92 N6 P488-500 (CHEM)-314 <CTS>
1962 V92 N8 P811-817 64-18602 <*> O
1962 V92 N9 P916-950 CHEM-333 <CTS>
1962 V92 N11 P1189-1197 64-10501 <*>
1962 V92 N2/3 P244-252 66-10343 <*>
1963 V93 N7 P916-918 65-13465 <*>
1963 V93 N1/2 P65-72 66-11299 <*>
1963 V93 N1/2 P73-80 66-11298 <*>
1963 V93 N1/2 P81-89 66-11297 <*>
1964 V94 P521-525 28T92I <ATS>
66-12719 <*>
1964 V94 P552-556 419 <CTS>
1964 V94 P797-803 39T92I <ATS>
1964 V94 P1278-1286 29T91I <ATS>
66-12714 <*>
1964 V94 N7 P797-803 66-12709 <*>
1964 V94 N10 P1028-1034 66-13246 <*>
1964 V94 N3/4 P340-355 65-12903 <*>
1965 V95 N3 P201-205 65-17072 <*> O
1965 V95 N1/2 P3-15 65-17078 <*> O

GAZZETTA INTERNAZIONALE DI MEDICINA E CHIRURGIA
1954 V58 P1376-1391 1088 <*>
1955 V59 P626-638 II-927 <*>
1957 V62 P1911-1918 58-1366 <*>
1957 V62 P2424-2428 64-16356 <*>
1958 V63 P803-817 59-15553 <*>
1959 V64 P259-265 60-14096 <*>
1960 V65 P2831-2837 62-16113 <*>
1961 V66 P2304-2310 63-10186 <*>

GAZZETTA MEDICA ITALIANA
1951 V110 N2 P43-45 65-14969 <*>
1953 V112 N4 P97-101 65-14970 <*>
1954 V113 P159-163 3055 <*>
1954 V113 P301-303 II-566 <*>
1955 V114 P69-72 57-1958 <*>
1956 V115 P71-74 57-1850 <*>
1957 V116 P149-153 58-826 <*>
1957 V116 P404-407 58-1181 <*>
1957 V116 N10 P492-494 64-18629 <*>
1958 V117 N5 P253-254 64-18710 <*>
1959 V118 P244-245 64-18735 <*>
72M42I <ATS>
1959 V118 N2 P64-74 W-61 <RIS>
1959 V118 N3 P123-124 64-18672 <*>
1960 V119 N1 P34-36 64-18656 <*>
1960 V119 N2 P57-62 61-10248 <*>
1961 V120 N9 P321-328 63-16503 <*>
1962 V121 P113-117 65-14971 <*>
1964 V123 P85-86 65-14284 <*> O

GAZZETTA MEDICA LOMBARDA. MILAN
1897 V48 P184 57-1493 <*>

GAZZETTA SANITARIA. MILANO
1960 V31 N4/5 P173-175 61-00910 <*>

GEBURTSHILFE UND FRAUENHEILKUNDE
1939 V1 P663-671 59-15859 <*> O
1951 V11 N4 P324-328 59-17012 <*>
1953 V13 P52 AL-23 <*>
1954 V14 P645-650 3028 <*>
1955 V15 P769-779 57-1962 <*>
1957 N17 P501-505 58-326 <*>
1957 N17 P581-591 58-325 <*>
1957 V17 N4 P381-387 61-00170 <*>
1957 V17 N6 P501-505 63-16414 <*>
1958 V18 N6 P824-828 59-17591 <*> P
1959 V19 N4 P346-352 65-14972 <*> O
1960 V20 P764-769 63-00068 <*>
1960 V20 N12 P1365-1371 61-16473 <*>

GEFLUEGELHOF
1949 V12 P339-340 65-12616 <*>

GELATINE, LEIM, KLEBSTOFFE
1944 V12 P25-33 T2079 <INSD>
1944 V12 P45-50 T2079 <INSD>

GEMATOLOGY I PERELIVANIE KROVI
1965 V1 P91-98 65-33461 <=$>
1965 V1 P132-137 65-33461 <=$>
1965 V1 P143-172 65-33461 <=$>
1965 V1 P248-250 65-33461 <=$>

GENEESKUNDIG TIJDSCHRIFT VOOR NEDERLANDSCH-INDIE

1940 V80 P1140-1151 C-2303 <NRC>
57-1279 <*>
1940 V80 P2693-2716 C-2322 <NRC>
57-2358 <*>
1941 V81 P477-485 C-2321 <NRC>
57-2357 <*>

GENEESKUNDIGE BLADEN UIT KLINIEK EN LABORATORIUM VOOR DE PRAKTIJK. HAARLEM
1932 V30 P181-210 57-1682 <*>

GENEESKUNDIGE GIDS
1954 V32 P89-94 3417-C <K-H>
1956 V34 N25 P477-480 57-63 <*>

GENETICA IBERICA
1950 V2 P295-322 2729-A <K-H>

GENIE CHIMIQUE. PARIS
1956 V76 N6 P172-177 64-16554 <*>
1957 V77 N2 P25-33 NP-TR-888 <*>
1957 V78 N2 P33-45 64-18093 <*>
1957 V78 N8 P46-53 61-10630 <*>
1958 V80 N4 P98-106 59-10296 <*>
1959 V82 N5 P117-126 S-2098 <RIS>
1960 V83 P170-181 65-12801 <*>
1960 V84 P109-119 65-14224 <*>
1960 V84 N3 P89-99 65-14189 <*>
1960 V84 N4 P109-119 64-30294 <*> P
1961 V85 N6 P252-262 62-14708 <*> 0
1961 V86 N1 P1-16 65-12800 <*> 0
1961 V86 N2 P39-44 65-10034 <*>
1962 V87 N1 P1-13 65-12799 <*>
1962 V87 N4 P108-113 65P64F <ATS>
1962 V88 P122-127 66-11530 <*> 0
1963 V89 P119-120 1027 <TC>
1963 V89 N5 P137-145 64-18773 <*>
1963 V89 N5 P163-166 NS-299 <TTIS>
1963 V90 N5 P127-134 65-11781 <*>
1964 V91 N6 P167-176 65-17312 <*>
1964 V92 P126-141 4402 <BISI>
1964 V92 N3 P64-68 65-13422 <*>
1965 V93 N4 P103-109 66-10362 <*>

GENIE CIVIL
1927 V90 P621-624 66-12367 <*>
1927 V91 P243-249 64-11814 <=>
1927 V91 P250-252 UCRL TRANS-1143(L) <*>
1928 V92 P353-355 57-2200 <*>
1931 V98 N17 P426-429 60-14427 <*> 0
1935 V107 N4 P93 66-12599 <*> 0
1943 V120 N18 P205-207 62-16802 <*> 0
1944 V121 P33-34 3337 <*>
1944 V121 N19/0 P156- 3336 <*>
1946 P327-329 T-2558 <INSD>
1947 N23 P459 3385 <*>
1948 V125 N2 P30-32 03328 <*>
3328 <*>
1948 V125 N13 P251-253 65-13004 <*>
1948 V125 N24 P472-474 3347 <*>
1950 V127 N5 P94 2276 <*>
1951 V128 N18 P345-347 AL-126 <*>
3384 <*>
1956 V133 N1 P7-10 64-16560 <*>
1956 V133 N12 P230-235 66-11351 <*> 0
1958 N24 P501-512 59-21037 <=> 0
1960 V137 P453-457 <LSA>

GENSHIRYOKU HATSUDEN
1960 V4 N3/4 P49-61 AEC-TR-4983 <*>

GENSHIRYOKU KOGYO
1963 V9 N1 P62-68 AEC-TR-6114 <*>
1963 V9 N11 P41-45 66-12134 <*>

GEOCHIMICA ET COSMOCHIMICA ACTA
1951 V2 N1 P62-75 57-3071 <*>
1960 V19 P70-79 37M46G <ATS>
1961 V21 P284-294 357 <TC>
1964 V28 N5 P683-693 65-12403 <*>

GEODETICKY A KARTOGRAFICKY OBZOR
1957 V3 N6 P113-114 62-19445 <=*>
1959 N6 P61-67 59-22150 <+*>
60-11165 <=>
1966 N12 P45-48 AD-641 276 <=$>

GEODEZIA ES KARTOGRAFIA
1958 V10 N1 P43-48 60-31358 <=>
1958 V10 N2 P152-158 60-31358 <=>
1958 V10 N3 P195 60-31358 <=>
1958 V10 N3 P222-225 60-31358 <=>
1958 V11 N1 P62-65 60-31358 <=>
1958 V11 N1 P167-169 60-31358 <=>
1958 V11 N1 P228-229 60-31358 <=>
1960 N3 P36-39 62-26031 <=$>
1960 N3 P55-62 62-26030 <=$>
1960 N4 P36-43 62-26029 <=$>
1960 N5 P33-37 62-26028 <=$>
1960 N6 P56-60 61-11712 <=>
1960 V12 N4 P252-255 62-19444 <*=>
1960 V12 N4 P271-274 62-19446 <*=>
1961 N6 P67 62-19528 <=*>
1961 V13 N1 P1-35 62-23780 <=*> P
1961 V13 N1 P25-34 62-19479 <*=>
1961 V13 N1 P55-56 62-19479 <*=>
1961 V13 N2 P73-87 62-19585 <=*>
1961 V13 N2 P91-93 62-19553 <=*>
1961 V13 N2 P112-115 62-19667 <=*>
1961 V13 N4 P286-288 62-15928 <*=>
1961 V13 N4 P291-292 62-15929 <*=>
62-15929 <=*>
1963 V15 N1 P62-63 63-31220 <=>
1963 V15 N2 P81-85 63-21999 <=>
1964 N12 P53-59 65-63415 <=$>
1964 V16 N1 P28-34 64-31301 <=>
1965 N4 P302-304 65-32805 <=*$>
1965 N6 P3-9 65-32230 <=*$>
1965 V17 N3 P161-166 65-32224 <=$*>
1965 V17 N3 P221-222 65-32224 <=$*>
1966 N2 P81-94 66-32737 <=$>
1966 N2 P98-106 66-32737 <=$>
1966 N2 P151 66-32737 <=$>
1966 V18 N3 P191-194 66-33986 <=$>
1966 V18 N3 P206-213 66-33986 <=$>
1966 V18 N3 P222-223 66-33986 <=$>
1966 V18 N4 P241-245 66-34363 <=$>

GEODEZIST. MOSKVA
1925 V1 N3 P46-48 60-11005 <=>
1925 V1 N3 P49-50 61-11589 <=>
1925 V1 N4/5 P45-48 60-11844 <=> 0
1925 V1 N4/5 P48-52 60-41121 <=>
1925 V1 N6/7 P6-20 60-41067 <=>
1925 V1 N6/7 P37-38 60-11278 <=>
1926 V2 N11 P14-33 60-11876 <=>
1926 V2 N11 P80-99 60-11866 <=>
1926 V2 N12 P1-11 60-11934 <=>
1926 V2 N1/2 P3-6 60-41120 <=>
1926 V2 N1/2 P13-16 60-41484 <=>
1926 V2 N1/2 P18-20 60-41485 <=>
1926 V2 N1/2 P34-36 60-11925 <=>
1926 V2 N3/4 P67-70 61-21178 <=>
1926 V2 N5/6 P6-7 59-13993 <=>
1926 V2 N5/6 P13-29 60-41118 <=>
1926 V2 N7/8 P2-8 61-11774 <=>
1926 V2 N7/8 P9-12 61-11728 <=>
1926 V2 N7/8 P46-49 61-21176 <=>
1926 V2 N7/8 P75-78 61-11730 <=>
1926 V2 N11/2 P65-73 59-13946 <=>
1926 V2 N13/4 P35-41 60-11880 <=>
1926 V2 N15/6 P31-38 60-11582 <=>
1926 V2 N15/6 P39-42 60-11377 <=>
1926 V2 N15/6 P48-49 59-13947 <=>
1926 V2 N17/8 P12-14 60-41483 <=>
1926 V2 N17/8 P35-38 60-41601 <=>
1926 V2 N17/8 P38-41 60-41011 <=>
1926 V2 N17/8 P42-43 59-13949 <=>
1926 V2 N19/0 P12-15 60-41597 <=>
1926 V2 N19/0 P34-40 60-11887 <=>

1926 V2 N21/2 P52-63 61-11010 <=>
1926 V2 N23/4 P26-30 60-11542 <=>
1926 V2 N23/4 P30-37 60-11891 <=>
1926 V2 N9/10 P3-7 60-41548 <=>
1926 V2 N9/10 P47-48 59-13945 <=>
1926 V2 N9/10 P66-69 61-11731 <=>
1927 V3 N1 P53-55 60-11277 <=>
1927 V3 N2 P12-21 60-41534 <=>
1927 V3 N2 P21-25 60-41007 <=>
1927 V3 N2 P28-37 60-11343 <=>
1927 V3 N2 P46-48 60-11267 <=>
1927 V3 N3 P1-12 61-21013 <=>
1927 V3 N3 P13-15 60-41547 <=>
1927 V3 N3 P22-30 61-21159 <=>
1927 V3 N3 P30-46 60-41437 <=>
1927 V3 N4 P14-23 61-21179 <=>
1927 V3 N4 P36-45 60-41472 <=>
1927 V3 N4 P48-50 60-41545 <=>
1927 V3 N6 P28-31 59-13950 <=>
1927 V3 N6 P35-37 59-13951 <=>
1927 V3 N7 P1-14 61-11914 <=>
1927 V3 N7 P59-65 60-41058 <=>
1927 V3 N8 P5-13 61-11913 <=>
1927 V3 N8 P13-23 60-11342 <=>
1927 V3 N8 P47-49 60-41757 <=>
1927 V3 N9 P1-13 60-41487 <=>
1927 V3 N9 P19-22 61-11771 <=>
1927 V3 N9 P30-35 61-11623 <=>
1927 V3 N9 P37-39 60-41486 <=>
1927 V3 N9 P41-43 60-11002 <=>
1927 V3 N9 P43 60-11004 <=>
1927 V3 N10 P1-4 60-41599 <=>
1927 V3 N10 P38-48 60-41206 <=>
1927 V3 N11 P30-39 60-41182 <=>
1927 V3 N11 P56-67 61-21025 <=>
1927 V3 N11 P99-112 60-11581 <=>
1927 V3 N11 P112-127 60-11962 <=>
1927 V3 N11 P130-138 61-21189 <=>
1927 V3 N12 P11-18 60-11910 <=>
1927 V3 N12 P18-23 60-11567 <=>
1927 V3 N12 P24-29 61-21012 <=>
1927 V3 N12 P40-41 59-13948 <=>
1927 V3 N12 P41-44 60-41544 <=>
1928 V4 N3 P28-37 60-11885 <=>
1928 V4 N4 P2-10 60-11977 <=>
1928 V4 N5 P27-32 60-11890 <=>
1928 V4 N5 P36-37 60-11922 <=>
1928 V4 N5 P37-40 60-41492 <=>
1928 V4 N6 P1-13 60-41056 <=>
1928 V4 N6 P41-49 60-41183 <=>
1928 V4 N7 P1-24 60-11573 <=>
1928 V4 N8 P1-8 60-11882 <=>
1928 V4 N8 P9-16 60-11937 <=>
1928 V4 N8 P30-31 60-11868 <=>
1928 V4 N9 P1-11 60-11935 <=>
1928 V4 N9 P12-16 60-11938 <=>
1928 V4 N9 P17-22 60-41133 <=>
1928 V4 N10 P5-12 60-11978 <=>
1928 V4 N12 P27-28 60-11960 <=>
1928 V4 N12 P53-57 61-21028 <=>
1928 V4 N2/3 P95 60-11373 <=>
1929 V5 N1 P8-18 60-11539 <=>
1929 V5 N1 P18-22 60-11576 <=>
1929 V5 N1 P41-46 60-11376 <=>
1929 V5 N2 P9-17 60-11336 <=>
1929 V5 N2 P38-44 60-41334 <=>
1929 V5 N2 P45-49 60-11855 <=>
1929 V5 N2 P53-57 60-41535 <=>
1929 V5 N3 P4-20 60-11577 <=>
1929 V5 N3 P30-35 61-11038 <=>
1929 V5 N3 P36-47 60-41065 <=>
1929 V5 N3 P52-53 60-11504 <=>
1929 V5 N4 P17-23 60-11500 <=>
1929 V5 N4 P24-40 60-11928 <=>
1929 V5 N4 P43-46 60-41140 <=>
1929 V5 N4 P51-52 60-11503 <=>
1929 V5 N5 P4-10 60-41335 <=>
1929 V5 N5 P22-24 60-41179 <=>
1929 V5 N5 P34-37 60-11979 <=>
1929 V5 N5 P1C3 60-11924 <=>
1929 V5 N6 P12-24 60-11501 <=>
1929 V5 N6 P35-43 60-41256 <=>
1929 V5 N6 P56-58 60-11578 <=>
1929 V5 N7 P10-17 60-11335 <=>
1929 V5 N7 P17-20 60-41010 <=>
1929 V5 N7 P21-27 60-11379 <=>
1929 V5 N7 P39-42 60-41084 <=>
1929 V5 N7 P43-44 60-41057 <=>
1929 V5 N8 P15-23 61-11037 <=>
1929 V5 N8 P30-34 60-41134 <=>
1929 V5 N9 P3-5 60-11378 <=>
1929 V5 N9 P6-13 60-41037 <=>
1929 V5 N9 P14-19 60-41181 <=>
1929 V5 N9 P20-22 60-41009 <=>
1929 V5 N9 P23-27 60-41083 <=>
1929 V5 N9 P27-30 60-11354 <=>
1929 V5 N10 P1-26 60-11857 <=>
1929 V5 N10 P36-40 60-11892 <=>
1929 V5 N10 P41-45 60-11909 <=>
1929 V5 N10 P45-51 60-11888 <=>
1929 V5 N10 P51-53 60-11502 <=>
1929 V5 N11 P13-20 60-11908 <=>
1929 V5 N11 P24-32 60-11883 <=>
1929 V5 N11 P32-42 60-11981 <=>
1929 V5 N11 P42-51 61-11857 <=>
1929 V5 N11 P51-52 61-21026 <=>
1929 V5 N11 P52-53 61-11587 <=>
1929 V5 N11 P54-58 61-21036 <=>
1929 V5 N12 P24-34 61-11753 <=>
1929 V5 N12 P35-40 60-11845 <=>
1929 V5 N12 P41-45 60-41336 <=>
1929 V5 N12 P45-57 61-11039 <=>
1929 V5 N12 P58-68 60-41598 <=>
1930 N4 P22-37 AD-634 929 <=$>
1930 V6 N1 P16-27 60-41467 <=>
1930 V6 N1 P28-35 61-11503 <=>
1930 V6 N1 P35-36 60-11374 <=>
1930 V6 N1 P37-39 60-41059 <*+>
1930 V6 N1 P45-55 60-41469 <=>
1930 V6 N1 P56-60 60-11927 <=>
1930 V6 N1 P60-62 60-41016 <=>
1930 V6 N1 P62-65 60-11375 <=>
1930 V6 N1 P65-68 60-41015 <=>
1930 V6 N1 P69-70 60-41038 <=>
1930 V6 N1 P72-73 60-41019 <=>
1930 V6 N1 P84-85 60-11541 <=>
1930 V6 N4 P9-15 60-41055 <=>
1930 V6 N4 P15-20 60-41539 <=>
1930 V6 N4 P42-43 60-41570 <=>
1930 V6 N4 P44-51 61-21040 <=>
1930 V6 N4 P52-56 60 41546 <=>
1930 V6 N5 P12-21 61-11734 <=>
1930 V6 N5 P30 60-41591 <=>
1930 V6 N5 P37-96 60-41613 <=>
1930 V6 N6 P13-25 60-41536 <=>
1930 V6 N6 P30-32 60-41537 <=>
1930 V6 N6 P33-46 61-19043 <+*>
1930 V6 N6 P46-54 61-21051 <=>
1930 V6 N6 P55 60-41337 <=>
1930 V6 N6 P63-85 61-11884 <=>
1930 V6 N6 P86-93 60-41135 <=>
1930 V6 N7 P15-16 60-11957 <=>
1930 V6 N10 P20-32 61-11817 <=>
1930 V6 N10 P35-39 61-19044 <+*>
1930 V6 N10 P57-67 61-11873 <=>
1930 V6 N10 P74-77 61-11887 <=>
1930 V6 N10 P77-79 61-19045 <+*>
1930 V6 N11 P53-56 61-11820 <=>
1930 V6 N11 P67-69 61-11834 <=>
1930 V6 N11 P88-89 61-21155 <=>
1930 V6 N12 P11-28 61-15949 <+*>
1930 V6 N12 P29-33 61-21187 <=>
1930 V6 N12 P34-38 61-11833 <=>
1930 V6 N12 P38-39 61-11933 <=>
1930 V6 N12 P40-43 61-11937 <=>
1930 V6 N12 P43-48 61-21138 <=>
1930 V6 N12 P52 61-11816 <=>
1930 V6 N2/3 P7-16 61-19042 <+*>
1930 V6 N2/3 P20 60-41020 <=>
1930 V6 N2/3 P30-35 60-41008 <=>

Citation	Number	Code
1930 V6 N2/3 P35-45	60-41125	<=>
1930 V6 N2/3 P46-57	60-41124	<=>
1930 V6 N2/3 P60-64	60-41021	<*+>
1930 V6 N2/3 P64-65	60-41018	<=>
1930 V6 N2/3 P109-141	66-11566	<=> 0
1930 V6 N8/9 P1-16	60-41470	<=>
1930 V6 N8/9 P16-25	61-11835	<=>
1930 V6 N8/9 P32-42	60-41603	<=>
1930 V6 N8/9 P43-47	60-11884	<=>
1930 V6 N8/9 P51-52	60-11787	<=>
1930 V6 N8/9 P60-62	60-41471	<=>
1930 V6 N8/9 P77-102	60-41612	<=>
1931 V7 N1 P13-23	61-11783	<=>
1931 V7 N1 P24-27	61-19308	<+*>
1931 V7 N1 P28-38	61-11951	<=>
1931 V7 N1 P39-40	61-11690	<=>
1931 V7 N1 P50-54	61-21045	<=>
1931 V7 N1 P54-56	61-11698	<=>
1931 V7 N4 P37-38	60-11623	<=>
1931 V7 N4 P39-41	60-11936	<=>
1931 V7 N4 P42	60-11958	<=>
1931 V7 N4 P57-58	61-21153	<=>
1931 V7 N4 P63-66	60-11617	<=>
1931 V7 N5 P41-42	60-41600	<=>
1931 V7 N5 P42-43	60-11923	<=>
1931 V7 N5 P47-54	61-11776	<=>
1931 V7 N5 P55-59	60-41122	<=>
1931 V7 N5 P67-70	60-11618	<=>
1931 V7 N6 P22-32	61-19046	<+*>
1931 V7 N6 P33-35	60-41538	<=>
1931 V7 N6 P35-37	61-19040	<+*>
1931 V7 N6 P41-42	60-11619	<=>
1931 V7 N6 P45-47	60-11856	<=>
1931 V7 N6 P52-56	60-11980	<=>
1931 V7 N7 P4-14	61-19047	<+*>
1931 V7 N7 P14-24	61-19048	<+*>
1931 V7 N7 P27-33	60-41017	<=>
1931 V7 N7 P37-41	61-11696	<=>
1931 V7 N7 P41-48	60-11926	<=>
1931 V7 N10 P5-12	60-41180	<=>
1931 V7 N10 P20-35	60-41060	<=>
1931 V7 N10 P36-43	61-11945	<=>
1931 V7 N1C P70	61-11751	<=*>
1931 V7 N2/3 P64-67	61-21154	<=>
1931 V7 N11/2 P17-19	61-11694	<=>
1931 V7 N11/2 P19-20	61-21152	<=>
1931 V7 N11/2 P21-23	61-21156	<=>
1931 V7 N11/2 P26-31	61-11971	<=>
1931 V7 N11/2 P31-34	61-21158	<=>
1931 V7 N11/2 P34-36	61-11700	<=>
1931 V7 N11/2 P56-59	61-21033	<=>
1931 V7 N11/2 P70-	61-11732	<=>
1932 V8 N4 P23-25	61-11005	<=>
1932 V8 N4 P26-30	61-19301	<=>
1932 V8 N4 P31-37	61-11755	<=>
1932 V8 N4 P55-58	61-21027	<=>
1932 V8 N5 P1-5	60-41436	<=>
1932 V8 N5 P13-31	61-11871	<=>
1932 V8 N5 P40-42	61-23098	<*=>
1932 V8 N5 P45-49	61-23097	<=*>
1932 V8 N5 P49-55	60-11959	<=>
1932 V8 N5 P56-61	61-11729	<=>
1932 V8 N5 P79-80	61-21034	<=>
1932 V8 N6 P15-18	61-21068	<=>
1932 V8 N6 P18-23	61-19167	<+*>
1932 V8 N2/3 P1-7	61-11752	<=>
1932 V8 N2/3 P7-13	61-19049	<+*>
1932 V8 N2/3 P20-22	61-19096	<+*>
1932 V8 N2/3 P22-32	60-41146	<=>
1932 V8 N2/3 P32-39	60-41085	<=>
1932 V8 N2/3 P80-100	60-11854	<=>
1932 V8 N2/3 P111-114	60-41039	<=>
1932 V8 N7/8 P94-115	61-21016	<=>
1932 V8 N11/2 P27-33	61-21038	<=>
1932 V8 N11/2 P86	60-11620	<=>
1932 V8 N11/2 P89-91	61-21032	<=>
1932 V8 N11/2 P110-113	61-21035	<=>
1932 V8 N9/10 P12-26	61-11947	<=>
1932 V8 N9/10 P26-31	61-21259	<=>
1932 V8 N9/10 P31-37	61-21260	<=>
1932 V8 N9/10 P45-48	61-11590	<=>
1932 V8 N9/10 P57-60	61-21088	<=>
1932 V8 N9/10 P60-74	61-11860	<=>
1932 V8 N9/10 P88-93	61-21030	<=>
1933 V9 N1/2 P13-16	61-19051	<+*>
1933 V9 N1/2 P21-24	61-21117	<=>
1933 V9 N1/2 P24-34	61-11966	<=>
1933 V9 N1/2 P54-71	60-11881	<=>
1933 V9 N3/4 P10-27	61-19052	<+*>
1933 V9 N3/4 P43-48	61-21115	<=>
1933 V9 N3/4 P48-65	61-19053	<+*>
1933 V9 N3/4 P94-98	61-21031	<=>
1933 V9 N3/4 P99-102	61-11775	<=>
1933 V9 N3/4 P103-104	61-21177	<=>
1933 V9 N5/6 P7-17	61-11965	<=>
1933 V9 N5/6 P17-26	60-11858	<=>
1933 V9 N5/6 P26-30	61-11926	<=>
1933 V9 N5/6 P38-42	61-11868	<=>
1933 V9 N5/6 P42-45	61-11920	<=>
1933 V9 N5/6 P46-55	61-11029	<=>
1933 V9 N5/6 P69-73	61-11909	<=>
1933 V9 N7/8 P14-27	61-21188	<=>
1933 V9 N7/8 P28-31	61-19054	<+*>
1933 V9 N7/8 P32-38	61-19738	<=*>
1933 V9 N7/8 P39-45	61-11856	<=>
1933 V9 N11/2 P23-28	61-11869	<=>
1933 V9 N11/2 P29-35	61-21087	<=>
1933 V9 N11/2 P36-40	61-11870	<=>
1933 V9 N11/2 P72-78	61-11989	<=>
1933 V9 N9/10 P41-43	61-11756	<=>
1933 V9 N9/10 P44-59	61-11087	<=>
1934 V10 N8 P14-19	61-11810	<=>
1934 V10 N8 P20-27	61-21020	<=>
1934 V10 N8 P28-32	61-21263	<=>
1934 V10 N8 P34-46	61-21014	<=>
1934 V10 N8 P51-52	61-21021	<=>
1934 V10 N8 P63-64	60-41590	<=>
1934 V10 N1/2 P45-50	60-41014	<=>
1934 V10 N1/2 P50-51	61-11893	<=>
1934 V10 N1/2 P52-53	61-19293	<+*>
1934 V10 N1/2 P71-72	60-11889	<=>
1934 V10 N1/2 P89	61-21112	<=>
1934 V10 N1/2 P90-99	61-11866	<=>
1934 V10 N3/4 P36-37	61-11500	<=>
1934 V10 N3/4 P38-42	61-11872	<=>
1934 V10 N3/4 P42-48	61-21113	<=>
1934 V10 N3/4 P51-55	61-21116	<=>
1934 V10 N5/6 P81-88	61-21029	<=>
1934 V10 N5/6 P92-96	61-21022	<=>
1934 V10 N11/2 P7-19	61-19415	<=*>
1934 V10 N11/2 P60-66	61-11921	<=>
1934 V10 N11/2 P67-72	61-11940	<=>
1934 V10 N9/10 P97-101	61-21080	<=>
1935 V11 N1 P29-32	60-41338	<=>
1935 V11 N1 P37-41	61-21086	<=>
1935 V11 N1 P51-59	61-21160	<=>
1935 V11 N2 P8-36	61-11996	<=>
1935 V11 N2 P46-48	61-21110	<=>
1935 V11 N2 P48-60	61-11952	<=>
1935 V11 N3 P8-17	61-11077	<=>
1935 V11 N3 P18-21	61-19291	<+*>
1935 V11 N3 P21-24	61-19290	<+*>
1935 V11 N3 P24-30	61-19289	<+*>
1935 V11 N3 P30-32	61-19088	<+*>
1935 V11 N3 P42-45	61-11983	<=>
1935 V11 N4 P34-46	61-11970	<=>
1935 V11 N4 P57-62	61-11702	<=>
1935 V11 N4 P62-70	61-11724	<=>
1935 V11 N5 P16-18	61-19055	<+*>
1935 V11 N5 P3C-32	61-19056	<+*>
1935 V11 N5 P46-48	61-11686	<=>
1935 V11 N7 P22-27	61-19160	<+*>
1935 V11 N7 P27-29	61-11943	<=>
1935 V11 N7 P29-34	61-21090	<=>
1935 V11 N7 P34-36	61-11944	<=>
1935 V11 N8 P17-22	61-19288	<+*>
1935 V11 N8 P29-32	61-19058	<+*>
1935 V11 N8 P33-43	61-21262	<=>
1935 V11 N8 P43-45	61-21111	<=>
1935 V11 N8 P45	61-11807	<=>

Citation	Translation
1935 V11 N9 P32	61-11806 <=>
1935 V11 N9 P45	61-11808 <=>
1935 V11 N9 P46	61-11804 <=>
1935 V11 N10 P5-14	61-11980 <=>
1935 V11 N10 P49-57	61-21018 <=>
1935 V11 N10 P71-76	61-11985 <=>
1935 V11 N12 P26-35	61-11910 <=>
1935 V11 N12 P73-77	61-21083 <=>
1936 V12 N1 P53-56	61-21089 <=>
1936 V12 N1 P57	61-11803 <=>
1936 V12 N2 P30-38	61-11928 <=>
1936 V12 N2 P39-43	61-11938 <=>
1936 V12 N2 P43-50	61-11927 <=>
1936 V12 N2 P70-72	61-19285 <+*>
1936 V12 N3 P13-32	61-21218 <=>
1936 V12 N4 P1-10	61-11707 <=>
1936 V12 N4 P10-21	61-21219 <=>
1936 V12 N4 P22-24	61-11802 <=>
1936 V12 N4 P25-26	61-21076 <=>
1936 V12 N4 P42-44	61-21091 <=>
1936 V12 N6 P27-28	61-21295 <=>
1936 V12 N7 P41-44	61-19281 <+*>
1936 V12 N7 P54-58	61-11705 <=>
1936 V12 N7 P61-62	61-11805 <=>
1936 V12 N7 P62-64	61-21079 <=>
1936 V12 N10 P58-62	61-21217 <=>
1936 V12 N11 P62-71	61-11878 <=>
1936 V12 N11 P72-80	61-21216 <=>
1936 V12 N8/9 P41-55	61-19066 <+*>
1936 V12 N8/9 P56-65	61-11708 <=>
1937 V13 N1 P46-54	61-19508 <=*>
1937 V13 N3 P42-45	60-13025 <+*>
1937 V13 N5 P49-52	61-19969 <=*>
1937 V13 N5 P53-54	61-19881 <=*>
1937 V13 N6 P51-57	61-19067 <+*>
1937 V13 N9 P32-35	61-19883 <=*>
1937 V13 N10 P9-32	61-11997 <=>
1937 V13 N10 P51-57	61-19274 <+*>
1938 V14 N4 P30-32	61-19970 <=*>
1938 V14 N4 P45-49	61-19268 <+*>
1938 V14 N7 P58-64	61-19981 <=*>
1938 V14 N7 P65-66	61-19068 <+*>
1938 V14 N8 P54-74	61-19976 <=*>
1938 V14 N12 P54-73	61-23484 <=*>
1939 V15 N1 P66-69	61-19069 <+*>
1939 V15 N3 P78-85	61-21010 <=>
1939 V15 N4 P35-47	61-19412 <+*>
1939 V15 N4 P48-56	61-19258 <+*>
1939 V15 N4 P84-87	61-19257 <+*>
1939 V15 N5 P38-45	61-19071 <+*>
1939 V15 N6 P51-54	61-19072 <+*>
1939 V15 N6 P68-70	61-21157 <=>
1939 V15 N7 P7-16	61-19464 <+*>
1939 V15 N7 P16-23	61-19423 <+*>
1939 V15 N7 P37-40	61-19073 <+*>
1939 V15 N9 P48-53	61-19075 <+*>
1940 V16 N1 P9-24	61-19150 <+*>
1940 V16 N2 P7-21	61-19990 <=*>
1940 V16 N2 P21-27	61-19499 <=*>
1940 V16 N5 P60-64	61-19146 <+*>
1940 V16 N5 P71-74	61-11809 <=>
1940 V16 N7 P9-21	61-19991 <=*>
1940 V16 N7 P35-39	61-19074 <+*>
1940 V16 N10 P5-12	61-19977 <=*>
1940 V16 N10 P13-20	61-19506 <=*>
1940 V16 N10 P27-30	61-19505 <=*>
1940 V16 N10 P56-61	61-19512 <=*>
1940 V16 N11 P26-29	61-23065 <=*>
1940 V16 N11 P48-57	61-19975 <=*>
1940 V16 N11 P62-64	61-19251 <=>
1940 V16 N12 P5-21	61-19988 <=*>
1940 V16 N12 P69-76	61-19979 <=*>

★GEODEZIYA I KARTOGRAFIYA

Citation	Translation
1957 N10 P36-44	IGR V1 N8 P73 <AGI>
	59-11422 <=>
1957 N10 P44-55	59-11421 <=>
1957 N11 P3-11	59-11309 <=>
1957 N11 P20-24	59-11310 <=>
1957 N11 P32-39	59-11311 <=>
1957 N11 P39-44	59-11312 <=>
1957 N11 P53-63	59-11313 <=>
1958 N3 P16-23	59-11377 <=> O
1958 N5 P42-45	64-13172 <=*$>
1958 N8 P3-79	60-19039 <+*>
1958 N11 P29-42	AD-621 736 <=*$>
1959 N6 P50-56	61-19449 <+*>
1960 N5 P6-10	60-31773 <=>
1961 N4 P77-80	62-15465 <*=>
1961 N6 P70-77	64-15158 <=*$>
1962 N1 P43-47	64-11602 <=>
1962 N4 P70-73	66-32382 <=$>
1962 N6 P37-46	62-33594 <=*>
1962 N8 P65-68	63-19255 <=*>
1963 N3 P8-14	63-31776 <=>
1963 N4 P38-42	63-31073 <=>
1963 N8 P62-66	63-31966 <=>
1964 N4 P75-76	AD-626 271 <=*$>
1964 N4 P77-78	AD-626 272 <=*$>
1964 N5 P16-17	AD-626 070 <=*$>
1964 N5 P42-48	AD-611 333 <=$>
1964 N6 P3-6	AD-608 933 <=>
1964 N6 P3-5	64-41042 <=>
1964 N6 P17-21	AD-610 472 <=$>
1964 N6 P23-26	AD-610 473 <=$>
1964 N6 P40-46	64-41051 <=>
1964 N6 P46-53	AD-614 107 <=$>
1964 N6 P53-60	AD-627 918 <=*$>
1964 N7 P3-10	AD-610 474 <=$>
1964 N7 P11-13	AD-610 475 <=$>
1964 N7 P25-28	AD-626 795 <=*$>
1964 N7 P34-46	AD-614 105 <=$>
	AD-627 917 <=$>
1964 N7 P46-48	AD-610 476 <=$>
1964 N7 P49-52	AD-614 106 <=$>
	AD-626 790 <=*$>
1964 N7 P52-55	AD-610 477 <=$>
1964 N7 P52-54	AD-626 079 <=*$>
1964 N7 P55-62	AD-610 875 <=$>
1964 N9 P21-24	64-51515 <=>
1964 N10 P3-8	AD-626 286 <=*$>
1964 N10 P15-17	AD-622 465 <=*>
1964 N10 P37-42	AD-622 465 <=*>
1964 N10 P42-45	AD-627 861 <=*$>
1964 N11 P67-77	65-30096 <=$>
1964 N12 P39-42	AD-627 117 <=$>
1964 N12 P46-52	AD-627 700 <=*$>
1965 N1 P11-23	AD-621 773 <=*$> M
1965 N6 P9-14	65-32231 <=$>
1965 N10 P3-7	AD-641 757 <=$>
1965 N11 P3-8	AD-634 285 <=$>
1965 N11 P25-29	AD-631 277 <=$>
1965 N11 P29-38	66-33803 <$=>
1965 N12 P3-17	AD-634 284 <=$>
1966 N1 P17-20	AD-641 274 <=$>
1966 N4 P10-14	66-34354 <=$>
1966 N6 P32-39	66-33803 <=$>
1966 N6 P48-51	66-33790 <=$>
1966 N6 P51-61	66-33689 <=$>

GEODEZIYA: SPRAVACHNOE RUKOVODSTVO

Citation	Translation
1939 V7 P186-198 PT2-4	63-13562 <=*>

GEODEZJA I KARTOGRAFIA

Citation	Translation
1953 N2 P117-126	R-462 <*>

GEOFISICA E METEOROLOGIA

Citation	Translation
1953 V1 P33-35	57M38I <ATS>
1955 V3 N1/2 P23-29	05H12I <ATS>
1958 V6 N1/2 P3-8	59-12697 <=*> O
1960 V8 N3/4 P80-82	64-14316 <*> O
1961 N9 P39-46	62-00394 <*>
1964 V13 N1/2 P12-14	64-20136 <*>
1964 V13 N1/2 P28-32	64-20154 <*>
1964 V13 N3/4 P65-66	65-12071 <*>
1964 V13 N3/4 P76-79	65-12072 <*>
1964 V13 N5/6 P107-108	65-14576 <*>
1965 V14 N1/2 P28-29	65-14574 <*>

GEOFISICA PURA E APPLICATA

1950 V16 P133-135 58M38I <ATS>
1950 V16 N4/5 P181-187 AL-51 <*>
1950 V17 P40-45 <ATS>
AL-52 <*>
1950 V18 P178-191 II-263 <*>
1951 V19 P19-32 C-2657 <NRC>
1952 V22 N1/2 P63-74 62-14275 <*>
1953 V25 P29-36 II-287 <*>
1955 V30 P155-159 57-3100 <*>
1955 V30 N1 P137-143 66-11354 <*>
1956 V34 P1-20 64-30670 <*>
1958 V40 P19-35 PT2 64-30661 <*>
1958 V41 N3 P133-140 62-11554 <=>
1961 V50 P235-242 AEC-TR-6422 <*>
1961 V50 P260-277 66-11754 <*>
1963 V56 P203-215 66-12033 <*>

GEOFIZICHESKAYA RAZVEDKA
1961 N4 P31- 18P63R <ATS>
1961 N5 P94-103 62-32881 <=*>
1961 N6 P47-59 63-21034 <=>
1961 N6 P93-95 63-21034 <=>
1961 N6 P118-124 63-21034 <=>
1961 N6 P124-129 63-21223 <=>
1962 N10 P34-44 64-15155 <=*$>
1963 N12 P48-60 06R80R <ATS>
1963 V13 P15-36 65-12724 <*>

GEOFIZICHESKIE METODY RAZVEDKA V ARKTIKI
1962 N4 P10-20 63-31894 <=>

GEOFIZICHESKII BYULLETEN. MOSCOW
1963 N13 P38-40 AD-628 037 <=$>
1963 N13 P80-93 AD-625 300 <=*$>
1964 N14 P15-25 TRANS-F-68 <NLL>
1964 N14 P89-91 65-33212 <=*$>

GEOFIZICHESKII BYULLETEN. LENINGRAD
1962 N12 P61-68 <GPRC> P
1962 N12 P74-80 <GPRC> P
1963 N13 P18-28 64-41000 <=>
1963 N13 P29-33 64-51255 <=>
1963 N13 P41-42 AD-609 170 <=>
1963 N13 P49-56 AD-609 929 <=$>
1963 N13 P67-73 AD-609 928 <=$>
1963 N13 P80-93 64-31977 <=>
1964 N14 P128-158 TRANS-325(M) <NLL>

GEOFIZICHESKII SBORNIK. AKADEMIYA NAUK SSSR. URALSKII FILIAL. INSTITUT GEOFIZIKI. SVERDLOVSK (SUBSERIES OF THE INSTITUTE'S TRUDY
1962 V2 N3 P211-219 64-19759 <=$>

GEOFIZICHESKII SBORNIK. AKADEMIYA NAUK UKRAINSKOI SSR. INSTITUT GEOFIZIKI. KIEV
1962 N2 P28-32 64-21707 <=>
1962 N2 P41-45 C-5583 <NRC>
1964 N8 P44-48 AD-632 486 <=$>

GEOFIZICHESKOE PRIBOROSTROENIE
1963 N16 P38-45 AD-638 977 <=$>
1963 N16 P94-101 AD-636674 <=$>

GEOFIZIKA I ASTRONOMIYA; INFORMATSIONNYI BYULLETEN
1965 N8 P103-105 66-32227 <=$>

GEOFIZIKAI KOZLEMENYEK
1955 V4 N2 P41-44 SCL-T-330 <*>
1956 V5 N1 P53-63 CSIR-TR-457 <CSSA>

GEOFYSIKALNI SBORNIK
1957 N62 P165-215 58-2697 <*>
59-12374 <=*>
1963 V10 P11-13 64-31289 <=>

GEOFYSISKE PUBLIKASJONER
1942 V12 N6 P1-15 C-2428 <NRC>

GEOGRAFICHESKII SBORNIK
1961 V14 ENTIRE ISSUE 62-33289 <=>
1962 V15 P198-207 64-13164 <=*$>
1963 N16 P47-64 SOV.GEOGR. V5 N6 P36 <AGS>

GEOGRAFICKY CASOPIS
1960 V12 N3 P224-225 62-19440 <=*>
1960 V12 N3 P225-226 62-19443 <*=>
1960 V12 N4 P247-252 62-19442 <*=>
1961 V13 N2 P147-151 62-19676 <*=>

GEOGRAFIYA
1957 V7 N7 P1-5 IGR V2 N9 P811 <AGI>
1958 V8 N1 P6-9 59-11143 <=*> 0

GEOGRAFIYA I KHOZYAISTVO
1958 N3/4 P47-54 SGRT V2 N3 P44 <AGS>
1958 N3/4 P64-70 SGRT V2 N3 P54 <AGS>
1960 N6 P32-33 SGRT V2 N3 P61 <AGS>
1960 N6 P34-44 SGRT V1 N8 P55 <AGS>
1960 N6 P69-70 SGRT V2 N3 P23 <AGS>
1960 N6 P71-80 SGRT V2 N3 P26 <AGS>
1961 N9 P16-20 SGRT V2 N10 P43 <AGS>
1961 N9 P60-64 SGRT V2 N10 P51 <AGS>
1961 N9 P74-81 SGRT V2 N10 P32 <AGS>
1961 N10 P19-23 SGRT V3 N4 P38 <AGS>
1961 N10 P44-49 SGRT V3 N4 P45 <AGS>
1961 N11 P6-8 SGRT V3 N7 P45 <AGS>
1961 N11 P92-95 SGRT V3 N7 P49 <AGS>
1961 N11 P96-102 SGRT V4 N1 P3 <AGS>
1963 N12 P3-8 SGRT V4 N9 P10 <AGS>
SGRT V4 N9 P3 <AGS>
1963 N12 P13-18 SGRT V4 N9 P16 <AGS>
1963 N12 P43-48 SGRT V4 N9 P27 <AGS>
1963 N12 P49-54 SGRT V4 N9 P34 <AGS>
1963 N12 P55-62 SGRT V4 N9 P43 <AGS>
1963 N12 P67-77 SGRT V4 N10 P3 <AGS>
1963 N12 P78-87 SGRT V4 N10 P19 <AGS>
1963 N12 P93 SGRT V4 N10 P71 <AGS>

GEOGRAFIYA V SHKOLE
1958 N5 P6-15 59-12691 <+*>
1959 N6 P28-41 SGRT V1 N4 P63 <AGS>
1959 N6 P41-46 SGRT V1 N4 P78 <AGS>
1960 N5 P37-41 SGRT V2 N2 P54 <AGS>
1960 N6 P7-15 SGRT V2 N4 P53 <AGS>
1961 N2 P5-15 SGRT V2 N7 P47 <AGS>
1961 N2 P16-20 SGRT V2 N7 P59 <AGS>
1961 N2 P64-65 SGRT V2 N7 P64 <AGS>
1961 N5 P5-13 SGRT V2 N9 P32 <AGS>
1962 N1 P10-18 SGRT V4 N5 P37 <AGS>
1962 N6 P17-20 SGRT V4 N5 P26 <AGS>
1963 N1 P19-23 SGRT V4 N5 P10 <AGS>
1964 N1 P13-25 65-63575 <=$>
1964 N1 P70-73 65-63575 <=$>
1964 N2 P8-15 AD-626 713 <=*$>
1964 N2 P16-18 AD-626 714 <=*$>
1964 N2 P18-23 AD-626 715 <=*$>
1964 N2 P70-74 AD-626 716 <=*$>
1964 N4 P10-13 SGRT V5 N10 P39 <AGS>
1965 N3 P85-86 65-64509 <=*$>
1965 N6 P2-5 SGRT V7 N7 P15 <AGS>
1965 N6 P6-12 SGRT V7 N7 P47 <AGS>

GEOGRAPHISCHE BERICHTE
1961 V6 N18 P1-9 62-19796 <=*> P
1961 V6 N18 P68 62-15957 <=*>

GEOGRAPHY. CHINESE PEOPLES REPUBLIC
SEE TI-LI

GEOGRAPHY AND INDUSTRIES. CHINESE PEOPLES REPUBLIC
SEE TI LI YU CH'AN YEH

★GEOKHIMIYA
1956 N1 P6-52 65-13783 <*> 0
1956 N1 P18-27 R-4646 <*>
1956 N2 P3-18 67M37R <ATS>
1956 N2 P68-73 62-14560 <=*>
1956 N2 P74-77 R-1660 <*>
R-1672 <*>
R-2023 <*>

65-13794 <*> 0
1956 N2 P94-96 R-4690 <*>
1956 N3 P3-8 R-1599 <*>
04J15R <ATS>
65-13793 <*>
1956 N4 P10-23 R-1661 <*>
1956 N5 P61-74 60-17467 <+*>
1956 N6 P3-24 04J19R <ATS>
1956 N6 P33-48 5J16R <ATS>
1956 N6 P84-94 R-1671 <*>
1956 N8 P3-9 65-13785 <*>
1957 N1 P36-45 61-15885 <=*>
1957 N2 P142-146 AEC-TR-4820 <=*>
1957 N2 P155-160 37K25R <ATS>
1957 N2 P161-165 76K20R <ATS>
1957 N3 P187-190 65-13781 <*>
78J18R <ATS>
1957 N3 P193-197 38K25R <ATS>
1957 N3 P204-213 R-5132 <*>
1957 N4 P304-311 60-10865 <*>
1957 N4 P337-345 58K21R <ATS>
1957 N5 P389-391 R-4637 <*>
1957 N6 P449-469 62-16666 <=*>
1957 N6 P481-492 60-23745 <+*>
1957 N6 P518-528 59K21R <ATS>
1957 N7 P559-565 28K26R <ATS>
1957 N7 P615-620 60K21R <ATS>
1957 N7 P626-637 R-4655 <*>
1958 N4 P296-306 25K27R <ATS>
1958 N5 P409-423 26K27R <ATS>
1959 N1 P3-5 59-13619 <=>
1959 N8 P679-695 61-15886 <*=> 0
1960 N1 P75-83 65N53R <ATS>
1960 N4 P362-370 19M47R <ATS>
1960 N4 P373-374 AEC-TR-4269 <+*>
1960 N7 P653-660 61-27662 <*=>
1961 N1 P30-39 83N52R <ATS>
1961 N2 P174-180 66N53R <ATS>
1961 N2 P187 62-11302 <=>
62-16146 <*=>
62-16146 <=*>
1961 N9 P812-820 61P61R <ATS>
1961 N10 P840-848 AEC-TR-5101 <=*>
1962 N5 P411-419 AEC-TR-5481 <=*>
1962 N5 P440-452 AEC-TR-5587 <*>
AEC-TR-5587 <=*>
1962 N6 P501-507 AEC-TR-5588 <*>
AEC-TR-5588 <*=>
1962 N9 P818-825 65-63968 <=$>
1962 N10 P851-871 65-23699 <$>
1962 N11 P989-992 63-19209 <=*>
1963 N1 P3-14 AEC-TR-5739 <*=>
1964 N1 P3- GI N1 1964 P1 <AGI>
1964 N1 P12- GI N1 1964 P3 <AGI>
1964 N1 P23- GI N1 1964 P8 <AGI>
1964 N1 P31- GI N1 1964 P13 <AGI>
1964 N1 P42- GI N1 1964 P22 <AGI>
1964 N1 P51- GI 64-1-6 <AGI>
1964 N1 P61- GI 64-1-7 <AGI>
1964 N1 P65- GI 64-1-8 <AGI>
1964 N1 P70- GI N1 1964 P29 <AGI>
1964 N1 P75- GI N1 1964 P32 <AGI>
1964 N1 P82- GI N1 1964 P37 <AGI>
1964 N1 P85- GI N1 1964 P40 <AGI>
1964 N2 P91- GI 64-1-13 <AGI>
1964 N2 P102- GI N1 1964 P44 <AGI>
1964 N2 P110- GI 64-1-15 <AGI>
1964 N2 P119- GI N1 1964 P50 <AGI>
1964 N2 P128- GI 64-1-17 <AGI>
1964 N2 P135- GI N1 1964 P57 <AGI>
1964 N2 P143- GI 64-1-19 <AGI>
1964 N2 P148- GI N1 1964 P63 <AGI>
1964 N2 P163- GI N1 1964 P73 <AGI>
1964 N2 P171- GI N1 1964 P78 <AGI>
1964 N2 P178- GI N1 1964 P83 <AGI>
1964 N2 P181- GI N1 1964 P86 <AGI>
1964 N2 P185- GI 64-1-25 <AGI>
1964 N3 P195- GI N2 1964 P193 <AGI>
1964 N3 P208- GI N2 1964 P198 <AGI>
1964 N3 P219- GI N2 1964 P205 <AGI>
1964 N3 P230- GI N2 1964 P213 <AGI>
1964 N3 P242- GI N2 1964 P221 <AGI>
1964 N3 P253- GI N2 1964 P229 <AGI>
1964 N3 P258- GI N2 1964 P232 <AGI>
1964 N3 P266- GI N2 1964 P238 <AGI>
1964 N3 P276- GI-2-10-64 <AGI>
1964 N3 P282- GI N2 1964 P246 <AGI>
1964 N3 P289- GI N2 1964 P251 <AGI>
1964 N3 P291- GI N2 1964 P253 <AGI>
1964 N4 P299- GI N2 1964 P256 <AGI>
1964 N4 P307- GI N2 1964 P262 <AGI>
1964 N4 P315- GI N2 1964 P268 <AGI>
1964 N4 P325- GI N2 1964 P275 <AGI>
1964 N4 P332- GI N2 1964 P279 <AGI>
1964 N4 P340- GI N2 1964 P285 <AGI>
1964 N4 P348- GI N2 1964 P290 <AGI>
1964 N4 P357- GI64-2-21 <AGI>
1964 N4 P362- GI N2 1964 P297 <AGI>
1964 N4 P368- GI64-2-23 <AGI>
1964 N4 P373- GI64-2-24 <AGI>
1964 N4 P375- GI N2 1964 P302 <AGI>
1964 N4 P380- GI N2 1964 P306 <AGI>
1964 N5 P395- GI N3 1964 P395 <AGI>
1964 N5 P399- GI N3 1964 P398 <AGI>
1964 N5 P404- GI N3 1964 P402 <AGI>
1964 N5 P414- GI N3 1964 P404 <AGI>
1964 N5 P420- GI N3 1964 P413 <AGI>
1964 N5 P431- GI N3 1964 P421 <AGI>
1964 N5 P441- GI N3 1964 P429 <AGI>
1964 N5 P457- GI N3 1964 P441 <AGI>
1964 N5 P468- GI N3 1964 P451 <AGI>
1964 N5 P477- GI N3 1964 P459 <AGI>
1964 N5 P481- GI N3 1964 P463 <AGI>
1964 N5 P483- GI N3 1964 P465 <AGI>
1964 N6 P491- GI N3 1964 P468 <AGI>
1964 N6 P491-499 65-20423 <$>
1964 N6 P500- GI N3 1964 P475 <AGI>
1964 N6 P505- GI N3 1964 P479 <AGI>
1964 N6 P518- GI N3 1964 P489 <AGI>
1964 N6 P529- GI N3 1964 P497 <AGI>
1964 N6 P542- GI N3 1964 P507 <AGI>
1964 N6 P549- GI N3 1964 P512 <AGI>
1964 N6 P564- GI N3 1964 P522 <AGI>
1964 N6 P573- GI N3 1964 P529 <AGI>
1964 N6 P578- GI N3 1964 P532 <AGI>
1964 N7 P587- GI N4 1964 P613 <AGI>
1964 N7 P601- GI N4 1964 P625 <AGI>
1964 N7 P610- GI N4 1964 P632 <AGI>
1964 N7 P623- GI N4 1964 P643 <AGI>
1964 N7 P635- GI N4 1964 P653 <AGI>
1964 N7 P641- GI N4 1964 P658 <AGI>
1964 N7 P646- GI N4 1964 P663 <AGI>
1964 N7 P650- GI N4 1964 P667 <AGI>
1964 N7 P654- GI N4 1964 P670 <AGI>
1964 N7 P660- GI N4 1964 P676 <AGI>
1964 N7 P672- GI N4 1964 P687 <AGI>
1964 N7 P678- GI N4 1964 P693 <AGI>
1964 N7 P683- GI N4 1964 P698 <AGI>
1964 N7 P684- GI N4 1964 P701 <AGI>
1964 N7 P687- GI N4 1964 P704 <AGI>
1964 N7 P692- GI N4 1964 P709 <AGI>
1964 N8 P715- GI N4 1964 P713 <AGI>
1964 N8 P744- GI N4 1964 P739 <AGI>
1964 N8 P754- GI N4 1964 P747 <AGI>
1964 N8 P758- GI N4 1964 P750 <AGI>
1964 N8 P766- GI N4 1964 P757 <AGI>
1964 N8 P780- GI N4 1964 P769 <AGI>
1964 N8 P788- GI N4 1964 P775 <AGI>
1964 N8 P795- GI N4 1964 P782 <AGI>
1964 N8 P802- GI N4 1964 P788 <AGI>
1964 N8 P811- GI N4 1964 P796 <AGI>
1964 N8 P817- GI N4 1964 P801 <AGI>
1964 N8 P825- GI N4 1964 P808 <AGI>
1964 N8 P828- GI N4 1964 P812 <AGI>
1964 N8 P831- GI N4 1964 P816 <AGI>
1964 N8 P836- GI N4 1964 P822 <AGI>
1964 N9 P843- GI N5 1964 P831 <AGI>
1964 N9 P849- GI N5 1964 P837 <AGI>
1964 N9 P855- GI N5 1964 P843 <AGI>
1964 N9 P866- GI N5 1964 P853 <AGI>

1964 N9 P872-	GI N5 1964 P855 <AGI>
1964 N9 P886-	GI N5 1964 P868 <AGI>
1964 N9 P893-	GI N5 1964 P875 <AGI>
1964 N9 P898-	GI N5 1964 P880 <AGI>
1964 N9 P908-	GI N5 1964 P889 <AGI>
1964 N9 P917-	GI 64-5-10 <AGI>
1964 N9 P926-	GI N5 1964 P898 <AGI>
1964 N9 P937-	GI N5 1964 P908 <AGI>
1964 N9 P945-	GI N5 1964 P915 <AGI>
1964 N9 P954-	GI N5 1964 P922 <AGI>
1964 N9 P956-	GI N5 1964 P925 <AGI>
1964 N9 P960-	GI N5 1964 P929 <AGI>
1964 N9 P962-	GI N5 1964 P933 <AGI>
1964 N10 P971-	GI N5 1964 P936 <AGI>
1964 N10 P980-	GI 64-5-20 <AGI>
1964 N10 P988-	GI N5 1964 P945 <AGI>
1964 N10 P995-	GI N5 1964 P951 <AGI>
1964 N10 P1015-	GI N5 1964 P970 <AGI>
1964 N10 P1022-	GI N5 1964 P977 <AGI>
1964 N10 P1028-	GI 64-5-25 <AGI>
1964 N10 P1037-	GI N5 1964 P983 <AGI>
1964 N10 P1043-	GI N5 1964 P989 <AGI>
1964 N10 P1048-	GI N5 1964 P994 <AGI>
1964 N10 P1055-	GI N5 1964 P1000 <AGI>
1964 N10 P1058-	GI N5 1964 P1003 <AGI>
1964 N10 P1060-	GI N5 1964 P1006 <AGI>
1964 N10 P1064-	GI N5 1964 P1011 <AGI>
1964 N10 P1067-	GI N5 1964 P1014 <AGI>
1964 N10 P1069-	GI N5 1964 P1017 <AGI>
1964 N10 P1070-	GI N5 1964 P1019 <AGI>
1964 N11 P1075-	GI 64-6-1 <AGI>
1964 N11 P1075-1085	65-23690 <$>
1964 N11 P1087-	GI 64-6-2 <AGI>
1964 N11 P1097-	GI N6 1964 P1057 <AGI>
1964 N11 P1116-	GI N6 1964 P1068 <AGI>
1964 N11 P1122-	GI 64-6-5 <AGI>
1964 N11 P1128-	GI N6 1964 P1075 <AGI>
1964 N11 P1138-	GI N6 1964 P1083 <AGI>
1964 N11 P1146-	GI 64-6-8 <AGI>
1964 N11 P1152-	GI N6 1964 P1091 <AGI>
1964 N11 P1157-	GI N6 1964 P1096 <AGI>
1964 N11 P1164-	GI N6 1964 P1103 <AGI>
1964 N11 P1182-	GI N6 1964 P1107 <AGI>
1964 N11 P1193-	GI N6 1964 P1115 <AGI>
1964 N11 P1199-	GI 64-6-15 <AGI>
1964 N11 P1200-	GI 64-6-16 <AGI>
1964 N11 P1202-	GI N6 1964 P1122 <AGI>
1964 N11 P1204-	GI 64-6-18 <AGI>
1964 N11 P1206-	GI 64-6-19 <AGI>
1964 N11 P1209-	GI 64-6-20 <AGI>
1964 N12 P1219-	GI N6 1964 P1129 <AGI>
1964 N12 P1219-1227	N65-19507 <=*$>
1964 N12 P1228-	GI N6 1964 P1137 <AGI>
1964 N12 P1240-	GI 64-6-23 <AGI>
1964 N12 P1244-	GI 64-6-24 <AGI>
1964 N12 P1256-	GI 64-6-25 <AGI>
1964 N12 P1261-	GI 64-6-26 <AGI>
1964 N12 P1267-	GI N6 1964 P1152 <AGI>
1964 N12 P1280-	GI N6 1964 P1164 <AGI>
1964 N12 P1286-	GI 64-6-29 <AGI>
1964 N12 P1293-	GI N6 1964 P1171 <AGI>
1964 N12 P1299-	GI 64-6-31 <AGI>
1964 N12 P1305-	GI 64-6-30 <AGI>
1964 N12 P1313-	GI 64-6-33 <AGI>
1964 N12 P1318-	GI N6 1964 P1179 <AGI>
1964 N12 P1319-	GI 64-6-35 <AGI>
1964 N12 P1324-	GI N6 1964 P1181 <AGI>
1964 N12 P1327-	GI N6 1964 P1185 <AGI>
1965 N1 P3-8	GI 65-1-1 <AGI>
1965 N1 P9-15	GI V2 N1 1965 P2 <AGI>
1965 N1 P16-24	GI 65-1-3 <AGI>
1965 N1 P25-31	GI V2 N1 1965 P10 <AGI>
1965 N1 P32-42	GI V2 N1 1965 P16 <AGI>
1965 N1 P43-55	GI V2 N1 1965 P25 <AGI>
1965 N1 P56-63	GI 65-1-7 <AGI>
1965 N1 P64-67	GI V2 N1 1965 P38 <AGI>
1965 N1 P68-72	GI V2 N1 1965 P42 <AGI>
1965 N1 P73-85	GI V2 N1 1965 P47 <AGI>
1965 N1 P86-91	GI 65-1-11 <AGI>
1965 N1 P92-96	GI V2 N1 1965 P60 <AGI>
1965 N1 P97-103	GI V2 N1 1965 P65 <AGI>
1965 N1 P104	GI V2 N1 1965 P71 <AGI>
1965 N1 P108-110	GI 65-1-16 <AGI>
1965 N1 P110-113	GI V2 N1 1965 P74 <AGI>
1965 N1 P113-114	GI V2 N1 1965 P77 <AGI>
1965 N1 P115-117	GI V2 N1 1965 P79 <AGI>
1965 N1 P117-121	GI V2 N1 1965 P83 <AGI>
1965 N1 P121-123	GI 65-1-21 <AGI>
1965 N1 P123-125	GI V2 N1 1965 P88 <AGI>
1965 N2 P131-158	GI V2 N1 1965 P92 <AGI>
1965 N2 P159-174	GI V2 N1 1965 P118 <AGI>
1965 N2 P175-179	GI 65-2-3 <AGI>
1965 N2 P180-187	GI V2 N1 1965 P133 <AGI>
1965 N2 P188-197	GI 65-2-5 <AGI>
1965 N2 P198-210	GI 65-2-6 <AGI>
1965 N2 P211-218	GI V2 N1 1965 P141 <AGI>
1965 N2 P219-226	GI V2 N1 1965 P148 <AGI>
1965 N2 P227-240	GI 65-2-9 <AGI>
1965 N2 P241-245	GI V2 N1 1965 P156 <AGI>
1965 N2 P245-247	GI 65-2-11 <AGI>
1965 N3 P259-268	GI 65-3-1 <AGI>
1965 N3 P269-272	GI V2 N2 1965 P192 <AGI>
1965 N3 P273-276	GI N2 N2 1965 P196 <AGI>
1965 N3 P277-290	GI 65-3-4 <AGI>
1965 N3 P291-301	GI V2 N2 1965 P202 <AGI>
1965 N3 P302-313	GI 65-3-6 <AGI>
1965 N3 P314-324	GI V2 N2 1965 P213 <AGI>
1965 N3 P325-336	GI V2 N2 1965 P223 <AGI>
1965 N3 P337-342	GI 65-3-9 <AGI>
1965 N3 P343-347	GI 65-3-10 <AGI>
1965 N3 P348-359	GI 65-3-11 <AGI>
1965 N3 P360-365	GI 65-3-12 <AGI>
1965 N3 P366-368	GI 65-3-13 <AGI>
1965 N3 P368-371	GI V2 N2 1965 P238 <AGI>
1965 N3 P371-372	GI V2 N2 1965 P242 <AGI>
1965 N3 P373-375	GI V2 N2 1965 P244 <AGI>
1965 N4 P387-389	GI V2 N2 1965 P249 <AGI>
1965 N4 P387-390	N65-29728 <=*$>
1965 N4 P390-405	GI V2 N2 1965 P254 <AGI>
1965 N4 P406-413	GI 65-4-3 <AGI>
1965 N4 P414-420	GI 65-4-4 <AGI>
1965 N4 P421-432	GI V2 N2 1965 P269 <AGI>
1965 N4 P433-442	GI 65-4-6 <AGI>
1965 N4 P443-455	GI 65-4-7 <AGI>
1965 N4 P456-465	GI 65-4-8 <AGI>
1965 N4 P466-473	GI V2 N2 1965 P286 <AGI>
1965 N4 P474-481	GI V2 N2 1965 P293 <AGI>
1965 N4 P482-489	GI V2 N2 1965 P301 <AGI>
1965 N4 P490-494	GI V2 N2 1965 P308 <AGI>
1965 N4 P494-496	GI V2 N2 1965 P313 <AGI>
1965 N4 P497-499	GI 65-4-14 <AGI>
1965 N5 P507-518	GI V2 N3 1965 P397 <AGI>
1965 N5 P519-527	GI 65-5-2 <AGI>
1965 N5 P528-533	GI V2 N3 1965 P408 <AGI>
1965 N5 P534-543	GI 65-5-4 <AGI>
1965 N5 P544-550	GI 65-5-5 <AGI>
1965 N5 P551-555	GI V2 N3 1965 P416 <AGI>
1965 N5 P556-561	GI V2 N3 1965 P421 <AGI>
1965 N5 P562-573	GI V2 N3 1965 P426 <AGI>
1965 N5 P574-581	GI V2 N3 1965 P437 <AGI>
1965 N5 P582-586	GI V2 N3 1965 P445 <AGI>
1965 N5 P587-601	GI V2 N3 1965 P449 <AGI>
1965 N5 P602-616	GI V2 N3 1965 P463 <AGI>
1965 N5 P617-619	GI V2 N3 1965 P476 <AGI>
1965 N5 P619-624	GI V2 N3 1965 P479 <AGI>
1965 N5 P624-627	GI 65-5-15 <AGI>
1965 N5 P627-629	GI 65-5-16 <AGI>
1965 N5 P629-632	GI V2 N3 1965 P488 <AGI>
1965 N6 P643-651	GI V2 N3 1965 P495 <AGI>
1965 N6 P652-660	GI 65-6-2 <AGI>
1965 N6 P661-667	GI 65-6-3 <AGI>
1965 N6 P668-673	GI V2 N3 1965 P506 <AGI>
1965 N6 P681-694	GI V2 N3 1965 P513 <AGI>
1965 N6 P695-706	GI 65-6-7 <AGI>
1965 N6 P707-723	GI V2 N3 1965 P527 <AGI>
1965 N6 P724-732	GI V2 N3 1965 P544 <AGI>
1965 N6 P733-738	GI V2 N3 1965 P553 <AGI>
1965 N6 P739-746	GI V2 N3 1965 P559 <AGI>
1965 N6 P747-756	GI V2 N3 1965 P566 <AGI>
1965 N6 P757-758	GI V2 N3 1965 P575 <AGI>

1965 N6 P758-762 GI V2 N3 1965 P577 <AGI>
1965 N6 P762-766 GI V2 N3 1965 P582 <AGI>
1965 N7 P779-790 GI 65-7-1 <AGI>
1965 N7 P791-800 GI V2 N4 1965 P608 <AGI>
1965 N7 P801-812 GI V2 N4 1965 P617 <AGI>
1965 N7 P813-826 GI V2 N4 1965 P629 <AGI>
1965 N7 P827-833 GI 65-7-5 <AGI>
1965 N7 P834-843 GI V2 N4 1965 P643 <AGI>
1965 N7 P844-850 GI V2 N4 1965 P653 <AGI>
1965 N7 P851-863 GI V2 N4 1965 P660 <AGI>
1965 N7 P864-869 GI 65-7-9 <AGI>
1965 N7 P870-873 GI 65-7-10 <AGI>
1965 N7 P874-878 GI 65-7-11 <AGI>
1965 N7 P878-879 GI V2 N4 1965 P675 <AGI>
1965 N7 P880-887 GI 65-7-13 <AGI>
1965 N7 P887-888 GI 65-7-14 <AGI>
1965 N7 P889-894 GI V2 N4 1965 P680 <AGI>
1965 N8 P909-917 GI 65-8-1 <AGI>
1965 N8 P918-935 GI 65-8-2 <AGI>
1965 N8 P936-943 GI 65-8-3 <AGI>
1965 N8 P944-960 GI 65-8-4 <AGI>
1965 N8 P961-976 GI V2 N4 1965 P692 <AGI>
1965 N8 P977-995 GI V2 N4 1965 P709 <AGI>
1965 N8 P996-1003 GI 65-8-7 <AGI>
1965 N8 P1004-1010 GI 65-8-8 <AGI>
1965 N8 P1011-1024 GI V2 N4 1965 P729 <AGI>
1965 N8 P1024-1037 GI V2 N4 1965 P741 <AGI>
1965 N8 P1056-1063 GI V2 N4 1965 P756 <AGI>
1965 N8 P1063-1064 GI V2 N4 1965 P765 <AGI>
1965 N9 P1067-1075 GI 65-9-1 <AGI>
1965 N9 P1076-1084 GI V2 N5 1965 P785 <AGI>
1965 N9 P1085-1092 GI V2 N5 1965 P794 <AGI>
1965 N9 P1093-1100 GI V2 N5 1965 P802 <AGI>
1965 N9 P1101-1104 GI V2 N5 1965 P810 <AGI>
1965 N9 P1106-1113 GI V2 N5 1965 P814 <AGI>
1965 N9 P1114-1118 GI V2 N5 1965 P822 <AGI>
1965 N9 P1120-1129 GI 65-9-8 <AGI>
1965 N9 P1130-1139 GI 65-9-9 <AGI>
1965 N9 P1140-1153 GI V2 N5 1965 P830 <AGI>
1965 N9 P1154-1164 GI V2 N5 1965 P843 <AGI>
1965 N9 P1165-1167 GI V2 N5 1965 P853 <AGI>
1965 N9 P1168-1170 GI V2 N5 1965 P857 <AGI>
1965 N9 P1171-1173 GI V2 N5 1965 P861 <AGI>
1965 N10 P1179-1185 GI V2 N5 1965 P865 <AGI>
1965 N10 P1186-1190 GI 65-10-2 <AGI>
1965 N10 P1191-1206 GI V2 N5 1965 P873 <AGI>
1965 N10 P1207-1211 GI 65-10-4 <AGI>
1965 N10 P1212-1221 GI V2 N5 1965 P886 <AGI>
1965 N10 P1223-1233 GI 65-10-6 <AGI>
1965 N10 P1234-1245 GI V2 N5 1965 P897 <AGI>
1965 N10 P1246-1256 GI V2 N5 1965 P908 <AGI>
1965 N10 P1257-1259 GI V2 N5 1965 P918 <AGI>
1965 N10 P1259-1260 GI V2 N5 1965 P921 <AGI>
1965 N10 P1261-1264 GI 65-10-11 <AGI>
1965 N10 P1265 GI 65-10-12 <AGI>
1965 N11 P1275-1312 GI V2 N6 1965 P947 <AGI>
1965 N11 P1313-1317 GI 11-2-65 <AGI>
1965 N11 P1318-1324 GI V2 N6 1965 P986 <AGI>
1965 N11 P1325-1334 GI V2 N6 1965 P993 <AGI>
1965 N11 P1335-1345 GI V2 N6 1965 P1001 <AGI>
1965 N11 P1346-1354 GI V2 N6 1965 P1012 <AGI>
1965 N11 P1355-1363 GI 65-11-7 <AGI>
1965 N11 P1364-1366 GI 65-11-8 <AGI>
1965 N11 P1367-1370 GI 65-11-9 <AGI>
66-12496 <*>
1965 N11 P1370-1372 GI V2 N6 1965 P1024 <AGI>
1965 N11 P1372 GI 65-11-11 <AGI>
1965 N11 P1373-1375 GI 65-11-12 <AGI>
1965 N11 P1376-1378 GI V2 N6 1965 P1030 <AGI>
1965 N12 P1395-1403 GI V2 N6 1965 P1034 <AGI>
1965 N12 P1404-1409 GI 65-12-2 <AGI>
1965 N12 P1410-1422 GI 65-12-3 <AGI>
1965 N12 P1423-1438 GI 65-12-4 <AGI>
1965 N12 P1439-1449 GI V2 N6 1965 P1045 <AGI>
1965 N12 P1450-1458 GI V2 N6 1965 P1057 <AGI>
1965 N12 P1459-1468 GI V2 N6 1965 P1066 <AGI>
1965 N12 P1469-1478 GI V2 N6 1965 P1076 <AGI>
1965 N12 P1479-1485 GI 65-12-9 <AGI>
1965 N12 P1486-1487 GI V2 N6 1965 P1086 <AGI>
1965 N12 P1487-1490 GI V2 N6 1965 P1088 <AGI>

1966 N1 P123-128 66-30701 <=$>

GEOLOGICAL REVIEW, CHINESE PEOPLES REPUBLIC
SEE TI CHIH LUN P'ING

GEOLOGICAL SCIENCE. CHINESE PEOPLES REPUBLIC
SEE TI CHIH K'O HSUEH

GEOLOGICHESKII SBORNIK. LENINGRAD
1955 V33 N1 P274-331 61-31073 <=>

GEOLOGICHESKII SBORNIK. LVOVSKOE GEOLOGICHESKOE OBSHCHESTVO
1957 N4 P187-195 IGR V1 N5 P24 <AGI>

GEOLOGICHNII ZHURNAL
SEE HEOLOHYCHNYI ZHURNAL

GEOLOGICKE PRACE
1962 N63 P99-108 65-14346 <*>
65-14346 <*>

GEOLOGIE
1952 V1 N1/2 P78-132 66-12821 <*>
1956 V5 N2 P65-100 62-16432 <*>
1964 V13 N10 P1191-1198 58S83G <ATS>

GEOLOGIE EN MIJNBOUW
1965 V44 P440-457 66-14178 <*>

GEOLOGISCHE RUNDSCHAU
1944 V34 P546-694 57-3190 <*>
1944 V34 N7/8 P545-694 57-3195 <*>
59-14262 <+*>
1948 V36 P10-29 AL-27 <*>
1949 V37 P100-101 AL-26 <*>
1950 V38 N1 P40-59 18J18G <ATS>
1955 V44 P3-25 65-12130 <*>
1957 V46 N1 P17-26 IGR V2 N7 P617 <AGI>
62-14569 <*>
1957 V46 N1 P26-29 62-14604 <*>
1958 V47 N1 P218-234 IGR V1 N9 P50 <AGI>
1960 V49 N1 P92-97 10N53G <ATS>
1961 V51 P530-546 63-14060 <*> 0
1961 V51 N2 P315-330 IGR V5 N8 P945 <AGI>
1965 V54 P208-224 93S88R <ATS>
1966 V55 N1 P1-20 66-14121 <*>
1966 V55 N1 P131-144 51T94G <ATS>

GEOLOGISCHES JAHRBUCH
1957 V74 N4 P39-62 57-3566 <*>
1958 V75 P183-196 AEC-TR-4099 <*>

GEOLOGISKA FORENINGENS I STOCKHOLM FORHANDLINGAR
1955 V77 P525-545 57-3290 <*>

GEOLOGIYA I GEOFIZIKA
1960 N1 P117-122 IGR V6 N12 P2127 <AGI>
1960 N3 P80-93 63M44R <ATS>
1960 N3 P94-102 64M44R <ATS>
1960 N3 P103-108 IGR V4 N6 P645 <AGI>
1960 N3 P128- 66-62215 <=*$>
1960 N4 P104-113 24N53R <ATS>
1960 N8 P100-105 61-21666 <=>
1960 N12 P120-123 96N52R <ATS>
1961 N1 P109-111 63-33493 <=*>
1961 N2 P106-110 64-15186 <=*$>
1961 N2 P111-116 64-15184 <=*$>
1961 N4 P13-21 01N56R <ATS>
1961 N6 P94-96 62-13852 <*=>
1961 N7 P3-15 64-15156 <=*$>
1961 N7 P96-98 64-15162 <=*$>
1961 N8 P82-92 64-19745 <=$>
1961 N9 P30-34 IGR V6 N6 P1076 <AGI>
IGR V6 N6 P1076-1079 <AGI>
1961 N10 P89-101 IGR V6 N2 P202 <AGI>
IGR V6 N2 P202-211 <AGI>

Citation	Reference
1961 N12 P70-79	IGR V6 N6 P1068 <AGI>
	IGR V6 N6 P1068-1075 <AGI>
1961 N12 P8C-94	63-20733 <=*$>
1962 N1 P88-99	48P63R <ATS>
1962 N5 P90-96	63-21117 <=>
1962 N5 P107-119	63-21120 <=>
1962 N6 P96-100	06P65R <ATS>
1962 N7 P8-15	IGR V6 N11 P1953 <AGI>
1962 N7 P46-52	63-27421 <=$>
1962 N7 P66-75	63-27420 <=$>
1962 N8 P104-106	63-21121 <=>
1962 N8 P133-134	63-13552 <=*>
1962 N9 P122-124	82Q67R <ATS>
1962 N10 P3-22	IGR V6 N12 P2132 <AGI>
1962 N10 P93-103	IGR V6 N3 P478 <AGI>
	IGR,V6,N3,P478-486 <AGI>
1962 N11 P114-124	AD-632 259 <=$>
1962 N12 P3-9	66R75R <ATS>
1962 N12 P109-113	AD-623 445 <=*$>
1963 N1 P134-136	63-21851 <=>
1963 N5 P59-67	47Q73R <ATS>
1963 N5 P130-133	64-19953 <=$>
1963 N7 P3-16	65-11089 <*>
1963 N7 P17-31	05R75R <ATS>
1963 N8 P3-12	NLLTB V6 N4 P349 <NLL>
1963 N8 P1CC-106	63-41031 <=>
1963 N8 P106-109	63-41032 <=>
1963 N9 P120-124	65-11119 <*>
1963 N9 P125-129	C-5064 <NRC>
1963 N9 P129-133	65-11125 <*>
1963 N10 P5-18	AD-623 447 <=*$>
1963 N10 P5-19	64-21518 <=>
1963 N10 P160-162	AD-611 055 <=$>
1963 N11 P3-17	64-21835 <=>
1963 N11 P106-114	AD-610 297 <=$>
1963 N11 P114-121	AD-637 062 <=$>
1963 N11 P128-131	64-21841 <=*$>
1963 N11 P132-134	64-21842 <=>
1963 N12 P3-10	65-11088 <*>
1964 N2 P48-62	64-31679 <=>
1964 N2 P93-104	64-31679 <=>
1964 N2 P125-129	64-31679 <=>
1964 N4 P96-107	65-12721 <*>
1964 N5 P32-37	66-10318 <*>
1964 N6 P48-57	GI N2 1964 P340 <AGI>
1964 N6 P130-135	GI N2 1964 P346 <AGI>
1964 N6 P158-163	GI N1 1964 P126 <AGI>
1964 N8 P57-65	65-14343 <*>
1964 N9 P95-99	65-14218 <*>
1964 N9 P112-115	65-30178 <=$>
1964 N10 P33-49	66-10306 <*>
1964 N10 P1C4-112	65-14217 <*>
1964 N11 P90-93	66-14242 <*$>
1964 N11 P94-110	66-14421 <*$>
1964 N12 P116-119	55S87R <ATS>
1964 N12 P119-123	11S85R <ATS>
1965 N2 P3-20	GI N5 1964 P1038 <AGI>
1965 N2 P92-98	66-33470 <=$>
1965 N2 P123-127	66-33469 <=$>
1965 N2 P157-159	27S89R <ATS>
1965 N2 P164-165	66-33468 <=$>
1965 N3 P3-24	IGR V8 N6 P716 <AGI>
1965 N3 P25-38	65-31473 <=$>
1965 N4 P3-18	IGR V8 N11 P1335 <AGI>
1965 N4 P66-67	IGR V8 N11 P1362 <AGI>
1965 N4 P103-113	IGR V8 N6 P643 <AGI>
1965 N5 P3-18	IGR V8 N12 P1428 <AGI>
1965 N5 P35-48	IGR V8 N10 P1141 <AGI>
1965 N5 P49-58	IGR V8 N10 P1208 <AGI>
1965 N6 P17-34	65-32876 <=*>
1965 N6 P55-65	IGR V8 N6 P739 <AGI>
1965 N6 P66-74	IGR V8 N11 P1372 <AGI>
1965 N6 P75-94	IGR V8 N5 P575 <AGI>
1965 N6 P85-98	IGR V8 N7 P757 <AGI>
1965 N6 P108-111	87T90R <ATS>
1965 N7 P52-59	IGR V8 N10 P1135 <AGI>
1965 N7 P68-81	85T90R <ATS>
1965 N7 P104-105	86T9CR <ATS>
1965 N7 P130-137	65-32926-A <=*$>
1965 N7 P14C	65-32925-A <=*$>
1965 N8 P3-15	IGR V8 N11 P1294 <AGI>
1965 N8 P50-64	IGR V8 N12 P1408 <AGI>
1965 N8 P66-67	IGR V8 N11 P1320 <AGI>
1965 N8 P87-96	65-33785 <=$>
1965 N8 P120-124	IGR V8 N11 P1316 <AGI>
1965 N8 P129-135	65-33785 <=$>
1965 N8 P148-149	65-33785 <=$>
1965 N9 P3-10	66-14216 <*$>
1965 N9 P127-130	66-30131 <=$>
1965 N9 P131-137	66-30131 <=$>
1965 N10 P12-22	AD-633 408 <=$>
1965 N1C P12-33	66-30514 <=*$>
1965 N10 P106-127	66-30514 <=*$>
1965 N10 P128-132	05T91R <ATS>
1965 N11 P66-79	66-23210 <*$>
1965 N11 P148-153	65-30781 <=$>
1965 N12 P15-2C	66-13936 <*$>
1966 N1 P10-20	AD-637 532 <=$>
1966 N1 P94-101	66-14123 <*$>
1966 N4 P3-14	IGR V8 N11 P1261 <AGI>
1966 N4 P107-119	66-33782 <=$>
1966 N8 P109-111	AC-651 470 <=$>
1967 N7 P3-10	IGR V8 N10 P1151 <AGI>

GEOLOGIYA MESTOROZHDENII REDKIKH ELEMENTOV

Citation	Reference
1959 N4 P1-54	60-11877 <=>
1962 N15 P98-100	IGR V6 N12 P2148 <AGI>

GEOLOGIYA NEFTI

Citation	Reference
1957 V1 N1 P64-73	63-26598 <=$>
1957 V1 N2 P32-36	42S88R <ATS>
1957 V1 N11 P13-29	63-16556 <*=>
1958 V2 N3 P35-37	65-13589 <*>
1958 V2 N3 P37-45	65-13590 <*>
1958 V2 N5 P37-48	62-33491 <=*> 0
1958 V2 N8 P39-44	22K27R <ATS>
1958 V2 N8 P49-52	23K27R <ATS>
1958 V2 N8 P55-65	24K27R <ATS>
1958 V2 N11 P25-27	54K28R <ATS>
1958 V2 N11 P50-60	76L28R <ATS>

★GEOLOGIYA NEFTI I GAZA

Citation	Reference
1959 N1 P18-25	59-11494 <=> 0
1959 V3 N1 P1-8	59-11491 <=>
1959 V3 N1 P9-13	59-11492 <=>
1959 V3 N1 P13-17	59-11493 <=>
1959 V3 N1 P33-38	59-11495 <=> 0
1959 V3 N1 P38-47	59-11496 <=> 0
1959 V3 N1 P47-51	29L31R <ATS>
1959 V3 N1 P52-55	28L31R <ATS>
1959 V3 N2 P49-52	71L31R <ATS>
1959 V3 N3 P22-29	73L32R <ATS>
1959 V3 N7 P33-39	64M37R <ATS>
1959 V3 N7 P39-44	96M38R <ATS>
1959 V3 N8 P44-47	94M38R <ATS>
1959 V3 N11 P1-6	60-11413 <=>
1959 V3 N12 P47-52	60-31186 <=>
1960 V4 N1 P41-48	61-16084 <=*>
1960 V4 N2 P41-45	68M42R <ATS>
1960 V4 N3 P42-44	98N51R <ATS>
1960 V4 N3 P60-63	61-21514 <=>
1960 V4 N4 P47-52	50M45R <ATS>
1960 V4 N5 P51-55	17M47R <ATS>
1960 V4 N6 P23-26	18M47R <ATS>
1960 V4 N6 P37-41	71M45R <ATS>
1960 V4 N7 P55-59	65-14830 <*>
1960 V4 N9 P35-39	18N51R <ATS>
1960 V4 N9 P41-43	25N51R <ATS>
1960 V4 N9 P43-47	19N51R <ATS>
1960 V4 N10 P46-47	72N53R <ATS>
1960 V4 N10 P48-51	26N51R <ATS>
1960 V4 N11 P9-10	69N51R <ATS>
1960 V4 N11 P11-14	70N51R <ATS>
1960 V4 N11 P18-21	51N52R <ATS>
1960 V4 N12 P21-24	11N50R <ATS>
	62-20295 <=*>
1960 V4 N12 P24-29	12N50R <ATS>
1961 V5 N1 P29-34	90Q66R <ATS>
1961 V5 N3 P29-34	72N55R <ATS>
1961 V5 N3 P35-40	37N55R <ATS>

1961 V5 N4 P38-39 71N55R <ATS>
1961 V5 N4 P40-43 38N55R <ATS>
1961 V5 N4 P43-45 39N55R <ATS>
1961 V5 N5 P50-54 79N55R <ATS>
1961 V5 N6 P46-50 02N56R <ATS>
1961 V5 N11 P41-43 11P61R <ATS>
1961 V5 N11 P50-51 23P60R <ATS>
1962 V6 N1 P48-52 87S82R <ATS>
1962 V6 N4 P46-51 02P63R <ATS>
1962 V6 N8 P50-52 07Q67R <ATS>
1962 V6 N8 P52-56 66P65R <ATS>
1962 V6 N11 P52-54 50Q68R <ATS>
1962 V6 N12 P54-58 42Q68R <ATS>
1963 V7 N6 P1-8 63-41252 <=>
1963 V7 N7 P25-29 31Q73R <ATS>
1963 V7 N11 P20-22 47R79R <ATS>
1963 V7 N12 P18-22 25R76R <ATS>
1964 V8 N1 P41-45 69S87R <ATS>
1964 V8 N7 P20-24 64-23311 <$>
1964 V8 N11 P26-30 16S83R <ATS>
1964 V8 N11 P51-55 87S81R <ATS>
1964 V8 N12 P1-8 53S84R <ATS>
1964 V8 N12 P18-23 66-10297 <*>
1964 V8 N12 P30-32 66-10295 <*>
1965 V9 N2 P5-12 91S84R <ATS>
1965 V9 N3 P43-48 66-14408 <*>
1965 V9 N4 P1-10 65-23805 <$>
1965 V9 N7 P33-35 75T93R <ATS>

★GEOLOGIYA RUDNYKH MESTORCZHDENII
1960 N2 P20-31 IGR V7 N2 P205 <AGI>
1960 N4 P123-126 61-11801 <=$>
1961 N1 P119-137 61-27761 <*=>
1961 N5 P64-79 63-24368 <=*$>
1962 N2 P109-115 62-32056 <=*>
1962 N3 P47-58 62-32552 <=*>
1962 N4 P3-6 62-33128 <=*>
1962 N4 P106-112 RTS-3497 <NLL>
1962 N6 P54-56 63-21511 <=>
1963 V5 N2 P76-90 RTS-2765 <NLL>
1963 V5 N6 P9-17 65-13144 <*>
1964 N2 P52-68 GI N1 1964 P114 <AGI>
1964 V6 N3 P7-15 66-13349 <*>
1965 N5 P18-30 GI V2 N6 1965 P1093 <AGI>
1966 N1 P41-54 GI V2 N4 1965 P766 <AGI>

GEOLOGIYA SSSR
1958 N5 P55-58 64-15168 <=*$>

GEOLOGY MONTHLY. CHINESE PEOPLES REPUBLIC
SEE CHUNG KUO TI CHIH

GEOLOGY AND PROSPECTING. CHINESE PEOPLES REPUBLIC
SEE TI CHIH YU K'AN T'AN

GEOLOSKI ANALI BALKANSKOGA POLUOSTRVA
1958 V25 ENTIRE ISSUE 60-21674 <=>

GEOLOSKI GLASNIK. SARAJEVO
1958 V2 ENTIRE ISSUE 60-21673 <=>

★GEOMAGNETIZM I AERONOMIYA
1961 N1 P21-29 AIAAJ V1 N2 P514 <AIAA>
1961 V1 N1 P34-40 65-14062 <*>
1961 V1 N1 P49-53 63-19849 <*=> 0
1961 V1 N2 P164-173 N65-22594 <=$>
1961 V1 N2 P174-177 N65-22595 <=$>
1961 V1 N2 P194-208 AIAAJ V1 N4 P994 <AIAA>
1961 V1 N3 P314-319 63-18717 <=*>
1961 V1 N3 P333-345 AIAAJ V1 N3 P738 <AIAA>
1961 V1 N3 P374-378 AD-610 996 <=$>
1961 V1 N3 P421-425 62-23349 <=*>
64-71574 <=>
1961 V1 N3 P436-440 62-23349 <=*>
1961 V1 N4 P453-478 62-23807 <=>
1961 V1 N4 P483-489 62-23807 <=>
1961 V1 N4 P494-499 62-23807 <=>
1961 V1 N4 P500-506 62-15648 <=>
1961 V1 N4 P507-530 62-23807 <=>
1961 V1 N4 P531-547 62-15648 <=>
1961 V1 N4 P548-551 62-23807 <=>
1961 V1 N4 P552-556 62-15648 <=>
1961 V1 N4 P557-560 62-23807 <=>
1961 V1 N4 P561-571 62-15648 <=>
1961 V1 N4 P572-587 62-23807 <=>
1961 V1 N4 P588-593 62-13615 <*=>
1961 V1 N4 P597-599 62-23807 <=>
1961 V1 N4 P599-605 62-15472 <*=>
1961 V1 N4 P606-620 62-23807 <=>
1961 V1 N4 P620-622 62-15648 <=>
1961 V1 N4 P623-626 62-13615 <*=>
1961 V1 N4 P627-628 62-23807 <=>
1961 V1 N5 P709-724 AIAAJ V1 N4 P1001 <AIAA>
1961 V1 N6 P832-834 N65-22857 <=$>
1961 V1 N6 P911-916 65-13655 <*>
1962 V2 N1 P38-40 N65-22601 <=*$>
1962 V2 N1 P41-47 N65-22590 <=$>
1962 V2 N1 P48-55 N65-23250 <=$>
1962 V2 N1 P56-57 N65-22584 <=$>
1962 V2 N1 P74-78 N65-22604 <=$>
1962 V2 N1 P80-85 N65-22856 <=$>
1962 V2 N1 P153-160 AIAAJ V1 N11 P2703 <AIAA>
63-23503 <=*$>
1962 V2 N1 P172-176 63-24076 <=*>
1962 V2 N1 P177-179 63-15231 <=*>
1962 V2 N2 P321-325 N65-22858 <=$>
1962 V2 N2 P363-364 N65-22589 <=$>
1962 V2 N2 P367-368 N65-22599 <=$>
1962 V2 N2 P368-369 N65-22600 <=$>
1962 V2 N3 P377-406 65-23226 <$>
1962 V2 N3 P541-544 64-13081 <=*$>
1962 V2 N3 P550-552 63-23200 <=*>
1962 V2 N3 P570-571 N65-23624 <=$>
1962 V2 N4 P674-687 63-01225 <*>
63-23764 <=*>
1962 V2 N4 P759-762 <GPRC>
1962 V2 N4 P777-781 65-13396 <*>
1962 V2 N5 P886-903 63-23709 <=*$>
1962 V2 N6 P1084-1094 66-20959 <$>
1962 V2 N6 P1091-1094 66-20960 <$>
1962 V2 N6 P1146-1147 65-23703 <=>
1963 V3 N1 P10-24 AD-611 121 <=$>
63-11696 <=>
63-21757 <=>
1963 V3 N1 P25-36 63-11684 <=>
1963 V3 N1 P94-103 63-23639 <=*>
1963 V3 N1 P121-126 63-11773 <=>
1963 V3 N1 P157-170 63-11651 <=>
63-21757 <=>
1963 V3 N2 P223-226 63-11712 <=>
63-24537 <=*$>
1963 V3 N2 P297-308 63-31932 <=>
1963 V3 N2 P354-361 64-11506 <=>
1963 V3 N2 P383-385 63-11757 <=>
1963 V3 N3 P408-416 63-11748 <=>
64-13173 <=*$>
1963 V3 N4 P626-634 63-20631 <=*$>
64-11511 <=>
64-11848 <=>
1963 V3 N4 P781-783 SC-T-535 <=*$>
1963 V3 N5 P816-822 63-41069 <=>
64-00454 <*>
64-19340 <=$>
1963 V3 N5 P858-867 63-41069 <=>
1963 V3 N5 P961-967 63-41069 <=>
1963 V3 N5 P968-975 63-41069 <=>
1963 V3 N5 P991-994 AD-609 794 <=>
1963 V3 N6 P1059-1064 64-00455 <*>
64-11818 <=>
1963 V3 N6 P1127-1128 N65-14855 <=*$>
64-21305 <=>
1964 V4 P668-674 N64-30857 <=$>
1964 V4 N2 P321-327 N65-17466 <=$>
1964 V4 N4 P626-634 N64-30849 <=$>
1964 V4 N4 P668-674 64-41604 <=>
1964 V4 N4 P722-731 64-31560 <=$>
1964 V4 N5 P825-831 N65-12020 <=$>
1964 V4 N5 P881-885 N65-11446 <=$>
1964 V4 N5 P886-891 N65-11447 <=$>
1964 V4 N5 P898-916 N65-18182 <=$>

1964 V4 N5 P917-923 N65-14428 <=$>
1964 V4 N5 P940-944 N65-14617 <=$>
1964 V4 N5 P945-947 N65-14616 <=$*>
1964 V4 N5 P946-947 N65-14619 <=*$>
1964 V4 N6 P977-986 N65-15162 <=$>
1964 V4 N6 P1015-1019 N65-15164 <=*$>
1964 V4 N6 P1020-1025 N65-15058 <=$>
1964 V4 N6 P1059-1063 N65-15890 <=*$>
1964 V4 N6 P1134-1136 65-31741 <=*$>
1965 V5 N1 P3-31 N65-21629 <=$>
1965 V5 N1 P55-69 N65-21635 <=$*>
1965 V5 N1 P97-103 N65-21638 <=$*>
1965 V5 N1 P104-112 N65-23251 <=*$>
1965 V5 N1 P148-154 65-63723 <=>
1965 V5 N1 P155-158 N65-22626 <=*$>
1965 V5 N4 P768-770 65-32661 <=*>
1965 V5 N4 P770-771 65-32662 <=$*>
1965 V5 N6 P1061-1067 66-33799 <=$>
1965 V5 N6 P1120-1123 N66-21688 <=$>
1966 V6 N3 P544-547 66-33217 <=$>
1966 V6 N3 P621-624 66-33168 <=$>

GEOPHYSICAL EXPLORATION. JAPAN
SEE BUTSURI TANKO

GEOPHYSICAL PROSPECTING. HAGUE
1954 V2 P203-226 64-30687 <*>
1955 V3 N2 P126-162 65-11309 <*>
1955 V3 N4 P339-349 65-10219 <*>
1956 V4 P394-423 65-10220 <*>
1957 V5 N3 P328-348 64-18095 <*>
1957 V5 N4 P421-448 64-30664 <*>
1960 V8 N2 P299-304 62-20287 <*> 0
1961 V9 P370-381 63-10281 <*>
1961 V9 N1 P60-73 64-16975 <*>
1961 V9 N2 P242-260 65-13017 <*>
1961 V9 N2 P276-295 66-12820 <*>
1962 V10 N1 P68-83 64P61F <ATS>
1962 V10 N3 P238-257 65-11569 <*>
1962 V10 N3 P352-402 65-11115 <*>
1963 V11 N2 P115-121 65-11086 <*>
1963 V11 N2 P131-163 64-10970 <*>
65-12478 <*>
1964 V12 N3 P258-282 65-11493 <*>
65-14206 <*>
1964 V12 N3 P283-289 65-13984 <*>
65-14207 <*>
1964 V12 N3 P308-324 65-14204 <*>
1965 V13 N1 P66-104 65-18072 <*>
66-10296 <*>
1965 V13 N3 P460-474 66-12692 <*>
66-15125 <*>

★ GEOTEKTONIKA. MOSCOW
1965 N6 P91-94 66-33471 <=$>

GEP; A GEPGYARTAS MUSZAKI FOLYOIRATA
1949 V1 N3 P93 58-921 <*>

GEPGYARTASTECHNOLOGIA
1964 N3 P86-90 NEL-TRANS-1533 <NLL>

GERBER
1926 V52 P63-65 60-10517 <*>
1935 V61 P43-44 60-10516 <*>
1935 V61 P49-51 60-10516 <*>
1937 V63 P1-3 60-10518 <*>

GERMANY. ELECTRICAL STANDARDS
1957 VDE-0750/10/7.7 66-11610 <*> 0
1957 VDE-0750/11/7.7 66-11610 <*> 0
1957 VDE-0750/14/7.7 66-11610 <*> 0
1957 VDE-0750/2/7.57 66-11610 <*> 0
1957 VDE-0750/3/7.57 66-11610 <*> 0
1957 VDE-0750/5/7.57 66-11610 <*> 0
1957 VDE-C750/6/7.57 66-11610 <*> 0
1957 VDE-0750/8/7.57 66-11610 <*> 0
1957 VDE-0750/9/7.57 66-11610 <*> 0
1959 VDE-0750/4/1.59 66-11610 <*> 0
1962 VDE-0107/12.62 66-11609 <*> 0
1962 VDE-0750/12/12. 66-11610 <*> 0
1963 VDE-0871/5.63 66-11611 <*> 0
1964 VDE-0750/1/8.64 66-11610 <*> 0

GERMANY, INDUSTRIAL STANDARDS
DIN 51,771 31R79G <ATS>
DIN 51,773 32R79G <ATS> P
DIN 51,754 33R79G <ATS> P
DIN 51,756 34R79G <ATS> P
DIN 51,759 35R79G <ATS> P
DIN 51,765 36R79G <ATS> P
DIN-106 CLAIRA T.1 <CLAI>
DIN-18,550 CLAIRA T.15 <CLAI>
DIN-1713 60-14656 <*>
DIN-107 60-18925 <*>
DIN-19532 66-11903 <*>
DIN-51,766 37R79G <ATS> P
DIN-51,769 38R79G <ATS> P
DIN-51,770 39R79G <ATS> P
DIN-52214 44T92G <ATS>
DIN-51,559 78Q71G <ATS>
1939 DIN-21501 66-10203 <*>
1941 DIN-52176 63-01666 <*>
1951 DIN-21191 66-10207 <*>
1953 DIN 19226 9H14G <ATS>
1953 DIN-4764 63-18329 <=*>
1955 DIN 51757 22N56G <ATS>
1956 DIN-53,422 SCL-T-322 <*>
1957 DIN-53,574 SCL-T-325 <*>
1957 DIN-52,322 62-18146 <=*>
1957 IV/33/57 61-10717 <*> 0
1958 DIN-53,420 SCL-T-319 <*>
1958 DIN-53,423 SCL-T-320 <*>
1958 DIN-53,572 SCL-T-321 <*>
1958 DIN-53,571 SCL-T-323 <*>
1958 DIN-53,421 SCL-T-324 <*>
1958 DIN-53,575 SCL-T-327 <*>
1958 DIN-7726 SCL-T-318 <*>
1959 DIN-53,573 SCL-T-326 <*>
1960 DIN-4762 63-18323 <=*>
1960 DIN-4762 SHEET3 63-18324 <*>
1960 DIN-4762 SHEET2 63-18325 <*>
1960 DIN-4762 SHEET1 63-18326 <*>
1960 DIN-4761 63-18327 <=*>
1960 DIN-4760 63-18328 <=*>
1961 DIN-53422 65-14093 <*>
1962 DIN-8061 66-11902 <*>
1964 DIN-8062 66-11922 <*>

GERONTOLOGIA
1959 V3 N4 P184-203 C-3715 <NRCC>

GESAMMELTE ABHANDLUNGEN ZUR KENNTNISS DER KOHLE
1923 V7 P213-224 59-10693 <*>
1937 V21 P204-216 60-10540 <*> P

GESETZBLATT DER DEUTSCHEN DEMOKRATISCHEN REPUBLIK
1958 N17 P161-162 PT.2 59-13881 <=>
1960 N7 P46-47 PT.3 62-19436 <=>
1961 N2 P 3-4 PT.2 62-19439 <=>
1961 N4 P11-12 PT.2 62-19441 <=>
1961 N4 P43-45 PT.3 62-23497 <=>
1962 N96 P817-820 PT2 63-21829 <=>
1965 N60 P409-413 65-31809 <=*$>
1965 N60 P413-414 65-31691 <*=$>
1966 N4 P45-51 66-30720 <=$>

GESUNDHEIT UND WOHLFAHRT
1951 V31 P89-99 3180-E <K-H>
1951 V31 P99-108 59-17030 <*>
1952 V32 P178-191 62-23012 <=*>

GESUNDHEITSINGENIEUR
1915 V38 N42 P477-496 C-2487 <NRC>
1915 V38 N42 P477-482 57-3033 <*>
1939 V62 P546-550 64-14131 <*> P
1941 V64 N1 P1-4 59-15130 <*>
1948 V69 N8 P224-228 3136 <*>
1949 V70 P259-261 63-20069 <*>
1953 V74 P1-3 II-807 <*>

1954 V75 P296-298 1779 <*>
1954 V75 P298-299 1487 <*>
1954 V75 N13/4 P224-226 75J15G <ATS>
1959 V80 N8 P225-235 C-3594 <NRCC>
1959 V80 N8 P235-238 C-3609 <NRCC>
1960 V81 N7 P207-210 63-14031 <*> 0
1963 V84 N2 P33-38 63-20385 <*>

GESUNDHEITWESEN UND DESINFEKTION
1962 V54 N6 P88-90 64-00429 <*>

GETREIDE, MEHL UND BROT
1950 V4 N5/6 P67-69 63-16311 <*>

GETREIDE UND MEHL
1951 V1 N8 P61-65 63-14973 <*>
1952 V2 N2 P2-16 63-16299 <*>
1955 V5 N1 P6-7 63-14968 <*> 0
1955 V5 N11 P82-84 63-14963 <*>
1957 V7 N9 P77-80 63-14865 <*>
1958 V8 N10 P73-76 60-10123 <*>
1958 V8 N11 P86-88 60-16577 <*>
1959 V9 N5 P49-52 63-16490 <*> 0
1959 V9 N10 P105-109 61-10172 <*> 0
1959 V9 N11 P117-119 NP-TR-515 <*>
1960 V10 N2 P13-16 PT.1 NP-TR-690 <*>
1960 V10 N2 P16-20 NP-TR-696 <*>
1960 V10 N5 P49-54 AEC-TR-5798 <*>
1960 V10 N11 P121-128 63-14975 <*>
1961 V11 N5 P49-53 61-20737 <*>

GETREIDEMUEHL
1957 V1 N7 P146-149 63-14864 <*>
1957 V1 N8 P170-173 63-14863 <*>
1958 V2 N8 P174-175 63-16395 <*>

GEWERBEFLEISS
1929 V108 N2/3 P35-42 60-10197 <*>

GIDROKHIMICHESKIE MATERIALY
1953 V20 P79-91 RJ-258 <*>
RJ-258 <ATS>
1955 V23 P31-35 59-22701 <+*>
1955 V23 P36-38 59-22700 <+*>
1955 V24 P23-27 90J19R <ATS>
1957 V26 P7-18 61-15911 <=*>
1957 V26 P243-245 61-13813 <*+>
1958 V27 P21-36 62-13088 <*=>
1964 V37 P133-143 1032 <TC>

GIDROLIZNAYA I LESOKHIMICHESKAYA PROMYSHLENNOST
1955 N3 P6-17 R-3178 <*>
1955 N4 P11-13 R-3169 <*>
1955 N6 P3 R-3165 <*>
1955 V8 N3 P1-3 R-3166 <*>
1955 V8 N4 P19-20 76K26R <ATS>
1955 V8 N8 P1-3 R-3164 <*>
1956 V9 N1 P26-27 R-3172 <*>
1956 V9 N5 P5-7 R-3154 <*>
1956 V9 N6 P3-6 R-3158 <*>
1956 V9 N6 P6-7 R-3168 <*>
1956 V9 N6 P27-28 R-1831 <*>
1957 V10 N1 P8-10 73J16R <ATS>
1957 V10 N1 P27-28 72K22R <ATS>
1957 V10 N2 P10-12 R-4538 <*>
1957 V10 N2 P28 R-3167 <*>
1957 V10 N7 P7-12 43L30R <ATS>
1957 V10 N7 P21-24 12119-B <K-H>
1957 V10 N8 P3-6 60-51161 <=>
1958 V11 N1 P4-6 60-51032 <=>
1959 V12 N3 P15-17 RTS-3305 <NLL>
1959 V12 N5 P23 63-11031 <=>
1960 V13 N1 P30 60-31687 <=>
1960 V13 N3 P6-9 62-14485 <=*> 0
1961 V14 N4 P29-30 62-13685 <*=>
1961 V14 N4 P30 62-13634 <*=>
1961 V14 N7 P1-3 62-23601 <=*>
1961 V14 N7 P4-8 64-13585 <=*$>
1961 V14 N8 P10 65-12898 <*>
1962 V15 N1 P7-9 62-20416 <=*> 0
1962 V15 N6 P19-22 63-27148 <=$>
1963 V16 N3 P1-3 63-31173 <=>
1963 V16 N3 P21-23 RTS-3334 <NLL>
1964 V17 N7 P1-3 65-30390 <=$>
1964 V17 N7 P7-10 66-13438 <*>
1964 V17 N7 P12-13 RTS-3003 <NLL>
1965 V18 N4 P18-19 84T90R <ATS>
1965 V18 N7 P18-19 65-34026 <=$>

★GIDROTEKHNICHESKOE STROITELSTVO
1933 V4 N2 P29-30 08P62R <ATS>
1946 V15 N9 P21 RT-358 <*>
1946 V15 N9 P26-29 RT-357 <*>
1946 V15 N10 P8-11 RT-365 <*>
1946 V15 N10 P25-26 RT-366 <*>
1946 V15 N11 P23-24 RT-367 <*>
1946 V15 N12 P13-16 RT-353 <*>
1946 V15 N12 P19-21 RT-368 <*>
1946 V15 N12 P21-22 RT-2313 <*>
1947 N9 P14-17 RT-375 <*>
1949 V18 N4 P7-10 62-25128 <=*>
1951 V20 N9 P35-38 60-21147 <=>
1951 V20 N12 P38-43 60-51077 <=>
1953 V22 N2 P6-11 61-13667 <+>
1953 V22 N5 P24-31 RT-1096 <*>
1953 V22 N7 P19-22 60-15843 <+*>
1953 V22 N7 P23-24 60-15844 <+*>
1955 N4 P30-33 64-00139 <*>
1956 V25 N5 P46-48 59-22689 <+*>
1956 V25 N9 P51-53 62-25127 <=*>
1956 V25 N10 P54-56 61-11426 <=>
1956 V25 N11 P34-36 62-25366 <=*>
1957 N11 P6-14 PB 141 292T <=>
1957 N11 P15-26 PB 141 221T <=>
1957 V26 N10 P50-51 62-25134 <=*>
1957 V26 N12 P38-39 74N48R <ATS>
1958 N4 P5-10 PB 141 173T <=>
1958 V27 N1 P20-26 59-11059 <=> 0
1958 V27 N11 P35-44 60-17251 <+*> 0
1959 V28 N3 P30-33 60-13824 <+*>
1959 V28 N6 P1-8 61-23927 <*=>
1959 V28 N6 P20-24 61-23928 <*=>
1959 V28 N11 P1-4 60-23537 <*+>
1959 V28 N11 P11-16 60-23539 <*+>
1959 V28 N11 P20-24 60-23538 <*+>
1960 V29 N4 P44-48 61-23028 <=*>
1960 V29 N5 P62-63 60-41327 <=>
1960 V29 N10 P56-58 61-11718 <=>
1960 V29 N11 P6-17 61-21128 <=>
1960 V30 N2 P45-47 61-15293 <*=>
1960 V30 N5 P38-39 62-25129 <=*>
1960 V30 N7 P13-17 61-31063/1-9 <=>
1960 V30 N7 P20-24 61-31063/1-9 <=>
1960 V30 N7 P27-35 61-31063/1-9 <=>
1960 V30 N7 P27-30 62-25368 <=*>
1960 V30 N7 P38-39 62-25367 <=*> P
1960 V30 N7 P39-46 61-31063/1-9 <=>
1960 V30 N10 P21-24 62-25124 <=*>
1960 V30 N10 P50-53 62-25123 <=*>
1960 V30 N10 P53-55 62-25126 <=*>
1960 V30 N11 P1-5 61-31155 <=>
1960 V30 N12 P5-22 61-31155 <=>
1961 V31 N2 P64-65 62-13349 <*=>
1961 V31 N6 P18-22 63-15187 <=*>
1962 V32 N4 P30-33 64-15891 <=*$>
1962 V32 N5 P13-16 62-33332 <=>
1962 V32 N5 P30-34 CEMBUREAU-TR.26 <CEM>
1962 V32 N5 P30-32 63-27827 <$>
1962 V32 N5 P34 63-27827 <$>
1962 V32 N9 P30-32 63-21787 <=>
1962 V32 N11 P39-41 RTS-3104 <NLL>
1963 V33 N2 P2-10 65-64512 <=*$>
1963 V33 N4 P44-46 64-19350 <=$>
1964 N1 P60-61 64-31108 <=>
1964 V34 N3 P11-18 65-31453 <=$>
1965 N11 P4-10 65-30218 <=*$>
1965 V35 N4 P27-30 65-64514 <=*$>
1965 V35 N11 P1-4 66-30465 <=$>

GIDROTEKHNIKA I MELIORATSIYA

Citation	Translation
1950 V2 N2 P3-12	51/3361 <NLL>
1952 V4 N5 P74-80	RT-2733 <*>
1952 V4 N7 P62-69	61-15374 <*+>
1954 V6 N12 P19-32	64-15826 <=*$> O
1956 V8 N1 P28-34	60-21107 <=>
1956 V8 N2 P33-42	60-21824 <=>
1956 V8 N4 P28-34	R-4436 <*>
1956 V8 N7 P46-50	60-21830 <=>
1956 V8 N8 P33-40	60-21153 <=>
1956 V8 N9 P13-19	63-11008 <=>
1956 V8 N9 P50-55	60-21828 <=>
1956 V8 N10 P9-18	60-21152 <=>
1956 V8 N10 P27-32	64-15828 <=*$> C
1957 V9 N3 P7-21	60-21826 <=>
1957 V9 N4 P20-27	65-11031 <*>
1957 V9 N10 P30-49	60-21827 <=>
1958 V10 N1 P22-29	60-21151 <=>
1958 V10 N1 P43-53	60-21862 <=>
1958 V10 N3 P18-25	60-21158 <=>
1958 V10 N8 P20-27	60-21108 <=>
1959 V11 N2 P7-13	61-11430 <=>
1959 V11 N4 P22-25	60-21829 <=>
1959 V11 N6 P3-19	61-11430 <=>
1959 V11 N6 P35-38	61-11430 <=>
1959 V11 N7 P10-16	61-11430 <=>
1960 V12 N2 P50-53	61-27249 <*=>
1960 V12 N10 P24-36	61-31071 <=>
1960 V12 N10 P41-47	61-31063/1-9 <=>
1960 V12 N10 P52-53	62-25131 <=*>
1960 V12 N11 P61-64	64-19349 <=*$>
1961 V13 N6 P47-50	65-29639 <$>
1963 V15 N9 P11-13	66-61874 <=*$>
1963 V15 N9 P19-24	AD-608 943 <=>
1964 V16 N4 P3-21	64-31589 <=>
1964 V16 N4 P56-59	65-63219 <=>
1964 V16 N7 P21-30	65-63750 <=>
	65-64369 <=*$>
1965 V17 N7 P26-29	66-62077 <=*$>
1966 N2 P1-6	66-31827 <=$>

GIDROTURBOSTROENIE. SBORNIK

Citation	Translation
1957 V4 P43-54	RTS-1528 <NLL>
1957 V4 P103-126	RTS-1529 <NLL>
1957 V4 P369-377	RTS-1536 <NLL>

GIESSEREI

Citation	Translation
1930 V17 N32 P770-774	60-10053 <*> O
1930 V17 N46 P1121-1122	57-3225 <*>
1930 V17 N51/2 P1211-1216	60-10054 <*> O
1931 P147-148	59-20223 <*> O
1931 V18 N5 P98-101	59-20441 <*> O
1935 V22 N13 P312-314	57-173 <*>
1936 V23 N24 P617-619	59-15440 <*>
1936 V23 N26 P674-685	59-20438 <*> O
1937 V24 P80-83	57-2188 <*>
1937 V24 P341-343	62-20068 <*> O
1937 V24 P490-494	60-18965 <*>
1937 V24 P519-521	60-18965 <*>
1937 V24 N21 P510-518	59-20439 <*> O
1938 V25 P242-247	60-18732 <*> O
1938 V25 P609-617	2658 <*>
1938 V25 N8 P177-183	59-20446 <*> O
1939 V26 P121-123	59-20226 <*>
1939 V26 P484-487	57-1053 <*>
1939 V26 P505-508	57-1053 <*>
1939 V26 N25 P597-604	58-1887 <*>
1939 V26 N26 P621-628	58-1887 <*>
1940 V27 N1 P12-14	59-20440 <*> O
1941 N10 P217-224	58-438 <*>
1941 V28 P491-492	61-18873 <*>
1941 V28 P501-503	61-18504 <*>
1943 V30 N11 P249-	62-10249 <*> OP
1951 V38 P51-52	2327 <*>
1951 V38 P476-482	62-10533 <*>
1952 V39 P381-387	64-10789 <*>
1952 V39 N13 P311-315	64-20835 <*>
1952 V39 N19 P467-472	64-18888 <*> O
1952 V39 N19 P491-498	63-14150 <*> O
1953 V40 N1 P16-24	64-18929 <*>
1953 V40 N3 P69-75	AL-765 <*>
1953 V40 N14 P354-359	64-10541 <*>
1953 V40 N19 P473-477	60-10782 <*> O
1953 V40 N26 P678-681	64-20376 <*> O
1954 V41 N11 P283-288	64-20377 <*> O
1954 V41 N12 P313-320	60-10241 <*> O
1954 V41 N19 P477-485	60-18300 <*> O
1955 V42 P433-440	TS-1370 <BISI>
1955 V42 N6 P121-123	59-20921 <*> O
1955 V42 N8 P187-191	63-14637 <*> C
1955 V42 N17 P433-440	62-10473 <*>
1955 V42 N24 P653-659	60-10279 <*>
1956 V43 P229-239	TS-1286 <BISI>
1956 V43 P305-315	3347 <BISI>
1956 V43 P540-547	66-10452 <*> O
1956 V43 N2 P33-40	C2362 <NRC>
	57-266 <*>
1956 V43 N18 P540-545	4050 <HB>
1956 V43 N22 P735-738	57-3301 <*>
1957 V44 P89-96	66-10453 <*> O
1957 V44 P301-305	8849 <IICH>
1957 V44 P909-917	TS-1378 <BISI>
1957 V44 N5 P113-120	3993 <HB>
1957 V44 N5 P129-133	6823 <HB>
1957 V44 N9 P216-227	59-20991 <*>
1957 V44 N9 P227-237	61-16222 <*>
1957 V44 N10 P277-290	59-20990 <*>
1957 V44 N24 P714-718	4666 <HB>
1958 V45 P229-239	61-20689 <*>
1958 V45 P409-417	TS-1298 <BISI>
1958 V45 N4 P87-92	61-18896 <*>
1958 V45 N14 P385-387	5710 <HB>
1958 V45 N15 P409-417	61-20588 <*>
1958 V45 N15 P417-420	4468 <HB>
1958 V45 N20 P615-623	61-10645 <*> O
	62-14495 <*>
1959 V46 P129-136	61-20832 <*>
1959 V46 P441-447	2964 <BISI>
1959 V46 P471-478	2964 <BISI>
1959 V46 P718-722	TS-1920 <BISI>
1959 V46 N1 P2-8	66-10451 <*> O
1959 V46 N4 P86-92	36L32G <ATS>
1959 V46 N11 P289-301	61-20829 <*>
1959 V46 N26 P977-980	4787 <HB>
1960 V47 P1-7	TS-1608 <BISI>
1960 V47 P608-614	2416 <BISI>
1960 V47 N1 P1-7	62-14537 <*>
1960 V47 N2 P94-98	1853 <BISI>
1960 V47 N6 P328-329	2039 <BISI>
1960 V47 N17 P447-451	5341 <HB>
1960 V47 N18 P497-500	6822 <HB>
1961 V48 P145-150	AD-632 672 <=$>
1961 V48 P737-742	4012 <BISI>
1961 V48 N3 P49-56	5200 <HB>
1961 V48 N5 P109-114	5175 <HB>
1961 V48 N10 P293-295	01R76G <ATS>
1961 V48 N15 P425-428	5222 <HB>
1961 V48 N17 P481-487	64-10544 <*>
1961 V48 N19 P533-540	5342 <HB>
1961 V48 N19 P545-558	5343 <HB>
1961 V48 N19 P545-548	62-14664 <*> O
1961 V48 N19 P579-586	5344 <HB>
1961 V48 N21 P639-647	5623 <HB>
1962 V49 N2 P43	5658 <HB>
1962 V49 N6 P133-136	5612 <HB>
1962 V49 N6 P136-142	5613 <HB>
1962 V49 N7 P153-156	6185 <HB>
1962 V49 N9 P235-243	5756 <HB>
1962 V49 N9 P243-247	5757 <HB>
1962 V49 N13 P377-378	5686 <HB>
1962 V49 N15 P414-417	5586 <HB>
1962 V49 N15 P417-421	5695 <HB>
1962 V49 N18 P605-610	5717 <HB>
1962 V49 N18 P619-623	5832 <HB>
1963 V50 P202-213	3366 <BISI>
1963 V50 P506-517	8900 <IICH>
1963 V50 N1 P3-6	3312 <BISI>
1963 V50 N3 P68-73	6050 <HB>
1963 V50 N15 P454-457	6784 <HB>
1963 V50 N20 P602-609	6084 <HB>

1963 V50 N23 P728-730 6783 <HB>
1964 V51 P113-117 4079 <BISI>
1964 V51 P227-234 4494 <BISI>
1964 V51 P289-291 4204 <BISI>
1964 V51 P403-410 4013 <BISI>
1964 V51 P442-446 4265 <BISI>
1964 V51 P634-641 4119 <BISI>
1964 V51 P655-665 4169 <BISI>
1964 V51 P819-820 4201 <BISI>
1964 V51 N1 P25-32 6196 <HB>
1964 V51 N4 P95-100 6781 <HB>
1964 V51 N5 P117-123 6225 <HB>
1964 V51 N8 P194-199 6293 <HB>
1964 V51 N8 P199-205 6386 <HB>
1964 V51 N8 P205-214 6292 <HB>
1964 V51 N17 P509-518 4184 <BISI>
1964 V51 N17 P518-527 4185 <BISI>
1964 V51 N19 P538-545 6357 <HB>
1964 V51 N22 P689-697 6399 <HB>
1964 V51 N22 P697-710 6450 <HB>
1965 V52 P161-171 4491 <BISI>
1965 V52 P319-320 4395 <BISI>
1965 V52 P375-382 4515 <BISI>
1965 V52 P461-470 4467 <BISI>
1965 V52 P813-820 4725 <BISI>
1965 V52 N1 P1-8 6466 <HB>
1965 V52 N2 P29-33 6463 <HB>
1965 V52 N2 P33-37 6544 <HB>
1965 V52 N3 P67-70 65-13845 <*>
1965 V52 N3 P71-76 6638 <HB>
1965 V52 N6 P153-160 4644 <BISI>
1965 V52 N7 P191-193 6523 <HB>
1965 V52 N7 P197-200 4517 <BISI>
1965 V52 N15 P457-461 6692 <HB>
1965 V52 N15 P471-479 6643-I <HB>
1965 V52 N16 P485-493 6643-2 <HB>
1965 V52 N19 P583-593 6769 <HB>
1966 V53 N1 P15-18 6802 <HB>
1966 V53 N6 P161-166 6852 <HB>
1966 V53 N8 P250-254 6864 <HB>

GIESSEREI. TECHNISCHE-WISSENSCHAFTLICHE BEIHEFTE. METALLKUNDE UND GIESSEREIWESEN
1950 N2 P57-63 59-20936 <*> 0
62-20211 <*>
1950 N4 P175-180 60-18823 <*>
1953 N10 P463-475 3605 <BISI>
1953 V10 P441-454 1252 <*>
1954 N14 P701-707 60-10094 <*> 0
1957 N17 P909-917 62-14424 <*>
1959 P1217-1246 TS 1344 <BISI>
1962 V14 P135-139 6132 <HB>
1963 V15 N3 P137-147 6091 <BISI>
1963 V15 N3 P147-159 6116 <HB>
1964 V16 N2 P87-90 5821 <HB>
1964 V16 N2 P99-106 6323 <HB>
1964 V16 N2 P106-110 6455 <HB>
1964 V16 N3 P127-134 6636 <HB>
1964 V16 N3 P143-148 6543 <HB>
1964 V16 N3 P149-154 385 <TC>

GIESSEREIPRAXIS
1936 V57 P213-214 59-20910 <*> 0
1936 V57 P289-290 59-17278 <*> 0
1936 V57 P313-314 59-20909 <*> 0
1936 V57 P480-482 59-20908 <*> 0
1937 V58 P136-140 59-20906 <*> 0
1937 V58 P494-497 59-20912 <*> 0
1937 V58 P513-518 59-20912 <*> 0
1937 V58 N23/4 P230-234 59-20911 <*> 0
1938 V59 P10-12 59-20912 <*> 0
1938 V59 P26-30 59-20912 <*> 0
1941 V61 N16 P213-217 59-15471 <*>
1957 V75 N8 P176-181 61-16223 <*>
1957 V75 N15 P345-346 58-1763 <*>
1958 V76 N11 P219-222 62-20086 <*>
1958 V76 N16 P309-311 62-10214 <*>
1959 N8 P145 66-12420 <*>
1959 V77 N8 P145 62-10216 <*>
1964 V82 P287-299 3997 <BISI>
1964 V82 N8 P149-154 4280 <BISI>
1964 V82 N14 P277-281 65-13859 <*> 0
1965 V83 P20-22 4274 <BISI>

GIESSEREITECHNIK
1956 V2 N1 P4-6 3935 <HB>
1956 V2 N9 P202-206 4297 <HB>
1957 V3 N4 P73-77 4008 <HB>
1957 V3 N7 P160-162 61-16221 <*>
1957 V3 N9 P193-197 4213 <HB>
1958 V4 P307-310 TS-1271 <BISI>
61-20686 <*>
1958 V4 N9 P225-233 62-10215 <*>
1960 V6 P3-11 2263 <BISI>
1960 V6 P314 2262 <BISI>
1960 V6 N9 P280-281 4981 <BISI>
1960 V6 N11 P331-338 5350 <HB>
1961 V7 P337-342 3136 <BISI>
1963 V9 N6 P173-176 6540 <HB>
1963 V9 N7 P204-208 6540 <HB>
1965 V11 N3 P69-79 4518 <BISI>

GIESSEREIZEITUNG
1929 V26 N16 P450-455 59-20480 <*>
1930 V27 N1 P14-18 59-20480 <*>
1930 V27 N8 P221-224 59-20480 <*>
1930 V27 N12 P328-330 59-20480 <*>
1930 V27 N14 P383-385 59-20481 <*> 0
1930 V27 N14 P388-391 59-20479 <*> 0
1930 V27 N19 P541-544 59-20479 <*> 0
1930 V27 N20 P557-564 57-3143 <*>
1930 V27 N21 P590-598 59-20478 <*> 0

★GIGIENA I SANITARIYA
1940 N2/3 P40-41 R-2589 <*>
1941 N1 P28-34 62-11691 <=>
1941 N1 P44-48 62-11753 <=>
1944 V9 N7 P1-7 65-60094 <=$>
1944 V9 N7 P33-34 65-60093 <=$>
1947 V12 N3 P1-9 65-63264 <=$>
1947 V12 N7 P39 RT-245 <*>
1947 V12 N7 P41 RT-244 <*>
1947 V12 N7 P42-45 RT-949 <*>
1947 V12 N11 P51-52 63-14301 <=*>
1948 V13 N6 P10-16 RT-1912 <*>
1949 V14 N3 P19-22 63-31380 <=>
1949 V14 N10 P38-44 61-13587 <*+>
1950 N5 P29-32 60-13705 <+*>
1950 N7 P45-48 RT-1867 <*>
1951 N3 P32-38 RJ-636 <ATS>
1951 N9 P11-16 RT-3835 <*>
1951 N12 P17-20 RT-2113 <*>
1951 V16 N1 P13-16 62-25392 <=*>
1951 V16 N8 P50-51 R-32 <*>
1951 V16 N10 P5-10 RT-3829 <*>
1951 V16 N11 P21-25 61-13666 <*+>
1952 N4 P49-50 C.T.S.NO.1 <NLL>
RT-2334 <*>
1952 N5 P49-51 RT-2340 <*>
1952 N7 P12-17 RJ-169 <ATS>
1952 N12 P8-11 RJ-400 <ATS>
1952 V17 N1 P15-20 RT-1275 <*> C
1952 V17 N1 P24-29 RT-3702 <*>
1952 V17 N4 P44-46 13M38R <ATS>
1952 V17 N4 P50-52 RT-2729 <*>
1952 V17 N6 P19-22 RT-120 <*>
61-28018 <*=>
1952 V17 N7 P12-17 RJ-169 <ATS>
59-15255 <+*>
1952 V17 N10 P31-33 63-31380 <=>
1952 V17 N11 P49-53 RT-121 <*>
1953 N1 P58-60 RT-1274 <*>
1953 N4 P3-11 RT-3081 <*>
1953 N4 P11-15 RT-1294 <*>
1953 N4 P22-28 RT-3077 <*> 0
1953 N4 P50-51 RT-3082 <*>
1953 N7 P3-11 RT-2559 <*>
1953 N8 P49 RT-2133 <*>
1953 N9 P23-28 RT-2613 <*>
1953 N10 P19-23 RT-2396 <*>

1953 N12 P36-39 RT-2227 <*>
1953 V18 N7 P52-54 60-23728 <+*>
1953 V18 N8 P49 64-10451 <=*$>
1954 N8 P11-15 RT-2697 <*>
1954 N8 P41-43 R-3506 <*>
1954 N9 P32-36 RT-2943 <*>
1954 N9 P36-40 RT-2944 <*>
1954 V19 N1 P24-29 62-18777 <=*>
1954 V19 N4 P13-16 1132 <TC>
1954 V19 N4 P46-47 64-10917 <=*$>
1954 V19 N5 P23-26 62-18767 <=*>
1954 V19 N6 P10 62-15306 <*=>
1954 V19 N8 P15-18 59-13203 <=> 0
1954 V19 N9 P46 62-25898 <=*>
1955 N1 P18-25 RT-3565 <*>
1955 N2 P46-47 R-148 <*>
1955 N3 P55-56 RT-3056 <*>
1955 N9 P16-18 R-390 <*>
RJ502 <ATS>
1955 N9 P19-24 RJ-428 <ATS>
1955 N9 P24-27 RT-4238 <*>
1955 N9 P53-54 R-388 <*>
1955 N11 P3-7 R-338 <*>
1955 N11 P7-9 R-337 <*>
1955 N11 P32-37 R-336 <*>
1955 N11 P38-42 R-335 <*>
1955 N11 P48-51 R-166 <*>
1955 N11 P52-56 RT-3690 <*>
1955 N11 P58-59 R-344 <*>
1955 N12 P12-16 RT-4424 <*>
1955 N12 P23-26 RT-4321 <*>
1955 N12 P45-47 RT-4320 <*>
1956 N1 P6-11 RJ-1501 <ATS>
1956 N1 P11-14 RJ-503 <ATS>
1956 N9 P32-37 R-4470 <*>
1956 V21 N1 P27-31 59-13706 <=> 0
1956 V21 N6 P87-89 62-33322 <=*>
1956 V21 N7 P26-30 65-13430 <*> 0
1956 V21 N8 P56-57 63-15709 <=*>
1956 V21 N9 P3- 62-15302 <=*>
1956 V21 N9 P85-86 60-17797 <+*>
1956 V21 N11 P20-23 59-17309 <+*> 0
1957 N9 P32-37 59-12063 <+*>
1957 V22 N1 P70-72 60-13049 <+*>
1957 V22 N2 P8-15 59-11191 <=> 0
1957 V22 N5 P30-37 R-3800 <*>
64-19438 <=$> 0
66-13690 <*>
1957 V22 N7 P86-88 62-16268 <=*>
1957 V22 N8 P35-41 64-13166 <=*$>
1957 V22 N8 P57-60 59-12640 <+*>
1957 V22 N9 P3-8 59-11195 <=> 0
1957 V22 N9 P85-86 59-11194 <=> 0
1957 V22 N11 P3-7 60-17796 <+*>
1958 N7 P22-26 59-13449 <=>
1958 N7 P72-79 59-13452 <=> 0
1958 N7 P85-87 59-13450 <=>
1958 N8 P6-9 59-13433 <=>
1958 N8 P57-59 59-13434 <=>
1958 N8 P71-72 59-13435 <=>
1958 V23 N1 P75-77 64-10775 <=*$>
1958 V23 N2 P77-83 59-10185 <+*> 0
1958 V23 N2 P83-85 62-16056 <=*>
1958 V23 N3 P88-89 59-10090 <+*>
1958 V23 N4 P75-77 63-31382 <=>
1958 V23 N7 P26-29 63-10923 <=*>
1958 V23 N7 P66-71 59-13451 <=>
1958 V23 N7 P84-85 59-22285 <+*>
1958 V23 N10 P3-92 59-13134 <=>
1958 V23 N10 P23-27 66-60636 <=*$>
1958 V23 N10 P32-37 62-15800 <=*>
1958 V23 N11 P53-57 61-13812 <*+>
1958 V23 N12 P6-9 10402-D <KH>
63-18472 <=*>
1959 V24 N1 P3-7 59-11695 <=>
1959 V24 N1 P25-30 59-11696 <=>
1959 V24 N1 P35-40 59-11697 <=>
1959 V24 N1 P44-50 59-11698 <=>
1959 V24 N1 P59-65 59-11699 <=>
1959 V24 N1 P86-87 59-11700 <=>
1959 V24 N1 P88-89 60-11836 <=>
1959 V24 N2 P11-15 59-11710 <=>
1959 V24 N3 P3-8 59-31030 <=>
1959 V24 N3 P9-14 59-11726 <=>
1959 V24 N3 P27-33 59-11727 <=>
1959 V24 N3 P71 60-11696 <=>
1959 V24 N3 P78-79 CSIR-TR-425 <CSSA>
1959 V24 N4 P3-7 59-13667 <=>
1959 V24 N4 P77-79 59-19270 <+*>
1959 V24 N6 P72-73 61-19644 <=*>
62-11692 <=*>
1959 V24 N6 P92 60-11329 <=>
1959 V24 N7 P3-5 59-11781 <=>
1959 V24 N7 P93-94 60-11421 <=>
1959 V24 N8 P3-11 59-13988 <=>
1959 V24 N8 P26-32 59-13980 <=>
1959 V24 N8 P83-84 59-13990 <=>
1959 V24 N9 P74-76 60-31709 <=>
1959 V24 N10 P71-72 60-31422 <=>
1959 V24 N11 P7-11 60-11550 <=>
1960 V25 N1 P11-15 61-27765 <*=>
1960 V25 N1 P90-93 61-27765 <*=>
1960 V25 N2 P3-12 62-33303 <=*>
1960 V25 N2 P27-33 60-11917 <=>
1960 V25 N2 P64-70 60-11919 <=>
1960 V25 N2 P86-88 60-11921 <=>
1960 V25 N2 P92-94 60-11580 <=>
1960 V25 N3 P6-12 60-11805 <=>
1960 V25 N4 P78-81 61-23470 <=*>
1960 V25 N5 P29-34 60-41615 <=>
1960 V25 N5 P77-82 60-41616 <=>
1960 V25 N5 P102 60-41617 <=>
1960 V25 N5 P103-106 61-11552 <=>
1960 V25 N7 P26-30 61-11022 <=>
1960 V25 N7 P56-60 61-19250 <+*>
1960 V25 N7 P63-64 61-11036 <=>
1960 V25 N8 P56-58 SMRE-TRANS-5111 <NLL>
1960 V25 N8 P74-79 AEC-TR-4428 <+*>
1960 V25 N9 P15-41 61-21254 <=>
1960 V25 N9 P50-53 61-21254 <=>
1960 V25 N9 P60-65 61-21254 <=>
1960 V25 N9 P70-88 61-21254 <=>
1960 V25 N9 P94-110 61-21254 <=>
1960 V25 N9 P114-116 61-21254 <=>
1960 V25 N10 P22-31 61-21610 <=>
1960 V25 N10 P61-62 61-15849 <+*>
1960 V25 N10 P72-73 61-15852 <+*>
1960 V25 N11 P32-36 61-21510 <=>
1960 V25 N11 P47-50 61-21719 <=>
1960 V25 N12 P20-25 AEC-TR-4978 <=*>
1960 V25 N12 P39-45 61-23553 <=*>
1961 V26 N1 P3-6 61-23471 <=*>
1961 V26 N1 P23-32 61-31468 <=>
1961 V26 N1 P44-50 61-31469 <=>
1961 V26 N1 P50-54 61-31470 <=>
1961 V26 N1 P100-104 63-13187 <=*>
1961 V26 N1 P117-119 61-23477 <=*>
1961 V26 N2 P47-53 61-23112 <=*>
1961 V26 N3 P25-31 61-28031 <*=>
1961 V26 N3 P96-97 03P63R <ATS>
1961 V26 N4 P9-14 61-27197 <*=>
1961 V26 N4 P21-24 61-27196 <*=>
1961 V26 N4 P71-76 61-31525 <=>
1961 V26 N4 P76-81 61-27208 <*=>
1961 V26 N5 P18-23 61-27729 <*=>
1961 V26 N5 P28-33 61-28306 <*=>
1961 V26 N5 P33-37 61-28299 <=*>
1961 V26 N5 P51-55 61-27730 <=*>
1961 V26 N6 P25-28 61-28266 <*=>
1961 V26 N6 P29-32 61-28270 <*=>
1961 V26 N6 P33-38 61-28277 <*=>
1961 V26 N6 P39-46 61-28247 <*=>
1961 V26 N6 P46-51 61-28268 <*=>
1961 V26 N6 P91-92 61-28283 <*=>
1961 V26 N6 P108-109 61-28272 <*=>
1961 V26 N7 P22-28 61-28865 <*=>
1961 V26 N7 P29-33 61-28930 <*=>
1961 V26 N7 P33-39 61-28931 <*=>
1961 V26 N7 P45-51 61-28856 <*=>
1961 V26 N7 P58-61 61-28902 <*=>

Citation	Translation
1961 V26 N8 P22-28	62-33141 <=*>
1961 V26 N8 P53-63	62-13343 <*=>
1961 V26 N8 P63-67	62-23615 <=*>
1961 V26 N9 P41-44	62-23403 <=*>
1961 V26 N9 P48-51	62-19791 <=*>
1961 V26 N9 P76-79	62-33143 <=*>
1961 V26 N9 P83-84	62-16746 <=*>
1961 V26 N10 P16-19	62-15487 <*=>
1961 V26 N10 P50-57	62-19804 <=*>
1961 V26 N11 P9-23	62-15675 <*=>
1961 V26 N12 P18-21	62-23815 <=*>
1961 V26 N12 P47-50	62-24387 <=*>
1961 V26 N12 P79-80	63-20985 <=*$>
1962 V27 N1 P51-53	66-12892 <*>
1962 V27 N1 P100-101	62-23627 <=*>
1962 V27 N2 P7-15	62-32442 <=*>
1962 V27 N2 P19-24	62-24732 <=>
1962 V27 N2 P34-37	62-24732 <=>
1962 V27 N2 P67-71	62-24732 <=>
1962 V27 N2 P100-102	62-24732 <=>
1962 V27 N3 P14-17	62-25622 <=*> 0
1962 V27 N3 P22-49	62-25622 <=*> 0
1962 V27 N3 P51-55	62-24745 <=*>
1962 V27 N3 P70-84	62-25622 <=*> 0
1962 V27 N4 P18-24	62-32545 <=$>
1962 V27 N4 P30-36	64-51287 <=>
1962 V27 N4 P56-60	62-32545 <=$>
1962 V27 N4 P67-71	62-32545 <=$>
1962 V27 N4 P88-91	62-32545 <=$>
1962 V27 N5 P3-8	62-32852 <=*>
1962 V27 N5 P13-17	62-33035 <=*>
1962 V27 N5 P17-21	62-33435 <=>
1962 V27 N5 P54-55	62-33435 <=>
1962 V27 N5 P68-72	62-33435 <=>
1962 V27 N5 P97-101	62-33435 <=>
1962 V27 N5 P104-106	62-32852 <=*>
1962 V27 N6 P53-55	62-32563 <=*>
1962 V27 N6 P81-86	62-33295 <=*>
1962 V27 N6 P90-93	62-33295 <=*>
1962 V27 N6 P93-95	62-33563 <=*>
1962 V27 N7 P8-11	62-33671 <=>
1962 V27 N7 P22-27	62-33671 <=>
1962 V27 N7 P27-30	62-33147 <=*>
1962 V27 N7 P33-42	62-33671 <=>
1962 V27 N7 P60-63	62-33671 <=>
1962 V27 N7 P79-87	62-33671 <=>
1962 V27 N7 P103-105	62-33671 <=>
1962 V27 N8 P60-63	63-13257 <=>
1962 V27 N8 P79-85	AEC-TR-5672 <=*>
	63-13257 <=>
1962 V27 N9 P30-34	63-15350 <=$>
1962 V27 N9 P82-85	63-16985 <=*>
1962 V27 N9 P85-90	63-15350 <=$>
1962 V27 N9 P98-99	63-15350 <=$>
1962 V27 N9 P103-109	63-15350 <=$>
1962 V27 N10 P25-28	63-19647 <=*>
1962 V27 N10 P39-62	66-60619 <=*$>
1962 V27 N10 P57-62	63-13485 <=>
1962 V27 N10 P73-74	63-13485 <=>
1962 V27 N10 P94-100	63-13485 <=>
1962 V27 N10 P122-124	63-13485 <=>
1962 V27 N11 P48-49	63-19462 <=*>
1962 V27 N12 P7-12	63-21755 <=>
1962 V27 N12 P13-19	64-10669 <=*$>
1962 V27 N12 P89-93	63-21755 <=>
1963 V28 N1 P31-36	65-17123 <*>
1963 V28 N1 P53-55	63-21475 <=>
1963 V28 N1 P83-89	63-21475 <=>
1963 V28 N1 P95-99	63-21475 <=>
1963 V28 N1 P105-108	63-21475 <=>
1963 V28 N2 P19-25	63-31020 <=>
1963 V28 N3 P104-105	64-21815 <=>
1963 V28 N4 P94-96	63-31060 <=>
1963 V28 N5 P24-29	63-31396 <=>
1963 V28 N6 P20-23	63-31657 <=>
1963 V28 N6 P35-38	63-31658 <=>
1963 V28 N8 P14-19	63-41040 <=>
1963 V28 N8 P61-63	63-41040 <=>
1963 V28 N8 P73-74	63-24035 <=*$>
1963 V28 N9 P5-11	63-41190 <=>
1963 V28 N9 P58-61	64-51287 <=>
1963 V28 N9 P97-98	63-41190 <=>
1963 V28 N9 P101-103	63-41190 <=>
1963 V28 N9 P124-125	63-41190 <=>
1963 V28 N10 P3-7	64-21180 <=>
	64-21591 <=>
1963 V28 N10 P45-48	64-21590 <=>
1963 V28 N10 P52-57	64-21589 <=>
1963 V28 N10 P73-78	64-21594 <=>
1963 V28 N12 P42-46	64-21556 <=>
1963 V28 N12 P62-65	64-21622 <=>
1963 V28 N12 P81-86	64-21557 <=>
1963 V28 N12 P87-96	64-19129 <=*$>
1964 V29 N1 P3-11	64-21911 <=>
1964 V29 N1 P28-34	64-21911 <=>
1964 V29 N1 P44-49	64-21911 <=>
1964 V29 N1 P56-65	64-19537 <=$>
1964 V29 N1 P74-77	64-21911 <=>
1964 V29 N1 P82-87	64-21911 <=>
1964 V29 N2 P23-27	64-31539 <=>
1964 V29 N2 P40-43	64-31050 <=>
1964 V29 N2 P102-104	64-21915 <=>
1964 V29 N3 P8-11	66-14882 <*$>
1964 V29 N3 P19-23	64-31366 <=>
1964 V29 N3 P56-58	64-31366 <=>
1964 V29 N3 P66-69	64-31366 <=>
1964 V29 N3 P79-83	64-31366 <=>
1964 V29 N3 P97-99	64-31435 <=>
1964 V29 N4 P3-8	64-41564 <=>
1964 V29 N4 P29-42	64-41564 <=>
1964 V29 N4 P58-64	64-41564 <=>
1964 V29 N4 P96-97	65-31363 <=*$>
1964 V29 N4 P98-100	65-31764 <=$>
1964 V29 N4 P103-112	65-30478 <=$>
1964 V29 N4 P114-117	65-30478 <=$>
1964 V29 N5 P3-6	64-41352 <=>
1964 V29 N5 P10-13	64-41352 <=>
1964 V29 N5 P53-57	64-41352 <=>
1964 V29 N5 P70-73	64-41352 <=>
1964 V29 N5 P80-82	64-41352 <=>
1964 V29 N5 P97-100	64-41352 <=>
1964 V29 N5 P104-105	64-41352 <=>
1964 V29 N5 P119-122	64-41352 <=>
1964 V29 N6 P3-9	64-41472 <=>
1964 V29 N6 P54-56	CSIR-TR.543 <CSSA>
1964 V29 N6 P72-74	64-41471 <=>
1964 V29 N7 P12-17	66-61746 <=*$>
1964 V29 N7 P25-30	64-51807 <=$>
1964 V29 N7 P39-44	AD-618 635 <=$>
	64-51809 <=$>
1964 V29 N7 P144-149	64-51808 <=$>
1964 V29 N8 P3-5	64-51455 <=$>
1964 V29 N8 P19-32	64-51455 <=$>
1964 V29 N8 P53-60	AD-614 396 <=$>
1964 V29 N8 P53-63	64-51455 <=$>
1964 V29 N8 P93-95	64-51455 <=$>
1964 V29 N8 P98-104	64-51455 <=$>
1964 V29 N8 P121-124	64-51455 <=$>
1964 V29 N9 P53-57	64-51425 <=$>
1964 V29 N11 P37-39	65-30205 <=$>
1964 V29 N12 P73-79	65-30476 <=$>
1965 V30 N2 P17-21	81S83R <ATS>
1965 V30 N2 P106-109	AD-638 189 <$=>
1965 V30 N3 P8-13	65-31077 <=$>
1965 V30 N4 P3-6	65-31491 <=$>
1965 V30 N4 P12-16	65-31491 <=$>
1965 V30 N4 P28-32	65-31491 <=$>
1965 V30 N4 P44-51	65-31491 <=$>
1965 V30 N4 P114-116	65-31491 <=$>
1965 V30 N6 P18-22	65-31777 <=$>
1965 V30 N6 P65-67	65-32930-B <=*$>
1965 V30 N6 P70-72	65-31778 <=$>
1965 V30 N6 P73-76	65-31958 <=*$>
	65-32930-B <=*$>
1965 V30 N7 P3-9	65-33436 <=*$>
1965 V30 N7 P60-61	65-32524 <=$>
1965 V30 N7 P68-69	65-32524 <=$>
1965 V30 N7 P125-126	65-33435 <=*$>
1965 V30 N8 P51-54	65-33069 <=*$>
1965 V30 N8 P83-86	65-33069 <=*$>

1965 V30 N8 P115-117 65-33069 <=*$>
1965 V30 N9 P111 66-30797 <=$>
1965 V30 N9 P116-117 66-30797 <=$>
1965 V30 N10 P107-109 66-30077 <=*$>
1965 V30 N11 P72-78 66-30726 <=$>
1965 V30 N12 P50-51 66-31104 <=$>
1965 V30 N12 P54-57 66-31104 <=$>
1965 V30 N12 P98-99 66-31104 <=$>
1965 V30 N12 P100-102 66-31104 <=$>

★ GIGIENA TRUDA I PROFESSIONALNYE ZABOLEVANIYA

1957 N3 P29-34 R-5369 <*>
1957 V1 N1 P10-14 59-13705 <=>
1957 V1 N4 P3-11 R-3150 <*>
1958 V2 N1 P9-16 59-11443 <=> 0
1958 V2 N1 P18-22 59-11437 <=>
1958 V2 N1 P44-48 63-31382 <=>
1958 V2 N2 P3-8 59-13315 <=>
1958 V2 N2 P12-16 62-33278 <=*>
1958 V2 N2 P36-41 59-13316 <=> 0
1958 V2 N2 P52-54 59-13317 <=> 0
1958 V2 N2 P56 59-13318 <=>
1958 V2 N4 P5-11 59-11382 <=>
1958 V2 N4 P23-30 59-11383 <=>
1958 V2 N5 P7-15 59-13462 <=>
1958 V2 N5 P15-20 61-11488 <=>
1958 V2 N6 P3-8 59-13468 <=>
1959 V3 N1 P3-6 59-11670 <=>
1959 V3 N1 P9-12 60-11629 <=>
1959 V3 N1 P22-28 60-41398 <=>
1959 V3 N1 P28-32 60-41399 <=>
1959 V3 N4 P29-34 66-10720 <*>
1959 V3 N5 P7-15 60-31022 <=>
1959 V3 N6 P3-8 60-11425 <=>
1959 V3 N6 P16-21 60-11425 <=>
1959 V3 N6 P22-28 60-11425 <=>
1960 V4 N3 P3-7 60-41216 <=>
1960 V4 N3 P8-11 60-41218 <=>
1960 V4 N3 P11-17 60-41219 <=>
1960 V4 N3 P28-32 60-41220 <=>
1960 V4 N3 P49-50 60-41221 <=>
1960 V4 N3 P57-58 60-41232 <=>
1960 V4 N5 P8-52 61-11043 <=>
1960 V4 N5 P61-63 61-11018 <=>
1960 V4 N6 P3-6 61-21103 <=>
1960 V4 N6 P7-11 61-21143 <=>
1960 V4 N6 P61 61-21182 <=>
1960 V4 N8 P10-13 61-11628 <=>
1960 V4 N8 P45-47 61-11478 <=>
1960 V4 N10 P35-40 61-21335 <=>
1960 V4 N12 P3-7 61-21621 <=>
1960 V4 N12 P55 61-21623 <=>
1961 V5 N1 P3-6 61-27108 <*=>
1961 V5 N1 P56-57 61-27089 <*=>
1961 V5 N2 P29-32 61-27073 <*=>
1961 V5 N2 P56 61-27073 <*=>
1961 V5 N3 P3-8 63-23217 <=*$>
1961 V5 N4 P3-65 61-31528 <=>
1961 V5 N10 P13-19 62-19762 <=*>
1961 V5 N11 P23-29 63-31382 <=>
1961 V5 N11 P55-57 64-51287 <=>
1961 V5 N12 P3-7 62-15790 <*=>
1961 V5 N12 P34-39 AEC-TR-4996 <=*>
1962 N11 P11-19 NLLTB V5 N3 P173 <NLL>
1962 V6 N1 P21-24 62-23626 <=*>
1962 V6 N1 P28-33 62-23626 <=*>
1962 V6 N2 P10-26 62-23838 <=*>
1962 V6 N2 P55-57 62-23838 <=*>
1962 V6 N3 P8-13 62-24998 <=*>
1962 V6 N4 P14-21 62-25522 <=>
1962 V6 N4 P26-35 62-25522 <=>
1962 V6 N4 P56-60 62-25522 <=>
1962 V6 N5 P8-13 62-32554 <=*>
1962 V6 N5 P22-27 62-32554 <=*>
1962 V6 N5 P43-46 62-32554 <=*>
1962 V6 N6 P5-37 62-33308 <=>
1962 V6 N7 P10-14 C-5105 <NRC>
1962 V6 N7 P19-22 62-33445 <=>
1962 V6 N7 P34-37 62-33445 <=>
1962 V6 N7 P48-50 62-32560 <=*>
1962 V6 N7 P54-58 62-33445 <=>
1962 V6 N9 P41-44 64-51287 <=>
1962 V6 N9 P57-58 63-24626 <=*$>
1962 V6 N10 P46-51 AEC-TR-5669 <=*$>
1962 V6 N12 P26-33 63-21332 <=>
1963 N4 P43- FPTS V23 N1 P.T51 <FASE>
1963 N6 P16- FPTS V23 N4 P.T743 <FASE>
1963 V7 N1 P4-13 63-21519 <=>
1963 V7 N1 P36-41 63-21519 <=>
1963 V7 N1 P41-44 64-51287 <=>
1963 V7 N2 P24-29 63-21756 <=>
1963 V7 N2 P27-30 64-13422 <=*$>
1963 V7 N2 P62-63 63-21756 <=>
1963 V7 N3 P7- 64-19267 <=*$>
1963 V7 N3 P33-37 64-19268 <=*$>
1963 V7 N4 P3-9 63-31118 <=>
64-13866 <=*$> 0
1963 V7 N4 P16-22 64-13482 <=*$>
1963 V7 N4 P27-31 64-13481 <=*$>
1963 V7 N4 P31-38 64-13622 <=*$>
1963 V7 N4 P38-43 64-13462 <=*$>
1963 V7 N4 P43- FPTS,V23,P.T51-T54 <FASE>
1963 V7 N5 P33-38 64-15405 <=*$>
1963 V7 N5 P38-43 63-31470 <=>
1963 V7 N5 P54-56 AD-617 074 <=$>
1963 V7 N5 P56-57 AD-617 163 <=$>
1963 V7 N5 P58-59 AD-617 076 <=$>
1963 V7 N5 P59-60 AD-617 075 <=$>
1963 V7 N5 P61 AD-616 697 <=$>
1963 V7 N6 P61-63 63-41103 <=>
1963 V7 N7 P3-8 63-41097 <=>
1963 V7 N7 P13-19 63-31735 <=>
64-15551 <=*$>
1963 V7 N7 P31-35 64-15552 <=*$>
1963 V7 N7 P35-40 64-15553 <=*$>
1963 V7 N9 P13-17 64-19242 <=*$>
1963 V7 N9 P45-47 63-41052 <=>
1963 V7 N10 P13-21 64-19243 <=*$>
1963 V7 N10 P38- FPTS V24 N1 P.T63 <FASE>
FPTS V24 P.T63-T64 <FASE>
1963 V7 N11 P3-8 64-21288 <=>
64-21468 <=>
1963 V7 N11 P17-20 64-21468 <=>
1963 V7 N11 P41-43 64-21468 <=>
1963 V7 N11 P51- FPTS V24 N1 P.T96 <FASE>
FPTS V24 P.T96-T98 <FASE>
1963 V7 N11 P60-64 64-19706 <=$>
64-21468 <=>
1963 V7 N12 P8-15 65-60160 <=$>
1963 V7 N12 P38-42 65-17043 <*>
1964 V8 N2 P9-14 65-31418 <=*$>
65-62347 <=$>
1964 V8 N2 P14-18 65-62344 <=$>
1964 V8 N2 P24-28 65-62372 <=$>
1964 V8 N3 P3-10 65-31366 <=*$>
1964 V8 N3 P3-7 65-62345 <=$>
1964 V8 N3 P7- FPTS V24 N2 P.T253 <FASE>
FPTS V24 P.T253-T254 <FASE>
1964 V8 N3 P38-42 65-61C55 <=$>
1964 V8 N4 P19-29 64-31498 <=>
1964 V8 N4 P25- FPTS V24 N4 P.T611 <FASE>
1964 V8 N4 P30- FPTS V24 N3 P.T551 <FASE>
1964 V8 N4 P30-35 79R80R <ATS>
1964 V8 N4 P40-46 65-62186 <=$>
1964 V8 N4 P57-62 64-31498 <=>
1964 V8 N5 P53-56 65-62373 <=$>
1964 V8 N6 P7-9 64-41338 <=>
1964 V8 N6 P10-13 64-41339 <=>
1964 V8 N6 P14-19 65-31545 <=$>
1964 V8 N6 P14-20 65-62350 <=$>
1964 V8 N7 P25- FPTS V24 N4 P.T608 <FASE>
1964 V8 N7 P39-43 65-31371 <=*$>
65-62355 <=$>
1964 V8 N7 P57-58 65-31182 <=*$>
1964 V8 N8 P3-7 64-51178 <=$>
1964 V8 N8 P18- FPTS V24 N5 P.T847 <FASE>
1964 V8 N8 P24-27 65-62354 <=$>
1964 V8 N8 P32-37 65-63599 <=$>
1964 V8 N8 P57-58 65-31460 <=$>
1964 V8 N9 P3-10 64-51538 <=$>

1964 V8 N9 P14- FPTS V24 N5 P.T877 <FASE>
1964 V8 N9 P14-19 65-31405 <=*$>
1964 V8 N9 P25-28 64-51538 <=$>
1964 V8 N9 P34-39 65-62353 <=$>
1964 V8 N9 P42-45 64-51538 <=$>
1964 V8 N9 P59-60 65-31406 <=*$>
1964 V8 N9 P62-63 64-51538 <=$>
1964 V8 N10 P3-9 65-30417 <=$>
1964 V8 N10 P18-41 65-30417 <=$>
1964 V8 N10 P26- FPTS V24 N6 P.T1121 <FASE>
1964 V8 N10 P42-44 65-62352 <=$>
1964 V8 N10 P45-49 65-30417 <=$>
1964 V8 N10 P52-56 65-30417 <=$>
1964 V8 N10 P57-58 65-31683 <=$>
1964 V8 N11 P22-25 65-30401 <=$>
1964 V8 N11 P29-36 65-64545 <=*$>
1964 V8 N11 P38-42 65-30401 <=$>
1964 V8 N11 P54-56 65-30401 <=$>
1964 V8 N11 P60-62 65-30401 <=$>
1964 V8 N12 P23-33 65-30902 <=$>
1965 N12 P3-7 66-30699 <=$>
1965 V9 N1 P3-6 65-30790 <=$>
1965 V9 N1 P12-17 65-64521 <=*$>
1965 V9 N1 P17-21 65-30791 <=$>
1965 V9 N1 P27-30 65-30792 <=$>
1965 V9 N1 P33- FPTS V25 N1 P.T91 <FASE>
1965 V9 N1 P36- FPTS V24 N6 P.T1123 <FASE>
1965 V9 N2 P29-33 65-64520 <=*$>
1965 V9 N2 P33- FPTS V24 N6 P.T951 <FASE>
1965 V9 N2 P58-60 65-31125 <=$>
1965 V9 N3 P3-15 65-30939 <=$>
1965 V9 N3 P20-24 66-60426 <=*$>
1965 V9 N3 P38-43 65-30939 <=$>
66-60427 <=*$>
1965 V9 N3 P60-63 65-30939 <=$>
1965 V9 N4 P42-49 65-31382 <=*$>
1965 V9 N4 P42-59 66-60880 <=*$>
1965 V9 N5 P20-23 65-32371 <=$>
1965 V9 N5 P28-43 65-32371 <=$>
1965 V9 N5 P39-44 66-60895 <=*$>
1965 V9 N5 P49-55 65-32371 <=$>
1965 V9 N5 P60-61 65-32371 <=$>
1965 V9 N5 P63-64 65-32371 <=$>
1965 V9 N6 P19-24 65-33615 <=$>
1965 V9 N6 P40-48 65-33615 <=$>
1965 V9 N6 P63-64 65-33615 <=$>
1965 V9 N7 P22- FPTS V25 N5 P.T851 <FASE>
1965 V9 N7 P32-36 65-33182 <=*$>
1965 V9 N7 P56 65-33182 <=*$>
1965 V9 N7 P60-62 65-33182 <=*$>
1965 V9 N8 P3-9 65-33183 <=$>
1965 V9 N8 P19-24 65-33183 <=$>
1965 V9 N8 P27-31 65-33183 <=$>
1965 V9 N8 P32-37 66-61324 <=*$>
1965 V9 N8 P37-41 65-33183 <=$>
1965 V9 N8 P61-63 65-33183 <=$>
1965 V9 N9 P18-21 65-33015 <=*$>
1965 V9 N9 P33- FPTS V25 N4 P.T657 <FASE>
1965 V9 N10 P7-11 66-61967 <=*$>
1965 V9 N11 P20- FPTS V25 N5 P.T843 <FASE>
1965 V9 N11 P26- FPTS V25 N5 P.T803 <FASE>
1965 V9 N11 P29- FPTS V25 N5 P.T854 <FASE>
1966 V10 N1 P5- FPTS V25 N6 P.T1063 <FASE>
1966 V10 N1 P5-6 66-31366 <=$>
1966 V10 N1 P26-34 66-31366 <=$>
1966 V10 N2 P3- FPTS V25 N6 P.T1066 <FASE>
1966 V10 N2 P23-27 66-61974 <=*$>
1966 V10 N12 P7-36 67-30567 <=>
1966 V10 N12 P40-42 67-30567 <=>
1966 V10 N12 P46-55 67-30567 <=>

GIGIENA TRUDA I TEKHNIKA BEZOPASNOSTI
1924 V2 N1 P79-80 65-60096 <=$>
1927 N7 P24-28 65-63255 <=>
1927 V5 N8 P12-19 64-23633 <=$>
1927 V5 N8 P45-52 768B <K-H>
1928 N2 P35-44 65-63243 <=>
1928 N6 P94-96 65-63256 <=>
1928 V6 N2 P44-47 1084C <K-H>
1936 V14 N1 P87-89 65-60091 <=$>

GINEKOLOGIA POLSKA
1960 V31 N2 P263-268 63-00852 <*>

GIORNALE DI BATTERIOLOGIA E IMMINOLOGIA
1934 V12 N2 P225-240 64-14167 <*> 0
1934 V13 P353-355 58-426 <*>
1935 V15 P864-871 58-425 <*>
1947 V36 P105-114 58-421 <*>
1954 V46 P381-397 60-13407 <=*> 0

GIORNALE DI BATTERIOLOGIA VIROLOGIA ET IMMUNOLOGIA
1958 V51 P149-153 64-18613 <*>
1958 V51 P590-595 64-18686 <*>
1959 V52 P134-141 64-18694 <*>
71M42I <ATS>

GIORNALE DI BIOCHIMICA
1955 V4 P33-41 58-1751 <*>
1955 V4 N1 P33-41 C-2591 <NRC>
1956 V5 P483-527 07K201 <ATS>
1956 V5 N4 P203-277 65-00090 <*>
1961 V10 N6 P393-410 62-01269 <*>

GIORNALE DI CHIMICA INDUSTRIALE ED APPLICATA
1921 V3 P197-199 2855 <*>
1924 V6 P10-11 2879 <*>
1924 V6 P323-325 2346 <*>
1924 V6 P571-575 1079 <*>
1925 V7 P333 AL-150 <*>
1932 V14 P69-75 II-647 <*>
1933 V15 P13-15 II-579 <*>
1933 V15 P494-499 II-675 <*>
1933 V15 N6 P273-281 87H11I <ATS>
1933 V15 N7 P350-351 I-1003 <*>
1934 V16 P538-552 64-18301 <*> 0
1934 V16 N12 P607 II-422 <*>

GIORNALE DI CLINICA MEDICA
1948 V29 P394-409 66-13462 <*> 0
1950 V31 P211-223 AL-167-II <*>
1951 V32 P1081-1110 65-11330 <*> 0
1952 V33 P964-976 57-145 <*>
1953 N6 P649- AL-989 <*>
1953 V34 N11 P1253-1347 3077 <*>
1957 V38 N4 P533-541 65-14963 <*> 0
1958 V39 N12 P1883-1891 66-11627 <*>
1960 V41 P76-84 10481-C <K-H>
63-16691 <*>
1960 V41 N10 P1303-1326 62-00420 <*>

GIORNALE DEL GENIO CIVILE
1917 V55 P341-343 I-371 <*>
1922 V60 P343-368 59-17098 <*> 0
1957 N6 P472-480 61-10166 <*>

GIORNALE DI GEOLOGIA
1940 V14 P17-38 50J16I <ATS>

GIORNALE DI GERONTOLOGIA
1957 N13 P175-180 59-12942 <+*>
1961 V9 N2 P129-132 62-00419 <*>
1963 V11 P47-55 64-18988 <*> 0
1964 V12 N6 P735-739 66-10934 <*>

GIORNALE ITALIANO DI CHEMIOTERAPIA
1954 V1 N2 P281-288 II-593 <*>
1954 V1 N3 P579-584 II-446 <*>
1955 V2 N1/2 P32-40 1081 <*>
1955 V2 N1/2 P63-70 II-844 <*>
1956 V3 P307-320 65-14964 <*> 0
1956 V3 P321-328 65-14965 <*> 0
1956 V3 N3/4 P569-571 59-15683 <*>
1963 V10 N1 P24-31 65-14542 <*>

GIORNALE ITALIANO DI CHIRURGIA
1958 V14 N8 P691-711 63-16949 <*>
8973-A <KH>

GIORNALE ITALIANO DI DERMATOLOGIA

1955 V96 N4 P359-371 57-2776 <*>
1955 V96 N5 P536-544 57-2777 <*>

GIORNALE ITALIANO DI OFTALMOLOGIA
1951 V4 N3 P226-231 2190 <*>
1952 V5 P22-29 65-13734 <*>
1955 V8 P42-52 59-18113 <=*> O
1956 V9 P429-443 66-12620 <*>

GIORNALE ITALIANO DELLA TUBERCOLOSI
1947 V1 N4 P207-208 2800 <*>
1949 V3 N5 P350- 2508 <*>
1950 V4 N3 P131-140 2803 <*>
1950 V4 N3 P143-146 2905 <*>

GIORNALE DI MALATTIE INFETTIVE E PARASSITARIE
1958 V10 P359-361 59-14056 <+*> O
1960 V12 N9/10 P522-524 62-00597 <*>
1961 V13 P614-619 63-20874 <*> O

GIORNALE DI MATEMATICHE DI BATTAGLINI PER IL PROGRESSO DEGLI STUDI NELLE UNIVERSITA ITALIANE
1894 V32 P209-291 N66-14059 <=$>
1947 V77 P119-144 AL-291 <*>

GIORNALE DI MEDICINA MILITARE
1938 V86 P355-374 57-1672 <*>

GIORNALE DI MICROBIOLOGIA
1955 V1 P127-143 58-1734 <*>

GIORNALE DI PSICHIATRIA E DI NEUROPATOLOGIA
1957 V85 P727-733 59-19211 <=*> O
1959 N4 P1-35 61-00353 <*>
1960 V88 P467-496 63-00782 <*>
1961 V89 P1527-1540 66-12166 <*>

GJUTERIET
1950 V40 P181-188 3537 <K-H>
1950 V40 N6 P81-88 I-735 <*>
1954 V44 P193-195 2658 <BISI>
1954 V44 N2 P19-28 60-18802 <*> O
1957 V47 N6 P96-98 66-10829 <*> O
1959 V49 N2 P33-38 66-11780 <*> O
1961 V57 N5 P69-74 2778 <BISI>

GLAS UND APPARAT
1924 V21 P164-165 BGIRA-556 <BGIR>

GLASERS ANNALEN FUER GEWERBE UND BAUWESEN
1953 V77 N1 P9-14 61-20940 <*> O
1953 V77 N8 P219-227 60-10233 <*>
1953 V77 N9 P264-281 59-20572 <*> P
1954 V78 P101-107 T-1414 <INSD>
1955 P42-43 T-2093 <INSD>
1955 P290-294 T2087 <INSD>
1955 V79 P150-154 T2132 <INSD>
1955 V79 P262-270 T2090 <INSD>
1955 V79 P336-338 T-2055 <INSD>
1955 V79 N7 P219-227 60-10235 <*>
1956 N4 P117-126 T-2377 <INSD>
1957 V81 N2 P36-49 60-10237 <*>
1957 V81 N3 P91-94 60-10236 <*>
1958 V82 P266-275 SCL-T-445 <*>
1961 V85 N3 P91-97 64-18074 <*>
1963 V87 N6/7 P391-393 66-11757 <*>

GLASHUETTE
1937 V67 P549-552 63-14504 <*> O

★GLASNIK HEMISKOG DRUSTVA. BEOGRAD
1949 V14 P229-231 2868 <*>
64-10461 <*>
1955 V20 P135-136 2820 <*>
1956 V21 P9-18 UCRL TRANS-673(L) <*>
1956 V21 P271-276 AEC-TR-3851 <*>
1962 V27 N1 P5-21 62-11757/1 <=>
1962 V27 N1 P23-35 62-11757/1 <=>
1962 V27 N1 P37-49 62-11757/1 <=>
1962 V27 N1 P51-61 62-11757/1 <=>
1963 V28 N1 P1-47 63-11452/1 <=>
1964 V29 N3/4 P81-93 64-11460/3-4 <=>
1964 V29 N3/4 P95-105 64-11460/3-4 <=>
1964 V29 N3/4 P107-113 64-11460/3-4 <=>
1964 V29 N3/4 P115-129 64-11460/3-4 <=>
1964 V29 N3/4 P131-137 64-11460/3-4 <=>
1964 V29 N3/4 P139-155 64-11460/3-4 <=>
1964 V29 N3/4 P157-193 64-11460/3-4 <=>

GLASNIK MATEMATICKO-FIZICKI I ASTRONOMSKI. HRVATSKO PRIRODOSLOVNO DRUSTVO
1959 V14 N4 P295-302 AEC-TR-4766 <*>

GLASNIK SUMARSKOG FAKULTET, UNIVERZITET U BEOGRADU
1955 N9 P311-315 60-21638 <=>

GLASNIK ZA SUMSKE POKUSE
1957 N13 P509-536 60-21653 <=>

GLASNIK. UNIVERZITET U BEOGRADU
1958 V2 N7 P175-178 60-31337 <=>
1958 V2 N7 P190-191 60-31337 <=>
1958 V2 N9 P243-266 60-31337 <=>
1958 V2 N10 P271-285 60-31337 <=>
1958 V2 N12 P315-318 60-31337 <=>
1958 V2 N12 P322-327 60-31337 <=>

GLASNIK ZEMALJSKOG MUSEJA U BOSNI I HERCEGOVINI
1945 P45-53 60-21689 <=>

GLAS-EMAIL-KERAMO-TECHNIK
1953 V4 N5 P172-175 62-18114 <*>
1953 V4 N7 P247-249 62-18114 <*>
1953 V4 N12 P437-439 62-18115 <*>
1954 P172-177 1774 <*>
1955 V6 N1 P1-5 62-18200 <*>
1957 V8 N6 P212-214 62-18132 <*>
1957 V8 N6 P214-215 61-14464 <*> O
1957 V8 N10 P390-391 62-14876 <*>
1959 V10 N3 P82-87 61N55G <ATS>
1959 V10 N5 P165-169 62-10094 <*>
1959 V10 N6 P219-224 60-16860 <*>
1959 V10 N6 P225-227 60-16841 <*>
1959 V10 N9 P337-340 63-00767 <*>
1961 V12 N1 P6-9 64-10564 <*>
1961 V12 N2 P37-40 BGIRA-570 <BGIR>
1961 V12 N5 P159-163 64-14073 <*>
1961 V12 N9 P322-327 62-18565 <*>
1961 V12 N10 P369-373 62-18564 <*>
1963 V14 N5 P161-168 65-13516 <*>
1963 V14 N5 P163-168 BGIRA-597 <BGIR>
1963 V14 N5 P174-180 BGIRA-596 <BGIR>
1963 V14 N7 P249-258 BGIRA-591 <BGIR>
1966 V17 N1 P1-7 66-14195 <*>

GLAS-UND HOCHVAKUUM-TECHNIK
1952 N7 P123-134 57-2757 <*>
60G8G <ATS>
1952 V1 N1 P9-13 BGIRA-562 <BGIR>
1952 V1 N1 P14-18 62-14878 <*>
1953 V2 P256-259 57-2001 <*>
1953 V2 N12 P256-259 62-16685 <*>

GLAS- UND INSTRUMENTEN-TECHNIK
SEE G-I-T

GLASTECHNISCHE BERICHTE
1928 V6 N8 P414-447 63-20138 <*>
1930 V8 N5 P257-265 63-20147 <*> O
1931 V9 N10 P529-545 BGIRA-616 <BGIR>
1932 V10 N1 P26-30 62-18138 <*>
1932 V10 N3 P126-130 62-18113 <*>
1932 V10 N8 P411-420 63-14505 <*> C
1935 V13 N7 P245-247 63-20139 <*>
1936 V14 N2 P60-66 60-10056 <*> O
1936 V14 N3 P89-103 64-10536 <*>
1937 V15 N7 P259-270 63-20141 <*> O
1937 V15 N7 P282-285 63-20140 <*> O
1938 V16 N7 P219-227 63-20145 <*> O
1938 V16 N7 P228-231 63-20144 <*> O

1938 V16 N7 P233-236 64-14085 <*>
1938 V16 N9 P296-304 63-20143 <*> O
1938 V16 N12 P389-391 63-20142 <*>
1939 V17 N10 P286-290 63-20146 <*> O
1939 V17 N12 P325-327 64-10528 <*>
1940 V18 N2 P33-45 65-13617 <*>
1940 V18 N3 P65-69 65-13617 <*>
1940 V18 N10 P267-273 64-14084 <*>
1941 V19 N7 P217-225 64-14035 <*>
1942 V20 N9 P258-262 62-14882 <*>
1943 V21 N12 P249-255 61-10180 <*>
1943 V21 N7/8 P149-170 61-10181 <*>
1949 V22 N11 P226-228 61-10186 <*>
1949 V22 N12 P262 62-18355 <*>
66-14806 <*>
1949 V22 N18 P420-423 61-10183 <*>
1949 V22 N9/10 P161-169 61-10184 <*>
1949 V22 N9/10 P173-177 62-18135 <*>
1949 V22 N9/10 P177-179 62-18136 <*>
1950 V23 P161-169 60-18795 <*>
1950 V23 N1 P1-10 60-16849 <*>
1950 V23 N3 P67-79 62-16468 <*>
1951 V24 P79-102 57-730 PT1 <*>
1951 V24 N12 P293-301 60-16859 <*>
1952 V25 N3 P71-83 60-16922 <*> O
1952 V25 N4 P101-106 61-10185 <*>
1952 V25 N9 P276-285 62-14887 <*>
1952 V25 N10 P307-324 60-18770 <*> P
60-18771 <*> P
63-18596 <*> P
1952 V25 N12 P387-392 66-10802 <*>
1952 V25 N12 P392-396 62-16131 <*>
1953 V26 N1 P5-12 62-16397 <*>
1953 V26 N7 P197-201 63-18580 <*>
1953 V26 N8 P232-238 60-14033 <*>
60-18777 <*>
1953 V26 N10 P300-306 60-10003 <*> O
62-10573 <*> O
1953 V26 N11 P333-341 66-10699 <*>
1954 V27 N2 P41-46 63-14059 <*>
1954 V27 N4 P105-116 60-16252 <*>
1954 V27 N5 P164-166 62-14498 <*>
1954 V27 N7 P239-247 61-20840 <*>
1954 V27 N10 P374-381 62-14883 <*>
1954 V27 N10 P381-392 62-14881 <*>
1954 V27 N11 P405-409 60-16845 <*>
1955 V28 P426-437 T-1383 <INSD>
1955 V28 N4 P131-136 99K25G <ATS>
1955 V28 N5 P185-190 60-18513 <*>
1955 V28 N7 P259-262 62-10445 <*>
1955 V28 N7 P265-272 62-18207 <*>
63-14034 <*> O
1955 V28 N9 P336-351 66-10671 <*>
1955 V28 N9 P359-368 63-18590 <*>
1955 V28 N10 P375-380 62-10945 <*>
1955 V28 N12 P455-467 64-16336 <*>
1956 V29 N2 P42-49 60-16858 <*>
1956 V29 N2 P49-51 62-18203 <*>
1956 V29 N3 P78-83 63-14065 <*>
1956 V29 N4 P109-119 62-14496 <*> O
1956 V29 N4 P120-128 59-10616 <*>
1956 V29 N4 P128-130 59-10618 <*>
1956 V29 N4 P131-137 62-14499 <=*> O
1956 V29 N4 P137-144 59-10617 <*>
1956 V29 N5 P169-174 59-10615 <*>
1956 V29 N5 P174-183 62-18112 <*>
1956 V29 N6 P233-238 59-10608 <*>
1956 V29 N6 P247-251 62-14863 <*>
1956 V29 N7 P269-275 63-14025 <*>
1956 V29 N8 P309-313 59-10620 <*>
1956 V29 N8 P314-318 62-14869 <*>
1956 V29 N9 P345-356 62-16461 <*>
1956 V29 N10 P386-392 57-3569 <*>
1956 V29 N10 P393-400 62-18121 <*>
1956 V29 N11 P417-426 59-10614 <*>
1956 V29 N11 P426-428 62-10949 <*>
1956 V29 N12 P453-459 64-14039 <*>
1956 V29 N12 P459-470 59-10619 <*>
1956 V29 N12 P483-487 66-10677 <*>
1957 V30 N1 P1-7 63-14047 <*>
1957 V30 N1 P8-14 59-10621 <*>
1957 V30 N2 P37-42 66-10676 <*>
1957 V30 N2 P42-52 59-10611 <*>
1957 V30 N3 P79-83 63-20370 <*>
1957 V30 N3 P88-94 59-10610 <*>
1957 V30 N4 P113-115 62-18134 <*>
1957 V30 N4 P115-116 62-14497 <*>
1957 V30 N4 P117-121 64-10571 <*>
1957 V30 N4 P122-129 59-10613 <*>
1957 V30 N4 P129-133 59-10609 <*>
1957 V30 N5 P157-163 62-10951 <*>
1957 V30 N5 P182-186 59-10612 <*>
1957 V30 N5 P186-188 59-10607 <*>
1957 V30 N6 P213-221 62-10952 <*>
1957 V30 N7 P282-287 62-18108 <*>
1957 V30 N7 P296-299 59-10928 <*>
1957 V30 N7 P299-307 62-14984 <*>
1957 V30 N7 P308-318 59-10926 <*>
1957 V30 N8 P332-335 59-10927 <*>
1957 V30 N9 P369-379 62-16414 <*>
1957 V30 N9 P379-386 59-10929 <*>
1957 V30 N10 P413-425 62-10948 <*>
1957 V30 N10 P425-434 59-10930 <*>
1957 V30 N10 P434-440 64-16592 <*>
1957 V30 N11 P463-470 61-20842 <*>
1957 V30 N11 P471-473 62-16136 <*>
1957 V30 N11 P473-475 62-16137 <*>
1958 V31 N1 P9-15 59-10931 <*>
66-11006 <*>
1958 V31 N2 P54-60 62-16139 <*>
1958 V31 N3 P81-84 59-10932 <*>
1958 V31 N3 P93-94 59-10933 <*>
1958 V31 N3 P94-95 59-10934 <*>
1958 V31 N4 P121-124 59-10935 <*>
1958 V31 N4 P133-137 62-18110 <*>
1958 V31 N5 P161-170 59-10936 <*>
1958 V31 N5 P176-179 62-18126 <*>
1958 V31 N5 P257-260 62-10935 <*>
1958 V31 N6 P221-228 59-10937 <*>
1958 V31 N7 P268-269 59-10938 <*>
1958 V31 N8 P311-315 59-10939 <*>
1958 V31 N9 P349-353 62-10946 <*>
1958 V31 N10 P377-381 62-16140 <*>
1958 V31 N10 P386-394 59-10940 <*>
1958 V31 N11 P422-428 36L35G <ATS>
1958 V31 N11 P428-431 62-14867 <*>
1958 V31 N11 P431-438 61-10507 <*>
1959 V32 N1 P1-9 62-14886 <*>
1959 V32 N1 P15-19 61-10508 <*>
1959 V32 N2 P41-47 61-10514 <*>
1959 V32 N2 P54-58 62-16138 <*>
1959 V32 N3 P81-88 65-14525 <*>
1959 V32 N3 P89-95 61-10512 <*>
1959 V32 N3 P103-106 60N55G <ATS>
1959 V32 N3 P107-120 62-16399 <*>
1959 V32 N4 P142-152 65-14526 <*>
1959 V32 N4 P153-157 61-10504 <*>
67N52G <ATS>
1959 V32 N4 P158-172 62-16398 <*>
1959 V32 N5 P181-185 61-10510 <*>
1959 V32 N5 P189-197 62-14249 <*> O
1959 V32 N6 P217-221 62-10944 <*>
1959 V32 N6 P221-231 61-10506 <*>
1959 V32 N6 P231-239 62-10932 <*>
1959 V32 N6 P243-246 62-14248 <*> O
1959 V32 N6 P247-251 62-14246 <*> O
1959 V32 N7 P278-281 62-14247 <*> O
65-14527 <*>
1959 V32 N8 P276-278 61-10503 <*>
1959 V32 N8 P307-313 61-10505 <*>
1959 V32 N9 P314-320 65-14528 <*>
1959 V32 N9 P357-361 61-10509 <*>
1959 V32 N9 P362-373 62-14244 <*> O
1959 V32 N9 P381-385 65-14529 <*>
1959 V32 N10 P397-401 62-14245 <*> O
68N52G <ATS>
1959 V32 N10 P402-413 61-10511 <*>
69N52G <ATS>
1959 V32 N10 P421-426 65-14530 <*>
1959 V32 N11 P437-442 61-10513 <*>

Citation	Translation
1959 V32 N11 P443-450	62-14243 <*> 0
1959 V32 N11 P450-459	62-18131 <*>
1959 V32 N11 P459-463	87M44G <ATS>
1959 V32K P II/49-II/53	62-10931 <*>
1959 V32K PT V P8-15	61-10518 <*>
1959 V32K PT.I P33-40	62-14250 <*> 0
1959 V32K PT.II P17-25	62-14252 <*>
1959 V32K PT.II P57-73	62-14506 <*>
1959 V32K PT.II P26-32	62-16410 <*>
1959 V32K PT.II P36-49	62-16411 <*>
1959 V32K PT.V P48-53	62-14251 <*> 0
1959 V32K VII/1-VII/5	62-16142 <*>
1959 V32K N4 P7-11	61-10517 <*> 0
1959 V32K N4 P30-34	61-10515 <*> 0
1959 V32K N4 P34-37	61-10516 <*>
1960 V33 N1 P1-7	62-14490 <*>
1960 V33 N1 P10-19	61-10696 <*>
1960 V33 N1 P20-24	62-14233 <*>
1960 V33 N2 P37-45	62-14232 <*> 0
1960 V33 N2 P45-47	62-16135 <*>
1960 V33 N2 P47-52	61-10695 <*> 0
1960 V33 N2 P52-55	60-18163 <*>
1960 V33 N3 P86-93	61-10694 <*>
1960 V33 N3 P93-101	64-16593 <*>
1960 V33 N4 P109-116	64-14047 <*>
1960 V33 N4 P117-120	62-14217 <*> 0
1960 V33 N4 P124-126	62-14218 <*> 0
1960 V33 N4 P127-132	61-10693 <*>
1960 V33 N5 P165-168	64-16522 <*>
1960 V33 N5 P169-173	61-10692 <*>
1960 V33 N6 P201-206	61-10691 <*> 0
1960 V33 N6 P206-213	62-14216 <*> 0
1960 V33 N6 P224-227	64-16530 <*>
1960 V33 N7 P250-252	61-10690 <*>
1960 V33 N7 P252-257	61-10689 <*>
1960 V33 N7 P257-261	62-14219 <*> 0
1960 V33 N8 P281-283	61-14727 <*>
1960 V33 N8 P283-285	64-16526 <*>
1960 V33 N8 P296-303	62-14220 <*>
1960 V33 N8 P303-304	64-10548 <*>
1960 V33 N9 P331-332	61-10688 <*>
1960 V33 N9 P332-338	61-14469 <*> 0
1960 V33 N9 P338-339	61-10687 <*>
1960 V33 N9 P340	62-14221 <*>
1960 V33 N10 P357-363	61-10686 <*>
1960 V33 N10 P370-376	63-14058 <*>
1960 V33 N10 P380-387	62-14234 <*> 0
1960 V33 N11 P401-411	62-14240 <*> 0
1960 V33 N11 P411-416	61-14728 <*>
1960 V33 N11 P416-421	63-14066 <*>
1960 V33 N11 P421-425	63-20366 <*>
1960 V33 N12 P441-449	61-14729 <*>
1960 V33 N12 P449-456	62-14222 <*> 0
1960 V33 N12 P469-475	65-12862 <*>
1961 N10 P489-500	66-11547 <*> 0
1961 V34 N1 P28-30	65-17335 <*>
1961 V34 N1 P30-37	66-10674 <*>
1961 V34 N1 P39	66-10064 <*>
1961 V34 N2 P49-55	BGIRA-544 <BGIR>
1961 V34 N2 P72-74	64-14048 <*>
1961 V34 N3 P91-101	65-14459 <*>
1961 V34 N3 P102-107	65-14460 <*>
1961 V34 N3 P107-120	65-14461 <*>
1961 V34 N3 P130-133	64-10547 <*>
1961 V34 N3 P146-152	61-20843 <*>
1961 V34 N4 P216-219	64-14042 <*>
1961 V34 N6 P311-320	66-10680 <*>
1961 V34 N6 P320-324	66-10065 <*>
1961 V34 N7 P363-369	64-14053 <*>
1961 V34 N9 P431-437	65-13619 <*>
1961 V34 N9 P456-460	66-11546 <*> 0
1961 V34 N10 P482-489	65-12881 <*>
1961 V34 N10 P489-500	65-13622 <*>
1961 V34 N11 P534-544	SCL-T-470 <*>
	65-11140 <*>
1961 V34 N11 P544-547	66-11550 <*> 0
1962 V35 N1 P8-13	65-13620 <*>
1962 V35 N1 P22-27	66-11545 <*> 0
1962 V35 N1 P37-43	66-11549 <*> 0
1962 V35 N1 P53-56	65-11147 <*>
1962 V35 N1 P56-60	65-11145 <*>
1962 V35 N1 P60-64	65-13518 <*>
1962 V35 N1 P65-71	66-11548 <*> 0
1962 V35 N4 P167-176	BGIRA-522 <BGIR>
1962 V35 N4 P177-181	65-11142 <*>
1962 V35 N5 P234-243	65-10659 <*>
1962 V35 N6 P272-278	65-18017 <*>
1962 V35 N6 P278-281	65-17338 <*>
	66-11675 <*> 0
1962 V35 N7 P312-316	66-13930 <*>
1962 V35 N7 P326-328	66-10670 <*>
1962 V35 N8 P347-354	BGIRA-546 <BGIR>
1962 V35 N8 P355-361	BGIRA-566 <BGIR>
	66-10660 <*>
1962 V35 N9 P394-399	66-10698 <*>
1962 V35 N11 P459-466	66-10068 <*>
1962 V35 N12 P495-500	BGIRA-547 <BGIR>
1962 V35 N12 P500-505	65-13513 <*>
1963 V36 P308-323	1134 <TC>
1963 V36 N1 P1-11	66-12441 <*>
1963 V36 N1 P16-22	66-12443 <*>
1963 V36 N2 P37-51	66-10681 <*>
1963 V36 N3 P73-86	BGIRA-573 <BGIR>
1963 V36 N3 P86-90	BGIRA-584 <BGIR>
	66-12412 <*>
1963 V36 N3 P91-92	66-10661 <*>
1963 V36 N4 P121-130	BGIRA-585 <BGIR>
1963 V36 N5 P158-169	66-12413 <*>
1963 V36 N5 P170-179	BGIRA-594 <BGIR>
1963 V36 N5 P183-193	66-10697 <*>
1963 V36 N6 P207-212	BGIRA-613 <BGIR>
	66-10682 <*>
1963 V36 N6 P217-225	66-10696 <*>
1963 V36 N7 P249-253	BGIRA-592 <BGIR>
1963 V36 N7 P253-258	BGIRA-590 <BGIR>
	65-14408 <*>
1963 V36 N7 P266-273	66-10678 <*>
1963 V36 N7 P285-288	66-10801 <*>
1963 V36 N7 P289-301	BGIRA-589 <BGIR>
1963 V36 N8 P323-326	66-10662 <*>
1963 V36 N9 P347-356	BGIRA-598 <BGIR>
1963 V36 N9 P371-376	66-10663 <*>
1963 V36 N10 P387-392	66-10679 <*>
1963 V36 N11 P425-444	66-10691 <*>
1963 V36 N11 P444-453	65-11742 <*> 0
1963 V36 N12 P461-467	BGIRA-606 <BGIR>
1963 V36 N12 P468-481	65-11743 <*>
1963 V36 N12 P481-483	BGIRA-605 <BGIR>
1963 V36 N12 P483-487	BGIRA-617 <BGIR>
1964 N2 P72-78	66-12406 <*>
1964 V37 N1 P1-15	BGIRA-615 <BGIR>
1964 V37 N2 P57-58	66-14826 <*>
1964 V37 N2 P102-115	BGIRA-624 <BGIR>
	66-14833 <*>
1964 V37 N2 P116-122	BGIRA-628 <BGIR>
1964 V37 N2 P122-125	BGIRA-610 <BGIR>
1964 V37 N2 P126-129	66-14829 <*>
1964 V37 N4 P206-218	BGIRA-627 <BGIR>
1964 V37 N5 P252-256	66-12409 <*>
1964 V37 N6 P275-281	BGIRA-629 <BGIR>
1964 V37 N6 P281-285	BGIRA-633 <BGIR>
1964 V37 N6 P286-292	BGIRA-635 <BGIR>
1964 V37 N6 P296-301	BGIRA-630 <BGIR>
1964 V37 N6 P301-305	BGIRA-632 <BGIR>
1964 V37 N6 P306-310	BGIRA-631 <BGIR>
1964 V37 N8 P381-385	66-13932 <*>
1964 V37 N9 P413-425	BGIRA-634 <BGIR>
1964 V37 N9 P432-435	66-14830 <*>
1964 V37 N10 P479-480	66-14828 <*>
1964 V37 N12 P553-562	BGIRA-642 <BGIR>
1965 V38 N2 P54-59	66-10081 <*>
1965 V38 N3 P98-103	66-13924 <*>
1965 V38 N4 P156-166	66-14824 <*>
1965 V38 N6 P229-232	66-14835 <*>
1965 V38 N8 P313-322	BIGRA-653 <BIGR>
1965 V38 N8 P334-335	66-14827 <*>
1966 V39 N1 P14-19	665 <BGRA>
1966 V39 N3 P81-89	66-12914 <*> 0
1966 V39 N3 P104-112	66-13469 <*>
1966 V39 N3 P126-130	66-12916 <*> 0

1966 V39 N3 P147-149 66-13470 <*>
1966 V39 N3 P149-155 66-14188 <*>
1966 V39 N3 P164-167 66-12915 <*> 0
1966 V39 N3 P167-169 66-12918 <*> 0
1966 V39 N3 P169-173 66-12917 <*> 0
1966 V39 N4 P175-176 66-13467 <*>
1966 V39 N4 P177-186 66-13468 <*>
1966 V39 N4 P186-190 66-14190 <*>
1966 V39 N4 P190-202 66-13466 <*>
1966 V39 N4 P203-217 66-13901 <*>
1966 V39 N5 P239-242 66-13900 <*>
1966 V39 N6 P279-283 66-14192 <*>
1966 V39 N6 P283-293 66-14191 <*>
1966 V39 N6 P294-302 66-14184 <*>
66-14504 <*>
1966 V39 N7 P319-323 66-14735 <*> 0
1966 V39 N7 P334-339 66-14737 <*> 0

GLASTEKNISK TIDSKRIFT
1954 V9 P131-148 T1586 <INSD>
1955 V10 N6 P151-159 62-14884 <*>
1955 V10 N6 P162-168 62-14884 <*>
1956 V11 N5 P127-136 62-18117 <*>
1956 V11 N6 P159-175 62-10928 <*>
1960 V15 N1 P5-8 88M44N <ATS>
1960 V15 N3 P63-68 63-14051 <*>
1960 V15 N5 P133-139 63-14026 <*>
1961 V16 N1 P5-18 64-16533 <*>
1961 V16 N1 P35 64-16533 <*>
1962 V17 N4 P91-98 63-18096 <*>
1963 V18 N2 P39-40 65-11023 <*>
1964 V19 N1 P9-20 BGIRA-625 <BGIR>
1965 V20 N2 P29-42 66-13995 <*>

GLOBEN
1961 V41 N1 P6-10 62-33343 <=*>

GLOBUS
1908 V94 P273- 57-1500 <*>

GLUECKAUF
1931 V67 N3 P86-94 00R76G <ATS>
1938 V74 P681-695 1940 <*>
1938 V74 N32 P681-695 97J17G <ATS>
1938 V74 N33 P705-715 97J17G <ATS>
1940 N8 P112- I-130 <*>
1940 V76 N41 P561-563 GB82 3927 <NLL>
1949 V85 N17/8 P287-294 66-10227 <*> 0
1950 V86 N3/4 P56-57 63-16849 <*>
1950 V86 N25/6 P498-509 1864 <*>
1950 V86 N37/8 P776-784 59-20559 <*>
1950 V86 N37/8 P798-802 T-1513 <INSD>
1952 V88 P346-350 01973 <*>
1973 <*>
1952 V88 N1/2 P8-13 1976 <*>
1952 V88 N15/6 P380-382 60-16798 <*>
1952 V88 N35/6 P868-875 T-1511 <INSD>
60-14525 <*>
1952 V88 N45/6 P1090-1094 1837 <*>
60-18801 <*> 0
1953 V89 N49/0 P1229-1239 66-10234 <*> 0
1954 V90 N1/2 P47-53 1984 <*>
1955 V91 P1266-1272 T1779 <INSD>
1955 V91 N27/8 P778-781 06R75G <ATS>
1956 V92 P320-328 T-2289 <INSD>
1956 V92 N11/2 P320-328 57-19 <*>
62-10009 <*>
1956 V92 N13/4 P380-388 C2206 <NRC>
1956 V92 N13/4 P389-397 C2209 <NRC>
1956 V92 N13/4 P397-411 C2211 <NRC>
1956 V92 N13/4 P412-414 C2212 <NRC>
1957 V93 N41/2 P1286-1290 58-944 <*>
1958 N3/4 P102-110 TS 1236 <BISI>
1958 N3/4 P110-121 TS 1237 <BISI>
1958 V94 P832-842 1593 <BISI>
1958 V94 N3/4 P102-110 61-20604 <*>
1958 V94 N3/4 P110-121 61-20622
1958 V94 N3/4 P125-128 58-2452 <*>
60-14031 <*>
1958 V94 N35/6 P1187-1189 60-14147 <*>

1959 V95 N11 P608-637 60-00016 <*>
1959 V95 N11 P701-705 62-10008 <*> 0
1959 V95 N20 P1237-1244 65-12297 <*>
1960 N5 P294-299 60-00197 <*>
1960 V96 N5 P294-299 60-19851 <*+>
1960 V96 N24 P1525-1534 55R75G <ATS>
1961 V97 N21 P1262-1271 66-10224 <*> 0
1962 V98 N2 P102-107 62-18019 <*>
1962 V98 N6 P348-357 3106 <BISI>
1963 V99 N19 P1048-1052 635 <TC>
1964 V100 N10 P549-561 82R79G <ATS>
1964 V100 N22 P1324-1334 65-00348 <*>
1964 V100 N25 P1473-1495 66-10251 <*> 0
1965 V101 P41-45 4315 <BISI>
1965 V101 N16 P961-970 66-10288 <*> 0

GODISEN ZBORNIK. FILOSOFSKI FAKULTET NA UNIVERZITET, SKOPJE
1954 V7 N4 P87-107 65-10886 <*>

GODISHNIK. KHIMIKO-TEKHNOLOGICHESKII INSTITUT SOFIA
1957 V4 P55-67 65-12937 <*>
1958 V5 N1 P111-126 65-17359 <*>
1959 V6 N2 P95-106 74Q69B <ATS>
1961 V8 N2 P1-10 65-12189 <*>
1961 V8 N2 P11-15 65-12188 <*>

GODISNIK. SUMARSKI NAUCNO OPITEN INSTITUT
1958 V3 P3-30 61-11207 <=>

GODISNJAK BIOLOGSKOG INSTITUTA U SARAJEVU
1950 V3 N1/2 P141-198 61-11205 <=>

GOLD UND SILBER
1961 V14 P20-22 58N56G <ATS>
1961 V14 P21-23 59N56G <ATS>
1961 V14 P28-31 60N56G <ATS>

GOMMA. MILANO
1939 V3 N2 P31-37 RCT V13 N1 P91 <RCT>
1939 V3 N4 P97-103 RCT V13 N1 P81 <RCT>
1940 V4 N1 P1-6 RCT V13 N1 P566 <RCT>
1940 V4 N1 P7-15 RCT V14 N1 P225 <RCT>

GORDIAN
1959 V58 N1402 P20-24 46L33G <ATS>

GORNO-OBOGATITELNOE DELO
1932 N1 P22-28 63-14597 <=*>
1932 N1 P29-30 63-16055 <=*>

GORNYI ZHURNAL
1880 V2 N4/6 P96-117 R-1147 <*>
1940 V116 N5/6 P65-68 63-15689 <=*$>
1947 V121 N5 P21-25 60-41435 <=>
65-10134 <*>
1948 V122 N1 P30-33 50/1808 <NLL>
1949 N8 P10-13 50/1338 <NLL>
1952 N11 P21-24 UCRL-TRANS-770(L) <=*>
1955 N6 P31-35 59-10439 <+*>
1955 N6 P51-54 RT-4298 <*>
1955 N8 P24-31 64-16846 <=*$> 0
1956 N2 P55-62 59-15821 <+*>
1956 N5 P29-32 AEC-TR-4005 <+*>
1957 N4 P10-14 60-23811 <+*>
1957 N4 P24-30 60-23625 <+*>
1957 N12 P26-32 60-13644 <+*>
1958 N6 P3-4 59-11355 <=>
1958 N7 P3-10 PB 141 202T <=>
1958 N9 P3-6 59-11123 <=>
1958 N9 P7-21 59-13135 <=> 0
1958 N10 P13-18 59-13235 <=>
1958 N11 P3-6 59-13214 <=>
1958 N11 P7-15 59-13214 <=>
1959 N2 P65-68 59-11539 <=> 0
1959 N5 P16 59-13865 <=>
1959 N5 P18 59-13865 <=>
1959 N5 P22 59-13865 <=>
1959 N8 P69-70 61-19107 <+*>

1959 N9 P44-47 63-11092 <=>
1959 N11 P75-76 60-41166 <=> 0
1959 N12 P5-6 59-13421 <=>
1959 N12 P9-10 59-13421 <=>
1959 N12 P15-17 59-13420 <=>
1959 N12 P18-23 59-13420 <=>
1960 N3 P1-8 61-19968 <=>
1960 N3 P11-14 60-41680 <=>
1960 N4 P54-59 E-751 <RIS>
1960 N4 P59-61 E-752 <RIS>
1960 N5 P8-18,75 60-41527 <=>
1960 N6 P6-11 61-19968 <=>
1960 N7 P62-65 61-11085 <=>
1960 N7 P66-69 61-11602 <=>
1960 N7 P69-71 61-11622 <=>
1960 N8 P70-74 61-11072 <=>
1960 N10 P19-21 61-19968 <=>
1960 N11 P28-30 61-19968 <=>
1961 N1 P3-5 61-27290 <=>
1961 N1 P18-22 61-23119 <=*>
1961 N1 P22-25 62-23477 <=*>
1961 N1 P59-65 61-27290 <=>
1961 N2 P5-9 61-27290 <=>
1961 N3 P70-73 63-19814 <=*>
1962 N1 P39-41 63-13109 <=>
1962 N1 P54-58 63-13109 <=>
1962 N3 P5-12 62-32461 <=>
1962 N3 P80 62-32461 <=>
1962 N4 P16-18 62-32210 <=>
1962 N6 P3-6 62-32846 <=*>
1962 N6 P6-11 63-13084 <=>
1962 N6 P12-14 62-32846 <=*>
1962 N6 P58-62 33T90R <ATS>
1962 N9 P26-27 63-15352 <=>
1962 N9 P54-55 63-15352 <=>
1962 N9 P74-75 63-15352 <=>
1962 N11 P36-41 63-21094 <=>
1962 N11 P46-51 63-21094 <=>
1962 N11 P58-64 63-21094 <=>
1962 N11 P70-74 63-21094 <=>
1962 N12 P8-10 63-21269 <=>
1963 N1 P3-15 63-21585 <=>
1963 N2 P55-61 63-21667 <=>
1963 N3 P13-15 63-21920 <=>
1963 N3 P27-29 63-21920 <=>
1963 N3 P37-40 63-21920 <=>
1963 N3 P52-53 63-21920 <=>
1963 N3 P65-66 63-21920 <=>
1963 N4 P3-12 63-31042 <=>
1963 N4 P20-25 63-31042 <=>
1963 N4 P26-31 63-31109 <=>
1963 N5 P5-8 63-31102 <=>
1963 N6 P61-62 63-31562 <=>
1963 N10 P36-39 33R79R <ATS>
64-18106 <=*$>
1964 N1 P3-5 NLLTB V6 N6 P541 <NLL>
TRANS.BULL.1964 P541 <NLL>
1964 N3 P76-77 64-41958 <=$>
1964 N7 P7-9 64-51015 <=*$>
1964 N11 P3-7 66-10278 <*> 0
1967 N3 P3-8 67-31453 <=$>

GORODSKOJE KHCZYAYSTVO MOSKVY
1953 V27 N12 P19-23 61-13499 <*+>
1962 V36 N5 P41-42 62-33332 <=>

GORYUCHIE SLANTSY: KHIMIYA I TEKHNOLOGIYA
1959 N3 P31-38 66-11517 <*>
1959 V3 P39-56 66-10987 <*>
1961 N4 P252-256 65-60403 <=$>

GOSPODARKA ADMINISTRACJA TERNOWA
1965 V6 N6 P35-38 65-31855 <=$>
1965 V6 N6 P42-44 65-31855 <=$>

GOSPODARKA MATERIALOWA. WARSAW
1964 N14 P471-475 64-41825 <=>
1967 N5 P167-171 67-31447 <=$>

GOSPODARKA LACZNOSCI
1957 N10 P7-10 59-11076 <=*>
1957 N12 P1-3 59-11029 <=*>
1958 N3 P1-4 59-11031 <=*>
1958 N3 P9-11 59-11032 <=*>
1958 N3 P11-12 59-11033 <=*>
1958 N3 P13-15 59-11030 <=*>
1958 N5 P5-8 PB 141 283T <=*>
1958 N7 P3-9 59-11481 <=*> 0

GOSPODARKA PLANOWA
1964 V19 N7 P26-31 64-41589 <=>
1964 V19 N7 P36-42 64-41589 <=>
1965 N7 P16-22 65-32304 <=$>

GOSPODARKA RYBNA
1955 V7 N11 P22 63-11395 <=>
1956 V8 N5 P9-10 63-11397 <=>
1958 V10 N5 P7-10 63-11355 <=>
1958 V10 N6 P7-12 63-11355 <=>

GOSPODARKA WODNA
1954 V14 N9 P347-350 60-21267 <=>
1955 V15 N6 P267-268 60-21547 <=>
1956 V16 N6 P263-266 60-21540 <=>
1958 V18 N1 P13-18 61-31311 <=>
1958 V18 N5 P194-198 61-31328 <=>
1958 V18 N8 P342-347 61-31300 <=>
1958 V18 N9 P390-396 61-31309 <*>
1959 V19 N7 P325 60-31183 <=>
1960 V20 N2 P77-89 64-41589 <=>
1960 V20 N3 P123-127 64-41589 <=>
1960 V20 N3 P141-144 64-41589 <=>
1960 V20 N3 P187-188 64-41589 <=>
1960 V20 N5 P198-207 64-41589 <=>
1960 V20 N5 P211-219 64-41589 <=>
1960 V20 N6 P250-263 64-41589 <=>
1960 V20 N6 P279-288 64-41589 <=>
1960 V20 N7 P296-301 64-41589 <=>
1960 V20 N7 P303-315 64-41589 <=>
1960 V20 N7 P333-336 64-41589 <=>
1960 V20 N8 P348-364 64-41589 <=>
1960 V20 N8 P377-384 64-41589 <=>
1960 V20 N9 P388-409 64-41589 <=>
1960 V20 N9 P412-419 64-41589 <=>
1960 V20 N9 P427-432 64-41589 <=>
1960 V20 N10 P434-451 64-41589 <=>
1960 V20 N10 P453-455 64-41589 <=>
1960 V20 N11 P477-489 64-41589 <=>
1960 V20 N11 P494-497 64-41589 <=>
1960 V20 N11 P507 64-41589 <=>
1960 V20 N11 P509-511 63-11363 <=>
1960 V20 N11 P513-520 63-11363 <=>
1960 V20 N11 P522-534 63-11363 <=>
1960 V20 N11 P537-544 63-11363 <=>
1960 V20 N11 P553 63-11363 <=>
1960 V20 N11 P557-564 63-11363 <=>
1964 V24 N7 P225-230 64-51407 <=>
1965 V25 N5 P189-192 65-31563 <=$>

GOUDRON POUR ROUTES
1958 P12-25 59-17472 <*>

GRADEVINAR
1963 V15 N7 P234-238 63-31996 <=>
1963 V15 N11 P436-440 64-21577 <=>
1965 N12 P467-474 66-30911 <=$>

GRAFISKA FORSKNINGSLABORATORIETS MEDDELANDE
1955 V36 P10 57-1716 <*>
1959 N42 P18-23 61-10412 <*> 0

GRASAS Y ACEITAS
1952 V3 P125-134 57-1977 <*>
1964 V15 N5 P247-252 65-14579 <*> 0

GRAVIMETRIIA: SBORNIK STATEI. MEZHDUVEDOMSTVENNYI KOMITET PO PROVEDENIIU MEZHDUNARODNOGO GEOFIZICHESKOGO GODA. AKADEMIYA NAUK SSSR. XIII. RAZDEL PROGRAMMY MGG
1960 N1 P5-62 61-21015 <=*>

1963 N3 ENTIRE ISSUE AD-611027 <=$>
1963 N4 P22-30 AD-611 122 <=$>
1963 N4 P44-60 AD-611 115 <=$>

GRAZHDANSKAYA AVIATSIYA
1956 V13 N4 P5 59-12387 <+*>
1956 V13 N4 P28-30 59-12387 <+*>
1956 V13 N4 P33-35 59-12387 <+*>
1956 V13 N6 P13-15 59-12396 <+*>
1956 V13 N6 P22-23 59-12396 <+*>
1956 V13 N6 P27-28 59-12396 <+*>
1956 V13 N9 P16-19 59-11984 <=>
1956 V13 N9 P22-23 59-12390 <+*>
1956 V13 N9 P27-28 59-12390 <+*>
1956 V13 N9 P32 59-12390 <+*>
1956 V13 N10 P22 59-12392 <+*>
1956 V13 N10 P23-25,28 59-12392 <+*>
1956 V13 N10 P30-34 59-12392 <+*>
1956 V13 N11 P4-5 59-12393 <+*>
1956 V13 N11 P15-17 59-12393 <+*>
1956 V13 N11 P18-21 59-12393 <+*>
1956 V13 N11 P23 59-12393 <+*>
1957 N12 P3-5 R-3461 <*>
1957 V14 N1 P3-5 59-16066 <+*> 0
1957 V14 N1 P8-9 59-16066 <+*> 0
1957 V14 N1 P11-13 59-16066 <+*> 0
1957 V14 N1 P15-19 59-16066 <+*> 0
1957 V14 N10 P13-14 63-16053 <*=>
63-16053 <=*>
1957 V14 N11 P17-20 59-11891 <=> 0
1958 N7 P4-5 59-16324 <+*>
1958 V15 N1 P1-4 60-19112 <+*> 0
1958 V15 N1 P5-11 59-19516 <+*>
1958 V15 N1 P19-24 59-19516 <+*>
1958 V15 N1 P30 59-19516 <+*>
1958 V15 N1 P32-37 59-19516 <+*>
1958 V15 N2 P17-18 60-19032 <+*>
1958 V15 N2 P33-36 59-13058 <=>
1958 V15 N3 P16-19 59-12165 <+*>
1958 V15 N3 P19-20 60-19136 <+*>
1958 V15 N3 P23 60-15066 <+*>
1958 V15 N4 P29-30 59-16035 <+*>
1958 V15 N5 P2-5 59-19514 <+*> 0
1958 V15 N5 P6-11 59-14111 <+*> 0
1958 V15 N5 P16-17 59-19514 <+*> 0
1958 V15 N5 P22-25 59-19514 <+*> 0
1958 V15 N6 P27-31 60-13108 <+*> 0
1958 V15 N7 P1-2 59-16325 <+*>
1958 V15 N7 P21-23 59-18417 <+*>
1958 V15 N7 P21-25 60-19003 <+*>
1958 V15 N7 P26-28 59-11508 <=>
1958 V15 N8 P3-4 60-19130 <+*>
1958 V15 N8 P10 59-19502 <+*>
1958 V15 N8 P14-21 60-19130 <+*>
1958 V15 N8 P26-27 60-19130 <+*>
1958 V15 N8 P35-37 59-16640 <+*> 0
1958 V15 N8 P38-39 60-19130 <+*>
1958 V15 N9 P22-26 60-21793 <=>
1958 V15 N9 P28-29 59-11371 <=>
1958 V15 N10 P24 59-19517 <+*>
1958 V15 N11 P3-7 60-15670 <+*>
1958 V15 N11 P34-35 60-15670 <+*>
1959 N3 P12 59-21063 <=>
1959 V16 N2 P27-28 61-23718 <=*>
1959 V16 N2 P33-35 61-23718 <=*>
1959 V16 N4 P6-9 59-22137 <+*>
1959 V16 N5 P22-23 61-23883 <*=>
1959 V16 N5 P30-31 61-23883 <*=>
1959 V16 N7 P22-23 59-22145 <+*>
1959 V16 N7 P32-34 60-17493 <+*> 0
61-23697 <=*>
1959 V16 N9 P16 61-27174 <*=>
1959 V16 N9 P18 61-27174 <*=>
1959 V16 N10 P13-15 61-28373 <*=>
1959 V16 N10 P27 61-28373 <*=>
1960 V17 N2 P22-23 61-19348 <+*>
1960 V17 N6 P24-25 61-28624 <*=>
1960 V17 N10 P30-31 61-23856 <*=>
1960 V17 N11 P12-13 61-23858 <*=>
1960 V17 N11 P20 61-27424 <*=>
1960 V17 N11 P28-30 61-27025 <*=>
1962 V19 N3 P18-20 62-32946 <=*>
1962 V19 N4 P10-13 62-25758 <=*>
1962 V19 N6 P5-6 62-33707 <=*>
1962 V19 N8 P25 63-13822 <=>
1962 V19 N9 P22-24 63-15565 <=*>
1962 V19 N11 P10-13 63-19474 <=*>
1962 V19 N11 P28-29 63-19474 <=*>
1963 V20 N3 P9 63-24049 <=*>
1963 V20 N4 P17 64-15249 <=*$>
1963 V20 N5 P10-11 63-24048 <=*>
1963 V20 N7 P18-19 64-15107 <=*$>
1963 V20 N11 P22-23 64-21213 <=>
1963 V20 N12 P14-15 64-11804 <=>
1964 V21 N1 P11-13 AD-610 790 <=$>
1964 V21 N8 P28-29 AD-620 974 <=$>
1965 V22 N11 P15 AD-640 958 <=$>
1966 N12 P14-19 67-30416 <=>
1966 V23 N3 P22-23 66-32872 <=$>

GROUND IMPROVEMENT
SEE TOCHI KAIRYO

GROZNENSKII NEFTYANIK
1936 V6 N3 P10-15 61-17283 <=$>
1936 V6 N1/2 P79-83 63-18396 <=*>

GRUDNAYA KHIRURGIYA
1960 V2 N3 P49-51 60-31801 <=>
1960 V2 N4 P3-5 61-11688 <=>
1960 V2 N4 P6-14 61-21097 <=>
1960 V2 N4 P15-18 61-11842 <=>
1960 V2 N4 P19-26 61-21161 <=>
1960 V2 N4 P26-32 61-11692 <=>
1960 V2 N4 P32-36 61-11704 <=>
1960 V2 N4 P37-46 61-11761 <=>
1960 V2 N4 P108-113 61-11811 <=>
1960 V2 N6 P18-30 61-21352 <=>
1961 V3 N5 P7-12 63-24383 <=*$>
1962 V4 N1 P12-17 62-23988 <=*>
1962 V4 N1 P52-58 62-23988 <=*>
1962 V4 N3 P13-17 62-32447 <=*>
1962 V4 N3 P40-47 62-32447 <=*>
1962 V4 N6 P45-50 63-21171 <=>
1962 V4 N6 P80-83 63-21171 <=>
1963 V5 N1 P8-18 63-21744 <=>
1963 V5 N1 P25-34 63-21744 <=>
1963 V5 N1 P116-128 63-21744 <=>
1963 V5 N3 P3-8 63-31469 <=>
1963 V5 N3 P54-57 63-31444 <=>
1963 V5 N5 P98-100 63-41159 <=>
1963 V5 N6 P34-39 64-21397 <=>
1964 V6 N1 P115-117 64-21916 <=>
1964 V6 N2 P3-9 64-31340 <=>
1964 V6 N5 P44-50 64-51980 <=$>
1964 V6 N6 P99-106 65-30156 <=$>
1965 V7 N2 P13-21 65-31702 <=$>
1965 V7 N2 P104-123 65-31702 <=$>
1965 V7 N4 P40-43 65-33219 <=$>
1965 V7 N4 P67-71 65-33219 <=$>
1965 V7 N4 P97-103 65-33219 <=$>

GRUNDLAGEN DER LANDTECHNIK
1953 N5 P81-86 CSIRO-3764 <CSIR>
1953 V5 P64-80 CSIRO-3763 <CSIR>
1956 N7 P11-27 61-19497 <+*>
1958 N10 P55-69 62-18736 <*>
1958 N10 P70-88 62-18737 <*>
1958 N10 P89-95 62-18738 <*>
1958 N10 P143-150 62-18739 <*>
1959 N11 P81-84 66-14625 <*>
1959 N11 P85-94 66-14747 <*>
1961 N13 P62-66 65-14153 <*> 0
1962 N14 P5-13 63-10599 <*>
1962 V14 P51-57 63-14400 <*>
1963 N16 P9-12 65-14151 <*> 0
1963 N16 P13-15 65-14152 <*> 0
1963 N17 P5-16 65-14196 <*>
1963 N17 P37-51 65-14083 <*>
1963 N18 P27-34 65-14197 <*>

1963 N18 P50-57 65-14082 <*> 0
1964 N21 P16-21 65-13926 <*>

GRUZLICA
1958 V26 N6 P473-478 64-18982 <*> 0
1961 V29 P70-72 65-29500 <$>

GUDOK
1959 N172 P1 61-23878 <=*>
1959 V38 N38 P1-8 60-19028 <+*>

GULLKORNET
1962 V12 N5 P27-35 63-27952 <$>

GUMMI UND ASBEST
1957 V10 N8 P456-460 59-10058 <*>
1958 V11 N5 P303-307 60-14009 <*>
1958 V11 N6 P344-347 60-14009 <*>
1958 V11 N11 P682-694 59-10758 <*>
1959 V12 N3 P146-156 59-15351 <*> 0
59-17214 <*>
1959 V12 N10 P722-726 60-18536 <*>

GUMMI, ASBEST UND KUNSTSTOFFE
1961 V14 P644-650 2315 <TTIS>
1961 V14 P653-655 2317 <TTIS>
1961 V14 P1024-1034 RCT V37 N4 P950 <RCT>
1961 V14 P1081-1094 RCT V37 N4 P950 <RCT>
1961 V14 N9 P806 62-14860 <*>
1961 V14 N9 P808 62-14860 <*>
1961 V14 N9 P812 62-14860 <*>
1961 V14 N9 P814 62-14860 <*>
1962 V15 P1182-1186 RCT V36 N4 P1148 <RCT>
1962 V15 N8 P741-751 RCT V37 N4 P1049 <RCT>
1962 V15 N10 P979-982 RCT V37 N4 P1049 <RCT>
1965 V18 N4 P425-432 1369 <TC>
1965 V18 N5 P540 1112 <TC>
1966 V19 N4 P356-360 66-12992 <*>

GUMMI UND ASBEST. STUTTGART
1955 V8 P68-76 RCT V30 N3 P962 <RCT>
1957 V10 P610-616 RCT V31 N5 P1035 <RCT>
1957 V10 P672-684 RCT V31 N5 P1035 <RCT>
1958 V11 P251-258 RCT V32 N4 P1039 <RCT>
1959 V12 P66- RCT V32 N4 P1050 <RCT>
1959 V12 P68- RCT V32 N4 P1050 <RCT>
1959 V12 P70- RCT V32 N4 P1050 <RCT>
1959 V12 P72- RCT V32 N4 P1050 <RCT>
1959 V12 P74- RCT V32 N4 P1050 <RCT>
1959 V12 P75- RCT V32 N4 P1050 <RCT>
1960 V13 P706-708 63-18368 <*>

GUMMI-BEREITUNG
1965 V41 N7 P13-30 66-13645 <*>

GUMMIZEITUNG UND KAUTSCHUK
1914 V28 P1580-1581 1615 <*>
1923 V37 P264-266 2866 <*>
1925 P94- 2891 <*>
1925 P892-895 2858 <*>
1925 P1044-1045 2888 <*>
1925 P1167-1168 2873 <*>
1925 P1740-1742 2936 <*>
1925 P2351-2352 2877 <*>
1925 V39 P434-435 AL-761 <*>
1927 P1112- 2937 <*>
1928 V42 P1303-1304 II-572 <*>
1932 V46 P497-498 2078 <*>
1935 P1360-1361 2874 <*>
1940 V54 N9 P141-142 RCT V14 N1 P489 <RCT>
1940 V54 N10 P167-168 RCT V14 N1 P489 <RCT>

GYERMEKGYOGYASZET
1956 V7 P102-104 57-3035 <*>

GYNAECOLOGIA
1953 V135 P87-92 AL-767 <*>
1956 V141 P255-260 57-1961 <*>
1956 V142 P300-303 57-2702 <*>
1959 V148 P1-8 62-01035 <*>
1961 V151 P301-313 63-01762 <*>

GYNECOLOGIE ET OBSTETRIQUE
1933 V27 P540-543 II-1033 <*>
1960 V59 N4 P514-518 65-13838 <*> 0

GYNECOLOGIE ET OBSTETRIQUE. SUPPLEMENT
1953 V5 P179-180 AL-530 <*>

GYNECOLOGIE PRATIQUE. PARIS
1964 V15 N3 P175-184 66-13176 <*>

GYOGYASZAT
1938 V78 P654-659 65-14973 <*>

GYOGYO KAGAKU SOSHO
1952 N4 P1-57 60-00620 <*>

GYOGYSZERESZET
1959 V3 N6 P222-225 60-11686 <*=>

GYOSEN KENKYO GIHO
1958 N12 P99-114 60-19855 <=*> 0
1958 N12 P149-155 60-13257 <=*> 0

GYPSUM
1952 V1 P225-229 1776 <*>

GYPSUM AND LIME. JAPAN
SEE SEKKO TO SEKKAI

HNO
1960 V8 N4 P121-123 62-20055 <*> 0
1960 V8 N11 P343-347 63-26243 <=$>
1961 V9 N7 P178-179 63-26241 <=$>
1963 V11 P249-253 65-12341 <*>

HACHINUCH
1961 V34 N2 P129-136 63-01241 <*>

HACIENDA
1944 V39 N6 P264-266 3363 <*>

HAEMATOLOGICA
1938 V19 P495-507 64-00438 <*>
1947 V30 P52-64 I-504 <*>
1954 V38 P263-282 1915 <*>
1954 V38 P891-898 AEC-TR-6079 <*>
1964 V49 N4 P253-272 65-12775 <*>

HAERTEREI-TECHNIK UND WAERMEBEHANDLUNG
1956 V2 N10 P89-90 4308 <HB>
1956 V2 N10 P92-96 4308 <HB>
1961 V7 P20-27 2387 <BISI>

HAERTEREI-TECHNISCHE MITTEILUNGEN
1960 V15 P89-93 PT.B 3469 <BISI>
1960 V15 N1 P14-19 1988 <BISI>
1960 V15 N2 P96-101 PTB 3230 <BISI>
1960 V15 N2 P72-77 3849 <BISI>
1960 V15 N4 P218-222 2767 <BISI>
1961 V16 P1-6 3055 <BISI>
1961 V16 N2 P75-79 PT A 64-16292 <*> 0
1961 V16 N3 P137-146 3149 <BISI>
1962 V17 P29-38 3384 <BISI>
1962 V17 P148-159 3535 <BISI>
1962 V17 N3 P131-140 3953 <BISI>
1963 V18 N4 P237-239 3969 <BISI>
1964 V19 P1-10 4186 <BISI>

HAI YANG YU HU CHAO
1958 V1 N2 P153-166 63-21405 <=>
1959 V2 N1 P1-9 61-28787 <*=>
1959 V2 N1 P11-14 61-21238 <=>
1959 V2 N1 P27-31 61-21238 <=>
1959 V2 N1 P35-60 61-21238 <=>
1959 V2 N3 P111-134 62-15490 <=>
1959 V2 N3 P136-144 62-15490 <=>
1959 V2 N3 P146-159 62-15229 <*=>
1959 V2 N3 P163-178 63-13674 <=>

1959 V2 N4 P203-212 62-13526 <*=>
1959 V2 N4 P244-261 63-13674 <=>
1959 V2 N4 P269-274 63-13674 <=>
1959 V2 N4 P278-282 63-13674 <=>

HAKKO KOGAKU ZASSHI
1953 V31 P39-42 57-2824 <*>
1953 V31 P72-74 57-2826 <*>
1953 V31 P86-90 57-2825 <*>
1953 V31 P126-130 57-2827 <*>
1953 V31 P230-234 57-2841 <*>
1953 V31 P271-275 PANSDOC-TR.120 <PANS>
1958 V36 P368-371 63-18904 <*>
1958 V36 P371-374 63-18905 <*>
1959 V37 N8 P295-298 63-16189 <*> 0
1959 V37 N8 P298-301 63-16190 <*> 0
1959 V37 N8 P301-304 63-16191 <*> 0
1959 V37 N8 P304-307 63-16192 <*> 0
1959 V37 N8 P307-309 63-16193 <*> 0
1959 V37 N8 P310-311 63-16194 <*>
1959 V37 N8 P311-313 63-16195 <*> 0
1959 V37 N8 P314-315 63-16196 <*> 0
1959 V37 N8 P315-318 63-16197 <*> 0
1959 V37 N8 P318-321 63-16198 <*> 0
1959 V37 N8 P322-324 63-16199 <*> 0
1959 V37 N8 P324-327 63-16200 <*> 0
1960 V38 N3 P131-135 80N56J <ATS>
1961 V39 N6 P328-332 62-14603 <*> 0
1961 V39 N12 P724-731 65-10469 <*> 0
1961 V39 N12 P736-742 65-10470 <*> 0
1962 V40 N12 P595-602 66-13673 <*>
1964 V42 N5 P288-293 09R792 <ATS>
1964 V42 N5 P294-297 08R79J <ATS>

HAKKO KYOKAISHI
1959 V17 N5 P201-206 62-18691 <*>
1959 V17 N6 P252-257 62-00331 <*>
1960 V18 N5 P7-12 62-16616 <*> 0
1961 V19 P401-404 22R75J <ATS>
1963 V21 P105-109 65-12487 <*>
1965 V23 N6 P267-269 1941 <TC>

HALBLEITERPROBLEME
1953 V1 P299-312 57-2049 <*>
1957 V4 P282-364 AEC-TR-4339 <=>
1958 P68-118 02L29G <ATS>

HANDBUCH DER BIOLOGISCHEN ARBEITSMETHODEN
1923 N2 P487-494 C6 PTB 64-10009 <*>
1923 N2 P494-500 C6 PTB 64-10010 <*>
1929 N2 P404-438 S4 P16 63-20406 <*>

HANDBUCH DER EXPERIMENTELLEN PHARMAKOLOGIE
1923 V1 P1-72 63-20424 <*> P
1923 V1 P73-121 63-20423 <*> OP
1923 V1 P122-132 63-20425 <*> P
1923 V1 P133-261 63-20419 P.178-190 <*> PO
63-20420 <*> PO
63-20421 <*> P
63-20422 P.253-254 <*> P
1923 V1 P262-388 63-20418 <*> PO
1923 V1 P504-639 63-20114 <*> P
63-20115 <*> P
63-20116 <*> OP
63-20117 <*> OP
63-20118 <*> P
1923 V1 P817-832 63-20098 <*> 0
1923 V1 P833-859 63-20093 <*> 0
1923 V1 P871-1048 63-20094 <*> P
1924 V2 P1748-1931 PT.2 63-20095 <*> OP
1927 V3 P568-619 PT.1 63-20112 <*> 0
1934 P1503-1534 63-18773 <*> P
1934 V3 P1575-1889 PT.3 63-20415 <*> P
1934 V3 P1747-1754 PT.3 63-20414 <*>
1934 V3 P2186-2187 PT.3 63-20097 <*>
1938 V7 P1-62 63-20100 <*> P

HANDBUCH DER NEUROCHIRURGIE
1957 V6 P177-218 UCRL TRANS-803 <*>

HANDBUCH DER PFLANZENANATOMIE
1929 S1A V3 P34-57 64-00324 <*>

HANDBUCH DER PFLANZENPHYSIOLOGIE
1956 V3 P215-246 CSIRO-3799 <CSIR>

HANDBUCH DER PHYSIK
V19 P895-899 59-20872 <*>
V26 P504-530 CHAP5 62-10724 <*>
1926 V10 P1-59 AEC-TR-4074 <*>
1926 V10 P1-52 RE 2-18-59W <NASA>
59-10953 <*>
1926 V10 P300-318 59-10952 <*>
1926 V10 P308-318 62-10527 <*>
1927 V18 P379-382 CH2C 61-14392 <*>
1928 V6 P416-422 AEC-TR-4614 <*>
1928 V20 P635-904 65-10374 <*> P
1933 V24 P499-579 PT2 1552 <*>
499-579 <*>
1953 V24 P499-579 PT2 1805 <*>
1955 V7 P290-316 PT 1 ORNL-TR-90 <*>
1955 V7 P513-563 60-14111 <*>
1955 V7 N1 P431-450 PT1 57-2977 <*>
1955 V7 N1 P563-629 62-10322 <*>
1955 V28 P79-145 63Q67G <ATS>
1956 V24 P555-645 63-14669 <*>
1957 V48 P734-774 65-14114 <*>

HANDLINGAR. CHALMERS TEKNISKA HOGSKOL
SEE CHALMERS TEKNISKA HOGSKOL. HANDLINGAR

HANDLINGAR OCH TIDSKRIFT K. KRIGSVETENSKAPS-AKADEMIEN
1952 V156 N9 P469-473 II-473 <*>
1579 <*>

HANG K'UNG CHI HSIEH YUEH K'AN
1959 N5 P23-24 61-27342 <*=>
1959 N9 P21 61-27342 <*=>

HANG KUNG CHIH SHIH
1964 N4 P10-13 AD-626 952 <=*$>
1964 N6 P3-6 64-51755 <=$>
1964 N10 P5-7 65-30286 <=$>
1964 N12 P22-23 66-35683 <=> 0

HANSA
1949 V86 P713-714 II-813 <*>
1949 V86 P717-718 II-813 <*>
1949 V86 P891-893 II-813 <*>
1951 N88 P1302- 2172 <BSRA>
1952 N33/4 P1090 60-13271 <=*>
1952 V89 N33/4 P1090-1092 1897 <*>
1952 V89 N46/7 P1588-1590 II-802 <*>
1953 V90 P1739-1744 II-589 <*>
1958 V95 N51/2 P2476-2478 60-17395 <=*>
1959 V96 N33/4 P1749-1761 60-17784 <=*>
1963 V100 N22 P2168 AD-627 507 <=*$>
1964 V101 P223- 2109 <BSRA>
1964 V101 N4 P331- 2241 <BSRA>
1964 V101 N18 P185- 2140 <BSRA>
1964 V101 N18 P1786- 2134 <BSRA>
1964 V101 N18 P1797- 2123 <BSRA>
1964 V101 N22 P2179- 2141 <BSRA>
1964 V101 N22 P2217- 2143 <BSRA>
1964 V101 N22 P2223- 2145 <BSRA>
1964 V101 N22 P2315- 2268 <BSRA>
1964 V101 N24 P2479- 2132 <BSRA>
1964 V101 N24 P2521- 2121 <BSRA>
1965 V102 N6 P499- 2227 <BSRA>
1965 V102 N6 P513- 2238 <BSRA>
1965 V102 N8 P663 2203 <BSRA>
1965 V102 N11 P1093 2251 <BSRA>
1965 V102 N12 P1134- 2262 <BSRA>
1965 V102 N12 P1178- 2251 <BSRA>
1965 V102 N18 P1675 2426 <BSRA>
1965 V102 N18 P1679 2423 <BSRA>
1965 V102 N18 P1751 2422 <BSRA>
1965 V102 N20 P1928 2430 <BSRA>
1965 V102 N20 P1967 2422 <BSRA>

1965 V102 N22 P2085 2408 <BSRA>

HAREFUAH
1956 V51 N4 P87 5758-D <K-H>
1958 V55 N6 P145-150 60-10943 <*>
1959 V56 P172-174 63-00953 <*>
1959 V57 P244-248 N66-11618 <=*$>
1962 N6 P201-204 63-01140 <*>
1965 V69 N2 P37-38 66-11832 <*>

HAROKEACH HAIVRI
1960 V8 N2 P77-83 62-32869 <=*>

HASLER MITTEILUNGEN
1952 V11 N2 P29-36 59-17892 <*> 0
1957 V16 N1 P1-4 66-14244 <*>

HASSADEH
1951 V32 P88-90 <CBPB>

HATANO TABAKO SHIKENJO HOKOKU, HOBUN
1952 N36 P1-22 58-1231 <*>
1952 V36 P63-75 13042 <KH>
1958 N39 P23-37 62-10919 <*> 0

HAUSMITTEILUNGEN AUS FORSCHUNG UND BETRIEB. FERNSEH A. G.
1939 V1 N5 P179-186 57-2423 <*>

HAUSMITTEILUNGEN. JOSEPH SCHNEIDER OPTISCHE WERKE
1962 V14 N5/6 P52-63 66-13762 <*>

HAUSZEITSCHRIFT DES V. A. W. U. D. ERFTWERK A. G. FUER ALUMINIUM
1930 P280 58-1861 <*>
1930 V2 N3 P75-81 60-10254 <*> 0
1931 P276 58-2020 <*>
1931 V3 P237 58-1884 <*>
1932 V4 P79 58-1988 <*>

HAUTARZT
1953 V4 P49-56 61-20188 <*> 0
1953 V4 N5 P210-212 63-00605 <*>
1956 V7 P168-171 57-1803 <*>
1957 V8 P58-65 63-14885 <*> 0
1957 V8 P397-405 58-1705 <*>
1957 V8 P486-488 64-18532 <*>
1957 V8 N4 P174-176 65-13730 <*> 0
1957 V8 N8 P352-359 58-1563 <*>
1958 V9 N1 P9-15 66-12712 <*> 0
1961 V12 N9 P399-402 62-14514 <*> 0
1963 V14 N11 P511 65-13960 <*> 0
1964 V15 N2 P51-57 65-17015 <*> 0
1964 V15 N6 P287-294 65-13601 <*>
1965 V16 N1 P1-6 65-14772 <*> 0

HAUTE PRECISION AOIP
1954 N3 P2-6 59-15586 <*> 0

HEALTH PHYSICS. NEW YORK
1964 V10 P453-468 66-12030 <*>

HEFTE ZUR UNFALLHEILKUNDE
1956 N55 P14-24 59-12431 <*=> 0

HEIDELBERGER BEITRAEGE ZUR MINERALOGIE UND PETROGRAPHIE
1950 V2 P216-234 UCRL TRANS-982(L) <*>
1955 V4 P434-442 63-18288 <*>

HEIKI TO GIJUTSU
1958 N8 P1 59-16010 <+*>
1958 N8 P22-31 59-14893 <+*> 0

HELGOLAENDER WISSENSCHAFTLICHE MEERESUNTERSUCHUNGEN
1942 V2 N3 P279-353 C-4251 <NRC> P
63-00275 <*>
1950 V3 P171-205 59-14292 <+*>

HELLENIKE IATRIKE
1961 V30 N7 P957-962 63-16532 <*>

HELVETICA CHIMICA ACTA
1921 V4 P914-920 58-204 <*>
1924 V7 P713-717 59-20983 <*>
1924 V7 P910-915 AEC-TR-6024 <*>
1925 V8 P314-322 64-16511 <*>
1925 V8 P900-923 63-14151 <*>
1926 V9 P1018-1049 64-20290 <*>
65-10716 <*>
1926 V9 P1066-1068 57-44 <*>
1927 V10 P140-167 64-20290 <*>
65-10716 <*>
1927 V10 P901-907 AEC-TR-5949 <*>
1928 V11 P416-425 1653 <*>
1928 V11 P1003-1026 63-16169 <*>
1928 V11 P1201-1209 5179-B <K-H>
1929 V12 P305-313 63-10557 <*>
1929 V12 P713-740 II-64 <*>
1929 V12 P806-817 64-16047 <*>
1930 V13 P650-666 65-12044 <*>
1930 V13 P842-864 37L29J <ATS>
1931 V14 P435-438 5179-A <K-H>
1932 V15 P619-634 65-11926 <*> 0
1932 V15 P1511-1520 II-982 <*>
8757 <IICH>
1933 V16 P169-181 60-10439 <*>
1933 V16 P1214-1225 58-1850 <*>
58-1876 <*>
1934 V17 P30-31 NP-TR-1022 <*>
1934 V17 P1226-1231 NP-TR-1023 <*>
1934 V17 P1530-1535 63-14035 <*> 0
1935 V18 P103-120 57-2258 <*>
58-1969 <*>
1935 V18 P166-178 I-716 <*>
1935 V18 P419-426 NP-TR-1024 <*>
1935 V18 P491-513 60-18149 <*>
1935 V18 P1028-1029 I-169 <*>
1936 V19 P29-68 60-10408 <*> P
1936 V19 P99-106 65-14975 <*> P
1936 V19 P223-234 60-10407 <*>
1937 V20 P1523-1528 57-208 <*>
1939 V22 P449-456 I-743 <*>
1939 V22 P1048-1058 58-1851 <*>
1939 V22 P1139-1143 2646 <*>
1940 V23 P1054-1062 58-1392 <*>
1940 V23 P1139-1146 62-16446 <*> 0
1940 V23 P1287-1297 I-872 <*>
1941 V24 P9-23 SPEC AL-623 <*>
1941 V24 P302-319 3270 <*>
1941 V24 P617-638 58-587 <*>
1941 V24 P986-988 65-14976 <*>
1941 V24 N2 P217-223 RCT V15 N1 P17 <RCT>
1942 V25 P252-295 NP-TR-1104 <*>
1942 V25 P353-361 58-707 <*>
1942 V25 P364-370 63-20497 <*>
1942 V25 P530-538 65-12011 <*>
1942 V25 P1116- I-763 <*>
1942 V25 P1421-1432 65-14977 <*>
1942 V25 P1543-1547 57-861 <*>
1943 V26 P1084-1098 CSIRO-2231 <CSIR>
1943 V26 P1185-1189 C-5042 <NRC>
1943 V26 P1741-1750 5629-E <K-H>
1943 V26 P1816-1828 1813 <*>
2843 <*>
1943 V26 N5/8 P1945-1965 60-10435 <*> 0
1944 V27 P268-273 II-130 <*>
1944 V27 P313-316 5562-B <K-H>
1944 V27 P317-319 5629-F <K-H>
1944 V27 P576-584 338 <TC>
1944 V27 P892-928 CSIRO-3389 <CSIR>
59-10822 <*>
1944 V27 P1480-1495 C-2402 <NRC>
63-20179 <*>
1944 V27 P1495-1501 64-18878 <*>
1944 V27 P1727-1735 65-14978 <*> P
1944 V27 N5 P1570-1583 59-15983 <*>
1945 V28 P129-148 99L36G <ATS>
1945 V28 P149-156 00L37G <ATS>

Citation	Translation
1945 V28 P248-250	59-20397 <*>
1945 V28 P257-274	57-2599 <*>
1945 V28 P1362-1370	63-14595 <*>
1945 V28 N6 P1638-1647	61-20487 <*>
1946 V29 P811-818	63-20177 <*>
1946 V29 P2006-2017	62-14299 <*>
1946 V29 N2 P364-370	AL-688 <*>
1946 V29 N4 P811-818	2884 <*>
1946 V29 N5 P1173-1783	C-5371 <NRC>
1946 V29 N7 P1842-1853	RCT V20 N2 P416 <RCT>
	RCT V21 N2 P314 <RCT>
1947 V30 P507-519	65-14979 <*> P
1947 V30 P539-543	65-14980 <*>
1947 V30 P1303-1320	I-629 <*> P
1947 V30 P1595-1599	62-00690 <*>
1947 V30 P1798-1804	I-630 <*>
1947 V30 N1 P307-328 PT1	RCT V27 N1 P36 <RCT>
1947 V30 N1 P64-78	64-10020 <*>
1947 V30 N2 P464-486 PT2	RCT V27 N1 P36 <RCT>
1947 V30 N2 P519-524	60-10398 <*>
1947 V30 N3 P839-859	RCT V28 N3 P694 <RCT>
1947 V30 N6 P1822-1836	72N57F <ATS>
1947 V30 N6 P1911-1927	59-17610 <*> O
1948 V31 P297-319	61-20292 <*>
1948 V31 P459-465	1521 <*>
1948 V31 P1029-1048	I-631 <*>
1948 V31 P1584-1602	62-18965 <*> O
1948 V31 N2 P459-465	49T94G <ATS>
1948 V31 N3 P737-748	61-16742 <*>
1948 V31 N4 P1029-1048	II-526 <*>
1949 V32 P1036-1040	5268-A <K-H>
1949 V32 P1314-1325	59-22396 <=*> O
1949 V32 N2 P436-	5062-B <K-H>
1949 V32 N2 P489-505	59-17609 <*> PO
1949 V32 N2 P538-544	65-12439 <*>
1949 V32 N3 P714-717	66-10259 <*>
1949 V32 N5 P1378-1390	65-10127 <*> P
1949 V32 N7 P2397-2399	2818-A <K-H>
1949 V32 N7 P2435-2439	60-10397 <*> O
1949 V32 N7 P2441-2444	61-00635 <*>
1950 V33 P184-198	58-2181 <*>
1950 V33 P1597-1605	61-20429 <*>
1950 V33 P2264-2267	62-25780 <=*>
1950 V33 N1 P25-36	AL-231 <*>
1950 V33 N3 P506-511	60-14526 <*>
1950 V33 N6 P1568-1581	4J16G <ATS>
1950 V33 N7 P2134-2152	AL-147 <*>
1951 V34 P1183-1202	I-494 <*>
1951 V34 P1419-1421	9188-C <K-H>
1951 V34 P1632-1634	5226-B <KH>
1951 V34 P1635-1641	I-914 <*>
1951 V34 P2182-2197	63-18823 <*> O
1951 V34 P2449-2457	62-25781 <=*>
1951 V34 N1 P210-221	59-17814 <*> O
1951 V34 N3 P952-958	II-403 <*>
1952 V35 P167-180	61-20654 <*> O
1952 V35 P276-283	65-14981 <*> P
1952 V35 P307-316	60-10742 <*>
1952 V35 P392-396	63-18245 <*>
1952 V35 P782-802	2910 <*>
1952 V35 P782-801	62-14943 <*> O
1952 V35 P2235-2237	61-14759 <*>
1952 V35 P2570-2573	5226-C <K-H>
1952 V35 N1 P53-60	60-18080 <*>
1952 V35 N1 P407-411	64-18880 <*>
1952 V35 N3 P1021-1030	62-14799 <*>
1952 V35 N5 P1486-1494	62-14801 <*>
1952 V35 N6 P1908-1913	61-10403 <*> O
1952 V35 N6 P1957-1970	62-14800 <*>
1952 V35 N7 P2248-2259	I-742 <*>
1952 V35 N7 P2547-2556	64-14842 <*>
1953 V36 P414-423	1600 <*>
1953 V36 P782-787	T-2042 <INSD>
	63-16062 <*>
1953 V36 P828-834	5472-D <K-H>
1953 V36 P1174-1177	I-96 <*>
1953 V36 P1436-1444	63-16871 <*>
1953 V36 P1649-1658	64-20351 <*>
1953 V36 N3 P597-605	65-12393 <*>
1953 V36 N4 P910-918	64-16448 <*> O
1953 V36 N6 P1352-1369	1075 <*>
1953 V36 N6 P1624-1630	64-10129 <*>
1953 V36 N144 P1160-1165	57-2327 <*>
1953 V36 N182 P1445-1454	58-2255 <*>
1953 V36 N247 P2009-2017	59-10152 <*>
1953 S4 V36 P875-886	57-1274 <*>
1954 V37 P332-334	3658 <K-H>
1954 V37 P647-649	I-635 <*>
1954 V37 P805-814	1620 <*>
1954 V37 P837-	II-975 <*>
1954 V37 P1767-1778	57-2721 <*>
1954 V37 P1901-1903	3504-C <K-H>
1954 V37 N2 P583-597	I-277 <*>
	2952-A <K-H>
	64-10153 <*>
1954 V37 N4 P937-957	60-14044 <*> O
1954 V37 N4 P1336-1338	60-14124 <*>
1954 V37 N5 P1392-1398	65-12528 <*>
1954 V37 N5 P1407-1422	59-17334 <*>
1954 V37 N5 P1423-1431	2443 <*>
1954 V37 N5 P1431-1436	57-1745 <*>
1954 V37 N5 P1548-1553	57-1919 <*>
1955 V38 P1749-1756	59-10438 <*>
1955 V38 P1931-1940	60-23459 <=*>
1955 V38 N1 P37-46	39R78G <ATS>
1955 V38 N1 P125-134	RCT V28 N3 P675 <RCT>
1955 V38 N1 P184-190	65-12564 <*>
1955 V38 N4 P935-942	65-12627 <*>
1955 V38 N7 P1786-1794	65-10240 <*>
1956 V39 P158-167	59-18959 <+*>
1956 V39 P304-317	57-3174 <*>
1956 V39 P596-606	1318 <*>
1956 V39 P944-950	T2237 <INSD>
1956 V39 N2 P505-513	61-10535 <*>
1956 V39 N3 P722-728	57-1790 <*>
1956 V39 N3 P767-775	57-733 <*>
1956 V39 N3 P812-827	64-20127 <*>
1956 V39 N4 P1071-1086	ORNL-TR-72 <*>
1956 V39 N6 P1654-1663	61-10307 <*> O
1956 V39 N16 P722-728	59-15745 <*>
1957 V40 P221-228 PT.1	UCRL TRANS-613(L) <*>
1957 V40 P228-233 PT.2	UCRL TRANS-615(L) <*>
1957 V40 P2322-2340 PT.4	UCRL TRANS-627(L) <*>
1957 V40 P234-236 PT.3	UCRL TRANS-616(L) <*>
1957 V40 P356-363	C-3855 <NRCC>
1957 V40 N1 P69-79	6287-B <K-H>
1957 V40 N1 P113-118	6124-A <KH>
1957 V40 N2 P350-355	60-18034 <*>
1957 V40 N3 P614-624	59-10447 <*>
1957 V40 N3 P717-721	6456-A <KH>
1957 V40 N4 P971-980	61-10482 <*>
1957 V40 N5 P1145-1157	63-00572 <*>
1957 V40 N7 P2322-2340	66-13231 <*> P
1957 V40 N7 P2410-2420	29K23G <ATS>
1957 V40 N7 P2411-2420	61-10662 <*>
1957 V40 N47 P387-394	57-2980 <*>
1958 V41 P1572-1581 PT.6	UCRL TRANS-609(L) <*>
1958 V41 P1771-1783 PT.7	UCRL TRANS-622(L) <*>
1958 V41 P824-843	UCRL TRANS-617(L) <*>
1958 V41 P1038-1046	59-15372 <*>
1958 V41 P1046-1052	59-15371 <*>
1958 V41 P1322-1332	63-10558 <*>
1958 V41 P1667-1692	59-17054 <*>
1958 V41 P2190-2199	1945 <TC>
1958 V41 N1 P275-288	29K23G <ATS>
1958 V41 N5 P1273-1286	60-10986 <*>
1958 V41 N5 P1287-1295	60-14178 <*>
1958 V41 N6 P132-152	60-14129 <*> O
1958 V41 N7 P1915-1932	59-15334 <*>
1958 V41 N7 P2135-2149	59-15335 <*>
1958 V41 N7 P2135-2148	59-17170 <*>
1958 V41 N7 P2308-2322	62-00281 <*>
1959 V42 P363-386	59-18075 <=*>
1959 V42 P628-660	61-13451 <=*> C
1959 V42 P1166-1170	61-10917 <*>
1959 V42 P1653-1657	62-10005 <*> O
1959 V42 N2 P417-425	59-15641 <*>
1959 V42 N3 P802-807	59-15951 <*>
1959 V42 N4 P1160-1165	61-10918 <*>
1960 V43 P623-628	60-16640 <*>

1960 V43 P889-1C04 AEC-TR-4297 <*>
1960 V43 P1019-1032 NP-TR-607 <*>
1960 V43 P1427-1431 10Q68G <ATS>
1960 V43 P1862-1863 62-10051 <*>
1960 V43 N1 P64-71 C-3490 <NRCC>
63-20269 <*> 0
1960 V43 N1 P433-438 61-00409 <*>
1960 V43 N2 P502-517 60-18164 <*>
1960 V43 N7 P2129-2138 64-16857 <*>
1960 V43 N214 P1745-1751 10607-D <KH>
1961 V44 P173-179 SC-T-64-49 <*>
1961 V44 P343-361 AEC-TR-4567 <*>
1961 V44 P667-674 62-10335 <*>
1961 V44 P977-985 71N54G <ATS>
1961 V44 P1162-1164 58P66G <ATS>
1961 V44 P1337-1349 NP-TR-858 <*>
1961 V44 P1829-1856 UCRL TRANS-851 <*>
1961 V44 N2 P447-456 66-14089 <*>
1961 V44 N2 P457-46C 66-14483 <*>
1961 V44 N3 P823-829 62-20442 <*> 0
1961 V44 N3 P859-865 AEC-TR-4819 <*>
1961 V44 N4 P969-976 62-10258 <*> 0
1961 V44 N4 P1123-1125 C-3976 <NRCC>
1961 V44 N5 P1287-1292 62-18725 <*>
1961 V44 N6 P1622-1645 <LSA>
1961 V44 N7 P1956-1966 63-10879 <*> 0
1961 V44 N7 P2141-2150 SCL-T-410 <*>
1962 V45 P675-685 62-18062 <*>
1962 V45 P1081-1089 64-15676 <=*$>
1962 V45 P1702-1721 AEC-TR-5791 <*>
1962 V45 N1 P212-224 63-18780 <*>
1962 V45 N1 P261-27C 63-10336 <*>
1962 V45 N2 P595-600 63-10399 <*> 0
65-12275 <*>
1962 V45 N2 P685-692 62-20120 <*>
1962 V45 N4 P1156-1161 65-11593 <*>
65-14470 <*> 0
1962 V45 N5 P1409-1415 63-10927 <*>
8719 <IICH>
1962 V45 N6 P2175-2185 65-12477 <*>
1962 V45 N6 P2266-2272 63-10516 <*> 0
1963 V46 P1339-1360 AEC-TR-5937 <*>
1963 V46 P2369-2388 66-10031 <*>
1963 V46 N4 P1385-1389 64-18141 <*>
1963 V46 N7 P2667-2676 66-14687 <*>
1963 V46 N49 P467-479 PT.2
AI-TR-44 <*>
1964 V47 P150-154 8923 <IICH>
1964 V47 N4 P992-1002 66-10896 <*>
1964 V47 N8 P2279-2288 697 <TC>
1965 V48 N3 P527-538 66-11108 <*>
1965 V48 N5 P1034-1039 66-14865 <*$>
1965 V48 N6 P1259-1270 1915 <TC>
1966 V49 N2 P1000-1002 66-14688 <*>
1966 V49 N3 P1119-1130 66-14873 <*>

HELVETICA CHIRURGICA ACTA
1953 SB V20 P175-195 SUP 7
66-12758 <*>
1965 V32 N1/2 P144-152 66-12159 <*>

HELVETICA MEDICA ACTA
1938 V5 P552-557 60-10671 <*> 0
1942 V9 P697-719 59-10977 <*> 0
1944 V11 P741-754 60-10666 <*> 0
1944 V11 P961-1010 60-10666 <*> 0
1947 V14 P212-230 59-14871 <+*> 0
1948 V15 P569-580 1317 <*>
1950 V17 P137-158 64-18995 <*> 0
1950 V17 P575-582 I-893 <*>
1952 V19 N4/5 P455-458 2664 <*>
1953 V20 P211-215 AL-234 <*>
1953 V20 P320-322 I-26 <*>
1953 V20 P340-345 I-953 <*>
1953 V20 P352-355 I-347 <*>
1953 V20 N4/5 P346-351 60-10606 <*> 0
1954 V21 P493-497 2016 <*>
1955 V22 P109-122 1399 <*>
1955 V22 P495-501 2990 <*>
1956 V23 N4/5 P515-522 65-14982 <*> 0
1957 V24 P453-458 58-16 <*>
1957 V24 P459-462 58-18 <*>
1957 V24 P463-471 58-23 <*>
1958 V25 N2 P101-152 65-13837 <*> P
1958 V25 N2 P153-183 61-00207 <*> P
1960 V27 P479-503 61-18277 <*>
1960 V27 P527-529 61-18276 <*> 0
1960 V27 P530-534 61-16101 <*>
1960 V27 P535-538 61-16109 <*>
1960 V27 P539-540 61-16103 <*>
1960 V27 P541-542 61-16105 <*>
1960 V27 P543-545 61-16099 <*>
1960 V27 P615-629 62-16037 <*>
1960 V27 P683-689 61-20046 <*>
1960 V27 N3 P245-263 62-01220 <*>
1960 V27 N5/6 P519-522 61-20907 <*>
1961 V28 P587-593 62-20032 <*>
1961 V28 N4 P487-495 64-16855 <*>
1962 V29 P543-549 63-18798 <*>
1962 V29 N5/6 P674-679 63-16346 <*>
1963 V30 P476-481 66-12037 <*>
1963 V30 N4/5 P512-516 64-16856 <*>
1964 V31 P432-436 65-13940 <*>

HELVETICA PAEDIATRICA ACTA
1948 V4 P334-337 59-14876 <+*>
1954 V9 N4 P298-310 63-00164 <*>
1955 V10 N4 P397-412 61-16561 <*> 0
1957 V12 P241-258 59-22261 <*+> 0
1957 V12 P596-604 59-22582 <*=> 0
1957 V12 N6 P569-595 61-16562 <*> 0
1958 V13 N3 P262-271 59-18960 <*=>
1961 V16 N5/6 P565-585 62-00836 <*>

HELVETICA PHYSICA ACTA
1930 N5/6 P335-390 57-1571 <*>
1934 V7 P298-330 AEC-TR-4356 <*>
1934 V7 P331-359 AEC-TR-4344 <*>
1934 V7 P620-627 62-20280 <*>
1934 V7 P676-683 62-01383 <*>
1934 V7 P878-883 62-20281 <*>
1935 V8 P674-681 59-10434 <*>
1936 V9 P63-83 63-18928 <*>
1936 V9 P265-284 T2228 <INSD>
1936 V9 P520-532 63-10092 <*>
1937 V10 P185-217 63-10091 <*>
1937 V10 P261 59-20435 <*>
1938 V11 P269-298 57-2394 <*>
1939 V12 N3 P270-276 RCT V13 N1 P285 <RCT>
1940 V13 N6 P435-450 66-12597 <*> 0
1941 V14 N3 P141-143 57-2115 <*>
1942 V15 P333-334 UCRL TRANS-582(L) <*>
1942 V15 P571-612 2273 <*>
1942 V15 N2 P127-161 57-820 <*>
1942 V15 N3 P199-220 57-820 <*>
1942 V15 N7 P645-684 57-2060 <*>
1943 V16 P207-209 57-2192 <*>
1943 V16 P214-216 63-01281 <*>
1943 V16 N4 P324-342 88S89G <ATS>
1945 V18 P251-262 2331 <*>
1945 V18 P369-388 UCRL TRANS-505(L) <*>
1945 V18 N7 P527-550 II-271 <*>
1946 V19 P77-90 II-22 <*>
1946 V19 P200-202 UCRL TRANS-581(L) <*>
1946 V19 P230-231 57-263 <*>
1946 V19 P404- 57-972 <*>
1946 V19 N3 P167-188 2532 <*>
1946 V19 N6/7 P463-492 57-966 <*>
59-10136 <*>
1947 V20 P96-104 57-1952 <*>
1947 V20 P139-152 I-376 <*>
1947 V20 P230-234 I-375 <*>
1947 V20 N1 P33-66 57-2373 <*>
1947 V20 N2 P207-221 57-308 <*>
1947 V20 N2 P225-228 57-981 <*>
1947 V20 N2 P307-340 57-2269 <*>
1948 V21 P170 UCRL TRANS-579(L) <*>
1948 V21 P197 AEC-TR-5962 <*>
1948 V21 P497-498 UCRL TRANS-580(L) <*>
1948 V21 N2 P151-168 60-18184 <*>

1948 V21 N2 P212-215	57-1197 <*>
1949 V22 P6C3-605	I-397 <*>
1949 V22 P606-609	II-13 <*>
1950 V23 P108-120	58-2510 <*>
1950 V23 P484-487	SCL-T-230 <*>
1950 V23 N4 P347-380	SCL-T-422 <*>
1950 V23 N4 P381-411	II-137 <*>
1950 V23 N6/7 P697-730	58-302 <*>
	94J7G <ATS>
1950 V23 N6/7 P795-844	65-12227 <*>
1951 V24 P307-309	UCRL TRANS-591(L) <*>
1951 V24 P310-314	I-54 <*>
1952 V25 P259-268	I-613 <*>
1952 V25 P469-	57-978 <*>
1952 V25 N4 P371-386	59-17964 <*>
1952 V25 N1/2 P142-152	NP-TR-743 <*>
1953 V26 P199-206	AEC-TR-5318 <*>
1953 V26 P578-583	II-298 <*>
1953 V26 P611-656	58-1917 <*>
1953 V26 N5 P545-562	57-215 <*>
1953 V26 N6 P611-656	64-16337 <*>
1954 V27 P3-44	I-377 <*>
	52F3G <ATS>
1954 V27 P99-124	C-2388 <NRC>
1954 V27 P241-248	UCRL TRANS-752(L) <*>
1954 V27 P259-282	17G7G <ATS>
1954 V27 P690-696	UCRL TRANS-746(L) <*>
1954 V27 N1 P74-80	F4G <ATS>
1954 V27 N2 P99-124	57-564 <*>
	59-15847 <*> 0
1954 V27 N3 P259-282	AEC-TR-3567 <*>
1954 V27 N4 P309-312	65-13504 <*>
1955 V28 P452-453	3262 <*>
1955 V28 N4 P3C0-303	CSIRO-3307 <CSIR>
1955 V28 N4 P389-394	NP-TR-1000 <*>
1955 V28 N56 P485-491	UCRL TRANS-531 <*>
1956 V29 P37-46	69M39G <ATS>
	79L29G <ATS>
1956 V29 P211-212	57-2303 <*>
1956 V29 P533-	AEC-TR-4114 <*>
1956 V29 N1 P37-46	AEC-TR-3853 <*>
1956 V29 N2 P147-186	NP-TR-1003 <*>
1956 V29 N5/6 P463-506	NP-TR-953 <*>
1957 V30 P315-330 PT.2	AEC-TR-3800 <*>
1957 V30 P157-182	C-3225 <NRCC>
1957 V30 P224-227	C-3765 <NRCC>
1957 V30 P254-257	T-2591 <INSD>
1957 V30 N1 P3-32	AEC-TR-3828 <*>
	65-12663 <*>
1957 V30 N1 P33-48	58-1647 <*>
1957 V30 N1 P91-92	AEC-TR-4002 <*>
1957 V30 N4 P297-315	AEC-TR-3761 <*>
1957 V30 N4 P331-346	AEC-TR-3801 <*>
	59-00797 <*>
1958 V31 P17-24	26K28G <ATS>
1958 V31 N1 P25-32	62-14561 <*>
1958 V31 N3 P173-204	62-16102 <*>
1958 V31 N7 P685-712	61-10202 <*> 0
1959 V32 P197-212	AEC-TR-5978 <*>
1959 V32 P567-600	62-10343 <*> 0
1959 V32 P615-654	63-14855 <*> 0
1959 V32 N2 P89-128	65-14523 <*>
1959 V32 N4 P293-296	62-00086 <*>
1959 V32 N4 P318-320	61-00673 <*>
1960 V33 P437-458	66-11152 <*>
1960 V33 P590-593	AEC-TR-5851 <*>
1960 V33 P6C8	UCRL TRANS-809(L) <*>
	62-00587 <*>
1960 V33 P933-940	63-18148 <*>
1960 V33 N9 P954-959	61-18691 <*>
1960 V33 N6/7 P657-666	61-16513 <*>
1961 V34 P370-373	62-16597 <*>
1961 V34 P391-392	SCL-T-440 <*>
1961 V34 P408-410	UCRL TRANS-921 <*>
1961 V34 P859-864	AEC-TR-4866 <*>
1962 V35 P500	AEC-TR-5925 <*>
1962 V35 N1 P3-33	NP-TR-1012 <*>
1962 V35 N4/5 P237-240	NP-TR-994 <*>
1964 V37 N3 P241-244	65-17309 <*> 0

HELVETICA PHYSIOLOGICA ET PHARMACOLOGICA ACTA

1945 V3 P269-289	I-1047 <*>
1945 V3 P437-443	2271 <*>
1946 V4 PC11-C12	63-18511 <*>
1947 V5 P169-177	62-20176 <*> 0
1947 V5 P348-360	62-00645 <*>
1949 V7 P210-229	AL-476 <*>
1949 V7 P382-405	61-14596 <*>
1950 V8 P525-543	58-231 <*>
1950 V8 N3 P272-285	65-14983 <*> 0
1952 V10 P1-19	62-00172 <*>
1952 V10 P184-206	58-223 <*>
1952 V10 P403-412	II-796 <*>
1953 V11 P20-29	65-00411 <*> 0
1953 V11 P142-156	NP-TR-983 <*>
1953 V11 N3 P239-250	60-10658 <*> P
1953 V12 N4 P284-292	II-322 <*>
1954 V12 N4 P64-66	II-304 <*>
1954 V12 N4 P75-77	II-321 <*>
1954 V12 N4 P83-84	II-362 <*>
1955 V12 P106-112	57-1832 <*>
1955 V13 P54-55	59-12343 <+*>
1955 V13 P113-120	II-1003 <*>
1955 V13 P331-353	62-01426 <*>
1955 V13 N1 P42-49	65-00412 <*> 0
1956 V14 P3-5	59-12343 <+*>
1956 V14 P27-29	57-2978 <*>
1956 V14 P45-47	57-2979 <*>
1956 V14 P53-54	59-12343 <+*>
1956 V14 N2 P207-221	3232 <*>
1956 V14 N2 PC34-C35	64-16212 <*>
1956 V14 N3 P279-288	57-120 <*>
1956 V14 N3 P353-362	57-143 <*>
1956 V14 N4 PC66-C70	58-801 <*>
	63-16396 <*>
1956 V14 N1028 P13-14	57-230 <*>
1957 V15 P14-24	57-3430 <*>
1957 V15 P38-54	61-01008 <*>
1957 V15 P177-183	58-2209 <*>
1957 V15 PC69-C71	58-780 <*>
	58-831 <*>
	63-16599 <*>
1957 V15 N1 P55-62	57-1808 <*>
1957 V15 N1 P105-116	57-1805 <*>
1957 V15 N2 P39-41	57-2659 <*>
1957 V15 N2 P284-292	57-2658 <*>
1957 V15 N3 P335-344	58-13 <*>
1957 V15 N3 P361-365	58-21 <*>
1957 V15 N4 P52-53	59-15051 <*>
1958 V16 P255-267	61-00848 <*>
1958 V16 P268-276	61-00847 <*>
1958 V16 N2 P127-145	62-00301 <*>
1958 V16 N2 P146-151	59-17596 <*>
1958 V16 N2 P152-162	59-17470 <*>
1958 V16 N2 PC7-C10	59-1758 <*>
1958 V16 N3 P208-221	59-10626 <*>
1958 V16 N5 P287-302	59-15681 <*>
1959 V17 P189-201	62-00317 <*>
1959 V17 P361-364	62-00535 <*>
1959 V17 N2 P18	63-00571 <*>
1960 V18 P394-403	61-10246 <*>
1960 V18 P482-490	62-01417 <*>
1960 V18 P545-562	65-17235 <*>
1960 V18 PC15-C16	63-00570 <*>
1960 V18 PC70-C73	63-01321 <*>
1960 V18 N1 PC9-C12	61-20576 <*> 0
1960 V18 N1 PC12-C15	60-18751 <*>
1960 V18 N2 P99-108	60-18558 <*>
1960 V18 N3 P259-265	61-10247 <*>
1960 V18 N3 P266-273	61-10245 <*>
1960 V18 N3 P274-295	61-14477 <*>
1960 V18 N3 P336-342	61-10816 <*>
1960 V18 N3 P357-365	61-10817 <*>
1960 V18 N3 P376-383	61-10815 <*>
1960 V18 N3 PC66-C67	61-20044 <*>
1961 V19 P42-57	61-16104 <*>
1961 V19 P335-343	63-18795 <*>
1961 V19 PC11-C13	62-00424 <*>
1962 V20 P135-147	63-20623 <*>

1962 V20 P291-293 63-10542 <*>
1962 V20 PC18-C20 63-10538 <*>
1962 V20 N4 PC64-C66 63-18745 <*>
1962 V20 N4 PC73-C76 64-20145 <*>
1963 V21 N3 PC3-C6 64-16081 <*>
1963 V21 N3 PC42-C45 64-16078 <*>
1963 V21 N4 P394-401 65-13991 <*>
1964 V22 PC34-C35 65-12166 <*>
1964 V22 PC80-C82 66-11669 <*>
1964 V22 N1 PC48 65-11294 <*>

HEMIKA HRONIKA
1955 V20A P127-135 57-745 <*>
1961 V26A P42-59 62-00683 <*>
1962 V27A P5-9 66-11576 <*>

HEMISKA INDUSTRIJA
1957 V7 P101-107 59-00771 <*>

HEMOSTASE
1962 V2 P305-307 65-17245 <*>

HEOLOHYCHNYI ZHURNAL
1957 V17 N3 P58-69 C-4658 <NRC>
1957 V17 N4 P64-69 66-14360 <*$>
1958 V18 N1 P39-45 IGR V1 N4 P26 <AGI>
1958 V18 N4 P61-70 AD-623 427 <=*$>
1958 V18 N6 P70-83 65-64317 <=>
1959 V19 N1 P57-67 64-13361 <=*$>
1962 V22 N5 P3-15 IGR V7 N8 P1521 <AGI>
1962 V22 N6 P10-21 IGR V7 N2 P297 <AGI>
1964 V24 N5 P16-29 N66-11604 <=*$>
1965 V25 N4 P3-12 66-12693 <*>
1965 V25 N6 P104-110 66-32970 <=$>

HIDROLOGIAI KOZLONY
1957 V37 P14-56 58-2107 <*>
1957 V37 N3 P245-248 13L34H <ATS>
1958 N38 P338-342 AEC-TR-4627 <*>
1959 V39 N4 P279-284 PANSDOC-TR.315 <PANS>
1961 V41 N3 P266-268 62-19316 <=*> P
1961 V41 N4 P284-288 62-19675 <=*> P
1961 V41 N4 P326-329 62-19564 <*=>
1961 V41 N4 P330-333 62-19540 <=*>
1961 V41 N4 P356-360 62-19578 <*=> P

HIDROTEHNICA
1963 V8 N11 P428-429 64-21745 <=>
1964 V9 N11 P585-590 90S84RU <ATS>

HIFU TO HINYO
1960 V22 N5 P65-67 63-16540 <*>

HIGH POLYMER
SEE KOBUNSHI

HIGHWAY. CHINESE PEOPLES REPUBLIC
SEE KUNG LU

HIGIENA
SEE IGIENA

HIGIJENA
1950 V2 N4 P319-321 AL-899 <*>
1956 V8 N4 P240-245 PB 141 291T <=>
1956 V8 N4 P250-267 PB-141-180T <=*>
1956 V8 N4 P273-276 59-13067 <=> 0
1960 V12 N4 P374-386 62-19706 <*=>
1960 V12 N2/3 P113-136 62-19434 <=>
1960 V12 N2/3 P145-152 62-19434 <=>
1960 V12 N2/3 P165-177 62-19434 <=>
1960 V12 N2/3 P217-224 62-19434 <=>

HINYOKIKA KIYO
1961 V7 N12 P1067-1073 63-16432 <*>
1963 V9 P472-477 65-14545 <*>
1964 V10 P812-821 66-13208 <*>
1964 V10 P822-828 66-13209 <*>

HIPPOKRATES
1936 V6 N9 P1-2 61-10441 <*>
1938 V9 N51 P1317-1325 T-2191 <INSD>
1956 V27 N18 P588-592 5796-C <K-H>
1956 V27 N18 P598-599 5796-B <K-H>
1957 V28 N3 P81-84 6445-C <K-H>
1959 V30 P246-248 60-16926 <*>
1961 V32 N12 P491-494 65-14984 <*>
1963 V34 N9 P383-384 66-10096 <*>
1964 V35 N1 P4-7 65-14778 <*> 0
1964 V35 N1 P31-32 66-10098 <*>

HIRADASTECHNIKA
1962 V13 N2 P70-74 62-33132 <=*>
1965 V16 P97-104 POED-TRANS-2219 <NLL>
1965 V16 N5 P156-157 65-31509 <=$>

HIROSAKI IGAKU
1960 V12 N1 P97-102 65-14737 <*>

HIROSHIMA DAIGAKU KOGAKUBU KENKYU HOKOKU
1955 V4 P95-101 II-1131 <*>
1955 V4 N1 P111-115 62-18129 <*>
1962 V11 P1-9 06Q73J <ATS>
1962 V11 P11-21 07Q73J <ATS>

HIROSAKI MEDICAL JOURNAL
SEE HIROSAKI IGAKU

HISPALIS MEDICA
1955 V12 N130 P155-161 62-00318 <*>

HISTOCHEMIE. BERLIN
1959 V1 P315-330 61-01050 <*>
1962 V3 P17-45 63-01131 <*>

HISTOIRE DE LA MEDECINE
1954 V4 P47-57 63-16628 <*>

HITACHI HYORON
1960 P80-86 SPEC.37 08N59J <ATS>
1963 V45 N5 P56-58 NEL-TRANS-1604 <NLL>
1964 V46 P512-519 4540 <BISI>

HITACHI ZOSEN GIHO
1963 N24 P24 2017 <BSRA>
1963 N24 P189- 2131 <BSRA>
1963 N24 P416- 2122 <BSRA>
1964 V25 P201 2420 <BSRA>

HITACHI ZOSEN TECHNICAL REVIEW
SEE HITACHI ZOSEN GIHO

HITACHI REVIEW. JAPAN
SEE HITACHI HYORON

HOAN EISEI GAKKAI
1954 V1 N4 P8-16 57-2563 <*>
1954 V1 N4 P17-26 II-498 <*>
1954 V1 N5 P7-17 65-00178 <*> 0
1954 V1 N7 P6-10 3059 <*>
1954 V1 N7 P11-12 1739 <*>

HOAT DONG KHOA HOC
1962 N6 P1-9 62-33148 <=*>

HOCHFREQUENZTECHNIK UND ELEKTROAKUSTIK
1933 V41 P116-128 AL-720 <*>
1933 V41 N4 P115-116 57-2139 <*>
57-3525 <*>
1933 V41 N5 P156-167 66-12602 <*> 0
1933 V42 N2 P64-67 57-1583 <*>
1934 V43 N4 P124-130 57-628 <*>
1934 V43 N5 P172-176 58-1308 <*>
1934 V43 N6 P195-199 66-13302 <*> 0
1934 V44 N4 P109-118 66-13299 <*> 0
1935 V45 N2 P42-50 57-5 <*>
1935 V45 N6 P187-198 66-12272 <*>
1935 V46 N2 P37-49 AL-288 <*>
1935 V46 N6 P187-192 57-922 <*>
1936 V47 N1 P8-12 57-939 <*>

1936 V47 N6 P177-181 57-604 <*>
1936 V48 N1 P1-7 2167 <*>
1936 V48 N4 P120-126 60-10101 <*> 0
1937 V50 P73-80 57-2123 <*>
1938 V52 N2 P37-44 66-12273 <*>
1938 V52 N2 P44-58 57-911 <*>
1938 V52 N2 P58-62 57-925 <*>
1938 V52 N3 P80-82 66-13337 <*>
1939 P116-121 59-15216 <*>
1939 V53 N4 P113-115 I-157 <*>
1939 V53 N4 P115-122 57-1353 <*>
1939 V53 N5 P145-146 66-13578 <*>
1939 V53 N5 P146-150 57-1335 <*>
1939 V54 N2 P62-69 I-832 <*>
1939 V54 N5 P153-156 57-758 <*>
1940 V55 N5 P141-143 66-11170 <*>
1940 V56 N1 P14-21 57-2248 <*>
1940 V56 N4 P104-111 57-2025 <*>
1940 V56 N4 P112-118 57-310 <*>
1941 P54-56 AL-87 <*>
1941 V57 N2 P40-47 66-12385 <*>
1941 V57 N2 P56-60 57-2160 <*>
1942 V56 N4 P97-105 57-2159 <*>
1942 V59 N1 P1-10 57-954 <*>
1942 V60 P67-73 66-14393 <*>
1942 V60 P155-160 AL-287 <*>
1942 V60 N1 P2-4 57-973 <*>
1943 V61 N1 P1-12 57-348 <*>
1943 V61 N6 P161-163 57-38 <*>
1943 V62 P14-15 AL-295 <*>
1953 V61 N2 P53-56 57-45 <*>
1954 V63 N5 P126-133 57-2595 <*>
1956 V65 N1 P4-15 65-12812 <*>
1956 V65 N3 P74-76 57-3140 <*>
1956 V65 N3 P77-86 C-2407 <NRC>
1959 V68 N3 P93-104 42N28G <ATS>
61-16149 <*>
1960 V69 N3 P94-103 62-18326 <*> 0
1960 V69 N4 P41-52 61-18768 <*>
1963 V72 N5 P155-163 64-71554 <=>
1963 V72 N5 P177-186 64-71554 <=>
65-10248 <*>
1964 V73 P117-120 65-17440 <*>

HODOWLA RCSLIN, AKLIMATYZACJA I NASIENNICTWO
1957 V1 N1 P31-57 61-31272 <=>
1957 V1 N2 P125-134 61-11371 <=>
1957 V1 N3 P415-430 61-31268 <=>
1957 V1 N4 P509-525 61-31277 <=>
1957 V1 N4 P527-539 61-31273 <=*>
1957 V1 N4 P559-561 61-31287 <=>
1958 V2 N1 P1-15 60-21417 <=>
1958 V2 N1 P21-66 61-31254 <=>
1958 V2 N1 P67-87 61-11399 <=>
1958 V2 N2 P157-180 61-31267 <=>
1958 V2 N2 P235-258 61-11327 <=>
1958 V2 N3 P313-332 61-31271 <=>
1958 V2 N4 P459-478 61-31275 <=>
1958 V2 N4 P549-559 61-11394 <=>
1958 V2 N4 P561-565 61-31253 <=>
1958 V2 N5 P567-577 61-11395 <=>
1958 V2 N5 P587-6C1 61-11396 <=>
1958 V2 N5 P623-631 61-31282 <=>
1959 V3 N2 P195-206 61-11398 <=>
1959 V3 N3 P277-319 61-31276 <=>
1959 V3 N3 P321-334 61-11397 <=>
1959 V3 N5 P597-636 61-31291 <=>
1959 V3 N6 P751-790 61-31274 <=>
1960 V4 N4 P449-476 11092-A <KH>
63-16936 <*>
1960 V4 N6 P636-649 61-31277 <=>
1961 V5 N3 P285-330 61-31276 <=>

HOHENKLIMAFORSCHUNGEN DES BASLER PHYSIOLOGISCHEN INSTITUTES
1945 P29-40 2008 <*>
1945 P41-49 1902 <*>
1948 P9-16 2006 <*>
1948 P41 1992 <*>
1948 P47-48 1992 <*>

HOJA TISIOLOGICA
1947 V7 N2 P69-73 2521 <*>
1948 V8 P33-47 2772 <*>
1950 V10 N1 P35-51 2495 <*>
1950 V10 N2 P159-168 2585 <*>
1950 V10 N2 P208-210 2505 <*>
1950 V10 N4 P421-425 2520 <*>
1950 V10 N4 P426-430 2916 <*>
1951 V11 N2 P134-138 2522 <*>
1952 V12 N1 P1-60 2543 <*>

HOKKAIDO DAIGAKU KOGAKUBU KENKYU HOKOKU
1953 N8 P50-65 T-1933 <INSD>

HOKKAIDO DAIGAKU KOGAKUBU KIYO
1947 V8 N1 P185-190 AL-631 <*>

HOKKAIDO DAIGAKU RIGAKUBU KIYO. DAI-2-RUI BUTSURIGAKU
SEE JOURNAL OF THE FACULTY OF SCIENCE, HOKKAIDO UNIVERSITY. SERIES 2. PHYSICS

HOKKAIDO DAIGAKU RIGAKUBU KIYO. DAI-3-RUI. KAGAKU
SEE JOURNAL OF THE FACULTY OF SCIENCE, HOKKAIDO UNIVERSITY. SERIES 3. CHEMISTRY

HOKKAIDO DAIGAKU RIGAKUBU KIYO. DAI-4-RUI. CHISHITSU-KOBUTSU GAKU
SEE JOURNAL OF THE FACULTY OF SCIENCE, HOKKAIDO UNIVERSITY. SERIES 4. GEOLOGY AND MINERALOGY

HOKKAIDO DAIGAKU RIGAKUBU KIYO. DAI-5-RUI. SHOKUBUTSUGAKU
SEE JOURNAL OF THE FACULTY OF SCIENCE, HOKKAIDO UNIVERSITY. SERIES 5. BOTANY

HOKKAIDO DAIGAKU RIGAKUBU KIYO. DAI-6-RUI. DOBUTSUGAKU
SEE JOURNAL OF THE FACULTY OF SCIENCE, HOKKAIDO UNIVERSITY. SERIES 6. ZOOLOGY

HOKKAIDO DAIGAKU RIGAKUBU KIYO. DAI-7-RUI. CHIKYUBUTSURIGAKU
SEE JOURNAL OF THE FACULTY OF SCIENCE, HOKKAIDO UNIVERSITY. SERIES 7. GEOPHYSICS

HOKKAIDO DAIGAKU SHOKUBAI KENKYUSHO KIYO
1966 N3 P196-208 ICE V6 N4 P730-736 <ICE>
1966 N3 P209-221 ICE V6 N4 P700-706 <ICE>

HOKKAIDO DAIGAKU SUISANGAKUBU KENKYU IHO
1951 V2 N1 P31-42 I-704 <*>
1952 V2 N4 P293-296 I-705 <*>
1952 V2 N4 P297-301 I-706 <*>
1956 V6 N4 P266-270 57-830 <*>
1956 V6 N4 P320-324 58-943 <*>
1956 V6 N4 P336-340 58-1209 <*>

HOKKAIDO IGAKU ZASSHI
1938 V16 N6 P1332-1342 57-118 <*>
1942 V20 P129-142 64-14655 <*>
1942 V20 P442-448 64-14658 <*>
1942 V20 P1620 64-14656 <*>

HOKKAIDO-KU SUISAN KENKYU-SHO KENKYU HOKOKU
1956 N14 P89-95 60-19847 <=*>
1958 N18 P11-22 59-15963 <*> 0

HOKKAIDO NOGYO SHIKENJO IHO
1956 N69 P103-114 66-14655 <*$>

HOKKAIDO MEDICAL JOURNAL
SEE HOKKAIDO IGAKU ZASSHI

HOKURIKU BYOGAICHU KENKYU KAIHO
1964 N12 P74-76 66-14207 <=$>

HOKURIKU NOGYO SHIKENJO HOKOKU
1961 N2 P1-16 66-11541 <*>

Citation	Reference
1964 N12 P57-58	66-13362 <*>

HOLZ ALS ROH-UND WERKSTOFF

Citation	Reference
1939 N5 P125-131	II-626 <*>
1940 V3 P397-407	64-20417 <*> 0
	8770 <IICH>
1940 V3 N2 P43-45	AL-258 <*>
1940 V3 N10 P305-321	3037 <*>
1951 N6 P232-242	2445 <*>
1951 V9 P260-270	8774 <IICH>
1951 V9 N1 P11-20	AL-756 <*>
1951 V9 N2 P56-62	I-165 <*>
1951 V9 N2 P62-71	I-51 <*>
1951 V9 N3 P84-97	I-145 <*>
1951 V9 N5 P173-175	59-10581 <*> 0
1951 V9 N5 P176-181	II-627 <*>
1951 V9 N6 P216-224	I-341 <*>
	II-627 <*>
1951 V9 N9 P333-334	I-185 <*>
1951 V9 N11 P427-431	60-10921 <*>
1952 P62-	1321 <*>
1952 V10 P121-134	I-330 <*>
1952 V10 P138-144	II-592 <*>
1952 V10 N3 P92-94	C-6145 <NRC>
1952 V10 N4 P134-138	I-312 <*>
1952 V10 N4 P157-170	I-164 <*>
1952 V10 N5 P201-207	II-791 <*>
1952 V10 N6 P229-238	AL-350 <*>
1952 V10 N9 P341-352	II-959 <*>
	1687 <*>
1952 V10 N11 P415-421	AL-371 <*>
1953 P389-392	II-216 <*>
1953 V10 N5 P187-197	I-328 <*>
1953 V10 N6 P239-244	I-328 <*>
1953 V10 N7 P269-278	I-328 <*>
1953 V11 P45-50	II-958 <*>
1953 V11 N4 P166-167	I-143 <*>
1953 V11 N5 P174-175	AL-459 <*>
1953 V11 N10 P377-382	64-20375 <*>
1953 V11 N11 P417-420	II-815 <*>
1954 P7-15	2077 <*>
1954 P213-223	II-482 <*>
1954 P312-316	I-973 <*>
1954 P334-342	1960 <*>
1954 V12 P233-241	2896 <*>
	8765 <IICH>
1954 V12 P267-270	2897 <*>
1954 V12 P308-312	57-2766 <*>
1954 V12 P329-334	63-01237 <*>
1954 V12 N2 P64-67	57-658 <*>
1954 V12 N4 P117-134	1777 <*>
1954 V12 N6 P223-226	64-20521 <*> 0
1954 V12 N6 P233-241	64-20515 <*> 0
1954 V12 N10 P391-402	2235 <*>
1954 V12 N11 P413-418	1489 <*>
1954 V12 N12 P453-454	1524 <*>
1954 V12 N12 P465-471	CSIRO-2837 <CSIR>
	1778 <*>
1955 P5-20	2405 <*>
1955 P245-	57-1724 <*>
1955 V13 P85-91	2360 <*>
1955 V13 P91-94	2362 <*>
1955 V13 P212-215	2367 <*>
	57-809 <*>
1955 V13 N2 P70-75	60-18450 <*>
1955 V13 N3 P85-91	2895 <*>
1955 V13 N5 P178-185	CSIRO-3039 <CSIR>
	57-194 <*>
1955 V13 N7 P245-249	57-2544 <*>
1955 V13 N8 P301-312	CSIRO-3059 <CSIR>
	57-655 <*>
1955 V13 N9 P323-338	II-1018 <*>
1955 V13 N12 P457-461	57-2723 <*>
1956 P162-172	57-191 <*>
1956 P172-181	57-198 <*>
1956 V14 P95-100	C-2224 <NRC>
	57-134 <*>
1956 V14 N1 P1-8	CSIRO-3521 <CSIR>
	57-3403 <*>
1956 V14 N2 P41-47	1343 <*>
1956 V14 N3 P92-93	57-1612 <*>
1956 V14 N3 P105-113	59-15652 <*> 0
1956 V14 N5 P161-162	C-2227 <NRC>
	57-2302 <*>
1956 V14 N5 P162-171	C-2250 <NRC>
1956 V14 N5 P172-181	C-2251 <NRC>
1956 V14 N6 P208-210	62-00075 <*>
1956 V14 N6 P218-222	C-2244 <NRC>
	57-197 <*>
1956 V14 N8 P285-295	59-15270 <*>
1956 V14 N9 P339-352	58-540 <*>
1956 V14 N10 P403-407	CSIRO-3811 <CSIR>
	58-2072 <*>
1956 V14 N11 P417-424	59-10100 <*>
1956 V14 N11 P424-439	C-2341 <NRC>
	57-1989 <*>
1956 V14 N11 P442-447	C-2305 <NRC>
1956 V14 N12 P466-472	C-2467 <NRC>
	57-3291 <*>
1956 V14 N12 P475-482	C-2465 <NRC>
	57-3102 <*>
1957 V15 P210-213	58-750 <*>
1957 V15 P247-252	60-00610 <*>
1957 V15 P449-453	57-3404 <*>
1957 V15 N2 P80-86	C-2489 <NRC>
	57-3042 <*>
1957 V15 N2 P96-109	58-541 <*>
1957 V15 N4 P159-170	C-2525 <NRC>
	57-3506 <*>
1957 V15 N5 P210-213	63-16554 <*>
1957 V15 N7 P281-297	58-1626 <*>
1957 V15 N10 P418-423	59-10174 <*>
1957 V15 N11 P449-453	58-2528 <*>
1957 V15 N11 P453-468	58-1735 <*>
1957 V15 N12 P500-511	59-20051 <*>
1958 V16 P138-145	58-2075 <*>
1958 V16 N4 P132-137	59-10477 <*> 0
1958 V16 N6 P215-220	C-3108 <NRCC> 0
1958 V16 N7 P274-276	63-10944 <*> 0
1958 V16 N9 P335-340	63-10945 <*>
1958 V16 N9 P340-346	59-15361 <*>
1958 V16 N11 P221-226	C-3099 <NRCC> 0
1958 V16 N12 P459-466	63-18519 <*>
1959 V17 P205-212	60-16641 <*> 0
1959 V17 P226-230	60-00813 <*>
1959 V17 N1 P1-9	60-16492 <*>
	61-20488 <*> 0
1959 V17 N2 P44-54	60-16493 <*>
1959 V17 N2 P54-61	60-00612 <*>
1959 V17 N3 P86	61-10761 <*>
1959 V17 N5 P165-171	61-10542 <*> 0
1959 V17 N5 P171-178	62-16630 <*> 0
1959 V17 N5 P205-212	63-14397 <*> 0
1959 V17 N8 P319-326	60-00809 <*>
1959 V17 N9 P341-351	60-00808 <*>
1959 V17 N9 P364-376	C-3491 <NRCC> 0
1959 V17 N10 P384-396	C-3395 <NRCC> 0
1960 V18 P325-331	C-3628 <NRCC> 0
1960 V18 N1 P19-25	61-00664 <*>
1960 V18 N5 P171-180	C-3489 <NRCC>
1960 V18 N8 P304-308	C-4456 <NRC> 0
1960 V18 N10 P391-399	61-20489 <*> 0
1960 V18 N11 P415-422	65-00378 <*>
1961 V19 P479-482	C-4998 <NRC>
1961 V19 N4 P159-167	62-01251 <*>
1961 V19 N10 P381-394	62-01577 <*>
1961 V19 N10 P420-421	62-01580 <*>
1961 V19 N11 P429-434	63-00127 <*>
1961 V19 N12 P489-494	C-3974 <NRC>
1962 V20 N1 P19-26	63-16784 <*>
1962 V20 N1 P27-38	63-10946 <*>
1962 V20 N5 P182-185	63-01234 <*>
1962 V20 N5 P189-195	63-01231 <*>
1962 V20 N7 P252-259	63-01561 <*>
1962 V20 N8 P292-303	63-01558 <*>
1962 V20 N8 P303-314	64-00089 <*>
1962 V20 N9 P347-351	63-01406 <*>
1962 V20 N9 P364-368	63-18523 <*>
1962 V20 N10 P381-392	64-00088 <*>
1962 V20 N10 P393-397	63-01663 <*>

1963 V21 N1 P14-19 64-00255 <*>
1963 V21 N1 P19-22 63-01564 <*>
1963 V21 N2 P47-61 C-5219 <NRC> 0
1963 V21 N2 P62-64 C-4497 <NRCC>
1963 V21 N4 P135-144 65-00104 <*>
1963 V21 N9 P337-345 65-00235 <*>
1963 V21 N11 P441-446 C-4927 <NRC> 0
1964 V22 N2 P47-51 C-5329 <NRC>
1964 V22 N2 P51-57 65-00106 <*>
1964 V22 N3 P107-113 C-4965 <NRC> 0
1964 V22 N4 P129-139 C-5408 <NRC>
1964 V22 N11 P413-418 C-5726 <NRC>
1965 V23 P271-284 C-6012 <NRC>
1965 V23 P394-396 C-6047 <NRC>
1965 V23 N10 P420-421 C-5985 <NRC>

HOLZFORSCHUNG
1948 V2 N4 P97-127 AL-681 <*>
1950 V4 N3/4 P79-107 63-18554 <*> 0
1953 V7 N1 P12-18 II-625 <*>
1954 V8 N4 P97-103 59-10131 <*>
1955 V9 P15-17 57-1785 <*>
1955 V9 N1 P5-10 1372 <*>
1955 V9 N2 P33-48 C-2248 <NRC>
57-113 <*>
1955 V9 N5 P129-140 2364 <*>
1955 V9 N6 P167-171 59-10127 <*>
1956 V10 P80-82 C2215 <NRC>
1956 V10 N1 P1-6 59-10129 <*>
1956 V10 N1 P12-18 59-10132 <*>
1956 V10 N3 P69-75 CSIRO-3318 <CSIR>
57-656 <*>
1957 V11 N1 P16-18 65-63439 <=$>
1957 V11 N2 P47-55 60-00812 <*>
1958 V11 N5/6 P169-174 59-15287 <*>
1958 V12 N3 P65-73 59-15360 <*>
1959 V13 N1 P16-20 60-14162 <*> 0
1959 V13 N5 P137-148 60-00609 <*>
1960 V14 N2 P52-56 61-00663 <*>
1962 V16 N3 P65-74 65-00337 <*>
1964 V18 P381- C-5372 <NRC>
1964 V18 N1/2 P33-38 65-12174 <*>

HOLZFORSCHUNG UND HOLZVERWERTUNG
1957 V9 N2 P21-24 58-1363 <*>
58-542 <*>
59-10855 <*>
1958 V10 N4 P57-65 C-2992 <NRCC> 0
1960 V12 N6 P115-119 C-3814 <NRC>
1961 V13 N1 P11-17 62-14828 <*>
1961 V13 N3 P41-49 62-16073 <*> 0
1963 V15 N5 P88-101 396 <TC>
1963 V15 N5 P110-111 65-00248 <*>
1963 V16 N4 P70-77 64-18170 <*>

HOLZINDUSTRIE
1953 V6 N10 P295-299 58-1277 <*>
1956 V9 N9 P238-241 59-10181 <*>
1957 V10 N7 P229-233 59-15251 <*> 0
1959 V12 N3 P77-79 59-21153 <=*> 0

HOLZSCHWELLE
1955 N17 P13-20 II-801 <*>
1959 N29 P1-23 60-10932 <*>
1962 V57 N39 P1-11 63-01410 <*>
1963 V58 N42 P1-15 65-00389 <*>

HOLZTECHNIK
1951 V31 N6 P164-167 I-148 <*>
1951 V32 N1 P10-12 I-1038 <*>

HOLZTECHNOLOGIE
1962 V3 N2 P145-149 63-01238 <*>
1963 V4 N1 P23-31 65-00147 <*>
1963 V4 N3 P222-228 65-00312 <*>
1964 V5 P66-68 SPEC NO NS-515 <TTIS>

HOLZZENTRALBLATT
1950 N86 P938-940 II-264 <*>
1952 N135 P1857-1858 59-15250 <*> 0
1953 V79 N105 P1127-1128 1065 <*>
1954 N117 P1371- 57-657 <*>
1954 V80 N108 P1266- 2236 <*>
1954 V80 N136 P1602- 2894 <*>
1954 V80 N136 P1602 60-14131 <*> 0
1956 N86 P1073- 57-2386 <*>
1957 V83 N125 P1521-1524 60-16513 <*> 0
1960 V86 N85 P1195-1197 63-16783 <*>
1960 V86 N101 P1407-1408 63-18524 <*> 0
1960 V86 N124 P1734 PT.1 62-00217 <*>
1960 V86 N127 P1791-1794 PT.2
62-00217 <*>
1960 V86 N130 P1828-1830 PT.3
62-00217 <*>
1960 V86 N155 P1-12 62-14811 <*>
1961 V87 N28 P435-438 C-5001 <NRC>
1961 V87 N32 P495-496 63-10164 <*>

HONVEDORVOS
1960 V12 N4 P335-342 62-19398 <*=>
1961 V13 N1 P73-79 62-19396 <*=>
1961 V13 N1 P80-84 62-19397 <*=>

HOPITAL. PARIS
1923 V11 P100-104 T-1951 <INSD>

HOPPE-SEYLER'S ZEITSCHRIFT FUER PHYSIOLOGISCHE CHEMIE
1884 V8 N4 P299-305 59-20763 <*>
1901 V33 P171-176 57-417 <*>
1901 V33 P412-416 57-418 <*>
1909 V61 P112-118 58-1349 <*>
1910 V65 N4 P318-322 II-946 <*>
1916 V97 P264-268 57-297 <*>
1916 V98 P1-10 61-00148 <*>
1924 V132 N4/6 P181-237 59-20759 <*> P
1927 V167 P91-114 57-383 <*>
1930 V187 P229-237 59-20823 <*> P
1930 V188 P219-224 59-20758 <*>
1931 V196 P169-186 59-20740 <*>
1931 V202 P116-127 59-20739 <*> P
1932 V205 N3/4 P99-114 57-2936 <*>
1932 V209 P207-210 57-294 <*>
1933 V222 P39-43 58-1353 <*>
1934 V224 P17-25 57-381 <*>
1934 V224 P26-33 57-382 <*>
1934 V229 P213-218 63-14870 <*>
1935 V233 N5/6 P257-264 63-20292 <*>
1935 V235 P8 58-1780 <*>
1935 V235 P235-245 63-14871 <*>
1936 V239 P167-168 59-15306 <*>
1936 V240 P43-45 3222 <*>
1937 V246 P181-193 59-20757 <*> P
1939 V257 P161-172 61-00638 <*>
1939 V258 P117-120 59-15417 <*>
1939 V258 N2/3 P57- 57-404 <*>
1939 V259 N2/4 P201-203 57-542 <*>
1939 V261 P240-248 57-96 <*>
1939 V262 N3/5 P4-6 57-281 <*>
1942 V273 P115-117 1594 <*>
1944 V280 N3/4 P76-87 64-20553 <*> 0
1948 V283 P27-30 64-10788 <*>
1948 V283 P31-34 63-14564 <*>
1949 V284 N1/6 P211-215 57-2965 <*>
1950 V285 P182-200 61-00905 <*>
1950 V285 P216-219 61-00614 <*>
1951 V287 P1-8 3121 <*>
1951 V287 P141-147 61-00988 <*>
1951 V288 N2/3 P123-124 2216 <*>
1951 V288 N4/6 P200-215 59-10589 <*>
1952 V290 P246-251 61-00615 <*>
1952 V291 P16-23 63-14964 <*>
1952 V291 P245 58-93 <*>
1953 V292 P32-50 CSIRO-2973 <CSIR>
1953 V293 P83-88 61-00911 <*>
1953 V293 P222-229 64-00440 <*>
1953 V295 P119-128 63-18478 <*> C
1953 V295 P164-173 II-480 <*>
1953 V295 P278-285 I-974 <*>
1954 V296 P67-73 64-00439 <*>

1954 V296 P246-256 30P60G <ATS>
1954 V298 P27-33 1601 <*>
1954 V298 N1/2 P1-26 65-12601 <*>
1954 V298 N3/5 P169-184 65-12569 <*>
1955 V299 P227-234 64-00437 <*>
1955 V299 N1/2 P66-73 62-00530 <*>
1955 V299 N1/2 P74-84 62-00606 <*>
1955 V300 P167-173 64-00430 <*>
1955 V301 P259-269 7G8G <ATS>
1955 V301 N1/2 P70-77 59-20753 <*> 0
1955 V301 N4/6 P210-223 65-12600 <*>
1955 V302 P10-19 57-102 <*>
1955 V302 P20-28 65-00116 <*>
1955 V302 P236-252 II-396 <*>
1956 V304 P166-172 1307 <*>
1956 V304 P221-231 58-2514 <*>
1956 V304 N1 P21-25 1316 <*>
1956 V305 P53-60 1481 <*>
1956 V305 P132-142 57-56 <*>
1956 V305 N4/6 P192-195 61-16532 <*>
1956 V306 P123-131 59-19003 <+*> 0
1956 V306 P143-144 9767-A <KH>
1957 V306 P145-153 58-2513 <*>
1957 V307 N2/6 P112-123 61-10598 <*>
1957 V308 P49-50 58-20 <*>
1957 V308 P188 58-2513 <*>
1958 V309 N46 P226-227 59-17092 <*>
1958 V311 P19-28 62-00676 <*>
1958 V311 N1/3 P41-45 60-14176 <*>
1958 V312 N4/6 P255-263 60-17504 <=*>
1958 V313 P97-108 66-12158 <*>
1959 V314 P262-275 62-00436 <*>
1959 V315 P86-89 61-00900 <*>
1959 V316 P45-60 65-00403 <*> P
1959 V316 N3/6 P157-163 65-14810 <*> 0
1959 V317 P131-143 62-00068 <*>
1959 V317 P276-280 63-20977 <*>
1960 V318 N1/2 P33-55 64-20033 <*> 0
1960 V320 P251-257 63-00202 <*>
1960 V321 P107-113 63-16483 <*>
1960 V322 P135-141 AEC-TR-5090 <*>
1960 V322 P258-266 63-00067 <*>
1961 V323 P116-120 62-00438 <*>
1961 V323 P129-144 65-10965 <*>
1961 V323 P199-210 63-01132 <*>
1961 V323 P236-248 11687 <K-H>
63-16889 <*>
1961 V323 P249-263 11547-B <KH>
63-16984 <*>
1961 V323 P278-284 AEC-TR-5089 <*>
1961 V324 P250-253 C-4724 <NRC>
1961 V324 P254-261 62-00636 <*>
1961 V325 P229-241 63-16657 <*>
1961 V325 P251-259 62-00299 <*>
1961 V326 N5/6 P177- AEC-TR-5418 <=*>
1962 V327 P135-143 NP-TR-1092 <*>
1962 V329 P31-39 63-00158 <*>
1962 V331 P118-123 63-01138 <*>
1963 V331 P95-104 63-01620 <*>
1963 V333 P133-139 66-12959 <*>
66-15718 <*> 0
1964 V335 N26 P277-279 65-14771 <*>

HORUMON TO RINSHO
1962 V10 N11 P843-849 70S89J <ATS>
1963 V11 N7 P658-664 87S83J <=$>

HOSHI YAKKA DAIGAKU KIYO
1958 V7 P17-25 C-4561 <NRC>

HOSPITAL. RIO DE JANEIRO
1946 V30 P729-754 2599 <*>
1952 V42 N1 P1-10 57-392 <*>
1953 V44 P21-32 AL-920 <*>
1954 V45 N5 P647-651 1941 <*>
1958 V54 P75-82 60-16924 <*>
1958 V54 N2 P251-256 59-10457 <*>
1959 V55 N4 P553-561 64-14600 <*>
9856-B <K-H>
1960 V57 N1 P73-80 60-17769 <*+> 0
1960 V58 N1 P11-22 63-18731 <*>
1960 V58 N1 P111-117 64-18655 <*>
1960 V58 N5 P959-964 61-18581 <*>
1961 V59 N1 P133-138 63-18490 <*>
1961 V60 N5 P581-594 63-00508 <*>
1962 V61 N2 P331-335 62-18049 <*>
1964 V66 N3 P625-633 66-10991 <*> 0

HOSPITALSTIDENDE
1932 V75 N1 P84-95 57-1878 <*>

HOUILLE BLANCHE
1946 V1 P257-263 I-775 <*>
1946 V1 P393-405 I-902 <*>
1947 V2 P41-47 I-902 <*>
1948 N5 P408-415 I-7 <*>
1948 V3 N6 P527-536 I-155 <*>
1949 P393-404 SPEC.A AL-583 <*>
1949 P393-404 3329 <*>
1951 N2 P147 II-836 <*>
1951 V6 P243-252 SPEC A II-842 <*>
1951 V6 P620-629 SPEC B II-1009 <*>
1951 V6 P59-64 I-90 <*>
1952 V6 P809-835 58-1216 <*>
1952 V7 P513-531 SPEC B I-708 <*>
1952 V7 P554-566 I-279 <*>
59-10552 <*>
1952 V7 P809-835 I-623 <*>
1953 P130-131 SPEC A 58-1000 <*>
1953 N4 P125-131 58-999 <*>
1953 V8 P346-359 II-10 <*>
1953 V8 N6 P815-829 II-434 <*>
1954 N4 P449-460 57-1165 <*>
1954 N4 P481-511 58-1218 <*>
1954 N5 P590-607 57-1165 <*>
1954 V9 N6 P811-822 58-1217 <*>
1955 V10 P332-339 SPEC A 332-339 <*>
1955 V10 P340-344 SP. NC 62-16757 <*>
1955 V10 N3 P392-407 1769 <*>
1956 P134 C-2245 <NRC>
1956 P144-172 C-2252 <NRC>
1956 V11 P575-607 T-2439 <INSD>
1956 V11 N6 P837-842 N65-17299 <=$>
1957 V12 N6 P903-919 87K23F <ATS>
1958 N2 P148-179 36L31F <ATS>
1958 V13 N3 P256-269 99L31F <ATS>
1960 N3 P247-267 64-00203 <*>
1961 V16 N1 P66-80 65-29159 <*>
1962 N1 P64-77 62-01366 <*>
1963 V18 N2 P143-158 69Q71F <ATS>

HOUTWERELD
1956 V10 N2 P3 58-2527 <*>

HRVATSKI SUMARSKI LIST
1952 V76 N7 P229-239 60-21635 <=>
1958 V82 N5/6 P163-176 60-21636 <=>

HSIN HUA PAN YUEH KAN
1957 N4 P119-122 63-15112 <=>
1957 N5 P57-58 63-15112 <=>

HUA HSUEH. FORMOSA
1961 P1-11 63-14030 <*>

*HUA HSUEH HSUEH PAO
1957 V23 P30-38 57-3134 <*>
62-18804 <=*> 0
1957 V23 P90-98 61-27606 <*=>
1957 V23 P330-339 29P62CH <ATS>
1957 V23 P438-446 66-10705 <*> 0
1958 V24 N5 P368-376 63-00619 <*>
63-19248 <=*>
1959 V25 N5 P1-4 60-11364 <=>
1960 V26 N3 P117-167 62-33040 <=>
1960 V26 N3 P169-177 62-33040 <=>
1962 V28 P75-79 66-11462 <*>
1963 V29 P307-312 66-11444 <*>
1964 V30 P211-213 66-11461 <*>
1964 V30 N4 P429-431 AD-637 099 <=$>

1964 V30 N5 P458-477 65-32030 <=$>

HUA HSUEH KUNG YEH
1957 N10 P49-51 61-20889 <*=>
62-16680 <=*>
1958 N5 P4-7 59-11460 <=>
1958 N5 P16-17 59-11460 <=>
1958 N5 P19-20 59-11460 <=>
1958 N5 P25-32 59-11460 <=>
1958 N12 P2-3 59-13324 <=>
1958 N12 P21-22 59-13324 <=>
1958 N13 P1-2 59-13324 <=>
1958 N13 P7-9 59-13324 <=>
1958 N13 P12-13 59-13324 <=>
1958 N13 P18-20 59-13324 <=>
1958 N13 P23-24 59-13324 <=>
1958 N13 P28-29 59-13324 <=>
1958 N13 P31-32 59-13324 <=>
1958 N14 P1-4 59-11466 <=>
1958 N14 P7-8 59-11466 <=>
1958 N15 P2-5 59-13496 <=>
1958 N15 P8-9 59-13496 <=>
1958 N15 P15-16 59-13496 <=>
1958 N15 P25-30 59-13496 <=>
1958 N16 P6-9 59-13496 <=>
1958 N16 P14-15,26-28 59-13496 <=>
1959 N1 P5-7 59-13418 <=>
1959 N1 P14-15 59-13418 <=>
1959 N1 P36-39 59-13418 <=>
1959 N1 P42-44 59-13418 <=>
1959 N1 P44-46 59-13418 <=>
1959 N1 P54-55 59-13418 <=>
1959 N1 P62-64 59-13418 <=>
1959 N2 P22-31 59-13694 <=>
1959 N3 P1-5 59-13694 <=>
1959 N3 P14-15 59-13694 <=>
1959 N3 P23-24 59-13694 <=>
1959 N17 P8-10 60-11357 <=>
1959 N17 P14-16 60-11357 <=>
1959 N18 P1-13 60-11357 <=>
1959 N18 P18-29 60-11357 <=>
1959 N18 P42 60-11357 <=>
1960 N2 P28-37 61-28872 <=*>
1960 N2 P42-46 61-28872 <=*>
1960 N4 P9-19 62-23480 <=> 0
1960 N4 P22-34 62-23480 <=> 0
1960 N4 P37-40 62-23480 <=> 0
1960 N5 P15 61-27936 <=>
1960 N5 P37-39 61-27936 <=>
1960 N5 P47 61-27936 <=>
1960 N6 P34 61-27936 <=>
1960 N6 P45-47 61-27936 <=>
1960 N7 P30-31 61-21680 <=>
1960 N7 P34 61-21680 <=>
1960 N7 P38-50 61-21680 <=>
1960 N8 P13 61-23138 <=>
1960 N8 P26 61-23138 <=>
1960 N8 P34 61-23138 <=>
1960 N8 P40-49 61-23138 <=>
1960 N10 P25 61-21602 <=>
1960 N10 P36-39 61-21602 <=>

HUA HSUEH SHIH CHIEH
1959 V14 P287-289 61-20518 <*=>
1959 V14 P292-293 33T93CH <ATS>
1959 V14 N3 P106-108 64-10591 <=*$>
1959 V14 N8 P398-399 63-23403 <=*$>
1959 V14 N10 P453 61-11592 <=>
1959 V14 N10 P455 61-11592 <=>
1959 V14 N10 P457 61-11592 <=>
1959 V14 N10 P461-464 61-11592 <=>
1959 V14 N10 P467-469 61-11592 <=>
1959 V14 N10 P476-479 61-11592 <=>
1959 V14 N10 P486 61-11592 <=>
1959 V14 N10 P492 61-11592 <=>
1959 V14 N10 P502 61-11592 <=>
1964 V19 N10 P435-448 65-30428 <=$>
1964 V19 N10 P472-474 65-30428 <=$>
1964 V19 N10 P486-487 65-30428 <=$>

HUA HSUEH TUNG PAO
1959 N10 P3-12 60-31096 <=>
1959 N10 P26 60-31096 <=>
1960 N2 P18-20 65-14895 <*>
1961 N5 P7-9 63-21 <=>
1962 N1 P1-2 62-33288 <=>
1962 N1 P31-38 64-11519 <=>
1962 N1 P39-44 62-33288 <=>
1962 N2 P49-54 62-33605 <=>
1962 N2 P97-102 65-11485 <*> 0
1962 N3 P52-53 62-33605 <=>
1962 N4 P49-52 62-33605 <=>
1962 N5 P50-53 65-14896 <*>
1962 N11 P27-34 65-13320 <*>
1962 N12 P41-47 63-31387 <=>
1963 N2 P26-33 64-21173 <=>
1963 N2 P38-40 AD-619 313 <=$>
1963 N2 P65 AD-625 794 <=*$>
1963 N3 P15-23 64-21073 <=>
1963 N5 P1-11 64-21407 <=>
1963 N5 P53-56 64-21407 <=>
1963 N5 P284-290 97S81CH <ATS>
1963 N5 P293-301 AD-618 007 <=$>
1964 N2 P1-7 AD-631 835 <=$>
1964 N2 P52-53 64-31475 <=>
1964 N2 P53-55 64-31475 <=>
1964 N2 P55-56 64-31475 <=>
1964 N3 P55-58 64-31535 <=>
1964 N3 P58-59 64-31535 <=>
1964 N3 P62 64-31535 <=>
1964 N4 P238-242 21S81CH <ATS>
1964 N5 P53-54 AD-637 066 <=$>
1964 N5 P53-57 64-31802 <=>
1964 N7 P11-19 65-30151 <=$>
1964 N7 P20-26 65-30151 <=$>
1964 N7 P49 65-30151 <=$>
1964 N7 P50-52 65-30151 <=$>
1964 N8 P36-44 ICE V5 N3 P498-507 <ICE>
1964 N8 P48-51 ICE V5 N2 P272-275 <ICE>
1964 N9 P61-63 65-30151 <=$>
1964 N10 P11-21 65-30944 <=$>
1964 N10 P44-49 65-30944 <=$>
1964 N10 P50-57 65-30944 <=$>
1964 N11 P22-23 65-33376 <=$>
1964 N11 P42-49 65-33376 <=$>
1964 N11 P50-54 ICE V5 N3 P440-445 <ICE>
1964 N12 P28-34 65-32934-B <=*$>
1965 N6 P1-9 65-33397 <=$>
1965 N6 P10-16 65-33472 <=$>
1965 N6 P34-44 65-33472 <=$>
1965 N8 P1-12 66-30011 <=$>
1965 N8 P13-19 65-33095 <=$>
1965 N8 P26-31 65-33095 <=$>
1965 N8 P55-57 66-30011 <=$>
1965 N9 P38-44 ICE V6 N2 P437-444 <ICE>
1965 N9 P38-56 65-33664 <=*$>
1965 N10 P24-34 65-34016 <=*$>
1965 N10 P35-38 65-33784 <=*$>

HUETTENWERK RHEINHAUSEN A.G. TECHNISCHE MITTEILUNGEN
1952 N5 P306-312 TRANS-51 <MT>

HULE MEXICANO Y PLASTICOS
1964 V19 P6-9 65-13633 <*>

HUMANITE NOUVELLE
1966 V2 N20 P1-6 66-33364 <=$>

HUTNICKE LISTY
1949 V4 N6 P169-174 60-18849 <*>
1950 V5 P102-106 60-18851 <*> 0
1950 V5 N3 P89-93 60-18850 <*>
1950 V5 N11 P454-457 60-18852 <*>
1954 V9 P272-274 AEC-TR-5010 <*>
1954 V9 N1 P2-6 63-16867 <*>
1954 V9 N3 P151-154 61-18424 <*>
1955 V10 P577-579 TS 1136 <BISI>
1955 V10 N3 P130-139 61-16230 <*> P
1955 V10 N6 P322-329 4596 <HB>

1955 V10 N9 P535-541 61-16231 <*>
1955 V10 N10 P600-605 4077 <HB>
1956 V11 N1 P1-9 <MT>
1956 V11 N3 P151-153 AEC-TR-4204 <*>
1956 V11 N4 P218-225 AEC-TR-4231 <*>
1956 V11 N4 P225- 60-10312 <*> 0
1956 V11 N5 P284-290 AEC-TR-4232 <*>
1956 V11 N6 P345-355 3861 <HB>
1956 V11 N7 P419-424 60-23015 <=*>
1956 V11 N9 P529-532 4207 <HB>
1956 V11 N12 P709-713 61-16228 <*>
1956 V11 N12 P715-721 3TM <CTT>
58-819 <*>
1957 V12 N1 P85-96 61-16229 <*>
1957 V12 N2 P125-130 5046 <HB>
1957 V12 N2 P125-131 58-1361 <*>
1957 V12 N2 P140-141 TS 1715 <BISI>
1957 V12 N3 P196-201 4023 <HB>
1957 V12 N4 P329-332 4042 <HB>
1957 V12 N6 P517-520 4560 <HB>
1957 V12 N7 P614-616 TRANS-97 <MT>
1957 V12 N7 P617-618 2081 <BISI>
1957 V12 N10 P590-600 4139 <HB>
1957 V12 N11 P989-1000 TS-1278 <BISI>
1957 V12 N12 P1077-1081 61-16232 <*>
1957 V12 N12 P1087-1093 TRANS-71 <MT>
1957 V12 N12 P1096-1103 62-14781 <*>
1958 V13 N1 P2-8 62-14527 <*>
1958 V13 N3 P199-205 TRANS-103 <MT>
1958 V13 N3 P213-220 TRANS-81 <MT>
1958 V13 N3 P226-229 TRANS-72 <MT>
1958 V13 N3 P233-241 TRANS-73 <MT>
1958 V13 N3 P287-298 1482 <BISI>
1958 V13 N5 P402-406 TRANS-09 <MT>
1958 V13 N6 P490-495 62-18211 <*>
1958 V13 N6 P512-517 TRANS-74 <MT>
1958 V13 N6 P517-526 <MT>
1958 V13 N7 P612-619 TS-1346 <BISI>
1958 V13 N7 P626-631 TS-1243 <BISI>
1958 V13 N8 P679-687 TS-1329 <BISI>
1958 V13 N8 P697-705 62-14782 <*>
1958 V13 N10 P878-881 TS 1505 <BISI>
62-14526 <*>
1958 V13 N10 P882-892 TS 1506 <BISI>
62-14174 <*>
1958 V13 N11 P1037-1050 TRANS-171 <MT>
1958 V13 N12 P1081-1086 TRANS-75 <MT>
1958 V13 N12 P1081-1087 28N54C <ATS>
1958 V13 N12 P1098-1105 TRANS-102 <MT>
1958 V13 N12 P1116-1123 TRANS-76 <MT>
1959 V14 P322-324 UCRL TRANS-774(L) <*>
1959 V14 N1 P47-54 1530 <BISI>
1959 V14 N1 P54-55 4495 <HB>
1959 V14 N1 P56-58 TRANS-77 <MT>
1959 V14 N2 P133-139 TRANS-78 <MT>
1959 V14 N2 P173-184 62-14656 <*>
1959 V14 N3 P211-214 <MT>
TS-1508 <BISI>
62-14185 <*>
1959 V14 N4 P316-319 61-20198 <*>
1959 V14 N5 P405-409 2255 <BISI>
1959 V14 N5 P469-472 TRANS-104 <MT>
1959 V14 N6 P478-484 TRANS-128 <MT>
1959 V14 N6 P489-493 2346 <BISI>
1959 V14 N6 P499-500 M.5818 <NLL>
1959 V14 N6 P507-511 TS 1509 <BISI>
1959 V14 N7 P569-573 TS 1539 <BISI>
1959 V14 N7 P598-602 TRANS-79 <MT>
1959 V14 N8 P688-692 TS-1510 <BISI>
62-14168 <*>
1959 V14 N8 P695-700 1511 <BISI>
1959 V14 N8 P743-748 TRANS-11 <MT>
1959 V14 N9 P777-786 TRANS-80 <MT>
1959 V14 N9 P829-840 62-14536 <*>
1959 V14 N10 P844-849 1948 <BISI>
1959 V14 N10 P903-904 1961 <BISI>
1959 V14 N11 P943-947 TS-1807 <BISI>
1959 V14 N11 P951-955 TRANS-17 <MT>
1959 V14 N12 P1073-1077 4773 <HB>
1959 V14 N12 P1097-1099 1669 <BISI>
1959 V14 N12 P1103-1105 1670 <BISI>
1959 V14 N12 P1105-1107 TS-1671 <BISI>
1959 V14 N12 P1108-1111 TS-1672 <BISI>
1959 V14 N12 P1115-1118 1673 <BISI>
1959 V14 N12 P1121-1124 1675 <BISI>
1959 V14 N12 P1169-1171 1676 <BISI>
1960 V15 P548-552 63-14467 <*>
1960 V15 N2 P109-119 TS 1668 <BISI>
1960 V15 N2 P120-125 TRANS-135 <MT>
1960 V15 N2 P157-166 TS-1774 <BISI>
1960 V15 N3 P179-188 TRANS-136 <MT>
1960 V15 N3 P200-207 3589 <BISI>
1960 V15 N4 P266-275 TRANS-114 <MT>
1960 V15 N4 P287-292 61-20080 <*> 0
1960 V15 N5 P355-366 2193 <BISI>
1960 V15 N5 P379-382 TRANS-153 <MT>
1960 V15 N6 P432-441 TRANS-173 <MT>
1960 V15 N6 P476-477 04N57C <ATS>
1960 V15 N6 P479-480 TRANS-154 <MT>
1960 V15 N7 P518-524 TRANS-137 <MT>
1960 V15 N7 P524-528 TRANS-138 <MT>
1960 V15 N7 P529-537 493 <MT>
1960 V15 N8 P609-615 TRANS-155 <MT>
1960 V15 N8 P619-625 B-68 <RIS>
1960 V15 N8 P657-669 TRANS-126 <MT>
1960 V15 N9 P679-686 2017 <BISI>
1960 V15 N9 P687-689 62-10333 <*> 0
1960 V15 N9 P705-710 TRANS-158 <MT>
1960 V15 N9 P720-724 TRANS-114 <MT>
1960 V15 N10 P755-762 TRANS-139 <MT>
1960 V15 N10 P778-781 TRANS-162 <MT>
1960 V15 N11 P839-842 5075 <HB>
1960 V15 N11 P842-851 5137 <HB>
1960 V15 N11 P864-867 TRANS-160 <MT>
1960 V15 N11 P867-876 TRANS-161 <MT> P
1960 V15 N12 P929-936 2301 <BISI>
1960 V15 N12 P936-945 TRANS-164 <MT>
1960 V15 N12 P945-950 2461 <BISI>
1960 V15 N12 P955-961 2464 <BISI>
1961 V16 N2 P94-102 TRANS-178 <MT>
1961 V16 N2 P120-128 TRANS-179 <MT>
1961 V16 N2 P129-135 TRANS-180 <MT>
1961 V16 N2 P135-138 63-10663 <*>
1961 V16 N3 P174-185 5433 <HB>
1961 V16 N3 P186-197 TRANS-188 <MT>
1961 V16 N5 P307-311 TRANS-198 <MT>
1961 V16 N5 P312-314 TRANS-189 <MT>
1961 V16 N5 P335-342 TRANS-200 <MT>
1961 V16 N5 P361-362 TRANS-343 <MT>
1961 V16 N6 P430-435 499 <MT>
1961 V16 N8 P565-572 TRANS-209 <MT>
1961 V16 N8 P587-588 TRANS-210 <MT>
1961 V16 N9 P637-645 TRANS-214 <MT>
1961 V16 N10 P704-710 5366 <HB>
1961 V16 N11 P803-806 TRANS-240 <MT>
1961 V16 N11 P806-810 TRANS-241 <MT>
1962 V17 N1 P19-23 TRANS-245 <MT>
1962 V17 N1 P53-58 AEC-TR-5826 <*>
1962 V17 N2 P101-110 TRANS-248 <MT>
1962 V17 N2 P111-114 TRANS-239 <MT>
1962 V17 N5 P311-318 TRANS-262 <MT>
1962 V17 N6 P383-390 TRANS-294 <MT>
1962 V17 N6 P390-395 3622 <BISI>
1962 V17 N7 P462-471 TRANS-268 <MT>
1962 V17 N7 P472-479 TRANS-269 <MT>
1962 V17 N7 P503-507 78Q67C <ATS>
1962 V17 N8 P581-582 TRANS-277 <MT>
1962 V17 N9 P617-626 5698 <HB>
1962 V17 N9 P626-629 TRANS-278 <MT>
1962 V17 N10 P705-711 TRANS-295 <MT>
1962 V17 N10 P720-724 TRANS-296 <MT>
1962 V17 N11 P761-766 5826 <HB>
1962 V17 N12 P853-856 TRANS-332 <MT>
1963 V18 P722-726 1066 <TC>
1963 V18 N1 P44-48 TRANS-306 <MT>
1963 V18 N1 P52-55 TRANS-307 <MT>
1963 V18 N2 P102-109 65-17177 <*> 0
1963 V18 N2 P850-858 TRANS-371 <MT>
1963 V18 N3 P157-166 TRANS-309 <MT>
1963 V18 N3 P185-193 4863 <BISI>

Citation	Translation
1963 V18 N3 P193-196	5921 <HB>
1963 V18 N4 P247-253	3556 <BISI>
1963 V18 N5 P349-350	TRANS-328 <MT>
1963 V18 N6 P401-403	6070 <HB>
1963 V18 N6 P425-428	TRANS-330 <MT>
1963 V18 N7 P489-499	66-14547 <*>
1963 V18 N9 P635-638	TRANS-347 <MT>
	64-14642 <*>
1963 V18 N11 P773-779	TRANS-366 <MT>
1964 V19 N1 P22-27	TRANS-388 <MT>
1964 V19 N2 P89-93	<SIC>
1964 V19 N2 P108-117	<SIC>
1964 V19 N2 P123-126	<SIC>
1964 V19 N2 P130-133	<SIC>
1964 V19 N3 P173-181	<SIC>
1964 V19 N5 P311-318	4199 <BISI>
1964 V19 N6 P382-388	3994 <BISI>
1964 V19 N6 P388-394	5880 <HB>
1964 V19 N6 P395-402	5816 <HB>
1964 V19 N8 P580-584	440 <MT>
1964 V19 N9 P633-641	441 <MT>
1964 V19 N11 P769-774	4646 <BISI>
1964 V19 N11 P794-799	TRANS-479 <MT>
1964 V19 N11 P809-815	66-11051 <*>
1964 V19 N12 P870-874	TRANS-464 <MT>
1965 V20 P473-478	NS-545 <TTIS>
1965 V20 N2 P87-92	6549 <HB>
1965 V20 N2 P111-116	6504 <HB>
1965 V20 N3 P163-169	4463 <BISI>
1965 V20 N3 P205-206	TRANS-546 <MT>
1965 V20 N4 P234-236	6550 <HB>
1965 V20 N4 P252-261	4566 <BISI>
1965 V20 N6 P411-418	4816 <BISI>
1965 V20 N7 P468-473	4610 <BISI>
1965 V20 N9 P630-636	4740 <BISI>
1965 V20 N9 P637-644	TRANS-66/23 <NLL>
1965 V20 N9 P651-658	TRANS-600 <MT>
1965 V20 N9 P658-662	4741 <BISI>
1965 V20 N11 P770-776	6790 <HB>

HUTNIK. PRAGUE

Citation	Translation
1955 V5 N7 P194-196	3869 <HB>
1956 V6 N10 P299-304	<MT>
	61-16227 <*>
1956 V6 N12 P366-371	<MT>
	TS-1179 <BISI>
1957 V7 N2 P47-50	61-16225 <*>
1957 V7 N8 P276-278	TRANS-125 <MT>
1958 V8 P171-173	TRANS-215 <MT>
1958 V8 P215	TRANS-215 <MT>
1958 V8 N2 P51-53	TS 1741 <BISI>
1959 V9 N4 P121-128	3338 <BISI>
1959 V9 N5 P163-166	1949 <BISI>
1959 V9 N6 P190-194	1823 <BISI>
1959 V9 N8 P273-276	1935 <BISI>
1959 V9 N12 P393-396	TRANS-15 <MT>
1960 V10 N1 P2-5	2063 <BISI>
1960 V10 N1 P21-24	TRANS-84 <MT>
	TS 1640 <BISI>
	62-14529 <*>
1960 V10 N2 P61-64	TRANS-96 <MT>
	5023 <HB>
1961 V11 N2 P59-74	2343 <BISI>
1961 V11 N3 P110-117	5183 <HB>
1961 V11 N4 P166-168	TRANS-190 <MT>
1961 V11 N6 P268-276	5215 <HB>
1961 V11 N9 P429-436	5224 <HB>
1961 V11 N9 P436-440	TRANS-364 <MT>
1962 V12 P586-590	3342 <BISI>
1962 V12 N1 P3-5	TRANS-272 <MT>
1962 V12 N1 P8-11	6048 <HB>
1962 V12 N2 P54-60	TRANS-244 <MT>
1962 V12 N3 P108-115	TRANS-252 <MT>
1962 V12 N3 P144-147	TRANS-253 <MT>
1962 V12 N4 P180-184	5699 <HB>
1962 V12 N5 P230-234	5714 <HB>
1962 V12 N6 P293-295	5701 <HB>
1962 V12 N7 P323-328	4649 <BISI>
1962 V12 N7 P331-335	3322 <BISI>
1962 V12 N7 P336-339	3290 <BISI>
1962 V12 N7 P339-347	3742 <BISI>
1962 V12 N9 P427-434	TRANS-370 <MT>
1962 V12 N11 P543-548	4214 <BISI>
1963 V13 N2 P53-58	TRANS-317 <MT>
1963 V13 N3 P129-133	3623 <BISI>
1963 V13 N4 P180-186	5098 <HB>
1963 V13 N5 P219-225	6071 <HB>
1963 V13 N5 P225-230	5870 <HB>
1963 V13 N6 P302-303	6095 <HB>
1963 V13 N7 P320-324	3687 <BISI>
1963 V13 N7 P328-334	TRANS-331 <MT>
1963 V13 N7 P334-337	424 <MT>
1963 V13 N8 P366-372	TRANS-335 <MT>
1963 V13 N9 P435-438	6140 <HB>
1963 V13 N9 P439-443	TRANS-354 <MT>
1963 V13 N10 P483-488	3888 <BISI>
1963 V13 N11 P544-547	6162 <HB>
1963 V13 N12 P579-583	3967 <BISI>
1964 V14 N1 P14-20	489 <MT>
1964 V14 N2 P76-81	4273 <BISI>
1964 V14 N4 P159-163	3929 <BISI>
1964 V14 N5 P229-233	TRANS-413 <MT>
1964 V14 N6 P278-283	TRANS-421 <MT>
1964 V14 N6 P293-298	422 <MT>
1964 V14 N8 P394-401	425 <MT>
1965 V15 N1 P4-7	4736 <BISI>
1965 V15 N2 P70-73	4737 <BISI>
1965 V15 N2 P79-82	6505 <HB>
1965 V15 N3 P113-116	6551 <HB>
1965 V15 N5 P234-236	TRANS-513 <MT>
1965 V15 N6 P232-233	6603 <HB>
1965 V15 N11 P481-485	6780 <HB>

HUTNIK, KATOWICE

Citation	Translation
1947 V14 N11 P519-534	3527 <BISI>
1951 V18 P393-397	II-511 <*>
	64-14137 <*>
1951 V18 P404-409	64-14149 <*> 0
1951 V18 N6 P225-227	TS 1733 <BISI>
1951 V18 N6 P229-233	TS 1734 <BISI>
1951 V18 N6 P234-235	TS 1725 <BISI>
1952 V19 N6 P197-201	60-18837 <*>
1954 V21 N10 P329-335	3625 <HB>
1955 V22 N10 P358-362	4650 <BISI>
1956 V23 N5 P202-207	57-3555 <*>
1956 V23 N11 P413-418	3948 <HB>
1957 N24 P222-227	TS-1557 <BISI>
1957 V24 N7/8 P266-271	4153 <HB>
1957 V24 N7/8 P277-282	4145 <HB>
1958 V25 P269-278	60-10076 <*> 0
1958 V25 N4/5 P142-156	59-11315 <=>
1960 V27 N3 P109-114	2064 <BISI>
1960 V27 N4 P127-137	5201 <HB>
1960 V27 N5 P182-188	2273 <BISI>
1960 V27 N5 P188-195	2397 <BISI>
1961 V28 N2 P44-49	5194 <HB>
1961 V28 N3 P85-91	3038 <BISI>
1961 V28 N4 P143-147	5704 <HB>
1961 V28 N5 P183-193	5245 <HB>
1961 V28 N5 P196-202	5252 <HB>
1961 V28 N6 P229-233	3099 <BISI>
1962 V29 N10 P378-381	5811 <HB>
1962 V29 N12 P451-457	65-00268 <*>
1963 V30 N9 P281-290	3903 <BISI>
1963 V30 N11 P359-367	4069 <BISI>
1964 V31 N3 P90-93	6347 <HB>
1964 V31 N6 P197-201	6391 <HB>
1964 V31 N12 P383-388	6552 <HB>
1965 V32 N2 P35-38	6584 <HB>
1965 V32 N4 P125-134	6640 <HB>
1965 V32 N4 P143-149	6641 <HB>
1965 V32 N5 P194-196	TRANS-576 <MT>
1965 V32 N9 P274-277	65-33622 <=$>
1965 V32 N10 P353-361	65-33800 <=$>
1965 V32 N12 P419-422	6836 <HB>

HUTOIPAR

Citation	Translation
1959 N2 P47-52	C-3466 <NRCC>
1961 V16 N5 P325-335	TRANS-199 <MT>

HWAHAK KWA HWAHAK KONGOP
1959 N5 P1-6 60-31077 <*=>
1960 V4 N2 P47-49 62-19454 <=*>
1965 N2 P85-93 ICE V6 N1 P57-63 <ICE>
1965 N6 P399-406 ICE V6 N4 P634-638 <ICE>

HYDROELECTRIC POWER. CHINESE PEOPLES REPUBLIC
SEE SHUI LI FA TIEN

HYDROGEOLOGY AND ENGINEERING GEOLOGY. CHINESE PEOPLES REPUBLIC
SEE SHUI WEN TI CHIH KUNG CH'ENG TI CHIH

HYDROGRAPHIC BULLETIN
SEE SUIRO YCHO

IGT NIEUWS
1956 V9 N10 P148- 57-1091 <*>
1956 V9 N10 P155- 57-1090 <*>
1956 V10 N9 P155-158 15 <HB>
1958 V11 N11/2 P176- 60-16914 <*> 0
1959 V12 N1 P11 60-14552 <*>
1960 V12 N13 P200-204 66-13756 <*>
1960 V12 N13 P204-206 66-13757 <*>

IRSID PUBLICATIONS
SEE PUBLICATIONS. INSTITUT DE RECHERCHES DE LA SIDERURGIE. ST. GERMAINE EN LAYE

IRSIC REPORT
SEE REPORT. INSTITUT DE RECHERCHES CE LA SIDURGIE. ST. GERMAIN-EN-LAYE

I. V. A. INGENIORSVETENSKAPSAKADEMIEN OCH DESS LABORATORIER
1943 V14 P160-165 SCL-T-263 <+*>
1946 V17 P9-16 I-585 <*>
1946 V17 P17-22 I-1053 <*>
1946 V17 N1 P5-9 I-586 <*>
1947 V18 N5 P143-149 60-10539 <*> 0
1948 V19 P134-139 <AC>
58-1783 <*>
59-10922 <*>
1956 V27 N3 P96-102 58-53 <*>

I-HSUEH SHIH YU PAO-CHIEN TSU-CHIH
1958 V2 N3 P192-197 59-13425 <=>

IDEGGYOGYASZAKI SZEMLE
1960 V13 N8 P225-234 64-21598 <=>
1961 V14 N2 P56-62 62-19604 <*=> P
1961 V14 N4 P119-123 62-19562 <=*>
1961 V14 N4 P124-127 62-19601 <=*> P
1961 V14 N8 P247-255 62-11086 <=>

IDENGAKU ZASSHI
1923 V1 N2 P81-85 65-17224 <*>
1929 V5 P140-144 65-17225 <*> 0

IDOJARAS
1950 V54 N8 P193-196 58-2469 <*>
1956 V60 P384-389 C-2654 <NRC>
1957 V61 P230-233 58-2111 <*>
1957 V61 P357-363 SCL-T-215 <=*>
1959 V63 N2 P125-128 60-31728 <*=>
1960 V64 N3 P143 62-19395 <=*>
1960 V64 N3 P182-184 62-19394 <=*>
1960 V64 N5 P264-268 62-19537 <=*>
1960 V64 N5 P276-280 AD-611 118 <=$>
1961 V65 N1 P23-29 62-19596 <=*>
1961 V65 N1 P31-34 62-19678 <*=>
1961 V65 N2 P65-79 M.5072 <NLL>
1961 V65 N2 P93-98 62-19149 <*=>
1961 V65 N3 P152-153 62-19558 <=*>
1962 V66 N1 P59-63 62-32213 <=*>
1962 V66 N6 P361-366 AD-636 572 <=$>
1963 V67 N6 P372-376 64-31781 <=>
1964 V68 N1 P45-46 65-12194 <*>
1964 V68 N5 P303-306 65-30969 <=*>
1964 V68 N5 P306-310 65-30968 <=*>
1964 V68 N6 P378-383 65-31079 <=$>
1965 V69 N1 P59-63 66-31463 <=$>
1965 V69 N2 P119-128 65-32347 <=$>
1965 V69 N5 P287-291 65-30954 <=$>
1965 V69 N4/5 P318-320 65-33918 <=*$> 0

IGAKU KENKYU
1953 V23 N9 P21-34 2743 <*>
1953 V23 N9 P35-45 1151 <*>
1954 V24 N8 P1531-1538 3439-A <K-H>
1954 V24 N8 P1539-1549 3490 <K-H>
1957 V27 N3 P603-606 58-265 <*>
1964 V34 P100-110 66-12955 <*>

IGAKU TO SEIBUTSUGAKU
1948 V14 P26-28 63-00544 <*>
1948 V14 P28-29 63-00542 <*>
1948 V14 P30-33 63-00543 <*>
1948 V14 P34-36 63-00541 <*>
1949 V14 N1 P26-28 63-26693 <$>
1949 V14 N1 P28-29 63-27096 <$>
1949 V14 N1 P30-33 63-27095 <$>
1949 V14 N1 P34-36 63-27094 <$>
1949 V14 N4 P221-222 61-00367 <*>
1950 V17 N5 P280-282 UCRL TRANS-113 <*>
1954 V32 N2 P81-83 55M38J <ATS>
1954 V32 N6 P279-281 56M38J <ATS>
1955 V35 N5 P147-151 59-15438 <*>
1964 V69 N4 P220-223 65-17383 <*>

IGIENA
1958 V7 N1 P51-59 59-11308 <=*>
1961 V9 N2 P113-119 83R74DV <ATS>
1961 V10 N1 P27-30 62-19408 <*=>
1961 V10 N1 P57-62 63-10625 <*>
1961 V10 N1 P89-91 62-19407 <=*>
1961 V13 N1 P39-48 62-19593 <=*>
1964 V13 N1 P1-10 64-31002 <=>
1966 N4 P193-201 66-32611 <=$>

IGIENE MENTALE
1959 V3 N1 P3-41 63-00614 <*>
63-27749 <$>

IGIENE MODERNA
1958 V51 N11/2 P801-820 63-16139 <*>

IKONOMICHESKA MISUL
1957 N6 P37-53 R-4325 <*>
1957 V2 N6 P69-86 R-4320 <*>
1966 V11 N8 P16-27 66-35552 <=>

IKONOMICHESKI ZHIVOT. SOFIA
1966 N5 P10 66-33247 <=$>
1966 N18 P6-7 67-30425 <=>

IKUSHUGAKU ZASSHI
1963 V13 N3 P194-195 66-12669 <*>

IMMUNITAETSFORSCHUNG
1950 V6 P106-111 65-00129 <*>
1950 V6 P130-139 65-00129 <*>

IMONO
1954 V26 N11 P581-588 64-10184 <*>
1955 V27 N1 P3-11 64-10185 <*>
1955 V27 N2 P81-91 64-10561 <*>
1955 V27 N3 P137-142 64-10187 <*>
1955 V27 N7 P380-388 66-12428 <*>
1956 V28 N1 P11-18 64-10524 <*>
1956 V28 N2 P63-70 64-10186 <*>
1956 V28 N4 P307-308 66-12429 <*>
1957 V29 N1 P11-21 66-12567 <*>
1957 V29 N6 P427-435 64-10520 <*>
1957 V29 N7 P508-514 62-11632 <=>
1958 V30 N1 P37-45 63-18623 <*>
1958 V30 N4 P231-232 4312 <HB>
1958 V30 N4 P247-249 64-10550 <*>
1958 V30 N4 P253-254 4811 <HB>
1958 V30 N4 P260-261 64-10578 <*>

1958 V30 N9 P667-668 64-10530 <*>
1958 V30 N10 P792-798 64-10533 <*>
1958 V30 N11 P880-886 64-10531 <*>
1959 V31 P97-98 66-12568 <*>
1959 V31 N1 P20-25 64-10545 <*>
1959 V31 N5 P498-508 64-10188 <*>
1959 V31 N7 P645-651 66-12433 <*>
1960 V32 N4 P261-267 5459 <HB>
64-10532 <*>
1960 V32 N9 P35-37 64-10577 <*>
66-12565 <*>
1961 V33 N3 P190-201 3652 <BISI>
1961 V33 N4 P266-271 66-12422 <*>
1961 V33 N4 P271-277 37N58J <ATS>
1961 V33 N6 P412-420 66-12425 <*>
1961 V33 N9 P644-655 66-12566 <*>
1963 V35 P220-229 3984 <BISI>
1963 V35 N4 P208-220 66-12437 <*>
1964 V36 P533-544 4388 <BISI>

L'INDUSTRIA MILANO
1935 V49 N5 P151-160 61-14933 <*>

INDUSTRIA ALIMENTARIA. PRODUSE ANIMALE.
1963 V11 N4 P111-113 63-31699 <=>

INDUSTRIA ALIMENTARIA. PRODUSE VEGETALE
1960 V11 P282-284 11816 <K-H>
1960 V11 P328-330 11249-B <K-H>
64-14538 <*>
1960 V11 N9 P282-284 64-14774 <*> 0

INDUSTRIA DELLA CARTA E DELLE ARTI GRAFICHE
1950 V4 P44-51 64-10783 <*>
1952 V6 P41-47 D-666 <RIS>
1952 V6 P77-83 D-667 <RIS>
1955 V9 P95-107 D-668 <RIS>
1955 V9 N2 P17- 57-676 <*>
1955 V9 N3 P31- 57-676 <*>
1956 V10 N1 P1-5 C-2185 <NRC>
2358 <*>
1956 V10 N1 P1- 57-3148 <*>
1956 V10 N6 P67- 57-2575 <*>
1957 V11 N10 P119 58-2121 <*>
1958 V12 N3 P25-29 59-10509 <*>
1959 V13 N12 P141-145 75N53I <ATS>
1962 V16 N11 P387-397 65-12097 <*>
1962 V16 N12 P425-432 66-10575 <*>
1963 V17 N1 P7-17 64-10167 <*> 0

INDUSTRIA DELLA CERAMICA E SILICATI
1951 V4 N4 P22-30 I-44 <*>

INDUSTRIA CHIMICA. ROMA
1926 V1 P237-240 63-10574 <*>
1929 V4 P990-999 AL-792 <*>
1931 V6 N2 P155- 1626 <*>
1932 V7 N5 P583- II-645 <*>
1934 V9 P1603-1615 I-369 <*>

INDUSTRIA CONSERVERA
1957 V32 N4 P277-280 58-2084 <*>

INDUSTRIA ITALIANA DEL CEMENTO
1934 N13 P312-320 64-16002 <*>
1952 V22 N2 P50-51 58-2546 <*>
1952 V22 N7/8 P170-179 I-74 <*>
1954 V24 N1 P14-16 AL-696 <*>
1955 V25 P145-151 57-614 <*>
1959 V29 N9 P219-223 60-14132 <*> 0
1959 V29 N10 P244-246 60-14132 <*> 0
1959 V29 N11 P276-280 60-14132 <*> 0
1959 V29 N11 P282 60-14132 <*> 0
1959 V29 N7/8 P185-189 60-14132 <*> 0
1960 V30 N9 P257-259 64-18000 <*> 0
1961 V31 N8 P385-396 63-14739 <*>
1963 V33 N1 P45-52 66-14100 <*>
1965 V35 N5 P275-286 66-10172 <*>

INDUSTRIA ITALIANA ELETTROTECNICA
1954 V3 P421-427 T-2077 <INSD>

INDUSTRIA LEMNULUI
1958 V7 N1 P5-8 62-00078 <*>
1959 V7 N12 P441-447 62-00920 <*>
1960 V9 N4 P140-143 61-28947 <=*> 0
1961 N6 P218-220 62-00921 <*>
1961 V9 N9 P331-340 62-00922 <*>
1961 V10 N12 P450-455 C-5055 <NRC> 0

INDUSTRIA LEMNULUI, CELULOZEI SI HIRTIEI
1955 N6 P237- 58-1326 <*>

INDUSTRIA MINERARIA. ROMA
1953 S11 V4 P270 AEC-TR-4287 <*>
1959 S11 V10 N9 P574- NP-TR-467 <*>

INDUSTRIA Y QUIMICA
1945 V7 P132-135 63-10986 <*>
1945 V7 P273-278 57-1928 <*>
1953 V15 N2/3 P62-63,76 AL-963 <*>
1955 V17 P78-82 58-1362 <*>
1955 V17 P104- 58-1362 <*>
1955 V17 N2 P78-82 58-731 <*>
1955 V17 N2 P104- 58-731 <*>
1957 V18 P163-165 C-2672 <NRC>
1961 V21 N4 P245-247 4435 <BISI>

INDUSTRIA SACCARIFERA ITALIANA
1939 V32 P408-410 64-20022 <*>
65-12146 <*>
1948 V41 P103-105 63-14881 <*>
1948 V41 P197-213 63-14869 <*>
1954 V47 P108-114 63-14857 <*>
1954 V47 P204-206 63-14880 <*> 0
1957 V50 P35-37 63-16356 <*>

INDUSTRIA TEXTILA
1956 V7 N1 P22-29 54N49R4 <ATS>
1957 V8 N2 P64-68 59-10242 <*>
60-18403 <*>
1958 V9 N6 P231-232 59-31054 <=>
1964 V15 N12 P673-676 42S86RU <ATS>

INDUSTRIA USOARA
1954 V1 P26-27 63-14072 <*>
1957 V4 N4 P172-175 59-15452 <*> 0
1958 V5 N2 P63-65 62-14866 <*>
64-23596 <=$>
1959 V6 N5 P177-180 62-14865 <*>
1961 V8 N1 P7-9 62-19468 <=*>

INDUSTRIA DELLA VERNICE
1950 V4 P190-193 58-1264 <*>
63-20638 <*>
1954 V8 P17-20 2108 <*>
1955 V9 P309-313 T-2336 <INSD>
1956 V10 P3-7 T-2336 <INSD>
1962 V16 P105-107 8016 <TTIS>

INDUSTRIAL PRODUCTION RESEARCH. JAPAN
SEE SEISAN KENKYU

INDUSTRIAL WATER. JAPAN
SEE KOGYO YOUSI

INDUSTRIE CERAMIQUE
1955 V463 P89-96 CSIRO-2940 <CSIR>
1957 N492 P351-352 AEC-TR-3652 <*>
1959 N510 P243-247 62-18217 <*>
1963 N549 P69-79 65-11285 <*>
1963 V555 P345-352 65-13248 <*> 0

INDUSTRIE CHIMIQUE
1940 P123-125 3323 <*>
1946 V33 N10 P177-181 63-14596 <*>
1949 V36 P116-119 T-1749 <INSD>
1957 V44 N475 P33-34 66J15F <ATS>
1959 V46 N499 P38-40 61-20892 <*>
1960 V47 P387 125-69 <STT>

1962 V49 N537 P1-5 98Q66F <ATS>

INDUSTRIE CHIMIQUE BELGE
1937 V8 P3-7 II-416 <*>
1951 V16 N3 P133-137 RCT V25 N2 P251 <RCT>
1952 P565-566 AL-171 <*>
1952 N1 P34-41 62-10411 <*> 0
1952 V17 P9-20 5309-I <K-H>
1952 V17 P652-659 II-854 <*>
1952 V17 N1 P9-20 64-16756 <*> 0
1953 V18 N8 P785-794 62-16547 <*>
1954 P29-32 TS 1575 <BISI>
1954 V20 P29-32 62-14522 <*>
1955 V20 P592-594 SPEC. 48H13F <ATS>
1955 V20 SPEC.NO. P289- 64-16502 <*>
1955 V20 P13-19 58-2124 <*>
1955 V20 N1 P35-53 57-1491 <*>
1956 V21 N10 P1053-1063 64-16823 <*>
1956 V21 N11 P1193-1202 64-16550 <*>
1956 V21 N12 P1309-1317 64-16551 <*>
1957 V22 N1 P39-51 64-16817 <*>
1957 V22 N5 P563-577 T-2428 <INSD>
1958 P608-613 SUP1 60-18508 <*> 0
1958 V23 P983-990 60-16228 <*>
1958 V23 N1 P3-14 61-16731 <*>
1959 V24 P481-485 SUP 1 <DIL>
1959 V24 P604-607 SUP 1 04M40F <ATS>
1959 V24 P608-613 SUP 1 09L37F <ATS>
1959 V24 P481-485 <DIL>
1959 V24 N7 P729-764 22M44F <ATS>
1959 V24 N7 P739-764 62-10381 <*> 0
1959 V24 N10 P1177-1188 61-10520 <*>
1960 V25 P1431-1436 AEC-TR-4850 <*>
1960 V25 N2 P127-132 61-18675 <*>
1960 V25 N4 P353-358 60-18430 <*>
1961 V26 N6 P780-787 3398 <BISI>
1962 V27 N8 P932-937 65-14117 <*>
1964 V29 N11 P1141-1164 1314 <TC>

INDUSTRIE CHIMIQUE, LE PHOSPHATE
1956 V43 P380-383 16K21F <ATS>

INDUSTRIE CHIMIQUE ET LE PHOSPHATE REUNIS
1939 P615- II-557 <*>

INDUSTRIE ELECTRIQUE
1918 V27 P444-448 57-1047 <*>
1930 V39 N922 P511-519 57-3264 <*>

INDUSTRIE DE LA PARFUMERIE
1947 V2 P83-84 5773-B <K-H>
1952 V7 P195-197 65-14985 <*>

INDUSTRIE DU PETROLE. BRUXELLES
1961 V29 N1 P36 69N50F <ATS>

INDUSTRIE DU PETROLE. PARIS
1956 V24 N7 P36-40 57-242 <*>
1956 V24 N12 P46-47 64-18376 <*>
1957 V26 N3 P63-64 70J17F <ATS>
1958 V26 N11 P6-16 65-10933 <*> P
1962 V30 P40-41 58P65F <ATS>
1962 V30 P43 58P65F <ATS>

INDUSTRIE DU PETROLE ET DE LA CHIMIE
1965 V33 N3 P83-84 66-10366 <*>
1966 V34 N364 P66-14566 66-14566 <*>

INDUSTRIE DES PLASTIQUES MODERNES
1953 V5 N3 P30-31 64-20063 <*>
1954 V6 N5 P33-39 64-20828 <*> 0
1955 V7 N7 P1-5 61-10524 <*> 0
1956 V8 N5 P48-51 57-3399 <*>
1956 V8 N6 P44-49 57-3399 <*>
1958 V10 P32-39 60-18548 <*>
1958 V10 P45-53 59-17738 <*>
1958 V10 N5 P45-53 8964 <K-H>
1958 V10 N6 P37-45 60-14073 <*>
1960 V12 N4 P18-20 65-10056 <*> 0
1961 V13 P25-30 61Q72F <ATS>
1962 V14 N3 P53-60 65-13302 <*> 0
1962 V14 N5 P49-52 66-10603 <*>
1962 V14 N9 P69-71 64-16286 <*>
1964 V16 N4 P117-123 64-18745 <*> 0

INDUSTRIE TEXTILE
1946 V63 P179 63-20537 <*>
1948 V65 N774 P333-336 61-14035 <*>
1950 V67 P106-108 64-10765 <*> 0
1950 V67 P197-208 61-16764 <*>
1952 V69 P656-658 61-14037 <*>
1952 V69 N786 P269-270 61-14036 <*>
1954 N813 P567-569 60-18396 <*> 0
1954 V71 N806 P41-42 61-14038 <*> 0
1954 V71 N809 P263-264 61-14039 <*>
1954 V71 N809 P275-276 61-14040 <*> 0
1954 V71 N812 P521-527 61-14508 <*>
1954 V71 N813 P563-565 61-14041 <*> 0
1955 V72 N820 P159-161 61-14509 <*>
1956 N838 P647-648 60-18398 <*>
1956 V73 N835 P421-424 61-14510 <*> 0
1957 V74 N843 P119-121 61-14511 <*> 0
1957 V74 N845 P277-280 61-14512 <*> 0
1958 V75 N855 P119-122 61-14513 <*> 0
61-16740 <*>
1958 V75 N857 P267-272 61-14514 <*> 0
1958 V75 N858 P339-343 61-14514 <*> 0
1958 V75 N861 P577-585 61-14515 <*> 0
1959 V76 N871 P533-537 61-14529 <*> 0
1960 V77 N884 P693-696 65-14183 <*>
1961 V78 N889 P245-246 62-14291 <*>
1961 V78 N890 P321-323 62-14269 <*>

INDUSTRIE-ANZEIGER
1953 V75 P906-914 2208 <BISI>
1953 V75 N12 P141-143 64-16547 <*> 0
1956 V78 N101 P1504-1507 61-16235 <*>
1957 P961-964 59-17784 <*>
1957 V79 N5 P21-23 61-16234 <*>
1957 V79 N81 P1217-1220 TS 1280 <BISI>
62-10500 <*>
1958 V80 P739-743 TS-1406 <BISI>
1958 V80 N51 P739-743 62-10502 <*>
1959 V81 P27-30 1763 <BISI>
1959 V81 P28-31 TS 1689 <BISI>
1959 V81 P410-412 TS 1625 <BISI>
1959 V81 N1 P410-412 62-14179 <*>
1959 V81 N8 P21-26 1762 <BISI>
1959 V81 N40 P28-31 62-14662 <*>
1959 V81 N43 P653-657 2176 <BISI>
1959 V81 N50 P18 AEC-TR-3983 <*>
1959 V81 N62 P13-21 1761 <BISI>
1959 V81 N76 P1223-1226 66-10719 <*> 0
1960 V82 P17-24 1951 <BISI>
1960 V82 N38 P583-588 TS 1776 <BISI>
1960 V82 N50 P775-780 4875 <HB>
1961 V83 N101 P29-33 5738 <HB>
1962 N60 P1511-1515 65-00188 <*>
1962 V84 N61 P1530-1533 65-14190 <*>
1963 V85 N91 P31-32 65-17179 <*>
1964 N74 P163-165 21S82G <ATS>
1964 V86 P1727-1732 4048 <BISI>
1964 V86 N6 P31-33 3913 <BISI>
1964 V86 N37 P683-689 66-14693 <*> 0
1964 V86 N80 P41-44 66-13459 <*>
1965 V87 P2485-2496 4733 <BIST>
1965 V87 P2497-2501 4734 <BIST>
1965 V87 N61 P197- NEL-TRANS-1745 <NLL>

INDUSTRIEBLATT
1959 P321-324 TS-1447 <BISI>
62-14397 <*>
1959 V59 P200-204 SCL-T-328 <*>
1961 V61 N1 P1-19 2831 <BISI>
1962 V62 N10 P623-626 3382, PT.I <BISI>
1963 V63 N4 P209-215 3532 <BISI>
1964 V64 N1 P8-12 3912 <BISI>

INDUSTRIE-LACKIER-BETRIEB
1959 V27 N3 P68-74 65-17457 <*> 0

1959 V27 N4 P97-101 65-17457 <*> 0
1960 V28 P207-216 2208 <TTIS>
1960 V28 P243-249 2208 <TTIS>
1960 V28 P249-250 2223 <TTIS>
1960 V28 P316-321 2244 <TTIS>
1960 V28 N5 P154-155 2193 <TTIS>
1961 V29 P191-194 FA56 <TTIS>
1962 V30 P109-113 2382 <TTIS>
1963 V31 P172-174 NS-165 <TTIS>
1964 V32 P69-77 NS 216 <TTIS>
1964 V32 P155-159 NS-250 <TTIS>
1964 V32 P381-391 NS-294 <TTIS>
1965 V33 P229-314 NS-504 <TTIS>

INDUSTRIELLE ORGANISATION
1959 V28 N7 P213-226 AEC-TR-4525 <*>

INDUSTRIES AGRICOLES ET ALIMENTAIRES
1953 V70 P387-394 T-1839 <INSD>
T1838 <INSD>
1953 V70 N9/10 P649-656 60-10449 <*>

INDUSTRIES ALIMENTAIRES ET AGRICOLES
1957 V74 P269-271 63-16353 <*>
1957 V74 P541-544 63-16300 <*>
1961 V78 P243-252 AEC-TR-5794 <*>

INDUSTRIES ATOMIQUES
1959 V3 N3/4 P37-46 62-32393 <=*>
1959 V3 N5/6 P69-74 59-11797 <=>
1959 V3 N5/6 P81-84 59-11798 <=>
1959 V3 N5/6 P91-95 59-11799 <=>
1959 V3 N9/10 P99-102 60-11239 <=>

INDUSTRIES DES CORPS GRAS
1945 V1 N11 P347-358 61-10292 <*>
1946 V2 P336-340 60-10519 <*>
1946 V2 P376-380 60-10520 <*>

INDUSTRIES DES PLASTIQUES
1947 V3 N4 P121-123 62-10546 <*> 0

INDUSTRIES ET TRAVAUX D'OUTRE MER
1957 V5 P711-720 59-14266 <=*> 0
1957 V5 P723-725 59-14267 <+*>

INDUSTRITIDNINGEN NORDEN
1947 V75 N8 P75-80 60-18302 <*>
1947 V75 N9 P90-94 60-18302 <*>
1947 V75 N11 P119-123 60-18302 <*>

INFORMACION DE QUIMICA ANALITICA, PURA Y APLICADA A LA INDUSTRIA
1950 V4 P127-139 I-496 <*>
1952 V6 P5-11 2495 <K-H>
1962 V16 N2 P42-51 65-12854 <*>
1963 V17 N2 P51-52 64-30061 <*>

INFORMACION DE MICROQUIMICA
1959 V3 N8 P29-37 46M46SP <ATS>

INFORMATION BULLETIN OF THE AGRICULTURAL SCIENCES
SEE NUNG YEH K'O HSUEH T'UNG HSUN

INFORMATION SNECMA
1962 N98 P73-83 AEC-TR-5390 <*>

INFORMATIONS CHIMIE. PARIS
1965 P6-7 66-14236 <*>
1965 P58 66-13436 <*>

INFORMATIONS THERAPEUTIQUES
1965 V3 P22-27 65-14986 <*>

INFORMATSIONNYI BYULLETEN ARKTICHESKII I ANTARKTICHESKII NAUCHNO-ISSLEDOVATELSKII INSTITUT
1961 N26 P36-38 62-24048 <=*> 0
1961 N28 P37-41 62-24045 <=*> 0
1961 N29 P27-29 62-24047 <=*> 0
1962 N27 P21-29 M-5601 <NLC>

INFORMATSIONNYI BYULLETEN INSTITUTA GEOLOGII ARKTIKI
1958 N12 P12-15 62-32963 <=*>
1958 N12 P16-25 63-15248 <=*>
1958 N12 P31-33 62-32962 <=*>
1959 N4 P30-33 IGR V2 N4 P327 <AGI>
1959 N16 P30-36 IGR V2 N10 P897 <AGI>

INFORMATSIONNYI BYULLETEN. MEZHDUNARODNYI GEOFIZICHESKII GOD
1958 N1 P3-111 59-11765 <=>
1958 N4 P56-57 59-15570 <+*>
61-17586 <=$>
1958 N4 P58-59 59-15571 <+*>
61-17583 <=$>
1958 N5 P3-81 59-12650 <+*> 0
1958 N5 P86-90 59-12650 <+*> C
1958 N5 P102-111 59-12650 <+*> 0
1959 N7 P3-45 60-11225 <=>
1959 N7 P42-45 61-11136 <=>
1959 N7 P57-64 60-11164 <=>
1959 N7 P65-73 60-11225 <=>
1959 N7 P68-102 61-11136 <=>
1959 N7 P103-104 60-11225 <=>
1960 N2 P37-42 63-11555 <=>
1960 N8 P7-15 61-21747 <=>
1960 N8 P42-46 61-19022 <=*>
1960 N8 P67-84 61-15262 <=*>
1961 N3 P19-23 63-19089 <=*>
1961 N3 P30-33 C-5160 <NRC>
1961 N9 P17 M.5103 <NLL>

*INFORMATSIONNYI BYULLETEN SOVETSKOI ANTARKTICHESKOI EKSPEDITSII 1955-58
1958 N1 P81-82 59-15220 <+*>
1958 N2 P31-35 63-13866 <=*>
1959 N4 P27-31 60-23059 <+*>
1959 N4 P33-36 60-23062 <+*>
1959 N4 P43-48 62-13507 <*=> 0
1959 N5 P2-34 60-23060 <+*>
1959 N5 P26-31 60-23061 <+*>
1959 N8 P8-11 63-19842 <=*$>
1959 V2 P115-156 61-23589 <=>
1960 N9 P6-9 C-3673 <NRC>
1960 N14 P24-28 61-21584 <=>
1960 N16 P323-328 63-21018 <=>
1960 N22 P32-35 64-00036 <*>
64-13754 <=*$>
1961 N25 P43-47 62-25359 <=*>
1961 N29 P49-52 62-01572 <*>
1961 N31 P44-47 AD-609 151 <=>
1962 N32 P5-9 63-15965 <=*>
65-61271 <=$>
1964 N50 P8-12 65-17375 <*>
1965 N57 P5-20 66-34227 <=$>
1965 V43 P58-91 66-32337 <=$>

INFORMATSIONNYI BYULLETEN. VYSTAVKA DOSTIZHENII NARODNOGO KHOZYAISTVA SSSR. MOSCOW
1962 N2 P8-10 62-25986 <=>
1962 N9 P20 63-13511 <=> C
1962 N9 P21 63-13648 <=>
1962 N11 ENTIRE ISSUE 63-13840 <=>
1962 N11 P17-18 63-13352 <=>
1962 N12 P10-11 63-21352 <=>
1963 N4 P3-7 63-31413 <=>
1963 N6 P4-6 63-31450 <=>
1963 N7 P5-6 63-31636 <=>
1963 N7 P8 63-31710 <=>
1963 N9 P3-5 63-41038 <=>
1963 N9 P21-22 63-41038 <=>
1963 V4 P3-7 63-31413 <=>
1964 N7 P5-8 64-41988 <=>
1964 N8 P25-26 64-51007 <=>
1964 N8 P28-29 64-51007 <=>
1964 N9 P32-34 64-51766 <=$>
1964 N11 P20 65-30160 <=$>
1965 N8 P13-14 65-32647 <=*$>
1965 V8 P33 65-32646 <=*$>

INFORMATSIONNYI SBORNIK TSENTRALNOGO NAUCHNO-ISSLEDOVATELSKOGO INSTUTA MORSKOGO FLOTA
1963 N98 P23-31 AD-637 627 <=$>

INFORMATSIONNYI SBORNIK VSESOYUZNOGO GEOLOGICHESKOGO INSTITUTA
1956 N3 P103-107 IGR V4 N7 P777 <AGI>
1956 V4 P145-146 65-14126 <*>

INFORMATSIONNYI UKAZATEL BIBLIOGRAFICHESKIKH SPISKOV I KARTOCHEK. SOSTAVLENNYKH BIBLIOTEKAMI SOVETSKOGO SOIZA. MOSCOW
1961 N9 P26-30 61-28071 <*=>

INFRARED PHYSICS
1962 V2 P141-153 63-18862 <*> 0
1962 V2 P175-181 63-18868 <*> 0
1963 V3 P117-127 66-33120 <=$>

INGEGNERE
1954 V28 P1374-1384 60-17588 <=*>
1964 V38 N8 P1-12 ORNL-TR-377 <*>

INGEGNERIA CHIMICA ITALIANA. SUPPLEMENT TO CHIMICA E L'INDUSTRIA
1965 V1 P51-58 65-18059 <*>
1965 V1 N2 P37-43 66-14679 <*>

INGEGNERIA FERROVIARIA
1952 V7 P273-276 SCL-T-335 <*>
1955 P917-926 T-2178 <INSD>
1955 V10 N9 P653-656 T2032 <INSD>
1956 V11 N7/8 P591-596 T-2170 <INSD>
1963 V11 P3-7 66-11473 <*> 0

INGEGNERIA MECCANICA
1954 V3 N11 P41-49 2290 <BISI>
1960 V9 N11 P19-29 152-181 <STT>

INGENIERIA. MEXICO
1965 V35 N2 P257-280 66-11939 <*>

INGENIERIA NAVAL
1955 V23 N245 P753-784 59-17165 <*>

INGENIEUR. GRAVENHAGE
1934 V49 N52 PE195-E200 61-18255 <*>
1935 V50 PE83-E90 59-15136 <*>
1947 V59 N4 PB1-B11 C-4285 <NRCC>
1949 V61 N10 BT 19-26 61-18963 <*>
1949 V61 N10 P19-26 58-2562 <*>
1949 V61 N13 BT 27-35 61-18963 <*>
1949 V61 N13 P27-36 58-2562 <*>'
1949 V61 N18 P.BT37-BT48 61-18999 <*>
1949 V61 N18 P37-48 58-2581 <*>
1950 V62 N21 P25-27 61-18998 <*>
1951 V63 N26 P33-35 43M38DU <ATS>
1952 N28 P92-101 58-2196 <*>
1952 V64 N6 P1-6 58-1144 <*>
1952 V64 N12 P33-40 AEC-TR-5109 <*>
1953 V65 PE20-E23 57-787 <*>
1953 V65 N10 PB43-B49 81R77DU <ATS>
1953 V65 N19 P23-28 58-2602 <*>
1953 V65 N19 PV23-V28 61-20233 <*> 0
1953 V65 N24 PO25-O27 I-240 <*>
1954 V66 N49 P91-94 CSIRO-3466 <CSIR>
1955 V67 N1 P1-7 58-524 <*>
1955 V67 N1 P7-11 58-524 <*>
1956 V68 N3 PL6-L12 C-2222 <NRC>
1956 V68 N25 PB77-B88 23Q73DU <ATS>
1956 V68 N40 P51-60 C-2661 <NRC>
1956 V68 N40 PG51-G57 58-1422 <*>
1957 V69 N3 P1-3 80N54DU <ATS>
1957 V69 N3 P3-7 58-755 <*>
81N54DU <ATS>
1957 V69 N3 P8-14 58-755 <*>
58-756 <*>
82N54DU <ATS>
1957 V69 N3 P14-15 83N54DU <ATS>
1957 V69 N3 P15-18 58-755 <*>
58-757 <*>
84N54DU <ATS>
1957 V69 N3 P18-19 85N54DU <ATS>
1957 V69 N15 P37-45 58-1463 <*>
1957 V69 N20 P(BT)45-(BT)53 61-18982 <*>
1957 V69 N20 PBT45-BT53 58-2559 <*>
1957 V69 N27 P87-92 44N5104 <ATS>
1957 V69 N28 PBT55-BT61 58-2559 <*>
1957 V69 N50 PE179-E181 43N51DU <ATS>
1959 V71 N25 P21-25 01N51DU <ATS>
1960 V72 PE67-E78 2298 <BISI>
1960 V72 PE79-E94 2298 <BISI>
1960 V72 PE103-E109 2406 <BISI>
1960 V72 N12 P75-80 60-18528 <*>
1960 V72 N25 PE42-E50 2297 <BISI>
1960 V72 N32 PCH99-CH100 65-12449 <*>
1960 V72 N38 PCH121-CH126 65-12455 <*>
1960 V72 N38 PCH126-CH128 65-12456 <*>
1960 V72 N38 PCH128-CH136 65-12457 <*>
1960 V72 N38 P121-126 12N52DU <ATS>
1961 V73 N2 CH1-7 65-13018 <*>
1962 V74 N21 PB143-B150 63-00101 <*>
1962 V74 N23 PB155-B160 63-00101 <*>
1962 V74 N32 PW135-W140 65-11576 <*>
1962 V74 N34 P72-78 75Q69G <ATS>
1962 V74 N34 PO72-O78 63-20350 <*>
1962 V74 N48 PE153-E156 63-20048 <*>
1963 V75 N33 P27CH-35CH 64-16683 <*>
1963 V75 N49 P85CH-91CH 64-30697 <*>
1964 V76 N16 P65-73 4305 <BISI>
1964 V76 N43 P58-62 65-13420 <*>
1965 V77 N3 PB1-B10 65-14922 <*> 0
1965 V77 N5 PB13-B21 65-14922 <*> C
1965 V77 N12 PCH15-CH21 66-10363 <*>
1965 V77 N15 PCH25-CH33 65-17126 <*>
1965 V77 N41 P225-229 66-14154 <*>

INGENIEUR DER DEUTSCHEN BUNDESPOST
1958 V1 N1 P11-17 51K28G <ATS>

INGENIEUR ELECTRICIEN
1935 V50 P83-90 59-15136 <*>

INGENIEUR-ARCHIV
1931 V2 P47-91 SCL-T-358 <*>
1931 V2 P359-371 63-14482 <*>
1932 V3 P454-462 57-1593 <*>
1933 V4 P384-393 T2179 <INSD>
1934 V5 P429-449 62-01111 <*>
1937 V8 P216-228 I-387 <*>
1939 V10 P125-132 65-13291 <*>
1941 V12 N2 P71-76 61-18756 <*>
1943 V14 P192-212 II-806 <*>
1944 V14 P286-305 AL-279 <*>
1947 V16 P39-44 T-2438 <INSD>
1947 V16 P51-71 65-12236 <*>
1949 V17 P94-106 65-12966 <*>
1949 V17 P129-141 58-2362 <*>
1949 V17 P199-206 AL-834 <*>
1949 V17 P418-419 62-18058 <*>
1949 V17 P450-480 CSIRO-2982 <CSIR>
1950 V18 P141-150 65-12589 <*>
1951 V19 P208-227 62-18393 <*> 0
1951 V19 N1 P42-65 II-932 <*>
1952 V20 P46-48 58-927 <*>
63-16647 <*>
1952 V20 P195-207 65-13178 <*>
1952 V20 N2 P67-72 62-16111 <*>
1952 V20 N2 P73-80 62-16165 <*>
1952 V20 N3 P170-183 64-10014 <*> P
1952 V20 N4 P211-228 2831 <*>
1953 V21 P20- 1763 <*>
1953 V21 P33- 3134 <*>
1953 V21 N2 P90- 3095 <*>
1953 V21 N4 P227-244 1559 <*>
2833 <*>
1953 V21 N5/6 P331-338 2972 <*>
1954 V22 N1 P21-35 65-12593 <*>
1954 V22 N3 P147-155 58-1194 <*>

1954 V23 N5 P295-322 65-13185 <*>
1955 V23 N3 P159-171 59-20146 <*>
1956 V24 P81-84 T1802 <INSD>
1956 V24 N4 P258-281 63-14484 <*>
65-13986 <*>
1957 V25 N1 P58-70 CSIRO-3833 <CSIR>
60-18392 <*> 0
1959 V27 N3 P137-152 6C-14066 <*>
1959 V27 N4 P201-226 66-11118 <*>
1960 V29 N6 P436-444 61-16126 <*>
1961 V30 N5 P293-316 C-4043 <NRCC>
1962 V31 N5 P317-342 64-11810 <=>
1963 V32 P201-213 N65-27675 <=*$>

INGENIEUR-CHIMISTE
1956 V38 P35-40 59-10657 <*>

INGENIEURS DE L'AUTOMOBILE
1963 V36 N6 P317-330 63-20571 <*> 0
1964 V38 N2 P87-96 65-11348 <*>

INGENIEURS ET TECHNICIENS
1955 V83 P8 57-1713 <*>

INGENIOREN
1951 V44 P733- AL-564 <*>
1958 V67 N3 P118-122 58-2592 <*>
1964 V73 N3 P136-141 64-18003 <*> P

INGENIOREN B
1958 V67 N3 P118-122 61-18973 <*>

INGENIORSVETENSKAPSAKADEMIENS MEDDELANDE
1950 N124 P53- 62-20474 <*> 0

INHISARLAR TUTUN ENSTITUSU RAPCRLARI
1943 V3 N1 P17-22 3006-F <K-H>
1943 V3 N1 P36-49 3006-E <K-H>

INSECT WCRLD
1937 V41 N2 P43-47 44J16J <ATS>

INSTITUT ZA OKEANOGRAFIJU I RIBARSTVO. FAUNA I FLORA JADRANA
1948 V1 P1-437 60-21661 <=>

INSTITUTO DEL HIERRO Y DEL ACERO. MADRID
1950 V3 P47-53 60-18824 <*> P
1950 V3 P115-131 60-18821 <*>
1951 V4 P52-61 I-428 <*>
1951 V4 P131-136 I-428 <*>
1954 V7 P464-467 TS-1531 <BISI>
62-16887 <*>
1955 V8 P136-148 62-16749 <*>
1956 V9 P478-480 SPEC 3878 <HB>
1956 V9 N44 P250-257 4553 <HB>
1957 V10 P166-170 61-00033 <*>
1958 V11 P177-179 TS-1457 <BISI>
1958 V11 P263-281 TS-1273 <BISI>
62-10487 <*> P
1958 V11 N57 P177-179 62-14392 <*>
1960 V13 N67 P332-335 4912 <HB>
1960 V13 N67 P432-435 4911 <HB>
1961 V14 P397-419 2347 <BISI>
1964 V17 P517-522 4779 <BISI>
1964 V17 N89 P59-72 6203 <HB>

INSTRUMENTS AND EXPERIMENTAL TECHNIQUES
1961 V5 P7-25 64-00248 <*>

INTERAVIA. REVUE DE L'AERONAUTIQUE MONDIALE. GENEVE
1965 V20 N2 P286-287 66-11177 <*>

INTERNAL MEDICINE. JAPAN
SEE NAIKA

INTERNAL MEDICINE AND PEDIATRICS. JAPAN
SEE RINSHO NAIKA SHONIKA

INTERNATIONAL ARCHIVES CF ALLERGY AND APPLIED IMMUNOLOGY
1955 V7 N2 P103-110 57-679 <*>
1957 V10 P348-354 58-2634 <*>
1957 V10 N2 P82-99 65-00344 <*> 0
1957 V10 N5 P276-284 58-550 <*>
1957 V10 N5 P3C5-316 58-544 <*>
1959 V14 N5/6 P339-362 65-00343 <*> 0

INTERNATIONAL JOURNAL OF APPLIED RADIATION AND ISOTOPES
1956 V1 P115-122 57-1866 <*>
57-936 <*>
1957 V1 N4 P237-245 57-2571 <*>
1959 V5 N3 P175-196 C-4608 <NRC>
1960 V8 P35-45 64-00230 <*>
1961 V11 P131-138 NP-TR-907 <*>
1961 V11 P161-173 AEC-TR-5096 <*>
1962 N13 P501-513 C-5527 <NRC>
1962 V13 P205 AEC-TR-5592 <*>
1962 V13 P501-513 AEC-TR-6308 <*>
1963 V14 P385-396 C-5561 <NRC>
1964 V15 P17-24 1049 <TC>
1964 V15 P25-41 HW-TR-70 <*>
1964 V15 N10 P587-598 65-12042 <*>
1965 V16 P301-318 66-11106 <*>

INTERNATIONAL JOURNAL OF HEAT AND MASS TRANSFER
1962 N5 P267-275 66-12356 <*>
1964 V7 P41-47 95R76G <ATS>
1964 V7 P517-527 UCRL TRANS-1115(L) <*>

INTERNATIONAL JOURNAL OF MECHANICAL SCIENCES. LONDON
1960 V1 N4 P322-335 61-00942 <*>

INTERNATIONAL NORTH PACIFIC FISHERIES COMMISSION, DOCUMENTS
1958 DOCUMENT 207 60-15538 <+*>

INTERNATIONALE ELEKTRONISCHE RUNDSCHAU
1964 V18 N9 P480-484 65-17210 <*>
1965 V19 N4 P199-260 66-11025 <*>
1965 V19 N8 P428-430 84T93G <ATS>
1965 V19 N10 P549-552 85T93G <ATS>
1965 V19 N10 P579-584 66-14624 <*>

INTERNATIONALE FACHSCHRIFT FUER DIE SCHOKOLADEINDUSTRIE
SEE REVUE INTERNATIONALE DE LA CHOCOLATERIE

INTERNATIONALE MITTEILUNGEN FUER BODENKUNDE. BERLIN, WIEN
1913 V3 P131-140 CSIRO-3274 <CSIR>

INTERNATIONALE ZEITSCHRIFT FUER ANGEWANDTE PHYSIOLOGIE, EINSCHLIESSLICH ARBEITSPHYSIOLOGIE
1956 V16 P237-249 65-12687 <*>
1956 V16 N6 P519-564 65-13902 <*>
1959 V18 N1 P1-12 N66-16574 <=$>
1959 V18 N1 P82-100 65-00364 <*> 0

INTERNATIONALE ZEITSCHRIFT FUER ELEKTROWAERME
1964 V22 N9 P331-339 56S86G <ATS>
66-11618 <*>
1965 V23 N2 P44-50 66-11439 <*> 0

INTERNATIONALE ZEITSCHRIFT DER LANDWIRTSCHAFT
1966 N6 P587-602 67-31227 <=$>

INTERNATIONALE ZEITSCHRIFT FUER METALLOGRAPHIE
1914 V5 N4 P278-287 57-2216 <*>
1914 V6 N1 P89-104 57-2217 <*>

INTERNATIONALE ZEITSCHRIFT FUER VITAMINFORSCHUNG
1949 V21 P151-161 57-1951 <*>
1955 V25 N4 P427-433 3154 <*>
1957 V27 N3 P4C5-413 61-00651 <*>

1957 V28 P157-160 22L30G <ATS>
1963 V33 P180-196 63-18943 <*>

INTERNIST. BERLIN
1961 V2 P403-412 62-20121 <*>

INTERNISTISCHE PRAXIS
1962 V4 N3 P414-415 65-12162 <*>

INVENTII SI INOVATII
1966 V2 P41-45 66-32288 <=$>

INVESTIGACION E INFORMACION TEXTIL
1959 V2 N4 P271-276 62-14307 <*>
1960 V3 N4 P285-293 62-16753 <*>

INZENYRSKE STAVBY
1962 V11 P412-415 CSIR-TR.326 <CSSA>

*INZHENERNO-FIZICHESKII ZHURNAL
1958 V1 N1 P65-73 63-14473 <=*>
1958 V1 N2 P75-88 59-17423 <+*>
61-13397 <+*>
61-23545 <=*>
1958 V1 N4 P31-39 62-11671 <=>
1958 V1 N6 P12-19 05L34R <ATS>
1958 V1 N7 P61-71 C-3436 <NRCC>
1958 V1 N8 P30-38 63-19417 <=*>
1958 V1 N9 P116-118 60-23570 <+*>
1958 V1 N10 P29-37 62-11609 <=>
1958 V1 N1C P88-93 4143 <BISI>
1958 V1 N11 P38-45 61-19781 <=*> 0
1958 V1 N11 P73-79 65-17379 <*>
1958 V1 N11 P80-91 C-3437 <NRCC>
1958 V1 N12 P54-58 65-17373 <*>
1958 V1 N12 P64-68 61-11175 <=>
1959 V2 N1 P93-95 62-33584 <=>
1959 V2 N1 P109-113 60-23562 <+*>
1959 V2 N2 P118-120 60-19038 <+*> 0
1959 V2 N3 P3-8 62-25049 <=*>
1959 V2 N3 P19-28 RTS-3179 <NLL>
1959 V2 N4 P8-14 62-20260 <=*>
1959 V2 N4 P78-86 62-24856 <=*>
1959 V2 N5 P3-7 60-21779 <=>
61-15624 <*+>
1959 V2 N5 P28-35 62-25402 <=*>
1959 V2 N5 P55-59 17M39R <ATS>
60-18738 <+*>
1959 V2 N7 P75-79 62-32473 <=*>
1959 V2 N8 P15-22 61-15091 <*+>
1959 V2 N8 P66-71 62-13985 <*=>
1959 V2 N8 P108-111 61-27988 <*=>
1959 V2 N8 P112-115 62-33551 <=*>
1959 V2 N9 P12-23 AEC-TR-4048 <+*>
1959 V2 N9 P24-29 61-10727 <=*>
1959 V2 N9 P83-91 60-11589 <=>
1959 V2 N9 P97-100 3939 <BISI>
1959 V2 N9 P111-112 60-11171 <=>
1959 V2 N1C P40-45 UCRL TRANS-694 <=*>
62-24852 <=*>
1959 V2 N10 P46-51 61-23178 <=*>
1959 V2 N11 P20-28 61-14959 <=*>
61-27779 <*=>
1959 V2 N11 P29-34 61-13332 <+*>
1959 V2 N11 P35-42 61-14964 <*=>
1959 V2 N11 P102-105 61-13765 <+*>
1959 V2 N12 P20-25 AEC-TR-4218 <+*>
1959 V2 N12 P64-67 61-23430 <=*>
1959 V2 N12 P105-109 60-31430 <=>
1960 V3 N1 P3-9 AEC-TR-4934 <*>
AEC-TR-4934 <=*>
1960 V3 N1 P94-97 61-31670 <=>
1960 V3 N1 P103-107 61-19387 <+*> 0
62-25394 <=*>
1960 V3 N1 P124-126 62-19148 <=*>
1960 V3 N2 P12-16 61-31539 <=>
64-11963 <=> M
1960 V3 N2 P41-45 62-14002 <=*>
1960 V3 N2 P71-73 61-19706 <=*> 0
1960 V3 N2 P83-85 62-13104 <*=>
1960 V3 N2 P105-110 62-13608 <*=>
1960 V3 N2 P111-114 62-23717 <=*>
1960 V3 N2 P128-132 60-31206 <=>
1960 V3 N3 P13-20 RTS-2990 <NLL>
1960 V3 N3 P69-73 61-15713 <+*>
1960 V3 N3 P117-122 ICE V2 N1 P3-5 <ICE>
62-15701 <*=>
1960 V3 N4 P18-22 61-15923 <+*>
1960 V3 N4 P44-48 61-15397 <*+>
1960 V3 N4 P54-64 61-21761 <=>
1960 V3 N4 P65-72 62-15705 <*=>
1960 V3 N5 P3-11 63-13621 <=*>
1960 V3 N5 P24-30 61-23838 <*=>
1960 V3 N5 P31-39 63-13622 <=*>
1960 V3 N5 P74-80 UCRL TRANS-677(L) <=*>
1960 V3 N5 P81-85 62-33640 <=*>
1960 V3 N5 P119-123 61-23332 <*=>
1960 V3 N5 P124- AEC-TR-4902 <*=>
1960 V3 N5 P138-144 61-27175 <*=>
1960 V3 N6 P9-16 62-15707 <*=>
1960 V3 N6 P52-54 61-27131 <*=> 0
1960 V3 N6 P62-65 61-31544 <=>
1960 V3 N6 P66-71 61-27161 <*=>
1960 V3 N6 P107-111 AEC-TR-6368 <=*$>
1960 V3 N6 P112-119 61-23850 <*=>
1960 V3 N7 P3-9 64-13808 <=*$>
1960 V3 N7 P88-94 63-23476 <=*$>
63-24316 <=*$>
1960 V3 N7 P106-111 63-13380 <=*>
1960 V3 N8 P39-46 61-31101 <=>
1960 V3 N9 P10-16 63-19508 <=*>
1960 V3 N9 P17-24 61-28242 <*=>
62-15516 <=*>
1960 V3 N10 P5-10 61-28379 <*=>
1960 V3 N10 P63-65 AEC-TR-4404 <+*>
1960 V3 N10 P139-143 61-11812 <=>
1960 V3 N11 P3-10 61-19801 <=*>
1960 V3 N11 P48-51 61-23851 <*=>
1960 V3 N11 P52-57 62-13004 <*=>
1960 V3 N11 P72-76 61-23852 <*=>
1960 V3 N11 P97-101 61-27471 <*=>
1960 V3 N12 P3-10 62-15808 <=*>
1960 V3 N12 P17-23 62-15703 <*=>
1960 V3 N12 P49-52 62-15810 <*=>
65-14041 <*>
1960 V3 N12 P72-77 62-15809 <=*>
1960 V3 N12 P111-113 62-15656 <*=>
1960 V4 N10 P44- ICE V2 N2 P149-153 <ICE>
1961 N12 P32-36 63-19198 <=*>
1961 V4 N1 P3-13 61-21722 <=>
1961 V4 N1 P14-21 UCRL TRANS-782 <=*>
62-19653 <=*>
1961 V4 N1 P37-43 62-19653 <=*>
1961 V4 N1 P58-62 62-15830 <=*>
63-18437 <=*>
1961 V4 N1 P98-103 62-13441 <*=>
1961 V4 N1 P104-108 62-15830 <=*>
1961 V4 N1 P116-119 62-25542 <=*>
1961 V4 N1 P131-148 61-23462 <=*>
1961 V4 N2 P18-26 62-19652 <=*>
1961 V4 N2 P70-76 64-21709 <=>
1961 V4 N2 P77-81 62-25831 <=*>
1961 V4 N2 P95-98 63-13209 <=*>
1961 V4 N2 P99-102 ANL-TRANS-148 <=$>
1961 V4 N2 P103-105 62-15404 <*=>
1961 V4 N2 P116-118 61-31110 <=>
1961 V4 N2 P119-130 62-19277 <=*>
1961 V4 N3 P10-17 62-19284 <=*>
64-23504 <=$>
1961 V4 N3 P39-45 ICE V1 N1 P64-67 <ICE>
61-31682 <*=>
62-19284 <=*>
1961 V4 N3 P72-82 51R78R <ATS>
64-11626 <=*$>
1961 V4 N3 P105-109 62-25819 <=*>
1961 V4 N4 P3-9 62-11676 <=> 0
1961 V4 N4 P16-24 NP-TR-812 <*>
1961 V4 N4 P32-37 64-11935 <=> M
1961 V4 N4 P43-48 63-13594 <=*>
1961 V4 N4 P49-54 62-14229 <=*> 0

Citation	Translation
1961 V4 N4 P123-125	AEC-TR-4635 <*>
	AEC-TR-4635 <=*>
1961 V4 N5 P33-37	62-25823 <=*>
1961 V4 N6 P3-12	ICE V2 N1 P25-30 <ICE>
1961 V4 N6 P33-41	62-32986 <=*>
1961 V4 N6 P42-50	62-13448 <=*>
1961 V4 N6 P64-69	63-19861 <=*>
1961 V4 N6 P70-77	62-32531 <=*>
1961 V4 N6 P90-100	62-32986 <=*>
1961 V4 N7 P3-10	ICE V2 N1 P31-35 <ICE>
1961 V4 N7 P19-24	62-32414 <=*>
1961 V4 N7 P84-90	62-33392 <=*>
1961 V4 N7 P96-104	AIAAJ V1 N5 P1264 <AIAA>
1961 V4 N7 P117-119	62-33402 <=*>
1961 V4 N8 P3-	ICE V2 N1 P35-38 <ICE>
1961 V4 N8 P30-35	62-32525 <=*>
1961 V4 N8 P57-62	62-13367 <=*>
1961 V4 N8 P93-	ICE V2 N1 P64-67 <ICE>
1961 V4 N8 P111-113	62-32525 <=*>
1961 V4 N9 P12-16	62-19275 <=*>
1961 V4 N9 P17-23	64-23618 <=$>
1961 V4 N9 P32-39	62-23669 <=*>
1961 V4 N9 P56-60	N65-26641 <=$>
1961 V4 N9 P86-89	AEC-TR-5471 <=*>
	AIAAJ V1 N6 P1497 <AIAA>
1961 V4 N9 P127-131	65-61382 <=$>
1961 V4 N10 P9-14	62-24353 <=*> 0
1961 V4 N10 P21-29	62-24353 <=*> 0
1961 V4 N10 P30-	ICE V2 N2 P161-164 <ICE>
1961 V4 N10 P36-	ICE V2 N2 P174-176 <ICE>
1961 V4 N10 P36-39	65-13657 <*>
1961 V4 N10 P44-51	62-23672 <=*>
1961 V4 N10 P52-63	62-32526 <=*>
1961 V4 N10 P64-70	64-11645 <=>
1961 V4 N10 P97-100	62-24353 <=*> 0
1961 V4 N10 P123-126	62-33369 <=*>
	63-15230 <=*>
1961 V4 N10 P127-128	62-32526 <=*>
1961 V4 N11 P10-	ICE V2 N2 P193-198 <ICE>
1961 V4 N11 P10-18	63-19662 <=*>
1961 V4 N11 P29-36	62-32306 <=*>
1961 V4 N11 P55-	ICE V2 N2 P182-184 <ICE>
1961 V4 N11 P105-108	62-11747 <=>
	62-32316 <=*>
1961 V4 N12 P11-15	63-19865 <=>
1961 V4 N12 P22-31	ICE V2 N2 P221-226 <ICE>
1961 V4 N12 P47-51	63-11510 <=>
1961 V4 N12 P118-130	NP-TR-1020 <=*$>
1962 N1 P7-	ICE V2 N3 P384-388 <ICE>
1962 N2 P3-	ICE V2 N3 P372-375 <ICE>
1962 N2 P15-	ICE V2 N3 P376-378 <ICE>
1962 N3 P80-	ICE V2 N3 P399-403 <ICE>
1962 N4 P3-	ICE V2 N4 P507-511 <ICE>
1962 N4 P10-	ICE V2 N4 P528-530 <ICE>
1962 N4 P15-	ICE V3 N1 P48-51 <ICE>
1962 N5 P5-14	ICE V2 N4 P455-459 <ICE>
1962 N5 P15-	ICE V2 N4 P460-463 <ICE>
1962 N6 P13-20	ICE V2 N4 P575-580 <ICE>
1962 N7 P11-	ICE V3 N1 P1-5 <ICE>
1962 N7 P99-	ICE V3 N1 P67-76 <ICE>
1962 N9 P3-	ICE V3 N1 P95-98 <ICE>
1962 N9 P53-	ICE V3 N1 P108-111 <ICE>
1962 N10 P26-	ICE V3 N2 P175-178 <ICE>
1962 N10 P70-	ICE V3 N2 P173-174 <ICE>
1962 N11 P12-	ICE V3 N2 P195-202 <ICE>
1962 N11 P41-	ICE V3 N2 P203-206 <ICE>
1962 N12 P16-	<ICE>
	ICE V3 N2 P251-255 <ICE>
1962 V5 N1 P33-41	64-71401 <=> M
1962 V5 N1 P52-54	63-15963 <=*>
1962 V5 N1 P115-129	62-33542 <=*>
1962 V5 N2 P42-46	62-24396 <=*>
1962 V5 N2 P47-51	62-24000 <=*>
1962 V5 N2 P108-112	AEC-TR-5454 <=*>
1962 V5 N3 P45-50	63-15534 <=$>
1962 V5 N3 P66-71	63-15534 <=$>
1962 V5 N3 P103-106	62-32982 <=*>
1962 V5 N4 P3-9	63-13208 <=*>
1962 V5 N4 P10-14	63-13921 <=*>
1962 V5 N4 P15-19	AEC-TR-5517 <=*>
	24P62R <ATS>
1962 V5 N4 P20-24	64-11932 <=> M
1962 V5 N4 P25-30	63-19505 <=*>
1962 V5 N4 P35-40	63-13172 <=*>
1962 V5 N4 P71-74	AEC-TR-5519 <=*>
	25P62R <ATS>
1962 V5 N5 P5-14	63-19436 <=*>
1962 V5 N5 P15-20	63-19435 <=*>
1962 V5 N5 P21-29	AD-635 274 <=$>
1962 V5 N5 P30-37	63-15535 <=*>
1962 V5 N5 P68-74	63-13586 <=*>
1962 V5 N5 P112-115	63-23705 <=*$>
1962 V5 N6 P3-7	63-19499 <=*>
1962 V5 N6 P21-26	63-19312 <=*>
	63-19565 <=*$>
1962 V5 N6 P55-	ICE V2 N4 P550-553 <ICE>
1962 V5 N7 P3-10	AEC-TR-5760 <=*$>
	25Q67R <ATS>
1962 V5 N7 P11-17	63-19074 <=*>
1962 V5 N7 P18-22	63-19500 <=*>
1962 V5 N7 P28-33	63-19500 <=*>
1962 V5 N7 P34-38	63-13207 <=*>
1962 V5 N7 P70-77	AD-609 161 <=>
1962 V5 N7 P78-82	CE TRANS-35866 <NLL>
1962 V5 N8 P10-16	63-19093 <=*>
1962 V5 N8 P138-139	62-33111 <=*>
1962 V5 N9 P9-	ICE V3 N3 P355-359 <ICE>
1962 V5 N9 P9-15	26Q67R <ATS>
	64-13804 <=*$>
1962 V5 N9 P25-32	63-19316 <=*>
1962 V5 N10 P47-52	64-11662 <=>
1962 V5 N10 P65-69	AEC-TR-5821 <=*$>
1962 V5 N10 P73-76	63-23198 <=*>
1962 V5 N11 P3-24	63-19336 <=*>
1962 V5 N11 P64-73	SC-T-530 <=*$>
1962 V5 N12 P16-22	63-18172 <=*>
	64-71345 <=> M
1962 V5 N12 P23-26	64-11954 <=> M
1962 V5 N12 P41-47	64-19110 <=*$>
1962 V5 N12 P80-83	63-11694 <=>
1963 N1 P3-	ICE V3 N3 P309-315 <ICE>
1963 N1 P19-	ICE V3 N3 P347-351 <ICE>
1963 N2 P31-	ICE V3 N3 P328-332 <ICE>
1963 N3 P69-	ICE V3 N4 P471-477 <ICE>
1963 N4 P3-	ICE V3 N4 P483-486 <ICE>
1963 N4 P27-	ICE V3 N4 P487-490 <ICE>
1963 N6 P20-30	ICE V3 N4 P571-576 <ICE>
1963 N6 P65-	ICE V3 N4 P567-571 <ICE>
1963 N7 P13-18	ICE V4 N1 P43-47 <ICE>
1963 N7 P19-25	ICE V4 N1 P50-54 <ICE>
1963 N8 P3-9	ICE V4 N1 P119-123 <ICE>
1963 N9 P40-45	ICE V4 N1 P132-136 <ICE>
1963 N9 P73-79	ICE V4 N1 P128-132 <ICE>
1963 N10 P27-32	ICE V4 N2 P212-215 <ICE>
1963 N10 P104-108	ICE V4 N2 P236-239 <ICE>
1963 N11 P20-25	ICE V4 N2 P320-324 <ICE>
1963 N11 P50-51	ICE V4 N2 P271-272 <ICE>
1963 N12 P27-34	ICE V4 N2 P324-328 <ICE>
1963 N12 P85-87	ICE V4 N2 P300-301 <ICE>
1963 V6 N1 P3-137	63-21746 <=>
1963 V6 N1 P59-65	64-19386 <=$>
1963 V6 N1 P105-108	65-10496 <*>
1963 V6 N2 P3-138	63-31047 <=>
1963 V6 N2 P15-19	AEC-TR-6403 <=$>
1963 V6 N2 P20-24	65-10398 <*>
1963 V6 N2 P52-59	64-11508 <=>
1963 V6 N2 P60-68	AEC-TR-5800 <=*$>
	63-24230 <=*$>
1963 V6 N2 P114-117	64-13160 <=*$>
1963 V6 N3 P51-57	RTS-2408 <NLL>
1963 V6 N3 P99-102	RTS-2762 <NLL>
1963 V6 N4 P3-138	63-31863 <=>
1963 V6 N4 P51-55	65-60501 <=$>
1963 V6 N5 P3-6	64-13731 <=*$>
1963 V6 N5 P50-54	64-19106 <=*$>
1963 V6 N6 ENTIRE ISSUE	64-21038 <=>
1963 V6 N6 P10-19	N66-23544 <=$>
1963 V6 N6 P94-99	64-71493 <=> M
1963 V6 N8 P22-27	ICE V4 N1 P110-113 <ICE>
1963 V6 N8 P37-40	64-15120 <=*$>

Citation	Translation
1963 V6 N8 P41-44	64-13699 <=*$>
1963 V6 N8 P82-87	64-11672 <=>
1963 V6 N8 P112-115	AEC-TR-6291 <=*$>
1963 V6 N8 P125-132	AD-611 861 <=$>
1963 V6 N10 P27-32	ICE V4 N2 P212-215 <ICE>
1963 V6 N10 P96-100	32Q74R <ATS>
1963 V6 N1C P104-108	ICE V4 N2 P236-239 <ICE>
1963 V6 N10 P129-136	64-15229 <=*$>
1963 V6 N11 P14-19	64-14129 <=*$>
1963 V6 N11 P20-25	ICE V4 N2 P320-324 <ICE>
1963 V6 N11 P50-51	ICE V4 N2 P271-272 <ICE>
1963 V6 N12 P3-10	AD-618 093 <=$>
1963 V6 N12 P3-126	64-21816 <=>
1963 V6 N12 P27-34	ICE V4 N2 P324-328 <ICE>
1963 V6 N12 P85-87	ICE V4 N2 P300-301 <ICE>
1963 V6 N12 P127-134	AD-609 136 <=>
1964 N1 P21-27	ICE V4 N P408-412 <ICE>
1964 N1 P55-58	ICE V4 N3 P499-501 <ICE>
1964 N2 P66-	ICE V4 N3 P405-408 <ICE>
1964 N3 P3-9	ICE V4 N4 P621-624 <ICE>
1964 N3 P68-71	ICE V4 N4 P603-605 <ICE>
1964 N4 P8-17	ICE V4 N4 P587-593 <ICE>
1964 N4 P111-113	ICE V4 N4 P598-599 <ICE>
1964 N5 P11-17	ICE V4 N4 P650-654 <ICE>
1964 N5 P96-99	ICE V4 N4 P600-602 <ICE>
1964 N6 P13-15	66-12345 <*>
1964 N6 P16-19	ICE V5 N1 P16-18 <ICE>
1964 N6 P40-43	ICE V5 N1 P77-79 <ICE>
1964 N6 P63-69	AD-636 665 <=$>
1964 N7 P15-19	ICE V5 N1 P55-57 <ICE>
1964 N7 P25-32	ICE V5 N1 P83-87 <ICE>
1964 N7 P91-95	AD-636 667 <=$>
1964 N7 P116-120	ICE V5 N1 P88-90 <ICE>
1964 N9 P19-24	ICE V5 N3 P393-396 <ICE>
1964 N9 P30-33	ICE V5 N2 P302-304 <ICE>
1964 N9 P71-77	ICE V5 N1 P112-115 <ICE>
1964 N10 P6-13	ICE V5 N2 P240-244 <ICE>
1964 N10 P37-44	ICE V5 N2 P213-217 <ICE>
1964 N10 P49-55	ICE V5 N2 P218-221 <ICE>
1964 N11 P16-21	ICE V5 N2 P341-343 <ICE>
1964 N11 P47-54	ICE V5 N2 P336-340 <ICE>
1964 N12 P8-12	ICE V5 N3 P432-435 <ICE>
1964 N12 P13-22	ICE V5 N3 P446-453 <ICE>
1964 N12 P45-53	ICE V5 N3 P461-466 <ICE>
1964 V7 P11-17	ICE V4 N4 P650-654 <ICE>
1964 V7 P96-99	ICE V4 N4 P600-602 <ICE>
1964 V7 N1 P3-11	AD-620 862 <=$>
1964 V7 N1 P21-27	ICE V4 N3 P408-412 <ICE>
1964 V7 N1 P55-58	ICE V4 N3 P499-501 <ICE>
1964 V7 N2 P3-9	AD-617 675 <=$>
1964 V7 N2 P21-24	64-71566 <=>
	65-64223 <=*$>
1964 V7 N2 P38-42	ICE V4 N3 P437-439 <ICE>
1964 V7 N2 P43-44	AD-631 436 <=$>
1964 V7 N2 P66	ICE V4 N3 P405-408 <ICE>
1964 V7 N2 P95-102	ICE V4 N3 P461-466 <ICE>
1964 V7 N3 P3-9	ICE V4 N4 P621-624 <ICE>
1964 V7 N3 P68-71	ICE V4 N4 P603-605 <ICE>
1964 V7 N3 P78-81	N65-29734 <=*$>
1964 V7 N3 P92-96	AD-640 302 <=$>
1964 V7 N4 P3-7	AD-637 388 <=$>
1964 V7 N4 P8-17	ICE V4 N4 P587-593 <ICE>
1964 V7 N4 P18-24	AD-630 868 <=$>
1964 V7 N4 P44-50	AD-634 806 <=$>
1964 V7 N4 P70-74	AD-634 806 <=$>
1964 V7 N4 P111-113	ICE V4 N4 P598-599 <ICE>
1964 V7 N4 P114-120	AD-634 806 <=$>
1964 V7 N5 P70-75	65-17232 <*>
1964 V7 N5 P105-111	RTS-2869 <NLL>
1964 V7 N6 P16-19	ICE V5 N1 P16-18 <ICE>
1964 V7 N6 P32-34	WAPD-TRANS-9 <*>
1964 V7 N6 P40-43	ICE V5 N1 P77-79 <ICE>
	66-12346 <*>
1964 V7 N6 P48-54	N66-14054 <=$>
1964 V7 N7 P15-19	ICE V5 N1 P55-57 <ICE>
1964 V7 N7 P25-32	ICE V5 N1 P83-88 <ICE>
1964 V7 N7 P116-120	ICE V5 N1 P88-90 <ICE>
1964 V7 N8 P59-63	AD-620 770 <=$>
1964 V7 N8 P124-127	AD-622 459 <=$*>
	N65-22623 <=$>
1964 V7 N9 P25-29	AD-651 090 <=$> M
1964 V7 N9 P30-33	WAPD-TRANS-17 <=$>
1964 V7 N9 P38-43	AD-651 090 <=$> M
1964 V7 N9 P44-51	WAPD-TRANS-19 <=$>
1964 V7 N9 P64-77	AD-651 090 <=$> M
1964 V7 N9 P71-77	ICE V5 N1 P112-115 <ICE>
1964 V7 N9 P83-91	AD-651 090 <=$> M
1964 V7 N9 P102-107	AD-651 090 <=$> M
1964 V7 N9 P108-112	65-63682 <=$>
1964 V7 N10 P62-66	87S86R <ATS>
1964 V7 N10 P112-116	AD-639 159 <=$>
1964 V7 N10 P117-120	66-61839 <=*$>
1964 V7 N10 P130-135	AD-633 659 <=$>
1964 V7 N11 P41-46	RTS-3180 <NLL>
1964 V7 N11 P55-61	AD-620 948 <=*$>
1964 V7 N11 P67-72	AD-625 066 <=*$>
1964 V7 N11 P73-76	AD-639 184 <=$>
1964 V7 N12 P85-89	ICE V5 N3 P418-420 <ICE>
1965 N1 P7-10	ICE V5 N3 P474-476 <ICE>
1965 N1 P93-97	ICE V5 N3 P552-554 <ICE>
1965 N1 P112-116	ICE V6 N1 P118-120 <ICE>
1965 N2 P187-195	ICE V6 N2 P228-232 <ICE>
1965 N2 P229-237	ICE V5 N4 P603-608 <ICE>
1965 N3 P290-293	ICE V5 N4 P666-668 <ICE>
1965 N3 P354-357	ICE V6 N2 P300-302 <ICE>
1965 N4 P430-431	ICE V5 N4 P613-615 <ICE>
1965 N4 P488-492	ICE V5 N4 P633-635 <ICE>
1965 N5 P563-567	ICE V6 N1 P1-4 <ICE>
1965 N5 P567-570	ICE V6 N3 P486-488 <ICE>
1965 N5 P579-585	ICE V5 N4 P720-723 <ICE>
1965 N5 P627-631	ICE V5 N4 P711-714 <ICE>
1965 N5 P647-653	ICE V6 N3 P502-506 <ICE>
1965 N5 P654-656	ICE V6 N3 P509-510 <ICE>
1965 N6 P703-7C6	ICE V6 N3 P495-497 <ICE>
1965 N6 P707-711	ICE V6 N1 P105-107 <ICE>
1965 N6 P722-728	ICE V6 N3 P498-501 <ICE>
1965 N6 P816-833	ICE V6 N4 P580-591 <ICE>
1965 V8 P30,294-299	RTS-3122 <NLL>
1965 V8 N1 P9-14	ICE V6 N1 P126-129 <ICE>
1965 V8 N1 P25-33	ICE V6 N1 P145-150 <ICE>
1965 V8 N1 P58-63	ICE V5 N3 P485-488 <ICE>
1965 V8 N1 P93-97	AD-622 359 <=$>
1965 V8 N2 P243-246	RTS-3066 <NLL>
1965 V8 N3 P369-374	ICE V5 N4 P662-665 <ICE>
1965 V8 N4 P432-438	67T90R <ATS>
1965 V8 N4 P447-450	AC-618 390 <=$>
1965 V8 N5 P568-573	N66-22301 <=$>
1965 V8 N5 P574-578	AD-640 325 <=$>
1965 V8 N5 P613-619	66-61840 <=*$>
1965 V8 N5 P684-686	AD-625 243 <=*$>
1965 V9 N2 P163-170	ICE V6 N2 P252-256 <ICE>
1965 V9 N2 P250-254	50T92R <ATS>
1965 V9 N3 P287	ICE V6 N2 P319-329 <ICE>
1965 V9 N4 P457-460	ICE V6 N3 P406-408 <ICE>
1965 V9 N4 P496-500	ICE V6 N3 P414-417 <ICE>
1965 V9 N4 P520-526	ICE V6 N3 P409-413 <ICE>
1965 V9 N6 P762-767	66-12323 <*>
	66-61303 <=$>
1965 V9 N6 P768-774	66-60598 <=*$>
1965 V9 N6 P783-787	66-61710 <=*$>
1966 V10 N5 P696-701	66-34466 <=$>

INZHENERNYI SBORNIK

Citation	Translation
1950 V6 P153-160	63-24043 <=*>
1951 V9 P143-166	AD-639 486 <=$>
1953 V15 P15-20	60-13816 <+*>
1953 V16 P119-148	63-19596 <=*>
1953 V17 P43-58	R-3921 <*>
1953 V17 P163-170	59-22607 <+*> C
1954 V19 P55-64	09Q69R <ATS>
1954 V19 P73-82	RT-3645 <*>
1954 V20 P49-54	60-13839 <+*>
1954 V20 P154-159	59-16552 <+*>
1955 V21 P79-96	T-2267 <INSD>
1955 V22 P48-52	R-4708 <*>
1956 V23 P36	59-17168 <+*>
1959 V25 P20-36	64-13643 <=*$>
1959 V25 P92-103	AD-449 227 <=$*> M
1959 V25 P92-100	62-18855 <=*>
1959 V25 P145-153	62-14005 <*>

1959 V25 P174-178 ARSJ V30 N11 P1034 <AIAA>
62-25886 <=*>
1959 V25 P179-187 ARSJ V30 N11 P1055 <AIAA>
1960 V27 P70-76 62-18856 <=*>
1960 V27 P87-91 62-23864 <=*>
1960 V28 P26-36 61-19712 <=*>
1960 V28 P36-43 AIAAJ V1 N11 P2689 <AIAA>
63-23478 <=*$>
1960 V28 P56-75 63-16632 <=*>
1960 V28 P87-96 SCL-T-392 <=*>
1960 V28 P134-144 63-13888 <=*>
1960 V28 P145-150 AD-610 795 <=$>
1960 V29 P124-135 63-23628 <=*$>
1960 V30 P139-148 62-33623 <=*>
1961 V31 P3-14 65-12814 <*>
1961 V31 P171-178 64-11931 <=> M
1961 V31 P206-216 AIAAJ V1 N11 P2692 <AIAA>

*INZHENERNYI ZHURNAL
1961 N1 P159-163 AIAAJ V1 N11 P2696 <AIAA>
1961 V1 N1 P60-83 AIAAJ V1 N6 P1473 <AIAA>
62-20138 <=*>
1961 V1 N1 P159-163 63-15545 <=*>
1961 V1 N1 P175-176 63-19663 <=*>
1961 V1 N2 P26-30 66-61861 <=*$>
1961 V1 N3 P22-39 64-13333 <=*$>
1961 V1 N3 P60-64 63-23635 <=*>
1961 V1 N3 P65-74 AD-632 285 <=$>
1961 V1 N3 P153-156 63-23635 <=*>
1961 V1 N3 P161-165 63-23635 <=*>
1961 V1 N3 P165-169 62-33632 <=*>
1961 V1 N4 P27-38 63-23624 <=*>
1961 V1 N4 P39-50 64-11675 <=>
1962 V1 N4 P18-26 63-23442 <=*> 0
1962 V1 N4 P51-58 63-23442 <=*> 0
1962 V2 N1 P79-86 63-19329 <=*>
1962 V2 N1 P163-170 63-19329 <=*>
1962 V2 N1 P170-174 AIAAJ V1 N10 P2449 <AIAA>
1962 V2 N1 P175-181 63-19329 <=*>
1962 V2 N2 P211-221 63-19166 <=*>
1962 V2 N2 P231-238 63-11709 <=>
63-19533 <=*>
1962 V2 N2 P365-368 AD-615 236 <=$>
1962 V2 N3 P14-42 64-15314 <=$*>
1962 V2 N3 P68-73 64-15314 <=*$>
1962 V2 N3 P119-125 64-15314 <=*$>
1962 V2 N3 P158-160 65-13656 <*>
1962 V2 N3 P161-162 64-15314 <=*$>
1962 V2 N4 P227-231 65-13393 <*>
1962 V2 N4 P341-343 63-23111 <=*>
1962 V2 N4 P343-344 65-13659 <*>
1962 V2 N4 P344-349 63-24225 <=*$>
1963 V3 N2 P203-206 63-20852 <=*$>
64-11964 <=> M
1963 V3 N2 P207-221 64-13162 <=*$>
1963 V3 N2 P215-221 64-15109 <=*$>
1963 V3 N2 P322-330 64-11964 <=> M
1963 V3 N2 P367-373 AD-610 354 <=$>
1963 V3 N3 P513-516 63-20961 <=*$>
1963 V3 N3 P568-572 64-14163 <=*$>
1963 V3 N4 P628-631 AD-617 681 <=$>
AD-625 787 <=*$>
1963 V3 N4 P694-699 64-14119 <=*$>
1964 V4 N1 P79-89 65-10375 <*>
1964 V4 N1 P90-100 65-12646 <*>
1964 V4 N1 P127-129 AD-620 811 <=$>
1964 V4 N1 P140-147 65-10408 <*>
1964 V4 N1 P147-155 65-12816 <*>
1964 V4 N2 P242-246 N65-11308 <=$>
1964 V4 N2 P321-325 N65-14564 <=*$>
1964 V4 N2 P330-336 66-14143 <=$>
1964 V4 N2 P392-423 N64-31340 <=$>
1964 V4 N3 P431-438 65-12644 <*>
65-62908 <=$>
1964 V4 N3 P446-450 AD-639 337 <=$>
1964 V4 N3 P475-485 65-12728 <*>
1964 V4 N3 P486-494 65-14353 <*>
1964 V4 N3 P504-509 AD-639 337 <=$>
1964 V4 N3 P527-532 65-12636 <*>
1964 V4 N3 P539-542 AD-609 148 <=>
1964 V4 N3 P566-570 N65-15159 <=$>
65-12727 <*>
1964 V4 N3 P571-572 65-12817 <*>
1964 V4 N4 P619-659 65-30823 <=$>
1964 V4 N4 P639-645 N65-23681 <=$>
1964 V4 N4 P673-784 65-30823 <=$>
1964 V4 N4 P743-750 AD-629 471 <=$>
1965 V5 N1 P1-278 65-30680 <*=$>
1965 V5 N1 P16-28 65-14555 <*>
65-63712 <=$>
1965 V5 N1 P29-34 65-14552 <*>
65-63188 <=$>
1965 V5 N1 P104-109 65-14554 <*>
65-63706 <=$>
1965 V5 N1 P189-192 65-14551 <*>
65-63192 <=$>
1965 V5 N1 P192-195 65-14046 <*>
65-63719 <=$>
1965 V5 N2 P203-210 65-14556 <*>
65-63709 <=$>
1965 V5 N2 P254-260 65-14557 <*>
65-63434 <=$>
1965 V5 N2 P282-286 65-14553 <*>
1965 V5 N3 P575-579 AD-635 777 <=$>
1965 V5 N4 P743-745 AD-635 834 <=$>
1965 V5 N5 P945-949 N66-23727 <=$>
1965 V5 N5 P996-999 66-61034 <=*$>
1965 V5 N6 P1087-1091 66-60593 <=*$>
1965 V5 N6 P1117-1121 66-12313 <*>
1965 V5 N6 P1127-1130 66-61981 <=*$>

INZYNIERIA I BUDOWNICTWO
1951 V8 N7/8 P276-287 61-18932 <*>
1958 N1 P11-16 59-17238 <*>

ION
1947 V7 N71 P375-381 62-18153 <*>
1951 V11 N117 P206-213 NP-TR-744 <*>
1951 V11 N117 P220 NP-TR-744 <*>
1951 V11 N121 P453-464 NP-TR-744 <*>
1954 V14 P599-607 T1826 <INSD>
1955 V15 P133-140 65-10020 <*> 0
1955 V15 P612-617 63-18874 <*> 0
1955 V15 P745-755 63-18880 <*> 0
1956 V16 P80-86 63-18881 <*> 0
1956 V16 N178 P285-297 65-10124 <*>
1958 V18 N202 P324-329 11K26SP <ATS>
1964 V24 P508-518 1220 <TC>
1964 V24 N227 P441-447 4125 <BISI>
1964 V24 N227 P456 4125 <BISI>

IPAREGESZSEGUGY
1960 V12 N10 P18-20 62-19222 <=*>

IRON AND STEEL. CHINESE PEOPLES REPUBLIC
SEE KANG T'IEH

IRRIGACION EN MEXICO
1940 V21 P17-36 60-16719 <*>
1941 V22 N5 P339-347 60-16725 <*>

IRYO
1956 V10 P616-620 58-2174 <*>
1956 V10 P856-857 58-1710 <*>
1963 V18 P8-15 66-13043 <*>

ISHIKAWAJIMA-HARIMA ENGINEERING REVIEW
SEE ISHIKAWAJIMA-HARIMA GIHO

*ISKUSSTVENNYE SPUTNIKI ZEMLI
1958 N2 P10-16 ARSJ V29 N10 P733 <AIAA>
1958 N2 P54-58 ARSJ V29 N10 P742 <AIAA>
1958 V1 P25-43 61-15042 <+*>
1958 V1 P44-49 60-31125 <=>
1958 V2 P3-9 62-14003 <*> 0
1958 V2 P10-16 61-15315 <=*>
1958 V2 P17-25 61-15316 <+*>
1958 V2 P26-31 61-15354 <=*>
1958 V2 P32-35 61-15357 <=*>
1958 V2 P36-49 61-19570 <*=>

Citation	Reference
	62-14011 <*> 0
1958 V2 P59-6C	C-3414 <NRCC>
1958 V2 P61-69	62-14012 <=*>
1959 P54-60	ARSJ V30 N7 P672 <AIAA>
1959 N3 P39-46	ARSJ V30 N7 P700 <AIAA>
1959 N3 P61-65	ARSJ V30 N4 P386 <AIAA>
1959 N3 P66-76	ARSJ V30 N7 P662 <AIAA>
1959 N3 P77-83	ARSJ V30 N4 P403 <AIAA>
1959 N3 P84-97	ARSJ V30 N4 P406 <AIAA>
1959 N3 P98-112	ARSJ V30 N7 P676 <AIAA>
1959 N3 P113-117	ARSJ V30 N7 P658 <AIAA>
1959 V3 P3-	60-21461 <=>
1959 V3 P3-12	61-27017 <*=>
1959 V3 P13-31	60-21607 <=>
1959 V3 P32-38	61-11110 <=>
	61-15118 <+*>
1959 V3 P39-47	61-11113 <=>
1959 V3 P39-46	63-24298 <=*$>
1959 V3 P47-53	61-11114 <=>
	64-16171 <=*$>
1959 V3 P54-60	61-1111 <=>
	63-24295 <=*$>
1959 V3 P66-76	60-21443 <=>
	62-24297 <=*$>
1959 V3 P77-	60-21466 <=>
1959 V3 P84-97	60-21465 <=>
	63-24296 <=*$>
1959 V3 P98-112	61-11198 <=>
1959 V3 P113-117	60-21442 <=>
1959 V3 P118-	60-21599 <=>
1959 V3 P118-124	62-14009 <*>
1960 N4 P18-30	ARSJ V31 N5 P706 <AIAA>
1960 N4 P31-34	ARSJ V32 N3 P485 <AIAA>
1960 N4 P56-85	ARSJ V31 N7 P976 <AIAA>
1960 N4 P86-117	ARSJ V32 N9 P1459 <AIAA>
1960 N4 P118-134	ARSJ V32 N1 P143 <AIAA>
1960 N4 P135-160	ARSJ V31 N9 P1329 <AIAA>
1960 N4 P161-164	ARSJ V31 N9 P1321 <AIAA>
1960 N4 P165-170	ARSJ V31 N9 P1341 <AIAA>
1960 N4 P184-194	ARSJ V31 N5 P715 <AIAA>
1960 N4 P195-204	ARSJ V31 N5 P699 <AIAA>
1960 N5 P16-23	ARSJ V31 N11 P1640 <AIAA>
1960 N5 P24-29	ARSJ V31 N7 P967 <AIAA>
1960 N5 P38-40	ARSJ V31 N5 P713 <AIAA>
1960 N5 P41-53	ARSJ V31 N11 P1624 <AIAA>
1960 N5 P573-576	ARSJ V31 N5 P698 <AIAA>
1960 V4 P3-17	60-27018 <=*>
	63-10312 <=*>
1960 V4 P18-30	61-27991 <*=>
	63-24322 <=*$>
1960 V4 P31-34	63-24320 <=*$>
1960 V4 P43-55	61-19531 <=*>
	63-10168 <=*>
1960 V4 P56-81	61-19039 <+*>
	63-10332 <=*>
1960 V4 P118-134	63-15941 <=*>
1960 V5 P41-53	61-31628 <=>
1960 V5 P60-65	62-14001 <=*>
1961 N6 P3-10	ARSJ V32 N5 P834 <AIAA>
1961 N6 P11-32	ARSJ V32 N11 P1762 <AIAA>
1961 N6 P33-47	ARSJ V32 N7 P1152 <AIAA>
1961 N6 P63-107	N65-15740 <=*$>
1961 N6 P127-131	ARSJ V32 N7 P1159 <AIAA>
1961 N6 P132-138	ARSJ V32 N5 P831 <AIAA>
1961 N7 P32-55	AIAAJ V1 N3 P748 <AIAA>
1961 N8 P5-45	AIAAJ V1 N8 P1985 <AIAA>
1961 N8 P46-56	AIAAJ V1 N2 P522 <AIAA>
1961 N9 P56-61	AIAAJ V1 N3 P744 <AIAA>
	IGR V4 N6 P631 <AGI>
1961 N9 P62-65	AIAAJ V1 N2 P520 <AIAA>
1961 N9 P71-77	AIAAJ V1 N2 P526 <AIAA>
1961 N10 P34-39	AIAAJ V1 N2 P516 <AIAA>
1961 V6 P3-10	TRANS.2 <STS>
	63-24332 <=*$>
1961 V6 P11-32	TRANS-3 <STS>
	63-24335 <=*$>
1961 V6 P48-62	62-14239 <=*>
	63-19733 <=*>
1961 V6 P63-10C	62-13589 <*=>
1961 V6 P101-107	TRANS.1 <STS>
1961 V6 P1C1-1C8	63-18705 <=*>
1961 V7 P3-22	62-11651 <=>
1961 V7 P23-31	62-11136 <=>
1961 V7 P32-55	62-24477 <=*>
1961 V7 P64-77	61-31577 <=>
1961 V7 P89-10C	61-31595 <=>
1961 V8 P5-45	AIAAJ,V1,N8 <AIAA>
	63-18706 <=*>
1961 V8 P46-56	63-13735 <=*>
1961 V8 P64-71	62-11687 <=>
	63-13381 <=*>
1961 V8 P87-89	63-20397 <=*>
1961 V8 P90-93	63-13381 <=*>
1961 V9 P20-29	62-18877 <=*>
	62-25066 <=*>
1961 V9 P30-40	TRANS-14 <STS>
	62-11679 <=>
1961 V9 P41-47	TRANS-15 <STS>
1961 V9 P56-61	TRANS-16 <STS>
1961 V9 P62-65	G6220 <STS>
1961 V9 P71-77	TRANS-17 <STS>
1961 V9 P78-85	TRANS-18 <STS>
1961 V9 P86-110	TRANS-19 <STS>
1961 V10 P10-21	63-10908 <=*>
1961 V10 P22-33	TRANS-4 <STS>
1961 V10 P34-39	TRANS-5 <STS>
1961 V10 P40-44	TRANS-6 <STS>
1961 V10 P45-47	TRANS-7 <STS>
1961 V10 P61-68	TRANS-8 <STS>
1961 V10 P69-71	TRANS-9 <STS>
1961 V10 P72-81	TRANS-10 <STS>
1961 V10 P93-95	TRANS-11 <STS>
1961 V10 P96-97	TRANS-12 <STS>
1961 V10 P98-101	TRANS-13 <STS>
1961 V10 P102-103	R6221 <STS>
	62-24051 <=*>
1961 V11 P15-22	N65-24036 <=*$>
1961 V11 P23-29	N65-24035 <=$>
1961 V11 P30-34	N65-24037 <=$>
1961 V11 P94-97	N65-24C38 <=$>
1961 V11 P98-107	N65-24106 <=$>
1962 N12 P1C5-118	65-13668 <*>
1962 N12 P145-150	AIAAJ V1 N9 P2209 <AIAA>
1962 N12 P151-158	AIAAJ V1 N9 P2212 <AIAA>
1962 N13 P61-66	AIAAJ V1 N11 P2700 <AIAA>
1962 N13 P81-84	AIAAJ V1 N5 P1254 <AIAA>
1962 N13 P1C7-1C9	AIAAJ V1 N9 P2216 <AIAA>
1962 V12 P3-5	TRANS-6228 <STS>
1962 V12 P31-34	TRANS-K6229 <STS>
1962 V12 P35-46	63-18701 <=*>
1962 V12 P47-50	TRANS-K6230 <STS>
1962 V12 P105-118	N65-23904 <=$>
1962 V12 P119-132	N65-23905 <=*$>
1962 V12 P145-150	63-20154 <=*$>
1962 V13 P1-66	63-19536 <=*>
1962 V13 P3-22	63-14894 <=*>
1962 V13 P23-52	63-18704 <=*>
1962 V13 P53-6C	63-18703 <=*>
1962 V13 P67-74	N65-23906 <=$>
1962 V13 P67-84	63-19534 <=*>
1962 V13 P75-81	TRANS-S6224 <STS>
1962 V13 P81-84	N65-23907 <=$>
1962 V13 P81-85	TRANS-S6225 <STS>
1962 V13 P85-88	TRANS-S6226 <STS>
1962 V13 P85-122	63-19536 <=*>
1962 V13 P89-96	N65-24673 <=$*>
1962 V13 P107-109	N65-24672 <=*$>
	TRANS-T6227 <STS>
1962 V13 P110-118	N65-24657 <=$*>
1962 V13 P123-133	63-19534 <=*>
1962 V14 P7-12	65-13658 <*>
1962 V14 P30-48	65-13672 <*>
1962 V14 P57-68	M.5391 <NLL> 0
1962 V14 P74-80	63-23135 <=*>
1962 V14 N12 P133-140	JAS V11,N4,P113-117 <AAS>
	N65-24669 <=$*>
1963 V15 P53-56	65-13381 <*>
1963 V15 P102-103	64-13104 <=*$>
1963 V16 P5-9	63-11774 <=>
1963 V16 P10-33	64-11512 <=>

1963 V17 P101-106	65-10751 <*>

ISOTOPEN-TECHNIK

1960 V1 N7 P200-203 PT.2	NP-TR-921 <*>
1960 V1 N5/6 P136-137 PT.1	NP-TR-921 <*>
1962 V2 P69-70	AEC-TR-6305 <*>
1962 V2 P278-279	65-11483 <*> 0
1962 V2 N3 P81-83	NP-TR-1105 <*>
1962 V2 N3 P83-85	NP-TR-1109 <*>
1962 V2 N8 P230-235	63-20990 <*> 0
1962 V2 N11 P341-343	63-21435 <=>
1962 V3 P66-69	AEC-TR-5689 <*>

ISOTOPES AND RADIATION. JAPAN
SEE DOITAI TO HOSHASEN

ISSLEDOVANIE SPLAVOV TSVETNYKH METALLOV

1959 V1 P111-116	60-13297 <+*>
1960 V2 P9-18	61-27523 <*=> 0
1960 V2 P197-204	5174 <HB>
	62-13736 <*=>
1962 V3 P143-148	63-11743 <=>
1963 V4 P17-24	64-21207 <=>
1963 V4 P108-116	64-21201 <=>
1963 V4 P117-129	64-21473 <=>
1963 V4 P185-203	64-21472 <=>
1963 V4 P211-223	65-64239 <=$*>
1963 V4 P266-278	64-21208 <=>

ISSLEDOVANIYA PO EKSPERIMENTALNOI I TEORETICHESKOI FIZIKE. FIZICHESKII INSTITUT IMENII P. N. LEBEDEVA. AKADEMIYA NAUK SSSR

1959 P62-70	62-33633 <=*>

ISSLEDOVANIYA FIZICHESKIKH PARAMETROV VESHCHESTVA ZEMNOI KORY

1964 N9 P3-13	65-33099 <=*$>
1964 N9 P114-117	65-33097 <=*$>
1964 N9 P118-123	65-33098 <=*$>

ISSLEDOVANIYA PO FIZIKE ATMOSFERY

1959 V1 P7-14	63-19658 <=*>
1960 V2 P67-114	63-10937 <=*>
1962 N3 P110-135	65-61544 <=$> 0
1962 V3 P5-22	65-61543 <=$>
1962 V3 P23-70	N65-11694 <=$>
	65-61328 <=$> 0
1962 V3 P72-84	65-61547 <=$>
1962 V3 P85-109	65-61545 <=$> 0
1963 V4 P111-119	65-61546 <=$> 0

ISSLEDOVANIYA PO GENETIKE

1961 V1 N1 P147-160	RTS-2870 <NLL>
1964 P3-20	65-32010 <=$>
1964 P46-85	65-32010 <=$>
1964 P125-139	65-32010 <=$>

ISSLEDOVANIYA IONOSFERY I METEOROV

1959 N1 ENTIRE ISSUE	60-21752 <=*>
	61-11158 <=>
1960 P106-113	62-15478 <=*> 0
1960 N2 P40-53	61-31578 <=>
1960 N3 P77-82	61-28216 <*=>
	62-23492 <=*>
1960 N5 P81-92	63-21212 <=>
1962 N8 P72-75	62-11750 <=>
1962 N10 P27-33	N65-24665 <=$>
1962 V8 P97-101	62-11748 <=>
1964 N12 P10-32	AD-613 982 <=$>
1964 N12 P74-88	AD-613 982 <=$>
1964 N12 P109-139	AD-613 982 <=$>
1964 N12 P152-159	AD-613 982 <=$>

ISSLEDOVANIYA I MATERIALY. INSTITUT PO IZUCENIJU SSSR. MUENCHEN. SERIYA 1

1959 S1 N49 P5-76	63-11087 <=>

ISSLEDOVANIYA PO PRIKLADNOI KHIMII. AKADEMIYA NAUK SSSR. OTDELENIE KHIMICHESKIKH NAUK

1955 P175-183	R-2238 <*>
1955 P184-191	R-2837 <*>

ISSLEDOVANIYA PROTSESSOV GORENIYA USSR

1958 N12 P78-80	62-13053 <=*>

ISSLEDOVANIYA SEREBRISTYKH OBLAKOV

1960 N1 P3-84	63-13877 <=*>

ISSLEDOVANIYA PO TEORII SOORUZHENII

1964 N13 P77-83	66-13359 <*>
1964 N13 P121-126	66-61736 <=*$>
1965 N13 P127-134	66-61719 <=$>

ISSLEDOVANIYA PO UPRUGOSTI I PLASTICHNOSTI

1961 N1 P46-51	64-16172 <*>
1963 N2 P48-58	64-10497 <*>
1963 N2 P59-65	64-10493 <=*$>
1963 N2 P66-73	RTS-2647 <NLL>
1963 N2 P74-80	64-10597 <=*$>
1963 N2 P81-90	64-10495 <=*$>
1963 N2 P90-104	64-10813 <=*$>
1963 N2 P121-131	64-10496 <*>
1963 N2 P212-215	64-10494 <=*$>
1963 N2 P347	AD-630 414 <=*$>
1963 N3 P52-61	RTS-2648 <NLL>
1963 N3 P225-231	RTS-2649 <NLL>
1964 P172-191	65-31966 <=$>
1964 N3 P107-113	65-10421 <*>
1964 N3 P114-123	65-10422 <*>
1964 N3 P265-270	65-10420 <*>
1965 N4 P65-71	66-61696 <=*$>

ISSLEDOVANIYA PO ZHAROPROCHNYM SPLAVAM

1956 P11-16	62-20249 <*>
1956 V1 P52-59	4235 <HB>
1957 V2 P3-8	M.5813 <NLL>
1957 V2 P52-56	60-16968 <*> 0
1957 V2 P131-134	60-16969 <*> 0
1957 V2 P135-140	60-16978 <*> 0
1957 V2 P141-147	60-16974 <*> 0
1957 V2 P148-157	60-16970 <*> 0
1957 V2 P158-162	60-16965 <*> 0
1957 V2 P171-180	60-14506 <+*> 0
1957 V2 P198-210	60-14507 <+*> 0
1957 V2 P234-245	60-16967 <*> 0
1957 V2 P246-250	60-16966 <*> 0
1957 V2 P251-256	60-16975 <*> 0
1957 V2 P257-265	60-16977 <*> 0
1957 V2 P266-274	60-16976 <*> 0
1957 V2 P320-329	60-13052 <+*>
1957 V2 P320-328	60-16979 <*> 0
1958 V3 P339-345	63-24146 P572-583 <*=$>
1958 V3 P381-383	63-24146 P643-646 <*=$>
1959 V4 P3-12	63-21017 <=>
1959 V4 P71-77	63-21017 <=>
1959 V4 P90-95	UCRL-TRANS-888(L) <=*>
	62-13161 <*=>
1959 V4 P170-175	62-13162 <*=>
1959 V4 P193-201	63-21017 <=>
1959 V4 P301-322	63-21061 <=>
1959 V4 P346-351	AD-636 626 <=$>
1959 V4 P352-359	63-21061 <=>
1959 V4 P360-366	61-27631 <*=> 0
1959 V5 P150-154	64-31371 <=>
1959 V5 P166-172	64-31371 <=>
1959 V5 P173-178	61-27522 <*=>
1959 V5 P210-219	61-27524 <*=> 0
1959 V5 P228-233	64-31371 <=>
1959 V5 P277-279	64-31371 <=>
1959 V5 P303-307	64-31371 <=>
1960 V6 P3-16	62-24902 <=>
1960 V6 P17-24	62-24908 <=>
1960 V6 P25-28	62-24909 <=*>
1960 V6 P29-33	62-24910 <=>
1960 V6 P64-70	62-24649 <=*>
1960 V6 P105-111	62-25295 <=*> 0
1960 V6 P120-129	63-21077 <=>
1960 V6 P140-145	62-24905 <=>
1960 V6 P169-173	62-15199 <*=>

1960 V6 P180-186 62-13170 <*=> O
1960 V6 P187-194 61-28810 <*=> O
1960 V6 P195-200 61-28809 <*=> O
1960 V6 P201-205 62-13171 <*=> O
1960 V6 P211-222 61-28808 <*=> O
63-23381 <=*$>
1960 V6 P240-250 62-23496 <*=>
1960 V6 P251-258 63-21077 <=>
1960 V6 P259-267 61-28806 <*=> O
1960 V6 P268-277 62-13172 <*=>
1960 V6 P278-283 63-19787 <=*> O
1960 V6 P308-311 62-23963 <=$>
1961 V7 P3-10 63-11590 <=>
1961 V7 P210-213 AD-615 256 <=$>
1961 V7 P309-316 63-19396 <=*>
1962 V8 P85-87 64-19007 <=*$> C
1962 V8 P88-126 63-13309 <=*>
1962 V8 P178-183 64-19136 <=$> O
1962 V8 P205-210 RTS-3240 <NLL>
1962 V8 P224-229 64-19006 <=*$> O
1962 V8 P242-250 64-19010 <=*$> C
1962 V9 P14-22 64-19145 <=$>
1962 V9 P42-46 AD-618 926 <=$>
65-61185 <=> O
1962 V9 P66-72 AD-618 926 <=$>
1962 V9 P87-88 AD-618 926 <=$>
1962 V9 P119-126 64-26148 <$>
65-61435 <=>
1962 V9 P133-139 AD-618 926 <=$>
1962 V9 P172-177 AD-615 220 <=$>
1962 V9 P187-189 AD-618 926 <=$>
1962 V9 P192-194 63-23215 <=*$>
1962 V9 P194-203 64-13712 <=*$>
1962 V9 P218-248 AD-618 926 <=$>
1963 V10 P27-31 AD-618 008 <=$>
1963 V10 P93-102 65-61338 <=$>
1963 V10 P116-123 65-61434 <=>
1963 V10 P123-130 M-5582 <NLL>
1963 V10 P130-137 65-61447 <=>
1963 V10 P214-218 65-61170 <=>
1963 V10 P239-246 65-61172 <=>
1963 V10 P252-257 M-5586 <NLL>

ISTANBUL UNIVERSITESI FEN FAKULTESI MECMUASI. SERI A
1951 V16A P30-32 AL-697 <*>
1952 V17 P159-177 65-12349 <*>

ISTANBUL UNIVERSITESI FEN FAKULTESI MECMUASI. SERI B. SCIENCES NATURELLES
1947 V12 P299-307 II-1528 <*>

ISTANBUL UNIVERSITESI FEN FAKULTESI MECMUASI. SERI C
1958 V23 P42-49 63-10159 <*>
1958 V23 N3 P197-200 89M43G <ATS>

ISTORIYA SSSR. MOSCOW
1966 N4 P232-233 AD-640 504 <=$>

ITALIA GRAFICA
1960 V15 N11 P9-10 64-16529 <*>
1960 V15 N13 P4-7 64-18351 <*>

ITOGI NAUCHNO-ISSLEDOVATELSKIKH RABOT VSESOYUZNOGO INSTITUTA ZASHCHITY RASTENII
1937 V1 P168-170 RT-1776 <*>

ITOGI NAUKI: BIOLOGICHESKIE NAUKI
1957 N1 P329-379 AEC-TR-3998 <+*>
1957 N1 P379-392 59-11758 <=>
1960 N4 P1-316 61-28566 <=*>

ITOGI NAUKI: FIZIKO-MATEMATICHESKIE NAUKI
1960 V3 P75-116 61-23146 <*=>

IUNYI TEKHNIK
1957 V2 N2 P34-36 59-12182 <+*>
1957 V2 N12 P46-48 59-11876 <=>
1958 V2 N4 P44-48 63-14682 <=*>
1960 V5 N6 P37-41 61-27266 <=*>
1961 V5 N3 P25-28 62-23459 <=*>
1961 V6 N2 P37-41 61-31625 <=>
1961 V6 N5 P8-13 62-19637 <=*> C
1962 V7 N3 P38 62-32556 <=*>
1962 V7 N3 P40-43 62-32556 <=*>

IWATE DAIGAKU NOGAKUBU HOKOKU
1953 V1 P187-191 74G8J <ATS>

IWATE IGAKU ZASSHI
1964 V16 N1 P15-21 66-11582 <*>

IZDANIYA. ZAVOD ZA RIBARSTVO NA N. R. MAKEDONIJA
1955 V1 N5 P135-147 60-21648 <=>

IZDANIYA. NOSOVSKAYA SELSKO-KHOZYAISTVENNAYA OPYTNAYA STANTSIYA. NOSOUKA. OTDEL AGO-KHIMICHESKII
1927 N47 ENTIRE ISSUE 66-51135/1 <=>

IZDANIIA. OBSHCHESTVO PO RASPROSTRANENIYU POLITICHESKIKH I NAUCHNYKH ZNANII. SERIIA 2
1961 S2 N1 P48 61-31108 <=>

IZDANIIA. OBSHCHESTVO PO RASPROSTRANENIYU POLITICHESKIKH I NAUCHNYKH ZNANII. SERIIA 3
1954 S3 N23 P1-31 RT-4275 <*>
1954 S3 N37 P3-24 RT-3899 <*>
1955 S3 N42 P3-32 RT-4253 <*> O

IZDANIYA VSESOYUZNOGO OBSHCHESTVA PO RASPROSTRANENIYU POLITICHESKIKH I NAUCHNYKA ZNANII. SERIYA 4. NAUKA I TEKHNIKA
1953 S4 N37 P31 61-13719 <=*> P
1954 S4 N40 P3-32 RT-4326 <*>
1957 N14 P1-39 59-16059 <+*>
1958 N11 P3-38 59-12183 <+*>
1958 N13 P16-24 62-19123 <=*>
1958 N17 P24-31 61-21951 <=> O
1958 S4 N34/5 P70 61-19676 <=*>
1959 N9 P1-31 62-13175 <*=>
1959 N10 P1-31 61-23680 <*=>
1959 S4 N28 P1-32 61-21652 <=>
1959 S4 N34 P21-29 61-19682 <=*>
1960 N8 P1-40 61-31152 <=>
1960 N12 P1-32 61-23625 <*=>
1960 N27 P1-46 62-13179 <*=>
1960 N35/6 P5-17 61-27263 <*=>
1960 S4 N29 P4 61-21388 <=> P
1960 S4 N29 P7-8 61-21388 <=> P
1960 S4 N29 P16-17 61-21388 <=> P
1960 S4 N29 P19 61-21388 <=> P
1961 S4 N4 P5-14 61-23128 <=>
1963 N2 P27-28 63-19701 <=*>

IZDANIIA. OBSHCHESTVO PO RASPROSTRANENIYU POLITICHESKIKH I NAUCHNYKH ZNANII. SERIIA 5. SELSKOE KHOZIAISTVO
1955 N14 P3-32 RT-4653 <*>
1958 S5 N4 ENTIRE ISSUE 59-16800 <+*>
1959 N18 P1-32 61-28127 <*=>

IZDANIIA. OBSHCHESTVO PO RASPROSTRANENIYU POLITICHESKIKH I NAUCHNYKH ZNANII. SERIIA 8. BIOLOGIIA I MEDITSINA
1958 N11 ENTIRE ISSUE 59-11549 <+> O
1958 N13 ENTIRE ISSUE 59-11549 <+> O
1960 N21 P1-31 61-21531 <=>
1960 S8 N4 P3-47 61-21764 <=>
1961 N9 P3-32 61-31585 <=>
1961 N16 P1-48 62-15456 <*=>
1962 N19 P1-64 63-21743 <=>
1963 N1 P1-48 64-13097 <=*$>

IZDANIIA. OBSHCHESTVO PO RASPROSTRANENIYU POLITICHESKIKH I NAUCHNYKH ZNANII. SERIIA 9. FIZIKA I KHIMIKA
1960 N20 P41-44 62-15821 <=*>
1960 S8 N19 P2-32 61-21257 <=>
1961 N3 P1-32 61-27689 <=*>
1961 N7 P33-45 62-19635 <=*>
1961 N15 P1-48 62-24017 <=*>
1963 S9 N2 ENRERE ISSUE 63-21383 <=>

IZDANIIA. OBSHCHESTVO PO RASPROSTRANENIYU POLITICHESKIKH I NAUCHNYKH ZNANII. SERIIA 10. MOLODEZHNAIA
1960 S10 N6 P1-38 61-23582 <=*>

IZDANIIA. OBSHCHESTVO PO RASPROSTRANENIYU POLITICHESKIKH I NAUCHNYKH ZNANII. SERIIA 11. PEDAGOGIKA
1960 N15 P3-47 61-23136 <=*>

IZDANIIA. OBSHCHESTVO PO RASPROSTRANENIYU POLITICHESKIKH I NAUCHNYKH ZNANII. SERIIA 12. GEOLOGIIA I GEOGRAFIIA
1963 N9 P7-16 63-31170 <=>
1963 N9 P37-42 63-31170 <=>

IZGRADNJA
1963 V17 N1 P50-51 63-31271 <=>

*IZMERITELNAYA TEKHNIKA
1950 N5 P52-54 R-2364 <*>
1955 P14-21 R-177 <*>
1955 P61- R-178 <*>
1955 N1 P3-5 R-47 <*>
1955 N1 P6-9 R-176 <*>
1955 N1 P22-27 R-57 <*>
1955 N1 P31-35 R-4060 <*>
63-15814 <=*>
1955 N1 P59- R-174 <*>
1955 N1 P62- R-691 <*>
1955 N2 P6-11 RT-4318 <*>
1955 N2 P20-23 RT-4350 <*>
1955 N2 P42-45 RT-4465 <*>
1955 N2 P61 RT-4317 <*>
1955 N3 P3-9 RT-4355 <*>
1955 N3 P10-19 RT-4272 <*>
1955 N3 P19-25 RT-4400 <*>
1955 N3 P40-42 RT-4401 <*>
1955 N4 P6-12 61-13538 <=*>
1955 N5 P3-6 86N49R <ATS>
1956 P3- R-4273 <*>
R-4274 <*>
1956 P6- R-4272 <*>
1956 P11- R-4271 <*>
1956 N1 P15-16 R-4059 <*>
63-15814 <=*>
1956 N1 P16-18 R-4061 <*>
1956 N1 P18-19 R-4063 <*>
1956 N1 P18-23 63-15814 <=*>
1956 N1 P20-23 R-4062 <*>
1956 N1 P43-46 R-4064 <*>
1956 N2 P12-16 R-4722 <*>
1956 N2 P57-61 59-18107 <+*>
1956 N4 P3-16 R-4276 <*>
1956 N4 P21-24 R-4721 <*>
1956 N5 P15-17 R-4723 <*>
1956 N5 P17-18 R-4724 <*>
1956 N5 P49-52 R-2343 <*>
1957 N2 P22-25 R-5101 <*>
1957 N4 P29-39 59-12145 <+*>
1957 N4 P43-47 66-11272 <*>
1957 N5 P29-32 AD-615 431 <=$>
1957 N5 P65-66 59-22536 <+*>
1957 N5 P67-68 59-22537 <+*>
1958 N2 P94-97 PB 141 480T <=>
1958 N4 P31-32 60-13846 <+*>
1958 N5 P24-25 60-21062 <=>
1958 N5 P65-66 60-21065 <=>
1958 N6 P15-17 59-18124 <+*>
1958 N6 P30-35 60-21785 <=>
1958 N6 P87-93 59-13445 <=> O
1959 N1 P58-62 60-17201 <+*> C
1959 N8 P27-28 62-33585 <=>
1959 N8 P29 61-13058 <=*>
1959 N8 P40-43 61-27635 <*=>
1959 N9 P3-5 59-31050 <=>
1959 N9 P5-8 59-31050 <=>
1959 N9 P8-11 59-31050 <=>
1959 N9 P12-14 59-31050 <=>
1959 N9 P57-58 59-31050 <=>
1959 N10 P11-13 93M45R <ATS>
1959 N11 P26-29 60-41266 <=>
1959 N11 P35-38 61-13055 <=*>
1959 N11 P67-68 60-11734 <=>
1959 N12 P28-32 61-13147 <=*>
1960 N4 P17-19 60-41342 <=>
1960 N4 P56-57 60-41342 <=>
1960 N4 P62-64 60-41342 <=>
1960 N5 P21-25 62-13120 <=*>
1960 N5 P64 60-41342 <=>
1960 N8 P41-43 61-23180 <=*>
1960 N12 P22-25 62-13607 <*=>
1960 N12 P40-42 61-28229 <=*>
1961 N1 P32-35 61-27836 <*=>
1961 N3 P25-28 62-13722 <*=>
1961 N4 P1-4 61-27276 <=*>
1961 N4 P58-64 61-27276 <=*>
1961 N4 P64 62-13374 <*=>
1961 N5 P17 62-24877 <=*>
1963 N5 P44-48 TIL/T-5616 <=$>
1964 N2 P30-31 65-14825 <*>
1964 N6 P7-9 64-51261 <=>
1964 N6 P21-22 RTS-2921 <NLL>
1964 N6 P33-36 64-51261 <=>
1965 N8 P51-52 66-30617 <=$>

IZOBRETATEL I RATSIONALIZATOR
1958 N11 P24-25 60-27013 <*+>
1959 N3 P45 NLLTB V1 N8 P41 <NLL>
1959 N4 P8-9 61-15652 <+*> O
1960 N5 P43-46 62-13274 <=*> C
1960 N8 P12-14 61-27234 <*=>
1960 N12 P12-19 61-28263 <*=>
1960 N12 P48-52 61-28263 <*=>
1961 N10 P20-21 62-32584 <=*>
1961 N10 P33 62-32584 <=*>
1962 N8 P29-30 62-11778 <=>
65-11171 <*>
1962 N10 P15-16 63-21324 <=>
1965 N1 P28-31 65-32115 <=$>
1965 N7 P9-13 65-33694 <=$>

IZVESTIYA (THE NEWSPAPER)
1954 N107 P2-3 RT-1883 <*>
1962 09/12 P3 NLLTB V5 N1 P34 <NLL>
1963 01/26 P1 NLLTB V5 N6 P494 <NLL>
1963 01/26 P3 NLLTB V5 N6 P494 <NLL>
1963 09/10 P2 NLLTB V6 N2 P144 <NLL>
1963 11/23 P3 NLLTB V6 N3 P236 <NLL>
1964 09/06 P4 NLLTB V7 N3 P201 <NLL>
1965 01/28 P3 NLLTB V7 N6 P522 <NLL>
1965 02/11 P4 NLLTB V7 N7 P618 <NLL>
1965 06/22 P4 NLLTB V7 N11 P941 <NLL>
1966 N37 P2 NLLTB V8 N6 P484 <NLL>

IZVESTIYA AKADEMII NAUK ARMYANSKOI SSR. EREVAN. BIOLOGICHESKIE NAUKI
1959 V12 N2 P3-15 60-18379 <=>
1959 V12 N2 P3-14 60-31460 <=>
1959 V12 N3 P85-93 62-25112 <=*>
1960 V13 N2 P3-16 62-13603 <=*>
1960 V13 N8 P35-58 61-27282 <*=>
1960 V13 N11 P3-9 61-23135 <=*>
1961 V14 N3 P97-99 62-13714 <*=>
1961 V14 N4 P7-18 63-23003 <=*>
1961 V14 N5 P85-91 62-23247 <=*>
1961 V14 N8 P3-21 64-15433 <=*$>
1962 V15 N1 P59-68 64-15439 <=*$>
1962 V15 N11 P85-93 63-21044 <=>
1962 V15 N12 P3-14 64-21643 <=>

1963 V16 N1 P3-7 64-19666 <=$>
1963 V16 N1 P23- FPTS V24 N2 P.T259 <FASE>
1963 V16 N2 P31-40 RTS-3107 <NLL>
1963 V16 N3 P19-26 65-60154 <=$>
1963 V16 N5 P3-6 64-19667 <=$>
1963 V16 N5 P7-13 65-60148 <=$>
1963 V16 N6 P11-26 65-62342 <=$>
1963 V16 N6 P69-74 65-63681 <=$>
1963 V16 N8 P21- FPTS V24 N2 P.T109 <FASE>
1964 V17 N1 P111-115 64-31307 <=>
1964 V17 N7 P45-54 AD-627 069 <=$>
1964 V17 N8 P47-51 65-63385 <=$>
1964 V17 N12 P47-53 RTS-3101 <NLL>

IZVESTIYA AKADEMII NAUK ARMYANSKOI SSR, EREVAN
BIOLOGICHESKIE I SELSKOKHOZYAISTVENNYE NAUKI
1950 V3 N4 P333-345 RT-3152 <*> 0
1954 V7 N4 P43-50 RT-4011 <*>
1954 V7 N4 P51-57 RT-4038 <*> 0
1954 V7 N5 P15-25 R-121 <*>
1954 V7 N10 P57-64 R-122 <*>
1955 V8 N5 P89-91 60-15528 <+*>
1956 V9 N2 P63-66 R-1528 <*>
1956 V9 N3 P123-130 RT-4610 <*>
1957 V10 N3 P85-88 59-10197 <+*>
1957 V10 N5 P41-56 61-28639 <*=>
1957 V10 N5 P65-67 60-15531 <+*>
1957 V10 N6 P25-34 62-14694 <=*>
1958 V11 N4 P33-43 63-11091 <=>
1958 V11 N5 P67-72 64-11006 <=>
1958 V11 N5 P97-99 63-11036 <=>
1958 V11 N6 P21-27 60-51159 <=>
1958 V11 N11 P13-17 62-15622 <=*>
1959 V12 N1 P69-73 59-11769 <=>
1959 V12 N12 P63-71 63-23064 <=*$>

IZVESTIYA AKADEMII NAUK ARMYANSKOI SSR, EREVAN.
GEOLOGICHESKIE I GEOGRAFICHESKIE NAUKI
1961 V14 N4 P37-43 IGR V4 N8 P925 <AGI>
1962 N1 P3-15 IGR V5 N11 P1432 <AGI>
1962 V15 N5 P35-46 IGR V7 N1 P1 <AGI>

IZVESTIYA AKADEMII NAUK ARMYANSKOI SSR, EREVAN.
KHIMICHESKIE NAUKI
1958 V11 N1 P3-11 62-18006 <=*>
1958 V11 N3 P201-209 62-24842 <=*>
1961 V14 N2 P147-149 63-10750 <=*>
1962 V15 P521-525 65-14899 <*>
1962 V15 N1 P33-37 65-23041 <$>
1963 V16 N3 P201-203 AD-622 388 <=*$>
1964 N6 P53-58 ICE V5 N3 P533-536 <ICE>
1964 V17 P301-305 65-10488 <*>
1964 V17 N5 P573-576 1399 <TC>
1964 V17 N5 P591-593 65-13956 <*>
1965 V18 P216-218 65-14901 <*>
1965 V18 P532-534 66-14177 <*$>

IZVESTIYA AKADEMII NAUK ARMYANSKOI SSR, EREVAN.
OBSHCHESTVENNYE NAUKI
1960 N3 P35-46 61-11002 <=>
1961 N11 P35-46 62-24979 <=*>

IZVESTIYA AKADEMII NAUK ARMYANSKOI SSR, EREVAN.
SERIYA FIZIKO-MATEMATICHESKIE, ESTESTVENNYE I
TEKHNICHESKIE NAUKI
1954 V7 P1-17 AMST S2 V35 P79-94 <AMS>
1955 V8 N3 P101-111 59-21144 <=> 0
1955 V8 N4 P73-78 59-15087 <+*>
1956 V9 N2 P3 60-14204 <+*>

IZVESTIYA AKADEMII NAUK ARMYANSKOI SSR. EREVAN.
SERIYA FIZIKO-MATEMATICHESKIKH NAUK
1958 V11 N1 P27-46 63-10198 <=*>
1958 V11 N2 P31-40 STMSP V4 P1-11 <AMS>
1958 V11 N5 P3-7 60-19015 <+*>
1958 V11 N6 P61-71 60-31855 <=>
1959 V12 N3 P95-99 62-15888 <=>
1959 V12 N5 P115-128 62-19934 <=*>
1960 V13 N1 P141-151 61-15940 <*=>
62-11566 <=>
1960 V13 N2 P105-108 63-18754 <=*>
1960 V13 N2 P123-129 61-28530 <*=>
1960 V13 N3 P89-95 61-28245 <*=>
1960 V13 N4 P65-68 62-23359 <=*>
1960 V13 N5 P55-64 62-15173 <*=>
1960 V13 N6 P377-383 61-28760 <*=>
1961 V14 N1 P41-49 62-15828 <*=>
1961 V14 N1 P79-86 62-19276 <=*>
1961 V14 N2 P7-16 63-11593 <=>
64-19993 <=$>
1961 V14 N2 P45-70 62-32284 <=*>
1961 V14 N5 P3-8 63-11614 <=>
65-60510 <=$>
1961 V14 N6 P117-123 UCRL TRANS-911(L) <=*>
1962 V15 N1 P123-134 UCRL-TRANS-1114(L) <=*$>
1962 V15 N2 P87-99 63-13055 <=*>
1962 V15 N2 P153-159 65-60505 <=$>
1962 V15 N4 P55-63 65-60487 <=$>
1962 V15 N5 P11-26 RTS-2642 <*>
1963 V16 N1 P55-61 64-19375 <=$>
1963 V16 N3 P65-82 RTS-2734 <NLL>
1963 V16 N5 P91-98 64-71527 <=>
1963 V16 N6 P117-124 AD-618 044 <*>
1964 V17 N1 P113-121 AD-621 054 <=*$>
1964 V17 N3 P29-53 65-14276 <*>
65-63198 <=$>
1964 V17 N3 P65-70 AD-640 259 <=$>
1964 V17 N4 P43-50 AD-631 766 <=$>
1964 V17 N5 P43-46 AD-618 014 <=$>
65-13346 <*>
1964 V17 N6 P57-64 65-13351 <*>
65-63732 <=*$>
1964 V18 N1 P53-60 65-63698 <=*$>
1965 V18 N1 P34-42 65-14659 <*>
65-63692 <=*$>
1965 V18 N1 P53-60 65-14658 <*>
1965 V18 N2 P32-38 65-14697 <*>

IZVESTIYA AKADEMII NAUK ARMYANSKOI SSR, EREVAN.
SERIYA GEOLOGICHESKIKH I GEOGRAFICHESKIKH NAUK
1957 V10 N4 P121-126 60-15887 <+*>
1958 V11 N2 P55-70 63-19958 <*=> 0

IZVESTIYA AKADEMII NAUK ARMYANSKOI SSR, EREVAN.
SERIYA KHIMICHESKIKH NAUK
1957 V10 N3 P203-212 31L35R <ATS>

IZVESTIYA AKADEMII NAUK ARMYANSKOI SSR, EREVAN.
SERIYA TEKHNICHESKIKH NAUK
1957 V10 N3 P21-34 61-19237 <+*>
1958 V11 N4 P69-74 NLLTB V1 N3 P30 <NLL>
59-13081 <=> 0
1959 V12 N5 P31-45 65-12908 <*>
1960 N6 P11-13 NLLTB V3 N7 P585 <NLL>
1960 V13 N1 P57-70 61-27472 <*=>
1960 V13 N4 P3-15 65-13383 <*>
1961 V14 N2 P7-11 62-25133 <=*>
1961 V14 N3 P43-57 64-19351 <=*$>
1963 V16 N2/3 P129-132 65-64513 <=*$>
1965 N3 P45-56 ICE V6 N1 P74-81 <ICE>
1965 N6 P57-63 ICE V6 N3 P445-448 <ICE>

IZVESTIYA AKADEMII NAUK AZERBAIDZHANSKOI SSR. BAKU
1936 N28 P25-32 61-18069 <=*>
1938 P75-79 61-18233 <=*>
1938 P133-150 61-18098 <=*>
1938 N55 P167-174 61-18070 <=*>
1954 N8 P89-94 R-4152 <*>
1955 N7 P3-25 35J14R <ATS>
1955 N8 P11-15 61-19325 <+*>
1956 N2 P12-27 C-2523 <*>
R-3061 <*>
1957 N9 P11-19 ICE V1 N1 P106-110 <ICE>
1957 V14 N3 P87-95 60-23743 <+*>
1957 V14 N9 P11-19 38N50R <ATS>

IZVESTIYA AKADEMII NAUK AZERBAIDZHANSKOI SSR. BAKU
SERIYA BIOLOGICHESKIKH I MEDITSINSKIKH NAUK
1960 N4 P105-112 62-25637 <=*>
1960 N4 P173-175 61-23020 <*=>
1961 N1 P15-23 63-13239 <=*>
1961 N10 P55-66 64-19547 <=$>
1963 N2 P61-68 64-15447 <=*$>
1963 N2 P117-126 64-15530 <=*$>

IZVESTIYA AKADEMII NAUK AZERBAIDZHANSKOI SSR. BAKU
SERIYA BIOLOGICHESKIKH NAUK
1964 N5 P95-101 66-60411 <=*$>
1965 N6 P53-57 66-32860 <=$>

IZVESTIYA AKADEMII NAUK AZERBAIDZHANSKOI SSR. BAKU
SERIYA FIZIKO-MATEMATICHESKIKH I TEKHNICHESKIKH
NAUK
1959 N3 P44 61-13308 <*+>
1959 N4 P23-32 62-14007 <=*>
1959 N4 P43-47 61-14961 <*=>
63M46R <ATS>
1959 N5 P127-134 ARSJ V31 N11 P1650 <AIAA>
1960 N1 P45-46 62-19861 <=>
1960 N2 P27-33 AIAAJ V1 N1 P258 <AIAA>
1960 N2 P101-106 61-21133 <=>
1960 N3 P23-29 ARSJ V32 N9 P1455 <AIAA>
1960 N4 P65-72 61-23920 <=*>
1960 N5 P91-98 19S86R <ATS>
1960 N6 P91-97 AD-617 673 <=$>
1962 N3 P69-75 65-10399 <*>
1962 N3 P95-97 78S84R <ATS>
1963 N4 P135-145 UCRL TRANS-1203(L) <=$>
1963 N5 P37-41 AD-611 533 <=$>
1963 N5 P51-54 65-10956 <*>
1963 N5 P79-84 65-10957 <*>
1963 V1 P73-79 RTS-2696 <NLL>

IZVESTIYA AKADEMII NAUK AZERBAIDZHANSKOI SSR. BAKU
SERIYA FIZIKO-TEKHNICHESKIKH I KHIMICHESKIKH NAUK
1958 N5 P7-13 12S85AZ <ATS>
1958 V15 N1 P37-45 60-23622 <+*>
1958 V15 N1 P49-63 33M39R <ATS>
1958 V15 N3 P201-205 60-17370 <+*>
1959 N1 P3-10 1994 <TC>

IZVESTIYA AKADEMII NAUK AZERBAIDZHANSKOI SSR. BAKU
SERIYA FIZIKO-TEKHNICHESKIKH I MATEMATICHESKIKH
NAUK
1964 N1 P91-93 65-10953 <*>
1964 N2 P81-84 29T95R <ATS>
1964 N2 P105-108 65-12709 <*>
1964 N2 P115-119 AD-615 869 <=$>
1964 N4 P83-87 66-11140 <*>
1964 N4 P95-99 66-11139 <*>
1964 N4 P101-108 66-11213 <*>
1964 N5 P69-72 AD-640 309 <=$>
1965 N1 P66-70 98S87R <ATS>
1965 N2 P83-86 19T92R <ATS>
1965 N3 P90-95 70T91R <ATS>

IZVESTIYA AKADEMII NAUK AZERBAIDZHANSKOI SSR. BAKU
SERIYA GEOLOGO-GEOGRAFICHESKIKH NAUK
1961 N3 P93-95 62-15950 <=*>
1961 N3 P99-101 62-15950 <=*>
1964 N1 P63-69 44S81R <ATS>
1964 N2 P69-76 58R79R <ATS>
1964 N6 P55-57 29S84R <ATS>
1965 N1 P80-83 22S87R <ATS>
1965 N1 P85-90 23S87R <ATS>

IZVESTIYA AKADEMII NAUK AZERBAIDZHANSKOI SSR. BAKU
SERIYA GEOLOGO-GEOGRAFICHESKIKH NAUK I NEFTI
1962 N4 P3-14 64-21459 <=>

IZVESTIYA AKADEMIYA NAUK AZERBAIDZHANSKOI SSR,
BAKU. SERIYA NAUK O ZEMLE
1966 N2 P115-121 66-33781 <=$>
1966 N6 P85-89 SHSP N6 1966 P656 <AGU>

IZVESTIYA AKADEMII NAUK BELORUSSKOI SSR
1948 N2 P114-124 63-16141 <=*>
1953 N3 P155-164 60-23001 <*+>
1955 N4 P93-102 65-50043 <=$>

IZVESTIYA AKADEMII NAUK ESTONSKOI SSR
SEE EESTI NSV TEADUSTE AKADEEMIA TOIMETISED

IZVESTIYA AKADEMII NAUK KAZAKHSKOI SSR, ALMA-ATA.
ASTROFIZICHESKII INSTITUT
1962 V14 P93-106 63-19354 <=*>

IZVESTIYA AKADEMII NAUK KAZAKHSKOI SSR. ALMA-ATA
SERIYA BIOLOGICHESKIKH NAUK
1965 N5 P64-71 66-30137 <=*$>
1966 N1 P93-95 66-32711 <=$>

IZVESTIYA AKADEMII NAUK KAZAKHSKOI SSR. ALMA-ATA.
SERIYA BOTANIKI I POCHVOVEDENIYA
1958 N1 P16-27 65-50107 <=>
1960 V7 N1 P89-100 61-11772 <=>
1961 N1 P52-56 60S85R <ATS>
1961 N3 P83-88 66-61565 <=*$>

IZVESTIYA AKADEMII NAUK KAZAKHSKOI SSR, ALMA-ATA.
SERIYA EKONOMIKI, FILOSOFII I PRAVA
1960 N2 P32-37 61-21495 <=>
1960 N2 P86-95 62-24782 <=>

IZVESTIYA AKADEMII NAUK KAZAKHSKOI SSR, ALMA-ATA.
SERIYA ENERGETICHESKAYA
1956 N11 P70-81 CSIR-TR.159 <CSSA>
1960 N2 P60-67 RTS-2766 <NLL>
1961 N1 P21-29 63-19498 <=*>

IZVESTIYA AKADEMII NAUK KAZAKHSKOI SSR. ALMA-ATA.
SERIYA FIZIKO-MATEMATICHESKIKH NAUK
1963 V16 N1 P61-73 N66-23610 <=$>

IZVESTIYA AKADEMII NAUK KAZAKHSKOI SSR, ALMA-ATA.
SERIYA GEOGRAFICHESKAYA
1948 V2 P78-101 62-33417 <=*>

IZVESTIYA AKADEMII NAUK KAZAKHSKOI SSR, ALMA-ATA.
SERIYA GEOLOGICHESKAYA
1957 N2 P67-93 64-13366 <=*$> O
1958 V30 P17-28 59-11336 <=> P
1958 V30 N1 P29-36 59-11335 <=> P
1958 V30 N1 P38-41 59-11333 <=>
1958 V30 N1 P47-51 59-11332 <=>
1959 V31 N2 P71-74 60-41166 <=> O
1960 V32 N4 P37-52 62-25184 <=*>
1962 N2 P24-31 63-13676 <=*>

IZVESTIYA AKADEMII NAUK KAZAKHSKOI SSR. ALMA-ATA.
SERIYA GORNOGO DELA
1959 V12 N1 P78-89 66-11018 <*>
1960 N2 P55-61 UCRL-TRANS-967(L) <=*$>

IZVESTIYA AKADEMII NAUK KAZAKHSKOI SSR, ALMA-ATA.
SERIYA GORNOGO DELA METALLURGII I STROIMATERIALOV
1957 V10 N4 P43-52 64-14949 <*> O

IZVESTIYA AKADEMII NAUK KAZAKHSKOI SSR, ALMA-ATA.
SERIYA IAZYKA I LITERATURY
1964 N1 P70-73 65-12269 <*>

IZVESTIYA AKADEMII NAUK KAZAKHSKOI SSR. ALMA-ATA.
SERIYA KHIMICHESKAYA
1953 N7 P70-74 RT-3489 <*>
1953 V5 P107-115 63-16205 <=*> O
1955 N8 P122-132 20M44R <ATS>
1956 N9 P23-32 RT-4637 <*>
1957 N1 P3-11 63-23245 <*=>
1958 V3 N1 P19-22 60-23673 <+*>
1958 V3 N1 P38-45 62-32957 <=*>
1958 V3 N1 P91-98 60-18919 <+*>
1958 V3 N1 P99-104 60-18918 <+*>
1958 V3 N1 P105-111 60-15816 <+*>
1960 V5 N1 P59-62 61-28852 <*=>
1960 V5 N1 P63-68 61-28780 <*=>

1960 V5 N2 P5-69 61-28741 <=>
1960 V5 N2 P81-91 62-32496 <=*>
1960 V5 N2 P112-113 61-28741 <=>
1962 V7 N1 P67-71 63-19326 <=*>
1962 V7 N2 P112-114 65-13698 <*>

IZVESTIYA AKADEMII NAUK KAZAKHSKOI SSR. SERIYA KHIMICHESKIKH NAUK
1965 N1 P42-51 ICE V6 N2 P221-227 <ICE>
1965 N2 P73-76 ICE V6 N2 P202-204 <ICE>

IZVESTIYA AKADEMII NAUK KAZAKHSKOI SSR, ALMA-ATA. SERIYA MATEMATIKI I MEKHANIKI
1962 N10 P56-59 62-33429 <=*>

IZVESTIYA AKADEMII NAUK KAZAKHSKOI SSR, ALMA-ATA. SERIYA MEDITSINSKIKH NAUK
1963 N1 P32-34 63-31871 <=$>
1963 N1 P37-52 63-31871 <=$>
1963 N2 P68-77 63-41091 <=>
1963 V11 N3 P50-53 64-51998 <=$>
1964 N1 P31-36 64-51141 <=>
1964 N1 P50-55 64-51145 <=$>
1964 N1 P84-90 64-51140 <=>

IZVESTIYA AKADEMII NAUK KAZAKHSKOI SSR, ALMA-ATA. SERIYA MEDITSINY I FIZIOLOGII
1960 N1 P7-13 61-28052 <*=>
1962 N2 P86-89 62-33334 <=*>

IZVESTIYA AKADEMII NAUK KAZAKHSKOI SSR, ALMA-ATA. SERIYA METALLURGII, OBOGASHCHENIYA I OGNEUPOROV
1959 N1 P80-84 61-27804 <*=>
1960 V3 P76-84 62-25462 <=*>

IZVESTIYA AKADEMII NAUK KAZAKHSKOI SSR, ALMA-ATA. SERIYA PARAZITOLOGICHESKAYA
1949 V7 P60-65 64-15633 <=*$> 0
1949 V7 N74 P49-54 R-4129 <*>

IZVESTIYA AKADEMII NAUK KIRGIZSKOI SSR, FRUNZE. SERIYA BIOLOGICHESKIKH NAUK
1960 V2 N7 P5-24 66-60976 <=$*>
1962 V4 N4 P65-69 64-15669 <=*$>
1964 V6 N2 P59-69 66-60982 <=*$>

IZVESTIYA AKADEMII NAUK KIRGIZSKOI SSR, FRUNZE. SERIYA ESTESTVENNYKH I TEKHNICHESKIKH NAUK
1959 V1 N3 P45-50 64-21033 <=>
1960 V2 N3 P67-73 63-24201 <=$>
1962 V4 N8 P117-124 65-63378 <=$>

IZVESTIYA AKADEMII NAUK LATVIISKOI SSR. RIGA
SEE LATVIJAS PSR ZINATNU AKADEMIJAS VESTIS, AND CORRESPONDING SERIES FOR LATER DATES

IZVESTIYA AKADEMII NAUK MOLDAVSKOI SSR, KISHINEV.
1961 N10 P57-63 65-10883 <*>
1961 N10 P91-94 65-10888 <*>
1961 N10 P95-97 65-10882 <*>
1962 N2 P13-22 63-21619 <=>

*IZVESTIYA AKADEMII NAUK SSSR. FIZIKA ATMOSFERY I OKEANA
1965 V1 P992-993 AD-634 302 <=$>
1965 V1 N4 P377-385 AD-640725 <=$> M
1965 V1 N5 P509-516 65-32990 <=*$>
1965 V1 N9 P920-951 65-33824 <=*$>
1965 V1 N9 P994-995 66-33819 <=$>
1965 V1 N11 P1151-1159 66-34283 <=$>
1965 V1 N12 P1270-1278 66-34324 <=$>
1966 V2 P523-533 66-33337 <=$>
1966 V2 N3 P263-271 66-32252 <=$>
1966 V2 N3 P305-307 66-32252 <=$>
1966 V2 N3 P312-315 66-32252 <=$>
1966 V2 N3 P320-323 66-32252 <=$>
1966 V2 N4 P367-422 66-32914 <=$>
1966 V2 N7 P695-704 66-34088 <=$>
1966 V2 N7 P714-720 66-34078 <=$>
1966 V2 N7 P721-728 66-34077 <=$>
1966 V2 N7 P729-739 66-34089 <=$>
1966 V2 N7 P740-757 66-34079 <=$>
1966 V2 N7 P762-765 66-34080 <=$>
1966 V2 N7 P772-774 66-34081 <=$>

*IZVESTIYA AKADEMII NAUK SSSR. FIZIKA ZEMLI
1965 N4 P9-22 AD-623 554 <=$>
1965 N5 P33-41 65-31717 <=$>
1965 N7 P9-21 65-32660 <=*$>
1965 N12 P1-20 66-32845 <=$>

IZVESTIYA AKADEMII NAUK SSSR. MEKHANIKA
1965 N1 P3-16 65-14559 <*>
1965 N1 P17-23 65-14454 <*>
1965 N1 P89-98 65-14415 <*>
65-63427 <=*$>
1965 N1 P99-108 65-14434 <*>
65-63417 <=$>
1965 N1 P200-203 44T90R <ATS>
1965 N2 P69-75 AD-634 515 <=$>
1965 N2 P88-94 N66-13478 <=$>
1965 N2 P131-134 65-64346 <=*$>
1965 N2 P152-154 65-14452 <*>
1965 N2 P152-159 65-14453 <*>
1965 N2 P152-154 65-63186 <=*$>
66-10924 <*>
1965 N2 P154-159 65-63187 <=*$>
1965 N2 P168-170 N66-13480 <=$>
1965 N2 P175-178 M.5796 <NLL>
1965 N3 P49-59 RTS-3288 <NLL>
1965 N3 P68-76 TP/T.3872 <NLL>
1965 N3 P114-118 65-14436 <*>
65-63421 <=$>
1965 N3 P124-130 65-14439 <*>
65-63424 <=$>
1965 N3 P131-135 65-14437 <*>
65-63429 <=$>
1965 N3 P149-153 65-14435 <*>
65-63761 <=*$>
1965 N3 P182-186 65-14438 <*>
65-63418 <=$>
1965 N4 P13-23 AD-638 204 <=$>
1965 N4 P129-130 65-64350 <=*$>
66-10473 <*>
1965 N4 P131-132 65-64347 <=*$>
66-10474 <*>
1965 N4 P133-139 66-10472 <*>
1965 N5 P7-10 66-10479 <*>
66-10480 <*>
1965 N5 P56-69 N66-21687 <=$>
1965 N5 P144-146 66-12664 <*>
66-61333 <=*$>
1965 N5 P146-148 AD-639 909 <=$>
1965 N5 P154-157 AD-636 505 <=$>
1965 N6 P3-9 66-12315 <*>
66-61366 <=*$>
1965 N6 P10-19 66-12316 <*>
66-61365 <=*$>
1965 N6 P84-86 N66-22282 <=$>
1965 N6 P86-87 N66-18452 <=$>
1965 V1 P3-16 65-63181 <=$>
1965 V1 P17-23 65-63223 <=$>
1965 V1 N7 P50-56 65-64096 <=*$>

*IZVESTIYA AKADEMII NAUK SSSR. MEKHANIKA ZHIDKOSTI I GAZA
1966 N1 P32-36 66-61355 <=*$>
1966 N1 P142-144 66-61335 <=*$>
1966 N1 P168-170 66-61334 <=*$>
1966 N2 P145-148 66-61707 <=*$>
1966 N2 P182-184 66-61693 <=*$>
1966 N2 P184-188 66-61705 <=*$>

*IZVESTIYA AKADEMII NAUK SSSR. METALLY
1965 N1 P59-61 6569 <BISI>
1965 N2 P180-186 M-5753 <NLL>
1965 N3 P123-127 6383 <HB>
1965 N4 P137-143 65-32807 <=$*>
1965 N4 P168-175 66-11199 <*>

*IZVESTIYA AKADEMII NAUK SSSR. NEORGANICHESKIE MATERIALY

1965 V1 N4 P449-459	65-31554 <=*$>
1965 V1 N4 P625-630	65-31555 <=*$>
1965 V1 N6 P873-876	66-12575 <*>
1965 V1 N7 P1229-1233	65-32991 <=*$>
1965 V1 N10 P1701-1709	66-31532 <=$>
1965 V1 N10 P1778-1786	66-31532 <=$>
1965 V1 N10 P1834-1837	66-31532 <=$>
1966 V2 N2 P223-228	66-13804 <*>

IZVESTIYA AKADEMII NAUK SSSR. OTDELENIE KHIMICHESKIKH NAUK

1938 P75-99	R-3365 <*>
1938 P101-109	R-2798 <*>
1938 P403-413	63-18112 <*>
1938 P925-940	R-2576 <*>
1938 N1 P101-109	64-20880 <*> 0
1938 N1 P137-146	R-3329 <*>
1938 N1 P177-184	R-2796 <*>
1938 N1 P195-201	R-2473 <*>
1938 N1 P203-216	R-3301 <*>
1940 P181-188	61-18024 <*=>
1940 P379-395	R-2940 <*>
1940 P511-528	R-2858 <*>
1940 P705-724	R-2949 <*>
1940 P795-810	R-2205 <*>
1940 P887-894	R-2942 <*>
1940 N1 P107-125	64-20157 <*>
1940 N1 P127-133	64-20158 <*> 0
1940 N1 P127-134	75N49R <ATS>
1940 N1 P135-142	61-16861 <*=>
1940 N1 P144-150	61-16863 <*=>
1940 N1 P153-160	61-16862 <*=>
1940 N3 P421-426	24R79R <ATS>
1940 N5 P617-626	62-14119 <*>
1940 N5 P681-689	62-14120 <*>
1940 N5 P727-738	RT-958 <*>
	63-20771 <=*$>
1940 N5 P825-831	RT-957 <*>
	63-20772 <=*$>
1940 N6 P997-1015	62-14122 <*>
1941 N1 P13-26	61-16875 <*=>
1941 N1 P27-33	61-16879 <*=>
1941 N1 P34-40	61-16884 <*=>
1941 N1 P41-48	61-16889 <*=>
1941 N1 P61-66	63-14391 <=*> 0
1941 N1 P97-106	61-18049 <*=>
1941 N1 P107-114	61-16888 <*=>
1941 N1 P115-119	63-14123 <=*>
1941 N1 P137-138	61-20795 <=*>
1941 N1 P177-190	61-16693 <*=>
1941 N1 P191-200	61-16695 <*=>
1941 N1 P201-205	61-18006 <*=>
1941 N1 P206-210	61-18007 <*=>
1941 N2 P297-302	62-14128 <*>
1941 N2 P327-331	62-14129 <*>
1941 N4/5 P533-543	61-18010 <*=>
1941 V1 P145-155	62-14126 <=*>
1942 P125-134	61-16658 <*=>
1942 P210-220	61-16665 <*=>
1942 P333-365	61-16661 <*=>
1942 N1 P8-13	62-15711 <*=>
1942 N1 P21-44	61-18058 <*=>
1942 N1 P45-53	61-16635 <*=>
1942 N4 P190-194	62-14616 <=*>
1942 N2/3 P87-97	RT-2142 <*>
1942 N2/3 P106-115	61-11056 <=>
1943 P65-72	61-16925 <*=>
1943 P145-151	61-16927 <*=>
1943 P198-204	61-20180 <*=>
1943 P259-263	R-2259 <*>
1943 P280-285	61-16910 <*=>
1943 P305-311	61-16912 <*=>
1943 P410-414	62-18772 <=*> 0
1943 P427-433	61-18190 <*=>
1943 P443-452	61-18189 <*=> 0
1943 N2 P99-107	31 <FRI>
1943 N3 P171-177	60-14164 <+*>
1943 N4 P271-279	RT-2038 <*>
1943 N4 P296-304	64-20552 <*> P
1943 N5 P381-388	RT-2037 <*>
1943 N6 P581-589	NP-TR-473 <*>
1944 P272-282	63-14098 <=*> 0
1944 P351-357	61-18228 <*=>
1944 P446-450	R-3340 <*>
1944 P729-733	61-18225 <*=>
1944 P763-773	61-18188 <*=>
1944 N1 P20-28	RT-104 <*>
	65-60040 <=$>
1944 N1 P65-69	62-14142 <=*> P
1944 N1 P122-128	61-18177 <*=>
1944 N1 P152-155	61-18176 <*=>
1944 N1 P238-242	61-18174 <*=>
1944 N4 P226-237	62-14138 <=*> P
1944 N4 P255-262	R-622 <*>
1944 N4 P255-261	61-18173 <*=>
1944 N5 P283-295	RJ-86 <ATS>
1944 N5 P320-324	RT-2144 <*>
1944 N5 P349-358	RT-616 <*>
1944 N5 P349-357	61-18186 <*=>
1944 N6 P416-431	RT-627 <*>
1944 N2/3 P137-142	62-14140 <=*> P
1944 N2/3 P156-159	62-14141 <=*> P
1945 P62-70	61-18236 <*=>
1945 P94-103	R-2642 <*>
1945 P163-166	61-18221 <*=>
1945 P271-278	61-18231 <*=>
1945 P359-363	61-18886 <*=>
1945 P375-383	61-18887 <*=>
1945 P394-408	62-14144 <=*> P
1945 P469-478	61-18892 <*=>
1945 P486-491	61-18891 <*=>
1945 P655-663	61-18888 <*=>
1945 P665-668	61-18895 <*=>
1945 P672-674	61-18885 <*=>
1945 N1 P35-43	61-18199 <*=>
1945 N1 P44-51	61-18205 <*=>
1945 N1 P53-60	61-18219 <*=>
1945 N1 P53-61	64-10277 <*>
1945 N1 P113-118	61-18220 <*=>
1945 N2 P104-112	64-11966 <=> M
1945 N2 P154-162	66-12058 <*>
1945 N3 P210-222	61-13013 <*+>
1945 N3 P251-	61-14473 <+*>
1945 N3 P279-281	RT-489 <*>
1945 N4 P330-338	62-14143 <=*> P
1945 N4 P375-384	63-18681 <=*>
	66-12057 <*>
1945 N5 P522-524	63-16035 <*>
1945 N6 P587-596	60-18538 <+*> 0
1945 N6 P609-616	84K26R <ATS>
1945 N6 P617-626	91N47R <ATS>
1946 P77-82	61-20110 <*=>
1946 P458-460	61-20111 <*=>
1946 N1 P47-56	RT-1909 <*>
1946 N1 P83-89	62-14147 <=*> P
1946 N1 P91-101	62-14148 <=*> P
1946 N3 P265-274	<INSD>
1946 N4 P381-389	RT-1908 <*>
1946 N4 P439-446	63-18252 <*>
	63-20468 <=*$>
1946 N4 P447-453	61-20115 <*=>
	63-18254 <=*> 0
1946 N5 P497-513	00N54R <ATS>
	62-20434 <=*>
1946 N5 P523-528	1439A <K-H>
	65F2R <ATS>
1946 N5 P557	1430 <KH>
	71F2R <ATS>
1946 N6 P587-600	<INSD>
1947 N1 P113-	60-17218 <*=>
1947 N5 P427-434	R.562 <RIS>
1947 N5 P435-442	R.563 <RIS>
1947 N5 P473-482	R.565 <RIS>
1947 N5 P483-493	R.566 <*>
1947 N5 P495-499	R.567 <RIS>
1947 N5 P501-508	R.568 <RIS>
1947 N5 P509-514	R.569 <RIS>
1947 N5 P515-522	63-20182 <=*$>

Citation	Translation
1947 N6 P605-616	62-11677 <=> 0
1947 N6 P636-647	1952B <KH>
1947 V6 P561-570	T-1750 <NLL>
1948 N1 P83-94	63-23553 <=*>
1948 N3 P290-301	62-11661 <=> 0
1948 N6 P568-580	NP-TR-472 <*>
	61-13389 <*+>
1948 N6 P581-586	NP-TR-473 <*>
1948 N6 P581-583	61-13390 <*+>
1948 N6 P631-641	RJ-358 <ATS>
	RJ-76G8R <ATS>
1949 N1 P44-55	TT.107 <NRC>
1949 N1 P110-114	RJ-24 <ATS>
	RT-617 <*>
1949 N3 P269-273	R-4034 <*>
1949 N3 P317-325	63-18674 <=*>
1949 N4 P379-385	RT-3977 <*>
1949 N4 P439-442	RT-659 <*>
	61-28015 <*=> 0
1949 N5 P504-521	RJ-359 <ATS>
1949 N9 P1320-	60-15761 <*=>
1950 P203-208	63-18214 <=*>
1950 N1 P15-26	59-10499 <+*>
1950 N1 P27-38	62-32474 <=*>
1950 N1 P66-76	RT-3607 <*>
1950 N1 P98-107	RT-1030 <*>
1950 N1 P108-113	RJ-36 <ATS>
	RT-3608 <*>
1950 N2 P152-161	RJ-42 <ATS>
1950 N2 P162-168	RJ-33 <ATS>
1950 N2 P178-184	RJ-39 <ATS>
1950 N3 P297-303	RT-2023 <*>
1950 N4 P418-425	1966A <KH>
1950 N5 P469-474	RT-1879 <*>
1950 N5 P521-530	63-18675 <=*>
1950 N6 P573-575	RT-1907 <*>
1950 N6 P576-581	RT-1906 <*>
1950 N6 P635-640	RT-3552 <*> 0
1951 N1 P32-41	R-4539 <*>
1951 N1 P86-94	63-18676 <=*>
1951 N1 P95-97	RT-3609 <*>
1951 N1 P485-491	RT-778 <*>
1951 N2 P100-114	RJ-224 <ATS>
1951 N2 P132-139	R-1866 <*>
	RJ-58 <ATS>
1951 N2 P133-139	RJ-58 <ATS>
1951 N2 P140-144	59-19107 <+*>
1951 N2 P145-149	RJ-392 <ATS>
1951 N2 P172-178	RT-3610 <*>
1951 N2 P201-204	RJ-57 <ATS>
1951 N3 P242-254	29 <FRI>
1951 N3 P268-272	RT-3611 <*>
1951 N3 P280-283	RJ-65 <ATS>
1951 N3 P317-320	RJ-66 <ATS>
1951 N3 P328-333	RJ-67 <ATS>
1951 N4 P341-349	RJ-99 <ATS>
1951 N4 P409-416	RT-3612 <*>
1951 N5 P634-652	61-13003 <*+>
1951 N6 P661-665	59-10278 <+*>
1951 N6 P661-666	61-17278 <=$>
1951 N6 P667-673	RT-4051 <*>
1951 N6 P722-727	RJ-130 <ATS>
1951 N6 P745-752	R-1901 <*>
	RJ-127 <ATS>
1951 N6 P782-794	63-11012 <=>
1951 N6 P806-808	65-10043 <*>
1951 N6 P809-818	R-1902 <*>
	RJ-129 <ATS>
1951 N6 P823-828	RT-306 <*>
1951 V2 P100-114	R-2097 <*>
1952 N1 P24-30	61-17228 <=$>
1952 N1 P40-43	63-20648 <=*$>
1952 N1 P44-52	R-1521 <*>
1952 N1 P59-63	63-20302 <=*>
1952 N1 P64-73	63-20304 <=*>
1952 N2 P236-237	59-15969 <+*>
1952 N3 P415-421	R-3626 <*>
	3290 <NLL>
1952 N3 P498-504	61-20958 <*=>
1952 N3 P547-555	RJ-101 <ATS>
1952 N3 P566-569	RJ-102 <ATS>
1952 N4 P603-615	R-1867 <*>
	RJ-116 <ATS>
	63-10395 <=*>
1952 N4 P671-681	RJ-111 <ATS>
1952 N4 P721-726	RJ-103 <ATS>
1952 N4 P727-735	RJ-106 <ATS>
1952 N4 P751-762	R-1173 <*>
	RJ-276 <ATS>
	RT-2819 <*>
1952 N5 P926-931	RJ-104 <ATS>
1952 N5 P932-939	RJ-105 <ATS>
1952 N6 P1041-1048	RT-1749 <*>
	64-15381 <=*$>
1952 N6 P1075-1081	62-25279 <=*>
1953 P121-125	R-684 <*>
1953 N2 P260-268	R-1874 <*>
	RJ-124 <ATS>
1953 N2 P298-302	RT-1852 <*>
	64-15377 <=*$>
1953 N3 P419-428	RT-3354 <*>
1953 N3 P442-447	RJ-160 <ATS>
	61-17209 <$=>
	63-10372 <=*>
1953 N3 P533-536	RT-2453 <*>
1953 N4 P607-614	CSIR-TR-444 <CSSA>
1953 N4 P615-622	<INSD>
1953 N4 P629-634	RJ-161 <ATS>
1953 N5 P889-900	RT-1456 <*>
1953 N5 P901-919	RT-1455 <*>
1953 N5 P945-950	RJ-205 <ATS>
1953 N6 P957-967	RJ-207 <ATS>
1953 N6 P1035-1042	49K26R <ATS>
1954 N1 P109-116	48K26R <ATS>
1954 N1 P117-123	61-13344 <+*>
1954 N1 P142-148	59-20292 <+*>
1954 N2 P217-224	R-1028 <*>
1954 N2 P266-277	RJ-225 <ATS>
1954 N2 P278-291	RJ-226 <ATS>
1954 N3 P478-483	61-17285 <=$>
1954 N4 P646-655	62-16177 <=*> 0
1954 N4 P663-669	TR.13 <ETHB>
1954 N4 P670-676	61-13203 <*+>
1954 N4 P694-706	RJ-229 <ATS>
1954 N5 P753-764	<CB>
1954 N5 P765-769	<CB>
1954 N5 P770-777	<CB>
1954 N5 P778-783	<CB>
1954 N5 P784-795	<CB>
1954 N5 P796-798	<CB>
1954 N5 P799-802	<CB>
1954 N5 P803-805	<CB>
1954 N5 P806-811	<CB>
1954 N5 P812-822	<CB>
1954 N5 P823-829	<CB>
1954 N5 P830-836	<CB>
1954 N5 P837-845	<CB>
1954 N5 P846-853	<CB>
1954 N5 P854-858	<CB>
1954 N5 P859-864	<CB>
1954 N5 P865-877	<CB>
1954 N5 P878-881	<CB>
1954 N5 P882-889	<CB>
1954 N5 P890-897	<CB>
1954 N5 P898-903	<CB>
1954 N5 P904-910	<CB>
1954 N5 P911-918	<CB>
1954 N5 P919-923	<CB>
	62-20098 <=*>
1954 N5 P924-930	<CB>
1954 N5 P931-935	<CB>
1954 N5 P936-944	<CB>
	60-15752 <+*>
1954 N5 P945-948	<CB>
1954 N6 P1008-1018	RT-3462 <*>
1954 N6 P1047-1052	RT-2876 <*>
1954 N6 P1119-1120	RJ-310 <ATS>
1954 N8 P128-136	<CP>
1954 V6 P1042-1046	C-2494 <NRC>
1955 P150-164	23N54R <ATS>

1955 P362-374	74G7R <ATS>
1955 N1 P3-8	<CB>
1955 N1 P9-16	<CB>
1955 N1 P17-20	<CB>
1955 N1 P21-30	<CB>
1955 N1 P31-39	<CB>
1955 N1 P40-47	<CB>
1955 N1 P48-53	<CB>
1955 N1 P54-61	<CB>
1955 N1 P62-70	<CB>
1955 N1 P71-77	<CB>
1955 N1 P163-171	RT-3461 <*>
1955 N3 P386-394	59-10003 <+*>
1955 N3 P570-572	RJ-337 <ATS>
1955 N4 P481-688	RT-4387 <*>
1955 N4 P624-638	RJ-13 <ATS>
	61-17378 <=$>
1955 N4 P665-668	RJ-333 <ATS>
1955 N4 P669-671	RJ-334 <ATS>
1955 N4 P672-675	RJ-336 <ATS>
1955 N4 P681-688	RJ-335 <ATS>
	64-19092 <=*$>
1955 N4 P689-695	RJ-341 <ATS>
1955 N4 P723-733	RJ-519 <ATS>
1955 N4 P766-767	RJ-512 <ATS>
1955 N4 P770-772	RJ-517 <ATS>
1955 N5 P789-792	59-22413 <+*>
1955 N5 P800-804	5866-A <K-H>
	64-18394 <*>
1955 N5 P863-868	<ATS>
1955 N5 P925-929	RJ-339 <ATS>
1955 N5 P934-941	RJ-338 <ATS>
1955 N5 P942-944	RJ-340 <ATS>
1955 N5 P952-953	RJ-576 <ATS>
1955 N5 P953-955	RJ-349 <ATS>
1955 N5 P1085-1089	62-20137 <=*>
1955 N6 P1112-1117	TR.15 <ETHB>
1955 N6 P1122-1124	RJ-487 <ATS>
1955 V4 P624-638	CSIRO-3210 <CSIR>
1955 V4 P681-688	1190 <*>
1956 P145-149	RJ-643 <ATS>
1956 P960-966	62-20101 <=*>
1956 P1144-1146	R-1132 <*>
1956 N1 P5-11	RJ-511 <ATS>
1956 N1 P43-49	62J14R <ATS>
1956 N1 P74-82	RJ-613 <ATS>
1956 N2 P145-149	61-17377 <=$>
1956 N2 P150-157	RJ-655 <ATS>
1956 N2 P253-254	64-30949 <*>
1956 N3 P265-269	RJ-18H11R <ATS>
1956 N3 P270-280	RJ-660 <ATS>
1956 N3 P282-286	RJ-506 <ATS>
1956 N3 P299-308	RJ-649 <ATS>
1956 N3 P322-326	59-10076 <+*>
	87H13R <ATS>
1956 N3 P368-372	RJ-624 <ATS>
1956 N3 P373-375	RJ-488 <ATS>
1956 N3 P375-376	88H13R <ATS>
1956 N3 P376	59-10075 <+*>
1956 N3 P376-377	89H13R <ATS>
1956 N3 P383-384	39H11R <ATS>
1956 N4 P443-450	RJ-622 <ATS>
1956 N4 P451-456	59-10074 <+*>
	90H13R <ATS>
1956 N4 P506-507	59-15776 <+*> 0
1956 N4 P506-508	65-29622 <$>
1956 N4 P508-509	91H13R <ATS>
1956 N5 P516-524	64-16307 <=*$>
1956 N5 P525-530	RJ-569 <ATS>
1956 N5 P538-543	RJ-620 <ATS>
1956 N5 P557-	RJ-663 <ATS>
1956 N6 P739-741	64-30950 <*>
1956 N6 P743-746	RJ-625 <ATS>
1956 N7 P784-789	RJ571 <ATS>
1956 N8 P902-912	77H13R <ATS>
1956 N8 P960-966	64-16557 <=*$>
	74J16R <ATS>
1956 N8 P986-991	35H13R <ATS>
1956 N9 P1148-1150	92J16R <ATS>
1956 N10 P1180-1184	53J14R <ATS>
	64-16900 <*>
1956 N10 P1197-1201	00L29R <ATS>
1956 N10 P1211-1229	74J14R <ATS>
1956 N10 P1256-1265	48J14R <ATS>
1956 N11 P1285-1293	R-3262 <*>
1956 N11 P1294-1303	72J14R <ATS>
	91L31R <ATS>
1956 N11 P1320-1328	73J14R <ATS>
1956 N11 P1370-1377	35J15R <ATS>
1956 N11 P1390-1398	70J14R <ATS>
1957 P166-173	R-1619 <*>
1957 P431-435	R-1616 <*>
1957 P485-489	R-2675 <*>
1957 P652-653	R-2671 <*>
1957 P1421-1428	R-4898 <*>
1957 N1 P100-111	63-16185 <=*>
1957 N1 P104-106	49K22R <ATS>
1957 N1 P118-120	65-29640 <$>
1957 N1 P125	82J16R <ATS>
1957 N2 P194-198	R-1618 <*>
1957 N2 P232-235	41K21R <ATS>
1957 N2 P236-237	R-1617 <*>
	24K23R <ATS>
1957 N2 P238-240	R-1595 <*>
	38J17R <ATS>
1957 N2 P242-243	23K23R <ATS>
1957 N3 P263-269	43K24R <ATS>
1957 N3 P270-273	59-15235 <+*>
	67J17R <ATS>
1957 N3 P303-309	46J17R <ATS>
1957 N3 P353-357	99M39R <ATS>
1957 N4 P397-407	60-14516 <+*>
1957 N4 P431-435	91K22R <ATS>
1957 N5 P589-597	60-15581 <+*>
1957 N5 P631-637	52J17R <ATS>
	63-16288 <=*> 0
	65-61307 <=$>
1957 N5 P638-640	64-30940 <*>
1957 N5 P646-648	60-15590 <+*>
1957 N5 P652-653	RJ-872 <ATS>
	59-15939 <+*>
	61-10463 <*+>
	62-10560 <*=>
	63-10633 <=*>
1957 N5 P652-658	65-14199 <*>
1957 N5 P924-928	59K20R <ATS>
1957 N6 P670-677	60-21745 <=> 0
1957 N6 P730-736	64J17R <ATS>
1957 N6 P737-745	79J17R <ATS>
1957 N7 P812-817	60-15591 <+*>
1957 N7 P878-879	TRANS-28 <MT>
1957 N8 P916-923	58K20R <ATS>
1957 N8 P972-975	28J19R <ATS>
1957 N8 P989-991	60-15592 <+*>
1957 N8 P1000-1001	62J19R <ATS>
1957 N8 P1002-1004	63J19R <ATS>
1957 N9 P1053-1058	52J19R <ATS>
1957 N9 P1064-1072	30K23R <ATS>
1957 N9 P1080	59-11868 <=>
1957 N9 P1080-1085	60-15593 <+*>
1957 N9 P1101-1104	81M37R <ATS>
1957 N9 P1123-1125	60-17144 <+*>
	67J19R <ATS>
1957 N9 P1125-1126	60-17145 <+*>
1957 N9 P1129-1132	86K22R <ATS>
1957 N9 P1134-1136	65J19R <ATS>
1957 N9 P1136-1138	66J19R <ATS>
1957 N10 P1218-1222	15K25R <ATS>
1957 N10 P1229-1234	63-16184 <=*> 0
1957 N10 P1245-1249	59-20867 <+*>
	60-16562 <+*>
1957 N10 P1271	66-12070 <*>
1957 N11 P63-68	86K24R <ATS>
1957 N11 P1349-1356	59-18752 <+*>
	65-61197 <=>
1957 N11 P1403-1404	94K22R <ATS>
1957 N11 P1408-1410	66K21R <ATS>
1957 N12 P1488-1489	93K23R <ATS>
1957 N12 P1497-1499	63-16380 <=*>
1957 N12 P1501	7K22R <ATS>

1957 V2 P230-232	NP-TR-37 <*>
1958 P242	65-14273 <*>
1958 P500-501	65-14275 <*>
1958 N1 P18-23	49K23R <ATS>
1958 N1 P59-63	55K23R <ATS>
1958 N1 P63-71	R-5129 <*>
1958 N1 P100-102	46K23R <ATS>
	63-14152 <=*>
1958 N1 P108-109	65-14274 <*>
1958 N1 P121-123	PB 141 169T <=>
1958 N2 P141-151	42K23R <ATS>
1958 N2 P157-165	38K23R <ATS>
1958 N2 P186-191	73K24R <ATS>
1958 N2 P192-199	72K24R <ATS>
1958 N2 P204-209	46K22R <ATS>
1958 N2 P217-220	44K22R <ATS>
	60-18385 <*+>
1958 N2 P221-226	47K22R <ATS>
1958 N2 P240-	54K23R <ATS>
1958 N2 P249-251	65K23R <ATS>
1958 N3 P277-284	66-11057 <*>
1958 N3 P328-334	59-17218 <+*>
	74K24R <ATS>
1958 N3 P353-356	14K24R <INSD>
1958 N3 P373-374	18K24R <ATS>
1958 N3 P375-376	61-10462 <*+>
	75K24R <ATS>
1958 N3 P378-380	19K24R <ATS>
1958 N3 P381	42K24R <ATS>
	65-11476 <*>
1958 N4 P219-424	94K26R <ATS>
1958 N4 P425-427	67K28R <ATS>
1958 N4 P504-506	22K26R <ATS>
1958 N4 P508-510	R-5154 <*>
1958 N5 P535-544	60-21745 <=> 0
1958 N5 P545-549	93K27R <ATS>
1958 N5 P570-574	59-20293 <+*>
	85L36R <ATS>
1958 N5 P575-583	85L32R <ATS>
1958 N5 P600-604	27K25R <ATS>
1958 N5 P618-625	26K25R <ATS>
1958 N5 P640-642	79K27R <ATS>
1958 N5 P646-648	87K26R <ATS>
1958 N6 P679-683	21K27R <ATS>
1958 N6 P726-729	63K27R <ATS>
1958 N6 P794-795	59-15598 <+*>
1958 N7 P880-885	92K25R <ATS>
1958 N7 P886-890	60-13572 <+*>
	92K26R <ATS>
1958 N7 P893-896	44K28R <ATS>
	59-10599 <+*>
1958 N7 P896-898	42K28R <ATS>
	59-10600 <+*>
1958 N7 P906	65-14365 <*> 0
1958 N8 P923-928	86K27R <ATS>
1958 N8 P929-936	37K27R <ATS>
1958 N8 P990-995	40K27R <ATS>
1958 N8 P1006-1008	59-10801 <+*>
1958 N9 P1070-1075	92K27R <ATS>
1958 N9 P1119-1122	19L30R <ATS>
	65-10329 <*>
	8999 <K-H>
1958 N9 P1133-1134	59-10524 <+*>
1958 N10 P1165-1174	61-23264 <=*>
1958 N10 P1175-1183	99L34R <ATS>
1958 N10 P1192-1198	47L29R <ATS>
	59-15451 <+*>
	61-11169 <=>
1958 N10 P1274-1275	61L29R <ATS>
1958 N11 P1345-1347	59-22538 <+*>
	60-10947 <+*>
1958 N11 P1383-1387	59-15260 <+*>
1958 V11 P1345-1347	90L29R <ATS>
1959 P1098-1101	1001 <TC>
1959 P1130-1133	1004 <TC>
1959 P1308-1311	1005 <TC>
1959 N1 P30-34	48L31R <ATS>
1959 N1 P35-40	49L31R <ATS>
1959 N2 P238-246	61-17595 <=$>
1959 N2 P263-266	12L33R <ATS>

	60-10987 <+*>
1959 N2 P365-366	61-10582 <*+>
	96L32R <ATS>
1959 N3 P417-424	16L33R <ATS>
1959 N3 P425-432	60-18079 <+*>
1959 N3 P444-449	98M37R <ATS>
1959 N3 P466-471	12L36R <ATS>
	63-24292 <=*$>
1959 N3 P491-498	43L34R <ATS>
1959 N3 P546-547	55L34R <ATS>
1959 N3 P550-552	99M37R <ATS>
1959 N4 P437-445	84L32R <ATS>
1959 N4 P586-593	61-10375 <*+>
1959 N4 P611-616	60-18573 <+*>
1959 N4 P721-726	03L37R <ATS>
1959 N4 P727-730	85L35R <ATS>
1959 N4 P736-738	60-19981 <+*>
1959 N4 P746-748	60-18572 <+*>
1959 N4 P755-756	60-21941 <=>
1959 N5 P797-805	60-15779 <+*> 0
1959 N6 P958	61-20767 <*=> 0
1959 N6 P971-974	<LSA>
	60-15777 <+*> 0
1959 N6 P1005-1010	23M43R <ATS>
1959 N6 P1058-1067	38L34R <ATS>
1959 N6 P1071-1078	60-16564 <+*>
	61-14968 <=*>
1959 N6 P1079-1087	01M38R <ATS>
	60-16901 <*+>
	61-10583 <*+>
1959 N6 P1088-1090	60-14335 <+*>
1959 N6 P1126-1127	13L36R <ATS>
1959 N6 P1134-1135	60-14478 <+*>
1959 N7 P1211-1215	40M38R <ATS>
1959 N7 P1346-1349	60-16597 <+*>
1959 N7 P1351	41M41R <ATS>
1959 N7 P1351-1352	60-10945 <+*>
1959 N8 P1397-1399	90L35R <ATS>
1959 N8 P1421-1424	61-23420 <*=>
1959 N8 P1433-1437	33M40R <ATS>
	60-18574 <+*>
1959 N8 P1453-1457	61-20683 <*=>
1959 N8 P1507	94M37R <ATS>
	96N47R <ATS>
1959 N9 P1515-1519	10R77R <ATS>
1959 N9 P1536-1545	61-10371 <*+>
1959 N9 P1562-1570	66-11506 <*>
1959 N9 P1571-1578	34M39R <ATS>
1959 N9 P1590-1594	15N48R <ATS>
	61-27510 <*=> 0
1959 N9 P1631-1635	73T94R <ATS>
1959 N9 P1640-1645	60-18187 <+*>
1959 N9 P1655-1662	60-18186 <+*>
1959 N9 P1670-1672	38M38R <ATS>
1959 N10 P1705-1715	60-16827 <*> 0
1959 N10 P1751-1759	28L37R <ATS>
1959 N10 P1760-1766	16N48R <ATS>
1959 N10 P1825-1829	65-13890 <*>
1959 N10 P1868	56M37R <ATS>
1959 N11 P1896-1904	56M40R <ATS>
1959 N12 P2080-2087	29M41R <ATS>
1959 N12 P2100-2111	85M40R <ATS>
1959 N12 P2142-2145	86M38R <ATS>
1959 N12 P2161-2164	94M39I <ATS>
1959 N12 P2171-2176	63-14089 <=*> 0
1959 N12 P2248-2250	97M43R <ATS>
1959 N12 P2251-2253	36M43R <ATS>
	61-18606 <*=>
1959 N12 P2257-2259	60-16797 <+*>
1959 N12 P2261	60-18143 <+*>
1960 P1053-1056	1002 <TC>
1960 N1 P31-38	84M43R <ATS>
1960 N1 P73-79	60-18571 <+*>
1960 N1 P94-97	60-16779 <+*>
1960 N1 P111-114	60-18440 <+*>
1960 N1 P115-119	61-10372 <*+>
1960 N1 P146	60-18142 <+*>
1960 N2 P182-187	62-23971 <=*>
1960 N2 P292-299	12M44R <ATS>
	62-16635 <=*>

Citation	Report
1960 N2 P361-363	61-18596 <*=>
1960 N2 P377	14M41R <ATS>
1960 N2 P378-381	60-41619 <=>
1960 N3 P427-434	52M45R <ATS>
1960 N3 P442-446	60-41490 <=>
1960 N3 P480-483	08M44R <ATS>
1960 N3 P505-512	61-11030 <=>
1960 N3 P561-563	66-11022 <*>
1960 N4 P588-598	61-21478 <=>
1960 N4 P605-613	99M45R <ATS>
1960 N4 P614-623	21M45R <ATS>
1960 N4 P645-650	61-11982 <=>
1960 N4 P686-692	96M42R <ATS>
1960 N4 P754-755	61-21294 <=>
1960 N5 P779-783	62-16619 <=*> 0
	91M43R <ATS>
1960 N5 P806-811	61-21999 <=>
1960 N5 P847-851	92M43R <ATS>
1960 N5 P944-945	10M44R <ATS>
	61-10160 <*+>
	61-11569 <=>
1960 N6 P1015-1021	61-27521 <*=> 0
1960 N6 P1044-1048	29M46R <ATS>
1960 N6 P1066-1072	28M44R <ATS>
1960 N6 P1073-1079	29M44R <ATS>
1960 N6 P1084-1093	01M47R <ATS>
1960 N6 P1098-1103	02M44R <ATS>
1960 N6 P1124-1126	61-27185 <*=>
1960 N6 P1126-1128	61-27182 <*=>
1960 N6 P1131-1133	07M46R <ATS>
1960 N7 P1147-1152	61-27532 <*=>
1960 N7 P1153-1161	61-27859 <*=>
1960 N7 P1178-1184	61-27520 <*=> 0
1960 N7 P1185-1190	61-27519 <*=>
1960 N7 P1191-1199	61-23053 <=*>
1960 N7 P1206-1214	31M46R <ATS>
	61-23054 <*=>
1960 N7 P1219-1223	69M45R <ATS>
1960 N7 P1224-1226	30M46R <ATS>
1960 N7 P1298-1300	47M47R <ATS>
1960 N7 P1313-1315	61-23064 <=*>
1960 N8 P1342-1347	08N48R <ATS>
1960 N8 P1354-1357	61-31676 <=>
1960 N8 P1445-1450	01M46R <ATS>
1960 N8 P1466-1470	02M46R <ATS>
1960 N9 P1529-1534	ICE V1 N1 P13-17 <ICE>
	39N50R <ATS>
1960 N9 P1535-1543	63-24326 <=*$>
1960 N9 P1668-1671	81N49R <ATS>
1960 N9 P1687-1692	62-24834 <=*>
1960 N9 P1716-1717	36M47R <ATS>
1960 N10 P1739-1750	61-16786 <*=> 0
	61-21561 <=>
1960 N10 P1751-1758	61-18597 <*=>
1960 N10 P1759-1762	61-18600 <*=>
1960 N10 P1772-1778	92N47R <ATS>
1960 N10 P1783-1786	28N48R <ATS>
1960 N10 P1838-1843	53N49R <ATS>
	62-10274 <*=>
1960 N10 P1844-1847	61-18602 <*=>
	97N48R <ATS>
1960 N10 P1883-1885	76N53R <ATS>
1960 N10 P1905-1908	61-21579 <=>
1960 N11 P1903-	61-10885 <+*>
1960 N11 P1974-1980	88N49R <ATS>
1960 N11 P1991-1997	49N49R <ATS>
1960 N11 P2019-2025	61-20685 <*=>
1960 N11 P2026-2031	48N49R <ATS>
1960 N12 P2156-2161	62-10441 <*=>
1960 N12 P2173-2177	27N50R <ATS>
	62-10432 <*=>
1960 N12 P2236-2237	61-16529 <*=>
	61-18601 <*=>
1961 P729-730	61-18824 <*=>
1961 P1947-1954	63-10236 <=*>
1961 N1 P164-166	62-16115 <=*>
1961 N1 P177-178	62-24848 <=*>
1961 N1 P184-186	73N53R <ATS>
1961 N2 P225-229	62-23883 <=*>
1961 N2 P360-363	11411-B <KH>
1961 N2 P360-362	63-18463 <=*>
1961 N2 P367-368	63-14084 <=*> 0
1961 N2 P379-380	62-13373 <*=>
1961 N3 P396-406	21N55R <ATS>
1961 N3 P430-436	62-14817 <=*>
1961 N3 P466-468	62-10428 <*=>
1961 N3 P468-473	65-13024 <*>
1961 N3 P524-525	62-14816 <=*>
1961 N4 P578-582	62-14815 <=*>
1961 N4 P653-657	62-14814 <=*>
1961 N4 P657-663	63-10227 <=*>
1961 N7 P1198-1205	63-15261 <=*>
1961 N7 P1240-1248	61-28967 <*=>
1961 N7 P1320-1325	62-16559 <=*>
1961 N7 P1326-1330	62-25321 <=*>
1961 N7 P1336-	ICE V2 N3 P348-353 <ICE>
1961 N7 P1336-1342	91N56R <ATS>
1961 N7 P1350-1352	62-16560 <=*>
	66-11454 <*>
1961 N7 P1357-1360	62-15218 <*=>
1961 N8 P1528	22P59R <ATS>
1961 N9 P1539-1542	35P59R <ATS>
1961 N10 P1742-1748	82P60R <ATS>
1961 N10 P1863-1870	42P59R <ATS>
	63-10238 <=*>
1961 N10 P1871-1874	43P59R <ATS>
1961 N10 P1874-1879	44P59R <ATS>
1961 N11 P1971-1976	62-24063 <=*>
1961 N11 P2020-2028	83P58R <ATS>
1961 N11 P2088-2090	62-14960 <=*>
1961 N12 P2204-2209	63-10241 <=*>
1961 N12 P2234-2235	62-14996 <=*>
1961 N12 P2254-2255	98P61R <ATS>
1962 N1 P19-22	62-P60R <ATS>
	65-12041 <*>
1962 N1 P117-128	65-63512 <=*$>
1962 N1 P155-157	65-12068 <*>
1962 N2 P195-201	AEC-TR-5117 <=*>
1962 N2 P236-243	06P62R <ATS>
1962 N2 P302-303	99Q70R <ATS>
1962 N2 P312-317	62-18064 <=*> 0
1962 N2 P324-327	AD-626 441 <=*$>
1962 N2 P358-359	62-32418 <=*>
1962 N2 P362-363	62-32418 <=*>
1962 N3 P517-518	63-15204 <=*>
1962 N3 P536-537	62-18060 <=*>
1962 N5 P789-799	63-23410 <=*>
1962 N5 P800-805	63-10085 <=*>
1962 N5 P930-931	81P64R <ATS>
1962 N6 P960-968	22P64R <ATS>
1962 N6 P996-998	63-18441 <*=>
1962 N6 P1120-1121	63-19688 <=*$>
1962 N7 P1169-1174	69P65R <ATS>
1962 N7 P1303-1305	43P65R <ATS>
1962 N7 P1305-1307	66-11455 <*>
1962 N8 P1322-1329	63-18834 <=*$>
1962 N8 P1365-1367	62-20448 <=*>
1962 N8 P1464-1467	63-18835 <=*>
1962 N9 P1546-1550	25Q69R <ATS>
1962 N9 P1584-1589	08P66R <ATS>
1962 N9 P1690-1692	63-21442 <=>
1962 N9 P1699	57Q67R <ATS>
1962 N10 P1748-1752	64-10873 <=*$>
1962 N10 P1753-1756	63-18334 <=*>
1962 N10 P1783-1788	66-12071 <*>
1962 N10 P1896-1899	63-15358 <=*>
1962 N11 P1945-1953	56Q68R <ATS>
1962 N11 P2002-2008	11Q70R <ATS>
1962 N12 P2251-2253	63-18353 <*=>
1962 V7 P1154-1163	63-20203 <=*>
1962 V8 P1357-1365	63-26203 <=$>
1963 N1 P17-21	51Q69R <ATS>
1963 N1 P166-170	63-20388 <=*$>
1963 N1 P186-187	67Q69R <ATS>
1963 N1 P190-192	63-11668 <=>
1963 N2 P328-332	39Q69R <ATS>
1963 N2 P359-364	47Q69R <ATS>
1963 N3 P414-425	64-14458 <=*$>
1963 N3 P481-487	64-15124 <=*$>
1963 N3 P528-531	64-14950 <=*$>

1963 N6 P1136-1139 64-15235 <=*$>
1963 N7 P1210-1215 27Q73R <ATS>
1963 N7 P1215-1219 28Q73R <ATS>
1963 N8 P1527-1528 33Q73R <ATS>

IZVESTIYA AKADEMII NAUK SSSR. OTDELENIE LITERATURY I YAZYKA
1959 V18 N3 P209-216 61-11046 <=>
1960 V19 N1 P60-77 61-11046 <=>
1962 V21 N1 P34-40 62-20408 <=*>
1962 V21 N2 P103-111 62-25499 <=*>

IZVESTIYA AKADEMII NAUK SSSR. OTDELENIE MATEMATICHESKIKH I ESTESTVENNYKH NAUK
1931 N7 P1123-1140 63-10360 <=*>
1932 N7 P857-905 R-5093 <*>
1933 N3 P373-386 T-456 <INSD>
1934 N10 P1501-1515 62-25018 <=*> O
1935 P1385-1397 64-14943 <*> P
1935 V2 N7 P1163-1168 61-31007 <=>

IZVESTIYA AKADEMII NAUK SSSR. OTDELENIE MATEMATICHESKIKH I ESTESTVENNYKH NAUK. SERIYA KHIMICHESKAYA
1936 P133-152 66-12065 <*> O
1936 P153-189 66-12066 <*> O
1936 P397-407 61-18065 <*>
1936 P408-422 61-18066 <*>
1936 P423-433 61-18067 <*>
1936 N2 P379-385 RT-2235 <*>
1937 P467-468 61-17992 <=$>
1937 P547-554 61-17247 <=$>
1937 P1119-1150 63-19523 <=$>
1937 N3 P529-538 R-5047 <*>
64-19521 <=$>
1937 N4 P1433-1450 R-3798 <*>
64-19426 <=$> O
1938 P403-413 66-12067 <*>
1938 P1225-1248 RT-2576 <*>
1938 P1249-1254 RT-2372 <*>
1938 N2 P489-497 R-4647 <*>
1938 V11 P1-109 RT-3034 <*>

IZVESTIYA AKADEMII NAUK SSSR. OTDELENIE TEKHNICHESKIKH NAUK
1937 N1 P49-70 RT-347 <*>
1937 N1 P125-149 RT-265 <*>
1937 N4 P479-488 R-2578 <*>
1938 N2 P3-54 RT-360 <*>
1938 N3 P43-60 61-28208 <=*$>
1938 N3 P61-69 62-16405 <=*>
1938 N4 P21-49 RT-293 <*>
1938 N10 P89-112 RT-1878 <*>
1939 N2 P53-67 63-15696 <=*$>
1939 N2 P107-128 RT-297 <*>
1939 N3 P85-109 61-16978 <=*$>
1939 N5 P33-48 66-12068 <*>
1940 N2 P41-48 2398 <BISI>
1940 N3 P27-34 61-16956 <*=>
1940 N4 P19-38 61-16832 <*=>
1940 N4 P29-35 61-16890 <=*>
1940 N5 P49-58 61-16831 <*=>
1940 N5 P135-136 RT-1665 <*>
1940 N8 P47-50 61-16885 <=*>
1940 N9 P99-108 61-16886 <=*>
1940 N10/1 P663-671 62-10875 <=*> P
1941 N4 P45-54 61-17991 <=$>
1941 N5 P3-10 61-17990 <=$>
1941 N7/8 P53-62 2K22R <ATS>
1942 N9 P54-59 61-16662 <=*>
1942 N10 P3-10 61-16663 <*=>
1942 N10 P11-25 61-16976 <*=>
1942 N3/4 P3-10 61-16645 <=*>
1942 N3/4 P11-19 61-16646 <=*>
1942 N3/4 P21-25 61-10143 <+*>
1942 N3/4 P21-26 61-16647 <*=>
62-18269 <=*>
1942 N3/4 P27-32 61-16654 <=*>
1942 N5/6 P3-16 61-16651 <=*>
1942 N5/6 P29-34 61-16652 <*=>
1942 N11/2 P3-10 61-16954 <=*>
1943 N7 P62-72 60-23027 <+*>
61-16944 <*=>
1943 N1/2 P10-22 61-16943 <*=>
1943 N5/6 P18-28 61-18893 <*=>
62-16441 <=*>
1943 N5/6 P71-77 61-17201 <=$>
1943 N11/2 P3-14 61-16945 <*=>
1943 N11/2 P37-49 61-18153 <*=>
1943 N9/10 P21-31 R-2245 <*>
1943 N9/10 P57-67 61-18350 <*=>
1943 N9/10 P68-69 61-16930 <*=>
1944 P275-287 61-20191 <*=>
1944 P440-445 61-18192 <*=>
1944 P672-684 61-18212 <*=>
1944 P685-689 61-18211 <*=>
1944 P690-694 61-18210 <*=>
1944 P695-708 61-18208 <*=>
1944 P709-715 61-18209 <*=>
1944 P716-723 61-18207 <*=>
1944 P724-728 61-18206 <*=>
1944 P734-739 61-18343 <*=>
1944 P740-744 61-18894 <*=>
1944 P745-751 61-18884 <*=> P
1944 N6 P337-345 62-10871 <=*> P
1944 N6 P346-350 62-10872 <=*> P
1944 N6 P405-412 RT-2618 <*>
1944 N6 P419-430 62-10873 <=*> P
1944 N1/2 P3-12 62-10868 <=*> P
1944 N1/2 P42-47 61-18170 <*=>
1944 N1/2 P48-50 62-10869 <=*>
1944 N1/2 P68-70 63-20994 <=*$>
1944 N7/8 P530-539 62-10874 <=*> P
1944 N10/1 P695-708 65-18162 <*>
1944 N10/1 P729-733 70K21R <ATS>
1945 P70-79 61-18890 <*=> P
1945 P185-189 61-18889 <*=> P
1945 N3 P190-202 62-24811 <=*>
1945 N3 P250-260 RT-294 <*>
1945 N9 P857-874 AEC-TR-5214 <=*>
1945 N1/2 P57-69 62-10877 <=*> P
1945 N4/5 P382-395 RT-628 <*>
1945 N10/1 P1105-1114 61-13665 <*+>
1945 N10/1 P1139-1144 <ATS>
R-1448 <*>
RT-1724 <*>
1946 N1 P107-114 RJ-463 <ATS>
2890 <HB>
518TT <CCT>
1946 N2 P215-218 RJ-464 <ATS>
2891 <HB>
1946 N3 P329-354 RT-349 <*>
1946 N4 P487-498 59-14628 <+*>
59-15095 <*>
1946 N4 P543-554 61-20108 <*=>
1946 N4 P555-559 61-20109 <*=>
1946 N5 P737-740 RT-83 <*>
1946 N6 P899-912 2053 <HB>
1946 N6 P1201-1210 RT-89 <*>
1946 N7 P969-974 59-15098 <+*>
1946 N9 P1219-1233 RT-805 <*>
1946 N9 P1243-1260 65-12307 <*>
1946 N10 P1355-1373 R-1110 <*>
RT-804 <*>
1946 N10 P1375-1384 RT-2799 <*>
1946 N10 P1459-1462 61-23173 <=*>
1946 N11 P1567-1580 RT-124 <*>
1946 V11 P1581-1589 RT-91 <*>
1947 P1369-1384 R-4458 <*>
1947 N1 P27-53 RT-361 <*>
1947 N1 P117-123 2033 <HB>
1947 N2 P143-150 R-1017 <*>
RT-525 <*>
60-27002 <+*>
1947 N3 P253-259 RT-1109 <*>
1947 N3 P271-300 RT-1868 <*>
1947 N3 P301-305 RT-350 <*>
1947 N3 P349-355 4806 <HB>
1947 N6 P735-747 65-23887 <$>
1947 N7 P805-808 63-20803 <=*$> O

Citation	Translation
1947 N7 P813-816	R-1909 <*>
1947 N7 P825-828	62-10308 <*=>
1947 N7 P863-870	63-19834 <=*>
1947 N7 P907-911	62-23877 <=*>
1947 N9 P1061-1068	59-14367 <+*>
1947 N9 P1107-1136	60-10099 <+*>
1947 N9 P1193-1206	61-13664 <*+> 0
1947 N10 P1271-1274	R-2353 <*>
1947 N12 P1639-1648	4681 <HB>
1947 V7 P895-900	T-1301 <INSD>
1948 N1 P79-86	20M46R <ATS>
1948 N2 P235-238	RTS-2673 <NLL>
1948 N3 P349-358	RT-1803 <*>
1948 N4 P433-448	C-4457 <NRC>
1948 N4 P449-458	RT-82 <*>
1948 N6 P789-800	RT-2976 <*>
1948 N6 P855-872	RT-264 <*>
1948 N6 P889-898	4609 <HB>
1948 N6 P899-906	2207 <HB>
1948 N7 P967-980	RT-2792 <*>
1948 N8 P1297-1312	RT-1216 <*>
1948 N9 P1389-1402	50/2799 <NLL>
1948 N9 P1457-1462	2246 <HB>
1948 N10 P1547-1560	61-23156 <=*>
1949 N1 P100-113	2329 <HB>
1949 N4 P508-513	50/2057 <NLL>
	60-13580 <+*>
1949 N5 P686-700	M.5667 <NLL>
	50/2714 <NLL>
1949 N5 P701-712	RT-2542 <*>
1949 N6 P601-606	59-19213 <+*>
1949 N7 P1067-1082	50/2712 <NLL>
1949 N8 P1231-	2391 <HB>
1949 N8 P1235-1241	74J19R <ATS>
1949 N9 P1284-1296	RT-81 <*>
1949 N9 P1365-1371	2457 <HB>
1949 N9 P1372-1377	59-19086 <+*>
1949 N11 P1620-1625	AEC-TR-981 <+*>
	AEC-TR-981 <*>
1949 N11 P1666-1674	2681 <HB>
1949 N11 P1685-1700	63-24581 <=*$>
1949 N11 P1719-1722	59-18545 <+*>
	60-19915 <+*>
1949 N12 P1753-1773	RT-624 <*>
	60-13873 <+*>
1949 N12 P1774-1787	RT-493 <*>
1949 N12 P1824-1831	RT-1355 <*>
	2928D <K-H>
	64-10144 <=*$> 0
1949 N12 P1848-1873	63-18216 <=*> P
1949 V9 P1352-1360	64-18400 <*> 0
1950 P1024-1033	62-24827 <=*>
1950 N1 P89-100	05R78R <ATS>
1950 N1 P108-125	RT-1999 <*>
1950 N2 P249-252	51/2245 <NLL>
1950 N2 P266-278	RT-909 <*>
1950 N3 P374-385	R-1982 <*>
1950 N4 P575-581	RT-1980 <*>
1950 N5 P689-694	61-17302 <=$>
1950 N6 P851-865	RT-4503 <*>
1950 N6 P888-	59-21130 <=>
1950 N7 P1040-1048	RTS 2679 <NLL>
1950 N7 P1071-1079	GB82 3931 <NLL>
1950 N8 P1105-	61-13663 <*+>
1950 N10 P1450-1462	RT-625 <*>
1950 N11 P1615-1644	R-3825 <*>
1950 N11 P1645-1647	62-10638 <=*>
1950 N12 P1784-1794	60-23942 <+*> 0
1950 N12 P1815-1826	62-23104 <=>
1951 P506-528	AEC-TR-1433 <+*>
1951 N1 P29-38	C-5614 <NRC>
1951 N1 P53-56	64-23529 <=>
1951 N2 P209-223	60-15767 <+*>
	61K20R <ATS>
1951 N3 P389-	59-21203 <=>
1951 N4 P506-528	RT-2107 <*>
1951 N4 P529-536	R-2412 <*>
	RT-1870 <*>
	64-16013 <=*$>
1951 N4 P565-575	2750 <HB>
1951 N6 P829-838	61-17288 <=$>
	63-13021 <=*>
1951 N6 P852-862	R-3772 <*>
1951 N7 P1015-1024	59-15131 <+*>
1951 N7 P1041-1045	RT-2176 <*>
1951 N8 P1188-1197	AEC-TR-3274 <+*>
1951 N8 P1230-1233	84J19R <ATS>
1951 N10 P1513-1521	RJ-97 <ATS>
	63-10396 <=*>
1951 N11 P1669-1681	53/0366 <NLL>
1951 N11 P1689-1695	RT-1984 <*>
1951 N11 P1744-1751	2851 <HB>
1951 N12 P1777-1785	5176 <HB>
1951 N12 P1792-1800	C-3721 <NRCC>
1951 N12 P1801-1811	N66-22302 <=$>
1952 N3 P367-373	59-22422 <+*>
1952 N3 P395-404	RT-2341 <*>
1952 N3 P423-432	RT-2354 <*>
1952 N6 P840-857	RTS-3031 <NLL>
1952 N6 P877-882	RT-1967 <*>
1952 N7 P1011-1025	RT-2802 <*>
1952 N8 P1226-1244	R-3819 <*>
1952 N9 P1360-1368	R-3822 <*>
1952 N10 P1448-1454	AEC-TR-3760 <+*>
1952 N10 P1472-1482	60-15826 <*+>
1952 N10 P1578-1581	60-17181 <*+>
1953 N1 P3-15	3288 <HB>
1953 N2 P283-298	62-19880 <=*>
1953 N3 P378-382	3194 <HB>
1953 N3 P432-440	RT-2636 <*>
1953 N4 P512-522	TT.514 <NRC>
1953 N4 P523-532	62-23051 <=>
	62-23051 <=*>
1953 N4 P562-598	61-13485 <*+>
1953 N5 P705-707	RT-3355 <*>
	60-23011 <+*>
1953 N5 P708-709	60-23012 <+*>
1953 N5 P708-729	63-14511 <*=>
	64-16121 <=*$> 0
1953 N5 P741-751	RT-3457 <*>
1953 N5 P762-766	R-3827 <*>
1953 N6 P905-909	RT-2279 <*>
1953 N6 P941-942	RT-1553 <*>
1953 N7 P969-991	61-13230 <*+>
1953 N7 P1022-1033	NP-TR-297 <*>
1953 N7 P1022-1034	RT-2241 <*>
1953 N7 P1022-1033	60-13004 <+*> 0
	60-13795 <+*> 0
1953 N7 P1035-1043	3274 <HB>
1953 N7 P1044-1057	3380 <HB>
1953 N8 P1110-1117	RT-2282 <*>
1953 N8 P1130-1136	RJ-168 <ATS>
1953 N9 P1241-1247	3224 <HB>
1953 N10 P1401-1416	RT-3726 <*>
1953 N10 P1444-1451	61-13280 <*+>
1953 N11 P1524-1531	59-18544 <+*>
1953 N11 P1630-1638	64-10355 <=*$> 0
1953 N12 P1790-1796	3417 <HB>
1953 N12 P1804-1812	3865 <HB>
1953 N12 P1819-1825	RT-3262 <*>
1953 N12 P1847-	59-21113 <=>
1954 N1 P54-60	RJ-232 <ATS>
1954 N1 P80-91	13J15R <ATS>
1954 N1 P114-127	62-24075 <=*>
1954 N1 P128-137	62-24093 <=*>
1954 N1 P162-164	R-1181 <*>
1954 N2 P52-59	3373 <HB>
1954 N2 P60-66	TT.575 <NRC>
1954 N2 P106-110	RT-2915 <*>
1954 N3 P72-82	3525 <HB>
1954 N3 P83-87	3390 <HB>
1954 N3 P88-90	UCRL TRANS-665L <*=>
1954 N3 P91-101	3858 <HB>
1954 N3 P102-109	3926 <HB>
1954 N3 P110-115	R-449 <*>
	3517 <HB>
1954 N3 P116-122	RJ-305 <ATS>
	RT-3165 <*>
	59-17495 <*>
	61-20199 <*=>

Citation	Translation
1954 N4 P80-85	64-14937 <=*$> 0
1954 N4 P140-146	RT-4614 <*>
	59-22359 <+*>
1954 N4 P147-151	R-3117 <*>
	RT-4615 <*>
1954 N5 P91-94	R-1004 <*>
1954 N6 P3-12	RT-2984 <*>
1954 N6 P25-36	62-23080 <=*>
1954 N6 P37-44	62-23200 <=*>
1954 N6 P47-52	62-11300 <=*>
1954 N6 P53-56	GB29 <NLL>
	R-3860 <*>
1954 N6 P137-144	RT-2924 <*>
	57-3127 <*>
1954 N7 P105-115	3449 <HB>
1954 N8 P29-36	RT-2962 <*>
1954 N8 P101-109	61-19440 <+*> 0
1954 N9 P62-79	61-23520 <=*>
1954 N9 P63-79	65-12264 <*>
1954 N9 P115-123	59-15043 <+*> 0
1954 N10 P23-30	3501 <HB>
1954 N10 P31-38	3509 <HB>
1954 N10 P39-46	3566 <HB>
1954 N10 P47-59	R-4476 <*>
1954 N10 P99-	59-21112 <=>
1954 N10 P99-111	62-23078 <=*>
1954 N12 P86-91	RT-3625 <*>
	59-16786 <+*>
1954 N12 P92-96	GB29 <NLL>
	R-3258 <=>
1954 N12 P102-119	61-18869 <*=> 0
1954 N12 P120-127	C-2122 <NLL>
1954 V12 P92-96	GB43 <NLL>
1955 P29-39	R-5065 <*>
1955 N1 P57-66	3537 <HB>
1955 N1 P80-95	3621 <HB>
1955 N1 P109-134	62-13753 <*=>
1955 N1 P155-156	62-23113 <=*>
1955 N1 P157-159	GB70 4193 <NLL>
1955 N2 P5-13	65K27R <ATS>
1955 N2 P98-104	GB43 <NLL>
1955 N2 P132-136	59-17491 <*>
	64-19987 <=$>
1955 N2 P139-140	RT-3672 <*>
1955 N3 P3-32	R-569 <*>
1955 N3 P33-68	60-17478 <+*>
1955 N3 P90-107	C-3482 <NRCC>
1955 N3 P139-149	59-12200 <+*>
1955 N4 P53-57	RT-3671 <*>
1955 N4 P110-119	CSIRO-3094 <CSIR>
	R-3633 <*>
1955 N4 P131-135	RJ-414 <ATS>
	RJ414 <ATS>
1955 N4 P136-140	R-15 <*>
	R-408 <*>
1955 N5 P3-41	49J15R <ATS>
1955 N5 P55-84	60-21429 <=>
1955 N5 P85-101	59-16723 <+*>
1955 N5 P109-113	R-3638 <*>
	3241 <*>
	3594 <HB>
1955 N5 P119-122	61-19439 <+*> 0
1955 N5 P123-128	GB70 4194 <NLL>
1955 N6 P101-108	C2356 <NRC>
1955 N6 P127-139	C-2500 <NRC>
	60-23778 <+*>
1955 N6 P140-146	C-2504 <NRC>
1955 N6 P140-148	RT-3951 <*>
1955 N7 P9-22	59-16710 <+*>
1955 N7 P23-33	R-1423 <*>
1955 N7 P70-74	NIAE-TRANS-170 <NLL>
1955 N7 P75-83	61-19910 <=*>
1955 N7 P84-88	60-14511 <+*>
1955 N7 P122-128	GB125 IB 1482 <NLL>
1955 N7 P145-149	CSIRO-3309 <CSIR>
	R-3917 <*>
1955 N8 P3-10	R-4073 <*>
1955 N8 P11-16	61-13081 <+*>
1955 N8 P17-21	NACA TM 1440 <NASA>
	R-4084 <*>
1955 N8 P22-36	63-20282 <=*> 0
1955 N8 P100-106	R-3115 <*>
	3635 <HB>
1955 N8 P145-147	GB43 T2150 <NLL>
	59-18665 <+*>
1955 N9 P3-13	61-17363 <=$>
1955 N9 P65-106	59-18708 <+*> 0
	62-25962 <=*>
1955 N9 P107-118	62-25923 <=*>
1955 N9 P125-136	63-24629 <=*$>
1955 N9 P160-166	R-1183 <*>
	59-21046 <=>
	61-13309 <*+>
1955 N10 P127-130	61-28159 <*=>
1955 N10 P131-137	RT-4303 <*>
1955 N10 P143-146	GB4 T1028 <NLL>
1955 N10 P147-151	M-5680 <NLL>
1955 N11 P13-24	64-20512 <*> 0
1955 N11 P58-61	59-19156 <+*>
1955 N11 P125-128	R-4976 <*>
1955 N11 P141-143	66-61518 <=*$>
1955 N12 P131-135	GB4 T1028 <NLL>
1955 N12 P139-143	T-2135 <INSD>
1956 N1 P21-29	22J15R <ATS>
1956 N1 P71-79	61-17358 <=$>
1956 N1 P108-118	CSIRO-3880 <CSIR>
1956 N1 P119-125	60-14510 <+*>
1956 N2 P35-42	46K21R <ATS>
1956 N2 P69-74	62-19817 <=>
1956 N3 P59-67	GB4 T1050 <NLL>
1956 N3 P68-76	4304 <HB>
1956 N4 P33-41	62-19850 <=*>
1956 N4 P85-93	R-4380 <*>
1956 N4 P115-121	62-19862 <=*>
1956 N4 P139-141	59-10362 <+*>
	62-11590 <=>
1956 N4 P151-152	62-11330 <*>
	62-11330 <=>
1956 N5 P29-39	61-20718 <*=>
1956 N5 P70-84	61-10820 <+*>
1956 N5 P113-121	42K22R <ATS>
1956 N5 P140-143	98M40R <ATS>
1956 N5 P147-151	4834 <HB>
1956 N5 P151-153	4686 <HB>
1956 N5 P156-163	70L34R <ATS>
1956 N6 P35-44	SCL-T-291 <+*>
	60-21604 <=>
1956 N6 P55-62	UCRL TRANS-785(L) <=*>
	UCRL-TRANS-785(L) <=*>
1956 N6 P63-76	59-14176 <+*>
	61-23861 <*=>
1956 N6 P77-88	60-00490 <*>
	60-13112 <+*>
1956 N6 P89-100	61-13033 <+*> 0
1956 N6 P113-118	62-11263 <=>
1956 N6 P136-143	63-19837 <*=>
1956 N7 P43-52	61-15381 <*+>
1956 N7 P86-97	60-15747 <+*>
1956 N7 P103-110	3912 <HB>
1956 N7 P114-117	62-11619 <=>
1956 N7 P121-123	R-3806 <*>
	59-18345 <+*>
1956 N8 P10-19	R-3655 <*>
1956 N9 P55-64	60-16981 <*> 0
1956 N9 P108-110	59-12639 <=>
1956 N9 P111-114	60-15772 <+*>
1956 N10 P37-47	R-5026 <*>
	60-15393 <+*>
1956 N10 P48-63	60-19976 <+*> 0
1956 N10 P112-116	62-11283 <=>
1956 N10 P116-117	62-19817 <=>
1956 N11 P28-39	R-3943 <*>
1956 N11 P52-57	4456 <HB>
1956 N11 P58-63	60-23022 <+*>
1956 N11 P77-81	59-18339 <*+>
1956 N11 P106-110	62-11214 <=>
1956 V11 P40-51	R-3186 <*>
1957 N1 P29-31	76K24R <ATS>
1957 N1 P77-84	90S87R <ATS>
1957 N1 P95-102	4833 <HB>

Citation	Translation
1957 N1 P123-135	3944 <HB>
	59-16986 <+*>
1957 N1 P136-138	3945 <HB>
1957 N1 P137-147	60-17362 <+*>
1957 N1 P153-156	59-22236 <+*> 0
1957 N2 P9-18	4030 <HB>
1957 N2 P19-26	4031 <HB>
1957 N2 P27-35	61-13101 <*+>
1957 N2 P120-122	60-16964 <*>
1957 N2 P122-123	61-13509 <=*>
1957 N2 P132-136	60-17237 <+*>
1957 N2 P159-162	4071 <HB>
1957 N3 P99-107	AMST S2 V28 P333-344 <AMS>
1957 N3 P180-182	RJ-944 <ATS>
1957 N4 P106-109	AD-620 120 <=$>
1957 N4 P115-120	61-21952 <=>
1957 N4 P172-173	96J16R <ATS>
1957 N5 P30-41	44K20R <ATS>
1957 N5 P48-55	<LSA>
	60-15836 <+*> 0
1957 N5 P77-84	4054 <HB>
1957 N5 P102-103	R-1584 <*>
	4074 <HB>
1957 N6 P32-36	4198 <HB>
1957 N6 P77-85	62-23227 <=*>
1957 N6 P86-92	63-13729 <=*>
1957 N6 P93-101	59-22532 <+*>
1957 N7 P24-28	R-4892 <*>
1957 N7 P35-40	AEC-TR-3158 <+*>
	R-3078 <*>
	R-3397 <*>
1957 N7 P89-93	62-15856 <=*>
1957 N7 P138-142	62-11205 <=> 0
1957 N7 P142-146	84L36R <ATS>
1957 N8 P13-19	R-4822 <*>
1957 N8 P20-34	61-28836 <*=>
1957 N8 P35-40	228TT <CCT>
1957 N8 P56-62	24M44R <ATS>
1957 N8 P72-77	60-14011 <+*> 0
1957 N8 P102-108	4081 <HB>
	4400 <HB>
1957 N8 P120-122	<HB>
1957 N8 P137-141	62-32933 <=>
1957 N8 P142-143	62-15882 <=*>
1957 N9 P37-44	4259 <HB>
1957 N9 P123-126	60-18054 <*> 0
1957 N10 P3-5	61-13357 <*+>
1957 N10 P68-70	59-18754 <+*>
1957 N10 P86-89	TS 1748 <BISI>
	61-13470 <*+>
1957 N10 P93-95	4422 <HB>
1957 N10 P111-112	83N48R <ATS>
1957 N11 P69-74	60-13818 <+*> 0
1957 N11 P186-188	60-13779 <+*>
1957 N12 P64-66	61-19345 <+*>
	62-15878 <=>
1957 N12 P71-76	4843 <BISI>
1957 N12 P76-77	4199 <HB>
	61-13146 <*+>
1957 N12 P80-82	62-13598 <*=>
1958 N1 P3-10	AEC-TR-3465 <+*>
1958 N1 P11-20	60-13848 <+*>
1958 N1 P26-34	91S87R <ATS>
1958 N1 P37-42	66L30R <ATS>
1958 N1 P44-51	61-13298 <*+>
1958 N1 P52-62	AEC-TR-3467 <+*>
1958 N1 P113-119	AEC-TR-3470 <+*>
1958 N1 P140-142	4905 <HB>
1958 N2 P7-14	4299 <HB>
1958 N2 P26-32	60-15649 <+*>
1958 N2 P123-125	AD-615 233 <=$>
1958 N3 P14-24	62-18954 <=*>
1958 N3 P25-32	61-31460 <=>
1958 N3 P33-41	62-16026 <=*>
	63-11525 <=>
	63-14662 <=*>
1958 N3 P42-50	64-18344 <*>
1958 N3 P104-108	59-17739 <+*>
1958 N3 P113-118	4508 <HB>
	60-15661 <+*>
1958 N3 P120-121	59-18125 <+*>
1958 N3 P121-125	N66-14870 <=*$>
1958 N3 P156-162	60-17380 <+*>
1958 N3 P163-164	60-23527 <+*>
1958 N3 P165-168	60-23529 <+*>
1958 N4 P41-53	60-17364 <+*>
1958 N4 P54-66	60-21750 <=>
1958 N4 P92-97	61-13004 <*+>
1958 N4 P102-109	60-21605 <=>
1958 N4 P114-117	60-17377 <+*>
1958 N4 P118-121	4340 <HB>
1958 N4 P122-123	4378 <HB>
1958 N5 P3-6	62-25329 <=*>
1958 N5 P7-15	62-24286 <=*>
1958 N5 P21-28	61-13531 <*+>
1958 N5 P46-50	AEC-TR-3928 <=*>
	AIAAJ V1 N7 P1729 <AIAA>
1958 N5 P46-53	61-20703 <*=>
1958 N5 P59-62	4339 <HB>
1958 N5 P63-68	61-19392 <+*>
1958 N5 P85-92	83K26R <ATS>
1958 N5 P110-115	63-14660 <=*>
1958 N6 P30-36	4494 <HB>
1958 N6 P96-98	59-10667 <+*>
1958 N6 P99-10C	59-22682 <+*>
1958 N6 P111-113	24N50R <ATS>
1958 N6 P134-136	39K28R <ATS>
1958 N6 P158-160	59-18772 <+*>
1958 N7 P3-9	SCL-T-326 <+*>
1958 N7 P15-23	60-17372 <+*>
1958 N7 P87-90	61-23638 <=*>
1958 N7 P98-101	CSIR-TR.173 <CSSA>
1958 N7 P102-105	ARSJ V32 N1 P130 <AIAA>
	59-20258 <+*>
	62-24830 <=>
1958 N7 P130-	59-11898 <=>
1958 N7 P136-139	4346 <HB>
1958 N7 P140-141	4344 <HB>
1958 N7 P149-150	61-23287 <*=>
1958 N8 P3-11	59-11929 <=>
1958 N8 P12-14	N66-23607 <=$>
1958 N8 P15-18	59-19155 <+*> 0
1958 N8 P26-31	59-22404 <+*>
1958 N8 P32-40	60-17923 <+*>
1958 N8 P81-87	61-27297 <*=>
1958 N8 P88-90	4401 <HB>
	59-22686 <+*>
1958 N8 P90-92	4402 <HB>
1958 N8 P124-127	49M42R <ATS>
1958 N8 P127-129	61-11190 <=>
1958 N8 P139-142	60-14337 <+*>
1958 N9 P17-24	4506 <HB>
1958 N9 P25	4417 <HB>
1958 N9 P25-30	4417 <HB>
1958 N9 P37-52	60-23209 <*+>
	61-10675 <*+>
1958 N9 P91-93	59-21114 <=>
1958 N9 P134-139	62-11694 <=>
1958 N10 P5-11	60-13684 <+*>
1958 N10 P153-155	AD-630 411 <=*$>
	24M38R <ATS>
	60-13252 <+*>
1958 N11 P15-24	60-23043 <*+>
1958 N11 P70-	59-21129 <=>
1958 N11 P83-86	59-16733 <+*>
1958 N11 P89-91	75L32R <ATS>
1958 N11 P95-97	67L30R <ATS>
1958 N11 P98-99	59-21202 <=>
1958 N11 P106-107	59-21116 <=>
1958 N11 P118-120	51L30R <ATS>
1958 N11 P136-138	62-10206 <*=>
1958 N12 P5-14	60-21770 <=>
1958 N12 P15-23	60-21772 <=>
1958 N12 P24-31	AEC-TR-4832 <=*>
1958 N12 P85-89	60-21763 <=>
1958 N12 P90-95	60-21066 <=>
1958 N12 P96-99	63-21264 <=>
1958 N12 P104-114	60-17468 <+*>
1958 N12 P115-119	60-16812 <+*>

IZVESTIYA AKADEMII NAUK SSSR. OTDELENIE TECHNICHESKIKH NAUK. ENERGETIKA I AVTOMATIKA

Citation	Translation
1959 N1 P20-38	59-13981 <=> 0
1959 N1 P20-25	62-10175 <=*>
1959 N1 P100-103	63-23955 <=*$>
1959 N1 P133-135	59-14683 <+*>
1959 N1 P138-143	62-25872 <=*>
1959 N1 P139-145	59-13981 <=> 0
1959 N2 P3-12	ARSJ V29 N10 P750 <AIAA>
1959 N2 P13-20	ARSJ V29 N10 P756 <AIAA>
	62-23108 <=>
1959 N2 P21-25	ARSJ V29 N1C P761 <AIAA>
	62-11464 <=>
1959 N2 P26-31	ARSJ V29 N10 P765 <AIAA>
	62-11539 <=>
1959 N2 P32-37	ARSJ V29 N10 P769 <AIAA>
1959 N2 P38-44	ARSJ V29 N10 P773 <AIAA>
1959 N2 P65-70	59-13981 <=> 0
	61-11180 <=>
1959 N2 P99-103	59-13981 <=> 0
1959 N2 P137-138	ARSJ V30 N1 P76 <AIAA>
1959 N2 P139-141	59-13981 <=> 0
1959 N2 P139	62-11357 <=>
1959 N2 P140-141	62-11311 <=>
1959 N3 P11-18	ARSJ V30 N1 P93 <AIAA>
	60-11059 <=>
1959 N3 P25-31	61-23248 <*=>
1959 N3 P32-42	ARSJ V30 N1 P86 <AIAA>
	60-11074 <=>
1959 N3 P43-49	62-14013 <*>
	62-23446 <=*>
1959 N3 P90-94	ARSJ V30 N4 P375 <AIAA>
1959 N3 P100-103	ARSJ V30 N1 P83 <AIAA>
	62-11246 <=>
1959 N3 P141-15C	60-11092 <=>
1959 N3 P151-154	60-11083 <=>
	60-17463 <+*>
1959 N3 P176-178	60-11054 <=>
1959 N3 P179-180	60-11082 <=>
1959 N3 P181-182	60-11056 <=>
1959 N3 P183-184	60-11085 <=>
1959 N3 P191-192	60-11078 <=>
	60-12159 <=>
1959 N4 P11-22	62-15248 <*=>
1959 N4 P59-78	62-15193 <*=>
1959 N4 P97-105	61-19793 <=*>
1959 N4 P116-125	61-27562 <*=>
1959 N4 P166-176	62-11567 <=>
1959 N4 P188-197	AWRE-TR-23 <*>
	62-25559 <=*>
1959 N4 P214-215	62-11204 <=>
1959 N4 P221	60-11171 <=>
1959 N4 P222-224	60-11171 <=>
1959 N4 P230-237	59-13981 <=> 0
1959 N5 P86-96	61-11191 <=>
	62-19815 <=>
1959 N5 P156-161	61-10762 <+*>
1959 N6 P16-21	61-27579 <*=>
1959 N6 P22-33	62-24654 <=*>
1959 N6 P79-89	ARSJ V30 N9 P834 <AIAA>
	60-31195 <=>
1959 N6 P99-107	60-31196 <=>
1959 N6 P108-117	ARSJ V30 N7 P685 <AIAA>
	60-31197 <=>
1959 N6 P134-140	61-10316 <*+>
1959 N6 P141-150	60-31198 <=>
1959 N6 P200-202	60-31631 <=>
1960 N1 P3-11	60-11948 <=>
1960 N1 P43-54	61-12258 <=>
	61-31459 <=> 0
1960 N1 P55-61	62-11255 <=>
1960 N1 P62-69	62-13061 <*=>
1960 N1 P70-75	60-19861 <+*>
	62-11344 <=>
1960 N1 P76-85	ARSJ V30 N11 P1036 <AIAA>
	60-31740 <=>
	62-15401 <=*>
1960 N1 P90-101	ARSJ V30 N9 P846 <AIAA>
1960 N1 P127-133	62-11293 <=>
1960 N1 P138-143	61-15222 <*+>
	62-11503 <=>
1960 N1 P165-168	ARSJ V30 N9 P843 <AIAA>
1960 N2 P20-30	64-15916 <=*$>
1960 N2 P38-43	62-11389 <=>
1960 N2 P44-53	62-23041 <=>
	62-23041 <=*>
1960 N2 P54-58	60-31414 <=>
1960 N2 P59-66	ARSJ V32 N7 P1130 <AIAA>
	60-31415 <=>
	61-11177 <=>
1960 N2 P73-82	60-31416 <=>
1960 N2 P83-97	61-27534 <*=> 0
1960 N2 P116-121	62-10176 <*=>
1960 N2 P122-131	ARSJ V31 N11 P1619 <AIAA>
	61-18811 <*=>
1960 N2 P137-152	ARSJ V31 N9 P1306 <AIAA>
1960 N2 P164-172	62-23125 <=*>
1960 N2 P177-180	RTS-3011 <NLL>
1960 N2 P185-191	ARSJ V32 N3 P467 <AIAA>
1960 N2 P195-197	ARSJ V31 N5 P691 <AIAA>
1960 N2 P205-2C9	62-13562 <*=>
1960 N3 P37-45	ARSJ V32 N7 P1143 <AIAA>
	62-23432 <=*>
1960 N3 P84-95	ARSJ V32 N5 P819 <AIAA>
1960 N3 P84-110	60-31850 <=>
1960 N3 P96-105	62-14264 <=*>
1960 N3 P138-144	ARSJ V32 N1 P133 <AIAA>
	62-10275 <=*>
1960 N3 P145-156	62-16716 <=*>
1960 N3 P180-182	61-21074 <=>
	62-11363 <=>
1960 N4 P36-47	AD-615 217 <=$>
1960 N4 P74-93	62-10177 <*=>
	62-13704 <=*>
1960 N4 P121-129	ARSJ V32 N5 P814 <AIAA>
1960 N4 P189-194	62-11189 <=>
1960 N5 P87-95	64-10800 <=*$>
1960 N5 P116-123	61-28992 <=*>
1960 N5 P159-165	61-28705 <*=>
1960 N5 P169-173	63-15910 <=*>
1960 N5 P174-178	AIAAJ V1 N4 P1014 <AIAA>
	62-13561 <=*>
1960 N5 P179-182	62-11222 <=>
	62-13699 <=*>
1960 N6 P107-112	62-15807 <=*>
1960 N6 P113-122	62-15198 <*=>
1960 N6 P123-132	61-27822 <*=>
1960 N6 P157-161	62-25454 <=*>
1960 N6 P162-164	62-15807 <=*>
1961 N1 P62-67	63-19895 <=*>
1961 N1 P78-80	62-24364 <=*>
1961 N1 P91-96	61-27215 <*=>
1961 N1 P97-109	61-27384 <*=>
1961 N1 P133-142	61-18799 <*=>
	61-27678 <*=>
1961 N1 P143-163	61-27655 <*=>
1961 N1 P166-170	61-27230 <*=>
	62-11367 <=>
1961 N1 P171-179	62-19860 <=*>
1961 N1 P180-184	61-27377 <*=>
1961 N1 P185-200	61-27762 <*=>
1961 N2 P62-64	62-25734 <=*>
1961 N2 P85-91	61-28558 <*=>
1961 N2 P92-105	62-25747 <=*>
1961 N2 P106-119	61-28558 <*=>
1961 N2 P134-143	61-27866 <*=>
1961 N2 P144-147	61-27953 <=*>
1961 N2 P148-158	61-27948 <*=>
	66-11258 <*>
1961 N2 P159-164	61-27950 <*=>
1961 N2 P165-170	61-31610 <=>
1961 N2 P173-175	61-27947 <*=>
1961 N3 P3-13	61-28869 <*=>
1961 N3 P46-57	61-28744 <*=>
1961 N3 P58-72	61-28749 <*=>
1961 N3 P73-81	61-28759 <*=>
1961 N3 P82-90	61-28859 <*=>
1961 N3 P91-96	61-28858 <*=>
1961 N3 P97-100	61-28735 <*=>
1961 N3 P105-1C9	63-19337 <=*>

1961 N3 P113-117	64-23520 <=>
1961 N4 P20-31	AD-614 629 <=$>
1961 N4 P37-42	62-32518 <=*>
1961 N4 P42-45	62-23668 <=*>
1961 N4 P46-54	62-15196 <*=>
1961 N4 P64-77	65-20425 <$>
1961 N4 P93-106	AD-615 215 <=$>
1961 N4 P191-196	62-32538 <=*>
1961 N4 P197-207	62-13569 <=*>
	63-10842 <=*>
	63-10902 <=*> 0
1961 N5 P7-12	63-13177 <=*>
1961 N5 P25-37	63-13177 <=*>
1961 N5 P51-71	62-24931 <=*>
1961 N5 P113-122	63-13177 <=*>
1961 N5 P123-135	62-20232 <=*>
1961 N5 P174-184	62-15812 <=*>
	63-15198 <=*>
1961 N6 P3-12	62-23756 <=*>
1961 N6 P13-20	64-23509 <=$>
1961 N6 P47-51	62-11723 <=>
1961 N6 P142-151	62-24936 <=*>
	62-32803 <=*>
1961 N6 P152-155	64-23562 <=$>
1961 V6 P93-100	976(L) <=$>
1962 N1 P37-44	63-23949 <=*$>
1962 N1 P45-53	56R77R <ATS>
1962 N1 P70-78	65-60706 <=> 0
1962 N1 P101-110	AEC-TR-5513 <=*>
1962 N1 P111-115	64-13271 <=$>
1962 N1 P116-125	62-32317 <=*>
1962 N1 P133-137	63-19202 <=*>
1962 N1 P152-165	63-19202 <=*>
1962 N2 P3-22	64-13010 <=*$>
1962 N2 P47-54	63-15544 <=*>
1962 N2 P104-114	65-13391 <*>
1962 N2 P124-129	AD-617 097 <=$>
1962 N3 P32-41	63-19180 <=*>
1962 N3 P114-121	63-21599 <=>
1962 N3 P163-172	NASA-TT-F132 <=>
1962 N3 P189-195	63-13836 <=*>
1962 N3 P196-200	63-11645 <=>
1962 N4 P66-68	63-15926 <=*>
1962 N4 P94-101	62-33470 <=*>
1962 N4 P102-108	62-33456 <=*>
1962 N4 P116-121	62-33467 <=*>
1962 N4 P122-129	62-33468 <=*>
1962 N4 P130-137	62-33469 <=*>
1962 N4 P145-150	AD-621 806 <=$>
	62-33457 <=*>
1962 N5 P89-95	AIAAJ V1 N10 P2445 <AIAA>
1962 N6 P34-38	63-21208 <=>
1962 N6 P39-49	63-21069 <=>
1962 N6 P76-81	63-20382 <=*>
	63-21207 <=>
1962 N6 P113-119	63-11720 <=>
1962 N6 P150-170	63-21098 <=>

IZVESTIYA AKADEMII NAUK SSSR. OTDELENIE TEKHNICHESKIKH NAUK. ENERGETIKA I TRANSPORT

1963 N1 P58-71	AEC-TR-6286 <=>
1963 N2 P138-141	65-60657 <=$>
1963 N2 P214-220	65-29514 <$>
1963 N3 P266-275	65-60740 <=>
1964 N1 P116-122	AD-610 410 <=$>
1964 N3 P370-377	AD-612 419 <=>
1964 N4 P466-473	AD-612 420 <=>
1964 N6 P743-749	AD-637 383 <=$>
1965 N2 P134-146	AD-632 675 <=$>
1965 N4 P98-110	AD-627 098 <=*$>
1965 N5 P159-172	AD-630 764 <=$>
1966 N1 P64-70	AD-639 908 <=$>
1966 N1 P131-134	AD-640 608 <=$>

IZVESTIYA AKADEMII NAUK SSSR. OTDELENIE TEKHNICHESKIKH NAUK. MEKHANIKA

1959 N1 P22-33	62-11167 <=>
1959 N1 P34-40	63-15791 <=>
	63-24263 <=*$>
1959 N1 P46-49	60-31882 <=>
	61-11165 <=>
	62-20251 <=*>
1959 N1 P89-96	61-20420 <*=>
1959 N1 P128-132	62-11572 <=>
1959 N1 P138-143	76L35R <ATS>
1959 N1 P154-157	AD-612 415 <=$>
1959 N1 P165-166	62-11443 <=>
	62-13388 <*=>
1959 N1 P170-173	61-15635 <+*>
1959 N2 P31-36	64-23534 <=$>
1959 N2 P81-87	62-11137 <=>
1959 N2 P151-155	ORNL-TR-19 <*>
	64-19118 <=*$>
1959 N2 P159-162	63-24333 <=*$>
1959 N2 P174-177	62-18854 <=*>
1959 N2 P205-207	62-32483 <=*>
1959 N3 P15-24	UCRL-TRANS-788(L) <=*>
1959 N3 P32-41	STMSP V4 P307-319 <AMS>
1959 N3 P59-64	65-12826 <*>
1959 N3 P78-83	ARSJ V31 N1 P98 <AIAA>
1959 N3 P84-95	62-25958 <=*>
1959 N3 P133-136	UCRL TRANS-729 <=*>
1959 N4 P3-10	63-24042 <=*>
1959 N4 P63-68	AD-607 964 <=>
1959 N4 P93-100	62-25383 <=*>
1959 N4 P130-131	59-22116 <=>
	62-11274 <=>
1959 N4 P131-133	ARSJ V30 N4 P414 <AIAA>
1959 N4 P134-135	59-22119 <=>
	61-23251 <*=>
	62-15385 <*=>
1959 N4 P140-141	AEC-TR-4092 <+*>
	ARSJ V30 N4 P374 <AIAA>
1959 N5 P9-15	ARSJ V30 N4 P416 <AIAA>
	60-12262 <=>
	62-11495 <=>
1959 N5 P58-63	61-11181 <=>
	62-20252 <=*> 0
1959 N5 P72-78	ARSJ V30 N7 P691 <AIAA>
1959 N5 P79-83	ARSJ V31 N7 P958 <AIAA>
1959 N5 P117-118	ARSJ V30 N9 P841 <AIAA>
	60-12519 <=>
	62-11449 <=>
1959 N5 P122-123	60-12265 <=>
	62-11227 <=>
1959 N5 P134-136	62-16099 <=*>
1959 N6 P3-6	60-31162 <=>
1959 N6 P7-13	62-32286 <=*>
1959 N6 P36-43	61-13334 <*+>
1959 N6 P93-99	60-31127 <=>
1959 N6 P110-112	60-12518 <=>
	62-11398 <=>
1959 N6 P127-131	61-27136 <*=>
1960 N1 P33-40	60-11812 <=>
1960 N1 P33-	61-21984 <=>
1960 N1 P41-46	60-11802 <=>
1960 N1 P60-69	60-11992 <=>
1960 N1 P70-78	60-11770 <=>
1960 N1 P106-112	ARSJ V30 N11 P1047 <AIAA>
1960 N1 P113-122	60-11782 <=>
1960 N1 P133-140	60-11991 <=>
1960 N1 P178-179	ARSJ V30 N9 P875 <AIAA>
1960 N1 P182-183	60-11983 <=>
	62-11221 <=>
1960 N2 P17-33	ARSJ V31 N7 P988 <AIAA>
1960 N2 P34-41	ARSJ V31 N7 P997 <AIAA>
	60-41169 <=>
1960 N2 P47-53	ARSJ V32 N7 P1135 <AIAA>
1960 N2 P128-129	65-64214 <=*$>
1960 N2 P135-138	60-41170 <=>
1960 N2 P141-144	60-41171 <=>
1960 N2 P144-148	00M74R <ATS>
1960 N2 P149-151	99M46R <ATS>
1960 N2 P175-176	63-23094 <=*>
1960 N3 P30-33	ARSJ V32 N1 P140 <AIAA>
	60-31796 <=>
1960 N3 P99-108	ARSJ V31 N11 P1644 <AIAA>
1960 N3 P117-120	ARSJ V31 N11 P1630 <AIAA>
1960 N3 P121-125	60-31797 <=>
	65-13175 <*>

Citation	Translation
1960 N3 P141-143	ARSJ V32 N7 P1139 <AIAA>
1960 N3 P143-146	ARSJ V32 N5 P805 <AIAA>
1960 N3 P147-151	75M46R <ATS>
1960 N3 P151-154	74M46R <ATS>
1960 N3 P163-164	97M45R <ATS>
1960 N4 P7-13	ARSJ V31 N11 P1635 <AIAA>
1960 N4 P14-18	ARSJ V32 N9 P1449 <AIAA>
1960 N4 P41-47	62-13743 <*=>
	87M47R <ATS>
1960 N4 P48-52	26M46R <ATS>
1960 N4 P84-92	ARSJ V32 N7 P1147 <AIAA>
1960 N4 P132	61-21998 <=>
	61-23838 <*=>
1960 N4 P163-165	ARSJ V32 N5 P809 <AIAA>
1960 N4 P166-167	62-25833 <=*>
1960 N4 P182-184	57N48R <ATS>
1960 N5 P3-8	61-31592 <=>
	62-15515 <=*>
1960 N5 P9-13	ARSJ V32 N11 P1788 <AIAA>
1960 N5 P33-39	63-10215 <=*>
	68N50R <ATS>
1960 N5 P40-48	89N49R <ATS>
1960 N5 P80-87	62-33540 <=*>
1960 N5 P118-121	61-27624 <*=>
1960 N5 P186	61-31537 <=>
1960 N5 P197-200	90N50R <ATS>
1960 N5 P200-203	29N50R <ATS>
1960 N6 P24-37	62-19255 <=*>
1960 N6 P38-46	62-15400 <*=>
1960 N6 P52-56	17N51R <ATS>
1960 N6 P57-62	62-15162 <*=>
1960 N6 P68-73	65-10425 <*>
1960 N6 P86-90	62-15286 <=*>
1960 N6 P105-108	62-19255 <=*>
1960 N6 P109-111	61-31626 <=>
	62-15400 <*=>
1960 N6 P117-120	AIAAJ V1 N1 P278 <AIAA>
1960 N6 P137-138	62-15286 <=*>
	62-24862 <=*>
1960 N6 P166-168	62-15286 <=*>
1960 N6 P176-179	65-13345 <*> 0
1961 N1 P64-69	AIAAJ V1 N1 P263 <AIAA>
	61-21637 <=>
1961 N1 P96-106	62-32279 <=*>
1961 N1 P114-120	62-32520 <=*>
1961 N1 P150-151	62-25856 <=*>
1961 N1 P152-153	AD-640 729 <=$>
1961 N1 P153-154	63-15706 <=*>
1961 N1 P175-177	64-13273 <=*$>
1961 N2 P17-23	66-61663 <=*$>
1961 N2 P55-60	62-32295 <=*>
1961 N2 P61-69	63-13879 <=*>
1961 N2 P70-77	62-15161 <*=>
1961 N2 P97-105	63-23477 <=*>
1961 N2 P106-111	63-19637 <=*>
1961 N2 P114-115	AD-624 913 <=*$>
1961 N2 P117-119	62-25834 <=*>
1961 N2 P148-152	65-64218 <=*$>
1961 N3 P5-17	62-32941 <=*>
1961 N3 P18-21	AD-607 970 <=> M
1961 N3 P40-57	63-19311 <=*>
1961 N3 P68-77	64-71467 <=> M
1961 N3 P102-112	63-23258 <=*>
	64-13020 <=*$>
1961 N3 P155-162	64-13542 <=*$>
1961 N3 P161-162	62-23958 <=>
1961 N4 P10-13	62-19282 <=*>
1961 N4 P26-32	62-33406 <=*>
1961 N4 P83-103	64-71356 <=> M
1961 N4 P170-172	62-15398 <*=>
1961 N4 P175-177	RCT V36 N1 P59 <RCT>
1961 N5 P11-15	63-19315 <=*>
1961 N5 P22-29	65-13388 <*>
1961 N5 P30-37	63-19315 <=*>
1961 N5 P159-162	96P59R <ATS>
1961 N5 P163-167	53P60R <ATS>
1961 N5 P174-177	00Q67R <ATS>
1961 N6 P3-10	62-33530 <=*>
1961 N6 P11-14	63-19530 <=*>
1961 N6 P15-25	62-33530 <=*>
	63-10829 <=*>
	63-10901 <=*> 0
1961 N6 P26-32	62-25547 <=*>
1961 N6 P33-36	63-13897 <=*>
1961 N6 P37-41	63-19532 <=*>
1961 N6 P167-169	AIAAJ V1 N5 P1263 <AIAA>
1962 N1 P25-31	N66-23608 <=$>
1962 N1 P32-41	63-19199 <=*>
1962 N1 P42-51	62-33625 <=*>
1962 N1 P66-72	63-15526 <=*>
1962 N1 P159-160	AIAAJ V1 N5 P1268 <AIAA>
1962 N2 P131-133	63-19525 <=*>
1962 N3 P19-23	63-19664 <=*>
1962 N3 P24-30	63-19479 <=*>
1962 N3 P31-37	65-13627 <*>
1962 N3 P106-113	63-19664 <=*>
1962 N3 P160-163	65-60357 <=$>
1962 N3 P173-175	63-19471 <=*>
1962 N4 P3-12	AD-618 334 <=$> M
1962 N4 P13-17	N65-1952 <=$>
1962 N4 P18-22	64-15090 <=*$>
1962 N4 P27-31	65-13386 <*>
1962 N4 P52-60	64-11974 <=> M
	65-62846 <=$>
1962 N4 P165-166	AIAAJ V1 N11 P2699 <AIAA>
1962 N5 P3-12	63-19334 <=*>
1962 N5 P26-31	63-23197 <=*>
1962 N5 P63-67	RTS-2646 <NLL>
1962 N5 P111-117	63-23445 <=*>
1962 N5 P184	65-13653 <*>
1962 N6 P3-5	AD-619 434 <=$>
1962 N6 P10-16	65-10397 <*>
1962 N6 P17-24	64-71494 <=> M
1962 N6 P33-49	63-23421 <*=>
1962 N6 P88-94	65-63566 <=*$>
1962 N6 P144-146	64-71494 <=> M
1962 N6 P146-147	AD-619 435 <=$>
1962 N6 P170-172	63-24542 <=*$>
1962 N6 P172-174	63-24543 <=*$>
1962 N16 P154-160	63-23711 <=*$>
1963 N3 P3-8	64-71488 <=> M
1963 N3 P149-155	64-11768 <=>

IZVESTIYA AKADEMII NAUK SSSR. OTDELENIE TEKHNICHESKIKH NAUK. MEKHANIKA I MASHINOSTROENIE. (TITLE VARIES)

Citation	Translation
1963 N1 P33-46	64-11845 <=>
1963 N1 P47-56	64-13620 <=*$>
1963 N1 P141-145	64-13384 <=*$>
1963 N2 P20-27	64-11968 <=> M
1963 N2 P28-32	64-15112 <=*$>
1963 N2 P102-119	65-60755 <=>
1963 N2 P120-125	64-11772 <=>
1963 N2 P146-148	65-60820 <=>
1963 N2 P149-152	64-71533 <=>
1963 N3 P16-28	AD-614 940 <=$>
1963 N3 P45-50	64-13690 <=*$>
1963 N3 P58-64	64-15307 <=*$>
1963 N3 P175-176	ICE V4 N2 P269-270 <ICE>
	48R74R <ATS>
1963 N4 P3-6	AD-615 228 <=$>
1963 N4 P7-9	64-71130 <=>
1963 N4 P10-17	64-19983 <=$>
1963 N4 P67-75	65-10811 <*>
1963 N4 P102-107	AD-625 060 <=*$>
1963 N4 P108-111	AD-609 788 <=>
1963 N4 P133-137	AD-638 883 <=$>
1963 N4 P162-166	64-10413 <=*$>
1963 N4 P167-171	64-14306 <=*$>
1963 N5 P29-41	64-71522 <=>
1963 N5 P60-66	AD-615 453 <=$>
1963 N5 P77-107	66-12713 <*>
1963 N5 P117-123	64-71323 <=>
1963 N5 P167-168	65-11172 <*>
1963 N5 P190-195	RTS-2653 <NLL>
1963 N6 P3-11	RTS-2735 <NLL>
1963 N6 P25-35	64-14806 <=*$>
	64-18425 <*>
1963 N6 P55-61	64-71375 <=>
1963 N6 P66-68	64-18426 <*>

Citation	Translation
1963 N6 P80-84	RTS-2716 <NLL>
	65-27638 <$>
1963 N6 P85-87	53R77R <ATS>
1963 N6 P116-118	AD-615 506 <=$>
	64-71539 <=>
1963 N6 P119-140	AD-624 800 <=*$>
1963 N6 P143	65-10827 <*>
1963 N6 P144-146	64-14725 <=*$>
1963 N6 P159-163	AD-633 820 <=$>
1963 N6 P169-173	RTS-2868 <NLL>
1963 V1 P68-75	RTS-2949 <NLL>
1963 V1 P197-200	64-71521 <=>
1963 V6 P150-154	64-71523 <=>
1964 N1 P115-120	AD-615 990 <=$>
1964 N1 P126-132	64R78R <ATS>
1964 N1 P133-136	ICE V5 N1 P91-94 <ICE>
	17R77R <ATS>
1964 N1 P141-142	AD-614 391 <=$>
1964 N1 P166-169	AD-625 052 <=*$>
	N64-25179 <=>
1964 N1 P170-173	N64-25173 <=>
1964 N1 P193-196	AD-621 050 <=*$>
1964 N1 P199-201	N64-25055 <=>
1964 N1 P202-205	66-11733 <*>
1964 N2 P9-18	AD-630 825 <=$>
1964 N2 P61-77	TP/T-3629 <NLL>
1964 N2 P123-129	AD-625 245 <=*$>
1964 N2 P168-171	64-18764 <*>
1964 N2 P179-182	04R79R <ATS>
1964 N2 P188-192	AD-611 859 <=$>
	64-51499 <=>
1964 N3 P16-28	AD-647 741 <=> M
1964 N3 P29-38	AD-632 460 <=$>
1964 N3 P137-141	65-10412 <*>
1964 N3 P142-144	AD-625 076 <=*$>
	65-10409 <*>
1964 N3 P144-146	65-10410 <*>
1964 N3 P178-182	65R80R <ATS>
1964 N4 P9-28	N65-16595 <=$>
	65-13176 <*>
1964 N4 P33-39	15R80R <ATS>
1964 N4 P63-76	AD-636 668 <=$>
1964 N4 P87-89	RTS-2920 <NLL>
1964 N4 P96-101	AD-636 668 <=$>
1964 N4 P142-143	N65-24559 <=*$>
1964 N4 P145-147	AD-618 059 <=$>
1964 N4 P148-150	AD-633 713 <=$>
1964 N4 P150-153	N65-15152 <=$>
1964 N4 P154-157	AD-625 061 <=*$>
1964 N4 P172-177	AD-636 668 <=$>
1964 N4 P180-184	06S81R <ATS>
1964 N5 P3-11	65-13174 <*>
1964 N5 P33-38	N66-22283 <=$>
1964 N5 P54-60	AD-632 308 <=$>
1964 N5 P108-111	65-14443 <*>
	65-63669 <=$>
1964 N5 P112-116	65-14561 <*>
	65-63677 <=$>
1964 N6 P10-19	65-30714 <=$>
1964 N6 P48-52	AD-629 044 <=*$>
1964 N6 P78-88	65-14456 <*>
	65-63690 <=$>
1964 N6 P96-105	65-14413 <*>
	65-63705 <=$>
1964 N6 P112-113	65-13158 <*>
	65-63734 <=$>
1964 N6 P113-116	65-13159 <*>
	65-63733 <=$>
1964 N6 P119-124	65-14455 <*>
	65-63694 <=$>
1964 V2 P25-29	64-71558 <=>
1964 V4 P40-46	N65-19509 <=$>
1964 V4 P116-119	65-12828 <*>

IZVESTIYA AKADEMII NAUK SSSR. OTDELENIE TEKHNICHESKIKH NAUK. METALLURGIYA I GORNOE DELO

Citation	Translation
1963 N1 P14-17	<SIC>
1963 N1 P33-40	RTS-2838 <NLL>
1963 N1 P41-58,69-75	<SIC>
1963 N1 P87-90	<SIC>
1963 N1 P91-96	3321 <BISI>
1963 N1 P97-99	6009 <HB>
1963 N1 P105-120	<SIC>
1963 N1 P129-175	<SIC>
1963 N1 P191-198	63-23919 <=*$>
1963 N1 P199-200	66-11293 <*>
1963 N2 P9-12	<SIC>
1963 N2 P13-16	6045 <HB>
1963 N2 P17-66	<SIC>
1963 N2 P17-20	6011 <HB>
1963 N2 P22-27	6012 <HB>
1963 N2 P79-85	64-13218 <=*$>
1963 N2 P96-99	<SIC>
1963 N2 P104-111	<SIC>
1963 N2 P116-129	<SIC>
1963 N2 P136-152	<SIC>
1963 N2 P169-172	<SIC>
1963 N2 P174-176	63-31079 <=>
1963 N3 P58-66	6000 <HB>
1963 N3 P127-146	3577 <BISI>
1963 N3 P191-192	63-41026 <=>
1963 N4 P151-153	64-19005 <=*$> 0
1963 N5 P3-12	<SIC>
	65-60811 <=>
1963 N5 P38-41	<SIC>
1963 N5 P50-57	<SIC>
1963 N5 P73-79	<SIC>
1963 N5 P98-113	<SIC>
1963 N5 P113-115	<SIC>
1963 N5 P116-120	<SIC>
1963 N5 P121-125	<SIC>
1963 N5 P126-128	<SIC>
1963 N5 P134-140	<SIC>
1963 N5 P145-151	<SIC>
1963 N6 P5-11	<SIC>
1963 N6 P12-23	<SIC> P
1963 N6 P52-60	<SIC>
1963 N6 P61-67	<SIC>
1963 N6 P68-73	<SIC>
1963 N6 P90-95	<SIC>
1963 N6 P116-119	<SIC>
1963 N6 P139-145	<SIC> P
1963 N6 P146-151	<SIC>
1963 N6 P152-154	<SIC>
1963 N6 P160-167	<SIC>
1963 N6 P168-173	<SIC>
1963 N6 P174-183	<SIC> P
1964 N1 P15-19	<SIC>
1964 N1 P41-44	<SIC>
1964 N1 P45-47	<SIC>
1964 N1 P67-77	<SIC>
	65-11417 <*>
1964 N1 P78-84	<SIC>
1964 N1 P85-91	<SIC>
1964 N1 P95-107	<SIC>
1964 N1 P113-119	<SIC>
1964 N1 P136-142	<SIC>
1964 N1 P143-150	65-61406 <=>
1964 N1 P166-169	<SIC>
1964 N1 P170-175	<SIC>
	6262 <HB>
1964 N1 P180-183	65-10868 <*>
1964 N2 P17-25	<SIC>
1964 N2 P26-30	4160 <BISI>
1964 N2 P45-50	64-25409 <SIC>
1964 N2 P69-74	64-25412 <SIC>
1964 N2 P85-91	<SIC>
1964 N2 P97-104	<SIC> P
1964 N2 P105-109	<SIC> P
1964 N2 P124-131	<SIC>
1964 N2 P137-142	<SIC> P
1964 N2 P149-155	<SIC> P
1964 N2 P164-166	64-25422 <SIC>
1964 N2 P172-176	<SIC> P
1964 N2 P188-191	<SIC>
1964 N3 P17-21	<SIC>
1964 N3 P32-36	<SIC>
1964 N3 P37-41	<SIC> P
1964 N3 P42-51	<SIC>
1964 N3 P52-57	<SIC> P

1964 N3 P58-62 <SIC>
1964 N3 P63-68 <SIC> P
1964 N3 P116-117 <SIC>
1964 N3 P118-121 <SIC> P
1964 N3 P122-124 <SIC>
1964 N3 P125-131 <SIC> P
1964 N3 P131-137 <SIC>
M-5594 <NLL>
1964 N3 P138-144 <SIC>
1964 N3 P145-147 <SIC> P
M-5584 <NLL>
1964 N3 P148-153 <SIC> P
M-5579 <NLL>
1964 N3 P154-157 <SIC>
1964 N3 P158-160 <SIC> P
1964 N3 P161-162 <SIC> P
1964 N3 P169-172 <SIC> P
66-10427 <*>
1964 N3 P179-182 64-41538 <=>
1964 N4 P6-16 <SIC>
1964 N4 P47-51 <SIC>
1964 N4 P68-75 <SIC>
1964 N4 P76-79 <SIC>
M-5610 <NLL>
1964 N4 P80-86 <SIC>
1964 N4 P91-105 <SIC>
1964 N4 P106-115 <SIC>
65-23680 <=$>
1964 N4 P123-126 <SIC> P
1964 N4 P127-130 <SIC> P
1964 N4 P143-146 AD-640 310 <=$>
1964 N4 P153-158 <SIC> P
1964 N4 P159-165 <SIC> P
M-5637 <NLL>
1964 N4 P166 <SIC>
66-10774 <*>
1964 N4 P167-171 <SIC>
1964 N5 P3-9 <SIC>
1964 N5 P18-22 <SIC>
1964 N5 P23-24 <SIC>
1964 N5 P35-44 <SIC>
1964 N5 P45-56 <SIC>
777 <TC>
1964 N5 P92-97 <SIC>
1964 N5 P98-100 <SIC>
1964 N5 P101-107 <SIC>
1964 N5 P108-111 <SIC>
1964 N5 P112-117 <SIC>
1964 N5 P118-120 <SIC>
1964 N5 P127-131 <SIC>
1964 N5 P132-136 <SIC>
1964 N5 P142-146 <SIC>
1964 N5 P147-149 <SIC>
1964 N5 P167-171 <SIC>
1964 N6 P3-9 M-5685 <NLL>
65-31350 <=*$>
1964 N6 P75-80 M-5686 <NLL>
1964 N6 P97-102 M-5687 <NLL>
1964 N6 P125-128 TP/T-3737 <NLL>
1964 V2 P51-57 <SIC>
1964 V2 P63-68 64-25411 <SIC>
1964 V2 P110-116 <SIC> P
1964 V2 P132-136 <SIC> P
1964 V2 P161-163 <SIC>

IZVESTIYA AKADEMII NAUK SSSR. OTDELENIE TEKHNICHESKIKH NAUK. METALLURGIYA I TOPLIVO

1959 N1 P9-12 60-11586 <=>
1959 N1 P44-49 63-21384 <=>
1959 N1 P75-77 61-31502 <=> 0
1959 N1 P106-109 TRANS-143 <FRI>
1959 N1 P117-119 60-15294 <+*>
1959 N2 P3-7 66-61667 <=*$>
1959 N2 P24-34 4702 <HB>
1959 N2 P40-42 4738 <HB>
1959 N2 P43-47 4736 <HB>
1959 N2 P62-69 1732 <BISI>
62-32801 <=*>
1959 N2 P69-77 63-24628 <=*$>
1959 N2 P78-81 72M39R <ATS>
1959 N2 P86-89 4752 <HB>
1959 N2 P90-95 40M39R <ATS>
1959 N2 P96-103 61-16115 <=*> 0
1959 N2 P104-108 62-23134 <=>
1959 N2 P111-112 61-16114 <=*>
1959 N2 P117-121 4611 <HB>
1959 N2 P124-138 1883 <BISI>
1959 N3 P25-28 4758 <HB>
1959 N3 P47-51 4756 <HB>
1959 N3 P62-69 3632 <BISI>
1959 N3 P83-87 AD-639 461 <=$>
1959 N3 P88-91 4776 <HB>
1959 N3 P92-98 61-19663 <=*> 0
1959 N3 P99-107 62-33186 <=*>
1959 N3 P108-112 4754 <HB>
1959 N3 P134-135 62-11267 <=>
1959 N3 P139-141 61-19663 <=*> 0
1959 N3 P144-145 91N55R <ATS>
1959 N3 P148-150 4755 <HB>
1959 N3 P150-151 60-11551 <=>
1959 N3 P184-186 60-23165 <+*>
1959 N4 P32-41 4826 <HB>
1959 N4 P42-45 4827 <HB>
1959 N4 P59-65 4829 <HB>
1959 N4 P73-81 61-23889 <=*>
1959 N4 P83-87 64-11584 <=>
1959 N4 P99-105 60-41565 <=>
1959 N4 P106-110 61-23889 <=*>
1959 N4 P114-123 65N50R <ATS>
1959 N4 P130-147 61-23889 <=*>
1959 N4 P135-142 61-28580 <*=>
1959 N4 P144-147 4709 <HB>
1959 N4 P153-155 62-11453 <=>
1959 N4 P156-158 62-11400 <=>
1959 N4 P190-199 60-31293 <=>
61-23889 <=*>
61-27505 <*=> 0
1959 N4 P202-204 61-23889 <=*>
1959 N5 P3-10 TRANS-168 <MT>
4862 <HB>
1959 N5 P42-44 4819 <HB>
1959 N5 P77-85 5663 <HB>
1959 N5 P86-90 62-11434 <=>
1959 N5 P123-126 AEC-TR-4202 <=>
1959 N5 P126-127 4784 <HB>
1959 N5 P127-130 4785 <HB>
1959 N5 P132-135 62-25891 <=*>
1959 N5 P135-138 64-41597 <=>
1959 N5 P138-141 AEC-TR-5323 <=*> 0
62-15709 <*=>
62-23130 <=*>
1959 N5 P174-184 62-24083 <=*>
64-14747 <=*$>
1959 N6 P3-7 2083 <BISI>
1959 N6 P121-126 60-31454 <=>
1959 N6 P127-129 61-19367 <+*> 0
1959 N6 P131-132 5103 <HB>
1959 N6 P133-137 60-11513 <=>
1959 N6 P137-141 61-13844 <*+> 0
1960 N1 P44-49 62-11245 <=>
1960 N1 P50-58 ARSJ V31 N7 P961 <AIAA>
1960 N1 P70-74 ARSJ V32 N1 P137 <AIAA>
1960 N1 P75-84 62-23001 <=*>
1960 N1 P85-89 ARSJ V31 N5 P694 <AIAA>
1960 N1 P93-100 62-14554 <=*>
1960 N1 P101-104 ARSJ V30 N11 P1062 <AIAA>
1960 N1 P111-122 UCRL TRANS-661(6) <=*>
1960 N1 P123-126 66-11480 <*>
1960 N1 P134-137 61-28232 <*=>
1960 N1 P138-144 4857 <HB>
1960 N1 P145-151 47M43R <ATS>
1960 N1 P174-175 4813 <HB>
1960 N2 P21-25 4805 <HB>
1960 N2 P53-55 62-19902 <=*>
1960 N2 P72-78 62-11439 <=>
1960 N2 P79-86 63-13718 <=*>
1960 N2 P104-109 4854 <HB>
1960 N2 P115-119 4853 <HB>
61-19711 <=*> 0
1960 N2 P146-150 80M43R <ATS>

Citation	Translation
1960 N2 P167-168	60-11939 <=>
1960 N3 P3-16	64-13809 <=*$>
1960 N3 P39-43	4906 <HB>
1960 N4 P81-84	62-26005 <=$>
1960 N4 P90-94	65-13363 <*>
1960 N4 P143-149	61-28814 <*=> 0
1960 N4 P150-155	62-11359 <=>
1960 N4 P185-186	62-11536 <=>
1960 N5 P70-85	62-33394 <=*>
1960 N5 P190-194	62-11365 <=>
1960 N6 P42-46	ARSJ V32 N11 P1804 <AIAA>
	61-27440 <*=> 0
1960 N6 P47-50	61-28236 <*=> 0
1960 N6 P56-60	61-27830 <*=> 0
1960 N6 P177-182	62-10208 <*=> 0
1961 N1 P24-30	6371 <HB>
1961 N1 P31-38	55R80R <ATS>
1961 N1 P58-73	62-33390 <=*>
1961 N1 P108-116	62-33390 <=*>
1961 N1 P142-145	AEC-TR-4865 <=*>
1961 N1 P144-148	6141 <HB>
1961 N1 P147-149	5181 <HB>
1961 N1 P152-157	62-32820 <=*>
1961 N2 P13-19	5179 <HB>
1961 N2 P43-48	63-15250 <=*>
1961 N2 P49-54	63-15251 <=*>
1961 N2 P55-59	63-15252 <=*>
1961 N2 P60-63	63-15253 <=*> 0
1961 N2 P64-67	63-15254 <=*>
1961 N2 P68-71	63-15255 <=*>
1961 N2 P72-76	63-15269 <=*>
1961 N2 P77-82	63-15256 <=*>
1961 N2 P83-87	63-15266 <=*>
1961 N2 P99-107	62-23149 <=>
1961 N2 P147-149	AEC-TR-4871 <*>
	AEC-TR-4871 <=*>
	63-10845 <=*>
1961 N2 P182-184	61-28514 <*=>
1961 N3 P40-54	64-13587 <=*$>
1961 N3 P50-54	63-19788 <=*> 0
1961 N3 P73-76	6040 <HB>
	62-13580 <*=>
1961 N3 P116-118	65-11657 <*>
1961 N3 P133-135	65-11668 <*>
1961 N3 P183-190	63-23982 <=*$>
1961 N4 P12-17	6034 <HB>
1961 N4 P39-42	62-10640 <=*>
1961 N4 P68-70	63-10129 <=*> 0
1961 N4 P90-94	5354 <HB>
1961 N4 P111-114	AD-624 908 <=*$>
1961 N4 P137-142	62-24951 <=*>
1961 N5 P19-21	63-10144 <=*>
1961 N5 P62-69	AD-624 909 <=*$>
1961 N5 P75-77	62-25309 <=*> 0
1961 N5 P96-100	63-10907 <=*>
1961 N5 P101-108	6036 <HB>
1961 N5 P109-112	5460 <HB>
1961 N5 P113-116	64-14843 <=*$>
1961 N6 P25-29	63-20249 <=*>
1961 N6 P45-51	6149 <HB>
	62-25307 <=*> 0
	63-21371 <=>
1961 N6 P56-73	AD-610 098 <=$>
1961 N6 P119-126	AEC-TR-5954 <=*$>
1961 N6 P130-136	63-19786 <=*> 0
1962 NC P125-130	<SIC>
1962 N1 P23-25	<SIC>
1962 N1 P46-53	<SIC>
1962 N1 P54-55	<SIC>
1962 N1 P56-59	<SIC>
1962 N1 P81-83	<SIC>
1962 N1 P100-104	<SIC>
1962 N1 P105-126	<SIC> P
1962 N1 P134-138	<SIC>
1962 N1 P139-146	<SIC>
1962 N1 P151-155	<SIC>
1962 N1 P156-159	<SIC>
	63-10108 <=*> 0
1962 N1 P160-162	<SIC>
	63-19926 <=*>
1962 N2 P15-21	<SIC>
1962 N2 P27-31	<SIC>
1962 N2 P42-48	C-4671 <NRC>
1962 N2 P56-62	AEC-TR-5884 <=*$>
1962 N2 P63-65	<SIC>
1962 N2 P71-77	<SIC>
1962 N2 P84-91	<SIC>
1962 N2 P92-97	<SIC>
1962 N2 P98-103	<SIC> P
1962 N2 P104-106	<SIC>
1962 N2 P107-112	<SIC>
1962 N2 P113-118	<SIC>
1962 N2 P119-125	<SIC>
	AEC-TR-5576 <=*$>
1962 N2 P126-130	<SIC> P
1962 N2 P131-133	<SIC> P
1962 N2 P134-135	<SIC>
1962 N2 P136-144	<SIC> P
1962 N2 P152-160	<SIC>
1962 N2 P174-175	<SIC>
1962 N2 P176	<SIC>
1962 N3 P3-9	<SIC> P
1962 N3 P10-19	<SIC>
1962 N3 P20-26	<SIC>
1962 N3 P56-57	<SIC>
1962 N3 P58-62	<SIC>
1962 N3 P63-70	<SIC>
1962 N3 P71-77	<SIC>
1962 N3 P78-80	<SIC>
1962 N3 P81-84	<SIC>
1962 N3 P85-87	<SIC>
	Z6222 <STS>
1962 N3 P88-93	<SIC>
	65-13379 <*>
1962 N3 P94-97	<SIC>
1962 N3 P98-101	<SIC> P
1962 N3 P102-106	<SIC> P
1962 N3 P107-113	<SIC>
1962 N4 P20-30	<SIC> P
1962 N4 P31-39	<SIC>
1962 N4 P44-50	<SIC>
1962 N4 P56-58	<SIC>
1962 N4 P62-69	<SIC>
1962 N4 P70-77	64-19933 <=$>
1962 N4 P78-81	<SIC>
1962 N4 P82-89	<SIC>
	65-13675 <*>
1962 N4 P90-102	<SIC>
1962 N4 P103-113	<SIC>
1962 N4 P114-118	<SIC>
1962 N4 P119-125	<SIC> P
1962 N4 P126-132	<SIC> P
1962 N4 P137-142	<SIC> P
1962 N4 P153-156	64-23571 <=>
1962 N5 P9-14	5800 <HB>
	63-13483 <=*>
1962 N5 P29-36	<SIC>
	5785 <HB>
1962 N5 P43-51	<SIC>
1962 N5 P74-80	<SIC>
1962 N5 P92-95	<SIC>
1962 N5 P110-116	<SIC>
1962 N5 P128-134	<SIC>
1962 N5 P135-138	<SIC>
1962 N5 P139-142	<SIC>
1962 N5 P143-150	<SIC>
1962 N5 P163-166	<SIC>
1962 N5 P167-180	6030 <HB>
1962 N5 P227-	ICE V3 N2 P277-280 <ICE>
1962 N5 P227-232	24Q67R <ATS>
1962 N6 P3-6	<SIC>
1962 N6 P12-26	<SIC>
1962 N6 P37-42	<SIC>
1962 N6 P61-66	<SIC>
1962 N6 P73-89	<SIC>
1962 N6 P73-80	<SIC> P
1962 N6 P81-89	<SIC> P
1962 N6 P90-97	M.5324 <NLL>
1962 N6 P90-111	63-21627 <=>
1962 N6 P103-106	<SIC>

1962 N6 P103-111 <SIC>
1962 N6 P103-106 M.5325 <NLL>
1962 N6 P107-111 <SIC>
1962 N6 P125-146 <SIC>
1962 N6 P131-136 <SIC>
1962 N6 P137-141 <SIC>
1962 N6 P142-146 <SIC>
1962 N6 P150-158 <SIC> P
1962 N6 P159-161 <SIC>
1962 N6 P159-166 <SIC>
1962 N6 P162-166 <SIC>
1962 N6 P167-170 AEC-TR-5935 <*>
AEC-TR-5935 <=*$>
1962 N6 P176-178 65-64596 <=*$>
1962 V1 P84-88 <SIC>
1962 V3 P114-115 <SIC>
1962 V5 P9-14 <SIC>
1962 V5 P52-65 <SIC>
1962 V5 P181-186 <SIC>

*IZVESTIYA AKADEMII NAUK SSSR. OTDELENIE TEKHNICHESKIKH NAUK. TEKHNICHESKAYA KIBERNETIKA
1963 N1 P1-208 63-31046 <=>
1963 N1 P41-47 63-20288 <=>
1963 N1 P42-50 AD-428 502 <=$>
1963 N1 P113-120 64-13614 <=*$>
1963 N1 P155-171 64-13607 <=*$>
1963 N2 P1-192 63-31222 <=>
1963 N2 P3-16 63-31172 <=>
1963 N2 P70-77 AD-611 521 <=$>
1963 N3 P1-208 63-31772 <=>
1963 N3 P81-93 74Q73R <ATS>
1963 N4 P3-25 63-41337 <=>
1963 N4 P26-43 63-41337 <=>
1963 N4 P44-45 63-41337 <=>
1963 N4 P56-59 63-41337 <=>
1963 N4 P60-79 63-41337 <=>
1963 N4 P80-83 63-41337 <=>
1963 N4 P84-93 63-41337 <=>
1963 N4 P94-99 63-41337 <=>
1963 N4 P100-106 63-41337 <=>
1963 N4 P107-114 63-41337 <=>
1963 N4 P115-126 63-41337 <=>
1963 N4 P127-139 63-41337 <=>
1963 N4 P140-146 63-41337 <=>
1963 N4 P147-156 63-41337 <=>
1963 N4 P157-163 63-41337 <=>
1963 N4 P164 63-41337 <=>
1963 N4 P165-170 63-41337 <=>
1963 N4 P171-179 63-41337 <=>
1963 N4 P180-183 63-41337 <=>
1963 N4 P184-195 63-41337 <=>
1963 N4 P196-201 63-20959 <=*$>
63-41337 <=>
1963 N4 P202-209 63-41337 <=>
1963 N5 P149 64-14164 <=*$>
1963 N5 P149-156 64-14317 <=*$>
1963 N6 P111-120 AD-610 796 <=$>
1963 N6 P128-130 65-64382 <=*$>
1963 V1 N2 P109-123 65-11692 <*>
1964 N3 P30-37 AD-621 001 <=*$>
1964 N3 P154-169 65-63650 <=$*>
1964 N5 P156-164 N65-21632 <=*$>
1964 N5 P165-172 N65-21633 <=$>
1964 N6 P10-22 66-11259 <*>
1965 N1 P3-200 65-31188 <=*$>
1965 N1 P114-117 EECL-LIB-TRANS-1272 <NLL>
1965 N2 P188-190 65-14699 <*>
65-63710 <=$>
1965 N5 P13-22 66-13958 <=$>

IZVESTIYA AKADEMII NAUK SSSR. SERIYA BIOLOGICHESKAYA
1936 N6 P1115-1172 65-11047 <*> 0
1937 V2 P907-911 RTS-2691 <NLL>
1938 N4 P975-981 65-60043 <=$> 0
1939 P349-361 61-11482 <=>
1939 N2 P159-170 61-11468 <=>
1939 N2 P215-227 61-11483 <=>
1939 N2 P362-370 61-11481 <=>
1941 N2 P278-282 61-28094 <*=> 0
1942 N5 P287-293 65-63232 <=>
1943 N2 P57-73 R-2377 <*>
1943 N2 P67-73 64-19452 <=$>
1945 N6 P654-663 64-15646 <=*$> 0
1945 V6 P654-663 63-01232 <*>
1947 N3 P409-422 <DIL>
R-4683 <*>
1948 N4 P493-504 60-18896 <+*> 0
1949 N2 P140-170 R-2376 <*>
1949 V6 P709-715 RTS-2891 <NLL>
1950 P85-101 51/3363 <*>
1950 N5 P102-109 59-19102 <+*> 0
1950 N5 P110-124 RT-2860 <*>
1950 N6 P30-41 2018-B <K-H>
1950 N6 P42-49 2018-A <K-H>
1951 N2 P44-52 2149G <K-H>
1951 N5 P120-135 RT-2848 <*>
1952 N4 P63-67 64-20427 <*>
1952 N6 P67-79 RT-3259 <*>
1952 V4 P63-67 66-12074 <*>
1953 N1 P96-104 R-1356 <*>
1953 N2 P14-33 RT-1160 <*>
1953 N6 P66-78 59-14057 <+*>
1953 N6 P111-116 RT-3380 <*>
1954 N1 P8-19 RT-2186 <*>
1954 N1 P53-58 RJ-425 <ATS>
1954 N1 P59-73 RT-2099 <*> P
1954 N2 P14-39 RT-3185 <*>
1954 N3 P49-61 RT-3028 <*>
1954 N6 P85-103 R-1360 <*>
1954 V18 N4 P473-488 <CTT>
1955 N2 P19-40 R-3927 <*>
1955 N2 P112-132 R-1358 <*>
1955 N3 P3-15 RT-3172 <*>
1955 N3 P49-53 R-1359 <*>
1955 N4 P3-13 RT-4360 <*>
1955 N6 P3-9 66-60913 <=*$>
1955 N6 P10-19 RT-3916 <*>
1956 N2 P18-28 R-322 <*>
1956 N5 P30-46 R-1849 <*>
1956 N6 P3-18 R-1850 <*>
1956 V6 P35-53 R-3437 <*>
1957 N2 P248-249 R-1851 <*>
1957 N6 P706-717 60-13457 <+*> 0
1957 V22 P41-54 61-19611 <*=>
1957 V22 P285-292 61-19575 <*=>
1957 V22 P357-359 CSIR-TR.295 <CSSA>
1957 V22 P649-673 61-28047 <*=>
1957 V22 N6 P770-772 PB-141234-T <=*>
1958 N1 P118-120 R-5216 <*>
1958 N2 P248-250 59-13359 <=>
1958 N3 P282-290 59-13308 <=> 0
1958 N3 P377-379 59-13309 <=>
1958 N4 P416-421 59-15404 <+*>
1958 N5 P592-596 59-12677 <+*>
1958 N6 P698-711 59-17366 <+*>
1958 N6 P758-760 59-13254 <=>
1958 V23 N1 P26-38 62-15064 <*=> 0
1958 V23 N1 P104-111 60-21885 <=>
1958 V23 N5 P579-583 59-20318 <+*>
1958 V23 N5 P592-596 59-18831 <+*> 0
1958 V23 N5 P618-622 59-16823 <+*>
1958 V23 N5 P625-628 59-16822 <+*>
1958 V23 N5 P628-631 59-16821 <+*>
1959 N2 P161-171 59-13625 <=>
1959 N2 P172-184 59-13626 <=>
1959 N2 P265-282 59-22322 <+*> 0
1959 N2 P316-319 60-11000 <=>
1959 N2 P319-320 60-11001 <=>
1959 V24 P428-430 61-23033 <*=>
1959 V24 N1 P103-110 59-16819 <+*> 0
1959 V24 N1 P151-154 59-16820 <+*>
1959 V24 N1 P155-158 59-16818 <+*>
1959 V24 N2 P186-192 63-21201 <=>
1959 V24 N2 P257-264 63-13420 <=*>
1959 V24 N3 P391-402 62-25664 <=*>
1959 V24 N3 P451-457 60-11808 <=>
1959 V24 N4 P552-557 61-19380 <+*>
1959 V24 N4 P599-611 60-11616 <=>

Citation	Translation
1959 V24 N4 P638-640	60-15356 <+*>
1959 V24 N5 P698-713	65-50125 <=>
1959 V24 N6 P814-831	60-17979 <+*>
1959 V24 N6 P855-864	60-18363 <+*>
	60-31079 <=>
1959 V24 N6 P935-943	60-31080 <=>
1959 V24 N6 P944-947	61-11015 <=>
1959 V24 N6 P947-949	61-11509 <=>
1959 V24 N6 P950-952	60-41726 <=>
1960 N2 P313-316	60-17936 <+*>
1960 V25 N1 P3-18	60-31170 <=>
1960 V25 N1 P19-38	60-11750 <=>
1960 V25 N1 P39-51	60-11751 <=>
1960 V25 N1 P52-62	60-11752 <=>
1960 V25 N1 P64-81	60-11753 <=>
1960 V25 N1 P82-97	60-17932 <+*> 0
1960 V25 N1 P122-128	60-17934 <+*>
1960 V25 N1 P129-131	60-17933 <+*> 0
1960 V25 N2 P161-176	60-17931 <+*>
1960 V25 N2 P177-196	60-17951 <+*>
1960 V25 N2 P230-239	60-17937 <+*>
1960 V25 N2 P291-293	60-17935 <+*>
1960 V25 N3 P326-337	61-13776 <*+> 0
1960 V25 N3 P355-363	61-13777 <*+> 0
1960 V25 N3 P451-458	61-15825 <+*> 0
1960 V25 N3 P470-472	60-23716 <+*>
1960 V25 N3 P472-473	60-23725 <+*>
1960 V25 N3 P474-478	61-15838 <+*>
1960 V25 N4 P519-531	61-11675 <=>
1960 V25 N4 P542-549	65-50058 <=>
1960 V25 N4 P601-606	63-11599 <=>
1960 V25 N4 P607-609	65-64361 <=$>
1960 V25 N4 P633-636	61-11631 <=>
1960 V25 N5 P762-767	63-11597 <=>
1960 V25 N5 P798-807	61-21345 <=>
1960 V25 N5 P814-815	61-21286 <=>
1960 V25 N6 P851-864	61-21748 <=>
1960 V25 N6 P926-934	61-21547 <=>
1961 N5 P664-668	NLLTB V4 N4 P339 <NLL>
1961 V26 N1 P117-157	61-27049 <*=>
1961 V26 N1 P158-162	61-27682 <*=>
1961 V26 N1 P163-169	61-19752 <=*>
	61-27343 <*=>
1961 V26 N1 P169-172	61-19753 <=*>
	61-27353 <*=>
1961 V26 N1 P173-175	61-19655 <=*>
	61-27952 <*=>
1961 V26 N2 P181-201	61-27506 <*=>
1961 V26 N2 P202-212	63-15925 <=*>
1961 V26 N2 P233-238	62-13451 <=*> 0
1961 V26 N2 P330-333	62-24401 <=>
1961 V26 N3 P377-385	61-27385 <*=>
1961 V26 N3 P386-410	61-28783 <*=>
1961 V26 N3 P412-423	61-28389 <*=>
1961 V26 N3 P425-440	61-27340 <*=>
1961 V26 N3 P472-482	61-28325 <*=>
1961 V26 N4 P561-573	62-32452 <=*>
1961 V26 N4 P574-581	55P64R <ATS>
1961 V26 N4 P620-629	63-11604 <=>
1961 V26 N4 P630-641	62-32452 <=*>
1961 V26 N5 P657-663	62-11085 <=>
1961 V26 N5 P828-830	62-11085 <=>
1961 V26 N6 P897-903	62-11129 <=>
1961 V26 N6 P926-930	63-13590 <=*>
1962 N1 P70-	FPTS V22 N4 P.T677 <FASE>
1962 N3 P52-73	62-33636 <=*>
1962 N3 P430-	FPTS V22 N3 P.T582 <FASE>
1962 N3 P442-	FPTS V22 N2 P.T326 <FASE>
1962 N3 P455-	FPTS V22 N3 P.T589 <FASE>
1962 N5 P782-	FPTS V22 N6 P.T1224 <FASE>
1962 N5 P791-	FPTS V23 N3 P.T511 <FASE>
1962 V27 P230-241	64-23530 <=>
1962 V27 N1 P3-12	62-24253 <=*>
1962 V27 N1 P84-95	62-24250 <=>
1962 V27 N1 P102-105	62-24250 <=>
1962 V27 N1 P117-121	63-19363 <=*>
1962 V27 N1 P122-130	62-24250 <=>
1962 V27 N1 P127-130	62-23993 <=*>
1962 V27 N1 P141-150	62-24250 <=>
1962 V27 N1 P151-152	62-24252 <=*>
1962 V27 N1 P207-219	65-63516 <=*>
1962 V27 N2 P153-162	62-33692 <=*>
1962 V27 N2 P260-270	63-19003 <=*>
1962 V27 N2 P308-311	62-32048 <=*>
1962 V27 N2 P314-316	62-32047 <=*>
1962 V27 N3 P100-113	IGR,V6,N3,P541-552 <AGI>
1962 V27 N3 P418-429	62-33596 <=*>
1962 V27 N3 P430-432	62-33681 <=*>
1962 V27 N3 P430-441	63-16427 <=*>
1962 V27 N3 P442-448	63-16454 <=*>
1962 V27 N3 P455-460	62-33596 <=*>
1962 V27 N4 P477-501	62-11794 <=>
1962 V27 N4 P502-522	63-24481 <=*$>
1962 V27 N4 P544-575	63-23045 <=*>
1962 V27 N4 P576-591	63-19381 <=*>
1962 V27 N4 P592-602	62-11794 <=>
1962 V27 N4 P603-613	63-13628 <=>
1962 V27 N4 P621-626	62-11794 <=>
1962 V27 N4 P636-643	62-11794 <=>
1962 V27 N4 P644-647	63-13478 <=*>
	63-13513 <=*>
1962 V27 N5 P79-	FPTS,V23,N3,P.T511 <FASE>
1962 V27 N5 P668-684	64-13308 <=*$>
1962 V27 N5 P718-729	65-60944 <=$>
1962 V27 N5 P804-812	63-21083 <=>
1962 V27 N5 P806-809	63-13848 <=*>
1962 V27 N6 P857-868	63-11660 <=>
1962 V27 N6 P869-884	63-21078 <=>
	63-21112 <=>
1962 V27 N6 P896-908	63-21112 <=>
1962 V27 N6 P925-928	64-41167 <=>
1962 V27 N6 P935-940	63-21385 <=>
1962 V27 N6 P948-950	63-21385 <=>
1963 N34 P59-	FPTS,V23,N4,P.T793 <FASE>
1963 V28 N1 P3-14	63-21498 <=>
1963 V28 N1 P15-23	63-11706 <=>
1963 V28 N1 P15-39	63-21723 <=>
1963 V28 N1 P15-23	63-24140 <=*$>
1963 V28 N1 P15-24	64-13441 <=*$>
1963 V28 N1 P24-	FPTS V23 N2 P.T410 <FASE>
1963 V28 N1 P99-105	64-13860 <=*$> 0
1963 V28 N1 P108-114	64-19549 <=$>
1963 V28 N1 P138-145	63-21723 <=>
1963 V28 N1 P152-156	63-21723 <=>
	63-21725 <=>
1963 V28 N2 P167-179	63-21988 <=>
	64-15542 <=*$>
1963 V28 N2 P180-200	64-15546 <=*$>
1963 V28 N2 P214-239	63-21988 <=>
1963 V28 N2 P232-239	64-19678 <=$>
1963 V28 N2 P249-260	63-31206 <=>
1963 V28 N2 P261-269	65-63548 <=*$>
1963 V28 N2 P270-282	64-13169 <=*$>
	64-19992 <=>
1963 V28 N2 P283-286	64-15322 <=*$>
1963 V28 N2 P331-334	63-23664 <=*$>
1963 V28 N3 P337-347	63-31684 <=>
1963 V28 N3 P348-365	63-31493 <=>
1963 V28 N3 P366-389	63-31882 <=*>
1963 V28 N3 P366-390	64-15500 <=*$> C
1963 V28 N3 P391-405	64-19209 <=*$>
1963 V28 N3 P405-418	63-31684 <=>
	64-15545 <=*$>
1963 V28 N3 P430-441	64-13824 <=*$> 0
1963 V28 N3 P441-444	64-15411 <=*$>
1963 V28 N3 P454-459	64-13832 <=*$>
1963 V28 N3 P459-	FPTS V23 N4 P.T793 <FASE>
1963 V28 N4 P497-513	63-31771 <=>
	63-41092 <=>
1963 V28 N4 P504-513	65-63518 <=*$>
1963 V28 N4 P514-525	AEC-TR-6112 <=$>
1963 V28 N4 P555-562	64-19691 <=$>
1963 V28 N4 P562-573	63-31963 <=>
1963 V28 N4 P613-620	63-31770 <=>
1963 V28 N4 P621-629	63-31769 <=>
1963 V28 N4 P630-634	63-31771 <=>
1963 V28 N5 P668-680	64-21249 <=>
1963 V28 N5 P719-723	64-13543 <=*$>
1963 V28 N5 P746-754	64-13543 <=*$>
	64-19282 <=*$> 0

1963 V28 N5 P763-767	63-41271 <=>
1963 V28 N6 P862-870	62-21744 <=>
1963 V28 N6 P871-879	64-19212 <=*$>
1963 V28 N6 P871-878	64-21398 <=>
1963 V28 N6 P880-891	64-15250 <=*$>
1964 N1 P95-109	66-61553 <=*$>
1964 V29 N1 P12-31	64-21924 <=>
1964 V29 N1 P132-136	64-21828 <=>
	65-10985 <*>
1964 V29 N1 P159-163	64-21919 <=>
1964 V29 N2 P177-196	64-41625 <=>
1964 V29 N2 P280-297	64-71531 <=>
1964 V29 N2 P306-311	64-71531 <=>
1964 V29 N2 P325-331	64-71531 <=>
1964 V29 N3 P341-387	64-41115 <=>
1964 V29 N3 P352-368	AD-615 432 <=$>
1964 V29 N3 P369-375	AD-610 348 <=$>
1964 V29 N3 P396-409	64-41438 <=$>
1964 V29 N3 P433-438	66-61561 <=*$>
1964 V29 N3 P482-489	AD-615 432 <=$>
	64-31831 <=>
1964 V29 N3 P492-494	64-31798 <=>
1964 V29 N4 P497-511	AD-615 525 <=$>
1964 V29 N4 P497-510	64-51250 <=$>
1964 V29 N4 P533-545	65-12754 <*>
1964 V29 N5 P677-689	64-51820 <=$>
1964 V29 N5 P690-694	65-31364 <=$>
1964 V29 N5 P721-726	65-63069 <=$>
1964 V29 N5 P779-788	65-61502 <=$>
1964 V29 N6 P936-941	65-30360 <=$>
1965 P231-242	RTS-3109 <NLL>
1965 N2 P274-	FPTS V25 N1 P.T31 <FASE>
1965 N2 P319-320	65-31064 <=$>
1965 N4 P507-	FPTS V25 N4 P.T675 <FASE>
1965 N6 P877-	FPTS V25 N5 P.T767 <FASE>
1965 V30 N1 P3-	FPTS V25 N1 P.T103 <FASE>
1965 V30 N1 P3-22	65-30527 <=$>
1965 V30 N1 P3-9	65-30653 <=$>
1965 V30 N1 P10-17	65-13231 <*>
	65-30652 <=$>
	65-63708 <=$>
1965 V30 N1 P10-18	66-60301 <=*$>
1965 V30 N1 P18-22	65-13270 <*>
	65-63700 <=$>
	66-60410 <=*$>
1965 V30 N1 P44-52	65-30745 <=$>
1965 V30 N1 P53-57	65-32548 <=*$>
	66-13159 <*>
1965 V30 N2 P169-187	65-31379 <=*$>
1965 V30 N2 P182-187	66-60412 <=*$>
1965 V30 N2 P188-200	65-32942 <=*$>
1965 V30 N2 P274-284	65-31379 <=*$>
1965 V30 N2 P298-303	65-32942 <=*$>
1965 V30 N3 P329-334	65-31370 <=$>
	65-32160 <=*$>
1965 V30 N3 P329-335	66-61059 <=*$>
1965 V30 N3 P335-359	66-60897 <=*$>
1965 V30 N3 P448-463	65-33584 <=*$>
1965 V30 N3 P467-470	65-33738 <=*$>
1965 V30 N4 P481-490	65-32403 <$=>
1965 V30 N4 P481-499	65-32452 <=*$>
1965 V30 N4 P491-499	65-32407 <=$>
1965 V30 N4 P491-500	66-61318 <=*$>
1965 V30 N4 P533-541	65-32756 <=*$>
	66-61319 <=*$>
1965 V30 N4 P576-580	65-32755 <=*$>
1965 V30 N5 P633-	FPTS V25 N4 P.T605 <FASE>
1965 V30 N5 P633-646	65-33692 <=*$>
1965 V30 N6 P884-	FPTS V25 N6 P.T981 <FASE>
1965 V30 N6 P935-936	66-62002 <=*$>
1966 V31 N1 P3-35	66-31715 <=$>
1966 V31 N2 P212-220	66-32660 <=$>
1966 V31 N3 P337-354	66-33164 <=$>
1966 V31 N3 P424-426	66-33164 <=$>
1966 V31 N4 P605-607	66-34867 <=$>

*IZVESTIYA AKADEMII NAUK SSSR. SERIYA FIZICHESKAYA

1937 V1 N6 P903-913	61-13508 <+*>
1938 P631-640	R-1957 <*>
1938 P757-759	RT-2262 <*>
1938 N3 P341-368	TT.443 <NRC>
1940 V4 N1 P111-113	RT-533 <*>
1941 P235-240	R-2994 <*>
1941 N2/3 P229-234	R-1730 <*>
1941 V5 P555-587	RT-3488 <*>
1941 V5 N2/3 P131-143	79N50R <ATS>
1941 V5 N2/3 P155-157	RT-514 <*>
1941 V5 N2/3 P222-228	RT-3236 <*>
1941 V5 N2/3 P229-234	RT-3244 <*>
1941 V5 N2/3 P258-265	RT-3230 <*>
	66-12069 <*> 0
1941 V5 N2/3 P266-270	RT-3231 <*>
1941 V5 N2/3 P272-276	RT-3232 <*>
1941 V5 N2/3 P277-279	RT-3233 <*>
1941 V5 N2/3 P280-283	RT-3234 <*>
1941 V5 N2/3 P284-288	RT-3245 <*>
1941 V5 N2/3 P293-295	RT-3246 <*>
1941 V5 N2/3 P308-312	RT-3247 <*>
1941 V5 N2/3 P328-330	RT-3239 <*>
1941 V5 N2/3 P366-375	RT-3248 <*>
1941 V5 N2/3 P376-386	RT-3237 <*>
1941 V5 N2/3 P387-390	RT-3238 <*>
1941 V5 N2/3 P391-392	RT-3235 <*>
1941 V5 N4/5 P457-466	63-14422 <=*>
1943 V7 N4 P114-133	59-20361 <+*>
1944 V8 N5 P275-279	63-14479 <=*>
1944 V8 N6 P352-356	RT-4578 <*>
1945 V9 P206-210	61-18355 <*=>
1945 V9 P222-224	61-18354 <*=>
1945 V9 P329-334	AEC-TR-5112 <=*>
1945 V9 P355-368	AEC-TR5110 <=*> 0
1945 V9 P699-702	63-14095 <=*> 0
1945 V9 N6 P691-698	R-1163 <*>
1945 V9 N1/2 P30-55	RT-103 <*>
	65-60048 <=$>
1945 V9 N4/5 P283-304	AEC-TR-5919 <=*$>
1946 N4 P415-424	RT-2794 <*>
1946 V10 P196-216	58-1289 <*>
1946 V10 N1 P37-48	R-4709 <*>
1946 V10 N1 P49-56	R-4467 <*>
1946 V10 N4 P403-414	RT-1003 <*>
1946 V10 N4 P533-539	R-4524 <*>
1946 V10 N5/6 P467-476	RT-1001 <*>
1946 V10 N5/6 P477-487	UCRL-TRANS-852(L) <=*>
1946 V10 N5/6 P505-508	RT-999 <*>
1947 V11 N2 P155-163	RT-383 <*>
1947 V11 N2 P165-182	RT-399 <*>
1947 V11 N3 P319-325	59-14680 <+*>
1947 V11 N5 P485-496	65-29992 <$>
1947 V11 N6 P667-675	RT-376 <*>
1948 V12 N1 P38-43	RT-857 <*>
1948 V12 N1 P44-48	RT-385 <*>
1948 V12 N2 P126-143	61-13908 <*+>
1948 V12 N2 P144-165	RT-1480 <*>
1948 V12 N2 P166-180	RT-1481 <*>
1948 V12 N3 P322-334	RT-386 <*>
1948 V12 N4 P358-361	RT-4516 <*>
1948 V12 N4 P372-375	RT-4514 <*>
1948 V12 N4 P382-385	RT-4515 <*>
1948 V12 N4 P455-456	R-4937 <*>
	RT-4133 <*>
1948 V12 N4 P475-476	61-13456 <+*>
1948 V12 N5 P527-532	64-18570 <*> 0
1948 V12 N5 P533-540	RT-3251 <*>
1948 V12 N5 P576-581	TT.496 <NRC>
1948 V12 N6 P695-710	RT-3449 <*>
1948 V12 N6 P711-723	62-19874 <=*>
1949 V13 N1 P18-32	RT-896 <*>
1949 V13 N1 P43-48	RT-941 <*>
1949 V13 N1 P67-74	RT-718 <*>
1949 V13 N1 P91-100	UCRL-TRANS-854(L) <=*>
1949 V13 N1 P149-160	RT-384 <*>
1949 V13 N1 P161-166	RT-632 <*>
1949 V13 N1 P188-202	59-17362 <+*>
1949 V13 N2 P218-223	RT-1699 <*>
1949 V13 N2 P254-256	11M42R <ATS>
1949 V13 N3 P331-339	61-17460 <=$>
1949 V13 N3 P392-406	RT-3450 <*>
1949 V13 N4 P409-420	RT-4437 <*>

Citation	Translation
1949 V13 N4 P473-485	RT-471 <*>
1949 V13 N5 P505-514	RT-4438 <*>
1949 V13 N5 P515-533	RT-4439 <*>
1949 V13 N5 P534-548	RT-4440 <*>
1949 V13 N6 P621-630	50/1807 <NLL>
1950 V14 P765-775	T-1574 <INSD>
1950 V14 P776-789	T-1575 <INSD>
1950 V14 N1 P70-94	59-19135 <+*>
1950 V14 N2 P145-173	60-23983 <+*>
1950 V14 N3 P257-262	RT-4002 <*>
1950 V14 N5 P630-633	59-10541 <+*>
1950 V14 N5 P665-669	21K21R <ATS>
1950 V14 N5 P689-692	C-2519 <NRC>
1950 V14 N5 P692-695	4650 <HB>
1950 V14 N6 P721-726	RT-1750 <*>
1951 V15 P637-648	UCRL-TRANS-855(L) <=*>
1951 V15 N1 P87-95	RT-310 <*>
1951 V15 N2 P164-169	RT-2096 <*>
1951 V15 N2 P225-238	RT-2155 <*>
1951 V15 N2 P239-242	RT-1772 <*>
1951 V15 N3 P283-284	59-19110 <+*>
1951 V15 N3 P285-293	59-19114 <+*>
1951 V15 N3 P294-305	59-19113 <+*>
1951 V15 N3 P306-316	59-19111 <+*>
1951 V15 N3 P317-322	RT-2218 <*>
	60-13578 <+*>
1951 V15 N3 P341-346	59-19115 <+*>
1951 V15 N3 P383-386	RT-189 <*>
1951 V15 N3 P387	59-19112 <+*>
1951 V15 N4 P409-412	RT-1869 <*>
1951 V15 N4 P477-486	RT-2384 <*>
1951 V15 N5 P533-541	TT.440 <NRC>
1951 V15 N5 P543-550	AEC-TR-4042 <+*>
1951 V15 N5 P628-636	09N49R <ATS>
	61-16147 <=*>
1952 V16 P139-153	R-4115 <*>
1952 V16 P155-168	R-3982 <*>
1952 V16 P728-738	R-4090 <*>
1952 V16 N1 P70-80	GB43 <NLL>
1952 V16 N2 P169-185	77S84R <ATS>
1952 V16 N3 P333-338	RT-2385 <*>
1952 V16 N3 P350-352	63-00763 <*>
1952 V16 N4 P398-411	62-23120 <=*>
1952 V16 N4 P412-419	62-15874 <=*>
1952 V16 N4 P432-448	R-186 <*>
1952 V16 N6 P653-663	R-1012 <*>
1952 V16 N6 P680-682	63-24398 <=*$>
1952 V16 N6 P683-687	R-878 <*>
1952 V16 N6 P728-738	R-889 <*>
1953 V17 N1 P51-64	RT-2820 <*>
1953 V17 N2 P219-223	25J17R <ATS>
	62-10578 <=*>
1953 V17 N2 P246-248	RT-1035 <*>
1953 V17 N3 P297-312	GB43 <NLL>
	R-792 <*>
1953 V17 N4 P503-506	RT-3414 <*>
1953 V17 N4 P507-510	RT-3415 <*>
1953 V17 N5 P523-530	59-22419 <+*>
	64-20479 <*> 0
	66-12076 <*>
1953 V17 N5 P538-545	C-2400 <NRC>
	R-2058 <*>
1953 V17 N5 P561-566	M-5264 <NLL> 0
1953 V17 N5 P596-603	R-4994 <*>
1953 V17 N6 P681-688	RJ-361 <ATS>
1954 N2 P263-	R-254 <*>
1954 V18 P79-87	R-5112 <*>
1954 V18 N1 P76-78	PB 141 260T <=> 0
1954 V18 N1 P88-92	RT-3340 <*>
1954 V18 N1 P148-154	RJ-236 <ATS>
1954 V18 N1 P155-160	RT-4505 <*>
1954 V18 N2 P161-172	TT-503 <NRC>
1954 V18 N2 P251-	R-249 <*>
1954 V18 N2 P256-	R-251 <*>
1954 V18 N2 P259-	R-255 <*>
1954 V18 N2 P275	R-252 <*>
1954 V18 N3 P307-311	<CTT>
1954 V18 N3 P312-318	<CTT>
1954 V18 N3 P319-327	<CTT>
1954 V18 N3 P328-338	<CTT>
1954 V18 N3 P339-349	<CTT>
1954 V18 N3 P350-359	<CTT>
1954 V18 N3 P360-367	<CTT>
1954 V18 N3 P368-377	<CTT>
1954 V18 N3 P378-381	RJ-240 <ATS>
1954 V18 N3 P382-399	<CTT>
1954 V18 N3 P400-402	<CTT>
1954 V18 N3 P403-405	RJ-239 <ATS>
1954 V18 N3 P406-408	<CTT>
1954 V18 N3 P409-411	<CTT>
1954 V18 N3 P412-416	<CTT>
	R-917 <*>
1954 V18 N4 P419-431	<CTT>
1954 V18 N4 P432-443	<CTT>
1954 V18 N4 P444-455	R-1439 <*>
1954 V18 N4 P444-445	R-1738 <*>
1954 V18 N4 P444-455	62-15857 <=*>
1954 V18 N4 P456-464	<CTT>
1954 V18 N4 P465-472	<CTT>
1954 V18 N4 P489-493	<CTT>
1954 V18 N4 P494-501	<CTT>
	R-912 <*>
	RT-4449 <*>
1954 V18 N4 P502-510	<CTT>
1954 V18 N4 P511-518	<CTT>
1954 V18 N4 P519-520	<CTT>
1954 V18 N5 P523-562	<CTT>
1954 V18 N5 P563-579	<CTT>
1954 V18 N5 P580-588	<CTT>
1954 V18 N5 P589-598	<CTT>
1954 V18 N5 P599-624	<CTT>
1954 V18 N6 P629-631	<CTT>
1954 V18 N6 P631-634	<CTT>
1954 V18 N6 P635-643	<CTT>
1954 V18 N6 P643-661	<CTT>
1954 V18 N6 P662-663	<CTT>
1954 V18 N6 P663	<CTT>
1954 V18 N6 P663-664	<CTT>
1954 V18 N6 P664-665	<CTT>
1954 V18 N6 P665-666	<CTT>
1954 V18 N6 P667-668	<CTT>
1954 V18 N6 P669-671	<CTT>
1954 V18 N6 P672	<CTT>
1954 V18 N6 P673	<CTT>
1954 V18 N6 P674-675	<CTT>
1954 V18 N6 P676-677	<CTT>
1954 V18 N6 P677-678	<CTT>
1954 V18 N6 P678-679	<CTT>
1954 V18 N6 P680	<CTT>
1954 V18 N6 P681-682	<CTT>
1954 V18 N6 P682-683	<CTT>
1954 V18 N6 P683-684	<CTT>
1954 V18 N6 P684-685	<CTT>
1954 V18 N6 P685-686	<CTT>
1954 V18 N6 P687	<CTT>
1954 V18 N6 P688-689	<CTT>
1954 V18 N6 P689-690	<CTT>
1954 V18 N6 P690-692	<CTT>
1954 V18 N6 P692-693	<CTT>
1954 V18 N6 P694-695	<CTT>
1954 V18 N6 P695-697	<CTT>
1954 V18 N6 P697-699	<CTT>
1954 V18 N6 P699-702	<CTT>
1954 V18 N6 P702-704	<CTT>
1954 V18 N6 P704-706	<CTT>
1954 V18 N6 P707	<TCT>
1954 V18 N6 P708-709	<CTT>
1954 V18 N6 P709-710	<CTT>
1954 V18 N6 P711-712	<CTT>
1954 V18 N6 P712-713	<CTT>
1954 V18 N6 P714-715	<CTT>
1954 V18 N6 P716-717	<CTT>
1954 V18 N6 P718-720	<CTT>
1954 V18 N6 P720-721	<CB>
1954 V18 N6 P721-723	<CTT>
1954 V18 N6 P723-725	<CTT>
1954 V18 N6 P725-726	<CTT>
1954 V18 N6 P726-728	<CTT>
1954 V18 N6 P728-729	<CTT>
1954 V18 N6 P729-730	<CTT>

Citation	Source
1954 V18 N6 P731-732	<CTT>
1954 V18 N6 P732-733	<CTT>
1954 V18 N6 P733-734	<CTT>
1954 V18 N6 P735-736	<CTT>
1954 V18 N6 P736-737	<CTT>
1954 V18 N6 P741-742	<CTT>
1955 N19 P387-394	<CTT>
1955 V19 P103-104	30M47R <ATS>
1955 V19 P149-150	R-3694 <*>
1955 V19 P395-403	<CTT>
1955 V19 P404-408	<CTT>
1955 V19 P409-428	<CTT>
1955 V19 P429-443	<CTT>
1955 V19 P444-446	<CTT>
1955 V19 P447-462	<CTT>
1955 V19 P462-473	<CTT>
1955 V19 P474-480	<CTT>
1955 V19 P481-489	<CTT>
1955 V19 P493-501	<CTT>
1955 V19 P502-507	<CTT>
1955 V19 P508-514	<CTT>
1955 V19 P515-518	<CTT>
1955 V19 P519-524	<CTT>
1955 V19 P525-532	<CTT>
1955 V19 P533-536	<CTT>
1955 V19 P537-540	<CTT>
1955 V19 P546-547	<CTT>
1955 V19 P547-548	<CTT>
1955 V19 P548-560	<CTT>
1955 V19 P561-572	<CTT>
1955 V19 P573-588	<CTT>
1955 V19 P589-603	<CTT>
1955 V19 P604-605	<CTT>
1955 V19 P605-	<CTT>
1955 V19 P611-616	<CTT>
1955 V19 P617-623	<CTT>
1955 V19 P624-628	<CTT>
1955 V19 P629-638	<CTT>
1955 V19 P639-650	<CTT>
1955 V19 P651-656	<CTT>
1955 V19 P657-660	<CTT>
1955 V19 P661-664	<CTT>
1955 V19 P663	<CTT>
1955 V19 P665	<CTT>
1955 V19 P666-676	<CTT>
1955 V19 P677-680	<CTT>
1955 V19 P681-686	<CTT>
1955 V19 P687-696	<CTT>
1955 V19 P697-699	<CTT>
1955 V19 P700-706	<CTT>
1955 V19 P707-710	<CTT>
1955 V19 P711-714	<CTT>
1955 V19 P715-719	<CTT>
1955 V19 P720-731	<CTT>
1955 V19 P732-736	<CTT>
1955 V19 P737-746	<CTT>
1955 V19 P747	<CTT>
1955 V19 P748	<CTT>
1955 V19 P750-752	<CTT>
1955 V19 P753-757	<CTT>
1955 V19 P758-761	<CTT>
1955 V19 N1 P5-6	<CTT>
1955 V19 N1 P7-8	<CTT>
1955 V19 N1 P8-9	<CTT>
1955 V19 N1 P10-11	<CTT>
1955 V19 N1 P11-14	<CTT>
1955 V19 N1 P15	<CTT>
1955 V19 N1 P16-17	<CTT>
1955 V19 N1 P18	<CTT>
1955 V19 N1 P18-	R-1473 <*>
	R-1694 <*>
1955 V19 N1 P19	<CTT>
1955 V19 N1 P20-21	<CTT>
1955 V19 N1 P21-22	<CTT>
1955 V19 N1 P23	<CTT>
1955 V19 N1 P24-25	<CTT>
1955 V19 N1 P25-27	<CTT>
1955 V19 N1 P27-28	<CTT>
1955 V19 N1 P27-	R-1466 <*>
1955 V19 N1 P28-29	<CTT>
1955 V19 N1 P30-31	<CTT>
1955 V19 N1 P31-34	<CTT>
1955 V19 N1 P34-35	<CTT>
1955 V19 N1 P35-36	<CTT>
1955 V19 N1 P36-38	<CTT>
1955 V19 N1 P38-40	<CTT>
1955 V19 N1 P40-42	<CTT>
1955 V19 N1 P42-44	<CTT>
1955 V19 N1 P44-45	<CTT>
1955 V19 N1 P45-47	<CTT>
1955 V19 N1 P48-49	<CTT>
1955 V19 N1 P49-52	<CTT>
1955 V19 N1 P52-53	<CTT>
1955 V19 N1 P54-55	<CTT>
1955 V19 N1 P55-56	<CTT>
1955 V19 N1 P56-57	<CTT>
1955 V19 N1 P57-58	<CTT>
1955 V19 N1 P58-59	<CTT>
1955 V19 N1 P58-	R-1467 <*>
1955 V19 N1 P60-61	<CTT>
1955 V19 N1 P60-	R-1468 <*>
1955 V19 N1 P61-64	<CTT>
1955 V19 N1 P61-	R-1469 <*>
1955 V19 N1 P64-65	<CTT>
1955 V19 N1 P66-67	<CTT>
1955 V19 N1 P67-70	<CTT>
1955 V19 N1 P70-72	<CTT>
1955 V19 N1 P72-73	<CTT>
	R-1470 <*>
1955 V19 N1 P74-75	<CTT>
1955 V19 N1 P75-76	<CTT>
1955 V19 N1 P75-	R-1471 <*>
1955 V19 N1 P77	<CTT>
1955 V19 N1 P78-79	<CTT>
1955 V19 N1 P78-	R-1472 <*>
1955 V19 N1 P79-80	<CTT>
1955 V19 N1 P81-82	<CTT>
1955 V19 N1 P82-84	<CTT>
1955 V19 N1 P84-86	<CTT>
1955 V19 N1 P86-87	<CTT>
1955 V19 N1 P87-88	<CTT>
1955 V19 N1 P89-93	<CTT>
1955 V19 N1 P94-96	<CTT>
1955 V19 N1 P97-98	<CTT>
1955 V19 N1 P98-100	<CTT>
1955 V19 N1 P100-101	<CTT>
1955 V19 N1 P102-103	<CTT>
1955 V19 N1 P103-104	<CTT>
1955 V19 N1 P104-106	<CTT>
1955 V19 N1 P106-113	<CTT>
1955 V19 N1 P113-114	<CTT>
1955 V19 N1 P115	<CTT>
1955 V19 N1 P115-116	<CTT>
1955 V19 N1 P116-117	<CTT>
1955 V19 N1 P117-119	<CTT>
1955 V19 N1 P119-120	<CTT>
1955 V19 N1 P120-121	<CTT>
1955 V19 N1 P122	<CTT>
1955 V19 N1 P122-123	<CTT>
1955 V19 N1 P128-129	<CTT>
1955 V19 N1 P131-132	<CTT>
1955 V19 N1 P132-133	<CTT>
1955 V19 N1 P135-136	<CTT>
1955 V19 N1 P136-138	<CTT>
1955 V19 N1 P138-139	<CTT>
1955 V19 N2 P147-148	RJ-611 <ATS>
1955 V19 N2 P149-150	RJ-612 <ATS>
	3884 <HB>
1955 V19 N2 P150	<ATS>
1955 V19 N2 P152-153	3885 <HB>
1955 V19 N2 P167-169	3619 <HB>
1955 V19 N2 P171-173	3617 <HB>
1955 V19 N2 P174-178	3624 <HB>
1955 V19 N2 P179-180	3620 <HB>
1955 V19 N4 P474-480	RJ-322 <ATS>
	RJ-382 <ATS>
1955 V19 N6 P664	<CTT>
1956 P736-739	<CTT>
1956 N8 P967-972	R-1752 <*>
1956 V20 P81-88	R-1904 <*>

R-444 <*>
1956 V20 P163-177 <CTT>
1956 V20 P178-184 <CTT>
1956 V20 P185-194 <CTT>
1956 V20 P195-198 <CTT>
1956 V20 P199-205 <CTT>
1956 V20 P206-210 <CTT>
1956 V20 P211-214 <CTT>
1956 V20 P215-218 <CTT>
1956 V20 P219-225 <CTT>
1956 V20 P226-230 <CTT>
1956 V20 P231-236 <CTT>
1956 V20 P237-250 <CTT>
1956 V20 P251-260 <CTT>
1956 V20 P261-267 <CTT>
1956 V20 P268-274 <CTT>
1956 V20 P275-289 <CTT>
1956 V20 P289-308 <CTT>
1956 V20 P308-312 <CTT>
1956 V20 P312-318 <CTT>
1956 V20 P318-328 <CTT>
1956 V20 P328-343 <CTT>
1956 V20 P343-347 <CTT>
1956 V20 P347-348 <CTT>
1956 V20 P348-354 <CTT>
1956 V20 P354-363 <CTT>
1956 V20 P363-367 <CTT>
1956 V20 P367-369 <CTT>
1956 V20 P369-371 <CTT>
1956 V20 P371-374 <CTT>
1956 V20 P374-376 <CTT>
1956 V20 P382-383 <CTT>
1956 V20 P383-384 <CTT>
1956 V20 P384-387 <CTT>
1956 V20 P388-391 <CTT>
1956 V20 P392-396 <CTT>
1956 V20 P397-409 <CTT>
1956 V20 P410-418 <CTT>
1956 V20 P419-423 <CTT>
1956 V20 P424-432 <CTT>
1956 V20 P433-442 <CTT>
1956 V20 P443-447 <CTT>
1956 V20 P448-454 <CTT>
1956 V20 P455-457 <CTT>
1956 V20 P458-463 <CTT>
1956 V20 P464-470 <CTT>
1956 V20 P471-475 <CTT>
1956 V20 P476-477 <CTT>
1956 V20 P477-478 <CTT>
1956 V20 P478-481 <CTT>
1956 V20 P482-487 <CTT>
1956 V20 P488-492 <CTT>
1956 V20 P493-501 <CTT>
1956 V20 P502-506 <CTT>
1956 V20 P507-513 <CTT>
1956 V20 P514-519 <CTT>
1956 V20 P520-523 <CTT>
1956 V20 P524-528 <CTT>
1956 V20 P529-532 <CTT>
1956 V20 P533-536 <CTT>
1956 V20 P537-539 <CTT>
1956 V20 P540-545 <CTT>
1956 V20 P553-563 <CTT>
1956 V20 P564-569 <CTT>
1956 V20 P570-573 <CTT>
1956 V20 P574-578 <CTT>
1956 V20 P579-582 <CTT>
1956 V20 P583-590 <CTT>
1956 V20 P591-595 <CTT>
1956 V20 P596-600 <CTT>
1956 V20 P601-604 <CTT>
1956 V20 P605-607 <CTT>
1956 V20 P611-613 <CTT>
1956 V20 P614-620 <CTT>
1956 V20 P621-623 <CTT>
1956 V20 P624-630 <CTT>
1956 V20 P631-635 <CTT>
1956 V20 P636-638 <CTT>
1956 V20 P639-640 <CTT>
1956 V20 P641-645 <CTT>
1956 V20 P646-649 <CTT>
1956 V20 P650-652 <CTT>
1956 V20 P653-658 <CTT>
1956 V20 P659-663 <CTT>
1956 V20 P664-670 <CTT>
1956 V20 P668-874 <CTT>
1956 V20 P671-675 <CTT>
1956 V20 P676-678 <CTT>
1956 V20 P679-683 <CTT>
1956 V20 P684-688 <CTT>
1956 V20 P689-692 <CTT>
1956 V20 P693-694 <CTT>
1956 V20 P695-699 <CTT>
1956 V20 P700-702 <CTT>
1956 V20 P703-705 <CTT>
1956 V20 P706-707 <CTT>
1956 V20 P708-713 <CTT>
1956 V20 P714-722 <CTT>
1956 V20 P723-728 <CTT>
1956 V20 P740-750 <CTT>
1956 V20 P751-754 <CTT>
1956 V20 P755-760 <CTT>
1956 V20 P761-763 <CTT>
1956 V20 P764-769 <CTT>
1956 V20 P770-779 <CTT>
1956 V20 P780-783 <CTT>
1956 V20 P784-789 <CTT>
1956 V20 P790-793 <CTT>
1956 V20 P794-797 <CTT>
1956 V20 P798-800 <CTT>
1956 V20 P801-808 <CTT>
1956 V20 P809-810 <CTT>
1956 V20 P811-814 <CTT>
1956 V20 P815-819 <CTT>
1956 V20 P820-823 <CTT>
1956 V20 P824-826 <CTT>
1956 V20 P827-829 <CTT>
1956 V20 P830-833 <CTT>
1956 V20 P834-837 <CTT>
1956 V20 P838-842 <CTT>
1956 V20 P843-844 <CTT>
1956 V20 P845-847 <CTT>
1956 V20 P848-852 <CTT>
1956 V20 P853-858 <CTT>
1956 V20 P859-867 <CTT>
1956 V20 P875-876 <CTT>
1956 V20 P877-882 <CTT>
1956 V20 P883-890 <CTT>
1956 V20 P891-895 <CTT>
1956 V20 P896-902 <CTT>
1956 V20 P903-908 <CTT>
1956 V20 P909-912 <CTT>
1956 V20 P913-924 <CTT>
1956 V20 P925-932 <CTT>
1956 V20 P933-940 <CTT>
1956 V20 P941-946 <CTT>
1956 V20 P951-955 <CTT>
1956 V20 P956-961 <CTT>
1956 V20 P967-974 <CTT>
1956 V20 P977-992 <CTT>
1956 V20 P994-1007 <CTT>
1956 V20 P1008-1022 <CTT>
1956 V20 P1025-1028 <CTT>
1956 V20 P1029-1037 <CTT>
1956 V20 P1039-1049 <CTT>
1956 V20 P1052-1064 <CTT>
1956 V20 P1071-1075 <CTT>
1956 V20 P1079-1085 <CTT>
1956 V20 P1085-1095 <CTT>
1956 V20 P1096-1104 <CTT>
1956 V20 P1105-1111 <CTT>
1956 V20 P1112-1119 <CTT>
1956 V20 P1120-1122 <CTT>
1956 V20 P1123-1126 <CTT>
1956 V20 P1127-1128 <CTT>
1956 V20 P1128-1134 <CTT>
1956 V20 P1135-1136 <CTT>
1956 V20 P1137-1141 <CTT>
1956 V20 P1142-1150 <CTT>
1956 V20 P1151-1152 <CTT>

1956 V20 P1153-1161 <CTT>
1956 V20 P1162-1164 <CTT>
1956 V20 P1165-1178 <CTT>
1956 V20 P1179-1188 <CTT>
1956 V20 P1189-1190 <CTT>
1956 V20 P1190-1194 <CTT>
1956 V20 P1195-1196 <CTT>
1956 V20 P1199-1206 <CTT>
1956 V20 P1207-1214 <CTT>
1956 V20 P1215-1219 <CTT>
1956 V20 P1220-1223 <CTT>
1956 V20 P1224-1225 <CTT>
1956 V20 P1226-1231 <CTT>
1956 V20 P1232-1235 <CTT>
1956 V20 P1236-1237 <CTT>
1956 V20 P1238-1244 <CTT>
1956 V20 P1245-1250 <CTT>
1956 V20 P1251-1254 <CTT>
1956 V20 P1255-1257 <CTT>
1956 V20 P1258-1261 <CTT>
1956 V20 P1262-1264 <CTT>
1956 V20 P1265-1266 <CTT>
1956 V20 P1267-1273 <CTT>
1956 V20 P1274-1278 <CTT>
1956 V20 P1279-1283 <CTT>
1956 V20 P1284-1298 <CTT>
1956 V20 P1299-1309 <CTT>
1956 V20 P1310-1317 <CTT>
1956 V20 P1318-1328 <CTT>
1956 V20 P1329-1335 <CTT>
1956 V20 P1336-1347 <CTT>
1956 V20 P1348-1358 <CTT>
1956 V20 P1359-1360 <CTT>
1956 V20 P1361-1364 <CTT>
1956 V20 P1365-1376 <CTT>
1956 V20 P1377-1386 <CTT>
1956 V20 P1387-1398 <CTT>
1956 V20 P1399-1406 <CTT>
1956 V20 P1407-1416 <CTT>
1956 V20 P1417-1418 <CTT>
1956 V20 P1419-1422 <CTT>
1956 V20 P1423-1429 <CTT>
1956 V20 P1430-1433 <CTT>
1956 V20 P1434-1437 <CTT>
1956 V20 P1438-1450 <CTT>
1956 V20 P1450-1454 <CTT>
1956 V20 P1455-1460 <CTT>
1956 V20 P1461-1468 <CTT>
1956 V20 P1469-1478 <CTT>
1956 V20 P1479-1483 <CTT>
1956 V20 P1484-1485 <CTT>
1956 V20 P1486-1490 <CTT>
1956 V20 P1491-1493 <CTT>
1956 V20 P1494-1495 <CTT>
1956 V20 P1496-1500 <CTT>
1956 V20 P1501-1508 <CTT>
1956 V20 P1509-1518 <CTT>
1956 V20 P1519-1520 <CTT>
1956 V20 P1521-1525 <CTT>
1956 V20 P1526-1532 <CTT>
1956 V20 P1533-1540 <CTT>
1956 V20 P1541-1547 <CTT>
1956 V20 P1550-1552 <CTT>
1956 V20 P1553-1559 <CTT>
1956 V20 P1560-1562 <CTT>
1956 V20 P1563-1568 <CTT>
1956 V20 P1569-1570 <CTT>
1956 V20 P1571-1580 <CTT>
1956 V20 N1 P65-75 R-943 <*>
RT-4009 <*>
1956 V20 N1 P76-80 R-1905 <*>
1956 V20 N2 P178-184 62-32905 <=*>
1956 V20 N2 P185-194 62-32646 <=*>
1956 V20 N2 P195-198 62-32647 <=*>
1956 V20 N2 P211-214 3860 <HB>
1956 V20 N12 P1571-1580 R-3759 <*>
1957 V21 P163-175 <CTT>
1957 V21 P176-178 <CTT>
1957 V21 P179-182 <CTT>
1957 V21 P183-191 <CTT>
1957 V21 P192-200 <CTT>
1957 V21 P201-205 <CTT>
1957 V21 P206-210 <CTT>
1957 V21 P211-219 <CTT>
1957 V21 P220-225 <CTT>
1957 V21 P226-232 <CTT>
1957 V21 P233-263 <CTT>
1957 V21 P264-274 <CTT>
1957 V21 P275-285 <CTT>
1957 V21 P286-288 <CTT>
1957 V21 P289-292 <CTT>
1957 V21 P295-304 <CTT>
1957 V21 P305-310 <CTT>
1957 V21 P311-321 <CTT>
1957 V21 P322-328 <CTT>
1957 V21 P329-333 <CTT>
1957 V21 P334-339 <CTT>
1957 V21 P340-351 <CTT>
1957 V21 P352-358 <CTT>
1957 V21 P359-367 <CTT>
1957 V21 P368-373 <CTT>
1957 V21 P374-378 <CTT>
1957 V21 P379-381 <CTT>
1957 V21 P382-389 <CTT>
1957 V21 P390-393 <CTT>
1957 V21 P394-396 <CTT>
1957 V21 P397-398 <CTT>
1957 V21 P399-401 <CTT>
1957 V21 P402-410 <CTT>
1957 V21 P411-422 <CTT>
1957 V21 P423-432 <CTT>
1957 V21 P433-438 <CTT>
1957 V21 P439-443 <CTT>
1957 V21 P444-449 <CTT>
1957 V21 P450-554 <CTT>
1957 V21 P455-465 <CTT>
1957 V21 P466-472 <CTT>
1957 V21 P475-482 <CTT>
1957 V21 P483- <CTT>
1957 V21 P484-493 <CTT>
1957 V21 P494- <CTT>
1957 V21 P495-498 <CTT>
1957 V21 P499-501 <CTT>
1957 V21 P502-503 <CTT>
1957 V21 P504- <CTT>
1957 V21 P505-506 <CTT>
1957 V21 P507- <CTT>
1957 V21 P508-509 <CTT>
1957 V21 P510- <CTT>
1957 V21 P511-520 <CTT>
1957 V21 P521-522 <CTT>
1957 V21 P523-534 <CTT>
1957 V21 P525-527 <CTT>
1957 V21 P528-529 <CTT>
1957 V21 P530- <CTT>
1957 V21 P531-533 <CTT>
1957 V21 P536-537 <CTT>
1957 V21 P538- <CTT>
1957 V21 P539-540 <CTT>
1957 V21 P541-543 <CTT>
1957 V21 P544- <CTT>
1957 V21 P545- <CTT>
1957 V21 P546-547 <CTT>
1957 V21 P548- <CTT>
1957 V21 P549- <CTT>
1957 V21 P550-551 <CTT>
1957 V21 P552- <CTT>
1957 V21 P553-554 <CTT>
1957 V21 P555-556 <CTT>
1957 V21 P557-569 <CTT>
1957 V21 P570-579 <CTT>
1957 V21 P580-586 <CTT>
1957 V21 P582-591 <CTT>
1957 V21 P587- <CTT>
1957 V21 P588- <CTT>
1957 V21 P589- <CTT>
1957 V21 P590- <CTT>
1957 V21 P593-594 <CTT>
1957 V21 P595-611 <CTT>
1957 V21 P612-618 <CTT>

Citation	Translation
1957 V21 P619-622	<CTT>
1957 V21 P623-631	<CTT>
1957 V21 P632-635	<CTT>
1957 V21 P636-642	<CTT>
1957 V21 P643-647	<CTT>
1957 V21 P655	<CTT>
1957 V21 P674	<CTT>
1957 V21 P675-677	<CTT>
1957 V21 P680	<CTT>
1957 V21 P682	<CTT>
1957 V21 P711-714	<CTT>
1957 V21 P715-	<CTT>
1957 V21 P716-720	<CTT>
1957 V21 P721-730	<CTT>
1957 V21 P731-747	<CTT>
1957 V21 P757-760	<CTT>
1957 V21 P763-770	<CTT>
1957 V21 P771-773	<CTT>
1957 V21 P774-778	<CTT>
1957 V21 P779	<CTT>
1957 V21 P781-782	<CTT>
1957 V21 P783	<CTT>
1957 V21 P787-789	<CTT>
1957 V21 P790-795	<CTT>
1957 V21 P796-800	<CTT>
1957 V21 P801	<CTT>
1957 V21 P802-816	<CTT>
1957 V21 P817-820	<CTT>
1957 V21 P821-823	<CTT>
1957 V21 P824-827	<CTT>
1957 V21 P828-832	<CTT>
1957 V21 P833-843	<CTT>
1957 V21 P844-848	<CTT>
1957 V21 P849-853	<CTT>
1957 V21 P862-864	<CTT>
1957 V21 P865-868	<CTT>
1957 V21 P869-878	<CTT>
1957 V21 P879-886	<CTT>
1957 V21 P887	<CTT>
1957 V21 P888-889	<CTT>
1957 V21 P890-903	<CTT>
1957 V21 P900-1001	<CTT>
1957 V21 P904-906	<CTT>
1957 V21 P907-908	<CTT>
1957 V21 P909-912	<CTT>
1957 V21 P913-917	<CTT>
1957 V21 P918-939	<CTT>
1957 V21 P940-953	<CTT>
1957 V21 P954-961	<CTT>
1957 V21 P962-965	<CTT>
1957 V21 P966-972	<CTT>
1957 V21 P973-977	<CTT>
1957 V21 P978-984	<CTT>
1957 V21 P985-986	<CTT>
1957 V21 P987-989	<CTT>
1957 V21 P1002-1003	<CTT>
1957 V21 P1004-1012	<CTT>
1957 V21 P1013-1016	<CTT>
1957 V21 P1017-1019	<CTT>
1957 V21 P1020-1024	<CTT>
1957 V21 P1025-1028	<CTT>
1957 V21 P1029-1033	<CTT>
1957 V21 P1034-1037	<CTT>
1957 V21 P1038-1046	<CTT>
1957 V21 P1047-1054	<CTT>
1957 V21 P1055-1063	<CTT>
1957 V21 P1064-1082	<CTT>
1957 V21 P1083-1087	<CTT>
1957 V21 P1088-1093	<CTT>
1957 V21 P1094-1104	<CTT>
1957 V21 P1105-1110	<CTT>
1957 V21 P1111-1115	<CTT>
1957 V21 P1116-1122	<CTT>
1957 V21 P1123-1130	<CTT>
1957 V21 P1131-1132	<CTT>
1957 V21 P1133-1139	<CTT>
1957 V21 P1140-1148	<CTT>
1957 V21 P1149-1161	<CTT>
1957 V21 P1162-1167	<CTT>
1957 V21 P1168-1169	<CTT>
1957 V21 P1170-1175	<CTT>
1957 V21 P1176-1177	<CTT>
1957 V21 P1177-1182	<CTT>
1957 V21 P1183-1184	<CTT>
1957 V21 P1184-1196	<CTT>
1957 V21 P1197-1204	<CTT>
1957 V21 P1205-1212	<CTT>
1957 V21 N2 P153-	59-22517 <*+> P
1957 V21 N5 P643-647	62-32644 <=*>
1957 V21 N5 P699-700	65-60709 <=>
1957 V21 N6 P844-848	R-3226 <*>
1957 V21 N6 P849-853	R-3211 <*>
1957 V21 N8 P1088-1093	TRANS-29 <MT>
1957 V21 N9 P1234-1238	59-17370 <+> 0
1957 V21 N9 P1293-1296	R-3722 <*>
1957 V21 N9 P1297-1301	R-3723 <*>
1958 V22 P561-565	R-5338 <*>
1958 V22 N3 P288-295	59-15878 <+*>
1958 V22 N3 P330-342	59-15403 <+*>
1958 V22 N3 P343-358	61-15613 <=*>
1958 V22 N3 P359-360	59-17068 <+*>
1958 V22 N4 P387-391	59-13329 <=> 0
	59-22513 <+*>
1958 V22 N6 P714-717	59-19890 <+*>
1958 V22 N6 P718-719	NP-TR-197 <+*>
1958 V22 N6 P720-724	NP-TR-194 <+*> 0
1958 V22 N6 P737-741	AEC-TR-3599 <+*>
1958 V22 N9 P1063-1067	62-32649 <=*>
1958 V22 N10 P1269-1272	61-13296 <+*> 0
1959 V23 P831	ORNL-TR-544 <=$>
1959 V23 N1 P122-125	60-18077 <+*>
1959 V23 N1 P126-130	60-16910
1959 V23 N1 P147-149	60-16561 <+*>
1959 V23 N4 P427-536	60-31144 <=>
1959 V23 N8 P941-947	62-18257 <=*>
1959 V23 N9 P1059-1061	60-12921 <=>
	62-19268 <=*>
1959 V23 N9 P1061-1063	60-12591 <=>
	62-11402 <=>
1959 V23 N9 P1096-1097	62-11342 <=>
1959 V23 N9 P1097-1099	60-12571 <=>
	62-11224 <=>
1959 V23 N9 P1105-1107	62-19846 <=>
1959 V23 N11 P1347-1350	33R78R <ATS>
1960 V24 N1 P2-104	<CTT>
1960 V24 N2 P114-253	<CTT>
1960 V24 N3 P258-377	<CTT>
1960 V24 N3 P261-271	UCRL TRANS-645 <*=>
1960 V24 N4 P354-361	AEC-TR-4266 <+*>
1960 V24 N5 P488-623	<CTT>
1960 V24 N6 P629-778	<CTT>
1960 V24 N7 P788	<CTT>
1960 V24 N7 P934	<CTT>
1960 V24 N8 P941-1018	<CTT>
1960 V24 N9 P1021-1169	<CTT>
1960 V24 N10 P1176-1304	<CTT>
1960 V24 N11 P1308-1440	<CTT>
1960 V24 N11 P1347-1349	61-19827 <*=>
	62-11428 <=>
1960 V24 N11 P1421-1435	61-21653 <=>
1960 V24 N12 P1444-1502	<CTT>
1961 V25 P237-256	63-00019 <*>
1961 V25 N1 P1-166	<CTT>
1961 V25 N1 P61-67	62-15817 <=*>
1961 V25 N1 P124-129	62-15817 <=*>
1961 V25 N1 P143-144	62-15817 <=*>
1961 V25 N3 P373-374	62-32619 <=*>
1961 V25 N4 P492-500	62-25735 <=*>
1961 V25 N6 P754-756	62-11695 <=>
1961 V25 N8 P910-1066	<CTT>
1961 V25 N8 P913-918	64-15839 <=*$>
1961 V25 N9 P1069-1188	<CTT>
1961 V25 N10 P1198-1308	<CTT>
1961 V25 N11 P1314-1430	<CTT>
1961 V25 N12 P1434-1524	<CTT>
1961 V25 N12 P1469-1472	63-16561 <=*>
1961 V25 N12 P1483-1486	63-19869 <=*>
1962 V26 N1 P14-20	62-33071 <=*> 0
1962 V26 N3 P322-330	62-32079 <=*>
1962 V26 N7 P846-855	62-33670 <=*>

Citation	Translation
1962 V26 N10 P1226-1230	63-15120 <=*>
1962 V26 N11 P1343-1348	63-19633 <=*>
1962 V26 N11 P1359-1365	63-19422 <=*>
1962 V26 N11 P1382-1385	63-01429 <*>
1963 V27 N1 P110-113	63S86R <ATS>
1963 V27 N2 P177-181	CSIR-TR-420 <CSSA>
1963 V27 N3 P319-321	65-10867 <*>
1963 V27 N4 P466-472	63-11766 <=>
1963 V27 N5 P634-637	64-15117 <=*$>
1963 V27 N7 P924-944	64-11787 <=>
1963 V27 N7 P1088-1093	64-11868 <=>
1963 V27 N8 P1040-1043	NP-TR-1086 <*>
1964 V28 N3 P533-536	AD-612 418 <=$>
1964 V28 N3 P540-544	AD-612 418 <=$>
1964 V28 N3 P572-579	64-10855 <*>
1964 V28 N3 P580-583	AD-612 418 <=$>
1964 V28 N5 P934-938	65-13122 <*>
1964 V28 N6 P969-976	AD-621 056 <=$>
1964 V28 N6 P977-979	AD-621 056 <=$>
1964 V28 N6 P993-995	AD-621 056 <=$>
1964 V28 N6 P996-999	AD-618 046 <=$>
1964 V28 N6 P1000-1001	AD-621 056 <=$>
1964 V28 N6 P1065-1068	AD-621 056 <=$>
1964 V28 N6 P1096-1099	AD-618 046 <=$>
1964 V28 N9 P1402-1408	65-30208 <=$>
1964 V28 N9 P1488-1490	65-30208 <=$>
1964 V28 N9 P1499-1503	65-30208 <=$>
1964 V28 N9 P1527-1533	65-30208 <=$>
1964 V28 N9 P1537-1540	65-30208 <=$>
1964 V28 N12 P2058-2074	AD-614 402 <=$>
1965 V29 N4 P604-614	65-31403 <=$>
1965 V29 N5 P782-786	66-11103 <*>
1966 V30 N7 P1150-1168	66-14560 <*>

*IZVESTIYA AKADEMII NAUK SSSR. SERIYA GEOFIZICHESKAYA

Citation	Translation
1951 N2 P40-42	R-5004 <*>
	R-5019 <*>
	59-12194 <+*>
1951 N2 P43-49	59-17173 <+*> 0
1951 N2 P50-55	59-17174 <+*> 0
1951 N3 P31-36	AEC-TR-3815 <*>
	62-14600 <=*>
1951 N3 P94-96	RT-4135 <*>
1951 N5 P1-30	60-18065 <+*>
	62-16404 <=*>
1951 N5 P93	RT-4175 <*>
1952 N1 P35-46	62-15979 <=*>
1952 N1 P47-56	<JBS>
1952 N1 P71-95	R-1355 <*>
1952 N2 P3-14	AD-610 903 <=$>
	RT-2539 <*>
1952 N2 P22-30	63-13565 <=*>
1952 N2 P31-45	60-15505 <+*>
1952 N2 P38-45	AD-633 409 <=$>
1952 N2 P56-74	RT-2405 <*>
1952 N2 P75-80	RT-3219 <*>
1952 N2 P81-84	RT-2798 <*>
1952 N3 P58-59	59-14103 <+*>
1952 N5 P55-56	62-15996 <=*>
1952 N2 P75-80	R-858 <*>
1953 N1 P26-32	59-14105 <+*>
1953 N1 P41-47	RT-2958 <*>
1953 N1 P69-77	RT-3395 <*>
1953 N1 P78-82	89K20R <ATS>
1953 N2 P155-165	R-3407 <*>
	64-71252 <*>
	64-71252 <=>
1953 N2 P191-192	RT-1585 <*>
	59-15573 <+*>
1953 N3 P201-208	62-15995 <=*>
1953 N3 P228-231	62-15995 <=*>
1953 N4 P318-323	RT-2019 <*>
1953 N4 P346-369	60-41097 <=>
1953 N5 P393-404	R-1168 <*>
1953 N5 P424-428	60-23741 <+*>
1953 N5 P445-450	<JBS>
1953 N5 P460-462	R-3911 <*>
1953 N5 P474-476	RT-2933 <*>
1953 N6 P523-525	AD-610 903 <=$>
	RT-2458 <*>
1953 N6 P546-560	RT-3222 <*>
1953 N5 P393-404	2313 <NRC>
1954 N1 P11-25	R-568 <*>
1954 N1 P26-48	R-674 <*>
1954 N1 P49-64	RT-4306 <*>
1954 N1 P65-76	RT-4217 <*>
1954 N1 P83-86	RT-2547 <*>
1954 N2 P108-113	CSIRO-3339 <CSIR>
	R-4581 <*>
1954 N2 P184-189	59-14102 <+*>
1954 N3 P209-222	RT-2717 <*>
1954 N3 P223-243	RT-2714 <*>
1954 N3 P244-263	RT-2718 <*>
1954 N3 P264-279	RT-2772 <*>
1954 N3 P288-292	61-23183 <=*> 0
1954 N3 P293-298	72K20R <ATS>
1954 N4 P349-359	62-15994 <=*>
1954 N5 P415-423	01M42R <ATS>
1955 N1 P45-56	<GPRC>
1955 N1 P57-59	65-11179 <*>
1955 N1 P69-79	RT-4609 <*>
1955 N1 P80-83	RT-3173 <*>
1955 N2 P137-155	RT-4277 <*>
1955 N2 P156-165	M.5142 <NLL>
	RT-3223 <*>
1955 N2 P181-184	62-15978 <=*>
1955 N2 P185-186	RT-4158 <*>
1955 N3 P234-248	56J15R <ATS>
	62-32824 <=*>
1955 N3 P267-269	62-14597 <=*>
1955 N3 P275-277	59-14114 <+*>
1955 N4 P323-331	65-64385 <=*$>
1955 N4 P332-338	60-19012 <+*>
1955 N4 P369-376	62-25426 <=*>
1955 N4 P381-383	59-17175 <+*> 0
1955 N5 P416-424	N65-16596 <=$>
1955 N5 P425-434	RT-4423 <*>
1955 N5 P464-467	RT-4422 <*>
1955 N5 P468-474	RT-4325 <*>
1955 N6 P514-520	RT-4581 <*>
1955 N6 P529-537	73K20R <ATS>
1955 N6 P538-540	RT-4417 <*>
1955 N6 P547-551	89K21R <ATS>
1955 N6 P561-565	59-20353 <+*> P
1955 N1 P45-56	RT-3991 <*>
1955 N1 P69-79	C-2207 <NLL>
1956 N1 P56-66	57J14R <ATS>
1956 N1 P87-91	33K21R <ATS>
1956 N1 P112-113	71K20R <ATS>
1956 N1 P114-116	74K20R <ATS>
1956 N2 P182-190	90K20R <ATS>
1956 N2 P202-209	61-23846 <*=>
1956 N2 P218-225	62-25424 <=*>
1956 N2 P232-235	75K20R <ATS>
1956 N3 P290-296	56J14R <ATS>
1956 N4 P410-418	62-18527 <=*>
1956 N6 P704-707	62-25425 <=*>
1956 N6 P712-721	62-25425 <=*>
1956 N6 P735-739	59-15470 <+*>
1956 N7 P755-775	60-23599 <+*>
1956 N7 P776-793	60-19131 <+*>
1956 N7 P835-837	R-203 <*>
1956 N7 P840-843	63-19907 <=*$>
1956 N7 P853-856	R-840 <*>
1956 N7 P862-864	R-655 <*>
	2285 <NRC>
1956 N8 P927-939	60-19011 <+*>
1956 N8 P997-998	62-25423 <=*>
1956 N8 P999-1000	AD-610 885 <=$>
	R-291 <*>
1956 N9 P1009-1020	59-10878 <+*>
1956 N9 P1021-1035	59-10598 <+*>
1956 N9 P1091-1098	SCL-T-251 <+*>
1956 N10 P1129-1144	59-15261 <+*>
1956 N10 P1161-1173	60-19010 <+*>
1956 N10 P1227-1231	90K21R <ATS>
1956 N10 P1236-1238	62-25422 <=*>
1956 N11 P1241-1257	59-15365 <+*>
1956 N11 P1332-1337	AD-610 962 <=$>

Citation	Translation
1956 N11 P1338-1353	59-15982 <+*> 0
1956 N11 P1358-1360	59-15981 <+*>
1956 N12 P1400-1410	R-5089 <*>
1956 N12 P1419-1426	59-12210 <+*>
1956 N12 P1461-1473	63-24007 <=*>
1956 N12 P1474-1483	59-15557 <+*>
1956 N7 P835-837	C-2261 <NRC>
1956 N7 P853-856	C-2298 <NRC>
1956 N8 P976-983	C-2585 <NRC>
	R-3452 <*>
1956 N8 P999-1000	C-2236 <NLL>
	R-14 <*>
1956 N11 P1332-1337	C-2498 <NRC>
1957 P1389-1392	61-13153 <*+>
1957 N3 P384-394	59-20354 <+*> PO
1957 N3 P404-406	59-20259 <+*> P
1957 N4 P426-439	R-4997 <*>
	65-13795 <*>
1957 N4 P449-457	65-13796 <*>
1957 N4 P512-515	AD-608 971 <=$>
1957 N4 P548-551	59-16423 <+*>
1957 N5 P569-574	59-15262 <+*>
1957 N6 P831-833	61-28821 <*=>
1957 N6 P834-837	R-4529 <*>
	59-22649 <+*>
1957 N8 P979-989	59-11407 <=> 0
1957 N8 P1000-1007	59-11408 <=> 0
1957 N8 P1045-1051	59-11413 <=>
1957 N8 P1060-1063	59-11409 <=> 0
1957 N8 P1064-1068	59-11410 <=> 0
1957 N8 P1069-1071	59-11411 <=> 0
1957 N11 P1313-1322	59-11634 <=>
1957 N11 P1347-1358	59-15122 <+*>
1957 N11 P1347-1348	62-14602 <=*>
1957 N11 P1347-1358	65-13797 <*>
1957 N11 P1366-1383	M.5138 <NLL> 0
1957 N11 P1393-1409	R-3475 <*>
1958 N1 P26-42	60-23748 <+*> 0
1958 N1 P46-53	60-17375 <+*>
	61-19331 <+*>
1958 N1 P54-64	59-11606 <=> 0
1958 N1 P93-104	R-3505 <*>
1958 N1 P146-152	59-11471 <=> 0
1958 N2 P157-164	59-12171 <+*>
1958 N2 P184-195	59-10877 <+*>
1958 N2 P244-254	PB 141 181T <=>
	R-5001 <*>
	59-12054 <+*>
1958 N2 P274-276	R-4951 <*>
	59-12196 <+*>
1958 N2 P277-279	59-16283 <+*>
1958 N3 P247-257	59-16022 <+*>
1958 N3 P337-346	59-12707 <+*>
1958 N4 P497-514	PB 141 272T <=>
1958 N4 P515-526	61-15087 <*+>
1958 N4 P527-535	59-16280 <+*>
1958 N4 P558-559	59-22213 <+*>
1958 N4 P560-563	59-22606 <+*>
1958 N5 P613-624	60-13782 <+*>
1958 N5 P648-654	59-13219 <=> 0
1958 N5 P655-663	59-13220 <=> 0
1958 N5 P673-677	60-17171 <+*>
1958 N6 P729-740	61-19010 <+*>
1958 N6 P741-751	59-22566 <+*>
	60-13086 <+*>
1958 N6 P752-764	59-13561 <=>
1958 N6 P791-795	63-24291 <=*$>
1958 N7 P891-902	60-00498 <*>
	60-13118 <+*>
1958 N7 P913-917	60-00498 <*>
1958 N7 P913-920	60-13118 <+*>
1958 N7 P917-920	60-00498 <*>
1958 N7 P923-926	61-13851 <*+>
1958 N8 P1040-1043	60-18926 <+*>
1958 N8 P1048-1051	59-12651 <+*>
	59-20260 <+*>
1958 N8 P1052-1053	59-12647 <+*>
1958 N9 P1099-1104	59-21119 <=>
1958 N9 P1105-1110	59-21120 <=>
	61-19326 <+*> 0
1958 N9 P1145-1150	59-11363 <=>
1958 N10 P1250-1253	59-13170 <=>
1958 N11 P1360-1373	59-13051 <=>
	59-16007 <+*>
	61-13535 <+*> 0
1958 N11 P1399-1401	59-11399 <=> 0
1958 N11 P1402-1405	59-11400 <=>
1958 N11 P1406-1417	60-14221 <+*>
1958 N12 P1477-1484	M.5112 <NLL>
1958 N12 P1517-1519	60-17422 <+*>
1958 N12 P1522-1529	59-13497 <=>
1958 V21 N10 P1153-1161	59-13161 <=>
1959 N1 P3-10	60-21036 <=>
1959 N1 P32-48	10L32R <ATS>
1959 N1 P108-112	59-11502 <=>
1959 N1 P122-130	59-11503 <=> 0
1959 N2 P262-275	57L34R <ATS>
1959 N2 P321-325	63-15918 <=*>
1959 N2 P326-329	60-31032 <=>
1959 N3 P344-360	24L32R <ATS>
1959 N3 P361-371	60-16897 <*+>
1959 N3 P389-409	59-11666 <=> 0
1959 N3 P432-444	60-41105 <=>
1959 N3 P480-488	59-13886 <=>
	59-13886 <=> 0
1959 N4 P505-515	20L34R <ATS>
1959 N4 P570-580	60-41095 <=>
1959 N4 P581-592	60-41113 <=>
1959 N5 P710-724	M.5120 <NLL>
1959 N5 P732-738	60-41108 <=>
1959 N6 P801-803	60-11337 <=>
1959 N6 P910-918	60-41214 <=>
1959 N7 P988-994	AEC-TR-4098 <=*>
1959 N7 P1079-1084	60-31866 <=>
1959 N8 P1157-1163	60-21073 <=>
1959 N8 P1164-1166	60-21043 <=>
1959 N8 P1167-1176	60-31854 <=>
1959 N8 P1238-1241	62-23277 <=*>
1959 N9 P1432-1433	M.5131 <NLL>
1959 N10 P1526-1527	60-31040 <=>
1959 N10 P1549-1552	61-11957 <=>
1959 N11 P1570-1578	60-11604 <=>
1959 N11 P1591-1598	61-15961 <*=> 0
1959 N11 P1656-1664	60-11631 <=>
1959 N11 P1665-1669	60-11632 <=>
	61-13067 <=*> 0
1959 N11 P1690-1693	60-19583 <+*>
1959 N11 P1699-1701	60-11605 <=>
1959 N11 P1721-1724	60-41111 <=>
1959 N12 P1799-1805	60-31112 <=>
1959 N12 P1821-1830	60-41109 <=>
1959 N12 P1889-1890	03M41R <ATS>
1960 N1 P98-106	60-11646 <=>
1960 N1 P107-114	60-19870 <+*>
1960 N2 P229-235	60-11813 <=>
1960 N2 P326-332	16M42R <ATS>
1960 N2 P341-345	60-11712 <=>
1960 N3 P449-458	60-23515 <+*>
1960 N4 P530-547	60-41549 <=>
1960 N4 P566-574	60-41550 <=>
	61-19757 <=*>
1960 N4 P607-616	06M43R <ATS>
1960 N5 P688-697	61-23530 <=*>
1960 N5 P698-706	60-31868 <=>
1960 N5 P707-713	60-41215 <=>
1960 N5 P714-719	61-11534 <=>
1960 N5 P743-745	60-41285 <=>
1960 N5 P753-755	60-41286 <=>
1960 N5 P756-757	60-41282 <=>
1960 N5 P758-761	60-41611 <=>
1960 N6 P836-846	60-41339 <=>
1960 N6 P882-885	61-21107 <=>
1960 N6 P892-897	61-21048 <=>
1960 N6 P898-902	61-11895 <=>
1960 N7 P1013-1021	61-11706 <=>
	61-27499 <*=>
1960 N7 P1042-1055	61-11706 <=>
1960 N7 P1056-1058	61-11711 <=>
1960 N8 P1216-1219	61-11523 <=>
1960 N8 P1229-1233	61-11523 <=>

1960 N8 P1271-1276	61-19937 <*=>
1960 N8 P1287-1288	61-11523 <=>
1960 N9 P1321-1327	UCRL TRANS-640(L) <*>
1960 N9 P1397-1406	61-11065 <=>
1960 N10 P1518-1528	61-21204 <=>
1960 N10 P1529-1533	61-21168 <=>
1960 N11 P1635-1641	61-19938 <*=>
1960 N11 P1642-1648	61-19934 <*=>
1961 N1 P25-39	61-23616 <*=>
1961 N1 P69-78	64N53R <ATS>
1961 N1 P160-161	62-24866 <=*>
1961 N2 P269-272	61-23043 <=*>
1961 N4 P541-543	94N53R <ATS>
1961 N6 P835-846	61-31681 <=> 0
1961 N6 P951-952	61-20095 <*=>
1961 N8 P1214-1223	62-13404 <=*>
1962 N6 P132-134	64-13677 <=*$>
1962 N8 P1083-1092	65-61152 <=> 0
1962 N10 P1425-1440	63-13547 <=>
1962 N10 P1450-1455	63-13547 <=>
1962 N12 P1825-1836	65-60794 <=> 0
1963 N1 P147-151	63-21439 <=>
1963 N1 P169-182	63-21439 <=>
1963 N5 P783-791	64-15999 <=*$>
1963 N5 P792-804	64-19000 <=*$>
1963 N5 P804-809	64-19001 <=*$> 0
1963 N8 P1270-1277	AD-611 308 <=$>
	65-61187 <=>
1963 N9 P1410-1416	64-11744 <=>
1964 N4 P604-614	N65-15156 <=$*>
1964 N11 P1697-1699	22S83R <ATS>
1964 N12 P1811-1818	23S83R <ATS>
1964 N12 P1849-1858	65-30292 <=$>
1964 N12 P1859-1868	65-30293 <=$>
1964 N12 P1869-1877	65-30294 <=$>

IZVESTIYA AKADEMII NAUK SSSR. SERIYA GEOGRAFICHES-KAYA

1951 N1 P74-79	62-33418 <=*>
1951 N4 P3-15	62-23503 <=*> 0
1952 N1 P3-13	59-15047 <+*> 0
1952 N2 P3-10	RT-1290 <*>
1952 N3 P52-54	62-15969 <=*>
1953 N3 P11-20	CSIRO-3218 <CSIR>
	R-3929 <*>
	62-10462 <*>
	65-12199 <*>
1953 N4 P64-74	M.5433 <NLL> 0
1953 N4 P75-76	M.5432 <NLL>
1954 N1 P3-10	RT-2321 <*>
1954 N1 P15-32	61-23686 <*=>
1954 N1 P91-93	RT-2482 <*>
1954 N2 P15-28	CSIRO-3844 <CSIR>
1954 N2 P29-44	R-4372 <*>
1954 N3 P17-41	RT-4363 <*>
1954 N3 P62-72	62-19040 <=*> 0
1954 N4 P18-28	RT-3171 <*>
	62-19048 <=*>
1954 N5 P3-16	RT-2570 <*>
1954 N2 P29-44	CSIRO-3265 <CSIR>
1955 N2 P5-15	CSIRO-3181 <CSIR>
1955 N3 P12-24	SGRT V1 N1-2 P3 <AGS>
1955 N4 P3-15	CSIRO-3181 <CSIR>
	R-4573 <*>
	RT-4262 <*>
1955 N4 P16-28	RT-4261 <*>
1956 N1 P115-143	RT-4181 <*>
1956 N3 P77-84	C-2569 <NRC>
1956 N4 P16-25	60-21094 <=>
1956 N7 P857-861	C-2294 <NRC>
	R-804 <*>
1957 N1 P19-35	60-21105 <=>
1957 N1 P36-55	PANSDOC-TR.43 <PANS>
1957 N1 P147-151	PANSDOC-TR.44 <PANS>
1957 N4 P449-457	59-17778 <+*>
1957 N5 P50-55	SGRT V1 N1-2 P21 <AGS>
1957 N5 P56-61	SGRT V1 N1-2 P33 <AGS>
1957 N5 P65-76	62-33418 <=*>
1958 N1 P28-47	60-51193 <=>
1958 N1 P129-133	SGRT V1 N1-2 P43 <AGS>
1958 N2 P9-21	60-21103 <=>
1958 N2 P144-145	SGRT V1 N1-2 P68 <AGS>
1958 N3 P22-36	60-21104 <=>
1958 N3 P147-150	SGRT V1 N1-2 P16 <AGS>
1958 N4 P24-32	C-3139 <NRCC>
1958 N4 P42-56	SGRT V1 N1-2 P48 <AGS>
1958 N5 P79-86	66-51099 <=>
1958 N5 P130-139	60-11431 <=>
1959 N1 P52-63	SGRT V1 N3 P20 <AGS>
1959 N1 P64-73	SGRT V1 N3 P9 <AGS>
1959 N1 P155-157	60-31155 <=>
1959 N2 P25-37	59-31032 <=>
1959 N2 P38-49	SGRT V1 N3 P42 <AGS>
1959 N2 P50-57	SGRT V1 N3 P33 <AGS>
1959 N2 P73-76	59-31033 <=>
1959 N2 P77-79	SGRT V1 N5 P85 <AGS>
1959 N2 P143-146	SGRT V1 N3 P3 <AGS>
	60-31057 <=>
1959 N3 P3-19	SGRT V1 N5 P3 <AGS>
1959 N3 P20-30	63-11077 <=>
1959 N3 P31-41	SGRT V1 N5 P24 <AGS>
1959 N3 P42-71	SGRT V1 N4 P30 <AGS>
1959 N6 P44-56	SGRT V1 N7 P59 <AGS>
1959 N6 P94-97	63-19840 <=*$>
1959 N6 P115-117	M.5127 <NLL>
1959 N6 P154-155	60-11288 <=>
1960 N1 P3-12	SGRT V1 N5 P37 <AGS>
1960 N1 P49-59	65-50033 <=$>
1960 N2 P42-54	SGRT V3 N2 P48 <AGS>
1960 N3 P3-11	63-11505 <=>
1960 N3 P55-61	SGRT V2 N2 P48 <AGS>
1960 N3 P100-107	SGRT V2 N1 P54 <AGS>
1960 N3 P135-137	SGRT V2 N1 P44 <AGS>
1960 N3 P138-141	SGRT V2 N4 P33 <AGS>
1960 N3 P138-142	60-31636 <=>
1960 N4 P3-10	60-41653 <=>
1960 N4 P138-139	60-41654 <=>
1960 N4 P139-142	60-41652 <=*>
1960 N4 P142-146	60-41655 <=>
1960 N5 P22-33	C-3588 <NRCC>
1960 N5 P53-59	65-50032 <=$>
1960 N5 P88-95	SGRT V2 N3 P35 <AGS>
1960 N5 P124-132	SGRT V2 N5 P13 <AGS>
1960 N6 P3-9	SGRT V2 N4 P40 <AGS>
1960 N6 P90-97	C-4220 <NRC>
1960 N6 P122-125	SGRT V2 N5 P23 <AGS>
1960 N6 P141-142	61-21354 <=>
1961 N1 P3-12	63-13880 <=*>
	63-19953 <=*$> 0
1961 N1 P158-161	SGRT V2 N5 P69 <AGS>
1961 N2 P3-12	SGRT V2 N8 P27 <AGS>
1961 N2 P37-45	SGRT V3 N10 P3 <AGS>
1961 N2 P66-76	SGRT V2 N9 P44 <AGS>
1961 N2 P123-130	61-27933 <*=>
1961 N3 P149-150	SGRT V3 N5 P57 <AGS>
1961 N5 P6-17	SGRT V3 N1 P27 <AGS>
1961 N5 P18-27	SGRT V3 N1 P49 <AGS>
1961 N5 P42-48	SGRT V3 N1 P59 <AGS>
1961 N5 P76-85	SGRT V5 N6 P11 <AGS>
	SOV.GEOGR. V5 N6 P11 <AGS>
1961 N5 P128-133	62-23437 <=*>
	65-61381 <=$>
1961 N5 P133-136	SGRT V3 N1 P67 <AGS>
1961 N5 P137-139	SGRT V3 N5 P41 <AGS>
1961 N5 P139-143	SGRT V3 N5 P60 <AGS>
1961 N6 P3-17	65-61270 <=$> 0
1961 N6 P36-46	SGRT V3 N10 P12 <AGS>
1961 N6 P117-125	SGRT V4 N3 P12 <AGS>
1961 N6 P150-153	SGRT V3 N2 P74 <AGS>
1962 N1 P6-16	SGRT V3 N5 P3 <AGS>
	63-19851 <=*> 0
1962 N1 P37-47	SGRT V3 N5 P30 <AGS>
1962 N1 P138-143	SGRT V3 N4 P65 <AGS>
1962 N2 P18-25	SGRT V4 N3 P30 <AGS>
1962 N3 P35-42	SGRT V3 N10 P26 <AGS>
1962 N3 P82-89	SHSP N4 1962 P441 <AGU>
1962 N4 P3-11	SGRT V3 N9 P3 <AGS>

1962 N4 P16-27 SGRT V3 N9 P14 <AGS>
1962 N4 P43-51 SGRT V3 N9 P29 <AGS>
1962 N4 P52-61 SGRT V3 N9 P39 <AGS>
1962 N4 P88-92 SGRT V3 N9 P49 <AGS>
1962 N5 P3-13 SGRT V3 N10 P37 <AGS>
1962 N5 P43-51 SGRT V4 N2 P3 <AGS>
1962 N5 P52-57 SGRT V4 N2 P12 <AGS>
1962 N5 P58-64 SGRT V4 N2 P18 <AGS>
1962 N5 P64-70 SGRT V4 N2 P25 <AGS>
1962 N5 P71-77 SGRT V4 N2 P32 <AGS>
1962 N5 P77-83 SGRT V4 N2 P39 <AGS>
1962 N5 P83-89 SGRT V4 N2 P46 <AGS>
63-13543 <=>
1962 N5 P89-94 SGRT V4 N2 P52 <AGS>
1962 N5 P95-101 SGRT V4 N5 P3 <AGS>
1962 N5 P127-137 63-13543 <=>
1962 N5 P184-187 TRANS-321(M) <NLL>
63-13543 <=>
1962 N6 P3-10 65-60715 <=> 0
1962 N6 P101-113 SGRT V4 N4 P3 <AGS>
1962 N6 P146-147 SGRT V4 N1 P60 <AGS>
1963 N1 P36-44 SGRT V5 N5 P13 <AGS>
SOV.GEOGR. V5 N5 P13 <AGS>
1963 N1 P45-49 63-21801 <=>
1963 N2 P3-13 SGRT V4 N6 P3 <AGS>
1963 N2 P14-24 SGRT V4 N6 P17 <AGS>
1963 N2 P35-45 SGRT V4 N6 P29 <AGS>
1963 N2 P79-87 65-61520 <=$> 0
1963 N3 P35-44 64-11566 <=>
1963 N3 P59-70 SGRT V5 N6 P20 <AGS>
1963 N3 P153-156 SGRT V4 N8 P25 <AGS>
1963 N3 P156-158 SGRT V4 N8 P31 <AGS>
1963 N4 P23-34 SGRT V5 N1 P26 <AGS>
63-11770 <=>
1963 N4 P35-46 SGRT V5 N1 P40 <AGS>
1963 N4 P60-67 SGRT V5 N5 P42 <AGS>
SOV.GEOGR. V5 N5 P42 <AGS>
1963 N4 P102-111 SGRT V4 N8 P3 <AGS>
1963 N4 P111-119 SGRT V5 N1 P15 <AGS>
1963 N4 P143-147 SGRT V5 N1 P53 <AGS>
1963 N5 P3-28 SGRT V5 N3 P3 <AGS>
64-15997 <=*$> 0
1963 N5 P45-50 SGRT V5 N5 P24 <AGS>
SOV.GEOGR. V5 N5 P24 <AGS>
1963 N5 P98-110 SGRT V5 N4 P3 <AGS>
64-15996 <=*$> 0
1963 N6 P15-26 65-60618 <=$> 0
1963 N6 P50-55 SGRT V5 N5 P33 <AGS>
SOV.GEOGR. V5 N5 P33 <AGS>
1963 N6 P122-123 SGRT V5 N2 P49 <AGS>
1963 N9 P1410-1416 64-00372 <*>
1963 N3 P59-70 SOV.GEOGR. V5 N5 P59 <AGS>
1964 N1 P44-51 SGRT V5 N6 P3 <AGS>
SOV.GEOGR. V5 N6 P3 <AGS>
1964 N1 P101-112 SGRT V5 N4 P18 <AGS>
64-21929 <=>
1964 N2 P3-5 64-31311 <=>
1964 N2 P168-169 SGRT V5 N6 P69 <AGS>
SOV.GEOGR. V5 N6 P69 <AGS>
1964 N3 P4-13 SGRT V5 N8 P23 <AGS>
SOV.GEOGR. V5 N8 P23 <AGS>
1964 N3 P89-101 SGRT V5 N8 P35 <AGS>
SOV.GEOGR. V5 N8 P35 <AGS>
1964 N4 P3-11 SGRT V5 N9 P3 <AGS>
1964 N4 P12-22 SGRT V5 N10 P61 <AGS>
1964 N4 P23-34 AD-628 841 <=*$>
1964 N4 P35-46 AD-628 842 <=*$>
SGRT V6 N10 P25 <AGS>
1964 N4 P47-54 SGRT V6 N10 P40 <AGS>
1964 N4 P67-78 AD-628 843 <=*$>
1964 N4 P158-162 TRANS-F.58. <NLL>
1964 N5 P13-15 SGRT V5 N10 P73 <AGS>
1965 N2 P3-22 TRANS-367(M) <NLL>
1965 N2 P3-33 65-31065 <=$>
1965 N2 P158-162 65-31065 <=$>
1965 N3 P53-63 SGRT V6 N8 P3 <AGS>
1965 N3 P64-72 SGRT V6 N8 P16 <AGS>
1965 N3 P110-119 SGRT V6 N8 P26 <AGS>
65-32279 <=$>
1965 N3 P122-125 AD-626 274 <=*$>
65-32393 <=$>
1965 N3 P136-138 AD-626 275 <=*$>
1965 N3 P139-142 AD-626 263 <=*$>
1965 N3 P139 AD-626 285 <=*$>
1965 N3 P142-149 AD-626 276 <=*$>
1965 N3 P145-147 AD-626 265 <=*$>
1965 N3 P147-149 AD-626 266 <=*$>
1965 N4 P31-39 SGRT V6 N8 P40 <AGS>
1965 N4 P61-69 SGRT V6 N8 P51 <AGS>
1965 N4 P113-126 SGRT V6 N9 P3 <AGS>
1965 N5 P3-15 SGRT V7 N6 P40 <AGS>
1965 N5 P37-49 SGRT V6 N10 P3 <AGS>
1965 N5 P50-54 SGRT V6 N10 P19 <AGS>
1965 N6 P17-22 AD-640234 <=$>
1965 N6 P47-55 SGRT V7 N2 P44 <AGS>
1966 N1 P27-33 66-31386 <=$>
1966 N1 P41-49 66-31386 <=$>
1966 N1 P82-88 66-31386 <=$>
1966 N2 P3-14 SGRT V7 N7 P3 <AGS>
1966 N2 P95-102 SGRT V7 N7 P19 <AGS>
1966 N2 P103-111 SGRT V7 N7 P28 <AGS>

IZVESTIYA AKADEMII NAUK SSSR. SERIYA GEOGRAFI-
CHESKAYA I GEOFIZICHESKAYA

1938 V2 N4 P307-315 62-15976 <=*>
1939 V3 N1 P3-29 62-15971 <=*>
1939 V3 N2 P155-173 R-2087 <*>
1939 V3 N3 P275-286 62-15974 <=*>
1940 N1 P33-118 RT-1090 <*>
1940 N3 P285-3C4 TT.489 <NRC>
1940 V4 N1 P3-16 62-15606 <=*>
1940 V4 N1 P133-152 62-15606 <=*>
1941 N4/5 P453-463 RT-910 <*>
1941 V5 P453-463 61-13005 <=*>
1941 V5 N1 P89-94 62-15975 <=*>
1941 V5 N3 P229-254 62-15967 <=*>
1943 N2 P70-82 TT.102 <*>
1943 V7 N6 P354-358 62-14596 <=*>
1944 V8 N1 P18-24 62-15970 <=*>
1944 V8 N6 P381-384 62-15986 <=*>
1945 V9 N2 P108-111 62-15985 <=*>
1945 V9 N3 P240-260 62-15984 <=*>
1945 V9 N4 P294-315 RT-4048 <*>
1945 V9 N5/6 P511-528 62-15598 <=*>
1945 V9 N5/6 P535-546 62-15598 <=*>
1946 V10 N2 P2C5-211 RT-329 <*>
1946 V10 N3 P285-290 RT-2002 <*>
1946 V10 N4 P3C1-310 RT-1144 <*>
59-15564 <+*>
1946 V10 N4 P313-316 RT-203 <*>
1946 V10 N5 P449-454 62-15990 <=*>
1947 V11 N1 P3-14 R-340 <*>
RT-2683 <*>
1947 V11 N1 P15-19 C-2219 <NLL>
R-50 <*>
RT-4639 <*>
1947 V11 N2 P137-140 RT-1241 <*>
1947 V11 N5 P395-408 RT-197 <*>
1947 V11 N5 P4C9-414 RT-195 <*>
1947 V11 N6 P465-487 RT-1356 <*>
1948 V12 P239-248 RT-122 <*>
1948 V12 N2 P137-146 RT-1235 <*>
1948 V12 N2 P155-168 62-14601 <=*>
1948 V12 N2 P169-176 R-1839 <*>
R-4945 <*>
1948 V12 N3 P249-254 RT-1239 <*>
1948 V12 N3 P271-282 RT-1206 <*>
1948 V12 N4 P289-305 RT-1203 <*>
62-19034 <=*>
1948 V12 N5 P377-386 R-1628 <*>
1948 V12 N5 P475-487 61-16188 <=*>
1948 V12 N6 P62-16548 62-16548 <=*>
1948 V13 N2 P108-114 63-15691 <=*$>
1949 N3 P50-68 56K20R <ATS>
1949 V13 N1 P33-57 RT-694 <*>
1949 V13 N4 P320-330 R-857 <*>
RT-3175 <*>
1949 V13 N4 P363-368 62-16549 <=*>
1949 V13 N6 P556-562 62-15992 <=*>
1950 V14 N3 P223-231 62-15991 <=*>

Citation	Translation
1950 V14 N5 P403-414	62-15993 <=*>
1950 V14 N5 P421-424	<JBS>
1950 V14 N5 P425-439	62-11190 <=>
1950 V14 N6 P473-500	RT-319 <*>
1950 V14 N6 P501-513	RT-313 <*>
1950 V14 N6 P562-570	R-3620 <*>
	R-4582 <*>
	3106 <NLL>
1951 V15 N1 P20-39	RT-1837 <*>
1951 V15 N1 P43-50	RT-1836 <*>
1951 V15 N1 P51-56	RT-1626 <*>

*IZVESTIYA AKADEMII NAUK SSSR. SERIYA GEOLOGICHESKAYA

Citation	Translation
1936 V1 N1 P3-33	62-33419 <=*>
1938 V3 N4 P659-666	62-33419 <=*>
1939 N1 P85-94	RT-1506 <*>
1942 N5/6 P102-114	RT-2797 <*>
1947 N2 P7-14	IGR V1 N1 P65 <AGI>
1947 N3 P19-38	RJ-557 <ATS>
1947 N4 P89-106	62-10426 <*=>
1947 N4 P117-125	R-4676 <*>
1948 N4 P3-50	R-1664 <*>
1948 V13 N3 P150-158	62-16675 <=*>
1949 N1 P35-52	66-14406 <*$>
1949 N6 P131-148	RJ-244 <ATS>
1950 N3 P103-113	CSIRO-3093 <CSIR>
1951 V16 N4 P99-101	62-16628 <=*> 0
1951 V16 N4 P102-105	62-16435 <=*>
1951 V16 N4 P106-126	62-16664 <=*>
1953 N2 P19-36	IGR V1 N1 P50 <AGI>
1953 N5 P12-49	50H12R <ATS>
1953 V18 N5 P12-49	RTS-2763 <NLL>
1954 N4 P3-37	IGR V6 N6 P1030 <AGI>
1954 N4 P38-49	RT-2747 <*>
1954 V19 N4 P3-37	IGR V6 N6 P1030-1056 <AGI>
1954 V19 N6 P3-14	RT-3089 <*>
1954 V19 N6 P85-93	91P60R <ATS>
1955 N4 P31-49	38H12R <ATS>
1955 N4 P50-56	41H12R <ATS>
1955 N4 P80-87	<ATS>
	43H12R <ATS>
1955 N5 P89-96	R-2664 <*>
1955 V3 P3-18	R-4923 <*>
1955 V20 N1 P34-51	62-18534 <=*> 0
1956 N2 P13-24	42H12R <ATS>
1956 N9 P44-50	IGR V2 N2 P138 <AGI>
1956 V21 N7 P49-60	62-25125 <=*>
1957 N3 P36-47	97K20R <ATS>
1957 N4 P108-112	R-4653 <*>
1957 N6 P3-13	IGR V1 N6 P79 <AGI>
1957 N11 P103-124	46L31R <ATS>
1957 V22 N3 P63-75	64-19831 <=$>
1957 V22 N5 P3-16	60-15560 <+*>
1957 V22 N5 P17-36	60-19557 <+*>
1957 V22 N8 P16-30	59-11530 <=> 0
1957 V22 N9 P3-18	59-11140 <=> 0
1957 V22 N11 P3-7	59-11487 <=>
1957 V22 N11 P8-14	59-11477 <=> 0
1957 V22 N11 P32-49	59-11489 <=>
1957 V22 N11 P50-57	59-10310 <+*>
1957 V22 N11 P58-81	59-11490 <=> 0
1957 V22 N12 P3-12	63-23601 <=*>
1957 V22 N12 P47-60	C-3756 <NRCC>
1958 N2 P85-93	59-10601 <+*> 0
1958 N7 P92-97	59-10597 <+*> 0
1958 N9 P3-24	59-10518 <+*> 0
1958 V23 N1 P62-73	60-23742 <+*>
1958 V23 N6 P58-68	61-15628 <=*>
1958 V23 N6 P95-100	43L29R <ATS>
1958 V23 N10 P3-20	59-13074 <=>
	60-13837 <+*>
1958 V23 N11 P3-19	59-13052 <=> 0
1959 V24 N1 P3-8	59-11459 <=>
1959 V24 N5 P3-15	61-15959 <*=> 0
1959 V24 N6 P3-15	59-31025 <=>
1959 V24 N6 P109-111	61-13946 <*+>
1959 V24 N6 P124-125	44L35R <ATS>
1959 V24 N8 P20-33	61-15957 <*=> 0
1959 V24 N9 P90-91	60-23533 <+*>
1959 V24 N10 P78-91	61-15910 <+*>
1959 V24 N12 P54-59	61-15199 <*+>
1960 V25 N1 P56-66	61-15888 <*=>
1960 V25 N1 P108-111	61-15014 <*+>
1960 V25 N2 P91-104	61-15277 <*+>
1960 V25 N3 P96-101	61-13496 <*+>
1960 V25 N3 P157-159	60-31772 <=>
1960 V25 N4 P89-95	61-15241 <*+>
1960 V25 N4 P115-117	60-41629 <=>
1960 V25 N7 P122-126	61-11000 <=>
1960 V25 N7 P127-	62-15270 <*=>
1960 V25 N10 P70-76	22N50R <ATS>
1960 V25 N11 P47-52	07N51R <ATS>
1962 N1 P10-25	IGR V5 N12 P1611 <AGI>
1962 N1 P66-84	IGR V5 N12 P1593 <AGI>
1962 N1 P85-94	IGR V5 N12 P1623 <AGI>
1962 N1 P99-108	IGR V5 N12 P1585 <AGI>
1962 N2 P3-24	IGR V6 N1 P99 <AGI>
1962 N2 P37-48	IGR V6 N1 P119 <AGI>
1962 N2 P49-63	IGR V6 N1 P129 <AGI>
1962 N2 P88-93	IGR V6 N1 P94 <AGI>
1962 N2 P94-97	IGR V6 N1 P139 <AGI>
1962 N3 P3-16	IGR V6 N3 P495 <AGI>
1962 N3 P17-29	IGR V6 N3 P507 <AGI>
1962 N3 P30-44	IGR V6 N3 P519 <AGI>
1962 N3 P72-87	IGR V6 N3 P531 <AGI>
1962 N3 P100-113	IGR V6 N3 P541 <AGI>
1962 N3 P120-121	IGR V6 N3 P553 <AGI>
1962 N4 P13-31	IGR V6 N4 P668 <AGI>
1962 N4 P40-	IGR V6 N4 P682 <AGI>
1962 N4 P100-101	IGR V6 N4 P690 <AGI>
1962 N4 P106-109	IGR V6 N4 P692 <AGI>
1962 N5 P1-26	IGR V6 N5 P860 <AGI>
1962 N5 P3-17	IGR V6 N5 P841 <AGI>
1962 N5 P27-40	IGR V6 N5 P850 <AGI>
1962 N5 P41-51	IGR V6 N5 P867 <AGI>
1962 N5 P63-75	IGR V6 N5 P875 <AGI>
1962 N5 P134-136	IGR V6 N5 P886 <AGI>
1962 N6 P3-11	IGR V6 N5 P889 <AGI>
	IGR,V6,N5,P889-894 <AGI>
1962 N6 P34-48	IGR V6 N5 P895 <AGI>
1962 N6 P49-61	IGR V6 N5 P904 <AGI>
1962 N6 P62-72	IGR V6 N5 P912 <AGI>
1962 N7 P59-69	IGR V6 N7 P1233 <AGI>
1962 N8 P3-13	IGR V6 N7 P1242 <AGI>
1962 N8 P14-20	IGR V6 N7 P1249 <AGI>
1962 N8 P60-77	IGR V6 N7 P1254 <AGI>
1962 N8 P115-120	IGR V6 N7 P1267 <AGI>
1962 N9 P3-11	IGR V6 N7 P1273 <AGI>
1962 N9 P27-47	IGR V6 N7 P1279 <AGI>
1962 N9 P75-89	IGR V6 N7 P1294 <AGI>
1962 N9 P90-96	IGR V6 N7 P1305 <AGI>
1962 N9 P109-111	IGR V6 N7 P1310 <AGI>
1962 N10 P3-19	IGR V6 N8 P1390 <AGI>
1962 N10 P46-54	IGR V6 N8 P1401 <AGI>
1962 N10 P69-73	IGR V6 N8 P1408 <AGI>
1962 N10 P86-89	IGR V6 N8 P1413 <AGI>
1962 N10 P90-96	IGR V6 N8 P1416 <AGI>
1962 N11 P3-17	IGR V6 N8 P1422 <AGI>
1962 N11 P18-31	IGR V6 N8 P1433 <AGI>
1962 N11 P50-57	IGR V6 N8 P1445 <AGI>
1962 N11 P78-83	IGR V6 N8 P1451 <AGI>
1962 N11 P84-94	IGR V6 N8 P1457 <AGI>
1962 N11 P110-111	IGR V6 N8 P1464 <AGI>
1962 N12 P5-94	IGR V6 N8 P1490 <AGI>
1962 N12 P23-31	IGR V6 N8 P1483 <AGI>
1962 N12 P95-101	IGR V6 N8 P1495 <AGI>
1962 V27 N3 P3-16	IGR,V6,N3,P495-506 <AGI>
1962 V27 N3 P17-29	IGR,V6,N3,P507-518 <AGI>
1962 V27 N3 P30-44	IGR,V6,N3,P519-530 <AGI>
1962 V27 N3 P72-87	IGR,V6,N3,P531-540 <AGI>
1962 V27 N3 P120-121	IGR,V6,N3,P553-554 <AGI>
1962 V27 N4 P40-	IGR,V6,N4,P682-689 <AGI>
1962 V27 N4 P100-101	IGR,V6,N4,P100-101 <AGI>
1962 V27 N4 P106-109	IGR,V6,N4,P692-696 <AGI>
1962 V27 N5 P3-17	IGR,V6,N5,P841-849 <AGI>
1962 V27 N5 P18-26	IGR,V6,N5,860-866 <AGI>
1962 V27 N5 P27-40	IGR,V6,N5,P850-859 <AGI>
1962 V27 N5 P41-51	867-874 <AGI>
1962 V27 N5 P63-75	IGR,V6,N5,P875-885 <AGI>

Citation	Translation
1962 V27 N5 P134-136	IGR,V6,N5,P886-888 <AGI>
1962 V27 N6 P34-48	IGR,V6,N5,P895-903 <AGI>
1962 V27 N6 P49-61	IGR,V6,N5,P904-911 <AGI>
1962 V27 N6 P62-72	IGR,V6,N5,P912-919 <AGI>
1962 V27 N7 P9-18	IGR,V6,N7,P1225-1232 <AGI>
1962 V27 N7 P59-69	IGR,V6,N7,P1233-1241 <AGI>
1962 V27 N8 P3-13	IGR,V6,N7,P1242-1248 <AGI>
1962 V27 N8 P14-20	IGR,V6,N7,P1249-1253 <AGI>
1962 V27 N8 P60-77	IGR,V6,N7,P1254-1266 <AGI>
1962 V27 N8 P115-120	IGR,V6,N7,P1267-1272 <AGI>
1962 V27 N9 P3-11	IGR,V6,N7,P1273-1278 <AGI>
1962 V27 N9 P27-47	IGR,V6,N7,P1279-1293 <AGI>
1962 V27 N9 P75-89	IGR,V6,N7,P1294-1304 <AGI>
1962 V27 N9 P90-96	IGR,V6,N7,P1305-1309 <AGI>
1962 V27 N9 P109-111	IGR,V6,N7,P1310-1311 <AGI>
1962 V27 N10 P3-19	IGR V6 N8 P1390-1400 <AGI>
1962 V27 N10 P33-45	65-60920 <=$>
1962 V27 N10 P46-54	IGR V6 N8 P1401-1407 <AGI>
1962 V27 N10 P69-73	IGR V6 N8 P1408-1412 <AGI>
1962 V27 N10 P86-89	IGR V6 N8 P1413-1415 <AGI>
1962 V27 N10 P90-96	IGR V6 N8 P1416-1421 <AGI>
1962 V27 N11 P3-17	IGR V6 N8 P1422-1432 <AGI>
	63-20678 <=*$>
1962 V27 N11 P18-31	IGR V6 N8 P1433-1444 <AGI>
1962 V27 N11 P50-57	IRG V6 N8 P1445-1450 <AGI>
1962 V27 N11 P78-83	IGR V6 N8 P1451-1456 <AGI>
1962 V27 N11 P84-94	IGR V6 N8 P1457-1463 <AGI>
1962 V27 N11 P110-111	
	IGR V6 N8 P1464-1465 <AGI>
1962 V27 N12 P3-22	IGR V6 N8 P1466-1482 <AGI>
1962 V27 N12 P23-31	IGR V6 N8 P1483-1489 <AGI>
1962 V27 N12 P85-94	IGR V6 N8 P1490-1494 <AGI>
1962 V27 N12 P95-101	IGR V6 N8 P1495-1499 <AGI>
1962 V27 N12 P117-118	
	IGR V6 N8 P1500-1501 <AGI>
1963 N1 P3-8	IGR V7 N1 P23 <AGI>
1963 N1 P19-45	IGR V7 N1 P27 <AGI>
1963 N1 P46-64	IGR V7 N1 P47 <AGI>
1963 N1 P66-76	IGR V7 N1 P62 <AGI>
1963 N1 P109-113	IGR V7 N1 P72 <AGI>
1963 N2 P3-8	IGR V7 N1 P78 <AGI>
1963 N2 P30-38	IGR V7 N1 P84 <AGI>
1963 N2 P50-59	IGR V7 N1 P98 <AGI>
1963 N2 P60-72	IGR V7 N1 P105 <AGI>
1963 N2 P80-89	IGR V7 N1 P116 <AGI>
1963 N2 P106-108	IGR V7 N1 P123 <AGI>
1963 N3 P3-9	IGR V7 N1 P126 <AGI>
1963 N3 P19-28	IGR V7 N1 P130 <AGI>
1963 N3 P29-39	IGR V7 N1 P136 <AGI>
1963 N3 P40-53	IGR V7 N1 P142 <AGI>
1963 N3 P54-62	IGR V7 N1 P151 <AGI>
1963 N3 P63-83	IGR V7 N1 P157 <AGI>
1963 N4 P8-23	IGR V7 N3 P385 <AGI>
1963 N4 P24-34	IGR V7 N3 P397 <AGI>
1963 N4 P52-66	IGR V7 N3 P405 <AGI>
1963 N4 P78-98	IGR V7 N3 P427 <AGI>
1963 N4 P114-124	IGR V7 N3 P416 <AGI>
1963 N5 P3-8	IGR V7 N3 P442 <AGI>
1963 N5 P23-46	IGR V7 N3 P446 <AGI>
1963 N5 P47-66	IGR V7 N3 P464 <AGI>
1963 N5 P67-81	IGR V7 N3 P476 <AGI>
1963 N5 P82-88	IGR V7 N3 P486 <AGI>
1963 N5 P89-95	IGR V7 N3 P491 <AGI>
1963 N5 P101-103	IGR V7 N3 P496 <AGI>
1963 N5 P121-124	IGR V7 N3 P501 <AGI>
1963 N6 P24-42	IGR V7 N5 P848 <AGI>
1963 N6 P58-72	IGR V7 N5 P862 <AGI>
1963 N6 P88-100	IGR V7 N5 P874 <AGI>
1963 N6 P112-115	IGR V7 N5 P889 <AGI>
1963 N7 P3-18	IGR V7 N7 P1157 <AGI>
1963 N7 P19-31	IGR V7 N7 P1168 <AGI>
1963 N7 P32-53	IGR V7 N7 P1176 <AGI>
1963 N7 P54-68	IGR V7 N7 P1194 <AGI>
1963 N8 P11-22	IGR V7 N7 P1207 <AGI>
1963 N8 P43-55	IGR V7 N7 P1217 <AGI>
1963 N8 P56-67	IGR V7 N7 P1227 <AGI>
1963 N8 P68-85	IGR V7 N7 P1237 <AGI>
1963 N9 P18-33	IGR V7 N7 P1259 <AGI>
1963 N9 P34-45	IGR V7 N7 P1271 <AGI>
1963 N9 P46-60	IGR V7 N7 P1281 <AGI>
1963 N9 P61-69	IGR V7 N7 P1252 <AGI>
1963 N9 P70-92	IGR V7 N7 P1293 <AGI>
1963 N9 P93-98	IGR V7 N7 P1313 <AGI>
1963 N10 P3-14	IGR V7 N11 P1907 <AGI>
1963 N10 P15-29	IGR V7 N11 P1917 <AGI>
1963 N10 P32-50	IGR V7 N11 P1949 <AGI>
1963 N11 P7-31	IGR V7 N11 P1928 <AGI>
1963 N11 P51-65	IGR V7 N11 P1963 <AGI>
1963 N11 P66-84	IGR V7 N11 P1977 <AGI>
1963 N11 P119-120	IGR V7 N11 P1992 <AGI>
1963 N12 P3-30	IGR V7 N11 P1994 <AGI>
1963 N12 P41-58	IGR V7 N11 P2033 <AGI>
1963 N12 P59-79	IGR V7 N11 P2017 <AGI>
1963 V28 N1 P3-8	IGR V7 N1 P23-26 <AGI>
1963 V28 N1 P9-18	63-21262 <=>
1963 V28 N1 P19-45	IGR V7 N1 P27-46 <AGI>
1963 V28 N1 P46-64	IGR V7 N1 P47-61 <AGI>
1963 V28 N1 P66-76	IGR V7 N1 P62-71 <AGI>
1963 V28 N1 P109-113	IGR V7 N1 72-77 <AGI>
1963 V28 N2 P3-8	IGR V7 N1 P78-83 <AGI>
1963 V28 N2 P30-38	IGR V7 N1 P84-89 <AGI>
1963 V28 N2 P39-49	IGR V7 N1 P90-97 <AGI>
1963 V28 N2 P50-59	IGR V7 N1 P98-104 <AGI>
1963 V28 N2 P60-72	IGR V7 N1 P60-72 <AGI>
1963 V28 N2 P80-89	IGR V7 N1 P116-122 <AGI>
1963 V28 N2 P106-108	IGR V7 N1 P123-125 <AGI>
1963 V28 N3 P3-9	IGR V7 N1 P126-129 <AGI>
1963 V28 N3 P19-28	IGR V7 N1 130-135 <AGI>
1963 V28 N3 P29-39	IGR V7 N1 P136-141 <AGI>
1963 V28 N3 P40-53	IGR V7 N1 P142-150 <AGI>
1963 V28 N3 P54-62	IGR V7 N1 P151-156 <AGI>
1963 V28 N3 P63-83	IGR V7 N1 P157-169 <AGI>
1963 V28 N5 P3-8	63-31134 <=>
1963 V28 N7 P19-31	65-60921 <=$>
1963 V28 N8 P23-42	65-60933 <=$>
1963 V28 N9 P61-69	67S86R <ATS>
1964 N5 P104-105	IGR V7 N3 P499 <AGI>
1964 N11 P90-94	GI N2 1964 P390 <AGI>
1964 N12 P3-17	IGR V7 N12 P2213 <AGI>
1964 N12 P18-32	IGR V7 N12 P2203 <AGI>
1964 N12 P42-61	IGR V7 N12 P2184 <AGI>
1964 V29 N1 P110-111	64-21647 <=>
1964 V29 N4 P20-37	RTS-2724 <NLL>
1964 V29 N9 P100-105	RTS-2855 <NLL>
1964 V29 N12 P3-17	<JBS>
1965 N1 P21-36	IGR V8 N4 P421 <AGI>
1965 N1 P37-43	IGR V8 N3 P266 <AGI>
1965 N1 P44-66	IGR V8 N5 P530 <AGI>
1965 N1 P67-79	IGR V8 N3 P346 <AGI>
1965 N1 P80-94	IGR V8 N2 P156 <AGI>
1965 N1 P130-133	IGR V8 N2 P168 <AGI>
1965 N2 P3-6	IGR V8 N3 P278 <AGI>
1965 N2 P7-15	IGR V8 N1 P36 <AGI>
1965 N2 P16-32	IGR V8 N6 P676 <AGI>
1965 N2 P33-43	IGR V8 N6 P731 <AGI>
1965 N2 P44-57	IGR V8 N3 P356 <AGI>
1965 N2 P83-101	IGR V8 N4 P391 <AGI>
1965 N2 P112-115	IGR V8 N4 P446 <AGI>
1965 N10 P15-27	IGR V8 N12 P1440 <AGI>
1965 N10 P89-101	IGR V8 N12 P1396 <AGI>
1965 N10 P126-131	IGR V8 N12 P1422 <AGI>
1965 N11 P3-13	IGR V8 N9 P1039 <AGI>
1965 N11 P14-16	IGR V8 N7 P768 <AGI>
1965 N11 P63-71	IGR V8 N7 P783 <AGI>
1965 N11 P92-97	IGR V8 N8 P908 <AGI>
1965 V30 N2 P7-15	65-30637 <=$>
1965 V30 N2 P83-101	RTS-3125 <NLL>
1965 V30 N2 P93-114	65-31297 <=*$>
1965 V30 N2 P148-150	65-30636 <=$>
1965 V30 N2 P154-160	65-30635 <=$>
1966 N1 P7-16	66-30975 <=$>
1966 N1 P57-72	IGR V8 N6 P631 <AGI>
1966 N1 P155-164	66-30975 <=$>
1966 N6 P3-9	GI V2 N6 1965 P1105 <AGI>
1966 N6 P10-20	GI V2 N6 1965 P1112 <AGI>
1966 N7 P75-85	AD-651 470 <=$>
1966 N7 P132-140	66-34925 <=$>
1966 V31 N6 P64	66-34738 <=$>

*IZVESTIYA AKADEMII NAUK SSSR. SERIYA KHIMICHESKAYA

Citation	Translation
1963 N1 P36-45	65-13028 <*>
1963 N3 P446-450	AD-609 785 <=>
1963 N6 P999-1003	65-17446 <*>
	66-10984 <*>
1963 N7 P1309-1310	63-24109 <=>
1963 N8 P1365-1367	64-18432 <*>
1963 N8 P1524	57R77R <ATS>
1963 N10 P1772-1775	14R75R <ATS>
1963 N10 P1857-1859	AD-609 787 <=>
1963 N10 P1957-1959	311 <TC>
1963 N11 P1905-1910	65-13486 <*>
1963 N11 P1920-1923	RTS-2779 <NLL>
	420 <TC>
1963 N11 P2032-2036	64-18976 <*> 0
1963 N11 P2061-2063	77R75R <ATS>
1963 N12 P62-69	64-16003 <*> 0
1963 N12 P2132-2136	65-60753 <=>
1963 V28 N8 P11-22	65-60516 <=$>
1964 N1 P155-157	65-12067 <*>
1964 N2 P209-215	75R77R <ATS>
1964 N2 P216-225	12R80R <ATS>
1964 N2 P225-234	43R76R <ATS>
1964 N2 P267-270	36R76R <ATS>
1964 N2 P344-346	M.5659 <NLL>
1964 N2 P363-365	65-10205 <*>
1964 N2 P379-381	ORNL-TR-259 <=$>
1964 N3 P543-548	65-10206 <*> 0
1964 N3 P569-572	64-51607 <=$>
1964 N3 P574-576	65-14894 <*>
1964 N3 P576-578	AD-620 064 <=$>
1964 N4 P604-613	22R79R <ATS>
1964 N4 P775	01R77R <ATS>
1964 N4 P775-776	02R77R <ATS>
1964 N5 P822-825	66-14474 <*> 0
1964 N5 P909-912	AD-618 645 <=$>
1964 N6 P985-990	74R79R <ATS>
1964 N6 P1108-1110	65-12460 <*>
1964 N7 P1225-1229	354 <TC>
1964 N7 P1296-1302	M-5716 <NLL>
1964 N7 P1359	ATS-34S84R <$=>
1964 N8 P1526-1528	65-30002 <=$>
1964 N8 P1555-1556	33S84R <ATS>
1964 N9 P1648-1653	65-11284 <*>
1964 N9 P1682-1685	65-11735 <*>
1964 N9 P1728-1729	65-13487 <*>
1964 N10 P1797-1801	59R80R <ATS>
1964 N10 P1801-1807	65-32982 <=*$>
1964 N11 P2051-2055	38S83R <ATS>
1964 V11 N3 P30-37	CSIR-TR-521 <CSSA>
1964 V29 N11 P3-8	<JBS>
1965 N1 P178-180	884 <TC>
1965 N2 P251-257	65-14310 <*>
1965 N2 P349-351	65-14806 <*>
1965 N3 P537-538	66-14348 <*$>
1965 N4 P720-721	66-14172 <*$>
1965 N4 P731-734	30S88R <ATS>
1965 N4 P759-760	70S84R <ATS>
1965 N4 P760	65-13798 <*>
1965 N4 P762	29S88R <ATS>
1965 N4 P763	28S88R <ATS>
1965 N5 P919-922	66-12207 <*>
1965 N5 P938-940	65-32328 <=*$>
1965 N6 P950-959	1138 <TC>
1965 N7 P1258-1260	66-10704 <*> 0
1965 N7 P1290-1292	66-12238 <*>
1965 N8 P1336-1345	66-14344 <*$>
1965 N8 P1370-1375	65-33166 <=*$>
1965 N8 P1416-1424	66-10886 <*> 0
1965 N8 P1466-1469	66-14345 <*$>
1965 N8 P1472-1474	66-14346 <*$>
1965 N8 P1508	66-10736 <*>
1965 N9 P1599-1606	65-33599 <=*$>
1965 N9 P1607-1613	66-13787 <*>
1965 N9 P1680-1682	66-14347 <*$>
1965 N9 P1696-1697	66-12736 <*>
1965 N10 P1899-1901	05S89R <ATS>
1965 N12 P2204-2206	54T92R <ATS>
1965 N12 P2224	53T92R <ATS>
1966 N1 P139-141	10T96R <ATS>

*IZVESTIYA AKADEMII NAUK SSSR. SERIYA MATEMATICHESKAYA

Citation	Translation
1938 P69-73	R-2981 <*>
1938 P613-623	AD-611 095 <=$>
1938 N5/6 P487-498	63-11763 <=>
1939 P603-626	RT-3929 <*>
1942 V6 P115-134	AMST S2 V1 P95-110 <AMS>
1942 V6 P309-330	63-20329 <=*>
1942 V6 N4 P309-330	64-11607 <=>
1944 V8 P3-48	AMST S2 V19 P1-46 <AMS>
1944 V8 P143-174	AMST S1 V11 P172-213 <AMS>
1944 V8 P225-232	AMST S1 V8 P186-194 <AMS>
1945 V9 P3-64	AMST S1 V8 P195-272 <AMS>
1945 V9 P291-300	AMST S1 V9 P214-227 <AMS>
1945 V9 P329-352	AMST S1 V9 P228-262 <AMS>
1945 V9 N2 P121-143	RT-629 <*>
1945 V9 N4 P291-300	RT-295 <*>
1945 V9 N5 P329-356	RT-1185 <*>
1946 V10 P97-104	R-1 <*>
1946 V10 P135-166	64-71352 <=> M
1946 V10 N1 P73-96	M.5365 <NLL>
1947 V11 P111-138	STMSP V2 P131-158 <AMS>
1947 V11 P363-400	AMST S2 V28 P1-35 <AMS>
1947 V11 P363-398	61-20660 <*>
1948 V12 P57-78	AMST S2 V19 P299-321 <AMS>
1948 V12 P279-324	AMST S1 V8 P305-364 <AMS>
1948 V12 P513-554	AMST S1 V2 P425-480 <AMS>
1948 V12 N4 P411-416	RT-1287 <*>
1948 V12 N5 P481-512	65-60804 <=>
1949 V13 P9-32	AMST S1 V9 P276-307 <AMS>
1949 V13 P97-110	AMST S1 V2 P170-186 <AMS>
1949 V13 P125-162	AMST S1 V7 P279-331 <AMS>
1949 V13 P193-200	AMST S1 V7 P332-345 <AMS>
1949 V13 P281-300	STMSP V2 P109-129 <AMS>
1949 V13 P389-402	AMST S1 V7 P399-417 <AMS>
1949 V13 P417-424	AMST S1 V1 P108-119 <AMS>
1949 V13 P425-446	AMST S1 V10 P55-83 <AMS>
1949 V13 P447-472	AMST S1 V1 P15-50 <AMS>
1949 V13 P483-494	AMST S1 V1 P211-227 <AMS>
1949 V13 N1 P9-32	RT-351 <*>
1949 V13 N2 P97-110	RT-343 <*>
1949 V13 N5 P389-402	RT-274 <*>
1949 V13 N5 P417-424	RT-275 <*>
1949 V13 N5 P425-446	RT-296 <*>
1949 V13 N5 P447-472	RT-271 <*>
1949 V13 N6 P483-494	RT-626 <*>
1950 V14 P7-44	AMST S1 V7 P346-398 <AMS>
1950 V14 P95-100	STMSP V1 P7-12 <AMS>
1950 V14 P123-144	AMST S1 V3 P79-106 <AMS>
1950 V14 P199-214	AMST S1 V2 P187-206 <AMS>
1950 V14 P261-274	AMST S2 V51 P317-332 <AMS>
1950 V14 P303-326	AMST S1 V11 P144-170 <AMS>
1950 V14 N1 P7-44	RT-272 <*>
1950 V14 N2 P123-144	RT-433 <*>
1950 V14 N3 P199-214	RT-342 <*>
1950 V14 N4 P303-326	RT-862 <*>
1951 V15 P17-46	AMST S2 V4 P31-58 <AMS>
1951 V15 P53-74	AMST S2 V59 P224-270 <AMS>
1951 V15 P109-130	AMST S1 V2 P207-232 <AMS>
1951 V15 P153-176	AMST S2 V59 P224-270 <AMS>
1951 V15 P205-218	STMSP V1 P41-53 <AMS>
1951 V15 P309-360	AMST S2 V1 P253-304 <AMS>
1951 V15 P463-476	AMST S2 V35 P1-14 <AMS>
1951 V15 N3 P199-230	RT-341 <*>
1951 V15 N3 P205-218	RT-784 <*>
1951 V15 N4 P361-383	63-16563 <=*>
1953 V17 P63-76	AMST S2 V3 P1-14 <AMS>
1953 V17 P189-248	AMST S2 V2 P147-205 <AMS>
1953 V17 P291-330	STMSP V1 997-134 <AMS>
1953 V17 P401-420	66-11240 <*>
1953 V17 N6 P525-538	RT-1282 <*>
1954 V18 P185-200	STMSP V2 P211-228 <AMS>
1954 V18 P261-296	AMST S2 V4 P107-142 <AMS>
1954 V18 P327-334	AMST S2 V4 P143-150 <AMS>
1954 V18 P389-418	AMST S2 V4 P151-183 <AMS>
1954 V18 P525-578	AMST S2 V4 P185-237 <AMS>
1955 V19 P81-96	STMSP V4 P105-121 <AMS>
1955 V19 P247-266	STMSP V1 P171-189 <AMS>
1956 V20 P3-16	AMST S2 V9 P217-231 <AMS>
1956 V20 P53-98	AMST S2 V10 P13-58 <AMS>

1956 V20 P99-136 AMST S2 V21 P51-86 <AMS>
1956 V20 P137-144 AMST S2 V27 P43-50 <AMS>
1956 V20 P145-166 AMST S2 V19 P87-108 <AMS>
1956 V20 P179-196 AMST S2 V10 P107-124 <AMS>
1956 V20 P307-324 AMST S2 V52 P75-94 <AMS>
1956 V20 P569-582 AMST S2 V39 P177-192 <AMS>
1956 V20 P673-678 AMST S2 V13 P1-7 <AMS>
1956 V20 P765-774 AMST S2 V14 P173-180 <AMS>
1956 V20 N1 P17-32 R-2084 <*>
1957 V21 P93-116 AMST S2 V14 P311-332 <AMS>
1957 V21 P171-198 AMST S2 V17 P173-200 <AMS>
1957 V21 P289-310 AMST S2 V45 P165-189 <AMS>
1957 V21 P515-540 AMST S2 V45 P191-220 <AMS>
1957 V21 P559-578 AMST S2 V14 P289-310 <AMS>
1957 V21 P605-626 AMST S2 V18 P295-319 <AMS>
1957 V21 P627-654 AMST S2 V18 P199-230 <AMS>
1957 V21 P655-680 AMST S2 V24 P79-106 <AMS>
1957 V21 N4 P501-514 65-11582 <*>
1957 V21 N6 P729-746 65-10272 <*>
1958 V22 P563-576 AMST S2 V59 P271-284 <AMS>
1958 V22 P737-756 AMST S2 V37 P59-78 <AMS>
1958 V22 N4 P449-474 61-11194 <=>
61-23169 <*=>
61-27192 <*=>
1959 V23 P3-34 AMST S2 V36 P63-99 <AMS>
1959 V23 P35-66 AMST S2 V22 P339-370 <AMS>
1959 V23 P115-134 AMST S2 V22 P139-162 <AMS>
1959 V23 P185-196 AMST S2 V21 P21-33 <AMS>
1959 V23 P197-212 AMST S2 V21 P35-50 <AMS>
1959 V23 P337-364 AMST S2 V46 P17-47 <AMS>
1959 V23 P489-502 AMST S2 V39 P193-206 <AMS>
1959 V23 P771-780 AMST S2 V27 P257-266 <AMS>
1959 V23 P781-808 AMST S2 V34 P375-404 <AMS>
1959 V23 P823-840 AMST S2 V27 P267-288 <AMS>
1959 V23 P841-870 AMST S2 V45 P105-137 <AMS>
1959 V23 N2 P157-164 65-11583 <*>
1959 V23 N3 P337-364 AMS TRANS V.46 P337 <AMS>
1959 V23 N3 P365-386 64-13564 <=*$>
1959 V23 N3 P387-420 62-24291 <=*>
1959 V23 N5 P643-660 60-25595 <=>
1959 V23 N6 P823-840 62-23000 <=>
1959 V23 N6 P913-924 62-19865 <=>
1960 V24 P3-42 AMST S2 V18 P341-382 <AMS>
1960 V24 P43-74 AMST S2 V40 P1-37 <AMS>
1960 V24 P75-92 AMS-TRANS V43 P11-30 <AMS>
AMST S2 V43 P11-30 <AMS>
1960 V24 P153-170 AMST S2 V52 P151-170 <AMS>
1960 V24 P357-368 AMST S2 V58 P110-124 <AMS>
1960 V24 P387-420 AMS TRANS V44 P155 <AMS>
AMST S2 V44 P155-191 <AMS>
1960 V24 P421-430 AMS TRANS V44 P1-11 <AMS>
AMST S2 V44 P1-11 <AMS>
1960 V24 P475-492 AMST S2 V58 P125-142 <AMS>
1960 V24 P493-510 AMST S2 V59 P1-22 <AMS>
1960 V24 P511-520 AMST S2 V45 P233-244 <AMS>
1960 V24 P531-548 AMS-TRANS V43 P31-50 <AMS>
AMST S2 V43 P31-50 <AMS>
1960 V24 P605-616 AMS-TRANS V43 P155 <AMS>
AMST S2 V43 P155-167 <AMS>
1960 V24 P629-706 AMS TRANS V.46 P65 <AMS>
AMST S2 V46 P65-148 <AMS>
1960 V24 P777-786 AMST S2 V58 P93-103 <AMS>
1960 V24 P825-864 AMS TRANS V.40 P39 <AMS>
AMST S2 V40 P39-84 <AMS>
1960 V24 P921-942 AMST S2 V39 P241-265 <AMS>
1960 V24 N1 P3-42 62-10147 <*=>
1960 V24 N1 P43-74 AMS TRANS V.40 P1 <AMS>
1960 V24 N2 P145-152 62-23156 <=*>
1960 V24 N2 P213-242 62-15889 <=*>
1960 V24 N3 P315-356 61-28520 <*=>
62-10145 <*=>
1960 V24 N5 P721-742 61-16454 <*=>
1961 N1 P164-165 61-27214 <*=>
1961 V25 P3-20 AMS-TRANS V42 P175 <AMS>
AMST S2 V42 P175-194 <AMS>
1961 V25 P21-86 AMS TRANS V.46 P213 <AMS>
AMST S2 V46 P213-284 <AMS>
1961 V25 P87-112 AMS TRANS V.40 P127 <AMS>
AMST S2 V40 P127-156 <AMS>
1961 V25 P153-172 AMST S2 V45 P245-264 <AMS>
1961 V25 P239-252 AMS TRANS V44 P12-28 <AMS>
AMST S2 V44 P12-28 <AMS>
1961 V25 P531-542 AMS TRANS V44 P115 <AMS>
AMST S2 V44 P115-128 <AMS>
1961 V25 P557-590 AMS TRANS V44 P29-66 <AMS>
AMST S2 V44 P29-66 <AMS>
1961 V25 P765-788 64-19999 <=$>
1961 V25 P789-796 AMST S2 V59 P73-81 <AMS>
1961 V25 P809-814 AMST S2 V45 P139-145 <AMS>
1961 V25 P815-824 AMST S2 V39 P287-298 <AMS>
1961 V25 P825-870 AMS TRANS V44 P192 <AMS>
AMST S2 V44 P192-240 <AMS>
1961 V25 P899-924 AMST S2 V39 P83-110 <AMS>
1961 V25 N4 P477-498 62-13646 <*=>
62-18955 <=*>
1962 V26 P5-52 AMST S2 V47 P217-267 <AMS>
1962 V26 P87-106 AMST S2 V39 P111-132 <AMS>
1962 V26 P107-124 AMS-TRANS-V43 P299 <AMS>
AMST S2 V43 P299-320 <AMS>
1962 V26 P281-292 AMST S2 V50 P127-140 <AMS>
1962 V26 P391-414 AMST S2 V48 P229-254 <AMS>
1962 V26 P427-452 AMST S2 V58 P216-244 <AMS>
1962 V26 P453-484 AMST S2 V55 P207-241 <AMS>
1962 V26 P495-512 AMST S2 V54 P125-144 <AMS>
1962 V26 P513-530 AMST S2 V39 P37-56 <AMS>
1962 V26 P605-624 STMSP V5 P285-307 <AMS>
1962 V26 P631-638 AMS TRANS V.46 P153 <AMS>
AMST S2 V46 P153-161 <AMS>
1962 V26 P677-686 AMST S2 V58 P143-154 <AMS>
1962 V26 P753-780 AMST S2 V47 P268-299 <AMS>
1962 V26 P797-824 AMST S2 V53 P221-252 <AMS>
1962 V26 P865-876 AMST S2 V55 P33-48 <AMS>
1962 V26 P877-910 AMST S2 V50 P141-177 <AMS>
1962 V26 P911-924 AMST S2 V58 P245-260 <AMS>
1962 V26 N3 P361-390 63-23443 <*=>
1963 V27 P3-8 AMST S2 V58 P104-109 <AMS>
1963 V27 P45-60 AMST S2 V56 P19-36 <AMS>
1963 V27 P161-240 AMST S2 V56 P103-192 <AMS>
1963 V27 P305-328 AMST S2 V54 P255-281 <AMS>
1963 V27 P561-576 AMST S2 V59 P111-127 <AMS>
1963 V27 P677-700 AMST S2 V53 P81-108 <AMS>
1963 V27 P855-882 AMST S2 V52 P171-200 <AMS>
1963 V27 P883-906 AMST S2 V54 P91-118 <AMS>
1963 V27 P937-942 AMST S2 V50 P59-65 <AMS>
1963 V27 P1055-1080 AMST S2 V53 P139-166 <AMS>
1963 V27 P1135-1164 AMST S2 V53 P253-284 <AMS>
1963 V27 P1181-1185 AMST S2 V49 P86-91 <AMS>
1963 V27 P1189-1210 AMST S2 V54 P153-176 <AMS>
1963 V27 P1343-1394 AMST S2 V54 P177-230 <AMS>
1963 V27 P1395-1440 AMST S2 V50 P189-234 <AMS>
1963 V27 N1 P61-66 63-18709 <=*>
1963 V27 N1 P101-160 63-31010 <=>
1964 V28 P91-122 AMST S2 V55 P270-304 <AMS>
1964 V28 P123-146 STMSP V6 P9 <AMS>
1964 V28 P249-260 STMSP V6 P257 <AMS>
1964 V28 P261-272 AMST S2 V48 P91-102 <AMS>
66-11173 <*>
1964 V28 P273-276 AMST S2 V48 P103-106 <AMS>
66-11172 <*>
1964 V28 P307-364 AMST S2 V52 P217-275 <AMS>
1964 V28 P365-474 AMST S2 V48 P271-396 <AMS>
1964 V28 P527-552 AMST S2 V53 P109-138 <AMS>
1964 V28 P571-582 AMST S2 V50 P66-76 <AMS>
1964 V28 P665-706 AMST S2 V56 P193-232 <AMS>
1964 V28 P1055-1082 AMST S2 V49 P130-161 <AMS>
1964 V28 P1363-1390 AMST S2 V59 P82-110 <AMS>
1964 V28 N2 P307-364 64-31491 <=>
1964 V28 N3 P481-514 64-41532 <=>
1965 V29 P379-436 AMST S2 V51 P215-272 <AMS>
1965 V29 P1147-1202 AMST S2 V58 P155-215 <AMS>

IZVESTIYA AKADEMII NAUK TADZHIKSKOI SSR. DUSHANBE(STALINABAD). OTDELENIE ESTESTVENNYKH NAUK
1956 V13 P65-75 AEC-TR-3867 <+*>

IZVESTIYA AKADEMII NAUK TADZHIKSKOI SSR. DUSHANBE(STALINABAD). OTDELENIE SELSKOKHOZYAISTVENNYKH I BIOLOGICHESKIKH NAUK
1961 N1 P87-93 65-60402 <=$>

IZVESTIYA AKADEMII NAUK TURKMENSKOI SSR. ASHKHABAD
1957 N3 P3-12 60-21075 <=>
1957 N6 P3-8 60-21014 <=>
1958 N4 P104-105 59-14129 <+*> 0
1958 N5 P108 60-13655 <+*>
1959 N1 P9-21 NLLTB V1 N9 P12 <NLL>
1959 N1 P45-52 65M42R <ATS>
1959 N1 P104-110 60-21015 <=>

IZVESTIYA AKADEMII NAUK TURKMENSKOI SSR. ASHKHABAD SERIYA BIOLOGICHESKIKH NAUK
1961 N2 P86-88 62-24401 <=>
1962 N1 P21-30 65-50030 <=*$>
1963 N2 P31-34 64-19676 <=$>
1963 N3 P38-42 65-60182 <=$>
1963 N5 P82- <CADA>
1964 N3 P94-95 64-41691 <=>
1966 N1 P94-95 66-33356 <=$>

IZVESTIYA AKADEMII NAUK TURKMENSKOI SSR. ASHKHABAD SERIYA FIZIKO-TEKHNICHESKIKH, KHIMICHESKIKH I GEOLOGICHESKIKH NAUK
1960 N6 P141 64-11775 <=>
1961 N1 P117-118 62-11672 <=>
1961 N2 P124 62-13224 <*=>
1961 N2 P125-126 62-13212 <*=>
1961 N2 P127 62-13219 <*=>
1963 N1 P30-37 64-14453 <=*$>
65-18181 <*>
1963 N4 P117-119 AD-620 946 <=*$>
1964 N2 P21-26 AD-622 422 <=$*>
1964 N3 P25-31 ICE V5 N2 P227-231 <ICE>
1964 N5 P3-11 65-30773 <=$>
1964 N5 P61-68 66-12780 <*> 0
1965 N4 P26 66-11145 <*>
1965 N6 P75-81 66-14412 <*>
1966 N1 P56-61 ICE V6 N3 P530-533 <ICE>
1966 N1 P128-129 66-33783 <=$>
1966 N3 P105 66-33169 <=$>

IZVESTIYA AKADEMII NAUK UZBEKSKOI SSR. TASHKENT
1956 N2 P55-63 59-17235 <+*>

IZVESTIYA AKADEMII NAUK UZBEKSKOI SSR. TASHKENT. SERIYA FIZIKO-MATEMATICHESKIKH NAUK
1957 N3 P11-26 AEC-TR-5668 <=*>
1957 N3 P47-62 61-27042 <*=>
1957 N3 P87-98 59-22669 <+*>
1957 N4 P39-54 65-61301 <=$>
1958 N1 P47-63 60-11769 <=>
1958 N2 P69-76 63-20074 <=*>
1958 N2 P107-115 61-23629 <*=>
1958 N5 P5-13 61-23688 <*=>
1958 N5 P15-22 61-23688 <*=>
1958 N6 P49-55 61-23631 <*=>
1959 N1 P25-30 59-00651 <*>
1959 N1 P45-53 62-11186 <=>
1959 N2 P14-25 62-19999 <=*>
1959 N2 P47-50 62-11349 <=>
1959 N4 P16-25 62-19905 <=>
1959 N6 P27-40 62-11545 <=>
1959 N6 P52-59 62-11188 <=>
1959 N6 P60-71 62-23027 <=*>
1960 N3 P38-43 61-11735 <=>
1960 N3 P52-55 62-11502 <=>
1960 N4 P31-37 AMS TRANS V.46 P57 <AMS>
AMST S2 V46 P57-64 <AMS>
1960 N4 P52-58 61-28698 <*=>
1960 N4 P89-92 62-11376 <=>
1960 N6 P78-92 62-33516 <=*> 0
1961 N5 P90-91 62-33631 <=*>
1961 N6 P23-33 63-31540 <=>
1961 N6 P34-39 62-32681 <=*>
1961 N6 P50-56 62-32681 <=*>
1961 N6 P57-64 63-19203 <=*>
1961 N6 P80-81 AD-618 049 <=$>
1962 N3 P75-80 63-19339 <=*>
1963 V7 N3 P29-36 64-71139 <=>
1963 V7 N4 P59-65 AEC-TR-6339 <=*$>
1963 V7 N6 P46-50 AD-610 420 <=$>
1964 V8 N1 P53-60 64-71132 <=>
1964 V8 N1 P92-94 AD-620 096 <=$*>
1964 V8 N3 P61-66 49T90R <ATS>
1964 V8 N4 P53-61 65-13692 <*>
1964 V8 N4 P62-72 N65-27701 <=*$>
1965 N1 P57-59 AD-634 814 <=$>
1965 V9 N1 P111-112 AD-625 294 <=*$>

IZVESTIYA AKADEMII NAUK UZBEKSKOI SSR. TASHKENT. SERIYA MEDITSINSKAYA
1959 N6 P3-6 60-31068 <=>

IZVESTIYA AKADEMII NAUK UZBEKSKOI SSR. TASHKENT. SERIYA TEKHNICHESKIKH NAUK
1959 N2 P48-56 61-13402 <+*>
1960 N5 P35-62 65-60488 <=$>
1961 N3 P66-70 62-15670 <=*>
1961 N4 P46-51 64-23543 <=$>
1961 N5 P30-39 62-32684 <=*>
1962 N5 P44-52 64-71496 <=> M
1963 N2 P34-40 64-19122 <=*$>
1963 N5 P35-45 RTS-3119 <NLL>
1964 N3 P43-53 AD-625 152 <=*$>
1964 V6 P66-73 AD-620 805 <=*>
1966 N3 P52-58 SHSP N1 1966 P93 <AGU>

IZVESTIYA AKADEMII PEDAGOGICHESKIKH NAUK. RSFSR
1954 N53 P108-123 RT-4468 <*>
1954 N53 P165-170 R-161 <*>
1954 N53 P171-180 R-162 <*>
1954 N53 P229-264 RT-4467 <*>
1954 V53 P202-228 RT-4466 <*>
1957 N85 P206-208 64-21024 <=>

IZVESTIYA AKADEMII STROITELSTVA I ARKHITEKTURY SSSR. MOSCOW
1959 N2 P22-40 NLLTB V2 N4 P256 <NLL>
1959 N2 P111-113 60-11208 <=>
1962 V4 N4 P3-8 NLLTB V5 N8 P710 <NLL>

IZVESTIYA ASTROFIZICHESKOGO INSTITUTA, AKADEMIYA NAUK KAZAKHSKOI SSR
1956 V3 N4 P89-98 61-19339 <+*> C
1957 V5 N7 P66-79 61-27031 <*=>
1957 V5 N7 P144-153 61-27031 <*=>
1958 V6 P3-72 60-21744 <=> 0
1958 V6 P92-102 60-19020 <+*>
1959 V8 P3-12 AIAAJ V1 N5 P1250 <AIAA>
1959 V8 P3-11 60-11477 <=>
1959 V8 P13-18 60-11477 <V>
1959 V8 P53-58 62-14778 <=*>
1959 V8 P68-81 60-11472 <=>
1959 V8 P82-97 60-11478 <=>
1959 V8 P98-107 65-13387 <*>
1959 V9 P10-20 ARSJ V31 N7 P970 <AIAA>
1960 N9 P3-9 60-41446 <=>
1960 N9 P35-39 60-41446 <=>
1960 N9 P53-55 60-41446 <=>
1960 V9 P86-117 60-41446 <=>
1961 N12 P37-55 62-23405 <=*>
1961 V12 P91-94 62-11724 <=>
1962 V14 P113-118 64-13165 <=*$> 0

IZVESTIYA ASTRONOMICHESKOI OBSERVATORII. PULKOVO SEE IZVESTIYA GLAVNOI ASTRONOMICHESKOI OBSERVATORII V PULKOVE

IZVESTIYA ASTRONOMICHESKOGO SOVETA KOMISSII PO FIZIKE PLANET
1959 N1 P81-84 ARSJ V32 N3 P488 <AIAA>
62-14795 <=*>
1959 N1 P85-92 AIAAJ V1 N10 P2433 <AIAA>
1959 V1 P55-58 AD-610 051 <=$>
1959 V1 P67-79 AD-607 724 <=>
1960 V2 P46-54 63-15568 <=*> 0
1960 V2 P65-72 63-15568 <=*> C
1961 V3 P16-30 N64-24832 <=>
64-19752 <=>
1961 V3 P31-40 AD-616 538 <=$>
1961 V3 P50-55 AD-612 878 <=$>

1961 V3 P68-73 63-19650 <=*>

IZVESTIYA BIOLOGO-GEOGRAFICHESKOGO NAUCHNO-ISSLEDOVATELSKOGO INSTITUTA
1933 V6 P109-133 61-28093 <=>

IZVESTIYA NA BULGARSKATA AKADEMIYA NA NAUKITE. FIZICHESKI INSTITUT
1962 V10 N2 P29-36 64-71166 <=>

IZVESTIYA DNEPROPETROVSKOGO GORNOGO INSTITUTA
1955 V24 N3 P191-200 62-24072 <=*>

IZVESTIYA ELEKTROPROMYSHLENNOSTI SLABOGO TOKA
1935 V4 N10 P2-12 RT-1864 <*>
1936 N1 P47-55 60-10220 <+*> 0
1940 N6 P1-16 R-4048 <*>

IZVESTIYA ELEKTROTEKHNICHESKOGO INSTITUTA. LENINGRAD
1964 N52 P141-150 AD-647 746 <=>
1964 N52 P202-204 AD-647 746 <=>

IZVESTIYA ENERGETICHESKOGO INSTITUTA
1940 V9 P23-29 61-16821 <=*> P

IZVESTIYA ESTESTVENNO-NAUCHNOGO INSTITUTA GOSUDARSTVENNYI UNIVERSITET. PERM
1957 V14 N1 P107-128 63-24210 <=>

IZVESTIYA NA FIZICHESKIYA INSTITUT S ANEB
1964 V12 N1/2 P5-12 N65-27718 <=*$>

IZVESTIYA FIZIKO-KHIMICHESKOGO NAUCHNO-ISSLEDOVATELSKOGO INSTITUTA PRI IRKUTSKOM GOSUDARSTVENNON UNIVERSITETE
1953 V2 N1 P5-9 25K22R <ATS>

IZVESTIYA FIZIKO-MEKHANICHESKOGO-MATEMATICHESKOGA FAKULTETA MOSKOVSKOGO UNIVERSITETA
1930 V3 N1A P103-127 64-19193 <=$>

IZVESTIYA GLAVNOI ASTRONOMICHESKOI OBSERVATORII. KIEV
1960 V3 N1 P94-104 62-25823 <=*>
1960 V3 N2 P27-76 63-21016 <=>
1960 V3 N2 P138-150 63-21016 <=>
1961 V4 N1 P3-11 62-33212 <=*>
1963 V5 N1 ENTIRE ISSUE N65-18507 <=$>

IZVESTIYA GLAVNOI ASTRONOMICHESKOI OBSERVATORII V PULKOVE
1954 V19 N5 P2-8 59-10120 <=*> 0
1955 V19 N6 P1-30 R-1464 <*>
1956 N4 P139-144 R-5351 <*>
1956 V20 N2 P22-45 63-23114 <=*>
1957 V21 N1 P116- 61-18839 <=*>
1960 V21 N4 P2- 61-18838 <=*>
1960 V21 N4 P73-87 63-15302 <=*>
1960 V21 N4 P96-105 63-15302 <=*>
1960 V21 N6 P2-11 63-15294 <=*>
1960 V21 N6 P83-113 63-13225 <=*>
1960 V22 N1 P3-122 66-11675 <=>
1960 V22 N1 P135-137 66-11675 <=>
1960 V22 N1 P151-156 66-11675 <=>
1960 V22 N1 P176-183 66-11675 <=>
1960 V22 N1 P184-196 63-15286 <=*>
1960 V22 N1 P197-208 66-11675 <=>
1961 V22 N3 P45-50 63-11676 <=>
1961 V22 N3 P56-64 63-11676 <=>
1961 V22 N3 P98-146 63-11676 <=>
1962 V23 N1 P169-178 AD-618 639 <*>
1962 V23 N2 P3-30 63-11677 <=>
1962 V23 N2 P41-55 63-11677 <=>
1962 V23 N2 P77-106 63-11677 <=>
1962 V23 N2 P117-167 63-11677 <=>
1963 V23 N2 P110-114 AD-614 395 <=$>
1963 V23 N2 P167-174 AD-615 222 <=$>
1964 V23 N3 P22-24 65-63510 <=$>
1964 V23 N5 P86-92 AD-639 485 <=$>
1964 V23 N5 P144-154 65-64609 <=*$>
1964 V24 N1 P38-56 AD-627 451 <=*$>
1964 V24 N177 P171-179 RTS-3148 <NLL>

IZVESTIYA GOSUDARSTVENNOGO INSTITUTA OPYTNOI AGRONOMII
1923 V1 N3 P106-107 60-13244 <=*>
1929 V6 N3/4 P113-114 60-13250 <=*>

IZVESTIYA GOSUDARSTVENNOGO RUSSKOGO GEOGRAFICHESKOGO OBSHCHESTVA
1882 V7 P187-188 62-25446 <=*>
1882 V7 P233-244 62-25446 <=*>
1884 V8 P225-243 62-15981 <=*>
1884 V8 P285-296 62-15981 <=*>
1884 V8 P332-335 62-15981 <=*>
1884 V8 P344-346 62-15981 <=*>
1886 V9 P68-89 63-13560 <=*>
1886 V9 P201-202 63-13560 <=*>
1886 V9 P205 63-13560 <=*>
1886 V9 P406-412 63-13560 <=*>
1896 V32 N1 P367-370 62-25434 <=*>
1896 V32 N6 P510-524 62-25435 <=*>
1901 V37 N3 P225-229 62-25451 <=*>
1903 V39 N4 P295-304 62-25453 <=*>
1903 V39 N5 P508-542 62-25433 <=*>
1908 V44 N6 P317-321 62-25437 <=*>
1908 V44 N1/2 P55-61 62-25449 <=*>
1909 V45 N10 P771-773 62-19039 <=*>
1912 V48 N1/5 P171-188 62-25448 <=*>
1912 V48 N1/5 P343-355 62-25448 <=*>
1914 V50 N7 P429-433 62-25447 <=*>
1915 V51 N3 P139-148 62-25452 <=*>
1918 V54 N1 P3-14 62-25436 <=*>
1918 V54 N1 P121-130 62-25436 <=*>
1928 V60 N2 P323-326 61-28846 <=*>
1931 V63 N5/6 P379-401 61-28844 <=*>
1936 V68 N3 P338-366 62-33413 <=*>

IZVESTIYA HOLOVNOI ASTRONOMICHNOI OBSERVATORII. AKADEMIYA NAUK URSR. KIEV
SEE IZVESTIYA GLAVNOI ASTRONOMICHESKOI OBSERVATORII. KIEV

IZVESTIYA IMPERATORSKOI AKADEMII NAUK
1897 S5 V7 P473-480 61-00596 <*>
1913 P153-162 II-1016 <*>
1913 N3 P153-162 RT-3399 <*>

IZVESTIYA NA INSTITUTA PO FIZIKOKHIMIYA. BULGARSKA AKADEMIIA NA NAUKITE. SOFIA
1962 V2 P93-100 64-71156 <=>

IZVESTIYA INSTITUTA PO OBSHTA I NEORGANICHNA KHIMIIA. BULGARSKA AKADEMIIA NA NAUKITE. SOFIA
1963 N1 P51-59 ICE V4 N1 P22-26 <ICE>
1964 V2 P131-139 ICE V5 N1 P31-36 <ICE>
1964 V2 P141-147 ICE V5 N1 P.19-22 <ICE>
ICE V5 N1 P19-22 <ICE>

IZVESTIYA INSTITUTA POCHVOVEDENIYA I GEOBOTANIKI SREDNE-AZIATSKOGO GOSUDARSTVENNOGO UNIVERSITETA TASHKENT
1924 V2 P495-496 R-2479 <*>
1924 V2 P496-498 R-2483 <*>
1933 V6 P159-168 R-2497 <*>

IZVESTIYA IRKUTSKOGO GOSUDARSTVENNOGO NAUCHNO-ISSLEDOVATELSKOGO PROTIVOCHUMNOGO INSTITUTA SIBIRI I DALNEGO VOSTOKA
1957 V15 P319-321 66-61742 <=$>
1959 V21 P108-121 63-23701 <=*$>
1959 V21 P122-127 63-23063 <=*>
1959 V21 P178-180 AD-616 767 <=$>

IZVESTIYA KARELSKOGO I KOLSKOGO FILIALOV AKADEMII NAUK SSSR
1959 N4 P157-159 60-11720 <=>

IZVESTIYA KAZANSKOGO FILIALA AKADEMII NAUK SSSR

1953 N3 P18-38 RJ-412 <ATS>
1957 N3 P57-61 60M38R <ATS>
1959 N7 P11-33 04Q69R <ATS>
1960 N14 P81-88 65-13676 <*>

IZVESTIYA NA KHIMICHESKIYA INSTITUT
1956 V4 P327-337 71K25B <ATS>

IZVESTIYA KHLOPCHATO-BUMAZHNOI PROMYSHLENNOSTI
1932 V2 N2 P19-22 59-15037 <*> O
1932 V2 N3 P36-39 59-15037 <*> O

IZVESTIYA KIEVSKOGO POLITEKHNICHESKOGO INSTITUTA
1956 V17 P130-133 60-23024 <+*>
1958 V28 P30-33 64-13743 <=*$>
1962 V41 P114-121 65-13056 <*>

IZVESTIYA KOMISSII PO FIZIKE PLANET
SEE IZVESTIYA ASTRONOMICHESKOGO SOVETA KOMISSII PO FIZIKE PLANET

IZVESTIYA KRYMSKOI ASTROFIZICHESKOI OBSERVATORII
1947 V1 P102-115 AD-621 197 <=$*>
1948 V3 P3-30 60-23863 <+*> O
1948 V3 P51-63 60-23863 <+*> O
1949 V4 P80-113 60-19069 <+*> O
1949 V4 N3 P3-22 60-13125 <+*>
1950 V5 P3-23 60-19070 <+*> O
1950 V5 P34-58 61-27011 <*=>
1950 V5 P109-134 59-10115 <+*>
61-15260 <*+>
1950 V5 N5 P86-99 59-19523 <+*>
1951 V7 P3-33 60-19071 <+*> O
1951 V7 P99-112 60-19071 <+*> O
1952 V8 P19-50 60-19072 <+*> O
1952 V8 P119-122 62-27425 <=*> O
1952 V9 P25-40 60-23864 <+*> O
1952 V9 P93-114 60-23864 <+*> O
1953 V10 P9-53 AD-637 808 <=$>
1954 V12 P3-32 R-478 <*>
R-4880 <*>
R-5090 <*>
1954 V12 P33-45 63-23113 <=*>
1954 V12 P93-116 RT-4524 <*>
1954 V12 P169-176 66-11070 <*>
1955 N13 P96-102 R-4399 <*>
1955 V13 P82-95 R-4453 <*>
R-4883 <*>
59-12191 <+*>
1955 V13 P96-102 59-12193 <+*>
1955 V15 P147-152 AD-610 116 <=$>
1956 N16 P159-161 64-00470 <*>
1956 V16 P3-11 60-14230 <+*>
1956 V16 P12-44 62-33205 <=*>
1956 V16 P80-99 61-19016 <=*>
1956 V16 P148-158 61-19008 <+*>
1956 V16 P159-161 64-11838 <=>
1957 V17 P129-161 60-14228 <+*>
1957 V17 P191-198 AD-621 193 <=$>
1957 V17 P232-241 61-19009 <+*>
1958 N18 P61-65 64-00474 <*>
1958 V18 P61-65 64-11825 <=>
1958 V18 P136-150 63-11687 <=>
1958 V19 P3-19 61-19379 <+*>
1958 V19 P20-45 61-27004 <*=>
1958 V19 P46-71 60-14231 <+*>
1958 V19 P72-99 60-14232 <+*>
1958 V19 P105-114 62-32232 <=*>
1958 V19 P126-139 61-27004 <*=>
1958 V19 P153-164 62-33376 <=*>
1958 V20 P12-21 61-23709 <=*>
61-27023 <*=>
1958 V20 P22-51 60-14229 <+*>
1958 V20 P52-66 60-14233 <+*>
1958 V20 P80-85 60-14234 <+*>
1958 V20 P123-155 62-32476 <=*>
1958 V20 P156-206 62-32951 <=*>
1959 V21 P24-39 62-33412 <=*>
1959 V21 P152-179 62-32954 <=*>
1959 V21 P180-189 RTS-2799 <NLL>
1959 V21 P190-197 AD-619 159 <=$>
1960 N22 P9-11 NASA-TT-F-137 <=>
1960 N22 P42-48 NASA-TT-F-138 <=>
1960 V22 P3-8 61-23978 <*=>
1960 V22 P12-41 61-28637 <*=>
63-13384 <=*>
1960 V22 P49-55 61-19680 <=*>
1960 V22 P67-74 61-23979 <*=>
1960 V22 P75-80 62-33066 <=*>
1960 V23 P184-211 63-16438 <=*>
1960 V23 P212-252 61-23144 <=*>
1960 V23 P253-276 61-23643 <=*>
1960 V23 P277-290 61-19674 <=*>
1960 V23 P291-298 62-24627 <=*>
1960 V23 P322-330 AD-621 196 <=$>
62-24626 <=*>
1960 V24 P3-15 N64-31082 <=>
1960 V24 P32-47 62-33399 <=*>
1960 V24 P52-77 RTS-2800 <NLL>
1960 V24 P235-257 AD-624 083 <=*$>
1960 V24 P281-292 62-13609 <*=>
1960 V24 P293-300 RTS 2801 <NLL>
1961 N25 P76-87 64-14752 <=*$>
1961 V25 P114-121 RTS-2802 <NLL>
1961 V25 P154-173 63-19147 <=*>
1961 V25 P174-179 RTS 2803 <NLL>
1961 V25 P180-233 AD-624 084 <=*$>
1961 V25 P234-248 AD-624 049 <=*$>
1961 V25 P277-280 RTS-2804 <NLL>
1961 V26 P41-44 AD-621 198 <=$*>
1962 N27 P120-139 64-13303 <=*$>
1962 N28 P166-193 64-13082 <=*$>
1962 N28 P194-223 NASA-TT-F-144 <=>
64-13067 <=*$>
1962 N28 P259-270 64-13068 <=*$>
1962 V27 P5-43 63-11572 <=>
1962 V27 P44-51 63-11561 <=>
1962 V27 P71-108 AD-624 076 <=*$>
1962 V27 P120-139 AD-621 195 <=$>
1962 V27 P148-161 RTS-2816 <NLL>
1962 V27 P167-177 N65-34019 <=$>
1962 V27 P309-317 AD-615 336 <=>
1962 V28 P159-165 63-24075 <=*>
1962 V28 P224-229 RTS-2813 <NLL>
1962 V28 P271-276 RTS-2815 <NLL>
1962 V28 N1 P166-193 65-60332 <=>
1962 V28 N1 P194-223 65-60340 <=$>
1963 V29 P3-14 64-15374 <=*$>
1963 V29 P15-67 65-63641 <=$>
1963 V29 P80-85 N64-24836 <=>
1963 V29 P141-145 AD-637 812 <=$>
1963 V29 P146-151 AD-637 811 <=$>
1963 V29 P152-159 AD-637 813 <=$>
1963 V30 P3-48 AD-611 043 <=$>
1963 V30 P19-24 65-30937 <=$>
1963 V30 P141-147 64-19897 <=$>
1963 V30 P161-184 RTS-3144 <NLL>
1963 V30 P267-272 RTS-2798 <NLL>
1964 V31 P118-125 65-60680 <=$>
1964 V31 P159-199 RTS-2806 <NLL>
1964 V31 P200-208 N65-11302 <=$>
RTS-2807 <NLL>
1964 V31 P209-215 RTS 2808 <NLL>
65-63525 <=$> C
1964 V31 P216-246 RTS-2809 <NLL>
1964 V31 P259-270 66-10804 <*>
1964 V31 P276-280 N65-11443 <=$>
1964 V31 P281-324 RTS-2876 <NLL>
1964 V32 P3-10 N65-17300 <=$>
1964 V32 P173-186 N65-17269 <=$>
1964 V32 P187-191 AD-630 280 <=*$>
1965 V33 P92-99 66-61567 <=$>

IZVESTIYA LENINGRADSKOGO ELEKTROTEKHNICHESKOGO INSTITUTA IM. V. I. ULYANOVA
1959 N38 P175-186 63-23346 <=*>
1963 N51 P94-106 AD-612 370 <=$>

IZVESTIYA NA MATEMATICHESKIYA INSTITUT. SOFIA
1960 V4 N2 P57-62 62-19467 <=*>

1960 V4 N2 P75-78 62-19467 <=*>

IZVESTIYA NA MIKROBIOLOGICHESKIYA INSTITUT. SOFIA
1951 V2 P3-28 57-1756 <*>

IZVESTIYA MOSKOVSKOGO ORDENA LENINA SELSKOKHOZYAISTVENNOI AKADEMII IM. K. A. TIMIRYAZEVA
SEE IZVESTIYA SELSKOKHOZYAISTVENNOI AKADEMII IMENI K. A. TIMIRYAZEVA

IZVESTIYA NA NARODNIYA SUBRANIE, BULGARIA
1960 V11 N63 P1 62-19292 <=*>
1960 V11 N70 P2-6 62-19292 <=*>
1960 V11 N96 P1 62-19401 <=*>
1960 V11 N105 P1-3 62-19402 <=*>
1961 V12 N10 P1-3 62-19403 <=*>
1961 V12 N11 P5-6 62-19404 <=*>
1962 V13 N80 P1-3 62-33690 <=*>
1962 V13 N90 P1-8 63-13521 <=*>

IZVESTIYA NAUCHNO-ISSLEDOVATELSKOGO INSTITUTA GIDROTEKHNIKI. LENINGRAD-MOSCOW
1934 V11 P75-94 RT-2893 <*>
1934 V14 P195-209 RT-259 <*>
1935 V15 P157-184 RT-1088 <*>
1936 V18 P44-49 RT-246 <*>
1936 V19 P5-24 RT-260 <*>
1936 V19 P25-48 RT-258 <*>

IZVESTIYA NAUCHNO-ISSLEDOVATELSKOGO INSTITUTA POSTOYANNOGO TOKA: SBORNIK. MOSCOW
1957 V2 P5-21 60-13790 <+*>
1957 V2 P84-95 60-13900 <*+>
1957 V2 P96-111 60-13901 <*+>
1958 V3 P5-19 61-13553 <=*>
1958 V3 P100-114 61-13513 <=*>
1958 V3 P129-142 61-13554 <=*>
1958 V3 P161-180 61-13555 <=*>
1958 V3 P181-196 60-13898 <*+>
1958 V3 P197-200 60-13897 <*+>
1958 V3 P201-209 60-13896 <*+>
1958 V3 P210-224 61-13556 <=*>
1958 V3 P234-254 61-28165 <*=>
1958 V3 P255-281 62-13602 <*=>
1959 V4 P5-18 62-13740 <*=>
1959 V4 P65-75 61-13759 <=*>
1959 V4 P97-113 62-13094 <*=>
1959 V4 P125-152 61-28464 <*=>
1960 V5 P23-63 62-13730 <*=>
1960 V6 P149-154 62-32775 <=*>
1960 V6 P155-163 C-4130 <NRCC>
62-32774 <=*>
1961 V7 P5-13 63-19210 <=*>
1962 V9 P5-27 65-63177 <=$>
1962 V9 P133-143 65-63724 <=*$>

IZVESTIYA OTDELENIYA MATEMATICHESKIKH I ESTESTVENNYKH NAUK AKADEMII NAUK SSSR
1934 N5 P711-732 60-14061 <*>
1935 P1224-1226 62-25063 <=$>

IZVESTIYA PETROGRADSKOGO LESNOGO INSTITUTA
1927 V35 P69-104 63-19969 <=*>

IZVESTIYA PETROGRADSKOGO NAUCHNOGO INSTITUTA IM. P. F. LESGAFTA
1923 V6 P71-81 RT-3435 <*>

IZVESTIYA. POLITEKHNICHESKOGO INSTITUTA. KIEV
SEE IZVESTIYA KIEVSKOGO POLITEKHNICHESKOGO INSTITUTA

IZVESTIYA POLITEKHNICHESKOGO INSTITUTA. TOMSK
1955 V80 P24-40 60-13637 <+*>
1956 V91 P103-107 62-25822 <=*>
1956 V91 P391-398 UCRL TRANS-947(L) <=*$>
1958 V95 P306-313 ORNL-TR-491 <=$>
1958 V101 P42-46 62-25817 <=*>
1962 V110 P79-86 RTS-2823 <NLL>

IZVESTIYA ROSSIISKOI AKADEMII NAUK
1917 N12 P883-890 RT-796 <*>

IZVESTIYA SEKTORA FIZIKO-KHIMICHESKOGO ANALIZA. INSTITUT OBSHCHEI I NEORGANICHESKOI KHIMII. AKADEMIYA NAUK SSSR
1926 V3 P14-41 64-30816 <*>
1926 V3 P461-462 64-30816 <*>
1933 V6 P169-184 R-2908 <*>
1936 V8 P135-140 RT-2103 <*>
1936 V9 P203-218 66-12063 <*> O
1936 V9 P219-253 66-12064 <*> O
1936 V10 P129-159 61-23515 <*=> O
1940 V13 P201-208 66-12072 <*>
1940 V13 P209-224 SCL-T-244 <*> P
66-12073 <*>
1940 V13 P229-320 63-20766 <=*$>
1940 V13 P299-320 RT-955 <*>
1940 V13 P331-353 R-2499 <*>
1941 V14 P205-210 R-3336 <*>
60-13882 <+*>
1941 V14 P373-386 R-3337 <*>
1941 V14 P411-430 63-14645 <=*>
1947 N15 P80-87 3 <FRI>
1947 V15 P47-57 65-60298 <=$>
1947 V15 P88-95 3980 C <K-H>
64-16445 <=*$> O
1947 V15 P157-199 R-2857 <*>
62-13111 <=*>
1947 V15 P200-233 R-3051 <*>
62-24898 <=*>
1947 V15 P234-265 76M40R <ATS>
1948 V16 N3 P108-126 R-2832 <*>
1949 V17 P209-219 3156 <HB>
1949 V17 P220-227 63-20435 <=*$>
1949 V17 P370-382 62-16438 <=*>
1949 V19 P33-40 RT-3752 <*>
1949 V19 P82-88 62-14847 <=*>
1949 V19 P113-119 R-2790 <*>
RT-2196 <*>
64-14139 <=*$> O
1949 V19 P120-125 RT-3753 <*>
1949 V19 P134-143 R-3041 <*>
1949 V19 P155-161 64-14411 <=*$> O
1949 V19 P419-428 64-71410 <=> M
1950 V20 P109-123 R-5055 <*>
1950 V20 P184-211 AEC-TR-1996 <+*>
1950 V20 P252-268 RJ-381 <ATS>
1950 V20 P317-325 AD-610 423 <=$>
1950 V20 P341-344 61-10290 <*+>
62-18910 <=*>
1950 V20 P383-388 R-2261 <*>
1952 V21 P159-171 R-3494 <*>
1952 V21 P228-249 5035-A <K-H>
64-16438 <=*$> O
1952 V21 P271-287 RJ-379 <ATS>
1953 V22 P92-103 R-4806 <*>
1953 V22 P93-103 514TT <CCT>
1953 V22 P104-110 AEC-TR-3629 <+*>
1953 V23 P110-117 UCRL TRANS-970 <=*$>
1953 V23 P176-182 64-18377 <*>
1953 V23 P233-240 64-14405 <=*$> O
1953 V23 P284-299 RJ-525 <ATS>
1954 V24 P59-123 65-62159 <=$>
1954 V24 P132-147 64-23560 <=>
1954 V24 P237-251 65-64501 <=*$>
1954 V24 P252-263 60-17717 <+*>
1954 V24 P280-298 65-60660 <=$>
79L32R <ATS>
1954 V25 P70-80 12J15R <ATS>
1954 V25 P89-93 59-10096 <+*>
1954 V25 P168-175 5866F <K-H>
64-18405 <*> O
1954 V25 P188-207 RT-3754 <*>
1309 <*>
1954 V25 P350-360 80L32R <ATS>
1954 V25 P375-380 65-63649 <=*$>
1955 V26 P56-61 60-13481 <+*>
1955 V26 P132-137 5866-D <K-H>

	64-18402 <*>
1955 V26 P147-155	RT-4209 <*>
1955 V26 P173-179	R-250 <*>
	RT-4238 <*>
1955 V26 P191-197	RT-4210 <*>
1955 V26 P229-241	65-60383 <=$>
1955 V26 P248-258	65L29R <ATS>
1955 V26 P266-269	18K27R <ATS>
1955 V26 P270-274	AEC-TR-3733 <+*>
	29L29R <ATS>
1956 V27 P126-132	R-1979 <*>
1956 V27 P198-208	59-16987 <+*>
	59-22078 <+*>
1956 V27 P255-267	AEC-TR-4233 <*+>
	64-20199 <*> 0
1956 V27 P402-411	60-17702 <+*>

IZVESTIYA SEKTORA PLATINY I DRUGIKH BLAGORODNYKH METALLOV. INSTITUT OBSHCHEI I NEORGANICHESKOI KHIMII. AKAJEMIYA NAUK SSSR

1948 N22 P145-148	TT-825 <NRCC>
1949 N23 P97-100	TT-826 <NRCC>
1949 V24 P15-25	95N55R <ATS>
1950 V25 P56-66	R-1760 <*>
1952 V27 P62-79	CSIR-TR-461 <CSSA>
1955 N29 P55-60	59-15625 <+*>
1955 V29 P163-182	65-61456 <=>
1955 V29 P197-206	M.5602 <NLL>
1955 V30 P109-119	AEC-TR-4109 <+*>

IZVESTIYA SELSKOKHOZYAISTVENNOI AKADEMII IM. K. A. TIMIRYAZEVA

1928 N2 P427-444	63-16078 <=*> 0

IZVESTIYA SIBIRSKOGO OTDELENIYA AKADEMII NAUK SSSR. NOVOSIBIRSK

1957 N1 P95-108	64-11066 <=>
1957 N8 P67-129	61-28048 <=>
1957 N8 P155-162	61-28048 <=>
1957 N9 P53-56	63-14315 <=*>
1958 N5 P43-52	C-3289 <NRCC>
1958 N7 P75-86	60-13115 <+*>
1958 N8 P3-18	60-13446 <+*>
1958 N9 P44-50	61-21118 <=>
1958 N10 P3-12	23L37R <ATS>
1958 N11 P83-94	63-19914 <=> 0
1958 N12 P51-60	59-13829 <=>
1958 N12 P93-94	63-23192 <=*>
1959 N1 P59-66	62-16145 <=*>
1959 N3 P18-29	61-13056 <*+>
1959 N3 P49-61	69N54R <ATS>
1959 N3 P133-134	60-11627 <=>
1959 N6 P70-77	61-27805 <*=>
1959 N6 P83-88	60-11311 <=>
1959 N7 P3-15	61-15499 <+*>
1959 N7 P126-127	60-11332 <=>
1959 N9 P92-97	60-41003 <=>
1959 N10 P5-14	60-11899 <=>
1959 N11 P50-56	64-13592 <=*$>
1959 N11 P103-104	60-31192 <=>
1959 N11 P105-107	60-31193 <=>
1959 N11 P107-108	60-31194 <=>
1959 N12 P11-24	C-3647 <NRCC>
1960 N1 P3-16	NLLTB V2 N11 P945 <NLL>
	60-11794 <=>
	63-23630 <=*$>
1960 N1 P41-55	62-24678 <=*>
1960 N2 P46-52	AEC-TR-4362 <=*>
	61-27882 <*=>
1960 N2 P71-75	79Q66R <ATS>
1960 N3 P129	61-11034 <=>
1960 N3 P129-132	61-11676 <=>
1960 N3 P131-132	61-11034 <=>
1960 N4 P132-135	60-41211 <=>
1960 N6 P3-16	61-27720 <*=>
1960 N6 P44-52	59P64R <ATS>
1960 N6 P53-58	61-27212 <*=>
1960 N6 P59-64	61-28255 <*=>
1960 N6 P124-126	61-28409 <*=>
1960 N7 P121-123	62-11562 <=>
1960 N8 P153-154	61-31462 <=> 0
1960 N11 P34-40	IGR V7 N2 P196 <AGI>
1960 N11 P144-145	61-21634 <=>
	62-11527 <=>
1960 N12 P67-77	ARSJ V32 N5 P825 <AIAA>
1960 N12 P135-137	61-21601 <=>
1961 N1 P25-30	5153 <HB>
1961 N1 P72-81	62-18620 <=*>
1961 N1 P123-126	61-28302 <*=>
1961 N3 P61-67	62-13750 <*=>
1961 N4 P116-122	61-31653 <=>
	62-13258 <=>
1961 N4 P120-122	62-25945 <=*>
1961 N5 P3-9	62-10642 <=*>
1961 N6 P14-22	63-13901 <=*>
1961 N6 P99-105	62-23763 <=*>
1961 N6 P130-135	62-24257 <=*>
1961 N7 P23-25	63-23347 <=*$>
1961 N8 P22-35	63-23347 <=*$>
1961 N8 P137-139	62-32893 <=*>
1961 N9 P21-27	62-32940 <=*>
1961 N9 P128-130	62-15451 <*=>
1961 N10 P3-9	63-31239 <=>
1961 N10 P137-141	64-31342 <=>
1961 N11 P40-	ICE V2 N4 P520-524 <ICE>
1961 N11 P130-132	AIAAJ V1 N12 P2915 <AIAA>
1961 N12 P30-36	62-18646 <=*>
	63-19483 <=*>
1961 V7 P126-129	62-24788 <=*>
1961 V8 P22-35	63-18011 <=*>
1962 N1 P3-4	62-11718 <=>
1962 N1 P5-10	63-23348 <=*$>
1962 N1 P97-101	62-25967 <=*>
1962 N1 P111-113	62-11718 <=>
	62-25967 <=*>
1962 N2 P37-42	62-25509 <=*>
1962 N2 P131-132	63-31581 <=>
1962 N3 P12-24	AIAAJ V1 N10 P2438 <AIAA>
1962 N4 P3-8	64-13058 <=*$>
1962 N4 P57-63	48S87R <ATS>
1962 N4 P108-114	62-33033 <=*>
1962 N5 P3-13	62-33709 <=*>
1962 N5 P14	ICE V3 N1 P36-44 <ICE>
1962 N5 P76-80	63-10736 <=*>
1962 N6 P46-	ICE V3 N1 P133-138 <ICE>
1962 N6 P99-102	63-19307 <=*>
1962 N8 P27-39	63-18702 <=*>
1962 N8 P103-109	63-21007 <=>
1962 N9 P13-	ICE V3 N2 P235-239 <ICE>
1962 N9 P20-	ICE V3 N2 P225-228 <ICE>
1962 N10 P51-59	63-21222 <=>
1962 N10 P125-138	63-21796 <=>
1962 N11 P59-65	337 <CTS>
1962 N11 P95-108	64-18979 <=*$> 0
1962 N11 P157-159	63-31581 <=>
1962 N12 P145-148	63-21838 <=>
1962 V10 P3-12	NLLTB V5 N7 P609 <NLL>

IZVESTIYA SIBIRSKOGO OTDELENIYA AKADEMII NAUK SSSR. NOVOSIBIRSK. GEOLOGICHESKII KOMITET

1922 V2 N6 P61-69	64-15160 <=*$>
1929 V10 N1 P1-5	64-15171 <=*$>

IZVESTIYA SIBIRSKOGO OTDELENIYA AKADEMII NAUK SSSR. NOVOSIBIRSK. SERIYA BIOLOGO-MEDITSINSKIKH NAUK

1963 N1 P23-46	64-41949 <=>
1963 N1 P53-61	64-41949 <=>
1963 N1 P105-108	64-41949 <=>
1963 N2 P34-43	64-21170 <=>
1963 N2 P62-66	64-21172 <=>
1963 N2 P79-84	64-21172 <=>
1963 N2 P85-90	64-21172 <=>
1963 N2 P99-107	63-41240 <=>
1963 N2 P134-136	63-41350 <=>
1963 N2 P136-139	63-41320 <=>
1963 N2 P139-141	63-41322 <=>
1964 N1 P56-58	64-41969 <=$>
1964 N1 P141-143	64-41969 <=$>
1964 N1 P145-147	64-41969 <=$>

*IZVESTIYA SIBIRSKOGO OTDELENIYA AKADEMII NAUK SSSR. NOVOSIBIRSK. SERIYA KHIMICHESKIKH NAUK
1963 N1 P25-31 63-24436 <=*$>

IZVESTIYA SIBIRSKOGO OTDELENIYA AKADEMII NAUK SSSR. NOVOSIBIRSK. SERIYA OBSHCHESTVENNYKH NAUK
1963 N1 P78-89 64-31123 <=>
1963 N3 P163-165 63-31529 <=>
1963 N5 P121-122 63-31886 <=>
1964 N2 P133-134 AD-626 711 <=*$>

IZVESTIYA SIBIRSKOGO OTDELENIYA AKADEMII NAUK SSSR. NOVOSIBIRSK. SERIYA TEKHNICHESKIKH NAUK
1963 N1 P7-11 64-21809 <=>
1963 N2 P121-123 64-71134 <=>
1964 N2 P30-33 65-30044 <=$>
1964 N2 P39-46 65-30043 <=$>
1964 N2 P93-98 AD-639 556 <=$>
1964 N3 P3-8 65-31544 <=$>
1964 N3 P18-26 65-31544 <=$>
1964 N3 P34-38 65-31544 <=$>
1964 N3 P45-50 65-31544 <=$>
1964 N3 P81-90 65-31544 <=$>
1964 N3 P103-109 65-31544 <=$>
1965 N2 P88-93 66-61787 <=$>

IZVESTIYA TADZHIKSKOGO FILIALA AKADEMII NAUK SSSR
1945 V6 P60-63 61-00459 <*>
64-15617 <=*$>

IZVESTIYA TEPLOTEKHNICHESKOGO INSTITUTA MOSCOW
1935 N7 P37-40 R-2526 <*>
147TM <CTT>
1935 N8 P43-47 R-2525 <*>
1935 V7 P37-40 64-20218 <*>

IZVESTIYA TIKHOOKEANSKOGO NAUCHNO-ISSLEDOVATELSKOGO INSTITUTA RYBNOGO KHOZYAISTVA I OKEANOGRAFII. VLADIVOSTOK
1940 V18 P5-105 C-4232 <NRC>
1946 V22 P259-261 C-2526 <NRC>
1947 V25 P33-51 60-51139 <=>
1948 V27 P29-42 60-51139 <=>
1948 V27 P115-137 60-51139 <=>
1948 V28 P3-27 R-5067 <*>
1948 V28 P43-101 C-4212 <NRC>
63-00874 <*>
1949 V31 P3-57 60-51139 <=>
1949 V31 P173-176 60-51139 <=>
1950 V32 P103-119 63-19734 <=*> 0
1950 V32 P158-160 C-3624 <NRCC>
1951 V34 P33-39 60-51139 <=>
1951 V34 P47-60 60-51139 <=>
1951 V34 P105-121 60-51139 <=>
1951 V34 P123-130 60-51139 <=>
1951 V35 P1-16 R-5075 <*>
1951 V35 P17-31 R-5069 <*>
1951 V35 P41-46 R-5074 <*>
1952 V37 P69-108 R-5076 <*>
1952 V37 P129-137 R-5073 <*>
1952 V37 P252-253 60-51139 <=>
1954 V39 P333-342 60-51139 <=>
1954 V41 P3-109 C-3832 <NRCC>
1954 V41 P111-195 60-21150 <=>
1954 V41 P319-322 63-23340 <=*$> 0
1954 V41 P333-336 <NRCC>
60-21097 <=>
1955 V43 P3-10 60-51139 <=>
1955 V43 P21-42 63-23341 <=*$> 0
1955 V43 P185-187 63-15225 <=*>
1957 V44 P97-110 C-3057 <NRCC>
60-21106 <=>
1957 V44 P253-255 63-23412 <=*$>
1957 V45 P3-16 60-51139 <=>
1957 V45 P29-35 59-15957 <+*>
1957 V45 P37-50 61-31037 <=>
1957 V45 P199-201 59-15955 <+*>
1957 V45 P201-202 59-15956 <+*>
1957 V45 P203-205 59-15958 <+*>
1964 V51 P141-150 65-63362 <=$>

IZVESTIYA TIMIRYAZEVSKOI SELSKO-KHOZYAISTVENNOI AKADEMII
1953 N2 P65-76 RT-4194 <*>
1953 S3 N2 P203-220 R-2228 <*>
1954 N3 P161-173 RT-4573 <*>
1955 V2 N9 P193-210 R-3643 <*>
1956 N2 P131-140 63-11016 <=>
1957 N3 P7-32 59-19600 <+*> 0
1957 N6 P51-60 63-11017 <=>
1958 V4 P89-98 60-51004 <=>
1958 V5 N24 P237-250 60-51023 <=>
1959 N2 P15-24 59-31015 <=> 0
1959 N2 P57-74 61-23031 <*=>
1959 N4 P7-22 C-5133 <NRC>
1959 N6 P239-250 NLLTB V2 N7 P545 <NLL>
60-11691 <=>
1960 N1 P77-88 61-15966 <+*>
1960 V33 N12 P137-156 63-00102 <*>
1961 N2 P7-19 62-13506 <*=>
1961 N2 P232-233 61-27147 <*=>
1961 N2 P234-236 61-27347 <*=>
1961 N2 P237-240 61-27389 <*=>
1962 N2 P165-183 64-11085 <=>
1962 N2 P188-189 63-15461 <=*>
1962 N2 P206-208 65-25351 <$>
1962 N3 P47-58 64-13343 <=*$>
1962 N3 P78-82 64-41080 <=>
1962 N4 P34-52 63-21252 <=>
1963 N3 P103-121 64-41208 <=>
1964 N2 P49-64 66-61551 <=*$>
1965 N4 P220-240 65-33112 <=*$>

IZVESTIYA TOMSKOGO INDUSTRIALNOGO INSTITUTA
1940 V60 N3 P61-78 R-2881 <*>

IZVESTIYA TOMSKOGO POLITEKHNICHESKOGO INSTITUTA
1962 V110 P87-94 RTS-2817 <NLL>
1962 V110 P95-102 RTS-2824 <NLL>

IZVESTIYA NA TSENTRALNATOI LABORATORII PO GEODEZIYA. SOFIA
1959 V2 P1-6 62-19406 <=*>
1959 V2 P127-157 62-19405 <=*>

IZVESTIYA UZBEKISTANSKOGO FILIALA GEOGRAFICHESKOGO OBSHCHESTVO SSSR
1956 V2 P86-99 65-50029 <=$>

IZVESTIYA VOSTOCHNYKH FILIALOV AKADEMII NAUK SSSR
1957 N1 P27-39 IGR V1 N9 P42 <AGI>
1957 N1 P27-33 59-11322 <=>

IZVESTIYA VSESOYUZNOGO GEOGRAFICHESKOGO OBSHCHESTVA LENININGRAD
1940 V72 N1 P3-14 62-19047 <=*>
1944 V76 N5 P226-276 62-19038 <=*>
1946 V78 N2 P171-182 62-19045 <=*>
1947 V79 N2 P189-198 RT-1364 <*>
1947 V79 N3 P317-324 RT-196 <*>
1947 V79 N6 P653-655 RT-2651 <*>
1950 V82 N2 P126-137 63-19128 <=*>
1950 V82 N3 P326-330 62-19037 <=*>
1952 V84 N2 P180-188 RT-2307 <*>
1953 V85 N3 P295-296 RT-2240 <*>
1954 V86 N5 P412-422 RT-2831 <*>
1956 V88 N4 P316-350 60-21825 <=>
1958 N5 P472-474 SGRT V1 N1-2 P40 <AGS>
1958 N6 P534-536 SGRT V1 N1-2 P28 <AGS>
1958 V90 P170-176 C-5467 <NRC>
1958 V90 N1 P25-38 IGR V2 N11 P925 <AGI>
1958 V90 N3 P288-293 PB 141 217T <=>
1959 N1 P51-59 SGRT V1 N4 P3 <AGS>
1959 N1 P109-119 SGRT V1 N4 P17 <AGS>
1959 V91 N1 P27-41 C-3608 <NRCC>
1959 V91 N3 P172-183 AD-631 658 <=$>
1959 V91 N3 P255-265 61-13118 <+*>
1959 V91 N5 P397-409 60-31054 <=>
1960 N1 P81-84 SGRT V1 N9 P67 <AGS>
1960 N3 P216-226 SGRT V2 N1 P63 <AGS>

1960 N6 P512-514 SGRT V2 N7 P42 <AGS>
1961 N1 P79-81 SGRT V2 N10 P57 <AGS>
1961 N3 P224-231 SGRT V2 N9 P20 <AGS>
1961 V93 N1 P69-76 65-50064 <=>
1961 V93 N5 P439-441 62-15413 <*=>
1962 N1 P3-14 SGRT V3 N6 P3 <AGS>
1962 N1 P15-25 SGRT V3 N7 P3 <AGS>
1962 N1 P51-55 SGRT V4 N3 P39 <AGS>
1962 N2 P159-167 SGRT V4 N1 P26 <AGS>
1962 N3 P193-201 SGRT V4 N1 P38 <AGS>
1962 N4 P295-303 SGRT V5 N3 P37 <AGS>
1962 N5 P405-413 SGRT V4 N5 P16 <AGS>
1962 N5 P414-424 SGRT V4 N4 P31 <AGS>
1962 N6 P465-473 SGRT V4 N3 P3 <AGS>
1962 V94 N1 P61-65 C-4055 <NRCC>
1962 V94 N4 P304-318 RTS-3000 <NLL>
1963 N1 P3-8 SGRT V5 N2 P41 <AGS>
1963 N3 P285-292 SGRT V4 N8 P34 <AGS>
1963 N4 P351-352 SGRT V4 N9 P52 <AGS>
1963 N4 P393-394 SGRT V4 N9 P55 <AGS>
1963 N6 P552-555 SGRT V5 N2 P51 <AGS>
1963 V95 N2 P107-114 SOU.HYDROL.1963 P175 <AGU>
1963 V95 N2 P134-142 SOU.HYDROL.1963 P134 <AGU>
1963 V95 N2 P143-153 SOU.HYDROL.1963 P143 <AGU>
1963 V95 N2 P164-168 65-61518 <=$> 0
1963 V95 N2 P168-173 65-61519 <=$> 0
1963 V95 N2 P174-177 SOU.HYDROL.1963 P201 <AGU>
1963 V95 N4 P367-368 C-4601 <NRC>
1963 V95 N5 P406-414 64-15998 <=*$> 0
1964 N1 P74-76 SGRT V5 N5 P52 <AGS>
1964 N2 P91-95 SGRT V5 N6 P85 <AGS>
1964 N5 P445-448 65-64602 <=*$>
1964 V96 N1 P74-76 SOV.GEOGR. V5 N5 P52 <AGS>
1964 V96 N2 P91-95 SOV.GEOGR. V5 N6 P85 <AGS>
1965 N2 P105-111 SGRT V6 N7 P3 <AGS>
1965 N3 P209-221 SGRT V6 N7 P11 <AGS>
1965 N4 P346-352 AD-626 712 <=*$>
1965 N6 P501-506 SGRT V7 N2 P3 <AGS>
1965 N6 P527-533 SGRT V7 N2 P36 <AGS>
1965 V97 N1 P87-92 65-63407 <=$>
1965 V97 N5 P427-437 66-31844 <=$>
1966 N2 P148-156 SGRT V7 N10 P3 <AGS>
1966 N3 P205-211 SGRT V7 N10 P15 <AGS>
1966 N3 P240-246 SGRT V7 N10 P24 <AGS>
1966 N3 P247-250 SGRT V7 N10 P46 <AGS>

IZVESTIYA VSESOYUZNOGO NAUCHNO-ISSLEDOVATELSKOGO INSTITUTA GIDROTEKHNIKI IM. V. E. VEDENEEVA. LENINGRAD
1949 V41 P98-109 63-10199 <=*>
1954 V52 P15-39 70J19R <ATS>
1954 V52 P203-222 60-17216 <+*>
1959 V62 P75-96 63-24102 <=*$> 0

IZVESTIYA VSESOYUZNOGO NAUCHNO-ISSLEDOVATELSKOGO INSTITUTA OZERNOGO I RECHNOGO RYBNOGO KHOZYAISTVA
1932 V14 P133-148 64-11098 <=>
1957 V42 ENTIRE ISSUE 60-51169 <=>
1957 V42 P304-314 60-17592 <+*>
1957 V42 P330 60-17581 <+*>
1958 V47 N1 P1-63 C-3451 <NRCC>
1959 V49 ENTIRE ISSUE 61-31056 <=>
1961 V50 ENTIRE ISSUE 66-51056 <=>
1961 V50 P51-61 RTS-3021 <NLL>
1961 V51 P37-46 63-19607 <=*$> 0
1961 V51 P47-51 63-19608 <=*$>
1963 V54 P119-129 65-60948 <=$>

IZVESTIYA VSESOYUZNOGO TEPLOTEKHNICHESKOGO INSTITUTA MOSCOW
1946 V15 N1 P16-23 60-13688 <+*>
1946 V15 N11 P23-25 RT-324 <*>
1947 N4 P1-5 60-13627 <+*>
1950 N2 P1-8 62-13145 <*=>
1950 N3 P17-22 59-14189 <+*>
1952 N8 P12-18 59-18342 <+*>

*IZVESTIYA VYSSHIKH UCHEBNYKH ZAVEDENII. AVIATSIONNAYA TEKHNIKA
1958 N1 P27-36 60-21611 <=>
1958 N1 P43-52 60-17356 <+*>
1958 N1 P53-60 60-17367 <+*>
1958 N1 P95-100 ARSJ V29 N10 P745 <AIAA>
1958 N1 P133-142 61-28500 <*=>
1958 N1 P143-149 62-24240 <=*>
1958 N1 P150-157 62-24858 <=*>
1958 N1 P165-170 62-24241 <=*>
1958 N1 P171-178 62-24239 <=*>
1958 N2 P9-15 60-23568 <+*>
1958 N2 P74-85 62-19258 <=*>
1958 N2 P93-111 62-19258 <=*>
1958 N3 P60-67 60-23518 <+*> 0
1958 N3 P78-88 63-23439 <=*>
1958 N4 P3-8 584-1 <AIS>
60-23520 <+*> 0
1958 N4 P22-29 60-13019 <+*>
1958 N4 P51-61 61-31535 <=>
1958 N4 P72-80 584-9 <AIS>
1958 N4 P101-108 AD-617 661 <=$>
1959 N1 P3-9 63-24308 <=*$>
1959 N1 P46-54 M.2828 <NLL>
1959 N1 P65-73 61-28492 <*=>
1959 N1 P125-133 61-28502 <*=>
1959 N2 P72-82 ARSJ V30 N1 P78 <AIAA>
1959 N2 P122-133 ARSJ V30 N4 P379 <AIAA>
1959 N2 P144-146 62-11708 <=>
62-14014 <*>
1959 N2 P147-150 61-28375 <*=>
1959 N2 P156-158 63-19863 <*=>
1959 N3 P3-8 60-21930 <=>
1959 N3 P26-45 62-25345 <=*>
1959 N3 P57-63 61-27419 <*=> 0
1959 N3 P64-71 61-27561 <*=>
1959 N3 P72-79 61-27780 <*=>
63-14658 <=*>
1959 N3 P91-100 ARSJ V31 N3 P408 <AIAA>
1959 N3 P119-129 ARSJ V30 N7 P695 <AIAA>
1959 N3 P152-154 61-27420 <*=>
1959 N4 P64-69 61-23693 <*=>
1959 N4 P161-165 62-13072 <*=>
1960 N1 P35-42 61-11934 <=>
1960 N1 P72-82 ARSJ V30 N11 P1041 <AIAA>
61-11892 <=>
1960 N1 P94-103 61-11961 <=>
62-14008 <*>
1960 N1 P104-110 61-11968 <=>
1960 N2 P3-13 61-28206 <*=>
1960 N2 P14-21 61-11090 <=>
1960 N2 P88-98 60-18763 <+*>
1960 N2 P99-104 61-23650 <*=>
1960 N3 P12-21 63-24601 <=*$>
1960 N3 P74-79 61-28966 <*=>
1960 N3 P80-86 62-18102 <=*>
1960 N3 P107-109 61-27838 <*=>
1960 N4 P10-30 62-15816 <=*>
1960 N4 P40-50 62-15811 <=*>
1960 N4 P51-60 63-24539 <=*$>
1960 N4 P61-71 62-13573 <=*>
1960 N4 P72-82 AIAAJ V1 N1 P267 <AIAA>
61-28501 <*=>
1960 N4 P83-92 61-28234 <*=>
1960 N4 P117-125 61-16083 <=*>
1960 N4 P126-131 61-19806 <*=>
1961 N1 P38-45 62-19254 <=*>
1961 N1 P61-73 62-24271 <=*>
1961 N1 P97-104 62-25739 <=*>
1961 N2 P3-16 62-32984 <=*>
1961 N2 P17-25 63-13620 <=*>
64-13158 <=*$>
1961 N2 P35-46 62-32984 <=*>
1961 N2 P147-157 AD-607 973 <=> M
1961 N2 P158-159 62-15393 <*=>
1961 N3 P3-13 64-13542 <=*$>
1961 N3 P14-20 64-13542 <=*$>
1961 N3 P21-30 64-13542 <=*$>
1961 N3 P31-37 62-25728 <=*>
64-16381 <=*$>
1961 N3 P38-55 64-13542 <=*$>
1961 N3 P56-68 64-13542 <=*$>
1961 N3 P69-77 62-15163 <*=>

1961 N3 P78-88	62-15283 <*=>
1961 N3 P89-99	64-13542 <=*$>
1961 N3 P100-112	61-28970 <*=>
1961 N3 P113-126	64-13542 <=*$>
1961 N3 P127-133	64-13542 <=*$>
1961 N3 P134-143	64-13542 <=*$>
1961 N3 P144-152	64-13542 <=*$>
1961 N3 P153-154	64-13542 <=*$>
1961 N3 P163-180	64-13542 <=*$>
1961 N4 P22-29	63-19494 <=*>
1961 N4 P75-81	AD-641 111 <=$>
1961 N4 P94-103	62-24928 <=*>
1961 N4 P111-119	63-11584 <=>
1961 N4 P128-131	62-24920 <=*>
1961 N4 P149-155	62-24920 <=*>
1962 N1 P82-91	63-15528 <=*>
1962 N2 P56-64	63-19164 <=*>
1962 N2 P86-94	63-19164 <=*>
1962 N2 P102-112	63-19335 <=*>
1962 N2 P130-137	63-19154 <=*>
1962 N2 P159-167	63-19335 <=*>
1962 N3 P166-176	63-19211 <=*>
1962 N4 P68-78	65-13654 <*>
1962 N4 P79-90	AD-619 557 <=$>
1962 N4 P91-102	AD-619 556 <=$>
1962 N4 P123-128	AD-619 555 <=$>
1962 V4 P3-11	AD-619 558 <=$>
1962 V5 N1 P19-31	63-19528 <=*>
1962 V5 N1 P66-74	63-19430 <=*>
1962 V5 N2 P65-71	63-19513 <=*>
1962 V5 N2 P113-123	AEC-TR-6247 <=*$>
1962 V5 N2 P113-133	63-19513 <=*>
1962 V5 N2 P138-151	63-19513 <=*>
1962 V5 N2 P168-170	AEC-TR-6251 <=*$>
1962 V5 N3 P25-33	63-23531 <=*>
1962 V5 N3 P79-86	63-19527 <=*>
1962 V5 N3 P157-165	AEC-TR-6250 <=*$>
1962 V5 N3 P177-186	63-19527 <=*>
1962 V5 N4 P12-19	AD-619 545 <=$>
1962 V5 N4 P18-26	63-23733 <=*$>
1962 V5 N4 P57-78	64-13322 <=*$>
1962 V5 N4 P103-110	64-71391 <=> M
1962 V5 N4 P111-122	64-71403 <=> M
1962 V5 N4 P129-138	64-71408 <=> M
1962 V5 N4 P139-144	64-71491 <=> M
1962 V5 N4 P145-150	63-24248 <=*$>
1962 V5 N4 P151-154	64-71402 <=> M
1963 N4 P186P	AD-629 441 <=*$> M
1963 V6 N1 P3-175	64-21747 <=>
1963 V6 N1 P33-47	NASA-TT-F-194 <=>
1963 V6 N1 P116-125	AEC-TR-6248 <=*$>
1963 V6 N2 P11-21	64-16021 <=*$>
1963 V6 N2 P70-77	AD-619 567 <=$>
1963 V6 N3 P11-20	63-20960 <=*$>
	64-10559 <=*$>
1963 V6 N3 P34-42	63-20962 <=*$>
	64-10558 <=*$>
1963 V6 N3 P92-99	64-21268 <=>
1963 V6 N4 P15-24	64-16006 <=*$>
1963 V6 N4 P25-32	64-16167 <*>
1963 V6 N4 P55-62	64-16005 <=*$>
1964 N1 P54-59	N65-16451 <=$>
1964 V7 N1 P47-53	64-18419 <*>
1964 V7 N1 P54-59	64-18422 <*>
1964 V7 N1 P64-74	64-18423 <*>
1964 V7 N1 P181-185	64-18555 <*>
1964 V7 N2 P68-80	RAE-LIB-TRANS-1103 <NLL>
1964 V7 N3 P3-10	65-12184 <*>
	65-63755 <=>
1964 V7 N3 P29-37	65-12729 <*>
1964 V7 N3 P50-57	TIL/T-5598 <NLL>
1964 V7 N3 P75-78	65-12732 <*>
1964 V7 N3 P79-86	65-12731 <*>
1964 V7 N3 P117-123	AD-639 183 <=$>
1964 V7 N4 P40-45	65-13285 <*>
1964 V7 N4 P106-110	N65-23715 <=$>
1964 V7 N11 P149-154	AD-625 293 <=*$>
1965 N1 P7-14	66-12743 <*>
1965 N2 P121-131	AD-630 288 <=*$>
1965 N3 P98-107	AD-635 071 <=$>
1965 N3 P157-162	N66-23542 <=$>
1965 N4 P3-7	66-61361 <=*$>
1965 N4 P29-37	66-61983 <=*$>
1965 N4 P97-102	66-12683 <*>
	66-61362 <=*$>
1965 V8 N2 P65-67	N65-32729 <=$>
1965 V8 N3 P108-118	AD-625 244 <=*$>
1966 N2 P38-43	66-61988 <=*$>

IZVESTIYA VYSSHIKH UCHEBNYKH ZAVEDENII. CHERNAYA METALLURGIYA

1958 N2 P113-121	60-13468 <+*>
1958 N3 P3-12	4701 <HB>
1958 N3 P21-26	5949 <HB>
1958 N3 P29-33	5952 <HB>
1958 N3 P42-51	5282 <HB>
	64-10365 <=*$>
1958 N4 P103-116	1845 <BISI>
1958 N7 P3-16	4632 <HB>
1958 N7 P29-35	4726 <HB>
1958 N7 P37-43	4610 <HB>
1958 N7 P163-165	4677 <HB>
1958 N7 P191-193	4678 <HB>
1958 N8 P3-6	5025 <HB>
1958 N8 P7-12	4640 <HB>
1958 N8 P172-173	60-13587 <+*>
1958 N9 P3-14	4737 <HB>
1958 N10 P3-13	4614 <HB>
1958 N10 P31-35	4619 <HB>
1958 N10 P37-42	4616 <HB>
	61-19548 <=*> O
1958 N10 P119-124	4617 <HB>
1958 N11 P3-8	4623 <HB>
	65-29156 <*>
1958 N11 P37-40	4621 <HB>
1958 N11 P95-98	4622 <HB>
1958 N12 P17-20	4704 <HB>
1958 N12 P45-50	4735 <HB>
1959 N1 P47-58	4675 <HB>
1959 N1 P81-88	63-23667 <=*>
1959 N1 P113-120	4627 <HB>
1959 N1 P145-153	4630 <HB>
1959 N2 P83-92	2698 <BISI>
1959 N2 P93-99	61-23164 <*>
1959 N2 P122-133	61-19457 <=*> O
1959 N3 P143-155	2019 <BISI>
1959 N4 P3-12	4693 <HB>
1959 N4 P53-55	4684 <HB>
1959 N4 P73-78	60-11476 <=>
1959 N4 P151-155	NLLTB V1 N12 P1 <NLL>
1959 N5 P19-27	4696 <HB>
1959 N5 P41-44	4689 <HB>
1959 N6 P31-36	4703 <HB>
1959 N6 P81-82	4763 <HB>
1959 N6 P145-154	4935 <HB>
1959 N7 P23-34	4759 <HB>
1959 N7 P35-39	4767 <HB>
1959 N7 P97-100	4766 <HB>
1959 N7 P101-103	4765 <HB>
1959 N8 P17-19	5119 <HB>
1959 N8 P39-44	2594 <BISI>
1959 N8 P54-55	4962 <HB>
1959 N8 P75-86	AEC-TR-4524 <*=>
1959 N9 P29-45	4871 <HB>
1959 N9 P189-194	60-11454 <=>
1959 N10 P17-21	4909 <HB>
1959 N10 P23-28	4855 <HB>
1959 N10 P29-41	61-10752 <+*>
	61-20197 <*=>
1959 N10 P43-48	61-19705 <=*> O
1959 N10 P49-56	2004 <BISI>
1959 N10 P89-98	2207 <BISI>
1959 N10 P191-193	4835 <HB>
1959 N11 P13-18	4858 <HB>
1959 N11 P31-37	4830 <HB>
1959 N11 P39-41	4876 <HB>
1959 N11 P43-46	4856 <HB>
1959 N11 P47-62	1989 <BISI>
1959 N11 P83-87	5041 <HB>
1959 N11 P113-118	4877 <HB>

Citation	Translation
1959 N12 P2-11	5002 <HB>
1959 N12 P13-30	4867 <HB>
1959 N12 P49-56	4872 <HB>
1959 N12 P57-64	4859 <HB>
1960 N1 P12-15	AEC-TR-4391 <*>
1960 N1 P28-34	61-11897 <=>
1960 N1 P33-40	2565 <BISI>
1960 N1 P104-114	61-31602 <=>
1960 N1 P183-190	4805 <HB>
1960 N2 P78-80	4843 <HB>
1960 N2 P151-156	4842 <HB>
1960 N2 P163-168	2078 <BISI>
1960 N3 P13-16	4914 <HB>
1960 N3 P91-95	AEC-TR-4441 <*=>
1960 N3 P106-109	61-13601 <*+> 0
1960 N3 P131-135	61-19707 <=*>
1960 N3 P140-145	4841 <HB>
1960 N3 P153-158	4902 <HB>
1960 N4 P5-18	2003 <BISI>
1960 N4 P19-22	5036 <HB>
1960 N4 P23-28	5037 <HB>
1960 N4 P29-36	5035 <HB>
1960 N4 P77-85	5040 <HB>
1960 N4 P190-193	4869 <HB>
1960 N5 P39-48	5081 <HB>
1960 N5 P49-54	E-727 <RIS>
	4882 <HB>
1960 N5 P55-60	60-31774 <=>
1960 N5 P61-67	60-31775 <=>
1960 N5 P68-71	60-31776 <=>
1960 N5 P91-92	4901 <HB>
1960 N5 P108-114	60-31777 <=>
1960 N5 P130-134	4900 <HB>
1960 N5 P137-142	61-28488 <*=>
1960 N5 P150-158	60-31778 <=>
1960 N5 P177-180	5602 <HB>
1960 N6 P106-113	5042 <HB>
1960 N7 P49-53	4927 <HB>
1960 N7 P54-59	4931 <HB>
1960 N7 P60-67	4926 <HB>
1960 N7 P129-134	4928 <HB>
	61-13896 <+*>
1960 N7 P156-162	62-23961 <=*>
1960 N7 P163-170	4929 <HB>
	61-20204 <*=> 0
1960 N7 P171-179	4930 <HB>
1960 N8 P15-21	64-11501 <=>
1960 N8 P22-28	5018 <HB>
1960 N8 P38-47	2136 <BISI>
1960 N8 P160-166	2391 <BISI>
1960 N9 P5-7	64-10563 <=*$>
1960 N9 P8-13	63-15948 <=*>
1960 N9 P18-28	5067 <HB>
1960 N9 P43-49	5074 <HB>
1960 N9 P157-159	5065 <HB>
1960 N9 P160-166	5068 <HB>
1960 N9 P167-168	5066 <HB>
1960 N10 P40-45	5085 <HB>
1960 N10 P79-83	5139 <HB>
1960 N10 P139-142	5251 <HB>
1960 N11 P20-26	5093 <HB>
1960 N11 P32-35	5092 <HB>
1960 N11 P50-60	5100 <HB>
1960 N11 P61-65	5091 <HB>
1960 N11 P113-120	5094 <HB>
1960 N12 P17-22	5105 <HB>
1960 N12 P23-30	5104 <HB>
1960 N12 P31-38	5082 <HB>
1960 N12 P111-113	5090 <HB>
1960 N12 P114-123	62-15274 <*=> 0
1960 N12 P173-182	5083 <HB>
1961 N1 P5-11	5134 <HB>
1961 N1 P21-30	5135 <HB>
1961 N1 P158-159	5131 <HB>
1961 N1 P166-170	62-25267 <=*>
1961 N1 P181-183	5237 <HB>
1961 N1 P184-190	2376 <BISI>
1961 N2 P5-9	5140 <HB>
1961 N2 P10-21	5141 <HB>
1961 N2 P22-33	5151 <HB>
1961 N2 P39-44	5150 <HB>
1961 N2 P45-50	5149 <HB>
1961 N2 P100-108	5145 <HB>
1961 N2 P122-127	5146 <HB>
1961 N2 P133-137	5147 <HB>
1961 N2 P191-196	5148 <HB>
1961 N3 P11-15	5165 <HB>
1961 N3 P16-22	5166 <HB>
1961 N3 P37-39	5167 <HB>
1961 N3 P82-90	AEC-TR-4958 <=*>
1961 N3 P120-125	64-31172 <=>
1961 N3 P148-153	3673 <BISI>
1961 N3 P197-205	5168 <HB>
1961 N4 P26-37	5203 <HB>
1961 N4 P31-37	5204 <HB>
1961 N4 P59-66	5208 <HB>
1961 N4 P67-78	2557 <BISI>
	62-25302 <=*> 0
1961 N4 P89-92	5240 <HB>
1961 N4 P107-113	5241 <HB>
1961 N4 P119-125	2966 <BISI>
1961 N4 P126-133	5238 <HB>
1961 N5 P28-36	63-18584 <=*>
1961 N5 P37-46	5244 <HB>
1961 N5 P50-57	5549 <HB>
1961 N5 P58-67	2359 <BISI>
1961 N5 P132-138	62-24820 <=>
1961 N5 P177-183	5243 <HB>
1961 N6 P11-19	5262 <HB>
1961 N6 P42-52	5265 <HB>
1961 N6 P129-133	5266 <HB>
1961 N6 P134-138	5267 <HB>
1961 N6 P139-144	3119 <BISI>
1961 N6 P157-163	5264 <HB>
1961 N6 P168-174	2400 <BISI>
1961 N7 P36-43	5345 <HB>
1961 N7 P44-54	5346 <HB>
1961 N7 P55-61	62-33370 <=*>
1961 N7 P62-66	2541 <BISI>
1961 N7 P78-87	5347 <HB>
1961 N7 P88-96	2997 <BISI>
1961 N7 P149-153	62-33370 <=*>
1961 N7 P180-194	5762 <HB>
1961 N7 P195-198	5661 <HB>
1961 N8 P24-31	5410 <HB>
1961 N8 P32-36	2504 <BISI>
1961 N8 P143-148	5369 <HB>
1961 N9 P32-43	62-25312 <=*> 0
	6361 <HB>
1961 N9 P44-53	62-25311 <=*> 0
	6240 <HB>
1961 N9 P54-58	5439 <HB>
1961 N9 P59-70	5440 <HB>
1961 N9 P71-78	2604 <BISI>
	62-25310 <=*> 0
1961 N9 P79-86	5438 <HB>
1961 N9 P129-137	62-32965 <=*>
1961 N9 P191-197	5437 <HB>
1961 N10 P29-36	5441 <HB>
1961 N11 P57-59	5471 <HB>
1961 N11 P60-70	5472 <HB>
1961 N11 P155-158	5133 <HB>
1961 N11 P170-175	3855 <BISI>
1961 N12 P27-30	5538 <HB>
1961 N12 P31-39	5539 <HB>
1961 N12 P50-53	62-25317 <=*>
1961 N12 P54-60	5541 <HB>
1961 N12 P114-116	5542 <HB>
1961 N12 P130-134	63-13913 <=*>
1961 N12 P135-143	5543 <HB>
1961 V4 N6 P67-74	M.5673 <NLL>
1962 N1 P33-40	5548 <HB>
1962 N1 P52-56	5552 <HB>
1962 N1 P61-69	5550 <HB>
1962 N1 P75-81	63-19082 <=*>
1962 N1 P78-89	63-19922 <=*>
1962 N1 P124-130	AEC-TR-6077 <=*$>
1962 N1 P160-168	5551 <HB>
1962 N1 P183-189	6152 <HB>
1962 N2 P8-14	63-19084 <=*>

Citation	Translation
1962 N2 P9-17	4014 <BISI>
1962 N2 P31-38	5565 <HB>
1962 N2 P44-50	5568 <HB>
1962 N2 P51-55	5567 <HB>
1962 N2 P101-103	5650 <HB>
	63-21442 <=>
1962 N2 P118-122	63-23917 <=*$>
1962 N2 P123-124	5564 <HB>
1962 N3 P12-23	5626 <HB>
1962 N3 P30-37	5628 <HB>
1962 N3 P53-59	5852 <HB>
1962 N3 P60-66	5600 <HB>
1962 N3 P67-76	5652 <HB>
1962 N3 P95-102	63-13875 <=*>
1962 N3 P111-121	C-5143 <NRC>
1962 N3 P122-124	63-16388 <=*>
1962 N4 P29-36	3842 <BISI>
1962 N5 P5-15	5662 <HB>
1962 N5 P37-43	6049 <HB>
1962 N5 P44-50	3188 <BISI>
1962 N5 P200-206	5924 <HB>
1962 N6 P16-20	3424 <BISI>
1962 N6 P26-31	5590 <HB>
1962 N6 P27-33	5263 <HB>
1962 N6 P39-50	5588 <HB>
1962 N6 P344-350	5592 <HB>
1962 N7 P37-45	5775 <HB>
1962 N7 P62-70	5884 <HB>
1962 N7 P71-77	5885 <HB>
1962 N7 P86-96	5886 <HB>
1962 N7 P97-102	5881 <HB>
1962 N7 P110-113	3449 <BISI>
1962 N7 P137-139	AEC-TR-6078 <=*$>
1962 N8 P21-28	5878 <HB>
1962 N8 P200-206	63-13322 <=*>
1962 N9 P17-25	5755 <HB>
1962 N9 P49-53	64-13385 <=*$>
1962 N9 P54-65	5795 <HB>
1962 N9 P92-98	5754 <HB>
1962 N9 P99-104	3441 <BISI>
1962 N9 P105-110	5753 <HB>
1962 N9 P160-164	63-15373 <=*>
1962 N9 P197-205	63-15373 <=*>
1962 N10 P31-41	5784 <HB>
1962 N10 P50-57	5778 <HB>
1962 N10 P107-110	5779 <HB>
1962 N10 P111-118	5780 <HB>
1962 N10 P125-130	5781 <HB>
1962 N10 P131-136	5782 <HB>
1962 N11 P57-60	5828 <HB>
1962 N11 P61-69	5941 <HB>
1962 N11 P76-79	63-20130 <=*>
1962 N11 P128-132	5835 <HB>
1962 N11 P133-139	3257 <BISI>
1962 N11 P170-174	5836 <HB>
1962 N11 P191-195	5837 <HB>
1962 N11 P202-208	5854 <HB>
1962 N12 P20-28	5882 <HB>
1962 N12 P29-40	5879 <HB>
1962 N12 P61-66	5819 <HB>
1962 N12 P117-119	5871 <HB>
1962 N12 P131-137	63-23923 <=>
1962 V5 N2 P15-22	63-23621 <=*>
1962 V5 N2 P39-47	63-23621 <=*>
1962 V5 N4 P123-128	63-20199 <=*> 0
1962 V5 N7 P78-85	6493 <HB>
1963 N1 P34-48	63-21578 <=>
1963 V6 P18-26	3671 <BISI>
1963 V6 N1 P34-41	5908 <HB>
1963 V6 N1 P42-48	5909 <HB>
1963 V6 N1 P111-116	5912 <HB>
1963 V6 N2 P23-30	5896 <HB>
1963 V6 N2 P31-38	5897 <HB>
1963 V6 N2 P68-76	5899 <HB>
1963 V6 N2 P95-103	AD-613 160 <=$>
1963 V6 N2 P115-119	5901 <HB>
1963 V6 N2 P120-128	5657 <HB>
1963 V6 N2 P129-132	5898 <HB>
1963 V6 N3 P26-33	AD-610 789 <=$>
1963 V6 N3 P46-52	3872 <BISI>
1963 V6 N3 P70-76	5553 <HB>
1963 V6 N3 P163-170	4358 <BISI>
1963 V6 N4 P20-26	5955 <HB>
1963 V6 N4 P27-33	5956 <HB>
1963 V6 N4 P37-49	5052 <HB>
1963 V6 N4 P109-114	5957 <HB>
1963 V6 N4 P156-162	RTS-3317 <NLL>
1963 V6 N5 P19-25	5982 <HB>
1963 V6 N5 P26-33	5983 <HB>
1963 V6 N5 P44-50	3350 <BISI>
1963 V6 N5 P58-64	5985 <HB>
1963 V6 N5 P65-69	5984 <HB>
1963 V6 N5 P81-89	4367 <BISI>
1963 V6 N5 P151-155	6297 <HB>
1963 V6 N5 P162-169	5986 <HB>
1963 V6 N6 P21-25	5582 <HB>
1963 V6 N6 P27-31	3599 <BISI>
1963 V6 N6 P32-34	6010 <HB>
1963 V6 N6 P43-48	6014 <HB>
1963 V6 N6 P49-50	6015 <HB>
1963 V6 N6 P58-67	6013 <HB>
1963 V6 N6 P96-99	6002 <HB>
1963 V6 N6 P130-137	AD-615 876 <=$>
1963 V6 N7 P33-38	6065 <HB>
1963 V6 N7 P39-46	6068 <HB>
1963 V6 N7 P70-78	6100 <HB>
1963 V6 N8 P72-81	64-21110 <=>
1963 V6 N8 P128-131	3875 <BISI>
1963 V6 N8 P161-163	6349 <HB>
1963 V6 N9 P5-10	6104 <HB>
1963 V6 N9 P28-32	6103 <HB>
1963 V6 N9 P39-44	6106 <HB>
1963 V6 N9 P50-54	6098 <HB>
1963 V6 N9 P62-69	6105 <HB>
1963 V6 N9 P88-91	AD-610 291 <=$>
	64-21508 <=>
1963 V6 N9 P142-144	6819 <HB>
1963 V6 N9 P145-147	65-61343 <=$>
1963 V6 N10 P166-170	5783 <HB>
1963 V6 N11 P5-10	AD-611 527 <=$>
1963 V6 N11 P22-29	6156 <HB>
1963 V6 N11 P161-167	66-60462 <=*$>
1963 V6 N11 P200-206	4385 <BISI>
1963 V6 N12 P28-34	6154 <HB>
1963 V6 N12 P45-53	4034 <BISI>
1963 V6 N12 P118-125	3965 <BISI>
1964 N5 P53-57	3821 <BISI>
1964 N9 P88-94	AD-636 663 <=$>
1964 V7 N1 P27-32	6229 <HB>
1964 V7 N1 P41-45	6247 <HB>
1964 V7 N1 P56-61	6228 <HB>
1964 V7 N1 P62-68	64-19141 <=$>
1964 V7 N1 P136-141	65-61327 <=$>
1964 V7 N1 P142-144	65-61410 <=>
1964 V7 N1 P174-177	3964 <BISI>
1964 V7 N1 P194-200	6208 <HB>
1964 V7 N1 P201-208	64-21788 <=>
1964 V7 N2 P5-12	6230 <HB>
1964 V7 N2 P13-18	3721 <BISI>
1964 V7 N2 P34-41	3972 <BISI>
1964 V7 N2 P49-55	6217 <HB>
1964 V7 N2 P58-63	6216 <HB>
1964 V7 N2 P64-72	4190 <BISI>
1964 V7 N2 P131-139	3722 <BISI>
1964 V7 N2 P183-188	6232 <HB>
1964 V7 N3 P16-22	6231 <HB>
1964 V7 N3 P47-52	64-41033 <=>
1964 V7 N3 P53-56	6233 <HB>
1964 V7 N3 P124-130	AD-621 737 <=*$>
1964 V7 N4 P11-19	4339 <BISI>
1964 V7 N4 P26-30	6286 <HB>
1964 V7 N4 P31-36	6288 <HB>
1964 V7 N4 P68-70	6289 <HB>
1964 V7 N4 P124-128	6291 <HB>
1964 V7 N4 P129-132	6290 <HB>
1964 V7 N4 P182-185	6346 <HB>
1964 V7 N5 P12-16	<SIC>
	6378 <HB>
1964 V7 N5 P22-25	<SIC>
1964 V7 N5 P40-45	6814 <HB>

Citation	Translation
1964 V7 N5 P90-93	M-5675 <NLL>
1964 V7 N5 P117	<SIC>
1964 V7 N5 P124-129	M-5581 <NLL>
1964 V7 N6 P17-22	6375 <HB>
1964 V7 N6 P23-25	6377 <HB>
1964 V7 N6 P26-28	6376 <HB>
1964 V7 N6 P47-51	M-5606 <NLL>
1964 V7 N6 P52-55	6335 <HB>
1964 V7 N6 P62-67	6337 <HB>
1964 V7 N6 P68-76	6336 <HB>
1964 V7 N6 P92-96	4007 <BISI>
1964 V7 N6 P125-130	6334 <HB>
1964 V7 N7 P10-18	4339 PT.2 <BISI>
1964 V7 N7 P29-35	3890 <BISI>
1964 V7 N7 P36-42	6329 <HB>
1964 V7 N7 P43-47	6330 <HB>
1964 V7 N7 P58-62	6338 <HB>
1964 V7 N7 P63-68	3891 <BISI>
1964 V7 N7 P162-164	LA-TR-65-5 <=*$>
1964 V7 N8 P44-49	6511 <HB>
1964 V7 N8 P188-194	AD-625 147 <=*$>
1964 V7 N8 P195-200	4006 <BISI>
1964 V7 N9 P195-201	6503 <HB>
1964 V7 N10 P59-63	M-5607 <NLL>
1964 V7 N1C P80-85	M-5608 <NLL>
1964 V7 N11 P34-40	6429 <HB>
1964 V7 N11 P120-123	6428 <HB>
1964 V7 N11 P131-136	6471 <HB>
1964 V7 N11 P137-141	6484 <HB>
1964 V7 N11 P155-157	1141 <TC>
1964 V7 N11 P180-184	M-5623 <NLL>
1964 V7 N12 P33-35	6568 <HB>
1965 N2 P161-167	4717 <BISI>
1965 N3 P10-14	4380 <BISI>
1965 N7 P206-208	4529 <BISI>
1965 V8 P36-42	6496 <HB>
1965 V8 P155-160	6601 <HB>
1965 V8 N1 P61-64	6726 <HB>
1965 V8 N1 P95-98	M-5743 <NLL>
1965 V8 N1 P99-103	M-5755 <NLL>
1965 V8 N1 P104-109	AD-634 834 <=$>
1965 V8 N1 P120-123	M-5751 <NLL>
1965 V8 N1 P136-142	6761 <HB>
1965 V8 N2 P29-35	6495 <HB>
1965 V8 N2 P43-48	6497 <HB>
1965 V8 N2 P58-64	6498 <HB>
1965 V8 N2 P65-67	6499 <HB>
1965 V8 N2 P99-102	6727 <HB>
1965 V8 N2 P106-110	6501 <HB>
1965 V8 N3 P29-35	6500 <HB>
1965 V8 N3 P45-52	6521 <HB>
1965 V8 N3 P53-58	6520 <HB>
1965 V8 N3 P59-63	6517 <HB>
1965 V8 N3 P77-80	M-5750 <NLL>
1965 V8 N3 P89-93	6516 <HB>
1965 V8 N3 P157-160	AD-627 121 <=$>
1965 V8 N3 P174-179	6519 <HB>
1965 V8 N4 P25-29	4593 <BISI>
1965 V8 N4 P53-58	6477 <HB>
1965 V8 N4 P72-74	6541 <HB>
1965 V8 N4 P91-95	6685 <HB>
1965 V8 N4 P123-125	6736 <HB>
1965 V8 N4 P164-168	65-31212 <=$>
	6571 <HB>
1965 V8 N4 P220-223	6587 <HB>
1965 V8 N5 P21-28	6574 <HB>
1965 V8 N5 P40-44	6507 <HB>
1965 V8 N5 P51-56	M.5814 <NLL>
1965 V8 N5 P157-161	6775 <HB>
1965 V8 N6 P16-21	6594 <HB>
1965 V8 N6 P27-32	6593 <HB>
1965 V8 N6 P50-53	6595 <HB>
1965 V8 N6 P60-63	6686 <HB>
1965 V8 N6 P64-67	6596 <HB>
1965 V8 N6 P127-130	6591 <HB>
1965 V8 N6 P149-155	6730 <HB>
1965 V8 N6 P178-179	6682 <HB>
1965 V8 N7 P5-10	6621 <HB>
1965 V8 N7 P11-15	6622 <HB>
1965 V8 N7 P20-22	6623 <HB>
1965 V8 N7 P28-31	6624 <HB>
1965 V8 N7 P37-42	6626 <HB>
1965 V8 N7 P50-55	6625 <HB>
1965 V8 N7 P74-77	6627 <HB>
1965 V8 N7 P113-115	M.5815 <NLL>
1965 V8 N7 P175-179	6629 <HB>
1965 V8 N8 P168-170	6732 <HB>
1965 V8 N9 P34-38	6671 <HB>
1965 V8 N9 P98-102	6853 <HB>
1965 V8 N9 P103-107	6772 <HB>
1965 V8 N9 P163-167	6731 <HB>
1965 V8 N10 P22-30	4612 <BISI>
1965 V8 N10 P37-41	6704 <HB>
1965 V8 N10 P49-51	6705 <HB>
1965 V8 N10 P59-61	6703 <HB>
1965 V8 N10 P90-93	6856 <HB>
1965 V8 N10 P116-119	6706 <HB>
1965 V8 N12 P36-39	6803 <HB>
1966 V9 N1 P10-14	6860 <HB>
1966 V9 N3 P15-21	6854 <HB>
1966 V9 N5 P31-35	4960 <BISI>
1966 V9 N5 P131-134	6907 <HB>

IZVESTIYA VYSSHIKH UCHEBNYKH ZAVEDENII. ELEKTRO-MEKHANIKA

Citation	Translation
1958 N11 P57-73	61-13330 <=*>
1959 N2 P131-134	62-32683 <=*>
1959 N5 P14-23	63-19167 <=*>
1959 N6 P34-40	61-15211 <=*>
1959 N10 P35-43	63-19190 <=*>
1959 N11 P132-141	60-31100 <=>
1959 N12 P12-17	60-11793 <=>
1959 N12 P88-95	61-28205 <*=>
1960 N1 P145-147	60-11971 <=>
1960 N2 P157-164	62-19993 <=*>
1960 N2 P168-170	60-11972 <=>
1960 N9 P62-72	ARSJ V32 N11 P1770 <AIAA>
1960 V3 N3 P140-143	61-28525 <*=>
1960 V3 N4 P20-26	62-15284 <*=>
1960 V3 N4 P129-133	63-19189 <=*>
1960 V3 N7 P3-8	61-11710 <=>
1960 V3 N7 P72-79	61-28381 <*=>
1960 V3 N9 P62-72	62-15834 <=*>
1960 V3 N11 P36-37	63-10189 <=*>
1960 V3 N11 P94-97	62-13071 <*=>
1960 V3 N12 P26-31	62-13073 <*=>
1961 V4 N1 P121-124	62-32422 <=*>
1961 V4 N7 P103-108	62-32291 <=*>
1961 V4 N9 P3-21	63-15907 <=*>
1961 V4 N10 P72-81	65-13376 <*>
1962 N10 P1097-1107	65-13670 <*>
1962 V5 N2 P130-140	64-71393 <=> M
1962 V5 N2 P140-167	63-15538 <=*>
1962 V5 N2 P177-195	63-15538 <=*>
1962 V5 N6 P632-649	63-19213 <=*>
1962 V5 N8 P909-918	63-19464 <=*>
1963 N8 P961-972	AD-635 272 <=$>
1963 V6 N5 P597-604	63-31989 <=>
1964 V7 N7 P848-857	AD-618 739 <=$>
1964 V7 N8 P958-970	65-30003 <=*$>

IZVESTIYA VYSSHIKH UCHEBNYKH ZAVEDENII. ENERGETIKA

Citation	Translation
1958 N2 P72-78	64-00318 <*>
1958 V1 N5 P100-108	62-24631 <=*>
1958 V1 N7 P84-89	64-13803 <=$>
1958 V1 N9 P61-64	50L34R <ATS>
1958 V1 N9 P65-67	52L34R <ATS>
1958 V1 N9 P77-89	64-71518 <=>
1958 V1 N10 P130-143	60-23804 <+*>
1959 N5 P93-101	64-21710 <=>
1959 V2 N4 P35-37	62-25406 <=*>
1959 V2 N4 P115-123	63-15242 <=*$>
1960 V3 N1 P78-88	61-27176 <*=>
1960 V3 N4 P127-132	62-15825 <*=>
1960 V3 N6 P136-144	62-23664 <=*>
1960 V3 N7 P51-55	63-19672 <=*>
1960 V3 N7 P97-107	M.5337 <NLL>
1960 V3 N8 P98-101	64-13386 <=*$>
1961 V4 N1 P73-81	62-32512 <=*>
1961 V4 N2 P27-33	62-13379 <*=>

Citation	Translation
1961 V4 N2 P103-108	62-32304 <=*>
1961 V4 N3 P87-92	62-15441 <*=>
1961 V4 N5 P97-104	62-15491 <*=>
1961 V4 N8 P37-46	62-32315 <=*>
1961 V4 N9 P55-61	62-24350 <=*>
1961 V4 N9 P76-82	62-24350 <=*>
1961 V4 N11 P77-83	AD-627 075 <=$>
1961 V4 N11 P101-104	63-23212 <=*>
1962 V5 N2 P85-91	65-60387 <=$>
1962 V5 N3 P67-70	AEC-TR-5410 <=*>
1962 V5 N4 P79-84	63-19177 <=*>
1962 V5 N5 P71-75	TP/T-3564 <NLL>
1962 V5 N5 P119-122	63-15783 <=*>
1962 V5 N7 P79-85	64-23531 <=$>
1962 V5 N8 P123-125	63-13365 <=>
1962 V5 N9 P86-93	RTS-2733 <NLL>
1962 V5 N12 P63-69	64-71421 <=> M
1963 V6 N1 P59-63	65-60745 <=>
1963 V6 N1 P99-102	63-24429 <=*$>
1963 V6 N5 P1-7	63-31638 <=>
1963 V6 N8 P80-89	AD-610 646 <=$>
1963 V6 N9 P74-78	65-64213 <=*$>
1964 N9 P49-53	AD-625 187 <=*$>
1964 V7 N2 P68-73	AD-615 429 <=$>
1964 V7 N5 P1-8	65-30073 <=$>
1964 V7 N5 P105-108	AD-620 065 <=$>
1964 V7 N6 P32-39	AD-612 416 <=$>
1964 V7 N8 P8-14	65-64209 <=*$>
1964 V7 N10 P58-63	66-60456 <=*$>
1965 N9 P38-46	AD-636 721 <=$>
1965 V8 N1 P105-109	AD-627 074 <=$>
1965 V8 N3 P61-66	AD-622 466 <=*$>

*IZVESTIYA VYSSHIKH UCHEBNYKH ZAVEDENII. FIZIKA

Citation	Translation
1957 N1 P103-110	62-18899 <=*>
	64-14028 <=*$>
1958 N1 P68-77	60-22093 <=>
	62-11492 <=>
1958 N2 P117-120	60-11547 <=>
1958 N2 P151-158	C-3641 <NRCC>
1958 N3 P91-99	61-27164 <*=>
1958 N4 P64-69	60-31148 <=>
1958 N4 P80-85	59-10501 <+*>
1958 N4 P127-134	R-5185 <*>
1958 N5 P102-107	59-14307 <+*>
1958 N5 P115-116	59-16744 <+*>
1958 N5 P133-134	59-14735 <+*>
1958 N6 P14-24	60-11913 <=>
1958 V1 N6 P145-151	35L35R <ATS>
1959 N1 P107-110	60-31159 <=>
1959 N1 P172-173	AEC-TR-4937 <=*>
	14N56R <ATS>
1959 N2 P36-38	10252 <K-H>
	60-18140 <+*>
1959 N2 P78-91	AD-621 783 <=*$>
1959 N2 P160-168	64-18598 <*>
1959 N2 P169-170	62-18090 <=*>
1959 N2 P171-172	60-11592 <=>
1959 N3 P16-22	60 22094 <=>
1959 N3 P68-77	62-19885 <=>
1959 N3 P84-94	62-23144 <=>
1959 N3 P95-101	62-25570 <=*>
1959 N4 P3-12	60-12543 <=>
	62-11487 <=>
1959 N4 P13-18	62-23040 <=>
1959 N4 P28-37	61-31638 <=> O
1959 N4 P43-47	62-11499 <=>
1959 N4 P52-58	60-22201 <=>
1959 N4 P59-63	62-11353 <=>
1959 N4 P64-67	61-15937 <*=>
	62-11260 <=>
1959 N4 P88-94	62-19997 <=>
1959 N4 P119-122	60-15457 <+*>
	62-11273 <=>
1959 N4 P135-139	68S83R <ATS>
1959 N4 P173-176	60 22082 <=>
	62-15608 <*=>
1959 N5 P109-118	62-23081 <=>
1959 N6 P14-20	61-13604 <+*>
1959 N6 P48-60	61-13604 <+*>
1959 N6 P90-94	62-11269 <=>
1959 N6 P95-101	60-12590 <=>
	62-11447 <=>
1959 N6 P145-151	61-31639 <=> O
	62-11285 <=>
1959 N6 P152-161	60-19859 <+*>
	62-11509 <=>
1959 N6 P168-169	61-21959 <=> O
	62-11455 <=>
1959 N6 P169-171	62-19840 <=>
1960 N1 P12-15	62-11412 <=>
1960 N1 P29-37	61-31637 <=>
1960 N1 P46-53	61-19808 <*=>
	62-11270 <=>
1960 N1 P54-56	62-11290 <=>
1960 N1 P57-59	61-31573 <=>
1960 N1 P64-67	62-11437 <*>
	62-11437 <=>
1960 N1 P80-86	62-11275 <*>
	62-11275 <=>
1960 N1 P90-92	61-31642 <=>
1960 N1 P93-103	62-11284 <=>
1960 N1 P135-138	62-11482 <=>
1960 N1 P145-154	TRANS-184 <FRI>
1960 N1 P197-202	63-13894 <=*>
1960 N1 P217-221	62-11462 <=>
1960 N1 P222-227	62-11461 <=>
1960 N1 P234-235	61-31644 <=>
1960 N1 P241-243	60-11964 <=>
1960 N1 P434-435	60-23575 <+*>
	62-11467 <=>
1960 N2 P75-80	62-11187 <=>
1960 N2 P99-102	61-15939 <+*>
	61-21995 <=>
	62-11500 <=>
1960 N2 P151-153	62-11484 <=>
1960 N2 P157-160	62-11483 <=>
1960 N3 P19-26	AD-633 537 <=$>
1960 N3 P32-38	62-23211 <=*>
1960 N3 P43-51	64-10580 <=*$>
1960 N3 P63-71	61-19823 <*=>
	62-11262 <=>
1960 N3 P97-102	61-27138 <*=>
1960 N3 P133-137	61-31678 <=>
1960 N3 P158-164	61-12575 <=>
	61-19825 <*=>
	62-11397 <=>
1960 N3 P212-217	64-21709 <=>
1960 N3 P223-229	62-11458 <=>
1960 N3 P233-234	AEC-TR-5177 <=*>
1960 N4 P13-21	62-23123 <=*>
1960 N4 P22-29	61-19826 <*=>
	62-11488 <=>
1960 N4 P56-59	62-13619 <*=>
	62-13619 <=>
1960 N4 P60-65	61-28537 <*=>
1960 N4 P107-109	62-13578 <*=>
1960 N4 P126-130	62-11315 <=>
	62-15609 <*=>
	62-24619 <=*>
1960 N4 P147-151	62-11578 <=>
1960 N4 P167-172	63-16547 <=*>
1960 N4 P173-177	61-28689 <*>
	62-11314 <=>
1960 N4 P183-189	62-11316 <=>
1960 N4 P206-216	62-23202 <=*>
1960 N4 P232-233	2828 <BISI>
1960 N4 P239	62-11247 <=>
1960 N4 P239-	62-11247 <=>
1960 N5 P26-34	62-15829 <=*>
1960 N5 P35-42	61-12474 <=>
	61-15936 <+*>
	62-11475 <=>
	65-13987 <=*$>
1960 N5 P77-81	61-28701 <*=>
1960 N5 P169-170	62-11579 <=>
	62-15610 <*=>
1960 N6 P13-19	62-13059 <*=>
1960 N6 P44-51	2249 <BISI>
1960 N6 P64-70	62-13577 <*=>

1960 N6 P170-171 64-13274 <=*$>
1961 N1 P20-23 62-15805 <=*>
1961 N1 P31-34 AIAAJ V1 N3 P758 <AIAA>
1961 N1 P63-70 62-10615 <=*>
1961 N1 P87-93 64-16668 <=*$>
1961 N1 P134-137 62-24267 <=*>
1961 N1 P138-142 62-15805 <=*>
1961 N2 P28-33 62-23679 <=*>
1961 N2 P46-51 62-25336 <=*>
1961 N2 P71-76 62-25338 <=*>
1961 N2 P85-91 64-11850 <=>
1961 N3 P28-34 61-28971 <*=>
1961 N4 P12-16 63-31540 <=>
1961 N4 P17-22 62-33539 <=*>
1961 N4 P52-55 62-14068 <=*> 0
1961 N4 P163-167 62-24923 <=*>
1961 N5 P26-29 62-24341 <=*>
1961 N5 P36-38 62-33077 <=*>
1961 N5 P170-171 64-19135 <=*$>
1961 N6 P22-26 63-13691 <=*>
1961 N6 P38-42 63-11516 <=>
1961 N6 P57-65 63-19521 <=*>
1962 N1 P25-35 63-15563 <=*>
1962 N1 P48-51 63-15563 <=*>
1962 N1 P84-87 62-32688 <=*>
1962 N1 P103-104 63-15549 <=*>
1962 N1 P111-117 63-19194 <=*>
1962 N1 P130-135 62-32242 <=*>
1962 N3 P55-61 63-19077 <*=>
1962 N3 P62-64 64-19988 <=$>
1962 N3 P65-73 63-19187 <=*>
1962 N3 P174-175 64-10952 <=*$>
1962 N4 P112-114 64-11500 <=>
66-13796 <*>
1962 N4 P118-122 63-11756 <=>
1962 N4 P163-170 63-11708 <=>
1962 N5 P34-45 63-23204 <=*>
1962 N5 P150-160 63-23204 <=*>
1962 N5 P156-160 63-23139 <=*$>
1962 N5 P177-178 64-11539 <=>
1962 N6 P90-98 63-23107 <*=>
1962 N6 P152-156 63-23720 <=*$>
1962 N6 P169-170 AD-614 026 <=$>
1962 N6 P174-176 AD-621 420 <=*$>
1962 N3 P82-92 SCL-T-467 <=*>
1963 N1 P21-30 AD-619 436 <=$>
1963 N2 P78-84 65-13428 <*>
1963 N2 P178-179 AD-620 088 <=$>
1963 N3 P7-11 65-10796 <*>
1963 N3 P18-22 64-71137 <=> M
1963 N3 P63-65 65-10802 <*>
1963 N5 P14-20 AD-611 523 <=$>
1963 N5 P27-31 65-10805 <*>
1963 N5 P32-38 65-10777 <*>
1963 N5 P75-81 AD-609 923 <=$>
1963 N5 P82-85 65-61191 <=>
1963 N5 P174-176 64-11572 <=>
1963 N6 P53-57 64-18759 <*>
1964 N1 P76-80 AD-639 182 <=$>
1964 N2 P110-115 64-41801 <=>
1964 N2 P126-130 64-41801 <=>
1964 N2 P147-148 AD-620 101 <=$>
1964 N2 P171-173 AD-619 326 <=$>
1964 N2 P173-175 AD-617 677 <=$>
1964 N3 P50-55 03S81R <ATS>
1964 N4 P3-5 N65-12274 <=$>
1964 N4 P17-21 65-17007 <*>
1964 N4 P22 66-10995 <*>
1964 N4 P68-71 65-12635 <*>
1964 N4 P72-75 AD-636 660 <=$>
1964 N4 P143-146 AD-636 660 <=$>
1964 N5 P28-33 AD-640 299 <=$>
1964 N5 P34-37 N65-23686 <=$>
1964 N5 P69-74 N65-17267 <=$>
1964 N5 P86-90 AD-639 482 <=$>
1964 N5 P135-138 65-13353 <*>
1964 N3 P3-6 AD-613 980 <=$>
1965 N1 P152-154 66-14880 <*$>
1965 N2 P106-111 48S89R <ATS>
1965 N3 P53-55 66-12185 <*>
1965 N3 P148-150 AD-638 456 <=$>
1965 N4 P46-49 66-11192 <*>
1965 N4 P76-81 66-11256 <*>
1965 N4 P116-118 66-11151 <*>
1965 N4 P140-143 66-11254 <*>
1965 N4 P151-155 68T91R <ATS>
1965 N4 P174-175 AD-631 586 <=$>
1965 N5 P128-133 66-12682 <*>
66-61363 <=*$>

*IZVESTIYA VYSSHIKH UCHEBNYKH ZAVEDENII. GEODEZIYA I AEROFOTOSEMKA

1957 N2 P37-54 64-11735 <=>
1957 N2 P37-55 64-19735 <=$>
1957 N2 P55-71 63-11617 <=>
64-19736 <=$>
1957 N2 P73-81 64-11658 <=>
1957 N2 P101-116 65-33203 <=*$>
1958 N5 P75-81 AEC-TR-6246 <=*$>
1959 N2 P7-13 61-18312 <*=>
1959 N3 P79-84 61-28941 <+*> 0
1959 N3 P109-117 59-31035 <=>
1959 N3 P144-146 60-11213 <=>
1959 N4 P25-33 60-19083 <+*>
1959 N5 P23-51 60-31177 <=> 0
1959 N6 P65-78 61-11976 <=>
1960 N1 P5-7 60-31320 <=>
1960 N1 P9-15 60-31320 <=>
1960 N1 P17-56 60-31320 <=>
1960 N1 P89-101 60-31320 <=>
1960 N2 P89-93 64-15172 <=*$>
1960 N3 P57-61 61-28498 <*=>
1960 N5 P73-78 62-15820 <*=>
1960 N6 P83-95 63-13404 <=*>
1960 N6 P121-132 61-28815 <*=>
1961 N2 P65-68 62-32292 <=*>
1961 N3 P103-106 62-11170 <=>
1961 N3 P127-138 63-13637 <=*>
63-15720 <=*>
1961 N3 P155-158 62-11170 <=>
1961 N4 P39-42 62-33211 <=*>
1961 N4 P67-74 62-23430 <=*>
1962 N6 P9-16 63-21577 <=>
1963 N6 P91-94 65-62435 <=$>
1963 V6 N5 P95-97 AD-633 406 <=$>
1964 N3 P27-31 AD-627 538 <=*$>
1964 N4 P137-140 65-64487 <=*$>
1965 N1 P81-90 65-31489 <=$>
1965 N2 P97-104 AD-623 435 <=*$>
1965 N2 P123-127 66-32627 <=$>
1965 N3 P3-11 AD-625 884 <=*$>
1966 N2 P27-48 66-34962 <=$>

IZVESTIYA VYSSHIKH UCHEBNYKH ZAVEDENII. GEOLOGIYA I RAZVEDKA

1958 N1 P107-123 59-11183 <=>
1958 N2 P3-17 IGR V1 N7 P30 <AGI>
1958 N4 P3-13 NLLTB V1 N3 P17 <NLL>
1958 N4 P121-132 IGR V4 N8 P971 <AGI>
1958 V1 N6 P83-94 65-11413 <*>
1959 N4 P124-136 IGR V2 N11 P936 <AGI>
1959 V2 N3 P87-88 61-19560 <*=>
1960 N12 P88-97 IGR V4 N3 P271 <AGI>
1960 V3 N1 P134-138 67M42R <ATS>
1960 V3 N1 P139-156 64-13591 <=*$>
1960 V3 N4 P103-111 61-11563 <=>
1960 V3 N4 P118-125 61-11563 <=>
1960 V3 N5 P61-66 37N50R <ATS>
1960 V3 N5 P128-136 61-11544 <=>
1960 V3 N6 P70-74 16N52R <ATS>
1960 V3 N10 P81-88 17N52R <ATS>
1961 N7 P3-21 IGR V5 N8 P985 <AGI>
1961 N8 P3-10 IGR V4 N9 P981 <AGI>
1961 N8 P60-73 IGR V5 N4 P379 <AGI>
1961 N9 P25-41 IGR V5 N8 P971 <AGI>
1961 N10 P66-71 IGR V5 N3 P253 <AGI>
1961 V4 N1 P83-93 05N54R <ATS>
1961 V4 N5 P97-101 47N56R <ATS>
1961 V4 N7 P68-74 98N57R <ATS>
1962 N1 P3-18 IGR V6 N2 P189 <AGI>

1962 N7 P95-101	IGR V6 N5 P799 <AGI>
1962 N8 P98-106	IGR V7 N4 P717 <AGI>
1962 V5 P120-129	66-14144 <*$>
1962 V5 N1 P3-18	IGR V6 N2 P189-201 <AGI>
1962 V5 N4 P134-142	43P63R <ATS>
1962 V5 N7 P95-101	IGR,V6,N5,P799-804 <AGI>
1962 V5 N9 P29-42	64-21364 <=>
1962 V5 N9 P73-85	03Q70R <ATS>
1962 V5 N10 P53-62	89Q69R <ATS>
1962 V5 N10 P113-119	AD-640 956 <=$>
1962 V5 N11 P36-50	63-19538 <=*>
1962 V5 N12 P35-43	RTS-2687 <NLL>
1963 N8 P19-25	IGR V7 N4 P724 <AGI>
1963 V6 N3 P13-29	88Q71R <ATS>
1963 V6 N10 P121-127	24R75R <ATS>
1964 N10 P112-140	IGR V8 N1 P1 <AGI>
1964 V7 N2 P39-55	64-21936 <=>
1964 V7 N2 P106-117	91R76R <ATS>
1964 V7 N2 P132-136	52R76R <ATS>
1964 V7 N3 P96-104	65-11099 <*>
1964 V7 N3 P110-119	09R78R <ATS>
1964 V7 N3 P120-123	64-31135 <=>
1964 V7 N6 P127-132	64-51363 <=$>
1964 V7 N9 P109-113	84S80R <ATS>
1964 V7 N10 P123-140	65-30034 <=$>
1964 V7 N11 P114-123	82S88R <ATS>
1964 V7 N12 P103-109	65-30324 <=$>
1964 V7 N12 P110-115	65-30323 <=$>
1965 N1 P18	ICE V6 N1 P55-56 <ICE>
1965 V8 N2 P132-141	28S84R <ATS>
1965 V8 N4 P102-106	66-13593 <*>
1965 V8 N5 P122-127	AD-637 536 <=$>
1965 V8 N7 P123-128	65-32757 <=$>
1965 V8 N8 P143-144	37T90R <ATS>
1965 V8 N9 P120-123	62T91R <ATS>
1965 V8 N10 P125-130	94T91R <ATS>
1966 N3 P155-156	IGR V8 N8 P883 <AGI>
1966 V9 N7 P30-31	AD-641 756 <=$>

IZVESTIYA VYSSHIKH UCHEBNYKH ZANEDENII. GORNYI ZHURNAL

1958 N1 P81-83	59-11351 <=>
1958 N3 P43-52	59-11648 <=> 0
1958 N6 P130-135	60-11311 <=>
1959 N10 P56-58	60-41560 <=>
1959 N10 P59-62	60-41563 <=>
1960 N2 P90-94	UCRL-TRANS-931(L) <=*$>
1960 N11 P69-72	UCRL-TRANS-936(L) <=*$>
1961 N1 P5-12	63-11092 <=>
1961 N10 P60-65	65-60881 <=$>
1962 V5 N3 P38-39	63-19898 <=*$>
1963 V6 N8 P54-60	CSIR-492 <CSSA>
1964 V7 N4 P119-124	ATS-79S85R <ATS>

IZVESTIYA VYSSHIKH UCHEBNYKH ZAVEDENI. KHIMIYA I KHIMICHESKAYA TEKHNOLOGIYA

1958 N1 P86-93	59-22619 <+*>
	60-13477 <+*>
1958 N2 P17-24	62-15391 <*=>
1958 N2 P142-146	64-10581 <=*$>
1958 N2 P152-159	60-13664 <+*>
1958 N3 P136-141	62-16679 <=*>
1958 N4 P122-127	60-13650 <+*>
	70M39R <ATS>
1958 N5 P52-57	66-13374 <*>
1958 N5 P70-75	60-13902 <+*>
	61-19540 <=*> 0
1958 V1 N1 P86-93	19L35R <ATS>
1959 V2 N1 P3-9	62-14347 <=*>
1959 V2 N1 P25-29	48M37R <ATS>
1959 V2 N1 P41-45	61-15220 <*+>
1959 V2 N1 P54-58	63-18917 <=*>
1959 V2 N1 P89-95	64-26381 <$>
1959 V2 N1 P96-101	61-20640 <*=>
1959 V2 N2 P196-199	62-15260 <*=>
1959 V2 N2 P231-237	63-24396 <=*$>
1959 V2 N2 P238-243	61-15866 <*=>
1959 V2 N2 P254-257	83P60R <ATS>
1959 V2 N3 P335-339	AEC-TR-4013 <+*>
1959 V2 N3 P420-423	62-18900 <=*>
1959 V2 N3 P430-436	RUB.CH.TECH.V36 N2 <ACS>
	62-15513 <=*>
1959 V2 N3 P437-442	61-27877 <*=>
1959 V2 N4 P475-479	27M42R <ATS>
1959 V2 N4 P485-489	61-27885 <*=>
1959 V2 N4 P516-521	60-17596 <+*> 0
1959 V2 N4 P582-588	61-27884 <*=>
1959 V2 N5 P651-656	62-13751 <*=>
1959 V2 N5 P685-692	63-20253 <=*>
1959 V2 N5 P797-802	AEC-TR-5574 <=*>
1959 V2 N6 P827-833	98M43R <ATS>
1959 V2 N6 P899-903	73M43R <ATS>
1959 V2 N6 P956-961	60-11746 <=>
1959 V2 N6 P974-977	C-4977 <NRC>
1960 V3 N1 P4-7	61-31627 <=>
1960 V3 N1 P8-13	77N51R <ATS>
1960 V3 N1 P29-32	62-13727 <*=>
1960 V3 N1 P36-40	14S83R <ATS>
1960 V3 N1 P104-108	RCT V35 N3 P794 <RCT>
1960 V3 N1 P186-190	60-31637 <=>
1960 V3 N2 P312-315	D-774 <RIS>
1960 V3 N2 P343-351	61-27133 <*=>
1960 V3 N3 P405-407	63-13398 <=*>
1960 V3 N3 P447-451	63-11620 <=>
1960 V3 N3 P550-559	66-13375 <*> 0
1960 V3 N4 P669-674	62-23272 <=*>
1960 V3 N4 P695-698	62-15424 <*=>
1960 V3 N4 P699-706	UCRL TRANS-745(L) <=*>
1960 V3 N4 P721-724	61-18677 <*=>
1960 V3 N4 P771-775	61-27374 <*=>
1960 V3 N5 P868-871	40N50R <ATS>
1960 V3 N5 P952-958	63-13399 <=*>
1960 V3 N6 P1040-1044	62-15806 <=*>
1960 V3 N6 P1079-1801	62-24295 <=*$>
1960 V3 N6 P1113-1116	61-28509 <*=>
1961 V4 P1041-1042	64-23647 <=$>
1961 V4 N1 P3-6	62-25933 <=*>
1961 V4 N1 P33-37	2571 <BISI>
1961 V4 N1 P38-44	63-13578 <=*>
1961 V4 N1 P138-141	62-25730 <=*>
1961 V4 N1 P142-	ICE V2 N2 P185-188 <ICE>
1961 V4 N1 P142-147	83N57R <ATS>
1961 V4 N2 P320-321	62-23581 <=*>
1961 V4 N2 P328-332	61-28550 <*=>
1961 V4 N3 P355-358	63-13422 <=*>
1961 V4 N3 P366-369	85Q69R <ATS>
1961 V4 N3 P437-445	66-10186 <*> 0
1961 V4 N3 P492-497	RUB.CH.TECH.V36 N2 <ACS>
1961 V4 N4 P545-549	62-14256 <=*>
1961 V4 N4 P550-553	63-13715 <=*>
1961 V4 N4 P636-638	62-20237 <=*>
1961 V4 N5 P854-858	RTS-2974 <NLL>
1961 V4 N5 P863-865	RTS-3142 <NLL>
1961 V4 N6 P992-997	63-10751 <=*>
1961 V4 N6 P1026-1029	RTS-2973 <NLL>
1962 N5 P691-693	N66-11617 <=*$>
1962 N6 P986-994	ICE V3 N3 P369-374 <ICE>
1962 V5 N1 P120-125	63-24091 <=*$>
1962 V5 N2 P322-325	64-16283 <*>
1962 V5 N2 P336-339	63-21012 <=>
1962 V5 N3 P474-476	RCT V37 N3 P770 <RCT>
1962 V5 N3 P496-501	64-23501 <=$>
1962 V5 N4 P529-532	63-19256 <=*>
1962 V5 N4 P621-624	81S81R <ATS>
1962 V5 N4 P675-680	63-11186 <=>
1962 V5 N5 P407-412	65-11010 <*>
1962 V5 N5 P808-814	RAPRA-1252 <RAP>
1962 V5 N5 P845	64-15151 <=*$>
1962 V5 N6 P911-915	AD-640 512 <=$>
1963 N2 P320-327	ICE V3 N4 P562-566 <ICE>
1963 N4 P659-	ICE V4 N2 P263-269 <ICE>
1963 N5 P774-780	ICE V4 N4 P581-585 <ICE>
1963 N5 P802-806	ICE V4 N2 P317-320 <ICE>
1963 N5 P807-	ICE V4 N2 P282-285 <ICE>
1963 N5 P816-	ICE V4 N3 P451-455 <ICE>
1963 N5 P879-881	ICE V4 N4 P573-574 <ICE>
1963 V6 P68-71	64-19855 <=$>
1963 V6 P408-415	65-64417 <=*$>
1963 V6 N1 P165-166	64-71187 <=>
1963 V6 N2 P257-259	64-13647 <=*$>

Citation	Translation
	64-14182 <=*$>
1963 V6 N4 P608-616	66-10279 <*>
1963 V6 N4 P659-	ICE V4 N2 P263-269 <ICE>
1963 V6 N5 P774-780	ICE V4 N4 P581-585 <ICE>
	67R75R <ATS>
1963 V6 N5 P802-806	ICE V4 N2 P317-320 <ICE>
1963 V6 N5 P807-810	ICE V4 N2 P282-285 <ICE>
1963 V6 N5 P816-	ICE V4 N3 P451-455 <ICE>
1963 V6 N5 P816	RTS-2856 <NLL>
1963 V6 N5 P879-881	ICE V4 N4 P573-574 <ICE>
	44R77R <ATS>
1964 N1 P141-147	ICE V4 N4 P609-613 <ICE>
1964 N2 P202-209	AD-638 669 <=$>
1964 N2 P313-319	ICE V5 N1 P157-161 <ICE>
1964 N3 P486-491	ICE V5 N1 P58-61 <ICE>
1964 N3 P497-500	ICE V5 N1 P109-111 <ICE>
1964 N4 P665-668	ICE V5 N2 P331-333 <ICE>
1964 V7 P691-693	92S85R <ATS>
1964 V7 N1 P106-110	AD-614 761 <=$>
1964 V7 N1 P137-140	65-63768 <=*$>
1964 V7 N1 P137-139	65-64765 <=*$>
1964 V7 N1 P141-147	ICE V4 N4 P609-613 <ICE>
1964 V7 N2 P269-273	4552 <BISI>
1964 V7 N3 P416-418	525 <TC>
1964 V7 N4 P651-654	M-5718 <NLL>
1964 V7 N5 P705-710	65-30298 <=$*>
1964 V7 N5 P797-800	65-13894 <*>
1964 V7 N5 P810-815	65-30298 <=$*>
1964 V7 N6 P989-992	65-14795 <*>
1965 N6 P1011-1013	ICE V6 N4 P592-594 <ICE>
1965 N6 P1014-1018	ICE V6 N4 P639-642 <ICE>
1965 V8 N1 P88-93	66-11275 <*>
1965 V8 N1 P131-134	22S89R <ATS>
1965 V8 N1 P142-150	ICE V5 N4 P656-661 <ICE>
1965 V8 N1 P151-	ICE V5 N4 P653-656 <ICE>
1965 V8 N2 P305-309	RAPA-1242 <RAP>

IZVESTIYA VYSSHIKH UCHEBNYKH ZAVEDENII. LESNOI ZHURNAL

Citation	Translation
1959 V2 N1 P112-125	62-00073 <*>
1959 V2 N4 P7-13	65-50047 <=>
1959 V2 N4 P40-45	65-50049 <=>
1960 V3 N1 P43-48	65-50040 <=>
1960 V3 N1 P77-82	66-11551 <*>
1960 V3 N1 P99-104	62-00082 <*>
1960 V3 N1 P130-133	62-00084 <*>
1960 V3 N1 P152-154	62-00083 <*>
1961 V4 N2 P102-111	64-13375 <=$>
1961 V4 N2 P112-120	64-13376 <=$>
1961 V4 N5 P16-20	66-51101 <=>
1961 V4 N5 P51-52	66-51101 <=>
1961 V4 N5 P147-155	M.5798 <NLL>
1961 V5 N1 P34-37	66-51101 <=>
1961 V5 N2 P152-160	66-51101 <=>
1961 V5 N3 P165-166	66-51101 <=>
1962 V5 N5 P37-42	66-51101 <=>
1962 V6 N2 P13-21	66-51101 <=>
1963 V6 N1 P127-130	65-63883 <=$>
1963 V6 N6 P9-13	C-5769 <NRC>
1963 V6 N6 P100-105	65-61571 <=$>

IZVESTIYA VYSSHIKH UCHEBNYKH ZAVEDENII. MASHINOSTROENIE

Citation	Translation
1958 N5 P73-81	60-15806 <=>
1958 N6 P75-80	62-25116 <=*>
1958 N3/4 P52-63	60-13783 <+*>
1958 N3/4 P134-143	61-15296 <=*> 0
1959 N12 P25-32	AD-619 544 <=$>
1960 N2 P19-30	63-19333 <=*>
1960 N2 P66-77	63-19333 <=*>
1960 N3 P43-46	60-41327 <=>
1960 N4 P24-31	4898 <HB>
1960 N4 P86-94	5339 <HB>
1960 N4 P103-113	4899 <HB>
1960 N5 P101-105	63-23214 <=*$>
1960 N5 P140-148	63-13213 <=*>
1960 N6 P139-148	63-13598 <=*>
1960 N7 P149-160	62-33639 <=*>
1960 N8 P115-119	63-11629 <=>
1960 N11 P103-107	AD-624 915 <=*$>
1960 N11 P130-134	61-21745 <=>
1960 N12 P46-54	AD-614 943 <=$>
1960 N12 P104-109	64-71398 <=> M
1961 N1 P32-38	62-13165 <*=>
1961 N4 P101-107	62-33395 <=*>
1961 N4 P161-167	62-13288 <*=>
1961 N5 P26-32	63-13180 <=*>
1961 N9 P137-142	64-19119 <=*$>
1961 N10 P114-123	AD-637 109 <=$> M
1961 N10 P168-179	75P61R <ATS>
1961 N12 P64-75	64-71353 <=> M
1962 N1 P7-13	62-32429 <=*>
1962 N1 P78-89	63-13254 <=*>
1962 N1 P95-104	62-32429 <=*>
1962 N2 P5-13	63-13388 <=*> 0
1962 N6 P64-70	63-23438 <*=>
1962 N6 P124-128	63-13831 <=*>
1962 N9 P5-14	63-21469 <=>
1962 N11 P193-198	64-19854 <=$>
1962 N12 P21-32	64-71406 <=> M
1962 N12 P158-167	64-11799 <=>
1963 N2 P208-223	AD-624 767 <=*$> M
1963 N3 P69-75	SMRE-TRANS-5114 <NLL>
1963 N3 P82-86	AD-621 003 <=$>
1963 N3 P194-199	63-31943 <=>
1963 N3 P209-217	63-31943 <=>
1963 N5 P92-99	64-71484 <=> M
1963 N9 P114-119	64-19856 <=$>
1963 N9 P130-133	AD-614 957 <=$>
1963 N9 P174-183	RTS-3318 <NLL>
1963 N10 P178-183	5923 <HB>
1963 N10 P205-208	65-63517 <=*$>
1963 N10 P216-223	65-60896 <=$>
1963 N12 P49-58	AD-610 352 <=$>
	65-10762 <*>
	65-63730 <=>
1963 N12 P193-202	6251 <HB>
1964 N3 P35-46	AD-620 965 <=$>
1964 N4 P138-146	65-63011 <=$>
1964 N7 P26-30	AD-615 871 <=$>
1964 N8 P42-48	AD-640 475 <=$>
1965 N3 P119-132	65-31123 <*>
1965 N10 P30-34	66-60592 <=$>
1965 V8 N9 P31-34	11TT93R <ATS>
1966 N2 P26-28	66-13489 <*>
	66-61691 <=*$>
1966 N2 P29-34	66-61722 <=$>
1966 N3 P22-30	66-13488 <*>
	66-61720 <=$>
1966 N3 P30-35	66-13477 <*>
	66-61733 <=*$>
1966 N3 P178-183	66-61709 <=*$>

IZVESTIYA VYSSHIKH UCHEBNYKH ZAVEDENII. MATEMATIKA

Citation	Translation
1958 N1 P143-151	AMST S2 V53 P1-12 <AMS>
1958 N1 P161-173	AMS TRANS V.46 P1-16 <AMS>
	AMST S2 V46 P1-16 <AMS>
1958 N4 P113-126	62-23038 <=*>
1958 N4 P218-221	63-15520 <=*>
1958 N5 P32-45	UCRL TRANS-1040(L) <=*$>
1958 N5 P46-51	NLLTB V1 N2 P33 <NLL>
1958 N5 P52-90	62-20210 <=*>
1958 N5 P126-157	61-16082 <=*>
1958 N5 P166-174	59-11770 <=> 0
1959 N2 P227-232	AMST S2 V32 P315-322 <AMS>
1959 N4 P21-26	62-19848 <=>
1959 N4 P27-37	62-23102 <=>
1959 N4 P56-63	62-11435 <=>
1959 N4 P89-93	62-11466 <=>
1959 N4 P122-125	62-19998 <=>
1959 N4 P141-149	AMST S2 V54 P75-84 <AMS>
1959 N4 P150-160	62-19882 <=>
1959 N5 P3-15	62-19849 <=>
1959 N5 P40-47	62-11355 <=>
1959 N5 P104-111	60-12517 <=>
1959 N5 P131-145	62-23216 <=>
	62-23216 <=*>
1959 N5 P208-218	62-19871 <=>
1959 N5 P219-221	62-23228 <=>
1959 N6 P38-43	62-11540 <=>

1959 N6 P118-130 61-19703 <=*>
1959 N6 P145-148 61-14925 <=*>
1960 N1 P166-174 ARSJ V31 N5 P682 <AIAA>
1960 N2 P223-235 C-124 <=>
1960 N3 P62-80 64-71476 <=> M
1960 N4 P36-48 AMST S2 V36 P337-350 <AMS>
1960 N4 P173-177 62-11291 <=>
1960 N4 P206-209 61-10636 <+*>
1960 N6 P62-73 AMST S2 V36 P383-395 <AMS>
1961 N1 P36-43 AMST S2 V36 P285-294 <AMS>
1961 N2 P41-53 61-27227 <=*>
1961 N3 P56-65 62-13703 <*=>
62-13703 <=*>
1961 N4 P78-92 66-13877 <*>
1961 N4 P112-118 63-15293 <=*>
1961 N6 P155-168 AMST S2 V54 P1-16 <AMS>
1961 V4 N9 P53-59 06P60R <ATS>
1962 N1 P172-177 STMSP V5 P308-314 <AMS>
1962 N2 P52-64 AMST S2 V47 P73-87 <AMS>
1962 N3 P48-58 62-32435 <=*>
1962 N3 P143-150 AMST S2 V53 P13-22 <AMS>
1962 N4 P146-151 62-33302 <=>
1963 N1 P66-74 64-13932 <=*$>
1963 N1 P158-168 63-20188 <=*>
1963 N1 P169-171 63-18371 <*=>
1963 N3 P173-184 64-15104 <=*$>
1964 N4 P111-117 AD-625 790 <=*$>
1964 N5 P3-7 65-12641 <*>
1964 N5 P91-94 65-12829 <*>
1964 N6 P59-66 AD-624 851 <=*$>
1964 N6 P159-167 AD-624 851 <=*$>
1965 N4 P91-99 66-11220 <*>
1965 N6 P176-179 AD-639 911 <=$>

IZVESTIYA VYSSHIKH UCHEBNYKH ZAVEDENII. NEFT I GAZ
1958 N1 P73-80 60-10988 <+*>
1958 N1 P81-85 60-10989 <+*>
1958 N6 P31-38 48L30R <ATS>
1958 N6 P73-76 59-18460 <+*>
1958 N7 P11-15 49L30R <ATS>
1958 N8 P17-23 50L30R <ATS>
1958 N8 P115-119 60-15730 <+*>
1958 V1 N1 P41-44 63-16989 <*=>
1958 V1 N2 P19-22 20R79R <ATS>
1958 V1 N5 P93-95 82M37R <ATS>
1958 V1 N7 P57-65 41N57R <ATS>
1958 V1 N8 P41-46 29N52R <ATS>
1958 V1 N10 P63-68 53N54R <ATS>
1959 N1 P27-33 23L32R <ATS>
1959 N2 P47-52 19L34R <ATS>
1959 N2 P59-65 28L34R <ATS>
1959 N3 P39-41 42L34R <ATS>
1959 N3 P63-66 04L34R <ATS>
1959 N4 P7-9 67L34R <ATS>
1959 N4 P41-48 45L34R <ATS>
1959 V2 N2 P67-74 61-28351 <*=>
1959 V2 N2 P99-105 04Q68R <ATS>
1959 V2 N3 P3-4 08M39R <ATS>
1959 V2 N3 P51-54 41L34R <ATS>
1959 V2 N4 P49-55 08L35R <ATS>
1959 V2 N4 P71-78 61-23426 <=*> O
1959 V2 N5 P31-36 85L34R <ATS>
1959 V2 N5 P37-43 63-10211 <=*>
1959 V2 N5 P75-78 88L35R <ATS>
1959 V2 N6 P43-51 83L36R <ATS>
1959 V2 N6 P103-108 04L37R <ATS>
1959 V2 N7 P9-12 79R78R <ATS>
1959 V2 N7 P17-24 68M37R <ATS>
1959 V2 N7 P33-40 51M39R <ATS>
1959 V2 N7 P41-48 63M37R <ATS>
1959 V2 N7 P91-95 89M37R <ATS>
1959 V2 N7 P97-102 45M37R <ATS>
1959 V2 N7 P103-106 94Q67R <ATS>
1959 V2 N8 P13-16 64-13688 <=*$>
1959 V2 N8 P25-26 68M39R <ATS>
1959 V2 N8 P51-53 14P63R <ATS>
1959 V2 N8 P55-61 14M39R <ATS>
1959 V2 N8 P109-113 51M43R <ATS>
1959 V2 N9 P11-18 IGR V3 N2 P141 <AGI>
1959 V2 N9 P31-35 12M39R <ATS>
1959 V2 N9 P37-40 11M39R <ATS>
1959 V2 N9 P41-48 41M39R <ATS>
65-18060 <*>
1959 V2 N9 P49-55 42M39R <ATS>
1959 V2 N9 P57-62 36M40R <ATS>
1959 V2 N9 P79-80 74M39R <ATS>
1959 V2 N9 P117-121 57M40R <ATS>
1959 V2 N10 P19-25 94M41R <ATS>
1959 V2 N10 P45-51 69Q73R <ATS>
1959 V2 N10 P73-77 62M43R <ATS>
1959 V2 N10 P87-92 60-41625 <=>
1959 V2 N11 P3-8 60M44R <ATS>
1959 V2 N11 P37-42 61M42R <ATS>
1959 V2 N12 P49-54 54N54R <ATS>
1959 V2 N12 P63-68 57M41R <ATS>
1960 V3 N1 P19-25 56M44R <ATS>
1960 V3 N1 P27-33 93P58R <ATS>
1960 V3 N1 P53-57 31M47R <ATS>
1960 V3 N1 P58-63 28M43R <ATS>
1960 V3 N1 P71-78 30M43G <ATS>
1960 V3 N1 P93-100 54R75R <ATS>
1960 V3 N2 P39-45 94P58R <ATS>
1960 V3 N2 P53-57 82M43R <ATS>
1960 V3 N2 P69-71 65-14629 <*>
1960 V3 N2 P81-85 60-11733 <=>
1960 V3 N3 P45-50 79M43R <ATS>
1960 V3 N3 P89-96 81M43R <ATS>
1960 V3 N4 P3-8 61-11977 <=>
1960 V3 N4 P49-53 47M44R <ATS>
1960 V3 N4 P67-71 73M46R <ATS>
1960 V3 N4 P91-98 61-28475 <=*>
1960 V3 N4 P124 61-11636 <=>
1960 V3 N5 P51-56 98M46R <ATS>
1960 V3 N5 P71-74 63N49R <ATS>
1960 V3 N6 P47-50 79N48R <ATS>
1960 V3 N6 P51-55 76M45R <ATS>
1960 V3 N6 P63-70 60N48R <ATS>
1960 V3 N6 P111-118 59N48R <ATS>
1960 V3 N7 P15-20 27N53R <ATS>
1960 V3 N7 P29-34 95P58R <ATS>
1960 V3 N7 P35-41 04N48R <ATS>
1960 V3 N7 P59-64 56N48R <ATS>
1960 V3 N8 P21-26 95N51R <ATS>
1960 V3 N8 P39-44 69M47R <ATS>
1960 V3 N8 P45-52 70M47R <ATS>
1960 V3 N8 P65-72 34M47R <ATS>
1960 V3 N8 P123-128 43M47R <ATS>
1960 V3 N9 P39-41 44N48R <ATS>
1960 V3 N9 P43-49 45N48R <ATS>
1960 V3 N9 P97-101 75N48R <ATS>
1960 V3 N9 P117-122 46N48R <ATS>
1960 V3 N10 P75-81 87N48R <ATS>
1960 V3 N10 P93-98 58N53R <ATS>
1960 V3 N11 P95-100 62-19265 <*=>
1960 V3 N12 P73-78 31N54R <ATS>
1961 V4 N1 P43-48 13N52R <ATS>
1961 V4 N1 P73-77 ICE V1 N1 P61-63 <ICE>
80N51R <ATS>
1961 V4 N2 P18 62-13248 <*=>
1961 V4 N2 P24 62-13248 <*=>
1961 V4 N2 P49-55 22N54R <ATS>
1961 V4 N2 P65-67 27N55R <ATS>
1961 V4 N2 P68 62-13247 <*=>
1961 V4 N2 P94 62-13236 <*=>
1961 V4 N2 P95-100 62-23675 <=*>
1961 V4 N2 P106 62-13236 <*=>
1961 V4 N2 P168-173 62-23770 <=>
1961 V4 N3 P43-46 40N55R <ATS>
1961 V4 N3 P75-79 92Q67R <ATS>
1961 V4 N4 P3-37 73N55R <ATS>
1961 V4 N4 P29-32 06N54R <ATS>
1961 V4 N4 P123-127 62-13231 <*=>
1961 V4 N5 P8 61-28281 <=>
1961 V4 N5 P18 61-28281 <=>
1961 V4 N5 P23-29 80N55R <ATS>
1961 V4 N5 P30 61-28281 <=>
1961 V4 N5 P39-46 54P60R <ATS>
63-27104 <=$>
1961 V4 N5 P52 61-28281 <=>
1961 V4 N5 P62 61-28281 <=>

1961 V4 N5 P68 61-28281 <=>
1961 V4 N5 P111-116 88N55R <ATS>
1961 V4 N6 P81-87 61-28475 <=*>
1961 V4 N7 P9-15 12N58R <ATS>
1961 V4 N7 P29-33 98P58R <ATS>
1961 V4 N7 P121-123 62-23770 <=>
1961 V4 N8 P37-44 44N58R <ATS>
1961 V4 N8 P45-50 45N58R <ATS>
1961 V4 N8 P51-57 38P59R <ATS>
1961 V4 N8 P59-62 46N58R <ATS>
1961 V4 N8 P69-73 81P58R <ATS>
1961 V4 N8 P75-81 47N58R <ATS>
1961 V4 N8 P83-86 29N58R <ATS>
1961 V4 N10 P3-10 19R79R <ATS>
1961 V4 N11 P23-27 44P60R <ATS>
1961 V4 N11 P47-52 11P60R <ATS>
1961 V4 N11 P53-58 01P65R <ATS>
1961 V4 N11 P65-70 05P61R <ATS>
1961 V4 N12 P61-62 87P60R <ATS>
1961 V4 N12 P85-94 81P60R <ATS>
1961 V4 N12 P95-98 02P61R <ATS>
1961 V4 N12 P113-116 06P61R <ATS>
1961 V5 N3 P25-30 57P62R <ATS>
1962 V5 N2 P7-10 01P62R <ATS>
1962 V5 N2 P15-18 26Q69R <ATS>
1962 V5 N2 P59-62 36P61R <ATS>
1962 V5 N3 P3-8 66P62R <ATS>
1962 V5 N4 P59-63 99P62R <ATS>
1962 V5 N5 P29-34 64P63R <ATS>
1962 V5 N5 P39-44 63P63R <ATS>
1962 V5 N5 P45-51 65P63R <ATS>
1962 V5 N5 P53-57 20P64R <ATS>
1962 V5 N6 P55-60 70P65R <ATS>
1962 V5 N6 P77-84 93Q67R <ATS>
1962 V5 N6 P93-96 16Q72R <ATS>
1962 V5 N7 P9-13 05Q67R <ATS>
1962 V5 N7 P31-36 91S82R <ATS>
1962 V5 N7 P37-41 59P66R <ATS>
1962 V5 N7 P95-99 17Q72R <ATS>
1962 V5 N9 P63-69 96Q66R <ATS>
1962 V5 N10 P35-40 95Q67R <ATS>
1962 V5 N10 P47-51 01Q68R <ATS>
1962 V5 N11 P13-16 63Q72R <ATS>
1962 V5 N11 P21-24 32Q69R <ATS>
1962 V5 N11 P39-45 11Q69R <ATS>
1962 V5 N11 P53-57 64-10982 <=*$>
1962 V5 N11 P71-74 64-10998 <=*$>
1962 V5 N12 P59- ICE V3 N3 P324-327 <ICE>
1962 V5 N12 P59-64 12Q69R <ATS>
1962 V5 N12 P111-115 63-21807 <=>
1963 N3 P59-64 ICE V4 N3 P519-524 <ICE>
1963 V6 N1 P35-40 48Q70R <ATS>
1963 V6 N1 P41-48 53Q70R <ATS>
1963 V6 N2 P23-28 70Q70R <ATS>
1963 V6 N2 P93-97 98Q70R <ATS>
1963 V6 N2 P101-104 C-5354 <NRC>
1963 V6 N3 P47-50 RTS-2717 <NLL>
1963 V6 N3 P59-64 ICE V4 N3 P519-524 <ICE>
92Q72R <ATS>
1963 V6 N3 P65-70 64-16422 <=*$>
1963 V6 N4 P19-24 89S82R <ATS>
1963 V6 N4 P25-28 41Q72R <ATS>
1963 V6 N4 P29-32 02Q72R <ATS>
1963 V6 N4 P93-98 64-16398 <=*$>
1963 V6 N5 P77-80 64-71380 <=>
1963 V6 N6 P43-48 69R77R <ATS>
1963 V6 N6 P77-82 88Q73R <ATS>
1963 V6 N7 P37-42 23R75R <ATS>
1963 V6 N7 P89-94 38R74R <ATS>
1963 V6 N8 P25-29 64-26106 <=$>
1963 V6 N8 P61-64 33R75R <ATS>
1963 V6 N9 P31-35 02R79R <ATS>
1963 V6 N9 P43-46 46R75R <ATS>
1963 V6 N10 P19-23 60R75R <ATS>
1963 V6 N10 P35-39 28R75R <ATS>
1963 V6 N10 P47-50 59R75R <ATS>
1963 V6 N11 P3-5 66-10308 <*>
1963 V6 N11 P23-27 84R75R <ATS>
1963 V6 N11 P97-99 82R75R <ATS>
1963 V6 N12 P3-7 65-12822 <*>
1963 V6 N12 P23-28 00S81R <ATS>
1963 V6 N12 P97-102 33R76R <ATS>
1964 V7 N1 P11-16 64-26429 <$>
1964 V7 N1 P23-28 51R76R <ATS>
1964 V7 N1 P35-38 65-27154 <$>
1964 V7 N1 P47-51 44R79R <ATS>
64-26500 <$>
1964 V7 N1 P61 66-10598 <*>
1964 V7 N1 P69-72 20R77R <ATS>
65-27479 <$>
1964 V7 N1 P73-75 66-10599 <*>
1964 V7 N1 P77-80 65-12634 <*>
65-27478 <$>
1964 V7 N2 P41-44 RTS-2700 <NLL>
65-27475 <$>
1964 V7 N2 P67-71 65-27476 <$>
1964 V7 N2 P103-108 65-27477 <$>
1964 V7 N3 P31-36 65-28104 <$>
1964 V7 N3 P36-41 RTS-2960 <NLL>
1964 V7 N3 P37-41 52R77R <ATS>
1964 V7 N3 P43-47 10R78R <ATS>
1964 V7 N3 P69-73 00S89R <ATS>
1964 V7 N4 P3-5 31R80R <ATS>
1964 V7 N4 P17-18 66-10309 <*>
1964 V7 N4 P23-27 64-23310 <$>
1964 V7 N4 P47-50 26R78R <ATS>
1964 V7 N6 P39-43 79R79R <ATS>
1964 V7 N6 P45-48 32S84R <ATS>
65-27474 <*>
1964 V7 N7 P71-75 16R80R <ATS>
1964 V7 N8 P31-37 51R80R <ATS>
1964 V7 N8 P47-52 98S82R <ATS>
1964 V7 N8 P53-56 48S81R <ATS>
1964 V7 N8 P83-87 52R80R <ATS>
1964 V7 N9 P69-71 72S81R <ATS>
1964 V7 N9 P77-81 65-14754 <*>
1964 V7 N9 P83-88 45S84R <ATS>
1964 V7 N11 P9-12 99S82R <ATS>
1964 V7 N11 P20-24 35S82R <ATS>
1964 V7 N11 P35-38 45T90R <ATS>
1964 V7 N11 P61-63 65-14753 <*>
1964 V7 N11 P64-68 RTS-2961 <NLL>
61S82R <ATS>
1964 V7 N11 P105-106 65S83R <ATS>
1964 V7 N12 P14-26 11S83R <ATS>
1964 V7 N12 P39-44 62S82R <ATS>
1964 V7 N12 P67-72 15S83R <ATS>
1965 N5 P25-27 24S87R <ATS>
1965 V8 N1 P18 44S85R <ATS>
1965 V8 N1 P29-34 44S83R <ATS>
1965 V8 N1 P35-39 80S82R <ATS>
1965 V8 N1 P83-86 10S83R <ATS>
1965 V8 N2 P82-88 66-11277 <*>
1965 V8 N3 P25-28 01S89R <ATS>
1965 V8 N3 P65-68 94S87R <ATS>
1965 V8 N3 P97-99 66-11278 <*>
66-14884 <*$>
1965 V8 N4 P3-8 43S86R <ATS>
1965 V8 N4 P41-44 22S88R <ATS>
1965 V8 N5 P47-49 66-10317 <*>
1965 V8 N5 P63-66 31S87R <ATS>
1965 V8 N6 P3-7 65-23806 <$>
1965 V8 N6 P28 66-20954 <$>
1965 V8 N6 P29-33 24S88R <ATS>
1965 V8 N6 P34 66-20954 <$>
1965 V8 N6 P111-115 65-25355 <$>
1965 V8 N7 P13-15 57T92R <ATS>
1965 V8 N7 P67-69 20S89R <ATS>
1965 V8 N7 P70 66-12208 <*>
1965 V8 N7 P71-78 66-11487 <*>
1965 V8 N8 P85-88 66-14237 <*$>
1965 V8 N9 P13-16 14T92R <ATS>
1965 V8 N9 P41-45 49T91R <ATS>
1965 V8 N10 P85-88 31T92R <ATS>
1965 V8 N10 P101-102 95T91R <ATS>
1965 V8 N11 P8 66-13781 <*>
1965 V8 N11 P45-48 05T94R <ATS>
1965 V8 N11 P69-74 66-14235 <*$>
1965 V8 N11 P93-96 66-14404 <*$>
1965 V8 N12 P38 44T93R <ATS>

1965 V8 N12 P42 44T93R <ATS>
1966 V9 N1 P69-74 72T94R <ATS>

IZVESTIYA VYSSHIKH UCHEBNYKH ZAVEDENII.
PISHCHEVAYA TEKHNOLOGIYA
1958 N2 P110-114 60-17920 <+*>
1958 N4 P131-135 59-22408 <+*>
1959 N4 P98-111 61-19221 <+*>
1959 N5 P157-164 61-28163 <*=>
1959 N6 P112-115 61-10554 <+*> 0
1960 N1 P144-145 61-27040 <*=>
1960 N4 P35-39 61-18830 <*=>
1960 N5 P91-96 11411-J <K-H>
1960 N5 P97-102 11401-H <KH>
61-20048 <*=>
1960 N5 P160-164 11401-A <K-H>
61-20248 <*=>
63-16921 <*=>
1960 N5 P165-169 11401-B <KH>
63-16689 <*=>
1961 N1 P12-16 11891 <K-H>
63-14965 <=*>
1961 N1 P45-49 65-11879 <*>
1961 N1 P140-146 63-14949 <=*>
1961 N2 P54-57 11455-F <KH>
64-14516 <=*$>
1961 N2 P58-63 63-16920 <=*>
1961 N3 P97-101 11538-C <KH>
64-14565 <=*$>
1961 N3 P164-168 61-28892 <*=>
1961 N4 P44-50 62-16515 <=*>
1961 N4 P51-57 63-10684 <=*>
1961 N4 P58-61 62-16516 <=*>
1961 N8 P24-25 62-13288 <*=>
1962 N2 P76-81 64-11593 <=>
1962 N4 P49-52 RTS 2347 <NLL>
1962 N4 P144-149 63-20135 <=*>
1962 N4 P150-153 63-18932 <=*> 0
1963 N1 P24-29 65-60949 <=$>
1963 N2 P47-51 RTS-2660 <NLL>
65-60198 <=$>
1963 N2 P77-82 64-19114 <=*$>
1963 N2 P130-137 64-19115 <=*$>
1964 N1 P13-19 AD-628 430 <=*$>
1964 N1 P40-43 AD-628 419 <=*$>
1964 N1 P61-63 AD-628 420 <=*$>
1964 N1 P79-81 AD-628 421 <=*$>
1964 N1 P87-94 AD-628 422 <=*$>
1964 N1 P167-171 AD-628 428 <=*$>
1964 N1 P174 AD-628 429 <=*$>
1964 N3 P77-79 76S85R <ATS>
1965 N1 P47-50 M-5779 <NLL>
1965 N4 P44-46 AD-626 068 <=*$>
1965 N4 P52-56 AD-630 855 <=$>
1965 N4 P57-59 AD-627 010 <=$>
1965 N4 P60-62 AD-627 698 <=$>
1965 N4 P68-73 AD-631 281 <=$>
1965 N4 P76-78 AD-630 854 <=$>
1965 N4 P78-81 AD-626 051 <=*$>
1965 N6 P17-20 AD-633 211 <=$>
1965 N6 P24-28 AD-634 415 <=$>
1965 N6 P39-42 AD-634 347 <=$>
1965 N6 P61-64 AD-634 346 <=$>

IZVESTIYA VYSSHIKH UCHEBNYKH ZAVEDENII.
PRAVOVEDENIE LENINGRAD
1959 V2 N3 P322-327 62-32956 <=*>
1960 N2 P42-50 60-41299 <=>

*IZVESTIYA VYSSHIKH UCHEBNYKH ZAVEDENII. PRIBO-
ROSTROENIE
1958 N3 P30-35 61-19316 <+*> 0
1958 N5 P37-45 60-21976 <=>
1958 N5 P69-83 61-23460 <=*>
1958 N6 P33-37 60-11426 <=>
1958 N6 P50-53 60-11400 <=>
1958 N6 P114-117 60-11401 <=>
1959 N2 P6-16 60-23039 <*+>
1959 N2 P56-58 ARSJ V30 N7 P661 <AIAA>
1959 N2 P67-75 63-13188 <=*>
1959 N3 P21-24 60-11614 <=>
61-13293 <=*>
1959 N3 P25-28 60-11615 <=>
1959 N3 P29-30 ARSJ V30 N7 P668 <AIAA>
1959 N3 P29-39 60-11815 <=>
1959 N3 P40-49 60-11798 <=>
1959 N3 P57- 60-11722 <=>
1959 N4 P91-96 61-21960 <=> 0
1959 N4 P97-105 61-21962 <=> 0
1959 N6 P17-22 61-31501 <=> 0
1959 N6 P29-37 61-31571 <=>
1959 V2 N1 P34-37 AD-608 604 <=>
1959 V2 N4 P38-45 63-23752 <=*$>
1960 N3 P59-65 61-11908 <=>
1960 N3 P66-73 61-21349 <=>
1960 N3 P74-76 61-21347 <=>
1960 N5 P88-94 AD-630 850 <=$>
1960 N5 P95-105 AD-630 851 <=$>
1960 N6 P43-49 AIAAJ V1 N9 P2222 <AIAA>
1960 V3 N1 P61-68 61-21162 <=>
1960 V3 N2 P116-118 61-23521 <=*>
1960 V3 N3 P54-58 61-21144 <=>
1960 V3 N5 P3-9 62-15195 <=*>
1960 V3 N5 P10-19 62-14000 <=*>
1960 V3 N5 P44-62 62-15195 <=*>
1960 V3 N5 P52-62 ARSJ V32 N9 P1473 <AIAA>
1960 V3 N5 P106-109 62-15195 <=*>
1960 V3 N6 P100-110 62-23380 <=*>
1961 N1 P94-98 ARSJ V32 N11 P1807 <AIAA>
1961 N2 P101-110 AD-630 852 <=$>
1961 V4 N1 P94-98 62-23375 <=*>
1961 V4 N2 P52-66 62-25738 <=*>
1961 V4 N2 P111-116 62-24866 <=*>
1961 V4 N3 P11-18 AD-630 929 <=$>
1961 V4 N3 P47-54 62-23374 <=*>
1961 V4 N3 P68-74 62-15180 <*=>
1961 V4 N3 P83-94 AD-630 929 <=$>
1961 V4 N3 P120-129 62-24354 <=*>
1961 V4 N3 P135-140 62-33062 <=*>
1961 V4 N4 P39-47 62-32945 <=*>
1961 V4 N4 P48-70 62-15194 <*=>
1961 V4 N4 P48-52 63-10914 <*>
1961 V4 N4 P78-84 63-13217 <=*>
1961 V4 N4 P101-108 AD-630 853 <=$>
1961 V4 N5 P43-46 63-19478 <=*>
1961 V4 N5 P76-83 62-33543 <=*>
1961 V4 N5 P94-104 AIAAJ V1 N6 P1491 <AIAA>
62-15833 <=*>
1961 V4 N6 P78-86 62-32299 <=*>
1961 V4 N6 P98-108 62-24947 <=*>
1962 V5 N1 P82-88 63-13697 <=*>
1962 V5 N3 P82-92 64-13007 <=*$>
1962 V5 N5 P140-147 63-13470 <=*>
1962 V5 N6 P55-57 64-13006 <=*$>
1963 V6 N1 P47-53 64-13605 <=*$>
1963 V6 N2 P89-98 64-21137 <=>
1963 V6 N3 P3-9 64-21371 <=>
1963 V6 N3 P10-18 64-21371 <=>
1963 V6 N3 P19-25 64-21371 <=>
1963 V6 N3 P26-35 64-21371 <=>
1963 V6 N3 P36-64 64-21371 <=>
1963 V6 N3 P45-54 63-20963 <=*$>
64-21371 <=>
1963 V6 N3 P55-62 64-21371 <=>
1963 V6 N3 P63-67 64-21371 <=>
1963 V6 N3 P68-76 64-16077 <=*$>
64-21371 <=>
1963 V6 N3 P77-84 64-21371 <=>
1963 V6 N3 P85-95 63-20958 <=*$>
64-21371 <=>
1963 V6 N3 P96-106 64-21371 <=>
1963 V6 N3 P107-114 64-21317 <=>
1963 V6 N3 P115-117 64-21371 <=>
1963 V6 N3 P118-123 64-21371 <=>
1963 V6 N3 P124-133 64-21371 <=>
1963 V6 N3 P134-142 64-21371 <=>
1963 V6 N3 P143-148 64-21371 <=>
1963 V6 N3 P149-160 64-21371 <=>
1963 V6 N3 P161-163 64-21371 <=>
1963 V6 N4 P30-37 AD-611 528 <=$>

1963 V6 N4 P90-97 64-10377 <=*$>
1963 V6 N4 P110-122 65-11005 <*>
1963 V6 N5 P3-10 64-21462 <=>
1963 V6 N5 P3-147 64-21462 <=>
1963 V6 N5 P11-19 64-21462 <=>
1963 V6 N5 P20-26 64-21462 <=>
1963 V6 N5 P27-33 64-21462 <=>
1963 V6 N5 P34-40 64-21462 <=>
1963 V6 N5 P41-50 64-21462 <=>
1963 V6 N5 P51-57 64-21462 <=>
1963 V6 N5 P58-68 64-21462 <=>
1963 V6 N5 P69-74 64-21462 <=>
1963 V6 N5 P75-83 64-21462 <=>
1963 V6 N5 P84-94 64-21462 <=>
1963 V6 N5 P95-102 64-21462 <=>
1963 V6 N5 P103-111 64-21462 <=>
1963 V6 N5 P112-125 64-21462 <=>
1963 V6 N5 P126-135 64-21462 <=>
1963 V6 N5 P136-138 64-21462 <=>
1963 V6 N5 P139-140 64-21462 <=>
1963 V6 N5 P141-145 64-21462 <=>
1965 V8 N1 P3-191 65-31005 <=*$>
1965 V8 N5 P1-164 66-30303 <=$>

*IZVESTIYA VYSSHIKH UCHEBNYKH ZAVEDENII. RADIO-FIZIKA

1958 V1 N1 P20-33 61-28688 <*=>
1958 V1 N1 P34-40 59-19488 <+*>
1958 V1 N1 P41-66 61-16438 <*=>
1958 V1 N1 P110-119 61-21285 <=>
1958 V1 N1 P120-140 64-13109 <=*$>
1958 V1 N1 P141-149 61-21389 <=>
1958 V1 N1 P150-152 59-19488 <+*>
1958 V1 N2 P19-26 60-17314 <+*>
1958 V1 N2 P36-50 60-19042 <+*>
1958 V1 N2 P59-65 59-14539 <=>
1958 V1 N2 P140-168 61-21389 <=>
1958 V1 N3 P3-12 59-13546 <=>
1958 V1 N3 P19-24 62-11442 <=>
62-11442 <=> O
1958 V1 N3 P25-29 59-13545 <=>
59-20695 <+*>
1958 V1 N3 P54-63 60-15983 <+*>
1958 V1 N3 P110-123 AEC-TR-4750 <*=>
1958 V1 N3 P158 60-17315 <+*>
1958 V1 N3 P159-161 60-17461 <+*>
1958 V1 N3 P161-162 59-13547 <=>
59-20689 <+*>
1958 V1 N4 P63-68 63-11723 <=>
1958 V1 N4 P81-89 61-27500 <=*>
1958 V1 N5/6 P17-28 N65-24627 <=$>
1958 V1 N5/6 P29-33 63-13899 <=*>
1958 V1 N5/6 P60-65 63-11717 <=>
1958 V1 N5/6 P66-74 64-11537 <=>
1958 V1 N5/6 P75-82 62-10641 <=*>
1958 V1 N5/6 P83-87 63-11713 <=>
1959 V2 N1 P3-146 60-41158 <=>
1959 V2 N1 P132-133 59-22734 <+*>
62-11226 <=>
1959 V2 N1 P145-146 62-11524 <=>
1959 V2 N2 P154-323 60-14050 <=>
1959 V2 N3 P355-369 61-11192 <=>
1959 V2 N3 P374-376 61-11167 <=>
1959 V2 N3 P508-509 60-15455 <+*>
62-11403 <=>
1959 V2 N4 P521-572 60-23781 <+*>
60-25488 <=>
1959 V2 N4 P521-673 60-41412 <=>
1959 V2 N4 P521-572 62-11560 <=>
1959 V2 N4 P654-655 60-12972 <=>
1959 V2 N4 P666-673 60-11851 <=>
1959 V2 N5 P677-837 60-41281 <=>
1959 V2 N5 P691-696 60-11382 <=>
1959 V2 N5 P827-829 60-11383 <=>
1959 V2 N6 P876-883 62-10614 <=*>
1959 V2 N6 P1005-1007 61-19373 <+*> O
1959 V2 N6 P1007-1009 61-28826 <*=>
62-00096 <*>
1960 V3 P393-404 ARSJ V32 N11 P1783 <AIAA>
1960 V3 N1 P5-32 61-00076 <*>
1960 V3 N1 P33-38 62-14004 <=*>
1960 V3 N1 P79-88 ARSJ V32 N3 P462 <AIAA>
1960 V3 N1 P97-101 ARSJ V32 N7 P1140 <AIAA>
1960 V3 N1 P146-148 ARSJ V32 N11 P1776 <AIAA>
1960 V3 N2 P192-198 60-23779 <+*>
62-11252 <=>
1960 V3 N2 P208-215 60-23780 <+*>
62-11183 <=>
1960 V3 N2 P257-268 62-15890 <=*>
1960 V3 N2 P316-327 62-10160 <*=>
1960 V3 N2 P341-342 60-31800 <=>
1960 V3 N3 P393-404 60-27006 <*+>
62-11185 <=>
1960 V3 N4 P551-583 1804 <RIA>
1960 V3 N4 P606-614 62-11476 <=>
1960 V3 N4 P619-630 61-31532 <=>
1960 V3 N4 P620-630 ARSJ V32 N11 P1777 <AIAA>
1960 V3 N4 P656-666 1802 <RIA>
1960 V3 N4 P722-723 ARSJ V32 N9 P1478 <AIAA>
63-19146 <=*>
1960 V3 N5 P737-745 61-31546 <=>
1960 V3 N5 P758-764 62-32632 <=*>
1960 V3 N5 P778-788 62-11501 <=>
1960 V3 N5 P901-903 61-31503 <=>
1960 V3 N5 P907-909 61-31540 <=>
1960 V3 N6 P943-948 62-11317 <=>
1960 V3 N6 P949-956 62-13627 <*=>
62-13627 <=>
63-15976 <=*>
1960 V3 N6 P1128-1129 62-11472 <=>
1961 V4 N1 P5-39 61-27848 <*=>
1961 V4 N1 P58-66 61-13624 <*=>
62-11571 <=>
1961 V4 N1 P74-89 62-13623 <*=>
62-13623 <=>
1961 V4 N1 P177-178 62-13622 <*=>
62-13622 <=>
1961 V4 N2 P244-252 62-25101 <=*>
1961 V4 N2 P253-258 62-25650 <=*>
1961 V4 N2 P282-292 62-23034 <=*>
1961 V4 N2 P293-305 62-32105 <=*>
1961 V4 N2 P376-377 62-13843 <*=>
1961 V4 N2 P377-379 62-13844 <*=>
1961 V4 N3 P415-424 N65-24670 <=*$>
1961 V4 N3 P444-475 62-32514 <=*>
1961 V4 N3 P484-495 63-11776 <=>
1961 V4 N3 P496-507 63-11718 <=>
1961 V4 N3 P508-514 64-11565 <=>
1961 V4 N4 P639-647 62-10634 <=*>
1961 V4 N5 P795-830 62-01373 <*>
63-19602 <=*$> OP
63-19603 <=*$> P
1962 V5 N1 P5-12 63-23350 <=*$>
1962 V5 N3 P464-467 N65-24662 <=*$>
1962 V5 N3 P468-472 N65-24663 <=*>
1962 V5 N3 P602-603 62-11786 <=>
1962 V5 N3 P604-606 62-11788 <=>
1962 V5 N4 P623-826 63-13673 <=>
1962 V5 N4 P687-696 63-11719 <=>
1962 V5 N4 P799-801 63-10510 <*>
1962 V5 N4 P801-802 63-10734 <=*>
1962 V5 N5 P998-1008 64-10799 <=*$>
1962 V5 N6 P1057-1061 63-11683 <=>
1962 V5 N6 P1187-1191 63-23611 <=*>
1963 V6 N2 P398-401 64-13304 <=*$>
1963 V6 N3 P431-436 66-11072 <*>
1963 V6 N3 P437-448 64-15082 <=*$>
1963 V6 N3 P616-623 63-20854 <=*$>
1963 V6 N4 P660-668 65-10892 <*>
1963 V6 N5 P877-1077 64-21748 <=>
1963 V6 N5 P897-903 AD-611 117 <=$>
1964 V7 N1 P51-58 AD-614 770 <=$>
1964 V7 N3 P395-398 N64-30858 <=$>
1964 V7 N3 P555-556 64-31561 <=$>
1964 V7 N4 P789-790 AD-634 812 <=$>
1964 V7 N5 P817-821 N65-16599 <=$>
1964 V7 N6 P1021-1031 N65-21005 <=$>
1964 V7 N6 P1041-1048 AD-635 891 <=*$>

*IZVESTIYA VYSSHIKH UCHEBNYKH ZAVEDENII.

RADIOTEKHNIKA. KIEV

Citation	Translation
1958 N1 P43-48	59-14251 <+*>
1958 N3 P271-287	59-16561 <+*>
1958 N4 P415-421	59-13110 <=>
1958 V1 N2 P214-221	60-21799 <=> O
1958 V1 N5 P551-554	61-23540 <=*>
1959 N3 P267-282	60-11104 <=>
1959 N4 P501-502	60-11189 <=>
1959 V2 N1 P38-47	60-21003 <=> O
1959 V2 N1 P80-85	61-27250 <*=>
1959 V2 N2 P155-164	62-23106 <=>
1959 V2 N3 P278-282	61-13549 <+*>
1959 V2 N4 P391-508	60-41031 <=>
1959 V2 N4 P424-430	62-11414 <=>
1959 V2 N4 P462-476	62-21923 <=>
1959 V2 N4 P480-485	60-12576 <=>
	62-23194 <=*>
1959 V2 N4 P490-491	62-19896 <=>
1959 V2 N5 P511-560	60-41341 <=>
1959 V2 N5 P511-533	62-19884 <=>
1959 V2 N5 P554-565	62-15891 <=>
1959 V2 N5 P581-588	62-11346 <=>
1959 V2 N5 P589-599	62-23191 <=>
1959 V2 N6 P658-671	62-23037 <=>
1959 V2 N6 P694-698	62-11465 <=>
1960 V3 N1 P3-12	60-31639 <=>
1960 V3 N1 P30-39	62-13093 <=*>
1960 V3 N1 P130-133	62-11526 <=>
1960 V3 N1 P133-134	62-11528 <=>
1960 V3 N2 P137-296	61-23934 <*=>
1960 V3 N3 P337-341	61-27634 <*=>
1960 V3 N4 P419-430	62-25760 <=*>
1960 V3 N4 P441-447	62-11535 <=>
1960 V3 N5 P425-434	62-13116 <*=>
1960 V3 N5 P517-518	62-11292 <=>
1960 V3 N5 P778-788	61-15938 <+*>
1960 V3 N6 P558-562	62-15611 <*=>
	62-15611 <=>
1960 V3 N6 P581-591	62-23142 <*=>
1960 V3 N6 P605-612	62-23146 <*=>
	62-23146 <=*>
1961 V4 N1 P3-124	62-13381 <=>
1961 V4 N1 P49-54	62-11570 <=>
1961 V4 N1 P64-76	62-11576 <=>
	62-13621 <*=>
1961 V4 N2 P127-236	62-13402 <*=>
1961 V4 N2 P140-147	62-11568 <=>
	62-13630 <*=>
1961 V4 N2 P192-197	62-11321 <=>
	62-13631 <*=>
1961 V4 N3 P1-366	62-13863 <=>
1961 V4 N3 P270-279	62-25942 <=*>
1961 V4 N3 P280-284	62-25105 <=*>
1961 V4 N3 P354-356	62-32106 <=*>
1961 V4 N4 P369-511	62-23789 <=*>
1961 V4 N6 P653-657	62-18528 <=*>
1962 V5 N2 P208-215	65-13669 <*>
1962 V5 N3 P301-413	63-13375 <=>
1962 V5 N4 P415-539	63-21101 <=>
1962 V5 N4 P469-475	63-15324 <=*>
1962 V5 N4 P523-527	63-10511 <=*>
	65-13384 <*>
1962 V5 N4 P538-539	63-15324 <=*>
1963 V6 N2 P197-199	64-11554 <=>
1963 V6 N2 P212-213	63-31965 <=>
1963 V6 N5 P455-583	64-21720 <=>
1963 V6 N5 P562-564	65-63775 <=$>
1964 V7 N2 P131-259	64-41750 <=>
1964 V7 N6 P1041-1048	NAS-5-3760 <=*$>

IZVESTIYA VYSSHIKH UCHEBNYKH ZAVEDENII. STROITELSTVO I ARKHITEKTURA NOUOSIBIRSK

Citation	Translation
1958 V1 N1 P87-97	63-19950 <=*>
1958 V1 N6 P17-27	63-24264 <=*$>
1958 V1 N6 P70-75	63-19698 <=*>
1959 V2 N1 P27-37	63-19564 <=*>
1962 N1 P7-13	AD-619 551 <=$>
1963 N1 P36-42	47S87R <ATS>
1964 N3 P100-109	LC-1239 <NLL>
1964 V7 N9 P80-85	RTS-3296 <NLL>

IZVESTIYA VYSSHIKH UCHEBNYKH ZAVEDENII. TEKHNOLOGIYA LEGKOI PROMYSHLENNOSTI

Citation	Translation
1958 N1 P129-137	59-21178 <=>
1959 N2 P28-38	RCT V34 N2 P588 <RCT>
1959 N2 P69-75	64-13957 <=*$>
1959 N6 P6-10	61-27665 <*=>
1959 N6 P11-17	61-27686 <*=>
1960 N1 P105-110	61-14539 <=*> O
1960 N4 P39-45	63-18385 <=*> O
1960 N4 P46-52	61-21612 <=>
1962 N2 P140-142	NLLTB V4 N10 P923 <NLL>
1963 N4 P46-53	RCT V37 N3 P714 <RCT>
1963 V3 N33 P70-75	64-19991 <=$>
1964 N1 P48-53	RTS-3173 <NLL>
1964 N2 P23-31	65-61562 <=*$>
1964 N6 P12-14	65-63770 <=$>

*IZVESTIYA VYSSHIKH UCHEBNYKH ZAVEDENII. TEKHNOLOGIYA TEKSTILNOI PROMYSHLENNOSTI

Citation	Translation
1958 V2 N2 P94-99	61-13769 <=*> O
1958 V2 N3 P80-86	59-19159 <+*>
1959 V3 N3 P11-19	61-28650 <=*>
1959 V3 N4 P71-74	61-13770 <=*>
1959 V3 N4 P75-85	61-13771 <=*> O
1959 V3 N5 P104-110	63-24363 <=*$>
1959 V3 N6 P104-106	16M46R <ATS>
1959 V3 N9 P37-45	61-15917 <+*>
1959 V3 N10 P19-25	60-23134 <=*>
1959 V3 N10 P88-95	61-15175 <+*>
1960 N1 P40-46	61-11827 <=>
1960 N3 P24-29	ICE V1 N1 P85-88 <ICE>
1960 V4 N1 P32-39	63-18375 <=*>
1960 V4 N1 P86-96	63-18268 <=*$>
	84M47R <ATS>
1960 V4 N1 P97-104	64-10928 <=*$>
1960 V4 N2 P9-16	61-23381 <*=>
	63-18335 <=*>
1960 V4 N2 P46-50	61-28646 <*=>
1960 V4 N3 P9-13	12N57R <ATS>
1960 V4 N3 P24-29	25N52R <ATS>
1960 V4 N3 P57-65	61-28647 <*=>
1960 V4 N3 P92-95	62-14260 <=*>
1960 V4 N4 P129-137	13N54R <ATS>
1961 V5 N1 P103-105	62-24847 <=*>
1961 V5 N2 P10-16	63-18363 <=*>
1962 N6 P31-37	65-12062 <*>
1962 V6 P42-49	65-61293 <=$>
1962 V6 N1 P42-49	63-10830 <=*>
1962 V6 N2 P85-90	63-24077 <=*>
1962 V6 N4 P69-72	65-61244 <=$>
1962 V6 N5 P55-62	64-10930 <=*$>
1963 N1 P34-38	65-12323 <*>
1963 N3 P98-102	65-63019 <=*$>
1963 N3 P118-125	65-63797 <=*$>
1963 V7 N1 P27-33	64-18748 <*>
1963 V7 N1 P118-122	64-13080 <=*$>
1963 V7 N4 P46-53	64-16847 <=*$> O
1965 N2 P28-31	66-10546 <*>

*IZVESTIYA VYSSHIKH UCHEBNYKH ZAVEDENII. TSVETNAYA METALLURGIYA

Citation	Translation
1958 N1 P107-115	PB 141 212T <=>
1958 N1 P116-120	60-13639 <+*>
1958 N1 P120-126	PB 141 213T <=>
1958 N1 P183-184	59-11330 <=>
1958 N2 P84-92	59-11709 <=>
1958 N2 P93-100	59-11687 <=>
1958 N3 P71-78	59-11709 <=>
1958 N3 P94-96	60-11311 <=>
1958 N6 P115-125	61-19550 <=*> O
1958 N6 P143-146	60-11118 <=>
1958 V1 N2 P77-83	60-15888 <+*>
1959 N1 P59-66	61-19551 <*=> O
1959 N1 P74-81	59-13913 <=>
1959 N1 P91-98	61-19551 <=*> O
1959 N1 P134-138	60-13646 <+*>
1959 N2 P54-57	4673 <HB>
1959 N3 P118-122	4714 <HB>
1959 N4 P151-152	60-11311 <=>

1959 N4 P152-155 60-11311 <=>
1959 N4 P158-160 60-11311 <=>
1959 V2 N1 P26-40 62-23932 <=*>
1959 V2 N1 P106-112 61-19543 <*=>
1959 V2 N1 P121-128 61-19543 <*=>
1959 V2 N2 P22-28 62-25270 <=*>
1959 V2 N4 P27-35 62-23933 <=*>
1959 V2 N4 P112-118 62-13159 <*=>
1959 V2 N5 P106-112 62-23715 <=*>
1959 V2 N5 P167-172 60-41450 <=>
1959 V2 N6 P47-51 60-11878 <=>
1959 V2 N6 P85-98 60-11878 <=>
1959 V2 N6 P121-125 61-19942 <=*>
1959 V2 N6 P126-132 61-27880 <*=>
1959 V2 N6 P134-141 60-11878 <=>
1959 V2 N6 P158-161 62-13160 <*=>
1959 V2 N6 P162-165 25N56R <ATS>
1959 V2 N6 P185-194 60-11878 <=>
1960 N2 P147-152 60-11843 <=>
1960 V3 N1 P29-34 61-13495 <*+>
1960 V3 N1 P57-63 60-11897 <=>
1960 V3 N1 P84-90 61-27290 <=>
1960 V3 N1 P165-166 60-11968 <=>
1960 V3 N2 P40-42 61-27559 <*=>
1960 V3 N2 P69-73 61-27535 <*=>
1960 V3 N2 P108-112 61-27543 <*=>
1960 V3 N2 P147-152 65-14577 <*> 0
1960 V3 N3 P54-61 63-23377 <=*>
1960 V3 N3 P155-160 61-27290 <=>
1960 V3 N4 P65-68 56M46R <ATS>
1960 V3 N4 P69-75 57M46R <ATS>
1960 V3 N4 P102-106 63-15133 <=*>
1960 V3 N4 P113-121 63-15314 <=*>
1960 V3 N4 P145-152 62-10082 <*=>
1960 V3 N5 P72-78 63-15457 <=*>
1960 V3 N6 P46-54 62-24943 <=*>
1960 V3 N6 P104-113 62-13595 <*=>
63-10858 <=*>
1960 V3 N6 P165-166 61-21708 <=>
1961 V4 N1 P5-6 61-27290 <=>
1961 V4 N1 P60-66 61-27290 <=>
1961 V4 N1 P86-88 61-27290 <=>
1961 V4 N1 P134-139 63-21348 <=>
1961 V4 N2 P43-52 AEC-TR-4867 <=*>
62-18258 <=*>
1961 V4 N4 P87-90 62-24650 <=*>
1961 V4 N4 P112-120 63-21243 <=>
1961 V4 N4 P139-144 62-13574 <*=>
1961 V4 N4 P145-148 63-24085 <=*>
1961 V4 N5 P81-89 62-25314 <=*> 0
1961 V4 N6 P58-64 31P60R <ATS>
1962 N4 P95-105 AEC-TR-5903 <*>
1962 V5 N1 P89-93 63R80R <ATS>
1962 V5 N1 P111-114 63-21243 <=>
1962 V5 N1 P142-149 AEC-TR-6411 <=$>
29Q71R <ATS>
1962 V5 N2 P118-123 CR-4026 <PS>
1962 V5 N2 P124-128 CR-4027 <PS>
1962 V5 N3 P129-134 63-21442 <=>
1962 V5 N3 P164-167 63-19080 <=*>
1962 V5 N4 P54-59 63-23613 <=*$>
1962 V5 N4 P95-105 AEC-TR-5903 <=*$>
1962 V5 N4 P122-131 62-33684 <=*>
1962 V5 N4 P143-148 62-33685 <=*>
1962 V5 N5 P113-122 63-19558 <=*>
63-21031 <=>
1962 V5 N6 P50-56 3682 <BISI>
1962 V5 N6 P136-139 65-60131 <=$>
1963 N1 P42-47 66-10249 <*>
1963 V6 N1 P117-120 6129 <HB>
1963 V6 N2 P120-126 TP/T-3577 <NLL>
1963 V6 N2 P133-138 22Q73R <ATS>
63-31658 <=>
1963 V6 N2 P162-166 6020 <HB>
1963 V6 N2 P172-173 6139 <HB>
1963 V6 N3 P106-110 6081 <HB>
1963 V6 N5 P28-32 66-10273 <*>
1963 V6 N5 P57-63 M-5619 <NLL>
1963 V6 N6 P120-123 65-10435 <*>
1964 N5 P119-122 65-13770 <*>
1964 N6 P125-130 AD-637 433 <=*>
1964 V7 N1 P61-65 M.5817 <NLL>
1964 V7 N4 P23-29 6433 <HB>
1964 V7 N4 P48-55 3813 <BISI>
1964 V7 N4 P56-61 3814 <BISI>
1964 V7 N4 P73-78 M.5816 <NLL>
1964 V7 N4 P114-118 AD-624 911 <=$>
1964 V7 N4 P124-129 N65-23716 <=$>
1964 V7 N5 P66-70 04S89R <ATS>
1964 V7 N6 P74-81 03S89R <ATS>
1965 V8 N1 P90-95 28S88R <ATS>
1965 V8 N2 P79-84 4568 <BISI>
1965 V8 N3 P77-81 66-12830 <*>
1965 V8 N4 P145-151 65-33585 <=*$>
1965 V8 N4 P152-161 65-33585 <=*$>
1965 V8 N6 P74-79 66-13046 <*>
1965 V9 N1 P90-95 1010 <TC>
1966 V9 N1 P82-84 66-14530 <*>

JAERI REPORTS
SEE REPORT. JAPAN ATOMIC ENERGY RESEARCH INSTITUTE

JAARBOEK VAN HET DEPARTEMENT VAN LANDBOUW, NIJVERHEID EN HANDEL IN NEDERLANDSCH-INDIE
1907 P226-298 57-3360 <*>
1908 P147-226 3049 <*>
1909 P216-258 57-3277 <*>
1910 P176-222 57-3276 <*>
1911 P140-182 57-3361 <*>

JAARBOEK. PLANTENZIEKTENKUNDIGE DIENST
1954 P238-243 CSIRO-3282 <CSIR>

JADERNA ENERGIE
1957 V3 N10 P290-292 UCRL TRANS-507 <*>
1957 V3 N12 P406-409 60-10798 <*>
1958 V4 N1 P19-26 ORNL-TR-81 <*>
1958 V4 N2 P34-39 AEC-TR-3593 <+*>
1958 V4 N2 P45-56 AEC-TR-3628 <+*>
1958 V4 N8 P210-215 59-13191 <=*>
1958 V4 N8 P237-238 59-13183 <=*>
1959 V5 N11 P373-377 62-25225 <=*>
1960 V6 P347 63-27139 <$>
1960 V6 P404-408 30P62C <ATS>
1960 V6 N3 P80-82 60-31272 <=*>
1960 V6 N3 P96-98 60-31273 <=*>
1960 V6 N5 P155-162 62-25226 <=*>
1960 V6 N7 P222-227 C-3571 <NRCC>
1960 V6 N8 P254-266 AEC-TR-4831 <*>
1960 V6 N10 P1-7 IA-666-TR <*>
1962 V8 P21 AEC-TR-5363 <*>
1962 V8 P43-50 AEC-TR-5330 <*>
1962 V8 P90-93 UCRL TRANS-1010(L) <*>
1962 V8 N2 P51 NP-TR-1077 <*>
1962 V8 N6 P205-207 AEC-TR-6101 <*>
1962 V8 N7 P221-224 AEC-TR-5924 <*>
1962 V8 N12 P427-428 AEC-TR-5923 <*>
1963 V9 N5 P146-155 66-12349 <*>
1963 V9 N10 P319-323 65-18075 <*>
1964 V10 P251-253 66-12468 <*>
1965 V11 N7 P263-267 66-10412 <*>
66-11111 <*>

JAHRBUCH. AKADEMIE DER WISSENSCHAFTEN UND DER LITERATUR IN MAINZ
1959 P404-408 62-24564 <=*>

JAHRBUCH DER DEUTSCHEN GESELLSCHAFT FUER CHRONOMETRIE
1956 P23-31 SC-T-517 <*>
1956 P49-64 SC-T-518 <*>

JAHRBUCH DER DEUTSCHEN LUFTFAHRTFORSCHUNG MUNICH
1937 P303-307 64-71578 <=>
1938 P206-219 65-13183 <*>
1938 P252-262 AL-462 <*>
1938 P515-516 59-15107 <*> 0
1938 S1 P303-305 65-12972 <*>
1941 S1 P364-366 65-12971 <*>

1942 P180-190 AL-863 <*>
1942 S1 P173-185 65-12968 <*>

JAHRBUCH DER DEUTSCHEN VERSUCHSANSTALT FUER LUFTFAHRT BERLIN
1930 P428-433 59-15348 <*>
1931 P233-245 AD-616 526 <=$>
1931 P591-593 66-12744 <*>

JAHRBUCH DER DRAHTLOSEN TELEGRAPHIE U. TELEPHONIE
1910 N4 P242-251 58-104 <*>
1910 V4 P176-187 57-2417 <*>
1914 V8 P1-34 57-1578 <*>
1914 V8 P132-139 57-1573 <*>
1919 P451- 57-1397 <*>
1919 V13 P552- 57-2121 <*>
1919 V13 N4 P280-286 66-12581 <*> O
1920 V15 P354- 57-2124 <*>
1920 V15 N5 P407-433 57-2429 <*>
1920 V16 P162- 57-2416 <*>
1923 V21 P77-100 63-14432 <*> O
1923 V21 P101-120 57-566 <*>
60-10193 <*>
1923 V22 P142-155 58-347 <*>
1925 V25 P56-61 58-1781 <*>
1925 V25 N2 P56-61 59-10002 <*>
1925 V25 N4 P111-114 58-359 <*>
1925 V26 N2 P29-37 57-2100 <*>
1927 V30 P32-33 I-200 <*>
1928 V31 P121-122 57-2281 <*>
1928 V32 N3 P77-83 57-2517 <*>
1929 V33 P9-15 57-2434 <*>
1929 V33 P47-52 57-2434 <*>
1929 V33 N2 P52-55 57-2418 <*>
1929 V33 N5 P166-175 57-646 <*>
1929 V33 N5 P176-180 58-147 <*>
1929 V33 N6 P219-223 59-20029 <*> O
1929 V34 P60-65 57-1590 <*>
1930 V35 N3 P109-115 57-2458 <*>
1930 V35 N5 P165-177 58-315 <*>
1930 V36 N1 P1-13 57-2436 <*>
1931 V37 N4 P123-125 57-1339 <*>
1931 V37 N4 P162-167 66-12587 <*>
1931 V37 N5 P175-187 66-12587 <*>
1931 V37 N6 P219-229 57-1576 <*>

JAHRBUCH DER GEOLOGISCHEN BUNDESANSTALT
1957 V100 N2 P269-298 10K24G <ATS>
1060-GJ <ATS>

JAHRBUCH DER HAMBURGISCHEN WISSENSCHAFTLICHEN ANSTALTEN
1909 V27 P1-31 59-17692 <*> O

JAHRBUCH FUER KINDERHEILKUNDE UND PHYSISCHE ERZIEHUNG
1888 V27 P1-7 65-11130 <*>
1918 V87 N2 P95-108 62-01202 <*>
1926 V120 P266-291 62-01267 <*>
1937 V149 P326-339 61-00625 <*>

JAHRBUCH DER LUFTFAHRTFORSCHUNG DER DEUTSCHEN DEMOKRATISCHEN REPUBLIK
1960 P118-129 64-18358 <*> P

JAHRBUCH DER MAX-PLANCK-GESELLSCHAFT ZUR FOERDERUNG DER WISSENSCHAFTEN
1954 P178-192 59-10658 <*> O

JAHRBUCH FUER PSYCHIATRIE UND NEUROLOGIE
1901 V20 P77-101 59-10706 <*>
1913 V34 P367-375 58-627 <*>

JAHRBUCH DER RADIOAKTIVITAET UND ELEKTRONIK
1912 V9 P355-418 66-12707 <*>
1915 V12 N12 P147-205 66-12279 <*>
1921 V17 P276-292 58-340 <*>

JAHRBUCH DER SCHIFFBAUTECHNISCHEN GESELLSCHAFT
1926 V27 P101-102 58-1452 <*>
1926 V27 P104- 58-1452 <*>
1933 V34 P205-227 1791 <*>
1950 V44 ENTIRE ISSUE 61-20682 <*>
1952 V46 P244-288 58-1150 <*>
1960 V54 P195-232 AD-630 685 <=$>
1964 V58 P113- 2234 <BSRA>
1964 V58 P169- 2230 <BSRA>
1964 V58 P268- 2226 <BSRA>

JAHRBUCH FUER WISSENSCHAFTLICHE BOTANIK
1927 V67 N2 P334-346 CSIRO-3150 <CSIR>
1935 V81 P758-766 CSIRO-3790 <CSIR>
1938 V86 P518-673 64-14305 <*> P

JAHRBUCH DER WISSENSCHAFTLICHEN GESELLSCHAFT FUER LUFTFAHRT
1954 P116-120 65-13184 <*>
1957 P117-134 60-13103 <=*>
1959 P288-291 SCL-T-450 <*>
1959 P292-298 65-18019 <*>
1960 P112-133 C-4897 <NRC>

JAHRESBERICHT DER DEUTSCHEN MATEMATIKERVEREINIGUNG
1942 V52 P83-95 UCRL TRANS-1226(L) <*>
1942 V52 P177-188 65-10833 <*>
1957 V60 P40-42 60-14227 <*>
1962 V65 N2 P45-71 65-11694 <*>
1963 V66 N2 P66-79 66-11210 <*>
1964 V66 N3 P106-118 66-11189 <*>

JAHRESBERICHT DES VEREINS SCHWEIZERISCHER ZEMENT-KALK- UND GIPS-FABRIKANTEN
1960 N50 ENTIRE ISSUE 65-13632 <*>
1960 N50 P8- 65-13632 <*>

JAHRSBERICHTE DER WIRKSAMKEIT DER AUGENKLINIK ZU BERLIN
1876 P50-54 57-1675 <*>

JAN LIAO HSUEH PAO
1956 V1 N2 P93-107 60-00703 <*>
1957 V2 N1 P39-51 59-18372 <+*>
1957 V2 N4 P311-322 59-10495 <*> O
1957 V2 N4 P333-340 59-10497 <*> O
1957 V2 N4 P341-351 59-10496 <*> O
1958 V3 P214-216 62-24664 <=*>
1959 V4 N2 P113-126 61-27718 <*=>
1959 V4 N2 P127-132 61-27112 <*=>
1959 V4 N2 P134-138 61-27715 <*=>
1959 V4 N2 P140-156 61-27939 <*=>
1959 V4 N2 P157-163 61-27874 <*=>
1959 V4 N2 P164-172 61-27206 <*=>
1959 V4 N2 P173-180 60-27940 <*=>
1959 V4 N2 P186-190 61-27279 <*=>
1959 V4 N4 P263-298 60-41439 <=> O

JAPAN ACCOUSTICAL SOCIETY. COLLECTION OF PAPERS
SEE NIPPON ONKYO GAKKAI KEON RONBUNSHU

JAPAN ANALYST
SEE BUNSEKI KAGAKU

JAPAN CHEMICAL FIBERS MONTHLY
SEE KASEN GEPPO

JAPAN INFORMATION CENTER OF SCIENCE AND TECHNOLOGY
SEE JOHO KANRI

JAPAN STEEL AND TUBE TECHNICAL REVIEW
SEE NIHON KOKAN GIHO

JAPAN STEEL WORKS TECHNICAL REVIEW
SEE NIHON SEIKO GIHO

JAPANESE AGRICULTURE
SEE NIPPON NOGYO

JAPANESE ARCHIVES OF INTERNAL MEDICINE
SEE NAIKA HOKAN

JAPANESE CIRCULATION JOURNAL
  SEE NIPPON JUNKANKI GAKUSHI

JAPANESE CLINICS
  SEE NIHON RINSHO

JAPANESE ECONOMIC YEARBOOK
  SEE NIPPON KEIZAI SHIMBUN

JAPANESE JOURNAL OF ALLERGY
  SEE ARERUGI

JAPANESE JOURNAL OF APPLIED ENTOMOLOGY AND ZOOLOGY
  SEE NIHON OYO DOBUTSU KONCHUGAKA ZASSHI

JAPANESE JOURNAL OF APPLIED PHYSICS
  1964 V3 N11 P728-732   CE-TRANS-4052 <NLL>

JAPANESE JOURNAL OF APPLIED ZOOLOGY
  SEE OYO DOBUTSUGAKU ZASSHI

JAPANESE JOURNAL OF BACTERIOLOGY
  SEE NIPPON SAIKINGAKU ZASSHI

JAPANESE JOURNAL OF BREEDING
  SEE IKUSHUGAKU ZASSHI

JAPANESE JOURNAL OF CANCER CLINICS
  SEE GAN NO RINSHO

JAPANESE JOURNAL OF CANCER RESEARCH
  SEE GANN

JAPANESE JOURNAL OF CLINICAL DERMATOLOGY
  SEE RINSHO HIFUKA

JAPANESE JOURNAL OF CLINICAL AND EXPERIMENTAL MEDICINE
  SEE RINSHO TO KENKYU

JAPANESE JOURNAL OF CLINICAL PATHOLOGY
  SEE RINSHO BYORI

JAPANESE JOURNAL OF DERMATOLOGY
  SEE NIHON HIFUKA GAKKAI ZASSHI

JAPANESE JOURNAL OF GENETICS
  SEE IDENGAKU ZASSHI

JAPANESE JOURNAL OF LEGAL MEDICINE
  SEE NIPPON HOIGAKU ZASSHI

JAPANESE JOURNAL OF LIMNOLOGY
  SEE RIKUSUI-GAKU ZASSHI

JAPANESE JOURNAL OF MEDICAL PROGRESS
  SEE NISSHIN IGAKU

JAPANESE JOURNAL OF NUTRITION
  SEE EIYOGAKU ZASSHI

JAPANESE JOURNAL OF PEDIATRICS
  SEE SHONIKA RINSHO

JAPANESE JOURNAL OF PHARMACY AND CHEMISTRY
  SEE YAKUGAKU KENKYU

JAPANESE JOURNAL OF PSYCHOLOGY
  SEE SHINRIGAKU KENKYU

JAPANESE JOURNAL OF SANITARY ZOOLOGY
  SEE EISEI DOBUTSU

JAPANESE JOURNAL OF THORACIC SURGERY
  SEE KYOBU GEKA

JAPANESE JOURNAL OF VETERINARY SCIENCE
  SEE NIHON JUIGAKU ZASSHI

JAPANESE JOURNAL OF ZOOLOGY
  SEE NIPPON GAKUJUTSU KAIGI

JAPANESE JOURNAL OF MEDICAL SCIENCE AND BIOLOGY
  SEE NIHON IJI SHINPO

JAPANESE SAFETY FORCES MEDICAL JOURNAL
  SEE HOAN EISEI GAKKAI

JAPANESE SCIENTIFIC MONTHLY
  SEE GAKUJUTSU GEPPO

JARMUVEK MEZOGAZDASAGI GEPEK
  1964 V11 N3 P90-97   TR.169 <NIAE>
  1964 V11 N4 P133-140   64-31744 <=>

JEMNA MECHANIKA A OPTIKA
  1958 N11 P363-369   60-18511 <*>
  1963 N8 P246   64-71144 <=>
  1964 N7 P201-205   AD-637 422 <=$>
  1965 N1 P5-6   AD-638 978 <=$>
  1965 N3 P75-76   AD-638 873 <=$>

JEN-MIN PAO-CHIEN
  1959 V1 N1 P79-83   60-11946 <=>
  1959 V1 N3 P272-277   60-11946 <=>
  1959 V1 N3 P281-287   60-11947 <=>
  1959 V1 N4 P373-375   60-11946 <=>
  1959 V1 N5 P472-474   60-11946 <=>
  1959 V1 N6 P563-566   60-41298 <=>
  1959 V1 N7 P663-669   60-11996 <=>
  1959 V1 N8 P764-768   60-41298 <=>
  1959 V1 N9 P855-860   60-11996 <=>
  1959 V1 N10 P896-903   60-11706 <=>
  1959 V1 N10 P965-968   60-41292 <=>
  1959 V1 N11 P1071   60-41298 <=>
  1959 V1 N12 P1100-1111   60-11467 <=>
  1959 V1 N12 P1123-1125   60-31387 <=>
  1959 V1 N12 P1156-1163   60-41298 <=>
  1960 V2 N3 P115-117   60-11869 <=>
  1960 V2 N3 P121-136   60-11869 <=>
  1960 V2 N3 P142-145   60-11869 <=>
  1960 V2 N3 P148-151   60-11869 <=>
  1960 V2 N3 P160-166   60-11869 <=>
  1960 V2 N5 P232-234   61-21003 <=>
  1960 V2 N5 P258-262   61-19201 <+>
  1960 V2 N5 P269-274   61-21245 <=>
  1960 V2 N5 P274   61-21003 <=>
  1960 V2 N6 P315-318   61-19201 <+>
  1960 V2 N6 P320-322   61-19201 <+>
  1960 V2 N6 P328-329   61-21245 <=>
  1960 V2 N6 P331   61-21245 <=>
  1960 V2 N6 P332-335   61-21003 <=>
  1960 V2 N6 P374-377   61-21245 <=>
  1960 V2 N7 P358-360   61-21003 <=>
  1960 V2 N7 P368-373   61-19201 <+>
  1960 V2 N7 P380-381   61-21003 <=>
  1960 V2 N7 P388   61-21003 <=>

JEN MIN YU TIEN
  1960 P5   61-11535 <=>
  1960 N1 P4-5   61-11081 <=>

JENA REVIEW
  1959 N2 P47-49   60-23951 <*=>

JENAER JAHRBUCH
  1952 P181-221   65-10291 <*> 0
  1956 P79-86   CSIRO-3858 <CSIR>
  1956 P87-93   CSIRO-3859 <CSIR>
  1959 P115-153   65-18092 <*>

JENAER UNIVERSITET. WISSENSCHAFTLICHE ZEITSCHRIFT. MATHEMATISCH-NATURWISSENSCHAFTLICHE REIHE
  SEE WISSENSCHAFTLICHE ZEITSCHRIFT DER FRIEDRICH SCHILLER-UNIVERSITET JENA. MATHEMATISCH-NATURUISSENSCHAFTLICHE REIHE

JENAISCHE ZEITSCHRIFT FUER MEDIZIN UND NATURWISSENSCHAFT. (TITLE VARIES)
  1902 V43 P729-798   59-10207 <*> 0

JERNKONTORETS ANNALER
1930 V114 N6 P304-316 57-3145 <*>
1932 V116 P1-19 58-998 <*>
1937 V121 N5 P219-231 2224 <BISI>
1940 V124 P179-212 2420 <*>
1940 V124 N5 P212-224 61-18472 <*> 0
1940 V124 N9 P511-531 61-18520 <*>
1941 V125 P333-422 I-304 <*>
1941 V125 N12 P663-694 61-18467 <*> 0
1942 V126 N4 P131-142 57-3018 <*>
1944 V128 N2 P77-80 57-3017 <*>
1944 V128 N4 P160-164 57-2790 <*>
1944 V128 N9 P457-520 63-16852 <*>
1944 V128 N10 P537-552 4901 <BISI>
1946 V130 P118-126 AL-965 <*>
1946 V130 N10 P477-552 60-18676 <*> 0
1947 V131 N1 P1-25 62-16803 <*> 0
1948 V132 N2 P27-41 63-10041 <*> 0
1948 V132 N4 P105-109 62-16928 <*> 0
1949 V133 N1 P1-28 66-10456 <*> 0
1950 V134 N3 P97-133 58-1167 <*>
1951 V135 N8 P403-494 61-16237 <*> 0
1953 V137 N1 P1-26 60-10272 <*>
1953 V137 N3 P100-114 59-20564 <*> 0
1953 V137 N4 P117-126 60-10299 <*>
1953 V137 N4 P128-138 60-10298 <*>
1953 V137 N7 P224-237 60-10301 <*>
1953 V137 N10 P725-743 3846 <HB>
1953 V137 N11 P767-784 61-18792 <*>
1954 V138 N7 P383-403 60-18662 <*> 0
1954 V138 N9 P539-572 60-18663 <*> 0
1954 V138 N12 P759-778 57-2000 <*>
1955 V139 N2 P78-134 60-10300 <*>
1955 V139 N4 P250-264 62-16986 <*> 0
1955 V139 N6 P412-438 50K22S <ATS>
1955 V139 N10 P805-816 61-18725 <*>
956 <BISI>
1955 V139 N10 P817-828 61-18400 <*>
954 <BISI>
1955 V139 N10 P829-846 955 <BISI>
1955 V139 N10 P847-852 61-16239 <*>
1956 V140 P612-619 T-2183 <INSD>
1956 V140 N1 P5-23 4353 <BISI>
1956 V140 N1 P75-80 3879 <HB>
1956 V140 N2 P116-129 3886 <HB>
1956 V140 N2 P130-136 3842 <HB>
1956 V140 N4 P314-344 2159 <BISI>
1956 V140 N5 P373-385 61-16236 <*>
1956 V140 N7 P467-492 4141 <HB>
1956 V140 N7 P512-519 3907 <HB>
61-10575 <*>
1956 V140 N11 P839-853 3831 <HB>
1956 V140 N12 P909-929 61-18359 <*>
1957 V141 N2 P90-94 4027 <HB>
1957 V141 N2 P90-99 4028 <HB>
61-16238 <*>
1957 V141 N4 P189-205 4025 <HB>
1957 V141 N4 P231-232 3879 <HB>
1957 V141 N5 P237-260 4000 <HB>
1957 V141 N5 P300-307 1865 <BISI>
1957 V141 N7 P380-389 61-18420 <*>
1957 V141 N9 P483-506 66-10145 <*> 0
1958 V142 N3 P128-164 TS-1331 <BISI>
62-14314 <*>
1958 V142 N4 P209-211 2951 <BISI>
1958 V142 N6 P289-319 4615 <HB>
1958 V142 N6 P319-355 4665 <HB>
1958 V142 N7 P401-428 4559 <HB>
59-17811 <*>
1958 V142 N7 P428-466 4638 <HB>
1958 V142 N10 P611-637 62-18289 <*>
1959 V143 N1 P1-28 TS-1495 <BISI>
62-14650 <*>
1959 V143 N6 P335-369 TRANS-86 <MT>
1959 V143 N8 P457-486 1884 <BISI>
1959 V143 N8 P487-507 TRANS-05 <MT>
1959 V143 N8 P508-522 TS-1703 <BISI>
66-10144 <*>
1960 V144 N8 P573-627 2311 <BISI>
1960 V144 N10 P739-756 62-14501 <*>
1960 V144 N11 P813-831 TRANS-177 <MT>
1960 V144 N11 P847-854 62-10281 <*> 0
1960 V144 N12 P859-967 TRANS-205 <MT>
1961 V145 N1 P25-46 61-20556 <*> 0
1961 V145 N8 P487-550 2553 <BISI>
1962 V146 P53-80 4028 <BISI>
1962 V146 N5 P383-402 TRANS-297 <MT>
1962 V146 N6 P438-452 3155 <BISI>
1962 V146 N6 P453-461 3309 <BISI>
1962 V146 N8 P549-679 3360 <BISI>
1962 V146 N9 P748-762 3140 <BISI>
1962 V146 N11 P862-868 5772 <HB>
1962 V146 N12 P924-934 5980 <HB>
1963 V147 N1 P52-56 5920 <HB>
1963 V147 N1 P116-132 5919 <HB>
1963 V147 N3 P319-334 5939 <HB>
1963 V147 N3 P347-348 5953 <HB>
1963 V147 N3 P348-351 5954 <HB>
1963 V147 N11 P913-916 6188 <HB>
1963 V147 N11 P917-930 6142 <HB>
1964 N5 P339-344 3894 <BISI>
1964 N6 P407-415 3894 <BISI>
1964 V148 N1 P51-63 3894 <BISI>
1964 V148 N2 P108-117 3894 <BISI>
1964 V148 N3 P181-194 3894 <BISI>
1964 V148 N4 P276-287 3894 <BISI>
1965 V149 N1 P1-11 6646 <HB>

JIBIINKOKA RINSHO
1961 V33 N6 P537-539 63-20224 <*>
1963 V35 P1120 65-14688 <*>

JIKKEN DOBUTSU
1961 V10 N3 P75-82 63-01121 <*>

JISHIN
1962 S2 V15 P238-254 AD-610 992 <=$>

JOHO KANRI
1964 V7 N1 P17-23 64-21868 <=>

JORNAL BRASILEIRO DE PSIQUIATRIA
1948 V1 P169-184 2040 <K-H>

JORNAL DOS CLINICOS
V24 N8 P7-24 2572 <*>

JORNAL DE PEDIATRIA
1955 V20 P264-308 4006-A <K-H>
1958 V23 P437-443 60-16943 <*>

JORNAL DE SOCIEDADE DES CIENCIAS MEDICAS DE LISBOA
1953 V117 P401-460 II-286 <*>
1962 V126 P275-294 64-18138 <*>

JOURNAL OF THE ACOUSTICAL SOCIETY OF JAPAN
SEE NIPPON ONKYO GAKKAISHI

JOURNAL OF THE ADHESION SOCIETY, JAPAN
SEE NIHON SECCHAKU KYOKAISHI

JOURNAL OF THE AGRICULTURAL CHEMICAL SOCIETY OF JAPAN
SEE NIHON NOGEIKAGAKU KAISHI

JOURNAL OF THE AGRICULTURAL ENGINEERING SOCIETY. JAPAN
SEE NOGYO DOBOKU KENKYU

JOURNAL OF AGRICULTURAL METEOROLOGY
SEE KISEICHUGAKU ZASSHI

JOURNAL OF AGRICULTURAL METEOROLOGY OF JAPAN
SEE NOGYO KISHO

JOURNAL DES AMERICANISTES DE PARIS
1950 SNS V39 P187-218 59-15608 <*>

JOURNAL OF ANCIENT VERTEBRATES. CHINESE PEOPLES REPUBLIC
SEE KU CHI CH'UI TUNG WU HSUEH PAO

JOURNAL OF APPLIED PHYSICS. LANCASTER
1965 V1 N4 P219-228 66-12685 <*>

JOURNAL OF APPLIED PHYSICS. JAPAN
SEE OYO BUTSURI

JOURNAL OF APPLIED POLYMER SCIENCE
1962 V6 N19 P98-102 62-18429 <*>
1962 V6 N19 P103-110 62-18428 <*>
65-1006 <*>

JOURNAL OF ARCHITECTURE. CHINESE PEOPLES REPUBLIC
SEE CHIEN CHU HSUEH PAO

JOURNAL OF ASTRONOMY, CHINESE PEOPLES REPUBLIC
SEE T'IEN WEN HSUEH PAO

JOURNAL OF ATMOSPHERIC AND TERRESTRIAL PHYSICS
1952 V2 P340-349 C-2396 <NRC>
1957 V11 N1 P14-22 C-2671 <NRC>

JOURNAL OF THE ATOMIC ENERGY SOCIETY OF JAPAN
SEE NIPPON GENSHI-RYOKO GAKKAI-SHI

JOURNAL BELGE DE RADIOLOGIE
1955 V38 P394-429 57-1880 <*>
1962 V45 N6 P728-741 AEC-TR-5889 <*>

JOURNAL OF BIOCHEMISTRY
SEE NIHON SEIKA-GAKKAI

JOURNAL OF BREWING. JAPAN
SEE JOZO GAKU ZASSHI

JOURNAL OF THE CANADIAN DENTAL ASSOCIATION
1958 V24 P615-628 59-15587 <*> 0

JOURNAL OF THE CERAMIC ASSOCIATION, JAPAN
SEE YOGYO KYOKAI-SHI

JOURNAL OF THE CHEMICAL SOCIETY. CHINESE PEOPLES REPUBLIC
SEE HUA HSUEH HSUEH PAO

JOURNAL OF THE CHEMICAL SOCIETY OF JAPAN
SEE NIPPON KAGAKU ZASSHI

JOURNAL OF THE CHEMICAL SOCIETY OF JAPAN. INDUSTRIAL CHEMISTRY SECTION
SEE KOGYO KAGAKU ZASSHI

JOURNAL OF THE CHEMICAL SOCIETY OF JAPAN. PURE CHEMISTRY SECTION
SEE NIPPON KAGAKU ZASSHI

JOURNAL FUER DIE CHEMIE, PHYSIK UND MINERALOGIE
1808 V5 N4 P690-712 63-16151 <*>

JOURNAL DE CHIMIE PHYSIQUE
1915 V13 P351-375 58-1945 <*>
1919 V17 P3-70 UCRL TRANS-979(L) <*>
1926 V23 P501-514 63-10983 <*> P
1926 V23 P521-544 62-10719 <*> 0
1929 V26 P219-223 AEC-TR-3659 <+*>
1930 V27 P119-162 I-1017 <*>
1930 V27 P169-170 AEC-TR-5942 <*>
1930 V27 P170-171 AEC-TR-5945 <*>
1930 V27 P401-442 98K21F <ATS>
1931 P163-173 AL-186 <*>
1931 V28 P605-621 64-00380 <*>
1932 V29 P453-473 II-329 <*>
1933 V30 P347-355 57-3271 <*>
58-2436 <*>
1933 V30 P556-559 57-3270 <*>
1938 V35 P387-394 53S85F <ATS>
1940 V37 P101-109 II-952 <*>
1942 V39 P57-72 3189-B <K-H>
1944 V41 P45-48 63-18492 <*>
1945 V42 P41-44 63-18399 <*> 0
1947 V44 P261-265 61-10784 <*> 0
1949 V46 P153-157 I-84 <*>
1949 V46 P337-343 1311 <*>
1949 V46 P480-484 T1476 <INSD>
1949 V46 N1/2 P30-34 64-16619 <*>
1950 V47 P223-228 28R76F <ATS>
1950 V47 P382-383 II-961 <*>
1950 V47 P704-707 63-20160 <*> 0
1950 V47 P747-763 63-18982 <*>
1950 V47 P776-794 58-825 <*>
1950 V47 P933-941 62-20339 <*>
1950 V47 N1/2 P11-20 60-10386 <*>
1950 V47 N1/2 P113-117 02S83F <ATS>
1950 V47 N5/6 P391-398 61-10410 <*> 0
1950 V47 N9/10 P805-806 II-686 <*>
1951 V48 P208-212 I-229 <*>
1951 V48 P368-371 I-359 <*>
1951 V48 P372- I-786 <*>
1951 V48 P385-398 58-78 <*>
63-16401 <*>
1951 V48 P429-437 AEC-TR-3560 <+*>
1951 V48 P579-581 92K22F <ATS>
1952 V49 P99-102 57-1873 <*>
1952 V49 P103-108 57-1871 <*>
1952 V49 P135-144 59-10171 <*>
1952 V49 P286-293 AEC-TR-4854 <*>
1952 V49 P302-307 58-1787 <*>
1952 V49 PC12-C13 1214 <*>
1952 V49 PC41-C46 61-18908 <*> 0
1952 V49 PC214-C218 61-14748 <*>
97N47F <ATS>
1952 V49 N3 P85-92 60-10385 <*> 0
1952 V49 N4 P239-244 60-10384 <*> 0
1952 V49 N6 P377-384 62-10563 <*>
1952 V49 N10 P527-536 59-10390 <*>
1953 V50 P53-70 58-2405 <*>
1953 V50 P501-506 2806 <*>
1953 V50 P507-511 1912 <*>
1953 V50 N6 P415-422 62-16911 <*> 0
1954 V51 P1-8 59-21054 <=>
1954 V51 P141-160 AEC-TR-4829 <*>
1954 V51 P161-164 II-962 <*>
1954 V51 P312-319 62-01380 <*>
1954 V51 P430-433 AEC-TR-4383 <*>
1954 V51 P474-481 3G6F <ATS>
1954 V51 P651-662 57-1762 <*>
1954 V51 P664-669 9652 <RIS>
1954 V51 P670-677 96953 <RIS>
1954 V51 P729-739 65-00243 <*>
1954 V51 N5 P255-259 60-18064 <*> 0
1955 V52 P41-47 58-2379 <*>
1955 V52 P181-200 64-00112 <*>
1955 V52 P201-206 UCRL TRANS-602 <*>
1955 V52 P246-258 HW-TR-31 <*>
1955 V52 P259-266 HW-TR-13 <*=>
60-10725 <*>
1955 V52 P267-271 HW-TR-14 <*>
1955 V52 P689-698 II-848 <*>
1955 V52 N1 P60-64 59-10263 <*> 0
1955 V52 N4 P327-330 61-00872 <*>
1955 V52 N10 P741-748 57-806 <*>
1955 V52 N7/8 P578- 58-1382 <*>
1956 V53 P536-541 AEC-TR-4863 <*>
1956 V53 P787-797 62-18897 <*>
65-11143 <*>
1956 V53 N4 P369-379 59-10553 <*>
63-27111 <$>
1956 V53 N4 P380-388 59-10554 <*>
1956 V53 N76 P705-713 58-502 <*>
1957 V54 P72-89 58-1658 <*>
1957 V54 P716-725 68P66F <ATS>
1957 V54 P931-937 58-2013 <*>
1957 V54 N1 P19-25 3452 <*>
1957 V54 N6 P483-484 61-10915 <*>
6673-A <K-H>
1957 V54 N11/2 P896-901 59-10421 <*>
1958 V55 P227 AEC-TR-5628 <*>

1958 V55 P315-319 60-19157 <*+>
1958 V55 P510-512 63-01069 <*>
1958 V55 N1 P5-8 59-10564 <*>
1958 V55 N3 P177-184 08K27F <ATS>
1958 V55 N3 P185-196 09K27F <ATS>
1958 V55 N5 P341-353 59-17445 <*>
1958 V55 N5 P377-382 64-14768 <*>
8650-C <K-H>
1958 V55 N9 P681-687 62-16526 <*> 0
1959 V56 P372-376 60-13176 <*+>
1959 V56 P532-547 NP-TR-446 <*>
1959 V56 P593-608 UCRL TRANS-625 <*>
1959 V56 P830-833 92M41F <ATS>
1959 V56 P1024-1035 88N53F <ATS>
1959 V56 N1 P94-102 65-10225 <*>
1959 V56 N4 P377-386 65-10221 <*>
1959 V56 N6 P593-608 61-10994 <*>
1960 V57 P33-37 93M41F <ATS>
1960 V57 P492-499 NP-TR-973 <*>
63-18931 <*>
1960 V57 P524-527 18N56F <ATS>
1960 V57 N3 P228-247 62-10096 <*>
1960 V57 N7/8 P600-605 61-20004 <*>
1960 V57 N7/8 P666-672 62-10976 <*>
1961 V58 P759 HW-TR-82 <*>
1961 V58 P926-933 AEC-TR-5868 <*>
1961 V58 N4 P418-441 62-18713 <*>
1962 V59 P675-680 66-12861 <*>
1962 V59 P1090-1092 66-14107 <*>
1962 V59 P1120-1121 AI-TR-26 <*>
1962 V59 N1 P15-26 65-12410 <*>
1962 V59 N2 P148-153 HW-TR-42 <*>
1962 V59 N4 P375-393 63-14741 <*>
1962 V59 N4 P394-411 63-14742 <*>
1962 V59 N11/2 P1142-1150 65-14417 <*>
1962 V59 N11/2 P1196-1206 40Q69F <ATS>
1962 V59 N11/2 P1207-1222 41Q69F <ATS>
1962 V59 N11/2 P1223-1232 42Q69F <ATS>
1963 V60 P397-401 66-13560 <*>
1963 V60 P659-666 NP-TR-1072 <*>
1963 V60 P891-898 66-13525 <*>
1963 V60 P1251 HW-TR-82 <*>
1963 V60 N3 P426-434 65-12471 <*>
1963 V60 N5 P676-683 65-12480 <*>
1963 V60 N10 P1219-1226 64-18776 <*>
1963 V60 N10 P1227-1230 64-16972 <*>
1963 V60 N11/2 P1409-1418 64-30711 <*>
65-10290 <*>
1964 V61 N3 P343-362 AD-612 653 <=$>
1964 V61 N3 P409-411 65-11087 <*>
1964 V61 N5 P748-750 65-11747 <*>
1964 V61 N11/2 P1477-1489 65-18066 <*>
1965 V62 N2 P224-225 66-11846 <*>
1965 V62 N3 P227-234 66-10369 <*>
1965 V62 N5 P475 1679 <TC>

JOURNAL OF CHINESE AGRICULTURE
SEE NUNG YEH HSUEH PAO

JOURNAL OF CHINESE METALLURGY
SEE CHIN SHU HSUEH PAO

JOURNAL OF CHINESE PHARMACEUTICS
SEE YAO HSUEH HSUEH PAO

JOURNAL DE CHIRURGIE
1930 V36 P697-711 61-14494 <*> 0
1958 V75 N3 P287-310 60-17652 <*=> 0
1961 V82 N1/2 P41-46 62-14319 <*>

JOURNAL OF CHROMATOGRAPHY
1958 V1 N1 P25-34 65-14987 <*> 0
1958 V1 N2 P122-165 65-14988 <*> PO
1959 V2 N1 P84-89 65-14989 <*>
1959 V2 N2 P218-220 60-18188 <*> 0
1960 V3 N3 P265-272 C-3462 <NRCC>
1961 V5 P351-355 NP-TR-1176 <*>
1961 V5 N5 P408-417 65-13544 <*> 0
1961 V6 P373-380 AEC-TR-5043 <*>
1961 V6 P505-513 64-10426 <*> 0
1961 V6 N1 P2-21 65-14990 <*>
1962 V7 P1-12 63-20932 <*> 0
1962 V7 N3 P392-399 63-14378 <*>
1962 V8 N3 P410-412 65-12905 <*>
1963 V10 P392-395 AEC-TR-6138 <*>
1963 V11 P389-393 AEC-TR-6375 <*>
1963 V11 P452-458 24Q73G <ATS>

JOURNAL OF CLINICAL DERMATOLOGY. JAPAN
SEE HIFUKA NO RINSHO

JOURNAL OF CLINICAL OPTHAMOLOGY. JAPAN
SEE RINSHO GANKA

JOURNAL OF THE COAL RESEARCH INSTITUTE. JAPAN
SEE TANKEN

JOURNAL OF COLLOID SCIENCE
1954 N1 P135-145 59-15777 <*>

JOURNAL OF CORRECTIONAL MEDICINE. JAPAN
SEE KYOSEI IGAKU

JOURNAL. DEPARTMENT OF AGRICULTURE, KYUSHU IMPERIAL UNIVERSITY
SEE KYUSHU DAIGAKU NOGAKUBU KIYO

JOURNAL DE L'ECOLE POLYTECHNIQUE
1889 V59 P97-128 AD-607 957 <=>

JOURNAL OF ECONOMIC STUDIES. JAPAN
SEE KEIZAI KENKYU

JOURNAL OF ELECTROANALYTICAL CHEMISTRY
1959 V1 P44-53 AEC-TR-6343 <*>
1963 V5 P186-194 AEC-TR-6263 <*>
1965 V10 P553-567 66-13355 <*>

JOURNAL OF THE ELECTROCHEMICAL SOCIETY OF JAPAN
SEE DENKI KAGAKU

JOURNAL OF THE ELECTROCHEMICAL SOCIETY OF JAPAN (NON-JAPANESE EDITION)
1949 V17 P52-53 59-17179 <*> 0
1949 V17 N6 P120-121 59-17181 <*> 0

JOURNAL OF ELECTRONMICROSCOPY. JAPAN
SEE DENSHI-KEMBIKYO GAKKAISHI

JOURNAL OF ELECTRONICS AND CONTROL
1958 V4 N2 P175-178 59-10807 <*>
1958 V5 N1 P15-18 59-15862 <*>
1959 V7 P285-288 SCL-T-349 <*>

JOURNAL OF EMBRYOLOGY AND EXPERIMENTAL MORPHOLOGY
1958 V6 N1 P171-177 61-20662 <*> 0

JOURNAL OF ENTOMOLOGY. CHINESE PEOPLES REPUBLIC
SEE K'UN CH'UNG HSUEH PAO

JOURNAL OF EXPERIMENTAL AND THEORETICAL PHYSICS
1949 V19 N11 P965-970 R-715 <*>
1952 V23 N6 P660-666 AEC-TR-2600 <*>

JOURNAL OF THE FACULTY OF AGRICULTURE, HOKKAIDO UNIVERSITY
1928 V24 N1 P25-38 63-14385 <*>

JOURNAL OF THE FACULTY OF AGRICULTURE. IWATE UNIVERSITY
SEE IWATE DAIGAKU NOGAKUBU HOKOKU

JOURNAL OF THE FACULTY OF ENGINEERING. UNIVERSITY OF TOKYO
SEE TOKYO DAIGAKU KOGAKUBU KIYO, A

JOURNAL OF THE FACULTY OF SCIENCE, HOKKAIDO UNIVERSITY. SERIES 2. PHYSICS
1938 V2 P29-53 66-11185 <*>

JOURNAL OF THE FERMENTATION ASSOCIATION. JAPAN
SEE HAKKC KYOKAISHI

JOURNAL OF FERMENTATION TECHNOLOGY. JAPAN
SEE HAKKO KOGAKU ZASSHI

JOURNAL CF FISHERIES
SEE SUISAN KENKYU-SHI

JOURNAL OF THE FOOD HYGIENIC SOCIETY OF JAPAN
SEE SHOKUHIN EISEIGAKU ZASSHI

JOURNAL CF FOOD SCIENCE AND TECHNOLOGY. JAPAN
SEE NIHON SHOKUHIN KOGYO GAKKAISHI

JOURNAL CF THE FCRMOSAN MEDICAL ASSOCIATION
SEE TAIWAN I HSUEH HUI TSA CHIH

JOURNAL DU FOUR ELECTRIQUE ET DE L'ELECTROLYSE ET DES INDUSTRIES ELECTROCHIMIQUES
1929 V38 P377-380 58-1985 <*>
1929 V38 P416-420 58-1985 <*>
1935 V44 P165-172 II-690 <*>
1935 V44 N9 P318-323 2279 <*>
1937 V46 P92-94 I-9 <*>
1946 P73-77 1099 <*>
1946 P90-96 1099 <*>
1948 V57 P26-27 63-14812 <*>
1950 V59 P112-113 3294 <*>
1953 V62 P45-47 I-557 <*>
1953 V62 N2 P45- 64-14389 <*>
1954 N3 P71-74 II-718 <*>
1955 V64 P13-15 2337 <*>
1964 V69 N1 P19-22 3741 <BISI>
1964 V69 N2 P51-57 3741 <BISI>
1964 V69 N8/9 P233-234 4327 <BISI>
1965 V70 N1 P7-9 66-11328 <*> 0
1965 V70 N3 P89-90 4726 <BISI>

JOURNAL FRANCAIS DE MEDECINE ET CHIRURGIE THORACIQUE
1954 V8 N3 P306-309 3359-B <KH>
1958 V12 N4 P388-402 60-10673 <*> 0
1958 V12 N5 P469-484 60-10674 <*>
1960 V14 P97-108 TC-1231 <TC>
1963 V17 P275-283 65-18003 <*>

JOURNAL FRANCAIS D'OTO-RHINO-LARYNGOLOGIE
1955 V4 N8 P851-858 11 <BIHR>
1958 V7 N5 P549-573 63-10718 <*> 0
1959 V8 N1 P103-111 65-11660 <*>
1960 V9 N8 P1067-1073 62-26242 <=$>

JOURNAL OF THE FUEL SOCIETY OF JAPAN
SEE NENRYO KYOKAI-SHI

JOURNAL OF GEOGRAPHY. CHINESE PEOPLES REPUBLIC
SEE TI-LI-HSUEH-PAC

JOURNAL CF GEOGRAPHY. JAPAN
SEE CHIGAKU ZASSHI

JOURNAL CF THE GEOLOGICAL SOCIETY OF JAPAN
SEE CHISTITSUGAKU ZASSHI

JOURNAL OF GEOLOGY. CHINESE PEOPLES REPUBLIC
SEE TI CHIH HSUEH PAO

JOURNAL OF GEOPHYSICAL PROSPECTING. CHINESE PEOPLES REPUBLIC
SEE TI CH'IU WU LI K'AN T'AN

JOURNAL OF GEOPHYSICAL RESEARCH
1949 V54 N1 P39-52 AEC-TR-6425 <*>
1951 V56 N4 P595-600 AL-114 <*>

JOURNAL OF GEOPHYSICS. CHINESE PEOPLES REPUBLIC
SEE TI CH'IU WU LI HSUEH PAO

JOURNAL FUER HIRNFORSCHUNG
1954 V1 N4/5 P281-325 57-1180 <*>

JOURNAL OF HOME ECONOMICS
SEE KASEIGAKU ZASSHI

JOURNAL OF HYDRAULIC ENGINEERING. CHINESE PEOPLES REPUBLIC
SEE SHUI LI HSUEH PAO

JOURNAL OF HYDROBIOLOGY. CHINESE PEOPLES REPUBLIC
SEE SHUI SHENG SHENG WE HSUEH CHI K'AN

JOURNAL OF HYDROLOGY
1965 V3 P73-87 66-12938 <*>

JOURNAL OF HYGIENE, EPIDEMIOLOGY, MICROBIOLOGY AND IMMUNOLOGY
1959 V3 P106-116 60-18556 <*> 0
1960 V4 N1 P6-11 61-00260 <*>
1960 V4 N4 P385-389 62-19433 <*=>

JOURNAL OF THE HYGIENIC CHEMICAL SOCIETY OF JAPAN
SEE NIHON EISEI KAGAKUKAI SHI

JOURNAL OF THE INDUSTRIAL EXPLOSIVES SOCIETY. JAPAN
SEE KOGYO KAYAKU KYOKAISHI

JOURNAL D'INFORMATIONS TECHNIQUES DES INDUSTRIES DE LA FONDERIE
1959 N102 P1-3 62-14594 <*>
1962 P11-21 2896 <BISI>

JOURNAL OF INORGANIC AND NUCLEAR CHEMISTRY
1958 V7 P129-132 AEC-TR-4015 <*>
1958 V8 P612-619 51L29I <ATS>
1960 V12 P297-303 AEC-TR-4975 <*>
1960 V14 P247-250 AEC-TR-4874 <*>
1960 V14 P251-254 AEC-TR-4875 <*>
1960 V15 P177-181 AEC-TR-4876 <*>
1960 V16 P87-99 NP-TR-893 <*>

JOURNAL OF INSECT PHYSIOLOGY
1957 V1 P143-149 60-18655 <*>
1957 V1 N1 P95-107 60-18017 <*>
1958 V1 N4 P346-351 64-30287 <*>
1960 V5 N3/4 P240-258 64-30298 <*>

JOURNAL OF THE INSTITUTE CF ELECTRICAL COMMUNICATION ENGINEERS OF JAPAN
SEE DENKI TSUSHIN GAKKAI ZASSHI

JOURNAL OF THE INSTITUTE OF ELECTRICAL ENGINEERS OF JAPAN
SEE DENKI GAKKAI ZASSHI

JOURNAL OF THE INSTITUTE CF TELEVISION ENGINEERS OF JAPAN
SEE TEREBIJON

JOURNAL OF INTERNAL MEDICINE
SEE NAIKA NO RYOIKI

JOURNAL OF THE IRON AND STEEL INSTITUTE OF JAPAN
SEE TETSU TO HAGANE

JOURNAL OF THE IWATE MEDICAL ASSOCIATION
SEE IWATE IGAKU ZASSHI

JOURNAL OF THE JAPAN FOUNDRYMEN'S SOCIETY
SEE IMONO

JOURNAL OF THE JAPAN GAS ASSOCIATION
SEE NIHON GASU KYOKAISHI

JOURNAL OF THE JAPAN INSTITUTE OF METALS
SEE NIPPON KINZOKU GAKKAISHI

JOURNAL OF THE JAPAN OIL CHEMISTS' SOCIETY
SEE YUKAGAKU

JOURNAL OF THE JAPAN PETROLEUM INSTITUTE
SEE SEKIYU GAKKAI-SHI

JOURNAL OF THE JAPAN SOCIETY OF AERONAUTICAL AND SPACE SCIENCES
SEE NIPPON KOKU GAKKAISHI

JOURNAL OF THE JAPAN SOCIETY OF CIVIL ENGINEERS
SEE DOBOKU GAKKAISHI

JOURNAL OF THE JAPAN SOCIETY OF COLOUR MATERIAL
SEE SHIKIGAI KYOKAI-SHI

JOURNAL OF THE JAPAN SOCIETY FOR LUBRICATION ENGINEERS
SEE JUNKATSU

JOURNAL OF THE JAPAN SOCIETY FOR MATERIALS
SEE ZAIRYO

JOURNAL OF THE JAPAN SOCIETY OF MECHANICAL ENGINEERS
SEE NIPPON KIKAI GAKKAISHI

JOURNAL OF THE JAPAN SOCIETY OF POWDER METALLURGY
SEE FUNTAI OYOBI FUNMATSU YAKIN

JOURNAL OF THE JAPAN WELDING SOCIETY
SEE YOSETSU GAKKAISHI

JOURNAL OF THE JAPAN WOOD RESEARCH SOCIETY
SEE MOKUZAI GAKKAISHI

JOURNAL OF THE JAPANESE ASSOCIATION FOR INFECTIOUS DISEASES
SEE NIHON DENSENBYO GAKKAI ZASSHI

JOURNAL OF THE JAPANESE ASSOCIATION OF MINERALO-GISTS, PETROLOGISTS AND ECONOMIC GEOLOGISTS
SEE GANSEKI KOBUTSU KOSHOGAKU ZASSHI

JOURNAL OF THE JAPANESE ASSOCIATION OF PETROLEUM TECHNOLOGISTS
SEE SEKIYU GIJUTSU KYOKAISHI

JOURNAL OF THE JAPANESE ASSOCIATION OF SNOW AND ICE
SEE SEPPYO

JOURNAL OF JAPANESE BIOCHEMICAL SOCIETY
SEE SEIKAGAKU

JOURNAL OF THE JAPANESE CERAMIC ASSOCIATION
SEE YOGYO KYOKAI-SHI

JOURNAL OF JAPANESE CHEMISTRY
SEE KAGAKU NO RYOIKI

JOURNAL OF THE JAPANESE FORESTRY SOCIETY
SEE NIPPON RINGAKKAI SHI

JOURNAL OF THE JAPANESE OBSTETRICAL AND GYNECOLOGICAL SOCIETY
SEE NIHON SANKA FUJINKA GAKKAI ZASSHI

JOURNAL OF THE JAPANESE ORTHOPAEDIC ASSOCIATION
SEE NIHON SEIKEI GEKA GAKKAI ZASSHI

JOURNAL OF THE JAPANESE SOCIETY OF FOOD AND NUTRITION
SEE EIYO TO SHOKURYO

JOURNAL OF THE JAPANESE SOCIETY FOR HORTICULTURAL SCIENCE
SEE ENGEI GAKKAI ZASSHI

JOURNAL OF THE JAPANESE SOCIETY FOR INTERNAL MEDICINE
SEE NIPPON NAIKA GAKKAI ZASSHI

JOURNAL OF THE JAPANESE SURGICAL SOCIETY
SEE NIPPON GEKAGAKKAI ZASSHI

JOURNAL OF THE JAPANESE TECHNICAL ASSOCIATION OF THE PULP AND PAPER INDUSTRY
SEE KAMI PARUPU GIJUTSU KYOKAI SHI

JOURNAL OF THE JAPANESE VETERINARY MEDICAL ASSOCIATION
SEE NIHON JUISHIKAI ZASSHI

JOURNAL OF THE KANSAI SOCIETY OF NAVAL ARCHITECTS, JAPAN
SEE KANSAI ZOSEN KYOKAI-SHI

JOURNAL OF THE KEIO MEDICAL SOCIETY
SEE KEIO IGAKU

JOURNAL OF THE KOREAN INSTITUTE OF CHEMICAL ENGINEERS
1963 V1 N1 P5-12 ICE V4 N2 P204-211 <ICE>
1966 N1 P9-14 ICE V6 N4 P668-673 <ICE>

JOURNAL OF THE KUMAMOTA MEDICAL SOCIETY. JAPAN
SEE KUMAMOTA I GAKKAI ZASSHI

JOURNAL OF KUSUNOKI AGRICULTURE
SEE KUSUNOKI NOHO

JOURNAL OF THE KYOTO MEDICAL ASSOCIATION
SEE KYOTO IGAKKAI ZASSHI

JOURNAL OF KYOTO PREFECTURAL MEDICAL UNIVERSITY
SEE KYOTO FURITSU IKADAIGAKU ZASSHI

JOURNAL OF THE KYUSHU HEMTOLOGICAL SOCIETY
SEE KYUSHU KETSUEKI KENKYU DOKOKAI-SHI

JOURNAL OF THE LESS-COMMON METALS
1959 V1 P439-455 AEC-TR-4320 <*>
1962 V4 N3 P252-265 TRANS-177 <FRI>
1963 V5 N1 P78-89 63-18308 <*> 0
63-18461 <*> 0
1964 V6 P132-151 66-12819 <*>
1965 V9 P133-151 66-11115 <*>

JOURNAL FUER MAKROMOLEKULARE CHEMIE
1943 V1 N7/9 P185-196 2012 <*>

JOURNAL OF THE MATHEMATICAL SOCIETY OF JAPAN
SEE NIHON SUGAKKAI

JOURNAL OF MATHEMATICS. CHINESE PEOPLES REPUBLIC
SEE SHU HSUEH HSUEH PAO

JOURNAL DE MATHEMATIQUES PURES ET APPLIQUEES
1940 V19 P27-43 I-356 <*>
1959 S9 V38 N4 P347-364 66-13798 <*>

JOURNAL DE MECANIQUE
1962 V1 N4 P367-394 66-13687 <*> 0

JOURNAL DE MECANIQUE ET DE PHYSIQUE DE L'ATMOSPHERE
1959 V1 N1 P43-47 SCL-T-362 <*>

JOURNAL OF THE MECHANICS AND PHYSICS OF SOLIDS
1957 V5 P95-114 3135 <BISI>
1961 P69-90 61-20274 <*>

JOURNAL DE MEDECINE DE BORDEAUX ET DE LA REGION DU SUD-OUEST
1918 V89 P185-188 57-1607 <*>
1948 V125 N7 P289-302 2784 <*>
1949 V126 P403-404 I-208 <*>
1950 V127 P461 I-208 <*>
1953 V130 N8 P1007-1010 57-2562 <*>
1957 V134 N4 P479-486 60-10095 <*>
1959 V136 N5 P632-634 60-14542 <*>
1960 V137 P346-359 61-16134 <*> P

1960 V137 N4 P392-396 66-13320 <*>
1962 V139 N7 P840-855 63-20991 <*> O

JOURNAL DE MEDECINE DE LYON
1948 V29 N695 P905-906 2799 <*>
1949 V30 N710 P585-587 2601 <*>
1949 V30 N719 P959-968 2771 <*>
1951 V32 P189-195 3250 <*>
1958 V39 P581-587 62-16612 <*>
1959 V40 N955 P987-997 10643-A <KH>
63-16576 <*>
1961 V42 P935-940 65-13742 <*>
1962 V43 P265-270 65-13747 <*>
1965 V46 P829-839 66-13217 <*>

JOURNAL DE MEDECINE DE PARIS
1949 V69 N4 P66-71 57-2335 <*>
1965 V85 N1 P3-17 66-12124 <*>

JOURNAL OF THE MEDICAL ASSOCIATION OF FORMOSA
SEE TAIWAN IGGAKAI ZASSHI

JOURNAL OF THE METAL FINISHING SOCIETY OF JAPAN
SEE KINZOKU HYOMEN GIJUTSU

JOURNAL OF METEOROLOGICAL RESEARCH. TOKYO
SEE KISHOCHO KENKYU JIHO

JOURNAL OF THE METEOROLOGICAL SOCIETY OF JAPAN
SEE KISHO SHUSHI

JOURNAL OF MICROBIOLOGY. CHINESE PEOPLES REPUBLIC
SEE WEI SHENG WU HSUEH PAO

JOURNAL DE MICROSCOPIE
1962 V1 P351-352 63-01066 <*>
1964 V3 N6 P579-588 66-11817 <*>
1964 V3 N6 P589-606 66-11818 <*>
1965 V4 N2 P253-264 66-11580 <*>
1965 V4 N5 P587-594 4765 <BISI>

JOURNAL OF THE MINING AND METALLURGICAL INSTITUTE OF JAPAN
SEE NIPPON KOGYO KAISHI

JOURNAL OF MOLECULAR BIOLOGY
1960 V2 P69-71 61-00359 <*>
1960 V2 P72-74 61-00351 <*>
1961 V3 N3 P787-789 62-00912 <*>
1961 V3 N5 P790-793 62-00911 <*>
1962 V4 P410-412 63-00151 <*>
1962 V4 P413-414 63-00064 <*>

JOURNAL OF THE NAGOYA CITY UNIVERSITY MEDICAL ASSOCIATION
SEE NAGOYA SHIRITSU DAIGAKU IGAKKAI ZASSHI

JOURNAL OF THE NAGOYA MEDICAL ASSOCIATION
SEE NAGOYA IGAKU

JOURNAL OF NARA GAKUGEI UNIVERSITY
SEE NARA GAKUGEI DAIGAKU KIYO

JOURNAL OF THE NARA MEDICAL ASSOCIATION
SEE NARA IGAKU ZASSHI

JOURNAL OF THE NATURAL HISTORY SOCIETY OF TAIWAN
1924 V14 N74 P4-21 61J8J <ATS>

JOURNAL OF NATURAL SCIENCES. CHI-LIN UNIVERSITY
SEE CHI-LIN TA HSUEH TZU JAN K'O HSUEH HSUEH PAO

JOURNAL OF THE NAUTICAL SOCIETY OF JAPAN
SEE NIHON KOKAI GAKKAI SHI

JOURNAL OF THE NIHON MEDICAL SCHOOL
SEE NIHON IKA DAIGAKU ZASSHI

JOURNAL OF NON-DESTRUCTIVE INSPECTION. JAPAN
SEE NIHON HIHAKAI KENSA KYOAKI SHI

JOURNAL OF NUCLEAR ENERGY
1955 V2 P101-109 62-10393 <*> O
1955 V2 N2 P110-111 AEC-TR-2701 <*>
1957 V4 P305-318 AEC-TR-5168 <*>
1957 V5 N3/4 P357-361 58-2316 <*>

JOURNAL OF NUCLEAR MATERIALS
1959 V1 P58-72 AEC-TR-3876 <*>
1959 V1 P113-119 AEC-TR-5840 <*>
1959 V1 P120-126 AEC-TR-4558 <*>
1959 V1 P203-209 66-11753 <*> O
1959 V1 P259-270 AEC-TR-5083 <*>
1959 V1 N3 P259-270 K-49 <RIS>
1960 V2 P69-74 AEC-TR-5779 <*>
1960 V2 N1 P43-50 AEC-TR-4522 <*>
1960 V2 N1 P75-80 AEC-TR-4364 <*>
1961 V3 N1 P60-66 61-20505 <*>
1961 V3 N3 P327-330 AEC-TR-5325 <*> O
1961 V4 P255-268 AEC-TR-5060 <*>
1961 V4 N2 P218-225 TRANS-186 <FRI>
64-14726 <*>
1961 V4 N3 P241-254 00Q72G <ATS>
1961 V4 N3 P269-271 ATS-01Q72G <ATS>
1962 V5 N1 P101-108 AEC-TR-5138 <*>
1962 V5 N1 P150-152 66-12866 <*>
1962 V5 N1 P153-155 AEC-TR-5104 <*>
1962 V5 N3 P287-300 AEC-TR-5301 <*>
1962 V6 P96-106 66-11098 <*>
1962 V6 P338-341 AEC-TR-5722 <*>
1962 V6 P342-345 AEC-TR-5717 <*>
1962 V6 N1 P96-106 NP-TR-992 <*>
1962 V6 N1 P137-138 AEC-TR-5742 <*>
1962 V6 N2 P203-212 HW-TR-43 <*>
1963 V8 P116-125 AEC-TR-6184 <*>
1963 V8 N1 P39-40 NP-TR-1070 <*>
1963 V10 P173-181 66-11749 <*>
1963 V10 N3 P215-223 NP-TR-1141 <*>
1964 V11 N1 P67-76 66-11620 <*> O
1964 V12 N2 P159-166 66-11086 <*>
1965 V16 P59-67 AEC-TR-6441 <*>
1965 V16 P190-207 65-14745 <*>

JOURNAL OFFICIEL DES COMMUNAUTES EUROPEENNES
1962 V5 P57 NP-TR-1161 <*>

JOURNAL OF THE OKAYAMA MEDICAL SOCIETY. JAPAN
SEE OKAYAMA IGAKKAI ZASSHI

JOURNAL D'OPHTHALMOLOGIE SOCIALE
1951 V12 P16-21 66-12622 <*>

JOURNAL OF THE ORGANIZATION OF HEALTH SERVICES AND THE HISTORY OF MEDICINE. CHINESE PEOPLES REPUBLIC
SEE I-HSUEH SHIH YU PAO-CHIEN TSU-CHIH

JOURNAL OF ORGANOMETALLIC CHEMISTRY
1964 V2 P197-205 545 <TC>
1965 V3 P16-24 8665 <IICH>

JOURNAL OF THE OSAKA CITY MEDICAL CENTER
SEE OSAKA SHIRITSU DAIGAKU IGAKU ZASSHI

JOURNAL OF THE OSAKA MEDICAL SOCIETY
SEE OSAKA IGAKKAI ZASSHI

JOURNAL OF THE OSAKA ODONTOLOGICAL SOCIETY
SEE SHIKA IGAKU

JOURNAL OF THE OTO-RHINO-LARYNGOLOGICAL SOCIETY OF JAPAN
SEE NIHON JIBI INKOKA GAKKAI KAIHO

JOURNAL OF PALEONTOLOGY. CHINESE PEOPLES REPUBLIC
SEE KU SHENG WU HSUEH PAO

JOURNAL OF PEDIATRIC PRACTICE. TOKYO
SEE SHONIKA SHINRYO

JOURNAL OF PEDOLOGY. CHINESE PEOPLES REPUBLIC

SEE TU JUNG HSUEH PAO

JOURNAL OF THE PEKING SCHOOL OF MINES
SEE PEI-CHING KUANG YEH HSUEH YUAN HSUEH PAO

JOURNAL OF PEKING UNIVERSITY, NATURAL SCIENCES
SEE PEI-CHING TA HSUEH HSUEH PAO-TZU JAN K'O HSUEH

JOURNAL OF THE PHARMACEUTICAL SOCIETY OF JAPAN
SEE YAKUGAKU ZASSHI

JOURNAL DE PHARMACIE DE BELGIQUE
1948 V3 P123-128 C-2278 <NRC>
65-14994 <*>
1954 V8 P44-58 II-112 <*>
1960 N7/8 P229-268 62-10265 <*>
1960 N9/10 P362-366 62-00500 <*>
1960 V15 N7/8 P229-268 62-10964 <*>
1961 V16 P286-300 62-00857 <*>
1962 V17 P267-271 63-01098 <*>

JOURNAL DE PHARMACIE ET DE CHIMIE
1917 S7 V15 P145-149 64-20023 <*>
1920 V22 N7 P321-323 3146 <*>
1927 S8 V5 P531-539 63-18664 <*>
1930 S8 V11 P46-47 63-14705 <*>
1934 S8 V20 P549-576 64-16208 <*>
1935 S8 V21 P5-23 64-16208 <*>
1939 S8 V29 P159-166 60-10605 <*> 0
1939 S8 V29 P167-175 60-10604 <*>
1939 S8 V30 P16-34 60-10602 <*> P

JOURNAL OF PHARMACY AND PHARMACOLOGY
1950 V2 P345-360 60-10857 <*>
1954 V6 P361-389 II-803 <*>
1954 V6 P361-384 2634 <*>
1954 V6 N6 P361-389 62-14953 <*> 0

JOURNAL OF PHYSICAL CHEMISTRY. ITHACA
1928 V32 P1263- 57-425 <*>

JOURNAL OF THE PHYSICAL SOCIETY OF JAPAN
SEE NIPPON BUTSURI GAKKAI

JOURNAL OF PHYSICS, CHINESE PEOPLES REPUBLIC
SEE WU LI HSUEH PAO

JOURNAL DER PHYSIK
1793 V7 P319-322 59-20034 <*> 0

JOURNAL OF THE PHYSIOLOGICAL SOCIETY OF JAPAN
SEE NIPPON SEIRIGAKU ZASSHI

JOURNAL DE PHYSIOLOGIE
1947 V39 P219-248 60-10864 <*> 0
1947 V39 P473-482 II-847 <*>
1948 V40 P89-110 59-20999 <*> P
1948 V40 P199-222 59-16053 <+*>
1950 V42 N3 P505-515 65-14995 <*> 0
1951 V43 P726-728 3170 <*>
1952 V44 P323-326 63-01628 <*>
1953 V45 N1 P363-365 2792 <*>
1954 V46 P414-418 63-01384 <*>
1956 V48 P608-612 63-01523 <*>
1957 V49 P215-217 63-20904 <*> 0
1957 V49 P667-675 57-3166 <*>
1957 V49 P1171-1200 59-16647 <+*>
1957 V49 P1201-1223 59-16648 <+*>
1958 V50 P435-438 59-10356 <*> 0
1958 V50 N2 P469-471 65-14996 <*>
1958 V50 N3 P587-661 59-12291 <+*> 0
1959 V51 P813-827 63-18737 <*>
1961 V53 N2 P260 62-14318 <*>
1961 V53 N2 P265-266 62-14915 <*>
1963 V55 P305-306 64-00190 <*>
1963 V55 N2 P249-250 64-00192 <*>
1964 V56 P213-233 66-11578 <*>

JOURNAL DE PHYSIOLOGIE ET DE PATHOLOGIE GENERALE
1919 V18 P70-82 61-14707 <*>
1919 V18 P83-94 61-14708 <*>
1919 V18 P295-304 61-14704 <*>
1919 V18 P885-894 61-14702 <*>

JOURNAL DE PHYSIQUE
1963 V24 P84-86 157 <TC>
1963 V24 N3 P39A-44A AI-TR-6 <*>
1965 V26 N2 P82A-83A 66-11037 <*>
1965 V26 N11 P684-688 66-14506 <*>

JOURNAL DE PHYSIQUE ET LE RADIUM
1926 S6 V6 N8 P225-229 AEC-TR-4083 <*>
1926 S6 V7 P25- II-464 <*>
1928 S6 V9 N6 P187-204 63-20345 <*>
1931 V2 N3 P86-100 2608F <ATS>
1932 S7 N3 P239-247 57-1055 <*>
1932 S7 V3 P590-613 66-14014 <*>
1934 V5 P257-261 52E2F <ATS>
1934 V5 N12 P628-634 34H10F <ATS>
1935 S7 V6 P257-262 66-14015 <*>
1936 V7 N4 P153-157 57-2400 <*>
1937 S7 V8 N11 P453-470 AL-137 <*>
1938 V7 N9 P157-158 58-2146 <*>
1938 S7 V9 P157-158 65-00108 <*>
1939 V10 P428-429 3312 <*>
1939 S7 V10 N10 P428-429 59-12934 <+*>
1940 V1 N8 P319-321 AL-141 <*>
1944 S8 V5 N6 P97-108 60-10827 <*>
62-16985 <*> 0
1946 V1 P7-10 2827 <*>
1946 V7 P203-208 I-246 <*>
1946 V7 N9 P259-265 AL-190 <*>
1946 S8 V7 N9 P259-263 31P66F <ATS>
1947 N6 P165-178 1568 <*>
1947 V8 N7 P200-211 1569 <*>
1947 V8 N12 P345-351 AL-854 <*>
1949 S7 V10 P177-188 AL-541 <*>
1949 S8 V10 P140-160 I-567 <*>
1950 V11 P105-119 UCRL TRANS-117 <*>
1950 V11 P208-212 58-889 <*>
1950 V11 P524-528 II-111 <*>
1950 V11 N1 P20 62-20274 <*> 0
1950 V11 N2 P49-61 61-10162 <*> 0
1950 V11 N2 P69-76 65-00185 <*>
1950 V11 N8 P2C- AL-534 <*>
1950 V11 N8/9 P524-528 AL-492 <*>
1951 V12 P563-564 UCRL TRANS-594(L) <*>
1951 V12 P602-606 58-2325 <*>
1951 V12 P751-755 I-1023 <*>
1951 V12 P893-899 AL-173 <*>
1951 V12 P941-949 I-113 <*>
1951 V12 N3 P252-255 65-12262 <*>
1951 V12 N4 P563-564 60-18042 <*>
1952 V13 P99A-104A SUP 7 2527 <*>
1952 V13 P247-248 I-768 <*>
1952 V13 P311-312 58-2425 <*>
1952 V13 P485-486 64-14116 <*>
1952 V13 P489-490 64-14123 <*>
1952 V13 P491 UCRL TRANS-597(L) <*>
1952 V13 P499-503 AEC-TR-3921 <*>
1952 V13 N5 P249-264 57-1004 <*>
1952 V13 N11 P499-505 AI-TR-14 <*>
1952 V13 N12 P645-649 I-669 <*>
1952 V13 N12 P651-657 I-669 <*>
1953 V14 P330- 59-22185 <+*> C
1953 V14 P374-380 66-11860 <*>
1953 V14 P501-509 II-255 <*>
1953 V14 P695-706 II-212 <*>
1953 V14 N1 P34-42 I-898 <*>
1953 V14 N5 P307-310 60-18085 <*>
1953 V14 N6 P424-425 59-11954 <=>
1953 V14 N10 P42S-43S II-1020 <*>
1953 V14 N10 P43S-44S II-926 <*>
1953 V14 N7/9 P494-495 59-11955 <=>
1954 V15 P117-121 I-125 <*>
1954 V15 P264-272 UCRL TRANS-577(L) <*>
1954 V15 P438-444 I-949 <*>
1954 V15 P477-482 AEC-TR-3712 <+*>
1954 V15 N2 P117-121 60-18045 <*>

1954 V15 N4 P225-239 57-967 <*>
1954 V15 N5 P101A-108A SUP
66-12656 <*>
1954 V15 N11 P776-777 64-16986 <*>
1955 V16 P155-158 57-802 <*>
1955 V16 P238 UCRL TRANS-590(L) <*>
1955 V16 P304-317 57-2994 <*>
58-571 <*>
1955 V16 P422-427 AEC-TR-4211 <*>
1955 V16 P704-706 T1796 <INSD>
1955 V16 P815-835 AEC-TR-4212 <*>
1955 V16 P849- 1253 <*>
1955 V16 N4 P334-338 63H9F <ATS>
1955 V16 N6 P444-445 C-3249 <NRCC>
1955 V16 N6 P489-490 59-11950 <=>
1955 V16 N6 P490-491 59-11951 <=>
1955 V16 N6 P491-492 59-11952 <=>
1956 V17 SUP.11 P113A- 58-1782 <*>
1956 V17 P57-59 64-16987 <*>
1956 V17 P347-349 64-16985 <*>
1956 V17 P379- 60-21185 <+*>
1956 V17 P384- 60 21469 <+*>
1956 V17 P420-425 91J15F <ATS>
1956 V17 P907-908 58-2204 <CTT>
1956 V17 P934-939 58-2187 <*>
1956 V17 P150A-158A 57-3047 <*>
1956 V17 N3 P169-306 63-14679 <*>
1956 V17 N3 P250-255 57-720 <*>
1956 V17 N5 P449 62-16662 <*>
1956 V17 N6 P104A-107A SUP
63-20723 <*>
1956 V17 N7 P585-586 57-1874 <*>
1956 V17 N8/9 P649 SCL-T-254 <+*>
1957 V18 P51- 60-21757 <+*>
1957 V18 P161-168 57-2593 <*>
1957 V18 P380-386 61-00221 <*>
1957 V18 P498-504 58-505 <*>
1957 V18 P585-589 58-2162 <*>
1957 V18 N2 P109-114 60-14396 <*>
1957 V18 N3 P41A-44A 59-15529 <*> 0
1957 V18 N4 P209-213 60-10095 <*>
1957 V18 N4 P260-279 62-16519 <*> 0
1957 V18 N8/9 P527-536 64-16569 <*>
1958 V19 P86-87 59-21187 <=>
1958 V19 P108 59-21186 <=>
1958 V19 P223-229 63-00591 <*>
63-27142 <$>
1958 V19 P573-581 AEC-TR-3809 <*>
1958 V19 P624-629 61-00222 <*>
1958 V19 N1 P38-91 61-00032 <*>
1958 V19 N2 P153-158 59-10966 <*>
1958 V19 N3 P262-269 63-14013 <*>
1958 V19 N11 P913-914 AEC-TR-5852 <*>
1958 V19 N12 P927-929 NP-TR-441 <*>
1959 V20 P160-168 AEC-TR-5726 <*>
1959 V20 P959-962 AEC-TR-4602 <*>
1959 V20 P494-544 SUP NP-TR-481 <*>
1959 V20 N5 P541-548 64-16618 <*>
1959 V20 N6 P633-646 63-01223 <*>
1959 V20 N7 P657-668 64-16617 <*>
1959 V20 N2/3 P160-168 65-12846 <*>
1960 V21 P261-263 62-25264 <=*>
1960 V21 P343-345 AEC-TR-4662 <*>
1960 V21 N2 P130-140 62-16558 <*> 0
1960 V21 N7 P97A-111A 61-14913 <*>
67M47F <ATS>
1960 V21 N7 P129A-133A SUP
62-10114 <*> 0
1960 V21 N11 P794-800 61-20825 <*>
1960 V21 N8/9 P679-680 NP-TR-820 <*>
NP-TR-830 <*>
1961 V22 P525-527 AEC-TR-4651 <*>
1961 V22 P615-617 66-11964 <*>
1961 V22 N2 P17A-26A SUP 76S87F <ATS>
1961 V22 N3 P135-141 65-18081 <*>
1961 V22 N6 P139A-140A 62-18912 <*>
1961 V22 N8/9 P472-480 65-18082 <*>
1962 V23 P672-676 AI-TR-64-3 <*>
1962 V23 P704-706 66-12186 <*>
1962 V23 N3 P173-183 63-14012 <*>
1962 V23 N5 P297-298 65-14648 <*>
1962 V23 N12 P113-118 C-5709 <NRC>
1962 V23 N12 P163A-165A SUP
64-10385 <*>
1962 V23 N8/9 P501-510 AI-TR-35 <*>
1962 V23 N8/9 P588-589 63-20922 <*> 0
65-14657 <*>

JOURNAL DE PHYSIQUE ET LE RADIUM. SUPPLEMENT
1957 V18 N12 P154A-157A 59-17456 <*>

JOURNAL DE PHYSIQUE THEORIQUE ET APPLIQUEE
1892 S3 V1 P265-285 66-14194 <*>
1901 P123-135 66-10917 <*>
1902 P151-156 66-10918 <*>
1910 V9 P129-134 66-13599 <*>
1910 V9 P457-468 57-3272 <*>
1917 S5 N7 P161 58-320 <*>

JOURNAL OF PLANT PROTECTION
SEE NIPPON SHOKUBUTSU AIGOKWAI

JOURNAL OF POLYMER SCIENCE. NEW YORK
1948 V3 P365-370 65-10315 <*>
1950 V5 P429-441 AL-466 <*>
1950 V5 N3 P333-353 65-12479 <*>
1951 V6 N5 P601-608 64-18581 <*> 0
1952 V8 N3 P289-311 65-17326 <*> 0
1952 V9 N6 P557-563 60-10764 <*>
1955 V16 P143-154 02G6F <ATS>
59-20985 <*>
1955 V16 P491-504 II-1015 <*>
64-16441 <*>
1956 V22 P409-422 64-16501 <*>
1958 V28 N118 P648-651 58-2388 <*>
59-17254 <*>
1958 V29 P191-217 60K25F <ATS>
1958 V29 P375-380 AI-TR-5 <*>
1958 V29 P584-594 59-15915 <*>
1958 V29 N120 P321-342 61-20721 <*> 0
1958 V29 N120 P367-374 60-18910 <*>
1958 V30 P131-147 61-10664 <*>
1958 V30 P363-374 RCT V32 N2 P588 <RCT>
1958 V30 P501-512 287 <TC>
1958 V30 P551-559 59-15832 <*> 0
1958 V30 N121 P131-147 62-14303 <*>
1959 V34 P551-568 03L31F <ATS>
61-10574 <*>
65-11969 <*>
1959 V34 P721-740 47L30G <ATS>
1959 V34 N127 P517-530 62-18278 <*> 0
1960 V48 P309-319 62-10088 <*>
1960 V48 P477-489 66-10066 <*>
1961 V54 P83-100 65-10073 <*>
1962 V59 N167 P13-27 65-13321 <*> 0
1962 V59 N167 P29-50 65-13323 <*> 0
1962 V59 N167 P51-70 65-13222 <*>
1963 N4 P507-519 66-12870 <*>
1963 N4 P649-672 65-18194 <*>
1964 N4 P103-107 65-12487 <*>
1964 V2 P1291-1299 8932 <IICH>
1964 V60 P83-91 296 <TC>

JOURNAL FUER PRAKTISCHE CHEMIE
1858 P41-62 64-20475 <*>
1877 V2 N16 P125-169 II-699 <*>
1890 V41 N2 P552-563 63-14158 <*> 0
1890 V42 N2 P110-126 63-14159 <*> 0
1892 V46 N2 P142-151 65-18022 <*>
1899 V59 N2 P10-11 63-16171 <*>
1900 V62 N2 P189-211 84Q71G <ATS>
1905 V71 P264-267 64-10787 <*>
1915 V92 P102-103 64-10433 <*>
1916 V93 P142-161 63-18556 <*>
1924 V108 N2 P61-74 376 <TC>
1927 V116 P163-174 65-12101 <*>
1927 V117 P245-261 T2075 <INSD>
1932 V135 P142-144 63-18114 <*>
1932 V135 N1/2 P15-35 12H10G <ATS>
1935 V143 P127-138 74J15G <ATS>

1935 V143 P139-142 72J15G <ATS>
1936 V147 P226-240 65-18153 <*>
1936 V147 N1/5 P99-109 60-10432 <*>
1939 V152 P237-266 57-412 <*>
1939 V153 P160-168 63-14036 <*>
1940 V157 N1/3 P19-88 RCT V15 N1 P473 <RCT>
1941 V157 P238-282 63-14160 <*>
1941 V157 P246-248 63-10214 <*>
1941 V157 P258-259 63-10214 <*>
1941 V157 P262-264 63-10214 <*>
1941 V157 N4/7 P158-176 RCT V15 N1 P523 <RCT>
1941 V158 N2 P275-294 63-20477 <*> 0
1941 V159 N8/10 P194-217 64-10515 <*>
1942 V160 N3/4 P95-119 RCT V17 N1 P331 <RCT>
1942 V160 N5/7 P176-194 60-18729 <*> 0
1942 V160 N10/2 P281-295 RCT V17 N1 P356 <RCT>
64-20408 <*>
1942 V160 N10/2 P296-314 RCT V18 N1 P267 <RCT>
1942 V161 P81-112 64-20467 <*>
1943 V161 P261-270 57-2189 <*>
1943 V161 N1 P55-75 63-14161 <*>
1943 V161 N11/2 P261-270 RCT V16 N1 P834 <RCT>
1943 V162 N1/7 P148-180 RCT V17 N1 P15 <RCT>
1944 V162 P224-236 62-16752 <*>
1955 V2 P105-120 II-933 <*>
1955 V2 N4 P250-260 1664 <*>
1955 S4 V2 P196-202 57G7G <ATS>
1955 S4 V2 N4 P233-242 66-12783 <*>
1956 S4 V4 P105-112 73L35G <ATS>
1957 V4 N5/6 P298-305 57-3392 <*>
1958 V6 N4 P46-57 12K26G <ATS>
1958 S4 V6 P103-114 68L30G <ATS>
1958 S4 V6 N5/6 P289-298 62-14213 <*> 0
65-10035 <*> 0
1959 S4 V8 P73-89 06L32G <ATS>
1959 S4 V8 P90-96 29L32G <ATS>
1959 S4 V8 P97-103 30L32G <ATS>
1959 S4 V8 P104-111 31L32G <ATS>
1959 S4 V8 P207-223 77L32G <ATS>
1959 S4 V8 P224-234 95L32G <ATS>
1959 S4 V8 N1/2 P1-16 JPC 10 <RIS>
1959 S4 V8 N1/2 P17-27 JPC-11 <RIS>
1959 S4 V8 N1/2 P28-30 JPC-12 <RIS>
1959 S4 V8 N1/2 P33-38 JPC-13 <RIS>
1959 S4 V8 N1/2 P39-43 JPC-14 <RIS>
1959 S4 V8 N1/2 P44-54 JPC-15 <RIS>
1959 S4 V8 N1/2 P55-63 JPC-16 <RIS>
1959 S4 V8 N1/2 P64-67 JPC-17 <RIS>
1959 S4 V8 N1/2 P68-72 JPC 18 <RIS>
1959 S4 V8 N1/2 P73-89 JPC-19 <RIS>
1959 S4 V8 N1/2 P90-96 JPC-20 <RIS>
1959 S4 V8 N1/2 P97-103 JPC-21 <RIS>
1959 S4 V8 N1/2 P104-111 JPC-22 <RIS>
1959 S4 V8 N1/2 P112-116 JPC-23 <RIS>
1959 S4 V8 N1/2 P117-120 JPC-24 <RIS>
1959 S4 V9 P232-236 65-10055 <*>
1959 S4 V9 N1/2 P3-6 60-10443 <*>
1959 S4 V9 N3/4 P97-103 65-13026 <*>
1961 V14 N4/6 P269-280 65-12428 <*>
1961 S4 V14 P131-138 74P59G <ATS>
1961 S4 V14 N3 P113-118 62-14451 <*> 0
1962 V4 N15 P205- NP-TR-1246 <*>
1962 S4 V15 P175-191 NS-40 <TTIS>
1962 S4 V17 N1/2 P107-112 63-10279 <*>
1962 S4 V18 N3 P150-155 63-18480 <*>
1964 V25 P135-140 740-A <TC>
1964 V25 N4 P150-159 740-C <TC>
1964 S4 V24 N1/2 P23-26 65-11407 <*>
1964 S4 V25 N4 P141-149 740-B <TC>

JOURNAL DES PRATICIENS
1897 V11 P809- 57-1681 <*>
1901 V15 P22-24 65-18028 <*>

JOURNAL FUER PSYCHOLOGIE UND NEUROLOGIE
1931 V41 N6 P383-420 61-14894 <*>

JOURNAL DE PSYCHOLOGIE NORMALE ET PATHOLOGIQUE
1925 V22 P625-666 61-14109 <*> OP
61-14684 <*>
1926 V23 P1003-1010 61-14147 <*>
1933 V30 P590-616 61-14576 <*>
1960 V57 N4 P421-430 65-11675 <*>

JOURNAL OF THE PUBLIC HEALTH ASSOCIATION OF JAPAN
SEE NIHON KOSKU HOKEN KYOKAI ZASSHI

JOURNAL OF QUANTITATIVE SPECTROSCOPY AND RADIATIVE TRANSFER
1963 V3 P221-245 65-63672 <=$>

JOURNAL DE RADIOLOGIE ET D'ELECTROLOGIE
1936 V20 P443-445 2816 <*>

JOURNAL DE RADIOLOGIE, D'ELECTROLOGIE ET ARCHIVES D'ELECTRICITE MEDICALE
1953 V34 N1/2 P18-27 63-01127 <*>
1953 V34 N3/4 P109-171 MF4 <*>
1953 V34 N3/4 P169-171 57-1705 <*>
1956 V37 N3/4 P164-169 57-3283 <*>
1956 V37 N7/8 P607-610 57-3257 <*>
1957 V38 N3/4 P282-284 57-3258 <*>

JOURNAL DE RADIOLOGIE, D'ELECTROLOGIE ET DE MEDECINE NUCLEAIRE
1960 V41 P242-246 AEC-TR-1080 <*>
1960 V41 N1 P731-745 NP-TR-1087 <*>
1960 V41 N3/4 P176 61-00168 <*>
1961 V42 P228-236 NP-TR-1081 <*>
1961 V42 P390-393 66-11653 <*> 0
1963 V44 P258-263 66-12652 <*> 0
1964 V45 N3/4 P133-148 82R77F <ATS>

JOURNAL OF RAILWAY ENGINEERING RESEARCH. JAPAN
SEE TETSUDO GIJUTSU KENYU SHIRYO

JOURNAL DE RECHERCHES ATMOSPHERIQUES
1964 V1 N2 P95-100 65-12004 <*>
1964 V1 N2 P101-108 65-12003 <*>
1964 V1 N3 P121-131 65-14567 <*>
1964 V1 N3 P151-157 65-11422 <*>

JOURNAL DES RECHERCHES DU CENTRE NATIONAL DE LA RECHERCHE SCIENTIFIQUE
1947 V1 P153-163 62-32334 <=*>
1948 V2 N6 P91-103 II-217 <*>
62-24775 <=*>
1950 N12 P113-122 57-2395 <*>
1950 V3 N10 P101-106 60-10092 <*> 0
1950 V3 N13 P169-175 16N56F <ATS>
1950 V10 P21-22 58-2519 <*>
1952 N18 P118-130 63-18976 <*> 0
1953 N24 P138-186 58-1675 <*>
1953 V24 P126-127 57-1178 <*>
58-2176 <*>
1954 N26 P277-289 58-2110 <*>
1954 V6 N29 P125-133 1490 <*>
1955 V6 N33 P404-407 60-18306 <*>
1957 V8 N38 P23-29 62-24056 <=*> 0
1958 N45 P287-304 60-00596 <*>
1958 N45 P287-303 65-17199 <*>
1960 N52 P237-248 62-10076 <*>
1960 V52 P231-236 62-10077 <*>
1961 N54 P23-59 65-10270 <*> P

JOURNAL FUER DIE REINE UND ANGEWANDTE MATHEMATIK
1883 V94 N1 P41-73 63-00931 <*>
63-27219 <$>
1901 V124 N1 P1-27 62-16022 <*>
1932 V167 P221-234 64-14714 <*>
1932 V168 N4 P233-252 64-18596 <*>
1961 V206 P61-66 65-11685 <*>

JOURNAL OF THE RESEARCH INSTITUTE FOR CATALYSIS HOKKAIDO UNIVERSITY
SEE HOKKAIDO DAIGAKU SHOKUBAI KENKYUSHO KIYO

JOURNAL OF THE RESEARCH INSTITUTE FOR METALS
SEE KINZOKU NO KENKYU

JOURNAL OF THE RESEARCH INSTITUTE OF TECHNOLOGY, NIPPON UNIVERSITY
SEE NIPPON DAIGAKU KOGAKU KENKYUJO IHO

JOURNAL OF SCIENCE OF LABOUR
SEE RODO KAGAKU

JOURNAL OF THE SCIENCE OF SOIL AND MANURE. JAPAN
SEE NIPPON DOJO HIRYOGAKU ZASSHI

JOURNAL DES SCIENCES MEDICALES DE LILLE

1955 V73 N3 P138-143 3710-A <K-H>
1958 V76 N8/9 P326-329 61-00158 <*>
1961 V79 N7 P288-293 63-18489 <*> O

JOURNAL OF THE SCIENTIFIC FISHERIES ASSOCIATION
SEE SUISAN GAKKAIHO

JOURNAL SCIENTIFIQUE DE LA METEOROLOGIE
1949 V1 P23-31 57-27 <*>
1958 V10 N38 P63-75 59-21121 <=> O

JOURNAL OF SEIBU ZOSENKAI
SEE SEIBU ZCSENKAI KAIHO

JOURNAL OF THE SEISMOLOGICAL SOCIETY OF JAPAN
SEE JISHIN

JOURNAL OF SERICULTURAL SCIENCE
SEE NIPPON SANSHIGAKU ZASSHI

JOURNAL OF THE SHIMONOSEKI UNIVERSITY OF FISHERIES
SEE SUISAN DAIGAKKO KENKYU HOKOKU

JOURNAL OF THE SHOWA MEDICAL ASSOCIATION
SEE SHOWA IGAKKAI ZASSHI

JOURNAL OF SILICATES. JAPAN
SEE KUEI SUAN YEH HSUEH PAO

JOURNAL DE LA SOCIETE DES INGENIEURS DE L'AUTOMOBILE
1951 V24 P135-138 I-196 <*>
58-1526 <*>
1952 V25 P119-129 60-10735 <*> O
1952 V25 P143-159 58-887 <*>
1953 V26 P33-40 59-17755 <*>
1954 V27 P292-298 60-10050 <*>
1955 V28 P337-342 59-20988 <*>
1957 P45-52 SPEC.NO. 58-824 <*>
1957 V30 N2 P102-106 58-1461 <*>
1957 V30 N12 P565-574 66-11744 <*> O

JOURNAL OF THE SOCIETY OF AGRICULTURAL MACHINERY, JAPAN
SEE NOGYO KIKAI GAKKAISHI

JOURNAL OF THE SOCIETY OF APPLIED MECHANICS OF JAPAN
SEE OYO-RIKIGAKU

JOURNAL OF THE SOCIETY OF CHEMICAL INDUSTRY OF JAPAN
SEE KOGYO KAGAKU ZASSHI

JOURNAL OF THE SOCIETY OF COSMETIC CHEMISTS. BRITISH EDITION
1963 V14 N11 P591-602 64-16188 <*>

JOURNAL OF THE SOCIETY OF NAVAL ARCHITECTS OF JAPAN
SEE ZOSEN KYOKAI RONBUNSHU

JOURNAL OF THE SOCIETY OF ORGANIC SYNTHETIC CHEMISTRY, JAPAN.
SEE YUKI GOSEI KAGAKU KYOKAI SHI

JOURNAL OF THE SOCIETY OF PRECISION MECHANICS OF JAPAN
SEE SEIMITSU KIKAI

JOURNAL OF THE SOCIETY OF THE RUBBER INDUSTRY OF JAPAN
SEE NIPPON GOMU KYOKAISHI

JOURNAL OF THE SOCIETY OF SCIENTIFIC PHOTOGRAPHY OF JAPAN
SEE NIHON SHASHIN GAKKAI KAI SHI

JOURNAL OF THE SOCIETY OF TEXTILE AND CELLULOSE INDUSTRY, JAPAN
SEE SEN-I-GAKKAISHI

JOURNAL OF THE SPECTROSCOPIAL SOCIETY OF JAPAN
SEE BUNKO KENKYU

JOURNAL OF THE STARCH SWEETENER TECHNOLOGICAL RESEARCH SOCIETY OF JAPAN
SEE DEMPUNTO GIJUTSU KENKYU KAIHO

JOURNAL SUISSE D'HORLOGERIE ET DE BIJOUTERIE
1948 V73 P93-114 59-20096 <*> O
1949 V74 P139-156 59-20096 <*> O

JOURNAL OF THE SULFURIC ACID ASSOCIATION OF JAPAN
SEE RYUSAN

JOURNAL OF SUN YAT-SEN UNIVERSITY. NATURAL SCIENCE EDITION. CHINESE PEOPLES REPUBLIC
SEE CHUNG SHAN TA HUEH PAO TZUJAN K'O HSUEH

JOURNAL OF SURVEYING AND MAPPING. CHINESE PEOPLES REPUBLIC
SEE T'SE LIANG CHI SHU HSUEH PAO

JOURNAL OF THE TAIHOKU SOCIETY OF AGRICULTURE AND FORESTRY
1936 V1 P171-175 AL-807 <*>
57-1828 <*>
1937 V1 P271-280 3758 <K-H>

JOURNAL OF THE TECHNOLOGICAL SOCIETY OF STARCH
SEE DENPUN KOGYO GAKKAISHI

JOURNAL OF THE TEISHIN MEDICAL SOCIETY. JAPAN
SEE TEISHIN IGAKU

JOURNAL DES TELECOMMUNICATIONS
1938 N9 P257-263 57-326 <*>
1940 V7 N2 P45-50 66-12527 <*>
1946 V13 N4 P73-79 57-2045 <*>
1946 V13 N7 P125-131 57-319 <*>
1946 V13 N8 P160-167 57-319 <*>

JOURNAL TELEGRAPHIQUE
1912 V36 P4-6 57-209 <*>
1912 V36 P25-29 57-209 <*>
1912 V36 P49-53 57-209 <*>
1913 V37 P97-101 57-929 <*>
1925 V49 N7 P121-126 57-909 <*>
1929 V53 N9 P196-198 58-139 <*>
1929 V53 N9 P203-204 57-2154 <*>
1930 V54 P306-309 60-10753 <*> O
1930 V54 N10 P265-266 66-13325 <*>
1930 V65 N4 P74-75 57-1549 <*>

JOURNAL OF THE TEXTILE MACHINERY SOCIETY OF JAPAN
SEE SEN-I KIKAI GAKKAISHI

JOURNAL OF THERAPY. JAPAN
SEE CHIRYO

JOURNAL OF THE TOKYO CHEMICAL SOCIETY
SEE TOKYO KAGAKU KAISHI

JOURNAL OF THE TOKYO DENTAL COLLEGE SOCIETY
SEE SHIKA GAKUHO

JOURNAL OF TRANSLATED PAPERS ON NONFERROUS METALS

SEE YU-SE CHEN SHU I SHU

JOURNAL OF TSING-HUA UNIVERSITY. CHINESE PEOPLES REPUBLIC
SEE CHING HUA TA HSUEH HSUEH PAO

JOURNAL OF ULTRASTRUCTURE RESEARCH
1959 V2 P444-452 62-00754 <*>
1959 V2 N4 P392-422 60-00614 <*>
1962 V6 N5/6 P437-448 63-01752 <*>

JOURNAL D'UROLOGIE MEDICALE ET CHIRURGICALE
1950 V56 P295-308 AL-428 <*>
1956 V62 P113-124 57-3159 <*>
1956 V62 N3 P132-140 61-16565 <*>
1959 V65 P248-257 60-17766 <+*> O
1959 V65 N10/1 P721-739 61-00598 <*>

JOURNAL D'UROLOGIE ET NEPHROLOGIE
1963 V69 N4/5 P223-233 64-00185 <*>
1964 V70 P789-797 66-13204 <*>

JOURNAL DES USINES A GAZ
1938 P32-40 57-630 <*>

JOURNAL OF THE VACUUM SOCIETY OF JAPAN
SEE SHINKU

JOURNAL OF VETERINARY MEDICINE. JAPAN
SEE JUI CHIKUSAN SHINPO

JOURNAL OF THE WAKAYAMA MEDICAL SOCIETY
SEE WAKAYAMA IGAKU

JOURNAL OF THE YONAKU MEDICAL ASSOCIATION
SEE YONAGO IGAKU ZASSHI

JOURNEES ANNUELLES DE DIABETOLOGIE DE L'HOTEL-DIEU
1962 V3 P141-149 63-18428 <*>

JOZO GAKU ZASSHI
1936 V14 P301-313 63-00509 <*>

JUGOSLOVENSKI PREGLAD
1961 P268-270 62-19702 <*=>
1961 V5 P177-179 62-19699 <*=>
1961 V5 P321-325 62-19708 <=*>
1961 V5 N2 P73-80 62-19700 <=*>
1962 V6 N3 P131-136 62-25525 <=*>
1963 V7 P31-42 63-21956 <=>
1964 V8 P23-25 65-31056 <=$>
1964 V8 P28-38 65-31056 <=$>
1964 V8 N3 P136-138 64-41491 <=>
1964 V8 N12 P475-482 65-31349 <=$>
1966 N10 P369-376 67-30957 <=$>
1966 V10 N4 P135-142 66-34564 <=$>

JUGOSLOVENSKO PRONALAZSTVO
1965 V6 N61 P13-16 65-31083 <=$>

JUI CHIKUSAN SHINPO
1960 V291 P2265-2270 63-01522 <*>
1960 V292 P2325-2330 63-01627 <*>

JUNKERS-NACHRICHTEN
1943 V14 N7/8 P53-59 58-466 <*>

JUSHI KOKO
1957 V6 P509-516 60K22J <ATS>
1957 V6 P565-569 61K22J <ATS>

JUSTUS LIEBIG'S ANNALEN DER CHEMIE
1875 V178 P153-155 AL-522 <*>
1875 V179 P68-70 66-13082 <*>
1875 V179 P72-73 66-13082 <*>
1878 V191 P261-285 66-13083 <*> P
1880 V203 P118-120 66-13084 <*>
1881 V209 P339-384 66-13085 <*> P
1885 V230 P1-42 66-13086 <*> OP
1885 V231 P152-195 66-13088 <*> P
1886 V232 P222-227 64-20877 <*>
1886 V232 P227-233 64-20878 <*> O
1887 V238 P137-219 66-13087 <*> P
1887 V241 P90-152 66-13089 <*> OP
1888 V249 P1-53 59-15871 <*>
1889 V250 P257-280 59-15870 <*>
1890 V259 N2/3 P253-276 59-20269 <*> P
1892 V270 N1 P1-63 II-722 <*>
1893 V272 N3 P271-288 60-10317 <*>
1893 V274 P184-186 1878 <*>
1893 V274 P185-186 66-13092 <*> P
1893 V274 P312-315 60-10533 <*>
1893 V278 P229-239 65-13449 <*>
1896 V291 P367-377 63-14207 <*>
1898 V300 P81-128 64-16064 <*>
1898 V303 P107-114 I-721 <*>
1899 V307 P329-331 58-2250 <*>
1899 V308 P333-343 59-20275 <*> P
1901 V315 P19-25 58-2268 <*>
1901 V315 N1 P104-137 60-10559 <*> P
1901 V316 P331-332 63-16023 <*>
1902 V326 P129-148 64-18894 <*> O
1904 V331 P46-57 AEC-TR-5677 <*>
1905 V343 N2/3 P207-266 60-10558 <*> P
1906 V348 P217 58-2104 <*>
1907 V351 P426-432 66-13091 <*>
1908 V358 P183-204 8861 <IICH>
1910 V371 N3 P32-68 60-10775 <*>
1912 V394 P362-384 66-13093 <*>
1913 V395 P362-377 2926 <*>
1918 V416 P230-233 T1857 <INSD>
1922 V429 P103-112 66-13094 <*>
1922 V429 P113-122 66-13095 <*>
1923 V434 P165-184 63-20531 <*> P
1924 V44D P121-139 64-14284 <*> O
1928 V462 P267-277 59-15937 <*>
1929 V473 P1-35 3559-A <K-H> O
64-12926 <*> O
1929 V473 P57-82 3559-B <K-H> O
64-14928 <*> O
1929 V474 P121-144 70P59G <ATS>
1929 V476 P113-150 63-18255 <*>
1930 V479 P123-134 3456 C <K-H>
64-14705 <*> O
1930 V479 P135-149 3456 D <K-H>
64-14703 <*>
1930 V480 P92-108 66-13096 <*>
1932 V495 P41-60 64-14767 <*> O
1932 V499 P188-200 3076 <*>
1933 V506 P171-195 90J17G <ATS>
1934 V508 P39-51 61-00400 <*>
1934 V509 P103-114 87J17G <ATS>
1934 V511 P13-44 3456-F <K-H> O
1934 V511 P13-31 65-11612 <*>
1934 V511 P41-44 65-11612 <*>
1934 V511 P45-63 64-14700 <*> O
1934 V511 P46-63 3456-E <K-H> O
1934 V511 P64-88 AL-844 <*>
64-10117 <*>
1934 V511 P64-77 65-11613 <*>
1934 V511 N3 P13-44 64-14896 <*> O
1935 V517 P73-104 63-18953 <*> O
1935 V517 P172-196 88J17G <ATS>
1935 V519 P113-114 61-16548 <*>
1935 V520 P144-150 89J17G <ATS>
1936 V522 P75-96 65-18145 <*>
1936 V524 P189-202 I-4 <*>
1937 V528 P101-113 AL-842 <*>
64-10118 <*>
1938 V535 P267-284 64-20876 <*>
1939 V540 P157-189 I-267 <*>
64-10142 <*>
1939 V542 P90-122 64-14906 <*> O
1941 V547 N2 P103-115 64-14017 <*> O
1942 V550 P99-133 18N52G <ATS>
1942 V551 N1/3 P80-119 60-10557 <*> P
1943 V554 P201-212 64-14766 <*>
1943 V554 P269-290 64-14765 <*>
1948 V560 P222-231 58-604 <*>
1949 V562 N2 P75-109 59-15983 <*>

Citation	Translation
	65-13562 <*>
1949 V563 N1/2 P110-126	59-10013 <*>
1949 V564 P35-43	59-19635 <**> O
1949 V565 P228	II-1170 <*>
1950 V566 P210-244	62-10017 <*>
1950 V567 P43-97	1718 <TC>
1950 V567 P179-184	64-14929 <*>
1950 V567 P185-195	3456-H <K-H> PO
	63-18952 <*> PO
	64-14702 <*> OP
1950 V567 P195-203	63-18956 <*> O
1950 V569 P161-183	2067 <*>
1951 V571 N3 P167-201	65-10085 <*> O
	65-13563 <*>
1951 V572 P23-82	64-18909 <*> O
1951 V572 P83-95	60-10151 <*>
1951 V572 N1 P23-82	60-16881 <*>
1951 V572 N1 P69-74	60-10136 <*>
1951 V573 N1 P60-84	II-491 <*>
1951 V573 N2 P142-162	65-14034 <*>
1951 V573 N3 P195-209	64-18086 <*>
1951 V574 P1-32	63-20535 <*> P
1952 V575 N3 P153-161	59-17051 <*>
1952 V576 P20-27	64-10762 <*> P
1952 V577 P60-68	63-18455 <*>
1952 V577 N3 P234-237	64-18916 <*>
1952 V578 P50-82	63-18685 <*> O
	65-12474 <*>
	66-11360 <*>
1952 V578 P94-100	66-13098 <*> P
1953 V579 P23-27	60-16880 <*>
1953 V581 P10-16	64-14764 <*> O
1953 V582 N1/2 P1-161	66-13125 <*>
1953 V582 N1/2 P87-116	65-10918 <*> O
1953 V584 N2/3 P156-176	59-15908 <*> O
1954 V585 P81-96	I-5 <*>
1954 V585 P114-124	CSIRO-3118 <CSIR>
1954 V585 N2 P81-90	61-10824 <*>
1954 V585 N2 P97-108	61-10819 <*>
1954 V587 N3 P177-194	57-2937 <*>
1954 V587 N3 P195-206	57-2939 <*>
1954 V589 P91-121	I-877 <*>
1954 V589 P157-162	2899 <*>
1954 V589 N2 P91-121	3215 <K-H> O
	64-14375 <*> O
1954 V589 N2 P122-156	64-16334 <*>
1955 V591 P108-117	58-2484 <*>
1955 V592 P38-53	64-14707 <*>
1955 V595 N1 P1-37	64-18476 <*>
1955 V596 P1-224	64-16159 <*> P
	65-13119 <*> P
1955 V596 P158-244	3858 <K-H>
1956 V601 N1/3 P81-138	40J15G <ATS>
	60-10713 <*>
1957 V601 N1/3 P81-138	62-16528 <*> PO
1957 V604 P104-110	56J17G <ATS>
1957 V604 N1/3 P168-178	65-11573 <*> O
1957 V605 N1/3 P93-97	61-10487 <*>
1957 V606 N13 P1-13	59-15942 <*> O
1957 V606 N1/3 P1-23	64-18386 <*> O
1957 V606 N1/3 P67-74	64-16887 <*>
1957 V607 P202-206	59-16556 <**>
1957 V607 N1/3 P35-45	61-10306 <*> O
1958 V611 P1-7	23K26G <ATS>
1958 V613 N1/3 P153-170	64-18802 <*>
1958 V614 N1/3 P1-2	65-14371 <*>
1958 V615 P1-14	62-16694 <*>
1958 V615 N1/3 P1-14	65-11447 <*> P
1958 V615 N1/3 P29-33	65-11446 <*>
1958 V617 P1-10	65-10190 <*>
1958 V618 P17-23	1440 <TC>
	65-10191 <*>
	88L31G <ATS>
1958 V618 P53-58	61-18637 <*>
1958 V618 P59-67	61-10533 <*>
1958 V618 P140-152	8695 <IICH>
1958 V618 N1/3 P31-43	60-14310 <*>
	65-11221 <*>
	9539 <K-H>
1958 V618 N1/3 P251-256	62-10456 <*> O

Citation	Translation
1959 V623 N1/3 P1-8	65-10909 <*> O
1959 V623 N1/3 P9-16	60-10132 <*>
1959 V623 N1/3 P176-183	65-10910 <*>
1959 V627 N1/3 P47-58	60-14521 <*>
1960 V629 P1-13	65-12463 <*>
1960 V629 P53-89	65-12923 <*> P
1960 V629 P121-166	65-12465 <*>
1960 V629 P172-198	65-12464 <*>
1960 V629 N1/3 P1-13	10024-11 <K-H>
	61-20449 <*>
	64-30976 <*> O
1960 V629 N1/3 P14-19	10024-12 <K-H>
	61-20450 <*>
1960 V629 N1/3 P20-22	10024-13 <K-H> O
	61-20451 <*>
1960 V629 N1/3 P23-33	61-20452 <*>
1960 V629 N1/3 P33-49	01Q67G <ATS>
	10024-9 <K-H> O
	61-20453 <*>
	61-20862 <*> O
1960 V629 N1/3 P50-52	10024-10 <K-H>
	61-20454 <*>
1960 V629 N1/3 P53-89	10024-4 <K-H> O
	61-20455 <*>
1960 V629 N1/3 P89-103	61-20456 <*>
1960 V629 N1/3 P104-121	10024 <K-H>
	61-20457 <*>
1960 V629 N1/3 P121-166	10024-2 <K-H> O
	61-20458 <*>
	65-11264 <*>
1960 V629 N1/3 P167-171	10024-16 <K-H>
	61-20459 <*>
1960 V629 N1/3 P172-198	10024-1 <K-H> C
	61-20460 <*>
	65-11260 <*>
1960 V629 N1/3 P198-206	10024-7 <KH>
	61-20504 <*>
1960 V629 N1/3 P207-209	61-20461 <*>
1960 V629 N1/3 P210-221	10024-3 <K-H>
	61-20462 <*>
	65-11262 <*>
1960 V629 N1/3 P222-240	61-20468 <*>
	65-12247 <*>
1960 V629 N1/3 P241-250	10024-6 <K-H> O
	61-20480 <*>
	65-11263 <*>
1960 V629 N1/3 P251-256	10024-18 <K-H>
	61-20463 <*>
1960 V629 N1/3 P251-254	65-12462 <*>
1960 V631 N1/3 P21-56	29M47G <ATS>
1960 V632 P1-7	AEC-TR-5205 <*>
1960 V632 N1/3 P38-52	10N49G <ATS>
1960 V632 N1/3 P104-115	61-10201 <*>
1960 V635 P1-21	34N51G <ATS>
	61-00569 <*>
1960 V636 N1/3 P1-18	61-10628 <*>
1960 V637 P73-93	80N50G <ATS>
1960 V637 P93-110	81N50G <ATS>
1960 V637 P111-118	82N50G <ATS>
1960 V638 P43-56	62-10047 <*>
1960 V638 P57-66	62-10039 <*>
1960 V638 P66-75	62-10038 <*>
1960 V638 P76-81	62-10037 <*>
1960 V638 P122-135	58Q67G <ATS>
1960 V638 N1/3 P122-135	65-12420 <*>
1961 V646 P10-17	65-13572 <*>
1961 V646 P65-77	9040 <IICH>
1962 V652 P1-7	11927 <K-H>
	65-14021 <*>
1962 V654 P23-26	65-14478 <*>
1962 V656 P1-9	63-16601 <*>
1962 V656 P15-17	63-16600 <*>
1962 V660 P1-23	63-16529 <*> O
1962 V660 P23-33	9003 <IICH>
1964 V677 N1 P67-83	65-11981 <*>
1965 V684 P37-57	66-14690 <*>
1965 V684 P112-115	1732 <TC>
1965 V684 P115-121	66-10418 <*> O
1965 V685 P89-96	1964 <TC>
1965 V685 P228-236	66-13434 <*>

JUSTUS LIEBIG'S ANNALEN DER CHEMIE UND PHARMACIE
1873 V169 P146-149 66-13080 <*>
1874 V172 P28-61 66-12338 <*>
1874 V174 P202-205 66-13081 <*>

KAELTETECHNIK
1954 V6 P267-269 59-16857 <+*>
1954 V6 N10 P271-274 CSIRO-3100 <CSIR>
1955 V7 N2 P34-38 59-17435 <*>
1955 V7 N3 P71-74 59-17435 <*>
1956 N5 P160-164 64-00027 <*>
1956 V8 N3 P90-93 65-10071 <*>
1956 V8 N5 P155-169 CSIRO-3214 <CSIR>
1956 V8 N5 P160-164 57-58 <*>
62-20182 <*>
1957 V9 N1 P7-12 58-2332 <*>
1957 V9 N2 P32-34 60J17G <ATS>
1957 V9 N3 P67-71 59J17G <ATS>
1958 V10 P3- 59-12947 <+*>
1958 V10 N9 P301-306 59-16997 <=*> 0
1959 V11 P202-203 9012 <IICH>
1960 V12 P334-339 AEC-TR-6241 <*>
1961 V13 N6 P229 62-25668 <=*>
1962 V14 N4 P98-105 95P64G <ATS>
1963 V15 N2 P39-44 62R74G <ATS>
1964 V16 N1 P5-11 4341 <BISI>

KAGAKU. KYOTO
1956 V11 N12 P874-883 63M39J <ATS>
1957 V12 P349-358 61-28940 <=*> 0
1959 V14 N9 P794-800 62M39J <ATS>
1960 V15 P290-298 18S81J <ATS>
1960 V15 N9 P719-724 83Q66J <ATS>
1960 V15 N9 P725-735 84Q66J <ATS>
1961 V16 N2 P151-159 44N57J <ATS>
1965 V20 N6 P552-562 66-14676 <*$>

KAGAKU. TOKYO
1943 V13 P128-129 380 <TC>
1949 V19 P379-380 64-14712 <*>
1949 V19 N10 P473-474 60-18533 <*>
1950 V20 P576-577 T-2447 <INSD>
1950 V20 N9 P419-420 88K24J <ATS>
1950 V20 N11 P514-515 87K24J <ATS>
1952 V22 N2 P96 59-15535 <*>
1953 V23 N11 P573-579 87F3J <ATS>
1955 V25 P137-138 AEC-TR-3511 <*>
1956 V26 N6 P313 58-1957 <*>
1957 V27 N9 P451-456 61-18841 <*> 0
1957 V27 N9 P541-546 59-15088 <*>
1957 V27 N12 P630-631 C-2593 <NRC>
58-607 <*>
60-22027 <NRCC>
1961 V31 P552 C-4090 <NRCC>
1962 V32 N7 P354-361 C-4367 <NRCC>

KAGAKU ASAHI. TOKYO
1964 N3 P91-96 64-31109 <=>
1964 V24 P23-28 65-30364 <=*$>
1965 P39-46 65-30515 <=$>

KAGAKU HYORON
1940 V6 P294-307 66-12881 <*>
1940 V6 P354-372 66-12882 <*>
1948 V18 N6 P277 AL-695 <*>

KAGAKU KEISATSU KENKYUJO HOKOKU
1963 V16 N4 P305-312 64-21851 <=>
1963 V16 N4 P319-347 64-21851 <=>

KAGAKU KENKYUSHO HOKOKU
SEE RIKAGAKU KENKYUSHO HOKOKU

★KAGAKU KOGAKU
1951 V15 N4 P164-172 57-3540 <*>
1951 V15 N5 P212-217 NP-TR-426 <*>
1953 V17 P438-447 NP-TR-633 <*>
1953 V17 N5 P176-184 64-20840 <*> 0
1953 V17 N7 P261-268 64-20841 <*> 0
1953 V17 N10 P382-386 07L29J <ATS>
1954 V18 N8 P369-372 87I8J <ATS>
1954 V18 N10 P467-473 65-11729 <*>
1954 V18 N12 P593-600 64-14701 <*> 0
1955 V19 P104-110 3626-B <K-H>
1955 V19 P155-161 II-816 <*>
1955 V19 N3 P104-110 64-16115 <*> 0
1955 V19 N12 P632-640 1426 <*>
1956 V20 P11-13 65H11J <ATS>
1956 V20 P272-279 52H12J <ATS>
1956 V20 N4 P148-155 NP-TR-982 <*>
1956 V20 N4 P156-162 NP-TR-1018 <*>
1956 V20 N6 P280-287 51H12J <ATS>
1956 V20 N12 P685-694 28K28J <ATS>
1957 N7 P408-412 66-13419 <*>
1957 N12 P780-783 66-13420 <*>
1957 V21 P32-36 R-1528 <*>
1957 V21 P721-726 62-18927 <*>
1957 V21 N1 P8-16 58-1501 <*>
1957 V21 N3 P139-146 27K28J <ATS>
1957 V21 N5 P278-286 32K28J <ATS>
1957 V21 N7 P420-425 NP-TR-1019 <*>
1957 V21 N8 P472-480 17TM <CTT>
1957 V21 N12 P784-790 66-13418 <*>
1957 V21 N12 P810-816 NP-TR-541 <*>
1958 V22 P7-15 58-1886 <*>
1958 V22 P200-207 58-2016 <*>
1958 V22 P476-482 59-15614 <*>
1958 V22 N9 P570-572 62-01584 <*>
1958 V22 N11 P680-686 AEC-TR-4225 <=>
1958 V22 N12 P758-763 NP-TR-922 <*>
1959 V23 N5 P284-290 18L35J <ATS>
1959 V23 N8 P502-505 61M37J <ATS>
1959 V23 N9 P589-594 NP-TR-987 <*>
1959 V23 N10 P647-654 AEC-TR-4225 <=>
1960 V24 N1 P12-19 05M41J <ATS>
1960 V24 N2 P70-80 NP-TR-646 <*>
1960 V24 N4 P198-203 NP-TR-887 <*>
1960 V24 N4 P219-225 NP-TR-889 <*>
1961 V25 N1 P34-40 64-11551 <=>
1961 V25 N6 P469-476 57P58J <ATS>
1961 V25 N8 P582-587 78P62J <ATS>
1961 V25 N12 P870-876 21P62J <ATS>
1962 N8 P856-863 ICE V4 N1 P165-173 <ICE>
1962 V26 N7 P800-806 12757 <K-H>
65-14845 <*>
1962 V26 N11 P1155-1160 26Q68J <ATS>
1963 N3 P156-161 ICE V4 N1 P173-178 <ICE>
1963 N6 P424-428 ICE V4 N1 P153-157 <ICE>
1963 N7 P477-483 ICE V4 N1 P179-185 <ICE>
1963 N11 P823-830 ICE V4 N2 P332-340 <ICE>
1963 V27 N11 P823-830 ICE V4 N2 P332-340 <ICE>
1963 V27 N12 P974-977 66-12872 <*>
1964 V28 N1 P26-32 1771 <TC>
1965 V29 N9 P681-687 66-12998 <*>

KAGAKU KOGYO
1960 V11 N10 P929-932 CODE-37 <SA>
1960 V11 N10 P933-936 CODE-38B <SA>
1960 V11 N10 P937-942 CODE-39 <SA>
1961 V14 N3 P204-212 63-18056 <*>

KAGAKU NANYO
1941 V4 N1 P64-75 AL-731 <*>
1942 V5 N1 P117-122 AL-469 <*>

KAGAKU NO RYOIKI
1948 V2 N5 P200-205 65-18080 <*>
1948 V2 N6 P262-270 61-10263 <*>
1950 V4 P557-561 59-21025 <=> 0
1950 V4 N11 P604-612 T-2411 <INSD>
1951 V5 P457-461 2464 <*>
1952 V6 P274-280 T-1966 <INSD>
1953 V7 N3 P167-169 77F4J <ATS>
1958 V12 N3 P186-192 06N53J <ATS>
1958 V12 N9 P659-669 62-00633 <*>
1959 V13 N10 P730-737 62-00973 <*>
1960 V14 P403-412 729 <TC>

1960 V14 N6 P379-393 98M45J <ATS>
1960 V14 N8 P578-581 62-00974 <*>
1961 V15 P936-942 63-16250 <*> O
1961 V15 N10 P21-29 62-14829 <*>
1963 V17 N5 P339-348 C-5535 <NRC>

KAGAKU RYOHO
1957 V5 N5 P222-223 61-10432 <*>
1957 V5 N7 P322 61-10427 <*>
1957 V5 N7 P322-323 61-10434 <*>
1958 V6 N6 P378-392 61-10433 <*>

KAGAKU TO KOGYO. OSAKA
1953 V27 P154-157 T1938 <INSD>
1953 V27 P180-183 T1939 <INSD>
1953 V27 P220-222 T-1939 <INSD>
1953 V27 P247-250 T1940 <INSD>
1953 V27 P250-253 T1941 <INSD>
1953 V27 P279-280 T1941 <INSD>
1954 V28 P267-271 5613-D <K-H>
1956 V30 N3 P73-76 00N48J <ATS>
1958 V32 P334-340 60-10944 <*>
1959 V33 P103-108 65-12901 <*>
1960 V34 P428-443 61-18694 <*>
1960 V34 P473-477 61-18695 <*> O
1962 V36 P129-132,142 12792 <K-H>
1962 V36 P129-132 65-14850 <*>
1964 V38 N12 P672-679 18S85J <ATS>

KAGAKU TO KOGYO. TOKYO
1955 V8 P467-477 57-3055 <*>
1958 V11 N7 P39-45 59-13083 <*=>
1960 V13 N4 P368-372 44M43J <ATS>
1962 V15 N4 P360-368 75S81J <ATS>
1964 V17 N12 P1344-1352 17S87J <ATS>
1965 V18 N6 P789-798 65-17171 <*>

KAGAKU TO SEIBUTSU
1963 V1 N2 P24-25 65-17223 <*>
1965 V3 N1 P36-38 66-10729 <*>

KAGOSHIMA DAIGAKU NOGAKUBU GAKUJUTSU HOKOKU
1964 N14 P1-50 65-00284 <*>

KAGOSHIMA DAIGAKU SUISAN GAKUBU KIYO
1959 V7 N2 P87-101 59-22355 <*+> O

KAIBOGAKU ZASSHI
1957 V32 N6 P271-582 60-15406 <+*> O

KAKU YUGO KENKYU
1958 V1 P542-563 NP-TR-542 <*>
1962 V9 N3 P267-304 AEC-TR-5564 <*>
1962 V9 N4 P351-377 AEC-TR-5612 <*>
1962 V9 N4 P351-375 NP-TR-1057 <*>
1962 V9 N5 P463-472 NP-TR-1034 <*>
1963 V10 N5 P373-386 AEC-TR-5985 <*>

KALI. BERLIN
1923 N13 P193-197 II-994 <*>

KALI UND STEINSALZ
1955 N8 P3-7 63-14625 <*>
1960 V3 N3 P98-100 63-16202 <*>
1961 N7 P234-241 64-16405 <*>
1963 V3 N10 P338-340 67S84G <ATS>
1965 V4 N4 P109-111 66-11961 <*>

KALI, VERWANDTE SALZE UND ERDOEL
1941 V35 N3 P37-41 63-18133 <*>

KALII
1932 N1 P26-27 63-14573 <=*>
1932 V1 N3 P20-23 64-10774 <=*$>
1932 V3 P20-23 R-3341 <*>
1933 N2 P24-28 R-2308 <*>
1933 N3 P19-29 64-10727 <=*$>
1933 N6 P17-21 64-18580 <=*$> O
1933 N8 P24-28 63-18108 <*>
1934 N7 P32-34 R-2789 <*>
1934 N8 P37-43 63-18109 <=*>
1935 N8 P18-19 63-14592 <=*>
1935 N8 P28-33 64-20397 <*> O
1935 V4 N9 P16-22 63-18113 <=*>
1936 N10 P37-41 R-3380 <*>
1936 V5 N6 P36-40 63-18302 <*>
1936 V5 N10 P19-36 63-18303 <=*>
1936 V10 P33-37 R-3384 <*>
1937 N1 P28-36 R-3385 <*>
1937 N1 P36-39 R-3381 <*>
1937 N2 P24-28 R-3383 <*>
1937 N2 P28-32 R-3379 <*>
1937 N3 P16-23 RT-2249 <*>
1937 V6 N3 P16-23 64-10137 <=*$> O

KAMI PARUPU GIJUTSU KYOKAI SHI
1951 V5 P563-572 57-2661 <*>
1956 V10 N61 P181-185 64-26110 <$>
1957 V11 N73 P239-244 2J18J <ATS>
1957 V11 N73 P245-248 1J18J <ATS>
1957 V11 N73 P261-265 3J18J <ATS>
1958 V12 N9 P607-612 60-18931 <*> O
1958 V12 N84 P169-173 59-17087 <*>
1958 V12 N88 P511-513 PANSDOC-TR.621 <PANS>
1958 V12 N93 P806-808 11L35J <ATS>
1959 V13 N2 P122-126 60-14235 <*>
1959 V13 N5 P304-308 60-14666 <*>
1959 V13 N95 P116-121 61-16577 <*>
1960 V14 P731-740 62-20209 <*> P
1960 V14 N115 P675-682 64-16030 <*>
1963 V17 P88-99 65-11365 <*> O
1963 V17 N8 P505-509 65-12135 <*>
66-11711 <*> O
1963 V17 N10 P655-664 CCL-65/25 <STC>
1965 V19 P27-32 20 <INT>
1965 V19 N7 P353-359 66-13885 <*$>

KANG T'IEH
1958 N9 P19-24 2038 <BISI>
1959 P602-608 62-24781 <=*>
1959 N21 P1033 61-11619 <=>
1959 N21 P1044-1055 61-11619 <=>
1959 N23 P1118-1125 61-11619 <=>
1959 N23 P1139 61-11619 <=>
1959 N23 P1147 61-11619 <=>
1959 N23 P1168 61-11619 <=>
1959 N24 P1197 61-11619 <=>
1959 N24 P1202 61-11619 <=>
1959 N24 P1208 61-11619 <=>
1959 N24 P1225 61-11619 <=>
1959 N24 P1235-1236 61-11619 <=>
1960 N1 P3-7,48-49 60-31852 <=>
1960 N2 P108-109 60-31852 <=>
1960 N3 P173 60-31852 <=>
1960 N9 P505-509 60-41631 <=>
1960 N9 P522-526 60-41631 <=>
1960 N9 P537 60-41631 <=>
1960 N9 P541 60-41631 <=>
1960 N9 P552 60-41631 <=>
1960 N9 P555,562-563 60-41631 <=>

KANTO TOSAN BYOGAICHU KENKYUKAI NENPO
1957 V7 P67-70 PANSDOC-TR.194 <PANS>
1964 N11 P59-60 66-13363 <*>

KANZO
1960 V1 N1 P17-36 61-00206 <*>

KAO FEN TZU T'UNG HSUN
1958 V2 N3 P195-196 59-13381 <=>
1959 V3 P374-375 NP-TR-777 <*>
1959 V3 N5 P193-195 60-11212 <=>
1959 V3 N5 P195- ICE V2 N1 P39-49 <ICE>
1959 V3 N5 P210-224 60-11193 <=>
1959 V3 N5 P297-300 60-11207 <=>
1959 V3 N6 P355-357 60-11761 <=>
1965 V7 N5 P304-314 66-30488 <=$>
1965 V7 N5 P322-335 66-30488 <=$>
1965 V7 N5 P355-372 66-30488 <=$>

KARAKULEVODSTVO I ZVEROVODSTVO

1953 V6 N2 P67-70	RT-3117 <*>
1955 V8 P51-55	CSIR-TR 337 <CSSA>
1955 V8 N5 P53-	R-2401 <*>

KARDIOLOGIYA

1961 V1 N3 P88-92	62-19808 <=*>
1962 V2 N1 P44-52	62-25999 <=*>
1962 V2 N1 P84-	FPTS V23 N1 P.T145 <FASE>
1962 V2 N1 P89-92	62-25999 <=*>
1962 V2 N2 P38-44	63-19357 <=*>
1962 V2 N2 P44-	FPTS V22 N4 P.T767 <FASE>
1962 V2 N2 P74-76	63-23008 <=*>
1962 V2 N3 P3-16	62-33297 <=*>
1962 V2 N3 P64-69	62-33297 <=*>
1962 V2 N4 P3-9	62-33597 <=*>
1962 V2 N4 P10-	FPTS V23 N1 P.T178 <FASE>
1962 V2 N4 P42-46	64-13844 <=*$>
1962 V2 N4 P75-78	64-13469 <=*$> 0
1962 V2 N4 P91-93	62-33339 <=>
1962 V2 N5 P18-27	64-13411 <=*$>
1962 V2 N5 P33-	FPTS V23 N1 P.T139 <FASE>
	FPTS,V23,P.T139-T141 <FASE>
1962 V2 N5 P69-70	63-15121 <=*>
1962 V2 N5 P90-94	63-13297 <=*>
1963 V2 N4 P46-52	64-13393 <=*$>
1963 V3 N1 P64-	FPTS V22 N6 P.T1154 <FASE>
	FPTS,V22,P.T1154-56 <FASE>
1963 V3 N2 P10-15	64-13846 <=*$>
1963 V3 N2 P23-27	64-13847 <=*$>
1963 V3 N2 P43-48	64-13434 <=*$>
1963 V3 N3 P14-	FPTS V23 N3 P.T601 <FASE>
	FPTS,V23,N3,P.T601 <FASE>
1963 V3 N3 P27-33	64-13855 <=*$>
1963 V3 N3 P33-	FPTS V23 N3 P.T590 <FASE>
	FPTS,V23,N3,P.T590 <FASE>
1963 V3 N3 P45-	FPTS V23 N3 P.T569 <FASE>
1963 V3 N3 P45-49	FPTS,V23,N3,P.T569 <FASE>
1963 V3 N3 P50-53	64-15528 <=*$>
1963 V3 N3 P62-68	64-15539 <=*$>
1963 V3 N4 P31-39	64-15513 <=*$> 0
1963 V3 N4 P57-	FPTS V23 N3 P.T583 <FASE>
	FPTS,V23,N3,P.T557 <FASE>
1963 V3 N4 P72-	FPTS V23 N3 P.T587 <FASE>
	FPTS,V23,N3,P.T587 <FASE>
1963 V3 N5 P33-	FPTS V23 N5 P.T1095 <FASE>
1963 V3 N5 P57-66	64-19653 <=$>
1963 V3 N6 P3-11	64-19701 <=$>
1963 V3 N6 P15-20	64-19702 <=$>
1964 V4 N1 P9-14	65-61107 <=$>
1964 V4 N1 P52-57	65-61108 <=$>
1964 V4 N1 P79-82	N66-12346 <=*$>
1964 V4 N2 P16-24	65-61113 <=$>
1964 V4 N2 P25-	FPTS V24 N3 P.T522 <FASE>
1964 V4 N2 P53-56	65-62187 <=>
1964 V4 N2 P65-67	65-61114 <=$>
1964 V4 N4 P27-31	65-63600 <=$>
1964 V4 N6 P30-	FPTS V24 N6 P.T971 <FASE>
1964 V4 N6 P62-69	65-63601 <=$>
1965 V5 N1 P13-18	66-60413 <=*$>
1965 V5 N1 P18-24	65-30904 <=$>
1965 V5 N1 P30-34	66-60414 <=*$>
1965 V5 N2 P3-21	65-31201 <=$>
1965 V5 N2 P28-	FPTS V25 N2 P.T350 <FASE>
1965 V5 N2 P61-66	66-60853 <=*$>
1965 V5 N2 P84-85	65-31201 <=$>
1965 V5 N3 P3-11,81	65-32175 <=$>
1965 V5 N4 P19-	FPTS V25 N3 P.T452 <FASE>
1965 V5 N4 P92-95	65-32937 <=*$>
1965 V5 N5 P8-	FPTS V25 N4 P.T615 <FASE>
1965 V5 N6 P41-	FPTS V25 N6 P.T993 <FASE>
1965 V5 N6 P47-	FPTS V25 N5 P.T775 <FASE>
1965 V5 N6 P78-84	66-30620 <=$>

KARTOFEL I OVOSHCHI

1960 V5 N1 P20-21	61-15865 <+*>
1960 V5 N1 P21-22	61-15862 <+*>
1961 V6 N1 P1-3	61-23476 <=>
1961 V6 N4 P1-3	61-27381 <=>
1966 N10 P2-4	67-30157 <=> 0

KARTOGRAPHISCHE NACHRICHTEN

1958 V8 N3 P69-70	59-11402 <=>
1958 V8 N3 P71-79	59-11403 <=>
1958 V8 N3 P79-82	59-11404 <=>
1958 V8 N3 P82-90	59-11405 <=>
1958 V8 N3 P90-98	59-11394 <=> 0
1958 V8 N3 P98-109	59-11393 <=>
1958 V8 N3 P109-116	59-11414 <=>

KASEIGAKU ZASSHI

1951 V1 N1 P14-17	57-1212 <*>
1956 V7 N3 P92-95	58-2340 <*>

★KAUCHUK I REZINA

1931 N4/5 P26-27	61-16689 <=*$>
1936 P912-917	61-18097 <=*$>
1936 P1114-1115	61-16992 <=*$>
1937 P37-48	RCT V14 N1 P786 <RCT>
1937 N2 P36-44	RT-3512 <*>
	63-18123 <=*> 0
1937 N4 P11-15	61-18053 <=*$>
1937 N4 P107	61-16673 <=*$>
1937 N7/8 P37-38	61-16630 <=*$>
1938 N2 P26-31	RT-300 <*>
1938 N2 P37-43	61-18335 <=*$>
1938 N11 P13-17	RT-299 <*>
1939 P25-34	RCT V13 N1 P361 <RCT>
1939 P36-37	RCT V13 N1 P263 <RCT>
1939 N7 P26-29	RT-317 <*>
1939 N8 P21-24	R-3972 <*>
1939 N8 P88-	R-3972 <*>
1939 N12 P22-25	RT-298 <*>
1940 P12-20	RCT V14 N1 P861 <RCT>
1940 N7 P16-19	61-18057 <*=>
1940 N10 P14-16	RCT V14 N1 P934 <RCT>
1940 N4/5 P1-5	61-20410 <*=>
1947 N1 P4-14	RCT V31 N1 P30 <RCT>
1957 N1 P22-31	RCT V31 N3 P526 <RCT>
1957 N2 P31-33	63-24324 <=*$>
1957 N3 P11-14	PB 141 285T <=>
	59-18713 <+*>
1957 N4 P9-19	RCT V31 N4 P907 <RCT>
1957 V16 P5-9	RCT V32 N1 P40 <RCT>
1957 V16 N1 P4-14	59-15624 <+*>
1957 V16 N1 P14-22	RCT V32 N1 P195 <RCT>
1957 V16 N2 P1-5	R-3469 <*>
1957 V16 N2 P33-35	C-2552 <NRC>
1957 V16 N4 P27-32	RCT V32 N2 P444 <RCT>
1957 V16 N5 P1-11	59-15623 <+*>
1957 V16 N6 P1-6	RCT V32 N1 P328 <RCT>
1957 V16 N7 P31-34	63-18753 <=*>
1957 V16 N8 P10-15	RCT V32 N2 P519 <RCT>
1957 V16 N9 P1-5	RCT V32 N2 P639 <RCT>
1957 V16 N11 P3	RAPRA-1243 <RAP>
1958 N3 P5-11	60-15562 <+*>
1958 V17 N2 P1-5	75K25R <ATS>
1958 V17 N2 P12-17	59-22719 <+*>
1958 V17 N3 P5-11	RCT V33 N4 P1019 <RCT>
1958 V17 N5 P14-21	61-19788 <=*> 0
1958 V17 N6 P2-3	59-16437 <+*> P
1958 V17 N7 P27-28	59-16437 <+*>
1958 V17 N7 P33-36	RCT V32 N3 P770 <RCT>
1958 V17 N8 P14-21	RCT V32 N4 P1199 <RCT>
	59-15620 <+*>
1958 V17 N8 P32-35	59-15619 <+*>
1958 V17 N9 P4-7	RCT V32 N3 P701 <RCT>
1958 V17 N9 P7-12	RCT V32 N4 P976 <RCT>
1958 V17 N9 P12-16	RCT V32 N3 P907 <RCT>
1958 V17 N9 P16-20	RCT V32 N2 P454 <RCT>
1958 V17 N10 P18-22	59-17452 <+*>
1958 V17 N11 P18-21	63-14418 <=*>
1958 V17 N12 P36-48	61-19332 <+*>
1959 N1 P16-21	RCT V32 N4 P983 <RCT>
1959 N1 P57-	NLLTB V1 N7 P36 <NLL>
1959 N11 P17-20	RCT V33 N3 P790 <RCT>
1959 V18 N1 P4-6	60-21007 <=>
1959 V18 N1 P27-30	RCT V32 N4 P1180 <RCT>
1961 V20 N1 P53-55	61-27964 <*=>
1961 V20 N1 P55-58	61-27967 <*=>

1961 V20 N3 P52-53	61-27225 <*=>
1961 V20 N5 P26-32	61-27938 <*=>
1961 V20 N5 P54-55	61-27966 <*=>
1961 V20 N5 P56-57	61-27934 <*=>
1961 V20 N6 P9-18	62-13432 <*=>
1961 V20 N6 P29-31	62-13326 <*=>
1961 V20 N6 P55-58	61-28307 <*=>
1961 V20 N8 P16-20	63-23376 <=*>
1961 V20 N8 P55-57	62-13762 <*=>
	62-15226 <*=>
1961 V20 N9 P13-	RCT V38 N3 P661 <RCT>
1961 V20 N12 P30-35	62-32678 <=*>
1962 V21 N8 P40-44	63-13079 <=>
1963 N10 P45-49	RAPRA-1228 <RAP>
1963 V22 N2 P32-36	64-13294 <=*$>
1963 V22 N3 P10-	SC-T-513 <=$>
1963 V22 N3 P55-57	64-71379 <=>
1964 V22 N6 P27-29	64-15264 <=*$>
1964 V23 N1 P32-35	66-10751 <*>
1965 N6 P19-24	ICE V6 N1 P108-112 <ICE>
1965 N7 P1-4	ICE V6 N1 P31-34 <ICE>
1965 V24 N2 P10-12	AD-627 124 <=$>
	1125 <TC>
1965 V24 N2 P16-19	AD-627 124 <=$>
1965 V24 N6 P54-56	66-11516 <*>
1965 V24 N9 P23-26	AD-630 744 <=$>
1966 N8 P2-4	66-34679 <=$>
1967 N1 P2-4	67-30810 <=$>

KAUCUK A PLASTICHE HMOTY

1962 N12 P429-431	RAPRA-1247 <RAP>
1963 N2 P37-43	RAPRA-1248 <RAP>

KAUTSCHUK

1925 P6-10	2872 <*>
1925 P10-13	2938 <*>
1927 P233-238	2871 <*>
1928 V4 P142-149	2852 <*>
1929 N2 P47-	57-2279 <*>
1931 P75	57-3085 <*>
1931 V7 N2 P26-32	57-3125 <*>
1938 V12 P231-232	57-2086 <*>
1938 V14 P23-25	49K21G <ATS>
1938 V14 P41-45	49K21G <ATS>
1938 V14 P203-210	64-18797 <*> 0
1938 V14 N11 P203-210	RCT V16 N1 P866 <RCT>
1939 V15 P160-166	63-16247 <*> 0
1939 V15 N9 P160-166	RCT V13 N1 P112 <RCT>
1939 V15 N10 P169-172	RCT V13 N1 P408 <RCT>
1939 V15 N11 P179-182	RCT V13 N1 P430 <RCT>
1940 V16 N1 P1-5	RCT V14 N1 P470 <RCT>
1940 V16 N2 P13-17	RCT V14 N1 P470 <RCT>
1940 V16 N3 P26-33	RCT V13 N1 P948 <RCT>
1940 V16 N5 P55-60	57-1588 <*>
1940 V16 N9 P109-116	RCT V14 N1 P899 <RCT>
1940 V16 N10 P121-125	57-332 <*>
1940 V16 N11 P138-139	RCT V14 N1 P372 <RCT>
1940 V16 N11 P140-145	57-332 <*>
1941 V17 N1 P1-7	RCT V14 N1 P920 <RCT>
1941 V17 N3 P31-33	RCT V15 N1 P874 <RCT>
1941 V17 N3 P33-36	RCT V15 N1 P854 <RCT>
1942 V18 P39-48	II-6 <*>
1942 V18 N11 P144-145	RCT V17 N1 P436 <RCT>
1943 V19 P47-49	RCT V18 N1 P62 <RCT>
1943 V19 P55-58	RCT V18 N1 P71 <RCT>
1944 V20 P1-2	RCT V18 N1 P401 <RCT>
1944 V20 P3-5	RCT V18 N1 P394 <RCT>
1944 V20 P15-19	RCT V18 N1 P646 <RCT>

KAUTSCHUK UND GUMMI KUNSTSTOFFE (TITLE VARIES)

1948 V1 N6 P149-152	RCT V23 N1 P292 <RCT>
1948 V1 N6 P153-156	63-16252 <*> 0
1948 V1 N9 P241-244	60-10168 <*> 0
1948 V1 N9 P250-251	65-11947 <*>
1949 V2 N11 P337-343	62-10123 <*> 0
1951 V4 P53-55	63-20000 <*>
1951 V4 P92-95	63-20001 <*>
1951 V4 P128-131	63-20002 <*>
1952 V5 N2 P17WT-22WT	RCT V27 N1 P1 <RCT>
1952 V5 N10 P157WT-161WT	RCT V28 N4 P1044 <RCT>
1953 V6 N7 P127WT-139WT	RCT V27 N4 P859 <RCT>
1953 V6 N8 P147WT-157WT	RCT V29 N4 P1373 <RCT>
1953 V6 N9 P171WT-177WT	RCT V27 N4 P839 <RCT>
1953 V6 N9 P178WT-186WT	RCT V27 N3 P549 <RCT>
1953 V6 N10 P204WT-205WT	60-18909 <*>
1953 V6 N11 P217WT-224WT	RCT V27 N3 P569 <RCT>
1954 V7 P122WT-127WT	RCT V30 N2 P728 <RCT>
1954 V7 N1 P1WT-7WT	RCT V28 N1 P153 <RCT>
1954 V7 N1 P8WT-12WT	I-293 <*>
1954 V7 N1 P8-12	57-1784 <*>
1954 V7 N2 P34WT-42WT	RCT V28 N1 P153 <RCT>
1954 V7 N3 P50WT-55WT	RCT V27 N4 P940 <RCT>
1954 V7 N4 P82WT-87WT	RCT V29 N2 P409 <RCT>
1954 V7 N5 P96WT-104WT	RCT V27 N4 P899 <RCT>
1954 V7 N6 P132WT-136WT	RCT V28 N3 P821 <RCT>
1954 V7 N7 P157WT-162WT	59-17558 <*>
1954 V7 N8 P170WT-178WT	65-11948 <*>
1954 V7 N9 P191WT-196WT	RCT V28 N4 P1082 <RCT>
1954 V7 N10 P218WT-221WT	65-17099 <*>
1954 V7 N11 P263WT-265WT	
	RCT V28 N2 P588 <RCT> 0
1954 V7 N12 P273WT-279 PT1	
	RCT V29 N1 P1 <RCT>
1955 V8 N1 P14WT-24WT	1335 <*>
1955 V8 N1 P15WT-24WT	RCT V30 N2 P470 <RCT>
1955 V8 N1 P2WT-8WT PT2	RCT V29 N1 P1 <RCT>
1955 V8 N2 P27WT-31WT PT3	RCT V29 N1 P1 <RCT>
1955 V8 N2 P31WT-34WT	RCT V30 N1 P354 <RCT>
1955 V8 N2 P35WT-39WT	RCT V29 N1 P166 <RCT>
1955 V8 N4 P85WT-90WT PT4	RCT V29 N1 P1 <RCT>
1955 V8 N5 P117WT-124WT	RCT V29 N3 P1082 <RCT>
1955 V8 N6 P145WT-152WT	RCT V29 N1 P207 <RCT>
1955 V8 N6 P157WT-160WT	RCT V29 N3 P1082 <RCT>
1955 V8 N9 P227WT-233WT	RCT V29 N2 P355 <RCT>
1955 V8 N9 P208-209	62-11327 <=>
1955 V8 N10 P251WT-257 PT5	
	RCT V29 N1 P1 <RCT>
1955 V8 N10 P251WT-257WT	59-15948 <*> 0
1955 V8 N10 P263WT-268WT	61-18771 <*>
1955 V8 N11 P273WT-280WT	RCT V29 N3 P901 <RCT>
1955 V8 N12 P302WT-306WT	RCT V30 N1 P180 <RCT>
1956 V9 P110WT-113WT	RCT V30 N1 P69 <RCT>
1956 V9 P197WT-205WT	RCT V30 N4 P1027 <RCT>
1956 V9 P2WT-9WT	T-1655 <INSD>
1956 V9 P269WT-272WT	RCT V30 N3 P903 <RCT>
1956 V9 P31WT-38WT	T-1655 <INSD>
1956 V9 N2 P27WT-30WT	RCT V29 N3 P894 <RCT>
1956 V9 N3 P56WT-62WT	RCT V29 N4 P1215 <RCT>
1956 V9 N3 P64WT-	RCT V29 N4 P1215 <RCT>
1956 V9 N3 P66WT-	RCT V29 N4 P1215 <RCT>
1956 V9 N6 P149WT-152WT	RCT V30 N1 P77 <RCT>
1956 V9 N6 P153WT-159WT	RCT V30 N1 P200 <RCT>
1956 V9 N12 P300WT-304WT	RCT V30 N4 P1017 <RCT>
1957 V10 P109WT-115WT	RCT V31 N3 P539 <RCT> 0
1957 V10 P185WT-194WT	RCT V31 N2 P286 <RCT> 0
1957 V10 P214WT-222WT	RCT V31 N2 P301 <RCT>
1957 V10 P23WT-30WT	RCT V30 N3 P911 <RCT>
1957 V10 P241WT-250WT	RCT V31 N2 P315 <RCT> 0
1957 V10 P273WT-277WT	RCT V31 N3 P548 <RCT>
1957 V10 P51WT-57WT	RCT V32 N1 P128 <RCT> 0
1957 V10 N2 P31WT-39WT	RCT V31 N1 P132 <RCT> 0
1957 V10 N3 P57WT-65WT	RCT V30 N4 P1078 <RCT> 0
1957 V10 N3 P57-65	30J18G <ATS>
1957 V10 N4 P81WT-88WT	RCT V31 N1 P117 <RCT> 0
1957 V10 N7 P161WT-167WT	RCT V31 N2 P262 <RCT>
1957 V10 N7 P168WT-172WT	RCT V31 N1 P105 <RCT>
	65-10260 <*>
1957 V10 N8 P204WT-	RCT V31 N3 P650 <RCT>
1957 V10 N8 P206WT-	RCT V31 N3 P650 <RCT>
1957 V10 N8 P208WT-	RCT V31 N3 P650 <RCT>
1957 V10 N10 P258WT-268WT	60-16486 <*>
	64-18516 <*>
1957 V10 N12 P302WT-307WT	RCT V31 N4 P925 <RCT>
1958 V11 P127WT-133WT	RCT V32 N1 P208 <RCT>
1958 V11 P210WT-214WT	RCT V32 N3 P759 <RCT> 0
1958 V11 P267WT-272WT	RCT V32 N2 P566 <RCT> 0
1958 V11 P278WT-281WT	RCT V32 N4 P1027 <RCT>
1958 V11 P281WT-292WT	RCT V32 N4 P1254 <RCT>
1958 V11 P3WT-8WT	RCT V32 N1 P295 <RCT>
1958 V11 P325WT-331WT	RCT V32 N4 P962 <RCT> 0

1958 V11 P51WT-56WT RCT V32 N1 P139 <RCT> 0
1958 V11 P57WT-62WT RCT V32 N2 P544 <RCT>
1958 V11 N6 P151WT-158WT 65-10230 <*>
1958 V11 N7 P185WT-190WT 65-10036 <*>
1958 V11 N9 P254WT-260WT 69L30G <ATS>
1958 V11 N9 P254-260 58-2492 <*>
1958 V11 N12 P332WT-335WT 61-16715 <*>
1959 V12 P1WT-4WT RCT V33 N2 P335 <RCT>
1959 V12 P122WT-128WT RCT V33 N3 P846 <RCT> 0
1959 V12 P233WT-239WT RCT V33 N4 P1051 <RCT> 0
1959 V12 P239WT-246WT RCT V33 N5 P1438 <RCT>
1959 V12 P284WT-296WT RCT V33 N5 P1438 <RCT>
1959 V12 P33WT-36WT RCT V33 N2 P326 <RCT>
1959 V12 P334WT-336WT RCT V33 N5 P1438 <RCT>
1959 V12 P59WT-68WT RCT V33 N4 P1029 <RCT>
1959 V12 P96WT-106WT RCT V33 N2 P282 <RCT>
1959 V12 N1 P5WT-10WT 59-15874 <*>
1959 V12 N3 P68WT-74WT 59-17213 <*> 0
1959 V12 N4 P83WT-95WT RCT V33 N3 P763 <RCT>
1959 V12 N5 P134WT-140WT 60-10799 <*>
1959 V12 N10 P270WT-283WT 61-10199 <*>
1959 V12 N10 P298WT-302WT 60-16589 <*>
1959 V12 N10 P298-302 61-18638 <*>
1960 V13 P1WT-12WT RCT V34 N3 P834 <RCT>
1960 V13 P49WT-59WT RCT V34 N2 P606 <RCT>
1960 V13 P336- 62-10374 <*>
1960 V13 P392-400 RCT V35 N1 P76 <RCT>
1960 V13 N2 P34WT-41WT 10N48G <ATS>
65-17122 <*>
1960 V13 N11 P362WT-364WT 64-10919 <*>
1960 V13 N11 P366WT-368WT 64-10919 <*>
1961 V14 P208WT-217WT RCT V35 N3 P776 <RCT>
1961 V14 P334WT- RCT V35 N4 P848 <RCT>
1961 V14 P347WT-358WT RCT V36 N1 P268 <RCT>
1961 V14 N2 P23WT-32WT 64-16693 <*> 0
1961 V14 N3 P54WT-60WT 77N55G <ATS>
1961 V14 N9 P261WT-268WT 65-10443 <*>
1961 V14 N10 P302WT-307WT 63-16251 <*> 0
1961 V14 N12 P364WT-372WT
RCT V37 N4 P910 <RCT> 0
1962 V15 P57WT-62WT RCT V36 N1 P236 <RCT>
1962 V15 N2 P37WT-41WT RCT V35 N3 P621 <RCT>
1962 V15 N6 P157WT-167WT RCT V36 N1 P158 <RCT>
1962 V15 N8 P215WT- RCT V38 N1 P176 <RCT>
1962 V15 N8 P251-254 65-11924 <*>
1962 V15 N12 P475WT-481WT 65-14859 <*> 0
1962 V15 N12 P475-481 12921 <K-H>
1963 V16 P313- RCT V39 N5 P1640 <RCT>
1963 V16 P426- RCT V38 N3 P581 <RCT> 0
1963 V16 N2 P84-87 65-17098 <*>
1963 V16 N5 P256-266 72Q73G <ATS>
1963 V16 N10 P55-558 RCT V37 N2 PT1 P477 <RCT>
1963 V16 N10 P553-555 RCT V37 N2 PT1 P446 <RCT>
1963 V16 N10 P571-582 RCT V37 N2 PT1 P408 <RCT>
1963 V16 N12 P655-659 RCT V37 N3 P698 <RCT>
1964 V17 N3 P115 65-18000 <*>
1964 V17 N5 P253-262 65-13936 <*>
1964 V17 N6 P310-314 65-11016 <*>
1964 V17 N7 P365-371 65-13937 <*>
1964 V17 N8 P434-437 65-14094 <*>
1964 V17 N9 P493-497 65-14171 <*>
1964 V17 N10 P572-574 65-14172 <*>
1964 V17 N11 P629-633 65-14219 <*>
1965 V18 P515-528 1802 <TC>
1965 V18 P750-752 66-11440 <*>
1965 V18 N2 P79-81 65-13938 <*>
1965 V18 N3 P138-145 65-13943 <*>
1965 V18 N7 P433-445 65-17463 <*>

KAZAN
1957 V1 N1 P47-57 IGR V3 N6 P518 <AGI>
1958 S2 V3 P1-16 SPEC ISS IGR V1 N6 P48 <AGI>
1959 V4 N2 P104-114 PT1 IGR V3 N8 P712 <AGI>
1959 V4 N2 P115-130 IGR V3 N9 P803 <AGI>
1959 V4 N3 P133-151 IGR V3 N10 P944 <AGI>

KAZANSKII MEDITSINSKII ZHURNAL
1930 V26 N11 P1139-1140 RT-792 <*>
1939 V35 N5/6 P49-52 R-1068 <*>
1941 V37 N1 P80-83 64-23638 <=$>
1959 N1 P5-7 59-13621 <=>
1959 V40 N2 P59-62 60-31184 <=>
1959 V40 N3 P70-73 59-13925 <=>
1959 V40 N3 P104-108 59-13936 <=>
1959 V40 N5 P63-65 60-11741 <=>
1959 V40 N5 P101-102 60-11742 <=>
1959 V40 N5 P104-106 60-11743 <=>
1959 V40 N5 P106-109 60-11744 <=>
1959 V40 N5 P109-110 60-11747 <=>
1959 V40 N5 P110-111 60-11748 <=>
1959 V40 N6 P8-14 60-11548 <=>
1959 V40 N6 P35-38 60-11548 <=>
1959 V40 N6 P99-101 60-11560 <=>
1960 V41 N1 P16-21 60-41457 <=>
1960 V41 N1 P71-73 60-41458 <=>
1960 V41 N1 P112-113 60-41459 <=>
1960 V41 N4 P82-83 61-11917 <=>
1960 V41 N4 P83-84 61-11916 <=>
1960 V41 N4 P84-87 61-11891 <=>
1960 V41 N5 P81-84 63-31382 <=>
1960 V41 N5 P100-104 61-28923 <*=>
1960 V41 N6 P76-78 61-21651 <=>
1960 V41 N6 P79-80 61-21626 <=>
1960 V41 N6 P80-82 61-21728 <=>
1960 V41 N6 P82-83 61-21718 <=>
1962 V43 N5 P3-8 63-21827 <=>
1963 V44 N3 P63-64 RTS-2701 <NLL>
1963 V44 N4 P36-38 RTS 2700 <NLL>

KEIJO JOURNAL OF MEDICINE
1939 V10 N2 P66-76 1901 <*>
57-1639 <*>

KEIKINZOKU
1952 N2 P22-27 II-1004 <*>
1953 V2 N7 P73-75 64-14280 <*>
1959 V9 N1 P53-63 30P61J <ATS>
1964 V14 N4 P19-32 AD-631 171 <=*$>
1965 V15 N2 P12-23 1583 <TC>

KEIO IGAKU
1957 V34 P345-352 64-00195 <*>
1959 V36 N10 P1287-1304 63-00310 <*>

KEIZAI KENKYU
1963 V14 N2 P114-122 64-21411 <=>

KEKKAKU
1953 V28 P800-802 57-2557 <*>
1955 V30 N10 P569-573 1361 <*>
1957 V32 P175-177 63-20944 <*>
1958 V33 N10 P671-673 60-15718 <=*> 0
1964 V39 P95-99 66-11594 <*>

KEMENTERIAN PENERANGAN R. I. INDONESIA
1951 P1-65 66-31631 <=$>

KEMIJA U INDUSTRIJI
1954 V3 N10 P277-282 1216 <*>
1956 V5 N1 P1-5 70J18CR <ATS>
1956 V5 N12 P55-58 58-1958 <*>
1961 V10 N7 PF71-F80 25P59CR <ATS>
1962 V11 N8 P476-477 64-14751 <*>
1964 V13 N8 P617-621 58S85CR <ATS>

KENCHIKU GIJUTSU
1956 N1 P72-76 CSIRO-3823 <CSIR>
1959 N95 P46-52 C-3401 <NRCC>

KENSHIN JIHO
1957 V22 N4 P5-18 SCL-T-346 <*>

KEP ES HANGTECHNIKA
1957 V3 N5/6 P109-113 92M39H <ATS>
1958 V4 N3 P77 93M39H <ATS>
1959 V5 N5 P141-143 47N48H <ATS>
1961 V7 N1 P6-8 62-19412 <*=>
1961 V7 N1 P30-31 62-19411 <=*>
1961 V7 N4 P102-106 66P58H <ATS>
1961 V7 N4 P109-112 80P59H <ATS>

1961 V7 N6 P161-165 79Q68H <ATS>
1962 V8 N1 P1-5 48Q72H <ATS>
1962 V8 N1 P17-18 27Q70H <ATS>
1963 V9 N4 P97-105 63-41194 <=>
1963 V9 N6 P164-171 55R76H <ATS>
1964 V10 N2 P45-49 11R79H <ATS>
1964 V10 N3 P69-73 12R79H <ATS>
1964 V10 N4 P97-105 64R80H <ATS>
1964 V10 N5 P110-118 42R80H <ATS>

KERAMICHESKII SBORNIK
1939 V1 N2 P49-50 R-769 <*>

KERAMIKA I STEKLO
1931 V7 N11/2 P36-41 63-20363 <=*$>

KERAMISCHE RUNDSCHAU UND KUNSTKERAMIK
1936 V44 P38-40 62-20200 <*> 0
1936 V44 N2C P231-234 62-18825 <*> 0
1936 V44 N20 P282-283 62-18825 <*> 0
1955 V79 N19 P297-302 57-2383 <*>

KERAMISCHE ZEITSCHRIFT
1957 V9 P438- 58-1283 <*>
1959 V11 N9 P476-478 63-14162 <*> 0
1959 V11 N10 P524-528 63-14163 <*> 0
1959 V11 N11 P475-480 63-14164 <*>
1959 V11 N11 P524-528 63-14164 <*>
1959 V11 N11 P570-573 63-14164 <*>
1961 V13 P617-622 65-13500 <*>
1963 V15 P192-193 UCRL TRANS-1043(L) <*>

KERAMOS
1930 V9 N13 P5-6 60-10251 <*>

KERNENERGIE
1958 V1 P921 61-18849 <*>
1959 V2 P893-899 NP-TR-478 <*>
1959 V2 P1148-1150 ORNL-TR-127 <*>
1960 V3 N4 P366-370 UCRL TRANS-650(L) <*>
1960 V3 N8 P707-717 62-24565 <=>
1960 V3 N8 P808-810 62-24565 <=>
1960 V3 N9 P816-821 62-24566 <=*>
1960 V3 N9 P866-868 AEC-TR-4985 <*>
1960 V3 N10/1 P973-978 NP-TR-861 <*>
1961 V4 P445-448 66-11983 <*>
1961 V4 P926-934 UCRL TRANS-952(L) <*>
1961 V4 N6 P435-439 ANL-TRANS-207 <*>
1962 V5 P118-119 AEC-TR-5605 <*>
1962 V5 P173-176 65-18104 <*>
1962 V5 P177-181 AEC-TR-5881 <*>
1962 V5 P472-473 AEC-TR-6182 <*>
1962 V5 P488 AEC-TR-6342 <*>
1962 V5 P515-533 AEC-TR-5754 <*>
1962 V5 P564-565 AEC-TR-6335 <*>
1962 V5 P839-845 ORNL-TR-309 <*>
1962 V5 P853-859 ORNL-TR-124 <*>
1962 V5 N6 P450-461 63-13491 <=>
1962 V5 N6 P462-471 NP-TR-956 <*>
1962 V5 N6 P474-480 63-13491 <=>
1962 V5 N6 P489-496 63-13491 <=>
1962 V5 N9 P685-689 63-21300 <=>
1962 V5 N10 P701-734 63-21225 <=>
1962 V5 N4/5 P264-266 AEC-TR-5907 <*>
1962 V5 N4/5 P267-269 AEC-TR-5909 <*>
1962 V5 N4/5 P329-331 AEC-TR-5913 <*>
1962 V5 N4/5 P438-442 63-13491 <=>
1963 V6 P37-39 AEC-TR-6186 <*>
1963 V6 P71-72 ORNL-TR-34 <*>
1963 V6 P73-76 AEC-TR-5908 <*>
1963 V6 P116-121 AEC-TR-5951 <*>
1963 V6 P121-122 AEC-TR-6163 <*>
1963 V6 P177-178 AEC-TR-6187 <*>
1963 V6 P243-251 AEC-TR-6267 <*>
1963 V6 P489-493 66-11752 <*>
1963 V6 P509-513 AEC-TR-6338 <*>
1963 V6 P514-516 AEC-TR-6351 <*>
1963 V6 N1 P43-46 NP-TR-1013 <*>
1963 V6 N4 P178-180 NP-TR-1103 <*>
1963 V6 N8 P390-409 NP-TR-1110 <*>
1963 V6 N9 P496-504 NP-TR-1206 <*>
1963 V6 N10 P546-552 NP-TR-1206 <*>
1963 V6 N11 P649-650 NP-TR-1125 <*>
1963 V6 N12 P685-691 ANL-TRANS-17 <*>
1963 V6 N12 P704-710 66-13643 <*>
1964 V7 P6-7 ORNL-TR-440 <*>
1964 V7 P406-411 ORNL-TR-440 <*>
1964 V7 N1 P11-18 66-12948 <*>
1964 V7 N2 P111-113 65-18129 <*>
1964 V7 N12 P838-845 65-30374 <*>
1965 V8 N5 P297-306 66-11117 <*>

KERNENERGIE IN DE LANDBOUW
1961 V2 N4 P1-5 AEC-TR-4806 <*>

KERNTECHNIK
1960 V2 P1-9 AEC-TR-4618 <*>
1960 V2 N4 P121-123 64-10568 <*>
1961 V3 P120-126 NP-TR-904 <*>
1961 V3 N1 P4-7 NP-TR-798 <*>
1961 V3 N2 P67-70 10P65G <ATS>
1961 V3 N6 P258-261 NP-TR-857 <*>
1961 V3 N11 P475-482 21Q73G <ATS>
1962 V4 P516-517 AEC-TR-6185 <*>
1962 V4 P517-518 AEC-TR-6181 <*>
1962 V4 N1 P1-7 NP-TR-894 <*>
1962 V4 N11 P485-491 AEC-TR-6319 <*>
1963 V5 P122-124 UCRL TRANS-1061(L) <*>
1963 V5 P293-295 ORNL-TR-28 <*>
1963 V5 N5 P207-212 3376 <BISI>
1963 V5 N5 P221-224 63-20849 <*> 0
63-20951 <*> 0
1963 V5 N7 P286-293 64-16166 <*>
1965 V7 P388-394 66-11133 <*>

KHIDROLOGIYA I METEOROLOGIYA. SOFIA
1958 N1 P14-20 59-11011 <=>
1960 N5 P28-36 62-19413 <=*>
1960 N5 P51-60 62-19413 <=*>
1961 V1 P35-88 C-4930 <NRC>
1963 N4 P3-8 64-21416 <=>
1964 V12 N2 P25 M.5780 <NLL>

KHIDROTEKHNIKA I MELIORATSII
1960 V5 N2 P56-59 62-19414 <=> 0
1960 V5 N4 P103-107 62-19414 <=> 0
1960 V5 N6 P165-168 62-19414 <=> 0
1964 V9 N1 P1-2 64-31246 <=>
1966 N9 P262-267 67-31017 <=$>

KHIGIENA EPIDEMIOLOGIYA I MIKROBIOLOGIYA
1960 V3 N6 P30-37 62-19399 <=*> 0
1961 V4 N2 P21-24 62-23600 <=*>
1961 V4 N2 P35 62-19147 <=*>
1961 V4 N3 P57-62 62-23447 <=*>
1961 V4 N4 P31-33 62-19776 <=*>
1961 V4 N4 P47-51 62-19774 <=*>
1962 V5 N5 P1-16 63-15383 <=>

KHIMICHESKAYA NAUKA I PROMYSHLENNOST
1956 V1 P317-324 61-23342 <*=>
1956 V1 P492-495 148TM <CTT>
1956 V1 N1 P32-37 65-61463 <=>
1956 V1 N1 P38-44 15H13R <ATS>
1956 V1 N2 P199-204 60-17440 <+*>
1956 V1 N3 P273-281 39M47R <ATS>
64-10101 <=*$>
65-17432 <*>
1956 V1 N3 P287-297 61-10974 <+*>
90M42R <ATS>
1956 V1 N3 P324-331 59-18848 <+*>
1956 V1 N4 P433-442 62-14912 <=*> P
1956 V1 N5 P505-511 60-31853 <=>
1956 V1 N5 P512-517 59-18530 <+*>
1956 V1 N5 P534-539 81J16R <ATS>
1956 V1 N5 P554-560 59-16981 <+*>
1956 V1 N6 P610-620 43K21R <ATS>
1957 N2 P299-305 8K20R <ATS>
1957 V2 P265-266 59-10328 <+*>
1957 V2 P274-279 R-5341 <*>

1957 V2 P339-347 65-26159 <$>
1957 V2 P392-393 59-15621 <+*>
1957 V2 P663 65-14132 <*>
1957 V2 P799-800 62-15683 <*=>
1957 V2 N1 P19-23 16J16R <ATS>
1957 V2 N1 P59-64 65N51R <ATS>
1957 V2 N1 P76-80 61-15373 <+*>
1957 V2 N1 P133 59-16409 <+*>
1957 V2 N2 P269-270 10L30R <ATS>
1957 V2 N3 P274-279 69K25R <ATS>
1957 V2 N3 P391-392 59-15233 <+*>
1957 V2 N3 P398-399 24K28R <ATS>
1957 V2 N4 P462-465 37J19R <ATS>
1957 V2 N5 P663 14K26R <ATS>
1958 V3 P191-221 62-10122 <*=>
1958 V3 P432-438 92M37R <ATS>
1958 V3 N1 P35-45 62-14505 <=*>
1958 V3 N1 P72-76 60-23985 <*+>
1958 V3 N1 P127 59-10463 <+*>
1958 V3 N2 P146-158 59-21177 <=>
1958 V3 N2 P174-190 62-10121 <*=>
1958 V3 N2 P284 65M38R <ATS>
1958 V3 N3 P443-448 62-19879 <=>
1958 V3 N4 P464-470 27P61R <ATS>
1958 V3 N4 P471-476 28P61R <ATS>
1958 V3 N4 P511-514 33L36R <ATS>
1958 V3 N4 P515-517 59-15833 <+*>
1958 V3 N4 P518-520 84L35R <ATS>
1958 V3 N5 P567-576 60-14064 <+*>
1958 V3 N5 P683 32L32R <ATS>
1958 V3 N6 P803-807 61-23893 <*=>
1958 V3 N6 P807-811 61-15271 <*+> 0
1959 V4 P281 662 <TC>
1959 V4 P287-288 N66-23549 <=$>
1959 V4 P472-478 AEC-TR-4851 <*>
1959 V4 P509-515 AEC-TR-4622 <=*>
1959 V4 P675-676 62-14022 <=*> 0
1959 V4 P747-756 62-10448 <=*>
1959 V4 P808 61-10401 <+*>
1959 V4 N1 P15-26 60-21746 <=> 0
1959 V4 N1 P35-41 60-10560 <+*>
60-21747 <=> 0
1959 V4 N1 P55-62 60-21748 <=> 0
1959 V4 N1 P100-111 60-21749 <=> 0
1959 V4 N1 P132-133 60-19893 <+*>
1959 V4 N1 P139-140 65-17005 <*>
1959 V4 N1 P139 65-17012 <*>
1959 V4 N2 P145-153 61-23072 <=*>
1959 V4 N2 P154-163 21L37R <ATS>
1959 V4 N2 P172-178 62-15272 <*=>
1959 V4 N2 P230-234 62-23176 <=>
1959 V4 N2 P235-241 62-13701 <*=>
62-13701 <=>
1959 V4 N2 P242-249 62-10695 <=*>
1959 V4 N2 P271-273 60-11392 <=>
1959 V4 N3 P407-408 RCT V33 N3 P796 <RCT>
1959 V4 N4 P423-434 60-41141 <=>
1959 V4 N4 P435-441 61-15729 <+*>
1959 V4 N4 P498-509 62-19758 <=*>
1959 V4 N4 P516-521 61-20678 <*=> 0
1959 V4 N4 P521-526 UCRL TRANS-551(L) <*+>
1959 V4 N4 P547-548 15M39R <ATS>
1959 V4 N5 P554-565 63-11146 <=>
1959 V4 N5 P617-622 61-19220 <+*>
1959 V4 N5 P655-661 90M41R <ATS>
1959 V4 N5 P686-687 60-18145 <*>

KHIMICHESKAYA PROMYSHLENNOST
1944 P438-440 AEC-TR-3302 <+*>
1944 N5 P10-12 RT-2413 <*>
1944 N1/2 P38-39 60-18497 <+*>
1945 N2 P1-5 61-18240 <*=>
64-30135 <*>
1945 N2 P5-6 61-18237 <*=>
1945 N2 P21 60-15850 <+*>
1945 N3 P8-9 63-14570 <=*> 0
1945 N3 P11-14 61-18238 <*=>
1945 N3 P14-16 62-10818 <=*> P
1945 N4 P1-3 62-10819 <=*> P
1945 N4 P7-9 61-18229 <*=>
1945 N6 P21-22 61-18230 <*=>
1945 N11 P11-14 RT-2979 <*>
1945 N12 P8-11 63-18677 <=*>
1946 N9 P8-9 R-3304 <*>
1947 N2 P16-18 61-13016 <=*>
1947 N4 P17-18 64-20345 <*>
1947 N51/3 P19-21 24K21R <ATS>
1954 P100-102 GB57 494 <NLL>
1954 P435-436 62-11116 <=>
1954 P480-485 62-19876 <*=>
1954 N1 P21-25 RT-2543 <*>
1954 N1 P56 RT-2239 <*>
1954 N2 P86-89 59-13831 <=>
1954 N4 P1-4 RT-2951 <*>
1954 N4 P213-216 63-16259 <*=> 0
1954 N6 P336-338 63-16258 <=*> 0
1955 P392-397 59-10460 <+*>
1955 N1 P27-34 RT-3525 <*>
1955 N1 P34-39 GB57 497 <NLL>
1955 N6 P344-345 RT-3563 <*>
1955 N7 P388-392 RJ-395 <ATS>
1955 N8 P1-5 RT-4630 <*>
1956 P156-161 34K26R <ATS>
1956 P324- 61-23252 <*=>
1956 P333-338 59-18535 <+*>
64-16513 <=*$>
1956 P347-358 60-23030 <+*>
1956 P405-407 781TM <CTT>
1956 P408-411 59-19616 <+*>
1956 N1 P26-31 64L30R <ATS>
1956 N1 P31-34 99J15R <ATS>
1956 N1 P35-41 RTS-2721 <NLL>
1956 N2 P69-77 63-20383 <=*$> 0
1956 N2 P78-89 59-16982 <+*>
1956 N2 P89-97 60-23630 <+*>
1956 N3 P16-22 172TM <CTT>
1956 N3 P358-363 <CTT>
1956 N4 P193-196 65-14296 <*>
1956 N7 P21-23 RCT V31 N1 P44 <RCT>
1956 N7 P405-407 R-2052 <*>
1956 N8 P469-474 63-16213 <=*> 0
1956 V5 P257-264 173TM <CTT>
1957 P129-132 60-15067 <+*>
1957 P193-201 60-17444 <+*>
1957 P349-352 R-5144 <*>
1957 P363-365 60-13665 <+*>
1957 P417-421 59-11617 <=>
1957 P489-493 62-19001 <=*>
1957 N1 P13-15 60-17710 <+*>
1957 N1 P15-19 89M41R <ATS>
1957 N2 P31-38 97K21R <ATS>
1957 N2 P77-79 510TT <CCT>
62-15356 <*=>
1957 N2 P108-116 48K21R <ATS>
1957 N3 P38-41 62-13130 <*=>
1957 N5 P54-61 61-23429 <=*>
1957 N5 P302-306 UCRL TRANS(L) <=*$>
65-10238 <*>
1957 N6 P22-26 35J19R <ATS>
1957 N7 P27-32 59-19445 <+*>
1958 P106-110 60-17371 <+*>
1958 P205-208 60-18986 <+*>
1958 P221-227 60-16960 <*+>
1958 P270-276 60-19565 <+*>
1958 P314 58L32R <ATS>
1958 P325-330 37L34R <ATS>
1958 P350-354 33L33R <ATS>
1958 P401-404 63L31R <ATS>
1958 P406-409 52L30R <ATS>
60-16949 <*+>
1958 P476-481 62-25388 <=*>
1958 N2 P42-46 60-17371 <+*>
1958 N2 P73-80 25M38R <ATS>
1958 N2 P80-84 62-10722 <=*>
1958 N3 P10-18 C-3578 <NRCC>
1958 N3 P58-59 59-18948 <+*>
1958 N3 P138-146 06T94R <ATS>
1958 N5 P261-267 66-14031 <*$>
1958 N5 P292-295 RTS 2386 <NLL>
1959 P62-68 32L36R <ATS>

Citation	Translation
1959 P134-139	61-27644 <*=>
1959 P243-250	A-124 <RIS>
1959 P288-290	87M43R <ATS>
1959 P385-387	31M40R <ATS>
1959 P388-390	61-27569 <*=>
1959 P404-408	60-31386 <=>
1959 P430-435	62-18104 <*>
1959 P436-441	17M46R <ATS>
	61-15474 <+*>
1959 P459	60-31084 <=>
1959 P566-573	62-20371 <=*>
1959 P577-580	61-27189 <*=>
1959 P622-624	61-27574 <*=>
1959 P674-680	83M42R <ATS>
1959 N1 P9-16	60-17320 <+*>
1959 N1 P79-84	60-17227 <+*>
1959 N2 P7-14	61-23421 <*=>
1959 N3 P25-48	60-11965 <=>
1959 N3 P32-34	61-13947 <+*>
1959 N3 P207-215	62M37R <ATS>
	66-14786 <*$>
1959 N4 P1-10	60-23868 <+*>
1959 N4 P10-12	61-15718 <+*>
1959 N4 P13-20	60-23868 <+*>
1959 N4 P16-20	61-10756 <+*>
1959 N4 P24-26	61-16597 <*=>
1959 N4 P88-89	61-27572 <*=>
1959 N4 P139-140	13077-B <K-H>
1959 N5 P419-425	AEC-TR-5528 <=*>
1959 N5 P426-429	65-64205 <=*$>
1959 N7 P86-88	60-41624 <=>
1959 N7 P586-591	63-19136 <=*>
1959 N8 P647-652	61-23432 <=*>
1959 N8 P652-658	66-11562 <*>
1959 N8 P672-674	61-23432 <=*>
1960 P186-192	62-10367 <*=> 0
1960 P193-201	61-27782 <=$>
1960 P265-272	24M47R <ATS>
1960 P287-293	08M46R <ATS>
1960 P362-367	62-25907 <=*>
1960 P375-382	61-21636 <=>
	61K118 <CTT>
1960 P401-406	36P62R <ATS>
1960 P411-413	62-33405 <=*>
	64-13302 <=*$>
1960 P444-452	70N49R <ATS>
1960 P488-492	62-10969 <=*>
1960 P496-499	62-10983 <=*>
1960 P529-537	48N58R <ATS>
1960 P613-614	61-21535 <=>
1960 P624-626	61-18705 <*=>
1960 P626-627	62-10248 <=*>
1960 P643-652	62-10666 <=*>
1960 N1 P25-31	65-18064 <*>
1960 N1 P34-41	62-24451 <=*>
1960 N1 P41-44	61-11898 <=>
1960 N1 P58-59	62-24308 <=*>
1960 N2 P89-95	10424 <K-H>
	65-12501 <*> 0
	78M46R <ATS>
1960 N4 P272-274	66-10002 <*>
1960 N4 P287-293	65-11482 <*>
1960 N6 P50-54	ICE V1 N1 P50-54 <ICE>
1960 N6 P488-492	63-14328 <=*>
1960 N7 P26-28	04N55R <ATS>
1960 N8 P15-23	ICE V1 N1 P39-49 <ICE>
1960 N8 P626-627	ICE V1 N1 P24-25 <ICE>
1960 N8 P642-643	63-16233 <=*> 0
	66-10619 <*>
1961 P105-108	62-15407 <*=>
1961 P185-190	62-32471 <=*>
1961 P204-210	03N54R <ATS>
1961 P213-214	75P59R <ATS>
1961 P258-263	62-13113 <*=>
1961 P335-338	62-14634 <=*>
1961 P339-340	62-00619 <*>
	62-25122 <=*>
1961 P416-420	62-16007 <=*>
1961 P420-424	62-24301 <=*>
1961 P431-432	62-32426 <=*>
1961 P498-502	38P63R <ATS>
1961 P564-566	39P63R <ATS>
1961 P672-677	62-18758 <=*>
1961 P711-714	66-10253 <*> 0
1961 P763-766	55P60R <ATS>
1961 P787-789	92Q66R <ATS>
1961 N1 P6-11	ICE V1 N1 P79-84 <ICE>
1961 N1 P51-56	ICE V1 N1 P68-73 <ICE>
	64-15877 <=*$>
1961 N1 P72-73	61-27949 <*=>
1961 N2 P31-34	61-28879 <*=>
1961 N2 P70-73	61-28602 <*=>
1961 N2 P70	61-28632 <*=>
1961 N2 P73	61-28632 <*=>
1961 N2 P122	ICE V1 N1 P74-79 <ICE>
1961 N3 P24-26	67P58R <ATS>
1961 N3 P56-62	61-28755 <*=>
1961 N3 P62-64	62-13922 <*=>
1961 N3 P65-66	61-28755 <*=>
1961 N3 P69-72	61-28755 <*=>
1961 N3 P185-190	NS-417 <TTIS>
1961 N3 P190-196	55Q69R <ATS>
1961 N4 P10-13	02P65R <ATS>
1961 N4 P13-15	61-28883 <*=>
1961 N4 P31-35	28N58R <ATS>
1961 N4 P45-	ICE V2 N2 P258-260 <ICE>
1961 N4 P45-47	59P58R <ATS>
1961 N4 P253-257	63-20044 <=*>
1961 N5 P29-30	65-14366 <*>
	66-14914 <*$>
1961 N5 P60-63	60P58R <ATS>
1961 N5 P70-73	61-28876 <*=>
1961 N5 P325-326	63-20014 <=*>
	66-14914 <*$>
1961 N5 P348-	ICE V2 N1 P61-64 <ICE>
1961 N6 P24-	ICE V2 N1 P98-104 <ICE>
1961 N6 P30-35	62-13540 <*=>
1961 N6 P54-	ICE V2 N1 P109-113 <ICE>
1961 N7 P6-8	62-13583 <*=>
1961 N7 P26-30	NLLTB V4 N1 P26 <NLL>
	36P59R <ATS>
1961 N7 P48-50	65-12750 <*>
1961 N7 P450-452	63-14165 <=*> 0
	63-16240 <=*>
1961 N7 P494-498	21Q71R <ATS>
1961 N8 P1-7	NLLTB V4 N2 P111 <NLL>
1961 N8 P535-537	64-15724 <=$>
1961 N9 P38-	ICE V2 N2 P210-215 <ICE>
1961 N9 P52-	ICE V2 N2 P282-289 <ICE>
1961 N9 P620-624	63-15959 <=*>
	65-12718 <*>
1961 N9 P644-650	65-60349 <=$>
1961 N11 P14-	ICE V2 N2 P251-258 <ICE>
1961 N11 P750-756	66-14503 <*$>
1961 N11 P787-	ICE V3 N2 P207-210 <ICE>
1961 N12 P4-	ICE V2 N3 P388-393 <ICE>
1961 N12 P28-	ICE V2 N3 P378-383 <ICE>
1961 N12 P837-839	64-16321 <=*$>
1961 V5 P313-316	65-10033 <*>
1962 P41-43	65-14012 <*>
1962 P121-122	AEC-TR-5379 <=*>
1962 P135-141	63-15516 <=*>
1962 P174-175	63-15518 <=*>
1962 P260-266	45P63R <ATS>
1962 P291-294	50P63R <ATS>
1962 P352-359	32P65R <ATS>
1962 P410-413	89Q67R <ATS>
1962 P562-566	44P66R <ATS>
1962 P695-697	64-14841 <=*$>
1962 N1 P30-37	64-11940 <=> M
1962 N1 P41-44	11911 <K-H>
1962 N1 P60-	ICE V2 N3 P408-412 <ICE>
1962 N1 P66-69	64-13714 <=*$>
1962 N1 P912-918	65-60133 <=$>
1962 N2 P29-	ICE V2 N3 P438-445 <ICE>
1962 N2 P100-105	66-14788 <$> 0
1962 N3 P43-	ICE V2 N4 P476-481 <ICE>
1962 N4 P30-36	ICE V2 N4 P567-575 <ICE>
1962 N4 P47-	ICE V2 N4 P585-590 <ICE>
1962 N4 P291-	ICE V3 N2 P157-160 <ICE>

Citation	Translation
1962 N5 P46-	ICE V3 N1 P24-32 <ICE>
1962 N5 P58-	ICE V3 N1 P60-64 <ICE>
1962 N5 P360-365	NS-416 <TTIS>
1962 N6 P24-	ICE V3 N2 P256-259 <ICE>
1962 N7 P53-	ICE V3 N2 P161-172 <ICE>
1962 N8 P592-	ICE V3 N1 P138-143 <ICE>
1962 N9 P14-	ICE V3 N2 P190-194 <ICE>
1962 N9 P625-630	64-19382 <=$>
1962 N9 P691-692	81Q68R <ATS>
1962 N10 P53-	ICE V3 N2 P259-263 <ICE>
1962 N11 P30-	ICE V3 N3 P333-338 <ICE>
1962 N11 P42-47	65-63515 <=$>
1962 N11 P50-56	65-63638 <=>
1962 N11 P74-	ICE V3 N3 P319-323 <ICE>
1962 N11 P798-801	NCB-A.2270/JG <NLL>
1962 N11 P819-822	65-11787 <*>
	65-60511 <=$>
1962 N11 P851-854	64-15730 <=$>
1962 N12 P1-	ICE V3 N3 P439-444 <ICE>
1962 N12 P859-863	<ICE>
	63-21657 <=>
1962 N12 P881-896	SC-T-512 <=$>
1962 V1 P43-47	63-23234 <=*>
1962 V11 P781-788	64-19896 <=$>
1963 P211-217	76Q76R <ATS>
1963 N1 P7-18	66-14790 <*$>
1963 N1 P29-34	ICE V4 N1 P10-16 <ICE>
	42Q72R <ATS>
1963 N1 P65-66	64-18134 <*>
1963 N2 P31-	ICE V3 N3 P365-369 <ICE>
1963 N2 P36-	ICE V3 N3 P425-432 <ICE>
1963 N2 P125-129	40R76R <ATS>
1963 N3 P51-57	ICE V3 N4 P496-502 <ICE>
1963 N3 P181-187	06T90R <ATS>
1963 N3 P190-192	71S81R <ATS>
1963 N3 P211-217	65-60492 <=$>
1963 N4 P32-36	ICE V3 N4 P502-506 <ICE>
1963 N4 P62	65-12099 <*>
1963 N5 P341-344	66-10250 <*>
1963 N5 P367-371	ICE V4 N1 P17-21 <ICE>
1963 N5 P389-390	65-18178 <*>
1963 N6 P45-49	ICE V4 N1 P80-84 <ICE>
1963 N6 P53-	ICE V4 N1 P136-141 <ICE>
1963 N6 P415-417	95R78R <ATS>
1963 N6 P419-424	65-60941 <=$>
1963 N6 P469-472	65-64414 <=*$>
1963 N7 P39-	ICE V4 N2 P245-253 <ICE>
1963 N7 P51-	ICE V4 N2 P254-257 <ICE>
1963 N7 P519-526	ICE V4 N2 P245-253 <ICE>
1963 N7 P531-533	ICE V4 N2 P254-257 <ICE>
1963 N8 P567-570	ICE V4 N2 P312-316 <ICE>
1963 N9 P35-37	65-11881 <*>
1963 N9 P641-	ICE V4 N2 P275-282 <ICE>
1963 N9 P641-646	ICE V4 N2 P275-282 <ICE>
1963 N9 P647-649	98R75R <ATS>
1963 N9 P662-665	64-18743 <*>
	66-10012 <*>
1963 N10 P50-55	ICE V4 N2 P239-245 <ICE>
1963 N10 P731-735	65-14793 <*>
	66-14353 <*$>
1963 N10 P742-743	65-12749 <*>
1963 N10 P752-754	84S87R <ATS>
1963 N10 P770-774	ICE V4 N2 P239-245 <ICE>
1963 N11 P823-830	4411 <BISI>
1963 N12 P9-14	ICE V4 N3 P492-498 <ICE>
1963 N12 P889-894	ICE V4 N3 P492-498 <ICE>
1964 N2 P14-20	ICE V4 N3 P473-480 <ICE>
1964 N2 P21-24	ICE V4 N3 P502-506 <ICE>
1964 N2 P81-86	64-31287 <=>
1964 N2 P94-101	ICE V4 N3 P473-480 <ICE>
1964 N2 P101-103	ICE V4 N3 P502-506 <ICE>
1964 N2 P134-141	65-11484 <*>
1964 N3 P20-23	AD-620 771 <=$>
1964 N3 P68-72	ICE V4 N4 P680-684 <ICE>
1964 N3 P171-174	66-11890 <*>
1964 N3 P196-198	82T94R <ATS>
1964 N3 P199-201	65-12459 <*>
1964 N3 P228-232	ICE V4 N4 P680-684 <ICE>
1964 N4 P307-310	AD-641 108 <=$>
1964 N5 P332-339	65-13885 <*>

Citation	Translation
1964 N5 P344-348	RTS 2854 <NLL>
1964 N6 P45-53	ICE V5 N2 P309-317 <ICE>
1964 N6 P422-425	65-13886 <*>
1964 N6 P429-431	RTS-2788 <NLL>
1964 N6 P442-445	AD-629 414 <=$>
1964 N6 P445-453	93R79R <ATS>
1964 N7 P28-34	ICE V5 N2 P201-208 <ICE>
1964 N7 P499-5C1	65-64406 <=*$>
1964 N8 P17-22	ICE V5 N2 P195-200 <ICE>
1964 N8 P597-605	RTS-2947 <NLL>
	905 <TC>
1964 N8 P605-610	RTS-2970 <NLL>
1964 N9 P28-31	ICE V5 N2 P355-359 <ICE>
1964 N9 P36-38	ICE V5 N2 P237-239 <ICE>
1964 N9 P49-53	ICE V5 N2 P231-236 <ICE>
1964 N9 P663-665	66-11894 <*>
1964 N9 P668-671	46R80R <ATS>
1964 N9 P683-684	66-10336 <*>
1964 N9 P689-693	45R80R <ATS>
	66-11913 <*>
1964 N10 P764-767	RTS-3281 <NLL>
1964 N11 P35-37	ICE V5 N2 P347-350 <ICE>
1964 N11 P801-804	65-30340 <=$>
1964 N11 P808-812	65-30739 <=*$>
1964 N11 P849-852	1398 <TC>
1964 N12 P881-887	23S82R <ATS>
	66-11912 <*>
1964 N12 P891-894	65-14759 <*>
1964 N12 P898-901	AD-636 639 <=$>
1964 N88 P569-574	44R80R <ATS>
1964 V4 P42-49	1233 <TC>
1964 V40 N8 P574-577	79S89R <ATS>
1964 V40 N8 P577-582	RTS-2969 <NLL>
1964 V40 N9 P643-649	31S84R <ATS>
1965 N1 P31-37	ICE V5 N3 P562-568 <ICE>
1965 N1 P60-64	1737 <TC>
1965 N2 P39-42	65-32966 <=$>
1965 N2 P137-139	66-10999 <*>
1965 N3 P46-51	ICE V5 N3 P555-561 <ICE>
1965 N3 P52-57	ICE V5 N4 P636-642 <ICE>
1965 N3 P186-188	66-12227 <*>
1965 N3 P206-211	36S83R <ATS>
1965 N5 P37-42	ICE V5 N4 P714-719 <ICE>
1965 N6 P12-15	ICE V6 N1 P51-55 <ICE>
1965 N6 P16-22	ICE V6 N1 P67-74 <ICE>
1965 N6 P28-37	ICE V6 N1 P4-15 <ICE>
1965 N7 P31-35	ICE V6 N1 P41-46 <ICE>
1965 N8 P16-17	66-14248 <*$>
1965 N9 P660-661	19T90R <ATS>
1965 N10 P40-42	ICE V6 N2 P297-300 <ICE>
1965 N10 P46-49	ICE V6 N2 P272-276 <ICE>
1965 V19 N11 P3-8	66-30670 <=$>
1965 V41 P57-59	1734 <TC>
1965 V41 P323-325	66-13981 <*>
1965 V41 P760-763	66-14878 <*>
1965 V41 N3 P191-196	66-14241 <*$>
1965 V41 N4 P251-254	66-13281 <*>
1965 V41 N4 P254-255	66-13982 <*>
1965 V41 N4 P255-257	75S87R <ATS>
1965 V41 N4 P263-264	66-13258 <*>
1965 V41 N5 P339-343	66-13620 <*>
1965 V41 N5 P369-373	66-13978 <*>
1965 V41 N5 P384-386	00S86R <ATS>
1965 V41 N6 P438-440	90S86R <ATS>
1965 V41 N7 P495-497	66-13621 <*>
1965 V41 N10 P50-	ICE V6 N2 P260-264 <ICE>
1965 V41 N11 P832-	ICE V6 N2 P264-266 <ICE>
1966 N1 P51-57	ICE V6 N3 P429-437 <ICE>
1966 N4 P1-2	NLLTB V8 N12 P1078 <NLL>
1966 N5 P9-13	ICE V6 N4 P650-655 <ICE>
1966 N5 P21-23	ICE V6 N4 P647-650 <ICE>
1966 N5 P322-329	41T95R <ATS>

KHIMICHESKIE VOLCKNA

Citation	Translation
1959 N1 P33-36	61-10236 <*+>
	78M42R <ATS>
1959 N1 P53	62-24891 <=*>
1959 N2 P21-24	60-18527 <+*>
1959 N2 P33-35	61-10234 <*+>
1959 N2 P44-47	61-10235 <*+>

Citation	Translation
1959 N2 P48-50	61-19805. <*>
1959 N3 P3-10	65-13198 <*>
1959 N3 P11-15	12M46R <ATS>
1959 N3 P39-43	61-19694 <=*>
1959 N3 P39-42	62-23871 <=*>
1959 N4 P23-26	61-18828 <*=>
1959 N4 P61-62	61-10223 <*+>
1959 N4 P62-64	61-27605 <*=>
1959 N4 P67-68	61-23667 <=*>
	65-10069 <*> 0
1959 N5 P3-12	63-19827 <=*>
1959 N5 P18-21	NASA-TT-F-191 <=>
1959 N6 P15-17	61-20709 <*=>
	89M46R <ATS>
1959 N6 P17-21	33M46R <ATS>
1959 N6 P31-33	87M46R <ATS>
1959 N6 P47-49	63-10786 <=*>
1960 N1 P23-26	61-13766 <+*>
1960 N1 P27-29	61-15466 <+*>
1960 N1 P32-33	61-15469 <+*>
1960 N1 P50-51	61-15472 <+*>
	63-10782 <=*>
1960 N1 P51-52	50M47R <ATS>
	61-15473 <+*>
1960 N1 P75	60-31843 <=>
1960 N2 P33-36	44N56R <ATS>
1960 N2 P53-56	<LSA>
	66-10540 <*>
1960 N2 P64-66	63-23399 <=*$>
1960 N3 P3-6	22P61R <ATS>
1960 N3 P15-18	63-19817 <*=>
	63M47R <ATS>
1960 N3 P46-47	63-18397 <=*> 0
1960 N3 P48-50	62-14259 <=*>
1960 N4 P1-2	61-21504 <=>
1960 N4 P10-13	C-143 <RIS>
1960 N4 P69-71	63-10785 <=*>
1960 N5 P37-40	62-23895 <=*>
1960 N6 P15-19	63-19793 <*=>
1960 N6 P57-58	62-15374 <*=>
	63-10812 <=*>
1961 N1 P1-2	61-23575 <=>
1961 N1 P17-18	64-13261 <=*$>
1961 N1 P19-21	62-23954 <=*>
1961 N1 P48-51	24P63R <ATS>
1961 N1 P60-67	63-10789 <=*>
1961 N1 P68-69	21P63R <ATS>
	63-10783 <=*>
1961 N2 P18-22	62-23946 <=*>
1961 N2 P23-24	62-20028 <=*>
1961 N2 P29-33	62-14257 <=*>
1961 N2 P33-37	62-24647 <=*>
	63-10807 <=*> 0
1961 N2 P40-46	63-18280 <*>
1961 N3 P1-2	61-31650 <=>
1961 N3 P21-24	62-24690 <=*>
1961 N3 P48-49	10P61R <ATS>
	65-60895 <=$>
1961 N3 P63-67	13Q70R <ATS>
1961 N3 P68	12Q70R <ATS>
1961 N4 P13-19	62-18969 <=*>
1961 N4 P33-34	63-10806 <=*> 0
	64-13258 <=*$>
1961 N4 P37-41	62-20027 <=*>
	64-13260 <=*$>
1961 N4 P42-46	54Q72R <ATS>
1961 N5 P13-17	62-10723 <*=>
	63-19777 <*=>
1961 N5 P24-27	62-14895 <=*>
1961 N5 P27-31	63-18294 <=*>
1961 N6 P19-22	63-19779 <*=>
1961 N6 P33-35	63-19946 <=*> 0
1961 N6 P39-41	63-18269 <=*>
1961 N6 P46-48	63-10808 <=*>
1961 N6 P56-57	44P64R <ATS>
	63-15180 <=*> 0
1961 N6 P58-59	63-15179 <=*>
1961 N6 P66	63-18400 <=*>
1961 V1 P37-39	63-18364 <=*>
1962 N1 P7-8	63-23365 <=*$>
1962 N1 P8-9	63-23364 <=*$>
1962 N1 P55-60	63-23363 <=*$>
1962 N1 P74-76	25P63R <ATS>
1962 N2 P20-23	64-14809 <=*$>
1962 N3 P2-12	62-33603 <=*>
1962 N3 P23-25	63-18291 <=*>
1962 N4 P1-3	63-18279 <=*>
1962 N4 P10-13	61T92R <ATS>
1962 N5 P5-15	63-23498 <=*$>
1962 N5 P25-27	13Q72R <ATS>
1962 N5 P30-32	65-12083 <*>
1962 N5 P37-44	63-23502 <=*$>
1962 N5 P45-46	63-23501 <=*$>
	64-19140 <=*$>
1962 N5 P49-51	63-23500 <=*$>
1962 N5 P52-54	63-23499 <=*$>
1962 N5 P55-56	RTS.2834 <NLL>
1962 N5 P60-62	RTS.2833 <NLL>
1962 N6 P14-21	M.5336 <NLL>
1962 N6 P25-27	64-10985 <=*$>
1962 N6 P28-30	64-10859 <=*$>
1962 N6 P31-34	65-12082 <*>
1962 N6 P53-55	64-13212 <=*$>
1962 N6 P57	63-16387 <*=>
	64-10986 <=*$>
1962 V4 P37-41	63-18270 <=*>
1963 N1 P19-23	10Q74R <ATS>
1963 N1 P33-38	M.5363 <NLL>
	63-23494 <=*$>
1963 N1 P38-41	M.5364 <NLL>
	63-23493 <=*$>
1963 N1 P47-50	47R74R <ATS>
	63-23496 <=*>
1963 N1 P60-63	63-23497 <=*$>
1963 N1 P64-65	34Q71R <ATS>
1963 N1 P69-70	35Q71R <ATS>
	63-18376 <=*>
1963 N1 P71-72	36Q71R <ATS>
1963 N2 P19-22	AD-636 623 <=$>
1963 N2 P25-29	64-10780 <=*$>
1963 N2 P37-40	65-12080 <*>
1963 N2 P57-58	64-13089 <=*$>
1963 N3 P5-8	63-31608 <=>
1963 N3 P9-11	AD-636 673 <=$>
1963 N3 P11-14	1846 <TC>
1963 N3 P15-18	65-12085 <*>
1963 N3 P20-24	AD-636 673 <=$>
1963 N3 P32-37	66-10562 <*>
1963 N3 P41-42	65-12081 <*>
1963 N3 P61-65	65-12098 <*>
1963 N3 P65-68	65-12078 <*>
1963 N3 P70-71	65-60752 <=>
1963 N4 P1-4	M-5720 <NLL>
1963 N4 P69-75	64-19137 <=*$>
1963 N5 P2-5	65-60750 <=>
1963 N5 P9-12	35R76H <ATS>
1963 N5 P35-39	65-12084 <*>
1963 N5 P40-45	M.5532 <NLL>
1963 N5 P59-64	37Q74R <ATS>
1963 N5 P64-67	65-60779 <=$>
1963 N6 P2-5	65-60776 <=>
1963 N6 P6-9,31-34	AD-615 982 <=$>
1963 N6 P10-13	64-18496 <*>
1963 N6 P16-18	64-21757 <=>
	65-60777 <=>
1963 N6 P18-23	65-60758 <=>
1963 N6 P38-40	65-60772 <=$>
1963 N6 P40-42	65-61280 <=$>
1963 N6 P64-65	64-16999 <*> 0
1964 P43-48	65-13135 <*>
1964 N1 P19-26	65-61183 <=>
1964 N1 P34-39	65-61180 <=>
1964 N2 P27-30	367 <TC>
1964 N3 P15-19	AD-621 057 <=$>
1964 N3 P26-29	M-5658 <NLL>
1964 N3 P34-35	66-10576 <*>
1964 N3 P36-39	66-10565 <*>
1964 N3 P61-62	M-5661 <NLL>
1964 N4 P14-17	65-10498 <*>
1964 N4 P20-24	66-10577 <*>

1964 N4 P24-28	66-10580 <*>
	66-11755 <*>
1964 N5 P2-8	M-5641 <NLL>
1964 N5 P36-38	66-10578 <*>
	66-11774 <*>
1964 N5 P38-41	66-10573 <*>
1964 N5 P41-47	66-10582 <*>
1964 N6 P14-18	77T91R <ATS>
1964 N6 P18-23	66-10563 <*>
	66-11777 <*>
1964 N6 P37-40	66-10574 <*>
1964 N6 P41-44	65-14374 <*>
1964 N6 P44-46	65-10559 <*>
	66-11776 <*>
1964 V6 P30-34	M-5719 <NLL>
1965 N1 P51-53	79S82R <ATS>
1965 N2 P12-15	66-10560 <*>
1965 N2 P57-58	66-10579 <*>
	66-11773 <*>
1965 N2 P58-59	66-10581 <*>
	66-11778 <*>
1965 N3 P65-67	66-10561 <*>
1965 N4 P6-8	RAPIA-1292 <RAP>
1965 N4 P22-25	66-14879 <*>
1965 N6 P33-35	40T94R <ATS>

★KHIMICHESKOE MASHINOSTROENIE

1959 N5 P22-26	ICE V1 N1 P88-92 <ICE>
1959 N5 P39-42	60-11652 <=>
1959 N6 P38-40	4727 <HB>
1959 V1 N2 P4-6	59-13865 <=>
1959 V1 N4 P1-3	60-11682 <=>
1959 V1 N4 P19-21	62-18967 <*=>
1959 V1 N5 P4-6	61-23092 <*=>
1959 V1 N5 P35-39	60-11651 <=>
1959 V1 N6 P1-3	60-11788 <=>
1959 V1 N6 P13-20	22N58R <ATS>
1959 V1 N6 P29-32	62-10970 <=*>
1960 V2 P8-10	61-21376 <=>
1960 V2 P24-28	61-21376 <=>
1960 V2 P37-42	61-21376 <=>
1960 V2 N1 P1-3	60-11860 <=>
1960 V2 N1 P17-20	23N58R <ATS>
1960 V2 N3 P9-11	64-13807 <=*$>
1960 V2 N4 P28-29	61-31657 <=>
1960 V2 N5 P1-3	61-21639 <=>
1960 V2 N5 P6-7	61-21639 <=>
1960 V2 N5 P42-43	61-21639 <=>
1960 V2 N5 P46-47	61-21639 <=>
1960 V2 N6 P1-3	61-21639 <=>
1960 V2 N6 P43-46	61-21639 <=>
1961 N3 P1-2	NLLTB V3 N11 P919 <NLL>
1961 N4 P1-4	NLLTB V4 N1 P38 <NLL>
1961 N5 P15-	ICE V2 N1 P94-97 <ICE>
1961 N5 P29-	ICE V2 N1 P73-77 <ICE>
1961 N6 P21-	ICE V2 N3 P315-318 <ICE>
1961 N6 P25-	ICE V2 N3 P337-341 <ICE>
1961 V3 P8-9	61-23556 <=>
1961 V3 P45-48	61-23556 <=>
1961 V3 N1 P1-2	62-24954 <=*>
1961 V3 N2 P1-6	61-27088 <=>
1961 V3 N2 P9-13	61-27088 <=>
1961 V3 N2 P17-22	62-25740 <=*>
1961 V3 N2 P38-40	61-27088 <=>
1961 V3 N2 P48-49	61-27088 <=>
1961 V3 N4 P12-15	64-19013 <=>
1961 V3 N5 P10-13	AD-617 66C <=$>
1962 N1 P14-	ICE V2 N3 P431-437 <ICE>
1962 N2 P13-17	ICE V2 N4 P580-584 <ICE>
1962 N3 P1-	ICE V2 N4 P536-538 <ICE>
1962 N3 P11-	ICE V2 N4 P546-549 <ICE>
1962 N5 P13-	ICE V3 N2 P185-190 <ICE>
1962 N6 P8-	ICE V3 N2 P215-219 <ICE>
1962 N6 P14-	ICE V3 N2 P220-225 <ICE>
1962 V4 N1 P7-9	65-10170 <*>
1962 V4 N3 P1-3	62-32209 <=>
1962 V4 N3 P23-25	AD-617 094 <=$>
1962 V4 N3 P38-40	62-32209 <=>
1962 V4 N3 P44-46	62-32209 <=>
1962 V4 N4 P22-25	64-23561 <=$>
1962 V4 N4 P35-37	65-60140 <=$>
1962 V4 N4 P39	65-60329 <=$>
1962 V4 N5 P1-3	63-13357 <=>
1962 V4 N5 P12	63-13357 <=>
1962 V4 N5 P17-20	63-19254 <=*>
1962 V4 N5 P42-44	63-13357 <=>
1962 V4 N6 P1-3	63-21218 <=>
1962 V4 N6 P5-6	63-21218 <=>
1962 V4 N6 P29-30	5844 <HB>
1962 V4 N6 P42-43	63-21218 <=>
1963 N2 P14-	ICE V3 N4 P455-459 <ICE>
1963 N3 P1-2	63-31344 <=>
1963 N3 P17-22	ICE V3 N4 P514-518 <ICE>
1963 N3 P22-25	ICE V3 N4 P519-522 <ICE>
1963 N3 P28-29	63-31344 <=>
1963 N3 P41-42	63-31344 <=>
1963 N3 P45-47	63-31344 <=>
1963 N4 P8-10	ICE V4 N1 P85-88 <ICE>
1963 N4 P19-23	ICE V4 N1 P88-92 <ICE>
1963 N4 P42-45	63-31857 <=>
1963 N5 P16-19	ICE V4 N1 P124-128 <ICE>
1963 N5 P19-23	ICE V4 N1 P114-118 <ICE>
1963 N5 P321-330	63-31725 <=>
1963 N5 P393	63-31725 <=>
1963 N6 P11-16	ICE V4 N2 P198-203 <ICE>
1963 N6 P20-22	ICE V4 N2 P189-192 <ICE>
1963 V5 N1 P1-5	63-21387 <=>
1963 V5 N1 P28-31	63-21387 <=>
1963 V5 N1 P38	63-21387 <=>
1963 V5 N1 P41	63-21387 <=>
1963 V5 N1 P43-45	63-21387 <=>
1963 V5 N2 P3-4	63-21843 <=>
1963 V5 N2 P44-46	63-21843 <=>
1963 V5 N4 P28-29	6109 <HB>
1963 V5 N4 P29-32	C-5161 <NRC>
1963 V5 N4 P35-36	6126 <HB>
1963 V5 N5 P28-30	6254 <HB>
1963 V5 N6 P11-16	ICE V4 N2 P198-203 <ICE>
1963 V5 N6 P20-22	ICE V4 N2 P189-192 <ICE>
1963 V5 N6 P27-31	64-31411 <=>
1963 V6 N3 P45-47	63-31344 <=>
1964 N2 P9-12	ICE V4 N3 P422-425 <ICE>
1964 N2 P17-21	ICE V4 N3 P487-491 <ICE>
1964 N3 P4-7	ICE V4 N4 P613-617 <ICE>
1964 V6 P17-21	ICE V4 N3 P487-491 <ICE>
1964 V6 N2 P9-12	ICE V4 N3 P422-425 <ICE>
1964 V6 N2 P30-32	AD-617 145 <=$>
1964 V6 N3 P4-7	ICE V4 N4 P613-617 <ICE>

★KHIMICHESKOE I NEFTYANOE MASHINOSTROENIE

1964 N2 P11-14	ICE V5 N1 P26-30 <ICE>
1964 N2 P15-20	ICE V5 N1 P95-101 <ICE>
1964 N3 P9-13	65-30038 <=$>
1964 N4 P35-37	AD-632 314 <=$>
1965 N1 P24-27	ICE V5 N3 P508-511 <ICE>
1965 N5 P20-25	ICE V6 N1 P35-40 <ICE>
1965 N5 P28-29	ICE V5 N4 P596-598 <ICE>
1965 N6 P15-18	ICE V6 N1 P47-50 <ICE>
1965 N6 P18-23	ICE V6 N1 P99-104 <ICE>
1965 N7 P4-	ICE V6 N1 P28-31 <ICE>
1965 N7 P33-34	4503 <BISI>
1965 N11 P11-13	ICE V6 N2 P285-287 <ICE>
1965 N11 P13-17	ICE V6 N2 P280-284 <ICE>
1965 N11 P23-25	ICE V6 N2 P308-310 <ICE>

KHIMICHNA PROMYSLOVIST UKRAINI (TITLE VARIES)

1963 N1 P43-44	66-11929 <*>
1963 N2 P3-5	63-31527 <=>
1963 N2 P52-56	ICE V4 N3 P525-529 <ICE>
	63R75V <ATS>
1963 N2 P56-59	63-31531 <=>
1963 N2 P70-72	63-31527 <=>
1963 N2 P76-78	63-31527 <=>
1963 N3 P46-48	64-21027 <=>
1963 N4 P20-24	AD-625 092 <=*$>
1964 N4 P7-8	AD-639 492 <=$>
1965 N4 P73-74	66-13953 <*$>

KHIMIYA I FIZIKO-KHIMIYA PRIRODNYKH I

SINTETICHESKIKH POLIMEROV
1962 N1 P23-28 AD-633 816 <=$>
1962 V1 P205-206 66-10337 <*>

★KHIMIYA GETEROTSIKLICHESKIKH SOEDINENII
1965 N1 P51-57 65-31202 <=$>
1965 N1 P148 1239 <TC>
1965 N2 P215-219 66-10275 <*>

KHIMIYA I INDUSTRIYA
1957 V29 P24-25 368 <TC>
1961 N6 P179- ICE V2 N3 P404-407 <ICE>
1961 V33 P171-175 72S83B <ATS>
1963 N2 P43-49 ICE V3 N4 P507-513 <ICE>
1963 N4 P133-137 ICE V4 N2 P192-197 <ICE>
1963 V35 N2 P65-68 64-10375 <*>
1963 V35 N4 P133-137 ICE V4 N2 P192-197 <ICE>
1964 N6 P207-209 ICE V5 N3 P415-417 <ICE>
1964 V35 N3 P81-82 64-41009 <=$>
1966 N7 P322-326 67-30951 <=$>

KHIMIYA I KHIMICHESKAYA TEKHNOLOGIYA
1963 V2 N3 P425-433 RTS-2662 <NLL>
1963 V6 N2 P233-235 65-63759 <=$>

KHIMIYA REDKIKH ELEMENTOV
1954 N1 P115-120 R-1374 <*>
1954 N1 P121-130 R-1372 <*>
1954 V1 P33-39 64-20856 <*> 0
1954 V1 P45-51 AEC-TR-4210 <*+>
1955 N2 P102-114 AEC-TR-4868 <=*>
1955 N2 P130-147 59-10051 <+*>
1955 V5 N2 P156-160 R-1784 <*>

KHIMIYA V SELSKOM KHOZYAISTVE
1964 N5 P8-11 64-31960 <=>
1964 N8 P2-6 64-51485 <=>
1964 N8 P35-68 64-51485 <=>
1964 V2 N7 P6-10 64-51876 <=$>
1964 V2 N7 P45-48 64-51876 <=$>
1964 V2 N7 P51-53 64-51876 <=$>
1964 V2 N7 P58-62 64-51876 <=$>
1964 V2 N8 P2-6 64-51485 <=>
1964 V2 N8 P35-68 64-51485 <=>
1964 V2 N9 P13-27 65-30509 <=$>
1964 V2 N9 P28-35 65-30570 <=$>
1964 V2 N9 P54-56 65-30509 <=$>
1964 V2 N9 P61-65 65-30509 <=$>
1964 V2 N11 P2-5 65-30384 <=$>
1964 V2 N11 P18-21 65-30384 <=$>
1964 V2 N11 P29-33 65-30384 <=$>
1964 V2 N11 P47-49 65-30384 <=$>
1965 V3 N5 P18-23 65-32075 <=*$>

KHIMIYA SERAOGANICHESKIKH SOEDINENII, SODERZHA-
SHCHIKHSYA V NEFTYAKH I NEFTEPRODUKTAKH.
AKADEMIYA NAUK SSSR. BASHKIRSKII FILIAL
1961 V4 P100-102 63-15522 <=*>
1961 V4 P231-235 63-15522 <=*>
1963 V5 P123-128 M-5578 <NLL>
1963 V5 P149-182 AD-611 071 <=$> M
1964 V6 P243-251 ICE V5 N1 P.62-67 <ICE>
ICE V5 N1 P62-67 <ICE>
1964 V6 P285-288 68S82R <ATS>
1964 V6 P301-307 ICE V5 N1 P22-26 <ICE>
1964 V7 P7-15 66-10982 <*>

KHIMIYA V SHKOLE
1954 V9 N3 P3-8 RT-2825 <*>
1963 V18 N3 P55-61 63-31528 <=>
1963 V18 N5 P18-27 64-21287 <=>
1965 V20 N6 P2-9 66-30657 <=$>

KHIMIYA I SOTSIALISTICHESKOE KHOZYAISTVO
1932 N8 P47-56 PANSDOC-TR.382 <PANS>

KHIMIYA I TEKHNOLOGIYA POLIMEROV
1960 V4 N7/8 P54-72 61-20639 <*=>
1960 V4 N7/8 P139-158 08Q67R <ATS>
1960 V4 N7/8 P139-157 61-16598 <*=>
1960 V4 N7/8 P174-186 61-16599 <*=>
1960 V4 N7/8 P196-221 61-23060 <=*>

KHIMIYA I TEKHNOLOGIYA TOPLIVA
1956 N2 P35-45 63K22R <ATS>
1956 N4 P13-19 86J19R <ATS>
1956 N4 P37-49 R-2001 <*>
53H12R <ATS>
1956 N4 P56-59 23H12R <ATS>
1956 N5 P43-47 RJ-578 <ATS>
1956 N5 P47-59 8K23R <ATS>
1956 N5 P64-69 RJ-577 <ATS>
1956 N6 P9-14 69H11R <ATS>
1956 N6 P14-20 70H11R <ATS>
1956 N6 P43-54 RJ-573 <ATS>
1956 N6 P57-60 71H11R <ATS>
72H11R <ATS>
1956 N8 P31-35 62K22R <ATS>
1956 V1 N1 P61-73 64-20204 <*> 0
1956 V1 N3 P54-58 59-22420 <+*>
1956 V1 N4 P13-19 60-13124 <+*>
1956 V1 N4 P37-49 61-15332 <*=>
1956 V1 N5 P1-7 59-14256 <+*>
1956 V1 N5 P47-59 62-19125 <*=>
64-20203 <*> 0
1956 V1 N6 P25-34 62-24665 <=*>
1956 V1 N7 P64-68 61-13163 <+*>
1956 V1 N8 P36-46 59-22519 <+*>
1956 V1 N9 P11-22 59-19180 <+*>
1956 V1 N10 P18-25 64-16562 <=*$>
65-11429 <*>
1956 V1 N12 P5-12 00M38R <ATS>
1956 V1 N12 P47-53 59-10662 <+*>

★KHIMIYA I TEKHNOLOGIYA TOPLIVA I MASEL
1957 N2 P18-24 92J19R <ATS>
1957 N2 P29-32 75K21R <ATS>
1957 N2 P52-56 4K22R <ATS>
1957 N3 P7-14 76K21R <ATS>
1957 N3 P19-24 77K21R <ATS>
1957 N3 P38-48 R-4782 <*>
1957 N3 P57-62 68J18R <ATS>
1957 N5 P19-27 93J19R <ATS>
1957 N5 P27-32 78K21R <ATS>
1957 N5 P32- R-3259 <*>
1957 N5 P40-48 79K21R <ATS>
1957 N6 P26-33 80K21R <ATS>
1957 N6 P66-68 63K20R <ATS>
1957 N7 P15-20 81K21R <ATS>
1957 N8 P51-57 R-4455 <*>
1957 N10 P6-9 72K21R <ATS>
1957 N10 P41-46 62K25R <ATS>
1957 N10 P66-71 73K21R <ATS>
1957 N10 P72- 83K21R <ATS>
1957 N11 P53-58 82K21R <ATS>
1957 N12 P38-44 R-5344 <*>
1957 N12 P52-56 61-17640 <=$>
1957 V2 N1 P3-13 62-25287 <=*>
1957 V2 N2 P3-11 59-10665 <+*>
1957 V2 N2 P25-29 74K21R <ATS>
1957 V2 N2 P33-40 61-10500 <+*>
1957 V2 N4 P38-41 59-19182 <+*>
1957 V2 N4 P47-53 65-10328 <*>
1957 V2 N5 P40-48 63-24568 <=*$>
1957 V2 N7 P39-42 60-15587 <+*>
1957 V2 N8 P1-4 57L30R <ATS>
1957 V2 N9 P41-49 59-19181 <+*>
1957 V2 N9 P63-66 60-15745 <+*>
1957 V2 N9 P66-70 61-27706 <=*>
1957 V2 N10 P41-46 60-17798 <+*>
1957 V2 N10 P72 60-23079 <+*>
1957 V2 N12 P1-12 60-19983 <*+>
1957 V2 N12 P45-52 59-18409 <+*>
1957 V2 N12 P64-69 60-15380 <+*>
1958 N3 P30-33 20K23R <ATS>
1958 N7 P66-70 63-00758 <*>
1958 V3 P48-52 66K23R <ATS>
1958 V3 N1 P9-17 96K24R <ATS>
1958 V3 N1 P40-46 59-17421 <+*>
1958 V3 N1 P47-52 57K23R <ATS>

Citation	Translation
1958 V3 N2 P22-28	60-19984 <*+>
1958 V3 N2 P34-41	95K24R <ATS>
1958 V3 N2 P41-46	60-18189 <+*>
	61-10489 <+*>
1958 V3 N2 P46-53	61-10488 <+*>
1958 V3 N3 P1-7	5K24R <ATS>
1958 V3 N3 P8-14	19K23R <ATS>
1958 V3 N3 P14-22	47K23R <ATS>
1958 V3 N3 P45-47	78K23R <ATS>
1958 V3 N3 P56-63	31K23R <ATS>
1958 V3 N3 P63-70	74K23R <ATS>
1958 V3 N3 P71-72	49K24R <ATS>
1958 V3 N4 P1-9	6K24R <ATS>
1958 V3 N4 P17-23	61-15377 <=*>
	7K24R <ATS>
1958 V3 N4 P23-25	78K25R <ATS>
1958 V3 N4 P39-46	59-16322 <+*>
1958 V3 N4 P46-51	60-23085 <+*>
	60K24R <ATS>
1958 V3 N4 P51-54	59-16478 <+*>
	60-23590 <+*>
1958 V3 N4 P55-61	8K24R <ATS>
1958 V3 N4 P62-66	61-15379 <+*>
	9K24R <ATS>
1958 V3 N5 P47-51	81K24R <ATS>
1958 V3 N6 P6-10	R-5237 <*>
1958 V3 N6 P52-57	60-14348 <+*>
1958 V3 N6 P65-70	74K25R <ATS>
1958 V3 N7 P1-8	61-13772 <*=>
1958 V3 N7 P8-14	60-15376 <+*>
1958 V3 N7 P45-53	60-10940 <+*>
1958 V3 N7 P53-62	90L32R <ATS>
1958 V3 N7 P62-66	61-19334 <+*>
1958 V3 N7 P66-70	61-15467 <+*>
1958 V3 N8 P9-15	59-22202 <+*>
1958 V3 N8 P38-44	57K27R <ATS>
	63-10400 <=*>
1958 V3 N8 P44-49	58K27R <ATS>
	59-10793 <+*>
	59-19447 <+*>
1958 V3 N8 P56-58	18L30R <ATS>
1958 V3 N8 P64-70	59K27R <ATS>
1958 V3 N9 P7-13	59-15009 <+*>
1958 V3 N9 P13-18	60-10077 <+*>
1958 V3 N9 P29-34	59-22416 <+*>
1958 V3 N9 P48-54	59-18018 <+*>
1958 V3 N10 P7-11	39L31R <ATS>
1958 V3 N10 P16-24	67L31R <ATS>
1958 V3 N10 P29-33	60-18916 <+*>
	96L28R <ATS>
1958 V3 N10 P44-49	41L31R <ATS>
1958 V3 N10 P50-55	08S81R <ATS>
	66-11316 <*>
1958 V3 N10 P64-69	52P63R <ATS>
1958 V3 N11 P10-12	60-10442 <+*>
	61-19317 <=*>
	93L29R <ATS>
1958 V3 N11 P13-15	94L29R <ATS>
1958 V3 N11 P15-20	34L36R <ATS>
1958 V3 N11 P23-27	59-19711 <+*>
1958 V3 N11 P33-35	53L30R <ATS>
1958 V3 N11 P62-66	60L29R <ATS>
1958 V3 N12 P9-15	44L30R <ATS>
	61-17600 <=$>
1958 V3 N12 P15-21	42L36R <ATS>
	61-13143 <+*>
1958 V3 N12 P22-26	60-13012 <+*>
1958 V3 N12 P27-32	45L30R <ATS>
1958 V3 N12 P32-36	46L30R <ATS>
1958 V3 N12 P41-46	99L33R <ATS>
1958 V3 N12 P46-48	12L31R <ATS>
1958 V3 N12 P52-55	65-14362 <*>
1959 V4 N1 P5-15	61-15256 <*+> O
1959 V4 N1 P15-19	63-16971 <=*> O
1959 V4 N1 P25-27	59-20295 <+*>
1959 V4 N1 P35-40	59-20294 <+*>
1959 V4 N1 P43-48	52L31R <ATS>
1959 V4 N1 P49-52	60-13797 <+*>
1959 V4 N1 P59-63	76L31R <ATS>
1959 V4 N1 P63-68	09L32R <ATS>
	61-13419 <*+>
1959 V4 N2 P18-24	61-17589 <=$>
1959 V4 N2 P25-27	59-18454 <+*> O
1959 V4 N2 P28-30	59-18431 <+*> O
1959 V4 N2 P71-72	59-18452 <+*>
1959 V4 N3 P10-12	63-15953 <=*>
1959 V4 N3 P17-20	60-14336 <+*>
1959 V4 N3 P30-32	96L33R <ATS>
1959 V4 N3 P60-64	07L35R <ATS>
1959 V4 N3 P64-68	47L33R <ATS>
1959 V4 N4 P7-12	59-19706 <+*> O
1959 V4 N4 P18-24	95L33R <ATS>
1959 V4 N4 P25-31	60-13538 <+*>
1959 V4 N4 P48-53	22L35R <ATS>
1959 V4 N5 P1-7	34N52F <ATS>
1959 V4 N5 P8-13	61-23284 <*=>
1959 V4 N5 P14-18	35N52R <ATS>
1959 V4 N5 P18-24	33N52R <ATS>
1959 V4 N5 P24-28	32N52R <ATS>
	60-13647 <+*>
	62-24854 <=*>
1959 V4 N5 P42-51	60-13537 <+*>
1959 V4 N6 P56-60	83M37R <ATS>
1959 V4 N7 P7-13	62-25268 <=*>
1959 V4 N7 P20-23	62-24850 <=*>
1959 V4 N8 P19-22	37L37R <ATS>
	60-16553 <+*>
1959 V4 N8 P22-25	64L36R <ATS>
1959 V4 N9 P8-15	60-11413 <=>
	61-28358 <*=>
1959 V4 N10 P10-16	33M38R <ATS>
1959 V4 N10 P28-31	61-27987 <*=>
1959 V4 N10 P31-34	21M40R <ATS>
1959 V4 N10 P34-	61-23268 <*=>
1959 V4 N10 P62-64	60-31038 <=>
	61-28335 <*=>
1959 V4 N11 P1-3	46M38R <ATS>
1959 V4 N11 P13-17	60-23000 <+*>
1959 V4 N11 P23-28	61-28338 <*=> O
	63-21183 <=>
1959 V4 N11 P36-38	60-16899 <*+>
1959 V4 N11 P39-41	60-16578 <+*>
1959 V4 N11 P41-48	A-126 <RIS>
	60-18067 <+*>
1959 V4 N12 P1-2	61-23234 <*=>
1959 V4 N12 P3-5	61-23235 <=*>
1959 V4 N12 P5-9	60-18324 <*> O
	61-23236 <=*>
1959 V4 N12 P9-13	61-23237 <=*>
1959 V4 N12 P12	61-27590 <*=>
1959 V4 N12 P14-19	61-23238 <=*>
1959 V4 N12 P19-22	61-23239 <=*>
1959 V4 N12 P22-24	61-23240 <=*>
1959 V4 N12 P25-29	61-23241 <=*>
	61-28943 <*=>
1959 V4 N12 P29-35	61-23242 <=*>
1959 V4 N12 P35-40	61-23243 <=*>
1959 V4 N12 P41-45	61-27593 <*=>
1959 V4 N12 P46-50	28Q71R <ATS>
	61-27592 <*=>
1959 V4 N12 P50-54	05M42R <ATS>
	61-27591 <*=>
1959 V4 N12 P54-59	61-19338 <+*>
	61-27590 <*=>
1959 V4 N12 P60-63	61-27589 <*=>
1959 V4 N12 P64-67	61-27588 <*=>
1959 V4 N12 P67-68	61-27587 <*=>
1960 N4 P66-69	NLLTB V3 N2 P79 <NLL>
1960 V5 N1 P8-12	20M40R <ATS>
1960 V5 N1 P32-35	06M42R <ATS>
	60-18739 <+*>
1960 V5 N1 P70-72	63-18698 <=*>
1960 V5 N2 P44-46	60-11732 <=>
1960 V5 N2 P52-54	A-125 <RIS>
	65-10074 <*>
1960 V5 N3 P1-5	54M44R <ATS>
1960 V5 N3 P17-21	62-01013 <*>
	64-23575 <=$>
1960 V5 N3 P22-28	60-18740 <+*>
1960 V5 N3 P28-35	29M43R <ATS>

1960 V5 N3 P35-42	12M43R <ATS>
	61-10365 <*+>
1960 V5 N3 P54-56	92M42R <ATS>
1960 V5 N4 P1-4	62-23365 <=*>
1960 V5 N4 P5-7	61-10364 <+*>
1960 V5 N4 P8-13	73M44R <ATS>
1960 V5 N4 P57-61	63-01692 <*>
1960 V5 N4 P66-69	61-21669 <=>
1960 V5 N5 P41-45	72M44R <ATS>
1960 V5 N6 P1-6	61-10363 <*+>
1960 V5 N6 P6-12	ICE V1 N1 P1-5 <ICE>
1960 V5 N6 P13-17	60-10362 <*+>
1960 V5 N6 P24-28	70Q73R <ATS>
1960 V5 N7 P24-29	96N49R <ATS>
1960 V5 N7 P33-38	06M46R <ATS>
	61-10726 <+*>
	61-23518 <*=>
1960 V5 N7 P52-55	61-10725 <+*>
1960 V5 N8 P15-21	25M47R <ATS>
1960 V5 N8 P29-34	61-10724 <+*>
1960 V5 N8 P63-66	61-27575 <*=>
1960 V5 N9 P57-61	61-27531 <*=>
1960 V5 N9 P71-72	61-11554 <=>
1960 V5 N10 P35-38	61-27706 <=*>
1960 V5 N1C P38-41	61-28176 <*=>
1960 V5 N11 P4-8	ICE V1 N1 P55-57 <ICE>
1960 V5 N11 P15-22	71N48R <ATS>
1960 V5 N11 P49-53	61-21549 <=>
1960 V5 N11 P57-64	61-28180 <*=>
	93N53R <ATS>
1960 V5 N11 P64-70	61-21599 <=>
1960 V5 N12 P18-24	02N51R <ATS>
1961 N2 P66-70	AIAAJ V1 N8 P2002 <AIAA>
1961 N4 P19-23	ARSJ V32 N11 P1800 <AIAA>
1961 N8 P1-	ICE V2 N1 P114-119 <ICE>
1961 V6 N1 P28-31	64-10971 <=*$> 0
1961 V6 N1 P52-54	96N51R <ATS>
1961 V6 N1 P54-57	59N51R <ATS>
1961 V6 N2 P25-28	13N53R <ATS>
1961 V6 N2 P46-52	62-13502 <=*> 0
1961 V6 N2 P66-70	AIAA V1 N08 <AIAA>
	62-11659 <=>
1961 V6 N3 P7-10	63-19954 <=*>
1961 V6 N3 P66-67	19N53R <ATS>
1961 V6 N4 P6-9	63-10240 <=*>
1961 V6 N4 P9-	ICE V2 N2 P189-192 <ICE>
1961 V6 N4 P9-14	81N57R <ATS>
1961 V6 N4 P23-27	88N54R <ATS>
1961 V6 N4 P27-31	87N54R <ATS>
1961 V6 N4 P56-57	62-14857 <=*>
1961 V6 N4 P64-67	86N54R <ATS>
1961 V6 N4 P70-72	03N56R <ATS>
1961 V6 N5 P11-14	62-16125 <=*>
1961 V6 N5 P15-17	76N56R <ATS>
1961 V6 N5 P17-21	53N56R <ATS>
	62-14845 <=*>
1961 V6 N5 P48-53	61-27706 <=*>
1961 V6 N6 P17-21	19N56R <ATS>
1961 V6 N6 P21-26	20N56R <ATS>
1961 V6 N6 P55-60	54N56R <ATS>
	62-16120 <=*>
1961 V6 N6 P61-66	62-16124 <=*>
	98N56R <ATS>
1961 V6 N7 P13-15	62-23376 <=*>
1961 V6 N7 P16-	ICE V2 N2 P164-168 <ICE>
1961 V6 N7 P16-21	17N57R <ATS>
1961 V6 N7 P56-61	62-33072 <=*>
1961 V6 N7 P62-63	62-25280 <=*>
1961 V6 N7 P63-70	79N56R <ATS>
1961 V6 N8 P47-53	62-19133 <=*>
1961 V6 N9 P32-37	42N58R <ATS>
1961 V6 N9 P51-54	43N58R <ATS>
1961 V6 N9 P65-69	62-18381 <=*>
	64-00085 <*>
1961 V6 N11 P23-27	92P58R <ATS>
1961 V6 N11 P55-59	62-25391 <=*>
1961 V6 N11 P64-66	65-13666 <*>
1961 V6 N12 P1-7	94P61R <ATS>
1961 V6 N12 P7-11	62-33626 <=*>
1961 V6 N12 P11-14	63-14325 <=*>
1961 V6 N12 P39-43	62-33626 <=*>
	63-10232 <=*>
	64-00249 <*>
1961 V6 N12 P53-57	63-20256 <=*>
1962 N1 P25-	ICE V2 N3 P360-364 <ICE>
1962 N1 P45-	ICE V2 N3 P353-356 <ICE>
1962 N3 P6-	ICE V2 N4 P501-503 <ICE>
1962 N4 P41-	ICE V2 N4 P539-543 <ICE>
1962 N5 P1C-	ICE V3 N1 P45-47 <ICE>
1962 N6 P41-	ICE V2 N4 P590-596 <ICE>
1962 N6 P61-64	65-14033 <*>
1962 N7 P38-	ICE V3 N2 P153-156 <ICE>
1962 N9 P1-	ICE V3 N3 P342-346 <ICE>
1962 N10 P1-	ICE V3 N2 P264-268 <ICE>
1962 N10 P9-	ICE V3 N2 P272-276 <ICE>
1962 N12 P1-	ICE V3 N2 P269-272 <ICE>
1962 V7 N1 P25-31	07P61R <ATS>
1962 V7 N1 P45-50	72P60R <ATS>
1962 V7 N1 P53-59	51P60R <ATS>
1962 V7 N2 P37-40	62-23836 <=*>
1962 V7 N3 P23-26	61P63R <ATS>
1962 V7 N3 P58-60	61Q71R <ATS>
	63-18411 <*>
1962 V7 N3 P61-64	62Q70R <ATS>
	63-18415 <=*>
1962 V7 N3 P64-66	AD-616 312 <=$>
1962 V7 N4 P11-14	62-32825 <=*>
1962 V7 N4 P15-21	59P63R <ATS>
1962 V7 N4 P21-24	60P63R <ATS>
1962 V7 N4 P53-56	63-26584 <=$>
	64-13000 <=*$>
1962 V7 N4 P60-66	63-26230 <=$>
1962 V7 N5 P10-13	56P63R <ATS>
1962 V7 N5 P23-26	10S88R <ATS>
	63-19972 <=*>
1962 V7 N5 P32-34	57P63R <ATS>
	63-19212 <=*>
1962 V7 N5 P43-49	58P63R <ATS>
1962 V7 N5 P70-73	27P63R <ATS>
1962 V7 N6 P61-64	63P65R <ATS>
1962 V7 N7 P38-43	28P65R <ATS>
	63-18421 <=*>
1962 V7 N8 P1-6	63-18887 <=*>
1962 V7 N8 P39-42	64-15274 <=*$>
1962 V7 N8 P53-58	63-18405 <=*>
1962 V7 N8 P58-66	63-26590 <=$>
	64-13001 <=*$>
1962 V7 N9 P1-8	00Q68R <ATS>
1962 V7 N10 P20-26	37Q68R <ATS>
	63-26583 <=*>
	65-13667 <*>
1962 V7 N10 P32-36	83Q67R <ATS>
1962 V7 N10 P56-59	AD-625 149 <=*$>
1962 V7 N11 P40-45	63-20257 <=*$>
1962 V7 N11 P66-67	89Q72R <ATS>
1962 V7 N12 P1-5	63-27415 <=$>
1962 V7 N12 P26-28	64-14255 <=*$>
	65-18195 <*>
1962 V7 N12 P44-47	64-10688 <=*$>
1962 V7 N12 P44-46	64-19857 <=$>
1962 V7 N12 P59-64	66Q70R <ATS>
1963 N1 P19-24	ICE V3 N4 P527-530 <ICE>
1963 N2 P10-14	ICE V4 N1 P1-4 <ICE>
1963 N3 P7-	ICE V3 N3 P375-378 <ICE>
1963 N6 P17-21	ICE V4 N2 P226-229 <ICE>
1963 N6 P46-62	ICE V3 N4 P523-527 <ICE>
1963 N8 P17-19	ICE V4 N1 P47-49 <ICE>
1963 N8 P19-23	ICE V4 N1 P40-42 <ICE>
1963 N10 P6-9	ICE V4 N3 P434-436 <ICE>
1963 N11 P11-15	ICE V4 N2 P216-219 <ICE>
1963 N11 P15-20	ICE V4 N2 P302-306 <ICE>
1963 V8 N1 P4-10	64-10689 <=*$>
1963 V8 N1 P19-24	38Q69R <ATS>
1963 V8 N1 P24-27	63-27101 <=$>
1963 V8 N1 P34-38	63-27100 <=$>
1963 V8 N2 P6-9	64-18434 <*>
1963 V8 N2 P10-14	18Q71R <ATS>
	66-14570 <*>
1963 V8 N2 P52-56	11S88R <ATS>

Citation	Reference
1963 V8 N3 P15-19	71R76R <ATS>
1963 V8 N4 P16-20	61R77R <ATS>
1963 V8 N4 P64-65	AD-611 529 <=$>
1963 V8 N5 P18-22	65-18196 <*>
1963 V8 N5 P34-38	64-11847 <=> M
1963 V8 N5 P42-46	60Q72R <ATS>
1963 V8 N5 P55-57	AD-625 143 <=*$>
1963 V8 N6 P12-16	64-14953 <=*$>
1963 V8 N6 P17-21	ICE V4 N2 P226-229 <ICE>
	01Q74R <ATS>
1963 V8 N6 P52-57	63-27423 <$>
	64-13002 <=*$>
	64-26099 <$>
1963 V8 N6 P60-65	AD-611 531 <=$>
1963 V8 N7 P58-62	64-71514 <=> M
1963 V8 N7 P64-67	13Q74R <ATS>
1963 V8 N7 P68-69	14Q74R <ATS>
1963 V8 N8 P49-54	AD-612 781 <=$>
1963 V8 N9 P11-16	66-60463 <=$>
1963 V8 N9 P16-20	64-16969 <=*$>
1963 V8 N9 P23-27	64-13709 <=*$>
1963 V8 N9 P31-38	64-71423 <=> M
1963 V8 N10 P6-9	ICE V4 N3 P434-436 <ICE>
	34R75R <ATS>
1963 V8 N10 P22-26	64-11578 <=>
1963 V8 N10 P48-53	64-26432 <$>
1963 V8 N11 P4-10	35R75R <ATS>
	65-11965 <*>
1963 V8 N11 P11-15	ICE V4 N2 P216-219 <ICE>
	42R74R <ATS>
	64-26320 <$>
1963 V8 N11 P15-20	ICE V4 N2 P302-306 <ICE>
	49R74R <ATS>
	66-10044 <*>
1963 V8 N11 P30-35	66-10045 <*>
1963 V8 N11 P52-57	64-15875 <=*$>
1963 V8 N11 P57-61	AD-608 734 <=>
1963 V8 N12 P7-12	83R75R <ATS>
1963 V8 N12 P62-65	AD-625 155 <=*$>
1964 N3 P22-26	ICE V5 N1 P12-16 <ICE>
1964 N6 P5-10	ICE V4 N4 P561-564 <ICE>
1964 N6 P14-18	ICE V4 N4 P606-608 <ICE>
1964 N6 P52-57	ICE V5 N3 P397-401 <ICE>
1964 N7 P8-12	ICE V5 N3 P402-404 <ICE>
1964 N8 P58-60	AD-633 714 <=$>
1964 N9 P1-7	ICE V5 N2 P208-212 <ICE>
1964 N9 P38-40	AD-631 448 <=$>
1964 N9 P53-56	AD-635 838 <=$>
1964 N12 P11-15	ICE V5 N3 P477-480 <ICE>
1964 V9 N1 P1-7	24S83R <ATS>
1964 V9 N1 P27-32	06R78R <ATS>
1964 V9 N1 P32-38	AD-617 947 <=$>
1964 V9 N1 P44-47	64-16695 <=*$>
1964 V9 N2 P5-11	94R78R <ATS>
1964 V9 N2 P22-28	66-10040 <*>
1964 V9 N2 P27-29	72R76R <ATS>
1964 V9 N2 P70-72	1017 <TC>
1964 V9 N3 P1-7	65-11964 <*>
1964 V9 N3 P22-26	ICE V5 N1 P12-16 <ICE>
	07R78R <ATS>
	65-13022 <*>
1964 V9 N3 P36-40	AD-620 077 <=$>
1964 V9 N3 P54-58	AD-629 415 <=$>
1964 V9 N3 P58-62	AD-640 307 <=$>
1964 V9 N4 P22-26	AD-618 045 <=$>
1964 V9 N4 P49-50	AD-633 849 <=$>
1964 V9 N4 P57-60	AD-641 107 <=$>
1964 V9 N4 P66-69	AD-618 045 <=$>
	47R77R <ATS>
1964 V9 N5 P6-12	64-26693 <$>
1964 V9 N5 P13-16	65-12452 <*>
1964 V9 N5 P61-65	64-27228 <$>
1964 V9 N5 P69-70	93R78R <ATS>
1964 V9 N6 P5-10	ICE V4 N4 P561-564 <ICE>
	65-27894 <=$>
1964 V9 N6 P14-18	ICE V4 N4 P606-608 <ICE>
1964 V9 N6 P48-52	68R78R <ATS>
1964 V9 N6 P52-57	06R79R <ATS>
	65-10468 <*>
1964 V9 N7 P6-8	65-11492 <*>
1964 V9 N7 P8-12	68R79R <ATS>
1964 V9 N7 P24-28	65-27896 <=$>
1964 V9 N7 P34-35	65-27898 <$>
1964 V9 N7 P36-39	65-11491 <*>
1964 V9 N7 P59-65	AD-636 596 <=$>
1964 V9 N8 P9-13	65-27895 <$>
1964 V9 N8 P50-53	AD-621 797 <=*$>
1964 V9 N8 P60-65	AD-621 797 <=*$>
	RTS-3039 <NLL>
1964 V9 N9 P18-22	65-13312 <*>
1964 V9 N9 P26-29	65-14286 <*>
1964 V9 N10 P14-19	65-13313 <*>
1964 V9 N10 P44-48	AD-618 011 <=$>
1964 V9 N11 P53-61	65-20551 <$>
1964 V9 N12 P11-15	19S82R <ATS>
1964 V9 N12 P39-43	67S82R <ATS>
1964 V9 N12 P44-47	AD-622 474 <=$*>
1964 V9 N12 P51-56	66-12776 <*>
1965 N1 P29-32	66-10646 <*>
1965 N1 P40-45	ICE V5 N4 P642-647 <ICE>
1965 N6 P1-4	66-12216 <*>
1965 N6 P56-60	66-10645 <*>
1965 N8 P25-27	66-12210 <*>
1965 N9 P14-17	ICE V6 N2 P239-241 <ICE>
1965 N9 P38-42	ICE V6 N2 P217-220 <ICE>
1965 N11 P41-45	AD-634 791 <=$>
1965 V10 P36-40	66-10735 <*>
1965 V10 N1 P2-6	77S86R <ATS>
1965 V10 N1 P12-16	85S86R <ATS>
1965 V10 N1 P17-20	81S86R <ATS>
1965 V10 N1 P27-29	75S86R <ATS>
1965 V10 N1 P29-32	83S86R <ATS>
1965 V10 N1 P40-45	ICE V6 N2 P267-271 <ICE>
	49S86R <ATS>
1965 V10 N2 P19-20	RTS-3210 <NLL>
1965 V10 N2 P47-48	29T90R <ATS>
1965 V10 N2 P48-52	59S88R <ATS>
1965 V10 N3 P14-16	76S86R <ATS>
	954 <TC>
1965 V10 N3 P19-22	08S86R <ATS>
1965 V10 N3 P23-25	ICE V6 N2 P257-259 <ICE>
	73S84R <ATS>
1965 V10 N3 P29-33	93S87R <ATS>
1965 V10 N3 P55-57	TRC-TRANS-1109 <NLL>
	65-20017 <$>
1965 V10 N3 P60-62	48S86R <ATS>
1965 V10 N4 P1-5	74S86R <ATS>
1965 V10 N4 P11-15	65-18062 <*>
	80S86R <ATS>
1965 V10 N4 P31-35	65-18063 <*>
	78S86R <ATS>
1965 V10 N4 P39-43	01S87R <ATS>
1965 V10 N4 P59-61	86S86R <ATS>
1965 V10 N5 P12-15	28S87R <ATS>
1965 V10 N5 P19-23	29S87R <ATS>
1965 V10 N5 P38-40	RTS-3432 <NLL>
1965 V10 N5 P49-52	AD-626 963 <=$>
1965 V10 N6 P13-18	30S87R <ATS>
1965 V10 N6 P34-37	66-13937 <*$>
1965 V10 N6 P56-60	60S88R <ATS>
1965 V10 N7 P32-35	1624 <TC>
1965 V10 N7 P41-45	02S88R <ATS>
1965 V10 N7 P63	30T90R <ATS>
1965 V10 N8 P19-24	ICE V6 N2 P242-246 <ICE>
	39S88R <ATS>
1965 V10 N9 P33-36	14T91R <ATS>
1965 V10 N9 P42-46	66-12219 <*> 0
1965 V10 N9 P53-57	66-14343 <*$>
1965 V10 N10 P27-29	97T90R <ATS>
1965 V10 N10 P46-50	98T90R <ATS>
1965 V10 N11 P50-52	AD-638 670 <=$>
1965 V10 N12 P15-18	96T91R <ATS>
1965 V10 N12 P24-31	70T92R <ATS>
1966 V11 N1 P54-57	79T92R <ATS>
1966 V11 N2 P38-43	66-23831 <=>
1966 V11 N2 P54-57	66-14518 <*>
1966 V11 N3 P9-13	90T94R <ATS>
1966 V11 N5 P5-8	52T95R <ATS>

KHIMIYA TVERDOGO TOPLIVA

1933 V4 P171-185	R-5134 <*>
1934 V5 P632-641	61-16608 <=*$>
1935 V6 P78-82	61-16606 <=*$>
1935 V6 P221-235	61-16628 <=*$>
1935 V6 P640-647	61-16609 <=*$>
1935 V6 P656-662	1978 <*>
1936 V7 P56-59	61-18081 <=*$>
1936 V7 P282-298	61-16624 <=*$>
1936 V7 P890-902	61-16611 <=*$>
1937 V8 P232-247	60-15295 <*+>
1937 V8 N10 P866-875	R-2106 <*>
	59-14126 <+*>
	62-20122 <=*>
1937 V8 N12 P1122-1137	R-2104 <*>

KHIMIYA I ZHIZN

1965 N4 P43-52	65-32508 <=$>

KHIMIZATSIYA SOTSIALISTICHESKOGO ZEMLEDELIYA

1934 N3 P78-82	R-2887 <*>
1936 N5 P3-16	R-2918 <*>
1937 V6 N9 P76-86	R-3320 <*>
1939 N10/1 P59-63	R-2251 <*>
1939 V8 N1 P33-43	R-2880 <*>
1940 V9 N5 P54-55	R-3312 <*>

KHIMSTROI

1933 V5 P2470-2473	R-2462 <*>
1933 V5 N2 P2053-2054	RJ-19 <ATS>
1933 V5 N43 P2038-2040	R-2808 <*>
1934 V6 P8-14	64-10760 <=*$> 0
1934 V6 P121-125	R-2463 <*>
1934 V6 P341-343	R-2505 <*>
1934 V6 P599-602	64-10759 <=*$> 0
1935 V7 P136-145	R-2280 <*>
1935 V7 P150-158	R-2305 <*>
1935 V7 P158-162	R-2464 <*>

KHIRURGIYA. MOSKVA

1937 N12 P61-63	R-1102 <*>
1938 N9 P117-119	R-1059 <*>
1938 N10 P6-23	RT-641 <*>
1938 V4 P45-49	R-1095 <*>
1939 N1 P30-32	R-1070 <*>
1940 N7 P3-13	R-1060 <*>
1943 N5/6 P15-22	RT-3867 <*>
1944 N2 P44-49	65-63250 <=>
1945 N6 P73-79	RT-1313 <*>
1945 N6 P79-82	RT-1844 <*>
1945 N6 P83-86	RT-1802 <*>
1947 N4 P53-56	60-23233 <+*>
1948 N8 P20-26	RT-370 <*>
1948 N11 P29-31	R-151 <*>
1948 V18 N8 P26-29	RT-1202 <*>
1949 N4 P22-26	RT-2700 <*>
1949 N8 P42-49	61-28100 <*=>
1950 N2 P80-	62-15301 <*=> P
1951 N2 P31-36	RJ-173 <ATS>
1951 N2 P31-37	RJ-173 <ATS>
1952 N5 P52-54	61-28157 <*=>
1952 N9 P11-17	59-12173 <+*>
1952 N9 P70-72	<CP>
1953 N3 P15-25	RT-3809 <*>
1953 N9 P81-84	RT-2534 <*>
1953 N10 P90-93	RT-2626 <*>
1954 N2 P65-68	RT-2473 <*>
1954 N9 P37-42	R-3223 <*>
1954 V30 N9 P37-42	64-19416 <=$> P
1955 N9 P48-54	R-707 <*>
	60-23182 <*+>
1955 N9 P89-93	R-703 <*>
1956 V32 N1 P24-31	62-15107 <=*> 0
1956 V32 N4 P6-9	62-15146 <=*>
1956 V32 N4 P17-24	62-15142 <*=> 0
1956 V32 N4 P33-41	62-15145 <*=> 0
1956 V32 N8 P80-86	62-23531 <=*> 0
1957 N10 P34-41	PB 141 288T <=>
1957 N11 P60-62	59-11607 <=>
1958 V34 N3 P66-71	63-16897 <=*> 0
	8921-A <K-H>
1959 N6 P3-12	59-11773 <=>
1959 N7 P3-20	59-13919 <=>
1959 N7 P16-20	60-15158 <+*>
1959 N7 P38-50	59-13919 <=>
1959 N7 P112-113	59-13919 <=>
1959 N7 P116-117	59-13919 <=>
	59-13924 <=>
1959 N11 P140-148	60-11264 <=>
1959 V35 N9 P33-37	62-15325 <=*>
1959 V35 N9 P133-138	60-31426 <=>
1960 N1 P133-137	60-11679 <=>
1960 N1 P137-139	60-31113 <=>
1960 V36 N2 P90-94	62-23571 <=*>
1960 V36 N5 P30-34	60-41370 <=>
1960 V36 N5 P77-81	60-41371 <=>
1960 V36 N5 P81-84	60-41372 <=>
1960 V36 N5 P85-87	60-41373 <=>
1960 V36 N5 P87-96	60-41374 <=>
1960 V36 N5 P96-104	60-41375 <=>
1960 V36 N5 P104-106	60-41376 <=>
1960 V36 N6 P93-97	60-41521 <=>
1960 V36 N6 P98-104	60-41522 <=>
1960 V36 N8 P3-10	61-11504 <=>
1960 V36 N8 P11-20	61-11097 <=>
1960 V36 N8 P31-36	61-21121 <=>
1960 V36 N8 P36-42	61-11086 <=>
1960 V36 N8 P42-49	61-11083 <=>
1960 V36 N8 P62-69	61-11088 <=>
1960 V36 N8 P85-91	61-11079 <=>
1960 V36 N9 P8-13	61-11896 <=>
1960 V36 N9 P68-72	61-21067 <=>
1960 V36 N9 P72-80	61-21142 <=>
1960 V36 N9 P81-86	61-21183 <=>
1960 V36 N9 P87-91	61-21181 <=>
1960 V36 N10 P8-15	61-21186 <=>
1960 V36 N10 P15-20	61-21148 <=>
1960 V36 N10 P21-32	61-11902 <=>
1960 V36 N10 P32-39	61-21042 <=>
1960 V36 N10 P39-42	61-21049 <=>
1960 V36 N10 P43-46	61-21093 <=>
1960 V36 N10 P54-61	61-21096 <=>
1960 V36 N10 P61-68	61-21171 <=>
1960 V36 N10 P72-79	61-21134 <=>
1960 V36 N10 P84-92	61-11930 <=>
1960 V36 N10 P93-100	61-21135 <=>
1960 V36 N10 P112-116	61-21190 <=>
1960 V36 N11 P28-34	61-21674 <=>
1960 V36 N11 P34-39	61-21678 <=>
1960 V36 N11 P40-42	61-21663 <=>
1960 V36 N11 P43-48	61-21567 <=>
1960 V36 N11 P48-54	61-21582 <=>
1960 V36 N11 P55-59	61-21583 <=>
1960 V36 N11 P82-86	61-21589 <=>
1960 V36 N11 P112-117	61-21587 <=>
1960 V36 N11 P117-122	61-21613 <=>
1960 V36 N12 P5-10	61-21416 <=>
1960 V36 N12 P11-18	61-21491 <=>
1960 V36 N12 P19-23	61-21432 <=>
1960 V36 N12 P24-27	61-21489 <=>
1960 V36 N12 P28-33	61-21490 <=>
1960 V36 N12 P50-52	65-13869 <*>
1960 V36 N12 P128-132	61-21744 <=>
1961 V37 N1 P16-25	61-21705 <=>
1961 V37 N1 P39-45	61-31150 <=>
1961 V37 N1 P53-58	61-31151 <=>
1961 V37 N1 P128-130	61-23008 <=*>
1961 V37 N3 P3-9	61-27223 <*=>
1961 V37 N3 P9-16	61-31530 <=>
1961 V37 N5 P7-12	61-27754 <*=>
1961 V37 N5 P12-15	61-27857 <*=>
1961 V37 N5 P70-75	61-27856 <*=>
1961 V37 N5 P113-115	61-27951 <*=>
1961 V37 N6 P37-44	61-31616 <=>
1961 V37 N6 P125-128	61-31617 <=>
1961 V37 N7 P39-46	62-13035 <=*>
1961 V37 N7 P130-132	62-13034 <=*>
1961 V37 N7 P145-147	62-13021 <*=>
1961 V37 N8 P28-30	62-13861 <=>
1961 V37 N8 P64-71	62-13861 <=>

1961 V37 N8 P75-79 62-13861 <=>
1961 V37 N8 P87-93 62-13861 <=>
1961 V37 N8 P121-130 62-13861 <=>
1961 V37 N11 P3-20 62-24786 <=*>
1961 V37 N11 P154-158 62-24786 <=*>
1961 V37 N12 P3-10 62-33038 <=*>
1961 V37 N12 P119-121 62-19753 <=*>
1962 V38 N1 P45-50 63-19367 <=*>
1962 V38 N6 P57-62 62-32577 <=*>
1962 V38 N7 P101-106 62-33009 <=*>
1962 V38 N8 P3-10 62-33311 <=*>
1962 V38 N10 P6-40 63-13657 <=>
1962 V38 N10 P117-119 63-13657 <=>
1962 V38 N11 P20-30 63-13633 <=>
1962 V38 N11 P81-100 63-13633 <=>
1962 V38 N12 P68-71 63-21127 <=>
1963 V39 P135-140 65-13833 <*>
1963 V39 N2 P137-141 63-21557 <=>
1963 V39 N2 P144-148 64-13203 <=*$>
1963 V39 N5 P6-18 63-31241 <=>
1963 V39 N5 P25-29 63-31241 <=>
1963 V39 N5 P36-40 63-31241 <=>
1963 V39 N5 P77-80 63-31241 <=>
1963 V39 N5 P98-111 63-31241 <=>
1963 V39 N5 P118-121 63-31241 <=>
1963 V39 N7 P13-19 63-41135 <=>
1963 V39 N7 P25-33 63-31737 <=>
1963 V39 N7 P39-43 63-31737 <=>
1963 V39 N7 P43-49 63-41135 <=>
1963 V39 N7 P80-82 63-31737 <=>
1963 V39 N7 P106-112 63-31737 <=>
1963 V39 N7 P115-122 63-31737 <=>
1963 V39 N9 P3-9 63-41223 <=>
1963 V39 N9 P10-14 63-41222 <=>
1963 V39 N9 P14-18 63-41221 <=>
1963 V39 N9 P19-22 63-41220 <=>
1963 V39 N11 P135-140 65-13443 <*> 0
1963 V39 N12 P29-46 64-21854 <=>
1964 V40 N1 P3-11 64-21862 <=>
1964 V40 N1 P93-98 65-60225 <=$>
1964 V40 N2 P6-11 65-60181 <=$>
1964 V40 N2 P81-86 64-21882 <=>
1964 V40 N2 P121-127 64-21882 <=>
1964 V40 N3 P9-15 64-31150 <=>
1964 V40 N3 P30- FPTS V24 N2 P.T295 <FASE>
1964 V40 N3 P30-38 FPTS V24 P.T295-T299 <FASE>
1964 V40 N3 P114-117 64-31151 <=>
1964 V40 N3 P135-140 64-31149 <=>
1964 V40 N5 P20-28 64-31858 <=>
1964 V40 N5 P52-58 64-31858 <=>
1964 V40 N5 P64-72 65-61109 <=$>
1964 V40 N5 P75-83 64-31858 <=>
1964 V40 N5 P149-153 66-10864 <*>
1964 V40 N7 P18-23 64-51217 <=$>
1964 V40 N7 P44-49 64-51217 <=$>
1964 V40 N7 P65-68 65-61220 <=$>
1964 V40 N7 P68-75 65-61221 <=$>
1964 V40 N8 P111-119 64-41904 <=$>
1964 V40 N9 P19-24 65-64517 <=*$>
1964 V40 N9 P64-67 65-62834 <=$>
1964 V40 N11 P80-85 65-63577 <=$>
1964 V40 N11 P103-108 65-63578 <=$>
1965 V41 N2 P77-80 65-30643 <=$>
1965 V41 N12 P3-9 66-30702 <=$>
1966 V42 N2 P3-8 66-31476 <=$>

KHIRURGIYA. SOFIA
1961 V14 P307-310 66-12165 <*>
1962 V15 N12 P1118-1122 63-21465 <=>

KHLEBOPEKARNAYA I KONDITERSKAYA PROMYSHLENNOST
1957 V1 N4 P12-14 63-16349 <=*>
1958 V2 N3 P13-19 66-12678 <*>
1958 V2 N6 P12-14 63-18052 <=*> 0
1959 V3 N9 P36-37 63-18034 <=*>
1959 V3 N11 P32-37 61-19108 <+*>
1960 V4 N5 P8-10 63-23402 <=*$>
1961 V5 N11 P5-9 CSIR-TR-432 <CSSA>
1963 V7 N10 P25-28 RTS-2780 <NLL>
1964 V8 N7 P12-16 M-5600 <NLL>
1965 V9 N4 P15-18 M-5735 <NLL>

KHLOPCHATO-BUMAZHNAYA PROMYSHLENNOST
1939 V9 N8/9 P27-31 61-15475 <=*> 0

KHLOPKOVODSTVO
1954 V4 N2 P42-46 RT-3327 <*>
1955 V5 N10 P38-40 RT-3889 <*>
1960 N7 P14-19 61-11958 <=>
1960 N9 P7-11 61-21244 <=>
1960 N10 P4-12 61-21533 <=>
1960 N11 P4-10 61-21721 <=>
1960 N11 P13-16 61-21721 <=>
1961 N2 P34-36 61-27130 <=>
1962 V12 N11 P59-60 65-61449 <=>
1963 V13 P42-44 66-60948 <=*$>

KHOA HOC THUAT
1962 N4 P22-25 62-32569 <=*>

KHOLODILNAYA PROMYSHLENNOST
1937 V15 N5 P19-26 TT-367 <NRC>

KHOLODILNAYA TEKHNIKA
1952 N3 P24-29 RT-3579 <*>
1952 V29 N3 P48-54 RT-3188 <*>
1952 V29 N3 P61-62 RT-3514 <*>
1953 V30 N1 P8-14 RT-2153 <*> 0
1953 V30 N1 P28-31 RT-3398 <*>
1953 V30 N1 P32-36 RT-2154 <*>
1953 V30 N1 P60-61 64-20471 <*> 0
1953 V30 N4 P53-55 RT-3317 <*>
1954 V31 N1 P33-38 RT-3260 <*>
1954 V31 N2 P45-51 R-3947 <*>
1954 V31 N2 P57-59 R-3922 <*>
1954 V31 N2 P60-61 R-4559 <*>
1954 V31 N2 P62-65 R-673 <*>
UCRL TRANS-280(L) <*>
1954 V31 N2 P68- R-4591 <*>
1954 V31 N2 P74-76 59-15152 <+*>
1954 V31 N3 P21-26 R-4577 <*>
1954 V31 N3 P27-34 59-15038 <+*>
1954 V31 N3 P38-44 R-4599 <*>
1955 V32 N3 P56-57 R-3616 <*>
1956 V33 N1 P62-63 R-440 <*>
1956 V33 N3 P55-61 R-1641 <*>
1957 V34 N3 P55-61 61-20878 <=*>
1957 V34 N4 P17-21 1757 <TC>
21S88R <ATS>
1958 V35 N1 P53-56 59-22422 <+*>
1958 V35 N2 P20-26 61-23091 <=*>
1958 V35 N4 P48-52 59-19598 <+*>
60-15426 <+*>
1959 N1 P52-55 NLLTB V1 N11 P18 <NLL>
1959 N2 P47-50 NLLTB V1 N8 P29 <NLL>
1959 N6 P17-19 NLLTB V2 N5 P364 <NLL>
1959 V36 N3 P51-54 60-19930 <+*> 0
1959 V36 N4 P10-14 02Q71R <ATS>
1959 V36 N4 P29-33 62-32935 <=*>
1959 V36 N5 P43-48 61-27610 <*=>
1959 V36 N6 P6-12 03Q71R <ATS>
1960 V37 N1 P20-24 62-23548 <=*>
1960 V37 N1 P39-41 61-15012 <+*>
1960 V37 N2 P47-49 61-15305 <+*>
1960 V37 N2 P77-79 60-41248 <=>
1960 V37 N3 P35-38 62-15819 <*=>
1960 V37 N4 P23-25 N66-11612 <=$>
1961 V38 N1 P1-4 62-23471 <=>
1961 V38 N1 P11-15 62-23471 <=>
1961 V38 N1 P35-38 62-23714 <=*>
1961 V38 N1 P64-65 61-27398 <*=>
1961 V38 N1 P66-68 62-23471 <=>
1961 V38 N1 P71 61-27398 <*=>
1961 V38 N3 P4-7 62-25731 <=*>
1961 V38 N3 P7-10 62-24624 <=*>
1961 V38 N3 P18-23 64-19957 <=$>
1961 V38 N4 P24-27 AEC-TR-5105 <=*>
1961 V38 N4 P39-42 <TIB>
1961 V38 N4 P66-68 62-23800 <=>
1961 V38 N5 P59-60 64-13737 <=*$>

1961 V38 N6 P31-33 63-15974 <=*>
1962 N6 P1-8 NLLTB V5 N6 P512 <NLL>
1962 V39 N1 P4-7 63-13238 <=*>
1962 V39 N1 P37-40 63-13237 <=*>
1962 V39 N1 P63-64 62-32571 <=>
1962 V39 N1 P68-70 63-15309 <=*>
1962 V39 N2 P54 63-13296 <=>
1962 V39 N3 P74-75 62-32565 <=*>
1962 V39 N4 P22-27 63-11576 <=> 0
63-23987 <=*>
1962 V39 N4 P34-39 64-19389 <=$>
1962 V39 N5 P35-36 28R80R <ATS>
64-10384 <=*$>
1963 V40 N1 P37-40 64-15364 <=*$>
1964 N4 P10-15 65-64103 <=*$>
1964 N4 P45-51 AD-631 769 <=$>
1964 N5 P28-30 NLLTB V7 N8 P706 <NLL>
1964 V40 N5 P28-30 TRANS.BULL.V7N8 <NLL>
1964 V41 N3 P45-47 AD-639 341 <=$>
1965 V42 N4 P5-6 65-33475 <=$>
1966 V43 N6 P18-20 66-34647 <=$>

KHRANITELNA PROMISHLENOST. SOFIA
1964 V13 N8/9 P1-4 64-51734 <=$>

KIANGSI MEDICAL JOURNAL. CHINESE PEOPLES REPUBLIC
SEE CHIANG HSI YAO TSA CHIH

KIANGSU TRADITIONAL MEDICINE
SEE CHIANG SU CHUNG I

KIKAI NO KENKYU
1956 V8 N3 P317-322 03K27J <ATS>
1956 V8 N4 P431-435 04K27J <ATS>
1956 V8 N5 P521-527 05K27J <ATS>
1956 V8 N6 P624-628 06K27J <ATS>
1958 V10 N5 P629-632 99K26J <ATS>
1958 V10 N6 P742-747 11K27J <ATS>

KIKAI TO KOGU
1958 P14-16 59-11368 <=> 0
1958 P43-48 59-11368 <=> 0
1958 P49-52 59-11368 <=> 0
1958 P59-66 59-11368 <=> 0

KINDERAERZTLICHE PRAXIS
1936 V7 P539-542 62-01237 <*>
1954 V22 N1 P1-4 59-18076 <=*>
1955 V23 N3 P132-134 II-915 <*>
1963 V31 P433-438 66-10956 <*>

★KINETIKA I KATALIZ
1960 V1 N1 P5-14 62-20381 <=*>
1960 V1 N1 P15-31 61-27601 <*=>
62-20398 <=*>
1960 V1 N1 P32-44 62-16620 <=*>
1960 V1 N1 P45-54 62-16621 <=*>
1960 V1 N1 P63-68 61-27600 <*=>
1960 V1 N1 P83-94 61-27599 <*=>
1960 V1 N1 P125-128 65M46R <ATS>
1960 V1 N1 P172-175 60-41478 <=>
1960 V1 N2 P221-228 61-28463 <*=>
1960 V1 N2 P229-236 61-27598 <*=>
66-13791 <*>
1960 V1 N2 P237-241 61-27597 <*=>
1960 V1 N2 P242-246 61-27596 <*=>
1960 V1 N2 P257-259 61-27602 <*=>
1960 V1 N2 P267-273 61-27595 <*=>
1960 V1 N3 P356-364 61-23088 <*=>
1960 V1 N3 P365-373 62-24960 <=*>
1960 V1 N3 P374-378 61-23640 <=*>
62-16566 <=*>
1960 V1 N3 P379-384 62-18965 <=*>
1960 V1 N3 P421-430 61-23658 <=*>
1960 V1 N3 P431-439 61-23659 <=*>
1961 N6 P838- ICE V2 N3 P425-430 <ICE>
1961 V2 P192-196 63-14319 <=*>
1961 V2 N1 P38-43 61-20740 <*=> 0
1961 V2 N1 P103-111 ICE V2 N1 P6-11 <ICE>
1961 V2 N1 P154-159 61-28576 <*=>
1961 V2 N2 P267-272 63-10228 <=*>
1961 V2 N2 P295-296 62-11081 <=>
1961 V2 N2 P297-302 61-27397 <*=>
1961 V2 N3 P305- ICE V2 N2 P227-235 <ICE>
1961 V2 N3 P319-339 62-18378 <=*>
1961 V2 N3 P340-349 62-11664 <=>
1961 V2 N3 P365-367 N66-21690 <=$>
1961 V2 N3 P374-377 62-16561 <=*>
1961 V2 N3 P386-393 66-14156 <*$>
1961 V2 N3 P408-417 86P61R <ATS>
1961 V2 N3 P429-434 62-16565 <=*>
1961 V2 N3 P435-439 62-16564 <=*>
1961 V2 N3 P440-445 62-16563 <=*>
1961 V2 N3 P454-461 62-16562 <=*>
1961 V2 N4 P481-489 63-10225 <=*>
1961 V2 N4 P507-508 63-10231 <=*>
1961 V2 N4 P525-528 63-10239 <=*>
64P59R <ATS>
1961 V2 N4 P562-566 41N58R <ATS>
1961 V2 N4 P584-589 63-10226 <=*>
1961 V2 N4 P637-638 62-23992 <=*>
1961 V2 N5 P643-647 NLLTB V4 N4 P356 <NLL>
62-24711 <=*>
1961 V2 N5 P803-805 62-23996 <=*>
1961 V2 N6 P847-853 63-10856 <=*>
1961 V2 N6 P940-941 62-11680 <=>
1962 V3 N1 P45-53 65-14097 <*>
1962 V3 N1 P114-117 63-18420 <*=>
1962 V3 N1 P139-144 63-18407 <=*>
1962 V3 N1 P145-153 62-18771 <=*>
65-14097 <*>
1962 V3 N2 P181-188 63-18840 <=*>
1962 V3 N2 P261-270 63-18838 <=*>
1962 V3 N2 P271-275 63-18837 <=*>
1962 V3 N3 P343-352 66-14350 <*$>
1962 V3 N3 P353-357 63-18404 <=*>
1962 V3 N3 P358-363 63-18406 <=*>
1962 V3 N3 P364-365 27P64R <ATS>
63-11627 <=>
1962 V3 N3 P421-426 28P64R <ATS>
1962 V3 N3 P427-430 63-18841 <=*>
1962 V3 N3 P445-448 42P65R <ATS>
1962 V3 N4 P509-517 63-18839 <=*>
1962 V3 N4 P518-519 63-18333 <=*>
1962 V3 N4 P529-540 65-13314 <*>
1962 V3 N4 P541-544 37P66R <ATS>
1962 V3 N4 P545-549 63-18832 <=*>
1962 V3 N5 P627-642 63-21244 <=>
1962 V3 N5 P661-673 13Q67R <ATS>
63-18408 <=*>
1962 V3 N5 P680-690 63-23824 <=*>
1962 V3 N5 P742-746 65-13315 <*>
1962 V3 N6 P870-876 57Q69R <ATS>
1962 V3 N6 P894-906 63-18485 <=*> 0
1962 V3 N6 P927-930 64-10103 <=*$>
1962 V3 N6 P937-941 64-13073 <=*$>
1962 V4 N3 P481-492 74P65R <ATS>
1963 N4 P508-516 ICE V4 N3 P467-472 <ICE>
1963 N4 P605-613 ICE V4 N1 P141-146 <ICE>
1963 V4 P244-251 63-18751 <=*>
1963 V4 N1 P53-59 64-14450 <=*$>
1963 V4 N1 P88-96 64-71336 <=> M
1963 V4 N3 P348-352 64-30136 <*>
1963 V4 N3 P467-474 02582R <ATS>
1963 V4 N4 P508-516 ICE V4 N3 P467-472 <ICE>
69R74R <ATS>
1963 V4 N4 P625-634 70R74R <ATS>
1963 V4 N4 P648-651 64-18778 <=*$>
1963 V4 N5 P688-697 68R74R <ATS>
1963 V4 N5 P779-782 64-18768 <*>
66-10008 <*>
1963 V4 N6 P823-828 66-10003 <*>
1963 V4 N6 P835-843 ANL-TRANS-39 <=>
1963 V4 N6 P930-932 20R75R <ATS>
1964 N4 P706-715 ICE V5 N3 P512-517 <ICE>
1964 V5 P350-354 65-13972 <*>
1964 V5 N1 P49-59 65-13029 <*>
1964 V5 N1 P144-153 65-13414 <*>
1964 V5 N1 P171-174 65-13415 <*>
1964 V5 N1 P192-193 65-13307 <*>

1964 V5 N1 P195-196 65-14726 <*>
1964 V5 N2 P201-210 N65-15888 <= $>
1964 V5 N2 P215-220 65-64113 <=*$>
1964 V5 N2 P347-350 65-10905 <*>
1964 V5 N2 P355-356 66-13792 <*>
1964 V5 N3 P388-398 TRC-TRANS-1101 <NLL>
65-28881 <$>
1964 V5 N3 P399-406 TRC-TRANS-1100 <NLL>
65-28880 <$>
65-63653 <=$>
1964 V5 N3 P434-440 65-12433 <*>
1964 V5 N3 P441-445 65-11642 <*>
1964 V5 N3 P460-468 66-10985 <*>
1964 V5 N3 P490-495 65-12453 <*>
1964 V5 N4 P609-615 65-14681 <*>
1964 V5 N4 P624-629 65-13609 <*>
1964 V5 N4 P649-657 67S83R <ATS>
1964 V5 N4 P706-715 32S81R <ATS>
1964 V5 N4 P751-752 65-13316 <*>
1965 V6 N1 P3-16 65-14747 <*>
1965 V6 N1 P23-30 AD-628 306 <=*$>
1965 V6 N1 P89-94 65-18054 <*>
1965 V6 N1 P121-127 65-18058 <*>
1965 V6 N2 P244-249 66-10368 <*>
1965 V6 N2 P313-319 65-18057 <*>
1965 V6 N2 P357-360 65-18353 <*>
1965 V6 N2 P365 65-17443 <*>
1965 V6 N3 P439-447 65-18055 <*>
1965 V6 N3 P559-562 66-12212 <*>
1965 V6 N4 P625-633 66-12211 <*>
1965 V6 N5 P864-868 66-11663 <*>
1965 V6 N5 P884-888 62S89R <ATS>
1965 V6 N5 P889-896 66-11273 <*>
1965 V6 N5 P909-915 66-11734 <*>
1965 V6 N5 P928-931 TRC-TRANS-1119 <NLL>
1965 V6 N6 P1121-1122 N66-23548 <=$>
1966 V7 N1 P21-26 66-24705 <$>

KINO-FOTO-KHIMICHESKAYA PROMYSHLENNOST
1937 V2 N5 P51-61 87L36R <ATS>

KINOMEKHANIK
1959 N4 P30-31 60-14173 <+*>
60-15098 <+*> 0
1960 N5 P19-22 61-11674 <=>
1961 N9 P21-29 53P62R <ATS>
1962 P21-23 63-21542 <=>
1962 N9 P21-22 63-13295 <=>
1963 N1 P38-39 63-21553 <=>
1963 N1 P40 63-21553 <=>
1963 N5 P28 63-31606 <=>
1963 N11 P22-25 64-18537 <*>

KINOTECHNIK. BERLIN
1959 N4 P25-29 62-10591 <*>
1961 V15 P255-256 66-14146 <*> 0
1961 V15 P259 66-14146 <*>
1964 N12 P294-298 65-12719 <*>
1964 V18 N4 P87-90 82R78G <ATS>
1964 V18 N4 P92 82R78G <ATS>
1964 V18 N4 P93-97 81R78G <ATS>

KINOTECHNIK. (DEUTSCHE KINOTECHNISCHE GESELLSCHAFT)
1930 V12 N15 P407-409 57-2494 <*>
1935 V17 P167-172 2000 <*>
1940 N22 P601 60-10196 <*>

KINZOKU
1951 V21 N2 P135-138 69P61J <ATS>
1955 V25 N5 P347-350 66-12592 <*>
1959 V29 P923-928 AEC-TR-5533 <*>
1960 V30 P28-32 2250 <BISI>
1963 V33 P36-45 3866 <BISI>

KINZOKU BUTSURI
1955 V1 N2 P69-70 57-2437 <*>
1957 V2 N2 P490-502 62-23213 <=>
1958 V4 N5 P193-290 59-15579 <*>
1958 V4 N5 P211-214 59-15578 <*>

KINZOKU HYOMEN GIJUTSU
1959 V10 N11 P408-412 83M45J <ATS>
1960 V11 P10-14 2137 <BISI>
1960 V11 N2 P41-44 63-18496 <*>
1964 V14 N12 P485-492 4537 <BISI>
1965 V16 N5 P210-214 66-11555 <*>

KINZOKU NO KENKYU
1934 V11 P305-316 2057 <*>
1934 V11 P549-560 II-520 <*>
64-14153 <*> 0
1934 V11 N6 P305-316 64-10444 <*> 0
1937 V14 P46-59 72G7J <ATS>

KISEICHUGAKU ZASSHI
1955 V4 N1 P12-18 59-16301 <=*> 0
1955 V4 N3 P258-261 59-16574 <=*>
1955 V4 N4 P388-393 59-14279 <+*> 0
1958 V7 N1 P51-55 59-14242 <=*> 0

KISERLETES ORVOSTUDOMANY
1951 V4 P60-65 62-00492 <*>
1952 V4 N1 P6-17 59-12872 <=*> 0
1957 V9 P172-178 63-00075 <*>
1958 V10 N1 P35-40 62-01207 <*>
1960 V12 N3 P296-305 62-00168 <*>
1962 V14 P269-272 12455 <K-H>
1962 V14 P298-305 63-00917 <*>

KISHO SHUSHI
1946 V24 P21-23 66-11228 <*>
1952 S2 V30 N11 P1-11 64-16995 <*>
1952 S2 V30 N11 P345-355 64-16995 <*>
1953 S2 V31 N6 P25-37 64-16994 <*>
1953 S2 V31 N6 P219-231 64-16994 <*>
1955 V33 N5 P217-219 66-11197 <*>
1961 S2 V39 N1 P29-44 C-4163 <NRCC>
1961 S2 V39 N5 P269-281 DSIS-61-210 <NRC>

KISHO TO TOKEI, OYO SUIKEIGAKU ZASSHI
1959 V10 N1 P23-31 60-21726 <=>

KISHOCHO KENKYU JIHO
1957 V9 P436-440 SCL-T-250 <=*>
1957 V9 N9 P652-655 SCL-T-231 <=*>
1957 V9 N11 P811-822 SCL-T-241 <=*>
1958 V10 P353-357 64-15677 <=*$>
1958 V10 P1053-1060 64-15675 <=*$>
1958 V10 N4 P358-360 64-15679 <=*$>
1958 V10 N11 P971-975 65-13688 <*> 0
1960 V12 N8 P509-517 C-3918 <NRCC>
1962 V14 P80-93 64-15678 <=*$>

KISLOROD
1944 N1 P17-31 RT-3062 <*>
1944 N1 P31-38 22F4R <ATS>
1944 N2 P1-12 RT-2843 <*>
1944 N2 P12-21 RT-2411 <*>
1944 N2 P21-29 RT-2412 <*>
1944 N2 P37-44 RT-2810 <*>
1944 N2 P45-53 RT-3001 <*>
1944 N3 P11-16 RT-3004 <*>
1944 N3 P20-28 RT-2900 <*>
65-10135 <*> 0
1944 N3 P39-43 RT-3006 <*>
1944 N4 P1-10 RT-3886 <*>
1944 N4 P10-16 RT-2989 <*>
1944 N4 P16-27 RT-3161 <*>
1944 N4 P27-38 RT-3002 <*>
1944 N4 P51-62 RT-3017 <*>
1944 V1 N1 P17-31 61-23300 <*>
1944 V1 N1 P26-31 R-1910 <*>
1944 V1 N3 P11-16 R-1908 <*>
1944 V1 N3 P43-51 R-749 <*>
1945 N1 P16-25 2025 <HB>
1945 V2 N1 P35-47 R-1907 <*>
1946 V3 N1 P1-14 61-23308 <*=>
1946 V3 N1 P12-14 R-1877 <*>
1946 V3 N2/3 P34-41 1952 <HB>

Citation	Translation
1946 V3 N2/3 P50-53	1969 <HB>
1946 V3 N2/3 P56-60	RT-3371 <*>
1947 N1 P5-14	771TM <CTT>
1947 V4 P5-14	R-1765 <*>
1947 V4 N4 P14-26	RT-1579 <*>
1947 V4 N5 P26-34	62-20158 <=*>
1947 V6 N194 P22-23	149TM <CTT>
1957 V10 N2 P27-33	61-23245 <=*>
1957 V10 N4 P28	60-13836 <+*>
1957 V10 N6 P1-11	TS-1282 <BISI>
	62-24220 <=*>
1958 V11 N3 P11-18	4459 <HB>
1958 V11 N3 P19-28	60-21718 <=> O
1958 V11 N4 P1-11	TS-1424 <BISI>
	62-25333 <=*>
1958 V11 N5 P1-10	59-11689 <=>
1958 V11 N5 P21-28	25L29R <ATS>
1958 V11 N5 P36-38	16L29R <ATS>
1958 V11 N5 P39-47	61-13069 <=*>
1958 V11 N5 P62-64	17L29R <ATS>
1958 V11 N5 P65-68	59-22612 <+*>
1959 V12 N1 P23-35	14L33R <ATS>
1959 V12 N1 P26	13L33R <ATS>
1959 V12 N2 P9-14	59-13985 <=>
1959 V12 N2 P37-38	16L34G <ATS>
1959 V12 N2 P38	61-15235 <*+>
1959 V12 N2 P39-44	59-13985 <=>
1959 V12 N2 P54	61-13757 <=*>
1959 V12 N3 P12-16	60-23040 <*+>
1959 V12 N3 P51-52	59-13985 <=>
1959 V12 N5 P1-6	57M37R <ATS>
1959 V12 N5 P58-59	60-31161 <=>
1959 V12 N6 P46-49	E-753 <RIS>

KITA-NIHON BYOGAICHU KENKYUKAI NENPO

Citation	Translation
1964 V15 P8-9	66-13966 <*$>
1964 V15 P82-83	66-13365 <*>
1964 V15 P86-87	66-13364 <*>

KITAI. PEKING

Citation	Translation
1963 N5 P6-9	63-31359 <=*>

KLEI EN KERAMIEK (TITLE VARIES)

Citation	Translation
1956 V6 N8 P331-376	CSIRO-3785 <CSIR>

KLEPZIG'S FACHBERICHTE

Citation	Translation
1960 P341-343	62-14530 <*>
1962 V70 P133-138	3545 <BISI>
1962 V70 P391-395	4044 <BISI>
1962 V70 P403-407	3316 <BISI>

KLIMAT I POGODA

Citation	Translation
1936 V65 N2 P12-14	RT-2308 <*>
1958 N1 P108-118	61-13313 <=$>
1958 N1 P162-171	61-13313 <=$>

KLIMAT SSSR

Citation	Translation
1959 N8 P1-220	AD-613 988 <=$> M
1961 V2 ENTIRE ISSUE	64-71329 <=> M
1962 N4 P1-360	AD-609 914 <=>

KLINCHESKAYA KHIRURGIYA

Citation	Translation
1962 N5 P3-22	62-33283 <=*>
1962 N5 P86	62-33283 <=*>
1962 N6 P47-52	62-32541 <=*>
1962 N9 P20-26	63-13504 <=*>
1962 V41 N1 P61-63	62-23817 <=*>
1962 V41 N2 P24-27	62-24405 <=*>
1962 V41 N4 P31-35	62-32074 <=*>

★KLINICHESKAYA MEDITSINA

Citation	Translation
1927 V5 P433-437	60-19607 <+*>
1938 V16 P386-	R-1054 <*>
1938 V16 N12 P1750-1756	65-60064 <=$>
1939 V17 N9/10 P90-93	RT-155 <*>
	65-60101 <=$>
1940 N5 P117-122	RT-1303 <*>
1940 V18 P91-98	R-1101 <*>
1940 V18 P117-122	65-63454 <=>
1940 V18 N7 P42-52	R-1096 <*>
1941 V19 P5-16	R-1090 <*>
1941 V19 N3 P17-25	R-1097 <*>
1942 V20 N10 P59-67	65-62962 <=>
	65-63248 <=>
1942 V20 N5/6 P83-87	65-60098 <=$>
1943 V21 N1/2 P3-6	RT-1818 <*>
1944 V22 N4 P44-52	RT-580 <*>
1944 V22 N4 P64-67	R-1091 <*>
1944 V22 N4 P64-66	RT-657 <*>
1944 V22 N4 P4452-	R-1065 <*>
1944 V22 N1/2 P25-31	RT-1825 <*>
1944 V22 N1/2 P36-42	RT-1830 <*>
1944 V22 N10/1 P30-34	RT-510 <*>
1946 V24 N6 P63-67	RT-157 <*>
1946 V24 N6 P67-69	RT-156 <*>
1948 V26 N11 P80-85	RT-4427 <*>
1949 V27 N8 P42-48	RT-158 <*>
1949 V27 N9 P24-29	60-10062 <+*>
1950 N4 P39-45	R-3839 <*>
1950 V28 N2 P3-10	61-28687 <*=>
1950 V28 N3 P15-24	AD-610 895 <=$>
	RT-1707 <*>
1950 V28 N6 P72-74	62-18254 <=*>
1951 V29 N7 P60-65	RT-2230 <*> P
	64-15579 <=*$> O
1952 V30 P44-53	66-60613 <=*$>
1952 V30 N2 P76-77	R-719 <RIS>
1952 V30 N6 P86-87	60-23193 <+*>
1952 V30 N7 P91	64-15070 <=*$>
1952 V30 N8 P76-78	R-3805 <*>
1952 V30 N10 P66-70	R-720 <RIS>
1953 V31 N2 P60-64	RT-3084 <*>
1953 V31 N5 P79-81	RT-3301 <*>
1953 V31 N5 P81-85	RT-3307 <*>
1953 V31 N8 P118-184	60-10855 <+*> O
1953 V31 N11 P44-48	RT-2524 <*>
	64-15380 <=*$>
1953 V31 N12 P57-61	RT-2525 <*>
	64-15371 <=*$>
1954 V32 P68-72	R-3802 <*>
1954 V32 N6 P9-20	60-23683 <+*> O
1954 V32 N9 P49-52	59-19198 <+*> O
1955 V33 N1 P88-89	62-15314 <=*>
1955 V33 N2 P38-46	62-15311 <=*> O
1955 V33 N3 P42-45	R-146 <*>
1955 V33 N6 P24-27	AEC-TR-3603 <+*>
	R-3547 <*>
1955 V33 N6 P37-40	AEC-TR-3604 <+*>
	R-3548 <*>
1955 V33 N6 P41-43	R-3549 <*>
1955 V33 N6 P44-48	RT-4434 <*>
1955 V33 N9 P12-17	59-10988 <+*>
	62-11146 <=>
1955 V33 N11 P54-56	R-579 <*>
1955 V33 N12 P3-12	62-23530 <=*>
1956 N8 P3-12	R-630 <*>
1956 V34 N1 P35-42	62-15141 <=*>
1956 V34 N3 P79-82	62-15313 <=*>
1956 V34 N4 P35-38	62-15335 <=*>
1956 V34 N6 P12-24	62-19809 <=*>
1956 V34 N6 P57-65	62-19809 <=*>
1956 V34 N6 P65-69	61-28684 <*=>
1956 V34 N7 P70-77	62-15734 <=*>
1957 V35 P21-34	65-63142 <=$>
1957 V35 N2 P3-15	59-11077 <=>
1957 V35 N7 P14-23	62-23547 <=*>
1957 V35 N7 P133-135	65-63688 <=$>
1957 V35 N8 P82-93	PB-141 225T <=> O
1957 V35 N9 P54-61	62-15108 <=*>
1957 V35 N9 P108-114	59-10342 <+> O
1957 V35 N9 P124-131	62-15956 <=*>
1957 V35 N10 P92-99	62-15084 <=*>
1957 V35 N11 P3	59-11507 <=>
1957 V35 N11 P5-12	PB 141 312T <=>
1957 V35 N11 P12-25	PB 141 313T <=>
1957 V35 N11 P31-36	PB 141 314T <=>
1958 V36 N5 P121-127	59-18853 <+*>
1958 V36 N9 P25-29	61-10871 <+*> O
1958 V36 N10 P460-462	63-10385 <*> O
1959 V37 N1 P3-12	59-13518 <=>

1959 V37 N1 P12-18 59-13519 <=>
1959 V37 N2 P3-7 59-13562 <=>
1959 V37 N4 P5-11 59-13622 <=>
1959 V37 N4 P11-18 59-13623 <=>
1959 V37 N4 P19-25 59-13624 <=>
1959 V37 N5 P28-32 62-15618 <=*>
1959 V37 N6 P6-14 60-31095 <=>
1959 V37 N7 P3-6 59-13954 <=>
1959 V37 N7 P7-11 63-16896 <=*>
9874-B <K-H>
1959 V37 N8 P14-19 59-31021 <=>
1959 V37 N9 P11-15 60-11072 <=>
1959 V37 N9 P104-110 62-25553 <=*>
1959 V37 N12 P3-12 60-31067 <=>
1960 V38 P122-124 66-11722 <=$>
1960 V38 N1 P10-18 60-31090 <=>
1960 V38 N1 P27-30 60-31104 <=>
1960 V38 N1 P103-107 63-19684 <=*>
1960 V38 N2 P143-145 62-15679 <=>
1960 V38 N5 P3-5 60-41686 <=>
1960 V38 N5 P20-26 60-41136 <=>
1960 V38 N5 P27-33 60-41687 <=>
1960 V38 N5 P34-36 60-41136 <=>
1960 V38 N5 P109-116 61-11014 <=>
1960 V38 N5 P121-126 60-41701 <=>
1960 V38 N5 P139-145 61-11076 <=>
1960 V38 N5 P149-151 61-11715 <=>
1960 V38 N6 P19-24 60-41381 <=>
1960 V38 N6 P41-44 60-41382 <=>
1960 V38 N6 P45-50 60-41383 <=>
1960 V38 N6 P50-53 60-41384 <=>
1960 V38 N6 P53-60 60-41388 <=>
1960 V38 N6 P60-66 60-41389 <=>
1960 V38 N6 P71-76 60-41392 <=>
1960 V38 N6 P95-100 60-41390 <=>
1960 V38 N6 P104-107 60-41391 <=>
1960 V38 N7 P3-12 60-41528 <=>
1960 V38 N7 P13-21 60-41529 <=>
1960 V38 N7 P36-43 60-41530 <=>
1960 V38 N8 P68-72 61-11530 <=>
1960 V38 N8 P72-74 61-11601 <=>
1960 V38 N8 P93-99 61-11630 <=>
1960 V38 N8 P137-138 61-11575 <=>
1960 V38 N9 P6-12 61-11531 <=>
1960 V38 N9 P43-47 61-11796 <=>
1960 V38 N9 P48-53 61-11795 <=> 0
1960 V38 N9 P90-95 61-21098 <=> 0
1960 V38 N9 P96-102 61-21207 <=>
1960 V38 N9 P137-139 61-21069 <=>
1960 V38 N9 P139 61-21102 <=>
1960 V38 N10 P105-109 10764-E <KH>
61-11836 <=>
1960 V38 N12 P24-29 61-21418 <=>
1960 V38 N12 P29-33 61-21420 <=>
1960 V38 N12 P34-38 61-21419 <=>
1960 V38 N12 P39-42 61-21383 <=>
1960 V38 N12 P125-128 61-21342 <=>
1961 V39 N1 P26-30 61-21659 <=>
1961 V39 N2 P3-12 61-23614 <=*>
1961 V39 N2 P148-151 61-23571 <=*>
1961 V39 N3 P3-11 61-27093 <*=>
1961 V39 N3 P12-18 65-13343 <*>
66-12404 <*>
1961 V39 N3 P19-29 61-27071 <*=>
1961 V39 N3 P94-99 61-27113 <*=>
1961 V39 N4 P41-48 61-27066 <*=>
1961 V39 N4 P123-127 61-31511 <=>
1961 V39 N5 P22-26 61-31553 <=>
1961 V39 N5 P65-73 61-31553 <=>
1961 V39 N5 P155-156 61-31553 <=>
1961 V39 N6 P7-10 61-31587 <=>
1961 V39 N6 P130-137 61-28057 <*=>
1961 V39 N7 P3-5 62-13647 <=>
1961 V39 N7 P110-124 62-13647 <=>
1961 V39 N8 P36-41 62-13549 <*=>
1961 V39 N8 P57-65 62-13518 <*=>
1961 V39 N8 P66-68 62-13552 <*=>
1961 V39 N9 P5-11 62-15943 <=*>
1961 V39 N10 P126-134 62-11165 <=>
1961 V39 N11 P43-56 62-23621 <=*>
1961 V39 N12 P132-138 62-23321 <=*>
1961 V40 N4 P37-41 63-23912 <=*$> 0
1962 V40 P752-754 63-10928 <*>
1962 V40 N1 P3-14 62-25475 <=*>
1962 V40 N1 P5-14 62-24018 <=*>
1962 V40 N1 P91-93 62-24018 <=*>
1962 V40 N2 P14-20 62-24404 <=*>
1962 V40 N2 P112- FPTS V22 N1 P.T105 <FASE>
1962 V40 N2 P119- FPTS V22 N1 P.T110 <FASE>
1962 V40 N3 P7-19 62-24721 <=*>
1962 V40 N3 P126- FPTS V22 N1 P.T138 <FASE>
1962 V40 N3 P126-129 62-24721 <=*>
1962 V40 N3 P137-143 63-10188 <*>
1962 V40 N5 P3-18 62-32887 <=*>
1962 V40 N5 P77-85 63-19015 <=*>
1962 V40 N5 P85- FPTS V22 N2 P.T353 <FASE>
1962 V40 N5 P96- FPTS V22 N2 P.T216 <FASE>
1962 V40 N6 P39- FPTS V22 N2 P.T383 <FASE>
1962 V40 N7 P3-14 62-33522 <=*>
1962 V40 N7 P15- FPTS V22 N3 P.T466 <FASE>
1962 V40 N7 P62-65 63-19369 <=*>
1962 V40 N7 P88-94 63-19358 <=*>
1962 V40 N8 P65- FPTS V22 N4 P.T664 <FASE>
1962 V40 N8 P73-77 63-19366 <=*>
1962 V40 N8 P96-100 63-23906 <=*>
1962 V40 N8 P107- FPTS V22 N3 P.T540 <FASE>
1962 V40 N8 P123-127 63-23009 <=*>
1962 V40 N9 P35-40 63-23010 <=*>
1962 V40 N9 P67-71 63-23665 <=*>
1962 V40 N9 P75-81 63-23012 <=*>
1962 V40 N9 P82-89 63-23013 <=*>
1962 V40 N9 P106- FPTS V22 N5 P.T869 <FASE>
1962 V40 N9 P106-111 1963,V22,N5,PT2 <FASE>
1962 V40 N9 P121-129 63-23014 <=*>
1962 V40 N9 P129-135 63-23015 <=*>
1962 V40 N10 P26-35 63-21052 <=>
1962 V40 N10 P33-35 63-23040 <=*$>
1962 V40 N10 P36- FPTS V22 N5 P.T861 <FASE>
1962 V40 N10 P36-42 1963,V22,N5,PT2 <FASE>
1962 V40 N10 P82-86 63-23041 <=*>
1962 V40 N10 P86-94 63-23042 <=*>
1962 V40 N11 P33-39 63-23052 <=*>
1962 V40 N11 P79- FPTS V22 N4 P.T778 <FASE>
1962 V40 N12 P71-75 63-24473 <=*$>
1962 V40 N12 P87 FPTS V22 N6 P.T1169 <FASE>
1962 V40 N12 P87-92 FPTS,V22,P.T1169-72 <FASE>
1962 V40 N12 P93-100 63-23865 <=*>
1962 V40 N12 P100- FPTS V22 N6 P.T1124 <FASE>
1962 V40 N12 P104-106 63-21071 <=>
1963 V41 N1 P19-24 63-23833 <=*$>
1963 V41 N1 P61-66 63-23866 <=*>
1963 V41 N1 P100-104 63-23867 <=*>
1963 V41 N2 P3-13 63-21458 <=>
1963 V41 N2 P30- FPTS V23 N2 P.T332 <FASE>
FPTS,V23,P.T332-T333 <FASE>
1963 V41 N2 P70- FPTS V23 N1 P.T194 <FASE>
FPTS,V23,P.T194-T196 <FASE>
1963 V41 N2 P78-83 64-13526 <=*$>
1963 V41 N2 P83- FPTS V23 N1 P.T7 <FASE>
1963 V41 N2 P83 FPTS,V23,N1,P.T7-T11 <FASE>
1963 V41 N2 P116- FPTS V23 N1 P.T5 <FASE>
1963 V41 N2 P116 FPTS,V23,P.T5-T6 <FASE>
1963 V41 N3 P3-10 63-21905 <=>
1963 V41 N3 P126-131 64-13463 <=*$>
1963 V41 N4 P59- FPTS V23 N2 P.T392 <FASE>
FPTS,V23,P.T392-T396 <FASE>
1963 V41 N5 P19-23 64-13848 <=*$>
1963 V41 N5 P45- FPTS V23 N2 P.T298 <FASE>
FPTS,V23,P.T298-T300 <FASE>
1963 V41 N5 P63-67 64-13873 <=*$>
1963 V41 N5 P67-70 64-13874 <=*$>
1963 V41 N5 P74- FPTS V23 N2 P.T304 <FASE>
FPTS,V23,P.T304-T306 <FASE>
1963 V41 N5 P79-83 64-13431 <=*$>
1963 V41 N5 P83-87 64-13432 <=*$>
1963 V41 N5 P92- FPTS V23 N4 P.T755 <FASE>
1963 V41 N5 P141- FPTS V23 N2 P.T307 <FASE>
1963 V41 N5 P141 FPTS,V23,N2,P.T307 <FASE>
1963 V41 N6 P35-38 64-15527 <=*$>
1963 V41 N6 P38-43 65-60157 <=$>

Citation	Translation
1963 V41 N6 P52-	FPTS V23 N3 P.T507 <FASE>
	FPTS,V23,N3,P.T507 <FASE>
1963 V41 N6 P79-85	64-13853 <=*$>
1963 V41 N6 P93-	FPTS V23 N3 P.T515 <FASE>
	FPTS,V23,N3,P.T515 <FASE>
1963 V41 N6 P101-105	64-15483 <=*$>
1963 V41 N6 P105-109	64-13842 <=*$>
1963 V41 N7 P23-27	64-15450 <=*$>
1963 V41 N7 P51-56	64-19256 <=*$>
1963 V41 N8 P126-128	64-15464 <=*$>
1963 V41 N9 P13-	FPTS V23 N5 P.T1161 <FASE>
	FPTS,V23,N5,P.T1161 <FASE>
1963 V41 N9 P37-44	64-19715 <=$>
1963 V41 N9 P61-67	N65-16593 <=$>
1963 V41 N10 P49-53	64-19231 <=*$>
1963 V41 N10 P73-87	64-19721 <=$>
1963 V41 N11 P78-84	64-19232 <=*$>
1963 V41 N12 P51-57	64-19642 <=$>
1964 V42 N2 P124-126	64-31013 <=>
1964 V42 N3 P5-9	64-31185 <=>
1964 V42 N3 P9-15	64-31186 <=>
1964 V42 N3 P20-25	64-31184 <=>
1964 V42 N3 P99-	FPTS V23 N6 P.T1311 <FASE>
	FPTS V23 P.T1311 <FASE>
1964 V42 N4 P113-	FPTS V24 N2 P.T237 <FASE>
	FPTS V24 P.T237-T238 <FASE>
1964 V42 N4 P139-144	64-31552 <=>
1964 V42 N5 P3-8	64-51180 <=$>
1964 V42 N5 P71-74	RTS-2993 <NLL>
1964 V42 N5 P74-77	64-51180 <=$>
1964 V42 N5 P77-	FPTS V24 N4 P.T632 <FASE>
1964 V42 N5 P87-93	64-51180 <=$>
1964 V42 N5 P108-114	64-51180 <=$>
1964 V42 N5 P126-128	64-51180 <=$>
1964 V42 N5 P148-150	64-51180 <=$>
1964 V42 N6 P3-8	64-51102 <=>
1964 V42 N6 P55-56	65-62330 <=$>
1964 V42 N6 P108-112	65-62329 <=$>
1964 V42 N6 P116-121	64-51320 <=$>
1964 V42 N7 P83-89	65-62324 <=$>
1964 V42 N7 P102-105	65-62326 <=$>
1964 V42 N9 P23-28	65-30081 <=$>
1964 V42 N9 P47-53	65-63579 <=$>
1964 V42 N9 P77-84	65-30081 <=$>
1964 V42 N9 P123-127	65-63580 <=$>
1964 V42 N10 P3-10	65-30001 <=$>
1964 V42 N11 P9-19	65-30344 <=$>
1964 V42 N11 P137-139	65-30344 <=$>
1965 V43 N1 P85-	FPTS V24 N6 P.T967 <FASE>
1965 V43 N1 P95-	FPTS V24 N6 P.T945 <FASE>
1965 V43 N2 P3-12	65-30839 <=$>
1965 V43 N2 P6-	FPTS V24 N6 P.T953 <FASE>
1965 V43 N2 P62-66	66-60415 <=*$>
1965 V43 N2 P78-83	66-60417 <=*$>
1965 V43 N3 P40-	FPTS V25 N1 P.T164 <FASE>
1965 V43 N3 P68-72	66-60418 <=*$>
1965 V43 N3 P72-75	66-60419 <=*$>
1965 V43 N3 P108-	FPTS V25 N1 P.T167 <FASE>
1965 V43 N4 P18-21	66-60844 <=*$>
1965 V43 N4 P124-126	65-31127 <=$>
1965 V43 N5 P50-54	66-60890 <=*$>
1965 V43 N5 P61-	FPTS V25 N2 P.T364 <FASE>
1965 V43 N5 P103-108	66-61082 <=*$>
1965 V43 N6 P98-104	66-61328 <=*$>
1965 V43 N6 P123-127	66-61330 <=*$>
1965 V43 N7 P5-19	65-32540 <=*$>
1965 V43 N7 P53-58	66-61329 <=*$>
1965 V43 N8 P14-18	65-33616 <=*$>
1965 V43 N8 P40-	FPTS V25 N3 P.T471 <FASE>
1965 V43 N8 P77-	FPTS V25 N4 P.T627 <FASE>
1965 V43 N8 P119-122	65-33045 <=*$>
	65-33616 <=*$>
1965 V43 N9 P29-	FPTS V25 N4 P.T619 <FASE>
1965 V43 N10 P117-121	66-61968 <=*$>
1965 V43 N10 P141-142	66-30076 <=*$>
1965 V43 N12 P107-109	66-30684 <=$>
1965 V43 N12 P121-125	66-30684 <=$>
1966 V44 N5 P157	66-33537 <=$>

KLINIKA OCZNA

Citation	Translation
1956 V26 N1 P1-18	58-1539 <*>
1962 V32 P393-402	65-17247 <*>

KLINISCHE MEDIZIN

Citation	Translation
1947 V2 P725-729	61-16131 <*> O
1949 V4 P517-522	59-15860 <*>
1950 V5 N2 P88-93	60-14090 <*> O
1955 P83-85	R-5040 <*>
1958 V13 N5 P196-200	61-00256 <*>

KLINISCHE MONATSBLAETTER FUER AUGENHEILKUNDE

Citation	Translation
1881 V19 P443-454	61-14161 <*>
1886 V24 P82-93	61-14156 <*>
1889 V27 P291-298	61-14103 <*>
1903 V41 P244-245	65-12007 <*>
1907 V45 P42-46	61-14860 <*>
1911 V49 N12 P772-775	61-14120 <*> OP
1912 V50 P273-278	61-14091 <*>
1914 V53 P240-246	61-14165 <*>
1914 V53 P415-417	61-14375 <*>
1920 V64 P593-606	61-14460 <*>
1920 V64 P846-847	61-14462 <*>
1921 V66 P474-476	61-14458 <*>
1921 V66 P759	61-14459 <*>
1922 V68 P588-597	64-14443 <*>
1922 V69 P514-515	61-14457 <*>
1923 V70 P1-16	61-14414 <*>
1923 V70 P534-537	61-14243 <*>
1923 V70 P537-539	61-14415 <*>
1927 V79 P17-42	61-14235 <*>
1927 V79 P528-530	61-14413 <*>
1929 V82 P358-360	61-14185 <*>
1932 V88 P514-517	61-14171 <*>
1932 V89 P493-499	57-1506 <*>
1933 V91 P667-672	61-14416 <*>
1934 V92 P1-10	61-14122 <*>
1937 V98 P195-205	61-14445 <*> OP
1937 V98 P728-734	57-1670 <*>
1938 V101 P97-99	61-14223 <*>
1953 V122 P225-226	AL-642 <*>
1954 V124 P386-392	3381-A <KH>
1954 V125 P57-61	II-93 <*>
1957 V130 N5 P666-676	59-14742 <+*> O
	65-00413 <*> O
1958 V133 N2 P276-278	59-10627 <*>
1958 V133 N5 P713-718	64-00353 <*>
1959 V134 N6 P833-835	60-14544 <*>
1959 V135 N3 P305-347	65-14942 <*>
1961 V139 P174-179	62-10913 <*>
1961 V139 P224-234	62-10915 <*>
1961 V139 P234-241	62-10730 <*>
1961 V139 N1 P26-29	62-16225 <*>
1963 V142 N4 P642-650	65-00401 <*>
1963 V142 N6 P982-1006	66-11810 <*> O
1964 V144 N2 P251-253	65-13954 <*> O
1964 V145 P107-123	66-12169 <*>
1965 V146 N1 P1-21	65-13957 <*> O
1965 V146 N3 P376-382	66-11702 <*>

KLINISCHE WOCHENSCHRIFT

Citation	Translation
1925 V4 N23 P1120-1121	9249-E <K-H>
1925 V4 N28 P1339-1343	60-10621 <*>
1926 V5 N15 P655-	58-1350 <*>
1926 V5 N15 P655	63-01092 <*>
1926 V5 N42 P1966-1967	59-17044 <*>
1927 V6 P1798-	57-419 <*>
1927 V6 P2182-	57-389 <*>
1928 V7 P1592-1596	57-127 <*>
1928 V7 N4 P163-165	59-17215 <*>
1928 V7 N14 P634-637	61-14674 <*>
1929 V8 N50 P2322-2323	60-10620 <*>
1930 V9 P58-63	58-1835 <*>
1930 V9 N52 P2433-2435	1319 <*>
1932 V11 N30 P1271-1272	60-10626 <*>
1933 V12 P1910	65-18032 <*>
1933 V12 N41 P1599-1601	60-10624 <*>
1934 V13 P393-399	64-18990 <*> P
1934 V13 P1082-1083	65-18033 <*>
1934 V13 N50 P1785-1786	60-10622 <*> O
1935 V14 N3 P79-83	60-10635 <*> O

1935 V14 N48 P1713-1716 60-10687 <*> 0
1936 V15 P19-21 65-18034 <*>
1936 V15 P1485-1488 65-18035 <*> 0
1936 V15 N51 P1875-1877 II-841 <*>
1937 V16 P65-66 64-14135 <*>
1937 V16 P88-90 57-397 <*>
1937 V16 N37 P1265-1268 60-10634 <*>
1938 V17 N24 P843-849 60-10690 <*> 0
1938 V17 N41 P1445-1446 60-10619 <*>
1938 V17 N44 P1550-1554 65-17159 <*>
1939 V17 P710-715 65-18037 <*> 0
1939 V18 N7 P225-231 57-150 <*>
1940 V19 P33-36 65-18038 <*> 0
1940 V19 P200-203 58-374 <*>
1941 V20 P1003-1004 64-20021 <*>
1941 V20 N14 P331-334 60-10618 <*>
1941 V20 N18 P437-440 60-10617 <*>
1941 V20 N20 P506-510 59-15647 <*> 0
1942 V22 P529-532 63-14579 <*>
1943 V22 N40/1 P624- 2210 <*>
1944 V23 N17/0 P169-172 57-667 <*>
1947 V24 P880-882 65-18039 <*> 0
1948 V26 N5/6 P70-76 2574 <*>
1949 V27 P511-512 65-18040 <*>
1950 V28 P96-99 22610 <K-H>
1950 V28 P582-584 AL-878 <*>
1950 V28 N11/2 P177-179 2563 <*>
1951 V29 P548-549 65-18041 <*> 0
1951 V29 N1/2 P1-9 2434 <*>
1951 V29 N11/2 P229-230 60-10633 <*>
1951 V29 N21/2 P375-377 59-17825 <*> 0
1951 V29 N23/4 P415-420 60-10632 <*>
1952 V30 P126-128 AL-47 <*>
1952 V30 P178-179 1692 <*>
1952 V30 P498-504 1436 <*>
1952 V30 P551-553 AL-151 <*>
1952 V30 P1009-1011 AL-396 <*>
1953 V31 P816- AL-69 <*>
1953 V31 P859-860 AL-904 <*>
1953 V31 P890-894 AL-95 <*>
1953 V31 P946-948 63-14599 <*>
1953 V31 N43/4 P1050-1051 II-782 <*>
1954 P732-734 57-846 <*>
1954 P1025-1030 2760 <*>
1954 V32 P49-57 I-594 <*>
1954 V32 P199-203 I-17 <*>
1954 V32 P220-224 I-626 <*>
1954 V32 P270-271 I-830 <*>
1954 V32 P304-310 2088 <*>
1954 V32 P410-416 II-924 <*>
1954 V32 P425-432 I-862 <*>
1954 V32 P445-450 I-404 <*>
1954 V32 P930-935 II-483 <*>
1954 V32 N15/6 P369-375 59-17037 <*> 0
1954 V32 N35/6 P856-863 62-10392 <*> 0
1955 V33 P124-126 1884 <*>
1955 V33 P372-378 2415 <*>
1955 V33 P562-567 57-3177 <*>
1955 V33 P624-625 57-1957 <*>
1955 V33 P900-903 1354 <*>
1955 V33 P956-958 65-18042 <*>
1955 V33 P1104-1105 II-411 <*>
1955 V33 P1113-1118 1717 <*>
1955 V33 N15/6 P372-378 59-15743 <*> 0
1955 V33 N17/8 P435-437 64-20024 <*> 0
1955 V33 N31/2 P750-758 57-2720 <*>
1956 V34 P366-371 1602 <*>
58-26 <*>
1956 V34 P624-630 59-10701 <*> 0
1956 V34 P929-941 57-2957 <*>
1956 V34 P953-957 57-201 <*>
1956 V34 P1079-1083 57-3206 <*>
1956 V34 P1105-1114 57-202 <*>
1956 V34 N1/2 P15-19 1350 <*>
1956 V34 N3/4 P61-69 3231 <*>
1956 V34 N23/4 P633-635 1520 <*>
1957 V35 P143- 58-11 <*>
1957 V35 P452-459 3067 <*>
1957 V35 P497-502 58-25 <*>
1957 V35 P540-550 58-184 <*>
1957 V35 P901-905 58-1577 <*>
1957 V35 N1 P50- 57-1085 <*>
1957 V35 N5 P225-236 59-20165 <*> P
1957 V35 N6 P308-310 61-20904 <*>
1957 V35 N12 P635-636 59-18486 <+*> 0
1957 V35 N15 P771-773 65-00269 <*> 0
1957 V35 N16 P812-814 62-14940 <*> 0P
1957 V35 N22 P1102-1105 62-00165 <*>
1958 V36 N8 P341-347 65-14221 <*>
1958 V36 N8 P379-382 62-01391 <*>
1958 V36 N12 P585 63-20661 <*>
1958 V36 N14 P677-678 59-15494 <*>
1958 V36 N15 P693-706 59-15688 <*>
1958 V36 N16 P772-773 59-10700 <*> 0
1958 V36 N17 P808-814 59-10456 <*>
1958 V36 N20 P960-963 60-10745 <*>
1958 V36 N22 P1061-1066 60-14177 <*>
1958 V36 N23 P1132-1138 AEC-TR-4782 <*>
1959 V37 P657-660 G-267 <RIS>
1959 V37 P997-1003 65-00237 <*> 0
1959 V37 N2 P71-76 59-15337 <*> 0
1959 V37 N7 P355-365 59-15689 <*>
1959 V37 N17 P918-926 63-00309 <*>
1959 V37 N17 P928-931 60-10984 <*>
1959 V37 N24 P1263-1278 63-00079 <*>
1960 V38 P126-134 62-18074 <*> 0
1960 V38 P673-679 AEC-TR-5498 <*>
1960 V38 P1088-1090 62-01194 <*>
1960 V38 N2 P69-71 61-10338 <*> 0
1960 V38 N3 P104-109 62-00538 <*>
1960 V38 N3 P123-126 61-00600 <*>
1960 V38 N11 P532-535 61-20568 <*> 0
1960 V38 N11 P552 60-16918 <*> 0
1960 V38 N13 P634-640 61-10240 <*>
1960 V38 N14 P707-716 61-20564 <*> 0
1960 V38 N15 P912-916 63-00951 <*>
1960 V38 N21 P1075-1080 61-00601 <*>
1961 V39 P599 65-18043 <*>
1961 V39 P881-884 62-10338 <*>
1961 V39 N1 P55-56 61-18582 <*>
1961 V39 N2 P100-101 61-00620 <*>
1961 V39 N5 P238-243 61-20582 <*> 0
1961 V39 N6 P293-298 61-20581 <*> 0
1961 V39 N6 P307-308 62-00834 <*>
1961 V39 N7 P369-371 61-18642 <*>
1961 V39 N11 P593-594 63-00170 <*>
1961 V39 N14 P759 62-10967 <*>
1961 V39 N15 P784-790 62-00687 <*>
1961 V39 N15 P790-795 62-16216 <*> 0
1961 V39 N17 P910-911 62-16217 <*> 0
1961 V39 N17 P924-925 62-10586 <*>
1961 V39 N19 P998-1006 62-16123 <*>
1961 V39 N19 P1006-1013 63-20628 <*>
1961 V39 N21 P1137-1141 63-20872 <*> 0
1962 V40 P1014 63-18009 <*>
1962 V40 N1 P54-56 62-20108 <*>
1962 V40 N1 P56-57 63-18744 <*>
1962 V40 N3 P149-151 62-18179 <*>
1962 V40 N10 P541-542 62-20038 <*>
1962 V40 N13 P701-702 64-10373 <*>
1962 V40 N20 P1048-1056 63-00924 <*>
1962 V40 N24 P1257-1258 AEC-TR-5785 <*>
1963 V41 P376-385 63-20228 <*> 0
1963 V41 P1186-1188 64-18147 <*>
1963 V41 N1 P43-44 64-10369 <*>
1963 V41 N2 P103-105 64-10370 <*>
1963 V41 N3 P130-138 63-01101 <*>
1963 V41 N5 P245-246 64-16670 <*>
1963 V41 N13 P633-636 64-16854 <*>
1963 V41 N14 P631-690 AEC-TR-6331 <*>
1963 V41 N19 P948-952 64-16079 <*>
1963 V41 N20 P1002-1006 65-12023 <*>
1964 V42 N4 P165-168 65-10974 <*> 0
1964 V42 N4 P196-198 65-10972 <*> 0
1964 V42 N18 P890-898 66-12721 <*>
1965 V43 N18 P975-980 66-10372 <*>

KLUBNYI KALENDAR
1962 N1 P46-47 62-25596 <=*>

KNIGA; ISSLEDOVANIYA I MATERIALLY
1964 V9 P371-386 PT3 64-31530 <=>

KNIHOVNA VEDECKO-TEORETICKY SBORNIK
1962 P479-515 64-30719 <*>

KNIZHNAYA LETOPIS
1955 N19 P78-89 RT-3699 <*>
1955 N20 P76-88 RT-3698 <*>
1955 N21 P84-98 RT-3697 <*>
1955 N23 P83-106 RT-4227 <*>
1955 N24 P89-105 RT-4228 <*>
1955 N25 P93-123 RT-4089 <*>
1955 N26 P89-127 RT-4459 <*>
1955 N46 P87-103 RT-4374 <*>
1955 N47 P79-91 RT-4373 <*>
1955 N48 P94-108 RT-4372 <*>
1955 N49 P105-116 RT-4371 <*>
1955 N50 P96-119 RT-4332 <*>
1955 N51 P103-118 RT-4331 <*>
1955 N52 P103-120 RT-4330 <*>
1956 N1 P102-122 R-622 <*>
1956 N4 P114-129 R-609 <*>
1956 N6 P111-125 R-606 <*>

KNOWLEDGE OF GEOGRAPHY. CHINESE PEOPLES REPUBLIC
SEE TI-LI CHIH SHIH

K'O HSUEH
1933 V17 P1018-1048 63-14650 <=*>

KO HSUEH HSIN WEN
1958 N2 P33-34 59-13402 <=>
1959 P609 SPEC. NO. 61-21426 <=>
1959 N3 P23-24 61-21426 <=>
1959 N13 P7-8 61-21234 <=>
1959 N13 P21-22 61-21234 <=>
1959 N19 P8-9 61-23944 <=>
1959 N19 P15 61-23944 <=>
1959 N20 P2-6 61-21346 <=>
1959 N21 P24 61-21426 <=>
1959 N22 P2-5 61-21458 <=>
1959 N22 P13 61-21458 <=>
1959 N23 P15 61-23944 <=>
1959 N23 P22 61-23944 <=>
1959 N24 P1-5 61-21428 <=>
1959 N24 P7-15 61-21428 <=>
1959 N26 P5-6 61-23944 <=>
1959 N26 P16 61-23944 <=>
1959 N26 P21 61-23944 <=>
1959 N28 P3-5 61-21272 <=>
1959 N28 P10-11 61-21272 <=>
1959 N28 P18 61-21272 <=>
1959 N28 P20 61-21272 <=>
1959 N28 P22 61-21272 <=>
1959 N29 P1-2 61-21424 <=>
1959 N29 P5 61-21424 <=>
1959 N29 P7-9 61-23904 <*=>
1959 N30 P1-8 60-31108 <=>
1959 N30 P8 61-11592 <=>
1959 N30 P8- 61-21506 <=>
1959 N30 P13-15 61-11592 <=>
1959 N30 P17 61-11592 <=>
1959 N30 P18-21 61-27203 <*=>
1959 N30 P22-25 61-11592 <=>
61-21506 <=>
1959 N30 P27-31 61-27203 <*=>
1959 N31 P8 61-21459 <=>
1959 N32 P3-4 60-11300 <=>
1959 N32 P5 60-11301 <=>
1959 N33 P1-4 61-21443 <=>
1959 N33 P9 61-21443 <=>
1959 N33 P12-13 61-21443 <=>
1959 N33 P16-21 61-21443 <=>
1959 N34 P8 61-21459 <=>
1959 N34 P10-14 60-11424 <=>
1959 N34 P15 61-21459 <=>
1959 N36 P8-12 61-21369 <=>
1959 N36 P15-16 61-21369 <=>
1959 N36 P19-20 61-21369 <=>
1959 N36 P23-24 61-21369 <=>
1959 N38 P2-3 61-21459 <=>
1959 N39 P8 61-21459 <=>
1959 N39 P10-12 61-21459 <=>
1959 N39 P18 61-21459 <=>
1959 N39 P20 61-21459 <=>
1959 N40 P4 61-21459 <=>
1959 N41 P2 61-21459 <=>
1959 N41 P4 61-21459 <=>
1959 N41 P7-8 61-21459 <=>
1959 N41 P10 61-21459 <=>
1959 N42 P9-10 61-21459 <=>
1959 N42 P15 61-21459 <=>
1959 N42 P22 61-21459 <=>

K'O HSUEH HUA PAO
1959 N3 P91 NLLTB V3 N5 P373 <NLL>
1962 N11 P422-423 AD-619 424 <=$>
1964 N2 P46-48 AD-639 443 <=$>
1965 P169-173 65-34128 <=$>
1965 N9 P297-301 65-33832 <=$>
1965 N10 P929-932 65-33898 <=$>

K'O HSUEH TA CHANG
1962 N1 P8-9 63-23185 <=*>
1962 N4 P99-100 63-19287 <=*>
1962 N11 P328-332 63-21700 <=>
1963 N7 P1-3 63-31766 <=>
1963 N7 P18-19 63-31766 <=>
1963 N9 P14-15 AD-608 180 <=>
1963 N11 P22-23 64-71314 <=>
1963 N12 P5 AD-637 064 <=$>
1963 N12 P22-23 64-71310 <=> M
1963 V11 P12-13 AD-610 373 <=$>
1964 N10 P366-368 65-30061 <=$>

K'O HSUEH T'UNG PAO
1954 N2 P37-41 57-774 <*>
1954 N12 P55-58 R-3 <*>
1955 V3 P36-39 57-608 <*>
1956 P37-41 59-13064 <=>
1956 N5 P93-94 63-15112 <=>
1956 N9 P90-91 63-15112 <=>
1956 N10 P72-76 63-15112 <=>
1956 V7 P73-75 C-2369 <*>
1957 N3 P95-96 59-13413 <=>
1957 N4 P97-104 61-27348 <*=>
1957 N4 P110 59-11498 <=>
1957 N9 P279-280 59-11511 <=> O
1957 N13 P385-394 66-31633 <=$>
1957 V1 P18 62-11307 <=>
1957 V4 P121-122 C2368 <NRC>
57-2573 <*>
1957 V11 P352 58-1552 <*>
1957 V14 P432 C-2503 <NRC>
58-584 <*>
1958 N3 P73-78 07N53CH <ATS>
1958 N7 P209-224 62-15383 <=>
1958 N8 P225-235 62-15351 <=>
1958 N8 P246-256 62-15351 <=>
1958 N14 P441-443 59-13340 <=>
1958 N14 P445-447 59-13340 <=>
1958 N15 P476-477 61-10551 <=*>
1958 N15 P477-478 61-10550 <+*>
1958 N16 P505-506 59-13489 <=>
1958 N18 P555-556 59-13424 <=>
1958 N21 P669-670 59-13399 <=>
1958 N23 P722-724 59-13528 <=>
1959 N2 P36-41 NP-TR-817 <*>
1959 N3 P98-99 59-11747 <=>
1959 N4 P108-109 59-11783 <=>
1959 N4 P109-112 59-11784 <=>
1959 N4 P112-115 59-11785 <=>
1959 N4 P116-117 59-11786 <=>
1959 N4 P118-120 59-11787 <=>
1959 N4 P120-122 59-11788 <=>
1959 N4 P122-123 59-11789 <=>
1959 N4 P124-125 59-11790 <=>
1959 N4 P126-128 59-11791 <=> O
1959 N16 P501-504 60-11388 <=>

Citation	Number
1959 N16 P505-507	60-31011 <=>
1959 N16 P508-510	60-11101 <=>
1959 N16 P511-513	60-11102 <=>
1959 N16 P514-516	60-11103 <=>
1959 N17 P561-562	60-31046 <=>
1959 N18 P565-566	60-11864 <=>
1959 N18 P568-571	60-11865 <=>
1959 N18 P572-576	60-41061 <=>
1959 N18 P586-593	60-11986 <=>
1959 N18 P594-602	60-11929 <=>
1959 N18 P6C7-612	60-11997 <=>
1959 N19 P627-629	60-31019 <=>
1959 N19 P637-641	60-31024 <=>
1959 N19 P650-654	60-31760 <=>
1959 N19 P659-660	60-31760 <=>
1959 N19 P695	60-31760 <=>
1959 N20 P666-673	60-11775 <=>
1959 N20 P674-677	60-11776 <=>
1959 N20 P678-681	60-11777 <=>
1959 N20 P691-693	60-31760 <=>
1959 N20 P695-708	60-31760 <=>
1959 N23 P773-776	60-11726 <=>
1959 N23 P776-779	60-11727 <=>
1959 N23 P789-790	60-31081 <=>
1959 N24 P830-832	63-01646 <*>
1959 N24 P832-833	60-31382 <=>
	60-31396 <+*>
	62-24033 <=*>
1960 P491-494	62-14830 <=*>
1960 N3 P133-135	60-11653 <=>
1960 N4 P117-118	63-31332 <=>
1960 N5 P146-149	60-11636 <=>
1960 N7 P193-197	62-13397 <=>
1960 N7 P2C6-208	62-13397 <=>
1960 N7 P216-223	62-13397 <=>
1960 N8 P247-252	61-11639 <=>
1960 N9 P270-282	61-23568 <=*>
1960 N9 P284-286	61-23568 <=*>
1960 N10 P291-310	61-21713 <=>
1960 N10 P313-316	60-41496 <=*>
1960 N10 P316-320	61-21713 <=>
1960 N15 P458-459	61-23565 <=*>
1961 N2 P2-6	61-27721 <*=>
1961 N2 P23-26	61-27201 <*=>
1961 N3 P7-11	61-27721 <*=>
1961 N3 P28-36	61-28517 <*=>
1961 N4 P11-16	61-27721 <*=>
1961 N11 P1-11	62-25468 <=>
1961 N11 P34-37	62-25468 <=>
1961 N11 P43-48	62-25468 <=>
1961 N12 P30-36	62-24966 <=>
1961 N12 P44-49	62-24966 <=>
1962 N1 P13-18	62-24966 <=>
1962 N1 P38-39	62-24966 <=>
1962 N1 P41-45	62-24966 <=>
1962 N1 P48	62-24966 <=>
1962 N2 P9-13	63-23654 <=*>
1962 N2 P28-33	63-13324 <=*>
1962 N2 P43-45	62-32374 <=>
1962 N3 P33-36	62-32374 <=>
1962 N3 P38-39	62-32374 <=>
1962 N4 P2-31	62-32374 <=>
1962 N4 P40-42	62-32374 <=>
1962 N5 P449-452	64-41012 <=>
1962 N6 P12-14	62-11735 <=>
1962 N6 P40	62-11735 <=>
1962 N7 P41-45	62-33444 <=>
1962 N9 P1-2	63-21279 <=>
1962 N9 P3-15	63-21527 <=>
1962 N9 P16-40	63-15355 <=*>
1962 N9 P42-49	63-21279 <=>
1962 N9 P52-54	63-21279 <=>
1962 N10 P27-49	63-15388 <=*>
1962 N11 P41-47	63-21150 <=>
1962 N11 P5C-57	63-21150 <=>
1962 N12 P1-25	63-21594 <=>
1962 N12 P48-52	63-21594 <=>
1962 N12 P54-62	63-21594 <=>
1963 N1 P21-27	63-31705 <=>
1963 N1 P69	64-15216 <=*$>
1963 N2 P60-63	63-21486 <=>
1963 N4 P1-6	63-31281 <=>
1963 N4 P7	63-31846 <=>
1963 N4 P32-38	63-31353 <=>
1963 N5 P28-34	63-31933 <=> 0
1963 N8 P28-44	AD-620 817 <=*$>
1963 N10 P1-2	64-21750 <=>
1963 N10 P15-22	64-21750 <=>
1963 N10 P29-33	64-21750 <=>
1963 N10 P40-49	64-21750 <=>
1963 N10 P57-58	64-21750 <=>
1963 N10 P65-67	64-21750 <=>
1963 N12 P4-21	64-31085 <=>
1964 N1 P27-34	64-31327 <=>
1964 N1 P35-48	64-31327 <=>
1964 N1 P48-51	64-31085 <=>
1964 N1 P52-56	64-31167 <=>
1964 N1 P56-64	64-31322 <=>
1964 N1 P71-73	64-31167 <=>
1964 N1 P76-77	64-31167 <=>
1964 N1 P78-81	64-31322 <=>
1964 N2 P46-47,64	AD-621 053 <=*$>
1964 N2 P151-154	AD-619 480 <=$>
1964 N3 P218-225	64-31451 <=>
1964 N3 P231-236	64-31451 <=>
1964 N3 P266-269	66-11460 <*>
1964 N3 P273	AD-626 951 <=$>
	64-31451 <=>
1964 N4 P295-305	64-31928 <=>
1964 N4 P306-334	64-31565 <=>
1964 N4 P335-340	64-31928 <=>
1964 N4 P340-342	64-31565 <=>
1964 N4 P343-344	64-31327 <=>
1964 N4 P344-345	64-31565 <=>
1964 N4 P346-359	64-31928 <=>
1964 N4 P361-363	64-31928 <=>
1964 N4 P365-370	64-31565 <=>
1964 N4 P374-375	AD-636 622 <=$>
1964 N5 P386-394	64-41012 <=>
1964 N5 P394-405	64-41012 <=>
1964 N5 P406-415	64-41012 <=>
1964 N5 P416-420	64-41012 <=>
1964 N5 P420-425	64-31834 <=>
1964 N5 P426-431	64-41012 P.14-26 <=>
1964 N5 P432	64-41012 <=>
1964 N5 P433	64-41012 <=>
1964 N5 P434	64-41012 <=>
1964 N5 P435-438	64-41012 <=>
1964 N5 P441-443	64-41012 <=>
1964 N5 P445-449	64-41090 <=>
1964 N5 P452-455	64-41012 <=>
1964 N5 P455-456	64-41012 <=>
1964 N5 P457-458	64-41012 <=>
1964 N5 P459-461	64-41012 <=>
1964 N5 P461-464	64-41012 <=>
1964 N5 P465	64-41012 P.81-83 <=>
1964 N5 P466	64-41012 P.11-13 <=>
1964 N5 P467	64-41012 P.84-86 <=>
1964 N5 P468	64-41012 <=>
	64-41012 <=>
1964 N6 P471-481	64-41388 <=>
1964 N6 P482-500	64-41452 <=$>
1964 N6 P510-514	64-41388 <=>
1964 N6 P518-521	64-41569 <=>
1964 N6 P526-529	64-41452 <=$>
1964 N6 P529-541	64-41569 <=>
1964 N6 P542-547	64-41388 <=>
1964 N6 P548-550	64-41452 <=$>
1964 N6 P551-563	64-41388 <=>
1964 N7 P565-595	64-51305 <=>
1964 N7 P607-619	64-51305 <=>
1964 N7 P619	64-51335 <=$>
1964 N7 P623	64-51335 <=$>
1964 N7 P630-632	64-51335 <=$>
1964 N7 P634-646	64-51335 <=$>
1964 N7 P651-656	64-51335 <=$>
1964 N7 P658	64-51335 <=$>
1964 N8 P667-674	64-41908 <=$>
1964 N8 P682-694	64-51672 <=$>
1964 N8 P727	64-51265 <=>

1964 N8 P733-736 AD-625 090 <=*$>
1964 N8 P737-739 AD-625 090 <=*$>
1964 N10 P854-861 65-30098 <=$>
1964 N10 P862-864 65-30079 <=$>
1964 N10 P865-868 65-30098 <=$>
1964 N10 P869-892 65-30079 <=$>
1964 N10 P900-907 65-30061 <=$>
1964 N10 P908-927 65-30061 <=$>
1964 N10 P928-941 65-30098 <=$>
1964 N10 P941-942 65-30079 <=$>
1964 N11 P943-962 65-30286 <=$>
1964 N11 P963-972 65-30431 <=$>
1964 N11 P991-1000 65-30274 <=$>
1964 N11 P1001-1007 65-30212 <=$>
1964 N11 P1007-1020 65-30286 <=$>
1964 N11 P1020-1023 65-30274 <=$>
1964 N11 P1024-1025 65-30212 <=$>
1964 N11 P1026-1031 65-30286 <=$>
1964 N11 P1036 65-30286 <=$>
1965 N1 P34-46 65-30619 <=$>
1965 N1 P82-85 65-30506 <=$>
1965 N2 P117-130 65-30673 <=$>
1965 N2 P177-179 65-30673 <=$>
1965 N3 P202-212 65-30794 <=$>
1965 N4 P339-361 65-31530 <=$>
1965 N4 P373-376 65-31530 <=$>
1965 N6 P472-493 65-33931 <=$>
1965 N7 P565-567 65-32676 <=$>
1965 N7 P634-648 65-32374 <=$>
1965 N8 P676-742 66-30180 <=$>
1965 N8 P729-732 65-33829 <=$>
1965 N9 P753-759 66-30049 <=$>
1965 N9 P760-774 66-30518 <=$>
1965 N9 P805-810 66-30049 <=$>
1965 N9 P831-832 66-30049 <=$>
1965 N10 P853-861 65-33813 <=*$>
1965 N10 P876-883 65-34030 <=*$>
1965 N10 P911-916 65-33813 <=*$>
1965 N10 P925-926 65-33813 <=*$>
1965 N11 P14-15 65-34128 <=$>
1965 N11 P941-1034 66-30202 <=$>
1965 N12 P1035-1042 66-30518 <=$>

KOBAYASHI RIGAKU KENKYUSHO HOKOKU
1959 V9 N1/2 P45-56 64-16490 <*> 0
1960 V10 N4 P159-162 65-10898 <*>

KOBUNSHI
1961 V10 N113 P703-708 66-13242 <*>
1965 V22 N239 P186-192 16S85J <ATS>

KOBUNSHI KAGAKU
1945 V2 P175-194 62-18939 <*>
1945 V2 P235-239 62-18940 <*>
1946 V3 P13-20 62-18938 <*>
1947 V4 P67-72 PT1 62-10419 <*> 0
1947 V4 P72-77 PT2 62-10419 <*> 0
1947 V4 P77-81 PT3 62-10419 <*> 0
1947 V4 P67-72 58-38 <*>
1948 V5 P57-73 CODE-H1 <SA> 0
62-18926 <*>
1949 V6 P497-501 57-1024 <*>
1949 V6 P502-504 57-3010 <*>
1950 V7 P122-128 57-3542 <*>
1950 V7 P188-193 9H11J <ATS>
1950 V7 P204-207 60Q67J <ATS>
1950 V7 P350-353 64-20355 <*> 0
1950 V7 N4 P129-131 62-14971 <*>
1950 V7 N6 P204-207 64-20354 <*> 0
1950 V7 N8 P292-294 64-20353 <*> 0
1951 V8 P128-133 33E1J <ATS>
57-2679 <*>
1951 V8 P133-136 57-2680 <*>
1951 V8 P317-320 61-20266 <*> 0
1953 V10 P247-252 97G5J <ATS>
1954 V11 N105 P14-17 01N48J <ATS>
1954 V11 N112 P337-343 60-18402 <*>
65-11434 <*> 0
1955 V12 P79-85 62-20092 <*>
1955 V12 P258-265 61-20090 <*>
1955 V12 P335-343 08Q74J <ATS>
1955 V12 P483-486 35S81J <ATS>
1955 V12 N126 P414-427 18TMS <CTT>
1955 V12 N126 P453-459 60-18461 <*>
65-11458 <*> 0
1955 V12 N128 P506-510 36S81J <ATS>
1956 V13 P491-495 61-16572 <*>
1956 V13 N129 P6-10 60-18458 <*>
1956 V13 N129 P11-17 60-18457 <*>
1956 V13 N129 P31-37 58P60J <ATS>
1956 V13 N130 P69-76 62-14970 <*>
1956 V13 N136 P323-329 60-10122 <*>
1956 V13 N137 P390-396 65-12879 <*>
1956 V13 N140 P531-539 65-18008 <*>
1957 V14 P359-362 60-18389 <*> 0
1957 V14 P430-434 82K24J <ATS>
1957 V14 P556-560 60-18328 <*> 0
1957 V14 N141 P14-19 59-17078 <*>
1957 V14 N142 P86-91 59-15268 <*>
1957 V14 N142 P101-106 60-18197 <*> 0
1957 V14 N143 P133-138 59-17079 <*>
1957 V14 N144 P176-183 59-15790 <*>
1957 V14 N144 P191-195 59-15791 <*>
1957 V14 N147 P363-366 59-15267 <*>
1957 V14 N150 P488-495 98L29J <ATS>
1957 V14 N152 P636-643 1407 <TC>
1958 V15 P60-64 62-20088 <*>
1958 V15 P83-88 62-20087 <*>
1958 V15 P160-164 62-20090 <*>
1958 V15 P165-169 62-20089 <*>
1958 V15 N1 P43-48 58-2491 <*>
1958 V15 N1 P65-70 58-2497 <*>
1958 V15 N153 P49-54 64K23J <ATS>
1958 V15 N153 P89-94 65-11433 <*> 0
1958 V15 N156 P228-232 59-10508 <*>
1958 V15 N156 P232-237 59-10507 <*>
1958 V15 N156 P238-242 59-10506 <*>
1958 V15 N159 P381-388 65-14367 <*>
1958 V15 N159 P412-416 59-17082 <*>
1958 V15 N159 P469-474 KOBU-159-1 <SA>
1958 V15 N161 P550-554 KOBU-161-1 <SA>
1958 V15 N162 P647-653 01L30J <ATS>
1958 V15 N162 P654-659 59-15240 <*>
1958 V15 N162 P664-670 59-15237 <*>
1959 V16 P239-243 REPT.61111 <RIS>
1959 V16 P260-266 09L33J <ATS>
1959 V16 P266-270 14L34J <ATS>
1959 V16 N165 P35-39 59-17081 <*>
1959 V16 N165 P45-48 REPT. 61112 <RIS>
1959 V16 N166 P129-132 REPT.61113 <RIS>
1959 V16 N167 P173-175 60-18390 <*>
1959 V16 N168 P247-280 M-5665 <NLL>
1959 V16 N169 P324-329 60-10116 <*>
1959 V16 N170 P333-336 60-10447 <*>
1959 V16 N171 P437-440 60-16677 <*>
1959 V16 N171 P453-455 60-14263 <*>
1959 V16 N171 P471-474 60-14264 <*>
1959 V16 N175 P693-698 60-18388 <*>
1959 V16 N176 P713-719 60-18129 <*>
1959 V16 N176 P720-723 60-18130 <*>
1960 V17 P197-201 NP-TR-807 <*>
1960 V17 P432-435 SC-T-64-1610 <*>
1960 V17 P478-481 66-14761 <*$>
1960 V17 N177 P13-16 60-18465 <*>
1960 V17 N177 P17-18 60-18464 <*>
1960 V17 N177 P18-20 60-18463 <*>
1960 V17 N178 P108-114 60-18387 <*>
1960 V17 N178 P115-119 60-18153 <*> 0
1960 V17 N180 P197-201 61-16765 <*>
1960 V17 N180 P210-215 98M44J <ATS>
1960 V17 N180 P222-226 61-16573 <*>
1960 V17 N180 P257-262 61118 <RIS>
1960 V17 N181 P207-210 99M44J <ATS>
1960 V17 N181 P305-311 61-16574 <*>
94M44J <ATS>
1960 V17 N181 P311-315 61-16570 <*>
95M44J <ATS>
1960 V17 N181 P325-328 96M44J <ATS>
1960 V17 N181 P329-332 97M44J <ATS>
1960 V17 N182 P333-336 32M47J <ATS>

1960 V17 N182 P359-363	65-14487 <*> 0
1960 V17 N182 P364-366	C-136 <RIS>
	61-10594 <*>
1960 V17 N183 P403-407	61-10596 <*>
1960 V17 N183 P407-412	61-10595 <*>
1960 V17 N183 P441-444	C-137 <RIS>
1960 V17 N183 P449-451	66-11384 <*>
1960 V17 N184 P475-477	C-134 <RIS>
1960 V17 N184 P489-492	61-10602 <*>
1960 V17 N184 P497-503	C-135 <RIS>
1960 V17 N185 P533-539	08N54J <ATS>
1960 V17 N186 P607-611	KOBU-186-1 <SA>
1960 V17 N186 P618-620	CODE-2 <SA>
1960 V17 N186 P621-627	61-16537 <*>
	81Q69J <ATS>
1960 V17 N186 P627-630	22Q69J <ATS>
1960 V17 N186 P635-640	61-16518 <*>
1960 V17 N186 P641-643	CODE-7 <SA>
1960 V17 N187 P672-675	1101 <FT>
	43N56J <ATS>
1960 V17 N188 P721-724	66-13288 <*>
1961 V18 P235-239	66-13889 <*>
1961 V18 N189 P22-25	CODE-47 <SA>
1961 V18 N191 P163-168	CODE-66 <SA>
	62-10224 <*>
1961 V18 N191 P175-182	62-10701 <*>
1961 V18 N191 P183-186	65-14368 <*> 0
1961 V18 N191 P187-190	<SA>
1961 V18 N192 P214-219	<SA>
1961 V18 N192 P240-251	66-13888 <*>
1961 V18 N192 P267-272	CODE-G81 <SA>
1961 V18 N194 P333-338	05P63J <ATS>
1961 V18 N194 P351-356	62-10690 <*>
	92N56J <ATS>
1961 V18 N196 P496-503	27P59J <ATS>
1961 V18 N197 P516-566	32Q70J <ATS>
1961 V18 N198 P589-595	62-18041 <*>
1961 V18 N198 P605-608	62-18036 <*>
1961 V18 N198 P645-652	62-16002 <*>
1961 V18 N198 P653-655	62-16001 <*>
1961 V18 N199 P667-673	62-16014 <*>
1962 N19 P261-266	ICE V4 N2 P346-351 <ICE>
1962 V19 P543-546	SC-T-529 <*>
1962 V19 N201 P19-24	56P64J <ATS>
1962 V19 N201 P25-30	KOBU-201-1 <SA>
1962 V19 N203 P148-153	29Q69J <ATS>
1962 V19 N203 P161-168	62-18983 <*>
1962 V19 N204 P224-228	KOBU-204-1 <SA>
1962 V19 N204 P239-244	63-10084 <*>
1962 V19 N204 P245-251	62-18982 <*>
1962 V19 N204 P251-255	62-18981 <*>
1962 V19 N204 P261-266	ICE V4 N2 P346-351 <ICE>
	KOBU-204-2 <SA>
	53Q73J <ATS>
1962 V19 N204 P276-280	21Q69J <ATS>
1962 V19 N207 P402-406	<LSA>
	KOBU-207-1 <SA>
1962 V19 N207 P431-435	66-11324 <*>
1962 V19 N207 P456-460	M-5652 <NLL>
1962 V19 N207 P461-466	M-5653 <NLL>
1962 V19 N208 P506-517	M-5731 <NLL>
1962 V19 N210 P571-574	KOBU-210-1 <SA>
1962 V19 N211 P682-690	65-10391 <*>
1962 V19 N211 P690-698	65-10390 <*>
1962 V19 N212 P715-721	KOBU-212-1 <SA>
1962 V19 N212 P728-733	12816 <K-H>
	65-14868 <*>
1963 N219 P461-466	ICE V4 N3 P552-557 <ICE>
1963 N220 P506-611	ICE V4 N3 P546-551 <ICE>
1963 V20 P587-590	65-17094 <*>
1963 V20 P591-595	65-17093 <*>
1963 V20 N213 P17-21	64-14265 <*>
1963 V20 N213 P22-26	64-10806 <*>
1963 V20 N213 P27-33	64-14266 <*>
1963 V20 N213 P49-57	ICE V4 N2 P352-360 <ICE>
1963 V20 N213 P49-64	ICE V4 N2 P352-366 <ICE>
1963 V20 N213 P49-57	97Q69J <ATS>
1963 V20 N213 P58-64	ICE V4 N2 P360-366 <ICE>
	98Q69J <ATS>
1963 V20 N213 P86-90	63-16418 <*>
1963 V20 N214 P102-107	65-10246 <*>
1963 V20 N215 P145-150	65-10245 <*>
1963 V20 N215 P180-184	KOBU-215-1 <SA>
1963 V20 N216 P251-256	KOBU-216-1 <SA>
1963 V20 N216 P262-267	KOBU-216-2 <SA>
1963 V20 N217 P289-296	KOBU-217-1 <SA>
1963 V20 N217 P303-311	KOBU-217-2 <SA>
1963 V20 N218 P357-363	KOBU-218-1 <SA>
1963 V20 N218 P364-368	64-16038 <*>
1963 V20 N219 P461-466	ICE V4 N3 P552-577 <ICE>
	KOBU-219-1 <SA>
	40R75J <ATS>
1963 V20 N220 P491-495	66-14848 <*$>
1963 V20 N220 P506-611	ICE V4 N3 P546-551 <ICE>
1963 V20 N220 P506-511	42R75J <ATS>
1963 V20 N221 P534-539	KOBU-221-1 <SA>
1963 V20 N222 P641-645	KOBU-222-1 <SA>
1964 N7 P403-408	ICE V5 N1 P169-174 <ICE>
1964 N8 P505-512	ICE V5 N1 P162-168 <ICE>
1964 N11 P673-677	ICE V5 N2 P379-383 <ICE>
1964 N226 P97-102	ICE V4 N3 P535-540 <ICE>
1964 V21 P657-665	66-14555 <*>
1964 V21 N226 P97-102	ICE V4 N3 P535-540 <ICE>
	20R76J <ATS>
1964 V21 N229 P300-304	KOBU-229-1 <SA>
	65-14796 <*> 0
1964 V21 N229 P304-311	KOBU-229-2 <SA>
1964 V21 N230 P337-346	65-10457 <*>
1964 V21 N231 P403-408	ICE V5 N1 P169-174 <ICE>
	71R79J <ATS>
1964 V21 N231 P467-472	66-11321 <*>
1964 V21 N232 P487-493	KOBU-232-1 <SA>
1964 V21 N232 P505-512	ICE V5 N1 P162-168 <ICE>
	65-11730 <*>
1964 V21 N232 P513-516	65-11349 <*>
1964 V21 N233 P564-567	KOBU-233-1 <SA>
1964 V21 N235 P673-677	13S81J <ATS>
1964 V21 N236 P729-736	65-13875 <*>
1965 N1 P64-68	ICE V5 N4 P756-760 <ICE>
1965 N2 P97-102	ICE V5 N3 P575-580 <ICE>
1965 N2 P128-134	ICE V5 N3 P569-575 <ICE>
1965 N3 P186-192	ICE V5 N4 P743-749 <ICE>
1965 N7 P394-404	ICE V6 N2 P330-340 <ICE>
	56S87J <ATS>
1965 N7 P410-416	ICE V6 N2 P340-346 <ICE>
1965 V22 N237 P1-8	65-13873 <*>
1965 V22 N237 P64-68	17S85J <ATS>
1965 V22 N238 P97-102	33S83J <ATS>
1965 V22 N238 P128-134	34S83J <ATS>
1965 V22 N239 P145-148	65-17222 <*>
1965 V22 N239 P148-151	65-17221 <*>
1965 V22 N243 P410-416	57S87J <ATS>
1965 V22 N247 P679-685	71T90J <ATS>
1966 N250 P103-106	ICE V6 N4 P707-710 <ICE>
1966 N251 P145-151	ICE V6 N4 P711-717 <ICE>
1966 N252 P222-228	ICE V6 N4 P751-757 <ICE>
1966 V23 P165-171	66-14536 <*>
1966 V23 N250 P103-106	27T94J <ATS>
1966 V23 N251 P145-151	28T94J <ATS>
1966 V23 N252 P222-228	75T95J <ATS>

KOGYO KAGAKU ZASSHI

1927 V30 P221-225	66-14548 <*>
1927 V30 P828-835	66-14549 <*>
1928 V31 P638-642	I-581 <*>
1929 V31 P638-642	98F3J <ATS>
1929 V32 P332-337	57-2668 <*>
1932 V35 P470 SUP	2431 <*>
1934 V37 N9 P1116-1175	65-12703 <*>
1936 V39 N1 P6-10	10K22J <ATS>
1936 V39 N7 P470-479	65-12702 <*>
1940 V43 N5 P366-371	CODE-G32 <SA>
1940 V43 N5 P1428-1448	RCT V13 N1 P856 <RCT>
1940 V43 N8 P534-538	AL-949 <*>
1940 V43 N11 P799-802	AL-949 <*>
1941 V44 P222-225	65-12141 <*>
	66-10810 <*>
1941 V44 P449-451	65-12141 <*>
	66-10813 <*>
1941 V44 P540-542	65-12141 <*>

Citation	Translation
	66-10812 <*>
1941 V44 P646-651	65-12141 <*>
	66-10811 <*>
1941 V44 P685-688	PANSDOC-TR.752 <PANS>
1941 V44 P1016-1019	II-475 <*>
	3000-E <K-H>
	64-10477 <*> 0
1941 V44 N8 P660-662	AL-949 <*>
1941 V44 N9 P737-740	60-10073 <*> 0
1941 V44 N10 P825-828	65-13638 <*> 0
1941 V44 N10 P860-862	10F4J <ATS>
1942 V45 P370-375	91H10J <ATS>
1942 V45 P486-491	92H10J <ATS>
1942 V45 P703-707	93H10J <ATS>
1942 V45 N1 P69-72	65-10263 <*>
1942 V45 N9 P971-974	I-11 <*>
1943 V46 P114-119	T-2063 <INSD>
1943 V46 P237-240	59-17884 <*> 0
1943 V46 P633-636	94H10J <ATS>
1943 V46 P706-711	57-2628 <*>
1943 V46 P879-884	52G5J <ATS>
1943 V46 P1037-1040	T1852 <INSD>
1943 V46 P1163-1165	95H10J <ATS>
1943 V46 P1165-1167	95H10J-B <ATS>
1943 V46 N2 P152-155	KKZ-46-2-1 <SA>
1943 V46 N9 P858-859	64-10730 <*>
1943 V46 N10 P1105-1108	65-12851 <*>
1943 V46 N11 P386-388	64-10731 <*> 0
1944 V47 P103-106	T1789 <INSD>
1944 V47 P516-518	65-13295 <*>
1944 V47 P849-853	63-14154 <*>
1944 V47 N1 P33-46	64-10732 <*> 0
1944 V47 N2 P202-203	61-16496 <*>
1944 V47 N5 P455-461	60-10035 <*> 0
1944 V47 N7 P627-629	60-10034 <*> 0
1944 V47 N8 P796-799	65-10262 <*>
1945 V48 P65-67	63-14707 <*>
1945 V48 N3 P40-41	60-10072 <*> 0
1945 V48 N10/2 P89-91	65-11275 <*>
1945 V48 N10/2 P91-92	65-11276 <*>
1946 V49 P121	63-14893 <*>
1946 V49 P169-170	61-20930 <*>
1946 V49 N1/2 P20-21	60-18888 <*>
1946 V49 N1/2 P22	60-18889 <*>
1946 V49 N1/2 P23-24	60-18891 <*>
1946 V49 N1/2 P24	60-18892 <*>
1947 V50 N10/1 P138-139	89S87J <ATS>
1948 V51 P3-4	64-14664 <*> 0
1948 V51 P129-132	1392 <*>
1948 V51 N1 P30-32	KKZ-51-1-1 <SA>
1948 V51 N1 P32-33	KKZ-51-1-2 <SA>
1949 V52 P37-39	57-249 <*>
1949 V52 P171-173	63-14155 <*> 0
1949 V52 P212-213	C-4933 <NRC>
1949 V52 P267-273	63-14156 <*> 0
1949 V52 N2 P52-53	66-10623 <*>
1949 V52 N4 P148-149	60-10027 <*> 0
1949 V52 N6 P261-263	65-10951 <*>
1949 V52 N7 P302-305	65-10952 <*>
1950 V53 P5-7	57-250 <*>
1950 V53 P152-154	99F2J <ATS>
1950 V53 P381-385	T1455 <INSD>
1950 V53 P409-411	2642 <K-H>
	2759J <K-H>
	63-20448 <*> 0
1950 V53 P578-579	57-251 <*>
1950 V53 N1 P42	CODE-G33 <SA>
1950 V53 N2 P57-58	98F2J <ATS>
1950 V53 N2 P69-70	65-13806 <*>
1950 V53 N3 P101-103	96F2J <ATS>
1950 V53 N4 P143-144	97F2J <ATS>
1950 V53 N5 P205-207	52R78J <ATS>
1950 V53 N7 P315-317	CODE-G34 <SA>
1950 V53 N8 P361-363	66-12649 <*> 0
1950 V53 N9 P339-401	66-12648 <*> 0
1950 V53 N9 P378-379	CODE.G35 <SA>
1950 V53 N10 P423-425	40N54J <ATS>
1951 V54 P17-18	57-252 <*>
1951 V54 P58-59	63-18284 <*>
1951 V54 P150-152	C-2234 <NRC>
1951 V54 P163-165	57-2672 <*>
1951 V54 P199-200	58-160 <*>
1951 V54 P239-242	C-2235 <NRC>
1951 V54 P268-270	2056 <*>
1951 V54 P358-361	2054 <*>
1951 V54 P524-525	34K24J <ATS>
1951 V54 P603-604	I-309 <*>
1951 V54 P750-752	57-2673 <*>
	59E1J <ATS>
1951 V54 P798-800	19D1J <ATS>
	57-2684 <*>
1951 V54 N1 P11-14	40K23J <ATS>
1951 V54 N1 P16-18	CODE.G36 <SA>
1951 V54 N2 P157-159	64-20385 <*> 0
1951 V54 N3 P213-215	64-16446 <*> 0
1951 V54 N4 P256-258	41N54J <ATS>
1951 V54 N6 P391-394	65-13858 <*>
1951 V54 N9 P592-594	62-26027 <=$>
1951 V54 N9 P597-599	66-13671 <*>
1951 V54 N9 P604-607	25R77J <ATS>
1951 V54 N12 P747-750	41E2J <ATS>
1951 V54 N12 P775-777	CODE-G37 <SA>
1952 V55 P3-6	UCRL TRANS-739 <*>
1952 V55 P120-121	89L33J <ATS>
1952 V55 P133-134	63-14157 <*> 0
1952 V55 P221-223	71J16J <ATS>
1952 V55 P275-276	72J16J <ATS>
1952 V55 P397-398	35K24J <ATS>
1952 V55 P419-422	G-257 <RIS>
1952 V55 P521-522	57-2683 <*>
	60E1J <ATS>
1952 V55 P589-591	64-14355 <*>
	65-18192 <*>
1952 V55 P673-674	61-20942 <*>
1952 V55 N1 P3-6	20D1J <ATS>
	57-2682 <*>
1952 V55 N1 P24-26	CODE-G38 <SA>
1952 V55 N1 P31-33	07K28J <ATS>
1952 V55 N2 P49-51	2050 <*>
1952 V55 N2 P83-84	61-10833 <*>
1952 V55 N2 P133-134	66-15297 <*>
1952 V55 N4 P219-221	60-14139 <*>
1952 V55 N4 P249-251	62-26026 <=$>
1952 V55 N5 P273-275	60-14139 <*>
1952 V55 N7 P455-457	64-20956 <*>
1952 V55 N7 P485-487	61-10603 <*>
1952 V55 N7 P492-494	40P66J <ATS>
	62-16184 <*>
1952 V55 N9 P607-608	57-1312 <*>
	90L33J <ATS>
1952 V55 N9 P628-629	64-20420 <*> P
1952 V55 N11 P718-720	91L33J <ATS>
1952 V55 N11 P724-725	KKZ-55-11-1 <SA>
1953 V56 P79-81	II-250 <*>
1953 V56 P100-101	66G6J <ATS>
1953 V56 P387-389	346 <TC>
1953 V56 P396-397	3293 <*>
1953 V56 P504-506	811 <TC>
1953 V56 P516-519	63-10926 <*> 0
1953 V56 P720-724	2633 <*>
1953 V56 N2 P92-94	25K23J <ATS>
	64-16039 <*>
1953 V56 N3 P159-160	64-20421 <*> P
1953 V56 N3 P184-185	62-10377 <*>
1953 V56 N3 P198-200	61-10599 <*>
1953 V56 N4 P281-284	57-1988 <*>
1953 V56 N5 P375-376	65-13330 <*>
1953 V56 N6 P387-389	65-11421 <*> 0
1953 V56 N6 P410-411	98G6J <ATS>
1953 V56 N6 P422-423	65-13329 <*>
1953 V56 N6 P440-441	61-10832 <*>
1953 V56 N7 P554-556	61-10600 <*>
1953 V56 N9 P704-707	12F3J <ATS>
	57-2677 <*>
1953 V56 N10 P727-731	NP-TR-438 <*>
1953 V56 N10 P800-801	91M38J <ATS>
1953 V56 N11 P824-826	NP-TR-444 <*>
1953 V56 N12 P970-971	II-822 <*>
1954 V57 P6-9	UCRL TRANS-862 <*>
1954 V57 P37-40	59-16158 <=*>

Citation	Translation
1954 V57 P161-164	1310 <TC>
1954 V57 P210-212	T1733 <INSD>
1954 V57 P257-258	57-253 <*>
1954 V57 P343-346	57-254 <*>
1954 V57 P369-370	59-16008 <=*>
1954 V57 P530-533	57-1930 <*>
1954 V57 P836-839	48K25J <ATS>
1954 V57 P932-934	12427-A <K-H>
1954 V57 P941-943	65-00380 <*>
1954 V57 N1 P18-20	42N54J <ATS>
1954 V57 N1 P53-56	66-10272 <*>
1954 V57 N2 P105-106	43N54J <ATS>
1954 V57 N2 P118-121	3871 A <K-H>
	64-16155 <*> 0
1954 V57 N2 P152-154	61-20934 <*>
1954 V57 N3 P191-193	KKZ-57-3-1 <SA>
1954 V57 N3 P212-214	61-10491 <*>
1954 V57 N3 P214-216	61-10490 <*>
	95H11J <ATS>
1954 V57 N3 P243-245	57-2674 <*>
1954 V57 N4 P261-264	63-16224 <*> 0
1954 V57 N5 P337-339	3763-D <K-H>
	5287 <HB>
	64-16142 <*> 0
1954 V57 N5 P339-341	CODE-G39 <SA>
	11Q67J <ATS>
1954 V57 N5 P357-359	64-30137 <*> 0
1954 V57 N6 P462-467	65-10243 <*>
1954 V57 N6 P471-474	G40 <SA>
1954 V57 N7 P479-481	60M40J <ATS>
	63-10475 <*>
1954 V57 N9 P641-643	59-17283 <*>
	65-17460 <*>
1954 V57 N9 P658-660	76S89J <ATS>
1954 V57 N10 P723-728	59-17283 <*>
1954 V57 N10 P725-728	65-17461 <*>
1954 V57 N10 P775-777	65-12160 <*>
1954 V57 N11 P797-800	49N54J <ATS>
1954 V57 N11 P800-801	50N54J <ATS>
1954 V57 N12 P897-898	64-16653 <*> 0
1954 V57 N12 P919-922	61-20933 <*> 0
1954 V57 N12 P932-934	65-14503 <*> 0
1954 V57 N658 P89-90	SCL-T-218 <+*>
	58-1383 <*>
1955 V58 P181-183	37P63J <ATS>
1955 V58 P286-291	57-2794 <*>
	67G7J <ATS>
1955 V58 P307-312	61-20939 <*> 0
1955 V58 P442-448	34G8J <ATS>
1955 V58 P805-806	709 <TC>
1955 V58 P946-	58-1281 <*>
1955 V58 N1 P76-77	20S88J <ATS>
1955 V58 N2 P83-86	64-30188 <*> 0
1955 V58 N2 P100-102	16G8J <ATS>
1955 V58 N2 P110-112	40H9J <ATS>
1955 V58 N2 P128-	2476 <*>
1955 V58 N2 P147-153	28M42J <ATS>
1955 V58 N3 P194-196	75S89J <ATS>
1955 V58 N3 P231-233	G41 <SA>
1955 V58 N8 P547-550	66P61J <ATS>
1955 V58 N8 P550-551	67P61J <ATS>
1955 V58 N8 P551-554	68P61J <ATS>
1955 V58 N10 P779-781	63-18583 <*>
1955 V58 N11 P820-824	C-2249 <NRC>
1955 V58 N12 P1001-1004	60-14142 <*>
1956 V59 P158-161	58-1374 <*>
1956 V59 P221-224	889 <TC>
1956 V59 P503-505	91J16J <ATS>
1956 V59 P564-567	RCT V33 N2 P416 <RCT>
1956 V59 P1129-1131	58-1184 <*>
1956 V59 P1132-1134	12427-B <K-H>
1956 V59 P1206-1209	5866-I <K-H>
1956 V59 P1230	62-10378 <*>
1956 V59 N1 P71-77	65-17435 <*>
1956 V59 N1 P82-87	60-18877 <*>
1956 V59 N1 P87-89	60-18999 <*>
1956 V59 N2 P148-151	44N54J <ATS>
1956 V59 N4 P468-470	64-30813 <*>
1956 V59 N6 P695-698	59-15275 <*>
1956 V59 N6 P698-700	60-18460 <*> 0
1956 V59 N7 P834-837	62-14973 <*>
1956 V59 N9 P1072-1074	61-16575 <*>
1956 V59 N10 P1104-1106	59-15241 <*>
1956 V59 N10 P1132-1134	65-14504 <*>
1956 V59 N10 P1206-1209	64-18392 <*>
1956 V59 N10 P1227-1228	10K26J <ATS>
1956 V59 N11 P1304-1308	63-01536 <*>
1956 V59 N11 P1369	KKZ-59-11-1 <SA>
1956 V59 N12 P1425-1429	G42 <SA>
1957 V60 P362-	RCT V31 N4 P800 <RCT>
1957 V60 P951-954	1692 <TC>
1957 V60 N3 P268-272	11949-A <K-H>
	65-14027 <*> 0
1957 V60 N4 P403-407	G43 <SA>
1957 V60 N4 P423-426	15L33J <ATS>
	59-20103 <*>
1957 V60 N5 P611-615	25L36J <ATS>
1957 V60 N6 P768-773	60-18902 <*> 0
1957 V60 N7 P840-843	59-15101 <*> 0
1957 V60 N8 P968-972	62-10937 <*>
1957 V60 N10 P1260-1262	45N54J <ATS>
1957 V60 N10 P1262-1265	46N54J <ATS>
1957 V60 N10 P1268-1271	64Q69J <ATS>
1957 V60 N10 P1294-1299	G44 <SA>
1957 V60 N11 P1434-1435	KKZ-60-11-1 <SA>
1958 V61 P106-109	12245-A <K-H>
1958 V61 P140-	RCT V33 N1 P211 <RCT>
1958 V61 P918-922	66-11113 <*>
1958 V61 P966-969	66-14738 <*$>
1958 V61 P1295-1298	61-16798 <*>
1958 V61 P1377-	RCT V33 N1 P217 <RCT>
1958 V61 P1636-1640	6117 <RIS>
1958 V61 N1 P13-16	90L36J <ATS>
1958 V61 N1 P106-109	65-14482 <*> 0
1958 V61 N2 P214-217	61-10597 <*>
1958 V61 N2 P241-243	58-2490 <*>
1958 V61 N2 P243-244	58-2503 <*>
1958 V61 N3 P384-387	60-18157 <*>
1958 V61 N4 P434-436	G45 <SA>
1958 V61 N4 P452-454	59-10493 <*>
1958 V61 N4 P466-469	59K24J <ATS>
1958 V61 N5 P598-600	38K24J <ATS>
1958 V61 N5 P614-619	45K24J <ATS>
1958 V61 N6 P638-641	59-10516 <*>
1958 V61 N6 P723-728	59-15496 <*>
	60-14065 <*> 0
1958 V61 N6 P728-734	59-15495 <*>
1958 V61 N7 P769-774	10L29J <ATS>
1958 V61 N7 P793-797	65-10078 <*>
1958 V61 N7 P862-865	56K27J <ATS>
1958 V61 N8 P946-948	59-10513 <*>
1958 V61 N8 P1024-1027	59-10514 <*>
1958 V61 N8 P1027-1030	59-10515 <*>
1958 V61 N8 P1046-1050	59-15500 <*>
1958 V61 N9 P1169-1172	59-15499 <*>
1958 V61 N10 P1312-1317	59-17126 <*>
1958 V61 N10 P1353-1359	57S88J <ATS>
1958 V61 N10 P1362-1366	KKZ-61-10-1 <SA>
	11796-A <K-H>
	65-13997 <*>
1958 V61 N11 P1430-1435	59-15498 <*>
1958 V61 N11 P1436-1439	59-15497 <*>
1958 V61 N11 P1508-1513	64Q67J <ATS>
1958 V61 N12 P1543-1546	67S88J <ATS>
1959 V62 P86-88	63-18285 <*>
1959 V62 P137-140	59-17125 <*>
1959 V62 P435-438	60-16743 <*>
1959 V62 P1254-	RCT V34 N2 P648 <RCT>
1959 V62 N1 P29-33	KKZ-62-1-1 <SA>
1959 V62 N1 P59-63	63-16147 <*>
1959 V62 N1 P89-90	65-14247 <*>
1959 V62 N1 P91-93	65-14248 <*>
1959 V62 N2 P268-273	KKZ-62-2-1 <SA>
1959 V62 N2 P273-276	91P64J <ATS>
1959 V62 N2 P276-278	07Q74J <ATS>
1959 V62 N3 P476-478	59-17130 <*> 0
1959 V62 N3 P485	KKZ-62-3-1 <SA>
1959 V62 N4 P538-542	66-11554 <*> 0
1959 V62 N5 P669-672	60-10445 <*>
1959 V62 N6 P825-828	60-10444 <*>

1959 V62 N6 P844-845 KKZ-62-6-1 <SA>
1959 V62 N6 P846-848 KKZ-62-6-2 <SA>
1959 V62 N7 P950-954 25M45J <ATS>
1959 V62 N7 P1048-1050 60-10802 <*>
1959 V62 N7 P1051-1054 KKZ-62-7-1 <SA>
1959 V62 N8 P1084-1087 36M39J <ATS>
1959 V62 N8 P1087-1089 37M39J <ATS>
1959 V62 N8 P1089-1091 38M39J <ATS>
1959 V62 N8 P1114-1116 KKZ-62-8-1 <SA> 0
1959 V62 N8 P1117-1119 KKZ-62-8-2 <SA> 0
60-10446 <*>
1959 V62 N9 P1269-1273 KKZ-62-9-1 <SA> 0
1959 V62 N9 P1313-1318 60-14246 <*>
1959 V62 N9 P1371-1373 60-14247 <*>
1959 V62 N9 P1449-1453 60-16672 <*>
1959 V62 N10 P1555-1559 42M38J <ATS>
1959 V62 N10 P1610-1612 60-18152 <*>
1959 V62 N11 P1677-1681 64-30139 <*> 0
1959 V62 N11 P1747-1752 KKZ-62-11-1 <SA>
1959 V62 N11 P1764-1766 66-13969 <*>
1959 V62 N11 P1781-1785 60-16673 <*>
1959 V62 N11 P1786-1788 61-10001 <*>
1959 V62 N12 P1875-1876 95M42J <ATS>
1959 V62 N12 P1897-1904 60-16678 <*>
1960 V63 P459-564 63-14038 <*> 0
1960 V63 N1 P114-118 51M47J <ATS>
1960 V63 N1 P173-175 10512-B <K-H>
1960 V63 N1 P176-178 60-18776 <*>
1960 V63 N1 P178-183 60-16674 <*>
1960 V63 N1 P186-188 60-16671 <*>
1960 V63 N2 P338-341 61-10007 <*>
1960 V63 N3 P418-427 64-14358 <*>
1960 V63 N3 P523-528 60-18146 <*>
1960 V63 N3 P536-540 66-13970 <*>
1960 V63 N4 P588-592 61-10682 <*> 0
1960 V63 N4 P608-611 UCRL TRANS-635(L) <*>
1960 V63 N4 P640-645 65-14080 <*>
1960 V63 N4 P645-652 65-14157 <*>
1960 V63 N5 P675-680 00M46J <ATS>
1960 V63 N5 P714-717 61-10830 <*>
1960 V63 N5 P734-737 KKZ-63-5-1 <SA>
69M44J <ATS>
1960 V63 N5 P797-799 60-18881 <*>
1960 V63 N5 P799-803 60-18882 <*>
1960 V63 N5 P851-854 61-10272 <*>
1960 V63 N5 P868-871 09M46J <ATS>
1960 V63 N5 P880-883 KKZ-63-5-2 <SA>
1960 V63 N5 P893-894 61-16523 <*>
1960 V63 N6 P903-906 61-10589 <*>
1960 V63 N6 P967-970 61-10601 <*>
1960 V63 N6 P1003-1010 KKZ-63-6-1 <SA>
1960 V63 N6 P1059-1061 61-16576 <*>
1960 V63 N7 P1166-1172 61-10831 <*>
1960 V63 N7 P1208-1211 CODE-1 <SA>
1960 V63 N7 P1214-1218 61-10593 <*>
1960 V63 N7 P1233-1235 62-14374 <*>
1960 V63 N8 P1364-1367 61-10605 <*>
1960 V63 N8 P1372-1376 63-20475 <*> 0
67P64J <ATS>
1960 V63 N8 P1390-1394 KKZ-63-8-1 <SA> 0
1960 V63 N8 P1427-1430 65-17436 <*>
1960 V63 N9 P1514-1516 M-5712 <NLL>
32N51J <ATS>
1960 V63 N9 P1527-1530 56N55J <ATS>
1960 V63 N9 P1586-1587 AEC-TR-4805 <*>
1960 V63 N9 P1631-1635 KKZ-63-9-1 <SA>
61-10840 <*>
1960 V63 N9 P1636-1639 KKZ-63-9-2 <SA>
61-10838 <*>
1960 V63 N9 P1639-1643 KKZ-63-9-3 <SA> 0
61-10839 <*>
1960 V63 N10 P1690-1694 CODE-G78 <SA>
1960 V63 N10 P1754-1757 CODE-92 <SA>
1960 V63 N10 P1769-1772 61-16125 <*>
1960 V63 N10 P1798-1801 66-11325 <*>
1960 V63 N10 P1807-1811 09Q74J <ATS>
1960 V63 N10 P1812-1817 CODE-107 <SA>
1960 V63 N10 P1817-1822 KKZ-63-10-3 <SA>
61-16519 <*>
1960 V63 N11 P1869-1875 66N51J <ATS>
1960 V63 N11 P1941-1944 61-16520 <*>
1960 V63 N11 P1945-1949 63-20358 <*> 0
1960 V63 N11 P2042-2045 61-16521 <*>
1960 V63 N11 P2059-2060 61-16522 <*>
1960 V63 N12 P2134-2140 63-20476 <*>
1960 V63 N12 P2177-2180 61-18630 <*>
1961 V64 P129-137 AEC-TR-6007 <*>
1961 V64 P137-141 63-18747 <*>
1961 V64 P380-382 CODE-G285 <SA>
1961 V64 P479- RCT V35 N2 P484 <RCT>
1961 V64 P483- RCT V35 N2 P491 <RCT>
1961 V64 P1013-1017 2480 <BISI>
1961 V64 P1291-1294 SC-T-64-2009 <*>
1961 V64 N1 P123-128 61-18681 <*>
1961 V64 N1 P213-218 55N55 <ATS>
61-16540 <*>
1961 V64 N1 P226-228 61-18676 <*>
1961 V64 N1 P229-231 61-18678 <*>
1961 V64 N2 P367-368 <SA>
KKZ-64-2-1 <SA>
1961 V64 N2 P369-372 CODE-G23B <SA>
1961 V64 N2 P378-379 CODE-G284 <SA> 0
1961 V64 N2 P392-395 26Q73J <ATS>
1961 V64 N2 P396-398 CODE-G290 <SA> 0
1961 V64 N2 P405-408 65-10052 <*> 0
1961 V64 N2 P405-411 65-13999 <*> 0
1961 V64 N2 P412-419 65-14000 <*> 0
1961 V64 N3 P467-468 CODE-G303 <SA>
1961 V64 N3 P539-540 28N56J <ATS>
1961 V64 N3 P541-543 27N56J <ATS>
1961 V64 N3 P564-567 CCL-64/8 <STC>
1961 V64 N3 P573-576 CODE-G26 <SA>
1961 V64 N4 P635-638 CODE-G346 <SA> 0
61-20681 <*> 0
1961 V64 N4 P643-653 04P63J <ATS>
1961 V64 N4 P745 CODE-G376 <SA>
1961 V64 N5 P787-791 63-18024 <*>
1961 V64 N5 P916-919 62-20254 <*> 0
1961 V64 N5 P925-928 62-10222 <*>
1961 V64 N5 P929-932 CODE-G419 <SA>
1961 V64 N6 P1017-1020 CODE-G441 <SA>
1961 V64 N6 P1031-1034 62-14289 <*>
1961 V64 N6 P1035-1040 62-14288 <*>
1961 V64 N6 P1043-1047 62-14287 <*>
1961 V64 N6 P1118-1121 CODE-38 <SA>
1961 V64 N6 P1130-1132 62-14968 <*>
1961 V64 N6 P1132-1134 62-14969 <*>
1961 V64 N6 P1134-1136 CODE-G469 <SA>
62-14965 <*>
1961 V64 N6 P1140-1145 CODE-43 <SA>
1961 V64 N7 P1218-1221 62-10686 <*>
1961 V64 N7 P1230-1233 CODE-G495 <SA>
62-14966 <*>
1961 V64 N7 P1234-1238 KKZ-64-7-2 <SA>
62-10223 <*>
1961 V64 N7 P1285-1289 CODE-G511 <SA>
1961 V64 N7 P1302-1304 65-14089 <*>
1961 V64 N7 P1321-1322 CODE-G524 <SA>
1961 V64 N8 P1397-1400 62-14874 <*> 0
1961 V64 N8 P1400-1403 62-16185 <*> 0
1961 V64 N8 P1452-1455 62-10684 <*>
1961 V64 N8 P1456-1460 62-10685 <*>
1961 V64 N8 P1489-1493 62-10689 <*>
1961 V64 N8 P1497-1501 CODE-G564 <SA>
1961 V64 N8 P1513-1514 CODE-G570 <SA>
33P62J <ATS>
62-20190 <*>
1961 V64 N9 P1515-1518 CODE-G572 <SA>
1961 V64 N9 P1523-1527 64-14353 <*> 0
1961 V64 N9 P1562-1567 CODE-G585 <SA>
1961 V64 N9 P1568-1573 CODE-G586 <SA>
1961 V64 N9 P1583-1588 65-14038 <*>
1961 V64 N9 P1585-1588 62-18748 <*>
1961 V64 N9 P1588-1592 64-10502 <*>
1961 V64 N9 P1620-1623 62-26136 <=$>
1961 V64 N9 P1662-1664 C-4405 <NRC> 0
63-18019 <*> 0
1961 V64 N9 P1668-1670 CODE-G610 <SA>
1961 V64 N9 P1676-1681 62-14967 <*>
1961 V64 N9 P1682-1686 CODE-G613 <SA>

62-18039 <*>
1961 V64 N10 P1859-1864 CODE-G646 <SA>
1961 V64 N10 P1864-1869 KKZ-64-10-2 <SA>
1961 V64 N11 P1939-1942 64-30140 <*> 0
1961 V64 N11 P1945-1948 85P61J <ATS>
1961 V64 N11 P1955-1957 63-10315 <*>
1961 V64 N11 P1995-1998 <INT>
1961 V64 N11 P2014-2017 63-10754 <*> 0
1961 V64 N11 P2049-2052 CODE-G691 <SA> 0
1961 V64 N11 P2072 CODE-G696 <SA>
1961 V64 N12 P2124-2126 62-14643 <*> 0
1961 V64 N12 P2126-2128 62-14642 <*> 0
1961 V64 N12 P2129-2131 CODE-G719 <SA>
1961 V64 N12 P2139-2142 63-16276 <*> 0
1961 V64 N12 P2142-2145 63-16277 <*> 0
1962 N8 P1207-1211 ICE V4 N4 P754-760 <ICE>
1962 V65 P880-884 KKZ-65-6-2 <SA>
1962 V65 P1041-1044 63-20245 <*>
1962 V65 P1139-1140 AEC-TR-5838 <*>
1962 V65 P1165-1167 UCRL TRANS-1044(L) <*>
1962 V65 P1168-1170 UCRL TRANS-1043(L) <*>
1962 V65 P1260-1265 63-20244 <*>
1962 V65 P1286-1290 SCL-T-498 <*>
1962 V65 P1388-1390 12641 <K-H>
1962 V65 P1622-1626 SCL-T-503 <*>
1962 V65 P1748-1753 63-14499 <*>
1962 V65 N1 P38-41 62-16190 <*> 0
1962 V65 N1 P82-88 65-13614 <*>
1962 V65 N1 P88-92 CODE-G21 <SA>
65-13520 <*>
1962 V65 N1 P93-95 65-12860 <*>
1962 V65 N1 P96-98 65-12868 <*>
1962 V65 N1 P132-136 60Q71J <ATS>
1962 V65 N2 P201-205 11949-B <K-H>
65-14026 <*>
1962 V65 N2 P206- 11949-C <K-H>
1962 V65 N2 P206-209 65-14020 <*> 0
1962 V65 N2 P209-213 65-14022 <*> 0
1962 V65 N2 P217-220 KKZ-65-2-1 <SA>
1962 V65 N2 P221-223 KKZ-65-2-2 <SA>
1962 V65 N2 P234-239 KKZ-65-2-3 <SA>
1962 V65 N2 P247-250 KKZ-65-2-4 <SA>
1962 V65 N2 P267-269 KKZ-65-2-5 <SA>
1962 V65 N2 P284-288 65-10244 <*>
1962 V65 N2 P289-292 63-14716 <*>
1962 V65 N2 P293 92P62J <ATS>
1962 V65 N3 P318-323 64-10189 <*>
1962 V65 N3 P368-370 62-18697 <*> 0
1962 V65 N3 P371-375 24S81J <ATS>
1962 V65 N3 P377-382 KKZ-65-3-1 <SA>
1962 V65 N3 P415-418 64-16184 <*>
1962 V65 N3 P419-422 64-16183 <*>
1962 V65 N3 P422-426 64-16186 <*>
1962 V65 N3 P427-431 64-16187 <*>
1962 V65 N4 P548-551 63-16274 <*> 0
1962 V65 N4 P552-556 KKZ-65-4-1 <SA>
63-16275 <*> 0
1962 V65 N4 P583-586 KKZ-65-4-2 <SA>
1962 V65 N4 P605-609 64-16185 <*>
1962 V65 N4 P613-615 KKZ-65-4-3 <SA>
1962 V65 N4 P642 <LSA>
KKZ-65-4-4 <SA>
65-13328 <*>
1962 V65 N5 P643-648 KKZ-65-5-1 <SA>
12425-A <K-H>
65-14511 <*>
1962 V65 N5 P649-657 65-14512 <*>
1962 V65 N5 P691-695 12425-C <K-H>
65-14508 <*>
1962 V65 N5 P695-698 <LSA>
KKZ-65-5-2 <SA> 0
65-14513 <*>
1962 V65 N5 P699-702 12425-E <K-H>
65-14514 <*>
1962 V65 N5 P712-716 12425-F <K-H>
65-14515 <*>
1962 V65 N5 P743-745 KKZ-65-5-3 <SA>
1962 V65 N6 P837-843 KKZ-65-6-1 <SA>
63-10083 <*>
1962 V65 N6 P945-949 KKZ-65-6-3 <SA>
1962 V65 N6 P1000-1003 66-11694 <*>
1962 V65 N7 P1049-1054 KKZ-65-7-1 <SA> 0
1962 V65 N7 P1054-1058 KKZ-65-7-2 <SA>
1962 V65 N7 P1059-1061 KKZ-65-7-3 <SA>
85Q66J <ATS>
1962 V65 N7 P1061-1063 KKZ-65-7-4 <SA>
1962 V65 N8 P1170-1174 KKZ-65-8-1 <SA>
1962 V65 N8 P1179-1182 63-16423 <*>
1962 V65 N8 P1207-1210 ICE V4 N4 P754-760 <ICE>
09Q71J <ATS>
62-20485 <*> 0
1962 V65 N8 P1211-1213 65-18018 <*>
1962 V65 N8 P1286-1290 K-K-Z-65-8-2 <NLL>
1962 V65 N9 P1388-1390 65-14845 <*>
1962 V65 N9 P1426-1430 63-20062 <*>
1962 V65 N10 P1603-1605 63-18870 <*>
1962 V65 N10 P1622-1626 02Q67J <ATS>
1962 V65 N11 P1733-1735 66-13886 <=$>
1962 V65 N11 P1779-1782 54R79J <ATS>
1962 V65 N11 P1861-1865 63-18514 <*> 0
1962 V65 N11 P1865-1868 63-18858 <*>
1962 V65 N11 P1869-1874 63-18857 <*> 0
1962 V65 N11 P1889-1896 M-5730 <NLL>
1962 V65 N12 P1911-1916 65-14252 <*>
1962 V65 N12 P1916-1918 65-14253 <*>
1962 V65 N12 P2005-2009 63-16644 <*>
1962 V65 N12 P2013-2017 63-18741 <*>
1962 V65 N12 P2027-2032 63-13512 <*>
1962 V65 N12 P2036-2043 40S81J <ATS>
1962 V65 N12 P2042-2046 41S81J <ATS>
1962 V65 N12 P2085-2086 62R75J <ATS>
1963 N10 P1433- ICE V4 N2 P340-345 <ICE>
1963 N10 P1466-1468 ICE V4 N3 P530-534 <ICE>
1963 V66 P100-103 1243 <TC>
1963 V66 P289-292 65-14856 <*>
1963 V66 P292-294 65-14855 <*>
1963 V66 P1538-1541 8940 <IICH>
1963 V66 N2 P188-191 66-11693 <*> 0
1963 V66 N2 P246-249 95Q69J <ATS>
1963 V66 N2 P249-252 96Q69J <ATS>
1963 V66 N2 P279-283 63-18909 <*>
1963 V66 N3 P324-326 64-18102 <*>
1963 V66 N3 P348-351 12963 <K-H>
65-14864 <*>
1963 V66 N3 P365-369 63-20687 <*>
1963 V66 N3 P382-386 64-10958 <*> 0
1963 V66 N4 P432-434 64-14987 <*> 0
87R74J <ATS>
1963 V66 N4 P435-437 64-10513 <*> 0
1963 V66 N4 P442-446 65-13327 <*>
1963 V66 N4 P446-450 65-13326 <*>
1963 V66 N4 P485-489 63-18902 <*>
1963 V66 N5 P577-580 64-10708 <*>
1963 V66 N5 P613-617 63-20685 <*>
1963 V66 N5 P618-620 63-20688 <*>
1963 V66 N5 P619-623 63-20686 <*>
1963 V66 N5 P621-624 63-20689 <*>
1963 V66 N5 P793-797 64-10702 <*>
1963 V66 N6 P797-803 63-20696 <*>
1963 V66 N6 P804-809 63-20695 <*>
1963 V66 N6 P824-827 64-16624 <*>
1963 V66 N7 P948-952 64-14357 <*>
1963 V66 N8 P1241-1244 64-10703 <*>
1963 V66 N9 P1287-1289 64-14986 <*> 0
1963 V66 N9 P1361-1364 65-12897 <*>
1963 V66 N9 P1404-1405 66-14843 <*>
1963 V66 N10 P1412-1416 66-10078 <*>
1963 V66 N10 P1433- ICE V4 N2 P340-345 <ICE>
1963 V66 N10 P1466-1468
ICE V4 N3 P530-534 <ICE>
78R74J <ATS>
1963 V66 N11 P1543-1559 64-31629 <=>
1963 V66 N11 P1601-1604 64-18129 <*> 0
1963 V66 N11 P1605-1609 N66-15637 <=$>
1963 V66 N11 P1610-1613 65-10484 <*>
1963 V66 N11 P1679-1682 65-11323 <*>
1963 V66 N11 P1694-1697 65-13332 <*> 0
1963 V66 N11 P1697-1702 65-13340 <*> 0
1963 V66 N11 P1703-1707 65-13331 <*> 0
1963 V66 N11 P1717-1720 64-20948 <*> 0

Citation	Translation
1964 N3 P411-414	ICE V5 N1 P175-179 <ICE>
1964 N3 P415-418	ICE V5 N1 P186-190 <ICE>
1964 N3 P456-460	ICE V5 N1 P180-185 <ICE>
1964 N10 P1612-1616	ICE V5 N2 P384-389 <ICE>
1964 N10 P1624-1629	ICE V5 N2 P367-373 <ICE>
1964 N11 P1798-1801	ICE V5 N2 P374-378 <ICE>
1964 V67 P289-292	GI N1 1964 P90 <AGI>
1964 V67 P292-297	GI N1 1964 P96 <AGI>
1964 V67 P576-578	65-11934 <*>
1964 V67 P626-629	733 <TC>
1964 V67 P714-719	65-18116 <*>
1964 V67 P1077-1081	84S83J <ATS>
1964 V67 P1144-1147	65-17437 <*>
1964 V67 P1670-1675	1311 <TC>
1964 V67 N1 P182-185	25R80J <ATS>
1964 V67 N1 P193-197	76S81J <ATS>
1964 V67 N1 P205-209	42R77J <ATS>
1964 V67 N2 P321-323	64-18122 <*> 0
1964 V67 N2 P362-366	64-18485 <*>
1964 V67 N3 P411-414	ICE V5 N1 P175-179 <ICE>
1964 V67 N3 P415-418	ICE V5 N1 P186-190 <ICE>
1964 V67 N3 P424-428	64-18493 <*>
1964 V67 N3 P429-432	65-10380 <*> 0
1964 V67 N3 P432-436	65-10379 <*>
1964 V67 N3 P456-460	ICE V5 N1 P180-185 <ICE>
1964 V67 N3 P474-478	65-13076 <*> 0
1964 V67 N4 P592-596	65-11544 <*> 0
1964 V67 N4 P604-607	65-10378 <*>
1964 V67 N5 P843-847	66-11503 <*>
1964 V67 N6 P880-883	66-10248 <*>
1964 V67 N6 P956-961	65-10393 <*>
1964 V67 N7 P1015-1018	66-11395 <*> 0
1964 V67 N7 P1018-1021	66-11396 <*> 0
1964 V67 N7 P1021-1025	66-11397 <*> 0
1964 V67 N7 P1026-1031	65-13685 <*>
	66-11398 <*> 0
1964 V67 N7 P1073-1076	65-13334 <*> 0
1964 V67 N7 P1081-1083	65-10489 <*>
1964 V67 N8 P1153-1157	66-10808 <*>
1964 V67 N9 P1396-1401	1946 <TC>
1964 V67 N9 P1473-1476	65-13684 <*>
1964 V67 N9 P1476-1478	65-13683 <*>
1964 V67 N9 P1479-1484	65-13682 <*>
1964 V67 N10 P1499-1501	77S81J <ATS>
1964 V67 N10 P1542-1545	63S84J <ATS>
1964 V67 N10 P1564-1566	65-13350 <*> 0
1964 V67 N10 P1612-1616	15S81J <ATS>
1964 V67 N10 P1624-1629	14S81J <ATS>
1964 V67 N10 P1652-1658	65-13333 <*> 0
1964 V67 N10 P1658-1660	65-14897 <*>
1964 V67 N10 P1661-1664	65-13335 <*> 0
1964 V67 N11 P1710-1713	N65-23679 <=$>
1964 V67 N11 P1744-1748	N65-27677 <=*$>
1964 V67 N11 P1778-1793	N65-27679 <=*$>
1964 V67 N11 P1931-1937	66-11556 <*>
	66-12778 <*>
1964 V67 N11 P1938-1941	95S81J <ATS>
1964 V67 N12 P2019-2023	66-13587 <*>
1964 V67 N12 P2023-2025	66-13588 <*>
1964 V67 N12 P2145-2149	66-11558 <*>
1964 V67 N12 P2150-2153	66-12572 <*>
1964 V67 N12 P2154-2156	66-12054 <*>
1964 V67 N12 P2163-2167	57S85J <ATS>
1965 N1 P63-67	ICE V5 N4 P737-742 <ICE>
1965 N1 P126-129	ICE V5 N4 P732-736 <ICE>
1965 N2 P283-286	ICE V5 N3 P580-583 <ICE>
1965 N2 P364-368	ICE V6 N1 P163-169 <ICE>
1965 N3 P419-423	ICE V5 N4 P749-756 <ICE>
1965 N3 P535-541	ICE V5 N4 P724-731 <ICE>
1965 N5 P904-908	ICE V6 N1 P170-176 <ICE>
1965 N5 P964-968	ICE V6 N1 P177-182 <ICE>
1965 N9 P1646-1651	ICE V6 N2 P351-358 <ICE>
1965 N9 P1651-1654	ICE V6 N2 P358-364 <ICE>
1965 N9 P1655-1661	ICE V6 N2 P364-373 <ICE>
1965 N9 P1729-1732	ICE V6 N2 P346-350 <ICE>
1965 V68 P126-129	82S82J <ATS>
1965 V68 P180-184	GI V2 N3 1965 P599 <AGI>
1965 V68 P1013-1016	1258 <TC>
1965 V68 N1 P63-67	83S82J <ATS>
1965 V68 N1 P67-72	65-14803 <*>
1965 V68 N1 P106-109	65-14804 <*> 0
1965 V68 N1 P174-179	65-14166 <*> 0
1965 V68 N2 P269-272	35S83J <ATS>
1965 V68 N2 P283-286	41S83J <ATS>
	66-15562 <*$>
1965 V68 N2 P300-303	66-14836 <*$>
	92S84J <ATS>
1965 V68 N2 P364-368	89S83J <ATS>
1965 V68 N2 P383-387	65-14167 <*> 0
1965 V68 N3 P419-423	89S84J <ATS>
1965 V68 N3 P481-485	66-11501 <*>
1965 V68 N3 P535-541	81S84J <ATS>
1965 V68 N3 P574-576	65-14168 <*> 0
1965 V68 N4 P654-658	65-14802 <*>
1965 V68 N4 P703-706	66-12577 <*>
1965 V68 N5 P881-885	66-11502 <*>
1965 V68 N5 P885-891	66-11492 <*>
1965 V68 N5 P904-908	58S87J <ATS>
1965 V68 N5 P957-960	66-10129 <*>
1965 V68 N5 P964-968	91S86J <ATS>
1965 V68 N5 P983-986	65-14907 <*>
1965 V68 N5 P991-997	1074 <TC>
1965 V68 N6 P1009-1012	1257 <TC>
1965 V68 N6 P1024-1026	1353 <TC>
1965 V68 N6 P1098-1102	66-11459 <*>
1965 V68 N6 P1132-1134	66-13879 <*>
1965 V68 N7 P1172-1175	1435 <TC>
1965 V68 N7 P1188-1194	66S88J <ATS>
1965 V68 N7 P1288-1291	66-13275 <*>
1965 V68 N7 P1292-1294	66-13278 <*>
1965 V68 N8 P1582-1586	66-12885 <*>
1965 V68 N9 P1675-1680	66-11458 <*>
1965 V68 N9 P1729-1732	17S89J <ATS>
1965 V68 N9 P1756-1761	66-1077 <*> 0
1965 V68 N10 P1682-1685	66-13881 <*>
1965 V68 N10 P1811-1815	66-14890 <*>
1965 V68 N10 P1815-1822	66-14889 <*>
1965 V68 N10 P1822-1827	66-14891 <*>
1965 V68 N10 P1842-1845	66-13882 <*>
1965 V68 N10 P1873-1877	66-12579 <*>
	66-13880 <*>
1965 V68 N10 P1910-1914	85T91J <ATS>
1965 V68 N10 P1941-1947	66-14517 <*>
1965 V68 N11 P2089-2092	66-13795 <*>
1965 V68 N11 P2186-2189	66-13987 <*>
1965 V68 N11 P2190-2195	66-13988 <*>
1965 V68 N11 P2196-2200	66-13989 <*>
1965 V68 N11 P2200-2204	66-13990 <*>
1965 V68 N11 P2205-2208	66-13991 <*>
1965 V68 N11 P2209-2212	66-13997 <*>
1965 V68 N11 P2213-2217	66-13992 <*>
1965 V68 N11 P2275-2277	66-14377 <*$> 0
1965 V68 N12 P2505-2508	66-12504 <*>
1966 N1 P36-41	ICE V6 N3 P546-553 <ICE>
1966 N1 P41-44	ICE V6 N3 P553-557 <ICE>
1966 N1 P45-51	ICE V6 N3 P558-566 <ICE>
1966 N3 P411-415	ICE V6 N4 P725-729 <ICE>
1966 N3 P426-431	ICE V6 N4 P717-724 <ICE>
1966 N4 P593-597	ICE V6 N4 P737-743 <ICE>
1966 N4 P690-694	ICE V6 N4 P744-751 <ICE>
1966 V69 N1 P6-8	66-13461 <*>
	66-13883 <*>
1966 V69 N1 P36-41	71T92J <ATS>
1966 V69 N1 P41-44	72T92J <ATS>
1966 V69 N1 P45-51	73T92J <ATS>
1966 V69 N1 P100-102	66-12574 <*>
1966 V69 N1 P103-105	66-12572 <*>
1966 V69 N3 P411-415	29T94J <ATS>
1966 V69 N3 P426-431	30T94J <ATS>
1966 V69 N4 P593-597	57T95J <ATS>
1966 V69 N4 P690-694	58T95J <ATS>

KOGYO KAYAKU KYOKAISHI

Citation	Translation
1953 V14 N3 P142-163	AD-620 222 <=$>
1955 V16 N2 P90-94	20TM <CTT>
1957 V18 N1 P64-66	82J17J <ATS>
1958 V19 N2 P118-121	69M38J <ATS>
1963 V24 N6 P318-329	66-12128 <*> 0
1963 V24 N6 P351-353	66-12127 <*>

1958 V3 N1 P20-25 60-21244 <=>
1958 V3 N1 P25-27 1765 <BISI>
1958 V3 N2 P47-50 61-11350 <=>
1958 V3 N4 P143-151 61-11350 <=>
1958 V3 N5 P184-189 61-11350 <=>
1959 V4 N4 P221-226 62-10024 <*>
1959 V4 N5/6 P211-215 60-17116 <=*$>
1959 V4 N5/6 P216-220 61-11355 <=>
1959 V4 N5/6 P238-240 61-11344 <=>
1960 V5 N1 P9-15 62-10128 <*>
1960 V5 N2 P49-58 61-11350 <=>
1960 V5 N2 P58-61 61-11350 <=>
1962 V7 N5 P181-184 3252 <BISI>
1964 V9 N3 P73-77 64-51256 <=$>
1964 V9 N3 P81-85 4374 <BISI>

KOKUDO KAIHATSU
1953 V2 P27-29 T-1949 <INSD>

KOKU UCHU NENKAN
1964 P91-103 65-30978 <=$>
1964 P126-131 65-31190 <=$>
1964 P215-230 65-31213 <=$>

KOKUSAI KAGAKU JOHO
1963 N5 P49-53 65-30026 <=$>
1963 N5 P57 65-30026 <=$>

KOLKHOZNCE PRCIZVODSTVO
1946 V6 N4 P42-43 RT-1778 <*>
1950 V10 N6 P43 RT-1036 <*>
1955 V15 N4 P24- R-1560 <*>
1956 V16 N5 P12- 1571 <*>
1959 N12 P24-25 60-31630 <=>
1959 V19 N4 P37-39 62-25673 <=*> 0

KCLKHOZNO SOVKHOZNOE PROIZVODSTVO TADZHIKISTANA
1963 V17 N1C P14-17 64-31106 <=>

KOLLOIDBEIHEFTE
1933 V37 P2-39 59-17342 <*>
1934 V40 P429-434 57-3256 <*>
1934 V40 P447-448 57-3256 <*>
1937 V46 P425-479 60-18726 <*>

KOLLOIDCHEMISCHE BEIHEFTE
1909 V1 P423-453 64-18230 <*>
1925 V21 P37-54 64P62G <ATS>
1930 V32 N1/4 P1-113 66-12384 <*>

KOLLOID GESELLSCHAFT
1958 V18 P122-131 59-10800 <*>
1958 V18 P156-168 59-15124 <*>

★ KOLLOIDNYI ZHURNAL
1937 V3 N3 P195-207 R-2266 <*>
1937 V3 N3 P273-280 AEC-TR-4530 <=*>
1938 V4 P699-704 62-16920 <=*> 0
1938 V4 N6/8 P483-495 59-19147 <+*>
1939 V5 N10 P925-932 RT-3964 <*>
1939 V5 N1/2 P105-11C 63-20566 <=*$>
1940 V6 P761-763 RCT V16 N1 P86 <RCT>
1940 V6 N2 P175-182 RCT V14 N1 P835 <RCT>
1947 V9 N4 P255-260 RCT V27 N3 P615 <RCT>
1947 V9 N5 P348-354 RCT V23 N1 P1 <RCT>
1948 V10 P122-124 59-15582 <+*>
1948 V10 N2 P94-102 RCT V23 N3 P553 <RCT>
1948 V10 N2 P148-154 RT-1541 <*>
1948 V10 N3 P241-244 RT-1545 <*>
1948 V10 N4 P268-280 RCT V23 N3 P563 <RCT>
1948 V10 N4 P289-304 TT.144 <NRC>
1948 V10 N6 P413-422 65-12392 <*>
1949 V11 P395-409 63-18998 <=*>
1949 V11 N1 P3-7 RJ-8 <ATS>
1949 V11 N1 P30-33 RJ-9 <ATS>
1949 V11 N3 P141-142 63-14569 <=*>
1949 V11 N3 P143-150 RCT V23 N1 P89 <RCT>
1949 V11 N3 P163-171 63-20814 <=*$>
1949 V11 N3 P172-175 RJ-1 <ATS>
RJ-2 <ATS>
1949 V11 N4 P209-210 65-13424 <*> 0
1949 V11 N4 P230-231 RJ-17 <ATS>
53/0199 <NLL>
1949 V11 N4 P280-282 RJ-18 <ATS>
1949 V11 N5 P308-310 RJ-15 <ATS>
1949 V11 N5 P311-313 RJ-16 <ATS>
1950 V12 N1 P50-61 RCT V25 N3 P596 <RCT>
1950 V12 N1 P62-66 RT-1745 <*>
59-19091 <+*>
1950 V12 N2 P102-111 RCT V24 N1 P99 <RCT>
1950 V12 N3 P184-193 RCT V24 N2 P344 <RCT>
1950 V12 N4 P241-247 RCT V24 N2 P328 <RCT>
1950 V12 N5 P347-351 RT-291 <*>
1950 V12 N5 P352-358 60-13875 <+*>
1950 V12 N6 P408-413 RCT V28 N1 P66 <RCT>
1950 V12 N6 P427-430 RT-676 <*>
1950 V12 N6 P437-447 65-60712 <=> 0
1950 V12 N6 P460-466 TT.267 <NRC>
1951 V13 P175-181 R-814 <*>
1951 V13 P401-407 65-11091 <*>
1951 V13 N1 P11-19 RCT V24 N4P810 <RCT>
1951 V13 N1 P38-45 62-25458 <=*>
1951 V13 N1 P55-63 RT-1952 <*>
1951 V13 N2 P83-88 TT.250 <NRC>
1951 V13 N4 P267-272 RCT V25 N1 P50 <RCT>
1951 V13 N4 P283-288 60-23571 <+*>
1951 V13 N5 P339-345 RCT V26 N1 P70 <RCT>
1951 V13 N5 P346-356 RT-3328 <*>
1951 V13 N5 P371-378 R-557 <*>
1951 V13 N5 P379-382 61-19688 <=*>
1951 V13 N5 P394-395 62-24373 <=*>
1952 V14 N1 P40-45 RCT V25 N4 P801 <RCT>
1952 V14 N1 P57-65 RT-2815 <*>
60-19923 <+*> 0
1952 V14 N1 P66-72 86K21R <ATS>
1952 V14 N2 P107-111 RT-3042 <*>
1952 V14 N2 P140-147 RCT V26 N1 P220 <RCT>
1952 V14 N3 P157-163 RCT V26 N3 P559 <RCT>
1952 V14 N3 P192-196 61-17249 <=$>
1952 V14 N3 P197-203 RCT V26 N3 P624 <RCT>
1952 V14 N4 P250-259 RCT V26 N4 P810 <RCT>
1952 V14 N4 P274-278 60-19922 <*+>
1952 V14 N5 P346-356 RCT V26 N4 P821 <RCT>
1952 V14 N5 P357-366 RCT V27 N4 P964 <RCT>
1952 V14 N6 P408-413 RT-3043 <*>
1952 V14 N6 P444-455 RCT V27 N2 P415 <RCT>
1953 V15 N1 P1-10 RCT V27 N2 P363 <RCT>
1953 V15 N1 P3-5 RJ-114 <ATS>
1953 V15 N1 P6-10 61-17195 <=$>
1953 V15 N1 P51-59 RCT V27 N4 P930 <RCT>
1953 V15 N1 P60-68 RCT V27 N1 P165 <RCT>
1953 V15 N2 P8-12 62-10424 <*=> 0
1953 V15 N2 P108-116 5309 <K-H> 0
59-15925 <+*>
64-16745 <=*$> 0
1953 V15 N3 P170-177 RCT V27 N3 P607 <RCT>
1953 V15 N4 P289-291 62-25393 <=*>
1953 V15 N5 P331-333 RJ-154 <ATS>
64-16759 <=*$>
1953 V15 N5 P347-360 RCT V27 N4 P883 <RCT>
1953 V15 N5 P361-364 RJ-203 <ATS>
1953 V15 N5 P365-370 20N49R <ATS>
1953 V15 N6 P448-454 RCT V28 N2 P527 <RCT>
1954 V16 P3-9 RCT V30 N2 P548 <RCT>
1954 V16 P36-43 RCT V28 N1 P84 <RCT>
1954 V16 N2 P89-93 RT-2770 <*>
1954 V16 N2 P126-133 RCT V28 N1 P57 <RCT>
1954 V16 N2 P141-149 R-448 <*>
1954 V16 N3 P171-178 RCT V28 N2 P494 <RCT>
1954 V16 N3 P211-219 RCT V28 N3 P684 <RCT>
1954 V16 N4 P287-296 RCT V28 N3 P838 <RCT>
1954 V16 N5 P313-321 C <CB>
1954 V16 N5 P322-324 <CB>
1954 V16 N5 P325-332 <CB>
1954 V16 N5 P333-339 <CB>
1954 V16 N5 P340-344 <CB>
1954 V16 N5 P345-349 <CB>

Citation	Translation
	RCT V29 N1 P296 <RCT>
1954 V16 N5 P350-357	<CB>
1954 V16 N5 P358-365	<CB>
1954 V16 N5 P366-375	<CB>
1954 V16 N5 P376-380	<CB>
1954 V16 N5 P381-386	<CB>
1954 V16 N5 P387-389	<CB>
1954 V16 N5 P390-395	<CB>
1954 V16 N5 P396-400	<CB>
1954 V16 N6 P412-420	RCT V29 N1 P135 <RCT>
1955 V17 N1 P18-23	RCT V29 N2 P391 <RCT>
1955 V17 N1 P24-30	RCT V29 N2 P463 <RCT>
1955 V17 N3 P168-170	R-104 <*>
	R-4600 <*>
1955 V17 N3 P171-172	R-105 <*>
	59-15041 <+*>
1955 V17 N3 P215-219	RCT V29 N3 P917 <RCT>
1955 V17 N3 P264	RT-3272 <*>
1955 V17 N6 P415-420	17J15R <ATS>
1955 V17 N6 P468-470	98J15R <ATS>
1956 V18 P167-179	RCT V31 N2 P329 <RCT>
1956 V18 P285-292	RCT V30 N3 P895 <RCT>
1956 V18 P404-412	RCT V30 N3 P837 <RCT>
1956 V18 P413-419	RCT V31 N2 P361 <RCT>
1956 V18 P528-535	RCT V31 N3 P655 <RCT>
1956 V18 N1 P3-6	45H13R <ATS>
1956 V18 N1 P7-12	RCT V30 N1 P54 <RCT>
1956 V18 N2 P135-144	84H13R <ATS>
1956 V18 N2 P167-179	<ATS>
1956 V18 N2 P233-236	CSIRO-3320 <CSIR>
	R-833 <*>
1956 V18 N3 P332-336	97J15R <ATS>
1956 V18 N4 P438-442	48J15R <ATS>
1956 V18 N5 P555-561	76P59R <ATS>
1956 V18 N6 P741-744	63-15201 <=*>
1956 V18 N6 P748-754	05M43R <ATS>
1957 V19 P72-77	60-17731 <+*>
1957 V19 P201-203	RCT V32 N2 P536 <RCT>
1957 V19 P261-267	RCT V31 N3 P592 <RCT>
1957 V19 P274-280	RCT V31 N4 P712 <RCT>
1957 V19 P287-292	RCT V31 N4 P756 <RCT>
1957 V19 P361-366	60-17728 <+*>
1957 V19 P376-383	RCT V31 N4 P691 <RCT>
1957 V19 P412-420	RCT V32 N1 P67 <RCT>
1957 V19 P421-429	RCT V32 N1 P184 <RCT>
1957 V19 P430-434	RCT V32 N2 P539 <RCT>
1957 V19 P587-591	RCT V31 N3 P513 <RCT>
1957 V19 P657-661	RCT V32 N2 P531 <RCT>
1957 V19 N1 P3-8	59-19178 <+*>
1957 V19 N1 P9-13	59-19179 <+*>
1957 V19 N1 P82-89	R-2314 <*>
1957 V19 N3 P311-318	59-10264 <+*>
1957 V19 N3 P375-382	62-14280 <=*>
1957 V19 N4 P459-464	63-18499 <=*>
1957 V19 N5 P562-571	76K22R <ATS>
1958 V20 P124-127	RCT V33 N2 P412 <RCT>
1958 V20 P260-271	RCT V32 N3 P785 <RCT>
1958 V20 P397-398	RCT V33 N4 P970 <RCT>
1958 V20 N1 P59-66	61-16704 <*=>
1958 V20 N2 P199-201	93L28R <ATS>
1958 V20 N3 P260-271	59-15236 <+*>
1958 V20 N3 P272-278	59-17599 <+*>
1958 V20 N3 P332-337	105 <LSB>
1958 V20 N3 P338-348	63-10176 <=*>
1958 V20 N4 P444-455	63-14709 <=*>
1958 V20 N5 P611-619	63-14713 <=*>
1958 V20 N6 P748-758	29L30R <ATS>
1959 V21 P244-245	RCT V33 N2 P398 <RCT>
1959 V21 P558-563	RCT V33 N3 P757 <RCT>
1959 V21 N1 P37-49	61-28533 <*=>
1959 V21 N2 P221-225	63-13871 <=*>
1959 V21 N3 P276-282	39M40R <ATS>
1959 V21 N3 P306-308	06M39R <ATS>
1959 V21 N6 P754-761	60-18796 <+*>
1960 V22 P168-175	RCT V35 N2 P326 <RCT>
1960 V22 P233-236	RCT V35 N2 P335 <RCT>
1960 V22 P277-281	62-15704 <*=>
1960 V22 P323-333	62-15700 <*=>
1960 V22 P334-339	RCT V35 N2 P311 <RCT>
1960 V22 N1 P16-22	61-10295 <*+>
1960 V22 N1 P57-62	61-16709 <*=>
1960 V22 N2 P176-185	04M44R <ATS>
1960 V22 N3 P334-339	62-14822 <=*>
1960 V22 N4 P385-392	64-10170 <=*$>
1960 V22 N4 P434-442	90N49R <ATS>
1960 V22 N4 P497-502	61-27777 <*=>
1960 V22 N4 P503-505	63-16979 <=*>
1960 V22 N4 P506-511	61-21668 <=>
1960 V22 N5 P599-605	62-15432 <*=>
1960 V22 N5 P639-640	61-10429 <+*>
1960 V22 N6 P663-670	61-18593 <*=>
1961 V23 P322-	RCT V35 N2 P449 <RCT>
1961 V23 N1 P112-117	62-14820 <=*>
1961 V23 N2 P157-162	AEC-TR-4789 <=*> 0
1961 V23 N4 P369-375	62-18379 <=*>
1961 V23 N4 P389-397	62-18380 <=*>
1961 V23 N4 P462-463	RCT V36 N1 P156 <RCT>
1961 V23 N5 P544-552	66-10701 <*> 0
1961 V23 N5 P582-591	60P60R <ATS>
1961 V23 N6 P646-651	87P59R <ATS>
1962 V24 P455-458	63-26597 <=$>
1962 V24 N2 P121-127	49P63R <ATS>
	63-23945 <=*>
1962 V24 N2 P141-151	63-14316 <=*>
1962 V24 N2 P185-194	66P63R <ATS>
1962 V24 N3 P289-292	33P63R <ATS>
	62-32839 <=*>
1962 V24 N3 P293-296	62-32858 <=*>
1962 V24 N3 P323-331	64-10100 <=*$>
1962 V24 N3 P355-356	22Q67R <ATS>
1962 V24 N6 P651-658	52Q68R <ATS>
1963 V25 P646-648	M-5574 <NLL>
1963 V25 N1 P82-85	64-14440 <=*$> 0
1963 V25 N2 P247-252	64-10881 <=*$>
1963 V25 N3 P329-333	64-71386 <M>
1963 V25 N4 P441-446	64-15955 <=*$>
1964 V26 P95-99	65-63009 <=$>
1964 V26 N1 P67-71	65-10203 <*>
1964 V26 N2 P207-214	AD-618 060 <=$>
1964 V26 N3 P341-349	65-63743 <=>
1964 V26 N3 P373-379	65-63772 <=>
1965 V27 N1 P106-112	65-30672 <=$>
1965 V27 N2 P250-253	18S87R <ATS>
1965 V27 N3 P374-378	RAPRA-1298 <RAP>
	66-11513 <*>
1965 V27 N3 P435-440	65-17464 <*>
1965 V27 N3 P446-452	19S88R <ATS>
	66-11515 <*>
1965 V27 N4 P624-626	AD-625 603 <=*$>
	66-13995 <*>
	66-14872 <*$>

KOLLOIDZEITSCHRIFT

Citation	Translation
1913 V13 P252-254	65-18044 <*>
1917 V20 P20-33	AEC-TR-6090 <*>
1920 V27 P18-27	57-311 <*>
1921 V29 P193-196	64-18245 <*>
1922 V30 P20-31	1470 <*>
1922 V31 P195-196	64-14425 <*>
1923 V33 P89-	57-538 <*>
1923 V33 P267-271	2870 <*>
1923 V33 P348-353	2867 <*>
1924 P367-374	2892 <*>
1924 V34 P117-119	64-18247 <*>
1924 V35 P166-169	I-975 <*>
1925 P205-214	2940 <*>
1925 P300-307	2876 <*>
1925 V37 P267-271	59-18269 <+*>
1925 V37 N1 P19-22	2859 <*>
1926 V38 P33-42	77S89G <ATS>
1926 V38 P259-	58-1471 <*>
1926 V38 P260-261	58-1472 <*>
1927 V42 P112-119	AL-521 <*>
1928 V45 P52-56	64-71519 <=>
1928 V46 N4 P337-345	60-10379 <*>
1928 V47 P65-76	64-14290 <*> 0
1928 V47 P241-246	57-3064 <*>
1930 V50 P217-228	58-1592 <*>
1930 V52 N1 P46-61	57-2036 <*>
1931 V54 N1 P278-284	59-20247 <*> 0

Citation	Translation
1931 V54 N3 P314-326	59-20455 <*> 0
1931 V55 P120-	57-942 <*>
1931 V55 N2 P129-143	63-16756 <*>
1931 V55 N2 P172-198	57-2245 <*>
1932 V59 N2 P217-226	58-1749 <*>
1933 V64 N2 P184-185	54L29G <ATS>
1933 V65 N2 P203-211	60-00611 <*>
1934 V68 N3 P289-298	NP-TR-966 <*>
1934 V69 P155-164	ORNL-TR-424 <*>
1935 V71 N1 P36-48	63-10038 <*>
1935 V71 N2 P172-176	62-00765 <*>
1935 V71 N3 P333-335	64-16163 <*> 0
1935 V72 P100-	I-46 <*>
1935 V72 N3 P336-345	CSIRO-3437 <CSIR>
	57-3292 <*>
1936 V74 N3 P253-265	66-12270 <*>
1936 V74 N3 P266-275	66-14234 <*>
1936 V75 N1 P80-88	66-12266 <*> 0
1936 V77 N2 P172-183	57-2511 <*>
1937 V79 N3 P257-273	63-14953 <*> 0
1938 V82 P87-99	58-1843 <*>
	59-18557 <*+>
1938 V83 N1 P120-128	60-14609 <*> 0
1938 V85 P74-87	2427 <K-H>
1939 V86 N2 P150-166	C-2424 <NRC>
1939 V86 N3 P313-339	50E1G <ATS>
	59-20615 <*> 0
1939 V87 N1 P21-36	63-16261 <*>
1939 V88 N2 P161-171	57-793 <*>
1939 V89 N2 P202-208	63-16753 <*>
1939 V89 N2 P237-238	63-01563 <*>
1940 V90 N1 P65-77	RCT V13 N1 P831 <RCT>
1940 V91 P287-294	AL-218 <*>
	58-1941 <*>
1941 V95 P212-	00H11G <ATS>
1941 V97 N1 P27-35	SCL-T-396 <*>
1942 V98 P90-92	66-14545 <*>
1942 V98 N2 P173-180	59-20110 <*> 0
1942 V99 P107-113	66-14883 <*>
1942 V100 P320-327	66-14544 <*>
1943 V102 N1 P1-14	CSIRO-313 <CSIR>
1943 V105 N3 P199-204	59-10695 <*> 0
1944 V106 N1 P50-67	II-51 <*>
1944 V107 N3 P201-205	62-14784 <*>
1948 V110 N2 P125-132	62-18400 <*> 0
1948 V110 N3 P214-221	43S89G <ATS>
1949 V112 N1 P21-26	55M46G <ATS>
1949 V112 N1 P34-60	II-808 <*>
1949 V112 N1 P60-66	61-16531 <*>
1949 V115 N1/3 P53-66	27K20G <ATS>
1949 V115 N1/3 P76-82	63-16474 <*> 0
1949 V115 N1/3 P183-189	63-16475 <*> 0
1950 V116 P1-9	60-23247 <*+>
1950 V117 P42-47	58-969 <*>
1950 V119 P69-73	2099 <*>
1950 V119 P157-160	I-271 <*>
	2100 <*>
1950 V119 N1 P23-38	65-11306 <*>
	71M37G <ATS>
1950 V119 N1 P42-45	61-14020 <*> 0
1951 V120 P24-34	8826 <IICH>
1951 V120 P57-65	8721 <IICH>
1951 V120 N1/3 P40-52	SCL-T-290 <*>
1951 V121 N3 P130-134	57-3052 <*>
1951 V121 N1/2 P1-20	60-10768 <*> 0
1951 V122 N1 P23-34	64-20561 <*> 0
1951 V123 P40-51	II-614 <*>
	3300 <*>
1951 V123 P66-83	SC-T-64-42 <*>
1951 V123 P92-99	63-20536 <*> 0
1951 V123 N1 P22-33	I-248 <*>
1951 V124 P41-43	II-221 <*>
1951 V124 P77-82	II-220 <*>
1951 V124 P116	SCL-T-276 <*>
1952 V125 N2 P106-108	38G6G <ATS>
1952 V125 N3 P174-	57-2228 <*>
1952 V126 P192-199	59H9G <ATS>
1952 V126 N2/3 P140-149	63-18981 <*>
1952 V127 P1-7	5241 <K-H>
1952 V127 N1 P1-7	64-16479 <*> 0
1952 V127 N1 P19-27	57-3066 <*>
1952 V128 P75-86	8973 <IICH>
1952 V128 P136-142	5309-G <K-H> 0
1952 V128 N3 P125-126	63-18987 <*> 0
1952 V128 N3 P136-142	64-16744 <*> 0
1952 V128 N3 P159-164	RCT V27 N1 P201 <RCT>
1952 V129 N2/3 P84-91	58-1054 <*>
1953 V130 P131-160	62-16534 <*>
1953 V130 N1 P64-65	59-17963 <*>
1953 V130 N2 P105-110	64-20318 <*> 0
1953 V132 N2/3 P84-99	62-10594 <*>
1953 V133 P26-32	SC-T-64-1633 <*>
1953 V133 P91-96	57-2503 <*>
1953 V133 N2/3 P91-96	2228 <*>
	64-20826 <*> 0
1953 V134 P149-189	63-14020 <*> P
	63-14021 <*> 0
1954 V136 N2/3 P84-99	63-14962 <*>
1954 V136 N2/3 P120-124	62-10412 <*> 0
1954 V137 P130-162	1040 <TC>
1954 V137 N1 P20-24	57-2727 <*>
1954 V137 N1 P24-26	57-2738 <*>
	58-247 <*>
1954 V137 N1 P26-28	57-2724 <*>
1954 V137 N2/3 P74-78	57-2504 <*>
1954 V137 N2/3 P93-103	62-18250 <*> 0
1954 V138 N3 P149-155	59-10148 <*>
1954 V139 N3 P146-150	62-18335 <*> 0
1954 V139 N3 P155-163	62-20063 <*>
1955 V140 P149-157	60-23433 <*+>
1955 V140 N2/3 P76-102	CSIRO-2870 <CSIR>
1955 V140 N2/3 P159-164	61-16182 <*> 0
1955 V141 N3 P146-159	72M45G <ATS>
1955 V141 N3 P165-173	CSIRO-3064 <CSIR>
1955 V141 N3 P177-187	61-10532 <*>
1955 V142 N1 P5-14	CSIRO-2870 <CSIR>
1955 V142 N2/3 P65-73	63-10929 <*>
1955 V142 N2/3 P132-150	57-2142 <*>
1955 V142 N2/3 P164-	57-1779 <*>
1955 V143 N1 P21-31	CSIRO-2870 <CSIR>
1955 V144 N1/3 P110-120	57-2726 <*>
1955 V144 N1/3 P125-148	65-11168 <*> 0
1956 V145 P17-46	T-1519 <INSD>
1956 V145 N2 P92-102	1675 <*>
1956 V145 N2 P119-125	59-20315 <*>
1956 V145 N3 P157-158	CSIRO-3323 <CSIR>
1956 V145 N3 P158-163	61-10320 <*>
1956 V146 N1/3 P44-48	45H10G <ATS>
1956 V147 N1/2 P78-79	65-14084 <*>
1956 V147 N1/2 P79-81	65-14085 <*>
1956 V149 N2/3 P67-72	64-16901 <*>
1956 V149 N2/3 P84-95	64-16328 <*>
1957 V150 P14-19	63-14412 <*> 0
1957 V150 N2 P128-134	63-16482 <*>
1957 V150 N2 P153-156	59-15949 <*> 0
	65-14087 <*>
1957 V151 N1 P18-24	59-15927 <*> 0
	61-20652 <*> 0
	64-16508 <*>
	65-14086 <*> 0
1957 V152 P53-57	8711 <IICH>
1957 V152 N1 P8-15	64-16812 <*>
1957 V152 N1 P18-23	59-10433 <*>
1957 V152 N1 P31-36	62-16706 <*> 0
1957 V152 N2 P116-121	64-16903 <*>
1957 V152 N2 P148-149	60-14008 <*>
1957 V153 N2 P128-155	61-16737 <*>
1957 V154 N2 P97-103	64-30673 <*>
1957 V154 N2 P130-141	63-16496 <*> 0
1957 V155 N1 P1-19	59-17516 <*>
1957 V155 N1 P45-55	26N48G <ATS>
	64-30674 <*>
1957 V155 N1 P55-64	27N48G <ATS>
	64-30675 <*>
1958 V156 N1 P8-14	62-20369 <*>
1958 V156 N1 P14-21	44K23G <ATS>
1958 V156 N1 P46-61	60-18798 <*> 0
1958 V156 N2 P102-107	59-10948 <*>
1958 V157 N2 P89-111	59-17253 <*>
1958 V157 N2 P111-123	59-15065 <*>

1958 V159 N2 P108-118 35K28G <ATS>
1958 V160 P97-106 62-10549 <*> 0
1958 V160 N1 P16-20 61-18834 <*>
1958 V160 N1 P21-26 82L31G <ATS>
1959 V162 N2 P138-140 59-20862 <*>
1959 V162 N2 P141-149 61-00223 <*>
1959 V163 P126-132 66-11736 <*> 0
1959 V163 N2 P106-115 60-14253 <*>
1959 V163 N2 P116-122 AEC-TR-4755 <*>
1959 V163 N2 P126-132 61-10644 <*> 0
1959 V164 N1 P8-13 32L33G <ATS>
1959 V164 N1 P34-37 60-18408 <*>
64-10923 <*>
1959 V165 N1 P3-15 59-20778 <*>
1959 V165 N1 P15-25 62-18810 <*> 0
1959 V166 N1 P10-14 65-12797 <*>
1959 V166 N1 P14-19 65-12798 <*>
1959 V167 N1 P55-62 62-18862 <*>
1959 V167 N2 P132-141 61-10214 <*>

1960 V166 N1 P25-38 54M43G <ATS>
1960 V168 N1 P37-49 60-18432 <*>
1960 V169 N1/2 P18-28 62-18665 <*>
1960 V169 N1/2 P34-41 61-10215 <*>
1960 V170 N2 P97-104 65-17396 <*>
1960 V171 N2 P119-122 8723 <IICH>
1961 V174 N1 P20-27 61-18847 <*> 0
1961 V174 N2 P134-142 66-11059 <*>
66-15398 <*>
1961 V175 P110-119 61-20503 <*>
1961 V176 N1 P49-62 63-20005 <*>
1961 V177 N2 P116-128 62-10480 <*>
1961 V177 N2 P149-153 33N56G <ATS>
62-14284 <*>
1961 V178 N2 P128-142 63-20488 <*> OP
1961 V179 N1 P11-29 62-18649 <*>
1961 V179 N2 P110-116 63-10817 <*>

KOLLOID-ZEITSCHRIFT UND ZEITSCHRIFT FUER POLYMERE
1962 V180 P11-26 8624 <IICH>
1962 V180 P163-164 SC-T-65-701 <*>
1962 V180 N1 P11-26 877 <TC>
1962 V180 N2 P87-108 63-10798 <*>
1962 V180 N2 P118-126 63-10824 <*> 0
1962 V180 N2 P150-160 8976 <IICH>
1962 V180 N2 P163-164 66-11060 <*>
1962 V182 N1/2 P75-85 63-10890 <*> 0
1962 V182 N1/2 P99-104 64-10680 <*>
1962 V182 N1/2 P140-145 65-14476 <*> 0
1962 V184 N1 P1-7 83S80G <ATS>
1962 V185 N2 P97-102 63-18367 <*>
1963 V187 N2 P107-109 66-14605 <*>
1963 V188 N2 P97-114 375 <CTS>
1963 V189 N1 P1-6 SC-T-65-711 <*>
1963 V189 N1 P14-22 8620 <IICH>
1963 V189 N1 P23-26 50R74G <ATS>
1963 V189 N1 P55-57 64-14281 <*>
1963 V190 P1-16 8622 <IICH>
1963 V190 N1 P1-16 66-10835 <*>
1963 V190 N1 P16-34 64-16192 <*>
1963 V191 N2 P123-130 66-10304 <*>
1963 V192 N1/2 P21-28 65-18016 <*>
1964 V195 N1 P1-8 876 <TC>
1964 V195 N1 P35-39 66-12931 <*>
1964 V196 N2 P97-125 66-12363 <*>
1964 V198 N1/2 P5- RCT V39 N4 PT1 P841 <RCT>
1964 V199 N1 P52-56 66-10305 <*>
1964 V199 N1 P63- RCT V39 N4 PT1 P863 <RCT>
1964 V199 N2 P125- RCT V39 N4 PT1 P858 <RCT>
1965 V201 N2 P111-116 1659 <TC>
1965 V201 N2 P143-147 66-10299 <*>
1965 V202 N1 P54-55 POED-TRANS-2197 <NLL>
1965 V202 N2 P97-107 916 <TC>
1965 V202 N2 P108-120 917 <TC>
1965 V202 N2 P127-132 993 <TC>
1966 V208 N2 P97-123 66-14674 <*>

KOLORADSKII ZHUK I MERY BORBY S NIM
1955 N1 P73-93 RT-4611 <*>

KOMMUNIST BELORUSSI
1959 V31 N5 P24-29 59-13909 <=>
1960 V32 P8-12 61-11863 <=>
1961 V33 N1 P43-47 61-23580 <=>
1962 V34 N7 P52-55 62-33601 <=>

KOMMUNIST ESTONII
1959 V15 N5 P41-49 59-13909 <=>
1959 V15 N7 P32-37 60-31018 <=>
1962 V18 N7 P13 63-13486 <=>
1963 V19 N4 P42-49 63-31406 <=>
1965 V21 N10 P18-26 66-30904 <=$>
1966 N5 P68-73 66-33360 <=$>

KOMMUNIST MOLDAVII
1962 V7 N9 P26-32 63-13664 <=>
1963 V8 N5 P56-58 63-31448 <=>
1966 N1 P21-26 66-31752 <=$>

KOMMUNIST. MOSCOW
1954 N15 P51-53 RT-4333 <*>
1954 V31 N9 P28-44 RT-2801 <*>
1957 V34 N7 P124-127 61-15695 <+*>
1958 V35 P35-46 59-19439 <+*>
1959 V36 N1 P53-69 59-13261 <*>
1959 V36 N5 P27-43 59-31029 <=>
1959 V36 N18 P10-19 60-31294 <=>
1960 V37 N8 P13-23 61-11186 <=>
1960 V37 N18 P48-54 61-27077 <=>
1962 V39 N9 P60-65 62-32365 <=*>
1962 V39 N15 P45-53 63-13342 <=>
1963 V39 N2 P56-59 63-21768 <=>
63-21885 <=>
1963 V39 N4 P14-26 63-21767 <=>
1963 V40 N10 P86 63-31900 <=>
1963 V40 N15 P114-123 64-21171 <=>
1964 V40 N2 P87-95 64-31040 <=>
1964 V41 P3 65-30167 <=$>
1964 V41 N10 P65-73 64-51694 <=$>
1965 V42 P2 65-31593 <=$>
1965 V42 N12 P48-73 65-32752 <=*$>
1965 V42 N16 P59-67 66-32682 <=$>
1966 N5 P3-128 66-31694 <=$>
1966 N9 P3-125 66-33535 <=$>

KOMMUNIST TADZHIKISTANA
1962 V18 N7 P35-37 62-33323 <=>
1962 V18 N7 P39-41 62-33323 <=>
1962 V18 N10 P16-19 63-13646 <=>
1963 V19 N2 P21 63-31449 <=>

KOMMUNIST UKRAINY
1960 P8-26 61-11819 <=>
1960 N6 P22-27 61-11743 <=>
1960 N7 P42-49 61-21598 <=>
1964 V39 N11 P10-18 65-30325 <=$>
1964 V39 N11 P47-54 65-30496 <=$>
1966 N5 P51-59 66-34285 <=$>

KOMMUNIST UZBEKISTANA
1964 V36 N1 P33-40 64-31561 <=>
1964 V36 N6 P63-69 64-51231 <=$>
1965 V37 N9 P31-37 66-30465 <=$>

KOMMUNIST. VILNA

1962 V45 N8 P33-36 63-13530 <=>

KOMMUNIST VOORUZHENNYKH SIL
1961 N9 P16-24 AD-629 859 <=$>
1962 V2 N15 P17-23 64-19368 <=$>
1963 V3 N12 P43-47 63-31711 <=>
1964 N5 P9-16 AD-626 077 <=*$>
1965 N13 P8-17 65-33670 <=*$>
1965 N15 P64-69 65-33635 <=*$>
1965 N17 P50-55 AD-626 073 <=*$>
1965 N17 P50-56 65-33680 <=*$>
1965 N19 P9-19 65-33761 <=*$>
1965 N19 P87-92 65-33827 <=*$>
1966 N22 P7-15 66-35696 <=>

1966 N22 P22-28 66-35696 <=>
1966 N22 P54-6C 66-35696 <=>
1966 N22 P73-77 66-35696 <=>
1966 N22 P87-89 66-35696 <=>
1966 N23 P73-78 67-30151 <=>
1966 N24 P24-31 67-30343 <=>
1967 N1 P48-55 67-30574 <=>
1967 N1 P87-88 67-30574 <=>
1967 N1 P90-95 67-30574 <=>

KOMSOMOLSKAYA PRAVDA (NEWSPAPER)
1965 P3 12/29 66-30521 <=*$>
1966 02 01/20 66-30635 <=*$>

KONEVODSTVO I KONNYI SPORT
1957 V27 N12 P29-31 60-15585 <+*>
1957 V27 N12 P31-32 60-15586 <+*>
1958 V28 N1 P30-35 64-19121 <=*$>
1960 V30 N9 P2-4 61-11969 <=>

KONSERVES
1955 V13 N7 P79-80 CSIRO-3882 <CSIR>

KONSERVNAYA I OVOSHCHESUSHILNAYA PROMYSHLENNOST
1958 V13 N5 P6-7 59-19608 <+*> 0
1958 V13 N12 P27-30 60-15383 <+*>
1961 V16 N7 P22-24 62-25640 <=*>
1961 V16 N9 P1-4 62-23985 <=*>
1961 V16 N9 P13-19 62-33100 <=*>
1962 V17 N4 P24-27 65-61618 <=$>
1962 V17 N11 P36-39 63-13667 <=*>
1963 V18 N8 P15-18 63-11768 <=>
1963 V18 N8 P20-22 64-11522 <=>
1963 V18 N9 P26-29 64-11559 <=>
1963 V18 N9 P33-34 64-11518 <=>
1963 V18 N10 P22-26 64-13629 <=*$>
1964 V19 N7 P33-34 M.5646 <NLL>
1965 N6 P20-24 AD-624 571 <=*>
1965 N6 P31-36 AD-624 571 <=*>
1965 N6 P41-43 AD-624 571 <=*>
1965 N6 P45-47 AD-624 571 <=*>
1965 V20 N6 P34-36 M.5803 <NLL>

KONSTRUKTII GRASHDANSKIKH ZDANII
1946 V3 P265-279 R-453 <*>
1946 V3 P28C-297 R-515 <*>
1946 V3 P326-349 R-454 <*>

KONSTRUKTION
1950 V2 N11 P321-325 58-713 <*>
1951 V3 N2 P5C-52 60-10239 <*>
1951 V3 N10 P302-308 58-1131 <*>
1953 V5 N2 P38-46 60-10238 <*>
1953 V5 N3 P85-90 60-10238 <*>
1954 V6 P105-108 T-1956 <INSD>
1954 V6 N3 P97-104 2731 <*>
1954 V6 N10 P384-389 61-20759 <*>
1954 V6 N575 P97-104 57-192 <*>
1955 V7 N2 P54-68 2420 <BISI>
1955 V7 N4 P157-163 62-18785 <*> 0
1956 V8 N8 P317-320 66-11785 <*> 0

KONSTRUKTICN IM MASCHINEN-, APPARATE- UND GERAETEBAU
1957 V9 N4 P147-153 61-00019 <*>
1959 N3 P82-89 TS-1392 <BISI>
1959 V11 N3 P82-89 62-10447 <*>
1960 V12 N4 P147-155 AEC-TR-4953 <*>
1960 V12 N5 P195-203 AEC-TR-4970 <*=>
1960 V12 N5 P210-218 AEC-TR-4959 <*>
1962 V14 N6 P234-239 62-20261 <*>
1964 V16 N4 P128-135 64-71479 <=>

KONTSENTRATSIYA NAPRYAZHENII. KIEV
1965 N1 P164-173 66-62006 <*$=>
1965 N1 P233-238 66-62011 <*$=>

KORA VYVETRIVANIYA
1956 V2 P77-84 <DIL>
R-1602 <*>
1956 V2 N2 P31-34 60-21836 <=>
1956 V2 N2 P45-60 60-21837 <=>

KORU TARU
1953 V5 P51-52 62-00110 <*>

KOROIDO TO KAIMEN-KASSEIZAI
1961 V2 P685-691 AEC-TR-6050 <*>

KORRESPONDENZBLATT FUER SCHWEIZER AERZTE
1906 V36 N31 P779-785 66-10974 <*>

KORROSION. WEINHEIM
1963 N16 P85-91 65-14722 <*>

KORROSION UND METALLSCHUTZ
1927 V3 P25-30 II-358 <*>
1928 V3 P49-53 57-164 <*>
1928 V4 P146-151 58-1846 <*>
1928 V6 P133-135 57-165 <*>
1930 V6 N1 P7-14 60-10751 <*> 0
1931 V7 N5 P108-111 58-2215 <*>
1932 N8 P253-260 57-1427 <*>
1933 V9 P268-273 I-339 <*>
1936 V12 P132-138 1866 <*>
1936 V12 P275-283 57-1040 <*>
1936 V12 P309- 57-865 <*>
1936 V12 N9 P257-260 60-10738 <*>
1937 V13 P95-97 59-15201 <*> 0
1937 V13 P144-157 1970 <*>
1937 V13 P274-275 57-1039 <*>
1937 V13 N6 P181-183 66-11790 <*>
1937 V13 N10/1 P380-383 58-1898 <*>
1938 V14 P173-174 57-1041 <*>
1938 V14 N1 P28-34 60-10754 <*> 0
1938 V14 N10/1 P350-353 57-1337 <*>
1939 V15 P105-121 60-10732 <*>
1939 V15 P225-241 63-01046 <*>
1939 V15 N4 P105-122 2650 <*>
1940 V16 P127-132 AEC-TR-6389 <*>
1940 V16 P236-246 58-1859 <*>
1940 V16 N9 P297-299 58-127 <*>
1940 V16 N10 P341-344 AL-276 <*>
58-350 <*>
1940 V16 N11 P331-338 58-1098 <*>
1941 N2 P52-55 II-1160 <*>
1941 N2 P56-65 AL-383 <*>
1941 V17 P117-123 I-603 <*>
1941 V17 P123- II-90 <*>
1941 V17 P377-380 II-50 <*>
1941 V17 P401-403 II-42 <*>
1941 V17 N1 P13-19 60-18712 <*> 0
1941 V17 N2 P52-55 II-1160 <*>
1941 V17 N3 P77-98 66-14618 <*>
1942 V18 P221-222 57-1038 <*>
1942 V18 P243-244 57-2786 <*>
1943 V19 P104-105 I-578 <*>
1943 V19 N1 P13-19 62-20478 <*> 0

KORROZIYA I BORBA S NEI
1939 V5 N1/2 P126-137 R-2575 <*>
1940 V6 N5/6 P26-31 R-2249 <*>

KORROZIYA METALLOV I SPLAVOV, SBORNIK
1963 P129-140 ORNL-TR-32 <=$>

KORUNK
1961 V20 N1 P39-49 62-19409 <*=>
1963 V22 N2 P154-160 63-21593 <=>
1963 V22 N2 P179-184 63-21593 <=>
1964 V23 N11 P498-503 65-30019 <=$>

KOSHU EISEIIN KENKYU HOKOKU
1953 V3 N1 P9 65-11013 <*> 0

★KOSMICHESKIE ISSLEDOVANIYA
1963 V1 N1 P98-112 65-13390 <*>
1963 V1 N1 P143-146 64-15588 <=*$>
1963 V1 N1 P169-171 65-13385 <*>
1963 V1 N1 P176-178 65-13389 <*>

1963 V1 N2 P209-215 64-23643 <=$>
1963 V1 N3 P339-480 64-13941 <=*$>
1963 V1 N3 P443-447 N65-11301 <=$>
1964 V2 N1 P64-70 66-61542 <=*$>
1964 V2 N2 P280-288 65-11173 <*>
1964 V2 N3 P441-454 65-12730 <*>
1964 V2 N5 P763-772 65-11282 <*>
1964 V2 N5 P779-782 65-11238 <*>
1964 V2 N6 P865-880 N65-15160 <=$>
1964 V2 N6 P909-916 N65-15161 <=$>
1964 V2 N6 P917-919 N65-14622 <=$>
1964 V2 N6 P928-932 N65-14620 <=$>
1964 V2 N6 P933-935 N65-14621 <=$>
1964 V2 N6 P948-951 N65-16600 <=$>
1965 V3 N2 P231-236 AD-618 465 <=$>
1965 V3 N2 P237-296 65-31099 <=*$>
1965 V3 N2 P340-342 66-61531 <=*$>
1965 V3 N3 P347-350 ICE V5 N4 P593-595 <ICE>
1965 V3 N3 P359-367 65-14442 <*>
65-63762 <=$>
1965 V3 N3 P374-379 65-14441 <*>
65-63795 <=$>
1965 V3 N3 P433-468 65-31705 <=$>
1965 V3 N3 P469-472 65-64380 <=*$>
1965 V3 N3 P495-496 66-61532 <=*$>
1965 V3 N4 P540-553 AD-625 882 <=*$>
1965 V3 N4 P595-603 AD-622 898 <=*$>
1965 V3 N4 P614-617 66-61530 <=*$>
1965 V3 N6 P811-952 AD-636 619 <=$>
1966 V4 N1 P162-164 66-61357 <=*$>

KOSMICHESKIE LUCHI: SBORNIK STATEI. MEZHDUVEDOMST-VENNYI KOMITET PO PROVEDENIIU MEZHDUNARODNOGO GEOFIZICHESKOGO GODA. AKADEMIYA NAUK SSSR. VII. RAZDEL PROGRAMMY MGG
1959 N1 P7-58 61-11525 <=$>
1960 N2 ENTIRE ISSUE N65-36039 <=$>

KOSMOS. STOCKHOLM
1956 V34 P180-208 62-19410 <=*>

KOSO KAGAKU SHIMPOJIUM
1957 V12 P201-207 59-18321 <=*> 0
1962 V14 P28-30 62-01513 <*>

KOTLOTURBOSTROENIE
1948 N5 P26-28 64-15706 <=*$>

KOTSU GIJUTSU
1954 N90 P42-47 T-2067 <INSD>

KOVOVE MATERIALY
1963 V1 N1 P21-41 TRANS-379 <MT>
1963 V1 N1 P44-61 TRANS-380 <MT>
1963 V1 N2 P196-214 TRANS-419 <MT>
1963 V1 N2 P238-249 TRANS-392 <MT>
1963 V1 N3 P321-327 420 <MT>
1963 V1 N3 P339-351 TRANS.377 <MT>
1964 V2 N1 P13-25 TRANS-431 <MT>
1964 V2 N1 P28-41 TRANS-384 <MT>
1964 V2 N1 P80-96 TRANS-385 <MT>
1964 V2 N1 P106-111 TRANS-393 <MT>
1964 V2 N2 P138-151 TRANS-401 <MT>
1964 V2 N2 P169-181 TRANS-402 <MT>
1964 V2 N2 P184-199 TRANS-403 <MT>
1964 V2 N2 P254-256 TRANS-541 <MT>
1964 V2 N3 P289-301 TRANS-433 <MT>
1964 V2 N3 P309-321 3983 <BISI>
1964 V2 N4 P333-342 M-5634 <NLL>
TRANS-451 <MT>
1964 V2 N5 P425-430 469 <MT>
1964 V2 N5 P433-443 TRANS-470 <MT>
1964 V2 N5 P433-444 66-11375 <*>
1964 V2 N6 P505-520 480 <MT>
1964 V2 N6 P549-556 483 <MT>
1965 V3 N1 P11-18 TRANS-502 <MT>
1965 V3 N1 P60-66 M.5820 <NLL>
1965 V3 N1 P161-169 M.5819 <NLL>
1965 V3 N2 P149-159 TRANS-527 <MT>
1965 V3 N3 P248-255 TRANS-635 <MT>
1965 V3 N4 P333-358 TRANS-569 <MT>

KOZAN CHISHITSU
1952 V2 N6 P216-219 57-2574 <*>
1962 V12 N51 P16-26 IGR V5 N5 P505 <AGI>

KOZARSTVI
1954 V4 N2 P38-39 58-2067 <*>
1954 V4 N4 P99-100 58-2067 <*>

KOZGAZDASAGI SZEMLE
1958 V5 N8/9 P885-903 59-11581 <=>
1961 V8 N7 P817-837 62-19535 <*=>
1964 V11 N5 P582-606 64-41215 <=$>
1967 N1 P108-115 67-30611 <=$>
1967 N2 P155-167 67-31023 <=$>

KOZHEVENNO-OBUVNAYA PROMYSHLENNOST
1959 N1 P4-8 60-15423 <+*>
1959 N1 P28-30 60-17908 <+*>
1959 N7 P28-33 RCT V33 N4 P1193 <RCT>
1960 N3 P16-20 61-15243 <*+>
1960 N11 P2-3 61-21697 <=>
1962 V4 N2 P24-27 63-23233 <=*>
1962 V4 N2 P27-31 63-23239 <*=>
1962 V4 N7 P30-33 64-19962 <=$>
1962 V4 N7 P33-34 64-19959 <=$>
1962 V4 N8 P4-7 64-13730 <=*$>
1962 V4 N8 P14-16 64-13728 <=*$>
1962 V4 N10 P14-15 64-16865 <=*$>
1963 V5 N10 P21-24 RTS-2758 <*>
1964 V6 N4 P33 RTS-3350 <NLL>
1964 V6 N6 P26-29 RTS-2958 <NLL>
1964 V6 N7 P1-3 64-51449 <=$>
1964 V6 N8 P36-38 RTS-3565 <NLL>

KOZLEKEDESI KOZLONY
1967 N4 P66-68 67-30857 <=$>

KOZLEMENYEI MUSZAKI ES GAZDASAGTUDOMANYI EGYETEM BANYA- KOHO ES ERDOMERNOKIKAR. MAGYAR KIRALYI JOSEF NADOR. BUDAPEST
1940 V12 P192-212 60-18833 <*>
1941 V13 P3-10 60-18832 <*>
1941 V13 P208-223 60-18831 <*>
1941 V13 P224-242 60-18830 <*>

KOZOGAKU ZASSHI
1934 V12 N4 P271-278 62-00983 <*>
1940 V18 N11 P830-833 62-00985 <*>

KRAFT OCH IJUS
1955 P231-236 57-760 <*>

KRAFTFAHRZEUGTECHNIK
1957 V7 N10 P365-369 59-14295 <=*>
1958 V8 N9 P321-325 59-21014 <+*> 0
1967 N4 P97-98 67-31416 <=$>
1967 N4 P121 67-31416 <=$>

KRAJ RAD
1963 N19 P8-11 63-31500 <=>
1963 N19 P24-25 63-31500 <=>

KRANKENHAUSARZT
1964 V37 N6 P169-171 66-11662 <*> 0

KRANKENHAUS-APOTHEKE
1960 V10 N4 P25-28 NP-TR-969 <*>

KRASNAYA ZVEZDA
1954 N118 P2 RT-2910 <*> 0
1954 V208 P3 RT-4039 <*>
1957 N87 P3 AEC-TR-3407 <*>
1957 N88 P4 AEC-TR-3407 <*>
1963 P3-4 AD-615 988 <=$>
1963 P1P AD-621 805 <=$>
1964 P2-3 66-33686 <=$>
1964 P3-4 65-13257 <*>
1964 P3 65-30551 <=$>

1964 P4 AD-615 993 <=$>
65-13256 <*>
65-13258 <*>
65-30210 <=$>
1965 P1 06/16 AD-636 609 <=$>
1965 P4 12/19 66-30535 <=$>
1965 P1 AD-629 473 <=$>
1965 P3 65-31086 <=*$>
1965 P5 65-33943 <=$>
1965 P6 65-31023 <=$>
65-33064 <=*$>
1966 P2 05/07 66-32021 <=$>
1966 P2 66-32185 <=$>
1966 P3 66-31039 <=$>
1966 P4 66-31492 <=$>
1966 2-3 09/01 66-34901 <=$>
1966 2-3 09/28 66-34901 <=$>
1966 2-3 09/29 66-34901 <=$>
1966 2-3 09/30 66-34901 <=$>
1966 N2 P2 C9/17 66-34294 <=$>

KRATKIE SOOBSHCHENIYA. INSTITUT NARODOV AZII. AKADEMIYA NAUK SSSR. MOSCOW
1961 N59 P69-84 AD-634 589 <=$>

KRATKIE SOOBSHCHENIYA INSTYTUT ARKHEOLOGII, AKADEMIYA NAUK URSR
1963 N93 P114-115 64-15434 <=$>

KREBSARZT
1949 V4 N5/6 P205-207 2208 <*>
1952 V7 P301-302 2026 <*>
1952 V7 N11/2 P352-353 3183-A <K-H>
1955 V10 P141-148 1349 <*>
1957 V12 N1 P1-4 57-1888 <*>
1957 V12 N3 P40 <RIS>
1957 V12 N3 P129-150 58-1611 <*>
1957 V12 N3 P15C-156 <RIS>
58-1607 <*>
1957 V12 N3 P157-166 <RIS>
58-1606 <*>
1957 V12 N4 P208-219 63-16677 <*>
8003-C <K-H>
1958 V13 N7 P318-322 63-16593 <*>
8500-A <KH>
1958 V13 N7 P322-33C 63-16580 <*>
8817-A <KH>
1959 V14 N1 P22-24 63-16901 <*>
9353-A <K-H>
1959 V14 N3 P81-84 61-00383 <*>
1959 V14 N5 P178-180 61-00382 <*>

KRIMINALSTIK
1956 V10 P434-440 59-13524 <=> 0

★KRISTALLOGRAFIYA
1956 V1 P209-213 59-18626 <+*> 0
1956 V1 P410-418 UCRL TRANS-836 <=*>
1956 V1 P634-643 R-1980 <*>
1956 V1 N1 P66-72 TS-1308 <BISI>
62-24222 <=*>
1956 V1 N1 P73-80 R-1384 <*>
1956 V1 N2 P166-170 R-442 <*>
1956 V1 N2 P2C5-2C8 R-443 <*>
1956 V1 N2 P209-213 63-14680 <=*>
1956 V1 N3 P291- AEC-TR-4557 <*=>
1956 V1 N3 P306-310 SC-T-511 <=$>
25Q72R <ATS>
1956 V1 N3 P370-372 R-709 <*>
1956 V1 N4 P393-402 R-4944 <*>
1956 V1 N5 P542-545 R-5179 <*>
63-14681 <=*>
1956 V1 N5 P572-576 65-64504 <=*$>
1956 V1 N5 P597-599 59-19607 <+*>
1956 V1 N6 P696-702 61-19377 <+*>
1957 V2 P400-407 R-5140 <*>
1957 V2 P414-418 R-5102 <*>
1957 V2 N1 P59-63 59-10196 <+*>
1957 V2 N1 P130-133 62-10357 <*=>
75L29R <ATS>
1957 V2 N1 P19C-192 60-15578 <+*> 0
1957 V2 N1 P195-197 60-15701 <+*>
1957 V2 N2 P226-232 CSIRO-3806 <CSIR>
1957 V2 N2 P287-288 59-10875 <+*> 0
1957 V2 N2 P307-308 60-15596 <+*>
1957 V2 N3 P424-427 65-61299 <=$>
1957 V2 N5 P613-617 59-10879 <+*>
1957 V2 N5 P700-702 R-3708 <*>
1957 V2 N6 P742-745 59-15015 <+*>
1958 V3 N1 P57-63 59-15083 <+*>
1958 V3 N1 P90-92 59-10883 <+*>
1958 V3 N1 P99-100 59-10881 <+*>
1958 V3 N2 P175-181 59-15081 <+*>
1958 V3 N2 P225-227 59-15082 <+*>
1958 V3 N2 P231-232 59-10882 <+*>
1958 V3 N2 P232-235 59-15080 <+*>
1958 V3 N4 P444-451 59-15876 <+*>
1958 V3 N5 P629-631 59-15877 <+*>
1958 V3 N5 P632-634 59-21206 <=>
1959 V4 N1 P3-12 60-10569 <+*>
1959 V4 N1 P13-19 59-20781 <+*>
1959 V4 N1 P20-24 59-20780 <+*>
1959 V4 N1 P121 59-17474 <+*>
1959 V4 N2 P235-238 60-10931 <+*>
1959 V4 N3 P399-409 60-14325 <+*>
1959 V4 N3 P414-417 60-10570 <+*>
1959 V4 N4 P526-533 61-13603 <*+> 0
1959 V4 N4 P554-562 62-19873 <=>
1959 V4 N4 P621-623 60-12574 <=>
62-11387 <=>
1959 V4 N4 P633-635 60-12573 <=>
62-11229 <=>
1959 V4 N5 P684-686 61-19342 <+*>
1959 V4 N6 P807-812 60-16954 <+*>
1960 V5 P962-976 62-23937 <=*>
1960 V5 N2 P273-281 62-10099 <*=>
1960 V5 N3 P476-477 61-10758 <+*>
1960 V5 N3 P477-478 61-13848 <=*>
1960 V5 N5 P726-731 61-18278 <*=>
1961 V6 P668-670 62-22464 <=*>
1961 V6 N3 P389-394 63-24341 <=*$>
1961 V6 N5 P714-726 62-32358 <=*>
1961 V6 N5 P80C-803 62-16598 <=*>
1962 V7 P316-318 20T92R <ATS>
1962 V7 N2 P271-275 62-18225 <=*>
1962 V7 N2 P309 65-12777 <*>
1962 V7 N2 P313-315 63-13596 <=*>
1962 V7 N3 P371-373 63-10657 <=*>
1962 V7 N5 P659-663 63-21377 <=>
1963 V8 P937-940 64-23527 <=$>
1964 V9 N4 P537-540 65-10741 <*>
1964 V9 N5 P722-726 29R80R <ATS>
1965 V10 P230-236 66-12512 <*>
1965 V10 N2 P268-270 66-11024 <*>

KRISTALLOGRAFIYA: SBORNIK STATEI
1955 N4 P3-46 62-23139 <=>

KROLIKOVODSTVO I ZVEROVODSTVO
1962 V5 N7 P25-27 AEC-TR-5849 <=*$>
1963 V6 N5 P18-19 65-60219 <=$>
1963 V6 N9 P18-22 65-60220 <=$>

KRUPPSCHE MONATSHEFTE
1921 V2 P117-126 63-20544 <*> 0
1929 V10 P184-185 57-2081 <*>

KRYLYA RODINY
1955 V6 N1 P9-10 RT-3389 <*>
1956 N5 P21 59-16957 <+*>
1957 N11 P7 59-12906 <+*>
1957 N12 P33 59-16050 <+*>
1957 V8 N1 P20-22 64-13700 <=*$> 0
1957 V8 N6 P20-22 64-13718 <=*$> 0
1957 V8 N7 P11-13 AD-430 014 <=$>
1958 N1 P30 59-12163 <+*>
1958 N2 P17 59-12168 <+*>
1958 N3 P9-10 59-12154 <+*> 0
1958 N6 P6-7 59-11973 <=> 0
1958 N6 P7-8 61-23876 <*=> 0

KUNSTSEIDE UND ZELLWOLLE

1949 V27 P195-202	T2076 <INSD>
1950 V28 N3 P84-87	63-18578 <*>

KUNSTSTOFFE

1911 P18-20	58-657 <*>
1913 V3 N20 P381-382	R-3422 <*>
1916 V6 N1 P4-7	R-3423 <*>
1918 V8 P26-29	57-2496 <*>
1926 V16 P41-43	AL-372 <*>
1926 V16 P69-70	AL-373 <*>
1926 V16 P167-168	AL-374 <*>
1926 V16 P199-201	AL-374 <*>
1927 V17 P269-272	AL-375 <*>
1930 N20 P1-2	65-11565 <*>
1930 V20 N6 P125-127	59-20431 <*>
1930 V20 N8 P173-175	59-20430 <*> 0
1933 V23 N12 P276-	57-906 <*>
1935 V25 N12 P305-308	59-20427 <*>
1936 V26 N9 P179-183	59-20426 <*> 0
1936 V26 N10 P207-209	59-20426 <*> 0
1937 V27 N6 P163-165	59-20429 <*> P
1937 V27 N7 P184-188	59-20248 <*> 0
1937 V27 N11 P287-290	59-15645 <*> 0
1938 V28 P161-170	58-861 <*>
1938 V28 N1 P5-8	59-20428 <*> 0
1940 V30 P337-341	62-10129 <*> 0
1940 V30 N6 P170-172	RCT V14 N1 P778 <RCT>
1941 V31 N4 P135-138	66-13233 <*>
1941 V31 N6 P223-225	63-18520 <*>
1941 V31 N11 P389-395	62-18420 <*> 0
1941 V31 N12 P417-421	59-15228 <*>
	63-10024 <*> 0
1942 V32 N2 P49-54	59-17678 <*>
1942 V32 N6 P180-182	59-17679 <*>
1942 V32 N7 P217-220	60-14412 <*>
1942 V32 N7 P221-223	62-18349 <*> P
1943 V33 N4 P97-102	63-18521 <*> 0
1944 V34 N2 P25-26	61-18305 <*>
1944 V34 N3 P51-57	62-18344 <*> 0
1944 V34 N6/7 P117-119	NP-TR-163 <*> 0
1947 V37 P25-29	I-468 <*>
1947 V37 P165-172	I-468 <*>
1947 V37 N8 P165-172	2084 <TC>
1947 V37 N2/3 P36-38	59-15849 <*>
1947 V37 N2/3 P43-45	<LSA>
1947 V37 N10/2 P213-218	61-10647 <*>
1948 V38 P74-76	2630 <K-H>
1949 V39 N1 P1-7	62-16975 <*> 0
1949 V39 N2 P47	66-11318 <*>
1950 V40 N7 P221-232	61-10646 <*> 0
1951 V41 P221-224	3302 <*>
1951 V41 N1 P1-6	64-16052 <*>
1951 V41 N3 P89-97	64-16065 <*>
1951 V41 N6 P186-188	57-3567 <*>
1951 V41 N12 P454-457	AL-973 <*>
	64-20313 <*>
1951 V41 N12 P457-462	57-986 <*>
1952 V42 N1 P9-12	64-10793 <*> 0
1952 V42 N3 P57-63	57-3573 <*>
1952 V42 N6 P45-46	61-10712 <*>
1952 V42 N12 P433-436	64-16050 <*> 0
1952 V42 N12 P445-449	5309-E <K-H> 0
	60-16280 <*>
	64-16763 <*> 0
1953 V43 P94-101	57-1191 <*>
1953 V43 P397-399	3324 <*>
	8832 <IICH>
1953 V43 P409-415	58-1225 <*>
1953 V43 P496-502	II-793 <*>
1953 V43 N1 P36-37	64-18883 <*> 0
1953 V43 N7 P266-270	AL-674 <*>
	57-1033 <*>
1953 V43 N12 P496-502	62-16533 <*> 0
	64-20352 <*> 0
1954 V44 P542-546	8834 <IICH>
1954 V44 P562-568	57-1327 <*>
1954 V44 N1 P3-5	62-18195 <*>
1954 V44 N2 P77-79	64-20337 <*> 0
1954 V44 N4 P131-133	62-10410 <*> 0
1954 V44 N4 P173-180	61-20946 <*> 0
1954 V44 N5 P221-226	61-20946 <*> 0
1954 V44 N7 P278-280	62-16449 <*>
1954 V44 N7 P281-284	64-20435 <*> 0
1954 V44 N7 P285-289	60-18441 <*>
1954 V44 N8 P341-347	2472 <*>
	57-3215 <*>
1954 V44 N8 P365-366	64-20821 <*>
1954 V44 N10 P429-430	64-16738 <*>
1954 V44 N10 P430-436	57-1194 <*>
1954 V44 N11 P525-528	63-10000 <*> 0
1954 V44 N11 P528-532	62-10420 <*> 0
1954 V44 N12 P105-111	57-2761 <*>
1954 V44 N12 P569-576	57-2696 <*>
1954 V44 N12 P601-607	12G6G <ATS>
	57-2664 <*>
1955 V45 P87-92	57-1944 <*>
1955 V45 P93-97	57-1328 <*>
1955 V45 N1 P9-12	344 <TC>
	5309-D <K-H>
	57-2754 <*>
	64-16743 <*>
1955 V45 N3 P99-103	57-1943 <*>
1955 V45 N4 P137-145	62K27G <ATS>
1955 V45 N7 P257-266	62-16999 <*> 0
1955 V45 N10 P410-414	1519 <*>
	57-1064 <*>
1955 V45 N11 P506-507	62-20131 <*>
1955 V45 N12 P101-104	61-10704 <*>
1955 V45 N12 P595-	61-10704 <*>
1956 V45 N10 P410-414	24J18G <ATS>
1956 V46 P55-58	T1994 <INSD>
1956 V46 P143-147	8787 <IICH>
1956 V46 P324-325	57-3402 <*>
1956 V46 P506-	<ES>
1956 V46 N1 P3-8	32H10G <ATS>
1956 V46 N1 P16-18	66-10131 <*> 0
1956 V46 N1 P18-20	15J15G <ATS>
	57-1789 <*>
	62-10132 <*>
	65-10441 <*>
1956 V46 N1 P442-450	67J18G <ATS>
1956 V46 N2 P81-86	5309 <K-H>
	64-16742 <*> 0
1956 V46 N3 P131-134	57-78 <*>
1956 V46 N4 P143-147	57-2585 <*>
1956 V46 N4 P176-180	62-16450 <*> 0
1956 V46 N5 P195-198	5309 A <K-H> 0
	64-16739 <*> 0
1956 V46 N6 P262-269	57-1063 <*>
	65-10712 <*>
	65-12243 <*>
1956 V46 N6 P274-280	48J19G <ATS>
1956 V46 N7 P341-344	59-15172 <*>
1956 V46 N8 P359-362	47L31G <ATS>
1956 V46 N8 P369-370	63-14169 <*> C
1956 V46 N10 P442-450	63-14168 <*> 0
	66-11334 <*> 0
1956 V46 N10 P450-459	44H130 <ATS>
1956 V46 N10 P460-466	64-16821 <*> 0
1956 V46 N10 P489-496	66-10130 <*> C
1956 V46 N11 P498-505	63-14170 <*> OP
	63-14176 <*> P
1956 V46 N12 P547-554	63-14171 <*> 0
1956 V46 N12 P555-556	63-14172 <*> 0
1956 V46 N12 P583-587	63-14173 <*> C
1957 V47 P510-512	58-1223 <*>
1957 V47 N1 P7-14	45K20G <ATS>
	65-17323 <*> 0
1957 V47 N3 P116-118	59-15654 <*> 0
1957 V47 N4 P153-156	60-16837 <*>
	62-10561 <*> 0
	63-14174 <*> 0
1957 V47 N4 P156-	62-10603 <*> C
1957 V47 N4 P180-182	62-14051 <*> 0
1957 V47 N4 P227-231	32J18G <ATS>
	64-16822 <*> 0
1957 V47 N5 P234-239	31J18R <ATS>
	62-10720 <*> C
1957 V47 N5 P250-254	23J18G <ATS>

65-10021 <*> 0
65-10442 <*>
1957 V47 N6 P299-302 63-14175 <*> 0
1957 V47 N6 P303-312 59-15697 <*>
61-16536 <*>
64-30663 <*>
1957 V47 N8 P409-412 59-15928 <*> 0
1957 V47 N8 P446-455 63-14177 <*> 0
1957 V47 N8 P510-512 65-11967 <*>
1957 V47 N8 P521-524 62-10599 <*>
1957 V47 N03 P116-118 66-11335 <*> 0
1957 V47 N10 P634-635 64-20117 <*> 0
1957 V47 N11 P651-655 59-10288 <*>
1957 V47 N12 P683-685 63-14178 <*>
65-18174 <*>
1957 V47 N12 P710-712 58-1258 <*>
1958 V48 P398-402 REPT. 61045 <RIS> 0
63-14039 <*> 0
1958 V48 N2 P65-66 65-11480 <*>
1958 V48 N3 P108-110 63-14179 <*> 0
65-18175 <*>
1958 V48 N3 P111-113 60-16884 <*>
1958 V48 N4 P165-166 66-11336 <*> 0
1958 V48 N5 P194-199 58-2144 <*>
63-14180 <*> 0
1958 V48 N6 P242-249 62-10519 <*> 0
1958 V48 N6 P257-261 78M39G <ATS>
1958 V48 N7 P350-352 59-15530 <*>
1958 V48 N8 P354-362 58-2327 <*>
59-15005 <*> 0
59-15592 <*>
65-10046 <*>
1958 V48 N8 P362-364 63-18526 <*>
1958 V48 N8 P391-393 62-10057 <*> 0
1958 V48 N8 P393 62-10372 <*> 0
1958 V48 N9 P406-408 63-18526 <*> 0
1958 V48 N11 P499-504 63-14181 <*> 0
1958 V48 N11 P525-530 59-20289 <*> 0
63-14182 <*> 0
1959 V49 P401-406 AEC-TR-4309 <*>
62-20455 <*>
1959 V49 P500-502 UCRL TRANS-529(L) <*>
1959 V49 P671-678 60-14535 <*>
1959 V49 N1 P9-14 60-10995 <*> 0
70L30G <ATS>
1959 V49 N1 P23-24 60-10923 <*>
1959 V49 N2 P50-55 63-14183 <*>
63-14183 <*> 0
1959 V49 N3 P109-113 61-20976 <*>
1959 V49 N4 P166-168 65-11966 <*>
1959 V49 N5 P213-216 63-14184 <*> 0
1959 V49 N5 P222-225 60-18787 <*> 0
1959 V49 N5 P230-231 60-16498 <*>
65-10192 <*>
1959 V49 N6 P264-268 64-30665 <*> 0
1959 V49 N7 P315-321 64-30686 <*>
1959 V49 N7 P321-322 60-16870 <+>
1959 V49 N8 P430-432 63-15587 <*> 0
1959 V49 N8 P432-434 63-15582 <*>
1959 V49 N8 P435 63-15588 <*>
1959 V49 N9 P455-458 63-14185 <*> 0
1959 V49 N10 P494-500 60-18433 <*>
1959 V49 N10 P500-502 60-14349 <*> 0
60-14545 <*>
65-10193 <*>
1959 V49 N10 P513-516 12798 <K-H>
63-14186 <*>
65-14869 <*>
1959 V49 N10 P516-525 64-30669 <*> 0
1959 V49 N10 P543-546 08L37G <ATS>
1959 V49 N10 P576-582 65-10194 <*> 0
1959 V49 N11 P616-621 63-14187 <*> 0
1959 V49 N12 P671-678 60-14535 <*>
1959 V49 N12 P679-683 62-10571 <*> 0
63-14188 <*> 0
1960 V50 P500-502 62-16730 <*>
1960 V50 P623-627 62-14032 <*>
1960 V50 N1 P23-26 66-11660 <*>
1960 V50 N1 P57-59 62-20024 <*>
1960 V50 N2 P137-140 63-18522 <*> 0
1960 V50 N3 P148-154 61-16725 <*>
64-16527 <*>
1960 V50 N3 P156-162 61-16722 <*>
1960 V50 N3 P163-165 63-14167 <*> 0
1960 V50 N3 P191-194 64-14049 <*>
1960 V50 N4 P227-234 65-18189 <*>
66-15530 <*> 0
1960 V50 N4 P255-256 61-10757 <*>
1960 V50 N6 P335-336 63-14189 <*> 0
1960 V50 N6 P360-365 11959 <K-H>
65-14025 <*> 0
1960 V50 N9 P485-490 61-20018 <*>
65-11590 <*> 0
1960 V50 N9 P500-502 61-10860 <*>
1960 V50 N10 P565-567 63-14190 <*> 0
1961 V51 N1 P18-20 65-17427 <*>
1961 V51 N2 P69-74 65-12924 <*>
1961 V51 N2 P104-105 64-10856 <*>
1961 V51 N3 P126-132 62-14292 <=*>
1961 V51 N3 P133-134 62-14293 <*>
1961 V51 N3 P137-144 61-20653 <*>
1961 V51 N6 P314-317 66-10163 <*> 0
1961 V51 N6 P338-340 62-14312 <*>
1961 V51 N9 P495-502 62-14441 <*>
1961 V51 N9 P503-508 SCL-T-481 <*>
1961 V51 N9 P512-517 SC-T-64-1623 <*>
66-11010 <*>
1961 V51 N9 P569- <ES>
1961 V51 N11 P698-702 63-14040 <*> 0
1961 V51 N11 P707-708 66-11058 <*>
1961 V51 N11 P708 63-14040 <*> 0
1961 V51 N11 P712-714 65-13739 <*>
1962 V52 N1 P15-18 06Q72G <ATS>
1962 V52 N8 P458-463 19Q67G <ATS>
1962 V52 N10 P599-603 SC-T-514 <*>
1963 V53 N2 P103-110 65-18152 <*>
1963 V53 N4 P210-217 63-20577 <*> 0
1963 V53 N7 P436 64-10857 <*>
1963 V53 N8 P502-509 34Q72G <ATS>
1963 V53 N8 P509-515 65-11637 <*>
1963 V53 N8 P541-546 65-11135 <*>
1963 V53 N10 P703-710 64-14985 <*> 0
1963 V53 N12 P941-943 65-13635 <*>
1964 V54 P432-435 333 <TC>
1964 V54 N3 P155-159 66-10747 <*>
1965 V55 P321-326 1437 <TC>
1965 V55 N2 P102-111 65-14957 <*>
66-12336 <*>
1965 V55 N3 P158-167 66-12791 <*>
1965 V55 N4 P263-268 66-12790 <*> 0
1965 V55 N5 P319-320 1130 <TC>
1965 V55 N5 P321-326 66-10376 <*>
1965 V55 N5 P329-332 1129 <TC>
66-10375 <*>
66-12810 <*>
1965 V55 N5 P335-338 1131 <TC>
1965 V55 N5 P346-350 1128 <TC>
1965 V55 N5 P372-374 65-14904 <*>
1965 V55 N6 P4380-4382 66-10537 <*>
1966 V56 N1 P15-23 66-14180 <*>
1966 V56 N2 P92-97 57T93G <ATS>
1966 V56 N3 P163-166 39T94G <ATS>

KUNST- UND PRESSTOFFE
1937 N1 P36-37 60-10211 <*>
1937 V2 P38- 57-2138 <*>

KUNSTSTOFFE-PLASTICS
1956 V3 N3 P247-254 59-17258 <*>
1957 V4 N3 P265-268 65-18025 <*>
1959 V6 P302-305 273 <TC>
1959 V6 N1 P32-37 19M46G <ATS>
1959 V6 N3 P297-301 11L32G <ATS>
1960 V7 N2 P179-192 88P62G <ATS>
1962 V9 N1 P32-33 3079 <BISI>

KUNSTSTOFF-RUNDSCHAU
1957 N7 P293-300 59-15968 <*>
1958 V5 N9 P385-389 95S86G <ATS>
1959 V6 P139-141 63-10579 <*>

65-10095 <*> 0
8626 <IICH>
1959 V6 N6 P217-220 62-10035 <*> 0
1960 V7 N1 P39-44 (CHEM)-168 <CTS>
61-16785 <*>
1961 V8 N1 P25-26 63-10497 <*>
1963 V10 P9-11 64-10925 <*>
1963 V10 N6 P277-284 66-12788 <*>
1963 V10 N7 P345-352 66-12789 <*>
1964 V11 P270-272 8906 <IICH>
1964 V11 N4 P209-214 64-30089 <*>
1964 V11 N5 P277-280 64-20958 <*>

KUZNECHNO-SHTAMPOVOCHNOE PROIZVODSTVO
1959 V1 N1 P25-31 TS 1541 <BISI>
62-14546 <=*>
62-32083 <=*>
1959 V1 N7 P38-40 TS 1738 <BISI>
62-32084 <=*>
1959 V1 N8 P18-21 61-27404 <*=> 0
1959 V1 N10 P30-36 1588 <BISI>
1959 V1 N11 P8-12 AEC-TR-4106 <+*>
1959 V1 N12 P1-5 AEC-TR-4624 <*=>
1959 V1 N12 P9-11 AEC-TR-4625 <*=>
1959 V1 N12 P11-14 AEC-TR-4626 <*=>
1960 N2 P13-15 60-11933 <=>
1960 V2 N1 P18-20 2172 <BISI>
1960 V2 N3 P3-8 TRANS-106 <MT>
1987 <BISI>
1960 V2 N3 P24-27 TRANS-133 <MT>
1960 V2 N3 P37-38 TRANS-134 <MT>
1960 V2 N5 P14-17 TRANS-163 <MT>
1960 V2 N7 P8-13 2281 <BISI>
1960 V2 N7 P17-21 2282 <BISI>
1960 V2 N7 P22 61-11972 <=>
1960 V2 N7 P39 61-11972 <=>
1960 V2 N7 P44 61-11972 <=>
1960 V2 N8 P19-21 TRANS-212 <MT>
1960 V2 N9 P1-5 2108 <BISI>
1960 V2 N9 P21-26 2287 <BISI>
1960 V2 N9 P26-33 2288 <BISI>
1960 V2 N10 P21-24 2454 <BISI>
1960 V2 N10 P30-35 2149 <BISI>
1960 V2 N11 P11-15 2413 <BISI>
1960 V2 N12 P5-9 2402 <BISI>
1960 V2 N12 P21 61-31115 <=>
1960 V2 N12 P44 61-31115 <=>
1961 V3 N1 P9-11 2612 <BISI>
1961 V3 N1 P14-18 62-23602 <=*> 0
1961 V3 N3 P1-2 61-23013 <=>
1961 V3 N3 P16-18 2528 <BISI>
5462 <HB>
1961 V3 N3 P18-19 2529 <BISI>
1961 V3 N3 P19-23 2530 <BISI>
1961 V3 N3 P27-30 61-23013 <=>
1961 V3 N4 P1-5 2696 <BISI>
1961 V3 N4 P46-48 62-13048 <*=>
1961 V3 N4 P48 61-28922 <*=>
1961 V3 N5 P4-5 2889 <BISI>
1961 V3 N5 P5-9 2673 <BISI>
1961 V3 N5 P10 2674 <BISI>
1961 V3 N5 P11-12 5593 <HB>
1961 V3 N5 P34-36 2656 <BISI>
1961 V3 N6 P7-13 2904 <BISI>
1961 V3 N6 P21-23 2903 <BISI>
1961 V3 N6 P38-43 2885 <BISI>
1961 V3 N7 P4-6 5744 <HB>
1962 V4 N5 P39-44 63-21569 <=>
1962 V4 N6 P26 62-33024 <=$>
1962 V4 N6 P29 62-33024 <=$>
1962 V4 N6 P30 62-33024 <=$>
1962 V4 N8 P8-11 63-19678 <=*>
1962 V4 N8 P27 62-33713 <=*>
1962 V4 N8 P30-33 62-33713 <=*>
1962 V4 N9 P22 63-13320 <=> P
1962 V4 N9 P25 63-13320 <=> P
1962 V4 N10 P39-42 63-13479 <=>
1962 V4 N11 P1-5 63-15725 <=*>
1963 N5 P27-31 AWRE-TRANS-49 <*>
1963 V5 N1 P29-32 3229 <BISI>
1963 V5 N2 P7-8 5915 <HB>
1963 V5 N2 P17-19 6033 <HB>
1963 V5 N4 P40-41 63-31015 <=>
1963 V5 N7 P31-34 63-31773 <=>
1963 V5 N8 P33-37 3683 <BISI>
1963 V5 N10 P32-36 3684 <BISI>
1963 V5 N12 P38 RTS-2674 <NLL>
1964 V6 N1 P5-10 6560 <HB>
1964 V6 N3 P4-8 RTS-2945 <BSRA>
1964 V6 N3 P11-16 AD-610 661 <=$>
1964 V6 N4 P37-38 6661 <HB>
1964 V6 N7 P16-18 64-51260 <=>
1964 V6 N8 P12-15 64-51880 <=$>
1964 V6 N8 P18-19 6369 <HB>
1964 V6 N8 P37-38 64-51880 <=$>
1964 V6 N9 P7-9 6566 <HB>
1964 V6 N9 P12-16 6298 <HB>
1964 V6 N9 P30-33 6556 <HB>
1964 V6 N10 P4-8 6425 <HB>
1964 V6 N11 P16-19 4250 <BISI>
1964 V6 N11 P37-39 6662 <HB>
1965 V7 N1 P21-24 6469 <HB>
1965 V7 N1 P25-26 6470 <HB>
1965 V7 N2 P24-28 6647 <HB>
1965 V7 N3 P7-9 6588 <HB>
1965 V7 N3 P9-10 6597 <HB>
1965 V7 N3 P13-14 6589 <HB>
1965 V7 N6 P1-4 6649 <HB>
1965 V7 N7 P1-5 6653 <HB>
1965 V7 N7 P5-8 6672 <HB>
1965 V7 N8 P1-6 6753 <HB>
1965 V7 N9 P1-8 6660 <HB>
1965 V7 N9 P36 6663 <HB>
1965 V7 N9 P37-38 6664 <HB>
1966 V8 N1 P3 6851 <HB>
1966 V8 N1 P46-47 6935 <HB>
1966 V8 N3 P3 6865 <HB>
1966 V8 N3 P8-10 6863 <HB>

KVASNY PRUMYSL
1955 V1 P132-134 64-14803 <*>
1956 V2 P59-61 64-16606 <*>
1957 V3 P1-5 64-16604 <*>
1958 N4 P79-80 59-15850 <*>
1960 V6 N8 P175-177 13N50C <ATS>
1960 V6 N9 P198-203 68N55C <ATS>

KWANGTUNG MEDICAL JOURNAL. CHINESE PEOPLES REPUBLIC
SEE KUANG TUNG I HSUEH HSUN TAI I HSUEH PAN

KWARTALNIK FILOZOFICZNY
1946 V16 P223-277 AMST S2 V29 P1-50 <AMS>

KYBERNETIK
1961 V1 N1 P1-6 SCL-T-401 <*>
61-18318 <*>
1961 V1 N1 P6 AEC-TR-4645 <*>
1961 V1 N2 P57-69 62-16587 <*>
1962 V1 N5 P200-208 64-14651 <*>
1964 V2 N2 P43-61 65-12216 <*>

KYOBU GEKA
1961 V14 P746-754 62-18080 <*>

KYODAI BOEN KENKYUSHO NENBO
1957 V127 P128-137 64-18475 <*> 0

KYOSEI IGAKU
1958 V7 N1 P40-52 63-01389 <*>
63-27742 <=$>

KYOTO DAIGAKU KAGAKU KENKYUJO
SEE BULLETIN OF THE INSTITUTE FOR CHEMICAL RESEARCH, KYOTO UNIVERSITY

KYOTO DAIGAKU KOGAKU KENKYUJO IHO

1954 V6 P9-15 19TM <CTT>
1954 V6 P57-60 II-1024 <*>

KYOTO DAIGAKU NIPPON KAGAKUSEN'I KENKYUSHO KOENSAU
1940 V5 P115-138 62-18943 <*>
1941 V6 P267-270 62-18947 <*> 0

KYOTO FURITSU IKADAIGAKU ZASSHI
1940 V28 P1041-1052 60-23118 <=*> P

KYOTO IGAKKAI ZASSHI
1908 V5 P115-117 1734 <*>

KYUSHU AGRICULTURAL RESEARCH
SEE KYUSHU NOGYO KENKYU

KYUSHU DAIGAKU KOGAKU SHUHO
1951 V24 P66-70 2051 <*>
1951 V24 N1 P16-20 66K26J <ATS>
1957 V29 N4 P229-234 97K26J <ATS>
1958 V30 N4 P263-268 98K26J <ATS>

KYUSHU DAIGAKU NOGAKOBU GAKUGII ZASSHI
1959 V17 P1-8 62-10301 <*>

KYUSHU DAIGAKU NOGAKUBU KIYO
1932 V3 N7 P149-178 60-16689 <*>

KYUSHU KETSUEKI KENKYU DOKOKAI-SHI
1959 V9 P722-753 63-00848 <*>
63-26695 <=$>

KYUSHU NOGYO KENKYU
1962 V24 P51-52 65-17226 <*>
1965 N27 P24-27 66-14018 <=$>
1965 N27 P29-30 66-13639 <*>
1965 N27 P31 66-13638 <*>
1965 N27 P36-37 66-13967 <*$>
1965 N27 P38 66-14026 <*$>
1965 V27 P15-17 66-13640 <*>

KYUSHU NOGYO SHIKENJO IHO
1960 V6 N4 P259-364 66-12889 <*> 0

LABORATORIO. GRANADA
1953 V16 N94 P301-320 1880 <*>

LABORATORNAYA PRAKTIKA
1940 V15 N5 P1-9 65-60102 <=$>
1940 V15 N7/8 P1-2 65-60028 <=$>

LABORATORNOE DELO
1956 V2 N2 P9-12 59-15384 <+*>
1957 V3 N2 P24-26 59-15892 <+*>
1958 V4 N2 P26-27 48Q69R <ATS>
1958 V4 N3 P44-46 61-19718 <*=>
1959 V5 N2 P45-46 C-3474 <NRCC>
60-00684 <*>
1959 V5 N5 P37-39 61-13796 <*+>
64-13562 <=*$>
1959 V5 N5 P42 61-13793 <*+>
1960 V6 N1 P40-43 60-11540 <=>
1960 V6 N2 P53-54 61-19650 <=*>
1960 V6 N3 P4-6 60-31765 <=>
1960 V6 N5 P22-24 61-11853 <=>
1960 V6 N5 P51-53 61-15856 <+*>
1960 V6 N6 P3-4 61-21564 <=>
1960 V6 N6 P12-18 61-21594 <=>
1960 V6 N6 P26-27 61-15854 <+*>
1960 V6 N6 P27-29 CSIR-TR 275 <CSSA>
61-15829 <+*>
1961 V7 N2 P10-12 61-21577 <=>
1961 V7 N3 P29-31 61-23018 <=*>
1961 V7 N3 P37-38 61-23938 <=*>
1961 V7 N3 P38-40 61-23113 <*=>
1961 V7 N3 P40-43 61-23938 <=*>
1961 V7 N6 P41-44 61-28901 <*=>
1961 V7 N6 P49-50 62-23627 <=*>
1962 V8 N4 P33-36 62-25050 <=*>
1962 V8 N4 P52-54 63-15576 <=*>
1962 V8 N6 P37-42 CSIR-TR.538 <CSSA>
1962 V8 N7 P3- FPTS V22 N3 P.T475 <FASE>
1962 V8 N7 P9-14 63-19382 <=*>
1962 V8 N7 P16-17 63-19004 <=*>
1962 V8 N7 P19- FPTS V22 N3 P.T483 <FASE>
1962 V8 N7 P22-24 63-19383 <=*>
1962 V8 N7 P26- FPTS V22 N4 P.T616 <FASE>
1962 V8 N7 P30-33 63-19384 <=*>
1962 V8 N7 P36-39 63-23044 <=*>
1962 V8 N7 P39- FPTS V22 N4 P.T642 <FASE>
1962 V8 N7 P47- FPTS V22 N3 P.T523 <FASE>
1962 V8 N8 P3-6 63-23007 <=*>
1962 V8 N8 P31- FPTS V22 N3 P.T528 <FASE>
1962 V8 N8 P35- FPTS V22 N5 P.T900 <FASE>
1962 V8 N8 P35-40 1963,V22,N5,PT2 <FASE>
1962 V8 N9 P5-9 63-21962 <=>
1962 V8 N9 P15-19 63-23898 <=*$> 0
1962 V8 N9 P26-29 64-13405 <=*$>
1962 V8 N9 P33-35 65-63147 <=$>
1962 V8 N9 P40-42 63-23850 <=*>
1962 V8 N9 P52-55 64-13914 <=*$>
1962 V8 N9 P55-59 63-24518 <=*$>
1962 V8 N10 P37-39 63-23828 <=*$> 0
1962 V8 N10 P49-51 63-23827 <=*>
1962 V8 N11 P15-17 63-24484 <=*$>
1962 V8 N11 P25-27 63-24485 <=*$>
1962 V8 N12 P15-18 64-13879 <=*$>
1962 V8 N12 P29-31 64-15409 <=*$>
1962 V8 N12 P41-45 64-15408 <=*$>
1963 N1 P35 63-24507 <=*$>
1963 N9 P3-5 63-24513 <=*$>
1963 N9 P44-48 63-24517 <=*$>
1963 V9 N1 P3-5 63-21392 <=>
1963 V9 N1 P33-34 64-19207 <=*$>
1963 V9 N1 P48- FPTS V23 N2 P.T383 <FASE>
1963 V9 N2 P9-11 63-21696 <=>
63-24602 <=*$>
1963 V9 N6 P61-63 63-20501 <=*$>
1963 V9 N9 P20-22 63-24514 <=*$>
1963 V9 N9 P22-25 RTS 2715 <NLL>
64-15388 <=*$>
1963 V9 N9 P29-34 63-24515 <=*$>
1963 V9 N9 P34-39 63-24516 <=*$>
1965 N2 P87-90 65-30759 <=$>
1965 N2 P110-112 65-30759 <=$>
1965 N5 P318-320 65-31372 <=*$>

LACZNOSC
1958 V9 N7 P3 60-31503 <=*>
1958 V9 N32 P5 60-21488 <=*>
1958 V9 N37 P1 60-31504 <=*>
1958 V9 N37 P4 60-31505 <=*>
1958 V9 N38 P4 60-31507 <=*>
1958 V9 N39 P3 60-31508 <=*>
1958 V9 N44 P2 60-31490 <=*>
1958 V9 N51/2 P1 60-31511 <=*>
1958 V9 N51/2 P2 60-31512 <=*>
1958 V9 N51/2 P4 60-31513 <=*>
1958 V9 N51/2 P5 60-31541 <=*>
1958 V9 N51/2 P12 60-31515 <=*>
1959 V10 N3 P3 60-31491 <=*>
1959 V10 N4 P2 60-315 <=*>
1959 V10 N4 P4 60-31517 <=*>
1959 V10 N4 P5 60-31518 <=*>
1959 V10 N5 P5 60-31482 <=*>
1959 V10 N10 P4 60-31519 <=*>
1959 V10 N10 P6 60-31520 <=*>
1959 V10 N16 P2 60-31492 <=*>
1959 V10 N48 P4 60-31493 <=*>
1959 V10 N51 P7 60-31529 <=*>
1959 V10 N51/2 P1-2 60-31527 <=*>
1959 V10 N51/2 P3 60-31528 <=*>
1960 V11 N1 P3 60-31521 <=*>
1960 V11 N1 P5 60-31522 <=*>
1960 V11 N4 P5 60-31494 <=*>
1960 V11 N7 P5 60-31495 <=*>
1960 V11 N9 P1 60-31496 <=*>
1960 V11 N10 P1 60-31523 <=*>
60-31524 <=*>
1960 V11 N12 P5 60-31497 <=*>

1960 V11 N13 P1 60-31525 <=*>
1960 V11 N13 P3 60-31526 <=*>

LAEKNABLADID
1961 V45 N1 P1-7 11279-E <KH>
63-18515 <*>

LAIT
1939 V19 P698-703 59-20653 <*>
1939 V19 P811-814 59-20653 <*>
1944 V24 P313-341 57-90 <*>
1945 V25 P27-50 II-213 <*>
1951 V31 P511-518 59-20651 <*>
1957 V37 N361 P9-20 65-12662 <*>

LAKOKRASOCHNYE MATERIALLY I IKH PRIMENENIE
1960 N1 P58-62 2248 <BISI>
1960 N1 P75-78 61-10558 <+*> 0
1960 N2 P3-5 60-31815 <=>
1960 N2 P95 60-31813 <=>
1960 N6 P21-26 64-30621 <*> 0
1961 N1 P23-29 64-15900 <=*$>
65-14664 <*> 0
1961 N1 P30-32 NS-408 <TTIS>
1961 N2 P1-2 61-27863 <*=>
1961 N3 P20-25 4015 <TTIS>
95P61R <ATS>
1961 N3 P25-26 4006 <TTIS>
1961 N3 P65-66 4012 <TTIS>
1961 N4 P2-5 4013 <TTIS>
1961 N4 P13-15 4011 <TTIS>
1961 N4 P30-34 4015 <TTIS>
1961 N5 P35-43 65-27480 <$> P
1961 N5 P46-53 65-17217 <*>
1961 N6 P31-35 NS 212 <TTIS>
1961 N6 P52-55 NS 118 <TTIS>
1961 N6 P76-80 NS-31 <TTIS> P
1962 N1 P17-20 4016 <TTIS>
1962 N1 P20-25 NS 261 <TTIS>
1962 N1 P44-49 NS-70 <TTIS>
1962 N1 P54-56 NS-69 <TTIS>
1962 N2 P4-7 4019 <TTIS>
1962 N2 P34-40 63-21180 <=>
1962 N2 P48-52 62-29004 <=$>
1962 N2 P52-54 4025 <TTIS>
1962 N3 P32-34 NS-72 <TTIS>
1962 N3 P42-44 NS-360 <TTIS>
1962 N3 P50-51 AD-620 075 <=$>
1962 N4 P30-32 NS-411 <TTIS>
1962 N4 P35-37 NS-286 <TTIS>
1962 N5 P30-37 TP/T-3598 <NLL>
1962 N5 P50-52 4027 <TTIS>
1962 N6 P15-19 63-20489 <=*> 0
1962 N6 P20-23 NS-103 <TTIS>
1962 N6 P52-53 NS-104 <TTIS>
1962 N6 P54-56 NS-106 <TTIS>
1962 V5 P27-30 NS-359 <TTIS>
1963 N1 P23-26 NS 129 <TTIS>
1963 N1 P26-30 NS-178 <TTIS>
1963 N1 P30-33 NS-179 <TTIS>
1963 N1 P36-38 NS-180 <TTIS>
1963 N1 P42-43 NS-166 <TTIS>
1963 N1 P52-54 NS 128 <TTIS>
1963 N1 P55-57 NS 130 <TTIS>
1963 N1 P70-71 NS 113 <TTIS>
1963 N1 P71-72 NS 112 <TTIS>
1963 N1 P72 NS-107 <TTIS>
1963 N2 P12-16 NS 120 <TTIS>
1963 N2 P25-29 NS 122 <TTIS>
1963 N2 P67-69 65-60742 <=>
1963 N3 P12-16 RAPRA-1274 <RAP>
1963 N3 P19-23 65-11278 <*>
1963 N3 P37-41 65-11342 <*>
1963 N3 P54-57 NS-409 <TTIS>
1963 N3 P58-60 NS-156 <TTIS>
1963 N3 P77-78 NS-167 <TTIS>
1963 N4 P5-10 NS-181 <TTIS>
1963 N4 P10-15 NS-182 <TTIS>
1963 N4 P23-26 NS-168 <TTIS>
1963 N4 P26-28 NS-285 <TTIS>
1963 N4 P47-50 65-14755 <*>
1963 N4 P51-52 NS-169 <TTIS>
1963 N4 P56-60 RTS-2775 <NLL>
1963 N5 P21-23 NS-190 <TTIS>
1963 N5 P24-26 NS-183 <TTIS>
1963 N5 P45-46 NS-191 <TTIS>
1963 N5 P47 NS-189 <TTIS>
1963 N5 P51-57 NS-295 <TTIS>
1963 N5 P52-55 NS-155 <TTIS>
1963 N5 P57-62 NS-296 <TTIS>
1963 N6 P4-7 NS 214 <TTIS>
1963 N6 P21-24 NS-200 <TTIS>
1963 N6 P57- NS-358 <TTIS>
1963 N6 P57 65-17459 <*>
1963 N6 P59-61 NS-337 <TTIS>
1963 V2 P1-5 NS 121 <TTIS>
1964 N1 P1-3 NLLTB V6 N8 P729 <NLL>
1964 N1 P28-31 NS-198 <TTIS>
1964 N1 P32-34 AD-616 308 <=$>
1964 N1 P35-39 NS 234 <TTIS>
1964 N1 P56-58 NS 225 <TTIS>
1964 N1 P64-71 NS 215 <TTIS>
1964 N1 P79-81 NS-281 <TTIS>
1964 N1 P82-85 NS-199 <TTIS>
1964 N2 P3-4 NS 247 <TTIS>
1964 N2 P13-15 NS 235 <TTIS>
1964 N2 P15-21 NS-248 <TTIS>
1964 N2 P21-23 NS 226 <TTIS>
1964 N2 P23-26 NS 227 <TTIS>
1964 N2 P33-36 NS 249 <TTIS>
1964 N3 P14-17 NS-282 <TTIS>
1964 N3 P18-19 NS-283 <TTIS>
1964 N3 P34-37 AD-626 432 <=*$>
C-5412 <NRC>
1964 N3 P49-51 NS-284 <TTIS>
1964 N4 P1-4 NS-506 <TTIS>
1964 N4 P8-15 65-14783 <*>
1964 N4 P22-25 NS-324 <TTIS>
533 <TC>
65-12909 <*>
1964 N4 P32-34 NS-507 <TTIS>
1964 N4 P42-45 NS-287 <TTIS>
1964 N4 P45-49 NS-288 <TTIS>
1964 N4 P47-50 NS-289 <TTIS>
1964 N4 P64-66 NS-290 <TTIS>
1964 N5 P9-11 NS-325 <TTIS>
1964 N5 P23-25 NS-353 <TTIS>
1964 N5 P31-33 NS-291 <TTIS>
1964 N5 P33-36 AD-628 185 <=*$>
1964 N5 P39-40 NS-292 <TTIS>
1964 N5 P41-44 NS-326 <TTIS>
1964 N5 P44-46 AD-629 863 <=*$>
1964 N5 P66-69 NS-293 <TTIS>
1964 N5 P77-79 NS-327 <TTIS>
1964 N6 P7-11 NS-450 <TTIS>
1964 N6 P34-39 NS-451 <TTIS>
1964 N6 P39-40 NS-410 <TTIS>
1964 N6 P45-50 NS-452 <TTIS>
1964 N6 P53-56 AD-628 194 <=*$>
1965 N1 P27-29 NS-449 <TTIS>
1965 N1 P56-63 M.5806 <NLL>
1965 N2 P22-25 NS-448 <TTIS>
1965 N3 P7-15 NS-508 <TTIS>
1965 N3 P18-21 NS-509 <TTIS>
66-12640 <*>
1965 N3 P22-25 NS-551 <TTIS>
1965 N4 P13-18 66-12741 <*>
1965 N4 P22-26 NS-510 <TTIS>
1965 N5 P25-30 NS-512 <TTIS>
1965 N5 P35-38 NS-552 <TTIS>
1965 N6 P1-6 NS-553 <TTIS>
1965 N6 P59-60 NS-554 <TTIS>
1965 V5 P1-4 NS-511 <TTIS>

LANCETTE FRANCAISE. (SUPPLEMENT TO GAZETTE DES HOPITAUX CIVILS ET MILITAIRES)
1960 V132 P673-676 SP.NO 64-14546 <*>

LAND DEVELOPMENT
SEE KOKUDO KAIHATSU

LANDARBEIT UND TECHNIK
1955 N20 P109-121 61-20425 <*>
1955 N20 P135-147 61-20424 <*>

LANDARZT
1957 V33 P789-791 58-1367 <*>
1959 V35 P470-475 9561-A <K-H>
1959 V35 N13 P470-475 64-16878 <$>
1960 V36 N11 P378-383 60-17659 <+*> 0
1963 V39 N25 P1100-1101 66-10095 <*>
1963 V39 N36 P1581-1582 66-10094 <*>

LANDBAUFORSCHUNG VOELKENRODE
1956 V6 N4 P91-93 CSIRO-3453 <CSIR>

LANDBOUWDOCUMENTATIE
1964 V20 N15 P516-523 65-14698 <*> 0

LANDBOUWKUNDIG TIJDSCHRIFT
1951 V63 P392-397 2173B <K-H>
58-1015 <*>
1955 V67 P345-356 CSIRO-3283 <CSIR>
1955 V67 P397-402 CSIRO-3072 <CSIR>
59-10827 <*>
1955 V67 P609-619 CSIRO-3191 <CSIR>
1955 V67 P620-628 CSIRO-3192 <CSIR>
1955 V67 P713-731 CSIRO-3184 <CSIR>
1955 V67 N4 P267-282 CSIRO-2902 <CSIR>
1955 V67 N4 P282-286 C-2283 <NRC>
1957 V69 P410-412 CSIRO-3747 <CSIR>

LANDTECHNIK
1953 V8 P376 59-17114 <*> 0
1953 V8 N17 P584-587 58-2184 <*>
1954 V9 P354-359 59-17844 <*> 0
1964 V19 P622-634 65-14092 <*> 0
1964 V19 N19 P700-702 65-14356 <*>
1964 V19 N19 P704-708 65-14399 <*>
1964 V19 N19 P708-711 65-14155 <*>
1964 V19 N19 P711-712 65-14357 <*>
1964 V19 N19 P713-716 65-14355 <*>

LANDTECHNISCHE FORSCHUNG
1951 V1 N1 P188- 59-17872 <*> 0
1953 V3 N1 P1-13 65-14146 <*> 0
1954 V4 N3 P79-81 65-14147 <*> 0
1955 V5 N4 P109-110 CSIRO-3234 <CSIR>
1955 V5 N4 P111-116 CSIRO-3231 <CSIR>
1958 V8 N4 P95-101 65-14149 <*>
1962 V12 N5 P125-128 63-14399 <*>
65-14148 <*> 0
1963 V13 N5 P142-150 65-14150 <*> 0
1964 V14 N5 P150-152 65-13927 <*>
1965 V15 N4 P105-110 66-11303 <*>
1965 V15 N4 P116-129 66-11304 <*>

LANDWIRTSCHAFTLICHE FORSCHUNG
1950 V2 P38-50 2565 <K-H>
1950 V2 N2 P97-110 58-2232 <*>
1955 V7 P113-116 II-891 <*>
1961 N15 P61-74 65-10458 <*> 0
1961 V14 N4 P246-254 63-10383 <*>

LANDWIRTSCHAFTLICHEN VERSUCHSSTATIONEN
1877 V20 P273-355 60-16717 <*>
1925 V103 P159-177 64-20477 <*>

LANDWIRTSCHAFTLICHES JAHRBUCH FUER BAYERN
1953 V30 P33-46 64-10333 <*>
1953 V30 N1/2 P55-63 63-16111 <*>

LANDWIRTSCHAFTLICHES JAHRBUCH DER SCHWEIZ
1918 V32 P221-248 59-20673 <*>
1945 V59 P915-927 59-15278 <*> 0
1948 V62 P853-875 CSIRO-3156 <CSIR>
1952 V66 P308-380 TT-790 <NRCC>
1954 V68 P291-368 C-3215 <NRCC>
1957 V71 N6 P143-164 TT-810 <NRCC>
1958 V72 N7 P163-182 C-3194 <NRCC>

LANGENBECK'S ARCHIV FUER KLINISCHE CHIRURGIE VEREINIGT MIT DEUTSCHE ZEITSCHRIFT FUER CHIRURGIE
1953 V274 N3 P225-236 62-20204 <*>
1955 V280 P592-608 5009-D <K-H>
1955 V280 N6 P641-647 1234 <*>
1957 V286 N2 P91-98 59-15680 <*> 0
1959 V292 P629-634 64-16937 <*> 0
1960 V295 P779-783 62-00856 <*>
1964 V308 P793-797 988 <TC>

LAO DONG
1960 N762 P3 60-41451 <*=>

LASTECHNIEK
1958 V24 P1-5 61-00035 <*>
1959 V25 P229-231 79M42OU <ATS>
1959 V25 P240-241 AEC-TR-4021 <*>
1959 V25 P248-250 AEC-TR-4024 <*>
1963 V29 N5 P94-99 3958 <BISI>
1964 V30 N7 P217-222 4053 <BISI>

LATTANTE
1960 V31 N1 P1-36 61-00984 <*>

LATVIJAS PSR ZINATNU AKADEMIJA. ELEKTRONIKAS UN SKATLOSONAS TEHNIKAS INSTITUTS. TRUDY. RIGA
1962 V3 P133-141 AD-617 672 <=$>
1963 V4 P21-33 AD-614 951 <=$>

LATVIJAS PADOMJU SOCIALISTISKAS REPUBLIKAS ZINATNU AKADEMIJA. TEHNISKO ZINATNU NODALA.SBORNIK STATEI
1956 N4 P95-121 SCL-T-332 <*>
1956 V4 P95-121 SCL-T-332 <=$>

LATVIJAS PSR ZINATNU AKADEMIJAS VESTIS
1948 N10 P100-110 R-2757 <*>
1951 N2 P333-346 AMST S2 V1 P239-252 <AMS>
1953 N11 P119-129 R-1999 <*>
61-15329 <+*>
1954 N10 P113-128 64-16735 <=*$>
1956 N1 P131-138 R-5117 <*>
1956 N5 P105-115 89N47R <ATS>
1956 V6 P80-83 64-19523 <=$>
1957 N7 P147-150 62-11765 <=>
1957 N9 P87-93 60-51163 <=>
1958 N3 P107-120 AMST S2 V18 P173-186 <AMS>
1958 N6 P101-107 60-41270 <=>
1959 N2 P85-90 70Q71R <ATS>
1959 N5 P73-76 90M46R <ATS>
1959 N9 P5-12 60-31008 <=>
1959 N9 P91-100 60-11444 <=>
1959 N11 P5-10 60-31228 <=>
1959 N11 P47-54 60-31226 <=>
1959 N11 P115-122 61-20716 <*=> 0
1959 N11 P169-177 60-31229 <=> 0
1959 N11 P183-190 60-31227 <=>
1959 N12 P83-90 40S87R <ATS>
64-10584 <=*$>
1960 N1 P5-16 60-11707 <=>
1960 N1 P73-76 62-13003 <*=>
1960 N2 P203-206 60-31689 <=*>
1960 N3 P11-30 60-41062 <=>
1960 N3 P57-64 AEC-TR-4633 <*>
1960 N5 P81-84 62-13238 <=*>
1960 N5 P89-96 62-18907 <=*>
1960 N5 P203-204 62-13237 <*=>
1960 N6 P199-206 61-11726 <=>
1960 N9 P79-84 63-10508 <=*>
1960 N12 P163-164 61-27242 <*=>
1960 N12 P165-168 61-27247 <*=>
1960 V7 P1-75 61-28756 <=>
1960 V7 P147-161 61-28756 <=>
1960 V7 P163-171 61-28756 <=>
1960 V7 P173-178 61-28756 <=>
1960 V7 P183-212 61-28756 <=>
1960 V7 P219-227 61-28756 <=>
1961 N1 P59-66 62-19650 <=*>
1961 N1 P180-184 62-24965 <=*>
1961 N2 P3-15 61-28916 <*=>
1961 N2 P17-29 61-28950 <*=>

1961 N2 P79-88 62-13038 <*=>
1961 N2 P89-93 62-32981 <=*>
1961 N3 P59-60 63-13018 <=*>
1961 N3 P67-76 96Q68R <ATS>
1961 N4 P85-92 64-14086 <=*$>
1961 N4 P151-165 62-13045 <*=>
1961 N5 P33-38 62-15779 <*=>
1961 N6 P43-49 62-33054 <=*>
1961 N8 P144 62-33006 <=>
1961 N8 P152 62-33006 <=>
1961 N11 P133-135 62-11718 <=>
1961 N12 P19-24 62-23605 <=*>
1962 N1 P131-142 62-33006 <=>
1962 N2 P51-57 62-24720 <=*>
1962 N2 P142 62-11718 <=>
1962 N2 P151-154 62-11718 <=>
1962 N3 P3-12 63-21100 <=>
1962 N3 P153 63-21100 <=>
1962 N6 P67-71 64-13117 <=*$>
1962 N6 P123-127 63-21100 <=>
1962 N6 P146 63-21100 <=>
1962 N7 P3-8 63-21671 <=>
1962 N10 P147-148 63-31695 <=>
1962 N12 P138 63-31695 <=>
1962 V8 P67-74 65-60679 <=$>
1963 N3 P3-14 63-31695 <=>
1963 N5 P51-56 64-11950 <=> M
1963 N6 P43-48 AD-610 346 <=$>
1963 N6 P49-57 64-10807 <=*$>
1963 N6 P75-84 64-10807 <=*$>
1963 N12 P49-59 AD-615 530 <=$>
1963 V12 P24-26 64-31986 <=>
1964 N1 P89-105 64-41035 <=>
1964 N2 P93-103 64-31537 <=>
1964 N4 P30-37 AD-640 257 <=$>
1964 N5 P43-53 65-30182 <=$>

LATVIJAS PSR ZINATNU AKADEMIJAS VESTIS. FIZIKAS TEHNISKO ZINATNU SERIJA
1964 N2 P46-52 65-64105 <=*$>
1964 N3 P77-87 64-41298 <=>
1964 N3 P95-119 64-41298 <=>
1964 N4 P23-30 66-14562 <*> 0
1964 N6 P101-106 65-30461 <=$>
1965 N2 P1-6 66-14561 <*> 0
1966 N2 P3-9 66-13684 <*>

LATVIJAS PSR ZINATNU AKADEMIJAS VESTIS. KIMIJAS SERIJA
1962 N3 P387-392 65-14686 <*>

LAVAL MEDICAL
1944 V9 P548-561 I-736 <*>
1949 V14 N7 P908-941 59-20972 <*>
1954 V19 P637-695 65-18046 <*> 0
1961 V31 N5 P604-610 11279-H <KH>
63-16929 <*>

LAVORI DELL'ISTITUTO DI ANATOMIA E ISTOLOGIA PATOLOGICA DELL' UNIVERSITA DEGLI STUDI DI PERUGIA
1954 V14 N3 P233-247 2706 <*>
1955 V15 N1 P5-31 2745 <*>
1955 V15 N1 P61-76 2746 <*>
1955 V15 N1 P77-85 2747 <*>
1963 V23 P167-202 65-17303 <*>
1963 V23 N2 P103-114 64-00355 <*>
1963 V23 N2 P115-128 64-00350 <*>

LAVORO UMANO
1957 V9 N9 P433-440 60-18181 <*>
1957 V9 N9 P545-552 60-18182 <*>

LA-YAARAN
1955 P7-11 60-51189 <=>

LEATHER CHEMISTRY. JAPAN
SEE HIKAKU KAGAKU

LEBENSMITTEL-INDUSTRIE
1964 V11 N7 P236-239 64-51156 <=>

LEDER
1950 V1 P3-12 PT1 2403 <*>
1950 V1 P257-259 60-10147 <*> 0
1951 V2 P241-242 T2211 <INSD>
1951 V2 N1 P4-8 59-20619 <*> 0
1953 V4 P289-292 62-14054 <*> 0
1953 V4 N10 P234-240 59-20620 <*>
1954 V5 N4 P73-82 62-10532 <*> 0
1954 V5 N5 P97-109 59-20613 <*>
1959 V10 P8-12 62-01075 <*>
1962 V13 N11 P253-262 66-10621 <*> 0

LEGKAYA ATLETIKA V SSSR
1965 N1 P15-16 65-30731 <=*$>

LEGKAYA PROMYSHLENNOST. KIEV
1944 N10/1 P21-23 R-2850 <*>
1946 N11/2 P43-45 RCT V21 N3 P727 <RCT>
1946 V6 N1 P37-39 62-16463 <=*>
1946 V6 N2 P39-41 62-16464 <=*>
1947 V7 N6 P23-25 62-10925 <=*>
1948 V8 N5 P22-24 62-10926 <=*>
1948 V8 N11 P18-20 63-23247 <=*>
1949 N7 P25-28 927TM <CTT>
1949 N9 P16-18 RT-2587 <*>
1949 N9 P24 50/2061 <NLL>
1949 N11 P24-26 50/2060 <*>

1950 V10 N2 P20-22 RCT V27 N3 P688 <RCT>
1950 V10 N9 P29-30 TT.290 <NRC>
1951 N5 P44-45 RCT V25 N2 P339 <RCT>
1951 V11 N8 P31-33 52/2595 <*>
1951 V11 N9 P29-31 59-18716 <+*>
1951 V11 N10 P27 52/2586 <NLL>
1952 V12 N2 P33-34 52/2589 <NLL>
1952 V12 N3 P26-27 52/2603 <NLL>
1952 V12 N4 P33 52/2583 <NLL>
1952 V12 N6 P17-18 52/2590 <NLL>
1952 V12 N7 P20-21 52/2591 <NLL>
1952 V12 N10 P36-39 64-16303 <=*$>
1952 V12 N10 P40-41 RCT V28 N3 P804 <RCT>
1954 V14 N1 P45-47 RCT V33 N4 P1010 <RCT>
1955 N9 P18-22 R-3785 <*>
1955 V15 P25-29 GB57 507 <NLL>
1955 V15 N1 P28-30 RCT V33 N4 P1015 <RCT>
1955 V15 N5 P10-18 R-301 <*>
R-3645 <*>
2971 <NLL>
1955 V15 N5 P39-40 R-2028 <*>
1955 V15 N9 P43-44 62-14899 <=*>
1956 N4 P38-44 RCT V32 N2 P628 <RCT>
1956 V16 N4 P32-38 60-21944 <=> 0
1956 V16 N6 P11-13 BGIRA-608 <BGIR>
1956 V16 N8 P22-24 65-29508 <$>
1956 V16 N10 P18-20 60-23138 <+*>
1957 V17 N3 P43-44 60-17788 <+*>
1957 V17 N5 P30-32 62-16406 <=*>
1957 V17 N9 P18-21 60-21809 <=> 0
1957 V17 N9 P26-27 60-21810 <=>
1957 V17 N9 P29-31 60-21811 <=> 0
1957 V17 N10 P14-17 60-19862 <+*>
1957 V17 N12 P16-18 59-18722 <+*>
1957 V17 N12 P32-35 63-10664 <=*>
1958 V18 N2 P13-14 60-21802 <=>
1958 V18 N2 P24-26 59-18019 <+*>
1958 V18 N2 P30-32 RCT V34 N1 P357 <RCT>
1958 V18 N3 P30-32 60-21794 <=>
1958 V18 N7 P20-22 60-21935 <=>
1958 V18 N11 P31-33 61-14519 <=*>
1958 V18 N12 P23-27 61-20704 <=*>

LEGKIE METALLY
1932 N7/8 P24-27 R-2586 <*>
1934 V3 N10 P1-9 R-2253 <*>
1935 V4 P16-23 R-2804 <*>
1935 V4 N1 P4-12 R-3287 <*>
1935 V4 N5 P22-27 R-2293 <*>
1935 V4 N5 P28-31 RT-2043 <*>

11920 <K-H>
1935 V4 N8 P24-34 62-19893 <=*>
1935 V4 N11 P1-4 65-10152 <*> 0
1935 V4 N11 P1-14 66-13984 <*>
1936 N8/9 P35-40 64-10729 <=*$> 0
1936 V5 N3 P1-15 R-2504 <*>
1936 V5 N4 P1-12 R-2504 <*>
1936 V5 N7 P52-55 5866-A <K-H>
64-18396 <*> 0
1936 V5 N10 P34-40 95J17R <ATS>
1936 V5 N12 P11-15 RT-2780 <*>
1937 N7/8 P9-14 RT-2175 <*>
1937 V6 N1 P21-22 64-14403 <=*$>
1937 V6 N1 P22-28 64-14660 <=*$> 0
1937 V6 N2 P13-15 RT-1556 <*>
1937 V6 N3 P17-24 64-14414 <=*$> 0
1937 V6 N3 P25-26 64-14907 <*>
1937 V6 N4 P10-11 RT-2786 <*>
1937 V6 N5/6 P6-9 R-2250 <*>
1937 V6 N5/6 P27-31 RT-1601 <*>
1937 V6 N7/8 P9-14 64-10453 <=*$> 0
1937 V6 N7/8 P20-23 R-2891 <*>
1938 N2 P12-16 R-4672 <*>
1938 V7 N2 P4-11 RT-1602 <*>
3322-A <K-H>
64-14415 <=*$>
1938 V7 N2 P12-16 61-16614 <=*$>

LEICHTBAU DER VERKEHRSFAHRZEUGE
1961 V5 N6 P255-268 65-12150 <*>
1963 V7 N2 P62-69 64-16029 <*>

LEITZ-MITTEILUNGEN FUER WISSENSCHAFT UND TECHNIK
1960 V1 P136-139 63-10859 <*>
1961 V2 N5 P146-147 64-14769 <*> 0

LEKARSKE LISTY
1953 V8 P523-526 2986-B <K-H>
1953 V8 N20 P468-472 PANSDOC-TR.484 <PANS>

LEKARSKY OBZOR
1961 N5 P305 62-23262 <*=>
1961 V10 N1 P51-59 62-19225 <=*>
1961 V10 N2 P125-127 62-19518 <*=>
1961 V10 N3 P135-139 62-19616 <=*>
1961 V10 N4 P193-196 62-19565 <=*>
1961 V10 N5 P306 62-19506 <*=>
1961 V10 N5 P307-308 62-19505 <=*>
1962 V11 N2 P105-112 62-24999 <=*>
1962 V11 N3 P181-182 62-24999 <=*>
1964 V13 N7 P353-359 64-41960 <=$>
1965 V14 N5 P257-265 65-31974 <=*$>

LEKARZ WOJSKOWY. EDINBURGH
1942 V34 N5 P263-277 57-1640 <*>

LEKARZ WOJSKOWY. WARSZAWA
1963 V39 P916-926 66-10969 <*>
1964 V40 N8 P577-585 64-51841 <=$> 0
1964 V40 N9 P666-674 64-51872 <=$>
1965 V41 N5 P344-348 65-31529 <=$>
1965 V41 N6 P417-422 65-32059 <=$>
1966 V42 N5 P393-397 66-33223 <=$>

LENZINGER BERICHTE
1962 V12 N9 P18-25 63-18277 <*>
1963 N13 P19-22 64-10874 <*>
1963 N13 P57-70 64-10860 <*> 0
1964 N16 P5-19 66-10882 <*>
1965 N18 P48-53 66-10881 <*>

LEPRO. OSAKA, KYOTO
1956 V25 N1 P2-14 57-2730 <*>

LEPROLOGIA
1963 V8 P155-159 65-17271 <*>
1963 V8 P175-179 65-17270 <*>
1963 V8 P186-188 65-17269 <*>

LES. LJUBLJANA
1957 V9 N4 P50-53 60-16514 <*> 0
1960 V12 N1 P4-8 C-3597 <NRCC>

LES I STEP
1950 V2 N2 P29-37 66-51081 <=>
1952 V4 N11 P48-52 R-1527 <*>
1953 V5 N2 P7-17 RT-4624 <*> 0

LESNAYA PROMYSHLENNOST
1947 N2 P10-12 60-51070 <=>
1950 V10 N10 P25-29 RT-149 <*>
1950 V10 N11 P22-27 RT-149 <*>
1950 V10 N12 P20-22 RT-1660 <*>
1951 N1 P7-8 RT-2594 <*>
1951 V11 N1 P29-32 60-14202 <+*> 0
1951 V11 N4 P27-30 RT-150 <*>
1951 V11 N5 P25-28 RT-1659 <*>
60-10936 <*>
1951 V11 N7 P16-19 RT-148 <*>
1951 V11 N7 P25-27 RT-151 <*>
1951 V11 N9 P10-13 RT-147 <*>
1951 V11 N9 P15-17 RT-146 <*>
1952 N4 P32-33 R-167 <*>
1952 V12 N4 P31-32 RT-1935 <*>
1952 V12 N4 P32-33 RT-4211 <*>
1952 V30 N1 P25-27 59-17420 <+*>
1953 N5 P26-29 RT-2168 <*>
1953 N6 P22-24 RT-2169 <*>
1953 V13 N8 P28-31 RT-2030 <*>
1954 V14 N5 P9-10 RT-4081 <*>
1954 V14 N11 P15-17 R-1045 <*>
RT-3316 <*>
1956 N2 P11-15 R-4560 <*>
R-675 <*>
1956 V16 N11 P18-19 C-3295 <NRCC>
1961 N9 P2-3 61-21697 <=>
1961 N10 P1-2 61-21697 <=>
1961 V39 N12 P13-14 62-01579 <*>
1963 V41 N7 P5 63-31927 <=>
1963 V41 N7 P6 63-31752 <=>
1963 V41 N12 P9-10 64-16905 <*>
1964 V42 N4 P17-18 C-5409 <NRC>
1964 V42 N6 P3-6 65-30168 <=$>
1965 N3 P4-8 65-31131 <=$>

LESNICKA PRACE
1954 P374-381 II-785 <*>
1963 V42 N7 P305-308 C-6050 <NRC>

LESNOE KHOZYAISTVO
1929 N8 P38-68 R-952 <*>
1929 N10/1 P19-44 R-953 <*>
1950 V3 N1 P27-31 RT-2207 <*>
1950 V3 N1 P38-45 60-21888 <=>
1951 P16-18 RT-169 <*>
1952 V5 N5 P23-29 RT-4234 <*>
1954 V7 N9 P72-73 63-11030 <=>
1955 V8 N2 P49-50 60-21867 <=>
1957 V10 N12 P46-48 63-11006 <=>
1958 V11 N6 P10-16 60-21866 <=>
1958 V11 N6 P15-19 60-21868 <=>
1958 V11 N9 P56-57 61-13804 <+*>
1959 V12 N4 P53-58 65-50034 <=>
1960 N8 P77-81 61-11815 <=>
1961 V14 N5 P25-27 66-51103 <=>
1962 V15 N1 P43-44 66-51103 <=>
1962 V15 N3 P27-28 66-51103 <=>
1962 V15 N6 P54-56 66-51103 <=>
1962 V15 N7 P32-34 66-51103 <=>
1962 V15 N9 P42-47 66-51103 <=>
1962 V15 N10 P36-37 66-51103 <=>
1962 V15 N11 P36-38 66-51103 <=>
1964 V17 N4 P28-29 C-5782 <NRC>

LESNOI ZHURNAL
SEE IZVESTIYA VYSSHIKH UCHEBNYKH ZAVEDENII. LESNOI ZHURNAL

LESOKHIMICHESKAYA PROMYSHLENNOST

1933 V2 N3 P11-15 63-10699 <=*> O
1935 V4 N3 P9-15 RT-666 <*>

LETECKY OBZOR
1962 V6 N12 P390-392 63-21949 <=>

LETECTVI
1965 N11 P350-352 AD-635 833 <=$>

LEYBOLD POLARGRAPHISCHE BERICHTE
1953 V1 N3 P49-52 AEC-TR-4001 <*>
1953 V1 N4 P55-58 AEC-TR-4196 <*>

LIBERATION ARMY MEDICAL JOURNAL. CHINESE PEOPLES REPUBLIC
SEE CHIEH FANG CHUN-I-HSUEH TSA CHIH

★LI HSUEH HSUEH PAO
1957 V1 N2 P141-151 62-13865 <=>
1957 V1 N2 P169-232 62-13865 <=>
1957 V1 N3 P250-349 62-13865 <=>
1957 V1 N4 P351-390 62-13865 <=>
1957 V1 N4 P416-422 62-13865 <=>
1958 V2 N1 P1-15 61-23487 <=>
1958 V2 N1 P17-27 61-23487 <=>
1958 V2 N1 P29-42 61-23487 <=>
1958 V2 N1 P43-88 61-23598 <=>
1958 V2 N2 P89-166 61-27124 <=>
1958 V2 N2 P167-180 61-27371 <=>
1958 V2 N3 P219-282 61-27371 <=>
1958 V2 N4 P283-306 61-19993 <=>
1958 V2 N4 P308-320 61-19993 <=>
1958 V2 N4 P334-342 61-19993 <=>
1958 V2 N4 P344-352 61-19993 <=>
1959 V3 N1 P29-86 61-23628 <=>
1959 V3 N2 P87-95 61-23628 <=>
1959 V3 N2 P111-143 61-23628 <=>
1959 V3 N2 P177-190 61-23628 <=>
1959 V3 N3 P217-268 61-27278 <=>
1959 V3 N4 P277-280 60-31367 <=>
1959 V3 N4 P281-297 62-23473 <=>
1959 V3 N4 P325-351 62-23473 <=>
1959 V3 N4 P356-363 62-23473 <=>
1960 V4 N1 P23-83 61-21682 <=>
1960 V4 N2 P85-93 61-23137 <=>
1960 V4 N2 P123-168 61-23137 <=>
1964 V7 N1 P12-28 AD-635 837 <=$>
1964 V7 N1 P63-80 AD-636 591 <=$>
1964 V7 N3 P184-194 65-32144 <=$>
1964 V7 N3 P196-210 AD-639 483 <=$>

LICHT UND LAMPE
1939 V28 P625-627 57-1471 <*>

LICHTTECHNIK
1950 V2 N3 P73-79 22N53G <ATS>
61-20527 <*> O
62-10060 <*> O
1950 V2 N4 P107-110 23N53G <ATS>
61-20528 <*>
1963 V15 N5 P266-268 65-10742 <*>
1963 V15 N6 P317-322 65-17211 <*>
1964 V16 P497-498 495 <TC>
1966 V18 N1 P8A-10A 66-13751 <*>

LIDE A ZEME
1965 V14 N2 P61-67 65-30912 <=*$>

LIETUVOS FIZIKOS RINKINYS
1961 V1 N1/2 P21-32 AD-612 378 <=$>
1961 V1 N3/4 P263-268 ANL-TRANS-10 <=>
1961 V1 N3/4 P271-281 ANL-TRANS-10 <=>
1962 V2 N3/4 P385-397 65-12241 <*>
1963 V3 P1-2 ANL-TRANS-123 <=$>
1963 V3 N1/2 P11-33 65-63763 <=*$>
1963 V3 N1/2 P47-72 AD-632 414 <=$>
1963 V3 N1/2 P159-166 65-63753 <=$>
1963 V3 N1/2 P205-226 AD-612 483 <=$>
1964 V4 P81-85 66-14111 <*$>
1964 V4 N1 P25-33 AD-632 807 <=$>
1964 V4 N2 P187-196 AD-628 969 <=*$>
1964 V4 N2 P198-212 ANL-TRANS-161 <=*$>
1964 V4 N4 P551-557 66-13772 <*>

LIETUVOS TSR MOKSLU AKADEMIJOS DARBAI. SERIJA B. VILNIUS
1955 N3 P9-16 61-15945 <+*>
1956 N2 P15-19 61-19029 <+*>
1956 SB N3 P85-95 64-31531 <=>
1956 SB N4 P101-113 64-31531 <=>
1957 N4 P123-131 CSIR-TR.530 <CSSA>
1957 SB N3 P17-28 59-10466 <+*>
1958 N1 P11-19 61-19031 <+*>
1958 N2 P3-16 61-19033 <+*>
1958 N2 P17-24 61-15944 <+*>
64-00462 <*>
1958 N4 P3-16 AD-616 679 <=$>
61-19032 <+*>
1958 N4 P17-29 61-20068 <*=>
1958 N4 P53-68 62-32486 <=*>
1958 N4 P71-80 62-32948 <=*>
1958 N4 P81-89 62-32489 <=*>
1960 SB N3 P141-151 65-60493 <=$>
1961 N3 P53-66 64-19342 <=$>
1961 N3 P99-105 64-00395 <*>
1961 SB N2 P281-290 61-28297 <=*>
1961 SB N4 P197-200 64-23608 <=$>
1962 N1 P161-168 65-64400 <=*$>
1963 P83-88 AD-622 464 <=*>
1963 N2 P9 65-12280 <*>
1963 SB N3 P99-105 AD-611 073 <=$>
1964 N1 P135-141 M-5710 <NLL>
1964 N1 P171-182 TRANS-329(M) <NLL>
1964 N3 P49-60 65-11929 <*>
1964 N3 P61-71 65-11928 <*>
1965 N4 P139-152 66-62010 <=*$>
1965 N4 P153-164 AD-639 185 <=$>
1965 N4 P153-163 66-62005 <=*$>

LIETUVOS TSR MOKSLU AKADEMIJOS DARBAI. SERIJA C. VILNIUS
1961 N1 P239-249 62-25478 <=*>

LIGHT INDUSTRY OF CHINA
SEE CHUNG-KUO CH'ING KUNG-YEH

LIGHT METALS. JAPAN
SEE KEIKINZOKU

★LIJECNICKI VIJESNIK
1961 V83 N5 P445-458 62-19629 <*=> P
1962 V84 N7 P698-701 49S87CR <ATS>
1963 V85 N7 P747-758 64-11895 <=>

LILLE CHIRURGICAL
1955 V11 P142-144 59-15354 <*>

LILLE MEDICAL
1960 V5 P850-857 66-11641 <*>
1962 S3 V7 N6 P500-506 65-10966 <*>

LIMNOLOGICA. BERLIN
1962 V1 P51-63 AEC-TR-5753 <*>

LIN YEH KO' HSUEH
1959 N2 P117-122 61-21695 <=>
1959 N5 P381-386 61-21695 <=>
1962 N1 P53-58 63-13537 <=*>
1962 N1 P67-72 63-13537 <=*>
1962 N1 P74-79 63-13537 <=*>
1963 N2 P194 63-31800 <=>

LINDE BERICHTE AUS TECHNIK UND WISSENSCHAFT
1958 N3 P25-32 61-20963 <*>
1959 N5 P3-14 NP-TR-417 <*>
1960 N9 P46-55 2080 <BISI>
62-20202 <*>
1960 N10 P12-16 61-20476 <*>
1960 N10 P17-22 61-20861 <*>
1961 N11 P40-41 61-20400 <*>

Citation	Translation
LISBOA MEDICA	
1944 V21 P410-	2922 <*>
LISTY CUKROVARNICKE	
1930 V49 P190-195	63-14574 <*>
1934 V53 P113-124	63-14601 <*>
1941 V60 P32-37	63-14547 <*> 0
1953 V69 P241-245	63-14874 <*>
1960 V76 P126-129	63-14076 <*> 0
1962 V78 N12 P277-281	44R75C <ATS>
1964 V80 N10 P265-269	66-12653 <*> 0
★LITEINOE PROIZVODSTVO	
1952 N1 P5	3410 <HB>
1952 N3 P9-10	5364 <HB>
1952 N5 P1-2	RT-2650 <*>
1952 N6 P16-18	46N56R <ATS>
1952 N9 P30-33	3242 <HB>
1952 V3 N2 P15-17	3395 <HB>
1952 V3 N5 P27	3429 <HB>
1952 V3 N8 P28	3063 <HB>
1952 V3 N9 P24-25	3299 <HB>
1953 N1 P21-22	3544 <HB>
1953 N2 P21-23	64-18380 <*>
1953 N5 P8-11	60-18839 <*>
1953 N6 P22-24	RT-3356 <*>
1953 N8 P17-18	3622 <HB>
1953 V4 N3 P18-19	3129 <HB>
1953 V4 N3 P20-21	3130 <HB>
1953 V4 N3 P21-23	3131 <HB>
1953 V4 N6 P9-15	3162 <HB>
1953 V4 N9 P25	3199 <HB>
1954 N1 P17-23	RT-1919 <*>
1954 N2 P18-19	RT-3626 <*>
	4097 <HB>
1954 N4 P18-22	08N56R <ATS>
1954 N4 P23-24	RT-4570 <*>
1954 N4 P25	RT-4357 <*>
1954 N5 P16-19	4458 <HB>
	59-14175 <+*>
1954 N7 P25-27	61-19429 <+*> 0
1954 N8 P14-16	R-986 <*>
	929TM <CTT>
1954 N9 P6-10	R-989 <*>
1954 V5 N1 P9-14	3386 <HB>
1954 V5 N1 P23-24	3372 <HB>
1954 V5 N1 P29-30	3383 <HB>
1954 V5 N1 P30-31	3384 <HB>
1954 V5 N2 P1-3	3364 <HB>
1954 V5 N2 P16-18	3374 <HB>
1954 V5 N6 P1-3	3452 <HB>
1955 N2 P17-20	61-15705 <+*>
1955 N4 P23-25	BGIRA-543 <BGIR>
1955 N5 P15-18	R-3831 <*>
1955 N6 P1-4	R-2336 <*>
1955 N9 P15-19	RT-3954 <*>
1955 N10 P10-13	R-3816 <*>
1955 N10 P22-23	R-3683 <*>
1955 N12 P1-2	3692 <HB>
1956 N1 P10-13	59-16542 <+*>
1956 N2 P15-16	06N56R <ATS>
1956 N2 P16-22	R-4464 <*>
1956 N3 P1-4	RT-4405 <*>
	59-10309 <+*> 0
1956 N3 P14-19	R-4462 <*>
1956 N5 P15-18	3815 <HB>
1956 N6 P7-11	TS1384 <BISI>
	62-25698 <=*>
1956 N7 P1-4	60-16496 <+*> 0
1956 N8 P25-27	62-24172 <=*>
1956 N9 P18-20	4078 <HB>
1956 N12 P19-20	59-19620 <+*>
1957 N1 P9-11	3946 <HB>
1957 N3 P19-20	4291 <HB>
1957 N4 P23-24	4096 <HB>
1957 N5 P17-20	60-17453 <+*>
1957 N5 P20-22	4910 <HB>
1957 N5 P22-25	60-13007 <+*>
1957 N6 P1-4	4272 <HB>
1957 N6 P5-7	60-13465 <+*>
1957 N6 P7-8	60-13050 <+*>
	61-23166 <*=> 0
1957 N6 P18-22	61-23166 <*=> C
1957 N8 P14-15	4341 <HB>
1957 N8 P27-29	4070 <HB>
1957 N10 P19-22	61-18911 <*=>
1957 N11 P18-19	4094 <HB>
1957 N11 P24-26	4170 <HB>
1957 N12 P1-3	60-17157 <+*>
1958 N1 P3-7	104 <LSB>
1958 N1 P7-8	4753 <HB>
1958 N2 P25	5275 <HB>
1958 N3 P10-11	4336 <HB>
1958 N3 P16-20	TRANS-30 <MT>
1958 N3 P20-24	4864 <HB>
1958 N4 P19-23	4865 <HB>
1958 N4 P24-25	107 <LSB>
1958 N4 P31-32	59-17623 <+*>
	8705 <K-H>
1958 N5 P19-21	TS-1415 <BISI>
	62-25331 <=*>
1958 N6 P22-26	TS 1538 <BISI>
1958 N7 P6-8	101 <LSB>
	4612 <HB>
1958 N7 P17-21	4633 <HB>
1958 N7 P25-26	4779 <HB>
	61-15295 <+*>
1958 N9 P32	60-19894 <+*>
1958 N10 P23-28	59-13202 <=> 0
1958 N11 P4-6	59-11395 <=>
1958 N11 P7-8	4917 <HB>
1958 N11 P14-15	60-18349 <*+>
1958 N11 P15-17	4518 <HB>
1958 N11 P18	4519 <HB>
1958 N11 P18-20	4520 <HB>
1958 N12 P8	4945 <HB>
1958 N12 P8-9	4945 <HB>
1958 N12 P10-12	4554 <HB>
1959 N1 P1	60-21203 <=>
1959 N1 P8	60-21204 <=>
1959 N1 P27-32	60-21205 <=> 0
	61-15410 <+*>
1959 N1 P48	60-21206 <=>
1959 N2 P42	4563 <HB>
1959 N3 P4-7	07N56R <ATS>
1959 N3 P12-14	60-14486 <*>
	61-15651 <+*>
1959 N3 P14-16	61-21958 <=> 0
1959 N3 P23-26	4849 <HB>
1959 N3 P29-30	5680 <HB>
1959 N3 P45-46	60-13752 <+*>
1959 N3 P46	4624 <HB>
1959 N5 P1	1986 <BISI>
1959 N5 P17-18	4651 <HB>
1959 N5 P28-30	60-16782 <+*>
1959 N5 P28-34	61-15294 <+*>
1959 N5 P36	4709 <HB>
1959 N6 P15-18	4713 <HB>
1959 N6 P22-27	4771 <HB>
1959 N7 P17-18	5363 <HB>
1959 N8 P13-14	61-10212 <*+>
	61-14975 <=*> 0
1959 N8 P23	TRANS-99 <MT>
1959 N9 P14-15	4815 <HB>
1959 N9 P38	4762 <HB>
1959 N10 P12-17	4847 <HB>
1959 N12 P1-2	61-15476 <+*>
1959 N12 P2-3	61-15477 <+*>
1959 N12 P3-4	61-15478 <+*>
1959 N12 P4-5	61-15479 <+*>
1959 N12 P5-7	61-15480 <+*>
1959 N12 P7-8	61-15481 <+*>
1959 N12 P8-9	61-15482 <+*>
1959 N12 P9-10	61-15483 <+*>
1959 N12 P11-12	4879 <HB>
	61-15484 <+*>
1959 N12 P13-14	61-15485 <+*>

1959 N12 P14-15 61-15486 <+*>
1959 N12 P15-16 61-15487 <+*>
1959 N12 P22-24 61-15488 <+*>
1959 N12 P24-27 61-15489 <+*>
1959 N12 P27-28 61-15490 <=*>
1959 N12 P28-31 61-15491 <=*>
1959 N12 P31-32 61-15492 <+*>
1959 N12 P32-35 61-15493 <=*>
1959 N12 P35-37 61-15494 <+*>
1960 N1 P10-12 66-14607 <*$>
1960 N1 P35-36 5044 <HB>
1960 N2 P36-37 4873 <HB>
1960 N3 P27 4818 <HB>
1960 N4 P42-44 4874 <HB>
1960 N5 P42-43 2150 <BISI>
1960 N6 P29-30 4884 <HB>
1960 N6 P30-31 1681 <TC>
1960 N6 P42-43 5187 <HB>
1960 N6 P47-48 5024 <HB>
1960 N7 P14-16 61-11673 <=>
1960 N7 P24-26 4907 <HB>
1960 N7 P26-32 09N56R <ATS>
1960 N7 P34-36 4908 <HB>
1960 N9 P12-13 5098 <HB>
1960 N9 P13-16 5202 <HB>
1960 N9 P25-28 5125 <HB>
1960 N10 P34-36 63-21902 <=>
1960 N10 P47 5029 <HB>
1960 N11 P1-5 61-21685 <=>
1960 N12 P17-23 146-205 <STT>
1960 N12 P33-34 5111 <HB>
1960 N12 P34-36 5180 <HB>
1961 N1 P43-48 38N58R <ATS>
1961 N6 P22-24 62-23874 <=*>
1961 N11 P22 5634 <HB>
1962 N5 P28-30 5694 <HB>
1962 N7 P26-27 5906 <HB>
1962 N11 P80-84 M-5390 <NLL>
1963 N1 P25-26 5916 <HB>
1963 N2 P7-8 5914 <HB>
1963 N3 P16-18 63-23931 <=*$>
1963 N8 P26-27 6279 <HB>
1963 N9 P28-29 6159 <HB>
1963 N10 P15 6153 <HB>
1963 N10 P45 6148 <HB>
1964 N1 P2-5 64-31512 <=>
1964 N1 P16-17 6222 <HB>
65-61339 <=$>
1964 N1 P42-44 64-31512 <=>
1964 N3 P27-31 64-31664 <=>
1964 N6 P19-22 AD-611 862 <=$>
1964 N8 P23-26 65-51690 <=$>
1964 N9 P5-6 M-5603 <NLL>
1964 N9 P8-10 M-5628 <NLL>
1964 N9 P27-31 M-5629 <NLL>
1964 N10 P7-8 6426 <HB>
1964 N10 P8-9 6234 <HB>
1966 N1 P11-12 6807 <HB>

LITERATURNAYA GAZETA
1966 P2,4 66-30576 <=$>

★LITOLOGIYA I POLEZNYE ISKOPAEMYE
1964 N1 P114-116 IGR V8 N5 P573 <AGI>
1964 N3 P5-19 IGR V8 N4 P467 <AGI>
1964 N3 P89-103 IGR V8 N4 P455 <AGI>
1964 N4 P21-42 IGR V8 N2 P197 <AGI>
1964 N4 P43-65 IGR V7 N12 P2135 <AGI>
1964 N4 P88-95 IGR V7 N12 P2197 <AGI>
1964 N5 P3-20 IGR V7 N12 P2161 <AGI>
1965 N1 P3-17 IGR V8 N5 P559 <AGI>
1965 N1 P18-30 IGR V8 N5 P549 <AGI>
1965 N2 P60-69 GI V2 N2 1965 P372 <AGI>
1965 N4 P18-49 IGR V8 N10 P1172 <AGI>
1966 N1 P129-134 66-33389 <=$>
1966 N2 P130-135 IGR V8 N11 P1270 <AGI>

LOKOMOTIVTECHNIK
1955 V79 P157-164 T-2089 <INSD>

LOODUS JA MATEMATIKA
1959 N1 P67-85 62-10972 <=*>

LOTTA CONTRO LA TUBERCOLOSI
1945 P41-57 2567 <*>
1945 P70-73 2567 <*>
1949 V19 N5 P422-435 2571 <*>
1950 V20 P68-73 2612 <*>
1950 V20 N1/2 P68-73 2621 <*>
1950 V20 N1/2 P74-81 2609 <*>
1951 V21 N3/4 P157-167 2776 <*>
1951 V21 N5/6 P282-287 2576 <*>
1953 V23 N10/1 P839-841 2780 <*>
1963 V33 P797-803 65-14631 <*> OP

LOW TEMPERATURE SCIENCE, SERIES A. PHYSICAL SCIENCE. JAPAN
SEE TEION KAGAKU, BUTSURI

LOW TEMPERATURE SCIENCE, SERIES B, BIOLOGICAL SCIENCE. JAPAN
SEE TEION KAGAKU, SEIBUTSU HEN

LOZANIA
1952 N4 P1-12 59-10330 <*>
1952 V1 P7-12 58-2474 <*>

LUCEAFARUL
1963 V6 N6 P6- 63-21749 <=>

LUCRARILE STIINTIFICE ALE INSTITUTULUI DE PATOLOGIE SI IGIENA ANIMALA
1959 V9 P59-69 60-00825 <*>
1959 V9 P71-87 60-00824 <*>

LUFTFAHRTFORSCHUNG
1928 V1 N4 P113- 57-2476 <*>
1928 V1 N4 P132-146 57-2088 <*>
1928 V1 N4 P153- 57-1556 <*>
1936 P20- 1749 <*>
1936 V13 P61-66 58-1991 <*>
1936 V13 N2 P67-70 58-462 <*>
1936 V13 N10 P31-39 65-13777 <*>
1936 V13 N12 P405-409 SCL-T-367 <*> O
1937 V14 P640-646 58-917 <*>
1937 V14 N2 P55-62 65-13006 <*>
1937 V14 N4/5 P215-223 2478 <*>
1938 V15 P60 58-1858 <*>
1938 V15 N4 P153-169 59-17729 <*>
1938 V15 N9 P445-462 63-20540 <*>
1938 V15 N1/2 P41-47 59-17689 <*>
61-00803 <*>
1938 V15 N10/1 P481-494 65-13002 <*> O
1939 V16 N1 P1-13 63-20539 <*> O
1939 V16 N1 P18-22 63-10462 <*>
1939 V16 N5 P251-275 58-474 <*>
59-17690 <*> O
1939 V16 N7 P370-383 63-20485 <*> O
1940 V17 N5 P154-160 59-17839 <*>
1940 V17 N11/2 P387-400 AL-647 <*>
1941 V18 N4 P135-141 63-20554 <*>
1941 V18 N4 P142-146 I-142 <*>
1941 V18 N8 P275-279 61-18855 <*> O
1941 V18 N10 P356-367 65-12974 <*>
1942 V19 P302-312 AEC-TR-5246 <*>
1942 V19 N4 P137-144 63-18822 <*>
1942 V19 N4 P153-156 58-486 <*>
1942 V19 N6 P201-209 58-500 <*>
1942 V19 N8 P282-291 AL-647 <*>
1942 V19 N9 P302-312 65-17403 <*>
1942 V19 N9 P326-330 58-472 <*>
1943 V20 P181-183 63-10985 <*>
1943 V20 N1 P16-21 58-900 <*>
1943 V20 N5 P137-146 59-20173 <*> O
1943 V20 N6 P102-106 58-860 <*>
1943 V20 N7 P217-219 AL-811 <*>
1943 V20 N8/9 P231-241 1764 <*>

LUFTFAHRTMEDIZIN
1939 V3 N4 P302-308 61-14455 <*>

LUFTFAHRTTECHNIK
1955 V1 N1 P14- 59-10474 <*>
1957 V3 N10 P227-228 58-1275 <*>
1958 V4 N3 P49-59 59-10056 <*>
1959 V5 P58-60 TS 1525 <BISI>
1959 V5 N2 P58-60 1525 <BISI>
62-14181 <*>
1961 V7 N5 P121-126 66-10710 <*> 0
1961 V7 N5 P132-139 66-10709 <*> 0

LUFTFAHRTTECHNIK-RAUMFAHRTTECHNIK
1964 V10 N4 P115-121 C-5081 <NRC>

LUFTVERUNREINIGUNG
1961 P1-6 65-13583 <*>

LUMIERE ELECTRIQUE
1891 V42 N52 P618-624 59-20524 <*> 0
59-20524 <*> 0
1912 V18 N24 P323-332 57-2118 <*>
1914 P289-297 57-2404 <*>
1914 P652-658 57-2404 <*>
1916 V32 N10 P242 57-2117 <*>
1916 V34 P145-150 57-2119 <*>
1916 V35 N41 P20- 57-1547 <*>
1916 V35 N51 P225-228 57-2034 <*>

LUNDS UNIVERSITETS ARSSKRIFT
SEE ACTA UNIVERSITATIS LUNDENSIS

LUPTA DE CLASA
1963 V43 N8 P23-35 63-31908 <=>
1963 V43 N9 P39-50 63-41066 <=>
1964 V44 N2 P27-40 64-21940 <=>
1964 V44 N4 P69-83 64-31542 <=>
1965 V45 N8 P35-46 65-32749 <=*>
1966 V46 N4 P95-99 66-32471 <=$>
1966 V46 N9 P36-44 66-34696 <=$>
1967 V47 N1 P16-30 67-30943 <=$>

LYMPHATOLOGIA
1952 N1 P82-87 I-437 <*>

LYON CHIRURGICAL
1931 V28 P560-570 61-14483 <*>
1936 V33 P187-190 61-14486 <*>
1937 V34 P20-26 61-14484 <*> 0
1941 V37 P37-39 61-16494 <*>
1941 V37 P177-178 61-16495 <*>
1941 V37 P207-210 61-14485 <*>
1956 V51 P157-183 <CP>

LYON MEDICAL
1957 V89 P545-552 62-16054 <*>
1957 V89 N3 P79-84 65-13748 <*>
1958 V200 N39 P385-406 64-14523 <*> 0
8599-B <K-H>
1960 N11 P687-707 61-10899 <*> 0
1961 V205 N16 P867-868 64-10364 <*>
1961 V205 N16 P871 64-10364 <*>

LYON PHARMACEUTIQUE
1960 V9 P3-16 61-20256 <*>
1963 N9 P403-410 66-12102 <*>

M.A.N. FORSCHUNGSHEFT
1953 P1-16 58-1055 <*>

MTZ
1939 V1 N6 P181-190 65-12970 <*>
1939 V2 P80-86 66-10814 <*>
1940 P227-229 63-20291 <*>
1940 P316-327 63-20788 <*>
1940 V2 P265-269 65-13010 <*>
1940 V2 P283-288 64-18180 <*>
1940 V2 N1 P7-15 65-12970 <*>
1940 V2 N12 P377-384 65-13005 <*>
1941 V3 P18-22 58-931 <*>
63-16653 <*>
1941 V3 N1 P11-16 65-13005 <*>
1942 V4 N9 P333-339 58-893 <*>
1943 V5 N8/9 P242-248 59-17724 <*>
1948 V9 N3 P33-36 63-20281 <*>
1949 P126-133 T-1622 <INSD>
1950 V11 N2 P29-35 63-18211 <*>
1950 V11 N3 P57-67 58-925 <*>
1950 V11 N5 P114-117 58-916 <*>
1950 V11 N6 P137-145 58-1466 <*>
1951 V12 P29-34 58-923 <*>
1952 N3 P38-40 58-884 <*>
1952 N3 P61-63 58-884 <*>
1952 V13 P41-44 58-888 <*>
1952 V13 P50-51 58-882 <*>
1952 V13 P213-221 58-881 <*>
1952 V13 P285-287 58-870 <*>
1952 V13 N2 P25-28 58-926 <*>
1952 V13 N2 P35-37 58-916 <*>
1952 V13 N8 P189-193 58-886 <*>
1952 V13 N10 P237-242 58-883 <*>
1953 V14 P29-39 58-869 <*>
1953 V14 P223-225 58-876 <*>
1953 V14 P254-256 58-875 <*>
1953 V14 N8 P223-225 63-16629 <*>
1953 V14 N8 P254-256 63-16440 <*>
1953 V14 N11 P317-323 61F2G <ATS>
1953 V14 N12 P359-363 58-1142 <*>
1954 V15 P3-11 58-877 <*>
1954 V15 N3 P61-69 05F4G <ATS>
1954 V15 N4 P105-106 58-897 <*>
63-16443 <*>
1954 V15 N6 P171-176 58-895 <*>
1954 V15 N7 P189-199 6F4G <ATS>
1954 V15 N11 P316-322 58-1101 <*>
1955 V16 P117-123 57-1936 <*>
1955 V16 N2 P32-42 62-18798 <*>
1955 V16 N3 P63-68 58-867 <*>
1955 V16 N3 P68-73 62-18798 <*>
1955 V16 N5 P117-123 36G6G <ATS>
1955 V16 N9 P245-254 62-18797 <*>
1956 V17 N1 P16-23 62-18742 <*>
1956 V17 N2 P53-58 62-18742 <*>
1956 V17 N3 P82-90 62-18742 <*>
1956 V17 N9 P306-313 5413 <K-H>
1956 V17 N12 P413-418 62-18794 <*>
1957 V18 N2 P46-48 58-1439 <*>
1957 V18 N5 P127-131 59-17621 <*>
1957 V18 N11 P363-365 62-18740 <*>
1959 V19 N9 P299-303 59-19499 <+*>
1959 V20 N1 P1-4 59-17519 <*> 0
65-12517 <*>
1959 V20 N1 P4-9 59-17518 <*>
65-12518 <*>
1959 V20 N7 P268-275 60-14500 <*>
1959 V20 N8 P313-316 60-10043 <*> 0
1960 V21 N1 P1-8 60-18111 <*>
1961 V22 N7 P261-265 64-00084 <*>
1961 V22 N9 P339-341 62-10728 <*>
1961 V22 N9 P342-343 62-10727 <*>
1962 V23 P177-183 63-14419 <*>
1962 V23 N5 P177-183 64-10920 <*>
1963 V24 N10 P333-338 64-14596 <*>
1963 V24 N10 P339-348 64-14597 <*>
1964 V25 P421- 2174 <BSRA>
1964 V25 N1 P28-33 64-30706 <*>
1964 V25 N1 P462- 2116 <BSRA>
1964 V25 N7 P289-291 65-14182 <*>
1964 V25 N11 P445- 2119 <BSRA>
1964 V25 N11 P465- 2115 <BSRA>
1964 V25 N12 P506- 2133 <BSRA>
1965 V26 P235 2197 <BSRA>
1966 V27 N7 P271-276 66-14556 <*> 0

MAANDBLAD VOOR DE GEESTELIJKE VOLKSGEZONDHEID
1962 V17 N7/8 P239-252 63-01756 <*>
63-26802 <=$>

MAANDBLAD VOOR DE LANDBOUWVOORLICHTINGSDIENST
1952 V9 N12 P454-455 63-16057 <*>

MAANDSCHRIFT VOOR KINDERGENEESKUNDE

1951 V19 N4 P141-144 62-00121 <*>
1954 V22 P280-287 II-885 <*>
1957 V25 P294-296 58-1579 <*>
1959 V27 N1 P26-32 60-17339 <*+>
1964 V32 P359-373 66-11801 <*>
1964 V32 P374 66-10965 <*>

MAANEDSSKRIFT FOR PRAKTISK LAEGEGERNING OG SOCIAL MEDICIN
1954 V32 P261-279 II-282 <*>

MACCHINE
1958 V13 N12 P1197-1199 2140 <BISI>
1960 V15 P1097-1107 175-180 <STT>
1960 V15 P1129-1147 186-181 <STT>
1963 V18 P481-487 3802 <BISI>

MACHINE DESIGN
1951 V23 N3 P121-123 59-17875 <*> 0

MACHINE INDUSTRY. CHINESE PEOPLES REPUBLIC
SEE CHI-HSIEH KUNG-YEH

MACHINE INDUSTRY WEEKLY. CHINESE PEOPLES REPUBLIC
SEE CHI HSIEH KUNG YEH CHOU-PAO

MACHINE MODERNE
1955 V49 N559 P17-23 59-17198 <*> OM
1956 P21-25 57-2026 <*>
1957 V51 N581 P9-11 59-10190 <*>
1960 V54 P24 122-97 <STT>
1960 V54 P41-45 146-305 <STT>

MACHINES ET METAUX
1947 V31 P333-339 64-71249 <=>
1947 V31 P341 64-71249 <=>
1948 V32 P227-234 62-18148 <*> 0

MADA
1963 V8 N1 P42-44 63-41011 <=>
1963 V8 N1 P53-56 63-41011 <=>
1963 V8 N1 P56 63-41011 <=>
1964 V9 P31-38 64-41848 <=>
1964 V9 P71-76 64-41848 <=>
1964 V9 P104 64-41848 <=>

MADAGASCAR; REVUE DE GEOGRAPHIE
1963 V2 P61-82 65-32658 <=$>

*MAGNITNAYA GIDRODINAMIKA
1965 N2 P3-10 66-13692 <*> 0
1965 V1 P73-79 N66-13294 <=$>

MAGYAR ALLATORVOSOK LAPJA
1957 V12 P35-37 59-21100 <=*$>
1963 V17 N3 P81-83 63-21456 <=>
1964 V19 N10 P451-452 65-30082 <=$>
1965 V20 N2 P86-88 65-30845 <=$>
1965 V20 N6 P241-256 65-32670.A <=$>
1965 V20 N7 P280-292 65-33567 <=*$>
1965 V20 N9 P390-396 66-14462 <*$>
1966 N2 P49-52 66-32193 <=$>
1966 N3 P97-100 66-32626 <=$>
1966 N3 P102-105 66-32584 <=$>

MAGYAR ASVANYOLAJES FOLDGAZ KISERLETI INTEZET KOSLEMENYII
1960 V1 P187-217 64-30142 <*> 0
1962 N3 P210- ICE V3 N1 P91-94 <ICE>
1962 V3 P184-188 64-30308 <*>
1962 V3 P189-197 64-30307 <*>
1964 N4 P6-12 ICE V5 N1 P72-76 <ICE>
1964 N5 P138-141 ICE V5 N4 P590-593 <ICE>
1964 V4 P6-12 ICE V5 N1 P72-76 <ICE>
1965 N6 P85-89 ICE V6 N3 P426-429 <ICE>

MAGYAR BELORVOSI ARCHIVUM ES IDEGGYOGYASZATI SZEMLE
1951 V4 P20-23 1346 <*>

MAGYAR BIOLOGIAI KUTATO INTEZET MUSKAI
1943 V15 P462-464 AL-326 <*>

MAGYAR FILOZOFIAI SZEMLE
1961 V5 N4 P538-552 62-19671 <=*>

MAGYAR FISIKAI FOLYOIRAT
1955 V3 P489-496 UCRL TRANS-468 <=>
1960 V8 P21-30 SCL-T-441 <*>
1960 V8 N5 P357-415 62-19495 <=*>
1961 V9 N1 P35-48 62-19669 <*=>
1961 V9 N1 P51-59 62-19617 <=*>
1961 V9 N1 P81-84 62-19670 <*=>
1961 V9 N4 P251-263 62-19157 <=*> P
1961 V9 N4 P265-268 62-19801 <=*>
1962 V10 N3 P183-187 872 <TC>

MAGYAR GYOGYSZERESTUDOMANYI TARSASAG ERTESITO
1931 V7 P125-130 5709-B <K-H>
1932 V8 P240-245 5472-B <K-H>

MAGYAR HIVADASTECHNIKA
1960 V11 N3 P104-107 62-19245 <*=>
1960 V11 N5 P165 62-19380 <=*>

MAGYAR KEMIAI FOLYOIRAT
1938 V44 P47-59 PT.I 58-1031 <*>
1939 V45 P19-30 PT.2II 58-1031 <*>
1950 V56 P201-203 60-18498 <*>
1951 V57 P68-73 HJ-669 <ATS>
57-2687 <*>
1953 V59 N7 P200-203 RJ-633 <ATS>
1953 V59 N7 P211-212 64-20348 <*>
1954 V60 N11 P347-348 4G6H <ATS>
57-2678 <*>
1955 V61 P33-42 11560A <K-H>
1955 V61 N2 P48-50 63-10816 <*>
1955 V61 N6 P176-182 11324 <K-H>
1955 V61 N10 P298-300 63-10795 <*>
1956 V62 P395-400 59-15987 <*> 0
1957 V63 N2/3 P95 61K25H <ATS>
1958 V64 P7-8 AEC-TR-5383 <*>
1958 V64 N5 P168-169 65-10264 <*>
1958 V64 N11 P417-428 60-31332 <*=>
1959 V65 N1 P31-36 59-13747 <=*>
1959 V65 N7 P245-249 AEC-TR-3959 <*>
1959 V65 N7 P280-281 18Q70H <ATS>
1960 V66 P321-324 AEC-TR-4429 <*>
1960 V66 P331-332 62-10023 <*> 0
1960 V66 P483-485 62-16688 <*>
1961 N7 P320- ICE V3 N3 P315-318 <ICE>
1961 V67 N1 P36-40 62-16263 <*> 0
1961 V67 N6 P257-259 62-19536 <=*>
1961 V67 N6 P259-266 RCT V35 N3 P599 <RCT>
1961 V67 N6 P266-268 RCT V35 N3 P611 <RCT>
1961 V67 N8 P360-364 64-00264 <*>
1962 N4 P167- ICE V3 N1 P55-59 <ICE>
1962 V68 P60-65 63-20710 <*>
1962 V68 N1 P5-9 63-20712 <*>
1962 V68 N1 P11-19 63-20718 <*>
1962 V68 N2 P54-59 63-20711 <*>
1962 V68 N7 P293-296 63-18873 <*> 0
1963 V69 N2 P56-60 13081 <K-H>
65-14874 <*>
1964 V70 P191-196 65-11752 <*>
1964 V70 N8 P361-365 65-17344 <*>
1964 V70 N12 P559-561 65-17424 <*>
1965 N9 P407-410 ICE V6 N2 P288-291 <ICE>
1965 V71 N2 P68-71 1425 <TC>
1965 V71 N10 P432-436 ICE V6 N2 P292-296 <ICE>
1966 N2 P77-78 ICE V6 N3 P507-508 <ICE>
1966 N3 P140-141 ICE V6 N4 P655-656 <ICE>

MAGYAR KEMIKUSOK LAPJA
1946 N4 P592-596 71K21H <ATS>
1947 V2 P317-350 1758 <K-H>
1949 V4 P524-531 1826 <TC>
1958 V13 P68-69 92L29H <ATS>
1959 V14 P314-317 NP-TR-600 <*>
1960 V15 P535-538 61-16770 <*>

1960 V15 N12 P566-567 61-20492 <*>
1961 V16 N1 P46-47 62-19216 <*=>
1961 V16 N4 P156-159 62-19547 <=*>
1961 V16 N10 P447- ICE V2 N2 P267-274 <ICE>
1962 N9 P405- ICE V3 N3 P303-308 <ICE>
1962 N11 P494- ICE V3 N2 P280-294 <ICE>
1962 V17 N1 P8-11 AEC-TR-5769 <=*$>
1962 V17 N2 P71-73 62-25609 <=*>
1962 V17 N3 P140-143 12377 <K-H>
65-14486 <*>
1962 V17 N4 P165-169 65-17365 <*>
1962 V17 N11 P488-493 63-21440 <=>
1963 N4 P152-156 ICE V3 N4 P557-562 <ICE>
1963 N5 P212-217 ICE V3 N4 P597-603 <ICE>
1963 N8 P395-403 ICE V4 N2 P285-299 <ICE>
1963 V18 P573-577 65-00148 <*>
1963 V18 N8 P395-403 ICE V4 N2 P285-299 <ICE>
1963 V18 N10 P461-469 66-13243 <*>
1963 V18 N12 P577-584 66-10018 <*>
1963 V18 N2/3 P53-65 63-21963 <=>
1964 N2 P89-92 ICE V4 N3 P395-399 <ICE>
1964 N9 P453-461 ICE V5 N2 P263-271 <ICE>
1964 N9 P470-474 ICE V5 N3 P421-425 <ICE>
1964 N10/1 P535-539 65-00219 <*>
1964 N10/1 P567-570 ICE V5 N2 P252-256 <ICE>
1964 V19 N2 P89-92 ICE V4 N3 P395-399 <ICE>
1964 V19 N10/1 P516-525 65-30046 <=$>
1964 V19 N10/1 P527-531 65-30046 <=$>
1965 N6 P304-312 ICE V6 N1 P150-159 <ICE>
1965 N7/8 P353-358 ICE V6 N2 P233-238 <ICE>
1965 V20 N1 P15-18 65-14801 <*>
1965 V20 N7/8 P341-352 65-32785 <=$>
1965 V20 N7/8 P359-374 65-32785 <=$>
1966 V21 N1 P47-49 66-31114 <=$>
1966 V21 N3 P150-153 66-32075 <=$>
1966 V21 N6 P331-332 66-33921 <=$>
1966 V21 N12 P620-626 67-30490 <=>

MAGYAR KOZLONY
1961 N5 P51-53 62-19218 <*=>
1964 N34 P3-7 64-51410 <=$>
1964 N54 P450-460 64-51410 <=$>
1965 P115-116 65-33198 <=*$>
1965 N69 P610-611 66-30252 <=*$>

MAGYAR LEGOLTALOM
1961 V3 N1 P20-21 62-19217 <*=>

MAGYAR MEZOGAZDASAG
1964 V19 N19 P1-16 64-31653 <=$>
1964 V19 N37 P2 64-51341 <=>
1964 V19 N37 P12-14 64-51341 <=>
1964 V19 N37 P18-19 64-51341 <=>

MAGYAR NOORVOSOK LAPJA
1954 V17 N1 P48-55 61-20005 <*>

MAGYAR ONKOLOGIA
1964 N3 P190-191 64-51676 <=*$>

MAGYAR PSZICHOLOGIAI SZEMLE
1960 V17 N1 P1-8 62-19211 <=*>
1960 V17 N3 P257-317 62-19210 <=*>
1961 V18 N1 P61-67 62-19213 <=*>
1961 V18 N1 P67-70 62-19212 <=*>
1961 V18 N4 P417-436 62-24981 <=*>
1964 V21 N3 P329-358 65-30196 <=$>

MAGYAR RADIOLOGIA
1961 V13 N1 P54-57 62-19244 <=*>

MAGYAR SZABVANYUGYI HIVATAL. MNOSZ. BUDAPEST
1957 P1-28 59-11314 <=>

MAGYAR TEXTILTECHNIKA
1955 N8 P288-293 63-10801 <*>
1956 V8 N8 P283-286 61-16721 <*>
1961 V13 N9 P405-407 <LSA>
1963 V15 N3 P111-114 59Q73H <ATS>

MAGYAR TUDOMANY
1959 V4 N11 P594-596 60-31233 <=*>
1959 V4 N11 P611-612 60-31233 <=*>
1959 V4 N12 P655-657 60-31233 <=*>
1959 V4 N12 P665 60-31233 <=*>
1960 N3 P135-151 60-31459 <*=>
1960 V5 P299-326 62-19181 <=>
1960 V5 P341-348 62-19181 <=>
1960 V5 N8 P486-489 62-19187 <*=> P
1960 V5 N8 P499-500 62-19186 <*=>
1960 V5 N9 P517-528 62-19192 <*=>
1960 V5 N9 P551-552 62-19193 <*=>
1960 V5 N9 P554-557 62-19194 <*=>
1960 V5 N9 P562-563 62-19189 <*=>
1960 V5 N11 P688-691 62-19188 <=*>
1960 V5 N11 P701-703 62-19190 <*=>
1960 V5 N12 P711-715 62-19182 <=>
1960 V5 N12 P718-721 62-19182 <=>
1960 V5 N12 P737-742 62-19182 <=>
1960 V5 N12 P749-751 62-19182 <=>
1960 V5 N12 P753-755 62-19182 <=>
1961 V6 N1 P43-47 62-19183 <=*>
1961 V6 N1 P48-50 62-19184 <*=>
1961 V6 N1 P58 62-19183 <=*>
1961 V6 N1 P59-60 62-19197 <=*>
1961 V6 N3 P153-159 62-19589 <*=>
1961 V6 N5 P269-294 62-19561 <=*>
1961 V6 N5 P311-313 62-19561 <=*>
1961 V6 N5 P313-317 62-19156 <=>
1961 V6 N6 P341-352 62-11164 <=>
1961 V6 N6 P359-360 62-23401 <=*>
1961 V6 N6 P361 62-19561 <=*>
1961 V6 N6 P362-363 62-19156 <=>
1961 V6 N6 P364-367 62-23400 <=*>
1961 V6 N6 P380- 62-19775 <=*>
1961 V6 N9 P529-536 62-24015 <=*>
1961 V6 N9 P537-544 62-19657 <*=>
1961 V6 N9 P551-553 62-23633 <=*> P
1961 V6 N9 P572-577 62-24004 <=*>
1961 V6 N12 P770-773 62-24403 <=*>
1961 V6 N7/8 P393-402 62-19600 <*=>
1961 V6 N7/8 P455-466 62-11090 <=>
1961 V6 N7/8 P470-472 62-19156 <=>
1962 V7 N2 P87-95 62-25533 <=*> P
1962 V7 N2 P112-113 62-25523 <=*>
1962 V7 N3 P181-183 62-32068 <=*>
1962 V7 N3 P191-192 62-32068 <=*>
1962 V7 N4 P203-224 62-25481 <=*>
1962 V7 N4 P269-271 62-25473 <=*>
1962 V7 N5 P281-306 62-32671 <=*>
1962 V7 N5 P327-333 62-32671 <=*>
1962 V7 N8 P515-518 63-15085 <=*>
1962 V7 N8 P521-525 63-21388 <=>
1962 V7 N8 P527-529 63-21259 <=>
1962 V7 N9 P588-590 63-21259 <=>
1962 V7 N10 P655-657 63-21436 <=>
1962 V7 N11 P683-696 63-21436 <=>
1962 V7 N12 P792-796 63-21271 <=>
1963 V8 N4 P288-292 63-31655 <=>
1963 V8 N5 P301-365 63-31796 <=>
1963 V8 N9 P573-577 64-21324 <=>
1963 V8 N9 P620-636 64-21320 <=>
1963 V8 N12 P830-831 64-21741 <=>
1963 V8 N12 P844-847 64-21742 <=>
1964 V9 N1 P45-46 64-13196 <=>
1964 V9 N1 P51-55 64-13196 <=>
1964 V9 N4 P257-261 64-41458 <=$>
1964 V9 N4 P263-265 64-31420 <=>
1964 V9 N5 P325-326 64-41394 <=>
1964 V9 N5 P326-328 64-41412 <=$>
1964 V9 N5 P329-330 64-41412 <=$>
1964 V9 N5 P335-336 64-31817 <=>
1964 V9 N6 P347-369 64-51174 <=$>
1964 V9 N7 P443-457 64-41994 <=$>
1964 V9 N7 P473-474 64-41994 <=$>
1964 V9 N11 P715-718 65-30159 <=$>
1964 V9 N11 P721-723 65-30159 <=$>
1964 V9 N12 P768-770 65-30861 <=$>
1964 V9 N12 P771-775 65-30862 <=$>
1964 V9 N12 P779-780 65-30861 <=$>

1964 V9 N8/9 P498-511 64-51890 <=$>
1964 V9 N8/9 P585-586 64-51889 <=$>
1965 V10 P638-653 66-30446 <=*$>
1965 V10 N1 P68-70 65-30536 <=$>
1965 V10 N4 P288-298 65-31645 <=$>
1965 V10 N5 P358-363 65-32861 <=$>
1965 V10 N5 P365-367 65-32849 <=*$>
1965 V10 N5 P370-372 65-32850 <=*$>
1965 V10 N9 P579-592 66-30084 <=*$>
1965 V10 N9 P609-613 66-30597 <=$>
1965 V10 N10 P623-627 66-30445 <=*$>
1965 V10 N11 P704-714 66-30455 <=$>
1965 V10 N11 P718-723 66-30454 <=$>
1965 V10 N12 P791,801 66-30785 <=*$>
1965 V10 N12 P797-798 66-30790 <=$>
1965 V10 N12 3-32 SUP 66-30928 <=*$>
1966 N2 P81-91 66-31767 <=$>
1966 N2 P123-126 66-31767 <=$>
1966 N2 P131-133 66-31767 <=$>
1966 N3 P196-207 66-31716 <=$>
1966 N4 P244-254 66-33387 <=$>
1966 N4 P255-259 66-33385 <=$>
1966 N4 P260-264 66-33386 <=$>
1966 N5 P280-286 66-33557 <=$>
1966 N5 P297-305 66-33556 <=$>
1966 N5 P314-328 66-33558 <=$>
1966 N5 P329-330 66-33559 <=$>
1966 N6 P351-352 66-33656 <=$>
1966 N6 P391-395 66-33656 <=$>
1966 N6 P399-401 66-33656 <=$>
1966 V11 N1 P48-52 66-31512 <=$>
1966 V11 N1 P59-62 66-31513 <=$>
1966 V11 N1 P66-67 66-31514 <=$>

MAGYAR TUDOMANYOS AKADEMIA AGRARTUDOMANYOK OSZTALYANAK KOZLEMENYEI
1964 V23 N3/4 P267-302 64-51435 <=$>

MAGYAR TUDOMANYOS AKADEMIA AKADEMIAI KOZLONYEI. BUDAPEST
1953 N3 P153-159 T-1918 <INSD>
1961 V12 N8 P61-63 62-19561 <=*> P

1963 V12 N4 P29-32 63-31219 <=>
1963 V12 N17 P151-154 64-21576 <=>
1964 V13 N3 P18-19 64-31350 <=>
1964 V13 N6 P37-60 64-31649 <=>
1964 V13 N7 P62-64 64-31704 <=>
1964 V13 N10 P108-115 64-51601 <=$>
1965 P147-151 66-30248 <=*$>
1965 N13 P128-130 65-33917 <=*$>
1965 V14 P153-156 65-30287 <=$>
1965 V14 N1 P1-4 65-30455 <=$>
1965 V14 N3 P26 65-30785 <=$>
1965 V14 N4 P33-36 65-30844 <=$*>
1965 V14 N5 P42-45 65-31031 <=*$>
1965 V14 N6 P53-54 65-31488 <=$>
1966 V15 N6 P41-44 66-33282 <=$>
1966 V15 N9 P63-83 66-34353 <=$>

MAGYAR TUDOMANYOS AKADEMIA ALKALMAZOTT MATEMATIKAI INTEZETENEK KOZLEMENYEI
1953 V3 N1/2 P109-127 STMSP V4 P203-218 <AMS>

MAGYAR TUDOMANYOS AKADEMIA. ATOMMAG KUTATO INTEZET. KOZLEMENYEK. DEBRECZEN
1962 V4 N3/4 P169-176 63-01512 <*>

MAGYAR TUDOMANYOS AKADEMIA KEMIAI TUDOMANYOK OSZTALYANAK KOZLEMENYEI
1960 V14 N2 P177-189 62-19195 <=>
1960 V14 N2 P191-199 62-19195 <=>
1960 V14 N2 P201-208 62-19195 <=>
1960 V14 N2 P213-231 62-19195 <=>
1960 V14 N2 P239-241 62-19195 <=>
1960 V14 N2 P243-244 62-19195 <=>
1960 V14 N3 P277-342 62-19541 <*=>
1960 V14 N3 P355-360 62-19591 <=*>
1960 V14 N3 P369-376 62-19590 <=*>
1960 V14 N4 P431-468 62-19191 <*=>
1961 V15 P18-28 ICE V1 N1 P93-99 <ICE>
1961 V15 N1 P1-15 61-27850 <*=>
1961 V15 N2 P247-250 62-19498 <=*>
1962 V18 N1 P23-36 ICE V3 N1 P6-11 <ICE>
1963 V19 N1 P83-85 63-21972 <=>
1963 V19 N1 P107-129 63-21972 <=>
1963 V19 N2 P277-278 63-31057 <=>
1963 V19 N2 P279-282 63-31070 <=>
1963 V19 N2 P283-285 63-31057 <=>
1963 V19 N3 P355-362 63-31070 <=>
1963 V19 N3 P363-369 63-31057 <=>
1963 V19 N3 P387-392 63-31057 <=>
1963 V19 N4 P477-478 63-31567 <=>
1963 V19 N4 P479-480 63-31560 <*>
1964 V21 N2 P109-183 64-51227 <=>
1964 V21 N3 P339-343 64-41892 <=$>
1964 V21 N4 P460-467 64-41892 <=$>
1965 N2 P159-172 66-31854 <=$>
1965 N4 P408-410 66-31596 <=$>
1966 V25 N2 P85-153 66-32036 <=$>
1966 V25 N3 P310-312 66-32460 <=$>
1966 V25 N3 P312-314 66-32461 <=$>
1966 V25 N4 P407-419 66-33324 <=$>
1966 V25 N4 P419-429 66-33325 <=$>

MAGYAR TUDOMANYOS AKADEMIA KOZPONTI FIZIKAI KUTATO INTEZETENEK KOZLEMENYEI
1959 V7 P391-398 AEC-TR-5384 <*>
1959 V7 N5 P296-301 55P65H <ATS>
1959 V7 N6 P366-373 56P65H <ATS>
1960 V8 N4 P189-193 54P65H <ATS>
1961 V9 N1/2 P57-62 NP-TR-902 <*>

MAGYAR TUDOMANYOS AKADEMIA MATEMATIKAI ES FIZIKAI OSZTALYANAK KOZLEMENYEI
1956 V6 P199-211 STMSP V4 P133-146 <AMS>
1960 V10 N4 P407-420 62-19196 <=*>
1961 V11 N1 P107-118 62-19586 <=*>
1961 V11 N3 P229-247 62-15944 <=*>
1961 V11 N3 P289-304 62-15943 <=*>
1962 V12 P7-14 STMSP V5 P373-380 <AMS>
1963 V13 N4 P313-340 64-21772 <=>
64-31271 <=>
1964 V14 N3 P225-274 64-51853 <=$>
1965 N8 P175-249 66-31791 <=$>

MAGYAR TUDOMANYOS AKADEMIA MATEMATIKAI KUTATO INTEZETENEK KOZLEMENYEI
1957 V2 P43-50 STMSP V4 P219-224 <AMS>
1958 V3 N1/2 P109-127 61-19903 <=*>
65-11557 <*>

MAGYAR TUDOMANYOS AKADEMIA MUSZAKI TUDOMANYOK OSZTALYANAK KOZLEMENYEI
1953 V9 N1/4 P57-70 66-14432 <*$>
1960 V25 N1/4 P239-240 62-19238 <=*>
1960 V25 N1/4 P404-405 62-19233 <=*>
1960 V26 N1/4 P5-7 62-19232 <*=>
1960 V26 N1/4 P9-23 62-19234 <*=>
1960 V26 N1/4 P235-253 62-19235 <=*>
1960 V26 N1/4 P269-295 62-19236 <=*>
1960 V26 N1/4 P301-319 62-19239 <*=>
1960 V26 N1/4 P321-333 62-19240 <*=>
1960 V27 N1/2 P1-33 62-19241 <*=>
1960 V27 N1/2 P80-82 62-19242 <=*>
1960 V27 N1/2 P103-104 62-19242 <=*>
1960 V27 N1/2 P114-130 62-19242 <=*>
1960 V27 N3/4 P393-427 62-19243 <=*>

1961 V29 N1/4 P5-32 62-24245 <=*>
1961 V29 N1/4 P39-42 62-24246 <=*>
1961 V29 N1/4 P207-235 ICE V2 N3 P319-333 <ICE>
1964 V34 N3 P187-210 65-30224 <=$>
1965 V36 N1/4 P11-34 66-31909 <=$>
1965 V36 N1/4 P57-69 66-31910 <=$>

MAGYAR TUDOMANYOS AKADEMIA V. ORVOSI TUDOMANYOK OSZTALYANAK KOZLEMENYEI
1960 V11 N1 P5-17 62-19198 <=*>

Citation	Translation
1961 N3 P344-349	66-31593 <=$>
1961 V12 N3 P245-261	62-19592 <=*>
1961 V12 N3 P271-284	62-19612 <=*>
1961 V12 N3 P311-315	62-19666 <=*>
1961 V12 N1/2 P157-165	62-19489 <=*>
1961 V12 N1/2 P167-175	62-19486 <=*>
1961 V12 N1/2 P177-202	62-19656 <=*>
1962 V13 N3 P201-218	63-21145 <=>
1963 V14 N2 P111-128	64-21146 <=$>
1964 N2 P107-124	64-51357 <=$>
1964 N2 P125-138	64-51379 <=$>
1965 V16 N1 P3-36	66-32135 <=$>
1965 V16 N1 P37-44	66-32137 <=$>
1965 V16 N1/2 P143-148	66-31630 <=$>
1966 V17 N1 P15-39	66-32136 <=$>

MAKROMOLEKULARE CHEMIE

Citation	Translation
1947 V1 P94-105	62-18688 <*> 0
1947 V1 P209-228	9011 <IICH>
1948 V2 P169-196	63-10009 <*> 0
1950 V4 P240-261	66-10832 <*>
1951 V5 P245-256	60-14121 <*> OP
1951 V6 P30-38	64-20058 <*>
1951 V6 P39-59	63-18543 <=*>
1951 V6 P71-77	64-14174 <*>
1951 V6 P292-317	63-18978 <*> 0
	65-12040 <*>
1951 V7 N1 P46-61	60-10866 <*>
1952 V7 P259-270	60-14122 <*>
1952 V8 N2 P147-155	64-20812 <*>
1953 V11 N2/3 P97-110	64-20314 <*>
1954 V12 P20-34	I-654 <*>
1954 V12 N1 P61-78	60-18409 <*>
1954 V13 N1 P53-70	<K-H>
	64-20229 <*>
1954 V13 N1 P76-89	63-14193 <*>
1954 V13 N1 P90-101	57-1787 <*>
	60-10769 <*>
1954 V13 N2/3 P210-222	I-311 <*>
1954 V14 P128-145	81L28G <ATS>
	8761 <IICH>
1954 V15 P169-178	8782 <IICH>
1954 V15 N2/3 P169-178	64-20516 <*>
1955 V15 P177-187	<K-H>
1955 V15 N2/3 P177-187	64-20241 <*>
1955 V16 P213-237	II-911 <*>
	58-243 <*>
1955 V16 N1 P71-73	59-15946 <*>
1955 V16 N1 P77-80	II-369 <*>
	36H9G <ATS>
	64-20557 <*> 0
1955 V16 N3 P213-237	64-16775 <*> 0
	64-20551 <*> 0
1955 V17 P62-73	82M39G <ATS>
1955 V17 P455-462	63-14194 <*> 0
1955 V17 N3 P201-218	61-10484 <*>
1955 V17 N3 P231-240	61-10483 <*>
1956 V20 P161-167	38M39G <ATS>
1956 V20 N2 P111-142	<ATS>
1956 V21 P13-36	RCT V30 N3 P805 <RCT>
1956 V21 P169-178	63-18305 <*>
1956 V21 P240-244	45K25G <ATS>
1956 V21 N3 P169-178	65-14088 <*>
1956 V21 N3 P240-244	57-1175 <*>
1956 V18/9 P37-47	8758 <IICH>
1956 V18/9 P151-165	63-18306 <*>
1956 V18/9 P186-	<ES>
1956 V18/9 P239-253	1678 <*>
1956 V18/9 P322-341	59-15947 <*>
	65-11635 <*>
1956 V18/9 P455-462	5388-A <K-H> 0
	64-16770 <*> 0
1957 V22 P59-80	60-14117 <*> 0
	65-10007 <*> 0
1957 V22 P237-239	59-15368 <*>
1957 V22 N1/2 P1-30	64-16335 <*>
1957 V22 N1/2 P59-80	64-16820 <*> P
1957 V22 N1/2 P131-146	59-15945 <*> 0
	63-18307 <*>
1957 V23 N1 P71-83	59-10751 <*>
1957 V23 N2/3 P175-179	63J17G <ATS>
1957 V24 P104-132	63-20934 <*> 0
1957 V24 P141-151	84M39G <ATS>
1957 V24 N1 P50-63	59-15917 <*> 0
1957 V24 N1 P64-75	64-18084 <*> 0
1957 V24 N2 P141-151	65-14186 <*>
1957 V24 N3 P173-204	64-18159 <*>
1957 V24 N3 P222-244	59-15379 <*>
1957 V24 N3 P245-257	59-15765 <*>
1957 V24 N3 P258-290	57-3574 <*>
	58-1265 <*>
	64-18151 <*>
1957 V25 N3 P159-175	59-15257 <*>
1958 V25 N3 P159-175	59-20328 <*>
1958 V25 N3 P199-204	59-15919 <*>
1958 V25 N3 P205-209	59-15920 <*>
1958 V26 N1/2 P61-66	59-15665 <*> 0
1958 V26 N1/2 P102-118	64-18813 <*> 0
1958 V27 N1/2 P1-22	50K26G <ATS>
	58-2495 <*>
1958 V28 N3 P221-235	33L31G <ATS>
1959 V29 P220-225	04L32G <ATS>
	60-14374 <*>
1959 V29 N1/2 P93-116	REPT. 61052 <RIS>
	17L31G <ATS>
1959 V30 N1 P23-38	62-10221 <*>
1959 V30 N1 P48-80	60-18412 <*>
1959 V30 N2/3 P123-153	59-20782 <*> 0
1959 V31 N1 P50-74	59-20804 <*>
1959 V32 N2/3 P170-183	61-10233 <*>
1959 V33 N2/3 P113-130	65-13146 <*>
1959 V34 P120-138	61-18875 <*>
1959 V34 P231-239	60-10706 <*>
1960 V37 P53-63	8686 <IICH>
1960 V37 P85-96	63-00603 <*>
1960 V37 N1/2 P71-84	65-14757 <*>
1960 V37 N1/2 P97-107	63-18354 <*>
1960 V39 N3 P101-117	62-10679 <*>
1960 V40 P39-54	61-20631 <*>
1960 V40 P161-171	97144 <RIS>
1960 V40 P172-188	97145 <RIS>
1960 V40 N3 P207-215	62-10331 <*>
1960 V40 N1/2 P25-38	65-11314 <*>
1960 V40 N1/2 P148-160	65-10040 <*>
1960 V41 P86-109	61-10989 <*>
1960 V41 P110-123	65-10147 <*> 0
1960 V41 P124-130	65-10072 <*> 0
1960 V41 P131-147	1042 <TC>
1960 V41 P148-173	1043 <TC>
1960 V42 P89-94	63-20263 <*>
1960 V42 N1 P1-11	38N53G <ATS>
	62-14304 <*>
1961 V43 P132-143	62-10547 <*> 0
1961 V47 P168-183	65-10093 <*> 0
1961 V47 P215-217	RCT V35 N2 P274 <RCT>
1961 V47 N2/3 P201-214	64-14171 <*>
1961 V48 P1-16	65-10102 <*> 0
1961 V48 P59-71	86N57G <ATS>
1961 V50 P253-256	RCT V35 N3 P615 <RCT>
1961 V44/6 P324-337	8815 <IICH>
1961 V44/6 P347-357	37N53G <ATS>
	62-18931 <*>
1961 V44/6 P358-387	C-3880 <NRCC>
1961 V44/6 P448-460	65-10101 <*> 0
1962 V51 P182-198	62-14338 <*>
1962 V51 P199-216	8956 <IICH>
1962 V52 P23-35	63-16565 <*>
1962 V52 P37-47	62-16618 <*>
1962 V52 P108-119	62-18670 <*>
	63-20063 <*>
1962 V52 P236-238	62-16963 <*>
	63-16598 <*>
1962 V54 P126-135	65-13411 <*>
1962 V55 P96-120	66-13905 <*>
1962 V56 P228-233	8793 <IICH>
1962 V56 P234-236	1215 <TC>
1962 V57 P105-108	65-14510 <*>
1962 V57 P220-240	65-13412 <*>
1962 V58 P1-17	65-12187 <*>
	65-12274 <*>

1962 V58 P43-64 63-14417 <*>
65-11538 <*>
1962 V58 P1C4-129 RCT V36 N3 P815 <RCT>
1962 V58 P247-250 8703 <IICH>
1962 V58 N1 P18-42 63-10948 <*>
1963 V61 P1-13 63-18699 <*>
1963 V61 P116-131 65-18051 <*>
1963 V62 P1-17 52Q70G <ATS>
1963 V62 P25-30 63-01664 <*>
1963 V62 P134-137 1222 <TC>
1963 V65 P1-15 8802 <IICH>
1963 V66 P19-30 65-10364 <*>
1963 V69 P1-17 8805 <IICH>
1963 V70 P222-259 1223 <TC>
1964 V70 P23-43 64-16981 <*>
1964 V70 P44-53 64-16977 <*>
64-30701 <*>
1964 V70 P222-259 65-12916 <*>
1964 V73 P85-108 167 <TC>
1964 V73 P128-140 8904 <IICH>
1964 V73 P168-176 64-18417 <*>
1964 V74 P29-38 65-11305 <*>
1964 V74 P55-7C 543 <TC>
65-13865 <*> 0
1964 V74 P71-91 8899 <IICH>
1964 V75 P35-51 64-30066 <*>
1964 V75 P113 <ES>
1964 V76 P183-189 65-13863 <*>
1964 V76 P190-195 65-13864 <*>
1964 V78 P24-36 04S88G <ATS>
1964 V78 P37-46 09S88G <ATS>
1964 V80 P36-43 544 <TC>
66-11672 <*>
1965 V82 P1-15 2054 <TC>
1965 V82 P169-174 66-11028 <*>
1965 V82 P175-183 05S88G <ATS>
1965 V82 P184-189 08S88G <ATS>
1965 V84 P36-50 65-17228 <*>
1965 V84 P261-273 1143 <TC>
1965 V84 P274-281 06S88G <ATS>
66-11571 <*>
1965 V84 P282-285 55S88G <ATS>
66-11569 <*>
1965 V84 P286-289 56S88G <ATS>
1965 V86 P89-97 1827 <TC>
1965 V87 P8- RCT V39 N5 P1411 <RCT>
1965 V88 P38-53 66-10838 <*>
1965 V88 P54-74 66-10833 <*>
1965 V88 P215-231 74S89G <ATS>
1966 V92 P149-169 59T93G <ATS>
1966 V92 P213-223 66-12576 <*>

MANUTENTION
1960 N11 P129-138 2318 <BISI>
1960 N6/7 P113-138 2318 <BISI>
1960 N8/9 P147-150 2318 <BISI>

MARBURGER SITZUNGSBERICHTE
1953 V76 P29-45 61-10673 <*> 0

MARCFES TROPICOUX DU MONDE
1957 N626 P2699-2701 59-14376 <=*>

MARINA ITALIANA
1951 V49 P238- 58-704 <*>
1955 V53 N12 P309-314 59-12968 <+*>
1965 V62 P261 2202 <BSRA>

MARKT INFORMATIENEN FUER INDUSTRIE UND AUSSENHANDEL
1965 P10 66-30030 <=>

MAROC MEDICAL
1952 V31 P469-471 I-881 <*>
1954 V33 N351 P780 3738-B <KH>
1957 V36 P456-458 61-16133 <*>
1957 V36 N390 P1084-1096 60-15088 <=*$> 0
1960 V39 P782-783 65-14597 <*>
1964 V43 P147-155 65-18125 <*>

MARSEILLE MEDICAL
1962 V99 P711-714 65-18197 <*> P

MASCHINENBAU
1960 V9 N12 P362-365 62-18923 <*>

MASCHINENBAU-DER BETRIEB
1929 V8 N2 P33-37 59-17682 <*>
1929 V8 N10 P318-323 59-20902 <*> 0
1929 V8 N13 P434-437 59-20902 <*> 0
1929 V8 N18 P611-618 59-20459 <*> 0
1929 V8 N22 P772-774 59-20458 <*> 0
1930 V9 N8 P257-262 59-20465 <*> 0
1930 V9 N11 P368-373 59-20464 <*>
1930 V9 N11 P375-379 59-20463 <*> 0
1930 V9 N13 P437-444 59-20462 <*> 0
1930 V9 N14 P480-483 59-20461 <*> 0
1930 V9 N14 P489-492 59-20460 <*> 0
1930 V9 N16 P533-540 57-3218 <*>
1930 V9 N16 P546-548 59-15633 <*> 0
1930 V9 N22 P739-744 59-20469 <*> P
1930 V9 N22 P745-749 59-20468 <*> 0
1930 V9 N24 P798-800 59-20467 <*> 0
1930 V9 N24 P800-803 59-20466 <*> 0
1931 P119-122 57-3099 <*>
1931 P265-270 57-3086 <*>
1931 V10 N2 P33-36 59-20496 <*> 0
1931 V10 N2 P37-39 59-15632 <*> 0
1931 V10 N2 P43-47 57-3229 <*>
1931 V10 N2 P47-50 59-20495 <*> 0
1931 V10 N15 P489-492 59-20494 <*> P
1935 V14 P510 57-1050 <*>
1935 V14 N15/6 P437-441 57-1074 <*>
1936 V15 N3/4 P63-66 59-20457 <*> 0
1936 V15 N7/8 P191-194 61-10085 <*> 0
1936 V15 N21/2 P623-625 59-20456 <*> 0
1937 V16 N9/10 P237-239 59-20492 <*> 0
1938 V17 N3/4 P77 60-18721 <*>
1939 V18 N1/2 P27-28 59-20491 <*>
1939 V18 N13/4 P331-333 59-20490 <*> 0
1939 V18 N15/6 P391-392 59-20489 <*> 0
1940 V19 P567-568 60-18982 <*>
1940 V19 N7 P3C3-304 59-20487 <*> 0
1940 V19 N9 P395-396 57-3079 <*>
59-15231 <*>
1940 V19 N11 P481-484 59-15142 <*>
1940 V19 N12 P519-520 59-15103 <*> 0
1940 V19 N12 P521-524 57-1119 <*>
1942 V21 N1 P11-14 62-20336 <*> 0
1942 V21 N12 P505-510 59-17020 <*> 0

MASCHINENBAU UND WAERMEWIRTSCHAFT
1952 V7 N8 P129-135 R-2272 <*>
1955 V10 P76-85 T-1936A <INSD>
1955 V10 N2 P37-44 58-863 <*>
1955 V10 N6 P177-184 58-863 <*>

MASCHINENBAUTECHNIK
1955 V4 P525- 2242 <BSRA>
1956 V5 P619- 2243 <BSRA>
1956 V5 N9 P455 58-2671 <*>
1959 V8 N5 P230-236 2691 <BISI>
1960 V9 N6 P283-290 127-303 <STT>
1960 V9 N7 P366-371 149-310 <STT>
1960 V9 N8 P405-412 154-299 <STT>
1960 V9 N8 P417-419 127-301 <STT>
1960 V9 N9 P452-456 127-315 <STT>
1960 V9 N9 P465-477 154-316 <STT>
1960 V9 N9 P477-487 145-317 <STT>
1960 V9 N12 P611-617 149-321 <STT>
1961 V10 N2 P61-65 151-601 <STT>
1961 V10 N5 P228-232 36N58G <ATS>
1963 V12 N5 P269-271 63-31395 <=>
1964 V13 N9 P471-479 4166 <BISI>
1965 V14 N7 P337-338 65-32304 <=$>
1965 V14 N7 P354 65-32304 <=$>

MASCHINEN-KONSTRUKTEUR-BETRIEBSTECHNIK
1930 V63 N22 P427-429 57-3220 <*>

MASCHINENMARKT

Citation	Translation
1963 V69 P16-25	3345 <BISI>

MASCHINENSCHADEN

Citation	Translation
1934 V11 P45-47	58-912 <*>
1937 N9 P136-142	58-940 <*>
1939 V16 N3 P37-43	58-903 <*>
1939 V16 N4 P54-58	59-20901 <*> 0
1939 V16 N5 P69-75	59-20900 <*> 0

MASCHINENWELT UND ELEKTROTECHNIK

Citation	Translation
1956 V11 N5/6 P63-67	59-20989 <*>
1961 V16 N3 P121-122	2596 <BISI>

MASHINNYI PEREVOD I PRIKLADNAIA LINGVISTIKA

Citation	Translation
1956 V5 N8 P34-38	85P65R <ATS>
1959 N2 P1-84	60-31699 <=>
1959 N3 P3-98	60-31817 <=>
1960 N4 P3-81	61-21711 <=>
1960 N4 P102-113	61-21711 <=>
1961 N5 P3-99	62-24713 <=*>
1961 N6 P3-38	62-24715 <=*>
1961 N6 P80-100	62-24715 <=*>
1962 N3 P79-81	63-21050 <=>
1962 N7 P3-87	63-31145 <=>
1962 N7 P102-110	63-31145 <=>
1962 V11 N5 P19-21	19Q68R <ATS>
1963 N1 P3-5	63-21919 <=>
1963 N2 P30-32	64-21052 <=>
1963 N4 P102-104	63-41096 <=>
1963 N6 P48-52	64-31523 <=>
1964 N3 P36-38	64-41954 <=*>
1964 N4 P54-57	AD-617 119 <=$>

MASHINOSTROENIE

Citation	Translation
1962 N1 P3-8	62-33145 <=*>
1962 N1 P36	62-33446 <=*>
1962 N2 P42	62-33307 <=>
1962 N2 P104-105	62-33333 <=> P
1962 N3 P3-6	62-33523 <=*>
1962 N3 P47-55	64-19984 <=$>
1962 N3 P56-60	64-13734 <=*$>
1962 N3 P87-90	62-33333 <=> P
1962 N3 P92-94	62-33333 <=> P
1962 N3 P99-100	62-33697 <=>
1962 N5 P32-35	63-13636 <=>
1962 N5 P86-87	63-13636 <=>
1962 N6 P3-6	63-21569 <=>
1962 N8 P35	M.5335 <NLL>
1963 N1 P57-60	63-31116 <=>
1963 N2 P6	63-31169 <=>
1963 N3 P88-90	63-31264 <=>
1963 N3 P93-94	63-31264 <=>
1963 N4 P92-94	63-41038 <=>
1963 N4 P102-104	63-41038 <=>
1963 N5 P78-81	64-21824 <=>
1963 N5 P95-98	64-21783 <=>
1963 N5 P104-112	64-21783 <=>
1963 N6 P37-40	64-31524 <=>
1963 N6 P80-84	64-21722 <=>
1963 N6 P84-86	64-21722 <=>
1964 N5 P3-4	65-30438 <=$>
1964 N5 P16-25	65-30438 <=$>
1964 N5 P35-36	AD-630 992 <=$>
1964 N12 P29-32	AD-639 597 <=$>
1965 N2 P38-44	65-31278 <=$>

MASHINOSTROENIE ZA RUBEZHOM

Citation	Translation
1955 N1 P177-181	61-13195 <*=>
1955 N1 P344-358	61-13584 <*=>

MASHINOSTROITEL

Citation	Translation
1938 N1 P44-45	60-10705 <+>
1957 N2 P1-11	60-17795 <+*>
1957 N2 P11-14	60-17452 <+*>
1958 N6 P41-42	64-71395 <=> M
1958 N6 P43-44	59-15828 <+*>
1959 N5 P3-6	60-11305 <=>
1959 N5 P33-36	61-21956 <=> 0
1959 N6 P45-46	60-31114 <=>
1959 N10 P40-41	60-11860 <=>
1959 N11 P40-42	60-11860 <=>
1960 N1 P2-4	60-31692 <=>
1960 N2 P1-4	60-31750 <=>
1960 N2 P4-9	60-31753 <=>
1960 N2 P10-11	60-31752 <=>
1960 N2 P12-21	60-31751 <=>
1960 N2 P22	60-31752 <=>
1960 N2 P23	60-31755 <=>
1960 N2 P25	60-31750 <=>
1960 N2 P26	60-31752 <=>
1960 N2 P31-33	60-31754 <=>
1960 N2 P42-43	60-31755 <=>
1960 N2 P45-48	60-31750 <=>
1960 N4 P2-5	61-21146 <=>
1960 N4 P6-19	61-11673 <=>
1960 N5 P11-13	61-11019 <=>
1960 N5 P32-34	62-11721 <=>
1960 N6 P14-15	61-11019 <=>
	61-21293 <=>
1960 N6 P30-31	61-21293 <=>
1960 N6 P43	61-11019 <=>
1960 N8 P3-7	61-21508 <=>
1960 N8 P10	61-21508 <=>
1960 N8 P13	61-21508 <=>
1960 N8 P25	61-21508 <=>
1960 N8 P27	61-21508 <=>
1960 N8 P37-38	61-21508 <=>
1960 N10 P3	61-31118 <=>
1960 N10 P11-12	61-31118 <=>
1960 N11 P13	61-21700 <=>
1960 N11 P15-16	61-21700 <=>
1960 N11 P22	61-21700 <=>
1960 N11 P32	61-21700 <=>
1960 N12 P6-7	61-21700 <=>
1960 N12 P27	61-21700 <=>
1960 N12 P29-30	61-21700 <=>
1960 N12 P33	61-27083 <=>
1960 N12 P34-35	61-21700 <=>
1960 N12 P36-37	AD-613 453 <=$>
1960 N12 P42	61-21700 <=>
1961 N1 P8-10	61-31157 <=>
1961 N1 P17	61-31157 <=>
1961 N1 P31-32	61-31157 <=>
1961 N1 P48	61-31157 <=>
1961 N2 P15-16	61-27074 <=>
1961 N2 P46	61-27074 <=>
1961 N3 P9-12	61-27232 <=>
1961 N3 P18	61-27232 <=>
1961 N9 P35	62-19647 <=*>
1962 N3 P17-20	62-33315 <=*>
1962 N3 P23-24	62-33315 <=*>
1962 N3 P25-27	62-32009 <=>
1962 N5 P24-25	62-32076 <=*>
1962 N6 P12-13	62-33526 <=*>
1962 N7 P44-46	62-33118 <=>
1962 N8 P42-44	62-33668 <=*>
1962 N9 P2	63-13352 <=>
1962 N9 P39-41	63-13524 <=>
1962 N10 P1-2	63-13664 <=>
1962 N10 P8-9	63-13479 <=>
1962 N10 P9	63-13847 <=>
1962 N12 P19	63-21204 <=>
1963 N1 P19	63-21979 <=>
1963 N1 P28	63-21426 <=>
1963 N1 P30	63-21426 <=>
1963 N1 P39	63-21387 <=>
1963 N2 P1	63-21619 <=>
1963 N2 P47	63-21788 <=>
1963 N3 P10	63-31101 <=>
	63-31108 <=>
1963 N3 P95-98	63-31287 <=>
1963 N3 P103-106	63-31287 <=>
1963 N4 P5-8	65-60020 <=$> P
1963 N4 P16-17	63-31169 <=>
1963 N4 P32	63-31169 <=>
1963 N4 P39	63-31169 <=>
1963 N5 P7	63-31231 <=>
	63-31440 <=>
1963 N5 P11	63-31440 <=>

1963 N5 P20	63-31295 <=>
	63-31440 <=>
1963 N5 P22	63-31231 <=>
	63-31233 <=>
	63-31606 <=>
1963 N5 P23	63-31440 <=>
	63-31732 <=>
1963 N5 P39	63-31440 <=>
1963 N5 P42	63-31440 <=>
1963 N5 P45-48	63-31451 <=>
1963 N6 P1-2	63-31553 <=>
1963 N6 P5	63-31686 <=>
1963 N6 P10	63-31686 <=>
1963 N6 P12	63-31855 <=>
1963 N6 P14	63-31499 <=> 0
1963 N6 P16-20	63-31686 <=>
1963 N6 P31	63-31686 <=>
1963 N6 P43	63-31797 <=>
1963 N7 P4	63-31636 <=>
	63-31927 <=>
1963 N7 P6-18	63-31843 <=>
1963 N7 P7	63-31752 <=>
1963 N7 P11	63-31636 <=>
	63-31752 <=>
1963 N7 P13	63-31752 <=>
	63-31927 <=>
1963 N7 P19	63-31843 <=>
1963 N7 P25	63-31782 <=>
1963 N7 P35	63-31782 <=>
1964 N2 P25	AD-613 178 <=$>
1964 N3 P6	64-51488 <=>
1964 N3 P34-35	AD-618 628 <=$>
1964 N3 P44	64-51488 <=>
1964 N4 P34-35	64-41226 <=>
1964 N7 P38-41	AD-622 409 <=*$>
1965 N6 P11-17	65-32215 <=$>
1965 N7 P6-7	65-32248 <=*$>
1965 N7 P34-36	65-32129 <=$>
1965 N9 P9-12	65-33487 <=*$>
1965 N10 P1-2	66-30464 <=$>
1965 N11 P78-79	66-30299 <=$>
1965 N12 P30-32	66-30859 <=$>
1966 N2 P17	NLLTB V8 N11 P933 <NLL>
	NLLTB V8 N8 P717 <NLL>
1966 N2 P41-43	66-31846 <=>
1966 N7 P35-36	66-35747 <=>
1966 N7 P37-40	66-35722 <=>

*MASHINOVEDENIE. MOSCOW

1965 V5 P55-58	65-33012 <=$>

MASLOBOINO-ZHIROVAYA PROMYSHLENNOST

1940 V16 N1 P12-13	1419 <K-H>
	66F2R <ATS>
1940 V16 N5/6 P30-32	61-18005 <*=> P
1940 V16 N5/6 P42-45	61-16891 <*=>
1953 V18 N1 P7-10	T.508 <*>
1953 V18 N7 P30-31	62-13122 <=*>
1953 V18 N8 P13-14	RJ-397 <ATS>
1953 V18 N8 P29-30	T-635 <INSD>
1953 V18 N9 P5-8	RT-1398 <*>
1953 V18 N9 P18-20	RT-1399 <*>
1953 V18 N10 P20-24	RT-1341 <*>
1954 N2 P35-36	<INSD>
1954 V19 N3 P34-35	T-2507 <INSD>
1954 V19 N6 P30-	RJ-505 <ATS>
1955 V21 N8 P5-9	T-2124 <INSD>
1955 V21 N8 P12-13	T-2123 <INSD>
1956 V21 P14-17	T-1841 <INSD>
1956 V21 N3 P26-29	59-16549 <+*>
1956 V22 N6 P18	78K22R <ATS>
1957 N9 P17-20	63K25R <ATS>
1957 V23 N6 P35-38	13M42R <ATS>
1957 V23 N7 P24-26	60-17212 <+*>
1957 V23 N9 P17-20	60-17458 <+*>
1958 V24 N1 P12-16	59-19158 <+*>
1958 V24 N1 P23-25	59-18129 <+*>
1958 V24 N3 P22-26	65-14194 <*>
1958 V24 N4 P19-22	59-18855 <+*>
1958 V24 N7 P1-9	60-15056 <+*>
1958 V24 N7 P26-30	59-18122 <+*>
1958 V24 N8 P10-12	59-18128 <+*>
1958 V24 N8 P23-29	61-15371 <+*>
1958 V24 N8 P45	61-15372 <+*>
1958 V24 N11 P17-19	59-19158 <+*>
1958 V24 N12 P17-20	05Q68R <ATS>
1959 V25 N5 P14-17	61-13491 <*+>
1959 V25 N5 P41	61-15709 <+*>
1959 V25 N7 P36-39	61-15391 <*+>
1959 V25 N7 P39-40	66-10624 <*>
1959 V25 N9 P41-43	61-15920 <+*>
1959 V25 N10 P22-24	23P63R <ATS>
1959 V25 N11 P30-32	07M44R <ATS>
1960 V26 N6 P27-29	72R79R <ATS>
1960 V26 N9 P20-22	98P62R <ATS>
1961 V27 N1 P36-38	62-32499 <=*>
1961 V27 N2 P29-30	736 <TC>
1961 V27 N2 P33-35	81N53R <ATS>
1961 V27 N4 P5-14	61-27383 <*=>
1961 V27 N4 P19-21	65-63398 <=$> 0
1961 V27 N5 P1-17	62-13040 <*=>
1961 V27 N9 P17-18	62-32490 <=*>
1961 V27 N10 P30-31	62-32498 <=*>
1961 V27 N12 P22-24	62-32967 <=*>
1961 V27 N12 P27	66-10620 <*>
1961 V27 N12 P31-32	65P60R <ATS>
1962 N2 P34-37	66-11566 <*>
1962 V28 N1 P16-21	63-18259 <=*>
	70Q69R <ATS>
1962 V28 N1 P30-32	64P60R <ATS>
1962 V28 N2 P14-17	65-60994 <=$>
1962 V28 N2 P23-26	64-19975 <=$>
1962 V28 N4 P17-19	65-60239 <=$>
1962 V28 N6 P11-13	64-19967 <=$>
1962 V28 N7 P15-22	63-23227 <=*>
1962 V28 N8 P10-15	65-60790 <=>
1962 V28 N8 P33-34	63-23226 <=*>
1963 V29 N1 P15-17	65-60950 <=$>
1963 V29 N1 P19-23	66-26105 <*$> 0
1963 V29 N2 P29-32	64-19963 <=$>
1963 V29 N3 P21-23	65-18183 <*>
1963 V29 N3 P36-37	64-13735 <=*$>
1963 V29 N4 P26-29	64-16970 <=*$>
	65-18182 <*>
1963 V29 N5 P9-11	65-63550 <=*$>
1963 V29 N5 P35-38	64-19381 <=$>
1963 V29 N6 P9-14	65-60935 <=$>
1963 V29 N6 P32-35	66-11568 <*>
1963 V29 N8 P22-24	RTS-2613 <NLL>
1963 V29 N10 P6-8	65-63400 <=$> 0
1963 V29 N10 P11-14	65-11311 <*>
1964 V30 N3 P17-19	M-5706 <NLL>
1964 V30 N6 P13-15	M.5599 <NLL>
1964 V30 N11 P26-29	AD-633 212 <=$>
1964 V30 N11 P30-32	65-13727 <*> 0
1965 V31 N8 P10-13	M.5792 <NLL>

MASLOBOINO-ZHIROVOE DELO

1935 P15-18	R-4901 <*>
1935 V11 P378-379	59-20632 <+*>
1936 V12 P546-547	64-11078 <=>
1938 V14 N1 P17-19	RT-1130 <*>
1938 V14 N3 P5-8	RT-4064 <*>
1938 V14 N6 P29-30	RJ-360 <ATS>

MASS SPECTROSCOPY. JAPAN
SEE SHITSURYO BUNSEKI

MASTE LESA

1960 N12 P4	61-21533 <=> P

MASTER UGLYA

1957 N12 P14-16	59-11301 <=>
1957 N12 P31-34	59-11301 <=>
1957 V6 N12 P14-16	60-17238 <+*>
1960 V9 N9 P9	63-11092 <=>

MATEMATICHE

1963 V18 P135-154	46S811 <ATS>
	65-13061 <*>

1964 V19 P108-115 47S811 <ATS>
65-13070 <*>

MATEMATICHESKIE ZAPISKI URALSKOGO MATEMATICHESKOGO OBSHCHESTVA. URALSKII GOSUDARSTVENNYI UNIVERSITET SVERDLOVSK
1964 V4 P30-35 AD-638 627 <=$>

*MATEMATICHESKII SBORNIK
1934 V41 N4 P561-574 64-13704 <=*$>
1936 V1 P815-844 AMST S2 V32 P1-35 <AMS>
AMST,S2,V32,P1-35 <AMS>
1937 V2 P467-499 AMST S2 V33 P1-40 <AMS>
1938 V3 N1 P47-100 RT-2863 <*>
1938 V3 N4 P47-100 AMST S1 V5 P175-241 <AMS>
1938 V4 P471-497 AMST S2 V34 P39-68 <AMS>
1939 V6 P95-138 AMST S2 V32 P37-81 <AMS>
AMST,S2,V32,P37-81 <AMS>
1940 V8 P205-237 AMST S1 V1 P228-282 <AMS>
1940 V8 P405-422 AMST S2 V45 P1-18 <AMS>
1943 V12 P99-108 AMST S2 V57 P113-122 <AMS>
1943 V13 P301-316 AMST S2 V36 P1-15 <AMS>
1944 V15 N3 P437-448 RT-737 <*>
1945 V17 P211-252 AMST S2 V22 P1-42 <AMS>
1946 V18 P3-28 AMST S1 V8 P273-304 <AMS>
1946 V19 P85-154 AMST S1 V8 P78-154 <AMS>
1946 V19 P165-174 AMST S1 V9 P263-275 <AMS>
1946 V19 P239-262 AMST S2 V13 P61-83 <AMS>
1946 V19 P311-340 AMST S1 V8 P155-185 <AMS>
1946 V19 N2 P165-174 RT-285 <*>
1947 V20 P351-363 AMST S2 V4 P59-72 <AMS>
1947 V21 P133-140 AMST S2 V37 P1-11 <AMS>
1947 V21 P233-284 AMST S1 V7 P149-219 <AMS>
1947 V21 P285-320 AMST S1 V2 P481-532 <AMS>
1947 V21 P405-434 AMST S1 V9 P1-41 <AMS>
1947 V21 N2 P285-320 RT-2864 <*>
1947 V21 N3 P405-434 RT-2865 <*>
1947 V21 N63 P233-284 60-13701 <+*>
1948 V22 P101-133 AMST S1 V1 P339-391 <AMS>
1948 V22 P391-424 AMST S1 V2 P284-336 <AMS>
1948 V22 P439-441 AMST S1 V1 P470-473 <AMS>
1948 V22 N1 P101-133 RT-308 <*>
1948 V22 N2 P319-348 RT-307 <*>
1948 V22 N31 P9-348 AMST S1 V1 P392-438 <AMS>
1948 V23 P3-52 AMST S1 V4 P24-101 <AMS>
1948 V23 P89-125 AMST S1 V6 P207-273 <AMS>
1948 V23 P187-228 AMST S1 V4 P102-158 <AMS>
1948 V23 P229-258 AMST S2 V32 P83-113 <AMS>
AMST,S2,V32,P83-113 <AMS>
1948 V23 P279-296 AMST S1 V2 P1-24 <AMS>
1948 V23 P361-382 AMST S2 V32 P115-137 <AMS>
AMST,S2,V32,P115-137 <AMS>
1948 V23 P399-418 AMST S1 V6 P1-26 <AMS>
1948 V23 N2 P187-228 RT-1180 <*>
1948 V23 N3 P399-418 RT-286 <*>
1949 V24 P129-162 AMST S1 V7 P220-278 <AMS>
1949 V24 P163-188 AMST S1 V7 P418-449 <AMS>
1949 V24 P227-235 AMST S1 V1 P1-14 <AMS>
1949 V24 P301-320 AMST S1 V4 P415-439 <AMS>
1949 V24 P321-346 AMST S1 V10 P378-407 <AMS>
1949 V24 P347-374 AMST S1 V4 P268-299 <AMS>
1949 V24 P385-389 AD-627 445 <=$>
1949 V24 P405-428 AMST S1 V6 P474-504 <AMS>
1949 V24 N2 P163-188 RT-288 <*>
1949 V24 N2 P227-235 RT-326 <*>
1949 V24 N2 P301-320 RT-558 <*>
1949 V24 N3 P321-346 RT-2866 <*>
1949 V24 N3 P347-374 RT-2867 <*>
1949 V24 N3 P405-428 RT-2868 <*>
1949 V25 P3-50 AMST S1 V1 P120-180 <AMS>
1949 V25 P51-94 AMST S2 V36 P17-62 <AMS>
1949 V25 P251-274 AMST S1 V1 P181-210 <AMS>
1949 V25 P275-306 AMST S1 V6 P430-473 <AMS>
1949 V25 P307-314 AMST S1 V2 P275-283 <AMS>
1949 V25 P321-346 AMST S1 V1 P439-469 <AMS>
1949 V25 P387-414 AMST S1 V8 P11-47 <AMS>
1949 V25 N1 P3-50 RT-567 <*>
1949 V25 N1 P107-150 RT-305 <*>
1949 V25 N2 P251-274 RT-566 <*>
1949 V25 N2 P275-306 RT-276 <*>
1949 V25 N2 P307-314 RT-277 <*>
1949 V25 N3 P321-346 RT-289 <*>
1949 V25 N3 P387-414 RT-352 <*>
1950 V26 P35-56 AMST S1 V4 P440-466 <AMS>
1950 V26 P103-112 AEC-TR-6133 <=*$>
1950 V26 P113-146 AMST S2 V4 P73-106 <AMS>
1950 V26 P215-223 AMST S1 V5 P498-510 <AMS>
1950 V26 P225-227 AMST S1 V8 P48-50 <AMS>
1950 V26 P228-236 AMST S1 V8 P51-61 <AMS>
1950 V26 P247-264 AMST S2 V13 P85-103 <AMS>
1950 V26 N1 P35-56 RT-284 <*>
1950 V26 N2 P215-223 RT-287 <*>
1950 V26 N2 P225-227 RT-352 <*>
1950 V26 N2 P228-236 RT-352 <*>
1950 V27 P47-68 AMST S1 V4 P300-330 <AMS>
1950 V27 P297-318 AMST S1 V8 P365-391 <AMS>
1950 V27 P427-454 AMST S2 V2 P89-115 <AMS>
1950 V27 N1 P47-68 RT-568 <*>
1950 V27 N2 P297-318 RT-559 <*>
1951 V28 P431-444 AMST S2 V2 P117-131 <AMS>
1951 V28 P567-588 AMST S2 V2 P1-21 <AMS>
1951 V28 N44 P5-452 AMST S2 V2 P133-140 <AMS>
1951 V29 P225-232 AMST S1 V3 P271-280 <AMS>
1951 V29 P233-280 AMST S2 V25 P199-248 <AMS>
1951 V29 P433-454 AMST S2 V15 P33-54 <AMS>
1951 V29 N1 P173-176 NASA-TT-F-193 <=>
1951 V29 N1 P225-232 RT-141 <*>
1952 V30 P181-196 AMST S1 V5 P396-413 <AMS>
1952 V30 P271-316 AMST S1 V6 P147-206 <AMS>
1952 V30 P349-462 AMST S2 V6 P111-244 <AMS>
1952 V30 N1 P181-196 RT-1055 <*>
1952 V30 N2 P271-316 RT-140 <*>
1952 V31 P497-507 AMST S2 V27 P9-18 <AMS>
1952 V31 P543-574 AMST S2 V38 P5-35 <AMS>
1952 V31 P645-674 AMST S2 V8 P209-241 <AMS>
1952 V31 P675-686 AMST S2 V8 P243-255 <AMS>
1952 V31 N3 P497-506 63-23133 <=*>
1953 V32 P353-364 AMST S2 V1 P37-48 <AMS>
1953 V32 P493-514 AMST S2 V37 P259-282 <AMS>
1953 V32 N1 P139-156 59-20498 <+*>
1953 V33 P73-100 AMST S2 V1 P67-93 <AMS>
1953 V33 P181-192 AMST S2 V28 P37-49 <AMS>
1953 V33 N2 P359-382 63-10121 <=*>
1954 V34 P3-54 AMST S2 V15 P245-295 <AMS>
1954 V34 P55-80 AMST S2 V6 P459-483 <AMS>
1954 V34 P145-199 AMST S2 V5 P67-114 <AMS>
1954 V34 P385-406 AMST S2 V10 P319-339 <AMS>
1954 V35 P3-20 AMST S2 V27 P125-142 <AMS>
1954 V35 P93-128 AMST S2 V17 P117-152 <AMS>
1954 V35 P129-173 AMST S2 V12 P301-342 <AMS>
1954 V35 P193-214 AMST S2 V18 P15-36 <AMS>
1954 V35 P317-356 AMST S2 V9 P155-193 <AMS>
1954 V35 P513-568 AMST S2 V35 P15-78 <AMS>
1954 V35 N3 P461-468 64-11959 <=> M
1954 V35 N77 P231-246 RTS-2579 <NLL>
1955 V36 P163-168 AMST S2 V27 P35-41 <AMS>
1955 V36 P281-298 AMST S2 V5 P285-304 <AMS>
1955 V36 P445-478 AMST S2 V5 P1-33 <AMS>
1955 V37 P21-68 AMST S2 V10 P125-175 <AMS>
1955 V37 P121-140 AMST S2 V9 P195-215 <AMS>
1955 V37 P141-196 AMST S2 V8 P87-141 <AMS>
1955 V37 P209-250 AMST S2 V10 P177-221 <AMS>
1955 V37 P385-434 AMST S2 V15 P297-349 <AMS>
1955 V37 P507-526 AMST S2 V17 P9-28 <AMS>
1955 V37 N2 P209-250 59-18401 <+*>
1956 V38 P51-92 AMST S2 V41 P1-48 <AMS>
1956 V38 P203-240 AMST S2 V36 P295-336 <AMS>
1956 V38 P283-302 AMST S2 V21 P1-20 <AMS>
1956 V38 N1 P51-92 AMS TRANS V.41 P1 <AMS>
62-18860 <=*>
1956 V39 P3-22 AMST S2 V12 P163-179 <AMS>
1956 V39 P51-148 AMST S2 V24 P173-278 <AMS>
1956 V39 P253-266 AMST,S2,V32,P273-287 <AMS>
1956 V39 P273-292 AMST S2 V17 P153-172 <AMS>
1956 V40 P23-56 AMST S2 V18 P117-149 <AMS>
1956 V40 P137-156 AMST S2 V15 P95-113 <AMS>
1956 V40 P415-452 AMST S2 V11 P115-153 <AMS>
1957 V41 P49-80 AMST S2 V13 P29-60 <AMS>
1957 V41 P159-176 AMST S2 V15 P15-32 <AMS>
1957 V41 P257-276 AMST S2 V32 P289-309 <AMS>

Citation	Translation
	AMST,S2,V32,P289-309 <AMS>
1957 V41 P415-416	AMST S2 V13 P29-60 <AMS>
1957 V41 N2 P221-230	64-11887 <=> M
1957 V42 P223-248	AMST S2 V14 P31-54 <AMS>
1957 V42 N1 P11-44	STMSP V1 P213-244 <AMS>
1957 V43 P257-276	AMST S2 V13 P9-27 <AMS>
1957 V43 P349-366	AMST S2 V37 P13-29 <AMS>
1957 V43 N2 P149-168	AMST S2 V14 P181-199 <AMS>
1958 V44 P3-52	AMST S2 V25 P283-334 <AMS>
1958 V44 P179-212	AMST S2 V52 P95-128 <AMS>
1958 V44 P481-508	AMST S2 V42 P41-69 <AMS>
1958 V44 N3 P353-408	62-23020 <=*>
1958 V44 N4 P409-456	62-23105 <=>
1958 V44 N4 P481-508	AMS-TRANS V42 P41-49 <AMS>
1958 V45 P233-260	AMST S2 V17 P251-275 <AMS>
1958 V45 N2 P195-232	64-19946 <=$>
1958 V46 P91-124	AMST S2 V21 P87-117 <AMS>
1958 V46 P229-258	AMST S2 V41 P121-154 <AMS>
1958 V46 N2 P229-258	AMS TRANS V.41 P121 <AMS>
1958 V46 N4 P389-398	62-20255 <=*>
1958 V47 N3 P401-414	63-20286 <=*$>
1959 V47 P3-16	AMST S2 V19 P179-192 <AMS>
1959 V47 P431-484	AMST S2 V41 P155-213 <AMS>
1959 V47 N3 P271-306	61-11573 <=>
	64-13651 <=*$>
	64-14614 <=*$>
1959 V47 N4 P431-484	AMS TRANS V.41 P155 <AMS>
1959 V48 P3-74	AMST S2 V28 P61-147 <AMS>
1959 V48 P75-91	AMST S2 V27 P159-177 <AMS>
1959 V48 P137-148	AMST S2 V35 P237-249 <AMS>
1959 V48 P191-212	AMST S2 V38 P95-118 <AMS>
1959 V48 P257-276	AMST S2 V19 P199-219 <AMS>
1959 V48 P335-376	AMST S2 V33 P189-231 <AMS>
1959 V48 P397-428	AMST S2 V28 P149-185 <AMS>
1959 V48 P499-508	AMST S2 V27 P179-189 <AMS>
1959 V48 N1 P105-116	64-71492 <=> M
1959 V49 P13-28	AMST S2 V30 P273-290 <AMS>
1959 V49 P109-132	AMST S2 V26 P273-297 <AMS>
1959 V49 P133-180	AMST S2 V34 P169-221 <AMS>
1959 V49 P331-340	AMST S2 V35 P251-262 <AMS>
1959 V49 P341-346	AMST S2 V19 P193-198 <AMS>
1959 V49 P381-430	AMST S2 V22 P289-337 <AMS>
1959 V49 P431-446	AMST S2 V34 P223-240 <AMS>
1959 V49 N2 P133-180	62-23151 <=>
1960 V50 P59-66	AMST S2 V45 P23-31 <AMS>
1960 V50 P101-108	AMS-TRANS V43 P139 <AMS>
	AMST S2 V43 P139-146 <AMS>
1960 V50 P233-240	AMS-TRANS V43 P147 <AMS>
	AMST S2 V43 P147-154 <AMS>
1960 V50 P241-246	AMST S2 V50 P288-294 <AMS>
1960 V50 P257-266	AMST S2 V45 P221-231 <AMS>
1960 V50 P299-334	AMST S2 V33 P233-275 <AMS>
1960 V50 P383-388	AMST S2 V45 P71-77 <AMS>
1960 V50 N3 P299-334	62-23131 <=*>
1960 V51 P3-26	AMST S2 V47 P131-156 <AMS>
1960 V51 P73-98	AMST S2 V55 P1-32 <AMS>
1960 V51 P99-128	AMS-TRANS V42 P199 <AMS>
	AMST S2 V42 P199-231 <AMS>
1960 V51 P217-226	AMST S2 V26 P299-309 <AMS>
1960 V51 P273-276	AMST S2 V50 P1-4 <AMS>
1960 V51 P427-458	AMST S2 V50 P235-266 <AMS>
1960 V51 P475-486	AMS-TRANS V42 P233 <AMS>
	AMST S2 V42 P233-245 <AMS>
1960 V51 P487-500	AMST S2 V59 P150-162 <AMS>
1960 V51 P515-536	AMST S2 V38 P119-140 <AMS>
1960 V52 P589-596	AMST S2 V27 P289-296 <AMS>
1960 V52 P597-628	AMST S2 V36 P351-381 <AMS>
1960 V52 P647-652	AMST S2 V36 P243-249 <AMS>
1960 V52 P661-700	AMST S2 V37 P157-196 <AMS>
1960 V52 P701-708	AMST S2 V37 P31-38 <AMS>
1960 V52 P739-788	AMST S2 V25 P77-130 <AMS>
1960 V52 P823-846	AMST S2 V26 P311-338 <AMS>
1960 V52 P847-862	AMST S2 V38 P141-158 <AMS>
1960 V52 P917-946	AMST S2 V45 P39-69 <AMS>
1960 V52 N4 P953-990	63-23177 <=*$> P
	63-23551 <=*>
1961 V53 P3-38	AMST S2 V37 P197-240 <AMS>
1961 V53 P73-136	AMS TRANS V.41 P49 <AMS>
	AMST S2 V41 P49-120 <AMS>
1961 V53 P207-218	AMS-TRANS V43 P267 <AMS>
	AMST S2 V43 P267-280 <AMS>
1961 V53 P261-263	AMST S2 V55 P1-32 <AMS>
1961 V53 P287-312	AMS TRANS V44 P241 <AMS>
	AMST S2 V44 P241-268 <AMS>
1961 V53 P353-366	AMST S2 V56 P233-248 <AMS>
1961 V53 P515-538	AMS-TRANS V43 P169 <AMS>
	AMST S2 V43 P169-193 <AMS>
1961 V53 N2 P195-206	61-28305 <*=>
1961 V54 P3-50	AMS-TRANS V43 P215 <AMS>
	AMST S2 V43 P215-266 <AMS>
1961 V54 P209-224	AMST S2 V59 P163-190 <AMS>
1961 V54 P381-384	AMST S2 V50 P183-187 <AMS>
1961 V55 P3-6	AMST S2 V45 P33-37 <AMS>
1961 V55 P35-100	AMST S2 V48 P161-228 <AMS>
1961 V55 P101-124	AMST S2 V45 P79-104 <AMS>
1961 V55 P125-174	AMST S2 V53 P23-80 <AMS>
1961 V55 P289-306	AMST S2 V39 P267-286 <AMS>
1961 V55 P329-346	AMST S2 V52 P129-150 <AMS>
1961 V55 P407-410	AMS-TRANS V42 P195 <AMS>
	AMST S2 V42 P195-198 <AMS>
1961 V55 N4 P449-472	63-31540 <=>
1962 V56 P3-42	AMST S2 V42 P247-288 <AMS>
1962 V56 P129-136	AMST S2 V55 P242-250 <AMS>
1962 V56 P353-374	AMS-TRANS V43 P51 <AMS>
	AMST S2 V43 P51-74 <AMS>
1962 V56 P433-468	AMST S2 V47 P89-129 <AMS>
1962 V56 N1 P3-42	AMS-TRANS V42 P247 <AMS>
1962 V56 N1 P77-93	62-11797 <=>
1962 V57 P13-44	AMST S2 V47 P157-191 <AMS>
1962 V57 P75-94	AMST S2 V37 P121-142 <AMS>
1962 V57 P95-104	AMST S2 V52 P32-41 <AMS>
1962 V57 P105-136	AMS TRANS V.40 P157 <AMS>
	AMST S2 V40 P157-192 <AMS>
1962 V57 P225-232	AMST S2 V45 P147-155 <AMS>
1962 V57 P319-322	AMS TRANS V44 P151 <AMS>
	AMST S2 V44 P151-154 <AMS>
1962 V57 P375-383	AMST S2 V55 P251-260 <AMS>
1962 V58 P3-16	AMS-TRANS V43 P75-89 <CSSA>
	AMST S2 V43 P75-89 <AMS>
1962 V58 P65-86	AMST S2 V47 P193-216 <AMS>
1962 V58 P707-748	AMS-TRANS V43 P91 <AMS>
	AMST S2 V43 P91-138 <AMS>
1962 V58 P785-791	AMST S2 V45 P157-163 <AMS>
1962 V58 N1 P65-86	AEC-TR-6004 <=*$>
1962 V59 P229-244	AMST S2 V56 P1-18 <AMS>
1963 V60 P17-28	AMST S2 V55 P141-152 <AMS>
1963 V60 P63-88	AMST S2 V39 P57-82 <AMS>
1963 V60 P89-119	AMST S2 V39 P133-164 <AMS>
1963 V60 P159-184	AMST S2 V58 P1-28 <AMS>
1963 V60 P270-292	AMST S2 V49 P107-129 <AMS>
1963 V60 P304-324	AMST S2 V54 P231-254 <AMS>
1963 V60 P366-392	AMST S2 V49 P241-268 <AMS>
1963 V60 P425-446	AMST S2 V56 P249-272 <AMS>
1963 V60 P447-485	AMST S2 V55 P153-194 <AMS>
1963 V60 N4 P393-410	64-71340 <=> M
1963 V61 P80-120	AMST S2 V51 P273-316 <AMS>
1963 V61 P147-174	AMST S2 V51 P82-112 <AMS>
1963 V61 P467-503	AMST S2 V59 P191-223 <AMS>
1963 V62 P53-74	AMST S2 V51 P132-154 <AMS>
1963 V62 P160-179	AMST S2 V50 P267-287 <AMS>
1963 V62 P186-248	AMST S2 V56 P37-102 <AMS>
1963 V62 P335-344	AMST S2 V59 P163-190 <AMS>
1964 N2 P169-214	AD-621 058 <=*$>
1964 V63 P23-42	AMST S2 V57 P123-143 <AMS>
1964 V63 P43-58	AMST S2 V58 P77-92 <AMS>
1964 V63 P554-581	AMST S2 V58 P29-56 <AMS>
1964 V64 P79-101	AMST S2 V49 P63-85 <AMS>
1964 V64 P102-114	AMST S2 V51 P201-214 <AMS>
1964 V64 P458-480	AMST S2 V53 P167-191 <AMS>
1964 V64 P481-496	AMST S2 V52 P201-216 <AMS>
1964 V64 P562-588	AMST S2 V57 P277-304 <AMS>
1964 V64 N1 P79-101	64-41427 <=$>
1964 V65 P198-211	AMST S2 V49 P92-106 <AMS>
1964 V65 P212-227	AMST S2 V57 P191-206 <AMS>

MATEMATICHESKOE PROSVESHCHENIE

Citation	Translation
1958 N3 P77-88	NLLTB V1 N6 P9 <NLL>
1958 N3 P221-227	NLLTB V1 N6 P1 <NLL>

MATEMATIKA V SHKOLE

1954 N3 P77-79 RT-2371 <*>

MATEMATIKAI ES TERMESZETTUDOMANYI ERTESITO
1911 V29 P165-192 PANSDOC-TR.157 <PANS>

MATEMATISK-FYSISKE MEDDELELSER
1933 V12 N8 P3-65 60-18880 <*>

MATERIA MEDICA NORDMARK
1956 V8 N2/3 P1-11 58-378 <*>

MATERIALE PLASTICE
1964 V1 N2 P77-81 94S88RU <ATS>
1964 V1 N3 P136-139 1413 <TC>
1965 V2 N4 P199-201 66-14647 <*$>

MATERIALKENNIS
1947 N1 P1-8 58-1147 <*>
1947 N3 P25-26 58-1147 <*>

MATERIALNO-TEKHNICHESKOE SNABZAENIE
1966 N2 P12-20 67-30083 <=>
1966 N2 P52-62 66-35673 <=>

MATERIALOVY SBORNIK. VYZKUMMY USTAV MATERIALU A TECHNOLOGIE
1961 N1 P61-72 3548 <BISI>

MATERIALPRUEFUNG
1959 V1 N1 P3-12 1260 <BISI>
61-20618 <*>
1959 V1 N5 P175-176 61-16422 <*>
1959 V1 N9 P297-302 C-5067 <NRC>
1960 V2 P56-64 TS 1704 <BISI>
1960 V2 P253-257 2619 <BISI>
1960 V2 N2 P45-50 1686 <BISI>
1960 V2 N2 P51-55 62-20294 <*>
1960 V2 N2 P65-68 1724 <BISI>
1960 V2 N3 P88-97 25N50G <ATS>
1960 V2 N6 P198-207 5838 <HB>
1960 V2 N7 P233-236 5349 <HB>
1960 V2 N8 P307-308 65-00161 <*>
1961 V3 P218-224 2428 <BISI>
1961 V3 P300-304 2493 <BISI>
1961 V3 N1 P1-4 5115 <HB>
1961 V3 N1 P15-17 65-17401 <*>
1961 V3 N5 P176-180 2970 <BISI>
1961 V3 N6 P228-233 3805 <BISI>
1961 V3 N11 P402-409 C-3957 <NRCC>
1961 V3 N11 P418-422 5506 <HB>
1962 V4 P21-25 4118 <BISI>
1962 V4 N2 P43-52 2838 <BISI>
1962 V4 N4 P117-128 64-18357 <*>
1962 V4 N7 P247-258 2979 <BISI>
1962 V4 N9 P331-337 5877 <HB>
1962 V4 N11 P407-410 5587 <HB>
1962 V4 N12 P463-469 3378 <BISI>
1962 V4 N12 P469-472 3228 <BISI>
1963 V5 N3 P107-113 64-10183 <*> 0
64-14891 <*>
1963 V5 N4 P144-153 3542 <BISI>
1963 V5 N5 P177-184 65-12136 <*>
1964 V6 P301-307 4299 <BISI>
1964 V6 P308-313 4531 <BISI>
1964 V6 P418-425 4455 <BISI>
1964 V6 N2 P37-42 6245 <HB>
1964 V6 N2 P56-62 6246 <HB>
1964 V6 N3 P100-105 6239 <HB>
1964 V6 N6 P196-200 6327 <HB>
1964 V6 N6 P201-204 6328 <HB>
1964 V6 N8 P261-265 4390 <BISI>
1964 V6 N8 P279-282 4384 <BISI>
1964 V6 N8 P283-285 3936 <BISI>
1964 V6 N9 P320-322 6390 <HB>
1965 V7 P361-365 4853 <BISI>
1965 V7 N7 P237-242 4829 <BISI>
1965 V7 N7 P243-250 6650-I <HB>
1965 V7 N8 P289-295 6650-II <HB>
1965 V7 N10 P379-383 6760 <HB>

MATERIALY BUDOWLANE
1956 V11 N1 P9-12 PB 141 199T <=>
1957 V12 N3 P65-72 59-11015 <=*>
1957 V12 N5 P150-152 59-11054 <=*>
1957 V12 N10 P289-298 PB 141 242T <=>
1957 V12 N12 P366-371 59-11138 <=*>
1958 V13 N2 P49-54 59-11128 <=*>

MATERIALY DESYATCGO VSESOYUZNOGO SOVESHCHANIYA PO SPEKTROSKOPII, LVOV. MOLEKULARNAYA SPEKTROSKOPIYA
1957 N1 P102-106 C-3435 <NRCC>
1957 N1 P107-111 C-3438 <NRCC>

MATERIALY PO EVOLYUTSIONNOI FIZIOLOGII
1956 V1 N1 P213-218 59-13877 <=> 0
1956 V1 N1 P219-230 59-13878 <=> 0
1956 V1 N1 P246-252 59-13879 <=>
1956 V1 N1 P333-348 59-13880 <=> 0
1957 V2 P86-101 59-13892 <=> 0
1957 V2 N2 P151-159 59-13893 <=>
1957 V2 N2 P160-171 59-13894 <=>
1957 V2 N2 P172-180 59-13895 <=>
1957 V2 N2 P186-195 59-13896 <=>
1960 V4 P77-82 61-28757 <*=>

MATERIALY PO GEOLOGII I POLEZNYM ISKOPAEMYM TSENTRALNYKH RAINOV EVROPEISKOI CHASTI SSR
1962 N5 P21-24 IGR V6 N7 P1174 <AGI>

MATERIALY PO GEOLOGII TSENTRALNOGO KAZAKHSTANA. MOSCOU
1960 V1 P9-26 63-19586 <=*>

MATERIALY DLYA IZUCHENIYA ESTESTVENNYKH PROIZVODITELNYKH SIL RCSSI, IZDAVAEMYYA KOMISSIEI PRI ROSSIISKOI AKADEMII NAUK
1926 N56 P14-20 RT-1296 <*>
1926 N56 P20-23 RT-1289 <*>
1926 N56 P23-33 RT-1293 <*>
1926 N56 P43-47 RT-1295 <*>
1926 N56 P47-49 RT-1297 <*>

MATERIALY KOMISSII EKSPEDITSIONNYKH ISSLEDOVANII
1929 V7 P1-28 62-15520 <=*>
1929 V7 P33-71 62-15520 <=*>
1930 V29 P181-208 62-25427 <=*>

MATERIALY PO KOMPLEKSNOMU IZUCHENIYA BELOGO MORYA
1957 V1 P44-73 63-15217 <=*> 0
1957 V1 P74-89 63-15228 <=*> 0
1957 V1 P90-104 63-15227 <=*> 0

MATERIALY PO LABORATORNYM ISSLEDOVANIYAM MERZLYKH GRUNTOV
1954 V2 P176-192 C-3476 <NRCC>
60-00706 <*>
1957 V3 P142-148 61-19831 <*=>
64-19047 <=*$>

MATERIALY PO MATEMATICHESKOI LINGVISTIKE I MASHINNOMU PEREVODU. (TITLE VARIES)
1963 V2 P2-195 64-41468 <=$>

MATERIALY K OSNOVAM PALEONTOLOGII
1958 N2 P57-67 61-28173 <*=> 0
1959 V3 P104-116 63-19689 <=*>

MATERIALY K OSNOVAM UCHENIYA O MERZLYKH ZONAKH ZEMNOI KORY
1955 V2 ENTIRE ISSUE C-3956 <NRCC>
1956 V3 P1-229 C-4543 <NRCC>

MATERIALY K POZNANIYU FAUNY I FLORY SSSR. OTDEL BOTANICHESKII
1947 V3 P110-145 61-23143 <=*> 0

MATERIALY K POZNANIYU FAUNY I FLORY SSSR. NOVAYA SERIYA. OTDEL ZOOLOGICHESKIE
1946 V5 P1-152 <BI>
64-20153 <*>

1950 V30 P30-43 R-506 <*>
1950 V30 P44-73 R-505 <*>
1950 V30 P85-105 R-504 <*>
1950 V30 P106-111 R-507 <*>
1950 V30 P166-187 R-503 <*>

MATERIALY PO TOKSIKOLOGII RADIOAKTIVNYKH VESHCHESTV
1964 V4 P3-9 64-51813 <=$>
1964 V4 P118 64-51813 <=$>

MATERIALY. TSENTRALNOGO NAUCHNO-ISSLEDOVATELSKOGO GEOLOGO-RAZVEDOCHNOGO INSTITUTA
1937 N3 P1-10 RT-55 <*>

MATERIALY VSESOYUZNOGO GEOLOGICHESKOGO INSTITUTA. LENINGRAD. GEOFIZIKA
1938 V7 P1-26 64-19751 <=$>
1941 N9/10 P3-55 RT-171 <*>

MATERIALY. VSESOYUZNOGO GEOLOGICHESKOGO INSTITUTA, LENINGRAD. NOVAYA SERIYA
1960 N40 P73-87 CSIR-TR.572 <CSSA>
1960 N40 P89-99 CSIR-TR.564 <CSSA>
1961 N46 P93-110 66-14362 <*$>

MATERIALY. VSESOYUZNOGO GEOLOGICHESKOGO INSTITUTA, LENINGRAD. NOVAYA SERIIA. GEOLOGII I POLEZNYM ISKOPANEMYM
1956 N8 P7-41 61-19218 <=*>

MATERIALY VSESOYUZNOGO NAUCHNO-ISSLEDOVATELSKOGO GEOLOGICHESKOGO INSTITUTA
1956 N12 P9-21 49P61R <ATS>

MATERIALY (I.E. VTORAGO) VSESOYUZNOGO SOVESHCHANIYA SPEKTROSKOPISTOV ANALITIKOV TSVETNOI METALLURGII
1955 P9-18 60-16927 <=*>

MATERIE PLASTICHE
1939 V6 P255-260 2073 <*>
1950 V16 N1 P215-219 57-2843 <*>
1950 V16 N5 P164-170 57-907 <*>
1956 P1023-1028 58-574 <*>
1956 N12 P1010-1022 <ATS>
1956 V21 P773-780 94K25I <ATS>
1956 V22 P361-370 69K23I <ATS>
1956 V22 P1010-1022 3247 <*>
1956 V22 N4 P288- 59-17314 <*> O
1956 V22 N5 P361-367 62-14305 <*> O
1956 V22 N12 P1010-1022 3435 <*>
57-3150 <*>
1957 V23 P247-250 58-1842 <*>
1957 V23 N2 P107-113 61-16732 <*> O
1958 N9 P839-846 62-10006 <*> O
1958 V24 P798-811 11901-B <K-H>
1958 V24 N5 P420-422 63-14103 <*> O
1958 V24 N9 P789-811 65-14013 <*> O
1958 V24 N12 P1081-1095 61-16726 <*>
1958 V24 N12 P1109-1116 60-14313 <*>
1959 V25 P21-33 11901-C <K-H>
1959 V25 P95-102 11901-D <K-H>
1959 V25 P231-247 NP-TR-474 <*>
1959 V25 P302-314 11901-G <K-H>
1959 V25 P357-364 54M47I <ATS>
1959 V25 P707-714 60-18782 <*> O
1959 V25 P729-736 (CHEM)-269 <CTS> O
1959 V25 P785-790 60-18783 <+*>
1959 V25 P1069-1076 11901-E <K-H>
1959 V25 N1 P21-33 65-14014 <*> O
1959 V25 N2 P95-102 65-14018 <*>
1959 V25 N4 P302-314 65-14019 <*>
1959 V25 N12 P1069-1076 65-14015 <*>
1960 V26 P723-730 74N521 <ATS>
1960 V26 N8 P723-730 62-14277 <*>
1960 V26 N12 P1140-1147 62-14306 <*>
1961 V27 N4 P321-326 62-14423 <*>
1961 V27 N8 P801-809 11901-F <K-H>
65-14016 <*>
1961 V27 N12 P1136-1145 47Q70I <ATS>
1961 V27 N12 P1156-1164 66-10572 <*>
1961 V27 N12 P1165-1170 65-18167 <*>
1962 V28 P466-471 63-18392 <*>
1962 V28 N1 P10-15 65-18184 <*>
1962 V28 N2 P138-146 65-13317 <*>
1962 V28 N3 P213-216 65-18172 <*>
1962 V28 N5 P563-569 64-10421 <*>
1962 V28 N11 P1345-1354 65-18185 <*>
1963 P56-64 RCT V36 N4 P1119 <RCT>
1963 V29 N7 P984-990 65-12061 <*>

MATERIE PLASTICHE ED ELASTOMERI
1963 V29 N11 P1375-1380 65-17096 <*>
1964 P1173-1181 898 <TC>
1964 V30 N2 P190-194 668 <TC>
1964 V30 N3 P278-282 66-10049 <*>
1964 V30 N4 P317-328 66-10594 <*>
8690 <IICH>
1964 V30 N4 P378-385 66-10595 <*>
1964 V30 N4 P386-389 1153 <TC>
1964 V30 N7 P643-653 RAPRA-1240 <RAP>
1964 V30 N7 P700-709 65-13815 <*>
1964 V30 N10 P969-976 66-10569 <*>
1964 V30 N10 P977-982 66-10567 <*>
1965 V31 N2 P111-124 66-12774 <*> O
1965 V31 N3 P268-279 66-10571 <*>
1965 V31 N5 P509-513 14S88I <ATS>
1965 V31 N8 P825-835 66-10878 <*>
66-13274 <*>
1965 V31 N10 P1045-1050 09T94I <ATS>

MATHEMATICS. JAPAN
SEE SUGAKU

MATHEMATISCHE ANNALEN
1869 V1 P225-252 65-10721 <*>
1904 V59 P383-397 66-12697 <*> O
1908 V66 P398-415 66-12764 <*> O
1909 V67 P355-386 1574 <*>
65-11273 <*>
1914 V75 P497-544 64-15896 <=*$>
1919 V79 P157-179 19K20G <ATS>
1931 V104 P415-458 62-10151 <*>
63-18360 <*>
1932 V106 P722-754 63-01049 <*>
1932 V107 P282-312 T-2318 <INSD>
1936 V112 P630-651 63-01539 <*>
1937 V114 P617-621 59-17244 <*>
1951 V123 N1 P96-124 65-10871 <*>
1952 V124 N3 P219-234 66-13943 <*>
1956 V131 N5 P411-428 65-12116 <*>
1958 V136 N5 P430-441 62-10143 <*>
1963 V150 N4 P317-324 65-12112 <*>

MATHEMATISCHE ZEITSCHRIFT
1944 V49 N4/5 P593- 1540 <*>
1955 V63 P39-52 T2164 <INSD>

MECANIQUE
1937 V21 N271 P69-83 59-17015 <*> O

MECANIZAREA SI ELECTRIFICAREA AGRICULTURII
1966 V11 N3 P6-13 66-32204 <=$>

MECHANIC--COLD PROCESSING
SEE CHI HSIEH KUNG JEN--LENG CHIA KUNG

MECHANIC--HEAT TREATMENT
SEE CHI HSIEH KUNG JEN--JE CHIA KUNG

MECHANIK
1946 V19 N5/6 P165-172 62-18156 <*> O
1949 V22 N1/2 P17-22 58-1119 <*>
1958 V31 N8/9 P406-411 60-18110 <*>
9905 <K-H>
1962 V35 N12 P670-672 63-21374 <=> O

1963 V36 N9 P437-443 NEL-TRANS- 1729 <NLL>
1964 V37 N7 P365-368 64-51051 <=>
1965 N9 P542-546 65-33656 <=$>
1965 N9 P555-558 65-33656 <=$>
1965 V38 N12 P687-689 66-31114 <=$>

MECHANIZACJA ROLNICTWA
1955 V2 N7 P15-16 60-21537 <=>
1956 V3 N10 P15-17 60-21545 <=>
1957 V4 N9 P11-16 60-21247 <=>
1958 V5 N7 P14-16 60-21238 <=>
1958 V5 N7 P28-29 60-21232 <=>
1964 V11 N13 P2-3 64-51350 <=>
1964 V11 N15 P2-3 64-51350 <=>
1967 N2 P15-17 67-30883 <=$>

MEDDELANDEN FRAN FINSKA KEMISTSAMFUNDET
1940 V49 P18-41 I-809 <*>
1950 V59 P40-43 64-14963 <*>
1954 V63 P22-41 TT-785 <NRCC>
1961 V70 N1 P33-39 17P60S <ATS>

MEDDELANDER. JERNKONTORETS TEKNISKA RAD
1954 V18 N203 P263-290 1643 <BISI>
62-14194 <*>
1958 N248 P155-164 62-16886 <*>
1959 N255 P405-430 62-16885 <*>
1960 N260 P711-722 1957 <BISI>
1961 N261 P723-732 2270 <BISI>

MEDDELANDEN FRAN STATENS CENTRALA FROKONTROLLAN-STALT
1963 V38 P41-44 65-12151 <*>

MEDDELANDEN FRAN STATENS FORSKNINGSANSTALT FOR LANTMANNABYGGNADER
1945 N5 P53-76 I-144 <*>

MEDDELANDEN. STATENS KOMMITTE FOR BYGGNADSFORSKNING
1945 N3 P17-37 64-18079 <*> 0
1945 N3 P38-85 64-18080 <*> 0

MEDDELANDEN FRAN STATENS PROVNINGSANSTALT
1942 N87 P3-35 61-18487 <*> 0
1957 N122 P11-32 61-10136 <*> 0

MEDDELANDEN FRAN STATENS SKOGSFORSKNINGSINSTITUT
1953 V30 P1-30 57-87 <*>

MEDDELANDEN FRAN STATENS UNDERSOKNINGS- OCH FORSOGSANSTALT FOR SOTVATTENSFISKET
1937 V13 N19 P19 C-5366 <NRC>
1943 N21 P1-48 FRBC TRANS SER 126 <NRCC>

MEDDELANDEN. STATENS VAGINSTITUT
1944 N69 P7-14 61-10137 <*>
1944 N69 P23-37 61-10137 <*>

MEDDELANDEN FRAN SVENSKA TEXTILFORSKNINGSINSTI-TUTET
1953 N33 P1-9 61-14005 <*> 0

MEDDELANDEN. SVENSKA TRAFORSKNINGSINSTITUTETS TRATEKNISKA AVDELNING
1952 N23 ENTIRE ISSUE I-329 <*>
1955 V67B P8- 1526 <*>

MEDDELELSER OM GRONLAND, AF KOMMISSIONEN FOR LEDELSEN AF DE GEOLOGISKE OG GEOGRAFISKE UNDERSOGELSER I GRONLAND
1959 V153 N3 P1-127 C-4482 <NRC>

MEDECINE ET HYGIENE. GENEVA
1953 V10 P206-207 AL-618 <*>
1953 V11 P198 AL-252 <*>
1954 V12 P429-431 3400-A <K-H>
1955 N13 P105-107 59-15679 <*>
1955 V13 P25- 2144 <*>
1956 V14 P342-343 58-2615 <*>
1957 V15 P159-161 65-18198 <*>
1959 V17 P478- 63-18509 <*> 0
1959 V17 P621-623 60-18768 <*>
1960 V18 P355 10413 <KH>
1960 V18 P505-507 62-10085 <*>
1960 V18 N5 P355 63-16891 <*>
1961 N506 P466 62-00276 <*>
1965 V23 P435 65-14758 <*>
1965 V23 N675 P1-15 17S86F <ATS>

MEDECINE INFANTILE
1963 V70 N9 P545-547 65-14547 <*>

MEDEDELINGEN VAN HET ALGEMEEN PROEFSTATION VOOR DEN LANDBOUW
1943 N82 P92 57-2896 <*>

MEDEDELINGEN. DIRECTEUR VAN DE TUINBOUW
1952 V15 N9 P811-815 2818-B <K-H>

MEDEDELINGEN VAN DE GEOLOGISCHE STICHTING
1948 N3 P75-80 47K27DJ <ATS>
1955 N8 P77-86 CSIRO-3129 <CSIR>

MEDEDELINGEN. INSTITUUT VOOR BIOLOGISCH EN SCHEIKUNDIG ONDERZOEK VAN LANDBOUWGEWASSEN
1960 N107 P139-141 62-14088 <*>

MEDEDELINGEN VAN HET INSTITUUT VOOR PLANTENZIEKTEN
1933 N81 P84 57-2851 <*>
1935 V84 P1-79 57-2839 <*>

MEDEDELINGEN VAN DE K. VLAAMSCHE ACADEMIE VOOR WETENSCHAPEN, LETTEREN EN SCHOONE KUNSTEN VAN BELGIE. KLASSE DER WETENSCHAPPEN
1957 V19 N5 P1-24 AEC-TR-3847 <*>
1958 V20 N8 P3-19 65-00300 <*>

MEDEDELINGEN VAN HET LABORATORIUM VOOR HOUTTECHNOLOGIE. RIJKSLANDBOUWHOGESCHOOL. GENT
1962 N16 P1-20 64-00253 <*>
1962 V15 P1755-1793 63-01075 <*>
1962 V17 P479-497 63-01235 <*>

MEDEDELINGEN VAN DE LANDBOUWHOOGESCHOOL EN DER OPZOEKINGSSTATIONS VAN DE STAAT TE GENT
1962 V27 N4 P1559-1571 64-14720 <*>

MEDEDELINGEN. PROEFSTATION VOOR DE AKKER- EN WEIDEBOUW. WAGENINGEN
1960 V39 P1-35 62-00601 <*>

MEDEDELINGEN DER VLAAMSCHE CHEMISCHE VEREENIGING
1958 V20 P102-113 59-17986 <*>
1958 V20 N1 P1-11 184 <TC>

MEDICAL JOURNAL OF OSAKA UNIVERSITY
SEE OSAKA DAIGAKU IGAKU ZASSHI

MEDICHNII ZHURNAL
1941 V11 N1 P169-174 60-19598 <=*$> P
1945 V14 P283-285 61-28021 <=*>
1950 V20 N4 P67-73 63-15162 <=*>
1953 V23 N3 P7-17 R-77 <*>
1953 V23 N3 P18-22 R-135 <*>
1953 V23 N3 P35-42 R-164 <*>
1953 V23 N3 P43-46 R-136 <*>
1953 V23 N3 P76-82 R-81 <*>
1954 V24 N1 P54-55 RT-3804 <*>

MEDICINA. BUENOS AIRES
1946 V6 N4 P389-404 59-20065 <*>
1964 V24 N3 P151-153 66-12609 <*>

MEDICINA. MADRID
1950 V18 P212-221 II-113 <*>
1960 V28 N1 P26-34 61-00189 <*>

MEDICINA. MEXICO
1947 V15 N2 P147-150 2591 <*>

1947 V27 P250-252 59-17882 <*>
1952 V32 P278-282 59-15741 <*>
1959 V39 N821 P234-239 61-00164 <*>
1961 V41 P25-38 63-16522 <*>
1961 V41 P190-191 63-16523 <*>
1961 V41 N41 P49-52 63-16538 <*>
1964 V44 P399-406 66-13410 <*> 0

MEDICINA. PARMA
1954 V4 P321-350 II-753 <*>
1955 V5 N3 P245-284 5436-B <K-H>

MEDICINA, CIRURGIA, FARMAGIA
1961 N291 P1-6 62-16119 <*> 0

MEDICINA CLINICA. BARCELONA
1947 V8 P104-107 II-261 <*>
1952 V18 P409-411 3024 <*>
1963 V41 N4 P302-306 65-10973 <*> 0

MEDICINA CONTEMPORANEA
1958 V76 N3 P101-127 61-00154 <*>

MEDICINA ESPANOLA
1956 V36 P91-101 59-12419 <**> 0

MEDICINA EXPERIMENTALIS
1960 V2 P132-137 61-20910 <*>
1960 V2 N2/4 P110-122 62-00635 <*>
1960 V2 N2/4 P141-147 62-00804 <*>
1960 V2 N2/4 P192-198 62-00480 <*>
1960 V3 N1 P45-52 65-14297 <*> 0
1962 V6 N2 P113-117 63-10594 <*>
1962 V7 P201-204 64-10480 <*>
1962 V7 P344-350 63-20642 <*>
1963 V8 N416 P228-236 64-16852 <*>
1963 V9 N5 P341-348 64-18983 <*> 0

MEDICINA INTERNA
1958 V10 P137-146 65-00095 <*>
1961 V13 P347-351 11548-E <KH>
64-10639 <*> 0
1961 V13 N1 P97-104 62-19208 <=*>
1961 V13 N2 P161-166 62-23263 <*=>
1961 V13 N2 P297-302 62-19575 <=*>
1966 V18 N3 P283-291 66-32153 <=$>

MEDICINA DEL LAVORO
1933 V24 P474-480 62-00674 <*>
1935 V26 P297-303 II-192 <*>
1935 V26 P376-378 II-191 <*>
1948 V39 P152-157 64-16202 <*>
1951 V42 N2 P49-68 MF-6 <*>
1951 V42 N11 P315-325 57-147 <*>
1954 V45 N10 P544-548 1G61 <ATS>
57-2695 <*>
1956 V47 P240-262 66-10070 <*>
1958 V49 N8/9 P504-511 62-01212 <*>
1959 V50 N3 P193-201 64-12332 <*>
1961 V52 N11 P653-657 64-00071 <*>
1964 V55 N3 P188-196 610 <TC>

MEDICINA Y SEGURIDAD DEL TRABAJO
1955 V3 P75-81 58-85 <*>
59-10849 <*> 0
1960 V8 N32 P36-42 64-14613 <*>

MEDICINA SPERIMENTALE. TORINO
1940 V6 P515-520 57-1669 <*>
1956 V30 P1-28 66-13218 <*>

MEDICINE. JAPAN
SEE SOGO IGAKU

MEDICINE AND BIOLOGY. JAPAN
SEE IGAKU TO SEIBUTSUGAKU

MEDICINISCHE JAHRBUCHER DES K. K. OESTERREICHISCHEN STAATES
1820 V6 P79-125 61-14679 <*>

MEDICINSKI GLASNIK
1961 V15 N3 P117-121 62-11092 <=*>
1961 V15 N4 P185-188 62-19559 <=*>
1961 V15 N6 P238-243 62-19580 <*=> P
1961 V15 N6 P260-263 62-19560 <=*>

MEDICINSKI PREGLED
1961 V14 N3 P135 62-19787 <=*>

MEDICO. MEXICO
1954 V5 P9-16 2489 <*>

MEDITSINSKAYA GAZETA
1963 V26 P3 63-31505 <=>
1964 V27 P1-4 65-30452 <=$>
65-30540 <=$>
1964 V27 P1 65-31721 <=$>
1964 V27 P3 65-30662 <=$>
1964 V27 09/08 P2 65-30338 <=$>
1964 V27 09/11 P1 65-30338 <=$>
1964 V27 09/11 P3 65-30338 <=$>
1964 V27 09/20 P2 65-31224 <=$>
1964 V27 10/06 P1 65-30338 <=$>
1964 V27 10/06 P2 65-30338 <=$>
1964 V27 12/01 P2 65-31736 <=$>
1964 V27 9/29 P2 65-31224 <=$>
1964 V27 N75 9/18 P3 64-51542 <=$>
1965 P3 07/27 AD-629 411 <=$>
1965 P3 09/21 AD-629 422 <=$>
1965 P3 09/07 AD-630 995 <=$>
1965 V28 P1-2 65-30763 <=$>
1965 V28 P2 65-30950 <=$>
65-31667 <=$>
1965 V28 P3 65-30778 <=$>
65-30796 <=$>
65-30828 <=*$>
65-30928 <=$>
65-31594 <*>
65-31959 <=$*>
65-32222 <=*$>
65-32463 <=$*>
65-32981 <=*$>
66-30365 <=$>
66-30969 <=$>
66-30977 <=$>
1965 V28 02/12 P1 65-31240 <=$>
1965 V28 02/19 P3 65-31240 <=$>
1965 V28 02/19 P2 65-31736 <=$>
1965 V28 02/26 P1 65-30919 <=$>
1965 V28 03/12 P3 65-31736 <=$>
1965 V28 03/25 P2 65-63335 <=*$>
1965 V28 05/28 P1 65-31427 <=$>
1965 V28 06/25 P3 65-31807 <=$>
1965 V28 06/29 P4 65-31807 <=$>
1965 V28 09/28 P4 65-33154 <=$>
1965 V28 10/26 P1 66-30888 <=$>
1965 V28 11/02 P3 66-30888 <=$>
1965 V28 11/05 P2 66-30888 <=$>
1965 V28 11/30 P2 66-30450 <=$>
1965 V28 12/07 P2 66-30888 <=$>
1966 P2 05/13 66-32639 <=$>
1966 P2-3 03/04 66-31475 <=$>
1966 12/05 P3 67-30590 <=> 0
1966 V29 P4 66-30837 <=$>
66-31757 <=$>
1966 V29 N55 P3 66-33952 <=$>

MEDITSINSKAYA PARAZITOLOGIYA I PARAZITARNYE BOLEZNI
1933 V2 P341-363 65-63235 <=>
1934 V3 N6 P460-479 RT-153 <*>
1936 V5 N5 P657-673 65-63268 <=>
1936 V5 N5 P937-941 64-23627 <=$>
1936 V5 N6 P1925-1926 63-15752 <=*>
1939 V8 N1 P89-108 63-24053 <=*$>
1939 V8 N1 P137-140 64-15698 <=*$>
1939 V8 N2 P191-206 61-28114 <=*$> 0
1940 V9 P44-53 R-1050 <*>

Citation	Translation
1940 V9 P583-588	60-19276 <+*> P
1940 V9 N3 P291-294	61-28116 <*=> 0
1940 V9 N3 P295-297	65-62999 <=>
1940 V9 N1/2 P44-53	60-18165 <+*>
1940 V9 N1/2 P93-105	61-00445 <*>
	64-15699 <=*$>
1941 V10 N3/4 P366-369	64-15832 <=*$>
1942 V11 N3 P11-14	64-23626 <=$>
1942 V11 N3 P44-52	61-19617 <*=>
1942 V11 N3 P94-99	65-63259 <=>
1942 V11 N3 P130	65-63457 <=>
1942 V11 N4 P112-117	65-60041 <=$>
1942 V11 N4 P125-126	65-63261 <=>
1942 V11 N1/2 P95-96	64-23637 <=$>
1943 V12 P40-48	65-63456 <=>
1943 V12 N1 P38-41	65-63452 <=>
1943 V12 N1 P41-44	65-60081 <=$>
1943 V12 N2 P3-8	65-63444 <=>
1943 V12 N2 P9-13	65-60055 <=$>
1943 V12 N3 P3-14	62-23771 <*=>
1943 V12 N3 P42-54	61-20411 <*=> P
	65-63234 <=> 0
1943 V12 N3 P83-84	RT-2102 <*>
1943 V12 N5 P33-36	RT-579 <*>
1944 V13 N5 P11-16	65-63267 <=>
1944 V13 N5 P16-23	65-63174 <=$> 0
1944 V13 N5 P89-90	63-00735 <*>
	63-19594 <=*>
1945 V14 N1 P12-18	59-15841 <+*> 0
1945 V14 N1 P18-24	61-00454 <*>
	64-15615 <=*$>
1945 V14 N2 P66-67	RT-1307 <*>
	65-63253 <=>
1945 V14 N4 P45-48	61-28110 <*=>
1945 V14 N6 P39-44	65-63271 <=$>
1945 V14 N6 P55-60	RT-263 <*>
1945 V14 N6 P60-66	RT-325 <*>
	64-15648 <=*$> 0
1945 V14 N6 P66-68	RT-262 <*>
1946 V15 N2 P68-75	RT-2116 <*>
1946 V15 N2 P76-83	RT-2117 <*>
1946 V15 N2 P84-85	RT-2115 <*>
	61-00448 <*>
	64-15607 <=*$>
1946 V15 N3 P58-62	61-00453 <*>
	64-15613 <=*$> 0
1946 V15 N4 P40-44	61-28121 <*=>
1946 V15 N4 P94-99	61-28221 <*=>
1946 V15 N5 P25-27	61-28113 <*=>
1946 V15 N6 P59-63	R-1055 <*>
	RT-373 <*>
	64-15614 <*>
	64-15614 <=*$>
1953 N3 P260-262	RT-2666 <*>
1953 N3 P279-280	RT-3725 <*>
1953 V22 P75-78	R-4096 <*>
1953 V22 N6 P559-560	63-24055 <=*$>
1954 N1 P43-45	RT-2998 <*>
1954 N1 P57-58	RT-2996 <*>
1954 N1 P60-61	RT-2995 <*>
1954 N1 P61-66	RT-2994 <*>
1954 N1 P71-77	RT-2990 <*>
1954 N1 P77-79	RT-2987 <*>
1954 N1 P79-83	RT-2988 <*>
1954 N1 P91-92	RT-2675 <*>
1954 N3 P281-284	RT-2225 <*>
1954 V23 N3 P270-271	59-15818 <+*>
1955 V24 P61-66	R-4574 <*>
	3017 <NLL>
1955 V24 N1 P71-72	R-143 <*>
	R-298 <*>
	64-19068 <=*$>
1956 V25 N1 P3-7	R-372 <*>
1956 V25 N1 P17-27	R-140 <*>
1956 V25 N1 P28-32	R-144 <*>
1956 V25 N3 P263-266	R-1860 <*>
1956 V25 N4 P318-323	59-18841 <+*>
1957 V26 P1-37	R-4128 <*>
1957 V26 P140-152	62-11162 <=>
	62-15072 <=*>
1957 V26 P643-650	61-27743 <*=>
1957 V26 N1 P56	59-15817 <+*>
1957 V26 N2 P167-172	R-5308 <*>
1957 V26 N3 P347-350	60-00626 <*>
1957 V26 N4 P458-463	61-19194 <+*>
1958 V27 P194-199	59-12740 <+*>
1958 V27 N1 P67-68	62-15636 <=*>
1958 V27 N6 P693-695	59-13923 <=> 0
1959 N1 P117	60-31155 <=>
1959 V28 P568-571	AD-638 582 <=$>
1959 V28 N3 P287-294	60-17968 <+*>
1959 V28 N3 P364-373	60-11387 <=>
1960 V29 P202-207	66-61173 <=*$>
1960 V29 N1 P66-72	61-28635 <*=>
1960 V29 N1 P105	61-27443 <*=> C
1960 V29 N2 P179-183	62-00440 <*>
	62-23688 <=*>
1960 V29 N3 P287-288	64-00285 <*>
	64-15710 <=*$>
	65-63987 <=$>
1960 V29 N4 P434-440	61-21546 <=>
1960 V29 N4 P466-474	61-21500 <=>
1960 V29 N5 P553-558	63-23131 <=*>
1961 V30 P6-11	AD-638 583 <=$>
1961 V30 N1 P3-5	61-21552 <=>
1961 V30 N1 P84-86	61-21552 <=>
1961 V30 N4 P389-394	62-19780 <=*>
1961 V30 N4 P475-476	63-23178 <=*$>
1961 V30 N5 P621	65-60619 <=$>
1961 V30 N6 P734-737	62-23765 <=*>
1962 V31 N1 P47-55	64-19538 <=$>
1963 V32 N1 P18-29	66-60946 <=*$>
1963 V32 N1 P54-61	66-60947 <=*$> 0
1963 V32 N1 P61-65	64-15641 <=*$> 0
1963 V32 N3 P313-319	64-15449 <=*$>
1963 V32 N5 P515-521	65-63159 <=$>
1963 V32 N5 P549-551	67-51227 <=>
1963 V32 N6 P738-739	AD-618 862 <=$>
1963 V32 N6 P739-740	AD-619 003 <=$>
1964 V33 N1 P25-31	65-63168 <=$>
1964 V33 N1 P47-53	64-21930 <=>
1964 V33 N1 P74-81	64-21899 <=>
1964 V33 N1 P82-86	64-21917 <=>
1964 V33 N2 P136-141	66-61046 <=*$>
1964 V33 N2 P141-144	66-61560 <=*$>
1964 V33 N2 P188-194	64-31452 <=>
1964 V33 N2 P233-234	64-31448 <=>
1964 V33 N3 P278-289	64-41305 <=>
1964 V33 N3 P376-379	64-51509 <=>
1964 V33 N4 P510	65-30727 <=$>
1964 V33 N6 P711-717	65-30697 <=$>
1964 V33 N6 P731-735	65-30698 <=$>
1965 V34 N1 P83-90	65-30936 <=$>
1965 V34 N2 P169-176	65-31439 <=$>
1965 V34 N2 P189-194	65-31430 <=$>
1965 V34 N4 P467-471	65-33068 <=*$>
1966 V35 N1 P107-109	66-32315 <=$>

MEDITSINSKAYA PROMYSHLENNOST SSSR

Citation	Translation
1954 N1 P39-42	64-10565 <=*$>
1956 V10 N1 P11-13	64-16310 <=*$>
1956 V10 N2 P29-31	62-10042 <*=>
1956 V10 N3 P22-27	00Q71R <ATS>
1957 V11 N4 P56-58	C-4469 <NRC>
1957 V11 N7 P23-26	47P66R <ATS>
1957 V11 N10 P13-18	59-11192 <=>
1957 V11 N10 P18-22	59-11193 <=>
1958 N4 P14-17	91K24R <ATS>
1958 N4 P36	92K24R <ATS>
1958 N4 P41-43	93K24R <ATS>
1958 V12 N3 P9-16	48P66R <ATS>
1958 V12 N7 P10-20	63-15462 <=*> 0
1958 V12 N7 P41-46	59-11470 <*> 0
1958 V12 N8 P3-5	59-13442 <=>
1958 V12 N12 P47-49	C-4470 <NRC>
1959 N1 P6-10	59-13579 <=>
1959 V13 N1 P3-5	59-13578 <=>
1959 V13 N1 P17-24	59-13580 <=> 0
1959 V13 N2 P3-10	59-13576 <=>

1959 V13 N2 P52-59 59-13577 <=> 0
1959 V13 N3 P6-13 64-21575 <=>
1959 V13 N3 P44-46 C-4480 <NRC>
1959 V13 N4 P3-5 59-13627 <=>
1959 V13 N4 P29-35 59-13914 <=>
1959 V13 N5 P3-5 59-13807 <=>
1959 V13 N5 P6-14 59-13809 <=>
1959 V13 N5 P48-52 61-15064 <+*> 0
1959 V13 N5 P54-57 59-13810 <=> 0
1959 V13 N6 P3-5 59-13834 <=>
1959 V13 N6 P5-7 59-13844 <=> 0
1959 V13 N6 P8-15 59-13859 <=>
1959 V13 N6 P15-31 45P66R <ATS>
1959 V13 N7 P3-6 59-13995 <=>
1959 V13 N7 P6-8 59-13996 <=>
1959 V13 N11 P55-58 60-11266 <=>
1959 V13 N12 P11-15 61-19724 <=*>
1959 V13 N12 P52-55 60-11460 <=>
1960 V14 N1 P3-65 60-41116 <=>
1960 V14 N3 P56-60 60-11850 <=>
1960 V14 N4 P3-6 60-41004 <=>
1960 V14 N4 P54-57 60-31794 <=>
1960 V14 N6 P3-6 60-41476 <=>
1960 V14 N6 P14-21 60-41417 <=>
1960 V14 N6 P63-65 60-41477 <=>
1960 V14 N10 P3-6 61-21413 <=>
1960 V14 N10 P7-9 61-21412 <=>
1960 V14 N10 P48-49 61-21411 <=>
1960 V14 N10 P50-56 61-21409 <=>
1960 V14 N11 P40-45 61-21662 <=>
1960 V14 N11 P54-57 61-15828 <+*> 0
1961 V15 N1 P33-38 61-21588 <=>
1961 V15 N1 P57-58 61-21566 <=>
1961 V15 N2 P63-64 61-21746 <=>
1961 V15 N3 P6-13 61-23122 <=*>
1961 V15 N3 P13-14 61-23123 <*=>
1961 V15 N3 P24-27 61-23587 <=*>
1961 V15 N3 P39-40 46P66R <ATS>
1961 V15 N3 P51-53 61-28588 <*=>
1961 V15 N3 P59-63 61-27382 <*=>
1961 V15 N5 P3-34 61-27928 <*=>
1961 V15 N5 P14-30 61-27724 <*=>
1961 V15 N6 P39-41 63-31040 <=>
1961 V15 N7 P3-10 62-13716 <*=>
1961 V15 N7 P43-45 62-13717 <*=>
1961 V15 N7 P46-50 62-13672 <=*>
1961 V15 N7 P50-53 62-13721 <*=>
1961 V15 N7 P55-58 62-13720 <*=>
1961 V15 N8 P39-43 62-13853 <=*>
1961 V15 N8 P45-48 62-13853 <=*>
65-63081 <=*$>
1961 V15 N8 P56-60 62-13853 <=*>
1961 V15 N9 P12-16 62-15653 <*=>
1961 V15 N9 P17-20 62-15663 <*=>
1961 V15 N9 P28-33 62-15649 <*=>
1961 V15 N9 P41-44 62-15654 <*=>
1961 V15 N9 P46-59 62-11131 <=>
1961 V15 N10 P56-61 62-15951 <=*>
1961 V15 N11 P3-12 62-23628 <=*>
1961 V15 N12 P19-25 62-19761 <=*>
1962 V16 N1 P3-6 62-23766 <=*>
62-25507 <=*>
1962 V16 N1 P60-63 62-25178 <=*>
1962 V16 N2 P15-19 62-25011 <=>
1962 V16 N2 P53-58 62-24746 <=*>
1962 V16 N2 P62 62-25011 <=>
1962 V16 N3 P55-57 62-24992 <=*>
1962 V16 N4 P7-11 62-32544 <=>
1962 V16 N4 P45-46 62-32544 <=>
1962 V16 N4 P52-57 62-23998 <=*>
1962 V16 N4 P57-59 62-23984 <=*>
1962 V16 N4 P60-63 62-32544 <=>
1962 V16 N6 P34-37 62-33005 <=*>
1962 V16 N6 P48-55 62-32853 <=*>
1962 V16 N7 P3-9 62-33455 <=>
1962 V16 N7 P34-39 62-33455 <=>
1962 V16 N7 P55-56 62-33455 <=>
1962 V16 N7 P58-62 62-33455 <=>
1962 V16 N8 P11-18 65-60481 <=$>
66-10218 <*> 0
1962 V16 N8 P18-25 62-33439 <=*>
64-13205 <=*$>
1962 V16 N8 P55-57 62-33439 <=*>
65-63079 <=$*>
1962 V16 N10 P39-42 75Q72R <ATS>
1963 V17 N2 P43-49 63-21673 <=>
1963 V17 N3 P23-31 63-31526 <=>
1963 V17 N4 P3-6 63-31215 <=>
1963 V17 N7 P3-8 63-31888 <=>
1963 V17 N7 P31-33 63-31887 <=>
1963 V17 N7 P37-38 <ES>
1963 V17 N8 P3-5 63-31930 <=>
63-31981 <=>
1963 V17 N8 P5-13 63-31953 <=>
1963 V17 N9 P8-15 63-31646 <=>
1963 V17 N9 P53-58 63-41213 <=> 0
1963 V17 N10 P6-14 64-21169 <=>
1963 V17 N11 P5-17 64-21529 <=>
1963 V17 N12 P3-8 64-21558 <=>
1964 V18 N1 P47-48 AD-614 390 <=$>
1964 V18 N2 P46-55 64-31805 <=>
1964 V18 N3 P3-7 64-31522 <=>
1964 V18 N3 P24-26 64-31522 <=>
1964 V18 N4 P6-15 64-31907 <=>
1964 V18 N4 P63-64 64-31724 <=>
1964 V18 N5 P3-10 64-41838 <=$>
1964 V18 N5 P46-51 65-31459 <=$>
1964 V18 N5 P51-53 AD-637 389 <=$>
1964 V18 N5 P55-56 64-41838 <=$>
1964 V18 N5 P64 64-41838 <=$>
1964 V18 N6 P14-20 64-41450 <=$>
1964 V18 N6 P20-22 64-41451 <=$>
1964 V18 N7 P3-9 65-31507 <=$>
1964 V18 N8 P3-11 64-51279 <=>
1964 V18 N8 P18-21 64-51279 <=>
1964 V18 N8 P29-30 64-51279 <=>
1964 V18 N8 P63-64 64-51279 <=>
1964 V18 N9 P3-5 65-30025 <=>
1964 V18 N9 P6-20 65-31660 <=$>
1964 V18 N9 P46-51 65-30025 <=>
1964 V18 N9 P54-55 65-30025 <=>
1964 V18 N11 P3-18 65-30280 <=$>
1964 V18 N11 P24-35 65-30280 <=$>
1964 V18 N11 P60-63 65-30280 <=$>
1964 V18 N12 P3-5 65-30929 <=$>
1964 V18 N12 P18-22 65-30929 <=$>
1964 V18 N12 P22-26 96S88R <ATS>
1964 V18 N12 P48-49 65-30929 <=$>
1964 V18 N12 P52-53 65-30929 <=$>
1965 N4 P3-9 NLLTB V7 N12 P1031 <NLL>
1965 N10 P48-52 NLLTB V8 N3 P208 <NLL>
1965 N11 P58-59 NLLTB V8 N4 P288 <NLL>
1965 V19 P48-54 65-30544 <=$>
1965 V19 N1 P3-6 65-30544 <=$>
1965 V19 N1 P7-16 65-31101 <=$>
1965 V19 N2 P17-23 65-30859 <=$>
1965 V19 N2 P58-63 65-30859 <=$>
1965 V19 N3 P3-6 65-30958 <=$>
1965 V19 N3 P7-11 65-31042 <=$>
1965 V19 N3 P12-13 93S84R <ATS>
1965 V19 N3 P58-61 65-30967 <=$>
1965 V19 N3 P63-64 65-30958 <=$>
1965 V19 N4 P28-30 65-32259 <=$>
1965 V19 N4 P40-44 65-32259 <=*$>
1965 V19 N4 P59-64 65-32259 <=*$>
1965 V19 N5 P58-63 65-32260 <=*$>
1965 V19 N7 P3-7 65-32341 <=$>
1965 V19 N7 P21-26 65-32341 <=$>
1965 V19 N7 P48-53 65-32341 <=$>
1965 V19 N8 P3-5 65-33483 <=*$>
1965 V19 N8 P49-53 65-33484 <=*$>
1965 V19 N9 P51-54 66-30290 <=*$>
1965 V19 N9 P63 66-30290 <=*$>
1965 V19 N10 P18-26 66-30344 <=*$>
1965 V19 N11 P62-64 66-30661 <=$>
1965 V19 N12 P3-7 66-30629 <=$>

*MEDITSINSKAYA RADIOLOGIYA

1956 N1 P27-35 R-4895 <*>
1956 N1 P43-49 R-1746 <*>

Citation	Translation
1956 N1 P80-87	R-4533 <*>
1956 N2 P29-32	R-1757 <*>
1956 V1 P66-68	R-4025 <*>
1956 V1 N1 P27-35	507TT <CCT>
1956 V1 N1 P49-59	R-1748 <*>
1956 V1 N2 P41-45	59-13708 <=>
1956 V1 N2 P46-51	61-15696 <+*>
1956 V1 N3 P42-52	59-13709 <=> 0
1956 V1 N3 P80-85	AEC-TR-3580 <+*> 0
	60-13681 <+*>
1956 V1 N4 P14-21	62-15298 <=*>
1956 V1 N5 P3-10	62-24958 <=*>
1956 V1 N5 P30-40	62-24959 <=*>
1956 V1 N6 P6-13	59-13710 <=>
1956 V1 N6 P36-40	60-13682 <+*>
1956 V1 N6 P61-65	509TT <CCT>
1956 V1 N6 P65-69	61-13170 <*+> 0
1957 N2 P13-18	R-4531 <*>
1957 N6 P3-12	PB 141 330T <=>
1957 N6 P19-25	PB 141 332T <=> 0
1957 N6 P26-36	PB 141 333T <=> 0
1957 N6 P61-64	PB 141 339T <=> 0
1957 N6 P69-72	59-11129 <=>
1957 V2 N1 P11-22	59-13711 <=>
1957 V2 N2 P3-12	62-24622 <=*> 0
1957 V2 N3 P3-8	PB 141 253T <=>
1957 V2 N3 P3-9	59-12788 <+*> 0
1957 V2 N3 P8-13	PB 141 254T <=>
1957 V2 N3 P19-23	PB 141 255T <=>
1957 V2 N3 P35-40	PB 141 497T <=>
1957 V2 N3 P40-47	59-11196 <=>
1957 V2 N3 P47-52	59-11007 <=>
1957 V2 N3 P52-54	PB 141 252T <=>
1957 V2 N3 P83-84	AEC-TR-3954 <+*>
1957 V2 N4 P11-17	AEC-TR-3622 <+*>
1957 V2 N4 P23-30	AEC-TR-3234 <+*>
1957 V2 N4 P38-43	AEC-TR-3624 <+*>
1957 V2 N4 P44-50	AEC-TR-3623 <+*>
1957 V2 N4 P68-74	R-4026 <*>
1957 V2 N5 P80-88	63-23089 <=*$>
1957 V2 N6 P12-18	PB 141 331T <=> 0
1957 V2 N6 P37-41	PB 141 334T <=>
1957 V2 N6 P41-43	PB 141 335T <=>
1957 V2 N6 P44-48	PB 141 336T <=>
1957 V2 N6 P49-55	PB 141 337T <=> 0
1957 V2 N6 P56-60	PB 141 338T <=>
1957 V2 N6 P65-69	PB 141 340T <=>
1958 N2 P42-46	R-5367 <*>
1958 V3 P80-84	R-2739 <*>
1958 V3 N1 P3-76	59-11127 <=> 0
1958 V3 N2 P3-10	59-13343 <=> 0
1958 V3 N2 P23-31	59-13341 <=> 0
1958 V3 N2 P32-36	59-13342 <=>
1958 V3 N2 P37-41	59-11027 <=>
1958 V3 N2 P49-52	59-13343 <=> 0
1958 V3 N2 P53-60	59-13343 <=> 0
1958 V3 N2 P72-85	59-13343 <=> 0
1958 V3 N3 P5-61	59-13139 <=> 0
1958 V3 N3 P65-67	59-13140 <=>
1958 V3 N3 P68-69	59-13141 <=> 0
1958 V3 N3 P70-77	59-13137 <=>
1958 V3 N3 P80-84	59-13142 <=>
1958 V3 N3 P85-90	59-13138 <=>
1958 V3 N4 P3-85	59-11520 <=>
1958 V3 N4 P34-41	62-13485 <=*>
1958 V3 N4 P79-85	AEC-TR-3569 <+*>
1958 V3 N4 P87-92	59-11520 <=>
1958 V3 N5 P14-19	59-13104 <=>
1958 V3 N5 P24-29	59-13105 <=>
1958 V3 N5 P37-49	59-13106 <=>
1958 V3 N5 P50-57	59-13107 <=>
1958 V3 N5 P58-64	59-18491 <+*> 0
1958 V3 N5 P95-96	59-13108 <=>
1958 V3 N8 P79-81	60-11066 <=>
1959 V4 N2 P1-9	60-11412 <=>
1959 V4 N2 P10-14	59-13581 <=> 0
1959 V4 N2 P15-19	AEC-TR-3923 <+*>
	59-13584 <=> 0
1959 V4 N2 P20-49	60-11412 <=>
1959 V4 N2 P50-53	59-13583 <=> 0
1959 V4 N2 P55-62	60-11412 <=>
1959 V4 N2 P63-66	59-13582 <=> 0
1959 V4 N2 P67-77	59-13596 <=>
1959 V4 N2 P78-79	60-11412 <=>
1959 V4 N2 P80-85	59-13597 <=>
1959 V4 N2 P86	60-11412 <=>
1959 V4 N2 P87-91	59-13598 <=>
1959 V4 N2 P91-96	60-11412 <=>
1959 V4 N3 P1-13	60-11530 <=>
1959 V4 N3 P14-21	59-13677 <=>
1959 V4 N3 P21-39	60-11530 <=>
1959 V4 N3 P39-42	59-13678 <=>
1959 V4 N3 P42-43	60-11530 <=>
1959 V4 N3 P44-48	59-13679 <=>
1959 V4 N3 P49-52	59-13680 <=> 0
1959 V4 N3 P52-57	59-13681 <=> 0
1959 V4 N3 P57-60	59-13682 <=>
1959 V4 N3 P61-70	60-11530 <=>
1959 V4 N3 P70-76	59-13683 <=> 0
1959 V4 N3 P77	59-13684 <=>
1959 V4 N3 P77-79	59-13685 <=>
1959 V4 N3 P80-81	AD-615 283 <=$>
	60-11530 <=>
1959 V4 N3 P81-82	59-13686 <=>
1959 V4 N3 P83-88	59-13687 <=>
1959 V4 N3 P89-91	60-11530 <=>
1959 V4 N3 P92-94	AEC-TR-3717 <+*>
1959 V4 N3 P92	59-13688 <=>
1959 V4 N3 P92-94	59-13689 <=>
1959 V4 N3 P95-96	60-11530 <=>
1959 V4 N4 P10-16	C-3602 <NRCC>
	62-00385 <*>
1959 V4 N4 P88-90	C-3649 <NRCC>
1959 V4 N5 P3-6	59-13811 <=>
1959 V4 N5 P7-22	60-11780 <=>
1959 V4 N5 P23-27	59-13812 <=>
1959 V4 N5 P28-47	60-11780 <=>
1959 V4 N5 P38-41	60-17644 <+*> 0
1959 V4 N5 P48-52	59-13813 <=>
1959 V4 N5 P52-71	60-11780 <=>
1959 V4 N5 P72-76	59-13814 <=>
1959 V4 N5 P77-83	60-11780 <=>
1959 V4 N5 P84-85	59-13815 <=>
1959 V4 N5 P85-90	60-11780 <=>
1959 V4 N5 P91-93	59-13816 <=>
1959 V4 N5 P93-95	60-11780 <=>
1959 V4 N5 P96	59-13817 <=>
1959 V4 N6 P3-55	60-11724 <=>
1959 V4 N6 P56-60	59-31016 <=>
	63-19555 <=*>
1959 V4 N6 P61-93	60-11724 <=>
1959 V4 N6 P73-75	AEC-TR-3938 <+*>
1959 V4 N6 P95-96	60-11724 <=>
1959 V4 N8 P3-9	60-11668 <=>
1959 V4 N8 P10-13	60-11060 <=>
1959 V4 N8 P13-17	60-11061 <=>
1959 V4 N8 P17-32	60-11668 <=>
1959 V4 N8 P32-37	60-11062 <=>
1959 V4 N8 P37-41	60-11063 <=>
1959 V4 N8 P42-48	60-11064 <=>
1959 V4 N8 P49-71	60-11668 <=>
1959 V4 N8 P72-78	60-1106K <=>
1959 V4 N8 P81-91	60-11668 <=>
1959 V4 N8 P81-82	64-13936 <=*$>
1959 V4 N8 P92-93	60-11067 <=>
1959 V4 N8 P93-95	60-11068 <=>
1959 V4 N9 P3-12	60-11105 <=>
1959 V4 N9 P13-17	60-11669 <=>
1959 V4 N9 P17-24	60-11140 <=>
1959 V4 N9 P24-28	60-11669 <=>
1959 V4 N9 P29-33	60-11141 <=>
1959 V4 N9 P45-51	AEC-TR-3878 <+*>
1959 V4 N9 P63-65	60-11138 <=>
1959 V4 N9 P66-90	60-11669 <=>
1959 V4 N9 P79-81	64-15087 <=*$>
1959 V4 N9 P91-92	60-11139 <=>
1959 V4 N10 P3-8	60-11333 <=>
1959 V4 N10 P9-13	60-11295 <=>
1959 V4 N10 P13-17	60-11296 <=>
1959 V4 N10 P17-46	60-11670 <=>

Citation	Translation
1959 V4 N10 P46-53	60-11334 <=>
1959 V4 N10 P54-58	60-11339 <=>
1959 V4 N10 P59-95	60-11670 <=>
1959 V4 N11 P10-14	64-15567 <=*$>
1959 V4 N11 P42-47	64-15905 <=*$>
1960 V5 N6 P39-42	62-15794 <=*>
1961 V6 N3 P17-21	62-14739 <=*> 0
1962 N1 P67-	FPTS V22 N1 P.T130 <FASE>
1962 V7 N1 P10-	FPTS V22 N1 P.T152 <FASE>
1962 V7 N1 P10-15	62-23775 <=*>
1962 V7 N1 P16-21	62-24381 <=*>
1962 V7 N1 P28-32	63-15423 <=*>
1962 V7 N1 P48-	FPTS V22 N1 P.T141 <FASE>
1962 V7 N1 P53-62	62-23775 <=*>
1962 V7 N1 P62-	FPTS V22 N1 P.T144 <FASE>
1962 V7 N1 P70-	FPTS V22 N1 P.T69 <FASE>
1962 V7 N2 P3-36	62-25008 <=> 0
1962 V7 N2 P42-48	62-25008 <=> 0
1962 V7 N2 P58-64	62-25008 <=> 0
1962 V7 N2 P68-71	62-25008 <=> 0
1962 V7 N2 P87-93	62-25008 <=> 0
1962 V7 N3 P8-13	63-23889 <=*$>
1962 V7 N3 P24-	FPTS V22 N6 P.T1191 <FASE>
1962 V7 N3 P24-27	FPTS,V22,P.T1191-93 <FASE>
1962 V7 N3 P31-35	63-23038 <*=>
1962 V7 N3 P35-	FPTS V22 N4 P.T700 <FASE>
1962 V7 N3 P39-45	63-23888 <=*>
1962 V7 N3 P45-	FPTS V22 N6 P.T1210 <FASE>
1962 V7 N3 P45-53	FPTS,V22,P.T1210-14 <FASE>
1962 V7 N3 P53-	FPTS V22 N4 P.T702 <FASE>
1962 V7 N3 P61-	FPTS V22 N6 P.T1194 <FASE>
1962 V7 N3 P61-66	FPTS,V22,P.T1194-96 <FASE>
1962 V7 N3 P67-	FPTS V22 N5 P.T836 <FASE>
1962 V7 N3 P67-72	1963,V22,N5,PT2 <FASE>
1962 V7 N4 P30-	FPTS V22 N1 P.T156 <FASE>
1962 V7 N4 P47-57	62-25538 <=>
1962 V7 N4 P54-	FPTS V22 N6 P.T1215 <FASE>
1962 V7 N4 P54-57	FPTS,V22,P.T1215-17 <FASE>
1962 V7 N4 P57-64	63-23887 <=*>
	64-71383 <=> M
1962 V7 N4 P64-66	62-25538 <=>
	63-23836 <=*>
1962 V7 N4 P66-	FPTS V22 N4 P.T694 <FASE>
1962 V7 N4 P71-75	62-25538 <=>
	63-23039 <=*>
1962 V7 N4 P79	FPTS V22 N1 P.T150 <FASE>
1962 V7 N4 P80-	FPTS V22 N1 P.T148 <FASE>
1962 V7 N4 P81-82	63-15416 <=*>
1962 V7 N4 P83-	FPTS V22 N1 P.T96 <FASE>
1962 V7 N4 P85-89	62-25538 <=>
1962 V7 N5 P3-	FPTS V22 N5 P.T855 <FASE>
1962 V7 N5 P3-6	1963,V22,N5,PT2 <FASE>
1962 V7 N5 P6-	FPTS V22 N5 P.T948 <FASE>
1962 V7 N5 P6-13	1963,V22,N5,PT2 <FASE>
1962 V7 N5 P27-31	63-23894 <=*>
1962 V7 N5 P31-	FPTS V22 N5 P.T833 <FASE>
1962 V7 N5 P31-37	1963,V22,N5,PT2 <FASE>
	62-32455 <=>
1962 V7 N5 P38-	FPTS V22 N5 P.T840 <FASE>
1962 V7 N5 P38-45	1963,V22,N5,PT2 <FASE>
1962 V7 N5 P45-48	63-23893 <=*>
1962 V7 N5 P49-53	62-32455 <=>
	63-23892 <=*>
1962 V7 N5 P91-93	62-32455 <=>
1962 V7 N6 P32-36	62-32862 <=*>
1962 V7 N6 P58-	FPTS V22 N5 P.T844 <FASE>
1962 V7 N6 P58-68	1963,V22,N5,PT2 <FASE>
1962 V7 N6 P58-76	62-32862 <=*>
1962 V7 N6 P80-87	62-32862 <=*>
1962 V7 N7 P45-	FPTS V22 N3 P.T485 <FASE>
1962 V7 N7 P50-	FPTS V22 N4 P.T690 <FASE>
1962 V7 N7 P57-62	63-19375 <=*>
1962 V7 N7 P62-	FPTS V22 N5 P.T814 <FASE>
1962 V7 N7 P62-67	1963,V22,N5,PT2 <FASE>
1962 V7 N7 P62-71	62-33314 <=>
1962 V7 N7 P68-	FPTS V22 N4 P.T687 <FASE>
1962 V7 N7 P72-74	63-19376 <=*>
1962 V7 N7 P75-90	62-33314 <=>
1962 V7 N7 P89-90	63-23897 <=*>
1962 V7 N7 P90-92	63-23896 <=*>
1962 V7 N7 P92-93	63-23895 <=*>
1962 V7 N8 P3-11	64-13525 <=*$>
1962 V7 N8 P22-29	63-23022 <=*>
1962 V7 N8 P40-46	64-23511 <=$>
1962 V7 N8 P52-59	63-23418 <*=>
1962 V7 N8 P59-68	63-13360 <=>
1962 V7 N8 P59-65	63-23419 <*=>
1962 V7 N8 P66-68	63-23905 <=*$>
1962 V7 N8 P74-	FPTS V22 N4 P.T697 <FASE>
1962 V7 N8 P91-97	63-23882 <=*>
1962 V7 N8 P100-101	63-13360 <=>
1962 V7 N9 P62-70	63-13534 <=*>
1962 V7 N9 P81-83	63-13534 <=*>
1962 V7 N10 P3-8	63-24447 <=*$>
1962 V7 N10 P8-15	63-24448 <=*$>
1962 V7 N10 P15-21	63-24449 <=*$>
1962 V7 N10 P32-35	64-13490 <=*$>
1962 V7 N10 P41-44	63-24450 <=*$>
1962 V7 N10 P44-	FPTS V23 N1 P.T65 <FASE>
	FPTS,V23,P.T65-T68 <FASE>
1962 V7 N10 P55-59	64-13489 <=*$>
1962 V7 N10 P59-63	64-13488 <=*$>
1962 V7 N10 P68-69	AD-616 696 <=$>
1962 V7 N11 P9-	FPTS,V22,P.1170-80 <FASE>
1962 V7 N11 P27-	FPTS V22 N6 P.T1204 <FASE>
1962 V7 N11 P27-31	FPTS,V22,P.T1204-06 <FASE>
1962 V7 N11 P36-39	63-24442 <=*$>
1962 V7 N11 P39-45	63-24443 <=*$>
1962 V7 N11 P45-50	63-24444 <=*$>
1962 V7 N11 P50-53	63-24445 <=*$>
1962 V7 N11 P53-59	64-13492 <=*$> 0
1962 V7 N11 P59-65	63-23857 <=*$>
1962 V7 N11 P65-	FPTS V22 N6 P.T1197 <FASE>
	FPTS,V22,P.T1197-201 <FASE>
1962 V7 N11 P77-82	63-23858 <=*$>
1962 V7 N11 P82-83	63-23859 <=*>
1962 V7 N11 P83-86	63-24446 <=*$>
1962 V7 N12 P3-7	63-24463 <=*$>
1962 V7 N12 P21-25	63-24462 <=*$>
1962 V7 N12 P32-37	63-24464 <=*$>
1962 V7 N12 P38-40	63-24471 <=*$>
1962 V7 N12 P40-43	63-24465 <=*$> 0
1962 V7 N12 P43-	FPTS V23 N2 P.T292 <FASE>
	FPTS,V23,P.T292-T295 <FASE>
1962 V7 N12 P49-55	63-21141 <=>
	63-24466 <=*$>
1962 V7 N12 P56-58	64-13496 <=*$>
1962 V7 N12 P68-77	63-21141 <=>
1963 V7 N10 P51-55	64-13491 <=*$>
1963 V7 N11 P9-	FPTS V22 N6 P.T1177 <FASE>
1963 V8 N1 P24-32	63-21676 <=>
1963 V8 N1 P36-46	63-21676 <=>
1963 V8 N1 P38-45	64-13922 <=*$>
1963 V8 N1 P45-46	63-24499 <=*$>
1963 V8 N1 P46-54	64-13530 <=*$> 0
1963 V8 N1 P54-	FPTS V23 N1 P.T62 <FASE>
	FPTS,V23,P.T62-T64 <FASE>
1963 V8 N1 P58-	FPTS V22 N6 P.T1202 <FASE>
1963 V8 N1 P58-61	FPTS,V22,P.T1202-03 <FASE>
1963 V8 N1 P62-64	63-21676 <=>
	64-13529 <=*$>
1963 V8 N1 P64-70	63-24500 <=*$>
1963 V8 N1 P71-76	64-19265 <=*$>
1963 V8 N2 P3-5	64-13918 <=*$>
1963 V8 N2 P5-10	64-13436 <=*$>
1963 V8 N2 P13-19	64-13867 <=*$>
1963 V8 N2 P23-25	64-19213 <=*$>
1963 V8 N2 P25-28	63-24249 <=*$>
	64-15515 <=*$>
1963 V8 N2 P28-35	64-13839 <=*$>
1963 V8 N2 P35-42	64-13840 <=*$>
1963 V8 N2 P42-47	64-13868 <=*$>
1963 V8 N2 P47-50	63-21805 <=>
	64-15704 <=*$>
1963 V8 N2 P58-66	64-13924 <=*$>
1963 V8 N3 P26-31	64-13865 <=*$>
1963 V8 N3 P34-38	64-13833 <=*$>
1963 V8 N3 P38-43	64-13407 <=*$>
1963 V8 N3 P50-57	64-13406 <=*$>
1963 V8 N3 P57-61	64-13834 <=*$>

Citation	Translation
1963 V8 N3 P66-71	64-13835 <=*$>
1963 V8 N3 P71-76	63-21695 <=>
	64-13836 <=*$>
1963 V8 N4 P62-68	64-13841 <=*$> 0
1963 V8 N4 P80-81	64-13817 <=*$>
1963 V8 N4 P81-92	63-31186 <=>
1963 V8 N4 P82-84	64-13869 <=*$>
1963 V8 N5 P29-32	64-15517 <=*$>
1963 V8 N5 P33-39	64-15518 <=*$>
1963 V8 N5 P39-43	63-31468 <=>
	64-15519 <=*$>
1963 V8 N5 P47-	FPTS V23 N3 P.T529 <FASE>
	FPTS,V23,N3,P.T529 <FASE>
1963 V8 N5 P79-82	63-31468 <=>
1963 V8 N5 P89-94	63-31830 <=>
1963 V8 N6 P27-73	63-31821 <=>
1963 V8 N6 P27-32	64-15534 <=*$>
1963 V8 N6 P32-42	64-15543 <=*$>
1963 V8 N6 P51-	FPTS V23 N4 P.T775 <FASE>
1963 V8 N6 P60-	FPTS V23 N4 P.T665 <FASE>
1963 V8 N6 P85-91	63-31821 <=>
1963 V8 N7 P3-11	63-31862 <=>
1963 V8 N7 P23-28	63-31862 <=>
1963 V8 N7 P29-34	64-15520 <=*$>
1963 V8 N7 P34-38	64-15521 <=*$>
1963 V8 N7 P52-56	63-31862 <=>
1963 V8 N7 P62-67	63-31862 <=>
1963 V8 N7 P68-	FPTS V23 N4 P.T783 <FASE>
1963 V8 N7 P71-	FPTS V23 N4 P.T785 <FASE>
1963 V8 N7 P82-87	63-31862 <=>
1963 V8 N8 P17-22	64-15522 <=*$>
1963 V8 N8 P25-31	64-19215 <=*$>
1963 V8 N8 P35-42	64-15523 <=*$>
1963 V8 N8 P42-48	64-11893 <=>
1963 V8 N8 P48-53	64-21200 <=>
1963 V8 N8 P54-	FPTS V23 N4 P.T780 <FASE>
1963 V8 N8 P54-58	64-11893 <=>
1963 V8 N8 P58-65	64-15524 <=*$>
	64-21219 <=>
1963 V8 N8 P77-87	64-21214 <=>
1963 V8 N9 P3-7	63-41138 <=>
1963 V8 N9 P7-14	64-19218 <=*$>
1963 V8 N9 P14-16	64-19680 <=$>
1963 V8 N9 P44-48	63-41277 <=>
1963 V8 N9 P48-52	63-41278 <=>
1963 V8 N9 P61-65	63-41279 <=>
1963 V8 N9 P66-68	63-41281 <=>
1963 V8 N10 P18-20	64-19684 <=$>
1963 V8 N10 P20-25	64-19685 <=$>
1963 V8 N10 P65-71	64-19686 <=$> 0
1963 V8 N11 P3-9	64-21376 <=>
1963 V8 N11 P9-13	64-19687 <=$>
1963 V8 N11 P20-24	64-19647 <=$>
1963 V8 N11 P30-33	64-19688 <=$> 0
	64-21376 <=>
1963 V8 N11 P33-	FPTS V23 N6 P.T1321 <FASE>
	FPTS V23 P.T1321 <FASE>
1963 V8 N11 P47-50	64-21376 <=>
1963 V8 N11 P55-59	64-21376 <=>
1963 V8 N11 P59-63	64-21376 <=>
1963 V8 N12 P25-30	64-19722 <=$>
1963 V8 N12 P38-42	65-60217 <=$>
1963 V8 N12 P81-83	64-71532 <=>
1964 V8 N6 P19-27	64-19214 <=*$>
1964 V9 N1 P20-	FPTS V24 N1 P.T126 <FASE>
1964 V9 N1 P20-24	FPTS V24 P.T126-T128 <FASE>
1964 V9 N1 P47-49	65-60150 <=$>
1964 V9 N1 P53-57	65-60190 <=$>
1964 V9 N2 P29-37	65-60162 <=$>
1964 V9 N2 P56-64	65-60191 <=$>
1964 V9 N2 P75-80	65-60163 <=$>
1964 V9 N3 P52-66	AD-610 300 <=$>
1964 V9 N3 P52-56	65-62202 <=$>
1964 V9 N3 P56-61	65-62203 <=>
1964 V9 N3 P61-66	65-62204 <=$>
1964 V9 N3 P84-89	AD-615 235 <=$>
1964 V9 N4 P36-	FPTS V24 N4 P.T685 <FASE>
1964 V9 N4 P41-46	64-31409 <=>
1964 V9 N5 P15-19	64-41475 <=>
1964 V9 N5 P24-34	64-41475 <=>
1964 V9 N5 P24-29	65-61125 <=$>
1964 V9 N5 P35-39	65-62205 <=>
1964 V9 N5 P45-62	64-41475 <=>
1964 V9 N5 P58-62	65-61218 <=$>
1964 V9 N6 P3-8	65-32130 <=$>
1964 V9 N6 P19-21	64-41800 <=>
1964 V9 N6 P34-43	65-62206 <=>
1964 V9 N6 P43-47	64-41800 <=>
1964 V9 N7 P7-17	64-51458 <=$>
	65-62207 <=$>
1964 V9 N7 P22-45	64-51458 <=$>
1964 V9 N7 P56-66	64-51458 <=$>
1964 V9 N7 P68-71	64-51458 <=$>
1964 V9 N8 P16-18	64-51216 <=$>
1964 V9 N8 P28-37	64-51216 <=$>
1964 V9 N8 P28-31	65-62340 <=$>
1964 V9 N8 P45-51	64-51216 <=$>
1964 V9 N8 P78-80	64-51216 <=$>
1964 V9 N9 P8-16	65-30018 <=$>
1964 V9 N9 P8-14	65-63616 <=$>
1964 V9 N9 P27-29	65-30018 <=$>
1964 V9 N9 P46-80	65-30018 <=$>
1964 V9 N9 P92-95	65-30018 <=$>
1964 V9 N11 P49-54	65-63581 <=$>
1964 V9 N12 P17-23	65-30641 <=$>
1964 V9 N12 P28-31	65-30641 <=$>
1964 V9 N12 P35-40	65-30641 <=$>
1964 V9 N12 P46-62	65-30641 <=$>
1964 V9 N12 P58-	FPTS V24 N6 P.T974 <FASE>
1965 N3 P3-7	NLLTB V7 N10 P864 <NLL>
1965 V10 N1 P33-39	66-60300 <=*$>
1965 V10 N1 P39-42	65-30642 <=$>
1965 V10 N1 P65-69	65-30642 <=$>
1965 V10 N2 P35-41	66-60881 <=*$>
1965 V10 N3 P3-7	65-31081 <=$>
1965 V10 N3 P44-48	65-31081 <=$>
1965 V10 N3 P61-	FPTS V25 N1 P.T95 <FASE>
1965 V10 N3 P66-	FPTS V25 N1 P.T99 <FASE>
1965 V10 N3 P90-92	65-31081 <=$>
1965 V10 N3 P92-94	65-31082 <=$>
1965 V10 N4 P42-49	65-31265 <=$>
1965 V10 N4 P53-59	65-31265 <=$>
1965 V10 N4 P62-65	65-31265 <=$>
1965 V10 N4 P62-66	66-60882 <=*$>
1965 V10 N4 P73-79	65-31265 <=$>
1965 V10 N4 P82-84	65-31265 <=$>
1965 V10 N5 P46-51	66-61325 <=*$>
1965 V10 N5 P55-59	66-61326 <=*$>
1965 V10 N5 P71-74	65-32644 <=*$>
1965 V10 N5 P91-92	65-32645 <=*$>
1965 V10 N6 P14-19	66-61075 <=*$>
1965 V10 N6 P83-85	65-32396 <=$>
1965 V10 N7 P56-	FPTS V25 N4 P.T637 <FASE>
1965 V10 N8 P32-39	65-33038 <=*$>
1965 V10 N8 P55-67	66-61308 <=*$>
1965 V10 N8 P67-73	66-61291 <=*$>
1965 V10 N9 P20-23	66-61623 <=*>
1965 V10 N9 P62-67	66-61624 <=*>
1965 V10 N10 P15-19	66-30401 <=$>
1965 V10 N10 P19-22	66-61597 <=*$>
1965 V10 N10 P33-34	66-61598 <=*$>
1965 V10 N10 P34-	FPTS V25 N5 P.T895 <FASE>
1965 V10 N10 P55-57	66-61599 <=*$>
1965 V10 N10 P75-80	66-30401 <=$>
1965 V10 N10 P89-93	66-30401 <=$>
1965 V10 N11 P34-	FPTS V25 N6 P.T1013 <FASE>
1965 V10 N11 P39-	FPTS V25 N6 P.T1079 <FASE>
1965 V10 N11 P66-71	66-61973 <=*$>
1965 V10 N12 P44-46	66-30990 <=$>
1965 V10 N12 P51-	FPTS V25 N6 P.T1059 <FASE>
1965 V10 N12 P51-57	66-30989 <=$>
1965 V10 N12 P63-66	66-30989 <=$>
1966 V11 N1 P74-77	66-31724 <=$>
1966 V11 N1 P88-93	66-31723 <=$>

MEDITSINSKAYA SESTRA

Citation	Translation
1947 N10 P1-6	RT-1189 <*>
1950 N8 P8-13	R-808 <*>
1960 V19 N4 P7-12	60-41702 <=>
1961 V20 N6 P23-26	62-24002 <=*>

1965 V24 N7 P3-7 65-32554 <=*$>

MEDITSINSKII RABCTNIK. MOSCOW
1952 N21 P2 RT-661 <*>
1952 V15 N38 P2-3 RT-1657 <*>
1952 V15 N58 P4 RT-1047 <*>
1952 V15 N64 P3 RT-1655 <*>
1952 V15 N65 P3-4 RT-1656 <*>
1952 V15 N65 P4 RT-1654 <*>
1953 N45 P3 RT-4035 <*>
1954 V17 N2 P3 RT-1559 <*>
1954 V17 N15 P4 RT-1371 <*>
1954 V17 N20 P4 RT-1365 <*>
1954 V17 N28 P4 RT-1554 <*>
1954 V17 N31 P3 RT-1568 <*>
1954 V17 N40 P3 RT-1734 <*>
1954 V17 N45 P4 RT-1733 <*>
1954 V17 N51 P3 RT-2048 <*>
1954 V17 N61 P3 RT-2047 <*>
1954 V17 N81 P2 RT-2920 <*>
1954 V17 N96 P2 RT-4584 <*>
1955 V18 N2 P2 RT-2827 <*>
1955 V18 N3 P4 RT-2919 <*>
1955 V18 N4 P3 RT-2785 <*>
RT-3661 <*>
1955 V18 N15 P4 RT-2973 <*>
1955 V18 N2C P3 RT-2972 <*>
1955 V18 N26 P3 RT-3072 <*>
1955 V18 N29 P2 RT-3055 <*>
1955 V18 N33 P2 RT-3142 <*>
1955 V18 N33 P3 RT-3288 <*>
1955 V18 N34 P3 RT-3068 <*>
1955 V18 N40 P3 RT-3071 <*>
1955 V18 N42 P2 RT-3054 <*>
1955 V18 N43 P2 RT-3169 <*>
1955 V18 N44 P4 RT-3214 <*>
1955 V18 N51 P2 RT-3170 <*>
1955 V18 N53 P2 RT-3097 <*>
1955 V18 N66 P3 RT-3132 <*>
1955 V18 N67 P3 RT-3216 <*>
RT-3337 <*>
1955 V18 N68 P3 RT-3133 <*>
1955 V18 N79 P3 RT-3662 <*>
1955 V18 N81 P3 RT-3335 <*>
1955 V18 N85 P1 RT-3539 <*>
1955 V18 N85 P2 RT-3498 <*>
1955 V18 N98 P2 RT-3497 <*>
1957 V20 N88 P1-2 60-19495 <+*>
1957 V20 N104 P4 61-19639 <=*>
1957 V2C N1C5 P3 61-27432 <*=>
1960 V23 P6 60-31199 <=>
1960 V23 N2 P4 60-31691 <=>
1960 V23 N4 P4 60-31185 <=>
1960 V23 N7 P1-2 60-31166 <=>
1960 V23 N8 P3 60-11900 <=>
1960 V23 N21 P1 60-31404 <=>
1960 V23 N22 P1-3 60-31404 <=>
1960 V23 N23 P1-2 60-31404 <=>
1960 V23 N33 P1 60-31175 <=>
1960 V23 N50 P2 60-31642 <=>
1960 V23 N72 P1-2 61-11506 <=>
1960 V23 N73 P1-2 61-11506 <=>
1960 V23 N75 P1 61-11506 <=>
1960 V23 N98 P1-3 61-21277 <=>
1960 V23 N43/4 P2-4 60-31788 <=>
1961 V24 N21 P1 61-27217 <*=>
1961 V24 N21 P4 61-27217 <*=>
1961 V24 N3C P3 61-27228 <*=>
1961 V24 N33 P4 61-28610 <*=>
1961 V24 N38 P4 61-28610 <*=>
1961 V24 N39 P2 61-27211 <*=>
1961 V24 N43 P3 61-28722 <*=>
61-28738 <*=>
61-28854 <*=>
1961 V24 N47 P3 61-27669 <*=>
1961 V24 N48 P3 61-27671 <*=>
1962 V25 P1-2 62-24726 <=*>
1962 V25 P2 62-25994 <=>
1962 V25 P3 62-23431 <=*>
62-24337 <=*>
62-25994 <=>
62-33460 <=*>
1962 V25 P4 62-23651 <=*>
1962 V25 N4 P4 62-23346 <=*>
1962 V25 N13 P3 62-25173 <=*>
1962 V25 N24 P3 62-24742 <=*>
1962 V25 N29 P4 62-25007 <=*>
1962 V25 N47 P2 63-13854 <=*>
1962 V25 N67 P3 62-32871 <=*>
1962 V25 N72 P3 63-13359 <=*>

MEDITSINSKII ZHURNAL UZBEKISTANA
1960 N9 P3-6 61-27717 <*=>
1960 N9 P7-10 61-27695 <*=>
1960 N9 P20-25 61-27680 <*=>
1960 N10 P8C-81 61-21436 <=>
1961 N7 P3-9 62-19805 <=*>
1962 N1 P3-12 62-25009 <=*>

MEDITSINSKOE OBOZRENIE SPRIMONA
1894 V42 N14 P97-118 R-802 <*>
RT-3849 <*>
64-19095 <=*$>
1897 V47 P1C2-104 60-19609 <*>
1897 V47 P440-448 60-19738 <=>
1898 N50 P317-320 RT-1843 <*>
1898 V50 P317-320 64-15587 <=*$>

MEDIZIN UND CHEMIE
1940 V4 P248-258 63-00513 <*>
1956 V5 P174-178 60-13456 <*+>
1956 V5 P181-184 60-13456 <*+>

MEDIZIN UND ERNAEHRUNG
1954 P553-555 1248 <*>
1961 V2 N10 P229-230 24Q70G <ATS>
1965 V6 N2 P30-35 66-12115 <*>

MEDIZINISCHE DOKUMENTATION
1960 V4 N3 P61-68 62-16228 <*> 0
1961 V5 P35-38 61-16450 <*> 0

MEDIZINISCHE GRUNDLAGENFORSCHUNG
1959 V2 ENTIRE ISSUE 63-16609 <*> 0
1959 V2 ENTIRE VOLUME 63-16609 <*> 0

MEDIZINISCHE KLINIK
1910 N44 P1741-1743 2394 <*>
1924 V20 P1237-1240 57-1666 <*>
1929 V25 N10 P390-392 22C9 <*>
1930 V26 P240-245 65-18199 <*>
1931 V27 N1C P359-361 2397 <*>
1931 V27 N23 P879-880 2754 <*>
1933 V28 P2-8 61-14179 <*>
1934 V30 P236-238 65-18200 <*>
1935 V31 P1561-1564 65-18201 <*>
1939 V35 P973-976 2957 <*>
1939 V35 P1169-1170 5147-A <K-H>
1939 V35 N29 P973-976 02957 <*>
1941 V37 P477-479 57-1495 <*>
1942 V38 P540-541 65-18202 <*>
1943 V39 P1302-1304 65-18203 <*>
1949 V44 P1530-1532 65-18204 <*>
1949 V44 P1540-1543 2220 <*>
1949 V44 N16 P506-510 2579 <*>
1950 V45 P1169-1171 65-18205 <*>
1951 V46 P1109-1110 I-372 <*>
1952 V47 P722- I-644 <*>
1952 V47 P991-994 65-18206 <*> 0
1953 V48 P932-934 AL-542 <*>
1953 V48 P1516-1518 AL-362 <*>
1953 V48 P1587-1589 AL-346 <*>
1953 V48 N44 P1632-1634 62-10262 <*>
1954 N23 P925-926 73F3G <ATS>
1954 V49 P295-298 I-406 <*>
1954 V49 P399-401 3058-B <K-H>
1954 V49 P481-482 I-612 <*>
1954 V49 P487- I-612 <*>
1954 V49 P959-962 3183-D <K-H>
1954 V49 N14 P559-560 64-10594 <*>

1954 V49 N17 P705-707 65-18207 <*> 0
1954 V49 N51 P2031-2033 63-10320 <*>
1955 N35 P1473-1474 94L30G <ATS>
1955 V50 P509-510 2142 <*>
1955 V50 P864-866 57-1837 <*>
1955 V50 P869- 3779-B <KH>
1955 V50 P1296-1301 57-1838 <*>
1955 V50 P1949-1950 II-728 <*>
1955 V50 N9 P357-359 59-10702 <*>
1955 V50 N10 P385-388 57-856 <*>
1955 V50 N24 P1022- 2681 <*>
1956 V51 P133-137 58-677 <*>
1956 V51 P761-764 5525-C <KH>
1956 V51 P1013-1017 57-207 <*>
1957 V52 N2 P46-49 63-01750 <*>
1957 V52 N5 P182-185 65-18208 <*> 0
1957 V52 N19 P820-822 <RIS>
1957 V52 N32 P1379-1381 65-18209 <*> 0
1958 V53 P1505-1506 58-2703 <*>
1958 V53 N29 P1272-1273 65-00253 <*>
1958 V53 N35 P1495-1497 59-10352 <*>
1958 V53 N35 P1507-1509 59-20699 <*> 0
1959 V54 N16 P800 59-20979 <*>
1959 V54 N35 P1541-1543 60-18160 <*> 0
1960 N24 P954-959 66-10373 <*> 0
1960 N43 P1932-1934 65-14348 <*> 0
1960 V55 P2064-2067 61-20566 <*> 0
1960 V55 N47 P2093-2100 62-00327 <*>
1960 V55 N48 P2141-2145 61-20575 <*> 0
1961 V56 P679-685 64-10418 <*>
1961 V56 P1497-1501 63-10567 <*>
1961 V56 N1 P26-27 11095 <K-H>
1961 V56 N3 P103-104 64-18618 <*>
1961 V56 N10 P377-380 63-00044 <*>
1961 V56 N15 P679-685 11363-C <K-H>
1961 V56 N37 P1584-1585 65-17416 <*>
1961 V56 N47 P2000-2003 62-14340 <*>
1962 V57 P749-752 63-10384 <*>
1962 V57 P1016-1017 63-10905 <*>
1962 V57 N32 P1370-1376 64-18148 <*>
1963 V58 P841-842 66-10100 <*>
1963 V58 N21 P882-884 66-10099 <*>
1963 V58 N30 P1234-1236 66-10097 <*>
1964 V59 P342-345 65-12190 <*>
1964 V59 N25 P1020-1023 65-14586 <*>
1964 V59 N35 P1399-1402 65-13236 <*>
1964 V59 N49 P1949-1951 66-13166 <*>
1965 V60 N3 P89-93 66-13034 <*>
66-14044 <*>
1965 V60 N19 P754-760 66-11606 <*>
1965 V60 N21 P847-851 03S86G <ATS>
1965 V60 N31 P1252-1253 66-11080 <*>
1965 V60 N35 P1398-1401 66-11704 <*>

MEDIZINISCHE MITTEILUNGEN. SCHERING-KAHLBAUM A.G.
1956 V17 N3 P63-73 57-272 <*>

MEDIZINISCHE MONATSSCHRIFT
1951 V5 N4 P239-244 59-20945 <*> 0
1953 V7 P165-167 65-18211 <*>
1953 V7 P515-516 AL-962 <*>
1953 V7 P577-580 AL-943 <*>
1953 V7 N3 P181-183 66-11698 <*>
1953 V7 N10 P652-654 3160-7 <K-H>
1954 V8 P36-37 II-373 <*>
1954 V8 P306-310 2389 <*>
1954 V8 P393-395 II-40 <*>
1955 V9 P808-811 2677 <*>
1956 V10 P162-165 1198 <*>
1956 V10 N10 P672-674 59-15755 <*>
1956 V10 N12 P794-796 61-00855 <*>
1957 V11 N5 P290-292 58-2666 <*>
1959 V13 N10 P651-653 64-18664 <*>
1963 V17 N2 P94-98 64-18965 <*> 0
1963 V17 N5 P309-310 66-10110 <*>
1963 V17 N5 P310-311 66-10111 <*>
1963 V17 N11 P691-694 65-18212 <*> 0
1964 V18 N3 P131-133 66-10109 <*>
1964 V18 N9 P420-422 66-10108 <*>
1964 V18 N11 P497-498 66-13168 <*>

MEDIZINISCHE WELT
1929 V3 P1321 2549 <*>
1929 V3 N8 P261-263 64-10665 <*>
1937 V11 P1554-1557 57-1637 <*>
1938 V12 P485-487 2835R <K-H>
1943 N8 P171- 2399 <*>
1952 P1202-1203 AL-555 <*>
1952 N37 P1157-1159 AL-793 <*>
1953 P918- AL-227 <*>
1953 N36 P1162-1164 65-18214 <*>
1953 N50 P1626-1629 I-183 <*>
1954 N1 P25-26 I-918 <*>
1954 N17 P593-595 3199-A <KH>
1955 P776-782 5322-A <KH>
1955 N5 P192-194 2070 <KH>
1955 N8 P290-291 65-18215 <*>
1955 N16 P606-608 2629 <*>
1955 N26 P955-957 II-820 <*>
1955 N49 P1704-1707 57-1849 <*>
1955 N50 P1750-1751 II-930 <*>
1955 N33/4 P1121-1126 2680 <*>
1956 P85-91 5223-B <KH>
1956 N7 P260-262 57-1097 <*>
1956 N7 P418-421 59-12327 <+*> 0
1956 N11 P391-393 06L31G <ATS>
1956 N33/4 P1143-1145 65-18216 <*> 0
1957 N1 P12-19 3248 <*>
1957 N2 P87-90 65-18217 <*>
1957 N3 P124-126 3394 <*>
1957 N45 P1659-1662 59-10057 <*>
1958 P543-549 59-12970 <+*> 0
1958 N23 P959-960 59-17066 <*>
1958 N36 P1400-1401 59-10840 <*>
1958 N36 P1401-1402 60-10068 <*> 0
1959 N21 P1034-1035 60-17650 <+*>
1959 N49 P2408-2412 77M42G <ATS>
1959 N51 P2528-2529 60-18161 <*> 0
1959 N51 P2530-2531 60-18162 <*> 0
1959 N52 P2575-2579 62-16230 <*> 0
1960 N8 P423-427 84M45G <ATS>
1960 N13 P672-676 62-00885 <*>
1960 N16 P841-843 62-00284 <*>
1960 N22 P1192-1205 61-18579 <*>
1960 N44 P2328-2329 62-10263 <*>
1960 N33/4 P1672-1676,1679
61-10335 <*> 0
1960 N33/4 P1679-1682 61-10333 <*>
1961 P921 63-20901 <*>
1961 N4 P195-197 61-20572 <*>
1961 N23 P1269-1271 62-14320 <*>
1961 N49 P2559-2563 63-00165 <*>
1961 N49 P2568-2571 63-16382 <*>
1962 N14 P763-765 72Q67G <ATS>
1962 N33 P1714 63-10932 <*>
1962 N38 P2000 64-00265 <*>
1962 N42 P2229-2231 63-20949 <*>
1962 N47 P2504-2508 UCRL TRANS-975 <*>
1963 N12 P643-645 64-18964 <*> 0
1963 N12 P662 64-18142 <*>
1963 N19 P1078-1081 64-18139 <*>
64-20151 <*>
1963 N20 P1138-1139 66-10089 <*>
1963 N20 P2235-2238 66-10090 <*>
1963 N37 P1879-1883 65-14424 <*>
1963 N42 P2144-2145 66-10085 <*>
1963 N51 P2617-2619 66-10084 <*>
1964 N9 P435-436 66-10088 <*>
1964 N11 P569-571 64-20138 <*>
1964 N15 P867-871 66-13041 <*>
1964 N20 P1125-1129 64-20147 <*> 0
1964 N29 P1592-1593 65-14773 <*> 0
1964 N31 P1655-1661 66-10086 <*>
1964 N42 P2231-2238 65-18213 <*>
1964 N42 P2251-2254 65-12787 <*>
1964 N48 P2583-2589 65-17209 <*>
1964 N50 P2705-2708 66-13199 <*>
1965 N14 P690-692 63S87G <ATS>
1965 N17 P943-947 65-14225 <*>
1965 N24 P1311-1320 66-13161 <*>

MEDIZINISCHE ZEITSCHRIFT
1944 N1 P19-23 64-16245 <*>
1955 N50 P1742-1744 II-1720 <*>

MEDIZIN-METEOROLOGISCHE HEFTE
1956 V11 P111-117 57-71 <*>

MEDLEMSBLAD FOR SVERIGES VETERINARFORBUND
1958 V10 N16 P289-290 64-14459 <*>

MEDUNARODNI TRANSPORT
1965 N4 P11-13 65-31450 <=$>
1965 N4 P38-41 65-31450 <=$>
1965 N4 P46-47 65-31450 <=$>

*MEDYCYNA DOSWIADCZALNA I MIKROBIOLOGIA
1951 V3 N1 P105-108 2233B <K-H>
1954 V6 N1 P51-62 3153 <*>
1955 V7 N3 P105-124 3161 <*>
1955 V7 N3 P323-329 3160 <*>
1957 V9 N1 P1-20 R-3230 <*>
70J15P <ATS>
1957 V9 N4 P437-440 58-625 <*>
1961 V13 N1 P35-41 62-19207 <=*>
1961 V13 N1 P47-52 62-19542 <=*>
1961 V13 N2 P127-133 62-19572 <=*>
1961 V13 N2 P135-140 62-19571 <=*>
1961 V13 N2 P173-181 65-10448 <*>
1962 V14 P545-548 65-00414 <*>
1965 V17 N2 P123-131 66-11682 <*>

MEDYCYNA PRACY
1960 V11 N2 P99-107 64-18992 <*> O
1960 V11 N6 P433-440 63-20889 <*> O
1961 V12 N5 P469-479 63-20888 <*> O
1963 V14 N5 P407-411 M.5744 <NLL>

MEDYCYNA WETERYNARYJNA
1954 V10 N3 P158-160 R-464 <*>
1954 V10 N3 P160-161 R-463 <*>
1957 V13 N2 P75-77 60-21515 <=>
1957 V13 N6 P333-334 60-00821 <*>
1957 V13 N7 P387-393 60-21222 <=>
1957 V13 N8 P481-484 65-00385 <*>
1957 V13 N9 P526-532 61-31265 <=>
1957 V13 N9 P533-534 60-21512 <=>
1957 V13 N11 P677-678 60-21240 <=>
1957 V13 N11 P694-698 60-21518 <=>
1958 V14 N2 P76-81 60-00806 <*>
1958 V14 N5 P275-280 61-11390 <=>
1959 V15 N6 P334-338 60-00820 <*>
1960 V16 N7 P385-389 62-19206 <=*>
1960 V16 N9 P533-534 62-19205 <=*>
1960 V16 N10 P577-581 62-19205 <=*>
1961 V17 N2 P65-70 62-19204 <=*>
1961 V17 N2 P73-74 62-19204 <=*>
1961 V17 N3 P136-142 62-19203 <*=>
1961 V17 N11 P687-691 62-25059 <=*>
1962 V18 N3 P131-137 63-00161 <*>
63-26694 <=*$>

MEGAMOT
1957 V8 N4 P411-418 65-11684 <*>

MEI K'UNG CHI SHU
1958 N2 P3-5 59-13721 <=>
1958 N2 P9-10 59-13721 <=>
1958 N4 P30-36 59-13721 <=>
1958 N5 P25 59-13721 <=>
1958 N5 P33-39 59-13721 <=>
1958 N8 P3-4 59-13721 <=>
1958 N8 P6-9 59-13721 <=>
1958 N9 P1-2 59-13721 <=>
1958 N9 P19-22 59-13721 <=>
1958 N10 P40-43 59-11452 <=>
1958 N11 P2 59-13721 <=>
1958 N11 P9-10 59-13721 <=>
1958 N11 P22-28 59-13721 <=>
1958 N11 P33-36 59-13721 <=>
1958 N11 P42-44 59-13721 <=>
1958 N12 P15 59-13721 <=>
1958 N12 P18-20 59-13721 <=>
1960 N5 P2 60-41519 <=>
1960 N5 P5-8 60-41519 <=>
1960 N5 P14 60-41519 <=>
1960 N5 P15 60-41519 <=> P

MEI-T'AN KUNG-YEH
1958 P11-12 59-11452 <=>
1958 N21 P7-12 59-11452 <=>
1958 N21 P24 59-11452 <=>
1958 N21 P26-29 59-11452 <=>
1958 N22 P12-13 59-11462 <=>
1958 N22 P18 59-11462 <=>
1958 N22 P28 59-11462 <=>

MEIJI SEIKA KENKYU NEMPO (YAKUHIN BUMON)
1959 N1 P33-36 64-14718 <*>

MEKHANICHESKAYA OBRABOTKA DREVESINY
1936 N11 P36-55 60-14130 <+*> O

MEKHANICHESKIE METODY ISPYTANIYA
1949 N5 P576-580 AD-255-398 <=$>

*MEKHANIKA POLIMEROV
1965 V1 N1 P93-99 160 <WFK>
1965 V1 N1 P124-127 159 <WFK>
1965 V1 N1 P151-158 158 <WFK>

MEKHANIKA TVERDOGO TELA
1966 N1 P67-73 66-61359 <=$>
1966 N2 P160-166 66-61700 <=$>

MEKHANIZATSIYA I AVTOMATIZATSIYA PROIZVODSTVA
1959 N6 P39 NLLTB V1 N10 P39 <NLL>
1959 N8 P3-5 60-11413 <=>
1959 V13 N4 P51 62-24826 <=*>
1959 V13 N11 P6-12 60-31124 <=> O
1960 V14 N2 P5-9 60-41002 <=>
1960 V14 N2 P9-10 60-41002 <=>
1960 V14 N2 P53-54 60-11933 <=>
1960 V14 N4 P1-6 61-11605 <=>
1960 V14 N4 P45-49 61-11605 <=>
1960 V14 N7 P59-60 61-21293 <=>
1960 V14 N8 P6-10 61-19912 <=*>
1960 V14 N8 P10-14 61-19795 <=*>
1960 V14 N8 P23-27 61-19911 <=*>
1960 V14 N10 P20-23 62-11108 <=>
1960 V14 N10 P42-46 61-31157 <=>
1960 V14 N11 P6-7 61-27083 <=>
1960 V14 N12 P52-54 61-21471 <=>
1961 N1 P1-4 NLLTB V3 N8 P645 <NLL>
1961 N2 P57-60 NLLTB V3 N8 P676 <NLL>
1961 N4 P1-5 NLLTB V3 N8 P661 <NLL>
1961 N6 P57-58 NLLTB V4 N4 P348 <NLL>
1961 N8 P1-7 NLLTB V4 N2 P135 <NLL>
1961 N8 P39-43 NLLTB V4 N3 P264 <NLL>
1961 V15 N2 P1-8 61-31527 <=>
1961 V15 N3 P4-10 62-13420 <=>
1961 V15 N3 P17-20 62-13420 <=>
1961 V15 N11 P10-11 62-33315 <=*>
1961 V15 N11 P39-41 62-33315 <=*>
1961 V15 N12 P36-40 2693 <BISI>
1962 N8 P28-32 NLLTB V5 N1 P41 <NLL>
1962 N10 P37-41 NLLTB V5 N4 P270 <NLL>
1962 N10 P41-43 NLLTB V5 N4 P317 <NLL>
1962 N11 P47-51 NLLTB V5 N4 P297 <NLL>
1962 N12 P25-28 NLLTB V5 N8 P683 <NLL>
1962 V16 N1 ENTIRE ISSUE 62-24734 <=*>
1962 V16 N2 P18-21 63-21702 <=>
1962 V16 N2 P54 62-25972 <=>
1962 V16 N3 P3-6 NLLTB V4 N11 P1015 <NLL>
1962 V16 N3 P48 62-33688 <=*>
1962 V16 N4 P59 62-25972 <=>
1962 V16 N6 P26-28 62-32879 <=*>
1962 V16 N6 P43-45 NLLTB V5 N1 P1 <NLL>
1962 V16 N6 P45-50 AD-622 462 <=*$>
1962 V16 N7 P57-59 NLLTB V4 N12 P1089 <NLL>
1962 V16 N7 P57-58 63-13340 <=> P

Citation	Translation
1962 V16 N8 P1-4	62-33718 <=*>
1962 V16 N8 P28-32	62-11793 <=>
1962 V16 N8 P45-46	62-11793 <=>
1962 V16 N8 P54	63-13102 <=*>
1962 V16 N9 P11-13	3906 <BISI>
1962 V16 N9 P55-56	63-13102 <=*>
1962 V16 N9 P56-58	63-13334 <=*>
1962 V16 N11 P26-33	63-21269 <=>
1962 V16 N11 P34-40	63-21156 <=>
1962 V16 N12 P54	63-21426 <=>
1963 N1 P40-42	NLLTB V5 N10 P914 <NLL>
1963 V17 P19-20	63-21694 <=>
1963 V17 N3 P1-5	63-21694 <=>
1963 V17 N3 P19-22	63-21694 <=>
1963 V17 N4 P3	63-31132 <=>
1963 V17 N4 P27-32	63-31132 <=>
1963 V17 N4 P56	63-31224 <=>
1963 V17 N7 P39-41	08582R <ATS>
1963 V17 N7 P53-54	63-31669 <=>
1963 V17 N10 P31-34	65-60356 <=$>
1963 V17 N11 P36-38	64-21239 <=>
1964 N5 P12-13	64-41078 <=>
1964 N5 P37-40	64-41078 <=>
1964 V18 N2 P14-15	64-19990 <=$>
1964 V18 N3 P1-4	NLLTB V6 N10 P896 <NLL>
	TRANS.BULL.1964 P896 <NLL>
1964 V18 N3 P5-11	64-51229 <=>
1964 V18 N5 P8-12	64-41027 <=>
1964 V18 N6 P22	RTS-2850 <NLL>
1964 V18 N9 P47-56	65-30290 <=$>
1964 V18 N11 P14-19	65-30650 <=$>
1964 V18 N11 P42-45	65-30429 <=$>
1964 V18 N12 P15-23	65-30573 <=$>
1965 N2 P39-42	NLLTB V7 N11 P946 <NLL>
1965 N5 P1-4	NLLTB V8 N1 P14 <NLL>
1965 N5 P35-39	65-31910 <=$>
1965 N9 P23-25	65-33058 <=*$>
1965 N10 P38-40	65-33857 <=$>
1965 N10 P50	65-33857 <=$>
1965 N10 P51	65-33857 <=$>
1965 N11 P41-42	66-30059 <=*$>
1965 V19 P52-55	65-30893 <=$>
1965 V19 N6 P5-7	RTS-3225 <NLL>
1965 V19 N6 P49-53	66-30979 <=$>
1966 N1 P25	NLLTB V8 N10 P833 <NLL>
1966 N3 P2-8	66-32129 <=$>
1966 N3 P5-8	NLLTB V8 N11 P943 <NLL>
1966 N3 P22-26	RTS-3672 <NLL>
1966 N4 P22-23	66-32873 <=$>
1966 N10 P56	66-35699 <=>
1966 N11 P49	67-30407 <=>
1966 V20 N4 P43-45	66-32975 <=$>
1966 V20 N6 P19-21	66-34586 <=$>
1966 V20 N6 P49-52	66-34502 <=$>
1966 V20 N6 P56-57	66-34502 <=$>
1966 V20 N7 P1-3	66-34877 <=$> 0
1966 V20 N8 P35-36	66-34846 <=$>

MEKHANIZATSIYA I ELEKTRIFIKATSIYA NA SELSKOTO STOPANSTVO

Citation	Translation
1966 N9 P31-32	66-35213 <=>

MEKHANIZATSIYA I ELEKTRIFIKATSIYA SOTSIALISTICHESKOGO SELSKOGO KHOZYAISTVA

Citation	Translation
1953 N3 P40-48	63-23001 <=*>
1957 N1 P24-28	63-19141 <=*>
1957 N4 P44-45	R-4435 <*>
1958 N2 P25-31	60-17910 <+*>
1959 V17 N3 P45-49	NIAE-TRANS-163 <NLL>
	TR-163 <NIAE>
1959 V17 N5 P51-53	61-19687 <=*>
1961 V19 N2 P4-7	61-27207 <=>
1961 V19 N2 P10-12	61-27207 <=>
1961 V19 N2 P25-29	61-27207 <=>
1961 V19 N2 P34-36	61-27207 <=>
1961 V19 N2 P51-54	61-27207 <=>
1962 N6 P3-7	63-21249 <=>
1962 V20 N1 P31-33	64-15879 <=*$>
1962 V20 N2 P29-32	TR.190 <NIAE>
1962 V20 N2 P32-33	TR.174 <NIAE>
1962 V20 N6 P38-40	TR.192 <NIAE>
1963 V21 N1 P14-17	TR.173 <NIAE>
1963 V21 N3 P20-23	NIAE-TRANS-162 <NLL>
1963 V21 N3 P49-50	65-14289 <*>
1964 V22 N2 P3-5	64-31324 <=>
1964 V22 N4 P1-3	64-51757 <=$>
1964 V22 N4 P21-24	64-51757 <=$>
1964 V22 N4 P42	64-51757 <=$>
1964 V22 N6 P11	65-30436 <=$>
1964 V22 N6 P32-34	65-30436 <=$>
1965 V23 N1 P30-36	65-31069 <=$>
1965 V23 N2 P1-4	65-31449 <=$>
1965 V23 N2 P14-16	65-31449 <=$>
1965 V23 N2 P28-30	65-31449 <=$>

MEKHANIZATSIYA SILSKOHO HOSPODARSTVA

Citation	Translation
1964 V15 N3 P32	64-51412 <=$>
1964 V15 N10 P17-18	65-30119 <=$>
1964 V15 N10 P30-31	65-30119 <=$>

MEKHANIZATSIYA STROITELSTVA

Citation	Translation
1950 V7 N3 P9-13	52/2587 <NLL>
	59-19097 <+*>
1951 V8 N2 P23-25	61-13656 <+*>
1951 V8 N4 P14-17	52/2582 <NLL>
	59-19095 <+*>
1951 V8 N4 P18-21	52/2596 <*>
	59-19098 <+*>
1951 V8 N6 P24-28	59-19100 <+*>
1951 V8 N12 P12-15	61-13372 <+*>
1952 V9 N1 P8-13	61-13659 <+*>
1952 V9 N2 P6-9	59-19282 <+*>
1952 V9 N4 P11-18	61-13658 <+*>
1952 V9 N8 P27-29	61-13660 <+*>
1952 V9 N8 P29-31	61-13661 <+*>
1953 V10 N7 P3-6	61-13662 <+*>
1953 V10 N8 P30-31	61-13657 <+*>
1954 V11 N10 P17-21	R-1629 <*>
1955 V12 N8 P3-8	59-17196 <+*> C
1956 V13 N1 P6-10	61-19830 <=*>
1957 V14 N1 P44-91	59-11136 <=> 0
1957 V14 N5 P6-8	61-15289 <+*>
1957 V14 N5 P12-16	61-15290 <+*>
1957 V14 N5 P18-23	61-15291 <+*>
1957 V14 N5 P29-30	61-15289 <+*>
1958 V15 N2 P8-14	59-11610 <=> 0
1958 V15 N4 P9-11	59-19281 <+*>
1958 V15 N7 P5-9	61-13283 <+*>
1958 V15 N8 P1-3	60-11086 <=>
1959 N12 P38-41	60-11682 <=>
1959 V16 N1 P5-12	59-13865 <=>
1959 V16 N1 P12-18	59-13865 <=>
1959 V16 N1 P18-24	59-13865 <=>
1959 V16 N9 P1-6	62-32809 <=*>
1959 V16 N9 P19-24	60-11305 <=>
1959 V16 N9 P26-28	60-17691 <+*>
1959 V16 N12 P1-3	60-11682 <=>
1960 V17 P3-8	60-41248 <=>
1960 V17 N1 P9-15	60-11860 <=>
1960 V17 N3 P1-19	60-41145 <=>
1960 V17 N5 P22-26	60-41327 <=>
1960 V17 N6 P8-12	61-23849 <*=>
1960 V17 N6 P30-31	61-23849 <*=>
1960 V17 N7 P21-23	61-11972 <=>
1960 V17 N9 P4-11	61-11861 <=>
1960 V17 N9 P16-18	61-11861 <=>
1960 V17 N10 P1-7	61-11919 <=>
1960 V17 N11 P3	61-21685 <=>
1960 V17 N11 P13	61-21685 <=>
1960 V17 N12 P1-2	61-21639 <=>
1960 V17 N12 P16-20	61-21639 <=>
1961 V18 N1 P26-27	61-23556 <=>
1961 V18 N2 P1-2	61-23013 <=>
1961 V18 N2 P9-13	61-23013 <=>
1961 V18 N2 P23	61-23013 <=>
1961 V18 N2 P25	61-23013 <=>
1961 V18 N3 P20-21	62-24275 <=*>
1961 V18 N3 P22	61-27074 <=>
1961 V18 N5 P7-11	61-27380 <=>
1962 V19 N10 P1-2	63-13347 <=>

1962 V19 N11 P5-8 63-13648 <=>
1962 V19 N11 P20-23 63-13648 <=>
1963 V20 N1 P1-2 63-21182 <=>
1963 V20 N7 P18-20 63-31797 <=>
1964 V21 N2 P22-24 LC-1287 <NLL>
1964 V21 N3 P22-24 LC-1287 <BRC>
1964 V21 N7 P29-30 64-51774 <=$>
1964 V21 N11 P3 65-30751 <=$>

MEKHANIZATSIYA TRUDOEMKIKH I TYAZHELYKH RABOT
1948 V2 N4 P28-31 RT-1432 <*>
1951 V5 N11 P9-10 61-13279 <=*>
1952 V6 N6 P29-33 60-15845 <+*>
1953 V7 N2 P10-12 63-19222 <=*>
1953 V7 N8 P42-45 61-13655 <+*>
1954 N3 P32-35 R-1229 <*>
1957 V11 N3 P23-24 60-17249 <+*>
1958 V12 N1 P18-23 60-17236 <+*>

MELDINGER FRA NORGES LANDBRUKSHOISKOLE
1953 N4 P78-84 58-1748 <*>

MELIORACJE ROLNE
1958 V3 P143-155 61-31312 <=>
1958 V3 N13 P175-182 61-31306 <=>

MELLIAND TEXTILBERICHTE
1934 V15 N7 P292-294 61-14049 <*>
1937 V15 P378-381 AL-355 <*>
1937 V18 P301-304 58-1492 <*>
1937 V18 P367-370 58-1494 <*>
1937 V18 P382-384 58-1059 <ATS>
1937 V18 P446-448 58-1494 <*>
1937 V18 P459-460 58-1059 <ATS>
1940 V21 N9 P441-443 61-14051 <*>
1941 V22 P21-28 20K21G <ATS>
1941 V22 P194-198 63-20509 <*> 0
1942 V23 P73-77 61-14052 <*>
1943 V24 P211-215 63-20504 <*>
1943 V24 N6 P249-254 61-14053 <*> 0
1943 V24 N7 P289-293 61-14054 <*> 0
1944 V25 N7 P234-236 59-10909 <*>
1947 V28 P150-153 61-14056 <*> 0
1949 V30 P359-363 58-626 <*>
1949 V30 P525-528 58-635 <*>
1950 V31 P47-48 58-633 <*>
1950 V31 P255-260 64-10229 <*>
1950 V31 P674-675 61-14058 <*>
1951 V32 P53-56 64-18875 <*>
1951 V32 P296-302 54H9G <ATS>
1951 V32 P314-317 57-803 <*>
1951 V32 P520-521 AL-110 <*>
1951 V32 P941-942 AL-727 <*>
1951 V32 P955-958 61-20702 <*>
1951 V32 N3 P205-209 CSIRO-3485 <CSIR>
1951 V32 N3 P210-212 57-3074 <*>
1951 V32 N6 P421-424 CSIRO-3487 <CSIR>
1952 V33 P619- I-316 <*>
1952 V33 P737-738 8784 <IICH>
1952 V33 P739-743 C-3286 <NRCC>
1952 V33 N1 P37-39 CSIRO-3055 <CSIR>
1952 V33 N6 P488-491 CSIRO-3488 <CSIR>
1952 V33 N6 P522-525 C-3223 <NRCC>
1952 V33 N6 P525-531 64-20056 <*>
1952 V33 N7 P598-601 61-14059 <*> 0
1952 V33 N7 P620-623 64-20527 <*> 0
8783 <IICH>
1952 V33 N7 P639-643 51P64G <ATS>
1952 V33 N8 P737-738 64-20528 <*> 0
1952 V33 N9 P823-828 58-963 <*>
1952 V33 N11 P1040-1043 64-10790 <*> 0
1953 V34 P971-972 I-676 <*>
1953 V34 P1076- I-676 <*>
1953 V34 N8 P749-752 1983 <*>
1953 V34 N9 P860-863 78H9G <ATS>
1953 V34 N5/6 P436-438 64-18934 <*>
1953 V34 N5/6 P527-530 64-18934 <*>
1953 V34 N8/9 P712-713 1849 <*>
1953 V34 N8/9 P822-823 1849 <*>
1953 V34 N9/11 P850-852 1865 <*>
1953 V34 N9/11 P951-953 1865 <*>
1953 V34 N9/11 P1065-1067 1865 <*>
1954 V35 P1322-1323 61-14523 <*> 0
1954 V35 N5 P533 63-10035 <*> 0
1954 V35 N6 P640-645 2456 <*>
1954 V35 N7 P725-727 1874 <*>
1954 V35 N10 P1084-1086 61-14521 <*> 0
1954 V35 N10 P1119-1121 60-10770 <*>
1954 V35 N11 P1227-1229 61-14522 <*> 0
1954 V35 N11 P1229-1230 61-14520 <*> 0
1955 V36 P873-877 2844 <*>
1955 V36 N1 P8 2999 <*>
1955 V36 N1 P55-58 61-14524 <*> 0
1955 V36 N2 P163-166 61-14524 <*> C
1955 V36 N3 P265-267 61-14524 <*> 0
1955 V36 N4 P356-360 57-2669 <*>
67G6G <ATS>
1955 V36 N4 P360-362 57-2739 <*>
68G6G <ATS>
1955 V36 N4 P362-367 3031 <*>
1955 V36 N5 P419 58-289 <*>
1955 V36 N5 P427-431 60-10771 <*>
1955 V36 N5 P466-469 57-2692 <*>
69G6G <ATS>
1955 V36 N7 P686-691 CSIRO-3113 <CSIR>
1955 V36 N7 P776-780 CSIRO-3113 <CSIR>
1955 V36 N8 P736- 57-2890 <*>
1955 V36 N8 P763- 57-1982 <*>
1955 V36 N10 P1028-1033 85H8G <ATS>
1955 V36 N10 P1033-1036 45H9G <ATS>
1955 V36 N11 P1108-1109 65-11584 <*>
1955 V36 N12 P1216-1217 60-18414 <*> C
1956 V37 P75-80 3438 <*>
1956 V37 P94-98 3442 <*>
1956 V37 P101-103 3440 <*>
1956 V37 P680-685 60-19901 <*+> 0
1956 V37 N1 P75-80 C2367 <NRC>
1956 V37 N1 P91-93 69H9G <ATS>
1956 V37 N1 P94-98 C2361 <NRC>
1956 V37 N1 P101-103 C2352 <NRC>
1956 V37 N7 P776-778 CSIRO-3302 <CSIR>
1956 V37 N7 P789-795 CSIRO-3301 <CSIR>
1956 V37 N8 P892-895 CSIRO-3321 <CSIR>
1956 V37 N10 P1142-1149 61-14525 <*>
1956 V37 N12 P1438-1442 77J14G <ATS>
1956 V37 N12 P1442-1448 78J14G <ATS>
1957 V38 P296-300 3448 <*>
1957 V38 P313-319 9038 <IICH>
1957 V38 N1 P78-82 37J16G <ATS>
1957 V38 N2 P181-188 37J16G <ATS>
1957 V38 N3 P321-322 37J16G <ATS>
1957 V38 N4 P428-431 24J16G <ATS>
1957 V38 N4 P442-448 61-14526 <*> 0
1957 V38 N6 P648-653 58-1495 <*>
1957 V38 N6 P654-655 47J17G <ATS>
1957 V38 N7 P777-783 59-15313 <*>
1957 V38 N7 P783-787 57J18G <ATS>
1957 V38 N8 P898-904 59-15744 <*>
61-20647 <*>
1957 V38 N9 P1039-1043 59-15286 <*>
1957 V38 N10 P1152-1157 24J19G <ATS>
1957 V38 N10 P1159-1163 23J19G <ATS>
1957 V38 N11 P1269-1273 54K10G <ATS>
65-14188 <*>
1958 V39 N1 P26-33 65 11585 <*>
1958 V39 N1 P55-60 58-1499 <*>
1958 V39 N1 P55-61 60-18410 <*>
1958 V39 N2 P135-136 59-15289 <*>
1958 V39 N2 P204-211 59-15067 <*> 0
59-17775 <*>
1958 V39 N3 P292-297 59-15288 <*>
1958 V39 N4 P408-414 59-17261 <*>
1958 V39 N5 P552-554 59-17407 <*>
1958 V39 N6 P619-621 61-14527 <*> 0
1958 V39 N8 P879-882 46L29G <ATS>
1958 V39 N9 P999-1001 59-17762 <*>
1958 V39 N10 P1121-1126 59-15285 <*>
1958 V39 N10 P1141-1145 00L35G <ATS>
1958 V39 N10 P1157-1160 60-14295 <*>
1958 V39 N11 P1240-1243 60-17505 <+*>

1958 V39 N12 P1327-1332 61-20650 <*>
1959 V40 P314-322 9035 <IICH>
1959 V40 P413-417 34L34G <ATS>
1959 V40 P1307-1315 62-10677 <*> 0
1959 V40 N1 P11-17 19L37G <ATS>
1959 V40 N1 P27-30 65-14177 <*>
1959 V40 N4 P355-357 60-15974 <+*>
1959 V40 N4 P403-408 49L33G <ATS>
1959 V40 N4 P413-417 61-10521 <*> 0
1959 V40 N11 P1304-1305 58M37G <ATS>
1959 V40 N11 P1315-1326 65-13209 <*>
1960 V41 P868-870 62-10678 <*> 0
1960 V41 N3 P330-338 C-3475 <NRCC>
1960 V41 N5 P527-528 63-10827 <*>
1960 V41 N5 P573-576 62-20022 <*>
1960 V41 N5 P596-600 C-3699 <NRCC>
1960 V41 N5 P611-613 63-18374 <*>
1960 V41 N6 P733-736 62-20022 <*>
1960 V41 N6 P738-741 48M43G <ATS>
1960 V41 N6 P770-779 63-10803 <*>
1960 V41 N7 P793 63-10828 <*>
1960 V41 N7 P804-812 64-10936 <*>
1960 V41 N9 P1067-1069 25N54G <ATS>
1960 V41 N9 P1121-1124 62-20023 <*>
1960 V41 N9 P1125-1129 C-3948 <NRCC>
1960 V41 N9 P1135-1144 62-10403 <*> 0
1961 V42 N1 P16-20 48N51G <ATS>
1961 V42 N1 P73-80 26N54G <ATS>
1961 V42 N7 P758-760 62-20124 <*>
1961 V42 N10 P1167-1172 C-3921 <NRCC>
1961 V42 N11 P1275-1279 63-16465 <*>
1961 V42 N11 P1292-1301 84P59G <ATS>
1962 V43 N1 P1144-1149 64-16189 <*> 0
1962 V43 N2 P156-164 63-16465 <*>
1962 V43 N3 P258-265 63-16465 <*>
1963 V44 N4 P365-366 64-10869 <*>
1963 V44 N8 P848-850 64-18498 <*>
1963 V44 N12 P1312-1317 64-18447 <*>
1963 V44 N5/6 P484-487 236 <TC>
1964 V45 N1 P24-25 65-12170 <*>
1965 V46 N1 P22-28 921 <TC>
1965 V46 N1 P67-72 65-14573 <*> 0
1965 V46 N5 P443-445 1706 <TC>
1965 V46 N5 P465-474 1701 <TC>
1966 V47 N5 P547-551 66-13581 <*>

MELYEPITESTUDOMANYI SZEMLE
1963 V13 N12 P546-550 96S87H <ATS>

MEMO TECHNIQUE. OFFICE NATIONAL D'ETUDES ET DE RECHERCHES AERONAUTIQUES
1958 N1 P1- 59-10324 <*>

MEMOIRES DE L'ACADEMIE DE CHIRURGIE
1940 V66 P136-140 61-16492 <*>
1940 V66 N10 P680-682 61-14489 <*>
1955 V81 P834-842 <CP>
1955 V81 P954-956 <CP>
1963 V89 P223-227 65-13604 <*>

MEMOIRES DE L'ACADEMIE R. DE BELGIQUE. CLASSE DES SCIENCES. COLLECTION IN -8 DEGREES
1961 V33 N3 P207-211 AEC-TR-5720 <*>

MEMOIRES DE L'ACADEMIE R. DE MEDECINE DE BELGIQUE
1961 V4 P69-148 64-16939 <*> 0

MEMOIRES DE L'ACADEMIE R. DES SCIENCES, DES LETTRES ET DES BEAUX-ARTS DE BELGIQUE
1898 V57 P1-225 3287 <*>

MEMOIRES ET COMPTES RENDUS DES TRAVAUX DE LA SOCIETE DES INGENIEURS CIVILS DE FRANCE
1928 P266-300 64-14029 <*>
1937 N90 P562-582 T1969 <INSD>
1945 V98 N113 P65-77 AD-608 539 <=>
1948 P155-178 C-3400 <NRCC>
1948 N3/4 P189-225 58-2552 <*>
61-18994 <*>
1955 V108 N2 P138-142 E-3159 <BISI>
1958 N3/4 P200-204 61-20211 <*>

MEMOIRES COURONNES ET AUTRES MEMOIRES P.P. L'ACADEMIE R. DE MEDECINE DE BELGIQUE
1865 V17 N3/5 P10-11 3078 <*>
1865 V17 N3/5 P119-120 3078 <*>
1865 V17 N3/5 P127 3078 <*>
1865 V17 N3/5 P149-153 3078 <*>
1865 V17 N3/5 P155-156 3078 <*>

MEMOIRES DE MATHEMATIQUES ET DE PHYSIQUE, ACADEMIE DES SCIENCES. PARIS
1776 V7 P343-382 59-10319 <*>

MEMOIRES SCIENTIFIQUES DE LA REVUE DE METALLURGIE
1959 V56 P393-402 HW-TR-24 <*>
1959 V56 P427-452 63-01277 <*>
1959 V56 P505-511 63-00533 <*>
1959 V56 N1 P22-29 1783 <BISI>
62-16848 <*>
1959 V56 N2 P144-150 SCL-T-359 <*>
1959 V56 N2 P163-171 62-18577 <*> 0
1959 V56 N2 P179-202 62-16897 <*>
1959 V56 N3 P301-306 UCRL TRANS-714(L) <*>
1959 V56 N5 P463-470 63-14685 <*>
1959 V56 N7 P131-143 1607 <BISI>
1959 V56 N7 P641-655 62-14550 <*>
1959 V56 N10 P211-220 62-16896 <*>
1959 V56 N12 P641-655 1563 <BISI>
1960 V57 P57-61 UCRL TRANS-657(L) <*>
1960 V57 P203-214 2121 <BISI>
1960 V57 P232-240 2122 <BISI>
1960 V57 P254-259 HW-TR-23 <*>
1960 V57 P338-344 AEC-TR-5276 <*>
1960 V57 P409-422 AEC-TR-4388 <*>
1960 V57 P423-434 HW-TR-48 <*>
1960 V57 P502-510 HW-TR-19 <*>
1960 V57 P511-519 HW-TR-18 <*>
1960 V57 P659-675 1981 <BISI>
1960 V57 P721-727 HW-TR-21 <*>
1960 V57 P943-948 2644 <BISI>
1960 V57 N1 P1-15 62-00512 <*>
95M46F <ATS>
1960 V57 N1 P23-34 61-16805 <*>
1960 V57 N2 P88-90 62-16892 <*>
1960 V57 N2 P153-158 1958 <BISI>
1960 V57 N3 P173-178 1934 <BISI>
1960 V57 N3 P215-231 1786 <BISI> P
1960 V57 N4 P241-253 2634 <BISI>
1960 V57 N4 P265-277 1786 <BISI> P
1960 V57 N5 P325-337 TRANS-179 <FRI>
62-18928 <*>
1960 V57 N5 P345-362 TRANS-180 <FRI>
63-16641 <*>
1960 V57 N6 P409-422 70P62F <ATS>
1960 V57 N6 P450-458 TRANS-150 <FRI>
61-16142 <*>
1960 V57 N7 P493-499 1910 <BISI>
1960 V57 N7 P527-534 1918 <BISI>
1960 V57 N7 P535-549 1911 <BISI>
1960 V57 N7 P550-556 AEC-TR-5324 <*>
1960 V57 N8 P643-648 3370 <BISI>
1960 V57 N9 P649-657 2040 <BISI>
1960 V57 N10 P729-740 2007 <BISI>
1960 V57 N10 P741-754 2008 <BISI>
1960 V57 N11 P829-844 3501-I <BISI>
1960 V57 N11 P863-875 2585 <BISI>
1960 V57 N12 P889-900 65-00155 <*>
1960 V57 N12 P949-971 3501, II. <BISI>
1961 V58 P1-10 AEC-TR-4760 <*>
1961 V58 P343-349 2603 <BISI>
1961 V58 P401-413 2609 <BISI>
1961 V58 P699-712 AEC-TR-5035 <*>
1961 V58 N1 P61-72 62-10205 <*>
1961 V58 N3 P176-182 TRANS-174 <FRI>
62-16233 <*>
1961 V58 N4 P241-260 AEC-TR-4904 <*>
1961 V58 N5 P383-387 NP-TR-988 <*>
1961 V58 N6 P440-448 62-10451 <*>
1961 V58 N7 P481-495 2602 <BISI>

1961 V58 N7 P496-502 2787 <BISI>
1961 V58 N7 P510-516 62-10304 <*>
1961 V58 N7 P517-534 2609(PT-2) <BISI>
1961 V58 N8 P557-573 2573 <BISI>
1961 V58 N9 P677-698 3017 <BISI>
1961 V58 N11 P869-880 2995 <BISI>
1961 V58 N12 P901-914 2690 <BISI>
1961 V58 N12 P915-926 2786 <BISI>
1961 V58 N12 P931-947 2609(PT-3) <BISI>
1962 V59 P713-734 63-20053 <*> 0
1962 V59 N1 P75-78 2784 <BISI>
1962 V59 N2 P147-159 2906 <BISI>
1962 V59 N4 P273-285 3015 <BISI>
1962 V59 N6 P405-415 3014 <BISI>
1962 V59 N6 P454-460 63-14078 <*>
1962 V59 N10 P629-642 3311 <BISI>
1963 V60 P513-530 3571 <BISI>
1963 V60 P551-563 3861 <BISI>
1963 V60 P625-636 4246 <BISI>
1963 V60 N1 P11-22 63-20572 <*>
1963 V60 N1 P41-46 4021 <BISI>
1963 V60 N2 P143-146 3369 <BISI>
1963 V60 N3 P165-17C 127 <TC>
1963 V60 N3 P177-188 3596 <BISI>
1963 V60 N3 P215-235 64-16290 <*>
1963 V60 N10 P681-690 64-30704 <*>
1963 V60 N11 P753-761 64-14307 <*> P
1963 V60 N11 P785-796 3753 <BISI>
1963 V60 N11 P801-818 3750 <BISI>
1964 V61 P123-128 4493 <BISI>
1964 V61 P221-228 4073 <BISI>
1964 V61 P271-282 4209 <BISI>
1964 V61 P361-377 4210 <BISI>
1964 V61 P525-542 3991 <BISI>
1964 V61 P657-676 4308 <BISI>
1964 V61 P753-760 AI-TRANS-88 <*>
1964 V61 N1 P33-42 3728 <BISI>
1964 V61 N3 P209-220 3733 <BISI>
1964 V61 N5 P389-397 4076 <BISI>
1964 V61 N6 P413-436 4523 <BISI>
1964 V61 N6 P577-585 4070 <BISI>
1964 V61 N10 P677-686 4063 <BISI>
1965 V62 P1-29 4304 <BISI>
1965 V62 P249-260 4275 <BISI>
1965 V62 P373-378 4591 <BISI>
1965 V62 P683-690 66-11954 <*>
1965 V62 N2 P129-134 4427 <BISI>
1965 V62 N3 P183-196 4406 <BISI>
1965 V62 N4 P313-322 65-17083 <*>
1966 V63 N1 P59-70 66-12307 <*>
1966 V63 N3 P249-252 66-13613 <*>

MEMOIRES DE LA SOCIETE DES INGENIEURS CIVILS DE FRANCE
SEE MEMOIRES ET COMPTES RENDUS DES TRAVAUX DE LA SOCIETE DES INGENIEURS CIVILS DE FRANCE

MEMOIRES DE LA SOCIETE R. DES SCIENCES DE LIEGE
1952 V12 N1/2 P71-86 58-2480 <*>
1961 V4 P146-178 63-10245 <*>

MEMOIRS OF THE FACULTY OF ENGINEERING, HOKKAIDO UNIVERSITY
SEE HOKKAIDO DAIGAKU KOGAKUBU KIYO

MEMOIRS OF THE FACULTY OF FISHERIES, KAGOSHIMA UNIVERSITY
SEE KAGOSHIMA DAIGAKU SUISAN GAKUBU KIYO

MEMOIRS OF THE FACULTY OF TECHNOLOGY. THE TOKYO METROPOLITAN UNIVERSITY
1962 V12 P895-902 3804 <BISI>

MEMOIRS OF GEOGRAPHY. CHINESE PEOPLES REPUBLIC
SEE TI-LI-HSUEH TZE-LIAO

MEMOIRS OF THE INSTITUTE OF HIGH SPEED MECHANICS, TOHOKU UNIVERSITY
SEE TOHOKU DAIGAKU KOSOKU RIKIGAKU KENKYUJO HOKOKU

MEMOIRS OF THE KAKIOKA MAGNETIC OBSERVATORY
SEE CHIJIKI KANSOKUJO YOHO

MEMORIAL DE L'ARTILLERIE FRANCAISE
1929 V8 N32 P837-902 AL-290 <*>
1935 V14 P127-152 AL-458 <*>
1935 V14 P177 AL-458 <*>
1948 V22 P595-611 64-30812 <*> 0
1950 V24 N94 P851-897 AL-882 <*>
1951 V25 N3 P625-633 59-20072 <*>
1955 V29 P333-345 PT2 1275 <*>
1955 V29 N2 P333-345 59-12436 <+*>
1955 V29 N2 P497-503 57-1203 <*>
59-19543 <+*>
1955 V29 N4 P801-816 57-893 <*>
1955 V29 N4 P810-816 59-12433 <+*>

MEMORIAL DES POUDRES
1926 V22 P180-190 AL-8 <*>
1928 V23 P158-177 89F3F <ATS>
1937 V27 P253-273 63-18562 <*>
1939 V29 P134-196 1090 <*>
71H9F <ATS>
1948 V30 P7-42 3251 <*>
1948 V30 P139-141 3289 <*>
1950 P107-120 AL-246 <*>
1952 V34 P167-177 57-2539 <ATS>
76F4F <ATS>
1953 V35 P213-222 59-19244 <*+>
1953 V35 P273-286 2973 <*>
1954 V36 P37-39 63-14891 <*>
1955 V37 P19-24 57-2806 <*>
59-12029 <+*>
1955 V37 P111-119 57-3101 <*>
59-12022 <+*>
1955 V37 P121-126 57-2059 <*>
1955 V37 P139-148 58-815 <*>
59-12027 <+*>
1955 V37 P153-162 57-2778 <*>
1955 V37 P197-206 57-2300 <*>
59-12303 <+*>
1955 V37 P217-223 59-18003 <+*>
1955 V37 P331-338 59-18004 <+*>
1955 V37 P351-363 13H14F <ATS>
1955 V37 P413-416 58-804 <*>
59-12018 <+*>
63-16422 <=*>
1956 V38 P267-299 58-1790 <*>
1961 V43 N3699 P1-84 64-30690 <*>

MEMORIAL DES SCIENCES PHYSIQUES
1929 V10 P1-82 AD-607 723 <=>

MEMORIAL DES SERVICES CHIMIQUES DE L'ETAT
1947 V33 P417-422 63-20522 <*>
1948 V34 P125-137 61-18260 <*> 0
1948 V34 P163-177 61-18257 <*> 0
1952 V37 P75-83 64-18445 <*>
1953 V38 N2 P109-124 65-14788 <*>
1956 V41 N7 P317-322 00L34F <ATS>
1957 V41 P243-251 73L31F <ATS>

MEMORIAS DO INSTITUTO BUTANTAN. SAO PAULO
1942 V16 P275-283 2667 <*>
1951 V23 P51-62 AL-706 <*>

MEMORIAS DO INSTITUTO OSWALDO CRUZ. RIO DE JANEIRO
1939 V34 P611-614 1139 B <K-H>
1953 V51 P485-492 57-1276 <*>

MEMORIAS. LABORATORIO NACIONAL DE ENGENHARIA CIVIL LISBOA
1961 N84 P1-20 C-3734 <NRCC>

MEMORIAS DE LA SOCIEDAD CUBANA DE HISTORIA NATURAL 'FELIPE POEY'
1941 V25 P327-335 T-2103 <INSD>

MEMORIE DELL'ISTITUTO LOMBARDO DI SCIENZE E

LETTERE. MILANO. CLASSE DI SCIENZE MATEMATICHE E NATURALI
1939 S3 V24/5 N2 P41-123 66-13776 <*> P

MEMORIE E NOTE DELL'ISTITUTO DI GEOLOGIA APPLICATA UNIVERSITA DEGLI STUDI. NAPOLI
1951 V4 P17-32 62-20299 <*>

MEMORIE DELLA R. ACCADEMIA D'ITALIA. ROMA
1930 V1 N6 P1-78 64-16711 <*> O

MEMORIE DELLA R. ACCADEMIA DELLE SCIENZE DELL'ISTITUTO DI BOLOGNA
1948 V6 P3-28 57-1127 <*>

MEMORIE DELLA SOCIETA ASTRONOMICA ITALIANA
1964 S3 V35 N2 P207-209 66-13611 <*>

MENSCH UND ARBEIT
1948 V1 N2 P11-20 64-10302 <*>
1951 V3 P27-39 58-2242 <*>
1952 V4 P23-31 58-2236 <*>
1953 V5 P1-17 58-2241 <*>
1953 V5 P1-10 65-12576 <*>
1956 V8 N3/4 P1-9 59-20885 <*>
1958 V10 N1 P9-10 59-20873 <*>
1958 V10 N1 P11-13 59-20877 <*>

MERCKS JAHRESBERICHT UEBER NEUERUNGEN AUF DEM GEBIET DER PHARMAKOTHERAPIE U. PHARMAZIE
1928 V42 P5-19 I-855 <*>
1949 V63 P5-14 T1450 <INSD>

MERES ES AUTOMATIKA
1959 V7 N8 P248-251 60-11305 <=>
1961 V9 N2 P57-61 62-19200 <*=>
1961 V9 N3 P91-92 62-19199 <=*>
1961 V9 N4 P123-125 62-19679 <*=>
1961 V9 N12 P370-371 62-23658 <=*>
1964 V12 N6 P189-195 AD-639 551 <=$>
1965 V13 N6 P163-164 65-32067 <=$>
1965 V13 N7 P197-205 65-33158 <=$>
1965 V13 N7 P216-219 65-33158 <=$>
1965 V13 N7 P228-229 65-33158 <=$>

MERZLOTNYE ISSLEDOVANIYA
1963 V3 P236-244 RTS-2789 <NLL>

MERZLOTOVEDENIE
1946 V1 P29-30 66-10320 <*>

MESSEN-STEUERN-REGELN
1963 V6 N12 P498-501 3784 <BISI>
1963 V6 N12 P501-504 3785 <BISI>
1963 V6 N12 P508-511 3786 <BISI>
1964 V7 N2 P63-72 AD-608 461 <=>

MESURES ET CONTROLE INDUSTRIEL
1963 V29 N313 P799-803 64-16844 <*>

METAALBEWERKING
1958 N22 P447-451 61-18785 <*>
1958 N22 P452-457 61-18407 <*>
1958 N23 P467-470 61-18407 <*>

METAL PHYSICS. JAPAN
SEE KINZOKU BUTSURI

METALEN
1950 V4 N7 P129-137 SC-T-64-903 <*>
1954 V9 N5 P91-98 57-2003 <*>
1957 V12 P148-153 61-16243 <*>
1957 V12 P169-174 61-16243 <*>
1958 V13 N15 P268-272 4505 <HB>
1960 V15 P277-279 2203 <BISI>
1961 V16 P56-63 2547 <BISI>
1963 V18 N11 P332-340 64-20951 <*>

METALL
1949 N22 P367- 2315 <*>
1949 N23/4 P418- 2315 <*>
1949 V3 P187-192 62-16924 <*> O
1949 V3 N7/8 P111-115 II-395 <*>
1949 V3 N13/4 P219-222 2303 <*>
1949 V3 N9/10 P150-154 II-395 <*>
1950 V4 P317-321 63-20507 <*> O
1950 V4 P365-369 63-20507 <*> O
1950 V4 P374-377 63-20516 <*> O
1950 V4 P407-416 AL-224 <*>
1950 V4 N5 P193-195 62-16960 <*> O
1950 V4 N19 P416-420 63-16775 <*>
1950 V4 N5/6 P85-87 65-11439 <*>
1950 V4 N7/8 P125-129 II-569 <*>
1950 V4 N17/8 P370-374 64-20050 <*>
1950 V4 N19/0 P407-416 58-2030 <*>
1950 V4 N9/10 P171-178 I-1049 <*>
1951 N19/0 P434-436 57-1968 <*>
1951 V5 P8-13 II-125 <*>
1951 V5 P63-67 1282 <*>
1951 V5 P141-145 58-1328 <*>
1951 V5 P444-446 2317 <*>
1951 V5 N3/4 P53-57 II-178 <*>
1951 V5 N5/6 P98-101 AL-710 <*>
1951 V5 N7/8 P135-141 58-2037 <*>
1951 V5 N13/4 P291-292 2320 <*>
1951 V5 N19/0 P434-436 2840 <*>
1952 P579- II-214 <*>
1952 V6 P513-515 2312 <*>
1952 V6 P530-534 63-10021 <*> O
1952 V6 N17 P519-522 62-18608 <*> O
1952 V6 N11/2 P285-291 1100 <*>
57-274 <*>
1952 V6 N13/4 P360-362 AL-461 <*>
1952 V6 N17/8 P504-509 58-1057 <*>
1952 V6 N21/2 P674-679 II-1006 <*>
1952 V6 N23/4 P744-753 60-18698 <*> O
1953 P875 II-1173 <*>
1953 V7 P25-29 I-1022 <*>
1953 V7 P155-161 I-689 <*>
1953 V7 P427-429 62-18406 <*> O
1953 V7 N1 P1-9 62-16965 <*> O
1953 V7 N3/4 P106-108 57-3541 <*>
62-18578 <*> O
1953 V7 N5/6 P155-161 I-690 <*>
1953 V7 N5/6 P171-182 57-3541 <*>
62-18639 <*> O
1953 V7 N5/6 P189-191 57-2512 <*>
1953 V7 N7/8 P250-254 62-16988 <*> O
1953 V7 N11/2 P436-441 II-169 <*>
1953 V7 N15/6 P602-603 I-385 <*>
64-20813 <*>
1953 V7 N15/6 P608-610 57-2635 <*>
1953 V7 N19/0 P751-754 57-153 <*>
1953 V7 N23/4 P1003-1006 64-20406 <*> O
1953 V7 N9/10 P343-347 1906 <*>
1954 V8 P459-462 62-16989 <*> O
1954 V8 P611-614 AEC-TR-4360 <*=>
1954 V8 N2 P83-88 58-1479 <*>
1954 V8 N9 P675-676 62-18150 <*> O
1954 V8 N19 P749-768 62-01081 <*>
1954 V8 N21 P850-852 62-18426 <*> O
1954 V8 N1/2 P25-29 SC-T-524 <*>
1954 V8 N5/6 P180-184 2305 <*>
1954 V8 N7/8 P280-282 63-10002 <*> O
1954 V8 N19/0 P749-758 57-00111 <*>
1954 V8 N23/4 P923-929 II-210 <*>
1955 N1/2 P14-22 4408 <HB>
1955 V9 P305-403 61-00122 <*>
1955 V9 P554-560 62-16969 <*> O
1955 V9 P652-655 63-16847 <*>
1955 V9 N1/2 P1-6 62-20095 <*>
1955 V9 N1/2 P7-13 5368G <ATS>
57-1370 <*>
1955 V9 N1/2 P14-22 4408 <HB>
1955 V9 N1/2 P27-33 62-14055 <*> O
1955 V9 N3/4 P104-109 58-1294 <*>
1955 V9 N5/6 P164-171 57-2570 <*>
62-16912 <*> O
1955 V9 N11/2 P466-471 66-11344 <*> O
1955 V9 N11/2 P472-473 57-1907 <*>

1955 V9 N13/4 P560-564 60-18076 <*> 0
1955 V9 N13/4 P593- 1806 <*>
1955 V9 N15/6 P686-689 AI-TRANS-52 <*>
1955 V9 N17/8 P758-763 60-19907 <=*> 0
1955 V9 N17/8 P776-779 64-20565 <*> 0
1955 V9 N21/2 P947-954 59-17772 <*>
1955 V9 N9/10 P358-366 57-2773 <*>
1955 V9 N9/10 P366-376 57-1942 <*>
1955 V9 N9/10 P382-386 60-19907 <=*> 0
1956 V9 N5/6 P198-199 57-3371 <*>
1956 V10 P523-527 59-15162 <*> 0
1956 V10 P916-920 58-2244 <*>
1956 V10 N1/2 P16-20 63-10016 <*>
1956 V10 N1/2 P36-38 62-16801 <*> 0
1956 V10 N3/4 P113-116 58-1485 <*>
1956 V10 N314 P106- 58-1476 <*>
1956 V10 N5/6 P205-211 59-20314 <*>
1956 V10 N11/2 P513-519 42N49G <ATS>
1956 V10 N11/2 P520-523 62-00374 <*>
1956 V10 N17/8 P795-800 58-1512 <*>
1956 V10 N1910 P921-925 43N49G <ATS>
1956 V10 N21/2 P1042-1044 65-17329 <*>
1956 V10 N9/10 P419-423 C-2403 <NRC>
1957 V11 P193-196 58-1418 <*>
1957 V11 P357-361 TS 1156 <BISI>
1957 V11 P756-757 AEC-TR-4036 <*>
1957 V11 N1 P1-7 63-16862 <*>
1957 V11 N1 P8-9 59-17625 <*>
8706 <K-H>
1957 V11 N5 P357-361 61-20680 <*> 0
1957 V11 N6 P49-498 58-1706 <*>
1957 V11 N7 P598-604 66-10149 <*> 0
1957 V11 N8 P676-677 R-5240 <*>
1957 V11 N9 P737-740 66-11339 <*> 0
1957 V11 N9 P769 R-3426 <*>
1957 V11 N10 P848-854 58-2542 <*>
66-11053 <*> 0
1957 V11 N10 P854-859 4241 <HB>
1957 V11 N12 P1029-1032 30M39G <ATS>
1957 V11 N12 P1038-1045 62-16839 <*> 0
1958 V12 P366-380 61-10809 <*> 0
1958 V12 P803-810 TS 1156 <BISI>
1958 V12 P817-821 TS-1401 <BISI>
1958 V12 N1 P12-20 61-16251 <*>
1958 V12 N1 P28-32 58-2377 <*>
1958 V12 N4 P262-268 61-10766 <*> 0
1958 V12 N6 P501-503 29M39G <ATS>
1958 V12 N6 P508-511 61-16166 <*> 0
1958 V12 N7 P585-593 62-18143 <*> 0
1958 V12 N7 P612-619 13L31G <ATS>
62-10942 <*>
1958 V12 N7 P619-622 61-16167 <*> 0
1958 V12 N7 P625-628 4649 <HB>
1958 V12 N7 P630-636 66-10148 <*> 0
1958 V12 N8 P707-713 62-10956 <*>
1958 V12 N8 P713-721 64-20950 <*>
1958 V12 N9 P803-810 61-20680 <*> 0
1958 V12 N9 P814-816 62-18185 <*>
1958 V12 N9 P817-821 62-14399 <*>
1958 V12 N10 P904-906 60-17102 <*+> 0
1958 V12 N11 P992-995 64-10621 <*>
1958 V12 N11 P1007-1014 C-3631 <NRCC>
1959 V13 N5 P379-385 61-16171 <*> 0
1959 V13 N5 P390-392 66-11340 <*> 0
1959 V13 N5 P392-397 01M39G <ATS>
1959 V13 N6 P547-549 3279 <BISI>
1959 V13 N6 P551 3279 (PT.III) <BISI>
1959 V13 N8 P750-752 74T91G <ATS>
1959 V13 N8 P752-759 62-14329 <*>
1959 V13 N9 P819-823 TRANS-145 <FRI>
61-10716 <*>
1959 V13 N10 P919-922 4722 <HB>
1960 N3 P183-186 AD-631 174 <=*$>
1960 V14 P25-36 AEC-TR-4101 <*>
1960 V14 P546-548 61-16163 <*> 0
1960 V14 N1 P23-25 64-11816 <=>
1960 V14 N3 P196-201 TRANS-111 <MT>
1960 V14 N7 P655-659 15M44G <ATS>
1960 V14 N7 P695-696 62-16492 <*>
1960 V14 N8 P782-784 65-17370 <*> 0
1960 V14 N9 P875-878 5845 <HB>
1960 V14 N10 P995-998 61-16096 <*> 0
1960 V14 N11 P1061-1072 66-12337 <*> 0
1961 N3 P211-214 AD-631 173 <=*$>
1961 V15 P97-101 63-10894 <*>
1961 V15 P425-432 8836 <IICH>
1961 V15 P675-679 63-01147 <*>
1961 V15 N1 P19-22 TRANS-175 <MT>
1961 V15 N5 P410-414 5725 <HB>
1961 V15 N5 P422-425 63-16309 <*>
1961 V15 N8 P761-763 4112 <BISI>
1961 V15 N9 P883-891 65-10901 <*>
1961 V15 N10 P1004-1013 65-10901 <*>
1961 V15 N11 P1076-1078 AEC-TR-4980 <*>
1961 V15 N12 P1194-1198 62-14838 <*> 0
1962 V16 N1 P6-10 64-16495 <*> 0
1962 V16 N3 P198-204 70P60G <ATS>
1962 V16 N3 P204-209 71P60G <ATS>
1962 V16 N4 P293-300 63-20402 <*> 0
1962 V16 N5 P403-407 3698 <BISI>
1962 V16 N5 P413-419 63-20403 <*> 0
1962 V16 N7 P639-642 00Q69G <ATS>
1962 V16 N7 P656-661 AEC-TR-5770 <*>
1962 V16 N10 P979-984 AEC-TR-5770 PT.2 <*>
1962 V16 N11 P1120-1122 NS-64 <TTIS>
1962 V16 N12 P1193-1195 AEC-TR-6021 <*>
1963 V17 N1 P21-23 5856 <HB>
1963 V17 N1 P24-26 5857 <HB>
1963 V17 N1 P36-38 63-18516 <*>
1963 V17 N5 P433-436 6435 <HB>
1963 V17 N8 P788-791 65-17178 <*> 0
1963 V17 N10 P989-996 64-14344 <*>
1963 V17 N11 P1108-1116 64-14855 <*>
1963 V17 N12 P1209-1212 3712 <BISI>
1964 V18 P466-468 388 <TC>
1964 V18 P1172-1177 65-18102 <*>
1964 V18 N1 P8-16 64-18522 <*>
1964 V18 N2 P171-172 65-17360 <*> 0
1964 V18 N6 P581-589 89R77G <ATS>
1964 V18 N7 P704-708 64-30037 <*>
1964 V18 N9 P908-918 66-13289 <*>
1964 V18 N9 P918-922 AD-636 600 <=$>
1964 V18 N12 P1300-1305 66-11893 <*>
1965 V19 P442-450 1671 <TC>
1965 V19 N3 P206-212 1687 <TC>
1965 V19 N5 P455-462 1305 <TC>
1965 V19 N7 P728-734 66-11759 <*>
1965 V19 N10 P1049-1052 64S88G <ATS>

METALL UND ERZ
1925 V22 P316-321 II-609 <*>
1926 P306-315 II-523 <*>
1927 V24 N19 P465-472 58-316 <*>
1928 V25 N14 P343-350 II-671 <*>
1928 V25 N14 P439- II-671 <*>
1928 V25 N17 P437-439 AL-202 <*>
63-20764 <*>
1929 V26 N3 P62-66 59-20658 <*> 0
1929 V26 N4 P88-92 AL-203 <*>
63-20763 <*>
1930 V27 P474-486 468 <TC>
1931 V28 P101-111 II-668 <*>
1931 V28 N9 P214-217 57-3137 <*>
1932 V29 P313-317 II-451 <*>
1934 V31 P290-293 II-488 <*>
1934 V31 P293-295 II-487 <*>
1934 V31 N8 P169-179 5241-A <K-H>
64-16736 <*> 0
1935 V32 P33-40 II-602 <*>
1935 V32 P511-519 1802 <*>
1935 V32 N24 P589- 1109 <*>
1936 V33 N6 P153-154 II-662 <*>
1937 V34 N17 P453-460 59-20657 <*> 0
1939 V36 P63-72 I-211 <*>
1939 V36 N3 P63-72 65-11295 <*> 0
1939 V36 N12 P325- 1130 <*>
1940 V37 N10 P194- I-129 <*>
1942 V39 P7-13 63-16853 <*>
1942 V39 P28-32 63-16853 <*>
1943 P81-86 1821 <*>

Citation	Translation
1943 P105-109	1821 <*>
1943 V40 N17/8 P246-252	60-16787 <*> O
	60-16789 <*> O
1944 V41 P124-133	58-878 <*>
1944 V41 N17 P203-205	62-18520 <*>
1944 V41 N9/10 P101-106	58-1877 <*>

METALL: WIRTSCHAFT, WISSENSCHAFT, TECHNIK

Citation	Translation
1948 P75-80	2307 <*>
1948 V2 N13 P229-230	63-10015 <*> O

METALLBOERSE

Citation	Translation
1928 V18 N93 P2581-2582	I-327 <*>
1929 V19 P2861-2862	63-10422 <*>
1929 V19 N1 P8	63-16164 <*>
1930 V20 N56 P1545-1546	59-17275 <*>
1930 V20 N58 P1603	59-17275 <*>
1931 V21 N7 P147-148	59-17274 <*>
1931 V21 N9 P195-196	59-17273 <*>
1931 V21 N29 P675-676	59-15638 <*>
1931 V21 N31 P723-724	59-15638 <*>
1931 V21 N33 P771-772	59-15638 <*>
1931 V21 N52 P1927-1928	59-20678 <*>
1931 V21 N54 P1275-1276	59-20677 <*> O
1931 V21 N58 P1372-1373	59-20677 <*> O
1931 V21 N60 P1421-1422	59-20677 <*> O
1933 P1118	II-677 <*>
1934 V24 P825-826	II-591 <*>
1934 V24 P859	II-591 <*>
1934 V24 N41 P645-647	II-639 <*>
1934 V24 N42 P714-	II-639 <*>
1935 V25 N15 P225-226	59-20676 <*>

METALLFORSCHUNG

Citation	Translation
1946 V1 P81-86	58-2109 <*>
1947 V2 P1-8	I-793 <*>
1947 V2 P213-225	58-1089 <*>

METALLHUETTENBETRIEBE: DIE VORGAENGE UND ERZEUGNISSE DER METALLHUETTENBEITRIEBE VOM STANDPUNKTE DER NEUESTEN FORSCHUNGSERGEBNISSE

Citation	Translation
1915 P390-437	II-423 <*>

METALLOBERFLAECHE

Citation	Translation
1947 V1 N2 P25-28	64-10203 <*>
1948 V1 N8 P161-184	58-705 <*>
1948 V2 P25-37	58-1122 <*>
1949 V3 N6 PA117-A125	1107 <*>
	57-2547 <*>
1949 V3 N7 PA113-A141	63-18213 <*> P
1950 V4 N2 P17-	AL-180 <*>
1951 V3 N5 PB72-B76	58-1125 <*>
1951 V5 N1 PA1-A7	64-20386 <*>
1951 V5 N12 P177A-185A	58-1506 <*>
1952 V4 P33B-36B	II-931 <*>
1952 V4 P162B-163B	I-332 <*>
1953 V5 N5 PB74-B76	62-10056 <*>
1953 V7A N8 P119-121	57-1924 <*>
1953 V7A N12 P177-183	57-1924 <*>
1954 V6 P69B-73B	2003 <*>
1954 V6 N4 PB51-B53	62-16978 <*> O
1954 V6 N4 PB53-B55	62-16979 <*> O
	65-17405 <*>
1954 V8 N4 P55-61	57-2009 <*>
1954 V8 N7 PA97-A103	57-1089 <*>
1954 V8 N10 P157-160	64-20849 <*>
1954 V8 N11 P172-176	64-20848 <*>
1955 V9 P129A-135A	II-562 <*>
1955 V9 P205A-207A	TS-1281 <BISI>
1955 V9 N1 P1-6	57-2006 <*>
1955 V9 N1 P28-31	57-2006 <*>
1955 V9 N2 P17A-22A	62-18614 <*> O
1955 V9 N3 P38B-39B	62-18397 <*>
1955 V9 N4 P54A-57A	65-17406 <*>
1955 V9 N4 PB52-B55	62-14073 <*> O
1955 V9 N5 P65A-68A	62-16571 <*> O
1955 V9 N5 P70B-73B	62-18607 <*> O
1955 V9 N7 P97A-103A	2974 <*>
	57-1908 <*>
1955 V9 N8 P118A-121A	1429 <*>
1955 V9 N9 P135A-140A	66-10440 <*> O
1955 V9 N12 P205A-207A	62-10482 <*>
1955 V9 N12 P208A-209A	65-17321 <*>
1955 V9A N12 P205A-207A	66-10914 <*> O
1955 SA V9 N1 P1-6	57-2293 <*>
	57-3547 <*>
1955 SA V9 N2 P28-31	57-2293 <*>
	57-3547 <*>
1956 V10 P230-233	T-2252 <INSD>
	8708 <IICH>
1956 V10 N2 P33-34	2475 <NRC>
	3862 <HB>
1956 V10 N5 P129-135	66-10438 <*> O
1956 V10 N8 P230-233	3826 <HB>
1957 V11 N2 P60-63	66-10159 <*> O
1957 V11 N3 P89-96	66-10075 <*> O
1957 V11 N8 P255-257	66-10439 <*> O
1957 V11 N9 P281-285	61-16242 <*>
1957 V11 N12 P377-379	59-17624 <*>
1957 V11 N12 P393-400	66-10169 <*> O
1958 V12 P150-151	97040 <RIS>
1958 V12 N3 P185-192	1834 <TC>
1958 V12 N12 P361-363	4513 <HB>
1959 V13 P167-168	63-01025 <*>
1959 V13 P229-233	TS 1573 <BISI>
1959 V13 N2 P33-35	66-10073 <*> O
1959 V13 N3 P84-87	60-14499 <*>
1959 V13 N6 P161-166	AD-638 506 <=$>
1959 V13 N8 P229-233	62-14169 <*>
1959 V13 N8 P242-245	AEC-TR-5203 <*>
1959 V13 N9 P269-273	60-18174 <*>
1959 V13 N10 P315-316	66-10441 <*> O
1959 V13 N11 P342-345	2131 <BISI>
1960 V14 N1 P1-6	4780 <HB>
1960 V14 N1 PB1-B2	4783 <HB>
1960 V14 N8 P229-235	5155 <HB>
1960 V14 N9 P266-269	5071 <HB>
	61-16514 <*> O
1961 V15 N1 P1-12	3863 <BISI>
1961 V15 N2 P38-41	61-20747 <*> O
1961 V15 N3 P71-72	5922 <HB>
1961 V15 N4 P113-117	61-20744 <*>
1961 V15 N5 P130-133	5987 <HB>
1961 V15 N5 P134-138	61-20746 <*> O
	62-14797 <*>
1961 V15 N5 P145-147	61-20743 <*> O
1961 V15 N6 P165-168	5988 <HB>
1961 V15 N7 P193-197	22N55G <*>
1961 V15 N9 P261-264	3333 <BISI>
1961 V15 N10 P293-301	3333 <BISI>
1961 V15 N12 P372-382	64-14645 <*>
1962 V16 N1 P11-14	62-14839 <*> O
1962 V16 N3 P65-69	62-18915 <*> O
1962 V16 N3 P81-84	63-10147 <*> O
1962 V16 N4 PB49-B56	62P61G <ATS>
1962 V16 N6 P181-182	2988 <BISI>
1962 V16 N8 P245-246	5862 <HB>
1963 V17 N7 P197-202	64-18353 <*> O
1964 V18 P701-704	4520 <BISI>
1964 V18 N2 P33-37	4713 <BISI>
1964 V18 N3 P70-75	65-13860 <*>
1964 V18 N4 P123	64-18523 <*>
	65-11403 <*>
1964 V18 N9 P263-267	09S86G <ATS>
	65-13074 <*> O
	66-11320 <*> O
1964 V18 N11 P325-329	66-11319 <*> O
1965 V19 P230-231	1829 <TC>
1965 V19 N1 P1-8	6476 <HB>
1965 V19 N1 P13-17	MFA/TS 711 <MFA>
1965 V19 N3 P84-87	MFA/TS 717 <MFA>
1965 V19 N4 P98-104	6575 <HB>
1965 V19 N6 P161-173	66-14759 <*> O
	6606 <HB>
1965 V19 N6 P187-191	6605 <HB>
1965 V19 N8 P257-262	6834 <HB>
1965 V19 N8 P263-264	6893 <HB>
1965 V19 N9 P281-283	MFA/TS-765 <MFA>
1965 V19 N10 P310-312	MFA/TS-766 <MFA>
1965 V19 N11 P339-344	MFA/TS-766 <MFA>

Citation	Translation
1965 V19 N11 P344-346	MFA/TS-764 <MFA>
1966 V20 N1 P10-13	MFA/TS-822 <MFA>
1966 V20 N2 P91-99	MFA/TS-821 <MFA>
1966 V20 N4 P173-174	6919 <HB>
METALLOVEDENIE: SBORNIK STATEI	
1959 V3 P58-73	66-34494 <=$>
1959 V3 P214-229	62-13169 <*=> 0
1959 V3 P358-366	62-13064 <*=>
METALLOVEDENIE I OBRABOTKA METALLOV	
1955 N1 P11-18	60-16982 <*> 0
1955 N1 P21-24	4119 <HB>
1955 N2 P4-8	R-3688 <*>
1955 N3 P17-20	3911 <HB>
1955 N3 P21-25	TS-1330 <BISI>
	62-10485 <*=>
1955 N4 P24-27	RT-4098 <*>
1955 N5 P28-32	4196 <HB>
1955 N6 P52-56	R-4072 <*>
1955 V1 P19-26	R-3142 <*>
1955 V1 N3 P54-57	3733 <HB>
1955 V1 N4 P19-24	3675 <HB>
1955 V1 N4 P48-51	3735 <HB>
1955 V1 N5 P9-14	3965 <HB>
1955 V1 N5 P17-23	3847 <HB>
1955 V1 N6 P3-9	3736 <HB>
1956 P50-56	4040 <HB>
1956 N1 P40-41	TS-1100 <BISI>
	62-24236 <=*>
1956 N1 P42-45	4551 <HB>
1956 N2 P23-33	3977 <HB>
1956 N3 P15-17	61-19435 <*=> 0
1956 N3 P23-27	868TM <CTT>
1956 N3 P52-55	R-4461 <*>
1956 N3 P2327	83J17R <ATS>
1956 N6 P2-16	R-641 <*>
1956 N6 P16-25	R-642 <*>
1956 N6 P36-47	<MT>
	62-24169 <=*>
1956 N7 P10-15	3645 <BISI>
1956 N7 P16-23	AEC-TR-5463 <=*>
1956 N8 P43-47	3970 <HB>
1956 N11 P4-10	4719 <HB>
1956 N11 P26-29	UCRL TRANS-647(L) <*=>
1956 N11 P40-49	3975 <HB>
1956 N12 P40-45	4013 <HB>
1956 V2 N3 P26-35	3812 <HB>
1956 V2 N4 P45-50	3809 <HB>
1956 V2 N4 P50-56	3805 <HB>
1956 V2 N6 P16-25	3811 <HB>
1956 V2 N7 P35-38	3813 <HB>
1956 V2 N7 P39-48	3814 <HB>
1956 V2 N8 P28-30	3854 <HB>
1956 V2 N9 P25-30	3855 <HB>
1956 V2 N10 P34-36	3856 <HB>
1957 N1 P2-15	R-4734 <*>
	59-18367 <+*>
1957 N1 P43-50	60-10232 <+*> 0
1957 N2 P46-48	CSIR-TR.167 <CSSA>
1957 N2 P49-54	00J16R <ATS>
	4046 <HB>
1957 N3 P2-8	3967 <HB>
1957 N3 P27-30	<HB>
1957 N3 P57-60	4202 <HB>
1957 N4 P2-9	4137 <HB>
1957 N6 P2-7	3979 <HB>
1957 N6 P24-31	3980 <HB>
1957 N6 P31-42	4043 <HB>
1957 N6 P43-47	5022 <HB>
	60-15594 <+*>
1957 N6 P54-58	3982 <HB>
1957 N7 P2-7	4289 <HB>
1957 N7 P45-48	4045 <HB>
1957 N8 P15-16	R-3431 <*>
1957 N8 P21-24	60-16994 <*> 0
1957 N8 P51-56	61-23162 <*=>
1957 N9 P2-4	60-17459 <+*>
1957 N9 P21-25	4015 <HB>
1957 N9 P25-27	4262 <HB>
1957 N9 P47-51	62-24119 <=*>
1957 N10 P19-21	4514 <HB>
1957 N10 P21-22	4044 <HB>
1957 N11 P18-42	59-11618 <=> 0
1957 N11 P56-65	59-11619 <=> 0
1957 N11 P77-80	4095 <HB>
1957 N12 P31-36	61-19442 <+*>
1957 N12 P53-61	60-17152 <+*>
1957 N12 P61-66	C-3179 <NRCC>
1958 N1 P2-6	4106 <HB>
1958 N1 P7-10	4107 <HB>
	64-71358 <=> M
1958 N1 P11-16	4108 <HB>
1958 N1 P17-20	4118 <HB>
1958 N1 P21-24	4119 <HB>
1958 N1 P26-29	4120 <HB>
1958 N1 P30-35	TRANS-32 <MT>
	4121 <HB>
1958 N1 P35-38	4122 <HB>
1958 N1 P38-42	4123 <HB>
1958 N1 P43-46	4124 <HB>
1958 N1 P46-49	4125 <HB>
1958 N1 P49-55	4126 <HB>
1958 N1 P55-57	4127 <HB>
1958 N2 P2-6	4140 <HB>
1958 N2 P6-11	4146 <HB>
1958 N2 P12-18	4147 <HB>
1958 N2 P19-22	TRANS-33 <MT>
	4148 <HB>
1958 N2 P23-28	4149 <HB>
1958 N2 P28-35	4150 <HB>
1958 N2 P35-37	4151 <HB>
1958 N2 P56-57	PB 141 294T <=>
1958 N3 P2-6	4171 <HB>
1958 N3 P6-10	4164 <HB>
1958 N3 P11-15	4172 <HB>
1958 N3 P24-29	4163 <HB>
1958 N3 P30-34	4174 <HB>
1958 N3 P35-38	4175 <HB>
1958 N3 P38-41	4166 <HB>
1958 N3 P42-48	4176 <HB>
1958 N3 P48-53	4177 <HB>
1958 N3 P53-56	4178 <HB>
1958 N3 P56-61	4179 <HB>
1958 N3 P60-61	4169 <HB>
1958 N4 P2-7	4180 <HB>
1958 N4 P7-9	4181 <HB>
1958 N4 P9-16	4182 <HB>
1958 N4 P16-25	4183 <HB>
1958 N4 P25-28	4184 <HB>
1958 N4 P29	59-22675 <+*> 0
1958 N4 P39-43	4186 <HB>
1958 N4 P44-45	4187 <HB>
1958 N4 P45-49	4188 <HB>
1958 N4 P49-51	4189 <HB>
1958 N4 P61-64	4191 <HB>
1958 N5 P4-8	4214 <HB>
1958 N5 P8-13	4215 <HB>
1958 N5 P13-16	4216 <HB>
1958 N5 P16-20	4217 <HB>
1958 N5 P20-23	4218 <HB>
1958 N5 P23-27	4219 <HB>
1958 N5 P27-30	4220 <HB>
1958 N5 P31-37	4221 <HB>
1958 N5 P37-40	4222 <HB>
1958 N5 P40-43	4223 <HB>
1958 N5 P43-48	4224 <HB>
1958 N5 P48-51	4225 <HB>
1958 N6 P2-6	4244 <HB>
	59-13152 <=> 0
1958 N6 P6-9	4245 <HB>
1958 N6 P10-14	4246 <HB>
1958 N6 P14-17	59-13153 <=> 0
1958 N6 P18-21	4248 <HB>
	59-13154 <=> 0
1958 N6 P21-25	4249 <HB>
1958 N6 P25-32	4250 <HB>
	59-13155 <=> 0
1958 N6 P33-37	4251 <HB>
	59-13156 <=> 0

1958 N6 P37-41 4252 <HB>
1958 N6 P41-47 4253 <HB>
1958 N6 P47-52 4254 <HB>
1958 N6 P52-56 4255 <HB>
1958 N7 P2-10 4274 <HB>
1958 N7 P15-18 4276 <HB>
1958 N7 P19-21 4277 <HB>
1958 N7 P22-27 4278 <HB>
1958 N7 P27-30 4279 <HB>
1958 N7 P30-35 4280 <HB>
1958 N7 P35-38 4281 <HB>
1958 N7 P39-42 4282 <HB>
1958 N7 P43-46 4283 <HB>
1958 N7 P46-50 4284 <HB>
1958 N7 P50-52 4285 <HB>
1958 N7 P52-53 4286 <HB>
1958 N7 P54-58 4287 <HB>
1958 N8 P2-13 4313 <HB>
1958 N8 P14-17 4314 <HB>
1958 N8 P18-21 4315 <HB>
1958 N8 P22-25 4316 <HB>
1958 N8 P26-29 4317 <HB>
1958 N8 P29-33 4318 <HB>
1958 N8 P34-37 4319 <HB>
1958 N8 P38-43 4320 <HB>
1958 N8 P43-46 4321 <HB>
1958 N8 P46-51 4322 <HB>
1958 N8 P57-58 4324 <HB>
1958 N8 P59-60 4325 <HB>
1958 N9 P2-12 4348 <HB>
1958 N9 P12-15 4349 <HB>
59-22624 <+*>
1958 N9 P15-19 4350 <HB>
1958 N9 P20-23 4351 <HB>
1958 N9 P27-28 4353 <HB>
1958 N9 P29-33 4354 <HB>
1958 N9 P39-42 4356 <HB>
1958 N9 P42-46 4357 <HB>
4383 <HB>
1958 N9 P46-49 4358 <HB>
1958 N9 P49-53 4359 <HB>
1958 N9 P54-55 4360 <HB>
1958 N10 P2-5 4387 <HB>
1958 N10 P5-10 4388 <HB>
1958 N10 P11-17 4389 <HB>
1958 N10 P17-22 4390 <HB>
1958 N10 P23-27 4391 <HB>
1958 N10 P28-33 4392 <HB>
1958 N10 P33-36 4393 <HB>
1958 N10 P36-40 4394 <HB>
1958 N10 P47-50 4396 <HB>
1958 N10 P56-57 4398 <HB>
1958 N10 P58-62 4399 <HB>
1958 N11 P6-19 4427 <HB>
1958 N11 P19-25 4428 <HB>
61-23383 <*=>
1958 N11 P25-32 4429 <HB>
1958 N11 P32-37 4430 <HB>
1958 N11 P39-44 4431 <HB>
1958 N11 P44-50 4432 <HB>
1958 N11 P51-56 4433 <HB>
1958 N11 P57-59 4434 <HB>
1958 N11 P60-62 4435 <HB>
1958 N12 P2-9 4442 <HB>
1958 N12 P10-16 4443 <HB>
1958 N12 P17-20 4444 <HB>
1958 N12 P21-28 4445 <HB>
1958 N12 P29-35 4446 <HB>
1958 N12 P35-41 4447 <HB>
1958 N12 P42-45 4448 <HB>
1958 N12 P45-52 4449 <HB>
1958 N12 P53-56 4450 <HB>
61-13135 <*=>
1958 N12 P56-57 4451 <HB>
1958 N12 P58-61 4452 <HB>

★METALLOVEDENIE I TERMICHESKAYA OBRABOTKA METALLOV
1959 N1 P2-61 4474-4485 <HB>
1959 N1 P6-10 4474 <HB>
1959 N1 P10-15 4475 <HB>
1959 N1 P15-19 4476 <HB>
1959 N1 P19-25 4477 <HB>
1959 N1 P26-30 4478 <HB>
1959 N1 P31-39 4479 <HB>
1959 N1 P39-41 4480 <HB>
1959 N1 P42-47 4481 <HB>
1959 N1 P47-49 4482 <HB>
1959 N1 P52-54 4483 <HB>
1959 N1 P55-57 4484 <HB>
1959 N1 P57-59 4485 <HB>
1959 N1 P64 4474-4485 <HB>
1959 N3 P2-6 4535 <HB>
1959 N3 P6-13 4536 <HB>
1959 N3 P13-16 4537 <HB>
1959 N3 P17-19 4538 <HB>
1959 N3 P19-24 4539 <HB>
1959 N3 P25-28 4540 <HB>
1959 N3 P28-32 4541 <HB>
1959 N3 P33-37 4542 <HB>
1959 N3 P38-43 4543 <HB>
1959 N3 P44-46 4544 <HB>
1959 N3 P44-48 61-19547 <=*> 0
1959 N3 P46-48 4545 <HB>
1959 N3 P48-51 4546 <HB>
1959 N3 P51-55 4547 <HB>
62-15710 <*=>
1959 N3 P56-59 4548 <HB>
1959 N4 P2-8 4564 <HB>
1959 N4 P8-14 4565 <HB>
1959 N4 P14-19 4566 <HB>
1959 N4 P19-22 4567 <HB>
1959 N4 P22-27 4568 <HB>
1959 N4 P27-33 4569 <HB>
1959 N4 P34-40 4570 <HB>
1959 N4 P41-44 4571 <HB>
1959 N4 P45-47 4572 <HB>
1959 N4 P48-51 4573 <HB>
1959 N4 P52-53 4574 <HB>
1959 N4 P53-55 4575 <HB>
1959 N4 P55-57 4576 <HB>
1959 N5 P7-14 4577 <HB>
1959 N5 P15-19 4578 <HB>
1959 N5 P19-23 4579 <HB>
1959 N5 P24-28 4580 <HB>
1959 N5 P28-30 4581 <HB>
1959 N5 P35-40 4582 <HB>
1959 N5 P40-44 4583 <HB>
1959 N5 P45-50 4584 <HB>
1959 N5 P51 4585 <HB>
1959 N5 P52-54 4586 <HB>
1959 N5 P54-57 4587 <HB>
1959 N5 P57 4588 <HB>
1959 N5 P58-59 4589 <HB>
1959 N5 P59-60 4590 <HB>
1959 N6 P2-6 4698 <HB>
1959 N6 P7-13 4740 <HB>
1959 N6 P13-17 4760 <HB>
1959 N6 P17-19 4742 <HB>
60-16484 <+*> 0
1959 N6 P19-23 4743 <HB>
1959 N6 P24-38 4744 <HB>
1959 N6 P38-41 4745 <HB>
1959 N6 P41-45 4746 <HB>
1959 N6 P46-50 4747 <HB>
1959 N6 P50-54 4748 <HB>
1959 N6 P55-58 4749 <HB>
1959 N6 P59-61 4750 <HB>
1959 N6 P61-62 4751 <HB>
1959 N7 P2-10 4652 <HB>
1959 N7 P10-15 4653 <HB>
1959 N7 P16-21 4654 <HB>
1959 N7 P22-30 4655 <HB>
1959 N7 P30-45 4656 <HB>
1959 N7 P31-34 5079 <HB>
1959 N7 P45-49 4657 <HB>
1959 N7 P50-57 4658 <HB>
1959 N7 P58-60 4659 <HB>
1959 N7 P61-64 4660 <HB>
1959 N8 P8-12 4948 <HB>

Citation	Translation
1959 N8 P12-14	4949 <HB>
1959 N8 P14-20	4950 <HB>
1959 N8 P20-24	4951 <HB>
1959 N8 P24-27	4952 <HB>
1959 N8 P27-31	4953 <HB>
1959 N8 P32-38	4954 <HB>
1959 N8 P39-41	4955 <HB>
1959 N8 P41-43	4956 <HB>
1959 N8 P43-45	4957 <HB>
1959 N8 P45-48	4958 <HB>
1959 N8 P49-50	4959 <HB>
1959 N8 P51-52	4960 <HB>
1959 N8 P52-54	4961 <HB>
1959 N9 P2-8	4963 <HB>
	61-19388 <+*>
1959 N9 P8-12	4964 <HB>
1959 N9 P12-17	4965 <HB>
1959 N9 P17-19	4966 <HB>
1959 N9 P19-22	4967 <HB>
	61-19388 <+*>
	62-25398 <=*>
1959 N9 P23-29	4968 <HB>
1959 N9 P30	4969 <HB>
1959 N9 P35-38	4969 <HB>
1959 N9 P39-41	4970 <HB>
1959 N9 P42-43	4971 <HB>
1959 N9 P44-46	4972 <HB>
	63-14700 <=*>
1959 N9 P47-48	4973 <HB>
1959 N9 P49-51	4974 <HB>
1959 N9 P52-55	4975 <HB>
1959 N9 P55-57	4976 <HB>
1959 N9 P58-59	4977 <HB>
1959 N9 P60-62	4978 <HB>
1959 N10 P2-5	4979 <HB>
1959 N10 P6-16	4980 <HB>
1959 N10 P16-19	4981 <HB>
1959 N10 P20-23	4982 <HB>
1959 N10 P24-27	4983 <HB>
1959 N10 P27-32	4984 <HB>
1959 N10 P33-37	4985 <HB>
1959 N10 P38-42	4986 <HB>
1959 N10 P42-44	4987 <HB>
	61-23161 <*=>
1959 N10 P44-47	4988 <HB>
1959 N10 P48-50	4989 <HB>
1959 N10 P51-60	4990 <HB>
1959 N10 P61-62	4991 <HB>
1959 N11 P5-13	4992 <HB>
1959 N11 P13-19	4993 <HB>
1959 N11 P19-23	4994 <HB>
1959 N11 P24-39	4995 <HB>
1959 N11 P31-34	5162 <HB>
1959 N11 P40-42	4996 <HB>
1959 N11 P43-44	TRANS-14 <MT>
	4997 <HB>
1959 N11 P45-48	4998 <HB>
1959 N11 P50	5161 <HB>
1959 N11 P51-53	4999 <HB>
1959 N11 P54-61	5001 <HB>
1959 N12 P12-18	5003 <HB>
1959 N12 P19-23	5004 <HB>
1959 N12 P24-31	5005 <HB>
1959 N12 P35-38	5007 <HB>
1959 N12 P39-43	5008 <HB>
1959 N12 P44-47	5009 <HB>
1959 N12 P47-50	5010 <HB>
1959 N12 P51-52	5011 <HB>
1959 N12 P53-54	5012 <HB>
1959 N12 P55-60	5013 <HB>
1960 N1 P2-5	5292 <HB>
1960 N1 P5-10	5293 <HB>
1960 N1 P10-13	5294 <HB>
1960 N1 P13-14	5295 <HB>
1960 N1 P15-17	5296 <HB>
1960 N1 P17-19	5297 <HB>
1960 N1 P20-24	5298 <HB>
1960 N1 P25-27	5299 <HB>
1960 N1 P28-31	5300 <HB>
1960 N1 P31-35	5301 <HB>
	62-11652 <=>
1960 N1 P36-38	5302 <HB>
1960 N1 P39-42	5303 <HB>
1960 N1 P43-44	5304 <HB>
1960 N1 P45-47	5305 <HB>
1960 N1 P47-48	5306 <HB>
1960 N1 P49-53	5307 <HB>
1960 N1 P52-53	5308 <HB>
1960 N1 P61-64	5309 <HB>
1960 N2 P2-6	5310 <HB>
1960 N2 P7-10	5311 <HB>
1960 N2 P11-13	5312 <HB>
1960 N2 P14-20	5313 <HB>
1960 N2 P20-31	5314 <HB>
1960 N2 P32-37	5315 <HB>
1960 N2 P37-42	5316 <HB>
1960 N2 P42-47	5317 <HB>
1960 N2 P48-52	5318 <HB>
1960 N2 P53-58	5319 <HB>
1960 N2 P58-61	5320 <HB>
1960 N2 P62-64	5321 <HB>
1960 N3 P2-7	5322 <HB>
1960 N3 P7-11	5323 <HB>
1960 N3 P13-17	5325 <HB>
1960 N3 P18-21	5326 <HB>
1960 N3 P22-29	5327 <HB>
1960 N3 P31-40	5328 <HB>
1960 N3 P41-47	5329 <HB>
1960 N3 P47-51	5330 <HB>
1960 N3 P52-58	5331 <HB>
1960 N3 P61-63	5332 <HB>
1960 N4 P2-15	5376 <HB>
1960 N4 P15-19	5377 <HB>
1960 N4 P19-22	5378 <HB>
1960 N4 P22-25	5379 <HB>
1960 N4 P26-29	5380 <HB>
1960 N4 P30	5381 <HB>
1960 N4 P31-34	5391 <HB>
1960 N4 P35-37	5381 <HB>
1960 N4 P38-41	5382 <HB>
1960 N4 P41-45	5383 <HB>
1960 N4 P46-47	5384 <HB>
1960 N4 P48-50	5385 <HB>
1960 N4 P50-51	5386 <HB>
1960 N4 P51-53	5387 <HB>
1960 N4 P53-54	5388 <HB>
1960 N4 P54-56	5389 <HB>
1960 N4 P56-58	5390 <HB>
1960 N5 P2-7	5392 <HB>
1960 N5 P7-10	5393 <HB>
1960 N5 P11-15	5394 <HB>
1960 N5 P15-18	5395 <HB>
1960 N5 P19-23	5396 <HB>
1960 N5 P24-30	5397 <HB>
1960 N5 P30	5398 <HB>
1960 N5 P31-34	5407 <HB>
1960 N5 P35-39	5398 <HB>
1960 N5 P40-41	5399 <HB>
1960 N5 P42-45	5400 <HB>
1960 N5 P45-48	5401 <HB>
1960 N5 P48-52	5402 <HB>
1960 N5 P53-54	5403 <HB>
1960 N5 P54-55	5405 <HB>
1960 N5 P55-57	5405 <HB>
1960 N5 P57-58	5406 <HB>
1960 N7 P3-7	5442 <HB>
1960 N7 P7-14	5443 <HB>
1960 N7 P15-16	5444 <HB>
1960 N7 P17-19	5445 <HB>
1960 N7 P19-24	5446 <HB>
1960 N7 P24-26	5447 <HB>
1960 N7 P27-30	5448 <HB>
1960 N7 P31-40	5449 <HB>
1960 N7 P40-42	5450 <HB>
1960 N7 P42-47	5451 <HB>
1960 N7 P48-52	5452 <HB>
1960 N7 P52-54	5453 <HB>
1960 N7 P55-57	5454 <HB>
1960 N7 P58-61	5455 <HB>
1960 N8 P3-9	5475 <HB>

1960 N8 P9-17	5476 <HB>
1960 N8 P17-19	5477 <HB>
	61-28183 <*=>
1960 N8 P20-25	5478 <HB>
1960 N8 P25-30	5479 <HB>
1960 N8 P30	5480 <HB>
1960 N8 P31-34	5484 <HB>
1960 N8 P35-38	5480 <HB>
1960 N8 P38-47	5481 <HB>
1960 N8 P48-53	5482 <HB>
1960 N8 P53-56	5483 <HB>
1960 N9 P2-6	5485 <HB>
	61-27633 <*=> 0
1960 N9 P7-11	5486 <HB>
	61-27632 <*=> 0
1960 N9 P12-16	5487 <HB>
	61-27628 <*=>
1960 N9 P17-19	5488 <HB>
1960 N9 P20-25	5489 <HB>
1960 N9 P25-29	5490 <HB>
1960 N9 P29-30	5491 <HB>
1960 N9 P31-34	5537 <HB>
1960 N9 P35-38	5491 <HB>
1960 N9 P38-41	5492 <HB>
1960 N9 P42-45	5493 <HB>
1960 N9 P46-51	5494 <HB>
1960 N10 P5-13	5495 <HB>
	61-27630 <*=> 0
1960 N10 P14-19	5496 <HB>
1960 N10 P19-21	5497 <HB>
1960 N10 P22-30	5490 <HB>
1960 N10 P31-36	5499 <HB>
1960 N10 P36-40	5501 <HB>
1960 N10 P41-46	5502 <HB>
	61-27993 <=*>
1960 N10 P46-54	5503 <HB>
1960 N10 P54-57	5504 <HB>
1960 N10 P57-6C	5505 <HB>
1960 N11 P2-7	5510 <HB>
1960 N11 P5	5522 <HB>
1960 N11 P7-12	5511 <HB>
1960 N11 P12-15	5512 <HB>
1960 N11 P16-20	5513 <HB>
1960 N11 P20-24	5514 <HB>
1960 N11 P25-28	5515 <HB>
	63-00769 <*>
1960 N11 P31-34	5524 <HB>
1960 N11 P37-38	5517 <HB>
1960 N11 P39-42	5518 <HB>
1960 N11 P42-44	5519 <HB>
1960 N11 P46	5520 <HB>
1960 N11 P46-47	5521 <HB>
1960 N11 P48	5522 <HB>
1960 N12 P2-7	5525 <HB>
1960 N12 P8-15	5526 <HB>
1960 N12 P16-17	5527 <HB>
1960 N12 P18-21	5528 <HB>
1960 N12 P21-25	5529 <HB>
1960 N12 P26-30	5530 <HB>
1960 N12 P31-33	5536 <HB>
1960 N12 P35-36	5530 <HB>
1960 N12 P36-39	5531 <HB>
1960 N12 P39-41	5532 <HB>
1960 N12 P42-43	5533 <HB>
1960 N12 P44-49	5534 <HB>
1960 N12 P49-52	5535 <HB>
1961 N1 P29-33	62-24837 <=*>
1961 N2 P2-11	63-19682 <=*>
1961 N2 P24-27	62-23872 <=*>
1961 N2 P59-61	61-31579 <=>
1961 N3 P27-30	5660 <HB>
1961 N3 P35	5660 <HB>
1961 N3 P40-42	5659 <HB>
1961 N3 P47-51	63-13925 <=*>
1961 N4 P60-64	61-28308 <=*>
1961 N6 P35-41	62-25350 <=*> 0
1961 N7 P23-28	62-33389 <=*>
1961 N7 P47-48	62-24778 <=*>
1961 N8 P29-36	62-24822 <=> P
1961 N9 P2-8	NLLTB V4 N3 P213 <NLL>
1961 N9 P33-43	62-32990 <=*> C
1961 N10 P13-16	62-25305 <=*> 0
1961 N11 P12-19	63-19289 <=*>
1961 N11 P25-33	63-13387 <=*>
1961 N11 P36-40	63-13387 <=*>
1961 N12 P2-12	62-23999 <=*>
1961 N12 P46-47	63-13904 <=*>
1962 N1 P38-40	63-13914 <=*>
1962 N1 P38-42	63-19790 <=*> C
1962 N1 P57-64	63-13914 <=*>
1962 N4 P2-6	63-13378 <=*>
1962 N4 P36-40	63-23357 <=*>
1962 N6 P2-5	63-19399 <=*>
1962 N6 P10-13	63-19400 <=*>
1962 N6 P14-17	63-19399 <=*>
1962 N7 P10-13	M.5329 <NLL>
1962 N7 P24-26	5681 <HB>
1962 N7 P52-57	64-23550 <=>
1962 N7 P64-65	62-33463 <=*>
1962 N9 P2-5	63-23206 <=*$>
	64-23549 <=>
1962 N9 P20-22	63-23210 <=*>
1962 N9 P22-23	5712 <HB>
1962 N9 P62-63	63-19288 <=*>
1962 N10 P5-8	63-16537 <=*> 0
1962 N10 P22-23	5749 <HB>
1962 N10 P194-201	63-13457 <=*>
1962 N12 P23-26	65-11763 <*>
1962 N12 P37-4C	M.5334 <NLL>
1962 N12 P40-43	M.5330 <NLL>
1962 N12 P49-50	5903 <HB>
1963 N1 P21-23	5904 <HB>
1963 N1 P25-29	63-23930 <=*$>
1963 N3 P59-61	63-41026 <=>
1963 N5 P61-63	5994 <HB>
1963 N8 P62-63	63-41026 <=>
1963 N11 P10-14	64-19160 <=$>
1964 N1 P2-5	64-19142 <=$>
1964 N1 P40-44	6179 <HB>
	65-60956 <=$>
1964 N1 P44-47	64-19143 <=$>
1964 N1 P48-49	6181 <HB>
1964 N1 P50-51	6182 <HB>
	65-60957 <=$>
1964 N1 P52-55	6183 <HB>
1964 N1 P55-56	6184 <HB>
1964 N3 P37-41	M-5696 <NLL>
1964 N3 P61-63	6220 <HB>
1964 N4 P15-18	65-61190 <=>
1964 N5 P1C-13	6284 <HB>
1964 N5 P15-17	4394 <BISI>
1964 N5 P52-54	M-5609 <NLL>
1964 N5 P55-56	M.5587 <NLL>
1964 N5 P56-58	65-30849 <=$>
1964 N7 P2-7	64-51968 <=$>
1964 N7 P47-48	M-5591 <NLL>
1964 N8 P9-11	M-5620 <NLL>
1964 N8 P39-41	M-5612 <NLL>
1964 N10 P57-58	6417 <HB>
1964 N11 P2-5	AD-620 773 <=$>
1964 N11 P2-24	65-30472 <=*$>
1964 N11 P39-41	64-51996 <=$>
1964 N12 P39-41	6565 <HB>
1965 N1 P12-15	M-5698 <NLL>
1965 N2 P47-48	M-5745 <NLL>
1965 N3 P2-5	66-32847 <=$>
1965 N4 P51-53	6481 <HB>
1965 N5 P15-21	65-31589 <=$>
1965 N5 P33-35	65-31589 <=$>
1965 N5 P45-50	65-31589 <=$>
1965 N5 P53-63	65-31589 <=$>
1965 N7 P25-29	M.5823 <NLL>
1965 N11 P37-38	6688 <HB>
1966 N2 P50-51	66-31855 <=$>

METALL-REINIGUNG+VORBEHANDLUNG

1960 V9 N1 P3-9	2378 <BISI>
1960 V9 N4 P53-56	1992 <BISI>
1960 V9 N5 P69-71	2280 <BISI>
1961 V10 N6 P89-93	5261 <HB>

Citation	Translation
1961 V10 N6 P108-113	5261 <HB>
1962 V11 N1 P1-8	3148 <BISI>
1963 V12 N1C P189-190	66-11745 <*>

METALLURG. MOSKVA, LENINGRAD

Citation	Translation
1933 V8 N3 P11-19	63-18111 <=*>
1934 V9 N4 P52-67	64-10235 <=*$>
1935 N6 P106-108	RT-1005 <*>
1935 V10 N3 P27-37	63-14290 <=*>
1935 V10 N4 P87-99	64-14900 <*> 0
	64-20839 <*>
1935 V10 N6 P82-105	RT-1761 <*>
	64-10446 <=*$> 0
1935 V10 N7 P100-113	63-18136 <=*>
	64-10458 <=*$> 0
1935 V10 N1C P1C0-113	RT-2145 <*>
1935 V10 N11 P85-98	RT-1517 <*>
	2928I <K-H>
	64-14158 <=*$> 0
1935 V10 N12 P67-73	RT-1518 <*>
	64-10158 <=*$> 0
1936 V11 N2 P54-72	R-3334 <*>
1936 V11 N2 P72-78	R-2417 <*>
1936 V11 N5 P19-33	3394 <HB>
1936 V11 N5 P48-56	R-2969 <*>
1936 V11 N12 P102-110	R-2583 <*>
1937 V12 N4 P27-41	R-2582 <*>
1937 V12 N4 P131-133	65-61192 <=>
1937 V12 N6 P31-36	3304 <HB>
1937 V12 N8 P85-92	RT-779 <*>
	63-207-65 <=*$>
1938 V13 N1 P96-99	5866-B <K-H> 0
	64-18500 <*> 0
1938 V13 N3 P15-19	R-2550 <*>
1938 V13 N11 P72-79	R-3350 <*>
1939 V14 N1 P15-27	R-3829 <*>
1939 V14 N9 P33-44	R-2247 <*>
1939 V14 N4/5 P26-34	R-2252 <*>
1940 V15 N1 P57-58	RT-2252 <*>
1940 V15 N2 P3-15	2461 <HB>
1940 V15 N2 P16-20	61-17187 <=$>
	62-23978 <=*> 0
1940 V15 N8 P3-8	66 <HB>
1940 V15 N9 P19-22	3485 <HB>

*METALLURG

Citation	Translation
1956 N2 P38-39	R-4980 <*>
1956 N4 P10-12	59-22477 <+*>
1956 N6 P36	59-18521 <+*>
1956 N7 P3-6	59-14178 <+*> 0
1956 N8 P12-15	3973 <HB>
	59-16544 <+*> 0
1956 N12 P16-19	4203 <HB>
1956 V1 P24-27	R-4465 <*>
1956 V1 N1 P10-11	3759 <HB>
1956 V1 N1 P20	3901 <HB>
1956 V1 N3 P1-7	61-23301 <*=>
1956 V1 N3 P14-16	4129 <HB>
1956 V1 N7 P18-21	3916 <HB>
1956 V1 N9 P19-22	4029 <HB>
1956 V1 N10 P5-7	3851 <HB>
1956 V1 N12 P23-24	4019 <HB>
1957 N5 P8-9	4333 <HB>
1957 N5 P16-18	3999 <HB>
1957 N5 P22-24	58-535 <*>
1957 N7 P3-5	4243 <HB>
1957 V2 N3 P6-8	61-13104 <*>
1957 V2 N3 P45-48	62-24138 <=*>
1957 V2 N5 P8	4083 <HB>
1957 V2 N5 P13-16	62-13205 <=*>
1957 V2 N5 P22-24	60-13048 <+*>
	761TM <CTT>
1957 V2 N6 P15-17	60-23084 <+*>
1957 V2 N9 P1-2	60-17159 <+*>
1957 V2 N9 P3-5	60-23075 <+*>
1957 V2 N9 P8-10	61-13116 <=*>
1957 V2 N9 P17-18	62-13196 <=*>
1957 V2 N10 P11-12	60-17162 <+*>
1957 V2 N10 P32-35	60-17163 <+*> 0
1958 N1 P15-17	84L31R <ATS>
1958 N2 P19-20	4242 <HB>
1958 N3 P16-21	4228 <HB>
1958 N4 P3-5	4205 <HB>
1958 N5 P12-16	4267 <HB>
1958 N6 P15-17	4343 <HB>
1958 N8 P11-14	4424 <HB>
1958 N1C P10-14	4423 <HB>
1958 N10 P15-18	4425 <HB>
1958 N11 P3-8	4550 <HB>
1958 V3 N2 P15-17	62-13209 <=*>
1958 V3 N3 P3-6	4728 <HB>
1958 V3 N5 P30-34	62-24089 <=*>
1958 V3 N6 P27-29	59-20085 <+*>
1958 V3 N6 P29-30	59-20084 <+*> 0
1958 V3 N9 P7-10	62-24203 <=*>
1959 N4 P3-7	4664 <HB>
1959 N4 P17-19	1600 <BISI>
1959 V4 N5 P4-6	5048 <HB>
1959 V4 N5 P11-13	4676 <HB>
1959 V4 N7 P17-21	TRANS-34 <MT>
1959 V4 N8 P17-20	TRANS-35 <MT>
1959 V4 N9 P7-9	TRANS-06 <MT>
1959 V4 N9 P17-19	TRANS-36 <MT>
1959 V4 N10 P4-6	TS-1838 <BISI>
1959 V4 N10 P32-36	63-16873 <=*> 0
1959 V4 N11 P14-16	TRANS-12 <MT>
1959 V4 N11 P17-20	TS 1651 <BISI>
	62-14704 <=*>
1959 V4 N12 P19-21	TRANS-18 <MT>
1959 V4 N12 P21-22	TRANS-87 <MT>
1960 V5 N1 P18-22	TRANS-172 <MT>
1960 V5 N2 P8-9	5016 <HB>
1960 V5 N2 P37-40	60-31115 <=>
1960 V5 N4 P4-8	TRANS-121 <MT>
1960 V5 N4 P14-17	TRANS-108 <MT>
1960 V5 N4 P21-24	TRANS-109 <MT>
1960 V5 N4 P28-29	TRANS-123 <MT>
1960 V5 N4 P30-31	TRANS-110 <MT>
1960 V5 N6 P9-11	1955 <BISI>
1960 V5 N6 P12-14	1956 <BISI>
1960 V5 N6 P22-24	TRANS-131 <MT>
1960 V5 N7 P5-7	TRANS-140 <MT>
1960 V5 N7 P7-9	TRANS-130 <MT>
1960 V5 N7 P21-23	TRANS-142 <MT>
1960 V5 N8 P4-6	TRANS-143 <MT>
	5124 <HB>
	62-23934 <=$>
1960 V5 N8 P7-8	TRANS-144 <MT>
1960 V5 N8 P18-19	TRANS-145 <MT>
1960 V5 N9 P7-8	TRANS-146 <MT>
1960 V5 N9 P9-10	TRANS-147 <MT>
1960 V5 N9 P16-17	5015 <HB>
1960 V5 N9 P19-20	TRANS-151 <MT>
1960 V5 N9 P21	TRANS-152 <MT>
1960 V5 N10 P3-7	2289 <BISI>
1960 V5 N10 P19-21	1866 <BISI>
1961 V6 N1 P18-20	2223 <BISI>
1961 V6 N1 P34	TRANS-181 <MT>
1961 V6 N2 P3-4	2154 <BISI>
1961 V6 N2 P4-6	TRANS-182 <MT>
1961 V6 N2 P7-11	TRANS-183 <MT>
1961 V6 N2 P23-24	TRANS-185 <MT>
1961 V6 N3 P6-9	TRANS-193 <MT>
1961 V6 N3 P9-12	2338 <BISI>
1961 V6 N3 P12-13	2337 <BISI>
1961 V6 N4 P4-6	TRANS-194 <MT>
	2457 <BISI>
1961 V6 N5 P10-12	TRANS-202 <MT>
1961 V6 N5 P25-26	5192 <HB>
1961 V6 N6 P3-7	TRANS-219 <MT>
1961 V6 N6 P7-9	TRANS-220 <MT>
1961 V6 N6 P13-16	5213 <HB>
1961 V6 N6 P17-18	TRANS-222 <MT>
1961 V6 N7 P4-8	TRANS-228 <MT>
1961 V6 N7 P11-13	TRANS-229 <MT>
1961 V6 N7 P13-17	TRANS-223 <MT>
1961 V6 N8 P4-6	TRANS-226 <MT>
1961 V6 N9 P3-4	TRANS-231 <MT>
1961 V6 N9 P5	TRANS-232 <MT>
1961 V6 N10 P10-12	TRANS-234 <MT>

1961 V6 N11 P3-8 TRANS-236 <MT>
1961 V6 N12 P2-4 TRANS-237 <MT>
1961 V6 N12 P6-8 TRANS-238 <MT>
1961 V6 N12 P12-15 2654 <BISI>
1962 V7 N1 P8-13 TRANS-242 <MT>
1962 V7 N2 P6-8 TRANS-249 <MT>
1962 V7 N2 P9-11 TRANS-250 <MT>
1962 V7 N2 P14-15 TRANS-251 <MT>
1962 V7 N2 P22-24 5560 <HB>
1962 V7 N2 P29-30 63-21347 <=>
1962 V7 N4 P1-8 63-13112 <=*>
1962 V7 N4 P9-10 TRANS-256 <MT>
1962 V7 N4 P13-14 TRANS-257 <MT>
1962 V7 N5 P1-2 62-32462 <=>
1962 V7 N5 P13-14 TRANS-266 <MT>
1962 V7 N6 P1-2 62-32547 <=>
1962 V7 N6 P3-5 TRANS-280 <MT>
62-33115 <=>
1962 V7 N7 P3 62-33520 <=>
1962 V7 N7 P7-8 62-33520 <=>
1962 V7 N7 P22 TRANS-279 <MT>
3073 <BISI>
1962 V7 N8 P2-5 TRANS-286 <MT>
1962 V7 N8 P19-23 62-33659 <=*>
1962 V7 N9 P1-2 62-33691 <=*>
1962 V7 N9 P3-7 5545 <HB>
1962 V7 N10 P4-7 63-15101 <=>
1962 V7 N10 P11-12 63-15101 <=>
1962 V7 N11 P5-10 63-13658 <=>
1962 V7 N11 P19-21 63-13860 <=*>
1962 V7 N12 P4-8 TRANS-318 <MT>
1963 V8 N1 P1-3 63-21616 <=>
1963 V8 N1 P34-35 63-21616 <=>
1963 V8 N2 P1-2 63-21693 <=>
1963 V8 N3 P3-5 TRANS-320 <MT>
5777 <HB>
1963 V8 N3 P10-20 63-31011 <=>
1963 V8 N4 P19-24 63-21921 <=> 0
1963 V8 N4 P22-24 TRANS-322 <MT>
1963 V8 N5 P1-2 63-31476 <=>
1963 V8 N5 P10-12 TRANS-324 <MT>
3323 <BISI>
1963 V8 N5 P19-21 63-31476 <=>
1963 V8 N5 P33-34 63-31476 <=>
1963 V8 N5 P36 63-31476 <=>
1963 V8 N5 P39 63-31476 <=>
1963 V8 N6 P1-3 63-31336 <=>
1963 V8 N6 P22-25 63-31336 <=>
1963 V8 N6 P34-35,40 63-31336 <=>
1963 V8 N7 P1-3 63-31880 <=>
1963 V8 N7 P8-9 63-31880 <=>
1963 V8 N7 P14-17 63-31880 <=>
1963 V8 N7 P34-36 63-31880 <=>
1963 V8 N8 P19-22 TRANS-360 <MT>
1963 V8 N11 P1-3 64-21378 <=>
1964 V9 N6 P1-3 64-41539 <=>
1964 V9 N9 P1-3 64-51707 <=$>
1965 V10 N1 P1-3 65-30893 <=$>
1965 V10 N1 P41 65-30893 <=$>
1965 V10 N2 P19-20 6437 <HB>
1965 V10 N3 P1-2 65-31238 <=$>
1965 V10 N9 P1-2 65-33116 <=$*>
1965 V10 N11 P1-24 66-34073 <=*$>
1966 V11 N1 P14-16 TRANS-673 <MT>
1966 V11 N6 P17-20 6920 <HB>
1966 V11 N6 P26-28 6929 <HB>

METALLURGIA ITALIANA
1931 V23 P265-291 59-20528 <*> 0
1935 V27 P563-569 59-20527 <*> 0
1935 V27 P570-575 59-20527 <*> 0
1935 V27 P703-706 59-20527 <*> 0
1938 V30 N11 P621-630 58-1904 <*>
1941 V33 N11 P478-492 66-15503 <*>
1942 V34 P5-29 66-15503 <*>
1942 V34 P50-64 66-15503 <*>
1942 V34 P90-106 66-15503 <*>
1947 V39 N5 P201-205 61-10807 <*> 0
1949 V41 N1 P1-6 58-2622 <*>
63-20153 <*>
1950 V42 N1 P22-29 59-20844 <*>
1950 V42 N12 P435-456 59-20561 <*> 0
1951 N3/4 P110-120 II-1041 <*>
1951 V43 P467-470 II-100 <*>
1952 N8/9 P424-431 II-990 <*>
1952 V44 P145-152 2526 <BISI>
1953 N9 P323-327 1846 <*>
1953 N12 P449-456 10TM <CTT>
1953 V45 N2 P41-46 64-20289 <*>
1953 V45 N5 P170-174 3388 <K-H>
1953 V45 N8 P273-283 02L34I <ATS>
1954 N9 P313-316 84J17I <ATS>
1954 V46 N11 P417-420 60-18697 <*>
1955 V47 P309-314 58-1547 <*>
1955 V47 P551-554 5256 <K-H>
1955 V47 N6 P251-258 57-2005 <*>
1956 N12 P449-456 57-2740 <*>
1957 N4 P275-289 58-1733 <*>
1957 V49 N3 P159-169 4109 <HB>
1957 V49 N6 P479-482 59-10097 <*>
1957 V49 N6 P483-487 2643 <BISI>
1957 V49 N9 P680-683 62-16021 <*>
1958 N8 P340-342 IS-TRANS-6 <*>
1958 V50 P173-180 63-14435 <*> 0
1959 V51 P257-273 TS 1702 <BISI>
1959 V51 P407-417 1977 <BISI>
1959 V51 N3 P89-92 4769 <HB>
1959 V51 N3 P94-100 TS 1619 <BISI>
62-16860 <*>
1959 V51 N4 P129-138 2052 <BISI>
1959 V51 N4 P139-144 2351 <BISI>
1959 V51 N7 P257-273 62-16868 <*>
1959 V51 N11 P509-514 AEC-TR-4811 <*>
68M44I <ATS>
1960 V52 P280-288 20[illegible] [illegible]I>
1960 V52 P661-668 2358 <BISI>
1960 V52 N7 P411-416 1897 <BISI>
1960 V52 N8 P551-560 61-18832 <*> 0
1960 V52 N8 P577-583 5575 <HB>
1960 V52 N11 P716-720 2990 <BISI>
1960 V52 N12 P786-794 5457 <HB>
1960 V52 N12 P795-800 5458 <HB>
1960 V52 N12 P835-841 NP-TR-929 <*>
1961 V53 P1-16 2360 <BISI>
1961 V53 P269-273 2527 <BISI>
1961 V53 P276 2527 <BISI>
1961 V53 N2 P53-59 3598 <BISI>
1961 V53 N4 P143-148 63-26201 <=$>
1961 V53 N8 P474-482 2748 <BISI>
1962 V54 N11 P485-496 5910 <HB>
1963 V55 P425-429 4283 <BISI>
1963 V55 N3 P107-111 3680 <BISI>
1963 V55 N4 P106-110 3390 <BISI>
1963 V55 N4 P129-144 NP-TR-1150 <*>
1963 V55 N8 P361-364 4282 <BISI>
1963 V55 N9 P449-455 3915 <BISI>
1963 V55 N9 P456-460 3699 <BISI>
1963 V55 N9 P467-473 6144 <HB>
1963 V55 N9 P477-478 6145 <HB>
1963 V55 N11 P560-562 3945 <BISI>
1963 V55 N12 P625-629 6215 <HB>
1963 V55 N12 P635-640 6214 <HB>
1964 V56 P54-60 4458 <BISI>
1964 V56 P343-349 4276 <BISI>
1964 V56 P450-456 4375 <BISI>
1964 V56 N1 P27-32 4151 <BISI>
1964 V56 N5 P175-181 6321 <HB>
1964 V56 N5 P181-187 6389 <HB>
1964 V56 N9 P443-449 6401 <HB>
1964 V56 N11 P539-544 6534 <HB>
1965 N4 P156-160 66-10495 <*>

METALLURGICAL JOURNAL. CHINESE PEOPLES REPUBLIC
SEE YEH CHIN PAO

METALLURGICHESKAYA I GORNORUDNAYA PROMYSHLENNOST
1962 N1 P3-4 62-32703 <=>
1962 N1 P58-60 62-32703 <=>
1962 N1 P67-68 62-32703 <=>
1962 N2 P19-23 62-32703 <=>

1962 N2 P48-51 62-32703 <=>
1962 N3 P3-6 62-32703 <=>
1962 N5 P3-7 63-15093 <=>
1962 N5 P12-19 63-15093 <=>
1962 N5 P48-53 63-21088 <=>
1962 N5 P73-75 63-15093 <=>
1962 N5 P86-87 63-15093 <=>
1962 N6 P8-14 63-21616 <=>
1963 N2 P50-53 63-31086 <=>
1963 N2 P61-66 63-31086 <=>
1963 N2 P70-71 63-31086 <=>
1963 N2 P81-84 63-31086 <=>
1963 N2 P87-88 63-31086 <=>
1963 N5 P37-41 59S82R <ATS>
758 <TC>
1964 N2 P24-26 64-31796 <=>
1964 N3 P77-79 64-41626 <=>
1964 N6 P8-12 65-30218 <=$>
1964 N6 P19-21 65-30218 <=$>
1964 N6 P75-76 65-30218 <=$>
1966 N3 P3-5 66-34792 <=$>
1966 N3 P89-92 66-34792 <=$>

METALLURGIE. BERLIN
1955 V5 N2 P50-60 TS 1691 <BISI>
62-16856 <*>
1955 V5 N3 P82-85 61-20096 <*>
1955 V5 N3 P85-102 2537 <BISI>

METALLURGIE. HALLE A. S.
1910 V7 N12 P396-402 AL-76 <*>
1910 V7 N23 P730-740 AL-944 <*>
1910 V7 N24 P755-770 AL-944 <*>
1911 N3 P72-77 58-456 <*>

METALLURGIE. MONS
1953 V1 N5 P127-138 2186 <BISI>
1960 V2 N1 P13-20 2107 <BISI>
1962 V3 N2 P27-38 64-14840 <*>
1963 V4 N4 P77-91 4277 <BISI>

METALLURGIE ET LA CONSTRUCTION MECANIQUE
1943 V75 N2 P3-6 62-16942 <*> 0
1943 V75 N3 P18 62-20272 <*>
1943 V75 N4 P15-16 62-20272 <*>
1944 V76 N2 P1-4 62-18242 <*> 0
1944 V76 N3 P1-4 62-18242 <*> 0
1945 V77 P22-23 58-922 <*>
1946 V78 N10 P15-17 62-20271 <*> 0
1947 V79 P17-21 64-71248 <=>
1953 V85 N1 P15-19 TS-1582 <BISI>
1956 V88 N4 P359-365 66-14643 <*> 0
1958 V90 N4 P263 TS-1375 <BISI>
1958 V90 N4 P263-271 62-10470 <*>
1958 V90 N4 P265 TS-1375 <BISI>
1958 V90 N4 P267 TS-1375 <BISI>
1958 V90 N4 P269-271 TS-1375 <BISI>
1959 V91 P541-543 2470 <BISI>
1959 V91 P545 2470 <BISI>
1959 V91 P547 2470 <BISI>
1959 V91 P665-673 2470 <BISI>
1960 V92 N3 P139-153 1954 <BISI>
1961 V93 N2 P131-133 64-10519 <*>
1963 V95 P443-450 3827 <BISI>
1965 V97 P795-797 4984 <BISI>

METALLURGIE UND GIESSEREITECHNIK
1952 N1 P14-17 TS 1232 <BISI>
1952 N1 P17-18 TS 1233 <BISI>
1952 V2 N1 P17-18 61-20669 <*>
1952 V2 N4 P109-112 5253 <HB>
1954 V4 P379-384 C-2242 <NLL>
57-258 <*>
1954 V4 N9 P379-384 64-16141 <*> 0
1954 V4 N11 P474-485 II-397 <*>

METALLURGIST. LONDON
1928 03/30 P47-48 AL-692 <*>

METALLURGIYA I METALLOVEDENIE CHISTYKH METALLOV
1959 N1 P5-43 60-11820 <=>
1959 N1 P44-62 60-11840 <=>
1959 N1 P63-69 60-41043 <=>
1959 N1 P78-90 60-41043 <=>
1959 N1 P91-105 60-11806 <=>
1959 N1 P106-143 60-11976 <=>
1959 N1 P144-178 60-41064 <=>
1959 N1 P179-212 60-41072 <=>
1959 N1 P213-243 60-41090 <=>
1960 N2 P1-335 61-23597 <*=>
1960 N2 P27-45 61-28235 <*=> 0
1961 N3 P82-95 AD-612 371 <=$>
1961 V3 P120-126 63-23619 <=*>
1961 V3 P137-151 4481 <BISI>
1963 N4 P47-57 AD-619 472 <=$>
1963 V4 P149-159 LA-TR-64-31 <=$*>

METALLURGIYA: SBORNIK STATEI
1959 V2 P3-21 61-28811 <*=>
1959 V2 P22-32 61-28812 <*=>
1959 V2 P89-114 62-13168 <*=> 0
1959 V2 P188-220 61-28990 <=*>
1959 V2 P221-235 62-15182 <*=> 0
1959 V2 P236-250 62-13176 <=*>
1959 V2 P251-268 62-13063 <*=> 0
1959 V2 P294-302 62-13177 <*=> 0

METALLWAREN-INDUSTRIE UND GALVANO-TECHNIK
1957 V48 N5 P194-204 5938 <HB>
1957 V48 N5 P214-215 4049 <HB>
1957 V48 N12 P528-535 62-18319 <*> 0

METALLWIRTSCHAFT, METALLWISSENSCHAFT, METALLTECHNIK
1925 N46 P197-198 63-16777 <*>
1930 V9 P499-502 59-20810 <*>
1930 V9 P1023-1028 66-10062 <*> 0
1930 V9 P1063-1066 59-15467 <*>
1930 V9 N4 P843-844 57-3230 <*>
1931 V10 P69-72 59-15867 <*> 0
1931 V10 P105-111 59-15713 <*>
1933 V12 P431- II-133 <*>
1933 V12 P431-434 64-20545 <*> 0
8777 <IICH>
1933 V12 P511 58-1975 <*>
1934 V13 P122- II-688 <*>
1934 V13 P655 58-1845 <*>
1934 V13 P725-731 196 <HB>
1934 V13 N23 P405-408 36 <HB>
1935 P1 58-1871 <*>
1935 V14 P265-267 57-2431 <*>
1935 V14 P545 58-1993 <*>
1935 V14 P605 58-1888 <*>
1935 V14 N2 P25-28 57-1052 <*>
1935 V14 N23 P445-449 59-17269 <*> 0
1935 V14 N23 P525-531 59-17269 <*> 0
1935 V14 N23 P581-587 59-17269 <*> 0
1935 V14 N23 P625-630 59-17269 <*> 0
1935 V14 N27 P525-531 57-2521 <*>
1935 V14 N30 P581-587 57-2521 <*>
1935 V14 N32 P625-630 57-2521 <*>
1935 V14 N46 P915-917 58-338 <*>
1936 N15 P27-32 2953D <KH>
1936 V15 P27-32 II-977 <*>
64-10442 <*>
1936 V15 P63-68 II-977 <*>
64-10442 <*>
1937 V16 P598-602 I-352 <*>
1937 V16 P721-725 II-542 <*>
1937 V16 P1038-1041 II-541 <*>
1937 V16 P1107-1112 57-3351 <*>
1937 V16 N22 P525-527 II-148 <*>
1938 N28 P755-757 II-171 <*>
1938 V17 P123-131 57-1151 <*>
1938 V17 P459-462 63-20970 <*>
1938 V17 P641-644 63-20973 <*>
1938 V17 P977-980 63-20974 <*>
1938 V17 N17 P459-462 57-2654 <*>
1938 V18 P833-835 64-10232 <*> 0
1939 V18 P576-584 63-18219 <*>

1939 V18 P945-950 61-10078 <*> O
1939 V18 N12 P249-254 62-18329 <*> O
1939 V18 N13 P271-276 62-18328 <*> O
1939 V18 N15 P315-320 62-18308 <*> O
1939 V18 N26 P563-567 59-20809 <*> O
1939 V18 N48 P963-968 57-1070 <*>
1940 V19 P141-143 57-1426 <*>
1940 V19 P404-407 64-10233 <*> O
1940 V19 P817-826 AL-569 <*>
1940 V19 P1005-1007 57-1076 <*>
1940 V19 P1029-1033 57-703 <*>
62-20082 <*>
1940 V19 N2/3 P247-251 60-10891 <*>
1940 V19 N12/3 P223-230 60-10891 <*>
1940 V19 N14/5 P267-276 AL-592 <*>
II-660 <*>
1940 V19 N14/5 P276-277 57-155 <*>
1942 V21 P131- I-554 <*>
1942 V21 N11/2 P158-162 58-454 <*>
1943 V22 P288-291 II-60 <*>
1943 V22 P401-405 II-94 <*>
1943 V22 P447-449 60-14422 <*>
1943 V22 P503-507 II-222 <*>
1943 V22 P543-545 57-705 <*>
1944 V23 N22/6 P221-227 58-2040 <*>
65-11443 <*>
1944 V23 N31/4 P20- I-364 <*>

METALS. JAPAN
SEE KINZOKU

METALS TECHNOLOGY REVIEW. TOHOKU UNIVERSITY
SEE TOHOKU DAIGAKU KINZOKU KENKYU-HOKOKU

METALURGIA. BUCURESTI
1964 V16 N10 P425-429 6479 <HB>

METALURGIA SI CONSTRUCTIA DE MASINI
1955 V7 N12 P41-45 2985 <BISI>
1957 V9 N12 P30-34 4711 <HB>
1959 V11 P928-933 4823 <HB>
1959 V11 N8 P671-682 4775 <HB>
1959 V11 N11 P934-939 2286 <BISI>
1959 V11 N12 P1033-1037 2286 <BISI>
1960 V12 N5 P390-395 5373 <HB>
1960 V12 N9 P831-836 5038 <HB>
1960 V12 N10 P865-867 5056 <HB>
1960 V12 N10 P868-870 5057 <HB>
1961 V13 N1 P1-9 5163 <HB>
1961 V13 N10 P847-857 5730 <HB>
1961 V13 N10 P858-862 5508 <HB>
1962 V14 P654 5771 <HB>
1962 V14 N1 P4-13 5622 <HB>
1962 V14 N10 P874-879 5891 <HB>
1963 V15 N4 P298-305 6427 <HB>
1963 V15 N6 P387-392 5724 <HB>
1963 V15 N8 P496-498 5726 <HB>
1963 V15 N8 P498-501 5727 <HB>
1963 V15 N12 P700-705 6207 <HB>

METALURGIA Y ELECTRICIDAD
1951 V15 N167 P36-42 64-20291 <*>
1951 V15 N168 P38-44 64-20291 <*>
1951 V15 N170 P40-46 64-20291 <*>

METANIERIBA DA TEKHNIKA
1959 N10 P4-7 AD-609 143 <=>
1959 N11 P14-17 AD-609 147 <=>
1960 N3 P14-18 AD-607 903 <=>
1961 N12 P25-28 AD-609 153 <=>

METANO
1954 V8 N5 P13-25 63-16846 <*>

METANO, PETROLIO E NUOVE ENERGIE
1957 V11 N2 P59-70 31K21I <ATS>

METAUX
1934 V9 P415-417 64-16723 <*> O
1936 V11 P31-37 II-420 <*>
1936 V11 N129 P116-117 60-14459 <*> O

METAUX ET CORROSION
1939 V14 P127-131 57-867 <*>

METAUX, CORROSION, INDUSTRIES
1951 V26 P126-130 I-81 <*>
1951 V26 P216-217 AEC-TR-4638 <*>
1951 V26 P235-249 I-498 <*>
1951 V26 N316 P417-496 64-10356 <*> O
1952 V27 N2 P312-317 57-885 <*>
1952 V27 N320 P143-149 62-16935 <*> O
1952 V27 N323 P312-317 60-18094 <*>
1954 V29 P88-89 62-18259 <*>
1954 V29 P399-403 67G7F <ATS>
1954 V29 N344 P151-166 2301 <*>
1955 V30 P78-87 61-16246 <*>
1955 V30 P294-303 62-18741 <*>
1955 V30 P476-483 TS-1348 <BISI>
1955 V30 N353 P34-36 II-966 <*>
1955 V30 N355 P134-138 62-16791 <*> O
1955 V30 N364 P476-483 62-10484 <*>
1956 P18-21 57-1776 <*>
1956 P274-288 58-2621 <*>
1956 N369 P219-232 57-3281 <*>
1956 N377 P18-21 C-2535 <NRC>
1956 V31 N367 P105-125 1815 <*>
1957 N379 P95-101 58-1276 <*>
1957 N379 P102-110 3426 <*>
1957 V32 P475-481 60-14126 <*>
1957 V32 N379 P122-131 63-14046 <*>
1957 V32 N388 P459-468 61-18450 <*>
1958 N397 P343-352 UCRL TRANS-713(L) <*>
1959 N401 P1-19 63-01695 <*>
1959 V34 P291-301 60-10527 <*>
1959 V34 P302-315 2592 <BISI>
1959 V34 N402 P48-71 2783 <BISI>
1959 V34 N402 P49-57 59-17157 <*> O
1959 V34 N406 P247-257 60-10527 <*>
1960 V35 N421 P336-343 05N56F <ATS>
1961 V36 N427 P107-111 62-14489 <*>
1961 V36 N427 P112-114 62-14488 <*>
1962 V37 N440 P127-141 5913 <HB>
1962 V37 N440 P141-153 5964 <HB>
1962 V37 N442 P227-231 63-20987 <*>
1965 N475 P91-109 65-14815 <*>
1965 V40 P91-109 4439 PT.1 <BISI>
1965 V40 P158-180 4439 PT.2 <BISI>
1965 V40 P202-219 4439 PT.3 <BISI>

METAUX, CORROSION-USURE
1941 V16 N189 P41-43 62-18235 <*> O
1943 V18 P209-213 I-219 <*>
1943 V18 N209 P1-21 AEC-TR-5247 <*>
1943 V19 N211 P43-48 61-16249 <*>
1944 V20 P92-93 62-16941 <*>
1945 V20 N236 P43-47 II-231 <*>
1947 V22 N259 P35-37 II-485 <*>
65-11414 <*> O
1948 V23 P15-18 II-992 <*>
1948 V23 P261- AL-256 <*>
1948 V23 N269 P1-4 62-18415 <*> O
1948 V23 N269 P9-11 62-16991 <*> O
1949 N287 P163-176 1083 <*>
1949 V24 N284 P87-117 63-10043 <*> O
1950 V25 P277-282 1218 <*>

METAUX ET MACHINES
1935 V19 P279-280 2653 <*>

*METEORITIKA
1951 N9 P71-101 RT-4128 <*>
1951 V9 P71-101 R-4938 <*>
1953 V11 P153-164 63-15313 <=*>
1956 V14 P38-53 R-4859 <*>
1956 V14 P62-69 59-20920 <+*>
62-10164 <=*>
1956 V14 P113-116 R-4638-A <*>
1958 N15 P115-135 IGR V2 N5 P380 <AGI>
1958 N15 P136-151 IGR V2 N4 P298 <AGI>

1959 V17 P85-92 62-33204 <=*>
1960 V18 P5-16 61-11932 <=>
1960 V18 P20-25 61-21296 <=>
1960 V18 P78-82 UCRL TRANS-533 <+*>
1960 V19 P103-104 61-23499 <*=>
1961 N20 P172-176 AD-622 241 <=*$>
1961 N21 P52-59 AD-622 241 <=*$>
1961 V20 P95-102 N65-14426 <=$>
1961 V20 P103-113 IGR V5 N7 P804 <AGI>
1961 V20 P124-136 64-13646 <=*$>
1961 V21 P15-31 64-71172 <=>
1962 N22 P71-73 AD-622 241 <=*$>
1962 N22 P74-82 AD-622 241 <=*$>
1962 N22 P104-109 AD-622 241 <=*$>
1962 V22 P83-93 64-71173 <=> M
1962 V22 P97-103 63-11547 <=>
1962 V22 P157-161 63-11548 <=>
1963 V23 P91-100 64-13358 <=*$>
64-21470 <=>
1963 V23 P103-130 64-13357 <=*$>
1964 V24 P5-15 N65-29731 <=*$>
1964 V24 P82-86 AD-614 018 <=$>
1964 V24 P108-111 65-63735 <=$>
1964 V25 P75-89 GI N2 1964 P350 <AGI>

METEOROLOGIA SI HIDROLOGIA
1956 N1 P5-10 PB 141 159T <=>
1957 V2 N2 P3-7 59-11567 <=>
1957 V2 N2 P39-43 59-11568 <=>

METEOROLOGIA, HIDROLOGIA SI GOSPODARINEA APELAR
1964 V9 N8 P377-382 64-51899 <=$>
1964 V9 N8 P432-438 64-51822 <=$> 0
1964 V9 N8 P439-445 64-51856 <=$>

METEORCLOGIA PRATICA. TURIN
1938 V19 P195-197 UCRL TRANS-513(L) <*>

METEOROLOGICAL MAGAZINE. CHINESE PEOPLES REPUBLIC
SEE CHI HSIANG HSUEH PAO

METEOROLOGICHESKIE ISSLEDOVANIYA
1963 N5 P100-107 66-21833 <$*>
1965 N8 P30-51 N66-12258 <=$>
1965 N9 P52-57 N66-12259 <=$>
1965 N9 P58-63 N66-12260 <=$>
1965 N9 P143-149 N66-12261 <=$>
1965 N9 P167-173 N66-13025 <=$>
1965 N9 P203-222 66-32844 <=$>

METEOROLOGICKE ZPRAVA
1958 V11 N3 P66-70 59-11436 <=*>
1958 V11 N4/5 P119-122 59 11637 <=> 0
1960 V13 N6 P140-146 62-19179 <=*>
1960 V13 N6 P153-156 62-19180 <=*>
1960 V13 N6 P159-160 62-19180 <=*>
1962 V15 N3/4 P104-105 65-12716 <*>
1963 V16 N3/4 P66-88 64-21309 <=> 0

METEOROLOGIE
1927 P420-425 CSIRO-3197 <CSIR>

METEOROLOGISCHE ABHANDLUNGEN. INSTITUT FUER METEOROLOGIE UND GEOPHYSIK, FREIE UNIVERSITAET BERLIN
1955 V2 N4 P39-52 58-122 <*>
1958 N5 P61-64 59-13151 <=>
1958 V4 N3 P1-55 59-00567 <*>
59-16278 <+*> 0
1963 V9 N6 P677-682 65-13281 <*>

METEOROLOGISCHE RUNDSCHAU
1949 V2 N3/4 P67-75 1583 <*>
1952 V5 P81-87 2417 <*>
1952 V5 P121-128 2417 <*>
1953 V6 N1/2 P1-6 58-2397 <*>
1954 V7 N12 P205-211 57-2586 <*>
58-2506 <*>
60-13213 <+*>
1954 V7 N3/4 P59-64 57-2588 <*>
1954 V7 N5/6 P100-101 2407 <*>
1954 V7 N11/2 P220-222 58-2399 <*>
1954 V7 N9/10 P161-166 59-18265 <+*>
1955 V8 N1/2 P12-16 58-1440 <*>
1956 V9 N1/2 P5-10 59-00480 <*>
1956 V9 N5/6 P89-92 CSIRO-3259 <CSIR>
1956 V9 N9/10 P171-173 CSIRO-3308 <CSIR>
1957 V10 N1 P11-20 CSIRO-3744 <CSIR>
1957 V10 N4 P119-124 58-2680 <*>
1957 V10 N6 P177-179 SCL-T-229 <+*>
1961 V14 N4 P125 SCL-T-474 <*>

METEOROLOGISCHE ZEITSCHRIFT
1895 P161-169 65-13370 <*>
1906 V9 P401-408 2159 <*>
1907 V24 P306-313 N65-13573 <=$>
1922 V39 N8 P241-242 57-625 <*>
1930 V47 P236-238 59-00588 <*>
1937 V54 P190-192 58-2505 <*>
1938 P415-417 CSIRO-3199 <CSIR>

METEOROLOGIYA I GIDROLOGIYA
1936 N1 P18-40 RT-569 <*>
1936 N6 P49-60 R-5003 <*>
1936 N8 P11-18 RT-570 <*>
1936 V1 N1 P91-93 RT-2015 <*>
1938 N3 P38-58 RT-729 <*>
1938 V4 N1 P21-28 RT-2461 <*>
1938 V4 N4 P123-131 RT-981 <*>
1938 V4 N6 P113-122 66-51077 <=$>
1939 N1 P36-46 RT-571 <*>
1939 V5 N1 P51-57 RT-1265 <*>
1939 V5 N5 P44-54 RT-4068 <*>
1939 V5 N5 P59-63 RT-509 <*>
1939 V5 N5 P85-89 RT-2299 <*>
1939 V5 N6 P34-45 RT-968 <*>
1939 V5 N9 P138-141 RT-2317 <*>
1939 V5 N10/1 P188-192 RT-467 <*>
1940 N4 P26-42 RT-971 <*>
1940 N8 P11-19 RT-931 <*>
1940 N9 P23-37 RT-972 <*>
1940 V6 N12 P33-45 R-756 <*>
1943 N6 P85-88 R-3296 <*>
1946 N1 P1-10 RT-2540 <*>
1946 N1 P33-38 RT-1519 <*>
1946 N1 P83-86 R-683 <*>
RT-4572 <*>
1946 N3 P28-35 RT-1520 <*>
1946 N3 P75-78 RT-725 <*>
1946 N3 P83-86 R-3212 <*>
1946 N4 P14-21 RT-1521 <*>
1946 N4 P59-70 65-11112 <*>
1946 N5 P21-31 RT-1475 <*>
1946 N5 P50-53 RT-1523 <*>
1947 N3 P30-33 RT-1476 <*>
1947 N6 P85-88 C-2497 <NRC>
1948 N2 P34-43 59-15998 <+*> 0
1949 N1 P119-122 59-15565 <+*> 0
1950 N2 P3-16 63-15311 <=*>
1950 N2 P34-45 63-15311 <=*>
1950 N2 P39-44 RT.1013 <*>
1952 N6 P8-11 65-17045 <*>
1952 N6 P29-33 87K21R <ATS>
1953 N1 P3-9 RT-2789 <*>
1953 N3 P7-10 RT-2790 <*>
1955 P121-125 60-41101 <=>
1955 N1 P3-7 RT-2971 <*>
1955 N1 P8-15 RT-2980 <*>
60-13841 <+*>
1955 N1 P23-28 RT-3031 <*>
1955 N1 P28-30 RT-3367 <*>
1955 N1 P31-32 RT-2981 <*>
1955 N1 P32-33 RT-2982 <*>
1955 N1 P42-43 RT-2983 <*>
1955 N1 P46-48 RT-2997 <*>
59-15572 <+*>
1955 N1 P52-58 RT-3003 <*>
1955 N1 P58-60 RT-2991 <*>
1955 N1 P66 RT-2992 <*>
1955 N1 P66-67 RT-3005 <*>

Citation	Translation
1955 N1 P67	RT-2993 <*>
1955 N1 P815-	R-2116 <*>
1955 N2 P30-32	R-4576 <*>
1955 N4 P21-24	R-27 <*>
1955 N4 P24-28	RT-3676 <*>
1955 N4 P28-31	R-53 <*>
1955 N4 P34-36	R-36 <*>
	60-15372 <+*>
1955 N4 P56-57	R-281 <*>
1955 N4 P57-58	R-48 <*>
1955 N5 P3-8	61-28623 <*=>
1955 N5 P15-21	RT-4370 <*>
1955 N5 P22-23	RT-4369 <*>
1955 N5 P24-26	RT-4368 <*>
1955 N5 P26-30	RT-4292 <*>
1955 N5 P30-35	RT-4191 <*>
1955 N5 P35-37	RT-4193 <*>
1955 N5 P40-44	RT-4198 <*>
1955 N5 P44-46	RT-4189 <*>
1955 N5 P46	RT-4188 <*>
1955 N5 P47-48	RT-4289 <*>
1955 N5 P49-52	RT-4290 <*>
1955 N5 P56-58	RT-4291 <*>
1955 N5 P58-60	RT-4362 <*>
1955 N5 P61-63	RT-4307 <*>
1955 N5 P63-64	RT-4293 <*>
1955 N5 P64-66	RT-4294 <*>
1955 N6 P3-6	R-44 <*>
1955 N6 P7-12	R-41 <*>
1955 N6 P13-18	R-52 <*>
1955 N6 P30-31	R-46 <*>
1955 N6 P34-35	R-30 <*>
1955 N6 P35-37	R-56 <*>
1955 N6 P37	R-29 <*>
1955 N6 P37-38	R-62 <*>
1955 N6 P40-43	R-310 <*>
1955 N6 P45-46	R-28 <*>
1955 N6 P48-49	R-55 <*>
1955 N6 P49-50	R-111 <*>
1955 N6 P53-54	R-61 <*>
1955 N6 P55-56	R-22 <*>
1955 N6 P58-59	R-42 <*>
1955 N6 P59-60	R-51 <*>
1955 N6 P60-61	R-31 <*>
1955 N6 P63-64	R-37 <*>
1955 N6 P66-67	R-63 <*>
1956 N1 P3-8	R-49 <*>
1956 N1 P9-12	R-34 <*>
1956 N1 P13-18	R-58 <*>
1956 N1 P25-29	R-40 <*>
1956 N1 P29-32	RT-4365 <*>
1956 N1 P32-34	CSIRO-3262 <CSIR>
1956 N1 P32-35	RT-4323 <*>
1956 N1 P50-52	RT-4322 <*>
1956 N1 P54-56	RT-4324 <*>
1956 N1 P56-	R-215 <*>
1956 N1 P57-58	R-221 <*>
1956 N1 P59-60	R-24 <*>
1956 N1 P60-61	R-54 <*>
1956 N1 P61-62	R-45 <*>
1956 N2 P3-9	RT-4413 <*>
1956 N2 P15-19	RT-4416 <*>
1956 N2 P19-21	RT-4415 <*>
1956 N2 P21-24	RT-4414 <*>
1956 N2 P32-34	R-3941 <*>
1956 N2 P35-37	RT-4408 <*>
1956 N2 P41-44	RT-4407 <*>
1956 N2 P46-47	RT-4444 <*>
1956 N2 P52-55	RT-4224 <*>
1956 N2 P58	RT-4410 <*>
1956 N2 P58-59	RT-4412 <*>
1956 N2 P60	RT-4409 <*>
1956 N3 P32-33	59-15558 <+*>
1956 N4 P12-18	59-17160 <+*> 0
1956 N4 P19-24	CSIRO-3260 <CSIR>
	62-23316 <=*>
1956 N4 P24-28	CSIRO-3261 <CSIR>
	R-4585 <*>
1956 N4 P35-36	59-17159 <+*> 0
1956 N4 P41-46	CSIRO-3263 <CSIR>
	R-4602 <*>
1956 N6 P41-44	59-15554 <+*> 0
1956 N6 P47-48	R-4388 <*>
1956 N6 P49-60	59-12055 <+*>
1956 N7 P8-13	59-15556 <+*> 0
1956 N7 P27-28	59-15559 <+*>
1956 N8 P9-13	C-2563 <NRC>
1956 N8 P20-23	CSIRO-3390 <CSIR>
	R-4374 <*>
1956 N8 P23-26	59-15882 <+*> 0
1956 N9 P12-21	C-2553 <NRC>
1956 N9 P32	60-17140 <+*>
1956 N9 P43-48	C-2537 <NRC>
1956 N10 P1-15	C-2560 <NRC>
1956 N10 P14-20	C-2724 <NRC>
1956 N11 P11-14	C-2714 <NRC>
1956 N11 P15-2C	59-15555 <+*> C
1956 N11 P21-28	59-18534 <+*>
1956 N11 P35-37	R-3640 <*>
1956 N12 P3-8	63-23798 <=*$>
1956 N12 P26-33	M.5117 <NLL>
1956 N12 P34-35	C-2555 <NRC>
1956 N12 P41-43	59-12211 <+*>
1956 N12 P46-48	C-2559 <NRC>
1957 N1 P8-18	C-2464 <NRC>
	R-5247 <*>
	59-12124 <+*>
1957 N1 P26-33	59-00684 <*>
	59-18995 <+*>
1957 N1 P36-37	59-22180 <+*>
1957 N1 P37-41	59-16282 <+*>
1957 N2 P3-9	59-15884 <+*>
	59-16021 <+*>
1957 N2 P10-18	64-13719 <=*$>
1957 N2 P24-26	59-16285 <+*>
1957 N2 P60-61	59-17158 <+*>
1957 N3 P34-37	59-15561 <+*>
	59-22183 <+*>
	62-10166 <=*>
1957 N3 P37-40	59-16020 <+*>
1957 N3 P44-45	59-22215 <+*>
1957 N3 P45-46	59-22212 <+*>
1957 N4 P20-24	59-16018 <+*>
1957 N4 P24-28	60-14286 <+*>
1957 N4 P28-30	59-16019 <+*>
1957 N4 P36-37	61-19231 <+*>
1957 N4 P37-40	61-15144 <*+>
1957 N4 P47-50	61-19529 <=*>
1957 N5 P3-11	59-16569 <+*>
1957 N5 P29-30	59-16570 <+*>
1957 N5 P43-45	59-22181 <+*>
1957 N5 P45-49	59-19503 <+*>
1957 N6 P3-11	59-22398 <+*>
	60-19137 <+*>
1957 N6 P12-20	59-11594 <=> 0
	59-16279 <+*>
1957 N6 P26-32	60-19137 <+*>
1957 N6 P32-33	59-16284 <+*>
1957 N6 P40-45	59-11595 <=>
	59-22230 <+*>
1957 N6 P46-47	59-11596 <=>
	59-19049 <+*>
	60-00260 <*>
1957 N6 P49-52	R-3193 <*>
	59-11597 <=> 0
1957 N6 P53-57	59-11598 <=>
	59-22214 <+*>
1957 N6 P59-61	59-11599 <=>
1957 N6 P63-64	59-11600 <=>
1957 N6 P64-65	59-11601 <=>
1957 N7 P3-11	59-11623 <=> 0
	60-14285 <+*>
1957 N7 P12-16	59-11624 <=>
1957 N7 P22-25	59-11625 <=> 0
1957 N7 P26-30	60-17433 <+*>
1957 N7 P39-40	59-11626 <=>
1957 N7 P40-45	60-17413 <+*>
1957 N7 P49-51	59-11627 <=> 0
1957 N7 P52-53	59-11628 <=>
1957 N7 P54-56	59-11631 <=> 0

Citation	Translation
1957 N7 P56-57	59-11629 <=>
1957 N7 P57-60	59-11630 <=>
1957 N8 P3-13	R-4736 <*>
1957 N8 P14-20	33K20R <ATS>
	60-17416 <+*>
1957 N8 P57-60	R-5025 <*>
1957 N8 P62-64	AEC-TR-3607 <+*>
1957 N9 P31-33	60-17419 <+*>
1957 N9 P58-59	R-4949 <*>
	59-12195 <+*>
1957 N10 P3-9	59-00660 <*>
	59-19002 <+*>
1957 N11 P3-	59-19061 <+*>
1957 N11 P7-16	59-00666 <*>
1957 N11 P17-25	59-00666 <*>
1957 N11 P26-31	59-00666 <*>
1957 N11 P32-40	R-5204 <*>
1957 N11 P32-39	59-00666 <*>
1957 N12 P16-21	60-13122 <+*>
1957 N12 P32-35	59-16281 <+*>
1957 N12 P36-39	59-10562 <+*>
	59-16568 <+*>
1957 N12 P44-45	R-5303 <*>
1958 N1 P22-27	R-5000 <*>
	59-12053 <+*>
1958 N1 P27-33	59-13294 <=> 0
1958 N1 P44-49	59-13294 <=> 0
1958 N1 P53-55	59-13294 <=> 0
1958 N1 P54-55	59-19000 <+*>
1958 N1 P61-62	59-13294 <=> 0
1958 N1 P65-67	59-13294 <=> 0
1958 N2 P3-9	R-5024 <*>
	59-12197 <+*>
1958 N2 P10-16	R-5202 <*>
	59-12198 <+*>
1958 N2 P24-27	59-16424 <+*>
1958 N2 P42	59-11113 <=>
1958 N2 P45-46	59-11112 <=>
1958 N2 P46	59-11111 <=>
1958 N2 P48-51	59-11110 <=>
1958 N2 P53-56	59-11109 <=>
1958 N2 P57-60	59-11108 <=>
1958 N2 P60	59-11114 <=>
1958 N2 P61	59-11115 <=>
1958 N2 P66-82	61-13311 <*=> 0
1958 N3 P3-12	59-13013 <=>
1958 N3 P13-20	59-13012 <=>
1958 N3 P21-27	59-13008 <=>
1958 N3 P28-33	59-13014 <=>
1958 N3 P34-36	59-11153 <=>
1958 N3 P56-58	60-13090 <+*>
1958 N3 P64-67	59-13007 <=>
1958 N3 P69	59-13000 <=>
1958 N3 P69-70	59-13016 <=>
1958 N3 P70-71	59-13002 <=>
1958 N4 P3-4	59-11178 <=>
1958 N4 P8-14	59-13148 <=>
1958 N4 P21-23	59-13172 <=>
1958 N4 P24-25	59-13171 <=>
1958 N4 P30-34	60-17414 <+*>
1958 N4 P40	59-11172 <=>
1958 N4 P42-44	59-11173 <=> 0
1958 N4 P47-49	59-11174 <=>
1958 N4 P53-56	59-11171 <=> 0
1958 N4 P61-62	59-11175 <=>
1958 N5 P3-11	59-13166 <=>
1958 N5 P12-19	R-5201 <*>
	59-12199 <+*>
1958 N5 P26-29	59-13165 <=> 0
1958 N5 P33-36	59-13164 <=>
1958 N5 P48-50	59-13150 <=>
1958 N5 P55-58	59-13163 <=>
1958 N6 P3-7	59-13127 <=>
1958 N6 P8-13	59-13128 <=>
1958 N6 P14-18	60-15975 <+*>
1958 N6 P19-22	59-13129 <=>
1958 N6 P25-29	59-13216 <=>
1958 N6 P32	61-15309 <+*>
1958 N6 P47-49	59-13130 <=>
1958 N6 P49-54	59-19065 <+*>
1958 N6 P55-58	SCL-T-261 <+*>
1958 N6 P61-63	59-13217 <=>
1958 N6 P64-65	59-13218 <=>
1958 N7 P3-10	59-11522 <=>
1958 N7 P11-17	59-11523 <=>
	62-15512 <=*>
1958 N7 P26-30	59-11524 <=>
1958 N7 P32-34	59-11525 <=>
1958 N7 P35-38	61-28211 <*=>
1958 N7 P41	59-16571 <+*>
1958 N7 P42-47	59-11526 <=>
1958 N7 P50-54	61-19230 <+*>
1958 N7 P59-60	59-11527 <=>
1958 N7 P60-64	59-11528 <=>
1958 N7 P65-66	59-11529 <=>
1958 N8 P3-10	59-13020 <=> 0
1958 N8 P11-16	59-13010 <=> 0
1958 N8 P29-35	59-13021 <=> 0
1958 N8 P33-36	59-13022 <=> 0
1958 N8 P47-48	59-13001 <=>
1958 N8 P51-57	59-11116 <=>
1958 N8 P59-62	59-13011 <=>
1958 N9 P22-24	59-13224 <=>
1958 N9 P27-31	59-13284 <=>
1958 N9 P32-34	59-13285 <=>
1958 N9 P44-46	59-13286 <=> 0
	61-15150 <*+>
1958 N9 P52-56	59-13225 <=>
1958 N10 P9-17	SCL-T-310 <+*>
	59-15563 <+*>
1958 N10 P18-23	60-51028 <=>
1958 N10 P24-29	60-17697 <+*>
1958 N10 P40-42	61-15153 <*+>
1958 N11 P3-10	59-13325 <=> 0
1958 N11 P30-31	60-17411 <+*>
1958 N11 P31	59-16567 <+*>
1958 N11 P31-35	60-21903 <=>
1958 N11 P41-46	59-13326 <=>
	60-23025 <+*>
1958 N11 P56-61	59-13327 <=> 0
1958 N11 P64-66	59-11506 <=>
1958 N11 P68-69	59-13328 <=>
1958 N12 P28-30	59-16425 <+*>
1958 N12 P30-32	61-15161 <*+>
1958 N12 P32-33	59-16426 <+*>
	59-18266 <+*>
1958 N12 P33-37	61-15157 <*+>
1958 N12 P44-45	60-17412 <+*>
1959 N1 P3-12	63-19806 <*=> C
1959 N1 P31-33	60-14290 <+*>
1959 N1 P37-40	60-21878 <=>
1959 N1 P40-44	60-51008 <=>
1959 N1 P44-46	62-11020 <=>
1959 N1 P47-49	60-51024 <=>
1959 N1 P54-57	60-51029 <=>
1959 N2 P7-14	61-15165 <*+>
1959 N2 P15-21	60-11025 <=>
	61-13583 <=*>
1959 N2 P22-27	60-17173 <+*>
1959 N2 P28-30	60-11026 <=>
1959 N2 P30-35	60-11027 <=>
	61-13582 <=*>
1959 N2 P46-51	61-15166 <*+>
1959 N2 P56-57	60-17169 <+*>
1959 N2 P62-63	60-11028 <=>
1959 N2 P63-66	60-11029 <=>
1959 N2 P70-71	60-11030 <=>
1959 N3 P3-9	60-11018 <=>
1959 N3 P27-29	60-11013 <=>
	61-15159 <*+>
1959 N3 P41-43	60-11019 <=>
1959 N3 P57-58	60-11014 <=> C
1959 N3 P58-62	60-11015 <=>
1959 N3 P63-64	60-11016 <=>
	61-15158 <*+>
1959 N3 P64-65	60-11017 <=>
1959 N4 P3-25	60-11044 <=>
1959 N4 P16-20	60-17696 <+*>
1959 N4 P21-25	61-15151 <*+>
1959 N4 P26-33	61-19011 <+*>

Citation	Number
1959 N4 P34-35	59-13907 <=>
1959 N4 P35-39	60-11046 <=>
	60-17699 <+*>
1959 N4 P39-42	61-15156 <*+>
1959 N4 P51	60 11047 <=>
1959 N4 P55-58	63-11021 <=>
1959 N4 P58-61	60-13893 <+*>
	60-17415 <+*>
1959 N4 P62-67	60-11053 <=>
	60-17420 <+*>
1959 N4 P70-73	60-11052 <=>
1959 N5 P3-8	60-11039 <=>
1959 N5 P9-17	60-11040 <=>
	61-13317 <=*> 0
	61-15136 <*+>
1959 N5 P25-30	60-17695 <+*>
1959 N5 P30-32	61-15162 <*+>
1959 N5 P45-48	60-51018 <=>
1959 N5 P50	61-23340 <*=>
1959 N5 P51-54	61-28006 <*=>
1959 N5 P54-55	60-11147 <=>
	60-17418 <+*>
1959 N5 P60-61	60-11041 <=>
1959 N5 P61-62	60-11042 <=>
1959 N5 P62-63	60-11043 <=>
1959 N6 P11-16	60-17421 <+*>
1959 N6 P17-20	60-21891 <=>
1959 N6 P39-41	61-15141 <*+>
1959 N6 P52-54	60-17417 <+*>
1959 N6 P55-58	59-31019 <=> 0
1959 N7 P3-13	60-11132 <=>
	61-15140 <*+>
1959 N7 P14-20	61-13322 <=*> 0
1959 N7 P26-29	61-15977 <+*>
1959 N7 P29-30	61-15147 <*+>
	61-15976 <+*>
1959 N7 P41-42	61-15154 <*+>
1959 N7 P48-49	61-15148 <*+>
	61-15298 <+*>
1959 N7 P49-50	61-15149 <*+>
	61-15299 <+*>
1959 N7 P54-57	60-11134 <=>
1959 N8 P8-12	60-11107 <=>
	60-17425 <+*>
1959 N8 P13-15	60-11144 <=>
1959 N8 P19-24	61-15142 <*+>
	62-10982 <=*>
1959 N8 P27-29	63-11022 <=>
1959 N8 P33	61-15135 <*+>
1959 N8 P35-37	60-11108 <=>
1959 N8 P47-50	60-11109 <=>
1959 N8 P50-54	60-11110 <=>
1959 N8 P55	61-15297 <+*>
1959 N9 P3-12	60-11215 <=>
	61-13312 <=*>
1959 N9 P20-24	60-11218 <=>
1959 N9 P24-29	60-11221 <=>
1959 N9 P33-34	60-11145 <=>
1959 N9 P39-40	60-11192 <=>
1959 N9 P41-45	60-11174 <=>
1959 N9 P46-49	60-11214 <=>
1959 N9 P54-57	60-11197 <=>
1959 N9 P58-59	60-11371 <=>
1959 N10 P3-7	61-19013 <+*>
1959 N10 P24-25	63-19839 <=*>
1959 N10 P38-39	60-11261 <=>
1959 N10 P45-49	60-11262 <=> 0
1959 N11 P3-15	60/11517 <=>
	61-28822 <*=>
1959 N11 P16-26	60-11518 <=>
	61-23705 <*=>
1959 N11 P27-28	60-17305 <+*>
1959 N11 P37-40	61-28828 <*=>
1959 N11 P58-60	61-28169 <*=>
1959 N11 P62-66	60-11535 <=>
1959 N12 P3-10	60-11635 <=>
1959 N12 P21-26	63-19737 <=*>
1959 N12 P41-48	60-11637 <=>
1959 N12 P55-58	60-11634 <=>
1959 N12 P58-59	60-11638 <=>
1959 N12 P64-69	60-11639 <=>
1960 N1 P3-9	60-41235 <=>
1960 N1 P34-37	65-13286 <*>
1960 N2 P3-10	60-11538 <=>
1960 N2 P9-10	63-19801 <=>
1960 N2 P35-38	60-11716 <=>
1960 N2 P53-57	60-11717 <=>
1960 N3 P11-17	61-19229 <+*>
1960 N3 P18-25	60-11789 <=>
1960 N3 P26-28	60-31433 <=>
1960 N3 P35-37	60-11791 <=>
1960 N3 P44-45	60-31434 <=>
1960 N3 P45-46	60-31435 <=>
1960 N3 P53-56	61-19228 <+*>
1960 N3 P57-61	60-31436 <=>
1960 N4 P3-9	60-11700 <=>
1960 N4 P10-16	60-11701 <=>
1960 N4 P25-26	61-19524 <=*>
1960 N4 P26-28	61-19523 <=*>
1960 N4 P28-31	60-11702 <=>
	61-19525 <=*>
1960 N4 P47-48	60-11703 <=>
1960 N4 P51-52	60-11705 <=>
1960 N4 P56-60	60-11704 <=>
1960 N5 P3-10	61-13981 <+*>
1960 N5 P56-58	61-11501 <=>
1960 N5 P58-59	61-11078 <=>
1960 N5 P65-66	61-11082 <=>
1960 N6 P3-8	60-41353 <=>
1960 N6 P9-12	65-61266 <=$> 0
1960 N6 P17-20	61-13502 <+*>
1960 N6 P20-24	60M45R <ATS>
	61-13501 <*+>
1960 N6 P35-37	61-23249 <*=>
1960 N6 P44-49	60-41316 <=>
1960 N6 P50	60-41283 <=>
1960 N7 P9-13	60-41664 <=>
1960 N7 P14-21	61-11839 <=>
1960 N7 P51-55	61-11564 <=>
1960 N8 P37-38	61-11598 <=>
1960 N8 P41-43	62-01464 <*>
1960 N8 P56-59	61-21191 <=>
1960 N8 P60-62	61-11541 <=>
1960 N9 P2-10	61-11582 <=>
	63-19755 <*=> 0
1960 N9 P16-20	61-28645 <*=>
1960 N9 P29-31	65-28932 <$>
1960 N9 P35-38	61-19569 <*=>
	61-21132 <=>
1960 N9 P41-43	61-28644 <*=>
1960 N9 P55-59	61-11582 <=>
1960 N10 P9-13	61-11825 <=>
1960 N10 P14-18	61-11826 <=>
1960 N10 P42-45	61-11929 <=>
1960 N10 P55-58	61-21073 <=>
1960 N11 P14-19	61-28975 <*=>
	63-19730 <=*> 0
1960 N11 P46-48	62-24809 <=*>
1960 N12 P3-10	61-23706 <*=>
1960 N12 P11-13	63-19881 <=*>
1960 N12 P14-18	61-21556 <=>
1960 N12 P37-42	61-19571 <=*>
	61-21555 <=>
	65-60765 <=>
1960 N12 P46-48	61-21554 <=>
1960 N12 P48-49	61-21558 <=>
1960 N12 P49-50	61-21559 <=>
1960 N12 P52-53	61-10858 <+*>
1960 N12 P53-58	61-21560 <=>
1961 N1 P3-9	61-23931 <*=>
1961 N1 P11-15	61-21641 <=>
1961 N1 P26-31	61-21644 <=>
1961 N1 P47-48	61-23930 <*=>
1961 N1 P61-64	61-27198 <*=>
1961 N2 P13-19	61-19994 <=>
1961 N2 P36-40	63-19951 <=*> 0
1961 N2 P42-43	61-19994 <=>
1961 N2 P47-48	61-19994 <=>
1961 N2 P49-52	61-21544 <=>
1961 N2 P55-59	61-19994 <=>

Citation	Translation
1961 N3 P3-10	63-19850 <=*>
1961 N3 P42-45	61-27287 <*=>
1961 N3 P43-45	AD-611 120 <=$>
1961 N3 P56-60	61-27287 <*=>
1961 N4 P35-36	61-27663 <=>
1961 N4 P47-49	61-27663 <=>
1961 N4 P51-54	61-27663 <=>
1961 N5 P24-26	61-28513 <*=>
1961 N5 P29-33	62-23296 <=*>
1961 N5 P42-43	62-15694 <=*>
1961 N5 P49-58	61-28513 <*=>
1961 N5 P60-61	61-28513 <*=>
1961 N6 P3-12	62-11110 <=>
	63-19846 <=*>
1961 N6 P21-27	65-61153 <=> 0
1961 N6 P57-61	62-11110 <=>
1961 N6 P65-66	62-11110 <=>
1961 N7 P16-21	65-63050 <=$>
1961 N7 P22-26	62-15793 <*=>
1961 N7 P52-55	62-13520 <=>
1961 N7 P58	62-13520 <=>
1961 N8 P3-10	63-23604 <=*$>
1961 N8 P11-19	62-15237 <=>
1961 N8 P26-32	62-15237 <=>
1961 N8 P38-41	62-15237 <=>
1961 N8 P52-55	62-15237 <=>
1961 N8 P57-60	62-15237 <=>
1961 N9 P3-22	62-15442 <*=>
1961 N9 P29-33	62-15442 <*=>
1961 N9 P34-36	62-11667 <=>
1961 N9 P46	62-15442 <*=>
1961 N9 P60	62-15442 <*=>
1961 N10 P3-10	65-61240 <=$> 0
1961 N10 P25-28	65-61441 <=>
1961 N10 P54-56	62-19159 <=*>
1961 N11 P3-6	66-11770 <*>
1961 N11 P7-14	62-23326 <=*> 0
1961 N11 P50-53	62-23326 <=*> 0
1961 N11 P66-68	62-23385 <=*>
1961 N12 P6-14	62-11684 <=>
	62-23808 <=*>
1961 N12 P15-22	62-11737 <=>
	65-60792 <=> 0
1961 N12 P23-25	62-11682 <=>
	62-23808 <=*>
1962 N1 P34-36	SHSP N1 1962 P2 <AGU>
1962 N1 P38-40	SHSP N1 1962 P5 <AGU>
1962 N1 P41-42	62-23802 <=>
1962 N1 P43-49	SHSP N1 1962 P7 <AGU>
1962 N1 P50-56	62-23802 <=>
1962 N1 P60-63	62-23802 <=>
1962 N2 P3-15	62-24378 <=>
1962 N2 P3-8	64-00478 <*>
	64-11836 <=>
	65-61374 <=$>
1962 N2 P16-22	SHSP N1 1962 P16 <AGU>
1962 N2 P23-27	SHSP N1 1962 P22 <AGU>
1962 N2 P23-30	62-24378 <=>
1962 N2 P32-37	SHSP N1 1962 P27 <AGU>
1962 N2 P37-40	SHSP N1 1962 P34 <AGU>
1962 N2 P42-45	SHSP N3 1962 P338 <AGU>
1962 N2 P46-47	62-24378 <=>
1962 N2 P62-67	62-24378 <=>
1962 N2 P69-71	62-24378 <=>
1962 N3 P39-43	SHSP N1 1962 P37 <AGU>
1962 N3 P55-57	SHSP N1 1962 P42 <AGU>
	63-15220 <=*>
1962 N3 P61-64	62-24722 <=*>
1962 N4 P14-21	62-25532 <=>
1962 N4 P27-33	SHSP N1 1962 P44 <AGU>
1962 N4 P43-46	64-15957 <=*$> 0
1962 N4 P46-47	62-25532 <=>
1962 N4 P47-51	SHSP N1 1962 P49 <AGU>
1962 N4 P66-73	62-25532 <=>
1962 N5 P29-33	M.5435 <NLL> 0
1962 N5 P40-43	62-20430 <=*>
1962 N5 P44-47	SHSP N1 1962 P54 <AGU>
1962 N6 P10-14	62-32199 <=*>
1962 N6 P20-27	SHSP N1 1962 P58 <AGU>
1962 N6 P35-38	SHSP N1 1962 P66 <AGU>

Citation	Translation
1962 N6 P58-62	62-32199 <=*>
1962 N7 P3-10	62-33004 <=>
1962 N7 P19-25	SHSP N2 1962 P111 <AGU>
1962 N7 P34-49	62-33004 <=>
1962 N7 P40-45	65-60963 <=$> 0
1962 N7 P55-56	SHSP N2 1962 P116 <AGU>
1962 N7 P64-68	62-33004 <=>
1962 N8 P3-15	63-13317 <=>
1962 N8 P3-10	65-60959 <=$> 0
1962 N8 P22-27	SHSP N2 1962 P119 <AGU>
1962 N8 P28-31	SHSP N2 1962 P124 <AGU>
1962 N8 P31-32	63-13317 <=>
1962 N8 P33-35	SHSP N2 1962 P128 <AGU>
1962 N8 P33	63-11609 <=>
1962 N8 P39-41	63-13308 <=*>
1962 N8 P42-46	63-13317 <=>
1962 N8 P54-62	63-13317 <=>
1962 N9 P4-6	63-13525 <=>
1962 N9 P13-18	65-60972 <=$> 0
1962 N9 P19-27	SHSP N4 1962 P448 <AGU>
1962 N9 P28-36	63-13525 <=>
1962 N9 P28-30	65-60774 <=$>
1962 N9 P40-44	SHSP N4 1962 P455 <AGU>
1962 N9 P44-45	SHSP N4 1962 P460 <AGU>
1962 N9 P49-51	SHSP N4 1962 P462 <AGU>
1962 N9 P52-55	TRANS-F-111 <NLL>
1962 N9 P58-64	63-13525 <=>
1962 N10 P12-21	63-11699 <=>
1962 N10 P28-36	63-13635 <=>
1962 N10 P34-36	64-13113 <=*$>
1962 N10 P44-46	65-60964 <=$> 0
1962 N10 P50-52	63-13635 <=>
1962 N10 P57-64	63-13635 <=>
1962 N11 P12-19	63-11612 <=>
	63-24395 <=*$> 0
1962 N11 P28-32	M.5361 <NLL> 0
1962 N11 P33-36	65-61514 <=$>
1962 N11 P45-47	SHSP N5 1962 P570 <AGU>
	65-24032 <$>
1962 N11 P47-50	SHSP N5 1962 P573 <AGU>
1962 N12 P3-8	ANL-TRANS-12 <=$>
1962 N12 P3-15	63-21328 <=>
1962 N12 P3-8	65-61524 <=$>
1962 N12 P9-15	65-61369 <=$> 0
1962 N12 P21-25	63-21328 <=>
1962 N12 P25-26	65-61476 <=> 0
1962 N12 P30-32	SHSP N6 1962 P679 <AGU>
1962 N12 P40-45	63-21328 <=>
1962 N12 P51-64	63-21328 <=>
1963 N1 P3-24	63-21397 <=>
1963 N1 P9-15	65-61521 <=$>
1963 N1 P31-37	AD-610 119 <=$>
1963 N1 P47	63-21397 <=>
1963 N1 P53-55	63-21397 <=>
1963 N1 P59-62	63-21397 <=>
1963 N2 P3-14	63-21635 <=> 0
1963 N2 P21-29	63-21635 <=> 0
1963 N3 P3-12	M.5412 <NLL>
	63-11753 <=>
1963 N3 P47-51	65-61259 <=$>
1963 N4 P3-15	63-31128 <=>
1963 N4 P34-39	63-31128 <=>
1963 N4 P42-44	63-31128 <=>
1963 N4 P54-57	63-31128 <=>
1963 N4 P59-63	63-31128 <=>
1963 N5 P3-8	63-20124 <=*>
1963 N5 P3-13	63-31267 <=>
1963 N5 P22-27	63-31267 <=>
1963 N5 P29-31	63-31267 <=>
1963 N5 P38-41	SOU.HYDROL.1963 P275 <AGU>
1963 N5 P38-50	63-31267 <=>
1963 N5 P42-44	65-60969 <=$> 0
1963 N5 P45-50	63-24598 <=*$>
1963 N5 P56-59	63-31267 <=>
1963 N5 P61-63	63-31267 <=>
1963 N6 P10-17	SOU.HYDROL.1963 P279 <AGU>
1963 N6 P18-30	63-31407 <=>
1963 N6 P30-34	SOV.HYDROL.1963 P284 <AGU>
1963 N6 P34-35	SOU.HYDROL.1963 P288 <AGU>
1963 N6 P35-40	63-31407 <=>

Citation	Source
1963 N6 P41-46	SOU.HYDROL.1963 P289 <AGU>
1963 N6 P47-50	63-31407 <=>
1963 N6 P51-53	65-61278 <=$> 0
1963 N6 P54-55	SOU.HYDROL.1963 P295 <AGU>
1963 N6 P61-64	63-31407 <=>
1963 N7 P3-17	63-31572 <=>
1963 N7 P18-24	65-61332 <=$> 0
1963 N7 P28-33	63-31572 <=>
1963 N7 P40-50	63-31572 <=>
1963 N7 P42-44	SOU.HYDROL.1963 P498 <AGU>
1963 N7 P51-56	63-11754 <=>
1963 N7 P58-61	63-31572 <=>
1963 N7 P64	63-31572 <=>
1963 N8 P3-10	63-31838 <=>
1963 N8 P21-28	SOV.HYDROL.1963 P409 <AGU>
1963 N8 P32-38	63-31833 <=>
1963 N8 P32-36	65-61260 <=$> 0
1963 N8 P43-46	65-61257 <=$>
1963 N8 P53	63-31833 <=>
1963 N8 P54-56	SOV.HYDROL.1963 P414 <AGU>
	65-61279 <=$> 0
1963 N8 P57-64	63-31833 <=>
1963 N9 P11-17	SOU.HYDRUL.1963 P501 <AGU>
1963 N9 P18-23	65-60762 <=> 0
1963 N9 P24-31	64-15959 <=*$> 0
1963 N9 P31-33	64-15956 <=*$>
1963 N9 P34-38	SOU.HYDROL.1963 P509 <AGU>
1963 N9 P38-42	SOU.HYDROL.1963 P514 <AGU>
1963 N9 P42-45	SOU.HYDROL.1963 P518 <AGU>
1963 N9 P46-48	SOV.HYDROL.1963 P522 <AGU>
1963 N10 P22-27	SCV.HYDROL.1963 P524 <AGU>
1963 N10 P33-36	65-63507 <=*$>
1963 N10 P36-37	SOU.HYDROL.1963 P529 <AGU>
1963 N10 P42-43	SCU.HYDROL.1963 P531 <AGU>
1963 N11 P47-49	SOV.HYDROL.1963 P618 <AGU>
1963 N11 P49-50	SOV.HYDROL.1963 P621 <AGU>
1963 N11 P53-54	64-21064 <=>
1963 N11 P57-61	64-21072 <=>
1963 N12 P3-7	TRANS-F.73 <NLL> 0
1963 N12 P14-21	SOV.HYDROL.1963 P623 <AGU>
1963 N12 P37-42	SOV.HYDROL.1963 P630 <AGU>
1964 N1 P12-22	N64-26783 <=>
1964 N2 P3-40	64-31083 <=>
1964 N2 P42-48	64-31083 <=>
1964 N2 P56-61	64-31083 <=>
1964 N2 P63-64	64-31083 <=>
1964 N3 P3-16	64-31253 <=>
1964 N3 P24-26	64-31253 <=>
1964 N3 P30-34	64-31253 <=>
1964 N3 P36-41	64-31253 <=>
1964 N3 P54-60	64-31253 <=>
1964 N3 P63-66	64-31253 <=>
1964 N4 P3-8	AD-617 948 <=$>
1964 N4 P3-15	64-31379 <=>
1964 N4 P9-15	TRANS-313(M) <NLL>
1964 N4 P33-35	64-31379 <=>
1964 N4 P38-39	TRANS-314(M) <NLL>
	64-31379 <=>
1964 N4 P43-47	TRANS-315(M) <NLL>
	64-31379 <=>
1964 N4 P54-57	64-31379 <=>
1964 N4 P59-64	64-31379 <=>
1964 N5 P4-9	64-18744 <*>
1964 N5 P45-49	TRANS-F.115. <NLL>
1964 N5 P51-54	TRANS-F-98 <NLL>
1964 N6 P3-10	TRANS-319 (M) <NLL>
1964 N6 P49-51	AD-613 979 <=$>
1964 N8 P23-29	64-51029 <=$>
1964 N8 P31-33	M.5784 <NLL>
1964 N8 P49-61	64-51029 <=$>
1964 N9 P3-9	TRANS-F-94 <NLL>
	64-51460 <=$>
1964 N9 P17-21	64-51460 <=$>
1964 N9 P22-28	AD-635 827 <=$>
1964 N9 P43-48	64-51460 <=$>
1964 N9 P51-59	64-51460 <=$>
1964 N10 P3-10	64-51914 <=$>
1964 N10 P22-27	64-51918 <=$>
1964 N10 P33-37	64-51917 <=$>
1964 N10 P45-46	64-51916 <=$>

Citation	Source
1964 N10 P48-51	64-51915 <=$>
1964 N10 P54-60	64-51911 <=$*>
1964 N11 P16-22	64-51863 <=$>
1964 N11 P23-26	64-51866 <=$>
1964 N11 P27-29	64-51867 <=$>
1964 N11 P53-54	64-51865 <=$>
1964 N11 P54-55	64-51869 <=$>
1964 N11 P55-57	64-51868 <=$>
	66-11216 <*>
1964 N11 P59-60	64-51871 <=$>
1964 N12 P3-8	65-30110 <=$>
1964 N12 P59	65-30109 <=$>
1965 N1 P3-13	65-30441 <=$>
1965 N1 P14-21	TRANS-F.116 <NLL> 0
1965 N1 P30-35	TRANS-F.119 <NLL> 0
1965 N1 P30-45	65-30441 <=$>
1965 N1 P40-45	TRANS-F.120. <NLL>
1965 N1 P53-58	65-30441 <=$>
1965 N1 P63-68	65-30441 <=$>
1965 N2 P13-19	TRANS-F.145 <NLL>
1965 N3 P21-23	M.5785 <NLL>
1965 N3 P36-41	65-30764 <=$>
1965 N3 P44-45	65-30764 <=$>
1965 N3 P49-60	65-30764 <=$>
1965 N6 P3-6	65-31423 <=$>
1965 N6 P14-29	65-31423 <=$>
1965 N6 P19-23	TRANS-F.137 <NLL> 0
1965 N6 P34-35	65-31423 <=$>
1965 N6 P46-48	65-14936 <*>
	65-31423 <=$>
1965 N6 P52-59	65-31423 <=$>
1965 N9 P3-26	65-33100 <=$>
1965 N9 P20-26	N66-13476 <=$>
1965 N9 P30-42	65-33100 <=$>
1965 N9 P39-42	66-11198 <*>
1965 N9 P48-55	65-33100 <=$>
1965 N9 P57-60	65-33100 <=$>
1965 N9 P64	65-33100 <=$>
1965 N12 P3-26	66-30447 <=$>
1965 N12 P31-40	66-30447 <=$>
1965 N12 P45-49	66-30447 <=$>
1965 N12 P57-58	66-30447 <=$>
1965 N12 P61-63	66-30447 <=$>
1966 N1 P43-45	SHSP N1 1966 P99 <AGU>
1966 N2 P13-18	66-30976 <=$>
1966 N2 P33-38	SHSP N1 1966 P101 <AGU>
1966 N2 P41-43	SHSP N1 1966 P107 <AGU>
1966 N4 P19-25	SHSP N2 1966 P211 <AGU>
1966 N6 P3-11	66-33708 <=$>
1966 N6 P12-20	66-33709 <=$>
1966 N6 P21-30	66-33710 <=$>
1966 N6 P31-35	66-33711 <=$>
1966 N6 P36-40	66-33712 <=$>
1966 N6 P40-44	SHSP N3 1966 P317 <AGU>
1966 N6 P45-48	66-33707 <=$>
1966 N6 P48-52	66-33703 <=$>
1966 N6 P53-56	66-33704 <=$>
1966 N7 P34-40	SHSP N4 1966 P435 <AGU>
1966 N8 P3-11	66-34835 <=$>
1966 N8 P12-15	66-34834 <=$>
1966 N8 P33-35	66-34833 <=$>
1966 N8 P35-37	66-34832 <=$>
1966 N8 P37-41	SHSP N4 1966 P441 <AGU>
1966 N8 P42-45	66-34831 <=$>
1966 N8 P46-49	66-34830 <=$>
1966 N8 P50-51	66-34829 <=$>
1966 N8 P52	66-34828 <=$>
1966 N8 P53-55	66-34827 <*$>
1966 N8 P55-58	66-34825 <=$>
1966 N8 P62-64	66-34826 <=$>
1966 N10 P48-49	SHSP N5 1966 P547 <AGU>
1966 N11 P46-49	SHSP N6 1966 P650 <AGU>
1966 N11 P49-53	SHSP N6 1966 P653 <AGU>
1967 N1 P1-20	67-31187 <=$>
1967 N1 P28-39	67-31187 <=$>
1967 N1 P49-66	67-31187 <=$>
1967 N2 P3-21	67-31495 <=$>
1967 N2 P29-65	67-31495 <=$>
1967 N2 P75-78	67-31495 <=$>
1967 N2 P82-84	67-31495 <=$>

1967 N2 P95-96 67-31495 <=$>
1967 N2 P100-101 67-31495 <=$>
1967 N2 P104-111 67-31495 <=$>
1967 N2 P116-128 67-31495 <=$>

METEOROLOGIYA: SBORNIK STATEI. MEZHDUVEDOMSTVENNYI KOMITET PO PROVEDENIYU MEZHDUNARODNOGO GEOFIZICHESKOGO GODA. AKADEMIYA NAUK SSSR. II. RAZDEL PROGRAMMY MGG
1960 N1 P124-140 64-13661 <=*$>
1960 N3 ENTIRE ISSUE 63-11142 <=>
1964 N7 ENTIRE ISSUE 65-60640 <=$> O

METEOROS
1951 V1 N1 P33-45 2370 <*>
1955 V5 N3 P185-206 CSIRO-3164 <CSIR>

METHODIK DER INFORMATION IN DER MEDIZIN
SEE METHODS OF INFORMATION IN MEDICINE

METHODS OF INFORMATION IN MEDICINE
1963 V2 N3 P90-94 64-00044 <*>
1963 V2 N3 P95-100 64-00042 <*>
1965 V4 N3 P107-111 65-17100 <*> O
1966 V5 N2 P75-80 66-13460 <*>

METODY I PROTSESSY KHIMICHESKOI TEKHNOLOGII
1955 V1 P45-64 R-1172 <*>

METODY VYCHISLENII
1963 N2 P29-38 AD-637 534 <=$>
1963 N2 P67-74 AD-620 765 <=$>
1963 N2 P95-131 AD-618 643 <=$>

MEXICO AGRARIO
1943 V5 N4 P237-245 57-2942 <*>

MEZHDUNARODNYI SELSKOKHOZYAISTVENNYI ZHURNAL
1959 N3 P62-72 NLLTB V2 N2 P97 <NLL>
1960 V4 N3 P118-128 60-31716 <=*>
1961 V5 N1 P130-133 61-27130 <=>
1961 V5 N5 P25-34 64-21975 <=>
1963 V7 N1 P55-61 64-21987 <=>

MEZHDUNARODNAYA ZHIZN
1958 N8 P68-71 R-4941 <*>

MEZODAZDASOGI ERTESTO
1967 V18 N8 P94-133 67-31227 <=$>

MEZOGAZDASAGI VILAGIRODALOM
1963 V5 N6 P113-118 65-14144 <*>

MEZSAIMNIECIBAS PROBLEMUINSTITUTA RAKSTII. RIHA
1956 N11 P121-162 65-50041 <=>
1957 V12 P79-90 18Q68R <ATS>
1959 V17 P93-103 17Q68R <ATS>

MICROBIOLOGIA ESPANOLA
1959 V12 N1 P65-84 60-10664 <*> O

MICROBIOLOGIA, PARAZITOLOGIA, EPIDEMIOLOGIA
1964 V9 N4 P277-278 64-41714 <=>

MIGRATSIY ZHIVOTNYKH
1959 N1 P27-40 C-3391 <NRC>
1962 N3 P10-20 64-13598 <=*$>
1962 N3 P87-91 64-13599 <=*$>
1962 N3 P92-105 64-13597 <=*$>

MIKROBIOLOGICHESKII ZHURNAL. KIEV
SEE MYKROBIOLOHICHNIYI ZHURNAL

*MIKROBIOLOGIYA
1932 V1 N3 P229- 59-19083 <+*>
1935 V4 N1 P24-43 C-4555 <NRCC>
1937 V6 P131-157 60-19661 <=*$> O
1937 V6 P1275-1292 65-11066 <*>
1937 V6 N3 P292-306 C-4556 <NRCC>
1937 V6 N3 P329-338 C-4410 <NRC>
1937 V6 N4 P449-464 C-3387 <NRCC>
1937 V6 N6 P754 61-17855 <=$>
1938 V7 N4 P466-484 AD-637 380 <=$>
1939 V8 P959-964 65-11080 <*> O
1939 V8 N6 P673-685 65-62904 <=>
1939 V8 N7 P828- 59-19084 <+*>
1939 V8 N8 P915-929 57K20R <ATS>
1940 V9 P217-231 65-11068 <*> O
1940 V9 N3 P267-281 RT-1685 <*>
1940 V9 N4 P361-376 RT-1073 <*>
1940 V9 N6 P608-614 RT-1074 <*>
1940 V9 N9/10 P879-887 64-15303 <=*$>
1940 V9 N9/10 P888-894 R-1547 <*>
64-19433 <=$>
1941 V10 P505-525 65-11079 <*> O
1941 V10 N3 P314-322 RT-1075 <*>
1941 V10 N5 P567-575 RT-2595 <*>
1942 V11 N4 P178-194 65-11048 <*> PO
1944 V13 N5 P251-255 64-15089 <=*$>
1945 V14 P177-190 65-11077 <*> O
1945 V14 P206-229 65-11076 <*> P
1945 V14 N3 P164-171 65-62907 <=>
1945 V14 N3 P172-176 65-60071 <=$>
1945 V14 N5 P347-352 RT-3850 <*>
1946 V15 P443-456 RJ-194 <ATS>
1946 V15 N3 P249-263 31E2R <ATS>
1946 V15 N4 P327-328 RT-2783 <*>
1946 V15 N4 P341-344 RT-2291 <*>
1946 V15 N5 P443-456 RJ-194 <ATS>
RT-3139 <*>
1946 V15 N6 P485-490 RT-2903 <*>
1947 V16 N5 P375-380 RT-1190 <*>
1948 V17 P77-81 59-10284 <+*>
1948 V17 N1 P76-81 63H12R <ATS>
1948 V17 N3 P271-281 61-32012 <=$>
1948 V17 N6 P463-468 RT-328 <*>
1949 V18 N1 P42-53 65-11071 <*> O
1949 V18 N2 P132-140 65-11043 <*> O
1949 V18 N2 P141-153 RT-3674 <*>
1949 V18 N4 P310-317 RJ-193 <ATS>
1949 V18 N5 P402-415 61-17070 <=$>
1949 V18 N6 P528-532 62-32149 <=*>
1950 V19 P294-298 51/0101 <NLL>
1950 V19 N1 P32-44 RT-3331 <*>
1950 V19 N1 P449-456 64-10039 <=*$> O
1950 V19 N2 P97-104 995 <TC>
1950 V19 N3 P193-202 32E2R <ATS>
61-17221 <=$>
1950 V19 N3 P203-210 30E2R <ATS>
1950 V19 N3 P275-279 RT-3820 <*>
1950 V19 N4 P308-316 51/0098 <NLL>
1950 V19 N4 P384-389 61-13654 <+*>
1951 V20 N1 P3-12 65-11053 <*> O
1951 V20 N1 P26-32 RT-397 <*>
1951 V20 N3 P217-222 RT-4314 <*>
1951 V20 N3 P245-255 RT-2977 <*>
1951 V20 N3 P256-265 RT-2904 <*>
1951 V20 N4 P324-329 R-2105 <*>
1951 V20 N5 P438-451 RT-2319 <*>
1951 V20 N6 P489-499 39L32R <ATS>
1952 V21 N1 P71-76 RT-2397 <*>
1952 V21 N1 P77-82 RT-1354 <*>
1952 V21 N1 P87-91 RT-1343 <*>
1952 V21 N2 P146-154 RAPRA-1237 <RAP>
1952 V21 N2 P219-225 60-51048 <=>
1952 V21 N2 P255-256 RT-1342 <*>
1952 V21 N2 P255 RT-1344 <*>
1952 V21 N3 P387-388 63-16127 <=*>
1952 V21 N4 P408-415 RT-2421 <*>
59-15123 <+*>
1952 V21 N6 P700-704 64-31531 <=>
1952 V21 N6 P710-717 65-64371 <=>
1952 V21 N6 P711-717 RT-1037 <*>
1953 V22 N1 P3-10 RT-2276 <*>
1953 V22 N1 P11-14 RT-1596 <*>
1953 V22 N2 P151-154 RT-1625 <*>
1953 V22 N3 P294-303 R-564 <*>

RT-2424 <*>
64-15065 <=*$> 0
1953 V22 N3 P308-310 RT-1340 <*>
1953 V22 N3 P311-315 RT-3357 <*>
1953 V22 N3 P316-324 RT-1558 <*>
1953 V22 N3 P325-337 RT-1339 <*>
1953 V22 N6 P656-662 62-32969 <=*>
1953 V22 N6 P714-718 RT-1684 <*>
1954 V23 N1 P3-14 RJ-191 <ATS>
1954 V23 N1 P15-21 RJ-192 <ATS>
1954 V23 N1 P22-26 RT-4082 <*>
1954 V23 N1 P49-52 C-4443 <NRC>
1954 V23 N2 P129-130 RT-2478 <*>
1954 V23 N2 P133-139 RJ-195 <ATS>
RT-2929 <*>
1954 V23 N2 P140-146 R-5208 <*>
1954 V23 N3 P249-251 59-11439 <=> 0
1954 V23 N3 P331-348 RJ-556 <ATS>
1954 V23 N5 P551-560 59-15717 <+*>
1955 V24 N1 P3-13 RJ-285 <ATS>
60-10648 <+*>
1955 V24 N1 P62-66 65-60675 <=$>
1955 V24 N1 P73-74 RJ-295 <ATS>
1955 V24 N1 P75-78 RJ-294 <ATS>
1955 V24 N2 P137-140 <CP>
R-4438 <*>
1955 V24 N2 P164-169 R-1508 <*>
RT-3531 <*>
64-19061 <=*$>
1955 V24 N2 P188-192 60-21077 <=>
1955 V24 N3 P315-320 65-61504 <=>
1955 V24 N3 P321-324 RJ-287 <ATS>
1955 V24 N3 P332-340 R-1591 <*>
1955 V24 N3 P371-381 R-5207 <*>
1955 V24 N3 P382-384 RJ-291 <ATS>
60-14338 <+*>
1955 V24 N4 P474-485 62-15120 <*=>
1955 V24 N5 P573-579 RJ-365 <ATS>
1955 V24 N5 P608-610 RJ-429 <ATS>
1955 V24 N5 P615-625 RJ-368 <ATS>
1955 V24 N6 P671-676 64-15294 <=*$>
1955 V24 N6 P681-684 C-4712 <NRC>
1955 V24 N6 P690-696 59-10086 <+*>
1955 V24 N6 P744-748 R-262 <*>
1956 V25 P261-267 T1893 <INSD>
1956 V25 N1 P66-71 RJ-606 <ATS>
1956 V25 N1 P629-638 61-13099 <=*>
1956 V25 N2 P156-163 59-12746 <+*> 0
1956 V25 N2 P191-194 62-10583 <=*> 0
1956 V25 N2 P195-199 R-4181 <*>
57-2926 <*>
1956 V25 N2 P200-207 R-5287 <*>
1956 V25 N2 P231-242 63-16125 <=*>
65-60681 <=$>
1956 V25 N3 P275-278 59-12743 <+*> 0
1956 V25 N3 P279-285 87K25R <ATS>
1956 V25 N3 P299-304 65-63436 <=$>
1956 V25 N3 P305-309 29H13R <ATS>
1956 V25 N4 P420-422 59-12747 <+*>
1956 V25 N4 P458-466 59-10866 <+*>
1956 V25 N4 P466-470 59-12755 <+*> 0
1956 V25 N5 P533-536 59-12745 <+*> 0
1956 V25 N5 P566-568 R-4188 <*>
59-12285 <+>
1956 V25 N5 P619-628 62-15117 <=*>
1956 V25 N6 P659-667 50J15R <ATS>
1956 V25 N6 P723-726 59-12751 <+*>
1956 V25 N6 P727-741 62-15118 <*=>
1956 V25 N6 P756-757 R-1818 <*>
1957 V26 N2 P255 R-3160 <*>
1957 V26 N3 P330-337 60-14092 <+*>
1957 V26 N4 P458-463 30K26R <ATS>
1957 V26 N5 P513-518 60-14091 <+*>
1957 V26 N5 P519-524 59-12060 <+*>
1957 V26 N5 P544-550 61K21R <ATS>
1957 V26 N5 P558-564 61J19R <ATS>
1957 V26 N5 P580-585 R-4886 <*>
1957 V26 N6 P651-658 60-14093 <+*>
98K22R <ATS>
1957 V26 N6 P745-749 60-10649 <+*>
1958 P4 29K27R <ATS>
1958 V27 N1 P104-109 29K26R <ATS>
1958 V27 N3 P387-389 66K25R <ATS>
1958 V27 N3 P390-392 67K25R <ATS>
1958 V27 N4 P478-483 28K27R <ATS>
1958 V27 N5 P626-633 40L37R <ATS>
1958 V27 N6 P740-752 15L30R <ATS>
1959 V28 N1 P3-6 59-13552 <=>
1959 V28 N1 P7-13 60-10951 <+*>
1959 V28 N1 P152-155 60-15956 <+*>
1959 V28 N2 P161-164 59-13663 <=>
1959 V28 N2 P165-167 59-13664 <=>
1959 V28 N2 P168-171 59-13665 <=>
1959 V28 N2 P172-174 59-13666 <=>
1959 V28 N2 P231-235 59-19261 <+*> 0
1959 V28 N3 P473-475 60-13486 <+*>
1959 V28 N4 P574-580 60-15945 <+*> 0
1959 V28 N5 P697-702 61-19721 <=*>
1959 V28 N5 P730-735 61-19722 <=*>
1959 V28 N5 P777-782 61-19720 <=*> 0
1959 V28 N5 P783-785 60-31150 <=>
1960 V29 N1 P73-78 63M40R <ATS>
1960 V29 N3 P408-414 15M47R <ATS>
1960 V29 N3 P415-418 16M47R <ATS>
1960 V29 N3 P428-432 55N48R <ATS>
1960 V29 N5 P710-714 23N51R <ATS>
1960 V29 N5 P715-720 24N51R <ATS>
1960 V29 N5 P738-744 61-19656 <=*>
1961 V30 N2 P280-285 34N55R <ATS>
1961 V30 N5 P809-817 62-13396 <*=>
1961 V30 N6 P985-989 80P60R <ATS>
1962 V31 N1 P111-120 63-31153 <=>
1962 V31 N1 P129-134 63-21568 <=>
1962 V31 N1 P189-191 63-31153 <=>
1962 V31 N2 P199-202 UCRL-TRANS-870 <=*>
1962 V31 N2 P209-215 62-33008 <=*>
1962 V31 N2 P350-356 62-32092 <=*>
64-13563 <=*$>
1962 V31 N3 P470-477 63-21912 <=>
1962 V31 N3 P571-573 62-33306 <=*>
1962 V31 N4 P608-615 54P64R <ATS>
1962 V31 N4 P745-757 62-33424 <=*>
1962 V31 N4 P763-764 63-13288 <=>
1962 V31 N4 P764-766 63-13469 <=*>
1962 V31 N6 P1102-1117 63-21960 <=>
1962 V31 N6 P1122-1128 63-21960 <=>
1963 V32 N2 P193-203 64-13171 <=*$>
1963 V32 N2 P362-370 64-13138 <=*$>
1963 V32 N3 P398-402 63-31596 <=>
1963 V32 N3 P403-404 63-31595 <=>
1963 V32 N3 P425-433 63-31597 <=>
1963 V32 N3 P551-557 AD-616 693 <=$>
1963 V32 N4 P582-597 64-13170 <=*$>
1963 V32 N4 P689-694 64-11858 <=>
64-21007 <=>
1963 V32 N4 P695-699 63-41329 <=>
1963 V32 N4 P727-731 64-13170 <=*$>
1963 V32 N4 P737-738 63-41335 <=>
1963 V32 N4 P740-741 64-13142 <=*$>
1963 V32 N5 P843-849 64-13637 <=*$>
1964 V33 N1 P3-6 64-31320 <=>
1964 V33 N1 P31-37 64-31320 <=>
1964 V33 N1 P91-96 AD-625 615 <=$>
1964 V33 N1 P118-133 64-31320 <=>
1964 V33 N1 P167-171 64-31076 <=>
1964 V33 N2 P356-363 N65-15057 <=*$>
1964 V33 N3 P454-458 AD-617 160 <=$>
1964 V33 N3 P472-476 AD-617 161 <=$>
1964 V33 N3 P508-515 AD-617 161 <=$>
1964 V33 N4 P565-568 64-41983 <=$>
1964 V33 N4 P610-612 64-41983 <=$>
1964 V33 N6 P992-995 65-30962 <=$>
1965 V34 N1 P91-100 65-13458 <*>
1965 V34 N5 P858-862 66-10900 <*>
1966 V35 N1 P188-189 66-32404 <=$>

MIKROCHEMICA ACTA
1937 V1 N1 P75-77 59-20618 <*> 0
1937 V2 N1/2 P80-84 60-31819 <=*>

MIKRCCHEMIE
1929 V7 N3 P305-313 66-14239 <*>
1932 V12 P133-136 63-20492 <*>
1932 V12 P315-320 58-1128 <*>
1934 V15 P95-98 58-1129 <*>
1937 V23 P9-16 59-10692 <*> 0
1938 V25 P228-233 II-1 <*>

MIKRCCHEMIE VEREINIGT MIT MIKROCHEMICA ACTA
1938 N3 P218-231 62-10449 <*> 0
1938 V24 P243-250 59-20672 <*>
1942 V30 N3 P241-258 61-16793 <*>
1943 V30 P279- II-731 <*>
1947 V33 P208-216 61-10229 <*>
1948 V34 P62-66 II-417 <*>
1948 V34 P319- 3036 <*>
1949 V34 P282-285 58-2403 <*>
1950 V35 P477-487 UCRL TRANS-916(L) <*>
8921 <IICH>
1951 V36 P793-800 63-10627 <*> 0
1951 V36/7 P379-392 58-2083 <*>
1951 V36/7 P420-424 58-1046 <*>
1951 V36/7 N2 P769-780 59-10073 <*>
1952 V39 P38-50 2970 <*>
1952 V39 P1C1-104 2969 <*>
1952 V39 P315-318 57-1284 <*>
1953 V40 P21-26 58-1605 <*>
1953 V40 P359-366 58-1443 <*>

MIKROCHIMICA (ET ICHNCANALYTICA) ACTA. TITLE VARIES
1953 N4 P414-420 57-2912 <*>
1953 N1/2 P79-88 65-12169 <*>
1954 P140-147 58-1720 <*>
1954 P376-387 58-1300 <*>
1954 N1 P140-147 3016 <*>
57-670 <*>
1954 N4 P349-365 II-615 <*>
1955 P187-201 290 <TC>
1955 P429-445 II-805 <*>
1955 N1 P29-36 62-20125 <*>
64-16133 <*> 0
1955 N1 P123-129 2849 <*>
58-2668 <*>
61-10486 <*>
63-14105 <*> 0
1955 N1 P187-202 65-13681 <*>
1955 N5 P845-849 II-865 <*>
1955 N2/3 P257-264 1121 <*>
1955 N2/3 P589-595 59-10128 <*>
1956 P667-681 58-1301 <*>
1956 P869-876 58-2541 <*>
1956 P1283-1306 65-00183 <*> 0
1956 N7 P1120-1135 59-10584 <*>
1956 N7 P1247-1263 57-2579 <*>
1956 N10 P1456-1472 57-1170 <*>
1956 N11 P1621-1648 58-749 <*>
1956 N11 P1705-1721 64-18366 <*>
1956 N11 P1735-1746 6048-A <K-H>
1956 N12 P1735-1746 64-20200 <*> 0
1956 N12 P1757-1761 62-20074 <*>
1956 N1/3 P171-178 63-27137 <$>
1956 N1/3 P263-267 1228 <*>
1956 N1/3 P474-483 61-16247 <*>
1956 N1/6 P71-90 62-00404 <*>
1956 N1/6 P171-178 63-00587 <*>
1956 N4/6 P869-876 12104 <K-H>
58-2669 <*>
59-10719 <*>
61-10485 <*>
63-14106 <*> 0
1956 N4/6 P949-954 64-10626 <*>
1957 N1 P113-124 66-14170 <*>
1957 N2 P150-158 57-2844 <*>
1957 N3/4 P313-317 64-20208 <*>
1957 N3/4 P607-612 63-14102 <*> 0
1958 P52-59 82L29G <ATS>
1958 N1 P9-27 63-10101 <*>
64-30114 <*> 0
1958 N1 P68-91 60-18878 <*>
1958 N1 P92-103 59-10082 <*>
1958 N1 P151-158 58-1667 <*>
1958 N3 P395-401 34L31G <ATS>
1958 N3 P415 62-00272 <*>
1958 N4 P545-552 AEC-TR-3630 <*>
1958 N5 P687-695 64-30954 <*> 0
1959 N2 P282-293 <DI>
1959 N2 P303-313 63-10100 <*>
1959 N3 P337-345 61-10387 <*> 0
1959 N6 P875-882 52M37G <ATS>
1959 N6 P883-890 54M37G <ATS>
1959 N6 P891-902 53M37G <ATS>
1959 N6 P903-907 55M37G <ATS>
1960 P294-298 63-10099 <*>
1960 P352-356 3094 <BISI>
1960 P394-404 AEC-TR-5292 <*>
1960 N1 P38-43 61-10313 <*> 0
1960 N1 P44-53 61-20748 <*>
1960 N1 P62-71 61-10314 <*> 0
1960 N2 P245-253 61-10312 <*> 0
1960 N2 P254-260 61-10311 <*>
1960 N4 P592-596 10467 <K-H>
65-12505 <*>
1960 N5/6 P641-649 65-00303 <*>
1960 N5/6 P670-674 65-12519 <*>
1960 N5/6 P854-862 61-20701 <*> 0
1961 P370-389 63-20981 <*>
1961 N2 P296-307 62-00637 <*>
1961 N3 P370-389 65-13539 <*> 0
1961 N4 P576-581 65-10366 <*> 0
1961 N5 P811-816 62-18998 <*>
1961 N6 P927-967 65-13996 <*> 0
1962 N4 P638-649 65-10367 <*>
1962 N5 P891-895 64-10712 <*>
1962 N5 P926-938 63-10584 <*>
63-18039 <*>
1963 N3 P456-466 64-16194 <*>
1963 N3 P499-505 65-18144 <*> 0
1963 N4 P759-768 64-16195 <*>
1963 N5/6 P831-850 4389 <BISI>
1963 N5/6 P1094-1108 64-14775 <*>
1964 N5 P778-783 66-14162 <*>
1964 N6 P1097-1105 66-10079 <*>
1964 N6 P1111-1114 540 <TC>
1964 N2/4 P196-201 65-10392 <*>
1964 N2/4 P272-297 142 <TC>
1965 N3 P206-208 4530 <BISI>
1965 N3 P471-478 66-11193 <*>
1965 N3 P503-514 4530 <BISI>
1965 N5 P842-851 66-14167 <*>
1966 N1/2 P357-369 66-14209 <*>

MIKROKOSMOS
1956 V45 P103-105 38L30G <ATS>

MIKROSKOPIE
1948 V3 N9/12 P257-309 57-1171 <*>
1950 V5 P101-116 65-18023 <*>
1953 V8 N7/8 P226-234 2018 <*>
1954 V9 P147-167 II-61 <*>
1959 V13 P289-304 61-20098 <*>
1961 V15 P11-12 63-00174 <*>
1961 V16 P291-293 62-01045 <*>
1961 V16 N5 P288-291 62-00914 <*>

MILCHWIRTSCHAFTLICHE FORSCHUNGEN
1934 V16 N4 P347-487 59-20671 <*> 0

MILCHWISSENSCHAFT
1947 V2 P235-247 2282 <NRC>
1947 V2 P335-347 58-1747 <*>
1950 V5 P13-17 57-1211 <*>
1950 V5 N2 P51-54 57-3246 <*>
1951 V6 N10 P351-354 58-1424 <*>
1951 V6 N11 P400-402 C-2640 <NRC>
58-1424 <*>
1956 V11 P381-384 58-561 <*>
1956 V11 N10 P381-384 C-2556 <NRC>
1961 V16 N1 P24-29 65-11674 <*>
1961 V16 N10 P523-531 65-11671 <*>

1964 V19 N6 P285-290	65-17024 <*>

M. LITAIRE SPECTATOR

1959 V128 N2 P64-71	59-21164 <=> 0

MINERACAO E METALLURGIA

1945 V8 N46 P263-268	63-14580 <*>
1945 V8 N46 P273-277	63-14032 <*> 0

MINERALNOE SYRE

1930 V5 P376-389	R-3298 <*>

MINERALNOE SYRE I TSVETNYE METALLY

1929 V4 P387-396	R-3376 <*>

MINERALNYE UDOBRENIYA I INSEKTOFUNGITSIDY

1935 V1 N1 P28-42	R-2605 <*>
1935 V1 N2 P24-40	R-2131 <*>
1935 V1 N2 P41-49	R-2585 <*>
1935 V1 N3 P11-19	R-3346 <*>
1935 V1 N3 P20-31	R-2869 <*>
1935 V1 N5 P7-20	R-2230 <*>
1935 V1 N5 P20-24	R-2204 <*>

MINERALOELE

1931 V4 P30-33	AL-242 <*>

MINERALOGICHESKII SBORNIK

1947 V1 P44-53	62-13082 <=*>
1950 V4 P167-168	R-1145 <*>
1951 N5 P211-218	IGR V2 N2 P129 <AGI>
1952 N6 P169-174	R-1652 <*>
1953 N7 P313-316	R-1157 <*>
1954 N8 P145-160	R-1144 <*>
1955 N9 P64-84	IGR V2 N3 P181 <AGI>
	R-1143 <*>
1955 V9 P321-323	62-16676 <=*>
1955 V9 N9 P309-312	65-64405 <=*$>
1956 N10 P132-134	35J16R <ATS>
1956 N10 P298-304	62-13108 <*=> 0
1957 N11 P303-321	IGR V1 N9 P1 <AGI>
	R-4688 <*>
1958 N12 P106-115	IGR V3 N6 P475 <AGI>
1958 N12 P116-128	IGR V4 N2 P127 <AGI>
1958 N12 P129-143	IGR V3 N7 P586 <AGI>
1958 N12 P196-224	IGR V3 N9 P784 <AGI>
1958 N12 P225-232	IGR V3 N8 P652 <AGI>
1958 N12 P255-261	IGR V3 N8 P658 <AGI>
1958 N12 P262-269	IGR V3 N8 P645 <AGI>
1959 N13 P139-148	IGR V3 N3 P195 <AGI>
1959 N13 P149-157	IGR V3 N2 P114 <AGI>
1959 N13 P158-177	IGR V3 N4 P337 <AGI>
1959 N13 P190-211	IGR V3 N5 P393 <AGI>
1959 N13 P220-234	IGR V3 N4 P325 <AGI>
1959 N13 P282-290	IGR V3 N3 P187 <AGI>
1959 N13 P349-362	IGR V3 N5 P385 <AGI>
1960 N14 P34-49	IGR V4 N7 P789 <AGI>
1960 N14 P50-79	IGR V4 N7 P799 <AGI>
1960 N14 P80-85	IGR V4 N6 P635 <AGI>
1960 N14 P86-104	IGR V4 N8 P929 <AGI>
1960 N14 P184-194	IGR V4 N6 P639 <AGI>
1960 N14 P195-207	IGR V5 N6 P663 <AGI>
1960 N14 P482-483	IGR V4 N6 P635 <AGI>
1961 V15 P182-188	62-33703 <=*>
1962 V16 P180-194	GI N1 1964 P132 <AGI>
	65-61576 <=$>

MINERVA ANESTESIOLOGICA

1954 V22 P110-114	57-732 <*>
1958 V24 N10 P425-427	64-16075 <*>
1958 V24 N12 P505-509	60-14195 <*>
1960 V26 P277-284	63-14770 <*>
1961 V27 N1 P22-24	64-20909 <*>
1963 V29 P133-135	65-00138 <*>
1963 V29 N2 P50-51	64-14780 <*>

MINERVA CARDIOANGIOLOGICA

1957 V5 N2 P597-600	63-16072 <*>
	8809-A <K-H>

MINERVA CHIRURGICA

1954 V9 N10 P488-490	62-20073 <*>
1954 V9 N23 P1132-1134	62-14403 <*>
1955 V10 P120-126	58-74 <*>
1956 V11 P663-667	58-678 <*>
1957 V12 P172-180	62-10267 <*> 0
1958 V13 N7 P403-404	64-18670 <*>
1958 V13 N8 P498-501	64-18730 <*>
1958 V13 N13 P790-794	64-18627 <*>
1960 V15 P347-351	63-14771 <*>
1960 V15 N5 P268-271	65-14780 <*> 0
1961 V16 P274-279	62-00881 <*>
1964 V19 N20 P693-695	66-11706 <*>

MINERVA DERMATOLOGICA

1953 V28 P221-229	62-10195 <*> P
1955 V30 N11 P340-348	57-57 <*>
1959 V34 P631-636	63-00546 <*>
1959 V34 N2 P121-122	63-00784 <*>
1959 V34 N3 P238-244	63-00552 <*>

MINERVA DIETOLOGICA

1962 V2 N4 P187-189	AEC-TR-5775 <*>

MINERVA FARMACEUTICA

1962 V10 P235-238	64-00356 <*>

MINERVA GINECOLOGICA

1952 V4 P128	AL-789 <*>
1952 V4 P666-669	II-981 <*>
1953 V5 P532-535	65-13831 <*>
1958 V10 P851-854	64-18707 <*>
1959 V11 N6 P221-230	64-18687 <*>
1961 V13 N1 P35-37	11548-F <K-H>
	64-14603 <*>
1962 V14 N5 P267-275	63-01285 <*>
1963 V15 N2 P121-126	66-12606 <*>
1964 V16 P592-594	66-12623 <*>

MINERVA MEDICA. ROME

1939 V30 N2 P207-209	313 <K-H>
1948 V39 P703-704	65-18219 <*>
1949 V40 N6 P162-164	62-00522 <*>
1950 V41 P1057	60-10008 <*>
1951 V42 N2 P1-4	60-10024 <*>
1951 V42 N3 P73-75	2124-F <K-H>
1952 V43 P1413-1416	65-18220 <*>
1953 V44 P331-341	57-1161 <*>
1953 V44 N6 P125-132	60-18166 <*> 0
1953 V44 N48 P1591-1594	57-2962 <*>
1953 V44 N86 P1202-1209	2374 <*>
1954 V45 P544-545	3471-A <K-H>
1954 V45 P1114-	II-960 <*>
1954 V45 P1403-1407	1858 <*>
1954 V45 P1660-1663	2462 <*>
1955 N101 P1864-1868	57-81 <*>
1955 V46 N1/13 P402-415	57-1273 <*>
1956 V47 P867-874	65-14325 <*> 0
1956 V47 P1378-1385	57-3447 <*>
1956 V47 P1831-	6405-B <K-H>
1956 V47 N76 P3-11	57-1221 <*>
1956 V47 N77 P855-865 SUP	65-11297 <*>
1956 V47 N77 P823-827 SUP	64-18736 <*>
1956 V47 N77 P828-836 SUP	64-18690 <*>
1956 V47 N77 P839-847 SUP	64-18731 <*>
1956 V47 N77 P847-850 SUP	64-18688 <*>
1956 V47 N77 P851-855 SUP	64-18689 <*>
1956 V47 N77 P865-866 SUP	64-18725 <*>
1956 V47 N1/2 P24-30	57-1818 <*>
1957 V48 P1352-1355	64-10663 <*>
1957 V48 P2165-2168	59-16015 <+*> 0
1957 V48 P2720-2723	65-18221 <*> 0
1957 V48 N18 P713-714	6492-A <K-H>
1958 V49 P77-82	65-18222 <*> 0
1958 V49 P1537-1538	65-14425 <*>
1958 V49 P4483-4488	63-14756 <*>
1958 V49 N19 P809-812	59-15684 <*> 0P
1958 V49 N32 P1523	62-14919 <*>
1958 V49 N32 P1527-1528	62-14980 <*>
1958 V49 N32 P1530	64-14716 <*>

1958 V49 N32 P1530-1531 64-16033 <*>
1958 V49 N97 P4476-4479 63-14761 <*>
1958 V49 N97 P4479-4482 64-18719 <*>
1958 V49 N97 P4488-4493 64-18727 <*>
1958 V49 N97 P4494-4497 64-18729 <*>
1958 V49 N97 P4498-4504 63-14757 <*> O
1958 V49 N97 P4504 64-18660 <*>
1958 V49 N97 P4508-4510 64-18639 <*>
1958 V49 N97 P4511-4512 63-14758 <*>
1958 V49 N97 P4512 63-14762 <*>
1958 V49 N97 P4513-4514 63-14760 <*>
1958 V49 N97 P4513 63-14899 <*>
1958 V49 N97 P4515 63-14763 <*>
1958 V49 N97 P4516-4518 63-14755 <*>
1958 V49 N97 P4518 64-18691 <*>
1958 V49 N97 P4519-4522 64-18728 <*>
1958 V49 N97 P4522-4523 64-18684 <*>
1958 V49 N97 P4524-4526 64-18666 <*>
1958 V49 N97 P4526-4527 64-18714 <*>
1958 V49 N97 P4530-4531 64-18726 <*>
1958 V49 N97 P4531 64-18657 <*>
64-18668 <*>
1958 V49 N97 P4532 63-14897 <*>
64-18628 <*>
1958 V49 N97 P4532-4533 64-18662 <*>
1958 V49 N97 P4533-4534 63-14896 <*>
1958 V49 N97 P4535-4536 64-18721 <*>
1958 V49 N97 P4536-4537 63-18495 <*>
1958 V49 N97 P4537-4538 63-14754 <*>
1958 V49 N97 P4537 64-18732 <*>
1958 V49 N97 P4538-4539 64-18708 <*>
1958 V49 N97 P4539-4542 64-18636 <*> O
1958 V49 N97 P4540-4541 63-14759 <*>
64-18640 <*>
1958 V49 N97 P4542 63-14895 <*>
1958 V49 N97 P4543-4545 64-18701 <*>
1958 V49 N97 P4543 64-18734 <*>
1958 V49 N97 P4545 64-18692 <*>
1958 V49 N97 P4546 64-18641 <*>
64-18661 <*>
64-18685 <*>
64-18720 <*>
1958 V49 N100 P4713-4724 58-2635 <*>
1958 V49 N100 P4726-4730 59-15880 <*>
1958 V49 N101 P4776-4779 64-20018 <*>
1959 V50 P1305 65-18223 <*>
1959 V50 N12 P377-380 60-14132 <*> O
1959 V50 N42 P1607 59-17075 <*>
1959 V50 N42 P1608-1613 59-17592 <*>
1959 V50 N42 P1613-1616 59-17069 <*>
1959 V50 N42 P1616-1620 59-17483 <*>
1959 V50 N42 P1620-1624 59-20962 <*>
1959 V50 N42 P1625-1628 59-17484 <*>
1959 V50 N42 P1628-1632 59-17076 <*>
1959 V50 N42 P1632-1638 59-20961 <*>
1959 V50 N57/8 P2349-2351 60-17482 <=*>
1959 V50 N59/0 P2406-2409 64-20016 <*>
1960 V51 P1154-1159 65-18224 <*>
1960 V51 P1163-1165 65-18225 <*> O
1960 V51 P3211-3215 61-16102 <*>
1960 V51 N23 P957-970 60-18759 <*>
1960 V51 N23 P971-977 61-10239 <*>
1960 V51 N23 P977-980 60-18757 <*>
1960 V51 N23 P980-982 60-18758 <*>
1960 V51 N23 P982-987 60-18761 <*>
1960 V51 N23 P987-990 61-10813 <*>
1960 V51 N23 P990-996 60-18169 <*>
1960 V51 N23 P996-998 60-18170 <*>
1960 V51 N39 P1816-1823 61-10812 <*>
1960 V51 N43 P2013-2023 61-01054 <*>
1960 V51 N77 P3208-3210 66M47I <ATS>
1960 V51 N85 P3586-3591 61-00581 <*>
1961 V52 P379-382 61-20435 <*>
1961 V52 P631-634 11095-C <KH>
63-16585 <*>
1961 V52 P2618-2623 63-26246 <=$>
1961 V52 N10 P363-366 61-16472 <*> O
1961 V52 N10 P382-383 61-16476 <*>
1961 V52 N10 P388-389 61-18580 <*>
1961 V52 N10 P393-394 61-16475 <*>
1961 V52 N29 P1352-1354 62-18051 <*>
1961 V52 N70 P2964-2968 11548-B <K-H>
63-16669 <*>
1962 V53 P1053-1055 63-01243 <*>
1962 V53 P1067-1069 65-00209 <*>
1962 V53 P1516-1521 12125 <K-H>
1962 V53 P1673-1676 63-14764 <*>
1962 V53 P2001-2007 66-12113 <*> O
1962 V53 P3800-3806 63-20946 <*>
1962 V53 N37 P1448-1449 63-00600 <*>
1962 V53 N38 P1466-1478 63-01623 <*>
1962 V53 N39 P1516-1521 63-16571 <*>
1962 V53 N41 P1583-1586 65-11335 <*> O
1963 V54 N49 P1849-1854 64-18146 <*>
1963 V54 N83 P3130-3136 64-18663 <*>
1963 V54 N88 P3341-3345 65-14302 <*> O
1963 V54 N88 P3345-3348 65-14299 <*> O
1963 V54 N88 P3348-3350 65-14300 <*>
1963 V54 N88 P3350-3352 65-14301 <*> O
1963 V54 N57/8 P2117-2122 64-18712 <*>
1964 V55 P2074-2081 65-12331 <*> P
1964 V55 P3651-3656 65-17419 <*>
1964 V55 N69 P2559-2562 SUP
66-10856 <*>
1964 V55 N100 P4005-4017 65-14719 <*>
1964 V55 N3/4 P75-77 64-18143 <*>
1965 V56 N6 P197-201 65-13536 <*>
1965 V56 N101 P4459-4464 66-12953 <*>

MINERVA MEDICA SICILIANA
1963 V8 N1 P16-18 64-16353 <*>

MINERVA MEDICOLEGALE
1951 V71 P136-139 62-16134 <*>
1956 V76 N1 P12-13 5367-A <K-H>
1960 V80 P32-34 65-10067 <*> O

MINERVA MEDICOPSICOLOGICA
1961 V2 P186-189 65-10987 <*>

MINERVA NEFROLOGICA
1964 V11 N4 P131-153 66-12757 <*> O
1964 V11 N4 P153-159 66-10860 <*> O

MINERVA NEUROCHIRURGICA
1962 V6 P101-103 66-12671 <*>
1962 V6 P153-157 66-12673 <*>
1962 V6 N4 P129-141 64-00434 <*>

MINERVA NIPIOLOGICA
1958 V8 N2 P44-52 61-16137 <*> O
1959 V9 P77-78 73M42I <ATS>

MINERVA NUCLEARE
1958 V2 P393-394 UCRL TRANS-954(L) <*>
1958 V2 N9 P238-244 60-19485 <=*> O
1963 V7 N6 P237-243 NP-TR-1112 <*>
1963 V7 N9 P353-362 65-18113 <*>
1963 V7 N11 P439-441 NP-TR-1197 <*>

MINERVA ORTOPEDICA
1953 N4 P29-35 57-2704 <*>
1958 V9 N1/2 P26-29 60-16605 <*> O

MINERVA OTORINOLARINGOLOGICA
1954 V4 N2 P91-93 II-310 <*>
1956 V6 P165-170 5811-A <K-H>
1958 V8 N3 P73-74 64-18669 <*>
1960 V10 P69-75 63-10719 <*> O
1961 V11 P430-437 65-14949 <*> OP
1961 V11 N10 P392-394 63-16504 <*>

MINERVA PEDIATRICA
1950 V2 N12 P593-597 2379 <*>
1952 V4 N16 P697-726 II-790 <*>
1952 V4 N16 P742-744 1293 <*>
1957 V9 N29/0 P785-792 62-14515 <*> O
1958 V10 P35-36 62-00437 <*>
1958 V10 P686-690 65-14328 <*>
1958 V10 P849-863 62-00437 <*>

1958 V10 N13 P356-359 64-18620 <*>
1958 V10 N37 P904-908 64-18709 <*> O
1960 V12 P127-128 65-18226 <*>
1960 V12 N49 P1623-1626 61-20908 <*>
1961 V13 P1619-1622 65-14642 <*> O
1962 V14 P1127-1140 65-17260 <*>
1964 V16 N20 P709-713 65-13974 <*>
1965 V17 P717-725 66-11584 <*>

MINERVA RADIOLOGICA, FIZIOTERAPIA E RADIOBIOLOGICA
1961 V6 N4 P187-190 62-00642 <*>

MINERVA UROLOGICA
1957 V9 N3 P109-110 59-15685 <*> O
1958 V10 N4 P101-106 64-18654 <*>
1960 V12 N5 P160-168 40N49I <ATS>

MINING GEOLOGY, JAPAN
SEE KOZAN CHISHITSU

MINNO DELO
1955 V10 N5 P85-91 59-13122 <=>
1963 V18 N12 P25-28 TRANS-556 <MT>

MINNO DELO I METALURGIYA
1961 V16 N9 P25-27 71583B <ATS>

MINZOKU EISEI
1963 V29 N1 P17-22 66-12108 <*>

MIROVAYA EKONOMIKA I MEZHDUNARODNYE OTNOSHENIYA
1958 N10 P16-25 59-16790 <+*>
1961 N2 P148-150 61-23646 <*=>
1962 N10 P3-14 63-13825 <=*>
1962 N10 P49-59 63-13825 <=*>
1965 N12 P42-59 66-30784 <=$>
1966 N1 P40-50 66-30642 <=$>

MISCELLANEOUS REPORTS OF THE RESEARCH INSTITUTE FOR NATURAL RESOURCES. JAPAN
SEE SHIGEN KAGAKU KENKYUSHO IHO

MISSILI. ASSOCIAZIONE ITALIANI RAZZI
1962 V4 P5-12 UCRL TRANS-1159(L) <*>

MITSUBISHI DENKI
1960 V33 N8 P1-11 SPEC SUP
62-19247 <=*>

MITSUBISHI DENKI GIHO
1959 N11 P1- 61-11123 <=>
1965 V39 N8 P7-12 66-11472 <*> O

MITSUBISHI ELECTRIC MANUFACTURING CO., TECHNICAL REPORTS
SEE MITSUBISHI DENKI GIHO

MITSUBISHI NIHON JUKO GIHO
1963 N2 P145 1907 <BSRA>
1963 V4 N2 P264- 1873 <BSRA>
1964 N1 P46- 2066 <BSRA>
1964 N1 P72- 2033 <BSRA>

MITSUI TECHNICAL REVIEW
SEE MITSUI ZOSEN GIHO

MITSUI ZOSEN GIHO
1963 N43 P2- 2195 <BSRA>

MITTEILUNGEN DER BEFA
1952 N11 ENTIRE ISSUE 62-20059 <*>
1954 N9 ENTIRE ISSUE 62-20058 <*>
1954 N11 ENTIRE ISSUE 62-20051 <*>
1955 N11 P1-13 T-1979 <INSD>
1955 N12/3 P1-5 T-2051 <INSD>

MITTEILUNGEN DER BIOLOGISCHEN BUNDESANSTALT FUER LAND-U. FORSTWIRTSCHAFT
1954 V80 P155-162 C-2310 <NRC>
1956 N86 P49-63 59-17785 <*> PO

MITTEILUNGEN DES CHEMISCHEN FORSCHUNGSINSTITUTS DER INDUSTRIE
1949 V3 P70-73 60-18327 <*>
63-14571 <*>
1949 V3 P91-94 71L33G <ATS>
1949 V3 P109-112 62-20075 <*>

MITTEILUNGEN DES CHEMISCHEN FORSCHUNGSINSTITUTS DER WIRTSCHAFT OESTERREICHS
1955 V9 N3 P57-58 57-2027 <*>
1958 V12 N5 P129-131 59-15534 <*>

MITTEILUNGEN DER DEUTSCHEN GESELLSCHAFT FUER HOLZFORSCHUNG
1958 N42 P72-75 60-16939 <*>
1960 V47 P61-71 62-01252 <*>
1961 V48 P9-13 63-01560 <*>
1961 V48 P18-28 63-01411 <*>
1961 V48 P45-52 63-18923 <*> O
1961 V48 P53-56 63-01559 <*>
1961 V48 P79-83 63-01409 <*>

MITTEILUNGEN DER DEUTSCHEN LANDWIRTSCHAFTS-GESELLSCHAFT
1957 P881-882 58-2518 <*>
1964 V79 N12 P389-390 65-14635 <*> P
1964 V79 N12 P392-394 65-14635 <*> P

MITTEILUNGEN DER DEUTSCHEN MATERIAL-PRUEFUNGS-ANSTALTEN
1930 N8 P98-99 60-10206 <*>
1930 N14 P38-40 59-15634 <*> O
1931 P67-69 SPEC. 57-985 <*>
1931 N15 P70-73 60-10216 <*> O

MITTEILUNGEN DER DEUTSCHEN PHARMAZEUTISCHEN GESELLSCHAFT
1963 V33 N5 P77-82 64-20916 <*>

MITTEILUNGEN DES DEUTSCHEN WETTERDIENSTES
1954 N8 P3-22 T2035 <INSD>
1955 V2 N12 P1-30 CSIRO-3037 <CSIR>

MITTEILUNGEN DES EIDGENOESSISCHEN INSTITUTS FUER SCHNEE-UND LAWINENFORSCHUNG
1961 N15 ENTIRE ISSUE 64-15902 <=*$>

MITTEILUNGEN UEBER FORSCHUNGSARBEITEN AUF DEM GEBIET DES INGENIEURWESENS
1908 N62 P31-67 65-12979 <*>

MITTEILUNGEN DER FORSCHUNGSGELLSCHAFT FUER BLECHVERARBEITUNG
1953 N13 P1-3 3760 <HB>
1955 N13 P153-160 2377 <BISI>
1957 N14 P4- 58-1325 <*>
1957 N14 P4 59-12036 <=*>
1960 N5 P50-55 66-10711 <*> O
1961 N17 P218-223 2622 <BISI>
1961 N17 P223-226 2651 <BISI>
1961 N17 P226-227 2652 <BISI>
1961 N21 P294-296 PN-82 <TTIS>
1962 P143-150 2422 <TTIS>
1962 P282-293 3425 <BISI>
1962 N16/7 P220-230 3158 <BISI>
1964 P274-283 4300 <BISI>

MITTEILUNGEN AUS DEM FORSCHUNGSINSTITUT FUER PHYSIK DER STRAHLANTRIEBE
1957 N12 P27-40 59-10297 <*>
8264 <K-H>

MITTEILUNGEN AUS DEM FORSCHUNGSINSTITUT. VEREINIGTE STAHLWERKE AKTIENGESELLSCHAFT
1933 V3 N8 P199-234 59-20588 <*> O
1933 V3 N8 P235-248 59-20594 <*>

MITTEILUNGEN DES FORSCHUNGSINSTITUTS UND PROBIERAMTS FUER EDELMETALLE DER STAATLICHEN

HOEHEREN FACHSCHULE SCHWAEB
1938 V12 N3 P17-29 59-17111 <*> 0
1938 V12 N1/2 P1-9 59-17112 <*>

MITTEILUNGEN AUS DEN FORSCHUNGSLABORATORIEN DER A G F A AKTIENGESELLSCHAFT FUER PHOTCFABRIKATION
1955 V1 P153-156 62-20482 <*>
1955 V1 P199-238 62-14405 <*> 0
1955 V1 P256-261 62-10993 <*> 0
1955 V1 P320-325 57-2903 <*>
1961 V3 P198-205 63-10780 <*>
1961 V9 P128-132 64-30648 <*> 0

MITTEILUNGEN AUS DEM FRAUNHOFER INSTITUT. FREIBURG
1959 V47 N3 P191-197 63-10668 <*> 0

MITTEILUNGEN AUS DEM GEBIET DER LEBENSMITTEL-UNTERSUCHUNG UND -HYGIENE
1930 V21 P312-314 5398-C <K-H>
1951 V42 P395-402 3036-E <K-H>
1953 V44 P472-474 3036-C <K-H>
1953 V44 N4 P371-377 58-1032 <*>
1953 V44 N6 P472-474 58-1211 <*>
1954 V45 N6 P473-476 3546-E <K-H>
58-1C27 <*>
1955 V46 P178-182 3799-A <K-H>
1956 V47 N1 P4-15 CSIRO-3501 <CSIR>
1956 V47 N1 P20-27 CSIRO-3382 <CSIR>
1956 V47 N1 P66-71 CSIRO-3498 <CSIR>
R-5099 <*>
59-10908 <*> 0
1956 V47 N2 P149-152 CSIRO-3499 <CSIR>
58-2370 <*>
1956 V47 N4 P221-231 59-10401 <*> 0
1957 V48 P94-116 58-1024 <*>
6673-B <K-H>
1959 V50 P18-39 63-18466 <*>
1959 V50 P159-165 9249 C <K-H>
9249-B <K-H>
1959 V50 P258-263 63-16679 <*>
9929-A <K-H>
1959 V50 N6 P523-531 60-16925 <*>
1960 V51 P69-74 10402-C <K-H>
63-16892 <*>
1960 V51 P3C3-320 10684-C <KH>
1960 V51 N1 P69-74 63-20127 <*>
1960 V51 N3 P159-165 65-13213 <*>
1960 V51 N5 P325-338 10969-F <KH>
62-20205 <*>,
1961 V52 N3 P135-244 62-18762 <*>
1962 V53 N3 P234-243 8702 <IICH>
1964 V55 N3 P154-181 65-17349 <*>
1964 V55 N4 P264-273 66-11081 <*>

MITTEILUNGEN AUS DEM GECLCGISCHEN STAATSINSTITUT IN HAMBURG
1949 N19 P89-109 49M38G <ATS>

MITTEILUNGEN DES HYDRAULISCHEN INSTITUTS DER TECHNISCHEN HOCHSCHULE, MUENCHEN
1926 N1 P75- 57-174 <*>
1934 N4 P70-93 57-174 <*>
1939 N9 P30-34 65-12976 <*>

MITTEILUNGEN. INSTITUT FUER AERODYNAMIK
1938 N6 ENTIRE ISSUE II-1464 <*>
1943 N8 P34-43 59-17727 <*>
1943 N8 P44-49 59-17713 <*>
1954 N21 P5-21 59-21109 <=>
1954 N21 P18-35 60-17352 <=*>

MITTEILUNGEN AUS DEM INSTITUT FUER ANGEWANDTE MATHEMATIK. EIDGENOSSISCHE TECHNISCHE HOCHSCHULE. ZURICH
1964 N9 P65-69 66-11017 <*>

MITTEILUNGEN AUS DEM INSTITUT FUER ENERGETIK LEIPZIG
1961 N34 P206-209 3528 <BISI>

MITTEILUNGEN AUS DEM INSTITUT FUER HOCHFREQUENZTECHNIK. EIDGENOESSISCHE TECHNISCHE HOCHSCHULE. ZURICH
1952 N16 P5-55 66-12600 <*> 0

MITTEILUNGEN DES KAELTETECHNISCHEN INSTITUTS UND DER REICHSFORSCHUNGSANSTALT FUER LEBENSMITTEL-FRISCHHALTUNG AN DER TECHNISCHEN HOCHSCHULE KARLSRUHE
1948 N2 P1-45 64-14402 <*>

MITTEILUNGEN AUS DEM KAISER-WILHELM-INSTITUT FUER EISENFORSCHUNG ZU DUESSELDCRF
1927 V9 P129-149 63-14386 <*>
1928 V10 P107-116 AL-483 <*>
1930 V12 N9 P115-124 59-20477 <*> 0
1930 V12 N11 P149-159 57-3122 <*>
1930 V12 N11 P161-164 57-3138 <*>
1930 V12 N12 P165-169 59-20476 <*> 0
1931 V13 N2 P29-41 59-20475 <*> 0
1931 V13 N13 P169-181 59-20474 <*> 0
1932 V14 P295-305 57-2621 <*>
1934 V16 N7 P77-91 60-18855 <*> 0
1936 V18 P1-14 59-20583 <*>
1937 V19 N2 P27-46 59-20473 <*> 0
1939 V21 P27-55 66-14850 <*> 0
1939 V21 N13 P201-212 TS-1399 <BISI>
62-14406 <*>
1940 V22 N7 P93-108 61-18557 <*> 0
1940 V22 N8 P109-119 61-18858 <*> 0
1940 V22 N8 P121-136 61-18852 <*=> 0
1940 V22 N14 P217-228 61-18566 <*> 0
1941 V23 N12 P195-245 61-18753 <*>

MITTEILUNGEN DES KOHLENFORSCHUNGSINSTITUTS IN PRAG
1935 V2 P142-149 NP-TR-278 <*>

MITTEILUNGEN FUER DIE LANDWIRTSCHAFT
1944 V59 N20 P435-44C 57-2952 <*>
1944 V59 N21 P467-471 57-2952 <*>

MITTEILUNGEN. MATERIALPRUFUNGSAMT TECHNISCHE HOCHSCHULE. BERLIN
1906 V24 P226-235 64-14244 <*>
1911 V29 N1 P2-28 60-10438 <*> 0
1919 V37 P12-18 64-14245 <*>

MITTEILUNGEN DER MATERIALPRUEFUNGSANSTALT AN DER TECHNISCHEN HOCHSCHULE, DARMSTADT
1933 N4 P1-102 62-01589 <*> P

MITTEILUNGEN. MAX-PLANCK GESSELLSCHAFT ZUR FORDERUNG DER WISSENSCHAFTEN
1957 N3 P143-163 59-16785 <+*>

MITTEILUNGEN AUS DEM MAX-PLANCK-INSTITUT FUER STROEMUNGSFORSCHUNG (UND DER AERODYNAMISCHEN VERSUCHSANSTALT). GOETTINGEN
1950 P1-54 62-00226 <*>
1952 N7 P1-48 65-13519 <*> 0

MITTEILUNGEN DER OESTERREICHISCHEN GESELLSHAFT FUER ANTHROPOLOGIE, ETHNOLOGIE UND PRAEHISTORIE. WIEN
1960 V90 P55-67 63-01626 <*>

MITTEILUNGEN DER OESTERREICHISCHEN GESELLSCHAFT FUER HOLZFORSCHUNG
1955 V7 N2 P11-17 2898 <*>

MITTEILUNGEN DER SCHWEIZERISCHEN ANSTALT FUER DAS FORSTLICHE VERSUCHSWESEN
1959 V35 N5 P373-409 63-C1562 <*>

MITTEILUNGEN UEBER TEXTIL-INDUSTRIE
1965 V72 P40-41 65-17307 <*>

MITTEILUNGEN DER VEREINIGUNG DER GROSSKESSEL-BESITZER
1951 P228-237 T2028 <INSD>

1953 N25 P533-538 56L31G <ATS>
1954 N29 P141-145 80L31G <ATS>
1954 N32 P360-363 64-20823 <*>
1955 N33 P413-415 57L31G <ATS>
1955 N37 P1-10 57-3204 <*>
1955 N37 P700-704 57-2858 <*>
1955 V38 P746-752 CSIRO-3233 <CSIR>
1956 N42 P155-158 61-16245 <*>
1956 N42 P159-166 T-2323 <INSD>
1956 N45 P409-426 62-01483 <*>
1956 V43 P235-240 58-2331 <*>
1956 V43 P241-252 58-1645 <*>
1956 V43 P253-258 58-2329 <*>
1956 V43 P258-273 58-2333 <*>
1957 N51 P417-421 59-18546 <+*> O
1957 N51 P421-430 59-18547 <=*> O
1957 V51 P417-421 58-2539 <*>
1957 V51 P421-430 58-2540 <*>
1958 P176-180 TS 1513 <BISI>
1958 P255-264 TS-1514 <BISI>
1958 P375-381 TS 1448 <BISI>
1958 P397-406 TS-1421 <BISI>
1958 P407-411 TS-1422 <BISI>
1958 P411-420 TS 1515 <BISI>
1958 N53 P92-98 AEC-TR-5108 <*>
1958 N54 P165-175 TS 1512 <BISI>
62-16844 <*>
1958 N54 P215-219 30L29G <ATS>
1958 N57 P375-381 62-14396 <*>
1958 N57 P393-397 TS-1420 <BISI>
1958 N57 P397-406 62-14444 <*>
1958 N57 P407-411 62-14398 <*>
1958 N57 P411-420 62-14680 <*>
1959 N60 P181-207 TS 1516 <BISI>
62-16845 <*>
1959 N62 P365-367 TS 1622 <BISI>
62-16851 <*>
1959 N63 P391- AEC-TR-4033 <*>
1959 N63 P391-401 93M40G <ATS>
1960 P230-242 2277 <BISI>
1961 N72 P155-158 62-18391 <*> O
1961 N74 P319-332 2692 <BISI>
1962 N76 P38-42 63-18494 <*>
1962 N78 P190-198 3224 <BISI>
1962 N79 P245-261 3224 <BISI>
1963 V84 P186-194 3352 <BISI>
1964 N88 P26-33 6226 <HB>
1964 N88 P33-39 6294 <HB>
1964 N88 P39-42 6324 <HB>
1964 N88 P42-45 6326 <HB>
1964 N88 P45-52 6340 <HB>
1964 N93 P394-399 4198 <BISI>

MITTEILUNGEN DER VEREINIGUNG SCHWEIZERISCHER VERSICHERUNGS-MATHEMATIKER
1945 V45 P97-163 58-770 <*>

MITTEILUNGEN DES VEREINS DEUTSCHER EMAILFACHLEUTE
1959 V7 P93-96 1752 <BISI>
1960 V8 N1 P1-10 AI-TR-41 <*>
1962 V10 P101-108 3244 <BISI>
1964 V12 P15-22 4114 <BISI>
1965 V13 N12 P85-90 66-12191 <*>

MITTEILUNGEN DER VERKAUFSVEREINIGUNG FUER TEERERZEUGNISSE. ESSEN
1956 V5 N1 ENTIRE ISSUE 61-20953 <*> O

MITTEILUNGEN DER VERSUCHANSTALT FUER WASSERBAU UND ERDBAU
1959 N44 P1- C-3483 <NRCC>

MITTEILUNGEN DES WOEHLER-INSTITUTS, BRAUNSCHWEIG
1938 N33 P55-65 65J15G <ATS>

MITTEILUNGEN AUS DER ZOOLOGISCHEN STATION ZU NEAPEL
1916 V22 N11 P329-366 AL-815 <*>

MITTEILUNGSBLATT FUER DIE AMTLICHE MATERIALPRUEFUNG IN NIEDERSACHSEN
1962 N2/3 P28-31 64-71436 <=>

MITTEILUNGSBLATT DER CHEMISCHEN GESELLSCHAFT IN DER DEUTSCHEN DEMOKRATISCHEN REPUBLIK
1963 V10 N3 P41-60 63-31187 <=>

MITTEILUNGSBLATT FUER MATHEMATISCHE STATISTIK UND IHRE ANWENDUNGSGEBIETE
1954 V6 N2 P164-169 64-16836 <*>

MLODY TECHNIK. WARSAW
1962 N7 P25-29 62-33669 <=>
1962 N7 P52-56 62-33669 <=>
1962 N8 P34-43 62-33669 <=>

MODERN MEDICINE. JAPAN
SEE SAISHIN IGAKU

MODERNE HOLZVERARBEITUNG
1963 N30 P176-180 C-5597 <NRC>

MOKSLINIAI PRANESIMAI
1960 V11 P48-68 AEC-TR-5891 <=$>
1960 V11 P73-80 AEC-TR-5871 <=$>

MOKSLINIAI PRANESIMAI. LIETUVOS TSR MOKSLU AKADEMIJA VILNA. GEOLOGIJOS IR GEOGRAFIJOS INSTITUTAS
1962 V13 ENTIRE VOLUME N64-24786 <=>
1962 V14 N1 P59-75 64-23609 <=$>

MOKUZAI GAKKAISHI
1957 V3 N2 P57-62 16J18 4 <ATS>
1959 V5 P41-44 61-17549 <=$>
1961 V7 N1 P19-23 57N55J <ATS>
65-12853 <*>
1961 V7 N5 P205-207 66-26093 <*$>
1962 V8 P133-138 63-20364 <*>
1962 V8 N5 P204-207 66-20965 <$>

MOKUZAI KENKYU
1949 N1 ENTIRE ISSUE II-1504 <*>
1952 N9 P42-62 C-3124 <NRCC> C
1953 N11 P1-2 AL-392 <*>

MOKUZAI KOGYO
1956 V11 N108 P23-30 C-3205 <NRCC>
1956 V11 N110 P236-243 CSIRO-3147 <CSIR>
57-3466 <*>
1958 V13 N8 P15-20 59-20086 <*>

MOLECULAR PHYSICS
1958 V1 N1 P23-43 NP-TR-976 <*>

MOLKEREIZEITUNG. HILDESHEIM
1935 V49 N23 P590-593 II-781 <*>
1936 V50 P1454-1455 2490 <*>
1936 V50 P1480-1482 II-757 <*>
1938 V52 P267-269 2491 <*>
1938 V52 P308-309 II-501 <*>
1938 V52 N11 P327-335 II-758 <*>

MOLKEREI- UND KAESEREI-ZEITUNG
1953 V7 P885- 59-15583 <*>
1959 V13 P243 64-14072 <*>

MOLOCHNAYA PROMYSHLENNOST
1951 V12 N1 P17-24 RT-731 <*>
1951 V12 N7 P29-31 R-112 <*>
1953 N2 P32-33 <INSD>
1957 V18 N1 P5-6 64-15432 <=*$>
1957 V18 N8 P32-34 64-19424 <=$> C
1957 V18 N9 P33-35 R-5274 <*>
1958 V19 N2 P16-18 59-18725 <+*>
1958 V19 N2 P43 59-18020 <+*>
1958 V19 N5 P7-8 61-15370 <+*>
1960 V21 N11 P14-17 61-23660 <=*>
1960 V21 N12 P21 61-21256 <=>

1961 V22 N2 P31-34 61-28768 <*=>
1961 V23 N3 P24-25 64-19546 <=$>
1962 V23 N4 P36-37 62-33410 <=*>
1962 V23 N5 P3-5 62-33050 <=*> 0
1962 V23 N5 P9-13 64-15437 <=*$>
1962 V23 N5 P43-45 65-63572 <=$*>
1963 V24 N4 P7-10 63-24067 <=*$>
1963 V24 N7 P29-30 64-15438 <=*$>
1963 V24 N8 P33-34 65-60678 <=$>
1964 V25 N7 P111-113 65-63010 <=$>
1965 V26 N5 P32-42 66-61558 <=*$>

MOLOCHNAYA PROMYSHLENNOST SSSR
1940 V7 N4 P5-6 50/1593 <NLL>
1940 V7 N4 P10-11 50/1592 <NLL>

MOLOCHNOE I MIASNOE SKOTOVODSTVO
SEE MOLOCHNOE I MIASNOE ZHIVOTNOVODSTVO

MOLOCHNOE I MYASNOE ZHIVOTNOVODSTVO
1959 N10 P49-51 60-31124 <=> 0
1960 N9 P1-6 61-11969 <=>
1960 V5 N10 P1-3 61-21598 <=>
1963 N1 P12-21 RTS-2811 <NLL>

MOLODOY KOMMUNIST
1954 V12 N9 P75-80 59-12975 <+*>
1960 V18 N2 P56-62 60-31173 <=>
1962 V20 N10 P8-17 63-13276 <=*>
1962 V20 N12 P45-50 63-21547 <=>
1962 V20 N12 P69-72 63-21326 <=>

MOMENTO
1959 V12 N1413 P2-37 59-15630 <*> 0

MONATSBERICHT DER DEUTSCHEN AKADEMIE DER WISSENSCHAFTEN ZU BERLIN
1959 V1 P400-404 PT.2 AEC-TR-5649 <*>
1959 V1 N1 P27-31 PT.1 AEC-TR-5649 <*>
1959 V1 N1 P32-34 62-32179 <=*>
1959 V1 N5 P290-296 61-15039 <*+> 0
1959 V1 N7/10 P400-404 AEC-TR-5649 <*>
1959 V1 N7/10 P446-451 5195 <HB>
1961 V3 P183-186 67R76G <ATS>
1961 V3 N5/6 P260-263 4567 <BISI>
1962 V4 N9 P596-597 66-12359 <*>
1962 V4 N10 P623-624 64-10492 <*>
1962 V4 N11/2 P761-766 63-18262 <*>
1963 V5 N3 P170-173 65-12088 <*>
1964 V6 N3 P220-224 6809 <HB>
1964 V6 N9 P641-643 66-10349 <*>

MONATSBERICHTE DES OESTERREICHISCHEN INSTITUTS FUER WIRTSCHAFTSFORSCHUNG. VIENNA
1959 V31 N9 P393-395 65-11765 <*>

MONATSBULLETIN. SCHWEIZERISCHER VEREIN VON GAS-U. WASSERFACHMAENNERN
1948 V28 N5 P105-113 66-11793 <*>
1949 V29 P59-62 AL-210 <*>
1952 V32 N4 P3-15 60-13243 <*=>
1963 V43 N1 P1-9 64-14947 <*>
1963 V43 N2 P34-45 64-14947 <*>
1963 V43 N4 P98-104 64-14947 <*>

MONATSHEFTE FUER CHEMIE
1881 V2 P398-409 63-14029 <*>
1885 V6 P989-996 64-10861 <*>
1887 V8 P180-186 63-20175 <*>
1893 V14 P685-698 60-14200 <*>
1896 V17 P232-235 63-20511 <*> P
1908 V29 P753-762 65-11614 <*>
1912 V33 P393-414 60-10081 <*> P
1912 V33 P859-871 61-18314 <*>
1912 V33 P1361-1377 MF-4 <*>
1912 V33 N10 P1407-1429 73H12G <ATS>
1914 V35 P391-406 65-17077 <*>
1926 V47 P17-38 59-10739 <*>
1926 V47 P39-56 59-10740 <*>
1927 V48 P113-121 62-18305 <*>
1927 V48 P673-687 62-18304 <*>
1928 V49 P283-315 63-14101 <*> 0
1933 V63 P186-200 1673 <*>
1936 V67 P241-247 63-14124 <*>
1947 V76 P398-405 57-2592 <*>
1947 V77 P65-72 1403A <K-H>
1948 V79 P311-315 SCL-T-472 <*>
1950 V81 P917-920 02M43G <ATS>
1950 V81 N5 P746-750 63-00739 <*>
1951 V82 P959-969 62-00670 <*>
1951 V82 N3 P489-493 59-20617 <*> 0
1952 V83 P802-817 AL-602 <*>
1952 V83 P1087-1089 63-10576 <*>
1952 V83 N3 P818-828 59-15511 <*>
1952 V83 N4 P988-1008 64-20447 <*> 0
1952 V83 N4 P1062-1068 I-1005 <*>
1953 V84 P250-256 59-15950 <*> 0
1953 V84 P406- AL-399 <*>
1953 V84 P579-584 58-2429 <*>
1953 V84 P677-685 58-1575 <*>
1953 V84 P996-1010 61-00301 <*>
1953 V84 P1119-1126 1480 <*>
1953 V84 P1146-1161 642 <TC>
1953 V84 N1 P99-101 64-20041 <*>
1953 V84 N4 P765-776 65-13052 <*>
1954 V85 P52-68 1650 <*>
1954 V85 P241-244 2103 <*>
58-2424 <*>
1954 V85 P245-254 2106 <*>
58-1546 <*>
1954 V85 P255-272 62-18663 <*> 0
1954 V85 P575-579 T-1463 <INSD>
1954 V85 P684-692 1963 <*>
1954 V85 P936-948 8766 <IICH>
1954 V85 P1024-1045 65-00379 <*> 0
1954 V85 P1251 AEC-TR-5482 <*>
1954 V85 N4 P896-905 3977 <K-H>
1954 V85 N4 P936-948 64-20820 <*>
1954 V85 N4 P957-971 CSIRO-2993 <CSIR>
1954 V85 N5 P67- I-1024 <*>
1954 V85 N5 P1015-1023 64-30141 <*>
1954 V85 N5 P1046-1054 I-1025 <*>
1954 V85 N6 P1276-1280 II-879 <*>
1955 V86 P637-642 1285 <*>
1955 V86 N1 P131-136 AEC-TR-3528 <+*>
1955 V86 N3 P474-484 67T95G <ATS>
1955 V86 N5 P718-734 62-14205 <*> 0
1955 V86 N6 P995-1003 5072-B <K-H>
1956 V87 P421-424 NP-TR-850 <*>
1956 V87 N1 P8-23 57-808 <*>
1956 V87 N1 P71-91 5562-C <K-H>
1957 V88 N3 P336-343 58-2460 <*>
59-10084 <*>
1957 V88 N4 P502-516 58-1396 <*>
1957 V88 N5 P739-748 58-1775 <*>
1958 V89 P618-624 63-20957 <*> 0
1958 V89 N1 P74-78 58-2467 <*>
1958 V89 N1 P79-82 AEC-TR-3735 <*>
1958 V89 N1 P88-95 C-3250 <NRCC>
213 <TC>
1958 V89 N6 P692-700 59-00745 <*>
1958 V89 N6 P701-707 59-00744 <*>
1958 V89 N4/5 P611-617 61-20838 <*>
1958 V89 N4/5 P618-624 64-16597 <*>
1959 V90 N1 P15-23 61-20837 <*>
1959 V90 N4 P443-457 65-11589 <*>
1960 V91 P176-187 AEC-TR-4343 <*>
1960 V91 P717-728 62-18539 <*>
1960 V91 N2 P249-262 C-3567 <NRCC>
1960 V91 N3 P400-405 65-12438 <*>
1960 V91 N6 P1020-1023 C-3653 <NRCC>
1961 V92 P841-855 AEC-TR-5000 <*>
1961 V92 P1176-1183 AEC-TR-5175 <*>
1961 V92 N1 P8-21 62-18567 <*>
1961 V92 N6 P1147-1154 62-18723 <*> 0
1961 V92 N6 P1279-1289 65-12306 <*>
1962 V93 N1 P74-77 RCT V36 N2 P558 <RCT>
62-18541 <*>
1962 V93 N5 P1000-1004 AEC-TR-5651 <*>
63-14451 <*>

1962 V93 N5 P1046-1054 63-14452 <*>
1962 V93 N5 P1176-1195 64-18609 <*> 0
1963 V94 P204-224 AEC-TR-6100 <*>
1963 V94 N2 P440-446 159 <TC>
1963 V94 N2 P473-476 63-01431 <*>
1963 V94 N2 P477-481 63-01430 <*>
1963 V94 N3 P507-517 64-18610 <*> 0
1963 V94 N4 P742-752 65-11536 <*>
1964 V95 P842-852 9002 <IICH>
1964 V95 N1 P89-93 66-14490 <*>
1964 V95 N1 P214-218 58R77G <ATS>
1964 V95 N1 P219-221 59R77G <ATS>
1964 V95 N3 P633-648 64-30647 <*> 0
1965 V96 N1 P95-97 66-12229 <*>

MONATSHEFTE FUER TIERHEILKUNDE
1953 V5 N1 P14-22 II-712 <*>
1961 V13 N5 P129-135 PT.1 64-00180 <*>
1961 V13 N6 P162-174 PT.2 64-00180 <*>
1961 V13 N7 P184-190 65-11412 <*>
1962 V14 N3 P77-90 64-00189 <*>
1962 V14 N9 P277-287 64-00078 <*>
1963 V15 N22 P191-199 64-00113 <*>

MONATSHEFTE FUER VETERINAERMEDIZIN
1947 V2 N10 P165-172 61-14502 <*>
1948 V3 N7 P121-127 61-18826 <*>
1952 V4 P145-161 61-10926 <*>
1955 V10 N17 P395-398 59-20973 <*> P
1960 V15 N24 P887-888 62-19379 <=*>
1961 V16 N9 P329-338 62-19723 <=*>
1961 V16 N9 P351-354 62-19733 <=*>
1961 V16 N10 P369-373 62-19724 <=*>
1961 V16 N10 P378-380 62-19734 <=*>
1961 V16 N10 P395-399 62-19735 <=*>
1961 V16 N11 P420-422 62-19744 <=*>
1961 V16 N11 P422-425 62-19743 <=*>
1961 V16 N11 P428-433 62-19730 <=*> P
1961 V16 N11 P433-437 62-19732 <=*>
1961 V16 N12 P449-453 62-19438 <=*>
1961 V16 N16 P605-609 62-19738 <=*>
1961 V16 N16 P624-626 62-19731 <=*>
1961 V16 N16 P638-639 62-23630 <=>
1961 V16 N17 P672 62-23630 <=>
1961 V16 N18 P714 62-23630 <=>
1961 V16 N19 P1-12 SUP 62-19793 <=*>
1961 V16 N20 P773-776 62-19739 <=*>
1961 V16 N21 P827 62-23630 <=>
1961 V16 N22 P867 62-23630 <=>
1961 V16 N23 P904 62-23630 <=>
1964 V19 N12 P449-452 65-17029 <*>

MONATSKURSE FUER DIE AERZTLICHE FORTBILDUNG
1957 V7 P153-159 57-2874 <*>

MONATSSCHRIFT FUER GEBURTSHILFE U. GYNAEKOLOGIE
1898 V7 P295-300 62-14510 <*> 0

MONATSSCHRIFT FUER KINDERHEILKUNDE
1907 V23 P181-187 66-10950 <*>
1932 V54 P203-211 II-108 <*>
1937 V68 P295-296 59-17848 <*> 0
1938 V73 P406-412 65-18227 <*>
1951 V99 N9 P325-328 59-20703 <*>
1954 V102 P65-67 I-187 <*>
1955 V103 N9 P401-407 62-00302 <*>
1958 V106 N9 P401-404 07M38G <ATS>
1960 V108 N3 P164-175 60-10863 <*>
1961 V109 N11 P477-481 63-14772 <*>
1962 V110 N1 P13-19 65-14550 <*>
1962 V110 N3 P217-220 64-16930 <*> 0
1963 V111 N4 P140-142 64-18697 <*>
1963 V111 N8 P297-299 65-12939 <*>

MONATSSCHRIFT FUER OHRENHEILKUNDE U. LARYNGO-RHINOLOGIE
1912 V46 P809-841 61-14600 <*>
1928 V62 N11 P1261-1270 61-16047 <*>
1929 V63 P1203-1207 61-16037 <*>
1936 V70 N1 P40-47 61-14302 <*>
1938 V72 P976-983 57-1533 <*>
1952 V86 P221-229 II-292 <*>
60-18551 <*>
1957 V91 P136-145 PT.3 59-14745 <=*> 0
1958 V92 N4 P193-196 63-16390 <*>
1962 V96 N7 P294-306 65-17417 <*>

MONATSSCHRIFT FUER PSYCHIATRIE UND NEUROLOGIE
1906 V19 P290-305 61-14182 <*> 0
1942 V105 N1/2 P116-117 57-852 <*>
1954 V128 P56-90 57-1185 <*>
1956 V131 P251-255 II-979 <*>
1956 V132 P81-95 57-203 <*>
1956 V132 P335-346 57-1852 <*>

MONATSSCHRIFT FUER TEXTILINDUSTRIE
1934 V49 P50-52 60-10146 <*>

MONATSSCHRIFT FUER UNFALLHEILKUNDE UND VERSICHERUNGSMEDIZIN
1961 V64 P429-433 64-14754 <*> 0

MONDE ILLUSTRE
1946 V90 N4381 P1113-1116 3085 <*>

MONDO DEL LATTE
1948 N8 P271-273 59-20679 <*>

MONITEUR BELGE
1959 V129 P8492-8496 60-25163 <ATS>

MONITEUR SCIENTIFIQUE
1925 S5 V15 P97-113 64-18217 <*>
1926 S5 V16 P97-101 64-18312 <*>

MONITEUR UNIVERSAL
1808 N148 P581-582 63-16152 <*>

MONITOR PETROLULUI ROMAN
1940 V41 N4 P205-210 60-18345 <*>

MONITORE OSTETRICO-GINECOLOGICO
1952 N23 P1-14 58-330 <*>
1960 V31 N4 P432-445 61-20915 <*>

MONITORE ZOOLOGICO ITALIANO
1948 V56 P188-191 2254 <*>
1948 V57 P3-11 SUP 1726 <*>

MONOGRAFIE SCIENTIFICHE DI AERONAUTICA
1946 N3 ENTIRE ISSUE 63-20681 <*>

MONOGRAFII PO PALEONTOLOGII SSSR. LENINGRAD, MOSCOW
1938 V47 N1 ENTIRE ISSUE 41R80R <ATS>

MONOGRAPHIEN ZU 'ANGEWANDTE CHEMIE' UND 'CHEMIE-INGENIEUR-TECHNIK'
1953 N64 P14-23 64-10288 <*>

MONOGRAPHIEN AUS DEM GESAMTGEBIET DER NEUROLOGIE U. PSYCHIATRIE
1937 V62 P14-46 II-828 <*>
1937 V62 P57-63 II-828 <*>
1937 V62 P74-116 II-828 <*>

MONOGRAPHIEN AUS DEM GESAMTGEBIET DER PHYSIOLOGIE DER PFLANZEN UND DER TIERE
1933 V30 P1-404 60-16715 <*>

MONTANISTISCHE RUNDSCHAU
1931 V23 N2 P3-9 60-10210 <*>

MONTAN-RUNDSCHAU
1959 V7 N9 P183-190 61-20728 <*>

MONTAZHNYE I SPETSIALIZIROVANNYE RABOTY V STROITELSTVE
1960 N11 P7-11 NLLTB V3 N5 P378 <NLL>
1960 V22 N11 P7-11 E-768 <RIS>

1960 V22 N11 P15-17 E-769 <RIS>
1962 V24 N7 P1-4 62-32565 <=*>
1962 V24 N11 P4-12 63-21088 <=>
1963 V25 N7 P18-22 63-31843 <=>

MONTHLY BULLETIN OF THE JAPAN FISHERIES RESOURCES CONSERVATION ASSOCIATION
SEE NIPPON SUISAN SHIGEN HOGO KYOKAI GEPPO

MONTHLY JOURNAL OF AVIATION MECHANICS. CHINESE PEOPLES REPUBLIC
SEE HANG K'UNG CHI HSIEH YUEH K'AN

MONTHLY JOURNAL OF THE ELECTRICAL COMMUNICATION LABORATORY
SEE TSUKEN GEPPO

MONTHLY JOURNAL OF HYDROGEOLOGY. CHINESE PEOPLES REPUBLIC
SEE SHUI WEN YUEH KAN

MONTHLY REPORT OF THE TRANSPORTATION TECHNICAL RESEARCH INSTITUTE. JAPAN
SEE UNYU GIJUTSU KENKYUSHO HOKOKU

MONTI E BOSCHI
V16 N3 P21-30 C-6272 <NRC>

MORNARICKI GLASNIK
1960 V10 N5 P541-563 62-19583 <=*>
1963 N2 P190-204 AD-632 869 <=$>

MORSKOI FLOT
1955 V15 N3 P9-10 RT-2917 <*>
1956 N2 P24-26 RT-4288 <*>
1956 V16 N4 P27 2099 <BSRA>
1957 V17 N5 P26- R-4142 <*>
1958 V18 N6 P21- 60-17672 <+*> 0
1958 V18 N9 P17-19 60-17466 <+*>
1958 V18 N12 P15-16 63-21169 <=>
1959 N11 P28 NLLTB V2 N3 P199 <NLL>
1959 V19 N1 P26-27 64-19944 <=$>
1959 V19 N3 P18-20 59-13850 <=>
1959 V19 N4 P21-23 60-17424 <+*>
1959 V19 N6 P5-7 64-19943 <=$>
1959 V19 N6 P10-14 61-19353 <+*>
1959 V19 N6 P40-41 61-19353 <+*>
1959 V19 N11 P28 62-33151 <=*>
1960 V20 N9 P24-26 62-11561 <=>
1960 V20 N10 P3-5 62-24783 <=>
1960 V20 N12 P34-35 61-21548 <=>
1961 V21 N3 P35-37 C-3760 <NRCC>
61-27870 <*=>
1961 V21 N7 P27-28 63-19420 <=*>
1962 N9 P31-32 NLLTB V5 N1 P56 <NLL>
1962 V22 N4 P29 65-61310 <=$>
1962 V22 N10 P30-32 63-13540 <=>
1963 V23 N3 P17-19 64-11558 <=>
1963 V23 N12 P31 1955 <BSRA>
1964 N1 P35-40 AD-633 832 <=$>
1964 N9 P36-40 AD-633 898 <=$>
1964 V24 N3 P35-39 AD-633 919 <=$>
1965 N7 P26-28 AD-633 926 <=$>
1965 N7 P33-34 AD-633 925 <=$>
1965 V25 N1 P18-19,23 65-30485 <=$>
1965 V25 N1 P20-21 65-31301 <=$>
1965 V25 N2 P24-27 65-31053 <=$>
1965 V25 N7 P13-15 AD-633 937 <=$>
1965 V25 N7 P16-18 65-32789 <=*$>
1965 V25 N7 P20-21 65-32797 <=*$>
1965 V25 N7 P40-41 AD-633 936 <=$>

MORSKOI SBORNIK
1937 N4 P105-132 R-4407 <*>
RT-4132 <*>
1961 N4 P65-71 AD-632 837 <=$>
1961 V44 N2 P33-34 65-30827 <=$> 0
1963 N2 P3-8 AD-632 863 <=$>
1963 V46 N1 P23-30 63-21904 <=>
1963 V46 N3 P67-74 64-21978 <=>
1963 V46 N11 P67-74 64-21385 <=>
1965 N7 P73-79 66-30064 <=*$> 0
1965 V48 N7 P62-68 66-30922 <=$>
1965 V48 N8 P31-37 65-32754 <=$*>
1965 V48 N11 P71-77 65-34106 <=*$>
1966 N5 P9-41 66-33097 <=$>

MOSKOVSKII PROPAGANDIST
1958 N1 P69-73 PB 141 270T <=>

MOSKVA
1960 N3 P178-182 60-31421 <=>

MOTOR. WARSAW
1959 V7 N23 P3 60-15626 <=*>
1959 V7 N23 P5 60-15626 <=*>
1959 V7 N23 P11 60-15626 <=*>

MOTOR-RUNDSCHAU
1948 N15 P171-172 59-19486 <*+>
1958 N21 P730-732 59-16295 <+*> 0
1962 V32 N16 P695 62-18826 <*>

MOTORTECHNISCHE ZEITSCHRIFT
SEE MTZ

MOTORYZACJA
1957 V7 N6 P197-199 59-11286 <=*> 0
1957 V7 N6 P199-203 59-11286 <=*> 0
1958 V14 N11 P265-266 62-25222 <=*>
1959 V15 N2 P31-36 62-25221 <=*>
1959 V15 N2 P49-51 62-25221 <=*>

MUANYAG
1965 V2 N5 P145-148 RAPRA-1295 <RAP>

MUEHLE
1955 V92 N44 P589- 63-14978 <*>
1956 V93 N21 P291-294 63-14786 <*>
1957 V94 P539-540 84L28G <ATS>
1957 V94 N42 P539-540 63-14960 <*>
1957 V94 N47 P67-68 63-16485 <*>
1958 V95 N12 P153-154 63-16340 <*>
1958 V95 N21 P276-277 85L28G <ATS>
1959 V96 N9 P111 63-16296 <*>
1959 V96 N17 P215-217 63-16488 <*>
1960 V97 N25 P317-318 63-16310 <*>
1960 V97 N29 P373 63-14974 <*>

MUEHLENLABORATORIUM
1932 V2 N14 P85-90 63-14954 <*>
1935 V5 P74-78 63-14955 <*>
1935 V5 N2 P19-26 63-14956 <*>
1937 V7 N12 P171-176 63-14957 <*>

MUELLEREI
1957 V10 N11 P23-24 63-16517 <*>

MUENCHNER BEITRAGE ZUR ABWASSER, FISCHEREI UND FLUSSBIOLOGIE
1958 V5 P193-209 61-20421 <*>
1962 V9 P163-183 65-12279 <*>

MUENCHENER MEDIZINISCHE WOCHENSCHRIFT
1904 V51 N18 P785-786 58-2652 <*>
1906 V53 N42 P2041-2047 63-16667 <*> 0
1910 V57 N45 P2358 66-11812 <*>
1911 V58 N31 P1680-1682 9249-F <K-H>
1918 V65 P579-581 59-18136 <+*>
1919 V66 N1 P589-591 59-10704 <*>
1919 V66 N51 P1463-1467 3367-A <KH>
1921 V68 P325 58-1821 <*>
1921 V68 N45 P1451-1452 61-14356 <*>
1922 V69 P1276-1277 58-1823 <*>
1923 V70 N23 P725-727 59-22173 <*+>
1927 V74 N24 P1019-1029 59-22171 <*+>
1929 V76 N48 P2014 62-16132 <*>
1930 V77 P496 62-16159 <*>
1930 V77 P682 62-16158 <*>
1932 V79 P1722-1723 65-18228 <*>

1935 V82 P869-871 64-20031 <*> 0
1935 V82 N46 P1823-1828 59-20704 <*>
1937 V84 N37 P1458-1460 59-20705 <*>
1938 V85 P252-253 63-14545 <*>
1939 V86 P65-67 65-18230 <*> 0
1939 V86 N22 P860 59-20706 <*>
1939 V86 N36 P1381-1383 2736-D <K-H>
1939 V86 N50 P1745- 3272 <*>
1940 V87 P453-454 65-18231 <*>
1942 P76- 57-1631 <*>
1942 P166-174 57-1677 <*>
1942 P174-175 57-1502 <*>
1943 V90 N1 P11-15 1691 <*>
1950 V92 N27/8 P1113-1122 2548 <*>
1951 V93 P1707-1710 65-18232 <*>
1951 V93 N4 P171- 2767 <*>
1951 V93 N10 P467-472 61-10265 <*>
1952 V94 P1105-1108 2697-H <K-H>
1953 V95 P654-655 65-18233 <*>
1953 V95 P1C49-1052 65-18234 <*>
1953 V95 P1207-1208 I-531 <*>
1953 V95 P1221-1225 AL-54 <*>
1953 V95 P1306-1307 AL-656 <*>
1954 V96 P296-297 I-12 <*>
1954 V96 P740-741 57-167 <*>
1954 V96 P979-980 II-493 <*>
1954 V96 P1336-1338 65-18235 <*>
1954 V96 P1366-1369 5352-A <KH>
1954 V96 N25 P724-726 2032 <*>
1954 V96 N25 P727-729 2029 <*>
1954 V96 N26 P756-759 II-389 <*>
1955 P872-875 T-2369 <INSD>
1955 V97 P551-552 63-16915 <*>
8025-A <K-H>
1955 V97 P595-597 65-18236 <*>
1955 V97 P1175- II-837 <*>
1955 V97 P1281-1283 65-18237 <*>
1955 V97 N3C P948-949 4006-C <KH>
1956 V98 P156-159 1355 <*>
1956 V98 P492-496 1339 <*>
1956 V98 P844-845 57-1257 <*>
1956 V98 N37 P1247-1250 66-12613 <*>
1958 V100 P1814-1817 65-00172 <*> 0
1958 V100 P1817-1819 65-00173 <*>
1958 V100 N1 P16-23 59-15857 <*>
1958 V100 N43 P1658-1660 64-20030 <*>
1958 V100 N44 P1700-1704 59-10818 <*>
1958 V100 N50 P1974-1976 59-15881 <*>
1958 V100 N52 P2025-2027 65-00345 <*>
1959 V101 P677-680 59-15891 <*>
1959 V101 P1911-1912 62-16036 <*>
1959 V101 N5 P184-186 60-14095 <*>
1959 V101 N16 P804-806 60-18168 <*>
1959 V101 N26 P1128-1132 60-16611 <*>
1960 V102 P276-279 63-18035 <*>
1960 V102 P1609-1611 65-14549 <*> 0
1960 V102 P2105-2106 62-00805 <*>
1960 V102 N14 P700-702 10401-D <K-H>
64-14582 <*>
1960 V102 N33 P1550-1553 67M46G <ATS>
1961 V103 P266-268 63-10515 <*> 0
1961 V103 P1174 61-20183 <*>
1961 V103 P1386 62-10733 <*>
1961 V103 P1628-1631 63-14765 <*>
1961 V103 P1752-1755 63-26235 <=$>
1961 V103 P1756-1757 63-26234 <=*>
1961 V103 P2288-2290 66-12042 <*> 0
1961 V103 P2339-2341 62-16081 <*>
1961 V103 P2485-2486 62-16059 <*>
1962 V104 P161 63-10717 <*> 0
1962 V104 P1490 63-10185 <*>
1962 V104 N14 P646-647 63-00857 <*>
1962 V104 N21 P1001-1002 64-18956 <*> 0
1962 V104 N21 P1003-1009 64-18972 <*> 0
1962 V104 N21 P1010-1016 64-18971 <*>
1962 V104 N21 P1016-1018 64-18970 <*> 0
1962 V104 N21 P1018-1022 64-18969 <*> 0
1962 V104 N21 P1022-1025 64-18968 <*> 0
1963 V105 P682-685 65-11762 <*>
1963 V105 P1273-1275 66-11615 <*> 0
1963 V105 N24 P1237-1242 64-00348 <*>
1963 V105 N24 P1242-1250 64-30652 <*>
1964 V106 N10 P1169-1173 66-10959 <*>
1964 V106 N18 P875 64-20140 <*>
64-20143 <*>
1964 V106 N18 P878 64-20141 <*>
1964 V106 N26 P1187-1188 65-11307 <*>
1964 V106 N45 P2033-2041 65-00251 <*> 0
1965 V107 N23 P1175-1176 65-17411 <*>
1965 V107 N43 P2124-2130 92T91G <ATS>

MUENCHENER TIERAERZTLICHE WOCHENSCHRIFT
1929 V80 N36 P497-500 66-12616 <*>
1930 V81 N23 P285-287 66-12617 <*>

MUKOMOLNO-ELEVATCRNAYA PRCMYSHLENNOST
1957 V23 N4 P8-9 R-4192 <*>
1959 V25 N12 P22-24 CSIR-TR 297 <CSSA>

MUNCA
1966 N5810 P2 66-33388 <=$>

MUNCITORUL SANITAR
1960 V11 N32 P4 62-19201 <=*>
1960 V11 N44 P2-3 62-19246 <=*>
1961 V11 N2 P3 62-19202 <*=>
1961 V12 N12 P2-4 62-19229 <=*> 0
1961 V12 N44 P4 62-19611 <=*>
1962 V13 N47 P3 63-13674 <=*>
1963 P2 64-21307 <=>
1963 V14 P4 63-31781 <=>
1963 V15 N4C P3 64-21103 <=>
1964 V15 02/08 P3 64-31089 <=>
1964 V15 N2 P4 64-21599 <=>
1964 V15 N9 P3 64-21964 <=>
1964 V15 N722 P1 64-31374 <=>
1966 P2 06/01 66-33070 <=$>
1966 P1 66-30872 <=$>
1966 P2 66-30872 <=$>
66-32074 <=$>

MUSZAKI ELET
1958 V13 P3 59-11268 <=*>
1958 V13 P8 59-11268 <=*>
1960 V15 N20 P11 62-19230 <=*>
1960 V15 N20 P16 62-19230 <=*>
1960 V15 N22 P11 62-19215 <=*>
1960 V15 N26 P4 62-19223 <=*>
1962 V17 N24 P3 63-21134 <=>
1963 V18 P3 63-21245 <=>
1963 V18 P6 63-21245 <=>
1966 N4 P4 66-31236 <=$>

MUSZAKI KONYOTORCSOK TAJEKOZTATOJA ORSZAGOS MUSZAKI KONYVTAR. BUDAPEST
1962 V9 N4 P1-12 63-21907 <=>

MUSZAKI KONYULAROS TAJEKOZTATOJA
1960 N5 P1-20 62-19222 <=*>

MYASNAYA INDUSTRIYA SSSR
1950 V21 N4 P85-87 TT.270 <NRC>
1950 V21 N6 P7C-71 TT.271 <NRC>
1952 V23 N2 P39-43 RT-3842 <*>
1952 V23 N2 P43-49 RT-3862 <*>
1952 V23 N4 P41-52 R-3499 <*>
1952 V23 N5 P33-41 RT-3743 <*>
1956 V27 N1 P53-56 R-3096 <*>
1957 V28 N2 P42-44 63-11048 <=$>
1957 V28 N5 P53-55 60-21136 <=>
1957 V28 N6 P46-47 6C-21082 <=>
1957 V28 N6 P48-49 60-21083 <=>
1957 V28 N6 P5C-51 60-21080 <=>
1957 V28 N6 P51-53 60-21081 <=>
1958 V29 N3 P52-53 63-11049 <=$>
1958 V29 N3 P54 59-18507 <+*>
1959 V30 N5 P6-10 60-19579 <+*> 0
1959 V30 N5 P42-43 60-19580 <+*>
1960 V31 N4 P1-6 63-13090 <=>
1960 V31 N6 P2-3 61-21697 <=>

1960 V31 N6 P9 61 21697 <=>
1961 V32 N3 P52-54 64-19964 <=$>
1962 V33 N1 P29-30 62-32571 <=>
1963 V34 N6 P43-50 64-26316 <$>
1964 V35 P29 M-5793 <NLL>
1964 V35 N4 P3-6 64-51739 <=$>
1964 V35 N4 P43-48 M.5647 <NLL>
1964 V35 N5 P52-55 M.5689 <NLL>
1965 V36 N2 P56-57 M.5770 <NLL>
1966 V37 N3 P6-9 66-34647 <=$>

MYKROBIOLOHICHNIYI ZHURNAL
1948 V9 N2/3 P71-79 63-15160 <=*>
1949 V11 N4 P43-49 63-15161 <=*>
1952 V14 N2 P40-45 RT-2128 <*>
1952 V14 N2 P47-53 RT-2129 <*>
1952 V14 N3 P68-72 RT-2130 <*>
1952 V14 N4 P24-27 RT-2132 <*>
1953 V15 N1 P27-33 RT-3135 <*>
1953 V15 N1 P59-65 R-5209 <*>
1953 V15 N1 P66-69 62-32970 <=*>
1953 V15 N2 P6-16 R-1987 <*>
1953 V15 N2 P72-80 RT-1777 <*>
1954 V16 N3 P90-95 RT-3473 <*>
1955 V17 N1 P5-10 R-199 <*>
1955 V17 N1 P46-50 R-223 <*>
1956 V18 N3 P59-61 59-12752 <+*>
1956 V18 N4 P38-43 59-13054 <=> O
1956 V18 N4 P47-56 62-15119 <*=>
1956 V18 N4 P60-62 59-13053 <=> O
1957 V19 N2 P11-13 59-12753 <+*>
1958 V20 N1 P34-39 59-13357 <=>
1958 V20 N1 P60-63 60-15957 <+*>
1958 V20 N3 P31-35 60-17952 <+*> O
1959 V21 N1 P3-4 59-11736 <=>
1959 V21 N1 P5-8 59-11737 <=>
1959 V21 N1 P9-12 59-11738 <=>
1959 V21 N1 P13-16 59-11739 <=>
1959 V21 N1 P17-20 59-11740 <=>
1959 V21 N1 P21-24 59-11741 <=>
1959 V21 N5 P58-65 64-11007 <=>
1960 V22 N2 P71-72 60-23151 <+*>
1960 V22 N3 P53-57 61-11099 <=>
1960 V22 N3 P69-74 61-11073 <=>
61-11633 <=*>
1960 V22 N4 P32-35 61-11613 <=>
1960 V22 N4 P45-50 61-11616 <=>
1961 V23 N2 P45-48 61-28040 <*=>
1961 V23 N2 P53-58 61-28041 <*=>
1961 V23 N2 P73-75 62-19163 <=*>
1961 V23 N5 P5-8 62-23567 <=*>
1962 V24 N1 P64-66 62-24738 <=*>
1962 V24 N2 P18-22 63-23899 <=*$>
1962 V24 N2 P48-54 63-24511 <=*$>
1962 V24 N3 P3- FPTS V22 N6 P.T1022 <FASE>
1962 V24 N3 P3-8 FPTS,V22,P.T1022-24 <FASE>
1962 V24 N3 P8- FPTS V22 N6 P.T1025 <FASE>
1962 V24 N3 P8-12 FPTS,V22,P.T1025-27 <FASE>
1962 V24 N3 P23-28 63-19553 <=*>
1962 V24 N4 P61-63 62-33562 <=*>
1962 V24 N5 P20-25 64-15410 <=*$>
1962 V24 N5 P30-34 63-13591 <=*>
1962 V24 N5 P35-38 63-15113 <=*>
1962 V24 N6 P32-36 63-24486 <=*$>
1962 V24 N6 P43-45 64-11755 <=>
1962 V24 N6 P49-52 64-13890 <=*$>
1962 V24 N6 P52-56 64-13517 <=*$>
1963 V25 N1 P3-8 64-13465 <=*$>
1963 V25 N1 P41-43 64-19283 <=*$>
1963 V25 N3 P10-17 64-15693 <=*$> O
1963 V25 N3 P23- FPTS V23 N4 P.T827 <FASE>
FPTS,V23,N4,P.T827 <FASE>
1963 V25 N3 P33-37 64-15703 <=*$>
1963 V25 N3 P53-58 64-15692 <=*$> O
1963 V25 N3 P64-70 64-15451 <=*$>
1963 V25 N4 P3-7 FPTS V23 N4 P.T829 <FASE>
1963 V25 N4 P3- FPTS,V23,N4,P.T829 <FASE>
1963 V25 N4 P34-39 63-41019 <=>
1963 V25 N5 P8-13 65-60173 <=$>
1963 V25 N5 P50-52 AD-617 392 <=$>
1963 V25 N6 P44-53 64-21718 <=>
1964 V26 N1 P8- FPTS V24 N1 P.T19 <FASE>
FPTS V24 P.T19-T22 <FASE>
1964 V26 N1 P42- FPTS V24 N1 P.T23 <FASE>
FPTS V24 P.T23-T24 <FASE>
1964 V26 N1 P45- FPTS V24 N1 P.T25 <FASE>
FPTS V24 P.T25-T26 <FASE>
1964 V26 N3 P31-36 65-62328 <=$>
1964 V26 N4 P3- FPTS V24 N5 P.T925 <FASE>
1964 V26 N4 P3-8 64-51720 <=$>
1964 V26 N4 P13-16 64-51720 <=$>
1964 V26 N4 P46-49 64-51720 <=$>
1964 V26 N4 P62-70 64-51720 <=$>
1964 V26 N4 P70- FPTS V24 N5 P.T933 <FASE>
1964 V26 N4 P80-85 66-12532 <*>
1964 V26 N5 P27-31 65-63613 <=$>
1964 V26 N6 P71-74 65-30549 <=$>
1965 V27 N1 P16- FPTS V25 N1 P.T135 <FASE>
1965 V27 N2 P73-78 66-61058 <=*$>
1965 V27 N3 P24-27 66-61057 <=$*>
1965 V27 N5 P70- FPTS V25 N4 P.T710 <FASE>
1965 V27 N6 P73- FPTS V25 N6 P.T1115 <FASE>
1965 V27 N6 P77- FPTS V25 N6 P.T1113 <FASE>

NHK GIJUTSU KENKYU
1959 V11 N5 P303-305 61-10357 <*>
1960 V12 N3 P199-223 62-18679 <*>
1960 V12 N5 P373-386 61-20230 <*>
1960 V12 N5 P393-411 62-14084 <*>
1965 V17 N3 P186-202 23T91J <ATS>
1965 V17 N3 P203-229 22T91J <ATS>

NA STROIKAKH ROSSII
1961 N4 P20-23 65-29613 <$>
1962 N5 P14-16 62-32210 <=>
1962 N8 P25 63-13340 <=> P
1962 N10 P17 63-15100 <=>
1962 N10 P37 63-13648 <=>
1962 N12 P10-11 63-21299 <=>
1962 V3 N10 P18 65-61621 <=$>
1962 V3 N10 P19 64-13630 <=*$>
1963 N2 P19-20 63-21789 <=>
1964 V5 N6 P18-19 64-41923 <=*$>
1964 V5 N7 P16-21 64-41923 <=*$>
1966 N11 P1-2 67-30511 <=>

NABLYUDENIYA ISKUSSTVENNYKH SPUTNIKOV ZEMLI. BYULLETEN STANTSII OPTICHESKOGO NABLYUDENIYA ISKUSSTVENNYKH SPUTNIKOV ZEMLI. SPETSIALNYI VYPUSK. AKADEMIYA NAUK SSSR. ASTRONOMICHESKII SOVET. MOSCOW
1963 N2 P19-24 AD-627 070 <=$>

NACHALNAYA SHKOLA
1966 V34 N6 P22-24 66-33188 <=$>

NACHRICHTEN. AKADEMIE DER WISSENSCHAFTEN IN GOETTINGEN
1946 S2A N1 P36-37 59-15018 <*>
1946 S2A N1 P52- I-1012 <*>
1949 S2A N1 P1-11 1338 <*>
1952 S2A N5 P21-30 63-00273 <*>
1953 S2A N7 P102-108 1541 <*>
1954 S2A N4 P71-93 AEC-TR-5892 <*>
1955 S2A N6 P87-120 26H9G <ATS>
1961 S2 N2 P27-37 62-00602 <*>
1963 S2 N23 P333-356 65-11372 <*>
1965 S2 N7 P95-114 66-11175 <*>

NACHRICHTEN. AKADEMIE DER WISSENSCHAFTEN IN GOETTINGEN. MATHEMATISCH-PHYSIKALISCHE KLASSE
1936 V1 N16 P161-173 66-11149 <*>

NACHRICHTEN AUS CHEMIE UND TECHNIK
1966 V14 N2 P29- 66-12874 <*>

NACHRICHTEN FUER DOKUMENTATION
1954 V5 P36-39 57-1311 <*>
1955 V5 P111-114 57-1172 <*>
1955 V5 P179-183 57-1172 <*>

1955 V6 P25-28 57-1172 <*>
1955 V6 P49-52 57-1172 <*>
1956 V7 P140-145 57-221 <*>
1957 V8 N1 P27-29 <ATS>
1960 V11 N3 P156-159 62-10544 <*>
1961 V12 P208-216 PB-166 544 <=>
1962 V13 N1 P13-19 55R74G <ATS>

NACHRICHTEN VON DER GESELLSCHAFT DER WISSENSCHAFTEN ZU GOETTINGEN
1903 V3 P126-131 AEC-TR-3805 <*>
1911 V5 P509-517 64-18120 <*> 0
1913 P582-592 59-20172 <*>
1918 P98-100 PT2 58-1459 <*>
1919 N2 P193-217 C-2635 <NRC>
58-1421 <*>
1923 N1 P1-15 66-14448 <*>
1925 N1 P49-69 I-835 <*>

NACHRICHTEN FUER DIE ZIVILE LUFTFAHRT
1964 V1 P2-3 64-41266 <=>

NACHRICHTENBLATT DES DEUTSCHEN PFLANZENSCHUTZDIENSTES. STUTTGART
1952 V32 P24-28 62-10300 <*>

NACHRICHTENTECHNIK
1952 V2 N6 P185-189 1780 <*>
1954 V4 N6 P254-258 66-14379 <*> 0
1955 N11 P481-489 57-3494 <*>
1955 V5 N4 P158-160 62-33213 <=*>
1955 V5 N8 P341-346 66-12978 <*> 0
1956 V6 P252-257 38J15G <ATS>
1956 V6 P500-502 58-200 <*>
1956 V6 N2 P63-70 62-18693 <*> 0
1956 V6 N6 P252-257 57-2997 <*>
1956 V6 N6 P266-273 62-18692 <*> 0
1956 V6 N7 P290-294 C-2666 <NRC>
1957 V7 N4 P148-152 58-1660 <*>
1957 V7 N5 P200-205 63-14663 <*>
1957 V7 N5 P210-215 03L32G <ATS>
1958 V8 P467- 61E1 <CTT>
1958 V8 N4 P154-158 63-14663 <*>
1959 V9 N3 P129-133 59-31006 <=> 0
1959 V9 N7 P306-309 65-10136 <*> 0
1959 V9 N8 P338-339 66-11331 <*>
1959 V9 N8 P343-346 66-11332 <*> 0
1959 V9 N8 P347-350 66-11333 <*> 0
1959 V9 N12 P543-545 60-31313 <=*$> 0
1959 V9 N12 P546-549 60-31314 <=*> 0
1960 V10 N10 P457-461 62-19231 <=*>
1961 V11 N2 P50-55 62-19178 <=*>
1961 V11 N2 P66-71 62-19176 <=*>
1961 V11 N2 P81-84 62-19177 <=*>
1961 V11 N6 P255-260 62-10631 <*>
1961 V11 N9 P422-425 62-18643 <*>
1962 V12 N7 P248-250 62-33686 <=*>
1962 V12 N11 P409-413 64-16950 <*> 0
1963 V13 N1 PU1-U3 63-21284 <=>
1963 V13 N3 P89-95 64-16282 <*>
1966 V16 N2 P46-49 66-32554 <=$>

NACHRICHTENTECHNISCHE FACHBERICHTE
1955 V2 P56-59 65-11460 <*>
1956 V5 P15-26 58-977 <*>
1957 V6 N2 P16-18 59-20918 <*>
1960 V19 P86-91 62-16595 <*>
1961 V22 P1-3 63-18354 <*>

NACHRICHTENTECHNISCHE ZEITSCHRIFT
1955 V8 N1 P8-13 36K28G <ATS>
1955 V8 N12 P641-646 18J18G <ATS>
58-133 <*>
1956 N10 P457-461 59-12160 <+*>
1956 V9 P424-430 3228 <*>
1956 V9 P532 59-22266 <=*>
1956 V9 N1 P1-18 62-25236 <=*> 0
1956 V9 N1 P21-28 62-25236 <=*> 0
1956 V9 N1 P34-39 58-249 <*>
62-25236 <=*> 0
1956 V9 N1 P39-42 59-12159 <+*>
1956 V9 N1 P47 62-25236 <=*> 0
1956 V9 N2 P55-59 62-16945 <*> 0
1956 V9 N3 P119-123 59-11416 <=> 0
1956 V9 N3 P124-128 58-1159 <*>
1956 V9 N10 P441-448 58-2133 <*>
1956 V9 N11 P493-498 60-17630 <=*>
1956 V9 N11 P513-518 60-17631 <=*>
1956 V9 N11 P519-532 60-17632 <=*>
1956 V9 N12 P561-565 3J19G <ATS>
1957 N12 P594-601 59-12158 <+*>
1957 V10 N1 P1-4 60-18053 <*>
1957 V10 N3 P120-124 58-1529 <*>
1957 V10 N6 P277-287 59-10865 <*>
1957 V10 N7 P335-343 82K27G <ATS>
1957 V10 N11 P551-558 59-19490 <*=>
1958 V11 P523-528 62-23849 <=*>
1958 V11 N4 P210-219 59-10964 <*>
1958 V11 N5 P225-237 64K28G <ATS>
1958 V11 N5 P238-244 59-10964 <*>
1958 V11 N6 P300-306 79L28G <ATS>
1958 V11 N8 P405-410 TT-813 <NRCC>
1958 V11 N8 P417-423 C-3440 <NRCC>
1958 V11 N9 P446-454 60-18100 <*>
1958 V11 N11 P557-560 64-16581 <*>
1959 V12 P374- 59-22220 <*=> 0
1959 V12 P464-466 62-00202 <*>
1959 V12 N1 P29-32 99L32G <ATS>
1959 V12 N3 P132-138 66-11239 <*>
1959 V12 N4 P181-186 60-18101 <*>
1959 V12 N9 P443-449 65-11461 <*>
1959 V12 N11 P561-565 62-10220 <*>
1959 V12 N12 P613-618 64-16620 <*>
1960 V13 N2 P57-63 61-10121 <*>
1960 V13 N3 P146 62-24041 <=*>
1960 V13 N4 P161-168 62-25233 <=*> 0
1960 V13 N6 P277-290 62-25232 <=*> 0
1960 V13 N11 P513-518 92N57G <ATS>
1960 V13 N11 P519-523 63-16704 <*>
93N57G <ATS>
1961 V14 N5 P242-248 63-10160 <*>
1961 V14 N8 P410-415 64-14235 <*>
1961 V14 N9 P436-440 62-16682 <*>
1961 V14 N10 P481-486 62-18707 <*>
1961 V14 N11 P555-559 POED-TRANS-2218 <NLL>
1962 N2 P71-78 63-01035 <*>
1962 V15 N7 P341-349 63-10730 <*>
1962 V15 N9 P438-441 64-16577 <*>
1963 V16 N4 P205-214 65-13642 <*>
1963 V16 N7 P353-357 65-11587 <*> 0
1963 V16 N11 P569-577 66-12358 <*>
1964 V17 N10 P515-519 15S82G <ATS>
65-13111 <*>
1964 V17 N12 P636-640 27S85G <ATS>
1965 V18 N2 P57-62 POED-TRANS-2009 <NLL>
1965 V18 N5 P268-274 POED-TRANS-2215 <NLL>
1965 V18 N6 P324-330 POED-TRANS-2220 <NLL>
1965 V18 N9 P503-510 POED-TRANS-2266 <NLL>
1966 V19 N2 P69-72 66-14500 <*>
1966 V19 N3 P139-142 66-12586 <*>

NAFTA. POLAND
1952 V8 N4 P99-102 PJ-466 <ATS>
1952 V8 N5 P125-128 PJ-467 <ATS>
64-16755 <*>
1952 V8 N6 P145-147 PJ-172 <ATS>
1952 V8 N6 P155-159 PJ-468 <ATS>
64-16754 <*>
1953 N9 P217-220 PJ246 <ATS>
1954 V10 N12 P285-286 57-1707 <*>
1955 V11 N10 P237-241 48N53P <ATS>
1956 V12 N1 P4-8 62-14599 <*>
1956 V12 N7/8 P186-191 59-15580 <*>
1957 N10 P275-278 IGR V2 N6 P522 <AGI>
1957 V13 N10 P275-278 PB 141 320T <=>
1958 V14 N1 P1-4 PB 141 164T <=*>
1958 V14 N4 P89-97 PB 141 323T <=*>
1958 V14 N4 P109-112 PB 141 168T <=*>
1958 V14 N7 P175-179 59-11582 <=>

1958 V14 N10 P273-276 59-21217 <=*> O
1959 V15 N11 P302-308 88M41P <ATS>

1961 V17 N5 P133-138 63-20703 <*>
1961 V17 N10 P281-284 65-11358 <*>
1964 V20 N5 P121-123 25R79P <ATS>
1964 V20 N6 P152-156 85R79P <ATS>
1964 V20 N7 P188-193 64-51223 <=>
1964 V20 N7 P193-195 64-51223 <=>
1964 V20 N8 P219-222 64-51641 <=$>
1964 V20 N9 P230-243 64-51704 <=$>
1964 V20 N9 P257-258 64-51704 <=$>
1965 V21 N6 P177-179 69T90P <ATS>
1965 V21 N7 P202-212 65-33622 <=$>
1966 V21 P163-166 1610 <TC>
1966 V22 N7 P213-217 66-34841 <=$>

NAFTA. YUGOSLAVIA
1956 V7 N11 P335-344 SC-T-40 <*>
1958 V9 N3 P71-76 59-10745 <*>
1959 V10 N4 P125-136 49M37CR <ATS>
1960 V11 N10 P233-236 11783 <*>
1964 V15 P110-120 66-14416 <*>

NAGASAKI IGAKKAI ZASSHI
1953 V28 N10 P1117-1121 58-1731 <*>
1959 V34 P834 62-01514 <*>
1959 V34 N2 P208-225 61-00190 <*>
1964 V39 P41-44 65-14714 <*>

NAGASAKI MEDICAL JOURNAL
SEE NAGASAKI IGAKKAI ZASSHI

NAGOYA JOURNAL OF MEDICAL SCIENCE
SEE NAGOYA IGAKKAI ZASSHI

NAGOYA KOGYA DAIGAKU GAKUHO
1955 V7 P165-171 59-20893 <*> O

NAGOYA KOGYO GIJUTSU SHIKEN-SHO HOKOKU
1954 V3 N3 P96-105 65-14700 <*>
1955 V4 N1 P24-29 65-14391 <*>
1958 V7 P743-752 PANSDOC-TR.325 <PANS>
1958 V7 N6 P429-435 65-14827 <*>
1958 V7 N9 P652-657 63L29J <ATS>
1958 V7 N11 P855-861 65-14828 <*>
1959 V8 N4 P311-316 60-15284 <*+>
1959 V8 N11 P793-798 65-10253 <*>
1960 V9 P453-457 AEC-TR-5912 <*>
1960 V9 N1 P33-40 65-10254 <*>
1960 V9 N2 P85-92 65-14829 <*>
1961 P626-635 AEC-TR-6386 <*>
1961 V10 N3 P199-205 65-11009 <*> O
1963 V12 N6 P296-302 66-10355 <*>
1963 V12 N9 P450-454 WAPD-TRANS-4 <*>

NAGOYA SANGYO KAGAKU KENKYU-SHU KENKYU HOKOKU
1952 V1 N3 P98-103 INT-6 <INT>

NAGOYA SHIRITSU DAIGAKU IGAKKAI ZASSHI
1952 V2 N4 P165-170 II-870 <*>

NAHRUNG
1960 V4 N4 P310-322 10573-D <K-H>
64-14532 <*>
1962 V6 P488-491 86S80G <ATS>
1962 V6 N2 P166-174 63-10722 <=*>
1962 V6 N2 P175-179 63-10201 <*>
1962 V6 N7/8 P732-743 64-10992 <*>

NAIKA
1959 V4 N2 P189-190 61-00587 <*>

NAIKA HOKAN
1958 V5 N2 P154-159 59-12739 <=*> O
1960 V7 N2 P176-183 15M41J <ATS>
60-00802 <*>

NANKAI-KU SUISAN KENKYUSHO KENKYU HOKOKU
1958 N7 P59-71 59-18378 <=*>
1958 N7 P105-128 60-15293 <=*>
1958 N7 P127-148 60-17121 <*=>
1958 N8 P31-48 59-18393 <+*>
1958 N9 P103-116 59-18551 <=*>
1959 N10 P72-87 59-18552 <+*>

NANYO SUISAN
1940 V6 N3 P12-19 AL-805 <*>
1940 V6 N4 P14-25 AL-805 <*>
1940 V6 N5 P9-15 AL-805 <*>
1941 V7 N9 P10-21 AL-32 <*>
1942 V8 N1 P29-41 AL-317 <*>

NANYO SUISAN JOHO
1937 N3 P2-6 AL-470 <*>
1938 N6 P2-12 AL-316 <*>
1941 V5 N1 P6-9 AL-318 <*>
1941 V5 N2 P2-6 AL-323 <*>
1941 V5 N3 P5-13 AL-314 <*>
1941 V5 N3 P13-17 AL-775 <*>
1941 V5 N4 P2- AL-864 <*>
1941 V5 N4 P9-12 AL-312 <*>
1942 V6 N1 P7-9 AL-322 <*>
1942 V6 N1 P10-13 AL-863 <*>

NARA IGAKU ZASSHI
1957 V8 P224-226 62-01512 <*>
1957 V8 N1 P76-83 61-01052 <*>
1958 V9 P36-47 8599-C <K-H>

NARODNA ARMIYA
1966 P12 66-33503 <=$>

NARODNA PROSVETA
1963 V19 N4 P71-80 63-31164 <=>
1964 V20 N6 P64-74 64-41164 <=>
1966 N4 P98-100 66-33222 <=$>
1966 N7 P18-20 66-34513 <=$>

NARODNA TVORCHYST TA ETNOHRAFIYA
1959 V3 N1 P14-21 60-11489 <=>

NARODNI SUMAR
1955 V12 N1/3 P48-58 60-21616 <=>

NARODNOE KHOZYASTVO KAZAKHSTANA
1959 N1 P82-85 59-13939 <=>
1959 N4 P28-31 60-11114 <=>
1959 N5 P26-27 60-11114 <=>
1959 N5 P35-40 60-11114 <=>
1959 N8 P21-26 60-11311 <=>
1959 N10 P39-43 60-11463 <=>
1959 N11 P14-21 60-41001 <=>
1959 N11 P25-29 60-41001 <=>
1959 N11 P57-58 60-11572 <=>
1959 V17 N11 P35-39 61-11664 <=>
1960 N2 P29 60-41002 <=>
1960 N3 P25-28 61-27273 <=>
1960 N4 P88-89 61-27273 <=>
1960 N5 P70-73 62-24777 <=*>
1960 N6 P41-44 61-11667 <=>
1960 N6 P45-47 61-11975 <=>
1960 N7 P11-12 61-11743 <=>
1960 N8 P28-31 61-11919 <=>
1960 N11 P50-51 61-23580 <=>
1960 N12 P19-22 61-21686 <=>
1960 N12 P33-34 61-21686 <=>
1960 V18 N1 P77-83 61-11664 <=>
1960 V18 N9 P9-10 61-21388 <=> P
1960 V18 N9 P38 61-21388 <=> P
1962 N3 P35-37 62-25985 <=>
1962 N3 P62-68 62-25983 <=>
1962 V20 N1 P25-34 63-13117 <=*>
1962 V20 N2 P3-8 62-32011 <=*>
1962 V20 N3 P11-15 63-13092 <=*>
1962 V20 N5 P3-10 62-33307 <=>
1962 V20 N5 P27-31 62-32848 <=*>

1962 V20 N5 P78-81 62-32848 <=*>
1962 V20 N7 P17-19 63-13645 <=>
1962 V20 N7 P37-41 63-13344 <=>
1962 V20 N7 P55-57 63-13292 <=>
1962 V20 N7 P58-59 63-13084 <=>
1962 V20 N8 P9-10 63-13342 <=>
1962 V20 N8 P18-21 63-13838 <=>
1962 V20 N8 P21-24 63-21253 <=>
1962 V20 N8 P29-31 63-15352 <=>
1962 V20 N8 P42-46 63-15352 <=>
1962 V20 N8 P47-50 63-21253 <=>
1962 V20 N8 P51-56 63-15352 <=>
1962 V20 N8 P65-67 63-15352 <=>
1962 V20 N8 P68-70 63-21253 <=>
1962 V20 N9 P14 63-21254 <=>
1962 V20 N9 P42-44 63-21363 <=>
1962 V20 N9 P50-56 63-13835 <=> O
1962 V20 N9 P66 63-21311 <=>
1962 V20 N9 P67-69 63-21363 <=>
1962 V20 N9 P71-72 63-21116 <=>
1962 V20 N10 P4-5 63-21501 <=>
1962 V20 N1C P9-12 63-21445 <=>
1962 V20 N10 P17-20 63-21445 <=>
1962 V20 N10 P28 63-21283 <=>
1962 V20 N1C P30-32 63-21467 <=>
1962 V20 N10 P61-63 63-21445 <=>
1962 V20 N11 P19-20 63-21770 <=>
1962 V20 N11 P20-22 63-21616 <=>
1962 V20 N11 P26-30 63-21770 <=>
1962 V20 N11 P68 63-21770 <=>
1963 V21 N1 P9-16 63-21792 <=>
1963 V21 N1 P32-33 63-21817 <=>
1963 V21 N1 P35-38 63-21792 <=>
1963 V21 N1 P65 63-31122 <=>
1963 V21 N1 P72-73 63-31122 <=>
1963 V21 N1 P74-75 63-31083 <=>
1963 V21 N2 P4-7 63-31083 <=>
1963 V21 N2 P7-8 63-21889 <=>
1963 V21 N2 P9 63-31083 <=>
1963 V21 N2 P12 63-31083 <=>
1963 V21 N2 P37-40 63-31122 <=>
1963 V21 N3 P3-6 63-21943 <=>
1963 V21 N3 P14-15 63-31453 <=>
1963 V21 N5 P21 63-31607 <=>
1963 V21 N6 P67 63-31752 <=>
1964 N8 P25-27 64-51833 <=$>
1964 V22 N9 P51-52 65-30120 <=$>
1965 V23 N1 P16-17 65-31238 <=$>
1965 V23 N11 P29-39 66-30568 <=*$>
1966 V24 N5 P49-52 66-33884 <=$>
1966 V24 N9 P10-13 66-35747 <=>

NARODNOE KHOZIAISTVO SREDNEI AZII. TASHKENT
1964 N8 P13-18 65-30155 <=$>

NARODNOE KHOZYAISTVO UZBEKISTANA
1961 N3 P9-12 61-27119 <=>
1961 N3 P21-24 61-27119 <=>
1961 N3 P34 61-23945 <=>
1961 V3 N4 P92-96 62-13508 <*=>
1962 V4 N1 P35-37 62-33348 <=>
1962 V4 N1 P67-68 62-32579 <=>
1962 V4 N1 P68-69 62-33332 <=>
1962 V4 N2 P47-48 62-32050 <=>
1962 V4 N4 P48-49 63-13850 <=>
1962 V4 N4 P72-77 62-32076 <=*>
1962 V4 N7 P41-44 62-33323 <=>
1962 V4 N7 P67-74 62-33601 <=>
1962 V4 N7 P81-83 62-33601 <=>
1962 V4 N8 P25-27 63-13329 <=> P
1962 V4 N8 P36-38 63-13515 <=>
1962 V4 N8 P90-92 63-13266 <=>
1962 V4 N9 P24-26 63-13658 <=>
1962 V4 N9 P64-65 63-13644 <=>
1962 V4 N9 P81-83 63-15100 <=>
1962 V4 N9 P84-85 63-13838 <=>
1962 V4 N11 P25-29 63-21295 <=>
1963 V5 N2 P21-23 63-21789 <=>
1963 V5 N2 P69-71 63-21789 <=>
1963 V5 N3 P64-68 63-21615 <=>
1963 V5 N5 P10-19 63-31608 <=>
1963 V5 N6 P3-9 63-31901 <=>
1963 V5 N6 P64-65 63-31599 <=>
1967 N1 P51-53 67-31062 <=$>

NARODNYE OBRAZOVANIYE
1959 N4 P107-111 60-11088 <=>

NARODNOE ZDRAVLJE
1961 V17 N5 P150-154 62-19482 <=*>
1961 V17 N5 P155-158 62-19497 <=*>
1961 V17 N6 P199-205 62-19496 <=*>
1961 V17 N6 P213 62-23457 <=*>
1961 V17 N7/8 P273-280 62-23457 <=*>

NARUCHNIK NA AGITATORA
1964 V19 N10/3 P49-71 64-31635 <=>

NASA VEDA
1961 V8 N2 P90-93 62-19219 <=*>

NASE GRADEVINARSTVO
1955 N5 P671-674 60-21703 <=>
1960 N3 P454-461 62-19326 <*=>

NASHA RODINA. SOFIA
1962 V9 N5 P8-10 62-32859 <=>

NASHONARU TEKUNIKARU REPOTO
1960 V6 N4 P364-371 69P59J <ATS>
1965 V11 N5 P359-372 66-12232 <*>

NATIONAL DEFENSE MEDICAL JOURNAL. JAPAN
SEE BOEI EISEI

NATIONAL MEDICAL JOURNAL OF CHINA
SEE CHUNG HUA I HSUEH TSA CHIH

NATIONAL TECHNICAL REPORT. JAPAN
SEE NASHONARU TEKUNIKARU REPOTO

NATTURUFROEDINGURINN
1949 V19 P20-26 65-12735 <*>
1949 V19 P27-32 65-12736 <*>

NATUR UND VOLK
1957 V87 N11 P399-401 58-1500 <*>

NATURA. BUCURESTI
1960 V12 N4 P94 62-19340 <*=>
1960 V12 N4 P95-97 62-19327 <*=>
1960 V12 N4 P1C0-108 62-19328 <*=>
1961 V13 N1 P12-19 62-19539 <*=> P
1961 V13 N2 P36-43 62-19490 <*=>
1961 V13 N3 P43-50 62-15942 <=*>
1961 V13 N4 P9-15 62-15930 <=*> P
1961 V13 N5 P61-65 62-24003 <=*>

NATURA. SERIA BICLOGIE
1962 V14 N4 P3-13 63-21555 <=>
1963 V15 N2 P3-12 64-71561 <=>
1964 V16 N4 P27-34 64-51708 <=$>

NATURA. SERIA GEOGRAFIE-GEOLOGIE
1964 V16 N4 P9-14 64-51476 <=$>
1964 V16 N4 P15-20 64-51424 <=$>
1965 N5 P14-23 65-33876 <=$>
1965 V17 N3 P1C-14 65-31482 <=$>

NATURA. SOCIETATU DE STIINTE NATURALE SI GEOGRAFIE DIN REPUBLICA POPULARA ROMINA. BUCURESTI
1962 V14 N6 P75-80 63-21422 <=>

NATURA E VITA
1954 V14 N1 P15-28 <ANSP>

NATURE. PARIS
1922 V50 P1C6-109 57-1577 <*>
1922 V50 P222-224 57-2104 <*>
1922 V50 P325-330 57-2103 <*>

Citation	Translation
1923 P364-366	57-1393 <*>
1923 V51 N2566 P364-366	60-10437 <*> 0
1926 V54 N2705 P81-83	63-18939 <*> 0
1936 P76-77	60-10131 <*> 0
1945 V73 N3081 P37-40	62-20291 <*> 0
1945 V73 N3093 P235-236	63-20178 <*>
1947 P177-180	II-425 <*>
1956 V84 N3250 P54-55	59-10266 <*>
1959 V87 N3291 P312-314	60-13139 <*+>
1961 V191 P238-240	62-01401 <*>
1963 N3343 P468-472	64-30083 <*>

NATUREN

Citation	Translation
1956 V80 N6 P323-333	<GSL>

NATURENS VERDEN

Citation	Translation
1951 V35 P1-2	MF-7 <*>
1951 V35 P19-33	M-7 <*>

NATURWISSENSCHAFTEN

Citation	Translation
1918 V6 N48 P698-702	C-2659 <NRC>
1924 V12 P619-620	57-1417 <*>
1924 V12 P1040-1046	64-14254 <*>
1926 V14 N46 P1005-1011	63-18999 <*> 0
1927 V15 N1 P49-51	65-10818 <*>
1928 V16 P1028-	I-863 <*>
1929 V17 P157-160	57-2457 <*>
1929 V17 N1 P13-	57-2030 <*>
1930 P756-766	I-67 <*>
1930 V18 N41 P867	59-20249 <*> 0
1930 V18 N20/1 P443-444	57-2322 <*>
1931 V19 N29 P626-634	64-16237 <*> 0
1931 V19 N23/5 P519-520	60-16831 <*> 0
1932 V20 N40 P732-738	57-3021 <*>
1933 V21 P465-	58-798 <*>
1933 V21 P787-788	AEC-TR-5850 <*>
1933 V21 N3 P39-43	61-14391 <*>
1933 V21 N21/3 P379-382	AEC-TR-5706 <*>
1934 V22 P420	66-11984 <*>
1934 V22 P648-	57-278 <*>
1934 V22 P838-	57-917 <*>
1934 V22 N22/4 P384-386	60-16748 <*>
1935 V23 P653-656	TRANS-206 <MT>
1938 V26 P188-	57-1895 <*>
1938 V26 P546-	II-565 <*>
1938 V26 P819-820	AEC-TR-6046 <*>
1938 V26 N24/5 P290-293	63-14481 <*>
1939 V27 N13 P214-215	57-541 <*>
1939 V27 N22 P390-	RCT V13 N1 P48 <RCT>
1940 V28 N26 P455-458	63-10089 <*>
1941 V29 P403-404	57-2860 <*>
1941 V29 P537-547	62-10002 <*> 0
1941 V29 P688-690	537 <TC>
1941 V29 P710-711	538 <TC>
1941 V29 P769-770	539 <TC>
1941 V29 N4 P49-61	66-12897 <*> 0
1941 V29 N10 P138-146	57-2317 <*>
1941 V29 N10 P150	59-10109 <*>
1941 V29 N21 P320-	63-16799 <*> 0
1942 V30 P63	60-10098 <*> 0
1942 V30 P107-108	61-20675 <*>
1942 V30 P242	59-20364 <*>
1942 V30 P577-582	AEC-TR-4770 <*>
1942 V30 N17/8 P260	59-10110 <*>
1943 N11/3 P147-148	58-693 <*>
1943 V31 P23	59-10536 <*>
1943 V31 P143-144	58-641 <*>
1943 V31 P147-148	58-649 <*>
1943 V31 P205-206	I-813 <*>
1943 V31 P487-490	66-11135 <*>
1943 V31 N1/2 P19-20	58-612 <*>
1943 V31 N1/2 P20-21	57-1403 <*>
1943 V31 N29/0 P335-344	3257 <*>
1944 N40/3 P299-300	58-1038 <*>
	58-907 <*>
1944 V32 P23	59-10535 <*>
1944 V32 P260-268	65-12486 <*>
1944 V32 N40/3 P260-268	CSIRO-3134 <CSIR>
	83L30G <ATS>
1946 V33 P221-222	63-20534 <*>
1946 V33 P251-	II-723 <*>
1946 V33 P280-	II-442 <*>
1946 V33 P281-282	II-54 <*>
1946 V33 N2 P37-40	AL-782 <*>
1946 V33 N4 P108-111	59-20612 <*> 0
1946 V33 N8 P239-243	AL-487 <*>
1946 V33 N10 P312-314	60-16551 <*> 0
1947 V34 P120-121	61-20007 <*>
1947 V34 P295-301	63-14561 <*>
1947 V34 N7 P194-201	AL-783 <*>
1947 V34 N7 P216	59-10022 <*>
1947 V34 N12 P372-	2332 <*>
1948 V35 P182-188	58-222 <*>
1948 V35 P255-256	AL-991 <*>
1948 V35 N6 P182-188	8J19G <ATS>
1948 V35 N7 P212-218	8J19G <ATS>
1949 V36 P327-333	8789 <IICH>
1949 V36 N1 P28-29	AEC-TR-3501 <*> 0
1949 V36 N2 P41-48	C-2442 <NRC>
1949 V36 N7 P218	58-2153 <*>
1949 V36 N9 P260-268	II-658 <*>
1949 V36 N10 P296-299	I-592 <*>
1949 V36 N11 P327-333	63-18568 <*>
1949 V36 N11 P359-362	63-18568 <*>
1950 V37 P138-139	64-10003 <*>
1950 V37 P164-165	I-989 <*>
1950 V37 P211-212	63-18234 <*>
1950 V37 P451-452	64-10026 <*>
1950 V37 P464-476	62-00689 <*>
1950 V37 P492-493	64-10002 <*>
1950 V37 N2 P43-44	65-11074 <*> 0
1950 V37 N16 P450	59-10876 <*>
1950 V37 N23 P544-545	AEC-TR-3475 <*> 0
1951 V38 P11-12	64-10001 <*>
1951 V38 P236-	2900 <*>
1951 V38 P288-290	64-10006 <*>
1951 V38 P306-307	I-815 <*>
1951 V38 P479-480	64-10000 <*>
1951 V38 P481-482	I-382 <*>
1951 V38 N19 P456	65-17426 <*> 0
1952 V39 P68	2545-D <K-H>
1952 V39 P88	3546-G <K-H>
1952 V39 P149-158	58-2676 <*>
1952 V39 N1 P20	64-14913 <*>
1952 V39 N6 P137-138	1210 <*>
1952 V39 N9 P209-210	66-11323 <*>
1952 V39 N11 P258-259	3309 <*>
1952 V39 N16 P378-379	1102 <*>
1952 V39 N17 P400-401	1646 <*>
	57-2273 <*>
	59-17918 <*>
1952 V39 N17 P401	59-17971 <*>
1952 V39 N24 P568-569	58-1766 <*>
1953 V40 P104-105	AL-286 <*>
1953 V40 P171	63-16293 <*>
1953 V40 P196-197	NP-TR-522 <*>
1953 V40 P460-	II-438 <*>
1953 V40 P625-	II-259 <*>
1953 V40 N4 P137-138	59-17962 <*>
1953 V40 N5 P163-164	62-01594 <*>
1953 V40 N11 P315	64-14276 <*> 0
1953 V40 N11 P315-316	66-11678 <*>
1953 V40 N17 P449-452	64-10340 <*>
1953 V40 N17 P460	64-10468 <*>
1953 V40 N19 P507-508	59-20840 <*>
1953 V40 N21 P550-551	I-80 <*>
1953 V40 N21 P557	66-11866 <*>
1953 V40 N22 P580-581	3160-D <K-H>
1953 V40 N22 P580	64-20112 <*>
1953 V40 N23 P608	66-11866 <*>
1954 V41 P1-3	T1896 <INSD>
1954 V41 P169	63-16294 <*>
1954 V41 P332-333	T2190 <INSD>
1954 V41 P436	T-2190 <INSD>
1954 V41 P447	6D16-B <K-H>
1954 V41 P472-473	2366 <*>
1954 V41 P481-482	57-1313 <*>
1954 V41 P570	17J16G <ATS>
1954 V41 N7 P162-	57-1032 <*>
	58-1595 <*>

Citation	Translation
1954 V41 N8 P183-184	58-2507 <*>
1954 V41 N12 P269-277	I-846 <*>
	57-779 <*>
1954 V41 N13 P301-302	1242 <*>
1954 V41 N19 P447	64-20197 <*>
1955 V42 P170-173	60-23447 <=>
1955 V42 P210	58-1235 <*>
1955 V42 P254-	3271 <*>
1955 V42 P256-	51Q71G <ATS>
1955 V42 P406-410	58-2620 <*>
1955 V42 P417	T2003 <INSD>
1955 V42 P490-	II-384 <*>
1955 V42 P628-629	58-224 <*>
1955 V42 N1 P21-22	59-17308 <*>
	63-16292 <*>
1955 V42 N2 P29-35	62-18333 <*> 0
1955 V42 N5 P126-	1191 <*>
1955 V42 N8 P210	5735-A <KH>
1955 V42 N9 P256-	1297 <*>
1955 V42 N11 P341-342	59-10958 <*>
1955 V42 N13 P399-	57-850 <*>
1955 V42 N15 P444-	57-1266 <*>
1955 V42 N15 P446-447	63-16291 <*>
1955 V42 N17 P478-482	57-2735 <*>
1955 V42 N18 P508-	II-681 <*>
1955 V42 N22 P6C4-605	63-16481 <*>
1956 V43 P39-	1401 <*>
	1446 <*>
1956 V43 P60-	57-765 <*>
1956 V43 P93-	1098 <*>
1956 V43 P126	58-585 <*>
1956 V43 P130-	II-937 <*>
1956 V43 P162-	1296 <*>
1956 V43 P180-181	1254 <*>
1956 V43 P208-	1061 <*>
1956 V43 P223-224	59-10642 <*>
1956 V43 N1 P12-13	57-3562 <*>
1956 V43 N2 P39	62-14936 <*>
1956 V43 N2 P39-40	62-14937 <*>
1956 V43 N4 P89-90	57-1436 <*>
1956 V43 N6 P21-28	58-2471 <*>
1956 V43 N9 P2C5-206	57-824 <*>
1956 V43 N11 P255-	1380 <*>
1956 V43 N13 P3C5	57-3421 <*>
1956 V43 N15 P351-352	59-10311 <*>
1956 V43 N15 P351	9061-C <K-H>
1956 V43 N17 P405	57-3417 <*>
	60-16817 <*> 0
1956 V43 N20 P472-473	57-70 <*>
1957 V44 P256-257	58-1573 <*>
1957 V44 P377-	57-2989 <*>
1957 V44 N1 P12	57-838 <*>
	62-14935 <*>
1957 V44 N2 P31	79R74G <ATS>
1957 V44 N2 P36	59-20611 <*>
1957 V44 N5 P108-109	09Q72G <ATS>
1957 V44 N8 P264-265	57-3433 <*>
1957 V44 N8 P265-	3070 <*>
1957 V44 N9 P283-284	59-17301 <*>
1957 V44 N9 P284-285	57-3446 <*>
1957 V44 N9 P284-	59-17302 <*>
1957 V44 N9 P285	62-01422 <*>
1957 V44 N11 P337-338	66-11322 <*>
1957 V44 N13 P371	60-18326 <*>
1957 V44 N14 P390	62-14056 <*>
1957 V44 N14 P399-400	59-15434 <*>
1957 V44 N14 P399	60-14045 <*>
1957 V44 N16 P453-454	CSIRO-395C <CSIR>
1957 V44 N18 P487	63-10358 <*>
1958 V45 P6-7	58-748 <*>
1958 V45 P137	58-1819 <*>
	58-1820 <*>
1958 V45 P166	59-17097 <*>
1958 V45 P192-193	58-1586 <*>
1958 V45 P294	61-00899 <*>
1958 V45 P415-416	AD-641 277 <$=>
1958 V45 P445	64-14537 <*>
	9621 <K-H>
1958 V45 P538-539	AEC-TR-3684 <+*>
1958 V45 N2 P34-35	60-00445 <*>
1958 V45 N2 P36	58-1679 <*>
1958 V45 N4 P73-80	64-10422 <*>
	8345-A <K-H>
1958 V45 N11 P263	59-15356 <*>
1958 V45 N11 P269-270	59-15357 <*>
1958 V45 N14 P327-329	59-10423 <*>
1958 V45 N15 P369-371	62-00972 <*>
1958 V45 N15 P373-374	59-17100 <*>
1958 V45 N15 P375-376	59-17099 <*>
1958 V45 N20 P490	65-10445 <*>
1958 V45 N21 P509-510	NP-TR-420 <*>
1958 V45 N21 P521	62-00756 <*>
1959 V46 P599	AEC-TR-4393 <*>
1959 V46 N2 P70-71	60-18434 <*>
1959 V46 N3 P108	64-10967 <*>
1959 V46 N6 P210	65-00162 <*>
1959 V46 N12 P399	65-14575 <*>
1959 V46 N12 P410-411	60-18800 <*>
1959 V46 N14 P457-458	63-00302 <*>
1959 V46 N15 P461-471	59M37G <ATS>
1959 V46 N17 P512	62-00755 <*>
1959 V46 N21 P607-608	60-14537 <*>
1960 V47 P13-14	AEC-TR-4016 <*>
1960 V47 P337-351	11558-B <K-H>
	64-14524 <*> 0
1960 V47 P477	AEC-TR-5535 <*>
1960 V47 N1 P15-16	62-00801 <*>
1960 V47 N4 P85	60-18068 <*>
1960 V47 N4 P89	63-01318 <*>
1960 V47 N6 P127-128	61-10863 <*>
1960 V47 N7 P155	63-00301 <*>
1960 V47 N9 P199-200	AEC-TR-4408 <*>
1960 V47 N12 P276	62-16606 <*>
1960 V47 N14 P313-317	63-18382 <*>
1960 V47 N18 P409-422	61-00366 <*>
1960 V47 N19 P446-447	62-00109 <*>
1960 V47 N20 P473	61-10283 <*>
1960 V47 N20 P474-475	61-10282 <*>
1960 V47 N21 P486-490	65-13271 <*>
1960 V47 N21 P492-493	61-20881 <*> 0
1960 V47 N23 P532-536	62-16557 <*> 0
1960 V47 N24 P605-606	10843-A <KH>
	63-16964 <*>
1961 V48 P441-445	62-18214 <*>
1961 V48 P497-498	AEC-TR-5048 <*>
1961 V48 P729-734	AEC-TR-5319 <*>
1961 V48 N5 P134-135	61-20569 <*>
1961 V48 N5 P135-136	61-01045 <*>
1961 V48 N7 P216	NP-TR-873 <*>
1961 V48 N9 P384-385	62-00824 <*>
1961 V48 N13 P474-475	66-11658 <*>
1961 V48 N16 P543-545	62-14798 <*> 0
1961 V48 N16 P548-549	AEC-TR-6124 <*>
1961 V48 N16 P549	62-20188 <*>
1961 V48 N18 P602-603	11647 <KH>
1961 V48 N20 P644-	NP-TR-967 <*>
1961 V48 N23 P712	62-18909 <*>
1961 V48 N23 P722-723	63-10353 <*>
1961 V48 N24 P737-738	AEC-TR-5389 <*>
1962 V49 P499-500	NP-TR-1028 <*>
1962 V49 N6 P128-129	65-12874 <*>
1962 V49 N7 P150-152	63-10162 <*>
	66-10690 <*>
1962 V49 N9 P201	63-18087 <*>
1962 V49 N9 P17A	64-14275 <*>
1962 V49 N10 P217-228	62-01510 <*>
1962 V49 N15 P350-351	63-20122 <*> 0
1962 V49 N17 P386-388	63-00783 <*>
1962 V49 N19 P449	63-10729 <*>
1962 V49 N19 P454-455	63-18925 <*>
1963 V50 P76-88	AEC-TR-5771 <*>
1963 V50 N2 P41	63-14444 <*>
1963 V50 N3 P90	63-18001 <*>
1963 V50 N3 P91	63-18169 <*>
	63-18170 <*>
1963 V50 N3 P102-103	65-00246 <*>
1963 V50 N8 P339-340	65-11065 <*>
1963 V50 N13 P470-471	AI-TR-18 <*>
1963 V50 N13 P471	158 <TC>
1963 V50 N20 P650-651	64-16362 <*>

1963 V50 N21 P660-661 66-11847 <*>
1964 V51 P460- GI N2 1964 P388 <AGI>
1964 V51 N1 P1-8 65-18114 <*>
1964 V51 N1 P12 NP-TR-1138 <*>
1964 V51 N13 P308 65-11725 <*> 0
1964 V51 N14 P332 8684 <IICH>
1964 V51 N17 P416-417 65-14398 <*>
1964 V51 N21 P497-503 66-11885 <*>
1964 V51 N21 P503-504 65-17153 <*>
1964 V51 N21 P504-505 65-17152 <*>
1964 V51 N23 P551-552 71S85G <ATS>
1965 V52 N7 P170-171 65-13535 <*>
1965 V52 N15 P461 65-18069 <*>
1965 V52 N21 P588 66-14507 <0>

NATURWISSENSCHAFTLICHE MONOGRAPHIEN UND LEHRBUCHER
1923 V5 P18-22 61-14257 <*>

NATURWISSENSCHAFTLICHE RUNDSCHAU. BRAUNSCHWEIG
1898 V13 P279- 2287 <*>

NATURWISSENSCHAFTLICHE RUNDSCHAU. STUTTGART
1948 V1 N3 P115-118 59-17201 <*> 0
1955 V8 P309-314 II-215 <*>
1958 V11 N7 P268-272 66-11661 <*>
1958 V11 N8 P299-310 66-11661 <*>
1960 V13 N10 P383-388 62-16070 <*>

NATURWISSENSCHAFTLICHER ANZEIGER
1822 V5 N9 P65-70 65-17283 <*>

NATUURKUNDE
V52 N9 P628-632 58-2077 <*>
1959 V68 N9 P142-147 62-01575 <*>

NATUURWETENSCHAPPELIJK TIJDSCHRIFT
1937 V19 P91-105 63-16289 <*>
1938 V20 P20-30 I-995 <*>
1940 V22 P249- 62-16291 <*>
1940 V22 P249-262 64-18582 <*>
1940 V22 P263-268 62-16267 <*>

NAUCHNI TRUDOVE. NAUCHNO-IZSLEDOVSATELSKI TEKHNOLOGICHESKI INSTITUT PO VINARSKA I PIVOVARNA PROMYSHLENNOST. SOFIA
1962 V5 P129-141 65-12781 <*>

NAUCHNOE SOOBSHCHENIE. TSENTRALNYI NAUCHNO-ISSLEDOVATELSKII INSTITUT STROITELNYKH KONSTRUKTSII
1960 N13 P1-60 64-13931 <=*$>

NAUCHNO-ISSLEDOVATELSKIE TRUDY TSENTRALNOGO NAUCHNO-ISSLEDOVATELSKOGO INSTITUTA KOZHEVENNO-OBUVNOI PROMYSHLENNOSTI. MOSCOW (TITLE VARIES)
1957 N28 P119-131 61-18261 <=*>

NAUCHNO-TEKHNICHESKII BYULLETEN. GOSUDARSTVENNOGO NAUCHNO-ISSLEDOVATEL'SKOGO INSTITUTA OZERNOGO I RECHNOGO RYBNOGO KHOZYAISTVA
1960 N11 P49-52 C-4196 <NRC>

NAUCHNO-TEKHNICHESKAYA INFORMATSIYA. TSENTRALNYI NAUCHNO-ISSLEDOVATELSKII INSTITUT TEKHNOLOGII I MASHINOSTROENIYA. MOSCOW
1963 N3 P12 66-13809 <*>
1963 N5 P24 66-13810 <*>
1963 N7 P12-20 64-16410 <=*$>
1963 N9 P17-21 64-21096 <=>
1963 N9 P32-37 64-21096 <=>
1963 N9 P38-43 64-21096 <=>
1963 N9 P49 64-21096 <=>
1964 N1 P23-24 65-12707 <*>
1964 N2 P20-27 AD-633 683 <=$>
1964 N2 P47-48 AD-633 683 <=$>
1964 N3 P22-32 66-10312 <*>
1964 N3 P47-51 38T91R <ATS>
1964 N3 P52-54 64-41034 <=>
1964 N5 P29-34 66-10311 <*>
1964 N5 P35-43 65-31091 <=$>
1964 N6 P37-42 65-31091 <=$>
1964 N7 P24-26 AD-639 490 <=$>
1964 N10 P10-14 AD-636 705 <=$>
1964 N10 P19-22 AD-636 705 <=$>
1964 N12 P38-41 65-31314 <=$>
1964 N12 P47 65-31269 <=$>
1965 N2 P35-42 65-33693 <=$>
1965 N2 P52-61 65-33693 <=$>
1965 N2 P69-75 65-33693 <=$>
1965 N2 P93-100 65-33693 <=$>
1965 N2 P114-120 65-33693 <=$>
1965 N3 P34-38 65-31526 <=*$>
1965 N4 P23-34 65-31852 <=$>
1965 N5 P22-25 AD-629 910 <=*$>
1965 N6 P20-22 65-33155 <=$>
1965 N7 P31-33 66-14222 <*$>
1965 N9 P3-9 66-30601 <=$>
1965 N9 P11-17 66-30601 <=$>
1965 N9 P21-35 66-30601 <=$>
1965 N9 P37-54 66-30601 <=$>
1965 N9 P63 66-30601 <=$>
1965 N10 P34-36 66-34230 <=$>
1966 N2 P3-46 66-33197 <=$>
1966 N3 P46 66-33198 <=$>

*NAUCHNO-TEKHNICHESKAYA INFORMATSIYA. VSESOYUZNYI INSTITUT NAUCHNOI I TEKHNICHESKOI INFORMATSIYA. MOSCOW
1963 N1 P3-6 63-31502 <=>
1963 N1 P7-10 63-31502 <=>
1963 N1 P14-17 63-31502 <=>
1963 N1 P18-23 63-31502 <=>
1963 N1 P24-30 63-31502 <=>
1963 N1 P31-33 63-31502 <=>
1963 N1 P38-39 63-31502 <=>
1963 N1 P48-49 63-31502 <=>
1963 N2 P3-6 63-31647 <=>
1963 N2 P23-28 63-31647 <=>
1963 N2 P36-41 63-31647 <=>
1963 N2 P42-45 63-31647 <=>
1963 N2 P46-49 63-31647 <=>
1963 N6 P6-9 63-24603 <=*$>
1963 N7 P3-4 NLLTB V6 N1 P53 <NLL>
1963 N10 P5-6 NLLTB V6 N4 P330 <NLL>
1963 V8 P3-5 NLLTB V6 N4 P338 <NLL>
1964 N6 P3-12 64-51881 <=$>
1964 N8 P6-9 65-31102 <=$>
1964 N9 P35-38 65-30479 <=$>
1964 N11 P13-19 65-30984 <=$>
1964 N11 P24-25 65-30479 <=$>
1964 N11 P35-40 65-30479 <=$>
1965 N1 P23-25 65-30918 <=$>
1965 N1 P39-49 65-30918 <=$>
1965 N4 P9-18 65-32316 <=$>
1965 N11 P1-60 66-31499 <=$>
1965 N12 P45-48 66-31501 <=$>
1965 N12 P64-65 66-30857 <=$>
1966 N7 ENTIRE ISSUE 66-35705 <=>
1966 N8 ENTIRE ISSUE 66-35564 <=>
1966 N11 P3-71 67-30780 <=$>
1966 N12 P3-14 67-31494 <=$>
1966 N12 P21-50 67-31494 <=$>
1966 N12 P52-67 67-31494 <=$>
1966 N12 P70-73 67-31494 <=$>
1966 N12 P75-76 67-31494 <=$>

NAUCHNO-TEKHNICHESKIE OBSHCHESTVA SSSR
1959 N6 P7-9 60-31316 <=>
1960 N3 P10-11 60-11933 <=>
1960 N4 P32-35 NLLTB V3 N1 P23 <NLL>
1960 N5 P29-32 61-21293 <=>
1960 N6 P23-27 60-31720 <=>
1960 N7 P36 61-21293 <=>
1960 N8 P17-23 62-13860 <*=>
1960 N10 P57-58 61-28392 <*=>
1960 N11 P11-15 61-21438 <=>
1960 N12 P23-28 61-23128 <=>
1961 N10 P24-27 62-11105 <=>
1961 V3 N6 P3-7 62-11172 <=>
1961 V3 N11 P40-42 62-11179 <=>

1962 N10 P34-35 NLLTB V5 N2 P89 <NLL>
1962 V4 N2 P56-57 62-25512 <=*>
1962 V4 N2 P60-61 62-25512 <=*>
1962 V4 N4 P43-44 63-13117 <=>
1962 V4 N7 P39 62-33299 <=*>
1962 V4 N9 P2-5 63-13548 <=*> 0
1962 V4 N9 P17-22 63-13102 <=*>
1962 V4 N9 P60-61 63-13548 <=*>
1962 V4 N10 P12-13 63-15101 <=>
1962 V4 N10 P36-39 63-13650 <=>
1962 V4 N10 P44-46 UCRL TRANS-1037 (L) <=$>
1962 V4 N11 P18-19 63-21142 <=>
1962 V4 N11 P20-21 63-21199 <=>
1962 V4 N11 P31-33 63-13824 <=>
1962 V4 N12 P18-19 63-21511 <=>
1962 V4 N12 P20-24 63-21585 <=>
1962 V4 N12 P40-42 63-21524 <=>
1963 V5 N1 P25 63-21582 <=>
1963 V5 N1 P29-30 63-21795 <=>
1963 V5 N1 P34-36 63-21768 <=>
1963 V5 N1 P39 63-21795 <=>
1963 V5 N1 P44 63-21423 <=>
1963 V5 N1 P46-48 63-21664 <=>
1963 V5 N2 P7-8 63-21814 <=>
1963 V5 N2 P28-29 63-21817 <=>
1963 V5 N3 P36 63-31029 <=>
1963 V5 N4 P8-11 63-21919 <=>
1963 V5 N4 P17-18 63-31216 <=>
1963 V5 N4 P29-32 63-31216 <=>
1963 V5 N5 P8 63-31295 <=>
1963 V5 N6 P10 63-31643 <=>
1963 V5 N6 P56 63-31524 <=>
1963 V5 N7 P24 63-31896 <=>
1963 V5 N10 P23-25 64-21704 <=>
1963 V5 N1C P31-34 64-21704 <=>
1963 V5 N10 P56-57 64-21704 <=>
1963 V5 N11 P60-61 NLLTB V6 N5 P458 <NLL>
TRANS.BULL.1964 P458 <NLL>
1964 V6 N7 P4-6 64-51832 <=$>
1964 V6 N8 P18-21 65-20548 <=$>
1965 N10 P35-36 66-30464 <=$>
1965 V7 N1 P25-29 65-30660 <=$>
1965 V7 N2 P52-55 65-31348 <=$>
1965 V7 N4 P43-45 AD-622 454 <=*$>
1965 V7 N4 P58-60 65-31767 <=$*> 0
1965 V7 N4 P61-63 65-31321 <=$>
1965 V7 N6 P18-23 65-32321 <=$>
1965 V7 N10 P5-10 65-33662 <=$> 0
1965 V7 N10 P17-19 65-33662 <=$> 0
1965 V7 N10 P21- 65-33921 <=$>
1966 N2 P8-9 66-31608 <=$>
1966 N6 P1-3 66-33997 <=$>
1966 N10 P8 67-30849 <=$>
1966 N10 P11 67-30849 <=$>
1966 N10 P17 67-30849 <=$>
1966 N10 P29 67-30849 <=$>
1966 N10 P39 67-30849 <=$>
1966 N12 P9 67-30849 <=$>
1966 N12 P15 67-30849 <=$>
1966 N12 P20 67-30849 <=$>

NAUCHNO-TEKHNICHESKIE PROBLEMY GORENIYA I VZRYVA. AKADEMIYA NAUK SSSR. SIBIRSKOE OTDELENIE. NOVOSIBIRSK
1965 N2 P22-34 ICE V6 N2 P393-401 <ICE>
1965 N2 P1C1-1C2 ICE V6 N2 P311-314 <ICE>

NAUCHNO-TEKHNICHESKII BYULLETEN POLYNARNYI NAUCHNO-ISSLEDOVATELSKII INSTITUT MORSKOGO RYBNOGO KHOZYAISTVA I OKEANOGRAFII. MURMANSK
1962 N4 P29-30 64-19880 <=$>
1962 N2/3 P37-38 63-00887 <*>

NAUCHNO-TEKHNICHESKII INFORMATSIONNYI BYULLETEN NAUCHNO-ISSLEDOVATELSKOGO INSTITUTA STEKLA. MOSCOW
SEE STEKLO. INFORMATSIONNYI BYULLETEN VSESOYUZNOGO NAUCHNO-ISSLEDOVATELSKOGO INSTITUTA STEKLA. MOSCOW

NAUCHNO-TEKHNICHESKII INFORMATSIONNYI BYULLETEN POLITEKHNICHESKII INSTITUT. LENINGRAD
1957 N10 P19-29 63-19151 <=*$>

NAUCHNO-TEKHNICHESKII SBORNIK PO DOBYCHE NEFTI
1962 N16 P78-81 65-12831 <*>
1962 V17 P66-68 65-12832 <*>

NAUCHNYE DOKLADY VYSSHEI SHKOLY. BIOLOGICHESKIE NAUKI
1958 N3 P193-194 59-16798 <+*>
1958 N4 P31- 59-20949 <+*>
1959 N1 P7-10 59-13998 <=>
1959 N1 P17-19 65-63163 <=$>
1959 N3 P59-65 C-3421 <NRCC>
1959 N4 P156-161 61-23654 <*=>
1960 N2 P208-209 61-11546 <=>
1960 N3 P97-106 MAFF-TRANS-NS-30 <NLL>
1960 N4 P59-62 63-11115 <=>
1960 N4 P77-81 63-11116 <=>
1960 V3 P136-141 65-50048 <=>
1961 N1 P11-12 63-11106 <=>
1961 N1 P58-61 63-11123 <=$>
1961 N1 P143- AEC-TR-4796 <=*>
1961 N1 P143-148 AEC-TR-5490 <=*>
1961 N4 P16-19 62-24976 <=*>
1961 N4 P29-33 C-4446 <NRC> 0
63-00881 <*>
1962 N1 P51-53 62-24969 <=*>
1962 N1 P57-61 62-24969 <=*>
1962 N1 P83-87 63-19038 <=*>
1962 N1 P88-92 62-24969 <=*>
1962 N1 P98-104 62-24969 <=*>
1962 N1 P159-164 62-24969 <=*>
1962 N2 P32-35 63-31021 <=>
1962 N2 P104-108 62-32459 <=*>
1962 N2 P147-151 63-13231 <=*>
1962 N2 P160-165 62-33564 <=*>
1962 N3 P63-65 63-23228 <=*$>
1962 N3 P90-93 62-33593 <=*>
1962 N4 P157-163 63-21309 <=>
1963 N1 P7-11 63-21897 <=>
1963 N1 P13-16 CSIR-495 <CSSA>
1963 N1 P68-71 63-21600 <=>
1963 N1 P92-95 63-21551 <=>
1963 N1 P96-99 63-21600 <=>
1963 N1 P10C-115 63-11726 <=>
1963 N1 P125-127 63-24535 <=*$>
1963 N2 P7-11 64-31095 <=>
1963 N2 P12-16 64-31036 <=>
1963 N3 P37-41 C-5471 <NRC>
1963 N3 P199-2C4 AD-610 374 <=$>
1963 N4 P84-89 64-21755 <=>
1964 N1 P7-17 64-31122 <=>
1964 N1 P56-61 64-31122 <=>
1964 N1 P65-72 64-31122 <=>
1964 N1 P92-98 64-31122 <=>
1964 N1 P171-179 64-31122 <=>
1964 N2 P11-14 64-41002 <=>
1964 N2 P55-59 64-41002 <=>
1964 N2 P80-84 64-51014 <=$>
1964 N2 P85-87 64-41002 <=>
1964 N2 P94-96 64-41002 <=>
1964 N2 P201-202 64-41001 <=>
1964 N6 P16-25 65-30462 <=$>
1965 N1 P22-25 65-20786 <=$>
1965 N3 P40-47 65-31622 <=$>
1965 N3 P50-51 65-32173 <=$>

NAUCHNYE DOKLADY VYSSHEI SHKOLY. EKONOMICHESKIYE NAUKI
1958 N3 P102-110 59-11649 <=>
1960 N3 P41-51 61-23096 <*=>
1960 N4 P25 61-21297 <=>
1960 V3 N11 P7-8 61-23096 <*=>
1962 N5 P3-10 63-21379 <=>

NAUCHNYE DOKLADY VYSSHEI SHKOLY. ELEKTROMEKHANIKA I AUTOMATIKA
1959 N1 P27-34 60-22099 <=>

62-11337 <=>

NAUCHNYE DOKLADY VYSSHEI SHKOLY. ENERGETIKA
1958 N3 P173-186 AEC-TR-5255 <=*>
1958 V2 P163-170 AD-619 012 <=$>
1959 N2 P229-239 AEC-TR-5518 <=*>
64-15212 <=*$>
65-60497 <$>
84P62R <ATS>

NAUCHNYE DOKLADY VYSSHEI SHKOLY. FILOLOGICHESKIYE NAUKI
1961 N1 P24-31 61-28866 <=*>
1962 N2 P33-41 62-25477 <=*>
1963 N1 P157-158 63-31161 <=>
1963 N4 P95-104 64-21743 <=>

NAUCHNYE DOKLADY VYSSHEI SHKOLY. FILOSOFSKIYE NAUKI
1958 N4 P218-221 59-13630 <=>
1959 N3 ENTIRE ISSUE 60-11282 <=>
1960 N4 P106-110 61-21356 <=>
1960 N4 P150-154 61-21357 <=>
1961 N3 P143-148 62-13261 <*=>
1961 N3 P189-193 62-13243 <*=>
1962 N2 P122-127 62-25160 <=*>
1962 N3 P3-21 62-33710 <=*>
1962 N3 P30-40 62-33710 <=*>
1962 N3 P68-74 62-33710 <=*>
1962 N5 P42-49 63-21356 <=>
1962 N5 P50-58 63-21356 <=>
1962 N5 P109-114 63-21194 <=>
1963 N1 P49-57 63-21834 <=>
1963 V6 N1 P58-65 63-31055 <=>
1963 V6 N1 P153-159 63-31146 <=>
1963 V6 N2 P37-47 63-31410 <=>
1964 V7 N5 P59-67 65-30227 <=$>
1964 V7 N6 P16-25 65-30788 <=$>
1965 V8 N1 P3-7 65-30606 <=$>
1965 V8 N3 P98-107 65-32159 <=$>
1965 V8 N3 P149-152 65-32159 <=$>

NAUCHNYE DOKLADY VYSSHEI SHKOLY. FIZIKO-MATEMATICHESKIYE NAUKI
1958 N1 P3-11 59-10993 <+*>
1958 N1 P84-88 59-10994 <+*>
1958 N1 P136-137 R-5114 <*>
1958 N1 P162-166 65-17444 <*>
1958 N2 P162-165 64-15088 <=*$>
1958 N2 P189-191 61-11166 <*>
1958 N2 P192-199 60-21787 <=>
1958 N2 P228-234 60-13833 <+*>
1958 N3 P196-202 65-63799 <=*$>
1958 N5 P29-30 62-11440 <=>
1958 N5 P38-39 64-71334 <=> M
1958 N5 P47-52 62-11582 <=>
1958 N5 P91-101 62-19933 <=>
1958 N5 P147-150 62-11341 <=>
1958 N5 P183-192 62-15887 <=>
1958 N6 P147-151 64-19976 <=$>
1958 N6 P219-224 62-11513 <=>
1959 N1 P141-145 61-19829 <*=>
62-11553 <=>
1959 N2 P86-94 AMST S2 V47 P47-57 <AMS>
1959 N2 P95-97 60-27008 <+*>
62-11259 <=>
1959 N2 P128-132 62-11243 <=>
1959 N2 P133-140 62-23190 <=>
1959 N2 P158-159 62-11580 <=>
1959 N2 P162-165 62-11286 <=>
1959 N3 P7-9 61-11183 <=>
61-15941 <+*>
62-11552 <=>
1959 N3 P126-131 62-11240 <=>
1959 N3 P132-140 62-11241 <=>
1959 N3 P141-144 UCRL TRANS-794(L) <=*>

NAUCHNYE DOKLADY VYSSHEI SHKOLY. GEOLOGOGEOGRAF-ICHESKIYE NAUKI
1958 N1 P124-132 59-11477 <=> O
1958 N1 P246-249 72L32R <ATS>
1958 N1 P269-270 59-11478 <=>
1958 N3 P116-123 63-10685 <=*> O
1958 N3 P178-184 59-11641 <=> O
1958 N3 P185-189 59-11642 <=> O
1958 N3 P190-194 59-11643 <=> O
1958 N3 P203-208 64-23482 <=*$>
1959 N2 P46-53 62-33094 <=*>
1959 N2 P62-65 IGR V2 N9 P769 <AGI>
1959 N2 P216-222 63-24348 <=*$>

NAUCHNYE DOKLADY VYSSHEI SHKOLY. KHIMIYA I KHIMICHESKAYA TEKHNOLOGIYA
1958 N1 P83-85 01L35R <ATS>
62-18977 <=*>
1958 N1 P110-114 61-20627 <*=>
1958 N1 P148-252 55L32R <ATS>
1958 N2 P229-232 65L32R <ATS>
1958 N2 P330-334 61-23880 <*=>
1958 N2 P339-341 61-23655 <*=>
1958 N2 P361-364 60-13571 <+*>
1958 N2 P385-387 60-11458 <=>
1958 N2 P388-391 60-11459 <=>
1958 N3 P408-412 61-19131 <*=>
1958 N4 P613-616 61-23956 <*=>
1958 N4 P621-623 UCRL TRANS-484(L) <*>
1958 N4 P635-639 60-15808 <+*>
1958 N4 P667-671 AEC-TR-6259 <=*$>
1958 N4 P676-679 61-19329 <+*> O
1958 N4 P779-781 65-10089 <*> O
1959 N1 P28-31 61-19103 <+*>
1959 N1 P62-66 62-11235 <=>
1959 N1 P70-73 62-25108 <=*>
1959 N1 P80-87 61-19408 <+*>
1959 N1 P162-165 61-15501 <=*>
1959 N1 P181-185 60-17476 <+*>
60-18988 <+*>
1959 N2 P244-246 AEC-TR-3960 <*>
1959 N2 P305-306 63-18891 <=*>
1959 N2 P350-353 61-13751 <+*>
1959 N2 P386-389 60-23593 <+*>
1959 N2 P402-405 61-27509 <*=>
1959 N2 P406-410 50M43R <ATS>
1959 N3 P449-453 60-15837 <+*>

NAUCHNYE DOKLADY VYSSHEI SHKOLY. LESOINZHENERNOE DELO
1959 V1 P217-222 62-01250 <*>

NAUCHNYE DOKLADY VYSSHEI SHKOLY. MASHINOSTROYENIYE PRIBOROSTROYENIYE
1958 N1 P75-86 62-32290 <=*>
1958 N2 P7-18 61-23075 <=*>
1958 N2 P68-71 62-15179 <*=>
1958 N2 P100-107 61-23075 <=*>
1958 N3 P15-24 61-28711 <*=>
1958 N3 P108-113 64-11944 <=> M
1958 N3 P122-131 61-23890 <*=> O
1958 N3 P141-148 61-23890 <*=> O
1958 N4 P73-81 64-15178 <=*$>
1958 N4 P134-143 61-13763 <=*>
1958 N4 P152-159 61-13761 <=*>
1958 N4 P160-169 61-13762 <=*>
1958 N4 P170-180 61-13764 <=*>
1959 N2 P110-119 60-41053 <=>

NAUCHNYE DOKLADY VYSSHEI SHKOLY. METALLURGIYA
1958 N1 P211-217 61-21970 <=>
1958 N1 P218-221 61-21971 <=>
1958 N1 P226-232 61-21972 <=> O
1958 N1 P233-238 61-21973 <=>
1958 N1 P244-246 61-21974 <=>
1958 N1 P252-255 61-21975 <=>
1958 N3 P226-230 UCRL-TRANS-908(L) <=*>
1958 N4 P28-33 2091 <BISI>
62-25266 <=*>
1958 N4 P169-173 61-21957 <=>
1958 N4 P178-183 61-31500 <=> O
1958 N4 P225-228 61-21961 <=>
1959 N1 P19-24 4715 <HB>

1959 N1 P37-41 4628 <HB>
1959 N1 P58-67 61-19428 <+*>
1959 N1 P68-72 4688 <HB>
1959 N1 P73-79 2077 <BISI>
1959 N1 P105-112 61-19428 <+*>
1959 N1 P146-150 UCRL TRANS-660(L) <*=>
1959 N1 P168-169 4687 <HB>
1959 N1 P182-188 61-19428 <+*>
1959 N1 P203-205 4716 <HB>
1959 N2 P5-8 4692 <HB>
1959 N2 P9-14 4683 <HB>
1959 N2 P20-26 4721 <HB>
1959 N2 P43-47 4695 <HB>
1959 N2 P48-56 4691 <HB>
1959 N2 P84-88 4699 <HB>
1959 N2 P123-130 4796 <HB>
1959 N2 P202-206 61-23635 <=>
65M41R <ATS>
1959 N2 P217-220 4690 <HB>
1959 N2 P221-223 4712 <HB>
1959 N2 P252-256 5171 <HB>

NAUCHNYE DOKLADY VYSSHEI SHKOLY. RADIOTEKHNIKA I ELEKTRONIKA
1958 N2 P13-19 59-16558 <+*>
1958 N3 P184-198 61-19542 <*=>
1959 N1 P83-90 62-11289 <=>
1959 N1 P105-116 62-11257 <=>
1959 N1 P182-187 62-11496 <=>

NAUCHNYE DOKLADY VYSSHEI SHKOLY. STROITELSTVO
1958 N2 P111-124 60-19573 <+*>

NAUCHNYE SOOBSHCHENIYA INSTITUTA FIZIOLOGII. AKADEMII NAUK SSSR. MOSCOW
1959 N1 P46-48 65-60645 <=$>

NAUCHNYE SOOBSHCHENIIA VSESOYUZOGO NAUCHNO-ISSLEDOVATELSKOGO INSTITUTA TSEMENTNOI PROMYSHLENNOSTI
1958 N2 P34-37 62-25568 <=*>
1960 N8 P19-23 65-11766 <*>
1961 N12 P1-7 64-13797 <=*$>
1961 N12 P30-35 64-14299 <=*$>

NAUCHNYE TRUDY. DNEPROPETROVSK METALLURGICHESKII INSTITUT
SEE SBORNIK TRUDOV. DNEPROPETROVSK METALLURGICHESKII INSTITUT

NAUGHNYE TRUDY. EKSPERIMENTALNYI NAUCHNO-ISSLEDOVATELSKII INSTITUT KUZNECHNO-PRESSOVOGO MASHINOSTROENIYA. MOSCOW
1964 N9 P5-14 AD-633 664 <=$>

NAUCHNYE TRUDY EREVANSKOGO GOSUDARSTVENNOGO UNIVERSITETA
1956 P85-94 R-4503 <*>
1956 V53 P85-94 <CTT>

NAUCHNYE TRUDY. GORNYI INSTITUT, KHARKOV
SEE SBORNIK NAUCHNYKH TRUDOV. GORNYI INSTITUT. KHARKOV. (TITLE VARIES)

NAUCHNYE TRUDY. INSTITUT INZHENEROV MORSKOGO FLOTA ODESSA
1958 V17 P3-11 22L37R <ATS>

NAUCHNYE TRUDY. LESOTEKHNICHESKII INSTITUT. LVOV
1957 V3 P88-89 62-01249 <*>

NAUCHNYE TRUDY. LESOTEKHNICHESKII INSTITUT. MOSCOW
1958 V9 P175-188 CSIR-TR.208 <CSSA>

NAUCHNYE TRUDY NAUCHNO-ISSLEDOVATELSKOGO INSTITUTA REDKIKH METALLOV. IRKUTSK
1961 N9 P21-30 66-60617 <=*$>

NAUCHNYE TRUDY TEKHNOLOGICHESKOGO INSTITUTA LEGKOI PROMYSHLENNOSTI. MOSCOW.
1957 N8 P103-107 65-10427 <*>
1961 N19 P54-58 RCT V36 N3 P803 <RCT>

NAUCHNYE TRUDY. TSENTRALNOGO NAUCHNO-ISSLEDOVATELSKOGO INSTITUT TSELLIULOGNOI I BUMAZHNOI PROMYSHLENNOSTI
1956 N41 P123-134 46L32R <ATS>

NAUCHNYE TRUDY UCHENYKH I PRAKITICHESKIKH VRACHEI UZBEKISTANA. SBORNIK. TASHKENT
1962 N3 P1-194 AD-636 618 <=$>

NAUCHNYE TRUDY. UKRAYINSKYI NAUKOVO DOSLIDNYI INSTYTUT ZAHHYSTU ROSLYN. KIEV
SEE NAUCHNYE TRUDY. UKRAINSKOGO NAUCHNO-ISSLEDOVATELSKOGO INSTITUTA ZASHCHITY RASTENII. KIEV

NAUCHNYE TRUDY UKRAINSKOGO NAUCHNO-ISSLEDOVATELSKOGO INSTITUTA ZASHCHITY RASTENII. KIEV
1957 V2 P41-47 61-23716 <=*$>
1959 V8 P5-15 61-23712 <=*>
1959 V8 P43-49 61-23035 <=*>
1959 V8 P50-56 61-23714 <=*>
1959 V8 P89-96 61-23034 <=*>
1959 V8 P131-136 61-23036 <=*>

NAUCHNYE TRUDY. VSESOYUZNOGO NAUCHNO-ISSLEDOVATELSKOGO INSTITUTA ELEKTRIFIKATSKOGO SELSKOGO KHOZYAISTVA. MOSCOW
1956 V2 P185- 61-13466 <*=>

NAUCHNYE TRUDY VYSSHIKH UCHEBNYKH ZAVEDENII LITOVSKOI SSSR; KHIMIYA I KHIMICHESKAYA TEKHNOLOGIYA
SEE LIETUVOS TSR AUKSTUJU MOKYKLU MOKSLO DARBAI: CHEMIYA IR CHEMINE TECHNOLOGIYA

NAUCHNYE ZAPISKI. GOSUDARSTVENNYI NAUCHNO-ISSLEDOVATELSKII I PROEKTNYI INSTITUT UGOLNOI, RUDNOI, NEFTYANOI I GAZOVOI PROMYSHLENNOSTI. KIEV
1962 N9 P24-28 64-71486 <=> M

NAUCHNYE ZAPISKI INSTITUTA MASHINOVEDENIYA I AVTOMATIKI
1961 V7 N7 P16-25 AD-611 819 <=$>
63-24151 <=*$>

NAUCHNYE ZAPISKI MOSKOVSKOGO INSTITUTA INZHENEROV VODNOGO KHOZYAISTVA
1958 V20 P321-329 62-15474 <=*>

NAUCHNYE ZAPISKI ODESSKOGO POLITEKHNICHESKOGO INSTITUTA
1955 V5 N2 P81-85 59-22570 <+*>

NAUCHNYE ZAPISKI. UZHGORODSKII GOSUDARSTVENNYI UNIVERSITET
1957 V18 P191-194 AMST S2 V18 P45-47 <AMS>

NAUCHNYI OTCHET. VSESOYUZNYI NAUCHNO-ISSLEDOVATELSKII INSTITUT AGROLESOMELIORATSII
1941 P82-90 60-51007 <=>

NAUHEIMER FORTBILDUNGSLEHRGAENGE
1935 V11 P14-33 58-1078 <*>

NAUKA I PEREDOVOI OPYT V SELSKOM KHOZYAISTVE
1956 N7 P4-7 R-1559 <*>
1957 V7 N8 P17-18 R-4185 <*>
1957 V7 N8 P65-66 R-4183 <*>

NAUKA POLSKA
1958 V6 N4 P80-84 63-21320 <=>
1959 V7 N2 P214-224 60-31732 <*=>
1959 V7 N2 P225-231 60-31733 <*=>
1959 V7 N3 P1-16 63-2133 <=>
1960 V8 N1 P53-77 62-19320 <*=>
1960 V8 N2 P107-114 62-19338 <=>
1960 V8 N2 P121-128 62-19338 <=>
1960 V8 N2 P153 62-19338 <=>

Citation	Translation	
1960 V8 N2 P155-182	62-19338 <=>	
1960 V8 N10 P1-52	60-31694 <*=>	
1960 V8 N10 P165-224	60-31694 <*=>	
1960 V8 N10 P247-255	60-31694 <*=>	
1960 V8 N10 P257-348	60-31694 <*=>	
1961 V9 N1 P1-16	62-19507 <*=>	
1961 V9 N1 P43-46	62-19504 <=*>	
1961 V9 N1 P59-70	62-19510 <=*>	
1961 V9 N1 P109-114	62-19563 <*=>	
1961 V9 N1 P115-122	62-19594 <*=>	
1961 V9 N1 P123-128	62-19609 <=*>	P
1961 V9 N1 P129-132	62-19568 <*=>	P
1961 V9 N1 P133-145	62-19502 <*=>	P
1961 V9 N1 P158-159	62-19548 <=*>	
1961 V9 N1 P160-171	62-19610 <*=>	P
1961 V9 N1 P229-248	62-19556 <*=>	P
1961 V9 N1 P249-276	62-23393 <=*>	
1961 V9 N1 P276-280	62-19579 <*=>	P
1961 V9 N2 P17-26	62-24406 <=*>	
1961 V9 N2 P261-267	62-25476 <=*>	
1961 V9 N4 P123-138	62-23442 <=*>	
1962 V10 N1 P3-40	62-33371 <=>	
1962 V10 N1 P63-86	62-33371 <=>	
1962 V10 N1 P121-132	62-33371 <=>	
1962 V10 N1 P139-149	62-33371 <=>	
1962 V10 N2 P3-47	62-32063 <=>	
1962 V10 N2 P49-66	62-33012 <=*>	
1962 V10 N2 P67-74	62-32063 <=>	
1962 V10 N2 P95-102	62-32063 <=>	
1962 V10 N2 P115-120	62-32063 <=>	
1962 V10 N2 P137-141	62-32063 <=>	
1962 V10 N2 P159-187	62-32063 <=>	
1962 V10 N2 P190-194	62-32063 <=>	
1962 V10 N2 P200-201	62-32063 <=>	
1962 V10 N4 P8-18	63-21360 <=>	
1962 V10 N5 P138-149	64-21290 <=>	
1963 V11 N3 P113-138	63-41036 <=>	
1964 V12 N4 P186-192	64-51857 <=$>	O

NAUKA I RELIGIYA

Citation	Translation
1964 N1 P30-44	65-32549 <=*$>
1965 N7 P21-24	65-32728 <=$>
1965 N7 P25-30	65-32727 <=$>
1965 N7 P89	65-32728 <=$>
1965 N8 P51-57,73	65-33102 <=*$>
1965 N8 P74-75	65-33103 <=*$>
1965 N11 P29-32,48-49	65-34070 <=*$>
1965 N11 P61-65	65-34070 <=*$>
1965 N12 P7-10	66-30266 <=*$>
1965 N12 P44-47	66-30266 <=*$>
1965 V6 N6 P36-38	65-32012 <=*$>
1965 V6 N6 P51	65-32012 <=*$>
1965 V6 N6 P73	65-32012 <=*$>
1966 N1 P10-14	66-30634 <=$>
1966 N1 P31-36	66-30634 <=$>
1966 N1 P72-75	66-30634 <=$>
1966 N2 P15-16	66-31015 <=$>
1966 N2 P26-28,95	66-31015 <=$>
1966 N4 P48-52	66-32139 <=$>
1966 N5 P2-3,13-23	66-33420 <=$>

NAUKA I SUSPILSTVO
SEE NAUKA I ZHYTTYA

NAUKA I TEKHNIKA, LATVIA

Citation	Translation
1962 N2 P28-29	63-31732 <=>
1962 N40 P3-4	64-21427 <=>
1963 N1 P3-4	63-31695 <=>
1963 N2 P4-7	63-31331 <=>
1963 N2 P16-18	63-21710 <=>
1963 N2 P28-29	63-31732 <=>
1963 N2 P38	63-31233 <=>
1963 N5 P3	63-31314 <=>
1963 N5 P28-31	63-31646 <=>
1963 N6 P4	63-31792 <=>
1963 N6 P14-18	63-31794 <=>
1963 N6 P19	63-31637 <=>
1963 N6 P20-22	63-31683 <=>
1963 N7 P8-19	63-31936 <=>
1963 V1 P35	63-31477 <=>
1963 V1 P38-39	63-31477 <=>
1963 V6 P26-27	63-31793 <=>
1963 V12 P39	AD-610 271 <=$>
1964 N4 P24-26	64-31846 <=>
1964 N4 P34-35	64-31763 <=>
1964 N7 P37-39	64-51310 <=>
1964 N7 P42-43	64-51324 <=$>
1964 N12 P5-8	65-31044 <=$>
1964 N12 P37	65-31045 <=$>
1965 N1 P37-40	AD-621 796 <=*$>
1965 N2 P7-9,31	AD-638 894 <=$>
1965 N2 P12-13	65-31612 <=$>
1965 N4 P20-23	65-31289 <=$>
1965 N6 P3-6	65-32446 <=*$>
1965 N10 P2-4	65-33981 <=*$>
1965 N10 P5-7	65-33982 <=*$>
1966 N5 P9	66-34964 <=$>
1966 N7 P13-15	66-34550 <=$>

NAUKA I ZHIZN

Citation	Translation	
1949 N5 P22-24	51/2152 <*>	
1951 V18 N6 P34-36	RT-1056 <*>	
1952 N1 P20-21	RT-468 <*>	
1952 N4 P24-25	RT-2597 <*>	
1952 V19 N1 P37	RT-1332 <*>	
1952 V19 N3 P13-15	RT-1292 <*>	
1953 V20 N1 P17-19	61-13722 <=*>	
1953 V20 N5 P36	62-25900 <=*>	
1953 V20 N8 P21-22	RT-1159 <*>	
1953 V20 N8 P33	RT-1161 <*>	
1953 V20 N11 P25-29	59-10233 <+*>	
1954 N6 P32	RT-2050 <*>	
	RT-2051 <*>	
1954 N12 P23-24	R-179 <*>	
	59-12044 <+*>	
1954 V21 N3 P24	RT-2569 <*>	
1954 V21 N3 P33	RT-2969 <*>	
1954 V21 N6 P9-11	RT-2804 <*>	
1954 V21 N9 P30-32	RT-2959 <*>	
1954 V21 N11 P26-29	RT-3776 <*>	
1954 V21 N12 P12-14	RT-4124 <*>	
1954 V21 N12 P25-26	R-1892 <*>	
1954 V21 N12 P27-29	RT-4180 <*>	
1955 N1 P7-10	RT-2885 <*>	
1955 N8 P17-20	PB 141 274T <=>	
1955 V22 N1 P43-45	RT-3777 <*>	
1955 V22 N4 P7-10	R-4568 <*>	
1955 V22 N11 P17-19	RT-3550 <*>	
1956 N4 P49-50	R-4331 <*>	
1956 V23 N7 P13-16	59-14172 <+*>	O
1956 V23 N7 P20	59-14173 <+*>	
1957 N2 P17-21	R-1779 <*>	
	61-15328 <+*>	
1957 N4 P14	PB 141 184T <=>	
1957 N6 P18-22	PB 141 269T <=>	
1957 V24 N5 P50	59-12213 <+*>	
1957 V24 N10 P63	60-12483 <=>	
1957 V24 N12 P11-16	61-23428 <=*>	O
1957 V24 N12 P33-36	60-17448 <+*>	O
1958 N7 P60-64	59-16451 <+*>	O
1958 V25 N1 P17-22	60-13815 <+*>	
1958 V25 N3 P3-5	59-16052 <+*>	
1958 V25 N3 P6-11	59-16052 <+*>	
1958 V25 N3 P12-16	59-16052 <+*>	
1958 V25 N3 P17-22	59-16052 <+*>	
1958 V25 N4 P67-68	60-13130 <+*>	
1958 V25 N4 P77-78	59-15576 <+*>	
1958 V25 N5 P68	63-14698 <=*>	
1958 V25 N7 P71	59-21001 <=>	
1958 V25 N8 P17-20	63-19502 <=*>	
1958 V25 N8 P37-38	59-16045 <+*>	O
1958 V25 N8 P69-70	60-15512 <+*>	
1958 V25 N9 P1-4,6	63-23529 <=*>	
1958 V25 N9 P7-12	59-19524 <+*>	O
1958 V25 N11 P17-21	61-23684 <=*>	O
1958 V25 N11 P23-27	61-19782 <=*>	O
1959 N11 P8	60-11448 <=>	
1959 V26 N1 P66	59-19481 <+*>	
1959 V26 N2 P2	59-21063 <=>	
1959 V26 N5 P20	59-19906 <+*>	

Citation	Translation
1959 V26 N5 P33-36	59-13757 <=>
1959 V26 N6 P17-21	60-11154 <=>
1959 V26 N8 P9-14	60-11055 <=>
1959 V26 N9 P2-8	60-31176 <=> 0
1959 V26 N9 P12	60-31176 <=> 0
1959 V26 N11 P67-68	61-13337 <=>
1959 V26 N12 P2-5	60-31237 <=> 0
	61-19323 <+*> 0
1959 V26 N12 P17-23	61-19323 <+*> 0
1960 V27 N2 P53-58	61-23522 <=*>
1960 V27 N3 P8-10	63-24438 <=*$>
1960 V27 N4 P58-64	61-23695 <=*>
1960 V27 N4 P66-67	61-23695 <=*>
1960 V27 N5 P14-16	60-31680 <=> 0
1960 V27 N5 P17-23	62-32285 <=*>
1960 V27 N6 P11-16	61-23431 <=*>
1960 V27 N6 P31-33	62-24351 <=*>
1960 V27 N9 P37-40	61-31462 <=> 0
	62-20194 <=*>
1960 V27 N9 P44-46	61-19870 <=*>
1960 V27 N9 P50	62-20194 <=*>
1960 V27 N10 P72	61-23857 <=*>
1960 V27 N11 P22-26	61-28210 <*=>
1960 V27 N11 P41-45	62-15172 <*=>
1960 V27 N11 P46-47	62-32637 <=*>
1960 V27 N11 P64	61-23932 <=>
1961 V28 N2 P8-10	61-27958 <*=>
	61-31111 <=>
1961 V28 N2 P28-32	61-27855 <*=>
1961 V28 N2 P33-39	61-27733 <*=>
1961 V28 N2 P60-63	61-27734 <*=>
1961 V28 N3 P7-11	62-24955 <=*>
1961 V28 N3 P17-32	62-24955 <=*>
1961 V28 N3 P72-73	62-13255 <*=>
1961 V28 N4 P8-11	62-25825 <=*>
1961 V28 N4 P37-41	62-25825 <=*>
1961 V28 N5 P17-22	63-13881 <=*>
1961 V28 N5 P27-28	62-24276 <=*>
1961 V28 N5 P29-31	62-23666 <=*>
1961 V28 N5 P43-45	62-15771 <=*>
1961 V28 N6 P1-5	61-27971 <*=>
1961 V28 N6 P6-9	61-28924 <*=>
1961 V28 N6 P50-54	62-13688 <=> 0
1961 V28 N6 P60-63	61-28059 <*=>
1961 V28 N6 P65-73	62-13033 <*=>
1961 V28 N7 P2-12	62-24789 <=*>
1961 V28 N7 P22-26	UCRL TRANS-742 <=*>
1961 V28 N7 P76	62-15784 <*=>
1961 V28 N7 P78-90	62-15784 <*=>
1961 V28 N9 P34-36	62-25824 <=*> 0
1961 V28 N9 P78-79	62-25824 <=*> 0
1961 V28 N10 P2-17	62-23831 <=*>
1961 V28 N10 P21-28	62-32301 <=*>
1961 V28 N10 P63-64	62-33537 <=*>
1961 V28 N11 P86-91	63-13883 <=*>
1961 V28 N12 P10-13	62-23460 <=*>
1961 V28 N12 P40-42	63-15409 <=*>
1962 V29 N1 P2-5	62-24982 <=>
1962 V29 N1 P7-19	62-24982 <=>
1962 V29 N1 P10-13	62-23860 <=*>
1962 V29 N1 P21-29	62-24982 <=>
1962 V29 N1 P32-69	62-24982 <=>
1962 V29 N1 P47-49	62-23850 <=*> 0
1962 V29 N1 P71	62-24982 <=>
1962 V29 N2 P13-19	62-33533 <=*>
1962 V29 N2 P20-24	62-25017 <=*>
1962 V29 N3 P18-23	62-25057 <=*>
1962 V29 N3 P24-32	AD-614 762 <=$>
1962 V29 N3 P80-83	63-13482 <=*>
1962 V29 N3 P96-97	62-25057 <=*>
1962 V29 N4 P1-112	63-13302 <=>
1962 V29 N5 P39-42	62-11719 <=>
1962 V29 N8 P76	64-71186 <=>
1962 V29 N9 P2-10	63-13487 <=*>
	63-15303 <=>
1962 V29 N9 P15-20	63-13487 <=*>
1962 V29 N9 P15-	63-15303 <=>
1962 V29 N9 P54-60	63-15303 <=>
1962 V29 N11 P28-29	63-23671 <=*>
1962 V29 N12 P28-33	66-30241 <=*$>
1962 V29 N12 P34-39	63-21322 <=>
1963 V30 N1 P9	63-31108 <=>
1963 V30 N1 P65-67	63-21528 <=>
1963 V30 N2 P12-17	63-21613 <=>
1963 V30 N2 P106-108	64-21012 <=> 0
1963 V30 N4 P32-33	63-31130 <=>
1963 V30 N4 P35-39	63-23643 <=*$>
1963 V30 N5 P29-34	63-31651 <=>
1963 V30 N5 P61-64	64-31431 <=>
1963 V30 N6 P24-29	63-24439 <=*$> 0
1963 V30 N6 P69	63-31855 <=>
1963 V30 N7 P78-82	63-31990 <=>
1963 V30 N8 P67-70	N65-16303 <=$>
	63-41231 <=>
1963 V30 N9 P24-25	64-15417 <=*$>
1963 V30 N9 P28-33	63-41232 <=>
1963 V30 N9 P39-41	64-21885 <=>
1963 V30 N10 P24-27	64-13640 <=*$> 0
1963 V30 N11 P33-35	64-21679 <=>
1963 V30 N11 P42-43	64-21684 <=>
1963 V30 N11 P46-49	64-21727 <=>
1963 V30 N11 P65-68	64-21719 <=>
1963 V30 N12 P40-44	64-15236 <=*$>
1964 N10 P97-104	JAS V12,N4,P159-164 <AAS>
1964 V31 N1 P26-31	AD-611 860 <=$>
1964 V31 N2 P54-57	AD-614 394 <=$>
1964 V31 N2 P64-66	64-31380 <=>
1964 V31 N2 P74-75	64-31380 <=>
1964 V31 N3 P82-89	AD-614 946 <=$>
1964 V31 N4 P60-62	64-41355 <=>
1964 V31 N4 P64-69	64-41355 <=>
1964 V31 N4 P78-82	64-41355 <=>
1964 V31 N4 P93-96	64-41355 <=>
1964 V31 N4 P130-131	AD-621 007 <=$>
1964 V31 N6 P16-18	64-41084 <=>
1964 V31 N8 P68-71	64-51181 <=>
1964 V31 N8 P124-125	AD-615 240 <=$>
1964 V31 N9 P16-21	AD-620947 <=*$>
	65-30814 <=$>
1964 V31 N9 P68-72	65-30815 <=$>
1964 V31 N9 P73	65-30813 <=$>
1964 V31 N10 P11-17	64-51995 <=$>
1964 V31 N10 P18-21	65-30695 <=$>
1964 V31 N10 P28-31	AD-625 277 <=*$>
1964 V31 N10 P35-40	64-51859 <=$>
1964 V31 N10 P97-104	64-51929 <=$>
1964 V31 N12 P15-18	65-30555 <=$>
1964 V31 N12 P56-57	65-30625 <=$>
1965 N6 P4-5	AD-624 569 <=*$>
1965 N10 P42-45	AD-635 568 <=$>
1965 V32 N1 P10-18	AD-618 319 <=$>
	65-31776 <=*>
1965 V32 N2 P80-85	65-31133 <=$>
1965 V32 N4 P14-18	65-31384 <=*$>
1965 V32 N5 P25-31	65-31475 <=$>
1965 V32 N5 P82-88	65-31997 <=$>
1965 V32 N6 P2-6	65-32241 <=*$>
1965 V32 N6 P10	AD-624 570 <=*$>
1965 V32 N6 P17-25	65-32241 <=*$>
1965 V32 N6 P26-28	AD-626 063 <=*$>
1965 V32 N6 P33-35	65-33216 <=$>
1965 V32 N6 P86-87	65-32381 <=$>
1965 V32 N8 P2-10	65-33107 <=*$>
1965 V32 N8 P63-64	AD-624 584 <=*$>
1965 V32 N8 P156-159	AD-626 069 <=*$>
1966 N3 P99-105	66-33593 <=$>
1966 N4 P16-22	66-33723 <=$>
1966 N4 P54-58	66-32403 <=*>
1966 N4 P59-65	66-32545 <=$>
1966 N6 P22-25	66-33560 <=$>
NAUKA I ZHYTTYA	
1959 N1 P50-53	66-31214 <=$>
1960 N1 P12-13	60-11760 <=>
1960 N1 P19-20	60-11760 <=>
1960 N1 P30	60-11760 <=>
1960 N1 P33	60-11760 <=>
1960 N1 P47	60-11760 <=>
1960 N1 P56	60-11760 <=>
1960 V10 N5 P16-20	61-21039 <=>

1960 V10 N7 P58-60 61-15996 <+*>
1960 V10 N10 P28-30 61-21415 <=>
1960 V10 N11 P42-44 61-23901 <=*>
1960 V10 N12 P42-44 61-31510 <=>
1961 V11 N1 P10-12 61-28275 <*=>
1961 V11 N1 P12-13 61-28273 <*=>
1961 V11 N1 P19-21 62-13383 <*=>
62-15189 <*=>
1961 V11 N1 P35-38 61-31596 <=>
1961 V11 N2 P41-44 62-13639 <*=>
1961 V11 N3 P43-46 61-28870 <*=>
1961 V11 N3 P46-48 61-28860 <*=>
1961 V11 N4 P44-47 61-28398 <*=>
1961 V11 N8 P13-17 62-33006 <=>
1961 V11 N9 P20-23 62-24398 <=*>
1961 V11 N10 P18-20 62-23674 <=*>
1961 V11 N12 P12-14 65-17404 <*>
1962 V12 N2 P12-13 62-25612 <=*>
1962 V12 N2 P20-23 63-21048 <=>
1962 V12 N5 P17-18 63-13630 <=*>
1962 V12 N12 P20-21 63-24040 <=*$>
1963 V13 N1 P29-31 63-31347 <=>
1963 V13 N1 P34-35 63-31347 <=>
1963 V13 N1 P39-41 63-31371 <=>
1963 V13 N7 P6-9 64-31090 <=>
1963 V13 N9 P36-38 64-21055 <=>
1965 N12 P18-25 66-30341 <=*$>

NAUKOVI POVIDOMLENNYA, FIZYKA. KIEV UNIVERSYTET
1956 N1 P11-12 61-19552 <=*>

NAUKOVI PRATSI. INSTYTUT LYUARNOHO VYROBNYTSTVA. AKADEMIYA NAUK URSR. KIEV
1962 V11 P80-84 63-23928 <=*$>

NAUKOVI ZAPYSKY. KREMENETSKYI DERZHAVNYI PEDAGOGICHNYI INSTYTUT
1961 V6 N1 P55-58 64-00197 <*>
64-15311 <=$>

NAUKOVI ZAPYSKY. KYYIVSSKYI DERZHAVNYI UNIVERSYTET
1957 V16 N17 P9-15 61-13480 <*+>
1957 V16 N18 P5-23 59-18230 <+*>

NAUKOVI ZAPYSKY. LVIVSKYI DERZHAVNYI UNIVERSYTET IM. IVANA FRANKA. SERIYA KHIMICHNA
1955 V34 N4 P79-83 R-3703 <*>
66-12402 <*>

NAUKOVI ZAPYSKY Z TSUKROVOYI PROMYSLOVOSTY
1930 V9 P57-62 R-2824 <*>
1930 V9 N1/2 P41-45 RT-959 <*>

NAUNYN-SCHMIEDEBERG'S ARCHIV FUER EXPERIMENTELLE PATHOLOGIE U. PHARMAKOLOGIE
1925 V109 P332-357 63-18452 <*>
1927 V122 P338-353 9767-B <K-H>
1928 V124 P334-342 60-10840 <*>
1928 V129 P133-149 I-1033 <*>
1928 V130 P323-325 60-10839 <*>
1928 V136 P129- 57-424 <*>
1929 V139 P120-128 277 <TC>
1929 V146 P113-128 64-14836 <*>
1929 V146 P327-346 59-15392 <*> 0
1930 V149 P336-342 63-18449 <*>
1930 V150 P257-284 64-14835 <*>
1930 V152 P250-256 64-14834 <*> P
1930 V154 P203-210 5470-B <K-H>
1930 V158 P233-246 66-11719 <*> 0
1930 V158 P247-253 66-11720 <*> 0
1932 V164 P501-508 63-18448 <*>
1932 V164 P685-694 64-14833 <*>
1932 V165 P520-537 64-14832 <*> P
1932 V168 P307-318 2919 <*>
1933 V169 P429-452 64-14831 <*> P
1933 V169 P453-458 66-11721 <*> 0
1933 V169 P687-723 57-1891 <*>
1933 V171 P329-339 64-14830 <*>
1934 V174 P405-415 66-11622 <*> 0
1934 V175 P401-405 57-289 <*>
1934 V175 P554-557 57-407 <*>
1935 V179 P504-523 62-01509 <*>
1936 V181 P317-324 60-10838 <*> 0
1936 V181 N1 P46-61 58-2494 <*>
59-10821 <*> 0
1936 V182 P141-159 AL-914 <*>
1936 V182 P160-163 AL-913 <*>
1936 V182 P390-400 64-14829 <*> 0
1936 V183 P87-105 57-1657 <*>
1936 V183 P571-586 64-14828 <*> 0
1936 V183 P587-594 64-14827 <*> 0
1937 V184 P645-658 57-399 <*>
1937 V185 P323-328 57-400 <*>
1937 V186 P428-433 57-279 <*>
1937 V186 P434-443 57-398 <*>
1938 V188 P226-246 57-144 <*>
1938 V188 P377-382 63-20482 <*>
1938 V188 P593-597 64-14826 <*>
1938 V189 P4-21 AEC-TR-4640 <*>
1938 V189 P581-599 59-10083 <*> 0
1938 V191 P465-481 08F4G <ATS>
1939 V191 P687-695 65-12384 <*>
1939 V192 P383-388 64-14825 <*> P
1939 V192 P405-413 2921 <*>
1939 V192 P472-485 64-14824 <*>
1939 V193 P619-621 64-14823 <*>
1939 V194 P621-628 58-157 <*>
1940 V195 P71-74 3164 <*>
1940 V195 P184-193 11N54G <ATS>
62-11390 <=>
1940 V196 P109-136 64-14822 <*> P
1940 V196 P266-273 60-10844 <*>
1940 V196 P505-520 I-296 <*>
1941 V197 P597-610 32Q73G <ATS>
1942 V199 P74-82 64-14821 <*>
1942 V199 P145-152 60-10496 <*> 0
1942 V199 P161-166 64-14820 <*>
1943 V201 P297-304 64-14819 <*>
1944 V203 P25-33 64-14818 <*> P
1947 V205 P33-35 1145-I <K-H>
1947 V205 P382-386 722 <TC>
1949 V207 P547-568 C-2114 <NRC>
C-2387 <NRC>
1949 V207 P696-702 64-14817 <*> 0
1950 V209 N4/5 P375-388 60-10495 <*> 0
1950 V212 P112-113 64-14816 <*> P
1950 V212 N1/2 P129-131 63-18940 <*>
1951 V212 N3/4 P331-338 66M45G <ATS>
1951 V213 P255-264 64-14815 <*> 0
1951 V213 N3/4 P207-234 59-17817 <*> 0
1952 V214 P165-173 64-14814 <*>
1952 V214 P176-184 I-255 <*>
1952 V214 P423-426 64-14813 <*> 0
1952 V214 N4 P374-380 60-10494 <*> 0
1952 V215 P19-24 58-158 <*>
64-14812 <*> 0
1952 V215 P217-230 63-18451 <*>
1952 V216 P323-326 3172 <*>
1953 V217 P293-311 58-180 <*>
1953 V218 P159-168 AL-122 <*>
1953 V218 P169-176 AL-121 <*>
1953 V218 P177-192 AL-123 <*>
1953 V219 P273-283 I-894 <*>
1953 V220 P153-154 57-1815 <*>
1953 V220 P268-289 I-747 <*>
1954 V221 P404-417 4520 <K-H>
1954 V222 P431-449 61-00617 <*>
1954 V222 P540-554 64-14811 <*> 0
1954 V222 N1/2 P214- 3199-E <K-H>
1954 V223 P169-176 II-524 <*>
1954 V223 P280-284 II-472 <*>
1954 V223 P338-347 II-361 <*>
1955 V224 P206-223 57-1916 <*>
1955 V224 P401-414 58-12 <*>
1955 V225 P160-162 3808-A <K-H>
1955 V225 P505-511 57-708 <*>
1955 V226 P319-327 57-1914 <*>
1955 V226 N2 P163-171 64-14990 <*> OP
1955 V226 N3 P207-218 64-14991 <*> 0
1955 V226 N5 P473-485 64-14992 <*>

1955 V227 P93-110 57-2133 <*>
62-18764 <*> 0
1955 V227 N3 P212-213 1482 <*>
1955 V227 N3 P234-238 64-14993 <*>
1956 V227 N5 P383-392 64-14994 <*> 0
1956 V228 P166-167 II-893 <*>
1956 V228 P474-481 57-2985 <*>
1956 V228 N3 P314-321 64-14997 <*> 0
1956 V228 N4 P340-346 58-381 <*>
1956 V228 N1/2 P166-167 64-14995 <*>
1956 V228 N1/2 P233-235 64-14996 <*>
1956 V229 P366-373 58-62 <*>
1956 V229 P381-388 57-3173 <*>
1956 V229 P432-440 10091 <K-H>
1956 V229 N2 P123-138 61-10340 <*> 0
1956 V229 N4 P338-347 57-2988 <*>
1956 V229 N4 P374-380 57-394 <*>
1957 V230 P26-44 57-1093 <*>
1957 V230 P73-79 57-1094 <*>
1957 V230 P559-593 58-1071 <*>
1957 V232 P319-320 48K20K <ATS>
1957 V232 N1 P137-161 65-13246 <*>
1958 V232 P481-486 61-00626 <*>
1958 V232 N2 P470-480 58-1727 <*>
1958 V233 P311-322 58-1173 <*>
1958 V233 P323-337 58-1170 <*>
1958 V234 P1-16 58-1632 <*>
1958 V234 P102-119 58-1701 <*>
1958 V234 P194-205 58-2415 <*>
1958 V234 N1 P35-45 65-11649 <*> 0
1958 V234 N5 P390-399 64-14999 <*> 0
1958 V234 N5 P426-431 59-10755 <*> 0
1958 V234 N6 P474-489 64-16000 <*>
1959 V235 P291-300 AEC-TR-4636 <*>
1959 V235 N2 P103-112 59-15489 <*>
1959 V235 N5 P400-411 59-17597 <*>
1959 V235 N5 P412-420 64-16001 <*>
1959 V235 N5 P437-463 60-10743 <*>
1959 V236 N2 P382-391 59-17594 <*> P
1959 V236 N3 P492-502 61-00213 <*>
1959 V237 P17-21 63-16516 <*>
1959 V237 P22-26 63-18020 <*>
1960 V237 P519-537 AEC-TR-4641 <*>
1960 V238 P66-67 63-16905 <*>
1960 V238 P486-501 70M42G <ATS>
1960 V238 N3 P339-347 60-16923 <*>
1961 V241 P236-253 11279-D <KH>
63-16900 <*>
1961 V241 P376-382 62-01520 <*>
1961 V241 N3 P317-334 65-14654 <*> 0
1961 V242 P90-95 62-18176 <*>
1961 V242 P188-200 62-10914 <*>
1961 V242 N1 P17-23 65-11755 <*>
1962 V243 P36-43 63-18734 <*>
1962 V243 P65-84 62-20240 <*> 0
1962 V243 P310-311 63-16341 <*>
1962 V243 P318-320 8720 <IICH>
1962 V243 P566-569 64-10817 <*>
1962 V243 N4 P359-360 64-16355 <*>
1962 V244 P97-108 63-14719 <*>
1962 V244 P270-282 63-01055 <*>
1962 V244 N2 P185-195 64-14619 <*> 0
1963 V244 P334-350 63-00864 <*>
1963 V244 N5 P442-456 65-17035 <*>
1963 V244 N6 P550-563 64-20152 <*>
1963 V245 N1 P10-28 63-01018 <*>
1963 V245 N1 P196-229 65-14661 <*>
1963 V245 N1 P283-284 64-16082 <*>
1963 V245 N4 P471-483 64-16850 <*>
1964 V247 N4 P298-299 65-11936 <*>
1964 V247 N4 P303-304 65-11937 <*>
1964 V249 N1 P71-84 65-17348 <*>
1965 V250 N2 P257-258 65-17412 <*>

NAUTICA
1962 V4 N12 P633 48Q67DU <ATS>
1962 V4 N12 P643 49Q67DU <ATS>

NAVIRES, PORTS ET CHANTIERS
1957 P350-352 58-1359 <*>

1957 N82 P176-181 59-16313 <=*>

NEDELYA
1961 P4-9 12/10 62-25177 <=*>
1961 P8-9 11/26 62-25480 <=*>
1961 N20 P10-11 62-23800 <=>
1961 N52 P18 62-23592 <=*>
1962 N1 P10 62-23800 <=>
1962 N16 P10-11 63-13107 <=*>
1962 N32 P6-7 63-13538 <=*>
1962 V45 P15 63-13455 <=*>
1963 N7 P11 63-21342 <=>
1964 P16 12/19 AD-626 960 <=$>
1964 P9 9/19 AD-615 535 <=$>
1965 N39 P22 66-30768 <=$>
1965 N48 P14-15 65-30158 <=*$>
1965 N50 P4-5 66-30243 <=*$>

NEDERLANDSCH LANCET
1850 S2 V6 P607-609 61-14155 <*>

NEDERLANDSCH MAANDSCHRIFT VOOR GENEESKUNDE
1929 V10 P1934-1946 57-1912 <*>

NEDERLANDSCH MELK-EN ZUIVELTIJDSCHRIFT
1952 V6 N2 P127-136 59-20645 <*> 0
63-18266 <*> OP
1956 V10 N3/4 P276-286 T2203 <INSD>

NEDERLANDS MILITAIR-GENEESKUNDIG TIJDSCHRIFT
1963 V16 P406-411 66-11626 <*> 0

NEDERLANDS TIJDSCHRIFT VOOR GENEESKUNDE
1912 N2A P656-660 PT2A 61-14297 <*>
1914 V50 P1846-1852 57-2949 <*>
1914 V50 N26 P2071-2075 57-1889 <*>
1921 V65 P285-287 PT2 61-14633 <*>
1921 V65 P2670-2673 61-14589 <*>
1925 P966-973 57-1933 <*>
1927 V71 P2400-2405 61-14634 <*>
1930 V74 P1156-1158 59-14239 <+*>
1933 V77 P3562-3579 63-20898 <*> 0
1936 V80 P304-311 784-B <K-H>
1936 V80 N33 P3715-3722 1319CQ <K-H>
1937 V81 P21-24 804 <K-H>
1937 V81 P1129-1139 784-C <K-H>
1937 V81 P1273-1275 66-12010 <*>
1939 V83 N1 P1340-1346 57-1512 <*>
1941 V85 P2189-2196 65-00198 <*>
1942 P1548-1552 65-00197 <*>
1942 V86 N3 P134-139 59-16653 <+*>
1946 V90 P1917-1927 65-00387 <*>
1948 V92 P2968-2973 AL-589 <*>
1951 V95 P120-124 AL-587 <*>
1951 V95 N17 P1309-1314 58-2661 <*>
1951 V95 N29 P2120-2122 59-17021 <*>
1951 V95 N33 P2420-2422 2265 <*>
1952 V96 P1842-1846 59-14549 <+*>
1952 V96 P1850-1855 63-16953 <*>
8003-A <KH>
1953 V97 P1412-1413 2923-E <K-H>
1953 V97 P2763-2764 57-2783 <*>
1953 V97 P3300-3302 1533 <*>
1953 V97 N18 P1118-1122 59-20975 <*>
1954 V98 P2025-2028 PT3 57-1267 <*>
1954 V98 N3 P2659-2665 61-16563 <*> 0
1955 V99 P2276-2278 PT.3 61-00633 <*>
1955 V99 P2276-2279 3869-C <K-H>
1956 V100 P2110-2116 57-3507 <*>
1956 V100 N16 P1207-1209 5258C <K-H>
1957 V101 P1695-1700 7036-B <K-H>
1957 V101 N10 P459-464 60-15704 <*+>
8003-F <K-H>
1957 V101 N12 P538-541 64-20003 <*>
1958 V102 P172-175 58-2617 <*>
1958 V102 P890-892 61-00990 <*>
1958 V102 P1808-1812 0992-B <K-H>
64-14544 <*>
1958 V102 P1858-1863 64-14543 <*>
9002-C <K-H>

1958 V102 N6 P274-278 62-18280 <*> 0
1958 V102 N24 P1144-1149 59-12108 <+*>
1958 V102 N38 P1851-1852 59-15858 <*>
1959 V103 N2 P1049-1057 85Q73DV <ATS>
1959 V103 N38 P1935-1936 63-00855 <*>
1960 V104 P1445-1447 66-11654 <*> 0
1960 V104 P2113-2123 10843-B <K-H>
1960 V104 P2113-2133 64-14512 <*>
1960 V104 N13 P617-619 65-00388 <*>
1960 V104 N44 P2202-2203 63-00861 <*>
1961 V105 P2458-2460 62-01271 <*>
1961 V105 N5 P222-223 62-00428 <*>
1961 V105 N34 P1673-1678 62-14572 <*>
1961 V105 N50 P2534-2538 65-14779 <*> 0
1962 V106 N16 P823-825 63-10753 <*>
1963 V107 P1598-1603 65-11475 <*>
1963 V107 N2 P74-78 63-00947 <*>
1963 V107 N15 P687-693 65-13368 <*> 0
1963 V107 N33 P1498-1500 65-11149 <*> 0
1963 V107 N36 P1614-1620 66-11588 <*>
1963 V107 N40 P1796-1800 64-20142 <*>
1964 V108 N23 P1133-1137 66-11595 <*>
1964 V108 N34 P1647-1648 65-12164 <*>
1965 V109 P1368-1369 66-13205 <*>
1965 V109 N15 P713-723 66-13186 <*>
1965 V109 N34 P1563-1569 66-11590 <*>

NEDERLANDS TIJDSCHRIFT VOOR NATUURKUNDE
1943 V10 N1 P1-16 59-10856 <*>
1950 V16 P145-152 1370 <*>
1951 V17 N7 P209-233 60-10793 <*> 0
1953 V19 P129-139 00N51DU <ATS>
1955 V21 P201-209 58-1004 <*>
1956 V22 P377-393 7TM <CTT>
1956 V22 P394-402 9TM <CTT>
1956 V22 N12 P377-393 58-296 <*>
1956 V22 N12 P394-402 58-250 <*>
1958 V24 P301-314 63-00470 <*>
1958 V24 N12 P301-314 62-18122 <*>
1959 V25 N10 P265-287 91N53DU <ATS>
1960 V26 N8 P225-238 62-18605 <*>
1962 V28 N10 P329-352 65-10825 <*>

NEDERLANDSCH TIJDSCHRIFT VOOR PSYCHOLOGIE EN HAAR GRENSGEBIEDEN
1955 P435-445 57-1892 <*>
1955 V10 P258-288 58-118 <*>

NEDERLANDS TIJDSCHRIFT VOOR VERLOSKUNDE EN GYNAECOLOGIE
1963 V63 P269-272 65-14630 <*> 0

NEDERLANDSE CHEMISCHE INDUSTRIE
1959 V14 N1 P11-14 63-10634 <*>

NEDERLANDSCH-INDISCHE BLADEN VOOR DIERGENEESKUNDE EN DIERENTEELT
1930 V42 N1 P56-72 64-00068 <*>
1930 V42 N2 P110-120 64-00072 <*>
1932 V44 P493-520 63-14829 <*>

NEFTEGAZOVAYA GEOLOGIYA I GEOFIZIKA. NAUCHNO-TEKHNICHESKII SBORNIK. MOSCOW
1965 N2 P31-34 66-20955 <*>
1965 N6 P37-40 66-14413 <*$>
1965 N6 P40-44 66-14411 <*>
1965 N8 P33-35 02T91R <ATS>
1965 N10 P29-32 06T91R <ATS>
1965 N11 P41-46 06T93R <ATS>
1965 N12 P44-49 07T93R <ATS>

*NEFTEKHIMIYA
1961 V1 N1 P60-64 64-14456 <=*$>
1961 V1 N1 P121-123 62-13678 <*=>
1961 V1 N2 P260-266 63-20215 <=*>
1961 V1 N3 P433-443 62-32800 <=*>
63-26586 <=$>
1961 V1 N4 P535-540 64-14457 <=*$>
1961 V1 N4 P573-575 NLLTB V4 N5 P438 <NLL>
1961 V1 N5 P613-623 63-15509 <=*>
1961 V1 N5 P669-674 63-15509 <=*>
1961 V1 N5 P675-682 64-10786 <=*$>
1961 V1 N6 P796-799 81Q66R <ATS>
1961 V1 N6 P828-835 63-15515 <=*>
1962 N2 P229-236 66-11570 <*>
1962 N5 P793-794 ICE V3 N4 P531-532 <ICE>
1962 V2 P68-70 AEC-TR-5861 <*>
AEC-TR-5861 <=$>
1962 V2 P280-287 63-18752 <=*>
1962 V2 P288-290 63-18846 <=*>
1962 V2 N1 P18-20 63-18847 <=*>
1962 V2 N1 P21-27 63-18842 <=*$>
1962 V2 N1 P54-67 AEC-TR-5900 <=*$>
1962 V2 N2 P160-163 63-23723 <=*$>
1962 V2 N2 P164-169 63-18750 <=*>
1962 V2 N2 P175-178 83R76R <ATS>
1962 V2 N2 P179-186 84R76R <ATS>
1962 V2 N2 P187-188 63-18848 <=*>
1962 V2 N2 P193-195 64-18478 <*>
1962 V2 N2 P237-241 34Q74R <ATS>
1962 V2 N2 P253-256 66-13627 <*>
1962 V2 N3 P298-304 64-14945 <=*$>
75R75R <ATS>
1962 V2 N3 P318-323 63-18845 <=*$>
63-19970 <=*>
1962 V2 N3 P362-367 64-16687 <=*$>
1962 V2 N3 P384-390 65-10232 <*>
1962 V2 N3 P391-397 65-10234 <*>
1962 V2 N3 P410-414 63-10665 <=*> 0
1962 V2 N3 P415-423 62-33525 <=*>
1962 V2 N3 P420-423 AD-610 648 <=$>
1962 V2 N4 P436-441 64-10776 <=*$>
1962 V2 N4 P442-447 63-18843 <=*>
1962 V2 N4 P448-456 63-20214 <=*>
1962 V2 N4 P457-466 64-10107 <=*$>
1962 V2 N4 P467-472 63-20211 <=*>
1962 V2 N4 P473-479 64-10102 <=*$>
1962 V2 N4 P487-494 63-20209 <=*>
66-10006 <*>
70R77R <ATS>
1962 V2 N4 P494-497 64-10094 <=*$>
1962 V2 N4 P531-535 63-20213 <=*>
1962 V2 N4 P592-599 63-20216 <=*>
1962 V2 N4 P604-610 63-20210 <=*>
1962 V2 N5 P681-687 64-10777 <=*$>
1962 V2 N5 P750-755 64-10778 <=*$>
1962 V2 N5 P788-792 64-14277 <=*$> P
64-16867 <=*$> 0
1962 V2 N5 P793-794 21Q70R <ATS>
64-14946 <=*$>
1962 V2 N6 P877-884 76R77R <ATS>
1962 V2 N6 P897-900 64-10105 <=*$>
1962 V2 N6 P901-905 64-10096 <=*$>
1963 N4 P541-547 ICE V4 N3 P400-404 <ICE>
1963 N4 P558-564 ICE V4 N3 P386-390 <ICE>
1963 N4 P565-571 ICE V4 N3 P391-395 <ICE>
1963 V3 P206-216 64-10162 <=*$> 0
1963 V3 P811-812 64-14454 <=*$>
1963 V3 N1 P10-12 64-16689 <=*$>
1963 V3 N1 P13-19 64-14455 <=*$>
1963 V3 N1 P20-27 65-10233 <*>
1963 V3 N1 P74-81 65-10237 <*>
1963 V3 N2 P198-200 66-13255 <*>
1963 V3 N2 P222-226 AD-615 983 <=$>
1963 V3 N2 P238-245 63-18844 <=*>
1963 V3 N3 P352-359 64-14955 <=*$>
1963 V3 N3 P417-424 AD-630 970 <=$>
1963 V3 N3 P430-435 66-10004 <*>
1963 V3 N4 P488-493 64-14954 <=*$>
1963 V3 N4 P541-547 ICE V4 N3 P400-404 <ICE>
64-18435 <*>
65R74R <ATS>
1963 V3 N4 P558-564 ICE V4 N3 P386-390 <ICE>
67R74R <ATS>
1963 V3 N4 P565-571 ICE V4 N3 P391-395 <ICE>
1963 V3 N5 P635-641 65-14680 <*>
1963 V3 N5 P683-689 188 <TC>
1963 V3 N5 P713-718 64-18772 <*>
1963 V3 N5 P719-724 65-14650 <*>
1963 V3 N5 P792-798 AD-621 008 <=$>

	66-10616 <*>
1963 V3 N6 P845-849	65-12454 <*>
1963 V3 N6 P850-852	65-12633 <*>
1963 V3 N6 P864-870	65-12748 <*>
1963 V3 N6 P905-910	65-63746 <=$>
1964 V4 P170-175	AD-622 942 <=$>
1964 V4 P707-712	65-13303 <*>
1964 V4 N1 P82-90	65-12458 <*>
1964 V4 N1 P96-99	65-11495 <*>
1964 V4 N2 P280-285	66-11937 <*>
1964 V4 N2 P286-289	66-12765 <*>
1964 V4 N2 P345-350	AD-615 245 <=$>
1964 V4 N3 P386-390	65-11496 <*>
1964 V4 N3 P399-405	65-13311 <*>
	65-14679 <*>
1964 V4 N3 P421-425	65-13306 <*>
1964 V4 N3 P426-430	65-13304 <*>
1964 V4 N3 P452-457	04S86R <ATS>
	65-13305 <*>
1964 V4 N3 P458-465	ICE V4 N4 P645-649 <ICE>
1964 V4 N3 P487-493	AD-640 250 <=$>
1964 V4 N3 P501-506	66-14355 <*$>
1964 V4 N3 P510-517	AD-618 010 <=$>
1964 V4 N4 P535-539	65-13310 <*>
1964 V4 N4 P584-590	66-10733 <*>
1964 V4 N4 P593-598	66-10643 <*>
1964 V4 N5 P722-726	ICE V5 N2 P305-308 <ICE>
	59S85R <ATS>
1964 V4 N5 P772-776	66-13619 <*>
1964 V4 N6 P829-833	ICE V5 N2 P351-354 <ICE>
	65-13308 <*>
1964 V4 N6 P834-838	ICE V5 N2 P344-347 <ICE>
1964 V4 N6 P844-849	66-12206 <*>
1964 V4 N6 P850-853	65-17124 <*>
1964 V4 N6 P906-915	65-14756 <*>
1965 V5 N1 P10-16	30S86R <ATS>
1965 V5 N1 P24-32	29S86R <ATS>
	66-13999 <*>
1965 V5 N1 P62-67	27S86R <ATS>
1965 V5 N1 P68-75	32S86R <ATS>
1965 V5 N1 P76-81	23S86R <ATS>
1965 V5 N1 P82-89	24S86R <ATS>
1965 V5 N1 P101-107	72S84R <ATS>
1965 V5 N1 P111-117	25S86R <ATS>
	66-10651 <*>
1965 V5 N1 P118-125	26S86R <ATS>
	65-17129 <*>
1965 V5 N1 P126-131	66-10734 <*>
1965 V5 N1 P136-140	28S86R <ATS>
1965 V5 N1 P149-152	46S88R <ATS>
1965 V5 N1 P153-159	47S88R <ATS>
1965 V5 N1 P191-194	73S86R <ATS>
1965 V5 N2 P187-190	82S86R <ATS>
1965 V5 N2 P195-203	84S86R <ATS>
1965 V5 N2 P211-216	79S86R <ATS>
1965 V5 N3 P303-312	66-12883 <*>
1965 V5 N3 P313-319	66-12884 <*>
1965 V5 N3 P340-346	66-14496 <*$>
1965 V5 N3 P388-393	38S87R <ATS>
1965 V5 N3 P406-409	42S87R <ATS>
1965 V5 N3 P417-424	71S87R <ATS>
1965 V5 N3 P438-444	72S87R <ATS>
1965 V5 N4 P493-497	ICE V6 N2 P276-279 <ICE>
	54S89R <ATS>
	66-12218 <*>
1965 V5 N4 P498-500	53S89R <ATS>
1965 V5 N4 P501-506	50S89R <ATS>
1965 V5 N4 P507-511	51S89R <ATS>
1965 V5 N4 P528-535	66-12217 <*>
1965 V5 N4 P545-548	66-12226 <*>
1965 V5 N4 P549-553	55S89R <ATS>
1965 V5 N4 P573-578	52S89R <ATS>
1965 V5 N4 P583-588	48T92R <ATS>
1965 V5 N5 P760-761	66-14363 <*$>
1965 V5 N6 P820-824	66-14356 <*$>
1966 N1 P131-138	ICE V6 N4 P575-580 <ICE>
1966 V6 N1 P3-8	66-13023 <*>
1966 V6 N1 P13-21	94T94R <ATS>
1966 V6 N1 P53-57	97T94R <ATS>
1966 V6 N1 P71-74	98T94R <ATS>
1966 V6 N1 P80-84	99T94R <ATS>
1966 V6 N1 P85-89	88T94R <ATS>

NEFTEPERERABOTKA I NEFTEKHIMIYA

1963 N1 P25-27	66-13595 <*>
1963 N11 P16-18	AD-630 878 <=$>
1963 N11 P22-27	64-31801 <=>
1964 N2 P22-26	AD-635 877 <=$>
1964 N6 P29-33	65-14290 <*>
1965 N5 P32-34	ICE V5 N4 P669-671 <ICE>

NEFTEPROMYSLOVOE DELO

1965 N7 P14-17	21S89R <ATS>
1965 N10 P19-22	50T91R <ATS>
1965 N10 P22-26	51T91R <ATS>
1965 N12 P17-20	98T92R <ATS>
1966 N2 P11-14	82T93R <ATS>

NEFTYANAYA PROMYSHLENNOST SSSR

1940 V21 N2 P46-53	R-4910 <*>

NEFTYANIK

1958 N12 P15-18	59-16320 <+*>
1958 V3 N1 P22-26	52K25R <ATS>
1958 V3 N10 P14-17	59-22159 <+*>
1959 V4 N1 P18	60-17930 <+*>
1959 V4 N3 P4-8	46L34R <ATS>
1959 V4 N6 P6-7	23M47R <ATS>
1959 V4 N7 P29-31	60-13185 <+*> C
1959 V4 N8 P5-6	79M40R <ATS>
1959 V4 N10 P1-2	60-11305 <=>
1959 V4 N10 P25	17M42R <ATS>
1959 V4 N12 P27	AD-610 794 <=$>
1960 V5 N1 P8-10	17M45R <ATS>
1960 V5 N8 P8-9	11M47R <ATS>
1960 V5 N9 P8-9	82M47R <ATS>
1960 V5 N11 P17-18	17N50R <ATS>
1960 V5 N12 P21-22	91Q66R <ATS>
1961 V6 N1 P8-10	72N52R <ATS>
1961 V6 N4 P21-22	61-28388 <*=>
1961 V6 N4 P28-29	99N53R <ATS>
1961 V6 N8 P14-16	51N56R <ATS>
1961 V6 N9 P23-24	45N57R <ATS>
1961 V6 N11 P16-17	62-33002 <=*>
1961 V6 N11 P23	62P58R <ATS>
1963 V8 N8 P30-31	15Q74R <ATS>

NEFTYANOE KHOZYAISTVO

1928 V14 P328-358	63-10419 <=*>
1928 V15 N11/2 P674-678	63-10363 <=*>
1929 V16 N2 P223-230	63-10418 <=*>
1929 V16 N3 P357-361	63-10411 <*>
1929 V17 P520-529	63-10341 <*>
1929 V17 N3 P362-363	63-10403 <=*>
1929 V17 N11/2 SUP. P813-834	
	62-16408 <=*>
	63-10217 <=*>
1930 V18 N2 P260-261	63-14087 <=*>
1930 V18 N4 P641-646	RT-1572 <*>
	63-14088 <=*>
	64-18557 <*>
1932 V23 P242-250	R-3782 <*>
1933 V25 N6 P29-35	61-16688 <=*$>
	63-10364 <=*>
1933 V25 N10 P27-35	63-14086 <*=> C
1933 V25 N10 P41-45	63-10412 <=*>
1934 V26 N3 P42-43	63-10365 <=*>
1934 V26 N9 P13-16	R-4905 <*>
1934 V26 N10 P37-39	61-16960 <=*$>
1935 V27 N1 P66-67	61-18094 <=*$>
1935 V27 N1 P74-81	61-16993 <=*$>
1936 V17 N6 P48-54	61-18079 <=*$>
1936 V17 N10 P60-63	61-16604 <=*$> P
1936 V17 N10 P70-71	61-18021 <=*$>
1936 V17 N11 P28-34	61-16603 <=*$>
1936 V17 N12 P25-26	61-18092 <=*$>
1936 V17 N12 P27-29	61-16988 <=*$>
1937 V18 N10 P49-59	61-16605 <=*$> P
1938 V19 N2 P36-41	63-20777 <=*$>
1938 V19 N7 P26-30	RT-3454 <*>

Citation	Translation
1939 V20 N2 P36-41	65-13194 <*>
1940 V21 N1 P19-21	61-16824 <*=>
1940 V21 N1 P22-25	61-18019 <*=>
1940 V21 N1 P25-29	61-18018 <*=>
1940 V21 N3 P28-32	61-16869 <*=>
1940 V21 N4 P86-88	62-14110 <*> P
1940 V21 N4 P89-92	62-14111 <*> PO
1940 V21 N5 P75-80	62-14112 <*> P
1940 V21 N6 P67-69	62-14114 <=*>
1940 V21 N6 P70-75	62-14115 <=*>
1941 V22 N5 P93-99	82L28R <ATS>
1946 V24 N1 P60-62	61-20153 <*=>
1946 V24 N2 P39-43	61-20163 <*=>
1946 V24 N2 P52-55	RT-2646 <*>
1946 V24 N2 P56-64	61-20169 <*=>
1946 V24 N5 P44-51	61-20162 <*=>
1946 V24 N3/4 P36-39	62-10813 <*=>
1946 V24 N3/4 P39-44	62-10814 <*=>
1946 V24 N3/4 P44-53	61-20154 <*=>
1946 V24 N3/4 P53-55	62-10815 <*=> P
1946 V24 N3/4 P55-57	62-10816 <*=> P
1946 V24 N3/4 P68-70	61-20151 <*=>
1947 N5 P42-47	RT-1472 <*>
1947 N6 P49-59	RT-1385 <*>
	RT-2604 <*>
1947 N7 P8-12	61-17838 <=$>
1947 N8 P45-50	61-17818 <=$>
1947 V25 N1 P33-42	61-17862 <=$>
1947 V25 N1 P45-50	61-17881 <=$>
1947 V25 N2 P6-15	61-32011 <=$>
1947 V25 N2 P42-46	RT-3893 <*>
1947 V25 N2 P46-50	<ATS>
1947 V25 N4 P26-32	61-17901 <=$>
1947 V25 N5 P42-47	61-17908 <=$>
1947 V25 N6 P31-35	61-17882 <=$>
1947 V25 N6 P49-54	61-32023 <=$>
1947 V25 N6 P54-59	61-17909 <=$>
1947 V25 N7 P13-17	33E2R <ATS>
1947 V25 N8 P16-20	61-32014 <=$>
1947 V25 N9 P30-39	TT.112 <NRC>
1947 V25 N9 P40-46	61-17977 <=$>
1947 V25 N10 P24-28	61-32013 <=$>
1947 V25 N10 P36-45	RT-2644 <*>
1947 V25 N10 P45-50	<ATS>
	RT-2645 <*>
1947 V25 N12 P8-15	63-20796 <=*$>
1948 N2 P56-57	RT-268 <*>
1948 N3 P36-38	RT-270 <*>
1948 N3 P44-45	RT-269 <*>
1948 N10 P16-18	61-17856 <=$>
1948 N12 P25-32	RT-3918 <*>
1948 V26 N1 P9-17	61-17986 <=$>
1948 V26 N1 P24-29	61-17955 <=$>
1948 V26 N2 P24	61-17848 <=$>
1948 V26 N2 P53-55	RT-237 <*>
1948 V26 N3 P15-18	63-20799 <=*$>
1948 V26 N3 P46-48	61-17907 <=$>
	63-20800 <=*$>
1948 V26 N4 P47-51	R.521 <RIS>
1948 V26 N5 P43-49	63-20802 <=*$>
1948 V26 N5 P52-53	61-17869 <=$>
1948 V26 N6 P43-50	RJ-424 <*>
1948 V26 N7 P18-28	61-17974 <=$>
1948 V26 N7 P34-35	61-17860 <=$>
1948 V26 N10 P16-18	63-20801 <=*$>
1948 V26 N10 P45-47	85T94R <ATS>
1948 V26 N10 P48-52	61-32028 <=$>
1948 V26 N12 P5	61-17871 <=$>
1948 V26 N12 P25-32	61-17906 <=$>
1948 V26 N12 P33-36	61-17870 <=$>
1954 N3 P1-4	RT-2306 <*>
1954 N4 P42-45	RT-2494 <*>
1954 N5 P19-30	RT-2709 <*>
1954 N9 P1-5	RT-2624 <*>
1954 N12 P57-60	<TCT>
1954 N12 P60-63	<TCT>
1954 N12 P64-67	<TCT>
1954 N12 P67-68	<TCT>
1954 N12 P69-73	<TCT>
1954 N12 P73-76	<TCT>
1954 N12 P76-80	<TCT>
1954 N12 P80-83	<TCT>
1954 N12 P87-88	<TCT>
1954 V32 N1 P33-38	62-14582 <=*>
	83K22R <ATS>
1954 V32 N2 P27-32	RJ-581 <ATS>
	59-22415 <+*>
1954 V32 N2 P33-38	61-17362 <=$>
	62-14595 <=*>
1954 V32 N3 P23-	RJ-568 <ATS>
1954 V32 N3 P38-41	AD-615 247 <=$>
1954 V32 N4 P9-15	62-16670 <=*>
1954 V32 N5 P6-12	RJ-563 <ATS>
1954 V32 N5 P70-73	R-2102 <*>
	RJ-221 <ATS>
	64-10472 <=*$>
1954 V32 N5 P81-85	R-4979 <*>
1954 V32 N7 P16-19	RJ-248 <ATS>
1954 V32 N7 P20-23	RJ-247 <ATS>
1954 V32 N7 P20-32	RJ-247 <ATS>
1954 V32 N8 P12-16	<TCT>
1954 V32 N8 P17-18	<TCT>
1954 V32 N8 P19-25	<TCT>
1954 V32 N8 P26	<TCT>
1954 V32 N8 P27-31	<TCT>
1954 V32 N8 P31-32	<TCT>
1954 V32 N8 P32-36	<TCT>
1954 V32 N8 P37-40	<TCT>
1954 V32 N8 P41-51	<TCT>
1954 V32 N8 P52-55	<TCT>
	03R77R <ATS>
1954 V32 N8 P61-63	<TCT>
1954 V32 N8 P67-75	<TCT>
1954 V32 N8 P76-82	<TCT>
1954 V32 N8 P82-85	<TCT>
1954 V32 N9 P9-12	<TCT>
1954 V32 N9 P13-17	<TCT>
1954 V32 N9 P17-19	<TCT>
1954 V32 N9 P19-24	<TCT>
	62-10566 <*=>
1954 V32 N9 P25-33	<TCT>
1954 V32 N9 P33-38	<TCT>
	65L31R <ATS>
1954 V32 N9 P38-41	<TCT>
1954 V32 N9 P41-43	<TCT>
1954 V32 N9 P43-44	<TCT>
1954 V32 N9 P45-55	<TCT>
1954 V32 N9 P56-59	<TCT>
1954 V32 N9 P59-61	<TCT>
1954 V32 N9 P62-69	<TCT>
1954 V32 N9 P70-74	<TCT>
1954 V32 N9 P74-76	<TCT>
1954 V32 N9 P77-84	<TCT>
1954 V32 N10 P1-5	<TCT>
1954 V32 N10 P5-9	<TCT>
	RJ-565 <ATS>
1954 V32 N10 P10-14	<TCT>
1954 V32 N10 P18-19	<TCT>
1954 V32 N10 P20-30	<TCT>
1954 V32 N10 P30-34	<TCT>
1954 V32 N10 P34-39	<TCT>
1954 V32 N10 P53-57	<TCT>
1954 V32 N10 P57-61	<TCT>
1954 V32 N10 P62-69	<TCT>
	49R80R <ATS>
1954 V32 N10 P70-73	<TCT>
1954 V32 N10 P73-79	<TCT>
1954 V32 N10 P79-84	<TCT>
1954 V32 N10 P84-88	<TCT>
1954 V32 N10 P94-96	RJ-245 <ATS>
1954 V32 N11 P10-15	<TCT>
1954 V32 N11 P15-19	<TCT>
1954 V32 N11 P20-22	<TCT>
1954 V32 N11 P22-26	<TCT>
1954 V32 N11 P26-30	<TCT>
1954 V32 N11 P31-33	<TCT>
1954 V32 N11 P38-39	<TCT>
1954 V32 N11 P39-42	<TCT>
1954 V32 N11 P43-45	<TCT>
1954 V32 N11 P46-49	<TCT>

1954 V32 N11 P50-51 <TCT>
1954 V32 N11 P51-52 <TCT>
1954 V32 N11 P53-55 <TCT>
1954 V32 N11 P56-62 <TCT>
1954 V32 N11 P62-67 <TCT>
1954 V32 N11 P68-70 <TCT>
1954 V32 N11 P71-79 <TCT>
1954 V32 N11 P79-83 <TCT>
1954 V32 N12 P7-9 <TCT>
1954 V32 N12 P9-13 <TCT>
1954 V32 N12 P13-14 <TCT>
1954 V32 N12 P15-16 <TCT>
1954 V32 N12 P15-53 34500 <HB>
1954 V32 N12 P17-19 <TCT>
1954 V32 N12 P19-21 <TCT>
1954 V32 N12 P21-25 <TCT>
1954 V32 N12 P25-31 <TCT>
1954 V32 N12 P32-36 <TCT>
1954 V32 N12 P36-39 <TCT>
RJ-374 <ATS>
1954 V32 N12 P39-44 <TCT>
1954 V32 N12 P44-49 <TCT>
1954 V32 N12 P49-51 <TCT>
1954 V32 N12 P53-57 <TCT>
60-14301 <+*>
1955 N6 P72-78 R-5290 <*>
1955 N7 P55-60 R-356 <*>
1955 N7 P71-74 R-2335 <*>
1955 N11 P59-62 61-17560 <=$>
1955 V33 N1 P1-12 <TCT>
1955 V33 N1 P22-29 <TCT>
1955 V33 N1 P30-36 <TCT>
1955 V33 N1 P37 <TCT>
1955 V33 N1 P38-46 <TCT>
1955 V33 N1 P46-52 <TCT>
1955 V33 N1 P53-57 <TCT>
1955 V33 N1 P57-63 <TCT>
1955 V33 N1 P64-70 <TCT>
1955 V33 N1 P71-76 <TCT>
1955 V33 N1 P76-80 <TCT>
1955 V33 N1 P80-83 <TCT>
1955 V33 N1 P83-88 <TCT>
1955 V33 N1 P95 <TCT>
1955 V33 N2 P1-9 <TCT>
1955 V33 N2 P10-12 <TCT>
1955 V33 N2 P13-22 <TCT>
1955 V33 N2 P23-25 <TCT>
1955 V33 N2 P26-28 <TCT>
1955 V33 N2 P28-30 <TCT>
1955 V33 N2 P31-38 <TCT>
1955 V33 N2 P39-42 <TCT>
1955 V33 N2 P43-49 <TCT>
1955 V33 N2 P49-54 <TCT>
1955 V33 N2 P55-61 <TCT>
1955 V33 N2 P61-66 <TCT>
1955 V33 N2 P67-71 <TCT>
1955 V33 N2 P71-78 <TCT>
1955 V33 N2 P79-84 <TCT>
1955 V33 N2 P85-86 <TCT>
1955 V33 N4 P39-43 62-14607 <=*>
66-14228 <*$>
1955 V33 N5 P1-12 <TCT>
1955 V33 N5 P12-15 <TCT>
1955 V33 N5 P17-20 <TCT>
1955 V33 N5 P20-23 <TCT>
1955 V33 N5 P23-28 <TCT>
1955 V33 N5 P28-32 <TCT>
1955 V33 N5 P33-36 <TCT>
1955 V33 N5 P37-41 <TCT>
1955 V33 N5 P41-44 <TCT>
1955 V33 N5 P45-48 <TCT>
1955 V33 N5 P48 <TCT>
1955 V33 N5 P49-57 <TCT>
1955 V33 N5 P58-62 <TCT>
1955 V33 N5 P63-69 <TCT>
62-25901 <=*>
1955 V33 N5 P69-77 <TCT>
1955 V33 N5 P78-85 <TCT>
1955 V33 N5 P85-87 <TCT>
1955 V33 N5 P91-93 <TCT>
1955 V33 N6 P1-4 <TCT>
1955 V33 N6 P5-7 <TCT>
1955 V33 N6 P7-9 <TCT>
1955 V33 N6 P10-13 <TCT>
1955 V33 N6 P13-14 <TCT>
1955 V33 N6 P15-20 <TCT>
1955 V33 N6 P21 <TCT>
1955 V33 N6 P21-23 <TCT>
1955 V33 N6 P24-27 <TCT>
1955 V33 N6 P31-32 <TCT>
1955 V33 N6 P32-35 <TCT>
1955 V33 N6 P35-42 <TCT>
1955 V33 N6 P42-50 <TCT>
1955 V33 N6 P51-61 <TCT>
1955 V33 N6 P61-63 <TCT>
1955 V33 N6 P63-72 <TCT>
1955 V33 N6 P72-78 <TCT>
1955 V33 N6 P78-79 <TCT>
RJ892 <ATS>
1955 V33 N6 P79-83 <TCT>
1955 V33 N6 P83-88 <TCT>
1955 V33 N6 P89 <TCT>
1955 V33 N6 P90 <TCT>
1955 V33 N6 P91-92 <TCT>
1955 V33 N7 P36-42 61-17602 <=$>
1955 V33 N7 P42-50 61-17383 <=$>
1955 V33 N7 P43-46 61-17347 <=$>
1955 V33 N7 P55-60 RJ-370 <ATS>
1955 V33 N9 P1-6 <TCT>
1955 V33 N9 P7-11 <TCT>
1955 V33 N9 P12-13 <TCT>
1955 V33 N9 P14-19 <TCT>
1955 V33 N9 P20-28 <TCT>
RJ-375 <ATS>
1955 V33 N9 P29-33 <TCT>
1955 V33 N9 P33-35 <TCT>
1955 V33 N9 P35-39 <TCT>
1955 V33 N9 P40-47 <TCT>
1955 V33 N9 P47-50 <TCT>
1955 V33 N9 P50-55 <TCT>
1955 V33 N9 P56-59 <TCT>
1955 V33 N9 P59-62 <TCT>
1955 V33 N9 P63-66 <TCT>
1955 V33 N9 P66-69 <TCT>
1955 V33 N9 P70-73 <TCT>
1955 V33 N9 P74-78 <TCT>
1955 V33 N9 P78-82 <TCT>
1955 V33 N9 P82-84 <TCT>
1955 V33 N9 P84-86 <TCT>
1955 V33 N9 P86-87 <TCT>
1955 V33 N9 P88-92 <TCT>
1955 V33 N10 P1-6 <TCT>
1955 V33 N10 P7-12 <TCT>
1955 V33 N10 P13-15 <TCT>
1955 V33 N10 P15-23 <TCT>
1955 V33 N10 P23-27 <TCT>
1955 V33 N10 P28-31 <TCT>
1955 V33 N10 P32-36 <TCT>
1955 V33 N10 P36-40 <TCT>
1955 V33 N10 P41-45 <TCT>
1955 V33 N10 P45-52 <TCT>
1955 V33 N10 P52-60 <TCT>
1955 V33 N10 P60-71 <TCT>
1955 V33 N10 P71-75 <TCT>
RT-4595 <*>
1955 V33 N10 P75-82 <TCT>
1955 V33 N10 P83-88 <TCT>
1955 V34 N5 P59-63 R-4783 <*>
1956 N4 P15-22 59-14186 <+*>
1956 N5 P59-63 61-17603 <=$>
1956 V34 N5 P20-23 RJ-564 <ATS>
1956 V34 N5 P31-36 RJ-567 <ATS>
1956 V34 N8 P49-53 64-16814 <=*$>
1956 V34 N8 P54-57 29J14R <ATS>
1956 V34 N9 P22-24 61-17380 <=$>
1956 V34 N11 P14-19 62-16719 <=*>
1956 V34 N12 P13-17 13J15R <ATS>
1957 N3 P30-36 61-17570 <=$>
1957 N7 P31-35 61-17361 <=$>
1957 N11 P51-54 61-17355 <=$>

Citation	Reference
1957 V35 N1 P24-30	64-16818 <*>
1957 V35 N1 P44-53	68L32R <ATS>
	719TM <CTT>
1957 V35 N2 P11-14	62-14558 <=*>
1957 V35 N2 P35-40	17J19R <ATS>
1957 V35 N5 P30-33	99J18R <ATS>
1957 V35 N5 P37-41	97J16R <ATS>
1957 V35 N7 P1-4	15L29R <ATS>
1957 V35 N7 P4-8	60-17142 <+*>
1957 V35 N7 P9-13	16K20R <ATS>
1957 V35 N7 P31-35	64-18507 <*>
1957 V35 N7 P35-38	85L31R <ATS>
1957 V35 N7 P45-46	64-18506 <*>
1957 V35 N8 P13-19	91J18R <ATS>
1957 V35 N8 P35-38	92J18R <ATS>
1957 V35 N9 P14-18	14K20R <ATS>
1957 V35 N9 P18-24	15K20R <ATS>
1957 V35 N9 P37-40	60-15599 <+*>
1957 V35 N9 P41-44	57K21R <ATS>
1957 V35 N10 P5-10	62-16658 <=*>
	64-18152 <*>
	85J19R <ATS>
1957 V35 N10 P11-16	45L36R <ATS>
1957 V35 N10 P17-20	17K20R <ATS>
1957 V35 N11 P55-58	52K22R <ATS>
1957 V35 N12 P42-51	36K26R <ATS>
1957 V35 N12 P51-55	37K26R <ATS>
1958 V36 N1 P9-13	31K26R <ATS>
1958 V36 N2 P7-13	38K26R <ATS>
	62-14579 <=*>
1958 V36 N2 P13-18	62-16439 <=*>
1958 V36 N2 P19-22	32K26R <ATS>
1958 V36 N3 P12-15	51K23R <ATS>
1958 V36 N3 P15-24	39K262 <ATS>
1958 V36 N3 P28-33	40K26R <ATS>
1958 V36 N3 P33-35	41K26R <ATS>
1958 V36 N3 P42-45	74L29R <ATS>
1958 V36 N4 P20-26	42K26R <ATS>
	61-15369 <+*>
1958 V36 N4 P31-32	43K26R <ATS>
1958 V36 N4 P44-50	35K26R <ATS>
1958 V36 N5 P19-22	44K26R <ATS>
1958 V36 N5 P28-37	97L30R <ATS>
1958 V36 N6 P32-36	91L30R <ATS>
1958 V36 N6 P36-39	12K27R <ATS>
1958 V36 N7 P6-13	38K27R <ATS>
1958 V36 N7 P13-16	39K27R <ATS>
1958 V36 N7 P52-55	90L30R <ATS>
1958 V36 N7 P55-57	17L30R <ATS>
1958 V36 N8 P38-46	35K27R <ATS>
	62-16659 <=*>
1958 V36 N8 P46-53	36K27R <ATS>
1958 V36 N9 P14-17	77L28R <ATS>
1958 V36 N9 P40-47	62-16660 <=*>
1958 V36 N9 P56-59	78L28R <ATS>
1958 V36 N10 P39-43	04L33R <ATS>
1958 V36 N10 P68-69	12M41R <ATS>
1958 V36 N11 P17-20	28M39R <ATS>
1958 V36 N11 P21-28	67L29R <ATS>
1958 V36 N11 P28-30	34L30R <ATS>
1958 V36 N11 P42-48	62-18533 <*>
1958 V36 N12 P26-28	45M38R <ATS>
1958 V36 N12 P33-35	47L35R <ATS>
1958 V36 N12 P40-45	48L35R <ATS>
1959 N4 P35-41	61-17575 <=$>
1959 V37 N1 P36-43	71L34R <ATS>
1959 V37 N1 P49-54	20L31R <ATS>
1959 V37 N1 P54-59	90M44R <ATS>
1959 V37 N1 P55-57	21L31R <ATS>
1959 V37 N2 P38-43	13L32R <ATS>
1959 V37 N3 P25-32	40L33R <ATS>
1959 V37 N3 P38-40	39L33R <ATS>
1959 V37 N3 P56-58	65L34R <ATS>
1959 V37 N4 P24-31	69L34R <ATS>
1959 V37 N4 P63-67	30L34R <ATS>
1959 V37 N4 P67-70	29L34R <ATS>
1959 V37 N5 P27-32	84L34R <ATS>
1959 V37 N6 P26-32	70L36R <ATS>
1959 V37 N6 P49-50	50L36R <ATS>
1959 V37 N6 P65-67	20L37R <ATS>
1959 V37 N7 P32-35	60-16950 <*+>
1959 V37 N8 P25-28	63R78R <ATS>
1959 V37 N9 P12-15	78M40R <ATS>
1959 V37 N9 P16-19	50M39R <ATS>
1959 V37 N9 P24-29	49M39R <ATS>
1959 V37 N10 P50-52	73M39R <ATS>
1959 V37 N11 P44-48	89M44R <ATS>
1959 V37 N12 P7-12	91M44R <ATS>
1959 V37 N12 P21-24	92M44R <ATS>
1959 V37 N12 P30-36	46M39R <ATS>
1960 N1 P36-38	NLLTB V2 N8 P679 <NLL>
1960 V38 N1 P38-43	48M39R <ATS>
1960 V38 N1 P56-59	45M39R <ATS>
1960 V38 N2 P41-49	46M43R <ATS>
1960 V38 N2 P49-54	75M44R <ATS>
1960 V38 N2 P55-59	27M43R <ATS>
1960 V38 N4 P14-20	63-10212 <=*>
	64-10108 <=*$>
1960 V38 N5 P39-43	77M45R <ATS>
1960 V38 N5 P43-48	78M45R <ATS>
1960 V38 N7 P24-28	15N52R <ATS>
1960 V38 N8 P9-13	12M47R <ATS>
1960 V38 N9 P18-21	42M47R <ATS>
1960 V38 N9 P24-25	97M46R <ATS>
1960 V38 N9 P32-36	40M47R <ATS>
1960 V38 N9 P36-39	41M47R <ATS>
1960 V38 N10 P17-20	15N49R <ATS>
1960 V38 N10 P20-25	29N49R <ATS>
1960 V38 N10 P28-30	14N50R <ATS>
1960 V38 N10 P48-52	46N49R <ATS>
1960 V38 N11 P30-33	49N50R <ATS>
1961 N5 P7-12	NLLTB V3 N12 P981 <NLL>
1961 V39 N1 P34-39	91Q72R <ATS>
1961 V39 N2 P1-5	62-13413 <*=>
1961 V39 N2 P5-9	62-13310 <*=>
1961 V39 N2 P67-68	62-13384 <*=>
1961 V39 N3 P1-6	62-13323 <*=>
1961 V39 N3 P16-23	62-13408 <*=>
1961 V39 N3 P35-40	33N55R <ATS>
1961 V39 N3 P43-46	62-13319 <*=>
1961 V39 N3 P51-54	80N53R <ATS>
1961 V39 N4 P1-6	62-13436 <*=>
1961 V39 N4 P33-37	53N53R <ATS>
	62-20282 <=*>
1961 V39 N4 P60-65	62-13425 <*=>
1961 V39 N5 P7-12	61-28785 <*=>
1961 V39 N6 P33-38	11N56R <ATS>
1961 V39 N6 P57-61	12N56R <ATS>
1961 V39 N8 P36-42	51P58R <ATS>
1961 V39 N9 P62-64	62-15350 <*=>
1961 V39 N11 P18-25	13P60R <ATS>
1961 V39 N11 P41-46	12P60R <ATS>
1961 V39 N11 P50-55	63-16986 <=*>
1961 V39 N12 P24-29	88S82R <ATS>
1961 V39 N12 P33-38	17Q73R <ATS>
1962 V40 N1 P1-6	63-26591 <=$>
1962 V40 N1 P40-44	63-20730 <=*$>
1962 V40 N3 P44-46	68R77R <ATS>
1962 V40 N4 P48-52	23P64R <ATS>
1962 V40 N4 P52-55	63-20705 <=*$>
1962 V40 N5 P49-50	55P63R <ATS>
1962 V40 N5 P62-63	54P63R <ATS>
1962 V40 N6 P43-47	07P65R <ATS>
	64-16417 <=*$>
1962 V40 N6 P47-50	64-16418 <=*$>
1962 V40 N8 P8-13	51P65R <ATS>
1962 V40 N8 P37-42	65-11094 <*>
1962 V40 N8 P42-49	21P66R <ATS>
1962 V40 N9 P51-56	06Q67R <ATS>
1962 V40 N10 P18-22	99Q66R <ATS>
1962 V40 N10 P43-48	12Q67R <ATS>
1962 V40 N10 P49-54	63-20706 <=*$>
1962 V40 N11 P47-52	06S83R <ATS>
1962 V40 N11 P56-60	53Q69R <ATS>
1963 V41 N1 P19-23	72Q70R <ATS>
1963 V41 N3 P8-12	49Q70R <ATS>
1963 V41 N3 P31-35	86Q70R <ATS>
1963 V41 N4 P14-19	46Q71R <ATS>
1963 V41 N4 P64-67	51R77R <ATS>
1963 V41 N6 P31-35	65-28105 <$>

1963 V41 N7 P13-17 63-27419 <=$>
1963 V41 N7 P51-56 TRC-TRANS-1104 <NLL>
1963 V41 N7 P69-70 49R77R <ATS>
1963 V41 N8 P1-5 63-31808 <=>
1963 V41 N9 P13-16 33Q74R <ATS>
1963 V41 N9 P68-70 16Q74R <ATS>
1963 V41 N10 P22-26 89R74R <ATS>
1963 V41 N10 P45-50 29R75R <ATS>
1963 V41 N11 P43-49 38R75R <ATS>
1963 V41 N12 P12-17 37R75R <ATS>
1964 N1 P50-57 ICE V5 N2 P318-322 <ICE>
1964 V42 N1 P29-31 50R76R <ATS>
1964 V42 N1 P37-42 11R76R <ATS>
1964 V42 N1 P50-57 80R79R <ATS>
1964 V42 N1 P57-63 81R79R <ATS>
1964 V42 N3 P35-38 14T93R <ATS>
1964 V42 N5 P7-11 59R79R <ATS>
1964 V42 N5 P44-49 40R78R <ATS>
1964 V42 N5 P49-54 08R80R <ATS>
1964 V42 N5 P58-59 43S81R <ATS>
1964 V42 N6 P8-11 26S81R <ATS>
1964 V42 N6 P12-17 27S81R <ATS>
1964 V42 N7 P5-9 45S81R <ATS>
1964 V42 N8 P51-55 RTS-2926 <NLL>
1964 V42 N11 P6-9 63S82R <ATS>
1964 V42 N11 P16-19 79S83R <ATS>
1964 V42 N11 P31-34 45S82R <ATS>
1964 V42 N12 P14-16 42S83R <ATS>
1964 V42 N12 P16-19 43S83R <ATS>
1964 V42 N12 P47-51 30S83R <ATS>
1964 V42 N9/10 P107-114 65-30035 <=$>
1964 V42 N9/10 P118-123 65-30035 <=$>
1965 N12 P62-65 67-31485 <=$>
1965 V43 N3 P22-27 06S86R <ATS>
1965 V43 N3 P52-57 03S85R <ATS>
1965 V43 N4 P36-39 02S89R <ATS>
1965 V43 N7 P25-30 25T92R <ATS>
1965 V43 N7 P33-36 19S89R <ATS>
1965 V43 N7 P56-59 73S88R <ATS>
1965 V43 N8 P26-28 73S89R <ATS>
1965 V43 N8 P43-46 72S89R <ATS>
1965 V43 N9 P28-31 43T90R <ATS>
1965 V43 N10 P31-36 24T91R <ATS>
1965 V43 N10 P49-53 99T91R <ATS>
1965 V43 N10 P54-57 03T91R <ATS>
1965 V43 N11 P41-46 77T93R <ATS>
1965 V43 N12 P58-62 83T93R <ATS>
1966 V44 N1 P47-52 19T94R <ATS>
1966 V44 N1 P52-54 76T93R <ATS>
1966 V44 N1 P55-56 55T93R <ATS>

NEFTYANOE I SLANTSEVOE KHOZYAISTVO
1923 V5 P636-646 R-4906 <*>
1925 N4 P640-646 61-17277 <=$>

NEHEZVEGYIPAII KUTATO INTEZET KOZLEMENYII UP SOROZAT
1963 N3/4 P187-193 ICE V4 N4 P625-628 <ICE>
1963 V3 N3/4 P187-193 ICE V4 N4 P625-628 <ICE>

NEMATOLOGICA
1957 V2 P424-433 SUP 66-10243 <*> P
1959 V4 P172-186 62-16713 <*> 0

NENRYC KYOKAI-SHI
1953 V32 P410-413 92L33J <ATS>
1953 V32 P414-419 93L33J <ATS>
1953 V32 N316 P447-460 58-2086 <*>
59-14283 <+*>
1953 V32 N319 P610-630 59-16355 <+*>
1954 V33 P134-143 59-15690 <*>
1954 V33 N325 P249-254 T-2122 <INSD>
1955 V34 P645-653 T-1498 <INSD>
1956 V35 P141-156 59-16358 <+*>
62-14565 <*>
1956 V35 P411-427 T1892 <INSD>
1956 V35 P647-649 T-2222 <INSD>
1956 V35 N353 P518-525 NEK-35-353-1 <SA>
1959 V38 N6 P384-387 60-17103 <*+>
1959 V38 N386 P361-373 63-16219 <*> 0

1960 V39 N402 P698-704 61-00786 <*>
1962 V41 N422 P539-544 NEK-41-422-1 <SA>
66-13266 <*>
1962 V41 N423 P615-623 NEK-41-423-1 <SA>
66-13267 <*>
1962 V41 N424 P706-711 NEK-41-424-1 <SA>
66-13268 <*>
1962 V41 N426 P860-863 NEK-41-426-1 <SA>
66-13269 <*>
1963 N1 P2-10 63-21512 <=>
1963 N1 P15 63-21512 <=>
1963 V42 P52-58 65-14760 <*>
1963 V42 N429 P52-58 66-13270 <*>
1963 V42 N431 P156-161 66-14220 <*$>
1964 V43 P118-124 66-12768 <*>
1965 N463 P744-750 ICE V6 N3 P540-546 <ICE>
1965 V44 N463 P744-750 74T92J <ATS>

NEOPLASIE
1950 V4 P303-310 57-828 <*>

NEOPLASMA
1961 V8 N1 P27-29 11092-C <K-H>
1961 V8 N1 P27-38 62-19330 <=*>
1961 V8 N1 P27-39 64-14514 <*> 0
1961 V8 N3 P315-321 62-19750 <=*>
1961 V8 N3 P323-329 62-19668 <=*>
1961 V8 N3 P331-335 62-19749 <=*>
1961 V8 N5 P509-522 62-01397 <*>
1962 V9 N5 P507-516 63-00850 <*>

NEOPLASMES. PARIS
1933 V12 P14-20 I-354 <*>

NEPEGESZSEGUGY
1949 V30 N8 P229-230 HJ-668 <ATS>
1950 V31 N3 P43 PANSDOC-TR.313 <PANS>
1954 V35 N7 P182-187 3384-A <KH>
1960 V41 N4 P87-90 62-19332 <*=>
1960 V41 N4 P102-108 62-19333 <*=>
1960 V41 N4 P108-111 62-19321 <*=>
1960 V41 N5 P118-131 60-31789 <*=>
1960 V41 N5 P140-142 60-31789 <*=>
1960 V41 N10 P338 62-19329 <=*>
1961 V42 N7 P193-214 62-19662 <=*>
1961 V42 N7 P220-221 62-19662 <=*>
1961 V42 N8 P225-253 62-19557 <=*>
1962 V43 N3 P65-68 62-25521 <=*>
1962 V43 N4 P97-101 62-25497 <=*>
1963 V44 N2 P58-60 13134A <KH>

NEPSZABADSAZ
1960 V18 N294 P11 62-19322 <*=>
1961 V19 N21 P2 62-19339 <*=>

NEPSYERU TECHNIKA
1961 V10 N7 P204-205 62-19680 <=*> P
1961 V10 N7 P213-215 62-19681 <=*> P

NERVENARZT
1928 V1 N1 P265-275 1077 <*>
1936 V4 P173-175 61S86G <ATS>
1947 V18 N11 P505-511 62-01213 <*>
1949 V20 P490-497 57-229 <*>
1954 V25 N1 P26-30 2387 <*>
1955 V26 P507-510 1716 <*>
1956 V27 P33- 1096 <*>
1956 V27 N5 P225-226 61-00631 <*>
1957 V28 P56-60 57-2888 <*>
1957 V28 N3 P122 58-226 <*>
1958 V29 P268-269 62-16902 <*>
1958 V29 P366-367 62-16610 <*>
1958 V29 N11 P520-522 59-15462 <*>
1959 V30 P516-518 62-18175 <*>
1959 V30 N7 P305-309 61-00163 <*>
1960 V31 N10 P471-472 62-20029 <*>

NERVNAYA SISTEMA
1960 N2 P32-36 62-20311 <=*>
1960 N2 P44-52 62-20303 <=*> C

1962 N3 P12-16 64-18754 <*> 0
1962 N3 P17-20 64-18750 <*>
1962 N3 P21-27 64-18751 <*>
1962 N3 P177-194 63-31522 <=>
1964 N4 P3-13 64-51234 <=$>
1964 N4 P146-148 64-51234 <=$>
1964 N4 P173-186 64-51234 <=$>
1964 N5 P105-122 65-30108 <=$>
1964 N5 P151-160 65-30108 <=$>

NETHERLANDS, STANDARDS
1957 NEN-1076 61-18779 <*>

NEUE BLAETTER FUER TAUBSTUMMENBILDUNG
1951 V5 N4/5 P151-158 64-10034 <*>

NEUE GIESSEREI
1949 V36 N4 P99-103 2299 <*>
1950 V37 N8 P145-148 I-614 <*>

NEUE HUETTE
1955 V1 N1 P6-10 62-14390 <*>
1955 V1 N1 P39-45 65-10214 <*>
1956 N6 P369 GB70 4212 <NLL>
1956 V1 N4 P235-237 3949 <HB>
1956 V1 N5 P303-307 6690 <HB>
1956 V1 N9 P561-562 3853 <HB>
1957 V2 P525-537 61-18412 <*>
1957 V2 N1 P35-39 58-138 <*>
1957 V2 N5 P280-285 4236 <HB>
1957 V2 N5 P289-299 4135 <HB>
1957 V2 N7 P404-409 61-18410 <*>
1957 V2 N8 P487-497 61-16397 <*>
1957 V2 N8 P497-502 4167 <HB>
1957 V2 N10 P601-604 4093 <HB>
1957 V2 N10 P6C5-606 4625 <HB>
1957 V2 N10 P621-626 4082 <HB>
1957 V2 N11 P671-678 4085 <HB>
1957 V2 N11 P692-702 63-15365 <=*>
1957 V2 N12 P764-767 62-16032 <*>
1957 V2 N2/3 P92-102 61-16398 <*>
1957 V2 N2/3 P103-111 1696 <BISI>
1957 V2 N2/3 P142-154 61-16396 <*>
1957 V2 N2/3 P157-168 61-16394 <*>
1958 P494-497 TS-1259 <BISI>
1958 V3 P85-93 TS-1164 <BISI>
1958 V3 P425-432 TS-1291 <BISI>
1958 V3 P475-482 TS 1104, PT.3 <BISI>
1958 V3 P608-615 TS-1487 <BISI>
1958 V3 P740-746 TS-1224 <BISI>
1958 V3 N1 P37-43 126 <TC>
1958 V3 N4 P225-232 4498 <HB>
1958 V3 N4 P233-235 4497 <HB>
1958 V3 N5 P300-302 61-16170 <*> 0
1958 V3 N6 P341-350 61-18435 <*>
1958 V3 N8 P462-474 TS 1163 <BISI>
1958 V3 N9 P523-532 59-11359 <=>
1958 V3 N9 P543-548 59-11360 <=*>
1958 V3 N10 P594-602 4851 <HB>
1958 V3 N10 P608-615 62-14445 <*>
1959 P244-246 TS-1334 <BISI>
1959 V4 P304-311 TS 1712 <BISI>
1959 V4 P352 TS-1393 <BISI>
1959 V4 P596-608 TS 1553 <BISI>
1959 V4 P725-730 TS-1727 <BISI>
1959 V4 P733-738 TS-1728 <BISI>
1959 V4 N2 P110-113 61-10654 <*> 0
66-11674 <*> 0
1959 V4 N2 P113-115 1571 <BISI>
1959 V4 N4 P218-231 TRANS-37 <MT>
1959 V4 N4 P250-252 4620 <HB>
1959 V4 N5 P304-311 62-16888 <*>
1959 V4 N6 P343-351 TS-1446 <BISI>
62-14337 <*>
1959 V4 N7 P417-419 66-10071 <*> 0
1959 V4 N7 P425-428 61-16183 <*> 0
1959 V4 N8 P493-501 61-16169 <*> 0
1959 V4 N10 P596-608 62-16891 <*>
1959 V4 N11 P663-668 3957 <BISI>
1959 V4 N12 P707-715 4791 <HB>

1960 V5 P63-69 TS 1705 <BISI>
1960 V5 P78-89 TS-1840 <BISI>
1960 V5 P224-232 TS 1759 <BISI>
1960 V5 P266-278 2163 <BISI>
1960 V5 P488-490 1980 <BISI>
1960 V5 N1 P31-36 1809 <BISI>
1960 V5 N2 P63-69 62-16869 <*>
1960 V5 N2 P99-105 62-00111 <*>
1960 V5 N3 P123-127 60-41662 <=*>
1960 V5 N3 P143-148 4831 <HB>
1960 V5 N4 P218-232 62-16867 <*>
1960 V5 N4 P238-244 1771 <BISI>
1960 V5 N5 P278-286 63-10294 <*> 0
1960 V5 N6 P348-354 1991 <BISI>
1960 V5 N7 P383-398 <MT>
1960 V5 N7 P420-425 1991 <BISI>
1960 V5 N10 P6C7-617 2057 <BISI>
1960 V5 N12 P720-729 146-296 <STT>
2669 <BISI>
1960 V5 N12 P730-733 152-297 <STT>
1960 V5 N12 P743-752 2056 <BISI>
1961 V6 P65-71 2324 <BISI>
1961 V6 P157-166 2616 <BISI>
1961 V6 P284-291 2304 <BISI>
1961 V6 P316-321 2590 <BISI>
1961 V6 P349-358 2479 <BISI>
1961 V6 P399-410 2410 <BISI>
1961 V6 P455-462 2481 <BISI>
1961 V6 P516-524 2607 <BISI>
1961 V6 P565-572 2608 <BISI>
1961 V6 N1 P17-22 5142 <HB>
1961 V6 N1 P56-60 2746 <BISI>
1961 V6 N2 P78-81 3048 <BISI>
1961 V6 N2 P81-87 2670 <BISI>
1961 V6 N2 P87-91 2671 <BISI>
65-17157 <*> 0
1961 V6 N2 P99-106 5178 <HB>
1961 V6 N2 P122-123 62-24555 <=*>
1961 V6 N3 P131-138 5143 <HB>
1961 V6 N3 P139-146 5144 <HB>
61-20760 <*>
1961 V6 N3 P151-156 2820 <BISI>
1961 V6 N3 P181-182 5188 <HB>
1961 V6 N5 P301-309 2589 <BISI>
1961 V6 N6 P333-342 5254 <HB>
1961 V6 N8 P500-506 2711 <BISI>
1961 V6 N9 P553-557 2708 <BISI>
1961 V6 N9 P588-591 2709 <BISI>
1961 V6 N12 P767-771 62-18695 <*>
1961 V6 N12 P795-796 2745 <BISI>
1962 V7 N1 P41-42 5610 <HB>
1962 V7 N1 P42-45 5611 <HB>
1962 V7 N2 P94-102 2836 <BISI>
1962 V7 N2 P111-123 2788 <BISI>
1962 V7 N3 P154-158 65-17361 <*>
1962 V7 N3 P164-166 3078 <BISI>
1962 V7 N3 P167-174 2918 <BISI>
1962 V7 N3 P180-181 5654 <HB>
1962 V7 N4 P206-213 2998 <BISI>
1962 V7 N5 P283-285 5667 <HB>
64-16112 <*> 0
1962 V7 N5 P299-306 5669 <HB>
1962 V7 N7 P395-402 3093 <BISI>
1962 V7 N7 P408-411 2949 <BISI>
1962 V7 N7 P411-416 3386 <HB>
1962 V7 N8 P504-505 5876 <HB>
1962 V7 N10 P623-628 3071 <BISI>
1962 V7 N11 P651-658 3317 <BISI>
1962 V7 N11 P658-664 3250 <BISI>
1962 V7 N11 P665-670 5829 <HB>
1962 V7 N11 P671-673 3885 <HB>
1962 V7 N11 P688-690 2947 <BISI>
1962 V7 N11 P695-698 5830 <HB>
1962 V7 N11 P698-699 5831 <HB>
1962 V7 N12 P729-736 5883 <HB>
1963 V8 P282-287 3768 <BISI>
1963 V8 P756-759 4492 <BISI>
1963 V8 N1 P2-5 3239 <BISI>
1963 V8 N1 P31-35 3267 <BISI>
1963 V8 N1 P35-39 3204 <BISI>

1964 V3 N1 P57-61 64-31454 <=>

NEVROPATOLOGIYA I PSIKHIATRIYA
1947 V16 N1 P49-57 RT-159 <*>
1947 V16 N2 P46-50 5K22R <ATS>
1948 V17 N5 P74-75 RT-4376 <*>
1949 V18 N2 P38-42 RT-4377 <*>
1949 V18 N2 P62-63 RT-4378 <*>
1949 V18 N4 P54-57 RT-3816 <*>
1950 V19 N3 P61-62 RT-2114 <*>
1950 V19 N4 P85-86 RT-1845 <*>
1951 V20 N2 P36-46 RT-161 <*>
1951 V20 N2 P56-60 RT-160 <*>

NEW CHINA SEMI-MONTHLY
SEE HSIN HUA PAN YUEH KAN

NEW CLINICAL MEDICINE. JAPAN
SEE SHIN RINSHO

NEW NIPPON ELECTRIC TECHNICAL REVIEW
SEE SHIN NIHON DENKI GIHO

NEW REMEDIES AND CLINIC. JAPAN
SEE SHINYAKU TO RINSHO

NICKEL-BERICHTE
1960 V18 N9/10 P291-296 63-18000 <*>

NIHON
SEE ALSO NIPPON

NIHON DENSENBYO GAKKAI ZASSHI
1958 V32 P756-757 66-13174 <*>
1958 V32 P947 66-13175 <*>
1958 V32 N6 P146-149 63-18032 <*>
1959 V33 P264-265 66-13201 <*>
1963 V37 P264-272 65-17019 <*>
1963 V37 P347-351 65-14358 <*>
1964 V38 P21-25 65-14359 <*>
1964 V38 P26-32 65-14670 <*>
1964 V38 P144-151 66-13042 <*>

NIHON GAKUSHIIN
1929 V5 N7 P284-286 C-3522 <NRC>
61-00067 <*>
1935 V11 P138-140 UCRL TRANS-847(L) <*>
1935 V11 P413-415 UCRL TRANS-848(L) <*>
1941 V17 P75-77 UCRL TRANS-849(L) <*>
1951 V27 N8 P493-500 00D1J <ATS>

NIHON GENSHIRYOKU KENKYUSHO CHOSA HOKOKU
SEE REPORTS OF THE JAPAN ATOMIC ENERGY RESEARCH INSTITUTE. TOKYO

NIHON GENSHIRYOKU KENKYUSHO KENKYU HOKOKU
SEE REPORTS OF THE JAPAN ATOMIC ENERGY RESEARCH INSTITUTE. TOKYO

NIHON GENSHIRYOKU KENKYUSHO NENPO
SEE REPORTS OF THE JAPAN ATOMIC ENERGY RESEARCH INSTITUTE. TOKYO

NIHON HIFUKA GAKKAI ZASSHI
1960 V70 N8 P815-834 61-00390 <*>

NIHON HIHAKAI KENSA KYOKAI SHI
1958 V7 N1 P30-31 59-17095 <*> O
1958 V7 N2 P79-85 59-17113 <*> O

NIHON HOSHASEN KOBUNSHI KENKYU KYOKAI NEMPO
1958 V1 P1-354 AEC-TR-6231 <=>
1963 V5 ENTIRE ISSUE AEC-TR-6565 <=$>

NIHON IGAKU HOSHASEN GAKKAI ZASSHI
1962 V22 N3 P199-203 63-01125 <*>

NIHON IJI SHINPO
1951 N1435 P17-19 63-18038 <*>
1957 N1710 P27-34 61-00205 <*>
1958 N1784 P39 61-00356 <*>

NIHON IKA DAIGAKU ZASSHI
1959 V26 N6 P513-528 63-00069 <*>

NIHON INSASTSU GAKKAI ROMBUNSHU
1962 V5 N9 P11-17 94Q72J <ATS>
1963 V6 N11 P74-78 66Q72J <ATS>

NIHON JIBI INKOKA GAKKAI KAIHO
1960 V63 N1 P109-121 62-14818 <*>
1962 V65 P662-671 65-12760 <*>
1962 V65 P998-1011 65-00340 <*> O
1962 V65 N5 P662-671 C-5115 <NRL>

NIHON JUI GAKKAI ZASSHI
1932 V9 P302-316 63-14853 <*>
1937 V16 N1 P11-16 66-12019 <*>

NIHON JUIGAKU ZASSHI
1954 V16 P53-64 64-14657 <*>
1962 V24 N3 P157-163 63-10314 <*> O

NIHON JUISHIKAI ZASSHI
1957 V10 N3 P125-127 61-00398 <*>

NIHON KAGAKU-KAI
1931 V6 P106- 2432 <*>
1931 V6 P152- 2081 <*>
1954 V27 N6 P386-388 64-16302 <*>

NIHON KIKAI GAKKAI
SEE BULLETIN OF J.S.M.E. TOKYO

NIHON KIKAI GAKKAI RONBUNSHU
1937 V3 N13 P334-344 AL-699 <*>
1938 V4 N4/5 P138-143 AL-734 <*>
1938 V4 N14/5 P86-93 AL-734 <*>
1950 V15 N51 P20-26 66-12762 <*>
1951 V17 N60 P119-125 64-26389 <$>
1952 V18 N65 P36-41 UCRL TRANS-691(L) <*>
1953 V19 N78 P32-39 2639 <*>
1953 V19 N84 P4-9 55K28J <ATS>
1953 V19 N88 P33-39 01K27J <ATS>
1958 V24 N140 P219-223 PT.2
NP-TR-629 <*>
1958 V24 N147 P873-879 66-12355 <*>
1962 V28 N195 P1489-1497 64-10821 <*> O
1964 V30 P413-420 21 <INT>

NIHON KOKAI GAKKAI SHI
1951 V158 P382-384 SCL-T-363 <*>
1961 N26 P49- 2224 <BSRA>
1961 N26 P55- 2223 <BSRA>
1965 N34 P19 2427 <BSRA>

NIHON KOKAN GIHO
1963 P36-49 3988 <BISI>
1963 N28 P425-432 4040 <BISI>

NIHON KOKUKA GAKKAI ZASSHI
1963 SB V16 N6 P387-391 66-12155 <*>

NIHON NOGEIKAGAKU KAISHI
1931 V7 P1036-1049 1304-A <K-H>
1932 V8 P404-410 5381-C <KH>
1932 V8 P515-518 63-14572 <*>
1934 V10 P374-378 62-00949 <*>
1934 V10 P1093-1103 57-222 <*>
1935 V11 P357-364 62-00980 <*>
1935 V11 P1089-1094 62-00997 <*>
1936 V12 N6 P497-502 57-225 <*>
1936 V12 N11 P1106-1116 62-00947 <*>
1937 V13 P89-93 62-00951 <*>
1937 V13 P444-453 432-A <KH>
1937 V13 P494-501 T-1394 <INSD>
62-16045 <*>
1937 V13 P499-501 2920 <*>
1938 V14 P342-348 T-1438 <INSD>
1938 V14 P505-506 62-00953 <*>

1940 V16 P293-298 AL-968 <*>
1940 V16 P910-916 T-2061 <INSD>
1940 V16 P917-924 T-2062 <INSD>
1940 V16 N6 P504-512 63-19395 <=*> 0
1940 V16 N16 P917-924 T-2062 <INSD>
1941 V17 N12 P1001-1004 3561-B <KH>
1943 V19 P809-815 62-10025 <*>
1948 V22 N2 P58-60 64-16707 <*>
64-18863 <*>
1950 V23 P468-477 59-15738 <*> 0
1950 V23 N10 P432-437 C-3850 <NRC>
61-01001 <*>
1950 V23 N10 P468-472 59-15799 <*>
1950 V23 N10 P473-477 59-15800 <*>
1951 V24 P412-416 AEC-TR-5845 <*>
1951 V24 N9 P399-402 59-15738 <*> 0
59-15798 <*>
1951 V25 P531-533 58-854 <*>
1951 V25 N2 P59-63 II-880 <*>
1952 V26 P24-27 63-20434 <*>
1952 V26 P159-162 3263-B <KH>
1952 V26 P490-493 3280-A <KH>
1952 V26 P528-533 63-14629 <*>
1952 V26 N1 P11-13 AL-768 <*>
1952 V26 N3 P151-154 05P59J <ATS>
1953 V27 P561-564 5575C <K-H>
1954 V28 P264-269 3580-A <KH>
1954 V28 P269-274 3580-B <KH>
1954 V28 P296-299 22G6J <ATS>
1954 V28 P387-391 3629-B <KH>
1954 V28 P618-621 3580-C <KH>
1954 V28 P621-629 3580-D <KH>
1954 V28 N10 P791-794 NNK-28-10-1 <SA>
1955 N29 P400-403 T1942 <INSD>
1955 N29 P404-407 T1943 <INSD>
1955 V29 P211-215 18Q67J <ATS>
1955 V29 P219-221 9188B <KH>
1955 V29 P222-225 6145 <KH>
1955 V29 P400-403 58-215 <*>
1955 V29 P404-407 58-216 <*>
1956 V30 P419-422 59-15615 <*>
1956 V30 P423-426 59-15616 <*>
1957 V31 P830-832 12138 <KH>
62-18832 <*>
1957 V31 N4 P272-275 64-16709 <*>
1957 V31 N4 P276-279 64-16725 <*>
1957 V31 N5 P297-299 62-20219 <*>
1957 V31 N6 P375-383 C-5205 <NRC>
1957 V31 N10 PA107-A118 63-20008 <*>
64-30961 <*>
8524 <K-H>
1958 V32 P26-29 58-1379 <*>
62-16046 <*>
1958 V32 N4 P321-324 58-1428 <*>
1958 V32 N6 P467-470 10521-B <KH>
62-18833 <*>
1958 V32 N7 P501-506 NNK-32-7-1 <SA>
1958 V32 N9 P667-670 NP-TR-680 <*>
1958 V32 N10 P778-783 59-20296 <*>
1959 V33 N3 P163-166 NNK-33-3-1 <SA>
1959 V33 N8 P707-710 NNK-33-8-1 <SA> 0
1960 V34 P404-405 61-10618 <*>
1960 V34 P440-447 61-20729 <*> P
1960 V34 P489-492 63-10862 <*>
1960 V34 N1 P100-103 NNK-34-1-1 <SA> 0
1960 V34 N5 P440-442 NNK-34-5-1 <SA> 0
1960 V34 N6 P475-479 65-17097 <*>
1960 V34 N6 P484-486 65-14179 <*>
1960 V34 N8 P662-665 11092-B <KH>
1960 V34 N12 P1043-1045 NNK-34-12-1 <SA> 0
1961 V35 P80-83 C-4294 <NRC>
1961 V35 P83-86 C-4388 <NRC>
1961 V35 N1 P40-45 58N55J <ATS>
62-18693 <*>
1961 V35 N2 P110-113 62-00377 <*>
1961 V35 N2 P113-118 62-00289 <*>
1961 V35 N2 P119-121 62-00321 <*>
1961 V35 N2 P155-159 62-00231 <*>
1961 V35 N2 PA19-A25 64-10527 <*>
1961 V35 N9 P868-870 11560-B <KH>
64-14612 <*> 0
1961 V35 N10 P908-915 62-18726 <*>
1961 V35 N14 P1378-1381 63-16248 <*>
1962 V36 N1 P18-23 63-16435 <*>
1962 V36 N1 P24-28 63-16436 <*>
1962 V36 N4 P374-377 63-16690 <*>
1962 V36 N5 P393-397 63-16437 <*>
1962 V36 N5 P398-402 63-16507 <*>
1962 V36 N7 P589-592 12240-C <K-H>
1962 V36 N12 P1013-1016 54R74J <ATS>
1963 V37 P611-614 65-63462 <=>
1965 V39 N1 P22-29 65-14180 <*>

NIHON SAKUMOTSU GAKKAI KIJI
1930 V2 P153-160 66-12555 <*>
1945 V14 P5-14 PANSDOC-TR.300 <PANS> P
1960 V29 N1 P51-54 66-11792 <*> 0
1962 V30 P237-240 66-13367 <*> 0

NIHON SEIKA-GAKKAI
1926 V6 P335-366 64-16196 <*>

NIHON SEISHIN SHINKEI GAKKAI
1954 V8 N1 P1-6 62-00452 <*>

NIHON SENBAI KOSHA CHUO KENKYUSHO KENKYU HOKOKU
1954 N90 P1-32 11436-A <KH>
64-14539 <*>
1954 N90 P33-38 11436-B <KH>
64-14541 <*>
1954 N90 P39-43 11436-C <K-H>
64-14540 <*>
1954 N90 P86-100 11436-D <KH>
64-14558 <*>
1954 N90 P107-117 11436-E <KH>
1954 N90 P131-141 11436-F <KH>
64-14557 <*>
1954 N90 P142-149 11436-G <KH>
64-14562 <*>
1954 N90 P150-162 11436-H <K-H>
64-14559 <*>
1954 V90 P107-117 63-18510 <*>
1956 N96 P48 PT2 6063C <K-H>
1956 N96 P25-29 5919-B <K-H>
1956 N96 P30-33 5919-C <K-H>
1956 N96 P34-38 5919-D <K-H>
1956 N96 P39-42 6063A <K-H>
1956 N96 P43-47 6063B <K-H>
1956 N96 P86-90 6063-F <K-H>
1956 N96 P91-105 6063-G <KH>
1957 N97 P27-39 58-1018 <*>
6445-E <KH>
1957 N97 P43-48 6445-D <KH>
1958 N99 P8-9 61-20231 <*>
8426 <K-H>
1960 N102 P6-9 10309-A <KH>
63-16120 <*>
1960 N102 P10-12 10309-B <KH>
64-10627 <*>
1960 N102 P13-15 10309-C <KH>
1961 N103 P19-24 11153-A <K-H>
61-20883 <*>
1961 N103 P45-49 11707 <KH>
63-16918 <*>
1961 N103 P56-62 64-14528 <*>
1962 N104 P37-41 12649 <K-H>
1962 N104 P71-75 12002-H <K-H>
1962 N104 P101-106 12002A <KH>
1962 N104 P115-121 12616 <KH>
1964 V106 P83-91 1375 <TC>

NIHON SENBAI KOSHA CHUO KENKYUSHO GYOTEI HOKOKU
1956 P79-105 NO 96 58-1199 <*>
1956 N96 P79-105 58-836 <*> P

NIHON SHASHIN GAKKAI KAI SHI
1954 V16 N3/4 P59-65 895 <TC>
1957 V20 P93-95 58-2413 <*>
1958 V21 P49-57 654 <TC>
1958 V21 P115-120 21L34J <ATS>

1958 V21 N1 P16-19 59-10300 <*>
1958 V21 N1 P20-23 40M40J <ATS>
1959 V22 N1 P18-25 60-16928 <*>
1959 V22 N1 P26-32 41M40J <ATS>
1959 V22 N3 P121-128 65-17C02 <*> 0
1959 V22 N3 P134-138 60-18483 <*>
1961 V24 N1 P36-40 90Q69J <ATS>
1962 V25 N2 P82-88 17Q67J <ATS>
1962 V25 N4 P161-174 77Q71J <ATS>
1962 V25 N4 P193-197 67Q72J <ATS>
1963 V26 N4 P160-166 1250 <TC>
1964 V27 P172-176 1358 <TC>
1964 V27 N1 P7-13 1251 <TC>
1964 V27 N3 P117-122 1252 <TC>
1965 V28 N1 P38-41 87S89J <ATS>
1965 V28 N3 P119-123 16T92J <ATS>

NIHON SHOKUBUTSU BYORIGAKU KAIHO
1934 V4 P66-68 66-14029 <*$>
1934 V4 P68-69 66-14028 <*$>
1955 V20 N1 P16-20 5709-D <K-H>
1957 V22 N1 P12 61-10868 <*>
6918-G <K-H>

NIHON SHOKUHIN KOGYO GAKKAISHI
1964 V11 N11 P499-501 65-25336 <$>

NIIMASH
1931 V1 N8 P363-390 60-10281 <=*> 0

NINETY-NINE. LONDON
1959 V121 N22 P881-886 9974-B <K-H>

NIPPON
SEE ALSO NIHON

NIPPCN ACTA RADIOLOGICA
SEE NIHON IGAKU HOSHASEN GAKKAI ZASSHI

NIPPON BUTSURI GAKKAI
1953 V8 N4 P545-548 66-14608 <*$>

NIPPON BUTSURI GAKKAI SHI
1961 V16 N7 P436-446 62-14812 <*>
65-10761 <*>
1962 V17 N3 P171-184 63-01227 <*>

NIPPON BYORIGAKKAI KAISHI
1922 V12 P1C9-110 II-484 <*>
1949 V38 N1/6 P108- II-636 <*>
1949 V38 N1/6 P111-112 II-678 <*>
1949 V38 N1/6 P113-114 II-608 <*>
1949 V38 N1/6 P120-121 57-108 <*>
1954 V43 P450-452 63-00300 <*>
1954 V43 P452-454 63-00282 <*>
1958 V47 N3 P612- 63-01249 <*>

NIPPON DAIGAKU KOGAKU KENKYUJC IHO
1954 N7 P16-20 T1932 <INSD>

NIPPON DOJC HIRYOGAKU ZASSHI
1951 V21 P253-260 66-11540 <*> 0
1953 V23 P117-120 CSIRO-2802 <CSTR>
1954 V24 P268-270 3263-A <K-H>
1954 V24 P331-333 3180-C <K-H>
1954 V25 P17-19 66-10732 <*>
1955 V26 N6 P7-14 4020-A <KH>
1957 V28 P181-184 11128-B <K-H>
1959 V30 P393-396 66-10731 <*>
1960 V31 P273-278 66-10842 <*>
1961 V32 N1 P15-18 11071-F <K-H>
1964 V35 N4 P115-118 66-12554 <*>
1964 V35 N11 P408 66-12935 <*>
1965 V36 N3 P45-48 66-13366 <*>

NIPPON GAKUJUTSU KAIGI
1922 V3 N3 P95-146 CSIR-3527 <CSSA>

NIPPON GANKA KIYO
1964 V15 P205-213 66-13195 <*>
1964 V15 N5 P1S0-194 65-12900 <*>

NIPPON GANKA GAKKAI ZASSHI
1957 V61 N8 P1325-1332 62-01517 <*>
1961 V65 N10 P2164-2170 65-17238 <*>
1962 V66 N9 P714-722 65-00163 <*> PC
1962 V66 N10 P1001-1009 66-12110 <*>
1963 V67 N9 P1223-1259 64-00275 <*>
1963 V67 N9 P1223-1239 65-29498 <$>
1963 V67 N10 P1323-1351 65-00239 <*>
65-29496 <$>

NIPPON GENSHI-RYCKO GAKKAI-SHI
1959 V1 P64-69 AEC-TR-4066 <*>
1959 V1 P319-329 AEC-TR-4172 <=>
1959 V1 P376-378 AEC-TR-4191 <=>
1959 V1 P405-411 AEC-TR-4939 <=>
63-27381 <$>
1959 V1 N1 P1-8 AEC-TR-4463 <=>
1959 V1 N1 P40-45 NEL-TRANS-1690 <NLL>
1959 V1 N1 P64-69 AEC-TR-4066 <=>
1959 V1 N3 P19C-195 AEC-TR-4467 <=> 0
1959 V1 N4 P259-271 AEC-TR-4065 <=>
1959 V1 N5 P308-318 ORNL-TR-200 <*>
1959 V1 N6 P359-362 50N56J <ATS>
1959 V1 N6 P363-369 AEC-TR-4225 <=>
1959 V1 N6 P370-375 AEC-TR-4464 <=>
1960 V2 P73-77 AEC-TR-4246 <*>
1960 V2 P89-95 AEC-TR-4380 <*>
1960 V2 P117-121 AEC-TR-4941 <=>
1960 V2 P182-189 AEC-TR-4381 <*>
1960 V2 P19C-195 AEC-TR-4465 <*>
1960 V2 P291-295 AEC-TR-5069 <*>
1960 V2 P598-602 AEC-TR-5070 <=>
1960 V2 N1 P6-14 AEC-TR-4466 <=> 0
1960 V2 N2 P73-77 AEC-TR-4246 <=>
1960 V2 N2 P89-95 AEC-TR-4380 <=>
1960 V2 N3 P136-146 AEC-TR-4468 <=>
1960 V2 N4 P182-189 AEC-TR-4381 <=>
1960 V2 N4 P19C-195 AEC-TR-4465 <=>
1960 V2 N5 P245-252 36P61J <ATS>
1960 V2 N5 P291-295 AEC-TR-5069 <=> C
1960 V2 N7 P389-393 AEC-TR-5073 <=>
1960 V2 N7 P406-411 AEC-TR-5076 <=>
1960 V2 N8 P451-459 AEC-TR-5078 <=>
1960 V2 N8 P474-477 AEC-TR-5077 <=>
1960 V2 N9 P511-517 AEC-TR-4517 <*>
1960 V2 N9 P518-522 AEC-TR-5074 <=>
1960 V2 N11 P659-670 AEC-TR-5071 <=>
1960 V2 N12 P731-735 AEC-TR-5072 <=>
1961 V3 P1-8 AEC-TR-4643 <*>
1961 V3 N3 P200-207 NP-TR-804 <*>
1961 V3 N4 P260-265 AEC-TR-5796 <*>
1961 V3 N5 P333-337 NP-TR-938 <*>
1961 V3 N5 P36C-364 NP-TR-864 <*>
1961 V3 N6 P457-461 NP-TR-901 <*>
1961 V3 N9 P705-710 C-4127 <NRCC>
1961 V3 N10 P763-766 NP-TR-919 <*>
1961 V3 N11 P868-873 AEC-TR-5135 <*>
1961 V3 N12 P918-922 AEC-TR-5136 <*>
1961 V3 N12 P923-928 AEC-TR-5134 <*>
1962 V4 P77-84 AEC-TR-5382 <*>
1962 V4 P94-99 AEC-TR-5378 <*>
1962 V4 N1 P30-36 AEC-TR-5466 <*>
42P63J <ATS>
1962 V4 N1 P37-44 NP-TR-944 <*>
1962 V4 N6 P355-361 NP-TR-933 <*>
1962 V4 N10 P7C3-707 AEC-TR-6054 <*>
1962 V4 N11 P797-807 AEC-TR-6085 <*>
1963 V5 P768-773 AEC-TR-6477 <*>
1963 V5 N6 P467-475 N66-13195 <=$>
1963 V5 N6 P497-503 66-12836 <*> 0
1963 V5 N12 P9S0-993 66-12859 <*>
1964 V6 N1 P15-20 66-12350 <*>
1964 V6 N2 P91-97 NSJ-TR-21 <*>
1964 V6 N7 P399-405 66-12133 <*>
1964 V6 N11 P646-655 N66-14687 <=$>
1965 N10 P554-562 ICE V6 N2 P373-381 <ICE>
1965 N11 P627-633 ICE V6 N2 P382-388 <ICE>
1965 N12 P680-686 ICE V6 N3 P534-539 <ICE>

Citation	Translation
1965 V7 N9 P480-484	66-11963 <*>
1965 V7 N9 P496-499	66-11966 <*>
1965 V7 N11 P627-633	60T90J <ATS>
1965 V7 N12 P68C-686	36T92J <ATS>

NIPPON GOMU KYOKAISHI

Citation	Translation
1948 V21 N4 P102-105	AL-972 <*>
1948 V21 N8 P171-172	76J18J <ATS>
1950 V23 P212-215	RCT V31 N3 P608 <RCT>
1950 V23 N8 P207-211	RCT V30 N3 P952 <RCT>
1951 V24 P130-132	SC-T-64-610 <*>
1951 V24 P133-138	SC-T-64-610 <*>
1951 V24 P263-265	RCT V31 N3 P612 <RCT>
1951 V24 P266-268	RCT V31 N3 P615 <RCT>
1952 V25 P267-270	1304 <*>
1952 V25 N9 P306-309	09G5J <ATS>
1953 V26 N5 P264-269	12TM <CTT>
1953 V26 N5 P269-278	13TM <CTT>
1953 V26 N7 P397-404	14TM <CTT>
1955 V28 P1-4	RCT V31 N3 P621 <RCT>
1955 V28 P141-145	RCT V31 N3 P618 <RCT>
1955 V28 P399-407	RCT V31 N3 P624 <RCT>
1955 V28 N4 P216-223	CODE-R1 <SA>
1958 V31 P961-965	61-16093 <*>
1959 V32 N7 P518-527	51P61J <ATS>
1959 V32 N7 P527-535	52P61J <ATS>
1960 V33 P882-892	RCT V35 N1 P182 <RCT>
1960 V33 N1 P29-35	23N52J <ATS>
1961 V34 P884-	RCT V38 N1 P219 <RCT>
1962 V35 N6 P404-413	CODE-R-0 <SA>
1962 V35 N12 P898-913	64-16828 <*>
1963 V36 P368-373	64-18186 <*>
1964 V37 N2 P81-86	65-11867 <*>
1964 V37 N2 P87-93	65-11866 <*>
1964 V37 N2 P99-120	66-13963 <*$>
1965 V38 N1 P13-22	66-13657 <*>
1966 V39 N3 P67-185	66-13962 <*$>
1966 V39 N3 P75-193	66-13962 <*$>

NIPPON HOIGAKU ZASSHI

Citation	Translation
1951 V5 N6 P75-84	59-19287 <=*>
1951 V5 N6 P85-93	59-19434 <=*>
1951 V5 N6 P94-100	59-19435 <=*>

NIPPON JUNKANKI GAKUSHI

Citation	Translation
1962 V26 P455-465	63-01286 <*>

NIPPON KAGAKU RYOHOGAKKAI ZASSHI

Citation	Translation
1955 V3 N4 P128-131	65-00194 <*>
1959 V7 N2 P109-127	62-01394 <*>

NIPPCN KAGAKU ZASSHI

Citation	Translation
1919 V40 P914-921	62-C0978 <*>
1930 V51 P138-150	57-1443 <*>
1931 V52 P668-672	59-20808 <*> 0
1931 V52 P685-690	AL-391 <*>
	1664 <K-H>
	63-14836 <*>
1932 V53 N6 P664-667	57-226 <*>
1934 V55 P11-14	AL-264 <*>
1934 V55 N6 P584-589	57-2742 <*>
1935 V56 N2 P192-195	1670 <*>
1935 V56 N4 P486-504	64-30147 <*> 0
1936 V57 N11 P1190-1194	57-2575 <*>
1936 V57 N11 P1205-1207	57-1844 <*>
	58-1654 <*>
1937 V58 N1 P1-3	AL-698 <*>
1937 V58 N8 P819-823	59-17085 <*>
1937 V58 N8 P824-825	59-17086 <*>
1937 V58 N10 P981-984	AEC-TR-5388 <*>
1938 V59 P673-674	1925 <*>
	64-14406 <*>
1938 V59 P1311-1320	64-10143 <*> 0
1938 V59 N10 P1145-1149	64-10791 <*> 0
1939 V60 P191-198	62-16284 <*>
1939 V60 P1173-1176	62-18215 <*>
1939 V60 P1287-1292	62-18216 <*>
1939 V60 N7 P625-631	53M42J <ATS>
1940 V61 P121-124	1442A <K-H>
1940 V61 P245-254	I-460 <*>
1940 V61 P889-903	1442B <K-H>
1940 V61 N3 P269-276	58-2654 <*>
1941 V62 P381-387	I-651 <*>
	64-14416 <*>
1941 V62 P509-515	71DJ <ATS>
1941 V62 P592-596	I-662 <*>
	64-10139 <*>
1941 V62 N2 P96-98	15TM <CTT>
1941 V62 N3 P259-266	26S84J <ATS>
1941 V62 N10 P990-994	I-764 <*>
1942 V63 P23-26	II-124 <*>
1942 V63 P634-643	59-15012 <*>
1942 V63 P760-762	<ATS>
1942 V63 P827-833	II-529 <*>
	64-14157 <*> 0
1942 V63 P1147-1150	1356A <K-H>
1942 V63 P1512-1515	1338B <K-H>
1942 V63 P1738-1742	I-652 <*>
	64-14383 <*> P
1942 V63 P1755-1758	1338C <K-H>
1942 V63 P1762-1765	59-15271 <*>
1942 V63 N5 P504-509	07N58J <ATS>
1943 V64 P338-340	65-11008 <*>
1943 V64 P883-886	II-531 <*>
	64-14150 <*> 0
1943 V64 P1211	1870 <*>
1944 V65 P148-153	II-530 <*>
	64-14155 <*> 0
1944 V65 P797-799	2013 <*>
	64-14404 <*> 0
1946 V67 N1 P29-34	2911B <K-H>
	64-10138 <*> 0
1948 V69 P16-17	63-18950 <*> 0
1948 V69 P45-52	3980 F <K-H>
1948 V69 P52-55	I-653 <*>
1948 V69 N1/3 P35-37	64-14282 <*>
1948 V69 N4/6 P45-52	64-16431 <*> 0
1948 V69 N4/6 P56-57	60-14012 <*>
1948 V69 N4/6 P72-75	64-14283 <*> C
1949 V70 P52-57	II-75 <*>
1949 V70 P103-104	4575 <BISI>
1949 V70 P253-257	58-162 <*>
1949 V70 P439-442	61-10002 <*>
1949 V70 N3 P45-47	II-504 <*>
	64-14152 <*> 0
1949 V70 N3 P47-52	II-505 <*>
1949 V70 N3 P52-57	64-10445 <*> 0
1949 V70 N4 P114-115	57-2762 <*>
1949 V70 N4 P134-136	61-10004 <*>
1949 V70 N7 P226-229	57-1572 <*>
1949 V70 N10 P373-376	61-10003 <*>
1949 V7C N1/2 P24-30	59-15801 <*>
1949 V7C N11/2 P447-448	64-20079 <*>
1950 V71 P40-42	63-16028 <*>
1950 V71 P212-214	63-16029 <*>
1950 V71 P627-629	I-392 <*>
1950 V71 N1 P29-32	64-16913 <*>
1950 V71 N1 P40-42	2160 <K-H>
1950 V71 N1 P59-60	18F4J <ATS>
	57-3013 <*>
1950 V71 N1 P77-80	61-10005 <*>
1950 V71 N2 P108-111	64-16912 <*>
1950 V71 N2 P145-148	57-2573 <*>
1950 V71 N5 P327-328	66-10803 <*>
1950 V71 N10 P494-496	45E2J <ATS>
1950 V71 N11 P556-557	64-20078 <*>
1950 V71 N11 P577-579	65-10079 <*>
1950 V71 N11 P580-584	1632 <*>
1950 V71 N6/7 P378-382	58-2670 <*>
1950 V71 N8/9 P464-466	58-1652 <*>
1951 V72 P124-127	AL-836 <*>
1951 V72 P236-237	UCRL TRANS-543(L) <*>
1951 V72 P398	58-1653 <*>
1951 V72 P423-426	UCRL TRANS-544(L) <*>
1951 V72 P764-765	27H10J <ATS>
1951 V72 N1 P101-104	61-10604 <*> 0
1951 V72 N1 P104-107	61-10606 <*>
1951 V72 N5 P431-434	AL-395 <*>
1951 V72 N5 P459-462	59-17009 <*>
1951 V72 N6 P495-498	NKZ-72-6-1 <SA>

Citation	Translation
1951 V72 N6 P498-501	NKZ-72-6-2 <SA>
1951 V72 N6 P532-535	PANSDOC-TR.483 <PANS>
1951 V72 N7 P641-644	II-110 <*>
1951 V72 N9 P812-815	AL-786 <*>
1951 V72 N9 P817-820	I-697 <*>
1951 V72 N11 P958-962	59-14734 <*+>
1951 V72 N11 P979-982	NKZ-72-11-1 <SA>
1951 V72 N12 P1022-1024	NKZ-72-12-1 <SA>
1951 V72 N12 P1067-1070	59-17005 <*>
1952 N2 P150-152	1300 <*>
1952 V73 P50-53	59-15916 <*> 0
1952 V73 P131-134	1497 <*>
1952 V73 P244-246	1497 <*>
1952 V73 P246-248	1662 <*>
1952 V73 P393-394	59-17869 <*>
1952 V73 P418-423	62-10751 <*> 0
1952 V73 P423-426	62-10752 <*> 0
1952 V73 P529	1617 <*>
1952 V73 P576-578	63-16566 <*>
1952 V73 P578-580	62-10753 <*> 0
1952 V73 P835-841	58-582 <*>
1952 V73 P889-891	58-580 <*>
1952 V73 N1 P59-63	35L29J <ATS>
1952 V73 N4 P223-225	59-15989 <*>
1952 V73 N4 P263-265	AL-324 <*>
1952 V73 N4 P267-270	I-577 <*>
1952 V73 N4 P278-279	1299 <*>
1952 V73 N6 P440-443	II-89 <*>
1952 V73 N7 P485-487	I-253 <*>
1952 V73 N8 P578-580	57-1755 <*>
1953 V74 P32-38	58-578 <*>
1953 V74 P298-305	58-581 <*>
1953 V74 P400-401	TS-1577 <BISI>
	58-588 <*>
1953 V74 P429-431	66-11260 <*>
1953 V74 P664-668	58-838 <*>
	58-849 <*>
	63-16393 <*> 0
	63-16447 <*> 0
1953 V74 P915-917	I-412 <*>
1953 V74 P1009-1011	60-14123 <*>
1953 V74 N1 P16-18	01L31J <ATS>
	64-20035 <*> 0
1953 V74 N1 P28-31	60-17431 <=*>
1953 V74 N2 P142-145	65-13737 <*>
1953 V74 N4 P256-258	1246 <*>
1953 V74 N4 P295-297	59-15943 <*> 0
1953 V74 N5 P349-352	86R76J <ATS>
1953 V74 N5 P383-385	I-434 <*>
1953 V74 N8 P642-644	59-14734 <*+>
1953 V74 N10 P832-834	C-4398 <NRC>
	63-00684 <*>
1954 V75 P94-96	59-15089 <*>
1954 V75 P182-189	I-580 <*>
1954 V75 P182-185	64-14395 <*> 0
1954 V75 P186-189	II-388 <*>
	64-14396 <*> 0
1954 V75 P348	63-14612 <*>
1954 V75 P378-380	T-1629 <INSD>
1954 V75 P380-383	T-1629 <INSD>
1954 V75 P499-502	UCRL TRANS-909(L) <*>
1954 V75 P586-588	T-1629 <INSD>
1954 V75 P596-602	58-218 <*>
1954 V75 P869-871	75J18J <ATS>
1954 V75 P931-933	63-27826 <$>
1954 V75 P1257-1259	62-25217 <=*>
1954 V75 N2 P218-222	65-13318 <*>
1954 V75 N3 P309-311	59-10213 <*>
1954 V75 N3 P324-327	59-17004 <*>
1954 V75 N4 P383-386	NKZ-75-4-1 <SA>
1954 V75 N5 P522-524	59-17010 <*>
1954 V75 N7 P703-707	59-17011 <*>
1954 V75 N7 P741-746	47N54J <ATS>
1954 V75 N8 P835-838	11G7J <ATS>
	57-2676 <*>
1954 V75 N9 P884-887	31G7J <ATS>
1954 V75 N10 P985-986	64-10576 <*>
1954 V75 N10 P990-993	C-3946 <NRCC>
	62-00646 <*>
1954 V75 N11 P1152-1155	57-76 <*>
1954 V75 N11 P1203-1208	2375 <*>
1955 V76 P14	62-25219 <=*> 0
1955 V76 P56-59	62-25218 <=*>
1955 V76 P60-63	57-742 <*>
1955 V76 P64-69	58-21 <*>
1955 V76 P220-222	61-20194 <*>
1955 V76 P402-406	8K22J <ATS>
1955 V76 P466-468	II-778 <*>
1955 V76 P1291-1293	60K28J <ATS>
1955 V76 P1361-1363	71H13J <ATS>
1955 V76 N3 P274-277	52S85J. <ATS>
1955 V76 N5 P545-548	61-20490 <*>
1955 V76 N5 P576-579	60-18154 <*>
1955 V76 N6 P687-691	59-15516 <*>
1955 V76 N7 P762-770	61-20490 <*>
1955 V76 N8 P944-951	57-1364 <*>
1955 V76 N9 P996-998	37H13J <ATS>
1955 V76 N11 P1291-1293	58-1732 <*>
	61-10291 <*>
1955 V76 N12 P1339-1345	59-10718 <*>
1956 V77 P238-244	62-25218 <=*>
1956 V77 P295-297	SCL-T-451 <*>
1956 V77 P401-402	63-16077 <*>
1956 V77 P1532-1536	57-3004 <*>
1956 V77 N1 P44-47	63-16206 <*> 0
1956 V77 N1 P48-51	63-16207 <*> 0
1956 V77 N1 P51-54	63-16208 <*> 0
1956 V77 N2 P385-388	62-14942 <*>
1956 V77 N6 P848-854	58-1692 <*>
1956 V77 N6 P962-964	63J14J <ATS>
1956 V77 N7 P1103-1105	60-14245 <*>
1956 V77 N10 P1525-1528	NKZ-77-10-1 <SA>
1956 V77 N10 P1568-1572	SCL-T-486 <*>
1956 V77 N12 P1811-1815	35K23J <ATS>
1956 V77 N12 P1815-1818	34K23J <ATS>
1957 V78 P129-131	R-5146 <*>
1957 V78 P280-282	1947 <TC>
1957 V78 P1171-1174	58-842 <*>
1957 V78 P1707-1709	PANSDOC-TR.195 <PANS>
1957 V78 N1 P131-138	65-13596 <*>
1957 V78 N2 P280-285	65-13586 <*>
1957 V78 N3 P313-316	61-10177 <*> 0
1957 V78 N5 P707-713	58-1393 <*>
1957 V78 N6 P754-759	NKZ-78-6-1 <SA>
1957 V78 N6 P814-817	51M41J <ATS>
1957 V78 N7 P981-982	PANSDOC-TR.265 <PANS>
	59-17268 <+*>
1957 V78 N8 P1174-1178	58-382 <*>
1957 V78 N8 P1178-1181	58-1391 <*>
1957 V78 N9 P1268-1272	59-14632 <=*$>
1957 V78 N10 P1517-1521	60-18459 <*>
1957 V78 N11 P1613-1617	64-16493 <*> 0
1958 V79 P32-50	AEC-TR-5222 <*>
1958 V79 P1492-1495	66-11968 <*>
1958 V79 N1 P17-21	69K24J <ATS>
1958 V79 N1 P21-26	69K24J <ATS>
1958 V79 N2 P131-134	58-2486 <*>
1958 V79 N2 P135-137	77L30J <ATS>
1958 V79 N2 P138-142	78L30J <ATS>
1958 V79 N3 P243-248	UCRL TRANS-539(L) <*>
1958 V79 N4 P499-504	58-2489 <*>
1958 V79 N5 P586-594	NP-TR-797 <*>
1958 V79 N7 P832-836	59-10491 <*>
1958 V79 N7 P843-848	59-10496 <*>
1958 V79 N10 P1169-1172	UCRL TRANS-540(L) <*>
1958 V79 N10 P1202-1204	62-32178 <=*$>
1958 V79 N12 P1513-1520	UCRL TRANS-541(L) <*>
1959 V80 P956-959	AEC-TR-5617 <=*>
1959 V80 P1369-1370	62-18945 <*>
1959 V80 N1 P21-25	35L32J <ATS>
1959 V80 N1 P25-27	59-17077 <*>
1959 V80 N1 P28-31	34L32J <ATS>
1959 V80 N1 P88-91	C-5566 <TC>
1959 V80 N2 P171-173	14L36J <ATS>
1959 V80 N3 P250-255	61L32J <ATS>
1959 V80 N3 P307-308	61-10006 <*>
1959 V80 N3 P326-328	59-17131 <*>
1959 V80 N9 P1038-1046	65-13228 <*>
1959 V80 N9 P1043-1046	60-14244 <*>

Citation	Translation
1959 V80 N9 P1061-1063	60-14243 <*>
1959 V80 N10 P1090-1094	62-18944 <*> 0
1959 V80 N10 P1094-1097	C-3950 <NRC>
1959 V80 N10 P1109-1112	UCRL TRANS-727(L) <*>
1959 V80 N10 P1112-1116	15M38J <ATS>
1959 V80 N11 P1203-1206	62-18942 <*> 0
1960 V81 P582	63-00250 <*>
1960 V81 P1266-1271	SCL-T-497 <*>
1960 V81 P1637-1642	SCL-T-394 <*>
1960 V81 P1643-1645	62-10625 <*>
1960 V81 N1 P175-179	60-16676 <*>
1960 V81 N3 P469-472	NKZ-81-3-1 <SA>
1960 V81 N4 P564-567	62-01054 <*>
1960 V81 N5 P694-698	66-11995 <*>
1960 V81 N6 P891-895	AEC-TR-4302 <*>
1960 V81 N6 P927-931	63Q70J <ATS>
1960 V81 N7 P997-1003	61-10837 <*>
1960 V81 N7 P1003-1007	61-10834 <*>
1960 V81 N7 P1016-1025	59S81J <ATS>
1960 V81 N7 P1034-1038	61-10835 <*>
1960 V81 N7 P1038-1041	61-10836 <*>
1960 V81 N9 P1414-1418	61-10829 <*>
1960 V81 N9 P1418-1421	61-10828 <*>
1960 V81 N11 P1648-1652	C-3634 <NRCC>
1960 V81 N12 P1871-1874	65-10261 <*>
1961 V82 P478-480	62-26045 <=$>
1961 V82 P545-549	66-14017 <*>
1961 V82 P550-554	20Q70J <ATS>
1961 V82 P841-845	62-16517 <*> 0
1961 V82 P880-882	62-18070 <*>
1961 V82 P1262-1265	12240A <K-H>
	63-18934 <*>
1961 V82 P1265-1267	12240B <K-H>
1961 V82 N3 P276-281	62-10696 <*>
1961 V82 N3 P281-285	62-10697 <*>
1961 V82 N3 P292-298	62-14964 <*>
1961 V82 N3 P339-343	64-16374 <*>
1961 V82 N3 P378-381	63Q69J <ATS>
1961 V82 N4 P417-419	62-10693 <*>
1961 V82 N4 P419-421	62-10692 <*>
1961 V82 N4 P421-423	62-10694 <*>
1961 V82 N4 P438-441	62-14963 <*>
1961 V82 N4 P483-486	14N53J <ATS>
1961 V82 N4 P486-490	15N53J <ATS>
1961 V82 N5 P630-632	62-00429 <*>
1961 V82 N6 P641-644	64-16373 <*>
1961 V82 N6 P713-715	63-18810 <*> 0
1961 V82 N6 P740-743	26N57J <ATS>
1961 V82 N7 P878-880	NKZ-82-7-1 <SA>
1961 V82 N7 P880-882	NKZ-82-7-2 <SA>
1961 V82 N9 P1272-1274	62-10954 <*> 0
1961 V82 N9 P1276-1279	62-14826 <*>
1961 V82 N1C P1287-1290	62-18035 <*>
1961 V82 N11 P1461-1463	63-00750 <*>
1961 V82 N11 P1479-1480	AEC-TR-5467 <*>
1961 V82 N11 P1485-1489	NP-TR-1015 <*>
1961 V82 N12 P1620-1624	63-20713 <*>
1961 V82 N12 P1672-1675	62-18037 <*>
1961 V82 N12 P1702-1708	62-14962 <*>
1962 V83 P528-532	62-20450 <*>
1962 V83 P532-536	62-20449 <*>
1962 V83 P836-838	63-10321 <*>
1962 V83 P1164-1167	SC-T-523 <*>
1962 V83 N1 P61-67	63-14017 <*> 0
1962 V83 N1 P67-72	63-14019 <*> 0
1962 V83 N1 P73-76	63-14018 <*> 0
1962 V83 N1 P92-95	65-13344 <*> 0
1962 V83 N4 P417-421	66-10199 <*> 0
1962 V83 N4 P421-425	66-10200 <*> 0
1962 V83 N5 P597-600	63-10915 <*>
1962 V83 N6 P693-695	63-18809 <*> 0
1962 V83 N6 P696-698	63-18715 <*> 0
1962 V83 N6 P729-731	62-18827 <*>
1962 V83 N7 P839-841	62-18930 <*>
1962 V83 N7 P850-851	62-18929 <*>
1962 V83 N8 P959	NKZ-83-8-1 <SA>
1962 V83 N9 P1023-1026	64-14808 <*> 0P
1962 V83 N9 P1042-1044	NKZ-83-9-1 <SA>
1962 V83 N1C P1142-1147	12557 A <K-H>
	65-14846 <*>
1962 V83 N10 P1158-1159	61Q71J <ATS>
1962 V83 N12 P1292-1294	63-16381 <*>
1963 V84 P48-50	12792 C <K-H>
1963 V84 P263-267	AEC-TR-6239 <*>
1963 V84 P402-404	66-12129 <*>
1963 V84 P897-902	65-18098 <*>
1963 V84 N1 P29-31	12792-A <K-H>
	65-14861 <*>
1963 V84 N1 P31-34	12792-B <K-H>
	65-14862 <*>
1963 V84 N1 P48-50	65-14847 <*>
1963 V84 N2 P115-119	63-20884 <*> 0
1963 V84 N2 P180-185	NKZ-84-2-1 <SA>
1963 V84 N5 P384-392	86R78J <ATS>
1963 V84 N5 P419-421	13147 <K-H>
1963 V84 N6 P496-499	31T91J <ATS>
1963 V84 N7 P547-552	64-14267 <*> 0
1963 V84 N11 P863-868	66-11421 <*>
1963 V84 N11 P868-871	66-11422 <*>
1963 V84 N11 P871-875	66-11423 <*>
1963 V84 N12 P968-972	41R75J <ATS>
1964 V85 N3 P152-155	66-13271 <*>
1964 V85 N3 P155-180	66-13272 <*>
1964 V85 N3 P159-168	65-10438 <*>
1964 V85 N3 P168-176	65-10467 <*>
1964 V85 N3 P227-231	66-13245 <*>
1964 V85 N3 P237-238	65-10394 <*>
1964 V85 N4 P247-252	66-13273 <*>
1964 V85 N4 P252-255	66-10617 <*>
1964 V85 N11 P753-756	65-13506 <*>
1965 V86 P172-176	1571 <TC>
1965 V86 N1 P35-39	1886 <TC>
1965 V86 N2 P131-150	65-17053 <*> 0
1965 V86 N7 P7C8-713	66-11465 <*>
1965 V86 N7 P737-74C	1713 <TC>
1965 V86 N8 P798-807	23T90J <ATS>
1965 V86 N9 P921-925	66-12515 <*>
1965 V86 N9 P950-954	66-12516 <*>

NIPPON KEIZAI SHIMBUN

Citation	Translation
1965 P11	65-33111 <=*$>

NIPPON KENCHIKU GAKKAI ROMBUN HOKOKU-SHU

Citation	Translation
1954 V27 P225-226	T-2065 <INSD>

NIPPON KETSUEKI GAKKAI ZASSHI

Citation	Translation
1955 V18 N5 P4C6-424	59-12768 <=*> C
1955 V18 N7 P647-658	58-201 <*>
1960 V23 P747-766	UCRL TRANS-652 <*>

NIPPON KIKAI GAKKAISHI

Citation	Translation
1932 V35 N183 P695-700	C-3477 <NRC>
1933 V36 N190 P127-130	C-3478 <NRC>
1934 V36 P367-374	62-25705 <=*>
1934 V37 N201 P15-21	C-3479 <NRC>
1952 V55 N397 P102-107	65-10369 <*> 0
1960 V63 N502 P1442-1451	52P59J <ATS>
1960 V63 N502 P1474-1482	65-14407 <*>
1962 V65 N520 P621-623	C-4564 <NRCC>
	63-01353 <*>
1964 V80 N918 P1047-1053	86S85J <ATS>

NIPPON KINZOKU GAKKAISHI

Citation	Translation
1940 V4 N7 P228-242	60-10040 <*> 0
1941 V5 P259-271	60-10033 <*> C
1950 V14B N2 P49-52	63-20461 <*>
1950 V14B N3 P30-33	62-14473 <*>
1951 P227-229	T1726 <INSD>
1951 V15 P528-531	3000B <K-H>
1951 V15B P202-205	57-1016 <*>
1951 V15B P528-531	II-477 <*>
	64-14140 <*> 0
1952 V16 N2 P121-124	59-17180 <*> 0
1952 V16 N3 P151-153	59-17182 <*> 0
1952 V16 N9 P486-492	60-10348 <*> C
1952 V16 N10 P547-551	6C-10368 <*> 0
1952 V16 N11 P607-610	60-10032 <*> 0
1953 V17 P191-194	60-15624 <=*> C
1953 V17 N1 P28-32	60-10349 <*> 0
1954 V18 N9 P549-551	61-18361 <*>

1956 V20 N5 P288-291 61-16526 <*>
1956 V20 N8 P460-465 57-2650 <*>
1957 V21 P536-540 TS-1316 <BISI>
1957 V21 N3 P176-180 AEC-TR-3456,REPT.1 <*>
1957 V21 N3 P180-183 AEC-TR-3456,REPT.2 <*>
1958 V22 P229-233 1848 <BISI>
1958 V22 N3 P120-123 4798 <HB>
1958 V22 N5 P225-229 1848 <BISI>
1958 V22 N12 P663-668 41L37J <ATS>
1959 V23 P696-698 1755 <BISI>
1959 V23 P717-721 1756 <BISI>
1959 V23 N8 P486-489 08M41J <ATS>
1959 V23 N8 P489-493 3344 <BISI>
1959 V23 N9 P508-511 65-10109 <*>
1959 V23 N11 P662-666 40P59 <ATS>
1960 V24 P710-714 5839 <HB> P
1960 V24 N1 P50-54 C-4050 <NRCC>
1960 V24 N2 P117-122 61-18297 <*>
1960 V24 N2 P130-134 4159J <ATS>
1960 V24 N6 P374-377 61-18298 <*>
1960 V24 N6 P377-379 61-20501 <*> 0
1960 V24 N11 P699-703 4334 <BISI>
1961 V25 N2 P116-120 64-10551 <*>
1961 V25 N11 P712-716 4115 <BISI>
1963 V27 N4 P191-195 4108 <BISI>
1963 V27 N10 P481-485 4041 <BISI>
1963 V27 N12 P599-604 66-60464 <=*$>
1963 V27 N12 P618 4546 <BISI>
1964 V28 P604-610 66-12832 <*>
1964 V28 N11 P717 4545 <BISI>

NIPPON KOGYO KAISHI
1957 V73 P837-840 61-10733 <*>
1958 V74 N838 P249-252 61-10737 <*>
1959 V75 P167-172 62-26046 <=$>
1959 V75 P313-317 65-00417 <*>
1959 V75 N858 P1105-1112 61-10738 <*>
1960 V76 N861 P179-195 61-10739 <*>
1960 V76 N866 P524-529 8939 <IICH>
1961 V77 P49-56 5114 <HB>
1963 V79 P525-530 NSJ-TR-2 <*>
1963 V79 P590- SC-T-64-904 <*>

NIPPON KOKU GAKKAISHI
1961 V9 N85 P37-42 62-20553 <*>
1962 V10 N101 P174-179 58P62J <ATS>

NIPPON NAIBUNPI GAKKAI ZASSHI
1936 V11 P62-63 1292-H <K-H>
1953 V29 N7/8 P155-188 64-14654 <*>
1957 V33 N1 P53-83 63-01614 <*>
1960 V36 N4 P499-518 61-18867 <*>
1964 V39 P897-900 65-17008 <*>

NIPPON NAIKA GAKKAI ZASSHI
1951 V40 P280- 58-1076 <*>
1954 V43 P539 66-13173 <*>
1957 V46 N3 P295-308 62-01003 <*>
1963 V52 N2 P172-184 64-18415 <*>
1963 V52 N5 P534-535 64-18416 <*>

NIPPON ONKYO GAKKAISHI
1958 V14 N2 P111-116 59-15995 <*>
1958 V14 N2 P164-169 59-15994 <*>
1958 V14 N2 P170-174 59-17245 <*>
1958 V14 N4 P281-290 16M41J <ATS>
1958 V14 N4 P291-299 17M41J <ATS>
60-14110 <*> 0
1959 V15 N1 P6-20 60-14519 <*>
1960 V16 N4 P277-279 62-18083 <*>
1960 V16 N4 P279-281 62-18084 <*>
1960 V16 N4 P281-283 62-14469 <*>
62-18081 <*>
1960 V16 N4 P283-285 62-14478 <*>
62-18082 <*>
1960 V16 N4 P286-288 62-18085 <*>
1962 V18 N3 P109 2434 <BSRA>

NIPPON RINGAKKAI SHI
1948 V30 N1/2 P1-11 I-238 <*>
1953 V35 P406-409 II-226 <*>
1953 V35 N12 P396-405 C-2229 <NRC>
1954 V36 P19-22 II-227 <*>
1954 V36 N1 P22-24 57-1132 <*>
1954 V36 N4 P106-112 2406 <*>
1955 V37 N4 P147-152 59-20087 <*>

NIPPON SAIKINGAKU ZASSHI
1955 V10 P823-827 66-13181 <*>
1955 V10 N9 P771-775 66-13180 <*> 0
1956 V11 N9 P803-812 AEC-TR-3524 <*> 0
1956 V11 N10 P879-884 59-19433 <+*> 0
1958 V13 N11 P1017-1022 PANSDOC-TR.671 <PANS>
1959 V14 P32-41 66-13172 <*>
1959 V14 N10 P868-872 68N57J <ATS>
1959 V14 N12 P1026-1029 69N57J <ATS>
1960 V15 N11 P1193-1199 63-01242 <*>
1962 V17 N8 P711-722 66-13040 <*>
1963 V18 N1 P44-51 64-00067 <*>

NIPPON SANSHIGAKU ZASSHI
1942 V13 N1 P1-3 61-00296 <*>
1953 V22 P78-80 37F4J <ATS>

NIPPON SEIBUTSU CHIRIGAKKAI KAIHO
1955 V16/9 P192-196 62-00118 <*>

NIPPON SEIRIGAKU ZASSHI
1942 V5 P79-88 3618 <K-H>
1951 V13 P203-210 9353-E <K-H>
1961 V23 P161-169 11401-F <KH>
64-10425 <*> 0
1961 V23 P527-546 63-20375 <*> 0
1962 V24 P443-450 65-00099 <*>
65-29497 <$>
1962 V24 P521-524 13161 <K-H>
64-10602 <*>

NIPPON SHONIKA GAKKAI ZASSHI
1961 V65 N2 P260-267 16N57J <ATS>
1961 V65 N2 P268-275 15N57J <ATS>

NIPPON SOCIETY OF APPLIED ENTOMOLOGY
SEE OYO-KONTYU

NIPPON SUGAKU BUTSURIGAKKWAI KIZI
1942 V24 P137-164 I-432 <*>
1943 V25 P540-552 2484 <*>
1943 S3 V25 P73-86 AD-621 881 <=$*>
66-10422 <*>

NIPPON SUISAN GEKKAISHI
1933 V2 N3 P107-111 AL-245 <*>
1934 V3 N4 P196-202 AL-245 <*>
1937 V6 N1 P13-21 AL-245 <*>
1937 V6 N2 P73-74 AL-354 <*>
1938 V7 N1 P79-88 AL-28 <*>
1940 V8 N6 P292-294 AL-491 <*>
1940 V9 N3 P100-102 AL-970 <*>
1940 V9 N3 P103-106 AL-774 <*>
1940 V9 N4 P145-148 AL-772 <*>
1941 V9 N6 P231-236 AL-771 <*>
1943 V11 N5/6 P179-183 AL-245 <*>
1948 V13 N5 P207-209 60-17430 <=*>
1950 V16 N2 P35-39 I-865 <*>
1952 V18 N6 P245-248 64T90J <ATS>
1953 V19 N7 P828-831 60-17428 <*=>
1953 V19 N7 P832-835 60-17429 <*=>
1953 V19 N8 P12-15 65-13853 <*>
1954 V19 P1012-1014 I-867 <*>
1954 V19 P1021-1027 I-710 <*>
1954 V20 N2 P136-139 65-13854 <*>
1954 V20 N3 P245-247 65-13855 <*>
1954 V20 N7 P610-612 65-13856 <*>
1955 V21 N1 P37-41 58-1707 <*>
1955 V21 N3 P187-189 65-13857 <*>
1955 V21 N8 P915-920 60-17427 <=*>
1957 V22 N12 P787-790 60-15282 <*+>
1957 V22 N12 P791-794 60-17105 <*+>
1957 V23 N5 P273-277 60-17107 <*+>

1957 V23 N5 P278-281 60-17106 <*+>
1958 V23 N11 P684-695 C-4701 <NRC>
59-18376 <*=>
1958 V24 N5 P334-337 60-13260 <=*>
1959 V24 N9 P735-738 60-17110 <*+>
1959 V24 N12 P957-960 59-19361 <*+>
1960 V26 N1 P45-48 80M47J <ATS>

NIPPON YAKURIGAKU ZASSHI
1927 V5 N3 P365-376 64-10676 <*> OP
1939 V27 P156-164 66-11723 <*> O
1954 V50 N1 P70-75 3178-H <K-H>
1955 V51 N1 P60-61 6445-F <K-H>
1955 V51 N1 P62-69 5668-B <K-H>
1956 V52 P113-120 63-00280 <*>
1956 V52 P429-435 61-10877 <*>
1957 V53 P1023-1028 63-18468 <*>
1957 V53 P1086-1118 64-14517 <*>
9061-B <K-H>
1957 V53 N3 P340-347 59-16161 <*=> O
1957 V53 N3 P553-565 59-14958 <*=> O
63-16914 <*>
6975-A <K-H>
1958 V54 P7-20 62-00853 <*>
1958 V54 P452-458 9428 <K-H>
1958 V54 P762-790 9477-C <K-H>
1958 V54 P825-837 9054-C <K-H>
1960 V56 N5 P1223-1234 11627-J <K-H>
63-16604 <*>
1961 V57 P353-362 66-12659 <*> O
1961 V57 P363-369 64-20010 <*> O
1964 V60 P563-568 66-12614 <*>

NISSHIN IGAKU
1955 V42 N10 P553-566 58-1739 <*>
1960 V47 P355-359 63-01521 <*>
1961 V48 P353-366 62-00867 <*>
1961 V48 N5 P328-330 64-18623 <*>

NITROCELLULOSE
1934 V5 N9 P159-162 AL-719 <*>
1934 V5 N10 P181-184 AL-719 <*>
1934 V5 N11 P203-206 AL-719 <*>
1939 V1C P1C9-110 AL-716 <*>
1939 V10 P128-130 AL-716 <*>
1940 N12 P227-229 59-18229 <+*>
1940 V11 P223-225 AL-233 <*>
1941 V12 N5 P83-88 AL-717 <*>

NO TO SHINKEI
1961 V13 N7 P539-547 66-10978 <*>
1963 V15 P91-95 66-12109 <*>
1964 V16 N1 P44-58 65-30183 <=$>

NOBEL HEFTE. SPRENGMITTEL IN FORSCHUNG UND PRAXIS
1955 V21 N1 P1-10 C-2438 <NRC>
1957 V23 P220-239 49J19G <ATS>
1957 V23 N4 P153-176 58-59 <*>
1958 V24 P161-170 60-10903 <*> O
1959 V25 N4 P149-159 SCL-T-406 <*>

NODONG SINMUN
1966 P2 66-33518 <=$>

NOGAKU KENKYU
1955 V43 P134-143 C-2587 <NRC>

NOGYO DOBOKU KENKYU
1962 V29 N7 P309-314 66-13584 <*>
1964 V32 N1 P15-23 66-14649 <*$>
1964 V32 N1 P24-30 66-13437 <*>

NOGYO DOKOKU KENKYU BESSATSU
1965 N12 P25-29 66-11538 <*>

NOGYO GIJUTSU
1964 V19 N5 P229-231 66-14653 <*$>

NOGYC KIKAI GAKKAISHI
1957 V19 N1 P13-17 66-10730 <*>
1963 V25 N2 P113-118 65-14142 <*> O

NOGYO KISHO
1954 V9 N3/4 P35-38 49H13J <ATS>
1957 V12 N4 P138-144 62-00100 <*>
1958 V14 N2 P45-53 62-00139 <*>
1960 V15 P41-45 62-01400 <*>
1960 V15 P93-96 62-01410 <*>
1960 V15 P123-129 62-01411 <*>

NOGYO OYOBI ENGEI
1943 V18 N2 P189-190 66-12657 <*>
1964 V39 N5 P827-828 66-14024 <*$>
1964 V39 N11 P1729-1730 66-14025 <=$> O

NORD INDUSTRIEL ET COMMERCIAL
1962 V29 P22-23 SPEC.NO. 3256 <BISI>
1962 V29 P26 SPEC.NO. 3256 <BISI>
1962 V29 P35 SPEC.NO. 3256 <BISI>

NORDISK BETONG
1957 V1 P117-127 62-10676 <*>
1959 V3 N2 P139-144 60-10350 <*>
1960 V4 N3 P231- 61-10805 <*> O
1961 V5 N1 P1-28 66-13784 <*> O
1963 V7 N1 P83-114 64-16389 <*>
1964 V8 N2 P147-170 65-11704 <*>
1964 V8 N2 P227-244 66-10171 <*> O
1965 V9 N1 P1-26 66-10170 <*> O

NORDISK HYGIENISK TIDSKRIFT
1956 V37 P1-9 58-1560 <*>
1959 V40 P71-88 11139 <K-H>
1960 V41 P199-216 10494-C <K-H>

NORDISK JORDBRUKSFORSKNING
1943 N6 P340-346 59-20646 <*>

NORDISK MEDICIN
1939 V4 P3805-3808 57-1661 <*>
1940 V8 P2324-2327 58-752 <*>
1941 V10 N14 P1046-1055 57-1621 <*>
1941 V10 N19 P1439-1451 57-1526 <*>
1941 V10 N22 P1748- 57-1649 <*>
1941 V11 N35 P2481-2484 57-1604 <*>
1941 V11 N35 P2484-2487 57-1536 <*>
1941 V12 P2924-2929 57-3426 <*>
1945 V26 P903 65-00365 <*>
1946 V30 P1415-1418 57-235 <*>
1947 V33 P1071- I-94 <*>
1947 V34 N15 P847-850 66-12021 <*>
1947 V35 P1586-1588 59-18140 <=*>
1949 V41 P451-455 3282 <K-H>
1949 V41 P506 1622 <K-H>
1949 V41 P627-632 57-292 <*>
1950 V43 N9 P363-366 59-20709 <*> O
1951 V46 N27 P1069-1072 62-18446 <*> O
1952 V47 P496 63-01057 <*>
1952 V47 N9 P271-275 59-16287 <+*>
1952 V48 N28 P967-969 63-00296 <*>
1952 V48 N39 P1321-1324 C-4475 <NRC>
1952 V48 N41 P1409-1411 58-1561 <*>
1952 V48 N51 P1754-1758 59-20708 <*>
1953 V49 P172-173 AL-2 <*>
1953 V49 P563-565 3061 <*>
1954 N52 P959 58-1841 <*>
1954 V51 N11 P376-377 62-14039 <*>
62-18224 <*> O
1955 V53 P951-956 3738-A <K-H>
1955 V54 N50 P1845-1848 4065-B <K-H>
1956 V55 P857- 59-22397 <*=>
1956 V55 N3 P85-87 57-234 <*>
1956 V56 N31 P1104-1105 58-373 <*>
1956 V56 N39 P1421-1423 62-00902 <*>
1956 V56 N42 P1511-1517 66-13157 <*> O
1957 V57 N12 P424-425 62-01303 <*>
1957 V57 N23 P822-825 58-1341 <*>
1957 V58 N34 P1268-1269 59-12870 <+*> O
1957 V58 N34 P1270-1271 58-1345 <*>

1957 V58 N42 P1582-1583 58-1354 <*>
1957 V58 N48 P1849-1852 62-00873 <*>
1957 V58 N51 P1970-1971 64-10416 <*>
8809-B <K-H>
1958 V59 N2 P55-61 60-15075 <*+>
1958 V59 N25 P827-832 NP-TR-195 <*> O
1958 V59 N25 P833-836 59-10455 <*> O
1959 V61 N1 P13-15 60-17651 <*+> O
1959 V61 N2 P41-47 66-13166 <*> O
1959 V61 N22 P819-823 60-15080 <*=> O
1959 V62 P1294-1296 63-16931 <*>
9531 <KH>
1959 V62 N49 P1752-1754 65-14309 <*> O
1960 V63 N2 P53-56 61-00844 <*>
1960 V63 N18 P565-566 63-18945 <*> O
1960 V64 P1288-1290 N65-18178 <=$>
1962 V67 N9 P280-283 53Q67D <ATS>
1962 V68 P930-931 63-01763 <*>
63-26697 <=$>
1962 V68 N43 P1379-1381 64-00433 <*>
1962 V68 N44 P1413-1416 63-20123 <*> O
1963 V69 P165-168 63-01767 <*>
1963 V69 N24 P694-697 66-12624 <*>
1964 V71 N4 P105-108 66-12157 <*>
1964 V71 N9 P271-273 85S89D <ATS>
1964 V72 N34 P1004-1005 65-12837 <*>
1965 V74 N33 P816-819 66-13167 <*>
1965 V74 N34 P841-843 65-17037 <*> O
1965 V74 N34 P843-844 65-17038 <*>

NORDISK MEDICINSK TIDSKRIFT
1931 V3 P561-567 63-16678 <*> O
1932 V4 P685-691 57-827 <*>
1932 V4 P1011-1014 58-1601 <*>
1934 V7 P744-747 63-16676 <*>
1935 V9 N1 P953-960 59-14910 <+*> O
1938 V15 P566-571 60-17573 <*+>

NORDISK MEJERITIDSSKRIFT
1951 V17 N7 P80-83 I-231 <*>
II-876 <*>
1952 V18 N11 P170-171 I-280 <*>
1952 V18 N11 P174-175 I-760 <*>
1962 V28 N5 P86-87 64-27257 <$>

NORDISK PSYKIATRISK TIDSSKRIFT
1962 V16 P99-102 66-11827 <*>
1962 V16 P103-108 66-12023 <*>
1962 V16 P108-111 66-12171 <*>
1962 V16 P111-116 66-12174 <*>
1963 V17 P169-177 66-12011 <*>,
1964 V18 P25-32 65-17300 <*>
1964 V18 P578-580 53T93D <ATS>
1965 V19 N1 P73-78 79S87N <ATS>

NORDISK TISSKRIFT FOR INFORMATIONSBEHANDLING
1963 V3 N1 P1-26 64-14635 <*>

NORDISK VETERINAERMEDICIN
1950 V2 P286-301 2570 <*>
1950 V2 N2 P906-915 2569 <*>
1951 V3 P109-126 II-556 <*>
1953 V5 P160-165 I-993 <*>
1953 V5 P663-669 MF-2 <*>
1954 V6 P533-546 61-20954 <*>
1955 V7 P1056-1062 58-363 <*>
63-16463 <*> O
1960 V12 P471-489 66-12018 <*>
1964 V16 P632-642 10R79D <ATS>

NORDISK VETERINAERMEDICIN. DANSK UDGAVE
1954 V6 P919-935 61-20422 <*>

NORDISKT MEDICINSKT ARKIV
1875 V7 N7 P1-33 66-12838 <*>

NORGES GEOLOGISKE UNDERSOGELSE
1953 V2 N164 P1044-1048 <GSL> O
1957 N203 P100-111 AEC-TR-4067 <*>

NORMALIZACJA
1956 V14 P330-332 T1955 <INSD>
1960 V28 N7/8 P360-365 61-11355 <=>

NORRLANDS SKOGSVARDSFORBUNDS TIDSKRIFT
1953 V4 P499-582 II-968 <*>
1956 N3 P293-388 C-3306 <NRCC>
1962 N1 P131-174 63-01063 <*>

NORSK FISKERITIDENDE
1912 V31 N11 P437-449 2845 <*>
1918 V37 P42-46 88P64N <ATS>

NORSK GALVANO-TEKNISK TIDSSKRIFT
1961 V4 N1 P6-8 2247 <BISI>

NORSK GEOLOGISK TIDSSKRIFT
1927 V9 P266-270 2293 <*>
1957 V37 N2 P1-7 17L33N <ATS>

NORSK MAGAZIN FOR LAEGEVIDENSKABEN
1901 V16 P1138-1145 66-10972 <*>

NORSK SKOGINDUSTRI
1950 V5 P119-136 AL-427 <*>
1951 V5 N10 P303-308 I-232 <*>
1952 V6 P226-241 I-1041 <*>
1952 V6 N4 P113-121 I-663 <*>
1952 V6 N10 P306-307 I-8 <*>
1952 V6 N11 P379-384 I-933 <*>
1953 V7 N6 P184-191 II-680 <*>
1953 V7 N7 P216-218 II-583 <*>
1953 V7 N7 P218-222 II-583 <*>
1953 V7 N7 P223 II-583 <*>
1953 V7 N7 P225-226 II-583 <*>
1953 V7 N10 P322-334 II-729 <*>
1953 V7 N10 P335-341 AL-714 <*>
1953 V7 N11 P424-427 I-450 <*>
1954 V8 N7 P236-240 CSIRO-2934 <CSIR>
1954 V8 N8 P269-270 2722 <*>
1954 V8 N11 P395-400 2234 <*>
1955 N5 P160- 57-1733 <*>
1956 P85- 57-1706 <*>
1956 N8 P292-294 10H13S-N <ATS>
1956 V10 N4 P123-136 59-10293 <*> O
1956 V10 N6 P204-213 C-2373 <NRC>
FPRB-TRANS-110 <NRC>
57-2629 <*>
1956 V10 N10 P344-348 C-2554 <NRC>
59-10292 <*>
1956 V10 N11 P389-394 59-10218 <*> O
1957 V11 N1 P14-21 C2372 <NRC>
1957 V11 N4 P124-130 58-1280 <*>
1957 V11 N11 P425-432 59-10217 <*> O
1957 V11 N12 P527-536 59-15913 <*>
1958 V12 N3 P81-86 59-10260 <*> O
1958 V12 N3 P87-94 59-10222 <*>
1958 V12 N3 P104-113 58-2685 <*>
1958 V12 N4 P147 58-2139 <*>
1958 V12 N9 P318-324 59-15234 <*>
1959 V13 N5 P146-157 60-18922 <*> O
1960 V14 N5 P167-178 61-10391 <*>
1960 V14 N5 P182-194 61-00666 <*>
97M42N <ATS>
1960 V14 N8 P281-291 61-10390 <*>
1960 V14 N10 P369-377 61-10850 <*>
1960 V14 N11 P441- 62-10231 <*> O
1960 V14 N11 P455-469 61-16116 <*>
61-16455 <*> O
1961 V15 P326-333 63-10179 <*> O
1961 V15 N3 P112-116 62-14480 <*> O
1961 V15 N3 P119-122 61-18863 <*>
1962 V16 N2 P58-64 62-16506 <*> O
1962 V16 N3 P87-96 63-10617 <*>
1962 V16 N11 P458-469 63-18695 <*> O
1963 V17 N11 P468-477 C-5049 <NRC>
1964 V18 P283-286 66-14433 <*>
1964 V18 N3 P92-97 65-10267 <*> O
1964 V18 N6 P210-211 65-00170 <*>
1964 V18 N9 P323-325 65-12917 <*>

1965 V19 N4 P155-163 65-14047 <*>
1965 V19 N5 P181-186 C-5587 <NRC>
66-10844 <*>
1966 V20 N3 P98-105 66-13911 <*>

NORSK TIDSSKRIFT FOR MILITAERMEDICIN
1930 V34 P121-126 2958 <*>

NOTAS DEL MUSEO DE LA PLATA
1947 V12 P207-216 I-546 <*>

NOTE. CEA. COMMISSARIAT A L'ENERGIE ATOMIQUE. FRANCE
CEA-NOTE-414 AEC-TR-6010 <=*$>
269 NP-TR-412 <*>
1957 222 AEC-TR-4504 <*>
1962 CEA-NOTE-381 AEC-TR-6009 <=*$>
1962 INT/SPR/62-298 UCRL TRANS-846 <*>

NOTE. INSTITUT DE PHYSIQUE DU GLOBE. PARIS
1964 N3 P1-27 N65-27703 <=*$>

NOTE RECENSION ENOTIZIE DI INSTITUTO SUPERIORE DELLE POSTE E DELLE TELECOMUNICAZIONI
1956 V5 N1 P3-13 R-5127 <*>

NOTES TECHNIQUES. OFFICE NATIONAL D'ETUDES ET DE RECHERCHES AERONAUTIQUES
NOTE NO 36 NACA TM-1410 <NASA>
1954 NO 3/1727/A 58-1274 <*>
1954 NOTE NO 22 3008 <*>
1957 NO 40 61-18373 <*>

NOTIZIARIO DELL'AMMINISTRAZIONE SANITARIA
1951 V4 N5 P137-155 2507 <*>
1952 V5 N4/5 P169-176 2620 <*>

NOTIZIARIO CHIMICO INDUSTRIALE
1927 V2 P122 58-2032 <*>
1928 P271-274 57-1927 <*>
1928 P346-352 57-1927 <*>

NOURISSON, REVUE D'HYGIENE ET DE PATHOLOGIE DE LA PREMIERE ENFANCE
1958 V46 N5 P179-186 61-00901 <*>
1959 V47 P91-102 60-17647 <+*>
1960 V48 P41-48 63-16502 <*>

NOUVEAUTES MEDICALES
1963 V12 N4 P230 64-18702 <*>

NOUVEAUTES TECHNIQUES MARITIMES
1965 V13 P88- 2194 <BSRA>
1965 V13 P152- 2205 <BSRA>
1965 V13 P197- 2210 <BSRA>

NOUVELLE REVUE FRANCAISE D'HEMATOLOGIE
1961 V1 P473-478 UCRL TRANS-779 <=*>
1961 V1 N3 P445-459 UCRL TRANS-778 <*>
1961 V1 N6 P872-879 64-00200 <*>
1963 V3 P1-4 63-18742 <*>
1965 V5 N4 P591-600 66-12654 <*> O

NOVA ACTA LEOPOLDINA
1957 SNS V19 N134 P55-75 63-00506 <*>

NOVA ACTA REGIAE SOCIETATIS SCIENTIARUM UPSALIENSIS
1927 S4 P3-23 61-16064 <*>

NOVA ADMINISTRACIJA. BELGRAD
1964 N5 P309-311 64-41825 <=>

NOVA MYSL
1962 N1 P48-57 62-24395 <=*>
1962 N7 P829-839 62-33672 <=*>
1962 N9 P1036-1046 62-33528 <=*>
1963 N7 P815-826 63-41049 <=>
1963 N8 P908-919 64-31132 <=>
1964 N7 P827-834 64-41449 <=>
1964 N9 P1051-1073 64-51281 <=>
1965 N3 P371-377 65-30843 <=$>
1965 N4 P434-444 65-30996 <=$>

NOVA TECHNIKA
1958 V3 N12 P539-550 59-21161 <=*> C

NOVAYA TEKHNIKA MONTAZHNYKH I SPETSIALNYKH RABOT V STROITELSTVE
1959 N3 P31-33 NLLTB V1 N9 P46 <NLL>
1959 V21 N4 P19-21 61-19681 <=*> C
1959 V21 N7 P14-16 59-13909 <=>
1959 V21 N9 P5-10 63-19342 <=*>
1959 V21 N12 P26-28 62-13280 <*=> O

NOVENYTERMELES
1952 V1 P51-66 3180-D <K-H>
1960 V9 N3 P247-250 81N56H <ATS>
1961 V10 N3 P269-274 62-23634 <=*>
1962 V11 N1 P3-16 65-22212 <$>
1962 V11 N4 P341-354 65-23697 <$>

NOVOE V NAUKA I TEKHNIKE VITAMINOY
1946 V1 P43-47 RT-3202 <*>

NOVOE V TEKHNOLOGII MASHINOSTROENIYA
1959 N1 P119-142 AD-259 240 <=$>

NOVOE V ZHIZNI, NAUKE, TEKHNIKE. SERIYA 2: FILOSOFIYA
1963 N9 P48 63-31339 <=>
1963 N24 P1-48 64-21998 <=>

NOVOE V ZHIZNI, NAUKE, TEKHNIKE. SERIYA 4: TEKHNIKA
1962 N9 P1-32 63-24533 <=*$>
1963 N9 P34-37 63-31555 <=>
1963 N21 P1-32 64-21584 <=>
1963 S4 N24 P1-39 64-41198 <=>
1965 N24 P3-48 66-33911 <=$>

NOVOE V ZHIZNI, NAUKE, TEKHNIKE. SERIYA 8: BIOLOGIYA I MEDITSINA
1963 N7 P3-54 63-31997 <=>
1963 N11 P1-48 64-21344 <=>
1964 N4 P1-32 64-31533 <=>
1964 N17/8 P1-80 65-30591 <=$>
1964 S8 N5 ENTIRE ISSUE 64-31449 <=>
1964 S8 N10 P1-40 64-41261 <=>
1964 S8 N16 P3-31 64-51983 <=$>
1964 S8 N19 P2-32 65-31962 <=$>
1964 S8 N1/2 P3-80 64-41490 <=>
1964 S8 N1/2 P36-79 64-31414 <=>

NOVOE V ZHIZNI, NAUKE, TEKHNIKE. SERIYA 9. FIZIKA MATEMATIKA ASTRONOMIYA
1963 N19 P3-31 64-21544 <=>
1963 N22 P1-24 64-21976 <=>

NOVOE V ZHIZNI, NAUKE, TEKHNIKE. SERIYA 12: GEOLOGIYA I GEOGRAFIYA
1964 S12 N7 P1-64 64-51146 <=>

NOVOSTI MEDITSINSKOI TEKHNIKI
1959 N1 P31-44 62-24525 <=*>
1959 N3 P42-63 61-28556 <*=>
1959 N3 P77-85 61-28586 <*=>
1959 N4 P50-70 61-28556 <*=>
1959 N4 P84-95 61-27668 <=*>
1960 N3 P27-45 63-13376 <=*>
1960 N6 P48-56 63-20578 <=*>
1961 N1 P47-59 64-10983 <=*$>

NOVOSTI MEDITSINY
1951 N24 P101-108 RT-1043 <*>
1952 V26 P65-68 62-15429 <*=>
1952 V26 P68-72 63-14100 <=*>

NOVOSTI NEFTYANOI I GAZAVOI TEKHNIKI. NEFTEPERERABOTKA I NEFTEKHIMIYA

1962 N5 P35-39 66-12871 <*>

NOVOSTI NEFTEPERERABOTKA
1934 N10 P5 61-18159 <*>
1934 N14 P6-7 61-18148 <*>
1934 N16 P1-2 61-16952 <*>
1934 N16 P3-4 61-16686 <*>
1934 N7/8 P1-2 61-18156 <*>
1934 N21/2 P6-7 61-16816 <*>
1935 N2 P4 61-16815 <*>
1935 N4 P1-2 61-16634 <*>
1935 N5 P2-4 61-16621 <*>

NOVOSTI NEFTYANOI TEKHNIKI. GEOLOGIYA
1958 N8 P3-7 IGR V1 N8 P79 <AGI>
1958 N8 P12-16 IGR V1 N9 P19 <AGI>
1958 N8 P34-41 IGR V1 N7 P24 <AGI>

NOVOSTI TEKHNIKI
1934 N26 P11 R-2617 <*>
1936 V5 N9 P34 61-16670 <*>
1939 N23/4 P50-51 R-3430 <*>
1940 N9 P32-33 RT-1431 <*>
1940 V9 N23 P20-21 59-17859 <+*>
1940 V9 N23 P33 61-20779 <*=>
1940 V9 N13/4 P38-40 61-20364 <*=> P
1941 V10 N2 P16-17 62-14641 <=*> P
1941 V10 N8 P13-15 62-14609 <=*>
1941 V10 N8 P38-39 62-14617 <=*>

NOVOSTI TORGOVOI TEKHNIKI
1957 N4 P32-35 60-21213 <=> 0

NOVYE KNIGI SSSR
1960 N40 P55-61 NLLTB V4 N3 P228 <NLL>
1962 N14 P57-60 NLLTB V5 N2 P104 <NLL>
1962 N17 P60-61 NLLTB V4 N10 P920 <NLL>
1962 N18 P51-53 NLLTB V4 N12 P1100 <NLL>
1962 N18 P56-63 NLLTB V5 N1 P19 <NLL>
1962 N35 P46-52 NLLTB V4 N12 P1111 <NLL>
1965 N42 P12-13 66-30090 <=$>
1965 N42 P46-47 66-30090 <=$>
1966 N6 P48-50 66-31845 <=$>

NOVYE METODY FIZIKO-KHIMICHESKIKH ISSLEDOVANII
1957 V2 P50-55 63-10156 <*>
1963 N4 P1-2 63-31065 <=>
1963 N4 P6 63-31065 <=>

NOVYE TOVARY
1959 N7 P6 64-13706 <=*$>
1959 N8 P4 64-13705 <=*$>
1961 N1 P6 61-23580 <=>
1961 N2 P11-12 61-23580 <=>
1962 N1 P4-5 62-32571 <=>
1962 N8 P4-5 63-21980 <=>
1962 N10 P7 63-15103 <=>
1962 N11 P18 63-21423 <=>
1962 N12 P1 63-21542 <=>
1962 N12 P6 62-32571 <=>
1963 N2 P1,5-7,11 63-21517 <=>
1963 N3 P1 63-31116 <=>
1963 N4 P3 63-31429 <=>
1963 N5 P8-9 63-31628 <=>
1963 N6 P2-3 63-31644 <=>
1963 N6 P6-7 63-31669 <=>
1963 N7 P3-4 63-31782 <=>

NOVYI KHIRURGICHESKII ARKHIV
1936 N36 P289-294 R-1082 <*>
1937 N38 P452-456 R-1057 <*>
1937 N39 P244-261 R-1087 <*>
1938 N3 P416 R-1094 <*>
1959 N4 P3-12 60-11087 <=>
1959 N4 P12-16 59-13983 <=>
1959 N6 P13-17 60-11465 <=>
60-11543 <=>
1960 N2 P18-24 60-41227 <=>
1960 N2 P58-63 60-41228 <=>
1960 N2 P64-68 60-41229 <=>
1960 N2 P89-96 60-41230 <=>
1960 N2 P96-101 60-41231 <=>
1960 V38 N2 P143-145 62-15679 <=*>
1960 V39 N4 P74-79 61-11640 <=>
1960 V39 N4 P79-81 61-11615 <=>
1960 V39 N4 P81-85 62-23484 <=*>
1961 V40 N1 P26-33 61-21740 <=>
1961 V40 N2 P45-49 61-31149 <=>
1961 V40 N4 P34-57 61-27676 <=*>
1961 V40 N4 P37-41 63-23912 <=*$> 0
1961 V40 N4 P66-67 61-27676 <=*>
1961 V40 N4 P75-81 61-27676 <=*>
1961 V40 N5 P3-11 61-28065 <*=>
1961 V40 N5 P11-19 61-28055 <*=>
1961 V40 N5 P31-36 61-28067 <*=>
1961 V40 N6 P35-41 61-28293 <*=>
1961 V40 N10 P26-32 62-15947 <=*>
1961 V40 N11 P51-57 62-23439 <=*>

NOVYI MIR
1960 V36 N8 P202-210 61-28577 <*=>
1965 N12 P194-213 66-30789 <=$>

NOWE ROLNICTWO
1957 V6 N18 P785-787 60-21535 <=>
1964 V13 N11 P41-43 64-31971 <=>
1966 N18 P14-15 67-30192 <=>
1966 N21 P4-7 67-30238 <=>
1966 N22 P7-9 67-30192 <=>
1967 N2 P4-8 67-30839 <=$>
1967 V16 N5 P1-4 67-31637 <=$>
1967 V16 N5 P10-12 67-31637 <=$>

NOWE DROGI
1961 V15 N2 P102-107 62-19323 <*=>
1961 V15 N2 P108-114 62-19324 <=*>
1961 V15 N10 P121-127 62-15784 <*=>
1964 V18 N8 P69-83 64-41514 <=>
1964 V18 N9 P94-103 64-51051 <=>
1964 V18 N10 P38-66 64-51481 <=>
1964 V18 N10 P120-131 64-51481 <=>
1966 V20 N10 P41-53 66-34801 <=$>
1967 N2 P3-18 67-30685 <=$>
1967 N2 P66-78 67-30685 <=$>

NOWOTWORY
1958 V8 P59-62 62-01425 <*>

NUCLEAR ENGINEERING. TOKYO
SEE GENSHIRYOKU KOGYO

NUCLEAR FUSION
1961 V1 N2 P82-100 63-23323 <=*>
1961 V1 N3 P189-194 64-14710 <*>
1961 V1 N3 P195-197 62-25557 <=*>
1962 V2 P259-263 SUP 65-17409 <*>
1962 V2 N3 P1045-1047 SUP 64-16004 <*>
1964 V4 P145-151 N65-27683 <=$>
1965 V5 N1 P7-16 AEC-TR-6669 <=$>
1965 V5 N1 P20-40 AEC-TR-6669 <=$>
1965 V5 N1 P85-86 AEC-TR-6669 <=$>
1965 V5 N2 P125-143 AEC-TR-6670 <=$>
1965 V5 N2 P150-155 AEC-TR-6670 <=$>
1965 V5 N3 P181-241 AEC-TR-6671 <=$>
1965 V5 N3 P249-250 AEC-TR-6671 <=$>
1966 V6 N3 P169-181 AEC-TR-6675 <=$>
1966 V6 N3 P188-199 AEC-TR-6675 <=$>
1966 V6 N3 P212-214 AEC-TR-6675 <=$>
1966 V6 N3 P228-230 AEC-TR-6675 <=$>

NUCLEAR FUSION RESEARCH REPORT. JAPAN
SEE KAKU YUGO KENKYU

NUCLEAR INSTRUMENTS AND METHODS
1959 V5 P300-311 NP-TR-691 <*>
1960 V9 P194-200 AEC-TR-4424 <*>
1961 V13 P282-286 NP-TR-947 <*>
1961 V13 P297-304 UCRL TRANS-866 <*>
1962 V15 P77-86 NP-TR-1042 <*>
1962 V15 P327-354 AERE-TR-928 <*>

Citation	Translation
	63-18781 <*> 0
1962 V16 P301-304	NP-TR-1014 <*>
1962 V17 P1-19	AEC-TR-5727 <*>
1962 V17 P20-30	63-00728 <*>
1963 V21 P65-74	66-12027 <*>
1963 V24 P197-212	66-12033 <*>
1964 V25 N2 P269-284	65-11150 <*>
1964 V29 P181-204	C-5672 <NRC>
1965 V34 P77-87	66-11973 <*>
1966 V39 P119-124	66-11869 <*>

NUCLEAR PHYSICS

Citation	Translation
1958 N5 P17-22	R-2782 <*>
1958 N7 P451-479	AEC-TR-3865 <=>
1958 N8 P91-105	AEC-TR-3865 <=>
1959 N10 P181-196	AEC-TR-3865 <=>
1959 N10 P509-526	AEC-TR-3865 <=>
1959 V13 N1 P136-139	NP-7814 <*>
1960 V15 N1 P89-91	UCRL TRANS-563(L) <=*>
1960 V17 N1 P153-162	62-01070 <*>
1961 V27 N2 P294-322	66-10058 <*> 0
1961 V28 P220-243	AEC-TR-5084 <*>
1962 V39 N2 P263-272	C-4427 <NRC>
1963 V44 P164-172	ORNL-TR-38 <*>
1963 V45 P529-554	66-14096 <*>
1964 V58 P593-600	UCRL TRANS-1081 <*>
1965 V73 P417-423	66-11132 <*>

NUCLEAR POWER. JAPAN
SEE GENSHIRYOKU HATSUDEN

NUCLEAR STUDY. JAPAN
SEE GENSHIKAKU KENKYU

NUCLEUS. REVUE SCIENTIFIQUE A L'AGE ATOMIQUE

Citation	Translation
1962 N2 P115-130	AEC-TR-5380 <*>
1963 V4 N2 P97-108	64-18133 <*>
1963 V4 N5 P384-391	SC-T-64-46 <*>

NUKLEONIK

Citation	Translation
1958 V1 P13-18	62-01197 <*>
1958 V1 P103-106	HW-TR-2 <*>
	HW-TR-2 <=*>
1958 V1 N1 P22-28	62-00192 <*>
1958 V1 N1 P29-40	SCL-T-271 <*>
1958 V1 N2 P41-48	62-00382 <*>
1958 V1 N2 P64-66	NP-TR-322 <*>
1958 V1 N2 P66-67	20L30G <ATS>
1958 V1 N2 P68-73	61-13737 <*=>
1959 V1 P197-208	AEC-TR-4351 <*>
1959 V1 P337-341	AEC-TR-4109 <*>
1959 V1 N5 P172	NP-TR-428 <*>
1959 V1 N8 P295-305	UCRL TRANS-631(L) <*>
1959 V1 N8 P319-324	UCRL TRANS-548(L) <*>
1959 V2 N4 P131-138	61-20508 <*> 0
1960 V2 N7 P271-276	AEC-TR-4630 <*>
1961 V3 P201-204	UCRL TRANS-842 <*>
1961 V3 N2 P61-76	AEC-TR-5039 <*>
	87N57G <ATS>
1961 V3 N3 P110-131	AEC-TR-5483 <*>
	56P62G <ATS>
1961 V3 N6 P257-267	AEC-TR-6053 <*>
1961 V3 N7 P295-301	62-01051 <*>
1962 V4 P1-9	AEC-TR-5856 <*>
1962 V4 P9-18	AEC-TR-5857 <*>
1962 V4 N1 P23-25	SCL-T-421 <*>
1962 V4 N1 P46-53	AEC-TR-5479 <*>
1962 V4 N2 P84-91	NP-TR-936 <*>
1962 V4 N4 P167-174	ORNL-TR-96 <*>
1962 V4 N6 P266-267	SCL-T-465 <*>
1962 V4 N7 P306-310	AEC-TR-5863 <*>
1962 V4 N7 P310-317	AEC-TR-5928 <*>
1962 V4 N8 P319-323	NP-TR-1041 <*>
1963 V5 N2 P74-82	AEC-TR-5846 <=*$>
1963 V5 N3 P115-120	65-13819 <*>
1963 V5 N3 P121-124	66-14454 <*>
1963 V5 N4 P154-159	66-12837 <*>
1963 V5 N6 P236-239	66-12141 <*>
1964 V6 P320-322	66-11839 <*>
1964 V6 N3 P153-157	65-17311 <*>
1964 V6 N4 P168-174	66-12031 <*>
1964 V6 N6 P288	66-11128 <*>
1965 V7 N1 P8-14	66-11876 <*>
1965 V7 N3 P130-144	66-11134 <*>
1966 V8 N3 P137-139	66-12142 <*>
1966 V8 N5 P273-282	66-11872 <*>

*NUKLEONIKA

Citation	Translation
1957 V2 N1 P1-30	AEC-TR-4140 <=>
1957 V2 N1 P1-8	62-16473 <*>
1957 V2 N1 P197-206	AEC-TR-4140 <=>
1957 V2 N1 P221-222	AEC-TR-4140 <=>
1957 V2 N2 P225-265	AEC-TR-4141 <=>
1957 V2 N2 P267-349	AEC-TR-4141 <=>
1957 V2 N2 P373-380	AEC-TR-4141 <=>
1957 V2 N3 P409-449	AEC-TR-4142 <=>
1957 V2 N3 P451-463	AEC-TR-4142 <=>
1957 V2 N3 P465	AEC-TR-4142 <=>
1957 V2 N3 P465-477	AEC-TR-4142 <=>
1957 V2 N3 P465-478	59-15169 <*> 0
1957 V2 N3 P489-505	AEC-TR-4142 <=>
1957 V2 N3 P507-523	AEC-TR-4142 <=>
1957 V2 N3 P525-527	AEC-TR-4142 <=>
1957 V2 N3 P529-534	AEC-TR-4142 <=>
1957 V2 N4 P605-615	AEC-TR-4112 <=*>
1957 V2 N4 P605-614	AEC-TR-4143 <=>
1957 V2 N4 P617-628	AEC-TR-4143 <=>
1957 V2 N4 P631-651	AEC-TR-4143 <=>
1957 V2 N4 P653-655	AEC-TR-4143 <=>
1957 V2 N4 P657-669	AEC-TR-4143 <=>
1957 V2 N4 P689-704	AEC-TR-4143 <=>
1958 V3 N1 P3-13	AEC-TR-4144 <=>
1958 V3 N1 P15-25	AEC-TR-4144 <=>
1958 V3 N1 P27-40	AEC-TR-4144 <=>
1958 V3 N1 P43-82	AEC-TR-4144 <=>
1958 V3 N1 P111-132	AEC-TR-4144 <=>
1958 V3 N3 P166-169	AEC-TR-4145 <=>
1958 V3 N3 P216-244	AEC-TR-4145 <=>
1958 V3 N3 P255-297	AEC-TR-4146 <=>
1958 V3 N3 P299-341	AEC-TR-4146 <=>
1958 V3 N3 P357-359	AEC-TR-4146 <=>
1958 V3 N4 P369-427	AEC-TR-4147 <=>
1958 V3 N4 P471-474	AEC-TR-4147 <=>
1958 V3 N5 P487-573	AEC-TR-4148 <=>
1958 V3 N5 P575-588	AEC-TR-4148 <=>
1958 V3 N5 P599-605	AEC-TR-4148 <=>
1958 V3 N6 P615-631	AEC-TR-4149 <=>
1958 V3 N6 P633-678	AEC-TR-4149 <=>
1958 V3 N6 P695-708	AEC-TR-4149 <=>
1959 V4 N1 P1-32	AEC-TR-4150 <=>
1959 V4 N1 P13-33	60-31667 <*=>
1959 V4 N1 P35-45	AEC-TR-4150 <=>
1959 V4 N1 P47-85	AEC-TR-4150 <=>
1959 V4 N1 P87-100	AEC-TR-4150 <=>
1959 V4 N1 P103-111	AEC-TR-4150 <=>
1959 V4 N2 P119-139	AEC-TR-4151 <=>
1959 V4 N2 P141-159	AEC-TR-4151 <=>
1959 V4 N2 P161-198	AEC-TR-4151 <=>
1959 V4 N2 P201-225	AEC-TR-4151 <=>
1959 V4 N2 P228-236	AEC-TR-4151 <=>
1959 V4 N3 P241-251	AEC-TR-4152 <=>
1959 V4 N3 P253-283	AEC-TR-4152 <=>
1959 V4 N3 P285-303	AEC-TR-4152 <=>
1959 V4 N3 P305-327	AEC-TR-4152 <=>
1959 V4 N3 P330-343	AEC-TR-4152 <=>
1959 V4 N4 P347-363	AEC-TR-4153 <=>
1959 V4 N4 P365-404	AEC-TR-4153 <=>
1959 V4 N4 P417-437	AEC-TR-4153 <=>
1959 V4 N4 P441-458	AEC-TR-4153 <=>
1959 V4 N4 P462-471	AEC-TR-4153 <=>
1959 V4 N5 P473-484	AEC-TR-4154 <=>
1959 V4 N5 P487-489	AEC-TR-4154 <=>
	AEC-TR-4213 <+*>
1959 V4 N5 P491-520	AEC-TR-4154 <=>
1959 V4 N5 P523-545	AEC-TR-4154 <=>
1959 V4 N5 P547-565	AEC-TR-4154 <=>
1959 V4 N5 P567-584	AEC-TR-4154 <=>
1959 V4 N6 P592-597	AEC-TR-4151 <=>
1959 V4 N6 P599-609	AEC-TR-4151 <=>
1959 V4 N6 P611-623	AEC-TR-4151 <=>

1959 V4 N6 P625-637 AEC-TR-4151 <=>
1959 V4 N6 P639-653 AEC-TR-4151 <=>
1959 V4 N6 P655-663 AEC-TR-4151 <=>
1959 V4 N6 P665-696 AEC-TR-4151 <=>
1960 V5 N12 P863-874 AEC-TR-5682 <*>
1960 V5 N1/2 P1-21 61-11508 <=*>
1960 V5 N1/2 P3-24 SUP 60-41657 <*=>
1961 V6 N1 P1-15 62-19628 <=*>
1961 V6 N1 P68-69 62-19569 <=*>
1961 V6 N2 P85-96 62-19686 <=*>
1961 V6 N2 P99-105 62-19663 <=*>
1961 V6 N2 P127-133 62-19538 <*=>
1961 V6 N3 P181-196 62-19509 <=*>
1961 V6 N3 P211-213 62-19508 <=*>
1961 V6 N4 P261-266 62-19683 <=*>
1961 V6 N4 P267-275 62-19685 <=*>
1961 V6 N4 P277-285 62-19685 <=*>
1961 V6 N4 P287-294 62-19685 <=*>
1961 V6 N5 P357-369 62-19684 <=*>
1962 V7 P389-406 UCRL TRANS-901 <=*>
1962 V7 P479-482 UCRL TRANS-910 <=*>

NUMERISCHE MATHEMATIK
1963 V5 P55-57 66-13626 <*>
1963 V5 N4 P353-370 66-12929 <*>
1963 V5 N5 P443-460 66-12930 <*>
1964 V6 N1 P55-58 64-18411 <*>

NUNG T'IEN SHUI LI
1965 N3 P12-18 65-31986 <=$>
1965 N3 P33 65-31986 <=$>

NUNG YEH CHI HSIEH CHI SHU
1965 N7 P4 65-34031 <=$>
1965 N7 P24-25 65-34031 <=$>
1965 N7 P28-29 65-34031 <=$>

NUNG YEH CHI SHU
1962 N12 P21-24 65-63773 <=$*>
1962 V11 P29-30 65-63741 <=>
1964 N8 P24-31 65-30391 <=$>

NUNG YEH HSUEH PAO
1959 V10 N1 P1-4 61-23592 <=*>
1959 V10 N3 P139-186 61-21715 <=>
1959 V10 N3 P189-219 61-21715 <=>
1959 V10 N4 P221-222 61-23464 <=>
1959 V10 N4 P240-242 61-23464 <=>
1959 V10 N4 P256-294 61-23464 <=>
1959 V10 N4 P296-302 61-23464 <=>
1959 V10 N4 P304-326 61-23464 <=>
1960 V11 N1 P30-40 63-23740 <=*>
1960 V11 N1 P41-46 63-23988 <=*$>
1960 V11 N2 P109-136 61-23468 <=*>
1960 V11 N2 P140-177 61-23925 <=>
1960 V11 N2 P179-197 61-23925 <=>
1960 V11 N2 P199-213 61-23925 <=>

NUNG YEH K'O HSUEH T'UNG HSUN
1958 N5 P236 59-11651 <=>
1958 N5 P242-245 59-11651 <=>
1958 N5 P255 59-11651 <=>
1958 N5 P285-286 59-11651 <=>
1959 N6 P183-185 61-23572 <=*>
1959 N7 P226-236 61-23572 <=*>
1959 N8 P280-282 61-23572 <=*>

NUNTIUS RADIOLOGICUS
1956 V22 P156-163 61-10584 <*> 0
1960 V26 P945-948 61-16132 <*>
1963 V29 N12 P935-939 66-11370 <*>

NUOVI ANNALI D'IGIENE E MICROBIOLOGIA
1953 V4 N4 P292-293 II-375 <*>
1954 V5 N4 P291-294 2737 <*>

NUOVO CIMENTO
1929 S8 V6 P141-142 66-13826 <*>
1932 V9 P43-50 I-379 <*>
1932 V9 P290-298 61-00542 <*>
1933 V10 P101-107 61-00423 <*>
1934 S8 V11 P34-47 63-14671 <*>
1934 S8 V11 P99-113 63-20836 <*>
1934 S8 V11 P157-166 II-580 <*>
1934 S8 V11 P621-634 63-20837 <*>
1935 V12 P423-425 AL-136 <*>
1935 V12 N4 P243-246 61-00362 <*>
1935 V12 N6 P348-357 57-874 <*>
1937 V14 P171-184 2455 <*>
1938 S8 V15 P88-99 58-1208 <*>
59-12028 <=*>
1938 S8 V15 P377-383 59-15345 <*>
1939 V16 N5 P253-260 2151 <*>
1943 V1 P120-125 57-1189 <*>
1946 S9 V3 P131-141 UCRL TRANS-632(L) <*>
1947 S9 V4 N3/4 P177-200 62-20433 <*> 0
1948 V5 P394-396 I-118 <*>
2183 <*>
1948 V5 N5 P416-446 2360 <*>
1948 S9 V5 N6 P589-590 I-734 <*>
1949 N4 P297-299 57-3014 <*>
1949 V6 N6 P327-335 57-971 <*>
1950 V7 P99-108 T1800 <INSD>
1950 V7 N2 P99-108 AL-164 <*>
1950 V7 N2 P159-160 3015 <*>
1950 S9 V7 N2 P159-160 59-17326 <*>
1951 V8 N6 P383-402 57-1966 <*>
1951 V8 N8 P552-568 II-117 <*>
1951 V8 N11 P896-897 AL-489 <*>
1952 V9 P169-183 T-1849 <INSD>
1952 V9 P282-290 II-553 <*>
1952 V9 P722-725 AL-37 <*>
1952 V9 P1242-1243 II-552 <*>
1952 V9 N7 P572-579 I-101 <*>
1953 V10 N1 P72-79 T-2417 <INSD>
1953 V10 N9 P1219-1260 I-678 <*>
1953 V10 N10 P1495 58-2127 <*>
1953 V10 N10 P1497 58-2127 <*>
1953 S9 V10 P268-280 UCRL TRANS-927(L) <*>
1953 S9 V10 N1 P55-82 C-85 <RIS>
1953 S9 V10 N1 P80-86 I-191 <*>
1954 V12 P5-17 2043 <*>
1954 V12 N1 P134-139 78F41 <ATS>
1954 S9 V12 P769-779 UCRL TRANS-804(L) <*>
1954 S9 V12 N1 P37-39 TT-795 <NRCC>
1955 V1 P657-668 II-696 <*>
1955 S10 V1 P501-503 UCRL TRANS-595(L) <*>
1955 S10 V1 N4 P273-294 63-14695 <*>
1956 S10 V4 N4 P922-928 TT-796 <NRCC>
1956 S10 V4 N5 P1133-1141 TT-797 <NRCC>
1957 V5 N5 P1316-1332 58-197 <*>
1957 V5 N5 P1374-1376 58-198 <*>
1957 V6 N10 P811-831 66-13561 <*>
1957 S10 V5 P243-266 SUP 1
UCRL TRANS-495(L) <*>
1957 S10 V5 N4 P842-852 61-16233 <*>
1959 S10 V11 P225-314 SUP 2
UCRL TRANS-495(L) <*>
1961 S10 V20 P361-383 57-2111 <*>
1961 S10 V20 N1 P28-58 UCRL TRANS-697 <*>
1962 S10 V23 N2 P433-439 62-18706 <*>
1963 V27 N3 P601-611 65-13115 <*>
1965 S10 V35 N2 P398-409 65-17103 <*>

NUOVO CIMENTO. SUPPLEMENT
1954 V12 P499-548 II-418 <*>
1957 N3 P1148-1167 C-4337 <NRC>
1960 N2 P168-180 61-18918 <*>

NUOVO ERCOLANI
1936 V14 P287-297 57-2054 <*>

NUOVO GIORNALE BOTANICO ITALIANO E BOLLETINO DELLA SOCIETA BOTANICA ITALIANA
1954 V61 P214-235 60-10703 <*>

NUTRITIO ET DIETA
1960 V2 P223-229 66-11619 <*>

OZE

1955 V8 N4 P121-125 C-2266 <NRC>

OBALY
1957 V3 N3 P75 62-10444 <*>

OBERFLAECHENTECHNIK. COBURG
1932 V9 N18 P191-193 64-20414 <*>
1936 V13 P101-102 59-15715 <*> 0
1937 V14 P59-61 60-10289 <*>
1940 V17 N1 P4-6 59-10255 <*>

OBOBSHCHENNYE FUNKTSII
1959 V1 P470- <API>

OBOGASHCHENIE RUD
1958 V3 N2 P32-34 63-10616 <=*>
1962 V7 N4 P30-32 3503 <BISI>
1963 V8 N6 P27-31 3968 <BISI>

OBRABOTKA METALLOV DAVLENIEM: SBORNIK STATEI
1952 V1 P17-41 2048 <BISI>
1952 V1 P231-237 2047 <BISI>
1953 V2 P76-92 2196 <BISI>
1958 V4 P84-92 4211 <BISI>
1959 V5 P62-72 61-27029 <*=>

OBROBKA PLASTYCZNA
1961 V3 N1 P9-49 NEL-TRANS-1630 <NLL>

OBSHCHESTVENNYE NAUKI V UZBEKISTANE
1964 N89 P92-96 65-30909 <=$>
1966 N3 P16-21 66-32546 <=$>

OBSTETRICIA Y GINECOLOGIA LATINO-AMERICANAS
1945 V3 P556-558 58-395 <*>

OBST-UND GEMUSEBAU
1927 P253-254 3879-C <K-H>

OCEANOLOGIA ET LIMNOLOGIA SINICA
SEE HAI YANG YU HU CHAO

OCEANOLOGY AND LIMNOLOGY. CHINESE PEOPLES REPUBLIC
SEE HAI YANG YU HU CHAO

OCHERKI PO OBSHCHIM VOPROSAM IKHTIOLOGII. IKHTIOLOGICHESKAYA KOMISSIYA. AKADEMIYA NAUK SSSR
1953 P47-51 R-5078 <*>
1953 P295-305 RT-3334 <*>

OCHRONA PRACY
1958 N7 P13-28 60-21211 <=*> 0

OCROTIREA SANATATII IN R.P.R.
1957 V2 P112-118 59-11279 <=*>

ODESSA POLITEKHNICHESKIY INSTITUT NAUCHNYYE ZAPISKI
1955 V2 N1 P87-96 60-19884 <+*>

ODONTOLOGISK TIDSKRIFT
1952 V60 P314-322 60-16934 <*>

OEFFENTLICHE GESUNDHEITSDIENST
1953 V15 N9 P305-313 5668-A <K-H>

OEL
1964 N5 P154-158 65-11002 <*> 0

OEL UND GAS
1962 N4 P28,30-32 66-14567 <*>
1962 N4 P35 66-14567 <*>

OEL UND KOHLE VEREINIGT MIT ERDOEL UND TEER
1935 N11 P499-503 58-1918 <*>
1937 V13 P799-802 58-2112 <*>
1937 V13 N40 P979-981 58-1051 <*>
1938 V14 N15 P299-309 58-1922 <*>
1938 V14 N15 P321-327 58-1922 <*>

OEL UND KOHLE VEREINIGT MIT PETROLEUM
1940 N40 P393-394 58-1042 <*>
1940 V36 N13 P122-128 63-10995 <*>
1940 V36 N45 P512-514 60-18318 <*>
1941 V37 P422-430 63-16043 <*>
1941 V37 N18 P327-329 58-640 <*>
1943 V39 P756-769 II-229 <*>
1943 V39 N9 P240-256 61-20941 <*>
1943 V39 N35/6 P788-792 61-20945 <*>
1943 V39 N43/4 P960-962 63-10994 <*>
1944 V40 P96- 58-914 <*>
1944 V40 N1/2 P15-19 58-859 <*>

OERLIKON SCHWEISSMITTEILUNGEN
1965 V23 N54 P4-15 4381 <BISI>

OESTERREICHISCHE APOTHEKERZEITUNG
1949 V3 P286- II-462 <*>
1951 V5 P540-541 AL-598 <*>
1952 V6 P248-250 AL-600 <*>

OESTERREICHISCHE BAUZEITSCHRIFT
1956 V11 N7/8 P137-155 T-2519 <INSD>

OESTERREICHISCHE BOTANISCHE ZEITSCHRIFT
1927 V76 P222-228 4078 <K-H>
1954 V101 P579-585 1948 <*>
1956 V103 P400-435 6272-A <K-H>

OESTERREICHISCHE CHEMIKERZEITUNG
1900 SNS V5 P441-444 AL-693 <*>
1919 V22 P66-67 I-32 <*>
1927 V30 N1 P1-5 59-10669 <*>
1935 V38 N10 P83-84 61-10783 <*>
1937 V40 P20-25 II-445 <*>
1937 V40 N10 P236-239 58-2057 <*>
1943 V46 P156-160 58-865 <*>
1944 V47 P52-58 65-11436 <*> 0
1948 V49 N1/2 P15-31 60-16785 <*> 0
63-20449 <*>
1948 V49 N3/4 P60-68 60-16784 <*> 0
1948 V49 N5/6 P102-114 60-16783 <*> 0
1950 V51 N2 P32-35 I-72 <*>
1950 V51 N5 P81-85 63-18220 <*>
1951 V52 N7 P125-131 62-10558 <*> 0
1952 V53 P158-160 58-1557 <*>
1952 V53 N5/6 P49-59 64-18926 <*>
1953 V54 P66-68 65-10006 <*> 0
1954 V55 N1/2 P11-21 63-10289 <*> 0
1954 V55 N5/6 P67-72 57-1770 <*>
1955 V55 N21/2 P301-305 RCT V29 N3 P880 <RCT>
1955 V56 P36-38 58-164 <*>
1955 V56 P66-71 63-16627 <*>
1957 V58 P8-13 PT3 57-3572 <*>
1957 V58 P195-208 AEC-TR-5544 <*>
1957 V58 N1/2 P2-8 65-12424 <*>
1959 V60 P65-69 AEC-TR-4797 <*>
1961 V62 N4 P102-115 64-16385 <*> 0
1962 V63 P49-51 64-14952 <*> 0
1962 V63 N1 P12-17 63-01073 <*>
1963 V64 N3 P76-77 NS-355 <TTIS>
1963 V64 N10 P301-312 377 <TC>
1964 V65 N11 P339-357 POED-TRANS-2010 <NLL>

OESTERREICHISCHE INGENIEUR-ZEITSCHRIFT
1958 V1 N1 P33-39 60-14003 <*>

OESTERREICHISCHE PAPIER-ZEITUNG
1959 V65 N1 P13 60-10119 <*>
1959 V65 N6 P9 60-18933 <*>

OESTERREICHISCHE ZEITSCHRIFT FUER BERG-U. HUETTENWESEN
1909 V57 N1 P1-5 59-15346 <*>

OESTERREICHISCHE ZEITSCHRIFT FUER PRAKTISCHE HEILKUNDE

1861 V7 N12 P179-180 66-10977 <*>

OESTERREICHISCHE ZEITSCHRIFT FUER STOMATOLOGIE
1963 V60 N4 P138-145 63-01095 <*>

OESTERREICHISCHE ZEITSCHRIFT FUER VERMESSUNGSWESEN
1959 V47 N5/6 P148-152 60-31759 <*=>

OESTERREICHISCHE ZOOLOGISCHE ZEITSCHRIFT
1951 V3 N3/4 P410-424 63-01093 <*>
1954 V5 P329-349 C-3690 <NRCC> 0
61-00594 <*>

OESTERREICHISCHES INGENIEURARCHIV
1948 V2 N5 P346-360 58-664 <*>
1949 V3 N1 P9-23 60-18028 <*>
1949 V3 N4 P336-344 T-2313 <INSD>
1950 V4 P376- 3130 <*>
1950 V4 P398- 3126 <*>
1950 V4 N1 P44- 3184 <*>
1952 V6 N3 P145-157 I-109 <*>
1952 V6 N3 P223-236 I-290 <*>
1953 V7 P273-284 II-739 <*>
1954 V8 N1 P1-10 65-12356 <*>
1955 V9 N2/3 P239-249 65-13773 <*>
1956 V10 P277-280 60-18652 <*>
1956 V10 N2/3 P155-160 59-20153 <*>
1961 V15 P199-214 65-12334 <*>
1962 V16 N3 P199-211 64-00315 <*>

OFFICIEL DES MATIERES PLASTIQUES
1959 V6 N54 P53-55 65-11758 <*>

OFTALMOLOGIA
1957 V2 P127-132 65-14178 <*>

OFTALMOLOGICHESKII ZHURNAL
1959 N1 P24-28 59-21218 <=>
1960 V15 N3 P187-192 60-41032 <=>
1965 N4 P316-319 65-32210 <=$>

OFVERSIGT AF FINSKA VETENSKAPSSOCIETETENS FORHANDLINGAR
1915 SA V57 N11 P1-13 58-2538 <*>

OFVERSIGT AF K. VETENSKAPSAKADEMIENS FORHANDLINGAR
1880 V37 N8 P42 62-11624 <=>
1880 V37 N8 P45 62-11624 <=>

*OGNEUPORY
1936 V4 P627- 60-13928 <+*>
1936 V4 P638-649 63-14043 <=*>
1937 V5 P809-812 R-2977 <*>
1940 V8 P205- 60-13920 <+*>
1940 V8 P295- 60-13914 <+*>
1940 V8 P326- 60-13915 <+*>
1941 V9 P29- 60-13927 <+*>
1946 V11 N3 P5- 60-15760 <+*>
1946 V11 N6 P35-37 63-18571 <*>
1946 V11 N7/8 P28-31 R-2162 <*>
1947 V12 P80 60-13942 <+*>
1947 V12 P206- 60-13956 <+*>
1948 V13 N8 P351-361 61-13241 <*+>
1949 V14 N3 P136-140 2406 <HB>
1950 V15 N5 P215-221 2765 <HB>
1950 V15 N7 P291-296 61-13223 <*+>
1950 V15 N8 P350-359 2728 <HB>
1950 V15 N10 P446-453 2729 <HB>
1950 V15 N11 P493-504 2845 <HB>
1951 V16 P51- 60-13949 <+*>
1951 V16 P60 60-13943 <+*>
1951 V16 P68-75 63-20436 <=*$>
1951 V16 P147- 60-13945 <+*>
1951 V16 P435- 60-13950 <+*>
1951 V16 P459- 60-13944 <+*>
1952 N12 P543-551 RT-3359 <*>
1952 V17 P22- 60-13951 <+*>
1952 V17 P169- 60-13955 <+*>
1952 V17 P206- 60-13947 <+*>
1952 V17 P211- 60-13953 <+*>
1952 V17 P262- 60-13948 <+*>
1952 V17 P364- 60-13952 <+*>
1952 V17 P465- 60-13946 <+*>
1952 V17 P507- 60-13954 <+*>
1952 V17 N3 P124-133 R-771 <*>
1952 V17 N8 P364-370 3271 <HB>
1955 N2 P64-69 R-341 <*>
1955 V20 P47 60-13957 <+*>
1955 V20 P72-79 58-1682 <*>
1955 V20 P72- 60-13959 <+*>
1955 V20 P243- 60-13961 <+*>
1955 V20 P263- 60-13960 <+*>
1955 V20 P305- 60-13958 <+*>
1955 V20 P315- 60-13968 <+*>
1955 V20 N4 P166-173 60-15817 <+*>
1955 V20 N7 P291 60-13962 <+*>
1956 V21 P37-39 871TM <CTT>
1956 V21 P193-202 T-2614 <INSD>
1956 V21 P220 T-2613 <INSD>
1956 V21 P221-226 T-2615 <INSD>
1956 V21 P233- 60-13969 <+*>
1956 V21 N1 P37- 60-13963 <+*>
1956 V21 N2 P73-75 3883 <HB>
1956 V21 N4 P166-170 59-14185 <+*> 0
1956 V21 N5 P207-211 59-14187 <+*> C
1956 V21 N6 P253-258 3849 <HB>
1957 V22 P38- 60-13965 <+*>
1957 V22 P42- 60-13966 <+*>
1957 V22 P120 60-13964 <+*>
1957 V22 P178- 60-13970 <+*>
1957 V22 P222- 60-13967 <+*>
1957 V22 P329- 60-13973 <+*>
1957 V22 N3 P105-108 59-19617 <+*>
1957 V22 N4 P145-152 60-13045 <+*>
1957 V22 N4 P169-173 59-19610 <+*> 0
1957 V22 N4 P186-188 4098 <HB>
1957 V22 N12 P529-533 TRANS-38 <MT>
1957 V22 N12 P549-556 4643 <HB>
1957 V22 N12 P557-562 4504 <HB>
60-13972 <+*>
1957 V22 N12 P562-566 4490 <HB>
59-15803 <+*>
1957 V22 N12 P568-571 4266 <HB>
1958 V23 N1 P5-11 61-15201 <*+>
1958 V23 N2 P82-87 60-13971 <+*>
1958 V23 N4 P145-150 59-18092 <+*> C
1958 V23 N5 P229-233 60-21948 <=> 0
1958 V23 N6 P257 60-13981 <+*>
1958 V23 N6 P284-285 61-13919 <*+>
1958 V23 N7 P303-307 TS 1267 <BISI>
62-24219 <=*>
1958 V23 N9 P385-395 60-13974 <+*>
1958 V23 N10 P454-461 60-13977 <+*>
1958 V23 N10 P476-478 60-13975 <+*>
1958 V23 N11 P481-493 60-13976 <+*>
1958 V23 N11 P498-504 3530 <BISI>
1958 V23 N12 P552-558 62-13166 <*=>
1959 V24 N2 P71-79 63-20376 <=*$> 0
1959 V24 N4 P157-161 AEC-TR-3988 <+*>
1959 V24 N4 P165-167 61-28770 <*=>
1959 V24 N5 P225-231 62-25271 <=*>
1959 V24 N5 P231-236 4883 <HB>
1959 V24 N7 P325-329 2278 <BISI>
1959 V24 N9 P419-423 2279 <BISI>
1959 V24 N10 P455-462 2216 <BISI>
1959 V24 N11 P510-517 2556 <BISI>
1959 V24 N11 P517-522 2611 <BISI>
1960 V25 P562-566 62-10305 <*=> 0
1960 V25 N1 P35-38 62-18903 <=*> C
1960 V25 N3 P132-137 C-3615 <NRCC>
1960 V25 N4 P171-175 2739 <BISI>
1960 V25 N4 P186-188 2217 <BISI>
1960 V25 N9 P386 61-11743 <=>
1960 V25 N9 P388 61-11743 <=>
1961 V26 P525-530 UCRL TRANS-863(L) <=*>
1961 V26 N1 P44-46 61-27935 <*=>
1961 V26 N2 P72-74 62-32996 <=*> 0
1961 V26 N4 P157-163 62-25952 <=*>
1961 V26 N4 P185-193 63-18004 <=*> C
1961 V26 N7 P335-338 62-23378 <=*>

1961 V26 N8 P385-386	62-15213 <*=>
1961 V26 N10 P465-469	62-23612 <=*> 0
1962 V27 N1 P40-42	65-13364 <*>
1962 V27 N5 P223-225	3216 <BISI>
1962 V27 N7 P297-299	63-23211 <*=>
1962 V27 N7 P332-336	63-13082 <=*>
1962 V27 N10 P449-453	63-13451 <=>
1962 V27 N10 P457-472	63-13451 <=>
1962 V27 N10 P480-481	63-13451 <=>
1962 V27 N11 P521	63-21311 <=>
1963 V28 N4 P163-165	5963 <HB>
1963 V28 N5 P224-231	65-60670 <=$>
1963 V28 N9 P400-407	6101 <HB>
1964 V29 N5 P197-200	6277 <HB>
1964 V29 N6 P264-269	4083 <BISI>
1964 V29 N9 P418-424	AD-623 208 <=$*>
1964 V29 N10 P1-2	65-30964 <=$>
1964 V29 N10 P442-444	6438 <HB>
1965 N2 P20-23	N65-27682 <=*$>
1965 V30 N1 P47	65-30774 <=$>
1965 V30 N3 P1-6	65-31134 <=$>
1965 V30 N4 P1-8	6529 <HB>
1965 V30 N4 P28-32	6680 <HB>
1965 V30 N7 P30-34	6689 <HB>
1965 V30 N11 P22-26	6838 <HB>
1965 V30 N12 P8-13	66-32799 <=$>
1966 V31 N3 P5-8	6875 <HB>

OGONEK

1955 P5-6	57-265 <*>
1955 N32 P5-8	C-2259 <NRC>
1957 N14 P7-8	59-16432 <+*> 0
1957 N29 P20-21	PB 141 223T <=>
1960 N27 P3	61-11743 <=>
1960 V38 N35 P2	61-11617 <=>
1960 V38 N35 P4-5	61-11617 <=>
1961 V39 P20-21	62-13515 <=*>
1961 V39 N14 P2-3	62-24268 <=*>
1961 V39 N17 P26-27	63-15529 <=*>
1961 V39 N49 P12-13	62-19170 <=*>
1961 V39 N49 P16-17	62-19170 <=*>
1962 V40 N9 P30-31	62-33664 <=*>
1962 V40 N22 P30-32	AEC-TR-5406 <=*>
	63-15226 <=*>
1962 V40 N25 P17-18	62-33664 <=*>
1962 V40 N32 P20-21	63-13477 <=>
1962 V40 N34 P4-5	AD-621 690 <=$>
	63-15556 <=*>
1962 V40 N36 P14-16	62-11784 <=>
1962 V40 N42 P14-15	63-13300 <=*>
1962 V40 N44 P10-11	63-13838 <=>
1962 V40 N44 P14	63-13832 <=*>
1962 V40 N47 P28-29	63-21001 <=>
1963 V41 N2 P5	63-21885 <=>
1965 N25 P4-8	AD-620 759 <=$>
1965 N38 P6	65-33065 <=*$>
1966 N12 P14-15	66-32756 <=$>

OITA DAIGAKU KYOIKUGAKUBU KENKYU KIYO, SHIGENKAKAGU

1954 V3 P11-18	58-1191 <*>

OKAJIMAS FOLIA ANATOMICA JAPONICA

1940 V19 P199-213	66-11815 <*>

OKAYAMA IGAKKAI ZASSHI

1934 V46 N3 P615-664	2023 <*>
1936 V48 N12 P2801-2808	994 <K-H>

*OKEANOLOGIYA

1961 V1 N1 P25-29	62-23382 <=*>
1961 V1 N1 P95-106	62-24071 <=*>
1961 V1 N2 P206-212	63-19882 <=*> 0
1961 V1 N2 P294-304	62-15505 <*=>
1961 V1 N2 P338-339	63-21526 <=>
1961 V1 N3 P450-455	62-15450 <*=>
1961 V1 N3 P522-530	62-24865 <=*>
1961 V1 N3 P543-549	62-11169 <*=>
1961 V1 N3 P554-555	64-13625 <=*$>
1961 V1 N4 P710-716	62-24867 <=*>
	63-24254 <=*> 0
1961 V1 N4 P741-743	62-32485 <=*>
1961 V1 N4 P743-744	C-3875 <NRC>
1961 V1 N4 P745-761	62-19789 <=*>
1961 V1 N4 P756-757	C-3874 <NRCC>
	C-3874 <NRCC>
1961 V1 N4 P757-761	62-24871 <=*>
1961 V1 N4 P761-763	62-24785 <=*>
1961 V1 N5 P886-887	64-13335 <=*$>
1961 V1 N6 P1034-1038	62-23797 <*=>
1961 V1 N6 P1039-1045	64-15416 <=*$>
1961 V1 N6 P1079-1084	62-25948 <=*>
1961 V1 N6 P1089-1096	62-25948 <=*>
1962 N5 P888-897	M.5317 <NLL>
1962 V2 N1 P118-125	63-15152 <=*>
1962 V2 N1 P134-138	63-15177 <=*>
1962 V2 N1 P139-143	63-19728 <=*$>
1962 V2 N1 P164-172	63-15137 <=*>
1962 V2 N1 P190-191	64-19138 <=*$>
1962 V2 N2 P205-209	63-15151 <=*>
1962 V2 N2 P251-256	63-15221 <*=>
1962 V2 N2 P293-304	63-21438 <=>
1962 V2 N2 P293-297	64-19138 <=*$> 0
1962 V2 N2 P305-310	63-15964 <=*>
1962 V2 N2 P334-345	64-13626 <=*$>
1962 V2 N2 P346-352	63-15222 <=*>
1962 V2 N2 P368-371	63-21438 <=>
1962 V2 N3 P385-392	63-15143 <=*>
1962 V2 N3 P393-409	63-10196 <=*>
1962 V2 N3 P457-463	64-13155 <=*$>
1962 V2 N4 P705-726	63-21646 <=>
1962 V2 N4 P743-745	64-13154 <=*$>
1962 V2 N5 P845-848	64-13153 <=*$>
1962 V2 N5 P849-863	64-11089 <=>
1962 V2 N5 P888-897	63-41000 <=>
1962 V2 N5 P898-903	63-23361 <=*$>
1962 V2 N6 P961-969	63-19974 <=*$>
	63-21481 <=>
	63-23506 <=*$>
1962 V2 N6 P999-1008	MAFF-TRANS-NS-64 <NLL>
1962 V2 N6 P1050-1059	63-31280 <=>
1962 V2 N6 P1093-1103	63-21162 <=>
1962 V2 N6 P1110-1112	63-23504 <=*$>
1962 V2 N6 P1115-1117	63-19975 <=*>
1963 V3 P752	65-12742 <*>
1963 V3 N1 P76-87	63-21453 <=>
1963 V3 N1 P123-126	63-21758 <=>
1963 V3 N1 P175-178	63-21758 <=>
1963 V3 N1 P182-184	63-21758 <=>
1963 V3 N2 P193-199	65-60930 <=$>
1963 V3 N2 P213-218	AD-620 000 <=$>
	C-5359 <NRC>
1963 V3 N2 P250-259	AD-620 001 <=$*>
	C-5347 <NRC>
1963 V3 N2 P308-312	63-31089 <=>
1963 V3 N3 P527-537	AD-636 219 <=$>
1963 V3 N4 P669-673	64-71429 <=>
1963 V3 N4 P697-705	63-41148 <=>
1963 V3 N4 P706-714	63-31918 <=>
1963 V3 N4 P720-730	C-5124 <NRC>
1963 V3 N4 P750-752	63-31918 <=>
1963 V3 N5 P840-847	AD-623 255 <=$*>
1963 V3 N5 P861-869	64-21657 <=>
1963 V3 N5 P876-885	64-21653 <=>
1963 V3 N5 P886-897	64-21656 <=>
1963 V3 N5 P898-906	64-21651 <=>
1963 V3 N5 P911-921	AD-630 741 <=$>
1963 V3 N5 P936-938	64-21652 <=>
1963 V3 N5 P938-940	64-21654 <=>
1963 V3 N5 P940	64-21649 <=>
1963 V3 N6 P1056-1060	64-21502 <=>
1963 V3 N6 P1079-1084	64-31136 <=>
1963 V3 N6 P1109-1114	64-31136 <=>
1963 V3 N6 P1119-1123	64-31136 <=>
1964 V4 P603-611	TRANS-336(M) <NLL>
1964 V4 P612-616	TRANS-337(M) <NLL>
1964 V4 N1 P68-73	64-31117 <=>
1964 V4 N1 P112-123	64-31112 <=>
1964 V4 N1 P112-124	65-61438 <=>
1964 V4 N1 P156-166	64-31138 <=>

1964 V4 N1 P167-174 64-31113 <=>
1964 V4 N1 P182-183 64-31111 <=>
1964 V4 N1 P183-186 64-31138 <=>
1964 V4 N2 P213-242 64-31473 <=>
1964 V4 N2 P325-339 64-31437 <=>
1964 V4 N2 P342-353 64-31473 <=>
1964 V4 N2 P359-361 64-31473 <=>
1964 V4 N3 P512-516 AD-625 850 <=*$>
64-41106 <=>
1964 V4 N3 P536-540 64-41107 <=>
1964 V4 N6 P1 65-30273 <=$>
1964 V4 N6 P6-10 65-30273 <=$>
1964 V4 N6 P939-953 65-30242 <=$>
1964 V4 N6 P1026-1029 65-30241 <=$>
1964 V4 N6 P1059-1061 65-30243 <=$>
1964 V4 N6 P1096-1100 65-30272 <=$>
1964 V4 N10 P115-120 66-23830 <$>
1965 V5 N1 P32-39 65-31874 <=$>
1965 V5 N1 P99-110 65-22211 <$>
1965 V5 N2 P210-221 TRANS-F.129 <NLL>
1965 V5 N3 P553-556 65-32881 <=*$>
1965 V5 N3 P566-568 65-32918 <=*$>
1965 V5 N4 P718-724 AD-633 271 <=$>
1965 V5 N5 P903-910 65-33934 <=*$>
1965 V5 N6 P1028-1042 66-30618 <=$>
1965 V5 N6 P1083-1084 66-30618 <=$>
1965 V5 N6 P1095-1099 66-30618 <=$>
1965 V5 N6 P1110-1112 66-30618 <=$>
1966 N1 P53-61 66-31577 <=$>
1966 N1 P159-161 66-31577 <=$>
1966 N1 P161-164 66-31577 <=$>
1966 N1 P172-175 66-31577 <=$>
1966 V6 N3 P513-519 66-33465 <=$>
1966 V6 N3 P529-530 66-33464 <=$>
1966 V6 N3 P535-542 66-33463 <=$>
1966 V6 N4 P599-607 66-34805 <=$>
1966 V6 N6 P1104-1105 67-31530 <=$>

OKEANOLOGICHESKIE ISSLEDOVANIYA
1961 N3 P30-51 65-20001 <$>
1961 N4 P18-24 64-15436 <=*$>
1965 N13 P61-65 AD-635 064 <=$>
1965 N13 P137-142 AD-632 739 <=$>
1965 N13 P181-188 AD-635 063 <=$>

OKEANOLOGIYA: SBORNIK STATEI. MEZHDUVEDOMSTVENNYI KOMITET PO PROVEDENIYU MEZHDUNARODNOGO GEOFIZICHESKOGO GODA. AKADEMIYA NAUK SSSR. X. RAZDEL PROGRAMMY MGG
1963 N8 P97-103 65-60594 <*>
1963 V8 P34-51 65-61354 <=$>

OKHRANA PRIRODY I ZAPOVEDNOE DELO V SSSR: BYULLETEN
1958 N3 ENTIRE ISSUE 61-11489 <=>

OKHRANA TRUDA I SOTSIALNOE STRAKHOVANIE
1965 N10 P24-26 66-30950 <=$>

OKTYABR
1965 N5 P149-162 65-33570 <=$>
1965 N11 P134-171 65-34111 <=$>

OLEAGINEAUX
1948 V3 N2 P57-64 59-20607 <*> PO
1949 V4 N2 P95-100 59-20608 <*> O
1951 V6 N1 P1-10 I-509 <*>
1952 V7 N1 P21-24 59-20609 <*>
1957 V12 N1 P1-6 59-20610 <*>
1962 V17 N6 P565-570 63-00872 <*>

OLEARIA
1947 V1 N6 P309-319 59-20680 <*> O
1952 V6 N5/6 P139-143 59-20656 <*>
1953 V7 N7/8 P183-187 59-20655 <*>

OLIVICOLTURA
1957 V12 N1 P1-8 58-1397 <*>
1958 V13 N4 P3-5 26L32I <ATS>

OMNIA MEDICA
1954 P99-106 57-1986 <*>

ONCOLOGIA. BASEL, NEW YORK
1953 V6 P1-26 58-2523 <*>
1955 V8 N2 P185-194 58-1690 <*>
1956 V9 P12-32 58-2308 <*>
1956 V9 P269 58-2303 <*>
1957 V10 P124-129 6251-A <K-H>
1957 V10 P157-186 6344 <K-H>
1957 V10 P281-294 58-2664 <*>
1957 V10 N2 P107-118 58-941 <*>
1957 V10 N2 P107-119 64-14555 <*> O
6609-B <K-H>
1957 V10 N2 P124-129 58-1317 <*>
1957 V10 N2 P137-155 58-1221 <*>
1957 V10 N2 P137-156 6251-B <KH>
1957 V10 N4 P307-329 65-00350 <*> O
1958 V11 P166-178 58-1372 <*>
1958 V11 N2 P138-147 62-01218 <*>
1958 V11 N3/4 P218-243 AEC-TR-3471 <*> O
1958 V11 N3/4 P244-253 61-00416 <*>
1958 V11 N3/4 P254-287 10332-A <K-H>
63-16933 <*>
1959 V12 N1 P22-27 64-10414 <*>
9249-A <K-H>
1960 V13 N2 P252-266 10573-C <K-H>
64-14556 <*>

ONDE ELECTRIQUE
1922 P101-123 57-10 <*>
1922 V1 P261-270 58-196 <*>
1924 P477-490 57-2140 <*>
1926 V5 P5-27 57-1058 <*>
1926 V5 P276-283 66-12261 <*>
1927 V6 P401-426 66-13142 <*>
1928 V7 P186-195 2170 <*>
1929 V8 N88 P160-170 66-13332 <*>
1930 V9 P229-244 66-13143 <*> O
1931 V10 N117 P369-424 57-2105 <*>
1932 V11 N12 P53-82 2187 <*>
57-3512 <*>
1938 V17 N197 P217-246 57-870 <*>
1938 V17 N198 P303-308 57-638 <*>
1947 V27 N249 P447-458 57-2532 <*>
1948 V28 N255 P236-242 66-13318 <*>
1949 V29 N262 P44-50 2178 <*>
1950 P458-461 I-703 <*>
1950 V30 N283 P433-437 1790 <*>
1950 V30 N285 P510-521 AEC-TR-3771 <+*>
1951 V31 N290 P205-209 61-14981 <*>
1952 V32 N303 P219-231 57-1011 <*>
1953 V33 N313 P217-234 57-2524 <*>
1953 V33 N317 P530-539 60-21427 <=*>
1954 V34 P242-244 57-553 <*>
1954 V34 P248-251 57-550 <*>
1954 V34 P431-440 38G8F <ATS>
57-1381 <*>
1954 V34 P838-841 66-12980 <*> O
1954 V34 N322 P7-13 II-431 <*>
1954 V34 N324 P282-291 57-260 <*>
1954 V34 N325 P323-338 63-14430 <*> O
1954 V34 N326 P413-417 57-1380 <*>
1954 V34 N328 P559-572 62-18425 <*> OP
1954 V34 N332 P883-896 60-21448 <=*>
63-10743 <=*>
1955 V35 P222-236 65-13612 <*> O
1955 V35 P336-337 57-3496 <*>
65-13612 <*> O
1955 V35 N335 P89-96 06H12F <ATS>
1955 V35 N336 P379-393 61-20853 <*> PO
1955 V35 N344 P1033-1047 NP-TR-948 <*>
1956 V36 P348- 58-1164 <*>
1956 V36 P801-814 63-16059 <*> O
1956 V36 N348 P194-213 58-1169 <*>
1956 V36 N355 P801-814 58-254 <*>
1957 V37 P28-35 AEC-TR-4175 <*>
1957 V37 P337-357 59-17450 <*> O
1957 V37 P671-678 59-17450 <*> O
1957 V37 P1083-1088 58-1794 <*>

Citation	Translation
1957 V37 N360 P259-261	66-10446 <*> 0
1957 V37 N364 P650-657	62-18819 <*> 0
1958 V38 P592-599	AEC-TR-4803 <*>
1958 V38 N372 P166-183	59-10872 <*>
1958 V38 N377 P606-616	60-14321 <*>
	61-10608 <*>
1958 V38 N377 P617-621	60-14320 <*>
1959 V39 N382 P40-45	AEC-TR-3849 <*>
1959 V39 N383 P74-87	60-18901 <*>
1960 V40 P751-761	61-18716 <*> 0
1960 V40 P825-830	62-25230 <=*$>
1960 V40 N402 P586-589	62-16186 <*> 0
1961 V41 P821-823	62-33345 <=*>
1961 V41 P1001-1006	UCRL TRANS-985 <*>
1961 V41 P1025-1028	UCRL TRANS-986 <*>
1961 V41 P1034-1041	UCRL TRANS-987 <*>
1961 V41 N417 P1042-1046	UCRL TRANS-988 <*>
1962 V42 N423 P541-553	NP-TR-997 <*>
1962 V42 N426 P746-753	65-11688 <*>
1962 V42 N426 P747-753	66-15375 <*>
1963 V43 P148-152	64-14633 <*>
1963 V43 N1 P56-58	64-16571 <*> 0
1963 V43 N434 P489-502	65-14068 <*>
1963 V43 N434 P540-555	65-11463 <*>
1963 V43 N439 P1003-1021	65-14067 <*>
1964 V44 N443 P127-146	65-14074 <*>
1964 V44 N445 P361-372	65-10900 <*>
	65-14071 <*>
1964 V44 N445 P378-387	65-14073 <*>
1964 V44 N451 P967-973	65-17150 <*>
1964 V44 N451 P1055-1063	65-14065 <*>
1965 V45 N463 P1204-1215	32T91F <ATS>
	66-11526 <*>

ONTODE

Citation	Translation
1953 V4 N4 P73-82	2467 <BISI>
1953 V4 N5 P97-103	2467 <BISI>
1954 V5 N12 P275-280	64-10552 <*>
1956 V7 N5 P112-115	64-10553 <*>
1959 V10 P111-113	TS-1814 <BISI>
1961 V12 N9 P197-203	5206 <HB>

OPHTHALMOLOGICA. BERLIN

Citation	Translation
1939 V97 N5 P320-321	61-14446 <*>
1940 V100 P351-354	62-10860 <*>
1949 V118 P751-763	2149-F <K-H>

OPPERVLAKTETECHNIEKEN VAN METALEN

Citation	Translation
1961 V5 P190-191	NS-28 <TTIS>

OPREDELITELI PO FAUNE SSSR, IZDAVAEMYE ZOOLOGICHESKIM MUZEEM AKADEMII NAUK

Citation	Translation
1938 N23 P84-	R-4123 <*>
1949 V29 P467-926	63-11056 <=>
1949 V30 P927-1382	63-11057 <=$>
1951 V38 P1-378 PT.1	63-11066/1 <=>
1951 V38 P149-183 PT.1	62-15621 <=*> 0
1951 V38 P26-29 PT.1	61-19199 <=*> PO
1951 V40 P385-667	62-15634 <=*> 0
	63-11066/2 <=>
1952 N44 ENTIRE ISSUE	61-3122 <=>
1952 N45 P3-104	65-50018 <=>
1954 V53 ENTIRE ISSUE	63-11160 <=>
1954 V53 P188-190	64-26258 <$>
1955 V59 P324-340	61-19191 <+*> 0
1956 V64 P79-83	64-15672 <=*$>
1960 V72 P3-130	62-15637 <=*> 0

OPSTINA

Citation	Translation
1966 V16 N6 P12-18	66-34564 <=$>

OPTICA ACTA

Citation	Translation
1956 V3 P97-99	66-11031 <*>
1956 V3 N4 P153-160	59-19765 <=*>
1957 V4 N4 P136-144	59-17750 <=>
1958 V5 P256-262	56Q71F <ATS>
1959 V6 N1 P52-76	UCRL TRANS-651(L) <*>
1959 V6 N4 P319-338	64-16390 <*>
1960 V7 P81-97	20M47F <ATS>
1960 V7 P173-178	62-10974 <*>
1960 V7 P243-261	57Q71F <ATS>
1960 V7 P385-398	58Q71F <ATS>
1961 V8 P161-168	59Q71F <ATS>
1962 V9 P121-148	65-10282 <*>
1962 V9 P335-364	65-00250 <*>
1963 V10 P1-19	64-00459 <*>
1964 V11 N3 P223-235	65-14602 <*>
	66-11160 <*>
1965 V12 N2 P161-166	71S88F <ATS>

OPTIK

Citation	Translation
1946 V1 N2 P134-143	AL-870 <*>
1947 V2 N2 P114-132	CSIRO-2915 <CSIR>
1947 V2 N4 P301-325	AL-109 <*>
1948 V3 P389-412	62-16518 <*> 0
1948 V3 N3 P201-220	66-12886 <*> 0
1948 V3 N1/2 P124-127	65-11165 <*>
1948 V3 N1/2 P128-136	1462 <*>
1948 V3 N5/6 P430-443	65-18093 <*>
1948 V3 N5/6 P495-498	AL-867 <*>
1948 V4 N1 P11-21	65-00137 <*>
1949 V5 N8/9 P469-478	61-20700 <*>
	62-16629 <*> 0
1949 V5 N8/9 P499-517	57-1900 <*>
1949 V5 N8/9 P518-530	62-16202 <*> 0
1950 V6 P56-58	58-2160 <*>
1950 V6 N6 P332-336	63-20493 <*>
1950 V7 N1 P1-12	AD-623 920 <=*$>
1951 V8 P550-560	AL-496 <*>
1951 V8 N7 P311-317	CSIRO-2813 <CSIR>
1952 V9 N4 P145-153	64-10038 <*>
1952 V9 N4 P174-179	63-24399 <=*$>
1953 V10 P426-438	AEC-TR-5398 <*>
1953 V10 N1/3 P116-131	83F4G <ATS>
1954 V11 P351-365	2229 <*>
1954 V11 N1 P13-17	65-11344 <*>
1954 V11 N5 P244-248	61-10124 <*>
1954 V11 N11 P509-510	63-10010 <*> 0
1955 V12 P377-384	63-18757 <*> 0
1955 V12 N2 P60-70	62-16028 <*>
1955 V12 N4 P166-172	CSIRO-2808 <CSIR>
1955 V12 N10 P467-475	II-896 <*>
1956 V13 N10 P437-462	66-11075 <*>
1957 V14 N7/8 P353-360	60-17401 <*=>
1958 V15 P116-126	UCRL TRANS-564(L) <*>
1958 V15 P242-260	61-18909 <*> 0
1958 V15 P372-381	AEC-TR-5725 <*>
1958 V15 N4 P242-260	62-14654 <*>
1958 V15 N11 P686-693	12T92G <ATS>
1959 V16 N5 P304-312	61-20717 <*>
1959 V16 N9 P527	AEC-TR-5274 <*>
1961 V18 P120	66-11880 <*>
1961 V18 N3 P147-156	64-00458 <*>
1961 V18 N10/1 P514-518	63-01724 <*>
1962 V19 N2 P122-131	ORNL-TR-145 <*>
1962 V19 N7 P357-368	63-18723 <*>
1962 V19 N9 P451-462	64-10906 <*>
1962 V19 N12 P640-651	N65-23794 <=$>
1963 V20 N8 P383-385	65-00127 <*>
1963 V20 N9/10 P475-480	65-12756 <*>
1964 V21 N7 P309-319	66-11183 <*>
1964 V21 N10 P550-566	65-17003 <*> 0

*OPTIKA I SPEKTROSKOPIYA

Citation	Translation
1956 V1 P403-406	AEC-TR-3570 <+*>
	R-3066 <*>
1956 V1 P650-657	AEC-TR-4319 <*>
1956 V1 P658-662	AEC-TR-4358 <=*>
1956 V1 P809-811	AEC-TR-2890 <*>
1956 V1 P972-982	60-13633 <+*>
1956 V1 P983-987	59-15139 <+*>
1956 V1 N1 P22-33	UCRL TRANS-706 <=*>
	62-18331 <=*> 0
1956 V1 N1 P34-40	59-11611 <=> 0
1956 V1 N1 P85-89	C2375 <NRC>
	R-1615 <*>
1956 V1 N1 P94-101	63-23389 <=*> 0
1956 V1 N1 P181-189	GAT-Z-4077 <+*>
1956 V1 N2 P113-124	63-20239 <=*>
1956 V1 N2 P175-180	R-4911 <*>

Citation	Translation
1956 V1 N2 P216-229	62-16522 <=*> 0
	63-11579 <=>
1956 V1 N3 P285-289	R-5320 <*>
	62-11615 <=>
1956 V1 N3 P302-320	R-2049 <*>
1956 V1 N3 P330-333	R-988 <*>
1956 V1 N3 P334-337	AEC-TR-4306 <+*>
1956 V1 N3 P338-347	59-15176 <+*>
1956 V1 N4 P469-477	62-25035 <=*>
1956 V1 N4 P49C-499	37J17R <ATS>
1956 V1 N4 P500-506	52J16 <ATS>
1956 V1 N4 P507-515	61-00429 <*>
1956 V1 N4 P516-522	59-14518 <+*>
1956 V1 N4 P536-545	10M38R <ATS>
1956 V1 N5 P627-635	R-4912 <*>
1956 V1 N5 P642-649	NP-TR-898 <*>
	63-23298 <=*> 0
1956 V1 N5 P672-684	N65-27702 <=*$>
1956 V1 N6 P738-746	59-15359 <+*>
1956 V1 N7 P821-832	62-15508 <*=>
1956 V1 N7 P863-866	AEC-TR-4314 <+*>
1956 V1 N8 P992-999	62-11557 <=>
	62-15205 <*=>
1957 N2 P158-161	R-3718 <*>
1957 V2 P282-284	AEC-TR-4596 <*>
	AEC-TR-4596 <*=>
1957 V2 P402-405	64S81R <ATS>
1957 V2 P645-65C	66-11023 <*>
1957 V2 P814-816	R-5002 <*>
1957 V2 N1 P43-48	63-23116 <=*>
1957 V2 N1 P62-74	64-16889 <=*$>
1957 V2 N1 P75-98	59-15899 <+*>
1957 V2 N1 P99-106	63-15933 <=*>
1957 V2 N1 P141-142	62-23722 <=*>
1957 V2 N2 P145-149	UCRL TRANS-707 <=*>
	62-16714 <=*>
1957 V2 N2 P210-219	63-11515 <=>
1957 V2 N2 P220-228	59-15487 <+*>
1957 V2 N2 P229-235	59-20505 <+*>
1957 V2 N2 P236-244	59-20506 <+*>
	62-10971 <=*> 0
1957 V2 N2 P245-253	62-10986 <=*>
1957 V2 N2 P254-262	61P101 <CTT>
1957 V2 N2 P263-268	59-20362 <+*>
1957 V2 N3 P298-303	R-5307 <*>
1957 V2 N3 P317-322	60-21768 <=>
1957 V2 N3 P323-329	R-3185 <*>
	26J18R <ATS>
1957 V2 N3 P361-370	59-16300 <+*>
1957 V2 N4 P480-487	AEC-TR-3568 <+*>
1957 V2 N4 P488-493	AEC-TR-3571 <+*>
1957 V2 N4 P510-513	65-10794 <*>
1957 V2 N4 P514-523	59-16300 <+*>
1957 V2 N5 P557-561	R-5318 <*>
1957 V2 N5 P568-577	AEC-TR-5967 <*>
	AEC-TR-5967 <=*$>
1957 V2 N5 P606-615	62-10163 <*=>
1957 V2 N5 P669-671	AEC-TR-3508 <+*>
1957 V2 N5 P673	62-14262 <=*>
1957 V2 N5 P674-676	65-10288 <*>
1957 V2 N6 P681-688	64-19965 <=$>
1957 V2 N6 P695-703	AEC-TR-3716 <+*>
1957 V2 N6 P710-716	60-13311 <+*>
1957 V2 N6 P710-723	60-17715 <+*>
1957 V2 N6 P717-723	60-13312 <+*>
1957 V2 N6 P809-811	59-22380 <+*>
	63-23706 <=*$>
1957 V2 N6 P814-816	10S81R <ATS>
1957 V2 N6 P816-818	AEC-TR-3530 <+*>
1957 V3 P77-81	R-4235 <*>
1957 V3 P246-250	25M39R <ATS>
1957 V3 N1 P3-8	AEC-TR-4519 <*=>
1957 V3 N1 P16-20	UCRL TRANS-766(L) <=*>
1957 V3 N1 P21-	59-11224 <=>
1957 V3 N1 P68-72	26M39R <ATS>
1957 V3 N1 P77-81	R-5275 <*>
1957 V3 N1 P94-95	R-5278 <*>
1957 V3 N2 P115-122	87K22R <ATS>
1957 V3 N2 P123-133	R-4532 <*>
1957 V3 N2 P158-161	62-18268 <=*>
	64-23576 <=$>
1957 V3 N2 P162-168	60-21472 <=>
1957 V3 N2 P180-181	R-3264 <*>
1957 V3 N3 P194-201	62-15509 <*=>
1957 V3 N3 P251-257	11M40R <ATS>
	66-15575 <*$>
1957 V3 N3 P289-293	61-27254 <=*> 0
1957 V3 N4 P391-393	59-22165 <+*>
1957 V3 N4 P401-402	60-15537 <+*>
1957 V3 N5 P417-433	62-01099 <*>
	64-23577 <=$>
1957 V3 N5 P457-472	60-23679 <*+>
1957 V3 N5 P473-479	R-5083 <*>
1957 V3 N5 P480-494	95M45R <ATS>
1957 V3 N5 P514-528	R-4921 <*>
1957 V3 N6 P560-	60-15984 <+*>
1957 V3 N6 P56C-567	60-19582 <+*> 0
1957 V3 N6 P560-	60-21471 <=>
1957 V3 N6 P61C-618	59-10777 <+*>
1957 V3 N6 P649-652	23M39R <ATS>
1958 V4 P539-541	66-11735 <*>
1958 V4 N1 P3-8	NP-TR-113 <*>
1958 V4 N1 P9-16	60-17905 <+*>
	61-19735 <*=>
	61-20836 <*=>
1958 V4 N1 P55-59	AEC-TR-4884 <=*>
1958 V4 N1 P82-86	59-15025 <+*>
1958 V4 N1 P112-113	59-18008 <+*>
1958 V4 N2 P189-195	59-10914 <+*>
1958 V4 N2 P196-202	AEC-TR-3549 <+*>
1958 V4 N2 P203-210	AEC-TR-6309 <=*$>
1958 V4 N2 P217-224	59-10913 <+*>
1958 V4 N2 P225-235	59-15022 <+*>
1958 V4 N2 P261-264	59-15026 <+*>
1958 V4 N2 P272-274	62-32638 <=*>
1958 V4 N2 P278-279	59-10912 <+*>
1958 V4 N2 P279-281	59-10911 <+*>
1958 V4 N3 P285-288	R-5064 <*>
1958 V4 N3 P328-334	60-23993 <+*>
1958 V4 N3 P348-353	59-20836 <+*>
1958 V4 N3 P402-404	62-13146 <*=>
1958 V4 N3 P407-409	60-23028 <+*>
1958 V4 N4 P421-429	AD-615 864 <=$>
1958 V4 N4 P474-480	60-23517 <+*> 0
1958 V4 N4 P486-493	59-17223 <+*>
1958 V4 N4 P494-500	59-17222 <+*>
1958 V4 N4 P501-505	59-17221 <+*>
1958 V4 N4 P521-523	59-20860 <+*>
1958 V4 N4 P523-525	59-17220 <+*>
1958 V4 N4 P532	59-20859 <+*>
1958 V4 N4 P539-541	AD-409 190 <=*$>
1958 V4 N4 P543-546	59-20858 <+*> C
1958 V4 N5 P580-585	59-11444 <=>
1958 V4 N5 P586-594	59-11445 <=>
1958 V4 N5 P595-601	59-17285 <+*>
1958 V4 N5 P602-619	21M39R <ATS>
1958 V4 N5 P62C-630	61-20484 <*>
	9681-B <K-H>
1958 V4 N5 P637-642	59-15991 <+*>
1958 V4 N5 P651-657	61-23378 <*=>
1958 V4 N5 P681-683	66-12691 <*>
1958 V4 N5 P692-695	59-15990 <+*>
1958 V4 N5 P696-697	59-15721 <+*>
1958 V4 N6 P715-718	UCRL TRANS-767(L) <=*>
1958 V4 N6 P725-733	AEC-TR-3681 <=*>
1958 V4 N6 P734-749	61-10905 <+*>
1958 V4 N6 P758-762	AEC-TR-3506 <+*>
1958 V4 N6 P779-790	59-18469 <+*>
1958 V4 N6 P795-797	59-20856 <+*>
1958 V4 N6 P799	59-20855 <+*>
1958 V4 N6 P803-805	59-15263 <+*>
1958 V5 N1 P15-22	61-10906 <+*>
1958 V5 N1 P88-90	60-31208 <+*>
1958 V5 N1 P90-92	59-11440 <=> 0
1958 V5 N1 P93-94	59-22647 <+*> C
1958 V5 N2 P147-155	27M39R <ATS>
1958 V5 N2 P184-190	59-17743 <+*>
1958 V5 N2 P219	59-18768 <+*>
1958 V5 N2 P222-224	09L36R <ATS>
1958 V5 N3 P236-237	AEC-TR-5321 <=*>

1958 V5 N3 P236-238 59-21085 <=>
1958 V5 N3 P256-263 34P62R <ATS>
1958 V5 N3 P290-296 RTS-2989 <NLL>
1958 V5 N3 P302-306 66-14640 <*>
1958 V5 N3 P334-337 AEC-TR-5262 <=*>
1958 V5 N3 P342-343 62-32640 <=*>
1958 V5 N4 P384-392 60-22594 <=>
62-11391 <=>
1958 V5 N4 P440-449 62-18621 <=*>
1958 V5 N4 P462-468 65-10851 <*>
1958 V5 N5 P490-499 61-11189 <=>
1958 V5 N5 P520-529 62-25567 <=*>
1958 V5 N5 P530-534 59-20857 <+*>
1958 V5 N5 P601-605 65-10850 <*>
1958 V5 N5 P614-617 59M47R <ATS>
1958 V5 N5 P617-619 65-10287 <*> O
1958 V5 N6 P692-698 59-20835 <+*>
1958 V5 N6 P709-711 24M39R <ATS>
62-10368 <*=> O
1959 V6 P564-565 60-00085 <*>
1959 V6 N1 P55-64 AEC-TR-3775 <+*>
1959 V6 N1 P65-69 AEC-TR-3776 <+*>
1959 V6 N1 P70-77 AEC-TR-3777 <+*>
1959 V6 N1 P98-101 AEC-TR-3778 <+*>
1959 V6 N1 P107-109 59-16560 <+*>
1959 V6 N1 P117-118 AEC-TR-3779 <+*>
1959 V6 N1 P122-124 AEC-TR-3780 <+*>
59-14903 <+*>
1959 V6 N2 P270-271 59-14906 <+*>
1959 V6 N3 P412-415 59N53R <ATS>
1959 V6 N4 P440-446 60-13643 <+*> O
1959 V6 N4 P450-456 60-15371 <+*>
1959 V6 N4 P565-566 AEC-TR-3824 <+*>
1959 V6 N5 P631-636 61-17550 <=$>
1960 V8 N1 P137-140 60-21764 <=>
1960 V8 N5 P585-593 60-41735 <=*>
1960 V9 N5 P631-634 61-15699 <+*>
1960 V9 N6 P784-786 61-15698 <+*> O
1961 V10 N3 P417 62-32623 <=*>
1961 V10 N4 P487-492 61-19926 <*=>
1961 V10 N5 P663-666 61-27253 <*=>
1961 V11 N4 P486-491 45P65R <ATS>
1961 V11 N4 P536-541 62-23319 <=*>
1962 V12 N3 P424-426 62-25065 <=*>
1966 V20 N4 P701-708 AD-641 140 <=$>

OPTIKA I SPEKTROSKOPIYA: SBORNIK STATEI
1963 V1 P51-57 65-14784 <*>

*OPTIKO-MEKHANICHESKAYA PROMYSHLENNOST
1940 V10 N1 P3-6 64-18564 <*> O
1940 V10 N11 P10-11 UCRL-TRANS-946(L) <=$>

OPUSCULA MEDICA
1957 V2 P62-64 57-1863 <*>
1957 V2 P70-76 57-1861 <*>

OPYT RABOTY PO TEKHNICHESKOI INFORMATSII I PROPAGANDE. MOSCOW
1963 N4 P7-8 64-41662 <=>
1963 N4 P46 64-41662 <=>

ORDNANCE AND TECHNOLOGY. JAPAN
SEE HEIKI TO GIJUTSU

ORELL FUESSLIS HEFTE ZUR SOZIAL-UND WIRTSCHAFTS-GESCHICHTE
1929 P499-509 65-13778 <*>

ORGAN FUER DIE FORTSCHRITTE DES EISENBAHNWESENS
1941 V96 P193-206 T-1936 <INSD>

ORGANISATION METEOROLOGIQUE MONDIALE
1954 N35 P11 58-1716 <*>

ORGANIZATSIYA I METODIKA NTI
1964 N2 P3-5 64-31876 <=>

ORIENTACION MEDICA
1961 V10 P102 61-16479 <*>

1964 N586 P606-608 C6S85SP <ATS>
1965 V14 P523 66-12676 <*>

ORIENTATION
1941 V1 N7 P93-99 61-14949 <*>

ORNITOLOGIYA
1959 N2 P153-156 61-31051 <=>
1962 N5 P177-182 RTS-3277 <NLL>

ORSZAGOS MUSZAKI KONYVTAI, BUDAPEST. TUDOMANYOS ES MUSZAKI TAJEKOZTATOS
1961 V8 N3 P4-6 64-30714 <*>
1963 P119-157 65-00039 <*>
1963 V10 N1 P3-13 64-30720 <*>
1963 V10 N1 P13-19 64-30717 <*>

ORTHOPEDIC SURGERY. TOKYO
SEE SEIKEIGEKA

ORTHOPEDICS. TOKYO
SEE SEIKEIGEKA

ORTOPEDIA E TRAUMATOLOGIA DELL'APPARATO MOTORE
1957 V25 N2 P279-335 60-15610 <=*> O

ORTOPEDIYA, TRAVMATOLOGIYA I PROTEZIROVANIE
1938 V12 N3 P52-62 60-23186 <*+>
1940 V14 N3 P50-61 R-1047 <*>
1940 V14 N5/6 P101-107 RT-1824 <*>
1956 V17 N6 P71-74 62-15157 <=*>
1957 V18 N2 P20-24 59-18149 <+*>
1957 V18 N4 P30-34 PB 141 296T <=> O
1958 V19 N6 P52-57 61-19732 <=*> C
1959 V20 P79-85 61-19733 <=*> O
1960 V21 N2 P76-79 60-31419 <=>
1960 V21 N5 P77-79 62-23936 <=*>
1960 V21 N6 P83-85 60-41566 <=>
1960 V21 N6 P85-86 60-41567 <=>
1960 V21 N6 P86-87 60-41568 <=>
1960 V21 N6 P87-94 60-41569 <=>
1961 V22 N5 P65-69 64-21467 <=>
1961 V22 N6 P42-46 62-13041 <=*>
1962 V23 P16-23 65-63738 <=>
1962 V23 P24-26 66-60612 <=*$>
1962 V23 N1 P64-68 62-24022 <=*>
1963 V24 N2 P3-9 63-21862 <=>
1963 V24 N2 P9-14 63-31575 <=>
1963 V24 N2 P18-21 63-21862 <=>
1963 V24 N2 P67-68 63-21862 <=>
1963 V24 N4 P49-66 63-31268 <=>
1963 V24 N6 P11-18 63-31822 <=>
1963 V24 N11 P21-23 64-21256 <=>
1963 V24 N12 P3-9 64-21638 <=>
1964 V25 N1 P3-7 64-31062 <=>
1964 V25 N1 P41-48 64-31062 <=>
1964 V25 N3 P88-92 64-31403 <=>
1964 V25 N4 P3-36 64-31991 <=>
1964 V25 N4 P50-56 64-31991 <=>
1964 V25 N6 P3-9 64-41731 <=>
1964 V25 N6 P61-62 64-41733 <=>
1964 V25 N6 P66 64-41734 <=>
1964 V25 N8 P53-55 64-51214 <=$>
1964 V25 N11 P18-24 65-30185 <=$>
1964 V25 N11 P48-51 65-30184 <=$>
1964 V25 N12 P49-52 65-30453 <=$>
1965 V26 N1 P52-60 65-30746 <=*>
1965 V26 N8 P28-32 65-33043 <=$>
1965 V26 N8 P62-65 65-33043 <=$>

ORVOSI HETILAP
1934 V78 N30 P684-687 63-16105 <*>
9641 <K-H>
1952 V93 N38 P1087-1093 57-1260 <*>
1953 V94 P690-692 1280 <*>
1955 V96 N8 P22-23 3759-B <KH>
1957 N47 P1290-1293 58-1178 <*>
1957 V98 P1018 59-15456 <*>
1960 V101 P258 85M46H <ATS>
1960 V101 P1211-1212 63-00781 <*>

1961 V102 P1707 63-00927 <*>
1961 V102 N23 P1063-1071 62-19682 <=*>
1961 V102 N26 P1227-1228 62-19615 <=*> P
1961 V102 N28 P1306-1313 62-19606 <*=>
1961 V102 N31 P1450-1452 62-19661 <=*>
1961 V102 N31 P1476-1477 62-19602 <=*>
1961 V102 N32 P1517 62-19688 <=*>
1961 V102 N32 P1534-1536 62-19687 <=*>
1961 V102 N35 P1655-1657 62-19622 <*=> P
1961 V102 N35 P1661-1662 62-19620 <=*>
1961 V102 N48 P2300-2302 62-19781 <=*>
1962 V103 N39 P1851-1854 63-15090 <=*>
1962 V103 N41 P1933-1935 64-11661 <=>
1963 V104 P1378-1379 66-12958 <*>
1964 V105 P1674-1679 64-51117 <=>
1964 V105 N21 P1005 64-31837 <=>
1964 V105 N21 P1008 64-31816 <=>
1964 V105 N38 P1817-1822 64-51364 <=$>
1965 N41 P1964-1965 65-33753 <=*$>
1965 V106 N3 P141-142 65-30403 <=$>
1965 V106 N4 P188 65-30550 <=$>
1965 V106 N16 P761-767 65-31276 <=$>
1965 V106 N17 P812-813 65-31276 <=$>
1965 V106 N30 P1436-1437 65-32345 <=$>
1965 V106 N33 P1569 65-32494 <=*$>
1965 V106 N41 P1964-1965 65-34075 <=*$>
1965 V106 N46 P2205-2206 65-34104 <=*$>
1966 N29 P1379-1381 66-33657 <=$>

ORVOSKEPZES
1961 V36 N1 P1-15 62-19664 <=*>
1961 V36 N1 P76-80 62-19603 <*=>
1961 V36 N4 P297-315 62-19674 <=*>

OSAKA CITY MEDICAL JOURNAL
SEE OSAKA SHI IGAKKAI

OSAKA DAIGAKU IGAKU ZASSHI
1955 V7 N4 P377-387 65-00418 <*> O
1958 V10 P1305-1309 66-12005 <*>
1958 V10 P1311-1316 66-12024 <*>
1958 V10 P1317-1323 66-12150 <*>

OSAKA FURITSU KOGYO SHOREIKAN HOKOKU
1961 N25 P58-62 ATS-75R80J <ATS>

OSAKA GANKASHU DANKAI
1963 P1-4 65-60312 <=$>

OSAKA IGAKKAI ZASSHI
1937 V36 P955-958 63-01680 <*>

OSAKA GIJUTSU SHIKEN SHO HOKOKU

1960 V11 N4 P229-234 65-17339 <*>

OSAKA SHI IGAKKAI
1959 V11 N11 P333-344 63-01524 <*>

OSAKA SHIRITSU DAIGAKU IGAKU ZASSHI
1955 V4 P376-380 C-3575 <NRCC>
1957 V6 P234-240 58-729 <*>
1959 V8 N9 P1335-1362 ANL-TR-6 <*>
79Q73J <ATS>
1962 V11 N5/8 P201-214 64-18682 <*>

★OSNOVANIYA, FUNDAMENTY I MEKHANIKA GRUNTOV
1960 V2 N1 P8-10 62-33614 <=*>
1960 V2 N1 P20-23 62-33615 <=*>
1961 V3 N3 P14-16 62-25130 <=*>
1961 V3 N4 P7-8 62-25132 <=*>
1962 V4 N1 P3-7 62-33616 <=*>
1962 V4 N6 P3-7 63-24213 <=*$>
1962 V4 N6 P11-13 66-61549 <=*$>
1963 V5 N1 P18-20 LC-1278 <NLL>
1963 V5 N2 P1-5 M-5707 <NLL>
1963 V5 N2 P8-12 66-61548 <=*$>
1963 V5 N5 P5-8 65-60990 <=$>
1963 V5 N6 P4 66-61547 <=*$>
1964 V6 N1 P24-26 66-61550 <=*$>

OSNOVY NEMATODOLOGII
1958 V7 ENTIRE ISSUE 65-50073 <=>

OSNOVY TSESTODOLOGII
1951 V1 P1-730 61-11490 <=>

OSPEDALE. TORINO
1962 V9 P1-3 63-01134 <*>

OSPEDALE MAGGIORE
1963 V58 P1013-1027 65-14193 <*>

OSPEDALE PSICHIATRICO
1947 V51 P97-104 58-414 <*>

OSSATURE METALLIQUE
1953 N10 P507-519 T2130 <INSD>

OSTEUROPA-NATURWISSENSCHAFT
1958 V1 N1/2 P31-39 62-19293 <*=>
1963 V7 N1 P16-43 64-21780 <=>

OTCHETNOST I KONTROL V SELSKOTO STOPANSTVO
1964 V9 N4 P181-184 64-31635 <=>
1964 V9 N9 P390-396 64-51734 <=$>
1966 N4 P153-156 66-31972 <=$>
1966 N4 P168-172 66-31972 <=$>
1966 N4 P178-182 66-31972 <=$>
1966 N4 P186-188 66-31972 <=$>
1966 N11 P452-455 66-35730 <=>
1966 N11 P475-482 66-35730 <=>

OTECHESTVEN FRONT
1966 03-4 05/15 66-32629 <=$>

OTOLARYNGOLOGIA POLSKA
1957 V11 N2 P170-174 64-10423 <*>
8043-A <K-H>
1957 V11 N2 P171-174 58-2459 <*>
1957 V11 N4 P371-378 8388-B <K-H>

OTO-RHINO-AND LARYNGOLOGICAL CLINIC
SEE JIBIINKOKA RINSHO

OTO-RHINO-LARYNGOLOGIE INTERNATIONALE
1940 V24 P46-63 61-14603 <*>

OTO-RINO-LARINGOLOGIA ITALIANA
1952 V20 P520-529 57-3570 <*>
1953 V21 P399-406 65-13958 <*>
1958 V26 N1 P68-84 62-00613 <*>
1959 V27 P261-286 63-10720 <*> O
1959 V28 P437-458 10265-D <K-H>
64-14580 <*>
1960 V29 N3 P165-178 65-17256 <*>

OUEST MEDICAL
1964 V17 N13 P789-791 66-13164 <*>

OVERSIGT AF FORHANDLINGAR. FINSKA VETENSKAPS-SOCIETEN, HELSINGFORS. SERIES A. MATEMATIK OCH NATURVETENSKAPER
1914 V57A N11 P1-13 61-10747 <*>

OVTSEVODSTVO
1959 N1 P10-14 59-31004 <=>
1961 N3 P24-28 61-27116 <=>
1964 V11 N9 P8-11 64-51438 <=$>

OYO BUTSURI
1948 V17 P104-109 SCL-T-378 <*>
1949 V18 P352-353 SC-T-64-915 <*>
1951 V20 N6/7 P237-239 65-10123 <*>
1951 V20 N6/7 P242-246 <ATS>
1951 V20 N8/9 P286-289 <ATS>

1952 V21 P403-408 T1727 <INSD>
1953 V22 N10 P358-359 II-640 <*>
1955 V24 P113-117 SC-T-64-916 <*>
1955 V24 N2 P49-56 C-4152 <NRCC>
1956 V25 P253-257 64-30106 <*> 0
1956 V25 N10 P389-395 62L30J <ATS>
1956 V25 N12 P475-479 63L30J <ATS>
1958 V27 N10 P577-584 60-14225 <*>
1958 V27 N1C P585-590 59-15276 <*>
1958 V27 N10 P590-595 60-10276 <*>
1958 V27 N10 P623-632 59-15722 <*>
1958 V27 N1C P633-634 59-15594 <*>
1958 V27 N10 P634-636 59-15593 <*>
1958 V27 N19 P595-599 60-14226 <*>
1959 V28 N4 P216-219 31R76J <ATS>
1959 V28 N8 P439-444 64-16496 <*> 0
1959 V28 N9 P531-534 66-12655 <*>
1959 V28 N11 P642-650 60-16956 <*>
1959 V28 N11 P668-669 61-10591 <*>
1959 V28 N12 P677-680 60-18991 <*>
1959 V28 N12 P681-687 60-18131 <*>
1959 V28 N12 P687-697 60-16956 <*>
1959 V28 N12 P711-720 60-18118 <*>
1960 V29 N3 P155-158 60-18989 <*>
1960 V29 N7 P438-442 SCL-T-403 <*>
1960 V29 N8 P509-514 SCL-T-415 <*>
1960 V29 N9 P636-641 62-10987 <*>
1960 V29 N12 P849-855 84N53J <ATS>
1961 V30 N1 P31-36 84N52J <ATS>
1961 V30 N1 P36-40 85N52J <ATS>
1961 V30 N7 P496-501 SC-T-65-626 <*>
1961 V30 N9 P647-653 SC-T-65-727 <*>
1961 V30 N9 P700-704 SC-T-65-731 <*>
1961 V30 N10 P778-785 60Q73J <ATS>
1961 V30 N12 P911-915 6.-14363 <*>
1962 V31 N3 P178-186 88R76J <ATS>
1962 V31 N9 P730-738 88Q68J <ATS>
1962 V31 N9 P739-745 C-4500 <NRCC>
1962 V31 N9 P749-752 C-4479 <NRCC>
1963 V32 P363-367 NSJ-TR-4 <*>
1963 V32 N2 P199-212 C-5310 <NRC>
1963 V32 N8 P562-567 40S85J <ATS>
1963 V32 N9 P673-676 00R75J <ATS>
1964 V33 N2 P82-86 65-11351 <*> 0
1964 V33 N3 P153-166 70S85J <ATS>
1964 V33 N9 P387-394 64-71517 <=>
1964 V33 N10 P721-726 41S82J <ATS>
1964 V33 N1C P738-740 99S88J <ATS>
1964 V33 N11 P823-825 72S86J <ATS>
1965 V34 N2 P97-107 803 <TC>

OYC DENKI KENKYUJO IHO
1955 V7 N1 P10-16 70S82J <ATS>
1956 V8 N3 P127-143 76S82J <ATS>

OYO DOBUTSUGAKU ZASSHI
1956 V21 N2 P53-62 64-00283 <*>

OYO-KONTYU
1950 V5 N4 P155-168 57-3279 <*>
1951 V7 N2 P59-60 C-2333 <NRC>
57-1775 <*>

OYO RIKIGAKU
1948 V1 P133-141 SCL-T-374 <*>

PTT TECHNISCHE MTTEILUNGEN
SEE TECHNISCHE MITTEILUNGEN PTT

PTT VESNIK. GENERALNE DIREKCIJE PTT
1959 V67 N16 P192 60-31321 <=*>
1959 V67 N22 P255-256 60-31325 <=*>

PTT-BEDRIJF
1950 V3 N2 P85-91 57-955 <*>
1951 V3 N4 P128-134 57-1020 <*>
1951 V4 N1 P34-41 57-1019 <*>
1952 V4 N3 P92-101 57-965 <*>
1952 V4 N3 P118-124 2526 <*>
1953 V5 N2 P84-88 57-1018 <*>
1953 V5 N3 P112-115 57-1017 <*>
1957 V18 N3 P74-77 60-31856 <=*>
1958 V19 N10 P294-295 60-31857 <=*>
1958 V19 N10 P307 60-31863 <=*>
1959 V2C N2 P52-54 60-31858 <=*>
1959 V20 N9 P268-269 6C-31646 <=*>
1959 V20 N9 P283 60-31647 <=*>

PVC ANC POLYMERS. JAPAN
SEE ENKA BINIIRU TC PCRIMA

PADAGOGIK
1963 V18 N1 P114-120 63-21503 <=>
1963 V18 N2 P159-167 63-21637 <=>

PAEDIATRISCHE FORTBILDUNGSKURSE FUER DIE PRAXIS. BASEL, NEW YORK
1964 V10 P53-62 66-10855 <*>

PALAECNTOGRAPHICA. CASSEL, STITTGART
1939 V84 N1/2 P1-20 57-3036 <*>

PALEONTOLOGICHESKII SBORNIK
1954 N1 P31-43 62-23744 <=*> C
1954 N1 P44-51 62-23743 <=*> 0
1954 N1 P69-80 89K27R <ATS>
1960 V3 N16 P133-173 63-32469 <=*>

*PALEONTOLOGICHESKII ZHURNAL
1959 N1 P25-36 C-3222 <NRCC>
1959 N3 P138-140 61-27881 <=*>
1959 N4 P48-64 61-15887 <=*> C
1959 N4 P85-89 RTS-3182 <NLL>
1959 N4 P9C-99 RTS-3183 <NLL>
1960 N1 P118-127 61-23968 <*=> 0
1960 N2 P97-109 RTS-3154 <NLL>
1960 N3 P28-42 63-19264 <=*>
1960 N3 P48-51 RTS-3168 <NLL>
1960 N4 P43-47 RTS-3167 <NLL>
1960 N4 P114-124 19N50R <ATS>
1960 N4 P125-128 20N50R <ATS>
1961 N1 P70-74 61-28663 <=*> C
1961 N1 P170-174 61-27100 <*=>
61-28606 <*=>
1961 N2 P140-141 61-28597 <*=>
1961 N2 P141-143 61-28600 <*=>
1962 N1 P3-6 IGR V5 N11 P1474 <AGI>
IGR,V5,N11,P1474-76 <AGI>
1962 N1 P7-18 IGR V5 N11 P1477 <AGI>
IGR,V5,N11,P.1477-86 <AGI>
1962 N1 P31-40 IGR V5 N11 P1487 <AGI>
IGR,V5,N11,P1487-95 <AGI>
1962 N1 P48-56 IGR V5 N11 P1501 <AGI>
1962 N1 P58-65 IGR,V5,N11,P1501-09 <AGI>
1962 N1 P105-110 IGR V5 N11 P1496 <AGI>
IGR,V5,N11,P1496-500 <AGI>
1962 N1 P157-158 IGR V5 N11 P1510 <AGI>
IGR,V5,N11,P1510-11 <AGI>
1962 N1 P166-168 IGR V5 N11 P1512 <AGI>
IGR,V5,N11,P1512-14 <AGI>
1962 N1 P173-174 IGR V5 N11 P1515 <AGI>
IGR,V5,N11,P1515-16 <AGI>
1962 N2 P9-20 IGR V5 N12 P1635 <AGI>
IGR,V5,N12,P1635-47 <AGI>
1962 N2 P21-33 IGR V5 N12 P1659 <AGI>
IGR,V5,N12,P1659-69 <AGI>
1962 N2 P34-44 IGR V5 N12 P1648 <AGI>
1962 N2 P45-48 IGR V5 N12 P1670 <AGI>
IGR,V5,N12,P1670-73 <AGI>
1962 N2 P49-70 IGR V5 N12 P1681 <AGI>
1962 N2 P117-121 IGR V5 N12 P1674 <AGI>
1962 N2 P160-162 63-24570 <=*$>
1962 N2 P166-167 IGR V5 N12 P1678 <AGI>
1962 N3 P40-46 IGR V6 N4 P700 <AGI>
1962 N3 P47-56 64-19961 <=$>
1962 N3 P57-60 IGR V6 N4 P711 <AGI>
1962 N3 P65-80 63-26962 <=$>
1962 N3 P102-110 IGR V6 N4 P716 <AGI>
1962 N3 P130-135 IGR V6 N4 P706 <AGI>
1962 N3 P136-137 IGR V6 N4 P724 <AGI>

1962 N4 P21-30 IGR V6 N9 P1596 <AGI>
IGR V6 N9 P1596-1603 <AGI>
1962 N4 P43-57 <JBS>
IGR V6 N9 P1617 <AGI>
IGR V6 N9 P1617-1629 <AGI>
1962 N4 P58-69 IGR V6 N9 P1630 <AGI>
IGR V6 N9 P1630-1641 <AGI>
1962 N4 P70-82 IGR V6 N9 P1604 <AGI>
IGR V6 N9 P1604-1616 <AGI>
1962 N4 P104-115 IGR V6 N9 P1642 <AGI>
IGR V6 N9 P1642-1651 <AGI>
1963 N1 P3-12 IGR V6 N10 P1814 <AGI>
1963 N1 P15-34 IGR V6 N10 P1827 <AGI>
1963 N1 P42-52 IGR V6 N10 P1858 <AGI>
1963 N1 P53-57 IGR V6 N10 P1868 <AGI>
1963 N1 P105-110 IGR V6 N10 P1822 <AGI>
1963 N1 P127-136 IGR V6 N10 P1847 <AGI>
1963 N1 P131-135 IGR V7 N6 P1116 <AGI>
1963 N1 P144-145 IGR V6 N10 P1855 <AGI>
1963 N2 P3-16 IGR V6 N12 P2204 <AGI>
1963 N2 P20-25 IGR V6 N12 P2229 <AGI>
1963 N2 P26-37 IGR V6 N12 P2235 <AGI>
1963 N2 P43-53 IGR V6 N12 P2214 <AGI>
IGR V6 N12 P2214 <IGR>
1963 N2 P144-147 IGR V6 N12 P2224 <AGI>
1963 N2 P148-150 IGR V6 N12 P2246 <AGI>
1963 N2 P151-154 IGR V6 N12 P2249 <AGI>
1963 N3 P3-9 44Q73R <ATS>
1963 N3 P10-17 IGR V7 N5 P910 <AGI>
1963 N3 P27-28 IGR V7 N5 P898 <AGI>
1963 N3 P39-48 IGR V7 N5 P918 <AGI>
1963 N3 P78-83 IGR V7 N5 P926 <AGI>
1963 N3 P110-112 IGR V7 N6 P1121 <AGI>
1963 N3 P120-123 IGR V7 N5 P931 <AGI>
1963 N4 P44-52 IGR V7 N6 P1075 <AGI>
1963 N4 P53-63 IGR V7 N6 P1084 <AGI>
1963 N4 P76-94 IGR V7 N6 P1094 <AGI>
1963 N4 P95-102 IGR V7 N6 P1110 <AGI>
1963 N4 P141- IGR V7 N6 P1074 <AGI>
1964 N1 P3-9 IGR V7 N8 P1410 <AGI>
1964 N1 P20-25 IGR V7 N8 P1423 <AGI>
1964 N1 P38-44 IGR V7 N8 P1429 <AGI>
1964 N1 P45-55 IGR V7 N8 P1435 <AGI>
RTS-2688 <NLL>
1964 N1 P63-70 IGR V7 N8 P1450 <AGI>
1964 N1 P82-87 IGR V7 N8 P1456 <AGI>
1964 N1 P97-106 IGR V7 N8 P1461 <AGI>
1964 N1 P107-113 IGR V7 N8 P1474 <AGI>
1964 N1 P114-119 IGR V7 N8 P1469 <AGI>
1964 N1 P131-132 IGR V7 N8 P1487 <AGI>
1964 N1 P132-134 IGR V7 N8 P1489 <AGI>
1964 N1 P139-141 IGR V7 N8 P1492 <AGI>
1964 N2 P22-31 IGR V7 N9 P1622 <AGI>
1964 N2 P85-98 IGR V7 N9 P1629 <AGI>
1964 N2 P99-114 IGR V7 N9 P1643 <AGI>
1964 N2 P115-120 IGR V7 N9 P1670 <AGI>
1964 N2 P121-124 IGR V7 N9 P1655 <AGI>
1964 N2 P125-131 IGR V7 N9 P1676 <AGI>
1964 N2 P132-142 IGR V7 N9 P1659 <AGI>
1964 N2 P149-152 IGR V7 N9 P1687 <AGI>
1964 N2 P152-154 IGR V7 N9 P1640 <AGI>
1964 N2 P169-173 IGR V7 N9 P1682 <AGI>
1964 N3 P4-14 IGR V7 N10 P1806 <AGI>
1964 N3 P23-29 IGR V7 N10 P1819 <AGI>
1964 N3 P35-46 IGR V7 N10 P1847 <AGI>
1964 N3 P52-57 IGR V7 N10 P1814 <AGI>
1964 N3 P58-72 IGR V7 N10 P1826 <AGI>
1964 N3 P111-114 IGR V7 N10 P1840 <AGI>
1964 N3 P121-123 IGR V7 N10 P1844 <AGI>
1964 N3 P125-126 65-63364 <=*$>
1964 N3 P127-131 IGR V7 N10 P1860 <AGI>
1964 N4 P3-9 IGR V7 N11 P2049 <AGI>
1964 N4 P10-22 IGR V7 N11 P2054 <AGI>
1964 N4 P23-31 IGR V7 N11 P2066 <AGI>
1964 N4 P101-103 IGR V7 N11 P2063 <AGI>
1964 N4 P104-110 IGR V7 N11 P2074 <AGI>
48S82R <ATS>
1965 N1 P13-25 IGR V7 N12 P2121 <AGI>
1965 N1 P26-38 IGR V7 N12 P2091 <AGI>
1965 N1 P54-59 IGR V7 N12 P2115 <AGI>
1965 N1 P133-144 IGR V7 N12 P2105 <AGI>
1965 N1 P136-141 IGR V8 N9 P1118 <AGI>
1965 N1 P148-150 IGR V7 N12 P2102 <AGI>
1965 N2 P3-17 IGR V8 N1 P48 <AGI>
1965 N2 P18-22 IGR V8 N1 P60 <AGI>
93S86R <ATS>
1965 N2 P55-62 IGR V8 N1 P64 <AGI>
1965 N2 P80-92 IGR V8 N1 P71 <AGI>
1965 N2 P126-128 IGR V8 N1 P81 <AGI>
1965 N2 P133-137 IGR V8 N1 P84 <AGI>
1965 N2 P144-146 IGR V8 N1 P89 <AGI>
1965 N2 P153-156 IGR V8 N1 P94 <AGI>
1965 N3 P16-22 IGR V8 N7 P795 <AGI>
1965 N3 P23-32 IGR V8 N7 P803 <AGI>
1965 N3 P59-72 IGR V8 N7 P811 <AGI>
1965 N3 P110-118 IGR V8 N7 P823 <AGI>
1965 N3 P119-122 IGR V8 N7 P838 <AGI>
1965 N3 P125-126 IGR V7 N10 P1858 <AGI>
1965 N3 P127-132 IGR V8 N7 P831 <AGI>
1965 N3 P133-136 66-14122 <*$>
1965 N3 P139-141 IGR V8 N7 P844 <AGI>
1965 N3 P145-147 IGR V8 N7 P848 <AGI>
1965 N4 P16-25 IGR V8 N8 P949 <AGI>
1965 N4 P33-40 IGR V8 N8 P958 <AGI>
1965 N4 P50-59 IGR V8 N8 P982 <AGI>
1965 N4 P68-74 IGR V8 N8 P976 <AGI>
1965 N4 P75-87 IGR V8 N8 P965 <AGI>
1965 N4 P88-91 IGR V8 N8 P991 <AGI>
1966 N1 P19-27 IGR V8 N9 P1050 <AGI>
1966 N1 P28-36 IGR V8 N9 P1058 <AGI>
1966 N1 P47-59 IGR V8 N9 P1067 <AGI>
1966 N1 P60-71 IGR V8 N9 P1078 <AGI>
1966 N1 P72-86 IGR V8 N9 P1089 <AGI>
1966 N1 P116-123 IGR V8 N9 P1102 <AGI>
1966 N1 P124-134 IGR V8 N9 P1109 <AGI>
1966 N1 P142-144 IGR V8 N9 P1123 <AGI>
1966 N1 P148-150 IGR V8 N9 P1126 <AGI>
1966 N1 P162-164 IGR V8 N9 P1129 <AGI>

PALEONTOLOGIYA I STRATIGRAFIYA BSSR
1955 V1 N1 P5-47 63-24366 <=*$>
1957 V2 P3-43 62-13737 <=*>

PALIVA
1950 V30 N11 P308-319 54Q70C <ATS>
1954 N5 P126-127 T1815 <INSD>
1954 V34 N5 P121-125 CSIRO-2825 <CSIR>
1955 V35 N10 P302-308 CSIRO-3203 <CSIR>
1956 V36 N5 P145-148 CSIRO-3222 <CSIR>
1956 V36 N6 P185-192 TS-1581 <BISI>
1581 <BISI>
62-14186 <*>
1956 V36 N6 P192-199 CSIRO-3298 <CSIR>
1956 V36 N11 P363-365 CSIRO-3419 <CSIR>
1957 V37 P33-36 21K20C <ATS>
1964 V44 N3 P75-81 4147 <BISI>
1964 V44 N8 P243-246 64-51704 <=$>

PALIVA A VODA
1946 V26 N8 P109-118 63-10703 <*>

PAMIETNIK INSTYTUTU ZOOTECHNIKI POLSCE
1956 P61-73 9J17P <ATS>

PAMIETNIK. PULAWSKI INSTYTUT UPRAWY, NAWOZENIA I GLEBOZNAWSTWA. WARSAW
1962 N7 P33-64 65-50334 <=>
1962 N8 P305-313 65-50337 <=>
1962 N8 P315-322 65-50338 <=>
1963 N9 P53-97 65-50320 <=>
1963 N10 P3-34 65-50339 <=>
1963 N10 P37-55 65-50340 <=>
1963 N10 P57-76 65-50341 <=>
1963 N10 P167-173 65-50346 <=>
1963 V9 P293-305 65-50322 <=>

PAPERI JA PUU
1941 V23 N15 P293-300 63-18181 <*> 0
1945 N7A P16-23 AL-565 <*>
1947 V29 P263-268 63-10756 <*> 0

Citation	Reference
1947 V29 P270-273	63-10756 <*> 0
1947 V29 N3 P38-44	60-18309 <*>
1951 V33 N2 P27-32	60-18191 <*> 0
1951 V33 N3 P53-56	65-12301 <*>
1952 V34 N9 P319-	AL-3 <*>
1954 V36 N1C P397-400	58-2684 <*>
1955 V37 N2 P23-28	57-2693 <*>
1956 N4 P231-	57-1983 <*>
1956 N4 P231	57-2893 <*>
1956 V38 P451-456	57-8C1 <*>
1956 V38 N9 P391-40C	C-2729 <NRC>
1956 V38 N9 P443-449	57-2846 <*>
1956 V38 N9 P451-454	57-2850 <*>
1956 V38 N9 P456	57-2850 <*>
1956 V38 N9 P457-461	C-2720 <NRC>
1956 V38 N10 P491	58-595 <*>
1956 V38 N4A P145-152	59-1C216 <*>
1957 V39 N3 P99-108	58-1625 <*>
1957 V39 N7 P352	58-2534 <*>
1958 V40 N11 P561-568	59-15723 <*>
	59-20801 <*> 0
1958 V40 N4A P159-160	59-10219 <*>
1958 V40 N4A P162-164	59-10219 <*>
1959 V41 N3 P97-99	65-10994 <*> 0
1959 V41 N9 P419	60-16582 <*> 0
1959 V41 N11 P559-560	60-16912 <*>
1959 V41 N12 P6C4-608,610	62-165C3 <*> 0
1960 V42 N3 P99-104	61-10920 <*>
1960 V42 N7 P4C1-409	64-10934 <*>
1960 V42 N8 P443-447	61-10609 <*> 0
1960 V42 N4A P235-237	61-10623 <*> 0
1961 V43 N2 P37-46	61-16553 <*>
1961 V43 N3 P83-92	61-20887 <*> 0
1961 V43 N4 P169-180	61-18800 <*> 0
1961 V43 N4 P169-175	63-10177 <*> 0
1961 V43 N4 P179-180	63-10177 <*> 0
1961 V43 N12 P757-761	62-16617 <*> 0
1961 V43 N4A P181-192	8691 <IICH>
1962 V44 N1 P3-16	C-3975 <NRC>
1962 V44 N9 P433-438	45Q69FN <ATS>
1962 V44 N9 P439-446	44Q69FN <ATS>
1962 V44 N9 P45C-457	44Q69FN <ATS>
1962 V44 N4A P159-162	63-10066 <*> 0
1962 V44 N4A P167-172	63-10066 <*> 0
1962 V44 N4A P217-222	63-10490 <*> 0
	63-18814 <*>
1963 V45 N4A P181-190	63-20564 <*> 0
	64-18593 <*> 0
1964 V46 P7-9	88S87FN <ATS>
1964 V46 P11-14	88S87FN <ATS>
1965 V47 N3 P109-121	C-5513 <NRC>
1965 V47 N9 P503-508	66-11004 <*>

PAPETERIE

Citation	Reference
1926 P722	59-20066 <*>
1926 P725	59-20066 <*>
1926 P769-770	59-20066 <*>
1926 P821-822	59-2C066 <*>
1926 P917-918	59-20066 <*>
1926 P921	59-20066 <*>
1949 V71 N2 P495-497	59-15117 <*> P
1949 V71 N4 P118-119	AL-584 <*>
1951 V73 P263-267	2169B <K-H>
1953 V75 N10 P659-673	61-20898 <*> 0
1955 P472-	57-1717 <*>
1955 P475-	57-1718 <*>
1955 N12 P811-	57-1734 <*>
1955 V77 N12 P803-	57-678 <*>
1958 V80 N7 P492-496	59-15518 <*>
1959 V81 N10 P705	63-10916 <*>
1959 V81 N1C P707	63-10916 <*>
1959 V81 N10 P709	63-10916 <*>
1959 V81 N1C P713	63-10916 <*>
1959 V81 N10 P715	63-10916 <*>
1959 V81 N10 P717	63-10916 <*>
1962 V84 N1 P26-29	62-20218 <*>
1962 V84 N2 P128-133	62-20102 <*>

PAPIER. DARMSTADT

Citation	Reference
1948 N15/6 P260-275	3280 <*>
1948 V2 N13/4 P225-228	65-13814 <*>
1948 V2 N15/6 P265-275	1793 <TC>
	65-14509 <*>
1948 V2 N15/6 P288-290	65-13814 <*>
1949 V3 P215-223	57-2992 <*>
1950 V4 N12 P249-254	62-10999 <*>
1951 V5 N9 P149-155	64-16368 <*> 0
1951 V5 N10 P2C9-212	I-207 <*>
1951 V5 N5/6 P75-83	I-819 <*>
1951 V5 N11/2 P256-260	I-207 <*>
1951 V5 N17/8 P361-371	I-69 <*>
1951 V5 N19/0 P411-417	I-291 <*>
1952 V6 P10-18	3301 <*>
1952 V6 N3/4 P47	58-251 <*>
1952 V6 N5/6 P75-80	65-14594 <*>
1952 V6 N5/6 P80-86	57-3244 <*>
1952 V6 N7/8 P115-119	65-14594 <*>
1952 V6 N7/8 P126-127	62-16500 <*>
1952 V6 N21/2 P443-449	61-20097 <*>
1952 V6 N23/4 P496-503	61-20097 <*>
1952 V6 N23/4 P504-510	64S82G <ATS>
1952 V6 N23/4 P523-524	64-20052 <*>
1952 V6 N9/10 P169-176	65-14594 <*>
1953 V7 N3/4 P41-46	C-2228 <NRC>
1953 V7 N9/10 P153-158	60-16758 <*>
1954 P205-	57-897 <*>
1954 V8 P43-48	I-292 <*>
	1618 <*>
1954 V8 P419-430	63-20136 <*> 0
1954 V8 P431-434	57-2350 <*>
1954 V8 P47C-479	3315 <*>
	57-1309 <*>
1954 V8 N7/8 P109-120	57-2331 <*>
1954 V8 N13/4 P247-258	65-10996 <*> 0
1954 V8 N17/8 P365-370	31L29G <ATS>
	64-16040 <*> 0
1954 V8 N19/0 P409-418	2718 <*>
1954 V8 N19/0 P419-430	65-10996 <*> 0
1954 V8 N9/10 P163-172	6C-14549 <*> 0
1955 P437-439	57-1728 <*>
1955 N11/2 P237-	2470 <*>
1955 V9 P13-14	57-3544 <*>
1955 V9 N3/4 P51-58	2469 <*>
1955 V9 N7/8 P133-	57-1731 <*>
1955 V9 N7/8 P153-157	3432 <*>
1955 V9 N7/8 P157-	57-2877 <*>
1955 V9 N11/2 P237-	57-1438 <*>
1955 V9 N13/4 P290-295	3769-A <K-H>
1955 V9 N13/4 P304-311	57-3311 <*>
1955 V9 N13/4 P311-	57-932 <*>
1955 V9 N13/4 P311-316	64-16041 <*>
1955 V9 N17/8 P429-437	57-3311 <*>
1955 V9 N23/4 P584-	57-1736 <*>
1955 V9 N23/4 P588-593	D-665 <RIS> C
1956 P109-120	2474 <*>
1956 P395-402	57-1720 <*>
1956 N11/2 P240-	57-1737 <*>
1956 V10 P264-270	8788 <IICH>
1956 V10 P405-	57-675 <*>
1956 V10 P406-409	63-18775 <*>
1956 V10 P540	57-675 <*>
1956 V10 N9 P4C9-412	57-1174 <*>
1956 V10 N78 P123-134	59-10156 <*> 0
1956 V10 N5/6 P88-90	C-3841 <NRCC>
1956 V10 N13/4 P264-270	65-12C02 <*>
1956 V10 N13/4 P295-300	59-10307 <*> 0
1956 V10 N17/8 P413-	57-1744 <*>
1956 V10 N19/0 P454-458	57-1721 <*>
1956 V10 N23/4 P535	58-258 <*>
1956 V10 N23/4 P54C-545	59-1C157 <*>
1956 V10 N23/4 P546-553	58-2639 <*>
	64-16043 <*>
1956 V10 N23/4 P564-	57-2905 <*>
1956 V10 N9/10 P183-189	64-16042 <*> 0
1957 V11 N1/2 P1-6	57-2873 <*>
	57-835 <*>
1957 V11 N1/2 P14-21	57-3191 <*>
1957 V11 N7/8 P125-133	45M46G <ATS>
	60-16801 <*> 0
1957 V11 N13/4 P274-	58-1616 <*>

Citation	Translation
1957 V11 N17/8 P391-	58-1618 <*>
1957 V11 N17/8 P396-398	58-2074 <*>
1957 V11 N17/8 P399-407	59-10281 <*>
1957 V11 N19/0 P443-445	58-2642 <*>
1957 V11 N19/0 P443	58-281 <*>
1957 V11 N23/4 P536-	58-1536 <*>
1957 V11 N23/4 P536-539	59-10279 <*>
1957 V11 N23/4 P553-562	59-10280 <*>
1957 V11 N23/4 P583-593	61-10767 <*> 0
1958 V12 N23 P624-632	60-16278 <*>
1958 V12 N1/2 P14-21	59-17409 <*>
1958 V12 N7/8 P126-136	59-20052 <*>
1958 V12 N11/2 P267-273	59-15320 <*>
1958 V12 N13/4 P318-334	64-16165 <*>
1958 V12 N13/4 P342-350	59-10985 <*>
1958 V12 N17/8 P464-467	59-15362 <*>
1958 V12 N19/0 P505-512	59-15629 <*> 0
1958 V12 N21/2 P553-567	59-15976 <*> 0P
1958 V12 N21/2 P568-578	60-18923 <*> 0
1958 V12 N23/4 P615-623	60-10916 <*> 0
1958 V12 N9/10 P196-200	32M44G <ATS>
1958 V12 N9/10 P196	58-1804 <*>
1958 V12 N9/10 P204-210	58-2682 <*>
1959 V13 N11 P237-244	61-10715 <*> 0
1959 V13 N1/2 P1-5	59-10986 <*> 0
1959 V13 N1/2 P5-12	59-17062 <*> 0
1959 V13 N1/2 P5-15	60-14797 <*>
1959 V13 N3/4 P37-44	60-10904 <*>
1959 V13 N3/4 P46-54	59-17048 <*> 0
	61-10796 <*>
1959 V13 N5/6 P85-92	59-17063 <*> 0
1959 V13 N7/8 P130-137	59-15918 <*> 0
1959 V13 N7/8 P141-149	59-20101 <*> 0
1959 V13 N11/2 P237-244	60-10128 <*> 0
1959 V13 N13/4 P293-301	60-14010 <*> 0
1959 V13 N17/8 P407-413	60-14284 <*> 0
	60-16572 <*>
1959 V13 N19/0 P459-469	61-10392 <*> 0
1959 V13 N19/0 P483-496	60-16271 <*> 0
1959 V13 N21/2 P519-530	60-10922 <*> 0
1959 V13 N23/4 P578-583	60-18503 <*> 0
	61-10769 <*> 0
1959 V13 N23/4 P583-592	61-10207 <*> 0
1959 V13 N9/10 P190-200	61-18266 <*> 0
1959 V13 N9/10 P201-207	59-17046 <*> 0
1960 V14 N1 P1-3	60-16637 <*>
1960 V14 N1 P5-11	60-18504 <*> 0
	61-10740 <*> 0
1960 V14 N2 P58-66	61-10742 <*> 0
1960 V14 N3 P85-91	60-18099 <*> 0
1960 V14 N3 P96-99	61-20667 <*>
1960 V14 N4 P123-130	60-18083 <*> 0
	61-16723 <*>
	66-10873 <*>
1960 V14 N6 P228-234	31M44G <ATS>
1960 V14 N7 P270-277	60-18098 <*> 0
1960 V14 N7 P278-283	61-18268 <*> 0
1960 V14 N7 P287-298	65-11215 <*> 0
1960 V14 N9 P399-410	61-20195 <*> 0
1960 V14 N9 P408-414	61-10015 <*>
1960 V14 N9 P414-422	61-16464 <*> 0
1960 V14 N10 P453-457	61-18571 <*> 0
1960 V14 N10 P458-461	61-20666 <*> 0
1960 V14 N12 P697-709	61-20485 <*> 0
1960 V14 N12 P718-723	61-14924 <*> 0
1960 V14 N12 P723-736	61-16120 <*> 0
1960 V14 N10A P535-541	61-10759 <*> 0
1960 V14 N10A P550-553	62-10276 <*> 0
1960 V14 N10A P554-564	61-10886 <*> 0
1960 V14 N10A P565-574	61-10931 <*> 0
1960 V14 N10A P575-580	61-20513 <*> 0
1960 V14 N10A P590-600	61-16804 <*> 0
1960 V14 N10A P600-609	62-14673 <*> 0
1960 V14 N10A P610-624	61-14905 <*> 0
1960 V14 N10A P625-630	61-10903 <*> 0
1960 V14 N10A P635-644	65-11217 <*> 0
1961 V15 P538-546	62-18917 <*>
1961 V15 N1 P10-16	61-20189 <*> 0
1961 V15 N3 P109-110	61-18805 <*>
1961 V15 N5 P44-51	<LSA>
1961 V15 N6 P222-229	62-10349 <*> 0
	62-16069 <*> 0
	62-18542 <*>
1961 V15 N6 P229-231	<LSA>
1961 V15 N6 P232-237	62-10277 <*> 0
1961 V15 N7 P295-301	62-20005 <*> 0
	65-10097 <*> 0
1961 V15 N7 P318-322	62-10398 <*>
	62-14904 <*> 0
	62-16494 <*>
	62-16708 <*>
1961 V15 N9 P407-410	64-16520 <*>
1961 V15 N9 P415-427	62-10309 <*> 0
	62-16200 <*> 0
	62-16495 <*> 0
1961 V15 N10A P522-529	63-10150 <*> C
1961 V15 N10A P610-625	62-16508 <*> 0
1961 V15 N10A P625-634	62-16505 <*> 0
1961 V15 N10A P635-643	62-16504 <*> 0
1962 V16 N1 P9-18	63-10791 <*>
1962 V16 N2 P53-61	62-18043 <*> C
1962 V16 N3 P95-102	62-18044 <*> 0
1962 V16 N4 P125-130	62-20469 <*> 0
	63-10060 <*> 0
1962 V16 N4 P131-138	62-18045 <*> 0
1962 V16 N6 P229-236	63-18272 <*>
1962 V16 N6 P236-243	63-10075 <*>
1962 V16 N9 P419-423	63-20576 <*>
1962 V16 N10 P457-462	63-01038 <*>
1962 V16 N11 P655-663	63-16642 <*>
1962 V16 N11 P664-670	63-20827 <*>
1962 V16 N12 P702-712	64-10935 <*>
1962 V16 N12 P724-728	64-10932 <*>
1962 V16 N10A P519-524	63-14470 <*> 0
1962 V16 N10A P568-574	63-18276 <*>
1962 V16 N10A P635-645	64-14348 <*>
1962 V16 N10A P645-654	63-20826 <*> 0
1963 V17 N1 P1-5	63-20574 <*>
1963 V17 N1 P14-19	64-00323 <*>
1963 V17 N2 P45-52	63-18366 <*> 0
1963 V17 N4 P141-147	64-14800 <*> 0
1963 V17 N4 P148-162	65-10296 <*> 0
1963 V17 N5 P191-196	64-10684 <*>
1963 V17 N5 P197-206	63-20825 <*> C
1963 V17 N6 P237-246	63-20727 <*>
1963 V17 N9 P448-450	64-14801 <*> 0
1963 V17 N11 P639-643	64-16363 <*>
1963 V17 N12 P687-696	64-14802 <*> 0
1963 V17 N10A P550-555	65-11216 <*> 0
1963 V17 N10A P612-622	65-11188 <*> 0
1964 V18 P437-442	484 <TC>
1964 V18 N2 P45-53	65-11183 <*> 0
1964 V18 N5 P207-213	65-11185 <*> 0
1964 V18 N6 P256-261	66-10754 <*>
1964 V18 N7 P308-314	65-17451 <*>
1964 V18 N9 P437-442	66-10182 <*>
1964 V18 N12 P741-746	66-10496 <*>
1964 V18 N12 P759-765	65-13358 <*>
1964 V18 N12 P765-771	66-10749 <*>
1964 V18 N10A P600-616	65-13813 <*>
1964 V18 N10A P652-657	1798 <TC>
1965 V19 N3 P93-96	65-14341 <*>
1965 V19 N3 P97-105	65-14397 <*> 0
	65-14580 <*>
1965 V19 N4 P145-150	57S84G <ATS>
1965 V19 N7 P352-357	66-12199 <*>
1965 V19 N7 P362-367	66-12200 <*>
1965 V19 N8 P452-459	66-10852 <*>
1965 V19 N9 P497-502	66-13463 <*>
1965 V19 N10 P557-570	66-13368 <*>
1965 V19 N10 P757-764	66-12181 <*>
1965 V19 N10A P649-654	55T92G <ATS>
1965 V19 N10A P711-719	66-10779 <*>
1965 V19 N10A P719-728	11T91G <ATS>
1965 V19 N10A P757-764	66-12377 <*>
1966 V20 N1 P1-4	66-13583 <*>
1966 V20 N4 P169-171	66-13775 <*>
1966 V20 N4 P171-173	66-13011 <*>

PAPIER, CARTON ET CELLULOSE

1955 P69- 57-1732 <*>
1958 V7 N1 P100-104 59-10093 <*> 0
1964 V13 N5 P119-125 NS-339 <TTIS>
1965 V14 N1 P96-98 86T93F <ATS>

PAPIERFABRIKANT
1927 V25 P76-85 AL-818 <*>
1928 V26 N8 P120-124 59-10208 <*> P
1929 V27 N28 P433-436 AL-449 <*>
1931 V29 N30 P485-490 63-14104 <*> 0
1932 P701-704 AL-942 <*>
1932 P725-729 AL-942 <*>
1932 P737-742 AL-942 <*>
1932 P749-751 AL-942 <*>
1932 V30 N43 P613-616 63-14108 <*>
1933 V31 N22/3 P317-321 57-183 <*>
1933 V31 N22/3 P331-336 57-183 <*>
1934 V32 N15 P169-170 63-14109 <*> 0
1935 V33 N37 P305-309 AL-704 <*>
1935 V33 N38 P313-319 AL-704 <*>
1937 V35 N9 P67-70 59-20665 <*> 0
1938 V36 P133-135 59-15244 <*>
1940 V38 N4 P17-22 60-18887 <*> 0
1940 V38 N48/9 P294-299 63-14107 <*>

PAPIERWERELD
1956 V10 N12 P275- 57-2351 <*>
1956 V11 N1 P7-10 57-3481 <*>
1957 V12 N5 P107-122 58-1407 <*>
1958 V12 N12 P303-309 60-14203 <*>
1958 V13 N2 P35- 60-14020 <*> 0
1960 V14 N11 P619-624 61-10452 <*>
1961 V15 N9 P243-248 62-18728 <*> 0
1961 V15 N9 P253 62-18728 <*> 0
1961 V16 N3 P63-65,68,73 62-16511 <*> 0
1961 V16 N4 P91-95,97 62-16511 <*> 0
1961 V16 N4 P100-103 62-16511 <*> 0
1963 V17 N8 P227-240 64-14347 <*>
1963 V17 N10 P289-302 65-11186 <*> 0

PAPIR A CELULOSA
1955 V10 N12 P260-264 63-18349 <*> 0
1957 V12 N9 P194-198 62-20467 <*> 0
1957 V12 N9 P206-208 01M40C <ATS>
1959 V14 N7 P149-150 28L36C <ATS>
1959 V14 N8 P176-181 65-11214 <*> 0
1959 V14 N10 P223-226 26M43C <ATS>
1960 V15 N6 P122-125 63-20823 <*> 0
1961 V16 P18-21 63-10525 <*>
1961 V16 N6 P121-124 62-00918 <*>
1962 V17 N11 P239-241 AD-619 814 <=$>

PAPIRIPAR
1959 V3 N4 P146-148 61-10308 <*>
1960 V4 N4 P171-172 61-20009 <*>

PAPIRJOURNALEN
1938 V31 P23-26 3236 <*>

PAPIR-ES NYOMDATECHNIKA
1955 V7 N2/3 P63-67 61-10024 <*> 0

PAPPER OCH TRA
SEE PAPERI JA PUU

PARASITICA
1950 V6 P98-106 60-10005 <*>
1952 V8 N1 P40-43 63-14110 <*>
1955 V11 N3 P74-80 59-10322 <*>

PARAZITOLOGICHESKII SBORNIK
1930 V1 P27-36 RT-1479 <*>
1930 V1 P75-96 RT-1478 <*>
1934 V4 P43-63 65-60046 <=$>
1940 V7 P7-44 64-15101 <=*$>
1940 V7 P100-133 64-15624 <=*$>
1947 V9 P183-190 63-23596 <=*$> 0
1947 V9 P191-222 61-28654 <*=>
1949 V11 P229-245 61-27452 <*=>
1950 V12 P3-12 AD-625 640 <=$>
1950 V12 P167-198 3019 <*>
59-15042 <+*> C
1950 V12 P225-271 63-23598 <=*$> 0
1950 V12 P272-274 64-15645 <=*$> 0
1951 V13 P343-354 CSIRO-2672 <CSIR>
R-4593 <*>
63-23704 <=*$>
1952 V14 P95-102 63-23595 <=*$> 0
1953 V15 P252-255 61-28653 <*=>
1953 V15 P275-296 61-28653 <*=>
1954 N172 P163-176 R-1857 <*>
1958 V18 P110-119 66-60943 <=*$>
1958 V18 P219-238 63-01283 <*>
63-24056 <=*$>
1958 V18 P239-254 63-23594 <=*$> C
63-23817 <=*$>
1958 V18 P255-282 64-13686 <=*$>
1960 V19 P26-31 AD-625-627 <=$>
1960 V19 P47-55 AD-625 633 <=$>
1960 V19 P333-353 64-13388 <=*$>
1961 V20 P148-182 63-15704 <=*>
1961 V20 P185-225 65-63664 <=$>
1961 V20 P226-247 65-61128 <=$>
1963 N21 P5-15 64-31377 <=>
1963 V21 P16-27 65-61129 <=$>
1963 V21 P39-43 65-60897 <=$>
1964 V22 P7-27 65-63152 <=$>

PARFUMERIE
1943 V1 P209-211 2483 <*>

PARFUEMERIE UND KOSMETIK
1953 V34 P15-16 59-20903 <*>
1955 V36 P160-162 58-962 <*>
1955 V36 P167- 58-962 <*>
1958 V39 N6 P348-350 64-10880 <*>
1959 V40 P677-679 62-00681 <*>
1962 V43 N4 P111-118 62-20041 <*>

PARFUMERIE, COSMETIQUE, SAVONS
1958 V1 P219-221 10K27F <ATS>
1958 V1 N1 P3-6 19L29F <ATS>
1958 V1 N9 P329-338 63-16967 <*>
1961 V4 P409-412 66-13226 <*>
1961 V4 N10 P401-404 66-13227 <*>
1963 V6 N4 P147-149 1245 <TC>

PARFUMERIE MODERNE
1923 V16 P260-262 64-10420 <*>
9915-B <K-H>
1934 V28 P31-33 63-20166 <*>

PARIS MEDICAL
1921 V83 N42 P2-5 59-15536 <*> 0
1934 V91 P249-251 57-1633 <*>
1934 V93 P386-389 57-1616 <*>

PARTIYNAYA ZHIZN
1955 N13 P19-25 R-170 <*>
1959 V29 N12 P31-36 60-41450 <=>
1959 V29 N12 P51-53 60-41450 <=>
1960 N2 P24-29 60-31692 <=>
1962 N9 P37-40 62-25985 <=>
1962 N11 P34-36 63-21623 <=>
1962 N12 P9-17 62-33019 <=>
1962 N14 P15-21 63-13320 <=> P
1962 N19 P15-19 63-13365 <=>
1962 N22 P37-40 63-13677 <=>
1962 N23 P27-33 63-21269 <=>
1962 N24 P16-20 63-21390 <=>
1963 N2 P21-22 63-21959 <=>
1963 N2 P28 63-21601 <=>
1963 N4 P52-54 63-31416 <=>
1963 N5 P73 63-31579 <=>
1963 N9 P35-40 63-31296 <=>
1963 N14 P34-37 63-31680 <=>
1963 V33 N4 P24-29 63-31554 <=>
1966 N2 P8-14 66-31700 <=$>
1966 N2 P73-77 66-31458 <=$>
1966 N4 P43-48 66-31457 <=$>

1966 N5 P30-35 66-31691 <=$>

PARTIYNAYA ZHIZN KAZAKSTANA
1960 V30 N8 P43-45 61-11743 <=>
1962 V32 N2 P34-36 62-25983 <=>
1962 V32 N3 P33-37 62-25975 <=>
1962 V32 N4 P70-72 62-25975 <=>
1962 V32 N5 P10-13 62-33307 <=>
1962 V32 N9 P34-36 63-21199 <=>
1962 V32 N9 P52-54 63-21261 <=>
1963 V33 N6 P55-57 63-31680 <=>
1963 V33 N11 P45-46 63-31400 <=>

PARUPU KAMI KOGYO ZASSHI
1959 V13 N6 P379-384 29L36J <ATS>
1960 V14 P166-169 61-20010 <*> 0

PASSOW-SCHAEFER BEITRAEGE ZUR PRAKTISCHEN UND THEORETISCHEN HALS-, NASEN-UND OHRENHEILKUNDE
1928 V27 N3 P21-59 61-00857 <*>
1931 V28 N3/4 P305-310 61-16040 <*>

PATHOLOGIA ET MICROBIOLOGIA
1960 V23 P344-350 62-16607 <*>
1961 V24 P768-773 63-10183 <*>
1962 V25 P340-342 63-10661 <*>
1962 V25 N5 P616-623 64-19677 <*>
1963 V26 N3 P303-312 64-20149 <*>
1963 V26 N3 P343-347 64-14312 <*>
1965 V28 P114-121 66-13037 <*>
1965 V28 N1 P50-57 65-13801 <*>
1965 V28 N3 P425-436 66-11716 <*>

PATHOLOGICA
1961 V53 P109-116 66-10944 <*>

PATHOLOGIE ET BIOLOGIE
1958 V6 P966-972 10764-D <KH>
64-14566 <*> 0
1958 V6 N9/10 P719-730 60-16600 <*>
1959 V7 N6 P1209-1217 61-10249 <*>
1959 V7 N9/10 P989-1015 63-00774 <*>
1960 V8 P1201-1210 66-11642 <*> 0
1960 V8 N5/6 P503-509 63-10523 <*> 0
1960 V8 N11/2 P1147-1154 62-14626 <*> 0
1960 V8 N11/2 P1193-1196 66-11643 <*> 0
1961 V9 N7/8 P825-830 11363-A <KH>
63-16583 <*>
1961 V9 N21/2 P2123-2125 63-16519 <*>
1963 V11 N3/4 P184-187 63-18450 <*>
1964 V12 P47-51 65-14711 <*>
1964 V12 N19/0 P988-989 815 <TC>
1965 V13 N9/10 P540-545 66-12746 <*> 0

PATOLOGIA POLSKA
1958 V9 N3 P215-220 63-16950 <*> 0
1959 V10 N1 P1-11 63-16951 <*>
9458 <K-H>

PATOLOGICHESKAYA FIZIOLOGIYA I EKSPERIMENTALNAYA TERAPIYA
1957 V1 N3 P14-21 <CP>
63-24186 <=>
1957 V1 N3 P39-44 62-33366 <=*>
1957 V1 N5 P3-12 62-15069 <*=>
1957 V1 N5 P60-67 60-23185 <+*> 0
1958 V2 N1 P3-11 PB 141 142T <=>
1958 V2 N1 P12-18 PB 141 143T <=>
1958 V2 N1 P19-21 PB 141 144T <=>
1958 V2 N1 P22-27 PB 141 145T <=>
1958 V2 N1 P27-33 PB 141 146T <=>
1958 V2 N1 P34-38 PB 141 147T <=>
1958 V2 N1 P39-44 PB 141 148T <=>
1958 V2 N1 P44-49 PB 141 149T <=>
1958 V2 N3 P10-13 AD-609 146 <=>
1958 V2 N4 P3-9 59-13040 <=>
1958 V2 N4 P38-42 59-13039 <=> 0
1958 V2 N4 P57-64 64-19829 <=$> 0
1958 V2 N5 P11-16 59-13319 <=>
1958 V2 N5 P17-20 59-13320 <=>
1958 V2 N5 P21-25 59-13321 <=> 0
1958 V2 N5 P29-33 59-13323 <=> 0
1959 V3 N1 P3-4 59-13638 <=>
1959 V3 N1 P12-20 59-13639 <=> 0
1959 V3 N1 P21-26 59-13640 <=> 0
1959 V3 N1 P27-29 59-13641 <=> 0
1959 V3 N1 P33-36 59-13789 <=>
1959 V3 N1 P36-39 59-13642 <=>
1959 V3 N1 P71 59-13643 <=>
1959 V3 N1 P72 59-13644 <=>
1959 V3 N1 P93-94 59-13645 <=>
1959 V3 N2 P74-80 61-27744 <*=>
1959 V3 N3 P22-27 60-18569 <+*>
1959 V3 N3 P77-78 59-13785 <=>
1959 V3 N3 P79-84 59-13803 <=>
1959 V3 N4 P26-31 63-24255 <=*$> 0
1959 V3 N4 P57-60 59-13965 <=>
1959 V3 N4 P76-82 59-13966 <=>
1959 V3 N5 P48-56 61-15715 <+*>
1959 V3 N5 P56-62 64-13040 <=*$> 0
1959 V3 N5 P65-69 59-31051 <=>
1959 V3 N6 P65-68 60-11205 <=>
1960 V4 N1 P3-13 60-11823 <=>
1960 V4 N1 P14-19 60-11824 <=>
1960 V4 N1 P20-24 60-11825 <=>
1960 V4 N1 P24-28 60-11826 <=>
1960 V4 N1 P28-32 60-11827 <=>
1960 V4 N1 P32-38 60-11828 <=>
1960 V4 N1 P39-44 60-11829 <=>
1960 V4 N1 P53-58 60-11830 <=>
1960 V4 N1 P61-66 63-19713 <=*> 0
1960 V4 N1 P71-73 60-11831 <=>
1960 V4 N1 P76-83 60-11832 <=>
1960 V4 N2 P54-58 62-14715 <=*>
1960 V4 N2 P89-91 60-31681 <=>
1960 V4 N3 P27-31 60-41596 <=>
1960 V4 N3 P31-35 60-41614 <=>
1960 V4 N3 P74-76 60-41607 <=>
1960 V4 N3 P76-77 60-41606 <=>
1960 V4 N3 P83-84 61-21053 <=>
1960 V4 N4 P32-38 61-11545 <=>
1960 V4 N4 P39-41 61-11556 <=>
1960 V4 N4 P42-46 61-11549 <=>
1960 V4 N4 P47-52 61-11559 <=>
1960 V4 N4 P61-67 61-11561 <=>
1960 V4 N4 P74 61-17555 <=>
1960 V4 N4 P75 61-11558 <=>
1960 V4 N5 P3-7 61-21289 <=>
1960 V4 N5 P56-57 61-21271 <=>
1960 V4 N5 P68-69 61-21313 <=>
1960 V4 N6 P17-23 61-21402 <=>
1960 V4 N6 P24-27 61-21403 <=>
1960 V4 N6 P49-53 61-21448 <=>
1960 V4 N6 P66-67 61-21445 <=>
1960 V4 N6 P71- 61-21447 <=>
1960 V4 N6 P77- 61-21385 <=>
1961 V5 N1 P38-40 61-23115 <*=>
1961 V5 N1 P63-65 61-23474 <*=>
1961 V5 N1 P69-70 61-23475 <=*>
1961 V5 N1 P74-82 63-23909 <=*> 0
1961 V5 N1 P83-88 61-23056 <=*>
1961 V5 N1 P92-93 61-23473 <=*>
1961 V5 N2 P3-9 61-31519 <=>
1961 V5 N2 P45-49 61-31519 <=>
1961 V5 N2 P61-64 61-31519 <=>
1961 V5 N2 P66-67 61-31519 <=>
1961 V5 N2 P70-71 65-63074 <=$>
1961 V5 N2 P78-82 62-19807 <=*>
1961 V5 N3 P54-59 62-15786 <*> 0
1961 V5 N3 P72-74 62-15786 <*> 0
1961 V5 N3 P84-85 62-15786 <*> 0
1961 V5 N4 P20-26 62-13361 <*=>
1961 V5 N4 P26-29 62-13371 <*=>
1961 V5 N4 P30-33 62-13411 <=*>
1961 V5 N4 P39-43 62-13366 <*=>
1961 V5 N4 P47-49 62-13316 <*=>
1961 V5 N4 P50-52 62-13363 <*=>
1961 V5 N4 P60-63 62-13365 <*=>
1961 V5 N4 P70-71 62-13416 <*=>
1961 V5 N4 P72-77 62-13328 <*=>

1961 V5 N4 P80-85 62-13313 <*=>
1961 V5 N4 P86-87 62-13327 <*=>
1961 V5 N5 P59-70 62-15366 <*=>
1961 V5 N5 P66 65-63053 <=$>
1962 V6 N1 P28-33 AD-622 350 <=*$>
62-24248 <=>
1962 V6 N1 P37-42 62-24248 <=>
1962 V6 N1 P49-53 62-24248 <=>
1962 V6 N1 P70-85 62-24248 <=>
1962 V6 N2 P18-35 62-32377 <=>
1962 V6 N2 P68-69 62-32377 <=>
1962 V6 N2 P76-82 62-32377 <=>
1962 V6 N3 P3-17 63-13355 <=>
1962 V6 N3 P37-40 63-13355 <=>
1962 V6 N3 P49-52 63-13355 <=>
1962 V6 N3 P57-67 63-13355 <=>
1962 V6 N3 P69-70 63-13355 <=>
1962 V6 N3 P85-93 63-13355 <=>
1962 V6 N4 P18-37 63-13274 <=>
1962 V6 N4 P57-62 63-19206 <=*>
1962 V6 N4 P87-91 62-33426 <=*>
1962 V6 N4 P89-92 63-13274 <=>
1962 V6 N5 P60-63 63-15396 <=*>
1962 V6 N5 P74-75 63-15396 <=*>
1962 V6 N6 P3-10 63-21206 <=>
1962 V6 N6 P41-56 63-21206 <=>
1963 V7 N1 P14- FPTS V23 N2 P.T284 <FASE>
FPTS,V23,P.T284-T286 <FASE>
1963 V7 N1 P14-31 63-21831 <=>
1963 V7 N1 P19-23 64-13416 <=*$>
1963 V7 N1 P32- FPTS V23 N3 P.T593 <FASE>
1963 V7 N1 P32-35 64-19205 <FASE>
1963 V7 N1 P39- FPTS V23 N3 P.T447 <FASE>
1963 V7 N1 P39-43 FPTS,V23,N3,P.T447 <FASE>
1963 V7 N1 P44-49 64-13858 <=*$>
1963 V7 N1 P49-52 64-13415 <=*$>
1963 V7 N1 P52- FPTS V23 N1 P.T82 <FASE>
FPTS,V23,P.T82-T84 <FASE>
1963 V7 N1 P65-71 64-13414 <=*$>
1963 V7 N2 P9-14 63-31356 <=>
1963 V7 N2 P9-15 64-13485 <=*$>
1963 V7 N2 P15- FPTS V23 N2 P.T417 <FASE>
FPTS,V23,P.T417-T419 <FASE>
1963 V7 N2 P19- FPTS V23 N3 P.T579 <FASE>
1963 V7 N2 P19-25 FPTS,V23,N3,P.T579 <FASE>
1963 V7 N2 P30-34 64-13484 <=*$>
1963 V7 N2 P38-41 64-13483 <=*$>
1963 V7 N2 P50- FPTS V23 N1 P.T12 <FASE>
FPTS,V23,P.T12-T14 <FASE>
1963 V7 N2 P58-62 63-31356 <=>
1963 V7 N2 P68-71 63-31356 <=>
1963 V7 N2 P78-80 64-41168 <=>
1963 V7 N3 P15-20 63-31688 <=>
1963 V7 N3 P30-33 63-31689 <=>
1963 V7 N3 P50-55 64-19643 <=$>
1963 V7 N3 P59- FPTS V23 N6 P.T1364 <FASE>
FPTS V23 P.T1364 <FASE>
1963 V7 N3 P74 63-31690 <=>
1963 V7 N3 P75-76 63-31691 <=>
1963 V7 N3 P77-82 63-31729 <=>
1963 V7 N3 P83-89 63-41287 <=>
1963 V7 N3 P90-91 AD-622 457 <=*$>
1963 V7 N4 P3-10 63-41311 <=>
1963 V7 N4 P27-30 63-41311 <=>
64-19229 <=*$>
1963 V7 N4 P31-34 63-41311 <=>
1963 V7 N4 P40- FPTS,V23,N5,P.T935 <FASE>
1963 V7 N4 P40-45 FPTS,V23,N5,P.T935 <FASE>
63-41311 <=>
1963 V7 N4 P50- FPTS,V23,N4,P.T807 <FASE>
1963 V7 N4 P60- FPTS,V23,N5,P.T939 <FASE>
1963 V7 N4 P60-63 FPTS,V23,N5,P.T939 <FASE>
1963 V7 N4 P64-65 63-41311 <=>
1963 V7 N4 P69-70 63-41311 <=>
1963 V7 N4 P71-72 63-41311 <=>
1963 V7 N4 P72-73 63-41311 <=>
1963 V7 N5 P9-15 63-41199 <=>
1963 V7 N5 P40-43 63-41214 <=>
1963 V7 N5 P45- FPTS V23 N5 P.T1166 <FASE>
1963 V7 N5 P45-48 FPTS,V23,N5,P.T1166 <FASE>
63-41215 <=>
1963 V7 N5 P49-53 64-19644 <=$>
1963 V7 N5 P55-58 64-19714 <=$>
1963 V7 N5 P63- FPTS V23 N5 P.T1059 <FASE>
FPTS,V23,N5,P.T1059 <FASE>
1963 V7 N5 P72-73 63-41216 <=>
1963 V7 N6 P3- FPTS V23 N6 P.T1343 <FASE>
FPTS V23 P.T1343 <FASE>
1963 V7 N6 P27-31 65-60206 <=$>
1963 V7 N6 P45- FPTS V23 N6 P.T1211 <FASE>
FPTS V23 P.T1211 <FASE>
1963 V7 N6 P50-53 65-60207 <=$>
1963 V7 N6 P57-61 65-61102 <=$>
1963 V7 N6 P61-65 65-61103 <=$>
1963 V7 N6 P65- FPTS V24 N2 P.T305 <FASE>
1963 V7 N6 P65-70 FPTS V24 P.T305-T308 <FASE>
1963 V7 N6 P72 64-21729 <=>
1964 V8 N1 P37-39 64-31127 <=>
65-62112 <=>
1964 V8 N1 P46-49 65-62113 <=>
1964 V8 N2 P17-22 64-31508 <=>
65-62114 <=$>
1964 V8 N2 P22-25 65-62115 <=$>
1964 V8 N2 P26-30 64-31509 <=>
1964 V8 N2 P30-34 64-31511 <=>
1964 V8 N2 P34-38 64-31510 <=>
1964 V8 N2 P79-81 64-31505 <=>
1964 V8 N3 P3-9 65-62116 <=>
1964 V8 N3 P30-35 65-62117 <=$>
1964 V8 N3 P35-48 64-41951 <=>
1964 V8 N3 P57-63 64-41951 <=>
1964 V8 N4 P33- FPTS V24 N5 P.T823 <FASE>
1964 V8 N5 P45-51 65-63614 <=$>
1964 V8 N5 P59- FPTS V24 N5 P.T861 <FASE>
1964 V8 N6 P65-70 65-64538 <=*$>
1964 V9 N1 P20-24 65-60168 <=$>
1964 V9 N2 P13-19 65-60145 <=$>
1965 V9 N1 P34- FPTS V24 N6 P.T1059 <FASE>
1965 V9 N2 P8-23 AD-638 187 <=$>
1965 V9 N2 P54-60 66-60896 <=*$>
1965 V9 N2 P79-81 65-31591 <=$>
1965 V9 N2 P84-86 65-31583 <=$>
1965 V9 N3 P34- FPTS V25 N4 P.T630 <FASE>
1965 V9 N3 P37-41 66-61629 <=*>
1965 V9 N3 P71-73 65-32258 <=*$>
1965 V9 N4 P32-38 66-61608 <=*>
1965 V9 N4 P59-64 66-61609 <=*>
1965 V9 N4 P88-89 65-33359 <=*$>
1965 V9 N4 P95-96 65-33360 <=*$>
1965 V9 N5 P49- FPTS V25 N5 P.T791 <FASE>

PCHELOVODSTVO
1947 V24 N6 P15-17 65-60021 <=$>
1949 V26 N7 P29-30 65-60084 <=$>
1954 V31 N12 P24-28 64-15697 <=*$> 0
1956 V33 P19-24 62-15688 <*=>
1956 V33 N11 P45-46 PANSDOC-TR.584 <PANS>
1957 V34 P31-32 62-15687 <*=>
1958 V35 N11 P46-48 65-60647 <=$>
1960 V37 N3 P43-45 64-15715 <=*$> 0
1963 N11 P14-16 65-63231 <$>
1963 V40 N7 P39 65-60644 <=$>
1965 N8 P20-21 66-62062 <=*$>
1965 N9 P21 66-62060 <=*$>

PEDIATRIA. NAPOLI
1936 V44 P1097-1108 58-416 <*>

PEDIATRIA DE LA AMERICAS
1945 V3 P663-675 58-430 <*>

PEDIATRIA INTERNAZIONALE
1959 V9 N4 P503-519 63-20612 <*>

PEDIATRIA PANAMERICANA
1962 V7 P169-171 64-18630 <*>

PEDIATRIA POLSKA
1955 V30 N9 P811-817 5409-B <K-H>
1956 V31 N1 P35-44 58-1339 <*>

1957 V32 N4 P377-385 62-00476 <*>
1960 V35 N8 P925-932 62-19581 <*=> P
1960 V35 N8 P933-934 62-19166 <=*>

PEDIATRICKE LISTY
1953 V8 P152-154 II-840 <*>

PEDIATRIE
1958 V13 N4 P455-458 59-10340 <*> 0
1959 V14 P541-542 64-18693 <*>

PEDIATRIYA
1943 N6 P45-49 64-10825 <=*$>
65-63273 <=>
1944 N5 P13-18 65-62966 <=>
1945 N5 P15-19 65-62906 <=*$>
1954 N4 P80-81 59-10848 <+*>
1957 N8 P39-44 AD-630731 <=*$>
1957 V35 N7 P63-65 59-14864 <+*>
1959 V37 N3 P73-77 63-19045 <=*>
1959 V37 N4 P42-45 05M47R <ATS>
1959 V37 N5 P93-95 59-13960 <=>
1959 V37 N6 P3-10 60-31653 <=>
1959 V37 N6 P11-15 60-11003 <=>
1959 V37 N6 P25-29 60-31654 <=>
1959 V37 N7 P31-34 61-10674 <+*>
1959 V37 N9 P75-79 60-11656 <=>
1959 V37 N10 P74-79 60-18376 <+*>
1959 V37 N11 P33-37 60-11200 <=>
1959 V37 N11 P38-41 60-11199 <=>
1959 V37 N12 P3-8 60-11384 <=>
1959 V37 N12 P39-41 64-13179 <=*$>
1959 V37 N12 P57 60-11384 <=>
1960 V38 N6 P62-66 61-28715 <*=>
1960 V38 N7 P65-70 63-19717 <=*>
1960 V38 N8 P84-87 64-13786 <=*$> 0
1960 V38 N9 P21-27 61-21230 <=>
1961 V39 N2 P32-42 61-21545 <=>
1961 V39 N3 P3-8 61-21760 <=>
1961 V39 N8 P3-6 62-13253 <*=>
1961 V39 N8 P11-15 62-13250 <*=>
1961 V39 N8 P52-55 62-13512 <=*>
1961 V40 N4 P29-34 62-23868 <=*>
1961 V40 N12 P8-11 62-23428 <=*>
1962 V41 N1 P67-73 63-15990 <=*>
1962 V41 N2 P3-10 62-24021 <=*>
1962 V41 N2 P11-16 63-13706 <=*>
1962 V41 N2 P94-95 62-24726 <=*>
1962 V41 N4 P35-39 62-25005 <=*>
1962 V41 N5 P9-20 63-13106 <=*>
1962 V41 N11 P3-16 63-21019 <=>
1963 V42 N1 P3-7 63-21828 <=>
1963 V42 N7 P59-62 97Q72R <ATS>
1963 V42 N8 P98-101 63-41018 <=>
1963 V42 N9 P89-90 65-61373 <=$> 0
1963 V42 N11 P33-37 64-21274 <=>
1964 V43 P32-35 AD-638 188 <=$>
1964 V43 N1 P37-43 65-60237 <=$>
1964 V43 N3 P3- FPTS V24 N3 P.T477 <FASE>
1964 V43 N3 P36-41 65-61118 <=$>
1964 V43 N4 P22-25 65-61119 <=$>
1964 V43 N8 P46-50 65-62835 <=$>
1965 V44 P8-12 66-60634 <=*$>
1965 V44 N2 P8-14 65-30857 <=$>
1965 V44 N2 P62-67 65-30857 <=$>
1965 V44 N9 P49-55 65-33651 <=*$>
1966 V45 N6 P7-10 66-33647 <=$>

PEDIATRIYA, AKUSHERSTVO I GINEKOLOGIYA
1955 N6 P54-57 62-11159 <=>
62-15074 <*=>
1956 N5 P49-53 62-11156 <=>
62-15075 <*=>
1957 N2 P46-48 62-11153 <=>
62-15077 <=*>
1957 N2 P49-54 60-13028 <+*> 0
62-11158 <=> 0
1957 N4 P43-45 62-11154 <=>
62-15068 <=*>
1963 N2 P54-55 RTS-2705 <NLL>

PEDOLOGICAL BULLETIN. CHINESE PEOPLES REPUBLIC
SEE T'U JANG T'UNG PAO

PEI-CHING KUANG YEH HSUEH YUAN HSUEH PAO
1959 N2 P10-18 63-15112 <=>

PEI-CHING TA HSUEH HSUEH PAO-TZU JAN K'O HSUEH
1956 V2 N4 P479-488 63-01282 <*>
63-24103 <=*>
1958 V4 N3 P305-313 63-24057 <=*$>

PEINTURES-PIGMENTS-VERNIS
1943 V19 N5 P142 62-18188 <*>
1944 V20 N4 P10 61-10616 <*>
1945 V21 N5 P152-153 59-15985 <*>
1956 V32 P419-431 58-793 <*>
1958 V34 P204-213 60-17761 <+*> 0
65-00283 <*> 0
1958 V34 P271-279 60-17761 <+*> 0
65-00283 <*> 0
1958 V34 P311-319 60-17761 <+*>
65-00283 <*> 0
1958 V34 N2 P68-72 60-14475 <*> 0
1958 V34 N6 P255-260 59-15655 <*>
1958 V34 N7 P309-310 59-17570 <*> 0
1958 V34 N11 P499-500 59-15667 <*>
1960 V36 P655-657 63-20936 <*>
1960 V36 N5 P264-266 61-10548 <*>
1960 V36 N10 P574-587 62-10707 <*> 0
1964 V40 N4 P184-188 65-14173 <*> 0
1965 V41 N9 P553-554 66-14437 <*>

PENISHIRIN SONO TA KOSEI BUSSHITSU
1950 V3 N7 P457-458 64-18614 <*> 0
1951 V4 P74 PANSDOC-TR.180 <PANS>
1952 V5 N11 P622-630 54E1J <ATS>
1952 V5B P559- 58-802 <*>
1953 SB V6 N5 P247-250 II-809 <*>
1954 SB N7 P81-84 2461 <*>
1954 SB V7 N3 P81-84 II-792 <*>
1955 V8 P138-145 58-384 <*>
1956 V9 N6 P297-302 58-365 <*>
1956 SB V9 P160-167 57-1431 <*>
1956 SB V9 P213-217 59-15716 <*>
1956 SB V9 P218-220 T-2016 <INSD>
1956 SB V9 N6 P297-302 75J14J <ATS>
1957 V10 N6 P255-259 64-18653 <*>
1959 V12 N5 P365-367 64-18634 <*>
1959 SB V12 N4 P300-304 62-00235 <*>
1961 SB V14 N2 P49-56 62-00976 <*>
1962 V15 N1 P7-10 64-18649 <*>
1963 V16B N5 P317-322 65-10982 <*>
1964 V17 P61-64 66-13187 <*>

PENTRU APARAREA PATRIEI
1961 V7 N11 P12-13 62-15349 <=*> 0

PENZUGYI SZEMLE
1967 N1 P1-14 67-31083 <=$>

PEOPLES HEALTH. CHINESE PEOPLES REPUBLIC
SEE JEN-MIN PAO-CHIEN

PEOPLES' POSTS AND TELECOMMUNICATIONS
SEE JEN MIN YU TIEN

PEREDOVOY NAUCHNO-TEKHNICHESKIY I PROIZVODSTVENNYY OPTY
1958 P3-45 60-31006 <=>

PEREMENNYE ZVEZDY
1949 V7 N3 P124-132 RT-3983 <*>
63-14918 <=*> 0
1950 V7 N4 P169-181 RT-3984 <*>
63-14923 <=*> 0
1950 V7 N4 P195-196 RT-3985 <*>
63-14913 <=*> 0
1951 V8 P192-201 63-14922 <=*> 0
1951 V8 P244- 63-14919 <=*> 0

1951 V8 N2 P83-120 RT-3986 <*>
63-14909 <=*> 0

PERICDICA POLYTECHNICA
1957 V1 N1 P27-51 AEC-TR-3723 <+*>
1958 V2 P59-63 65-10257 <*>
1958 V2 P183-187 65-10258 <*>
1958 V2 N4 P333-354 61-18917 <*>
1959 V3 P167-176 62-18757 <*> 0
62-20134 <*>
1961 V5 N21 P65-87 AD-631 595 <=$>

PERIODICO DI MINERALOGIA
1955 V24 P49-83 30K28I <ATS>

PERMANENT WAY. JAPAN
SEE TETSUDO SENRO

PESTICIDE AND TECHNIQUE. JAPAN
SEE NOYAKU SEISAN GIJUTSU

PETERMANNS GEOGRAPHISCHE MITTEILUNGEN
1948 V92 P161-162 57-3214 <*>
1959 V103 P225-232 62-20203 <*>
1959 V103 N3 P257-272 66-34338 <=$> 0
1961 N90 P16-19 62-19478 <=> 0
1962 V106 N4 P286 63-21505 <=>
1963 V107 N1 P14-19 63-31635 <=> 0

PETITE REGION AGRICOLE
1952 P41-48 65-12357 <*>

PETROL SI GAZE
1958 V9 N3 P116-119 62-18523 <*>
1958 V9 N10 P467-468 02N50RU <ATS>
1959 V10 N1 P28-36 65-12432 <*>
1959 V10 N5 P181-189 59M38RU <ATS>
1961 V12 N1 P33-43 66-10356 <*>
1961 V12 N10 P454-457 63-20716 <*>
1962 V13 N10 P453-458 64-16401 <*>
1963 V14 N1 P10-15 64-10830 <*> P
1963 V14 N2 P57-59 64-16416 <*>
1964 V15 P542-549 8790 <IICH>
1964 V15 N5 P232-237 35R80RU <ATS>
1964 V15 N7 P342-347 03R80RU <ATS>
1964 V15 N8 P381-386 64-51233 <=$>
1964 V15 N8 P423-426 64-51271 <=>
1964 V15 N8 P438-452 64-51407 <=>
1965 V16 N11/2 P605-608 92T92RU <ATS>
1965 V16 N11/2 P650-658 38T93RU <ATS>

PETROLEUM. BERLIN
1924 V20 N34 P1887-1891 63-10477 <*>
1930 V26 N15 P5 60-10263 <*> 0
1930 V26 N15 P451-466 59-17282 <*> P
1930 V26 N21 P3-9 59-20367 <*>
1930 V26 N50 P12-13 59-20366 <*>
1932 N19 P1-14 58-1155 <*>
1935 V31 N15 P3-5 SUP 581 <TC>
1938 V34 N37 P1-3 58-1045 <*>

PETROLEUM REFINING. CHINESE PEOPLES REPUBLIC
SEE SHIH-YU LIEN-CHIH

PETROLEUM SURVEY. CHINESE PEOPLES REPUBLIC
SEE SHIH-YU K'AN T'AN

PETROLEUM ZEITSCHRIFT
1928 V24 P898-902 63-10698 <*>
1931 V27 N3 P3-4 ASPHT. SEC
63-10457 <*>
1931 V27 N4 P53-56 63-10480 <*> 0
1932 V28 P2-5 58-2164 <*>
1932 V28 N44 P7-8 58-1896 <*>
1932 V28 N50 P1-6 58-1930 <*>
1933 V29 P5 SUP5 59-10010 <*>
1933 V29 P1-3 58-1929 <*>
1933 V29 N4 P1-7 SUP 58-1925 <*>
1933 V29 N18 P1-4 58-1900 <*>
58-2161 <*>
1933 V29 N31 P2-5 58-1899 <*>
1933 V29 N40 P1-14 58-1924 <*>
1933 V29 N48 P4-8 58-1926 <*>
1935 V31 N5 P4-8 58-2119 <*>
1935 V31 N15 P1-7 58-1910 <*>
1937 V33 N10 P1-2 63-10700 <*>
1937 V33 N16 P5-6 58-2173 <*>
1937 V33 N45 P5-10 63-10482 <*> 0
1938 V34 N32 P33-34 58-2165 <*>

PFLANZENSCHUTZBERICHTE
1950 V4 N3/4 P33-46 1340 <*>

PFLUEGERS ARCHIV FUER DIE GESAMTE PHYSIOLOGIE DES MENSCHEN UND DER TIERE
1878 V16 P272-292 63-18528 <*>
1879 V20 N4/5 P210-214 62-20184 <*>
1883 V30 P312-347 61-14644 <*>
1891 V48 P195-306 61-14683 <*>
1895 V59 P16-42 61-14607 <*>
1897 V67 P299-344 66-11816 <*>
1897 V68 P596-598 61-14598 <*>
1898 V70 P494-507 61-14718 <*>
1898 V70 P507-510 61-14718 <*>
1899 V77 P311-330 61-14586 <*>
1900 V81 P328-348 61-14425 <*>
1900 V90 P1-40 61-14238 <*>
1905 V109 P63-72 61-14191 <*>
1907 V119 P29-38 61-14432 <*>
1909 V129 P35-45 58-951 <*>
1910 V133 P305-312 62-00631 <*>
1911 V137 P515-544 62-00630 <*>
1913 V150 P128-138 57-3487 <*>
1913 V151 P57-64 2262 <*>
1915 V162 P261-281 57-138 <*>
1918 V170 P646-676 57-128 <*>
1918 V170 P646- 57-416 <*>
1921 V186 P61-81 61-14641 <*>
1921 V191 P234-257 N65-27676 <=*$>
1922 V194 P629-646 61-14240 <*> 0
1922 V196 P185-199 C-2446 <NRC>
1922 V196 P331-344 DSIS-T-91-G <NRCC>
1922 V196 P540-559 57-413 <*>
1923 V198 P421-426 61-14437 <*>
1923 V200 P374- 57-414 <*>
1924 V203 P186-198 61-14199-1 <*>
1924 V204 P177-202 61-14429 <*>
1924 V204 P203-233 61-14177 <*>
1924 V204 P234-246 61-10961 <*>
1924 V204 P247-260 61-14178 <*>
1924 V205 P328-337 C-2670 <NRC>
1924 V205 P669-686 61-14433 <*> 0
1925 V210 P514-520 61-14903 <*>
1927 V215 P291-328 57-637 <*>
1927 V215 P588-607 61-14409 <*>
1930 V226 P559-577 57-422 <*>
1931 V226 P578-584 57-387 <*>
1931 V228 P213-224 61-14248 <*>
1931 V228 P234-257 61-14671 <*>
1931 V228 P322-328 61-14669 <*>
1931 V228 P724-730 61-14186 <*>
1931 V228 P742-750 57-283 <*>
1932 P26-32 MF-1 <*>
1932 V229 N4/5 P439-440 61-14824 <*>
1932 V230 P401-411 I-206 <*>
1932 V231 P26- 57-384 <*>
1934 V234 P13-28 61-14136 <*>
1934 V235 P538-544 61-14665 <*>
1936 V238 P319-326 61-14329 <*>
1937 V238 P78-90 57-665 <*>
1937 V238 P279-289 57-79 <*>
1937 V239 P290-292 61-14326 <*>
1938 V240 P263-281 57-146 <*>
1941 V245 P112-120 61-14285 <*>
1942 V245 P511-523 61-14655 <*>
1943 V246 N6 P757-789 62-00862 <*>
1943 V247 P145-148 61-10965 <*>
1943 V247 P149-159 61-14421 <*>
1944 V247 P576-592 64-16297 <*>
1950 V252 P331-344 61-16058 <*>

1951 V253 P435-458 II-265 <*>
1951 V253 P477-502 63-18865 <*>
1952 V256 N2 P87-95 2198 <*>
1953 V257 P308-317 II-415 <*>
1953 V257 P329-342 58-2226 <*>
1954 V259 N3 P212-225 1215 <*>
1954 V260 N2 P170-176 59-17469 <*>
1955 V261 P62-77 57-262 <*>
1956 V262 P377-394 57-3113 <*>
1956 V262 P431-442 1458 <*>
1956 V262 N6 P573-594 59-15195 <*> P
59-15248 <*>
1956 V263 P603-614 61-20000 <*> 0
1957 V264 P245-259 57-3114 <*>
1957 V264 P314-324 57-3115 <*>
1957 V264 P325-334 58-1585 <*>
1957 V264 N3 P217-227 59-10637 <*>
59-15551 <*> OP
1957 V265 N4 P314-327 61-00150 <*>
1958 V266 P569-585 62-00883 <*>
1958 V266 P611-627 62-00835 <*>
1958 V266 P628-641 62-00906 <*>
1958 V268 P177-180 61-00853 <*>
1959 V269 N2 P114-129 61-00263 <*>
1960 V271 P634-654 62-00320 <*>
1960 V271 P776-781 62-00479 <*>
1960 V271 P782-796 61-00300 <*>
1960 V272 N1 P41 61-00912 <*>
1960 V272 N1 P68 62-00316 <*>
1960 V272 N2 P187-190 61-00616 <*>
1961 V273 P562-572 62-00876 <*>
1961 V273 P575-578 62-00878 <*>
1961 V274 P227-251 62-01058 <*>
1961 V274 P252-261 62-01006 <*>
1961 V274 P311-317 62-00877 <*>
1962 V274 P553-566 63-00142 <*>
1962 V274 P567-580 62-01204 <*>
1962 V274 P593-607 63-00858 <*>
1962 V274 N6 P608-614 63-00159 <*>
1962 V274 N6 P615-623 63-00153 <*>
1962 V275 P12-22 62-01265 <*>
1963 V276 P336-356 63-00865 <*>
1965 V284 N1 P1-17 66-12715 <*>

PHARMACEUTICA ACTA HELVETIAE
1943 V18 P44-46 61-20253 <*> 0
1943 V18 P6C-65 61-20254 <*> 0
1947 V22 P612-613 60-10019 <*>
1949 V24 N3 P77-84 2924 <*>
1949 V24 N12 P430-443 59-17818 <*> 0
1951 V26 N3 P77-91 59-20879 <*>
1951 V26 N7 P243-249 65-12670 <*>
1952 V27 N7 P179-187 59-20707 <*>
1952 V27 N9 P235-250 59-20869 <*>
1952 V27 N2/3 P54-70 65-12694 <*>
1956 V31 N9 P409-417 62-14877 <*>
1959 V34 N2 P65-78 60-14539 <*> 0
1961 V36 N4 P232-237 65-11399 <*>
1963 V38 N6 P358-370 65-14587 <*> 0
1963 V38 N9 P641-654 64-14311 <*>
1964 V39 N12 P741-751 65-11424 <*>

PHARMACEUTICAL BULLETIN. CHINESE PEOPLES REPUBLIC
SEE YAO HSUEH T'UNG PAO

PHARMACEUTISCH WEEKBLAD VOOR NEDERLAND
1903 V40 N16 P309-313 62-00942 <*>
1924 V61 N31 P841-846 65-13297 <*>
1928 V65 N31 P731-738 1500-B <K-H>
1941 V78 P1277-1282 T-2474 <INSD>
1946 V81 P103-104 I-234 <*>
1950 V85 P937-950 2570A <K-H>
1952 V87 P861-865 992 <TC>
1955 V90 P241-251 58-1381 <*>
1957 V92 P1-24 T-2560 <INSD>
1957 V92 P775-781 61-20245 <*>
8388-A <K-H>
1965 V1C0 N21 P659-661 66-11703 <*>

PHARMAKOTHERAPIA
1963 V1 P65-75 64-18959 <*> 0

PHARMAZEUTISCHE INDUSTRIE
1950 V12 P39-44 63-18559 <*> 0
1955 V17 P523-524 2675 <*>
1955 V17 N9 P370 60-16612 <+*>
1957 V19 P88-93 3316 <*>
1962 V24 N6 P261-268 63-10514 <*>
1962 V24 N8 P370-372 65-12524 <*>
1963 V25 N11 P674-676 65-12498 <*>
1963 V25 N12 P733-735 65-12499 <*>
65-17034 <*>

PHARMAZEUTISCHE MONATSHEFTE
1934 V15 P161-162 3167 <*>

PHARMAZEUTISCHE PRAXIS
1963 N3 P37-43 64-18975 <*> 0

PHARMAZEUTISCHE ZEITUNG. BERLIN
1918 V63 P241 85H11G <ATS>
1918 V63 N33 P197-198 86H11G <ATS>
1937 V82 N101 P1189-1194 58-2102 <*>
1947 V83 P423-425 63-20169 <*>
1954 V90 N34 P941-942 62-16157 <*>
1955 V100 P70-71 <INSD>
1963 V108 N40 P1374-1379 65-17413 <*>

PHARMAZEUTISCHE ZENTRALHALLE FUER DEUTSCHLAND
1927 V69 P337-345 2545C <K-H>
1928 V69 P305-307 2627F <K-H>
1934 V75 P646-647 2627D <K-H>
1942 V83 N41 P481-484 63-10537 <*>
1956 V95 P91-96 5470-C <K-H>
1956 V95 P143-145 1178 <*>
1956 V95 P309-312 <INSD>
1957 V96 P68-71 T-2559 <INSD>
1961 V100 P213-221 11411-H <K-H>
64-14579 <*>
1961 V100 N7 P321-328 11627-E <K-H>
64-14567 <*> 0

PHARMAZIE
1950 V5 P276-278 63-20168 <*>
1952 V7 P133-143 64-10017 <*>
1953 V8 P251- 3138-C <K-H>
1953 V8 N3 P221-223 2682 <*>
1953 V8 N12 P1005-1010 66-11680 <*>
1954 V9 P806-812 58-1404 <*>
1954 V9 N8 P643-654 3928-B <K-H>
1954 V9 N9 P719-734 3963-E <K-H>
1954 V9 N10 P834-843 3963-F <K-H>
1955 V10 P102-103 T-1732 <INSD>
1955 V10 P141-157 3738-C <K-H>
1955 V10 P806-812 3471-B <K-H>
1955 V10 N8 P490-493 65-12613 <*>
1956 V11 P39-42 5339-C <KH>
1956 V11 P476-478 T-2550 <INSD>
1956 V11 P638-652 CSIRO-3468 <CSIR>
1956 V11 N1 P1-12 8480 <K-H>
1956 V11 N7 P476-478 65-12614 <*>
1957 V12 P391- <ES>
1957 V12 N1 P24-30 64-14530 <*> 0
9353-C <K-H>
1957 V12 N5 P262-264 57-2774 <*>
1958 V13 N5 P266-276 66-12720 <*> 0
1958 V13 N10 P628-631 59-15872 <*> 0
1959 V14 N8 P445-447 60-14543 <*>
1959 V14 N8 P466-473 64-14526 <*>
9856-C <K-H>
1960 V15 P374-377 63-10504 <*> 0
1960 V15 N4 P158-161 60-18552 <*> 0
1961 V16 P308-310 63-20879 <*> 0
1961 V16 N1 P26-30 63-20890 <*> 0
1961 V16 N4 P54-58 62-23450 <=*>
1961 V16 N4 P217-222 11296-A <KH>
64-14527 <*>
1962 V17 N1 P36-41 12057-B <K-H>
64-14548 <*> 0
1963 V18 N10 P704-708 64-18985 <*> 0

1964 V19 N5 P334-335 65-11406 <*>
1964 V19 N11 P708-715 65-14600 <*>
1965 V20 N11 P252-255 66-14138 <*>
1966 V21 N1 P58 66-14656 <*>

PHILIPS RESEARCH REPORTS
1949 V4 P291-315 66-11683 <*>
1950 V5 P250-261 1057 <*>
1950 V5 P262-269 58-1303 <*>
1951 V6 P54-74 1058 <*>

PHILOSOPHISCHE STUDIEN. LEIPZIG
1889 V5 P601-617 61-14296 <*>

PHOSPHORSAEURE
1934 V4 P429-440 1068 <*>
1955 V15 P47-49 63-14623 <*>
1955 V15 P50-63 3424A <K-H>
63-14628 <*>
1962 V22 P303-315 66-10192 <*>
1962 V22 P316-328 66-10193 <*> 0

PHOTOGRAMMETRIA
1951 V8 N1 P20- 3129 <*>

PHOTOGRAPHIE UND WISSENSCHAFT
1958 V7 N1 P5-8 SCL-T-249 <=*>

PHOTOGRAPHISCHE INDUSTRIE
1930 V28 P680-681 60-10214 <*> 0
1930 V28 P821-823 60-10198 <*> 0
1930 V28 P1156-1157 60-10213 <*>
1933 V31 P997-998 65-13071 <*>
1937 V35 P778-780 65-13069 <*> 0
1939 V37 P490-491 65-13068 <*> 0
1940 V38 P159-160 66-13754 <*>
1940 V38 P174-175 66-13754 <*>
1940 V38 N11 P159-160 59-17325 <*>
1940 V38 N12 P174-175 59-17325 <*>

PHOTOGRAPHISCHE KORRESPONDENZ
1910 N601 P469- 63-18297 <*> 0
1910 V47 P219-222 531 <TC>
1935 V71 N5 P65-68 64-16494 <*> 0
1935 V71 N5 P73-76 64-16494 <*> 0
1937 V73 P17-24 1467B <K-H> 0
1937 V73 P37-40 1467B <K-H> 0
1937 V73 P57-62 1467B <K-H> C
1939 V75 N10/2 P138-145 64-10616 <*>
1957 V93 N2 P17-25 63-14096 <*>
1958 V94 N1 P3-11 59-15988 <*>
1958 V94 N2 P19-26 59-15988 <*>
1958 V94 N5 P69-73 59-15355 <*>
1958 V94 N6 P83-91 59-15355 <*>
1958 V94 N8 P115-121 60-16559 <*>
1958 V94 N9 P131-137 60-16559 <*>
1958 V94 N10 P147-153 64-16664 <*>
1959 V95 N1 P7-13 59-17722 <*>
1959 V95 N2 P28-30 59-17722 <*>
1959 V95 N2 P35-42 59-20098 <*>
1959 V95 N8 P115-124 60-16560 <*>
1959 V95 N9 P131-134 60-16560 <*>
1959 V95 N10 P149-159 60-18481 <*>
1959 V95 N11 P165-171 60-18482 <*>
1960 V96 N1 P3-9 60-16955 <*>
1960 V96 N5 P67-74 58M46G <ATS>
1960 V96 N7 P99-104 09M47G <ATS>
1960 V96 N6/8 P83-88 NP-TR-868 <*>
1960 V96 N6/8 P99-102 NP-TR-868 <*>
1960 V96 N6/8 P115-125 NP-TR-868 <*>
1963 V99 N7 P102-105 166 <TC>
1964 V100 N3 P39-45 65-11944 <*>
1964 V100 N6 P99-102 43R78G <ATS>
1965 V101 N5 P69-73 55S84G <ATS>
1965 V101 N6/7 P85-92 66-10409 <*>

PHOTO-REVUE
1949 V61 N5 P79-80 AL-22 <*>

PHYSICA. NEDERLANDSCH TIJDSCHRIFT VOOR NATUURKUNDE
EINDHOVEN
1924 V4 P286-301 AEC-TR-4800 <*>
1926 V6 P361-365 UCRL TRANS-887(L) <*>
1928 V8 N1 P13-23 66-13320 <*>
1928 V8 N1 P34-37 58-36 <*>
1929 V9 P65-80 1326 <*>
1930 V10 P267-269 63-20860 <*>
1930 V10 N9 P273-286 57-883 <*>
1931 V11 N5 P146-149 57-587 <*>

PHYSICA
1934 V1 P53-59 I-461 <*>
1934 V1 P273-280 63-14460 <*>
1934 V1 P1003-1006 63-14459 <*>
1937 V4 P1190-1199 UCRL TRANS-585(L) <*>
1938 V5 N8/9 P384-388 65-14690 <*>
1939 V6 N8 P737-763 64-18838 <*> 0
1940 V7 N6 P552-562 64-30082 <*>
1941 V8 P1-22 1784 <*>
1942 V9 N9 P923-924 AL-292 <*>
1943 V10 P81-89 AL-293 <*>
1944 V11 N2 P78-90 59-15283 <*> 0
1944 V11 N3 P129-143 SCL-T-353 <*>
1944 S2 V11 N3 P129-143 60-15637 <=*>
1946 S2 V12 P589-594 59-17337 <*>
1950 V16 P239-248 UCRL TRANS-102 <*>
1951 V17 N8 P717-736 57-2505 <*>
1951 S2 V17 N10 P885-898 64-30286 <*>
1954 S2 V20 N11 P1110-1114
62-18813 <*>
1955 V21 P867-876 58-591 <*>
1964 V30 N8 P1513-1528 65-17343 <*>

PHYSICA STATUS SOLIDI
1961 V1 P366-385 63-18168 <*>
1961 V1 P636-649 UCRL TRANS-907(L) <*>
1961 V1 P704-715 UCRL TRANS-907(L) <*>
1961 V1 P739-757 65-17081 <*> 0
1961 V1 N5 P421-423 63-10255 <*>
1962 V2 P160-163 65-13131 <*>
1962 V2 P481-516 NP-TR-963 <*> OP
1962 V2 P668-691 AEC-TR-5568 <*>
1962 V2 P734-766 63-01031 <*>
1962 V2 P1005-1020 AEC-TR-5842 <*>
1962 V2 P1393-1402 66P66G <ATS>
1962 V2 N2 PK27-K30 63-14751 <*>
1962 V2 N4 P411-416 62-20229 <*>
1962 V2 N8 P187-191 AEC-TR-5876 <*>
1963 V3 P383- 63-20882 <*> 0
1963 V3 P744-759 AEC-TR-6016 <*>
1963 V3 P1153-1200 66-11257 <*>
1963 V3 P1480-1490 66-12353 <*>
1963 V3 N4 PK159-K162 65-11470 <*>
1963 V3 N5 P932-949 NP-TR-1102 <*>
1963 V3 N6 P1059-1071 64-16927 <*>
1963 V3 N6 P1098-1106 66-13568 <*> 0
1963 V3 N7 P1247-1251 64-16110 <*> 0
1963 V3 N9 P1619-1628 64-16946 <*>
1964 V4 P439-451 AI-TRANS-59 <*>
1964 V4 N3 P509-520 65-12158 <*>
1964 V4 N3 P685-696 AD-637 436 <=$>
1964 V5 P279-301 26R80G <ATS>
1964 V5 P421-434 66-11376 <*>
1964 V6 P441-460 21S84G <ATS>
1964 V6 N1 P185-205 65-27884 <=$>
1964 V6 N3 P615-625 65-63725 <=$>
1964 V7 P189-203 19S81G <ATS>
1964 V7 P577-590 AI-TRANS-75 <*>
1964 V7 P701-710 66-11989 <*>
1964 V7 P833-849 CE-TRANS-4013 <NLL>
1964 V7 P863-868 88S83G <ATS>
1964 V7 P937-952 66-11378 <*>
1964 V7 P953-971 AI-TRANS-90 <*>
1965 V8 P533-542 00S85G <ATS>
1965 V8 P831-840 13T91G <ATS>
1965 V8 PK5-K8 65-17156 <*>
1965 V8 N2 P613-618 66-11179 <*>
1965 V8 N3 P881-896 AD-636 649 <=$>
1965 V8 N3 PK163-K166 65-14660 <*>
1965 V10 P269-282 66-12136 <*>

1965 V10 PK131-K133 66-13615 <*>
1965 V11 P339-354 66-11875 <*>
1965 V11 P819-829 66-11871 <*>
1965 V11 N1 PK73-K76 66-10426 <*>
1965 V11 N2 P635-650 66-14389 <*>
1965 V12 P235-250 66-11050 <*>
1965 V12 P251-264 66-11481 <*>
1965 V12 P333-339 66-14364 <*>
1965 V12 P405-419 66-11867 <*>
1965 V12 N1 P115-123 66-14494 <*>
1966 V14 N2 PK181-K182 66-13235 <*>
1966 V16 N1 P237-246 66-14505 <*>

PHYSICS BULLETIN. CHINESE PEOPLES REPUBLIC
SEE WU LI T'UNG PAO

PHYSICS AND CHEMISTRY OF SOLIDS
1956 V1 P175-178 AEC-TR-4386 <*>
1958 V4 P71-77 4009 <BISI>
1958 V4 N4 P283-305 C-3174 <NRCC>
1958 V6 P136-143 AEC-TR-4661 <*>
1958 V6 P144-154 65-17076 <*> 0
1958 V6 P155-168 62-10417 <*> 0
63-16973 <*>
1958 V6 N2/3 P155-168 65-11478 <*>
1958 V7 P22-51 62-16961 <*>
1958 V7 P218-227 UCRL TRANS-829(L) <*>
1958 V7 P295-300 72N49G <ATS>
1959 V9 N2 P181-182 60-14072 <*> 0
1959 V10 P174-181 65-11575 <*>
1959 V10 N2/3 P126-137 C-3379 <NRCC>
1959 V10 N2/3 P174-181 C-3382 <NRCC>
1959 V11 P310-314 62-14087 <*> 0
1959 V12 N1 P74-88 62-16203 <*> 0
62-16735 <*> 0
1960 V12 P233-244 ANL-TRANS-206 <*>
1960 V12 P298-313 64-10412 <*>
1960 V16 N3/4 P253-264 66-12354 <*>
1960 V16 N3/4 P265-278 62-14461 <*> 0
62-16207 <*> 0
62-18279 <*> 0
1961 V18 P129-138 AEC-TR-4739 <*>
1961 V18 P196-202 AEC-TR-4274 <*>
1961 V21 P119-122 66-14108 <*>
1961 V21 N1/2 P33-39 SCL-T-402 <*>
1961 V21 N3/4 P156-171 62-16206 <*> 0
62-18644 <*>
1962 V23 P639-658 AI-TR-12 <*>
1962 V23 P975-983 64-18345 <*> 0
1962 V23 P1621-1629 66-11097 <*>
1963 V24 P1617-1624 66-12944 <*>
1963 V24 N7 P969-973 AEC-TR-5977 <*>
1963 V24 N11 P1285-1289 26S83G <ATS>
1964 V25 P559-564 66-11087 <*>
1965 V26 N7 P1143-1145 66-10348 <*>

PHYSICS LETTERS
1962 V3 N1 P52-55 NP-TR-1050 <*>
1963 V4 N3 P191-194 63-01513 <*>
1963 V4 N5 P302-304 AEC-TR-5848 <*>
1963 V5 P33-35 63-18821 <*> 0
1963 V5 N3 P179-181 64-10401 <*> 0
64-16615 <*>
64-18588 <*>
1964 V8 N1 P14 65-00320 <*>
1964 V8 N1 P15-17 ORNL-TR-66 <*>
1964 V13 N3 P215-216 65-14321 <*>
1965 V18 P304-307 65-22511 <=$>

PHYSIK DER KONDENSIERTEN MATERIE
1963 V1 P78-104 C-4588 <NRC>
1963 V1 N2 P105-124 3574 <BISI>
1963 V1 N4 P296-315 AI-TRANS-34 <*>
1964 V2 P133-138 C-5590 <NRC>
1964 V2 N5 P367-376 65-13143 <*>
1964 V3 P1-17 65-14072 <*>

PHYSIK IN REGELMAESSIGEN BERICHTEN
1936 V4 N1 P17-33 63-18567 <*> 0
1940 V8 N4 P127-148 63-20150 <*>

PHYSIKALISCHE BERICHTE
1920 N14 P885 58-136 <*>

PHYSIKALISCHE BLAETTER
1947 V3 N5 P159-160 AL-673 <*>
1951 V7 N5 P205-214 MF-1 <*>
1952 V8 N11 P493-500 57-719 <*>
1954 V10 N12 P565-577 58-523 <*>
1958 V14 P207-212 62-11332 <=>
62-15384 <*=>
1962 V18 N6 P255-263 NP-TR-1056 <*>
1964 V20 N12 P559-562 66-10919 <*>

PHYSIKALISCHE ZEITSCHRIFT
1901 V2 N22 P329-334 57-1591 <*>
1907 V8 N25 P923-924 57-2233 <*>
1909 V10 P168- 57-323 <*>
1910 V11 P294-311 58-1911 <*>
1910 V11 P430-433 59-10595 <*>
1910 V11 P708-709 63-14410 <*>
1910 V11 N7 P257- 57-1548 <*>
1911 V12 P614-620 58-84 <*>
59-10638 <*>
1911 V12 N14 P561-568 64-10372 <*>
1911 V12 N15 P609- 57-2219 <*>
1911 V12 N21 P920-924 59-15221 <*>
1912 V13 P118-120 62-11494 <=>
1912 V13 P864-870 I-188 <*>
1912 V13 N13 P577-583 59-10639 <*> 0
1913 V14 P324-332 AEC-TR-4650 <*>
1913 V14 P343-349 57-1342 <*>
1913 V14 P832-835 63-00584 <*>
63-27134 <$>
1913 V14 P1269-1271 3237 <*>
1913 V14 N13 P561-562 65-10821 <*>
1913 V14 N21 P1042-1045 57-01581 <=>
1914 V15 P362-363 58-202 <*>
1915 V16 P59-62 57-2074 <*>
1917 V18 P261-270 58-307 <*>
1917 V18 N19 P445-453 66-12378 <*>
1917 V18 N21 P509-515 63-01753 <*>
1919 V20 P104-114 58-129 <*>
1919 V20 P245-251 58-128 <*>
1919 V20 P371-375 66-13309 <*>
1919 V20 P401-403 57-2209 <*>
1919 V20 N8 P183-188 I-693 <*>
1920 V21 P168 61-14344 <*>
1920 V21 P264-270 58-188 <*>
1920 V21 P499-500 61-14880 <*>
1920 V21 N8 P187-192 58-348 <*>
1921 V22 P218-224 1435 <*>
1921 V22 P282-286 1435 <*>
1921 V22 N12 P345-352 NP-TR-845 <*>
1923 V24 P232 58-189 <*>
1923 V24 P261-265 UCRL TRANS-891(L) <*>
1923 V24 P344- 58-772 <*>
1923 V24 N16 P344-350 59-15222 <*>
64-71265 <=>
1924 V25 P166-167 UCRL TRANS-891(L) <*>
1926 V27 P115-133 60-18974 <*> 0
1926 V27 P787-789 58-2154 <*>
1926 V27 N11 P361-366 57-3025 <*>
1927 V28 N3 P153- 57-1460 <*>
1927 V28 N23 P838-841 66-13141 <*>
1928 V29 P2-5 61-14271 <*>
1928 V29 P18 61-14271 <*>
1928 V29 P34-41 57-565 <*>
1928 V29 P793-810 57-2084 <*>
1928 V29 N18 P654-655 57-1467 <*>
1929 V30 P115-125 57-364 <*>
1929 V30 P467-473 62-18223 <*>
1929 V30 N7 P177-196 66-12260 <*> 0
1929 V30 N16 P489-493 11E2G <ATS>
1929 V30 N21 P721-745 57-365 <*>
1929 V30 N22 P839-846 66-13305 <*>
1929 V30 N23 P849-856 NP-TR-437 <*>
1930 V31 P419-428 57-1023 <*>
1930 V31 P626-640 57-1365 <*>
1930 V31 P824-835 57-2068 <*>

57-3475 <*>
1930 V31 P857-868 57-2068 <*>
1930 V31 P964-969 57-3096 <*>
1930 V31 N12 P561-574 57-3226 <*>
1931 V32 P48-56 58-2149 <*>
1931 V32 P517-520 62-18303 <*> O
1931 V32 N7 P286-288 57-315 <*>
1931 V32 N21 P833-842 66-12394 <*>
1931 V32 N23 P942-945 57-937 <*>
1932 V33 P32-38 58-2194 <*>
1932 V33 P727-729 64-16247 <*> O
1932 V33 P844-847 65-13362 <*>
1932 V33 N21 P835-841 66-13648 <*>
1933 V34 P1-24 NP-2042 <*>
1933 V34 P478-482 66-11854 <*>
1933 V34 P756-761 63-18147 <*>
1935 V36 P61-66 63-14458 <*>
1935 V36 P552-558 C-2411 <NRC>
1935 V36 N4 P142-144 59-17681 <*>
1936 V37 P569-578 66-13564 <*>
1936 V37 P708-720 T2229 <INSD>
1936 V37 P737-753 AD-631 200 <=$>
1936 V37 P757-763 59-20820 <*>
1936 V37 N6 P185-203 66-13656 <*> O
1936 V37 N11 P414-415 57-330 <*>
1936 V37 N22/3 P912-914 61-14272 <*>
1937 V38 P112-122 58-2195 <*>
61-10846 <*> O
1937 V38 P951- II-360 <*>
1937 V38 P964-965 I-905 <*>
1937 V38 N7 P202-224 66-12909 <*>
1938 V39 P105-109 UCRL TRANS-502 <*>
1938 V39 P673-687 57-884 <*>
1938 V39 N11 P460-462 58-2508 <*>
1938 V39 N14 P546-559 66-12265 <*>
1939 V40 P159-161 AL-554 <*>
1939 V40 P337-345 T-1368 <INSD>
1939 V40 P461-466 UCRL TRANS-586(L) <*>
1939 V40 P645-663 T-1369 <INSD>
1939 V40 N12 P416-428 II-289 <*>
1940 V41 P308-325 63-10656 <*>
1940 V41 P399-401 64-10201 <*>
1940 V41 P434-442 E-634 <RIS>
1940 V41 P475-480 57-1028 <*>
1940 V41 N11/2 P285-290 57-6 <*>
1941 V42 P55-57 T-2083 <INSD>
1941 V42 P349-360 66-12282 <*>
1942 V43 P91-101 I-801 <*>
1943 V44 P296-302 58-501 <*>
1943 V44 N3/4 P63-77 60-18586 <*>
1944 V45 P199-205 66-11574 <*>

PHYSIKALISCHE ZEITSCHRIFT DER SOWJETUNION
1932 V1 P733-746 SCL-T-443 <*> P
1933 V4 P197-211 62-18902 <*>
1933 V4 P397-419 62-20352 <*> P
1933 V4 P501-515 NP-TR-442 <*>
1933 V4 P723-734 17K24G <ATS>
1934 V5 N5 P687-705 R-2656 <*>
57-913 <*>
1934 V6 P224-243 AEC-TR-5956 <*>
1935 V8 N5 P489-500 66-10712 <*> O
1936 V9 N1 P57-71 AL-856 <*>
1936 V10 N1 P34-43 60-10823 <*>
1937 V11 P18-25 66-11873 <*>
1937 V11 P445-457 UCRL TRANS-784 <*>
1937 V11 N18 P25 63-01726 <*>
1937 V12 P389-403 RJ-182 <*>
RJ-182 <ATS>
62-18766 <*>
1938 V13 P198- 57-812 <*>

PHYSIKERTAGUNG. VERBAND DEUTSCHER PHYSIKALISCHER GESELLSCHAFTEN
1955 V3 P73-88 AEC-TR-5926 <*>

PHYSIOLOGIA BOHEMOSLOVENICA
1960 V9 P376 61-16123 <*> O
1963 V12 N1 P150-155 65-60850 <=$>

PHYSIOLOGIA PLANTARUM
1954 V7 N3 P538-547 64-30283 <*>
1965 V18 P1037-1043 66-11865 <*>

PHYSIOLOGY AND ECOLOGY. JAPAN
SEE SEIRI SEITAI

PHYSIOLOGY TODAY CHINESE PEOPLES REPUBLIC
SEE SHENG-LI K'O-HSUEH CHIN-CHAN

PHYTON. ANNALES REI BOTANICAE
1950 V2 N1/3 P182-192 59-17190 <*>

PHYTOPATHOLOGISCHE ZEITSCHRIFT
1934 V7 N3 P255-258 14F3G <ATS>
1950 V17 P218-228 65-12538 <*>
1951 V17 P371-373 3238 <*>
1952 V18 N4 P404-415 II-986 <*>
1954 V21 N4 P395-406 C-2269 <NRC>
1954 V22 N10 P449-453 58-2476 <*>
1955 V23 N3 P328-334 C-2231 <NRC>
1957 V28 P319-328 60-18637 <*>
1957 V29 N3 P299-304 59-15259 <*>
1961 V41 P257-264 63-10725 <*>
1961 V42 P362-374 65-11678 <*>
1962 V44 P179-188 65-00377 <*>
1962 V45 N1 P53-56 63-01037 <*>

PIRELLI, RICERCA E SVILUPPO
1954 N2 P1-20 RCT V28 N4 P1054 <RCT>
1956 N4 P2-10 RCT V29 N3 P743 <RCT>
1956 N4 P12-24 RCT V29 N3 P753 <RCT>

PISHCHEVAYA PROMYSHLENNOST SSSR
1945 N1 P24-26 60-10885 <+*>

PISHCHEVAYA TEKHNOLOGIYA
SEE IZVESTIYA VYSSHIKH UCHEBNYKH ZAVEDENII. PISHCHEVAYA TEKHNOLOGIYA

PITTURE E VERNICI
1961 V37 P449-452 8012 <TTIS>
1962 V38 P6-8 8019 <TTIS>
1962 V38 N6 P203-212 NS-101 <TTIS>
1963 V39 P341-343 66-11746 <*> O
1965 V41 N2 P56-63 66-11358 <*>
1965 V41 N6 P227-234 65-17468 <*>

PLANOVA STOPANSTVO
1960 V15 N3 P77-80 62-19302 <=*>

PLANOVANE HOSPODARXTVI
1958 V11 N3 P165-176 PB 141 305T <=>

PLANOVOE KHOZYAISTVO
1939 V16 N2 P73-80 RT-1856 <*> O
1945 N2 P57-67 RT-1855 <*>
1951 N4 P4-20 61-13653 <=*>
1958 N7 P20-21 59-22196 <+*>
1958 N7 P32-44 NLLTB V1 N1 P13 <NLL>
59-22638 <+*>
1958 N8 P29-39 NLLTB V1 N1 P32 <NLL>
59-22639 <+*>
1958 N12 P16-27 59-11472 <=>
1959 N1 P53-64 59-11722 <=>
1959 N6 P79-81 60-11764 <=>
1959 N9 P26-37 60-11413 <=>
1959 N9 P59-69 60-11572 <=>
1959 N12 P30-34 61-23015 <=>
1960 N1 P38-58 60-11816 <=>
1960 N1 P82-86 60-11816 <=>
1960 N5 P60-65 61-23015 <=>
1960 N5 P71-73 61-23015 <=>
1960 N6 P74-75 61-21293 <=>
1960 N6 P77-80 61-23015 <=>
1960 N7 P49-57 61-11062 <=>
1960 N8 P9-18 61-21497 <=>
1960 N8 P19-29 61-21139 <=> P
1960 N9 P92-95 62-23487 <=*>
1960 N11 P12 61-21697 <=>

1960 N11 P75-78 61-31155 <=>
1960 N12 P59-62 61-21697 <=>
1960 V37 N12 P78-82 61-23932 <=>
1961 N7 P31-39 SGRT V3 N1 P39 <AGS>
1962 V39 N1 P14-26 65-32856 <=*$>
1962 V39 N9 P65-71 63-13344 <=>
1962 V39 N11 P72-78 63-21033 <=>
1963 V40 N1 P70-72 63-21542 <=>
1963 V40 N2 P45-46 63-21569 <=>
1963 V40 N3 P1-10 63-21930 <=>
1963 V40 N3 P63-66 63-21674 <=>
1963 V40 N6 P75-81 63-31448 <=>
1963 V40 N7 P54-75 63-31677 <=>
1963 V40 N7 P76-81 63-31665 <=>
1963 V40 N7 P87-88 63-31665 <=>
1964 V41 N2 P43-50 64-31352 <=>
1964 V41 N3 P69-74 64-31310 <=>
1964 V41 N3 P90-92 64-31310 <=>
1964 V41 N5 P24-32 64-31763 <=>
1964 V41 N6 P11-21 64-41139 <=>
1964 V41 N6 P29-38 64-31670 <=>
1965 N7 P45-57 65-32733 <=$>
1965 V42 N4 P39-49 65-31441 <=$>
1965 V42 N4 P64-72 65-31441 <=$>
1965 V42 N5 P59-64 65-31706 <=$>
1965 V42 N5 P83-86 65-32672 <=*$>
1965 V42 N8 P9-16 65-33819 <=*$>
1965 V42 N9 P86-88 65-33199 <=*$>
1965 V42 N10 P18-29 66-30187 <=*$>
1965 V42 N11 P50-55 65-34025 <=*$>
1965 V42 N12 P20-29 66-30578 <=$>
1965 V42 N12 P56-72 66-30893 <=$>
1966 N2 P29-35 66-31660 <=$>
1966 V43 N1 P24-32 66-31049 <=$>
1966 V43 N1 P88-89 66-30856 <=$>
1966 V43 N1 P93-95 66-30856 <=$>
1966 V43 N4 P55-66 66-32354 <=$>
1966 V43 N7 P31-41 66-34914 <=$>
1966 V43 N7 P60-69 66-34914 <=$>
1966 V43 N11 P13-19 67-30031 <=>
1966 V44 N6 P30-40 66-32958 <=$>
1967 N2 P37-53 67-31084 <=$>
1967 N2 P72-79 67-31057 <=$>

PLANOVO STOPANSTVO I STATISTIKA
1964 V19 N6 P27-44 64-41121 <=>
1965 V20 N2 P3-8 65-30826 <=$>
1967 N2 P14-23 67-31219 <=$>

PLANSEEBERICHTE FUER PULVERMETALLURGIE
1956 V4 P7-9 8773 <IICH>
1956 V4 N1 P2-6 59-10525 <*> 0
1956 V4 N1 P7-9 65-10010 <*> 0
1957 V5 P2-19 58-2375 <*>
1957 V5 N3 P104-120 59-10780 <*> 0
1958 V6 P17-21 61-20076 <*>
1959 V7 P6-17 AEC-TR-3822 <*>
1959 V7 N2 P67-78 61-16165 <*> 0
1961 V9 P65-76 AEC-TR-4834 <*>
1962 V10 P42-64 AEC-TR-5459 <*>
1962 V10 P137-143 AEC-TR-6084 <*>
1962 V10 P168-177 AEC-TR-5952 <*>
1963 V11 N1 P18-19 NP-TR-1076 <*>
1963 V11 N3 P146-157 65-10383 <*> 0

PLANT PROTECTION, JAPAN
SEE SHOKUBUTSU BOEKI

PLANT AND SOIL
1959 V11 N2 P157-169 64-30296 <*>
1961 V15 N4 P284-290 63-10488 <*>
63-16249 <*>

PLANTA
1927 V4 P467-475 II-948 <*>
1928 V5 P563-615 1062 <K-H>
1934 V22 P800 58-2294 <*>
1934 V23 P264-283 57-2736 <*>
1937 V26 P6-18 878 <K-H>
1937 V27 P392-398 CSIRO-2701 <CSIR>
1938 V28 N2 P344-351 CSIRC-3160 <CSIR>
1939 V30 P118-128 AL-116 <*>
1942 V33 P278-289 65-12351 <*>
1950 V38 P1-11 AEC-TR-6018 <*>
1952 V40 P199-253 T1902 <INSD>
1954 V44 P269-285 65-12599 <*>
1957 V49 N1 P1-10 60-18010 <*>
1958 V50 N5 P461-497 59-10144 <*>
1960 V55 P480-495 61-20673 <*> C
1960 V55 N5 P480-495 11499-F <KH>
1963 V59 P535-562 AEC-TR-6261 <*>
1963 V60 P178-204 AEC-TR-6281 <*>
1965 V65 P102-104 66-11881 <*>

PLASTE UND KAUTSCHUK
1954 V1 N8 P178-180 65-18169 <*>
1955 V2 N4 P79-81 65-11437 <*> C
1955 V2 N11 P249-252 59-20922 <*>
1957 V4 P84-87 RCT V31 N4 P681 <RCT>
1957 V4 N1 P3-6 59-17780 <*>
1958 V5 N1 P3-5 60-14145 <*>
1958 V5 N1 P20-21 58-2355 <*>
1958 V5 N1 P37-40 62-10028 <*> 0
1958 V5 N2 P43-48 AEC-TR-3987 <*>
1958 V5 N2 P52-54 60-14144 <*>
1958 V5 N2 P83-84 62-10028 <*> 0
1958 V5 N3 P93-96 65-18173 <*>
1958 V5 N3 P96-103 60-14141 <*>
1958 V5 N3 P110-112 59-18720 <+*> 0
1958 V5 N11 P428-433 60-10431 <*>
1958 V5 N11 P434-435 59-15701 <*>
1958 V6 N1 P23-24 58-2356 <*>
1959 V6 N7 P325-326 60-18301 <*>
1959 V6 N8 P361-364 RCT V34 N2 P474 <RCT>
1960 V7 P228-230 2216 <TTIS>
1960 V7 P426-428 2220 <TTIS>
1960 V7 N4 P208-211 62-10029 <*> C
1960 V7 N9 P445-449 146-111 <STT>
1960 V7 N11 P528-533 63-14016 <*> 0
1961 V8 P282- 2297 <TTIS>
1961 V8 P296-300 63-10820 <*> 0
1961 V8 P332-333 2314 <TTIS>
1961 V8 P387-391 2312 <TTIS>
1961 V8 P435-440 2330 <TTIS>
1961 V8 P482-486 2345 <TTIS>
1961 V8 P529-531 2345 <TTIS>
1961 V8 P591-593 2361 <TTIS>
1961 V8 N5 P282-287 64-10196 <*> 0
1961 V8 N6 P304-310 63-14478 <*>
78P60G <ATS>
1961 V8 N8 P427-429 62-14532 <=*>
1962 N4 P186-188 65-13529 <*>
1962 V9 N3 P121-123 65-13529 <*>
1962 V9 N6 P282-287 RCT V36 N2 P365 <RCT>
RUB.CH.TECH.V36 N2 <ACS>
1962 V9 N7 P343-344 NS-39 <TTIS>
1962 V9 N8 P407-409 NS-53 <TTIS>
1962 V9 N9 P425-428 NS-55 <TTIS>
1962 V9 N10 P495-496 NS-61 <TTIS>
1962 V9 N11 P516-519 <TTIS>
1963 V10 N1 P22-24 64-10929 <*>
1963 V10 N1 P25-27 RCT V37 N1 P146 <RCT>
1963 V10 N3 P140-145 NS-81 <TTIS>
1963 V10 N3 P153 NS-82 <TTIS>
1963 V10 N6 P373-376 NS 133 <TTIS>
1963 V10 N7 P390-391 64-71158 <=>
1963 V10 N9 P552-556 64-16682 <*>
1963 V10 N11 P660-665 RAPRA-1249 <RAP>
1963 V10 N12 P728-730 77R80G <ATS>
1964 V11 P131-133 2115 <TC>
1964 V11 N4 P248-250 NS 237 <TTIS>
1964 V11 N5 P269-270 NS-345 <TTIS>
1964 V11 N5 P270-275 NS-346 <TTIS>
66-10877 <*>
1964 V11 N5 P319- NS-347 <TTIS>
1964 V11 N6 P333-335 NS-348 <TTIS>
1964 V11 N6 P343-349 66-13865 <*>
1964 V11 N10 P580-586 66-11355 <*> 0
1964 V11 N10 P609-613 AD-636 651 <=$>
1964 V11 N10 P624-627 698 <TC>

1964 V11 N10 P632-635 NS-349 <TTIS>
1964 V11 N11 P670-673 1738 <TC>
1965 V12 N1 P16-20 56T93G <ATS>
1965 V12 N2 P122-124 65-17469 <*>
1965 V12 N4 P215-218 RAPRA-1299 <RAP>
1965 V12 N5 P370-374 NS-455 <TTIS>
1966 V13 N1C P577-578 66-35492 <=> 0
1966 V13 N10 P586 66-35492 <=> 0
1967 N2 P73-78 67-31017 <=$>

PLASTICA
1951 V4 P55-58 58-114 <*>
1951 V4 P188-189 2674 <*>
1952 V5 N1 P19-24 60-18351 <*>
1956 V9 N8 P448-454 63-14111 <*> 0
1959 V12 N1 P24-36 61-10907 <*>
1960 V13 N12 P1216-1219 62-16592 <*>
1962 V15 P666-670 RAPRA-1224 <RAP>
1963 V16 N6 P272-275 RAPRA-1225 <RAP>
1963 V16 N1C P500-511 RAPRA-1227 <RAP>
1963 V16 N11 P539-543 RAPRA-1229 <RAP>
1963 V16 N12 P596-6C3 78R79DU <ATS>
1964 V17 N9 P44C-447 65-11619 <*>
1964 V17 N12 P624-627 65-14883 <*>
1965 V18 N1 P12-16 70S86DU <ATS>
1966 V19 N4 P133-143 66-13454 <*>
1966 V19 N5 P175-183 66-14366 <*>

★ PLASTICHESKIE MASSY
1935 N3 P4-8 RT-892 <*>
1959 N1 P3-9 61-27320 <*=> 0
1959 N1 P10-16 61-23370 <=*>
1959 N1 P17-20 61-23369 <=*>
1959 N1 P21-25 61-23368 <*=>
1959 N1 P26-29 61-23367 <=*>
1959 N1 P30-31 61-23366 <=*>
1959 N1 P32-34 61-23365 <=*>
1959 N1 P34-40 61-23364 <=*>
1959 N1 P34-49 61-27320 <*=> 0
1959 N1 P40-49 61-23363 <=*>
1959 N2 P25-29 61-13161 <=*>
1960 N1 P36-38 61-27643 <*=>
1960 N1 P55-59 61-11828 <=>
1960 N2 P17-19 63-18895 <=*>
1960 N3 P20-23 61-31656 <=>
1960 N3 P58-63 62-19126 <*=>
1960 N4 P71-73 60-41675 <=>
1960 N5 P15-17 64-26380 <$>
1960 N5 P22-25 63-16977 <=*>
1960 N5 P64-65 C-5270 <NRC>
1960 N5 P65-66 63-18892 <=*>
1960 N6 P2-5 27M46R <ATS>
1960 N6 P11-13 AD-613 451 <=$>
1960 N6 P19-22 61-31063/1-9 <=>
1960 N7 P17-20 62-11595 <=> 0
1960 N8 P3-6 35R78R <ATS>
1960 N8 P69-71 61-21299 <=>
1960 N9 P4-7 62-33618 <=*>
1960 N9 P61-68 63-24202 <=*$>
1960 N9 P77-78 62-13256 <*=>
1960 N9 P78-79 62-13234 <*=>
1960 N9 P79 62-13233 <*=>
1960 N10 P4-5 62-13150 <*=>
1960 N10 P6-8 62-13155 <*=>
1960 N10 P42-46 62-13152 <*=>
1960 N10 P64-68 62-13151 <*=>
1960 N10 P76-78 61-21423 <=>
1960 N12 P2-3 61-19205 <+*>
1960 N12 P14-15 62-14045 <=*>
1960 N12 P19-22 62-19128 <*=>
1961 N1 P35-37 63-24338 <=*$>
1961 N1 P63-65 61-27963 <*=>
1961 N2 P1-2 62-13228 <*=>
1961 N2 P9-12 63-18862 <=*>
1961 N3 P35-36 12166 <K-H>
65-14036 <*> 0
1961 N4 P14-17 62-10607 <=*>
1961 N5 P26-28 62-00076 <*>
62-13426 <*=>
69N55R <ATS>
1961 N5 P72-75 62-13433 <*=>
1961 N6 P13-20 62-13437 <*=>
1961 N6 P34-40 62-13428 <*=>
1961 N7 P59-65 ICE V2 N1 P18-24 <ICE>
24N55R <ATS>
1961 N8 P69-75 62-13686 <*=>
1961 N9 P26-29 62-25930 <=*> 0
1961 N9 P30-35 63-18290 <=*>
1961 N11 P47-48 62-14987 <=*>
1961 N12 P3- ICE V2 N2 P242-246 <ICE>
1961 N12 P3-7 62-33090 <=*>
1961 N12 P7-10 62-14988 <=*>
1961 N12 P15-19 65-61457 <=>
1961 N12 P22-26 62-20020 <=*>
63-19899 <*=>
1961 N12 P7C 62-32102 <=*>
1962 N2 P77-80 63-15176 <=*> C
1962 N3 P8- ICE V2 N3 P445-449 <ICE>
1962 N3 P44-51 57P61R <ATS>
1962 N3 P77 62-32102 <=*>
1962 N4 P1-2 62-32102 <=*>
1962 N4 P76-78 63-23351 <=*>
1962 N5 P45-47 M-5723 <NLL>
1962 N6 P62-63 63-18356 <=*>
1962 N7 P11-14 13327A <K-H>
63-18650 <=*>
65-18149 <*>
1962 N7 P32-35 63-24394 <=*$>
1962 N8 P38-40 63-10094 <=*>
1962 N9 P24-28 75Q70R <ATS>
1962 N9 P61-64 06Q70R <ATS>
1962 N10 P71-73 63-235C5 <=*$>
1962 N11 P33-35 63-19641 <=*>
1962 N11 P43-44 50Q73R <ATS>
1962 N12 P18-21 63-23511 <=*$>
1962 N12 P40-45 ANL-TRANS-15 <=$>
1963 N1 P56-64 SCL-T-487 <=*$>
1963 N1 P74-75 63-21382 <=>
1963 N1 P78-80 63-21382 <=>
1963 N2 P6-11 ICE V3 N3 P383-388 <ICE>
1963 N2 P14-15 NS 115 <TTIS>
1963 N3 P17 63-27149 <=$>
1963 N3 P19-23 64-18135 <*>
1963 N3 P39-42 AD-624 752 <=*$>
64-14839 <=*$>
66-14791 <*$>
1963 N4 P7-11 ICE V3 N4 P577-581 <ICE>
1963 N4 P70-72 40Q72R <ATS>
1963 N5 P24-27 63-20621 <=*$>
1963 N7 P24-28 64-15115 <=*$>
1963 N8 P3-7 ICE V4 N1 P75-80 <ICE>
1963 N8 P13-16 64-10719 <=*$>
1963 N8 P2C-22 64-10718 <=*$>
1963 N8 P27-33 ICE V4 N2 P230-236 <ICE>
1963 N9 P50-51 64-18487 <*>
1963 N10 P3-7 ICE V4 N2 P307-312 <ICE>
1963 N10 P17-21 64-15213 <=*$>
1964 N1 P71-72 64-21848 <=>
1964 N2 P3-6 ICE V5 N3 P519-523 <ICE>
60S82R <ATS>
1964 N3 P5-9 ICE V4 N4 P593-597 <ICE>
1964 N3 P61-62 66-10237 <*>
1964 N4 P14-20 09R77R <ATS>
1964 N7 P17-19 M-5575 <NLL>
1964 N7 P19-20 65-29619 <$>
1964 N7 P21-23 708 <TC>
1964 N7 P24-26 65-10459 <*>
1964 N7 P65-66 AD-618 058 <=$>
1964 N8 P13-16 ICE V5 N1 P36-39 <ICE>
1964 N8 P18-19 M-5573 <NLL>
1964 N8 P20-23 AD-619 482 <=$>
1964 N8 P31-33 AD-619 482 <=$>
1964 N8 P33-36 M-5592 <NLL>
1964 N8 P36-40 64R79R <ATS>
1964 N9 P18-20 AD-622 355 <=*$>
1964 N9 P23-26 14S84R <ATS>
1964 N9 P41-43 AD-622 355 <=*$>
1964 N10 P49-51 1632 <TC>
66-11564 <*>
1964 N10 P62-64 13S84R <ATS>

1964 N11 P1-9 AD-629 165 <=*$>
ICE V5 N3 P405-414 <ICE>
1964 N11 P37-39 AD-629 165 <=*$>
1964 N11 P42-45 AD-629 165 <=*$>
1964 N12 P20-26 1018 <TC>
1965 N3 P14-16 73S87R <ATS>
1965 N4 P63-65 74S87R <ATS>
1965 N4 P72-74 AD-626 049 <=$>
1965 N5 P57-58 15S85R <ATS>
1965 N7 P13-14 97T91R <ATS>
1965 N7 P17-20 98T91R <ATS>
1965 N8 P53-55 ICE V6 N2 P247-249 <ICE>
72S88R <ATS>
1965 N9 P20-22 AD-625 613 <=$>
1965 N10 P35-36 46S89R <ATS>
1965 N11 P5-8 96T90R <ATS>
1965 N12 P10-12 AD-639 442 <=$>
1966 N1 P47-50 66-14671 <*$>
1966 N3 P15-17 94T93R <ATS>
1966 N3 P59-61 95T93R <ATS>
1966 N5 P9-12 66-14221 <*$>

PLASTICKE HMOTY A KAUCUK
1964 V1 N10 P289 RAPRA-1246 <RAP>
1965 V2 N3 P67-71 2103 <TC>
1965 V2 N11 P323-326 66-13806 <*>
1966 N2 P33-38 ICE V6 N3 P511-517 <ICE>

PLASTIQUES
1944 V2 N3 P80-86 58-2481 <*>

PLASTISCHE MASSEN IN WISSENSCHAFT UND TECHNIK
1937 V7 P172-174 57-1550 <*>

PLASTVARLDEN
1963 V13 N5 P476-481 <CCA>
65-13629 <*>

PLASTVERARBEITER
1956 V7 N9 P342-348 61-15991 <+*> 0
1957 V8 N1 P19-21 65-12807 <*>
1957 V8 N4 P149-151 64-18808 <*>
1957 V8 N6 P214-215 64-18809 <*>
1957 V8 N7 P260-261 64-18809 <*>
1957 V8 N9 P337-338 64-18810 <*>
1957 V8 N10 P381-383 64-18811 <*>
1957 V8 N12 P463-465 64-18812 <*>
1959 V10 N4 P137-143 60-18520 <*>
1960 V11 P241-249 65-12751 <*>
1960 V11 N10 P441-447 61-14476 <*> 0
1961 V12 N10 P454-458 63-18347 <*>
1961 V12 N11 P504-511 63-18390 <*>
1961 V12 N12 P559-563 63-14748 <*>
1962 V13 N1 P14-21 63-18388 <*>
1962 V13 N10 P527-529 64-16690 <*>
1962 V13 N12 P643-644 65-12096 <*>
1962 V13 N12 P651-656 63-18769 <*> 0
1963 V14 P91-98 63-18763 <*> 0
1963 V14 N10 P624-628 43S84G <ATS>
1963 V14 N10 P642-649 394 <TC>
1963 V14 N11/2 P689-696 125 <TC>
1964 V15 N3 P149-152 1566 <TC>
1965 V16 N5 P251-258 65-14955 <*>

★ POCHVOVEDENIE
1932 N2 P178-211 61-11499 <=>
1934 N3 P311-325 60-51072 <=>
1937 N6 P775-782 65-50081 <=$>
1937 N8 P1139-1159 65-50082 <=>
1938 N4 P595-605 R-4701 <*>
1938 V33 N2 P163-179 RT-1353 <*>
1939 N2 P39-41 66S82R <ATS>
1939 N2 P42-50 63-10065 <=*>
1939 N2 P81-82 63-10064 <=*>
1939 N7 P10-43 60-21832 <=>
1940 N7 P3-21 RJ-85 <ATS>
1943 N7 P17-23 RT-1359 <*>
1943 N8 P13-19 RT-2364 <*>
1943 N9/10 P30-36 60-21840 <=>
1944 N9 P393-409 60-21852 <=>
1944 N9 P426-432 60-21115 <=>
1944 N10 P491-499 RT-4213 <*>
1944 N4/5 P159-179 60-21841 <=>
1945 N2 P122-130 65-20005 <$>
1945 N7 P327-339 65-50056 <=>
1945 N8 P381-402 RT-1028 <*>
1945 N9 P539-549 RT-4636 <*>
1945 N3/4 P146-151 60-21842 <=>
1945 N3/4 P152-161 60-21850 <=>
1945 N3/4 P189-198 60-21854 <=>
1945 N3/4 P199-208 60-21114 <=>
1945 N5/6 P242-249 60-21849 <=>
1946 N6 P379-384 60-21846 <=>
1947 N3 P183-187 60-21130 <=>
1947 N5 P265-276 65-50106 <=$>
1947 N9 P533-548 RT-255 <*>
1947 N10 P617-624 RT-3663 <*>
1947 N12 P697-703 60-21847 <=>
1948 N1 P3-13 65-50057 <=>
1948 N4 P217-226 60-21839 <=>
1948 N4 P252-259 63-20818 <=*$>
1949 N9 P505-517 51/3476 <NLL>
1949 N10 P611-618 60-21848 <=>
1949 N11 P684-687 50/2800 <NLL>
1949 N12 P394-399 60-21078 <=>
1949 V11 P638-651 CSIRO-3394 <CSIR>
R-4583 <*>
1949 V11 P682-683 50/2804 <NLL>
1950 N3 P151-157 N65-27685 <=*$>
1951 N1 P57-59 RT-2039 <*>
1951 N8 P489-491 R-1534 <*>
1952 N1 P21-27 60-21831 <=>
1952 N2 P117-123 RT-961 <*>
1952 N2 P132-144 63-11082 <=>
1952 N7 P611-627 60-21845 <=>
1952 N7 P668-670 RT-2737 <*>
1952 N9 P851-859 61-31161 <=>
1952 N10 P936-944 63-11037 <=>
1952 N11 P1019-1026 65-50060 <=>
1953 N2 P52-59 RT-2094 <*>
1953 N5 P55-66 RT-1673 <*>
1953 N10 P79 RT-2414 <*>
1954 N3 P55-59 R-1578 <*>
1954 N3 P77-80 RT-4041 <*>
1954 N3 P94-96 RT-3998 <*>
1954 N7 P113-125 RT-4086 <*>
1954 N8 P52-64 61-11493 <=>
1954 N9 P23-34 RT-4205 <*>
1954 N9 P52-63 RT-3677 <*>
1954 N9 P64-71 RT-3678 <*>
1954 N10 P52-61 RT-3396 <*>
1954 N11 P71-79 R-4205 <*>
1955 N1 P12-15 RT-3840 <*>
1955 N1 P83-87 59-22521 <+*>
1955 N2 P73-78 61-11494 <=>
1955 N2 P85-87 R-4207 <*>
1955 N2 P92-93 R-3156 <*>
1955 N2 P94-96 RT-4621 <*>
1955 N4 P62-68 R-4206 <*>
1955 N4 P69-81 60-21139 <=>
1955 N5 P29-43 R-3646 <*>
1955 N6 P60-65 60-21872 <=>
1955 N6 P74-82 60-21887 <=>
1955 N7 P1-12 RT-4232 <*>
1955 N7 P74-83 R-4696 <*>
1955 N7 P92-93 R-3617 <*>
1955 N9 P17-24 CSIRO-3165 <CSIR>
R-4564 <*>
1955 N9 P25-36 R-4208 <*>
1955 N9 P25-35 R-4601 <*>
1955 N9 P37-44 R-3903 <*>
1955 N9 P49-55 CSIRO-3087 <CSIR>
R-4209 <*>
R-4590 <*>
1955 N9 P70-72 R-3937 <*>
1955 N10 P1-10 R-4385 <*>
1955 N11 P26-35 RT-4358 <*>
1955 N11 P86-90 63-23240 <=*>
1955 N11 P100-103 RT-4625 <*>
1956 N1 P50-53 59-22074 <+*> 0

Citation	Translation
1956 N1 P70-88	61-11496 <=>
1956 N2 P27-41	CSIRO-3264 <CSIR>
	R-3908 <*>
1956 N2 P75-89	29J17R <ATS>
1956 N3 P18-30	61-19850 <=*>
1956 N4 P1-23	60-21134 <=>
1956 N4 P36-43	60-21838 <=>
1956 N4 P59-69	60-51128 <=>
1956 N5 P17-24	65-50068 <=$>
1956 N5 P25-30	02N55R <ATS>
1956 N6 P76-81	61-31001 <=>
1956 N6 P82-102	61-11495 <=>
1956 N8 P49-56	60-21121 <=>
1956 N8 P57-73	R-1562 <*>
	65-63365 <=*$>
1956 N10 P69-90	60-21122 <=>
1956 N12 P73-75	R-3625 <*>
1957 N1 P76-81	R-4157 <*>
1957 N1 P107-110	60-21900 <=>
1957 N1 P124-128	60-51183 <=>
1957 N3 P19-31	65-50067 <=>
1957 N3 P40-47	60-21908 <=>
1957 N3 P110-116	R-3934 <*>
1957 N6 P24-34	R-4391 <*>
1957 N6 P74-80	65-50059 <=>
1957 N6 P102-107	60-21899 <=>
1957 N7 P91-98	62-32139 <=*>
1957 N8 P46-53	60-21883 <=>
1957 N8 P60-65	R-4149 <*>
1957 N9 P10-19	61-11497 <=>
1957 N9 P37-48	63-11080 <=>
1957 N9 P70-78	R-4881 <*>
1957 N10 P1-15	60-21141 <=>
1957 N10 P16-32	60-51026 <=>
1957 N11 P1-13	60-21138 <=>
1957 N12 P1-19	60-21113 <=>
1957 N12 P45-51	60-21140 <=>
1957 N12 P62-71	62-13926 <*=>
1957 N12 P90-97	60-21877 <=>
1958 N4 P61-66	59-14996 <+*>
1958 N10 P28-37	60-15332 <+*> 0
1958 N12 P28-35	60-17503 <+*>
1959 N9 P12-21	61-27150 <*=>
1959 N9 P118-123	61-27195 <*=>
1960 N4 P125-127	60-11987 <=>
1960 N5 P21-29	62-25138 <=*>
1960 N8 P84-86	65-29151 <$>
1960 N9 P114-115	61-11814 <=>
1962 N1 P74-83	63-19152 <=*>
1962 N4 P103-108	62-32893 <=*>
1962 N11 P109-114	66-51088 <=>
1963 N1 P75-83	64-23524 <=$>
1965 N8 P81-88	65-33762 <=*$>

PODSHIPNIK

Citation	Translation
1953 N5 P13-19	62-20457 <=*>
1953 N8 P1-5	60-10260 <+*> 0

PODZEMNAYA GAZIFIKATSIYA UGLEI

Citation	Translation
1957 N1 P3-15	R-5165 <*>
1957 N1 P46-51	59-15134 <+*>
1957 N1 P57-60	59-18326 <+*>
1957 N2 P27-31	59-22338 <+*>
1957 N2 P31-38	62-13052 <*=>
1957 N2 P65-67	R-5152 <*>
1957 N2 P79-86	59-19318 <+*>
1957 N3 P13-16	59-10204 <+*>
1957 N3 P16-21	59-10203 <+*>
1957 N3 P38-43	59-10202 <+*>
1957 N3 P59-65	60-23031 <+*>
1957 N4 P3-7	59-10201 <+*>
1957 N4 P11-15	59-16380 <+*>
1957 N4 P17-19	59-10205 <+*>
1957 N4 P24-28	59-16377 <+*>
1957 N4 P31-39	59-19319 <+*>
1957 N4 P39-41	59-16379 <+*> 0
1957 N4 P49-50	R-4480 <*>
1957 N4 P67-69	59-22475 <+*>
1957 V1 P65-67	R-5167 <*>
1957 V1 N1 P65-67	59-18333 <+*>
1957 V1 N1 P88-89	59-18344 <+*>
1957 V1 N2 P65-67	59-18336 <+*>
1957 V1 N2 P67-74	59-18327 <+*>
1957 V1 N4 P15-	59-18355 <+*>
1957 V1 N4 P51-55	59-18380 <+*>
1957 V1 N4 P63-67	59-10206 <+*>
1958 N1 P31-34	59-16737 <+*>
1958 N2 P25-31	59-16378 <+*>
1958 N2 P31	59-16736 <+*>
1958 N2 P32-35	59-10798 <+*>
1958 N2 P38-43	59-10797 <+*>
1958 N2 P51-	59-15133 <+*>
1958 N3 P43-49	62-33774 <=*>
1958 V2 N1 P41-43	59-18375 <+*>
1958 V2 N1 P43-47	59-18364 <+*>
1958 V2 N1 P51-54	59-18391 <+*> 0
1958 V2 N5 P43-45	59-22550 <+*>
1959 N1 P3-9	60-16588 <+*>
1959 N2 P73-77	61-10728 <+*>
1959 N3 P8-10	60-16587 <+*>
1959 N4 P69-72	60-16959 <+*>
1966 N37 P9	66-34788 <=$>

POGLED

Citation	Translation
1966 N36 P12	66-34788 <=$>

POKROKY MATEMATIKY, FYSIKY A ASTRONOMIE

Citation	Translation
1957 V2 P534-543	AEC-TR-4548 <*>
1957 V2 P668-674	AEC-TR-4548 <*>

POKROKY PRASKOVE METALURGIE VUPM

Citation	Translation
1963 N1 P14-24	5731 <HB>
1963 V1 N1 P47-59	6572 <HB>

POLAND, STANDARDS

Citation	Translation
PN-53-B-03150	70 <DTS>
PN-63/C-84123	65-11292 <*>

POLICLINICO. SEZIONE MEDICA

Citation	Translation
1946 V53 P197-205	64-16010 <*> 0
1956 V63 P24-34	1222 <*>

POLICLINICO. SEZIONE PRATICA

Citation	Translation
1950 V57 N32 P1051-1052	2323A <K-H>
1954 V61 N29 P847-851	3306-C <K-H>
1955 V62 P1229-1232 PT2	57-205 <*>
1961 V68 P37-44	62-20424 <*>
1961 V68 N4 P109-113	61-01006 <*>
1963 V70 P1038-1041	65-14269 <*>

POLIGRAFICHESKOE PROIZVODSTVO

Citation	Translation
1957 N8 P5-8	9069 <IICH>
1958 N1 P18-20	60-14041 <+*>
1961 N8 P14-16	64-10911 <=*$>
1962 N1 P11-14	63-18863 <=*$>
1962 N1 P23-25	63-20563 <=*>
1962 N2 P18-19	62-18876 <=*>
1962 N2 P19-20	62-18880 <=*>
1962 N4 P20-21	64-10984 <=*$>
1962 N4 P21-23	62-32867 <=>
1962 N4 P25-26	64-10912 <=*$>
1962 N9 P11-13	14P66R <ATS>
1962 V38 N4 P21-23	62-18677 <=*>
1963 N1 P10-13	NLLTB V5 N10 P926 <NLL>
1963 N6 P3	63-31504 <=>
1963 N7 P4-6	63-31853 <=>
1963 N10 P22-23	66-13746 <*>
1963 N11 P13-14	66-12458 <*>
1963 N11 P18-19	66-13747 <*>
1963 N12 P13-14	66-12448 <*>
1963 N12 P14-16	66-13737 <*>
1964 N4 P19-21	66-13739 <*>
1964 N10 P38-39	66-13738 <*>

POLIGRAFIYA

Citation	Translation
1964 N2 P12-14	66-12543 <*>
1964 N4 P18-19	66-12544 <*>
1964 N4 P21-23	66-12551 <*>
1964 N5 P18-19	66-12545 <*>
1964 N6 P16-18	66-12542 <*>

1964 N6 P18 66-12546 <*>
1964 N7 P15-16 66-12547 <*>
1964 N7 P16-17 66-12548 <*>
1964 N7 P19-20 66-13759 <*>
1964 N9 P28-30 66-12549 <*>
1964 N9 P33-35 66-13758 <*>
1964 N10 P31-33 66-13761 <*>
1964 N10 P33-35 66-12449 <*>
1965 N6 P25-26 66-12550 <*>

POLIGRAFIYA ZA RUBEZHOM
1964 N10 P138-139 50T90R <ATS>

POLIMERY (TWORZYWA WIELKOCZASTEZKOWE) TITLE VARIES
1962 V7 P199-206 65-10268 <*>
1962 V7 P210-215 NS-54 <TTIS>
1962 V7 P245-248 6001 <TTIS>
1962 V7 N4 P125-128 43Q68P <ATS>
1962 V7 N11 P410-414 66-10047 <*>
88Q72P <ATS>
1962 V7 N11 P415-418 NS 213 <TTIS>
1962 V7 N7/8 P255-256 64-10779 <*>
1963 V8 P287-290 NS-193 <TTIS>
1963 V8 N1 P1-3 63-31061 <=>
1963 V8 N2 P69-72 RAPRA-1268 <RAP>
1963 V8 N4 P151-154 2039 <TC>
1963 V8 N11 P409-411 NS-202 <TTIS>
1963 V8 N11 P420-423 NS-210 <TTIS>
1963 V8 N7/8 P302-303 49S81P <ATS>
1964 V9 P15-17 65-10474 <*>
1964 V9 P315-317 2133 <TC>
1964 V9 N1 P15-17 NS-228 <TTIS>
1964 V9 N2 P48-49 66-11426 <*>
1964 V9 N3 P103-107 66-11772 <*>
1964 V9 N6 P221-225 61S85P <ATS>
1964 V9 N9 P353-356 66-12904 <*>
1964 V9 N9 P393-396 65-30244 <=$>
1964 V9 N7/8 P289-293 66-10825 <*>
1965 V10 P97-101 NS-453 <TTIS>
1965 V10 P101-103 NS-454 <TTIS>
1965 V10 P199-202 NS-555 <TTIS>
1965 V10 N2 P63-66 NS-516 <TTIS>

POLIPLASTI E PLASTICI RINFORZATI
1962 V10 N57 P5-7,16 1912 <TC>
1962 V10 N57 P5-7 66-13785 <*>
1962 V10 N57 P16 66-13785 <*>
1964 V12 P36-39 65-13988 <*>
1964 V12 N76 P5-10 66-10543 <*>

POLITEKHNICHESKOYE OBUCHENIYE
1957 N6 P66-79 61-19182 <+*> 0
1959 N7 P75-80 60-11645 <=>

POLITICHESKOE SAMOOBRAZOVANIE
1959 V3 N4 P75-86 60-31483 <=>
1959 V3 N12 P57-60 60-31204 <+*>
1965 V9 N2 P73-80 65-32556 <=$>
1965 V9 N7 P12-22 65-33059 <=$>
1966 V10 N7 P99-102 66-34515 <=$>

POLITYKA
1965 N36 P1 65-32523 <=$>

POLIZEIBEAMTE
1959 V20 P308-312 60-31069 <=*> 0

POLIZEI TECHNIK VERKEHR
1959 N5 P1-8 60-18125 <*>
9892 <K-H>

POLJOPRIVREDA
1955 V3 N4 P5-9 61-11211/1-9 <=>
1955 V3 N7/8 P17-23 61-11211/1-9 <=>
1957 V5 N3 P14-25 61-11211/1-9 <=>
1957 V5 N10 P43-50 61-11211/1-9 <=>
1958 V6 N2 P54-74 61-11211/1-9 <=>
1958 V6 N10 P13-26 61-11211/1-9 <=>
1964 V12 N8 P4-11 64-51300 <=$>
1966 V14 N12 P49-52 67-30943 <=$>

POLJOPRIVREDNA TEHNIKA
1964 V2 N2 P31-37 TR.176 <NIAE>

POLSKA GAZETA LEKARSKA
1935 V14 N17 P309-312 2419 <*>

POLSKI PRZEGLAD RADIOLOGICZNY
1959 V23 N1 P15-18 63-10716 <*> 0

POLSKI TYGODNIK LEKARSKI
1950 V5 P1654-1657 2104 <K-H>
1951 V6 N1/2 P18-22 II-834 <*>
1952 V7 N44 P1405-1407 2411 <*>
1953 V8 N15 P553-556 3166-C <K-H>
1953 V8 N41 P1413-1416 II-835 <*>
1955 V10 N1 P6-7 3596-B <K-H>
1955 V10 N6 P191 59-17024 <*>
1956 V11 N2 P65-68 1206 <*>
1956 V11 N16 P705-717 5538-A <K-H>
1956 V11 N49 P2084-2087 6424 <K-H>
1956 V11 N52 P2183-2188 58-1802 <*>
1957 V12 N31 P1181-1184 63-16944 <*> 0
6999-A <K-H>
1957 V12 N49 P1881-1885 64-14518 <*>
8535-C <K-H>
1958 V13 N12 P428-430 62-00536 <*>
1958 V13 N26 P994-998 65-00367 <*> 0
1958 V13 N45 P1788-1793 59-20717 <*> 0
1959 V14 N43 P1904-1908 10022 <K-H>
1959 V14 N47 P2064-2067 10148 <K-H>
64-10415 <*>
1959 V14 N52 P2279-2282 64-10607 <*>
1960 V15 P1106-1110 10643-C <K-H>
1960 V15 N8 P3-15 60-31880 <=*>
1960 V15 N9 P326-330 60-31290 <=*>
1960 V15 N33 P1271-1276 62-19305 <*=>
1960 V15 N34 P1321-1327 62-19312 <*=>
1960 V15 N35 P1366-1369 62-19312 <*=>
1960 V15 N36 P1382-1384 62-19300 <=*>
1960 V15 N49 P1898-1900 62-19311 <=*>
1961 V16 P287-292 11170-B <K-H>
1961 V16 N3 P87-94 62-19299 <=*>
1961 V16 N13 P641-643 62-19477 <=*>
1961 V16 N14 P517-522 62-19475 <=*>
1961 V16 N15 P566-570 62-19474 <=*>
1961 V16 N16 P592-595 62-19476 <*=>
1962 V17 N1 P1-7 20Q71P <ATS>
1962 V17 N13 P489-490 64-20915 <*>
1962 V17 N28 P1093-1098 65-14946 <*>

POLSKIE ARCHIWUM HYDROBIOLOGII
1956 V3 P166-173 65-50370 <$>
1957 V4 P221-250 63-11399 <=>
1959 V6 P117-124 65-50369 <=>

POLSKIE ARCHIWUM MEDYCYNY WEWNETRZNEJ
1955 V25 P149-152 57-3508 <*>
1956 V26 N3 P393-402 6518-A <K-H>
1956 V26 N4 P547-552 61-16129 <*>
1960 V30 N7 P923-926 63-18476 <*>
1960 V30 N7 P1004-1005 10809-A <K-H>
63-16940 <*>

POLSKIE PISMO ENTOMOLOGICZNE
1926 V5 N1/2 P104-108 61-31261 <=>
1927 V6 N1/2 P100-102 61-31260 <=>
1955 V25 N14 P213-226 60-21405 <=>
1958 V27 N2 P21-36 60-21562 <=>
1958 V27 N4 P39-69 60-21402 <=>
1958 V27 N5 P71-73 60-21564 <=>
1958 V27 N11 P109-113 60-21563 <=>

POLUPROVODNIKOVYE PRIBORY I IKH PRIMENENIE: SBORNIK STATEI
1960 N6 P63-91 62-13006 <*=>
1960 N6 P102-124 61-28494 <*=> 0
1960 V4 P298-307 63-23103 <=*$>

POLYARNYE SIVANIYA I SVECHENIE NOCHNOGO NEBA:

SBORNIK STATEI
1960 N4 ENTIRE ISSUE 62-13339 <=>
1960 N2/3 ENTIRE ISSUE 61-27714 <=>
1961 N5 P29-31 61-31630 <=>
1963 N9 P59-60 AD-631 580 <=$>

POLYMER BULLETIN. CHINESE PEOPLES REPUBLIC
SEE KAO FEN TZU T'UNG HSUN

POLYTECHNISCH TIJDSCHRIFT
1954 V9A N31/2 P644-650 C-2514 <NRC>
57-3478 <*>
1963 V18A N24 P1101-1103 65-14665 <*> O

POLYTECHNISCHES ZENTRALBLATT
1875 S3 V29 P49-53 63-10428 <*>

POMIARY, AUTOMATYKA, KONTROLA
1960 V6 N1 P20-21 62-23538 <=*>
1961 V7 N1 P19-20 64-11949 <=>

POMORSTVO
1966 N11/2 P318 67-30446 <=>
1967 N1 P5-6 67-31082 <=$>
1967 N1 P13-15 67-31082 <=$>

POPULAR SCIENCE. CHINESE PEOPLES REPUBLIC
SEE K'O HSUEH TA CHANG

POPULATION. PARIS
1946 P623-642 2268 <*>
1946 V1 P91-98 59-17249 <*>
1957 V12 N3 P413-444 62-00638 <*>
1961 V16 N1 P27-48 62-00668 <*>
1961 V16 N2 P261-282 62-00639 <*>

PORAROGUAFII. JAPAN
1957 V5 P41-59 58-1292 <*>

★ POROSHKOVAYA METALLURGIYA
1961 V1 N1 P20-29 AEC-TR-5604 <=*>
62-32494 <=*>
1961 V1 N1 P43-49 5579 <HB>
1961 V1 N1 P61-67 5580 <HB>
63-19783 <*=> O
1961 V1 N1 P68-74 5581 <HB>
1961 V1 N1 P82-91 5995 <HB>
1961 V1 N2 P3-13 AEC-TR-5572 <=*>
63-23224 <=*$>
1961 V1 N2 P101-107 63-21846 <=> O
1961 V1 N2 P108-115 64-13382 <=*$>
1961 V1 N3 P40-46 64-41271 <=>
1961 V1 N4 P63 63-19398 <=*>
1961 V1 N4 P80-85 63-18486 <=*> O
1961 V1 N4 P86-102 63-19398 <=*>
1961 V1 N5 P69-73 5589 <HB>
62-25652 <=*>
1961 V1 N6 P55-60 5940 <HB>
1961 V1 N6 P70-74 5555 <HB>
1962 V2 N1 P50-55 63-19301 <=*>
1962 V2 N1 P88-91 5601 <HB>
1962 V2 N2 P27-37 63-19925 <=*>
1962 V2 N3 P80-85 64-15827 <=*$>
1962 V2 N4 P10-19 63-13500 <=*>
1962 V2 N5 P60-67 5999 <HB>
1962 V2 N5 P84-88 5847 <HB>
1962 V2 N5 P89-98 63-21927 <=>
64-23548 <=>
1962 V2 N6 P3-11 64-00465 <*>
64-11832 <=>
1962 V2 N6 P12-13 64-00466 <*>
64-11826 <=>
1962 V2 N6 P50-53 5975 <HB>
1962 V2 N6 P72-80 AD-610 349 <=$>
1963 N1 P59-76 AD-619 568 <=$>
1963 V3 N1 P3-12 63-31343 <=>
1963 V3 N1 P17-25 1107 <TC>
1963 V3 N1 P33-34 64-13223 <=*$>
1963 V3 N1 P42-47 5981 <HB>
1963 V3 N1 P42-48 63-23921 <=*$>
1963 V3 N1 P49-53 6238 <HB>
1963 V3 N1 P54-59 63-23920 <=*$>
1963 V3 N1 P76-78 6102 <HB>
1963 V3 N1 P79-82 5996 <HB>
1963 V3 N1 P93-111 63-31343 <=>
1963 V3 N2 P3-7 64-15256 <=*$>
1963 V3 N2 P22-25 64-13091 <=*$>
1963 V3 N2 P26-30 5997 <HB>
1963 V3 N2 P65-79 65-10875 <*>
1963 V3 N3 P25-29 64-13214 <=*$>
1963 V3 N3 P30-36 6073 <HB>
1963 V3 N3 P52-56 64-13217 <=*$>
1963 V3 N3 P99-103 6115 <HB>
64-21018 <=>
1963 V3 N3 P11C 63-41026 <=>
1963 V3 N4 P3-5 64-21149 <=*$>
1963 V3 N4 P40-48 65-60816 <=>
1963 V3 N4 P49-53 6090 <HB>
1963 V3 N5 P3-9 64-21237 <=>
1963 V3 N5 P15-20 6186 <HB>
65-61344 <=$>
1963 V3 N5 P21-27 6187 <HB>
1963 V3 N5 P43-46 64-21237 <=>
1963 V3 N5 P68-76 64-21237 <=>
1963 V3 N6 P24-30 6249 <HB>
1963 V3 N6 P36-37 6252 <HB>
1963 V3 N6 P46-50 64-19146 <=$>
1963 V3 N6 P54-56 6257 <HB>
1963 V3 N6 P88-93 6256 <HB>
1964 V4 N1 P34-41 65-61319 <=$>
1964 V4 N1 P81-89 65-61433 <=>
1964 V4 N1 P91-95 6264 <HB>
1964 V4 N1 P111-112 65-61186 <=>
1964 V4 N3 P23-28 6331 <HB>
1964 V4 N5 P16-21 65-11279 <*>
6564 <HB>
1964 V4 N5 P52-56 M-5630 <NLL>
1964 V4 N5 P98-101 6509 <HB>
1964 V4 N6 P12-16 M-5678 <NLL>
6462 <HB>
1964 V4 N6 P96-97 6461 <HB>
1965 N1 P60-70 N65-27706 <=*$>
1965 N2 P41-49 AD-637 381 <=$>
1965 N5 P54-57 NLLTB V8 N3 P194 <NLL>
1965 N9 P65-68 NLLTB V8 N3 P194 <NLL>
1965 N10 P85-90 N66-23728 <=$>
1965 V5 N2 P1-3 6530 <HB>
1965 V5 N2 P3-8 6531 <HB>
1965 V5 N2 P27-31 M-5699 <NLL>
1965 V5 N2 P104-107 6539 <HB>
1965 V5 N4 P9-11 M-5747 <NLL>
1965 V5 N5 P9-16 6570 <HB>
1965 V5 N5 P17-19 6615 <HB>
1965 V5 N6 P1-4 6616 <HB>
1965 V5 N7 P67-73 65-17425 <*>
1965 V5 N7 P97-99 6759 <HB>
1965 V5 N7 P100-107 65-14762 <*>
6684 <HB>
1965 V5 N8 P19-22 6656 <HB>
1965 V5 N8 P82-86 6655 <HB>
1965 V5 N9 P104-109 65-23201 <$>
1965 V5 N10 P75-79 AD-641 930 <=$>
1965 V5 N11 P52-54 6812 <HB>
1965 V5 N11 P83-86 6813 <HB>
1966 N1 P81-84 6808 <HB>
1966 N2 P31-39 6872 <HB>
1966 N2 P97-99 6871 <HB>

POSEBNA IZDANJA. BIOLOSKI INSTITUT N. R. SRBIJE. BEOGRAD
1958 V3 ENTIRE ISSUE 61-11203 <=>

POSEBNA IZDANJA. INSTITUT ZA FIZIOLOGIJU RAZVICA, GENETIKU I SELEKCIJU. SRPSKA AKADEMIJA NAUKA
1951 V182 ENTIRE ISSUE 60-21630 <=>

POSEBNA IZDANJA. MATEMATICKI INSTITUT, SRPSKA AKADEMIJA NAUKA
1953 V5 P45-52 60-21707 <=>

POSEBNE PUBLIKACIJE. JUGOSLOVENSKA DRUSTVO ZA PROUCAVANJE ZEMLJISTA
1956 N4 P9-73 60-21616 <=>

POSTARCHIV
1940 V68 N4 P645-684 66-13295 <*> 0

POSTEPY BIOCHEMII
1959 V5 N1 P47-65 60-31017 <=>
1960 V6 P43-58 UCRL TRANS-626 <*>
1961 V7 N2 P289-298 85N56P <ATS>
1965 V11 N2 P247-263 66-23826 <$>

POSTEPY FIZYKI
1954 V5 N2 P212-219 62-01133 <*>
1958 V9 N3 P261-279 65N57P <ATS>
1961 V12 N1 P71-87 62-19513 <=*>
1961 V12 N1 P89-97 01P61P <ATS>
1961 V12 N4 P443-466 62-01371 <*>
1961 V12 N4 P467-476 CSIR-TR.258 <CSSA>
1961 V12 N4 P486-488 62-19584 <*=>
1964 V15 N3 P287-322 AD-631 767 <=$>
1965 V16 N1 P25-40 AD-637 058 <=$>
1965 V16 N3 P257-278 66-13526 <*>

POSTEPY HIGIENY I MEDYCYNY DOSWIADCZALNEJ
1958 V12 N3 P241-262 61-11321 <=>
1959 V13 N3 P315-317 65-00227 <*>
1959 V13 N6 P787-803 60-41088 <*=>
1960 V14 N4 P381-403 61-19425 <*=>
1961 V15 N3 P285-296 62-19544 <=*>

POSTEPY NAUKI ROLNICZEJ
1956 V3 N2 P73-76 60-21538 <=>
1956 V3 N6 P11-20 62-00829 <*>
1959 V6 N2 P61-73 60-21568 <=>
1960 V7 N1 P3-20 60-31748 <*=>
1962 V9 N2 P109-116 12596 <K-H>

POUMON
1953 V9 P77-88 2053 <*>

POVRATAK U ZIVOT
1956 V4 N4 P8-14 57-110 <*>

POZHARNOE DELO
1957 V3 N6 P9 62-24899 <=*>
1962 V8 N10 P12 63-18504 <=*>
1963 V9 N11 P8-11 RTS-2564 <NLL>
1963 V11 P21-23 65-60351 <=$>
1964 N10 P12-13 RTS-3026 <NLL>

PRACE BADAWCZE INSTYTUTU BADAWCZEGO LESNICTWA. WARSAW
1946 N47 P128- 61-11339 <=>
1953 N94 P1-96 60-21235 <=>
1961 N210 P103-162 65-50351 <=>
1963 N259 P61-69 65-50357 <=>

PRACE BRNENSKE ZAKLADNY, CESKOSLOVENSKE AKADEMIE VED
1959 V31 P148-156 AEC-TR-5495 <*>

PRACE GIMO
1950 V2 P93-99 65-14794 <*> 0
1950 V2 P101-103 65-14808 <*>
1951 N3 P173-181 60-18853 <*>
1951 N4 P267-277 60-18854 <*>
1951 V3 P17-22 59-17496 <*>
1951 V3 P115-148 58-1148 <*>
1952 V4 N2 P99-108 60-18838 <*>

PRACE GEOFYSIKALNIHO USTAVU CESKOSLOVENSKE AKADEMIE VED
1962 N162 P11-23 64-21174 <=>

PRACE GEOGRAFICZNE
1962 N34 P1-185 65-50312 <=>

PRACE GLOWNEGO INSTYTUTU GORNICTWA
1947 N17 P1-9 60-21296 <=>
1947 N19 ENTIRE ISSUE 60-21280 <=>
1947 N19 P1-19 60-21280 <=>
1947 N20 P1-42 60-21299 <=>
1948 N23 P1-82 60-21291 <=>
1948 N29 P3-32 60-21266 <=>
1948 N31 ENTIRE ISSUE 60-21280 <=>
1948 N33 P1-30 60-21276 <=>
1948 N34 P1-6 60-21293 <=>
1948 N37 ENTIRE ISSUE 60-21530 <=>
1949 N42 ENTIRE ISSUE 60-21528 <=>
1949 N43 ENTIRE ISSUE 60-21529 <=>
1949 N45 P3-10 60-21282 <=>
1949 N46 P1-16 60-21302 <=>
1949 N53 P1-9 6021300 <=>
1949 N59 P1-15 60-21353 <=>
1950 N62 P1-37 60-21278 <=>
1950 N68 ENTIRE ISSUE 60-21368 <=>
1950 N69 P1-14 60-21355 <=>
1951 N78 P1-4 60-21288 <=>
1951 N79 P1-16 60-21298 <=>
1951 N88 P1-21 60-21365 <=>
1951 N93 P1-28 60-21275 <=>
1951 N100 P1-19 60-21363 <=>
1951 N101 ENTIRE ISSUE 61-11352 <*>
1951 N103 P1-12 60-21356 <=>
1951 N106 ENTIRE ISSUE 60-21362 <=>
1952 N85 P1-10 60-21301 <=>
1952 N110 P1-12 60-21292 <=>
1952 N111 P1-20 60-21366 <=>
1952 N116 ENTIRE ISSUE 60-21369 <=>
1952 N119 P1-12 60-21277 <=*>
1952 N121 P3-21 60-21290 <=>
1952 N123 ENTIRE ISSUE 60-21303 <=>
1952 N123 P1-9 60-21303 <=>
1952 N125 ENTIRE ISSUE 60-21272 <=>
1952 N126 P1-9 60-21367 <=>
1953 N136 P1-26 60-21371 <=>
1953 N137 ENTIRE ISSUE 60-21268 <=>
1953 N138 P1-39 60-21360 <=>
1953 N139 P1-38 60-21354 <=>
1953 N144 ENTIRE ISSUE 60-21264 <=>
1954 N153 P1-16 60-21357 <=>
1954 N156 P1-10 60-21304 <=>
1954 N159 P1-34 61-11346 <=>
1954 N161 P1-16 60-21361 <=>
1954 N162 P3-20 61-11351 <*>
1954 N163 P1-15 60-21370 <=>
1954 N164 ENTIRE ISSUE 60-21359 <=>
1954 N167 P3-32 63-11384 <=>
1954 SB N142 P3-10 T-1922 <INSD>
1954 SB N152 P2-16 T-1923 <INSD>
1954 SB N162 ENTIRE ISSUE 59-18366 <=>
1955 N169 P3-30 63-11384 <=>
1955 N172 ENTIRE ISSUE T-1628 <NLL>
1956 N183 P1-15 60-21358 <=>
1956 SB N189 P1-10 58-1737 <*>
1957 N197 P3-11 63-11384 <=>
1957 N198 P2-24 63-11384 <=>
1958 N208 ENTIRE ISSUE 60-21270 <=>
1958 N211 ENTIRE ISSUE 60-21271 <=>
1958 N213 P3-11 60-21269 <=>
1958 N214 P1-12 60-21295 <=>
1958 N215 P1-18 60-21274 <=>
1958 N220 P3-28 61-11350 <=>
1958 N226 P3-44 61-11356 <=>
1959 N227 P3-12 63-11384 <=>
1959 N228 P3-7 63-11384 <=>
1959 N231 P3-23 63-11384 <=>
1959 N241 P1-15 61-11349 <=>
1960 N242 P3-12 61-11350 <=>
1960 N244 P1-8 61-11349 <=>
1960 N245 P3-12 63-11384 <=>
1960 N247 P2-16 63-11384 <=>
1960 N255 P3-15 63-11384 <=>
1960 N261 P1-17 61-31322 <=>

PRACE. GLOWNY INSTYTUT PRZEMYSLU ROLNEGO I SPOZYWCZEGO
1958 V8 N2 P105-111 60-21570 <=>

PRACE INSTYTUTOW HUTNICZYCH. KATOWICE
1953 V5 P89-104 5866-C <KH>
1953 V5 N2 P89-104 64-18395 <*> 0
1955 N7 P1C1-105 3707 <HB>
1955 V7 P30-34 3706 <HB>
1955 V7 N5/6 P270-275 3601 <HB>
1956 V8 N2 P85-88 4436 <HB>
1956 V8 N8 P207-214 3890 <HB>
1958 V10 P333-347 AEC-TR-6159 <*>
1958 V10 N2 P120-123 1259 <BISI>
1958 V10 N3 P180-183 1202 <BISI>
1959 V11 P229-234 61-31319 <=>
1959 V11 N1 P1-9 63-11382 <=>
1959 V11 N1 P11-17 1818 <BISI>
1959 V11 N2 P71-74 5214 <HB>
1961 V13 N1 P27-36 2969 <BISI>
1961 V13 N3 P153-172 3009 <BISI>
1962 V14 P75-80 UCRL TRANS-953(L) <*>
1962 V14 P285-289 6038 <HB>
1964 V16 N3 P113-125 460C <BISI>
1965 V17 P219-223 4824 <BISI>

PRACE INSTYTUTOW MINSTERSTWA HUTNICTWA
SEE PRACE INSTYTUTOW HUTNICZYCH

PRACE INSTYTUTU CELULOZOWO-PAPIERNCZEGO
1955 V4 N2 P1-10 61-11381 <=>
1955 V4 N2 P55-60 61-11389 <=>
1957 V6 N1 P16-30 61-31270 <=>

PRACE INSTYTUTOW I LABORATORIOW BADAWCZYCH PRZEMYSLU SPOZYWCZEGO
1953 V3 P52-54 588 <TC>
1958 V8 N5 P55-66 61-11383 <=>
1960 V10 N2 P69-83 61-11369 <=>

PRACE INSTYTUTU LOTNICTWA
1964 N23 P36-54 66-12148 <*>

PRACE. INSTYTUTU MASZYN PRZEPLYWOWYCH. POLSKA AKADEMIA NAUK
1963 N13 P19-36 AD-608 452 <=>

PRACE INSTYTUTU MECHANIKI PRECYZYJNEJ
1960 V8 N29 P1-21 AD-617 550 <=$>
1961 V9 N34 P11-24 AD-610 050 <=$>
1961 V9 N34 P37-50 64-71576 <=>
1962 V10 N11 P16-31 3688 <BISI>
1962 V10 N35 P1-15 M-5411 <NLL>
1965 V13 N1 P22-24 NEL-TRANS-1724 <NLL>
1965 V13 N1 P25-28 NEL-TRANS-1723 <NLL>

PRACE. INSTYTUTU METALURGII
1951 V3 P297-304 6459 <HB>
1961 N3 P149-160 6458 <HB>

PRACE INSTYTUTU NAFTOWEGO. KATOWICE
1963 N84 P3-10 67R77P <ATS>

PRACE INSTYTUTU OCLEWNICTWA. WARSAW
1955 V5 N1 P4-14 64-10586 <*>
1955 V5 N3 P84-88 64-10587 <*>
1956 V6 N1/2 P10-21 62-10212 <*>

PRACE INSTYTUTU PRZEMYSLU WLOKIEN LYKOWYCH
1958 V6 P27-43 61-11392 <=>
1958 V6 P45-65 61-11338 <=>
1958 V6 P91-105 61-31269 <=>
1959 V7 P153-158 61-11380 <=>
1962 V9 P11-20 61-11392 <=>

PRACE INSTYTUTU TELE- I RADIOTECHNICZNEGO
1961 V5 N4 P85-90 65-18013 <*>

PRACE INSTYTUTU WLOKIENNICTWA
1959 V9 N2 P39-58 60-21577 <=*>

PRACE KOMISJI MATEMATYCZNO-PRZYRODNICZEJ
1938 SA P290-303 63-14441 <*>
1959 V7 N9 P79-96 20Q74P <ATS>

PRACE. LODZKIE TOWARZYSTWO NAUKOWE. WYDZIAL 3. NAUKI MATEMATYCZNO-PRZYRODNICZE
1954 N29 P1-30 60-21486 <=>

PRACE MATEMATYCZNE. POLSKIE TOWARZYSTWO MATEMATYCZNE. WARSZAWA
1960 V4 P7 66-12361 <*>

PRACE MORSKIEGO INSTYTUTU RYBACKIEGO W GDYNI
1957 N9 P79-102 60-21520 <=>
1957 N9 P549-563 60-21306 <=>
1957 N9 P597-632 60-21373 <=>
1957 N9 P633-679 60-21534 <=>
1957 N9 P681-703 60-21294 <=>
1957 V17 N9 P45-78 61-11366 <=>
1957 V17 N9 P103-150 61-11364 <=>
1957 V17 N9 P151-173 60-21283 <=>
1957 V17 N9 P221-246 61-11361 <=>
1957 V17 N9 P313-379 60-21297 <=>
1957 V17 N9 P427-437 65-50374 <=>
1957 V17 N9 P565-596 60-21364 <=>
1957 V17 N9 P705-738 61-11363 <=>
1959 N10A P441-459 61-11365 <=>
1959 N10/A P35-52 65-50368 <=>
1959 V19 N10A P53-67 61-11368 <=>
1959 V19 N10A P361-374 61-11359 <=>
1959 V19 N10A P383-402 61-11360 <=>
1959 V19 N10/A P351-359 61-11358 <=>

PRACE NAUKOWE INSTYTUTU OCHRONY ROSLIN
1959 V1 N1 P7-44 60-21574 <=>
1959 V1 N1 P107-134 60-21574 <=>
1959 V1 N1 P231-273 60-21574 <=>
1959 V1 N2 P7-167 60-21573 <=>
1959 V1 N3 P5-27 60-21574 <=>
1959 V1 N3 P37-72 60-21574 <=>
1959 V1 N3 P75-91 60-21574 <=>
1959 V1 N3 P95-131 60-21574 <=>
1960 V2 N1 P57-86 60-21533 <=>
1962 V4 N1 P53-100 65-50325 <=>
1962 V4 N1 P155-176 65-50327 <=>

PRACE P.I.T. PRZEMYSLOWY INSTYTUT TELEKOMUNIKACJI
1957 V8 N22 P11-13 58-1533 <*>
1958 V8 N24 P41-46 66-12403 <*>
1963 V13 N30 P1-8 64-71603 <=>
1965 V15 N5 P129-150 65-31805 <=$>

PRACE RADY NAUKOWO-TECHNICZNEJ HUTY. KRAKOW
1964 N14 P3-12 6674 <HB>
1965 N17 P4753 6820 <HB>
1965 V16 P3-7 06676 <HB>

PRACE SLOVENSKEJ AKADEMIE VIED. II SEKCIE. SERIA BIOLOGICKA
1955 V1 N5 P5-24 83P62C <ATS>

PRACE I STUDIA KOMITET GOSPODARKI WODNEZ POLSKA AKADEMIIA NAUK
1956 V1 P161-181 60-21372 <=>

PRACE USTAVU PRO NAFTOVY VYZKUM PUBLIKACE. BRNO
1955 N4/8 P111-121 62-14566 <*>
1956 N17/1 P73-77 09N52C <ATS>
1958 N9 P29-46 65-12256 <*>
1958 N9 P89-100 65-12257 <*>
1958 V9 N9 P29-44 48M40C <ATS>
1958 V9 N37 P57-59 16L37C <ATS>

PRACE USTAVU PRO VYZKUM A VYUZITI PALIV
1957 V8 N17 P110-131 59-11139 <=>
1959 V10 P293-303 62-10965 <*>

PRACE VYZKUMNEHO USTAVU CESKOSLOVENSKYCH NAFTOVYCH DOLU. PUBLIKACE. BRNO
1965 N4/8 P111-121 84S4C <ATS>

PRACE I WYDAWNICTWA. SZKOLA GLOWNA GOSPODURSTWA WIEJSKIEGO

1934 V1 P1-21 61-11342 <=>

PRACE WYDZIALU NAUK PRZYRODNICZO-ROLNICZYCH. SZCZECINSKIE TOWARZYSTWO NAUKOWE
1959 V1 N1 P1-17 65-50332 <=>

PRACE ZAKLODA APARATOW MATEMATYCZNYCH. PRACE A. POLSKA AKADEMIA NAUK. WARSAW
1959 SA N1 P11 61-11303 <=>
1959 SA N2 P1-11 61-11303 <=>
1959 SA N3 P1-11 61-11303 <=>

PRACOVNI LEKARSTVI
1950 V2 P49-61 63-18772 <*> 0
1950 V2 P187-189 65-12000 <*>
1953 N5 P191-198 II-503 <*>
1953 V5 P268-270 C-3689 <NRCC>
1954 V6 P99-101 62-32390 <=*$>
1956 V8 P433-435 59-22093 <=*>
1956 V8 N4 P262-265 62-11789 <=>
1959 V11 P206-215 AEC-TR-4417 <*>
1959 V11 N8 P432-433 62-19304 <*=>
1963 V15 N4 P145-149 65-10497 <*> 0

PRACTICA OTO-RHINO-LARYNGOLOGICA
1957 V19 P339-343 58-609 <*>
63-16521 <*>

PRACTICAL PHARMACY. JAPAN
SEE YAKKYOKU

PRAKTICKY ZUBNI LEKAR
1954 V2 N10 P223-225 3653 <K-H>

PRAKTIKA TES AKADEMIAS 'ATHENON
1927 V2 P179 58-2320 <*>
1928 V3 P250-253 59-10889 <*>
1928 V3 P538 58-2317 <*>
1950 V25 P42-47 59-10888 <*>
1954 V29 P255-273 2227 <*>

PRAKTISCHE ARZT
1927 V12 P410-413 65-11313 <*> 0
1953 N7 P226-232 AL-586 <*>

PRAKTISCHE ARZT. WIEN
1953 V7 P293-295 I-891 <*>

PRAKTISCHE CHEMIE
1952 V3 N2 P31-33 64-20430 <*>
1952 V3 N3 P56-58 63-14022 <*>
1956 V7 N10 P323-334 14K21G <ATS>
1957 V8 N1 P6-8 14K21G <ATS>

PRAKTISCHE ENERGIEKUNDE
1959 V7 N2 P107-141 TS-1758 <BISI>

PRAKTISCHE METALLOGRAPHIE
1964 V1 N1 P19-24 3908 <BISI>
1965 V2 P64-66 4833 <BISI>
1965 V2 P251-263 4715 <BISI>
1965 V2 P264-269 66-13604 <*>

PRATIQUE DES INDUSTRIES MECANIQUES
1933 V15 P392-394 59-17017 <*> 0
1937 V20 N1 P3-10 II-766 <*>
1941 N23 P107-109 62-18681 <*> 0
1941 V23 P159-160 62-20289 <*>
1946 V29 N1 P3-7 AD-617 193 <=$> 0
1946 V29 N2 P31-36 AD-617 193 <=$> 0
1964 V47 N10 P247-252 6654 <HB>

PRATIQUE DU SOUDAGE
1959 N2 P25-29 2058 <BISI>

PRATSI UKRAINSKYYI RESPUBLIKANKA NAUKOVA KONFERENTSIYA MOLODYKH UCHENYCH V HALUZI FIZIOLOHII ROSLYN
1963 V1 P127-130 65-13294 <*>

PRAVDA
1960 03/15 P2 NLLTB V2 N7 P579 <NLL>
1960 05/30 P1 NLLTB V2 N9 P783 <NLL>
1960 V48 P2 60-11933 <=>
1961 05/05 P4 NLLTB V3 N8 P637 <NLL>
1961 06/15 P3 NLLTB V3 N9 P750 <NLL>
1962 09/22 P2 NLLTB V5 N2 P79 <NLL>
1962 10/06 P2 NLLTB V5 N5 P380 <NLL>
1962 11/28 P2 NLLTB V5 N4 P261 <NLL>
1962 12/9 P3 NLLTB V5 N4 P283 <NLL>
1962 N23 P1-2 NLLTB V4 N5 P397 <NLL>
1963 02/05 P2 NLLTB V5 N5 P396 <NLL>
1963 03/22 P2 NLLTB V5 N7 P603 <NLL>
1963 04/21 P3 NLLTB V5 N9 P819 <NLL>
1963 05/19 P3 NLLTB V5 N9 P810 <NLL>
1963 10/17 P4 NLLTB V6 N2 P165 <NLL>
1963 11/07 P6 NLLTB V6 N4 P315 <NLL>
1964 02/04 P1-2 NLLTB V6 N7 P630 <NLL>
1964 02/06 P2-3 NLLTB V6 N7 P613 <NLL>
1964 05/31 P3 NLLTB V6 N9 P823 <NLL>
1964 06/07 P2 NLLTB V6 N9 P829 <NLL>
1964 08/30 P4 NLLTB V7 N3 P221 <NLL>
NLLTB V7 N4 P346 <NLL>
1964 11/22 P3 NLLTB V7 N3 P227 <NLL>
1965 P2-3 NLLTB V8 N1 P1 <NLL>
1965 02/04 P2 NLLTB V7 N7 P609 <NLL>
1965 10/20 P2 NLLTB V8 N2 P122 <NLL>
1965 12/19 P3 NLLTB V8 N4 P303 <NLL>
1966 P3 06/25 66-33005 <=$>
1966 P4 05/14 66-32639 <=$>
1966 02/03 P1-2 NLLTB V8 N5 P363 <NLL>
1966 04/22 P3 NLLTB V8 N8 P707 <NLL>

PRAXIS
1949 V38 P1134-1136 57-356 <*>
1953 V42 P806-812 I-532 <*>
1953 V42 N46 P968-972 65-14637 <*> 0
1954 V43 P9-10 I-561 <*>
1954 V43 P177-181 2568 <*>
1954 V43 P228- I-547 <*>
1954 V43 P338-343 I-950 <*>
1955 V44 P138-139 2110 <*>
1955 V44 P223-224 1861 <*>
1956 V45 P56-58 II-760 <*>
1956 V45 P339-343 1063 <*>
1956 V45 P571- 57-731 <*>
1956 V45 P966-968 62-18888 <*> 0
1956 V45 P993-998 3319 <*>
1956 V45 N10 P219-223 59-20946 <*>
1957 V46 P465-466 58-1174 <*>
1957 V46 P660-662 58-1175 <*>
1957 V46 P767-770 58-830 <*>
1957 V46 P906-915 58-2695 <*>
1957 V46 P991-992 62-14998 <*>
1957 V46 P1037-1039 58-601 <*>
1960 V49 P187-188 60-16944 <*>
1960 V49 P506-511 G-294 <RIS>
1960 V49 N39 P930-932 62-25707 <=> 0
1961 V50 P200-201 61-18780 <*>
1961 V50 N12 P249-254 64-00116 <*>
1961 V50 N21 P548-552 61-20664 <*> 0
62-10284 <*>
1962 V51 P279-283 66-11636 <*>
1962 V51 N27 P704-707 66-12040 <*>
1964 V53 N24 P846-850 65-12167 <*>
1964 V53 N50 P1695-1698 65-13975 <*> 0
1965 V54 N18 P547-549 65-14226 <*>

PRAXIS DER KINDERPSYCHOLOGIE UND KINDERPSYCHIATRIE
1960 V9/10 P12-14 65-12115 <*>

PRAXIS DER PNEUMOLOGIE
1964 V18 P798-802 65-12747 <*>

PRECONTRAINTE
1952 V2 N1 P31-40 60-18192 <*> 0

PREDELNO DOPUSTIMYE KONTSENTRATSII ATMOSFERNYKH ZAGRIAZNENII
1964 V8 P5-21 65-30671 <=$>

PREDVARITELNYE ITOGI NAUCHNYKH ISSLEDOVANII S POMOSHCHIIU PERVYKH SOVETSKIKH ISKUSSTVENNYKH SPUTNIKOV ZEMLI I RAKET

Citation	Accession
1958 N1 P13-21 PT1	AD-610 989 <=$>
1958 N1 P40-108 SECT.2	AD-610 974 <=$>

PREOVOJ

Citation	Accession
1960 V4 N24 P4-5	62-19690 <*=>

PREHLED VEDECKE CINNOSTI USTAVU PRO PECI O MATKU A DITE. CZECHOSLOVAKIA

Citation	Accession
1961 P20-23	64-00179 <*>

PRENSA MEDICA. LA PAZ

Citation	Accession
1964 V29 N5/6 P169-172	65-13941 <*>

PRENSA MEDICA ARGENTINA

Citation	Accession
1935 V22 P539-546	64-10834 <*> 0
1936 V23 P972-977	57-1667 <*>
1941 V28 P1213-1222 PT1	57-1673 <*>
1949 V36 P435-441	T-2634 <INSD>
1949 V36 N36 P2383-2390	65-17262 <*>
1950 V37 P2461-2466	1708 <*>
1951 V38 N23 P1434-1435	2433 <*>
1954 V41 P742-748	I-997 <*>
1954 V41 P1502-1508	II-71 <*>
1954 V43 P480-483	1710 <*>
1956 V43 N2 P166-179	65-00296 <*>
1957 V44 P1273-	<RIS>
1957 V44 P1385-1390	65-17286 <*>
1957 V44 N17 P1273-1280	OA 2 <RIS>
1957 V44 N30 P2368-2371	64-10623 <*>
	8133-A <K-H>
1957 V44 N34 P2642-2645	59-15686 <*> 0
1958 V45 N1 P152-156	66-10939 <*>
1958 V45 N16 P1613-1617	62-14511 <*>
1958 V45 N37 P3035-3036	66-12122 <*>
1959 V46 P1970-1976	28M45SP <ATS>
1959 V46 P2005-2007	65-00423 <*>
1959 V46 P2867-2869	62-16903 <*>
1959 V46 N29 P1853-1859	10265-C <K-H>
	64-10617 <*>
1960 V47 P386-388	64-10487 <*>
1960 V47 P2797-2800	11339 <K-H>
1960 V47 P2918-2923	61-20772 <*>
1960 V47 N1 P179-186	61-10902 <*>
1961 V48 P903-968	63-16526 <*>
1962 V49 N23 P1286-1291	64-14178 <*> 0
1963 V50 N32 P2290-2292	65-11761 <*>
1964 V51 N8 P467-470	65-14790 <*> 0

PRENSA MEDICA MEXICANA

Citation	Accession
1960 V25 N7/9 P432-433	26N50SP <ATS>
1961 V26 P289	63-16524 <*>

PRESLIA

Citation	Accession
1951 V29 P125-131	65-12246 <*>

PRESSE MEDICALE

Citation	Accession
1899 V6 N102 P362	62-01044 <*>
1901 V9 P229-233	59-15524 <*>
1907 N15 P641-644	58-327 <*>
1910 V18 P421-423	62-00806 <*>
1921 V29 P294-296	65-11442 <*>
1925 V33 N80 P1330-1332	62-00166 <*>
1935 V43 N50 P1004-1008	61-14498 <*>
1936 V44 P1779	61-14496 <*>
1936 V44 N69 P1356-1357	61-14495 <*> 0
1939 V47 N8 P150	61-16485 <*>
1939 V47 N74 P1365-1366	59-20702 <*>
1940 V48 N47/8 P533-535	61-16846 <*> 0
1940 V48 N62/3 P667-669	61-16488 <*>
1940 V48 N91/2 P935-937	61-14497 <*>
1941 V49 N37 P449-451	65-17285 <*>
1941 V49 N5/6 P44-47	61-16847 <*>
1942 V50 N42 P588	61-16483 <*>
1942 V50 N15/6 P169-171	61-16848 <*>
1943 V51 N11 P138-139	61-20259 <*>
1946 N2 P499-500	57-2610 <*>
1946 V54 N63 P862-863	66-10990 <*> 0
1947 V55 P778-779	59-17828 <*> 0
1947 V55 N24 P17 SUP	2586 <*>
1949 V57 N1 P4-5	2795 <*>
1949 V57 N36 P505-506	2589 <*>
1949 V57 N60 P837-838	59-17819 <*> 0
1949 V57 N61 P852-854	2566 <*>
1950 V58 N16 P271-273	2588 <*>
1950 V58 N22 P392-393	2550 <*>
1950 V58 N38 P684-687	57-3260 <*>
1951 V59 P1240-1242	58-1073 <*>
1951 V59 N17 P325-327	2553 <*>
1952 V60 P1719-1721	58-172 <*>
1952 V60 N66 P623-637	60-10020 <*> 0
1952 V60 N66 P1045-1047	60-10020 <*> 0
1953 V61 P825-826	AL-770 <*>
1953 V61 P1059-1060	I-746 <*>
1953 V61 N28 P582-584	60-18937 <*>
1953 V61 N34 P701-703	62-00453 <*>
1953 V61 N49 P1029-1030	59-17593 <*> 0
1953 V61 N80 P1646	59-17868 <*>
1954 V62 P759-760	II-460 <*>
1954 V62 P939-941	2637 <*>
1954 V62 P1125-1126	1879 <*>
1954 V62 P1529	3390-E <K-H>
1954 V62 P1751-1752	57-1162 <*>
1954 V62 P1775-1777	62-18098 <*> 0
1954 V62 N5 P79-81	2923 <*>
1954 V62 N6 P123-124	54K22F <ATS>
1954 V62 N31 P654-657	60-10657 <*> 0
1954 V62 N33 P712-715	3084-C <KH>
1954 V62 N44 P939-941	62-14934 <*> 0
1954 V62 N46 P972-976	61-16564 <*>
1955 V63 P171-173	T-1636 <INSD>
1955 V63 P375-376	63-10317 <*> 0
1955 V63 P1015	II-895 <*>
1955 V63 P1478-	II-568 <*>
1955 V63 N12 P221-223	57-857 <*>
1955 V63 N37 P765	3779-H <K-H>
1955 V63 N54 P1105-1107	66-13411 <*> 0
1955 V63 N62 P1247-1248	57-1946 <*>
1956 V64 P1857-1858	57-2715 <*>
1956 V64 P2189-2192	57-1853 <*>
1956 V64 N47 P1097-1098	5525-D <K-H>
1956 V64 N64 P1474-1475	57-2970 <*>
1956 V64 N77 P1765-1768	58-156 <*>
1957 P1353-1354	58-948 <*>
1957 V65 P485-487	57-2775 <*>
1957 V65 P1263-	57-2655 <*>
1957 V65 P1353-1354	63-16664 <*> 0
1957 V65 P1930-1932	58-1770 <*>
1957 V65 P2060-2063	58-606 <*>
1957 V65 N25 P571-573	62-14588 <*> 0
1957 V65 N52 P1230-1235	62-01073 <*>
1957 V65 N86 P1943-1945	66-11800 <*>
1957 V65 N88 P1991-1996	61-10210 <*> 0
1958 V66 P131-134	58-827 <*>
1958 V66 N1 P121-125	58-321 <*>
1958 V66 N24 P529-530	61-10679 <*> 0
1958 V66 N52 P1220	60-14872 <*>
1958 V66 N57 P1307-1309	62-14606 <*> 0
1958 V66 N59 P1337-1340	9149-C <K-H>
1958 V66 N61 P1369-1371	8933 <K-H>
1958 V66 N78 P1745-1746	59-10625 <*> 0
1958 V66 N80 P1792-1793	64-18733 <*>
1959 V67 P891-895	66-12041 <*>
1959 V67 N4 P129-130	60-14183 <*> 0
1959 V67 N17 P684-685	59-17589 <*>
1959 V67 N34 P1386-1388	62-14952 <*> 0
1959 V67 N38 P1488-1490	61-10922 <*>
1960 V68 N23 P860-863	62-00326 <*>
1960 V68 N37 P1397	61-00355 <*>
1960 V68 N38 P1421-1424	62-14921 <*>
1960 V68 N39 P1441-1444	62-14939 <*>
1960 V68 N49 P1827-1829	61-10921 <*>
1960 V68 N50 P1867-1869	65-14323 <*>
1960 V68 N54 P2091-2092	64-18739 <*>
1960 V68 N58 P2318-2321	10938-C <KH>

64-14529 <*>
1961 V69 P230-232 62S87F <ATS>
1961 V69 P1044 61-20182 <*>
1961 V69 P1527 61-20249 <*>
1961 V69 N13 P587-590 62-00497 <*>
1961 V69 N20 P842-844 63-16512 <*>
1961 V69 N21 P912-914 62-14938 <*>
1961 V69 N23 P1042 62-14631 <*>
1961 V69 N26 P1189 64-18633 <*>
1961 V69 N33 P1511-1514 62-01524 <*>
1961 V69 N34 P1546-1548 62-01519 <*>
1961 V69 N35 P1581-1583 62-14528 <*>
1961 V69 N38 P1638 62-16015 <*>
1961 V69 N47 P2096 62-14324 <*>
1962 V70 P765-766 63-18023 <*>
1962 V70 P2833-2836 63-16345 <*>
1962 V70 N27 P1365 62-20146 <*>
1963 V71 P199 63-14720 <*>
1963 V71 P1141 63-20624 <*>
1963 V71 P1905-1908 64-10969 <*>
1963 V71 N14 P710-712 65-13245 <*> O
1963 V71 N14 P736 64-16851 <*>
1963 V71 N18 P911-914 66-12119 <*>
1963 V71 N18 P919-920 65-10954 <*>
1963 V71 N27 P1403 64-16849 <*>
1963 V71 N28 P1439 64-18652 <*>
1963 V71 N31 P1555-1556 63-20984 <*>
1963 V71 N31 P1572 64-20912 <*>
1963 V71 N43 P2025-2027 64-18738 <*>
1963 V71 N55 P2688-2689 66-11644 <*> O
1964 V72 P999-1002 65-17316 <*>
1964 V72 P1895-1896 65-14426 <*>
1964 V72 N2 P75-79 65-17030 <*>
1964 V72 N8 P472 64-16848 <*>
1964 V72 N12 P661-665 65-13615 <*>
1964 V72 N13 P793-797 65-00109 <*>
1964 V72 N19 P1139-1141 64-20148 <*>
1964 V72 N27 P1579-1582 65-13244 <*> O
1964 V72 N27 P1631 64-20914 <*>
1964 V72 N30 P1840 64-20908 <*>
1964 V72 N36 P2089-2093 65-14706 <*>
1964 V72 N37 P2135-2136 66-12952 <*>
1964 V72 N37 P2141-2142 65-13233 <*>
1964 V72 N40 P2325-2330 65-17244 <*>
1964 V72 N42 P2481 65-12165 <*>
1964 V72 N47 P2829-2830 65-13235 <*>
1964 V72 N51 P3097-3101 65-13234 <*> O
1964 V72 N54 P3263-3265 66-13193 <*>
1965 V73 P457-459 66-12625 <*>
1965 V73 P893-894 66-13196 <*>
1965 V73 P1415-1416 66-13035 <*>
1965 V73 P1529-1534 66-13206 <*>
1965 V73 N6 P304 66-13228 <*>
1965 V73 N13 P725-726 904 <TC>
1965 V73 N20 P1157-1162 65-14814 <*>
1965 V73 N35 P2003-2006 66-11597 <*>
1965 V73 N39 P2215-2220 66-12722 <*>
1965 V73 N40 P2247-2250 66-13183 <*>
1965 V73 N47 P2700 66-13184 <*>

PRESSE DER SOWJETUNION
1955 03/11 P603 57-607 <*>
1960 N124 P2561 65-10105 <*>

★PRIBOROSTROENIE
1956 N4 P31 59-14181 <+*>
1957 N2 P9-11 R-1775 <*>
1957 N3 P1-7 59-18526 <+*>
1957 N4 P26-27 R-1494 <*>
1957 N4 P51-64 R-4893 <*>
1957 N12 P25-26 60-13589 <+*>
1958 N3 P1-19 59-21183 <=>
1958 N3 P22-25 59-21183 <=>
1958 N3 P27-31 59-21183 <=>
1958 N4 P1-5 60-13476 <+*>
1958 N5 P22-26 60-21722 <=> O
1958 N5 P26-29 59-15864 <+*>
1958 N5 P30-31 60-21722 <=> O
1958 N7 P4-8 59-11187 <=>
1958 N10 P28-30 59-13096 <=>
1958 N11 P1-3 59-11467 <=>
1958 N11 P4-7 59-11467 <=>
1958 N12 P1-2 60-15503 <+*>
1958 N12 P16-18 59-11509 <=>
1959 N1 P24-25 60-13832 <+*>
1959 N2 P25-27 49L36R <ATS>
1959 N3 P3-6 62-24667 <=*> O
1959 N4 P20-21 60-19059 <+*>
1959 N6 P23-25 61-13365 <+*>
1959 N8 P22-25 60-11305 <=>
1959 N10 P26-30 60-11734 <=>
1959 N11 P29-31 60-11860 <=>
1959 N12 P1-2 60-11860 <=>
1959 N12 P24-26 60-11860 <=>
1960 N1 P1-2 60-11860 <=>
1960 N1 P24-25 60-11860 <=>
1960 N2 P24-32 60-41173 <=>
1960 N3 P27-28 60-11933 <=>
1960 N4 P23-26 62-13279 <=*>
1960 N4 P29 60-41342 <=>
1960 N5 P29-32 60-41342 <=>
1960 N6 P9-12 61-11127 <=> O
1960 N6 P30-31 61-11674 <=>
1960 N7 P16-19 61-21150 <=>
1960 N8 P9-15 61-21150 <=>
1960 N8 P26-27 61-21150 <=>
1960 N9 P19-20 E-772 <RIS>
1960 N11 P1-2 61-21687 <=>
1960 N11 P3-5 ARSJ V32 N5 P811 <AIAA>
1960 N11 P13-17 61-21679 <=>
1960 N12 P1-5 61-21687 <=>
1960 V4 N2 P22-24 61-19675 <=*>
1961 N2 P30-31 61-23110 <=*>
1961 N3 P24 66-14157 <*$>
1961 N4 P8-11 62-15725 <=*>
1961 N9 P9-12 62-13653 <*=> O
1961 N9 P12-14 62-13555 <*=>
1962 N1 P4-5 62-24974 <=*>
1962 N2 P1-3 62-24249 <=*>
1962 N2 P3-5 62-24399 <=*>
1962 N2 P32 62-25944 <=*>
1962 N10 P30-31 63-19496 <=*>
1962 N12 P24 63-23790 <=*>
1962 V11 P6-8 63-24020 <=*>
1963 N4 P29 63-23670 <=*>
1963 N8 P3-6 63-41169 <=>
1963 N8 P21-23 65-61346 <=$>
1963 N9 P18-19 64-11803 <=>
1963 N11 P1-2 64-21084 <=>
1963 N11 P3-8 64-31417 <=>
1964 N1 P5-17 64-31087 <=>
1964 N1 P25-26 64-31087 <=>
1964 N1 P30-31 64-21676 <=>
64-31087 <=>
1964 N5 P41-46 AD-640 308 <=$>
1965 N3 P23-24 AD-625 145 <=*$>
1965 N8 P176-181 AD-631 224 <=$>
1965 N9 P1-4 65-34095 <=*$>

★PRIBORY I TEKHNIKA EKSPERIMENTA
1956 N1 P38-43 R-1750 <*>
1956 N1 P38-42 61-13858 <+*>
1956 N1 P83 59-11244 <=>
1956 N2 P26-28 87J14R <ATS>
1956 N2 P45-49 59-15017 <+*>
1956 N2 P70-71 88J14R <ATS>
1956 N2 P86-87 89J14R <ATS>
1956 N2 P99-104 90J14R <ATS>
1956 N2 P116-122 91J14R <ATS>
1956 N2 P122-124 92J14R <ATS>
1956 N3 P9- R-2658 <*>
R-3103 <*>
1957 N1 P15-21 28J16R <ATS>
1957 N1 P33-36 AEC-TR-5540 <=*>
1957 N1 P49-55 29J16R <ATS>
60-23586 <+*>
1957 N1 P55-57 30J16R <ATS>
1957 N1 P64-71 31J16R <ATS>
1957 N1 P71-77 32J16R <ATS>
1957 N1 P77-82 33J16R <ATS>

1957 N1 P97-98	62-14484 <=*>
1957 N1 P106-110	62-25560 <=*>
1957 N1 P116	34J16R <ATS>
1957 N2 P15-18	59-14763 <+*>
1957 N2 P18-22	NP-TR-193 <+*> 0
	R-4172 <*>
1957 N2 P26-29	R-4268 <*>
1957 N2 P48-53	63-23145 <=*>
1957 N2 P60-63	E-536 <RIS>
1957 N2 P103-105	60-23082 <+*>
1957 N3 P115-117	61-13510 <=*>
	66-14384 <*$>
1957 N4 P20-24	AEC-TR-3860 <*=>
1957 N4 P25-27	R-5053 <*>
1957 N4 P39-43	AEC-TR-3701 <+*> 0
1957 N4 P66-67	61-31574 <=>
1957 N5 P54-58	AEC-TR-4018 <+*>
1957 N5 P114-116	59-10325 <+*>
1957 N5 P116-117	59-22411 <+*>
1957 N5 P122-123	61-20050 <*=>
1957 N6 P22-30	AEC-TR-3935 <+*>
1957 N6 P45-49	60-10129 <*>
1957 N6 P50-54	59-19175 <+*>
1957 N6 P57-67	62-18530 <=*>
1957 N6 P72-74	AEC-TR-3932 <+*>
1957 N6 P80-82	60-19146 <+*>
1957 N6 P92-94	61-20049 <*=>
1957 N6 P113-114	NP-TR-1054 <=$>
1958 N1 P3-16	59-22509 <+*>
1958 N1 P31-34	AEC-TR-3663 <+*>
1958 N1 P38-41	AEC-TR-3664 <+*>
1958 N1 P41-46	AEC-TR-3651 <+*>
1958 N1 P59-62	59-22505 <+*>
1958 N1 P62-68	59-10299 <+*>
1958 N1 P116-119	60-13584 <+*>
1958 N1 P138-139	63-14064 <=*>
1958 N2 P21-24	59-10791 <+*>
1958 N2 P24-28	AEC-TR-3916 <+*>
1958 N2 P86-91	PB 141 182T <=>
	R-5182 <*>
1958 N2 P103-104	59-11957 <=>
1958 N3 P21-26	AEC-TR-3912 <+*>
1958 N3 P39-45	59-22545 <+*>
	60-23981 <*+>
1958 N3 P58-61	59-14383 <+*> 0
1958 N3 P100-101	AEC-TR-3542 <+*>
	59-22546 <+*>
	60-14302 <+*>
1958 N4 P37-39	59-21003 <=>
1958 N4 P107	59-22648 <+*>
1958 N5 P99-101	59-10298 <+*>
1958 N6 P68-71	61-13859 <+*>
1958 N6 P89-93	60-13849 <+*>
	61-19446 <+*> 0
1959 N1 P121-125	61-19328 <+*> 0
1959 N1 P135-136	AEC-TR-3933 <+*>
1959 N1 P142-145	62-16756 <=*>
1959 N2 P8-11	62-32768 <=*>
1959 N2 P12-15	62-32757 <=*>
1959 N2 P15-18	62-32766 <=*>
1959 N2 P19-24	60-21193 <=> 0
	62-32765 <=*>
1959 N2 P24-26	60-21194 <=> 0
	62-32764 <=*>
1959 N2 P27-30	62-32763 <=*>
1959 N2 P31-33	62-32762 <=*>
1959 N2 P33-35	62-32761 <=*>
1959 N2 P36-38	62-32760 <=*>
1959 N2 P38-40	62-32759 <=*>
1959 N2 P41-45	62-32758 <=*>
1959 N2 P45-49	62-32757 <=*>
1959 N2 P49-52	62-32756 <=*>
1959 N2 P53-56	62-32755 <=*>
1959 N2 P57-58	62-32754 <=*>
1959 N2 P59-61	62-32753 <=*>
1959 N2 P62-64	62-32752 <=*>
1959 N2 P65-67	62-32751 <=*>
1959 N2 P68-75	62-32750 <=*>
1959 N2 P75-78	62-32749 <=*>
1959 N2 P78-80	62-32748 <=*>
1959 N2 P81-82	62-32742 <=*>
1959 N2 P83-85	62-32746 <=*>
1959 N2 P86-90	62-32745 <=*>
1959 N2 P91-94	62-32744 <=*>
1959 N2 P95-102	62-32743 <=*>
1959 N2 P103-107	62-32742 <=*>
1959 N2 P108-110	62-32741 <=*>
1959 N2 P111-112	62-32740 <=*>
1959 N2 P113-114	62-32739 <=*>
1959 N2 P115-116	62-32737 <=*>
1959 N2 P117	62-32728 <=*>
1959 N2 P118-119	62-32736 <=*>
1959 N2 P120	62-32735 <=*>
1959 N2 P121-123	62-32734 <=*>
1959 N2 P124-127	62-32733 <=*>
1959 N2 P128-131	62-32732 <=*>
1959 N2 P132-133	62-32731 <=*>
1959 N2 P134-135	62-32729 <=*>
1959 N2 P136-138	62-32730 <=*>
1959 N2 P139-141	62-32728 <=*>
1959 N2 P142	62-32727 <=*>
1959 N2 P143-145	62-32726 <=*>
1959 N2 P145-146	62-32725 <=*>
1959 N2 P147	62-32724 <=*>
1959 N2 P148-150	62-32723 <=*>
1959 N2 P150-151	62-32722 <=*>
1959 N2 P152-153	62-32721 <=*>
1959 N2 P154	62-32720 <=*>
1959 N2 P155	62-32719 <=*>
1959 N2 P156	62-32718 <=*>
1959 N2 P157	62-32717 <=*>
1959 N2 P158	62-32714 <=*>
1959 N2 P159	62-32715 <=*>
1959 N2 P160	62-32716 <=*>
1959 N3 P71-76	61-13131 <+*>
1959 N3 P140-142	61-23701 <*=> 0
1959 N4 P161-163	60-11499 <=>
1959 N5 P134-137	61-21964 <=>
1959 N6 P10-13 PT.2	AEC-TR-4137 <*>
1960 N1 P48-50	61-10151 <+*>
1960 N1 P50-57	61-28152 <*=>
1960 N1 P118-122	60-41340 <=>
1960 N2 P146-148	61-23673 <=*>
1960 N2 P150-152	61-23673 <=*>
1960 N5 P64-66	61-31534 <=>
1960 N5 P106-108	62-24029 <=*> 0
1960 N5 P145-146	61-28200 <*=>
1961 N1 P55-57	62-13055 <=*>
1961 N1 P101-102	62-19251 <*=>
1961 N1 P120-121	62-13055 <=*>
1961 N1 P129	UCRL TRANS-704 <=*>
1961 N1 P130-132	62-13055 <=*>
1961 N2 P72-	UCRL TRANS-705 <=*>
1961 N2 P118-119	62-15726 <*=>
1961 N2 P166-169	62-13074 <*=>
1961 N3 P22-25	AEC-TR-5137 <=*>
1961 N3 P63-66	62-24274 <=*>
1961 N4 P89-91	62-24278 <=*>
1961 N5 P120-126	62-24064 <=*> 0
1961 N6 P41-44	AEC-TR-5239 <=*>
1961 N6 P80-83	62-25949 <=*>
1962 N1 P65-77	62-32440 <=*>
1962 N1 P96-98	62-32534 <=*> 0
1962 N1 P110-117	63-15218 <=*>
1962 N1 P142-144	63-13915 <=*>
1962 N3 P89-92	63-15405 <=*>
1962 N3 P119-122	AEC-TR-5622 <=*>
1962 N5 P106-112	SC-T-525 <=*$>
1962 N5 P174-176	63-21242 <=>
1962 N6 P132	81S80R <ATS>
1963 N1 P152-154	63-23150 <*=>
1963 N4 P90-94	64-11953 <=> M
1963 N4 P103-106	64-11806 <=>
1963 N5 P123-124	64-11859 <=>
1963 N5 P203-207	TRC-TRANS-1093 <NLL>
1963 N6 P124-127	N65-16309 <=*$>
1964 N1 P30-33	64-71129 <=>
1964 N3 P94-102	N65-12270 <=$>
1964 N3 P102-107	N65-12271 <=$>
1964 N4 P171-175	65-10743 <*>

1964 V9 N1 P121-123 AD-617 908 <=$>
1964 V9 N3 P88-94 N65-12269 <=$>
1965 N1 P169-173 AD-633 896 <=$>
1965 N2 P146-149 AD-625 157 <=*$>
1965 N2 P160-163 AD-619 864 <=$>
1965 N5 P177-182 65-33754 <=*$>
1965 V10 N1 P54-59 65-14653 <*> 0

PRIKLADNA MEKHANIKA
SEE PRYKLADNA MEKHANIKA

★PRIKLADNAYA BIOKHIMIYA I MIKROBIOLOGIYA
1965 V1 N2 P167-174 65-31857 <=$>
1965 V1 N4 P466-468 66-61545 <=$*>

PRIKLADNAYA GEOFIZIKA
1954 N11 P152-162 60-13441 <+*>
1955 N12 P93-106 71J17R <ATS>
1955 N12 P157-176 60-15665 <+*>
1956 N14 P97-114 62-14580 <=*>
1956 N14 P115-129 60-19009 <+*>
1956 N15 P91-102 60-19008 <+*>
1956 N15 P135-139 AEC-TR-4228 <*+>
1957 N16 P3-36 60-19562 <+*>
1957 N16 P37-49 71L29R <ATS>
1957 N16 P50-84 61-13238 <*+>
70L29R <ATS>
1957 N16 P130-144 66-61864 <=*$>
1957 N17 P33-66 59-15843 <+*>
1957 N17 P76-92 59-15845 <+*>
1958 N19 P3-22 60-17918 <+*>
1958 N19 P57-108 63-20720 <=*$>
1958 N19 P230-244 IGR V2 N10 P874 <AGI>
1958 N20 P3-25 72L29R <ATS>
1958 N20 P26-45 73L29R <ATS>
1958 N20 P141-154 63-19724 <=*$>
1959 N23 P193-201 IGR V2 N10 P867 <AGI>
63-16996 <=*>
1960 N24 P213-221 62-32482 <=*>
1960 N25 P55-65 61-28207 <*=>
1960 N28 P23-24 62-13446 <*=> 0
62-25908 <=*>
1960 N28 P70-91 63-13192 <=*>
1961 N29 P120-122 AD-623 436 <=*$>
1961 N29 P136-137 62-23786 <=*>
1961 N30 P92-102 64-10977 <=*$>
1961 N30 P103-114 64-10976 <=*$>
1961 N31 P230-247 83Q69R <ATS>
1961 V29 P20-38 63-26599 <=$>
1962 N32 P142-154 51Q68R <ATS>
1962 N33 P190-204 64Q71R <ATS>
1962 N33 P206-212 65Q71R <ATS>
1962 N33 P213-224 64Q73R <ATS>
1962 N34 P3-22 63-27103 <=$>
1962 N34 P76-86 RTS-2531 <NLL>
1962 N35 P142-156 92Q71R <ATS>
1963 N36 P167-180 65-62427 <=$>
1963 N36 P181-186 65-62428 <=$>
1963 N36 P187-194 65-62429 <=$>
1963 N37 P135-146 AD-635 573 <=$>
1964 N39 P75-90 IGR V8 N4 P379 <AGI>
65-14293 <*>
1964 N39 P94-106 IGR V8 N6 P656 <AGI>
11S81R <ATS>
1964 N39 P107-113 IGR V8 N4 P416 <AGI>
12S81R <ATS>
1964 N39 P123-135 IGR V8 N4 P480 <AGI>
1964 N39 P153-166 IGR V8 N1 P28 <AGI>
1964 N39 P179-196 IGR V8 N1 P17 <AGI>
1964 N40 P120-126 65-31117 <=$>
1964 N40 P174-180 IGR V7 N12 P2151 <AGI>
1965 N41 P43-52 19S87R <ATS>

★PRIKLADNAYA MATEMATIKA I MEKHANIKA
1938 V2 P209-210 63-18506 <=*>
1940 V4 N1 P3-32 RT-1083 <*>
1940 V4 N2 P3-6 RTS-3161 <NLL>
1940 V4 N3 P31-36 RT-239 <*>
1940 V4 N3 P37-42 RT-354 <*>
1940 V4 N5/6 P19-34 RT-278 <*>
1940 V4 N5/6 P123-127 RT-283 <*>
1941 V5 N1 P57-70 RT-355 <*>
1941 V5 N2 P165-192 RT-980 <*>
1941 V5 N2 P193-222 RT-1466 <*>
1941 V5 N3 P431-438 RT-282 <*>
1941 V5 N3 P439-452 RT-348 <*>
1941 V5 N3 P453-470 RT-1838 <*>
1942 V6 P187-196 RT-1954 <*>
1942 V6 P287-316 RT-1087 <*>
1942 V6 P411-448 AMST S1 V5 P242-290 <AMS>
RT-1081 <*>
1942 V6 N5 P381-394 RT-1797 <*>
1942 V6 N2/3 P241-246 RT-1955 <*>
1943 V7 P81-96 63-20579 <=*$> 0
1943 V7 P273-292 R-761 <*>
1943 V7 P431-438 65-60775 <=>
1943 V7 N2 P109-130 RT-1080 <*>
1943 V7 N3 P193-222 RT-557 <*>
1943 V7 N3 P226-230 RT-1078 <*>
1943 V7 N4 P245-272 RT-1082 <*>
1943 V7 N4 P273-292 RT-1077 <*>
1943 V7 N4 P316-320 RT-2590 <*>
1943 V7 N5 P331-340 62-16104 <=*>
1943 V7 N5 P389-392 RT-1119 <*>
1943 V7 N6 P405-412 RT-1079 <*>
1944 V8 P241-245 AMST S1 V5 P291-297 <AMS>
1944 V8 N1 P3-14 RT-2322 <*>
1944 V8 N1 P15-24 RT-556 <*>
1944 V8 N1 P84-87 RT-3923 <ATS>
63-20148 <=*>
1944 V8 N2 P109-140 RT-546 <*>
1944 V8 N3 P169-186 RT-1397 <*>
1944 V8 N3 P201-224 RT-1214 <*>
1944 V8 N4 P273-286 RT-518 <*>
1944 V8 N5 P337-360 RT-526 <*>
1944 V8 N5 P361-394 RT-547 <*>
1945 V9 P67-78 R-3843 <*>
1945 V9 N1 P67-78 RT-248 <*>
1945 V9 N1 P101-110 RT-415 <*>
1945 V9 N1 P111-128 RT-247 <*>
1945 V9 N3 P207-218 RT-414 <*>
1945 V9 N4 P293-311 R-3588 <*>
RT-4380 <*>
1945 V9 N4 P347-352 T.767 <INSD>
1945 V9 N5 P389-412 T.780 <INSD>
1946 V10 P21-32 59-14261 <+*>
1946 V10 P75-92 59-12969 <+*>
1946 V10 P365-368 R-4525 <*>
1946 V10 P397-406 62-13252 <*=>
1946 V10 N1 P153-164 RT-861 <*>
1946 V10 N2 P39-50 AD-625 517 <=*$>
1946 V10 N2 P251-272 RT-911 <*>
1946 V10 N3 P397-405 62-25370 <=*>
1946 V10 N4 P449-474 RT-528 <*>
1946 V10 N4 P481-502 RT-322 <*>
65-13255 <*>
1946 V10 N4 P503-512 RT-323 <*>
1946 V10 N4 P513-520 RT-321 <*>
1946 V10 N4 P521-524 RT-330 <*>
RT-516 <*>
1946 V10 N5/6 P623-638 RT-320 <*>
1947 V11 P171-176 AMST S1 V11 P339-347 <AMS>
1947 V11 P215-222 AMST S1 V11 P325-338 <AMS>
1947 V11 P313-328 AMST S1 V4 P1-23 <AMS>
RT-143 <*>
1947 V11 P370-376 RT-249 <*>
1947 V11 P371-376 RT-497 <*>
RT-997 <*>
1947 V11 P459-464 AMST S1 V11 P348-355 <AMS>
1947 V11 P495-496 R-00562 <*>
R-562 <*>
1947 V11 P527-532 61-20031 <*=>
1947 V11 P565-592 AMST S1 V11 P178-230 <AMS>
1947 V11 N1 P69-84 50/1894 <NLL>
1947 V11 N1 P105-118 RT-251 <*>
1947 V11 N1 P119-128 RT-421 <*>
1947 V11 N1 P129-146 RT-1184 <*>
1947 V11 N1 P147-164 RT-2946 <*>
1947 V11 N1 P165-170 RT-423 <*>
1947 V11 N1 P171-176 RT-453 <*>

Citation	Translation
1947 V11 N1 P177-192	RT-428 <*>
1947 V11 N1 P193-198	RT-250 <*>
1947 V11 N1 P199-202	RT-417 <*>
1947 V11 N2 P205-214	RT-422 <*>
1947 V11 N2 P215-222	RT-2945 <*>
1947 V11 N2 P223-230	RT-426 <*>
1947 V11 N2 P231-236	RJ-197 <ATS>
1947 V11 N2 P293-296	RT-998 <*>
1947 V11 N2 P297-300	RT-420 <*>
1947 V11 N3 P301-312	RT-314 <*>
1947 V11 N3 P363-370	RT-419 <*>
1947 V11 N3 P371-376	64-71460 <=> M
1947 V11 N3 P383-386	RT-2947 <*>
1947 V11 N3 P391-394	RT-418 <*>
1947 V11 N3 P395-396	RT-374 <*>
1947 V11 N4 P449-458	RT-1464 <*>
1947 V11 N4 P459-464	RT-403 <*>
1947 V11 N4 P465-474	RT-427 <*>
1947 V11 N4 P493-494	61-19573 <*=>
	62-24799 <=*>
1947 V11 N4 P495-496	RT-416 <*>
1948 V12 P69-74	61-20030 <*=>
1948 V12 P109-128	AMST S1 V11 P231-257 <AMS>
1948 V12 P301-328	61-15317 <+*>
1948 V12 P561-596	AMST S1 V5 P298-359 <AMS>
1948 V12 P673-690	AMST S1 V5 P360-388 <AMS>
1948 V12 N1 P47-52	64-19346 <=$>
1948 V12 N1 P53-62	RT-3201 <*>
1948 V12 N1 P63-68	RT-515 <*>
1948 V12 N2 P165-180	RT-964 <*>
1948 V12 N3 P339-344	RT-675 <*>
1948 V12 N6 P757-760	RT-2897 <*>
1949 V13 P3-40	AMST S1 V4 P159-206 <AMS>
1949 V13 P381-390	60-13819 <+*>
1949 V13 P561-596	R-00566 <*>
	R-566 <*>
1949 V13 N1 P55-78	RT-1154 <*>
1949 V13 N2 P171-186	RT-344 <*>
1949 V13 N2 P209-216	RT-720 <*>
1949 V13 N2 P217-218	RT-2803 <*>
1949 V13 N3 P257-266	R-00559 <*>
	RT-698 <*>
1949 V13 N3 P267-276	RT-695 <*>
1949 V13 N3 P307-316	T.766 <INSD>
1949 V13 N4 P449-456	R-565 <*>
	RT-697 <*>
	00565 <*>
1949 V13 N5 P513-525	RT-290 <*>
1949 V13 N5 P537-542	RT-721 <*>
1949 V13 N5 P543-546	RT-905 <*>
1949 V13 N5 P551-556	T-2367 <INSD>
1949 V13 N6 P561-596	RT-292 <*>
1949 V13 N6 P597-608	RT-696 <*>
1949 V13 N6 P655-658	RT-915 <*>
1949 V13 N6 P659-662	RT-914 <*>
1950 V14 P197-202	R-3840 <*>
1950 V14 P405-414	AMST S1 V11 P258-272 <AMS>
	RT-142 <*>
1950 V14 P670-671	R-3841 <*>
1950 V14 N1 P93-98	63-10252 <=*>
1950 V14 N2 P139-170	RT-551 <*>
1950 V14 N2 P171-182	RT-552 <*>
1950 V14 N2 P183-192	RT-1232 <*>
1950 V14 N2 P193-196	RT-555 <*>
	52/1397 <NLL>
1950 V14 N2 P197-202	RT.554 <*>
1950 V14 N2 P203-208	RT-553 <*>
1950 V14 N2 P209-214	RT-442 <*>
1950 V14 N2 P215-217	RT-443 <*>
1950 V14 N2 P218-224	RT-444 <*>
1950 V14 N3 P265-276	RT-401 <*>
1950 V14 N3 P316-318	62-18562 <=*>
1950 V14 N3 P321-339	52/1398 <NLL>
1950 V14 N4 P441-443	RT-907 <*>
1950 V14 N6 P651-658	RT-761 <*>
1950 V15 P295-302	RT-1439 <*>
1951 V15 P120	60-23938 <+*>
1951 V15 P167-174	65-17305 <*>
1951 V15 P323-348	AMST S2 V1 P163-187 <AMS>
1951 V15 P371-372	<EEUP>
1951 V15 P765-770	77L35R <ATS>
1951 V15 N1 P3-26	RT-240 <*>
1951 V15 N2 P167-174	RTS-2944 <NLL>
1951 V15 N2 P227-236	AD-618 127 <=$>
1951 V15 N6 P751-761	60-13756 <+*>
1952 V16 P365-368	62-10146 <*=>
1952 V16 P569-574	65-12630 <*>
1952 V16 P617-619	R-4398 <*>
1952 V16 P706-710	AEC-TR-4041 <+*>
1952 V16 N1 P67-78	61-17229 <=$>
1952 V16 N1 P116-118	S-2104 <RIS>
1952 V16 N1 P116	59-15414 <+*>
1952 V16 N1 P116-118	62-23229 <*=>
1952 V16 N2 P255-256	R-1118 <*>
	RT-3114 <*>
	TT.383 <NRC>
	64-14915 <=*$>
1952 V16 N3 P319-322	RT-2387 <*>
1952 V16 N3 P341-344	62-18563 <=*>
1952 V16 N3 P345-348	AD-619 554 <=$>
1952 V16 N3 P369-374	65-11034 <*>
1952 V16 N4 P465-486	RT-3047 <*>
1952 V16 N4 P495-499	61-15210 <+*>
1952 V16 N5 P569-574	59-19140 <+*>
1952 V16 N5 P629-632	RT-3588 <*>
1952 V16 N5 P635-648	64-14269 <=*$>
1952 V16 N6 P719-722	S-2105 <RIS>
1953 V17 P3-16	22L29R <ATS>
1953 V17 P73-86	61-13532 <=*>
1953 V17 P165-178	59-14582 <+*>
1953 V17 P261-274	61-23311 <*=>
1953 V17 P455-460	UCRL TRANS-989(L) <*>
	66-13897 <*$> O
1953 V17 P485-490	RT-1488 <*>
1953 V17 P529-540	63-16987 <=*>
1953 V17 P615-618	59-18663 <+*>
1953 V17 N3 P361-368	RT-2556 <*>
	61-10443 <=*>
1953 V17 N3 P369-372	62-10181 <=*>
1953 V17 N3 P382-386	AD-632 286 <=$>
1953 V17 N4 P389-400	AD-618 128 <=$>
1953 V17 N5 P517-528	RT-3647 <*>
1953 V17 N5 P579-592	S-2106 <RIS>
	62-10182 <*=>
1953 V17 N6 P705-726	S-2107 <RIS>
	62-10183 <=*>
1953 V17 N6 P743-744	R-1119 <*>
	64-14916 <=*$>
1953 V17 N6 P755-760	RT-3646 <*>
1954 V18 P35-42	62-26064 <=$>
1954 V18 P75-94	63-20238 <=*>
1954 V18 P163-166	62-10185 <=*>
1954 V18 P257-264	R-3740 <*>
1954 V18 P303-312	64-18556 <*>
1954 V18 P345-350	62-10184 <=*>
1954 V18 P469-510	AMST S2 V1 P189-237 <AMS>
1954 V18 P512	63-16987 <=*>
1954 V18 N1 P35-42	RT-3643 <*>
	62-23112 <=*>
	63-15671 <=*>
1954 V18 N2 P181-186	RT-3644 <*>
1954 V18 N2 P212-214	61-27041 <*=>
1954 V18 N3 P257-264	64-71240 <=>
1954 V18 N3 P371-378	62-15870 <=*>
1954 V18 N4 P452-453	63-18316 <*=>
1954 V18 N5 P631-636	S-2108 <RIS>
1954 V18 N6 P667-674	NACA TM1433 <NASA>
	T2194 <INSD>
1955 V19 P41-54	R-981 <*>
	875TM <CTT>
1955 V19 P251-254	62-26065 <=$>
1955 V19 P509-512	R-4401 <*>
	64-71367 <=>
1955 V19 P681-692	876TM <CTT>
1955 V19 N1 P121-126	64-16022 <=*$>
1955 V19 N2 P179-210	62-20253 <=*>
1955 V19 N2 P249-250	RT-3641 <*>
1955 V19 N2 P251-254	RT-3642 <*>
1955 V19 N3 P279-286	62-20396 <=*>
1955 V19 N3 P295-314	61-10404 <=*>

1955 V19 N3 P353-358	S-2109 <RIS>
1955 V19 N4 P453-462	61-10402 <+*>
1955 V19 N4 P463-470	64-16382 <=*$>
1955 V19 N5 P516-530	61-10773 <+*>
1955 V19 N5 P531-540	61-23272 <=*>
1955 V19 N5 P599-616	62-20394 <=*>
1955 V19 N5 P623-624	61-14763 <=*>
1955 V19 N6 P644-680	61-10842 <+*>
1956 V20 P203-210	62-15919 <=*>
1956 V20 P325-327	T1909 <INSD>
1956 V20 P449-474	AD-639 553 <=$>
1956 V20 P487-499	59-11986 <=>
1956 V20 P513-518	R-2331 <*>
1956 V20 P532-544	66-14392 <*$>
1956 V20 N1 P3-20	60-21468 <=>
1956 V20 N1 P21-38	60-17360 <+*>
1956 V20 N1 P67-72	NACA TM 1413 <NASA>
1956 V20 N1 P73-86	61-27046 <*=>
1956 V20 N1 P133-135	61-23273 <=*>
1956 V20 N2 P211-222	RE 11-22-58W <NASA>
1956 V20 N2 P284-288	R-3623 <*>
1956 V20 N2 P293-294	64-13758 <=*$>
1956 V20 N3 P297-308	59-12142 <+*>
	60-18644 <+*>
1956 V20 N3 P395-401	60-17214 <+*>
1956 V20 N4 P475-486	62-14583 <=*>
1956 V20 N4 P552-554	60-10311 <+*>
1956 V20 N5 P599-605	59-17702 <+*>
1956 V20 N5 P663-665	62-11193 <=>
1956 V20 N6 P761-763	62-11326 <=>
1956 V20 N9 P671-672	64-71419 <=> M
1957 V21 P175-183	62-11277 <=>
1957 V21 P309-319	AMST S2 V18 P275-287 <AMS>
1957 V21 P658-669	AMST S2 V26 P339-351 <AMS>
1957 V21 P670-677	AMST S2 V28 P323-332 <AMS>
1957 V21 N1 P3-14	59-18093 <+*>
1957 V21 N1 P83-88	60-23076 <+*>
1957 V21 N1 P119-120	61-16124 <=*>
1957 V21 N1 P143-144	63-18317 <*=>
1957 V21 N2 P157-168	61-19347 <+*>
1957 V21 N2 P175-183	RE 3-10-59W <NASA>
	59-11996 <=>
1957 V21 N2 P189-194	RE 5-1-59W <NASA>
1957 V21 N2 P195-206	59-10317 <+*>
	62-23023 <=>
1957 V21 N2 P207-212	60-17349 <+*>
1957 V21 N2 P213-220	60-17350 <+*>
1957 V21 N2 P244-252	59-16782 <+*>
1957 V21 N2 P253-261	59-16779 <+*>
1957 V21 N2 P262-271	60-17347 <+*>
1957 V21 N3 P320-329	61-10621 <+*>
1957 V21 N3 P347-352	M.5143 <NLL>
1957 V21 N3 P440-444	64-13387 <=*$>
1957 V21 N3 P445-448	64-11957 <=> M
1957 V21 N4 P449-458	61-11988 <=>
	61-13364 <=*>
1957 V21 N4 P525-532	S-2110 <RIS>
1957 V21 N4 P585-590	59-16058 <+*>
1957 V21 N4 P593-595	64-11946 <=> M
1957 V21 N5 P597-605	60-13561 <+*>
	60-21612 <=>
	64-71399 <=> M
1957 V21 N5 P606-614	60-17361 <+*>
1957 V21 N5 P615-623	60-21609 <=>
1957 V21 N5 P644-657	NACA TM-1438 <NASA>
1957 V21 N5 P670-677	60-18806 <+*>
	63-14455 <*=>
1957 V21 N5 P693-695	61-27298 <*=>
1957 V21 N5 P696-700	59-12164 <+*>
	59-22534 <+*>
	61-17582 <=$>
	64-71347 <=>
1957 V21 N6 P725-739	59-14155 <+*>
1957 V21 N6 P769-774	60-17381 <+*>
1957 V21 N6 P801-814	65-10783 <*>
1957 V21 N6 P815-822	59-21004 <=>
1957 V21 N6 P856-859	62-19818 <=*>
1958 V22 N2 P155-166	60-13101 <+*>
1958 V22 N2 P173-178	60-13101 <+*>
1958 V22 N2 P245-249	60-13101 <+*>
1958 V22 N3 P350-358	59-21060 <=>
1958 V22 N3 P412-414	CSIR-TR.221 <CSSA>
1958 V22 N4 P565-568	59-15483 <+*>
	62-23229 <*=>
1958 V22 N6 P766-780	59-19458 <+*>
1959 V23 N2 P292-298	61-15273 <*+>
1959 V23 N2 P412-413	61-14760 <=*>
1959 V23 N5 P862-878	60-21192 <=>
1959 V23 N5 P885-892	62-23064 <=>
1959 V23 N6 P1115-1123	60-31152 <=*>
1959 V23 N6 P1124-1128	60-31151 <=*>
1960 V24 N2 P303-308	60-23792 <+*>
1960 V24 N4 P732-733	61-10668 <*>
1960 V24 N4 P746-749	61-10750 <=*>
1961 V25 N1 P17-23	62-10148 <*=>
1961 V25 N2 P381-382	61-27818 <*=>
1961 V25 N3 P413-419	62-15939 <=*>
1961 V25 N3 P498-502	RTS-2651 <NLL>
1961 V25 N5 P845-850	62-15405 <*=>
1961 V25 N5 P965-968	62-15405 <*=>
1962 V26 N1 P15-21	63-19245 <=*$>
1962 V26 N1 P22-28	62-25408 <=*>
1962 V26 N1 P29-38	63-31540 <=>
1962 V26 N1 P62-79	63-31540 <=>
1962 V26 N2 P247-258	62-33545 <=*>
1962 V26 N2 P328-334	62-11717 <=>
1962 V26 N2 P365-369	62-33545 <=*>
1962 V26 N3 P431-443	62-18958 <=*>
1962 V26 N3 P431-433	62-32549 <=*>
1962 V26 N3 P444-448	63-14398 <*=>
1962 V26 N6 P992-1002	63-20287 <=*>
1963 V27 N1 P126-134	63-31760 <=>
1963 V27 N1 P175-177	63-18711 <=*>
1963 V27 N1 P179-182	ICE V5 N1 P8-11 <ICE>
	83R77R <ATS>
1963 V27 N2 P193-202	63-23196 <=*>
1963 V27 N2 P211-217	63-18712 <=*>
1963 V27 N2 P255-264	64-71307 <=> M
1963 V27 N2 P377-378	63-18815 <=*>
1963 V27 N2 P383-389	64-71505 <=> M
1963 V27 N3 P393-417	64-11962 <=> M
1963 V27 N3 P428-435	64-11962 <=> M
1963 V27 N3 P450-458	64-71387 <=> M
1963 V27 N3 P496-508	AD-613 574 <=$>
1963 V27 N3 P541-546	63-31707 <=>
1963 V27 N4 P593-608	63-20851 <=*$>
	64-11937 <=> M
1963 V27 N5 P918-923	64-21054 <=>
1964 V27 N3 P548-553	65-63647 <=$*>
1964 V28 N2 P193-215	AD-618 650 <=$>
1964 V28 N2 P351-355	65-10384 <*>
1964 V28 N2 P366-372	64-41145 <=>
1964 V28 N2 P375-380	64-41145 <=>
1964 V28 N5 P949-951	AD-620 775 <=$>
1964 V28 N5 P974-976	N65-11304 <=$>
1964 V28 N6 P1008-1014	N65-15734 <=$>
1964 V28 N6 P1135-1137	65-13336 <*>
	65-63230 <=*$>
1965 N6 P122-125	AD-630 168 <=*$>
1965 V29 P334-336	AD-640 526 <=$>
1965 V29 N1 P106-113	AD-629 416 <=$>
1965 V29 N1 P141-155	AD-629 416 <=$>
1965 V29 N4 P616-634	AD-628 303 <=*$> M
1965 V29 N4 P745-750	65-64355 <=*$>
	66-10471 <*>
1965 V29 N4 P751	65-64354 <=*$>

★PRIKLADNAYA MEKHANIKA

1965 V1 N1 P52-61	AD-645 819 <=>
1965 V1 N1 P92-97	AD-645 819 <=>
1965 V1 N2 P48-55	N66-13049 <=*$>
1965 V1 N2 P78-85	AD-625 297 <=*$>
1965 V1 N2 P104-109	N66-14056 <=$>
1965 V1 N2 P134-138	N66-12262 <=*$>
1965 V1 N3 P123-127	65-64358 <=*$>
1965 V1 N5 P11-20	65-14431 <*>
	65-63715 <=$>
1965 V1 N5 P21-28	65-14540 <*>
	65-63704 <=$>
1965 V1 N5 P100-104	65-14538 <*>

65-63707 <=$>
1965 V1 N5 P127-128 65-14541 <*>
65-63711 <=$>
1965 V1 N5 P133-137 N66-12502 <=$>
N66-23546 <=$>
1965 V1 N6 P71-77 65-63785 <=$>
66-10466 <*>
1965 V1 N6 P78-84 65-63786 <=*$>
66-10469 <*>
1965 V1 N6 P92-96 66-10462 <*>
66-61706 <=*$>
1965 V1 N7 P101-106 65-64095 <=*$>
66-10465 <*>
1965 V1 N8 P17-22 66-10461 <*>
1965 V1 N8 P57-62 66-10464 <*>
1965 V1 N9 P14-19 66-12485 <*>
1965 V1 N9 P20-25 66-12484 <*>
66-61033 <=*$>
1965 V1 N9 P115-118 66-10463 <*>
1965 V1 N11 P1-6 66-12481 <*>
66-60597 <=$>
1965 V1 N11 P12-19 66-12482 <*>
66-60590 <=$>
1965 V1 N11 P20-27 66-60589 <=$>
1965 V1 N11 P106-112 66-12483 <*>
1965 V1 N12 P82-86 66-61980 <=*$>
1966 V2 N2 P100-103 66-61331 <=*$>
1966 V2 N2 P110-116 66-61337 <=*$>
1966 V2 N2 P130-135 66-61336 <=*$>
1966 V2 N5 P65-70 66-13474 <*>
66-61726 <=$>
1966 V2 N6 P105-111 66-62296 <=*$>
1966 V2 N6 P112-121 66-62297 <=*$>
1966 V2 N6 P122-124 66-62298 <=*$>

PRIMENENIE MATHEMATICHESKIKH METODOV V BIOLOGII
1963 V2 P5-11 64-21569 <=>
1963 V2 P124-131 64-21569 <=>
1963 V2 P140-145 64-21569 <=>
1963 V2 P146-151 64-21569 <=>
1963 V2 P161-169 64-21569 <=>
1963 V2 P234-238 64-21569 <=>

PRIMENENIE ULTRAAKUSTIKI K ISSLEDOVANIYA VESHCHESTVA
1960 N12 P121-123 65-12740 <*>
1960 N12 P141-145 64-71397 <=> M

PRIRODA
1929 V12 N8/9 P321-329 61-01058 <*>
1937 V26 N12 P100-102 RT-3049 <*>
1938 N4 P109-112 R-796 <*>
1940 N1 P79-81 <ATS>
1940 N2 P65 <ATS>
1940 N6 P115-116 R-1482 <*>
1940 V29 N8 P21-27 RT-3634 <*>
1944 V33 N1 P3-11 60-27015 <*+> O
1944 V33 N5/6 P107-109 60-13263 <+*>
1945 V34 N6 P29 1537 <K-H>
1946 V35 N4 P5-8 62-14804 <=*>
1946 V35 N4 P58-60 RT-2425 <*>
1946 V35 N5 P7-23 RT-3061 <*>
1946 V35 N5 P24-28 RT-1286 <*>
1946 V35 N8 P33-38 R-3157 <*>
1947 N11 P3-12 R-922 <*>
1947 N11 P53-54 C-2214 <NLL>
R-39 <*>
RT-4605 <*>
1947 V36 P19-23 R-4800 <*>
1947 V36 N1 P37-40 R.626 <RIS>
1947 V36 N1 P46 R.627 <RIS>
1947 V36 N1 P47-50 R.642 <RIS>
1947 V36 N1 P60-61 R.632 <RIS>
1947 V36 N1 P61-62 R.630 <RIS>
1947 V36 N1 P62-63 R.641 <RIS>
1947 V36 N1 P74-84 R.631 <RIS>
1947 V36 N2 P7-14 R.625 <RIS>
1947 V36 N2 P43-44 R.637 <RIS>
1947 V36 N2 P44-45 R.622 <RIS>
1947 V36 N2 P45-48 R.623 <RIS>
1947 V36 N2 P48-49 R.363 <RIS>
1947 V36 N2 P48 R.635 <RIS>
1947 V36 N3 P44-45 R.620 <RIS>
1947 V36 N3 P45-46 R.618 <RIS>
1947 V36 N6 P13-17 R.624 <RIS>
1947 V36 N6 P49 R.639 <RIS>
1947 V36 N6 P49-50 R.640 <RIS>
1947 V36 N6 P49 R-614 <RIS>
1947 V36 N6 P50-52 RT-2432 <*>
1947 V36 N7 P16-20 618 <RIS>
1947 V36 N9 P25-32 61-11193 <=>
1947 V36 N11 P3-12 RT-4029 <*>
1947 V36 N11 P53-54 62-18267 <=*>
1947 V36 N11 P75-78 RT-538 <*>
RT-538 <*> O
64-15631 <=*$> O
1948 V37 N4 P62-63 RT-760 <*>
1948 V37 N5 P33-34 RT-3558 <*>
1948 V37 N6 P46-48 RT-2433 <*>
1948 V37 N7 P16-24 60-19019 <+*>
1948 V37 N12 P55-56 CSIR-TR 270 <CSSA>
1949 V38 N1 P3-14 RT-1188 <*>
1949 V38 N2 P61-62 R-271 <*>
1949 V38 N3 P61-62 61-28118 <*=>
1949 V38 N7 P59-60 RT-3151 <*> O
1949 V38 N10 P23-26 RT-3425 <*>
1949 V38 N10 P51-53 51/0103 <NLL>
1949 V38 N11 P10-23 RT-2108 <*>
1950 N1 P25-32 RT-2197 <*>
1950 N6 P35-46 50/2802 <NLL>
1950 N11 P5-10 RT-2199 <*>
1950 V39 N1 P25-32 61-13177 <*+>
1950 V39 N4 P64-65 R-432 <*>
64-19504 <=$>
1950 V39 N6 P75 C-5286 <NRC>
1950 V39 N6 P79-80 RT-1791 <*>
1950 V39 N10 P67-68 RT-537 <*>
1950 V39 N11 P5-10 61-13177 <*+>
1951 N1 P60-62 RJ-53 <ATS>
1951 N2 P54- 61-17620 <=$>
1951 V40 N1 P60-62 RJ-53 <ATS>
1951 V40 N1 P73-75 R-1778 <*>
1951 V40 N2 P40-42 RT-2084 <*>
1951 V40 N2 P54 RT-2105 <*>
1951 V40 N5 P64 52/2588 <NLL>
1951 V40 N9 P48-52 RT-1689 <*>
1951 V40 N11 P39-46 61-13652 <*=>
1951 V40 N11 P48-52 61-23419 <=*>
1951 V40 N12 P20-28 59-19108 <+*>
1951 V40 N12 P44 RT-1564 <*>
1952 V41 N4 P37-46 RT-1038 <*> O
1952 V41 N5 P49-59 R-5018 <*>
RT-4015 <*>
1952 V41 N7 P92-95 C-4274 <NRC>
1952 V41 N8 P107-111 R-3095 <*>
1952 V41 N9 P94-99 RT-1393 <*>
1953 P102-104 R-5342 <*>
1953 N1 P35-43 R-1544 <*>
1953 N1 P109-112 C-2105 <NLL>
1953 N2 P114 RT-2713 <*>
1953 N3 P101-104 R-1454 <*>
1953 N8 P33-41 RT-2374 <*>
1953 V42 N1 P60-63 59-19146 <+*>
1953 V42 N1 P90-91 RT-3048 <*>
1953 V42 N2 P15-22 RT-2902 <*>
1953 V42 N2 P107-110 62-25064 <=>
1953 V42 N2 P128 RT-3295 <*>
1953 V42 N3 P37-47 RT-2152 <*>
1953 V42 N3 P101-104 64-19482 <=$>
1953 V42 N6 P20-25 RT-3981 <*>
1953 V42 N6 P85-87 RT-3025 <*>
1953 V42 N7 P13-21 64-11945 <=> M
1953 V42 N10 P97-102 60-21909 <=>
1953 V42 N12 P88-90 AD-607 722 <=>
1954 N1 P75-76 RT-2375 <*>
1954 N2 P23-29 RT-4470 <*>
1954 N2 P75-77 RT-2536 <*>
1954 N2 P103-106 RT-4577 <*>
1954 N5 P67-73 R-2337 <*>
1954 N5 P87-92 RT-4366 <*>

Citation	Translation
1954 N6 P63-66	RT-3477 <*>
1954 N6 P121-123	RT-2303 <*>
1954 N10 P36-42	IGR V2 N4 P311 <AGI>
1954 N10 P91-95	84K21R <ATS>
1954 N12 P3-12	RT-3360 <*>
1954 V43 N1 P45-53	T.898 <INSD>
1954 V43 N1 P54-63	RT-3221 <*>
1954 V43 N2 P30-40	91K21R <ATS>
1954 V43 N3 P11-20	RT-3296 <*> 0
1954 V43 N5 P45-51	R-4565 <*>
	2985 <NLL>
1954 V43 N6 P63-66	AD-610 907 <=$>
1954 V43 N6 P112-113	RT-2899 <*>
1954 V43 N8 P116	SCL-T-233 <*> P
1954 V43 N10 P36-42	R-4425 <*>
1954 V43 N12 P51-56	RT-3447 <*>
1955 N5 P37-44	760TM <CTT>
1955 N5 P81-85	RT-2960 <*>
1955 N7 P96-98	RT-3306 <*>
1955 N8 P3-19	R-1764 <*>
1955 N8 P28-34	R-3685 <*>
1955 N12 P79-82	RT-4484 <*>
1955 V44 N1 P10-15	RT-4392 <*>
1955 V44 N1 P76-77	RT-3227 <*>
1955 V44 N1 P84-85	60-11409 <=>
1955 V44 N2 P37-45	R-5092 <*>
1955 V44 N4 P9-23	RT-2942 <*>
1955 V44 N5 P80-81	RT-3046 <*>
1955 V44 N6 P89-93	RT-3640 <*>
1955 V44 N7 P13-22	RT-3338 <*>
1955 V44 N7 P65-72	C-2273 <NRC>
	R-4812 <*>
1955 V44 N7 P92-96	RT-3339 <*>
1955 V44 N7 P96-98	AD-610 904 <=$>
1955 V44 N9 P61-68	R-4815 <*>
1955 V44 N10 P86-88	RT-3476 <*>
1955 V44 N11 P3-12	RT-4266 <*>
1955 V44 N12 P79-82	R-1039 <*>
1955 V44 N12 P121-122	RT-4225 <*>
1956 N2 P23-37	RT-4173 <*> 0
1956 N5 P41-48	60-00254 <*>
1956 V45 N1 P24-34	RT-4267 <*>
1956 V45 N4 P43-51	R-4418 <*>
1956 V45 N4 P85-89	61-23403 <=*>
1956 V45 N4 P96-97	RT-4620 <*>
1956 V45 N5 P17-26	R-1813 <*>
1956 V45 N5 P41-48	60-17692 <+*>
1956 V45 N5 P71-73	60-15947 <+*>
1956 V45 N6 P13-18	59-13963 <=>
1956 V45 N6 P33-41	63-23695 <=*>
1956 V45 N7 P89-91	C-3303 <NRCC>
	60-17426 <+*>
1956 V45 N8 P15-21	64-10964 <=*$>
1956 V45 N8 P114-115	60-21116 <=>
1956 V45 N9 P94-95	C-2258 <NRC>
	R-404 <*>
1956 V45 N1C P94-95	59-18515 <+*>
1956 V45 N11 P43-47	R-1594 <*>
1956 V45 N12 P95-97	61-19714 <=*> 0
1957 N1 P87-90	IGR V2 N4 P346 <AGI>
1957 N4 P27-33	R-3151 <*>
1957 N5 P11-20	IGR V2 N1 P43 <AGI>
1957 N5 P21-30	R-3138 <*>
1957 V46 N1 P43-46	64-13944 <=*$>
1957 V46 N1 P117	C-2339 <NRC>
	R-1459 <*>
1957 V46 N2 P14-24	59-13599 <=> 0
1957 V46 N3 P31-37	59-15778 <+*>
1957 V46 N3 P98-100	R-2399 <*>
1957 V46 N4 P71-73	60-13009 <+*>
	62-11486 <=>
1957 V46 N4 P105-106	64-00473 <*>
	64-11822 <=>
1957 V46 N5 P21-30	AD-610 910 <=$>
	C-2515 <NRC>
	59-14106 <+*>
1957 V46 N5 P117-119	59-19522 <+*>
1957 V46 N6 P3-8	59-11295 <=>
1957 V46 N6 P3-12	59-16069 <+*>
1957 V46 N6 P3-8	60-15589 <+*>
1957 V46 N6 P26-28	PB 141 247T <=>
1957 V46 N6 P71-73	61-23279 <*=>
1957 V46 N6 P1C7-108	R-3797 <*>
	64-19422 <=$>
1957 V46 N7 P19-30	59-11985 <=> 0
1957 V46 N7 P95-97	62-33216 <=*>
1957 V46 N8 P35-44	59-11987 <=>
1957 V46 N8 P88-89	R-5210 <*>
1957 V46 N8 P90-92	60-14002 <+*> 0
1957 V46 N8 P97-98	R-5214 <*>
1957 V46 N8 P113-114	R-5213 <*>
1957 V46 N9 P120 INSERT	AD-610 963 <=$>
1957 V46 N9 P3-12	60-17357 <+*>
1957 V46 N9 P87-91	63-19572 <=*>
1957 V46 N10 P73-78	59-16041 <+*>
1957 V46 N10 P89-92	<ANSP>
1957 V46 N10 P126-127	60-15595 <+*> 0
1957 V46 N11 P79-88	60-41663 <=>
1957 V46 N12 P15-20	61-19346 <+*>
1957 V46 N12 P27-34	AD-610 966 <=$>
	C-2639 <NRC>
	R-4020 <*>
1957 V46 N12 P49-54	59-15954 <+*> C
1957 V46 N12 P87-88	AD-610 911 <=$>
	C-2584 <NRC>
	R-3454 <*>
1957 V46 N12 P88-91	59-11327 <=>
1957 V46 N12 P107-111	63-31577 <=>
1957 V46 N12 P113	61-13794 <+*>
1958 N1 P87-89	C-2691 <NRC>
1958 N1 P95-99	IGR V2 N7 P623 <AGI>
1958 N2 P46-52	R-4744 <*>
1958 N5 P75-77	PB 141 457T <=>
1958 N9 P15-24	IGR V1 N8 P40 <AGI>
1958 N10 P112	NLLTB V1 N2 P19 <NLL>
1958 N12 P11-14	NLLTB V1 N5 P35 <NLL>
1958 V47 N1 P19-25	AD-610 967 <=$>
	C-2623 <NRC>
	R-3742 <*>
1958 V47 N1 P87-89	R-4264 <*>
1958 V47 N1 P95-99	PB 141 235T <=>
1958 V47 N3 P17-25	PB 141 455T <=>
1958 V47 N3 P26-31	60-18760 <+*>
1958 V47 N3 P32-35	61-13979 <+*> 0
1958 V47 N3 P88-90	AD-610 972 <=$>
	C-2690 <NRC>
	R-4356 <*>
1958 V47 N4 P78-79	61-23401 <=*>
1958 V47 N4 P79-80	59-19172 <+*>
1958 V47 N4 P87-90	AD-610 873 <=$>
	R-4457 <*>
1958 V47 N4 P111-112	59-19173 <+*>
1958 V47 N4 P113-114	66-51096 <=>
1958 V47 N5 P49-	59-11933 <=>
1958 V47 N5 P77-79	59-18849 <+*>
1958 V47 N5 P86-87	61-23392 <*=>
1958 V47 N6 P85-87	R-4953 <*>
	T 304 R <NRC>
1958 V47 N7 P84-85	59-16046 <+*>
1958 V47 N7 P85-88	62-24699 <=*>
1958 V47 N7 P107-108	62-25064 <=>
1958 V47 N8 P3-12	59-11667 <=> 0
1958 V47 N8 P14	60-13026 <+*>
1958 V47 N8 P50-55	59-20256 <+*>
1958 V47 N8 P66-73	59-19174 <+*>
1958 V47 N8 P83-85	59-11668 <=> 0
1958 V47 N8 P1C9-111	RTS-2810 <NLL>
1958 V47 N8 P113-115	AEC-TR-3526 <+*> 0
1958 V47 N9 P38-44	59-11999 <=>
1958 V47 N9 P68-71	59-15281 <+*>
1958 V47 N9 P85-88	59-11429 <=> 0
1958 V47 N9 P97-99	62-25663 <=*> 0
1958 V47 N9 P107-108	62-10435 <*=>
1958 V47 N10 P71-77	59-19487 <+*>
1958 V47 N10 P90-91	61-10354 <*+>
1958 V47 N11 P85-87	59-15279 <+*>
	61-17587 <=$>
1958 V47 N11 P96-98	11311 <K-H>
	61-14969 <+*>
1958 V47 N12 P71-78	60-17176 <+*>

Citation	Translation
	61-13536 <+*>
1958 V47 N12 P88-90	59-15280 <+*>
	61-17584 <=$>
1958 V47 N13 P33-38	59-13529 <=> 0
1959 N2 P17-	59-21063 <=>
1959 N3 P84-88	IGR V1 N10 P40 <AGI>
1959 N6 P19-26	IGR V1 N12 P72 <AGI>
1959 V48 N1 P43-45	60-13812 <+*>
1959 V48 N1 P103-108	AEC-TR-3769 <+*>
1959 V48 N1 P108-111	59-13821 <=>
1959 V48 N2 P7-11	59-16824 <+*> 0
1959 V48 N2 P98-100	59-16825 <+*> 0
1959 V48 N2 P123	C-3415 <NRCC>
1959 V48 N3 P35-43	59-13670 <=> 0
1959 V48 N3 P115	59-16425 <+*>
1959 V48 N4 P5-8	62-14015 <=*>
1959 V48 N4 P97-99	60-11071 <=>
1959 V48 N5 P74-76	60-13894 <+*>
	62-32784 <=*>
1959 V48 N5 P77-81	62-32783 <=*>
1959 V48 N5 P108-109	60-11114 <=>
1959 V48 N6 P19-26	61-15965 <*=> 0
1959 V48 N6 P62-68	60-11023 <=>
1959 V48 N6 P75-	60-21200 <=> 0
1959 V48 N6 P84-87	62-24699 <=*>
1959 V48 N6 P98-100	C-3242 <NRCC>
1959 V48 N7 P13-18	60-21927 <=>
1959 V48 N7 P27-32	61-13233 <*+>
1959 V48 N7 P97-99	61-28151 <*=> 0
1959 V48 N8 P3-10	60-31498 <=>
1959 V48 N8 P11-18	60-31003 <=>
	61-15815 <+*>
1959 V48 N8 P86-89	60-21010 <=>
1959 V48 N8 P93-94	AD-617 146 <=$>
1959 V48 N9 P3-10	60-11778 <=>
1959 V48 N10 P1-7	60-11536 <=>
	60-31250 <=>
1959 V48 N10 P19-26	62-10144 <*=>
1959 V48 N10 P27-34	C-3285 <NRCC>
	60-11508 <=>
1959 V48 N10 P35-44	60-11516 <=>
	65-62708 <=>
1959 V48 N10 P51-54	60-11427 <=>
1959 V48 N10 P57-59	60-11358 <=>
1959 V48 N10 P67-69	60-11902 <=>
1959 V48 N10 P82-85	60-11498 <=>
1959 V48 N10 P85-86	60-11369 <=>
1959 V48 N10 P87-89	60-11509 <=>
1959 V48 N10 P89-91	61-10810 <+*>
1959 V48 N10 P95-97	60-10893 <+*>
1959 V48 N10 P107-108	60-11903 <=>
1959 V48 N11 P16-24	60-23145 <+*>
1959 V48 N12 P22-28	60-41173 <=>
1959 V48 N12 P35-40	61-13756 <=*>
1959 V48 N12 P45-50	60-17958 <+*> 0
1959 V48 N12 P98-100	60-17957 <+*>
1960 N6 P75-77	IGR V3 N3 P227 <AGI>
1960 V49 N1 P9-19	60-11596 <=>
1960 V49 N1 P20-25	60-11597 <=>
1960 V49 N1 P75-82	61-27566 <*=>
1960 V49 N2 P33-38	60-11683 <=>
1960 V49 N2 P75-77	60-41267 <=>
1960 V49 N2 P124-128	61-27496 <*=>
1960 V49 N3 P5-10	60-31456 <=>
1960 V49 N3 P16-22	62-13120 <=*>
1960 V49 N3 P33-38	60-31756 <=>
1960 V49 N4 P3-11	62-25989 <=>
1960 V49 N4 P64-70	ARSJ V31 N5 P678 <AIAA>
	60-41205 <=>
1960 V49 N4 P93-94	61-27565 <*=>
1960 V49 N4 P104-105	62-25989 <=>
1960 V49 N5 P11-18	60-41047 <=>
1960 V49 N5 P50-53	60-31738 <=>
1960 V49 N5 P54-56	60-31739 <=>
1960 V49 N6 P3-6	61-21681 <=>
1960 V49 N6 P7-14	64-11069 <=>
1960 V49 N6 P23-26	61-21681 <=>
1960 V49 N6 P27-34	61-11859 <=>
1960 V49 N6 P35-42	61-28008 <*=>
1960 V49 N6 P100	63-13026 <=*>
1960 V49 N6 P104-105	61-21681 <=>
1960 V49 N6 P125	61-21681 <=>
1960 V49 N7 P49-54	62-33085 <=*>
1960 V49 N8 P19-24	61-11973 <=>
	61-23209 <*=> 0
1960 V49 N8 P68-69	61-11013 <=>
1960 V49 N8 P91-93	61-11697 <=>
1960 V49 N9 P3	61-28700 <*=>
1960 V49 N9 P5	61-28700 <*=>
1960 V49 N9 P27-32	61-27376 <*=>
1960 V49 N9 P92-94	61-27650 <*=>
1960 V49 N9 P94-96	61-27653 <*=>
1960 V49 N10 P8-15	61-23952 <*=>
1960 V49 N10 P16-20	61-21401 <=>
1960 V49 N10 P21-26	62-23269 <=*> 0
1960 V49 N10 P34-41	61-21434 <=>
1960 V49 N10 P42-45	61-28208 <*=>
1960 V49 N10 P76-	61-21425 <=>
1960 V49 N10 P78-80	61-27360 <*=>
1960 V49 N10 P90-	61-21392 <=>
1960 V49 N11 P14-21	62-19285 <=*>
1960 V49 N11 P68	62-33001 <=*>
1960 V49 N11 P69-72	62-10186 <=*>
1960 V49 N12 P10-16	62-15275 <*=>
1960 V49 N12 P32-39	61-27139 <*=> 0
1961 N1 P41-50	IGR V4 N3 P253 <AGI>
1961 V50 N1 P7-16	61-23926 <=*>
1961 V50 N1 P17-24	61-27392 <*=>
1961 V50 N1 P51-55	61-23617 <*=>
1961 V50 N1 P61-65	61-23626 <=*>
1961 V50 N1 P94-96	61-21488 <=>
1961 V50 N2 P23-30	61-28697 <*=>
1961 V50 N2 P55-59	62-15934 <=*>
1961 V50 N3 P9-14	61-27151 <*=>
1961 V50 N3 P61-66	61-27082 <*=>
1961 V50 N3 P67-70	61-27087 <*=>
1961 V50 N3 P71-72	61-27115 <*=>
1961 V50 N4 P1-4	61-28269 <*=>
1961 V50 N4 P3-8	61-28039 <*=>
1961 V50 N4 P9-16	61-28690 <*=>
1961 V50 N5 P3-6	61-31545 <=>
1961 V50 N5 P65-66	62-25500 <=*>
1961 V50 N5 P126	62-32235 <=*>
1961 V50 N6 P1-2	61-31599 <=>
1961 V50 N6 P3-5	62-15933 <=*>
1961 V50 N7 P3-4	62-13314 <*=>
1961 V50 N7 P5-12	62-13335 <*=>
1961 V50 N7 P13-18	62-13661 <*=>
1961 V50 N7 P25-32	61-28522 <*=> 0
1961 V50 N7 P40-44	62-13336 <*=>
1961 V50 N7 P81-87	61-31648 <=> 0
1961 V50 N8 P7-15	62-32420 <=*>
1961 V50 N8 P16-20	62-13708 <*=>
	62-24869 <=*>
1961 V50 N8 P33-41	63-15277 <=*>
1961 V50 N8 P88-93	62-19797 <=*>
1961 V50 N8 P90-91	63-16246 <=*>
1961 V50 N9 P9-10	62-15207 <*=>
1961 V50 N9 P24-31	62-15452 <*=>
1961 V50 N9 P41-47	62-13858 <=*>
1961 V50 N9 P54-60	62-15207 <*=>
1961 V50 N9 P75-77	62-15207 <*=>
1961 V50 N9 P105-109	CSIR-TR.200 <CSSA>
1961 V50 N10 P23-31	DSIS T 366 R <NRCC>
	62-14557 <=*>
1961 V50 N10 P23-37	62-24949 <=*>
1961 V50 N10 P82-87	62-24949 <=*>
1961 V50 N11 P8-9	63-24342 <=*$>
1961 V50 N11 P44-52	62-23777 <=*>
1961 V50 N11 P72-78	62-23387 <=*>
1961 V50 N11 P79-83	62-25941 <=*>
1961 V50 N11 P93-97	AD-611 038 <=$>
	63-13393 <=*>
1961 V50 N11 P102-103	C-4511 <NRC>
1961 V50 N11 P103-104	C-4468 <NRC>
1961 V50 N11 P112	62-15413 <*=>
1961 V50 N12 P26-34	62-24397 <=*>
1961 V50 N12 P41-47	62-11673 <=>
1961 V50 N12 P73-75	CSIR-TR.215 <CSSA>
1961 V50 N12 P109-110	62-24000 <=*>

1962 N6 P14-18 IGR V4 N10 PT1 P1097 <AGI>
1962 N9 P105-108 IGR V6 N3 P435 <AGI>
1962 N11 P51-58 AD-637 807 <$=>
1962 V51 N2 P33-39 62-24251 <=*>
1962 V51 N2 P40-48 62-25845 <=*>
1962 V51 N3 P3-8 61-23138 <=>
1962 V51 N3 P58-61 62-32441 <=*> 0
1962 V51 N3 P105-107 AD-617 685 <=$>
1962 V51 N3 P115-118 62-24729 <=*>
62-33362 <=*> 0
1962 V51 N4 P12-14 65-50078 <=>
1962 V51 N4 P27-34 62-25181 <=*>
1962 V51 N4 P42-47 62-33031 <=*>
1962 V51 N4 P55-67 62-33428 <=*>
1962 V51 N4 P104-105 62-33081 <=*>
1962 V51 N5 P25-33 62-33032 <=*>
1962 V51 N5 P47-51 62-33032 <=*>
1962 V51 N5 P52-60 63-13413 <=*> 0
1962 V51 N5 P104-107 63-13029 <=*> 0
1962 V51 N6 P3-13 63-13502 <=*> 0
1962 V51 N6 P54-59 62-33773 <=*>
1962 V51 N6 P84-91 IGR V6 N10 P1735 <AGI>
1962 V51 N7 P3-12 62-32873 <=*>
1962 V51 N7 P13-36 62-33020 <=>
1962 V51 N7 P106-108 63-21925 <=>
1962 V51 N7 P113-116 62-33020 <=>
1962 V51 N8 P3-13 UCRL-TRANS-890(L) <=*>
1962 V51 N8 P14-23 63-10071 <=*> 0
1962 V51 N8 P93-95 63-21103 <=>
1962 V51 N9 P3-22 63-15304 <=*>
1962 V51 N9 P32-49 63-13654 <=>
1962 V51 N9 P77-86 63-13545 <=*>
1962 V51 N9 P89-91 63-13654 <=>
1962 V51 N9 P105-108 IGR,V6,N3,P435-438 <AGI>
1962 V51 N9 P111-114 62-33680 <=*>
1962 V51 N9 P114-115 63-11631 <=>
1962 V51 N10 P9-11 63-21321 <=>
64-13466 <=*$>
1962 V51 N10 P69-71 63-21137 <=>
1962 V51 N10 P101-103 63-13833 <=*> 0
1962 V51 N11 P3-4 64-71535 <=>
1962 V51 N11 P24-40 63-21131 <=>
1962 V51 N11 P106-107 63-23708 <=*$>
1962 V51 N12 P19-25 63-24482 <=*$> 0
1962 V51 N12 P26-33 C-4502 <NRC>
1962 V51 N12 P46-49 63-21168 <=>
1962 V51 N12 P68-73 C-4306 <NRCC>
1962 V51 N12 P94-95 63-15703 <=*>
1962 V51 N12 P100 C-4255 <NRCC>
63-23647 <=*$>
1962 V51 N12 P102 63-24497 <=*$>
1962 V51 N12 P105-108 63-19149 <=*>
1963 N4 P90-93 66-11065 <*>
1963 N5 P36-43 AD-632 864 <=$>
1963 N6 P25-33 AD-630 327 <=*$>
1963 V52 N1 P3-15 63-21651 <=>
1963 V52 N1 P35-43 63-21468 <=>
1963 V52 N1 P54-60 63-21651 <=>
1963 V52 N1 P100-101 C-4384 <NRC>
63-21651 <*>
1963 V52 N1 P102-104 C-4399 <NRC>
1963 V52 N1 P104-106 C-4408 <NRC>
1963 V52 N2 P3-13 63-21662 <=>
1963 V52 N2 P14-26 63-31730 <=>
1963 V52 N2 P27-32 64-19201 <=*$>
1963 V52 N2 P99-100 65-63553 <=$>
1963 V52 N2 P116 C-4382 <NRC>
1963 V52 N3 P24-35 65-60643 <=$>
1963 V52 N3 P36-39 C-4721 <NRC>
64-19240 <=*$>
64-26256 <$>
1963 V52 N3 P102-103 63-11700 <=>
1963 V52 N3 P103-104 AD-624 640 <=*$>
1963 V52 N4 P29-35 63-21914 <=>
64-19668 <=$>
1963 V52 N4 P90-93 63-31315 <=>
1963 V52 N5 P30-35 AD-615 241 <=$>
1963 V52 N5 P61-67 FPTS V23 N4 P.T701 <FASE>
1963 V52 N5 P61- FPTS,V23,N4,P.T701 <FASE>
1963 V52 N5 P61-67 63-31199 <=> 0
1963 V52 N6 P9-16 63-31861 <=>
1963 V52 N6 P25-33 65-63534 <$=> 0
1963 V52 N6 P65-70 63-31861 <=>
1963 V52 N6 P109-111 63-31787 <=>
1963 V52 N6 P115-116 N65-18336 <=$>
1963 V52 N7 P11-28 63-31785 <=>
1963 V52 N7 P11-18 64-15903 <=*$>
1963 V52 N7 P25-31 63-31715 <=>
1963 V52 N7 P32-37 64-13317 <=*$>
1963 V52 N8 P3-17 63-41120 <=>
1963 V52 N8 P20-30 63-41113 <=>
1963 V52 N8 P38-44 63-41120 <=>
1963 V52 N8 P45-52 63-41113 <=>
1963 V52 N8 P62-66 63-41120 <=>
1963 V52 N8 P73-75 C-5443 <NRC>
1963 V52 N8 P85-88 63-41120 <=>
1963 V52 N8 P88-90 63-41120 <=>
1963 V52 N9 P17-21 63-41193 <=>
1963 V52 N9 P27-31 AD-621 799 <=*$>
1963 V52 N9 P50-55 64-15199 <=*$>
1963 V52 N9 P112-114 63-41282 <=>
1963 V52 N10 P3-11 64-21026 <=>
1963 V52 N10 P87-89 C-4830 <NRC>
1963 V52 N10 P102-104 63-41349 <=>
1963 V52 N11 P79-83 C-5460 <NRC>
1963 V52 N11 P85-92 C-5080 <NRC>
1963 V52 N11 P97-99 64-51688 <=$>
1963 V52 N11 P102-104 64-21633 <=>
1963 V52 N11 P113-114 C-5102 <NRC>
1963 V52 N12 P25-29 64-31128 <=>
1963 V52 N12 P62-68 64-31129 <=>
1963 V52 N12 P75-79 64-31130 <=>
1964 N3 P87-89 65-10795 <*>
1964 N9 P2-12 NLLTB V7 N5 P400 <NLL>
66-11202 <*>
1964 V53 N1 P95-99 64-31427 <=>
1964 V53 N1 P107-110 64-31427 <=>
1964 V53 N1 P121-122 64-41022 <=>
1964 V53 N2 P10-17 64-31060 <=>
1964 V53 N2 P114-116 64-31169 <=>
1964 V53 N2 P128 64-31057 <=>
1964 V53 N3 P9-18 64-31634 <=>
1964 V53 N3 P44-51 64-31432 <=>
1964 V53 N3 P52-58 64-31428 <=>
1964 V53 N3 P64-67 64-31341 <=>
1964 V53 N3 P87-89 AD-608 976 <=$>
1964 V53 N3 P87-97 64-31447 <=>
1964 V53 N5 P20-25 64-41036 <=>
1964 V53 N5 P20-26 65-62111 <=$>
1964 V53 N5 P26-33 64-51119 <=>
1964 V53 N5 P33 64-41036 <=>
1964 V53 N5 P110-111 64-51795 <=$>
1964 V53 N6 P29-32 64-51247 <=>
65-62325 <=$>
1964 V53 N6 P44-49 64-41718 <=>
1964 V53 N6 P50-55 64-41718 <=>
1964 V53 N6 P56-64 64-51247 <=>
1964 V53 N6 P89-95 64-51247 <=>
1964 V53 N6 P119-124 64-51247 <=>
1964 V53 N7 P9-12 64-51683 <=>
1964 V53 N7 P14-23 66-30504 <=*$>
1964 V53 N7 P24-31 64-51103 <=>
1964 V53 N7 P32-38 64-51100 <=>
1964 V53 N7 P39-53 64-41721 <=>
1964 V53 N7 P87-91 C-5101 <NRC>
1964 V53 N7 P106-108 AD-615 287 <=$>
1964 V53 N8 P10-20 66-30504 <=*$>
1964 V53 N8 P44-50 64-51026 <=$>
1964 V53 N9 P2-12 TRANS. BULL,V7 N5 <NLL>
1964 V53 N9 P90-95 64-51815 <=$>
1964 V53 N9 P113-114 64-51811 <=$>
1964 V53 N10 P32-35 64-71675 <=$>
1964 V53 N11 P2-9 65-30192 <=$>
1964 V53 N11 P10-22 N65-17298 <=$>
1964 V53 N11 P34-44 65-30192 <=$>
1964 V53 N11 P45-52 65-30362 <=$>
1964 V53 N11 P111-112 65-30192 <=$>
1964 V53 N12 P31-38 65-31063 <=$>
1964 V53 N12 P31-39 65-64543 <=*$>
1965 N8 P71-74 AD-626 792 <=*$>

1965 N10 P86-94 AD-626 706 <=*$>
1965 V54 P114-115 65-31443 <=$>
1965 V54 N1 P12-34 65-30628 <=$>
1965 V54 N1 P25-34 N65-18183 <=$>
1965 V54 N1 P124-126 65-30628 <=$>
1965 V54 N2 P26-32 65-30820 <=$>
1965 V54 N2 P68-71 65-63413 <=$>
1965 V54 N2 P78-79 AD-615 768 <=$>
1965 V54 N3 P16-24 65-31383 <=$>
1965 V54 N3 P50-58 N66-13293 <=$>
1965 V54 N3 P87-93 65-32603 <=*$>
1965 V54 N3 P113-114 65-31353 <=$>
1965 V54 N4 P9-16 65-31443 <=$>
1965 V54 N4 P17-24 N66-13295 <=$>
1965 V54 N4 P25-31 65-32048 <=*$>
1965 V54 N4 P46-53 65-31443 <=$>
1965 V54 N4 P54-63 65-32048 <=*$>
1965 V54 N4 P78-81 65-31443 <=$>
1965 V54 N4 P108-110 65-31443 <=$>
1965 V54 N4 P114-115 65-31443 <=$>
1965 V54 N5 P17-28 65-32094 <=*$>
1965 V54 N5 P32-36 65-32094 <=*$>
1965 V54 N5 P51-55 65-32094 <=*$>
1965 V54 N6 P24-34 65-32207 <=$>
1965 V54 N6 P48-57 65-33571 <=*$>
1965 V54 N6 P58-62 65-32208 <=$>
1965 V54 N6 P109-110 65-32447 <=*$>
1965 V54 N7 P2-10 65-33387 <=*$>
1965 V54 N7 P65-66 C-5717 <NRC>
1965 V54 N7 P95-98 AD-627 128 <=$>
1965 V54 N8 P11-19 65-33070 <=*$>
1965 V54 N8 P20-27 65-33011 <=*$>
1965 V54 N8 P103-109 N66-14057 <=$>
1965 V54 N8 P119-120 AD-626 050 <=*$>
1965 V54 N9 P12-23 AD-630 846 <=$>
1965 V54 N9 P45-50 65-33873 <=*$>
1965 V54 N10 P12-21 65-33896 <=*$>
1965 V54 N10 P34-39 65-33911 <=*$>
1965 V54 N11 P14-17 66-31365 <=$>
1965 V54 N11 P17-24 AD-633 209 <=$>
1965 V54 N11 P76-79 65-34020 <=*$>
1965 V54 N11 P88-101 66-30700 <=$>
1965 V54 N12 P32-38 66-32025 <=$>
1965 V54 N12 P72-75 66-32025 <=$>
1965 V54 N12 P93-95 66-31117 <=$>
1966 N1 P2-6 66-31043 <=$>
1966 N2 P65-66 AD-635 022 <=$>
1966 N2 P102-105 NLLTB V8 N10 P823 <NLL>
1966 N2 P120-122 66-33187 <=$>
1966 N4 P2-3 66-33912 <=$>
1966 N4 P10-17 66-33912 <=$>
1966 N4 P93-97 66-33245 <=$> O
1966 N4 P114-115 66-33250 <=$>
1966 N5 P93-98 66-33284 <=$>
1966 N5 P113-114 66-33283 <=$>
1966 N6 P5 66-34651 <=$>
1966 N6 P6-18 66-34568 <=$>
1966 N6 P18-19 66-34651 <=$>
1966 N7 P24-34 66-33472 <=$>
1966 N7 P41-49 66-34902 <=$>

PRIRODA. SOFIYA
1960 V9 N6 P11-16 62-19307 <*=> P
1962 V11 N4 P106-107 63-13632 <=*>
1962 V11 N6 P108-112 63-21970 <=>
1963 V12 N1 P4-6 63-31168 <=> P
1963 V12 N4 P119-124 64-21358 <=>
1963 V12 N5 P106-107 64-21662 <=>
1964 V13 N4 P3-6 64-51571 <=$>
1964 V13 N4 P6-12 64-51572 <=$>
1964 V13 N4 P12-16 64-51573 <=$>
1964 V13 N4 P16-20 64-51574 <=$>
1964 V13 N4 P20-23 64-51575 <=$>
1967 V20 N2 P54-67 67-31330 <=$>

PRIRODA I ZNANIE
1962 V15 N10 P23-24 63-21292 <=>
1964 V17 N5 P1-3 64-51441 <=$>

PRIRODNAYA OCHAGOVOST BOLEZNEI I VOPROSY
PARAZITOLOGII
1955 P239-243 R-142 <*>
1955 P248-252 R-141 <*>

PRIVOREDNI PREGLAD
1960 V10 P10 SPEC.SUP. 60-31881 <=>
1960 V10 P11 SPEC.SUP. 60-31881 <=>
1960 V10 P2 SPEC.SUP. 60-31881 <=>
1960 V10 P3 SPEC.SUP. 60-31881 <=>
1960 V10 P6 SPEC SUP. 60-31881 <=>
1961 N21 P4 62-19702 <*=>
1961 V11 N14 P8 62-19701 <=*>
1963 P1 63-21672 <=>
1963 P6-7 63-21672 <=>
1963 V13 P12 63-31622 <=>
1963 V13 N24 P8 63-31272 <=>
1963 V13 N2245 P3 63-31621 <=>
1964 V14 P6 64-51443 <=$>

PRIX NOBEL
1956 P177-181 63-01313 <*>

PROBLEME AGRICOLE
1964 V16 N2 P46-54 64-21953 <=>
1964 V16 N3 P14-18 64-31120 <=>
1964 V16 N8 P8-14 64-51169 <=>
1964 V16 N8 P41-49 64-51300 <=$>
1966 N3 P4-12 66-32518 <=$>
1966 N5 P4-13 66-33836 <=$>

PROBLEME ECONOMICE
1964 V17 N4 P13-27 64-31635 <=>
1964 V17 N8 P32-47 64-51151 <=$>
1964 V17 N8 P117-128 64-51151 <=$>
1965 V18 N7 P24-55 65-32042 <=$>
1966 N11 P3-19 67-30238 <=>
1966 V19 N8 P3-16 66-34946 <=$>
1966 V19 N10 P22-34 67-30410 <=>
1966 V19 N11 P54-68 67-30322 <=>
1967 N1 P92-108 67-30860 <=$>

PROBLEME DE EPIZOOTOLOGIE. INSTITUTUL DE PATOLOGIE SI IGIENA ANIMALA. BUBURESTI. (TITLE VARIES)
1958 N8 P11-19 60-00827 <*>

PROBLEME DE TERAPEUTICA
1955 V2 P74-81 5513-C <K-H>
1955 V2 P83-108 5497-B <K-H>
1958 V9 N2 P17-27 64-10674 <*> O
9331-C <K-H>
1959 V10 N3 P21-24 64-10625 <*> O

PROBLEME ZOOTEHNICE SI VETERINARE
1958 N8 P19-30 60-31002 <=>
1961 V11 N4 P58-60 62-19526 <=*>
1961 V11 N5 P3-9 62-19525 <=*>
1961 V11 N5 P71-73 62-19524 <*=>
1961 V11 N6 P57-61 62-23265 <=*>
1961 V11 N7 P33-42 62-19597 <=*>
1961 V11 N7 P44-49 62-19501 <=*> P

PROBLEMES & TECHNIQUES
1958 N43 P19-33 59-15056 <*>

PROBLEMI SPOLJNE TRGOVINE I KONJUNKTURE. INSTITUT ZA SPOLJNA TRGOVINA. BELGRAD
1966 N1 P37-41 67-31047 <=$>
1966 N1 P57-61 67-31097 <=$>

PROBLEMY
1955 V11 N10 P681-694 59-11635 <=> O
1958 V14 N6 P394-402 PB 141 498T <=*> O
1961 V17 N3 P180-191 62-19470 <*=> O
1962 V18 N7 P488-492 62-32859 <=>

PROBLEMY ARKTIKI
1937 N5 P123-124 RT-1308 <*>
1938 N4 P151-154 RT-3076 <*>
1938 N5/6 P9-25 RT-1151 <*>
64-15217 <=>

1940 N1 P36-43 RT-3941 <*>
1940 N2 P51-68 PB 129 433T <+*>
1940 N2 P69-85 PB 130 600T <+*>
1940 N5 P7-12 60-13216 <+*>
1940 N9 P41-45 AD-610 919 <=$>
60-00676 <*>
1940 V2 P51-68 R-1836 <*>
1940 V2 P69-85 R-1837 <*>
1943 N2 P68-74 RT-4195 <*>
1944 V2 P122- R-1121 <*>
1945 N2 P51-58 R-4116 <*>
R-5020 <*>
RT-4479 <*>
1945 N2 P122 RT-4049 <*>
1957 N1 P19-27 61-15950 <+*>
1957 N1 P29-33 C-3560 <NRCC>
1957 N1 P111-113 63-15688 <=*$>
1957 N2 P5-17 60-17177 <+*>
61-15034 <+*>
1957 N2 P19-31 61-13879 <*+>
1957 N2 P41-51 59-18150 <+*>
1957 N2 P53-58 61-15038 <+*>
1957 N2 P59-71 60-14289 <+*>
61-15954 <+*>
1957 N2 P85-91 61-13869 <*+>
1957 N2 P93-96 61-13867 <*+>
1957 N2 P127-132 60-31036 <=>
1957 N2 P133-139 61-13880 <*+>
1957 N2 P149-159 AD-610 969 <=$>
1957 N2 P161-170 61-15955 <+*>
1957 N2 P193-204 60-14288 <+*>
61-15036 <+*>
1957 N2 P205-218 61-13870 <*+>
1957 V2 P141-147 C-2664 <NRC>
R-4243 <*>
1957 V2 P149-159 C-2732 <NRC>
R-4551 <*>
1958 N3 P35-40 59-18998 <+*>
62-24065 <=*>
1958 N3 P53-59 AD-610 980 <=$>
59-15569 <+*>
1958 N3 P61-67 59-18994 <+*>
1958 N3 P69-78 61-13882 <*+>
1958 N3 P79-82 65-12497 <*>
1958 N3 P95-101 61-19012 <+*>
1958 N3 P106-108 C-3131 <NRCC>
1958 N3 P109-110 64-11820 <=>
1958 N3 P116-119 60-17168 <+*>
1958 N3 P124-127 64-19774 <=$>
1958 N4 P5-13 61-13874 <*+>
1958 N4 P23-28 59-22216 <+*>
1958 N4 P45-49 C-3132 <NRCC>
1958 N4 P65-77 C-3463 <NRCC>
59-19001 <+*>
1958 N4 P103-110 59-19632 <+*>
1958 N5 P19-26 61-19226 <+*>
1958 N5 P27-31 61-19017 <+*>
1958 N5 P33-44 61-19020 <*=>
1958 N5 P73-80 C-3473 <NRCC>
61-15232 <=*>
1958 N5 P121-124 61-13974 <=*>
1958 N5 P125-127 61-23703 <*=>
1958 N5 P127 61-19244 <+*>
1959 N6 P13-21 61-19526 <=*>
1959 N6 P37-41 61-19227 <*=>
62-24694 <=*>
1959 N6 P117-119 C-3293 <NRCC>
1959 N6 P121-137 61-13977 <=*>
1959 N7 P5-14 61-19021 <=*>
1959 N7 P79-89 61-15233 <*=>
1960 N6 P14-26 63-19754 <=*> 0

PROBLEMY ARKTIKI I ANTARKTIKI
1959 N1 P5-10 63-23138 <=*$>
1959 N1 P11-23 63-23584 <=*>
1959 N1 P25-31 60-41054 <=>
1959 N1 P33-39 61-15227 <=*>
1959 N1 P49-58 61-13864 <*=>
1959 N1 P59-63 61-19025 <=*>
1959 N1 P65-71 61-19014 <=*>
1959 N1 P73-80 61-13853 <*=>
1959 N1 P81-85 61-19026 <*=>
1959 N1 P116-118 63-23585 <=*>
1960 N2 P5-16 61-19037 <+*>
1960 N2 P125-127 61-19036 <+*>
1960 N3 P5-15 61-19521 <=*>
1960 N3 P65-76 62-23493 <=*>
1960 N3 P112-115 61-28829 <*=>
62-00091 <*>
1960 N3 P118-119 C-3467 <NRCC>
60-00675 <*>
1960 N5 P53-58 63-19727 <=*$>
1960 N5 P63-66 63-19852 <=*> 0
1961 N7 P5-10 65-61267 <=$> 0
1961 N9 P37-61 62-32070 <=*>
1961 N9 P63-65 62-32070 <=*>
1961 N9 P85-87 62-32070 <=*>
1961 N9 P89-92 62-32070 <=*>
1961 N9 P104 62-32070 <=*>
1963 N12 P21-32 64-19994 <=$>
1963 N13 P5-12 64-23502 <=$>
1963 N13 P67-78 AD-626 071 <=*$>
1964 N15 P41-51 64-41303 <=>
1964 N16 P75-82 64-41207 <=>

PROBLEMY BIOKHIMII I MICHURINSKOI BIOLOGII
1949 N1 P169-187 2443-B <K-H>

PROBLEMY BOTANIKI
1962 V6 P7-17 64-19942 <=>
1962 V6 P83-94 64-13085 <=>

PROBLEMY DIFRAKTSII I RASPROSTRANENIIA VOLN
1962 V2 ENTIRE ISSUE 64-51381 <=$>
1964 V3 N3 ENTIRE ISSUE 64-41745 <=$>

PROBLEMY ENDOKRINOLOGII I GORMONOTERAPII
1955 V1 P121-126 R-705 <*>
1955 V1 N1 P5-12 R-704 <*>
1955 V1 N1 P71-77 59-15742 <+*>
1955 V1 N5 P95-109 62-15127 <=*>
1955 V1 N5 P121-126 62-15124 <*=>
1956 V2 N1 P111-117 62-15123 <*=>
1956 V2 N3 P49-53 53K24R <ATS>
1956 V2 N3 P100-109 R-439 <*>
64-19479 <=$>
1956 V2 N5 P61-71 62-15106 <=*>
1956 V2 N6 P90-98 62-15104 <*=>
1956 V2 N6 P108-116 62-15125 <*=>
1957 V3 N2 P82-90 62-15126 <*=>
1957 V3 N3 P25-34 62-15122 <*=>
1957 V3 N3 P72-74 38M43R <ATS>
1957 V3 N4 P3-9 59-12059 <+*>
1957 V3 N5 P12-26 62-15082 <=*>
1957 V3 N5 P27-37 62-15086 <=*>
1957 V3 N5 P38-48 62-15081 <*=>
1957 V3 N5 P49-56 62-15091 <=*>
1957 V3 N5 P62-73 59-11620 <=>
1957 V3 N5 P110-113 59-11621 <=>
1957 V3 N5 P114-117 59-11622 <=>
1958 V4 N2 P14-22 59-13337 <=> 0
1958 V4 N2 P26-30 59-13338 <=>
1958 V4 N2 P115-122 59-13339 <=>
1958 V4 N3 P22-27 62-24325 <=*> 0
1958 V4 N3 P120-123 PB 141 496T <=>
1958 V4 N4 P21-24 59-13046 <=>
1958 V4 N4 P31-37 59-13047 <=> 0
1958 V4 N4 P114-118 59-13048 <=>
1958 V4 N5 P15-24 59-20993 <+*>
1959 V5 N1 P70-79 AD-607 542 <*>
1959 V5 N2 P3-9 60-18568 <+*>
1959 V5 N3 P80-87 61-23899 <*=>
1959 V5 N6 P112-115 60-31155 <=>
1960 V6 N1 P3-9 60-11657 <=>
1960 V6 N1 P33-37 60-11658 <=>
1960 V6 N1 P46-51 63-16719 <=*> 0
1960 V6 N1 P102-105 60-11659 <=>
1960 V6 N2 P120-125 60-41006 <=>
1960 V6 N3 P3-8 60-41293 <=>
1960 V6 N3 P18-21 60-41294 <=>

1960 V6 N3 P22-26 60-41295 <=>
1960 V6 N3 P27-31 60-41296 <=>
1960 V6 N3 P32-45 63-19721 <=*>
1960 V6 N3 P52-61 60-41297 <=>
1960 V6 N3 P86-90 60-41317 <=>
1960 V6 N3 P91-94 60-41318 <=>
1960 V6 N3 P95-103 60-41319 <=>
1960 V6 N4 P3-7 61-21120 <=>
1960 V6 N4 P28-36 61-21119 <=>
1960 V6 N4 P54-57 61-11855 <=>
1960 V6 N5 P3-6 61-11703 <=>
1960 V6 N5 P20-23 63-19046 <=*> 0
1960 V6 N5 P24-26 61-11739 <=>
1960 V6 N6 P28-34 61-21248 <=>
1960 V6 N6 P34-42 61-21224 <=>
1960 V6 N6 P112-115 63-19707 <*=> 0
1961 V7 N2 P104-117 61-31473 <=>
1961 V7 N3 P22-31 62-13443 <*=>
1961 V7 N3 P117-124 62-13538 <=*>
1961 V7 N4 P26-32 62-13415 <*=>
1961 V7 N4 P82-88 62-13311 <*=>
1961 V7 N4 P119-121 62-13409 <*=>
1961 V7 N6 P15-19 62-23327 <=>
1961 V7 N6 P43-46 62-23327 <=>
1961 V7 N6 P105-116 62-23327 <=>
1962 V8 N2 P57-64 62-24712 <=*>
1962 V8 N3 P11-14 62-32851 <=*>
1962 V8 N4 P34-38 62-33114 <=*>
1962 V8 N4 P104-106 62-33114 <=*>
1962 V8 N5 P15-20 63-15401 <=*>
1963 V9 P87-92 65-13452 <*>
1963 V9 N1 P3-7 64-13493 <=*$>
1963 V9 N1 P7-11 63-23877 <=*$> 0
1963 V9 N1 P12-18 64-13494 <=*$>
1963 V9 N1 P23- FPTS V22 N5 P.T945 <FASE>
1963,V22,N5,PT2 <FASE>
1963 V9 N1 P34- FPTS V22 N5 P.T961 <FASE>
1963,V22,N5,PT2 <FASE>
1963 V9 N1 P44- FPTS V22 N4 P.T628 <FASE>
1963 V9 N1 P111-118 64-51168 <=>
1963 V9 N2 P17- FPTS V22 N6 P.T1184 <FASE>
FPTS,V22,P.T1184-86 <FASE>
1963 V9 N2 P26- FPTS V22 N6 P.T1187 <FASE>
FPTS,V22,P.T1187-90 <FASE>
1963 V9 N2 P36- FPTS V23 N1 P.T197 <FASE>
FPTS,V23,P.T197-T201 <FASE>
1963 V9 N2 P51-56 64-13915 <=*$> 0
1963 V9 N2 P73-76 63-21754 <=>
1963 V9 N3 P3-7 63-31211 <=>
64-15454 <=*$> 0
1963 V9 N3 P40- FPTS V23 N5 P.T969 <FASE>
FPTS,V23,N5,P.T969 <FASE>
1963 V9 N4 P39- FPTS V23 N2 P.T399 <FASE>
1963 V9 N4 P39 FPTS,V23,P.T399-T400 <FASE>
1963 V9 N4 P50- FPTS V23 N4 P.T769 <FASE>
1963 V9 N5 P3-12 64-19266 <=*$>
1963 V9 N5 P12-17 64-15460 <=*$>
1963 V9 N5 P17-19 64-15537 <=*$>
1963 V9 N5 P35-40 64-15538 <=*$>
1963 V9 N5 P40-46 64-15501 <=*$> 0
1963 V9 N5 P81- FPTS V23 N4 P.T767 <FASE>
1964 V10 P19- FPTS V23 N6 P.T1253 <FASE>
1964 V10 N1 P9- FPTS V23 P.T1253 <FASE>
1964 V10 N1 P20-24 64-21754 <=>
1964 V10 N1 P32-37 64-19727 <=$>
64-21754 <=>
1964 V10 N1 P57- FPTS V23 N6 P.T1275 <FASE>
1964 V10 N1 P66- FPTS V23 N6 P.T1271 <FASE>
FPTS V23 P.T1271 <FASE>
1964 V10 N1 P73- FPTS V23 N6 P.T1296 <FASE>
1964 V10 N1 P77- FPTS V24 N1 P.T183 <FASE>
1964 V10 N1 P81- FPTS V23 N6 P.T1293 <FASE>
1964 V10 N1 P91-97 64-19697 <=$>
1964 V10 N1 P97-103 65-60186 <=$>
1964 V10 N2 P73- FPTS V23 N6 P.T1264 <FASE>
FPTS V23 P.T1264 <FASE>
1964 V10 N2 P103- FPTS V24 N1 P.T48 <FASE>
1964 V10 N3 P62-65 64-31573 <=>
1964 V10 N3 P76-79 64-31573 <=>
1964 V10 N3 P105- FPTS V24 N3 P.T379 <FASE>
1964 V10 N5 P20- FPTS V24 N4 P.T593 <FASE>
1964 V10 N5 P41-48 65-30152 <=$>
1964 V10 N5 P52- FPTS V24 N4 P.T603 <FASE>
1965 V11 N1 P71- FPTS V25 N1 P.T63 <FASE>
1965 V11 N1 P87- FPTS V25 N1 P.T59 <FASE>
1965 V11 N1 P109- FPTS V25 N1 P.T51 <FASE>
1965 V11 N2 P50- FPTS V25 N1 P.T55 <FASE>
1965 V11 N2 P72- FPTS V25 N1 P.T69 <FASE>
1965 V11 N2 P89-93 66-60862 <=*$>
1965 V11 N2 P106-114 65-31369 <=$>
1965 V11 N3 P71- FPTS V25 N2 P.T268 <FASE>
1965 V11 N3 P88-95 66-61068 <=*$>
1965 V11 N4 P105- FPTS V25 N4 P.T672 <FASE>
1965 V11 N4 P125-126 65-32329 <=*$>
1965 V11 N5 P68- FPTS V25 N3 P.T496 <FASE>
1965 V11 N5 P100- FPTS V25 N4 P.T669 <FASE>
1965 V11 N6 P120-121 66-30740 <=$>
1966 V12 N1 P54- FPTS V25 N6 P.T1034 <FASE>
1966 V12 N1 P102- FPTS V25 N6 P.T1019 <FASE>
1966 V12 N1 P125-126 66-31260 <=$>

PROBLEMY FIZICHESKOI KHIMII
1959 N2 P27-38 61-28303 <*=>
1959 N2 P118-131 61-27814 <*=>
61-28607 <*=>
1959 N2 P132-145 61-28238 <*=>
1963 N3 P66-76 AD-637 053 <=$>
1963 N3 P76-85 AD-637 053 <=$>

PROBLEMY FIZIKI ATMOSFERY
1963 V1 P60-70 64-71331 <=>
1963 V2 P28-47 N64-24833 <=>
1963 V2 P48-66 N64-22676 <=>
1963 V2 P67-86 N64-24835 <=>
1963 V2 P87-112 N64-24834 <=>
1963 V2 P113-126 N64-22677 <=>

PROBLEMY FIZIOLOGICHESKOI AKUSTIKI
1949 V1 P122-127 62-25749 <=*>
1950 V2 P123-128 62-15322 <*=>
1950 V2 P129-138 62-15316 <*=>
1955 V3 P5-17 62-23067 <=>
1955 V3 P18-26 R-681 <*>
1955 V3 P27-33 R-1140 <*>
31K28R <ATS>
1955 V3 P60-66 R-1449 <*>
1955 V3 P67-74 59-11542 <=> 0
1955 V3 P122-125 62-11525 <=>

PROBLEMY FIZIOLOGICHESKOI OPTIKI
1941 N1 P77-79 60-13098 <+*> 0
1941 V1 P77-79 RT-2769 <*>
1949 V7 P25-33 R-4735 <*>
65-61609 <=$> 0
1949 V7 P34-38 R-4730 <*>
65-61584 <=$> 0
1953 N8 P47-54 59-10059 <+*>

★PROBLEMY GEMATOLOGII I PERELIVANIYA KROVI
1956 V1 N1 P5-9 RT-4283 <*>
1956 V1 N1 P41-46 R-204 <*>
64-19507 <=$>
1956 V1 N2 P10-14 62-15105 <=*>
1956 V1 N3 P38-42 R-131 <*>
R-300 <*>
64-19070 <=*$> 0
1956 V1 N3 P45-48 R-130 <*>
R-296 <*>
64-19071 <=*$>
1956 V1 N3 P47-52 R-133 <*>
R-293 <*>
64-19069 <=*$> 0
1956 V1 N4 P49-52 R-785 <*>
64-19516 <=$>
1956 V1 N5 P49-54 R-623 <*>
1960 V5 N6 P3-8 60-31814 <=>
1960 V5 N9 P3-7 62-20309 <=*>
1961 V6 N5 P44-47 61-28256 <*=>
1961 V6 N10 P3-8 62-18174 <=*>
1962 V7 N3 P53-55 63-00847 <*>

63-19687 <=*$>
1962 V7 N8 P55-56 66-61759 <=*$>
1963 V8 N1 P17-21 63-21712 <=>
1963 V8 N1 P21-24 63-23868 <=*$>
1963 V8 N1 P34-42 63-21712 <=>
1963 V8 N1 P38- FPTS V22 N5 P.T903 <FASE>
1963 V8 N1 P38-42 1963,V22,N5,PT2 <FASE>
1963 V8 N1 P46-47 63-21712 <=>
1963 V8 N2 P16-20 63-23837 <=*$>
1963 V8 N2 P28-34 63-21552 <=>
63-23836 <=*$> 0
1963 V8 N2 P34- FPTS V22 N5 P.T891 <FASE>
1963 V8 N2 P37-44 63-21552 <=>
1963 V8 N2 P37-41 63-23835 <=*$>
1963 V8 N3 P10-14 63-24519 <=*$>
1963 V8 N3 P20-26 63-21702 <=>
63-41136 <=>
64-13403 <=*$>
1963 V8 N3 P26-32 64-13402 <=*$>
1963 V8 N3 P42- FPTS V23 N1 P.T175 <FASE>
FPTS,V23,P.T175-T177 <FASE>
1963 V8 N4 P9-15 63-31346 <=>
1963 V8 N4 P19-22 63-31346 <=>
1963 V8 N4 P47-51 63-31346 <=>
1963 V8 N5 P3-37 63-31355 <=>
1963 V8 N5 P3-16 64-13859 <=*$> 0
1963 V8 N5 P37-41 64-13395 <=*$>
1963 V8 N5 P41-47 64-13394 <=*$>
1963 V8 N5 P47-52 63-31355 <=>
1963 V8 N6 P3- FPTS V23 N2 P.T322 <FASE>
FPTS,V23,P.T322-T325 <FASE>
1963 V8 N6 P18-21 64-13851 <=*$>
1963 V8 N6 P21-27 64-13852 <=*$>
1963 V8 N6 P45-48 63-31471 <=>
1963 V8 N6 P54 63-31439 <=>
1963 V8 N7 P29-32 64-15529 <=*$>
1963 V8 N7 P32- FPTS V23 N3 P.T605 <FASE>
FPTS,V23,N3,P.T587 <FASE>
1963 V8 N8 P3-8 63-41160 <=>
1963 V8 N8 P22- FPTS V23 N3 P.T491 <FASE>
1963 V8 N8 P22-25 FPTS,V23,N3,P.T491 <FASE>
63-41160 <=>
1963 V8 N8 P46-49 64-15462 <=*$> 0
1963 V8 N8 P53-56 63-41160 <=>
1963 V8 N8 P57-58 63-41160 <=>
1963 V8 N9 P32-34 64-15452 <=*$>
1963 V8 N9 P44-45 64-15453 <=*$>
1963 V8 N10 P3- FPTS V23 N6 P.T1226 <FASE>
FPTS V23 P.T1226 <FASE>
1963 V8 N10 P12- FPTS V23 N5 P.T1054 <FASE>
FPTS,V23,N5,P.T1054 <FASE>
1963 V8 N10 P26-29 63-41280 <=>
1963 V8 N10 P40-42 63-41117 <=>
64-19645 <=$>
1963 V8 N11 P26- FPTS V23 N5 P.T1015 <FASE>
FPTS,V23,N5,P.T1015 <FASE>
1963 V8 N11 P33- FPTS V23 N5 P.T1063 <FASE>
FPTS,V23,N5,P.T1063 <FASE>
1963 V8 N11 P50-53 64-19646 <=$>
1963 V8 N12 P3-12 65-63064 <=$>
1963 V8 N12 P13-17 64-19625 <=$>
1963 V8 N12 P27-30 64-19626 <=$>
1963 V8 N12 P30- FPTS V23 N6 P.T1219 <FASE>
FPTS V23 P.T1219 <FASE>
1964 V9 N1 P3-10 64-21775 <=>
1964 V9 N1 P11-17 65-60143 <=$>
1964 V9 N1 P31-40 64-21775 <=>
1964 V9 N1 P40- FPTS V24 N1 P.T180 <FASE>
1964 V9 N1 P40-43 FPTS V24 P.T180-T182 <FASE>
1964 V9 N1 P44-46 64-21595 <=>
1964 V9 N1 P52-55 64-21775 <=>
1964 V9 N2 P3-8 64-31174 <=>
1964 V9 N2 P27-32 65-60226 <=$>
1964 V9 N2 P32- FPTS V23 N6 P.T1247 <FASE>
1964 V9 N2 P32-36 FPTS V23 P.T1247 <FASE>
64-31174 <=>
1964 V9 N2 P37-55 64-31034 <=>
1964 V9 N3 P26- FPTS V24 N2 P.T309 <FASE>
1964 V9 N3 P26-31 FPTS V24 P.T309-T312 <FASE>
1964 V9 N3 P49-53 65-62107 <=$>
1964 V9 N4 P12-15 65-62343 <=$>
1964 V9 N4 P29-43 64-31650 <=>
1964 V9 N4 P31-37 65-62315 <=$>
1964 V9 N4 P42-43 09S81R <ATS>
1964 V9 N4 P49-50 64-31650 <=>
1964 V9 N5 P3-7 64-31866 <=>
1964 V9 N5 P27-31 65-61095 <=$>
1964 V9 N5 P52-55 65-63204 <=*$>
1964 V9 N6 P18-23 64-41057 <=>
1964 V9 N6 P34-37 64-41057 <=>
1964 V9 N6 P51-52 64-41057 <=>
1964 V9 N7 P32-37 65-63201 <=*$>
1964 V9 N8 P44-49 64-51358 <=$>
1964 V9 N8 P53-54 64-51358 <=$>
1964 V9 N9 P44- FPTS V24 N4 P.T748 <FASE>
1964 V9 N10 P36- FPTS V24 N5 P.T841 <FASE>
1964 V9 N10 P47-49 66-61758 <=*$>
1964 V9 N11 P12-17 65-63615 <=$>
1964 V9 N12 P3-8 65-63666 <=$>
1964 V9 N12 P26-33 65-30197 <=$>
1964 V9 N12 P33- FPTS V25 N1 P.T153 <FASE>
1965 V10 N1 P16-19 65-63671 <=$>
1965 V10 N1 P40-50 65-30529 <=$>
1965 V10 N1 P50-52 65-64525 <=*$>
1965 V10 N1 P53-54 65-30529 <=$>
1965 V10 N2 P3-14 66-60305 <=*$>
1965 V10 N2 P23-26 65-63667 <=$>
1965 V10 N2 P35-36 66-60314 <=*$>
1965 V10 N2 P51-54 65-30595 <=$>
1965 V10 N2 P56-58 65-30595 <=$>
1965 V10 N3 P30-34 66-60856 <=*$>
1965 V10 N3 P55-57 65-30970 <=$>
1965 V10 N4 P14-20 66-60857 <=*$>
1965 V10 N5 P28-34 66-60879 <=*$>
1965 V10 N5 P54-56 66-60315 <=*$>
1965 V10 N6 P23- FPTS V25 N2 P.T370 <FASE>
1965 V10 N6 P27-30 66-61067 <=*$>
1965 V10 N6 P44-47 65-32226 <=$>
1965 V10 N7 P18- FPTS V25 N3 P.T469 <FASE>
1965 V10 N7 P38-42 66-60565 <=*$>
1965 V10 N7 P42- FPTS V25 N3 P.T455 <FASE>
1965 V10 N8 P30-36 65-33389 <=*$>
1965 V10 N8 P32- FPTS V25 N4 P.T645 <FASE>
1965 V10 N8 P50-53 65-33389 <=*$>
1965 V10 N9 P28- FPTS V25 N4 P.T633 <FASE>
1965 V10 N10 P23-34 66-30209 <=*$>
1965 V10 N10 P28- FPTS V25 N5 P.T805 <FASE>
1965 V10 N10 P38-43 66-61045 <=*$>
1965 V10 N11 P28- FPTS V25 N4 P.T640 <FASE>
1965 V10 N11 P39-44 66-61600 <=$*>
1965 V10 N12 P3-13 66-31065 <=$>
1965 V10 N12 P13-18 66-61999 <=*$>
1965 V10 N12 P18-25 66-61975 <=*$>
1965 V10 N12 P32- FPTS V25 N5 P.T809 <FASE>
1966 N5 P3-7 66-33167 <=$>
1966 V11 N1 P3-9 66-31409 <=$>
1966 V11 N1 P54-57 66-31408 <=$>
1966 V11 N2 P50- FPTS V25 N6 P.T978 <FASE>
1966 V11 N2 P60- FPTS V25 N6 P.T975 <FASE>
1966 V11 N3 P3-7 66-61763 <=$>
1966 V11 N3 P24-31 66-32112 <=$>
1966 V11 N3 P43-48 66-32112 <=$>

★PROBLEMY KIBERNETIKI
1958 N1 P5-23 59-11770 <=> 0
1958 N1 P190-202 60-41030 <=>
1959 N2 P7-38 60-11347 <=>
1959 N2 P51-68 60-11428 <=>
1959 N2 P69-71 60-11394 <=>
1959 N2 P73-74 60-11393 <=>
1959 N2 P75-122 60-11837 <=>
1959 N2 P123-138 60-11353 <=>
1959 N2 P181-184 60-11291 <*>
1959 N2 P185-189 60-11338 <=>
1959 N2 P191-201 60-11303 <=>
1959 N2 P203-212 60-11255 <=>
1959 N2 P213-228 60-11289 <=>
1959 N2 P229-282 60-11286 <=>
1960 N3 P181-272 61-11784 <=>
1960 N4 P5-22 61-27688 <=*> 0

1960 N4 P23-26 61-28515 <*=>
1960 N4 P37-52 61-21461 <=>
1960 N4 P53-58 61-28781 <*=>
1960 N4 P59-68 61-21194 <=>
1960 N4 P96-110 61-21273 <=>
1960 N4 P111-120 61-27691 <*=>
1960 N4 P121-149 61-11649 <=>
1960 N4 P151-181 61-11649 <=>
1960 N4 P183-195 61-11649 <=>
1960 N4 P197-205 61-21037 <=>
1960 N4 P207-257 61-21037 <=>
1960 V4 P69-93 61-21131 <=>
1961 N5 P7-15 62-13410 <*=>
1961 N5 P17-29 62-13419 <*=>
1961 N5 P31-48 62-13412 <*=>
1961 N5 P49-60 62-13613 <*=>
1961 N5 P97-103 62-13529 <*=>
1961 N5 P105-121 63-23785 <=*$>
1961 N5 P123-136 63-23785 <=*$>
1961 N5 P165-182 62-13523 <*=>
1961 N5 P183-197 61-28731 <=*>
1961 N5 P199-215 61-28733 <*=>
1961 N5 P217-243 61-28742 <*=>
1961 N5 P271-277 62-13534 <*=>
1961 N5 P283-285 62-13536 <*=>
1961 N5 P289-294 62-13539 <*=>
1961 N6 P15-43 62-23983 <=>
1961 N6 P45-67 62-23983 <=>
1961 N6 P83-100 62-19158 <=*>
1961 N6 P101-160 62-23426 <=*>
1961 N6 P161-181 62-19752 <=*>
64-21817 <=>
1961 N6 P207-287 62-23983 <=>
1961 N6 P289-297 62-23983 <=>
1961 N6 P298-302 62-19757 <=*>
1962 N8 P5-356 64-21800 <=>
1962 V8 P293-308 63-31764 <=>
1962 V8 P309-336 64-11956 <=> M
1963 N8 P253-291 63-31022 <=>
1963 N9 P241-264 63-23816 <=*$>
1963 N9 P317-319 64-14270 <=*$>
1964 V11 P5-24 64-51913 <=$>
1964 V11 P147-151 64-51913 <=$>
1964 V11 P153-187 64-51913 <=$>
1964 V11 P189-198 64-51913 <=$>
1964 V11 P215-244 64-51913 <=$>
1964 V11 P276-279 64-51913 <=$>
1965 V12 P165-168 66-10721 <*>
1965 V12 P169-179 66-11764 <*>
1965 V15 P1-292 66-30863 <=$>

PROBLEMY KINETIKI I KATALIZA
1940 V4 P28-46 61-18112 <*>
1940 V4 P125-133 61-18090 <*>
1940 V4 P143-166 61-18089 <*>
1940 V4 P167-184 61-18234 <*> P
1948 V5 P258-273 934TM <CTT>
1949 N6 P223-231 RJ-364 <ATS>
1949 N6 P426-431 RT-4655 <*>
1949 V6 P202-205 375 <TC>
1949 V6 P223-231 RJ-364 <ATS>
1955 V8 P17-33 59-10253 <+*>
1955 V8 P53-60 60-10100 <+*> 0
1955 V8 P180-188 RJ-583 <ATS>
1957 N9 P76-83 60-13825 <+*> 0
1957 V9 P5-30 89L31R <ATS>
1957 V9 P45-60 64-10701 <=*$>
1957 V9 P61-75 60-13777 <+*>
1957 V9 P84-90 59-15324 <+*>
1957 V9 P107-116 34M41R <ATS>
60-13778 <+*>
1957 V9 P129-132 60-13641 <+*>
1957 V9 P152-161 60-13805 <+*>
60L32R <ATS>
1957 V9 P168-172 60L32R <ATS>
1957 V9 P218-233 60-41273 <=>
1957 V9 P242-244 60-11713 <=>
1957 V9 P264-266 60-11714 <=>
1957 V9 P267-273 60-11772 <+*>
61-15981 <+*>
1957 V9 P363-368 60-11715 <=>
1957 V9 N1 P97-103 33M41R <ATS>
1957 V9 N9 P162-167 64L34R <ATS>
1960 V10 P62-66 62-10992 <=*> 0
1960 V10 P88-89 61-27558 <*=>
1960 V10 P90-94 63-26815 <$>
1960 V10 P102-107 AEC-TR-4979 <=*>
1960 V10 P108-110 AEC-TR-4857 <=*>
1960 V10 P279-284 65-17114 <*>
1960 V10 P285-290 C-5153 <NRC>
1960 V10 P410-414 62-33222 <=*>

PROBLEMY KOSMICHESKOI BIOLOGII
1962 V1 P1-462 63-19218 <=> M
1964 V3 P3-490 64-31578 <=>

PROBLEMY METALLOVEDENIYA I FIZIKI METALLOV
1954 V5 P128-137 3929 <HB>
1955 P103-112 4033 <HB>
1955 P461-464 3950 <HB>
1955 P577-594 4001 <HB>
1955 V4 P377-387 62-19901 <=*>
1955 V4 P425-431 62-11281 <=>
1958 V5 P210-234 RAE-LIB-TRANS-1080 <NLL>
65-63888 <=$>
1959 V6 P259-292 65-25086 <$>
1962 V7 P34-92 63-15345 <=*>
1962 V7 P156-174 63-15345 <=*>
1962 V7 P231-245 63-15345 <=*>
1962 V7 P281-306 63-15345 <=*>

PROBLEMY METALLOVEDENIY I TERMICHESKOI OBRABOTKI: SBORNIK STATEI
1956 V1 P63-99 61-27163 <=*> 0

★PROBLEMY PEREDACHI INFORMATSII
1960 N5 P1-125 61-28538 <*=>
1960 N6 P24-33 62-32082 <=*>
1960 N6 P34-45 63-23641 <=*$>
1960 N7 P1-204 63-24050 <=*>
1961 N8 P5-66 62-23994 <=*>
1961 N8 P121-131 62-23994 <=*>
1961 N9 P79-82 63-15732 <=*>
1961 N10 P5-23 63-13197 <=*>
1961 N10 P35-56 63-13197 <=*>
1961 N10 P49-56 62-25858 <=*>
1963 N13 P1-172 64-41124 <=>
1963 N14 P43-58 AD-612 374 <=$>
1963 N15 P61-70 AD-620 785 <=$>
1963 N15 P71-74 AD-620 785 <=$>
1963 V15 P77-79 AD-617 656 <=$>
1964 V17 P17-70 AD-637 417 <=$>
1965 SNS V1 N1 P113-116 66-11236 <*>
1965 SNS V1 N1 P125-126 66-11191 <*>
1965 SNS V1 N4 P33-40 66-30222 <=*$>

PROBLEMY PROCHNOSTI V MASHINOSTROENII
1959 N2 P54-71 61-27008 <=*> 0
1959 V4 P47-60 62-24907 <=*>
1959 V5 P10-33 65-61413 <=>

PROBLEMY PROJEKTOWE HUTNICTWA
1960 V8 N2 P53-59 5177 <HB>
1965 V13 P37-40 4896 <BISI>
1967 V15 N2 P50-52 67-31604 <=$>

PROBLEMY PSIKHOLOGII SPORTA
1960 N1 P254-260 65-31871 <=$>

PROBLEMY REGULIROVANIYA RECHNOGO STOKA
1956 V6 P5-93 61-11431 <=>
1956 V6 P107-153 61-11431 <=>
1956 V6 P165-174 61-11431 <=>
1956 V6 P230-247 61-11431 <=>

PROBLEMY REVMATIZMA
1961 V1 N1 P3-10 61-31588 <=>
1961 V1 N1 P43-51 61-31588 <=>
1961 V1 N1 P57-66 61-31588 <=>
1961 V1 N1 P72-86 61-31588 <=>

★PROBLEMY SEVERA
1958 N1 P5-29 60-23766 <+*>
1958 N1 P42-51 61-15035 <+*>
1958 N1 P52-64 61-15951 <+*>
1958 N1 P330-336 61-15033 <+*>
1958 N1 P337-340 61-15952 <+*>
1958 N1 P341-345 60-17307 <+*>
1958 N1 P346-353 60-23767 <+*>
1958 N2 P1-16 60-23764 <+*>
1958 N2 P47-79 61-27999 <*=>

PROBLEMY SOVETSKOGO PCCHVCVEDENIYA
1939 N7 P149-162 R-2418 <*>

PROBLEMY SOVREMENNOI NEIRCKHIRURGII
1962 N4 P36-43 65-63205 <=*$>
1962 V6 N4 P25-35 66-60560 <=*$>

PROBLEMY TUBERKULEZA
1936 N7 P1146-1147 63-15753 <=*>
1940 N10 P5-14 R-893 <*>
1942 N3 P36- R-1072 <*>
1944 N2 P41-44 RT-1468 <*>
1944 N3 P9-15 RT-658 <*>
1949 N1 P50-54 R-835 <*>
1950 N6 P42-44 R-892 <*>
1951 N1 P44-48 63-20940 <=*$>
1952 N2 P25-30 R-891 <*>
1952 N3 P3-13 R-839 <*>
1952 N3 P22-28 R-890 <*>
1954 N3 P78-81 RT-2949 <*>
1954 N3 P81-82 RT-2950 <*>
1955 N4 P72-76 62-15155 <*=>
1960 N2 P69-73 60-31458 <=>
1961 V39 N4 P93-94 64-51287 <=>
1961 V39 N5 P105-110 64-13648 <=*$> O
1962 N1 P122-125 62-24726 <=*>
1963 P34-37 AD-638 190 <=$> O
1963 V41 N9 P3-7 64-21198 <=>

PROBLEMY VRACHEBNCGC KCNTRCLYA
1960 V5 P301-316 62-23813 <=*>
1960 V5 P332-343 62-23813 <=*>

PROBLEMY YUNOSHESKOGO SPORTA
1958 N1 P230-241 60-11033 <=>

PROBLEMY ZHIVOTNOVODSTVA
1933 N4 P95-100 RT-2702 <*>
1937 N10 P53-72 RT-2704 <*> O
1938 N11 P131-134 RT-2703 <*>
1938 N12 P33-42 RT-2701 <*>
1938 V7 N3 P175-178 R-2151 <*>

PROCEEDINGS OF THE ASSOCIATION FOR PLANT PRCTECTION OF HOKURIKU. JAPAN
SEE HOKURIKU BYOGAICHU KENKYU KAIHO

PROCEEDINGS OF THE CROP SCIENCE SOCIETY OF JAPAN
SEE NIHON SAKUMOTSU GAKKAI KIJI

PROCEEDINGS OF THE JAPAN ACADEMY. TOKYO
SEE NIHON GAKUSHIIN

PROCEEDINGS OF THE JAPAN CEMENT ENGINEERING ASSOCIATION
SEE SEMENTO GIJUTSU NEMPO

PROCEEDINGS OF THE KANTO TOSAN PLANT PROTECTION SOCIETY
SEE KANTO TOSAN BYOGAICHU KENKYUKAI NENPO

PROCEEDINGS OF THE PHYSICAL SOCIETY. LONDON
1950 V63 N2 P126-128 AL-604 <*>
1950 V63 N362B P122-126 PT2 AL-113 <*>

PROCEEDINGS OF THE PHYSICAL SOCIETY CF JAPAN
SEE NIPPON BUTSURI GAKKAI SHI

PROCEEDINGS OF THE PHYSICC-MATHEMATICAL SOCIETY OF JAPAN
SEE NIPPON SUGAKU BUTSURIGAKKWAI KIZI

PROCEEDINGS OF THE RESEARCH SOCIETY OF JAPAN SUGAR REFINERIES TECHNOLOGY
SEE SEITO GIJUTSU KENKYUKAISHI

PROCEEDINGS OF THE ROYAL SOCIETY. LONDON
1958 V246 N1245 P240-247 SCL-T-221 <+*>

PROCEEDINGS OF THE SECTION OF SCIENCES. K. NEDERLANDSE AKADEMIE VAN WETENSCHAPPEN
1912 V15 P52-54 1667 <*>
1930 V33 P972-984 66-11101 <*>
1936 V39 P190-200 I-269 <*>
1937 V40 P164-173 58-2101 <*>
1939 V42 P468-475 AL-152 <*>
1954 V57 P524-539 63-01686 <*>

PROCHNOSTI I USTCICHIVOSTI ELEMENTCV TONKOSTENNYKH KONSTRUKTSII
1963 N1 P97-109 AS-619 548 <=$> M

PRODUITS PHARMACEUTIQUES
1957 V12 P33-37 3318 <*>
1961 V16 N5 P216-220 62-01268 <*>

PRODUITS ET PROBLEMES PHARMACEUTIQUES
1963 V18 N2 P51-66 64-14618 <*>

PROFESSIONALNO-TEKHNICHESKOE OBRAZOVANIE
1960 V17 N5 P20-21 65-17391 <*>

PROGRES MEDICAL
1909 P306-308 57-1597 <*>
1959 V87 N21 P380-384 10085-A <K-H>
64-10633 <*>
1961 N6 P123-125 65-10969 <*>

PROGRESELE STIINTEI
1966 V2 N4 P116-121 66-33784 <=$>

PROGRESS IN EXPERIMENTAL TUMOR RESEARCH. BASEL, NEW YORK
1960 V1 P112-161 62-00135 <*>

PROGRESS IN MATHEMATICS. CHINESE PEOPLES REPUBLIC
SEE SHU HSUEH CHIN CHAN

PROGRESS OF PHYSICAL SCIENCES
V54 N2 P358-360 58-03 <*>

PROGRESS IN PHYSICAL CHEMISTRY, JAPAN
SEE BUTSURI KAGAKU NO SHINPO

PROGRESSO MEDICO. NAPOLI
1953 V9 N16 P489-494 59-17042 <*> P
1960 V16 N9 P265-274 62-14981 <*>
1960 V16 N21 P734-736 11279-A <KH>
64-14575 <*>
1961 V17 N3 P86-92 63-26247 <=$>

PROIZVODSTVO TRUB
1962 N6 P37-46 3770 <BISI>
1965 N15 P62-66 66-33110 <=$>
1965 N15 P80-89 66-33110 <=$>
1965 N15 P95-100 66-33110 <=$>

PRO-METAL
1965 N106 P236-239 66-13602 <*>

PROMYSHLENNAYA AERODINAMIKA
1958 N10 P36-42 59-19441 <+*>
1958 N10 P43-60 60-15659 <+*>
1958 N10 P77-110 62-25956 <=*>
1958 N10 P78-110 60-13178 <+*>
1958 N11 P40 59-11467 <=>
1960 N18 P54-107 63-13686 <=>
1960 N18 P65-79 63-15167 <=*>

1960 N19 P9-20	64-13935 <=*$>
1961 N20 P57-81	63-15980 <=*>
1962 N23 P66-71	AD-646 499 <=>
1962 N23 P72-79	AD-610 323 <=$>
1962 N23 P99-106	65-62706 <=$>
1962 N23 P107-165	AD-646 499 <=>
1962 N23 P174-199	AD-614 760 <=$>
1962 N24 P34-47	AD-619 474 <=$>
1962 N24 P63-73	AD-619 474 <=$>
1963 N25 P121-183	RTS-3297 <NLL>

PROMYSHLENNAYA ENERGETIKA

1947 V4 N2 P13-14	78T95R <ATS>
1950 V7 N4 P14-16	RT-1641 <*>
1950 V7 N5 P6-7	RT-1642 <*>
1952 N1 P1-3	RT-2537 <*>
1952 N6 P4-9	RT-2744 <*>
1952 V9 N1 P4-7	R-190 <*>
1958 V13 N3 P12-13	62-32936 <=*>
1958 V13 N9 P1-7	59-19134 <+*>
1958 V13 N11 P34-35	59-11467 <=>
1958 V13 N11 P36-38	59-11467 <=>
1959 V14 N6 P42	60-11086 <=>
1959 V14 N8 P6-8	60-11086 <=>
1959 V14 N8 P8-10	60-11086 <=>
1959 V14 N11 P52-54	60-11757 <=>
1959 V14 N11 P64	60-11757 <=>
1960 V15 P50-	61-11791 <=>
1960 V15 P6C-	61-11791 <=>
1960 V15 N3 P56-57	60-41247 <=>
1960 V15 N4 P47-49	61-11789 <=>
1960 V15 N5 P23-25	62-13559 <*=>
1960 V15 N5 P61	60-41342 <=>
1960 V15 N5 P64	60-41342 <=>
1960 V15 N6 P15-16	62-13298 <*=>
1960 V15 N6 P48	61-21307 <=>
1960 V15 N8 P9-13	61-11537 <=> P
1960 V15 N10 P1-3	61-21139 <=> P
1960 V15 N1C P6-8	61-21139 <=> P
1960 V15 N1C P16-18	61-21139 <=>
1960 V15 N11 P3-5	61-21538 <=>
1961 N2 P38-44	62-13337 <*=>
1961 V16 P54-55	61-31582 <=>
1961 V16 N2 P4-8	61-27119 <=>
1961 V16 N3 P6-9	61-31550 <=>
1961 V16 N3 P9-11	61-27119 <=>
1961 V16 N3 P62-63	62-23476 <=>
1961 V16 N5 P54	61-31584 <=>
1961 V16 N5 P59-60	61-27385 <=>
1961 V16 N5 P62-63	61-31584 <=>
1961 V16 N6 P30-33	62-25343 <=*>
1962 N10 P1-5	NLLTB V5 N2 P110 <NLL>
1962 V17 N2 P41	62-33324 <=>
1962 V17 N2 P44	63-21719 <=>
1962 V17 N2 P54-56	63-21719 <=>
1962 V17 N5 P1-5	62-33332 <=>
1962 V17 N5 P49	62-25972 <=>
1962 V17 N5 P61	62-25972 <=>
1962 V17 N6 P54	62-33664 <=*>
1962 V17 N7 P49-50	62-33115 <=>
1962 V17 N7 P62-63	62-33115 <=>
1962 V17 N8 P41	63-13310 <=>
1962 V17 N8 P53-54	63-13519 <=>
1962 V17 N10 P52-53	63-13486 <=>
1962 V17 N1C P54-55	63-13365 <=>
1962 V17 N11 P61-62	63-13843 <=*>
1963 V18 N3 P47,63	63-31452 <=>
1963 V18 N5 P53	63-31065 <=>
	63-31097 <=>
1963 V18 N5 P6C	63-31097 <=>
1963 V18 N6 P13-21	63-31619 <=>
1963 V18 N6 P49-50	63-31618 <=>
1963 V18 N6 P63-64	63-31561 <=>
1963 V18 N7 P47-48	63-31710 <=>
1964 V19 N1 P42-43	AD-610 278 <=$>
1964 V19 N7 P2-6	64-51219 <=>
1967 V22 N2 P2-4	67-31207 <=$>
1967 V22 N2 P50-51	67-31207 <=$>

PROMYSHLENNC-EKONOMICHESKAYA GAZETA

1959 P2	60-31063 <=>
1959 P4	60-31064 <=>
	60-31178 <=>
1959 N58 P2	NLLTB V1 N9 P6 <NLL>
1959 N58 P4	NLLTB V1 N9 P6 <NLL>
1959 N77 P3	59-22138 <+*>
1959 N96 P4	60-19082 <+*> 0

PROMYSHLENNOE STROITELSTVO

1958 V36 N1 P14-19	61-15365 <*+>
1958 V36 N7 P16-18	61-28670 <*=>
1958 V36 N12 P21-23	59-19284 <+*>
	61-13892 <*+>
1959 V37 N6 P30-33	59-13909 <=>
1959 V37 N7 P4	59-13939 <=>
1959 V37 N7 P44-46	60-17432 <+*>
1959 V37 N11 P6-9	60-11572 <=>
1960 V38 N1 P1-2	60-11569 <=>
1960 V38 N2 P43-45	61-27301 <*=>
1960 V38 N7 P8	61-11743 <=>
1960 V38 N7 P64	61-28136 <*=>
1960 V38 N11 P2-4	61-21388 <=> P
1960 V38 N12 P2-5	61-21536 <=>
1961 V39 N2 P2-5	61-23575 <=>
1961 V39 N2 P49-53	62-18616 <=*> C
1962 V40 N7 P2-3	62-33144 <=*>
1962 V40 N9 P2-6	63-21088 <=>
1962 V40 N10 P11-14	63-13650 <=>
1962 V40 N11 P31-34	63-21352 <=>
1963 V41 N1 P22-26	64-13760 <=*$>
1963 V41 N1 P32-36	64-13760 <=*$>
1963 V41 N2 P10-12	63-31115 <=>
1963 V41 N3 P5-10	63-21817 <=>
1963 V41 N5 P2-3	63-31321 <=>
1965 V43 N2 P29-30	65-64605 <=*$>
1965 V43 N12 P6-9	66-61834 <=*$>

PROMYSHLENNOE STROITELSTVC I INZHENERNYE SOORUZHENIIA

1962 V4 N5 P28-32	63-15103 <=>
1962 V4 N6 P11-15	63-21390 <=>
1962 V4 N6 P24-27	63-21390 <=>
1963 V5 N1 P1-3	63-21789 <=>

PROMYSHLENNOST ARMENII

1961 N2 P8-10	61-31527 <=>
1961 N3 P24-28	61-31584 <=>
1961 N3 P39	61-27385 <=>
1961 N3 P39-40	61-31584 <=>
1961 V4 N1 P23-24	61-27290 <=>
1961 V4 N9 P30-34	63-13116 <=*>
1962 N3 P31-36	62-32065 <=>
1962 V5 N1 P21-22	63-21959 <=>
1962 V5 N2 P7-10	AD-620 766 <=$>
1962 V5 N4 P41-45	63-13645 <=>
1962 V5 N5 P27-29	62-32461 <=>
1962 V5 N6 P3-5	62-32579 <=>
1962 V5 N7 P33-35	63-13079 <=>
1962 V5 N7 P35-39	63-13310 <=>
1962 V5 N7 P46-47	63-13079 <=>
	63-13477 <=>
1962 V5 N9 P14-18	63-13645 <=>
1962 V5 N9 P34-35	63-13310 <=>
1962 V5 N9 P49-53	63-13477 <=>
1962 V5 N9 P58-59	63-13477 <=>
1962 V5 N10 P25-29	63-21363 <=>
1962 V5 N10 P35	63-21156 <=>
1962 V5 N10 P43-47	63-21363 <=>
1962 V5 N1C P55-56	63-21352 <=>
1962 V5 N10 P59	63-21156 <=>
1962 V5 N11 P8-19	63-21224 <=>
1962 V5 N11 P14-15	63-21269 <=>
1962 V5 N11 P44-46	63-21224 <=>
1962 V5 N11 P5C-51	63-15351 <=*>
1962 V5 N11 P52-54	63-21224 <=>
1962 V5 N11 P58-60	63-21381 <=>
1962 V5 N12 P18-20	63-21682 <=>
1963 V6 N1 P65	63-21817 <=>
1963 V6 N1 P70-72	63-21582 <=>
1963 V6 N2 P55-59	63-31224 <=>

1963 V6 N2 P69-70 63-31102 <=>
1963 V6 N2 P70-71 63-31132 <=>
1963 V6 N2 P70 63-31472 <=>
1963 V6 N2 P70-72 63-31472 <=>
1963 V6 N2 P71 63-31352 <=>
1963 V6 N2 P72 63-31169 <=>
63-31472 <=>
63-31495 <=>
1963 V6 N3 P69-70 63-31451 <=>
1963 V6 N3 P72 63-31303 <=>
1963 V6 N4 P3-11 63-31384 <=>
1963 V6 N4 P32-34 63-31399 <=>
1963 V6 N4 P37-41 63-31425 <=>
1963 V6 N4 P43-44 63-31579 <=>
1963 V6 N4 P50-51 63-31398 <=>
1963 V6 N4 P54-56 63-31399 <=>
1963 V6 N4 P65-74 63-31398 <=>
1963 V6 N4 P67 63-31399 <=>
1963 V6 N4 P70-72 63-31384 <=>
1963 V6 N4 P74 63-31384 <=>
63-31399 <=>
63-31420 <=>
63-31797 <=>
1963 V6 N5 P6-9 63-31636 <=>
1963 V6 N5 P10-12 63-31733 <=>
1963 V6 N5 P16-18 63-31599 <=>
1963 V6 N5 P27-30 63-31710 <=>
1963 V6 N5 P3C-34 63-31636 <=>
1963 V6 N5 P66 63-31599 <=>
1963 V6 N6 P9-13 63-41088 <=>
1963 V6 N6 P13-14 63-41088 <=>
1963 V6 N6 P15 63-31906 <=>
63-31924 <=>
1963 V6 N6 P34-38 498 <TC>
64-16388 <=*$>
1964 V7 N1 P46-49 6670 <HB>
1965 N12 P30-32 66-32467 <=$>
1966 P62-64 67-31207 <=$>
1966 N12 P6-9 67-31161 <=$>
1966 N12 P22-23 67-31161 <=$>
1966 V9 N2 P33-35 66-32069 <=$>
1966 V9 N5 P28-29 67-30085 <=>
1966 V9 N5 P41-42 67-30085 <=>

PROMYSHLENNOST ORGANICHESKOI KHIMII
1936 N2 P79-87 R-4939 <*>
1936 V1 P350-353 R-4635 <*>
1936 V1 P537-540 61-16820 <=*$>
1936 V2 P203-205 61-16819 <=*$>
1936 V2 P382-387 61-16697 <=*$>
1937 V3 P162-163 <ATS>
1937 V4 N14 P104-109 RT-1796 <*>
1938 V5 N7 P489-492 RJ-73 <ATS>
RJ73 <ATS>
1938 V5 N8/9 P551-553 RJ-352 <ATS>
1939 V6 P101-103 5917279 <+*> 0
1939 V6 P434-437 R-3965 <*>
1939 V6 N2 P93-95 RT-2450 <*>
1939 V6 N8 P457-459 RT-2359 <*>
1939 V6 N9 P519-521 RT-960 <*>
1939 V6 N12 P670-671 RT-919 <*>
1939 V6 N4/5 P203-207 63-18611 <=*>
1940 V7 P24-25 <ATS>
1940 V7 P210-214 60-18494 <+*>
1940 V7 P240-241 63-18977 <=*>
1940 V7 P296-304 R-4866 <*>
62-14071 <=*>
1940 V7 P300-301 R-2771 <*>
1940 V7 P314-317 65-10148 <*>
1940 V7 N10 P552-562 62-10898 <=*>
1940 V7 N10 P562-571 62-10899 <*=> P
1940 V7 N11 P593-599 61-18107 <*=>
1940 V7 N11 P618-621 62-14701 <=*> P
1940 V7 N12 P660-661 63-18116 <=*>
1940 V7 N12 P662-663 62-14067 <=*>
1940 V7 N4/5 P221-223 RT-1373 <*>
1940 V7 N4/5 P223-225 RT-1400 <*>
1940 V7 N4/5 P225-227 RT-1401 <*>

PROPHYLAXIE SANITAIRE ET MORALE
1954 V26 N1 P263-264 3881-A <K-H>

PROTEIN, NUCLEIC ACID AND ENZYME. TOKYO
SEE TANPAKUSHITSU KAKUSAN KOSO

PROTEZIROVANIE I PROTEZCSTRCENIE
1962 N6 P5-13 63-31246 <=>
1962 N6 P43-49 63-31246 <=>

PROTOKOLL DER VERHANDLUNGEN DES VEREINES DEUTSCHER PORTLAND-ZEMENT FABRIKANTEN
1904 V27 P86- 64-14247 <*>
1909 V32 P206-230 64-18215 <*> 0
1909 V32 P244-259 64-18215 <*> C
1910 V33 P238 65-17421 <*>
1910 V33 P246 65-17421 <*>
1912 V35 P217-249 64-18296 <*> 0
1913 V36 P273-301 64-18276 <*> PO
1913 V36 P313-344 64-14252 <*> C
1913 V36 P347-373 64-18288 <*> 0
1914 V37 P145-160 64-18290 <*> 0
1914 V37 P16C-180 64-18283 <*>
1914 V37 P180-189 64-18294 <*>
1914 V37 P221-233 64-14433 <*>
1914 V37 P245-269 64-14251 <*> 0
1921 V44 P109-125 64-18234 <*> P
1922 V45 P76 64-18268 <*>
1922 V45 P98-116 64-18233 <*>

PROTOPLASMA
1936 V26 P557-576 II-46 <*>
1952 V41 P258-260 2709B <K-H> C
1956 V46 N1/4 P194-197 57-3420 <*>
1956 V46 N1/4 P711-742 58-950 <*>
1956 V47 N1/2 P217-235 58-2343 <*>
1959 V51 N3 P373-376 63-00141 <*>

PRUMYSL POTRAVIN
1953 V4 N4 P159-161 CJ399 <ATS>
RJ-399 <ATS>
1958 V9 N12 P644-645 12818 <K-H>
60-18480 <*>
1959 V10 N5 P258-265 62-10743 <*> 0
1959 V10 N6 P300-301 59-21216 <=*>
1963 V14 N2 P70-73 RAPRA-1259 <RAP>
1963 V14 N4 P211-212 65-11984 <*>
1963 V14 N5 P262-265 80R75C <ATS>

PRYKLADNA MEKHANIKA
1955 V1 N4 P378-390 59-16727 <+*>
1957 N1 P108-112 R-2103 <*>
1957 V3 P225-228 UCRL-TRANS-876(L) <=*>
1957 V3 N1 P13-18 61-19546 <*=>
1958 V4 N4 P369-375 59-11484 <=>
1959 V5 N1 P106-113 AD-625 805 <=*$>
1959 V5 N3 P344-348 60-11593 <=>
1959 V5 N4 P391-401 65-63643 <=*$>
1959 V5 N4 P455-456 60-11389 <=>
1960 V6 N2 P224-228 AD-630 413 <=*$>
1960 V6 N2 P233-234 60-31823 <=>
1960 V6 N3 P241-249 AD-633 817 <=$>
1960 V6 N3 P272-280 63-15338 <=*>
1960 V6 N4 P375-384 63-23109 <=*$>
1960 V6 N4 P385-392 AD-633 818 <=$>
1960 V6 N4 P445-448 61-27784 <*=>
1961 V7 N1 P11C-112 63-15547 <=*>
1961 V7 N2 P228-230 61-28387 <*=>
1961 V7 N3 P253-257 62-32529 <=*>
1961 V7 N3 P266-271 AD-634 840 <=$>
1961 V7 N3 P326-331 64-13157 <=*$>
1961 V7 N4 P459-462 62-32843 <=*>
1961 V7 N5 P496-502 63-15273 <=*>
63-15580 <=*>
1961 V7 N6 P601-608 62-32694 <=*>
1961 V7 N6 P649-655 64-13163 <=*$>
1962 V8 N3 P342-346 63-13348 <=>
1962 V8 N5 P489-499 AD-609 786 <=>
1963 V9 N2 P126-132 65-60352 <=$>
1963 V9 N3 P249-258 AD-610 799 <=$>
64-16049 <=*$>

1963 V9 N3 P259-263 64-16831 <=*$>
1963 V9 N5 P465-472 AD-611 422 <=$>
1963 V9 N5 P466-472 64-10329 <=*$>
1963 V9 N5 P473-479 64-10828 <=*$>
1963 V9 N5 P520-528 64-14723 <=*$>
1963 V9 N6 P612-618 64-14784 <=*$>
1963 V9 N6 P619-626 64-14724 <=*$>
1963 V9 N6 P649-658 64-14785 <=*$> O
1963 V9 N6 P659-669 64-71528 <=>
1963 V9 N6 P677-682 64-16008 <=*$>
1964 V10 N1 P24-31 64-11894 <=>
64-16007 <=*$>
1964 V10 N1 P32-39 AD-610 766 <=$>
64-16829 <=*$>
1964 V10 N1 P72-76 64-16830 <=*$>
1964 V10 N2 P149-157 64-18421 <*>
1964 V10 N2 P205-215 64-18424 <*>
1964 V10 N2 P222-225 64-18420 <*>
1964 V10 N3 P237-246 65-14043 <*>
1964 V10 N3 P254-262 AD-625 162 <=*$>
1964 V10 N5 P503-507 65-12642 <*>
1964 V10 N5 P547-551 AD-639 160 <=$>
1964 V10 N6 P594-599 AD-633 582 <=$>
1964 V10 N6 P600-607 AD-625 799 <=*$>
1964 V10 N6 P639- AD-625 799 <=*$>
1964 V10 N6 P654-659 65-14045 <*>
1964 V10 N6 P660-663 AD-625 799 <=*$>
65-14044 <*>

PRZEGLAD BIBLIOTECZNY
1959 V27 N1/2 P29-44 65-12048 <*>

PRZEGLAD ELEKTRONIKI
1960 V1 N1 P2-6 62-19306 <*=>
1962 V3 N3 P145 34Q7CP <ATS>
1963 V4 N5/6 P329-338 65-10810 <*>
1964 V5 N5 P252-255 AD-640 306 <=$>
1964 V5 N11 P597-602 66-14646 <*$>
1965 N1 P39-41 AD-636 678 <=$>

PRZEGLAD DROBNIJ WYTWOROZOSCI. (TITLE VARIES)
1965 N22 P1-4 66-30296 <=$>

PRZEGLAD ELEKTROTECHNICZNY
1958 V34 N10 P541-549 1846 <BISI>
1958 V34 N2/3 P45 PB 141 318T <=>
1958 V34 N2/3 P46-50 59-11277 <=> O
1958 V34 N2/3 P51-61 PB 141 488T <=>
1958 V34 N2/3 P61-66 PB 141 319T <=>
1958 V34 N2/3 P66-80 PB 141 487T <=>
1958 V34 N2/3 P80-89 59-11003 <=>
1958 V34 N2/3 P89-92 PB 141 325T <=>
1958 V34 N2/3 P92-99 PB 141 326T <=>
1958 V34 N2/3 P99-104 PB 141 327T <=>
1958 V34 N2/3 P104-114 PB 141 328T <=> C
1958 V34 N2/3 P115-121 59-11073 <=>
1958 V34 N2/3 P121-127 59-11074 <=> O
1958 V34 N2/3 P128-133 59-11090 <=> O
1958 V34 N2/3 P133-136 59-11023 <=>
1958 V34 N2/3 P136-140 59-11022 <=>
1958 V34 N2/3 P140-150 59-11021 <=>
1958 V34 N2/3 P150-157 59-11080 <=>
1958 V34 N2/3 P157-163 59-11046 <=> O
1958 V34 N2/3 P163-170 59-11047 <=> O
1958 V34 N2/3 P170-176 59-11060 <=>
1958 V34 N2/3 P176-188 59-11001 <=>
1958 V34 N2/3 P188-192 59-11002 <=>
1958 V34 N2/3 P193-197 59-11006 <=>
1958 V34 N2/3 P197-202 59-11005 <=>
1958 V34 N2/3 P202-204 59-11004 <=> O
1960 V36 P382-384 SCL-T-409 <*>
1961 V37 N2 P53-55 62-23402 <=*>
1961 V37 N9 P381-385 62-32341 <=*$>
1961 V37 N10 P397-403 SCL-T-408 <*>
1962 V38 P264-265 64-00035 <*>
1965 N8 P315-319 AD-638 668 <=$>
1965 N12 P501-503 66-30482 <=$>

★PRZEGLAD EPIDEMIOLOGICZNY
1961 V15 P179-187 63-00042 <*>
1961 V15 N1 P67-74 62-19309 <=*>
1961 V15 N1 P77-85 62-19310 <=*>

PRZEGLAD GEODEZYJNY
1960 V32 N7 P242-244 62-19291 <*=>
1960 V32 N7 P258-261 62-19318 <*=>
1960 V32 N12 P431-445 62-19512 <*=> O
1960 V32 N12 P447-448 62-19319 <*=>
1960 V32 N12 P449-450 62-19512 <*=> O
1960 V32 N12 P458-459 62-19319 <*=>
1961 V33 N1 P24-26 62-19317 <=*>
1961 V33 N3 P82-95 62-19587 <=>
1961 V33 N3 P97-100 62-19587 <=>
1961 V33 N3 P102-112 62-19587 <=>
1961 V33 N3 P116-120 62-19587 <=>
1961 V33 N4 P125I 62-19484 <=*>
1961 V33 N5 P191-195 62-23443 <=> P
1961 V33 N7 P241-249 62-23443 <=>
1961 V33 N7 P252-254 62-23443 <=>
1961 V33 N7 P274-278 62-23443 <=> P
1961 V33 N8 P308-311 62-23443 <=> P
1962 V34 N11 P454-455 63-21246 <=>
1964 V36 N3 P103-104 64-31364 <=>
1964 V36 N7 P274-276 64-51974 <=$>
1965 V37 N8 P316-319 65-33834 <=*$>

PRZEGLAD GEOFIZYCZNY
1959 V4 N3/4 P272-274 60-31847 <*=>
1959 V4 N3/4 P275-276 63-11391 <=>
1960 V5 N1 P77-82 62-19287 <=*>
1964 V9 N2 P245-284 65-50376 <=>

PRZEGLAD GEOGRAFICZNY
1960 V32 N3 P303-320 65-50321 <=$>
1961 V33 N3 P569-580 62-23596 <=*>
1964 V36 N12 P475-476 AD-631 582 <=$>

PRZEGLAD GEOLOGICZNY
1955 N9 P426-429 R-342 <*>
1958 N5 P229-232 IGR V1 N8 P67 <AGI>
1958 V6 N3 P132-134 79L31P <ATS>
1958 V6 N5 P229-232 PB-141-282T <=*>
1958 V6 N11 P469-474 62-19308 <*=>
1959 V7 N1 P27-29 59-11638 <=>
1961 V9 N4 P210-214 16P61P <ATS>

PRZEGLAD GORNICZY
1954 V10 N10 P357-359 60-10052 <*>
1957 V13 N10 P504-510 TRANS-568 <MT>
1957 V13 N12 P594-601 60-21378 <=>
1958 V14 N9 P21-24 61-11350 <=>
1959 V15 N10/1 P487-501 61-11356 <=>
1960 V16 N10 P502-506 63-11386 <=>
1961 V17 P209-214 UCRL TRANS-768(L) <*>
1963 V19 N6 P258-260 94R77P <ATS>
1963 V19 N10 P410-413 01R79P <ATS>
1963 V19 N10 P18-21 SUP 16R79P <ATS>

PRZEGLAD KOLEJOWY
1956 N7 P158-160 59-11548 <=>
1958 V10 N7 P241-246 59-11329 <=*>
1958 V10 N11 P425-531 59-11501 <=*>
1964 V16 N8/9 P25-57 64-51338 <=>

PRZEGLAD KOMUNIKACYJNY; MIESIECZNIK EKONOMICZNO-TECHNICZNY
1964 N11 P412-420 AD-636 586 <=$>

PRZEGLAD LEKARSKI
1955 V11 P155-157 R-3790 <*>
1955 S2 V11 N10 P314-315 12057-A <K-H>
63-18513 <*>
1961 S2 V17 N6 P227-230 62-19545 <=*> P

PRZEGLAD MECHANICZNY
1957 V16 N2 P58-64 60-21011 <=*>
1960 V19 N17 P510-516 62-19314 <*=>
1961 V20 N9 P286-288 62-19315 <*=>
1963 V22 N20 P626-628 NEL-TRANS-1688 <NLL>

PRZEGLAD ODLEWNICTWA
1953 V3 N12 P341-346 60-18829 <*>
1958 V8 P221-224 64-10556 <*>
1960 V10 N4 P106-111 2020 <BISI>
1960 V10 N9 P263-266 3526 <BISI>
1962 V12 N7 P213 M-5389 <NLL>
1963 V13 P197-201 3961 <BISI>
1966 N4 P116-121 66-32532 <=$>
1967 N1 P2-4 67-30894 <=$>

PRZEGLAD PAPIERNICZY
1954 V10 P328-332 63-10062 <*> 0
1954 V10 P341 63-10062 <*> 0
1954 V10 N6 P168-172 56T90P <ATS>
1955 V11 N9 P263-267 C-2188 <NRC>
2359 <*>
1958 V14 N1 P2-4 61-31266 <=>
1958 V14 N12 P1-6 61-31292 <=>
1959 V15 P241-244 62-10608 <*>
1959 V15 N7 P208-210 62-14483 <*> 0
1960 V16 N1 P1-3 61-11382 <=>
1960 V16 N5 P139-145 64-18497 <*> 0
1960 V16 N7 P193-198 63-20992 <*> 0
1960 V16 N8 P225-228 61-20897 <*> 0
1961 V17 N1 P3-6 63-10529 <*> 0
1961 V17 N1 P18-20 64-18499 <*> 0
1961 V17 N10 P301-306 64-16600 <*> 0
1962 V18 N2 P33-38 62-20468 <*> 0
1962 V18 N11 P337-342 64-16602 <*> 0

PRZEGLAD SAMOCHODOWY
1958 N2 P23-29 59-21013 <=*>

PRZEGLAD SPAWALNICTWA
1955 V7 N1 P7-12 TS-1337 <BISI>

PRZEGLAD TECHNICZNY
1958 N5 P166-170 59-11574 <=> 0
1958 N10 P438-441 59-11612 <=>
1960 N38 P6 62-19296 <*=>
62-19297 <=*>
1960 V81 N48 P7 62-19298 <=*>

PRZEGLAD TELEKOMUNIKACYJNY
1957 V30 N1 P5-8 62-11531 <=>
1958 V31 N7 P200-207 62-23046 <=>
62-23046 <=*>
1961 V34 N1 P1-36 62-19295 <*=>
1961 V34 N1 P1-4 SUP 62-19295 <*=>
1961 V34 N3 P78-84 62-19472 <=*>
1962 N6 P171-175 66-11478 <*>
1962 V35 N11 P324-326 63-21362 <=>
1964 V37 N3 P84-85 64-31468 <=>
1964 V37 N6 P161-165 64-51366 <=>
1965 V37 N4 P97-100 65-31759 <=$>

PRZEGLAD WLOKIENNICZY
1962 V16 N1 P32-34 62-18906 <*>

PRZEGLAD WOJSK LADOWYCH
1963 V5 N2 P125-140 AD-637 416 <=$>

PRZEGLAD ZBOZOWO-MLYNARSKI
1959 V3 N5 P145 60-13692 <=*>
1959 V3 N6 P171-173 60-13006 <=*>

PRZEMYSL CHEMICZNY
1927 V11 P146-182 63-14028 <*> 0P
1928 V12 P40-48 57-1627 <*>
1928 V12 P501-525 63-10697 <*> 0
1928 V12 N7 P333-341 63-10450 <*> 0
1929 V13 P185-195 63-10433 <*> 0
1929 V13 P209-220 63-10434 <*> 0
1929 V13 P455-460 63-10435 <*>
1934 V18 P458-464 63-16104 <*> 0
1934 V18 P464-470 63-16103 <*> 0
1934 V18 P628-633 64-20822 <*> 0
1935 V19 P122-125 63-16109 <*> 0
1936 V20 P56-69 63-16102 <*> 0
1937 V21 P228-237 63-14614 <*>
1937 V21 P279-289 63-14568 <*> 0
1938 V22 P293-296 34H12P <ATS>
58-356 <*>
1938 V22 N11/2 P444-448 RCT V13 N1 P435 <RCT>
1939 P181-188 57-1619 <*>
1948 V27 P22-29 R-672 <*>
1950 V4 P194-197 58-2694 <*>
1950 V6 P227-280 63-18662 <*>
1950 V6 N4 P194-197 55Q67P <ATS>
1950 V6 N2/3 P133-134 40K21P <ATS>
1950 V29 P277-280 64-20335 <*>
1950 V29 N4 P166-170 61-10848 <*> 0
1950 V29 N4 P170-173 63-20438 <*> 0
1952 V8 P285 58-2658 <*>
1952 V8 P508-510 T-1844 <INSD>
1952 V8 N7/8 P318-320 64-20343 <*> 0
1952 V31 N6/7 P365-368 59-15327 <*>
1952 V31 N7/8 P358-360 59-20675 <*>
1953 V32 P72-76 61-20952 <*> 0
1953 V32 P589-596 61-20854 <*> 0
1954 V10 P392-396 PJ-476 <ATS>
1955 V2 P295-301 T-1550 <INSD>
1955 V34 P630-635 60-14143 <*>
1955 V34 N1 P36-39 61-31285 <=>
1955 V34 N7 P375-380 61-11384 <=>
1956 V12 N8 P433-442 65-10212 <*>
1956 V12 N9 P520-522 65-10168 <*>
1956 V35 N9 P520-522 62-16451 <*>
1957 V13 P721-723 63-16187 <*> 0
1957 V13 N7 P401-405 61-11388 <=>
1957 V13 N7 P405-410 61-11340 <=>
1957 V13 N8 P463-467 61-11340 <=>
1957 V13 N9 P532-536 61-11340 <=>
1957 V36 P26-30 60-13230 <+*>
1957 V36 P273-276 60-13231 <+*>
1957 V36 P291-293 61-10298 <*>
1957 V36 P607-609 59-00761 <*>
1957 V36 N2 P82-87 C-3142 <NRC>
59-17783 <*>
1957 V36 N8 P445-449 61-31284 <=>
1957 V36 N9 P494-495 60-21230 <=>
1958 V14 N2 P98-102 61-31283 <=>
1958 V37 P30-34 43L33P <ATS>
1958 V37 P340-343 62-14089 <*>
1958 V37 P408-411 60-19158 <=>
1958 V37 P411-415 62-14090 <*>
86L29P <ATS>
1958 V37 P468-469 62-14091 <*> 0
1958 V37 P525-529 62-14092 <*> 0
1958 V37 P590-592 9033 <K-H>
1958 V37 P630-634 63-20242 <*>
1958 V37 N9 P590-592 65-10914 <*>
1958 V37 N10 P637-639 60-21037 <=>
1959 V15 P599-603 64-16536 <*>
1959 V38 P87-88 60-17100 <*=>
1959 V38 P358-362 13N55P <ATS>
1959 V38 P551-554 60-17744 <=*>
1959 V38 N3 P168-175 61-11373 <=>
1959 V38 N4 P216-220 61-11378 <=>
1959 V38 N6 P329-331 66-10274 <*>
1959 V38 N8 P481-483 NP-7808 <*>
1960 V16 N7 P436-438 65-13196 <*>
1960 V39 P768-772 AEC-TR-4846 <*>
1960 V39 N4 P205-210 60-41559 <*=>
1960 V39 N4 P228-231 61-11376 <=>
1960 V39 N8 P496-502 65-10119 <*>
1960 V39 N8 P519-520 62-19294 <=*>
1960 V39 N10 P599-603 03N55P <ATS>
1960 V39 N10 P629-633 51N50P <ATS>
1961 V17 N4 P203-206 63-14113 <*>
1961 V17 N7 P375-379 63-14114 <*> 0
1961 V17 N9 P529-533 65-12632 <*>
1961 V17 N12 P694-695 62-20235 <*> 0
1961 V40 P15-17 NP-7810 <*>
1961 V40 P48-51 ICE V1 N1 P124-129 <ICE>
1961 V40 N4 P203-206 65-10051 <*> 0
1961 V40 N4 P206-209 81S87P <ATS>
1961 V40 N5 P253-255 65-10116 <*>
1961 V40 N7 P396- ICE V2 N1 P132-134 <ICE>
1961 V40 N8 P433- ICE V2 N1 P105-108 <ICE>

1961 V40 N11 P646- ICE V2 N2 P290-296 <ICE>
1961 V40 N11 P651- ICE V2 N3 P342-344 <ICE>
1961 V40 N12 P684-687 46P65P <ATS>
1962 N2 P64- ICE V2 N4 P504-506 <ICE>
1962 N2 P95- ICE V2 N4 P543-545 <ICE>
1962 N4 P186- ICE V2 N4 P511-514 <ICE>
1962 N9 P527- ICE V3 N2 P178-185 <ICE>
1962 N10 P590- ICE V3 N3 P339-341 <ICE>
1962 V41 N1 P40-43 09P62P <ATS>
1962 V41 N10 P563-565 59Q68P <ATS>
1963 N1 P35- ICE V3 N3 P351-355 <ICE>
1963 N2 P99- ICE V3 N3 P433-438 <ICE>
1963 N3 P135- ICE V3 N4 P467-471 <ICE>
1963 N6 P313-317 ICE V4 N1 P66-71 <ICE>
1963 N10 P563-566 ICE V4 N2 P258-262 <ICE>
1963 N11 P614-617 ICE V4 N3 P413-417 <ICE>
1963 N11 P633-635 ICE V4 N3 P447-450 <ICE>
1963 N4/5 P242-245 ICE V3 N4 P582-585 <ICE>
1963 V19 N3 P176 95R74P <ATS>
1963 V19 N8 P419-421 64-16067 <*>
1963 V19 N8 P439-441 64-16173 <*>
1963 V19 N10 P563-566 ICE V4 N2 P258-262 <ICE>
96R74P <ATS>
1963 V42 P629-632 65-12492 <*>
1963 V42 N3 P139-140 64-14289 <*> 0
1963 V42 N7 P349-354 82S87P <ATS>
1963 V42 N8 P430-432 ICE V4 N1 P72-75 <ICE>
1963 V42 N8 P435-439 ICE V4 N1 P61-66 <ICE>
1963 V42 N9 P508-511 83S87P <ATS>
1963 V42 N11 P578-581 ICE V4 N3 P456-460 <ICE>
1963 V42 N11 P614-617 ICE V4 N3 P413-417 <ICE>
1963 V42 N11 P633-635 ICE V4 N3 P447-450 <ICE>
1963 V42 N12 P691-693 65-12491 <*>
1964 N4 P190-193 ICE V4 N4 P695-700 <ICE>
1964 N4 P201-203 ICE V4 N4 P628-631 <ICE>
1964 N4 P226-229 ICE V4 N4 P701-705 <ICE>
1964 N5 P275-279 ICE V4 N4 P655-660 <ICE>
1964 N6 P328-331 ICE V5 N1 P40-44 <ICE>
1964 N9 P484-490 ICE V5 N2 P323-330 <ICE>
1964 N9 P508-515 ICE V5 N2 P280-288 <ICE>
1964 N11 P630-633 ICE V5 N2 P222-226 <ICE>
1964 N11 P633-638 ICE V5 N2 P360-366 <ICE>
1964 V20 P445-449 66-12230 <*>
1964 V20 N5 P255-259 65-13409 <*>
1964 V20 N10 P534-537 66-11921 <*>
1964 V43 P9-12 1431 <TC>
1964 V43 P153-157 NS-323 <TTIS>
1964 V43 P538-540 1421 <TC>
1964 V43 P569-572 1430 <TC>
1964 V43 P594-596 1429 <TC>
1964 V43 P605-607 1422 <TC>
1964 V43 P608-609 NS-414 <TTIS>
1964 V43 N4 P190-193 ICE V4 N4 P628-631 <ICE>
1964 V43 N4 P201-203 65-11409 <*>
1964 V43 N4 P203-207 ICE V4 N4 P575-580 <ICE>
1964 V43 N4 P226-229 ICE V4 N4 P701-705 <ICE>
1964 V43 N5 P275-279 ICE V4 N4 P655-660 <ICE>
1964 V43 N6 P328-331 ICE V5 N1 P40-44 <ICE>
1964 V43 N7 P361-365 64-41514 <=>
1964 V43 N8 P409-419 64-51343 <=$>
1964 V43 N9 P481-484 66-11920 <*>
1964 V43 N11 P610-614 66-14224 <*$>
1964 V43 N12 P684- 1730 <TC>
1965 N1 P24-28 ICE V5 N3 P547-551 <ICE>
1965 N6 P305-308 ICE V6 N1 P63-67 <ICE>
1965 N7 P361-363 ICE V6 N1 P160-162 <ICE>
1965 N10 P573-576 ICE V6 N2 P303-307 <ICE>
1965 N12 P689-692 ICE V6 N3 P422-425 <ICE>
1965 V21 N6 P323-324 1851 <TC>
1965 V44 N2 P79-81 66-13580 <*>
1966 N3 P121-127 ICE V6 N4 P618-625 <ICE>
1966 N4 P179-183 ICE V6 N4 P626-630 <ICE>
1966 N5 P254-256 ICE V6 N4 P631-633 <ICE>
1966 N5 P276-278 ICE V6 N4 P642-646 <ICE>
1966 V21 N3 P142-145 66-14542 <*>
1966 V45 N2 P85-91 93T93P <ATS>
1967 N4 P203-207 ICE V4 N4 P575-580 <ICE>

PRZEMYSL SPOZYWCZY
1953 V7 N2 P58-62 60-21413 <=>
1956 V10 N1 P22-25 59-11686 <=>
1956 V10 N7 P280-282 60-21415 <=>
1957 V11 N1 P26-30 60-21510 <=>
1958 V12 N4 P132-135 60-21511 <=>
1958 V12 N5 P169-174 60-21532 <=>
1958 V12 N9 P339-344 59-21075 <=*> 0

PRZEMYSL WLOKIENNICZY
1957 V11 N6 P273-275 65-29150 <$>

PSICOTECNIA
1944 V5 P103-114 62-16804 <*>

PSYCHE. BERLIN-PANKOW
1958 V12 N7 P401-407 65-00314 <*>
1958 V12 N7 P408-414 60-13096 <+*>
65-00229 <*>

PSYCHIATRIA ET NEUROLOGIA
1957 V134 P224-235 58-1182 <*>
1958 V135 P361-377 62-01222 <*>
1959 V137 P40-48 62-18052 <*>

PSYCHIATRIA ET NEUROLOGIA JAPONICA
SEE SEISHIN SHINKEIGAKU ZASSHI

PSYCHIATRIE, NEUROLOGIE AND MEDIZINISCHE PSYCHOLOGIE
1950 V2 P116-117 60-10006 <*>
1958 V10 P260-264 60-16930 <*>
1963 V15 N6/7 P215-226 64-31661 <=>

PSYCHOLOGIA WYCHOWAWCZA
1963 V6 P2-16 64-00278 <*>

PSYCHOLOGISCHE BEITRAEGE
1961 V6 N2 P196-207 63-00849 <*>

PSYCHOLOGISCHE FORSCHUNG
1922 V2 P1-4 61-14268 <*>
1924 V6 P113-120 61-14233 <*>
1925 V6 P121-126 61-14284 <*>
1926 V8/9 P318-335 61-14664 <*>
1930 V13 P135-144 61-14190 <*>
1932 V16 P166-170 59-20878 <*>
1938 V22 P238-266 61-14339 <*>
1949 V23 P1-9 64-10015 <*>
1949 V23 P10-24 64-10008 <*>
1950 V23 P263-286 65-12530 <*>
1951 V23 P399-408 64-10048 <*>
1953 V24 P215-229 65-12363 <*>
1964 V27 N4 P377-402 66-11727 <*>

PSYCHOLOGISCHE RUNDSCHAU
1951 V2 P39-42 57-3039 <*>
1954 V5 N4 P291-297 59-12939 <+*>

PSYCHOLOGISCHE STUDIEN
1907 V2 P129-202 61-14363 <*>
1916 V10 N3 P239-259 61-14254 <*>

PSYCHOTECHNISCHE ZEITSCHRIFT
1932 V7 N5 P139-147 61-14672 <*>

PTITSEVODSTVO
1958 V8 N11 P26-29 60-51050 <=>
1960 V10 P95 66-61879 <=*$>
1960 V10 N9 P3-5 61-11969 <=>
1960 V10 N10 P6-9 61-21533 <=>
1960 V10 N12 P6-8 61-27077 <=>

PTITSY SOVETSKOGO SOYUZA
1951 V2 P75-79 64-19347 <=$>

PUBBLICAZIONI. ISTITUTO DI CHIMICA, UNIVERSITA DI TRIESTE
1958 N21 P1-17 59K261 <ATS>

PUBBLICAZIONI DELLA STAZIONE ZOOLOGICA DI NAPOLI
1933 V18 N1 P80-85 <DI>

PUBBLICAZIONI DELL'UNIVERSITA CATTOLICA DEL SACRO CUORE. SERIES 8. STATISTICA. MILAN
1926 S8 V2 P34-53 PT.2 63-00169 <*>
1926 S8 V2 P5-33 PT.1 63-00059 <*>

PUBLICACIONES DE LA COMISION NACIONAL DE LA ENERGIA ATOMICA, REPUBLICA ARGENTINA. SERIE FISICA
1954 V1 N3 P77-82 GAT-T-646 <+*>

PUBLICACIONES DE LA COMISION NACIONAL DE LA ENERGIA ATOMICA, REPUBLICA ARGENTINA. SERIE QUIMICA
1955 V1 N3 P19-28 1563 <*>
1955 V1 N4 P29-41 1562 <*>

PUBLICACIONES. INSTITUTA DE INVESTIGACIONES MICROQUIMICAS. UNIVERSIDAD NACIONAL DEL LITORA
1943 V7 P47-52 3149 <*>
1953 V17 P87-93 1591 <*>
1953 V17 P94-103 1591 <*>
1953 V17 P170-184 1591 <*>
1957 N23 P75-80 75L34SP <ATS>

PUBLICACIONES. LABORATORIO DE ENSAYO DE MATERIALES E INVESTIGACIONES TECNOLOGICAS
1947 N19 P1-48 63-20515 <*>
1959 N9 P49-60 61-18777 <*>

PUBLICACOES. LABORATORIO NACIONAL DE ENGENHARIA CIVIL. LISBON
1954 N48 P38 II-497 <*>

PUBLICATIES BECETEL. GHENT RIJKSUNIVERSITEIT. TECHNICUM
1961 N21 ENTIRE ISSUE 65-18148 <*>

PUBLICATIES. CENTRAAL INSTITUUT VOOR LANDBOUWKUNDIG ONDERZOEK. DROOGTECHNISCH LABORATORIUM. EAGENINGEN
1954 P70-78 CSIRO-3371 <CSIR>
1954 P219-228 CSIRO-3374 <CSIR>

PUBLICATIONS. BANYASZATI KUTATO INTEZET. HUNGARIAN RESEARCH INSTITUTE FOR MINING. BUDAPEST
1957 N1 P3-16 61-16224 <*>

PUBLICATIONS DU CENTRE NATIONAL DE RECHERCHES METALLURGIQUES. BRUXELLES
1964 N1 P15-22 4002 <BISI>
1964 N1 P45-63 4001 <BISI>
1965 P19-25 4307 <BISI>
1965 P53-63 4919 <BISI>
1965 N2 P3-8 4107 <BISI>
1965 N2 P9-17 6542 <HB>
1965 N5 P73-81 4707 <BISI>

PUBLICATIONS DE L'INSTITUT NATIONAL POUR L'ETUDE AGRONOMIQUE DU CONGO BELGE
1939 N24 P51 57-2840 <*>

PUBLICATIONS. INSTITUT DE RECHERCHES DE LA SIDE-RURGIE. ST. GERMAINE EN LAYE
1958 SA N174 P1-83 1835 <BISI>
1958 SA N194 P1-47 TS-1647 <BISI>

PUBLICATIONS DE L'INSTITUT DE STATISTIQUE DE L'UNIVERSITE DE PARIS
1957 V6 N3 P227-240 66-11169 <*>

PUBLICATIONS MATHEMATIQUES. INSTITUT DES HAUTES ETUDES SCIENTIFIQUES. PARIS
1963 N18 P295-316 AMST S2 V59 P128-149 <AMS>

PUBLICATIONS. OFFICE NATIONALE D'ETUDES ET DE RECHERCHES AERONAUTIQUES
1954 N71 ENTIRE ISSUE NACA TM 1435 <NASA>
1956 N83 ENTIRE ISSUE 58-1012 <*>

PUBLICATIONS SCIENTIFIQUES ET TECHNIQUES DU MINISTERE DE L'AIR
1948 N128 P1-82 58-2402 <*>
1948 N218 P1-82 AEC-TR-3386 <*>
1950 N237 ENTIRE ISSUE 2985 <*>
1953 N283 P15-35 58-1507 <*>
1953 N283 P118- 2438 <*>

PUBLICATIONS DE LA STATION FEDERALE D'ESSAIS VITICOLES, ARBORICOLES ET DE CHIMIE AGRICOLE
1946 N349 P16 57-137 <*>

PUBLICATIONS TECHNIQUES. CENTRE D'ETUDES ET DE RECHERCHES DE L'INDUSTRIE DES LIANTS HYDRAULIQUES
1949 N14 P1-23 66-10183 <*> P
1951 N43 P3-4 1195 <*>

PUBLIKACIJE. ELEKTROTEHNICKI FAKULTET, UNIVERZITET U BEOGRADU
1956 N1 P1-20 62-24060 <*=>

PUBLIKATIONER. GROENLANDSKE TEKNISKE ORGANISATION
1957 N1 P1-20 C-3973 <NRCC>
1958 N2 P1-30 C-3962 <NRCC>

PUBLIKATSII KIEVSKOI ASTRONOMICHESKOI OBSERVATORII
1950 N4 P17-31 61-15269 <*+> 0
1953 N5 P143-145 60-19018 <+*>
1953 N5 P147-154 60-19018 <+*>
1954 N6 P139-152 60-19007 <+*>

PUT I PUTEVOE KHOZYAISTVO
1959 N2 P46-48 60-21004 <=> 0
1959 N5 P42-48 61-28035 <*=> 0
1960 N8 P14 61-11743 <=>
1965 V9 P16-18 65-17193 <*>
1965 V9 P19 65-17192 <*>

Q. S. DIGEST
1959 V10 P367 64-18717 <*> 0
1959 V10 P433 64-18646 <*>

Q. S. T. FRANCAIS ET RADIOELECTRICITE
1930 V11 P28-30 60-10199 <*>

QUADERNI DI CLINICA OSTETRICA E GINECOLOGICA
1955 V10 N8 P717-723 61-00414 <*>
1957 V12 N2 P112-128 57-3278 <*>
1960 V15 P475-487 61-18868 <*>

QUADERNI DI GEOFISICA APPLICATA
1963 V24 P1-25 65-14215 <*>
1963 V24 P27-38 65-14214 <*>

QUADERNI DELLA NUTRIZIONE
1936 V3 P472-479 59-20674 <*>

QUALITA. ITALY
1959 V9 P28-35 64-30678 <*>
1959 V9 P36-45 64-30679 <*>
1959 V9 P45-48 64-30680 <*>

QUARTERLY JOURNAL OF THE ROYAL METEOROLOGICAL SOCIETY
1936 P9-11 AL-576 <*>

QUARTERLY JOURNAL OF SEISMOLOGY. TOKYO
SEE KENSHIN JIHO

QUARTERLY REPORT
SEE REPORT

QUATERNARIA SINICA
SEE CHUNG KUO TI SSU CHI YEN CHIU

QUATERNARY RESEARCH. JAPAN
SEE DAI-4-KI KENKYU

QUIMICA Y INDUSTRIA
1935 V12 N143 P277-278 60-14432 <*> 0

RCN BULLETIN
1958 V2 P142-146 59-00767 <*>

R.I.L.E.M. BULLETIN
1959 N4 P66-75 60-18355 <*>
1962 N17 P37-61 64-18824 <*>
64-30618 <*>

RABOTNICHESKO DELO
1967 P2 67-30775 <=$>
1967 P4 67-30839 <=$>

RABOTY PO KHIMI RASTVOROV I KOMPLEKSNYKH SOEDINENII
1954 P29-49 3737-E <K-H>
61-23652 <*=>
64-16160 <*> O
1954 P113-131 3950 <K-H>
64-16427 <*$>

RABOTY VOLZHSKOI BIOLOGICHESKOI STANTSII
1929 V10 N4 P149-158 RT-2044 <*> O
64-15658 <=*$> O

RACE HYGIENE. JAPAN
SEE MINZOKU EISEI

RAD JUGOSLAVENSKE AKADEMIJE ZNANOSTI I UMJETNOSTI
1953 V294 P245-268 60-21697 <=>
1957 N5 P5-24 60-21655 <=>
1962 P11 62-25187 <=*>

RADEX RUNDSCHAU
1949 N6 P208-219 57-1216 <*>
1950 N2 P79-84 63-16735 <*>
1951 P178-187 TS 1597 <BISI>
1951 N1 P13-24 3233 <*>
65-10025 <*>
1951 N5 P178-187 62-16870 <*>
1951 N6 P257-260 59-17500 <*>
1952 N3 P120-123 46G6G <ATS>
57-2670 <*>
1952 N4 P181-186 60-18861 <*>
1952 N6 P267-269 64-18322 <*> O
1953 N1 P28-35 62-20187 <*>
1954 N4/5 P107-121 60-18847 <*>
1954 N4/5 P152-164 TS-1598 <BISI>
1956 N3 P91-104 T1775 <INSD>
1956 N7 P354-362 66-10152 <*> O
1956 N4/5 P147-152 57-2611 <*>
1956 N4/5 P153-162 57-2894 <*>
58-87 <*>
1956 N4/5 P163-172 57-2943 <*>
1957 P727-737 2006 <BISI>
1957 P776-778 TS 1596 <BISI>
1957 N2 P501-507 61-18374 <*>
1957 N7 P861-867 59-17230 <*>
1957 N9 P711-724 2002 <BISI>
1957 N5/6 P693-707 61-20845 <*>
1957 N5/6 P725-726 2502 <BISI>
1957 N5/6 P738-753 2992 <BISI>
1957 N5/6 P754- 2993 <BISI>
1957 N5/6 P768-770 2503 <BISI>
1957 N5/6 P776-783 62-16898 <*>
1958 N1 P3-29 61-18797 <*>
1958 N6 P253-276 61-18796 <*>
1958 N6 P277-281 59-15706 <*>
1958 N7 P323-347 61-18795 <*> O
1958 N7 P348-353 61-20601 <*>
1958 N3/4 P146-153 61-18813 <*>
1958 N3/4 P154-172 61-18387 <*>
1959 P517-532 1394 <BISI>
1959 P533-545 TS-1395 <BISI>
1959 P546-555 TS-1396 <BISI>
1959 P629-639 TS-1601 <BISI>
1959 P714-721 1682 <BISI>
62-16882 <*>
1959 N12 P704-713 1896 <BISI>
1960 N2 P86-99 5031 <HB>
1960 N2 P100-103 61-10182 <*>
1960 N4 P71-85 1767 <BISI>
1960 N6 P373-388 63-19123 <=*>
1960 N6 P400-405 64-14063 <*>
1960 N10 P257-261 1993 <BISI>
1961 P543-545 2893 <BISI>
1961 P623-640 2389 <BISI>
1961 P641-646 2390 <BISI>
1961 P657-663 2415 <BISI>
1961 N1 P449-484 1180 <BISI>
1961 N1 P485-491 2164 <BISI>
1961 N1 P506-525 TRANS-186 <MT>
1961 N3 P579-589 2317 <BISI>
1961 N3 P607-615 2368 <BISI>
1961 N4 P647-656 3129 <BISI>
1961 N6 P739-746 4235 <BISI>
1961 N6 P747-773 2887 <BISI>
64-18005 <*> O
1961 N6 P802-809 2587 <BISI>
1962 P82-88 2835 <BISI>
1962 P123-132 2864 <BISI>
1962 N3 P133-156 GI N3 1964 P578 <AGI>
1962 N4 P157-173 GI N3 1964 P578 <AGI>
1962 N4 P199-210 <LSA>
1962 N5 P250-270 3069 <BISI>
63-10864 <*>
1962 N6 P291-302 3120 <BISI>
1962 N6 P311-331 BGIRA-552 <BGIR>
3294 <BISI>
1963 N1 P335-350 3278 <BISI>
1963 N2 P405-412 3259 <BISI>
1963 N2 P415-420 3268 <BISI>
1963 N3 P441-461 3340 <BISI>
1963 N4 P519-534 ANL-TRANS-150 <*>
1963 N5 P573-584 3517 <BISI>
1963 N6 P635-646 3572 <BISI>
1964 P251-265 4200 <BISI>
1964 N1 P3-19 3737 <BISI>
1964 N1 P20-41 3738 <BISI>
1964 N1 P55-70 3739 <BISI>
1964 N4 P226-232 4058 <BISI>
1964 N5 P277-289 6442 <HB>
1964 N5 P290-298 6443 <HB>
1964 N5 P299-312 4453 <BISI>
1964 N6 P341-356 4266 <BISI>
1965 P439-446 4293 <BISI>
1965 P468-474 4294 <BISI>
1965 P672-686 4604 <BISI>
1965 N1 P369-377 6527 <HB>
1965 N1 P388-392 6525 <HB>
1965 N1 P406-417 BGIRA-646 <BGIR>
1965 N2 P425-431 6554 <HB>
1965 N2 P432-438 6555 <HB>
1965 N2 P447-460 6657 <HB>
1965 N2 P461-467 6558 <HB>
1965 N2 P483-486 6559 <HB>
1965 N4 P545-558 4525 <BISI>
1965 N4 P559-576 4526 <BISI>
1965 N5 P623-646 4670 <BISI>
1965 N5 P647-657 6773 <HB>
1965 N6 P722-732 6816 <HB>
1965 N6 P759-767 6817 <HB>
1966 N1 P51-59 6867 <HB>

RADIATION BOTANY. LONDON, NEW YORK
1965 V5 P403-416 66-12139 <*>

RADIO. CHINESE PEOPLES REPUBLIC
SEE WU-HSIEN-TIEN

RADIO. MOSCOW
1948 N5 P6-8 RT-1280 <*> O
1948 N5 P9 RT-1279 <*>
1948 N5 P10-13 RT-1278 <*>
1952 N5 P14-15 RT-1046 <*>
1952 N6 P37-40 RT-1931 <*> O
1952 N6 P44 RT-1691 <*> O
1952 N7 P7 RT-1930 <*>
1952 N8 P51 RT-1929 <*> O
1952 N6 P21-24 R-4108 <*>
1953 N1 P17-19 59-15996 <+*> O

Citation	Number
1953 N11 P55-56	RT-1726 <*>
1953 N11 P57-60	62-13350 <*=>
1954 N2 P13-14	RT-2561 <*>
1954 N6 P22-23	RT-2237 <*>
1954 N6 P34-37	RT-2743 <*>
1954 N10 P12-14	RT-2955 <*>
1955 N6 P26-27	R-359 <*>
1955 N6 P28-29	R-358 <*>
1955 N6 P30-	R-360 <*>
1955 N7 P58	RT-3166 <*>
1955 N8 P50-53	63-13242 <=*>
1955 N12 P27-28	R-4981 <*>
1956 N2 P23	61-27438 <*=> 0
1956 N2 P32-33	R-604 <*>
1956 N9 P13-14	61-19783 <=*> 0
1956 N11 P15-16	59-12189 <+>
1956 N11 P18	59-12189 <+>
1956 N11 P25	59-12189 <+>
1957 N5 P20-21	R-3270 <*>
	R-3480 <*>
1957 N5 P22-23	62-23486 <=*>
1957 N6 P14-17	R-2361 <*>
	R-3059 <*>
	R-4874 <*>
1957 N6 P17-19	R-2439 <*>
	R-3060 <*>
1957 N7 P17-20	R-2761 <*>
1957 N7 P17-23	R-3261 <*>
1957 N7 P17	R-3273 <*>
1957 N7 P17-23	R-3477 <*>
1957 N7 P17	R-3481 <*>
1957 N7 P17-20	59-11242 <=>
1957 N7 P21-23	R-2762 <*>
	R-3269 <*>
	R-3479 <*>
	59-11243 <=>
1957 N7 P24-25	R-2440 <*>
	R-3058 <*>
	R-3272 <*>
	R-3478 <*>
1957 N8 P17-19	R-2387 <*>
1957 N8 P17-20	R-2760 <*>
1957 N8 P17-19	R-2764 <*>
	R-3209 <*>
	59-11240 <=>
1957 N8 P17-20	59-14248 <+*>
1957 N8 P19-20	R-3206 <*>
	59-11241 <=>
1957 N11 P7-10	59-11877 <=> 0
1957 N11 P34-35	R-4432 <*>
1957 N12 P24-29	59-11991 <=> 0
1957 V8 P19-20	R-2386 <*>
1958 N1 P3-5	60-19147 <+*>
1958 N1 P35-36	R-5299 <*>
1958 N2 P27-31	59-12166 <+*>
1958 N3 P30	59-12162 <+*>
1958 N5 P45	59-21157 <=> 0
1958 N5 P51-53	59-21158 <=> 0
1958 N5 P54-55	59-21159 <=>
1958 N5 P61	59-21160 <=> 0
1958 N7 P3-5	59-19492 <+*>
1958 N7 P17	59-19569 <+*>
1958 N7 P19	59-19569 <+*>
1958 N7 P22	59-19569 <+*>
1958 N8 P19	60-21739 <=>
1958 N8 P20-22	60-21708 <=> 0
1958 N8 P29-30	59-21047 <=>
	60-21741 <=>
1958 N8 P31-33	59-13174 <=> 0
	60-21742 <=> 0
1958 N10 P35	59-11863 <=>
1958 N11 P12-13	59-13530 <=>
1958 N11 P17	59-13531 <=>
1958 N11 P59-60	59-13532 <=>
1958 N12 P30-33	61-23184 <*=>
1958 N12 P32-33	59-16434 <+*>
1958 N12 P45-46	61-23184 <*=>
1959 N1 P18-20	60-13784 <+*>
1959 N2 P6-7	59-16157 <+*> 0
	59-16410 <+*>
	59-21063 <=>
	60-13110 <+*> 0
1959 N3 P27-31	60-19062 <+*>
1959 N11 P8-11	60-31097 <=>
1960 N4 P29-30	60-41131 <=>
1960 N7 P34-38	62-15693 <*=>
1960 N9 P12-13	61-28554 <*=>
1960 N10 P3	61-28240 <*=>
1960 N11 P9-12	62-11325 <=>
1960 N11 P37-39	62-24272 <=*>
1961 N1 P4-7	61-21638 <=>
1961 N1 P11-17	61-27327 <*=> 0
1961 N3 P10	61-31464 <=>
1961 N3 P12-13	61-31464 <=>
1961 N3 P33-34	61-27824 <*=>
1961 N4 P53-54	61-28309 <*=>
1961 N5 P3-5	62-23472 <=*>
1961 N7 P6-7	61-28712 <*=>
1961 N8 P27-28	62-25846 <=*>
1961 N10 P16-17	62-11106 <=>
1961 N10 P39	62-32332 <=*>
1961 N10 P59-61	62-32332 <=*>
1961 N11 P3-7	62-32297 <=$> 0
1961 N12 P3-5	62-24342 <=*>
1962 N2 P3-5	62-24001 <=*>
1962 N2 P9	62-32879 <=*>
1962 N10 P29	63-21531 <=>
1962 N10 P64	63-21531 <=>
1963 N1 P2-5	63-21553 <=>
1963 N1 P39	63-21553 <=>
1963 N1 P49	63-21553 <=>
1963 N1 P51	63-21553 <=>
1963 N2 P36-37	63-21636 <=>
1963 N2 P41	63-21636 <=>
1963 N3 P3-6	63-31375 <=>
1963 N3 P10-12	63-15521 <=*>
1963 N4 P2	63-21967 <=>
1963 N4 P5-6	63-21967 <=>
1963 N4 P55-58	63-21969 <=>
1963 N5 P2	63-31324 <=>
1963 N5 P7-9	63-31324 <=>
1963 N5 P29-32	63-31324 <=>
1963 N6 P13	63-31565 <=>
1963 N6 P35-38	63-31565 <=>
1963 N7 P3	63-31725 <=>
1963 N7 P4-6	63-31818 <=>
1963 N11 P28-31	64-21437 <=>
1964 N1 P18-20	AD-611 526 <=$>
1964 N4 P57-58	64-31367 <=>
1964 N7 P50-53	64-41719 <=>
1965 N2 P53-56	65-31220 <=$>
1965 N3 P10-11	65-31390 <=$>
1965 N5 P50-52	65-31375 <=$>
1965 N5 P54	65-31375 <=$>
1966 N12 P9	67-31016 <=$>

RADIO UND FERNSEHEN

Citation	Number
1957 V6 N11 P368-	59-14250 <+*>
1958 V7 N15 P488-489	61-27357 <=*> 0
1958 V7 N24 P717-719	60-23949 <=*>
1959 V8 N23 P722	60-31263 <=*>
1959 V8 N24 P749-750	60-31673 <=*>
1960 V9 N1 P2	60-31230 <=>
1960 V9 N1 P3-4	60-31232 <=> 0
1960 V9 N3 P66	60-31267 <=*>
1960 V9 N3 P71-73	60-31268 <=*> 0
1960 V9 N3 P85	60-31269 <=*>
1960 V9 N4 P98	60-31444 <=*>
1960 V9 N4 P99-102	60-31445 <=*>
1960 V9 N16 P513-514	62-25207 <=*> 0
1960 V9 N17 P550-553	62-25208 <=*> 0
1960 V9 N21 P665-666	62-25206 <=*> 0
1961 V10 N4 P118-119	62-25209 <=*> 0
1963 V12 N18 P557-560	64-71178 <=>

RADIO FRANCAISE

Citation	Number
1950 P1-7	57-763 <*>

RADIO MENTOR

Citation	Number
1956 V22 N11 P702-704	PB 141 183T <=>

1960 N11 P887-893 UCRL TRANS-724 <*>

RADIO I TELEVISIYA
1967 V16 N2 P34-35 67-31575 <=$>

RADIOACTIVE ISOTOPE IN KLINIK UND FORSCHUNG
1963 V5 P416-427 65-13438 <*> 0

RADIOBIOLOGIA, RADIOTHERAPIA
1963 V4 N4 P491-498 ANL-TR-1 <*>
1963 V4 N4 P499-508 ANL-TR-2 <*>

★RADIOBIOLOGIYA
1961 V1 N1 P86-92 65-30434 <=$>
1961 V1 N3 P321-462 62-19174 <=*>
1961 V1 N3 P437-439 66-61747 <=*$>
1961 V1 N4 P535-542 63-15291 <=*>
1961 V1 N4 P550-554 63-15291 <=*>
1961 V1 N4 P583-590 63-15291 <=*>
1961 V1 N5 P663-667 63-19691 <=*>
1962 V2 N1 P105- FPTS V22 N6 P.T1036 <FASE>
FPTS,V22,P.T1036-41 <FASE>
1962 V2 N1 P115-120 63-23890 <=*$>
1962 V2 N1 P134- FPTS V22 N5 P.T818 <FASE>
1963,V22 N5,PT2 <FASE>
1962 V2 N2 P211-215 AEC-TR-5721 <=*>
1962 V2 N3 P390-394 63-24244 <=*$>
1962 V2 N3 P442-449 63-19253 <=*>
1962 V2 N4 P569-572 AD-619 546 <=*$>
1962 V2 N4 P611-615 66-61762 <=$>
1962 V2 N5 P641-646 AEC-TR-5432 <=>
1962 V2 N5 P647-653 AEC-TR-5432 <=>
1962 V2 N5 P654-661 AEC-TR-5432 <=>
1962 V2 N5 P662-666 AEC-TR-5432 <=>
1962 V2 N5 P667-673 AEC-TR-5432 <=>
1962 V2 N5 P674-680 AEC-TR-5432 <=>
1962 V2 N5 P681-684 AEC-TR-5432 <=>
1962 V2 N5 P685-689 AEC-TR-5432 <=>
1962 V2 N5 P690-694 AEC-TR-5432 <=>
1962 V2 N5 P695-699 AEC-TR-5432 <=>
1962 V2 N5 P700-704 AEC-TR-5432 <=>
1962 V2 N5 P705-708 AEC-TR-5432 <=>
1962 V2 N5 P709-712 AEC-TR-5432 <=>
1962 V2 N5 P713-714 AEC-TR-5432 <=>
1962 V2 N5 P715-718 AEC-TR-5432 <=>
1962 V2 N5 P719-725 AEC-TR-5432 <=>
1962 V2 N5 P726-731 AEC-TR-5432 <=>
1962 V2 N5 P732-740 AEC-TR-5432 <=>
1962 V2 N5 P741-748 AEC-TR-5432 <=>
1962 V2 N5 P749-757 AEC-TR-5432 <=>
1962 V2 N5 P758-762 AEC-TR-5432 <=>
1962 V2 N5 P763-767 AEC-TR-5432 <=>
1962 V2 N5 P768-772 AEC-TR-5432 <=>
1962 V2 N5 P773-779 AEC-TR-5432 <=>
1962 V2 N5 P780-784 AEC-TR-5432 <=>
1962 V2 N5 P785-789 AEC-TR-5432 <=>
1962 V2 N5 P790-798 AEC-TR-5432 <=>
1962 V2 N6 P801-952 AEC-TR-5433 <*>
1963 V3 N4 P545-548 65-60931 <=$>
1963 V3 N5 P633- FPTS V23 N6 P.T1315 <FASE>
1963 V3 N5 P667-670 64-19711 <=$>
1963 V3 N5 P703- FPTS V24 N1 P.T175 <FASE>
1963 V3 N5 P703-710 FPTS V24 P.T175-T179 <FASE>
1963 V3 N5 P711-716 65-60187 <=$>
1963 V3 N5 P766- FPTS V24 N1 P.T133 <FASE>
1963 V3 N5 P766-769 FPTS V24 P.T133-T134 <FASE>
1963 V3 N6 P877-880 64-19689 <=$> 0
1964 V4 N1 P18-25 65-60147 <=$>
1964 V4 N1 P29- FPTS V24 N1 P.T129 <FASE>
1964 V4 N1 P29-35 FPTS V24 P.T129-T132 <FASE>
1964 V4 N1 P47-52 65-60195 <=$>
1964 V4 N1 P102-108 65-60194 <=$>
1964 V4 N1 P114-117 65-60211 <=$>
1964 V4 N1 P167-169 65-61106 <=$>
1964 V4 N2 P313- FPTS V24 N4 P.T688 <FASE>
1964 V4 N3 P337-343 65-62108 <=$>
1964 V4 N3 P344-348 65-62109 <=$>
1964 V4 N3 P402-409 65-62341 <=$>
1964 V4 N4 P578-582 65-62339 <=$>
1964 V4 N4 P594-599 65-62841 <=$>
1964 V4 N4 P637-638 65-62840 <=$>
1964 V4 N5 P708-715 65-63619 <=$>
1964 V4 N6 P804-810 65-64550 <=*$>
1964 V4 N6 P840-843 65-64549 <=*$>
1964 V4 N6 P843-847 65-63620 <=$>
1964 V4 N6 P878- FPTS V24 N6 P.T1038 <FASE>
1965 V5 N4 P612-615 65-33218 <=*$>

RADIOCHIMICA ACTA
1963 V1 N3 P117-123 AI-TR-25 <*>
1964 V3 P169-185 N66-15683 <=$>

RADIOELECTRICITE
1920 P292- 57-2098 <*>
1922 V3 P147-152 60-10096 <*> 0
1924 V5 P65-73 57-2021 <*>
1924 V5 N57 P35-37 2172 <*>
1926 P142-144 2129 <*>
1926 P275-278 2154 <*>

RADIOELECTRICITE ET Q. S. T. FRANCAIS
1931 P46-47 57-695 <*>
1931 V12 P70-73 58-39 <*>

RADIO-ISOTOPES. TOKYO
1959 V8 N1 P28-31 AEC-TR-5789 <*>
1959 V8 N2 P185-189 AEC-TR-6280 <*>
78Q73J <ATS>
1960 V9 N1 P6-16 NP-TR-742 <*>
1961 V10 N1 P8-18 63-20886 <*> 0
1961 V10 N1 P27-36 63-20867 <*> 0
1961 V10 N2 P181-185 AEC-TR-5582 <*>
1961 V10 N4 P488-492 NP-TR-1093 <*>
1962 V11 N1 P9-16 AEC-TR-5457 <*>
1962 V11 N1 P16-22 AEC-TR-5581 <*>
1963 V12 N3 P311-317 NSJ-TR-11 <*>

★RADIOKHIMIYA
1959 V1 N1 P43-51 60-11588 <=>
1959 V1 N2 P131-135 64-18822 <*>
65-60338 <=$>
1959 V1 N2 P147-154 61-28181 <*=>
1959 V1 N2 P155-161 62-25561 <=*>
1959 V1 N2 P196-203 AEC-TR-4555 <*=>
50M42R <ATS>
1959 V1 N2 P208-211 63-13872 <=*>
1959 V1 N3 P257-269 61-27048 <*=>
1959 V1 N3 P336-345 AEC-TR-4554 <*=>
51M42R <ATS>
1959 V1 N5 P545-547 63-23287 <=*>
1959 V1 N5 P567-572 63-24360 <=*$>
1959 V1 N5 P573-580 64-13378 <=*$>
1959 V1 N5 P596-602 62-10478 <*=>
1959 V1 N6 P660-664 62-25395 <=*>
1959 V1 N6 P691-693 ORNL-TR-526 <=*$>
1959 V1 N6 P712-716 NP-TR-554 <*>
62-23991 <=*>
1960 V2 P451-457 AEC-TR-4598 <*=> 0
1960 V2 P495-499 AEC-TR-4518 <=*>
1960 V2 N1 P13-19 AEC-TR-4852 <=*>
1960 V2 N3 P274-280 62-25397 <=*>
1960 V2 N3 P351-356 AEC-TR-4239 <*+>
1960 V2 N5 P509-632 AEC-TR-4578 <=>
1960 V2 N5 P541-548 61-21696 <=>
1960 V2 N6 P637-755 AEC-TR-4578 <=>
1961 V3 P121-128 AEC-TR-4840 <=*>
1961 V3 P155-164 AEC-TR-4830 <*>
AEC-TR-4830 <=*>
1961 V3 P165-172 AEC-TR-4841 <*=>
1961 V3 P396-402 AEC-TR-4982 <=*>
1961 V3 N1 P7-9 AEC-TR-4842 <=*>
1961 V3 N1 P101-113 AEC-TR-4623 <*=> 0
1961 V3 N2 P195-198 AEC-TR-4787 <=*>
1961 V3 N3 P348-355 63-23300 <=*>
65-60736 <=>
1961 V3 N4 P417-421 ORNL-TR-209 <=$>
1961 V3 N4 P486-489 62-18730 <=*>
1962 V4 N1 P44-49 AEC-TR-5253 <=*>
1962 V4 N1 P59-66 64-13199 <=*$>
1963 V5 P157-159 AEC-TR-6136 <=*$>

Citation	Translation
1963 V5 N2 P189-197	66-20968 <$>
1963 V5 N2 P198-205	66-26098 <*$>
1963 V5 N2 P244-248	64-71474 <=>
1963 V5 N3 P299-304	66-12369 <*>
1963 V5 N3 P335-342	AEC-TR-6311 <=*$>
1963 V5 N3 P351-355	ANL-TRANS-29 <=$>
1964 V6 P419-425	ORNL-TR-457 <=*$>
1964 V6 P440-444	ORNL-TR-450 <=*$>
1964 V6 P445-448	ORNL-TR-452 <=$>
1964 V6 P500-502	ORNL-TR-542 <*>
1964 V6 N1 P130-132	AD-618 638 <=$>
1964 V6 N2 P237-241	M.5631 <NLL>
1964 V6 N3 P377	ORNL-TR-453 <=$>
1965 V7 P125-126	ORNL-TR-625 <=$>

RADIOLOGIA AUSTRIACA

Citation	Translation
1961 V12 N1/2 P3-8	65-13731 <*> 0

RADIOLOGIA CLINICA

Citation	Translation
1947 V16 P73-77	I-410 <*>
1949 V18 P300-305	1149 <*>
1950 V19 N3 P170-173	2606 <*>
1953 V22 P130-139	UCRL TRANS-610 <*>
1956 V25 P371	58-2298 <*>
1959 V28 N2 P88-101	61-10588 <*>
1962 V31 N4 P247-254	NP-TR-996 <*>

RADIOLOGIA MEDICA

Citation	Translation
1938 V25 N7 P583-597	66-11635 <*>
1950 V36 N3 P194-217	II-649 <*>
1951 V37 N5 P389-411	AL-528 <*>
1951 V37 N6 P471-488	AL-528 <*>
1952 V38 N3 P261-272	II-58 <*>
1961 V47 N1 P10-34	62-00970 <*>
1963 V49 P238-273	65-00093 <*>
1963 V49 N2 P97-126	63-01611 <*>
	64-26246 <$>

RADIOTEKHNICHESKOE PROIZVODSTVO

Citation	Translation
1957 V8 P29-33	63-18598 <=*>

★RADIOTEKHNIKA

Citation	Translation
1948 V3 N2 P11-20	R-3992 <*>
1948 V3 N4 P47-55	RT-170 <*>
1949 V4 N3 P21-35	R-1113 <*>
1949 V4 N3 P57-68	RT-4025 <*>
1949 V4 N5 P13-27	RT-3994 <*>
1951 V6 N5 P38-46	63-23542 <=*$>
1952 V7 N6 P55-66	63-23542 <=*$>
1954 V9 N1 P3-10	61-19904 <=*>
1954 V9 N2 P5-12	1448TM <CTT>
	59-16688 <+*>
1954 V9 N2 P21-30	62-23204 <=>
1954 V9 N3 P3-11	62-23183 <=*>
1954 V9 N3 P19-32	RT-3956 <*>
1954 V9 N3 P33-37	59-16713 <+*>
1954 V9 N5 P3-7	RT-3944 <*>
1954 V9 N6 P78-80	59-16690 <+*>
1954 V9 N6 P81-83	61-23642 <=*>
1955 N10 P44-57	R-3393 <*>
1955 N12 P3-10	R-357 <*>
1955 V10 N1 P23-26	59-16720 <+*>
1955 V10 N1 P44-52	62-15868 <=*>
1955 V10 N1 P72-73	61-19797 <=*>
1955 V10 N1 P74-77	61-19796 <=*>
1955 V10 N2 P14-20	62-23167 <=*>
1955 V10 N2 P50-65	R-426 <*>
1955 V10 N3 P14-24	RT-4103 <*>
	62-14042 <=*>
	62-23092 <=*>
1955 V10 N3 P29-32	62-11239 <=>
1955 V10 N3 P63-67	R-740 <*>
1955 V10 N4 P7-25	RT-4265 <*>
1955 V10 N4 P26-35	RT-3393 <*>
1955 V10 N4 P48-55	R-2783 <*>
	59-16715 <+*>
1955 V10 N4 P56-58	62-23062 <=*>
1955 V10 N6 P16-20	62-23231 <=*>
1955 V10 N6 P52-57	1446TM <CTT>
	59-16697 <+*>
1955 V10 N7 P3-7	62-23014 <=*>
1955 V10 N7 P52-57	R-734 <*>
1955 V10 N7 P69-73	62-11279 <=>
1955 V10 N8 P3-21	R-864 <*>
1955 V10 N8 P44-57	57-2584 <*>
1955 V10 N9 P3-13	RT-4190 <*>
	1444TM <CTT>
	59-16721 <+*>
1955 V10 N9 P77-80	R-697 <*>
1955 V10 N10 P3-14	R-742 <*>
1955 V10 N10 P15-22	62-23177 <=*>
1955 V10 N10 P39-50	62-19900 <=*>
1955 V10 N10 P74-75	62-23114 <=*>
1955 V10 N11 P3-11	62-15923 <=*>
1955 V10 N11 P65-79	62-19851 <=*>
	831TM <CTT>
1956 V11 N1 P3-6	62-11586 <=>
1956 V11 N1 P7-16	R-975 <*>
	62-15905 <=*>
1956 V11 N1 P57-60	R-473 <*>
	62-23030 <=*>
1956 V11 N2 P14-28	R-4622 <*>
	R-977 <*>
1956 V11 N2 P43-53	TT-829 <NRCC>
1956 V11 N2 P60-63	59-17789 <+*>
1956 V11 N2 P74-76	62-23058 <=*>
1956 V11 N3 P3-69	59-14139 <+*>
1956 V11 N3 P23-40	R-1441 <*>
1956 V11 N3 P34-40	62-10158 <*=>
1956 V11 N3 P51-62	TT-829 <NRCC>
1956 V11 N3 P70-80	62-23180 <=*>
1956 V11 N4 P5-14	62-15909 <=*>
1956 V11 N4 P15-30	62-24411 <=*>
1956 V11 N4 P31-35	62-23208 <=*>
1956 V11 N4 P36-43	R-973 <*>
1956 V11 N4 P36-80	62-24411 <=*>
1956 V11 N5 P3-20	R-2068 <*>
	62-15896 <=*>
1956 V11 N5 P21-25	62-11637 <=>
1956 V11 N5 P26-80	59-14140 <+*>
1956 V11 N6 P71-74	R-1582 <*>
1956 V11 N7 P1-80	59-14152 <+*>
1956 V11 N7 P57-59	SCL-T-287 <+*>
	62-11407 <=>
1956 V11 N7 P60-62	R-1015 <*>
1956 V11 N8 P31-36	R-424 <*>
	R-437 <*>
1956 V11 N8 P64-70	R-1382 <*>
1956 V11 N9 P3-7	R-1954 <*>
1956 V11 N9 P8-11	R-659 <*>
1956 V11 N9 P12-20	R-3891 <*>
1956 V11 N9 P28-38	R-670 <*>
1956 V11 N9 P39-45	R-730 <*>
1956 V11 N9 P59-71	R-733 <*>
1956 V11 N11 P5-6	R-1583 <*>
1956 V11 N12 P3-14	R-664 <*>
1956 V11 N12 P37-52	62-14438 <=*>
1956 V11 N12 P53-54	R-3183 <*>
1957 V12 N1 P3-11	R-874 <*>
	59-12049 <+*>
1957 V12 N1 P76-77	R-1586 <*>
1957 V12 N2 P10-21	R-3084 <*>
	59-12050 <+*>
1957 V12 N2 P22-27	R-2384 <*>
1957 V12 N2 P51-58	62-23084 <=*>
1957 V12 N2 P65-70	62-32906 <=*>
1957 V12 N3 P25-30	66-12797 <*> 0
1957 V12 N3 P62-	60-14343 <+*>
1957 V12 N4 P13-23	R-5191 <*>
1957 V12 N5 P62-66	96K20R <ATS>
1957 V12 N6 P1-81	59-16083 <+*>
1957 V12 N9 P12-19	86TM <CTT>
1957 V12 N10 P17-30	R-3153 <*>
1957 V12 N10 P40-46	R-3055 <*>
1957 V12 N11 P50-61	60-14345 <+*>
1957 V12 N12 P10-18	R-4820 <*>
1957 V12 N12 P19-28	59-12156 <+*>
1958 V13 N3 P48-60	R-5242 <*>
1958 V13 N4 P15-25	59-10199 <+*>
1958 V13 N4 P53-62	R-5243 <*>

1958 V13 N4 P77-79	66-14599 <*> 0
1958 V13 N5 P3-6	59-22678 <+*>
1958 V13 N7 P55-62	60-14344 <+*>
1958 V13 N8 P47-49	60-21931 <=>
1958 V13 N8 P63-70	61-11179 <=>
1958 V13 N8 P71-79	59-22463 <+*>
1958 V13 N11 P5-10	60-13843 <+*>
	60-15802 <+*>
1958 V13 N11 P27-38	59-17458 <+*>
1958 V13 N11 P39-43	60-21775 <=>
1958 V13 N12 P3-10	59-21017 <=>
1959 V14 N3 P9-21	59-00689 <*>
1959 V14 N3 P76-77	60-13146 <+*>
	60-21466 <=>
1959 V14 N5 P12-22	61-19177 <+*>
1959 V14 N5 P44-48	60-21771 <=>
1959 V14 N6 P3-16	59-31001 <=> 0
1959 V14 N6 P63	59-31001 <=> 0
1959 V14 N7 P3-7	60-12521 <=>
	60-15454 <+*>
	62-11329 <=>
1959 V14 N7 P56-70	60-19087 <+*>
1959 V14 N8 P3-7	60-12542 <=>
	62-11180 <=>
1959 V14 N8 P8-13	62-11223 <=>
1959 V14 N9 P13-16	60-12269 <=>
	62-11469 <=>
1959 V14 N10 P3-14	60-31838 <=> 0
1959 V14 N10 P15-22	60-12520 <=>
	60-15453 <+*>
1959 V14 N11 P35-42	62-11392 <=>
	62-11392 <=*>
1959 V14 N12 P44	62-11463 <=>
1959 V14 N12 P50-57	62-11504 <=>
	62-11514 <=>
1959 V14 N12 P58-68	62-23170 <=>
	62-23170 <=*>
1960 V15 N1 P35-37	61-23539 <=*>
1960 V15 N1 P68-71	61-23539 <=*>
1960 V15 N3 P45-52	61-23382 <*=>
1960 V15 N5 P5-8	62-11174 <=>
1960 V15 N6 P10-17	61-20539 <*=>
1960 V15 N9 P10-20	62-10187 <=*>
1960 V15 N9 P47-53	62-10188 <*=>
1960 V15 N10 P11-13	61-27613 <=*>
1960 V15 N12 P10-12	62-32654 <=*>
1961 V16 N3 P3-12	62-11394 <=>
	62-13633 <*=>
1961 V16 N3 P13-21	62-13629 <*=>
1961 V16 N6 P40-44	62-25913 <=*>
1961 V16 N10 P3-9	62-23659 <=*>
1962 V17 N2 P3-9	63-15229 <=*>
1962 V17 N12 P48-51	63-21618 <=>
1963 V18 N2 P37-42	63-23112 <*=>
1963 V18 N4 P3-7	64-71425 <=> M
1963 V18 N5 P59-65	63-24600 <=*$>
1963 V18 N5 P66-71	63-31189 <=>
1963 V18 N11 P20-26	64-71320 <=>
	65-17233 <*>
1963 V18 N11 P62-70	AD-610 355 <=$>
	64-21270 <=>
1964 V19 N1 P13-17	AD-610 353 <=$>
	65-11586 <*>
1964 V19 N8 P70-74	64-51750 <=$>
1964 V19 N11 P37-41	AD-627 904 <=*$>
1964 V19 N11 P65-70	65-30207 <=$>
1965 V20 N5 P5-9	65-32486 <=*$>

★RADIOTEKHNIKA I ELEKTRONIKA

1956 P869-872	C-2540 <NRC>
1956 V1 P38-50	62-11429 <=>
1956 V1 P205-212	R-1766 <*>
1956 V1 P447-468	R-854 <*>
1956 V1 P593-600	R-1810 <*>
1956 V1 P638-646	R-853 <*>
1956 V1 P739-742	62-11352 <=>
1956 V1 P1127-1134	UCRL TRANS-528 <+*>
1956 V1 N1 P5-22	62-19122 <=*>
1956 V1 N1 P23-33	62-15904 <=*>
1956 V1 N1 P71-78	64-11507 <=>
1956 V1 N1 P79-87	R-2327 <*>
1956 V1 N1 P88-97	R-2330 <*>
1956 V1 N1 P623-626	R-1554 <*>
1956 V1 N2 P131-142	62-23077 <=*>
1956 V1 N2 P143-161	63-23757 <=*$>
1956 V1 N2 P205-212	62-23181 <=*>
1956 V1 N2 P233-244	R-4393 <*>
1956 V1 N3 P269-273	R-1772 <*>
1956 V1 N3 P281-292	60-14028 <+*>
1956 V1 N3 P293-308	59-12068 <=>
1956 V1 N3 P309-312	62-11282 <=>
1956 V1 N3 P344-357	R-4394 <*>
1956 V1 N3 P358-369	64-71450 <=>
1956 V1 N3 P370-376	64-71448 <=>
1956 V1 N4 P407-417	59-22685 <+*>
1956 V1 N4 P418-427	62-23129 <=*>
1956 V1 N4 P447-468	66-12641 <*>
1956 V1 N4 P469-477	59-10886 <+*>
1956 V1 N4 P497-511	62-23098 <=*>
1956 V1 N5 P560-574	R-1155 <*>
1956 V1 N5 P575-592	R-1139 <*>
	59-19136 <+*>
1956 V1 N5 P593-600	R-680 <*>
1956 V1 N5 P601-612	R-1490 <*>
1956 V1 N5 P638-646	66-12642 <*>
1956 V1 N5 P683-	R-949 <*>
1956 V1 N6 P695-703	R-330 <*>
1956 V1 N6 P704-719	66-11174 <*>
1956 V1 N6 P720-731	60-15524 <+*>
1956 V1 N6 P743-746	R-328 <*>
1956 V1 N6 P747-751	R-1234 <*>
1956 V1 N6 P752-757	R-198 <*>
1956 V1 N6 P794-797	62-11444 <=*> 0
1956 V1 N6 P805-808	62-11613 <=>
1956 V1 N6 P831-837	62-11201 <=>
1956 V1 N6 P864-868	C-2534 <NRC>
	R-3056 <*>
1956 V1 N6 P869-872	58-563 <*>
1956 V1 N6 P873-877	R-3082 <*>
1956 V1 N8 P1052-1057	R-1773 <*>
1956 V1 N8 P1135-1143	62-15892 <=*>
1956 V1 N9 P1284-1287	62-11203 <=>
1956 V1 N10 P1364-1373	70P66R <ATS>
1956 V1 N11 P1435-1443	62-10108 <*=> 0
1956 V1 N12 P1503-	59-11997 <=> 0
1956 V1 N12 P1515-1519	62-23022 <=*>
1956 V1 N12 P1525-1526	R-1587 <*>
1957 N3 P311-316	39J16R <ATS>
1957 N3 P317-322	<ATS>
1957 N3 P328-333	42J16R <ATS>
1957 N5 P566-578	763TM <CTT>
1957 V2 N1 P34-37	59-18773 <+*>
1957 V2 N1 P65-74	59-10885 <+*>
1957 V2 N2 P136-143	R-3069 <*>
1957 V2 N2 P157-172	R-1590 <*>
1957 V2 N2 P219-221	15J16R <ATS>
	62-16529 <=*>
1957 V2 N3 P323-327	41J16R <ATS>
	59-00506 <*>
1957 V2 N5 P631-636	62-20139 <=*>
1957 V2 N6 P705-713	59-17247 <+*>
1957 V2 N6 P714-726	UCRL TRANS-470(L) <*=>
1957 V2 N7 P833-842	R-2046 <*>
1957 V2 N7 P935-937	R-2383 <*>
1957 V2 N7 P937-938	R-2047 <*>
1957 V2 N7 P943-	R-2379 <*>
1957 V2 N7 P944-	R-2380 <*>
1957 V2 N8 P947-950	59-10795 <+*>
1957 V2 N8 P1053-1061	R-3068 <*>
1957 V2 N9 P1200-1209	R-3892 <*>
1957 V2 N11 P1344-1389	59-19518 <+*> 0
1957 V2 N11 P1413-1434	59-19518 <+*> 0
	60-13039 <+*>
1958 V3 P288-290	R-3720 <*>
1958 V3 N1 P38-45	59-10880 <+*>
1958 V3 N1 P155-	R-4029 <*>
1958 V3 N2 P172-179	59-20690 <+*>
1958 V3 N2 P276-277	R-3737 <*>
1958 V3 N2 P290-291	R-3719 <*>
1958 V3 N2 P291-292	R-3736 <*>

Citation	Translation
1958 V3 N2 P297-298	R-3733 <*>
1958 V3 N2 P298-299	R-3734 <*>
1958 V3 N3 P415-420	60-13659 <+*>
1958 V3 N3 P428-429	60-13660 <+*>
1958 V3 N4 P459-466	59-20688 <+*>
1958 V3 N4 P478-486	59-18739 <+*>
1958 V3 N4 P552-554	59-21155 <=>
1958 V3 N4 P570-571	59-11235 <=>
1958 V3 N4 P587-591	R-4713 <*>
1958 V3 N5 P615-627	59-11270 <=>
1958 V3 N5 P628-633	59-11271 <=>
1958 V3 N5 P641-648	59-11272 <=>
1958 V3 N5 P675-689	R-4715 <*>
1958 V3 N5 P690-697	R-4469 <*>
1958 V3 N7 P873-881	59-11858 <=>
1958 V3 N7 P882-889	59-16563 <+*>
1958 V3 N7 P954	60-17211 <+*>
1958 V3 N7 P970-971	60-10704 <+*>
1958 V3 N7 P971-972	59-20687 <+*>
1958 V3 N9 P1107-1121	59-12938 <+*> 0
	59-14252 <+*>
	60-13658 <+*>
1958 V3 N9 P1218-1219	59-22684 <+*>
1958 V3 N9 P1221-1222	59-22681 <+*>
1958 V3 N10 P1274-1279	61-23670 <*=>
1958 V3 N11 P1379-1383	59-13221 <=> 0
1958 V3 N11 P1404-1405	59-21050 <=> 0
1958 V3 N11 P1407	59-16746 <+*>
1958 V3 N12 P1411-1429	59-16749 <+*>
1958 V3 N12 P1441-1449	59-14905 <+*>
1958 V3 N12 P1451-1462	59-14901 <+*>
1958 V3 N12 P1485-1494	59-14904 <+*>
1958 V3 N12 P1495-1500	59-14736 <+*>
1958 V3 N12 P1501-1515	59-10812 <+*>
1959 V4 N1 P17-20	59-21200 <=>
1959 V4 N1 P21-27	60-21063 <=>
1959 V4 N1 P88-96	60-21473 <=>
1959 V4 N1 P131	60-14341 <+*>
1959 V4 N1 P142-144	60-14342 <+*>
1959 V4 N2 P161-180	59-13887 <=>
1959 V4 N2 P161-171	60-17491 <+*>
1959 V4 N2 P195-201	59-18394 <+*>
1959 V4 N2 P202-	59-16014 <+*>
1959 V4 N2 P202-211	61-15813 <+*>
1959 V4 N3 P463-467	62-11181 <=>
1959 V4 N4 P592-598	61-19459 <+*>
1959 V4 N4 P681-687	62-19127 <*=>
1959 V4 N5 P894-896	61-16146 <=*>
	65M39R <ATS>
1959 V4 N5 P897-900	61-16145 <*=>
	64M39R <ATS>
1959 V4 N6 P936-941	61-13515 <=*>
1959 V4 N6 P960-965	AEC-TR-5950 <=*$>
1959 V4 N6 P966-971	27M40R <ATS>
	60-17490 <+*>
1959 V4 N6 P972-979	28M40R <ATS>
1959 V4 N6 P980-987	26M40R <ATS>
1959 V4 N8 P1238-1243	AEC-TR-4277 <+*>
1959 V4 N8 P1339-1358	60-11499 <=>
1959 V4 N8 P1381-1386	61-11150 <=> 0
1959 V4 N8 P1393-1394	61-13516 <=*>
1959 V4 N9 P1427-1433	68M40R <ATS>
1959 V4 N9 P1475-1479	70M40R <ATS>
1959 V4 N9 P1499-1504	60-14346 <+*>
1959 V4 N9 P1505-1512	60-14347 <+*>
1959 V4 N9 P1538-1942	28M40R <ATS>
1959 V4 N9 P1563-1565	60-11183 <=>
1959 V4 N10 P1585-1593	47M41R <ATS>
1959 V4 N10 P1602-1608	48M41R <ATS>
1959 V4 N10 P1747-1749	00M42R <ATS>
1959 V4 N11 P1765-1773	61-23528 <=*>
1959 V4 N11 P1816-1820	99M41R <ATS>
1959 V4 N11 P1821-1830	26M41R <ATS>
	61-20417 <*=>
1959 V4 N11 P1869-1877	25M41R <ATS>
1959 V4 N11 P1929-1931	61-10854 <+*>
1959 V4 N11 P1943-1944	60-21052 <=>
1959 V4 N12 P1951-1956	61-23537 <=*>
1959 V4 N12 P1957-1966	61-13336 <=*>
1959 V4 N12 P2094-2095	62-32909 <=*$>
1959 V4 N12 P2097-2100	61-10853 <=*>
1960 V5 N1 P150-161	62-19926 <=>
1960 V5 N1 P162-166	62-11468 <=>
1960 V5 N1 P167-169	62-11546 <=>
1960 V5 N3 P221-222	62-32616 <=*>
1960 V5 N3 P355-375	61-23536 <*=>
1960 V5 N4 P551-561	61-23541 <=*>
1960 V5 N7 P1052-1064	62-32664 <=*>
1960 V5 N7 P1065-1071	62-15728 <*=>
1960 V5 N7 P1158-1164	63-16645 <=*>
1960 V5 N9 P1359-1369	61-27839 <*=>
1960 V5 N9 P1380-1386	61-27840 <*=>
1960 V5 N9 P1475-1477	61-28707 <*=>
1960 V5 N10 P1576-1591	62-32631 <=*>
1960 V5 N11 P1751-1763	61-28209 <*=>
1960 V5 N12 P1919-1924	61-21690 <=>
	62-32666 <=*>
1960 V5 N12 P1974-1985	61-31538 <=>
	62-19932 <=*>
1960 V5 N12 P2057-2059	61-27172 <*=>
1960 V5 N12 P2069-2073	61-21377 <=>
1961 V6 N1 P9-13	62-15697 <=*>
1961 V6 N1 P14-21	62-15695 <*=>
1961 V6 N1 P125-136	64-13152 <=*$>
1961 V6 N3 P355-362	62-11320 <=>
	62-13620 <*=>
1961 V6 N3 P422-429	62-24410 <=*>
1961 V6 N3 P422-436	62-24874 <=*>
	62-32665 <=*>
1961 V6 N4 P536-544	61-28329 <*=>
1961 V6 N4 P676-678	62-13118 <*=>
1961 V6 N5 P707-715	65-11659 <*>
1961 V6 N6 P867-885	62-23318 <=*>
1961 V6 N9 P1411-1419	62-15514 <=*>
1961 V6 N9 P1420-1431	62-19273 <=*>
1961 V6 N9 P1540-1544	62-22471 <=*>
1961 V6 N10 P1707-1717	62-25541 <=*>
1961 V6 N11 P1888-1893	62-23676 <=*>
1961 V6 N12 P1961-1973	62-16690 <=*>
1961 V6 N12 P2084-2092	62-25737 <=*>
1962 V7 N1 P133-141	62-24476 <=*>
1962 V7 N2 P187-194	62-25409 <=*>
1962 V7 N2 P352-353	62-25756 <=*>
1962 V7 N3 P542-546	62-25571 <=*>
1962 V7 N3 P557-565	62-18859 <=*>
1962 V7 N5 P866-873	SCL-T-449 <=*>
1962 V7 N7 P1253-1254	62-11779 <=>
1962 V7 N11 P1896-1900	64-13663 <=*$>
1963 V8 N1 P8-23	63-21566 <=>
1963 V8 N3 P416-424	63-23794 <=*>
1963 V8 N4 P723-724	63-18718 <=*>
	65-10758 <*>
1963 V8 N6 P942-949	RPQ,V1,N1,P136-142 <IPIX>
1963 V8 N6 P950-958	63-31885 <=>
1963 V8 N7 P1130-1138	64-15106 <=*$>
1963 V8 N7 P1179-1186	64-11800 <=>
1963 V8 N8 P1451-1461	15R75R <ATS>
1963 V8 N9 P1577-1586	64-11852 <=>
1963 V8 N10 P1783-1786	64-15105 <=*$>
1963 V8 N11 P1862-1871	64-21469 <=>
1964 V9 P1179-1187	65-13755 <*>
1964 V9 N1 P114-117	64-21689 <=>
1964 V9 N4 P563-570	65-10411 <*>
1964 V9 N4 P616-624	65-30829 <=$>
1964 V9 N6 P943-948	AD-633 272 <=$>
1964 V9 N6 P1076-1079	AD-633 272 <=$>
1964 V9 N6 P1099-1113	64-41082 <=>
1964 V9 N8 P1327-1337	N64-30853 <=$>
1964 V9 N8 P1494-1495	N64-30855 <=$>
1964 V9 N8 P1530-1533	N64-30856 <=$>
1964 V9 N10 P1735-1739	N65-11444 <=$>
1964 V9 N10 P1875-1877	N65-14603 <=$>
1964 V9 N10 P1893-1897	65-12810 <*>
1964 V9 N11 P1903-1919	AD-613 466 <=$>
1964 V9 N11 P1933-1937	N65-14613 <=$>
1965 V10 P195-198	N65-27690 <=*$>
1965 V10 P201-203	N65-27719 <=*$>
1965 V10 N1 P102-111	N65-27717 <=$>
1965 V10 N2 P228-234	65-30715 <=$>
1965 V10 N2 P235-244	N65-21003 <=*$>

1960 V40 N16 P936-940 63-00615 <*>
1960 V40 N16 P955-961 66-11602 <*> 0

RASSEGNA INTERNAZIONALE DI STOMATOLOGIA PRATICA
1961 V12 P273-280 62-20109 <*>

RASSEGNA ITALIANA D'OTTALMOLOGIA
1933 V2 P519-552 58-431 <*>
1960 N11/2 P1-40 66-11601 <*>
1962 N7/8 P1-35 66-11596 <*>

RASSEGNA MEDICA SARDA
1950 V52 P222-232 58-418 <*>

RASSEGNA DI MEDICINA APPLICATA AL LAVORO INDUSTRIALE
1935 V6 P211-239 64-16211 <*>
1935 V6 N6 P435-440 64-16216 <*> 0

RASSEGNA DI MEDICINA INDUSTRIALE
1940 V11 P372-380 63-18762 <*>
1943 V14 P113-123 64-16370 <*> 0
1957 V26 P16-24 57-3152 <*>
58-1040 <*>

RASSEGNA DI NEUROLOGIA VEGETATIVA
1955 V11 P233-246 60-16935 <*>

RASSEGNA DI NEUROPSICHIATRIA E SCIENZE AFFINI
1952 V6 P401-420 1257 <*>

RASSEGNA DI STUDI PSICHIATRICI
1937 V26 P797-805 57-1689 <*>
1956 V45 N5 P1094-1096 61-10705 <*>
6999-C <K-H>
1962 V51 N1 P117-123 12159-C <K-H>

RASTENIEVUDNI NAUKI
1964 V1 N3 P159-166 64-31439 <=>
1964 V1 N8 P3-12 64-51558 <=$>

RASTITELNOST KRAINEGO SEVERA SSSR I EE OSVOENIE
1958 V3 P154-244 65-50044 <=$>

RASTITELNYE RESURSY
1965 V1 N1 P102-107 66-60316 <=*$>
1965 V1 N4 P507- FPTS V25 N6 P.T1069 <FASE>

RATIONALIZATSIYA I STANDARTIZATSIYA
1966 N6 P19-21 66-33981 <=$>
1966 N8 P5-7 66-34958 <=$>

RAYON REVUE
1949 V3 P142-148 61-14013 <*>
1953 V7 P45-53 61-16717 <*> 0
1954 V8 P54-77 61-16716 <*> 0
1954 V8 P178-183 61-14506 <*>
1954 V8 N4 P125-134 61-14014 <*>

RAYONNE ET FIBRES SYNTHETIQUES. BRUSSELLS
1952 V8 N4 P51-55 61-14021 <*>
1952 V8 N4 P57-67 61-14022 <*>
1956 V12 P1444-1456 60-18517 <*>
1957 V13 N3 P289-298 61-14530 <*>
1958 V14 N2 P205-214 38L29F <ATS>
1958 V14 N8 P1157-1182 60-10121 <*>
1959 V15 N10 P1113-1120 62-14647 <*> 0
1959 V15 N10 P1125-1132 62-14294 <*>
1959 V15 N11 P1285-1293 62-14295 <*>
1960 V16 N1 P79-88 62-14296 <*>
1960 V16 N2 P141-144 61-10284 <*>
61-16728 <*>
1961 V17 N8 P797-807 63-15583 <*> 0
1962 V18 N6 P635-640 63-14497 <*>
1963 V19 P935-940 64-14015 <*> 0
1963 V19 N6 P579-589 64-10376 <*> 0

RAZPRAVE. SLOVENSKA AKADEMIJA ZNANOSTI IN UMETNOSTI
1955 V3 P3-65 61-11242 <=>
1958 V4 P5-38 61-11243 <=>
1958 V4 P85-124 61-11282 <=>
1959 V5 P5-22 61-11241 <=>
1959 V5 P141-182 61-11244 <=>

RAZVEDKA NEDR
1937 V7 N3 P15-18 62-23148 <=*>
1938 V8 N3 P6-10 R-2465 <*>
1939 V10 N1 P38-41 R-4442 <*>
RT-168 <*>
1940 V10 N12 P32-43 26E2R <ATS>
1940 V10 N12 P32-42 61-18078 <*=>
1946 N2 P34-39 R-4674 <*>
1946 V12 N2 P34-39 RT-56 <*>
1946 V12 N2 P51-55 RT-57 <*>
1946 V12 N5 P41-42 RT-165 <*>
1947 N5 P67-71 RT-267 <*>
1947 V13 N3 P20-24 24E2R <ATS>

★RAZVEDKA I OKHRANA NEDR
1955 V21 N6 P59 62-16718 <=*>
1957 N4 P31-33 IGR V1 N3 P31 <AGI>
1957 N5 P43-51 59-11329 <=> 0
1957 N5 P57-58 IGR V1 N9 P39 <AGI>
59-11376 <=> 0
1957 V23 N1 P6-12 26K21R <ATS>
1957 V23 N1 P12-16 28K21R <ATS>
1957 V23 N1 P25-32 27K21R <ATS>
1957 V23 N4 P13-17 39J18R <ATS>
1957 V23 N4 P17-23 AEC-TR-4472 <=*>
1957 V23 N9 P11-16 99K27R <ATS>
1958 N2 P8-9 IGR V1 N10 P45 <AGI>
59-11356 <=>
1958 N5 P38-44 IGR V1 N9 P33 <AGI>
PB 141 271T <=>
1958 N9 P11-14 IGR V1 N7 P71 <AGI>
1958 V24 N2 P42-45 30K27R <ATS>
1958 V24 N3 P46-47 85K24R <ATS>
1958 V24 N6 P41-48 31K27R <ATS>
1958 V24 N8 P9-14 <RIA>
1958 V24 N9 P11-14 59-13279 <=>
1958 V24 N9 P22-25 21P61R <ATS>
59-13280 <=> 0
1958 V24 N9 P26-32 59-13281 <=> 0
1958 V24 N9 P33-35 59-13282 <=> 0
1958 V24 N9 P55-56 59-13283 <=>
1958 V24 N10 P1-4 74L28R <ATS>
1958 V24 N10 P18-22 59-16370 <+*>
1958 V24 N10 P45-58 75L28R <ATS>
1958 V24 N11 P62-64 59-13617 <=>
1958 V24 N12 P44-47 81L31R <ATS>
1959 V25 N1 P19-22 07M39R <ATS>
1959 V25 N1 P25-27 57L32R <ATS>
1959 V25 N1 P28-30 93L31R <ATS>
1959 V25 N2 P5-10 41L32R <ATS>
1959 V25 N2 P12-25 59-18354 <+*>
1959 V25 N4 P4-6 60-11118 <=>
1959 V25 N6 P1-4 60-13241 <+*>
1959 V25 N6 P3-6 R-1-1 <RIA>
1959 V25 N7 P5-9 R-1-2 <RIA>
1959 V25 N7 P25-30 29M40R <ATS>
1959 V25 N7 P34-36 R-1-3 <RIA>
1959 V25 N11 P60-63 60-11740 <=>
1959 V25 N12 P54-59 60-31103 <+*>
1960 V26 N2 P39-42 66M42R <ATS>
1960 V26 N3 P43-48 60-31883 <=>
1960 V26 N9 P1-6 61-27290 <=>
1960 V26 N10 P1-4 61-27290 <=>
1960 V26 N11 P6-10 61-28881 <*=>
1961 V27 N1 P18-23 62N53R <ATS>
1961 V27 N1 P27-32 63N53R <ATS>
1961 V27 N3 P17-26 70N55R <ATS>
1961 V27 N5 P8-15 37P59R <ATS>
1961 V27 N5 P21-25 78N55R <ATS>
1961 V27 N5 P36-40 63-19514 <=*>
1961 V27 N9 P48-50 43P60R <ATS>
1962 N5 P36-42 IGR V6 N3 P400 <AGI>
1962 N12 P28-30 63-21770 <=>
1962 V28 N2 P1-7 63-21920 <=>
1962 V28 N2 P30-34 63-21920 <=>

1962 V28 N4 P8-15 63-13118 <=*>
1962 V28 N5 P34-36 62-32461 <=>
1962 V28 N5 P36-42 IGR,V6,N3,P400-404 <AGI>
98Q68R <ATS>
1962 V28 N5 P47-50 05P65R <ATS>
1962 V28 N6 P21-24 42P64R <ATS>
1962 V28 N8 P62-63 62-33664 <=*>
1962 V28 N9 P41-45 86S82R <ATS>
1962 V28 N1C P17-20 63-15100 <=>
1962 V28 N10 P57-60 61Q67R <ATS>
1962 V28 N12 P28-30 63-21511 <=>
1963 V29 N1 P1-8 63-21616 <=>
1963 V29 N1 P9-13 63-21664 <=>
1963 V29 N2 P52-54 17Q71R <ATS>
1963 V29 N3 P8-23 63-31009 <=>
1963 V29 N4 P36-39 22Q72R <ATS>
1963 V29 N5 P1- 63-31420 <=>
1963 V29 N9 P27-32 63-41208 <=>
1964 V30 N4 P30-34 64-31494 <=>
1964 V30 N4 P35-39 64-31485 <=>
1964 V30 N5 P22-25 60R79R <ATS>
1964 V30 N5 P27-30 61R79R <ATS>
1964 V30 N5 P63-64 64-41958 <=$>
1964 V30 N6 P15-17 64-41633 <=>
1964 V30 N7 P30-33 64-41262 <=>
1964 V30 N10 P34-36 78S82R <ATS>
1964 V30 N1C P39-42 64-51902 <=$>
1964 V30 N12 P1-11 65-30582 <=$>
1964 V30 N12 P22-27 93S82R <ATS>
1965 N1 P44-48 67-31485 <=$>
1965 V31 N2 P25-29 65-30703 <=$>
1965 V31 N2 P29-31 65-30702 <=$>
1965 V31 N2 P48-49 05S86R <ATS>
1965 V31 N5 P32-45 65-31422 <=$>
1965 V31 N6 P36-40 65-32497 <=*$>
1965 V31 N6 P40-45 65-32498 <=$*>
1965 V31 N9 P23-25 79T91R <ATS>
1966 V32 N6 P62-64 66-33688 <=$>

RAZVEDOCHNAYA I PROMYSLOVNYA GEOFIZIKA
1957 V17 P12-19 PANSDOC-TR.277 <PANS>
1957 V17 P21-32 PANSDOC-TR.278 <PANS>
1957 V17 P32-36 PANSDOC-TR.279 <PANS>
1957 V17 P36-40 PANSDOC-TR.280 <PANS>
1958 V21 P82-90 39M45R <ATS>
1958 V22 P3-76 60-23615 <+*>
1958 V22 P76-97 17M40R <ATS>
1959 V29 P78-82 38M45R <ATS>
1959 V31 P3-8 62-13081 <=*>
1960 V35 P3-12 63-18443 <=$>
1960 V36 P3-7 64-15192 <=*$>
1960 V36 P8-13 64-15192 <=*$>
1960 V36 P14-23 UCRL TRANS-992(L) <=*$>
64-15192 <=*$>
1960 V36 P24-30 64-15192 <=*$>
1961 V42 P72-76 63-13096 <=*>
1962 V43 P52-64 39Q70R <ATS>
1962 V43 P72-79 63-15110 <=*>
1962 V43 P118-120 17Q70R <ATS>
1962 V44 P111-116 63-13464 <=*>
1962 V44 P117-122 C-5375 <NRC>
1962 V44 P123-126 63-13464 <=*>
1963 N47 P29-34 66-14414 <*>
1963 N48 P15-17 66-10321 <*>
1963 N49 P59-64 65-12823 <*>
1963 V47 P46-50 RTS-2693 <NLL>
1963 V47 P51-58 RTS-2698 <NLL>
1963 V48 P45-47 64-31976 <=>
1963 V48 P63-65 64-31976 <=>
1963 V48 P87-91 64-31976 <=>
1963 V49 P12- 65-12824 <*>
1963 V49 P86-87 48R79R <ATS>
65-12825 <*>
1963 V50 P3-11 64-71503 <=> M
1964 V51 P49-58 65-62426 <=$>
1964 V51 P59-67 AD-623 444 <=*$>

REAKSTII I METODY ISSLEDOVANIYA ORGANICHESKIKH SOEDINENII
1957 V6 P343-367 62-10564 <=*> O

1964 V14 P173-298 RTS-3302 <NLL>

RECHERCHE AERONAUTIQUE
1949 N9 P33-42 3009 <*>
1951 N20 P61-67 59-10338 <*>
1951 N23 P41-5C 62-18622 <*> O
1953 N31 P37-44 1457 <*>
1954 N40 P3-5 C-3664 <NRCC>
1954 N42 P17-21 1491 <*>
1955 N47 P11-14 C-3686 <NRCC>
1955 N47 P27-37 65-12999 <*>
1955 V45 P45-51 3898 <HB>
1956 N49 P31-39 57-749 <*>
1956 N49 P39-4C 60-10351 <*>
1956 N50 P39-44 59-14138 <+*> O
1956 N52 P3-11 C-2469 <NRC>
1956 N54 P3-8 C-4135 <NRCC>
1956 V52 P3-11 57-3031 <*>
1957 N56 P3-12 59-16778 <+*>
1957 N58 P29-35 59-10320 <*>
1957 N59 P21-26 62-18648 <*>
1957 N61 P35-52 62-18187 <*>
1958 N65 P19-27 101 <LS>
1958 N67 P11-19 64-30684 <*>
1958 V63 P41-52 N66-13278 <=$>
1958 V63 P41-51 66-11153 <*>
1959 N68 P9-19 AEC-TR-4026 <*>
1959 N71 P11-28 02L37F <ATS>
1960 N76 P52-53 SCL-T-375 <*>
1960 N79 P35-44 65-17234 <*>
1960 N79 P36-37 62-18647 <*>
1961 N80 P37-45 62-01365 <*>
1961 N82 P13-26 65-17001 <*>
1961 N84 P31-38 C-4593 <NRC>

RECHERCHE AEROSPATIALE
1963 N94 P3-7 65-11029 <*>
1963 N95 P7-15 66-10488 <*>
1963 N96 P39-49 RAPRA-1241 <RAP>
1964 N99 P11-16 N65-33065 <=$>
66-11565 <*>
1964 N102 P43-49 N65-24371 <=$>
1965 N105 P3-9 N66-13050 <=*$>
1965 N107 P25-27 66-11671 <*>
1965 N108 P57-61 N66-23726 <=$>

RECHNOI TRANSPORT
1950 V10 N2 P20-22 50/3173 <NLL>
1960 N9 P1-4 NLLTB V4 N3 P252 <NLL>
1960 V19 N9 P1-4 62-24783 <=>
1961 V20 N10 P26-28 AD-631 681 <=$>
1961 V20 N12 P29 63-19487 <=*>
1962 V21 N4 P34-36 62-33089 <=*> O
1962 V21 N9 P5-10 63-13540 <=>
1962 V21 N9 P27-28 63-13540 <=>
1962 V21 N9 P31-32 63-13540 <=>
1964 N7 P42-43 AD-639 489 <=$>
1964 N11 P47-48 AD-629 861 <=*$>
1964 V23 N10 P29-30 AD-639 439 <=$>

RECORD OF THE ELECTRICAL ANC COMMUNICATION ENGINEERING CONVERSATION. TOHCKU UNIVERSITY
SEE TOHOKU DAIGAKU DENTSU DANWAKAI KIROKU

RECUEIL DE MEDECINE VETERINAIRE DE L'ECOLE D'ALFORT. PARIS
1941 V117 P116-125 I-1011 <*>
1941 V117 P329-338 I-535 <*>
1953 V129 N1 P7-15 MF-2 <*>
1953 V129 N2 P86-89 I-497 <*>
1953 V129 N3 P167-179 I-1001 <*>
1955 V131 N2 P73-85 58-389 <*>
1956 V132 N9 P657-673 65-12661 <*>

RECUEIL DES TRAVAUX BOTANIQUES NEERLANDAIS
1915 V15 P115-135 ANL-TRANS-199-B <*>

RECUEIL DES TRAVAUX CHIMIQUES DES PAYS-BAS ET DE LA BELGIQUE
1894 V13 P36 63-18669 <*>

1904 V23 N6 P357-359 357-359 <*> P
1905 V24 P165-175 AL-877 <*>
1909 V28 P42-66 59-10749 <*>
1910 V29 P85-112 63-14115 <*>
63-14116 <*> P
1910 V29 P173-184 63-14117 <*>
1915 V34 P78-95 63-20513 <*> O
1919 V38 P1C1-105 63-18246 <*>
1920 V39 P704-710 57-55 <*>
1921 V40 P452-453 I-268 <*>
64-10157 <*>
1923 V42 P821 4410 <BISI>
1924 V43 P643-644 63-10628 <*>
1925 V44 P271-274 63-10629 <*> O
1927 V46 P173-176 2842 <*>
1931 V50 N12 P239-256 5876-B <K-H>
1934 V53 P186-190 63-20330 <*> O
1935 V54 P847-852 I-774 <*>
1936 V55 P1000-1006 63-14118 <*> O
1937 V56 P351-355 59-20604 <*>
1937 V56 N9/10 P985-999 63-18043 <*> O
1938 V57 P217-224 I-771 <*>
64-10479 <*>
1938 V57 P445-455 63-14119 <*>
1938 V57 P1075-1086 I-366 <*>
1939 V58 P411-422 63-14643 <*>
1939 V58 P778-784 63-14120 <*>
1940 V59 P1117-1122 58-909 <*>
63-16391 <*>
1940 V59 P1141-1155 63-14121 <*> O
1940 V59 P1206-1219 61-20540 <*>
1942 V61 P500-512 62-10100 <*>
1945 V64 P239-249 AEC-TR-4224 <*>
1946 V65 P193-202 63-20270 <*>
1949 V68 P789-806 59-17133 <*> O
1949 V68 P817-826 57-2566 <*>
1949 V68 P1106-1122 63-18921 <*>
1949 V68 N9/10 P789-806 64-20360 <*> O
1950 V69 P1495-1503 SCL-T-397 <*>
1951 V70 P696 66-10338 <*> O
1951 V70 P730-732 2907 <*>
1951 V70 N8 P638-646 59-17771 <*> P
1956 V75 P194-198 5756-A <K-H>
1956 V75 P648-657 AEC-TR-3706 <*>
1962 V81 P5-18 63-00706 <*>

RECUEL DES TRAVAUX DE L'INSTITUT DE RECHERCHES SUR LA STRUCTURE DE LA MATIERE
SEE BULLETIN INSTITUTA ZA NUKLEARNE NAUKE, BORIS KIDRIC. BELGRAD

REDIA
1955 V40 P197-212 60-15412 <=*> O

REDKIE METALLY
1934 V3 N2 P25-33 TT.311 <NRC>
1934 V3 N3 P40-43 64-18378 <*> O
64-20523 <*> O

REFORM OF THE WRITTEN SCRIPT
SEE WEN TZU KAI KO

REFRACTORIES. JAPAN
SEE TAIKABUTSU

REFRIGERATION. JAPAN
SEE REITO

REGELUNGSTECHNIK
1953 V1 N1 P13-17 64-10343 <*>
1953 V1 N1 P17-20 65-12525 <*>
1953 V1 N4 P74-78 64-10341 <*>
1954 V2 P157-162 2014 <*>
1954 V2 N8 P177-181 65-12602 <*>
1955 V3 N11 P266-268 59-17178 <*> O
1957 V5 N10 P339-342 61-20725 <*>
1958 V6 N12 P442-446 61-20696 <*>
1959 V7 P248-253 TS-1485 <BISI>
1959 V7 N5 P160-165 SCL-T-347 <*>
1959 V7 N7 P248-253 62-14651 <*>
1960 V8 N1 P15-18 1697 <BISI>
1960 V8 N8 P261-266 61-20723 <*>
1961 V9 N4 P149-153 63-00247 <*>
1965 V13 N2 P62-68 66-14033 <*> O

REGELUNGSTECHNISCHE PRAXIS
1964 V6 N1 P15-20 4447-I <BISI>
1964 V6 N2 P60-68 4447-II <BISI>

REICHSBERICHE FUER CHEMIE
1944 V1 N2 P121-140 66-12589 <*>

REICHSBERICHTE FUER PHYSIK
1945 V1 N5 P154-158 AEC-TR-6081 <*>

RENDICONTI DELL'ACCADEMIA DELLE SCIENZE FISICHE E MATEMATICHE
1909 S3 V15 P15-17 3438 <KH>
64-14689 <*>
1927 S3A V33 N9/12 P240-243
AD-620 604 <=$>
AL-158 <*>
1928 S3 V34 N5/8 P149-151 AL-159 <*>

RENDICONTI. ASSOCIAZIONE ELETTROTECNICA ITALIANA
1954 P3-10 57-1006 <*>

RENDICONTI DEL CIRCOLO MATEMATICO DI PACERMO
1907 V24 P111-117 T-1638 <INSD>
1907 V24 P117-136 T-2671 <INSD>
1907 V24 P137-140 T-1638 <INSD>

RENDICONTI DELL'ISTITUTO LOMBARDO DI SCIENZE E LETTERE
1957 V91 P347-370 AEC-TR-3583 <=*>
1957 V91 P378-385 AEC-TR-3583 <=*>
1957 V91 N1 P68-75 90L28I <ATS>
1958 V92B P424-430 63-00283 <*>
1959 V93 P695-706 CHEM-247 <CTS>
1959 V93A P707-714 62N51I <ATS>
1961 V95A P83-91 <LSA>
1962 V96A P513-522 66-12732 <*>
1965 V99A N1 P21-29 66-12734 <*>

RENDICONTI DELL'ISTITUTO SUPERIORE DI SANITA. ROME
1944 V7 P604-615 58-394 <*>
1947 V10 N2 P265-277 64-10843 <*>
1949 V12 P106-137 II-560 <*>
1949 V12 P138-157 1589 <*>
1950 V13 P3-19 AL-79 <*>
1950 V13 P85-92 58-424 <*>
1951 P509-529 UCRL TRANS-593(L) <*>
1951 V14 P779-791 PT10 I-962 <*>
1951 V14 P171-178 3142 <*>
1951 V14 P779-791 58-1515 <*>
1952 V15 P188-194 3150 <*>
1954 V17 P326-332 II-1031 <*>
1955 V18 P367-375 62J17I <ATS>
1955 V18 P1312-1321 64-16177 <*>
1955 V18 P1322-1330 64-16180 <*>
1955 V18 P1331-1338 64-16178 <*>
1962 V25 P473-486 64-00181 <*>

RENDICONTI E MEMORIE DELL'ACCADEMIA DI SCIENZE, LETTERE ED ARTI DEGLI ZELANTI
1912 S3 V7/8 P35-36 3902-B <K-H>

RENDICONTI DEL SEMINARIO MATEMATICO, UNIVERSITA E POLITECNICO DI TORINO
1950 V9 P309-324 66-12520 <*>
1955 V15 P59-64 66-14164 <*>

RENTGENOGRAFIYA MINERALNOGO SYRIA
1964 N4 P135-143 GI V2 N2 1965 P388 <AGI>
1964 N4 P164-169 GI V2 N2 1965 P382 <AGI>

REPERTOIRE GENERAL D'ANATCMIE ET DE PHYSIOLCGIE PATHOLOGIQUES ET DES CLINIQUES CHIRURGICALS
1826 V2 P100-107 62-18778 <*> O

REPORT. AKTIEBOLAGET ATOMENERGIE. STOCKHOLM

1955 AEF-49	58-2205 <*>
1956 AEF-68	58-1914 <*>
1956 AEF-69	58-2070 <*>
1956 AEF-65	58-2206 <*>
1957 AEF-54	66-12860 <*>
1957 AEF-71	58-1913 <*>

REPORT. ANSALDO S.P.A. GENOA

1960 LAB/STU-1416	EURAEC-49 <=*>
1961 LAB/STU-1469	EURAEC-209 <=*>
1961 LAB/STU-1451	EURAEC-99 <=*>

REPORT. BADISCHE ANILIN-UND SODA-FABRIK. A.G. LUDWIGSHAFTEN. GERMANY

1930 KUKO-9	64-10241 <=*$>
1931 KUKO-26	64-10243 <=*$>
1931 KUKO-21	64-10268 <=*$>
1932 KUKO-31	64-10244 <=*$>
1932 KUKO-27	64-10267 <=*$>
1932 KUKO-33	64-10269 <=*$>
1932 KUKO-28	64-10721 <=*$>
1932 KUKO-32	64-10722 <=*$>
1933 KUKO-44	64-10245 <=*$>
1933 KUKO-50	64-10246 <=*$>
1933 KUKO-46	64-10263 <=*$>
1933 KUKO-46C	64-10264 <=*$>
1933 KUKO-46B	64-10266 <=*$>
1933 KUKO-45	64-10270 <=*$>
1933 KUKO-46A	64-10284 <=*$>
1934 KUKO-16	64-10242 <=*$>
1934 KUKO-52C	64-10247 <=*$>
1934 KUKO-54	64-10248 <=*$>
1934 KUKO-52A	64-10265 <=*$>
1934 KUKO-53	64-10271 <=*$>
1934 KUKO-58	64-10272 <=*$>
1935 KUKO-64	64-10249 <=*$>
1935 KUKO-70	64-10250 <=*$>
1935 KUKO-74	64-10251 <=*$>
1935 KUKO-66	64-10273 <=*$>
1936 KUKO-79	64-10255 <=*$>
1936 KUKO-83	64-10257 <=*$>
1936 KUKO-86	64-10258 <=*$>
1936 KUKO-93	64-10259 <=*$>
1936 KUKO-92	64-10260 <=*$>
1936 KUKO-89	64-10274 <=*$>
1936 KUKO-90	64-10275 <=*$>
1936 KUKO-75	64-10723 <=*$>
1936 KUKO-84	64-10724 <=*$>
1937 KUKO-103	64-10253 <=*$>
1937 KUKO-106	64-10254 <=*$>
1937 KUKO-97	64-10262 <=*$>
1963 KUKO-87	64-10725 <=*$>

REPORT OF THE CENTRAL RESEARCH INSTITUTE. JAPAN MONOPOLY CORPORATION
SEE NIHON SENBAI KOSHA CHUO KENKYUSHO GYOTEI HOKOKU

REPORT. CENTRE D'ETUDE DE L'ENERGIE NUCLEARE. BRUSSELS
SEE RAPPORT. CENTRE D'ETUDE DE L'ENERGIE NUCLEARE. BRUSSELS

REPORT. CENTRE EUROPEAN DE RECHERCHE NUCLEARE GENEVA

MPS/INTLIN-62-5	66-15624 <*>
1955 CERN-PS/MM-14	58-1490 <*>
1961 CERN-61-12	UCRL-TRANS-864 <=*$>
1962 CERN-62-24	66-15067 <*>
1964 CERN-123-63	65-18086 <*>
1965 AR/INT.SG/65-28	66-13524 <*>
1965 MPS/MU/S-65-1	66-15365 <*>

REPORT. CENTRO INFORMAZIONI STUDII ESPERIENZE MILAN

CISE-R-40	EURAEC-151 <=*>
CISE-R-30	EURAEC-44 <=*>
CISE-R-34	EURAEC-98 <=*>
1960 CISE-R-18	EURAEC-36 <=*>
1960 CISE-R-24	EURAEC-38 <=*>
1960 CISE-R-28	EURAEC-39 <=*>
1960 CISE-R-23	EURAEC-41 <=*>
1960 CISE-R-27	EURAEC-42 <=*>
1960 CISE-R-26	EURAEC-43 <=*>
1961 CISE-R-47	EURAEC-173 <=*>
1961 CISE-R-32	EURAEC-40 <=*>
1961 CISE-R-31	EURAEC-45 <=*>
1961 CISE-R-37	EURAEC-97 <=*>
1962 CISE-R-51	EURAEC-222 <=*>
1962 CISE-R-61	EURAEC-332 <=*>
1962 CISE-R-60	EURAEC-401 <=*$>
1962 CISE-R-43	EURAEC-445 <=*$>
1962 CISE-R-64	EURAEC-475 <=*$>
1962 CISE-R-65	EURAEC-476 <=*$>
1963 CISE-R-71	EURAEC-537 <=*$>

REPORT. COMITATO NAZIONALE PER L'ENERGIA NUCLEARE. ITALY

1960 CNI-35	AEC-TR-4935 <*>
1961 CNI-95	UCRL TRANS-998(L) <*>
1962 CNEN-158	AEC-TR-6253 <*>
1962 ISR-87	AEC-TR-5309 <*>
1962 LTEC/PR/16	AEC-TR-5641 <*>
1963 LNF-63/26	AI-TR-19 <*>
1963 RT/EL(63)-3	UCRL TRANS-1021(L) <*>
1964 CEC-85	ORNL-TR-85 <*>
1964 CEC-94	ORNL-TR-349 <*>

REPORT. COMITATO NAZIONALE PER LE RICERCHE NUCLEARE. ITALY

1959 CNI-24	AEC-TR-4171 <=>
	61-18292 <*> P
1960 CNTR-3	AEC-TR-4616 <*>
1960 CNTR-9	AEC-TR-4617 <*>
1961 CNI-95	UCRL TRANS-998(L) <*>
1963 RT/EL(63)-3	UCRL TRANS-1021(L) <*>

REPORT. COMMISSARIAT A L'ENERGIE ATOMIQUE. FRANCE
SEE RAPPORT COMMISSARIAT A L'ENERGIE ATOMIQUE. FRANCE

REPORT. COMMISSIE VOOR UITUOENING VON RESEARCH. NETHERLANDS

CUR-23	<CCA>

REPORT. COMPAGNIE FRANCAISE THOMSON-HOUSTON

1961 R/211	62-10629 <*>

REPORT. DEUTSCHES ELEKTRONEN-SYNCHROTRON HAMBURG, GERMANY

1959 DESY-AL.5	UCRL-TRANS-816 <=*>
1964 DESY-64/11	ORNL-TR-547 <*>
1964 DESY-64/7	66-14105 <*>
1964 DESY-64/13	66-15402 <*>

REPORT OF THE ELECTRICAL TESTING INSTITUTE. JAPAN
SEE DENKI SHIKEN-SHO IHO

ANNUAL REPORT OF THE ENGINEERING RESEARCH INSTITUTE. UNIVERSITY OF TOKYO
SEE SOGO SHIKENJO NENPO

REPORT OF THE EUROPEAN ATOMIC ENERGY COMMUNITY. EURATOM

1960 EUR-C-434-60 I	EURAEC-11 <=*>
1960 EUR-C-1323-3-59	EURAEC-13 <=*>
1960 EUR-3081-60.F	EURAEC-52 <=*>
1961 EUR-C-3283-61 E	EURAEC-116 <=*>
1961 EUR-C-262-2-60D	EURAEC-137 <=*>
1961 EUR-C-1356-60-F	EURAEC-216 <=*>
1961 EUR-C-355-1-60F	EURAEC-31 <=*>
1961 EUR-C-1658-603I	EURAEC-56 <=*>
1961 EUR-C-434-2-60I	EURAEC-61 <=*>
1961 EUR-C-1928-1-60	EURAEC-66 <=*>
1961 EUR-025-60-10	EURAEC-50 <=*>
1961 EUR-3081-60.F	EURAEC-195 <=*>
1962 EUR-C-1356-60.F	EURAEC-115 <=*>
1962 EUR-C-262-2-60D	EURAEC-201 <=*>
1962 EUR-C-659-62-E	EURAEC-232 <=*>

1962 EUR-118.F 65-17091 <*>
1962 EUR-103.F 66-11974 <*>
1963 CEX-R-3 NP-TR-1121 <*>
1963 CEX-R-4 NP-TR-1122 <*>
1963 EUR-14CF HW-TR-50 <*>
1963 EUR-204I NP-TR-1091 <*>
1963 EUR-294I NP-TR-1142 <*>
1963 EUR-272-F 66-11373 <*>
1963 EUR-364I NP-TR-1134 <*>
1964 CETIS-31 66-15422 <*> 0
1964 EUR-1888D ORNL-TR-548 <*>
1964 EUR-2211.D N66-14379 <=*$>
1964 EUR-2158F 66-11372 <*>
1964 EUR-365.F 66-15422 <*> 0
1964 R.2290 ANL-TRANS-74 <*>
1964 RD71/6713/01/MS ORNL-TR-539 <*>
1965 EUR-2449.F 66-15641 <*>

REPORTS OF THE FACULTY OF ENGINEERING. YAMANASHI UNIVERSITY
SEE YAMANASHI DAIGAKU KOGAKUBU KENKYU HOKOKU

ANNUAL REPORT OF THE FACULTY OF PHARMACY, KANAZAWA UNIVERSITY
SEE KANAZAWA DAIGAKU YAKUGAKUBU KENKYU NEMPO

REPORT. FIAT. SEZIONE ENERGIA NUCLEARE. TURIN
1962 FN-E-19 EURAEC-268 <=*>
1962 FN-E-21 EURAEC-310 <=*>
1962 FN-E-22 EURAEC-311 <=*>
1962 FN-E-23 EURAEC-321 <=*>
1962 FN-E-27 EURAEC-406 <=*$>
1962 FN-E-29 EURAEC-500 <=*>
1963 FN-E-28 EURAEC-492 <=*$>
1963 FN-E-30 EURAEC-568 <=*$>
1963 FN-E-31 EURAEC-584 <=*$>

REPORT OF THE FOOD RESEARCH INSTITUTE. TOKYO
SEE SHOKURYO KENKYUSHO KENKYU HOKOKU

REPORT. GOSUDARSTVENNYI KOMITET PO ISPOLZOVANIYU ATOMNOI ENERGII
1964 NO 79 66-11125 <*>

REPORTS OF THE GOVERNMENT CHEMICAL INDUSTRIAL RESEARCH INSTITUTE. TOKYO
SEE TOKYO KOGYO SHIKENSHO HOKOKU

REPORT OF THE GOVERNMENT INDUSTRIAL RESEARCH INSTITUTE. NAGOYA
SEE NAGOYA KOGYO GIJUTSU SHIKENSHO KENKYU HOKOKU

REPORTS OF THE GOVERNMENT INDUSTRIAL RESEARCH INSTITUTE. OSAKA
SEE OSAKA GIJUTSU SHIKEN SHO HOKOKU

REPORT OF THE GOVERNMENT MECHANICAL LABORATORY TOKYO
SEE KIKAI SHIKENJO HOKOKU

REPORT. HAHN-MEITNER INSTITUT FUER KERNFORSCHUNG. BERLIN
EU-13-18 EURAEC-385 <=*>
1959 HMI-B 7 66-11877 <*>
1961 EUR-710-26-D EURAEC-121 <=*>

REPORTS OF THE HIMEJI TECHNICAL UNIVERSITY
SEE HIMEJI KOGYO DAIGAKU KENKYU HOKOKU

REPORT. HOKKAIDO FORESTS PRODUCTS RESEACH INSTITUTE
SEE HOKKAIDORITSU RINSAN SHIKENJO KENKYU HOKOKU

ANNUAL REPORT OF THE HOSHI COLLEGE OF PHARMACY
SEE HOSHI YAKKA DAIGAKU KIYO

REPORT. I.G. FARBENINDUSTRIE. WEST GERMANY
1937 NO 502 58-1874 <*>
1937 NO 522 58-2017 <*>
1938 NO 1541 57-2632 <*> P
1938 NO 1542 SEC10.1 57-2634 <*>
1938 NO 73 58-1885 <*>
1940 NO 581 58-2018 <*>
1942 NO 353 2830 <*>
1942 NO 638 58-1849 <*>

REPORTS OF THE INDUSTRIAL RESEARCH INSTITUTE. OSAKA PREFECTURE UNIVERSITY
SEE OSAKA FURITSU KOSYO SHOREIKAN HOKOKU

REPORT. INSTITUT ATOMNOI ENERGII. AKADEMIYA NAUK SSSR. GOSUDARSTVENNYI KOMITET PO ISPOLZOVANIYU ATOMNOI ENERGII
1962 IAE-23/236 AEC-TR-5804 <=>
1963 IAE-517 AEC-TR-6340 <=$>
1963 IAE-70 64-23312 <$>
1964 IAE-562 ANL-TRANS-68 <=$> P
1964 IAE-733 ANL-TRANS-156 <=$>

REPORT. INSTITUT FIZIKI. AKADEMIYA NAUK SSSR
A-40 64-18760 <*>
1961 A-133 62-24796 <=*>
1962 A-210 UCRL-TRANS-945(L) <=*$> P
63-10654 <=*> P
1962 A-61 UCRL-TRANS-833(L) <=*>
1962 A-7 64-14622 <=*$>
1962 A-80 63-10860 <=*>
1962 A-83 63-10861 <=*>
1963 A-15 64-10802 <=*$>
1963 A-25 63-20697 <=*> P
1963 A-29 64-13589 <=*$>
1964 A-53 N66-10450 <=*$> P
1965 A-16 65-13905 <*>
1965 A-23 65-13906 <*>
1965 A-95 66-11208 <*> P

REPORT. INSTITUT FRANCO-ALLMAND DE RECHERCHES DE SAINT-LOUIS. FRANCE. (DEUTSCH-FRANZOESISCHES FORSCHUNGSINSTITUT)
1950 ISL-8/50 65-60832-1 <=>
1953 ISL-1/53 65-60832-2 <=>

REPORT. INSTITUT NATIONAL DE SECURITE. PARIS
NO 269-26-62 9070 <IICH>
NO 330-32-63 9029 <IICH>
NO 371-34-64 9071 <IICH>

REPORT. INSTITUT TEORETICHESKOI I EKSPERIMENTALNOI FIZIKI. AKADEMIYA NAUK SSSR. GOSUDARSTVENNYI KOMITET PO ISPOLZOVANIYU ATOMNOI ENERGII
ITEF-161 64-26323 <$>
ITEF-165 64-26324 <$>
ITEF-191 64-26384 <$>
1962 ITEF-61 AEC-TR-5625 <=$>
1963 ITEF-121 ANL-TRANS-44 <=$>
1963 161 64-00149 <*>
1964 IAE-N-310 65-14520 <*>
1964 ITEF-251 AEC-TR-6581 <=$>
1965 ITEF-333 AEC-TR-6807 <=>
1965 ITEF-363 65-23420 <$>
1965 ITEF-375 66-24790 <=$>

REPORTS OF THE INSTITUTE OF JAPANESE CHEMICAL FIBERS. KYOTO UNIVERSITY
SEE KYOTO DAIGAKU NIPPON KAGAKUSENI KENKYUSHO KOENSHU

ANNUAL REPORT (INSTITUTE FOR NUCLEAR STUDY, UNIVERSITY OF TOKYO)
SEE TOKYO DAIGAKU GENSHIKOKU KENKYUSHO ANNUAL REPORT

REPORT. INSTITUTE OF NUCLEAR RESEARCH. WARSAW POLISH ACADEMY OF SCIENCES
INR-588/9/R LA-TR-66-54 <*>

REPORTS OF THE INSTITUTE OF PHYSICAL AND CHEMICAL RESEARCH. JAPAN
SEE RIKAGAKU KENKYUSHO HOKOKU

REPORT OF THE INSTITUTE OF SCIENCE AND TECHNOLOGY. UNIVERSITY OF TOKYO
SEE TOKYO DAIGAKU RIKOGAKU KENKYUSHO HOKOKU

REPORT. INTERNATIONAL COMMISSION ON IRRIGATION AND DRAINAGE. INDIA
1966 NO 10 67-61145 <=$>
1966 NO 11 67-61144 <=$>
1966 NO 12 67-61007 <=>
1966 NO 24 67-61009 <=>
1966 NO 8 67-61015 <=>

REPORT. INTERNATIONAL NORTH PACIFIC FISHERIES COMMISSION
1958 N207 ENTIRE ISSUE 59-18361 <+*> 0

REPORTS OF THE JAPAN ATOMIC ENERGY RESEARCH INSTITUTE. TOKYO
JAERI-107I 66-11845 <*>
1960 JAERI-1012 AEC-TR-5079 <=>
1960 JAERI-5002(2) NP-TR-813 <*>
1960 JAERI-5002 P97- NP-TR-824 <*>
1960 JAERI-5002 P243 NP-TR-833 <*>
1961 JAERI-1015 ANL-TR-26 <*>
1962 JAERI-5006 ANL-TRANS-76 <*>
1962 JAERI-1018 NP-TR-1047 <*>
1962 JAERI-1020 NP-TR-1060 <*>
1962 JAERI-1018 63-27380 <$>
1962 JAERI-1020 63-27399 <$>
63-27400 <$>
1962 JAERI-4022 64-26318 <$>
1963 JAERI-1045 AEC-TR-5979 <*>
1963 JAERI 1048 NSJ-TR-10 <*>

ANNUAL REPORT OF THE JAPANESE ASSOCIATION FOR RADIATION RESEARCH ON POLYMERS
SEE NIHON HOSHASEN KOBUNSHI KENKYU KYOKAI NEMPO

REPORT. JOINT INSTITUTE FOR NUCLEAR RESEARCH. DUBNA. SSSR
P-351 UCRL-TRANS-604 <*=>
P-710 UCRL TRANS-702 <=*>
P-802 UCRL-TRANS-819(L) <=*>
P-845 AEC-TR-5115 <=*>
1957 P-110 AEC-TR-4205 <+*>
1957 P-28 R-2370 <*>
1957 P-41 R-4894 <*>
1959 P-335 NP-TR-334 <=*>
1959 P-344 UCRL-TR-508 <+*>
1959 P-346 61-10415 <+*>
1959 P-373 UCRL-TR-509 <+*>
1959 P-384 UCRL-TR-1041(L) <=$>
1960 P-470 UCRL TRANS-664(L) <*=>
1960 P-509 UCRL TRANS-566 <+*>
1961 P-759 UCRL-TRANS-759 <=*>
1961 P-765 62-10923 <=*>
62-26003 <=$>
1961 P-769 66-15384 <*>
1961 P-777 UCRL TRANS-798 <=*>
1961 P-787 UCRL-TRANS-843 <=*>
1961 P-841 UCRL-TRANS-818 <=*>
1961 P-848 UCRL-TRANS-810 <=*>
1961 P-849 UCRL TRANS-813 <=*>
1961 P-855 UCRL TRANS-814 <=*>
1962 P-1010 AEC-TR-5569 <=*>
1962 P-1070 UCRL TRANS-918 <=*$>
1962 P-1072 UCRL TRANS-919 <=*$>
1962 P-1075 UCRL TRANS-922 <=*$>
1962 P-1072 UCRL-TRANS-919 <=*$>
1962 P-1118 UCRL-TRANS-940 <=*>
1963 P-1174 UCRL-TRANS-973 <=*$>
1963 P-1277 AEC-TR-6103 <=*$>
1963 P-1299 AEC-TR-6102 <=*$>
1963 P-1343 64-19307 <=$>
1963 P-1357 P1-38 ANL-TRANS-5 <*>
1963 P-1383 P1-26 AEC-TR-6369 <=*$>
1963 P-1486 UCRL-TRANS-1080 <=$>
1964 P-1513 ANL-TRANS-32 <=$>
1964 P-1568 64-27258 <$>
1964 P-1634 P1-80 ANL-TRANS-83 <*>
1964 P-1689 65-29501 <$>
1964 P-1853 65-13980 <*>
1964 P-2037 65-13981 <*>
1964 P-2089 65-13978 <*>
1965 P-1949 AD-627 443 <=$>
1965 P-2037 66-11850 <*>
1965 P-2202 M.5801 <NLL>
65-17105 <*>
1965 P-2231 66-23810 <$>
1965 P-2495 66-61338 <=*$>

REPORT. K. FORTIFIKATSIONSFORVALTNINGEN, FORSKNINGSEKTIONEN. STOCKHOLM
1960 NO 109:15 AEC-TR-5026 <=*>
1961 NO 103:24 AEC-TR-5025 <=*>

REPORT. KERNREAKTOR BAU- UND BETRIEBS-GESELLSCHAFT. KERNFORSCHUNG ZENTRUM KARLSRUHE. WEST GERMANY
1964 KFK-225 66-14711 <*>

REPORT. LABORATORIE DE RECHERCHES SUR LA PHYSIQUE DES PLASMA. LAUSANNE
1963 LRP-5-63 65-18006 <*>

REPORT. LILIENTHAL-GESELLSCHAFT FUER LUFTFAHRTFORSCHUNG
NO 129 3132 <*>
3180 <*>
3214 <*>
3220 <*>
3221 <*>
1939 106 PT.1P.9-14 AL-267 <*>
1941 NO 135 65-13776 <*> P
1941 N135 P61-74 1755 <*>
1942 NO 156 65-13775 <*> P
1942 S13/1 P.40-68 AL-668 <*>

REPORT. METALLWERK PLANSEE GESELSCHAFT M,B.H., REUTTE, AUSTRIA
1951 NO 05/83/51 62-18312 <*>
1952 NO 05/94/52 62-18314 <*>

REPORT OF THE NANKAI REGIONAL FISHERIES RESEARCH LABORATORY
SEE NANKAI-KU SUISAN KENKYUSHO KENKYU HOKOKU

REPORT. OESTERREICHISCHE STUDIENGESELLSCHAFT FUER ATOMENERGIE. SEIBERSDORF REAKTOR ZENTRUM
1964 SGAE-CH-4 66-14443 <*>
1964 SGAE-CH-12 66-15366 <*>

REPORT OF THE OHARA INSTITUTE FOR AGRICULTURAL BIOLOGY
SEE NOGAKU KENKYU

REPORT. POLSKA AKADEMIS NOUK. INSTYTUT BADAN JADROWYCH
1960 175/111 NP-TR-649 <*>
1963 NO 440/IX AEC-TR-6545 <=> P

REPORT. RADIOTEKHNICHESKII INSTITUT. AKADEMIYA NAUK SSSR
1958 NT.150-03-58 63-10251 <*>
1960 NT.2460-34 63-23292 <=*$>
1960 NT.2460-33 63-23293 <=*$>
1960 NT.2460-35 63-23309 <=*$>
1961 NT.3461-60 63-26184 <=$>
65-61445 <=>
1962 NT.4162-63 63-23489 <=*$>

REPORTS OF THE RESEARCH INSTITUTE OF DENTAL MATERIALS
SEE SHIKA ZAIRYO KENKYUSHO HOKOKU

REPORTS OF THE RESEARCH LABORATORY, ASAHI GLASS COMPANY
SEE ASAHI GARASU KENKYU HOKOKU

REPORT. RUHRCHEMIE A.G. WEST GERMANY

1939 R-460 64-18869 <*>
1943 R-656 64-18868 <*>

REPORTS OF THE SCIENTIFIC POLICE RESEARCH INSTITUTE
SEE KAGAKU KEISATSU KENKYUJO HOKOKU

REPORT OF THE SCIENTIFIC RESEARCH INSTITUTE. TOKYO
SEE RIKAGAKU KENKYUSHO HOKOKU

REPORT OF SHIONOGI RESEARCH LABORATORY
SEE SHIONOGI KENKYUSHO NEMPO

REPORTS OF THE SHIP RESEARCH INSTITUTE. JAPAN
SEE SENPAKU GIJUTSU KENKYUJO HOKOKU

REPORT. SOCIETE BELGE POUR L'INDUSTRIE NUCLEAIRE BRUXELLS
1962 BN-6204-01 EURAEC-301 <=*$>
1962 BN-6204-08 EURAEC-307 <=*$>
1962 BN-6207-02 EURAEC-408 <=*$>
1962 BN-6207-03 EURAEC-410 <=*$>
1962 BN-6206-01 EURAEC-431 <=*$>
1962 BN-6210-01 EURAEC-490 <=*$>

REPORT. SOCIETE GRENOBLOISE D'ETUDE ET APPLICATION HYDRAULIQUES FRANCE
SOGREAH-R-7884 EURAEC-160 <=*>
1962 SOGREAH-R-8107 EURAEC-286 <=*>

REPORT. SOCIETE NATIONALE D'ETUDE ET DE CONSTRUCTION DE MOTEURS D'AVIATION. PARIS
1963 YTO-93 65-17048 <*>

ANNUAL REPORT OF THE SOCIETY OF PLANT PROTECTION OF NORTH JAPAN
SEE KITA-NIHON BYOGAICHU KENKYUKAI NENPO

REPORT. TNO INSTITUTE FOR STRUCTURAL MATERIALS AND BUILDING STRUCTURES
1959 BI-59-22 61-10123 <*>

ANNUAL REPORT OF TAKAMINE LABORATORY. JAPAN
SEE TAKAMINE KENKYUSHO NEMPO

ANNUAL REPORT OF THE TAKEDA RESEARCH LABORATORIES
SEE TAKEDA KENKYUSHO NEMPO

REPORTS OF THE TEXTILE RESEARCH INSTITUTE. JAPAN
SEE SEN-I KOGYO SHIKENSHO IHO

ANNUAL REPORT OF TOHOKU COLLEGE OF PHARMACY
SEE TOHOKU YAKKA DAIGAKU KIYO

REPORT. TOKYO METROPOLITAN ISOTOPE CENTRE. ANNUAL
1964 V3 P57-64 66-15334 <*>

REPORT. USTAV JADERNEHO VYZKUMU. CESKOSLOVENSKA AKADEMIE VED
1962 NO 666 CE.TRANS.3574 <CEGB>
1963 REZ-UJ-929163 AEC-TR-6377 <=$>

REPORT. ZENTRALE FUER WISSENSCHAFTLICHES BERICHTWESEN DER LUFTFAHRTFORSCHUNG DES GENERALLUFTZEUGMEISTERS. BERLIN-ADLERSHOF
FB1008 1765 <*>
FB1437 3107 <*>
M1278 1759 <*>
1935 FB430 1761 <*>
1940 FB1209 AL-620 <*>
1940 FB1320 2046 <*>
1942 FB1665 1760 <*>
1942 M690 AL-182 <*>
1943 M1023/3 AL-540 <*>
1943 M788 AL-388 <*>
1944 M1293 1758 <*>
1944 M94 AL-225 <*>

REPROGRAPHIE. WEST GERMANY
1963 V3 P199-204 34R78G <ATS>
1965 V4 N1 P9-12 1033 <TC>
1965 V4 N1 P13-14 1034 <TC>
1965 V5 N1 P13-15 49S83G <ATS>
1965 V5 N2 P31-34 40S84G <ATS>

REPULES. BUDAPEST
1963 N6 P16 64-11975 <=>
1963 N8 P14-15 AD-608 152 <=>
1965 V18 N5 P14-15 AD-639 347 <=$>

RESEARCH BULLETIN OF THE COLLEGE EXPERIMENTAL FORESTS. COLLEGE OF AGRICULTURE. HOKKAIDO UNIVERSITY
SEE HOKKAIDO DAIGAKU NOGAKUBU ENSHURINKENKYU HOKOKU

RESEARCH BULLETIN OF THE FACULTY OF AGRICULTURE, GIFU UNIVERSITY
SEE GIFU DAIGAKU NOGAKUBU KENKYU HOKOKU

RESEARCH BULLETIN OF THE FACULTY OF EDUCATION. OITA UNIVERSITY. NATURAL SCIENCE
SEE OITA DAIGAKU KYOIKUGAKUBU KENKYU KIYO, SHIGENKAKAGU

RESEARCH BULLETIN, GIFU COLLEGE OF AGRICULTURE
SEE GIFU NORIN SEMMON GAKKO GAKUJUTSU HOKOKU

RESEARCH BULLETIN. HOKKAIDO NATIONAL AGRICULTURAL EXPERIMENT STATION
SEE HOKKAIDO NOGYO SHIKENJO IHO

RESEARCH FILM
1957 V2 N5 P233-242 58-377 <*>

RESEARCH JOURNAL OF THE HINDI SCIENCE ACADEMY
1961 V4 P149-160 UCRL TRANS-938(L) <*>
1961 V4 P173-178 AEC-TR-5552 <*>

RESEARCH REPORTS OF THE FACULTY OF TECHNOLOGY OF CHIBA UNIVERSITY. JAPAN
SEE CHIBA DAIGAKU KOGAKUBU KENKYU HOKOKU

RESEARCH REPORT OF THE NAGOYA INDUSTRIAL SCIENCE RESEARCH INSTITUTE
SEE NAGOYA SANGYO KAGAKU KENKYU-SHU KENKYU HOKOKU

RESEARCHES OF THE ELECTROTECHNICAL LABORATORY. JAPAN
SEE DENKI SHIKEN-SHO KENKYU HOKOKU

RESENHA CLINICO-CIENTIFICA
1955 V24 N7 P179-183 5062-A <K-H>
1956 V25 N2 P35-40 5538-B <K-H>

RESIN FINISHING AND APPLICATIONS
SEE JUSHI KOKO

REUMATISIMO. MILAN
1960 V12 P367-374 SUP 1 61-20906 <*>

REUMATOLOGIA POLSKA
1964 V2 N2 P97-100 66-10980 <*>

REVIEW OF THE KOBE UNIVERSITY OF MERCANTILE MARINE PART 2. NAVIGATION, MARINE ENGINEERING AND SCIENTIFIC SECTION
SEE KOBE SHOSEN DAIGAKU KIYO. DAI-2-RUI. KOBAI, KIKAN, RIGAKU-HEN

REVIEW OF PHYSICAL CHEMISTRY OF JAPAN
1957 V27 N2 P59-67 44K24J-E <ATS>

REVIEW OF POLAROGRAPHY. JAPAN
SEE PORAROGUAFII

REVISTA DE LA ACADEMIA DE CIENCIAS EXACTAS, FISICO-QUIMICAS Y NATURALES DE ZARAGOZA
1951 S2 V6 N2 P15-23 64-18513 <*>

1954 S2 V9 P117-123   64-16706 <*>

REVISTA DE LA ADMINISTRACION NACIONAL DEL AGUA
1946 V108 P429-435   I-22 <*>

REVISTA ARGENTINA DE AGRONOMIA
1946 V13 P256-276   I-588 <*>
1952 V19 P171-178   1508 <*>
1956 V23 P53-70   CSIRO-3455 <CSIR>

REVISTA ARGENTINA DE DERMATO-SIFILOGIA
1942 V26 P1030-1042   593 <TC>

REVISTA ARGENTINA DE QUIMICA E INDUSTRIA
1937 V6 P126-132   63-10496 <*> 0
1959 V10 N2 P107-108   64-10374 <*> 0

REVISTA ARGENTINO-NORTEAMERICANA DE CIENCIAS MEDICAS
1943 V1 N2 P145-161   59-20715 <*>

REVISTA DE LA ASOCIACION BIOQUIMICA ARGENTINA
1941 V8 P15-18   60-10859 <*>
1941 V8 N23 P3-18   60-10858 <*> 0
1945 V12 P91-92   2333 <*>
1953 V18 P259-273   2998-C <K-H>
1956 V21 P207-214   C-2557 <NRC>

REVISTA DE LA ASOCIACION MEDICA ARGENTINA
1949 V63 P647-648   2559 <*>
1955 V69 P65-72   64-20028 <*> 0
1955 V69 N789 P1-5   64-20004 <*> 0
1957 V71 P57-64   60-14194 <*> 0
1957 V71 P161-163   8378 <K-H>
1957 V71 N7 P208-212   64-14584 <*>
   8619-F <K-H>
1959 V73 N4 P125-130   64-10393 <*>
1960 V74 P82-83   62-10285 <*>
1960 V74 N9 P531-533   05N51SP <ATS>
   62-14352 <*>
1965 V79 N6 P271-274   66-12610 <*>

REVISTA DA ASSOCIACAO MEDICA BRASILEIRA
1963 V9 N5 P153-156   65-14789 <*>

REVISTA DE BIOLOGIA MARINA
1951 V4 N1/3 P184-185   59-18323 <+*>
1951 V4 N1/3 P239-243   59-19347 <+*>

REVISTA BRASILEIRA DE BIOLOGIA
1941 V1 N4 P431-434   59-20994 <*> 0
1948 V8 N1 P57-60   2927 <*>
   62-16035 <*>
1951 V11 P29-31   T-2033 <INSD>

REVISTA BRASILEIRA DE CHIMICA (SCIENCIA E INDUSTRIA)
1938 V6 P72-73   63-10118 <*>
1945 V20 P283-286   1540 <K-H>

REVISTA BRASILEIRA DE CIRURGIA
1959 V38 P153-159   60-17562 <=*>
1960 V39 N6 P518-522   63-16575 <*>
1963 V46 P329-348   66-11535 <*>

REVISTA BRASILEIRA DE ENTOMOLOGIA
1954 V1 N205 P207-211   57-2053 <*>

REVISTA BRASILEIRA DE GASTROENTEROLOGIA
1950 V2 N6 P595-608   66-10994 <*> 0
1957 V9 P5-18   61-00849 <*>

REVISTA BRASILEIRA DE MALARIOLOGIA E DOENCAS TROPICAIS
1956 V8 N2 P397-404   59-15600 <*>
1956 V8 N3 P527-534   59-15599 <*>
1956 V8 N4 P583-587   59-15602 <*> 0

REVISTA BRASILEIRA DE MEDICINA
1946 V3 N1 P64-65   2604 <*>
1950 V7 N12 P787-788   1259 <*>
1954 V11 N9 P633-635   3449-A <KH>
1954 V11 N12 P831-833   5225-A <K-H>
1955 V12 N12 P879-880   5796-H <K-H>
1956 V13 N5 P380-382   5691-C <K-H>
1956 V13 N7 P553-555   6049-A <K-H>
1956 V13 N9 P687-688   5920-C <K-H>
1957 V14 N4 P252-253   6950 <K-H>
1958 V15 N8 P547-550   59-15682 <*>
1960 V17 N10 P899-901   11006-B <KH>

REVISTA BRASILEIRA DE OTO-RINO-LARINGOLOGIA
1955 V23 P159-200   65-13871 <*>

REVISTA BRASILEIRA DE TUBERCULOSE E DOENCAS TORACICAS
1949 V17 N123 P229-252   2796 <*>
1949 V17 N125 P451-464   2788 <*>
1950 V18 N128 P157-172   2906 <*>
1951 V19 N133 P39-111   2524 <*>
1952 V20 N141 P165-170   2915 <*>
1952 V20 N141 P181-188   2626 <*>
1952 V20 N143 P583-592   2502 <*>
1957 V25 P7-14   59-14872 <=*> C

REVISTA CAILOR FERATE. BUCHAREST
1965 V4 N1 P82   65-33879 <=$>
1965 V4 N1 P87-128   65-33879 <=$>

REVISTA DE CALCULO AUTOMATICO Y CIBERNETICO
1957 V6 N16 P1-7   SCL-T-500 <*>

REVISTA DEL CENTRO ESTUDIANTES DE FARMACIA Y BIOQUIMICA
1943 V33 P116-144   60-18158 <*>

REVISTA DEL CENTRO DE QUIMÍCOS INDUSTRIALES
1948 V3 N12 P11-22   61-10297 <*>

REVISTA CERES
1948 V5 N29 P313-317   57-2950 <*>

REVISTA CHILENA DE HIGIENE Y MEDICINA PREVENTIVA
1942 V5 P39-45   57-1499 <*>

REVISTA DE CHIMICA PURA E APPLICADA
1950 S4 V1 P24-41   UCRL TRANS-558(L) <*>

REVISTA DE CHIMIE
1954 V3 P537-545   64-16761 <*>
1955 V6 P125-132   57-2058 <*>
1956 V7 N1 P12-30   93M46RU <ATS>
1956 V7 N3 P159-162   48K24RU <ATS>
1956 V7 N11 P657-658   CSIRO-3788 <CSIR>
1957 V8 P93-99   65-11440 <*>
1957 V8 N1 P52-54   62-10673 <*>
1957 V8 N6 P406-409   65-11450 <*>
1957 V8 N10 P625-633   65-10063 <*> 0
1957 V8 N11 P688-691   44N49RU <ATS>
1958 V9 N2 P84-88   CSIR-TR.567 <CSSA>
1958 V9 N3 P150   60-10816 <*>
1958 V9 N7/8 P361-366   63-16140 <*> 0
1958 V9 N7/8 P370-373   65-10038 <*>
1958 V9 N7/8 P415-419   64-30681 <*>
1958 V9 N7/8 P447-450   62-14503 <*>
1959 V10 N2 P103-104   65-10960 <*>
1959 V10 N3 P149-150   65-10217 <*>
1959 V10 N6 P320-326   64-30146 <*> 0
1959 V10 N10 P562-566   35R77RU <ATS>
1960 V11 N1 P49   65-14223 <*>
1960 V11 N2 P107-108   65-10959 <*> 0
1960 V11 N3 P144-147   62-16489 <*>
1960 V11 N3 P151-155   65-11566 <*>
1960 V11 N4 P204-206   68P65RV <ATS>
1960 V11 N10 P569-575   63-18586 <*>
1960 V11 N12 P688-690   62-19336 <=*> P
1961 N12 P706-   ICE V2 N3 P357-359 <ICE>
1961 V12 P275-281   65-10266 <*>
1961 V12 N2 P96-97   61N53RU <ATS>
1961 V12 N5 P275-281   ICE V1 N1 P110-116 <ICE>

34N57RU <ATS>
1961 V12 N7 P387- ICE V2 N1 P68-72 <ICE>
1961 V12 N9 P544- ICE V2 N2 P156-160 <ICE>
1961 V12 N10 P614-615 37S82RU <ATS>
1961 V12 N11 P649-651 65-14202 <*> 0
1962 N1 P9- ICE V2 N4 P464-471 <ICE>
1962 N1 P16- ICE V2 N4 P472-475 <ICE>
1962 N3 P141- ICE V3 N1 P19-24 <ICE>
1962 N4 P205- ICE V3 N1 P80-91 <ICE>
1962 N8 P465- ICE V3 N2 P210-215 <ICE>
1962 V13 P486-489 66-10490 <*>
1962 V13 P675-678 RAPRA-1269 <RAP>
1962 V13 N3 P153-157 64-18121 <*> 0
1962 V13 N4 P231-233 64-11664 <=>
1962 V13 N6 P345-350 66-10609 <*>
1962 V13 N7 P383-384 63-13509 <=*>
1962 V13 N7/8 P471-473 53Q68RU <ATS>
1963 N4 P217- ICE V3 N3 P445-452 <ICE>
1963 N6 P341-345 ICE V4 N1 P26-31 <ICE>
1963 V14 P385-390 RAPRA-1256 <RAP>
1963 V14 N6 P324-327 4202 <BISI>
1964 N1 P18-22 ICE V4 N3 P506-510 <ICE>
1964 N5 P265-273 ICE V5 N1 P138-151 <ICE>
1964 N7 P381-385 ICE V5 N1 P102-108 <ICE>
1964 N7 P404-408 ICE V5 N2 P257-262 <ICE>
1964 N7 P409-411 ICE V5 N2 P276-280 <ICE>
1964 V15 N1 P18-22 ICE V4 N3 P506-510 <ICE>
1964 V15 N5 P265-273 ICE V5 N1 P.138-151 <ICE>
1964 V15 N7 P381-385 ICE V5 N1 P102-108 <ICE>
1964 V15 N8 P443-458 64-51478 <=>
1964 V15 N8 P489-496 64-51478 <=>
1964 V15 N9 P546-550 2134 <TC>
1965 N6 P322-325 ICE V6 N3 P417-421 <ICE>
1965 V16 N3 P134-138 66-12205 <*>
1965 V16 N9 P428-433 66-13592 <*>
1965 V16 N9 P443-446 65-33964 <=$>
1966 V17 N1 P3-7 61-31777 <=$>
1966 V17 N1 P48-49 66-13591 <*>
1966 V17 N2 P69-72 66-13961 <=$>
1966 V17 N2 P94-97 66-13778 <*>

REVISTA DE CIENCIA APLICADA
1951 P403-407 T-1393 <INSD>
1953 V7 N31 P152-160 T-2420 <INSD>
1962 V16 N1 P36-41 64-16963 <*>

REVISTA DE CIENCIA VETERINARIA
1956 V51 N358 P191-229 CSIRO-3535 <CSIR>

REVISTA CLINICA ESPANOLA
1949 V34 N5 P315-325 58-375 <*>
1954 V53 N6 P360-363 58-371 <*>
1957 V64 N3 P196-199 62-00910 <*>
1957 V67 N4 P225-257 8500-B <K-H>
1957 V67 N4 P255-257 63-16578 <*>
1959 V75 P83-88 65-17032 <*>
1960 V77 P252-256 65-00126 <*>
1960 V78 N5 P268-275 61-00403 <*>
1962 V86 N5 P337-347 64-18651 <*>
1963 V91 N1 P11-17 64-20919 <*>

REVISTA COLOMBIANA DE OBSTETRICIA Y GINECOLOGIA
1964 V15 N5 P413-419 65-13729 <*>

REVISTA CONSTRUCTIILOR SI A MATERIALELOR DE CONSTUCTII
1964 N10 P501-509 65-30744 <=$>

REVISTA CUBANA DE CIRUGIA
1964 V12 N8 P419-450 64-51338 <=>
1964 V12 N8 P460 64-51338 <=>

REVISTA CUBANA DE LABORATORIO CLINICO
1948 V2 N1 P165-179 2501 <*>

REVISTA CUBANA DE TUBERCULOSIS
1944 V9 N7/9 P158-160 2790 <*>
1946 V10 N4 P361-376 2778 <*>
1946 V10 N11/2 P755-760 2789 <*>
1948 V12/3 P95-107 2779 <*>
1950 V14 N7/8 P676-684 2539 <*>
1950 S5 V14 N5/6 P495-500 2584 <*>
1951 V15 N3 P197-203 2580 <*>
1951 V15 N12 P1117-1118 2913 <*>
1951 V15 N12 P1120-1150 2583 <*>
1951 V15 N12 P1150-1154 2635 <*>
1951 V15 N12 P1175-1187 2912 <*>
1951 V15 N12 P1193-1199 2590 <*>
1951 V15 N4/5 P432-434 2565 <*>
1951 V15 N4/5 P459-461 2793 <*>
1951 V15/6 P90-92 2801 <*>
1952 V16 N7/8 P695-700 2544 <*>
1952 S5 V16 N12 P1149-1151 2594 <*>

REVISTA ESPANOLA DE LAS ENFERMEDADES DEL APARATO DIGESTIVO Y DE LA NUTRICION
1959 V18 P1671-1674 61-00412 <*>

REVISTA ESPANOLA DE FISIOLOGIA
1951 V7 P221-224 60-13218 <*+>
65-00357 <*>
1956 V12 N1 P17-20 57-2607 <*>
1960 V16 P163-173 SUP 3 64-16841 <*> 0
1965 V21 N3 P85-90 66-11834 <*>

REVISTA ESPANOLA DE MEDICINA Y CIRUGIA
1927 V10 P210-212 57-2982 <*>

REVISTA ESPANOLA DE MEDICINA Y CIRUGIA DE GUERRA
1941 V6 P113-122 57-6100 <*>

REVISTA ESPANOLA DE OBSTETRICIA Y GINECOLOGIA. VALENCIA
1959 V18 P13-24 62-01223 <*>
1960 V19 P230-248 10860-C <K-H>
62-00311 <*>
64-14577 <*>
1963 V22 P510-512 65-11317 <*>

REVISTA ESPANOLA DE OTO-NEURO-OFTALMOLOGIA Y NEUROCIRUGIA
1959 V18 P381-385 62-00307 <*>
1961 V20 P329-332 65-10970 <*>

REVISTA ESPANOLA DE PEDIATRIA
1958 V14 N82 P545-552 59-1421 <+*> 0
1960 V16 N95 P695-701 39N49SP <ATS>
1963 V19 N114 P765-773 65-14548 <*>

REVISTA DE LA FACULTAD DE AGRONOMIA, UNIVERSIDAD NACIONAL DE LA PLATA
1940 V25 P21-54 I-121 <*>

REVISTA DE LA FACULTAD DE AGRONOMIA Y VETERINARIA, UNIVERSIDAD DE BUENOS AIRES
1948 V12 P51-67 60-14280 <*> 0

REVISTA DE LA FACULTAD DE CIENCIAS QUIMICA Y FARMACIA, UNIVERSIDAD NACIONAL DE LA PLATA
1945 V20 P181-191 66-10597 <*> 0
1946 V21 P47-54 59-20964 <*>
1950 V25 P103-110 3169-B <K-H>
1957 V30 P93-99 63-20783 <*>

REVISTA DE LA FACULTAD DE HUMANIDADES Y CIENCIAS, UNIVERSIDAD DE LA REPUBLICA, URUGUAY
1958 N3 P65-73 60-18519 <*>

REVISTA DE LA FACULTAD DE INGENIERIA QUIMICA UNIVERSIDAD NACIONAL DEL LITORAL. SANTA FE, ARGENTINA
1939 V8 P129-131 I-530 <*>

REVISTA DE LA FACULTAD DE MEDICINA, UNIVERSIDAD NACIONAL DE COLOMBIA
1942 V11 P183-194 57-1650 <*>
1960 V28 P197-203 63-16525 <*>

REVISTA DE LA FACULTAD NACIONAL DE AGRONOMIA.

MEDELLIN, COLOMBIA
1944 V5 N21 P267-281 57-2916 <*>

REVISTA FARMACEUTICA
1952 P60-72 58-1548 <*>

REVISTA FONOAUDIOLOGICA
1961 V7 N2 P128-164 64-18595 <*>

REVISTA DE GEODEZIE SI ORGANIZAREA TERITORULUI
1961 V5 N3 P33-46 62-19658 <=*> P

REVISTA GEOGRAFICA. CUBA
1960 V30 N1 P5-18 62-33277 <=*>
1960 V30 N2 P33-48 62-33277 <=*>

REVISTA DE GINECOLOGIA E D'OBSTETRICIA
1954 V48 P681-692 II-795 <*>

REVISTA DE HIGIENE
1942 V23 N4 P64-87 57-1519 <*>

REVISTA DO HOSPITAL DAS CLINICAS DE FACULDADE DE MEDICINA DA UNIVERSIDADE DE SAO PAULC
1960 V15 N3 P201-206 61-10678 <*>

REVISTA DEL HOSPITAL DE NINOS
1962 V4 N14 P137-143 63-20895 <*>
1962 V4 N15 P230-231 63-20894 <*>

REVISTA IBERICA DE ENDOCRINOLOGIA
1956 V3 P565-584 57-1255 <*>
1963 V10 P149-156 64-18429 <*>
1963 V10 P311-328 64-18066 <*> 0

REVISTA IBERICA DE PARASITOLOGIA
1950 V10 N2 P187-203 60-15255 <=*>
1959 V19 N4 P417-425 63-01771 <*>
1960 V20 N1 P23-30 63-01116 <*>
1960 V20 N1 P39-52 63-01060 <*>
1961 V21 N2 P193-196 62-00832 <*>

REVISTA INDUSTRIAL Y AGRICOLA DE TUCUMAN
1940 N30 P221-226 1883 <K-H> 0

REVISTA DE INGENIERIA QUIMICA. CHILE
1943 V2 N2 P17-28 63-14551 <*>

REVISTA DO INSTITUTO ADOLFO LUTZ
1952 V2 P141-158 AL-956 <*>

REVISTA DEL INSTITUTO MALBRAN
1954 V16 N3 P169-172 57-239 <*>

REVISTA DEL INSTITUTO DE SALUBRIDAD Y ENFERMEDADES TROPICALES
1942 V3 P117-130 57-1538 <*>
1950 V11 N2/4 P167-173 1894 <*>
1956 V16 N4 P1-5 61-20054 <*> 0

REVISTA DE INVESTIGACIONES AGRICOLAS
1948 V2 N2 P93-96 60-18488 <*>
1951 N5 P288-294 64-10698 <*>
1953 V7 P131-145 61-10557 <*> 0
1956 V10 N1 P5-34 59-10257 <*>
1956 V10 N4 P349-372 60-17313 <*+>

REVISTA KUBA DE MEDICINA TROPICAL Y PARASITOLOGIA
1950 V6 P30-32 57-1885 <*>

REVISTA LATINAMERICAN DE SIDERUGIA
1965 V5A P25-32 4477 <BISI>

REVISTA MEDICA. MEXICO
1952 V32 N654 P278-282 58-95 <*>

REVISTA MEDICA DE CHILE
1942 V70 P679-685 57-1655 <*>
1957 V85 N4 P193-195 62-20425 <*>
1961 V89 P366-368 65-13361 <*>
1961 V89 P840-843 63-10178 <*>
1961 V89 P1082-1084 63-00062 <*>
1961 V89 P1085-1086 63-00160 <*>
1961 V89 N10 P767-771 62-20426 <*>
1964 V92 N3 P183-186 65-17267 <*>

REVISTA MEDICA DE COSTA RICA
1962 V19 N334 P101-108 63-10518 <*> 0

REVISTA MEDICA CUBANA
1958 V69 N6 P248-262 60-13142 <*=>
1959 V70 N9 P441-460 64-10606 <*>

REVISTA MEDICALA
1962 V8 P251-253 65-12780 <*>

REVISTA DE MEDICINA EXPERIMENTAL
1960 V1 N2 P77-83 65-17304 <*>

REVISTA DE MEDICINA MILITAR
1940 V29 P178- 57-1599 <*>

REVISTA DE MEDICINA VETERINARIA Y PARASITOLOGIA. CARACAS
1953 V12 P63-89 I-796 <*>
1953 V12 N1/4 P3-62 I-795 <*>

REVISTA DE MEDICINA VETERINARIA Y DE ZOOTECNIA
1953 V5 P1-8 57-1811 <*>

REVISTA MEDICO-CIRURGICA DO BRAZIL
1944 V52 P119-130 <K-H>
1945 P171-179 2610 <*>

REVISTA MEXICANA DE CIENCIAS MEDICAS Y BIOLOGICAS
1960 S2 V1 P1-6 62-01055 <*>
1961 S2 V2 P1-32 62-01055 <*>

REVISTA MEXICANA DE CIRUGIA, GINECOLOGIA Y CANCER
1939 V7 P375-385 57-1652 <*>

REVISTA MEXICANA DE FISICA
1954 V3 P107-114 UCRL TRANS-596(L) <*>

REVISTA MEXICANA DE TUBERCULOSIS Y ENFERMEDADES DEL APARATO RESPIRATORIC
1944 V6 N29 P31-46 2582 <*>
1945 V7 N39 P377-389 2794 <*>
1956 V17 N6 P522-526 65-12006 <*>

REVISTA MILITAR, LISBCN
1957 V9 N10 P565-595 59-12010 <+*>
1957 V9 N8/9 P469-502 59-12011 <+*> 0

REVISTA DE MINAS E HIDROCARBUROS
1951 V2 N3 P165-184 59-10982 <*>

REVISTA MINELOR
1956 V7 N2 P77-78 8K26RU <ATS>
1957 V8 P392-394 63-16160 <*>
1960 V11 N1 P16-20 AEC-TR-4366 <*>
1960 V11 N3 P93-100 AEC-TR-4367 <*>
1966 V17 N8 P335-345 67-30279 <=>
1966 V17 N8 P351-367 67-30279 <=>

REVISTA DEL MUSEO DE LA PLATA. SECCION BOTANICA
1937 SNS V1 P211-250 I-470 <*>

REVISTA NACIONAL DE AERONAUTICA
1958 P33 59-16294 <+*> C
1958 V18 N200 P46-52 59-16303 <+*> C

REVISTA DE OBRAS PUBLICAS. MADRID
1952 V100 N2845 P173-179 T-1385 <INSD>

REVISTA DE OBSTETRICIA Y GINECOLOGIA DE VENEZUELA
1954 V14 P28-41 2202 <*>

REVISTA ODONTOLOGICA. BUENOS AIRES
1945 V33 N3 P87-105 64-00194 <*>

REVISTA OTO-LARINGOLOGICA DE SAO PAULC
1938 V6 P437-442 61-14578 <*>

REVISTA DE OTORINOLARINGOLOGIA DEL LITORAL
1941 V1 N1 P65-75 66-13317 <O>

REVISTA PADURILOR
1961 V76 N5 P282-284 65-23230 <$>
1963 V78 N8 P431-434 63-41021 <=>
1965 N7 P349-350 65-32400 <=$>
1965 N7 P353-358 63-32400 <=$>

REVISTA DE PARASITOLOGIA, CLINICA Y LABORATORIO
1935 P117-120 T1911 <INSD>

REVISTA PAULISTA DE MEDICINA
1953 V43 N4 P291-304 1968 <*>

REVISTA DE PEDAGOGIE
1963 V12 N10 P5-14 64-21278 <=>

REVISTA DE PLASTICOS
1955 V6 P4-9 57-1329 <*>
1955 V6 N33 P140-146 62-25669 <=*> O
1960 V11 N66 P409-414 AEC-TR-5050 <*>

REVISTA DE PLASTICOS MODERNOS
1963 V14 P49-52 SC-T-64-45 <*>

REVISTA PORTUGUESA DE ZOOLOGIA E BIOLCGIA GERAL
1959 V2 P117-152 66-12000 <*>

REVISTA DE PSICOANALISIS
1957 V14 N4 P368-374 63-16615 <*>
8619-C <K-H>

REVISTA DE PSICOLOGIA GENERAL Y APLICADA
1951 V6 N19 P537-562 64-10035 <*>
1955 V10 N34 P366-380 65-12572 <*>

REVISTA DE PSIHOLOGIE. BUCURESTI
1963 V9 N4 P519-541 64-21873 <=>

REVISTA DE QUIMICA INDUSTRIAL. RIO DE JANEIRO
1946 V15 P340-344 AL-277 <*>

REVISTA QUIMICO-FARMACEUTICA
1957 V22 P19-30 58-552 <*>
1957 V22 P33-39 58-554 <*>

REVISTA DE LA R. ACADEMIA DE CIENCIAS EXACTAS, FISICAS Y NATURALES DE MADRID
1954 V48 P53-63 58-1229 <*>
1956 V50 N1 P135-140 61-20550 <*>

REVISTA DE SANIDAD Y ASISTENCIA SOCIAL
1942 P745-748 57-1531 <*>
1946 V11 P387-390 2455B <K-H>
1955 V20 P339-364 65-00200 <*> O
1955 V20 N5/6 P339-364 59-14580 <=*> O

REVISTA DE LA SOCIEDAD ARGENTINA DE BIOLOGIA
1947 V23 P219-225 C-2493 <NRC>
57-3482 <*>
1949 V25 N3/4 P91-112 59-20995 <*>
1952 V28 P219-224 3036-G <K-H>
1954 V30 P57-61 57-238 <*>
1954 V30 N6/8 P188-193 57-73 <*>
1955 V31 N7/8 P222-241 57-119 <*>
1962 V38 P391-399 66-12607 <*>

REVISTA DE LA SOCIENDAD MEXICANA DE HISTORIA NATURAL
1947 V8 P29-46 I-952 <*>
1953 V14 N1/4 P23-33 57-3197 <*>

REVISTA. SOCIEDAD QUIMICA DE MEXICO
1957 V1 P33-38 <CP>
1963 V7 N5 P191-198 64-18987 <*> O

REVISTA DE STATISTICA
1964 V13 N1 P25-31 64-31175 <=>
1964 V13 N9 P33-39 64-51746 <=$>
1966 V15 N1 P29-34 66-31627 <=$>
1966 V15 N9 P32-41 66-35669 <=>

REVISTA STIINTELOR MEDICALE
1934 V23 P1539-1549 60-19793 <*>

REVISTA SUDAMERICANA DE ENDOCRINOLOGIA, IMMUNOLOGIA Y QUIMIOTERAPIA
1935 V20 P45-46 60-10025 <*>

REVISTA TECNICA IEM
1958 V3 N22 P19-21 SCL-T-248 <*>

REVISTA TELEFONICA INTERNACIONAL
1922 V5 P7-9 57-2194 <*>

REVISTA TELEGRAFICA ELECTRONICA
1937 V25 P247-249 57-2132 <*>
1941 V29 P27-30 57-1077 <*>

REVISTA TRANSPORTURILOR
1961 V8 N3 P121-127 62-19577 <=*>
1963 V10 N9 P416-427 63-41295 <=>

REVISTA TRIMESTRAL MICROGRAFICA
1896 V1 P131-167 57-1083 <*>

REVISTA DE TUBERCULOSIS. HABANA
1939 V5 N10 P1234-1240 2511 <*>

REVISTA DE TUBERCULOSIS DEL URUGUAY
1945 V13 N1 P1-184 2775 <*>

REVISTA UNIVERSIDAD CENTRAL FACULTAD DE AGRONOMIA, CARACAS
1954 V1 P219-222 3180-A <K-H>

REVISTA DE LA UNIVERSIDAD INDUSTRIAL DE SANTANDER
1962 V4 N1 P43-48 <JTBC>
1964 V6 P70-78 65-10454 <*>

REVISTA DE LA UNIVERSIDAD NACIONAL DE CORDOBA
1944 V31 P1706-1709 61-20948 <*>
1944 V31 P1710-1714 61-20947 <*>

REVISTA YPF. BUENOS AIRES
SEE BOLETIN DE INFORMACIONES PETROLERAS. BUENOS AIRES

REVUE DES ACCIDENTS DU TRAVAIL ET DE MALADIES PROFESSIONELLES
1954 V47 P163-176 II-257 <*>
1958 V52 N3 P221-231 1517 <BISI>

REVUE D'ACOUSTIQUE
1935 V4 N3 P57-85 57-872 <*>

REVUE AERONAUTIQUE INTERNATIONALE
1936 V6 N19 P68-74 63-18828 <*> O

REVUE DE L'ALUMINUM ET DE SES APPLICATIONS
1938 P1189-1190 II-180 <*>
1938 V15 P1196-1214 I-48 <*>
1947 V24 P259-264 62-18094 <*>
1948 V25 P335-338 62-18237 <*> O
1949 V26 P315-319 57-100 <*>
1950 N169 P314-315 2322 <*>
1950 V27 N166 P175-182 62-18354 <*> O
1951 V28 P230-232 57-214 <*>
1951 V28 N182 P393-395 62-18439 <*> O
1952 V29 N194 P431-437 2158 <*>
1953 V30 P299-306 64-20347 <*> O
1953 V30 P347-374 1903 <*>
1953 V30 N195 P5-11 2158 <*>
1953 V30 N196 P45-54 2158 <*>
1953 V30 N197 P87-95 2158 <*>
1953 V30 N202 P307-318 1363 <*>

1954 V31 N208 P96-97 57-1698 <*>
1954 V31 N210 P184-188 II-62 <*>
1955 N227 P1117- 1200 <*>
1955 V32 P701-712 1332 <*>
1955 V32 N220 P367-372 66-10155 <*> O
1955 V32 N223 P713-715 II-698 <*>
1955 V32 N226 P1011-1014 II-925 <*>
1955 V32 N227 P1150-1154 1324 <*>
1956 N236 P923-930 57-2292 <*>
1956 V33 P233 615-619 <*>
1956 V33 P1167-1170 3984 <HB>
1956 V33 N233 P615-619 C2348 <NRC>
1957 N243 P525-530 57-3563 <*>
1958 V35 N256 P751-757 63-10280 <*> O
1959 N269 P1051-1055 60-15999 <*+>
1964 N318 P275-277 8905 <IICH>

REVUE. ASSOCIATION FRANCAISE DES AMIS DES CHEMINS DE FER
1959 P164-166 TS 1653 <BISI>
62-16853 <*>
1959 P176-186 TS 1654 <BISI>
62-16852 <*>

REVUE DE L'ATHEROSCLEROSE
1959 V1 P239-255 10402-2 <K-H>
63-16947 <*>

REVUE BELGE DES MATIERES PLASTIQUES
1963 N4 P309-317 66-10604 <*> O
1963 V4 N1 P1-19 65-14782 <*> O

REVUE BELGE DE PATHOLOGIE ET DE MEDECINE EXPERIMENTALE
1952 V22 N2 P113-125 61-00387 <*>
1956 V25 N6 P491-497 58-1600 <*>

REVUE DU BOIS ET DE SES APPLICATIONS
1956 V10 N3 P5-10 C-2164 <NRC>
1313 <*>
1956 V11 N1 P26-31 C-2240 <NRC>
1789 <*>
1957 P26-28 C-2326 <NRC>
57-1860 <*>
1957 P38-39 C-2308 <NRC>
57-1859 <*>
1958 V13 N1 P13-18 58-1736 <*>
1959 V14 N4 P47-51 TRANS-131 <NRCC> O

REVUE DE BOTANIQUE APPLIQUEE ET D'AGRICULTURE
1932 V12 N128 P261-282 57-3060 <*>
1932 V12 N128 P347-358 57-3060 <*>

REVUE DE LA BRASSERIE
1952 V2 P171-178 AL-533 <*>

REVUE CANADIENNE DE BIOLOGIE
1942 V1 N4 P354-365 63-18099 <*> O
1943 V2 N2 P168-243 57-1530 <*>
59-15538 <*> OP
1948 N7 P254-292 57-2711 <*>
1952 V11 N2 P180-184 64-10325 <*>
1952 V11 N2 P185-189 64-10326 <*>
1952 V11 N2 P190-194 64-10324 <*>
1957 V16 N4 P434-444 59-18958 <+*> O
1963 V22 N1 P107-111 63-20945 <*> O

REVUE DE CHIMIE. ACADEMIA REPUBLICII POPULARE ROMINE.
1954 V2 P73-95 UCRL TRANS-894(L) <*>

REVUE DE CHIMIE. BUCAREST
1959 V4 N2 P133-139 62-19335 <*=>
1962 V7 N2 P921-928 65-14731 <*>

REVUE DE CHIMIE INDUSTRIELLE ET LE MONITEUR SCIENTIFIQUE QUESNEVILLE REUNIS
1927 N36 P51-58 9874-D <K-H>
1927 V36 P12-15 9793-B <K-H>

REVUE DE CHIRURGIE ORTHOPEDIQUE ET REPARATRICE DE L'APPAREIL MOTEUR
1951 V37 P395-407 59-10707 <*> O
1955 V41 P763-766 59-18142 <+*>
1957 V43 N1 P29-37 66-10861 <*> O

REVUE CLINIQUE D'UROLOGIE
1912 V1 P601-613 57-3259 <*>

REVUE DU CORPS DE SANTE MILITAIRE
1952 V8 N1 P27-39 1115 <*>

REVUE DE DE'FENSE NATIONALE
1958 V14 P120-126 59-16791 <+*>
1958 V14 P1567-1575 59-18314 <+*> O

REVUE DE LA DOCUMENTATION
1962 V29 N1 P8-13 62-16755 <*>

REVUE DE DOCUMENTATION MILITAIRE
1958 N102 P78-94 59-14894 <=*>

REVUE DOMINICAINE
1955 V61 N1 P230-237 60-18654 <*>

REVUE DES EAUX ET FORETS
1930 V68 P92-100 57-1239 <*>
1932 S7 V70 P203-209 57-1238 <*>
1932 S7 V70 P471-481 57-1236 <*>
1932 S7 V70 P561-572 57-1236 <*>
1932 S7 V70 P647-654 57-1236 <*>
1937 S7 V75 N7 P681-685 I-744 <*>

REVUE D'ECONOMIE POLITIQUE
1957 N67 P208-215 59-20158 <*>
1958 V68 N1 P256-263 60-18012 <*>
1958 V68 N6 P1026-1035 60-18639 <*>
1962 V72 N6 P877-890 65-12107 <*>

REVUE ECONOMIQUE FRANCAISE PUBLIEE PAR LA SOCIETE DE GEOGRAPHIE COMMERCIALE
1957 N5 P841-850 60-18033 <*>

REVUE DE L'EDUCATION PHYSIQUE
1957 N181 P1-12 59-20870 <*>

REVUE DE L'ELECTROTECHNIQUE ET D'ENERGETIQUE
1960 V5 P73-88 66-12684 <*>
1962 V7A P237-247 UCRL TRANS-1039(L) <*>
1962 SB V7 N2 P261-272 63-20926 <*>

REVUE D'ELEVAGE ET DE MEDECINE VETERINAIRE DES PAYS TROPICAUX
1957 V10 P357-368 PT.1 63-00289 <*>
1959 V12 P369-379 63-01755 <*>
1964 V17 N1 P23-33 65-17248 <*>

REVUE DE L'EMBOUTEILLAGE ET DES INDUSTRIES CONNEXES
1963 P52-56 BGIRA-611 <BGIR>

REVUE DE LA FEDERATION DES INGENIEURS DES TELECOMMUNICATIONS DE LA COMMUNANTE EUROPEENNE (REVISTA, TIJDSCHRIFT, ZEITSCHRIFT)
1963 P31-37 66-11471 <*>

REVUE DES FERMENTATIONS ET DES INDUSTRIES ALIMENTAIRES
1954 V9 N3 P117-119 I-181 <*>
1955 V10 P103-118 3963-G <K-H>

REVUE DE FONDERIE MODERNE
1937 V31 P23-25 60-10752 <*> O

REVUE FRANCAISE D'ALLERGIE
1962 V2 N3 P127-132 63-18507 <*> O

REVUE FRANCAISE DES CORPS GRAS
1956 V5 P336-351 58-263 <*>
1961 V8 P37-46 SPEC NO 486-B <TC>

Citation	Translation
1961 V8 P403-404	486-A <TC>
1961 V8 N1 P15-24	62-14753 <*>
1961 V8 N2 P85-100	62-14754 <*>
1961 V8 N3 P152-165	62-14752 <*>

REVUE FRANCAISE D'ENDOCRINOLOGIE

Citation	Translation
1924 V2 P301-325	59-22163 <+*>

REVUE FRANCAISE DE L'ENERGIE

Citation	Translation
1959 V104 P133-141	65-11753 <*>

REVUE FRANCAISE D'ETUDES CLINIQUES ET BIOLOGIQUES

Citation	Translation
1956 V1 P29-38	58-177 <*>
1956 V1 P175-186	59-12336 <+*> 0
1956 V1 P619-630	1595 <*>
1956 V1 N1 P67-78	57-2938 <*>
1956 V1 N5 P531-553	58-379 <*>
1957 V2 N2 P161-177	59-12411 <+*> 0
1957 V2 N10 P1025-1037	59-12292 <+*> 0
1958 V3 P763-767	65-18106 <*>
1958 V3 P943-944	SCL-T-275 <*>
1958 V3 N1 P57-58	59-12290 <+*> 0
1958 V3 N3 P257-258	59-19184 <+*>
1958 V3 N3 P263-267	59-10898 <*> 0
1958 V3 N6 P558-584	60-17667 <*+> 0
1958 V3 N9 P977-981	59-20321 <=>
1958 V3 N10 P1101-1105	62-00869 <*>
1959 V4 P239-241	AEC-TR-4237 <*>
1959 V4 P442-446	59-22717 <+*> 0
1959 V4 N2 P146-150	59-20316 <*> 0
1959 V4 N3 P210-225	AEC-TR-3774 <+*> 0
	60-13270 <*=> 0
1959 V4 N3 P226-230	SCL-T-280 <*>
1959 V4 N3 P226-238	60-13415 <*+>
	61-10928 <+*> 0
1959 V4 N3 P239-241	60-13288 <*+>
1959 V4 N5 P423	60-13141 <*+>
1960 V5 P254-261	63-20896 <*> 0
1961 V6 P155-160	62-10320 <*>
1961 V6 P1034-1043	UCRL TRANS-884 <*>
1961 V6 N1 P66-68	61-20524 <*>
1961 V6 N1 P88-92	61-01047 <*>
1961 V6 N3 P227-231	62-00908 <*>
1961 V6 N6 P549-552	62-00427 <*>
1961 V6 N6 P553-559	62-00433 <*>
1961 V6 N7 P717-718	62-16082 <*>
1961 V6 N9 P916-917	62-16065 <*>
1962 V7 N5 P543-552	63-00297 <*>
1963 V8 P292-302	64-18671 <*>
1964 V9 P203-206	66-12173 <*>
1965 V10 N2 P198-211	66-12117 <*>

REVUE FRANCAISE DE GYNECOLOGIE ET D'OBSTETRIQUE

Citation	Translation
1942 V37 P51-61	62-10585 <*> 0
1962 V57 P831-834	64-00260 <*>

REVUE FRANCAISE DE MECANIQUE

Citation	Translation
1962 N2/3 P157-163	66-14385 <*>

REVUE FRANCAISE D'ODONTO-STOMATOLOGIE

Citation	Translation
1958 V7 P1-32	65-00221 <*> 0

REVUE FRANCAISE DE RECHERCHE OPERATIONNELLE

Citation	Translation
1962 V6 N25 P323-333	66-12112 <*>

REVUE FRANCAISE DE TRAITEMENT DE L'INFORMATION CHIFFRES

Citation	Translation
1963 V6 P231-253	66-13240 <*>

REVUE GENERALE DE BOTANIQUE

Citation	Translation
1950 V57 P65-77	57-93 <*>
1953 V60 N710 P239-249	64-10304 <*>
1955 V62 P238-242	1688 <*>

REVUE GENERALE DU CAOUTCHOUC

Citation	Translation
1939 N9 P329-335	RCT V14 N1 P676 <RCT>
1939 V16 N8 P301-306	RCT V14 N1 P696 <RCT>
1939 V16 N10 P365-374	RCT V13 N1 P451 <RCT>
1940 V17 N2 P49-50	RCT V13 N1 P931 <RCT>
	RCT V19 N1 P482 <RCT>
1941 V18 P198-208	59-17679 <*>
1941 V18 P223-235	59-17679 <*>
1941 V18 N2 P39-50	RCT V19 N1 P329 <RCT>
1941 V18 N3 P90-93	RCT V19 N1 P494 <RCT>
1941 V18 N8 P268-271	RCT V19 N1 P948 <RCT>
1941 V18 N9 P289-302	RCT V19 N1 P876 <RCT>
1942 V19 N1 P13-20	RCT V19 N1 P296 <RCT>
1942 V19 N3 P79-84	RCT V19 N1 P319 <RCT>
1942 V19 N3 P85-90	RCT V19 N1 P466 <RCT>
1942 V19 N3 P91-	RCT V19 N1 P349 <RCT>
1942 V19 N5 P165-166	RCT V19 N1 P478 <RCT>
1942 V19 N7 P207-211	RCT V19 N1 P1051 <RCT>
1942 V19 N10 P273-276	RCT V20 N1 P63 <RCT>
1943 V20 N6 P111-115	RCT V19 N1 P938 <RCT>
1943 V20 N6 P126-	RCT V19 N1 P400 <RCT>
1943 V20 N8 P155-157	RCT V19 N1 P1061 <RCT>
1943 V20 N9 P177-182	RCT V19 N1 P1061 <RCT>
1944 V21 P89-93	RCT V20 N4 P962 <RCT>
1944 V21 N1 P3-8	RCT V20 N4 P949 <RCT>
1944 V21 N2 P39-40	RCT V19 N1 P1085 <RCT>
1944 V21 N3 P50-56	RCT V20 N2 P479 <RCT>
1944 V21 N4 P83-84	RCT V20 N1 P249 <RCT>
1944 V21 N6 P122-124	RCT V20 N3 P769 <RCT>
1944 V21 N8 P155-160	RCT V19 N1 P176 <RCT>
1944 V21 N8 P168-	RCT V19 N1 P123 <RCT>
1944 V21 N9 P177-180	RCT V20 N2 P409 <RCT>
1944 V21 N9 P189-191	RCT V19 N1 P933 <RCT>
1944 V21 N12 P243-245	RCT V20 N4 P972 <RCT>
1944 V21 N12 P250-252	RCT V19 N1 P486 <RCT>
1945 V22 N1 P3-5	RCT V21 N1 P247 <RCT>
1945 V22 N3 P49-55	RCT V20 N2 P457 <RCT>
1945 V22 N3 P63-66	RCT V21 N1 P141 <RCT>
1945 V22 N5 P93-95	RCT V20 N1 P308 <RCT>
1946 V23 N2 P28-33	RCT V21 N1 P168 <RCT>
1946 V23 N5 P101-106	RCT V21 N2 P505 <RCT>
1947 V24 P81-86	SCL-T-509 <*>
1947 V24 N1 P4-7	RCT V21 N1 P60 <RCT>
1947 V24 N12 P425-431	62-18234 <*> 0
1947 V24 N12 P436-445	RCT V21 N3 P684 <RCT>
1948 V25 N1 P20-24	57-634 <*>
1948 V25 N2 P56-62	57-634 <*>
1948 V25 N3 P85-90	57-634 <*>
1949 V26 N1 P10-14	RCT V22 N3 P690 <RCT>
1949 V26 N2 P85-90	RCT V22 N4 P1000 <RCT>
1949 V26 N3 P167-172	RCT V22 N4 P912 <RCT>
1949 V26 N3 P172-176	RCT V22 N4 P1028 <RCT>
1949 V26 N4 P273-278	RCT V23 N1 P281 <RCT>
	62-18353 <*> 0
1949 V26 N5 P341-345	RCT V25 N1 P124 <RCT>
1949 V26 N6 P426-432	RCT V25 N1 P74 <RCT>
1949 V26 N7 P505-508	RCT V25 N1 P124 <RCT>
1949 V26 N11 P740-744	RCT V23 N2 P352 <RCT>
1950 V27 N4 P205-208	RCT V23 N4 P897 <RCT>
1950 V27 N7 P409-414	RCT V24 N1 P199 <RCT>
1950 V27 N8 P473-474	RCT V24 N1 P195 <RCT>
1950 V27 N12 P731-732	RCT V24 N4 P921 <RCT>
1951 V28 N1 P39-42	RCT V25 N2 P303 <RCT>
1951 V28 N2 P105-107	RCT V25 N2 P303 <RCT>
1951 V28 N8 P563-566	RCT V25 N1 P99 <RCT>
1951 V28 N8 P570-576	RCT V26 N1 P207 <RCT>
1952 V29 N2 P114-117	RCT V25 N3 P549 <RCT>
1952 V29 N4 P278-282	RCT V25 N3 P621 <RCT>
1952 V29 N7 P506-510	RCT V26 N1 P143 <RCT>
1952 V29 N9 P660-663	RCT V26 N1 P136 <RCT>
1952 V29 N12 P894-899	RCT V27 N2 P481 <RCT>
1953 V30 N1 P42-45	RCT V26 N2 P411 <RCT>
1953 V30 N4 P262-264	RCT V27 N1 P271 <RCT>
1953 V30 N8 P559-562	RCT V27 N1 P157 <RCT>
1953 V30 N9 P654-659	RCT V27 N1 P147 <RCT>
1954 N3 P216-	RCT V27 N4 P1005 <RCT>
1954 V31 P49-53	58-2521 <*>
1954 V31 N1 P46-48	RCT V27 N4 P1041 <RCT>
1954 V31 N2 P123-131	RCT V28 N2 P438 <RCT>
1954 V31 N5 P393-396	RCT V28 N3 P814 <RCT>
1954 V31 N9 P724-727	RCT V28 N2 P420 <RCT>
1954 V31 N11 P898-900	RCT V28 N3 P788 <RCT>
1954 V31 N12 P977-982	RCT V28 N4 P1175 <RCT>
1955 V32 N2 P133-137	RCT V28 N4 P1175 <RCT>
1955 V32 N3 P229-236	RCT V29 N1 P302 <RCT>
1955 V32 N10 P889-892	RCT V29 N3 P1026 <RCT>

1956 V33 P427-430 65-18166 <*>
1956 V33 P973-984 RCT V31 N1 P166 <RCT>
1956 V33 N4 P335-341 RCT V29 N4 P1509 <RCT>
1956 V33 N6 P516-517 64-16808 <*>
1956 V33 N10 P860-863 64-16304 <*>
1957 V34 P366-370 RCT V30 N4 P1175 <RCT>
1957 V34 P1122-1126 RCT V31 N2 P387 <RCT>
1957 V34 N11 P1122-1126 58-1336 <*>
1957 V34 N12 P1260-1267 RCT V31 N3 P631 <RCT>
1958 V35 N3 P305-308 65-11240 <*> 0
1958 V35 N5 P615-619 64-20118 <*> 0
1958 V35 N11 P1349-1354 59-15519 <*>
1958 V35 N11 P1379-1384 59-15358 <*>
1958 V35 N11 P1387-1395 59-15058 <*>
1959 V36 N3 P369-372 66-13282 <*>
1962 V39 N2 P217-222 RCT V35 N4 P848 <RCT>
1963 V40 N3 P406-412 RCT V37 N3 P720 <RCT>
1963 V40 N5 P748-753 RCT V37 N1 P88 <RCT>
1964 V41 N3 P371-375 65-12343 <*>
1964 V41 N3 P429-433 65-12344 <*>
1964 V41 N6 P989-992 22S81F <ATS>
1964 V41 N7/8 P1155-1159 65-12912 <*>

REVUE GENERALE DES CHEMINS DE FER ET DES TRAMWAYS
1939 V58 P65-102 T1662 <INSD>
1953 P1-21 T1834 <INSD>
1954 V73 P173-191 62-16594 <*>
1955 N7 P504-507 60-21027 <=*> 0
1955 V74 P761-783 T-1953 <INSD>
1955 V74 P813-822 T1827 <INSD>
1955 V74 P823-826 T1878 <INSD>
1955 V74 P872-879 T1880 <INSD>
1959 P684-686 TS-1536 <BISI>
62-14184 <*>
1959 P687-697 TS-1537 <BISI>
62-14180 <*>
1960 N7/8 P401-403 1830 <BISI>

REVUE GENERALE DES COLLOIDES
1923 V1 P33-39 2939 <*>
1925 P289-293 2932 <*>
1925 P321-324 2931 <*>

REVUE GENERALE DE L'ELECTRICITE
1917 V1 N9 P331-337 57-2177 <*>
1917 V2 N8 P288-292 57-2180 <*>
1918 V4 P740-742 UCRL TRANS-1144(L) <*>
1919 V6 P131- 57-2145 <*>
1919 V6 P163- 57-2035 <*>
1919 V6 P195-199 57-01458 <*>
1919 V6 P195- 57-1458 <*>
1919 V7 P137- 57-940 <*>
1921 V9 P138-141 57-2165 <*>
1922 V11 P663-666 58-100 <*>
1922 V12 P755-763 57-2112 <*>
1922 V12 N13 P479-483 59-20526 <*> 0
1923 V13 P653-654 57-2039 <*>
1923 V14 P963-965 66-12905 <*>
1925 V17 P337-349 57-2243 <*>
1925 V17 P363-368 57-2109 <*>
1925 V17 N7 P243-245 2127 <*>
1925 V17 N8 P312-315 59-20470 <*>
1926 V19 N14 P523-529 57-342 <*>
1926 V20 N8 P270-278 57-1045 <*>
1927 V21 P195-198 57-2212 <*>
1927 V21 P539-547 57-2131 <*>
1927 V21 P1030-1036 57-1580 <*>
1927 V21 N1 P3-13 66-12259 <*>
1927 V21 N2 P43-59 66-12259 <*>
1927 V22 P329-330 66-12906 <*>
1927 V22 P871-875 57-1338 <*>
1927 V22 P879 57-1338 <*>
1928 V23 P73-76 UCRL TRANS-1141(L) <*>
1928 V23 P169-170 65-18118 <*>
1928 V24 P7-11 58-1873 <*>
1929 V25 N3 P87-98 57-648 <*>
1930 V28 N4 P127-135 59-20497 <*> 0
1930 V28 N8 P273-282 57-3144 <*>
1932 V31 N16 P519-534 66-13818 <*>
1933 V34 N26 P894-895 57-1402 <*>
1934 V35 N2 P47-48 57-2453 <*>
1934 V35 N19 P665-666 57-1078 <*>
1934 V35 N21 P709-719 57-2197 <*>
1936 V40 N8 P227-239 57-2424 <*>
1938 V44 N8 P241-242 57-2163 <*>
1939 V45 P303-316 2188 <*>
1939 V45 P333-350 2188 <*>
1940 V47 N17/8 P317-321 2165 <*>
1945 V54 N6 P183-191 59-17948 <*> 0
1945 V54 N8 P239-241 62-18348 <*> 0
1945 V54 N10 P298-302 62-18347 <*> 0
1946 V55 N4 P143-151 62-16467 <*>
1949 V58 N11 P478-480 57-2759 <*>
1950 V59 N3 P114-116 57-2614 <*>
1950 V59 N3 P133-136 C-4483 <NRCC>
1950 V59 N4 P169-174 NP-TR-1039 <*>
1950 V59 N11 P479-501 60-21715 <=>
1950 V59 N12 P519-524 58-293 <*>
1951 N60 P279-291 UCRL TRANS-578 <*>
1951 N60 P317-328 UCRL TRANS-578 <*>
1951 V60 P107-121 T1739 <INSD>
1952 V61 N9 P383-387 62-18162 <*> 0
1952 V61 N12 P551-559 57-2419 <*>
1953 V62 P442-448 T-1740 <INSD>
1953 V62 N6 P277-285 64-10352 <*>
1953 V62 N6 P293 64-10352 <*>
1956 V65 N9 P513-536 3123 <*>
1957 V66 P207-225 58-517 <*>
58-57 <*>
1957 V66 N3 P187-188 58-2069 <*>
1958 V67 N7 P354-364 62-32391 <=*>
1959 V68 N12 P693-695 66-13237 <*>
1960 V69 N1 P3-18 63-14988 <*> 0
1960 V69 N11 P579-590 64-18200 <*> P
1960 V69 N11 P601-611 62-14653 <*> 0
1963 V72 N3 P155-163 65-14069 <*>
1963 V72 N3 P165-172 65-14070 <*>
1964 V73 N4 P217-227 66-13872 <*>
1964 V73 N5 P281-294 66-13239 <*>
1965 V74 N2 P177-190 66-13238 <*> 0P

REVUE GENERALE D'ELECTRONIQUE (FORMERLY ELECTRONIQUE 1946-1960)
1963 N195 P29-33 66-12530 <*>
1964 N206 P20-24 66-12493 <*>
1964 N207 P15-19 66-12494 <*>

REVUE GENERALE DE L'ETANCHEITE ET DE L'ISOLATION
1963 N53 P3-9 9068 <IICH>

REVUE GENERALE DU FROID ET DES INDUSTRIES FRIGORIFIQUES
1953 V30 P1167-1170 65-10450 <*> 0
1958 N7 P681-685 AEC-TR-3691 <=>

REVUE GENERALE DES MATIERES COLORANTES, DE LA TEINTURE, DE L'IMPRESSION ET DES APPRETS
1930 V34 P55-56 58-2100 <*>
1939 V43 P45-48 58-1923 <*>

REVUE GENERALE DES MATIERES PLASTIQUES
1938 V14 P3-5 65-10149 <*>
1940 V16 P54-57 62-16934 <*> C

REVUE GENERALE DE MECANIQUE
1955 V39 N74 P39-45 65-18083 <*>
1955 V39 N79 P247-254 65-18083 <*>
1955 V39 N83 P395-400 65-18083 <*>
1955 V39 N84 P445-451 65-18083 <*>
1956 P429-433 58-131 <*>
1956 N92 P309-313 57-3172 <*>
1956 V40 P315-320 64-16514 <*>
1956 V40 N86 P81-87 65-18083 <*>
1956 V40 N88 P159-164 65-18083 <*>
1956 V40 N95 P429-433 65-12015 <*>
1957 V41 P302-306 73J19F <ATS>
1957 V41 N96 P13-18 61-16405 <*>
1957 V41 N97 P63-67 61-16405 <*>

REVUE GENERALE DES ROUTES ET DES AERODROMES

1955 V25 P59-74 T-1668 <INSD>
1956 V26 N292 P53-72 62-14225 <*>
1958 V28 N321 P37-82 65-13371 <*>
1960 V30 N341 P61-84 65-13372 <*>

REVUE GENERALE DES SCIENCES PURES ET APPLIQUEES
1900 V11 N23 P1261-1271 66-14342 <*>
1919 V30 P5-17 64-10121 <*> 0
1925 V36 P365-370 57-2426 <*>
1959 V66 P39 63-16971 <*>

REVUE DE GEOGRAPHIE PHYSIQUE ET DE GEOLOGIE DYNAMIQUE
1957 V1 P189-198 64-14285 <*>

REVUE DE GEOLOGIE ET DE GEOGRAPHIE
1959 V3 N1 P139-150 63-20732 <*>

REVUE DE GEOMORPHOLOGIE DYNAMIQUE
1951 V2 N3 P128-134 2716F <ATS>

REVUE H.F. TIJDSCHRIFT. (TITLE VARIES)
1962 V5 N8 P255-275 63-01434 <*>

REVUE DES HAUTES TEMPERATURES ET DES REFRACTAIRES
1964 V1 N2 P97-106 65-14826 <*>
1965 V2 P47-54 65-14900 <*>
1965 V2 N2 P163-172 66-11457 <*>

REVUE D'HEMATOLOGIE
1949 V4 N2 P173-176 58-1346 <*>
1953 V8 P282-298 I-550 <*>
1955 V10 N4 P745-752 59-14528 <+*> 0
1956 V11 P437-450 57-3437 <*>
1956 V11 N2 P190-215 57-148 <*>
1956 V11 N3 P324-326 57-3326 <*>
1956 V11 N5 P477-485 57-2922 <*>
1957 V12 N2 P211-221 59-14527 <+*> 0
1958 V13 N1 P61-77 59-14526 <+*> 0
1959 V14 N1 P75-89 59-18800 <+*> 0
1959 V14 N2 P97-117 63-00943 <*>
1960 V15 P52-71 62-10749 <*> 0
1960 V15 N1 P3-9 SCL-T-424 <*>
1960 V15 N2/3 P115-161 63-01748 <*>
1960 V15 N2/3 P162-173 62-00234 <*>

REVUE D'HYGIENE ET DE MEDECINE SOCIALE
1957 V5 N5 P423-449 58-2336 <*>

REVUE D'IMMUNOLOGIE ET DE THERAPIE ANTIMICROBIENNE
1939 V5 P279-284 57-1494 <*>
1939 V5 P299-316 57-1622 <*>
1939 V5 P535-536 57-1647 <*>
1941 V6 P363-380 MF-2 <*>
1942 V7 P237-240 93F2F <ATS>
1946 V10 P71-81 61-10449 <*>
1951 V15 N2 P93-157 63-00741 <*>
63-27748 <$>
1951 V15 N1/2 P47-67 <ATS>
1952 V16 N4/5 P257-273 <ATS>
1954 V18 N3 P199-205 2645 <*>
1956 V20 N1/2 P27-36 57-1262 <*>
1960 V24 P226-232 63-20219 <*> 0
1962 V26 N5/6 P297-311 63-18358 <*>
1963 V27 N1/2 P87-96 64-00358 <*>

REVUE DE L'INDUSTRIE MINERALE
1934 N35 P228-239 59-14721 <+*>
1937 P565-570 57-605 <*>
1947 N520 P191-199 AL-532 <*>
3331 <*>
1948 N530 P173-194 60-10807 <*>
1951 N3 P556-567 59-17310 <*>
1956 V38 N639 P193-199 CSIRO-3310 <CSIR>
1956 V38 N639 P200-204 CSIRO-3311 <CSIR>
1957 V39 N7 P629-643 59-18560 <+*>
1958 V40 N3 P215-229 58-2543 <*>
59-22344 <=*> 0
1960 V42 N6 P542-551 NP-TR-849 <*>
1962 V44 N12 P869-884 3642 <BISI>
1963 V45 N7 P509-536 65-11090 <*>

REVUE DE L'INSTITUT FRANCAIS DU PETROLE ET ANNALES DES COMBUSTIBLES LIQUIDES
1946 V1 N3 P145-151 64-20383 <*>
1947 V2 N6 P288-292 60-10344 <*> 0
1949 V4 P107-118 57-3077 <*>
1949 V4 P653-660 63-18217 <*> 0
1951 V6 N8 P305-311 AL-77 <*>
1952 V7 P170-180 57-3072 <*>
1952 V7 P183-197 3235 <*>
1953 V8 P491-503 64-10115 <*> 0
1953 V8 N4 P129-151 60-14047 <*> 0
1953 V8 N5 P193-222 60-14047 <*> 0
1953 V8 N6 P248-276 60-14047 <*> 0
1953 V8 N8 P423-435 79F2F <ATS>
1954 V9 P133-143 59-18472 <*=>
1954 V9 N2 P51-66 64-14677 <*> 0
1954 V9 N5 P202-213 61-16457 <*>
1954 V9 N6 P296-307 65-12422 <*>
1955 V10 P164-169 3759-C <K-H>
1955 V10 N2 P103-114 62-16657 <*>
1955 V10 N3 P181-213 65-12431 <*>
1955 V10 N5 P466-469 73G7F <ATS>
1955 V10 N5 P470-476 79G7F <ATS>
1955 V10 N5 P477-486 C-5485 <NRC>
1955 V10 N5 P487-499 59-10814 <*>
1955 V10 N8 P886-911 62-16656 <*>
1955 V10 N10 P1280-1283 64-16437 <*> 0
1956 V11 P1298-1312 59-10660 <*> 0
1956 V11 N1 P134-157 65-12675 <*>
1956 V11 N2 P150-196 62-16655 <*>
1956 V11 N4 P501-548 64-18409 <*> 0
1956 V11 N10 P1269-1277 59-15011 <*>
1956 V11 N10 P1298-1312 64-30111 <*> 0
1956 V11 N10 P1313-1324 64-30127 <*> 0
1956 V11 N11 P1485-1488 64-16309 <*>
98K24F <ATS>
1956 V11 N11 P1489-1495 64-16553 <*>
1956 V11 N12 P1611-1615 93K21F <ATS>
1957 V12 N3 P304-329 64-16813 <*>
1957 V12 N5 P576-598 58-521 <*>
64-16884 <*>
1957 V12 N6 P715-759 60-10790 <*>
1957 V12 N9 P956-969 62-18751 <*>
1957 V12 N9 P971-983 62-18751 <*>
1957 V12 N11 P1161-1167 65-12930 <*>
1957 V12 N12 P1236-1240 65-12434 <*>
1958 V13 N3 P213-221 59K25F <*>
1958 V13 N3 P267-288 64-30095 <*> 0
1958 V13 N6 P1040-1063 64-30962 <*>
68L31F <ATS>
1958 V13 N9 P1247-1252 65-12273 <*>
1958 V13 N7/8 P1157-1196 64-30978 <*> 0
1958 V13 N7/8 P1197-1216 61-10702 <*> 0
65-10299 <*> 0
1958 V13 N7/8 P1217-1242 61-10570 <*>
1959 V14 P55-71 60-18102 <*>
1959 V14 N10 P1295-1306 64-30683 <*>
1959 V14 N12 P1595-1614 65-12270 <*>
1959 V14 N12 P1615-1636 60-16478 <*>
1959 V14 N4/5 P519-534 59M39F <ATS>
1959 V14 N4/5 P535-548 18L34F <ATS>
1959 V14 N4/5 P599-619 39L35F <ATS>
60-14303 <*>
1959 V14 N4/5 P647-667 65-14187 <*>
1960 V15 N1 P3-68 18Q74F <ATS>
63-16545 <*>
1960 V15 N2 P419-430 65-11254 <*> 0
1960 V15 N3 P529-566 83M43F <ATS>
1960 V15 N10 P1384-1400 65-10475 <*>
1960 V15 N11 P1567-1574 16N49F <ATS>
65-12272 <*>
1960 V15 N12 P1731-1740 65-12921 <*>
1960 V15 N12 P1847-1880 58N51F <ATS>
1961 V16 N2 P140-149 65-11572 <*>
1961 V16 N2 P150-160 65-10471 <*> 0
73N56F <ATS>
1961 V16 N3 P263-274 65-12268 <*>
1961 V16 N3 P332-362 36N54F <ATS>

1961 V16 N4 P367-381 63-10222 <*>
1961 V16 N4 P468-484 65-12755 <*>
66-10325 <*> 0
1961 V16 N6 P659-677 65-12436 <*>
1961 V16 N6 P678-700 65-12437 <*>
1961 V16 N11 P1255-1272 24P60F <ATS>
1961 V16 N11 P1307-1329 65-12313 <*>
1961 V16 N7/8 P886-893 65-12435 <*>
1962 V17 P379-394 63-14314 <*>
1962 V17 P852- 12710 <K-H>
1962 V17 N4 P478-490 65-12296 <*>
1962 V17 N4 P537-557 63-20196 <*> 0
1962 V17 N4 P585-595 65-12289 <*>
1962 V17 N5 P669-713 65-11528 <*> 0
1962 V17 N6 P830-841 57P65F <ATS>
1962 V17 N9 P1107-1116 65-11120 <*>
1962 V17 N9 P1149-1180 65-12271 <*>
1962 V17 N10 P1214-1231 65-10213 <*>
1962 V17 N1C P1232-1259 65-12629 <*>
1962 V17 N11 P1372-1377 64-10176 <*> 0
1962 V17 N12 P1454-1472 64-16421 <*>
1962 V17 N12 P1508-1519 65-18014 <*>
1963 V18 P56-77 SPEC.IS. 15R76F <ATS>
1963 V18 P78-96 SPEC NC 65-11526 <NLL>
1963 V18 P7-16 1744 <TC>
1963 V18 P229-240 202 443 <ATS>
1963 V18 P258-283 95S82F <ATS>
1963 V18 P17-31 SUP 66-10315 <*>
1963 V18 N1 P41-49 14Q72F <ATS>
1963 V18 N4 P611-619 43R75F <ATS>
64-18433 <*> 0
1963 V18 N5 P679-696 65-18010 <*>
1963 V18 N5 P724-759 64-18766 <*>
1963 V18 N11 P1648-1672 65-18193 <*> 0
1963 V18 N7/8 P996-1011 65-10218 <*>
65-10478 <*> 0
1964 V19 N3 P297-334 65-12282 <*>
1964 V19 N5 P613-626 66-10324 <*>
1964 V19 N9 P941-957 65-12395 <*>
1964 V19 N1C P1067-1092 47S82F <ATS>
1964 V19 N11 P1244-1263 65-17115 <*> 0
65-18050 <*>
1964 V19 N12 P1391-1404 616 <TC>
1964 V19 N7/8 P872-900 65-13750 <*>
1964 V19 N7/8 P921-937 66-13940 <*>
1965 V20 P181-190 66-10367 <*>
1965 V20 N1 P27-93 66-10359 <*>
1965 V20 N1 P160-180 66-15150 <*>
797 <TC>
1965 V20 N5 P755-803 66-13789 <*>
1965 V20 N5 P866-876 66-12215 <*>
1965 V20 N11 P1610-1622 66-13941 <*>
1965 V20 N7/8 P1135-1201 66-13614 <*>
1966 V21 N2 P190-211 66-13939 <*>
1966 V21 N2 P227-238 66-13938 <*>

REVUE DE L'INSTITUT INTERNATIONAL DE STATISTIQUE
1961 V29 N3 P44-56 63-00176 <*>

REVUE INTERNATIONALE D'HEPATOLOGIE
1953 V3 N4 P461-492 59-15531 <*> P
1953 V3 N4 P497-510 60-18936 <*>
1953 V3 N4 P519-523 60-18935 <*>
1953 V3 N4 P525-536 59-15528 <*>
1953 V3 N5 P579-601 60-18934 <*> 0
1959 V9 P301-310 60-16606 <*>
1959 V9 P329-352 60-16603 <*> 0
1959 V9 P353-357 60-16607 <*>
1959 V9 P359-370 60-16602 <*> 0

REVUE INTERNATIONALE DES PRODUITS COLCNIAUX
1929 V4 N44/5 P290-295 57-2842 <*>

REVUE INTERNATIONALE DES PRODUITS TROPICAUX
1960 V35 N372 P111 11411-A <K-H>
1960 V35 N372 P181 63-16912 <*>
1960 V35 N372 P191 11411-A <K-H>
63-16912 <*>

REVUE INTERNATIONALE DES TABACS
1948 V23 P99-1C1 3169-A <K-H>
1949 V24 P123 58-1220 <*>
1949 V24 P125 58-1220 <*>
1949 V24 P127-128 58-1220 <*>
1949 V24 P130-132 58-1220 <*>
1949 V24 P183,189 2278B <K-H>
1950 V25 P17-24 2141 <K-H>
1950 V25 P99-1C2 2447B <K-H>
1950 V25 P151-156 2447C <K-H>
1950 V25 P196- 2012B <K-H>
1951 V26 P37-41 2447A <K-H>
1953 V28 P223-225 3036-B <K-H>
1954 V29 P33 3046-A <K-H>
1954 V29 P35-36 3046-A <K-H>
1954 V29 P37 3160-H <K-H>
1955 V30 P171-180 5629-A <K-H>
1955 V30 P195-206 5629-B <K-H>
1955 V30 N273 P171-180 58-843 <*>
1955 V30 N274 P195-206 58-843 <*>
1960 V35 N330 P147 10833-B <K-H>
1960 V35 N330 P149 10833-B <K-H>
1960 V35 N330 P158 10833-B <K-H>
1961 V36 N335 P4,5,23 10897-A <K-H>
64-10610 <*>

REVUE INTERNATIONALE DU TRACHCME
1963 V40 P173-184 66-13202 <*>

REVUE JURIDIQUE ET POLITIQUE, INDEPENDANCE ET COOPERATION
1965 V19 P179-193 65-31973 <=$>

REVUE DE LARYNGOLOGIE, D'CTCLCGIE ET CE RHINOLOGIE
1920 V41 N4 P91-112 61-16075 <*>
1958 V79 P1125-1135 3641 <K-H>
1960 V81 N1/2 P119-123 65-13732 <*> P

REVUE; LITTERATURE, HISTCIRE, ARTS, ET SCIENCES
1957 N11 P385-394 60-18650 <*>

REVUE LYONNAISE DE MEDECINE
1955 P28-35 57-2981 <*>
1955 V4 P93-100 59-15677 <*> 0
1961 V10 P1085-1093 63-16542 <*>

REVUE M.B.L.E. MANUFACTURE BELGE DE LAMPES ET DE MATERIEL ELECTRONIQUE
1961 V4 N4 P222-249 NP-TR-920 <*>
1963 V6 N4 P5-29 UCRL TRANS-1007(L) <*>

REVUE MARITIME
1955 N110 P737-752 2847 <*>
62-00561 <*>
1957 N139 P1503-1513 59-12015 <+*>

REVUE DES MATERIAUX DE CCNSTRUCTION ET DE TRAVAUX PUBLICS. EDITION C. INDUSTRIES: CEMENT CHAUX, PLATRE, AGGLOMERES
1925 N186 P57 64-18203 <*> C
1925 N187 P91 64-18203 <*> 0
1931 P1-5 64-18207 <*> 0
1931 P45-49 64-18207 <*> C
1931 N267 P485-490 64-18206 <*>
1932 N268 P9-15 64-18206 <*>
1949 N405 P179-187 61-18979 <*>
1949 N405 P188-194 C-2480 <NRC>
57-3476 <*>
1949 N406 P219-224 C-2480 <NRC>
57-3476 <*>
1949 N407 P259-266 58-2563 <*>
1949 N407 P267-272 C-2480 <NRC>
57-3476 <*>
1949 N408 P295-303 58-2563 <*>
1949 N408 P304-308 C-2480 <NRC>
57-3476 <*>
1950 N414 P71-74 I-443 <*>
1950 N415 P148-150 58-2601 <*>
61-18996 <*>
1950 N417 P187-189 64-14349 <*> 0
1951 P302-304 I-102 <*>

1951 P331-334 I-102 <*>
1951 N430 P236-237 57-615 <*>
1952 P24-28 I-102 <*>
1952 N440 P132-136 57-619 <*>
1952 N441 P173-175 57-619 <*>
1953 N457 P282-288 57-618 <*>
1953 N458 P324-328 57-618 <*>
1953 N459 P329-331 I-177 <*>
1953 N459 P343-345 57-618 <*>
1954 N460 P7-12 AL-636 <*>
1954 N462 P55-67 AL-351 <*>
1955 N474 P59 II-1141 <*>
1955 N474 P59-69 II-1141 <*>
62-18414 <*>
1955 N474 P79-86 I-270 <*>
1955 N477 P153-164 II-371 <*>
64-18248 <*> 0
1955 N478 P181-192 II-372 <*>
64-18248 <*> 0
1956 N490 P155-172 57-2265 <*>
1956 N492 P191-209 57-2265 <*>
1957 N496 P12-18 57-2267 <*>
1959 N523 P89-95 60-14557 <*> 0
1959 N526 P173-179 60-14021 <*> 0
1962 N566 P301-314 64-18001 <*>
1963 N569 P37-47 66-12813 <*> 0
1963 N570 P67-78 66-12814 <*>
1963 N573 P190-203 66-12815 <*>

REVUE DES MATERIAUX DE CONSTRUCTION ET DE TRAVAUX PUBLICS. CIMENTS ET BETONS
1960 N532 P1-5 60-18540 <*>
1960 N533 P31-42 60-17536 <=*>
61-10786 <*>
1960 N537 P139-148 63-18497 <*> 0
1960 N538 P169-182 63-18502 <*> 0
1960 N538 P183-189 63-18497 <*> 0
1960 N540 P215-231 63-18502 <*> 0
1960 N540 P231-232 61-10785 <*>
1960 N540 P233-243 63-18497 <*> 0
1960 N541 P251-276 63-18502 <*> 0
1960 N542 P295-308 66-13232 <*> P
1960 N542 P309-315 61-18848 <*>
1960 N543 P327-331 62-16841 <*>
1961 N546 P189-200 62-14230 <*> 0
1962 N561 P168-180 63-10197 <*> 0
1963 N570 P79-81 63-20312 <*>
1963 N572 P157-161 66-11941 <*>
1963 N576 P271-275 64-18007 <*>
65-11293 <*>
1965 N594 P130-141 66-10778 <*>

REVUE DE MATHEMATIQUES PURES ET APPLIQUES
1963 V8 P611-645 65-17434 <*>

REVUE DE MECANIQUE APPLIQUEE
1956 V1 N1 P71-88 61-31567 <=>
1956 V1 N1 P141-155 57-2932 <*>
1962 V7 N1 P173-183 63-27395 <$> P
1963 V8 N2 P193-216 64-71543 <=>

REVUE DE MEDECINE. PARIS
1910 V30 P785-801 64-00074 <*>

REVUE MEDICALE DE L'EST. NANCY
1915 V35 N11 P605-615 59-15548 <*>

REVUE MEDICALE FRANCAISE
1959 V40 N2 P161-171 64-18715 <*> 0

REVUE MEDICALE FRANCAISE D'EXTREME-ORIENT
1942 V20 N8/10 P1077-1084 2700 <*>

REVUE MEDICALE DE LIEGE
1949 V4 N19 P573-576 3160-N <K-H>
1952 V7 N21 P713-717 58-1839 <*>
1954 V9 N9 P271-276 II-868 <*>
1960 V15 N7 P234-242 65-13182 <*>
1961 V16 N20 P569-576 64-14313 <*>
1964 V19 N3 P88-90 65-14712 <*>

REVUE MEDICALE DU MOYEN-ORIENT
1958 V15 N3 P195-208 59-15453 <*>

REVUE MEDICALE DE NANCY
1950 V70 P19-28 59-15475 <*>
1951 V76 P408-410 59-15424 <*>
1954 V79 P173-178 3199-B <KH>
1957 V82 P743-755 6871-A <KH>
1957 V82 N7 P743-755 63-16998 <*>

REVUE MEDICALE DE LA SUISSE ROMANDE
1939 N59 P738-748 57-2714 <*>
1952 V72 P27-44 10177-B <KH>
63-16582 <*>
1953 V73 P891-899 1847 <*>
1954 V74 N1 P39-44 59-15432 <*>
1955 V75 N12 P822-840 64-18994 <*> 0
1961 V81 N11 P797-800 59T92F <ATS>
1964 V84 N11 P811-822 41S84F <ATS>

REVUE DE METALLURGIE
1913 V10 P944-947 59-15198 <*> 0
1917 P16-21 II-545 <*>
1922 V19 P298-302 57-2444 <*>
1928 V25 P397-404 57-1557 <*>
1929 V26 N5 P238-247 59-20853 <*> 0
1929 V26 N8 P435-443 59-20419 <*> 0
1930 V27 P362-377 58-2120 <*>
1930 V27 N10 P522-534 59-20433 <*> 0
1930 V27 N10 P560-562 59-20432 <*> 0
1930 V27 N10 P563-569 59-20454 <*> 0
1931 V28 N3 P162-164 59-10650 <*> 0
1933 P511-519 II-437 <*>
1935 V32 N10 P494-500 59-15442 <*>
1936 P638-640 57-9 <*>
1936 V33 P295-302 <CSIR>
1938 V35 N8 P363-378 AD-614 282 <=$>
1938 V35 N9 P407-424 AD-614 282 <=$>
1938 V35 N10 P448-474 AD-614 282 <=$>
1939 V36 P21-29 59-20230 <*> 0
1939 V36 N12 P497-508 AEC-TR-5534 <*>
1942 V39 P54-60 I-91 <*>
1943 V40 P33-37 3249 <BISI>
1943 V40 P65-72 3249 <BISI>
1943 V40 P143-155 I-931 <*>
1943 V40 P175-182 I-931 <*>
1943 V40 P202-208 I-930 <*>
1943 V40 P252-256 I-930 <*>
1944 V41 N11 P378-388 59-17928 <*> 0
1944 V41 N11 P403-408 59-17928 <*> 0
1945 V24 N1/3 P1-10 63-14657 <=*>
1945 V42 N3 P79-92 60-18293 <*>
1945 V42 N4 P125-132 60-18293 <*>
1945 V42 N5 P156-167 60-18293 <*>
1945 V42 N6 P194-202 60-18293 <*>
1945 V42 N7 P230-239 60-18293 <*>
1945 V42 N10 P333-335 62-18440 <*> 0
1947 V44 N5/6 P174-179 62-18442 <*> 0
1947 V44 N5/6 P180-186 62-18441 <*> 0
1947 V44 N5/6 P187-192 62-18352 <*> 0
1948 V45 P49-59 8717 <IICH>
1948 V45 N7 P205-210 63-10026 <*> 0
1948 V45 N12 P481-489 60-16883 <*> 0
1948 V45 N1/2 P9-18 59-10071 <*>
63-20828 <*>
1948 V45 N1/2 P49-59 I-10 <*>
65-10140 <*> 0
1949 V46 P321-329 61-20498 <*> 0
1949 V46 P329-338 61-18907 <*> 0
1949 V46 P661-675 64-10230 <*> 0
1949 V46 N2 P79-83 63-10014 <*> 0
1949 V46 N5 P309-314 57-12 <*>
1949 V46 N9 P594-615 II-845 <*>
1949 V46 N11 P719-726 2097 <*>
1950 V47 P589-600 58-2045 <*>
1950 V47 N4 P299-305 61-10592 <*>
1950 V47 N4 P317-323 14M44F <ATS>
65-17229 <*>
1950 V47 N5 P375-387 61-10590 <*>

Citation	Translation
1950 V47 N7 P547-557	3914 <BISI>
1950 V47 N11 P856-862	59-20796 <*>
1950 V47 N12 P873-888	64-18575 <*> O
1951 V48 P1-16	SC-T-64-911 <*>
1951 V48 P363-368	I-43 <*>
1951 V48 N2 P73-84	60-18601 <*>
1951 V48 N4 P262-266	59-17297 <*> P
1951 V48 N4 P303-313	60-18602 <*>
1951 V48 N12 P923-928	II-127 <*>
1951 V48 N12 P970-974	II-91 <*>
1952 V49 P1-19	II-392 <*>
1952 V49 N4 P267-282	62-18106 <*> O
1952 V49 N5 P339-363	I-287 <*>
1952 V49 N5 P374-378	58-1505 <*>
1952 V49 N6 P439-452	I-287 <*>
1952 V49 N9 P613-622	1431 <*>
1953 V50 N4 P229-247	60-18297 <*>
1953 V50 N4 P275-290	60-18298 <*>
1953 V50 N4 P291-296	65-13177 <*>
1953 V50 N5 P317-327	64-20367 <*>
1953 V50 N6 P424-426	60-10176 <*>
1953 V50 N9 P629-634	SCL-T-356 <*>
1953 V50 N10 P727-736	59-21002 <=>
1953 V50 N11 P775-780	60-10165 <*> O
1953 V50 N12 P817-828	60-10164 <*> O
1954 V51 P1-12	57-1971 <*>
1954 V51 P173-178	II-461 <*>
1954 V51 P617-623	II-365 <*>
1954 V51 N1 P13-16	II-128 <*>
1954 V51 N2 P101-107	58-1960 <*>
	58-2309 <*>
1954 V51 N3 P179-190	61-16254 <*>
1954 V51 N3 P192-202	62-18951 <*> O
1954 V51 N6 P385-400	60-10183 <*> O
1954 V51 N7 P482-488	2302 <*>
1954 V51 N7 P503-513	TS-1352 <BISI>
	62-10507 <*>
1954 V51 N9 P589-597	II-381 <*>
1954 V51 N9 P597-603	II-410 <*>
1954 V51 N9 P603-613	59-17340 <*> P
1954 V51 N9 P614-616	II-366 <*>
1954 V51 N9 P623-657	61-16412 <*> O
1954 V51 N10 P603-701	57-2541 <*>
1954 V51 N10 P665-673	60-10181 <*> O
1954 V51 N10 P693-701	58-976 <*>
1954 V51 N10 P723-734	61-16257 <*>
1954 V51 N11 P749-757	II-465 <*>
1955 N8 P710-715	61-18739 <*>
1955 V52 P369-374	58-1571 <*>
1955 V52 P477-484	2329 <BISI>
1955 V52 P559-568	64-10537 <*>
1955 V52 P982-994	TS-647 <BISI>
1955 V52 N2 P85-93	3565 <HB>
1955 V52 N2 P121-134	57-3448 <*>
	63-01041 <*>
1955 V52 N7 P569-578	SCL-T-354 <*>
1955 V52 N9 P676-690	1201 <*>
1955 V52 N9 P731-734	II-383 <*>
1955 V52 N10 P757-763	57-3325 <*>
1955 V52 N11 P869-886	TS 1469 <BISI>
1955 V52 N12 P961-964	65-14940 <*>
1955 V52 N12 P982-994	61-16256 <*>
1956 V53 P461-470	TS-1353 <BISI>
	62-10508 <*>
1956 V53 P575-583	61-18398 <*>
1956 V53 N4 P255-262	58-51 <*>
1956 V53 N8 P584-618	61-18626 <*>
1956 V53 N8 P638-644	AD-613 694 <=$>
1956 V53 N9 P665-681	61-18396 <*>
1956 V53 N9 P689-700	61-16284 <*>
1956 V53 N12 P897-914	59-20986 <*>
1957 V54 P175-180	TS-836 <BISI>
1957 V54 N2 P101-106	61-18393 <*>
1957 V54 N3 P175-180	61-16253 <*>
1957 V54 N7 P529-537	61-16258 <*>
1957 V54 N7 P537-554	61-16255 <*> ⊙
1957 V54 N8 P597-610	58-304 <*>
1957 V54 N9 P733-736	61-16252 <*>
1957 V54 N10 P803-812	3047 <BISI>
1958 P448-452	HW-TR-3 <*>

Citation	Translation
1958 V55 P407-416	59-17777 <*>
1958 V55 P453-458	58-2293 <*>
1958 V55 P601-612	TS-1146 <BISI>
1958 V55 P1188-1194	HW-TR-1 <*>
	HW-TR-1 <=*>
1958 V55 N1 P9-16	61-18397 <*> O
	919 <BISI>
1958 V55 N1 P24-33	4591 <HB>
1958 V55 N1 P39-52	90M40F <ATS>
1958 V55 N1 P53-60	65-17330 <*>
1958 V55 N2 P107-122	TS-1451 <BISI>
1958 V55 N2 P186-200	61-18667 <*>
1958 V55 N5 P486-494	61-18668 <*>
1958 V55 N6 P524-530	AEC-TR-3531 <=*>
1958 V55 N6 P573-594	2455 <BISI>
1958 V55 N7 P601-612	62-10490 <*>
1958 V55 N7 P656-678	61-18740 <*>
1958 V55 N7 P679-695	60-10748 <*>
1958 V55 N9 P815-828	NP-TR-828 <*>
	72L28F <ATS>
1958 V55 N9 P829-839	73L28F <ATS>
1958 V55 N10 P913-917	60-15286 <=*> O
1958 V55 N11 P1023-1041	61-10254 <*>
	65-10026 <*> P
1958 V55 N11 P1091-1109	62-10506 <*>
1958 V55 N12 P1113-1125	59-15837 <*>
1959 P445-460	TS 1693 <BISI>
1959 N3 P237-241	65-12325 <*>
1959 V56 P163-170	TS-1380 <BISI>
1959 V56 P175-180	TS 1565 <BISI>
1959 V56 P179-202	TS 1693 <BISI>
1959 V56 P211-220	TS 1546 <BISI>
1959 V56 P237-246	UCRL TRANS-656(L) <*>
1959 V56 P395-406	TS 1403 <BISI>
1959 V56 N1 P11-24	TS-1263 <BISI>
	62-14372 <*>
1959 V56 N1 P40-48	63-14661 <*>
1959 V56 N1 P49-54	61-18967 <*>
1959 V56 N1 P55-60	HW-TR-6 <=*>
1959 V56 N1 P83-103	TS-1264 <BISI>
	62-14059 <*>
1959 V56 N2 P105-121	HW-TR-5 <=*>
1959 V56 N2 P163-170	62-10512 <*>
1959 V56 N2 P171-178	62-14342 <*>
1959 V56 N4 P395-406	62-14373 <*>
1959 V56 N5 P409-417	64-14007 <*> O
1959 V56 N5 P487-503	65-10098 <*> O
1960 P911-917	AEC-TR-5272 <*>
1960 P919-923	AEC-TR-5273 <*>
1960 V57 P88-90	TS-1719 <BISI>
1960 V57 P481-490	TS-1850 <BISI>
1960 V57 P815-826	2067 <BISI>
1960 V57 P911-917	2167 <BISI>
1960 V57 P925-933	2166 <BISI>
1960 V57 P1091-1105	2373 <BISI>
1960 V57 P1125-1132	2374 <BISI>
1960 V57 P1133-1142	AEC-TR-5278 <*>
1960 V57 P1133-1141	2375 <BISI>
1960 V57 P1143-1157	2265 <BISI>
1960 V57 N2 P107-114	62-14458 <*> O
1960 V57 N2 P117	1824 <BISI>
1960 V57 N2 P135-148	1816 <BISI>
1960 V57 N4 P297-303	61-10626 <*>
1960 V57 N4 P337-346	3238 <BISI>
1960 V57 N4 P347-353	1796 <BISI>
1960 V57 N5 P379-386	1895 <BISI>
1960 V57 N5 P397-400	61-10625 <*>
1960 V57 N5 P401-407	61-10633 <*>
1960 V57 N5 P423-436	2074 <BISI>
1960 V57 N5 P437-444	1873 <BISI>
1960 V57 N6 P491-499	1851 <BISI>
1960 V57 N6 P507-519	1889 <BISI>
1960 V57 N7 P573-587	2892 <BISI>
1960 V57 N7 P589-600	2024 <BISI>
1960 V57 N7 P607-612	1670 <BISI>
1960 V57 N7 P623-630	61-16463 <*> O
1960 V57 N8 P715-724	1978 <BISI>
1960 V57 N9 P797-803	2206 <BISI>
1960 V57 N9 P801-803	61-10624 <*>
1960 V57 N10 P919-923	2010 <BISI>

1960 V57 N1C P947-954 5290 <HB>
1960 V57 N11 P985-989 1968 <BISI>
1960 V57 N11 P991-997 2106 <BISI>
1960 V57 N11 P1003-1018 3341 <BISI>
1960 V57 N11 P1019-1031 2055 <BISI>
1960 V57 N12 P1081-1089 2153 <BISI>
1960 V57 N12 P1105-1106 62-16204 <*>
1961 V58 P39-43 2327 <BISI>
1961 V58 P92-116 4296 <BISI>
1961 V58 P176-182 AEC-TR-4795 <*>
1961 V58 P187-193 2174 <BISI>
1961 V58 P429-436 61-00953 <*>
1961 V58 P716-724 HW-TR-37 <*>
1961 V58 N1 P13-24 2210 <BISI>
1961 V58 N2 P131-136 2702 <BISI>
1961 V58 N2 P165-176 2719 <BISI>
1961 V58 N3 P249-256 2703 <BISI>
1961 V58 N3 P257-266 2738 <BISI>
1961 V58 N3 P271-275 5235 <HB> P
1961 V58 N6 P465-482 2944 <BISI>
1961 V58 N6 P527-535 2747 <BISI>
1961 V58 N7 P589-598 2700 <BISI>
1961 V58 N8 P655-660 62-14425 <*> 0
1961 V58 N8 P661-665 62-10742 <*>
1961 V58 N9 P723-734 2615 <BISI>
1961 V58 N9 P751-763 3142 <BISI>
1961 V58 N10 P807-821 62-14223 <*> 0
1961 V58 N10 P849-856 2653 <BISI>
1961 V58 N11 P981-990 2717 <BISI>
1961 V58 N12 P1039-1048 2830 <BISI>
1962 V59 P547-551 HW-TR-47 <*>
1962 V59 N2 P93-109 2917 <BISI>
1962 V59 N3 P223-236 2940 <BISI>
1962 V59 N3 P236-241 62-16838 <*> 0
1962 V59 N3 P237-241 2928 <BISI>
1962 V59 N3 P249-254 2961 <BISI>
1962 V59 N4 P319-322 64-14013 <*> P
1962 V59 N5 P445-450 3110 <BISI>
1962 V59 N6 P511-518 3053 <BISI>
1962 V59 N9 P751-756 3254 <HB>
1962 V59 N10 P817-827 63-16466 <*> 0
1962 V59 N11 P815-951 3182 <BISI>
1962 V59 N12 P1043-1054 3182 <BISI>
1962 V59 N7/8 P631-637 3107 <BISI>
1962 V59 N7/8 P651-660 4261 <BISI>
1963 P1221-1232 ORNL-TR-234 <*>
1963 V60 P359-365 4898 <BISI>
1963 V60 P535-544 4361 <BISI>
1963 V60 P631-643 4259 <BISI>
1963 V60 P1221-1232 4262 <BISI>
1963 V60 P1233-1240 4847 <BISI>
1963 V60 N1 P9-22 3380 <BISI>
1963 V60 N1 P23-37 3679 <BISI>
1963 V60 N1 P49-58 3701 <BISI>
1963 V60 N2 P105-113 3283 <BISI>
1963 V60 N2 P127-140 3375 <BISI>
1963 V60 N3 P215-236 3649 <BISI>
1963 V60 N3 P237-249 3457 <BISI>
1963 V60 N3 P269-272 3371 <BISI>
1963 V60 N5 P371-376 6019 <HB>
1963 V60 N5 P421-425 3792 <BISI>
1963 V60 N5 P439-456 3396 <BISI>
1963 V60 N5 P477-485 3629 <BISI>
1963 V60 N9 P763-773 3662 <BISI>
1963 V60 N9 P807-814 6138 <HB>
1963 V60 N10 P845-849 6113 <HB>
1963 V60 N10 P855-858 6130 <HB>
1963 V60 N10 P859-861 3711 <BISI>
64-16018 <*>
1963 V60 N1C P863-868 3714 <BISI>
1963 V60 N10 P869-870 6131 <HB>
1963 V60 N10 P889-903 3730 <BISI>
1963 V60 N11 P944-947 3998 <BISI>
1963 V60 N11 P948-957 3999 <BISI>
1963 V60 N12 P1241-1247 64-20954 <*>
1963 V60 N7/8 P623-629 3529 <BISI>
1963 V60 N7/8 P663-668 3458 <BISI>
1963 V60 N7/8 P684-691 66-14469 <*>
1964 V61 P493-503 4017 <BISI>
1964 V61 P781-789 65-11728 <*>
1964 V61 P1043-1060 4302 <BISI>
1964 V61 N1 P1-26 3706 <BISI>
1964 V61 N1 P53-63 3852 <BISI>
1964 V61 N1 P63-69 65-11727 <*>
1964 V61 N1 P71-85 3850 <BISI>
1964 V61 N1 P87-91 3811 <BISI>
1964 V61 N2 P121-145 3809 <BISI>
1964 V61 N2 P205-208 4075 <BISI>
1964 V61 N3 P233-255 3858 <BISI>
1964 V61 N3 P281-296 3816 <BISI>
1964 V61 N3 P303-309 3800 <BISI>
1964 V61 N4 P379-388 3856 <BISI>
1964 V61 N4 P389-394 3857 <BISI>
1964 V61 N5 P475-491 3996 <BISI>
1964 V61 N9 P769-775 6416-I <HB>
1964 V61 N9 P775-780 6416-2 <HB>
1964 V61 N9 P781-789 65-11728 <*>
1964 V61 N10 P831-838 4597 <BISI>
1964 V61 N12 P1061-1063 6486 <HB>
1964 V61 N12 P1093-1097 6465 <HB>
1964 V61 N7/8 P623-634 3971 <BISI>
1964 V61 N7/8 P639-646 4068 <BISI>
1964 V61 N7/8 P653-659 3919 <BISI>
1964 V61 N7/8 P661-670 3920 <BISI>
1965 V62 P37-45 4712 <BISI>
1965 V62 P197-200 4363 <BISI>
1965 V62 P299-309 4628 <BISI>
1965 V62 P687-710 4513 <BISI>
1965 V62 P1161-1179 4786 <BISI>
1965 V62 N1 P15-35 4404 <BISI>
1965 V62 N1 P49-57 6514 <HB>
1965 V62 N2 P87-114 4230 <BISI>
1965 V62 N3 P243-248 4364 <BISI>
1965 V62 N4 P339-341 4432 <BISI>
1965 V62 N4 P351-357 66-12834 <*>
1965 V62 N12 P1187-1191 6832 <HB>
1965 V62 N7/8 P711-719 6652 <HB>
1966 V63 P23-26 4845 <BISI>
1966 V63 P242-246 4970 <BISI>
1966 V63 P285-295 4948 <BISI>

REVUE METAPSYCHIQUE
1958 P1091-1109 TS-1225 <BISI>

REVUE DE MICROPALEONTOLOGIE
1962 V4 N4 P237-248 11P63F <ATS>

REVUE MILITAIRE D'INFORMATION
1957 N6 P13-17 59-11210 <=>

REVUE NEUROLOGIQUE
1902 V10 P394-401 1139 <*>
1908 V16 P2028-2029 61-10242 <*>
1943 V75 P154 61-16499 <*>
1946 V78 P581-584 57-3165 <*>
1947 V79 P683-687 57-3164 <*>
1947 V79 N4 P273-276 57-02699 <*>
1948 V80 P321-337 65-13934 <*>
1952 V86 N2 P168-171 64-10321 <*>
1952 V86 N2 P171-174 64-10320 <*>
1952 V86 N4 P319-327 59-20319 <*> 0
1954 V91 P428-444 57-2337 <*>
1955 V93 N5 P783-787 60-18941 <*> 0
1956 V95 N5 P381-387 61-10241 <*>
1957 V97 N6 P466-481 60-17646 <*+>
1958 V98 N4 P416-419 61-10652 <*>
1959 V100 P786-789 62-16066 <*>
1959 V100 N6 P786-789 65-17415 <*>
1959 V101 N3 P463-467 61-10043 <*> 0
1959 V101 N4 P501-523 63-01016 <*>
1959 V101 N4 P576-582 61-00262 <*>
1960 V103 N3 P276-283 65-12803 <*>
1962 V106 N2 P89-105 66-11809 <*>
1962 V107 N3 P204-210 65-12035 <*>
1962 V107 N3 P269-271 63-01096 <*>
1963 V108 N2 P199-201 66-12375 <*> 0
1964 V110 N4 P377-393 65-17266 <*>

REVUE DU NICKEL
1958 P39-46 60-141112 <*>

1960 V26 N5 P132-139 22N48F <ATS>
1964 V30 N3 P53-67 4020 <BISI>

REVUE D'OKA, AGRONOMIE, MEDECINE, VETERINAIRE
1940 V14 N3 P69-78 2396-F <K-H>
1956 V30 P30-36 65-12654 <*>

REVUE D'OPTIQUE THEORIQUE ET INSTRUMENTALE
1922 V1 N1 P13-22 58-40 <*>
1922 V1 N9 P397-412 58-37 <*>
1927 V6 P31-33 57-309 <*>
1928 V7 P2C5-214 61-14150 <*>
1929 V8 P59-69 64-71233 <=>
66-11950 <*> 0
1930 V9 N4 P164-168 61-14201 <*>
1930 V9 N10 P414-420 61-14108 <*>
1932 V11 N3 P97-104 AD-441 397 <=>
1945 P27-102 I-435 <*>
1945 V24 N1/3 P1-10 AL-289 <*>
59-19005 <=*>
1950 V29 P499-512 3011 <*>
1951 V30 N11 P453-470 57-2406 <*>
1952 V31 N10 P457-458 62-18419 <*> 0
1953 V32 N5 P257-268 65-12340 <*>
1954 V33 N10 P497-501 57-1899 <*>
1954 V33 N10 P513-518 57-1898 <*>
1955 V34 N3 P150-163 64-18504 <*>
1956 V35 N1 P37-42 60-16811 <*>
1956 V35 N7 P414-419 62-10136 <*>
1956 V35 N11 P569-589 61N49F <ATS>
1956 V35 N12 P642-656 62N49F <ATS>
1957 V36 P281-284 UCRL TRANS-955(L) <*>
1961 V40 N5 P213-230 65-13080 <*>
1961 V40 N8 P407-422 32P60F <ATS>
1964 V43 N11 P573-593 66-13755 <*>

REVUE D'ORTHOPEDIE ET DE CHIRURGIE DE L'APPAREIL MOTEUR
1910 V30 P729-767 59-15523 <*>
1947 V33 P3-6 66-10866 <*>

REVUE D'OTO-NEURO-OPHTALMOLOGIE
1929 V7 P255-262 61-14647 <*>
1956 V31 P425-447 57-3151 <*>

REVUE DE PATHOLOGIE COMPAREE ET D'HYGIENE GENERALE
1951 V51 N629 P411-414 61-20734 <*> 0

REVUE DE PATHOLOGIE GENERALE ET COMPAREE
1955 V55 N667 P527-556 3779-I <K-H>

REVUE DE PATHOLOGIE GENERALE ET DE PHYSIOLOGIE CLINIQUE
1956 V56 N674 P80-92 59-15433 <*>
1960 V60 P643-680 63-14387 <*>

REVUE PETROLIFERE. PARIS
1927 P22- 63-10451 <*>
1930 P469-472 63-10465 <*>
1938 N784 P573 63-10423 <*>

REVUE DE PHILOLOGIE DE LITTERATURE ET D'HISTOIRE ANCIENNES
1904 SNS V28 N57 P462-492 61-14691 <*>
1906 SNS V30 N61 P526-529 61-14690 <*>

REVUE PHILOSOPHIQUE DE LA FRANCE ET DE L'ETRANGER
1893 V33 P334-336 AL-61 <*>

REVUE DE PHYSIQUE. BUCHAREST
1959 V4 P317-326 00Q73R <ATS>
1959 V4 N2 P235-243 64-15594 <=*$>
1961 V6 P211-217 65-10799 <*>
1961 V6 N1 P73-79 65-13616 <*>
1961 V6 N3 P363-383 NP-TR-952 <*>
1963 V8 P269-275 66-11238 <*>
1963 V8 P379-382 66-11155 <*>

REVUE DE PRATICIEN
1958 V8 N1C P1C39-1040 60-10665 <*>
1960 V10 N9 P917-930 65-17273 <*>
1960 V10 N20 P2204-2206 61-10821 <*>

REVUE DES PRODUITS CHIMIQUES ET L'ACTUALITE-SCIENTIFIQUE REUNIES
1934 V37 P705-712 2066 <*>
1934 V37 P737- 2064 <*>
1935 P33- 2065 <*>
1935 P65- 2063 <*>
1935 P385- 2061 <*>
1935 P417- 2060 <*>
1936 P449- 2062 <*>
1954 V57 P4C1-404 II-1030 <*>
1954 V57 N1207 P401-404 59-15166 <*>
1954 V57 N1207 P451-454 59-15166 <*>
1957 V60 P305-309 R-4778 <*>
1960 V63 N1 P3-7 61-20888 <*>
1960 V63 N2 P53-55 61-20888 <*>
1962 V65 P3-7 659 <TC>
1965 V68 P50-52 65-14929 <*>

REVUE DE PSYCHOLOGIE APPLIQUEE
1951 V1 N2 P11-17 64-10332 <*>
1951 V1 N2 P61-72 65-12542 <*>
1951 V1 N2 P73-80 64-10332 <*>
1953 V3 N1 P1-16 65-12558 <*>
1953 V3 N3 P295-298 58-2237 <*>
1953 V3 N3 P301-308 65-12378 <*>
1954 V4 P55-63 65-12596 <=$>
1954 V4 N1 P64-74 65-12358 <*>
1954 V4 N3 P317-340 65-12376 <*>
1955 V5 N1 P13-28 65-12565 <*>
1956 V6 P15-27 65-12664 <*>
1957 V7 N3 P177-185 60-18014 <*>
1957 V7 N3 P221-227 59-20169 <*>
1958 V8 N1 P25-35 60-18013 <*>
1958 V8 N3 P189-198 61-20433 <*>
1958 V8 N4 P257-264 60-18649 <*>
1960 V10 N2 P93-100 62-18063 <*>
1960 V10 N3 P139-150 64-30303 <*>
1961 V11 P353-360 SPEC 64-00182 <*>

REVUE DES QUESTIONS SCIENTIFIQUES
1958 S5 V19 P58-89 AEC-TR-4776 <*>
1963 V24 P343-366 64-41835 <=$>

REVUE DU RHUMATISME ET DES MALADIES OSTEOARTICULAIRES
1958 V25 N2 P93-102 59-10623 <*>
1961 V28 N12 P637-642 64-16933 <*> 0
1964 V31 N9 P479-485 65-14765 <*> 0
1965 V32 N10 P561-565 66-11534 <*> 0

REVUE ROMANDE D'AGRICULTURE, DE VITICULTURE ET D'ARBORICULTURE
1952 V8 N5 P36-37 2817-B <K-H>

REVUE ROUMAINE DE CHIMIE
1965 V16 P94-96 1427 <TC>

REVUE ROUMAINE DE METALLURGIE
1963 V8 N1 P67-77 3948 <BISI>

REVUE DES SCIENCES MEDICALES
1954 V2 P106 <CSIR>

REVUE SCIENTIFIQUE. PARIS
1899 S4 V11 N7 P207-208 63-01065 <*>
1922 V50 P838-840 57-1574 <*>
1929 V67 P65-69 57-899 <*>
1946 N84 P259-262 AL-178 <*>
1952 V90 N1 P120-134 63-00039 <*>
1953 V91 N2 P1C1-115 62-01521 <*>

REVUE. SOCIETE R. BELGE DES INGENIEURS ET DES INDUSTRIELS
1956 V5 P191-220 AEC-TR-3581 <*> 0
1960 N3 P101-126 AEC-TR-4399 <*>

REVUE DE LA SOUDURE
1955 V11 N4 P214-225 62-20126 <*>
1956 V12 N1 P50-70 62-20126 <*>
1956 V12 N2 P130-138 62-20126 <*>

REVUE DE LA SCUDURE AUTOGENE
1939 V31 P678-684 2666 <*>

REVUE DE STOMATOLOGIE
1954 V55 N4 P209-227 2744 <*>
1955 V56 N4 P257-269 2758 <*>
1955 V56 N2/3 P150-159 2759 <*>
1960 V61 N6 P326-330 11455-G <K-H>

REVUE SUISSE DE ZOOLOGIE
1957 V64 P236-246 63-00547 <*>
1958 V65 N37 P779-792 63-00612 <*>

REVUE TECHNIQUE C.F.T.H.
1960 N32 P7-39 62-14842 <*>
1960 N32 P41-58 62-14843 <*>
1960 N32 P59-72 62-14840 <*>
1960 N32 P73-82 62-14841 <*>

REVUE TECHNIQUE LUXEMBOURGEOISE
1952 V44 P171-180 3604 <BISI>
1958 V50 P157-159 TS-1841 <BISI>
1961 V53 N4 P238-248 62-18863 <*>

REVUE DES TELEPHONES, TELEGRAPHES ET T.S.F.
1935 P625-643 57-1071 <*>
1935 V13 N136 P366-385 57-2235 <*>

REVUE TEXTILE-TIBA-RUSTA-RAYONNE
1955 V54 P172-177 61-14531 <*> 0

REVUE TEXTILIS
1959 V15 N11 P35-48 61-14532 <*> 0

REVUE TRIMESTRIELLE CANADIENNE
1952 P14-16 59-17232 <*>

REVUE DE LA TUBERCULOSE ET DE PNEUMOLCGIE
1949 V13 N9/10 P746-761 2600 <*>
1950 V14 N3/4 P315-317 2597 <*>
1951 V15 P1188-1193 2596 <*>
1951 V15 N3 P215-228 2558 <*>
1954 S5 V18 P567-576 3359-A <K-H>

REVUE UNIVERSELLE DES MINES
1925 S7 V5 N5 P246-257 63-14044 <*>
1931 S8 V5 P106-109 59-15636 <*> 0
1931 S8 V5 N4 P102-106 60-10240 <*> 0
1937 S8 V13 N7 P310-312 59-15637 <*>
1945 P118-121 TS 1229 <BISI>
1945 P128-137 TS-1231 <BISI>
1945 S9 V1 N2/3 P118-121 61-20670 <*>
1945 S9 V1 N2/3 P122-128 TS 1230 <BISI>
61-20671 <*>
1945 S9 V1 N2/3 P128-137 62-14335 <*>
1948 S9 V4 N1 P64-66 64-71508 <=>
1948 S9 V4 N5 P335-345 62-10706 <*> 0
1951 V7 N3 P85-106 I-346 <*>
1951 S9 V7 P410-412 T-1444 <INSD>
1953 S9 V96 P563-567 60-10162 <*>
1953 S9 V96 N8 P644-657 60-10161 <*> 0
1955 S9 V11 N3 P111-120 57-1204 <*>
1955 S9 V98 N2 P68-75 3504 <BISI>
1955 S9 V98 N2 P86-97 3810 <HB>
1956 S9 V12 P454-465 4052 <BISI>
1956 S9 V12 P641-652 61-16406 <*>
1956 S9 V12 P652-660 2330 <BISI>
1956 S9 V12 N12 P641-652 58-536 <*>
1957 P165-212 TS-664 <BISI>
1957 S9 V13 N5 P165-212 61-16411 <*> 0
1957 S9 V13 N12 P699-707 61-14465 <*>
1958 S9 V14 P80-105 61-18365 <*>
1958 S9 V14 P261-269 61-18378 <*>
1958 S9 V14 P270-279 61-18745 <*>
1959 V15 P430-436 AEC-TR-5270 <*>
1959 S9 V15 P281-289 2087 <BISI>
1959 S9 V15 P783-795 1579 <BISI>
1959 S9 V15 N11 P783-795 62-16858 <*>
1960 S9 V16 P177-200 66-12858 <*>
1960 S9 V16 N6 P272-294 1952 <BISI>
1960 S9 V16 N7 P309-321 66-11068 <*> 0
1960 S9 V16 N10 P436-443 2098 <BISI>
1960 S9 V16 N11 P463-480 2157 <BISI>
1961 V17 N3 P99-110 3088 <BISI>
1961 V104 P431-437 8709 <IICH>
1961 S9 V17 N3 P110-134 2266 <BISI>
1961 S9 V17 N7 P401-408 5249 <HB>
1961 S9 V17 N10 P514-524 2953 <BISI>
1962 V18 N3 P287-303 2986 <BISI>
1962 V18 N6 P426-436 3114 <BISI>
1962 V18 N9 P559-577 63-20605 <*>
1962 V18 N12 P680-690 3353 <BISI>
1962 S9 V18 N3 P304-310 2763 <BISI>
1962 S9 V18 N3 P317-324 2934 <BISI>
1962 S9 V18 N6 P451-461 5688 <HB>
1962 S9 V18 N7 P482-499 3639 <BISI>
1962 S9 V18 N9 P559-577 63-10940 <*>
1962 S9 V105 P284-287 5690 <HB>
1962 S9 V105 P310-316 2764 <BISI>
1962 S9 V105 N6 P465-470 5689 <HB>
1963 V106 N1 P3-15 3643 <BISI>
1963 S9 V19 N6 P226-237 6076 <HB>
1963 S9 V19 N9 P367-383 3538 <BISI>
1963 S9 V19 N10 P403-418 3595 <BISI>
64-14257 <*>
1963 S9 V106 N4 P163-167 5950 <HB>
1963 S9 V106 N7 P289-303 3591 <BISI>
1963 S9 V106 N8 P325-341 3494 <BISI>
1964 V107 P233-241 4258 <BISI>
1964 V107 N2 P41-58 3787 <BISI>
1964 S9 V20 N2 P27-33 3661 <BISI>
1964 S9 V20 N4 P1-8 3704 <BISI>
1965 V108 P291-299 4474 <BISI>

REYON, SYNTHETICA, ZELLWOLLE
1951 V29 P362-367 61-16714 <*>

REYON, ZELLWOLLE UND ANDERE CHEMIEFASERN
1952 V30 N7 P343-346 I-324 <*>
60-10795 <*>
1952 V30 N10 P524-530 I-829 <*>
59-10220 <*> 0
1953 P27-28 I-484 <*>
1953 V31 P628-629 I-661 <*>
1954 V32 P69-71 1907 <*>
1954 V32 N6 P356-360 61-14023 <*> 0
1954 V32 N7 P403-413 61-14006 <*> 0
1954 V32 N8 P481-482 61-14024 <*>
62-10600 <*>
1955 N3 P142-155 CSIRO-2994 <CSIR>
1955 N5 P321-326 61-16781 <*>
1955 N9 P626-631 61-16782 <*>
1955 N11 P751-757 61-16783 <*>
1955 N12 P818-819 61-16784 <*>
1956 P88-93 57-1786 <*>
1956 P242-246 60-18516 <*>
1956 N9 P612-617 61-14533 <*>
1956 N12 P853-860 13K25G <ATS>
1956 V34 N1 P27-29 59-17101 <*>
1958 V8 N1 P28-32 58-1510 <*>
1958 V8 N1 P46-50 4K23G <ATS>
1958 V8 N2 P108-110 58-1493 <*>
1958 SNS V8 P666-670 37L29G <ATS>
1959 V9 N12 P786-787 63-10810 <*>
1959 SNS V9 P85-94 59-20992 <*>
1959 SNS V9 N7 P431-436 61-10226 <*>

REZULTATY ISSLEDOVANII PO MEZHDUNARODNYM GEOFIZICHESKII PROEKTAM
1963 N1 P25-32 AD-611 114 <=$>
1963 N1 P33-36 AD-611 116 <=$>
1963 N1 P37-63 AD-611 123 <=$>
1963 N1 P73-87 AD-611 036 <=$>
1963 N1 P88-94 AD-611 037 <=$>

RHEOLOGICA ACTA
1958 V1 P274-280 66-15572 <*> O
1958 V1 N2/3 P274-280 63-14374 <*> O
1958 V1 N2/3 P318-321 AEC-TR-4757 <*>
1958 V1 N4/6 P608-617 63-10799 <*>
1961 V1 N4/6 P361-370 C-5615 <NRC>
1962 V2 N4 P273-280 3393 <BISI>
1964 V3 N3 P178-180 38T90G <ATS>

RICERCA SCIENTIFICA
1934 V1 N8 P452-453 59-12933 <+*>
1934 V2 N3/4 P87-88 63-10114 <*>
1934 S1 V5 N5 P283- II-49 <*>
1935 S1 V6 N2 P123-125 II-2 <*>
1935 S1 V6 N11/2 P581-584 II-3 <*>
1936 V7 N2 P13- 59-22332 <+*>
1937 V8 P413-421 62-16457 <*> O
1939 V10 P28-31 66-10823 <*>
1939 V10 P905-914 63-14584 <*> O
1939 V10 P1143-1144 58-1288 <*>
1940 V11 P269-270 63-14834 <*>
1945 V15 P264-266 T-1454 <INSD>
1946 P1459-1462 57-1351 <*>
1947 V17 N12 P1998-2005 63-10618 <*>
1948 V18 N6 P831-839 <LSA>
1949 V19 N8 P851-862 11K20I <ATS>
1949 V19 N8 P870-874 I-110 <*>
1949 V19 N9 P1007-1009 I-110 <*>
1950 V20 N11 P1658-1661 I-485 <*>
1951 V21 N3 P372-378 I-870 <*>
1952 V22 N5 P864-893 2398 <*>
1953 V23 N1 P98-106 64-00460 <*>
1953 V23 N7 P1234-1237 59-10559 <*>
1954 V24 P73-75 1612 <*>
1954 V24 N9 P1858-1861 65-13120 <*>
66S811 <ATS>
1955 V25 P695-706 61-17634 <=$>
1955 V25 N10 P2834-2837 57-616 <*>
1955 V25 N12 P3244-3268 62-14570 <*>
1956 V26 P2-12 29J19I <ATS>
1956 V26 N3 P779-802 59-15119 <*>
1956 V26 N6 P1883-1885 5875 <K-H>
1956 V26 N9 P2792-2797 58-1724 <*>
1956 V26 N11 P3337-3341 65-11754 <*>
1957 V27 N5 P1448-1455 NP-TR-432 <*>
1957 V27 N5 P1509-1525 NP-TR-427 <*>
1957 V27 N5 P1546-1548 58-1364 <*>
1957 V27 N6 P1853-1864 NP-TR-431 <*>
1957 V27 N7 P2163-2172 58-1789 <*>
1957 V27 N8 P2468-2474 58-560 <*>
1958 V28 P1473-1479 59-17059 <*>
59-17760 <*> O
1958 V28 N3 P507-515 60-18407 <*>
1958 V28 N6 P1168-1173 59-22049 <*+>
1958 V28 N7 P1435-1442 78L31I <ATS>
1958 V28 N7 P1444-1450 61-16528 <*>
1958 V28 N8 P1611-1625 59-22583 <*+> O
1958 V28 N10 P2135-2137 61-10701 <*>
65-10333 <*>
1959 V29 N1 P100-105 62-16129 <*>
1959 V29 N3 P484-495 65-10222 <*>
1959 V29 N4 P804-809 61M43I <ATS>
1959 V29 N11 P2301-2313 C-3667 <NRCC>
1959 V29 N11 P2420-2426 62-16456 <*> O
1960 V30 N5 P680-685 63-15590 <=*>
1960 V30 N11 P1675-1679 11005 <K-H>
65-13541 <*>
1960 V30 N12 P1963-1968 11171-A <K-H>
65-13561 <*>
1960 V30 N12 P2030-2037 61-20559 <*>

RICERCA SCIENTIFICA. PARTE II. RENDICONTI. SEZIONE A: ABIOLOGICA
1962 V2 N5 P449-461 64-10908 <*>
1964 SA V6 N1 P37-46 1641 <TC>

RICERCA SCIENTIFICA. PARTE 11B. RENDICONTI SEZIONE BIOLOGICA
1961 SB V1 N1 P30-36 62-20242 <*>

RICERCHE DI INGEGNERIA
1937 V15 P117-121 10735 <KH>

RICERCHE E STUDI. ISTITUTO SPERIMENTALE STRADALE
1938 V2 P135-142 62Q73I <ATS>

RICHTLINIEN DES VEREINS DEUTSCHER INGENIEURE
SEE VDI RICHTLINIEN

RIECHSTOFFE UND AROMEN
1960 V10 P158-159 44N55G <ATS>

RIFORMA MEDICA
1934 V50 P1604-1611 59-22714 <=*> O
1935 V51 P93-94 60-13031 <=*>
1950 V64 N15 P405-408 59-17835 <*>
1954 V68 P1269-1270 1207 <*>
1956 V70 P417-423 65-11321 <*>
1956 V70 P936-940 65-11322 <*> O
1957 V71 P1421-1422 59-12134 <+*>
1958 V72 N1 P3-15 59-12429 <+*>
1958 V72 N32 P889-892 64-20017 <*>
1959 V73 P1347-1348 64-10618 <*>
1960 V74 P35 10764-C <K-H>
1960 V74 P604-606 64-20019 <*>
1960 V74 P989 10764-C <K-H>
1960 V74 P996 10764-C <K-H>
1960 V74 N35 P989-996 64-14573 <*>
1961 V75 P321-324 62-10916 <*>
1961 V75 P575-576 62-00831 <*>
1961 V75 P1392-1396 63-10569 <*>
1961 V75 N9 P1020 63-16518 <*>
1961 V75 N35 P1028-1029 62-01063 <*>
1963 V77 P34-39 64-18980 <*>
1964 V78 P29-32 65-14662 <*>

RIKAGAKU KENKYUSHO HOKOKU
1929 V8 P552-561 14G5J <ATS>
1932 N2 P124-134 5525-F <K-H>
1940 V19 P167-169 64-14409 <*> O
1941 V20 P201-208 15G5J <ATS>
1941 V20 N8 PT.1 P489-513 AL-896 <*>
1941 V20 N8 PT.2 P489-513 AL-897 <*>
1942 V21 P843-848 16G5J <ATS>
1942 V21 P992-999 I-252 <*>
1943 V22 N1 P112-114 63-14437 <*>
1944 V23 P788-790 80K24J <ATS>
1944 V23 N5 P281-283 RKK-23-5-1 <SA>
1947 V24 P14-16 66-20977 <$>
1948 V24 P454-457 UCRL TRANS-559(L) <*>
1948 V24 N4 P121-124 64-18862 <*> O
1950 V25 N4/5 P165-167 UCRL TRANS-603(L) <*>
1950 V26 P255-257 63-18954 <*> O
1950 V26 N1 P1-6 RKK-26-1-1 <SA> O
1950 V26 N10/1 P255-257 64-18890 <*>
1951 V27 P423-430 II-1148 <*>
1951 V27 P495-502 T-1540 <INSD>
1952 V28 P275-288 I-436 <*>
1954 V30 N4 P233-248 AD-614 278 <=$>
1954 V30 N6 P306-328 65-12233 <*>
1957 V33 N2 P53-57 62-20484 <*> O
1957 V33 N6 P350-352 58-2412 <*>
1958 V34 P330-338 63-00434 <*>
1958 V34 N4 P283-288 80M40J <ATS>
1958 V34 N4 P289-293 81M40J <ATS>
1959 V35 N6 P426-428 10607-C <KH>
63-16687 <*>
1960 V36 P386-398 64-10500 <*> O
1961 V37 P429-431 63-20671 <*>
1961 V37 N4 P271-275 NP-TR-1027 <*>
1961 V37 N5 P290-296 62-16842 <*>
1962 V38 P90-94 65-12074 <*>
1962 V38 N1 P81-89 RKK-38-1-1 <SA>
78P64J <ATS>
1962 V38 N3 P288-300 SCL-T-502 <*>
1963 V39 N1 P27-34 64-10499 <*> O
68Q72J <ATS>
1965 V41 P194-200 66-11709 <*> O

RIKUSUI-GAKU ZASSHI

1958 V19 N3/4 P118-129 59-19343 <+*>

RINASCENZA MEDICA
1927 V4 P424-426 57-1540 <*>

RINGYO SHIKENJO KENKYU HOKOKU
1952 N53 P59-68 I-866 <*>
1954 N68 P229-243 2058 <*>
1957 N97 P61-72 58-2529 <*>
1957 V97 P64-71 CSIB-3786 <CSSA>
1959 N116 P75-83 C-3255 <NRCC> O
1960 N119 P95-166 C-3497 <NRCC> P

RINSHO BYORI
1953 V1 N3 P284-292 1187 <*>

RINSHO FUJINKA SANKA
1962 V16 P615-617 63-16489 <*>

RINSHO GANKA
1957 V11 N3 P524-525 57-3436 <*>
1963 V17 P875-877 65-14546 <*>

RINSHO GEKA
1963 V18 P1319-1326 65-17264 <*>
1964 V19 N10 P1348-1349 66-11805 <*>

RINSHO NAIKA SHONIKA
1948 V3 N11 P429-434 64-14653 <*>

RIVISTA DI ANATOMIA PATOLOGICA E DI ONCOLOGIA
1955 V9 N6 P131-138 62-00319 <*>
1955 V9 N6 P661-696 II-1043 <*>

RIVISTA DI AUDIOLOGIA PRATICA
1953 V3 N1/3 P21-52 60-18555 <*>
1956 V6 P65-88 57-3149 <*>
66J16I <ATS>

RIVISTA DI BIOLOGIA
1955 V47 P181-189 1467 <*>
1958 V50 N4 P345-360 66-12840 <*>
1961 V54 P155-167 62-00875 <*>

RIVISTA DI CLINICA MEDICA
1940 V41 P61-85 1856 <*>

RIVISTA DI CLINICA PEDIATRICA
1957 V57 P238-249 62-00371 <*>
1957 V60 N5 P474-480 61-00196 <*>

RIVISTA DEI COMBUSTIBILI
1951 V5 P556-567 64-14371 <*> O
1954 P417-428 II-953 <*>
1954 V8 P491-512 03G51 <ATS>
1954 V8 N1 P5-14 63G51 <ATS>
1954 V8 N5 P449-453 14F4I <ATS>
1954 V8 N7/8 P526-544 61-10150 <*> O
1954 V8 N7/8 P545-550 61-10153 <*> O
1955 V9 P417-434 64-16134 <*> O
1955 V9 N7 P595-606 64-14958 <*>
1956 N2 P77-99 T-1664 <INSD>
1956 V10 N2 P77-100 65-10950 <*> O
1956 V10 N11 P803-829 65-12418 <*>
1957 V11 P245-259 61-10576 <*>
1957 V11 P308-318 58-1188 <*>
1957 V11 N3 P157-166 64-20209 <*> O
1957 V11 N12 P829-840 65-10229 <*>
1958 V12 N3 P187-195 64-18798 <*>
1958 V12 N6 P474-493 59-15532 <*>
61-10577 <*>
65-11449 <*>
1959 V13 N3 P187-196 65-11757 <*>
1959 V13 N6 P443-449 86M42I <ATS>
1959 V13 N7/8 P493-522 65-11444 <*>
1962 V16 P340-349 CHEM-332 <CTS>
1962 V16 N9 P371-378 66-11910 <*>
1962 V16 N9 P379-383 63-26860 <=$>
1962 V16 N10 P423-428 65-12266 <*>
1964 V18 N4 P140-142 61R78I <ATS>
1964 V18 N5 P177-182 65-12717 <*>
1964 V18 N6 P221-232 66-10358 <*>
1964 V18 N9 P331-344 66-10357 <*>
1965 V19 N9 P413-425 1861 <TC>

RIVISTA CRITICA DI CLINICA MEDICA
1957 V57 P51-58 60-14193 <*>

RIVISTA DI EMOTERAPIA ED IMMUNOEMATOLOGIA
1955 V2 P190-197 58-264 <*>
1957 V4 N2 P97-112 65-00288 <*> O
1957 V4 N4 P265-276 65-00263 <*>

RIVISTA DEL FREDDO
1937 V23 P45-48 59-20769 <*>

RIVISTA DI GASTRO-ENTEROLOGIA
1959 V11 P409-415 61-10900 <*> O

RIVISTA DI GEOFISICA APPLICATA
1953 V1 P3-13 2886 <*>

RIVISTA DE HIDROCARBUROS Y MINAS
1953 V4 N13 P45-77 59-10979 <*> O

RIVISTA DI INGEGNERIA
1954 V4 N2 P165-171 SCL-T-389 <*>
1962 V12 N7 P721-733 3685 <BISI>

RIVISTA DELL'ISTITUTO SIEROTERAPICO ITALIANO
1948 V23 P204-222 58-401 <*>
1949 V24 P157-167 58-408 <*>
1950 V25 P1-8 58-410 <*>
1950 V25 P175-183 58-432 <*>
1951 V26 P185-192 58-427 <*>
1954 V29 P361-365 58-1814 <*>
1954 V29 N4 P309-315 2222 <*>
1955 V30 N2 P105-122 1445 <*>
1955 V30 N6 P414-423 1212 <*>
1960 V35 N2 P320-324 62-00328 <*>
1965 V40 P73-81 66-11589 <*>

RIVISTA DI ISTOCHIMICA NORMALE E PATOLOGICA
1954 V1 N1 P15-28 57-863 <*>
1956 V2 P449-460 66-11824 <*>
1957 V3 P95-104 66-11823 <*>
1964 V10 N1/2 P53-131 65-00165 <*>

RIVISTA ITALIANA ESSENZE, PROFUMI, PIANTE OFFICINALI, OLLI VEGETALE, SAPONI
1960 V42 N5 P203-209 10880 B <K-H>

RIVISTA ITALIANA DI GINECOLOGIA
1952 V35 P128-140 58-773 <*>

RIVISTA ITALIANA D'IGIENE
1941 V1 N12 P3-5 AL-135 <*>

RIVISTA ITALIANA DELLA SALDATURA
1955 V7 P55-67 62-20193 <*>
1955 V7 P115-134 62-20193 <*>
1962 V14 P70-73 3612 <BISI>
1964 V16 P63-73 4149 <BISI>

RIVISTA ITALIANA DI STOMATOLOGIA
1947 V2 N9 P773-780 63-16965 <*>
1947 V2 N11 P922-935 63-16966 <*>

RIVISTA DI MALARIOLOGIA
1928 V7 P690-712 57-1508 <*>
1930 V9 P224-231 63-01091 <*>
1937 V16 P419-433 SECT.1 63-01115 <*>

RIVISTA DI MECCANICA. MILANO
1956 V7 N130 P27-33 58-1430 <*>

RIVISTA DI MEDICINA AERONAUTICA E APAZIALE
1963 V26 P478-508 N65-18180 <=$>

RIVISTA DI METEOROLOGIA AERONAUTICA

1963 V23 N1 P14-30 64-15599 <*>
1966 V26 N1 P11-27 66-13155 <*>

RIVISTA DI NEUROBIOLOGIA
1961 V7 P463-480 62-20427 <*>

RIVISTA DI NEUROLOGIA
1930 V3 P515-539 65-00169 <*> O
1952 V22 P720-733 59-14863 <+*>
1953 V23 N4 P357-363 59-17169 <*>
1958 V28 N2 P216-226 64-14570 <*>
8650-B <K-H>
1959 V29 P151-178 60-14871 <*>

RIVISTA DI OSTETRICIA E GINECOLOGIA PRATICA
1959 V41 P719-731 61-20247 <*>

RIVISTA OTO-NEURO-OFTALMOLOGICA
1952 V27 P197-205 3207-C <K-H>
1961 V36 N1 P14-27 65-17274 <*>

RIVISTA DI PARASSITOLOGIA
1937 V1 N4 P273-300 57-1265 <*>
1938 V2 N3 P233-241 57-1258 <*>
1939 V3 N1 P39-55 57-1429 <*>
1952 V15 N4 P513-540 60-15973 <*+>
1954 V15 N4 P285-304 57-1159 <*>
1958 V19 N1 P67-72 62-00310 <*>

RIVISTA DI PATOLOGIA E CLINICA
1955 V10 N7 P259-282 5796-D <K-H>
1955 V10 N7 P283-302 5796-E <K-H>
1955 V10 N12 P673-684 5796-G <K-H>
1955 V10 N12 P685-694 66-12046 <*>

RIVISTA DI PATOLOGIA NERVOSA E MENTALE
1940 V56 P555-605 T-1525 <INSD>
1958 V79 N1 P149-188 59-16348 <+*>

RIVISTA DI PATOLOGIA SPERIMENTALE
1932 V7 N7 P73-98 60-10849 <*>
1932 V8 N3/5 P327-335 I-649 <*>

RIVISTA SPERIMENTALE DI FRENIATRIA E MEDICINA LEGALE DELLE ALIENAZIONI MENTALI
1940 V64 P653-707 58-2660 <*>
1955 V79 P465-483 57-1913 <*>
1958 V82 N1 P205-220 59-15053 <*> O

RIVISTA TESSILE ARACNE
1957 V12 N11 P1111-1118 59-15272 <*>

RIVISTA TESSILE-TEXTILIA. MILANO
1962 V38 N3 P83-93 1444 <TC>
65-18056 <*>

RIVISTA TRIMESTRALE DI DIRITTO E PROCEDURA CIVILE
1960 V14 P169-177 61-18624 <*>

RIVISTA DI VITICOLTURA E DI ENOLOGIA
1953 V6 N4 P98-102 CSIRO-3699 <CSIR>
1953 V6 N5 P130-135 CSIRO-3701 <CSIR>
1953 V6 N6 P176-178 CSIRO-3702 <CSIR>
1954 V7 P247-255 3051 <*>
62-14955 <*>

ROCZNIK AKADEMII MEDYCZNEJ IM. JULIANA MARCHLEWSKIEGO W BIALYSTOKU
1958 P5-62 SUP2 10401B <K-H>
1961 N16 P1-84 SUP AD-636 585 <=$>

★ROCZNIK CHEMII
1926 V6 P404-414 65-14761 <*>
1931 V11 P664-669 3274 <*>
1933 V13 P5-15 MF-6 <*>
1934 V14 P430-450 AD-633 414 <=$>
1936 V16 P104-112 58-1287 <*>
1938 V18 P434-438 AL-143 <*>
1938 V18 P614-624 63-16800 <*>
1949 V23 N5 P361-379 64-20311 <*> O
1950 V24 P88-117 64-20130 <*>
1950 V24 P229-237 64-20129 <*>
1950 V24 N1/6 P144-166 60-16795 <*>
1951 V25 P35-45 64-20121 <*>
1951 V25 P388-391 63L35P <ATS>
1952 V26 P281-286 R-1807 <*>
1952 V26 P494-495 57-2787 <*>
1952 V26 P688-689 II-440 <*>
2956 <*>
1953 V27 P161-166 T1821 <INSD>
1953 V27 P207-208 47Q71P <ATS>
1953 V27 P218-226 AEC-TR-4631 <*>
1953 V27 P311-313 57-653 <*>
1953 V27 P482-493 II-441 <*>
2648 <*>
2885 <*>
1953 V27 N4 P527-528 57-536 <*>
57-652 <*>
1954 V28 P3-10 1255 <*>
1954 V28 P109-123 21P59P <ATS>
1954 V28 P611-628 SC-T-64-1605 <*>
1954 V28 P673-675 PJ-662 <ATS>
1954 V28 N12 P12-20 117TM <CTT>
1955 V29 N2/3 P921-925 61-11372 <=>
1955 V29 N2/3 P943-944 1144 <*>
1956 V30 P185-194 118TM <CTT>
58-790 <*>
1956 V30 P195-199 116TM <CTT>
1956 V30 P311-314 63-20128 <*> O
1956 V30 P323-325 <LSA>
1956 V30 P385-397 33M47P <ATS>
1956 V30 P655-658 57-1990 <*>
1956 V30 P723-732 17K21P <ATS>
1956 V30 N1 P29-38 995TT <CCT>
1956 V30 N2 P569-586 61-31278 <=$>
1956 V30 N2 P587-606 61-31279 <=$>
1956 V30 N3 P981-983 63-11390 <=>
1957 V31 P319-321 1031-TT <CCT>
59-11922 <=*>
1957 V31 P323-324 1030TT <CCT>
1957 V31 P327-328 59-11920 <=*>
953TT <CCT>
1957 V31 P349-350 61-18717 <*>
1957 V31 P497-515 64-20111 <*>
1957 V31 P621-626 58-1179 <*>
1957 V31 P707-710 AD-626 822 <=*$>
1957 V31 P915-925 AEC-TR-6145 <*>
1957 V31 N2 P543-551 65-12927 <*>
1957 V31 N3 P793-800 02Q68P <ATS>
1957 V31 N3 P1065-1066 58-1351 <*>
1957 V31 N4 P1167-1176 59-15301 <*>
1957 V31 N4 P1327-1329 63-11390 <=>
1958 V32 P29-38 59-11919 <=*>
1958 V32 P545-552 59-15378 <*>
1958 V32 P837- 60-16592 <*>
1958 V32 P1257-1268 61-11387 <=>
1958 V32 N5 P1189-1190 NP-TR-604 <*>
1958 V32 N6 P1301-1309 61-10009 <*>
1959 V33 P549-550 75M37P <ATS>
1959 V33 P849-852 65-11238 <*>
9757 <K-H>
1959 V33 N1 ENTIRE ISSUE AEC-TR-4292 <=>
1959 V33 N1 P133-144 61-11374 <=>
1959 V33 N1 P145-156 61-11375 <=>
1960 V34 P385-389 AEC-TR-4827 <*>
1960 V34 P1149-1153 45N51P <ATS>
1960 V34 N5 P1211-1538 AEC-TR-4609 <=>
1960 V34 N3/4 P1155-1160 65-50328 <=>
1961 V35 P365-367 63-10242 <*>
1961 V35 N2/3 P671-678 62-23268 <=*>
1962 V36 P365-366 63-16594 <*>
1962 V36 N2 P203-213 64-16583 <*>
1962 V36 N2 P365-366 65-14376 <*>
1962 V36 N3 P403-409 66-10205 <*> O
1962 V36 N4 P929-936 10P66P <ATS>
1963 V37 P1525-1531 65-00274 <*>
1964 N6 P1007-1013 ICE V5 N1 P1-4 <ICE>
1964 N8 P1691-1697 ICE V5 N4 P599-602 <ICE>
1964 N9 P1347-1353 ICE V5 N2 P289-293 <ICE>
1964 N9 P1355-1360 ICE V5 N2 P293-296 <ICE>

1964 N9 P1361-1366 ICE V5 N3 P426-428 <ICE>
1964 N10 P1539-1547 ICE V5 N2 P297-301 <ICE>
1964 V38 N1 P35-42 695 <TC>
1964 V38 N1 P43-46 696 <TC>
1964 V38 N1 P105-113 RTS-2862 <NLL>
1964 V38 N3 P437-445 65-12889 <*>
1964 V38 N5 P899-901 65-11736 <*> 0
1964 V38 N6 P1007-1013 ICE V5 N1 P.1-4 <ICE>
1965 N3 P425-434 ICE V5 N4 P616-622 <ICE>
1965 V39 P1491-1497 ICE V6 N2 P315-319 <ICE>
1965 V39 N1 P101-106 1432 <TC>
1965 V39 N2 P141-149 65-14272 <*>
1965 V39 N2 P263-271 ICE V5 N4 P608-612 <ICE>
1966 V40 P1083-1090 ICE V6 N4 P663-668 <ICE>

ROCZNIK NAUK LESNYCH
1954 N5 P137-160 60-21523 <=>
1955 V11 P111-129 60-21391 <=>
1955 V13 P45-55 61-11328 <=>
1956 V14 P223-245 60-21383 <=>
1956 V14 P247-266 60-21382 <=>

ROCZNIK NAUK ROLNICZYCH
1937 V38 N1 P134-140 60-21485 <=>
1949 V51 P281-285 60-21489 <=>

ROCZNIK NAUK ROLNICZYCH. SERIA A. ROSLINNE
1954 V70 N1 P77-96 60-21526 <=>
1957 V74 N2 P359-372 60-21411 <=>
1957 V76 N2 P217-290 61-31255 <=>
1957 SA V74 N2 P287-314 60-21412 <=>
1959 V79 N3 P911-925 61-11326 <=>
1960 V81 N4 P975-990 11203 <KH>
1960 V81 N4 P1073-1C84 11151 <KH>
1960 V81 N4 P1097-1107 11095-H <K-H>
64-14569 <*>
1960 V81 N4 P1109-1117 11085-G <K-H>
64-14568 <*>
1961 V84 N4 P563-592 61-31325 <=>

ROCZNIK NAUK ROLNICZYCH. SERIA B. ZOOTECHNIKA
1954 V67 N4 P473-486 60-21571 <=>
1955 V69 N3 P399-439 60-21568 <=>
1955 V7C N1 P1-5 60-21418 <=>
1957 V71 N2 P223-249 60-21568 <=>
1957 V71 N3 P399-420 60-21568 <=>
1957 V71 N3 P469-533 60-21578 <=>
1958 V72 N4 P545-563 60-21568 <=>
1958 V73 N2 P119-148 61-11357 <=>
1959 V74 N1 P199-208 60-21572 <=>
1959 V74 N2 P375-378 60-21571 <=>
1960 V75 N1 P135-154 60-21572 <=>
1960 V75 N3 P283-299 60-21571 <=>
1962 V80 N1 P5-22 65-50502 <=>
1962 V80 N2 P115-125 65-50502 <=>
1962 V80 N3 P333-346 65-50502 <=>
1962 V81 N2 P151-305 65-50503 <=>

ROCZNIK NAUK ROLNICZYCH. SERIA C. MECHANIZACJA, KHRAKOV
1953 V66 N2 P97-123 61-11341 <=>
1957 V67 N1 P1-100 60-21569 <=>
1958 V67 N3 P357-381 60-21569 <=>
1960 V67 N4 P419-509 60-21569 <=>

ROCZNIK NAUK ROLNICZYCH. SERIA D. MONOGRAFIE
1954 V69 P5-171 60-21496 <=>

ROCZNIK NAUK ROLNICZYCH. SERIA E. WETERYNARIA
1960 V69 N3 P413-427 61-20902 <*>
1960 V70 N1/4 P86-88 62-01415 <*>

ROCZNIK NAUK ROLNICZYCH. SERIA F. MELIORACJI I UZYTKOW ZIELCNYCH
1955 V71 N1 P21-42 60-21544 <=>
1955 V71 N1 P57-71 60-21542 <=>
1955 V71 N1 P73-85 60-21536 <=>
1955 V71 N1 P90-109 60-21539 <=>
1958 V72 N3 P1-6 60-21416 <=>
1959 V73 N4 P679-712 60-21541 <=>

ROCZNIK PANSTWOWEGO ZAKLADU HYGIENY
1951 V2 P139-160 3759-A <K-H>
1957 V8 N5 P481-493 63-11402 <=>
1959 V10 N3 P197-216 61-00182 <*>
1959 V10 N4 P395-402 63-11402 <=>
1959 V10 N5 P413-421 61-00151 <*>
1960 V11 N4 P329-334 63-11402 <=>
1960 V11 N6 P541-550 11279-B <K-H>
1962 V13 N2 P175-182 19Q71P <ATS>

ROCZNIK POLSKIEGC TOWARZYSTWA GEOLOGICZNEGC
1937 V13 P194-250 60-21506 <=>
1962 V32 N1 P107-114 51Q70R <ATS>
1964 V34 P425-445 IGR V8 N4 P435 <AGI>

ROCZNIK POLSKIEGC TOWARZYSTWA MATEMATYCZNEGO
1948 V20 P347-372 T-2275 <INSD>

ROCZNIK POMORSKIEJ AKADEMII MEDYCZNEJ IM. GEN. KAROLA SWIERCZEWSKIEGO W SZCZECINIE
1958 V4 P213-226 64-18989 <*> 0

ROCZNIK SEKCJI DENDROLOGICZNEJ POLSKIEGO TOWARZYSTWA BOTANICZNEGC
1955 V1C P53-96 61-11331 <=>
1955 V10 P165-189 60-21381 <=>
1955 V10 P275-304 60-21228 <=>
1957 V12 P413-418 60-21394 <=>
1957 V12 P413-420 60-21394 <=>
1957 V12 P421-429 60-21393 <=>

ROCZNIKI CHEMII
SEE ROCZNIK CHEMII

ROCZNIKI GLEBOZNAWCZE
1960 V9 P3-25 61-31323 <=>
1960 V9 P49-86 65-50323 <=>
1960 V9 P87-101 61-31324 <=>
1960 V9 P137-141 61-31325 <=>
1960 V9 N1 P67-68 65-50318 <=>
1960 V9 N2 P57-66 61-11348 <=>
1963 V13 N1 P51-56 65-50317 <=>
1963 V13 N1 P67-78 65-50319 <=>

RODOVIA
1958 V20 N215 P17-2C 59-14581 <+*>
1958 V20 N215 P25-29 59-14581 <+*>

RODO KAGAKU
1956 V32 N8 P620-625 57-1881 <*>
1956 V32 N9 P682-691 57-3201 <*>

RODRIGUESIA
1942 V6 N15 P83-85 3351 <*>

ROENTGEN-BLAETTER
1957 V10 P315-318 58-1772 <*>
1959 V12 N4 P1-8 AEC-TR-3782 <*>
1959 V12 N4 P1C4-111 59-00727 <*>
1960 V13 P166-172 61-18670 <*>
1963 V16 N1C P305-314 18R75G <ATS>
1963 V16 N10 P321-328 19R75G <ATS>

ROHDE UND SCHWARZ MITTEILUNGEN
1952 N1 P4-15 II-873 <*>
1955 N7 P441-456 57-2314 <*>
1956 12/08 P1-12 57-3495 <*>
1957 N9 P73-129 59-14104 <+*>
1958 N10 P145-154 59-20864 <*> 0
1958 N11 P185-205 SCL-T-253 <+*>
61-19212 <+*>
1960 N14 P386-390 48R75G <ATS>
64-16961 <*>
1961 N15 P11-14 65-11602 <*>

ROHRE, ROHRLEITUNGSBAU, RCHRLEITUNGSTRANSPCRT
1963 V2 P313-315 4003 <BISI>

ROLNICTWO

1958 V5 N17 P23-33 61-31256 <=>
1959 V9 N25 P127-134 61-31257 <=>
1960 V10 N29 P35-45 61-31258 <=>

ROPA A UHLIE
1965 V7 N2 P35-43 65-18047 <*>
1965 V7 N6 P171-173 66-12879 <*>

RORSCHACHIANA
1952 V1 N3 P249-267 1888 <*>

ROST KRISTALLCV
1964 V4 P61-67 AD-613 985 <=$>
1964 V4 P68-73 AD-613 986 <=$>
59S86R <ATS>
1964 V4 P74-80 AD-613 984 <=$>
60S86R <ATS>

ROUMAINES DE PATHOLOGIE EXPERIMENTALE ET DE MICROBIOLOGIE
1933 V6 N1/2 P5-134 65-12366 <*> 0

ROUTE ET LA CIRCULATICN RCUTERIE
1938 N10 P165-172 AL-931 <*>
3340 <*>
1938 N1C P2C6-2C8 AL-931 <*>
1946 N23/4 P311-316 3334 <*>

ROUTES ET DES AERODROMES
1957 N11 P339-349 59-14889 <=*> 0
1957 N11 P373- 59-14899 <=*> 0

ROZHLEDY V TUBERKULOSE (A V NEMOCECH PLICNICH)
1959 V19 N8 P607-611 10997-A <KH>
1960 V20 N2 P116-121 10177-D <KH>
1960 V20 N6 P482-488 10501 <K-H>
63-16666 <*>
1962 V22 N1 P51-57 62R77C <ATS>

ROZPRAVY CESKE AKADEMIE VED A UMENI
1928 V37 N24 P1-83 64-11535 <=>

ROZPRAWY ELEKTROTECHNICZNE
1958 V4 N3 P287-317 59-18239 <=>
62-19869 <=*>
1961 V7 N1 P17-67 62-16919 <=*>

ROZPRAWY HYDRCTECHNICZNE
1957 N3 P3-83 60-21566 <=>
1958 N4 P97-120 60-21582 <=>
1958 N4 P169-180 61-31313 <=>
1959 N5 P91-106 61-31303 <=>
1959 N6 P95-108 60-21566 <=>
1959 N7 P89-110 61-31299 <=>

ROZPRAWY INZYNIERSKIE
1957 V5 N1 P119-134 65-64502 <=*$>
66-10484 <*>
1960 V8 N2 P203-210 63-16631 <*>
1962 V10 N3 P459-496 64-15595 <=*$>

ROZPRAWY MATEMATYCZNE
1957 V13 P3-41 STMSP V3 P131-166 <AMS>

ROZPRAWY WYDZIALU NAUK MEDYCZNYCH, POLSKA AKADEMIA NAUK. WARSZAWA
1957 V2 N1 P1-288 61-11335/2 <=>
1957 V2 N1 P137-192 60-21576 <=>
1958 V3 N1 P1-84 61-11335/3 <=>
1958 V3 N2 P1-67 61-11335/4 <=>
1958 V3 N3 P1-224 61-11335/5 <=>
1959 V4 N1 P1-254 61-11335/6 <=>
1960 V5 N1 P1-196 61-11335/7 <=>
1960 V5 N2 P1-232 61-11335/8 <=>

RUBRIEK VOOR HANDEL EN INDUSTRIE
1934 V31 P889-893 60-10781 <*>

★ RUDARSKO-METALURSKI ZBORNIK
1959 N1 P41-62 5113 <HB>

1962 N1 ENTIRE ISSUE 62-19467-1 <=>
1962 N3 P235-243 63-11450/1 <=>
1962 N4 P313-339 62-19467-4 <=>
1962 N4 P341-351 62-19467-4 <=>
1962 N4 P353-364 62-19467-4 <=>
1963 N1 P1-13 63-11450/1 <=>
1963 N1 P15-31 63-11450/1 <=>
1963 N1 P33-47 63-11450/1 <=>
1963 N1 P49-68 63-11450/1 <=>
1963 N2 P89-101 63-11450-2 <=>
1963 N2 P103-110 63-11450-2 <=>
1963 N2 P111-123 63-11450-2 <=>
1963 N2 P125-137 63-11450 <=>
1963 N2 P139-142 63-11450-2 <=>
1963 N3 P193-285 64-11450-3 <=>
1963 N4 P345-357 63-11450/4 <=>
1963 N4 P359-393 63-11450/4 <=>
1963 N4 P395-408 63-11450/4 <=>
1964 N3 P225-3C5 64-11458-3 <=$>
1965 N2 P137-207 65-50402/2 <=>
1965 N2 P223-228 65-50402/2 <=>

RUDE PRAVO
1963 P4 63-21237 <=>

RUDY. PRAGUE
1957 V5 N1 P341-346 PB 141 295T <=> 0
1960 V8 N12 P407-414 66-10263 <*>

RUDY I METALE NIEZELAZNE
1957 V2 N3 P79-84 59-11580 <=>
1957 V2 N4 P105-112 59-11580 <=>
1965 V10 N7 P383-386 CSIR-TR.576 <CSSA>
1967 N2 P79-82 67-31219 <=$>

RUMANIA STANDARDS
STAS 4741-55 63-10747 <*>
1955 STAS 4743-55 59-17473 <*> 0

RUNDFUNKTECHNISCHE MITTEILUNGEN
1959 V3 N1 P51-55 61-18607 <*>
1960 V4 N2 P66-73 33M44G <ATS>
62-14482 <*>
1962 V6 P102-1C3 65-11631 <*>
1962 V6 P111-113 64-14853 <*>
1962 V6 N1 P7-14 62-20215 <*>
1965 V9 N3 P145-156 59T91G <ATS>

RUSSIA STANDARDS
SEE USSR STANDARDS

RUSSKAYA KLINIKA
1924 V2 P307-317 59J16R <ATS>

RUSSKII GIDROBIOLOGICHESKII ZHURNAL
1923 V2/3 N11/2 P280-290 62-13757 <=*>
1925 V4/5 N1/2 P1-10 62-13757 <=*>
1926 V5 N1/2 P2-10 R-247 <*>
RT-2348 <*>
64-15211 <=*$>

RUSSKII OFTALMOLOGICHESKII ZHURNAL
1929 V10 P646-649 61-14424 <*>

RUSSKII VRACH
1906 V5 P135 59-10839 <*>
1911 V10 N48 P1829-1831 RT-3851 <*>
60-19629 <*> 0

RUSSKII ZHURNAL TROPICHESKOI MEDITSINY
1924 V2 N3 P7-14 64-15570 <=*$>
1927 N5 P5C2-512 65-60C66 <=$>
1927 V5 P654-663 60-19286 <=*$>
1929 V7 P563-565 R-1099 <*>
1929 V7 N9 P577-581 61-28115 <=> 0

RUSSKI ZOOLOGICHESKII ZHURNAL
1928 V8 N3 P39-65 62-11019 <=>
1928 V8 N4 P41-88 62-11019 <=>

RUSSKOE ENTCMOLOGICHESKOE OBOZRENIE
1926 V20 N3/4 P296-307 T.522 <INSD>
1930 V24 N3/4 P135-155 60-16642 <+*> O

RYBNOE KHOZYAISTVO
1946 N12 P31-34 FRBC-TRANS-70 <NRC>
1954 V30 N9 P19-23 63-15144 <=*>
1954 V30 N11 P41-43 63-41321 <=>
1955 V31 N2 P51-53 C-3024 <NRCC>
1955 V31 N2 P53-55 C-2997 <NRCC>
1956 V32 N2 P22-29 61-19746 <=*>
1956 V32 N2 P55-59 60-21098 <=>
1956 V32 N4 P54-59 60-51139 <=>
1957 V33 N6 P52-54 59-22305 <+*>
1957 V33 N7 P70-71 C-3900 <NRCC>
60-51139 <=>
1957 V33 N7 P74-75 65-61182 <=>
1958 V34 N1 P12-15 63-11103 <=*$>
1958 V34 N1 P22-24 61-23391 <*=>
1958 V34 N1 P36-42 65-61300 <=$>
1958 V34 N3 P12-16 61-28150 <*=> O
1958 V34 N5 P8-13 60-21100 <=>
1958 V34 N7 P22-24 59-11664 <=>
1958 V34 N7 P32-34 61-19943 <=*>
1958 V34 N7 P43-49 60-21146 <=>
1958 V34 N9 P6-9 60-13242 <+*>
1958 V34 N9 P29-31 59-14722 <+*>
1958 V34 N9 P40-49 60-51195 <=>
1958 V34 N9 P94 60-13246 <+*>
1958 V34 N10 P11-13 64-15983 <=*$>
1958 V34 N10 P13-16 60-13245 <+*>
1958 V34 N10 P62-64 60-21096 <=>
1958 V34 N11 P21-26 C-3231 <NRCC>
1958 V34 N12 P13-15 C-3402 <NRCC>
1959 V35 N1 P52-59 60-51190 <=>
1959 V35 N2 P5-8 60-17590 <+*>
1959 V35 N2 P33-39 60-51129 <=>
1959 V35 N2 P39-42 60-51180 <=>
1959 V35 N2 P43-46 60-51080 <=>
1959 V35 N2 P61-64 60-19181 <+*>
1959 V35 N2 P65-69 60-15881 <+*>
60-21132 <=>
1959 V35 N3 P20-22 60-31793 <=>
1959 V35 N4 P13-14 60-17589 <+*>
1959 V35 N4 P16-18 C-3392 <NRCC>
1959 V35 N4 P32-33 C-3622 <NRCC>
1959 V35 N6 P15-22 C-3393 <NRC> O
60-25017 <*>
1959 V35 N6 P23-24 C-3418 <NRCC>
1959 V35 N7 P7-16 C-3701 <NRCC>
61-19742 <*=>
1959 V35 N7 P16-21 C-3696 <NRCC>
1959 V35 N7 P53-58 C-4455 <NRC>
60-17117 <+*>
1959 V35 N7 P59-65 64-15985 <=*$>
1959 V35 N8 P8-9 C-3424 <NRCC>
1959 V35 N9 P22-24 C-3381 <NRCC>
1959 V35 N9 P39-46 61-19947 <=*>
1959 V35 N12 P42-47 C-4777 <NRC>
1959 V35 N12 P62-64 64-21286 <=>
1960 N7 P8-14 NLLTB V3 N2 P94 <NLL>
1960 V36 N1 P12-15 C-3635 <NRCC>
1960 V36 N7 P15-43 63-21559 <=>
1960 V36 N8 P20-25 C-3632 <NRCC>
61-00998 <*>
1960 V36 N9 P7-16 63-21863 <=> O
1960 V36 N9 P69-73 61-19916 <=*>
1960 V36 N11 P27-33 63-15126 <=*>
1960 V36 N11 P62-65 61-23639 <*=>
1961 N7 P17-22 NLLTB V4 N2 P153 <NLL>
1961 V37 N1 P3 61-23580 <=>
1961 V37 N1 P9-11 C-4460 <NRC>
63-00886 <*>
1961 V37 N1 P27-31 MAFF-TRANS-NS-41 <NLL>
1961 V37 N5 P40-43 63-41204 <=>
1961 V37 N5 P62-64 63-41204 <=>
1961 V37 N7 P43-49 MAFF-TRANS-NS-67 <NLL>
1961 V37 N9 P27-30 C-4451 <NRC>
63-00891 <*>
1961 V37 N10 P37-41 64-15984 <=*$>
1961 V37 N10 P46-50 62-23276 <=*>
1961 V37 N11 P13-18 C-4229 <NRC> O
1961 V37 N11 P45-47 62-32127 <=*> P
1961 V37 N11 P76-81 RTS 2778 <NLL>
1961 V37 N12 P6-11 62-32126 <=*>
1961 V37 N12 P53-57 62-32124 <=*>
1962 V38 N1 P3-8 62-32123 <=*>
1962 V38 N1 P9-13 63-31096 <=> O
1962 V38 N1 P26-36 63-31096 <=> O
1962 V38 N1 P37-40 MAFF-TRANS-NS-32 <NLL> O
1962 V38 N3 P44-48 64-51963 <=$>
1962 V38 N3 P62-63 RTS-2827 <NLL>
1962 V38 N4 P47-55 65-61371 <=$> O
1962 V38 N6 P3-7 63-21982 <=>
1962 V38 N6 P15-23 63-21982 <=>
1962 V38 N8 P32-41 AD-619 269 <=$>
1962 V38 N11 P92-94 63-21280 <=>
1963 V39 N1 P85 63-21708 <=>
1963 V39 N1 P87-90 63-21708 <=>
1963 V39 N2 P3-9 63-21606 <=>
1963 V39 N4 P2-6 63-31627 <=>
1963 V39 N5 P27-36 64-15982 <=*$>
1963 V39 N8 P3-9 63-41127 <=>
1963 V39 N8 P35-38 63-41127 <=>
1963 V39 N8 P44-47 MAAF-TRANS-NS-61 <NLL>
1963 V39 N8 P83-85 63-41127 <=>
1963 V39 N8 P86-87 63-41127 <=>
1964 V40 N1 P3-9 64-31877 <=>
1964 V40 N1 P18-20 64-31877 <=>
1964 V40 N1 P23-25 64-31877 <=>
1964 V40 N1 P31-36 64-31877 <=>
1964 V40 N1 P56-58 MAFF-TRANS-NS-62 <NLL>
1964 V40 N2 P16-33 64-31383 <=>
1964 V40 N2 P82-84 64-31383 <=>
1964 V40 N3 P3-7 64-41114 <=>
1964 V40 N3 P63-65 64-41114 <=>
1964 V40 N3 P79-83 64-41114 <=>
1964 V40 N3 P90-92 64-41114 <=>
1964 V40 N4 P3-7 64-31853 <=>
65-33805 <=*$>
1964 V40 N4 P13-16 64-31853 <=>
1964 V40 N4 P51-53 64-31853 <=>
1964 V40 N4 P68-70 64-31853 <=>
1964 V40 N4 P75-76 64-31853 <=>
1964 V40 N4 P78-93 64-31853 <=>
1964 V40 N5 P3-11 64-31942 <=>
1964 V40 N5 P17-18 64-31942 <=>
1964 V40 N5 P76-78 64-31942 <=>
1964 V40 N5 P87-88 64-31942 <=>
1964 V40 N8 P15-18 C-5439 <NRC>
1964 V40 N8 P70-73 RTS-2965 <NLL>
1964 V40 N11 P3-6 67-60363 <=>
1964 V40 N11 P15-20 67-60363 <=>
1964 V40 N11 P24-26 67-60363 <=>
1965 V41 N1 P77-79 65-30645 <=$>
1965 V41 N2 P25-27 65-31183 <=$>
1965 V41 N3 P33-35 MAFF-TRANS-NS-68 <NLL>
1965 V41 N4 P57-58 M.5757 <NLL>
1965 V41 N7 P7-10 67-31530 <=$>
1966 V42 N5 P69-70 66-34923 <=$>
1966 V42 N6 P43-49 67-31530 <=$>
1966 V42 N8 P38-42 66-34723 <=$>

RYBNOE KHOZYAISTVO DALNEGO VOSTOKA
1934 N1/2 P65-68 C-5448 <NRC>

RYBOVODSTVO I RYBOLOVSTVO
1961 V4 N5 P20 64-13715 <=*$>
1961 V4 N5 P26-27 64-13716 <=*$>
1963 V6 N2 P2-3 63-31507 <=>
1963 V6 N2 P13-16 63-31508 <=>
1964 V7 N3 P2-8 64-41378 <=>
1964 V7 N3 P19-22 64-41378 <=>

RYUSAN
1964 V17 N11 P239-243 1385 <TC>

SEL NACHRICHTEN
1962 S1 V10 P9-12 POED-TRANS-2200 <NLL>

SVF FACHORGAN FUER TEXTILVEREDLUNG
1952 N7 P157-168 T2150 <INSD>
1952 V7 N1 P8-12 T2149 <INSD>
1952 V7 N2 P49-55 T2149 <INSD>
1953 V8 N1 P1-8 58-2182 <*>
1953 V8 N2 P53-58 T-2146 <INSD>
1953 V8 N3 P93-99 T-2146 <INSD>
1954 V9 P24-30 T2138 <INSD>
1954 V9 P170-171 T2136 <INSD>
1955 V10 P538-543 T-2144 <INSD>
1957 V12 N9 P602-612 61-14538 <*>
1960 V15 N1 P22-25 C-3656 <NRCC>
1960 V15 N1 P62-66 C-3657 <NRCC>
1960 V15 N3 P212-215 64-10915 <*>
1962 V17 N1 P67-71 62-20201 <*>
1965 V20 P722-730 66-12180 <*>
1965 V20 N2 P115-124 66-11305 <*>

SAD I OGOROD
1946 N3 P73-74 RT-1792 <*>
1947 N7 P25-30 RT-3155 <*>
1947 N11 P72-77 RT-1793 <*>
1947 N12 P65-66 RT-1794 <*>
1948 N3 P59-62 RT-1785 <*>
1948 N6 P63-64 RT-1932 <*>
1948 N8 P70-72 RT-3194 <*>
1948 N9 P10-11 RT-1784 <*>
1948 N9 P12-13 RT-3156 <*>
1949 N5 P72-73 RT-3157 <*>
1949 N11 P23-24 RT-1786 <*>
1950 N2 P25-26 RT-3291 <*>
1950 N2 P35-37 RT-3292 <*>
1950 N2 P78-79 RT-3378 <*>
1950 N6 P61-64 RT-3190 <*>
1950 N7 P74-75 RT-3193 <*>
1950 N8 P32-34 RT-3191 <*>
1950 N8 P34-36 RT-3192 <*> O
1950 N12 P7-9 <CBPB>
1950 N12 P26-27 <CBPB>
1950 N12 P30-33 RT-3293 <*>
1950 N12 P55-56 <CBPB>
1952 N12 P28-31 <CBPB>
1953 N2 P63-67 RT-2365 <*>
1954 N1 P45-48 RT-1690 <*>
1955 N8 P27-29 RT-3875 <*>
1955 N10 P88-90 RT-4139 <*> O
1956 N3 P37-40 RT-4628 <*>
1956 N3 P78-81 RT-4629 <*>
1958 V96 N7 P21-23 60-51049 <=>
1959 V97 N9 P23-29 61-23290 <*=> O

SADOVODSTVO
1960 V98 N10 P37-39 64-23581 <=$>

SADOVODSTVO VINOGRADARSTVO I VINODELIE MOLDAVII
1957 V12 N3 P34-35 63-19597 <=*>
1960 V15 N4 P27-32 62-32972 <=*>

SAGLIK DERGISI
1963 V37 N3/4 P9-22 66-30249 <=$>

SAISHIN IGAKU
1957 V12 N2 P332-340 3263 <*>
1957 V12 N2 P341-351 58-2179 <*>
1960 V15 P450-455 65-12134 <*>
1961 V16 N1 P104-111 AEC-TR-5778 <*>

SAKHARNAYA PROMYSHLENNOST
1948 V22 N4 P22-25 63-16375 <*=>
1949 V23 N10 P8-10 63-14867 <=*>
1954 V28 N3 P11-14 63-16364 <*=> O
1954 V28 N4 P45-46 63-14862 <=*>
1954 V28 N5 P14-15 63-16368 <=*>
1957 V31 N7 P20-23 63-16357 <=*>
1958 V32 N7 P12-17 63-16377 <=*> O
1958 V32 N9 P12-16 85M39R <ATS>
1960 V34 N11 P1-3 61-21256 <=>
1960 V34 N12 P4 61-21697 <=>
1961 V35 N3 P16-18 61-18698 <*=>
1963 V37 N1 P54-58 63-21553 <=>
1963 V37 N2 P22-23 63-21929 <=>

SAKHARNAYA SVEKLA
1959 N8 P43-47 NLLTB V2 N4 P247 <NLL>
1959 N10 P18-23 60-31294 <=>
1959 N11 P39-44 60-31294 <=>
1960 N8 P30-34 61-11923 <=>
1960 N9 P1-3 61-21258 <=>
1960 N10 P29-31 61-11923 <=>
1962 V7 N9 P42-44 63-13315 <=*> PO
1962 V7 N12 P22-24 64-13532 <=*$>

SALUD PUBLICA DE MEXICO
1962 V4 N3 P385-401 64-00427 <*>

SAMMLUNG CHEMISCHER UND CHEMISCH-TECHNISCHER VORTRAEGE
1920 V26 N2 P46-63 59-10717 <*>
1926 V28 P347-408 I-1026 <*>

SAMMLUNG VON VERGIFTUNGSFALLEN
1934 V5 P23-26 62-11549 <=>
1934 V5 P57-60 60-12491 <=>
62-11529 <=>
1937 V8 P13-18 62-11194 <=>

SAMOLET
1936 V13 N10 P28-29 RT-1498 <*>
1936 V13 N10 P30-33 RT-786 <*>
1936 V13 N10 P33 RT-1498 <*>
1936 V14 N11 P24-26 RT-1499 <*>
1937 V15 N1 P35-38 RT-1500 <*>
1937 V15 N4 P35-36 RT-1501 <*>
1937 V15 N4 P47 RT-1501 <*>
1937 V15 N5 P39-41 RT-1502 <*>
1937 V15 N6 P24-25 RT-1503 <*>
1937 V15 N8 P37-38 RT-1504 <*>

SAMOLETNOE ELEKTROOBORUDOVANIE
1960 V1 N1 P63-69 62-13016 <*=> O
63-13222 <=*>

SANATATEA
1963 N2 P12-13 63-21463 <=>
1963 V11 N3 P19 63-21727 <=>

SANG. BIOLOGIE ET PATHOLOGIE
1948 V19 N7 P417-419 59-15544 <*>
1949 V20 N7 P417-421 66-12627 <*>
1950 V21 P76-87 65-17265 <*>
1954 V25 P683-706 57-64 <*>
1954 V25 N8 P769-776 II-1010 <*>
1954 V25 N8 P777-787 II-1011 <*>
1955 V26 N7 P633-649 65-00206 <*> O
1957 V28 N6 P553-568 65-00091 <*> O
1958 V29 N4 P292-297 65-00322 <*> O
1958 V29 N9 P764-789 65-00271 <*> O
1958 V29 N9 P796-804 NP-TR-299 <*>
1959 V30 N7 P762-765 61-00167 <*>

SANITARNAYA OKHRANA VODOEMOV OT ZAGRYAZNANCYA PROMYSHLENNYMI STOCKNYMI VODAMI
1964 V6 P3-29 65-30684 <=$>
1964 V6 P357-358 65-30684 <=$>

SANSHI SHIKENJO HOKOKU
1960 V16 N4 P141-164 62-00903 <*>

SAOPSTENJA. HIDROTEHNICKI INSTITUT. BEOGRAD
1954 N1 P35-42 60-21642 <=>
1955 N2 P3-14 60-21647 <=>
1956 N3 P5-10 60-21642 <=>
1956 N5 P19-26 60-21642 <=>
1957 N8 P11-21 60-21644 <=>

SAOPSTENJA. INSTITUT ZA VODOPRIVREDU "JAROSLAV CERNI"
1957 N6 P1-18 60-21651 <=>
1957 N6 P37-45 60-21650 <=>
1962 V9 N22 ENTIRE NO. 62-19470/1 <=>

1962 V9 N23 ENTIRE NO. 62-19470/2 <=>
1963 V10 N28 ENTIRE ISSUE 63-11451/3 <=>
1964 V11 N30 P1-94 64-11459/1 <=>
1965 V12 N34 P1-11 65-50401/3 <=>
1965 V12 N34 P13-17 65-50401/3 <=>
1965 V12 N34 P19-44 65-50401/3 <=>

SAPPORO IGAKU ZASSHI
1952 V3 N5 P209-212 1499 <*>
1952 V3 N5 P224-229 57-180 <*>
1954 N4 P239-241 57-2600 <*>
1954 V5 N2 P90-94 57-1813 <*>
1954 V5 N3 P175-181 57-1261 <*>
1954 V5 N5 P282-286 57-1217 <*>
1954 V5 N5 P287-290 57-1883 <*>

SAPPORO MEDICAL JOURNAL
SEE SAPPORO IGAKU ZASSHI

SAVREMENA TEHNIKA
1963 N15 P288-289 64-11663 <=>

SBORNIK CESKOSLOVENSKE AKADEMIE ZEMEDELSKYCH VED. RADA B: ZEMEDELSKA EKONOMIKA
1959 N5 P655-658 60-31731 <=*>

SBORNIK CESKOSLOVENSKE AKADEMIE ZEMEDELSKYCH VED. RADA C. ROSLINNA VYROBA
1956 V2 P193-202 AEC-TR-3803 <*>
1957 V3 N11 P1141-1160 CSIR-TR.328 <CSSA>

SBORNIK INSTITUTA ETNOGRAFII
1951 V13 P125-153 AOTN N4 P124 <AINA>

SBORNIK INSTITUTA STALI. MOSKVA
1951 P147-180 2906 <HB>
1951 V30 P70-84 2878 <HB>
1955 V33 P154-192 2050 <BISI>
1956 V35 P283-289 4794 <HB>
1956 V35 P290-297 4209 <HB>
1958 P196-208 59-00683 <*>
1958 N38 P185-195 4674 <HB>
1958 N38 P209-225 4626 <HB>
1958 N38 P405-419 4634 <HB>
1958 V38 P5-195 61-27034 <=*>
1958 V38 P79-87 63-21589 <=>
1958 V38 P427-432 63-21589 <=>

SBORNIK LEKARSKY
1952 V54 P28-47 84M38C <ATS>
1960 V62 P185-188 10643-B <KH>
63-16656 <*>
1963 V65 N2 P33-37 64-20037 <*> O

SBORNIK LENINGRADSKOGO ORDENA LENINA INSTITUTA INZHENEROV ZHELEZNODOROZHNOGO TRANSPORTA IMENI AKADEMIKA V. N. OBRAETSOVA.
1954 V146 P294-312 60-15773 <+*>

SBORNIK. MEZINARODNI POLAROGRAFICKY SJEZD
1951 V1 P611-618 PT1 I-331 <*>

SBORNIK MOSKOVSKOGO INSTITUTA STALI I SPLAVOV
SEE SBORNIK INSTITUTA STALI. MOSCOW

SBORNIK NARODNIHO MUSEA V PRAZE
1959 V15 N3/4 P156-160 61-00656 <*>

SBORNIK NAUCHNO-ISSLEDOVATELSKIKH RABOT TASHKENTSKOGO TEKSTILNOGO INSTITUTA
1956 N3 P40-54 61-15002 <*+>

SBORNIK NAUCHNO-TEKHNICHESKIKH PROISVODSTVENNYKH STATEY PO GEODEZII, KARTOGRAFII, TOPOGRAFII, AEROSYEMKE I GRAVIMETRII. MOSKVA
1941 V1 P3-19 62-15532 <=*>
1941 V1 P25-43 62-15532 <=*>
1941 V1 P46-49 62-15532 <=*>
1941 V1 P119-131 62-15532 <=*>
1944 N4 P3-11 60-11672 <=>
1944 V3 P3-40 62-15595 <=*>
1944 V3 P68-69 62-15595 <=*>
1944 V5 P24-65 62-15528 <=*>
1944 V7 P50-56 63-24262 <=*$>
1948 V21 P58-76 64-19561 <=$>
1949 V23 P21-28 62-25654 <=*>
1949 V23 P61-79 62-25654 <=*>
1949 V24 P8-20 62-15524 <=*>
1949 V24 P35-38 62-15524 <=*>
1949 V24 P68-72 62-15524 <=*>
1949 V24 P76-82 62-15524 <=*>

SBORNIK NAUCHNO-TEKHNICHESKIKH TRUDOV CHELYABINSKII NAUCHNO-ISSLEDOVATELSKII INSTITUT METALLURGII
1960 N1 P120-129 63-15278 <=>
1960 N2 P163-167 AEC-TR-5306 <=*>

SBORNIK NAUCHNYKH RABOT BELORUSSKOGO POLITEKHNICHESKOGO INSTITUTA. MINSK
1957 V56 P316-332 62-25352 <=*>
1957 V61 P175-190 62-24683 <=*>
1958 V63 P27-40 63-20368 <=*>
1960 V82 P3-15 64-15310 <=*$>
1960 V82 P86-93 64-13319 <=*$>
1960 V82 P94-99 64-13319 <=*$>
1960 V82 P100-111 64-13319 <=*$>
1960 V82 P115-118 64-13319 <=*$>
1960 V82 P116-119 64-13319 <=*$>
1960 V82 P120-125 64-13319 <=*$>
1960 V82 P126-136 64-13319 <=*$>
1960 V82 P137-143 64-13319 <=*$>
1960 V82 P144-148 64-13319 <=*$>
1960 V82 P149-154 64-13319 <=*$>
1960 V82 P155-163 64-13319 <=*$>

SBORNIK NAUCHNYKH RABOT ELISTINSKOI PROTIVOCHUMNOI STANTSII
1959 V1 ENTIRE ISSUE 62-33449 <=>
1959 V1 P193-203 62-23304 <=*>
1959 V1 N1 P19-29 61-28155 <*=>

SBORNIK NAUCHNYKH RABOT INSTITUTA FIZIKO-ORGANICHESKOI KHIMII
1960 N8 P119-125 65-17130 <*>

SBORNIK NAUCHNYKH RABOT INSTITUTA KHIMII. MINSK
1958 N6 P83-91 61-15100 <*+>
1959 V7 P185-187 28P59R <ATS>

SBORNIK NAUCHNYKH RABOT INSTITUTA METALLOFIZIKI. AKADEMIYA NAUK UKRAINSKOI SSR. KIEV
1952 N3 P114-121 42L30R <ATS>
1955 V6 P85-91 60-16650 <+*> O
1957 N8 P187-198 TRANS-140 <FRI>
1957 V8 P163-169 A-128 <RIS>
1959 N9 P22-26 2371 <BISI>
1959 N10 P111-120 62-11491 <=>
1959 N10 P121-129 62-11490 <=>
1960 V11 P3-21 62-18524 <=*>
1960 V11 P121-128 2832 <BISI>
1961 V12 P21-36 63-13171 <=*>
1961 V12 P46-60 63-13171 <=*>
1961 V12 P73-87 62-25308 <=*> O
1961 V12 P93-97 63-10865 <=*>
1961 V12 P98-101 63-10866 <=*>
1961 V12 P124-134 63-13171 <=*>
1961 V13 P35-43 64-15835 <=*$>
1961 V13 P177-180 63-18454 <=*>
1962 N15 ENTIRE ISSUE AD-635 407 <=$>
1962 N16 P39-43 M-5402 <NLL>
64-13094 <=*$>
1962 N16 P168-177 64-13093 <=$>
1962 V14 P116-120 63-19427 <=*>
1962 V14 P121-125 AD-617 156 <=$>
1962 V14 P147-151 63-19427 <=*>
1962 V16 P3-15 64-21709 <=>
1962 V16 P128-131 AD-610 764 <=$>
1962 V16 P132-135 AD-610 325 <=$>
1963 N17 P60-63 65-61416 <=>

1963 N17 P98-137 64-31362 <=>
1963 V17 P68-71 AD-639 335 <=$>
1963 V17 P98-110 65-61415 <=>
1963 V17 P209-210 AD-630 971 <=$>
1964 N19 P51-53 66-10996 <*>
1964 N19 P196-205 AD-632 071 <=$>
1964 V18 P69-73 AD-617 157 <=$>

SBORNIK NAUCHNYKH RABOT LENINGRADSKOGC INSTITUTA SOVETSKOI TORGOVLI IMENI F. ENGELSA
1959 V13 P8-13 62-25294 <=*>
1959 V13 P14-20 62-25293 <=*>
1959 V13 P133-138 61-20496 <*>

SBORNIK NAUCHNYKH RABOT MOSKOVSKOGO INZHENERNO-FIZICHESKOGO INSTITUTA
1960 N2 P137-143 63-15289 <=*>
1962 N3 P1-192 AD-617 098 <=$>

SBORNIK NAUCHNYKH RABOT SVERDLOVSKOGO OTDELENIYA VSESOYUZMOGO OBSHCHESTVA ANATOMOV, GISTOLOGOV I EMBRIOLOGOV
1957 V1 N1 P47-50 62-11148 <=>
62-15078 <=*>

SBORNIK NAUCHNYKH RABOT VSESOYUZNOGO NAUCHNO-ISSLEDOVATELSKOGO INSTITUTA POLIGRAFICHESKOI PROMYSHLENNOSTI I TEKHNIKI
1956 V7 P3-54 59-13120 <=>
1956 V7 P103-119 59-13364 <=> O
1956 V7 P120-131 59-13365 <=> O
1956 V7 P183-215 59-13366 <=> O
1956 V7 P216-228 59-13367 <=> O
1962 N14 P125-129 65-14603 <*>
1962 V14 P125-129 66-10332 <*>

SBORNIK NAUCHNYKH TRUDOV. BELORUSSKII INSTITUT MEKHANIZATSII SELSKOGO KHOZYAISTVA. MINSK
1959 N2 P221-229 64-71344 <=> M

SBORNIK NAUCHNYKH TRUDOV BELORUSSKOGO NAUCHNO-ISSLEDOVATELSKOGO INSTITUTA ZEMLEDELIYA
1961 V7 P299-305 64-13342 <=>

SBORNIK NAUCHNYKH TRUDOV FIZIKO-TEKHNICHESKOGO INSTITUTA AKADEMII NAUK BELORUSSKOI SSR. MINSK
1958 V5 N4 P196-212 60-19889 <+*>
1958 V5 N4 P213-219 60-23809 <+*>
1958 V5 N4 P220-224 60-17929 <+*>
1961 V8 N7 P56-59 M.5323 <NLL>
1961 V8 N7 P60-64 M.5322 <NLL>
1961 V8 N7 P135-140 M.5321 <NLL>
1961 V8 N7 P150-156 M.5326 <NLL>
1961 V8 N7 P157-160 M.5327 <NLL>

SBORNIK NAUCHNYKH TRUDOV. GOSUDARSTVENNYI MEDITSINSKII INSTITUT. ROSTOV-ON-THE-DON
1947 N6 P96-102 63-23088 <=*>

SBORNIK NAUCHNYKH TRUDOV GOSUDARSTVENNOGO NAUCHNO-ISSLEDOVATELSKOKO INSTITUTA TSVETNYKH METALLOV. MOSCOW
1956 V12 P45-51 C-3384 <NRCC>
1956 V12 P150-162 AEC-TR-5129 <=*>
1957 V13 P115-122 61-15627 <+*>
1961 V18 P259-274 65-61179 <=>
1962 N19 P795-799 AD-638 869 <=$>
1963 N20 P144-148 66-12202 <*>

SBORNIK NAUCHNYKH TRUDOV. INZHENERNO-STROITELNYI INSTITUT. LENINGRAD
1956 V23 P183-189 63-24315 <=*$>

SBORNIK NAUCHNYKH TRUDOV. INZHENERNO-STROITELNYI INSTITUT. MOSCOW
V29 P258-280 64-00150 <*>
1958 V22 P59-89 RTS-2454 <NLL>

SBORNIK NAUCHNYKH TRUDOV. KUIBYSHEVSKOGO INDUSTRIALNOGO INSTITUTA
1955 N5 P196-203 61-13082 <=*>
61-23292 <*=>
1956 V6 P287-296 65-14647 <*>
1956 V6 N2 P3-22 60-13626 <=*> O

SBORNIK NAUCHNYKH TRUDOV LENINGRADSKOGO INSTITUTA FARMATSEVCHESKII
1957 V2 P38-55 RTS-2729 <NLL>

SBORNIK NAUCHNYKH TRUDOV MOSKOVSKOGO INSTITUTA TSVETNYKH METALLOV I ZOLOTA
1955 V25 P209-225 62-13756 <*=>
1958 V29 P323-329 62-13749 <=*>
1958 V29 P339-348 61-19217 <=*>
1958 V31 P281-297 64-13595 <=*$>
1958 V31 P305-307 62-24374 <=*>

SBORNIK NAUCHNYKH TRUDOV TASHKENTSKOGO INSTITUTA INZHENEROV ZHELEZNODOROZHNOGO TRANSPORTA
1956 N151 P301-313 59-18520 <+*>

SBORNIK NAUCHNYKH TRUDOV TOMSKOGO INZHENERNO-STROITELNOGO INSTITUTA
1957 V2 P97-123 61-11433 <=>

SBORNIK NAUCHNYKH TRUDOV TSENTRALNOGO NAUCHNO-ISSLEDOVATELSKOGO INSTITUTA SVYAZI. MOSCOW
1961 N2 P91-101 64-16835 <*>

SBORNIK NAUCHNYKH TRUDOV. UKRAINSKII NAUCHNO-ISSLEDOVATELSKII INSTITUT OGNEUPOROV
1960 N3 P274-281 AEC-TR-5872 <=$>
1960 V3 P129-152 65-63747 <=$>

SBORNIK PRACI. VYZKUMMY USTAV ZEMEDELSKYCH STROJU. PRAGUE
1961 V1 N1 P133-153 TR.160 <NIAE>

SBORNIK RABOT PO GIDROLOGII
1966 N6 P3-11 SHSP N3 1966 P293 <AGU>
1966 N6 P55-62 SHSP N3 1966 P300 <AGU>
1966 N6 P63-68 SHSP N3 1966 P306 <AGU>
1966 N6 P69-77 SHSP N3 1966 P310 <AGU>
1966 N6 P108-113 SHSP N5 1966 P528 <AGU>
1966 N6 P114-118 SHSP N5 1966 P533 <AGU>
1966 N6 P119-126 SHSP N5 1966 P537 <AGU>
1966 N7 P14-27 SHSP N6 1966 P557 <AGU>
1966 N7 P28-40 SHSP N6 1966 P568 <AGU>

SBORNIK RABOT INSTITUTA PRIKLADNOI ZOOLOGII I FITOPATOLOGII
1953 V2 P33-39 60-17282 <+*> O

SBORNIK RABOT LENINGRADSKOGO VETERINARNOGO INSTITUTA
1954 V14 P9-24 61-23977 <*=>
1955 V15 P133-136 RTS-2940 <NLL>
1959 V22 P435-441 RTS-3187 <NLL>

SBORNIK RABOT PO LESNOMU KHOZYAISTVU. VSESOYUZNYI NAUCHNO-ISSLEDOVATELSKII INSTITUT LESOVODSTVA I MEKHANIZATSII LESNOGO KHOZYAISTVA
1958 N30 P144-160 66-51075 <=>
1958 V34 P140-157 65-50035 <=$>
1958 V37 P106-122 62-33098 <=*>

SBORNIK RABOT PO MEZHDUNARODNOMU GEOFIZICHESKOMU GODU
1961 N1 P26-36 63-21724 <=>
1961 N1 P49-51 63-21724 <=>
1963 N2 P90-97 C-5175 <NRC>

SBORNIK RABOT PO SILIKOZU. URALSKII FILIAL, AKADEMIYA NAUK SSSR
1960 V2 P71-78 64-13276 <=$>

SBORNIK RABOT TSENTRALNOGO MUZEYA POCHVOVEDENIYA IM. V. V. DOKUCHAEVA. MOSCOW
1954 V1 P100-159 65-50066 <=$>

SBORNIK RABOT TSIMLIANSKOI GIDROMETEOROLOGICHESKOI
1958 V1 P135-148 63-11145 <=>

SBORNIK RABOT PO VOPROSAM ELEKTROMEKHANIKI
1960 N4 P189-201 62-32686 <=*>
1961 N5 P3-28 62-23804 <=>
1961 N5 P39-84 62-23804 <=>
1961 N5 P84-94 62-25713 <=*>
1961 N5 P149-161 62-24020 <=>
1961 N5 P162-176 63-15290 <=*>
1961 N5 P176-188 62-24020 <=>
1961 N5 P201-266 62-24020 <=>
1961 N5 P276-281 62-24020 <=>
1963 N9 P38-45 AD-617 158 <=$>
1963 N9 P87-101 AD-633 661 <=$>
1963 N9 P131-145 AD-636 610 <=$>

SBORNIK REKOMENDUEMYKH TERMINOV. AKADEMIYA NAUK SSSR. MOSCOW
1953 N16 P6-26 RT-2533 <*>

SBORNIK STATEI PO GEODEZII I KARTOGRAFII
1953 V4 P43-55 61-28839 <*=>
1955 V9 P20-41 60-15507 <+*>
1960 V11 P63-71 AD-635 923 <=$>

SBORNIK STATEI PO KARTOGRAFII
1954 V6 P17-23 62-25421 <=*>

SBORNIK STATEI LABORATORII AEROMETODOV. AKADEMIYA NAUK SSSR
1953 P119-132 62-25439 <=*>

SBORNIK STATEI LENINGRADSKOGO INSTITUTA TOCHNOI MEKHANIKI I OPTIKI
1957 V24 P127-133 60-10895 <*>
1957 V24 P134-139 60-10896 <*>

SBORNIK STATEI. NAUCHNO-ISSLEDOVATELSKII INSTITUT PO BEZOPASNOSTI RABOT V GORNI PROMYSHLENNOSTI
1960 V14 P10-12 64-13370 <=*$>
1960 V14 P23-26 63-19955 <=*>
1961 V17 P10-13 63-19780 <=*$>

SBORNIK STATEI PO OBSHCHEY KHIMII. AKADEMIYA NAUK SSSR
1953 N1 P179-185 3737-D <KH>
1953 N1 P246-251 02L36R <ATS>
1953 N2 P878-881 59-22676 <+*>
1953 V1 P179-185 64-16117 <=*$> O
1953 V1 P311-314 R-3712 <*>
1953 V2 P944-948 88M47R <ATS>
1953 V2 P1227-1231 64-16147 <=*$> O
1953 V2 P1249-1260 T-1687 <INSD>
1953 V2 P1366-1369 R-5353 <*>

SBORNIK STATEI. OTDELENIE TEKHNICHESKIKH NAUK. AKADEMIYA LATVIISKOI SSR. RIGA
SEE LATVIJAS PADOMJU SOCIALISTISKAS REPUBLIKAS ZINATNU AKADEMIJA. TEHNISKO ZINATNU NODALA. SBORNIK STATEI. RIGA

SBORNIK STATEI PO PALEONTOLOGII I BIOSTRATIGRAFII
1958 V12 P77-79 61-23961 <=*$> O
64-15121 <=*$> O

SBORNIK STATEI. RASCHETY NA PROCHNOST. TEORETICHESKIE EKSPERIMENTALNYE ISSLEDOVANIYA PROCHNOSTI MASHINOSTROITELNYKH KONSTRUKTSII
1958 V3 P310-354 RTS-3024 <NLL>
1963 V9 N9 P173-195 RTS-2771 <NLL>
1963 V9 N9 P270-279 RTS-2793 <NLL>
1964 N10 P211-260 AD-618 317 <=$>

SBORNIK STATEI. RAZVEDDOCHNAYA I PROMYSLOVAYA GEOFIZIKA
1957 V17 P12-19 60-19892 <+*>

SBORNIK STATEI VSESOYUZNOGO NAUCHNO-ISSLEDOVATELSKOGO INSTITUTA KHIMICHESKOGO MASHINOSTROENIYA
1958 V25 P75-86 2293 <BISI>
63-23380 <=*$>

SBORNIK STATEI VSESOYUZNOGO NAUCHNO-ISSLEDOVATELSKOGO I KONSTRUKTORSKOGO INSTITUTA KHIMICHESKOGO MASHINOSTROENIYA. MOSCOW
1958 V24 P107-125 62-24042 <=*>

SBORNIK STATNIHO VYZKUMNEHO USTAVU TEPELNE TECHNIKY. PRAGUE
1956 P144-162 AEC-TR-4540 <*>
1958 P103-125 SCL-T-298 <*>

SBORNIK TRUDOV PO AGRONOMICHESKOI FIZIKE. FIZIKO-AGRONOMICHESKII INSTITUT VSESOYUZNOI AKADEMII SELSKOKHOZYAISTVENNYKH NAUK IMENI V. I. LENINA
1962 V9 P102-105 65-61216 <=$>

SBORNIK TRUDOV. DNEPROPETROVSK METALLURGICHESKII INSTITUT
1955 N1 P143-155 60-13809 <+*>
1955 V33 P442-448 59-19622 <+*>
1955 V33 P449-456 59-19623 <+*>
1965 V49 P332-342 4923 <BISI>
1965 V49 P343-359 4924 <BISI>

SBORNIK TRUDOV INSTITUTA ELEKTRTEKHNIKI. AKADEMIYA NAUK UKRAINSKOI SSR
1955 N12 P115-117 R-677 <*>
1956 V13 P5-34 59-22502 <+*>
1956 V13 P35-41 59-22503 <+*>

SBORNIK TRUDOV INSTITUTA OSNOVANII / PODZEMNYKH SOORUZHENII. AKADEMIYA STROITELSTVA / ARKHITEKTURY. MOSCOW (TITLE VARIES)
1958 V32 P93-103 62-18745 <=*>

SBORNIK TRUDOV INSTITUTA STALI TSENTRALNOGO NAUCHNO-ISSLEDOVATELSKOGO INSTITUTA CHERNOI METALLURGII
1960 V17 P184-203 2126 <BISI>
62-15369 <*=>
1960 V17 P204-227 2123 <BISI>
62-15370 <*=>
1960 V17 P228-246 2125 <BISI>
62-15368 <*=>
1960 V17 P295-310 2124 <BISI>
62-15367 <*=>
1960 V17 P386-397 61-27627 <*=> O
1960 V23 P23-33 62-25651 <=*>

SBORNIK TRUDOV INSTITUTA TEPLO ENERGETIKI. AKADEMIYA NAUK UKRAINSKOI SSR
1949 V1 P118-124 64-19981 <=>
1953 V9 N10 P24-31 61-13032 <=>
1955 V12 P21-53 AD-608 537 <=>
1958 V14 P167-173 63-19884 <=*$>
1958 V14 P174-185 63-19885 <=*$>

SBORNIK TRUDOV KHARKOVSKOGO NAUCHNO-ISSLEDOVATELSKOGO INSTITUTA VAKTSIN I SYVOROTOK IMENI MECHNIKAVA
1954 V20 P285-291 60-15638 <+*>
1954 V20 P293-296 60-15602 <+*> O

SBORNIK TRUDOV PO KHIMICHESHKOI TEKHNOLOGII MINERALNOGO SYRYA KOLSKOGO POLUOSTROVA
1959 V1 P129-147 63-23398 <=*>

SBORNIK TRUDOV LABORATORII HIDRAVLICHNYKH MASHYN
1956 N6 P74-85 59-22672 <+*>
1956 N6 P123-134 60-21463 <=>
1956 N6 P142-158 61-13984 <*+>
1962 N10 P27-37 66-61986 <=*$>
1964 N11 P3-12 65-12726 <*>
1964 N11 P54-61 65-12725 <*>

SBORNIK TRUDOV LENINGRADSKOGO NAUCHNOGO OBSHCHESTVA NEVROPATOLOGOV I PSIKHIATROV
1958 V3 ENTIRE ISSUE 60-21173 <=>

1958 V4 ENTIRE ISSUE 60-21174 <=>

SBORNIK TRUDOV MOSKOVSKOGO INZHENERNO-STROITELNOGO INSTITUTA
1957 V20 P5-24 CSIR-TR-523 <CSSA>
1963 V44 P103-118 AD-637 430 <=$>

SBORNIK TRUDOV RESPUBLIKANSKOGO NAUCHNO-ISSLEDOVATELSKOGO INSTITUTA MESTNYCH STROITELNYKH MATERIALOV ROSNIIMS. MOSCOW
1959 N16 P90-99 67N51R <ATS>
1962 N25 P54-62 GI N2 1964 P331 <AGI>

SBORNIK TRUDOV TSENTRALNOGO NAUCHNO-ISSLEDOVATELSKOGO INSTITUTA CHERNOI METALLURGII. MOSCOW
1960 V17 P327-357 61-27321 <=*> O
1960 V17 P398-418 61-27321 <=*> O
1960 V17 P472-488 61-27321 <=*> O
1960 V23 P23-33 5138 <HB>
1962 V24 P105-111 4121 <BISI>
1962 V25 P5-32 62-33725 <=>
1962 V25 P344-360 62-33725 <=>
1962 V28 P138-145 4538 <BISI>
1963 N30 P7-16 AD-619 334 <=$>
1963 N35 P11-23 64-31518 <=>
1963 N35 P31-45 64-31518 <=>
1963 N35 P69-77 64-31518 <=>
1964 N38 P123-135 AD-635 856 <=$>

SBORNIK TRUDOV TSENTRALNOGO NAUCHNO-ISSLEDOVATELSKOGO LESOKHIMICHESKOGO INSTITUTA. KHIMKI
1957 N12 P136-144 CSIR-TR.327 <CSSA>

SBORNIK TRUDOV VSESOYUZNOGO NAUCHNO-ISSLEDOVATELSKOGO INSTITUTA GIDROLIZNOI I SULFITNO-SPIRTOVOI PROMYSHLENNOSTI. MOSCOW. (TITLE VARIES)
1964 V12 P189-194 84T92R <ATS>

SBORNIK TRUDOV. VSESOYUZNOGO NAUCHNO-ISSLEDOVATELSKOGO INSTITUTA TVERDYKH SPLAVOV. MOSCOW
1960 V2 P15-23 63-15344 <=*>
1960 V2 P24-36 63-19668 <=*>

SBORNIK TRUDOV PO ZEMLEDLCHESKOI MEKHANIKE
1954 V2 P3-12 62-25346 <=*>
1954 V2 P61-72 62-25347 <=*>

SBORNIK VEDECKYCH PRAC. CESKE VYSOKE UCENI TECHNICKE V PRAZE
1958 V1 P109-121 62-14075 <*>

SBORNIK VEDECKYCH PRACI LEKARSKE FAKULTY V HRADCI KRALOVE
1963 V6 P147-153 C-5131 <NRC>
1964 V7 N1/3 P275-281 66-12153 <*>

SBORNIK VEDECKYCH PRACI VYSOKE SKOLY BANSKE. MORAVSKA OSTRAVA
1959 V5 N3 P249-261 TRANS.83 <MT>
1961 V7 N2 P247-250 456 <MT>
1961 V7 N6 P569-606 3365 <BISI>
1961 V7 N7 P731-733 2955 <BISI>
1961 V7 N7 P735-748 2956 <BISI>
1961 V7 N7 P799-803 2959 <BISI>
1963 V9 N3 P439-451 3578 <BISI>

SBORNIK VEDECKYCH PRAC VYSOKEJ SKOLY TECHNICKEJ V KOSICIACH
1957 V1 N1 P65-85 TRANS-39 <MT>

SBORNIK VYSOKE SKOLY CHEMICKO-TECHNOLOGICKE V PRAZE. ODDIL TECHNOLOGIE VODY
1962 V5 N1 P7-15 CSIR-TR.573 <CSSA>

SBORNIK VYSOKE SKOLY ZEMEDELSKE A LESNICKE V BRNE RADA C. SPISY FAKULTY LESNICKE
1955 SC V4 P275-289 63-00528 <*> P
1959 SC V7 N1/3 P199-217 60-00823 <*> P
63-00823 <*> P

SBORNIK VYSKUMNYCH PRAC Z ODBORU CELULOZY A PAPIERA
1964 V9 P218-226 64-10498 <*> OP
1965 V10 P39-61 66-14036 <*$>
1965 V10 P79-89 66-14035 <*$>

SBORNIK VYZKUMNYCH PRACI V PRUMYSLU KOZEDELNEM
1955 P207-217 64-16599 <*>
1958 V3 P97-116 65-10276 <*> O
1960 V5 P108-110 62-19527 <*=>
1960 V5 P117-131 62-19527 <*=>

SCALPEL ET LIEGE MEDICALE
1932 V85 N18 P525-531 63-00035 <*>
1959 V112 N29 P689-695 65-14655 <*>
1961 V114 N13 P289-295 65-14656 <*> O

SCHALLTECHNIK
1930 V3 N1 P1-5 57-1409 <*>

SCHETOVODSTVA I KONTROL
1964 N584 P4-5 64-51730 <=$>

SCHIFF UND HAFEN
1952 V4 P83-84 AL-278 <*>
1956 V8 N3 P167-176 60-17397 <+*>
1956 V8 N3 P176-177 60-17433 <=*>
1958 P545-547 63-01071 <*>
1958 V10 N6 P464-468 62-00209 <*>
1958 V10 N6 P492-497 63-00181 <*>
1959 N1 P62-66 65-12743 <*>
1961 V13 P607- 2175 <BSRA>
1961 V13 N2 P146-152 62-25863 <=*>
1962 V14 N8 P701-706 65-00402 <*>
1964 V16 P112- 2142 <BSRA>
1964 V16 P720- 2118 <BSRA>
1964 V16 P893- 2129 <BSRA>
1965 V17 P10- 2101 <BSRA>
1965 V17 P20- 2108 <BSRA>
1965 V17 P21- 2117 <BSRA>
1965 V17 P29- 2113 <BSRA>
1965 V17 P101- 2225 <BSRA>
1965 V17 P115- 2236 <BSRA>
1965 V17 P188- 2235 <BSRA>
1965 V17 P206- 2228 <BSRA>
1965 V17 P289 2204 <BSRA>
1965 V17 P559 2387 <BSRA>
1965 V17 P690 2411 <BSRA>
1965 V17 P939 2436 <BSRA>

SCHIFFBAU, SCHIFFAHRT UND HAFENBAU
1929 V30 P336-338 63-20549 <*>
1934 V35 N4 P49-53 59-15135 <*>

SCHIFFBAUFORSCHUNG
1964 V3 N3/4 P141- 2149 <BSRA>
1965 V4 N3/4 P97 2432 <BSRA>
1965 V4 N3/4 P150 2417 <BSRA>
1965 V4 N3/4 P157 2416 <BSRA>
1965 V4 N3/4 P183 2421 <BSRA>
1965 V4 N5/6 P193 2419 <BSRA>
1965 V4 N5/6 P264 2418 <BSRA>

SCHIFFBAUTECHNIK
1956 V6 N4 P79-84 59-10195 <*>
1959 V9 N7 P337-340 62-25706 <=*>
1963 V13 P184 2401 <BSRA>
1964 V14 P456- 2167 <BSRA>
1964 V14 P482 2425 <BSRA>
1964 V14 P591 2424 <BSRA>
1964 V14 N5 P449-450 64-51226 <=>
1964 V14 N12 P640-642 NS-356 <TTIS>
1965 V15 P9 2209 <BSRA>
1965 V15 P466 2404 <BSRA>
1965 V15 N2 P68- 2260 <BSRA>

1946 V12 N9 P278-288 62-20357 <*> 0
1947 V13 N1 P9-14 62-20359 <*> 0
1947 V13 N8 P232-238 62-16936 <*> 0
1947 V13 N9 P268-275 62-16937 <*> 0
1948 P330-336 I-152 <*>
1948 V14 N9 P257-274 62-20358 <*> 0
1948 V14 N12 P353-362 57-1110 <*>
1949 V15 P75-84 57-1905 <*>
61-16404 <*>
1949 V15 P308-316 I-602 <*>
1949 V15 N4 P97-116 62-20356 <*> 0
1949 V15 N9 P273-278 58-2596 <*>
61-18951 <*>
1949 V15 N10 P299-307 63-14655 <*>
1949 V15 N10 P308-316 63-14656 <*>
1950 V16 P97-114 AL-700 <*>
1950 V16 N8 P225-243 65-14265 <*> 0
1952 V18 P351-362 63-10030 <*> 0
1952 V18 N6 P178-189 RCT V26 N2 P465 <RCT>
1952 V18 N11 P379-380 57-2619 <*>
1952 V18 N12 P395-404 63-18033 <*>
1953 V19 P316-322 57-1026 <*>
1953 V19 N1 P1-6 RCT V27 N2 P494 <RCT>
1954 N3 P75-80 57-162 <*>
1954 V20 N2 P56-58 62-18447 <*> 0
1954 V20 N6 P198-200 58-1718 <*>
1954 V20 N10 P313-319 57-622 <*>
1955 V21 N5 P165-168 T-2274 <INSD>
1955 V21 N6 P169-178 64-16153 <*> 0
1955 V21 N7 P209-222 58-1149 <*>
60-15719 <=*>
1955 V21 N8 P251-257 64-16469 <*> 0
1955 V21 N10 P337-342 62-00113 <*>
1955 V21 N12 P392-404 64-16732 <*> 0
1956 V22 P178-182 61-16403 <*>
1956 V22 P258-260 TS 1379 <BISI>
1956 V22 N1 P18-22 59-17622 <*>
8533 <K-H>
1956 V22 N6 P178-182 66-10146 <*> 0
1956 V22 N8 P258-260 62-14331 <*>
1956 V22 N10 P334-338 65-17325 <*> 0
1957 V23 N1 P1-7 64-20196 <*> 0
1957 V23 N1 P14-19 4072 <HB>
1957 V23 N4 P97-104 61-16401 <*>
1957 V23 N4 P121-127 61-16402 <*>
1957 V23 N8 P243-248 64-10286 <*>
1957 V23 N8 P249-258 61-16282 <*>
1957 V23 N9 P292-304 61-16282 <*>
1959 V25 P426-439 AEC-TR-5281 <*>
1959 V25 N6 P201-210 66-10909 <*> 0
1959 V25 N11 P415-418 1972 <BISI>
1959 V25 N12 P426-439 4795 <HB>
1960 V26 P157-162 2043 <BISI>
1960 V26 P163-170 2044 <BISI>
1960 V26 P304-311 2249 <TTIS>
1960 V26 P455-477 95N56G <ATS>
1961 V27 P388-389 2341 <TTIS>
1961 V27 P461-463 2962 <BISI>
1961 V27 N3 P108-114 07P59G <ATS>
1961 V27 N8 P344-351 2929 <BISI>
1962 V28 N3 P125-126 5653 <HB>
1962 V28 N11 P473-476 5849 <HB>
1962 V28 N12 P485-490 5864 <HB>
1963 V29 P341-351 LA-TR-64-35 <*>
1963 V29 N3 P91-102 NS 127 <TTIS>
1963 V29 N12 P413-427 3689 <BISI>
1964 V30 N3 P66-74 66-10335 <*> 0
1964 V30 N11 P345-354 4547 <BISI>

SCHWEIZER ARCHIV FUER NEUROLOGIE, NEUROCHIRURGIE UND PSYCHIATRIE
1960 V86 N1/2 P34-48 63-01770 <*>

SCHWEIZER ARCHIV FUER NEUROLOGIE UND PSYCHIATRIE
1921 V8 P215-232 65-11156 <*>
1921 V9 P42-64 58-947 <*> P
1922 V10 P48-79 58-1704 <*>
1937 V40 P164- 57-1521 <*>
1953 V71 P360-371 57-206 <*>
1953 V71 P360-377 62-01208 <*>
1954 V74 P148-163 57-3318 <*>
1955 V75 P67-76 63-10755 <*>

SCHWEIZER ARCHIV FUER TIERHEILKUNDE
1949 V91 N4 P232-237 65-12391 <*>
1950 V92 N12 P737-755 2497 <*>
1951 V93 35-71 SP. ISS. 59-20959 <*> PC
1952 V94 N3 P197-198 59-20970 <*>
1954 V96 N3 P127-143 51F3G <ATS>
1954 V96 N3 P127-142 58-165 <*>
1959 V101 N12 P600-605 60-14184 <*>

SCHWEIZER BRAUEREI RUNDSCHAU
SEE REVUE DE LA BRASSERIE

SCHWEIZER ELEKTRORUNDSCHAU
1936 V28 N7 P45-48 60-14437 <*> 0

SCHWEIZERISCHE APOTHEKERZEITUNG
1954 V92 P524-540 1294 <*>
1963 V101 P364-365 63-20999 <*>

SCHWEIZERISCHE BAUZEITUNG
1927 V90 N23 P291-294 64-11813 <*>
1933 V101 N22 P260-264 3023 <*>
1935 V53 N15 P175-176 59-15350 <*> 0
1942 V119 N10 P107- I-390 <*>
1945 V125 N23/4 P269-273 58-983 <*>
1945 V125 N23/4 P278-284 58-983 <*>
1946 V127 N11 P127-130 64-15592 <=$>
1947 V65 N41 P557-561 59-22385 <*=>
1949 V67 P193-198 AL-835 <*>
1949 V67 P212-215 AL-835 <*>
1949 V67 P225-228 AL-835 <*>
1949 V67 N38 P543-545 62-18399 <*> 0
1949 V67 N28/9 P383-387 II-918 <*>
1949 V67 N28/9 P395-398 II-918 <*>
1950 V68 N19 P253-257 64-71232 <=>
1950 V68 N20 P265-267 64-71232 <=>
1952 V70 N40 P580-582 1957 <*>
63-20790 <*>
1953 V71 N29 P422-425 II-512 <*>
1954 V72 N26 P371-375 1078 <*>
1955 V73 N14 P200-203 59-22390 <*=>
1958 V76 N28 P411-419 TT-792 <NRCC>
1961 V79 N46 P801-809 65-18109 <*>
1961 V79 N46 P841-844 65-12183 <*>

SCHWEIZERISCHE BRAUEREI-RUNDSCHAU
1953 V64 N1 P1-4 I-605 <*>

SCHWEIZERISCHE LANDWIRTSCHAFTLICHE MONATSHEFTE
1952 V30 N4 P145-152 II-36 <*>
1954 V32 N6 P225-238 65-12577 <*>
1955 V33 N9/10 P407-415 65-12669 <*>

SCHWEIZERISCHE MEDIZINISCHE WOCHENSCHRIFT
1935 V65 P363-366 58-812 <*>
1937 V67 N19 P422 59-20977 <*> P
1937 V67 N23 P505-508 65-10004 <*>
1938 V68 N25 P711-713 60-10023 <*> 0
1939 V69 N34 P757-762 61-14308 <*>
1941 V71 N49 P1526-1535 64-16234 <*> 0
1942 V72 P544-547 60-10022 <*> 0
1942 V72 P755-761 57-1539 <*>
1942 V72 P1099-1102 66-11804 <*>
1942 V72 N11 P321- 57-1524 <*>
57-1645 <*>
1942 V72 N11 P322- 57-1644 <*>
1942 V72 N14 P385-388 57-1501 <*>
1943 V73 P442-445 58-2618 <*>
1948 V78 P282-284 62-10269 <*>
1948 V78 P1151 AL-294 <*>
1948 V78 P1155 AL-723 <*>
1948 V78 P1155- 57-1641 <*>
1948 V78 N1 P10-13 1292I <K-H>
1948 V78 N13 P293-294 64-20025 <*>
1949 V79 P144 AL-722 <*>
1949 V79 P978-980 57-736 <*>
1949 V79 P986-990 57-355 <*>

Citation	Translation
1949 V79 N6 P119-120	2622 <*>
	2909 <*>
1951 N3 P61-65	I-990 <*>
1951 V81 P89-9C	T-2315 <INSD>
1951 V81 P817-837	66-11822 <*>
1951 V81 P970-974	65-00097 <*> 0
1952 V82 P848-850	66-11617 <*> 0
1952 V82 N5 P104-106	59-17865 <*> 0
1953 V83 P25-30	I-47 <*>
1953 V83 P425-427	58-1838 <*>
1953 V83 P971-975	AL-802 <*>
1953 V83 P1C12-1015	I-367 <*>
1953 V83 P1127-1135	II-119 <*>
1953 V83 N1 P4-7	1247 <*>
1953 V83 N50 P1199-1202	65-13238 <*>
1953 V83 N50 P1208-1209	65-12604 <*>
1954 P1272-1273	I-880 <*>
1954 V84 P1-20	II-492 <*>
1954 V84 P406-416	AL-453 <*>
1954 V84 P741-743	II-303 <*>
1954 V84 P968-970	I-826 <*>
1954 V84 P1392-1394	II-685 <*>
1954 V84 P1415-1416	1862 <*>
1954 V84 N2 P25-27	65-13237 <*> 0
1954 V84 N5 P167-169	2380 <*>
1954 V84 N28 P765-766	3174 <*>
1954 V84 N4C P1147-	2224 <*>
1955 V85 P53-55	3417-B <K-H>
1955 V85 P167-170	1889 <*>
1955 V85 P244-248	58-1420 <*>
1955 V85 P305-309	II-323 <*>
1955 V85 P309-310	2487 <*>
1955 V85 P390-393	57-2709 <*>
1955 V85 P420-423	57-2708 <*>
1955 V85 P439-440	2124 <*>
1955 V85 P440-442	2122 <*>
1955 V85 P442-443	2123 <*>
1955 V85 P443-444	2126 <*>
1955 V85 P612-614	2125 <*>
1955 V85 P631-634	2416 <*>
1955 V85 P660-	2135 <*>
	2137 <*>
1955 V85 P661-	2136 <*>
1955 V85 P661	2138 <*>
1955 V85 P662-	2140 <*>
1955 V85 P664-	2128 <*>
1955 V85 P664	2141 <*>
1955 V85 P665	2139 <*>
1955 V85 P688-	II-973 <*>
1955 V85 P751-752	2148 <*>
1955 V85 P952-	II-1035 <*>
1955 V85 P991-994	2966 <*>
1955 V85 P994-1000	57-1332 <*>
1955 V85 P1000-1001	2975 <*>
1955 V85 P1003-1005	57-1331 <*>
1955 V85 P1085-1088	II-1029 <*>
	57-1835 <*>
1955 V85 P1218-1220	II-861 <*>
	II-978 <*>
1955 V85 N6 P128-131	2673 <*>
1955 V85 N17 P387-	1265 <*>
1955 V85 N46 P1120-1122	65-11333 <*> 0
1955 V85 N49 P1190-1196	60-17335 <+*> 0
1955 V85 N52 P1274-1276	60-13073 <+*> 0
1955 V85 N38/9 P942-945	57-2929 <*>
1956 N10 P252-254	1725 <*>
1956 V86 P94-96	1080 <*>
1956 V86 P122-123	1219 <*>
1956 V86 P162-165	II-635 <*>
1956 V86 P249-251	1718 <*>
1956 V86 P413-415	58-2701 <*>
1956 V86 P659-	57-1095 <*>
	57-1096 <*>
1956 V86 P688-691	2679 <*>
1956 V86 P691-694	1274 <*>
1956 V86 P1262-1263	57-2986 <*>
1956 V86 N21 P623-625	60-13059 <*+> 0
1956 V86 N23 P669-675	5497-C <K-H>
1957 N14 P406-408	3068 <*>
1957 V87 P163	58-19 <*>
1957 V87 P269-272	57-2959 <*>
1957 V87 P307-315	63-00529 <*>
1957 V87 P787-790	3066 <*>
1957 V87 P1087-1089	58-71 <*>
1957 V87 P1228-1229	58-2636 <*>
1957 V87 P1318-	0A-13 <RIS>
1957 V87 P1427-1430	57-3577 <*>
1957 V87 N7 P155-159	59-15730 <*>
1957 V87 N26 P822-828	64-00065 <*>
1957 V87 N26 P828-833	60-17329 <*+>
1957 V87 N26 P881-885	66-12012 <*>
1957 V87 N34 P1076-1098	59-15454 <*>
1957 V87 N39 P1210-1218	AEC-TR-4783 <*>
1958 V88 P184-185	58-1176 <*>
1958 V88 P474-476	58-1631 <*>
1958 V88 P813-817	58-2614 <*>
1958 V88 P1160-1164	58-2707 <*>
1958 V88 N6 P127-132	65-00202 <*> 0
1958 V88 N15 P349-353	63-16909 <*>
1958 V88 N20 P488-491	62-00136 <*>
1958 V88 N26 P634-635	63-00575 <*>
1958 V88 N30 P740-742	59-10024 <*>
1958 V88 N34 P835-839	58-2443 <*>
1958 V88 N35 P858-863	58-2443 <*>
1958 V88 N45 P1132-1136	63-16675 <*>
	8973-B <K-H>
1959 P1232-1234	62-10729 <*>
1959 V89 P1313-1318	60-10966 <*>
	61-20766 <*>
1959 V89 N7 P181-187	59-15879 <*>
1959 V89 N12 P325-331	59-15597 <*>
1959 V89 N12 P331-334	59-15687 <*>
1959 V89 N13 P353-361	59-17071 <*>
1959 V89 N14 P363-377	61-00388 <*>
1959 V89 N15 P405-407	59-17587 <*>
1959 V89 N15 P407-408	66-12519 <*>
1959 V89 N19 P510-512	59-20965 <*>
1960 V90 P1163-1164	63-10925 <*>
1960 V90 P1265-1269	63-20909 <*> 0
1960 V90 N5 P113-118	60-14875 <*>
1960 V90 N17 P471-473	60-16688 <*> 0
1960 V90 N44 P1246-1249	63-20871 <*> 0
1960 V90 N44 P1256-1257	63-20891 <*>
1960 V90 N46 P1315-1320	61-20663 <*> 0
1960 V90 N48 P1379-1383	61-10901 <*>
	61-20911 <*>
1960 V90 N51 P1458-1467	61-00418 <*>
1961 V91 P774-778	62-18001 <*>
1961 V91 P914-918	62-10086 <*>
1961 V91 P1245-1249	AEC-TR-5499 <*>
1961 V91 N3 P87-93	61-16474 <*>
1961 V91 N32 P939-943	65-14838 <*>
1962 V92 P639-647	63-10190 <*>
1962 V92 P684-691	63-18735 <*>
1962 V92 P1127-1132	63-10724 <*>
1962 V92 P1295-1306	63-14721 <*>
1962 V92 N2 P50-53	62-16077 <*>
1962 V92 N5 P125-130	63-01122 <*>
1962 V92 N33 P1007-1009	63-10115 <*>
1963 V93 P826-830	63-20941 <*> 0
1963 V93 P959-973	63-20625 <*>
1963 V93 P1011-1016	64-16074 <*>
1963 V93 P1811-1815	64-16360 <*>
1963 V93 N2 P81-84	65-10967 <*>
1963 V93 N2 P9C-92	64-20910 <*>
1963 V93 N5 P223-227	64-20139 <*> 0
1963 V93 N20 P737-739	64-14585 <*>
1963 V93 N25 P914-918	64-16832 <*>
1963 V93 N30 P1027-1030	64-16080 <*>
1963 V93 N33 P1061-1065	65-17241 <*>
1963 V93 N42 P1511-1512	64-18144 <*>
1964 V94 N6 P198-201	65-14703 <*>
1964 V94 N7 P235-240	64-20917 <*>
1964 V94 N7 P262-266	64-18140 <*>
1964 V94 N34 P1158-1164	65-12324 <*>
1964 V94 N47 P1652-1655	65-12805 <*>
1965 V95 P387-395	65-18140 <*>
1965 V95 P517-520	66-13200 <*>
1965 V95 N4 P134-137	65-13939 <*>
1965 V95 N4 P1591-1596	74T90G <ATS>

1965 V95 N14 P460-466 66-1070 <*>
1965 V95 N17 P571-578 66-13170 <*> O
1965 V95 N19 P628-636 65-17135 <*>
1965 V95 N20 P667-672 89S88G <ATS>
1965 V95 N21 P702-705 66-13171 <*>
1965 V95 N28 P944-950 66-13169 <*>
1965 V95 N35 P1151-1154 66-12723 <*> O
1965 V95 N35 P1155 66-13221 <*>
1965 V95 N35 P1157-1160 66-12750 <*> O

SCHWEIZERISCHE MILCHZEITUNG
1956 V82 N37 P281-285 C-2538 <NRC>
57-3556 <*>
1958 V85 N31 P244-248 C-2641 <NRC>

SCHWEIZERISCHE MINERALOGISCHE UND PETROGRAPHISCHE MITTEILUNGEN
1948 V28 N1 P456-467 AI-TRANS-54 <*>
1952 V32 N1 P111-159 2860A <K-H>
63-20740 <*>
1959 V39 P1-84 PT1 IGR V3 N2 P119 <AGI>
1959 V39 P1-84 PT2 IGR V3 N3 P202 <AGI>
1961 V41 P335-369 65-12804 <*>
1963 V43 N1 P259-263 65-11471 <*>

SCHWEIZERISCHE MONATSSCHRIFT FUER ZAHNHEILKUNDE
1946 V56 P1-23 61-18819 <*> P
1952 V62 N7 P712-715 II-751 <*>
1955 V65 P253-262 57-2605 <*>
1965 V75 N2 P168-180 65-14834 <*>

SCHWEIZERISCHE RUNDSCHAU FUER MEDIZIN
SEE REVUE SUISSE DE MEDECINE

SCHWEIZERISCHE TECHNISCHE ZEITSCHRIFT
1958 V55 P974-976 92L31G <ATS>
1960 N52 P1085-1089 127-337 <STT>

SCHWEIZERISCHE VIERTELJAHRSSCHRIFT FUER ZAHNHEILKUNDE
1910 V20 P204-208 65-00160 <*>

SCHWEIZERISCHE ZEITSCHRIFT FUER ALLGEMEINE PATHOLOGIE UND BAKTERIOLOGIE
1949 V12 P289-305 65-00175 <*>
1949 V12 P313-350 62-19852 <=*>
1950 V13 N5 P570-574 I-597 <*>
1952 V15 P458-461 60-10007 <*> O
1952 V15 N4 P517-525 65-12353 <*>
1953 V16 N2 P165-189 2738 <*>
1953 V16 N3 P484-490 65-00297 <*>
1953 V16 N6 P987-994 3081 <*>
1954 V17 P520-524 II-351 <*>
1954 V17 P703-719 1606 <*>
1955 V18 P564-582 <CP>
1955 V18 P907-918 63-16963 <*>
6918-D <KH>
1955 V18 N1 P32-37 62-00594 <*>
1956 V19 P217-243 65-00158 <*> O
1956 V19 P620-624 58-1832 <*>
1956 V19 P738 65-00330 <*> O
1956 V19 N2 P129-149 62-00916 <*>
1956 V19 N3 P331-350 1386 <*>
1956 V19 N5 P582-597 65-12625 <*>
1956 V19 N5 P639-646 57-2987 <*>
1957 V20 P17-22 57-1256 <*>
1957 V20 P544-547 58-566 <*>
1957 V20 P716-723 58-608 <*>
1957 V20 N5 P614-618 58-565 <*>
1957 V20 N6 P703-710 62-00994 <*>
1958 V21 P1014-1017 58-2689 <*>
1958 V21 N3 P688-706 63-00466 <*>
1958 V21 N4 P773-820 63-00609 <*>
1958 V21 N5 P1018-1023 59-15488 <*>
1959 V22 N5 P607-608 60-14181 <*>

SCHWEIZERISCHE ZEITSCHRIFT FUER FORSTWESEN
1927 V78 P145-153 57-1229 <*>
1927 V78 P307-311 57-1226 <*>
1927 V78 N11 P345-357 R-958 <*>
1927 V78 N11 P390-397 R-958 <*>
1953 P508-516 I-790 <*>

SCHWEIZERISCHE ZEITSCHRIFT FUER HYDROLOGIE
1958 V20 P218-254 63-01510 <*>

SCHWEIZERISCHE ZEITSCHRIFT FUER PILZKUNDE
1955 V33 N5 P69-74 CSIRO-3196 <CSIR>
57-1186 <*>

SCHWEIZERISCHE ZEITSCHRIFT FUER PSYCHOLOGIE UND IHRE ANWENDUNGEN
1947 P95-113 60-18643 <*>
1949 V8 P32-60 60-13409 <*+>
1956 V15 P34-50 65-11696 <*>

SCHWEIZERISCHE ZEITSCHRIFT FUER STRASSENWESEN UND VERWANDTE GEBEITE
SEE REVUE SUISSE DE LA ROUTE

SCHWEIZERISCHE ZEITSCHRIFT FUER TUBERKULOSE UND PNEUMOLOGIE
1950 V7 P3-7 3305 <*>
1950 V7 N1 P17-26 2552 <*>
1958 V15 N4 P198-201 59-10836 <*>

SCHWEIZERISCHES ARCHIV FUER VERKEHRSWISSENSCHAFT UND VERKEHRSPOLITIK
1957 N3 P213-233 59-11887 <=>

SCIENCE. CHINESE PEOPLES REPUBLIC
SEE K'O HSUEH

SCIENCE (CHEMISTRY). TOKYO
SEE KAGAKU. TOKYO

SCIENCE BULLETIN OF THE FACULTY OF AGRICULTURE, KYUSHU UNIVERSITY
SEE KYUSHU DAIGAKU NOGAKOBU GAKUGII ZASSHI

SCIENCE EDUCATION, CHINESE PEOPLES REPUBLIC
SEE K'O HSUEH CHIAO YU

SCIENCE AND ENGINEERING REVIEW OF DOSHISHA UNIVERSITY
SEE DOSHISHA DAIGAKU RIKOGAKU KENKYU HOKOKU

SCIENCE ET INDUSTRIE. PARIS
1932 V16 N225 P401-408 63-16808 <*>
1932 V16 N226 P459-469 63-16808 <*>

SCIENCE ET INDUSTRIES PHOTOGRAPHIQUES
1931 V2 N9 P321-332 57-3059 <*>
1945 S2 V16 N5/6 P129-137 65-13067 <*> O
1947 V18 N11 P321-332 66-11157 <*>
1947 V18 N12 P353-359 66-11158 <*>
1952 V23 N2 P354-358 58-1744 <*>
1953 S2 V24 N1 P1-21 II-1047 <*>
1954 S2 V25 P225-230 1415 <*>
1954 S2 V25 N11 P425-432 1785 <*>
1955 V26 N8 P289-304 65-14882 <*>
1955 V26 N12 P465-471 62-10103 <*> O
1957 S2 V28 N1 P36-37 62-10264 <*> O
1958 V29 P262-265 65-14958 <*>
1958 S2 V29 N10 P361-364 60-14140 <*>
1959 S2 V30 P3-12 62-16736 <*> O
1960 S2 V31 N9 P346-353 62-10988 <*>
1963 V34 N12 P341-349 66-13774 <*>
1964 V35 N2 P113-118 936 <TC>

SCIENCE AND INDUSTRY. OSAKA
SEE KAGAKU TO KOGYO. OSAKA

SCIENCE INFORMATION. JAPAN
SEE KOKUSAI KAGAKU JOHO

SCIENCE OF THE LIVING BODY. JAPAN
SEE SEITAI NO KAGAKU

SCIENCE OF MIND. JAPAN

SEE KIKAI NO KENKYU

SCIENCE PICTORIAL. CHINESE PEOPLES REPUBLIC
SEE K'O HSUEH HUA PAO

SCIENCE RECORD. ACADEMIA SINICA. CHINESE PEOPLES REPUBLIC
1957 V1 N6 P383-384 94M42CH <ATS>

SCIENCE REPORTS OF THE TOHOKU UNIVERSITY. FIRST SERIES. PHYSICS, CHEMISTRY, ASTRONOMY
SEE TOHOKU DAIGAKU RIHA HOKOKU. DAI-I-SHU. BUTSUNGAKU, KAGAKU, TENMONGAKU

SCIENCE OF THE SEAS
SEE KAIYO NO KAGAKU

SCIENCE OF TELECOMMUNICATION. CHINESE PEOPLES REPUBLIC
SEE TIEN HSIN K'O HSUEH

SCIENCE ET VIE
1949 P113-117 II-122 <*>
1949 V76 P3-8 64-16897 <*> 0

SCIENCES ET AVENIR
1958 N12 P614-617 60-15993 <*+>
1958 N139 P478-483 59-16290 <=*> 0

SCIENCES DE LA TERRE
1954 V2 N3 P73-97 1512 <*>

SCIENCIA MEDICA
1929 V1 P371-372 T-1910 <INSD>

SCIENTIA. BOLOGNA, MILANO
1959 V53 N6 P1-4 60-17773 <*=>

SCIENTIA. SERIE PHYSICO-MATHEMATIQUE. PARIS
1911 N31 P1-93 60-16720 <*>
1912 N33 P3-81 60-16722 <*>

SCIENTIA ELECTRICA
1958 V4 N3 P92-107 08L34G <ATS>

SCIENTIA PHARMACEUTICA
1952 V20 N2 P69-75 64-10036 <*>
1964 V32 N2 P111-121 65-14335 <*>
1964 V32 N2 P154-158 65-11428 <*> 0
1964 V32 N2 P162-169 65-11427+<*> 0

SCIENTIA SILVAE SINICA
SEE LIN YEH KO' HSUEH

SCIENTIA SINICA
1955 V4 N2 P307-312 57-775 <*>
1958 V7 N2 P217-249 59-22198 <+*> 0
1961 V10 N1 P70- ICE V2 N4 P493-500 <ICE>
1961 V10 N2 P225-230 61-28557 <*=>
1961 V10 N6 P653-657 N66-13058 <=$>
1963 V12 N4 P575-585 65-14140 <*>
1963 V12 N8 P1197-1211 63-41046 <=>
1964 V13 N3 P525-527 65-13803 <*>
1964 V13 N5 P813-821 65-13602 <*>

SCIENTIFIC ASAHI. TOKYO
SEE KAGAKU ASAHI. TOKYO

SCIENTIFIC INSECT CONTROL. JAPAN
SEE BOCHU KAGAKU

SCIENTIFIC INSTRUMENTS. CHINESE PEOPLES REPUBLIC
SEE K'O HSUEH I CHI

SCIENTIFIC NEWS. CHINESE PEOPLES REPUBLIC
SEE KO HSUEH HSIN WEN

SCIENTIFIC PAPERS OF THE CENTRAL RESEARCH INSTITUTE. JAPAN MONOPOLY CORPORATION
SEE NIHON SENBAI KOSHA CHUO KENKYUSHO KENKYU HOKOKU

SCIENTIFIC PAPERS OF THE INSTITUTE OF PHYSICAL AND CHEMICAL RESEARCH
1926 V4 N50 P85-101 64-18884 <*> 0

SCIENTIFIC PAPERS OF THE ODAWARA SALT EXPERIMENT STATION, JAPAN MONOPOLY CORPORATION
SEE SHIKEN HOKOKU (NIHON SENBAI KOSHA ODAWARA SEIEN SHIKENJO)

SCIENTIFIC PUBLICATIONS OF THE FUJI PHOTO FILM CO. LTD
SEE FUJI SHASHIN FUIRUMII KENKYU HOKOKU

SCIENTIFIC REPORTS OF THE HOKKAIDO SALMON HATCHERIES. JAPAN
SEE HOKKAIDO SAKE, MASU, FUKAJO KENKYU HOKOKU

SCIENTIFIC REPORTS MEIJI SEIKA K. K., PHARMACEUTICALS DIVISION
SEE MEIJI SEIKA KENKYU NEMPO (YAKUHIN BUMON)

SCIENTIFIC REPORTS OF THE SAIKYO UNIVERSITY. NATURAL SCIENCE AND LIVING SCIENCE
SEE SAIKYO DAIGAKU GAKUJUTSA HOKOKU (RIGAKU OYOBI KASEIGAKU)

SCIENTIFIC REPORT OF THE TOYO SODA MANUFACTURING COMPANY, LTD
SEE TOYO SODA KENKYU HOKOKU

SCIENTIFIC RESEARCHES. FINLAND'S INSTITUTE OF TECHNOLOGY
SEE TIETEELLISIA TUTKIMUKSIA

SDELOVACI TECHNIKA
1964 V12 N7 P248-252 AD-641 105 <=$>

SEEVERKEHR
1965 N7 P402-407 65-32671 <=$>
1966 N12 P495-496 67-30335 <=>
1967 V7 P25-26 67-30495 <=>
1967 V7 N1 P5-6 67-30446 <=>

SEEWART; NAUTISCHE ZEITSCHRIFT
1954 V15 N4 P125-131 1882 <*>
1955 V16 N6 P166-168 1516 <*>

SEIFEN-OELE-FETTE-WACHSE
1949 V75 P481-482 II-832 <*>
1950 V76 P145-147 59-17874 <*>
1951 V77 P203-206 63-20643 <*>
1951 V77 P407-408 AL-687 <*>
1951 V77 P494-496 AL-363 <*>
1951 V77 N13 P315-316 79M37G <ATS>
1952 V78 P537-538 5710-E <K-H>
1952 V78 N1 P18 59-20852 <*>
1952 V78 N10 P212-214 3306 <*>
65-10389 <*>
1952 V78 N10 P267-269 3306 <*>
1952 V78 N11 P267-269 65-10389 <*>
1952 V78 N15 P373-375 II-942 <*>
1952 V78 N16 P395-396 II-942 <*>
1952 V78 N17 P409-410 59-20794 <*>
1953 V79 N9 P228 64-20832 <*>
1953 V79 N21 P546 59-17866 <*>
1954 V80 P239- 59-15482 <*>
1954 V80 P436-437 T-2422 <INSD>
1954 V80 N7 P158-159 59-17842 <*>
1954 V80 N8 P170-172 48P64G <ATS>
1954 V80 N9 P225-227 49P64G <ATS>
1954 V80 N10 P245-247 5CP64G <ATS>
1955 V81 P721-723 8785 <IICH>
1956 V82 P493-494 58-958 <*>
1956 V82 P517-520 58-959 <*>
1956 V82 N9 P245-246 58-1537 <*>
1956 V82 N20 P573-574 58-960 <*>
1957 V83 N13 P361-367 58-961 <*>
1958 N2 P47-48 58-1210 <*>

1959 V85 N25 P785-788 60-16777 <*>
1961 V87 N21 P677-678 62-14745 <*>
1962 V88 N4 P93-97 63-10008 <*>
1962 V88 N8 P191-195 62-20376 <*>
1962 V88 N25 P853-855 63-16553 <*> 0
1963 V89 N7 P183-186 66-11383 <*> 0
1963 V89 N12 P367-370 65-12209 <*>
1963 V89 N19 P573-578 64-18137 <*>
1963 V89 N20 P605-608 65-12204 <*>
1964 V90 P105-110 507 <TC>

SEIFENSIEDERZEITUNG
1914 V41 N14 P394-395 60-18966 <*>
1922 V49 N29 P507-508 59-20789 <*>
1929 V56 N3 P19-20 59-20788 <*> P
1935 V62 P189 64-10651 <=*$>
1935 V62 P197 64-10660 <=*$>
1935 V62 P205 64-10650 <=*$>
1935 V62 P209 64-10654 <=*$>
1935 V62 P218 64-10662 <=*$>
1935 V62 P229 64-10659 <=*$>
1935 V62 P230 64-10656 <=*$>
1935 V62 P249 64-10648 <=*$>
1935 V62 P272 64-14090 <*>
1935 V62 P292 64-10657 <=*$>
1935 V62 P311 64-14093 <*>
1935 V62 P353 64-10653 <=*$>
64-14094 <*>
1935 V62 P365 64-10655 <=*$>
1935 V62 P373 64-14096 <*>
1935 V62 P375 64-14091 <*>
1935 V62 P376 64-14092 <*>
1935 V62 P384 64-10658 <=*$>
1935 V62 P395 64-10661 <=*$>
1935 V62 P396 64-14089 <*>
1935 V62 P416 64-10228 <*>
1935 V62 P498 64-14095 <*>
1935 V62 P954 64-14101 <*>
1935 V62 P1058 64-14108 <*>
1935 V62 N9 P161-163 59-20623 <*>
1935 V62 N10 P195-196 59-20623 <*>
1936 V63 P69 64-14111 <*>
1936 V63 P89 64-14107 <*>
1936 V63 P401 64-14484 <*>
1936 V63 P972 64-10227 <*>
1936 V63 N13 P265 64-10206 <*>
1936 V63 N41 P833-835 59-20786 <*> 0
1937 V64 P4 64-14503 <*>
1937 V64 P26 64-14506 <*>
64-14508 <*>
1937 V64 P50 64-14495 <*>
1937 V64 P87 64-14510 <*>
1937 V64 P93 64-14507 <*>
1937 V64 P103 64-14505 <*>
1937 V64 P111 64-14856 <*>
1937 V64 P116 64-14509 <*>
1937 V64 P168 64-14860 <*>
1937 V64 P229 64-14100 <*>
1937 V64 P238 64-14858 <*>
1937 V64 P248 64-14863 <*>
1937 V64 P274 64-14864 <*>
64-14867 <*>
1937 V64 P310 64-14862 <*>
1937 V64 P348 64-14870 <*>
1937 V64 P357 64-14874 <*>
1937 V64 P358 64-14106 <*>
1937 V64 P364 64-14872 <*>
1937 V64 P368 64-14869 <*>
1937 V64 P387 64-14873 <*>
1937 V64 P404 64-14865 <*>
64-14868 <*>
1937 V64 P458 64-14871 <*>
1937 V64 P592 64-14883 <*>
1937 V64 P598 64-14876 <*>
1937 V64 P602 64-14498 <*>
1937 V64 P654 64-14877 <*>
1937 V64 P665 64-14878 <*>
1937 V64 P675 64-14875 <*>
1937 V64 P832 64-14885 <*>
1946 V72 N3 P45-46 3523-A <K-H>
1947 V73 P41-42 60-10080 <*>

SEIKAGAKU
1947 V19 N4/5 P85-91 57-1817 <*>
1947 V19 N4/5 P92-93 57-1816 <*>
1956 V28 P70-74 62-00971 <*>
1956 V28 P218-223 62-00954 <*>
1957 V29 N2 P10-69 58-181 <*>
1957 V29 N5 P290-293 59-15437 <*>
59-15774 <*> 0
1957 V29 N5 P293-295 60-14085 <*> 0
1958 V29 N12 P915-918 59-14277 <+*> 0
1958 V29 N12 P918-921 59-14317 <=*>
1958 V30 N6 P444-446 59-15771 <*>
1958 V30 N6 P446-448 59-15772 <*> 0
1958 V30 N6 P449-452 59-15773 <*> 0
1958 V30 N6 P452-456 60-14080 <*> 0
1961 V33 P794-798 C-4164 <NRCC>
1962 V34 N3 P79-87 62-01396 <*>
1962 V34 N8 P319-330 63-00862 <*>
63-27746 <=$>
1963 V35 P278-281 23 <*>
1963 V35 N4 P187-197 63-01319 <*>
1964 V36 N10 P735-747 65-17026 <*>

SEIMITSU KIKAI
1947 V13 P5-7 61-10660 <*>

SEIRI SEITAI
1953 V5 N3/4 P97-103 60-15416 <*+> 0
1955 V6 N2 P127-144 <TIDC>
1957 V7 N2 P134-144 C-5315 <NRC>

SEISAN KENKYU
1958 V10 N12 P400-401 53P58J <ATS>
1960 V12 N12 ENTIRE ISSUE 63-11595 <=>

SEISHIN IGAKU
1963 V4 N10 P741-754 65-10984 <*>

SEISHIN SHINKEIGAKU ZASSHI
1949 V51 N1 P25-29 57-2334 <*>
1951 V52 P204-215 63-01387 <*>
1955 V56 N10 P37-41 57-2545 <*>
1958 V60 N4 P109-120 63-01386 <*>
63-27743 <=$>
1962 V64 P85-89 63-00776 <*>

SEISMOLOGIIA: SBORNIK STATI. MEZHDUVEDOMSTVENNYI KOMITET PO PROVEDENIIU MEZHDUNARODNOGO GEOFIZICHESKOGO GODA. AKADEMIYA NAUK SSSR. XII. RAZDEL PROGRAMMY MGG
1960 N4 P1-13 64-19737 <=$>
1960 N4 P7-77 62-15348 <=*>
1960 N4 P32-41 64-19737 <=$>
1965 N6 P23-30 NIOT-88 <NLL>
1965 N6 P31-36 NIOT-89 <NLL>
1965 N13 P163-171 NIOT-87 <NLL>

SEITAI NO KAGAKU
1955 V6 N4 P154-159 60-15415 <=*> 0
1956 V7 N8 P408-411 61-00183 <*>
1958 V9 N3 P235-240 63-00846 <*>
1959 V10 N1 P30-38 63-00851 <*>

SEITETSU KENKYU
1960 P2841-2862 3325 <BISI>

SEITO GIJUTSU KENKYUKAISHI
1959 V8 P22-29 88Q66J <ATS>

SEKIYU GAKKAI-SHI
1958 V1 N1 P33-40 SEKI-1-1-1 <SA>
1958 V1 N2 P128-132 SA-SEKI-1-2-1 <SA>
74L32J <ATS>
1959 V2 N1 P8-14 SEKI-2-1-1 <SA> 0
1959 V2 N3 P216-221 SEKI-2-3-1 <SA> 0
1959 V2 N5 P520-521 SA-SEKI-2-5-1 <SA>
1959 V2 N6 P580-588 SA-SEKI-2-6-1 <SA>
1960 V3 N2 P116-120 56P58J <ATS>

1960 V3 N7 P552-556 SEKI-3-7-1 <SA>
1960 V3 N8 P636-640 60N48J <ATS>
1960 V3 N9 P722-726 SEKI-3-9-1 <SA>
1960 V3 N10 P791-796 SEKI-3-10-1 <SA>
1961 V4 N1 P24-28 SEKI-4-1-1 <SA>
1961 V4 N1 P29-33 87N55J <ATS>
1961 V4 N3 P197-200 SEKI-4-3-1 <SA>
1961 V4 N6 P440-447 SEKI-4-6-1 <SA>
1961 V4 N6 P452-457 SEKI-4-6-2 <SA>
1961 V4 N7 P509-514 SEKI-4-7-1 <SA>
1961 V4 N7 P538-545 SEKI-4-7-2 <SA>
1961 V4 N8 P592-595 SEKI-4-8-1 <SA>
1961 V4 N9 P670-673 SEKI-4-9-1 <SA>
1961 V4 N9 P674-677 SEKI-4-9-2 <SA>
1961 V4 N9 P678-682 SEKI-4-9-3 <SA>
1961 V4 N10 P761-765 SEKI-4-10-1 <SA>
1961 V4 N11 P834-839 SEKI-4-11-1 <SA>
1961 V4 N11 P840-844 17P66J <ATS>
1962 N4 P238-242 ICE V4 N1 P158-164 <ICE>
1962 N4 P243-248 ICE V4 N2 P367-374 <ICE>
1962 N8 P578-581 ICE V4 N4 P748-754 <ICE>
1962 V5 P18-22 65-13284 <*>
1962 V5 P720-724 63-20006 <*>
1962 V5 N1 P18-22 SEKI-5-1-1 <SA>
1962 V5 N1 P31-37 SEKI-5-1-2 <SA>
1962 V5 N2 P78-83 SEKI-5-2-1 <SA>
1962 V5 N4 P238-242 20Q68J <ATS>
1962 V5 N4 P243-248 ICE V4 N2 P367-374 <ICE>
87P65J <ATS>
1962 V5 N6 P384-390 SEKI-5-6-1 <SA>
1962 V5 N6 P398-403 SEKI-5-6-2 <SA>
1962 V5 N8 P564-567 SEKI-5-81 <SA>
1962 V5 N8 P578-581 ICE V4 N4 P748-754 <ICE>
37R76J <ATS>
1962 V5 N11 P821-826 SEKI-5-11-1 <SA>
1962 V5 N12 P885-890 SEKI-5-12-1 <SA>
1963 V6 N1 P20-25 SEKI-6-1-1 <SA>
1963 V6 N3 P191-196 SEKI-6-3-1 <SA>
65-14263 <*> 0
1963 V6 N8 P588-592 SEKI-6-8-1 <SA>
1963 V6 N9 P666-670 SEKI-6-9-1 <SA>
1963 V6 N11 P864-867 SEKI-6-11-1 <SA>
1963 V6 N12 P919-924 SEKI-6-12-1 <SA>
1964 V7 P20-24 66-13861 <*>
1965 V8 N1 P8- 2083 <TC>
1965 V8 N9 P697-703 07T91J <ATS>

SEKIYU GIJUTSU KYOKAISHI
1964 V29 N2 P72-76 66-13934 <*$>

SEKKO TO SEKKAI
1954 N13 P591-595 CSIRO-3822 <CSIR>
1955 N18 P841-945 63-16119 <*>
1957 N26 P1269-1275 44Q71R <ATS>
1957 N27 P1315-1318 42Q71J <ATS>
1958 N33 P1627-1636 62-26044 <=$>
1959 N40 P337-342 43Q71J <ATS>
1959 N42 P465-469 45Q71J <ATS>
1963 N62 P21-26 73Q69J <ATS>
1963 N64 P120-125 45R79J <ATS>
1963 N65 P155-157 46R79J <ATS>
1964 N70 P96-99 56R79J <ATS>

SELECTA. NETHERLANDS.
1963 V5 P1598-1599 64-00188 <*>
1963 V43 P1330-1332 65-10977 <*> 0

SELEKTSIYA I SEMENOVODSTVO
1947 V14 N10 P26-30 RT-3158 <*> 0
1947 V14 N10 P58-63 RT-3159 <*>
1947 V14 N11 P41-52 RT-766 <*>
1947 V14 N11 P75-79 RT-763 <*>
1947 V14 N12 P27-32 RT-765 <*>
1948 V15 N2 P67-68 RT-1787 <*>
1948 V15 N11 P31-41 RT-2773 <*>
1949 V16 N10 P69-70 R-1576 <*>
1950 V17 N5 P35-39 RT-2726 <*>
1950 V17 N7 P31-37 RT-3148 <*>
1951 N4 P67-70 <CBPB>
1951 N7 P59-62 <CBPB>
1951 N9 P3-14 <CBPB>
1952 V19 N12 P56-59 RT-3294 <*>
1953 V20 N3 P37-42 RT-3322 <*>
1953 V20 N5 P31-33 RT-1522 <*>
1956 V21 N4 P7-12 60-21892 <=>
1956 V21 N5 P47-48 61-15853 <+*>
1957 V22 N1 P8-13 59-16796 <+*>
1957 V22 N4 P23-27 61-19645 <=*>
1957 V22 N6 P75-77 R-5233 <*>
1958 V23 N2 P22-26 60-23147 <+*>
1958 V23 N2 P66-69 60-23148 <+*>
1958 V23 N2 P69 60-23149 <+*>
1958 V23 N3 P43-46 60-23153 <+*>
1958 V23 N3 P59-60 60-23154 <+*>
1958 V23 N5 P25-30 59-19254 <+*> 0
1959 V24 N1 P3-7 60-17961 <+*>
1959 V24 N1 P49-50 60-17962 <+*>
1959 V24 N1 P51-52 59-16827 <+*>
1959 V24 N1 P60-65 60-23146 <+*> 0
1959 V24 N4 P70-72 61-27299 <*=>
1959 V24 N5 P13-14 60-17953 <+*> 0
1959 V24 N5 P15-19 60-17954 <+*>
1959 V24 N5 P29-31 60-17955 <+*>
1959 V24 N6 P31-33 60-23155 <+*>
1959 V24 N6 P54-58 60-23156 <+*>
1959 V24 N6 P60-66 60-23157 <+*>
1960 V25 N1 P28-30 62-23568 <=*>
1962 V27 N2 P22-26 64-13533 <=*$>
1963 V28 N3 P32-38 64-13571 <=*$> 0
1963 V28 N5 P53-59 64-31384 <=>
1963 V28 N6 P71-72 RTS-2791 <NLL>

SELKHOZMASHINA
1950 N8 P1-6 RT-4212 <*>
1951 N10 P10-13 60-21155 <=>
1952 N3 P5-10 60-21119 <=>
1952 N10 P10-13 59-17115 <+*> 0
1954 N5 P3-9 RT-2233 <*>
1954 N10 P3-7 RT-3131 <*>
1954 V10 P21-23 R-1911 <*>
1955 N5 P14-17 R-128 <*>
1955 N9 P10-13 RT-4310 <*>
1956 N2 P13-16 CSIRO-3107 <CSIR>
1956 N3 P22-24 60-21084 <=>
1956 N11 P103 59-18533 <+*>
1956 V2 P13-16 R-3913 <*>
1957 N3 P10-12 60-17791 <+*>

SELSKAYA ZHIZN
1960 N195 P2 61-21129 <=>
1960 N205 P2 61-21129 <=>
1961 N41 P2 61-27391 <=>
1961 N48 P1-2 61-27391 <=>
1961 N63 P2 61-27391 <=>
1963 N46 P4 63-21546 <=>

SELSKII MEKHANIZATOR
1960 N8 P3-4 61-11736 <=>
1961 N2 P22-23 61-27153 <=>
1963 N4 P29 63-31143 <=>
1963 N5 P22-23 63-31209 <=>
1964 N1 P5-7 64-31994 <=>
1964 N3 P10-11 64-31994 <=>
64-41104 <=>
1966 N8 P1-5 66-34871 <=$>
1966 N8 P28-29 66-34871 <=$>
1966 N9 P32-33 67-30153 <=>

SELSKOE KHOZYAISTVO KAZAKHSTANA
1958 N1 P83-84 62-26023 <=$>
1959 N12 P51-55 60-31294 <=>
1959 N12 P76-78 60-31294 <=>
1959 V7 N10 P82-85 63-11023 <=>
1960 N8 P47-48 61-11923 <=>
1960 N9 P76-79 61-11923 <=>
1961 V8 N1 P66-68 61-27482 <*=>
1964 V12 N4 P32-35 65-60905 <=$>

SELSKOE KHOZYAISTVO KIRGIZII
1960 V6 N10 P18-20 62-24957 <=*>

SELSKOE KHOZYAISTVO POVOLZHYA
1956 V1 N5 P45-46 61-15864 <+*>
1957 V2 N6 P67-69 61-28469 <*=>
1958 V3 N5 P55-58 60-21894 <=>
1958 V3 N8 P64-65 63-11047 <=>
1958 V4 N12 P43-46 61-15861 <+*>
1959 V5 N6 P87-89 61-21146 <=>
1960 V6 N2 P91-93 63-11055 <=>
1960 V6 N12 P43-45 66-51094 <=>
1961 V7 N2 P82-84 61-27153 <=>
1962 V8 N2 P73-76 65-61500 <=$>

SELSKOE KHOZYAISTVO PODMOSKOVIA
1960 V9 N7 P23-24 61-21339 <=>

SELSKOE KHOZYAISTVO SEVERNOGO KAVKAZA
1959 V2 N1 P20-23 59-31015 <=> O
1960 V3 N1 P34-37 60-31696 <=>
1960 V3 N12 P47 61-31515 <=>
1961 V4 N3 P1-2 61-27116 <=>
1966 N9 P36 67-30153 <=>

SELSKOE KHOZYAISTVO SEVERO-ZAPADNOY ZANY
1959 V2 N3 P65-68 61-31004 <=>
1959 V2 N12 P52-54 60-31294 <=>
1960 V3 N3 P68-70 60-31499 <=>
1961 V4 N3 P12-15 61-27116 <=>
1962 V5 N5 P69-71 62-33348 <=> P

SELSKOE KHOZYAISTVO SIBIRI
1957 N6 P85-90 R-2002 <*>
1958 N6 P93-94 62-25136 <=*>
1959 N4 P21-27 64-15673 <=*$>
1959 N8 P9-13 60-31018 <=>
1960 N9 P3-5 61-21100 <=>
1961 N2 P44-46 61-31146 <=>
1961 V6 N4 P51-52 64-13798 <=*$>
1961 V6 N10 P42-43 62-24985 <=*>
1961 V6 N10 P53-54 62-24985 <=*>

SELSKOE KHOZYAISTVO TADZHIKISTANA
1960 N12 P51 61-23011 <=>

SELSKOE KHOZYAISTVO TURKMENISTANA
1960 N6 P10-12 61-23476 <=>
1960 N6 P51-54 61-23476 <=>

SELSKOE STROITELSTVO
1960 N4 P21 60-11814 <=>
1961 N3 P21 61-23945 <=>

SELSKOKHOZYAISTVENNOE PROIZVODSTVO SIBIRI I DALNEGO VOSTOKA
1964 N5 P3-6 AD-611 054 <=$>
1965 N5 P42-43 65-33000 <=*$>

SELSKOKHOZYAISTVENNOE PROIZVODSTVO URALA
1965 V3 N6 P24-25 65-32220 <=$>

SELSKOSTOPANSKA NAUKA
1964 V3 N1 P53-57 64-31446 <=>
1964 V3 N1 P77-86 64-31444 <=>
1966 N2 P3-7 66-34708 <=$>
1966 N2 P27-38 66-34708 <=$>
1966 N2 P75-90 66-34693 <=$>
1966 N2 P103-109 66-34694 <=$>

SEMAINE DES HOPITAUX DE PARIS
1946 V22 N44 P2011-2016 66-10707 <*> P
1947 V23 P895-896 65-14609 <*>
1948 V24 N9 P247-255 2554 <*>
1949 V25 N7 P296-300 2805 <*>
1949 V25 N86 P3592-3603 2787 <*>
1949 V25 N96 P4040-4042 2555 <*>
1950 V26 P3929-3936 61-20762 <*> P
1950 V26 P3936-3946 61-20763 <*> P
1951 P344-351 2581 <*>
1952 V28 P1062-1070 61-00373 <*>
1953 V29 P385-392 II-33 <*>
1953 V29 P1745-1748 T-1676 <INSD>
3148 <*>
1953 V29 P1959-1968 AL-947 <*>
1953 V29 P1994- AL-349 <*>
1953 V29 N37 P1849-1854 59-20971 <*>
1954 V30 P1692-1701 65-14322 <*> O
1955 V31 P67-70 2662 <*>
1955 V31 N3 P1-16 58-1762 <*>
1956 V32 P1051-1057 5226-A <K-H>
1957 V33 N62 P3630-3643 6871-D <K-H> P
1957 V33 N51/2 P3043-3044 59-15425 <*>
1958 N13/4 P1083-1104 AEC-TR-4352 <*>
1958 V34 N13/4 P1083-1104 59-18798 <+*> O
1959 V35 N60/1 P2294-2301 60-14548 <*>
1961 V37 P29-40 61-16560 <*> O
1962 V38 P1593-1596 63-10386 <*>
1962 V38 N1/2 PE28-E34 64-18665 <*>
1963 V39 P2459-2463 65-14710 <*>
1963 V39 N15 P694-695 64-18644 <*>
1964 V40 N1 P17-20 66-12747 <*> O

SEMAINE MEDICALE
1894 V14 P89-92 59-15610 <*> O
1901 V21 P241-243 II-987 <*>
1907 V27 P37-40 59-15609 <*> O
1908 N29 P345 58-324 <*>
1911 V31 N7 P232-235 57-107 <*>

SEMAINE THERAPEUTIQUE. (TITLE VARIES)
1960 V36 N10 P909-912 63-10491 <*>
1963 V39 P258-261 66-12047 <*>
1963 V39 N3 P175-178 64-18667 <*>
1964 V40 N5 P335-338 66-13182 <*>

SEMANA MEDICA. BUENOS AIRES
1936 V43 N52 P1883-1887 63-16680 <*>
6960 <K-H>
1937 V44 P1430-1434 57-1664 <*>
1938 V45 P386-388 57-1596 <*>
1938 V45 P1019-1021 57-1517 <*>
1943 V50 P121-130 57-1634 <*>
1943 V50 P1420-1425 57-1503 <*>
1952 N28 P650-653 AL-187 <*>
1952 V100 P807-811 58-2197 <*>
1956 V108 N14 P446-449 62-00309 <*>
1960 N117 P1142-1147 55N51SP <ATS>
1960 V116 N16 P466-468 62-14351 <*>
1960 V117 P1142-1147 62-16229 <*>
1960 V117 N18 P678-683 62-14350 <*>
1960 V117 N36 P1373-1384 65-17240 <*>
1961 V119 P1171-1174 63-10662 <*>
1961 V119 N41 P1738-1740 65-11540 <*>
1962 V120 N20 P734-737 63-10920 <*>
1962 V121 P1705-1710 64-18737 <*>
1963 V122 P189 65-17287 <*>
1963 V122 N6 P161-164 65-17287 <*>
1963 V123 N27 P1052-1057 65-12926 <*>
1964 V124 P1744-1750 65-12319 <*>
1964 V124 P1748-1750 65-13599 <*>
1964 V124 N7 P409-412,423 65-12163 <*>
1964 V125 P1285-1286 65-13597 <*>
1964 V125 P1436-1438 65-13598 <*>

SEMANA MEDICA DE MEXICO
1960 V27 N9 P319-322 62-16083 <*>
1961 V30 P290-291 62-14321 <*>

SEMENTO GIJUTSU NEMPO
1955 V9 N41 P25-28 61-20742 <*>
1955 V9 N41 P46-49 61-20742 <*>
1956 V10 P51-56 59-17896 <*>
1957 V11 P48-54 59-15073 <*>
1958 V12 P309-311 60-18117 <*>
1962 V16 P70-77 66-26096 <$*>
1963 V17 P359-365 1884 <TC>

SEMIA I SHKOLA
1952 V7 N9 P32-33 RT-4538 <*>

SENCKENBERGIANA
1938 V20 N1/2 P1-42 11K24G <ATS>

SEN-I-GAKKAISHI
1953 V9 N2 P81-85 T-2199 <INSD>
1953 V9 N9 P446-448 59-17083 <*>
1956 V12 P412-427 T-2200 <INSD>
1956 V12 N7 P479-486 62-14961 <*>
1956 V12 N11 P823-827 C-2518 <NRC>
58-49 <*>
1957 V13 P292-304 61-16772 <*>
1957 V13 P801-807 SCL-T-267 <*>
1957 V13 N8 P553-559 60-18456 <*>
1957 V13 N12 P808-812 SCL-T-264 <=*>
59-15807 <*>
1957 V13 N12 P813-816 SCL-T-265 <=*> O
1957 V13 N12 P861-865 61-16583 <*>
1958 V14 P92-95 18L29J <ATS>
1958 V14 N3 P133-141 61-16582 <*>
1958 V14 N6 P481-483 02M45J <ATS>
1958 V14 N7 P484-487 64-18131 <*>
1958 V14 N8 P557-562 60-18455 <*>
1958 V14 N11 P814-817 61-16539 <*>
1958 V14 N11 P818-821 61-16538 <*>
1959 V15 P527-531 60-10150 <*>
1959 V15 P532-536 60-10149 <*>
1959 V15 P863-867 02M41J <ATS>
1959 V15 N1 P9-12 61-16579 <*>
1959 V15 N1 P13-15 61-16578 <*>
1959 V15 N2 P90-94 61-16581 <*>
1959 V15 N4 P254-259 61-16571 <*>
1959 V15 N5 P368-372 61-16569 <*>
1959 V15 N9 P708-715 61-16580 <*>
1959 V15 N10 P814-823 65-13445 <*>
1959 V15 N12 P951-959 60-18453 <*>
61-16517 <*>
1959 V15 N12 P985-991 60-18454 <*>
1960 V16 N1 P7-15 61-16525 <*>
1960 V16 N1 P15-20 61-16524 <*>
1960 V16 N6 P458-469 00M45J <ATS>
1960 V16 N10 P810-812 CODE-S110 <SA>
1960 V16 N10 P839-848 C-140 <RIS>
1960 V16 N11 P955-961 62-14267 <*>
1961 V17 P325-329 76N57J <ATS>
1961 V17 N1 P27-30 62-16011 <*>
94N51J <ATS>
1961 V17 N3 P243-252 79N53J <ATS>
1961 V17 N10 P997-1001 CODE-S265 <SA> O
SEG-7-10-1 <SA>
1961 V17 N11 P1088-1093 CODE-S277 <SA> O
1962 V18 P147-152 998 <TC>
1962 V18 P1076-1081 65-63423 <=$>
1962 V18 N3 P189-193 SEG-18-3-1 <SA>
1962 V18 N3 P194-198 SEG-18-3-2 <SA>
1962 V18 N3 P198-206 SEG-18-3-3 <SA>
1962 V18 N3 P236-240 12P66J <ATS>
1962 V18 N3 P240-242 13P66J <ATS>
1962 V18 N4 P356-360 34Q68J <ATS>
1962 V18 N4 P361-370 SEG-18-4-1 <SA>
1962 V18 N8 P741-748 64-18491 <*>
1963 V19 P369-372 94Q73J <ATS>
1963 V19 N3 P230-242 22R78J <ATS>
1963 V19 N10 P828-832 SEG-19-10-1 <SA>
1963 V19 N12 P953-961 M-5654 <NLL>
1964 V20 N5 P311-318 999 <TC>
1964 V20 N6 P356-364 37S81J <ATS>
1964 V20 N7 P448-453 1075 <TC>
1964 V20 N8 P519-524 94S80J <ATS>
1964 V20 N10 P662-665 45S87J <ATS>

SEN-I KAI
1957 V16 N174 P64-68 62-20094 <*>

SEN-I KIKAI GAKKAISHI
1958 N11 P758-761 04M42J <ATS>
1959 V12 N2 P7-13 60-18485 <*>
1959 V12 N3 P163-168 60-18486 <*>
1959 V12 N10 P16-21 C-145 <RIS>
1960 V13 N2 P104-108 10M46J <ATS>
1960 V13 N6 P53-58 C-144 <RIS>
1960 V13 N6 P435-440 62-14301 <*>
1960 V13 N9 P637-643 C-138 <RIS>
1961 V14 P518-524 1144 <TC>
1964 V17 N7 P513-519 M-5733 <NLL>

SEN-I KOGYO SHIKENSHO IHO
1958 N45 P87-98 61-16778 <*>
1959 N48 P1-6 63-10683 <*>
1959 N49 P65-68 62-10718 <*> O
1959 N49 P69-76 62-10716 <*> O
1959 N49 P77-82 62-10717 <*> O
1960 N52 P29-37 14M46J <ATS>
1960 N53 P1-8 62-14020 <*> O
1960 N53 P9-16 15M46J <ATS>
1960 N54 P1-4 07N55J <ATS>

SEN-I-SO KOGYO
1941 V17 P81-89 65-12869 <*>

SEPPYO
1941 V3 P109-121 II-374 <*>
1941 V3 P225-236 1954 <*>
1941 V3 P260-263 II-1046 <*>
1941 V3 P264-270 2449 <*>
1941 V3 P291-294 II-539 <*>
1941 V3 P303-307 1951 <*>
1941 V3 P333-342 2452 <*>
1941 V3 P343-349 II-1001 <*>
1941 V3 P414-418 I-707 <*>
1941 V3 P503-506 57-727 <*>
1941 V3 N7 P1-4 II-540 <*>
2837 <*>
1942 V4 N8 P229-234 I-1029 <*>
1943 V5 P249-256 2242 <*>
1959 V21 P178-181 63-00685 <*>
1959 V21 N6 P178-181 C-4389 <NRC>

SETTIMANA MEDICA
1965 V53 N22 P1153-1160 66-12951 <*>
1965 V53 N22 P1177-1183 66-12949 <*>
1965 V53 N22 P1185-1194 66-12950 <*>

SETTIMANA DEGLI OSPEDALI
1964 V6 P446 65-13971 <*>

SEVERNYI MORSKOI PUT
1939 V11 P62-63 RT-2760 <*>

SHAKHTNOE STROITELSTVO
1958 N5 P23-25 60-17242 <+*>
1959 N3 P6-9 60-11114 <=>
1959 N4 P18-19 62-15266 <*=>
1959 N11 P16-17 61-27594 <*=>
1960 V4 N7 P17-20 63-19942 <*=>
1961 V5 N3 P8-10 64-13293 <=*$>
1961 V5 N10 P26-28 63-24571 <=*$>
1961 V5 N11 P27-30 63-24572 <=*$>
1962 V6 N11 P28-29 63-21053 <=>
1963 V7 N5 P33 63-31535 <=>
1963 V7 N7 P31-34 63-31797 <=>
1964 N4 P1-3 NLLTB V6 N9 P809 <NLL>

SHA MO TI CHU TI TSUNG HO T'IAO CH'A YEN CHIU PAO KAO
1958 N2 P1-82 63-21341 <=>

SHANGHAI CHI HSIEH
1962 N8 P43 63-21279 <=>
1963 N4 P4-9 64-31200 <=>
1964 N3 P13-16 64-41581 <=>
1964 N7 P29-31 66-30011 <=$>

SHANGHAI MACHINERY. CHINESE PEOPLES REPUBLIC
SEE SHANGHAI CHI HSIEH

SHENG HSUEH HSUEH PAO
1964 V1 N2 P76-83 67-30064 <=>
1965 V2 N1 P24-28 66-35530 <=>
1965 V2 N1 P39-40 66-35530 <=>
1965 V2 N2 P102-103 66-35530 <=>
1966 V3 N1 P1-7 66-35607 <=>
1966 V3 N1 P14-26 66-35607 <=>
1966 V3 N1 P34-39 66-35607 <=>

1966 V3 N1 P45-46 66-35607 <=>
1966 V3 N1 P52-54 66-35607 <=>

SHENG LI HSUEH PAO
1956 V20 N1 P37-49 63-18529 <=*> O
1960 V24 N2 P85-94 61-28608 <*=>
1960 V24 N2 P95-104 62-13309 <*=>
1960 V24 N2 P105-109 62-13214 <=>
1960 V24 N2 P110-120 61-28549 <*=>
1960 V24 N2 P121-128 61-28574 <*=>
1960 V24 N2 P129-135 62-13214 <=>
1960 V24 N2 P136-140 61-28567 <*=>
1960 V24 N2 P141-148 62-13214 <=>
1961 V24 N3/4 P220-226 AD-620 967 <=*$>

SHENG-LI K'O-HSUEH CHIN-CHAN
1957 V1 N1 P1-9 59-11318 <=>
1957 V1 N1 P11-21 59-11320 <=>
1957 V1 N1 P23-30 59-11319 <=>
1957 V1 N1 P31-37 59-11317 <=>
1957 V1 N1 P39-51 59-11321 <=>

SHENG WU HSUEH T'UNG PAO
1959 N10 P436-440 60-11317 <=>
1959 N10 P441-448 60-11365 <=>
1959 N10 P455 60-11365 <=>
1959 N10 P462-467 60-11352 <=>
1959 N10 P468-471 60-31025 <=>
1959 N10 P475 60-31025 <=>
1960 N3 P97-106 61-21581 <=>
1960 N3 P113 61-21581 <=>
1964 N2 P1-8 AD-636 599 <=$>
64-41090 <=>
1964 N2 P9-13 64-41012 P.129-156 <=>
1964 N2 P14-16 64-41012 P.121-128 <=>
1965 N1 P16-23 65-30648 <=$>

SHENG WU HUA HSUEH YU SHENG WU WU LI HSUEH PAO
1964 V4 N1 P1-5 64-14278 <*> O

SHIGEN KAGAKU KENKYUSHO IHO
1950 N17/8 P61-68 80E1J <ATS>
1954 N36 P104-107 61-00455 <*>
1954 V36 P108-119 62-00116 <*>
1955 V37 P127-128 62-00117 <*>
1958 V48 P28-38 62-00115 <*>
1959 V51 P22-23 62-00627 <*>

SHIH-YU K'AN T'AN
1958 N18 P4-5 59-11535 <=>
1958 N18 P7-9 59-11535 <=>
1958 N18 P12-28 59-11535 <=>
1958 N18 P37 59-11535 <=>
1958 N18 P43 59-11535 <=>

SHIH-YU LIEN-CHIH
1958 N5 P1-5 59-11535 <=>
1958 N5 P20-22 59-11535 <=>
1958 N5 P23-27 59-11535 <=>
1958 N5 P48-49 59-11535 <=>

SHIKA GAKUHO
1960 V60 P91-93 63-16431 <*>

SHIKA ZAIRYO KENKYUSHO HOKOKU
1962 V2 N4 P375-387 SCL-T-533 <*>

SHIKEN HOKOKU (NIHON SENBAI KOSHA ODAWARA SEIEN SHIKENJO)
1961 N6 P35-46 NEL-TRANS-1693 <NLL>

SHIKOKU ACTA MEDICA. TOKUSHIMA
SEE SHIKOKU IGAKU ZASSHI

SHIKOKU AGRICULTURAL RESEARCH
SEE SHIKOKU NOGYO KENKYU

SHIKOKU IGAKU ZASSHI
1959 V14 N5 P897-932 11216 <K-H>

SHIN RINSHO
1947 V2 P12-15 64-14659 <*>

SHINDAN TO CHIRYO
1954 V42 P824-828 65-14447 <*>

SHINKU
1960 V3 P304-308 65-11641 <*>
1960 V3 N9 P341-348 32N53J <ATS>
61-20534 <*>
1960 V3 N11 P416-429 56N53J <ATS>
61-20849 <*>
1962 V5 N8 P310-322 N65-23675 <=*$>

SHINRIGAKU KENKYU
1937 V12 N5 P43-44 1478 <*>
1937 V12 N5 P465-482 1477 <*>
1951 V22 N2 P77-88 61-00624 <*>
1957 V28 N4 P210-231 60-19057 <=*>
1958 V28 N5 P37-43 59-11134 <=>
1962 V32 N6 P381-387 63-01312 <*>

SHINRYO
1960 V13 N11 P1411-1418 61-00410 <*>

SHINYAKU TO RINSHO
1956 V5 N5 P357-359 62-01526 <*>
1958 V7 P1003-1009 66-13213 <*>
1958 V7 P1011-1016 64-18722 <*>
1962 V11 P549-554 66-13163 <*>

SHIONOGI KENKYUSHO NEMPO
1964 N14 P100-108 65-14596 <*> O

SHIPBUILDING IN CHINA
SEE CHUNG KUO TSAO CHUAN

SHITSURYO BUNSEKI
1964 V12 N25 P25-36 65-14749 <*>

SHKOLA I PROIZVODSTVO
1960 V4 N9 P50-53 61-11974 <=>
1961 V5 N9 P84-85 62-24988 <=*>

SHOKUBAI. TOKYO
1961 V3 N2 P183-187 62-10306 <*>
1962 V4 N1 P5-7 22P66J <ATS>
1962 V4 N1 P88-92 30Q70J <ATS>

SHOKUBUTSU BOEKI
1956 V10 N4 P7-10 C-2247 <NRC>
57-103 <*>

SHOKOBUTSU KAGAKU ZASSHI
1929 V4 P327-340 T-2425 <INSD>
1931 V5 P271-287 65-13640 <*>

SHOKUBUTSU OYOBI DOBUTSU
1934 V2 P1212 65-11057 <*>

SHOKUBUTSUGAKU ZASSHI
1929 V43 P227-229 65-11045 <*> O
1937 V51 N605 P306-317 892-E <K-H>
1949 V62 P9-13 3140-A <K-H>
1949 V62 P49-52 3140-B <K-H>

SKOKURYO KENKYUSHO KENKYU HOKOKU
1957 N12 P71-72 66-10843 <*>

SHOWA IGAKKAI ZASSHI
1954 V14 N3 P223-245 65K28J <ATS>
1958 V17 N6 P481-489 66K28J <ATS>

SHU HSUEH CHIN CHAN
1957 V3 P434-444 AEC-TR-5118 <=*>
1964 V7 N1 P38 AD-637 431 <=$>
1965 V8 N1 P105-107 65-32374 <=*$>

*SHU HSUEH HSUEH PAO
1953 V2 P203-229 AMST S2 V38 P235-258 <AMS>

1953 V2 P288-323 AMST S2 V32 P163-194 <AMS>
AMST,S2,V32,P163-194 <AMS>
1954 V4 P201-221 AMST S2 V38 P171-188 <AMS>
1954 V4 P323-346 AMST S2 V11 P155-172 <AMS>
1954 V4 P365-379 AMST S2 V38 P159-170 <AMS>
1955 V5 P1-25 AMST S2 V32 P195-220 <AMS>
AMST,S2,V32,P195-220 <AMS>
1955 V5 P193-204 AMST S2 V33 P47-58 <AMS>
1955 V5 P205-242 AMST S2 V32 P221-263 <AMS>
AMST,S2,V32,P221-263 <AMS>
1955 V5 P347-368 STMSP V4 P17-38 <AMS>
1955 V5 P369-379 AMST S2 V37 P241-252 <AMS>
1955 V5 P401-410 AMST S2 V38 P259-268 <AMS>
1955 V5 P463-470 AMST S2 V32 P265-272 <AMS>
AMST,S2,V32,P265-272 <AMS>
1956 V6 P233-241 AMST S2 V38 P291-300 <AMS>
1956 V6 P631-637 AMST S2 V38 P189-194 <AMS>
1957 V7 P167-179 AMST S2 V38 P277-289 <AMS>
1957 V7 P235-241 AMST S2 V38 P269-276 <AMS>
1958 V8 P1-11 AMST S2 V37 P143-156 <AMS>
1958 V8 P102-130 AMST S2 V38 P195-234 <AMS>
1958 V8 P210-221 STMSP V3 P191-203 <AMS>
1958 V8 P272-275 AMST S2 V37 P253-257 <AMS>
1958 V8 P333-347 STMSP V3 P205-224 <AMS>

SHUI LI FA TIEN
1959 N19 P9-11 59-11742 <=>

SHUI LI HSUEH PAO
1959 V6 N3 P35-40 62-25376 <=*>
1963 V10 N1 P77 63-31766 <=>

SHUI LI SHUI TIEN CHIEN SHE
1958 N1 P12-17 59-11497 <=>
1959 N17 P7-20 60-11446 <=>
1959 N17 P28-35 60-11446 <=>
1959 N18 P18-20 60-11446 <=>
1959 N18 P23-25 60-11446 <=>
1959 N18 P32-33 60-11446 <=>
1959 N18 P41 60-11446 <=>

SHUI LI YU TIEN LI
1958 N1 P6-8 59-11742 <=>
1958 N2 P22-27 59-11742 <=>

SHUI WEN TI CHIH KUNG CH'ENG TI CHIH
1959 N1 P17-32 61-28575 <=> 0
1959 N1 P33-46 61-28861 <=>
1959 N1 P48 61-28861 <=>
1959 N5 P3-8 61-28310 <=>
1959 N5 P10-15 62-13362 <*=>
1959 N5 P15-18 62-13718 <*=>
1959 N5 P19-25 62-13334 <*=>
1959 N5 P26-37 62-13020 <*=>
1959 N5 P41 62-13020 <*=>
1959 N6 P3-17 61-27872 <=>
1959 N6 P18-22 62-15232 <*=>
1959 N6 P22-24 61-28789 <*=>
1959 N6 P25-33 62-13670 <*=>
1959 N6 P34-36 61-27962 <=>
1959 N6 P37-40 61-28855 <*=>
1959 N6 P49-60 61-27962 <=>
1959 N7 P1-8 61-28563 <=>
1959 N7 P13-41 61-28563 <=>
1959 N8 P1-37 61-23919 <=>
1959 N9 P1-42 61-23919 <=>
1959 N10 P1-6 62-13222 <*=>
1959 N10 P7-8 61-28603 <*=>
1959 N10 P11-15 62-13713 <=*>
1959 N10 P18-21 61-28572 <=>
1959 N10 P23 61-28572 <=>
1959 N10 P28-29 61-28572 <=>
1959 N10 P30-32 62-13713 <=*>
1959 N10 P32-33 62-13656 <*=>
1959 N10 P34-37 61-28570 <*=>
1959 N11 P13-28 61-27957 <=>
1959 N11 P29-31 62-13656 <*=>
1959 N12 P10-37 61-23124 <=>
1959 N12 P39-41 61-23124 <=>
1960 N1 P3-14 61-27157 <*=>
1960 N1 P14-17 62-13257 <=*>
1960 N1 P21-22 62-13435 <*=>
1960 N2 P1-23 61-27122 <=>
1960 N2 P28-42 61-27122 <=>
1960 N3 P4-42 61-27094 <=>
1960 N4 P3-8 61-27149 <=>
1960 N4 P10-35 61-27149 <=>
1960 N4 P37-39 61-27149 <=>
1960 N5 P3-23 62-15439 <=>
1960 N5 P25-39 62-15439 <=>
1960 N5 P44-48 62-15439 <=>
1960 N6 P1-5 61-28304 <=>
1960 N6 P15-18 61-28304 <=>
1960 N6 P26-28 62-13848 <=>
1960 N6 P29-30 61-28045 <=>
1960 N6 P31-33 61-28631 <*=>
1960 N6 P34-41 61-28045 <=>
1960 N6 P42-44 62-13848 <=>

SHUI WEN YUEH KAN
1959 N5 P6 61-15747 <+>
1959 N5 P12-21 61-15747 <+>
1959 N5 P22 62-13556 <=>
1959 N5 P24-25 62-13556 <=>
1959 N5 P27-29 62-13556 <=>
1959 N5 P35-37 62-13556 <=>
1959 N5 P43 62-13556 <=>
1959 N9 P7-15 61-21247 <=>
1959 N9 P20-21 61-21247 <=>
1959 N11 P5-8 61-21109 <=>
1959 N11 P8-17 62-13649 <=>
1959 N11 P17-25 62-15240 <=>
1959 N11 P34-36 62-15240 <=>
1959 N12 P14-21 62-13711 <=>
1959 N12 P24-27 62-13711 <=>
1959 N12 P34-35 62-13711 <=>
1959 N12 P41 62-13711 <=>
1960 N2 P1-3 61-11954 <=>
1960 N2 P4-31 61-28910 <=>
1960 N2 P36-39 61-28910 <=>
1960 N3 P1-25 61-28910 <=>
1960 N3 P28-34 61-28910 <=>
1960 N3 P38-39 61-28910 <=>
1960 N4 P8-20 61-28910 <=>
1960 N4 P23-27 61-28910 <=>
1960 N4 P33-35 61-28910 <=>
1960 N5 P6-18 61-28910 <=>
1960 N5 P26-30 61-28910 <=>
1960 N5 P37-39 61-28910 <=>
1960 N10 P4-10 61-21278 <=>
1960 N10 P11-14 61-21440 <=>

SHUI YUN
1957 N6 P11-13 61-11103 <=> 0
1957 N6 P15-17 61-11103 <=> 0
1957 N6 P35 61-11103 <=> 0
1957 N6 P38-40 61-11103 <=> 0
1957 N8 P5 61-11103 <=> 0
1958 N5 P6 61-11103 <=> 0
1959 N7 P7-17 60-11673 <=>
1959 N7 P25-26 60-11673 <=>
1965 N9 P8-9 66-30104 <=$>
1965 N9 P14-15 66-30104 <=$>
1965 N9 P22 66-30104 <=$>
1965 N9 P25-26 66-30104 <=$>

SIBIRSKIE OGNI
1962 V41 N5 P146-160 62-32877 <=>

SIBIRSKII GEOGRAFICHESKII SBORNIK
1962 N1 P85-95 SGRT V4 N4 P17 <AGS>

*SIBIRSKII MATEMATICHESKII ZHURNAL
1960 V1 P14-44 AMST S2 V36 P251-284 <AMS>
1960 V1 P71-77 AMST S2 V48 P36-44 <AMS>
1960 V1 P117-138 STMSP V5 P103-125 <AMS>
1960 V1 P242-247 AMST S2 V54 P85-90 <AMS>
1960 V1 P427-455 AMST S2 V32 P323-357 <AMS>
AMST,S2,V32,P323-357 <AMS>
1960 V1 N2 P198-204 62-19813 <=>

1960 V1 N2 P238-241 62-19936 <=>
1960 V1 N4 P609-610 62-11399 <=>
1961 V2 P89-99 AMST S2 V36 P231-242 <AMS>
1961 V2 P129-137 STMSP V5 P191-200 <AMS>
1961 V2 P237-260 AMST S2 V57 P144-170 <AMS>
1961 V2 P622-638 AMST S2 V53 P202-220 <AMS>
1961 V2 P639 AMST S2 V48 P36-44 <AMS>
1961 V2 N1 P59-67 62-15614 <*=>
62-15614 <=>
1961 V2 N5 P746-758 63-31540 <=>
1961 V2 N5 P767-788 63-31540 <=>
1961 V2 N5 P792-800 63-31540 <=>
1961 V2 N6 P904-912 63-31540 <=>
1962 V3 P103-131 AMST S2 V59 P23-55 <AMS>
1962 V3 P645-694 STMSP V5 P315-372 <AMS>
1962 V3 N3 P313-332 44S82R <ATS>
1962 V3 N4 P481-499 62-33312 <=*>
1962 V3 N5 P701-709 63-13456 <=*>
1962 V3 N5 P768-796 63-19549 <=*$>
1963 V4 P120-137 AMST S2 V54 P29-48 <AMS>
1963 V4 P138-144 AMST S2 V51 P74-81 <AMS>
1963 V4 P295-302 STMSP V6 P189 <AMS>
1963 V4 P951-955 AMST S2 V50 P178-182 <AMS>
1963 V4 P1137-1149 AMST S2 V54 P49-62 <AMS>
1963 V4 P1263-1270 AMST S2 V55 P261-269 <AMS>
1963 V4 P1365-1375 AMST S2 V54 P63-74 <AMS>
1963 V4 N2 P295-302 63-31103 <=>
1963 V4 N2 P426-445 63-31354 <=>
1963 V4 N3 P622-631 64-11891 <=> M
1963 V4 N6 P1328-1341 64-21836 <=*$>
1964 V5 P86-93 STMSP V6 P198 <AMS>
1964 V5 P253-289 STMSP V6 P207 <AMS>
1964 V5 P319-336 AMST S2 V58 P57-76 <AMS>
1964 V5 P377-386 AMST S2 V51 P189-200 <AMS>
1964 V5 P509-531 AMST S2 V56 P273-295 <AMS>
1964 V5 P549-556 AMST S2 V52 P42-49 <AMS>
1964 V5 P750-767 STMSP V6 P218 <AMS>
1964 V5 P1282-1304 AMST S2 V52 P9-31 <AMS>
1964 V5 N1 P34-38 AD-610 650 <=$>
1964 V5 N1 P109-123 AD-616 309 <=$>
1964 V5 N2 P290-309 AD-615 226 <=$>
1964 V5 N5 P996-1006 64-51900 <=$>
1964 V5 N5 P1163-1180 AD-636 595 <=$>

SIEMENS-ZEITSCHRIFT
1924 V4 P6-14 58-351 <*>
1926 V6 P351-354 57-2466 <*>
1929 V9 N3 P135-144 57-645 <*>
1929 V9 N5/6 P389-390 66-14614 <*> 0
1930 V10 N10 P562-566 66-13306 <*>
1931 V11 N1 P26-32 59-20941 <*> 0
1934 V14 N3 P76-83 57-3026 <*>
1934 V14 N9 P302-308 57-3024 <*>
1936 V16 N1 P1-4 59-15443 <*>
1937 V17 N6 P259-269 59-20533 <*> 0
1938 V18 N5 P249-254 59-20532 <*> 0
1938 V18 N7 P329-338 66-12264 <*>
1939 V19 N2 P70-73 58-655 <*>
1939 V19 N4 P166-172 58-303 <*>
1939 V19 N8 P357-368 62-18370 <*> 0
1939 V19 N11 P534-537 66-13649 <*>
1952 V26 N3 P3-10 62-18873 <*> 0
1952 V26 N5 P213-219 63-10044 <*> 0
1952 V26 N8 P377-382 2528 <*>
1953 V27 P62-73 I-301 <*>
1953 V27 N1 P1-8 2530 <*>
1954 V28 N5 P224-228 45J15G <ATS>
57-2998 <*>
1954 V28 N7 P299-302 58-1094 <*>
1954 V28 N8 P370-376 58-1092 <*>
59-12033 <+*>
1955 V29 N10 P434-440 57-2313 <*>
1955 V29 N5/6 P206-208 62-18874 <*>
1956 V30 N1 P29-35 57-2047 <*>
1956 V30 N8 P390-398 51J15G <ATS>
57-2797 <*>
1956 V30 N9 P543-548 61-16400 <*>
1957 V31 N4 P177-178 47K28G <ATS>
1957 V31 N4 P179-181 52K28G <ATS>
1957 V31 N4 P182-186 53K28G <ATS>
1957 V31 N4 P190-191 58-1581 <*>
1957 V31 N4 P227-230 58-1543 <*>
1957 V31 N8 P400-404 59-10963 <*>
1957 V31 N8 P404-409 59-10997 <*>
1957 V31 N9 P469-472 59-16345 <+*>
1958 V32 P109 59-10789 <*>
1958 V32 P128-134 60-16662 <*>
1958 V32 N2 P83-87 60-16663 <*>
1958 V32 N3 P110-115 61-16179 <*>
1958 V32 N3 P122-128 61-16177 <*> 0
1958 V32 N10 P692-701 62-25201 <=*> 0
1958 V32 N10 P719-725 62-25201 <=*> 0
1958 V32 N10 P732-735 62-25201 <=*> 0
1958 V32 N12 P845-847 62-14388 <*>
1959 V33 N5 P305-310 62-25202 <=*> 0
1959 V33 N8 P509-516 61-20876 <*>
1959 V33 N11 P685-690 61-18618 <*>
1959 V33 N11 P698-703 60-31782 <*=>
1959 V33 N12 P763-766 61-18265 <*>
1960 V34 N1 P26-31 63-10674 <*>
1960 V34 N2 P96-104 3113 <BISI>
1960 V34 N6 P385-390 62-16020 <*>
1960 V34 N7 P397-404 62-16092 <*>
1960 V34 N10 P653-656 61-20726 <*> 0
1960 V34 N11 P777-786 61-20543 <*>
92N52G <ATS>
1961 V35 N4 P251-252 65-10037 <*>
1961 V35 N6 P493-499 62-14787 <*>
1961 V35 N8 P574-578 62-14788 <*>
1962 V36 N11 P808-811 63-18743 <*>
1962 V36 N12 P822-829 63-01428 <*>
1963 V37 N4 P329-330 64-14011 <*> 0
1963 V37 N4 P339-341 66-12711 <*> 0
1963 V37 N7 P532-536 64-16587 <*>
1963 V37 N9 P667-669 64-18604 <*> 0
1963 V37 N11 P809-815 65-11465 <*> 0
1964 V38 N2 P97-100 65-10382 <*> 0
1964 V38 N7 P558-562 65-12843 <*>
1964 V38 N8 P640-646 65-10381 <*> 0
1965 V39 N2 P138-145 65-17134 <*> 0
1965 V39 N4 P320-322 66-12281 <*>
92S87G <ATS>
1965 V39 N12 P1304-1306 66-12177 <*>

SIGLO MEDICO
1856 V2 P211 64-00064 <*>
1926 V78 P205-208 58-423 <*>
1926 V78 P233-236 58-423 <*>

SIGNAL UND DRAHT
1955 P10-14 T1976 <INSD>

SIGNAL UND SCHIENE
1962 V6 N11 P407-409 63-15108 <=*> 0
1967 N1 P6-8 67-30860 <=$>

SILICATES INDUSTRIELS
1949 V14 N5 P95-97 65-12316 <*> 0
1949 V14 N6 P125-133 64-14032 <*>
1950 V15 N10 P1-3 64-10542 <*>
1951 V16 P138-143 64-16925 <*>
1951 V16 P296-299 T-1398 <INSD>
1952 V17 P326-329 63-18435 <*> 0
65-11157 <*> 0
1952 V17 N7 P243-246 66-10685 <*>
1953 V18 P151 T-1778 <INSD>
1953 V18 N1 P17-19 62-20238 <*>
1954 V19 N4 P141-154 62-16236 <*> 0
64-10395 <*>
1954 V19 N5 P207-209 61P64F <ATS>
1955 V20 P244-254 C-2291 <NRC>
57-1277 <*>
1956 V21 P271-279 57-2286 <*>
64-14321 <*> 0
1957 P291-299 57-3407 <*>
1957 V22 P532 SC-T-527 <*>
1957 V22 N5 P533-540 60-18147 <*>
1957 V22 N12 P675-681 59-17648 <*>
1958 V23 P241-247 TS-1558 <BISI>
1958 V23 P327-332 TS 1559 <BISI>

1958 V23 N5 P253-260 61-14470 <*>
63-18694 <*>
1958 V23 N12 P643-649 61-18801 <*>
1959 V24 N1 P7-13 60-16844 <*>
1961 V26 P284-289 65-14893 <*>
1961 V26 N1 P9-16 64-10582 <*>
1961 V27 N10 P451-468 BGIRA-588 <BGIR>
1962 V27 N1C P469-481 66-12584 <*>
1963 V28 P89-99 66-10684 <*>
1963 V28 P123-134 8814 <IICH>
1963 V28 N1 P9-17 65-14416 <*>
1963 V28 N1 P19-27 BGIRA-572 <BGIR>
1963 V28 N2 P59-70 BGIRA-569 <BGIR>
1963 V28 N5 P223-229 577 <BGIR>
1963 V28 N7/8 P323-331 BGIRA-600 <BGIR>
1963 V28 N7/8 P345-354 BGIRA-599 <BGIR>
1964 V29 N1 P15-23 BGIRA-623 <BGIR>
1965 V30 N9 P515-516 66-13903 <*>
1966 V31 N4 P165-167 66-13896 <*>

SILIKATTECHNIK
1951 V2 N3 P80- 63-16056 <*>
1952 V3 N6 P265-267 57-2393 <*>
1953 V4 P483-485 1419 <*>
1953 V4 N4 P147-150 62-10947 <*>
1953 V4 N5 P195-199 62-10930 <*>
1953 V4 N10 P435-442 63-20373 <*>
1953 V4 N12 P531-535 57-752 <*>
64-16532 <*>
1954 V5 P243-247 4570 <BISI>
1954 V5 P446-447 I-160 <*>
1954 V5 N4 P147-150 59-15326 <*>
1955 V6 P99-104 64-10481 <*> 0
1955 V6 P473-475 T-2009 <INSD>
1955 V6 N4 P159-160 I-59 <*>
1955 V6 N6 P260-261 II-343 <*>
1955 V6 N8 P338 64-18337 <*> 0
1955 V6 N9 P372-377 62-16133 <*>
1955 V6 N10 P415-422 63-14063 <*> 0
1955 V6 N11 P476-477 63-14710 <*>
1956 V7 N5 P179-182 BGIRA-548 <BGIR>
1956 V7 N10 P38C-390 62-16413 <*>
1956 V7 N10 P399 62-16413 <*>
1956 V7 N10 P436-448 65-11701 <*> 0
1956 V7 N11 P451-465 63-20396 <*> 0
1956 V7 N12 P505-510 57-3338 <*>
1956 V7 N12 P537-547 CSIRO-3652 <CSIR>
1957 V8 N1 P24-27 59-10733 <*>
1957 V8 N6 P231-232 62-18119 <*>
1957 V8 N8 P343-344 62-16624 <*>
1957 V8 N11 P467-469 59-15328 <*>
1957 V8 N11 P470-473 62-18624 <*>
1957 V8 N11 P474-476 R-3388 <*>
1958 V9 P74-77 AEC-TR-5711 <*>
1958 V9 P443-448 9046 <IICH>
1958 V9 N2 P51-62 62-14486 <*>
1958 V9 N4 P162-164 60-15982 <*+>
1958 V9 N12 P552-555 62-18183 <*>
1959 V10 P190-192 TS 1483 <BISI>
1959 V10 N3 P105-119 62-18123 <*>
1959 V10 N3 P119-122 62-14487 <*>
1959 V10 N4 P190-192 62-14361 <*>
1959 V10 N8 P39C-400 65-13048 <*>
1959 V10 N8 P401-404 64-10567 <*>
1959 V10 N8 P4C5-408 62-14873 <*>
1959 V10 N11 P530-536 60-16868 <*>
1959 V10 N11 P537-538 63-14067 <*>
1959 V10 N12 P605-609 60-14556 <*>
1960 V11 N2 P52-55 CLAIRA T.10 <CLAI>
1960 V11 N4 P149-151 01N57G <ATS>
1960 V11 N4 P181-183 64-14263 <*>
1960 V11 N6 P273-275 CLAIRA T.11 <CLAI>
1960 V11 N9 P401-404 62-10680 <*>
1960 V11 N9 P404-407 64-14075 <*>
1960 V11 N10 P455-459 64-14083 <*>
1960 V11 N11 P506-511 64-14065 <*>
1960 V11 N11 P539-540 64-14060 <*>
1960 V11 N12 P575-576 64-14262 <*>
1961 V12 N1 P9-14 64-10546 <*>
1961 V12 N1 P17-21 2845 <BISI>
1961 V12 N2 P67-70 62-10670 <*>
1961 V12 N5 P218-219 63-14074 <*>
1961 V12 N5 P219-220 63-20838 <*>
1961 V12 N6 P278-280 62-18011 <*>
1961 V12 N6 P292-294 62-18010 <*>
1961 V12 N8 P313-316 578 <BGIR>
1961 V12 N8 P368-376 BGIRA-587 <BGIR>
1961 V12 N9 P408-421 65-11148 <*>
1962 V13 N11 P382-389 BGIRA-567 <BGIR>
1963 V14 N5 P145-150 BGIRA-643 <BGIR>
1964 V15 N3 P78-82 BGIRA-618 <BGIR>
1964 V15 N4 P120-123 BGIRA-621 <BGIR>
1964 V15 N5 P144-146 BGIRA-622 <BGIR>
1964 V15 N12 P383-387 65-12468 <*>
1965 V16 P431-433 SPEC. 66-14740 <*>
1965 V16 N4 P109-110 1672 <TC>

SILIKATY
1957 V1 P322-329 65-13825 <*>
1958 V2 P176-181 TS-1289 <BISI>
1958 V2 N3 P269-272 62-18128 <*>
1959 V3 N2 P168-176 60-14119 <*>
1959 V3 N3 P139-153 AD-649 477 <=$>
1959 V3 N3 P192-201 AD-649 477 <=$>
1960 V4 N4 P307-319 CLAIRA T.14 <CLAI>
1961 V5 P189-2C2 62-18714 <*>
1962 V6 N3 P258-272 65-13511 <*>
1962 V6 N3 P281-290 63-14737 <*> 0
1963 V7 P1-18 45T93C <ATS>
1963 V7 N4 P278-283 65-11025 <*>
1964 V8 N1 P27-44 65-11C26 <*>
1965 V9 N3 P177-205 66-13814 <*>

SILVA FENNICA
1955 N85 P32 2357 <*>

SILVAE GENETICA
1962 V11 N2 P29-39 63-01236 <*>

SINOPSIS MEDICA INTERNACIONAL
1955 V3 P13-17 62-16057 <*>
1955 V3 P19-20 62-16057 <*>

SINTETICHESKII KAUCHUK
1932 V1 N2 P5-12 61-18095 <=*$>
1932 V1 N3 P12-14 61-18027 <=*$>
1933 V2 N3 P5-12 61-18052 <=*$>
1933 V2 N3 P13-18 61-16808 <=*$>
1933 V2 N4 P11-14 61-18030 <=*$>
1933 V2 N5 P5-9 61-16696 <=*$>
1933 V2 N6 P12-16 61-16625 <=*$>
1934 V3 N1 P13-22 61-18093 <=*$>
1934 V3 N1 P31-34 61-16629 <=*$>
1934 V3 N1 P36-40 61-16683 <=*$>
1934 V3 N2 P15-29 61-16680 <=*$>
1934 V3 N3 P7-12 61-16674 <=*$>
1934 V3 N3 P12-19 61-16677 <=*$>
1934 V3 N3 P37-41 61-16699 <=*$>
1934 V3 N4 P3-7 61-16951 <=*$>
1934 V3 N4 P7-11 61-16814 <=*$>
1934 V3 N4 P20-35 61-16950 <=*$>
1934 V3 N5 P29-36 61-16837 <=*$>
1934 V3 N5 P36-38 61-16682 <=*$>
64-20805 <*>
1934 V3 N6 P13-18 61-16997 <=*$>
1934 V3 N6 P19-29 61-16998 <=*$>
1935 V4 N1 P8-27 61-18200 <=*>
1935 V4 N1 P37-42 61-18048 <=*$>
1935 V4 N2 P6-16 61-16994 <=*$>
1935 V4 N2 P23 61-16678 <=*$>
1935 V4 N3 P11-12 61-16995 <=*$>
1935 V4 N4 P13-23 61-18047 <=*$>
1935 V4 N4 P31-35 61-16827 <=*$>
1935 V4 N5 P6-10 61-18042 <=*$>
1935 V4 N5 P11-18 61-16600 <=*$>
1935 V4 N6 P22-29 61-16828 <=*$>
1936 N4 P8 R-3386 <*>
1936 V5 N2 P17-19 61-16676 <=*$>
1936 V5 N2 P27-29 61-16684 <=*$>
1936 V5 N2 P29-31 61-16685 <=*$>

1936 V5 N3 P3-7 61-18029 <=*$>
1936 V5 N4 P12-14 61-16691 <=*$>
1936 V5 N4 P18-22 61-16949 <=*$>
1936 V5 N4 P27-28 61-16622 <=*$>
1936 V5 N4 P29-31 61-18044 <=*$>
1936 V5 N4 P31-33 61-18043 <=*$>
1936 V5 N5 P20-24 61-16627 <=*$>
1936 V5 N6 P2-7 61-18028 <=*$>
1936 V5 N6 P17-21 61-16986 <=*$>
1936 V5 N9 P6-12 61-16672 <=*$>
1936 V5 N7/8 P3-19 61-16620 <=*$>
1936 V5 N7/8 P30-32 61-16623 <=*$>
1936 V5 N11/2 P6-9 61-16987 <=*$>

*SINTEZY ORGANICHESKIKH SOEDINENII: SBORNIK
1950 V1 P23-28 64-20160 <*>
1950 V1 P144-145 64-20159 <*>
1952 V2 P39-43 30N57R <ATS>

SISTEMA NERVOSA
1950 V1 P1-14 2197 <*>
1955 V4 P301-310 II-319 <*>
1959 V11 P418-439 63-00738 <*>
1961 V13 P167-175 63-01619 <*>
1962 V14 P389-408 63-20899 <*> 0

SITZUNGSBERICHTE DER AKADEMIE DER WISSENSCHAFTEN IN WIEN. MATHEMATISCH-NATURWISSENSCHAFTLICHE KLASSE
1870 V61 N2 P257-262 61-14612 <*>
1872 V66 P117 66-12078 <*>
1873 V68 P124-140 PT3 61-14722 <*>
1874 V69 P121-135 PT2 61-14721 <*>
1876 S1 N1/5 P272-284 28K24G <ATS>
1881 S2 V83 P943-954 66-62015 <=*$>
1913 V122 P845-877 65-12563 <*>
1913 S1 V122 P845-875 65-11054 <*> 0
1927 V136 P271-306 I-335 <*>
1932 S2A V141 P533-537 II-386 <*>
1935 V144 P393-396 I-256 <*>
1940 S2A V149 P31-58 AEC-TR-3548 <+*>

SITZUNGSBERICHTE DER BAYERISCHEN AKADEMIE DER WISSENSCHAFTEN ZU MUENCHEN. MATHEMATISCH-PHYSIKALISCHE KLASSE
1948 P255-261 62-00844 <*>
1955 P303-307 57-748 <*>

STISUNGSBERICHTE. CESKA SPOLECNOST NAUK. PRAGUE
1872 N6 P65-74 61-14621 <*>

SITZUNGSBERICHTE DER DEUTSCHEN AKADEMIE DER LANDWIRTSCHAFTWISSENSCHAFTEN ZU BERLIN
1959 V8 N4 ENTIRE ISSUE 61-10559 <*>

SITZUNGSBERICHTE DER DEUTSCHEN AKADEMIE DER WISSENSCHAFTEN ZU BERLIN
1877 P144-163 57-2751 <*>
1919 N34 P579-584 UCRL TRANS-871(L) <*>
1920 V24 P426-447 64-30982 <*> 0
1920 V24 P426-442 8529-D <K-H>
1933 N9 P380-401 UCRL TRANS-981 <*>
1935 P449-472 II-708 <*>
1937 N9/12 P118-124 65-11655 <*>

SITZUNGSBERICHTE DER GESELLSCHAFT NATURFORSCHENDER FREUNDE ZU BERLIN
1891 P131- 65-17279 <*>

SITZUNGSBERICHTE DER PHYSIKALISCH-MEDIZINISCHEN GESELLSCHAFT ZU WUERZBURG
1870 V16 P573-574 64-18791 <*>

SKANDINAVISCHES ARCHIV FUER PHYSIOLOGIE
1918 V35 P220-237 57-718 <*>
1936 V64 P299-316 1268 <*>
1936 V74 P60-61 61-14110 <*>
1936 V74/5 P1-46 62-00258 <*>
1938 V78 P218-219 61-14113 <*> P
1938 V79 N14 P15-26 59-14241 <+*> 0
1939 V82 N3/4 P185-192 59-18225 <+*> 0
62-20465 <=*> 0
1939 V82 N3/4 P193-200 59-18228 <+*> 0
1939 V82 N3/4 P201-211 59-18227 <+*> C
1939 V82 N3/4 P212-220 59-18226 <+*> 0
1939 V83 P56-57 61-14111 <*> P

SKANDINAVISK VETERINAERTIDSKRIFT
1942 V32 P257-289 60-18553 <*>
1944 V34 P129-136 61-14501 <*>
1948 V38 N9 P506-519 3152 <*>

SKIPSTEKNIKK
1965 V3 N5 P5- 2257 <BSRA>
1965 V3 N6 P25- 2250 <BSRA>
1965 V3 N6 P35- 2249 <BSRA>

SKLAR A KERAMIK
1954 V4 P77-80 64-10554 <*>
1956 V6 P12-15 6TM <CTT>
1961 V11 P188-190 AEC-TR-5516 <*>
1961 V11 N6 P159-160 SCL-T-489 <*>
1961 V11 N8 P211-214 65-14419 <*>
1961 V11 N9 P247-250 65-14418 <*>
1962 V12 P193-195 NP-TR-1149 <*>

SKOGEN
1952 V39 N6 P141- 2365 <*>
1953 V40 N6 P113- 2365 <*>
1961 N18 P322-323 C-3958 <NRCC>
1961 N19 P344-345 C-3958 <NRCC>
1965 V52 N22 P443 66-11624 <*>

SKRZYDLATA POLSKA
1959 N18 P8-9 60-00493 <*>
1959 N18 P13 60-00493 <*>
1960 N31 P5 NP-TR-681 <*>
1961 V17 N27 P13 62-19552 <=*>
1961 V17 N28 P13 62-19552 <=*>
1962 V18 N25 P9 62-33430 <=*>
1962 V18 N30 P9 62-33430 <=*>
1964 V20 N30 P9 64-71565 <=>
1966 N33 P10-11 08/14 66-34358 <=$>

SLABOPROUDY OBZOR
1954 V15 N12 P579-586 59-20923 <*> 0
1955 V16 N6 P323-327 3974 <HB>
1957 V18 N2 P91-106 34TM <CTT>
58-1295 <*>
1958 V19 N3 P140-143 62-25862 <=*>
1958 V19 N7 P464-466 62-11271 <=>
1960 V21 N7 P398-401 64-16922 <*>
1960 V21 N10 P596-602 62-25249 <=*> 0
1961 N12 P671-674 64-00456 <*>
1961 V22 N1 P2-5 62-25250 <=*> 0
1961 V22 N2 P66-70 62-32339 <=*$>
1961 V22 N3 P153-159 27Q67C <ATS>
1961 V22 N6 P338-342 62-11790 <=>
1961 V22 N8 P450-452 63-01433 <*>
1962 V23 N3 P155-163 71P66C <ATS>
1963 V24 N6 P356-363 AD-636 650 <=$>
1964 V25 N1 P21-24 AD-641 110 <=$>
1964 V25 N6 P313-318 AD-610 130 <=$>

SLEVARENSTVI
1954 V2 N2 P34-43 2466 <BISI>
1954 V2 N3 P72-74 2466 <BISI>
1954 V2 N9 P265-267 64-20537 <*>
1954 V2 N10 P292-296 3889 <HB>
1958 V6 N12 P384-390 TS-1532 <BISI>
1959 V7 P191-194 TS 1585 <BISI>
1959 V7 N11 P433-438 TRANS-129 <MT>
1960 V8 N9 P331-332 5358 <HB>
1961 V9 N2 P41-46 2601 <BISI>
1961 V9 N4 P126-131 2665 <BISI>
1961 V9 N4 P143-145 TRANS-363 <MT>
176-650 <STT>
1963 V11 N4 P144-149 TRANS-326 <MT>
1963 V11 N10 P413-418 TRANS-396 <MT>
1963 V11 N11 P471-477 TRANS-373 <MT>

1964 V12 N4 P152-153 6683 <HB>
1964 V12 N5 P173-178 TRANS-406 <MT>
1964 V12 N6 P213-216 TRANS-417 <MT>
1964 V12 N6 P226-227 TRANS-423 <MT>
1964 V12 N8 P293-299 4297 <BISI>
1964 V12 N8 P316-318 AD-638 900 <=$>

SLUZBA DROWIA
1960 V12 N43 P1 62-19343 <=*>
1960 V12 N43 P4 62-19343 <=*>
1960 V12 N46 P4 62-19349 <=*>
1961 V13 N27 P1-4 62-19624 <=*>
1961 V13 N27 P6 62-19624 <=*>

SLUZBENI LIST. BELGRADE
1960 N8 P205-206 62-19350 <=*>
1960 N22 P453-456 62-19350 <=*>
1960 V16 N23 P469-471 62-19352 <=*>
1960 V16 N2[illegible] P485-487 62-19353 <=*>
1960 V16 N26 P534-536 62-19351 <=*>
1962 V18 N31 P646-651 63-13077 <=*>
1962 V18 N53 P937-941 63-21720 <=>
1963 V19 N11 P154-155 63-21722 <=>
1963 V19 N12 P163 63-21705 <=>
1963 V19 N23 P483 63-31411 <=>
1964 V20 N18 P349-354 64-31674 <=>
1965 V21 N5 P97-99 65-30656 <=$>
1965 V21 N15 P636-639 65-31030 <=$>
1965 V21 N21 P926-928 65-31266 <=$>
1965 V21 N25 P1076-1086 65-31854 <=$>
1965 V21 N30 P1186-1191 65-32059 <=$>
1965 V21 N31 P1198-11205 65-32487 <=*$>
1965 V21 N32 P1213-1227 65-32360 <=$>
1965 V21 N35 P140[illegible]-1404 65-32344 <=$>
1965 V21 N35 P1417 65-32344 <=$>
1965 V21 N36 P1421-1426 65-32689 <=*$>
1966 N49 P955-958 67-30547 <=>
1967 N3 P11-19 67-31031 <=$>

SMENA
1960 V37 N21 P3 61-23932 <=>
1961 V38 N7 P3-4 63-15524 <=*>
1961 V38 N13 P22-23 63-15510 <=*>
1963 V40 N11 P10-12 63-31562 <=>
1966 N14 P23-25 66-33805 <=$>

SOCIALT ARBEID
1964 V38 N2 P34-43 65-17282 <*>

SOCIJALIZAM
1964 V7 N6 P749-770 64-41384 <=>
1965 N5 P626-649 65-32219 <=$>
1965 N10 P1288-1310 66-30163 <=*$>

SOCIJALNA POLITIKA
1963 V13 N11 P1059-1066 64-21617 <=>
1965 V15 N2 P149-166 65-31486 <=$>

SOCIOLOGIE DU TRAVAIL
1961 V3 N1 P18-29 65-11658 <*>

SOGO IGAKU
1954 V11 N2 P143-146 62-01522 <*>

SOGO RINSHO
1957 V6 N10 P208-215 61-00175 <*>
1962 V11 P1360-1366 64-18594 <*>
1963 V12 P1590-1594 66-13215 <*> 0

SOGO SHIKENJO NENPO
1956 P11-17 61-18719 <*>
1964 V23 N1 P40-47 66-13914 <*$>

SOKKO JIHO
1957 N24 P365-370 59-16685 <=>

SOLDAT UND TECHNIK
1960 V3 N9 P464-470 62-33126 <=*>

SOLID-STATE ELECTRONICS. LONDON
1960 V1 P39-45 73N49G <ATS>
1960 V1 P172-175 AEC-TR-5827 <*>
1960 V1 N2 P123-130 AEC-TR-6322 <*>
1962 V5 P71-84 65-00189 <*>
1962 V5 P249-259 63-18756 <*>
1963 V6 P605-610 66-12344 <*>
1966 V9 N1 P83-85 66-13661 <*>

SOLNECHNYE DANNYE
1959 N1 P36 60-23677 <*+>
1959 N4 P79-80 61-15649 <+*>
1959 N4 P82-83 61-15648 <+*>
1959 N5 P83-85 61-15230 <*+>
1959 N9 P83-85 61-19533 <=*>
1959 N9 P86-89 61-19372 <+*>
1959 N10 P72-74 61-19371 <+*>
1959 N10 P75-77 61-19374 <+*>
1959 N10 P78-80 62-24629 <=*>
1960 N4 P72-75 62-33400 <*>
62-33400 <=*>
1960 N6 P62-64 61-27454 <*=> 0
1960 N6 P65-66 61-27455 <*=>
1960 N7 P65-68 63-13908 <=*>
1960 N7 P68-73 61-27453 <*=> C
1960 N7 P74-77 61-27456 <*=> 0
1961 N1 P61-65 62-32357 <=*>
1961 N1 P66-69 62-24628 <=*>
1961 N3 P68-69 63-13600 <=*>
1961 N3 P73 63-19114 <=*>
1961 N4 P62-64 63-13601 <=*>
1961 N4 P65-67 63-13240 <=*>
1961 N10 P58-62 63-19143 <=*>
1961 N10 P63 63-19102 <=*>
1961 N12 P55-57 AD-619 263 <=$>
1962 N2 P53-58 AD-619 264 <=$>
64-19835 <=$>
1962 N3 P72-76 63-23404 <=*> C
1962 N12 P50-57 64-19834 <=$>
1963 N1 P55-67 64-19884 <=$>
1963 N2 P43-50 64-19884 <=$>
1963 N3 P46-55 N65-32726 <=$>
1963 N3 P52-58 64-19858 <=$>
1963 N3 P64-70 N65-32966 <=$>
1963 N4 P67-77 N65-32967 <=$>
65-60899 <=$>
1963 N4 P77-81 N65-64255 <=$*>
1963 N6 P53-59 AD-610 301 <=$>
1963 N11 P58-61 AD-613 978 <=$>
1963 N12 P60-70 65-60666 <=$>
1963 V11 P71-76 65-64104 <=$*>
1964 N1 P71-78 65-64384 <=*$>
1964 N8 P45-49 65-63766 <=$>
1964 N11 P60-64 66-61027 <=*$>
1964 N12 P72-73 66-61025 <=*$>
1965 N3 P54-58 66-61850 <=*$>
1965 N3 P58-67 66-61332 <=*$>
1965 N5 P58-62 66-61546 <=*$>
1965 N5 P69-74 66-61571 <=$>

SOLS ET FONDATIONS
1952 N58 P966-971 3388 <*>

SONDERHEFT. KOELNER ZEITSCHRIFT FUER SOZIOLOGIE UND SOZIALPSYCHOLOGIE
1962 N6 P36-49 65-12113 <*>

SOOBSHCHENIYA AKADEMII NAUK GRUZINSKOI SSR, TIFLIS
1941 V2 N6 P499-504 RT-2882 <*>
1942 V3 N1 P73-79 61-00463 <*>
64-15628 <=*$>
1943 V4 N1 P3-9 RJ-603 <ATS>
1944 V5 N10 P975-981 <ATS>
1951 V12 N7 P397-401 62-11421 <=>
1952 V13 N4 P241-247 RT-2072 <*> 0
1952 V13 N5 P273-280 RT-4017 <*>
1953 P3-38 RT-2823 <*>
1953 P88-89 RT-2823 <*>
1954 V15 N1 P3-6 63-13730 <=*>
1954 V15 N1 P13-20 62-20324 <=*>
1954 V15 N9 P575-581 62-15897 <=*>

1955 V16 P509-516 R-4512 <*>
1955 V16 N4 P277-279 R-401 <*>
64-19502 <=$>
1956 V17 N6 P489-494 62-11427 <=>
1957 V18 N4 P393-400 61-27992 <*=>
1957 V19 N1 P33-36 64-11092 <=>
1958 V20 N1 P21-25 R-5115 <*>
1958 V21 N2 P183-186 60-15534 <+*>
1959 V23 N1 P27-34 75L35R <ATS>
1960 V24 N5 P619-623 61-23006 <*=>
1960 V25 N4 P417-424 63-19111 <=*>
1961 V26 P385-387 AMS TRANS V.46 P162 <AMS>
AMST S2 V46 P162-164 <AMS>
1961 V26 N1 P87-94 62-13348 <*=>
1961 V26 N2 P161-165 62-25496 <=*>
1961 V26 N6 P681-686 AD-633 218 <=$>
1961 V27 N3 P299-305 65-60489 <=$>
1961 V27 N5 P529-536 62-24258 <=*>
62-24980 <=*>
1962 V28 N1 P33-39 62-32899 <=*>
1962 V28 N3 P375-380 62-32367 <=*>
1962 V28 N4 P409-416 82P64R <ATS>
1962 V28 N5 P553-560 AD-608 541 <=>
1962 V28 N5 P593-606 62-33335 <=*>
1962 V29 N1 P101-106 62-33716 <=*>
1962 V29 N2 P229-234 63-13278 <=*>
1962 V29 N3 P365-371 63-13369 <=*>
1962 V29 N6 P765-772 63-21562 <=>
1963 V30 N2 P213-220 63-31095 <=>
1963 V30 N5 P601-606 <JBS>
1963 V31 N2 P277-281 CSIR-491 <CSSA>
1963 V31 N3 P577-581 64-21417 <=>
1964 V35 N1 P37-44 65-10401 <*>
1964 V36 N1 P18-25 RTS-3266 <NLL>
1964 V36 N2 P279-286 AD-625 303 <=*$>

SOOBSHCHENIYA ASTRONOMICHESKOGO INSTITUTA. MOSCOW UNIVERSITETA
1949 V16 P71-92 63-14914 <=*> 0
1957 V100 P36-50 63-23250 <=*>

SOOBSHCHENIYA BYURAKANSKOI OBSERVATORII
1958 V25 P15-32 63-14910 <=*> 0
1958 V25 P35-42 59-20919 <+*>
1958 V25 P67-73 63-14908 <=*> 0
1959 V27 P35-41 61-19679 <*=>

SOOBSHCHENIYA GOSUDARSTVENNOGO ASTRONOMICHESKOGO INSTITUTA IMENI P. K. SHTERNBERGA
1951 V80 P3-17 AD-607 958 <=>
1957 V100 P3-35 63-23251 <=*>
1961 V114 P13-32 62-32937 <=*>
1963 N126 P66-71 AD-629 412 <=$>
1964 N131 P37-41 65-63529 <=$*>
1964 N133 P3-9 65-63771 <=$>

SOOBSHCHENIYA LABORATORII LESOVEDENIYA
1959 V1 P41-56 62-20266 <=*> 0
1959 V1 P70-81 65-50050 <=>
1961 N5 P53-61 66-51083 <=>
1962 N1 P3-96 63-31091 <=>

SOOBSHCHENIYA O NAUCHNYKH RABOTAKH CHLENOV VSESOYUZNOGO KHIMICHESKOGO OBSHCHESTVA IM. D. I. MENDELEEVA
1955 N3 P2-4 61-19130 <+*>
1955 N3 P43-45 61-13162 <+*>
1955 N3 P46-49 60-17197 <+*>

SOOBSHCHENIYA OTDELE MEKHANIZATSI I AVTOMATIZATSII INFORMATSIONNYKH RABOT. INSTITUT NAUCHNOI INFORMATSII. AKADEMIA NAUK SSSR
1960 V1 P1-250 62-13845 <=*>
1961 V2 P1- 62-23835 <=*>

SOPROTIVLENIE MATERIALOV I TEORIYA SORRUZHENIE
1965 N1 P7-17 65-64351 <=*$>
1965 N1 P17-23 65-64359 <=*$>

SOTILASLAAKETIETEELLINEN AIKAKAUSLEHTI
1959 V34 P59-69 63-01747 <*>
63-26698 <=$>
1960 V35 P220-227 63-16033 <*>
1961 V36 P5-15 11340 <KH>
63-16106 <*>
1961 V36 P62-68 11517-B <KH>
63-16903 <*>
1961 V36 P69-78 11433-B <KH>
63-16906 <*>

SOTSIALISTICHESKAYA ZAKONNOST
1965 V42 N6 P70-72 65-31859 <=$>

SOTSIALISTICHESKOE SELSKOE KHOZYAISTVO
1941 V12 N4 P70-75 RT-2652 <*>
1946 V17 N3 P70-75 RT-1781 <*>
1947 V18 N12 P57-61 RT-762 <*>
1953 V24 N11 P27-35 64-10353 <=*$>
1954 V25 N6 P37-47 RT-2581 <*>
1954 V25 N6 P37- 61-13718 <+*>
1956 V27 N6 P51-60 65-12665 <*>

SOTSIALISTICHESKOE SELSKOE KHOZYAISTVO AZERBAIDZHANA
1960 V9 N7 P22-23 61-21339 <=>
1961 V10 N5 P49-51 63-13030 <=*>

SOTSIALISTICHESKII TRUD
1956 N2 P26-37 RT-4650 <*>
1958 N8 P65-72 59-31004 <=>
1959 N1 P42-55 59-13374 <=>
1959 N5 P104-108 60-31042 <=>
1959 N7 P115-120 60-31042 <=>
1960 N3 P108-109 61-11975 <=>
1960 N6 P11-20 61-21737 <=>
1960 N6 P58-64 61-21737 <=>
1960 N11 P115-120 61-23011 <=>
1962 V7 N1 P50-52 62-33124 <=*>
1962 V7 N6 P70-75 62-32878 <=*>
1962 V7 N10 P129-131 63-21269 <=>
1962 V7 N12 P23 63-21788 <=>
1962 V7 N12 P34-38 63-21792 <=>
1962 V7 N12 P73-77 63-21616 <=>
1963 V8 N5 P65-70 63-31676 <=>
1963 V8 N7 P71-78 63-41015 <=>
1963 V8 N7 P79-84 63-31673 <=>
1964 V9 N4 P18-24 64-31434 <=>
1964 V9 N5 P65-70 64-41154 <=>
1964 V9 N6 P42-47 64-41206 <=>
1965 V10 N9 P60-66 66-30081 <=*$>
1965 V10 N12 P23-29 66-30444 <=*$>
1965 V10 N12 P30-36 66-30430 <=*$>

SOTSIALISTICHESKOE ZEMLEDELIE
1939 P3 RT-2338 <*>
1939 P4 RT-2336 <*>

SOUDAGE ET TECHNIQUES CONNEXES
1957 V11 P277-289 61-16280 <*>
1957 V11 N9/0 P323-332 61-16279 <*>
1957 V11 N9/10 P333-337 61-16281 <*>

1959 V13 P433- 2107 <BSRA>
1959 V13 N1/2 P33-42 AEC-TR-4813 <*>
95M40F <ATS>
1959 V13 N5/6 P165-179 60-14197 <$*> 0
1959 V13 N9/10 P349-357 62-14875 <*>
1963 V17 P253 2389 <BSRA>
1963 V17 N11/2 P426-438 3745 <BISI>
1963 V17 N9/10 P357-366 65-11337 <*>
1965 V19 P297-310 4586 <BISI>
1965 V19 N3/4 P134-135 65-17195 <*>

SOUDURE ET TECHNIQUES CONNEXES
1954 V8 N7/8 P187-193 66-10067 <*> 0

SOUTH CHINA JOURNAL OF AGRICULTURAL SCIENCE
SEE HUA NAN NUNG YEH KO HSUEH

SOUTH SEA FISHERIES. JAPAN

SEE NANYO SUISAN

SOUTH SEA FISHERY NEWS. JAPAN
SEE NANYC SUISAN JOHO

SOUTH SEA SCIENCE
SEE KAGAKU NANYO

SOUTHEAST ASIAN STUDIES
SEE TONAN AJIA KENKYU. KYOTO

SOVETSKAYA AGRONOMIYA
1939 N12 P46-53 CSIR-TR 152 <CSSA>
1940 V2 N10 P64-65 60-21870 <=>
1946 V4 N7 P25-32 66-10245 <*> O
1947 V5 N2 P42-48 65-61288 <=$> O
1947 V5 N4 P72-75 RT-3147 <*>
1947 V5 N6 P92-93 RT-3146 <*>
1947 V5 N11 P10-17 RT-769 <*>
1947 V5 N11 P70-76 RT-768 <*>
1949 V7 N9 P75-79 RT-1789 <*>
1949 V7 N10 P60-68 R-3025 <*>
1950 N6 P51-56 RT-4155 <*>
1950 N9 P72-78 <CBPB>
1952 N7 P62-66 R-1535 <*>
1952 V10 N5 P43-50 R-1530 <*>
1952 V10 N6 P60-67 60-51056 <=>
1952 V10 N11 P41-45 RT-1960 <*>
1952 V10 N12 P19-26 RT-4059 <*>
1953 V11 N3 P48-54 60-51047 <=>
1953 V11 N4 P48-54 RT-3619 <*>
1953 V11 N5 P75-80 60-51046 <=>

SOVETSKAYA ANTARKTICHESKAYA EKSPEDITSIYA
1959 V2 P157-162 64-71319 <=>

SOVETSKAYA ANTROPOLOGIYA
1957 V1 N1 P7-30 60-21175 <=>

SOVETSKAYA ARKHEOLOGIYA
1956 V25 P64-86 42L29R <ATS>

SOVETSKAYA ARKTIKA
1938 N2 P48-50 RT-2271 <*>
1938 N7 P98-101 R-797 <*>

SOVETSKAYA AVIATSIYA
1955 11/22 P2- 60-13539 <+*>
1957 N152 P2 AEC-TR-3537 <+*>
1957 N173 P1 59-16432 <+*> O
1957 N176 P4 62-18255 <=*>
1957 N233 P2 61-23844 <*=>
1957 N261 P2 60-19016 <+*>
1957 N277 P3 4208 <HB>
1957 N282 P3 59-16339 <+*> PO
1958 04/05 P4 63-19507 <=*>
1958 N49 P2 59-16308 <+*> P
1958 N136 P2 59-16452 <+*>
1958 N141 P3 60-23526 <+*>
1958 N147 P2 60-13550 <+*>
1958 N183 P3 60-19034 <+*>
1958 N213 P4 59-22057 <+*>
1958 N228 P2 59-18417 <+*>
1958 N232 P2 59-22207 <+*> P
1958 N247 P2 59-12711 <+*> P
1958 N271 P3 59-22204 <+*>
1958 N274 P3- 61-10859 <+*>
1958 N278 P3 59-12692 <+*>
60-27021 <*+> O
1958 N280 P4 59-12687 <+*> P
1958 N289 P3 59-12689 <+*> P
1958 N290 P2 59-22201 <+*> P
1958 N291 P1 59-22203 <+*> P
1958 N293 P3 59-12686 <+*> P
1959 N37 P2 59-19479 <+*>
1959 N47 P2 60-15502 <+*>
1959 N65 P2 60-19024 <+*>
1959 N65 P3 60-19033 <+*>
1959 N67 P21 61-23175 <=*>
1959 N72 P2 59-19702 <+*> PO
60-19031 <+*>
1959 N86 P4 59-16339 <+*>
1959 N107 P1 60-19052 <+*> O
1959 N112 P1 60-13303 <+*> O
1959 N119 P2 59-19695 <+*> P
60-19051 <+*>
1959 N121 P4 60-15516 <+*>
1959 N126 P2 61-23181 <=*>
1959 N126 P4 61-23181 <=*>
1959 N130 P2 60-19055 <+*>
1959 N142 P1 61-23698 <*=>
1959 N142 P4 62-19259 <=*>
1959 N177 P1 61-23443 <=*>
1959 N180 P1 61-23700 <*=>
1959 N187 P3 61-23698 <*=>
1959 N189 P4 60-19080 <+*>
1959 N198 P4- 61-19454 <+*> O
1959 N215 P4 60-19081 <+*>
1959 N218 P2 61-23690 <=*>
1959 N218 P4 61-23690 <=*>
1959 N221 P3 61-23690 <=*>
1959 N227 P4 60-13202 <+*>
1959 N244 P3 60-13197 <+*> O
1959 N268 P4 60-13203 <+*>

SOVETSKAYA BIBLIOGRAFIYA
1958 N52 P65-73 61-21074 <=>
1959 N2 P116-118 NLLTB V1 N8 P1 <NLL>
1959 N3 P25-34 NLLTB V2 N2 P134 <NLL>
1959 N3 P102-104 NLLTB V2 N2 P129 <NLL>
1959 N5 P3-10 NLLTB V2 N4 P302 <NLL>
1959 N5 P11-27 NLLTB V2 N5 P387 <NLL>
1959 N6 P10-21 62-25581 <=*>
1960 N1 P50-53 NLLTB V2 N6 P507 <NLL>
1961 N6 P20-25 NLLTB V3 N6 P504 <NLL>
1965 N2 P6-12 NLLTB V7 N10 P849 <NLL>
1965 V93 N5 P28-38 NLLTB V8 N8 P670 <NLL>

SOVETSKAYA BOTANIKA
1940 V8 N5/6 P144-153 63-23126 <=*$>
1944 N4/5 P81-84 65-20014 <$>

SOVETSKAYA ETNOGRAFIYA
1950 N4 P150-168 AOTN N2 P3 <AINA>
1950 N4 P169-173 AOTN N2 P24 <AINA>
1952 N1 P36-50 AOTN N2 P207 <AINA>
1952 N2 P69-72 AOTN N2 P29 <AINA>
1952 N2 P73-85 AOTN N2 P33 <AINA>
1952 N2 P86-91 AOTN N2 P47 <AINA>
1953 N1 P38-63 AOTN N2 P73 <AINA>
1953 N2 P37-52 AOTN N2 P102 <AINA>
1955 N2 P16-26 AOTN N2 P119 <AINA>
1956 N3 P35-51 AOTN N2 P144 <AINA>
1958 N4 P9-17 AOTN N2 P197 <AINA>
1959 N1 P29-37 AOTN N2 P62 <AINA>
1959 N2 P38-46 AOTN N2 P54 <AINA>
1960 N1 P72-82 61-31223 <=>
1961 N2 P122-123 61-28407 <*=>
1961 N5 P9-26 SGRT V3 N4 P3 <AGS>
1965 N5 P45-56 66-30381 <=$>
1966 N2 P50-58 66-32417 <=$>
1966 N4 P38-51 66-34663 <=$>

*SOVETSKAYA GEOLOGIYA (SBORNIK STATEI. TITLE VARIES)
1939 V9 N2 P98-105 R-1159 <*>
R-1215 <*>
1939 V9 N7 P71-77 R-1146 <*>
1939 V9 N12 P68-73 AD-610 891 <=$>
RT-2620 <*>
1940 N5/6 P182-184 RT-2707 <*>
1948 N28 P133-138 64-15164 <=*$> O
1955 N42 P156-163 RJ-762 <ATS>
61-13417 <=*> C
1957 N55 P93-113 60-19872 <+*>
1957 N58 P138-149 AD-623 429 <=*$>
1958 N1 P3-24 IGR V1 N4 P1 <AGI>
1958 N1 P86-113 IGR V2 N5 P397 <AGI>
1958 N1 P145-149 IGR V1 N5 P51 <AGI>
1958 N1 P150-155 IGR V1 N4 P75 <AGI>

1958 N4 P3-17	IGR V1 N3 P39 <AGI>
1958 N4 P18-	IGR V1 N5 P41 <AGI>
1958 N4 P33-42	IGR V1 N4 P18 <AGI>
1958 N4 P43-52	IGR V1 N6 P66 <AGI>
1958 N4 P53-72	IGR V1 N4 P31 <AGI>
	TT-806 <NRCC>
	59-15575 <+*>
1958 N4 P73-80	IGR V1 N6 P60 <AGI>
1958 N4 P81-	IGR V1 N5 P29 <AGI>
1958 N4 P124-143	IGR V1 N3 P17 <AGI>
1958 N4 P162-165	IGR V1 N9 P24 <AGI>
1958 N4 P165-	IGR V1 N3 P34 <AGI>
1958 N5 P141-145	PB 141 495T <=>
1958 N6 P3-23	61-23981 <*=>
1958 N7 P138-147	IGR V2 N10 P888 <AGI>
1958 N8 P114-136	IGR V2 N10 P851 <AGI>
1958 N11 P111-130	61-18583 <*=>
1958 N12 P3-12	IGR V2 N12 P1071 <AGI>
	NLLTB V1 N10 P1 <NLL>
	59-13921 <=>
1958 N12 P36-42	59-13903 <=>
1958 N12 P69-97	59-13922 <=> O
1958 V1 N6 P3-23	64-19742 <=$>
1958 V1 N11 P16-25	63-27413 <=$>
1959 N1 P163-165	59-13657 <=>
1959 N4 P50-66	61-28032 <*=>
1959 N4 P67-82	61-27785 <*=>
1959 N5 P11-24	IGR V1 N11 P28 <AGI>
1959 N5 P25-44	IGR V1 N11 P58 <AGI>
1959 N5 P46-65	IGR V1 N12 P1 <AGI>
1959 N5 P66-72	IGR V2 N2 P103 <AGI>
1959 N5 P73-80	IGR V1 N12 P56 <AGI>
1959 N5 P81-95	IGR V1 N11 P37 <AGI>
1959 N5 P96-108	IGR V2 N2 P93 <AGI>
1959 N5 P119-129	IGR V2 N1 P60 <AGI>
1959 N5 P131-141	IGR V2 N1 P52 <AGI>
1959 N6 P13-29	60-19853 <+*>
1959 N8 P3-13	60-11250 <=>
1959 N8 P160-165	60-11251 <=>
1959 N9 P3-19	60-11574 <=>
1959 N10 P152-154	60-31116 <=>
1959 N11 P3-15	IGR V2 N12 P1085 <AGI>
1959 V2 N1 P92-110	64-19738 <=$>
1959 V2 N8 P61-80	IGR V3 N2 P100 <AGI>
1959 V2 N10 P146-152	IGR V3 N1 P71 <AGI>
1960 N1 P10-33	IGR V3 N10 P839 <AGI>
1960 N1 P72-74	IGR V3 N1C P859 <AGI>
1960 N1 P126-129	IGR V3 N10 P861 <AGI>
1960 N1 P129-134	IGR V3 N10 P864 <AGI>
1960 N1 P140-146	IGR V3 N1C P871 <AGI>
1960 N2 P48-56	IGR V3 N10 P878 <AGI>
1960 N2 P57-73	IGR V3 N10 P885 <AGI>
1960 N2 P83-86	IGR V3 N6 P482 <AGI>
1960 N2 P115-124	IGR V3 N10 P900 <AGI>
1960 N2 P125-138	IGR V2 N12 P1017 <AGI>
1960 N3 P28-39	IGR V3 N10 P907 <AGI>
1960 N3 P48-60	IGR V3 N10 P917 <AGI>
1960 N3 P113-119	IGR V3 N10 P927 <AGI>
1960 N3 P120-121	IGR V3 N10 P931 <AGI>
1960 N3 P130-135	IGR V3 N1C P936 <AGI>
1960 N3 P136-140	IGR V3 N1C P939 <AGI>
1960 N4 P57-65	IGR V3 N11 P989 <AGI>
1960 N4 P120-123	IGR V3 N11 P995 <AGI>
1960 N5 P32-46	IGR V3 N11 P998 <AGI>
1960 N5 P74-87	IGR V3 N5 P373 <AGI>
	62-13107 <*=>
1960 N5 P107-115	IGR V3 N11 P1019 <AGI>
1960 N6 P3-25	IGR V3 N11 P1027 <AGI>
1960 N6 P78-92	IGR V3 N11 P1048 <AGI>
1960 N6 P111-118	IGR V3 N11 P1060 <AGI>
1960 N6 P119-128	IGR V3 N11 P1068 <AGI>
1960 N6 P137-139	IGR V3 N11 P1076 <AGI>
1960 N7 P3-27	IGR V3 N12 P1126 <AGI>
1960 N7 P72-81	IGR V3 N12 P1143 <AGI>
1960 N7 P82-94	IGR V3 N12 P1150 <AGI>
1960 N7 P103-115	IGR V3 N12 P1159 <AGI>
1960 N8 P66-74	IGR V3 N12 P1168 <AGI>
1960 N8 P75-86	IGR V3 N12 P1174 <AGI>
1960 N8 P102-114	IGR V3 N12 P1185 <AGI>
1960 N9 P57-72	IGR V4 N1 P6 <AGI>

1960 N9 P103-112	IGR V4 N1 P17 <AGI>
1960 N9 P113-121	IGR V4 N1 P24 <AGI>
1960 N9 P128-130	IGR V4 N1 P30 <AGI>
1960 N9 P134-137	IGR V4 N1 P33 <AGI>
1960 N9 P162-165	IGR V4 N1 P40 <AGI>
1960 N9 P197-203	IGR V4 N1 P37 <AGI>
1960 N10 P3-23	IGR V3 N7 P557 <AGI>
1960 N10 P24-41	IGR V4 N1 P43 <AGI>
1960 N10 P42-59	IGR V4 N1 P55 <AGI>
196C N10 P81-98	IGR V4 N1 P67 <AGI>
1960 N10 P99-111	IGR V4 N1 P79 <AGI>
1960 N10 P129-132	IGR V4 N1 P89 <AGI>
1960 N11 P94-1C8	IGR V4 N2 P199 <AGI>
1960 N11 P156	IGR V4 N2 P211 <AGI>
1960 N12 P3-44	IGR V4 N2 P166 <AGI>
1960 N12 P22-32	IGR V4 N2 P212 <AGI>
1960 N12 P33-43	IGR V4 N2 P221 <AGI>
1960 N12 P52-64	IGR V4 N2 P229 <AGI>
1961 N1 P49-64	IGR V4 N4 P415 <AGI>
1961 N1 P98-108	IGR V4 N4 P427 <AGI>
1961 N1 P121-127	IGR V4 N4 P435 <AGI>
1961 N1 P142-145	IGR V4 N4 P442 <AGI>
1961 N2 P39-54	IGR V4 N5 P561 <AGI>
1961 N2 P55-67	IGR V4 N5 P570 <AGI>
1961 N2 P92-107	IGR V4 N5 P578 <AGI>
1961 N2 P177-178	IGR V4 N5 P588 <AGI>
1961 N3 P56-68	IGR V4 N9 P1008 <AGI>
1961 N3 P79-87	IGR V4 N9 P1017 <AGI>
1961 N3 P88-96	IGR V4 N9 P1023 <AGI>
1961 N3 P113-114	IGR V4 N9 P1030 <AGI>
1961 N3 P115-119	IGR V4 N9 P1032 <AGI>
1961 N3 P120-126	IGR V4 N9 P1035 <AGI>
1961 N3 P132-143	IGR V4 N9 P1040 <AGI>
1961 N3 P144-148	IGR V4 N9 P1050 <AGI>
1961 N4 P7-24	IGR V4 N10 PT1 P1105 <AGI>
	9OP6OR <ATS>
1961 N4 P25-46	IGR V4 N10 PT1 P1118 <AGI>
1961 N4 P25-45	61-27335 <*=>
1961 N4 P47-59	IGR V4 N10 PT1 P1135 <AGI>
1961 N4 P60-85	IGR V5 N1 P5 <AGI>
1961 N4 P114-126	IGR V4 N10 PT1 P1144 <AGI>
1961 N4 P140-145	IGR V4 N10 PT1 P1154 <AGI>
1961 N5 P16-38	IGR V4 N11 P1214 <AGI>
1961 N5 P59-76	IGR V4 N11 P1235 <AGI>
1961 N5 P96-109	IGR V4 N11 P1252 <AGI>
1961 N5 P134-138	IGR V4 N11 P1263 <AGI>
1961 N6 P3-7	IGR V4 N12 P1304 <AGI>
1961 N6 P89-101	IGR V4 N12 P1326 <AGI>
1961 N6 P102-117	IGR V4 N12 P1337 <AGI>
1961 N6 P130-134	IGR V4 N12 P1350 <AGI>
1961 N6 P134-138	IGR V4 N12 P1354 <AGI>
1961 N6 P138-140	IGR V4 N12 P1363 <AGI>
1961 N7 P34-45	IGR V5 N1 P29 <AGI>
1961 N7 P68-94	IGR V5 N1 P38 <AGI>
1961 N7 P95-106	IGR V5 N1 P63 <AGI>
1961 N7 P134-141	IGR V5 N1 P71 <AGI>
1961 N8 P31-44	IGR V5 N2 P171 <AGI>
1961 N8 P45-57	IGR V5 N2 P180 <AGI>
1961 N8 P113-115	IGR V5 N2 P189 <AGI>
1961 N8 P119-122	IGR V5 N2 P192 <AGI>
1961 N8 P127-137	IGR V5 N2 P196 <AGI>
1961 N8 P150-154	IGR V5 N2 P206 <AGI>
1961 N9 P8-56	IGR V5 N3 P289 <AGI>
1961 N9 P57-70	IGR V5 N3 P321 <AGI>
1961 N9 P108-114	IGR V5 N3 P331 <AGI>
1961 N9 P127-135	IGR V5 N3 P339 <AGI>
1961 N9 P148-152	IGR V5 N3 P335 <AGI>
1961 N10 P42-57	IGR V5 N4 P425 <AGI>
1961 N10 P100-107	IGR V5 N4 P433 <AGI>
1961 N10 P118-136	IGR V5 N4 P437 <AGI>
1961 N11 P9-20	IGR V5 N5 P528 <AGI>
1961 N11 P21-36	IGR V5 N5 P539 <AGI>
1961 N11 P86-108	IGR V5 N5 P557 <AGI>
1961 N11 P109-120	IGR V5 N5 P577 <AGI>
1961 N11 P131-137	IGR V5 N5 P534 <AGI>
1961 N11 P137-145	IGR V5 N5 P585 <AGI>
1961 N11 P158-164	IGR V5 N5 P593 <AGI>
1961 N12 P3-13	IGR V5 N6 P698 <AGI>
1961 N12 P14-2C	IGR V5 N6 P706 <AGI>
1961 N12 P55-77	IGR V4 N7 P757 <AGI>

62-24024 <=*>
1961 N12 P107-113 IGR V5 N6 P716 <AGI>
1961 N12 P117-121 IGR V5 N6 P722 <AGI>
1961 N12 P128-131 IGR V5 N6 P727 <AGI>
1961 N12 P131-132 IGR V5 N6 P731 <AGI>
1961 V4 N6 P130-134 66-14354 <*$>
1962 N1 P35-53 IGR V5 N7 P815 <AGI>
1962 N1 P54-70 IGR V5 N7 P830 <AGI>
1962 N1 P93-103 IGR V5 N7 P842 <AGI>
1962 N1 P131-145 IGR V5 N7 P850 <AGI>
1962 N1 P160-164 IGR V5 N7 P859 <AGI>
1962 N2 P3-16 IGR V5 N9 P1077 <AGI>
1962 N2 P61-79 IGR V5 N9 P1087 <AGI>
1962 N2 P100-121 IGR V5 N9 P1099 <AGI>
1962 N3 P67-82 IGR V5 N9 P1123 <AGI>
1962 N3 P83-95 IGR V5 N9 P1114 <AGI>
1962 N3 P96-107 IGR V5 N9 P1137 <AGI>
1962 N3 P108-118 IGR V5 N9 P1147 <AGI>
63-21986 <=>
1962 N3 P128-135 IGR V5 N9 P1156 <AGI>
1962 N4 P3-14 IGR V5 N10 P1290 <AGI>
1962 N4 P9-10 63-21525 <=>
1962 N4 P10-14 63-15367 <=*>
1962 N4 P40-52 IGR V5 N10 P1280 <AGI>
1962 N4 P77-92 IGR V5 N10 P1297 <AGI>
1962 N5 P104-112 IGR V5 N11 P1450 <AGI>
1962 N5 P114-130 IGR V5 N11 P1457 <AGI>
1962 N6 P3-16 IGR V6 N1 P68 <AGI>
1962 N6 P17-32 IGR V6 N1 P75 <AGI>
1962 N6 P33-42 IGR V6 N1 P85 <AGI>
1962 N6 P43-51 IGR V6 N1 P58 <AGI>
1962 N6 P52-76 IGR V6 N1 P39 <AGI>
1962 N7 P3-7 63-15367 <=*>
1962 N7 P8-25 IGR V6 N2 P263 <AGI>
1962 N7 P26-40 IGR V6 N2 P277 <AGI>
1962 N7 P41-63 IGR V6 N2 P287 <AGI>
1962 N7 P94-113 IGR V6 N2 P303 <AGI>
1962 N8 P3-15 IGR V6 N4 P624 <AGI>
63-23159 <=*>
1962 N8 P16-24 IGR V6 N4 P633 <AGI>
1962 N8 P33-39 IGR V6 N4 P639 <AGI>
1962 N8 P79-91 SHSP N5 1962 P558 <AGU>
1962 N8 P104-121 IGR V6 N4 P644 <AGI>
1962 N8 P151-157 IGR V6 N4 P656 <AGI>
1962 N9 P8-23 IGR V6 N5 P810 <AGI>
1962 N9 P45-60 IGR V6 N5 P820 <AGI>
1962 N9 P113-121 IGR V6 N5 P830 <AGI>
1962 N10 P16-27 IGR V6 N6 P1085 <AGI>
1962 N10 P39-56 IGR V6 N6 P1099 <AGI>
1962 N11 P15-36 IGR V6 N8 P1370 <AGI>
1962 N11 P36-37 IGR V6 N8 P1341 <AGI>
1962 N11 P88-97 IGR V6 N8 P1336 <AGI>
1962 N12 P16-29 IGR V6 N9 P1532 <AGI>
1962 N12 P80-99 IGR V6 N9 P1557 <AGI>
1962 N12 P133-139 IGR V6 N9 P1585 <AGI>
1962 N12 P140-141 IGR V6 N9 P1591 <AGI>
1962 V5 N5 P104-112 IGR,V5,N11,P1450-56 <AGI>
1962 V5 N5 P114-130 IGR,V5,N11,P1457-69 <AGI>
1962 V5 N6 P43-51 IGR V6 N1 P58-67 <AGI>
1962 V5 N7 P8-25 IGR V6 N2 P263-276 <AGI>
1962 V5 N7 P26-40 IGR V6 N2 P277-286 <AGI>
1962 V5 N7 P41-63 IGR V6 N2 P287-302 <AGI>
1962 V5 N7 P94-113 IGR V6 N2 P303-316 <AGI>
1962 V5 N8 P3-15 IGR,V6,N4,P624-632 <AGI>
1962 V5 N8 P16-24 IGR,V6,N4,P633-638 <AGI>
1962 V5 N8 P33-39 IGR,V6,N4,P639-643 <AGI>
1962 V5 N8 P104-121 IGR,V6,N4,P644-655 <AGI>
1962 V5 N8 P151-157 64-15358 <=*$>
1962 V5 N8 P168-172 IGR,V6,N4,P662 <AGI>
1962 V5 N8 P656-661 IGR,V6,N4,P656-661 <AGI>
1962 V5 N9 P8-23 IGR,V6,N5,P810-819 <AGI>
1962 V5 N9 P45-60 IGR,V6,N5,P820-829 <AGI>
1962 V5 N9 P113-121 IGR,V6,N5,P830-835 <AGI>
1962 V5 N10 P16-27 IGR V6 N6 P1085-1098 <AGI>
1962 V5 N10 P39-56 IGR V6 N6 P1099-1109 <AGI>
1962 V5 N10 P140-143 IGR V6 N6 P1110-1112 <AGI>
1962 V5 N11 P6-14 63-27414 <$>
1962 V5 N11 P15-36 IGR V6 N8 P1370-1383 <AGI>
1962 V5 N11 P36-69 IGR V6 N8 P1341-1369 <AGI>
1962 V5 N11 P88-97 IGR V6 N8 P1336-1340 <AGI>
1962 V5 N11 P110-112 64-26095 <=$>
1962 V5 N12 P16-29 IGR V6 N9 P1532-1540 <AGI>
1962 V5 N12 P70-79 IGR V6 N9 P1541-1556 <AGI>
1962 V5 N12 P80-99 IGR V6 N9 P1557-1572 <AGI>
1962 V5 N12 P100-114 IGR V6 N9 P1573-1584 <AGI>
1962 V5 N12 P133-139 IGR V6 N9 P1585-1590 <AGI>
1962 V5 N12 P140-141 IGR V6 N9 P1591-1592 <AGI>
1963 N1 P11-28 IGR V6 N10 P1750 <AGI>
1963 N1 P40-52 IGR V6 N10 P1761 <AGI>
1963 N1 P70-81 IGR V6 N10 P1773 <AGI>
1963 N1 P97-109 IGR V6 N10 P1782 <AGI>
1963 N1 P110-118 IGR V6 N10 P1792 <AGI>
1963 N1 P119-128 IGR V6 N10 P1798 <AGI>
1963 N1 P129-134 IGR V6 N10 P1805 <AGI>
1963 N2 P3-16 63-21850 <=>
1963 N2 P17-24 IGR V6 N11 P2009 <AGI>
1963 N2 P25-33 IGR V6 N11 P2015 <AGI>
1963 N2 P34-44 IGR V6 N11 P2020 <AGI>
1963 N2 P58-71 IGR V6 N11 P2027 <AGI>
1963 N2 P72-81 IGR V6 N11 P2036 <AGI>
1963 N2 P110-117 IGR V6 N11 P2042 <AGI>
1963 N2 P143-150 IGR V6 N11 P2046 <AGI>
1963 N2 P150-152 IGR V6 N11 P2053 <AGI>
1963 N3 P25-42 IGR V7 N3 P507 <AGI>
1963 N3 P43-56 IGR V7 N3 P518 <AGI>
1963 N3 P61-81 IGR V7 N3 P526 <AGI>
1963 N3 P82-93 IGR V7 N3 P543 <AGI>
1963 N3 P124-133 IGR V7 N3 P551 <AGI>
1963 N4 P3-23 IGR V7 N4 P576 <AGI>
1963 N4 P25-53 IGR V7 N4 P592 <AGI>
1963 N4 P85-98 IGR V7 N4 P621 <AGI>
1963 N4 P99-108 IGR V7 N4 P630 <AGI>
1963 N4 P123-126 IGR V7 N4 P638 <AGI>
1963 N4 P129-133 IGR V7 N4 P614 <AGI>
1963 N4 P133-140 IGR V7 N4 P642 <AGI>
1963 N4 P141-144 IGR V7 N4 P655 <AGI>
1963 N4 P145-154 IGR V7 N4 P649 <AGI>
1963 N5 P7-18 IGR V7 N4 P659 <AGI>
1963 N5 P19-37 IGR V7 N4 P668 <AGI>
1963 N5 P63-75 IGR V7 N4 P681 <AGI>
1963 N5 P104-114 IGR V7 N4 P696 <AGI>
1963 N5 P115-118 IGR V7 N4 P702 <AGI>
1963 N5 P125-129 IGR V7 N4 P705 <AGI>
1963 N5 P129-133 IGR V7 N4 P709 <AGI>
1963 N5 P133-138 IGR V7 N4 P713 <AGI>
1963 N6 P3-19 IGR V7 N5 P764 <AGI>
1963 N6 P20-31 IGR V7 N5 P777 <AGI>
1963 N6 P32-50 IGR V7 N5 P788 <AGI>
1963 N6 P51-64 IGR V7 N5 P816 <AGI>
1963 N6 P75-93 IGR V7 N5 P803 <AGI>
1963 N6 P112-119 IGR V7 N5 P826 <AGI>
1963 N6 P134-145 IGR V7 N5 P832 <AGI>
1963 N7 P24-51 IGR V7 N6 P1013 <AGI>
1963 N7 P52-76 IGR V7 N6 P1039 <AGI>
1963 N7 P77-89 IGR V7 N6 P1030 <AGI>
1963 N7 P90-104 IGR V7 N6 P1057 <AGI>
1963 N7 P130-132 IGR V7 N6 P1068 <AGI>
1963 N8 P17-26 IGR V7 N8 P1361 <AGI>
1963 N8 P40-48 IGR V7 N8 P1368 <AGI>
1963 N8 P78-94 IGR V7 N8 P1387 <AGI>
1963 N8 P118-125 IGR V7 N9 P1569 <AGI>
1963 N8 P141-148 IGR V7 N8 P1401 <AGI>
1963 N8 P143-149 IGR V7 N9 P1577 <AGI>
1963 N8 P156-161 IGR V7 N9 P1583 <AGI>
1963 N9 P73-89 IGR V7 N9 P1550 <AGI>
1963 N10 P3-13 IGR V7 N10 P1718 <AGI>
1963 N10 P38-46 IGR V7 N9 P1588 <AGI>
1963 N10 P47-62 IGR V7 N9 P1594 <AGI>
1963 N10 P102-113 IGR V7 N9 P1606 <AGI>
1963 N10 P150-154 IGR V7 N9 P1608 <AGI>
1963 N11 P26-39 IGR V7 N10 P1724 <AGI>
1963 N11 P40-50 IGR V7 N10 P1733 <AGI>
1963 N11 P116-121 IGR V7 N10 P1742 <AGI>
1963 N12 P19-35 IGR V7 N10 P1747 <AGI>
1963 N12 P36-57 IGR V7 N10 P1756 <AGI>
1963 N12 P68-88 IGR V7 N10 P1771 <AGI>
1963 N12 P123-129 IGR V7 N10 P1784 <AGI>
1963 N12 P133-138 IGR V7 N10 P1789 <AGI>
1963 N12 P142-145 IGR V7 N10 P1794 <AGI>
1963 N12 P150-156 IGR V7 N10 P1797 <AGI>

1963 V6 N1 P11-28	IGR V6 N10 P1750 <AGI>
1963 V6 N1 P40-52	C-5317 <NRC>
	IGR V6 N10 P1761 <AGI>
1963 V6 N1 P70-81	IGR V6 N10 P1773 <AGI>
1963 V6 N1 P97-109	IGR V6 N10 P1782 <AGI>
1963 V6 N1 P110-118	IGR V6 N10 P1792 <AGI>
1963 V6 N1 P119-128	IGR V6 N10 P1798 <AGI>
1963 V6 N1 P129-134	IGR V6 N10 P1805 <AGI>
1963 V6 N2 P17-24	C-4925 <NRC>
	IGR V6 N11 P2009 <AGI>
1963 V6 N2 P25-33	IGR V6 N11 P2015 <AGI>
1963 V6 N2 P34-44	IGR V6 N11 P2020 <AGI>
1963 V6 N2 P58-71	IGR V6 N11 P2027 <AGI>
1963 V6 N2 P72-81	IGR V6 N11 P2036 <AGI>
1963 V6 N2 P110-117	IGR V6 N11 P2042 <AGI>
1963 V6 N2 P143-150	IGR V6 N11 P2046 <AGI>
1963 V6 N2 P150-152	IGR V6 N11 P2053 <AGI>
1963 V6 N3 P113-123	C-4925 <NRC>
1963 V6 N7 P3-13	64-21380 <=>
1963 V6 N9 P62-72	64-26267 <$>
1964 N9 P3-28	IGR V8 N7 P851 <AGI>
1964 N9 P29-46	IGR V8 N5 P591 <AGI>
1964 N9 P77-94	IGR V8 N10 P1241 <AGI>
1964 N10 P122-128	IGR V8 N12 P1451 <AGI>
1964 N11 P108-113	AD-623 423 <=*$>
1964 N12 P70-89	IGR V8 N2 P171 <AGI>
1964 N12 P115-118	IGR V8 N5 P570 <AGI>
1964 N12 P125-132	IGR V7 N12 P2156 <AGI>
1964 V7 N1 P123-129	65-27471 <=>
1964 V7 N4 P3-23	64-51059 <=$>
1964 V7 N6 P144-146	66S86R <ATS>
1964 V7 N11 P3-17	66-13351 <*>
1964 V7 N11 P61-71	AD-623 421 <=*$>
1964 V7 N11 P99-105	AD-623 449 <=*$>
1964 V7 N11 P117-120	AD-623 422 <=*$>
1965 N1 P3-20	IGR V8 N5 P505 <AGI>
1965 N1 P17-34	IGR V8 N3 P317 <AGI>
1965 N1 P35-53	IGR V8 N3 P290 <AGI>
1965 N1 P54-75	IGR V8 N2 P127 <AGI>
1965 N1 P76-94	IGR V8 N5 P515 <AGI>
1965 N1 P95-104	IGR V8 N3 P309 <AGI>
1965 N1 P132-137	IGR V8 N3 P331 <AGI>
1965 N2 P3-15	IGR V8 N3 P336 <AGI>
1965 N2 P29-44	IGR V8 N4 P404 <AGI>
1965 N2 P45-59	IGR V8 N3 P253 <AGI>
1965 N2 P138-140	IGR V8 N6 P702 <AGI>
1965 N2 P140-145	IGR V8 N1 P98 <AGI>
1965 N3 P10-22	IGR V8 N6 P665 <AGI>
1965 N3 P23-34	IGR V8 N2 P144 <AGI>
1965 N3 P35-42	IGR V8 N3 P271 <AGI>
1965 N3 P113-115	IGR V8 N2 P154 <AGI>
1965 N3 P118-124	IGR V8 N3 P281 <AGI>
1965 N3 P132-134	IGR V8 N3 P288 <AGI>
1965 N3 P134-137	IGR V8 N3 P306 <AGI>
1965 N4 P3-18	IGR V8 N9 P1017 <AGI>
1965 N4 P49-62	IGR V8 N8 P930 <AGI>
1965 N4 P63-73	IGR V8 N9 P1009 <AGI>
1965 N4 P74-89	IGR V8 N8 P896 <AGI>
1965 N5 P38-56	IGR V8 N11 P1347 <AGI>
1965 N9 P63-73	IGR V8 N10 P1199 <AGI>
1965 N10 P94-96	IGR V8 N9 P1029 <AGI>
1965 N10 P97-109	IGR V8 N10 P1226 <AGI>
1965 N10 P127-131	IGR V8 N9 P1032 <AGI>
1965 N10 P136-137	IGR V8 N10 P1197 <AGI>
1965 N10 P145-146	IGR V8 N9 P1037 <AGI>
1965 N11 P3-18	IGR V8 N11 P1305 <AGI>
1965 N11 P91-103	IGR V8 N10 P1218 <AGI>
1965 N11 P125-131	IGR V8 N10 P1237 <AGI>
1965 N11 P137-143	IGR V8 N11 P1329 <AGI>
1965 N11 P153-156	IGR V8 N10 P1215 <AGI>
1965 N12 P17-26	IGR V8 N7 P874 <AGI>
1965 N12 P37-51	IGR V8 N8 P918 <AGI>
1965 N12 P52-68	IGR V8 N8 P885 <AGI>
1965 N12 P93-99	IGR V8 N8 P940 <AGI>
1965 N12 P100-119	IGR V8 N10 P1157 <AGI>
1965 V8 N1 P76-94	80S83R <ATS>
1965 V8 N6 P165	65-22518 <$>
1965 V8 N8 P157-159	66-14410 <*$>
1965 V8 N12 P93-99	AD-633 407 <=$>
1965 V8 N12 P153-155	66-32969 <=$>
1966 N6 P7-19	IGR V8 N12 P1387 <AGI>
1966 V9 N5 P47-57	AD-641 754 <=$>

SOVETSKAYA KHIRURGIYA

1933 V5 N2 P2053-2054	RJ-19 <ATS>
1934 V7 P462-472	R-1074 <*>
1934 V7 N2/3 P462-472	RT-1706 <*>
1936 V6 N10 P707-711	R-1507 <*>
	RT-3421 <*>

SOVETSKAYA KNIGA

1948 N2 P23-25	RT-2361 <*>
1948 N2 P25-28	RT-4093 <*>
1948 N9 P31-34	RT-3789 <*>
1948 N9 P34-38	RT-3791 <*>
1948 N10 P19-22	RT-3786 <*>
1948 N11 P29-30	RT-3788 <*>
1948 N11 P31-32	RT-3790 <*>
1948 N12 P15-20	RT-3758 <*>
1948 N12 P20-23	RT-3787 <*>
1949 N11 P19-21	RT-4040 <*>
1950 N5 P23-25	RT-2909 <*>
1953 N8 P28-33	RT-1091 <*>
1953 N8 P40-42	RT-1349 <*>
1953 N8 P42-43	RT-1350 <*>

SOVETSKAYA LITVA

1961 03/31 P4	63-15511 <=*>
1961 07/23 P2	61-28721 <=*>
1964 11/04 P3/4	64-51955 <=$>

SOVETSKAYA MEDITSINA

1938 N10 P35-45	R-2460 <*>
1938 N22 P38-40	R-643 <*>
1940 V4 N1 P48	63-21962 <=>
1940 V4 N16 P34-36	RT-634 <*>
1940 V4 N20 P13-14	62-01042 <*>
	62-32587 <=*>
1940 V4 N13/4 P5-8	R-1048 <*>
1940 V4 N13/4 P16-18	RT-1828 <*>
1941 N15 P16-	R-1098 <*>
1941 N13/4 P41-42	R-944 <*>
1941 V5 N13/4 P41-42	64-19513 <=$>
1941 V5 N17/8 P18-19	RT-794 <*>
1942 N9 P15-	R-1089 <*>
1943 N11/2 P12-15	RT-1821 <*>
1943 N11/2 P15-	R-1062 <*>
1943 N11/2 P18-	R-1106 <*>
1943 V7 N4 P12-14	63-19100 <=*>
1943 V7 N11/2 P15-16	RT-633 <*>
1944 N6 P12-15	65-63450 <=>
1944 N12 P14-15	RT-795 <*>
1944 V8 N6 P9-10	65-63455 <=>
1944 V8 N12 P14-15	60-19546 <=*$>
1947 N6 P19-21	R-4497 <*>
	RT-1861 <*>
1949 N4 P7-8	RT-2357 <*>
1949 V13 N4 P7-8	R-303 <*>
	64-15069 <=*$>
1949 V13 N11 P23-24	61-28101 <*=>
1950 V14 N6 P24-26	60-13861 <+*>
1951 V15 N2 P21	RT-2112 <*>
1953 N2 P34-36	RT-3079 <*>
1953 N2 P36-38	RT-3078 <*>
1953 N4 P3-7	R-702 <*>
1953 N7 P33	RT-4425 <*>
1953 V17 N5 P40-41	RT-2009 <*>
1953 V17 N9 P25-26	RT-2968 <*>
1953 V17 N10 P41-44	RT-2265 <*>
1954 N3 P36-38	RT-2696 <*>
1954 N7 P31-33	RT-2742 <*>
1954 N10 P39-40	RT-2918 <*>
1954 V18 N1 P41-44	RT-2226 <*>
1954 V18 N3 P41-42	59-18138 <+*>
1954 V18 N7 P31-33	R-788 <*>
	65-63257 <=*$>
1954 V18 N12 P3-5	R-635 <*>
1955 V19 N1 P73-77	R-1536 <*>
	64-19064 <=*$>
1955 V19 N2 P36-40	62-15312 <=*>

Citation	Translation
1955 V19 N2 P62-65	62-15310 <=*>
1955 V19 N8 P51-56	R-413 <*>
	R-484 <*>
	64-19481 <=$>
1955 V19 N11 P57-58	62-15339 <=*>
1956 V20 N1 P82-90	RT-4268 <*>
1956 V20 N3 P41-45	65-13442 <*>
1956 V20 N11 P8-18	62-15148 <*=> 0
1956 V20 N11 P23-29	62-15307 <*=>
1957 N11 P85-89	R-4446 <*>
1957 V21 N1 P110-112	64-19528 <=$>
1957 V21 N2 P35-41	62-11145 <=>
	62-15076 <=*>
1957 V21 N3 P69-73	59-13516 <=>
1957 V21 N9 P18-26	59-11561 <=>
1957 V21 N10 P10-24	62-15085 <=*>
1957 V21 N11 P10-15	60-17978 <+*>
1957 V21 N11 P85-89	64-19454 <=$>
1957 V21 N11 P89-94	61-19537 <=*>
1958 V22 N1 P69-74	59-10030 <+*>
1958 V22 N2 P94-103	59-13661 <=> 0
1958 V22 N2 P109-112	59-13495 <=>
1958 V22 N3 P37-42	59-13335 <=>
1958 V22 N3 P100-102	59-13336 <=> 0
1958 V22 N6 P86-90	64-19820 <=$> 0
1958 V22 N6 P135-141	59-13208 <=>
1958 V22 N7 P133-134	62-19263 <*=>
1958 V22 N10 P100-102	64-19822 <=$>
1958 V22 N10 P116-119	59-13251 <=>
1958 V22 N10 P129-131	59-13252 <=>
1958 V22 N10 P153-158	59-13253 <=>
1958 V22 N12 P125-130	60-41474 <=>
1959 V23 N1 P3-7	59-11719 <=>
1959 V23 N1 P8-17	59-11714 <=>
1959 V23 N1 P100-106	59-11712 <=>
1959 V23 N1 P152-153	59-11715 <*>
1959 V23 N2 P6-12	59-11723 <=>
1959 V23 N2 P141-149	59-11724 <=>
1959 V23 N2 P149-151	59-11725 <=>
1959 V23 N5 P28-34	59-20716 <+*>
1959 V23 N5 P144-151	60-11228 <=>
1959 V23 N7 P3-12	60-11093 <=>
1959 V23 N7 P54-62	61-27745 <*=>
1959 V23 N11 P3-14	62-15397 <=*>
1959 V23 N12 P3-7	60-11362 <=>
1959 V23 N12 P8-25	60-11419 <=>
1959 V23 N12 P26-33	60-11386 <=>
1959 V23 N12 P61-67	60-11363 <=>
1959 V23 N12 P124-128	60-11391 <=>
1960 V24 N1 P22-23	61-11047 <=>
1960 V24 N1 P42-48	61-11978 <=>
1960 V24 N1 P48-53	61-11699 <=>
1960 V24 N1 P63-65	61-21206 <=>
1960 V24 N1 P100-103	61-21205 <=>
1960 V24 N1 P123-126	61-21203 <=>
1960 V24 N1 P136-137	61-11723 <=>
1960 V24 N1 P138-143	60-11966 <=>
1960 V24 N3 P3-10	60-41033 <=>
1960 V24 N3 P52-55	60-41034 <=>
1960 V24 N3 P114-119	63-19118 <=*> 0
1960 V24 N3 P123-126	60-41035 <=>
1960 V24 N3 P150-155	60-41036 <=>
1960 V24 N4 P34-42	60-41427 <=>
1960 V24 N4 P43-47	60-41428 <=>
1960 V24 N4 P54-61	62-15768 <*=>
1960 V24 N4 P70-76	60-41429 <=>
1960 V24 N4 P105-109	60-41430 <=>
1960 V24 N4 P121-126	60-41431 <=>
1960 V24 N5 P116-118	60-31792 <=>
1960 V24 N6 P76-81	60-41422 <=>
1960 V24 N6 P81-85	60-41423 <=>
1960 V24 N6 P106-107	60-41424 <=>
1960 V24 N6 P123-126	60-41425 <=>
1960 V24 N6 P128-134	60-41426 <*=>
1960 V24 N6 P149-153	60-41462 <=>
1960 V24 N7 P95-103	60-41448 <=>
1960 V24 N7 P119-123	61-28028 <*=>
1960 V24 N7 P131-132	63-19712 <=*>
1960 V24 N8 P30-39	61-11924 <=>
1960 V24 N8 P39-46	61-11942 <=>
1960 V24 N8 P94-97	61-21044 <=>
1960 V24 N8 P131-135	61-11185 <=>
1960 V24 N8 P148-150	61-21041 <=>
1960 V24 N9 P26-31	61-11642 <=>
1960 V24 N9 P92-97	61-11642 <=>
1960 V24 N10 P3-12	61-21532 <=>
1960 V24 N10 P33-40	61-21532 <=>
1960 V24 N10 P111-112	61-21532 <=>
1960 V24 N10 P151-154	61-21505 <=>
1960 V24 N11 P52-55	61-21391 <=>
1960 V24 N11 P56-60	61-21308 <=>
1960 V24 N11 P61-64	61-21287 <=>
1960 V24 N11 P100-106	61-21303 <=>
1960 V24 N11 P120-123	61-21279 <=>
1960 V24 N11 P132-134	61-21292 <=>
1960 V24 N11 P138-143	61-21394 <=>
1960 V24 N11 P144-148	61-21397 <=>
1960 V24 N12 P20-28	61-21660 <=>
1960 V24 N12 P74-81	61-21677 <=>
1960 V24 N12 P122-126	61-21675 <=>
1961 V25 N2 P101-104	61-27051 <*=>
1961 V25 N2 P114-118	62-23474 <=*>
1961 V25 N3 P71-79	61-23554 <=*>
1961 V25 N3 P122-126	61-23586 <=*>
1961 V25 N4 P3-8	62-19152 <*=>
1961 V25 N4 P9-16	61-28524 <*=>
1961 V25 N5 P107-114	61-28857 <*=>
1961 V25 N6 P3-13	62-15361 <*=>
1961 V25 N6 P55-66	62-15361 <*=>
1961 V25 N7 P8-19	62-13331 <*=>
1961 V25 N7 P134-138	62-13270 <=*>
1961 V25 N8 P29-33	62-15492 <=>
1961 V25 N8 P41-54	RTS-2980 <NLL>
1961 V25 N8 P63-76	62-15492 <=>
1961 V25 N8 P86-95	62-15492 <=>
1961 V25 N8 P123-127	62-15492 <=>
1961 V25 N8 P129-131	62-15492 <=>
1961 V25 N9 P139-142	62-23757 <=*>
1961 V25 N9 P150-151	62-23610 <=>
1962 V25 N1 P137-139	63-19390 <=*>
1962 V25 N1 P145-147	63-19389 <=*>
1962 V25 N3 P3-8	62-25045 <=>
1962 V25 N3 P113-117	62-25045 <=>
1962 V25 N4 P22-30	64-51287 <=>
1962 V25 N4 P57-60	63-15232 <=*>
1962 V25 N4 P107-112	63-20907 <=*$>
1962 V26 N2 P75-81	63-15564 <=*>
1962 V26 N3 P58-65	63-15558 <=*>
1962 V26 N4 P53-56	62-25502 <=*>
1962 V26 N4 P126-128	62-25502 <=*>
1962 V26 N5 P3-9	63-13653 <=*>
1962 V26 N6 P5-13	62-33146 <=*>
1962 V26 N6 P115-118	62-33146 <=*>
1962 V26 N6 P145-151	62-33146 <=*>
1962 V26 N8 P72-78	63-15557 <=*>
1962 V26 N8 P80-84	63-15569 <=*>
1962 V26 N9 P151-152	63-21827 <=>
1962 V26 N10 P3-13	63-21086 <=>
1962 V26 N10 P42-45	63-21086 <=>
1962 V26 N10 P110-114	63-21086 <=>
1962 V26 N11 P98-103	63-21075 <=>
1962 V26 N12 P13-16	63-21915 <=>
1963 V26 N1 P17-23	63-24508 <=*$>
1963 V26 N1 P30-37	63-23869 <=*$>
1963 V26 N1 P60-65	63-24476 <=*$>
1963 V26 N1 P79-84	63-24477 <=*$>
1963 V26 N1 P85-90	64-13519 <=*$> 0
1963 V26 N1 P91-92	63-23870 <=*$>
1963 V26 N1 P99-103	63-23832 <=*$>
1963 V26 N1 P109-111	63-24478 <=*$>
1963 V26 N1 P111-113	63-24479 <=*$>
1963 V26 N1 P115-	FPTS V22 N6 P.T1220 <FASE>
1963 V26 N1 P115-116	FPTS,V22,P.T1220-21 <FASE>
1963 V26 N1 P120-122	63-24480 <=*$> C
1963 V26 N1 P129-	FPTS V22 N5 P.T797 <FASE>
1963 V26 N1 P129-133	1963,V22,N5,PT2 <FASE>
1963 V26 N2 P8-17	64-13421 <=*$>
1963 V26 N2 P72-78	64-13420 <=*$>
1963 V26 N2 P109-111	64-13419 <=*$>
1963 V26 N2 P116-119	64-13418 <=*$>

1963 V26 N2 P134-138 64-13417 <=*$>
1963 V26 N2 P154-159 63-41025 <=>
1963 V26 N3 P35-38 64-13456 <=*$>
1963 V26 N3 P52-55 64-13623 <=*$>
1963 V26 N3 P65-69 64-13454 <=*$>
1963 V26 N3 P69-72 64-13453 <=*$> O
1963 V26 N3 P104-106 64-13452 <=*$>
1963 V26 N3 P106-109 64-13451 <=*$>
1963 V26 N3 P110-114 64-13450 <=*$>
1963 V26 N3 P114-117 64-13449 <=*$>
1963 V26 N3 P118-124 64-13448 <=*$>
1963 V26 N3 P127-131 64-13447 <=*$>
1963 V26 N3 P131-133 64-13446 <=*$>
1963 V26 N4 P41-47 AD-630 845 <=$>
1963 V26 N4 P91-95 64-13872 <=*$>
1963 V26 N4 P110-113 64-13435 <=*$>
1963 V26 N5 P8-13 64-13875 <=*$>
1963 V26 N5 P42-46 64-13911 <=*$>
1963 V26 N5 P49-54 64-13876 <=*$>
1963 V26 N5 P54-60 64-13925 <=*$>
1963 V26 N5 P74-79 10Q72R <ATS>
1963 V26 N5 P74-80 64-13885 <=*$>
1963 V26 N5 P80-85 64-13919 <=*$>
1963 V26 N5 P102-106 64-13822 <=*$>
1963 V26 N5 P118-120 64-13877 <=*$>
1963 V26 N5 P127-132 64-13920 <=*$>
1963 V26 N5 P132-134 64-13878 <=*$>
1963 V27 N6 P82-87 64-15533 <=*$>
1963 V27 N6 P136-138 63-31817 <=>
1963 V27 N7 P56- FPTS V23 N3 P.T441 <FASE>
1963 V27 N7 P56-62 FPTS,V23,N3,P.T441 <FASE>
63-31833 <=>
1963 V27 N7 P74-78 63-31832 <=>
64-15407 <=*$>
1963 V27 N8 P3-8 64-21023 <=>
1963 V27 N8 P77-81 64-15463 <=*$>
1963 V27 N8 P81-84 64-15486 <=*$> O
1963 V27 N9 P21- FPTS V23 N5 P.T984 <FASE>
FPTS,V23,N5,P.T984 <FASE>
1963 V27 N10 P25-32 64-19281 <=*$>
1963 V27 N10 P46-49 64-19692 <=$> O
1963 V27 N10 P58-62 64-19253 <=*$>
1963 V27 N10 P62-66 64-19254 <=*$>
1963 V27 N10 P82-87 64-19237 <=*$>
1963 V27 N11 P25- FPTS V23 N6 P.T1261 <FASE>
FPTS V23 P.T1261 <FASE>
1963 V27 N11 P59-62 64-19640 <=$>
1963 V27 N11 P62-68 64-19699 <=$>
1963 V27 N11 P68-70 64-19629 <=$>
1963 V27 N11 P78-82 64-19206 <=*$>
1963 V27 N11 P92-96 64-19641 <=$> O
1963 V27 N11 P99-104 64-19622 <=$>
1963 V27 N12 P15-22 64-19627 <=$>
1963 V27 N12 P22-30 65-60204 <=$>
1963 V27 N12 P30-33 65-60205 <=$>
1964 V27 N1 P6-14 65-60228 <=$>
1964 V27 N1 P41-45 65-60208 <=$>
1964 V27 N1 P66-71 65-62618 <=$>
1964 V27 N2 P42- FPTS V24 N3 P.T525 <FASE>
1964 V27 N3 P38-42 65-60184 <=$>
1964 V27 N3 P48-53 65-61110 <=$>
1964 V27 N4 P37-41 65-60231 <=$>
1964 V27 N7 P17-22 65-63621 <=$>
1964 V27 N7 P39-44 65-63617 <=$>
1964 V27 N8 P16-20 65-62110 <=$>
1964 V27 N11 P50-54 65-63618 <=$>
1964 V28 N8 P88-92 65-31504 <=$>
1964 V28 N9 P13- FPTS V24 N4 P.T597 <FASE>
1965 V28 N1 P3-9 66-60317 <=*$>
1965 V28 N5 P46-51 66-61295 <=*$>
1965 V28 N5 P80-84 66-60869 <=*$>
1965 V28 N6 P70-74 66-61353 <=*$>
1965 V28 N7 P21-24 66-61294 <=*$>
1965 V28 N8 P52- FPTS V25 N4 P.T683 <FASE>
1965 V28 N11 P60- FPTS V25 N5 P.T800 <FASE>
1966 V29 N5 P156-157 66-33103 <=$>

SOVETSKAYA METALLURGIYA
1936 V8 N12 P11-20 RT-1682 <*>
1937 V9 N4 P56-64 R-3361 <*>

SOVETSKAYA MOLDAVIYA
1961 09/15 P4 66-34903 <=$>
1966 07/07 P3-4 66-33690 <=$>
1966 08/05 P3,4 66-33996 <=$>

SOVETSKAYA NAUKA
1940 V5 P108-123 R-2421 <*>

SOVETSKAYA NEVROPATOLOGIYA, PSIKHIATRIYA I PSIKHOGIGIENA
1935 V4 N1 P140-142 58J16R <ATS>

SOVETSKAYA PECHAT
1961 N12 P20-22 62-32868 <=>
1962 N12 P13-14 63-21885 <=>
1966 N10 P39-41 66-34682 <=$>

SOVETSKAYA PEDAGOGIKA
1952 N9 P52-62 AD-610 959 <=$>
1952 N9 P52-56 C-2669 <NRC>
1952 N9 P52-62 R-4241 <*>
1954 V18 N8 P102-112 RT-3756 <*>
1955 V19 N1 P41-54 RT-3220 <*>
1957 V21 N6 P151-155 R-3487 <*>
64-19421 <=$>
1958 V22 N4 P95-109 59-11512 <=>
1958 V22 N4 P121-128 59-11513 <=>
1960 V24 N2 P16-44 60-31118 <=>
1960 V24 N7 P39-50 63-18926 <=*>
1960 V24 N9 P156-160 61-11829 <=>
1960 V24 N10 P151-152 61-11668 <=>
1961 V25 N4 P151-160 61-31597 <=>
1961 V25 N11 P64-70 62-24988 <=*>
1962 V26 N4 P45-55 62-25603 <=*>
1962 V26 N6 P70-76 62-33007 <=*>
1965 V29 N3 P14-20 65-32339 <=*>
1965 V29 N4 P149-153 65-32426 <=*$>
1966 N2 P142-145 66-31753 <=$>

SOVETSKAYA PEDIATRIYA
1934 N8/9 P220-224 R-1104 <*>

SOVETSKAYA PSIKHONEVROLOGIYA
1937 V13 N7 P101-104 R-1077 <*>

SOVETSKAYA ROSSIYA
1964 12/05 P1 65-30270 <=$>
1965 01/17 P2 65-30270 <=$>
1966 P3 07/06 66-33316 <=$>
1966 N169 P4 07/23 66-34134 <=$>

SOVETSKAYA TORGOVLYA
1961 N2 P3-4 61-23580 <=>
1962 V35 N12 P45-46 63-19416 <*=>
1963 N9 P37-40 AD-625 856 <=*$>
1963 V36 N5 P47-48 63-31310 <=>
1963 V36 N6 P18-21 63-31559 <=>
1964 V37 N8 P8-12 64-51739 <=$>
1965 N10 P22-25 65-34024 <=*$>
1965 V38 N7 P40-42 AD-626 797 <=*$>
1965 V38 N10 P1-4 65-34067 <=*$>
1965 V38 N10 P19-21 65-33764 <=*$>

SOVETSKAYA UKRAINA
1959 N8 P127-137 60-11244 <=>

SOVETSKAYA VETERINARIYA
1934 V11 P40-81 R-3079 <*>
1939 V16 N5 P45-46 59-15897 <+*>
1940 V17 N1 P53-55 R-1076 <*>
1940 V17 N5 P61-68 RT-301 <*>

SOVETSKAYA VRACHEBNAYA GAZETA
1934 N12 P889-894 R-1056 <*>

SOVETSKAYA ZOOTEKHNIYA

1950 V5 N5 P37-46 RT-4606 <*>

SOVETSKIE PROFSOIUZY
1963 N1 P5-6 63-21682 <=>
1963 N5 P15 63-31116 <=>
1966 V22 N3 P12-13 66-31979 <=$>
1966 V22 N3 P41-43 66-31979 <=$>

SOVETSKIE SUBTROPIKI
1939 N2/3 P51-53 R-4884 <*>
1940 N5 P38-43 R-2254 <*>

SOVETSKIE KAZAKHSTAN
1959 N5 P125-126 60-41622 <=>
1959 N11 P99-105 60-31124 <=> 0

SOVETSKII KAUCHUK
1934 N6 P24-27 RT-4598 <*>

SOVETSKII KRASNYI KREST
1960 V10 N1 P12-13 AD-620810 <=*$>
62-24912 <=*>
1961 N1 P1-4 62-13249 <*=>
1961 N2 P1-3 61-27731 <*=>
1961 N2 P21-23 61-27390 <*=>
1963 V13 N2 P20-21 64-31133 <=>

SOVETSKII SOYUZ
1955 N2 P12-13 RT-3118 <*>
1959 N10 P12-14 60-13191 <+*>

SOVETSKII VRACHEBNYI ZHURNAL
1935 V39 N7 P583-586 65-60097 <=$>
1939 N10 P547-551 R-1093 <*>
1939 V43 N7 P391-402 RT-1817 <*>
1940 V44 P835-840 60-17653 <+*>
1940 V44 N6 P439-444 60-19758 <=$> 0

SOVETSKOE FOTO
1959 V19 N7 P73 64-13116 <=*$>
1960 V20 N6 P31 61-10285 <*+>
1961 V21 N5 P40-42 68N56R <ATS>
1961 V21 N10 P26-29 63-18071 <=*>
1961 V21 N12 P26-27 62-18676 <=*>
1962 V22 N7 P39 63-18059 <=*>
1962 V22 N10 P30-33 63-18070 <=*>
1963 V23 N4 P37 63-31113 <=>
1963 V23 N6 P32 63-31855 <=>
1963 V23 N8 P30-31 65-11362 <*>
1963 V23 N9 P30-33 65-11363 <*>
1963 V23 N10 P32-35 65-11361 <*>
1963 V23 N11 P34-35 65-11360 <*>
1963 V23 N12 P40-41 64-20088 <*>
1964 V24 N1 P38 65-13822 <*>
1964 V24 N3 P38-39 65-11359 <*>
1964 V24 N12 P34 65-14531 <*>
1964 V24 N12 P38 65-14572 <*>
1965 V25 N1 P34 65-14614 <*>
1965 V25 N1 P37 65-14611 <*>
1965 V25 N1 P40-41 65-14613 <*>
1965 V25 N2 P32 65-14615 <*>

SOVETSKOE GOSDARSTVO I PRAVO
1958 N1 P27-34 63-31751 <=>
1958 N7 P52-58 64-71374 <=>
1959 N11 P25-37 60-11579 <=>
1960 N3 P34-45 60-41045 <=>
1961 N10 P116-125 62-13715 <*=>
1962 V32 N7 P3-24 63-21037 <=>
1964 V34 N1C P81-90 30459 <=$>
1964 V34 N11 P98-103 65-31191 <=$>
1965 V35 N4 P90-101 65-31299 <=$>
1965 V35 N6 P83-92 65-32095 <=*$>
1965 V35 N7 P121-125 65-33042 <=*$>
1966 N12 P24-33 67-30304 <=>

SOVETSKOE KOTLOTURBOSTROENIE
1940 N1 P30-32 RT-359 <*>
1940 N3 P107-111 62-24184 <=*>
1940 N8 P261-269 RT-4036 <*>
1940 N10 P362-366 59-19151 <+*>

*SOVETSKOE ZDRAVOOKHRANENIE
1945 N4/5 P45-49 RT-1841 <*>
1947 N4 P52-54 RT-1177 <*>
1954 V13 N1 P21-25 RT-2472 <*>
1955 V14 N1 P20-24 62-15337 <*=>
1955 V14 N1 P61-62 R-165 <*>
1957 V16 N9 P14-18 59-13179 <=> 0
1957 V16 N9 P55-62 59-13178 <=>
1958 V17 N2 P30-34 60-23729 <+*>
1958 V17 N4 P8-12 63-19042 <=*>
1958 V17 N12 P55-56 59-13888 <=>
1959 V18 N1 P11-15 59-13955 <=>
1959 V18 N1 P16-20 59-13956 <=>
1959 V18 N2 P3-4 59-13632 <=>
1959 V18 N2 P25-30 59-13633 <=>
1959 V18 N2 P30-33 62-11139 <=>
1959 V18 N4 P3-6 59-13653 <=>
1959 V18 N4 P6-11 59-13654 <=>
1959 V18 N4 P11-18 59-13655 <=>
1959 V18 N5 P3-9 59-11719 <=>
1959 V18 N5 P9-14 59-11720 <=>
1959 V18 N5 P34-38 59-11721 <=>
1959 V18 N6 P3-7 59-13790 <=>
1959 V18 N7 P3-9 59-13905 <=>
1959 V18 N7 P57-64 59-13858 <=>
1959 V18 N8 P19-25 59-31020 <=>
1960 V19 N1 P3-8 60-31487 <=>
1960 V19 N1 P36-40 60-31486 <=>
1960 V19 N1 P83-87 60-31410 <=>
1960 V19 N2 P25-32 60-11684 <=>
1960 V19 N2 P82-87 61-11851 <=>
1960 V19 N2 P102-104 60-11685 <=>
1960 V19 N4 P51-54 60-41005 <=>
1960 V19 N7 P3-6 61-11762 <=>
1960 V19 N7 P14-19 61-27972 <*=>
1960 V19 N7 P91-93 61-11762 <=>
1960 V19 N8 P3-6 61-21149 <=>
1960 V19 N8 P6-13 61-21145 <=>
1960 V19 N8 P18-21 61-11660 <=>
1960 V19 N9 P3-8 61-11992 <=>
1960 V19 N9 P48-68 61-11992 <=>
1960 V19 N9 P75-80 61-11992 <=>
1960 V19 N11 P13-15 61-21690 <=>
1960 V19 N11 P15-20 61-21656 <=>
1960 V19 N11 P20-26 61-21511 <=>
1960 V19 N11 P84-86 61-21512 <=>
1960 V19 N11 P90-91 61-21689 <=>
1960 V19 N12 P3-7 61-21630 <=>
1960 V19 N12 P12-16 61-31109 <=>
1960 V19 N12 P50-54 64-71315 <=>
1960 V19 N12 P59-62 62-13642 <*=>
1961 V20 N1 P21-25 61-21543 <=>
1961 V20 N2 P10-16 61-31156 <=>
1961 V20 N2 P16-20 61-27712 <*=>
1961 V20 N2 P25-29 61-27711 <*=>
1961 V20 N2 P65-68 61-27709 <*=>
1961 V20 N2 P68-72 61-27710 <*=>
1961 V20 N3 P4-29 61-27052 <*=>
1961 V20 N4 P3-8 61-27244 <=*>
1961 V20 N4 P9-15 61-27241 <=*>
1961 V20 N4 P63-65 61-27213 <*=>
1961 V20 N5 P3-9 61-28784 <*=>
1961 V20 N5 P63-67 61-28730 <*=>
1961 V20 N6 P3-8 61-31652 <=>
1961 V20 N6 P45-51 62-13517 <*=>
1961 V20 N6 P52-58 62-13317 <*=>
1961 V20 N6 P58-66 62-13510 <*=>
1961 V20 N6 P81-88 62-13346 <*=>
1961 V20 N8 P13-30 62-11089 <=>
1961 V20 N8 P35-47 62-11089 <=>
1961 V20 N8 P89-90 62-11089 <=>
1961 V20 N11 P3-11 62-23632 <=*>
1961 V20 N11 P39-43 62-23632 <=*>
1961 V20 N12 P6-24 63-15106 <=*>
1962 V21 N1 ENTIRE ISSUE 62-23760 <=*>
1962 V21 N3 P34-39 62-25467 <=>
1962 V21 N3 P43-51 62-25467 <=>
1962 V21 N4 P3-10 62-33613 <=*>

1962 V21 N4 P44-48 62-33613 <=*>
1962 V21 N6 P9-14 62-33044 <=*>
1962 V21 N6 P25-37 62-33044 <=*>
1962 V21 N6 P53-59 62-33044 <=*>
1962 V21 N8 P47-50 62-33279 <=*>
1962 V21 N9 P4-17 63-13641 <=*>
1962 V21 N9 P58-59 63-13095 <=>
1962 V21 N9 P61-67 63-13641 <=*>
1962 V21 N10 P47-58 63-21082 <=>
1963 V22 N6 P7-12 63-31714 <=>
1963 V22 N6 P13-16 63-31718 <=>
1963 V22 N6 P16-17 63-31783 <=>
1963 V22 N7 P3-8 63-31917 <=>
1963 V22 N7 P71-76 63-31916 <=>
1963 V22 N10 P34-41 64-21222 <=>
1963 V22 N10 P49-52 64-21013 <=>
1964 V23 N1 P59-63 64-21655 <=>
1964 V23 N1 P93-96 64-21648 <=>
1964 V23 N2 P3-13 64-21921 <=>
1964 V23 N4 P3-9 64-31391 <=>
1964 V23 N5 P3-17 64-41153 <=>
1964 V23 N5 P22-25 64-41153 <=>
1964 V23 N5 P29-32 64-41153 <=>
1964 V23 N5 P56-58 64-41153 <=>
1964 V23 N6 P3-19 64-41254 <=>
1964 V23 N6 P27-30 64-41254 <=>
1964 V23 N6 P41-45 64-41254 <=>
1964 V23 N7 P3-12 64-41679 <=>
1964 V23 N7 P34-36 64-41679 <=>
1964 V23 N7 P50-54 64-41956 <=$>
1964 V23 N7 P58-61 64-41956 <=$>
1964 V23 N8 P3-31 64-41952 <=$>
1964 V23 N8 P62-66 64-41952 <=$>
1964 V23 N8 P71 64-41952 <=$>
1964 V23 N12 P3-33 65-30232 <=$>
1964 V23 N12 P40-43 65-30232 <=$>
1965 N5 P3-11 66-30314 <=$>
1965 V24 N1 P46-48,93-95 65-30866 <=$>
1965 V24 N2 P23-27 65-30612 <=$>
1965 V24 N2 P34-40 65-30612 <=$>
1965 V24 N2 P91-94 65-31204 <=$>
1965 V24 N5 P26-30 65-31411 <=$>
1965 V24 N12 P15-20 66-31099 <=$>
1965 V24 N12 P70-73 66-31127 <=$>
1965 V24 N12 P74-77 66-31128 <=$>
1966 N12 P3-8 67-31554 <=$>
1966 N12 P13-18 67-31554 <=$>
1966 N12 P27-30 67-31554 <=$>
1966 N12 P43-49 67-31554 <=$>
1966 N12 P67-70 67-31554 <=$>
1966 V25 N11 P3-26 67-30385 <=>
1966 V25 N11 P92-93 67-30385 <=>

SOVETSKOE ZDRAVOOKHRANENIE KIRGIZII
1962 N2 P34-49 62-32466 <=*>
1962 N6 P3-7 63-21459 <=>
1962 N6 P27-29 63-21459 <=>
1964 N3 P17-31 64-51658 <=$>
1964 N3 P42-45 64-51658 <=$>
1964 N3 P50-56 64-51658 <=$>
1964 N5 P24-27 65-30144 <=$>
1965 N2 P21-26 65-32276 <=*$>

SOVETY DEPUTATOV TRUDIASHCHIKASIA. MOSCOW
1963 N5 P49-51 63-31110 <=>

SOVKHOZNOE PROIZVODSTVO
1946 V6 N2/3 P30 RT-3145 <*>
1954 N1 P88-89 RT-3308 <*> 0
1957 N8 P31-36 61-19919 <*=>
1959 N10 P52-55 60-31124 <=> 0
1960 N6 P14-18 60-31718 <=>
1960 N6 P27-30 60-31718 <=>
1960 N7 P53-56 61-21339 <=>
1960 N8 P62-63 61-11625 <=>
1960 N10 P54-56 61-21339 <=>
1960 N12 P4-23 61-23584 <=*>
1961 N2 P2-4 61-27130 <=>
1961 N2 P13-18 61-27130 <=>
1961 N3 P2-4 61-27078 <=>
1961 N3 P16-21 61-27078 <=>
1961 N3 P33-36 61-27078 <=>

SOVREMENNA MEDITSINA
SEE SUVREMENNA MEDITSINA

SOVREMENNA MEDITSINA, USSR
1862 V3 N20 P377-380 59-22172 <+*>
1862 V3 N22 P417-423 59-22172 <+*>
1862 V3 N23 P433-439 59-22172 <+*>
1862 V3 N24 P452-458 59-22172 <+*>

SOVREMENNAYA PSIKHONEVROLOGIYA
1928 V7 P272-277 60-21756 <=>
1929 V8 P531-537 R-1086 <*>

SOVREMENNYE PROBLEMY ONKOLOGII
1955 V7 N4 P3-13 62-15113 <=*>
1955 V7 N5 P3-9 62-15114 <=*>
1956 V8 N1 P3-14 62-15111 <*=>
1956 V9 N4 P3-25 60-23231 <+*>

SOVREMENNYE PROBLEMY TUBERKULEZA
1955 V6 N1 P3-8 62-15731 <*=>
1955 V6 N5 P3-8 62-15730 <*=>
1955 V6 N6 P3-8 62-23545 <=*>
1956 V7 N1 P3-10 62-15732 <*=>
1956 V7 N4 P3-13 62-23501 <=*>

SOWJETWISSENSCHAFT.
1949 V2 N3 P113-149 62-25671 <=*>

SPARWIRTSCHAFT
1930 V8 N12 P563-568 57-3082 <*>

SPECIAL REPORT ON PRE-EXAMINATION OF BLIGHTS AND HARMFUL INSECTS. JAPAN
SEE BYOGAICHU HASSEI YOSATSU TOKUBETSU HOKOKU

SPECTROCHIMICA ACTA
1939 V1 P249-269 I-488 <*>
1939 V1 N3 P239-248 62-18664 <*> 0
1940 V1 N4 P318-322 59-15190 <*>
1940 V1 N5 P400-402 I-874 <*>
1941 V1 P548-559 I-642 <*>
1941 V2 P92-97 65-14928 <*> 0
1944 N2 P269-290 12J17I <ATS>
1944 V2 P333-339 I-82 <*>
1944 V2 P396-416 AEC-TR-2870 <*>
1947 V3 P18-39 62-16939 <*> 0
1947 V3 N1 P40-67 62-18359 <*> 0
1948 V3 P214-232 I-190 <*>
1948 V3 N2 P214-232 58-97 <*>
1950 V4 P85-92 I-621 <*>
59-10570 <*> 0
1950 V4 P237-251 I-361 <*>
1950 V4 P237 58-1662 <*>
1952 V5 P114-123 TRANS-567 <MT>
1952 V5 N4 P322-326 64-20317 <*> 0
1954 V6 P288-301 59-10592 <*> 0
1955 V7 P25-31 3005 <*>
1955 V7 P32-44 3006 <*>
1955 V7 P128-133 64-14930 <*>
1957 V10 N1 P61-69 61-10527 <*>
1958 V12 N4 P305-320 64-30966 <*> 0
1959 N8 P549-556 TS-1569 <BISI>
1959 V15 P454-460 AEC-TR-5285 <*>
1961 V17 P352-355 UCRL TRANS-994(L) <*>
1961 V17 P523-529 63-00206 <*>
1962 V18 P549-560 63-10291 <=>
1962 V18 N2 P183-199 62-16684 <*>
1963 V19 N2 P533-540 3346 <BISI>
1964 V20 P1815-1828 707 <TC>
1964 V20 N5 P785-798 140 <TC>
1965 V21 N1 P141-153 65-14081 <*> 0

SPELEON
1950 V1 N2/3 P3-22 61-16516 <*>

SPERIMENTALE

1953 V103 P131-135 57-84 <*>
1956 V106 N2 P183-188 58-1709 <*>
1957 V107 N1 P12-23 47P59I <ATS>
1957 V107 N1 P24-32 46P59I <ATS>
1959 V109 N4 P383-394 61-20909 <*>
1962 V112 N6 P446-456 63-01107 <*>
1965 V115 N3 P127-140 66-12677 <*> 0

SPIEGEL. HANOVER, HAMBURG
1959 N12 P89-91 65-11233 <*> 0
1960 V14 N47 P42-57 63-16622 <*> 0
1963 V17 N5 P32-44 63-00860 <*>

SPINNER UND WEBER
1956 V74 N10/1 P455-458 61-14534 <*> 0
1957 V75 N6 P240-243 61-14535 <*> 0
1957 V75 N18 P949-950 61-14536 <*> 0
1957 V75 N24 P1323-1325 61-14537 <*>

SPINNER UND WEBER + TEXTILVEREDLUNG
1960 V78 P398-403 C-3739 <NRCC>
1961 V79 N10 P944-948 31P63G <ATS>

SPIRTOVAYA PROMYSHLENNOST
1956 V22 N3 P36-37 71K22R <ATS>
1956 V22 N4 P10-11 718TM <CTT>
1961 N4 P21-25 61-28260 <*=>
1961 N5 P42-43 61-28584 <*=>
1963 N1 P27-28 63-21623 <=>

SPISANIE NA BULGARSKATA AKADEMIYA NA NAUKITE
1960 V5 N3 P3-23 62-19347 <=*>
1961 V6 N1 P95-98 62-19614 <=*>
1961 V6 N2 P15-20 62-23806 <=*>
1962 V7 N4 P50-57 63-21874 <=>
1963 V8 N4 P30-34 64-31966 <=>
1964 V9 N1/2 P4-46 64-51848 <=$>

SPISY LEKARSKE FAKULTY MASARYKOVY UNIVERSITY. BRNO
1957 V30 N5 P193-198 62-01038 <*>
1961 V34 N4 P125-137 65-14641 <*> 0

SPISY VYDAVANE PRIRODOVEDECKOU FAKULTOU KARLOVY UNIVERSITY
1926 N56 P3-15 34J15C <ATS>

SPISY VYDAVANE PRIRODOVEDECKOU FAKULTOU MASARYKOVY UNIVERSITY
1951 N331 P231-268 AL-958 <*>

SPOOR- EN TRAMWEGEN
1951 V24 N18 P295-296 57-983 <*>
1955 V28 P193-195 T-1879 <INSD>

SPORTAERZTLICHE PRAXIS
1960 V3 P12-19 63-26249 <=*$>

SPRACHE IM TECHNISCHEN ZIETALTU
1963 N7 P547-556 65-12114 <*>

SPRAWOZDANIA Z POSIEDZEN TOWARZYSTWA NAUKOWEGO WARSZAWSKIEGO
1939 V32 P102-109 66-11251 <*>

SPRECHSAALL FUER KERAMIK, GLAS, EMAIL
1932 V65 P8-9 II-618 <*>
1932 V65 N52 P925-926 62-18120 <*>
1933 V66 P573-574 63-14352 <*> 0
1933 V66 P591-594 63-14352 <*> 0
1937 V70 P29-30 60-10264 <*>
1938 V71 P185-187 60-10265 <*>
1939 V72 N35 P453-454 60-18600 <*>
1940 V73 P153-157 63-18129 <*>
1941 V74 N25 P246-248 64-10403 <*>
1941 V74 N26 P255-256 64-10402 <*>
1942 V75 N13/4 P120-125 60-18769 <*>
63-20365 <*> 0
1942 V75 N9/10 P82-85 63-18427 <*>
1950 V83 N2 P21-23 62-16470 <*> P
1950 V83 N7 P121-122 62-14885 <*>
1950 V83 N21 P436 65-12865 <*>
1952 V85 N3 P53-54 57-730 PT3 <*>
1953 V86 N16 P389-392 62-10555 <*>
1954 V87 N2 P25-26 BGIRA-609 <BGIR>
1954 V87 N9 P479-481 II-882 <*>
1955 V88 P265-266 NS-457 <TTIS>
1955 V88 N4 P72 62-10557 <*>
1955 V88 N9 P188-191 62-10924 <*>
1956 V89 P138-140 NS-456 <TTIS>
1956 V89 N1 P3-4 62-10918 <*>
1956 V89 N14 P329-330 62-10943 <*>
1957 V90 N1 P26-28 64-14025 <*>
1957 V90 N15 P365-367 60-14029 <*> 0
1958 V91 N6 P112-114 65-11290 <*>
1959 V92 N12 P310-312 60-16838 <*>
1959 V92 N24 P615-619 64-14022 <*>
1960 V93 N6 P159-163 UCRL TRANS-644(L) <*>
1961 V94 N7 P130-132 64-14066 <*>
1962 V95 N11 P316-326 AI-TR-30 <*>
1962 V95 N12 P347-348 63-10615 <*>
64-14286 <*>
1962 V95 N17 P464-467 BGIRA-557 <BGIR>
1962 V95 N18 P484-487 BGIRA-558 <BGIR>
1962 V95 N19 P518-520 BGIRA-559 <BGIR>
1962 V95 N24 P683-688 63-18804 <*>
1962 V95 N24 P691-694 BGIRA-560 <BGIR>
1963 V96 N2 P36-39 BGIRA-561 <BGIR>

SPRECHSAAL FUER KERAMIK, GLAS, EMAIL, SILIKATE
1964 V97 N1 P15-16 BGIRA-619 <BGIR>
1964 V97 N6 P123-126 BGIRA-620 <BGIR>
1964 V97 N16 P451-453 NS-357 <TTIS>
65-13049 <*>
1964 V97 N18 P538-539 65-10460 <*>
1964 V97 N18 P555-557 65-10903 <*> 0
1965 V98 N4 P77-78 65-14138 <*>
1965 V98 N12 P338-344 66-11528 <*>

SPRENGTECHNIK
1952 N12 P221-227 59-12440 <+*>

SREDEN MEDITSINSKI RABOTNIK
1962 N6 P3-13 63-15383 <=>

SREDNEE SPETSIALNOE OBRAZOVANIE
1963 V10 N8 P23-30 64-21792 <=>
1963 V10 N8 P34-38 64-21792 <=>
1966 V13 N1 P40-44 66-32492 <=*>

STAAT UND RECHT
1963 V12 N4 P593-636 63-21878 <=>
1965 V14 N10 P1609-1622 65-33735 <=*$>
1965 V14 N10 P1622-1636 65-33736 <=*$>

STAEDTEHYGIENE
1954 V5 N1 P5-6 NP-TR-738 <*>
1959 V10 N9 P178-180 63-20457 <*> 0

STAERKE
1951 V3 N5 P112-115 60-10115 <*>
1951 V3 N7 P202-210 63-14970 <*>
1953 V5 N3 P65-69 59-20793 <*> 0
1954 V6 N5 P87-90 64-20449 <*> 0
1958 V10 P79-86 58-2643 <*>
1958 V10 P119 58-2627 <*>
1958 V10 N2 P39-41 60-14518 <*> 0
1960 V12 N7 P197-201 NP-TR-689 <*>
1960 V12 N8 P237-243 74M47G <ATS>
1960 V12 N9 P257-265 75M47G <ATS>
1960 V12 N12 P351-358 62-18761 <*> 0
1961 V13 N5 P174-181 62-16498 <*> 0
1964 V16 P351-359 TT-664 <TC>
1965 V17 P284-289 1436 <TC>

STAHL UND EISEN
1907 V27 N37 P1309-1315 AL-891 <*>
1907 V27 N37 P1347-1353 AL-891 <*>
1912 V32 P816-822 58-211 <*>
1912 V32 P863-867 58-211 <*>
1917 V37 N19 P442-448 AL-892 <*>

Citation	Translation
1917 V37 N19 P474-479	AL-892 <*>
1917 V37 N19 P497-499	AL-892 <*>
1918 V38 P316-317	63-14600 <*>
1919 V39 N49 P1497-1506	63-16076 <*> O
1923 V43 P1191-1199	57-3248 <*>
	57-3461 <*>
1923 V43 N33 P1073-1075	63-16737 <*>
1924 V44 P590-592	65-14764 <*>
1924 V44 P1250	63-16778 <*>
1924 V44 P1687-1694	TS 1245 <BISI>
1924 V44 N12 P309-311	66-12381 <*>
1925 V45 N28 P1110-	57-3350 <*>
1925 V45 N3/4 P79-86	57-2064 <*>
1925 V45 N3/4 P109-114	57-2064 <*>
1926 V46 P1423-1428	58-337 <*>
1926 V46 N23 P782-784	66-13816 <*>
1926 V46 N52 P1886	66-12392 <*>
1928 V48 N39 P1372-1373	2655 <*>
1928 V48 N49 P1705-1712	59-20053 <*> O
1930 V50 P223-238	57-166 <*>
1930 V50 P1741-1744	57-2639 <*>
1930 V50 N43 P1501	59-20412 <*> O
1930 V50 N46 P1611-1616	59-20411 <*> O
1931 V51 P104-106	57-3126 <*>
1931 V51 P189-196	57-3091 <*>
1931 V51 P1033-1034	3049 <BISI>
1931 V51 N30 P944-945	59-20414 <*> O
1931 V51 N40 P1221-1228	35 <HB>
	35 REISSUE <HB>
1931 V51 N41 P1256-1263	35 <HB>
	35 REISSUE <HB>
1932 V52 P220-231	58-1893 <*>
1932 V52 P625-631	II-692 <*>
1932 V52 N3 P71	63-16754 <*>
1932 V52 N20 P490-492	59-20592 <*>
1932 V52 N22 P529-539	II-603 <*> P
1932 V52 N26 P625-633	II-691 <*>
1932 V52 N35 P845-849	57-1062 <*>
1933 V53 P1049-1052	63-16771 <*>
1933 V53 N33 P849-856	59-20224 <*> O
1933 V53 N51 P1330-1332	63-18826 <*>
1934 V54 N19 P462-466	AL-886 <*>
1934 V54 N30 P773-775	II-644 <*>
1934 V54 N31 P797-801	58-1908 <*>
1934 V54 N43 P1110-1111	62-18487 <*> O
1934 V54 N50 P1289-1291	27G6G <ATS>
1935 V55 N13 P349-351	336 REISSUE <HB>
	366 <HB>
1935 V55 N16 P444-448	58-2118 <*>
1935 V55 N21 P557-564	59-20898 <*>
1935 V55 N22 P586-589	59-20898 <*>
1935 V55 N23 P616-623	59-20581 <*>
1935 V55 N24 P648-653	59-20582 <*>
1935 V55 N25 P670-680	59-20897 <*>
1936 V56 P490-492	4499 <BISI>
1936 V56 N15 P444-446	59-20577 <*>
1937 V57 P261-269	1973A <BISI>
1937 V57 P296-300	1973B <BISI>
1937 V57 P889-899	AL-950 <*>
1937 V57 P1245-1248	62-10291 <*>
1937 V57 P1269-1279	62-10291 <*>
1937 V57 N21 P593-601	58-2163 <*>
1937 V57 N26 P732-735	63-16824 <*>
1937 V57 N32 P889-899	59-10684 <*> O
1937 V57 N41 P1178-1179	58-1932 <*>
1938 V58 P157-165	61-18454 <*>
1938 V58 P265-275	62-14462 <*>
1938 V58 P546-549	2688 <BISI>
1938 V58 P1239-1250	AL-951 <*>
1938 V58 N12 P313-316	65-13179 <*>
1939 V59 P1-8	57-1152 <*>
1939 V59 P537-548	58-1895 <*>
1939 V59 P1155-1157	AEC-TR-6394 <*>
1939 V59 N1 P54-63	61-18406 <*>
1939 V59 N27 P785-790	II-156 <*>
1939 V59 N38 P1057-1067	60-18203 <*>
1939 V59 N38 P1067-1069	59-20413 <*> O
1939 V59 N44 P1197-1203	60-18595 <*> O
1940 V60 P245-252	2134 <BISI>
1940 V60 P634-640	61-18533 <*> O
1940 V60 P655-658	61-18533 <*> O
1940 V60 P724-727	57-1043 <*>
1940 V60 P877-880	61-18540 <*>
1940 V60 P921-929	61-18562 <*>
1940 V60 P948-955	61-18562 <*>
1940 V60 P1021-1027	61-18535 <*>
1940 V60 P1069-1075	61-18854 <*>
1940 V60 P1075-1083	61-18530 <*> O
1940 V60 P1107-1113	61-18854 <*>
1940 V60 N38 P844-	57-163 <*>
	57-169 <*>
1940 V60 N46 P1027-1037	57-933 <*>
1941 V61 P53-63	61-18538 <*>
1941 V61 P73-83	61-18537 <*>
1941 V61 P100-107	61-18537 <*>
1941 V61 P129-136	61-18521 <*> O
1941 V61 P164-170	61-18521 <*> O
1941 V61 P185-187	61-18552 <*> O
1941 V61 P465-473	61-18561 <*>
1941 V61 P529-535	61-18525 <*> O
1941 V61 P597-606	61-18529 <*> O
1941 V61 P606-609	61-18563 <*>
1941 V61 P624-630	61-18529 <*> O
1941 V61 P649-653	61-18528 <*> O
1941 V61 P654-658	61-18565 <*> O
1941 V61 P671-680	61-18524 <*> O
1941 V61 P756-759	61-18567 <*> O
1941 V61 P769-776	61-18559 <*> O
1941 V61 P777-778	61-18552 <*> O
1941 V61 P785-791	61-18560 <*> O
1941 V61 P801-806	61-18564 <*> O
1941 V61 P909-919	61-18469 <*> O
1941 V61 P929-937	I-569 <*>
	61-18513 <*> O
1941 V61 P949-956	I-569 <*>
	61-18513 <*> O
1941 V61 P1033-1035	61-18493 <*>
1941 V61 P1073-1078	61-18460 <*>
1941 V61 P1105-1109	61-18512 <*>
1942 V62 P81-89	61-18553 <*> O
1942 V62 P115-116	61-18553 <*> O
1942 V62 P301-307	61-18502 <*> C
1942 V62 P774-779	61-18550 <*> O
1942 V62 P921-923	61-18492 <*>
1942 V62 P983-986	61-18471 <*>
1942 V62 P997-1001	61-18483 <*>
1942 V62 P1022-1033	61-18484 <*> C
1942 V62 P1083-1090	61-18479 <*> O
1942 V62 N2 P21-30	63-20155 <*>
1942 V62 N17 P347-352	60-10190 <*> O
1942 V62 N27 P576	I-761 <*>
1943 V63 P21-30	61-18475 <*>
1943 V63 P85-94	61-18459 <*> O
1943 V63 P110-113	61-18481 <*>
1943 V63 P113-114	61-18470 <*>
1943 V63 P189-199	61-18458 <*> O
1943 V63 P217-220	61-18516 <*> O
1943 V63 P229-236	61-18453 <*>
1943 V63 N36 P653-659	57-3000 <*>
1943 V63 N38 P695-700	60-10169 <*> O
1944 V64 P285-290	61-18545 <*>
1944 V64 P399-404	61-18547 <*>
1944 V64 P413-419	61-18548 <*>
1945 V65 N9/10 P118-121	60-18211 <*> O
1947 V66 P87-90	62-16915 <*> C
1947 V66 N11 P171-180	62-20338 <*> O
1947 V66 N15 P244-250	62-18401 <*> O
1947 V67 P307-312	II-617 <*>
1947 V67 N25/6 P411-416	60-18219 <*>
1947 V66/7 P250-255	57-2834 <*>
1947 V66/7 P255-257	57-2626 <*>
1947 V66/7 N25/6 P411-416	63-16299 <*> O
1948 V68 N17/8 P287-294	TRANS-120 <MT>
1949 V69 N1 P1-8	60-18603 <*>
1949 V69 N1 P19-22	62-18230 <*> O
1949 V69 N2 P49-53	63-16780 <*>
1949 V69 N9 P306-308	60-18604 <*>
1949 V69 N10 P319-325	63-16740 <*>
1949 V69 N13 P443-450	60-18226 <*>
1949 V69 N14 P464-465	60-18227 <*>

Citation	Translation
1949 V69 N14 P468-475	2485 <*>
1949 V69 N14 P468-474	60-18227 <*>
1949 V69 N14 P468-475	62-18232 <*> O
1949 V69 N15 P503-508	60-18228 <*>
1949 V69 N22 P759-762	60-18605 <*>
1949 V69 N22 P762-764	60-18606 <*>
1949 V69 N23 P813-819	60-18231 <*> O
1949 V69 N23 P827-835	60-18232 <*>
1950 V70 P1208-1211	1816 <*>
1950 V70 N2 P41-51	62-18356 <*> O
1950 V70 N11 P452-459	59-20565 <*> O
1950 V70 N13 P541-543	60-18607 <*>
1950 V70 N13 P552-561	60-18608 <*>
1950 V70 N13 P561-565	60-18609 <*> O
1950 V70 N14 P582-596	62-18155 <*>
1950 V70 N15 P633-640	60-18610 <*>
1950 V70 N18 P765-767	4445 <BISI>
	63-16825 <*> O
1950 V70 N21 P925-929	59-20566 <*>
1950 V70 N22 P995-1004	60-18611 <*> O
1950 V70 N22 P995-1003	63-16763 <*>
1950 V70 N25 P1166-1174	60-18615 <*> O
1951 V71 P283-287	3299 <HB>
1951 V71 P568-575	2160 <BISI>
1951 V71 N3 P125-128	57-3313 <*>
	58-1460 <*>
1951 V71 N4 P157-170	60-18612 <*>
	62-18436 <*> O
1951 V71 N12 P597-605	60-18613 <*>
1951 V71 N12 P619-	57-2627 <*>
1951 V71 N13 P664-669	60-18613 <*>
1951 V71 N13 P669	58-996 <*>
1951 V71 N16 P836-839	5070 <HB>
1951 V71 N20 P1040-1044	AL-275 <*>
1951 V71 N21 P1081-1090	60-18614 <*> O
	61-16354 <*>
1951 V71 N21 P1090-1097	61-16164 <*> O
1951 V71 N21 P1103-1114	63-16779 <*>
1951 V71 N21 P1114-1115	62-16051 <*>
1951 V71 N23 P1199-1204	60-18616 <*>
1951 V71 N23 P1212-1218	60-18617 <*>
1951 V71 N23 P1219-1225	62-16981 <*> O
1952 V72 P243-245	63-16829 <*>
1952 V72 P425-426	2349 <BISI>
1952 V72 P459-466	63-16834 <*>
1952 V72 P898-904	63-16835 <*>
1952 V72 P1217-1221	2313 <BISI>
1952 V72 P1261-1267	TS 599 <BISI>
	61-16346 <*>
1952 V72 P1268-1277	TS 463 <BISI>
1952 V72 P1609-1610	T1924 <INSD>
1952 V72 N1 P10-12	60-18618 <*> O
1952 V72 N2 P66-69	63-10005 <*> O
1952 V72 N2 P70-75	59-17132 <*> O
	60-18619 <*> O
1952 V72 N2 P75-79	60-18620 <*>
1952 V72 N4 P176-185	AL-464 <*>
	60-18621 <*> O
1952 V72 N6 P284-287	60-18622 <*>
1952 V72 N7 P341-345	60-18623 <*> O
1952 V72 N7 P345-347	3354 <HB>
1952 V72 N9 P466-475	60-18624 <*> O
1952 V72 N10 P561-569	60-18625 <*>
1952 V72 N10 P597-605	61-16363 <*>
1952 V72 N10 P611-616	60-18626 <*>
1952 V72 N12 P663-668	60-18233 <*>
1952 V72 N12 P683-687	60-18627 <*> O
1952 V72 N16 P935-941	60-18234 <*> O
1952 V72 N20 P1193-1195	58-1010 <*>
1952 V72 N21 P1268-1277	60-18235 <*>
1952 V72 N21 P1278-1285	61-20602 <*>
1952 V72 N21 P1285-1298	60-18236 <*>
1952 V72 N23 P1418-1426	60-18237 <*> O
1952 V72 N24 P1509-1513	60-18238 <*>
1952 V72 N25 P1577-1579	60-18239 <*> O
1952 V72 N25 P1583-1587	60-18240 <*>
1952 V72 N25 P1587-1595	62-18437 <*> OP
1952 V72 N26 P1633-1642	60-18241 <*> O
1953 V73 P81-84	TS 1253 <BISI>
	61-20614 <*>
1953 V73 P84-91	2312 <BISI>
1953 V73 P1169-1174	63-16841 <*>
1953 V73 P1404-1409	2395 <BISI>
1953 V73 P1426-1428	II-95 <*>
1953 V73 P1706-1717	63-16842 <*>
1953 V73 N1 P6-22	62-18936 <*> O
1953 V73 N2 P65-80	62-16946 <*> O
1953 V73 N4 P219-222	59-17041 <*> O
	60-10175 <*> O
1953 V73 N5 P257-266	60-18242 <*>
	62-20186 <*> O
1953 V73 N5 P279-283	57-2612 <*>
1953 V73 N5 P283-292	60-18243 <*> O
1953 V73 N6 P360-364	63-16839 <*>
1953 V73 N8 P461-469	62-18935 <*> O
1953 V73 N8 P485-492	AEC-TR-6141 <*>
1953 V73 N8 P492-494	60-18244 <*>
1953 V73 N10 P621-629	62-18438 <*> O
1953 V73 N11 P721-727	60-18245 <*> O
1953 V73 N14 P895-902	60-18246 <*> O
1953 V73 N18 P1169-1174	CSIRO-3851 <CSIR>
1953 V73 N21 P1342-1349	60-18247 <*>
1953 V73 N23 P1446-1452	61-18659 <*>
1953 V73 N23 P1453-1457	60-18248 <*> O
1953 V73 N23 P1457-1462	60-18254 <*> O
1953 V73 N23 P1463-1464	60-18255 <*> O
1953 V73 N23 P1468-1472	60-18249 <*> O
1953 V73 N23 P1496-1503	60-18250 <*> O
1953 V73 N25 P1654-1657	57-2638 <*>
	60-18251 <*>
1953 V73 N26 P1717-1720	60-18252 <*>
1954 V74 P1054-1062	T2029 <INSD>
1954 V74 P1402-1413	T1773 <INSD>
1954 V74 P1492-1502	TS-1176 <BISI>
1954 V74 N2 P89-94	3666 <HB>
1954 V74 N3 P133-145	59-20940 <*>
1954 V74 N7 P396-402	59-20939 <*>
1954 V74 N14 P881-888	3445 <HB>
1954 V74 N17 P1045-1054	60-18253 <*> O
1954 V74 N18 P1062-1069	2001 <BISI>
1954 V74 N23 P1492-1502	61-20661 <*>
1954 V74 N23 P1502-1509	60-18256 <*> O
1954 V74 N23 P1515-1521	1184 <*>
	60-10167 <*> O
1954 V74 N24 P1659-1661	3019 <BISI>
1955 V75 P9-24	1805 <BISI>
1955 V75 P70-75	3273 <*>
1955 V75 P478-485	TS 1700 <BISI>
	62-16855 <*>
1955 V75 P587-590	T-1624 <INSD>
1955 V75 P633-640	2355 <BISI>
1955 V75 P682-690	1874 <BISI>
1955 V75 P958-974	TS-1808 <BISI>
1955 V75 P1273-1274	T-2469 <INSD>
1955 V75 P1300-1310	T-2010 <INSD>
1955 V75 P1324-1330	57-3303 <*>
1955 V75 P1627-1640	GB41/509 <NLL>
1955 V75 N2 P70-75	60-18257 <*>
	62-20069 <*>
1955 V75 N3 P141-144	64-20513 <*> O
1955 V75 N3 P144-162	61-16415 <*>
1955 V75 N8 P502-513	58-1165 <*>
	59-12126 <=*>
1955 V75 N9 P549-559	61-16357 <*>
1955 V75 N9 P559-570	60-18259 <*> O
1955 V75 N11 P691-693	4069 <HB>
	60-10038 <*> O
1955 V75 N11 P709-718	60-18260 <*> O
1955 V75 N12 P779-784	61-16415 <*>
1955 V75 N14 P900-906	60-18258 <*>
1955 V75 N17 P1101-1106	62-14834 <*> O
1955 V75 N19 P1252-1263	C-2430 <NRC>
	57-1767 <*>
1955 V75 N20 P1295-1300	60-18261 <*>
1955 V75 N20 P1317-1324	3705 <HB>
1955 V75 N22 P1452-1460	66-10718 <*> O
1955 V75 N22 P1480-1492	61-16298 <*>
1955 V75 N22 P1494-1501	61-16355 <*>
1955 V75 N23 P1553-1557	63-16865 <*>
1955 V75 N23 P1571-1582	57-117 <*>

	59-17304 <*> 0
1955 V75 N24 P1627-1640	60-18264 <*>
1955 V75 N25 P1691-1695	5391 <HB>
1955 V75 N25 P1696-1701	5352 <HB>
1955 V75 N25 P1701-1705	5353 <HB>
1955 V75 N25 P1705-1710	58-738 <*>
	61-16331 <*> 0
1955 V75 N25 P1710-1718	61-16415 <*>
1955 V75 N26 P1774-1784	57-2533 <*>
1956 P971-976	57-3139 <*>
1956 P1690-1698	57-1545 <*>
1956 V76 P133-144	T1830 <INSD>
1956 V76 P398-402	61-16348 <*>
1956 V76 P968-970	61-16380 <*>
1956 V76 P1032-1040	61-18449 <*>
1956 V76 P1085-1099	2314 <BISI>
1956 V76 P1290-1292	2315 <BISI>
1956 V76 P1453-1456	TS 1215 <BISI>
1956 V76 N1 P1-13	61-16367 <*>
1956 V76 N1 P41-44	3795 <HB>
1956 V76 N2 P61-68	GB70 4247 <NLL>
	59-20981 <*>
1956 V76 N2 P68-78	60-18263 <*>
1956 V76 N3 P125-133	61-18385 <*>
1956 V76 N5 P257-261	66-10160 <*> 0
	66-11786 <*> 0
1956 V76 N7 P393-397	60-10185 <*>
1956 V76 N10 P588-595	63-16850 <*>
1956 V76 N13 P789-799	61-16356 <*>
1956 V76 N13 P799-805	60-10242 <*> 0
1956 V76 N15 P968-970	62-20118 <*> 0
1956 V76 N16 P1028-1032	4048 <HB>
1956 V76 N16 P1040-1049	3807 <HB>
1956 V76 N16 P1049-1052	3808 <HB>
1956 V76 N17 P1107-1116	6846 <HB>
1956 V76 N17 P1138	3829 <HB>
1956 V76 N19 P1229-1231	61-16359 <*>
1956 V76 N19 P1231-1246	60-18264 <*>
1956 V76 N22 P1410-1416	61-16338 <*>
1956 V76 N22 P1426-1441	60-18265 <*> 0
1956 V76 N22 P1442-1454	61-16288 <*>
1956 V76 N22 P1453-1456	<MT>
	61-20600 <*>
1956 V76 N23 P1570-1573	59-20847 <*>
1956 V76 N24 P1616-1628	4065 <BISI>
1956 V76 N25 P1665-1668	61-16365 <*>
1956 V76 N25 P1698-1700	61-16364 <*>
1956 V76 N26 P1721-1728	3900 <HB>
	59-10265 <*>
	60-18266 <*>
1956 V76 N26 P1728-1734	59-20987 <*>
1956 V76 N26 P1734-1740	61-16293 <*>
1957 P24-32	TS-923 <BISI>
1957 P359-362	TS-1404 <BISI>
1957 P435-438	TS-1404 <BISI>
1957 P512-525	TS-1404 <BISI>
1957 P562-567	TS-1234 <BISI>
1957 P837-845	58-2499 <*>
1957 P853-859	57-3049 <*>
1957 P1476-1482	58-1630 <*>
1957 V77 P15-23	TS 922 <BISI>
	61-18423 <*>
1957 V77 P24-28	61-18403 <*>
1957 V77 P133-143	63-16855 <*>
1957 V77 P294-296	61-16384 <*>
1957 V77 P359-362	62-10498 <*>
1957 V77 P421-426	61-18793 <*>
1957 V77 P435-438	62-10498 <*>
1957 V77 P496-503	61-16295 <*>
1957 V77 P512-525	62-10498 <*>
1957 V77 P805-813	61-16332 <*>
1957 V77 P867-881	61-16417 <*>
1957 V77 P917-926	61-16337 <*>
1957 V77 P926-931	61-16334 <*>
1957 V77 P1006-1018	61-16361 <*>
1957 V77 P1660	59-10489 <*>
1957 V77 P1795-1802	61-18392 <*>
1957 V77 N1 P11-15	2486 <BISI>
1957 V77 N1 P33-36	61-16369 <*>
1957 V77 N1 P36-43	61-16299 <*>
1957 V77 N2 P69-78	61-16330 <*>
1957 V77 N2 P84-91	61-16339 <*>
1957 V77 N2 P92-95	61-16379 <*>
1957 V77 N2 P95-100	61-16352 <*>
1957 V77 N4 P197-204	59-20938 <*>
1957 V77 N4 P204-215	61-16383 <*>
1957 V77 N6 P324-334	61-16296 <*>
1957 V77 N7 P394-408	61-16345 <*>
	62-18042 <*> 0
1957 V77 N7 P409-421	3466 <BISI>
1957 V77 N8 P469-476	61-16290 <*> 0
1957 V77 N8 P483-486	61-16285 <*>
1957 V77 N8 P487-491	61-16289 <*>
1957 V77 N9 P562-567	62-10466 <*> P
1957 V77 N9 P567-576	61-16297 <*>
1957 V77 N9 P576-581	61-16333 <*>
1957 V77 N9 P591-593	61-16276 <*>
1957 V77 N10 P643-651	61-16377 <*>
1957 V77 N11 P685-693	61-16368 <*>
1957 V77 N11 P693-701	61-16366 <*>
1957 V77 N11 P715-727	61-16353 <*>
1957 V77 N11 P727-734	61-16418 <*>
1957 V77 N12 P773-783	61-16362 <*>
1957 V77 N13 P859-867	59-15077 <*>
1957 V77 N14 P926-931	45J18G <ATS>
1957 V77 N15 P988-998	61-16414 <*>
1957 V77 N15 P998-1006	61-16413 <*>
1957 V77 N16 P1064-1069	61-16419 <*>
1957 V77 N16 P1070-1074	61-16375 <*>
1957 V77 N17 P1117-1122	61-16386 <*>
1957 V77 N17 P1126-1135	61-16385 <*>
1957 V77 N17 P1146-1160	61-16341 <*>
1957 V77 N18 P1181-1195	61-16347 <*>
1957 V77 N18 P1204-1209	61-16382 <*>
1957 V77 N18 P1215-1220	59-10241 <*>
	61-16381 <*>
1957 V77 N18 P1233-1244	61-16389 <*>
1957 V77 N19 P1284-1296	61-16388 <*>
1957 V77 N19 P1296-1303	61-16387 <*>
1957 V77 N19 P1303-1308	61-16376 <*>
1957 V77 N19 P1315-1329	61-16373 <*>
1957 V77 N20 P1356-1362	61-18388 <*>
1957 V77 N20 P1368-1374	4117 <HB>
1957 V77 N20 P1376	5998 <HB>
1957 V77 N21 P1442-1450	4079 <HB>
1957 V77 N21 P1451-1459	61-16360 <*>
1957 V77 N21 P1464-1476	61-16378 <*>
1957 V77 N21 P1483-1487	4110 <HB>
1957 V77 N21 P1488-1496	61-16370 <*>
1957 V77 N21 P1497-1499	61-16358 <*>
1957 V77 N22 P1577-1582	61-16392 <*>
1957 V77 N22 P1583-1593	61-16391 <*>
1957 V77 N22 P1607-1610	61-16390 <*>
1957 V77 N23 P1675-1686	61-18448 <*>
1957 V77 N23 P1686-1690	4471 <HB>
1957 V77 N23 P1692-1698	61-16434 <*>
	892 <BISI> P
1957 V77 N23 P1698-1699	4079 <HB>
1957 V77 N24 P1717-1728	61-16336 <*>
1957 V77 N24 P1729-1733	61-16226 <*>
1957 V77 N24 P1740-1747	1595 <BISI>
1957 V77 N24 P1752-1759	<MT>
1957 V77 N25 P1817-1830	4100 <HB>
1957 V77 N26 P1863-1867	61-16350 <*>
1957 V77 N26 P1868-1873	61-16351 <*>
1957 V77 N26 P1873-1877	61-16349 <*>
1957 V77 N26 P1885-1887	61-16372 <*>
1957 V77 N26 P1895-1896	59-10270 <*>
1958 P178-180	58-1262 <*>
1958 P1678-1690	TS-1355 <BISI>
1958 V78 P955-960	61-18377 <*>
1958 V78 P1033-1038	TS-1359 <BISI>
1958 V78 P1045-1064	TS 1170 <BISI>
1958 V78 P1108-1109	61-18382 <*>
1958 V78 P1165-1169	61-18376 <*>
1958 V78 P1169-1175	61-18378 <*>
1958 V78 P1191-1200	TS 1071 <BISI>
1958 V78 P1389-1395	UCRL TRANS-780(L) <*>
1958 V78 P1456-1462	61-18743 <*>
1958 V78 P1536-1546	TS 1143 <BISI>

Citation	Translation
1958 V78 P1564-1574	2189 <BISI>
1958 V78 P1662-1670	1152 <BISI>
1958 V78 P1670-1676	TS 1153 <BISI>
1958 V78 P1678-1690	62-10499 <*>
1958 V78 P1734-1745	TS-1167 <BISI>
1958 V78 P1793-1798	TS-1275 <BISI>
	61-20620 <*>
1958 V78 P1798-1808	TS-1276 <BISI>
	61-20607 <*>
1958 V78 P1812-1815	TS 1476 <BISI>
1958 V78 P1881-1891	TS 1241 <BISI>
1958 V78 N1 P14-21	58-1584 <*>
1958 V78 N1 P21-27	61-16343 <*>
1958 V78 N1 P27-34	61-16371 <*>
1958 V78 N1 P35-39	61-18880 <*>
1958 V78 N1 P40-46	59-15820 <*>
1958 V78 N2 P79-87	61-18727 <*>
1958 V78 N2 P87-92	61-18656 <*>
1958 V78 N2 P94-100	61-16342 <*>
1958 V78 N2 P100-103	61-16335 <*>
1958 V78 N3 P141-148	61-18109 <*>
1958 V78 N3 P149-152	61-18390 <*>
1958 V78 N3 P156-160	61-16344 <*>
1958 V78 N3 P160-167	61-16393 <*>
1958 V78 N4 P215-220	61-18394 <*>
1958 V78 N4 P235-239	61-16340 <*>
1958 V78 N5 P229-303	61-18425 <*>
1958 V78 N5 P273-284	61-18372 <*>
1958 V78 N5 P284-291	61-18370 <*>
1958 V78 N5 P291-298	61-18369 <*>
1958 V78 N6 P333-343	4230 <HB>
1958 V78 N6 P358-364	59-10373 <*>
	61-18415 <*>
1958 V78 N6 P364-367	59-20712 <*>
1958 V78 N6 P368-377	TS 1681 <BISI>
	62-14542 <*>
1958 V78 N7 P397-406	61-18360 <*>
1958 V78 N7 P412-418	61-18647 <*>
1958 V78 N7 P419-429	4200 <HB>
	61-16173 <*> 0
1958 V78 N8 P465-475	61-18432 <*>
1958 V78 N8 P475-483	61-18663 <*>
1958 V78 N9 P600-606	4680 <HB>
1958 V78 N10 P646-654	61-18446 <*>
1958 V78 N10 P658-663	61-18413 <*>
1958 V78 N11 P736-743	61-18732 <*>
	66-11690 <*> 0
1958 V78 N12 P792-798	61-18441 <*>
1958 V78 N12 P799-804	61-18657 <*>
1958 V78 N12 P804-812	61-18658 <*>
1958 V78 N12 P812-820	61-18442 <*>
1958 V78 N13 P889-890	4270 <HB>
1958 V78 N14 P937-947	61-18972 <*>
1958 V78 N15 P1020-1027	61-18437 <*>
1958 V78 N15 P1028-1032	4330 <HB>
1958 V78 N15 P1033-1038	62-10494 <*>
1958 V78 N15 P1041-1058	61-18748 <*>
1958 V78 N16 P1100-1107	61-18436 <*>
1958 V78 N16 P1119-1126	59-10161 <*> 0
1958 V78 N17 P1191-1200	61-20605 <*>
1958 V78 N18 P1225-1229	1287 <BISI>
	59-17642 <*>
1958 V78 N18 P1251-1262	1070 <BISI>
	59-10159 <*> P
	61-18383 <*>
1958 V78 N19 P1313-1320	61-18746 <*>
1958 V78 N19 P1320-1326	61-20593 <*>
1958 V78 N19 P1326-1332	61-18747 <*>
1958 V78 N20 P1383-1389	61-18786 <*>
1958 V78 N21 P1472-1475	59-20849 <*>
1958 V78 N22 P1493-1505	61-18731 <*>
1958 V78 N22 P1505-1513	61-18741 <*>
1958 V78 N22 P1514-1525	4648 <HB>
1958 V78 N22 P1525-1536	<MT>
	61-18730 <*>
1958 V78 N22 P1536-1546	<MT>
1958 V78 N22 P1546-1556	TRANS-40 <MT>
1958 V78 N22 P1556-1563	3467 <BISI>
1958 V78 N22 P1575-1585	3986 <BISI>
1958 V78 N22 P1609-1611	61-18781 <*>
1958 V78 N22 P1611-1613	61-18782 <*>
1958 V78 N23 P1657-1662	61-18734 <*>
1958 V78 N23 P1662-1670	61-18738 <*>
1958 V78 N24 P1745-1748	59-20050 <*>
1958 V78 N25 P1808-1811	61-18791 <*>
1958 V78 N25 P1812-1815	62-14176 <*>
1958 V78 N25 P1822-1827	61-10322 <*>
1958 V78 N25 P1832-1834	59-15066 <*>
1958 V78 N26 P1861-1865	61-20599 <*>
1958 V78 N26 P1865-1867	61-20589 <*>
1958 V78 N26 P1891-1898	1187 <BISI>
	61-18783 <*>
1959 V79 P8-17	TS 1208 <BISI>
1959 V79 P23-32	TS 1198 <BISI>
1959 V79 P46-51	TS-1410 <BISI>
1959 V79 P80-88	TS 1228 <BISI>
1959 V79 P89-94	TS 1252 <BISI>
1959 V79 P129-134	TS-1256 <BISI>
1959 V79 P135-137	TS 1254 <BISI>
1959 V79 P137-141	TS 1255 <BISI>
1959 V79 P205-210	TS-1265 <BISI>
1959 V79 P210-215	TS-1296 <BISI>
1959 V79 P291-294	TS 1294 <BISI>
	61-20590 <*>
1959 V79 P325-331	TS-1306 <BISI>
1959 V79 P349-356	TS-1300 <BISI>
1959 V79 P356-361	TS 1307 <BISI>
	61-20659 <*> 0
1959 V79 P419-426	TS-1312 <BISI>
1959 V79 P457-463	TS-1376 <BISI>
	62-10492 <*>
1959 V79 P477-485	TS-1318 <BISI>
1959 V79 P485-494	TS-1321 <BISI>
1959 V79 P494-500	TS-1322 <BISI>
1959 V79 P500-514	TS-1323 <BISI>
1959 V79 P629-634	TS-1385 <BISI>
	62-10504 <*>
1959 V79 P669-674	TS-1386 <BISI>
1959 V79 P674-683	TS-1369 <BISI>
1959 V79 P683-693	TS-1366 <BISI>
1959 V79 P694-703	TS 1550 <BISI>
1959 V79 P703-711	TS 1439 <BISI>
1959 V79 P777-785	TS-1367 <BISI>
1959 V79 P786-797	TS-1365 <BISI>
	TS-1365A <BISI>
1959 V79 P797-802	TS-1368 <BISI>
1959 V79 P846-854	TS-1471 <BISI>
1959 V79 P969-976	TS-1411 <BISI>
1959 V79 P977-989	TS 1412 <BISI>
1959 V79 P996-1002	1772 <BISI>
1959 V79 P1065-1075	TS-1426 <BISI>
1959 V79 P1075-1079	TS 1548 <BISI>
1959 V79 P1087-1090	TS 1458 <BISI>
1959 V79 P1120-1129	TS-1431 <BISI>
1959 V79 P1273-1280	1623 <BISI>
1959 V79 P1295-1297	TS-1773 <BISI>
1959 V79 P1352-1356	TS-1490 <BISI>
1959 V79 P1398-1411	TS 1561 <BISI>
1959 V79 P1430-1431	TS-1560 <BISI>
1959 V79 P1461-1471	TS 1492 <BISI>
1959 V79 P1471-1478	TS 1493 <BISI>
1959 V79 P1479-1483	TS 1494 <BISI>
1959 V79 P1483-1491	2407 <BISI>
1959 V79 P1545-1554	TS 1518 <BISI>
1959 V79 P1581-1590	1677 <BISI>
1959 V79 P1627-1637	TS 1690 <BISI>
1959 V79 P1715-1722	TS 1527 <BISI>
1959 V79 P1801-1803	TS 1630 <BISI>
	62-14195 <*>
1959 V79 P1862-1867	TS 1634 <BISI>
1959 V79 P1872-1875	1891 <BISI>
1959 V79 P1912-1923	TS 1555 <BISI>
1959 V79 N1 P8-22	61-20598 <*>
1959 V79 N1 P23-32	61-20597 <*>
1959 V79 N1 P32-36	TS-1309 <BISI>
1959 V79 N1 P46-51	62-10505 <*>
1959 V79 N2 P65-73	TS-1313 <BISI>
1959 V79 N2 P74-80	4499 <HB>
1959 V79 N2 P81-88	61-20665 <*> 0
1959 V79 N2 P89-94	61-20612 <*>

Citation	Translation
1959 V79 N3 P129-134	61-20616 <*>
	65-10169 <*>
1959 V79 N3 P135-137	61-20615 <*>
1959 V79 N3 P137-140	61-20617 <*>
1959 V79 N4 P205-210	62-10496 <*>
1959 V79 N4 P210-215	62-10489 <*>
1959 V79 N5 P267-272	4552 <HB>
1959 V79 N5 P272-276	4561 <HB>
1959 V79 N5 P276-282	4606 <HB>
1959 V79 N6 P325-331	62-10481 <*> 0
1959 V79 N6 P337-349	60-16658 <*>
1959 V79 N6 P349-356	61-20611 <*>
1959 V79 N6 P365-366	66-10717 <*> 0
1959 V79 N7 P405-407	4599 <HB>
1959 V79 N7 P408-410	4600 <HB>
1959 V79 N7 P410-414	4601 <HB>
1959 V79 N7 P414-419	4602 <HB>
1959 V79 N7 P419-426	62-10483 <*>
1959 V79 N8 P463-468	4603 <HB>
1959 V79 N8 P468-472	4604 <HB>
1959 V79 N8 P472-477	4605 <HB>
1959 V79 N8 P477-485	<MT>
1959 V79 N8 P485-494	62-10493 <*>
1959 V79 N8 P494-500	62-10524 <*>
1959 V79 N9 P634-638	4606 <HB>
1959 V79 N10 P669-674	62-14400 <*>
1959 V79 N10 P674-683	62-10467 <*>
1959 V79 N10 P694-703	62-14548 <*>
1959 V79 N10 P703-711	62-14649 <*>
1959 V79 N10 P711-718	4618 <HB>
1959 V79 N11 P786-797	62-10486 <*>
	62-14363 <*> OP
1959 V79 N11 P797-802	62-10503 <*>
1959 V79 N12 P846-854	62-14401 <*>
1959 V79 N13 P905-917	TRANS-41 <MT>
	4662 <HB>
1959 V79 N13 P917-926	TRANS-42 <MT>
	4663 <HB>
1959 V79 N13 P926-933	TS-1434 <BISI>
	62-14402 <*>
1959 V79 N13 P940-948	60-16832 <*>
1959 V79 N14 P977-989	62-16857 <*>
1959 V79 N15 P1041-1057	1587 <BISI>
1959 V79 N15 P1058-1064	4661 <HB>
1959 V79 N15 P1075-1079	62-16859 <*>
1959 V79 N15 P1087-1090	5021 <HB>
	62-14541 <*>
1959 V79 N16 P1120-1129	62-10511 <*>
1959 V79 N16 P1129-1141	62-14364 <*>
1959 V79 N17 P1183-1186	4679 <HB>
1959 V79 N17 P1187-1196	5375 <HB>
1959 V79 N17 P1187-1201	61-20075 <*>
1959 V79 N18 P1258-1263	2718 <BISI>
1959 V79 N18 P1263-1268	1529 <BISI>
1959 V79 N18 P1268-1273	<MT>
	62-14325 <*>
1959 V79 N19 P1334-1344	4700 <HB>
1959 V79 N20 P1385-1391	TS-1743 <BISI>
	60-10294 <*>
1959 V79 N20 P1398-1411	62-14652 <*>
1959 V79 N20 P1412-1419	62-14326 <*>
	66-10912 <*> 0
1959 V79 N20 P1430-1431	62-14678 <*>
1959 V79 N21 P1461-1471	62-16881 <*>
1959 V79 N21 P1471-1478	62-16846 <*>
1959 V79 N21 P1479-1483	62-14175 <*>
1959 V79 N22 P1545-1554	62-14193 <*>
1959 V79 N22 P1591-1601	1552 <BISI>
1959 V79 N22 P1601-1615	62-14540 <*>
1959 V79 N22 P1627-1637	62-16862 <*>
1959 V79 N23 P1715-1722	62-14518 <*>
1959 V79 N23 P1722-1730	1833 <BISI>
1959 V79 N23 P1730-1742	1834 <BISI>
1959 V79 N25 P1852-1862	2984 <BISI>
	66-13464 <*>
1959 V79 N25 P1862-1867	62-16849 <*>
1959 V79 N25 P1869-1872	4733 <HB>
1959 V79 N26 P1912-1923	62-14172 <*>
1959 V79 N26 P1924-1932	1749 <BISI>
1959 V79 N26 P1933-1938	1751 <BISI>
1959 V79 N26 P1954-1958	4790 <HB>
1960 V80 P12-19	TS-1730 <BISI>
1960 V80 P20-27	TS 1576 <BISI>
1960 V80 P36-45	TS-1692 <BISI>
1960 V80 P79-90	1621 <BISI>
1960 V80 P90-101	TS-1745 <BISI>
1960 V80 P129-136	TS-1645 <BISI>
1960 V80 P150-159	1726 <BISI>
1960 V80 P159-164	TS 1628 <BISI>
1960 V80 P165-169	TS-1646 <BISI>
1960 V80 P220-233	TS 1652 <BISI>
1960 V80 P272-276	TS 1784 <BISI>
1960 V80 P282-285	TS-1742 <BISI>
1960 V80 P290-296	TS 1679 <BISI>
1960 V80 P325-337	TS-1684 <BISI>
1960 V80 P356-365	1924 <BISI>
	61-18315 <*>
1960 V80 P372-374	1995 <BISI>
1960 V80 P397-407	TS 1695 <BISI>
1960 V80 P417-428	1710 <BISI>
1960 V80 P473-483	TS 1706 <BISI>
1960 V80 P483-491	1831 <BISI>
1960 V80 P514-520	1798 <BISI>
1960 V80 P520-535	TS-1799 <BISI>
1960 V80 P616-619	TS-1760 <BISI>
1960 V80 P659-669	TS 1544 <BISI>
1960 V80 P669	TS 1544A <BISI>
1960 V80 P751-756	1785 <BISI>
1960 V80 P756-763	1913 <BISI>
1960 V80 P788-796	1929 <BISI>
1960 V80 P801-809	TS 1789 <BISI>
1960 V80 P882-888	TS-1813 <BISI>
1960 V80 P1000-1006	1852 <BISI>
1960 V80 P1018-1023	2307 <BISI>
1960 V80 P1067-1072	1885 <BISI>
1960 V80 P1072-1083	1909 <BISI>
1960 V80 P1117-1135	2060 <BISI>
1960 V80 P1136-1147	1890 <BISI>
1960 V80 P1148-1160	1931 <BISI>
1960 V80 P1185-1194	1926 <BISI>
1960 V80 P1207-1211	1963 <BISI>
1960 V80 P1261-1268	2423 <BISI>
1960 V80 P1275-1283	1953 <BISI>
1960 V80 P1318-1321	1947 <BISI>
1960 V80 P1321-1336	1965 <BISI>
1960 V80 P1347-1348	1937 <BISI>
1960 V80 P1348-1349	2034 <BISI>
1960 V80 P1398-1403	1997 <BISI>
1960 V80 P1411-1414	1999 <BISI>
1960 V80 P1443-1448	2011 <BISI>
1960 V80 P1492-1496	1975 <BISI>
1960 V80 P1501-1507	2188 <BISI>
1960 V80 P1517-1524	2076 <BISI>
1960 V80 P1524-1531	2456 <BISI>
1960 V80 P1531-1540	4359 <BISI>
1960 V80 P1540-1549	2012 <BISI>
1960 V80 P1551-1553	1974 PT.I <BISI>
1960 V80 P1554-1564	1974 PT.4 <BISI>
1960 V80 P1564-1573	1974 PT.5 <BISI>
1960 V80 P1573-1574	1974,PT.6 <BISI>
1960 V80 P1618-1623	2433 <BISI>
1960 V80 P1681-1689	1985 <BISI>
1960 V80 P1711-1727	1996 <BISI>
1960 V80 P1753-1759	2113 <BISI>
1960 V80 P1770-1775	2036 <BISI>
1960 V80 P1776-1779	2068 <BISI>
1960 V80 P1787-1798	1974 <BISI>
1960 V80 P1803-1805	2037 <BISI>
1960 V80 P1838-1851	2026 <BISI>
1960 V80 P1852-1863	2027 <BISI>
1960 V80 P1864-1877	2028 <BISI>
1960 V80 P1878-1890	2035 <BISI>
1960 V80 P1952-1954	2162 <BISI>
1960 V80 N1 P1-12	1729 <BISI>
1960 V80 N1 P47-49	2975 <BISI> P
1960 V80 N2 P65-73	62-14666 <*>
1960 V80 N2 P197-206	1843 <BISI>
1960 V80 N3 P159-164	1628 <BISI>
	62-14190 <*>
1960 V80 N4 P220-233	62-16854 <*>

Citation	Number
1960 V80 N5 P272-276	62-16895 <*>
1960 V80 N5 P277-281	4800 <HB>
1960 V80 N5 P282-285	62-16863 <*>
1960 V80 N5 P285-289	1678 <BISI>
1960 V80 N5 P290-296	62-16872 <*>
1960 V80 N6 P325-336	63-16558 <*>
1960 V80 N7 P397-407	62-16879 <*>
1960 V80 N7 P407-416	4828 <HB>
1960 V80 N7 P429-437	1711 <BISI>
1960 V80 N7 P444-446	4814 <HB>
1960 V80 N8 P473-483	62-16873 <*>
1960 V80 N8 P503-507	1717 <BISI>
1960 V80 N8 P508-513	1718 <BISI>
1960 V80 N8 P534-536	3126 <BISI>
1960 V80 N8 P539-541	1821 <BISI>
1960 V80 N8 P541-546	TS-1822 <BISI>
1960 V80 N9 P616-621	62-16877 <*> P
1960 V80 N10 P641-652	<MT>
	1778 <BISI>
1960 V80 N10 P659-669	<MT>
1960 V80 N11 P737-744	4848 <HB>
	63-14856 <*> O
1960 V80 N11 P744-751	1800 <BISI>
1960 V80 N12 P781-788	4852 <HB>
1960 V80 N12 P801-809	62-16889 <*>
1960 V80 N12 P821-822	<MT>
1960 V80 N13 P854-861	5156 <HB>
1960 V80 N13 P863-877	1827 <BISI>
1960 V80 N13 P878-882	4878 <HB>
1960 V80 N14 P913-918	4881 <HB>
1960 V80 N14 P918-925	4880 <HB>
1960 V80 N15 P981-987	4889 <HB>
1960 V80 N19 P1288-1289	4938 <HB>
1960 V80 N20 P1337-1338	5028 <HB>
1960 V80 N21 P1417-1423	5014 <BISI>
1960 V80 N22 P1453-1457	4942 <HB>
1960 V80 N22 P1469-1477	4941 <HB>
1960 V80 N22 P1477-1481	4940 <HB>
1960 V80 N22 P1481-1486	5000 <HB>
1960 V80 N26 P1952-1954	61-18699 <*> O
1961 N19 P1253-1263	2639 <BISI>
1961 V81 P1-12	2184 <BISI>
1961 V81 P195-199	2436 <BISI>
1961 V81 P217-220	2323 <BISI>
1961 V81 P284-294	2132 <BISI>
1961 V81 P303-311	2130 <BISI>
1961 V81 P361-366	2192 <BISI>
1961 V81 P408-421	2326 <BISI>
1961 V81 P422-431	2322 <BISI>
1961 V81 P431-445	2182 <BISI>
1961 V81 P629-640	2306 <BISI>
1961 V81 P676-684	2591 <BISI>
1961 V81 P725-728	2331 <BISI>
1961 V81 P787-794	2362 <BISI>
1961 V81 P795-800	2333 <BISI>
1961 V81 P849-858	2468 <BISI>
1961 V81 P870-880	2452 <BISI>
1961 V81 P909-920	2364 <BISI>
1961 V81 P924-933	2363 <BISI>
1961 V81 P977-987	2381 <BISI>
1961 V81 P987-990	2382 <BISI>
1961 V81 P1006-1014	2401 <BISI>
1961 V81 P1017-1021	2446 <BISI>
1961 V81 P1047-1056	2424 <BISI>
1961 V81 P1093-1094	2425 <BISI>
1961 V81 P1172-1179	2561 <BISI>
1961 V81 P1180-1182	2642 <BISI>
1961 V81 P1192-1202	2453 <BISI>
1961 V81 P1274-1283	2463 <BISI>
1961 V81 P1284-1287	2464 <BISI>
1961 V81 P1305-1308	2543 <BISI>
1961 V81 P1313-1321	2574 <BISI>
1961 V81 P1322-1329	2484 <BISI>
1961 V81 P1410-1415	2570 <BISI>
1961 V81 P1457-1463	2550 <BISI>
1961 V81 P1472-1473	2647 <BISI>
1961 V81 P1497-1502	2554 <BISI>
1961 V81 P1510-1519	2577 <BISI>
1961 V81 P1519-1526	2578 <BISI>
1961 V81 P1527-1529	2579 <BISI>
1961 V81 P1599-1609	2580 <BISI>
1961 V81 P1618-1632	2582 <BISI>
1961 V81 P1804-1809	2620 <BISI>
1961 V81 N1 P12-22	5054 <HB>
1961 V81 N1 P46-50	2185 <BISI>
1961 V81 N1 P54-57	5073 <HB>
1961 V81 N1 P58-60	2072 <BISI>
1961 V81 N2 P73-88	2082 <BISI>
1961 V81 N2 P88-95	5063 <HB>
1961 V81 N2 P96-102	2090 <BISI>
1961 V81 N2 P103-110	2094 <BISI>
1961 V81 N3 P149-154	5121 <HB>
1961 V81 N3 P155-163	5088 <HB>
1961 V81 N3 P163-172	5089 <HB>
1961 V81 N3 P172-183	5045 <HB>
1961 V81 N3 P184-194	2195 <BISI>
1961 V81 N4 P220-228	2109 <BISI>
1961 V81 N4 P235-238	2112 <BISI>
1961 V81 N4 P239-248	2151 <BISI>
1961 V81 N4 P249-251	61-16097 <*>
1961 V81 N5 P295-302	5110 <HB>
1961 V81 N5 P311-320	5109 <HB>
1961 V81 N6 P337-349	3468 <BISI>
1961 V81 N6 P361-366	63-14887 <*>
1961 V81 N7 P448-449	5189 <HB>
1961 V81 N8 P498-501	5172 <HB>
1961 V81 N9 P529-537	2294 <BISI>
1961 V81 N9 P552-558	5157 <HB>
1961 V81 N9 P559-561	5158 <HB>
1961 V81 N9 P562-566	5159 <HB>
1961 V81 N9 P566-571	5276 <HB>
1961 V81 N9 P572-578	2261 <BISI>
1961 V81 N9 P579-589	2253 <BISI>
1961 V81 N10 P641-645	2799 <BISI>
1961 V81 N10 P661-669	2253 <BISI>
1961 V81 N10 P670-675	2718 <BISI>
1961 V81 N11 P701-707	5221 <HB>
1961 V81 N11 P712-719	5205 <HB>
1961 V81 N11 P720-724	2417 <BISI>
1961 V81 N11 P737-738	4087 <BISI>
1961 V81 N12 P778-782	5193 <HB>
1961 V81 N13 P885-893	5207 <HB>
1961 V81 N15 P1001-1005	5218 <HB>
1961 V81 N16 P1041-1047	5239 <HB>
1961 V81 N16 P1057-1062	2636 <BISI>
1961 V81 N16 P1063-1069	2637 <BISI>
1961 V81 N16 P1069-1072	2638 <BISI>
1961 V81 N17 P1101-1107	2641 <BISI>
1961 V81 N17 P1107-1116	5190 <HB>
1961 V81 N17 P1116-1122	5256 <HB>
1961 V81 N17 P1123-1138	2564 <BISI>
1961 V81 N18 P1183-1187	5268 <HB>
1961 V81 N18 P1187-1192	5432 <HB>
1961 V81 N19 P1264-1270	5280 <HB>
1961 V81 N19 P1270-1274	2640 <BISI>
1961 V81 N19 P1297-1301	2710 <BISI>
1961 V81 N21 P1381-1388	5277 <HB>
1961 V81 N21 P1388-1392	5348 <HB>
1961 V81 N21 P1395-1403	2983 <BISI>
1961 V81 N21 P1404-1409	2855 <BISI>
1961 V81 N22 P1437-1449	2646 <BISI>
1961 V81 N22 P1449-1456	5365 <HB>
1961 V81 N22 P1464-1472	2569 <BISI>
1961 V81 N23 P1503-1510	2803 <BISI>
1961 V81 N23 P1530-1536	2731 <BISI>
1961 V81 N24 P1565-1574	2679 <BISI>
1961 V81 N24 P1574-1581	2680 <BISI>
1961 V81 N24 P1581-1592	2681 <BISI>
1961 V81 N24 P1592-1598	5425 <HB>
1961 V81 N24 P1610-1618	2581 <BISI>
1961 V81 N24 P1643-1649	5426 <HB>
1961 V81 N24 P1647-1649	5465 <HB>
1961 V81 N24 P1649-1661	2682 <BISI>
1961 V81 N24 P1665-1672	3127 <BISI>
1961 V81 N24 P1693-1696	5424 <HB>
1961 V81 N25 P1729-1739	2894 <BISI>
1961 V81 N25 P1739-1745	5434 <HB>
1961 V81 N26 P1810-1815	5468 <HB>
1961 V81 N26 P1816-1820	2649 <BISI>
1961 V81 N26 P1820-1824	2858 <BISI>

Citation	Reference
1962 V82 P137-146	2687 <BISI>
1962 V82 P1028-1035	4264 <BISI>
1962 V82 N1 P1-11	5466 <HB>
1962 V82 N1 P11-18	5467 <HB>
1962 V82 N1 P18-23	2714 <BISI>
1962 V82 N1 P23-30	2895 <BISI>
1962 V82 N2 P77-90	3647 <BISI>
1962 V82 N2 P97-105	5474 <HB>
1962 V82 N3 P147-154	5523 <HB>
1962 V82 N4 P203-206	5540 <HB>
1962 V82 N5 P253-260	5597 <HB>
1962 V82 N5 P261-268	5598 <HB>
1962 V82 N5 P268-275	2890 <BISI>
1962 V82 N5 P276-282	48Q71G <ATS>
	64-18355 <*> P
1962 V82 N6 P313-337	2761 <BISI>
1962 V82 N6 P338-347	2973 <BISI>
1962 V82 N7 P394-401	2865 <BISI>
1962 V82 N7 P401-419	2876 <BISI>
1962 V82 N7 P420-428	2852 <BISI>
1962 V82 N8 P449-457	5614 <HB>
1962 V82 N9 P505-513	3124 <BISI>
1962 V82 N9 P513-517	2991 <BISI>
1962 V82 N9 P518-525	3125 <BISI>
1962 V82 N9 P539-546	2877 <BISI>
1962 V82 N10 P597-603	5630 <HB>
1962 V82 N1C P604-612	5631 <HB>
1962 V82 N11 P665-671	5646 <HB>
1962 V82 N11 P687-692	5647 <HB>
1962 V82 N11 P692-696	5648 <HB>
1962 V82 N12 P762-771	2911 <BISI>
1962 V82 N14 P945-952	5664 <HB>
1962 V82 N14 P952-957	3133 <BISI>
1962 V82 N14 P957-963	3042 <BISI>
1962 V82 N14 P963-966	5702 <HB>
1962 V82 N15 P1009-1017	2965 <BISI>
1962 V82 N15 P1017-1026	3041 <BISI>
1962 V82 N15 P1036-1041	5671 <HB>
1962 V82 N15 P1041-1044	5861 <HB>
1962 V82 N15 P1044-1053	3008 <BISI>
1962 V82 N16 P1077-1092	2974 <BISI>
1962 V82 N16 P1093-1099	3003 <BISI>
1962 V82 N16 P1099-1105	5691 <HB>
1962 V82 N16 P1105-1113	2982 <BISI>
1962 V82 N17 P1149-1155	5696 <HB>
1962 V82 N17 P1176-1186	3246 <BISI>
1962 V82 N17 P1187-1196	2987 <BISI>
1962 V82 N18 P1213-1222	3046 <BISI>
1962 V82 N18 P1222-1232	3063 <BISI>
1962 V82 N18 P1232-1236	3128 <BISI>
1962 V82 N19 P1273-1278	5715 <HB>
1962 V82 N19 P1279-1287	5709 <HB>
1962 V82 N19 P1287-1298	5700 <HB>
1962 V82 N20 P1335-1341	5718 <HB>
1962 V82 N20 P1341-1345	6077 <HB>
1962 V82 N20 P1345-1348	5719 <HB>
1962 V82 N20 P1349-1355	5720 <HB>
1962 V82 N20 P1356-1366	3036 <BISI>
1962 V82 N20 P1367-1371	5721 <HB>
1962 V82 N20 P1372-1375	5722 <HB>
1962 V82 N21 P1410-1422	3062 <BISI>
1962 V82 N21 P1423-1432	3172 <BISI>
1962 V82 N21 P1432-1436	5752 <HB>
1962 V82 N21 P1445-1449	3045 <BISI>
1962 V82 N22 P1476-1485	5743 <HB>
1962 V82 N22 P1500-1508	5745 <HB>
1962 V82 N22 P1508-1511	5892 <HB>
1962 V82 N22 P1512-1520	3315 <BISI>
1962 V82 N22 P1527-1540	3081 <BISI>
1962 V82 N22 P1552-1560	5746 <HB>
1962 V82 N22 P1561-1565	5747 <HB>
1962 V82 N22 P1579-1584	5768 <HB>
1962 V82 N23 P1641-1647	5742 <HB>
1962 V82 N23 P1655-1661	3097 <BISI>
1962 V82 N24 P1720-1726	5794 <HB>
1962 V82 N24 P1727-1739	5869 <HB>
1962 V82 N25 P1775-1781	6078 <HB>
1962 V82 N25 P1783-1790	3134 <BISI>
1962 V82 N25 P1790-1796	5802 <HB>
1962 V82 N25 P1796-1801	5803 <HB>
1962 V82 N25 P1801-1809	3219 <BISI>
1962 V82 N26 P1839-1846	63-18013 <*>
1962 V82 N26 P1846-1855	3141 <BISI>
1962 V82 N26 P1855-1857	5804 <HB>
1962 V82 N26 P1865-1873	5889 <HB>
1963 V83 N1 P10-17	3153 <BISI>
1963 V83 N1 P30-36	3137 <BISI>
1963 V83 N2 P65-75	3167 <BISI>
1963 V83 N2 P75-80	5863 <HB>
1963 V83 N2 P81-92	3173 <BISI>
1963 V83 N2 P93-99	3217 <BISI>
1963 V83 N3 P145-154	3255 <BISI>
1963 V83 N3 P154-162	3195 <BISI>
1963 V83 N3 P162-166	5859 <HB>
1963 V83 N3 P166-172	3213 <BISI>
1963 V83 N4 P213-219	3209 <BISI>
1963 V83 N5 P257-265	5900 <HB>
1963 V83 N5 P265-270	6035 <HB>
1963 V83 N5 P270-281	3210 <BISI>
1963 V83 N5 P282-290	3211 <BISI>
1963 V83 N5 P290-298	3212 <BISI>
1963 V83 N5 P298-304	5707 <HB>
1963 V83 N6 P317-327	3241 <BISI>
1963 V83 N6 P328-335	6488 <HB>
1963 V83 N6 P336-344	3351 <BISI>
1963 V83 N6 P345-35C	5893 <HB>
1963 V83 N6 P356-358	3335 <BISI>
1963 V83 N7 P377-381	5937 <HB>
1963 V83 N7 P382-387	3258 <BISI>
1963 V83 N7 P388-396	3271 <BISI>
1963 V83 N7 P398-406	3440 <BISI>
1963 V83 N7 P4C6-415	3272 <BISI>
1963 V83 N8 P441-449	3354 <BISI>
1963 V83 N8 P449-457	3310 <BISI>
1963 V83 N9 P489-495	3301 <BISI>
1963 V83 N9 P496-504	3302 <BISI>
1963 V83 N9 P5C4-509	3303 <BISI>
1963 V83 N9 P509-514	3304 <BISI>
1963 V83 N9 P517-522	3305-I <BISI>
1963 V83 N9 P522-526	3306-II <BISI>
1963 V83 N9 P526-532	3307 <BISI>
1963 V83 N9 P532-539	5918 <HB>
1963 V83 N10 P569-577	3318 <BISI>
1963 V83 N10 P578-585	3319 <BISI>
1963 V83 N10 P585-593	3336 <BISI>
1963 V83 N10 P594-597	5961 <HB>
1963 V83 N1C P602-604	3692 <BISI>
1963 V83 N11 P669-675	3326 <BISI>
1963 V83 N12 P702-715	3349 <BISI>
1963 V83 N12 P715-723	3557 <BISI>
1963 V83 N14 P859-864	6211 <HB>
1963 V83 N15 P907-911	3428 <BISI>
1963 V83 N15 P921-929	3446 <BISI>
1963 V83 N16 P961-969	6031 <HB>
1963 V83 N16 P969-971	6079 <HB>
1963 V83 N16 P972-978	3431 <BISI>
1963 V83 N16 P979-986	3381 <BISI>
1963 V83 N17 P1025-1034	6039 <HB>
1963 V83 N17 P1035-1051	3583 <BISI>
1963 V83 N17 P1051-1058	3456 <BISI>
1963 V83 N17 P1058-1066	3392 <BISI>
1963 V83 N17 P1067-1070	6066 <HB>
1963 V83 N18 P1099-1106	3430 <BISI>
1963 V83 N18 P1107-1116	3432 <BISI>
	64-14438 <*>
1963 V83 N18 P1117-1125	3560 <BISI>
1963 V83 N19 P1153-1159	6052 <HB>
1963 V83 N19 P1162-1169	3564 <BISI>
1963 V83 N19 P1169-1176	3453 <BISI>
1963 V83 N19 P1180-1184	6069 <HB>
1963 V83 N20 P1209-1226	3523, PT.I <BISI>
1963 V83 N20 P1226-1234	3452 <BISI>
1963 V83 N20 P1235-1250	3448 <BISI>
1963 V83 N21 P1277-1286	3506 <BISI>
1963 V83 N21 P1294-1302	3490 <BISI>
1963 V83 N21 P1302-1315	3523, PT.II <BISI>
1963 V83 N22 P1337-1348	6092 <HB>
1963 V83 N22 P1345-1348	6093 <HB>
1963 V83 N22 P1348-1356	3508 <BISI>
1963 V83 N23 P1397-1407	6122 <HB>

Citation	Number
1963 V83 N23 P1408-1413	3576 <BISI>
1963 V83 N23 P1414-1425	3536 <BISI>
1963 V83 N23 P1426-1432	3650 <BISI>
1963 V83 N23 P1433-1440	3552 <BISI>
1963 V83 N23 P1440-1451	3620 <BISI>
1963 V83 N23 P1451-1458	3824 <BISI>
1963 V83 N23 P1467-1477	3579 <BISI>
1963 V83 N23 P1477-1484	3923 <BISI>
1963 V83 N23 P1485-1490	6123 <HB>
1963 V83 N23 P1490-1497	6124 <HB>
1963 V83 N23 P1501-1506	3553 <BISI>
1963 V83 N23 P1513-1518	5833 <HB>
1963 V83 N24 P1541-1546	3558 <BISI>
1963 V83 N24 P1546-1553	3551 <BISI>
1963 V83 N24 P1553-1561	3533 <BISI>
1963 V83 N24 P1561-1565	3534 <BISI>
1963 V83 N25 P1605-1616	3719 <BISI>
1963 V83 N25 P1626-1640	3608 <BISI>
1963 V83 N25 P1640-1646	6112 <HB>
1963 V83 N26 P1691-1695	64-18343 <*>
1963 V83 N26 P1695-1697	6143 <HB>
1963 V83 N26 P1697-1698	3536 <BISI>
1964 V84 P434-436	4047 <BISI>
1964 V84 P625-632	4887 <BISI>
1964 V84 P681-684	4023 <BISI>
1964 V84 P722-728	3798 <BISI>
1964 V84 P932-946	3925 <BISI>
1964 V84 P1105-1120	4512 <BISI>
1964 V84 P1344-	4133 <BISI>
1964 V84 P1437-1444	4222 <BISI>
1964 V84 P1585-1592	4094 <BISI>
1964 V84 P1853-1858	4191 <BISI>
1964 V84 N1 P9-15	3637 <BISI>
1964 V84 N1 P15-24	3585 <BISI>
1964 V84 N1 P31-34	NS-192 <TTIS>
1964 V84 N1 P36-38	3586 <BISI>
1964 V84 N1 P38-42	6213 <HB>
1964 V84 N2 P57-62	6197 <HB>
1964 V84 N2 P63-73	3587 <BISI>
1964 V84 N2 P82-88	6212 <HB>
1964 V84 N3 P117-137	3653 <BISI>
1964 V84 N4 P181-190	3667 <BISI>
1964 V84 N5 P168-169	3672 <BISI>
1964 V84 N5 P237-244	3853 <BISI>
1964 V84 N5 P256-259	3674 <BISI>
1964 V84 N5 P259-264	3697 <BISI>
	65-12019 <*>
1964 V84 N6 P327-349	3717 <BISI>
1964 V84 N6 P361-368	3717A <BISI>
1964 V84 N7 P389-403	3723 <BISI>
1964 V84 N7 P403-410	3724 <BISI>
1964 V84 N7 P411-421	3725 <BISI>
1964 V84 N7 P421-429	6494 <HB>
1964 V84 N8 P453-460	3749 <BISI>
1964 V84 N8 P460-469	3778 <BISI>
1964 V84 N8 P479-482	3717A <BISI>
1964 V84 N9 P513-520	3879 <BISI>
1964 V84 N9 P520-528	3795 <BISI>
1964 V84 N9 P528-536	3796 <BISI>
1964 V84 N9 P552-556	3869 <BISI>
1964 V84 N10 P632-639	3772 <BISI>
1964 V84 N11 P660-667	3797 <BISI>
1964 V84 N11 P685-692	4091 <BISI>
1964 V84 N12 P713-718	6299 <HB>
1964 V84 N12 P718-722	3843 <BISI>
	65-11961 <*>
1964 V84 N12 P728-733	6301 <HB>
1964 V84 N14 P837-848	3724 <BISI>
1964 V84 N14 P859-868	3819 <BISI>
1964 V84 N14 P868-876	3820 <BISI>
1964 V84 N15 P909-913	6322 <HB>
1964 V84 N15 P913-919	3836 <BISI>
1964 V84 N15 P920-925	3837 <BISI>
1964 V84 N16 P979-986	4018 <BISI>
1964 V84 N16 P987-998	3887 <BISI>
1964 V84 N16 P1006-1013	3928 <BISI>
1964 V84 N16 P1013-1017	6360 <HB>
1964 V84 N17 P1041-1046	6342 <HB>
1964 V84 N17 P1046-1052	6454 <HB>
1964 V84 N17 P1062-1070	6343 <HB>
1964 V84 N17 P1070-1075	6344 <HB>
1964 V84 N18 P1128-1135	4011 <BISI>
1964 V84 N19 P1169-1174	4082 <BISI>
1964 V84 N19 P1180-1187	4163 <BISI>
1964 V84 N19 P1187-1197	3922 <BISI>
1964 V84 N19 P1197-1202	4084 <BISI>
1964 V84 N20 P1243-1250	3951 <BISI>
1964 V84 N21 P1297-1303	6381 <HB>
1964 V84 N21 P1304-1313	4116 <BISI>
1964 V84 N21 P1322-1328	3992 <BISI>
1964 V84 N22 P1353-1359	6393 <HB>
1964 V84 N22 P1359-1365	6467 <HB>
1964 V84 N22 P1382-1392	4000 <BISI>
1964 V84 N22 P1392-1398	6394 <HB>
1964 V84 N22 P1399-1410	4015 <BISI>
1964 V84 N22 P1410-1424	4016 <BISI>
1964 V84 N23 P1497-1500	6418 <HB>
1964 V84 N23 P1500-1505	6419 <HB>
1964 V84 N23 P1505-1511	4064 <BISI>
1964 V84 N24 P1561-1568	4236 <BISI>
1964 V84 N24 P1569-1576	4050 <BISI>
1964 V84 N24 P1576-1585	4077 <BISI>
1964 V84 N25 P1641-1647	4150 <BISI>
1964 V84 N25 P1656-1658	6528 <HB>
1964 V84 N27 P1866-1872	4192 <BISI>
1964 V84 N27 P1873-1877	6456 <HB>
1964 V84 N27 P1877-1884	6508 <HB>
1965 V85 P29-36	4102 <BISI>
1965 V85 P149-155	4483 <BISI>
1965 V85 P180-182	4221 <BISI>
1965 V85 P183-184	4729 <BISI>
1965 V85 P189-197	4312 <BISI>
1965 V85 P202-203	4599 <BISI>
1965 V85 P243-250	4220 <BISI>
1965 V85 P331-341	4457 PT 2 <BISI>
1965 V85 P342-353	4378 <BISI>
1965 V85 P397-400	4442 <BISI>
1965 V85 P406	2177 <BSRA>
1965 V85 P526-535	4318 <BISI>
1965 V85 P535-546	4344 <BISI>
1965 V85 P546-550	4365 <BISI>
1965 V85 P550-555	4366 <BISI>
1965 V85 P675-688	4343 <BISI>
1965 V85 P738-745	4460 <BISI>
1965 V85 P963-970	4671 <BISI>
1965 V85 P970-977	4444 <BISI>
1965 V85 P1040-1046	4914 <BISI>
1965 V85 P1061-1063	4501 <BISI>
1965 V85 P1198-1199	4812 <BISI>
1965 V85 P1229-1240	5006 <BISI>
1965 V85 P1267-1280	5006 <BISI>
1965 V85 P1335-1340	4549 <BISI>
1965 V85 P1361-1371	4638 <BISI>
1965 V85 P1378-1387	4920 <BISI>
1965 V85 P1412-1417	4702 <BISI>
1965 V85 P1441-1446	4801 <BISI>
1965 V85 P1479-1481	4817 <BISI>
1965 V85 P1517-1525	4634 <BISI>
1965 V85 P1713-1722	4789 <BISI>
1965 V85 N1 P1-11	4369 <BISI>
1965 V85 N1 P25-28	6457 <HB>
1965 V85 N1 P36-39	6474 <HB>
1965 V85 N2 P61-71	4124 <BISI>
1965 V85 N3 P117-124	6472 <HB>
1965 V85 N3 P125-127	6473 <HB>
1965 V85 N3 P127-136	4231 <BISI>
1965 V85 N4 P173-180	6402 <HB>
1965 V85 N4 P198-202	6460 <HB>
1965 V85 N5 P233-243	4219 <BISI>
1965 V85 N5 P257-261	6513 <HB>
1965 V85 N5 P262-266	1210 <TC>
	6464 <HB>
1965 V85 N6 P297-307	4243 <BISI>
1965 V85 N6 P308-316	6515 <HB>
1965 V85 N6 P317-331	4457-I <BISI>
1965 V85 N6 P353-364	6404 <HB>
1965 V85 N6 P364-372	6600 <HB>
1965 V85 N7 P385-391	6537 <HB>
1965 V85 N7 P391-397	6538 <HB>
1965 V85 N7 P400-405	4407 <BISI>

1965 V85 N8 P456-464 4387 <BISI>
1965 V85 N8 P464-471 4284 <BISI>
1965 V85 N8 P472-479 4426 <BISI>
1965 V85 N9 P513-519 6545 <HB>
1965 V85 N9 P519-526 6546 <BISI>
1965 V85 N1C P619-622 4446 <BISI>
1965 V85 N12 P713-721 6577 <HB>
1965 V85 N12 P721-724 6578 <HB>
1965 V85 N13 P785-794 6590 <HB>
1965 V85 N13 P804-810 6634 <HB>
1965 V85 N14 P857-865 6607 <HB>
1965 V85 N15 P897-901 6632 <HB>
1965 V85 N15 P902-907 6533 <HB>
1965 V85 N16 P977-981 6658 <HB>
1965 V85 N16 P982-987 6659 <HB>
1965 V85 N16 P987-990 6776 <HB>
1965 V85 N17 P1025-1032 6648 <HB>
1965 V85 N17 P1033-1039 6403 <HB>
1965 V85 N19 P1173-1182 6582 <HB>
1965 V85 N19 P1187-1195 6665 <HB>
1965 V85 N20 P1240-1247 6681 <HB>
1965 V85 N21 P1297-1307 4636 <BISI>
1965 V85 N21 P1297-1306 66-11544 <*> 0
1965 V85 N21 P1308-1311 4613 <BISI>
1965 V85 N21 P1311-1320 6687 <HB>
1965 V85 N21 P1320-1327 6839 <HB>
1965 V85 N22 P1398-1403 4669 <BISI>
1965 V85 N22 P1446-1451 6788 <HB>
1965 V85 N23 P1525-1532 4615 <BISI>
1965 V85 N23 P1546-1548 6787 <HB>
1965 V85 N24 P1588-1595 6777 <HB>
1965 V85 N25 P1686-1691 6789 <HB>
1965 V85 N26 P1751-1754 6796 <HB>
1966 V86 P8-16 01/13 4788 <BISI>
1966 V86 P89-99 01/27 4775 <BISI>
1966 V86 P525-532 4886 <BISI>
1966 V86 N2 P65-73 6827-I <HB>
1966 V86 N2 P74-76 6827-2 <HB>
1966 V86 N2 P77-81 6828 <HB>
1966 V86 N2 P81-88 6829 <HB>
1966 V86 N3 P129-137 6868 <HB>
1966 V86 N3 P137-141 4852 <BISI>
1966 V86 N3 P150-160 4822 <BISI>
1966 V86 N7 P413-416 6861 <HB>
1966 V86 N9 P548-552 6878 <HB>

STAHLBAU
1937 V10 N10 P73-76 58-470 <*>
1955 V24 N9 P202-206 59-20842 <*>
1956 V25 P2C5-210 TS-1432 <BISI>
1956 V25 N8 P181-184 66-10799 <*> 0
1956 V25 N9 P205-21C 62-14389 <*>
1959 V28 N6 P156-159 1922 <BISI>
1960 V29 P186-191 1881 <BISI>
1960 V30 N1 P24-27 C-3851 <NRCC>
1961 V30 N1 P16-23 63-20346 <*> 0
1962 V31 N5 P129-136 63-18850 <*> 0
1963 P199-203 64-00320 <*>

STAHLBAU-RUNDSCHAU
1957 V3 N2 P29-31 61-16291 <*>

*STAL
1935 N4 P345-348 R-187 <*>
1938 V8 N1C P33-37 R-1883 <*>
1938 V8 N8/9 P28-31 R-1846 <*>
1939 N12 P18-20 R-3807 <*>
1940 V10 N2 P17-22 62-24181 <=*> 0
1940 V10 N5/6 P57-59 R-2062 <*>
1941 V1 N5 P47-53 R-1927 <*>
1946 V6 P627-631 CSIRO-3232 <CSIR>
1946 V6 N2 P99-104 3163 <HB>
1946 V6 N11/2 P697 2794 <HB>
1947 N1 P3-6 RT-1067 <*>
1947 N1 P7-10 RT-400 <*>
1947 V7 N1 P39-48 1826 <BISI>
1947 V7 N2 P149-151 R-2077 <*>
1947 V7 N5 P395-399 2111 <HB>
1947 V7 N6 P485-489 2294 <HB>
1947 V7 N6 P511-518 TS-1311 <BISI>
62-24224 <=*>
1947 V7 N10 P903-910 3405 <HB>
1948 V8 N1 P28-36 2467 <HB>
1948 V8 N3 P232-240 61-23304 <*=>
1948 V8 N1C P911-916 2795 <HB>
1949 V9 N12 P18-20 59-18328 <+*>
1952 V12 P330-336 TS-1364 <BISI>
1952 V12 N4 P330-336 62-24228 <=*>
1952 V12 N12 P1134- 62-24157 <=*>
1954 V14 N11 P975-983 3245 <BISI>
1954 V14 N12 P1106-1107 1939 <BISI>
1955 P34-50 GB66 <NLL>
1955 P371-372 GB66 <NLL>
1955 N4 P329-333 R-189 <*>
1955 N5 P472- GB12 <NLL>
1955 N7 P635-637 R-3786 <*>
1955 V15 P398-407 63-16874 <=*>
1955 V15 P637-639 TS-1583 <BISI>
1955 V15 P771-776 59-19321 <+*>
1955 V15 P1120-1123 R-4404 <*>
1955 V15 N1 P11-18 4384 <HB>
1955 V15 N1 P63-68 3597 <HB>
1955 V15 N1 P69-70 3598 <HB>
1955 V15 N1 P86-88 3599 <HB>
1955 V15 N3 P199-204 62-25273 <=*>
1955 V15 N3 P2C8-215 3836 <HB>
1955 V15 N4 P295-301 60-17207 <+*>
1955 V15 N5 P391-396 3844 <HB>
1955 V15 N5 P431-438 62-24128 <=*>
1955 V15 N5 P449-454 3669 <HB>
1955 V15 N6 P488-497 GB70 4197 <NLL>
3676 <HB>
1955 V15 N7 P583-591 1592 <BISI>
61-13753 <=*>
1955 V15 N7 P637-639 62-25876 <=*>
1955 V15 N8 P692-698 1871 <BISI>
61-27555 <*=>
1955 V15 N8 P7C9-713 3673 <HB>
1955 V15 N8 P714-719 3636 <HB>
1955 V15 N8 P720-727 3637 <HB>
1955 V15 N9 P8C1-806 3643 <HB>
1955 V15 N9 P815-820 62-23966 <=*>
1955 V15 N10 P887-891 3645 <HB>
1955 V15 N10 P891-894 3646 <HB>
1955 V15 N10 P922-930 62-24153 <=*>
1955 V15 N11 P963-968 4014 <HB>
1955 V15 N11 P984-989 60-13308 <+*>
1955 V15 N11 P994-1000 3691 <HB>
1955 V15 N11 P1037-1038 62-21458 <=*>
1955 V15 N12 P1063-1073 RT-4255 <*>
61-13107 <=*> 0
63-16876 <*=>
1956 P62- GB70 T 4079 <NLL>
1956 N3 P200-2C3 GB70 4238 <NLL>
1956 N4 P356 GB70 4182 <NLL>
1956 N9 P790-793 936TM <CTT>
1956 V16 N1 P62-66 R-4839 <*>
1956 V16 N1 P68-69 R-4337 <*>
1956 V16 N2 P1C8-114 3875 <HB>
1956 V16 N2 P115-124 T-1896 <INSD>
1956 V16 N2 P157-160 59-10316 <+*>
1956 V16 N2 P181-182 3881 <HB>
1956 V16 N3 P200-203 3868 <HB>
1956 V16 N3 P212-214 59-19611 <+*>
1956 V16 N3 P263-265 R-1906 <*>
1956 V16 N4 P327-330 3825 <HB>
1956 V16 N4 P333-337 62-13186 <=*>
1956 V16 N4 P343-347 3838 <HB>
1956 V16 N4 P348-351 4016 <HB>
1956 V16 N4 P367-368 2238 <BISI>
1956 V16 N5 P402-408 62-25272 <=*>
1956 V16 N5 P450-452 5C78 <HB>
1956 V16 N5 P465-467 4018 <HB>
1956 V16 N6 P514-518 3843 <HB>
1956 V16 N6 P545-548 3848 <HB>
64-19390 <=$>
1956 V16 N7 P582-585 3867 <HB>
62-24156 <=> P
1956 V16 N7 P585-586 3882 <HB>
1956 V16 N8 P675-682 4092 <HB>

Citation	Reference
1956 V16 N8 P682-689	62-23968 <=*>
1956 V16 N8 P727-734	4010 <HB>
1956 V16 N8 P734-737	4764 <HB>
1956 V16 N8 P746-748	2232 <BISI>
1956 V16 N9 P771-773	62-24117 <=*>
1956 V16 N9 P780-782	62-24147 <=*>
1956 V16 N9 P790-793	R-1993 <*>
1956 V16 N9 P798-802	62-24125 <=*>
1956 V16 N9 P815-817	4017 <HB>
1956 V16 N10 P883-890	3852 <HB>
1956 V16 N10 P890-894	R-1542 <*>
1956 V16 N10 P909-915	3931 <HB>
1956 V16 N10 P927-933	62-24096 <=*>
1956 V16 N11 P975-976	4309 <HB>
1956 V16 N11 P977-983	4778 <HB>
1956 V16 N11 P1006-1015	60-15575 <+*>
1956 V16 N11 P1042-1048	3888 <HB>
1956 V16 N12 P1097-1098	3908 <HB>
1956 V16 N12 P1099-1103	62-24201 <=*> P
1956 V16 N12 P1103-1105	62-24155 <=*> P
1956 V16 N12 P1135	60-17153 <+*>
1957 V17 P431-435	61-16202 <=*>
1957 V17 P521-522	R-4047 <*>
	63-16875 <*=>
1957 V17 P602-608	TS 1528 <BISI>
1957 V17 P779-787	TS 1578 <BISI>
1957 V17 N1 P7-15	4066 <HB>
1957 V17 N1 P16-20	62-13207 <=*>
1957 V17 N1 P20-24	4087 <HB>
1957 V17 N1 P64-69	62-13191 <=*>
1957 V17 N1 P69-71	62-24120 <*>
1957 V17 N2 P99-	62-24134 <=*>
1957 V17 N2 P103-105	4090 <HB>
1957 V17 N2 P106-114	62-24173 <=*> P
1957 V17 N2 P124-129	62-23969 <=*>
1957 V17 N2 P152-157	62-24161 <=*>
1957 V17 N2 P185-187	62-24166 <=*>
1957 V17 N3 P195-199	4385 <HB>
1957 V17 N3 P209-213	4004 <HB>
	62-24154 <=*>
1957 V17 N3 P228-232	4113 <HB>
1957 V17 N3 P232-238	62-24159 <=*>
1957 V17 N3 P243-253	4642 <HB>
	62-24146 <=*>
1957 V17 N3 P261-263	62-24114 <=*>
1957 V17 N3 P268-	62-24160 <=*> P
1957 V17 N3 P275-276	4195 <HB>
1957 V17 N3 P280	4063 <HB>
1957 V17 N4 P320-322	4022 <HB>
1957 V17 N4 P320-321	62-24162 <=*>
1957 V17 N4 P322-325	3985 <HB>
	60-21023 <=> 0
1957 V17 N4 P326-328	3995 <HB>
1957 V17 N4 P333-340	62-13188 <=*>
1957 V17 N4 P347-351	62-24127 <=*>
1957 V17 N4 P358-361	62-24164 <=*>
1957 V17 N4 P362-365	62-24163 <=*>
1957 V17 N4 P374-	62-24141 <=*>
1957 V17 N5 P389-391	62-24131 <=*>
1957 V17 N5 P402-405	62-13194 <=*>
1957 V17 N5 P406-411	4035 <HB>
1957 V17 N5 P411-413	62-24139 <=*>
1957 V17 N5 P441-444	62-24152 <=*>
1957 V17 N5 P453-456	4134 <HB>
1957 V17 N5 P464-465	4076 <HB>
1957 V17 N5 P468-469	2236 <BISI>
1957 V17 N6 P481-488	62-24133 <=*>
1957 V17 N6 P483-486	4051 <HB>
1957 V17 N6 P487-492	4052 <HB>
1957 V17 N6 P493-495	62-24132 <=*>
1957 V17 N6 P496-500	62-13195 <=*>
1957 V17 N6 P507-511	4047 <HB>
1957 V17 N6 P512-513	4708-B <HB>
1957 V17 N6 P521-522	4080 <HB>
1957 V17 N6 P569-570	62-24145 <=*>
1957 V17 N6 P571-573	62-13203 <=*>
1957 V17 N7 P580-584	62-24135 <=*>
1957 V17 N7 P602-608	<MT>
	62-14173 <=*>
	62-25870 <=*>
1957 V17 N7 P611-615	1921 <BISI>
1957 V17 N7 P616-621	62-24136 <=*>
1957 V17 N7 P636-640	4088 <HB>
1957 V17 N7 P658-662	62-24130 <=*>
1957 V17 N7 P661-662	3989 <HB>
1957 V17 N7 P663-665	60-15598 <+*>
1957 V17 N8 P673-684	62-13183 <=*>
1957 V17 N8 P685-690	60-17901 <+*>
1957 V17 N8 P693-700	60-17454 <+*>
1957 V17 N8 P701-707	62-13182 <=*>
1957 V17 N8 P707-713	62-13181 <=*>
1957 V17 N8 P718	62-23879 <=*>
1957 V17 N8 P728-730	62-13184 <=*>
1957 V17 N8 P754-	4021 <HB>
1957 V17 N9 P772-778	62-13187 <=*>
1957 V17 N9 P779-787	62-14549 <=*>
	62-25875 <=*>
1957 V17 N9 P800-804	4062 <HB>
1957 V17 N9 P819-822	62-13192 <=*$>
1957 V17 N9 P823-828	4060 <HB>
1957 V17 N9 P858-860	4061 <HB>
1957 V17 N10 P868-873	4068 <HB>
1957 V17 N10 P884-887	4056 <HB>
1957 V17 N10 P899-901	4271 <HB>
1957 V17 N10 P929-933	60-13222 <+*>
1957 V17 N10 P934-935	4058 <HB>
1957 V17 N10 P936-940	62-24110 <=*>
1957 V17 N10 P941-943	62-13190 <=*>
1957 V17 N10 P948	4053 <HB>
1957 V17 N11 P961-964	62-24167 <=*>
1957 V17 N11 P965-968	62-24123 <=*>
1957 V17 N11 P968-976	59-11602 <=> 0
1957 V17 N11 P982-987	PB 141 279T <=>
1957 V17 N11 P987-991	59-11585 <=>
1957 V17 N11 P992-996	59-11588 <=> 0
1957 V17 N11 P997-1005	62-13193 <=*$>
1957 V17 N11 P1006-1010	103 <LSB>
1957 V17 N11 P1032-1033	59-11589 <=>
1957 V17 N11 P1034-1038	59-11590 <=>
1957 V17 N11 P1038-1041	59-11591 <=>
1957 V17 N11 P1042-1045	59-11592 <=>
1957 V17 N11 P1046-1055	59-11593 <=> 0
1957 V17 N12 P1068-1071	4142 <HB>
1957 V17 N12 P1071-1073	62-24111 <=*>
1957 V17 N12 P1086-1093	62-24112 <=*>
1957 V17 N12 P1094-1096	62-24122 <=*>
1957 V17 N12 P1097-1098	5853 <HB>
1957 V17 N12 P1099-1103	62-24124 <=*>
1957 V17 N12 P1103-1107	62-23967 <=*>
1957 V17 N12 P1128-1130	62-24121 <=*>
1957 V17 N12 P1140-1141	62-24118 <=*>
1958 P202-205	TS-1416 <BISI> P
1958 N4 P289-294	<PS>
1958 N5 P434-441	<PS>
1958 N6 P506-509	<PS>
1958 N12 P1112-1117	62-24227 <=*>
1958 V18 P16-25	4731 <HB>
1958 V18 P44-59	1870 <BISI>
1958 V18 P176-206	62-24104 <=*>
1958 V18 P207-226	62-13204 <=*>
1958 V18 P227-242	62-24087 <=*>
1958 V18 P243-258	62-24103 <=*>
1958 V18 P247-250	TS 1570 <BISI>
	62-14677 <=*>
	62-25874 <=*>
1958 V18 P248	4201 <HB>
1958 V18 P248-	4201 <HB>
1958 V18 P259-275	62-24200 <=*>
1958 V18 P285-313	TS 1572 <BISI>
1958 V18 P295-313	62-24198 <=*>
1958 V18 P314-337	62-24199 <=*$>
1958 V18 P338-351	62-13200 <=*>
1958 V18 P520-525	4405 <HB>
1958 V18 P89-114 SUP	3251 <BISI>
1958 V18 P276-294 SUP	1041 <BISI>
1958 V18 N1 P21-22	62-24113 <=*>
1958 V18 N1 P22-23	4509 <HB>
1958 V18 N1 P43-48	4165 <HB>
1958 V18 N1 P57-60	62-24106 <=*>
1958 V18 N1 P60-66	62-24107 <=*>

Citation	Translation
1958 V18 N1 P67-70	4707 <HB>
1958 V18 N1 P75-81	4256 <HB>
	60-13617 <+*>
1958 V18 N2 P97-104	4229 <HB>
1958 V18 N2 P105-109	4161 <HB>
1958 V18 N2 P110-113	62-13202 <=*>
1958 V18 N2 P114-120	62-24092 <=*>
1958 V18 N2 P126-130	TS 1268 <BISI>
	62-24216 <=*>
1958 V18 N2 P131-137	PB 141 166T <=>
	4193 <HB>
	60-13620 <+*>
1958 V18 N2 P138-144	62-24225 <=*>
1958 V18 N2 P144-151	62-24193 <=*>
1958 V18 N2 P171-178	911 <BISI>
1958 V18 N2 P179-185	60-3984 <+*>
1958 V18 N3 P193-202	60-13622 <+*>
1958 V18 N3 P202-205	62-24238 <=*> P
1958 V18 N3 P214-218	4608 <HB>
1958 V18 N3 P218-223	4240 <HB>
1958 V18 N3 P246-248	4260 <HB>
	60-13623 <+*>
1958 V18 N3 P253-256	62-13206 <=*>
1958 V18 N3 P271-280	62-24101 <=*>
1958 V18 N4 P289-294	4233 <HB>
1958 V18 N4 P295-297	60-15297 <*+>
1958 V18 N4 P311-316	4708 <HB>
	59-20079 <+*> 0
1958 V18 N4 P335-339	62-24098 <=*>
1958 V18 N4 P358-363	PB 141 172T <=>
1958 V18 N4 P379-383	62-24100 <=*>
1958 V18 N5 P385-390	62-24099 <=*>
1958 V18 N5 P391-397	TRANS-43 <MT>
	4403 <HB>
1958 V18 N5 P398-402	4298 <HB>
1958 V18 N5 P411-414	4453 <HB>
1958 V18 N5 P417-425	4300 <HB>
1958 V18 N5 P428-433	TS 1200 <BISI>
	62-24209 <=*>
1958 V18 N5 P442-446	62-24094 <=*>
1958 V18 N5 P446-448	4263 <HB>
1958 V18 N6 P481-485	PB 141 289T <=>
1958 V18 N6 P486-488	59-11500 <=>
1958 V18 N6 P489-495	4301 <HB>
1958 V18 N6 P495-502	61-13329 <=*>
1958 V18 N6 P509-511	4310 <HB>
1958 V18 N6 P520-525	4405 <HB>
1958 V18 N6 P561-568	PB 141 290T <=>
	62-24090 <=*>
1958 V18 N7 P586-593	4386 <HB>
1958 V18 N7 P599-604	4337 <HB>
1958 V18 N7 P617-620	4562 <HB>
1958 V18 N7 P629-633	62-24088 <=*>
1958 V18 N7 P642	59-20080 <+*>
1958 V18 N7 P643-647	62-24086 <=*>
1958 V18 N7 P652-653	62-24194 <=*>
1958 V18 N7 P668-672	4342 <HB>
1958 V18 N8 P722-726	3181 <BISI>
1958 V18 N8 P742-744	5080 <HB>
1958 V18 N9 P796-799	62-24192 <=*>
1958 V18 N9 P823-824	62-24204 <=*>
1958 V18 N10 P867-869	4414 <HB>
1958 V18 N10 P877-882	TRANS-44 <MT>
1958 V18 N10 P890-893	62-24188 <=*>
1958 V18 N10 P894-899	TRANS-45 <MT>
1958 V18 N10 P931-938	<PS>
1958 V18 N10 P942-946	4886 <HB>
1958 V18 N11 P980-983	4667 <HB>
1958 V18 N11 P983-987	4470 <HB>
	59-00682 <*>
1958 V18 N11 P987-992	TS 1649 <BISI>
	62-14681 <=*>
	62-25878 <=*>
1958 V18 N11 P990-1002	4511 <HB>
1958 V18 N11 P1012-1017	TS 1219 <BISI>
	62-24211 <=*>
1958 V18 N11 P1018-1020	TS-1466 <BISI>
	62-25326 <=*>
1958 V18 N11 P1029-1035	2869 <BISI>
1958 V18 N11 P1046-1055	TS-1186 <BISI>

Citation	Translation
	62-24207 <=*>
1958 V18 N12 P1-4 ADD	62-24205 <=*>
1958 V18 N12 P1057-1065	4500 <HB>
1958 V18 N12 P1066-1071	<PS>
	4613 <HB>
1958 V18 N12 P1071-1077	4635 <HB>
1958 V18 N12 P1089-1095	4720 <HB>
1958 V18 N12 P1095-1102	TS 1194 <BISI>
	62-24226 <=*>
1958 V18 N12 P1103-1107	60-14665 <+*>
1958 V18 N12 P1112-1117	TS-1235 <BISI>
1958 V18 N12 P1130-1132	4503 <HB>
1959 V19 P22-23	TS-1456 <BISI>
	62-25332 <=*>
1959 V19 P95-103	TS 1680 <BISI>
1959 V19 P133-148	TS-1480 <BISI>
1959 V19 P133-138	62-25325 <=*>
1959 V19 P151-159	TS 1499 <BISI>
	62-25328 <=*>
1959 V19 P171-179	TS 1481 <BISI>
	62-25327 <=*>
1959 V19 P180-187	TS 1491 <BISI>
1959 V19 N5 P444-447	60-10030 <+*> 0
1959 V19 N9 P802-807	60-11179 <=>
1959 V19 N9 P812-817	60-11179 <=>
1960 V20 N8 P691-694	61-11618 <=>
1960 V20 N8 P701-703	61-11577 <=>
1960 V20 N8 P741-745	61-11637 <=>
1960 V20 N10 P911-914	61-11684 <=>
1960 V20 N11 P1004-1007	61-21707 <=>
1961 P395-411	62-33398 <=*> 0
1961 P424-435	62-33398 <=*> 0
1961 P441-446	62-33398 <=*> 0
1961 P455-461	62-33398 <=*> 0
1961 P487-489	62-33398 <=*> 0
1961 V21 P73-84	2730 <BISI>
1961 V21 P335-353	2727 <BISI>
1961 V21 P365-372	2729 <BISI>
1961 V21 P455-461	5609 <HB>
1961 V21 N4 P382-383	61-27117 <*=>
1962 V22 N11 P1001-1005	63-13839 <=*>
1962 V22 N11 P1035-1039	63-13839 <=*>
1963 V23 N2 P136	63-23296 <=*>
	63-23925 <=*$>
1963 V23 N2 P157	64-13013 <=*$>
1963 V23 N4 P374-378	63-31120 <=>
1963 V23 N6 P523-528	63-31336 <=>
1963 V23 N6 P533-536	63-31378 <=>
1963 V23 N6 P544-546	64-13012 <=*$>
1963 V23 N7 P577-580,623	63-31602 <=>
1963 V23 N9 P835-838	63-41151 <=>
1963 V23 N10 P889-892	64-21202 <=>
1963 V23 N10 P908-910	63-41300 <=>
1963 V23 N11 P1024-1027	64-21113 <=>
1964 V24 N1 P5-9	65-30774 <=$>
1964 V24 N4 P346-347	64-31311 <=>
1964 V24 N5 P385-391	64-31788 <=>
1964 V24 N7 P642-645	64-51215 <=$>
1964 V24 N8 P673-675	64-51817 <=$>
1964 V24 N9 P845-848	M-5621 <NLL>
1964 V24 N11 P961-963	65-30412 <=$>
1964 V24 N11 P1037-1040	M-5622 <NLL>
1964 V24 N11 P1041-1045	65-30412 <=$>
1964 V24 N11 P1051	65-30412 <=$>
1964 V24 N12 P1120-1122	M-5676 <NLL>
1965 N6 P573-574	65-31771 <=*$>
1965 N8 P673-675	66-33143 <=$>
1965 V25 N1 P1-4	65-30774 <=$>
1965 V25 N2 P163-168	65-30766 <=$>
1966 N9 P834-836	67-30934 <=$>

STANDARDISIERUNG

Citation	Translation
1959 V5 P2/1153-2/1168	2994 <BISI>

STANDARDIZAREA

Citation	Translation
1962 V14 N5 P244-247	65-20008 <$>
1962 V14 N8 P421-425	65-20015 <$>

STANDARTIZATSIYA

Citation	Translation
1954 N3 P41-45	RT-2471 <*>

Citation	Translation
1954 N6 P46-49	RT-3350 <*>
1955 N5 P37-44	NS-31 <TTIS> P
1957 N6 P70-71	59-18402 <+*>
1958 N4 P23-28	61-23543 <*=>
1959 N3 P13-18	60-31028 <=>
	61-13037 <=*>
1959 N8 P41-48	60-11405 <=>
1959 N10 P22-24	61-13335 <*=>
1959 N11 P51-52	60-31062 <=>
1959 N12 P56-62	60-31102 <=>
1960 N2 P11-13	62-15285 <*=>
1960 N4 P9-10	60-41342 <=>
1960 N4 P58	60-41342 <=>
1960 N5 P6-9	61-11033 <=>
1960 N5 P17-19	61-11033 <=>
1960 N8 P8-11	61-21343 <=>
1961 N3 P53-56	61-27080 <*=>
1961 N3 P57	62-13374 <*=>
1961 V25 N5 P20-23	64-13374 <=*$>
1961 V25 N6 P16-22	64-13745 <=*$>
1961 V25 N8 P45-46	64-13373 <=*$>
1962 N2 P50-52	62-25002 <=*>
1962 N3 P17-20	62-32449 <=>
1962 N4 P32-34	62-32009 <=>
1962 N4 P46-49	62-32081 <=*>
1962 N5 P52-54	62-32449 <=>
1962 V26 N3 P43	62-32565 <=*>
1962 V26 N3 P57-59	62-11669 <=*$>
1962 V26 N6 P65-67	62-33024 <=$> P
1962 V26 N9 P26	63-13268 <=>
1962 V26 N11 P23-27	63-21186 <=>
1962 V26 N11 P54-55	63-21186 <=>
1963 V27 N1 P39-40	63-21979 <=>
1963 V27 N1 P59	63-21387 <=>
1963 V27 N2 P24-27	63-21654 <=>
1963 V27 N3 P8-14	63-21633 <=>
1963 V27 N5 P18-22	65-63576 <=$>
1963 V27 N6 P51-52	63-31727 <=>
1963 V27 N6 P52-53	63-31710 <=>
1963 V27 N7 P44-45	63-31856 <=>
1963 V27 N7 P51-54	63-31853 <=>
1963 V27 N11 P3-6	64-21616 <=>
1963 V27 N11 P46-48	64-21616 <=>
1964 V28 N1 P17-23	64-31228 <=>
1964 V28 N3 P3-8	64-31883 <=>
1964 V28 N3 P55	64-31883 <=>
1964 V28 N4 P45-54	64-41317 <=>
1964 V28 N5 P3-6	64-41619 <=>
1964 V28 N5 P8-18	64-41619 <=>
1964 V28 N5 P47-48	64-41619 <=>
1964 V28 N7 P14-16	AD-625 301 <=*$>
1965 V29 N2 P48-51	65-31218 <=$>
1965 V29 N6 P7-10	66-30598 <=$>

STANDARTY I KACHESTVO

Citation	Translation
1966 N1 P42-45	66-31938 <=$>
1966 N2 P1-4	66-32968 <=$>

★STANKI I INSTRUMENT

Citation	Translation
1941 V12 N2 P20-22	60-13906 <+*>
1949 V20 N5 P21-22	63-15268 <=*>
1949 V20 N10 P5-7	50/3392 <NLL>
1949 V20 N11 P15-16	R-3842 <*>
1950 V21 N3 P6-9	50/2862 <NLL>
1950 V21 N4 P4-9	50/2307 <NLL>
1950 V21 N4 P17-18	50/2308 <NLL>
1950 V21 N4 P19-20	50/2312 <*>
1950 V21 N4 P24-25	50/2309 <NLL>
1950 V21 N10 P3-6	51/2156 <NLL>
1951 V22 N1 P34-36	51/2153 <NLL>
1951 V22 N2 P28	4090 <BISI>
1951 V22 N5 P20-22	2821 <HB>
1952 V23 N3 P22-27	62-23097 <=*>
1952 V23 N4 P24-25	RT-1048 <*>
1952 V23 N6 P7-11	61-15366 <+*>
1952 V23 N7 P12-16	61-15367 <+*>
1952 V23 N7 P17-19	3133 <HB>
1953 N7 P12-14	R-4077 <*>
1953 V24 N1 P30-31	RT-3348 <*>
1953 V24 N8 P21-23	RT-4120 <*>
1953 V24 N8 P23-27	4672 <HB>
1953 V24 N9 P18-20	3418 <BISI>
1953 V24 N10 P9-15	61-15368 <+*>
1953 V24 N10 P25-27	62-10365 <*=>
1954 N6 P33-34	RT-2373 <*>
1954 V25 N5 P16-20	3484 <HB>
1954 V25 N5 P29-30	3479 <HB>
1954 V25 N5 P33-34	3480 <HB>
1954 V25 N7 P21-23	60-13298 <+*> O
1954 V25 N11 P12-16	R-1233 <*>
1955 N2 P28-30	R-4463 <*>
1955 V26 N2 P11-17	3688 <HB>
1955 V26 N2 P17-19	3595 <HB>
1955 V26 N2 P26-27	R-1890 <*>
1955 V26 N2 P33-34	R-1891 <*>
1955 V26 N4 P21-22	62-20047 <=*>
1955 V26 N7 P28-29	3725 <HB>
1955 V26 N8 P17	3633 <HB>
1955 V26 N10 P24-26	R-3149 <*>
1955 V26 N11 P18-23	R-4841 <*>
1956 N1 P1-7	R-4330 <*>
1956 N1 P7-13	R-4333 <*>
1956 N5 P10-14	R-4846 <*>
1956 V27 N2 P21-23	59-10194 <+*> O
1956 V27 N4 P1-8	60-23560 <+*>
1956 V27 N5 P27-30	3839 <HB>
1956 V27 N6 P36-37	59-14188 <+*>
1956 V27 N8 P23-27	65-60504 <=$>
1956 V27 N11 P1-7	59-12391 <+*>
1956 V27 N11 P5-7	59-16553 <+*> O
1956 V27 N12 P9-12	59-15846 <+*>
1956 V27 N12 P23-25	59-14113 <+*>
	59-22571 <+*>
1957 N5 P24-25	4334 <HB>
1957 V28 N2 P1-9	61-13478 <*+>
1957 V28 N3 P38-39	R-3773 <*>
1957 V28 N4 P29-30	3998 <HB>
1957 V28 N4 P34-35	3988 <HB>
1957 V28 N6 P1-4	60-19558 <+*>
1957 V28 N6 P26	3966 <HB>
1957 V28 N7 P26-28	62-24105 <=*>
1957 V28 N9 P1-9	60-13437 <+*>
1957 V28 N9 P10-14	59-11992 <=> O
1958 N10 P20-22	59-18856 <+*>
1958 V29 N4 P28-30	62-32241 <=*>
1958 V29 N10 P32-33	4641 <HB>
1958 V29 N12 P27-29	4922 <HB>
1959 N1 P1-2	NLLTB V1 N6 P42 <NLL>
1959 N5 P33-34	NLLTB V1 N9 P1 <NLL>
1959 V30 N3 P17-19	AD-615 229 <=$>
1959 V30 N5 P1-2	61-31527 <=>
1959 V30 N7 P1-2	61-31527 <=>
1959 V30 N11 P1-2	61-31527 <=>
1960 V31 N5 P42	62-23485 <=*>
1963 V34 N1 P43	5855 <HB>
1963 V34 N3 P1-3	63-21769 <=>
1963 V34 N3 P32	5907 <HB>
1963 V34 N4 P1-2	63-21966 <=>
1963 V34 N5 P1-2	63-31232 <=> P
1963 V34 N6 P1-2	63-31524 <=>
1963 V34 N7 P42-44	63-31673 <=>
1963 V34 N8 P30	6118 <HB>
1964 V35 N1 P26-28	64-71530 <=>
1964 V35 N3 P1-2	64-31528 <=>
1964 V35 N3 P8-17	64-31528 <=>
1964 V35 N8 P1-3	64-51819 <=$>
1964 V35 N11 P1-2	65-30568 <=$>
1965 N10 P38-40	65-33765 <=$>
1965 V36 N1 P1-2,41	65-31096 <=$>
1965 V36 N4 P1-2	65-31613 <=$>
1966 N2 P45-46	66-31342 <=$>

STARSHINA SERZHANT

Citation	Translation
1963 N1 P24-25	63-31711 <=>
1963 N2 P17,32	AD-632 521 <=$>
1964 N11 P2-3	65-30181 <=$>

STATISTICA. THE HAGUE

Citation	Translation
1953 V7 N1 P15-22	2537 <*>
1953 V7 N1 P23-40	57-1025 <*>

1953 V7 N4 P193-198 57-2646 <*>
1953 V7 N4 P209-221 57-2615 <*>

STATISTICKY OBZOR
1958 V38 N6 P243-250 59-11069 <=>

STATISTIQUES. OFFICE INTERNATIONAL DES EPIZOOTIES
1954 V42 P267-277 57-1154 <*>

STATISTISCHE NACHRICHTEN. AUSTRIA
1948 V3 N2 P242-245 3050 <*>

STATISTISCHE PRAXIS
1963 V18 N1 P15-19 63-21522 <=>
1965 N2 P82-85 65-30798 <=$>
1965 V20 N5 P214-217 65-32069 <=*$>

STATISZTIKAI SZEMLE
1961 V39 N5 P529-537 63-15432 <=*>
1964 V42 N8/9 P877-888 64-51396 <=>
1965 V43 N5 P491-504 65-31989 <=$*>
1965 V43 N12 P1238-1246 66-30296 <=$>
1966 V44 N8/9 P842-854 66-35301 <=>

*STAUB (REINHALTUNG DER LUFT. TITLE VARIES)
1955 P436-467 58-14631 <+*> 0
1955 N46 P481-488 58-1714 <*>
1956 N44 P159-173 58-1713 <*>
1958 V18 P15-17 <ES>
1958 V18 N1 P3-14 62-16695 <*> 0
1959 P291-296 2257 <BISI>
1959 V19 P253-255 2353 <BISI>
1959 V19 N12 P413-416 66-10902 <*>
1960 V20 P393 AEC-TR-5531 <*>
1960 V20 N3 P69-100 62-10270 <*>
1961 V21 P298-300 UCRL TRANS-1045 <*>
1961 V21 N5 P212-215 5617 <HB>
1962 V22 P1C5-108 UCRL TRANS-1046 <*>
1962 V22 N7 P270-275 3277 <BISI>
1962 V22 N9 P343-390 AEC-TR-5607 <*>
1963 V23 P69-76 UCRL TRANS-1047 <*>
1963 V23 P92-94 66-12364 <*>
1963 V23 N2 P64-69 65-17090 <*>
1963 V23 N6 P304-309 64-10810 <*> 0
1963 V23 N9 P424-430 NP-TR-1151 <*>
1963 V23 N10 P443-451 3810 <BISI>
1964 V24 N5 P175-182 66-11960 <*>
1964 V24 N6 P2C1-205 4611 <BISI>
1964 V24 N6 P205-210 3889 <BISI>
1964 V24 N6 P223-228 66-11958 <*>
1964 V24 N9 P353-359 3993 <BISI>
1964 V24 N10 P396-400 66-12144 <*>
1964 V24 N11 P444-448 4285 <BISI>
1964 V24 N12 P525-528 66-11959 <*>
1965 V25 N1 P15-21 4326 <BISI>
1965 V25 N5 P175-179 4759 <BISI>
1966 V26 N2 P65-69 4938 <BISI>

STAVEBNICKY CASOPIS
1962 V10 N4 P193-212 CSIR-TR.299 <CSSA>
1962 V10 N5 P275-281 CSIR-TR.300 <CSSA>
1962 V10 N6 P378-382 CSIR-TR.289 <CSSA>

STAVIVO
1956 V34 N5 P165-171 R-3920 <*>
1963 V41 P52-54 66-10233 <*> 0

STAZIONI SPERIMENTALI AGRARIE ITALIANE
1916 V49 P405-421 3893-A <K-H>
1920 V53 P81-96 5484-B <K-H>

*STEKLO I KERAMIKA
1948 V5 N12 P20 60-13982 <+*>
1949 V6 N2 P15 60-15000 <+*>
1949 V6 N3 P10-14 BGIRA-549 <BGIR>
1949 V6 N5 P17-18 60-15006 <+*>
1949 V6 N8 P10-14 62P64R <ATS>
1949 V6 N8 P17-20 R-3196 <*>
1949 V6 N9 P10-14 RT-1631 <*>
1949 V6 N10 P20 60-15001 <+*>
1949 V6 N11 P11-17 64-14036 <=*$>
1950 V7 N2 P16-18 RT-977 <*>
1950 V7 N3 P7-12 65-61165 <=> 0
1950 V7 N9 P13 6C-15002 <+*>
1951 V8 N3 P17 60-15003 <+*>
1951 V8 N4 P9-12 65-12872 <*>
1951 V8 N7 P4-10 62-10936 <=*>
1952 N5 P6-10 T-2078 <INSD>
1952 V9 N8 P8-9 RT-3039 <*>
1952 V9 N8 P17 60-15005 <+*>
1952 V9 N11 P14 60-15004 <+*>
1953 N9 P13-15 T2081 <INSD>
1953 N10 P6-11 T-2080 <INSD>
1953 N11 P4-7 RJ-175 <ATS>
1953 V10 N2 P4-7 62-16520 <=*>
1953 V10 N11 P4-7 RJ-175 <ATS>
62-10565 <=*> 0
1953 V10 N12 P4-11 <INSD>
1954 N9 P9-12 RT-3014 <*>
1954 V11 N3 P16-18 AD-625 196 <=*$>
1954 V11 N4 P4-6 59-10975 <+*>
1954 V11 N6 P3-11 62-14900 <=*>
1954 V11 N8 P3-5 60-10152 <+*>
1954 V11 N8 P23-25 60-10155 <+*> C
1954 V11 N9 P8 60-13986 <+*>
1954 V11 N10 P24-26 97L34R <ATS>
1955 V12 N12 P16-18 CSIRO-3325 <CSIR>
R-3916 <*>
60-10153 <+*>
1955 V12 N12 P22-23 CSIRO-3326 <CSIR>
R-3924 <*>
1955 V12 N12 P25-27 CSIRO-3327 <CSIR>
R-3953 <*>
1956 N3 P9-13 R-4840 <*>
1956 N7 P79-83 132TM <CTT>
1956 V13 N1 P18 60-13998 <+*>
1956 V13 N4 P25 60-13987 <+*>
1956 V13 N4 P29 60-13988 <+*>
1956 V13 N9 P1 60-13990 <+*>
1956 V13 N9 P15 6C-13989 <+*>
1956 V13 N10 P26 60-13991 <+*>
1956 V13 N12 P5-9 62-16418 <=*>
1956 V13 N12 P21-22 R-3942 <*>
1957 V14 N1 P15 60-13993 <+*>
1957 V14 N1 P19-21 60-23080 <+>
1957 V14 N1 P25 60-13992 <+*>
1957 V14 N2 P8 60-13994 <+*>
1957 V14 N2 P11-18 64-14051 <=*$>
1957 V14 N3 P12 60-13995 <+*>
1957 V14 N4 P9-13 60-18535 <+*>
1957 V14 N4 P19-22 65-13515 <*>
1957 V14 N8 P22-23 62-14591 <=*>
1958 V15 N1 P4-9 62-10953 <=*>
1958 V15 N1 P28 60-13996 <+*>
1958 V15 N3 P9-13 62-14898 <=*>
1958 V15 N3 P13-16 61-19549 <=*>
64-16525 <=*$>
1958 V15 N4 P11-16 62-18197 <=*>
1958 V15 N5 P25-29 62-16148 <=*>
1958 V15 N6 P4-6 60-18912 <+*>
1958 V15 N6 P43-45 60-13997 <+*>
1958 V15 N8 P22-25 59-17239 <+*>
1958 V15 N10 P22-25 60-15396 <+*>
1958 V15 N10 P31-34 61-14468 <+*>
1959 V16 N3 P8-11 62-10950 <=*>
1959 V16 N3 P44 60-23009 <+*>
1959 V16 N5 P1-4 61-19426 <+*>
1959 V16 N7 P12-15 61-23867 <*=>
1959 V16 N8 P1-3 59-13939 <=>
1959 V16 N8 P13-16 62-16417 <=*>
1959 V16 N8 P21-22 61-19432 <+*>
1959 V16 N9 P25-29 ICE V1 N1 P99-105 <ICE>
14N51R <ATS>
64-14050 <=*$>
1959 V16 N10 P1-4 60-11208 <=>
1959 V16 N10 P5-6 63-18621 <=*>
1959 V16 N10 P9-14 15N51R <ATS>
1959 V16 N10 P14-20 62-10933 <=*>
1959 V16 N11 P1-2 60-11359 <=>
1959 V16 N11 P4-7 61-27568 <*=>

1959 V16 N11 P48 67M43R <ATS>
1960 V17 N1 P1-3 60-11569 <=>
1960 V17 N1 P7-12 65-11191 <*>
1960 V17 N2 P28-31 61-27406 <*=>
61-28944 <*=> 0
1960 V17 N2 P47 61-27406 <*=>
1960 V17 N5 P1-8 60-41063 <=>
1960 V17 N5 P23-24 64-14024 <=*$>
1960 V17 N6 P18-21 61-23674 <*=>
1960 V17 N6 P29-33 89N53R <ATS>
1960 V17 N7 P1 60-41507 <=>
1960 V17 N7 P21-244 62-13156 <*=>
1960 V17 N8 P1-3 61-11020 <=>
1960 V17 N8 P4-7 62-13154 <*=>
1960 V17 N8 P7-9 64-10566 <=*$>
1960 V17 N8 P22-25 64-10555 <=*$>
1960 V17 N9 P1-5 61-11919 <=>
1960 V17 N1C P1-4 63-18615 <=*>
63-23512 <=*>
1960 V17 N11 P1-4 61-21388 <=> P
1960 V17 N12 P40 61-23019 <=>
1961 V18 N2 P42-43 61-23949 <*=>
1961 V18 N2 P46-47 61-23949 <*=>
1961 V18 N3 P1-3 61-23063 <=>
1961 V18 N11 P46-47 62-24011 <=*>
1962 N12 P8-10 AD-631 433 <=*$>
1962 V19 N1 P4 62-24919 <=*>
1962 V19 N4 P37-38 63-13110 <=*>
1962 V19 N6 P1-3 62-32847 <=*>
1962 V19 N7 P47-48 62-32847 <=*>
1963 V20 N1 P23-25 63-21788 <=>
1963 V20 N1 P43-44 63-21623 <=>
1963 V20 N5 P39-41 63-31371 <=>
1964 V21 N1 P22-26 AD-615 227 <=$>
1964 V21 N3 P12-16 AD-618 641 <=$>
1965 N10 P27 NLLTB V8 N6 P496 <NLL>
1965 V22 N1 P27-30 CSIR-TR.522 <CSSA>

STEKOLNAYA I KERAMICHESKAYA PROMYSHLENNOST
1946 V3 N1/2 P11-13 64-16537 <=*$>
1947 N9 P16 R-529 <RIS>
1947 N9 P17-18 R.530 <RIS>
1947 V4 N7 P9-11 62-16152 <=*>

STERNE UND WELTRAUM
1962 N9 P193-195 65-10844 <*>
1965 N4 P87-88 66-11150 <*>

STIINTA SOLULUI
1966 V4 N2 P4-11 66-33737 <=$>

STIINTA SI TEHNICA. ACADEMIA REPUBLICII POPULARE ROMINE. BUCURESTI
1961 V13 N9 P16-17 62-19665 <=*>
1962 V14 N4 P30-33 62-33029 <=*>
1962 V14 N9 P9-10 63-21514 <=>
1962 V14 N11 P8-9 63-21514 <=>
1962 V14 N11 P14-16 63-21514 <=>
1962 V14 N11 P24-25 63-21514 <=>
1962 V14 N11 P45 63-21514 <=>
1963 V15 N1 P19-21 64-71164 <=>
1963 V15 N2 P40-41 64-11770 <=>
1963 V15 N3 P3-5 63-21873 <=>
1963 V15 N3 P10-11 63-21721 <=>
1963 V15 N12 P3-5 64-21505 <=>
1963 V15 N12 P26-29 64-21434 <=>
64-71180 <=>
1964 V16 N4 P20-21 64-31476 <=>
1964 V16 N6 P3-5 64-41204 <=>
1964 V16 N9 P36-37 64-51583 <=*$>
1964 V16 N11 P12-15 65-30366 <=$>
1965 V17 N2 P24-25 65-30578 <=$>
1965 V17 N3 P3-5 65-30842 <=$>
66-32463 <=$>
1965 V17 N3 P12 66-32463 <=$>
1965 V17 N4 P8-9 65-31259 <=$>
1965 V17 N5 P6-7 65-31396 <=$>
1965 V17 N8 P7-9 65-32604 <=$>
1966 V18 N6 P8-9 66-33580 <=$>

STOMATOLOGIYA. MCSCOW
1937 N4 P56-58 RT-1301 <*>
1938 N4 P3-5 RT-1300 <*>
1951 V29 N1 P25-27 AD-622 475 <=*$>
1951 V29 N2 P12-13 2237A <K-H>
1955 N6 P32 62-15304 <*=>
1960 V39 N1 P75-79 60-31140 <=>
1961 V40 N4 P27-28 62-32136 <=*>
1961 V40 N6 P14-19 AD-619 318 <=$>
1963 N3 P7-10 AD-637 420 <=$>
1963 V42 N3 P29-33 AD-618 899 <=$>
1963 V42 N4 P11-15 AD-627 123 <=$>

STOMATOLOGIYA. SOFIYA
1954 N2 P40-44 RT-2309 <*> 0
1954 N2 P53-54 RT-2305 <*>
1963 V13 N6 P2-9 64-21686 <=>

STRADE
1955 V35 P135-146 T-1730 <INSD>
1955 V35 P188-194 T-1730 <INSD>
1956 V36 P297-301 T-2005 <INSD>

STRAHLENTHERAPIE
1928 V29 N3 P367-374 57-277 <*>
1929 V33 P362-374 57-126 <*>
1930 V38 P521-542 57-537 <*>
1932 V45 P700-710 I-452 <*>
1934 V50 P357- 57-433 <*>
1935 V52 P282-298 57-282 <*>
1935 V52 P531-536 57-434 <*>
1935 V52 P537-544 57-385 <*>
1936 V55 P498-523 I-841 <*>
II-409 <*>
1940 V67 P487-499 58-2306 <*>
1947 V77 P91-106 65-00210 <*>
1948 V77 P573-584 66-10938 <*>
1950 V81 P177-186 58-2524 <*>
1950 V81 P273-280 58-2648 <*>
1950 V81 N2 P187-192 59-10451 <*> 0
1950 V83 N4 P654-662 C-2440 <NRC>
57-3028 <*>
1952 V86 P227-240 58-2400 <*>
1952 V88 N2 P261-275 I-520 <*>
1953 V90 P78-87 58-2525 <*>
1953 V90 N4 P546-552 59-17357 <*>
1953 V91 N1 P149-153 59-10452 <*> 0
1953 V91 N4 P551-554 9382 <K-H>
1953 V92 P555-562 AEC-TR-3732 <+*>
1953 V92 P576-589 2040 <*>
1953 V92 N4 P612-620 NP-TR-872 <*>
1953 V92 N4 P649-653 AEC-TR-3473 <*> 0
1954 V93 P89-93 I-445 <*>
1954 V94 P527-538 II-121 <*>
1954 V94 N1 P72-78 59-10454 <*> 0
1954 V94 N3 P455-459 NP-TR-906 <*>
1954 V95 P302-311 65-00360 <*>
1955 V96 N2 P169-200 1262 <*>
1955 V96 N2 P241-249 57-683 <*>
1955 V97 P549-567 58-2305 <*>
1955 V98 P453-463 59-10323 <*>
1955 V98 N4 P570-575 1433 <*>
1956 V99 P94-105 65-00265 <*> 0
1956 V99 P290-300 II-702 <*>
1956 V100 N2 P259-268 62-16221 <*> 0
1957 V102 P590-595 58-1185 <*>
1957 V102 N1 P65-72 62-16222 <*> 0
1957 V103 P472-476 58-2299 <*>
1957 V104 P169-181 58-1708 <*>
1957 V104 P338-340 59-15052 <*> 0
1957 V104 P345-354 65-00103 <*> 0
1957 V104 P494-506 G-262 <RIS>
1958 V105 P278-295 65-00366 <*> 0
1958 V105 N1 P39-44 64-14581 <*> 0
8500-D <K-H>
1958 V105 N1 P138-142 29M45G <ATS>
1958 V106 P44-57 65-00167 <*> 0
1958 V106 N4 P606-626 G-263 <RIS>
1958 V107 P298-308 G-264 <RIS>
1958 V107 P437-443 G-265 <RIS>

1959 V108 P63-72 G-266 <RIS>
1959 V108 P594-601 AEC-TR-4312 <*>
1959 V108 N1 P8-16 TT-817 <NRCC>
1959 V108 N1 P57-62 60-18546 <*>
1959 V108 N2 P257-261 NP-TR-429 <*>
1959 V108 N2 P296-300 NP-TR-424 <*>
1959 V109 P412-425 62-01049 <*>
1959 V109 P464-482 G-268 <RIS>
1959 V109 N2/3 P464-482 AEC-TR-3854 <*>
1959 V110 N2 P248-259 C-3360 <NRCC>
1960 V111 P266-272 AEC-TR-4131 <*>
1960 V111 N1 P65-74 62-00611 <*>
1961 V116 P251-258 63-18719 <*>
1961 V116 P420-425 AEC-TR-5381 <*>
1961 V116 N1 P85-96 62-16305 <*> 0
1963 V122 N3 P463-471 66-12376 <*> 0
1964 V123 P132-138 AEC-TR-6521 <*>

STRAHLENTHERAPIE-SONDERBAND
1957 V37 P128-132 64-14547 <*>
8003-D <K-H>
1959 V43 P373-378 61-00298 <*>

STRASBOURG MEDICAL
1956 V7 N3 P171-179 65-13735 <*>
1960 V11 N6 P429-435 66-11645 <*> 0

STRASSE
1965 V5 N1 P23-25 AD-638 209 <=$>
1965 V5 N1 P25-30 AD-638 209 <=$>

STRASSE UND AUTOBAHN
1950 V1 P9-11 T-1451 <INSD>
1950 V1 P20-22 T-1401 <INSD>
1950 V1 N9 P1-5 T-1445 <INSD>
1951 V2 P67-71 T-1416 <INSD>
1951 V2 P85-89 T-1436 <INSD>
1951 V2 N8 P250-252 58-2600 <*>
61-18933 <*>
1953 V4 N5 P161-167 64-10323 <*>
1953 V4 N7 P223-227 I-959 <*>
1955 V6 N7 P246-255 58-2572 <*>
61-28960 <*>
1955 V6 N8 P278-282 58-2570 <*>
61-18995 <*>
1956 N3 P78-82 C-2576 <NRC>
58-786 <*>
1956 V7 N11 P388-393 T-2515 <INSD>
1961 V12 N6 P197-202 105 <CCA>
65-13502 <*>

STRASSE UND VERHEHR
SEE ROUTE ET LA CIRCULATION ROUTERIE

STRASSEN- UND TIEFBAU
1948 V2 P292-298 T-1964 <INSD>
1950 V4 P181-184 T1447 <INSD>
1950 V4 P190-192 T-1448 <INSD>
1950 V4 P211-213 T-1453 <INSD>
1950 V4 P248-250 T-1453 <INSD>
1950 V4 P286-287 T-1446 <INSD>
1950 V4 P294-297 T-1453 <INSD>
1952 V6 N4 P126-128 I-695 <*>
1954 V8 N11 P575-576 1771 <*>
1954 V8 N11 P578 1771 <*>
1957 N11 P658-661 58-2245 <*>
1957 V11 N8 P454-461 R-2591 <*>
61-18981 <*>
1960 V14 N9 P674 104 <CCA>
1960 V14 N9 P674-693 65-13365 <*>
1960 V14 N9 P676 104 <CCA>
1960 V14 N9 P678 104 <CCA>
1960 V14 N9 P680 104 <CCA>
1960 V14 N9 P682 104 <CCA>
1960 V14 N9 P684 104 <CCA>
1960 V14 N9 P686 104 <CCA>
1960 V14 N9 P688 104 <CCA>
1960 V14 N9 P690 104 <CCA>
1960 V14 N9 P692-693 104 <CCA>
1962 V16 N12 P1197-1209 RAPRA-1263 <RAP>

STRIDES OF MEDICINE. JAPAN
SEE IGAKU NO AYUMI

STROIKAKH ROSSII
1963 N1 P3-7 63-21929 <=>
1963 N2 P19-20 63-21789 <=>
1963 N2 P27 63-21789 <=>

STROITEL. MOSCOW
1959 V5 N10 P14 60-11359 <=>
1959 V5 N11 P1 60-11470 <=>
1960 V6 N8 P27-28 61-11861 <=>
1963 N8 P3-6 NLLTB V6 N2 P127 <NLL>
1963 V9 N8 P3-6 NLLTB,V6,N2,P127-139 <HMSO>
1966 N6 P6-7 NLLTB V8 N12 P1063 <NLL>
1966 V12 N11 P3 67-30302 <=>

STROITELNAYA GAZETA
1959 V23 12/23 P1 60-11470 <=>
1960 V24 08/19 P4 61-11020 <=>
1962 04/20 P2 62-32064 <=>
1962 05/18 P2 62-32064 <=>
1964 N55 P86-91 AC/66/II-105 <CEMB>

STROITELNAYA MEKHANIKA I RASCHET SOORUZHENII. MOSKVA
1959 N2 P9-16 63-00124 <*>
1963 V5 N3 P7-11 65-64367 <=*$>

STROITELNAYA PROMYSHLENNOST
1938 V16 N2 P43-44 C-2243 <*>
R-292 <*>
1939 V17 N6 P47-50 63-15695 <=*$>
1943 V21 N9 P11-13 RT-1683 <*>
60-17740 <+*>
1950 V28 N12 P11-12 60-13874 <+*>
1951 V29 N1 P12-14 59-19144 <+*>
1954 V32 N2 P38-41 T.1672 <*>
1954 V32 N9 P23-25 RT-2793 <*>
1954 V32 N10 P14-17 RT-3401 <*>
60-17741 <+*>
1954 V32 N11 P30-31 TT-830 <NRCC>
1954 V32 N12 P22-23 TT-830 <NRCC>
1955 N3 P34-38 R-2351 <*>
1955 N9 P39-40 R-3681 <*>
1955 N12 P8-12 RT-4287 <*>
1955 V33 N4 P23-26 C-3564 <NRCC>
1956 N3 P29-31 59-16989 <+*>
1956 N4 P27-28 R-5286 <*>
1956 V34 N6 P25-27 TT-830 <NRCC>
1957 N11 P2-4 PB 141 209T <=>
1957 V35 N1 P50 TT-830 <NRCC>
1957 V35 N6 P23-27 60-17150 <+*> 0
1957 V35 N7 P18-21 TT-830 <NRCC>
1958 N4 P11-13 60-13160 <+*>
1958 V36 N2 P28-31 60-17633 <+*>
1958 V36 N7 P37-39 60-15980 <+*>

STROITELNOE I DOROZHNOE MASHINOSTROENIE
1957 V2 N9 P15-18 61-17566 <=$>
1958 V3 N1 P3-5 PB 141 222T <=> 0
1958 V3 N1 P5-7 PB 141 222T <=> 0
1958 V3 N1 P17-20 PB 141 222T <=> 0
1958 V3 N6 P5-11 59-11604 <=> 0
1958 V3 N6 P11-14 59-11556 <=> 0
1958 V3 N6 P14-20 59-11557 <=> 0
1958 V3 N6 P20-26 59-11558 <=> 0
1958 V3 N6 P26-32 59-11559 <=> 0
1958 V3 N6 P33-39 59-11605 <=> 0
1958 V3 N8 P16-18 59-22561 <+*>
1958 V3 N10 P13-17 59-22563 <+*>
1958 V3 N12 P12-14 59-22562 <+*>
1959 V4 N2 P5-7 61-13544 <+*> 0
1959 V4 N2 P13-14 61-13545 <+*>
1959 V4 N10 P3-5 60-11305 <=>
1960 N8 P3-8 61-11972 <=>
1960 N8 P35-36 61-11972 <=>
1960 V5 N1 P25-26 CSIR-TR.185 <CSSA>
1960 V5 N3 P12-13 60-41145 <=>

Citation	Reference
1960 V5 N3 P34	60-41145 <=>
1960 V5 N4 P8-12	60-41248 <=>
1960 V5 N4 P33	60-41248 <=>
1960 V5 N7 P1-5	61-11861 <=>
1960 V5 N7 P17-20	61-11861 <=>
1960 V5 N7 P24-25	61-11861 <=>
1960 V5 N7 P27	61-11861 <=>
1960 V5 N7 P35	61-11861 <=>
1960 V5 N8 P22-24	61-21374 <=>
1960 V5 N9 P3-4	61-21374 <=>
1960 V5 N9 P12	61-21374 <=>
1960 V5 N9 P14-15	61-21374 <=>
1960 V5 N9 P21	61-21374 <=>
1960 V5 N9 P31-34	61-21374 <=>
1960 V5 N10 P9	61-21374 <=>
1960 V5 N10 P12	61-21374 <=>
1960 V5 N10 P21	61-21374 <=>
1960 V5 N10 P36-38	61-21374 <=>
1960 V5 N12 P3-11	61-31115 <=>
1960 V5 N12 P23-27	61-31115 <=>
1960 V5 N12 P34-36	61-31115 <=>

STROITELNYE I DOROZHNYE MASHINY

Citation	Reference
1958 V3 N8 P13-16	62-18743 <=*>
1960 V5 N1 P20-25	62-18744 <=*>
1961 V6 N1 P3-5	61-23556 <=>
1961 V6 N1 P15-16	61-23556 <=>
1961 V6 N1 P19	61-23556 <=>
1961 V6 N1 P21-22	61-23556 <=>
1961 V6 N1 P31-32	61-23556 <=>
1961 V6 N2 P3	61-23013 <=>
1961 V6 N2 P14-15	61-23013 <=>
1961 V6 N2 P31	61-23013 <=>
1961 V6 N3 P3-9	61-27074 <=>
1961 V6 N3 P22-23	61-27074 <=>
1961 V6 N3 P27-28	61-27074 <=>
1961 V6 N3 P31-32	61-27074 <=>
1962 V7 N3 P1-8	63-13093 <=>
1962 V7 N3 P12-15	63-13093 <=>
1962 V7 N3 P35-37	63-13093 <=>
1962 V7 N4 P1-7	62-32057 <=>
1962 V7 N4 P33-37	62-32057 <=>
1962 V7 N5 P1-11	62-32214 <=> P
1962 V7 N6 P1-4	62-32215 <=*>
1962 V7 N6 P9-10	66-11019 <*>
1962 V7 N7 P1-8	62-32845 <=>
1962 V7 N7 P14-16	62-32845 <=>
1962 V7 N8 P1-3	63-13347 <=>
1962 V7 N8 P7	63-13073 <=>
1962 V7 N8 P16-19	63-13073 <=>
1962 V7 N8 P38	63-13073 <=>
1962 V7 N9 P1-4	63-13347 <=>
1962 V7 N9 P10	63-13532 <=> P
1962 V7 N9 P16-18	63-13532 <=> P
1962 V7 N9 P31-36	63-13532 <=> P
1962 V7 N10 P1	63-13336 <=> P
1962 V7 N10 P3-6	63-13336 <=> P
1962 V7 N10 P11	63-13336 <=> P
1962 V7 N10 P15-16	63-13336 <=> P
1962 V7 N10 P28-29	63-13336 <=> P
1962 V7 N10 P33	63-13336 <=> P
1963 V8 N1 P1-3	63-21369 <=> P
1963 V8 N1 P7-13	63-21369 <=> P
1963 V8 N1 P17	63-21369 <=> P
1963 V8 N1 P19-20	63-21369 <=> P
1963 V8 N1 P27-28	63-21369 <=> P
1963 V8 N2 P1-2	63-21484 <=>
1963 V8 N5 P18	63-31264 <=>
1963 V8 N5 P30-31	63-31264 <=>
1963 V8 N6 P7	63-31797 <=>
1963 V8 N6 P14	63-31797 <=>
1963 V8 N6 P35-36	63-31797 <=>
1963 V8 N7 P1-13	63-31660 <=>
1963 V8 N7 P16-20	63-31660 <=>
1963 V8 N7 P32	63-31660 <=>
1963 V8 N8 P6-7	63-41149 <=>
1963 V8 N8 P10-11	63-41149 <=>
1963 V8 N8 P13-14	63-41149 <=>
1963 V8 N8 P18-21	RTS-2573 <NLL>
1963 V8 N8 P22-23	63-41149 <=>
1963 V8 N8 P27-29	63-41149 <=>
1963 V8 N8 P31-32	63-41149 <=>
1963 V8 N8 P32	63-41149 P7 <=>
	63-41149 P8 <=>
1963 V8 N8 P35-36	63-41149 <=>
1963 V8 N9 P1-5	64-11583 <=>
1963 V8 N12 P1-2	64-21723 <=>
1963 V8 N12 P11-14	64-21723 <=>
1963 V8 N12 P27-29	64-21723 <=>
1964 V9 N11 P1-3	65-30524 <=$>
1965 V10 N1 P1-3	65-31653 <=$>
1965 V10 N1 P16-17	65-31653 <=$>
1965 V10 N1 P22-25	65-31653 <=$>
1965 V10 N8 P1-2	66-30156 <=$>
1965 V10 N11 P1	66-30156 <=$>
1966 N4 P1-2	NLLTB V8 N11 P936 <NLL>
1966 N8 P3-4	66-34931 <=$>
1967 N2 P1-3	67-31431 <=$>
1967 N2 P33	67-31431 <=$>

STROITELNYE MATERIALY

Citation	Reference
1955 N9 P13-15	63-13700 <=*>
1955 V1 N9 P11-13	64-19376 <=$>
1956 V2 N10 P31-32	R-5325 <*>
1958 V4 N6 P8-11	59-22227 <+*>
1959 V5 N2 P23-26	60-15739 <+*>
1959 V5 N5 P12-16	60-15741 <+*>
1959 V5 N7 P1-8	59-13939 <=>
1959 V5 N7 P21-22	61-13951 <=*>
1959 V5 N7 P25	61-13951 <=*>
1959 V5 N9 P1-3	60-11208 <=>
1959 V5 N10 P1-3	60-11208 <=>
1959 V5 N11 P1-3	60-11470 <=>
1960 V6 N1 P1-3	60-11569 <=>
1960 V6 N2 P29-31	61-28137 <*=>
1960 V6 N3 P1-3	60-11764 <=>
1960 V6 N3 P16-18	60-11764 <=>
1960 V6 N4 P3-6	60-11814 <=>
1960 V6 N4 P7-12	61-28140 <*=>
1960 V6 N4 P34-37	61-28139 <*=>
1960 V6 N5 P1-3	60-41063 <=>
1960 V6 N5 P8-9	62-14532 <=*>
1960 V6 N6 P1-3	60-41507 <=>
1960 V6 N7 P1-2	60-41507 <=>
1960 V6 N8 P1-3	61-11722 <=>
1960 V6 N11 P1-2	61-21388 <=> P
1960 V6 N11 P32-34	62-25939 <=*>
1961 N1 P1-3	61-21686 <=>
1961 V7 N1 P15-16	<CLAI>
1961 V7 N4 P7	61-23945 <=>
1961 V7 N4 P30-33	<CLAI>
1962 V8 N2 P36-39	63-24204 <=*$>
1962 V8 N6 P34-36	63-24206 <=*$>
1963 V9 N1 P31-34	65-20426 <$>
1963 V9 N5 P1-2	TR-17 <JLRD>
1964 N12 P35-36	<JLRD>
1964 V10 N2 P27	<APC>
1964 V10 N6 P8-9	<APC>
1965 N2 P1-3	NLLTB V7 N9 P777 <NLL>
1965 V11 N10 P30-32	66T91R <ATS>
1966 N1 P1-4	NLLTB V8 N8 P692 <NLL>
1966 N8 P9-13	66-34931 <=$>
1966 N11 P10	67-30302 <=>
1967 N1 P1-4	67-31308 <=$>
1967 N1 P9-10	67-31308 <=$>
1967 N1 P12	67-31308 <=$>
1967 N1 P18-20	67-31308 <=$>
1967 N2 P4-5	67-31431 <=$>

STROITELSTVO. BULGARIA

Citation	Reference
1959 N1 P9-13	59-11586 <=> 0

STROITELSTVO I ARKHITEKTURA. KIEV

Citation	Reference
1959 V7 N1 P81-94	60-21792 <=>
1960 V8 N4 P34-35	61-11972 <=>
1960 V8 N7 P2-4	61-11020 <=>
1961 V9 N1 P1-4	61-23063 <=*>
1961 V9 N6 P135-145	62-33619 <=*>
1963 V11 N5 P10-12	<BRS>

STROITELSTVO I ARKHITEKTURA MOSKVY
1959 V7 N12 P6-7 60-11569 <=>
1960 N11 P21-22 61-21538 <=>
1960 V9 N9 P23-26 61-11919 <=>
1960 V9 N10 P1-6 61-21388 <=> P

STROITELSTVO I ARKHITEKTURA SREDNEI AZII
1964 N5 P32-34 AD-629 862 <=*$>

STROITELSTVO TRUBOPROVODOV
1958 V3 N110 P1 60-13551 <+*>
1958 V3 N114 P2 60-13551 <+*>
1958 V3 N116 P2 60-13551 <+*>
1958 V3 N118 P1 60-13551 <+*>
1958 V3 N119 P2 60-13551 <+*>
1959 V4 N2 P6-9 62-16288 <=*>
1959 V4 N4 P15-17 58L35R <ATS>
1959 V4 N7 P28-30 59-13984 <=>
1959 V4 N10 P29-30 60-11572 <=>
1960 V5 N12 P12-15 62-10602 <*=>
1961 V6 N3 P16-18 95N57R <ATS>
1961 V6 N7 P14-16 99N56R <ATS>
1961 V6 N12 P10-11 66P59R <ATS>
1966 V11 N4 P15-18 66-32472 <=>

STROJE NA ZPRACOVANI INFORMACI. CESKOSLOVENSKA AKADEMIE VED. USTAV MATEMATICKYCH STROJU. PRAGUE
1957 V5 P9-37 60-12608 <=>

STROJIRENSKA VYROBA
1959 V7 N2 P68-71 64-14058 <*>

STROJIRENSTVI
1955 V5 P219-221 AEC-TR-4541 <*>
1955 V5 N3 P194-198 R-3796 <*>
1958 V8 N12 P923-928 95M41C <ATS>
1959 V9 N1 P33-40 430 <MT>
1959 V9 N3 P163-169 93M38C <ATS>
1959 V9 N4 P303-308 62-25199 <=*> O
1959 V9 N5 P329-335 62-23002 <*=>
1959 V9 N12 P939-942 3202 <BISI>
1960 V10 N2 P139-149 62-25200 <=*> O
1960 V10 N5 P374-377 62-19812 <*=>
1960 V10 N10 P778-781 TRANS-169 <MT>
1960 V10 N11 P819-829 144-70 <STT>
1960 V10 N11 P870-873 189-73 <STT>
1960 V10 N12 P901-906 145-288 <STT>
1960 V10 N12 P920-924 175-290 <STT>
1960 V10 N12 P927-932 145-291 <STT>
1960 V10 N12 P941-943 154-293 <STT>
1961 V11 N4 P275-282 TRANS-191 <MT>
1961 V11 N6 P449-453 TRANS-204 <MT>
1961 V11 N11 P843-847 TRANS-230 <MT>
1961 V11 N12 P915-920 TRANS-350 <MT>
1962 V12 N1 P39-43 3364 <BISI>
1962 V12 N4 P283-287 TRANS-289 <MT>
1962 V12 N8 P608-616 490 <MT>
1962 V12 N9 P677-684 63-16487 <*>
1963 V13 P844-846 4061 <BISI>
1963 V13 N1 P46-51 TRANS-315 <MT>
1963 V13 N3 P163-171 65-11329 <*>
1964 V14 N1 P39-43 4194 <BISI>
1964 V14 N2 P94-100 NEL-TRANS-1549 <NLL>
1964 V14 N3 P360-363 66-13385 <*>
1964 V14 N9 P667-673 NEL-TRANS-1608 <NLL>
1964 V14 N11 P834-842 NEL-TRANS-1635 <NLL>
1964 V14 N12 P883-887 TRANS-1651 <NLL>
1964 V14 N12 P888-898 NEL-TRANS-1680 <NLL>
1965 V15 N1 P3-12 NEL-TRANS-1664 <NLL>
1965 V15 N2 P144-151 487 <MT>
1965 V15 N4 P254-259 NEL-TRANS-1702 <NLL>
1965 V15 N4 P260-269 TRANS-1714 <NLL>
1965 V15 N4 P278-283 TRANS-538 <MT>
1965 V15 N4 P307-311 TRANS-540 <MT>
1965 V15 N5 P333-340 NEL-TRANS-1720 <NLL>
1965 V15 N6 P433-439 NEL-TRANS-1730 <NLL>
1965 V15 N7 P491-497 NEL-TRANS-1783 <NLL>
1965 V15 N7 P506-511 NEL-TRANS-1772 <NLL>
1965 V15 N10 P723-730 NEL-TRANS-1788 <NLL>
1965 V15 N10 P757-762 TRANS-638 <MT>
1965 V15 N11 P814-822 NEL-TT-1795 <NLL>

STROJNOELEKTROTECHNICKY CASOPIS
1953 V4 N1 P5-28 57-1766 <*>
1958 V9 N4 P240-246 61-16158 <*>
63-18616 <*>
1958 V9 N9 P523-530 AEC-TR-4542 <*>

STUDI DI MEDICINA E CHIRURGIA DELLO SPORT
1950 V4 P12-23 59-10968 <*> O

STUDI SASSARESI
1962 V40 N1/2 P68-73 64-20906 <*>

STUDI TRENTINI DI SCIENZE NATURALI
1959 V36 N1 P10-48 88N50I <ATS>

STUDIA FILOZOFICZNE. WARSAW
1965 N2 P231-237 65-32457 <=$>

STUDIA GEOPHYSICA ET GEODAETICA
1958 V2 N1 P40-43 61-28462 <=*>
1960 V4 P111-118 62-19286 <=*>
1962 V6 N4 P369-399 63-21273 <=*>
1962 V6 N4 P400-406 63-21274 <=>
1962 V6 N4 P407-409 63-21270 <=>
1963 V7 P146-154 16R75G <ATS>
1964 V8 P109-119 AD-623 451 <=*$>
1964 V8 P274-286 65-13287 <*>
1964 V8 N3 P239-246 AD-631 837 <=$>
1965 V9 N9 P185-200 65-31894 <=$*>

STUDIA SOCIETATIS SCIENTIARUM TORUNENSIS. SECTIO B. CHEMIA
1961 SB V3 N2 P33-62 64-14079 <*>

STUDIA UNIVERSITATIS BABES-BOLYAI. BUCHAREST
1959 S2 N2 P179-181 65-29148 <$>
1960 S2 N1 P275-276 65-29147 <$>
1961 V6 N1 P157-162 64-71301 <=>
1963 N5 P57-67 755 <TC>
1963 N5 P69-78 756 <TC>

STUDIES ON SOLID STATE PHYSICS AND CHEMISTRY. JAPAN
SEE BUSSEIRON KENKYU

STUDII SI CERCETARI DE ASTRONOMIE. BUCURESTI
1964 V9 N1 P89-92 AD-637 435 <=$>
1964 V9 N1 P101-111 AD-637 435 <=$>
1964 V9 N1 P113-120 AD-637 435 <=$>

STUDII SI CERCETARI. BAZA DE CERCETARI STIINTIFICE TIMISOARA, ACADEMIA RPR. STIINTE CHEMICE.
1959 V6 N3/4 P9-19 62-14448 <*>
1959 V6 N3/4 P27-32 33N49RU <ATS>
1960 V7 P317-319 64-10398 <*>

STUDII SI CERCETARI DE CHIMIE. BUCURESTI
1954 V2 N1/2 P15-26 CSIR-TR.296 <CSSA>
1959 V7 P461 AEC-TR-5106 <*>
1960 V8 P509-517 65-12899 <*>
1962 V10 N3/4 P291-294 63-21691 <=>
1963 V11 N3/4 P325-330 66-11904 <*>

STUDII SI CERCETARI DE CHIMIE. FILIALA CLUJ, ACADEMIA RPR
1957 V8 N3/4 P199-206 66-10265 <*> O
1960 V11 N1 P55-66 66-10266 <*> O
1962 V13 N2 P157-170 66-10267 <*> O
1963 V14 N1 P93-101 66-10269 <*> O
1963 V14 N1 P103-110 66-10270 <*> O
1963 V14 N1 P111-124 66-10271 <*> O
1964 V15 N2 P281-287 ICE V4 N4 P664-666 <ICE>

STUDII SI CERCETARI DE ENDOCRINOLOGIE. BUCURESTI
1963 V14 N4/6 P433-436 64-21555 <=>
1964 V15 N4 P289-292 64-51849 <=$>

STUDII SI CERCETARI DE ENERGETICA

1955 V5 N3/4 P287-316 58-744 <*>
1959 V9 N1 P119-135 65-12245 <*>
1959 V9 N4 P647-680 65-12287 <*>
1959 V10 N3 P601-612 65-12288 <*>
1962 V12 P513-531 66-12245 <*>

STUDII SI CERCETARI DE FIZICA. BUCURESTI
1955 V6 P373-376 UCRL TRANS-1034(L) <*>
1956 V7 N1 P63-66 AEC-TR-3609 <+*>
1956 V7 N4 P567-577 NP-TR-317 <*>
1957 V8 N1 P41-53 65-17170 <*>
1958 V9 N3 P317-322 80M39RU <ATS>
1958 V9 N4 P459-463 78M38RJ <ATS>
1958 V9 N4 P489-496 79M38RU <ATS>
1959 V10 N1 P63-73 30M40R4 <ATS>
1960 V11 N1 P83-96 NP-TR-648 <*>
1960 V11 N1 P117-128 NP-TR-538 <*>
1960 V11 N2 P351-356 AEC-TR-4963 <=*>
74N55RU <ATS>
1961 V12 P825-838 ORNL-TR-228 <*>
1961 V12 N4 P753-763 62-18913 <*>
1961 V12 N4 P801-803 62-18768 <*>
1961 V12 N4 P815-820 62-18769 <*>
1962 V13 P29-35 AEC-TR-5906 <*>
1962 V13 P947-957 ORNL-TR-116 <*>
1962 V13 N1 P29-37 64-16704 <*>
1962 V13 N4 P651-665 53R79RU <ATS>
1963 V14 P731-733 66-11156 <*>

STUDII SI CERCETARI DE FIZIOLOGIE
1960 V5 N2 P321-325 62-19543 <=>
1960 V5 N2 P341-345 62-19543 <=>
1960 V5 N2 P349-352 62-19543 <=>
1960 V5 N2 P355-367 62-19543 <=>
1960 V5 N2 P389-396 62-19543 <=>
1960 V5 N2 P405-408 62-19543 <=>
1960 V5 N2 P421-427 62-19543 <=>
1960 V5 N2 P429-432 62-19543 <=>
1960 V5 N2 P435-441 62-19543 <=>
1960 V5 N3 P466-471 62-19543 <=>
1960 V5 N3 P515-519 62-19543 <=>
1960 V5 N3 P565-569 62-19543 <=>
1960 V5 N3 P587-596 62-19543 <=>
1960 V5 N3 P601-609 62-19543 <=>
1963 V8 N3 P335-337 63-41042 <=>
1964 V9 N5/6 P459-465 65-33617 <=*$>

STUDII SI CERCETARI DE GEOFIZICA
1962 V13 N4 P1001-1020 3461 <BISI>
1965 N2 P137-143 65-33127 <=*$>

STUDII SI CERCETARI DE INFRAMICROBIOLOGIE, MICROBIOLOGIE SI PARAZITOLOGIE. BUCURESTI
1960 V11 P111-115 NP-TR-579 <*>
1960 V11 P133-140 61-00653 <*>
1961 V12 N1 P9-16 62-19514 <=*>
1961 V12 N1 P19-25 62-19515 <=*>
1963 V14 P7-16 66-10932 <*>
1963 V14 P261-267 66-10955 <*>
1964 V15 N4 P321-323 64-51752 <=$>

STUDII SI CERCETARI DE MATEMATICA. FILIALA CLUJ, ACADEMIA RPR
1958 V9 P439-480 STMSP V3 P239-276 6 <AMS>
1958 V9 P481-490 STMSP V4 P321-329 <AMS>
1961 V12 N1 P29-40 64-71302 <=>

STUDII SI CERCETARI DE MECANICA APLICATA
1958 V9 P625-629 02P59RU <ATS>
1960 V11 P739-743 62-10725 <*>
1961 V12 N3 P633-640 66R77RU <ATS>
1963 V14 N4 P805-816 64-71161 <=>
1963 V14 N5 P1073-1087 N65-23687 <=$>
1964 V16 N4 P815-820 65-30214 <=$>

STUDII SI CERCETARI DE MEDICINA INTERNA. BUCURESTI
1961 V2 P237-250 11548-D <K-H>
64-10638 <*> 0

STUDII SI CERCETARI DE METALURGIE. BUCURESTI
1957 V2 P499-522 NP-TR-319 <*>
1958 N1 P39-52 TS-1251 <BISI>
1958 N2 P161-202 TS-1754 <BISI>
1959 V4 N4 P561-586 4724 <HB>
1960 V5 N1 P7-27 64-10588 <*>
1960 V5 N3 P371-390 5283 <HB>
1961 V6 N1 P103-112 5258 <HB>
1961 V6 N2 P161-169 5428 <HB>
1961 V6 N2 P171-183 5429 <HB>
1962 V7 P405-414 53Q72RU <ATS>
1962 V7 N1 P73-87 5703 <HB>
1963 V8 N2 P191-199 6199 <HB>
64-30132 <*> 0
1963 V8 N3 P277-296 6204 <HB>
1963 V8 N4 P409-424 6205 <HB>

STUDII SI CERCETARI DE NEUROLOGIE
1963 V8 N3 P325-335 64-21562 <=>

STUDII SI CERCETARI STIINTIFICE. BAZA DE CERCETARI STIINFICE, TIMISOARA, ACADEMIA RPR. SERIA I. STIINTE MATEMATICE, FIZICE, CHIMICE SI TEHNICE
1955 V2 P133-142 58-1437 <*>

STUDII SI CERCETARI STIINTIFICE. BAZA DE CERCETARI STIINTIFICE, TIMISOARA, ACADEMIA RPR. SERIA STIINTE CHIMICE
1961 V8 N1/2 P151-159 65-13522 <*>

STUDII SI CERCETARI STIINTIFICE. FILIALA CLUJ, ACADEMIA RPR
1952 V3 N3/4 P99-102 59-10654 <*>

STUDII SI CERCETARI STIINTIFICE. FILIALA CLUJ, ACADEMIA RPR. SERIA I. STIINTE MATEMATICE, FIZICE CHIMICE SI TEHNICE
1955 V3 N3/4 P225-232 63-14075 <*>

STUDII SI CERCETARI STIINTIFICE. FILIALA IASI, ACADEMIA RPR
1952 V3 P71-81 AEC-TR-4047 <*>
1954 V5 N1/2 P87-94 AEC-TR-3993 <=*>
AEC-TR-3994 <*>
1954 V5 N1/2 P215-220 AEC-TR-4102 <*>

STUDII SI CERCETARI STIINTIFICE. FILIALA IASI. ACADEMIA RPR. CHIMIE. BUCHAREST
1956 V7 N2 P1-4 PT.2 AEC-TR-4117 <*>
1956 V7 N2 P149-190 96L29RU <ATS>
1962 N2 P319-331 ICE V4 N2 P219-225 <ICE>
1964 N2 P281-287 ICE V4 N4 P664-666 <ICE>

SUCHASNIST
1965 N10 P93-105 65-33867 <=*$>
1965 N12 P82-89 66-31282 <=$>
1965 N12 P90-99 66-30988 <=$>
1965 N12 P113-116 66-30788 <=$>
1966 N1 P27-39 66-31396 <=$>
1966 N1 P87-97 66-31348 <=$>
1966 N1 P98-108 66-31343 <=$>
1966 N2 P67-82 66-31523 <=$>
1966 N2 P114-117 66-31444 <=$>

SUCRERIE BELGE
1956 V75 N10 P418-437 63-14858 <*> 0
1956 V76 P225-251 63-16351 <*> 0
1956 V76 N3 P101-103 63-14868 <*> 0
1960 V80 P49-65 63-14062 <*> 0

SUCERERIE FRANCAISE
1950 V91 P218-220 63-16355 <*>
1955 V96 N7 P227-230 63-16371 <*>
1960 V101 N7 P163-165 63-16534 <*>

SUD MEDICAL ET CHIRURGICAL
1958 V91 P7235-7236 64-10509 <*>

SUDEBNO-MEDITSINSKAYA EKSPERTISA
1959 N2 P59-60 60-31155 <=>
1959 V2 N2 P3-6 59-13800 <=>

Citation	Translation
1959 V2 N3 P47-49	60-11330 <=>
1959 V2 N4 P3-5	60-11299 <=>
1960 V3 N1 P3-8	60-31141 <=>
1961 V4 N1 P48-53	61-23469 <=*>
1961 V4 N3 P3-7	62-23819 <=*>
1962 V5 N3 P3-8	62-33677 <=*>
1962 V5 N4 P16-23	63-15115 <=*>
1964 V7 N3 P31-32	64-51375 <=$>
SUDOSTROENIE	
1938 V8 N2 P92-97	RT-945 <*>
1938 V8 N6 P411-417	RT-946 <*>
1956 V22 N2 P10-15	65-34045 <=*$>
1956 V22 N3 P4-10	60-13062 <+*>
1956 V22 N7 P1-5	64-19018 <=*$>
1957 N1 P11-14	C-2574 <NRC>
	R-3453 <*>
1957 V23 N3 P5-11	60-13555 <+*> 0
1957 V23 N3 P25-29	60-19976 <+*> 0
1957 V23 N3 P47-51	65-33807 <=*$>
1957 V23 N5 P20-25	66-31074 <=$>
1957 V23 N5 P26-30	65-34049 <=*$>
1957 V23 N6 P18-21	59-16651 <+*> 0
1957 V23 N6 P29-31	62-13083 <*=> 0
1957 V23 N8 P24-29	61-15932 <+*>
1957 V23 N9 P24-31	61-15733 <+*>
1957 V23 N10 P9	59-12409 <+*>
1957 V23 N10 P32-35	2008 <BSRA>
1958 V24 N1 P6-	59-19819 <+*>
1958 V24 N3 P10-16	C-3738 <NRCC>
1958 V24 N4 P1-4	AD-610 924 <=$>
	60-21034 <=>
1958 V24 N4 P9-12	59-22588 <+*>
1958 V24 N5 P1-5	60-21033 <=>
1958 V24 N5 P12-15	65-34047 <=*$>
1958 V24 N6 P1-7	59-19440 <+*>
1958 V24 N9 P25-28	60-21724 <=>
1958 V24 N9 P28-33	60-21725 <=>
1958 V24 N9 P78-80	59-11378 <=>
1958 V24 N10 P7-9	61-31570 <=>
1958 V24 N11 P5-9	62-23876 <=*>
1958 V24 N11 P29-34	62-13919 <*=>
1959 N1 P26-33	NLLTB V1 N7 P39 <NLL>
1959 N5 P33-37	NLLTB V2 N1 P25 <NLL>
1959 V25 N1 P26-33	59-13527 <=> 0
1959 V25 N1 P37-42	89R75R <ATS>
1959 V25 N2 P24-28	60-13828 <+*>
1959 V25 N3 P38-42	61-13847 <+*>
1959 V25 N4 P1-7	61-19683 <=*>
1959 V25 N7 P20-24	60-23592 <+*>
1959 V25 N7 P68-69	88R75R <ATS>
1959 V25 N8 P9-11	61-15312 <*+>
1959 V25 N8 P30-33	61-13552 <=*>
1959 V25 N8 P39-42	90R75R <ATS>
1959 V25 N8 P51-52	65-34046 <=*$>
1959 V25 N9 P1-3	60-15783 <+*>
1959 V25 N12 P71-73	61-13469 <=*>
1960 N3 P72	NLLTB V2 N7 P587 <NLL>
1960 N10 P71-73	NLLTB V3 N6 P479 <NLL>
1960 V26 N1 P52-56	60-31190 <=>
1960 V26 N2 P5-7	61-13994 <*+>
1960 V26 N2 P29-33	65-30852 <=$>
1960 V26 N3 P15-18	64-13278 <=*$>
1960 V26 N3 P75	61-28302 <*=>
1960 V26 N5 P1-3	61-27570 <*=>
1960 V26 N5 P6-9	61-27570 <*=>
1960 V26 N5 P18-	2171 <BSRA>
1960 V26 N5 P25-26	64-13390 <=*$>
1960 V26 N7 P9-12	60-21974 <=>
1960 V26 N7 P36-38	61-27640 <*=>
1960 V26 N8 P5-7	61-15693 <+*>
1960 V26 N8 P48-50	61-15694 <+*>
1960 V26 N8 P58-59	61-28573 <*=>
1960 V26 N8 P74	62-23489 <=*>
1960 V26 N8 P75-76	62-23488 <=*>
1960 V26 N9 P1-5	62-24782 <=>
1960 V26 N9 P11-13	61-27611 <*=>
1960 V26 N10 P1-5	62-24782 <=>
1960 V26 N10 P18-24	61-19133 <+*>
1960 V26 N10 P31-32	61-28202 <*=>
1960 V26 N12 P56-59	61-19520 <=*>
1961 V27 N1 P15-	64-19024 <=*$>
1961 V27 N2 P25-30	62-13119 <*=>
1961 V27 N3 P24-27	65-30981 <=$>
1961 V27 N3 P45-47	62-13157 <*=>
1961 V27 N3 P48-51	64-13265 <=*$>
1961 V27 N4 P80-83	61-28302 <*=>
1961 V27 N6 P36-38	62-15241 <*=>
1961 V27 N7 P55-60	63-19957 <=*$>
1961 V27 N7 P62-63	64-13267 <=*$>
1961 V27 N8 P11-14	62-11111 <=>
1961 V27 N8 P21-38	62-11111 <=>
1961 V27 N8 P48-53	64-13263 <=*$>
1961 V27 N10 P3-4	62-19171 <=*>
1961 V27 N10 P22-24	64-13264 <=*$>
1961 V27 N10 P24	2275 <BSRA>
1961 V27 N10 P33-36	65-60682 <=$>
1961 V27 N10 P41-43	64-13266 <=*$>
1961 V27 N10 P76-78	62-19171 <=*>
1961 V27 N11 P19-23	62-25946 <=*>
	65-60684 <=$>
1961 V27 N11 P24	65-60683 <=$>
1961 V27 N11 P28-35	64-23521 <=$>
1961 V27 N12 P4-8	63-15200 <=*>
1962 V28 N1 P51-53	65-61309 <=$>
1962 V28 N2 P17-19	64-13286 <=*$>
1962 V28 N4 P10-12	65-33706 <=*$>
1962 V28 N4 P13-14	C-4124 <NRCC>
1962 V28 N5 P74	65-33705 <=*$>
1962 V28 N6 P14-18	64-15262 <=*$>
1962 V28 N6 P27-29	64-15259 <=*$>
1962 V28 N6 P32-	64-15257 <=*$>
1962 V28 N7 P65-	64-13027 <=*$>
1962 V28 N8 P1-3	C-4269 <NRCC>
1962 V28 N8 P15-16	63-24565 <=*$>
1962 V28 N8 P41-44	63-24664 <=*$>
1962 V28 N9 P7-12	63-24358 <=*$>
1962 V28 N10 P4-7	64-19025 <=*$>
1962 V28 N10 P36-40	63-24566 <=*$>
1962 V28 N10 P48-54	64-19019 <=*$>
1962 V28 N12 P43-49	64-19026 <=*$>
1962 V28 N12 P49-	64-19015 <=*$>
1962 V28 N12 P59-	63-24567 <=*$>
1962 V28 N12 P61-	64-19016 <=*$>
1963 V29 N1 P10-12	64-19014 <=*$>
1963 V29 N1 P12-16	64-13240 <=*$>
1963 V29 N2 P5-	64-13238 <=*$>
1963 V29 N2 P58-59	64-13237 <=*$>
1963 V29 N3 P19-21	64-15258 <=*$>
1963 V29 N3 P29-34	64-13008 <=*$>
	64-31357 <=>
1963 V29 N3 P34-38	64-13608 <=*$>
1963 V29 N4 P6-9	63-31368 <=>
	64-31357 <=>
	65-61349 <=$>
1963 V29 N4 P13-	65-61351 <=$>
1963 V29 N4 P26-29	64-13239 <=*$>
1963 V29 N4 P32	65-60761 <=>
1963 V29 N4 P48-50	65-61350 <=$>
1963 V29 N5 P1-3	64-19020 <=*$>
1963 V29 N5 P4-	64-19017 <=*$>
1963 V29 N5 P9-12	65-33707 <=*$>
1963 V29 N5 P18-21	64-19028 <=*$>
1963 V29 N5 P21-22	65-61534 <=$>
1963 V29 N5 P28-30	65-61536 <=$>
1963 V29 N6 P1-4	65-61324 <=$>
1963 V29 N6 P4-6	65-61325 <=$>
1963 V29 N6 P8-11	65-61375 <=$>
1963 V29 N6 P15-18	65-61322 <=$>
1963 V29 N6 P25-27	65-61323 <=$>
1963 V29 N6 P28-29	65-61285 <=$>
1963 V29 N6 P50-51	65-60739 <=>
1963 V29 N7 P1-4	65-61370 <=$>
1963 V29 N7 P9-14	64-19023 <=*$>
1963 V29 N7 P20-22	64-19021 <=*$>
1963 V29 N7 P56-59	64-19022 <=*$>
1963 V29 N8 P54-60	65-30321 <=$>
1963 V29 N9 P28-30	1924 <BSRA>
1963 V29 N9 P36-38	57R76R <ATS>
1963 V29 N10 P7-11	65-61252 <=$>

1963 V29 N10 P11-13 65-61250 <=$>
1963 V29 N10 P22-25 1855 <BSRA>
1963 V29 N10 P26-29 65-60787 <=>
1963 V29 N1C P30-33 65-61251 <=$>
1963 V29 N11 P43 1981 <BSRA>
1963 V29 N11 P48-52 1984 <BSRA>
1963 V29 N11 P48 65-61321 <=$>
1963 V29 N12 P32- 1963 <BSRA>
1964 N6 P57-59 AD-633 086 <=$>
1964 N11 P14-16 65-33600 <=*$>
1964 V30 N1 P42-44 1929 <BSRA>
1964 V30 N2 P8-9 1945 <BSRA>
1964 V30 N2 P21-25 2015 <BSRA>
1964 V30 N2 P26-27 65-30318 <=$>
1964 V30 N3 P5 1970 <BSRA>
1964 V30 N3 P7 1964 <BSRA>
1964 V30 N3 P32 1931 <BSRA>
1964 V30 N3 P45 1957 <BSRA>
1964 V30 N4 P1 1971 <BSRA>
1964 V30 N4 P22 1949 <BSRA>
1964 V30 N4 P25 1956 <BSRA>
1964 V30 N5 P7-8 1958 <BSRA>
1964 V30 N5 P9-10 1961 <BSRA>
1964 V30 N5 P21-24 1977 <BSRA>
1964 V30 N6 P12 2048 <BSRA>
1964 V30 N6 P42- 2049 <BSRA>
1964 V30 N7 P5- 2007 <BSRA>
1964 V30 N7 P6- 2004 <BSRA>
1964 V30 N7 P8 2021 <BSRA>
1964 V3C N7 P9- 2062 <BSRA>
1964 V30 N7 P33- 2169 <BSRA>
1964 V30 N8 P9-49 AD-627 803 <=$>
1964 V30 N8 P43-47 AD-634 058 <=$>
1964 V30 N9 P23- 2050 <BSRA>
1964 V3C N9 P49- 2072 <BSRA>
1964 V30 N11 P29-32 RTS-3175 <NLL>
1964 V30 N11 P33-34 AD-640 291 <=$>
1965 N6 P76-77 AD-633 895 <=$>
1965 N7 P12-15 65-33575 <=*$>
1965 N9 P44-51 AD-629 509 <=*$>
1965 N9 P51-56 AD-629 510 <=*$>
1965 N9 P61-62 AD-629 167 <=*$>
1965 N11 P15 2414 <BSRA>
1965 N12 P61 2428 <BSRA>
1965 N12 P65-69 AD-633 897 <=$>
1965 N12 P7C 2410 <BSRA>
1965 V31 N1 P1-4 AD-627 251 <=$>
1965 V31 N1 P18-20 65-31341 <=$>
1965 V31 N1 P21-23 AD-641 649 <=$>
1965 V31 N2 P9-17 65-31317 <=$>
1965 V31 N2 P17-20 65-31342 <=$>
1965 V31 N3 P15-17 65-31614 <=$>
1965 V31 N3 P26-30 AD-625 806 <=*$>
1965 V31 N3 P35- 2248 <BSRA>
1965 V31 N5 P12-18 AD-626 703 <=*$>
1965 V31 N6 P17-19 65-34048 <=*$>
1965 V31 N6 P24-27 AD-625 847 <=*$>
1965 V31 N6 P78-81 AD-633 947 <=$>
1965 V31 N7 P15-17 65-34050 <=*$>
1965 V31 N7 P22-36 AD-630 742 <=$>
1965 V31 N8 P18-19 AD-633 826 <=$>
1965 V31 N8 P19-20 AD-633 638 <=$>
1965 V31 N8 P53 2397 <BSRA>
1965 V31 N9 P19 2405 <BSRA>
1965 V31 N11 P21 2403 <BSRA>
1965 V31 N11 P35-38 AD-634 110 <=$>
1966 N1 P3-8 66-31566 <=$>
1966 N3 P9 2435 <BSRA>
1966 N3 P40-45 66-33579 <=$>
1966 N3 P46 2429 <BSRA>

SUELO ARGENTINO
1944 V3 N31 P10-11 3361 <*>
1944 V3 N31 P51 3361 <*>
1944 V3 N31 P56 3361 <*>

SUGAKU
1953 V5 N2 P65-72 AMST S2 V8 P1-12 <AMS>

SUISAN GAKKAIHO
1915 V1 N1 P1-24 AL-370 <*>
1917 V2 N1 P106-108 AL-364 <*>
1921 V3 N3 P196-204 1794 <*>
1924 V4 N2 P87-92 AL-367 <*>
1926 V4 N3 P125-137 AL-368 <*>
1929 V5 N2 P234-257 61-00065 <*>

SUISAN KAGAKU
1957 V6 N3/4 P2-6 58-1755 <*>
1957 V6 N3/4 P6-12 58-1753 <*>
1957 V6 N3/4 P13-20 58-1752 <*>
1957 V6 N3/4 P20-24 58-1754 <*>
1957 V6 N3/4 P25-28 58-1756 <*>
1957 V6 N3/4 P44 58-1754 <*>

SUISAN KOZA
1949 V6 P17-94 AL-320 <*>

SUISAN KENKYU-SHI
1943 V38 N10 P186-187 AL-736 <*>
1943 V38 N12 P222-224 AL-736 <*>

SUIYOKAI-SHI. KYOTO
1954 V12 N6 P265-270 73K25J <ATS>

SULZER TECHNICAL REVIEW. (TECHNISCHE RUNDSCHAU SULZER)
1945 N1 P6-24 58-442 <*>

SUMARSTVO
1955 V8 N9 P534-545 60-21639 <=>
1957 V10 N3/4 P245-257 60-21637 <=>
1958 V11 N7/8 P444-448 60-21640 <=>

SUMITOMO DENKI
1962 N79 P70-74 20S86J <ATS>
1965 N90 P44-52 66-14554 <*>

SUMITOMO ELECTRIC REVIEW
SEE SUMITOMO DENKI

SUMITOMO KEIKINZOKU GIHO
1962 V3 N1 P73-81 66-20963 <$>

SUMITOMO KINZOKU
1955 V7 N4 P98-109 1801 <BISI>
1956 V8 N1C P55-60 61-16399 <*>
1957 V9 N7 P143-157 TS 1351 <BISI>
1958 V10 N1 P1-4 61-18384 <*> P
1958 V10 N4 P258-264 66-11665 <*>
1959 V11 N4 P252-260 2500 <BISI>
1960 V12 N4 P8C-90 2144 <BISI>
1964 V16 P238-247 4485 <BISI>

SUMITOMO LIGHT METAL TECHNICAL REPORT
SEE SUMITOMO KEIKINZOKU GIHO

SUMITOMO METALS
SEE SUMITOMO KINZOKU

SUOMALAISEN ELAIN- JA KASVITIETEELLISEN SEURAN VANAMON TIEDONANNOT. HELSINKI
1954 V16 N2 P1-48 1924 <*>
1962 V17 N1 P25-38 63-01059 <*>

SUOMALAISEN TIEDEAKATEMIAN TOIMITUKSIA. SARJA A
1927 V29 N12 P1-7 62-20214 <*>
1930 V30 P1-156 I-709 <*>
1957 SA V6 N5 P1-17 AI-TR-17 <*>
1957 SA V6 N7 P1-12 AI-TR-16 <*>
1958 SA V6 N9 P3-16 AI-TR-15 <*>

SUOMALAISEN TIEDEAKATEMIAN TOIMITUKSIA. SARJA AI. MATHEMATICA PHYSICA
1954 ENTIRE ISSUE 58-995 <*>

SUOMEN ELAINLAAKARILEHTI
SEE FINSK VETERINARTIDSKRIFT

SUOMEN GEODEETTISEN LAITOKSEN JULKAISUJA

1955 N46 P81-87 62-25629 <=*>

SUOMEN KEMISTILEHTI
1943 V16B P13-14 2893 <*>
1945 V18A P60-64 2487 <K-H>
1945 V18A P199-207 872B <K-H>
1945 V18B N1/2 P6 63-14582 <*>
1957 V30 P92-98 31J17FN <ATS>
57-3147 <*>
1957 V30 P98-100 32J17FN <ATS>
57-3146 <*>
1957 VA30 N4 P85-92 69K27FN <ATS>
1959 VA32 P6-10 19L36FN <ATS>
1963 V36 N3 PA76-A77 64-14805 <*> 0
1963 V36A N3 P76-77 65-11367 <*> 0

SUOMEN LAAKARILEHTI
1958 V13 P545-547 63-20004 <*>

SUOMEN PAPERI- JA PUUTAVARALEHTI
SEE PAPERI JA PUU

SURFACE SCIENCE
1964 V1 N1 P22-41 65-10869 <*>
1964 V1 N1 P42-53 65-11873 <*>

SURVEYING AND CARTOGRAPHY BULLETIN. CHINESE PEOPLES REPUBLIC
SEE T'SE HUI T'UNG PAO

SUVREMENNA MEDITSINA
1953 V4 N10 P6-22 R-4999 <*>
1954 V5 N3 P60-68 3381-B <K-H>
1958 V9 N1 P43-49 39M46B <ATS>
1958 V9 N3 P88-90 8973-C <K-H>
1959 V10 N9/10 P46-53 60-11611 <*=>
1961 V12 N1 P3-11 62-19554 <*=>
1961 V12 N1 P51-60 62-19659 <=*>
1961 V12 N1 P51-61 97Q66B <ATS>
1961 V12 N2 P31-35 62-19660 <=*>
1961 V12 N2 P105-108 63-16916 <*>
1961 V12 N10 P29-34 63-10058 <*>
1963 V14 N10 P33-37 65-14634 <*> 0
1964 V15 N8 P3-22 65-30033 <=$>

SVARKA: SBORNIK STATEI
1958 N1 P38-48 66-33951 <=$>
1958 N1 P144-155 66-33951 <=$>
1959 V2 P156-162 64-21076 <=>
1959 V2 P163-173 61-27002 <=*>
1959 V2 P195-213 61-27002 <=*>

*SVARCCHNCE PRCIZVCDSTVO
1955 N1 P14-17 62-20046 <=*>
1955 N1 P17-19 3611 <HB>
1955 N4 P9-13 62-20045 <=*>
1955 N6 P9-10 5209 <HB>
1955 N7 P7-9 3605 <HB>
1955 N7 P13-16 RT-3528 <*>
1955 N7 P25-26 3681 <HB>
1955 N8 P6-8 59-22670 <+*>
1955 N9 P23-27 R-4079 <*>
1955 N11 P8-13 3715 <HB>
1955 N12 P4-6 59-22660 <+*>
1955 V1 P14-17 3610 <HB>
1955 V7 P25-26 GB70 4210 <NLL>
1956 N2 P13-15 62-20048 <=*>
1956 N3 P1-4 4840 <HB>
1956 N3 P20-22 62-13185 <=*>
1956 N4 P4-12 3823 <HB>
1956 N4 P9-12 3670 <HB>
1956 N5 P20-22 3779 <HB>
65-12020 <*>
1956 N6 P1-5 3894 <HB>
1956 N6 P5-11 62-32500 <=*>
1956 N6 P20-21 <HB>
1956 N6 P22-26 62-20133 <=*>
1956 N6 P27-29 3906 <HB>
1956 N7 P10-13 62-20021 <=*>
1956 N9 P23-24 4492 <HB>
1956 N10 P9-13 62-20132 <=*>
1957 N1 P18-20 62-20043 <=*>
1957 N1 P20-24 62-13597 <*=>
1957 N3 P1-3 4426 <HB>
1957 N5 P1-3 4426 <HB>
1957 N5 P24-25 59-15806 <+*>
1957 N7 P9-10 4116 <HB>
1957 N8 P6-10 4115 <HB>
1957 N8 P31-34 4463 <HB>
62-24195 <=*>
1957 N9 P19-23 4502 <HB>
1957 N9 P23-24 4937 <HB>
1957 N10 P1-7 61-28029 <*=>
1957 N10 P16-18 59-22659 <+*>
1957 N12 P30-31 62-13099 <*=>
1957 V1 P15-17 3947 <HB>
1958 N1 P19-21 1377 <BISI>
62-24232 <=*>
1958 N1 P27-28 4454 <HB>
1958 N1 P32 4669 <HB>
1958 N1 P34-36 4294 <HB>
1958 N2 P1-6 61-19940 <*=>
1958 N2 P7-10 60-23588 <+*>
1958 N2 P14-18 62-13746 <*=>
1958 N3 P26-29 TRANS-46 <MT>
1958 N4 P1-5 59-16462 <+*> P
1958 N4 P5-10 102 <LSB>
60-14135 <+*> 0
1958 N4 P22-25 TRANS-47 <MT>
1958 N5 P1-5 4487 <HB>
1958 N5 P8-11 TRANS-48 <MT>
59-22658 <+*>
1958 N5 P12-14 5047 <HB>
1958 N5 P27-29 TRANS-49 <MT>
1958 N5 P34-36 5053 <HB>
1958 N6 P24-28 60-19569 <+*>
1958 N6 P36-41 33L30R <ATS>
1958 N6 P41-49 61-13483 <=*>
1958 N7 P10-14 61-15405 <*+>
1958 N7 P14-17 61-15723 <+*>
1958 N7 P35-38 4517 <HB>
1958 N8 P10-14 4821 <HB>
1958 N9 P19-22 4670 <HB>
1958 N9 P24 59-22268 <+*>
1958 N9 P26-36 TS-1304 <BISI>
1958 N9 P26-32 62-24223 <=*>
1958 N10 P14-18 4812 <HB>
1958 N10 P18-21 60-15556 <+*>
1958 N10 P26-29 59-22252 <+*> 0
1958 N10 P29-30 60-15057 <+*>
1958 N10 P35-37 59-22596 <+*>
1958 N11 P7-10 60-15559 <+*>
1958 N11 P10-12 4913 <HB>
60-13470 <+*>
1958 N11 P15-16 73L30R <ATS>
1958 N11 P19-23 60-15062 <+*>
1958 N11 P29-32 5136 <HB>
60-17915 <+*>
1958 N11 P33-34 60-15428 <+*>
1958 N11 P36 60-13430 <+*>
1958 N11 P37-38 60-17474 <+*>
1958 N12 P1-5 60-19177 <+*>
1958 N12 P9-13 60-17909 <+*>
1958 N12 P13-17 60-13431 <+*>
1958 N12 P21-23 60-23670 <+*>
1959 N1 P16-20 60-19570 <+*>
1959 N2 P6-8 1489 <BISI>
60-17927 <+*>
1959 N2 P8-12 60-17926 <+*>
1959 N2 P19-21 4734 <HB>
1959 N2 P23-25 2348 <BISI>
1959 N3 P12-15 TRANS-50 <MT>
1959 N3 P16-18 61-23542 <*=>
1959 N3 P35-37 60-21731 <=>
1959 N4 P1-48 <BWRA>
1959 N5 P1-5 63-14487 <=*>
1959 N5 P17-20 64-14059 <=*$>
1959 N5 P35-36 TS-1405 <BISI>
62-24237 <=*>
1959 N6 P9-13 59-13940 <=> 0

1959 N7 P16-19	60-11155 <=>
1959 N9 P1-4	42M43R <ATS>
1959 N9 P5-6	43M43R <ATS>
1959 N9 P27-30	TS 1629 <BISI>
	62-14189 <=*>
1959 N11 P12-14	60-15561 <+*>
1960 N11 P26-28	146-121 <STT>
1961 N1 P34-37	62-13173 <*=> 0
1961 N1 P38-40	62-23362 <*=>
1961 N1 P41-43	61-28486 <*=> 0
1961 N2 P41-43	61-31579 <=>
1961 N3 P5-8	62-16944 <=*> 0
1961 N11 P7-13	62-15411 <*=>
	63-21349 <=>
1961 N11 P30-32	66-11079 <*>
1961 V7 N4 P6-10	62-16943 <=*> 0
1962 N2 P8-12	63-15506 <=*>
1962 N3 P13-16	63-19308 <=*>
1962 N5 P3-6	63-11600 <=>
1962 N6 P18-20	62-32066 <=*> 0
1962 N7 P25-28	5728 <HB>
1962 N8 P14-16	5692 <HB>
1962 N9 P42-43	63-21402 <=>
1962 N11 P13-15	63-21465 <=>
1962 N12 P25-28	60Q69R <ATS>
1962 N12 P30-34	63-21659 <=>
1963 N4 P19-21	63-31058 <=>
1963 N4 P28-31	63-31058 <=>
1963 N5 P6-9	63-41152 <=>
1963 N5 P9-13	63-41152 <=>
1963 N5 P21-24	63-31935 <=>
1963 N5 P41-44	63-41152 <=>
1963 N6 P42-43	64-41225 <=>
1963 N7 P28-29	63-31584 <=>
1963 N8 P22-28	63-31955 <=>
1963 N10 P10-12	64-21205 <=>
1963 N10 P15-17	63-41266 <=>
1963 N12 P10-12	64-31115 <=>
1964 N5 P36-40	AD-610 408 <=$>
	64-41306 <=>
1964 N5 P45	AD-610 408 <=$>
	64-41306 <=>
1964 N6 P1-4	64-51050 <=>
1964 N6 P16-19	AD-609 157 <=>
1964 N7 P13-19	64-51371 <=>
1964 N7 P41-45	64-51705 <=$>
1964 N9 P16-19	64-51860 <=$>
1964 N9 P30-31	64-51860 <=$>
1964 N10 P6-9	AD-616 281 <=$>
1964 N10 P39-40	65-30246 <=$>
1964 N10 P42	65-30245 <=$>
1965 N3 P29-30	AD-619 319 <=$>
1965 N4 P15-18	AD-637 374 <=$>
1965 N4 P25-28	AD-637 374 <=$>
1965 N8 P1-3	65-32864 <=*$>
1965 N8 P37-38	65-32864 <=*$>
1965 N8 P39-42	65-33380 <=*$>
1966 N7 P47	66-34736 <=$>

SVENSK BOTANISK TIDSKRIFT

1913 V7 P97-188	II-777 <*>

SVENSK FARMACEUTISK TIDSKRIFT

1942 V46 P551-552	64-10677 <*>
	9827-B <K-H>
1954 V58 N31 P745-751	58-362 <*>

SVENSK KEMISK TIDSKRIFT

1926 V38 P111-112	64-18314 <*>
1929 V41 P249-257	64-18877 <*>
1929 V41 N6 P141-152	64-18334 <*> 0
1933 V45 P141-150	CRNL-TR-253 <*>
1933 V45 P151-152	58-1779 <*>
1937 V49 N2 P29-52	UCRL TRANS-679 <*>
1940 V52 P241-246	I-587 <*>
1941 V53 P233-241	AL-448 <*>
1943 V55 P135-144	1568C <K-H>
1944 V56 P156-158	57-2993 <*>
1945 V57 P54	AL-824 <*>
1945 V57 N6 P158-168	517 <MT>
1947 V59 P14-18	AEC-TR-4007 <*>
1954 V66 N5/6 P143-157	2756 <*>
1957 V69 N10 P476-486	59-10160 <*>
1957 V69 N6/7 P328-342	56K23G <ATS>
1958 V70 N4 P182-196	59-10135 <*> 0
1959 V71 N8 P367-386	62-18517 <*>
1960 V72 N2 P69-87	66M46N <ATS>
1960 V72 N2 P93-106	91M45S <ATS>
1960 V72 N3 P151-165	62-18069 <*>
1962 V74 N12 P609-617	AEC-TR-6015 <*>
1964 V76 N11 P617-627	66-13669 <*>

SVENSK PAPPERSMASSE-TIDNING

1959 V5 P275-280	59-17129 <*> 0

SVENSK PAPPERSTIDNING

1927 V30 P651-653	63-16052 <*> 0
1936 V39 N17 P316-319	I-34 <*>
1938 V41 P267-271	63-14079 <*>
1939 V42 N22 P554-557	1842 <*>
1941 N19 P427-429	57-2921 <*>
1941 V44 P427-429	66-10413 <*>
1942 V45 P421-428	AL-481 <*>
1943 V46 N23 P565-569	<ATS>
1944 V47 N2 P29-33	<ATS>
1944 V47 N18 P439-444	60-16282 <*> 0
1946 V49 N3 P51-61	60-18308 <*> 0
1946 V49 N9 P204	C-2226 <NRC>
1946 V49 N10 P215-234	58-1633 <*>
1947 P215-221	58-1310 <*>
1947 V50 P145-148	II-129 <*>
1947 V50 P159-168	63-16024 <*> 0
1947 V50 P196-204	63-16038 <*> C
1947 V50 N16 P363-369	<ATS>
1948 V51 P531-538	3321 <*>
1948 V51 N2 P23-30	58-1402 <*>
1948 V51 N3 P45-49	59-10916 <*>
1948 V51 N10 P230-232	58-1186 <*>
1948 V51 N19 P447-453	74F3S <ATS>
1949 V52 N20 P493-504	2473 <*>
1950 V53 P287-294	8827 <IICH>
1950 V53 P321-326	8828 <IICH>
1950 V53 N7 P179-182	57-1304 <*>
1950 V53 N11 P287-294	2721 <*>
1950 V53 N12 P321-326	2720 <*>
1950 V53 N20 P638-643	I-315 <*>
1950 V53 N21 P694-697	59-17816 <*>
1951 V54 N8 P267-274	I-202 <*>
1951 V54 N14 P469-476	C-2189 <NRC>
1951 V54 N19 P671-678	I-805 <*>
1952 V55 N1 P15-18	<ATS>
	57-2667 <*>
1952 V55 N4 P134-138	2719 <*>
1952 V55 N5 P153-167	I-166 <*>
1952 V55 N15 P515-517	C-2233 <NRC>
1953 V56 N15 P590-597	2080 <*>
1953 V56 N16 P625-633	2079 <*>
1953 V56 N23 P893-899	57-2332 <*>
1954 V57 P437-440	57-1303 <*>
1954 V57 N1 P1-8	2471 <*>
1954 V57 N1 P1-	57-896 <*>
1954 V57 N1 P19-22	58-994 <*>
1954 V57 N3 P92-	57-898 <*>
1954 V57 N8 P307-308	2475 <*>
1954 V57 N15 P542-548	C-2184 <NRC>
1954 V57 N16 P583-590	II-797 <*>
1954 V57 N23 P867-871	57-2552 <*>
1955 P635-	57-1738 <*>
1955 V58 P392-394	3649 <K-H>
1955 V58 N6 P185-195	C-2324 <NRC>
	57-2359 <*>
1955 V58 N9 P332-	57-6C <*>
1955 V58 N10 P392-394	57-1714 <*>
1955 V58 N12 P452-	58-759 <*>
1955 V58 N13 P483-	57-2878 <*>
1955 V58 N17 P621-	57-1739 <*>
1955 V58 N19 P706-712	59-10861 <*> 0
1955 V58 N20 P758-763	1768 <*>
	2361 <*>
1956 V59 P395-401	57-1735 <*>

Citation	Translation
1956 V59 N2 P61-65	62-10362 <*> 0
1956 V59 N3 P98-103	C-2510 <NRC>
	57-3480 <*>
	59-10154 <*>
1956 V59 N4 P157-171	59-10155 <*>
1956 V59 N4 P172-174	59-10225 <*>
1956 V59 N5 P172-	57-61 <*>
1956 V59 N8 P296-300	62-18222 <*>
1956 V59 N9 P329-334	58-1278 <*>
1956 V59 N21 P747	58-285 <*>
1957 V60 N2 P50-55	CSIRO-3471 <CSIR>
	57-3409 <*>
1957 V60 N5 P59-	57-1984 <*>
1957 V60 N7 P252-	58-1401 <*>
1957 V60 N11 P425-428	59-15374 <*> 0
	59-15698 <*> 0
1957 V60 N12 P447	58-280 <*>
1957 V60 N12 P460-466	58-1408 <*>
	58-2188 <*>
	62-10364 <*> 0
1957 V60 N13 P484-488	C-2615 <NRC>
1957 V60 N13 P484-	58-761 <*>
1957 V60 N16 P584-594	59-17053 <*>
1957 V60 N19 P713-719	C-2616 <NRC>
1957 V60 N19 P713	58-597 <*>
1957 V60 N19 P713-	58-760 <*>
1957 V60 N24 P905-910	59-10153 <*>
1958 V61 N1 P10-18	59-10134 <*>
1958 V61 N3 P55-60	59-10163 <*>
1958 V61 N4 P91-95	58-1538 <*>
	59-15164 <*>
1958 V61 N5 P133-139	59-10164 <*>
	59-17753 <*>
1958 V61 N14 P445-453	59-15179 <*>
	59-17515 <*> 0
	82K25S <ATS>
1958 V61 N17 P531-539	59-17714 <*>
1958 V61 N18 P568-580	59-10983 <*> 0
1958 V61 N18 P625-632	60-16756 <*>
1958 V61 N18 P794-802	59-15050 <*> P
	63-18373 <*>
1958 V61 N19 P834-843	59-10478 <*> 0
1958 V61 N19 P844-850	59-17057 <*> 0
1958 V61 N20 P871-880	30L36S <ATS>
1958 V61 N18B P625-632	61-00665 <*>
1958 V61 N18B P648-655	59-17152 <*>
1958 V61 N18B P718-725	59-10098 <*> 0
1958 V61 N18B P768-774	60-10917 <*> 0
1958 V61 N18B P815-826	59-10162 <*> 0
1959 V62 N6 P198	60-10135 <*>
1959 V62 N7 P241-244	59-17650 <*> 0
1959 V62 N7 P245-250	60-14274 <*> 0
1959 V62 N8 P284-288	60-14039 <*> 0
1959 V62 N9 P308-317	59-17740 <*> 0
1959 V62 N10 P345-350	59-17449 <*> 0
1959 V62 N11 P381-389	91L36S <ATS>
1959 V62 N12 P425-433	60-14068 <*> 0
1959 V62 N15 P524-533	60-14292 <*>
1959 V62 N17 P612-620	60-16272 <*>
1959 V62 N18 P631-639	60-10709 <*> 0
1959 V62 N18 P640-645	60-14661 <*>
1959 V62 N18 P652-655	60-16273 <*>
1959 V62 N19 P692-699	79L36S <ATS>
1959 V62 N22 P829-833	60-14383 <*> 0
	60-14663 <*>
1960 V63 N1 P1-3	60-18138 <*> 0
1960 V63 N2 P15-23	51M44S <ATS>
	61-10389 <*> 0
1960 V63 N2 P15-	61-10448 <*> 0
1960 V63 N11 P356-368	34M44S <ATS>
	61-20497 <*> 0
	61-20510 <*> 0
1960 V63 N13 P425-430	61-10670 <*>
1960 V63 N14 P447-448	70M44G <ATS>
1960 V63 N18 P609-	61-10908 <*>
1960 V63 N19 P658-664	61-18568 <*> 0
1961 V64 P533-544	62-10611 <*> 0
1961 V64 N4 P101-108	61-20042 <*> 0
1961 V64 N4 P130-137	61-16460 <*>
1961 V64 N5 P175-179	61-20041 <*>
1961 V64 N12 P461-468	61-20486 <*> 0
	61-20690 <*> 0
1961 V64 N23 P870-871	66-10872 <*> P
1962 V65 N5 P164-172	62-18709 <*> 0
1962 V65 N6 P222-223	62-16615 <*> 0
1962 V65 N8 P322-325	62-18883 <*> 0
1962 V65 N12 P488-493	62-18709 <*> 0
1962 V65 N18 P681-684	63-10768 <*>
1962 V65 N18 P711-713	63-10061 <*>
1962 V65 N20 P785-789	63-14471 <*> 0
1962 V65 N20 P817-818	63-01039 <*>
1962 V65 N24 P991-1000	63-01253 <*>
1963 V66 N2 P37-41	63-14472 <*> 0
1963 V66 N3 P51-54	63-18811 <*>
1963 V66 N5 P159-167	64-10795 <*>
1964 V67 N6 P246-248	65-10277 <*> 0
1964 V67 N10 P421-4311	C-5293 <NRC>
1964 V67 N19 P784-789	65-14170 <*>
1965 V68 N2 P25-33	65-14396 <*>
1965 V68 N7 P230-248	65-13895 <*> 0
1965 V68 N10 P369-377	65-14588 <*>
1965 V68 N10 P378-383	65-17197 <*>
1965 V68 N14 P477-481	65-17176 <*>
1965 V68 N18 P618-627	65-17453 <*>
1966 V69 N6 P191-198	66-12921 <*>

SVENSK TANDLAEKARE TIDSKRIFT

Citation	Translation
1956 V49 N4 P195-204	65-11312 <*>
1956 V49 N10 P687-693	58-559 <*>

SVENSKA FARAVELSFORENINGENS TIDSKRIFT

Citation	Translation
1955 V35 P81-86	CSIRO-3163 <CSIR>

SVENSKA INSTITUTET FOR KONSERVERINGSFORSKNING PUBLIKATION

Citation	Translation
1956 N115 P79-94	62-18872 <*>

SVENSKA LAEKARTIDNINGEN

Citation	Translation
1939 V36 N2 P73-101	66-12604 <*>
1946 V43 P3109-3136	63-16579 <*>
1947 V44 P785-804	63-16913 <*> 0
	6918-B <KH>
1948 V45 P1-17	8003-B <K-H>
1948 V45 N1 P1-17	63-16684 <*> 0
1949 V46 N45 P2413-2426	59-16142 <=*> 0
1951 V48 N40 P2340-2348	58-1342 <*>
1953 V50 P690-692	II-810 <*>
1953 V50 P925-930	2998 <*>
1954 V51 P3267-3288	1735 <*>
1956 V53 P1514-1519	8TM <CTT>
1956 V53 P2140-2145	5613-B <K-H>
1956 V53 P2634-2653	5668-C <K-H>
1956 V53 N16 P989-1003	57-681 <*>
1956 V53 N16 P1003-1009	57-1435 <*>
1956 V53 N16 P1009-1016	57-1272 <*>
1957 V54 P1479-1484	6404-B <K-H>
1957 V54 N32 P2355-2361	57-3552 <*>
1957 V54 N36 P2646-2654	62-00228 <*>
1957 V54 N36 P2655-2661	62-00143 <*>
1958 V55 P608-615	AEC-TR-6268 <*>
1959 V56 P90-97	60-13140 <=*>
1959 V56 P983-986	63-16668 <*>
	9372-D <KH>
1959 V56 P1899-1911	63-16681 <*>
	9477-A <K-H>
1959 V56 N11 P3106-3111	61-10822 <*> C
1959 V56 N26 P1843-1851	63-16942 <*>
	9955 <KH>
1959 V56 N27 P1899-1911	62-00483 <*>
1959 V56 N27 P2080-2089	64-10634 <*>
	9477-B <K-H>
1959 V56 N28 P1975-1986	62-00444 <*>
1959 V56 N29 P2025-2034	62-00450 <*>
1959 V56 N31 P3139-3144	62-00430 <*>
1959 V56 N33 P2254-2262	63-16683 <*>
	9513 <K-H>
1960 V57 P398-406	62-00330 <*>
1960 V57 P1568-1579	64-10624 <*>
1960 V57 N6 P398-406	10401-A <K-H>
	63-16894 <*>

1960 V57 N22 P1669- V-97 <RIS>
1960 V57 N41 P2821-2833 61-00415 <*>
1960 V57 N41 P2833-2845 61-00393 <*>
1960 V57 N41 P2846-2852 61-00644 <*>
1961 V58 P1722-1725 66-11520 <*>
1961 V58 P3379-3381 64-14845 <*>
1961 V58 N31 P2153-2163 11499-G <KH>
1963 V60 P3325-3333 66-11831 <*>
1963 V60 N1C P681-690 63-01383 <*>
1964 V61 N49 P3794-3807 86S83S <ATS>
1964 V61 N50 P3945-3947 65-14832 <*> 0

SVENSKA SKOGSVARDSFORENINGENS TIDSKRIFT
1954 N2 P157-158 C-3229 <NRCC>
1955 V53 N2 P201-227 C-3126 <NRCC>
1955 V53 N4 P337-394 C-3512 <NRCC>

SVENSKA TROFORSKNINGSINSTITUTET TRATEKNIK
1957 V86B P1-16 CSIRO-3856 <CSIR>
1957 V86B P16 58-2532 <*>

SVENSKA VAGFORENINGENS TIDSKRIFT
1949 V36 N3 P112-115 T-2045 <INSD>
1955 N9 P345-350 T-1543 <INSD>

SVENSKA VETENSKAPSAKADEMIENS HANDLINGAR
SEE KUNGLIGA SVENSKA VETENSKAPAKADEMIENS HANDLINGAR

SVETOTEKHNIKA
1955 V1 N2 P8-13 TT-799 <NRCC>
1955 V1 N4 P7-9 C-3366 <NRCC>
1957 V3 N2 P16-19 60-23078 <+*>
1957 V3 N2 P19-21 60-13325 <+*>
1957 V3 N8 P1-6 C-2642 <NRC>
R-4294 <*>
1957 V3 N8 P20-23 C-2645 <NRC>
R-4296 <*>
1957 V3 N8 P25-27 C-2636 <NRC>
R-4293 <*>
1958 V4 N5 P3-12 60-13667 <+*>
1958 V4 N5 P13-20 60- 13668 <+*>
1958 V4 N7 P26-27 59-19438 <+*>
1958 V4 N8 P1-5 59-22560 <+*>
1958 V4 N12 P11-16 60-15427 <+*>
1959 V5 N7 P13-17 61-13952 <=*>
1959 V5 N8 P5-13 61-15506 <=*>
1959 V5 N8 P14-17 E-529 <RIS>
1959 V5 N8 P18-20 61-15506 <=*>
1959 V5 N8 P19-20 E-527 <RIS>
1959 V5 N9 P1-7 61-28945 <*=> 0
1959 V5 N11 P8-13 E-531 <RIS>
1959 V5 N11 P13-15 E-532 <RIS>
1960 V6 N3 P22-25 62-32795 <=*>
1960 V6 N3 P32 60-41173 <=>
1960 V6 N4 P26-30 61-15638 <+*>
1960 V6 N5 P26-29 60-41498 <=>
1960 V6 N5 P32 60-41498 <=>
1960 V6 N8 P22-25 M-5053 <NLL> 0
1960 V6 N8 P26-28 62-32795 <=*>
65-61238 <=$> 0
1960 V6 N8 P30 62-32795 <=*>
1960 V6 N8 P32 61-21307 <=>
1960 V6 N9 P22-24 65-60710 <=>
1960 V6 N9 P31-32 61-21307 <=>
1960 V6 N10 P1-8 65-61150 <=>
1960 V6 N11 P19-21 62-32798 <=$> 0
65-61151 <=>
1960 V6 N12 P8-12 62-25282 <=*>
1960 V6 N12 P17-20 62-32798 <=$> 0
65-60708 <=>
1961 N11 P13-17 63-13405 <=*>
1961 V7 N1 P29-30 62-32798 <=> 0
1961 V7 N2 P1-9 62-32796 <=*>
1961 V7 N2 P12-18 65-10843 <*>
1961 V7 N2 P23-25 62-32797 <=*>
1961 V7 N8 P1-7 62-25337 <=*>
1961 V8 N8 P8-17 63-19227 <=*>
1962 V8 N2 P1-5 63-19469 <=*>
1962 V8 N3 P1-6 63-24205 <=*$>
1962 V8 N3 P32 62-32867 <=>
1962 V8 N5 P7-12 64-71534 <=>
1962 V8 N7 P1-5 62-32860 <=*>
1962 V8 N8 P26-27 63-24223 <=*$>
1962 V8 N9 P1-10 64-15883 <=*$>
1962 V8 N11 P3-9 63-15103 <=>
1962 V8 N11 P28-30 63-15103 <=>
1962 V8 N12 P31 63-21381 <=>
1963 V9 N2 P32 63-21768 <=>
1963 V9 N5 P1-5 AD-611 773 <=$>
1963 V9 N5 P19-24 63-31224 <=>
1963 V9 N7 P27-29 63-31687 <=>
1963 V9 N7 P32 63-31687 <=>
1963 V9 N10 P7-10 RTS-2790 <NLL> 0
1963 V9 N11 P13-17 RTS-3073 <NLL>
1964 V10 N1 P8-13 RTS-2919 <NLL>
1964 V10 N7 P3-5 RTS-3074 <NLL>
1964 V1C N10 P7-13 LC-1260 <NLL>
1965 V11 N2 P5-10 RTS-3510 <NLL>
1966 N11 P3-5 63-13403 <=*>
1966 V12 N1 P16-19 66-62078 <=*$>

SVINOVODSTVO
1959 N9 P18-26 60-31124 <=> 0
1960 N9 P2-7 61-21258 <=>
1961 N2 P27-29 61-31146 <=>
1961 N3 P15-16 61-27116 <=>
1961 N3 P22-24 61-27116 <=>
1963 N9 P41 RTS-3016 <NLL>
1963 V17 N2 P34-35 RTS-2783 <NLL>
1964 N2 P28-29 RTS-3053 <NLL>
1964 N3 P24 RTS-3054 <NLL>
1964 N9 P40 RTS-3055 <NLL>
1965 N1 P37-38 RTS-3060 <NLL>

SWEDISH STANDARD
1963 SEMKO 14A APP10 66-11363 <*> P

SYDOWIA
1951 V5 N1/2 P196-197 59-16195 <+*>

SYLWAN
1952 V96 N3 P283-304 60-21241 <=>
1954 V98 N1 P332-344 C-2168 <NRC>
1954 V98 N3 P193-199 60-21250 <=>
1954 V98 N4 P332-344 1196 <*>
1955 V99 N6 P5C1-504 60-21380 <=>
1956 N3 P89-96 61-01042 <*>
1956 V100 N1 P29-35 CSIRO-3303 <CSIR>
57-1710 <*>
1957 V101 N6 P1-24 60-21385 <=>
1957 V101 N10 P5-10 65-00299 <*>
1959 V103 N4 P5-22 60-21392 <=>
1959 V103 N4 P31-39 60-21386 <=>
1959 V103 N5 P7-19 60-21395 <=>
1959 V103 N5 P21-31 60-21395 <=>
1959 V103 N5 P33-38 60-21395 <=>
1959 V103 N5 P73-90 65-50329 <=>
1960 N10 P61-66 61-01041 <*>
1960 V104 N1 P1-13 65-50331 <=>
1961 V105 N2 P49-56 62-00919 <*>
1961 V105 N3 P13-24 62-00823 <*>
1961 V105 N4 P1-9 62-01361 <*>
1961 V105 N5 P53-61 65-50330 <=>
1962 V106 N2 P23-30 62-01418 <*>
1962 V106 N3 P25-38 63-01072 <*>
1962 V106 N3 P55-61 65-50348 <=>

SYMPOSIA ON ENZYME CHEMISTRY
SEE KOSO KAGAKU SHIMPOJIUM

SZKLO I CERAMIKA
1961 N9 P266-269 66-12415 <*>
1961 V12 N4 P97-103 NP-TR-1249 <*>
1961 V12 N12 P361-364 65-11357 <*>
1963 V14 N1 P1-4 ATS-71S86P <ATS>
1963 V14 N11 P294-297 BGIRA-607 <BGIR>
1964 V15 P202-204 66-13813 <*>
1964 V15 N4 P90-92 66-14383 <*$>
1965 V16 P69-71 4673 <BISI>

TIBA. REVUE GENERALE DE TEINTURE, IMPRESSION, BLANCHIMENT, APPRET ET DE CHIMIE TEXTILE ET TINCTORIALE
1926 V4 P331-335 60-10884 <*>

TZ FUER PRAKTISCHE METALLBEARBEITUNG
1934 P339-343 II-763 <*>
1937 V47 N19/0 P893-899 60-10207 <*> 0
1939 V49 N19 P703-704 60-18592 <*>
1939 V49 N19 P750-752 60-18592 <*>
1939 V49 N7/8 P317-318 60-10204 <*> 0
1939 V49 N7/8 P320 60-10204 <*> 0
1939 V49 N7/8 P322 60-10204 <*> 0
1965 V59 N7 P454-458 4807 <BISI>

TABACCO
1947 V51 N573 P6-15 2239 <K-H>
1948 V52 N582 P3-15 11517-A <KH>
62-18004 <*>
1950 V54 P14-23 2124-E <K-H>
1950 V54 P96-98 2150 B <KH>
1950 V54 P128-157 2203C <K-H> 0
1950 V54 N606 P3-13 2203A <K-H>
1950 V54 N607 P55-61 2170C <K-H>
1950 V54 N607 P62-64 2124G <K-H>
1950 V54 N613 P260-265 58-845 <*>
1950 V54 N615 P330-332 2170B <K-H>
1950 V54 N616 P343-353 2186C <K-H>
1951 V55 P83-88 2261B <K-H>
1951 V55 P343-406 2650F <K-H> 0
1951 V55 P1003-1012 2650C <K-H> 0
1951 V55 N620 P89-93 2203B <K-H> 0
1951 V55 N624 P215-234 2650E <K-H>
1952 V56 N631 P50-52 3374-B <K-H>
1952 V56 N635 P178-179 3280-B <K-H>
1952 V56 N637 P227-245 3160-D <K-H>
1953 V57 N651 P354-357 2974B <K-H>
1953 V57 N653 P399-408 I-53 <*>
1954 V58 P39-55 T-1667 <INSD>
1954 V58 P227-239 3270-C <K-H>
1954 V58 P264-278 3384-B <K-H>
1954 V58 N654 P3-33 5629-D <K-H>
1955 V59 N660 P17-31 58-846 <*>
1955 V59 N672 P257-272 CSIRO-2936 <CSIR>
1956 V60 N678 P3-70 58-1035 <*>
1956 V60 N681 P347-351 11999-A <K-H>
63-16881 <*>
1957 V61 P349-357 11860 <KH>
1957 V61 N682 P26-48 11558-A <KH>
62-18005 <*> 0
1958 N686 P46-52 61-10878 <*>
1958 V62 N686 P46-52 8650-E <K-H>
1959 V63 N692 P285-292 10281-B <K-H>
63-16573 <*>
1959 V63 N693 P431-443 10693-D <K-H>
64-10630 <*>
1960 V64 N695 P194-199 61-10770 <*>
1961 V65 P219-228 62-10395 <*> 0
1961 V65 P238-244 62-10396 <*> 0
1961 V65 N699 P219-228 11537-B <K-H>
63-16574 <*>
1961 V65 N699 P229-237 11627-I <KH>
1961 V65 N699 P238-244 11537-A <K-H>
63-16572 <*>

TABAK
1938 V9 N4 P36-39 2339I <K-H>
1939 V10 N5 P16-18 2339G <K-H>
1939 V10 N5 P21-23 2339H <K-H>
1951 V12 N3 P39-41 61-10330 <*+>
1952 V13 N1 P56-58 2821A <K-H>
1953 V14 N2 P16-18 3084-A <K-H>
1953 V14 N3 P26-29 63-23222 <=*>
1953 V14 N5 P26-29 3160(I) <K-H>
1954 V15 N1 P23-25 3504-A <K-H>
1954 V15 N3 P40-44 3439-B <K-H>
63-15958 <=*>
1955 V16 N1 P46-49 5613-C <K-H>
1955 V16 N4 P57- R-129 <*>
1956 V17 N1 P23-24 60-18512 <+*>
1956 V17 N4 P53-55 PANSDOC-TR.283 <PANS>
60-19574 <+*>
1957 N2 P32-35 R-4005 <*>
1957 N2 P47-52 R-3769 <*>
1957 V18 N1 P40-42 64-19895 <=$>
1957 V18 N1 P43-44 63-16922 <*=>
1957 V18 N1 P52-53 R-3794 <*>
1959 V20 N1 P14-17 60-10813 <+*>
1959 V20 N2 P41-43 61-15172 <*+>
62-10908 <=*>
9900-B <K-H>
1959 V20 N2 P44-45 61-13960 <*+>
1959 V20 N3 P22-23 10177-A <K-H>
63-16926 <*=>
1959 V20 N3 P28-29 60-10812 <+*>
1959 V20 N3 P55-59 11249-A <KH>
63-16924 <=*>
1959 V20 N4 P25-27 11071-C <KH>
64-10697 <=*$>
1959 V20 N4 P41-43 61-13959 <*+>
1959 V20 N4 P50-52 11433-C <KH>
64-14760 <=*$>
1960 V21 N1 P40-42 10789 <K-H>
63-16923 <*=>
1960 V21 N1 P49-51 64-10601 <=*$>
1960 V21 N1 P54-57 61-15975 <+*>
1960 V21 N2 P21-23 11053-B <K-H>
61-20016 <*=>
1960 V21 N2 P34-46 11480-B <K-H>
1960 V21 N2 P43-46 62-18836 <=*>
1960 V21 N3 P35-39 61-20584 <=*>
1960 V21 N4 P17-20 61-10698 <+*>
1960 V21 N4 P117-120 10997-B <K-H>
1962 V23 N2 P47-52 64-13084 <=*$>
1962 V23 N3 P49-53 63-10669 <=*>
1963 V24 N4 P37-40 65-10904 <*>
1964 V25 N1 P38-39 RTS-2858 <NLL>
1964 V25 N1 P43-46 RTS-2755 <NLL>
1964 V25 N2 P48-53 90S89R <ATS>
1964 V25 N2 P57-60 RTS-2754 <NLL>
1965 V26 N2 P55-56 RTS-3315 <NLL>

TABAK-FORSCHUNG
1950 N3 P1 58-1197 <*>
1950 V4 P31-32 3180-F <K-H>
1953 N10 P1-3 58-979 <*>
1953 V10 P1-3 3517-B <KH>
1953 V10 P20-22 3625 <K-H>
1954 V11 P4 3523-B <K-H>
1954 V13 P9-12 3471-C <K-H>
1955 V15 P29-32 6690-A <K-H>
1955 V16 P25-28 6518-B <K-H>
1956 V18 P33-38 61-20017 <*>
1957 V19 P42-45 62-20233 <*>
1957 V21 P53-60 62-20234 <*> 0
63-16997 <*>
1957 V22 P61-64 62-20239 <*>

TAEGLICHE PRAXIS. (1960-)
1963 V4 N1 P70-73 64-18658 <*>

TAGUNGSBERICHTE DER ZEMENTINDUSTRIE
1954 N10 P25-49 T-1737 <INSD>
1960 N19 P27-45 63-14752 <*>

TAIRIKU MONDAI
1952 V1 N3 P37-42 62-16096 <*>

TAIWAN SUISAN ZASSHI
1935 N241 P8-10 AL-900 <*>

TAIWAN IGAKKAI ZASSHI
1925 N238 P95-98 61-10973 <*>

TAKAMINE KENKYUSHO NEMPO
1955 N7 P202-208 61J17J <ATS>
1958 V10 P44-51 61-16558 <*>

TAKEDA KENKYUSHO NEMPO
1945 N8 P22-29 58-1358 <*>

1959 N18 P37-43 61-00643 <*>

TALANTA
1960 V4 N1 P8-12 AEC-TR-4398 <*>
1961 V8 P629-640 AEC-TR-5011 <*>
1961 V8 N6 P446-449 63-20604 <*>
1962 V9 P997-1002 66-13253 <*>
1963 V10 P493-497 63-18157 <*>
1963 V10 P1229-1233 66-11205 <*>

AL-TALIAH. CAIRO
1965 N7 P68-83 65-31995 <=*$>

TALLINNA POLUTEHNILISE INSTIDUDI TOIMETISED SERIA A
1953 SA N45 P1-43 62-24618 <=*>
1962 N197 P155-165 AD-618 316 <=$>
1963 N206 P45-53 RTS-3606 <NLL>

TANDLAEGEBLADET
1955 V59 N2 P55-56 2703 <*>

TANKEN
1950 V1 P67-73 62-10125 <*> 0
1957 V8 N3 P65-70 65-10048 <*>

TANKIST
1950 N9 P14-19 RT-1891 <*>
1950 N9 P57-60 RT-1890 <*>

TANSO
1953 V3 P96-99 65S88J <ATS>
1954 V4 P9-12 60-10255 <*>
1957 V6 N1 P8-12 21TM <CTT>
1958 V7 P82-84 PANSDOC-TR.326 <PANS>
1959 V7 N3 P70-73 AEC-TR-4135 <*>
1960 V8 N1 P2-6 64N49J <ATS>
1962 N32 P22-27 65-12217 <*>

TAP CHI Y HOC VIET-NAM. HANOI (Y HOC VIET NAM)
1962 N2 P11-18 63-15098 <=>
1962 N2 P30-37 63-15098 <=>
1962 N2 P62-73 63-15098 <=>
1962 N2 P78-87 63-15098 <=>

TARSADALMI SZEMLE
1964 V19 N5 P20-32 64-31744 <=>
1967 N2 P39-47 67-30962 <=$>

TARTU ASTRONOOMIA OBSERVATOORIUMI PUBLIKATSIOONID
1955 V32 N1 P5-41 60-15274 <+*>
1955 V33 N1 P3-34 60-15373 <+*>

TAVRUAH
1956 N8 P30-34 PANSDOC-TR.496 <PANS>

TECHNICA. BASEL
1953 V2 N21 P4-8 62-20178 <*> 0
1959 V8 P1333-1338 2221 <BISI>
1960 V9 P1295-1298 2116 <BISI>
1961 V10 P682-687 62-10388 <*> 0
1964 V13 P350-351 3959 <BISI>
1965 V14 N10 P803-806 65-13908 <*>
1965 V14 N11 P961-964 65-13908 <*>
1965 V14 N12 P1037-1051 65-13908 <*>

TECHNICAL JOURNAL OF JAPAN BROADCASTING CORPORATION
SEE NHK GIJUTSU KENKYU

TECHNICAL PHYSICS OF THE U.S.S.R.
1935 V1 N5/6 P555-590 65-12198 <*>

TECHNICAL REPORT OF FISHING BOAT. JAPAN
SEE GYOSEN KENKYO GIHO

TECHNICAL REPORT OF FUJI IRON AND STEEL CO., LTD.
SEE FUJI SEITETSU GIHO

TECHNICAL REPORT OF THE NATIONAL AEROSPACE LAB. JAPAN
SEE KOKU UCHU GIJUTSU KENKYUSHO HOKOKU

TECHNICAL REPORTS OF OSAKA UNIVERSITY
SEE OSAKA DAIGAKU KOGAKUBU KOGAKU HOKOKU

TECHNICAL REVIEW, MITSUBISHI NIHON HEAVY INDUSTRIES
SEE MITSUBISHI NIHON JUKO GIHO

TECHNICKA PRACA
1962 N3 P210- ICE V2 N4 P526-527 <ICE>
1963 V15 N8 P606 64-71169 <=>

TECHNICKE ZPRAVY
1962 N4 P2-14 63-31351 <=> 0
1962 N4 P15-20 64-21102 <=>
1962 N4 P21-26 63-31351 <=> 0
1964 V1 P1-5 ORNL-TR-402 <*>
1964 V1 P6-9 ORNL-TR-403 <*>
1964 V1 P10-14 ORNL-TR-404 <*>
1964 V1 P15-17 ORNL-TR-405 <*>
1964 V1 P18-26 ORNL-TR-406 <*>
1964 V1 P27-32 ORNL-TR-407 <*>
1964 V1 P33-39 ORNL-TR-408 <*>
1964 V1 P48-61 ORNL-TR-409 <*>
1964 V1 P71-84 ORNL-TR-410 <*>

TECHNIK. BERLIN
1946 V1 N3 P114-118 AL-702 <*>
1947 V2 N1 P25-28 64-10011 <*>
1948 V3 N3 P133-134 62-20293 <*> 0
1948 V3 N4 P170-174 64-10298 <*>
1948 V3 N9 P381-386 62-16995 <*> 0
1948 V3 N11 P473-475 62-16995 <*> 0
1949 V4 P229-234 TS 1556 <BISI>
1949 V4 N1 P14-19 62-16927 <*> 0
1949 V4 N5 P229-234 62-14171 <*>
1949 V4 N8 P357-360 62-16967 <*> 0
1950 V5 N4 P170-172 TS 1734 <BISI>
1950 V5 N9 P460-466 AEC-TR-5020 <*>
1951 V6 N1 P13-14 AL-229 <*>
1951 V6 N1 P21-23 58-1133 <*>
1953 V8 N9 P609-612 60-18694 <*>
1954 V9 N3 P152-156 60-18693 <*>
1954 V9 N11 P615-622 62-18299 <*> 0
1956 V11 P695-700 57-890 <*>
1956 V11 N9 P645-650 59-20558 <*>
1956 V11 N10 P709-716 59-20563 <*>
1956 V11 N11 P767-772 58-1470 <*>
1956 V11 N12 P845-847 94K27G <ATS>
1957 V12 N11 P733-735 58-2416 <*>
1958 V13 P489-492 TS-1244 <BISI>
1958 V13 P718-721 59-21035 <=> 0
1958 V13 N2 P94-99 2114 <BISI>
1959 P53-54 59-16154 <+*> 0
1959 V14 P484-485 60-21045 <=> 0
1959 V14 N5 P333-336 60-23946 <=*>
1959 V14 N5 P337-341 60-23947 <=*>
1960 V15 N3 P192-195 60-31343 <=*>
1960 V15 N3 P196-198 60-31444 <=*> 0
1960 V15 N3 P205-208 60-31345 <=*> 0
1960 V15 N3 P216-220 60-31346 <=*> 0
1960 V15 N3 P249-250 60-31347 <=*> 0
1960 V15 N3 P254 60-31348 <=*>
1960 V15 N9 P586-592 62-25198 <=*> 0
1962 V17 N6 P440-443 63-16943 <*> 0
1963 V18 N4 P277-281 63-21797 <=>
1963 V18 N7 P469-473 3569 <BISI>
1964 V19 N1 P42-45 64-21559 <=>
1964 V19 N11 P719-724 65-30006 <=$>
1965 V20 N1 P16-22 65-30503 <=$>
1965 V20 N2 P89-91 65-30503 <=$>
1965 V20 N11 P725-727 66-30508 <=*$>
1966 N5 P307-313 66-32696 <=$>
1966 V21 N3 P176-177 66-34355 <=$>
1966 V21 N3 P180-182 66-34355 <=$>

1966 V21 N4 P257-263 66-32696 <=$>
1966 V21 N10 P523-524 66-35492 <=> O

TECHNIK UND BETRIEB. WIEN
1962 V14 N3 P42-43 3489 <BISI>

TECHNIK UND FORSCHUNG
1960 08/31 P498-500 1969 <BISI>
1961 N181 P758-760 3284 <BISI>

TECHNIK UND KULTUR
1931 V22 N2 P17-24 57-3098 <*>

TECHNIKA. BUDAPEST
1961 V5 N7 P2 62-19516 <=*> O

TECHNIKA LOTNICZA
1955 V10 N5 P144-148 57-2015 <*>
1963 N7 P183-188 M.5825 <NLL>
1964 N12 P302-305 AD-639 063 <=$>

TECHNIKA MOTORYZACYJNA
1958 V8 N5 P169-178 59-11917 <=*> O

TECHNIQUE:REVUE INDUSTRIELLE
1946 V21 N2 P119-120 63-10017 <*>

TECHNIQUE MODERNE
1946 V38 N23/4 P281-286 62-18422 <*> O
1947 V39 N1/2 P17-21 62-18422 <*> O
1955 V47 N11 P470-472 6247-C <K-H>
1956 P497-503 57-3282 <*>
1957 V49 N7 P63-65 61-16268 <*>
1957 V49 N7 P66-69 61-16267 <*>
1957 V49 N7 P70-72 61-16266 <*>
1957 V49 N7 P73-77 61-16265 <*>
1957 V49 N7 P87-93 61-16264 <*>
1957 V49 N7 P103-104 61-16263 <*>
1957 V49 N7 P108-110 61-16262 <*>
1957 V49 N7 P114-118 61-16261 <*>
1957 V49 N7 P366-369 61-18430 <*>
1958 V50 N9 P395-401 59-17106 <*> O
1959 V51 P197-206 TS 1460 <BISI>
1959 V51 N4 P197-205 62-14367 <*>
1960 V52 N6 P312-318 2276 <BISI>
61-18300 <*>
62-10251 <*> PO
1960 V52 N11 P513-517 177-420 <STT>
1961 V53 N9 P321-328 63-20487 <*> P
1962 V54 N12 P41-46 3214 <BISI>
1962 V54 N12 P57-59 3215 <BISI>
1963 V55 P198-199 3960 <BISI>

TECHNIQUE MODERNE - CONSTRUCTION. PARIS
1954 V9 N6 P171-178 62-10093 <*>
1954 V9 N7 P249-255 62-10093 <*>
1955 V10 N6 P218-224 CSIRO-2821 <CSIR>

TECHNIQUE DU PETROLE
SEE TECHNIQUES ET APPLICATIONS DU PETROLE ET AUTRES ENERGIES

TECHNIQUE ROUTIERE
1956 V2 P9-17 T-2477 <INSD>
1959 V4 N1 P4-25 59-17480 <*>
1960 V5 N1 P1-18 61-20558 <*>

TECHNIQUE SANITAIRE ET MUNICIPALE
1953 V48 P139-148 57-3384 <*>
1955 V50 N2 P25-30 II-563 <*>

TECHNIQUE ET SCIENCES AERONAUTIQUES
1951 N5 P268-278 AL-559 <*>
1955 V5 P297-302 1812 <*>
1956 N3 P118-124 57-2367 <*>
1956 N5 P221-225 57-3452 <*>
1957 N2 P49-79 59-16065 <*=>
1959 N3 P127-138 C-3742 <NRCC>
1959 N3 P177-183 61-11176 <=>
1959 N4 P231-235 61-11122 <=>
1960 N2 P79-91 C-3984 <NRCC>

TECHNIQUE ET SCIENCES AERONAUTIQUES ET SPATIALES
1962 V1 P31-45 AD-613 334 <=$>
1964 P489-495 4429 <BISI>
1965 V5 P431-445 66-12644 <*>

TECHNIQUES ET APPLICATIONS DU PETROLE ET AUTRES ENERGIES
1956 P4303-4304 62-16653 <*>
1956 P4305-4307 62-16652 <*>
1956 P4308-4310 62-16654 <*>
1956 P4733-4735 36J15F <ATS>
1961 V16 N189 P8524 62-18066 <*>
1964 N1062 P39-43 65-14344 <*>

TECHNIQUES ET ECONOMIQUES ET ECONOMIE INDUSTRIELLES
1965 P65-77 66-10150 <*> O

TECHNIQUES ET SCIENCES MUNICIPALES
1961 V36 N1 P9-13 65-11652 <*>

TECHNISCHE BERICHTE. ZENTRALE FUER TECHNISH-WISSEN SCHAFTLICHES BERICHTSWESEN UBER LUFTFAHRFORSCHUNG
1942 V9 N4 P101-105 64-10405 <*>

TECHNISCHE HAUSMITTEILUNGEN. NORDWESTDEUTSCHER RUNDFUNK
1954 V6 P4-5 1741 <*>
1954 V6 P5-7 1745 <*>
1954 V6 P8-15 1744 <*>
1954 V6 P16-18 1743 <*>
1954 V6 P18-23 1428 <*>
1954 V6 P24-27 1746 <*>
1954 V6 P27-29 1740 <*>
1954 V6 P29-39 1746 <*>
1954 V6 P40-41 1454 <*>
1954 V6 P42-46 1742 <*>
1954 V6 P46-51 1783 <*>
1954 V6 P52-54 1453 <*>
1955 V7 P77-78 91H8G <ATS>
1955 V7 P101-110 92H8G <ATS>
1955 V7 N3/4 P77-87 57-2804 <*>
1955 V7 N5/6 P101-110 57-2753 <*>

TECHNISCHE MECHANIK UND THERMO-DYNAMIK
1930 V1 P309-315 58-1906 <*>
1930 V1 P349-357 58-1906 <*>
1930 V1 N7 P268-276 65F4G <ATS>

TECHNISCHE MITTEILUNGEN. ESSEN
1952 V45 P3-18 I-679 <*>
1954 V47 N3 P137-145 65-12553 <*>
1956 V49 P41-49 T-1914 <INSD>
1956 V49 P2-12 4055 <BISI>
1957 V50 N4 P162-174 3956 <HB>
1957 V50 N4 P174-181 3957 <HB>
1958 V51 N3 P95-106 3776 <BISI>
1958 V51 N8 P349-356 62-18750 <*>
1958 V51 N8 P356-363 62-18754 <*>
1958 V51 N8 P370-374 62-18753 <*>
1958 V51 N10 P513-519 1764 <BISI>
1959 V52 N3 P109-113 62-10491 <*>
1960 V53 N2 P92-98 62N55G <ATS>
1961 V54 P77-78 2383 <BISI>
1961 V54 P80-82 2384 <BISI>
1961 V54 P82-87 2385 <BISI>
1961 V54 P88-98 2386 <BISI>
1961 V54 N11 P448-451 3043 <BISI>
1963 V56 P199-205 4818 <BISI>

TECHNISCHE MITTEILUNGEN DES FERNMELDEWERKS, ABTEILUNG FUER FERNSPRECHGERAET
1939 V2 P22-27 57-3522 <*>

TECHNISCHE MITTEILUNGEN DES INSTRUMENTENWESENS DES DEUTSCHEN WETTERDIENSTES
1955 V1 P3-18 57-1179 <*>

TECHNISCHE MITTEILUNGEN KRUPP
1934 V2 P44-46 57-170 <*>
1937 V5 N1 P19-21 59-20584 <*>
1938 V6 N7 P151-166 60-10252 <*> 0
1954 V12 N1 P5-12 62-14474 <*>
1955 V13 N2 P23-38 1405 <*>
1955 V13 N2 P39-43 1366 <*>
1955 V13 N2 P44-47 1365 <*>
1957 V15 N1 P13-22 58-1535 <*>
64-16582 <*>
1957 V15 N6 P130-144 66-11350 <*> 0
1958 V16 N2 P27-33 60-16792 <*>
1958 V16 N2 P34-38 60-16791 <*>
1958 V16 N2 P39-40 60-16786 <*>
1958 V16 N2 P41-46 60-16790 <*>
1959 V17 P330-342 2158 <BISI>
1959 V17 N1 P44-52 60-14220 <*>
1959 V17 N2 P1-15 NP-TR-739 <*>
1959 V17 N2 P82-89 61-16174 <*> 0
1959 V17 N4 P179-183 10156 <K-H>
60-18112 <*>
1959 V17 N6 P310-317 4808 <HB>
1959 V17 N6 P318-829 4807 <HB>
1960 V18 N1 P1-8 5069 <HB>
1960 V18 N2 P64-80 2485 <BISI>
1961 V19 P17-31 2179 <BISI>
1961 V19 P173-181 AEC-TR-5213 <*>
1961 V19 N1 P1-7 5127 <HB>
1961 V19 N1 P7-16 5359 <HB>
1961 V19 N1 P32-35 5812 <HB>
1961 V19 N2 P37-39 5271 <HB>
1961 V19 N2 P73-77 2396 <BISI>
1961 V19 N3 P107-121 5436 <HB>
1961 V19 N3 P122-129 5435 <HB>
1961 V19 N3 P130-153 62-16910 <*> P
1961 V19 N4 P208-213 5554 <HB>
1961 V19 N4 P213-218 5618 <HB>

TECHNISCHE MITTEILUNGEN KRUPP. FORSCHUNGSBERICHTE
1938 N2 P37-46 66-13145 <*> 0
1939 V2 N2 P5-14 60-10219 <*> 0
1941 V4 P31-36 I-294 <*>
1962 V20 P1-9 3064 <BISI>
1962 V20 P44-49 3072 <BISI>
1962 V20 N3 P65-72 5788 <HB>
1963 V21 P1-4 3696 <BISI>
1963 V21 P134-138 3752 <BISI>
1963 V21 P207-213 3707 <BISI>
1963 V21 N2 P37-43 3593 <BISI>
1963 V21 N2 P44-56 3594 <BISI>
1963 V21 N4 P123-127 6236 <HB>
1963 V21 N4 P139-146 6237 <HB>
1964 V22 P93-100 4242 <BISI>
1964 V22 N1 P24-40 4067 <BISI>
1964 V22 N2 P55-62 6382 <HB>
1964 V22 N2 P63-64 6209 <HB>
1964 V22 N3 P65-82 4879 <BISI>
1964 V22 N4 P125-142 4879 <BISI>
1965 V23 P146-156 4891 <BISI>
1965 V23 P157-178 4744 <BISI>

★TECHNISCHE MITTEILUNGEN KRUPP. WERKSBERICHTE
1943 V10 N2 P59-61 1762 <*>
1961 V20 N1 P29-38 2584 <BISI>
1962 V20 P165-179 3081 <BISI>
1962 V20 N2 P73-77 3374 <BISI>
1962 V20 N5 P263-267 5868 <HB>
1962 V20 N5 P267-272 6125 <HB>
1963 V21 P1-10 3540 <BISI>
1963 V21 N11 P125-136 4208 <BISI>

TECHNISCHE MITTEILUNGEN PTT
1951 N5 P7- 58-719 <*>
1951 V29 N5 P161-167 II-715 <*>
1951 V29 N8 P281-297 2903 <*>
1951 V29 N10 P390-392 63-10080 <*> 0
1951 V29 N12 P445-466 57-2487 <*>
1952 V30 N12 P363-379 2535 <*>
1952 V30 N12 P379-383 2717 <*>
1954 V32 P362-363 62-20290 <*> 0
1955 V33 N4 P143-155 61-16427 <*>
1956 N4 P172-179 T-2099 <INSD>
1956 N6 P250-257 T2100 <INSD>
1956 V34 N1 P1-26 11J17F <ATS>
57-3184 <*>
1956 V34 N6 P259-268 57-3194 <*>
1956 V34 N7 P303-309 66-11349 <*> 0
1956 V34 N9 P370-376 57-3185 <*>
58-1008 <*>
1957 V35 N9 P387-395 59-10853 <*>
1957 V35 N11 P441-455 60-31835 <=*>
1958 V36 N1 P13-32 98K25G <ATS>
1958 V36 N2 P61-62 60-31845 <=*>
1958 V36 N2 P67-82 60-31846 <=*>
1958 V36 N7 P261-276 65-11473 <*>
1959 V37 P109-113 TS-1350 <BISI>
1959 V37 N3 P89-94 1879 <BISI>
1959 V37 N6 P194-197 60-31461 <=*>
1959 V37 N7 P241-252 60-31469 <=*>
1959 V37 N8 P283-303 65-12234 <*>
1959 V37 N9 P355-370 60-31887 <=*>
1959 V37 N9 P381-392 60-31888 <=*>
1959 V37 N9 P400-405 60-31889 <=*>
1959 V37 N11 P503-508 60-31501 <=*>
1960 V38 N1 P1-20 60-31806 <=*>
1960 V38 N1 P20-25 60-31807 <=*>
1960 V38 N2 P51-61 60-31429 <=*>
1961 V39 N7 P217-243 62-14470 <*>
1961 V39 N9 P316-319 62-16604 <*>
1962 V40 N10 P354-363 66-14600 <*> 0
1963 V41 N3 P85-89 64-16279 <*>
1965 V43 N4 P97-107 POED-TRANS-2264 <NLL>
1965 V43 N11 P441-450 POED-TRANS-2265 <NLL>

TECHNISCHE MITTEILUNGEN. POST-, TELEGRAPHEN- UND TELEPHONVERWALTUNG
1947 V25 N3 P92-96 57-2082 <*>
1950 V28 N2 P41-50 57-585 <*>

TECHNISCHE MITTEILUNGEN. TELEGRAPHEN- UND TELEPHON-VERWALTUNG
1925 V3 P145-147 57-2028 <*>
1927 V5 N5 P174-176 57-2207 <*>
1928 V6 P101-106 66-13822 <*>
1932 V10 N2 P95-103 66-13815 <*>
1938 V16 N3 P92-99 58-355 <*>
1939 V17 P38-39 66-12688 <*>
1939 V17 P180-186 57-919 <*>
1940 V18 N5 P183-189 57-1103 <*>
1940 V18 N6 P205-215 60-14113 <*>
1941 V19 N2 P58-66 57-2454 <*>
1946 V24 N2 P55-69 57-880 <*>

TECHNISCHE RUNDSCHAU UND ALLGEMEINE INDUSTRIE -UND HANDELSZEITUNG
1932 04/27 P20-21 63-10446 <*>
1959 N52 P25-27 79M41G <ATS>
1962 N4 P13 3226 <BISI>
1962 N4 P15 3226 <BISI>
1963 N2 P29-30 65-18015 <*> C

TECHNISCHE UEBERWACHUNG. ESSEN
1960 V1 N4 P156-158 61-00029 <*>
1961 V2 N1 P12-16 873 <TC>
1961 V2 N12 P464-466 65-10451 <*> C
1963 V4 N3 P77-84 6074 <HB>
1963 V4 N10 P371-373 64-18823 <*>
65-18141 <*>

TECHNISCHE UEBERWACHUNG. MUENCHEN
1955 N12 P423-429 58-2092 <*>
1955 V7 N12 P423-429 62-18963 <*>
1957 N5 P107-116 58-1321 <*>
1959 V11 N1 P16-17 SCL-T-333 <*>

TECHNISCHE-WETENSCHAPPELIJKTIJDSCHRIFT
1951 V10 P222-241 T-1548 <INSD>

TECHNISCH-WISSENSCHAFTLICHE ABHANDLUNGEN AUS DER OSRAM-GESELLSCHAFT (KONZERN)

1958 V7 P175-184 66-11863 <*>

TECHNISCH-WISSENSCHAFTLICHES SCHRIFTUM DES TABAK TECHNIKUM. HAMBURG
1963 N35 P1-7 65-10738 <*>

TECHNOLCGY REPORTS OF THE FACULTY OF ENGINEERING, KYUSHU IMPERIAL UNIVERSITY
SEE KYUSHU DAIGAKU KOGAKU SHUHO

TECNICA ITALIANA
1953 V8 N5 P297-300 66-10284 <*>
1955 P93-98 57-185 <*>
1961 V26 P247-261 2575 <BISI>
1963 V28 P361- 2102 <BSRA>
1963 V28 N11 P4-23 65-00193 <*>
1964 V29 PT.1 P573- 2138 <BSRA>
1964 V29 PT.2 P717- 2138 <BSRA>
1964 V29 P223 2437 <BSRA>
1964 V29 P729 2196 <BSRA>
1965 V3C P6C5 2438 <BSRA>

TECNICA METALURGICA
1957 V13 N1C9 P1-43 61-18735 <*>

TECNICA ED ORGANIZZAZIONE
1954 V15 P8-15 T-2047 <INSD>

TECNICA VETRARIA
1956 V1 N6 P1-19 T-2285 <INSD>
1960 V5 N3 P13-28 64-10583 <*>

TECNOLOGIA CHIMICA
1942 V20 P197- 1071 <*>

TEER UND BITUMEN
1930 V28 N26 P421-424 59-20785 <*>
1930 V28 N27 P440-443 59-20785 <*> P

TEHNICA NOUA
1962 V9 N52 P1 63-21483 <=>
1962 V9 N52 P3 63-21483 <=>

TEHNICKI PREGLAD
1952 V4 P46-51 SPEC NO. 99Q67CR <ATS>
1962 V14 N2 P55-60 63-13105 <=*>

TEHNICKI VJESNIK
1943 V60 N1/2 P28-34 72J19CR <ATS>

TEHNIKA
1956 V11 N7 P1069-1071 63-16373 <*> O
1958 N5 P0B81-0B83 59-21077 <=*> O
1958 N6 P0B116-0B118 59-21102 <=*>
1960 V15 N3 P394-397 62-19344 <*=>
1960 V15 N3 P483-486 62-19345 <=*>
1962 V17 N5 P837-842 62-25971 <=*>
1962 V17 N1C P1846-1847 63-13361 <=*>
1962 V17 N12 P2341-2344 63-15364 <=*>
1963 V18 N3 P426-428 AD-636 657 <=$>
1963 V18 N4 P779-782 63-21974 <=>
1963 V18 N5 P805-810 63-31340 <=>
1963 V18 N5 P819 63-31340 <=>
1963 V18 N8 P1431-1432 63-31879 <=>
1964 V19 N1 P25-32 64-21586 <=>
1964 V19 N4 P623-625 64-31254 <=>
1964 V19 N6 P1011-1014 64-31964 <=>
1964 V19 N6 P1112-1117 64-41009 <=$>
1964 V19 N7 P1211-1222 64-41531 <=>
1964 V19 N7 P1339-1344 64-41531 <=>
1964 V19 N10 P1781-1786 64-41898 <=$>
1964 V19 N10 P1821-1822 64-51790 <=$>
1965 V2C N2 P228-233 65-30572 <=*>
1965 V20 N4 P637-644 65-31326 <=$>
1965 V20 N4 P644-649 65-31327 <=$>
1965 V20 N4 P650-652 65-31325 <=$>
1965 V20 N5 P877-884 65-31461 <=$>
1965 V20 N6 P1021-1027 65-31945 <*$=>
1965 V20 N6 P1053-1056 65-31965 <=*$>
1965 V20 N6 P1158-1165 65-31945 <*$=>
1965 V20 N6 P1183-1184 65-31945 <*$=>
1965 V20 N7 P1256-1260 65-32162 <=*$>
1965 V20 N8 P1453-1463 65-32750 <=*$>
1965 V20 N9 P1656-1660 65-33420 <=*$>
1965 V20 N9 P1736-1744 65-33421 <=*$>
1965 V20 N10 P1852-1857 65-33391 <=*$>
1966 V21 N8 P1397-1402 66-34413 <=$>

TEHNIKA JA TOOTMINE. TALLINN
1959 N6 P33 NLLTB V1 N9 P52 <NLL>

TEINTEX
1937 V2 P139-149 3290 <*>
1948 V13 P399-412 2032 <TC>
1950 V15 P271-281 T-2109 <INSD> P
1950 V15 P329-337 T-2109 <INSD> P
1950 V15 P353-357 T-2109 <INSD>
1952 V17 P71-99 T2110 <INSD>
1954 V19 P3-13 55N49F <ATS>
1954 V19 P811-838 T-2111 <INSD>
1954 V19 N2 P83-99 64-20422 <*> O
1954 V19 N11 P811-838 62-14083 <*> O
1956 V21 N12 P955-973 59-15315 <*>
1957 V22 P759-777 CHEM-107 <CTS>
1962 V27 N8 P535-553 63-18278 <*>
1962 V27 N12 P873-885 63-10939 <*> O
1963 V28 N10 P759-774 64-20942 <*> O

TEINTURE ET APPRETS
1960 N59 P110-115 35N54R <ATS>

TEION KAGAKU, BUTSURI
1949 N2 P119-128 2959 <*>
1954 V12 P94-111 64-16998 <*> O
1954 SA V12 P113-119 57-2953 <*>
1954 SA V13 P1C5-109 C-2297 <NRC>
1956 V15 P34-42 60-17344 <=*>
1956 SA V15 P171-183 58-2407 <*>
1958 V17 P147-166 62-00395 <*>
1959 V18 P97-114 C-3757 <NRCC>
61-00920 <*>
1959 V18 P115-129 62-00397 <*>
1959 V18 P131-147 62-00396 <*>
1960 V19 P175-186 62-00087 <*>
1960 V19 P188-201 62-00737 <*>
1960 V19 P203-213 62-00736 <*>
1963 V21 P173 1987 <BSRA>

TEION KAGAKU, SEIBUTSU HEN
1947 SB P187-192 II-724 <*>
1949 V2 P119-128 AL-955 <*>
1954 V12 P39-61 AEC-TR-6176 <*>
1954 SB V12 P95-111 57-3067 <*>
1962 SB V20 P57-67 AEC-TR-5885 <*>
1962 SB V20 P109-119 AEC-TR-5886 <*>
1964 V22 P109-118 66-11838 <*>

TEISHIN IGAKU
1954 V6 N12 P953-959 62-16602 <*>
1962 V14 P643-645 64-18718 <*>

TEKEL ENSTITULERI RAPCRLARI
1954 V6 N2 P256-261 3779-A <K-H>

TEKHNICHESKIE BIBLIOTEKI SSSR. MOSCOW
1964 V30 N8 P3-7 66-20987 <$>

TEKHNICHESKIE SOVETY KOLKHOZAM, RTS, SOVKHOZAM
1959 V20 N4 P2-6 60-23612 <*=>

TEKHNICHESKO DELC
1964 V15 06/13 P2 64-41155 <=>
1964 V15 06/20 P3 64-41155 <=>
1965 V16 N569 P1-2 65-31071 <=$>
1966 N12 P3 66-33984 <=$>

TEKHNIKA, SOFIYA
1964 V13 N2 P27-31 64-31321 <=>
1964 V13 N7 P7-9 64-51487 <=>
1964 V13 N7 P12 64-51487 <=>

Citation	Translation
TEKHNIKA KINO I TELEVIDENIYA	
1957 N1 P50	60-16233 <+*>
1957 N6 P74-78	60-14174 <+*>
1958 N2 P13-21	60-19560 <+*>
1958 N8 P27-29	60-23177 <+*> 0
1958 N9 P10-19	59-15829 <+*>
1958 N9 P36-41	61-11101 <=>
1958 N9 P57-63	59-17248 <+*>
1958 N11 P27-37	60-17386 <+*>
1958 N11 P62-65	59-20833 <+*>
1958 N12 P32-38	63-11686 <=>
1959 V3 N4 P11-19	60-16747 <+*>
1959 V3 N6 P11-16	60-14120 <+*>
1959 V3 N6 P47-54	60-14019 <+*>
1959 V3 N6 P60-67	60-16568 <+*>
1959 V3 N7 P19-22	60-14277 <+*>
1959 V3 N10 P1-12	60-16566 <+*>
1959 V3 N10 P47-50	60-16296 <+*>
1959 V3 N10 P58-62	60-16569 <+*>
1959 V3 N11 P16-18	60-16565 <+*> 0
1959 V3 N11 P25-33	60-31359 <=>
1959 V3 N12 P1-11	60-11860 <=>
1959 V3 N12 P12-17	60-16567 <+*>
1959 V3 N12 P18-25	60-16754 <+*>
1959 V3 N12 P32-34	60-18993 <+*>
1959 V3 N12 P55-56	60-31130 <=>
1960 V4 N2 P9-18	60-18993 <+*>
1960 V4 N3 P93-94	60-41247 <=>
1960 V4 N4 P70-72	60-41445 <=>
1960 V4 N4 P95-96	60-41445 <=>
1960 V4 N5 P1-31	60-41275 <=>
1960 V4 N5 P4-12	61-18287 <*=>
1960 V4 N5 P24	61-11674 <=>
1960 V4 N5 P44-51	61-20067 <*=>
1960 V4 N5 P56-61	61-11674 <=>
1960 V4 N5 P73-74	61-10321 <+*>
1960 V4 N6 P1-8	60-31766 <=>
1960 V4 N6 P38-44	62-19260 <*=>
1960 V4 N6 P66-70	61-21114 <=>
1960 V4 N6 P78	61-21114 <=>
1960 V4 N8 P1-7	61-21114 <=>
1960 V4 N8 P8-11	61C102 <CTT>
1960 V4 N9 P1-9	61-11791 <=> P
1960 V4 N9 P96	61-11791 <=>
1960 V4 N10 P1-7	61-11843 <=>
	61-11964 <=>
	61-21985 <=> 0
1960 V4 N10 P58-62	61-11964 <=>
1960 V4 N11 P36-40	61-11864 <=>
	74Q68R <ATS>
1961 V5 N1 P7-10	61-21638 <=>
1961 V5 N1 P95	61-21638 <=>
1961 V5 N3 P1-3	61-31464 <=>
1961 V5 N3 P4-13	60Q68R <ATS>
1961 V5 N3 P13-19	61Q68R <ATS>
1961 V5 N4 P29-34	61K117 <CTT>
1961 V5 N5 P23-27	75Q68R <ATS>
1961 V5 N5 P52-58	95Q72R <ATS>
1961 V5 N7 P10-20	81P59R <ATS>
1961 V5 N7 P43-50	76Q68R <ATS>
1961 V5 N8 P15-24	77Q68R <ATS>
1961 V5 N10 P38-41	63-15282 <=*>
1961 V5 N10 P51-56	62-25754 <=*>
1961 V5 N12 P3-10	63-19165 <=*>
1962 V6 N1 P8	62-32868 <=>
1962 V6 N2 P37-48	AD-616 268 <=$>
1962 V6 N4 P17-18	09P66R <ATS>
1962 V6 N5 P34-39	78Q68R <ATS>
1962 V6 N6 P10-17	35Q69R <ATS>
1962 V6 N9 P3-19	63-13642 <=>
1962 V6 N9 P49-50	71Q68R <ATS>
1962 V6 N9 P51-54	72Q68R <ATS>
1962 V6 N10 P1-11	64-10507 <=*$> 0
1962 V6 N10 P45-53	73Q68R <ATS>
1962 V6 N10 P66	63-13642 <=>
1962 V6 N10 P68	63-13642 <=>
1962 V6 N11 P5-15	63-13666 <=*>
1962 V6 N11 P40-42	56R75R <ATS>
1962 V6 N11 P65-68	63-21311 <=>
1963 V7 N1 P38-47	63-20573 <=*>
1963 V7 N1 P66-68	63-21428 <=>
1963 V7 N1 P92-94	63-21428 <=>
1963 V7 N2 P1-3	63-21736 <=>
1963 V7 N2 P21-26	64-10409 <=*$>
1963 V7 N2 P93-94	63-21736 <=>
1963 V7 N3 P73-74	63-18720 <=*>
1963 V7 N3 P73-75	63-21885 <=>
1963 V7 N5 P42	63-31230 <=>
1963 V7 N5 P92-93	63-31230 <=>
1963 V7 N5 P95	63-31230 <=>
1963 V7 N6 P66-67	63-31576 <=>
1963 V7 N6 P74-76	63-31576 <=>
	65-13823 <*>
1963 V7 N6 P94-95	63-31576 <=>
1963 V7 N7 P92-94	63-31727 <=>
1963 V7 N9 P25-29	AD-611 746 <=$>
1963 V7 N10 P35-38	64-18533 <*>
1963 V7 N11 P16-23	65-11213 <*>
1963 V7 N12 P5-12	64-16696 <=*$>
1964 V8 N1 P67-69	64-16697 <=*$>
1964 V8 N3 P1-6	64-20098 <*>
1964 V8 N3 P7-14	355 <TC>
1964 V8 N8 P1-6	66-13623 <*>
1964 V8 N8 P18-21	66-13618 <*>
1964 V8 N9 P9-13	64-51101 <=>
1964 V8 N9 P33-35	65-13917 <*>
1964 V8 N9 P36-47	65-13914 <*>
1964 V8 N10 P50-52	65-13916 <*>
1964 V8 N10 P53-58	65-13915 <*>
1964 V8 N11 P55-59	793 <TC>
1964 V8 N12 P28-34	66-12450 <*>
1965 V9 N1 P1-3	65-14616 <*>
1965 V9 N2 P1-12	66-13741 <*>
1965 V9 N2 P54-60	66-13742 <*>
1965 V9 N3 P38-42	86T92R <ATS>
1965 V9 N4 P56-59	66-13743 <*>
1965 V9 N7 P12-14	66-13744 <*>
1965 V9 N11 P40-44	66-13745 <*>
1966 V10 N2 P21-26	66-13740 <*>
1966 V10 N3 P1-6	66-13720 <*>
1966 V10 N3 P7-12	66-13719 <*>
1966 V10 N4 P48-51	66-13808 <*>
TEKHNIKA MOLODEZHI	
1953 N2 P9-12	UCRL TRANS-358(L) <*>
1954 N9 P2-6	UCRL TRANS-358(L) <*>
1957 N7 P7	UCRL TRANS-358(L) <*>
1957 V25 N8 P26	59-18796 <+*>
1957 V25 N9 P13	59-11450 <=>
1957 V25 N9 P26-27	60-19068 <+*>
1957 V25 N12 P16-19	61-19863 <=*>
1958 V26 N2 P7-8	60-19067 <+*>
1958 V26 N4 P3-4	60-19056 <+*>
1958 V26 N4 P5-7	63-19170 <=*> 0
1958 V26 N4 P17-18	60-19144 <+*> 0
1958 V26 N9 P37	59-11875 <=>
1958 V26 N10 P20-22	61-27269 <*=> 0
1958 V26 N11 P8-10	59-31015 <=> C
1958 V26 N12 P35-36	60-15065 <+*>
1959 V27 N1 P15-16	62-13057 <*=>
1959 V27 N2 P5-6	59-11770 <=> 0
1959 V27 N3 P22-23	60-13440 <+*>
1959 V27 N3 P37-39	59-19694 <+*>
1959 V27 N6 P7-9	61-19352 <+*> 0
1959 V27 N6 P16-17	61-23070 <*=>
1959 V27 N7 P34-36	59-22211 <+*>
1959 V27 N10 P5-6	60-31074 <=>
1959 V27 N10 P18-21	61-19784 <=*>
1959 V27 N10 P37-38	60-13528 <+*>
1960 V28 N1 P37-38	61-19350 <+*> 0
1960 V28 N6 P3-7	61-21343 <=>
1960 V28 N7 P37-39	61-18674 <*=>
1960 V28 N8 P14-15	61-27367 <*=>
1960 V28 N9 P22-23	64-13323 <=*$>
1960 V28 N10 P7-10	63-19516 <=*> 0
1960 V28 N11 P12-13	61-23932 <=>
1960 V28 N11 P23	61-23932 <=>
1961 V29 N1 P2-4	62-13567 <*=>
1961 V29 N1 P10-12	62-11080 <*=>

1961 V29 N1 P18-19 62-19261 <*=>
1961 V29 N1 P22-23 62-19261 <*=>
1961 V29 N1 P37 62-15277 <*=>
1961 V29 N2 P30-33 62-24277 <=*>
1961 V29 N10 P22-23 62-32421 <=*>
1961 V29 N11 P23-25 63-11509 <=>
1961 V29 N12 P12-13 62-32999 <=*>
1961 V29 N12 P28-30 63-21367 <=>
1961 V29 N12 P34-36 63-21367 <=>
1962 V30 N1 P30-31 62-32868 <=>
1962 V30 N5 P9 62-32867 <=>
1962 V30 N6 P28-30 62-33464 <=*>
1962 V30 N8 P27-29 63-13827 <=*>
1962 V30 N9 P25-26 63-13505 <=*>
63-15532 <=*>
1962 V30 N9 P27 63-13494 <=*>
1962 V30 N9 P36-38 63-21062 <=>
1963 V31 N1 P37 63-31675 <=>
1963 V31 N3 P6-7 63-23631 <=*$>
1963 V31 N3 P10-13 63-21981 <=>
1963 V31 N3 P35-37 63-24231 <=*$> 0
1963 V31 N6 P13 64-71170 <=>
1963 V31 N6 P30 63-11769 <=>
1963 V31 N9 P3-4 AD-620 975 <=$>
1963 V31 N12 P29 AD-610 415 <=$>
1964 V32 N4 P5 65-62124 <=*$>
1964 V32 N4 P22 64-71665 <=$>
1964 V32 N8 P20-21 AD-611 409 <=$>
1964 V32 N8 P36 64-51170 <=>
1964 V32 N9 P17-21 65-31733 <=*$>
1964 V32 N11 P21 AD-627 854 <=$>
1964 V32 N11 P22 AD-627 991 <=$>
1965 N4 P8 AD-637 438 <=$>
1965 N12 P14-15 66-30954 <=$>
1965 V33 N3 P14-16 65-31244 <=$> 0
1965 V33 N4 P33-36 65-31642 <=$>
1965 V33 N5 P5-9 65-31712 <=$>
1965 V33 N5 P36 65-31712 <=$>
1965 V33 N6 P18-19 65-31747 <=$>
1965 V33 N8 P12-13 65-32825 <=*$>
1965 V33 N11 P2-3 66-30750 <=$>
1965 V33 N11 P7-11 66-30750 <=$>
1965 V33 N11 P26-28 65-34105 <=*$>
1965 V33 N12 P22-23 66-31042 <=$>
1965 V33 N12 P35-36 66-30765 <=$>
1966 V34 N3 P10-13 66-31793 <=$>

TEKHNIKA V SELSKOM KHOZYAISTVE
1958 V18 N9 P1-4 59-11406 <=>
1958 V18 N9 P10-15 59-11406 <=>
1958 V18 N10 P24-27 59-11406 <=>
1959 V19 N7 P47-49 60-31001 <=>
1960 V20 N1 P15-17 60-51063 <=>
1960 V20 N1 P25-27 60-51064 <=>
1960 V20 N1 P30-31 60-51065 <=>
1960 V20 N1 P32-34 60-51038 <=>
1960 V20 N1 P43-44 63-11050 <=>
1960 V20 N1 P47-50 63-11010 P.1-5 <=>
1960 V20 N1 P50-52 63-11010 P.6-10 <=>
1960 V20 N4 P74 61-11518 <=>
1960 V20 N4 P77 61-11709 <=>
1960 V20 N5 P74-75 65-63551 <=$>
1960 V20 N5 P80-81 61-11709 <=>
1960 V20 N7 P73-75 61-11625 <=>
1960 V20 N10 P79-81 63-11051 <=>
1960 V20 N11 P57-59 61-23011 <=>
1960 V20 N11 P76-77 61-23011 <=>
1960 V20 N12 P1-5 61-27379 <=>
1960 V20 N12 P18-24 61-27379 <=>
1960 V20 N12 P46-49 61-27379 <=>
1961 V21 N1 P85-87 63-23750 <=*>
1961 V21 N2 P39-40 61-27153 <=>
1962 N2 P35-38 NLLTB V4 N7 P593 <NLL>
1962 V22 N9 P36-39 63-13347 <=>
1962 V22 N9 P48-52 63-13844 <=>
1962 V22 N9 P52 63-13486 <=>
1962 V22 N10 P47-50 63-13844 <=>
1963 V23 N4 P37-40 64-19551 <=$> 0
1963 V23 N6 P11-14 63-31574 <=>
1963 V23 N6 P38-40 63-31574 <=>
1963 V23 N9 P1-3 64-21160 <=>
1963 V23 N9 P4-9 64-21160 <=>
1963 V23 N9 P42-46 64-21160 <=> P
1963 V23 N9 P57-60 64-21160 <=> P
1963 V23 N9 P71 64-21160 <=>
1963 V23 N12 P41-43 64-31324 <=>
1964 V24 N9 P1-3 64-51693 <=$>
1964 V24 N9 P27-30 64-51693 <=$>
1964 V24 N9 P59-63 64-51693 <=$>
1964 V24 N9 P85 64-51693 <=$>
1965 V25 N1 P1-15 65-30835 <=*$>
1965 V25 N2 P39-42 65-30787 <=$>
1965 V25 N2 P47-48 65-30787 <=$>
1965 V25 N2 P50-51 65-30787 <=$>
1965 V25 N2 P69-71 65-30787 <=$>
1965 V25 N6 P1-4 65-32733 <=$>
1965 V25 N6 P42-45 65-32733 <=$>
1965 V25 N11 P1-4 66-30890 <=$>
1965 V25 N11 P35-40 66-30890 <=$>
1965 V25 N12 P72 66-30761 <=$>
1965 V25 N12 P85-93 66-30761 <=$>
1966 V26 N10 P1-6 66-35718 <=>
1966 V26 N10 P52-55 66-35718 <=>
1966 V26 N10 P86-87 66-35718 <=>
1966 V26 N10 P92-93 66-35718 <=>

TEKHNIKA I VOORUZHENIE
1963 N3 P40-43 64-31308 <=>
1964 N3 P2-87 64-71616 <=$>
1964 N3 P94-96 64-71616 <=$>
1964 N9 P82-84 AD-614 404 <=$>
65-60242 <=$>
1965 N8 P72-79 66-30132 <=*$>
1966 N3 P1-81 AD-643 646 <=>
1966 N3 P94-96 AD-643 646 <=>
1966 N6 P3-6 66-33754 <=$>
1966 N7 P1-83 AD-646 220 <=>
1966 N7 P94-96 AD-646 220 <=>
1966 N8 ENTIRE ISSUE AD-645 520 <=> 0

TEKHNIKA VOZDUSHNOGO FLOTA
1934 V8 N5 P39-66 R-3823 <*>
1935 V9 N1 P25-56 RT-1107 <*>
1944 V18 N8/9 P18-19 RT-1197 <*>
1946 V20 N10 P7-12 AD-613 461 <=$>

TEKHNOLOGIYA I KONSTRUIROVANIE GIROPROBOROV
1964 N59 ENTIRE ISSUE AD-634 796 <=$>

TEKHSOVETY MTS
1953 V14 N19 P1-6 RT-2040 <*> 0
1954 V15 N24 P1-23 RT-3902 <*>

TEKNILLISEN KEMIAN AIKAKAUSILEHTI
1962 V19 N16 P659-669 RUB.CH.TECH.V36 N2 <ACS>

TEKNISK TIDSKRIFT
1921 V51 N6 P84-89 57-1478 <*>
1930 V60 N8 P57-64 59-15649 <*> 0
1930 V60 N9 P65-74 59-15648 <*> 0
1932 V62 N30 P285-288 66-12700 <*>
1933 V63 N9 P98- 60-10693 <*>
1935 V65 P81-85 1638 <*>
1935 V65 P97-104 1638 <*>
1938 V68 N24 P295-296 63-14849 <*>
1939 N1 P1-4 57-976 <*>
1940 V70 N27 P114-118 57-344 <*>
1941 V71 P549-551 61-18474 <*> 0
1942 N5 P33-39 AEC-TR-5626 <*>
1943 V73 PE20-E28 62-18494 <*> 0
1943 V73 N7 PE10-E125 64-16201 <*> 0
1943 V73 N27 PE109-E116 57-43 <*>
1944 V74 P947-950 66-10226 <*> 0
1944 V74 P1043-1046 66-10687 <*>
1945 P271-275 59-15325 <*>
1945 V75 P279-284 64-16138 <*> 0
1947 V77 N44 P823-826 57-338 <*>
1948 V78 P160-162 58-1521 <*>

1949 V79 P141-147 UCRL TRANS-959(L) <*>
1950 V80 P451-455 AL-418 <*>
1950 V80 P1157-1164 36E1S <ATS>
1950 V80 N21 P553-554 61-16471 <*> 0
1950 V80 N31 P741-750 62-20344 <*> 0
1951 V81 P657-667 62-18496 <*> 0
1952 V82 P1095-1096 63-16400 <=*>
1952 V82 N10 P221-229 66-12666 <*>
1952 V82 N23 P1095-1096 58-711 <*>
1953 V83 P789-791 65-12045 <*>
1953 V83 P945-947 58-1079 <*>
1953 V83 N15 P303-309 60-10280 <*>
1954 V84 N38 P903-907 57-152 <*>
1954 V84 N44 P1043-1047 64-20569 <*> 0
1955 P469-471 57-151 <*>
1955 V85 N2C P457-461 59-20562 <*>
1956 V86 P512-521 57-3154 <*>
86J16S <ATS>
1956 V86 P1051-1057 37L32S <ATS>
1957 P333-336 T-2566 <INSD>
1957 V87 N21 P491-496 63-14676 <*>
1959 V89 P363-370 1007 <TC>
65-10115 <*>
1959 V89 P721-728 1683 <BISI>
1960 V90 P551-556 62-18327 <*>
1960 V90 P983-988 63-16214 <*> 0
1960 V90 N10 P221-226 5469 <HB>
1961 V91 N19 P529-536 185-664 <STT>
1962 V92 N1C P197-200 64-10484 <*> 0
1964 V94 N45 P1229-1232 6815 <HB>
1965 V95 P531- 2266 <BSRA>
1965 V95 P1077 2413 <BSRA>

TEKNISK UKEBLAD
1921 V39 N7 P77-79 59-20622 <*> 0
1921 V39 N8 P85-89 59-20622 <*> 0
1940 V87 P289-291 57-1147 <*>
1940 V87 P305-307 57-1147 <*>
1951 V98 N32 P631-638 I-582 <*>
1957 V104 N27 P567-571 AEC-TR-3424 <*>
1958 V105 P323-331 TS-1479 <BISI>
1958 V105 P1011-1018 TS 1435 <BISI>
1958 V105 N28 P651-658 60-14007 <*> 0
1958 V105 N43 P1011-1018 62-14328 <*>
1960 V107 N28 P616-617 61-10804 <*>
1962 V109 P351-360 63-14080 <*> 0

TEKNISKA FORENINGENS I FINLAND FORHANDLINGAR
1948 N1 P3-11 62-18495 <*> 0

TEKNISKA MEDDELANDEN FRAN K. TELEGRAFSTYRELSEN
1941 N7/9 P97-115 66-12254 <*>
1947 V1 P1-15 66-12280 <*> 0

TEKNISKT FORUM
1961 N20 P543-546 62-18617 <*>

TEKSTIL
1964 V13 N2 P69-78 65-17202 <*>
66-11557 <*>

TEKSTILNAYA PROMYSHLENNOST
1945 V5 N7 P18-22 61-20407 <*=>
1945 V5 N7 P25-27 61-20405 <*=>
1947 V7 N12 P18-21 27J15R <ATS>
59-19104 <+*>
1948 N6 P22-23 61-17125 <=$>
1949 V9 N3 P27-28 66-61555 <=*$>
1950 V10 N2 P35-36 50/3261 <NLL>
1951 N3 P31-32 T2145 <INSD>
1952 N7 P17-19 RT-2342 <*>
1952 N12 P21-23 RT-2500 <*>
1952 V12 N1 P31-33 TT.434 <NRC>
1952 V12 N4 P10 RT-2366 <*>
1952 V12 N4 P31-32 60-15374 <+*>
1952 V12 N7 P34-35 RT-4044 <*>
1952 V12 N10 P28-31 RT-3349 <*>
1952 V12 N11 P25-29 RT-3349 <*>
1952 V12 N12 P23-24 T2145 <INSD>
1954 V14 N2 P40-41 R-3553 <*>
1954 V14 N3 P31-34 R-3114 <*>
1954 V14 N4 P4-6 R-3552 <*>
1954 V14 N6 P41-44 T-2140 <INSD>
1954 V14 N9 P11 M-6262 <NLL>
1954 V14 N9 P16-19 59-15151 <+*>
1954 V14 N10 P31-33 R-3113 <*>
1954 V14 N12 P40-42 R-990 <*>
1955 N2 P34-38 T2141 <INSD>
1955 V15 N1 P36-37 65-61453 <=>
1955 V15 N3 P27-30 60-21798 <=>
1955 V15 N4 P40-42 M-6242 <NLL>
1955 V15 N10 P42-43 T-2134 <INSD>
61-14540 <=*>
1955 V15 N11 P38-39 CSIRO-3279 <CSIR>
1955 V15 N12 P46-47 CSIRO-3280 <CSIR>
1956 V16 N1 P12-14 R-5292 <*>
1956 V16 N1 P33-34 65-61540 <=$>
1956 V16 N2 P26-31 CSIRO-3508 <CSIR>
R-3650 <*>
1956 V16 N6 P33-36 62-32478 <=*$>
1956 V16 N6 P51-52 60-18489 <*+>
1956 V16 N7 P15-18 65-61537 <=$>
1956 V16 N7 P49-52 R-5049 <*>
59-16988 <+*> 0
1956 V16 N8 P20-24 CSIRO-3508 <CSIR>
R-3650 <*>
1956 V16 N9 P19-20 65-61538 <=$>
1956 V16 N9 P47 R-5277 <*>
1956 V16 N10 P43-44 59-16990 <+*>
1956 V16 N12 P34-36 61-14541 <=*>
1957 V17 N4 P22-26 62K24R <ATS>
1957 V17 N5 P6-7 59-18719 <+*>
62-16743 <=*> C
1957 V17 N5 P11-14 60-17455 <+*>
1957 V17 N5 P19-21 61-14542 <=*> 0
1957 V17 N7 P22-26 60-17780 <+*>
1957 V17 N8 P36-40 63-23753 <=*$>
1957 V17 N12 P16-17 59-17430 <+*>
1958 V18 N1 P16-17 65-61305 <=$>
1958 V18 N1 P25-27 63-16978 <=*> 0
1958 V18 N2 P15-16 60-10266 <+*>
60-10794 <*>
1958 V18 N2 P28-31 59-22668 <*>
1958 V18 N2 P61 66-60916 <=*$>
1958 V18 N2 P62 61-14543 <=*>
1958 V18 N3 P37-40 59-18089 <+*>
1958 V18 N3 P67 65-64603 <=*$>
1958 V18 N4 P9-11 63-18438 <=*>
1958 V18 N4 P58-59 59-22618 <+*>
1958 V18 N5 P17-19 64-10858 <=*$>
1958 V18 N5 P44-47 61-15238 <*+>
1958 V18 N5 P53-55 61-13075 <*+>
1958 V18 N6 P56-60 61-14544 <=*>
1958 V18 N7 P49-50 61-14545 <=*>
1958 V18 N8 P1-4 60-21218 <=>
1958 V18 N8 P10-14 60-21439 <=>
1958 V18 N8 P15-17 60-21432 <=>
1958 V18 N9 P16-17 59-21220 <=>
1958 V18 N9 P20-22 60-15533 <+*>
1958 V18 N9 P44-47 59-21221 <=>
1958 V18 N9 P51-53 60-13570 <+*>
1958 V18 N9 P58-60 66-60915 <=*$>
1958 V18 N9 P63-65 59-21196 <=>
1958 V18 N11 P9-12 60-13802 <+*>
63-18343 <=*>
1958 V18 N11 P36 60-15970 <+*>
1958 V18 N12 P32-36 61-13169 <*+>
1959 N4 P1-3 NLLTB V1 N11 P26 <NLL>
1959 N4 P4-9 NLLTB V1 N11 P30 <NLL>
1959 V19 N1 P1-9 60-15058 <+*> 0
1959 V19 N1 P10-12 60-15058 <+*> 0
60-15059 <+*>
1959 V19 N1 P13-16 60-15433 <+*>
1959 V19 N1 P16-19 60-15656 <+*>
1959 V19 N1 P20-24 60-15434 <+*>
1959 V19 N1 P25-27 60-15436 <+*>
1959 V19 N1 P27-31 60-15060 <+*>
1959 V19 N1 P31-33 60-15061 <+*>
1959 V19 N1 P34-37 60-15655 <+*>
60-17312 <+*>

1959 V19 N1 P38-41 60-15880 <+*>
1959 V19 N1 P41-44 60-15679 <+*>
1959 V19 N1 P44-47 60-15678 <+*>
1959 V19 N1 P48-49 60-15893 <+*>
1959 V19 N1 P50-51 60-15737 <+*>
1959 V19 N1 P52-55 60-15892 <+*>
1959 V19 N1 P55-58 60-15891 <+*>
1959 V19 N1 P58 60-15651 <+*>
1959 V19 N1 P59-60 60-17471 <+*>
1959 V19 N1 P61-63 60-15890 <+*>
1959 V19 N1 P63-65 60-15889 <+*>
1959 V19 N1 P65-69 60-17924 <+*>
1959 V19 N1 P69-71 60-15989 <+*>
1959 V19 N1 P71-73 60-15677 <+*>
1959 V19 N1 P74-77 60-15988 <+*>
1959 V19 N1 P79-81 60-15987 <+*> 0
1959 V19 N1 P82-86 60-15676 <+*>
1959 V19 N1 P86 60-15652 <+*>
1959 V19 N1 P87 60-17319 <+*>
1959 V19 N1 P88-89 60-17472 <+*>
1959 V19 N1 P89-91 60-15986 <+*>
1959 V19 N1 P92-93 60-15736 <+*>
1959 V19 N1 P93-94 60-15735 <+*>
1959 V19 N1 P94 60-15734 <+*>
1959 V19 N1 P95 60-15985 <+*>
1959 V19 N2 P56 61-13275 <*+>
1959 V19 N3 P45-53 60-10148 <+*>
1959 V19 N4 P55-58 60-13853 <+*>
1959 V19 N5 P73-77 26M44R <ATS>
60-18518 <+*>
61-15011 <*+> 0
1959 V19 N8 P40-41 60-23142 <+*>
1959 V19 N9 P28- 60-23013 <+*> 0
1959 V19 N10 P33-39 61-13031 <*+>
1960 V20 N2 P23-28 61-19792 <=*>
1960 V20 N2 P96-97 61-31462 <=> 0
1960 V20 N3 P36-42 62-23939 <=*>
1960 V20 N3 P46-49 61-13387 <*+>
1960 V20 N4 P84-86 61-19914 <=*>
1960 V20 N10 P18-24 CSIR-TR.255 <CSSA>
1960 V20 N11 P32-35 62-32955 <=*>
1960 V20 N12 P53-55 62-13745 <*=>
1960 V20 N12 P76-77 <FS>
1961 V21 N1 P14-19 62-33553 <=*>
1961 V21 N1 P43-47 58N54R <ATS>
1961 V21 N1 P47-49 65-63016 <=$>
1961 V21 N2 P54-56 64-13596 <=*$>
1961 V21 N2 P70 65-60665 <=$>
1961 V21 N4 P38-40 62-23719 <=*>
1961 V21 N5 P33-35 62-24304 <=*>
1961 V21 N5 P43-45 63-13602 <=*>
1961 V21 N5 P57-59 63-15213 <=*>
1961 V21 N6 P45-47 62-25285 <=*>
1961 V21 N6 P48-50 62-24698 <=$>
1961 V21 N7 P11-15 62-24646 <=*>
1961 V21 N7 P22-25 63-15263 <=*>
1961 V21 N7 P82-85 63-19129 <=*>
1961 V21 N8 P48-50 63-19773 <=*>
1961 V21 N1C P47-49 63-15212 <=*>
1961 V21 N11 P13-14 63-15264 <=*>
1961 V21 N11 P19-23 63-15972 <=*>
1961 V21 N11 P35-41 63-15265 <=*>
1961 V21 N11 P41-46 63-15307 <=*>
1961 V21 N11 P58-60 62-32952 <=*>
1961 V21 N11 P72-76 63-15973 <=*>
1961 V21 N12 P12-16 <LSA>
1961 V21 N12 P22-26 63-15216 <=*>
1961 V21 N12 P48-50 63-24625 <=*$>
1962 N3 P1-4 NLLTB V4 N10 P927 <NLL>
1962 N8 P8-9 NLLTB V5 N4 P293 <NLL>
1962 N8 P39-40 NLLTB V5 N4 P288 <NLL>
1962 V22 N3 P42-44 63-23517 <=*>
1962 V22 N4 P65 64-13593 <=*$>
1962 V22 N10 P41-44 65-60937 <=$>
1962 V22 N10 P51-53 63-18378 <=*> 0
1962 V22 N10 P57-60 63-24074 <=*>
1962 V22 N11 P6C-64 64-13594 <=*$>
1962 V22 N11 P72-77 63-18377 <=*>
1962 V22 N12 P13-19 65-60939 <=$>
1962 V22 N12 P24-29 64-19995 <=$>
1963 V23 N2 P45-48 64-23506 <=$>
1963 V23 N4 P31-32 65-60922 <=$>
1963 V23 N4 P37-43 64-23508 <=$>
1963 V23 N7 P5-15 63-31906 <=>
1963 V23 N7 P64-66 64-19878 <=$>
1963 V23 N7 P75-81 64-23507 <=$>
1963 V23 N9 P58-61 RTS-2746 <NLL>
1963 V23 N10 P18-21 65-60667 <=$>
1963 V23 N10 P26-28 65-63015 <=$>
1963 V23 N11 P83-84 65-60668 <=$>
1964 N4 P68-69 65-63798 <=*$> 0
1964 N10 P5-8 NLLTB V7 N5 P430 <NLL>
1964 V24 N1 P86-87 64-19373 <=$>
1964 V24 N4 P58-61 65-64604 <=*$>
1964 V24 N4 P67-68 65-64383 <=*$>
1964 V24 N7 P70-73 65-63017 <=$>
1964 V24 N8 P84-87 66-60999 <=*$>
1964 V24 N9 P61-63 65-63014 <=$>
1964 V24 N10 P5-8 TRANS BULL,V7,N5 <NLL>
1964 V24 N12 P36-44 RTS-3247 <NLL>
1964 V24 N12 P44-47 RTS-3216 <NLL>
1964 V24 N12 P53-56 65-63765 <=>
1964 V24 N12 P56-59 66-10564 <*>
1965 N12 P1-6 NLLTB V8 N6 P469 <NLL>
1965 N12 P8C-81 NLLTB V8 N7 P614 <NLL>
1965 N12 P89 NLLTB V8 N8 P716 <NLL>
1965 V25 N2 P31-33 66-61166 <=*$>
1965 V25 N3 P12-14 66-61554 <=$*>
1965 V25 N5 P79-81 66-61552 <=$*>
1966 V26 N1 P21-24 RTS-3691 <NLL>
1966 V26 N1 P27-30 RTS-3692 <NLL>

TELE
1958 N3 P1-5 POED-TRANS-2216 <NLL>

TELECOMUNICATII
1960 V4 N6 P261-267 62-23264 <*=>
1962 V6 N4 P145-150 63-13101 <=*>
1962 V6 N6 P285 63-21365 <=>
1963 V7 N2 P45-50 AD-613 458 <=$>
1964 V8 N2 P73-78 AD-641 112 <=$>
1964 V8 N5 P193-198 64-51298 <=$>
1964 V8 N5 P209 64-51298 <=$>
1966 V10 N1 P3-14 66-32640 <=$>

TELEFUNKEN-ROEHRE
1935 N3 P103-112 66-14621 <*>
1939 N16 P198-209 AL-898 <*>
1940 N18 P50-57 AL-749 <*>
1953 P110-116 SPEC. 57-1192 <*>
1953 V50 P1-22 57-2259 <*>
1953 V50 P1-9 57-995 <*>
1961 N39 P121-159 64-18418 <*>
1961 N40 P169-186 62-18097 <*>

TELEFUNKENZEITUNG
1929 V10 N53 P54-60 57-2153 <*>
1935 V16 N70 P35-43 66-13134 <*> 0
1935 V16 N71 P17-36 66-13134 <*> 0
1953 V26 P123-127 57-2492 <*>
1953 V26 N99 P95-101 59-17756 <*>
1953 V26 N99 P111-120 57-1369 <*>
62-18962 <*> 0
1954 V27 N105 P172-186 58-30 <*>
64J18G <ATS>
1954 V27 N106 P237-245 57-2801 <*>
9J8G <ATS>
1955 V28 N10 P222-226 AEC-TR-4039 <*>
1955 V28 N107 P15-22 28H9G <ATS>
57-2785 <*>
1956 V29 N113 P182-191 CSIRO-3396 <CSIR>
1956 V29 N114 P256-266 58-1604 <*>
58-35 <*>
1957 V30 P55-61 64-16959 <*> 0
1957 V30 N117 P207-214 58-1273 <*>
1958 V31 P124-130 64-14711 <*>
1958 V31 P179-187 63-01436 <*>
1958 V31 N120 P90-97 C-3397 <NRCC>
1958 V31 N120 P97-99 C-3287 <NRCC>
1958 V31 N120 P115-123 62-10282 <*>

1958 V31 N121 P150-161 59-11504 <=> O
1958 V31 N122 P232-239 47R75G <ATS>
64-16911 <*>
1961 V34 N131 P5-12 65-13644 <*>
1961 V34 N133 P221-231 62-18095 <*>
1963 V36 N3/4 P119-126 64-18597 <*>
1963 V36 N3/4 P146-151 65-10274 <*> O
1964 V37 N3/4 P194-209 66-11027 <*>

TELEGRAAF EN TELEFOON
1958 V59 P19-28 59-17790 <*>

TELEGRAFI E TELEFONI
1922 V3 P37-41 66-12249 <*>
1924 V5 N6 P316-324 57-2044 <*>

TELEGRAPHEN, FERNSPRECH-, FUNK- UND FERNSEH TECHNIK
1920 V9 P93-95 57-2090 <*>
1921 V10 P36-41 66-13336 <*> O
1921 V10 P55-60 66-13336 <*> O
1922 N4 P32-33 58-152 <*>
1925 V14 N4 P93-98 57-1468 <*>
1925 V14 N7 P189- 57-2092 <*>
1925 V14 N8 P205-210 57-2094 <*>
1925 V14 N10 P274-283 57-02425 <*>
1926 V15 P13-18 60-10097 <*>
1926 V15 P214-222 66-13308 <*> O
1926 V16 P288-293 57-2072 <*>
1927 V16 N4 P91- 57-2093 <*>
1928 V17 N3 P61-71 57-2175 <*>
1928 V17 N6 P178- 57-1481 <*>
1929 V18 P129-139 66-12599 <*>
1929 V18 P235-241 57-2161 <*>
1929 V18 N5 P140-142 66-13820 <*>
1930 V19 P133-141 66-11009 <*>
1930 V19 P180-185 57-3097 <*>
1930 V19 N1 P1-7 57-2101 <*>
1930 V19 N1 P7-17 57-1059 <*>
1930 V19 N1 P26-28 57-1401 <*>
1930 V19 N6 P167-180 57-2432 <*>
1930 V19 N9 P265-275 57-2172 <*>
1930 V19 N11 P348-357 66-12395 <*> O
1931 V20 N5 P154-155 57-328 <*>
1931 V20 N6 P171-180 57-2116 <*>
1932 V21 N1 P8-10 57-2206 <*>
1933 V22 N1 P3-13 57-2208 <*>
1934 V23 N4 P98-99 57-1391 <*>
1934 V23 N5 P107-117 57-1486 <*>
1934 V23 N7 P165-170 58-319 <*>
1934 V23 N8 P191-198 58-319 <*>
1935 V24 N1 P5-10 57-2166 <*>
57-3379 <*>
1935 V24 N2 P40-43 66-13650 <*> O
1935 V24 N10 P245-250 57-2255 <*>
57-3375 <*>
1935 V24 N12 P316-319 66-12585 <*>
1936 V25 N5 P117-118 63-16405 <*>
1936 V25 N7 P187-192 66-12268 <*>
1936 V25 N8 P217-218 57-2067 <*>
1936 V25 N10 P273-279 66-12267 <*>
1937 V26 N12 P275-280 57-592 <*>
1938 V27 P157- 57-1352 <*>
1938 V27 P158-166 66-12262 <*> O
1938 V27 P518-524 57-582 <*>
1938 V27 N2 P68-69 66-13651 <*>
1938 V27 N11 P395-404 2186 <*>
1938 V27 N12 P575-583 66-12271 <*>
1939 V28 N2 P63-68 57-369 <*>
1939 V28 N7 P249-257 57-2516 <*>
1939 V28 N7 P264-267 66-13341 <*> O
1939 V28 N8 P298-310 57-950 <*>
1939 V28 N8 P311-318 57-920 <*>
1939 V28 N11 P403-407 57-584 <*>
1939 V28 N12 P433-446 59-15165 <*>
1940 N9 P258-270 57-1454 <*>
1940 V29 N1 P9-16 61-10031 <*> O
1940 V29 N2 P49-54 57-1456 <*>
1940 V29 N4 P122-125 58-210 <*>
1940 V29 N5 P151-157 58210 <*>
1941 V30 N2 P52-57 63-14438 <*>
1941 V30 N12 P347-352 57-1480 <*>

TELEKOMUNIKACIJE
1960 V9 N4 P1-47 62-19360 <=*> P
1966 N2 P1-2 66-31210 <=$>

TELEKTRONIKK
1959 V1/2 P19-28 63-18922 <*>

TELE-RADIO
1958 V3 N12 P607-629 62-25204 <=*$>
1959 V4 N3 P113-123 62-25197 <=*> O

TELETEKNIK
1950 N3 P95-102 57-2508 <*>
1951 V2 N2 P203-206 57-2507 <*>
1952 V3 N1 P43-48 57-2484 <*>
1953 P153-159 57-2271 <*>
1957 V8 N3/4 P171-175 59-10956 <*>

TELLUS
1951 V3 N4 P240-247 2164 <*>

TEMAS DE TISIOLOGIA. UNIVERSIDAD NACIONAL. CORDOBA
1948 V2 N2/3 P232-251 2802 <*>

TEMATICHESKII SBORNIK. AKADEMIYA NAUK TADZHIKSKOI SSR. OTDEL FIZIOLOGII I BIOFIZIKII RASTENII
1963 N3 P29-40 64-18831 <*>
65-64386 <=*$>

TENSIDE
1964 V1 N1 P7-18 387 <TC>
1964 V1 N1 P18-26 66R79G <ATS>
1964 V1 N2 P50-59 879 <TC>
1964 V1 N3 P1-8 98S86G <ATS>
1964 V1 N4 P112-115 1118 <TC>
1964 V1 N4 P116-125 1299 <TC>
1965 V2 N11 P365-367 66-14125 <*>
1965 V2 N11 P368-373 1848 <TC>

TEOLLISUUDEN KESKUSLABORATORION. TIEDONANTOJA
1959 N235 P217-223 64-10921 <*>

TEORIYA I PRAKTIKA FIZICHESKOY KULTURY
1953 V16 N11 P785-792 RT-4312 <*>
1955 V18 N2 P131-142 RT-3391 <*>
1960 V23 N4 P307-318 60-41463 <=>
1960 V23 N11 P805-810 61-11880 <=>
1961 V24 N1 P7-13 61-31114 <=>
1961 V24 N1 P26-32 62-25524 <=*>
1961 V24 N1 P67-69 62-25524 <=*>
1961 V24 N4 P246-249 61-28393 <*=>
1961 V24 N4 P291-297 62-25474 <=*>
1961 V24 N4 P305-317 61-28296 <*=>
1961 V24 N8 P627-630 62-25001 <=*>
1961 V24 N9 P701-706 62-25536 <=*>
1961 V24 N10 P757-760 62-25051 <=*>
1961 V24 N10 P773-775 62-23593 <=>
1961 V24 N10 P778-779 62-23593 <=>
1961 V24 N10 P798-799 62-23593 <=>
1961 V24 N11 P867-870 62-24990 <=*>
1961 V24 N12 P907-909 62-24972 <=*>
1962 V25 N5 P69-79 62-32369 <*>
1962 V25 N7 P78-79 62-32854 <=*>
1963 V26 N3 P1-6 65-33499 <=$*>
1963 V26 N5 P77-78 63-31980 <=>
1963 V26 N5 P78-79 63-31979 <=>
1964 V27 N5 P6-10 64-41799 <=>
1964 V27 N5 P23-29 64-41799 <=>
1964 V27 N5 P31-37 64-41799 <=>
1964 V27 N5 P76-79 64-41799 <=>
1964 V27 N6 P1-14 64-51211 <=$>
1964 V27 N6 P45-49 64-51211 <=$>
1964 V27 N6 P57-61 64-51211 <=$>
1964 V27 N6 P67-70 64-51211 <=$>
1965 V28 N4 P12-20 65-31315 <=$>
1965 V28 N5 P22-30 65-31315 <=$>

TEORIYA I PRAKTIKA METALLURGII
1938 N4 P20-32 61-17185 <=$>
62-24182 <=*>
1939 V10 N1 P23-27 700 <HB>

*TEORIYA VEROYATNOSTEI I EE PRIMENENIE
1956 V1 N1 P3-17 R-1154 <*>
1956 V1 N3 P320-327 R-967 <*>
1957 V2 N1 P106-116 59-10884 <+*>
1957 V2 N4 P417-443 62-19883 <=>
1957 V2 N4 P473-475 64-11947 <=> M
1958 V3 P99-103 AMST S2 V12 P247-250 <AMS>
1958 V3 N1 P84-96 60-12610 <=>
1958 V3 N1 P99-102 R-4171 <*>
1958 V3 N2 P205-211 TS 1753 <BISI>
62-25696 <=*>
1958 V3 N3 P285-317 66-34283 <=$>
1958 V3 N4 P395-412 51M46R <ATS>
60-21789 <=>
1960 V5 N1 P134-136 60-13840 <=>
1960 V5 N2 P237-243 61-10416 <+*>
1961 V6 N2 P164-181 62-15444 <*=>
1961 V6 N2 P222-228 62-15669 <*=>
1961 V6 N2 P234-242 62-15444 <*=>
1961 V6 N4 P377-391 63-15327 <=*>
1962 V7 N2 P208-213 62-32051 <=*>
1962 V7 N3 P283-311 63-23728 <=*$>
1962 V7 N7 P438-446 RTS-3336 <NLL>
1963 V8 N3 P324-330 64-21952 <=>

*TEPLOENERGETIKA
1954 V1 N2 P49 RTS-2623 <NLL>
1954 V1 N2 P50 RTS-2624 <RTS>
1954 V1 N4 P49-54 R-2527 <*>
1955 N3 P34-37 NP-TR-173 <*>
1955 N6 P16-18 R-3687 <*>
1955 N9 P43-48 R-4978 <*>
1955 V2 N1 P12-17 62-23164 <=*>
1955 V2 N1 P17-23 60-12482 <=>
1955 V2 N1 P27-31 61-15200 <*+>
1955 V2 N2 P3-10 RTS-3272 <NLL>
1955 V2 N5 P17-21 R-3143 <*>
1955 V2 N5 P34-37 64-15272 <=*$>
1955 V2 N5 P38-44 36K25R <ATS>
1955 V2 N7 P30- 59-21059 <=>
1955 V2 N9 P37-42 UCRL TRANS-501(L) <*+>
1955 V2 N10 P34-78 61-15909 <+*>
1955 V2 N10 P38-45 RTS-2981 <NLL>
1955 V2 N11 P19-23 R-3885 <*>
1955 V2 N12 P32-36 62-11459 <=>
1956 N6 P51-56 NP-TR-175 <*>
1956 V3 N1 P16-21 62-24366 <=*>
1956 V3 N1 P22-26 59-16543 <+*>
1956 V3 N3 P39-47 61-15717 <+*>
1956 V3 N4 P36-39 59-22499 <+*>
1956 V3 N6 P39-44 60-19877 <+*>
1956 V3 N7 P16-23 62-24074 <=*>
1956 V3 N7 P40-45 63-15922 <=*>
1956 V3 N8 P3-10 64-15927 <=*$>
1956 V3 N8 P10-13 64-15928 <=*$>
1956 V3 N9 P18-24 59-22473 <+*>
1956 V3 N9 P49-52 65-60490 <=$>
1956 V3 N10 P47-51 61-19402 <+*> 0
1956 V3 N10 P62-63 R-4392 <*>
1956 V3 N11 P3-10 59-19627 <+*>
1956 V3 N11 P19-25 59-19626 <+*>
1956 V3 N11 P25-29 59-16539 <+*>
1956 V3 N11 P37-42 59-22476 <+*>
1956 V3 N12 P10-14 59-22486 <+*>
1956 V3 N12 P14-20 3K25R <ATS>
1956 V3 N12 P35-37 59-22691 <+*>
1957 V4 N1 P3-8 62-24674 <=*> 0
1957 V4 N1 P39-41 65-12330 <*>
1957 V4 N1 P42-45 65-60390 <=$>
1957 V4 N2 P3-7 59-22523 <+*>
60-23587 <+*>
1957 V4 N2 P20-23 59-22524 <+*>
1957 V4 N3 P17-22 61-15388 <+*>
1957 V4 N3 P22-26 59-19625 <+*>
1957 V4 N3 P50-54 61-20714 <*=>
1957 V4 N3 P55-57 61-23395 <*=>
1957 V4 N4 P3-6 61-13299 <*+>
1957 V4 N4 P6-10 61-13212 <*+>
1957 V4 N4 P10-16 62-24673 <=*>
1957 V4 N4 P16-21 60-13221 <+*>
1957 V4 N4 P28-33 59-22522 <+*>
1957 V4 N5 P24-30 2K25R <ATS>
1957 V4 N5 P44-48 60-13223 <+*>
1957 V4 N6 P12-16 60-15574 <+*>
1957 V4 N6 P21-25 61-15461 <+*>
1957 V4 N6 P43-50 60-23029 <*+>
1957 V4 N7 P57-59 59-22495 <+*>
1957 V4 N7 P68-72 60-13326 <+*>
1957 V4 N7 P72-79 60-15566 <+*>
1957 V4 N8 P15-18 60-17149 <+*>
1957 V4 N8 P18-23 60-17793 <+*>
1957 V4 N8 P48-53 60-13327 <+*>
1957 V4 N8 P65-68 R-4702 <*>
59-22677 <+*>
60-15582 <+*>
1957 V4 N8 P68-72 91M40R <ATS>
1957 V4 N9 P7-12 62-24672 <=*>
1957 V4 N9 P12-16 60-17151 <+*>
1957 V4 N9 P37-40 60-15577 <+*>
1957 V4 N9 P45-48 60-15576 <+*>
1957 V4 N9 P58-63 60-17143 <+*>
1957 V4 N9 P64-67 60-15583 <+*>
1957 V4 N9 P73-75 60-15584 <+*>
1957 V4 N10 P27-30 60-13310 <+*>
1957 V4 N10 P73-77 60-13850 <+*> C
1957 V4 N11 P70-73 AEC-TR-6104 <*>
AEC-TR-6104 <=*$>
03P61R <ATS>
1957 V4 N11 P96 NP-TR-120 <*>
1957 V4 N12 P16-20 62-24102 <=*>
1957 V4 N12 P61-65 60-17720 <+*>
1957 V4 N12 P69-72 SCL-T-220 <+*>
1958 V5 N1 P3-8 60-15700 <+*>
1958 V5 N1 P22-24 60-16543 <+*> 0
60-17799 <+*>
1958 V5 N1 P31-33 59-10923 <+*>
1958 V5 N2 P47-51 62-15408 <*=>
1958 V5 N2 P56-61 1441 <BISI>
1958 V5 N2 P63-69 61-15462 <+*>
1958 V5 N2 P72-75 63-15922 <=*>
1958 V5 N2 P76-77 60-17906 <+*>
1958 V5 N2 P77-80 59-18091 <+*>
1958 V5 N3 P47-50 T62-2 <MUL>
1958 V5 N3 P51-54 60-23077 <+*>
1958 V5 N3 P55-60 61-23884 <*=>
1958 V5 N3 P67-72 C-3599 <NRCC>
61-15630 <+*>
1958 V5 N4 P34-41 60-17446 <+*>
1958 V5 N4 P63-68 61-15364 <+*>
61-23678 <*=>
1958 V5 N4 P79-80 61-23678 <*=>
1958 V5 N5 P25-31 59-22474 <+*>
1958 V5 N5 P38-43 TRANS-52 <MT>
1958 V5 N5 P44-46 59-19448 <+*>
1958 V5 N5 P49-54 59-22666 <+*>
1958 V5 N5 P57-61 59-18120 <+*>
1958 V5 N6 P63-70 61-15363 <+*>
1958 V5 N6 P71-76 59-19444 <+*>
62-13131 <*=>
1958 V5 N7 P3-6 62-25461 <=*>
1958 V5 N7 P7-9 61-13909 <*+>
1958 V5 N7 P10-13 61-15409 <+*>
1958 V5 N7 P13-17 61-15396 <+*>
1958 V5 N7 P18-21 61-13998 <*+>
1958 V5 N7 P26-30 62-24365 <=*>
1958 V5 N7 P38-43 62-24372 <=*>
1958 V5 N7 P55-63 59-19446 <+*>
1958 V5 N7 P68-74 60-13803 <+*>
1958 V5 N8 P18-23 62-25928 <=*>
1958 V5 N8 P44-48 59-19314 <+*>
59-22436 <+*>
1958 V5 N8 P48-51 65-60491 <=$>
1958 V5 N8 P56-60 59-19813 <+*>
1958 V5 N8 P61-65 62-24371 <=*>
1958 V5 N8 P66-74 60-13775 <+*>

Citation	Report
1958 V5 N8 P74-78	62-25357 <=*>
1958 V5 N9 P30-33	59-18412 <+*>
1958 V5 N9 P57-60	62-13744 <*=>
1958 V5 N9 P68-70	64-24591 <=*$>
1958 V5 N9 P74-79	TS-1714 <BISI>
	62-25700 <=*$>
1958 V5 N9 P87-89	59-22651 <+*>
1958 V5 N10 P30-34	61-23281 <=*>
1958 V5 N10 P35-42	61-15237 <*+>
1958 V5 N10 P42-46	61-15203 <*+>
1958 V5 N10 P46-51	61-15401 <+*>
1958 V5 N10 P51-53	61-13490 <*+>
1958 V5 N10 P61-65	AEC-TR-477 <=*>
1958 V5 N10 P90-95	63-19888 <*=>
1958 V5 N11 P3-9	60-13776 <+*>
1958 V5 N11 P9-13	59-19315 <+*>
1958 V5 N11 P14-20	59-18860 <+*>
1958 V5 N11 P33-41	64-15947 <=*$>
1958 V5 N11 P64-69	59-18858 <+*>
1958 V5 N12 P56-62	62-24772 <=*>
1958 V5 N12 P79-85	59-19247 <+*>
1959 V6 N1 P28-37	60-13467 <+*>
1959 V6 N1 P49	60-13568 <+*>
1959 V6 N1 P50-56	61-13985 <*+>
1959 V6 N1 P62-65	61-13057 <*+>
1959 V6 N1 P68-72	60-13013 <+*>
1959 V6 N1 P72-80	64-23532 <=>
1959 V6 N1 P80-83	61-13986 <*+>
1959 V6 N2 P23-28	61-13211 <*+>
1959 V6 N2 P33-39	60-19939 <*+>
1959 V6 N2 P50-53	AEC-TR-5049 <=*>
	61-13206 <*+>
	61-13996 <*+>
	63-15146 <=*> 0
1959 V6 N2 P53-57	63-24590 <=*$>
1959 V6 N2 P79-83	60-19888 <+*>
1959 V6 N2 P83-88	74L35R <ATS>
1959 V6 N3 P26-29	61-13300 <*+>
	63-24553 <=*$>
1959 V6 N3 P30-31	61-13020 <*+>
1959 V6 N3 P32-35	62-25358 <=*>
1959 V6 N3 P40-44	59-22590 <+*>
1959 V6 N3 P45-49	59-19249 <+*>
1959 V6 N3 P66-72	AD-615 428 <=$>
1959 V6 N4 P49-55	64-19927 <=$>
1959 V6 N4 P72-79	61-13243 <*+>
1959 V6 N5 P50-55	63-19941 <*=>
1959 V6 N6 P35-39	61-13089 <*+> 0
1959 V6 N7 P41-45	61-15236 <*+>
1959 V6 N7 P59-65	61-13232 <*+>
1959 V6 N7 P65-69	61-13231 <+*>
1959 V6 N7 P74-84	RTS-2727 <NLL>
	65-60374 <=$>
1959 V6 N7 P84-86	60-23038 <*+>
1959 V6 N8 P9-11	60-19178 <+*>
1959 V6 N8 P11-14	60-19572 <+*>
1959 V6 N8 P30-33	60-23807 <+*>
1959 V6 N8 P33-37	60-15682 <+*>
1959 V6 N8 P38-43	60-15650 <+*>
1959 V6 N8 P43-48	60-23568 <+*>
1959 V6 N8 P51-53	60-23073 <+*>
1959 V6 N9 P3-6	60-12194 <=>
1959 V6 N9 P3-7	60-17230 <+*>
1959 V6 N9 P7-15	62-19935 <=>
	62-19935 <=*>
1959 V6 N9 P15-21	61-13529 <*+> 0
	61-13529 <=>
	61-23397 <*=>
	62-33587 <=*>
1959 V6 N9 P22-27	62-11335 <=>
1959 V6 N9 P32-39	61-15330 <+*>
1959 V6 N9 P39-46	RTS-2301 <NLL>
1959 V6 N9 P46-50	60-23757 <+*>
1959 V6 N9 P50-56	60-23626 <+*>
1959 V6 N9 P57-62	61-23871 <*=> 0
1959 V6 N9 P74-79	AEC-TR-4812 <=*>
	61-13224 <*+>
	61-28187 <*=>
	92M40R <ATS>
1959 V6 N10 P3-8	59-31049 <=>
1959 V6 N10 P8-17	60-17221 <+*>
	61-13122 <+*>
1959 V6 N10 P67-70	62-11333 <=>
1959 V6 N10 P70-74	62-23150 <=>
	62-23150 <=*>
	62-25348 <=*>
1959 V6 N10 P74-77	60-12188 <=>
	62-11404 <=>
1959 V6 N11 P34-38	61-15078 <*+>
1959 V6 N11 P37-41	61-15001 <*+>
1959 V6 N11 P45-47	61-15216 <+*>
1959 V6 N11 P48-52	65-60659 <=$>
1959 V6 N11 P53-57	61-13492 <*+>
1959 V6 N11 P58-65	61-13124 <*+> 0
	63-13425 <=*>
1959 V6 N12 P3-13	61-28767 <*=>
1959 V6 N12 P13-18	61-15017 <*+>
1959 V6 N12 P19-26	64-13801 <=*$>
1959 V6 N12 P43-46	60-23005 <*+>
1959 V6 N12 P46-51	61-15626 <+*>
	61-27178 <*=>
1959 V6 N12 P51-55	61-15400 <+*>
1959 V6 N12 P77-80	62-24302 <=*>
1959 V6 N12 P80-83	61-15261 <*+>
1960 V7 N1 P30-33	65-60658 <=$>
1960 V7 N1 P33-37	61-15406 <+*>
1960 V7 N1 P63-69	62-24370 <=*>
1960 V7 N1 P69-75	62-24369 <=*>
1960 V7 N1 P75-79	61-13132 <*+> 0
1960 V7 N1 P79-85	AEC-TR-5602 <=*>
1960 V7 N1 P85-87	62-13127 <*=>
1960 V7 N2 P3-11	62-24689 <=*>
1960 V7 N2 P18-24	61-13527 <*+>
1960 V7 N2 P34-40	RTS-2484 <NLL>
1960 V7 N2 P41-47	66-61527 <=*$>
1960 V7 N2 P60-66	61-15006 <*+>
1960 V7 N2 P70-77	62-24303 <=*>
1960 V7 N2 P77-80	62-25697 <=*>
1960 V7 N2 P78-87	61-27262 <*=>
1960 V7 N3 P3-8	63-15148 <=*> 0
1960 V7 N3 P8-12	63-15154 <=*> C
1960 V7 N3 P13-17	61-19794 <=*>
	62-23357 <=*>
1960 V7 N3 P20-24	61-15639 <+*>
1960 V7 N3 P83-87	62-24368 <=*>
1960 V7 N4 P18-24	61-15714 <+*>
1960 V7 N4 P33-38	62-24658 <=*>
1960 V7 N4 P42-47	63-15155 <=*> 0
1960 V7 N4 P62-67	61-15922 <+*>
1960 V7 N4 P72-73	62-25109 <=*>
1960 V7 N4 P74-81	60-11771 <=>
1960 V7 N5 P12-16	61-15399 <+*>
1960 V7 N5 P19-24	61-27619 <*=>
1960 V7 N5 P27-32	62-13128 <*=>
1960 V7 N5 P76-81	64-13151 <=*$>
1960 V7 N5 P81-88	AEC-TR-4740 <*=>
	61-15867 <+*>
1960 V7 N6 P3-7	61-13497 <*+>
1960 V7 N6 P47-49	62-25938 <=*>
1960 V7 N6 P58-62	61-27607 <*=>
1960 V7 N6 P63-66	62-19216 <=*>
1960 V7 N6 P67-69	61-15007 <+*>
1960 V7 N6 P69-72	64-13802 <=*$>
1960 V7 N6 P72-76	64-13668 <=*$>
	64-13810 <=*$>
1960 V7 N7 P3-4	61-14566 <=*>
1960 V7 N7 P4-12	61-14567 <=*>
1960 V7 N7 P12-16	61-14568 <=*>
1960 V7 N7 P16-23	61-14569 <=*>
1960 V7 N7 P23-27	61-16855 <*=>
1960 V7 N7 P32-34	65-31209 <=$>
1960 V7 N7 P44-51	62-13114 <*=>
1960 V7 N7 P59-64	63-15141 <=*> 0
	63-15193 <=*>
1960 V7 N7 P80-86	62-00138 <*>
	62-15293 <*=>
1960 V7 N8 P29-32	62-25922 <=*>
1960 V7 N8 P57-60	TR-101 <INFO>
	61-19905 <=*>
1960 V7 N8 P71-74	61-23656 <=*>

Citation	Report
1960 V7 N8 P74-78	65-60375 <=$>
1960 V7 N9 P6-12	61-27549 <*=>
1960 V7 N9 P30-34	63-15153 <=*> 0
1960 V7 N9 P44-49	61-27188 <*=>
1960 V7 N9 P60-62	61-23374 <*=>
1960 V7 N9 P67-71	62-19143 <*=>
1960 V7 N9 P71-75	61-19918 <=*>
1960 V7 N10 P6-13	62-24888 <=*>
1960 V7 N10 P14-22	62-24887 <=*>
1960 V7 N10 P22-27	62-23941 <=*>
1960 V7 N10 P63-67	62-25460 <=*>
1960 V7 N10 P67-68	TR-102 <INFO>
	61-19697 <=*>
1960 V7 N10 P69-73	61-21094 <=>
1960 V7 N10 P80-89	61-19799 <=*>
1960 V7 N10 P95	61-28694 <*=>
1960 V7 N11 P3-8	61-21139 <=> P
1960 V7 N11 P9-15	61-19941 <=*>
1960 V7 N11 P46-48	61-27538 <=*>
	62-25957 <=*>
1960 V7 N11 P59-64	61-19803 <=*>
1960 V7 N11 P64-66	61-27184 <*=>
1960 V7 N11 P66-69	61-19698 <=*>
1960 V7 N11 P90-91	61-21139 <=>
1960 V7 N12 P3-7	61-21538 <=>
1960 V7 N12 P14-18	2065 <BISI>
1960 V7 N12 P19-23	62-25888 <=*>
1960 V7 N12 P27-33	62-15427 <*=>
1960 V7 N12 P33-38	62-15197 <*=>
1960 V7 N12 P67-71	62-23896 <=*>
1960 V7 N12 P90-91	61-23637 <*=>
1961 N3 P63-66	62-32811 <*=>
1961 N3 P100-107	63-19328 <=*>
1961 N3 P116-128	63-19328 <=*>
1961 N3 P137-163	63-19328 <=*>
1961 V8 N1 P3-7	61-31550 <=>
1961 V8 N1 P18-22	61-28915 <*=>
1961 V8 N1 P30-34	63-19847 <=*>
1961 V8 N1 P37-44	62-24644 <=*>
1961 V8 N1 P56-58	62-15174 <*=>
	62-24863 <=*>
1961 V8 N1 P58-65	62-13917 <=*>
1961 V8 N1 P72-78	62-24288 <=*>
1961 V8 N2 P24-28	62-24676 <=*>
	62-24940 <=*>
1961 V8 N2 P67-71	62-33770 <=*>
1961 V8 N2 P72-75	62-25929 <=*>
1961 V8 N2 P75-79	AEC-TR-5710 <=*>
1961 V8 N3 P36-40	62-24935 <=*>
1961 V8 N3 P46-48	62-25955 <=*>
1961 V8 N3 P49-52	63-15140 <=*> 0
1961 V8 N3 P53-56	63-13882 <=*>
1961 V8 N3 P62-64	63-13882 <=*>
1961 V8 N3 P64-66	62-32805 <=*>
1961 V8 N3 P67-70	62-32470 <=*>
1961 V8 N3 P74-75	65-60751 <=>
1961 V8 N3 P91-94	61-31550 <=>
1961 V8 N4 P12-15	62-25935 <=*>
1961 V8 N4 P25-27	62-25954 <=*>
1961 V8 N4 P28-30	62-24677 <=*>
1961 V8 N4 P37-41	62-24666 <=*>
1961 V8 N4 P41-43	AD-611 108 <=$>
	62-24669 <=*>
1961 V8 N4 P44-48	62-25121 <=*>
	62-25921 <=*>
	63-19853 <=*>
1961 V8 N4 P60-63	AD-610 326 <=$>
1961 V8 N4 P76-81	61-27786 <*=>
1961 V8 N4 P85-89	63-24081 <=*>
1961 V8 N5 P3-6	601 <FT>
	61-27766 <*=>
	62-24670 <=*>
1961 V8 N5 P6-11	605 <FT>
	62-13923 <*=>
	62-24682 <=*>
1961 V8 N5 P11-17	614 <FT>
1961 V8 N5 P17-24	604 <FT>
1961 V8 N5 P25-28	609 <FT>
1961 V8 N5 P28-34	2451 <BISI>
	606 <FT>
	62-24634 <=*>
1961 V8 N5 P35-39	612 <FT>
1961 V8 N5 P39-45	608 <FT>
1961 V8 N5 P45-52	607 <FT>
1961 V8 N5 P52-57	AD-610 327 <=$>
	610 <FT>
1961 V8 N5 P57-59	613 <FT>
1961 V8 N5 P66-70	611 <FT>
1961 V8 N5 P71-76	602 <FT>
1961 V8 N5 P76-78	603 <FT>
1961 V8 N5 P79-81	616 <FT>
	63-19797 <=*$>
1961 V8 N5 P81-85	61-27766 <*=>
	615 <FT>
1961 V8 N6 P5-8	61-28913 <*=>
1961 V8 N6 P9-11	62-24630 <=*>
1961 V8 N6 P16-20	AEC-TR-4943 <*>
1961 V8 N6 P25-28	63-13247 <=*>
1961 V8 N6 P42-47	62-32305 <=*>
1961 V8 N6 P92-93	65-60728 <=>
1961 V8 N7 P24-29	63-23387 <=*$>
1961 V8 N7 P48-49	62-24640 <=*>
1961 V8 N7 P57-60	62-25736 <=*>
1961 V8 N7 P60-67	RTS-3038 <NLL>
1961 V8 N7 P73-76	64-13667 <=*$>
1961 V8 N8 P3-12	M.5024 <NLL>
1961 V8 N8 P12-17	M.5025 <NLL>
1961 V8 N8 P18-23	65-60734 <=>
1961 V8 N8 P23-27	63-24554 <=*$>
1961 V8 N8 P32-37	63-23385 <=*$>
	64-71500 <=> M
1961 V8 N8 P48-49	62-24687 <=$>
1961 V8 N8 P56-60	AEC-TR-4853 <=*>
	26N55R <ATS>
1961 V8 N9 P33-36	62-32810 <=*>
1961 V8 N9 P44-49	62-32479 <=*>
1961 V8 N9 P50-55	49P60R <ATS>
	62-25463 <=*>
1961 V8 N9 P56-60	62-32480 <=*>
1961 V8 N9 P60-65	62-24686 <=*>
1961 V8 N9 P65-68	62-32530 <=*>
1961 V8 N9 P68-72	62-32425 <=*>
1961 V8 N9 P73-77	AD-619 570 <=$>
1961 V8 N9 P78-83	63-19735 <=*>
1961 V8 N9 P92-93	62-32495 <=*>
1961 V8 N10 P5-9	65-61156 <=>
1961 V8 N10 P40-43	62-32812 <=*>
1961 V8 N10 P55-60	63-23947 <=*$>
1961 V8 N10 P65-68	65-60714 <=>
1961 V8 N11 P5-12	65-60705 <=>
1961 V8 N11 P27-29	62-32524 <=*>
1961 V8 N11 P30-37	65-60732 <=>
1961 V8 N11 P73-79	NP-TR-884 <*>
1961 V8 N11 P81-83	62-24307 <=*>
1961 V8 N12 P3-7	65-61157 <=>
1961 V8 N12 P21-26	63-19890 <*=>
1961 V8 N12 P45-50	M.5021 <NLL>
1961 V8 N12 P45-51	62-32303 <=*>
1961 V8 N12 P55-59	M.5022 <NLL>
	62-32303 <=*>
1961 V8 N12 P60-63	M.5023 <NLL>
1961 V8 N12 P70-77	62-24358 <=*>
1962 N2 P60-62	AEC-TR-5414 <*>
1962 V9 N1 P14-18	64-13021 <=*$>
1962 V9 N1 P28-31	62-32307 <=*>
	63-24551 <=*$>
1962 V9 N1 P32-36	65-61405 <=>
1962 V9 N1 P36-39	64-13024 <=*$>
1962 V9 N1 P40-44	63-15192 <=*>
1962 V9 N1 P52-57	63-19795 <=*$>
1962 V9 N1 P57-65	63-19794 <=*$>
1962 V9 N1 P75-77	64-18790 <*>
1962 V9 N1 P80-82	65-17402 <*>
1962 V9 N1 P83-86	63-15693 <=*>
1962 V9 N2 P48-54	63-15195 <=*>
1962 V9 N2 P60-62	AEC-TR-5414 <=*>
1962 V9 N2 P77-79	63-13216 <=*>
1962 V9 N2 P79-85	63-23002 <=*>
1962 V9 N3 P18-21	63-24096 <=*>
1962 V9 N3 P25-29	63-24047 <=*>

Citation	Translation
1962 V9 N3 P34-37	62-32682 <=*>
1962 V9 N3 P41-45	64-15910 <=*$>
1962 V9 N3 P48-50	63-15700 <=*>
	63-23965 <=*$>
1962 V9 N3 P51-56	63-13693 <=*>
	64-13025 <=*$>
1962 V9 N3 P64-67	63-15519 <=*>
1962 V9 N3 P79-83	AEC-TR-6431 <=$>
1962 V9 N3 P82-86	62-33532 <=*>
1962 V9 N4 P7-12	63-23950 <=*$>
1962 V9 N4 P12-17	63-24083 <*>
1962 V9 N4 P48-57	62-33332 <=*>
1962 V9 N5 P15-17	64-19938 <=>
1962 V9 N5 P17-19	63-24088 <=*$>
1962 V9 N5 P32-35	64-13018 <=*$>
1962 V9 N5 P35-38	63-15541 <=*>
	63-23957 <=*$>
1962 V9 N5 P38-43	63-19504 <=*>
	63-19864 <=*>
1962 V9 N5 P43-46	63-24552 <=*$>
1962 V9 N5 P47-50	63-23953 <=*$>
1962 V9 N5 P65-70	63-15541 <=*>
1962 V9 N5 P70-72	63-15188 <=*>
	63-15541 <=*>
1962 V9 N6 P3-9	63-24097 <=*$>
1962 V9 N6 P24-31	63-19497 <=*>
1962 V9 N6 P35-37	65-60730 <=>
1962 V9 N6 P44-47	ORNL-TR-107 <=>
1962 V9 N6 P72-78	AD-619 561 <=$>
1962 V9 N6 P79-82	62-33332 <=>
1962 V9 N6 P88-93	63-19497 <=*>
1962 V9 N7 P37-42	63-23968 <=*>
1962 V9 N7 P42-46	64-23563 <=$>
1962 V9 N7 P46-49	64-23624 <=$>
1962 V9 N7 P50-53	63-19162 <=*>
1962 V9 N7 P59-64	65-60625 <=$>
1962 V9 N7 P77-80	63-19162 <=*>
1962 V9 N7 P81-83	AEC-TR-5538 <=*>
1962 V9 N8 P2-6	3108 <BISI>
1962 V9 N8 P21-23	63-24090 <=*$>
1962 V9 N8 P23-31	64-13023 <=*$>
1962 V9 N8 P32-39	64-13231 <=*$>
1962 V9 N8 P42-47	63-24555 <=*$>
1962 V9 N8 P47-55	63-19341 <=*>
1962 V9 N8 P55-59	64-13039 <=*$>
1962 V9 N8 P69-72	63-20391 <=*> 0
	63-23932 <=$>
1962 V9 N8 P77-81	AEC-TR-5539 <=*>
	63-23966 <=*>
1962 V9 N9 P32-36	65-61535 <=$>
1962 V9 N9 P36-42	64-15929 <=*$>
1962 V9 N9 P42-	65-61229 <=$>
1962 V9 N9 P46-49	63-15701 <=*>
1962 V9 N9 P56-62	63-15698 <=*>
1962 V9 N9 P58-62	63-23975 <=*>
1962 V9 N9 P68-70	64-15911 <=*$>
1962 V9 N9 P71-77	65-60480 <=$>
1962 V9 N9 P81-85	63-24092 <=*>
1962 V9 N10 P21-26	63-24597 <=*$>
1962 V9 N10 P26-31	63-23978 <=*>
1962 V9 N10 P32-35	63-24586 <=*$>
1962 V9 N10 P35-40	63-23980 <=*$>
1962 V9 N10 P41-47	64-23566 <=$>
1962 V9 N10 P47-52	63-19158 <=*>
1962 V9 N10 P55-59	AD-611 108 <=$>
	63-19340 <=*>
1962 V9 N10 P59-64	63-24587 <=*$>
1962 V9 N10 P65-72	N66-18447 <=$>
1962 V9 N10 P77-83	64-11882 <=>
	65-60503 <=$>
1962 V9 N11 P10-13	64-15946 <=*$>
1962 V9 N11 P23-27	63-23096 <*=>
1962 V9 N11 P34-36	63-23981 <=*$>
1962 V9 N11 P41-45	63-24089 <=*$>
1962 V9 N11 P47-51	64-15279 <=*$>
1962 V9 N11 P63-67	63-23972 <=*>
1962 V9 N11 P69-74	63-24584 <=*$>
1962 V9 N12 P2-9	63-24585 <=*$>
	64-19931 <=$>
1962 V9 N12 P9-13	64-19108 <=*$>
1962 V9 N12 P19-21	64-15277 <=*$>
1962 V9 N12 P31-37	63-24095 <=*>
1962 V9 N12 P54-56	RTS-2951 <NLL>
1962 V9 N12 P57-63	ONRL-TR-108 <=>
1962 V9 N12 P64-66	63-23248 <=*>
1962 V9 N12 P83-85	63-21293 <=>
1963 N3 P58-62	ICE V4 N3 P426-430 <ICE>
1963 V10 N1 P2-8	63-24098 <=*>
1963 V10 N1 P26-28	64-15914 <=*$>
1963 V10 N1 P34-40	64-15944 <=*$>
1963 V10 N1 P61-64	64-15940 <=*$>
1963 V10 N1 P68-70	65-60661 <=$>
1963 V10 N2 P2-9	64-15912 <=*$>
1963 V10 N2 P9-11	64-15270 <=*$>
1963 V10 N2 P11-14	64-15913 <=*$>
1963 V10 N2 P19-26	64-13059 <=*$>
1963 V10 N2 P23-26	64-15945 <=*$>
1963 V10 N2 P26-30	64-15948 <=*$>
1963 V10 N2 P30-35	65-61465 <=>
1963 V10 N2 P41-50	64-71499 <=> M
1963 V10 N2 P64-69	TP/T-3683 <NLL>
1963 V10 N2 P78-82	65-60386 <=$>
1963 V10 N3 P8-12	64-15269 <=*$>
1963 V10 N3 P12-16	64-15271 <=*$>
1963 V10 N3 P18-22	64-15943 <=*$>
1963 V10 N3 P22-25	63-24583 <=*$>
1963 V10 N3 P30-33	65-61455 <=>
1963 V10 N3 P51-53	64-15915 <=*$>
1963 V10 N3 P58-62	ICE V4 N3 P426-430 <ICE>
	61R75R <ATS>
1963 V10 N3 P66-69	AEC-TR-6401 <=$>
1963 V10 N3 P78-81	TP/T-3416 <NLL>
1963 V10 N3 P82-87	RTS-2794 <NLL>
1963 V10 N4 P2-7	65-60814 <=>
1963 V10 N4 P8-14	65-60813 <=>
1963 V10 N4 P18-21	65-60789 <=>
	65-60812 <=>
1963 V10 N4 P28-30	64-19027 <=*$>
1963 V10 N4 P41-44	64-15931 <=*$>
1963 V10 N4 P49-52	65-60388 <=$>
1963 V10 N4 P55-57	64-15919 <=*$>
1963 V10 N4 P61-66	65-60376 <=$>
1963 V10 N4 P66-71	65-60373 <=$>
1963 V10 N5 P35-38	63-24574 <=*$>
1963 V10 N5 P49-52	AEC-TR-6252 <=*$>
1963 V10 N5 P64-69	64-19929 <=$>
1963 V10 N5 P69-70	65-60331 <=>
1963 V10 N5 P71-75	65-60370 <=$>
1963 V10 N5 P75-79	64-15273 <=*$>
	65-60484 <=$>
1963 V10 N5 P80-83	AD-610 299 <=$>
1963 V10 N5 P91-93	AD-619 520 <=$>
1963 V10 N6 P6-10	64-15941 <=*$>
1963 V10 N6 P10-15	64-15942 <=*$>
1963 V10 N6 P16-20	M-5742 <NLL>
1963 V10 N6 P25-29	AEC-TR-6402 <=$>
1963 V10 N6 P29-33	65-60379 <=$>
1963 V10 N6 P33-35	65-60393 <=$>
1963 V10 N6 P35-40	64-15920 <=*$>
1963 V10 N6 P40-45	TP/T-3542 <NLL>
1963 V10 N6 P46-48	AD-619 521 <=$>
1963 V10 N6 P49-52	64-15930 <=*$>
1963 V10 N6 P58-61	AD-619 521 <=$>
1963 V10 N7 P16-21	TP/T-3541 <NLL>
	64-15260 <=*$>
	64-19924 <=$>
1963 V10 N7 P42-47	RTS-2509 <NLL>
1963 V10 N7 P47-51	64-15918 <=*$>
1963 V10 N7 P64-67	64-13741 <=*$>
1963 V10 N7 P68-73	64-19937 <=$>
1963 V10 N7 P75-76	64-13389 <=*$>
1963 V10 N7 P77-82	63-13744 <=*$>
1963 V10 N8 P10-16	RTS-2591 <NLL>
1963 V10 N8 P16-19	RTS 2792 <NLL>
1963 V10 N8 P24-28	64-15261 <=*$>
1963 V10 N8 P38-40	65-60378 <=$>
1963 V10 N8 P40-43	64-15939 <=*$>
1963 V10 N8 P46-50	65-60382 <=$>
1963 V10 N8 P50-54	65-60385 <=$>

1963 V10 N8 P61-64 64-19111 <=*$>
1963 V10 N8 P64-69 64-19978 <=$>
1963 V10 N8 P76-78 AEC-TR-6290 <=*$>
65-10354 <=$>
1963 V10 N9 P15-19 65-64232 <=*>
1963 V10 N9 P19-25 66-61674 <=*$>
1963 V10 N9 P54-57 64-19980 <=$>
1963 V10 N9 P57-60 RTS-2443 <NLL>
1963 V10 N9 P61-65 64-19940 <=>
1963 V10 N9 P71-76 64-19979 <=>
1963 V10 N9 P76-80 65-60139 <=>
1963 V10 N10 P2-10 TP/T-3544 <NLL>
1963 V10 N10 P2-12 64-23500 <=$>
1963 V10 N10 P18-22 65-60655 <=$>
1963 V10 N10 P23-28 TP/T-3548 <NLL>
1963 V10 N10 P28-29 TP/T-3549 <NLL>
1963 V10 N10 P30-35 65-60621 <=$>
1963 V10 N10 P45-51 CE-TRANS-3732 <NLL>
1963 V10 N10 P51-56 65-60654 <=$>
1963 V10 N10 P63-69 65-60138 <=>
65-60622 <=$>
1963 V10 N10 P69-72 ANL-TR-50 <*>
CE-TRANS-3621 <NLL>
64-19388 <=$>
1963 V10 N10 P72-75 AEC-TR-6393 <=$>
1963 V10 N10 P76-78 AEC-TR-6414 <=$>
CE-TRANS-3622 <NLL>
1963 V10 N10 P82-87 AD-624 801 <=*$>
1963 V10 N11 P2-12 AD-618 313 <=$>
1963 V10 N11 P12-18 65-61256 <=$>
1963 V10 N11 P22-28 CE-TRANS-3782 <NLL>
1963 V10 N11 P38-39 TP/T-3575 <NLL>
1963 V10 N11 P55-60 65-61533 <=$>
1963 V10 N11 P60-66 65-64224 <=*>
1963 V10 N11 P74-80 65-60788 <=>
1963 V10 N11 P81-85 AD-618 313 <=$>
1963 V10 N12 P20-26 65-61255 <=$>
1963 V10 N12 P52-56 65-60960 <=$>
1963 V10 N12 P78-82 64-19120 <=*$>
1964 N9 P2-6 AD-634 815 <=$>
1964 N9 P19-21 AD-634 815 <=$>
1964 V11 N1 P40-43 WAPD-TRANS-10 <=$>
66-12347 <*>
1964 V11 N2 P81-87 AEC-TR-6443 <=$>
1964 V11 N3 P23-28 65-63528 <=$>
1964 V11 N3 P86-88 AEC-TR-6442 <=$>
1964 V11 N5 P54-57 AD-618 047 <=$>
1964 V11 N6 P20-22 AD-614 769 <=$>
1964 V11 N7 P75-77 65-12481 <*>
1964 V11 N8 P33-36 AD-614 015 <=$>
1964 V11 N8 P51-57 64-51931 <=*$>
1964 V11 N9 P16-19 AD-615 533 <=$>
1964 V11 N9 P71-73 AD-618 634 <=$>
1964 V11 N11 P72-74 AD-630 849 <=$>
1964 V11 N12 P2-31 65-30314 <=$>
1965 N3 P81-85 AD-625 289 <=*$>
1965 N7 P54-58 AD-629 474 <=$>
1965 N11 P29-34 AD-636 298 <=$>
1965 V12 N3 P41-47 AD-622 363 <=*$>
1965 V12 N4 P2-13 TP/T-3773 <NLL>
1965 V12 N5 P40-44 TP/T-3797 <NLL>
1965 V12 N5 P59-62 TP/T-3798 <NLL>
1965 V12 N5 P95-96 TP/T-3831 <NLL>
1965 V12 N7 P33-35 TP/T-3854 <NLL>
1965 V12 N8 P2-6 65-32787 <=*$>
1965 V12 N8 P47-50 65-32298 <=$*>
1965 V12 N10 P71-74 AD-641 856 <=$>
1965 V12 N11 P25-29 AD-635 17C <=$>
1966 N1 P2-5 66-31013 <=$>
1966 N1 P16-20 66-31013 <=$>
1967 N3 P1-4 67-31602 <=$>
1967 N4 P2-5 67-31602 <=$>

*TEPLOFIZIKA VYSOKIKH TEMPERATUR
1963 V1 N2 P281-290 AD-602 314 <=>
1964 V2 N1 P53-57 AD-613 161 <=$>
1964 V2 N1 P130-131 N65-21002 <=$>
1964 V2 N4 P525-534 AD-620 804 <=$>
1964 V2 N4 P573-582 N65-16308 <=$>
1964 V2 N5 P736-741 AD-629 228 <=*$>
1965 N3 P409-420 ICE V6 N1 P137-144 <ICE>
1965 N3 P421-426 ICE V6 N1 P81-85 <ICE>
1965 V3 N2 P186-191 AD-624 835 <=*$>
1965 V3 N2 P285-293 AD-618 512 <=$>
1965 V3 N3 P340-353 N66-11616 <=*$>
1965 V3 N3 P401-408 AD-638 667 <=$>

TEPLOVYE NAPRIAZHENIYA V ELEMENTOV KONSTRUKTSII
1962 V2 P141-148 RTS-2737 <NLL>
1964 N4 P355-358 AD-637 379 <=$>

TEPLOVYE NAPRIAZHENIIA V ELEMENTAKH TURBOMASHIN
1961 V1 P121-137 AD-610 417 <=$>
1962 N2 P171-175 64-19956 <=$>
1962 V2 P162-170 AD-610 371 <=$>

TERAPEVTICHESKII ARKHIV
1940 V18 P595-611 65-63789 <=>
1950 V22 N4 P20-29 RT-2320 <*>
1951 V23 N4 P30-37 RT-107 <*>
1951 V23 N4 P37-40 RT-108 <*>
1953 V25 N1 P88-89 RT-3080 <*>
1953 V25 N6 P12-24 RT-1993 <*>
64-15556 <=*$>
1954 V26 N5 P3-6 RT-3797 <*>
1956 V28 N4 P73-77 62-15394 <=*> 0
1957 V29 N1 P3-19 61-13888 <+*> 0
1957 V29 N7 P26-32 62-15937 <=*>
1958 V30 N2 P60-64 PB 141 150T <=>
1958 V30 N2 P64-67 PB 141 151T <=>
1958 V30 N3 P70-76 59-13358 <=> 0
1958 V30 N4 P37-40 59-13331 <=>
1958 V30 N4 P49-57 59-13332 <=> 0
1958 V30 N4 P94-95 59-13333 <=>
1959 V31 N1 P3-12 59-13517 <=>
1959 V31 N1 P46-52 2820 <K-H>
63-16132 <*=>
1959 V31 N4 P17-30 61-13850 <+*> 0
1959 V31 N4 P62-69 59-13780 <=>
1959 V31 N7 P53-60 62-15959 <=*>
1959 V31 N7 P93-95 59-13889 <=>
1959 V31 N8 P3-11 59-13974 <=>
1959 V31 N8 P12-17 59-13975 <=>
1959 V31 N11 P75-77 62-13509 <=*>
1959 V31 N11 P90-93 60-11781 <=>
1960 V32 N1 P5-10 60-11556 <=>
1960 V32 N2 P86-95 60-31287 <=>
1960 V32 N4 P3-8 60-41196 <=>
1960 V32 N4 P8-12 60-41197 <=>
1960 V32 N4 P13-18 60-41198 <=>
1960 V32 N4 P18-26 60-41199 <=>
1960 V32 N4 P26-29 60-41200 <=>
1960 V32 N4 P36-38 60-41201 <=>
1960 V32 N4 P38-43 60-41202 <=>
1960 V32 N5 P3-6 60-41640 <=>
1960 V32 N5 P7-12 60-41641 <=>
1960 V32 N5 P12-18 60-41642 <=>
1960 V32 N5 P19-26 60-41643 <=>
1960 V32 N5 P26-34 60-41644 <=>
1960 V32 N5 P47-53 60-41645 <=>
1960 V32 N5 P53-66 60-41646 <=>
1960 V32 N5 P66-77 60-41647 <=>
1960 V32 N5 P77-85 60-41648 <=>
1960 V32 N7 P3-10 60-41571 <=>
1960 V32 N7 P17-21 60-41572 <=>
1960 V32 N7 P21-24 60-41573 <=>
1960 V32 N7 P25-28 60-41574 <=>
1960 V32 N7 P29-30 60-41575 <=>
1960 V32 N7 P40-42 60-41576 <=>
1960 V32 N7 P58-61 60-41577 <=>
1960 V32 N7 P62-65 60-41578 <=>
1960 V32 N7 P78-81 60-41579 <=>
1960 V32 N7 P81-83 60-41580 <=>
1960 V32 N7 P84-85 60-41581 <=>
1960 V32 N8 P3-14 61-11585 <=>
1960 V32 N8 P44-49 61-11576 <=>
1960 V32 N8 P50-57 61-11603 <=>
1960 V32 N8 P62-65 61-11572 <=>
1960 V32 N8 P65-71 61-11557 <=>
1960 V32 N8 P72-74 61-11621 <=>

1960 V32 N8 P75-83 61-11586 <=>
1960 V32 N8 P83-86 61-11579 <=>
1960 V32 N8 P89-90 61-11580 <=>
1960 V32 N9 P15-19 61-21140 <=>
1960 V32 N9 P20-27 61-11935 <=>
1960 V32 N9 P28-32 61-11687 <=>
1960 V32 N9 P64-65 61-11749 <=>
1960 V32 N10 P14-22 61-21060 <=>
1960 V32 N10 P22-30 61-21435 <=>
1960 V32 N1C P31-36 61-21228 <=>
1960 V32 N10 P84-89 61-21236 <=>
1960 V32 N11 P3-9 61-21226 <=>
1960 V32 N11 P57-64 61-31467 <=>
1960 V32 N11 P93-95 61-23114 <=*>
1960 V32 N12 P34-38 61-21348 <=>
1960 V32 N12 P38-42 61-21311 <=>
1960 V32 N12 P42-47 61-21316 <=>
1960 V32 N12 P47-53 61-21449 <=>
1960 V32 N12 P63-71 61-21446 <=>
1960 V32 N12 P72-78 61-21444 <=>
1961 V33 N1 P3-9 61-21592 <=>
1961 V33 N1 P14-18 61-31125 <=>
1961 V33 N1 P19-26 61-31126 <=>
1961 V33 N1 P26-29 61-21565 <=>
1961 V33 N1 P29-35 61-21749 <=>
1961 V33 N1 P36-40 61-21605 <=>
1961 V33 N1 P79-84 61-21755 <=>
1961 V33 N2 P58-71 61-23605 <=*>
1961 V33 N3 P46-53 61-23613 <*=>
1961 V33 N3 P97-101 61-23552 <=*>
1961 V33 N4 P13-17 61-27354 <*=>
1961 V33 N4 P18-26 61-27355 <*=>
1961 V33 N4 P87-88 61-27344 <*=>
1961 V33 N4 P112-116 62-25166 <=*>
1961 V33 N6 P23-27 62-13025 <=*>
1961 V33 N6 P28-32 62-13269 <=*>
1961 V33 N6 P41-46 62-13026 <*=>
1961 V33 N6 P72-75 61-28920 <*=>
1961 V33 N6 P75-84 62-13051 <*=>
1961 V33 N6 P92-94 61-28778 <*=>
1961 V33 N6 P95-96 62-13050 <*=>
1961 V33 N7 P3-9 62-13857 <=>
1961 V33 N7 P35-38 62-13857 <=>
1961 V33 N7 P46-53 62-13857 <=>
1961 V33 N7 P58-67 62-13857 <=>
1961 V33 N7 P91-99 62-13857 <=>
1961 V33 N7 P110-113 62-13857 <=>
1961 V33 N7 P123-126 62-13857 <=>
1962 V34 N1 P25- FPTS V22 N1 P.T50 <FASE>
1962 V34 N1 P31- FPTS V22 N1 P.T165 <FASE>
1962 V34 N1 P39- FPTS V22 N1 P.T76 <FASE>
1962 V34 N1 P95- FPTS V22 N1 P.T71 <FASE>
1962 V34 N2 P31-38 63-15418 <=*>
1962 V34 N2 P96- FPTS V22 N1 P.T177 <FASE>
1962 V34 N2 P101- FPTS V22 N2 P.T346 <FASE>
1962 V34 N2 P121-125 62-33460 <=*>
1962 V34 N3 P28- FPTS V22 N2 P.T212 <FASE>
1962 V34 N3 P46-52 63-15419 <=*>
1962 V34 N3 P81-87 63-15420 <=*>
1962 V34 N4 P36-46 62-25607 <=*>
63-15128 <=*>
1962 V34 N4 P53- FPTS V22 N1 P.T135 <FASE>
1962 V34 N5 P10- FPTS V22 N1 P.T91 <FASE>
1962 V34 N5 P22-25 63-15417 <=*>
1962 V34 N5 P37-42 63-15421 <=*>
1962 V34 N5 P42- FPTS V22 N1 P.T63 <FASE>
1962 V34 N5 P53- FPTS V22 N1 P.T67 <FASE>
1962 V34 N5 P57- FPTS V22 N1 P.T74 <FASE>
1962 V34 N7 P3C- FPTS V22 N2 P.T232 <FASE>
1962 V34 N7 P40- FPTS V22 N3 P.T494 <FASE>
1962 V34 N7 P49- FPTS V22 N2 P.T229 <FASE>
1962 V34 N7 P58- FPTS V22 N2 P.T247 <FASE>
1962 V34 N7 P72-78 63-15987 <=*>
1962 V34 N8 P17-22 64-13527 <=*$>
1962 V34 N8 P29-37 63-23016 <=*>
1962 V34 N8 P46-52 63-23017 <=*>
1962 V34 N8 P53-59 63-23018 <=*>
1962 V34 N8 P96-102 63-23019 <=*>
1962 V34 N8 P1C2-109 63-23020 <=*>
1962 V34 N8 P109-113 63-23021 <=*>

1962 V34 N9 P38- FPTS V22 N5 P.T857 <FASE>
1963,V22,N5,PT2 <FASE>
1962 V34 N9 P45-53 63-23023 <=*>
1962 V34 N9 P84-89 63-13325 <=*>
1962 V34 N9 P95-99 63-23024 <=*>
1962 V34 N9 P102-106 63-13325 <=*>
63-23025 <=*>
1962 V34 N10 P3-10 63-13280 <=*>
1962 V34 N1C P20-26 63-23031 <=*>
1962 V34 N10 P26-31 63-23032 <=*>
1962 V34 N10 P36-40 63-23033 <=*>
1962 V34 N10 P44-49 63-23034 <=*>
1962 V34 N10 P50-55 63-23035 <=*>
1962 V34 N1C P55-58 63-23036 <=*>
1962 V34 N10 P64-69 63-23037 <=*>
1962 V34 N10 P69-76 63-23849 <=*>
1962 V34 N10 P84- FPTS V22 N5 P.T911 <FASE>
1963,V22,N5,PT2 <FASE>
1962 V34 N11 P19-24 63-23050 <=*>
1962 V34 N11 P41- FPTS V22 N4 P.T634 <FASE>
1962 V34 N11 P46-51 63-15360 <=*>
63-23851 <=*$>
1962 V34 N11 P63-68 63-23059 <=*>
1962 V34 N11 P68- FPTS V22 N4 P.T613 <FASE>
1962 V34 N11 P73-79 63-23852 <=*$>
1962 V34 N11 P80-88 63-23855 <=*$>
1962 V34 N11 P56- FPTS V22 N4 P.T782 <FASE>
1962 V34 N11 P101-103 63-15360 <=*>
1962 V34 N11 P114-117 63-15360 <=*>
1962 V34 N12 P8-14 63-23854 <=*$>
1962 V34 N12 P14- FPTS V22 N5 P.T882 <FASE>
1963,V22,N5,PT2 <FASE>
1962 V34 N12 P20- FPTS V22 N6 P.T1143 <FASE>
FPTS,V22,P.T1143-46 <FASE>
1962 V34 N12 P27-32 N64-25054 <=>
63-23853 <=*$>
1962 V34 N12 P38- FPTS V22 N5 P.T893 <FASE>
1963,V22,N5,PT2 <FASE>
1962 V34 N12 P48-52 63-23856 <=*$>
1962 V34 N12 P52- FPTS V22 N5 P.T897 <FASE>
1963,V22,N5,PT2 <FASE>
1962 V34 N12 P74-80 64-13206 <=*$>
1962 V34 N12 P84-87 63-23840 <=*>
1962 V34 N12 P92-95 63-23839 <=*>
1963 V35 N1 P3C-34 345 <TC>
63-24488 <=*$> 0
1963 V35 N1 P43-49 63-24489 <=*$>
1963 V35 N1 P65-70 63-24490 <=*$>
1963 V35 N1 P71- FPTS V22 N6 P.T1173 <FASE>
FPTS,V22,P.T1173-76 <FASE>
1963 V35 N1 P78-83 63-24491 <=*$>
1963 V35 N1 P84- FPTS V22 N6 P.T1181 <FASE>
1963 V35 N1 P84-88 FPTS,V22,P.T1181-83 <FASE>
1963 V35 N1 P84-94 63-21575 <=>
1963 V35 N1 P89-94 63-24492 <=*$> 0
1963 V35 N2 P16-22 64-13511 <=*$>
1963 V35 N2 P22-30 64-13843 <=*$>
1963 V35 N2 P37-42 64-13510 <=*$>
1963 V35 N2 P54-58 64-13509 <=*$>
1963 V35 N2 P58-60 64-13856 <=*$>
1963 V35 N2 P6C- FPTS V23 N1 P.T142 <FASE>
FPTS,V23,P.T142-T144 <FASE>
1963 V35 N2 P66- FPTS V23 N2 P.T334 <FASE>
1963 V35 N3 P12-15 64-13473 <=*$>
1963 V35 N3 P30-34 64-13472 <=*$>
1963 V35 N3 P34-38 64-13917 <=*$>
1963 V35 N3 P38-43 64-13471 <=*$>
1963 V35 N3 P43-46 64-13470 <=*$>
1963 V35 N3 P46-51 64-13445 <=*$>
1963 V35 N3 P51-58 64-13444 <=*$>
1963 V35 N3 P70-78 64-13443 <=*$>
1963 V35 N5 P10-15 64-13430 <=*$>
1963 V35 N5 P21-28 64-15526 <=*$>
1963 V35 N5 P34-39 64-13849 <=*$>
1963 V35 N5 P39-44 64-13850 <=*$>
1963 V35 N5 P46- FPTS V23 N2 P.T301 <FASE>
FPTS,V23,P.T301-T303 <FASE>
1963 V35 N5 P71-78 64-13428 <=*$> 0
1963 V35 N6 P33-37 64-19900 <=$>
1963 V35 N6 P38-46 64-15487 <=*$> 0

1963 V35 N6 P46-50 64-15488 <=*$>
1963 V35 N6 P56-62 64-19257 <=*$>
1963 V35 N6 P67-73 64-15489 <=*$>
1963 V35 N6 P85-90 64-15465 <=*$>
1963 V35 N8 P3-9 63-41186 <=>
1963 V35 N8 P39-44 64-15583 <=*$>
1963 V35 N8 P56- FPTS,V23,N4,P.T901 <FASE>
1963 V35 N8 P56-62 63-41171 <=>
1963 V35 N8 P93-99 64-15531 <=*$>
1963 V35 N8 P118 64-21056 <=>
1963 V35 N9 P14-19 FPTS,V23,N3,P.T562 <=*$>
1963 V35 N9 P67-76 64-19234 <=*$>
1963 V35 N9 P98-106 64-19654 <=$>
1963 V35 N10 P22-27 64-19217 <=*$>
1963 V35 N10 P77-82 64-19235 <=*$>
1963 V35 N10 P131-137 FPTS,V23,N3,P.T541 <=*$>
1963 V35 N11 P64-68 64-19236 <=*$>
1963 V35 N11 P95-99 64-19273 <=*$>
1963 V35 N12 P37- FPTS V23 N6 P.T1361 <FASE>
1963 V35 N12 P37-41 FPTS V23 P.T1361 <FASE>
1963 V35 N12 P42-47 64-19677 <=$>
1963 V35 N12 P48- FPTS V23 N6 P.T1257 <FASE>
FPTS V23P.T1257 <FASE>
1963 V35 N12 P68-73 65-60221 <=$>
1963 V35 N12 P73719 64-19704 <=$>
1964 V36 N1 P36-43 64-19650 <=$>
1964 V36 N1 P40- FPTS V24 N2 P.T255 <FASE>
1964 V36 N1 P51- FPTS V23 N6 P.T1237 <FASE>
FPTS V23 P.T1237 <FASE>
1964 V36 N1 P88-100 64-19703 <=$>
1964 V36 N2 P50-58 65-61121 <=$>
1964 V36 N2 P80-87 65-61122 <=$>
1964 V36 N2 P95-100 65-62121 <=$>
1964 V36 N3 P15-19 65-62122 <=$>
1964 V36 N3 P40-46 FPTS V24 P.T255-T258 <FASE>
1964 V36 N3 P60-65 65-60575 <=$>
1964 V36 N3 P71-78 65-61117 <=$>
1964 V36 N3 P105-111 65-60151 <=$>
1964 V36 N4 P28-32 65-62103 <=$>
1964 V36 N4 P41- FPTS V24 N3 P.T455 <FASE>
1964 V36 N4 P54- FPTS V24 N2 P.T291 <FASE>
FPTS V24 P.T291-T294 <FASE>
1964 V36 N4 P112-119 65-60152 <=$>
1964 V36 N5 P3-8 65-61219 <=$>
1964 V36 N6 P71-78 65-62364 <=$>
1964 V36 N7 P26-32 65-12193 <*>
1964 V36 N7 P51-57 65-62104 <=>
1964 V36 N9 P7- FPTS V24 N5 P.T845 <FASE>
1964 V36 N9 P10-14 65-62363 <=$>
1964 V36 N9 P14-18 65-62362 <=$>
1964 V36 N10 P76-81 66-13409 <*> O
1964 V36 N10 P107-113 65-64515 <=*$>
1964 V36 N11 P37-40 65-63625 <=$>
1964 V36 N11 P59-63 65-63622 <=$>
1964 V36 N12 P59-64 65-64540 <=*$>
1965 V37 N1 P63-67 65-52855 <=*$>
1965 V37 N1 P83-89 65-64531 <=*$>
1965 V37 N1 P93-100 66-60845 <=*$>
1965 V37 N1 P100-107 65-64530 <=*$>
1965 V37 N2 P3-13 65-30986 <=$>
1965 V37 N2 P22-28 65-30987 <=$>
1965 V37 N2 P28-31 65-30988 <=$>
1965 V37 N2 P31-41 65-64519 <=*$>
1965 V37 N2 P46-51 66-60318 <=*$>
1965 V37 N2 P75-80 66-60319 <=*$>
1965 V37 N2 P102-104 65-30989 <=$>
1965 V37 N3 P43-48 66-30841 <=*$>
1965 V37 N3 P56- FPTS V25 N2 P.T367 <FASE>
1965 V37 N3 P93-97 66-60840 <=*$>
1965 V37 N4 P37-41 66-61019 <=*$>
1965 V37 N4 P66-67 65-31506 <=$>
66-60866 <=*$>
1965 V37 N5 P51-54 65-32535 <=*$>
1965 V37 N7 P18- FPTS V25 N3 P.T493 <FASE>
1965 V37 N7 P67-71 66-61080 <=$>
1965 V37 N8 P3-8 65-33497 <=*$>
1965 V37 N9 P66- FPTS V25 N4 P.T704 <FASE>
1965 V37 N10 P47- FPTS V25 N6 P.T1005 <FASE>
1965 V37 N10 P106- FPTS V25 N5 P.T781 <FASE>
1965 V37 N11 P96-100 66-30941 <=$>
1965 V37 N11 P100- FPTS V25 N5 P.T797 <FASE>
1966 V38 N1 P19-25 66-61971 <=*$>
1966 V38 N1 P41- FPTS V25 N6 P.T1041 <FASE>
1966 V38 N1 P59-65 66-61970 <=*$>
1966 V38 N1 P83- FPTS V25 N6 P.T972 <FASE>
1966 V38 N1 P118-119 66-31690 <=$>

TERAPIA. INSTITUTO SIEROTERAPICO MILANESE
1963 V48 N361 P83-93 64-00202 <*>

TERMESZETTUDOMANYI KOZLONY
1939 P1-11 PANSDOC-TR.345 <PANS>
1962 V6 N11 P526-527 63-21209 <=>

TERMOFIZIKA VYSOKIKH TEMPERATUR
1964 V2 N3 P337-343 AD-634 228 <=$>

TERMOTECNICA
1951 V5 N2 P64-66 AL-238 <*>
1960 V14 N6 P275-280 62-00191 <*>
1961 V15 N2 P89-100 C-4323 <NRCC>
1963 V17 P734-738 1103 <TC>

TERRESTRIAL MAGNETISM AND ATMOSPHERIC ELECTRICITY
1932 V37 N3 P273-277 C-3260 <NRCC>

TESLA. BEOGRAD
1961 V8 N1 P2-6 62-19478 <=> O
1961 V8 N1 P51 62-19478 <=> O
1962 V9 N3 P2-6 62-32841 <=*>

TETRAHEDRON
1958 V3 P91-93 289 <TC>
1960 V9 P67-75 9087 <IICH>
1961 V14 N3/4 P190-200 65-13576 <*>
1962 V18 P1023-1028 55P66G <ATS>
1965 V21 P1855-1879 1374 <TC>

TETRAHEDRON LETTERS
1959 N10 P13-15 65-10937 <*>
1960 V8 N3/4 P336-339 10444 <K-H>
1961 N4 P156-160 61-20270 <*>
1961 N18 P624-627 65-13580 <*>
1962 N20 P913-916 63-20016 <*>
1964 N8 P383-386 8675 <IICH>
1964 N19 P1137-1142 66-10048 <*>
1964 N33 P2267-2272 65R78G <ATS>
1965 N15 P969-972 1063 <TC>
1966 N1 P129-133 2057 <TC>

TETRAHEDRON. SUPPLEMENT
1963 V19 N2 P223-241 66-11029 <*>
1963 V19 N2 P315-335 183 <TC>

TETSU TO HAGANE
1950 V36 N12 P21-25 1966 <BISI>
1951 V37 N1 P23-29 1966 <BISI>
1951 V37 N5 P23-30 60-10037 <*> O
1951 V37 N10 P25-30 4158 <HB>
1952 V38 P283-288 I-418 <*>
1952 V38 P531-536 64-14384 <*> P
1952 V38 P532-536 I-419 <*>
1952 V38 N5 P51-56 60-10295 <*>
1952 V38 N6 P29-33 60-10296 <*> O
1952 V38 N7 P41-47 60-10297 <*>
1952 V38 N11 P33-36 4157 <HB>
1955 P861-869 TS-1240 <BISI>
1955 P1102-1107 58-1163 <*>
1955 V41 P593-601 TS 1430 <BISI>
1955 V41 P1258-1264 TS-1147 <BISI>
1955 V41 N3 P193-195 3850 <HB>
1955 V41 N6 P593-601 62-14170 <*>
1955 V41 N8 P861-869 62-10510 <*>
1956 V42 P342-343 61-16260 <*> O
1956 V42 P864-865 61-18883 <*>
1956 V42 P928-930 TS 1665 <BISI>
1956 V42 P1102-1105 TS-1121 <BISI>
1956 V42 N1 P49-52 4112 <HB>
1956 V42 N3 P306-307 1936 <BISI>
1956 V42 N9 P935-936 61-16259 <*>

Citation	Translation
1956 V42 N9 P936-938	61-16408 <*>
1957 V43 P24-28	61-16407 <*>
1957 V43 P628-632	TS-1123 <BISI>
1957 V43 P663-668	TS-1666 <BISI>
1957 V43 P695-699	TS 1124 <BISI>
1957 V43 P699-703	TS 1125 <BISI>
1957 V43 P894-897	TS-1257 <BISI>
1957 V43 P1194-1199	61-18366 <*> P
1957 V43 P1222-1228	2399 <BISI>
1957 V43 N2 P117-121	TS-1122 <BISI>
1957 V43 N3 P45-46	4789 <HB>
1957 V43 N3 P265-266	4101 <HB>
1957 V43 N3 P351-352	4102 <HB>
1957 V43 N9 P899-901	4155 <HB>
1957 V43 N9 P1015-1017	4156 <HB>
1957 V43 N9 P1017	4064 <HB>
1957 V43 N9 P1073-1074	4160 <HB>
1957 V43 N9 P1074-1076	4159 <HB>
1958 V44 P260-262	TS 1477 <BISI>
1958 V44 P386-388	TS-1667 <BISI>
1958 V44 P552-559	TS-1631 <BISI>
1958 V44 P660-668	TS-1632 <BISI>
1958 V44 N2 P151-157	4510 <HB>
1958 V44 N3 P222-223	4768 <HB>
1958 V44 N3 P260-262	62-14658 <*>
1958 V44 N7 P733-739	5096 <HB> P
1958 V44 N9 P1055-1056	5621 <HB>
1958 V44 N9 P1056-1058	5269 <HB>
1958 V44 N9 P1058-1060	5270 <HB>
1958 V44 N9 P1060-1062	5372 <HB>
1958 V44 N9 P1062-1063	5507 <HB>
1958 V44 N9 P1132-1134	TS 1568 <BISI>
1959 V45 N3 P177-178	4710 <HB>
1959 V45 N3 P183-185	4723 <HB>
1959 V45 N3 P189-192	4770 <HB>
1959 V45 N3 P248-249	5620 <HB>
1959 V45 N5 P32-37	64-31478 <=>
1959 V45 N7 P34-38	64-31401 <=>
1959 V45 N9 P883-885	5672 <HB>
1959 V45 N9 P893-895	5285 <HB>
1959 V45 N9 P895-897	5286 <HB>
1959 V45 N9 P957-959	4943 <HB>
1959 V45 N9 P959-961	4947 <HB>
1959 V45 N9 P961-962	4925 <HB>
1959 V45 N9 P966-967	5117 <HB>
1959 V45 N9 P1089-1092	2105 <BISI>
1959 V45 N10 P1145-1151	4774 <HB>
1959 V45 N12 P1341-1345	5223 <HB>
1960 V46 N2 P123-129	5058 <HB>
1960 V46 N3 P153-155	4936 <HB>
1960 V46 N3 P240-241	4848 <HB>
1960 V46 N3 P241-243	4839 <HB>
1960 V46 N3 P283-285	5122 <HB>
1960 V46 N3 P311-313	2111 <BISI>
1960 V46 N3 P358-361	5257 <HB>
1960 V46 N3 P377-379	5216 <HB>
1960 V46 N3 P379-381	5120 <HB>
1960 V46 N3 P414-416	5273 <HB>
1960 V46 N3 P575-577	5064 <HB>
1960 V46 N7 P753-757	5219 <HB>
1960 V46 N9 P1091-1095	4653 <BISI>
1960 V46 N10 P1103-1105	5017 <HB>
1960 V46 N10 P1130-1132	5942 <HB>
1960 V46 N10 P1132-1134	5943 <HB>
1960 V46 N10 P1134-1136	5944 <HB>
1960 V46 N10 P1136-1137	5945 <HB>
1960 V46 N10 P1138	5946 <HB>
1960 V46 N10 P1140-1142	5947 <HB>
1960 V46 N10 P1142-1144	5948 <HB>
1960 V46 N10 P1180-1181	5101 <HB>
1960 V46 N10 P1180-1182	5247 <HB>
1960 V46 N10 P1182-1185	5097 <HB>
1960 V46 N10 P1185-1187	5096 <HB>
1960 V46 N10 P1187-1189	5095 <HB>
1960 V46 N10 P1209-1211	5217 <HB>
1960 V46 N10 P1219-1221	2780 <BISI>
1960 V46 N10 P1233-1234	5102 <HB>
1960 V46 N10 P1259-1261	2779 <BISI>
1960 V46 N10 P1275-1276	5072 <HB>
1960 V46 N10 P1280-1282	2781 <BISI>
1960 V46 N10 P1286-1289	5556 <HB>
1960 V46 N10 P1349-1352	5107 <HB>
1960 V46 N12 P1733-1740	5108 <HB>
1961 V47 N2 P116-124	5220 <HB>
1961 V47 N3 P286-288	63-20071 <*>
1961 V47 N3 P341-343	5226 <HB>
1961 V47 N3 P343-345	5227 <HB>
1961 V47 N3 P345-348	5228 <HB>
1961 V47 N3 P348-350	5229 <HB>
1961 V47 N3 P350-352	5230 <HB>
1961 V47 N3 P353-355	5232 <HB>
1961 V47 N3 P390-391	5019 <HB>
1961 V47 N3 P519-521	6063 <HB>
1961 V47 N3 P548-550	5284 <HB>
1961 V47 N3 P552-554	5170 <HB>
1961 V47 N3 P554-556	5173 <HB>
1961 V47 N3 P556-558	5291 <HB>
1961 V47 N3 P558-560	5336 <HB>
1961 V47 N3 P561-562	5337 <HB>
1961 V47 N3 P563-564	5338 <HB>
1961 V47 N6 P828-833	5594 <HB>
1961 V47 N7 P891-896	5278 <HB>
1961 V47 N10 P1346-1348	5355 <HB>
1961 V47 N10 P1348-1350	5356 <HB>
1961 V47 N10 P1350-1352	5374 <HB>
1961 V47 N10 P1353-1354	5673 <HB>
1961 V47 N10 P1355-1357	5599 <HB>
1961 V47 N10 P1357-1359	5685 <HB>
1961 V47 N10 P1360-1362	5461 <HB>
1961 V47 N10 P1362-1363	5629 <HB>
1961 V47 N10 P1516-1518	5546 <HB>
1961 V47 N10 P1518-1520	5578 <HB>
1962 V48 N4 P361-362	5682 <HB>
1962 V48 N4 P362-363	5683 <HB>
1962 V48 N4 P363-365	6021 <HB>
1962 V48 N4 P418-419	5684 <HB>
1962 V48 N4 P452-453	63-20072 <*>
1962 V48 N4 P453-455	63-20070 <*>
1962 V48 N4 P464-465	5674 <HB>
1962 V48 N4 P465-467	5675 <HB>
1962 V48 N4 P467-468	5705 <HB>
1962 V48 N4 P468-470	5676 <HB>
1962 V48 N4 P470-471	5678 <HB>
1962 V48 N4 P472-473	5677 <HB>
1962 V48 N4 P473-475	5991 <HB>
1962 V48 N4 P475-477	5992 <HB>
1962 V48 N4 P477-478	5711 <HB>
1962 V48 N4 P513-514	5760 <HB>
1962 V48 N4 P515-517	5761 <HB>
1962 V48 N4 P644-645	5951 <HB>
1962 V48 N4 P645-647	6016 <HB>
1962 V48 N6 P747-752	5633 <HB>
1962 V48 N7 P845-849	5666 <HB>
1962 V48 N8 P933-940	5670 <HB>
1962 V48 N11 P1219-1221	5806 <HB>
1962 V48 N11 P1237-1238	5865 <HB>
1962 V48 N11 P1251-1252	5962 <HB>
1962 V48 N11 P1273-1274	5867 <HB>
1962 V48 N11 P1294-1295	5842 <HB>
1962 V48 N11 P1297-1299	5965 <HB>
1962 V48 N11 P1321-1322	6165 <HB>
1962 V48 N11 P1322-1324	6146 <HB>
1962 V48 N11 P1338-1340	5807 <HB>
1962 V48 N11 P1340-1342	5805 <HB>
1962 V48 N11 P1353-1354	5770 <HB>
1962 V48 N11 P1354-1356	5841 <HB>
1962 V48 N11 P1356-1357	5739 <HB>
1962 V48 N11 P1358-1359	5769 <HB>
1962 V48 N11 P1359-1362	5741 <HB>
1962 V48 N11 P1362-1363	5765 <HB>
1962 V48 N11 P1363-1364	5766 <HB>
1962 V48 N11 P1365-1366	5767 <HB>
1962 V48 N11 P1366-1368	5851 <HB>
1962 V48 N11 P1368-1371	5850 <HB>
1962 V48 N11 P1371-1372	5759 <HB>
1962 V48 N11 P1372-1374	5758 <HB>
1962 V48 N11 P1381-1383	5958 <HB>
1962 V48 N11 P1383-1385	6056 <HB>
1962 V48 N11 P1394-1395	5843 <HB>
1962 V48 N11 P1410-1411	5791 <HB>

1962 V48 N11 P1411-1413 5792 <HB>
1962 V48 N11 P1413-1415 5793 <HB>
1962 V48 N11 P1438-1439 5808 <HB>
1962 V48 N11 P1454-1456 5798 <HB>
1962 V48 N11 P1456-1458 5799 <HB>
1962 V48 N11 P1458-1460 5732 <HB>
1962 V48 N11 P1485-1487 5874 <HB>
1962 V48 N11 P1487-1488 5875 <HB>
1962 V48 N11 P1488-1490 5873 <HB>
1962 V48 N11 P1495-1496 6027 <HB>
1962 V48 N11 P1496-1498 6029 <HB>
1962 V48 N11 P1500-1502 6028 <HB>
1962 V48 N11 P1511-1513 5968 <HB>
1962 V48 N11 P1515-1517 6085 <HB>
1962 V48 N11 P1517-1519 6058 <HB>
1963 V49 P22-29 3670 <BISI>
1963 V49 P394-395 5927 <HB>
1963 V49 N3 P294-297 6017 <HB>
1963 V49 N3 P329-33C 6054 <HB>
1963 V49 N3 P364-366 5989 <HB>
1963 V49 N3 P366-368 6072 <HB>
1963 V49 N3 P390-392 5925 <HB>
1963 V49 N3 P392-393 5926 <HB>
1963 V49 N3 P395-397 5298 <HB>
1963 V49 N3 P397-399 5929 <HB>
1963 V49 N3 P399-400 5930 <HB>
1963 V49 N3 P400-403 5931 <HB>
1963 V49 N3 P403-404 5932 <HB>
1963 V49 N3 P4C5-406 5933 <HB>
1963 V49 N3 P407-408 5934 <HB>
1963 V49 N3 P408-410 5935 <HB>
1963 V49 N3 P410-412 5936 <HB>
1963 V49 N3 P418-419 6062 <HB>
1963 V49 N3 P431-433 6258 <HB>
1963 V49 N3 P433-435 6259 <HB>
1963 V49 N3 P435-438 6260 <HB>
1963 V49 N3 P438-439 6261 <HB>
1963 V49 N3 P461-463 6057 <HB>
1963 V49 N3 P463-465 6046 <HB>
1963 V49 N3 P465-468 5970 <HB>
1963 V49 N3 P468-469 5971 <HB>
1963 V49 N3 P469-471 5972 <HB>
1963 V49 N3 P471-472 5973 <HB>
1963 V49 N3 P525-526 6047 <HB>
1963 V49 N3 P527-528 6053 <HB>
1963 V49 N3 P536-538 5969 <HB>
1963 V49 N3 P538-54C 6018 <HB>
1963 V49 N3 P540-551 6055 <HB>
1963 V49 N3 P583-585 5974 <HB>
1963 V49 N3 P612-615 6306 <HB>
1963 V49 N3 P620-623 5976 <HB>
1963 V49 N3 P623-625 5977 <HB>
1963 V49 N3 P625-626 5978 <HB>
1963 V49 N3 P627-629 5979 <HB>
1963 V49 N3 P629-631 6059 <HB>
1963 V49 N3 P631-632 5967 <HB>
1963 V49 N3 P634-636 6051 <HB>
1963 V49 N3 P638-64C 6088 <HB>
1963 V49 N3 P641-643 6060 <HB>
1963 V49 N3 P643-645 6089 <HB>
1963 V49 N9 P1151-1162 4025 <BISI>
1963 V49 N9 P1163-1175 4025 PT.2 <BISI>
1963 V49 N9 P1176-1199 4025 PT.3 <BISI>
1963 V49 N9 P1200-1227 4025 PT.4 <BISI>
1963 V49 N1C P1261-1262 6158 <HB>
1963 V49 N1C P1262-1264 6150 <HB>
1963 V49 N10 P1281-1282 6111 <HB>
1963 V49 N1C P1363-1365 6166 <HB>
1963 V49 N10 P1365-1367 6167 <HB>
1963 V49 N10 P1367-1369 6168 <HB>
1963 V49 N10 P1369-1371 6169 <HB>
1963 V49 N10 P1371-1373 6170 <HB>
1963 V49 N1C P1373-1375 6171 <HB>
1963 V49 N10 P1375-1376 6172 <BISI>
1963 V49 N10 P1377-1378 6173 <HB>
1963 V49 N1C P1379-1380 6174 <HB>
1963 V49 N10 P1380-1382 6175 <HB>
1963 V49 N1C P1382-1384 6176 <HB>
1963 V49 N1C P1384-1386 6177 <HB>
1963 V49 N10 P1386-1388 6178 <HB>
1963 V49 N10 P1417-1419 6134 <HB>
1963 V49 N10 P1419-1421 6135 <HB>
1963 V49 N10 P1421-1423 6136 <HB>
1963 V49 N10 P1423-1425 6137 <HB>
1963 V49 N1C P1459-1461 5823 <HB>
1963 V49 N10 P1461-1463 6608 <HB>
1963 V49 N10 P1463-1465 6304 <HB>
1963 V49 N10 P1468-147C 6548 <HB>
1963 V49 N10 P1507-1508 6261 <HB>
1963 V49 N1C P1508-1510 5822 <HB>
1963 V49 N10 P1541-1543 6110 <HB>
1963 V49 N10 P1543-1545 6107 <HB>
1963 V49 N1C P1549-1551 6180 <HB>
1963 V49 N10 P1551-1553 6223 <HB>
1963 V49 N10 P1553-1555 6224 <HB>
1963 V49 N1C P1561-1563 6303 <HB>
1963 V49 N10 P1565-1567 6305 <HB>
1963 V49 N1C P1569-1571 6069 <HB>
1963 V49 N10 P1589-1590 6083 <HB>
1963 V49 N10 P1598-1600 6191 <HB>
1963 V49 N1C P1604-1605 6192 <HB>
1963 V49 N10 P1605-1607 6194 <HB>
1963 V49 N1C P1607-1609 6193 <HB>
1963 V49 N10 P1617-1618 6082 <HB>
1963 V49 N10 P1627-1629 6221 <HB>
1963 V49 N1C P1629-1631 6097 <HB>
1963 V49 N10 P1635-1637 5824 <HB>
1963 V49 N1C P1699-1701 6319 <HB>
1964 V50 P46-62 6384 <HB>
1964 V50 P911-917 4469 <BISI>
1964 V50 N2 P174-182 3726 <BISI>
1964 V50 N3 P364-366 PT28 4177 <BISI>
1964 V50 N3 P325-327 6392 <HB>
1964 V50 N3 P327-33C 6397 <HB>
1964 V50 N3 P341-343 6387 <HB>
1964 V50 N3 P366-369 4178 <BISI>
1964 V50 N3 P369-371 4179 <BISI>
1964 V50 N3 P371-373 4180 <BISI>
1964 V50 N3 P373-376 4181 <BISI>
1964 V50 N3 P393-396 6307 <HB>
1964 V50 N3 P396-398 6308 <HB>
1964 V50 N3 P398-400 6309 <HB>
1964 V50 N3 P400-403 6310 <HB>
1964 V50 N3 P4C3-406 6311 <HB>
1964 V50 N3 P406-409 6312 <HB>
1964 V50 N3 P4C9-412 6314 <HB>
1964 V50 N3 P412-414 6315 <HB>
1964 V50 N3 P414-417 6316 <HB>
1964 V50 N3 P417-42C 6317 <HB>
1964 V50 N3 P420-422 6318 <HB>
1964 V50 N3 P435-436 6253 <HB>
1964 V50 N3 P439-442 6362 <HB>
1964 V50 N3 P447-450 6300 <HB>
1964 V50 N3 P456-458 6724 <HB>
1964 V50 N3 P494-496 6385 <HB>
1964 V50 N3 P496-498 6405 <HB>
1964 V50 N4 P565-567 6388 <HB>
1964 V50 N4 P603-605 6320 <HB>
1964 V50 N4 P614-617 6365 <HB>
1964 V50 N4 P626-628 6380 <HB>
1964 V50 N4 P641-644 6364 <HB>
1964 V50 N4 P684-686 6406 <HB>
1964 V50 N4 P722-724 6363 <HB>
1964 V50 N5 P743-752 5796 <HB> P
1964 V50 N11 P1697-1699 6431 <HB>
1964 V50 N11 P1700-1702 6432 <HB>
1964 V50 N11 P1702-1705 6430 <HB>
1964 V50 N11 P1712-1715 6445 <HB>
1964 V50 N11 P1715-1718 6709 <HB>
1964 V50 N11 P1718-1720 6721 <HB>
1964 V50 N11 P1720-1723 6734 <HB>
1964 V50 N11 P1723-1725 6407 <HB>
1964 V50 N11 P1725-1727 6408 <HB>
1964 V50 N11 P1727-1730 6409 <HB>
1964 V50 N11 P1733-1735 6412 <HB>
1964 V50 N11 P1735-1738 6413 <HB>
1964 V50 N11 P1738-1740 6414 <HB>
1964 V50 N11 P1740-1742 6415 <HB>
1964 V50 N11 P1859-1862 6763 <HB>
1964 V50 N11 P1870-1873 6791 <HB>

1964 V50 N12 P2004-2006 5825 <HB>
1964 V50 N12 P2006-2008 6287 <HB>
1964 V50 N12 P2011-2013 6441 <HB>
1964 V50 N12 P2044-2046 6725 <HB>
1964 V50 N12 P2056-2C57 4974 <BISI>
1964 V50 N12 P2080-2082 4976 <BISI>
1964 V50 N12 P2083-2086 4977 <BISI>
1964 V50 N12 P2088-2091 6723 <HB>
1964 V50 N12 P2091-2092 4979 <BISI>
1964 V50 N12 P2095-2097 4357 <BISI>
1964 V50 N12 P2097-2099 4980 <BISI>
1964 V50 N12 P2099-2102 4356 <BISI>
1964 V50 N12 P2139-2140 6447 <HB>
1964 V50 N13 P2166-2175 6436 <HB>
1965 V51 N4 P717-720 6768 <HB>
1965 V51 N4 P720-722 6739 <HB>
1965 V51 N4 P722-725 6740 <HB>
1965 V51 N4 P725-727 6741 <HB>
1965 V51 N4 P727-729 6742 <HB>
1965 V51 N4 P729-732 6743 <HB>
1965 V51 N4 P732-735 6744 <HB>
1965 V51 N4 P735-737 6745 <HB>
1965 V51 N4 P737-740 6746 <HB>
1965 V51 N4 P740-742 6747 <HB>
1965 V51 N4 P743-744 6748 <HB>
1965 V51 N4 P752-755 6628 <HB>
1965 V51 N4 P774-777 6764 <HB>
1965 V51 N4 P786-788 6729 <HB>
1965 V51 N4 P807-810 6696 <HB>
1965 V51 N4 P813-816 6694 <HB>
1965 V51 N4 P816-819 6695 <HB>
1965 V51 N4 P855-857 6849 <HB>
1965 V51 N4 P870-873 6698 <HB>
1965 V51 N4 P873-876 6699 <HB>
1965 V51 N4 P876-878 6483 <HB>
1965 V51 N10 P1722-1725 6844 <HB>
1965 V51 N10 P1733-1736 6701 <HB>
1965 V51 N10 P1834-1837 6711 <HB>
1965 V51 N10 P1860-1863 6918 <HB>
1965 V51 N10 P1881-1883 6915 <HB>
1965 V51 N1C P1901-1904 6708 <HB>
1965 V51 N10 P1904-1906 6700 <HB>
1965 V51 N10 P1909-1911 6712 <HB>
1965 V51 N10 P1911-1914 6713 <HB>
1965 V51 N10 P1920-1922 6735 <HB>
1965 V51 N1C P1922-1925 6714 <HB>
1965 V51 N10 P1925-1927 6715 <HB>
1965 V51 N10 P1928-1930 6716 <HB>
1965 V51 N1C P1930-1933 6717 <HB>
1965 V51 N10 P1935-1937 6848 <HB>
1965 V51 N10 P1937-1939 6719 <HB>
1965 V51 N10 P1959-1962 6710 <HB>
1965 V51 N10 P1970-1973 6869 <HB>
1965 V51 N11 P1987-1990 6901 <HB>
1965 V51 N11 P2021-2025 6720 <HB>
1965 V51 N11 P2031-2033 6899 <HB>
1965 V51 N11 P2101-2103 6722 <HB>
1965 V51 N11 P2144-2146 6842 <HB>
1966 V52 N2 P113-119 4858 <BISI>
1966 V52 N3 P389-391 6880 <HB>
1966 V52 N3 P469-471 6904 <HB>

TETSUDO GIJUTSU KCNYU SHIRYO
1956 V13 P2-32 T-1556 <NLL>
1956 V13 P289-296 61-17554 <=$>

TEX. ENSCHEDE
1958 V17 P816-821 61-16776 <*>

TEXTIELWEZEN
1945 V1 N3 P9-11 61-14028 <*>
1946 V2 P601 61-14029 <*>
1952 V8 N5 P19-25 61-14030 <*>

TEXTIL
1959 V14 P11-12 62-14310 <*>
1959 V14 N1 P28-31 92L35C <ATS>
1960 V15 N11 P415-418 10Q67C <ATS>
1961 V16 P331-332 64-16068 <*> 0
1961 V16 N3 P109-111 84N57C <ATS>
1961 V16 N9 P347-348 19P62SK <ATS>
1963 V18 N8 P298-300 M-5724 <NLL>
1963 V18 N9 P347-350 M-5725 <NLL>

TEXTIL OCH KONFEKTION
1947 V4 N9 P376-377 61-14007 <*> 0

TEXTILES. (TEXTILES NOUVEAUTES, TEXTILE NOUVEAUTES, VETEMENTS CREATIONS) PARIS
1947 N3 P127-137 61-14008 <*> 0

TEXTILES PANAMERICANOS
1952 V12 P27-28,71 61-14071 <*>

TEXTILINDUSTRIE
1965 V67 N2 P67-73 1113 <TC>

TEXTILIS
1961 V17 N9 P26-35 C-5370 <NRC>

TEXTIL-PRAXIS
1952 P220-223 57-1793 <*>
1952 V7 P967-971 60-18515 <*>
1952 V7 N8 P570-573 I-452 <*>
1952 V7 N8 P574-576 I-660 <*>
1953 V8 N2 P118-121 CSIRO-3114 <CSIR>
1953 V8 N5 P375-377 CSIRO-3486 <CSIR>
1953 V8 N7 P587-590 61-14060 <*>
1953 V8 N9 P795-800 61-14061 <*>
1953 V8 N12 P1C30-1033 61-14063 <*>
1953 V8 N12 P1077-1079 61-14062 <*>
1954 V9 N2 P174-175 1836 <*>
1954 V9 N3 P264-268 3032 <*>
1954 V9 N4 P364-367 3032 <*>
1954 V9 N5 P423-427 61-14064 <*> 0
1954 V9 N5 P439-440 61-14065 <*> 0
1954 V9 N6 P538 61-14066 <*>
1954 V9 N7 P607 60-10787 <*>
1954 V9 N10 P920-923 61-14067 <*> 0
1954 V9 N10 P923-925 61-14546 <*> 0
1955 V10 P220-223 60-10871 <*>
1955 V10 N2 P133-140 CSIRO-3125 <CSIR>
1955 V10 N4 P335-340 CSIRO-3126 <CSIR>
1955 V10 N6 P561-566 61-14547 <*> 0
1955 V10 N11 P1134-1137 61-14548 <*> 0
1956 V11 P1112-1115 C-2983 <NRCC>
1956 V11 N2 P163-170 CSIRO-3067 <CSIR>
1956 V11 N3 P240-244 CSIRO-3119 <CSIR>
58-2372 <*>
1956 V11 N5 P477-480 CSIRO-3248 <CSIR>
1956 V11 N7 P652-654 61-14549 <*>
1956 V11 N10 P999-1002 CSIRO-3792 <CSIR>
1956 V11 N11 P1072-1C74 CSIRO-3512 <CSIR>
1957 N3 P219- 58-1491 <*>
1957 V12 P171-176 RJ-577G <ATS>
1957 V12 N1 P3-5 61-14550 <*> 0
1957 V12 N1 P6-11 CSIRO-3483 <CSIR>
1957 V12 N1 P17-25 61-14551 <*> 0
1957 V12 N1 P62-69 76J17G <ATS>
1957 V12 N2 P121-125 61-14552 <*> 0
1957 V12 N4 P337-338 CSIRO-3553 <CSIR>
1957 V12 N9 P851-858 63K24G <ATS>
1958 V13 N2 P115-120 61-16724 <*>
1958 V13 N11 P1157-1160 59-20827 <*>
1959 V14 P1152-1158 62-10071 <*>
1959 V14 N1 P35-41 61-14553 <*> C
1959 V14 N2 P139-141 65-11505 <*>
1959 V14 N3 P276-281 65-11506 <*>
1959 V14 N4 P383-388 65-11507 <*>
1959 V14 N5 P435-443 61-14554 <*>
1959 V14 N5 P468-472 65-11508 <*>
1959 V14 N7 P698-700 65-11509 <*>
1959 V14 N8 P759-764 C-3365 <NRCC>
1959 V14 N8 P793-796 65-11510 <*>
1959 V14 N10 P1003-1005 60-17506 <*=>
1959 V14 N10 P1020-1024 65-11511 <*>
1959 V14 N11 P1133-1137 65-11512 <*>
1959 V14 N12 P1236-1240 65-11513 <*>
1960 V15 P480-484 65-11514 <*>
1960 V15 P824-827 65-11515 <*>

1960 V15 P913-917 65-11516 <*>
1960 V15 P1030-1034 65-11517 <*>
1960 V15 P1143-1145 65-11518 <*>
1960 V15 N5 P448-452 46N51G <ATS>
1960 V15 N6 P573-579 64-10937 <*>
1960 V15 N6 P590-594 47N51G <ATS>
1960 V15 N8 P801-810 63-15586 <*>
1960 V15 N12 P1293-1299 C-4261 <NRCC>
1961 V16 N1 P39-40 C-3705 <NRCC>
1961 V16 N1 P50-51 C-3682 <NRCC>
1961 V16 N5 P447-451 C-3917 <NRCC>
1961 V16 N6 P610-615 C-5039 <NRC>
1961 V16 N7 P737-738 C-3994 <NRCC>
1961 V16 N9 P875-877 C-3955 <NRCC>
66-10545 <*>
1961 V16 N10 P1032-1037 C-4084 <NRCC>
1963 V18 N8 P764-767 C-4941 <NRC>
1964 V19 P627-632 65-11019 <*> 0
1964 V19 N11 P1073-1075 66-10743 <*>
1964 V19 N12 P1169-1178 1057 <TC>
65-14595 <*>
1965 V20 P633-634 1622 <TC>
1966 V21 N6 P432-439 66-14498 <*>
1966 V21 N6 P440-448 66-14131 <*>
1966 V21 N6 P448-451 66-14499 <*>

TEXTIL-RUNDSCHAU
1946 V1 N4 P105-112 65-10995 <*> 0
1949 V4 N6 P199-211 58-1498 <*>
1950 V5 N4 P144-147 61-14069 <*>
1951 V6 N3 P111-116 64-18901 <*>
1951 V6 N4 P169-175 3433 <*>
1953 V8 N3 P131-133 I-859 <*>
57-1782 <*>
1953 V8 N4 P168-176 61-10329 <*> 0
1953 V8 N5 P233-242 61-10301 <*> 0
1954 V9 N1 P23-24 59-15310 <*>
1954 V9 N10 P509-514 65-14268 <*>
1955 V10 N4 P179-187 57-2330 <*>
1955 V10 N7 P353-359 C-2312 <NRC>
57-2353 <*>
1956 V11 N1 P1-15 66H11G <ATS>
1956 V11 N2 P70-82 66H11G <ATS>
1956 V11 N3 P131-136 66H11G <ATS>
1957 V12 P124-129 C-3505 <NRCC>
1957 V12 N2 P59-77 CSIRO-3480 <CSIR>
1957 V12 N10 P551-560 CSIRO-3832 <CSIR>
1958 V13 N2 P61-66 65-17310 <*> 0
1958 V13 N3 P163-166 36L29G <ATS>
1958 V13 N6 P323-335 42K25G <ATS>
1958 V13 N6 P336-346 C-3118 <NRCC> 0
1958 V13 N7 P396-411 C-3118 <NRCC> 0
1958 V13 N12 P694-704 59-20220 <*> 0
1959 V14 N3 P1-15 60-17565 <*+>
1959 V14 N6 P316-328 60-18406 <*>
1960 V15 N2 P82-85 62-20117 <*>
1960 V15 N3 P109-121 44M41G <ATS>
61-16712 <*>
1960 V15 N5 P233-238 32N48G <ATS>
61-16121 <*>
1960 V15 N11 P586-591 C-3852 <NRCC>
1961 V16 P531-560 <LSA>
96P62G <ATS>
1961 V16 P593-603 63-18649 <*> 0
1961 V16 N9/10 P531-560 62-10407 <*> 0
1961 V16 N9/10 P580-585 65-12191 <*>
1962 V17 N1 P1-12 77P62G <ATS>
1962 V17 N4 P200 63-16313 <*>
1962 V17 N4 P201 63-16314 <*>
1962 V17 N4 P201-203 63-16315 <*>
1963 V18 N3 P117-124 64-10918 <*>
1963 V18 N11 P589-593 65-12254 <*>
1964 V19 N5 P245-254 65-12195 <*>

TEXTIL-UND FASERSTOFFTECHNIK
1953 V3 P269-273 T-2324 <INSD>
1954 V4 P172-174 T-2325 <INSD>
1954 V4 N1 P27-37 61-14031 <*> 0
1954 V4 N6 P339-345 61-14032 <*> 0
1954 V4 N6 P358-363 61-14033 <*> 0
1954 V4 N6 P363-364 61-14034 <*> 0
1954 V4 N11 P641-642 61-14555 <*> 0
1956 V6 N10 P443-445 61-14556 <*> 0

TEYSMANNIA
1908 V19 N7 P431-435 3401 <*>
1908 V19 N8 P494-498 57-2889 <*>
1909 V20 N7 P409-417 3398 <*>
1910 V21 N12 P777-780 3400 <*>
1912 V23 N10/1 P610-612 3399 <*>
1915 V26 N1/2 P54-57 3402 <*>

THARANDTER FORSTLICHES JAHRBUCH
1936 V87 P369-412 64-16014 <*>

THEE
1926 N2 P52-53 57-3357 <*>
1926 V7 N3 P87-91 57-2835 <*>
1926 V7 N4 P156-158 57-3539 <*>

THEORETICA CHIMICA ACTA
1964 V2 N1 P63-74 66-10723 <*>

THERAPEUTISCHE BERICHTE
1959 V31 P67 65-00393 <*> 0
1959 V31 P71-75 62-00669 <*>

THERAPEUTISCHE UMSCHAU UND MEDIZINISCHE BIBLIOGRAPHIE
1956 V13 N9 P165-172 65-12720 <*>
1957 V14 P248-251 58-1368 <*>
1962 V19 P571-575 63-18488 <*>
1965 V22 N4 P184-185 88S84G <ATS>
1965 V22 N4 P194 89S84G <ATS>
1965 V22 N8 P410-417 66-10858 <*>

THERAPIE
1952 V7 P108-113 3304 <*>
1954 V9 N4 P450-455 3440-A <KH>
1955 V10 N6 P907-926 60-10002 <*>
1956 V11 N2 P300-306 65-14283 <*>
1959 V14 P68-69 62-14983 <*>
1960 V15 P967-973 65-14385 <*> 0
1960 V15 N2 P297-305 10454-A <K-H>
63-16954 <*>
1962 V17 P629-633 65-13933 <*> 0
1963 V18 P670-673 64-18145 <*>
1964 V19 P9-17 66-13162 <*>
1964 V19 P419-429 65-14280 <*> 0

THERAPIE DER GEGENWART
1941 V83 N1 P46-47 60-10017 <*> 0
1950 V89 N6 P169-177 59-17880 <*> 0
1952 P167-168 AL-14 <*>
1953 V92 P127 AL-171 <*>
1953 V92 P305- I-848 <*>
1954 V93 P184-187 II-984 <*>
1955 V94 P44-49 2121 <*>
1955 V94 N3 P92-95 57-715 <*>
1956 V95 P260-261 57-1806 <*>
1956 V95 P374-376 62-18890 <*> 0
1958 V97 N8 P312-314 59-10831 <*>
1960 V99 N8 P403 10764-B <K-H>
63-16618 <*>
1963 V102 P579-583 66-11612 <*>
1964 V103 P1427-1430 66-13165 <*>

THERAPIEWOCHE
1955 V5 N11/2 P275-282 62-00858 <*>
1957 V7 P137- 57-2867 <*>
1962 V12 N18 P816-823 65-11340 <*>
1963 V13 P448-450 66-11731 <*>
1963 V13 N12 P464-469 63-01244 <*>
1964 V14 P433-434 66-11614 <*>

THONG BAO THOI TIET
1961 N6 P1-3,5 62-19533 <=*>
1961 N7 P1-3 62-19533 <=*>

THORAXCHIRURGIE

1956 V4 P125-135 57-2701 <*>
1958 V5 N6 P528-532 59-10892 <*> 0
1960 V8 P436-449 62-10587 <*> 0

THORAXCHIRURGIE UND VASKULAERE CHIRURGIE
1964 V12 N3 P202-207 989 <TC>

THROMBOSIS ET DIATHESIS HAEMORRHAGICA
1961 V6 N1 P172-176 63-20893 <*>
1962 V7 N5/6 P444-456 63-10505 <*> 0
1962 V8 N1/2 P112-120 63-20947 <*>

TI CHIH HSUEH PAO
1954 V34 N4 P339-365 PT1 IGR V6 N6 P953 <AGI>
1954 V34 N4 P371-398 PT2 IGR V6 N7 P1177 <AGI>
1954 V34 N4 P339-365 IGR V6 N6 P953-978 <AGI>
1954 V34 N4 P371-398 IGR,V6,N7,P1177-1216 <AGI>
1956 V36 N3 P315-367 PT1 IGR V2 N3 P197 <AGI>
1956 V36 N3 P315-367 PT2 IGR V2 N4 P273 <AGI>
1956 V36 N4 P535-543 IGR V2 N10 P833 <AGI>
1959 V39 N2 P115-134 IGR V1 N11 P73 <AGI>
1959 V39 N2 P167-184 IGR V1 N12 P40 <AGI>
1959 V39 N2 P229-234 IGR V2 N1 P1 <AGI>
1960 V40 N1 P70-87 62-25529 <=*>
1960 V40 N1 P89-119 62-25529 <=*>
1962 V42 N1 P1-14 63-31191 <=> 0
1962 V42 N1 P93-101 63-31191 <=> 0
1962 V42 N1 P102-104 62-33676 <=>
1962 V42 N3 P15-24 67-30741 <=$>
1963 V43 N1 P90-96 63-31998 <=>
1965 V45 N2 P131-141 AD-635 925 <=$>
1965 V45 N2 P153-164 AD-635 371 <=$>

TI CHIU HUA HSUEH
1961 P325-333 65-30306 <=$>

TI CHIH LUN P'ING
1957 V17 N4 P381-382 63-21684 <=>
1957 V17 N4 P430-437 63-21684 <=>
1957 V17 N4 P437-440 66-33071 <=$>

TI CHIH YU K'AN T'AN
1960 N7 P11-14 61-11875 <=>

TI CH'IU WU LI HSUEH PAO
1958 V7 N2 P103-107 AD-631 062 <=$>
1959 V8 N1 P55-59 63-20721 <=*$>
1959 V8 N2 P123-131 AD-631 120 <=$>
1960 V9 N1 P38-46 AD-623 113 <=*>
1963 V12 N1 P12-31 AD-611 029 <=$>
1963 V12 N1 P32-40 AD-611 031 <=$>
1963 V12 N1 P100-117 AD-631 061 <=$>
1964 V13 N1 P20-23 AD-632 408 <=$>
1964 V13 N3 P234-242 65-31336 <=$>
1965 V14 N1 P33-44 AD-635 370 <=$>
1965 V14 N3 P173-180 AD-636 970 <=$>
C-6135 <NRC>

TI CH'IU WU LI K'AN T'AN
1957 V5 N3 P328-348 26N49G <ATS>
1958 N2 P43-44 64-31016 <=>
1958 N6 P5-8 64-31016 <=>
1959 V7 N9 P8-12 60-31023 <=>
1961 V9 N3 P427-443 77N57F <ATS>

TI-LI
1961 N4 P156-157 63-21741 <=>
1961 N4 P168-169 63-21741 <=>
1961 N4 P172-177 63-21741 <=>
1961 N5 P200-205 63-21741 <=>
1961 N5 P224-225 63-21741 <=>
1961 N6 P250 63-31192 <=>
1961 N6 P258-267 63-31192 <=>
1961 N6 P280-281 63-31192 <=>
1962 N4 P135-139 63-31257 <=>
1962 N4 P148 63-31257 <=>
1962 N4 P157-160 63-31257 <=>
1962 N5 P172-177 63-31074 <=>
1963 N1 P1-20 63-31433 <=>
1963 N1 P12-17 63-31951 <=>
1963 N1 P21-23 66-31138 <=$>
1963 N1 P28-32 66-31138 <=$>
1963 N1 P33-38 63-31433 <=>
1963 N1 P41 63-31433 <=>
1963 N1 P44-45 63-31433 <=>
1963 N1 P49-50 66-31138 <=$>
1963 N3 P143-144 63-31951 <=>
1965 N2 P49-51 65-32786 <=$>
1965 N2 P60-76 65-32786 <=$>
1965 N2 P84-89 65-32786 <=$>
1965 N2 P94-96 65-32786 <=$>
1965 N3 P106-110 65-32617 <=$>
1965 N3 P132-139 65-32617 <=$>
1965 N3 P141-144 65-32617 <=$>
1965 N4 P147-156 65-33833 <=$>
1965 N4 P166-170 65-33833 <=$>
1965 N5 P203-206 66-30563 <=$>
1965 N5 P215-221 203-206 <=$>
1965 N5 P225-226 203-206 <=$>

TI-LI CHI K'AN
1957 N1 P38-76 60-31052 <=>
1964 N8 P1-27 66-34410 <=$>

TI-LI CHIH SHIH
1958 V9 P204-207 60-11037 <=>
1958 V9 P342-344 60-11314 <=>
1958 V9 N2 P49-53 60-11076 <=>
1958 V9 N3 P109-114 60-11076 <=>
1958 V9 N10 P451-455 60-11348 <=>
1958 V9 N10 P476-478 60-11177 <=>
1958 V9 N11 P484-487 60-11455 <=>
1958 V9 N11 P488-493 60-11447 <=>
1958 V9 N12 P539-543 60-11126 <=>
1958 V9 N12 P547-549 60-11125 <=>
1959 V10 N2 P53-57 60-11075 <=>
1959 V10 N10 P433-440 60-11561 <=>
1959 V10 N10 P441-443 60-11526 <=>
1959 V10 N10 P446 60-11526 <=>
1966 N3 P140-143 67-30229 <=>

TI-LI-HSUEH-PAO
1935 V2 N1 P53-62 61-23612 <=*>
1935 V2 N2 P75-90 61-23612 <=*>
1954 N3 P313-329 59-11734 <=> 0
1954 N3 P333-344 59-11735 <=>
1954 V20 N3 P255-266 59-11545 <=>
1955 V21 N3 P245-257 59-13553 <=> 0
1956 V22 N4 P295-322 60-11575 <=>
1956 V22 N4 P325-337 60-11553 <=>
1956 V22 N4 P339-351 60-11553 <=>
1957 V23 N2 P145-159 59-13662 <=>
1957 V23 N3 P235-242 59-11534 <=>
1958 V24 N1 P33-46 60-31053 <=> 0
1958 V24 N1 P79-82 60-31053 <=> 0
1959 V25 N1 P33-39 60-31089 <=>
1959 V25 N1 P47-89 60-31089 <=>
1960 V26 N1 P1-8 63-21507 <=>
1960 V26 N1 P13-22 63-21507 <=>
1960 V26 N1 P34-60 63-21507 <=>
1960 V26 N1 P71-72 63-21507 <=>
1962 V28 N2 P111-122 63-21617 <=>
1962 V28 N2 P137-147 63-21617 <=>
1962 V28 N2 P149-174 63-21617 <=>
1962 V28 N3 P175-185 63-21670 <=>
1962 V28 N3 P189 63-21670 <=>
1962 V28 N3 P191 63-21670 <=>
1962 V28 N3 P196 63-21670 <=>
1962 V28 N3 P200 63-21670 <=>
1962 V28 N3 P205 63-21670 <=>
1962 V28 N3 P241-256 63-21670 <=>
1962 V28 N4 P257-305 63-31403 <=>
1962 V28 N4 P320-334 63-31403 <=>
1963 V29 N1 P1-12 63-31432 <=>
1963 V29 N1 P14-23 63-31432 <=>
1963 V29 N1 P25-32 63-31432 <=>
1963 V29 N1 P36-47 63-31432 <=>
1963 V29 N1 P63-86 63-31432 <=>
1963 V29 N2 P109-125 64-21271 <=>
1963 V29 N2 P156-168 64-21271 <=>

1965 V31 N2 P170-178 66-34251 <=$>
1966 V32 N1 P1-35 67-30089 <=>
1966 V32 N1 P37-94 67-30089 <=>

TI-LI-HSUEH TZE LIAO
1957 N1 P8-15 59-11674 <=>
1957 N1 P40-46 59-11675 <=>
1957 N1 P46-56 59-11676 <=>
1957 N1 P81-88 59-11677 <=>
1957 N1 P88-94 59-11678 <=>
1957 N1 P97-108 59-11731 <=>
1957 N1 P123-138 59-11733 <=>
1957 N1 P138-145 59-11732 <=>
1957 N1 P153-164 59-11729 <=>
1957 N1 P165-174 59-11730 <=>

TI LI YU CH'AN YEH
1956 V1 N1 P48-49 PB 141 219T <=>

TIDSIGNAL
1966 P4-5 66-33792 <=$>
1966 P12-14 66-33792 <=$>
1966 P22-28 66-33792 <=$>

TIDSKRIFT FOR DOKUMENTATION
1956 V12 N6 P69-72 62-14078 <*>
1959 V15 N2 P13-16 62-10063 <*> 0

TIDSKRIFT. KUNGLIGA LANTBRUKSAKADEMI
SEE KUNGLIGA LANTBRUKSAKADEMIENS TIDSKRIFT

TIDSKRIFT SVENSKA VAGFORENINGEN
1958 V45 N6 P259-263 60-15411 <*+> 0

TIDSKRIFT FOR VARME-, VENTILATIONS- OCH SANITETSTEKNIK
1948 V19 P67 AL-343 <*>

TIDSSKRIFT FOR HERMETIKINDUSTRI
1941 V26 P255-266 AL-331 <*>
1952 V38 P249-250 64-20381 <*> 0
1952 V38 P253-258 64-20381 <*> 0
1952 V38 P261-262 64-20381 <*> 0
1962 V48 N9 P394-402 TRANS-329 <MT>

TIDSSKRIFT FOR KEMI OG BERGVESEN
1933 V13 N2 P18-22 II-611 <*>
1933 V13 N2 P37-41 II-611 <*>
1938 V18 N8 P127- 1175 <*>

TIDSSKRIFT FOR KJEMI, BERGVESEN OG METALLURGI
1941 N10 P177-180 AL-959 <*>
1942 N2 P67-68 3980 <K-H>
1942 V2 N2 P67-68 64-16430 <*>
1943 N5 P55-59 62-20223 <*>
1943 V3 N5 P55-59 07R80N <ATS>
1943 V3 N8 P98-99 57-1200 <*>
1944 V4 N1 P2-4 57-1378 <*>
1944 V4 N4 P26-29 57-1375 <*>
1944 V4 N5 P40-43 57-1377 <*>
1944 V4 N6 P50-52 57-1367 <*>
1944 V4 N9 P81-83 57-1366 <*>
1945 V5 N6 P77-83 57-1196 <*>
1946 N3 P32-36 62-20222 <*>
1946 N5 P59-62 62-20224 <*>
1951 N6 P78-82 57-2405 <*>
57-3526 <*>
1951 V11 N6 P78-82 II-194 <*>
1954 V14 N5 P73-78 58-2301 <*>
1956 V16 N5 P77-82 59-16641 <*=> 0
1958 V18 N5 P75-79 59-15980 <*>
1958 V18 N6 P92-98 59-15980 <*>
1961 V21 N3 P62-68 65-13108 <*>
1964 V24 N10 P178-187 34S81N <ATS>

TIDSSKRIFT FOR MATHEMATIK
1886 S5 V4 P130-137 66-13954 <*>

TIDSSKRIFT FOR DEN NORSKE LAEGEFORENING
1954 N15 P1- I-583 <*>
1954 V74 N17 P549-554 3359-C <K-H>
1956 V76 N3 P67-72 5073 <K-H>
1956 V76 N16 P549-552 5470-D <K-H>
1958 V78 N13/4 P634-635 66-12541 <*> 0
1959 V79 N7 P394-396 61-00180 <*>
1959 V79 N19 P1127-1128 62-16016 <*>
1962 V82 P895-897 65-14951 <*>
1963 V83 P1449-1452 64-00349 <*>
1963 V83 P1721-1724 66-17299 <*>
1964 V84 P617-619 32S85N <ATS>
1965 V85 N2 P557-558 66-11579 <*>

TIDSSKRIFT FOR DET NORSKE LANDBRUK
1955 V62 P59-70 1683 <*>

TIDSSKRIFT FOR SKOGBRUK
1951 V59 N6 P176-185 63-20660 <*>

TIDSSKRIFT FOR TEXTILTEKNIK
1952 V10 N2 P25-39 61-14009 <*>
1959 V17 N1 P7-11 61-14557 <*>
1960 V18 P55-57 75P64D <ATS>
1960 V18 P64-65 33N48N <ATS>

TIEN-CHI YUEH-K'AN
1958 N9 P1-6 59-13503 <=>
1959 N9 P47 60-31762 <=>

TIEN HSIN CHI SHU T'UNG HSUN
1964 V1 P23-30 AD-625 261 <=$>

TIEN HSIN K'O HSUEH
1957 N2 P64-65 60-11603 <=>
1959 N1 P25-28 61-11026 <=>
1959 N7 P1-13 62-19630 <=*>
1959 N7 P18-22 62-19630 <=*>
1959 N7 P46- 62-19630 <=*>
1959 N7 P52-62 62-19630 <=*>

TIEN SHIH CHIEH
1963 V16 N8/11 P323-324 65-6400 <=*$> 0
1963 V16 N8/11 P325-326 65-63981 <=$> 0

TIEN TZU CHI SHI
1964 N4 P31-33 AD-631 768 <=$>

TIEN TZU HSUEH PAO
1960 V4 N1 P59-68 61-20879 <*>
1963 N2 P109-119 66-31195 <=$>
1964 N2 P3-15 65-32306 <=*$>
1964 N2 P67-90 65-32306 <=*$>

T'IEN WEN HSUEH PAO
1959 V7 N1 P1-3 62-24467 <=*>
1959 V7 N2 P163-174 63-23417 <=*>
1959 V7 N2 P169-174 63-01033 <*>
1962 V10 N1 P24-33 64-71199 <=>
1962 V10 N2 P208-210 63-31442 <=>
1963 V11 N1 P41-47 65-17044 <*>
1963 V11 N1 P74-78 AD-611 047 <=$>
1964 V12 N2 P169-179 65-33546 <=*$>
1965 V13 N1 P1-21 N66-14911 <=*$>

TIERAERZTLICHE RUNDSCHAU
1933 V39 N3 P1-8 66-14402 <*>
1941 V47 P58-62 61-10255 <*> P

TIERAERZTLICHE UMSCHAU
1961 V16 N2 P46-48 63-14409 <*>
1963 V18 N6 P288-290 65-11940 <*>

TIERERNAEHRUNG UND TIERZUCHT
1930 V2 P517-522 57-98 <*>
1930 V3 P631-638 59-15274 <*> 0

TIERZUCHT. BERLIN
1949 V5 P4217 82G7G <ATS>

TIERZUECHTER
1957 V9 N22 P554-557 58-2517 <*>

TIJCSCHRIFT VCOR DIERGENEESKUNDE
1944 V69 N15 P505-514 58-2078 <*>
1947 V72 P550-557 58-2076 <*>
1950 V75 P589-590 3178 <*>
1955 V80 N1 P13-16 58-91 <*>
1956 V81 N9 P430-435 CSIRO-3135 <CSIR>

TIJCSCHRIFT VAN HET K. NEDERLANDSCH AARDRIJKS-KUNDIG GENOOTSCHAP
1950 S2 V67 N3 P303-325 44K27DU <ATS>
1950 S2 V67 N3 P326-333 53K27DU <ATS>
1950 S2 V67 N3 P334-335 45K27D4 <ATS>

TIJDSCHRIFT. NEDERLANDS MILITAR GENEESKUNDIG
1957 V10 N9 P264-274 60-19086 <*+>
1963 V16 P200-215 65-00218 <*>

TIJDSCHRIFT VOOR OPPERVLAKTETECHNIEKEN VAN METALEN
1963 V7 P93-94 PN-425 <TTIS>

TIJERETAZOS SOBRE MALARIA
1942 V6 N2 P45-49 57-1692 <*>

TIMONE
1953 N10/7 P3 58-1145 <*>

TIN TUC HOAT DONG KHOA HOC
1960 N7 P1-2 62-19356 <=*>
1960 N7 P32-33 62-19356 <=*>
1960 N8 P8-13 62-19357 <=*>
1960 N9 P21-27 62-19358 <=*>
1960 N11 P25-33 62-19359 <=>
62-19359 <=*>
1960 N11 P44-46 62-19359 <=>
62-19359 <=*>
1960 N11 P49 62-19359 <=>
62-19359 <=*>
1961 N11 P54-55 62-19607 <=*>
1962 N5 P1-9 62-32058 <=*>
1962 N5 P10-15 62-32193 <=>
1962 N5 P34 62-32193 <=>
1962 N5 P58 62-32193 <=>
1962 N10 P5-7 63-21176 <=>

TINCTORIA
1953 V50 P1-6 T-2161 <INSD>
1954 V51 P43-49 T-2158 <INSD>
1954 V51 P321-325 T-2150 <INSD>
T-2160 <INSD>
1954 V51 P359-364 T-2159 <INSD>
1954 V51 P397-405 T-2159 <INSD>
1954 V51 N11 P406-408 61-14558 <*>
1958 V55 P237-238 72L34I <ATS>
1960 V57 N3 P89-96 61-16775 <*>

TITAN I EGO SPLAVY
1959 V2 P133-144 60-31825 <=>
1961 V5 P135-142 65-13296 <*>
1961 V5 P225-232 51Q73R <ATS>
1962 V7 P130-139 62-33609 <=*>
1962 V8 P114-118 65-11768 <*>
1963 V9 P166-171 64-18549 <*>
1963 V10 P151-167 64-31378 <=>
1963 V10 P218-223 64-31378 <=>
1963 V10 P254-261 64-31378 <=>
1963 V10 P317-321 64-11862 <=> M
1963 V10 P317-331 64-31378 <=>
1963 V10 P357-361 64-31378 <=>

TLUSZCZE I SRODKI PIORACE
1963 V7 P260-263 64-18951 <*>
1965 V9 N6 P360-368 66-13428 <*>

TCCHI KAIRYO
1954 N4/5 P156-162 T-1950 <INSD>

TOHOKU DAIGAKU DENTSU DANWAKAI KIROKU
1954 V23 N1 P39-40 53N50J <ATS>
1954 V23 N1 P43-44 54N50J <ATS>
1955 V24 N1 P45-46 56N50J <ATS>
1955 V24 N2 P43-44 55N50J <ATS>
1959 V28 N1 P39-40 57N50J <ATS>

TOHOKU DAIGAKU HISUI YOEKI KAGAKU KENKYU-SHO HOKOKU
1959 V8 P41-46 65-14581 <*>

TOHOKU DAIGAKU IGAKUBU
1942 V44 P130-157 61-00852 <*>
1944 V46 P295-322 63-18813 <*>
1949 V51 N1/2 P9-16 2575 <*>

TOHOKU DAIGAKU KAGAKU KEISOKU KENKYUSHO HOKOKU
1958 V7 N2 P125-140 59-17899 <*>
1958 V7 N2 P175-184 59-17900 <*>
1962 V11 P121-139 66-14453 <*$>
1962 V11 P141-160 66-14452 <*$>
1962 V11 P161-188 66-14455 <*$>
1962 V11 P227-234 66-14456 <*$>

TOHOKU DAIGAKU KINZOKU KENKYU-HOKOKU
1963 V1 N1 P30-34 66-11737 <*>
1963 V1 N1 P44-49 66-11739 <*>
1963 V1 N1 P50-53 66-11740 <*>
1963 V1 N1 P54-62 66-11741 <*>
1963 V1 N1 P63-71 66-11742 <*>
1963 V1 N1 P94-97 66-11743 <*>
1963 V1 N1 P98-102 66-11738 <*>

TOHOKU DAIGAKU KOSOKU RIKIGAKU KENKYUJO HOKOKU
1952 V8 N76 P63-76 AD-618 204 <=$>

TOHOKU DAIGAKU SENKO SEIREN KENKYUSHU IHO
1953 V9 N2 P235-240 79M39J <ATS>
1955 V11 N2 P155-158 57-2583 <*>

TOHOKU IGAKU ZASSHI
1954 V50 P246-251 63-00562 <*>

TOHOKU JOURNAL OF EXPERIMENTAL MEDICINE
SEE TOHOKU DAIGAKU IGAKUBU

TOHOKU KAIKU SUISAN KENKYUSHO KENKYU HOKOKU
1955 N4 P62-82 59-19338 <+*>
1955 N4 P83-100 59-19342 <+*>
1955 N4 P101-119 59-19341 <+*>
1955 N5 P43-52 59-18563 <*=> 0

TOHOKU MEDICAL JOURNAL
SEE TOHOKU IGAKU ZASSHI

TOHOKU NOGYO SHIKENJO KENKYU HOKOKU
1959 N29 P53-74 66-10840 <*>
1959 V14 P16-20 66-11539 <*>
1964 N30 P1-149 66-14435 <*$> 0
1965 N32 P109-144 66-13964 <*$>

TOKAI-KU SUISAN KENKYUSHO KENKYU HOKOKU
1958 N21 P15-24 59-19340 <*=> 0
1962 N32 P49-118 C-4761 <NRC>

TOKSIKOLOGIYA NOVYKH PROMYSHLENNYKH KHIMICHESKIKH VESHCHESTV
1961 N3 P108-112 64-11665 <=>
1965 N7 P3-4 65-33186 <=$>
1965 N7 P122-138 65-33186 <=$>
1965 N7 P187 65-33186 <=$>

TOKYO DAIGAKU JISHIN KENKYUSHO IHO
1938 V16 N3 P670-675 LC-1250 <NLL>
1962 V40 P357-369 PT2 IGR V6 N3 P420 <AGI>
1962 V40 P333-355 IGR V6 N11 P1903 <AGI>
1962 V40 N2 P357-369 IGR,V6,N3,P420-429 <AGI>

TOKYO DAIGAKU KOGAKUBU KIYO
1938 V21 N3 P115-168 63-16827 <*>
1958 N1 P153-163 AEC-TR-3713 <*>

TOKYO DAIGAKU NOGAKUBU ENSHURIN HOKOKU

1955 N49 P205-216 C-2195 <NRC>
2356 <*>

TOKYO DAIGAKU RIKOGAKU KENKYUSHO HOKOKU
1949 V3 N11/2 P316-322 AEC-TR-6099 <*>
1950 V4 P236-239 2213 <KH>
63-18237 <*>
1950 V4 N3/4 P31-36 62-25703 <=*>
1952 V6 P191-195 62-01273 <*>
1952 V6 N3 P111-116 38F4J <ATS>
1952 V6 N6 P367-371 AEC-TR-6098 <*>
1953 V7 P23-24 68P63J <ATS>
1954 V8 P39-48 3012 <*>

TOKYO IGAKU ZASSHI
1957 V61 P343-354 57-3435 <*>

TOKYO IJI SHINSHI
1953 V70 N10 P51-73 63-00604 <*>
1954 V71 N4 P17-18 61-00858 <*>

TOKYO JOURNAL OF MEDICAL SCIENCE. JAPAN
SEE TOKYO IGAKU ZASSHI

TOKOYO KAGAKU KAISHI
1918 V39 P1116-1121 62-00955 <*>

TOKYO KOGYO SHIKEN-SHO HOKOKU
1950 V45 P355-368 57-2694 <*>
70E1J <ATS>
1951 V46 P1-10 57-2671 <*>
63E1J <ATS>
1951 V46 P11-19 57-2665 <*>
64E1J <ATS>
1951 V46 P103-113 57-2662 <*>
61E1J <ATS>
1951 V46 P114-124 57-2675 <*>
62E1J <ATS>
1952 V46 N9 P341-349 64-14142 <*> 0
1952 V46 N9 P364-371 64-14154 <*> 0
1952 V46 N9 P385-391 64-14162 <*> 0
1952 V47 P258-263 67N54J <ATS>
1952 V47 P264-269 66N54J <ATS>
1952 V47 P327-336 94L32J <ATS>
1953 V48 P12-26 64-20340 <*>
1953 V48 N4 P133-138 60-18156 <*>
1953 V48 N4 P139-157 60-18155 <*>
1953 V48 N8 P311-324 91M39J <ATS>
1954 V49 P278-279 II-1044 <*>
1954 V49 P280-282 II-940 <*>
1954 V49 P283-286 II-941 <*>
1955 V50 N6 P185-195 11L30J <ATS>
1956 V51 N11 P397-405 12L30J <ATS>
1956 V51 N11 P406-414 13L30J <ATS>
1957 V52 N1 P30-34 40M46J <ATS>
1957 V52 N3 P364-367 60-00594 <*>
1957 V52 N8 P281-284 81K23J <ATS>
1957 V52 N8 P285-288 45K21J <ATS>
1958 V53 N12 P429-432 62-16458 <*> 0
1960 V55 N9 P346-350 NP-TR-794 <*>
1961 V56 N2 P59-63 61-18680 <*>
1961 V56 N6 P241-252 65-11773 <*>
1962 V57 N1 P50-57 CODE-T1 <SA>
63-20007 <*>
1962 V57 N4 P187-193 65-14902 <*>
1963 V58 N6 P268-280 06S82J <ATS>

TOKYO MEDICAL JOURNAL
SEE TOKYO IJI SHINSHI

TONAN AJIA KENKYU. KYOTO
1964 V3 P1-113 65-31780 <=$>

TONINDUSTRIEZEITUNG
1897 V21 P1148-1151 64-18304 <*>
1897 V21 P1157-1159 64-18304 <*>
1914 V38 P537-539 64-18237 <*>
1924 V48 N22 P221-223 64-18293 <*>
1925 V49 N19 P881 64-14335 <*>
1926 V50 P299 64-18226 <*> 0
1927 V51 N22 P344-347 64-18239 <*>
1928 V52 P418-421 64-14323 <*>
1928 V52 N38 P757-760 64-18325 <*> 0
1928 V52 N39 P782-786 64-18325 <*> 0
1928 V52 N65 P1318-1321 64-18328 <*> 0
1928 V52 N97 P1939-1940 57-3178 <*>
1929 V53 P681-684 64-18330 <*>
1934 V58 N57 P689-690 64-18277 <*>
1934 V58 N57 P700-701 64-18277 <*>
1934 V58 N57 P713-715 64-18277 <*>
1935 V59 N25 P316-317 64-14846 <*>
1935 V59 N70 P849-853 64-18329 <*> 0
1936 V60 N21 P270-273 60-14431 <*> 0
1936 V60 N21 P282-285 60-14431 <*> 0
1936 V60 N62 P761-763 64-14298 <*> 0
1936 V61 N3 P27-29 I-818 <*>
1936 V61 N4 P4C-41 I-818 <*>
1936 V61 N5 P51-53 I-818 <*>
1937 V61 N6 P59-61 57-2170 <*>
1937 V61 N7 P74-76 57-2170 <*>
1937 V61 N28 P318-319 64-14848 <*>
64-18327 <*>
1940 V64 P47-48 1281 <*>
1940 V64 P55-57 1281 <*>
1940 V64 P95- 1281 <*>

TONINDUSTRIE-ZEITUNG UND KERAMISCHE RUNDSCHAU
1951 N9/10 P135-136 3392 <*>
1951 V75 N3/4 P33-36 60-18692 <*> 0
1951 V75 N9/10 P135-136 II-426 <*>
1953 V77 P83-88 63-14603 <*>
1953 V77 P165-170 T-1491 <INSD>
1953 V77 P234- 57-2256 <*>
1954 V78 N1/2 P9-11 AEC-TR-4647 <*>
1955 V79 P91-94 I-360 <*>
1955 V79 P239-241 II-328 <*>
1955 V79 N1/2 P2-3 62-16471 <*>
1955 V79 N11/2 P167-169 CLAIRA T.4 <CLAI>
1956 V80 N1/2 P1-7 CLAIRA T.2 <CLAI>
1957 V81 N19/0 P325-332 58-2387 <*>
58-787 <*>
1958 V82 N16 P341-348 CLAIRA T.7 <CLAI> P
1958 V82 N3/4 P38-41 61-18445 <*>
1959 V83 P275-281 2539 <BISI>
1959 V83 N2 P30-33 18N55G <ATS>
1959 V83 N10 P219-238 62-18198 <*>
1959 V83 N14 P334-341 TRANS-07 <MT>
1959 V83 N18 P432-441 3231 <BISI> P
1959 V83 N22 P535-536 60-14258 <*>
1959 V83 N24 P579-586 65-12883 <*>
1960 V84 P77-88 2046 <BISI>
1960 V84 P215-219 2540 <BISI>
1960 V84 N14 P335-338 65-11141 <*>
1960 V84 N17 P414-417 61-10794 <*>
1960 V84 N24 P585-598 62-14276 <*> 0
1962 V86 N6 P138-141 62-18824 <*>
1962 V86 N10 P234-236 64-10175 <*>
1962 V86 N20 P497-504 BGIRA-553 <BGIR>
1962 V86 N24 P606-612 63-16983 <*>
1962 V86 N22/3 P586-589 3597 <BISI>
1964 V88 N2 P25-32 BGIRA-644 <BGIR>
1964 V88 N4 P73-80 BGIRA-645 <BGIR>
1964 V88 N10 P225-231 BGIRA-649 <BGIR>
1964 V88 N7/8 P153-159 BGIRA-648 <BGIR>
1964 V88 N17/8 P389-396 65-11706 <*>
1965 V89 N5/6 P124-130 65-18071 <*> 0

TOOL ENGINEER. JAPAN
SEE KIKAI TO KOGU

TORAX
1957 V6 N1 P35-52 64-14452 <*> PO

TORFYANAYA PROMYSCHLENNCST
1946 V23 N2 P29-32 RT-119 <*>
65-63246 <=>
1949 N8 P23-25 59-22379 <+*>
1950 V26 N1 P2-6 61-13445 <*+>
1953 V29 N10 P24-25 60-13571 <*+>
1954 N1 P22-25 R-4427 <*>

1960 V37 N7 P22-24 64-15925 <=>
1961 V38 N3 P24-26 63-15279 <=*>
1962 V39 N2 P9-11 63-15280 <=*>
1962 V39 N5 P10-14 63-23244 <=*>
1962 V39 N5 P14-17 63-23243 <=*>
1962 V39 N5 P17-19 63-15967 <=*>
1962 V39 N5 P28-29 63-15956 <=*>
1962 V39 N8 P3-4 63-24362 <=*$>
1962 V39 N8 P5-7 64-13380 <=*$>
1962 V39 N8 P13-14 64-13379 <=*$>
1963 V40 N1 P13-14 63-21553 <=>
1963 V40 N6 P28-29 64-19954 <=$>

TOSHIBA REBYU
1950 V5 N4/5 P33-38 59-17451 <*>
1956 V11 N12 P1349-1356 57N53J <ATS>
61-20848 <*>
1957 V12 N3 P265-272 66-12506 <*>
1958 V13 N8 P826-831 60-14520 <*>
1958 V13 N10 P1001-1006 66-12508 <*>
1959 V14 N12 P1238-1245 97M41J <ATS>
1961 V16 N9 P1161-1166 62-14760 <*> 0
1963 V18 N2 P1-9 UCRL TRANS-984 <*>
1964 V19 N4 P451-458 65-11630 <*>

TOSHIBA REVIEW
SEE TOSHIBA REBYU

TOULOUSE MEDICAL
1949 V50 P127-135 3044 <*>
1953 V54 N9 P651-662 3058-C <K-H>
1957 V58 N1 P1-9 6471 <K-H>

TOUTE LA RADIO ET RADIO CONSTRUCTEUR
1949 N137 P218-220 57-2481 <*>
1960 N242 P39-41 60-16951 <*>

TOWARZYSTWO NAUK SCISLYCH W PARYZU PAMNIETNIK
1874 V5 N2 P1-15 UCRL TRANS-885(L) <*>

TOYAMA DAIGAKU KOGAKUBU KIYO
1960 V11 N1/2 P86-91 64-10526 <*>
1961 V12 N1 P1-3 64-10525 <*>

TRABAJOS DEL INSTITUTO CAJAL DE INVESTIGACIONES BIOLOGICAS. MADRID
1960 V52 P153-167 63-00779 <*>

TRABAJOS DEL LABORATORIO DE INVESTIGACIONES BIOLOGICAS DE LA UNIVERSIDAD DE MADRID
1910 V8 P1-26 1323 <*>
1910 V8 P63-134 66-11806 <*>
1913 V11 P81-102 66-12163 <*>

TRADUCTION AUTOMATIQUE
1960 N2 P3-19 62-19355 <=>
1960 N2 P34 62-19355 <=>
1960 N5 P2-13 62-19354 <=>
1960 N5 P17-19 62-19354 <=>
1960 N5 P24- 62-19354 <=>
1960 N5 P29-33 62-19354 <=>

TRAE
1958 V5 N10 P19-23 61-01039 <*>
1958 V5 N10 P33 61-01039 <*>

TRAEIDUSTRIEN. STOCKHOLM
1957 V7 N5 P67-74 58-1266 <*>

TRAITEMENT THERMIQUE
1965 V2 P23-35 4505-3 <BISI>

TRAKTORIST KOMBAINER I SHOFER
1959 P30-31 60-17388 <+*>

TRAKTORY I SELKHOZMASHINY
1958 V28 N7 P27-29 NIAE-TRANS-166 <NLL>
1959 N1 P10-13 60-15976 <+*>
1959 N2 P40 59-19609 <+*>
1959 N3 P4- 60-17206 <*+>
1959 N4 P48-49 60-21734 <=>
1959 N6 P47-48 60-23574 <+*>
1959 N8 P35-38 60-31018 <=>
1959 N9 P1-3 60-31018 <=>
1959 N11 P19- 61-13301 <*+>
1959 N12 P1-2 60-31499 <=>
1959 N12 P39-40 60-31499 <=>
1959 V2 N11 P1-3 60-31630 <=>
1959 V2 N11 P15-19 61-28638 <*=>
1959 V2 N11 P24-26 60-31630 <=>
1959 V2 N11 P32-34 60-31630 <=>
1959 V2 N11 P38-41 60-31630 <=>
1959 V2 N12 P18-19 61-15863 <+*>
1960 N1 P1-2 60-31736 <=>
1960 N4 P1-2 NLLTB V2 N8 P670 <NLL>
60-31737 <=>
1960 N4 P25-26 60-31737 <=>
1960 N5 P31-32 61-11625 <=>
1960 V3 N1 P33-36 61-11625 <=>
1960 V3 N1 P33 61-11709 <=>
1960 V3 N1 P36-37 61-11709 <=>
1960 V3 N2 P1-2 60-31787 <=>
1960 V3 N2 P35 61-11574 <=>
1960 V3 N3 P1-3 60-31787 <=>
1960 V3 N3 P19-24 61-11734 <=>
1960 V3 N3 P36 61-11709 <=>
1960 V3 N3 P38 61-11709 <=>
1960 V3 N4 P32-34 61-11574 <=> P
1960 V3 N5 P32-33 61-11709 <=>
1960 V3 N6 P29-30 61-11709 <=>
1960 V3 N6 P32 61-11574 <=> P
1960 V3 N7 P31-33 61-11574 <=> P
1960 V3 N7 P33 61-11709 <=>
1960 V3 N9 P33-36 61-11709 <=>
1960 V3 N9 P37-38 61-11574 <=>
1960 V3 N10 P1-4 61-11736 <=>
1960 V3 N10 P47 61-11736 <=>
1960 V3 N11 P1-2 61-21468 <=>
1960 V3 N11 P30 61-21468 <=>
1960 V3 N11 P35-37 61-21468 <=>
1960 V3 N12 P1-4 61-23623 <=>
1960 V3 N12 P20-24 61-23623 <=>
1960 V3 N12 P28-34 61-23623 <=>
1960 V30 N1 P19-22 62-32487 <=*$>
1960 V30 N4 P12-14 63-11010 P.11-16 <=>
1960 V30 N5 P17-20 63-11010 P.17-24 <=>
1960 V30 N6 P22-24 TR.195 <NIAE>
1960 V30 N7 P37-43 63-11010 <=>
1960 V30 N8 P1-2 61-11518 <=>
1960 V30 N8 P38-39 61-11518 <=>
1960 V30 N9 P1-2 61-11518 <=>
1961 V4 N2 P1-3 61-27153 <=>
1961 V4 N2 P33-34 61-27153 <=>
1961 V4 N2 P38-39 61-27153 <=>
1962 V32 N3 P1-3 62-32890 <=*>
1962 V32 N3 P7 62-32890 <=*>
1962 V32 N3 P1C 62-32890 <=*>
1962 V32 N3 P34-36 62-32890 <=*>
1962 V32 N6 P22-28 62-33018 <=> P
62-33018 <=*> P
1962 V32 N6 P33-34 62-33018 <=> P
62-33018 <=*> P
1962 V32 N6 P37-38 62-33024 <=$>
1962 V32 N7 P28-32 63-13331 <=>
1962 V32 N8 P1-5 64-15865 <=*$>
1962 V32 N9 P33-34 63-13265 <=>
1962 V32 N10 P42-44 63-13823 <=*>
1962 V32 N11 P1-2 63-21049 <=>
1962 V32 N11 P33-36 63-21049 <=>
1962 V32 N11 P38-43 63-21049 <=>
1962 V32 N12 P32 65-63552 <=$>
1963 V32 N1 P1-4 63-21433 <=>
1963 V32 N1 P29-30 63-21433 <=>
1963 V32 N1 P32 63-21433 <=>
1963 V32 N1 P34-35 63-21433 <=>
1963 V32 N9 P36-38 63-13265 <=>
1963 V33 N2 P24-27 63-21711 <=>
1963 V33 N2 P41-43 63-21711 <=>
1963 V33 N5 P4-5 63-31296 <=> P
1963 V33 N5 P35 63-31296 <=> P

1963 V33 N5 P37-39 63-31296 <=> P
1963 V33 N5 P41 63-31296 <=> P
1963 V33 N6 P1-5 63-31499 <=> O
1963 V33 N6 P42-44 63-31499 <=> O
1963 V33 N7 P22-24 63-31631 <=>
1963 V33 N7 P32-34 63-31631 <=>
1963 V33 N7 P48 63-31631 <=>
1963 V33 N8 P36-38 63-41038 <=>
1963 V33 N8 P38 63-41038 <=>
1963 V33 N8 P40-41 63-41038 <=>
1963 V33 N9 P4-6 64-19876 <=$>
1964 V34 N1 P11-15 65-63520 <=$*>
1964 V34 N1 P15-18 65-60995 <=$>
1964 V34 N6 P24-27 TR.180 <NIAE>
1964 V34 N12 P1-2 65-30505 <=$>
1965 V35 N1 P1-11 65-30609 <=$>
1965 V35 N2 P1-2 65-30818 <=$>
1965 V35 N2 P17-19 65-30818 <=$>
1965 V35 N4 P3-6 TR.177 <NIAE>
1965 V35 N7 P1-2 65-32614 <=$>
1965 V35 N7 P17-19 65-32614 <=$>
1965 V35 N7 P26-28 65-32614 <=$>
1965 V35 N7 P49 65-32614 <=$>
1966 N1 P1-3 66-31125 <=$>
1966 N1 P17-18 66-31125 <=$>
1966 N1 P39 66-31125 <=$>
1966 N1 P40 66-31125 <=$>
1967 N1 P1-2 67-31452 <=$>

TRANSACTIONS OF THE AGRICULTURAL ENGINEERING SOCIETY OF JAPAN
SEE NOGYO DOKOKU KENKYU BESSATSU

TRANSACTIONS OF THE ARCHITECTURAL INSTITUTE OF JAPAN
SEE NIPPON KENCHIKU GAKKAI ROMBUN HOKOKU-SHU

TRANSACTIONS OF THE JAPAN SOCIETY OF CIVIL ENGINEERS
SEE DOBOKU GAKKAI RONBUNSHU

TRANSACTIONS OF THE JAPAN SOCIETY OF MECHANICAL ENGINEERS
SEE NIHON KIKAI GAKKAI RONBUNSHU

TRANSACTIONS OF THE JAPANESE PATHOLOGICAL SOCIETY
SEE NIPPON BYORIGAKKAI KAISHI

TRANSACTIONS OF THE MINING AND METALLURGICAL ALUMNI ASSOCIATION. KYOTO
SEE SUIYOKAI-SHI, KYOTO

TRANSACTIONS OF THE NATIONAL RESEARCH INSTITUTE FOR METALS. JAPAN
SEE KINZOKU ZAIRYO GIJUTSU KENKYUSHO OBUN HOKOKU

TRANSACTIONS OF THE NATURAL HISTORY SOCIETY OF TAIWAN
SEE TAIWAN HAKUBUTSU GAKKAI

TRANSPORT I KHRANENIE NEFTI I NEFTEPRODUKTOV. NAUCHNO-TEKHNICHESKII SBORNIK (TITLE VARIES)
1961 N1 P29-41 AD-610 372 <=$>

TRANSPORTNOE STROITELSTVO
1958 V8 N9 P30 60-13471 <+*>
1959 V9 N11 P12-15 61-11128 <=>
1962 V12 N10 P46-48 64-71417 <=> OM
1963 V13 N2 P9-14 64-19132 <=*$>
1963 V13 N2 P23-27 64-11585 <=>
1963 V13 N2 P29-32 65-60326 <=$>
1963 V13 N4 P5-9 64-11582 <=>
1963 V13 N4 P9-11 64-11582 <=>
1963 V13 N4 P11-12 64-11582 <=>
1963 V13 N4 P13-18 64-11582 <=>
1963 V13 N4 P22-24 64-11582 <=>
1963 V13 N4 P34-46 64-11582 <=>
1963 V13 N4 P37-38 64-11582 <=>
1963 V13 N4 P68-70 64-11582 <=>
1966 V16 N1 P1-5 66-31880 <=$>

TRASPORTI PUBBLICI
1959 P1126-1128 1832 <BISI>

TRATTAMENTI DEI METALLI
1964 V7 P19-22 4290 <BISI>
1964 V7 P24-26 4409 <BISI>

TRAVAIL HUMAIN
1952 V15 N3/4 P256-264 64-10319 <*>
1957 V20 N3/4 P339-349 59-20866 <*>
1958 V21 N1/2 P141-146 60-18011 <*>

TRAVAIL ET METHODES
1949 N3 P28-37 58-2234 <*>
1951 N38 P13-16 64-10043 <*>
1956 N97 P181-184 65-12566 <*>
1956 N104 P24-28 65-12623 <*>
1956 N107 P16-20 65-12676 <*>
1956 N7/8 P30-33 60-18030 <*>
1958 N1 P173-174 60-18657 <=>
1958 N1 P229-232 60-18658 <*>
1958 N124 P109-115 60-18646 <*>

TRAVAUX: ARCHITECTURE, CONSTRUCTION, TRAVAUX PUBLICS
1948 N164 P378-382 3332 <*>
1948 V164 P386-387 3348 <*>
1949 P2-22 62-20093 <*>
1949 P47-64 62-20093 <*>
1949 P347-351 60-15706 <*+> O
1951 V35 N204 P559-566 I-929 <*>
1953 V37 N9 P437-451 61-16432 <*>
1954 P463-474 58-2575 <*>
1954 V38 P463-474 61-20517 <*> O
1955 V39 N238 P225-232 1773 <*>
1955 V39 N248 P499-515 61-18991 <*> P
1958 P45-47 58-1001 <*>
1958 P951-957 59-14531 <+*> O
1963 V47 N339 P24-30 63-18040 <*>

TRAVAUX DE L'INSTITUT SCIENTIFIQUE CHERIFIEN. SERIE SCIENCES PHYSIQUES. TANGER
1959 N4 ENTIRE ISSUE 62-00393 <*>

TRAVAUX DE LA SOCIETE DE PHARMACIE DE MONTPELLIER
1946 V6 P89-91 63-14546 <*>
1962 V22 P49-54 65-14707 <*>

*TRENIE I IZNOS V MASHINAKH
1953 V7 P12-33 64-71181 <=>
1953 V8 P107-204 62-20917 <=*>
1958 V12 P288-294 59-22710 <+*>
1959 V13 ENTIRE ISSUE <ASME>
63-12522 <*>
1960 V14 P63-92 63-19304 <=*>
1962 N15 P227-253 AD-635 278 <=$>

TRIBUNA. RUMANIA
1963 V7 N47 P1 64-21128 <=>
1963 V7 N47 P9 64-21128 <=>

TROPICHESKAYA MEDITSINA I VETERINARIYA
1930 N8/9 P303-304 RT-1497 <*>

TROPISCHE NATUUR
1924 V8 P113-122 T-2650 <INSD>

TRUD I TSENI
1964 V6 N7 P86-91 64-51109 <=>
1966 N1 P53-60 66-31918 <=$>
1966 V4 N13 P1 66-31918 <=$>
1966 V4 N13 P4 66-31918 <=$>

TRUDOVE OT CHERNOMORSKATA BIOLOGICHNA STANTSIYA V GR. VARNA
1933 V1 P26-33 II-1026 <*>

TRUDY AKADEMII MEDITSINSKIKH NAUK SSSR. MOSKVA
1952 V20 P56-75 60-10106 <+*> O

62-13292 <*=>
1952 V20 P103-107 62-15795 <=*>
1952 V22 N1 P47-50 R.714 <*>
1954 V31 P168-189 60-16921 <*+>

TRUDY AKADEMII NAUK BSSR
1939 P135-155 61-16835 <*>

TRUDY AKADEMII NAUK LITOVSKOI SSR.
SEE LIETUVOS TSR MOKSLU AKADEMIJOS DARBAI AND CORRESPONDING SERIES FOR LATER DATES

TRUDY AKADEMII NAUK TADZHIKSKOI SSR. STALINABAD
1956 V54 P41-56 SCL-T-306 <+*>
1956 V54 P57-77 SCL-T-300 <+*>
1956 V54 P103-114 SCL-T-309 <+*>
61-19396 <+*>
1957 V71 P39-45 SCL-T-304 <+*>
1958 V73 P233-257 65-50124 <=$>

TRUDY ALMA-ATINSKOGO BOTANICHESKOGO SADA. ALMA-ATA
1956 V3 P102-104 61-15394 <*=>

TRUDY ALTAISKOGO GORNOMETALLURGICHESKOGO NAUCHNO-ISSLEDOVATELSKOGO INSTITUTA. ALMA-ATA
1956 V3 P110-121 59-11401 <=>
1957 V4 P69-83 60-23628 <+*>
1957 V4 P160-171 59-11357 <=>
1957 V5 P158-170 60-11316 <=>
1958 V6 P157-164 60-15437 <+*>
1961 V11 P48-55 64-19356 <=$> 0
1963 V14 P66-74 RTS-2573 <NLL>
4195 <BISI>

TRUDY ARALO-KASPIISKOI KOMPLEKSNOI EKSPEDITSII
1956 V7 P197-221 61-31047 <=>

TRUDY ARKTICHESKOGO I ANTARKTICHESKOGO NAUCHNO-ISSLEDOVATELSKOGO INSTITUTA. LENINGRAD
1936 V30 P71-136 40R80R <ATS>
1936 V44 P7-18 62-32129 <=*> 0
1938 V110 P43-55 63-15594 <=*>
1959 N226 P76-98 60-00277 <*>
1959 V226 P5-18 61-19038 <+*>
1959 V226 P19-29 60-17694 <+*>
1959 V226 P30-41 61-13863 <*+>
61-23341 <*=> 0
1959 V226 P42-47 61-13855 <*+>
61-23337 <*=> 0
1959 V226 P48-60 61-13852 <*+>
1959 V226 P61-65 61-19019 <+*>
1959 V226 P66-75 61-13862 <*+>
1959 V226 P76-98 60-17698 <+*>
1959 V226 P99-108 61-13861 <*+>
1959 V226 P109-112 61-19024 <+*>
1959 V226 P113-122 60-41690 <=*>
61-19027 <+*>
1959 V226 P123-135 61-13975 <*+>
1959 V226 P136-141 61-13976 <*+>
1959 V226 P142-150 61-19023 <*=>
1959 V226 P151-161 61-13860 <*+>
1959 V228 P14-37 62-32931 <=*>
1959 V228 P68-86 62-33767 <=*>
1959 V228 P162-167 61-13854 <*+>
1959 V228 N1 P5-13 18M41R <ATS>
61-23707 <*=>
1959 V228 N1 P55-67 19M41R <ATS>
61-28830 <*=>
1959 V228 N1 P68-86 27M45R <ATS>
1959 V228 N1 P87-99 61-28827 <*=>
1959 V228 N1 P100-112 20M41R <ATS>
1959 V228 N1 P124-134 61-23704 <*=>
1959 V228 N1 P135-145 21M41R <ATS>
61-28824 <*=>
1959 V228 N1 P146-148 22M41R <ATS>
61-28825 <*=>
1959 V228 N1 P149-154 23M41R <ATS>
1959 V228 N1 P155-161 24M41R <ATS>
61-28823 <*=>
1959 V228 N1 P168-174 61-19035 <+*>
1960 V218 P200-208 C-4375 <NRC>
1960 V223 P5-20 62-25666 <=*>
1960 V223 P110-149 62-18849 <=*>
1960 V230 P17-28 63-19810 <=*$> 0
1961 V210 P106-110 64-31125 <=>
1961 V240 P4-23 65-60797 <=> 0
1961 V240 P24-32 65-61241 <=$> 0
1961 V240 P52-67 65-60616 <=$> 0
1961 V240 P68-87 65-60966 <=$> C
1961 V240 P95-146 65-60763 <=> 0
1962 V239 N2 P5-10 64-51936 <=$>
1962 V239 N2 P64-74 64-51937 <=$>
1962 V239 N2 P95-103 64-51938 <=$>
1962 V239 N2 P128-133 64-51939 <=$>
1962 V239 N2 P134-138 64-51940 <=$>
1962 V239 N2 P139-143 64-51941 <=$>
1963 V253 P122-131 65-60971 <=$> 0
1963 V253 P132-137 65-61383 <=$> 0
1963 V255 P16-46 TRANS-338(M) <NLL>
1963 V255 P108-118 TRANS-340(M) <NLL>
1963 V255 P156-168 TRANS-341(M) <NLL>
1963 V255 P169-183 TRANS-342(M) <NLL>
1963 V263 P148-231 TRANS-F.57 <NLL> 0
1964 V267 P13-18 AD-631 236 <=$>
1964 V267 P150-152 AD-626 446 <=$>
1964 V271 P5-18 TRANS-F-95 <NLL>
1964 V271 N1 P31-44 AD-630 669 <=*$>
1964 V271 N1 P100-114 65-32848 <=$*>
1965 V273 P46-63 TRANS-F-151 <NLL>

TRUDY ARKTICHESKOGO NAUCHNO-ISSLEDOVATELSKOGO INSTITUTA. LENINGRAD
1933 V1 P109-123 RT-975 <*>
1933 V10 N1 P5-37 RT-739 <*>
1936 V40 ENTIRE ISSUE R-4252 <*>
1936 V44 P7-59 RT-3229 <*>
1938 V110 P5-13 RT-1104 <*>
1938 V110 P33-38 RT-1102 <*>
1938 V110 P39-41 RT-1103 <*>
1938 V110 P43-55 R-4406 <*>
RT-4027 <*>
1938 V110 P57-81 RT-1132 <*>
1938 V110 N1 P83-100 RT-2272 <*>
1938 V110 N1 P101-108 RT-740 <*>
1946 V187 P1-442 RT-1369 <*>
1946 V187 P1-70 RT-2853 <*>

TRUDY ARMYANSKOGO NAUCHNO-ISSLEDOVATELSKOGO INSTITUTA ZHIVOTNOVODSTVA I VETERINARII. EREVAN
1956 V1 N9 P35-41 R-3503 <*>
1956 V1 N9 P93-99 61-15978 <+*>
1956 V1 N9 P107-112 61-15980 <+*>
1956 V1 N9 P113-117 61-15979 <+*>

TRUDY ASTROFIZICHESKOGO INSTITUTA. AKADEMIYA NAUK KAZAKHSKOI SSR. ALMA-ATA
1962 V3 P62-65 63-23809 <=*$>
1962 V3 P144-148 63-23587 <=*$>
1962 V3 P163-170 64-13749 <=*$>

TRUDY ASTRONOMICHESKOGO SEKTORA INSTITUTA FIZIKI
1961 V12 P167-175 63-19661 <=*> 0

TRUDY ASTRONOMICHESKOI OBSERVATORII KHARKOVSKOGO GOSUDARSTVENNOGO UNIVERSITETA. KHARKOV
1952 V2 P48-172 66-10414 <*>

TRUDY AVIATSIONNOGO INSTITUTA. KHARKOV
1954 N15 P133-148 R-2536 <*>

TRUDY. AVIATSIONNYI INSTITUT. UFA
1957 V3 P41-62 61-27412 <=*> 0
1957 V3 P169-180 61-27412 <=*> 0

TRUDY AZERBAIDZHANSKOGO FILIALA AKADEMII NAUK SSSR
1936 P59-70 61-18096 <*>

TRUDY. AZERBAIDZHANSKII GOSUDARSTVENNYI UNIVERSITET. BAKU. SERIIA KHIMICHESKAYA
1959 P97-105 COR80R <ATS>

TRUDY. AZERBAIDZHANSKOGO NAUCHNO-ISSLEDOVATELSKOGO INSTITUTA PO DOBYCHE NEFTI. BAKU
1956 N4 P244-270 IGR V6 N3 P405 <AGI>
1956 V4 P244-270 IGR,V6,N3,P405-419 <AGI>

TRUDY AZERBAIDZHANSKOGO NAUCHNO-ISSLEDOVATELSKOGO VETERINARNAYA OPYTNYA STANTSIYA. BAKU
1957 V6 P79-85 64-13603 <=>
1962 V13 P5-15 64-15320 <=*$>

TRUDY AZOVSKOGO NAUCHNO-ISSLEDOVATELSKOGO INSTITUTA RYBNOGO KHOZYAISTVA. ROSTOV-ON-THE-DON
1963 N6 P111-117 RTS-3112 <NLL>

TRUDY BALTIISKOGO NAUCHNO-ISSLEDOVATELSKOGO INSTITUTA MORSKOGO RYBNOGO KHOZYAISTVA I OKEANOGRAFII. KALININGRAD
1960 V6 P124 63-11108 <=>

TRUDY BARABINSKOGO OTDELENIYA GOSUDARSTVENNOGO NAUCHNO-ISSLEDOVATELSKOGO INSTITUTA OZERNOGO I RECHNOGO RYBNOGO KHOZYAISTVA
1953 V6 N2 P31-51 65-60479 <=$>

TRUDY BASHKIRSKOGO NAUCHNO-ISSLEDOVATELSKOGO INSTITUTA PO PERERABOTKE NEFTI
1959 N1 P5-19 60-13530 <*=>
1962 V5 P238-250 63-15919 <=*>

TRUDY BELORUSSKOGO NAUCHNO-ISSLEDOVATELSKOGO INSTITUTA MELIORATSII I VODNOGO KHOZYAISTVA
1958 V8 P35-54 63-11018 <=>
1958 V8 P142-154 63-11019 <=>
1958 V8 P313-327 63-11020 <=>

TRUDY BIOGEOKHIMICHESKOI LABORATORII. AKADEMIYA NAUK SSSR. LENINGRAD
1939 P90-111 R-4662 <*>
1939 P201-204 R-3488 <*>
1939 V5 P201-204 64-19437 <=$>
1944 V6 P98-105 R-4670 <*>
1944 V7 P130-135 R-4678 <*>
1954 V10 P98-115 RJ-369 <ATS>

TRUDY BIOLOGICHESKOGO INSTITUTA. SIBIRSKOE OTDELENIE, AKADEMIYA NAUK SSSR. NOVOSIBIRSK
1961 V7 P7-21 64-23505 <=$>
1961 V7 P103-107 64-13736 <=*$>
1962 V8 P33-134 63-21257 <=>

TRUDY BIOLOGICHESKOI STANTSII 'BOROK'. MOSKVA
1955 V2 P5-23 C-4098 <NRCC>

TRUDY BOTANICHESKOGO INSTITUTA. AKADEMIYA NAUK SSSR. SERIIA 3. GEBOTANIKA. LENINGRAD, MOSKVA
1952 V8 P40-69 63-23127 <=*$>
1963 V15 P94-105 65-60404 <=$>

TRUDY BOTANICHESKOGO INSTITUTA. AKADEMIYA NAUK SSSR. SERIIA 4. EKSPERIMENTALNOGO BOTANIKA
1933 S4 V1 P205-222 RT-1857 <*>
1953 S4 V9 P123-131 R-5260 <*>
1958 N12 P268-269 AEC-TR-4818 <=*>
1958 N12 P290-298 64-11006 <=>
1960 N14 P54-72 63-24624 <=*$>
1960 N14 P73-88 64-13725 <=*$>
1960 N14 P258-282 63-15249 <=*>
1962 N15 P193- AEC-TR-5488 <=*>

TRUDY BOTANICHESKOGO INSTITUTA. AKADEMIYA NAUK SSSR. SERIYA 5. RASLITELNOE SYRE. LENINGRAD, MOSCOW
1949 N2 P292-325 61-15722 <+*>
1949 N2 P470-478 61-15722 <+*>
1955 N5 P85-89 61-15746 <+*>

TRUDY BOTANICHESKOGO INSTITUTA. AKADEMIYA NAUK SSSR. SERIIA 6. INTRODUKTSIYA RASTENII I ZELENOE STROITELORSTVO. LENINGRAD, MOSKVA
1955 N4 P310-316 60-51013 <=>

TRUDY BOTANICHESKOGO SADA. TASHKENT
1954 V4 P136-155 65-50042 <=>

TRUDY BYURO KOLTSEVANIYA. MOSKVA
1955 N8 P3-21 60-21160 <=>
1955 N8 P179-184 60-21160 <=>
1957 V9 P5-45 62-11001 <=>
1957 V9 P144-207 62-11001 <=>
1957 V9 P215-222 61-13963 <*+>

TRUDY CHKALOVSKOGO GOSUDARSTVENNOGO MEDITSINSKOGO INSTITUTA.
1950 V2 P91-98 64-31531 <=>
1950 V2 P103-107 64-31531 <=>

TRUDY DALNEVOSTOCHNOGO FILIALA, AKADEMII NAUK SSSR
1956 V3 P265-268 64-15725 <=*$>
1956 V3 N6 P265-268 64-00286 <*>
1961 V5 P46-98 01Q70R <ATS>

TRUDY DALNEVOSTOCHNOGO NAUCHNO-ISSLEDOVATELSKOGO GIDROMETEOROLOGICHESKOGO INSTITUTA VLADIVOSTOK
1958 N6 P30-43 60-23762 <+*>
1959 N7 P33-45 60-31784 <=>
1961 N12 P106-110 63-11504 <=>
1964 N17 P64-68 65-31703 <=$>
1965 N20 P110-141 66-34809 <=$>

TRUDY DNEPROPETROVSKOGO INSTITUTA INZHENEROV-ZHELEZNODOROZHNOGO TRANSPORTA. DNEPROPETROVSK
1954 N24 P115-135 90J18R <ATS>
1958 V26 P322-335 62-11773 <=>

TRUDY DNEPROPETROVSKOGO KHIMIKO-TEKHNOLOGICHESKOGO INSTITUTA
1955 V4 P25-34 62-24457 <=*>
1958 V6 P99-109 62-24466 <=*>

TRUDY DREIFUYUSHCHEI STANTSII 'SEVERNYI POLYUS'
1940 V1 P209-245 RT-1709 <*>

TRUDY ELBRUSSKOGO VYSOKOGORNOGO EKSPEDITSIYA
SEE TRUDY VYSOKOGORNOGO GEOFIZICHESKOGO INSTITUTA ELBRUSSKOGO VYSOKOGORNOGO EKSPEDITSKIYA

TRUDY ENERGETICHESKOGO INSTITUTA IM. I. G. ESMANA. AKADEMIYA NAUK AZERBAIDZHANSKOI SSR. BAKU
1962 V15 P180-194 64-13930 <=*$>

TRUDY. FARMATSEUTICHESKII INSTITUT. TASHKENT
1957 V1 P359-364 65-60494 <=$>

*TRUDY FIZICHESKOGO INSTITUTA IMENI P. N. LEBEDEVA. AKADEMIYA NAUK SSR. MOSKVA
1950 V5 P339-386 AEC-TR-4600 <*=>
1950 V5 P340-386 R-3741 <*>
1955 V6 P199-268 63-18469 <=*>
1956 V8 P13-64 60-21603 <=>
1960 V12 P3-53 AEC-TR-5067 <=*>
1960 V13 P3-173 62-24725 <=*>
1961 V15 P178-229 AEC-TR-5377 <=>
1962 V17 P3-41 63-13460 <=*>
1962 V18 P105-116 64-15917 <=*$>
1962 V27 P42-83 AIAAJ V2 N1 P193 <AIAA>
1964 V28 P104-115 65-31568 <=$>

TRUDY PO FIZIKE POLUPROVODNIKOV
1962 N1 P15-18 AD-636 602 <=$>

TRUDY. FIZIKO-TEKHNICHESKII INSTITUT. AKADEMIYA NAUK UZBEKSKOI SSR. TASHKENT
1955 V6 P43- 61-23826 <=*>

TRUDY FIZIOLOGICHESKOGO INSTITUTA IM. I. P. PAVLOVA. AKADEMIYA NAUK SSSR. MOSCOW, LENINGRAD
1949 V4 P113-116 62-15318 <=*>

TRUDY FIZIOLOGICHESKIKH LABORATORII IMENI I. P. PAVLOVA. MOSKVA, LENINGRAD

1941 V10 P51-155 RT-4452 <*> 0

TRUDY GELMINTOLOGICHESKOI LABORATORII. AKADEMIYA NAUK SSSR. MOSKVA, LENINGRAD
1948 V1 P198-201 T-2583 <INSD>
1951 V5 P318-322 RTS-2748 <NLL>
1952 V6 P72-73 C-2710 <NRC>
1952 V6 P235-250 C-2715 <NRC>
1952 V6 P251-258 C-2680 <NRC>
1954 V7 P290-291 C-2726 <NRC>
1954 V7 P392-393 RTS 2749 <NLL>
1961 V11 P340-352 RTS 2867 <NLL>
1962 V12 P5-8 RTS-3325 <NLL>

*TRUDY GEOFIZICHESKOGO INSTITUTA. AKADEMIYA NAUK SSSR. MOSKVA, LENINGRAD
1948 N1 P71-73 61-28004 <*=>
62-00569 <*>
1949 N4 ENTIRE ISSUE 65-64443 <=$>
1949 N4 P3-102 RT-4187 <*>
1949 N5 P76-133 62-33416 <=*>
1949 N5 P94-99 64-13340 <=*$>
1949 N7 P1-93 62-33416 <=*>
1950 N12 P3-21 62-33416 <=*>
1953 N19 P146- R-638 <*>
1953 N20 P3-19 60-15506 <+*>
1953 N20 P20-36 62-16437 <=*>
1954 N22 P3-18 59-14149 <+*>
1954 N22 P50-58 59-14098 <+*>
1954 N22 P102-110 23L35R <ATS>
1954 N22 P111-116 60-13442 <+*>
1954 N23 P65-91 RT-4549 <*>
1954 N23 P92-105 RT-4548 <*>
1954 N24 P1-88 59-16082 <+*>
1954 N24 P163-187 R-3899 <*>
59-15097 <+*>
1954 N25 P162-180 24L35R <ATS>
61-19183 <+*>
65-61337 <=$> 0
1954 N25 P181-191 25L35R <ATS>
1955 N26 P153-159 64-13359 <=*$>
1955 N26 P208-210 59-22217 <+*>
1955 N27 P1-168 RT-4274 <*>
1955 N30 P240-271 60-19013 <+*>
1955 N30 P272-277 59-16547 <+*>
1956 N34 P5-73 63-23603 <=*$>
1956 N34 P243-268 61-19233 <+*> 0
1957 N16 P39-51 60-14027 <+*>
1957 N38 P1-90 <CB>
1957 N40 P1-67 <CB>
59-11307 <=>

TRUDY GEOLOGICHESKAGO I MINERALOGICHESKAGO MUZEYA IMENI PETRA VELIKAGO IMPERATORSKOI AKADEMII NAUK
1923 V3 N1 P1-24 33K22R <ATS>

TRUDY GEOLOGIYA INSTITUTA. AKADEMIIA NAUK ESTONSKOI SSR. TALLININ
1959 V4 ENTIRE ISSUE 61-28454 <=*> 0

TRUDY GEOLOGICHESKOGO INSTITUTA. AKADEMIYA NAUK SSSR. MOSCOW, LENINGRAD
1956 N1 P5-49 PT2 IGR V4 N11 P1175 <AGI>
1956 V1 P5-102 PT.2 62-23845 <=*$> 0
1957 N6 P13-38 IGR V4 N3 P310 <AGI>
1959 N23 P5-17 IGR V4 N4 P379 <AGI>
1959 V25 P1-62 79N57R <ATS>
1959 V32 P7-21 61-27735 <=*> 0
1959 V32 P45-78 61-27736 <=*> 0
1959 V32 P97-114 61-27737 <=*>
1959 V32 P122-137 61-27738 <=*> 0
1961 V48 P1-56 65-60650 <=$> 0
1963 N80 P5-33 66-10310 <*>
1963 N80 P128-151 66-10316 <*>
1963 V50 N88 P245 RTS-2572 <NLL>

TRUDY GEOLOGICHESKOGO MUZEYA AKADEMII NAUK SSR. LENINGRAD
1931 V1 P1-14 62-33414 <=*>
1931 V1 P15-60 64-19740 <=$>

TRUDY GLAVNOGO BOTANICHESKOGO SADA. PUSHKINSKOYE, MOSCOW
1959 V6 P7-48 65-50053 <=>

TRUDY GLAVNOGO BOTANICHESKOGO SADA. AKADEMIYA NAUK SSSR. LENINGRAD-PUSHKIN
1960 V7 P32-54 63-23261 <=*>
1960 V7 P55-66 62-32492 <=*>
1961 V8 P113-140 62-32493 <=*>
1961 V8 P162-195 64-11080 <=>

TRUDY GLAVNOI GEOFIZICHESKOI OBSERVATORII IMENI A. I. VOEIKOVA. LENINGRAD, MOSCOW
1940 V31 N8 P42-52 59-22434 <+*>
1950 V19 P122-132 TT.395 <NRC>
1950 V19 N81 P122-132 RT-1675 <*>
1950 SNS V19 P14-24 TT.282 <NRC>
1951 N25 P20-26 RT-1224 <*>
1951 N87 P84-87 RT-1201 <*>
1955 N55 P3-12 60-41103 <=>
1955 N116 P78-80 88K21R <ATS>
1956 N56 P69-78 SCL-T-266 <+*>
1956 N58 P8-16 61-11152 <=>
1956 N58 P23-30 61-11152 <=>
1956 N59 P3-8 63-11149 <=>
1956 N60 P32-39 60-23758 <+*>
1956 N60 P51-52 61-13817 <=*>
1956 N60 P53-59 61-23087 <*=>
1956 N60 P60-66 60-15361 <+*>
1956 N60 P67-79 60-15364 <+*>
1956 N60 P86-91 61-13816 <*+>
1956 N63 P32- 62-32360 <=*>
1956 N63 P138-167 61-11874 <=>
1956 N63 P168-171 SCL-T-247 <+*>
1956 N119 P44-49 14K22R <ATS>
1956 N119 P88-100 13K22R <ATS>
1956 V57 N119 P101-107 12K22R <ATS>
1956 V59 N121 P37-39 AEC-TR-3527 <+*>
1956 V119 N57 P3-18 37K21R <ATS>
1956 V119 N57 P19-35 15K22R <ATS>
1956 V119 N57 P111-112 11K22R <ATS>
1957 N67 P3-32 32M38R <ATS>
1957 N67 P33-58 58L34R <ATS>
1957 N67 P59-103 68L36R <ATS>
1957 N67 P104-113 59L34R <ATS>
1957 N67 P114-120 60L34R <ATS>
1957 N67 P121-128 31M38R <ATS>
1957 N67 P129-130 29M38R <ATS>
1957 N67 P131-143 30M38R <ATS>
1957 N67 P144-152 61L34R <ATS>
1957 N68 P164-180 PB 141 228T <=>
1957 N69 P36-40 61-13562 <+*> 0
1957 N69 P41-44 61-13563 <+*> 0
1957 N70 P60-91 62-25866 <=*>
1957 N71 P112-128 63-11144 <=>
1957 N71 P229-238 61-28010 <*=>
1957 N72 P33-38 60-23874 <+*>
1957 N72 P110-138 60-23874 <+*>
1957 N72 P110-117 61-14921 <=*>
1957 N72 P118-126 61-14923 <=*>
1957 N72 P127-133 61-14922 <=*>
1957 N74 P3-21 60-15362 <+*>
1957 N74 P41-60 61-13500 <*+>
1957 N74 P71-102 59-11055 <=>
1957 V74 N74 P41-60 63-11697 <=>
1958 N73 P50-53 64-13666 <=*$>
1958 N77 P3-6 61-13315 <+*> 0
1958 N77 P7-19 60-23064 <+*>
1958 N77 P20-33 61-28372 <*=>
1958 N77 P34-42 61-23084 <*=>
1958 N77 P57-64 61-28765 <*=>
1958 N77 P65-71 61-23083 <*=>
1958 N77 P72-75 61-23086 <*=>
1958 N77 P79-83 61-23081 <*=>
1958 N77 P84-94 61-23087 <*=>
1958 N77 P95-98 60-15526 <+*>
1958 N77 P99-103 60-19869 <+*>
1958 N82 P26-35 61-23297 <*=>
1958 N82 P36-40 AD-610 115 <=$>

1958 N82 P41-44 74M37R <ATS>
1958 N82 P68- 61-23298 <*=> 0
1958 N83 P20-24 TRANS-317(M) <NLL>
1959 N80 P88-96 60-23761 <+*>
1959 N87 P10-31 63-19915 <=*$> 0
1959 N87 P66-85 63-19803 <*=> 0
1959 N92 P27-49 M.5392 <NLL> 0
1959 N92 P102-126 61-23077 <*=>
1959 N93 P3-20 50N52R <ATS>
1959 N93 P70-80 62-24970 <=*>
1959 N93 P88-94 99N57R <ATS>
1959 N96 P84-100 62-24801 <=*>
1959 N96 P101-106 62-24803 <=*>
1959 N98 P3-16 64-11969 <=> M
1959 N99 P105-111 63-19841 <*=> 0
1959 N101 ENTIRE ISSUE 62-33065 <=>
1959 V80 P51-69 63-11698 <=>
1959 V87 P40-45 63-19804 <*=> 0
1959 V89 P8-29 63-19917 <*=> 0
1959 V89 P53-59 63-19809 <*=> 0
1960 N88 P3-15 65-61474 <=> 0
1960 N90 P11-26 61-31122 <=>
1960 N90 P27-42 61-31113 <=>
62-24895 <=*> 0
1960 N90 P43-62 61-31123 <=>
1960 N90 P63-78 61-31124 <=>
62-24893 <=*> 0
1960 N91 P56-59 62-24806 <=*>
1960 N91 P62-70 62-24808 <=*>
1960 N94 P3-7 62-33223 <=*>
1960 N94 P8-28 62-24802 <=*>
1960 N94 P29-32 61-23079 <*=>
1960 N94 P33-38 62-00729 <*>
62-24804 <=*>
1960 N94 P39-41 62-32901 <=*>
1960 N94 P138-144 62-32902 <=*>
1960 N94 P145-148 62-24807 <=*>
1960 N94 P149-155 62-24805 <=*>
1960 N94 P156-162 AD-625 913 <=*$>
1960 N97 P3-4 54N53R <ATS>
1960 N97 P5-15 48N52R <ATS>
1960 N97 P16-33 49N52R <ATS>
1960 N97 P48-50 39N48R <ATS>
1960 N100 P53-57 63-19744 <=*> 0
1960 N102 P3-20 63-19813 <=*> 0
1960 N102 P50-57 82M46R <ATS>
1960 N102 P58-62 47N52R <ATS>
1960 N102 P63-93 84M46R <ATS>
1960 N102 P94-103 83M46R <ATS>
1960 N103 P85-92 62-15231 <*=>
1960 N103 P93-102 65-60687 <=$> 0
1960 N103 P103-112 65-61265 <=$> 0
1960 N104 P3-13 37N52R <ATS>
1960 N104 P3-95 62-13579 <=*>
1960 N104 P14-24 38N52R <ATS>
1960 N104 P25-38 39N52R <ATS>
1960 N104 P39-45 40N52R <ATS>
1960 N104 P46-52 41N52R <ATS>
1960 N104 P53-67 42N52R <ATS>
1960 N104 P68-74 43N52R <ATS>
1960 N104 P75-78 44N52R <ATS>
1960 N104 P85-94 46N52R <ATS>
1960 N106 P10-18 M.5082 <NLL>
1960 N106 P19-36 65-61154 <=> 0
1960 N106 P62-68 65-61383 <=$>
1960 N108 P59-63 62-13285 <=*>
1960 N114 P90-103 63-19750 <*=>
1960 N115 P77-94 62-13286 <=*>
1960 V88 P96-110 63-19909 <=*> 0
1960 V90 P95-115 63-19843 <=*> 0
1960 V100 P58-64 63-19811 <=*> 0
1960 V114 P104-113 63-19749 <*=> 0
1961 N109 P38-52 62-11706 <=>
1961 N109 P61-75 63-19956 <*=>
1961 N109 P76-89 64-00469 <*>
64-11827 <=>
1961 N111 P71-76 64-15958 <=*$>
1961 N111 P81-98 62-25471 <=*>
1961 N111 P136-178 62-25471 <=*>
1961 N118 P26-41 64-16287 <*>
1961 N120 P3-14 02N58R <ATS>
1961 N120 P27-36 03N58R <ATS>
1961 N120 P37-44 04N58R <ATS>
1961 N120 P45-51 05N58R <ATS>
1961 N120 P52-59 06N58R <ATS>
1961 N121 P109-124 AD-624 759 <=*$> M
1961 N121 P125-140 63-15371 <=>
1961 N122 P48-60 63-19848 <=*>
66-11206 <*>
1961 N122 P68-74 63-19031 <=*>
1962 N125 P54-57 63-19694 <=*>
1962 N125 P62-75 M.5438 <NLL> 0
1962 N126 P3-7 12P65R <ATS>
1962 N126 P3-89 63-21573 <=>
1962 N126 P8-9 13P65R <ATS>
1962 N126 P10-15 14P65R <ATS>
1962 N126 P16-21 15P65R <ATS>
1962 N126 P22-24 16P65R <ATS>
1962 N126 P33-39 18P65R <ATS>
1962 N126 P40-56 19P65R <ATS>
1962 N126 P57-61 20P65R <ATS>
1962 N126 P62-69 21P65R <ATS>
1962 N126 P70-78 22P65R <ATS>
1962 N126 P79-89 23P65R <ATS>
1962 N127 P3-13 AD-628 972 <=*$>
1962 N127 P26-34 AD-630 552 <=$>
1962 N127 P35-47 AD-631 965 <=$>
1962 N127 P57-68 AD-630 671 <=$>
1962 N127 P82-87 65-60685 <=$> 0
1962 N129 P31-39 65-61477 <=> 0
1962 N129 P101-110 65-61513 <=$>
1962 N129 P111-117 65-61515 <=$>
1962 N129 P118-121 65-61196 <=> 0
1962 N130 P3-10 SHSP N1 1962 P71 <AGU>
1962 N130 P11-28 SHSP N1 1962 P77 <AGU>
1962 N130 P51-64 SHSP N1 1962 P95 <AGU>
1962 N131 P3-9 M.5439 <NLL> 0
1962 N131 P29-36 SHSP N4 1962 P426 <AGU>
1962 N131 P37-44 SHSP N4 1962 P434 <AGU>
1962 N131 P45-51 M.5434 <NLL> 0
63-13313 <=*>
1962 N133 P3-25 M.5415 <NLL> 0
1962 N134 P38-74 65-61470 <=> 0
1962 N134 P102-112 65-61469 <=> C
1962 N134 P119-122 65-60965 <=$> 0
1962 N135 P55-59 65-61194 <=>
1962 N135 P120-128 63-23622 <=*$>
1962 N135 P147-157 65-60798 <=> 0
1962 N136 P83-95 65-60783 <=>
1962 V126 P25-32 65-61155 <=> 0
1962 V134 P38-74 63-21857 <=>
1962 V134 P102-112 63-21857 <=>
1962 V135 P135-146 63-23781 <=*> 0
1963 N95 P3-12 SOV.HYDROL.1963 P315 <AGU>
1963 N95 P13-18 SOV.HYDROL.1963 P322 <AGU>
1963 N95 P60-65 SOV.HYDROL.1963 P326 <AGU>
1963 N112 P107-115 SOU.HYDROL.1963 P423 <AGU>
1963 N112 P191-198 65-60717 <=> 0
1963 N138 P3-55 64-51535 <=*$>
1963 N138 P73-81 65-60613 <=$> 0
1963 N139 P3-15 AD-626 065 <=*$>
1963 N139 P43-60 65-30772 <=$>
1963 N139 P93-107 SOV.HYDROL.1963 P597 <AGU>
1963 N140 P65-70 AD-637 466 <=$>
1963 N142 P3-12 65-60637 <=$> 0
1963 N143 P147 AD-628 789 <=*$> M
1963 N144 P202-208 SOV.HYDROL.1963 P612 <AGU>
1963 N145 P3-12 64-51800 <=$>
1963 N145 P20-22 64-51802 <=$>
1963 N145 P23-29 64-51803 <=$>
1963 N145 P35-48 64-51801 <=$>
1963 N145 P36-48 TRANS-250(M) <NLL>
1963 N148 P59-67 65-61385 <=$> C
1963 N148 P77-89 AD-610 113 <=$>
65-61387 <=$>
1963 N148 P122-132 65-60614 <=$> 0
1963 N149 P29-34 AD-611 014 <=$>
1963 N149 P38-47 SOU.HYDROL.1963 P430 <AGU>
1963 V148 P38-58 65-60631 <=$>
1964 N150 P14-20 65-30099 <=$>

1964 N150 P21-35 65-30100 <=$>
1964 N150 P63-68 65-30101 <=$>
1964 N150 P78-84 65-30102 <=$>
1964 N150 P85-98 65-30103 <=$>
1964 N150 P152-160 65-30104 <=$>
65-30104 <=$> O
1964 N151 P17-40 64-51697 <=$>
1964 N151 P60-76 64-51697 <=$>
1964 N151 P77-80 TRANS-F-97 <NLL>
1964 N152 P3-15 N65-20986 <=$>
1964 N152 P16-30 N65-20987 <=$>
1964 N152 P68-80 N65-21434 <=$>
1964 N152 P81-89 N65-21435 <=$>
1964 N152 P90-95 N65-21227 <=$>
1964 N152 P96-109 N65-21228 <=$>
1964 N152 P121-141 N65-21649 <=$>
1964 N152 P142-159 64-51823 <=$>
1964 N152 P160-167 N65-20988 <=$>
1964 N152 P168-175 64-51823 <=$>
1964 N152 P176-185 N65-21230 <=$>
1964 N152 P186-191 N65-21436 <=$>
1964 N152 P192-211 N65-20989 <=$>
1964 N153 P11-17 AD-628 308 <=*$>
TRANS-F-90 <NLL>
1964 N153 P18-23 TRANS-F.91 <NLL>
1964 N153 P56-74 TRANS-F-92 <NLL>
1964 N153 P75-77 AD-628 308 <=*$>
1964 N153 P89-92 TRANS-F-93 <NLL>
1964 N153 P93-101 TRANS-F.113 <NLL>
1964 N153 P102-110 TRANS-F.114 <NLL> O
1964 N154 P3-10 65-30060 <=$>
1964 N154 P20 M.5782 <NLL>
1964 N154 P20-29 65-30052 <=$>
1964 N154 P46-57 65-30059 <=$>
1964 N154 P65-77 65-30058 <=$>
1964 N154 P78-84 65-30057 <=$>
1964 N154 P85-89 65-30056 <=$>
1964 N154 P99-104 65-30055 <=$>
1964 N154 P105-108 AD-630 668 <=$>
1964 N156 P31-45 65-30189 <=$>
1964 N156 P60-82 65-30190 <=$>
1964 N156 P101-117 65-30191 <=$>
1964 N157 P22-30 64-51787 <=$>
1964 N157 P31-38 64-51788 <=$>
1964 N157 P48-53 AD-623 191 <=*$>
64-51785 <=$>
1964 N157 P54-58 64-51781 <=$>
1964 N157 P70-72 N65-21231 <=$>
1964 N157 P73-75 N65-20884 <=$>
1964 N158 P3-21 TRANS-F.82. <NLL>
1964 N158 P3-40 65-30148 <=$>
1964 N158 P22-31 TRANS-F.83 <NLL>
1964 N158 P41-45 TRANS-F-84 <NLL>
1964 N158 P46-101 65-30148 <=$>
1964 N158 P46-49 65-30148 <=$>
1964 N158 P50-55 TRANS-F-86 <NLL>
1964 N158 P56-68 TRANS-F.87. <NLL>
1964 N158 P77-83 TRANS-F.88. <NLL>
1964 N158 P84-87 TRANS-F-89 <NLL>
1964 N158 P109-119 65-30148 <=$>
1964 N158 P123-135 65-30148 <=$>
1964 N159 P5-17 AD-645 788 <=>
1964 N159 P33-50 AD-645 788 <=>
1964 N159 P35-39 AD-628 967 <=*$>
1964 N159 P48-58 TRANS-F-106 <NLL>
1964 N159 P59-64 TRANS-F-107 <NLL>
1964 N159 P65-69 AD-629 010 <=*$>
1964 N159 P79-84 TRANS-F-108 <NLL>
1964 N159 P85-88 AD-628 971 <=*$>
1964 N159 P111-194 AD-645 788 <=>
1964 N159 P273-282 AD-645 788 <=>
1964 N159 P289-300 AD-645 788 <=>
1964 N160 P138-143 TRANS-F-112 <NLL>
1964 N161 P46-59 65-30090 <=$>
1964 N161 P60-63 TRANS-F.65 <NLL>
1964 N163 P76-86 TRANS-F-102 <NLL>
1964 N163 P93-103 TRANS-F-104 <NLL>
1964 N164 P3-20 TRANS-F.76 <NLL>
1964 N164 P21-28 TRANS-F-77 <NLL>
1964 N164 P29-42 TRANS-F-78 <NLL>
1964 N166 P282-294 TRANS-F.56. <NLL>
1964 V152 P110-125 N65-21229 <=$>
1964 V152 P212-221 N65-20883 <=$>
1965 N172 P3-22 66-31158 <=$>
1965 N172 P35-41 66-31158 <=$>
1965 N172 P58-73 66-31158 <=$>
1965 N172 P79-85 66-31158 <=$>
1965 N172 P183-212 66-31158 <=$>
1965 N173 P3-8 AD-630 676 <=$>
1965 N173 P9-18 AD-631 968 <=$>
1965 N173 P19-25 AD-629 008 <=*$>
1965 N173 P26-33 AD-630 675 <=$>
1965 N173 P63-70 AD-630 674 <=$>
1965 N173 P71-75 AD-630 554 <=$>
1965 N173 P76-90 AD-631 967 <=$>
1965 N175 P5-23 TRANS-F.140 <NLL>
1965 N175 P24-30 TRANS-F.141 <NLL>
1965 N175 P31-58 TRANS-F-142 <NLL>
1965 N175 P59-66 TRANS-F-143 <NLL>
1965 N176 P60-68 TRANS-F.128 <NLL>
1965 N177 P42-45 AD-631 246 <=$>
1965 N177 P55-58 AD-631 246 <=$>
1965 N177 P59-63 AD-631 246 <=$>
1965 N177 P64-65 AD-631 246 <=$>
1966 N195 P133-137 SHSP N5 1966 P543 <AGU>

TRUDY GORNO-GEOLOGICHESKOGO INSTITUTA. URALSKII FILIAL, AKADEMIYA NAUK SSSR. MOSKVA
1955 N23 P3-8 63-13867 <=*>
1959 N42 P193-201 64-15179 <=*$>
1960 N54 P125-130 UCRL TRANS-820(L) <=*>
1962 N58 P129-153 IGR V8 N2 P234 <AGI>

TRUDY GOSUDARSTVENNOGO ASTRONOMICHESKOGO INSTITUTA IM. P. K. SHTERNBERGA. MOSCOW
1936 V7 N1 P5-125 62-18953 <=*>
1939 V9 N2 P5-45 65-17088 <*>
1940 V14 N1 P7-41 61-23881 <*=>
1940 V14 N1 P69-71 63-10322 <=*$>
1949 V16 P47-70 RT-3979 <*>
63-14917 <=*> O
1949 V16 P71-92 RT-3980 <*>
1952 V21 P3-18 ARSJ V31 N3 P399 <AIAA>
1954 V24 P3-9 ARSJ V31 N3 P387 <AIAA>
1954 V24 P10-16 ARSJ V31 N3 P390 <AIAA>
1954 V24 P17-39 N65-24105 <=$>

TRUDY GOSUDARSTVENNOGO GIDROLOGICHESKOGO INSTITUTA. LENINGRAD
1937 N5 P3-15 RT-1477 <*>
1956 V54 P5-74 63-11015 <=>
1956 V54 P80-91 63-11147 <=>
1956 V54 P119-125 60-21881 <=>
1957 V62 P3-23 59-18556 <+*>
1959 V69 P43-69 62-32147 <=*>
1960 V73 P90-140 65-60802 <=>
1961 V91 P5-13 AD-608 948 <=>
1962 N82 P3-33 SHSP N2 1962 P131 <AGU>
1962 N82 P76-101 SHSP N2 1962 P159 <AGU>
1962 N82 P102-114 SHSP N2 1962 P182 <AGU>
1962 N85 P3-16 SHSP N3 1962 P227 <AGU>
1962 N85 P37-51 SHSP N3 1962 P238 <AGU>
1962 N85 P87-107 SHSP N4 1962 P345 <AGU>
1962 N87 P3-26 SHSP N5 1962 P469 <AGU>
1962 N87 P27-45 SHSP N5 1962 P488 <AGU>
1962 N87 P46-54 SHSP N6 1962 P581 <AGU>
1962 N94 P3-21 SHSP N3 1962 P249 <AGU>
1962 N94 P22-86 SHSP N3 1962 P267 <AGU>
1962 N94 P115-128 SHSP N3 1962 P325 <AGU>
1962 N95 P101-155 SHSP N4 1962 P362 <AGU>
1962 N95 P156-172 SHSP N4 1962 P411 <AGU>
1962 N96 P99-108 SHSP N5 1962 P503 <AGU>
1962 N96 P109-112 SHSP N5 1962 P512 <AGU>
1962 N96 P123-130 SHSP N5 1962 P515 <AGU>
1962 N97 P204-242 SHSP N5 1962 P521 <AGU>
1962 N98 P56-98 SHSP N6 1962 P588 <AGU>
1962 N98 P247-262 SHSP N6 1962 P624 <AGU>
1962 N99 P261-273 SHSP N6 1962 P637 <AGU>
1962 V83 P60-67 64-41331 <=>
1962 V85 P52-68 64-41146 <=>

1963 V100 P88-95 SOV.HYDROL.1963 P330 <AGU>
1963 V100 P96-121 SOV.HYDROL.1963 P336 <AGU>
1963 V100 P122-135 SOV.HYDROL.1963 P356 <AGU>
1963 V101 P5-18 SOU.HYDROL.1963 P116 <AGU>
1963 V101 P19-43 SOU.HYDROL.1963 P105 <AGU>
1963 V101 P44-50 SOU.HYDROL.1963 P134 <AGU>
1963 V101 P51-55 SOU.HYDROL.1963 P140 <AGU>
1963 V102 P209-226 SOU.HYDROL.1963 P211 <AGU>
1963 V103 P41-56 SOU.HYDROL.1963 P144 <AGU>
1963 V104 P37-59 SOU.HYDROL.1963 P156 <AGU>
1963 V105 P45-79 SOU.HYDROL.1963 P226 <AGU>
1963 V106 P146-173 SOV.HYDROL.1963 P366 <AGU>
1963 V106 P182-196 SOV.HYDROL.1963 P388 <AGU>
1963 V107 P112-135 SOV.HYDROL.1963 P537 <AGU>
1963 V107 P136-151 SOV.HYDROL.1963 P555 <AGU>
1963 V108 P89-114 SOU.HYDROL.1963 P253 <AGU>
1966 N132 P18-45 SHSP N1 1966 P1 <AGU>
1966 N132 P57-67 SHSP N1 1966 P26 <AGU>
1966 N132 P90-109 SHSP N1 1966 P47 <AGU>
1966 N132 P110-125 SHSP N1 1966 P63 <AGU>
1966 N132 P126-138 SHSP N1 1966 P75 <AGU>
1966 N133 P22-41 SHSP N2 1966 P117 <AGU>
1966 N133 P82-89 SHSP N2 1966 P135 <AGU>
1966 N133 P148-175 SHSP N2 1966 P141 <AGU>
1966 N134 P80-114 SHSP N3 1966 P225 <AGU>
1966 N134 P151-164 SHSP N3 1966 P252 <AGU>
1966 N135 P3-26 SHSP N4 1966 P331 <AGU>
1966 N135 P27-51 SHSP N4 1966 P353 <AGU>
1966 N135 P121-129 SHSP N4 1966 P375 <AGU>
1966 N135 P130-149 SHSP N5 1966 P455 <AGU>
1966 N135 P150-157 SHSP N4 1966 P382 <AGU>
1966 N136 P36-44 SHSP N5 1966 P471 <AGU>
1966 N136 P45-56 SHSP N5 1966 P477 <AGU>
1966 N136 P82-91 SHSP N5 1966 P487 <AGU>
1966 N136 P129-147 SHSP N5 1966 P495 <AGU>
1966 N136 P148-159 SHSP N5 1966 P507 <AGU>
1966 N136 P214-230 SHSP N5 1966 P515 <AGU>
1966 N137 P125-180 SHSP N4 1966 P389 <AGU>

TRUDY GOSUDARSTVENNOGO INSTITUTA EKSPERIMENTALNOI VETERINARII. MOSCOW
1926 V3 P82-85 T-2209 <INSD>
1926 V3 P99-113 T2208 <INSD>
1957 V20 P12-22 59-19602 <+*>
1959 V22 P112-119 61-28448 <*=>
1959 V22 P120-125 62-13477 <*=>
1959 V22 P177-184 61-23004 <=*> 0
1959 V22 P189-194 62-13465 <*=>
1959 V22 P195-201 62-13464 <=*>
1959 V22 P213-218 61-28219 <*=>
1959 V23 P3-28 62-23564 <=*>
1959 V23 P29-48 62-23285 <=*>
1959 V23 P49-58 62-23286 <=*> 0
1959 V23 P160-174 62-23563 <=*>
1959 V23 P202-214 62-23562 <=*>
1959 V23 P215-244 62-32134 <=*> P
1959 V23 P245-256 62-32132 <=*>
1959 V23 P257-275 62-32133 <=*>
1959 V23 P282-294 62-32135 <=*>
1959 V23 P321-324 62-13471 <*=>
1959 V23 P324-330 62-32090 <=*>
1959 V23 P330-337 62-13450 <*=>
1959 V23 P337-340 62-13473 <*=>
1959 V23 P352-354 62-13472 <*=>
1959 V23 P354-355 62-32094 <=*>
1959 V23 P356-358 62-32095 <=*>
1959 V23 P358 62-13462 <*=>
1959 V23 P371 62-13463 <*=>
1959 V23 P371-373 62-32096 <=*>
1959 V23 P374-375 62-32097 <=*>
1959 V23 P384-388 62-32348 <=*>
1959 V23 P397-398 62-13459 <*=>
1959 V23 P405 62-13460 <*=>
1959 V23 P406 62-13461 <*=>
1959 V23 P458-474 62-32352 <=*>
1961 V24 P53-61 62-24526 <=*> 0
1961 V25 P78-88 62-32091 <=*>
1961 V25 P89-98 63-13234 <=*>
1961 V25 P99-102 63-13024 <=*>
1961 V25 P103-108 62-33558 <=*> 0
1961 V25 P113-121 63-13229 <=*>
1961 V25 P384-398 63-13411 <=*>
1961 V25 P399-410 63-13412 <=*>
1961 V27 P80-82 64-15639 <=*$>
1962 V26 P46-55 63-19610 <=*>
1962 V29 P177-199 65-60902 <=$> 0
1962 V29 P200-208 65-60906 <=$>
1962 V29 P209-216 65-60903 <=$> 0
65-61585 <=$> 0
1963 V28 P40-50 64-19887 <=$>
1963 V28 P124-128 64-19850 <=$>
1963 V28 P215-218 64-23583 <=$>
1964 V30 P288-293 66-61557 <=*$>

TRUDY GOSUDARSTVENNOGO INSTITUTA PRIKLADNOI KHIMII
1927 V8 P5-13 65-60760 <=>
1932 N15 P3-7 R-551 <*>
1932 N15 P8-36 RJ-396 <ATS>
1932 V16 P16-32 RT-1581 <*>
1932 V16 P106-122 R-3030 <*>
1934 V22 P7-24 64-20054 <*> 0
1934 V22 P35-40 64-20053 <*> 0
1934 V25 P9-19 64-10276 <=*$>
1934 V25 P20-39 64-10238 <=*$>
1934 V25 P39-69 64-10240 <=*$> 0
1934 V25 P69-78 63-18536 <=*>
1934 V25 P78-81 63-18137 <=*>
1934 V25 P81-84 63-18558 <=*>
1934 V25 P84-90 64-10769 <=*$>
1935 N24 P54-66 T-2395 <INSD>
1935 N24 P67-77 T-2396 <INSD>
1935 N24 P77-80 RT-1927 <*> 0
1935 V23 P100-109 63-18055 <=*>
1935 V24 P21-31 RT-1576 <*>
64-18561 <*>
1935 V24 P32-47 RT-1632 <*>
63-14123 <=*>
64-18560 <*>
1939 V30 N5 P56-73 63-16177 <=*>
1940 N32 P147-166 R-2291 <*>
1940 V32 P25-41 59-16384 <+*>
1940 V32 P95-114 59-16381 <+*>
1940 V32 P166-179 59-16382 <+*>
1960 V46 P324-328 66-14552 <*>
1962 V49 P170-182 AD-637 434 <=$>

TRUDY GOSUDARSTVENNOGO MEDITSINSKOGO INSTITUTA. VORONEZH
1942 V21 P143-154 60-13730 <+*>
1957 V29 P33-39 65-63148 <=$>
1957 V29 P77-81 65-61210 <=$>

TRUDY GOSUDARSTVENNOGO NAUCHNO-ISSLEDOVATELSKOGO INSTITUTA GORNO-KHIMICHESKOGO SYRYA. MOSCOW
1950 N1 P23-50 66-10235 <*>
1962 N7 P280-289 66-10262 <*> 0

TRUDY GOSUDARSTVENNOGO NAUCHNO-ISSLEDOVATELSKOGO INSTITUTA KHIMICHESKOI PROMYSHLENNOSTI. MOSCOW
1958 N7 P61-67 <TIB>

TRUDY GOSUDARSTVENNOGO NAUCHNO-ISSLEDOVATELSKOGO INSTITUTA PSIKHIATRII. MOSKVA
1962 V35 P1-26 63-41237 <=>

TRUDY GOSUDARSTVENNOGO NAUCHNO-ISSLEDOVATELSKOGO PROEKTNOGO INSTITUTA AZOTNOY PROMYSHELNNOSTI
1957 V7 P33-37 63-24289 <=*$>

TRUDY GOSUDARSTVENNOGO NAUCHNO-KONTROLNOGO INSTITUTA VETERINARNYKH PREPARATOV. MOSCOW
1956 V6 P194-197 59-14678 <+*>
1956 V6 P197-211 59-16145 <+*> 0
1956 V6 P256-262 59-19257 <+*>
1956 V6 P262-264 59-14677 <+*>
63-19230 <=*>
1956 V6 P264-266 59-14669 <+*>
1957 V7 P3-11 59-16146 <+*>
1957 V7 P20-25 59-16147 <+*>
1957 V7 P177-193 59-16149 <+*> 0

1957 V7 P194-200 59-22304 <+*>
1957 V7 P284-289 59-19678 <+*>
1961 V9 P50-57 63-13410 <=*>

TRUDY GOSUDARSTVENNOGO OKEANOGRAFICHESKOGO INSTITUTA. MOSCOW
1932 V1 N4 P3-66 63-11108 <=> P
1956 N2 P59-68 R-4009 <*>
1957 N36 P87-127 63-01635 <*>
1957 N38 P3-10 AD-631 014 <=$>
1957 V34 P5-72 65-63051 <=$>
1957 V36 P63-86 M.5141 <NLL>
1957 V40 P50-56 C-3276 <NRCC>
1957 V40 N52 P147-155 61-28002 <=*>
1958 N42 P105-114 AD-633 635 <=$>
1959 N37 P3-13 61-27995 <=*>
1959 V37 P14-33 61-28001 <*=>
1959 V37 P34-41 61-27998 <*=>
1959 V37 P42-52 62-00565 <*>
1959 V37 P73-78 60-23776 <*+>
1959 V37 P205-230 62-33569 <=*>
1959 V37 N49 P42-52 61-28005 <=*>
1959 V37 N49 P79-84 61-27997 <=*>
1960 N50 P54-144 63-23149 <=*>
1960 V49 P189-197 M.5126 <NLL>
1960 V50 P27-32 AD-636 218 <=$>
1960 V52 P28-48 AD-633 637 <=$>
1960 V54 P35-60 61-28371 <*=>
63-19752 <=*$>
1960 V58 P252-267 62-33571 <=*>
1961 V61 P153-158 64-19532 <=*>
1961 V63 P95-103 64-19531 <=$>
1961 V64 P103-111 62-25465 <=*>
1961 V64 P112-117 62-32481 <=*>
1961 V64 P118-127 62-24376 <=*>
1965 V86 P75-94 66-32687 <=$>
1965 V86 P112-123 66-32688 <=$>
1965 V86 P144-152 66-32689 <=$>
1965 V87 P166-173 66-32917 <=$>
1965 V87 P174-183 66-32916 <=$>

TRUDY GOSUDARSTVENNOGO OPTICHESKOGO INSTITUTA
1955 V24 N144 P1-42 59-10187 <+*>

TRUDY GOSUDARSTVENNOGO VSESOYUZNOGO NAUCHNO-ISSLEDOVATELSKOGO INSTITUTA TSEMENTNOI PROMYSHLENNOSTI. MOSCOW
1952 V5 P59-81 62-25569 <=*>

TRUDY GROZNENSKOGO NEFTYANOGO NAUCHNO-ISSLEDOVATELSKOGO INSTITUTA
1963 V12 P61-68 AD-610 290 <=$>

TRUDY INSTITUTA AGROKHIMII I POCHVOVEDENIYA. AKADEMIYA NAUK AZERBAIDZHANSKOI SSR
1955 V7 P215-223 60-21079 <=>
1956 V7 P225-232 CSIRO-3276 <CSIR>

TRUDY INSTITUTA BETONA I ZHELEZOBETONA AKADEMIYA STROITELSTVA I ARKHITEKTURY PEROVO. MOSCOW
1959 V5 P29-53 63-10200 <=*>
1959 V5 P54-77 63-20310 <=*$>
1961 V23 P74-126 63-24203 <=*$>

TRUDY INSTITUTA BIOLOGII. URALSKII FILIAL, AKADEMIYA NAUK SSSR. SVERDLOVSK
1955 N6 P119-144 60-21890 <=>
1957 N4 P59-69 61-28175 <*=>
1957 N4 P135-139 PB 141 261T <=>
1960 N12 P46-118 61-27666 <=*>
1960 N12 P189-193 63-15760 <=*>
1962 N22 P7-16 63-24139 <=*$>
1962 N28 ENTIRE ISSUE 64-15232 <=$>
1963 N30 P1-79 63-41112 <=>

TRUDY INSTITUTA BIOLOGII VODOKHRANILISHCH. AKADEMIYA NAUK SSSR. MOSKVA
1959 V1 P309-323 61-13958 <+*>
1959 V1 P324-331 63-24385 <=*$>
1960 V3 P248-265 64-13310 <=*$>
1960 V3 P266-272 63-11111 <=>
1963 V5 P10-20 C-5148 <NRC>

TRUDY INSTITUTA CHERNOI METALLURGII. AKADEMIYA NAUK UKRAINSKOI SSR. KIEV
1953 V6 P91-100 59-13616 <=> 0
1960 V12 P37-67 4622 <BISI>
1960 V12 P163-168 3962 <BISI>
1960 V13 P51-61 3950 <BISI>
1960 V13 P85-101 M-5694 <NLL>
1960 V13 P126-139 4170 <BISI>
1961 V14 P60-65 3966 <BISI>

TRUDY INSTITUTA CHISTYKH KHIMICHESKIKH REAKTIVOV
1930 V9 P161-167 63-14126 <=*>
1939 N17 P65-69 54L32R <ATS>
1941 V18 P3-10 61-18149 <*=>
1941 V18 P80-82 61-18140 <*=>

TRUDY INSTITUTA DVIGATELEI. AKADEMIIA NAUK SSSR, MOSCOW
1960 V5 P27-33 AD-627 071 <=$>
1962 N6 P138-139 AD-619 437 <=*$>
1962 V6 P61-75 AD-611 086 <=$>
1962 V6 P76-81 64-71175 <=> M
1962 V6 P110-117 64-71175 <=> M

TRUDY INSTITUTA EKSPERIMENTALNOI MEDITSINY. AKADEMIYA NAUK LATVIISKOI SSR. RIGA
1962 N4 P83-86 64-13934 <=*$>
1962 N4 P117-128 64-13934 <=*$>
1962 N8 P125-130 63-31305 <=>
1962 N8 P209-216 63-31305 <=>
1962 N8 P283-284 63-31305 <=>
1962 V28 N4 P171-243 63-31999 <=>
1962 V28 N8 P255-256 63-31999 <=>

TRUDY INSTITUTA EKSPERIMENTALNOI MEDITSINY. AKADEMIYA NAUK LITOVSKOI SSR. VILNIUS
SEE EKSPERIMENTINES MEDICINOS INSTITUTO DARBAI. VILNIUS

TRUDY INSTITUTA EKSPERIMENTALNOI METEOROLOGII
1937 V1 P102-108 59-12698 <+*>

TRUDY INSTITUTA ELEKTROKHIMII. URALSKII FILIAL, AKADEMIYA NAUK SSSR. SVERDLOVSK
1964 V5 P7-16 ORNL-TR-578 <=$>

TRUDY INSTITUTA ELEKTRONIKI, AVTOMATIKI I TELEMEKHANIKI. AKADEMIYA GRUZINSKOI USSR. TIFLIS
1960 V1 P14-15 61-28908 <=>
1960 V1 P17-23 61-19966 <=*>
1960 V1 P27 61-28908 <=>
1960 V1 P41-63 61-28908 <=>
1962 V3 P3-33 AD-623 207 <=*$>

TRUDY INSTITUTA ELEKTRONIKI I VYCHISLITELNOI TEKHNIKI. AKADEMIYA NAUK LATVIISKOI SSR. RIGA
SEE LATVIJAS PSR ZINATNU AKADEMIJA. ELEKTRONIKAS UN SKATLOSONAS TEHNIKAS INSTITUTS. TRUDY. RIGA

TRUDY INSTITUTA ENERGETIKI. AKADEMIYA NAUK BELORUSSKOI SSR. MINSK
1955 V2 P162-177 AD-614 948 <=$>
1960 V11 P31-39 64-11925 <=> M

TRUDY INSTITUTA ENERGETIKI. AKADEMIYA NAUK KAZAKHSKOI SSR. ALMA-ATA
1960 V2 P244-251 65-60369 <=$>

TRUDY INSTITUTA ENERGETIKI. AKADEMIYA NAUK LATVIISKOI SSR. RIGA.
1961 V12 P5-22 AD-612 417 <=$>
1961 V12 P23-48 AD-614 941 <=$>

TRUDY INSTITUTA EPIDEMIOLOGII, MIKROBIOLOGII I GIGIENY IM. PASTERA I INSTITUTA EKSPERIMENTALNOI MEDITSINY AKADEMII MEDITSINSKIKH NAUK SSSR
1959 V20 P106-123 64-19844 <=$>

1959 V20 P124-129 64-19837 <=$>
1963 V25 P26-31 64-19888 <=$>
1963 V25 P92-100 65-60848 <=$>
1963 V25 P102-122 65-60846 <=$>
1963 V25 P123-134 65-60845 <=$>
1963 V25 P135-152 64-19893 <=$>
1963 V25 P178-184 64-19885 <=$>
1963 V25 P185-191 65-60847 <=$>
1963 V25 P210-218 64-19886 <=$>
1963 V25 P219-224 65-60414 <=$>
1963 V25 P245-250 64-23645 <=$>
1963 V25 P251-259 64-19894 <=$>
1963 V25 P334-345 64-19892 <=$>

TRUDY INSTITUTA ETNOGRAFII. AKADEMIYA NAUK SSSR
1951 SNS V14 P95-108 ACTN N2 P300 <AINA>
1952 SNS V18 P5-87 ACTN N2 P220 <AINA>
1952 SNS V18 P199-238 AOTN N4 P84 <AINA>
1958 SNS V36 ENTIRE ISSUE AOTN N3 <AINA>
1959 SNS V51 P114-156 AOTN N4 P3 <AINA>
1959 SNS V51 P157-192 AOTN N4 P46 <AINA>

TRUDY INSTITUTA EVOLYUTSIONNOI FIZIOLOGII I PATOLOGII VYSSHEI NERVNOI DEYATELNOSTI IM. I. P. PAVLOVA. MOSKVA
1947 V1 P325-333 RT-2622 <*>
1947 V1 P365-368 RT-2663 <*>

TRUDY INSTITUTA FIZICHESKOI KHIMII. AKADEMIYA NAUK SSSR. MOSCOW
1951 N2 P62-68 62-18853 <=*>
1951 V2 N1 P136-145 RJ-315 <ATS>
1951 V2 N1 P146-165 RJ-607 <ATS>
63-20529 <=*$>
64-20193 <*>
1951 V2 N3 P79-82 64-14895 <=*$> 0
1952 V3 N4 P3-60 59-17788 <+*> P
1955 N5 P237-266 1442 <BISI>
1955 V4 P172-197 AD-610 957 <=$>
1955 V4 P172-179 C-2650 <NRC>
1955 V4 P172-197 R-4021 <*>
1957 V6 N2 P69-78 63-24361 <=*$>
1959 N7 P5-10 AEC-TR-5009 <=*>
1959 N7 P11-21 AEC-TR-5008 <=*>
62-18773 <=*>
1959 N7 P22-40 62-32239 <=*>
1959 N7 P51-53 AEC-TR-5005 <=*>
1959 N7 P78-84 AEC-TR-5007 <=*>
1959 N7 P85-95 AEC-TR-5006 <=*>
1959 N7 P96-100 AEC-TR-5004 <=*>
1959 N7 P139-140 62-32802 <=*>
1960 V8 P155-172 64-11930 <=>
1960 V8 P333-344 64-71405 <=> M

TRUDY INSTITUTA FIZIKI. AKADEMIYA NAUK GRUZINSKOI SSR. TIFLIS
1956 V4 P3-95 65-13051 <=$>
1960 V7 P193-199 64-11928 <=> M
1962 V8 P277-285 AD-615 221 <=$>
1962 V8 P313-321 AD-610 787 <=$>
1963 V9 P159-169 ANL-TRANS-93 <*>

TRUDY INSTITUTA FIZIKI, AKADEMIYA NAUK LATVIISKOI SSR
SEE FIZIKAS INSTITUTA RAKSTI

TRUDY INSTITUTA FIZIKI. AKADEMIYA NAUK UKRAINSKOI SSR. KIEV
1952 V3 P36-51 61-31685 <=>
1953 V4 P11-27 64-15586 <=*$>
1953 V4 P71-92 64-13299 <=*$>
1954 N5 P105-181 AEC-TR-4255 <*>
1954 V5 P94-104 64-15584 <=*$>
1954 V5 P119-136 AEC-TR-4256 <*>
1956 N7 P17-23 R-2012 <*>
1956 N7 P24-34 AEC-TR-4176 <*>
1956 V7 P3-16 10115 <KH>
60-18883 <+*>
1956 V7 P35-43 64-71447 <=>

TRUDY INSTITUTA FIZIKI I ASTRONOMII AKADEMIYA NAUK ESTONSKOI SSR
SEE EESTI NSV TEADUSTE AKADEEMIA FUUSIKA JA ASTRONOOMIA INSTITUUDI UURIMUSED

TRUDY INSTITUTA FIZIKI ATMOSFERY. AKADEMIYA NAUK SSSR. MOSKVA
1958 N1 ENTIRE ISSUE <CB>
1958 N2 P23-49 60-41106 <=>
61-13564 <+*>
1958 N2 P50-53 61-13574 <*+>
1958 N2 P54-65 61-13576 <*+>
1958 N2 P66-104 60-41092 <=>
61-13568 <+*>
61-13568 <*+>
1958 N2 P105-113 61-13565 <*+>
1958 N2 P114-118 61-13567 <*+>
1958 N2 P119-141 61-13566 <*+>
1958 N2 P142-159 65-13377 <*>
1958 N2 P160-166 60-41094 <*>
60-41094 <=>
1962 N4 P5-20 AD-625 621 <=*$>
AD-628 973 <=*$>
63-31788 <=>
1962 N4 P21-29 64-19004 <=*$> 0
1962 N4 P101-136 AD-630 667 <=$>
1962 N4 P144-146 65-12820 <*>
1962 N4 P147-202 63-41057 <=>
65-61331 <=$>
1962 N4 P203-256 63-23650 <=*$>

TRUDY INSTITUTA FIZIKI I GEOFIZIKI. AKADEMIYA NAUK TURKMENSKOI SSR. ASHKHABAD
1958 V5 P47-74 63-15790 <=*> C

TRUDY INSTITUTA FIZIKI I MATEMATIKI. AKADEMIYA NAUK AZERBAIDZHANSKOI SSR. BAKU
1958 V9 P5-9 60-15882 <+*>
1958 V9 P10-19 AD-627 073 <=$>
1958 V9 P20-25 60-15683 <+*>
1958 V9 P27-39 60-19561 <+*>
1958 V9 P42-57 60-17916 <+*>
1958 V9 P60-68 60-19175 <+*>
1963 V11 P52-107 66-11249 <*>

TRUDY INSTITUTA FIZIKI METALLOV. URALSKII FILIAL, AKADEMIYA NAUK SSSR. SVERDLOVSK
1954 N15 P65-69 62-11411 <=>
1955 N16 P91-96 R-4335 <*>
1956 N17 P111-118 61-19390 <+*>
1956 N18 P66-71 2018 <BISI>
1958 N19 P58-64 61-27516 <*=> 0
1958 N19 P95-100 59-13940 <=> 0
1958 N20 P5-11 AEC-TR-4081 <+*>
1958 N20 P5-40C AEC-TR-4513 <=>
1958 N20 P53-69 66-11076 <*>
1958 N20 P111-124 61-13295 <+*> C
1958 N20 P163-168 AEC-TR-4080 <+*>
1958 N20 P265-272 61-27027 <*=>
1958 N20 P273-282 AEC-TR-4377 <*=>
1958 N20 P329-338 07P62R <ATS>
1959 N21 ENTIRE ISSUE AEC-TR-4512 <=>
1959 N22 P27-31 E-759 <RIS>
1959 N22 P59-66 65-10745 <*>

TRUDY INSTITUTA FIZIKI ZEMLI. AKADEMIYA NAUK SSSR. MOSCOW
1958 N1 P44-64 60-31191 <=>
1959 N4 P8-21 64-13364 <=*$>
1959 N6 P104-105 63-10220 <=*>
1959 N6 P120-135 22N57R <ATS>
1959 N8 P3-77 60-11122 <=>
1960 N9 P1-18 62-23704 <=*>
1960 N9 P1-327 63-19325 <=>
1960 N9 P75-114 62-19113 <*=>
1960 N11 ENTIRE ISSUE 63-11178 <=>
1960 N11 P121-132 61-21433 <=>
1962 N23 P97-100 65-11131 <*>
1962 N24 P86-115 63-21108 <=>
1962 N24 P116-123 63-23634 <=*$>

1964 N31 P3-48 65-62425 <=*$>
1964 N35 P3-143 65-30925 <*>

TRUDY INSTITUTA FIZIOLOGII. AKADEMIYA NAUK GRUZINSKOI SSR. TIFLIS
1961 V12 P153-161 62-14732 <*> P
1961 V12 P163-173 62-14776 <*>
1961 V12 P193-198 62-14765 <*>

TRUDY INSTITUTA FIZIOLOGII IMENI I. P. PAVLOVA. AKADEMIYA NAUK SSSR. MOSCOW
1952 V1 P43-49 RT-1969 <*>
1952 V1 P50-60 RT-1970 <*>
1952 V1 P196-204 RT-1971 <*>
1952 V1 P228-236 RT-1972 <*>
1952 V1 P237-242 RT-1973 <*>
1952 V1 P243-250 R-711 <*>
65-63054 <=$>
1952 V1 P266-271 RT-1757 <*>
1952 V1 P369-375 RT-1974 <*>
1953 V2 P5-15 RT-3955 <*>
1953 V2 P31-51 RT-3934 <*>
1953 V2 P52-63 RT-3931 <*>
1953 V2 P69-75 RT-3947 <*>
1953 V2 P276-286 RT-3935 <*>
1953 V2 P306-315 RT-4032 <*>
1953 V2 P321-334 RT-4047 <*>
1953 V2 P433-448 RT-4003 <*>
1954 V3 P19-31 RT-3283 <*>
1954 V3 P105-126 RT-3285 <*>
1954 V3 P171-192 RT-3286 <*>
1954 V3 P278-286 RT-3284 <*>
1954 V3 P289-302 RT-4342 <*>
1954 V3 P303-315 RT-4343 <*>
1954 V3 P316-322 RT-4344 <*>
1954 V3 P349-354 RT-4345 <*>
1954 V3 P355-368 RT-4346 <*>
1954 V3 P369-376 RT-4347 <*>
1954 V3 P395-403 RT-3287 <*>
1954 V3 P412-418 RT-4348 <*>
1954 V3 P447-453 RT-4349 <*>
1954 V3 P454-460 RT-3281 <*>
1954 V3 P490-505 RT-3279 <*>
1954 V3 P607-611 RT-3280 <*>
1954 V3 P615-618 RT-3282 <*>
1954 V3 P619-621 62-24757 <=*>
1955 V4 P5-16 60-13479 <+*>
1955 V4 P17-21 60-13478 <+*>
1955 V4 P22-23 60-17912 <+*>
1955 V4 P34-50 60-17913 <+*>
1955 V4 P51-57 60-19176 <+*>
1955 V4 P58-62 60-19559 <+*>
1955 V4 P63-67 60-23800 <+*>
1955 V4 P68-74 60-17914 <+*>
1955 V4 P75-80 61-13240 <*+>
1955 V4 P81-92 61-13997 <*+>
1955 V4 P93-102 61-15646 <+*>
1955 V4 P103-108 61-15218 <*+>
1955 V4 P109-122 61-15244 <*+>
1955 V4 P123-131 61-15739 <*+>
1955 V4 P132-135 61-15647 <+*>
1955 V4 P136-144 61-19944 <=*>
1955 V4 P213-216 61-19101 <+*>
1955 V4 P217-229 61-27781 <*=>
1956 V5 P267-277 62-15044 <*=>
1956 V5 P307-316 61-00087 <*>
1956 V5 P409-415 R-5362 <*>
1957 V6 P254-259 65-63171 <=$>
1957 V6 P470-476 59-13205 <=> 0
1959 V8 P77-80 61-27186 <*=>
1959 V8 P421-425 64-13751 <=*$> 0
1959 V8 P426-432 64-13635 <=*$> 0
1959 V8 P433-440 64-13634 <=*$> 0
1959 V8 P476-478 64-13177 <=*$> 0
1959 V8 P501-508 62-14731 <=*>
1959 V8 P526-531 63-16457 <=*>
1960 V9 ENTIRE ISSUE 63-11172 <=>
1962 V10 ENTIRE ISSUE 63-11173 <=>
1962 V10 P245-254 65-60646 <=$>
1962 V10 P294-302 62-20360 <=*>
1962 V18 P159-200 65-60367 <=$>

TRUDY INSTITUTA FIZIOLOGII RASTENII IMENI K. A. TIMIRYAZEVA. MOSKVA
1955 N10 P150-155 R-1593 <*>
1955 V10 P81-90 R-1363 <*>
1955 V10 P91-100 63-11034 <=>
1955 V10 P129-138 R-1364 <*>
1955 V10 P150-154 64-15864 <=*$>
1955 V10 P156-160 R-1365 <*>
1955 V10 P210-249 R-1366 <*>
1955 V10 P257-264 R-1367 <*>
1955 V10 P265-271 R-1368 <*>
1956 V11 P97-115 R-1362 <*>

TRUDY INSTITUTA GENETIKI. AKADEMIYA NAUK SSSR. LENINGRAD
1935 V10 P227-232 61-11475 <=>
1950 V18 P247-259 61-11474 <=>
1958 V24 P372-384 63-11046 <=>
1958 V24 P426-434 61-28665 <*=> P
1960 V27 P154-173 08N51R <ATS>
1961 V28 P153-157 65-50024 <=>
1961 V28 P277-282 64-19947 <=>

1962 V29 P185-193 65-61217 <=$>
1963 V30 P321-356 RTS-2768 <RTS>
1964 V31 P182-194 RTS-3306 <NLL>

TRUDY INSTITUTA GEOFIZIKI. AKADEMIYA NAUK GRUZINSKOI SSR. TIFLIS
1939 V4 P97-108 62-33421 <=*>
1947 V10 P125-191 62-33421 <=*>
1955 V14 P61-75 60-23735 <+*> 0
1955 V14 P79-90 62-16672 <=*>
1955 V14 P219-228 64-13367 <=*$>
1955 V14 P229-236 64-13365 <=*$>
1959 V18 P23-27 AD-623 430 <=*$>
1959 V18 P31-42 65-64323 <=>
1959 V18 P97-107 AD-623 432 <=*$>
1959 V18 P129-143 65-64326 <=>

TRUDY INSTITUTA GEOGRAFII. AKADEMIYA NAUK SSSR. LENINGRAD
1946 V37 P331-348 R-4426 <*>

TRUDY INSTITUTA GEOLOGICHESKIKH NAUK. AKADEMIYA NAUK BELORUSSKOI SSR. MINSK
1958 V1 P108-118 61-28843 <=*>

TRUDY INSTITUTA GEOLOGICHESKIKH NAUK. AKADEMIYA NAUK KAZAKHSKOI SSR. ALMA-ATA.
1962 N6 P5-27 IGR V7 N2 P241 <AGI>
1962 N6 P170-180 IGR V7 N2 P233 <AGI>

TRUDY INSTITUTA GEOLOGICHESKIKH NAUK. AKADEMIYA NAUK SSSR. MOSCOW
1940 V39 N8 P37-40 R-1609 <*>
1946 N24 P1-83 PT1 IGR V2 N7 P551 <AGI>
1946 N24 P1-83 PT2 IGR V2 N8 P645 <AGI>
1950 V114 N40 P22-64 R-1208 <*>
1951 N135 P263-265 63-24627 <=*$>
1951 V124 N45 P371- R-2770 <*>
1951 V132 N53 P158-171 60-23750 <+*>
1952 N115 ENTIRE ISSUE 61-27674 <=>
1954 V25 P124-133 64-19755 <=$>
1955 N155 P156-164 RJ-561 <ATS>
1955 V155 N66 P3-115 PT1 ONLY
IGR V1 N5 P1 <AGI>
1955 V155 N66 P3-115 PT2 ONLY
IGR V1 N6 P11 <AGI>
1955 V155 N66 P3-115 PT3 ONLY
IGR V1 N7 P40 <AGI>
1955 V163 P53 63-13869 <=*>
1955 V163 P59-61 63-13869 <=*>
1955 V163 P76-77 63-13869 <=*>
1957 N64 P13-58 <DI>
1957 N64 P72-92 59-21091 <=>
1957 N152 P13-58 IGR V1 N10 P1 <AGI>
R-4645 <*>

57K26R <ATS>
1957 N152 P72-92 87L28R <ATS>
1957 V152 N64 P59-71 R-4643 <*>

TRUDY INSTITUTA GEOLOGICHESKIKH NAUK. AKADEMIYA NAUK UKRAINSKOI SSR. KIEV
1939 V12 P163-178 64-19741 <=$>

TRUDY INSTITUT GEOLOGICHNYKH NAUK. AKADEMIYA NAUK URSR. SERIYA GEOFIZICHESKAYA. KIEV
1958 V2 P3-14 62-33415 <=*>
1958 V2 P27-42 62-33415 <=*>
1958 V2 P62-68 62-33415 <=*>
1958 V2 P79-97 62-33415 <=*>
1958 V2 P120-129 64-11091 <=$>

TRUDY INSTITUTA GEOLOGII. AKADEMIYA NAUK UZBEKSKOI SSR. TASHKENT
1953 N9 P109-114 5J19R <ATS>

TRUDY INSTITUTA GEOLOGII I GEOFIZIKI. NOVOSIBIRSK.
1960 N1 P16-25 63-24346 <=*$>
1960 N5 P8C-96 62-32145 <=*>
1961 N7 P25-28 65-14232 <*>
1961 N7 P29-30 65-14233 <*>
1963 N6 P106-108 03S87R <ATS>
1965 N4 P134-137 67S87R <ATS>
1965 N10 P106-117 04T91R <ATS>

TRUDY INSTITUTA GEOLOGII I RAZRABOTKI GORIUCHIK-SHOGI ISKOPAEMYKH
1960 V1 P240-246 62-20301 <=*>
1960 V1 P247-259 63-10224 <*>

TRUDY INSTITUTA GEOLOGII RUDNYKH MESTOROZHDENII, PETROERAFII, MINERALOGII I GEOKHIMII. AKADEMIYA NAUK SSSR. MOSKVA
1956 N6 P57-72 62-16421 <=*>
1958 N11 P1-62 61-27123 <*=>
1958 N16 P1-76 61-15963 <*=>
1959 N28 P5-137 61-19564 <=*> 0
1959 N28 P142-148 61-19564 <=*> 0
1959 V28 P142-147 IGR V3 N1 P5 <AGI>
1960 N2 P83-91 07T90R <ATS>
1960 N42 P3-20 62-01087 <*>
1961 N41 P5-14 IGR V6 N4 P589 <AGI>
IGR,V6,N4,P589-595 <AGI>
1961 N41 P15-36 IGR V6 N4 P596 <AGI>
IGR,V6,N4,P596-611 <AGI>
1961 N41 P37-47 IGR V6 N4 P612 <AGI>
1961 N41 P37-41 IGR,V6,N4,P612-619 <AGI>
1961 N41 P48-66 IGR V6 N5 P768 <AGI>
IGR,V6,N5,P768-783 <AGI>
1961 N41 P67-72 IGR V6 N5 P784 <AGI>
IGR,V6,N5,P763-767 <AGI>
1961 N41 P73-87 IGR V6 N5 P789 <AGI>
IGR,V6,N5,P789-798 <AGI>
1961 N41 P86-98 IGR V6 N6 P1003 <AGI>
IGR V6 N6 P1003-1013 <AGI>
1961 N41 P99-105 IGR V6 N6 P1014 <AGI>
IGR V6 N6 P1014-1019 <AGI>
1961 N41 P106-118 IGR V6 N6 P1020 <AGI>
IGR V6 N6 P1020-1029 <AGI>
1961 N41 P119-133 IGR V6 N7 P1143 <AGI>
IGR,V6,N7,P1143-1153 <AGI>
1961 N41 P134-140 IGR V6 N7 P1154 <AGI>
IGR,V6,N7,P1154-1159 <AGI>
1961 N41 P140-157 IGR V6 N7 P1160 <AGI>
IGR,V6,N7,P1160-1173 <AGI>
1961 N41 P158-170 IGR V6 N11 P1977 <AGI>
1961 N41 P171-177 IGR V6 N11 P1987 <AGI>
1961 N41 P178-183 IGR V6 N11 P1993 <AGI>
1961 N41 P184-188 IGR V6 N11 P1998 <AGI>
1961 N41 P189-193 IGR V6 N11 P2002 <AGI>
1961 N41 P194-198 IGR V6 N12 P2165 <AGI>
1962 N59 P18-29 64-19833 <=$>
1962 N59 P7C-103 64-19833 <=$>
1963 P152-159 IGR V7 N2 P218 <AGI>
1963 P160-180 IGR V7 N2 P272 <AGI>
1963 P182-195 IGR V7 N2 P307 <AGI>

TRUDY INSTITUTA GORNOGO DELA. AKADEMIYA NAUK SSSR
1955 V2 P193-205 62-13734 <*=>
1956 V3 P98-116 60-15285 <+*>
1963 N21 P195-205 NCB-A.2391/SEH. <NLL>

TRUDY INSTITUTA GORYUCHIKH ISKOPAEMYKH. AKADEMIYA NAUK SSSR. MOSCOW
1939 V1 N1 P15-22 91R75R <ATS>
1939 V1 N1 P23-28 92R75R <ATS>
1939 V1 N1 P29- 24R76R <ATS>
1954 V3 P52-58 C-2385 <NRC>
1954 V3 P86-94 C-2349 <NRC>
1954 V3 P124-139 C-2410 <NRC>
1955 V5 P11-35 R-4700 <*>
1955 V5 P61-72 CSIRO-3541 <CSIR>
R-3950 <*>
1955 V5 P115-126 CSIRO-3542 <CSIR>
R-4369 <*>
1955 V5 P144-155 R-4698 <*>
1955 V5 N6 P3-15 59-18347 <+*>
1955 V6 P3-15 R-5166 <*>
1955 V6 P35-42 R-3653 <*>
1955 V6 P178-187 R-4382 <*>
1957 V7 P61-65 59-16374 <+*>
1959 V8 P14-20 61-15962 <=$>
63-19943 <=*>
1959 V8 P21-30 61-15962 <=$>
62-24901 <=*>
1959 V8 P45-50 61-15962 <=$>
1959 V8 P239-253 63-19856 <=*>
1959 V10 P7-44 62-24823 <=*>
1959 V10 P45-50 62-24824 <=*>
1959 V11 P30-38 64-13870 <=*$>
1959 V11 P82-90 65-60389 <=$>
1959 V11 P127-132 62-25400 <=*>
1959 V11 P215-225 61-28491 <*=>
1960 V13 P1-166 63-11063 <=>
1960 V14 P3-20 65-61292 <=$>
1960 V14 P91-97 62-33771 <=*>
1961 V12 P30-36 63-24175 <=*$>
1961 V12 P82-93 64-15965 <=*$>
1961 V12 P167-170 63-19934 <*=>
1962 V18 P135-148 63-11663 <=>
1962 V18 P254-258 65-14687 <*>
1962 V19 P35-45 N65-14944 <=$>
1962 V19 P46-58 63-23646 <=*>
1962 V19 P160-163 65-63509 <=$>
1963 V20 P198-207 BRITCOKE-TRANS-80 <NLL>
1963 V21 P179-184 66-14219 <*$>
1963 V21 P185-189 66-11443 <*>

TRUDY INSTITUTA IKHTIOLOGII I RYBNOGO KHOZYAISTVA. AKADEMIYA NAUK KAZAKHSKOI SSR. ALMA-ATA
1963 V4 P3-18 64-21480 <=>
1963 V4 P33-46 64-21480 <=>
1963 V4 P69-74 64-21480 <=>
1963 V4 P152-187 64-21480 <=>
1963 V4 P211-233 64-31071 <=>

TRUDY INSTITUTA INZHENEROV VODNOGO TRANSPORTA. LENINGRAD
1957 V24 P162-174 NP-TR-859 <*>
1962 V32 P32 65-60747 <=>
1962 V33 P8 65-61348 <=$>
1962 V33 P20-31 65-60784 <=>
1962 V33 P69 65-61379 <=$>
1964 N62 P20-26 2412 <BSRA>

TRUDY INSTITUTA ISPOLZOVANIYA GAZA V KOMMUNALNOM KHOZYAISTVE I PROMYSHLENNOSTI. AKADEMIYA NAUK UKRAINSKOI SSR. KIEV
1955 V3 P105-113 60-19876 <+*>

TRUDY INSTITUTA KHIMICHESKIKH NAUK. AKADEMIYA NAUK KAZAKHSKOI SSR. ALMA-ATA
1958 V2 P70-76 M-5604 <NLL>
1958 V2 P182-187 62-18071 <=*>
1958 V2 N2 P150-157 60-13807 <+*>
1958 V3 P64-71 61-15926 <+*>

1958 V3 P72-81 61-15701 <+*>
1960 V6 P178-183 62-10616 <=*>
1962 V8 P109-114 65-13821 <*>
1964 V11 P56- ICE V5 N1 P116-137 <ICE>
1964 V11 P56-88 ICE V5 N1 P116-137 <ICE>
1964 V12 P18-25 CSIR-TR-574 <CSSA>
1965 V13 P67-117 ICE V6 N3 P449-480 <ICE>

TRUDY INSTITUTA KHIMII. AKADEMIYA NAUK AZERBAIDZHANSKOI SSR. BAKU
1954 V13 P38-47 R-1626 <*>

TRUDY INSTITUTA KHIMII. AKADEMIYA NAUK GRUZINSKOI SSR. TBILISKI
1956 V12 P157-168 00J15R <ATS>
1957 V13 P145-164 60-16822 <+*>
60-23801 <+*>
61-13475 <*=>

TRUDY INSTITUTA KHIMII. AKADEMIYA NAUK KIRGIZSKOI SSR. FRUNZE
1956 V7 P69-77 63-16352 <=*> 0

TRUDY INSTITUTA KHIMII. AKADEMIYA NAUK UZBEKSKOI SSR. TASHKENT
1949 N2 P94-112 UCRL TRANS-568(L) <=*>

TRUDY INSTITUTA KHIMII. URALSKII FILIAL, AKADEMIYA NAUK SSSR. MOSKVA, SVERDLOVSK
1958 N2 P45-49 AEC-TR-4303 <+*>
1958 N2 P51-56 AEC-TR-4304 <+*>
1958 N2 P57-62 AEC-TR-4305 <+*>
1959 N3 P33-41 66-51116 <=>
1960 N4 P21-32 63-20198 <=*>
1960 N4 P85-94 63-24240 <=*$>
1963 N7 P3-6 RTS-2895 <NLL>
1963 N7 P7-11 RTS-2896 <NLL>
1963 N7 P13-17 RTS-2897 <NLL>

TRUDY INSTITUTOV KOMITETA STANDARTOV MER I IZMERITELNYKH PRIBOROV. MOSCOW
1960 N48 P3-4 64-31631 <=>
1960 N48 P7-23 64-31631 <=>
1960 N48 P24-41 64-31631 <=>
1960 V39 P130-142 61-15930 <=*>
1960 V46 P5-29 62-24292 <=*>
1960 V46 P30-42 62-24293 <=*>
1960 V46 P43-54 62-25110 <=*>
1960 V46 P55-61 62-13755 <=*>
1960 V46 P62-67 62-13747 <=*>
1960 V46 P68-80 62-13748 <=*>
1960 V46 P81-95 62-24377 <=*>
1960 V46 P96-106 62-24294 <=*>
1960 V46 P107-116 62-13921 <=*>
1961 N47 P193-198 65-60502 <=$>
1961 N55 P42-54 62-23692 <=*>
1961 V55 P35-41 62-23691 <=*>
1961 V55 P55-65 62-23690 <=*>
1961 V55 P90-97 62-23689 <=*>
1962 N61 P47-57 63-21457 <=>
1962 N65 P13-20 64-31206 <=>
1965 N76 P218-228 66-61857 <=$>

TRUDY INSTITUTA KRISTALLOGRAFII. AKADEMIYA NAUK SSSR. MOSKVA
1948 V4 P179-184 AEC-TR-4032 <+*>
1953 N8 P5-12 R-3254 <*>
1953 N8 P21-26 R-3249 <*>
1953 N8 P35-40 R-3684 <*>
1953 N8 P99-110 61-16148 <=*>
61M47R <ATS>
1953 N8 P237-246 R-3137 <*>
1953 N8 P315-328 R-3245 <*>
1953 N8 P341-354 R-3253 <*>
1953 V8 P13-20 63-11507 <=>
1953 V8 P27-34 R-2530 <*>
62-11796 <=>
1953 V8 P77-88 R-2531 <*>
1953 V8 P165-188 63-15946 <=*> 0
1953 V8 P233-236 R-2532 <*>
1953 V8 P261-273 R-4078 <*>
1953 V8 P273-282 R-2533 <*>
1953 V9 N8 P111-128 60M47R <ATS>
1954 N9 P165-210 66-60627 <=*$>
1954 N9 P211-220 R-1032 <*>
1954 N10 P99-116 R-4007 <*>
1954 V10 P76-83 GB4 T/1030 <NLL>
61-18844 <*=>
1955 N11 P33-67 R-1456 <*>
1955 N11 P121-123 C-3261 <NRCC>
1955 N11 P206-211 R-980 <*>
1955 N11 P221-222 64J19R <ATS>
1955 V11 P121-123 26Q72R <ATS>
1955 V11 P239-242 62-16428 <=*>

TRUDY INSTITUTA LESA. AKADEMIYA NAUK SSSR
1951 N1 P252-264 R-5205 <*>
1951 V1 P5-37 61-11428 <=>
1951 V8 P5-35 60-21910 <=>
1953 V9 P29-38 64-11070 <=>
1953 V9 P70-88 R-3632 <*>
1953 V9 P89-114 61-11423 <=>
1953 V9 P108-204 C-2166 <NRC>
1953 V9 P121-126 II-1264 <*>
1953 V9 P186-204 II-1192 <*>
1953 V9 P229-235 II-1479 <*>
1953 V9 P236-249 C-2210 <NLL>
1953 V9 P315-331 60-51131 <=>
1953 V9 P332-336 61-11420 <=>
1953 V9 P337-346 61-11421 <=>
1953 V9 P347-370 61-11424 <=>
1954 V16 P5-9 60-51020 <=>
1954 V16 P110-128 60-51011 <=>
1954 V16 P336-346 60-51006 <=>
1955 V24 P167-194 66-51125 <=>
1957 V34 P345-375 60-51166 <=>
1957 V36 ENTIRE ISSUE 65-50054 <=>
1958 V37 P46-66 63-11026 <=>
1958 V37 P120-141 60-51021 <=>
1958 V37 P171-174 60-21145 <=>
1958 V41 P111-121 65-50051 <=>
1958 V43 P138-151 64-11067 <=>
1959 V49 ENTIRE ISSUE 61-31225 <=>
1962 V51 P91-106 64-00086 <*>
1962 V51 P107-119 63-01655 <*>
1965 V65 P95-103 C-5282 <NRC>

TRUDY INSTITUTA LESA I DREVESINY. SIBIRSKOE OTDELENIE, AKADEMIYA NAUK SSSR
1961 V56 P89-103 63-31237 <=>
1962 V51 P124-134 63-15961 <=*>
1963 V65 P48-65 66-61876 <=*$>

TRUDY INSTITUTA LESOKHOZYAISTVENNYKH PROBLEM. AKADEMIYA NAUK LATVIISKOI SSR. RIGA
1957 V12 P5-24 60-00615 <*>
1957 V12 P91-100 60-51162 <=>

TRUDY INSTITUTA LESOKHOZYAISTVENNYKH PROBLEM I KHIMII DREVESINY. AKADEMIYA NAUK LATVIISKOI SSR. RIGA
1959 V17 P3-14 61-20833 <=*>
1959 V17 P105-114 61-20834 <=*>
64-10585 <*>
1960 V19 P65-70 C-4462 <NRC>

TRUDY INSTITUTA MASHINOVEDENIYA. SEMINAR PO PROCHNOSTI DETALEI MASHIN. AKADEMIYA NAUK SSSR
1953 V1 N2 P3-30 R-1886 <*>
1953 V1 N2 P78-84 R-1232 <*>

TRUDY INSTITUTA MASHINOVEDENIYA. SEMINAR PO TEORII MASHIN I MEKHANIZMOV. AKADEMIYA NAUK SSSR. MOSKVA
1948 V5 N17 P5-39 60-18641 <+*>
1953 V13 N51 P33-53 R-1894 <*>
1954 V14 N56 P5-19 62-23226 <=*>
1954 V14 N56 P48-58 62-23187 <=*>
1955 V15 N8 P35-51 61-16756 <*=>
62-195875 <=*>
1958 V18 N71 P22-42 62-25889 <=*>

1959 V19 N75 P16-30 61-13088 <*+> 0
1961 V21 N83 P5-28 62-23824 <=*>

TRUDY INSTITUTA MASHINOVEDENIYA. SEMINAR PO TOCHNOSTI V MASHINOSTROENII I PRIBOROSTROENII. AKADEMIYA NAUK SSSR. MOSKVA
1963 N17 P3-11 AD-609 142 <=>
1963 N17 P89-97 AD-609 142 <=>

TRUDY INSTITUTA MATEMATIKI I MEKHANIKI. AKADEMIYA NAUK UZBEKSKOI SSR. TASHKENT
1953 N12 P71-75 R-286 <*>

TRUDY INSTITUTA MELIORATSII VODNOGO I BOLOTNOGO KHOZYAISTVA. AKADEMIYA NAUK BELORUSSKOI SSR
1956 V6 ENTIRE ISSUE 60-51054 <=>

TRUDY INSTITUTA MERZLOTOVEDENIYA IM. V. A. OBRUCHEVA. MOSKVA, LENINGRAD
1944 V4 P125-204 RT-3959 <*>
1946 V1 P137-166 RT-1875 <*>
1947 V5 P15-17 RT-1406 <*>
1955 N2 P5-25 47J19R <ATS>
1959 N15 P132-143 65-14292 <*>
1960 V16 P36-45 63-19126 <=*>
1960 V16 P60-71 63-15275 <=*>

TRUDY INSTITUTA METALLURGII IM. A. A. BAIKOVA. AKADEMIYA NAUK SSSR. MOSCOW
1957 N1 P34-38 59-22372 <+*>
1957 N1 P60-66 59-00822 <*>
60-13251 <+*>
1957 N1 P158-161 R-4745 <*>
1957 N2 P54-64 64-30975 <*> 0
1957 N2 P164-166 60-14509 <+*>
1958 N3 ENTIRE ISSUE 60-51087 <=>
1958 N3 P63-68 4924 <AB>
1958 N3 P160-164 62-25304 <=*> 0
1958 N3 P181-190 AEC-TR-4093 <*+>
1958 N3 P191-194 60-21938 <=>
1958 N3 P203-215 62-13136 <*=>
1958 N3 P216-230 62-15712 <*=>
1958 N3 P295-306 2497 <BISI>
1960 N4 P3-8 5182 <HB>
1960 N4 P58-66 5087 <HB>
1960 N5 P22-35 5154 <HB>
1960 N5 P22-49 61-28816 <*=> 0
1960 N5 P85-94 60-41354 <=>
1960 N5 P95-99 60-41350 <=>
62-13101 <*=>
1960 N5 P127-132 63-19828 <=*>
1960 N5 P139-144 60-41356 <=>
1960 N5 P151-155 60-41368 <=>
63-13363 <=*>
1960 N5 P156-161 60-41369 <=>
61-28816 <*=> 0
1960 N5 P202-237 60-41357 <=>
1960 N6 P32-40 61-27418 <*=> 0
1960 V6 P41-53 64-31580 <=>
1961 N6 P187-195 CE-TRANS-3888 <NLL>
1961 N6 P209-218 M-5636 <NLL>
1961 N8 P54-106 63-13174 <=*>
1961 N8 P120-127 63-13174 <=*>
1961 N8 P149-165 63-13174 <=*>
1961 N8 P269-277 63-13174 <=*>
1962 N9 P82-86 65-61167 <=>
1962 N9 P105-108 6448 <HB>
1962 N9 P109-114 6449 <HB>
1962 N9 P141-146 AD-640 287 <=$>
1962 N10 P152-154 AEC-TR-6075 <=*$>
1962 N10 P181-187 63-13473 <=*>
1962 N10 P202-208 AD-615 875 <=$>
63-13463 <=*>
1962 N10 P215-216 64-15992 <=*$>
1962 N11 P16-30 3631 <BISI>
1962 N11 P36-53 64-13037 <=*$>
1962 N11 P78-82 AD-610 798 <=$>
1962 N11 P114-119 AD-618 636 <=$>
1963 N12 P118-120 6163 <HB>
1963 N13 P64-86 64-51840 <=$>
1963 N13 P108-121 AD-610 350 <=$>
1963 N13 P157-162 AD-614 017 <=$>
1963 N14 P50-57 64-19149 <=$>
1963 N14 P68-77 3839 <BISI>
1963 N14 P78-89 64-31330 <=>
1963 N14 P90-100 64-19148 <=$>
1963 N14 P120-138 64-31330 <=>
1963 N14 P147-154 64-31330 <=>
1963 N15 P79-87 64-21197 <=>
1963 V12 P45-48 64-19996 <=$>
1964 P28-31 AD-639 484 <=$>
1964 P63-74 M-5769 <NLL>
1964 P75-78 M-5768 <NLL>
1964 P108-113 M-5762 <NLL>
1964 P132-138 AD-636 676 <=$>
1964 P179-184 M-5763 <NLL>
1964 P185-193 M-5764 <NLL>
1964 P209-215 M-5765 <NLL>
1964 P262-266 M-5761 <NLL>
1964 P304-309 M-5766 <NLL>
1964 P345-351 M-4767 <NLL>
1964 P358-366 M-5760 <NLL>

TRUDY INSTITUTA METALLURGII. AKADEMIYA NAUK GRUZINSKOI SSR. TBILISI
1962 V12 P159-165 AD-610 293 <=$>
1962 V13 P255-263 M.5826 <NLL>

TRUDY INSTITUTA METALLURGII I OBOGASHCHENIYA. AKADEMIYA NAUK KAZAKHSKOI SSR. ALMA-ATA
1959 V1 P27-30 61-19946 <*=>
1959 V1 P46-52 61-28166 <*=>
1959 V1 P65-75 62-13731 <*=>
1959 V1 P115-126 62-32958 <=*>
1962 V4 P43-47 65-17366 <*>
1962 V5 P3-11 63-41188 <=>
1963 V6 P22-29 34R78R <ATS>
1963 V6 P78-90 65-61174 <=>
1963 V6 P91-105 TRANS-559 <MT>
1963 V7 P24-29 65-61281 <=$>
1963 V7 P36-37 65-61184 <=>
1963 V8 P6-7 65-10763 <*>
1963 V8 P13-18 66-11536 <*>

TRUDY INSTITUTA METALLURGII. URALSKII FILIAL, AKADEMIYA NAUK SSSR. SVERDLOVSK
1957 N1 P74-79 63-20294 <=*> 0
1957 V1 P39-50 60-13323 <+*>
1957 V1 P78-84 60-13320 <+*>
1957 V1 P85-92 60-15580 <+*>
1957 V2 P135-138 AEC-TR-3920 <+*>
1958 N2 P95-105 61-27788 <*=>
1958 N2 P169-180 61-27787 <*=>
1958 N4 P87-95 66-12305 <*>
1959 N3 P93-101 65-11998 <*> 0
1960 N4 P53-57 5086 <HB>
1960 N4 P240-242 5129 <HB>
1960 N4 P254-256 5130 <HB>
1960 N5 P145-150 60-41366 <=>
1960 N5 P166-173 60-41367 <=>
1962 N11 P209-220 63-11705 <=>
1963 N12 P214-240 64-13707 <=*$>

TRUDY INSTITUTA MIKROBIOLOGII. AKADEMIYA NAUK LATVIISKOI SSR
1956 N5 P27-45 59-12750 <+*>
1956 N5 P47-55 59-12744 <+*>
1958 N7 P67-75 63-11006 <=>

TRUDY INSTITUTA MIKROBIOLOGII. AKADEMIYA NAUK SSSR
1952 N2 P3-32 RJ-170 <ATS>
1952 N2 P51-63 RT-4340 <*>
1952 N2 P64-77 RT-4257 <*>
1952 N2 P89-99 RT-4258 <*>
1952 N2 P121-129 RJ-171 <ATS>
1952 N2 P139-149 RT-4259 <*>
1952 N2 P150-156 RT-4642 <*>
1954 N3 P144-153 RT-4389 <*>
1954 N3 P221-223 RT-4391 <*>
1954 N3 P224-229 RT-4390 <*>

1958 V5 P6-17 59-19265 <+*>
1959 V6 P20-37 62-18851 <=*>
1961 V9 P45-56 63-20250 <=*>
1961 V9 P81-85 63-20251 <=*>
1961 V9 P101-104 57R80R <ATS>

TRUDY INSTITUTA MIKROBIOLOGII I VIRUSOLOGII. AKADEMIYA NAUK KAZAKHSKOI SSR. ALMA-ATA
1956 V1 N1 P87-95 60-23174 <+*> 0

TRUDY INSTITUTA MINERALNYKH RESURSOV. AKADEMIYA NAUK UKRAINSKOI SSR. KIEV
1960 V2 P53-58 65-64001 <=$>

TRUDY INSTITUTA MINERALOGII, GEOKHIMII I KRISTALLOKHIMI REDKIKH ELEMENTOV. MOSCOW
1957 P93-119 59-10318 <+*>
1957 N1 P23-34 AEC-TR-4186 <*+>
1957 N1 P35-37 AEC-TR-4185 <*+>
1959 N2 P1-84 61-15871 <*=> 0
1959 N2 P74-77 60-19866 <+*>
1959 N2 P211-242 61-15870 <*=>
1959 V3 P3-25 IGR V3 N1 P9 <AGI>
1961 N6 P111-117 65-63984 <=>
1963 N18 P53-59 65-63989 <=>

TRUDY INSTITUTA MORFOLOGII ZHIVOTNYKH IM. A. N. SEVERTSOVA. AKADEMIYA NAUK SSSR. MOSCOW, LENINGRAD
1950 V2 P85-139 64-15126 <=*$>
1952 N6 P38-42 RT-1884 <*>
1959 V13 P118-128 C-4970 <NRC> 0
1959 V27 P175-186 61-28425 <*=>
1959 V27 P221-230 61-15436 <+*>
1961 N34 P35-53 NIOT-81 <NLL>
1961 V33 P151-172 63-19101 <=*>

TRUDY INSTITUTA NEFTI. AKADEMIYA NAUK KAZAKHSKOI SSR. ALMA-ATA
1958 V2 P53-60 91N50R <ATS>
1958 V2 P93-99 59-18952 <+*>

TRUDY INSTITUTA NEFTI. AKADEMIYA NAUK SSSR. MOSKVA
1950 V1 N2 P285-294 34Q73R <ATS>
1952 V2 P39-46 C-2342 <NRC>
1952 V2 P53-72 RJ-580 <ATS>
1952 V2 P110-132 53L36R <ATS>
1954 N4 P101-115 C-2311 <NRC>
1954 V4 P176-198 59-10327 <+*>
1955 V6 P30-34 81J14R <ATS>
1955 V6 P71-78 14N58R <ATS>
1955 V6 P199-206 94M45R <ATS>
1956 V8 P94-99 78K20R <ATS>
1956 V8 P107-113 23H13R <ATS>
1956 V8 P114-119 65J14R <ATS>
1956 V8 P131-133 20H13R <ATS>
501TM <CTT>
1956 V8 P180-184 60-13321 <+*>
92J15R <ATS>
1958 N12 P213-227 NP-TR-832 <*>
1958 V11 P46-52 93N53R <ATS>
1958 V11 P85-110 52N54R <ATS>
1958 V11 P209-227 76N55R <ATS>
1958 V11 P228-239 IGR V3 N1 P60 <AGI>
59-00748 <*>
1958 V11 P283-293 71N53R <ATS>
1958 V12 P76-82 63-18054 <=*>
1958 V12 P267-371 04M41R <ATS>
1959 V9 P68-80 53L32R <ATS>
1959 V13 P80-96 61-28158 <*=>
1959 V13 P224-240 65-60499 <=$>

TRUDY INSTITUTA OKEANOLOGII. AKADEMIYA NAUK SSSR. MOSCOW
1951 V5 P11-13 C-4176 <NRCC>
1951 V5 P14-16 C-4169 <NRCC>
1951 V5 P46-49 C-3898 <NRCC>
1953 V7 P91-110 RT-2670 <*>
1953 V7 P198-213 59-14723 <+*>
1953 V7 P252-258 59-14723 <+*>
1953 V7 P320-326 RT-2456 <*>
1954 V8 P18-40 65-61566 <=$>
1954 V9 P5-22 R-3649 <*>
1954 V10 P5-20 RT-3070 <*>
1954 V10 P35-43 RT-3069 <*>
1954 V10 P169-178 RT-3120 <*>
1954 V11 P56-61 60-51040 <=>
1954 V11 P206-219 60-21110 <=>
1954 V11 P246-257 62-25386 <=*>
1955 V12 P161-176 61-15101 <*+>
1955 V12 P177-183 M.5110 <NLL>
1955 V12 P298-310 M.5140 <NLL>
1955 V13 P77-82 62-32834 <=*>
1955 V18 P28-47 60-23775 <+*>
1955 V18 P95-99 M.5122 <NLL>
1955 V18 P100-112 M.5121 <NLL>
1955 V18 P134-141 R-4977 <*>
1956 V17 P18-41 59-11614 <=> 0
1956 V17 P148-175 M.5137 <NLL>
1956 V17 P176-191 59-11615 <=> 0
1956 V17 P192-203 61-15102 <*+>
1956 V19 P117-128 59-15869 <+*> 0
1956 V19 P324-329 M.5113 <NLL>
1957 V20 P186-227 C-3243 <NRCC> 0
1957 V22 P252-259 M.5111 <NLL>
1957 V24 P173-199 62-24795 <=*> 0
1957 V25 P7-24 63-20133 <=*>
1957 V25 P78-87 63-24187 <=*$>
1957 V25 P134-142 60-11147 <=>
1957 V25 P143-152 M.KUOI <NLL>
65-60951 <=$>
1957 V25 P153-170 61-11407 <=>
1957 V25 P171-197 60-17119 <+*> 0
1958 V26 P3-7 AD-633 318 <=$>
1958 V26 P243-246 C-3942 <NRCC>
1958 V27 P219-257 C-3626 <NRCC>
1958 V28 P37-55 60-15448 <+*> 0
1958 V28 P56-58 59-10047 <+*>
1958 V28 P93-99 M.5129 <NLL>
1959 V29 P17-64 64-00463 <*>
64-11837 <=>
1959 V30 P148-155 RTS-3020 <NLL>
1959 V33 P114-125 61-19562 <*=>
1959 V35 P33-44 M.5128 <NLL>
1959 V35 P78-85 61-27618 <*=>
1959 V35 P113-117 61-27618 <*=>
1959 V35 P190-224 61-27618 <*=>
1959 V35 P238-244 61-27618 <*=>
1960 V31 P205-285 61-31031 <=>
1960 V39 P10-24 65-33114 <=*$>
1960 V39 P33-38 65-64363 <=*$>
1960 V39 P96-100 65-64364 <=*$>
1960 V41 P31-41 62-15291 <*=>
1960 V41 P42-47 62-15290 <*=>
1960 V41 P146-152 64-15125 <=*$>
1960 V41 P175-191 61-19745 <=*>
1960 V41 P192-197 63-23825 <=*$>
1960 V41 P257-265 63-23142 <=*$>
1961 V42 P40-91 65-60321 <=$> 0
1961 V43 P225-269 63-23143 <=*$>
1961 V45 P223-239 50P60R <ATS>
63-21042 <=>
1961 V45 P259-278 65-60320 <=$> 0
1961 V46 P3-84 63-21079 <=>
1961 V47 P199-200 62-33624 <=*>
1961 V47 P203-216 65-61302 <=$>
1961 V49 P60-64 63-24191 <=*$>
1961 V49 P118-136 63-24616 <=*$>
1961 V49 P137-146 63-23759 <=*$>
1961 V49 P147-155 63-23697 <=*$>
1961 V49 P162-179 63-23760 <=*$>
1961 V49 P200-204 63-23761 <=*$>
1961 V49 P205-223 63-24189 <=*$>
1961 V49 P266-273 63-24188 <=*$>
1961 V51 P57-81 63-19929 <=*$>
1961 V53 P112-122 63-19726 <=*$>
1961 V53 P123-148 AD-622 379 <=*$>
1962 V36 P15-22 <JBS>
1962 V55 P3-22 AD-627 256 <=$>
1962 V55 P37-44 M.5318 <NLL>

1962 V55 P62-66 65-25334 <$>
1962 V57 P156-199 64-15995 <=*$> 0
1962 V58 P67-79 63-23698 <=*$>
1962 V58 P8C-111 64-15863 <=*$>
1962 V59 P183-203 63-21510 <=>
1963 V62 P62-67 65-62670 <=$>
1963 V62 P68-95 63-31088 <=>
1963 V62 P211-223 65-64362 <=*$>
1963 V66 P29-44 AD-623 811 <=*$>
1963 V66 P46-58 AD-628 147 <=*$>
1963 V66 P91-124 AD-631 421 <=$>
1963 V70 P279 AD-619 257 <=$>
1963 V72 P167-177 MAFF-TRANS-NS-66 <NLL>
1964 V64 P136-156 <JBS>
1964 V64 P154-157 TRANS-320(M) <NLL>
1964 V64 P158-181 65-61353 <=$>
1964 V64 P182-201 <JBS>
65-61439 <=>
1964 V64 P214-225 66-23209 <*$>
1964 V64 P265-270 64-23622 <=$>
1964 V73 ENTIRE ISSUE 65-50120 <=>
1964 V73 P93-138 65-64106 <=*$> 0
1964 V75 P3-61 65-30971 <=*$>
1964 V75 P62-68 65-30972 <=$*>
1964 V75 P69-72 65-30973 <=$>
1964 V75 P73-98 65-30959 <=*$>
1964 V75 P99-131 65-30974 <=$>
1964 V75 P157-170 66-30095 <=*$>
1965 V75 P69-72 TRANS-36 <NLL>
1965 V79 P49-59,87-96 65-33842 <=*$>

TRUDY INSTITUTA ONKOLOGII. AKADEMIYA MEDITSINSKIKH NAUK SSSR. MOSKVA
1958 V2 P256-276 59-18223 <=*> 0

TRUDY INSTITUTA POCHVOVEDENIYA I AGROKHIMII. AKADEMIYA NAUK AZERBAIDZHANSKOI SSR. BAKU
1956 N7 P225-232 R-4366 <*>

TRUDY INSTITUTA POCHVOVEDENIYA. AKADEMIYA NAUK KAZAKHSKOI SSR. ALMA-ATA
1953 V2 P84-149 60-51179 <=>

TRUDY INSTITUTA POLEVODSTVA. AKADEMIYA NAUK GRUZINSKOI SSR. TBILISKI
1951 V6 P79-97 R-546 <*>
1954 V8 P105-106 R-542 <*>

TRUDY INSTITUTA PRIKLADNOI MINERALOGII I METALLURGII. MOSKVA
1924 N8 P1-16 R-2289 <*>
1933 N59 P1-47 59-17503 <*>

TRUDY INSTITUTA SEISMOLOGII. AKADEMIYA NAUK TADZHIKSKOI SSR
1957 V71 P47-58 SCL-T-308 <*>

TRUDY INSTITUTA SELSKOKHOZYAISTVENNOGO MASHINOSTROENIYA. ROSTOV-ON-DON (TITLE VARIES)
1957 V8 P247-262 PT1 62-23521 <=*>

TRUDY INSTITUTA STROITELSTVA I STROIMATERIALOV. AKADEMIYA NAUK KAZAKHSKOI SSR. ALMA-ATA
1959 V2 P163-166 63-19669 <=*>
1959 V2 P179-182 63-19669 <=*>

TRUDY INSTITUTA TEORETICHESKOI ASTRONOMII. AKADEMIYA NAUK SSSR. MOSKVA
1952 N1 P87-222 62-15966 <=*>
1952 V1 P5-86 62-16112 <=*>
1956 N6 P5-66 60-15511 <+*>
1961 V8 ENTIRE ISSUE 63-14443 <=*>
1961 V8 P3-134 63-14442 <=*>

TRUDY INSTITUTA TEORETICHESKOI GEOFIZIKI. AKADEMIYA NAUK SSSR. MOSKVA, LENINGRAD
1947 V2 P108-111 62-33420 <=*>
1947 V3 P3-45 62-33420 <=*>

TRUDY INSTITUTA TEPLOENERGETIKI. AKADEMIYA NAUK URSR. KIEV
1952 N8 P34-42 RJ-413 <ATS>

TRUDY INSTITUTA TOCHNOI MEKHANIKI I VYCHISLITELNOI TEKHNIKI. AKADEMIYA NAUK SSSR. MOSKVA
1961 V2 ENTIRE ISSUE 62-24260 <=*>

TRUDY INSTITUTA TONKOI KHIMICHESKOI TEKHNOLOGII. MOSCOW
1953 N4 P30-36 RCT V29 N4 P1496 <RCT>
1955 V5 P50-58 65-11011 <*>

TRUDY INSTITUTA TORFA. AKADEMIYA NAUK BELORUSSKOI SSR. MINSK (TITLE VARIES)
1957 V6 P38-51 61-13217 <*+>
1957 V6 P52-67 60-19881 <+*>
1957 V6 P68-79 60-23619 <+*>
1957 V6 P274-290 60-23620 <+*>
1959 V7 P36-49 64-11072 <=>

TRUDY INSTITUTA VETERINARII KAZAKHSKOGO FILIALIA VSESOYUZNAYA AKADEMIYA SELSKOKHOZYAYSTVENNYKH NAUK. ALMA ATA
1955 V7 P52-53 R-2027 <*>
1955 V7 P154-166 61-15179 <*+>

*TRUDY INSTITUTA VYSSHEI NERVNOI DEYATELNOSTI. AKADEMIYA NAUK SSSR. SERIYA PATOFIZIOLOGICHESKAYA
1956 V2 P88-98 62-13794 <=*>
1956 V2 P129-135 62-13793 <*=>
1956 V2 P209-228 62-15101 <*=>
1956 V2 P315-334 59-12006 <+*>
1957 V3 ENTIRE ISSUE 60-21086 <=>
1958 V5 ENTIRE ISSUE 61-31022 <=>
1960 V5 ENTIRE ISSUE 61-31020 <=>
1961 V6 ENTIRE ISSUE 63-11167 <=>
1961 V9 ENTIRE ISSUE 63-11170 <=>

TRUDY INSTITUTA YADERNOI FIZIKI. AKADEMIYA NAUK KAZAKHSKOI SSR. ALMA-ATA
1958 V1 P288-295 61-27955 <*=>
1962 V5 P111-116 63-11737 <=>
1962 V5 P147-154 63-19510 <=*>
1963 V6 P105-111 64-71294 <=> M

TRUDY INSTITUTA ZEMNOGO MAGNETIZMA, IONOSFERY I RASPROSTRANENIYA RADIOVOLN. MOSKVA
1960 N17 P27-49 63-15517 <=*>
1960 N17 P187-202 63-15546 <=*>
1960 N17 P216-281 63-19095 <=*>
1961 N18 P77-86 C-4976 <NRC>
1961 N19 P31-43 C-4393 <NRC>
1961 N19 P48-51 C-4341 <NRC>
1961 N19 P71-84 63-21726 <=>
1961 N19 P131-139 64-51944 <=$>
1961 N19 P140-150 63-21842 <=>
1961 V18 P14-26 AD-623 437 <=*$>

TRUDY. INSTITUTA ZHIVOTNOVODSTVA KAZAKHSKII FILIAL ALMA-ATA
1955 N3 P16-28 R-3094 <*>

TRUDY INSTITUTA ZOOLOGII. AKADEMIYA NAUK GRUZINSKOI SSR. TBILISI
1956 V14 P357 R-514 <*>

TRUDY INSTITUTA ZOOLOGII. AKADEMIYA NAUK KAZAKHSKOI SSR. ALMA-ATA
1953 P84-101 R-1861 <*>
1953 V1 P30-36 59-15840 <+*>
1955 V3 P161-163 63-10935 <=*> 0
1956 V2 P3-20 62-15326 <*=> 0
1956 V2 P149-174 62-13791 <*=>
1956 V2 P335-346 62-13792 <*=>
1957 V7 P252-257 60-16932 <=> 0
1957 V7 P290 62-13931 <*=>
1958 V9 P167-175 62-15639 <=*> 0
1958 V9 P183-186 60-16931 <*+>
63-10934 <=*>
1958 V9 P187-197 63-10936 <=*> 0

1960 V12 P227-235 62-15736 <*=>
1962 V16 P15-22 63-01251 <*>
64-15636 <=*$>
1962 V16 P183-185 64-15635 <=*$> O
1963 V19 P220-225 63-41270 <=>
1963 V24 P171-177 66-61001 <=*$>

TRUDY INSTITUTA ZOOLOGII. AKADEMIYA NAUK UKRAINSKOI SSR. KIEV
SEE TRUDY INSTYTUTU ZOOLOHIYI. AKADEMIYA NAUK UKRAYINSKOYI RSR. KIEV

TRUDY INSTITUTA ZOOLOGII I PARAZITOLOGII. AKADEMIYA NAUK KIRGIZSKOI SSR. FRUNZE
1955 N4 P79-87 66-61187 <=$*>
1955 N5 P121-127 59-15767 <+*>
1956 N5 P121-127 66-60944 <=*$>

TRUDY INSTYTUTU KHEMIYI. KHARKOV
SEE TRUDY INSTITUTA KHIMII. KHARKOV

TRUDY INSTYTUTU ZOOLOHIYI. AKADEMIYA NAUK UKRAYINSKOYI RSR. KIEV
1955 P91- R-1859 <*>
1955 V21 P365-367 R-508 <*>

TRUDY KARADAHSKOYI NAUCHNOYI STANTSIYI IMENI T. I. VYAZEMSKOHO. SIMFEROPOL
1952 V12 P50-67 60-23773 <+*>
1952 V12 P68-77 60-23772 <+*>
1952 V12 P78-95 60-23771 <+*>
1952 V12 P111-115 61-15010 <*+>
1957 V14 P155-157 60-17118 <+*>

TRUDY KARELSKOGO FILIALLA. AKADEMIYA NAUK SSSR
1959 V14 P5-12 62-00154 <*>
1959 V14 P5-13 63-10933 <=*>
1959 V14 P5-12 64-15625 <=*$> O
1959 V14 P14-32 65-60932 <=$>
1959 V14 P72-83 63-01252 <*>
1959 V17 P5-37 63-23544 <=*$>

TRUDY KAZAKHSKOGO NAUCHNO-ISSLEDOVATELSKOGO GIDROMETEOROLOGICHESKOGO INSTITUTA. ALMA-ATA
1956 V7 P57-65 60-21898 <=>
1959 V10 P37-72 63-19738 <=*>
1959 V10 P73-81 63-19919 <*> O
1960 V15 P11-26 61-28402 <*=>
1960 V15 P32-40 61-28400 <=*>
1960 V15 P73-79 63-19812 <=*> O

TRUDY KAZAKHSTANSKOGO FILIALA, AKADEMII NAUK SSSR
1947 V6 P3-13 60-15950 <+*>

TRUDY KAZANSKOGO AVIATSIONNOGO INSTITUTA
1937 N7 ENTIRE ISSUE AMST S1 V5 P1-174 <AMS>
1959 V45 P45-62 63-24037 <=*>
1961 V66 P3-17 AD-637 052 <=$> M
1961 V66 P49-62 AD-635 280 <=$>
1963 V75 P61-69 AD-612 412 <=$>
1963 V75 P91-101 AD-632 260 <=$> M
1963 V77 P35-48 AD-630 860 <=$>
1963 V77 P130-147 AD-624 758 <=*$> M

TRUDY KAZANSKOGO KHIMIKO-TEKHNOLOGICHESKOGO INSTITUTA IM. S. M. KIROVA. KAZAN
1950 N15 P26-31 72TM <CTT>
1951 N16 P3-10 99L29R <ATS>
1951 N16 P133-150 61-14960 <=*> O
1953 N18 P8-21 00L30R <ATS>
1957 N23 P201-204 63-19452 <=*$>
1959 N26 P15-18 66-10613 <*>
1959 N26 P63-66 63-15566 <=*>
1961 N27 P193-207 RTS-2723 <NLL>

TRUDY KHARKOVSKOGO INSTITUTA INZHENEROV ZHELEZNODOROZHNOGO TRANSPORTA IM. S. M. KIROVA. MOSKVA
1962 N55 P88-96 64-31084 <=$>

TRUDY KHARKOVSKOGO POLITEKHNICHESKOGO INSTITUTA IM. V. I. LENINA. KHARKOV
1958 V14 N2 P23-41 62-24681 <=*>
1963 V45 P1-10 64-26392 <$>
1963 V45 P11-2C 64-26393 <$>

TRUDY PO KHIMII I KHIMICHESKOI TEKHNOLOGII
1958 N1 P487-496 65-61164 <=>
1958 V1 P187-189 54M40R <ATS>
1958 V1 P426-430 65-10439 <*>
1959 V2 P549-656 64-10812 <=*$>
1960 V3 P538-544 AEC-TR-5300 <=*>
1960 V3 N3 P501-504 63-10805 <=*>
1963 N1 P8-11 AD-630 412 <=*$>
1963 N1 P26-30 AD-634 813 <=$>
1963 N1 P43-46 AD-630 874 <=$>
1963 V6 N1 P186-187 65-12159 <*>

TRUDY KHIMICHESKOGO INSTITUTA IMENI L. YA. KARPOVA MOSKVA
1924 N3 P54-65 66-11936 <*>
1926 N5 P81-89 61-18141 <=*>
1927 P157-168 61-18133 <=*>

TRUDY PO KHIMII PRIRODNYKH SOEDINENII. KISHINEV UNIVERSITET
1959 N2 P3-18 66-10287 <*> O

TRUDY KIEVSKOGO TEKHNOLOGICHESKOGO INSTITUTA PISHCHEVOI PROMYSHLENNOSTI
1962 N25 P36-41 99R78R <ATS>

TRUDY KOMISSII PO AKUSTIKE. AKADEMIYA NAUK SSSR. MOSCOW
1951 N6 P82-85 62-23110 <=>
1953 N7 P36-39 62-23111 <=>
1953 N7 P40-42 62-19826 <=>
1953 N7 P43-47 62-23109 <=>
1953 N7 P53-60 62-11518 <=>
62-11581 <=>
1955 N8 P21-45 61-31541 <=>

TRUDY KOMISSII ANALITICHESKOI KHIMII. AKADEMIYA NAUK SSSR. MOSCOW
1949 V2 P35-53 R-5131 <*>
1951 V3 P5-22 T.797 <INSD>
1951 V3 P101-115 T-2273 <INSD>
1952 V4 N7 P282-299 R-1425 <*>
1955 V6 P182-190 61-23377 <*=>
1955 V6 P351-364 UCRL TRANS-898 (L) <=*$>
1955 V6 P478-482 R-3654 <*>
1955 V6 N9 P343-350 R-4412 <*>
1956 V7 P64-76 62-24524 <=*>
1956 V7 P89-95 AEC-TR-3481 <+*>
1958 V8 N11 P52-74 PB 141 317T <=>
1958 V8 N11 P178-182 60-11483 <=>
1958 V9 P361-366 AEC-TR-4037 <+*>
1960 V10 P91-96 63-13911 <=*>
1960 V10 P97-1C2 762-6 <MUL>
1960 V10 P117-121 63-13911 <=*>
1960 V10 P129-136 C-3969 <NRCC>
1960 V11 P261-272 63-19131 <=*>
1960 V11 P309-322 66-12188 <*>
1960 V12 P227-235 CSIR-TR.272 <CSSA>
1963 V13 P3-7 63-20610 <=*$>
1963 V13 P15-19 65-10490 <*>
1963 V13 P20-23 63-20609 <=*$>
1963 V13 P24-27 65-10491 <*>
1963 V13 P33-35 63-20608 <=*$>
1963 V13 P187-191 66-12869 <*>
1963 V13 P277-283 516 <TC>
1963 V13 P315-319 66-11000 <*>
1963 V13 P359-366 89S89R <ATS>
1963 V13 P383-388 64-71378 <=>
1963 V14 P31-46 65-64419 <=*$>
1963 V14 P47-58 65-64409 <=*$>
1963 V14 P191-201 N66-23534 <=$>

TRUDY KOMISSII PC BORBE S KORROZIEI METALLOV. AKADEMIYA NAUK SSSR. MOSKVA
1956 N2 P11-21 R-3673 <*>

TRUDY KOMISSII PO IZUCHENIYU CHETVERTICHNOGO PERIODA. AKADEMIYA NAUK SSSR. LENINGRAD
1959 V14 P1-128 64-11074 <=>

TRUDY KOMMISSII PO IZUCHENIIU VECHNOI MERZLODY
1934 N3 P13-19 RT-3948 <*>
1935 N4 P170-187 RT-3933 <*>
1940 N10 P5-36 RT-3966 <*>
1940 N10 P85-107 RT-3957 <*>

TRUDY KOMPLEKSNOI NAUCHNOI EKSPEDITSII PO VOPROSAM POLEZASHCHITNOGO LESORAZVEDENIYA. AKADEMIYA NAUK SSSR. MOSKVA
1952 V2 P231-239 PT2 63-11006 <=>

TRUDY KRASNODARSKOGO INSTITUTA PISHCHEVOI PROMYSHLENNOSTI
1956 N8 P61-69 59-17797 <+*>
1957 N16 P83-86 61-15202 <+*>
1958 V19 P23-29 65-64098 <=*$>

TRUDY KRASNODARSKOGO SELSKO-KHOZYAISTVENNOGO INSTITUTA
1935 V1 P65-79 63-11041 <=>

TRUDY KUIBYSHEVSKOGO AVIATSIONNOGO INSTITUTA
1962 V15 N1 P37-45 RTS-3248 <NLL>

TRUDY. L. P. I.
SEE TRUDY LENINGRADSKOGO POLITEKHNICHESKOGO INSTITUTA IM. M. I. KALININA

TRUDY LABORATORII AEROMETODOV. AKADEMIYA NAUK SSSR MOSKVA
1958 V6 P146-176 63-26687 <=$>
1959 N7 P58-69 61-10882 <=*> 0
1959 V7 P25-31 62-10968 <=*>
1959 V7 P114-120 62-15503 <=*>

TRUDY LABORATORII DVIGATELEI. AKADEMIYA NAUK SSSR. MOSKVA
1956 V2 P91-107 59-19788 <+*>
1957 V3 P164-193 64-13372 <=*$>
1958 V4 P79-84 64-13033 <=*$>

TRUDY LABORATORII GEOLOGII DOKEMBRIYA. AKADEMIYA NAUK SSSR. MOSKVA
1959 V8 P353-369 61-19099 <*=>
1961 V12 P123-132 71P61R <ATS>

TRUDY LABORATORII GEOLOGII UGLYA. AKADEMIYA NAUK SSSR. MOSKVA
1956 P.121-130 PT.6 T-2463 <INSD>
1956 N4 P5-102 R-4872 <*>
1956 N6 P9-17 IGR V1 N9 P12 <AGI>
1956 V6 P9-17 59-16373 <+*>
1956 V6 P137-138 61-20951 <*=>
1956 V6 N4 P5-102 60-23746 <*+>
1958 V8 P314-322 76M39R <ATS>

TRUDY LABORATORII GIDROGEOLOGICHESKIKH PROBLEM IM. F. P. SAVARENSKOGO
1955 V12 P71-88 62-14598 <=*>
1961 V35 ENTIRE VOLUME <CB>
1962 V45 P27-43 SHSP N6 1962 P649 <AGU>
1962 V45 P49-61 SHSP N6 1962 P663 <AGU>
1962 V45 P62-66 SHSP N6 1962 P674 <AGU>

TRUDY LABORATORII OZEROVEDENIYA. AKADEMIYA NAUK SSSR. MOSKVA
1959 V8 P3-22 61-28414 <*=>
1960 V10 P36-44 63-13083 <=*>

TRUDY LABORATORII VULKANOLOGII. AKADEMIYA NAUK SSSR. MOSKVA
1961 V19 P5-11 IGR V6 N9 P1523-1527 <AGI>
1962 N21 P83-99 IGR V7 N2 P325 <AGI>
1962 N22 P131-142 IGR V7 N2 P272 <AGI>

TRUDY LENINGRADSKOGO GIDROMETEOROLOGICHESKOGO INSTITUTA
1961 V10 N1 P43-51 62-23956 <*> 0
1961 V10 N1 P52-55 62-23957 <=*>
1961 V10 N1 P157-172 62-24362 <=*>
1963 V18 P3-128 65-30092 <=$>
1963 V18 P131-166 65-30092 <=$>
1963 V18 P181-183 65-30092 <=$>
1963 V19 P3-30 AD-612 366 <=$>

TRUDY LENINGRADSKOGO INDUSTRIALNOGO INSTITUTA
1938 N1 P3-26 RT-777 <*>
1938 N1 P36-49 RT-776 <*>
1938 V1 P81-97 RT-1525 <*>
1939 N1 P49-58 RT-833 <*>
1939 N1 P60-73 R-2827 <*>
RT-1524 <*>
1940 N4 P54-58 RT-1133 <*>
1940 N4 P60-74 RT-2203 <*>

TRUDY LENINGRADSKOGO INSTITUTA KINOINZHENEROV
1955 N3 P179-187 60-14279 <+*>
1955 N3 P244-247 60-23671 <+*>
1959 N5 P165-176 61-18865 <=*>
1959 N5 P177-181 61-18866 <*=>
1959 N5 P183-189 61-20201 <*=>
1959 N5 P200-209 61-18285 <*=>
1961 N6 P91-98 66-13763 <*>
1962 N8 P37-42 02Q74R <ATS>

TRUDY LENINGRADSKOGO INSTITUTA VODNOGO TRANSPORTA
1957 V24 P162-174 63-19348 <=*>

TRUDY LENINGRADSKOGO INZHENERNO-EKONOMICHESKOGO INSTITUTA
1958 N24 P81-108 60-11662 <=>

TRUDY LENINGRADSKOGO KORABLESTROITELNOGO INSTITUTA
1959 V29 P257 TRANS-201 <MT>
1960 V32 P35- 64-13292 <=*$>
1961 V33 P57-63 63-23225 <=*>
1962 N38 P15 2192 <BSRA>
1962 N39 P69 2207 <BSRA>
1962 V36 P103-105 64-71600 <=>
1962 V36 P121-133 AD-635 852 <=$> M
1962 V38 P27 65-60754 <=>
1962 V38 P89-96 65-60786 <=>
1962 V38 P97 65-61380 <=$>
1962 V38 P187 65-61347 <=$>
1964 N43 P27 2211 <BSRA>
1964 N43 P53- 2222 <BSRA>
1964 N44 P125- 2237 <BSRA>

TRUDY LENINGRADSKOGO MEDITSINSKOGO INSTITUTA
1935 N6 P13-16 R-1069 <*>

TRUDY LENINGRADSKOGO METALLICHESKOGO ZAVODA IMENI STALINA
1957 V5 P345-350 62-25890 <=*>

TRUDY LENINGRADSKOGO NAUCHNO-ISSLEDOVATELSKOGO INSTITUTA GIGIENY TRUDA I PROFESSIONALNYKH ZABOLEVANII
1948 V11 P130-131 PT2 61-13805 <*+> 0
1948 V11 P146-156 PT2 61-13805 <*+> 0
1948 V11 P159-170 PT2 61-13805 <*+> 0
1948 V11 P39-67 PT2 61-13805 <*+> 0
1948 V11 P72-107 PT2 61-13805 <*+> 0

TRUDY LENINGRADSKOGO OBSHCHESTVA ESTESTVOISPYTATELEI
1899 V28 N5 P97 65-61580 <=$> 0
1899 V28 N5 P134-136 65-61580 <=$> 0
1914 V45 P197-210 RTS-3069 <NLL>
1936 V65 P254-261 AEC-TR-4826 <=*$>
1938 V67 N4 P96-108 AEC-TR-5051 <=*>
1952 V71 N4 P73-81 61-27449 <*=>
1954 V72 N4 P124-159 65-63997 <=>
1957 V73 N4 P229-233 62-15628 <=*>
1961 V72 N1 P134-137 63-21015 <=>

TRUDY LENINGRADSKOGO POLITEKHNICHESKOGO INSTITUTA
IM. M. I. KALININA
1938 N1 P3-26 63-20752 <=*$>
1938 N1 P36-48 63-20758 <=*$>
1939 N1 P49-58 63-20753 <=*$>
1940 N4 P3-12 64-16436 <=*$> 0
1940 N4 P36-46 64-16426 <=*$> 0
1940 N4 P54-59 63-20821 <=*$>
1940 N4 P60-74 64-14156 <=*$>
1942 N1 P124-138 63-18655 <=*> 0
1948 N3 P105-114 R-1614 <*>
1948 V3 P105-114 C-2374 <NRC>
1955 N176 P175-188 61-13165 <=*> 0
1955 N176 P191-200 61-15015 <*+>
1955 N176 P201-213 61-15906 <+*>
1955 N178 P133-149 62-23283 <=*>
1955 N180 P32-37 62-13164 <*=>
1955 N180 P38-43 62-13163 <*=>
1957 N188 P40-44 94R74R <ATS>
1957 N189 P23-33 59-22631 <+*>
1957 N189 P34-42 59-22630 <+*>
1957 N189 P43-50 59-22629 <+*>
1957 N189 P51-57 59-22628 <+*>
1957 N189 P58-67 59-22627 <+*>
1957 N189 P111-122 59-22626 <+*>
1958 N193 P38-50 60-17921 <+*>
1958 N193 P66-74 60-17922 <+*>
1958 N194 P143-153 UCRL TRANS-783 <=*>
1958 N199 P26-47 65-60359 <=$>
1958 N199 P48-52 64-19939 <=$>
1958 N199 P64-74 64-13739 <=*$>
1959 N201 P24-34 4141 <BISI>
1959 N201 P102-109 6201 <HB>
1959 N202 P43- 62-13158 <*=>
1960 N210 P7-22 63-15731 <=*>
1961 N217 P7-16 63-19200 <=*>
1961 N217 P27-36 63-19657 <*=>
1961 N217 P176-179 AD-614 020 <=$>
CSIR-TR.462 <CSSA>
1962 N219 P108-114 AEC-TR-5953 <=*$>
1962 N221 P5-46 63-19173 <=*>
1962 N221 P103-109 63-19173 <=*>
1963 N222 P8-14 AD-609 916 <=> 0
1963 N222 P68-70 64-21752 <=>
1963 N222 P71-72 64-21752 <=>
1963 N224 P24-59 M.5834 <NLL>
1963 N224 P61-83 M.5832 <NLL>
1963 N224 P97-112 64-31426 <=>
1963 N224 P113-123 M5831 <NLL>
1963 N224 P124-132 M.5830 <NLL>
1963 N224 P133-141 M.5829 <NLL>
1963 N224 P142-152 M.5827 <NLL>
1963 N224 P153-175 M.5833 <NLL>
1963 N224 P177-202 64-31426 <=>
1963 N224 P217-228 M.5828 <NLL>
1963 N225 P143-148 5820 <HB>
1963 N228 P44-54 AD-627 113 <=$>
1963 V222 P20-27 AD-611 088 <=$>
1963 V223 P43-48 378 <TC>
1963 V223 P49-54 379 <TC>
1964 N230 P7-12 AD-622 452 <=$>
1964 N230 P7-13 65-64352 <=*$>
1964 N232 P42-46 AD-640 476 <=$>
1964 N233 P77-84 AD-633 699 <=$>
1964 N234 P89-95 AD-622 468 <=$*>
1964 N239 P39-56 AD-635 276 <=$>
1964 N239 P70-81 AD-635 276 <=$>
1965 N248 P7-13 66-10470 <*>
1965 N248 P56-58 66-61730 <=*$>
1965 N248 P59-64 66-12475 <*>
66-60594 <=*$>
1965 N248 P65-73 66-60595 <=*$>
1965 N248 P74-81 66-60596 <=*$>

TRUDY LENINGRADSKOGO SANITARNO-GIGIENICHESKOGO
MEDITSINSKOGO INSTITUTA
SEE TRUDY SANITARNO-GIGIENICHESKOGO
MEDITSINSKOGO INSTITUTA. LENINGRAD

TRUDY LENINGRADSKOGO TEKHNOLOGICHESKOGO INSTITUTA
IMENII LENINGRADSKOGO SOVETA (LENSOVETA)
1946 N12 P141-151 64-10430 <=*$>
1946 V12 P141-151 RT-1586 <*>
1946 V12 P152-164 3789 <K-H>
64-16149 <*> 0
1955 V33 P38-47 62-15916 <=*>
1957 N37 P78-9C 64-15102 <=*$>
1957 N38 P228-237 RCT V33 N3 P868 <RCT>
1957 N40 P155-162 64-15103 <=*$>
1957 N42 P30-38 26L30R <ATS>
1957 V37 P154-158 60-10894 <+*> 0
1959 N51 P46-51 64-10994 <=*>
1960 N56 P54 61-27056 <=*>
1960 N56 P65 61-27056 <=*>
1960 N56 P81 61-27056 <=*>
1960 N56 P84 61-27056 <=*>
1960 N56 P93 61-27056 <=*>
1960 N56 P93-98 62-14258 <=*>
1960 N56 P113 61-27056 <=*>

TRUDY LENINGRADSKOGO TEKHNOLOGICHESKOGO INSTITUTA
KHOLODILNOI PROMYSHLENNOSTI (TITLE VARIES)
1956 V10 P33 59-19603 <+*>
1956 V11 P109-114 62-13141 <*=>
1956 V11 P115-124 62-13139 <*=>
1959 V13 P41-45 63-13248 <=*>

TRUDY LENINGRADSKOGO TEKHNOLOGICHESKOGO INSTITUTA
PISHCHEVOI PROMYSHLENNOSTI
1956 V16 P225-226 61-23398 <=*>
1958 V15 P55-77 AD-625 430 <=*$>

TRUDY PO LESNOMU OPYTNOMU DELU UKRAINY. KHARKOV
1930 V7 P18-20 R-950 <*>

TRUDY LESOTEKHNICHESKOI AKADEMII IM. S. M. KIROVA.
LENINGRAD
1946 V8 P85-94 63-24290 <=*>

TRUDY MAKEEVSKOGO NAUCHNO-ISSLEDOVATELSKOGO
INSTITUTA PO BEZOPASNOSTI RABOT V GORNOI
PROMYSHLENNOSTI
1956 V6 N2 P22-26 62-25579 <=*>
1960 V11 P3-9 62-25455 <=*>
1960 V11 P239-248 62-25466 <=*>
1961 V12 P12-18 63-15244 <=*>

TRUDY MASHINOSTROITELNYY INSTITUT. OMSK
1956 N1 P107-118 61-19921 <=*>

TRUDY MATEMATICHESKOGO INSTITUTA IMENI V. A. STEK-
LOVA. AKADEMIYA NAUK SSSR. LENINGRAD AND MOSCOW
1945 V12 P1-106 RT-2312 <*>
1945 V12 P106 R-3409 <*>
1949 P25- RT-364 <*>
1949 N25 ENTIRE ISSUE AMST S2 V11 P82-143 <AMS>
1949 V28 P104-144 63-20748 <=*$>
1950 N32 ENTIRE ISSUE AMST S1 V3 P197-270 <AMS>
1950 V32 P3-99 RT-1550 <*>
1951 V38 P42-67 AMST S1 V2 P337-366 <AMS>
RT-2870 <*>
1951 V38 P176-189 AMST S2 V15 P1-14 <AMS>
1951 V38 P190-243 AMST S2 V3 P39-89 <AMS>
1951 V38 P279-316 AMST S2 V29 P51-89 <AMS>
1951 V38 P321-344 AMST S2 V3 P15-38 <AMS>
1954 V42 P375- 60-51085 <=>
1955 N44 ENTIRE ISSUE AMST S2 V9 P1-122 <AMS>
1955 N45 ENTIRE ISSUE AMST S2 V11 P1-114 <AMS>
1955 N46 ENTIRE ISSUE AMST S2 V7 P1-98 <AMS>
AMST S2 V7 P98-134 <AMS>
1955 N47 ENTIRE ISSUE AMST S2 V7 P135-321 <AMS>
1955 V48 P3-110 AMST S2 V30 P1-102 <AMS>
1957 V50 P1-65 60-41520 <=>
1958 V51 P5-141 60-11974 <=>
1958 V51 P143-157 60-11970 <=>
62-23019 <=>
1958 V51 P158-173 60-11613 <=>
1958 V51 P174-185 60-11804 <=>
1958 V51 P186-225 60-41027 <=>
1958 V51 P226-269 60-41028 <=>

1958 V51 P270-360 60-41069 <=>
1958 V52 P7-65 AMST S2 V29 P91-151 <AMS>
1958 V52 P66-74 AMST S2 V29 P153-161 <AMS>
1958 V52 P75-139 AMST S2 V23 P15-81 <AMS>
1958 V52 P226-311 AMST S2 V23 P109-189 <AMS>
1958 V52 P315-348 AMST S2 V29 P163-195 <AMS>
1959 N53 P387-391 63-00895 <*>
1959 V53 P186-192 61-15073 <+*>
62-23036 <=>
1959 V53 P387-391 63-23144 <=*$>
1959 V54 P1-136 AMST S2 V30 P103-233 <AMS>
1960 V57 P1-84 63-23095 <=*>
1960 V59 P37-86 AMST S2 V35 P297-350 <AMS>
1960 V59 P115-173 63-23100 <=*>
1961 V60 P5-21 AMS-TRANS V43 P281 <AMS>
AMST S2 V43 P281-297 <AMS>
1961 V60 P42-81 AMS TRANS V.40 P85 <AMS>
1961 V60 P82-95 63-14733 <=*$>
63-19475 <=*>
1961 V60 P101-144 AMST S2 V48 P107-150 <AMS>
1961 V60 P181-194 AMST S2 V35 P351-363 <AMS>
1961 V60 P238-261 STMSP V4 P245-269 <AMS>
1961 V60 P262-281 AMS TRANS V44 P129 <AMS>
AMST S2 V44 P129-150 <AMS>
1961 V60 P304-324 AMS TRANS V44 P67-88 <AMS>
AMST S2 V44 P67-88 <AMS>
1961 V62 P3-47 AMST S2 V28 P221-268 <AMS>
1961 V62 P48-60 AMST S2 V28 P269-282 <AMS>
1961 V62 P61-97 AMST S2 V28 P283-322 <AMS>
1961 V64 P9-27 AMS-TRANS V43 P195 <AMS>
AMST S2 V43 P195-213 <AMS>
1961 V64 P52-6C STMSP V5 P201-209 <AMS>
1961 V64 P61-78 AMS TRANS V44 P89 <AMS>
AMST S2 V44 P89-107 <AMS>
1961 V64 P79-89 AMST S2 V55 P195-206 <AMS>
1961 V64 P165-172 AMST S2 V52 P1-8 <AMS>
1961 V64 P185-210 STMSP V5 P210-239 <AMS>
1961 V64 P316-346 AMST S2 V37 P85-114 <AMS>
1962 V66 P4-15 63-23638 <=*>
1962 V66 P37-44 63-23638 <=*>
1962 V66 P113-135 AMS TRANS V.40 P229 <AMS>
AMST S2 V40 P229-255 <AMS>
65-10373 <*>
1962 V66 P136-146 AMS TRANS V.40 P257 <AMS>
AMST S2 V40 P257-269 <AMS>
1962 V66 P147-165 AMS TRANS V.40 P147 <AMS>
AMST S2 V40 P271-290 <AMS>
1962 V67 P385-457 AMST S2 V57 P1-84 <AMS>
1962 V68 P1-25 64-10720 <=*$>
1962 V68 P3-88 AMS TRANS V.41 P215 <AMS>
AMST S2 V41 P215-317 <AMS>
1963 V69 P3-122 AERE-TRANS-1002 <*>
1964 N71 P3-16 STMSP V6 P37 <AMS>
1964 N71 P17-20 STMSP V6 P51 <AMS>
1964 N71 P21-25 STMSP V6 P55 <AMS>
1964 N71 P26-34 STMSP V6 P61 <AMS>
1964 N71 P35-45 STMSP V6 P71 <AMS>
1964 N71 P46-50 STMSP V6 P84 <AMS>
1964 N71 P51-61 STMSP V6 P89 <AMS>
1964 N71 P62-77 STMSP V6 P100 <AMS>
1964 N71 P78-81 STMSP V6 P118 <AMS>
1964 N71 P82-87 STMSP V6 P123 <AMS>
1964 N71 P87-101 STMSP V6 P129 <AMS>
1964 N71 P102-103 STMSP V6 P144 <AMS>
1964 N71 P104-112 STMSP V6 P147 <AMS>
1964 N71 P113-117 STMSP V6 P157 <AMS>

TRUDY I MATERIALY URALSKOGO NAUCHNO-ISSLEDOVATELSKOGO I PROEKTNOGO INSTITUTA MEDNOI PROMYSHLENNOSTI. SVERDLOVSK
1957 V2 P316-321 62-13725 <=*>
1957 V2 P322-328 62-13726 <=*>

TRUDY MINERALOGICHESKOGO INSTITUTA. AKADEMIYA NAUK SSSR. LENINGRAD
1959 V11 N9 P172-175 60-23736 <*+>

TRUDY MINERALOGICHESKOGO MUZEYA AKADEMII NAUK SSSR. LENINGRAD, MOSKVA
1954 N6 P117-121 R-4684 <*>
1955 V7 P70-75 R-1601 <*>
1955 V7 P132-150 IGR V1 N8 P52 <AGI>
R-1222 <*>
1957 V8 P25-28 61-19235 <=*>
1957 V8 P61-76 62-15740 <*=>
1957 V8 P128-131 61-19236 <=*>
1957 V8 P164-167 61-19239 <=*> 0
1959 V9 P138-145 61-19238 <=*>
1961 N12 P235-238 65-63370 <=>
1965 N8 P29-36 TRANS-F-139 <NLL>

TRUDY MIROVOI METEOROLOGICHESKII TSENTR. LENINGRAD
1964 N4 P3-26 65-30777 <=$>
1964 N4 P44-49 65-30777 <=$>
1965 N5 P35-40 TRANS-F.150 <NLL>
1965 N8 P3-14 TRANS-F-138 <NLL>
1965 N8 P45-67 67-30248 <=>
1965 N8 P76-86 67-30248 <=>

TRUDY MONGOLSKOI KOMISSII. AKADEMIYA NAUK SSSR
1951 V41 ENTIRE ISSUE 64-11073 <=>

TRUDY MORSKOGO GIDROFIZICHESKOGO INSTITUTA. AKADEMIYA NAUK SSSR. MOSKVA
1950 V4 P15-22 59-22614 <+*> 0
1954 V4 P31-71 CSIR-TR.545 <CSSA>
1956 V8 P44-62 59-22613 <+*> 0
1956 V9 P141 AD-637 960 <=$>
1957 V10 P106-118 ORNL-TR-515 <=$>
1957 V11 P84-96 63-20639 <=*$>
1958 V13 P54-64 AD-627 151 <=$>
1958 V13 P73-77 60-11172 <=>
1959 V15 P80-85 65-61503 <=$>
1959 V15 P144-166 61-13310 <*+> 0
1959 V18 P94-113 62-23560 <=*>
1959 V18 P113-116 63-23762 <=*$>
1960 V19 P3-20 62-24073 <=*>
1960 V20 P36-43 65-61269 <=$>
1960 V22 P3-4 62-33572 <=*>
1960 V22 P26-32 62-33573 <=*>
1961 V23 P139-147 65-60803 <=> 0
1961 V23 P148-150 63-23629 <=*$>
1962 V25 P142-186 64-19360 <=$> 0

TRUDY MOSKOVSKOE VYSSHEE TEKHNICHESKOE UCHILISHCHE
1953 V26 P221-238 RTS-3088 <NLL>
1955 V32 P75-99 64-71333 <=> M
1961 V101 P5-107 62-32410 <=*>
1961 V101 P186-216 62-32410 <=*>
1961 V101 P236-252 62-32410 <=*>

TRUDY MOSKOVSKOGO AVIATSIONNOGO INSTITUTA
SEE TRUDY MOSKOVSKOGO ORDENA LENINA AVIATSIONNOGO INSTITUTA IM. SERGO ORDZHONIKIDZE

TRUDY MOSKOVSKOGO AVIATSIONNOGO TEKHNOLOGICHESKOGO INSTITUTA
1956 N27 ENTIRE ISSUE 59-18109 <+*>
1959 V37 P5-32 61-27171 <*=> 0
1959 V39 P74-90 65-17407 <*>
1959 V40 P21-54 63-19441 <=*>
1959 V42 P105-113 62-32939 <=*>
1960 V43 P68-85 62-25905 <=*>
1960 V43 P86-9C 62-19134 <*=>
1961 V49 ENTIRE ISSUE AD-617 093 <=$>
1961 V49 P5-46 62-23681 <=*>
1961 V50 ENTIRE ISSUE AD-646 733 <=>
1961 V52 P20-28 63-15357 <=*>
1961 V52 P33-44 63-19413 <=*>
1961 V52 P52-60 63-19317 <=*>
1961 V52 P61-72 63-11581 <=>
1962 V53 P5-22 63-19676 <=*$>
1963 V57 P95-98 63-41333 <=>
1963 V57 P99-1C3 63-41333 <=>
1963 V57 P104-109 63-41333 <=>
1963 V57 P110-113 63-41333 <=>
1963 V57 P114-119 63-41333 <=>
1963 V57 P120-126 63-41333 <=>
1963 V57 P127-134 63-41333 <=>
1963 V57 P135-139 63-41333 <=>

1963 V58 P5-20 64-31337 <=>
1964 N59 P74-82 AD-638 902 <=$>
1964 V60 P72-107 AD-623 211 <=*$>
1964 V61 P73-85 RTS-3035 <NLL>

TRUDY MOSKOVSKOGO AVTOMOBILNO-DOROZHNOGO INSTITUTA
1955 V16 P79-106 61-15358 <+*> 0
1955 V16 P120-133 61-17567 <=$>
1955 V17 P85-100 59-22417 <+*>

TRUDY MOSKOVSKOGO ENERGETICHESKOGO INSTITUTA
1953 V11 P20-39 63-19798 <=*$>
1956 V28 N18 P135-146 59-15617 <+*>
1963 N48 P75-83 RTS-3243 <NLL>
1963 N48 P147-149 CSIR-TR-459 <CSSA>

TRUDY MOSKOVSKOGO FIZIKO-TEKHNICHESKOGO INSTITUTA
1958 V2 P13-29 63-21485 <=>
1959 V4 P49-61 62-32772 <=*> 0
1959 V4 P85-89 65-11617 <*>
1961 V7 P110-113 63-13219 <=*>
1961 V7 P136-151 63-23651 <=*>
1961 V7 P152-157 AD-630 287 <=*$>
1962 V8 P67-72 63-15402 <=*>
1962 V8 P73-76 63-23188 <=*>
1962 V10 P67-79 64-71135 <=>
1962 V10 P80-86 64-15394 <=*$>

TRUDY MOSKOVSKOGO GEOLOGO-RAZVEDOCHNOGO INSTITUTA IM. ORDZHONIKIDZE
1956 V29 P203-213 56K26R <ATS>
1961 V37 P190-204 66-24685 <$>
1963 V39 P92-100 IGR V7 N2 P343 <AGI>

TRUDY MOSKOVSKOGO INSTITUTA INZHENEROV GEODEZIL, AEROFOTOSEMKI I KARTOGRAFII
1953 V16 P25-30 62-15547 <=*>
1953 V16 P37-50 62-15545 <=*>
1953 V16 P51-57 62-15546 <=*>
1955 V20 P23-34 62-25414 <=*>
1955 V21 P33-48 62-15548 <=*>
1955 V21 P57-83 62-15549 <=*>
1955 V22 P3-25 62-15550 <=*>
1957 V24 P3-42 62-15551 <=*>
1957 V24 P57-59 62-15552 <=*>
1957 V24 P105-110 62-15553 <=*>
1957 V25 P23-42 62-15554 <=*>
1957 V25 P43-50 62-15555 <=*>
1957 V25 P51-59 62-15556 <=*>
1957 V25 P67-75 62-15557 <=*>
1957 V27 P45-51 62-15983 <=*>
1957 V27 P65-78 62-15983 <=*>
1957 V27 P91-93 62-15983 <=*>
1957 V28 P3-26 60-19113 <+*> 0
1957 V28 P27-40 62-15558 <=*>
1957 V28 P41-52 62-15559 <=*>
1957 V28 P61-70 62-15560 <=*>
1957 V28 P71-76 62-15561 <=*>
1957 V28 P77-104 62-15562 <=*>
1957 V29 P27-32 62-15576 <=*>
1957 V29 P57-68 62-15576 <=*>
1957 V29 P81-108 62-15576 <=*>
1958 V30 P5-11 62-25417 <=*>
1958 V32 P37-60 62-15573 <=*>
1958 V32 P107-112 62-15573 <=*>
1958 V33 P15-18 AD-633 403 <=$>
1959 N31 P123-131 66-13773 <*>
1959 V34 P129-130 63-21035 <=>
1960 V39 P89-110 C-4472 <NRC>
1960 V40 P103-113 63-13221 <=*>
1961 V44 P85-88 63-19407 <=*>

TRUDY MOSKOVSKOGO INSTITUTA INZHENEROV ZHELEZNODOROZHNOGO TRANSPORTA
1950 V74 P334-356 99P61R <ATS>
1961 N139 P134-158 63-19332 <=*>
1961 N139 P206-209 64-19109 <=*$>
1961 V139 P31-44 64-71213 <=>
1961 V139 P105-110 64-71575 <=>
1961 V139 P122-130 AD-609 160 <=>
1962 V155 P176-180 64-23503 <=$>
1963 N165 P82-91 AD-621 784 <=$*>
1963 V164 P85-91 64-30054 <*>

TRUDY MOSKOVSKOGO INSTITUTA KHIMICHESKOGO MASHINOSTROENIYA
1950 V10 P19-31 63-24621 <=$>
1963 V25 P172-180 AD-617 658 <=$>
1964 V27 P194-200 AD-625 286 <=*$>

TRUDY MOSKOVSKOGO INSTITUTA MEKHANIZATSII
1958 V26 P94-100 66-61717 <=$>

TRUDY MOSKOVSKOGO INSTITUTA MEKHANIZATSII I ELEKTRIFIKATSII SELSKOGO KHOZYAISTVA
1956 N2 P102-122 60-23032 <+*>

TRUDY MOSKOVSKOGO INSTITUTA NEFTEKHIMICHESKOI I GAZOVOI PROMYSHLENNOSTI
1939 P129-142 61-18113 <*>
1946 V4 P123-127 63-19962 <*=>
1955 V14 P115-134 95R79R <ATS>
1956 V17 P101-111 07R77R <ATS>
1957 V18 P157-167 63-10283 <=*>
1958 V22 P115-125 24Q69R <ATS>
1958 V23 P9-21 39R77R <ATS>
1959 V24 P95-106 63-10304 <=*>
1959 V25 P268-284 23Q69R <ATS>
65-11400 <*>
1960 N31 P64-67 IGR V5 N6 P692 <AGI>
1960 V28 P44-55 AD-619 476 <=$> M
1960 V28 P56-67 AD-619 476 <=$> M
1960 V29 P63-69 80R76R <ATS>
1960 V31 P68-77 IGR V7 N1 P11-16 <AGI>
1961 V33 P122-130 32Q71R <ATS>
1961 V33 P296-304 31Q71R <ATS>
1963 N41 P84-87 65-12737 <*>
1963 N44 P96-100 ICE V4 N4 P618-620 <ICE>
1963 V42 P213-221 66-10301 <*>
1963 V44 P92-95 ICE V4 N4 P643-645 <ICE>
48R76R <ATS>
1963 V44 P96-100 ICE V4 N4 P618-620 <ICE>
49R76R <ATS>
1963 V44 P251-257 AD-636 662 <=$>
1964 N50 P224-231 66-13862 <*>
1964 N51 P48-53 AD-639 575 <=$>
1964 V51 P240-250 RTS-3340 <NLL>

TRUDY MOSKOVSKOGO KHIMIKO-TEKHNOLOGICHESKOGO INSTITUTA IM. D. I. MENDELEEVA
1940 N6 P31-47 R-3299 <*>
1949 V16 N2 P43-52 51R79R <ATS>
1949 V16 N2 P87-101 T.764 <INSD>
1956 V21 P177-193 62-13109 <*=>
1956 V22 P113-115 40N51R <ATS>
1957 V22 N24 P96-110 61-15226 <*+>
1957 V22 N24 P318-323 46M37R <ATS>
59-15810 <+*>
1957 V22 N24 P389-404 61-13749 <=*>
1957 V22 N24 P413-427 61-23511 <*=>
1959 N29 P11-14 63-10595 <=*>
1959 N29 P23-33 63-10826 <=*>
1961 N32 P259-265 63-19875 <=*>
1961 N35 P48-59 65-64397 <=*$>
1961 N35 P60-66 65-64395 <=*$>
1961 V32 P266-271 64-23558 <=>
1961 V35 P60-66 65-11345 <*>
1961 V35 P101-107 SNSRC-169 <SNSR>
1962 N38 P52-58 65-64418 <=*$>
1963 N41 P197-210 65-12955 <*>

*TRUDY MOSKOVSKOGO MATEMATICHESKOGO OBSHCHESTVA
1952 V1 P39-166 AMST S2 V6 P245-378 <AMS>
1952 V1 P187-246 AMST S2 V24 P107-172 <AMS>
1954 V3 P181-270 AMST S2 V16 P103-193 <AMS>
1954 V3 P271-306 AMST S2 V38 P37-73 <AMS>
1955 V4 P237-290 AMST S2 V20 P1-53 <AMS>
1955 V4 P375-404 AMST S2 V9 P123-154 <AMS>
1955 V4 P421-438 AMST S2 V38 P75-93 <AMS>
1955 V4 N33 P3-374 STMSP V4 P345-387 <AMS>

1956 V5 P3-80	AMST S2 V11 P173-254 <AMS>
1956 V5 P89-201	AMST S2 V16 P195-314 <AMS>
1956 V5 P311-351	AMST S2 V21 P193-238 <AMS>
1956 V5 P353-366	AMST S2 V14 P201-216 <AMS>
1956 V5 P367-432	AMST S2 V13 P105-175 <AMS>
1957 V6 P4-133	AMST S2 V11 P255-385 <AMS>
1957 V6 P371-463	AMST S2 V21 P239-339 <AMS>
1958 V7 P1-62	AMST S2 V35 P167-235 <AMS>
1958 V7 P391-405	AMST S2 V29 P197-215 <AMS>
1958 V7 P407-412	AMST S2 V23 P7-13 <AMS>
1959 V8 P71-81	AMST S2 V25 P161-172 <AMS>
1959 V8 P83-120	AMST S2 V34 P241-281 <AMS>
1959 V8 P121-153	AMST S2 V36 P101-136 <AMS>
1959 V8 P199-257	AMST S2 V25 P11-76 <AMS>
1959 V8 P283-320	AMST S2 V27 P191-230 <AMS>
1959 V8 P321-390	AMST S2 V37 P351-429 <AMS>
1959 V8 P391-412	AMST S2 V27 P231-255 <AMS>
1959 V8 P413-496	AMST S2 V34 P283-373 <AMS>
1959 V8 P497-518	STMSP V3 P291-313 <AMS>
1960 V9 P81-120	64-19949 <=$>
	65-17111 <*>
1960 V9 P121-128	64-19950 <=$>
1960 V9 P143-189	STMSP V5 P127-178 <AMS>
1960 V9 P237-282	AMST S2 V36 P137-187 <AMS>
1960 V9 P425-453	AMST S2 V39 P207-239 <AMS>
1960 V9 P507-535	AMST S2 V32 P359-393 <AMS>
	AMST,S2,V32,P359-393 <AMS>
1961 V10 P181-216	AMST S2 V36 P189-229 <AMS>
1961 V10 P217-272	AMST S2 V55 P49-140 <AMS>
1961 V10 P297-350	AMST S2 V49 P1-62 <AMS>
1962 V11 P3-35	AMS TRANS V.40 P193 <AMS>
	AMST S2 V40 P193-228 <AMS>
1962 V11 P99-126	AMST S2 V55 P49-140 <AMS>
1962 V11 P199-242	AMST S2 V50 P5-58 <AMS>
1963 V12 P3-466	<AMS>

TRUDY MOSKOVSKOGO NEFTYANOGO INSTITUTA IM. I. M. GUBKINA

1953 V12 P80-91	60-18917 <+*>
1956 V16 P70-81	53L34R <ATS>
1957 V18 P39-75	60-23672 <+*>
1957 V18 P108-127	60-19882 <+*>
1957 V18 P168-183	60-19883 <+*>
	62-18526 <=*>
1958 V22 P142-157	60-18748 <+*>
1958 V22 P158-169	61-10358 <*+>
1958 V22 P198-206	60-16961 <+*>
1958 V23 P78-83	95L36R <ATS>
1958 V23 P84-94	20M42R <ATS>
1960 N31 P68-77	IGR V7 N1 P11 <AGI>

TRUDY MOSKOVSKOGO OBSHCHESTVA ISPYTATELEI PRIRODY OTDEL BIOLOGICHESKII

1963 V7 N2 P30-41	AD-610 805 <=$>

TRUDY MOSKOVSKOGO ORDENA LENINA AVIATSIONNOGO INSTITUTA IM. SERGO ORDZHONIKIDZE

1956 V61 P20-29	64-11890 <=>
	64-11890 <=> M
1956 V61 P30-36	64-71412 <=> M
1956 V61 P37-40	64-11951 <=> M
1956 V61 P58-66	64-11874 <=> M
1956 V68 P42-60	61-13006 <*=>
1957 N93 P5-20	59-19477 <+*> O
1957 N93 P39-63	59-19477 <+*> O
1958 V95 P77-119	62-15395 <*=>
1958 V97 P1-238	61-27317 <*=>
1959 V104 P12-17	64-14122 <=*$>
1959 V114 P1-69	62-13563 <=*>
1959 V117 P82-102	RTS-2682 <RTS>
1960 V119 P1-174	62-13568 <*=>
1960 V119 P43-71	AD-640 728 <=$>
1960 V119 P162-173	61-28377 <*=>
1960 V123 P17-34	62-15273 <=*> O
1960 V123 P45-52	62-15273 <=*> O
1960 V123 P65-68	62-15273 <=*> O
1960 V129 P100-111	62-32697 <=*> O
1960 V130 P19-56	63-19172 <=*>
1960 V130 P110-132	63-19172 <*=>
1961 N139 P71-107	63-13196 <=*>
1961 V132 ENTIRE ISSUE	62-24263 <=*>
1961 V133 ENTIRE ISSUE	64-13945 <=*$>
1961 V134 ENTIRE ISSUE	AD-646 932 <=> M
1961 V138 ENTIRE ISSUE	63-13205 <=*>
1961 V138 P28-47	AD-611 013 <=$>
1961 V139 P119-128	63-15281 <=*>
1961 V139 P129-133	63-13903 <=*>
1961 V140 P29-36	64-19982 <=$>
1961 V142 P5-141	62-33534 <=*>
1962 V145 P60-98	63-19492 <=*>
1962 V147 P1-139	63-23737 <=*>
1964 N157 P5-16	AD-651 086 <=$> M
1964 N157 P51-90	AD-651 086 <=$> M
1964 N158 P29-34	AD-625 148 <=*$>
1964 N159 P257-272	AD-637 432 <=$>
1964 V160 P1-284	AD-637 105 <=$>

TRUDY MOSKOVSKOGO TEKHNOLOGICHESKOGO INSTITUTA PISHCHEVOI PROMYSHLENNOSTI

1957 N7 P7-20	R-5234 <*>
1957 N8 P37-41	C-3882 <NRC>
1962 N25 P41-44	98R78R <ATS>

TRUDY MOSKOVSKOGO TEKHNOLOGICHESKOGO INSTITUTA RYBNOI PROMYSHLENNOSTI I KHOZYAISTVA

1939 P163-172	C-4631 <NRC>
1953 N5 P201-206	60-17919 <+*>

TRUDY MOSKOVSKOGO VYSSHOGO TEKHNICHESKOGO UCHILISHCHA

1959 V92 P66-84	63-24252 <=*$>
1962 N110 ENTIRE ISSUE	65-62713 <=>

TRUDY MOSKOVSKOGO ZOOPARKA

1941 V2 N2 P94-101	R-803 <*>
	RT-4117 <*>
	64-19097 <=$>

TRUDY MOSKOVSKOI ORDENA LENINA SELSKOKHOZYAISTVENNOI AKADEMII IM. K. A. TIMIRYAZEVA

1940 N4 P29-50	R-2854 <*>
1956 V12 P116-122	59-10089 <+*>

TRUDY MOSKOVSKOI VETERINARNOI AKADEMII

1956 V12 P54-61	59-10087 <+*>
1956 V12 P73-81	R-3612 <*>
1956 V17 P120-123	R-2388 <*>

TRUDY MURGABSKOI GIDROBIOLOGICHESKOI STANTSII

1957 V3 P114-129	62-23863 <=*> O

TRUDY MURMANSKOGO MORSKOGO BIOLOGICHESKOGO INSTITUTA AKADEMII NAUK SSSR

1960 V1 P3-27	61-28133 <=*>
1960 V2 P68-113	63-11106 <=>
1960 V2 P172-185	61-28156 <*=>
1960 V2 P186-202	63-11124 <=>
1960 V2 P245-252	63-11164 <=>
1961 V3 P147-169	64-15987 <=*$>
1962 V4 P244-249	64-19345 <=$>

TRUDY MURMANSKOI BIOLOGICHESKOI STANTSII

1948 V1 P215-241	61-15918 <+*>
1958 V4 P1-196	60-41300 <=>

TRUDY NAUCHNO GORNYY INSTITUT. KHARKOV

1955 V2 P127-135	77L34R <ATS>

TRUDY NAUCHNOGO INSTITUTA PO UDOBRENIYAM I INSEKTOFUNGITSIDAM IM. YA. V. SAMOILOVA. MOSCOW

1922 N64 P5-61	R-2859 <*>
1929 N59 P5-160	R-3052 <*>
1932 N92 P59-67	R-2878 <*>
	720TM <CTT>
1932 N92 P117-127	R-2958 <*>
1933 N101 P5-11	R-2203 <*>
1933 N101 P12-20	R-2199 <*>

1933 N101 P21-26 R-2426 <*>
1933 N101 P33-38 R-2202 <*>
1933 N101 P38-45 R-2915 <*>
1933 N101 P45-53 R-2200 <*>
1933 N101 P73-91 R-2198 <*>
1933 N101 P91-98 R-2425 <*>
1933 N101 P113-122 R-799 <*>
1933 N110 P4-11 R-1555 <*>
R-798 <*>
1933 N110 P23-34 R-2490 <*>
1933 N113 P64-67 R-3288 <*>
1933 N113 P74-78 R-2271 <*>
1933 N113 P78-79 R-2886 <*>
1935 N123 P162-167 R-2591 <*>
1935 N123 P176-179 R-2246 <*>
R-545 <*>
1935 N123 P215-226 R-3335 <*>
1936 N127 P62-66 R-3047 <*>
1937 N137 P10-35 <LSA>
1937 N137 P10-86 R-2927 <*>
1937 N137 P102-109 R-2456 <*>
1937 N137 P110-126 R-2457 <*>
98L30R <ATS>
1937 N137 P127-144 R-2897 <*>
1937 N139 P3-73 R-2270 <*>
1938 N140 P128-164 R-3360 <*>
1938 N141 P70-76 R-2264 <*>
1938 N141 P87-96 R-3357 <*>
1938 V141 P98-104 R-118 <*>
1938 V141 P256-261 R-2176 <*>
1940 N153 P12-42 R-2987 <*>
RJ-87 <ATS>
1940 N153 P103-110 R-2843 <*>
1940 N153 P143-192 R-3014 <*>
1940 N153 P215-227 R-2618 <*>
1940 N153 P228-241 70J16R <ATS>
1940 V153 P111-121 R-2482 <*>
1940 V153 P193-201 R-2992 <*>
1940 V153 P202-214 R-2991 <*>
1940 V153 P228-241 63-14607 <=*>
1958 N160 P5-8 11T90R <ATS>
1961 N171 P52-60 78R75R <ATS>

TRUDY. NAUCHNO-ISSLEDOVATELSKOGO FIZIKO-KHIMICHESKOGO INSTITUTA
1963 N3 P66-75 65-10830 <*>
1963 N3 P130-133 65-12938 <*>

TRUDY. NAUCHNO-ISSLEDOVATELSKII GIDROMETEOROLOGICHESKII INSTITUT. TIFLIS
1962 V8 P113-116 64-41781 <=$>
1962 V10 P117-120 63-24145 <=*$>
1963 N10 P124-134 TRANS-F.124. <NLL>

TRUDY NAUCHNO-ISSLEDOVATELSKIKH UCHREZHDENII. SERIYA 1: METEOROLOGIYA. GLAVNOE UPRAVLENIE GIDROMETEOROLOGICHESKOI SLUZHBY. USSR
1946 N34 P69-76 RT-1402 <*>
1947 P3-10 RT-2054 <*> 0
1947 P68-86 RT-2584 <*>
1947 N39 P77-85 RT-1854 <*>

TRUDY NAUCHNO-ISSLEDOVATELSKOGO INSTITUTA AEROKLIMATOLOGII. MOSCOW
1958 N5 P35-41 61-13094 <=*> 0
1958 N6 P52-63 C-4041 <NRCC>
1963 N9 P86-94 TRANS-F.125. <NLL>
1963 N21 P95-133 TRANS-362(M) <NLL>
1963 N21 P134-150 TRANS-363(M) <NLL>
1964 N27 P123-128 65-31313 <=$>
1964 N28 P3-27 65-31309 <=$>
1964 N28 P89-95 65-31311 <=$>
1964 N28 P96-111 65-31312 <=$>
1965 N30 P119-132 66-31053 <=$>

TRUDY NAUCHNO-ISSLEDOVATELSKOGO INSTITUTA BETONA I ZHELEZOBETONA
1958 N3 P140-162 01L29R <ATS>

TRUDY NAUCHNO-ISSLEDOVATELSKOGO INSTITUTA PO BEZOPASNOSTI RABOT I GORNOI PROMYSHLENNOSTI
1957 N3 P18-20 64-19154 <=*$> P
1959 V9 P330-348 63-24563 <=*$>
1959 V9 N2 P3-28 61-15740 <*=>
1959 V9 N2 P94-115 61-15919 <*=>
1959 V9 N2 P127-141 61-15390 <*=>
1960 V11 P214-238 64-19150 <=*$>
1961 V12 P324-338 62-32950 <=*>
1962 V13 P219-240 63-24562 <=*$>
1962 V14 P36-40 SMRE-TRANS-5083 <NLL>
1962 V14 P55-67 SMRE-TRANS-5082 <NLL>
1962 V14 P136-145 SMRE-TRANS-5081 <NLL>
1962 V14 P146-155 SMRE-TRANS-5080 <NLL>
1962 V14 P156-166 SMRE-TRANS-5079 <NLL>
1962 V14 P249-265 SMRE-TRANS-5090 <NLL>
1962 V14 N5 P3-9 SMRE-TRANS-5085 <NLL>
1962 V14 N5 P13-35 SMRE-TRANS-5084 <NLL>
1962 V14 N5 P209-219 SMRE-TRANS-5091 <NLL>
1962 V14 N5 P250-293 SMRE-TRANS-5089 <NLL>
1962 V14 N5 P294-301 SMRE-TRANS-5088 <NLL>
1963 P261-300 SMRE-TRANS-5060 <NLL>
1963 V15 P29-42 SMRE-TRANS-5057 <NLL>
1963 V15 P42-58 SMRE-TRANS-5058 <NLL>
1963 V15 P58-68 SMRE-TRANS-5059 <NLL>
1963 V15 P356-374 SMRE-TRANS-5062 <NLL>

TRUDY NAUCHNO-ISSLEDOVATELSKOGO INSTITUTA GEODEZII I KARTOGRAFII. MOSCOW
1931 N1 P20-24 62-15593 <*=>
1931 N1 P25-41 62-15594 <*=>
1931 N3 P3-72 62-15588 <=*>
1931 N4 P187-198 62-25418 <=*>
1931 N4 P272-286 62-25418 <=*>
1936 N11 P2-7 62-25420 <=*>
1936 N11 P59-72 62-25420 <=*>
1937 N19 P5-34 62-25416 <=*>
1949 N66 P75-83 63-13564 <=*>
1949 V68 ENTIRE ISSUE 62-15563 <=*>
1950 N73 P77-112 63-13563 <=*>
1950 N73 P153-173 63-13563 <=*>
1950 V73 P11-77 AD-639 111 <=$>
1951 V86 P96 AD-641 753 <=$>
1952 N93 P5-50 62-15564 <*=>
1953 P105-154 59-11370 <=> 0
1955 N107 P219-226 60-16962 <*+>
1956 N111 P1-339 62-15572 <*=>
1956 V112 P1-8 AD-639 113 <=$>
1956 V112 P9-12 AD-639 112 <=$>
1956 V112 P13-21 AD-640 380 <=$>
1957 N114 P17-125 62-15518 <*=>
1957 N120 P1-47 62-15582 <*=>
1957 N122 P3-32 62-15544 <*=>
1957 N122 P33-70 62-15602 <*=>
1958 N123 P1-294 62-15571 <*=>
1958 V118 P2-28 63-13566 <=*>
1958 V118 P141-147 63-13566 <=*>
1959 N129 P1-259 62-19786 <*=>
1960 N131 ENTIRE ISSUE 61-31207 <=>
1960 N137 ENTIRE ISSUE 63-26229 <=$>
1960 V139 P3-59 62-25987 <=>
1960 V139 P61-75 AD-612 375 <=$>
1960 V139 P77-81 62-25987 <=>
1960 V139 P83-87 62-25987 <=>
1960 V139 P89-101 62-25987 <=>
1960 V139 P103-145 62-25987 <=>
1961 N138 P1-11 64-15170 <=*$>
1962 N147 P17-33 63-13462 <=*>
1962 N156 P38-47 63-24353 <=*$>
1962 N156 P55-58 63-24353 <=*$>
1962 N156 P144-149 63-24353 <=*$>
1962 V145 P3-82 63-19244 <=*>
1962 V145 P3-21 64-11090 <=$>
1962 V145 P23-41 64-11090 <=$>
1962 V145 P43-59 64-11090 <=$>
1962 V146 P3-15 64-11876 <=>
1962 V156 P55-58 63-24353 <=*$>
1962 V156 P144-149 63-24353 <=*$>
1963 V154 P125-148 64-31314 <=>
1964 N149 P70-92 AD-636 597 <=$>
1964 V159 P29-59 AD-632 485 <=$>

1965 V157 P1-46 AD-638 421 <=$>
1965 V157 P69-84 AD-634 927 <=$>
1965 V157 P85-100 AD-634 930 <=$>
1965 V157 P116-124 AD-633 404 <=$>
1965 V157 P125-134 AD-633 405 <=$>
1965 V157 P135-138 AD-637 537 <=$>
1965 V157 P139 AD-634 934 <=$>

TRUDY NAUCHNO-ISSLEDOVATELSKOGO INSTITUTA GEOLOGII ARKTIKI. LENINGRAD
1956 V86 P9-112 62-18719 <=*>
1957 V81 P11-20 61-19134 <+*>
1957 V81 P21-22 61-19018 <+*>
1959 V65 P4-15 64-19747 <=$>
1959 V65 P57-72 64-19748 <=$>
1959 V91 ENTIRE ISSUE 61-15879 <=*> 0
1959 V98 N1 P63-72 65-14291 <*>
1960 V109 ENTIRE ISSUE 63-23495 <=*> 0
1961 V115 P1-222 63-24376 <=*$>
1961 V119 P23-37 66-14417 <*$>
1961 V120 P18-46 53P65R <ATS>
1961 V120 P258-262 53P65R <ATS>
1961 V125 P103-112 63-24352 <=*$>
1961 V125 P135-159 65-60341 <=$>
1962 V121 P154-164 63-23252 <=*$>

TRUDY NAUCHNO-ISSLEDOVATELSKOGO INSTITUTA GIDROMETEOROLOGICHESKOGO PRIBOROSTROENIYA. MOSCOW
1953 N3 P42-58 RT-4254 <*>
1957 N4 P75-79 SCL-T-258 <=*>
1957 N5 P3-16 SCL-T-226 <+*>
59-11418 <=> 0
1957 N5 P17-40 SCL-T-238 <+*>
59-11419 <=>
1957 N5 P138-144 SCL-T-224 <+*>
1959 N8 P11-22 61-23339 <*=>
1959 N8 P23-29 61-23338 <*=>
1959 V7 P3-24 64-13295 <=*$>
1959 V7 P3-51 64-13295 <=*$>
1959 V7 P25-35 64-13295 <=*$>
1959 V7 P36-51 64-13295 <=*$>
1961 N10 P99-100 63-23821 <=*>
1963 V11 P49-61 65-30874 <=$>
1964 V12 P84-88 65-31184 <=$>

TRUDY NAUCHNO-ISSLEDOVATELSKOGO INSTITUTA MATEMATIKI I MEKHANIKI. SEMINAR PO VEKTORNOMU I TENZORNOMU ANALIZU
1949 N7 P233-259 RT-1094 <*>

TRUDY NAUCHNO-ISSLEDOVATELSKOGO INSTITUTA OSNOVNOI KHIMII. LENINGRAD
1961 V13 N1 P147-154 65-11415 <*>

TRUDY NAUCHNO-ISSLEDOVATELSKOGO INSTITUTA PTITSEVODSTVA. (TITLE VARIES)
1954 V24 P5-12 58J14R <ATS>
1954 V24 P126-155 41J14R <ATS>
1954 V24 P156-169 38J14R <ATS>
1954 V24 P170-175 37J14R <ATS>
1954 V24 P176-184 40J14R <ATS>
1954 V24 P202-210 39J14R <ATS>
1954 V24 P211-221 36J14R <ATS>

TRUDY NAUCHNO-ISSLEDOVATELSKOGO INSTITUTA SINTETICHESKIKH ZHIROZAMENITELEI I MOIUSHCHIKH SREDSTV. BELGOROD
1961 N2 P5-13 66-11491 <*>
1961 V2 P5-13 65-10226 <*>
1961 V2 P25-29 65-10227 <*>
1961 V2 P30-33 65-10228 <*>
1961 V2 P33-38 63-20003 <=*>
1961 V2 P78-79 65-13288 <*>
1962 N3 P71-74 66-10634 <*>

TRUDY NAUCHNO-ISSLEDOVATELSKOGO INSTITUTA VAKTSIN I SYVOROTOK. TOMSK
1954 V20 P253-257 60-15419 <+*> 0
1954 V20 P259-264 60-15400 <+*> 0
1954 V20 P273-278 60-15417 <+*> 0
1954 V20 P279-284 60-15421 <+*> C
1956 V8 P125-132 61-18646 <*=>
1958 V9 P5-14 59-19680 <+*>
1958 V9 P15-18 59-19674 <+*>
1958 V9 P19-22 59-19676 <+*>
1958 V9 P23-26 59-19682 <+*>
61-18645 <*=>
64-15689 <=*$>
1958 V9 P27-32 59-19683 <+*>
61-18644 <*=>
64-15695 <=*$>
1958 V9 P53-61 59-19684 <+*> 0
1958 V9 P101-106 59-19679 <+*>
1958 V9 P117-120 59-19677 <+*>

TRUDY. NAUCHNO-ISSLEDOVATELSKII INSTITUT ZASHCHITY RASTENII. ALMA-ATA
1961 V6 P254-255 64-19875 <=$>
1962 V7 P191-196 CSIR-TR.322 <CSSA>

TRUDY NAUCHNO-ISSLEDOVATELSKOGO INSTITUTA ZEMNOGO MAGNETIZMA, IONOSFERY I RASPROSTRANENIYA RADIOVOLN. LENINGRAD
1953 V10 P159 60-23227 <+*>
1957 V11 P162-173 61-27021 <*=>
1960 N17 P173-186 ARSJ V32 N7 P1161 <AIAA>
1960 N17 P187-202 ARSJ V32 N11 P1794 <AIAA>
1960 V17 P3-26 62-11569 <=>
62-13628 <*=>
1960 V17 P173-186 62-15178 <*=>
1960 V17 P208-215 62-11366 <=>
62-13841 <=*>
1960 V17 P282-286 62-13625 <*=>
1960 V17 P287-291 62-13626 <*=>
62-13626 <=>

TRUDY NAUCHNO-ISSLEDOVATELSKOGO KINO-FOTO-INSTITUTA. MOSKVA
1947 N8 P75-94 R-4788 <*>
1947 N8 P95-104 R-4790 <*>

TRUDY NAUCHNO-ISSLEDOVATELSKOGO VETERINARNOGO INSTITUTA. MINSK
1960 V1 P105-109 64-13684 <=*$>

TRUDY NAUCHNO-TEKHNICHESKOGO OBSHCHESTVA SUDOSTROITELNOI PROMYSHLENNOSTI.
1955 V6 N3 P50-65 60-13808 <+*> C

TRUDY NOVOCHERKASSKOGO POLITEKHNICHESKOGO INSTITUTA IM. S. ORDZHONIKIDZE
1955 V26 P112-119 59-18737 <+*>
1955 V31 P73-77 61-15207 <*+>
1956 V27 N41 P151-165 59-17637 <+*>
1956 V33 P223-238 N65-28915 <=*$>
1956 V33 P239-248 N65-20990 <=$>
1957 V39 N53 P81-86 60-17199 <+*>
1959 V79 P95-107 62-13918 <*=>
1960 V107 P3-22 63-23237 <*=>

TRUDY NOVOSIBIRSKOGO GOSUDARSTVENNOGO MEDITSINSKOGO INSTITUTA
1957 V29 P3-6 65-63146 <=*$>
1957 V29 P7-32 65-61203 <=$>
1957 V29 P40-46 65-61204 <=$>
1957 V29 P47-52 65-61205 <=$>
1957 V29 P53-57 65-61206 <=$>
1957 V29 P58-64 65-61207 <=$>
1957 V29 P65-67 65-61208 <=$> C
1957 V29 P68-76 65-61209 <=$> 0
1957 V29 P82-87 65-61211 <=$> C
1957 V29 P88-91 65-61212 <=$>
1957 V29 P92-97 65-61213 <=$>

TRUDY OBSHCHESTVA PSIKHIATROV V S.-PETERBURGE
1912 V79 P137-147 65-60893 <=$>

TRUDY OBSHCHESTVA RUSSKIKH VRACHEI V S-PETERBURGIE
1885 V11 P184-202 65-62994 <=$>
1911 V78 P221-233 60-10667 <*>

TRUDY ODESSKOGO DERZHAVNOGO UNIVERSITETA
1953 V3 P61-65 T-2400 <INSD>
1958 V148 N6 P25-27 28P63V <ATS>
1959 V149 P9-23 62-33645 <=*>
1959 V149 P31-36 62-33648 <=*>

TRUDY ODESSKOGO GIDROMETEOROLOGICHESKOGO INSTITUTA ODESSA
1958 V17 P119-129 63-11148 <=>

TRUDY ODESSKOGO NAUCHNO-ISSLEDOVATELSKOGO INSTITUTA EPIDEMIOLOGII I MIKROBIOLOGII
1960 V4 P157-161 AD-638 578 <=$>
1960 V4 P163-164 AD-638 577 <=$>

TRUDY OKEANOGRAFICHESKOI KOMISSII. AKADEMIYA NAUK SSSR. MOSKVA
1958 V3 P118-121 59-22377 <+*> 0
1958 V3 P138- 59-13907 <=>
1959 V5 P63-78 65-60664 <=$>
1960 V7 P3-22 62-24068 <=*>
1960 V7 P23-26 62-24070 <=*>
1960 V7 P37-51 64-13662 <=*$>
1960 V7 P52-97 64-15316 <=*$>
1960 V7 P98-115 62-24069 <=*>
1960 V9 ENTIRE ISSUE 62-33084 <=*>
1960 V10 N2 P61-68 M.5104 <NLL>
1960 V10 N4 P3-7 64-11100 <=>
1960 V13 N1/2 P1-44 64-00317 <*>

1961 V8 P3-32 IGR V5 N6 P671 <AGI>
1961 V8 P45-59 IGR V6 N2 P228-237 <AGI>
1961 V8 P109-113 IGR,V6,N3,P487-490 <AGI>
1961 V8 P125-128 IGR V4 N11 P1206 <AGI>
1961 V8 P129-135 IGR V5 N11 P1440 <AGI>
IGR,V5,N11,P.1440-45 <AGI>
1961 V8 P136-157 IGR V6 N2 P212-227 <AGI>
1962 V14 P5-6 63-31934 <=>
1962 V14 P7-23 63-31798 <=>
1962 V14 P39-42 64-19534 <=$>
1962 V14 P69-73 64-19773 <=$>
1962 V14 P73-75 64-19535 <=$>
1962 V14 P76-107 63-31934 <=>
1962 V14 P116-130 63-31934 <=>
1962 V14 P116-122 64-13711 <=*$>
1962 V14 N6 P45-68 63-31798 <=>

TRUDY OPYTNO-ISSLEDOVATELSKOGO ZAVADA 'KHIMGAZ' SEE TRUDY VSESOYUZNOGO NAUCHNO-ISSLEDOVATELSKOGO INSTITUTA KHIMICHESKIKH PERERABOTKI GAZOV

TRUDY PALEONTOLOGICHESKOGO INSTITUTA. AKADEMIYA NAUK SSSR. LENINGRAD
1948 V15 N3 P36-116 <AGI>
1948 V15 N8 P5-37 IGR V4 N8 P845 <AGI>
1955 V56 P118-128 61-15077 <*+>
1956 V60 P127-133 61-15092 <*+>
1956 V60 P150-155 61-15092 <*+>
1958 V73 P1-146 61-15956 <*=> 0
1960 V76 ENTIRE ISSUE 63-19217 <=*>
1960 V83 P107-136 PT4 IGR V5 N10 P1308 <AGI>
1960 V83 P3-25 PT2 IGR V5 N7 P789 <AGI>
1960 V83 P64-106 PT3 IGR V5 N8 P915 <AGI>
IGR V5 N9 P1164 <AGI>
1961 V89 P14-40 38P60R <ATS>
1962 V88 P57-61 64-23512 <=$>

TRUDY PERVOGO MOSKOVSKOGO GOSUDARSTVENNOGO MEDITSINSKOGO INSTITUTA
1965 V37 P185-191 65-33737 <=$>
65-33737 <=*$>

TRUDY POCHVENNOGO INSTITUTA IMENI V. V. DOKUCHAEVA AKADEMIYA NAUK SSSR
1949 V30 P253-261 63-11078 <=>
1950 V34 P5-27 62-24296 <=*>
1950 V34 P110-142 65-50084 <=>
1952 V35 P27-31 59-14995 <+*>
1952 V36 P11-105 60-51167 <=>
1954 V43 P5-128 60-51173 <=>
1954 V43 P129-198 60-51174 <=>
1954 V44 P79-156 60-51184 <=>
1954 V44 P211-233 60-17777 <+*>
1956 V49 P7-72 61-19384 <+*>
1956 V49 P73-85 61-13277 <*+>
1956 V49 P152-181 61-15225 <*+>
1958 V53 P104-112 66-62219 <=*$>

TRUDY POLYARNOGO NAUCHNO-ISSLEDOVATELSKOGO INSTITUTA MORSKOGO RYBNOGO KHOZYAISTVA I OKEANOGRAFII IMENI N. M. KNIPOVICHA. MURMANSK
1956 N9 P5-61 C-2646 <NRC>
1956 N9 P62-123 C-2647 <NRC>
1956 N9 P226-233 C-ILKO <NRC>
1957 N10 P5-29 C-3114 <NRCC> 0
1957 N10 P30-53 C-3113 <NRCC> 0
1957 N10 P54-77 C-3093 <NRCC> 0
1957 N10 P88-105 C-3077 <NRCC> 0
1957 N10 P106-124 C-3075 <NRCC> 0
1957 V10 P78-87 65-60808 <=>
1957 V10 P125-144 62-25580 <=*> 0
1957 V10 P145-160 65-61235 <=$>
1957 V10 P161-171 65-60809 <=>
1957 V10 P172-185 65-60807 <=>
1957 V10 P198-211 65-60810 <=>
1957 V10 P212-229 65-61231 <=$>
1957 V10 P230-243 65-61230 <=$>
1959 V11 P35-53 64-15209 <=*$> 0
1959 V11 P142-148 64-15208 <=*$> 0
1960 N12 P5-70 63-00890 <*>
1960 V12 P5-70 C-4439 <NRC> PO
1960 V12 P71-87 64-19155 <=*$>

TRUDY PO PRIKLADNOI BOTANIKE, GENETIKE I SELEKTSII. LENINGRAD
1923 V13 P525-559 63-11040 <=>
1925 V14 P143-146 T-1862 <INDS>
1928 V21 N1 P241-292 CSIRO-2839 <CSIR>
1957 V30 N3 P271-277 59-16811 <+*> 0
1959 V32 N3 P299-303 61-28447 <*=>
1960 V32 N2 P259-270 64-15210 <=*$>
1964 V36 N3 P158-170 RTS-3673 <NLL>

TRUDY PROBLEMNYKH I TEMATICHESKIKH SOVESHCHANII. ZOOLOGICHESKII INSTITUT, AKADEMIYA NAUK SSSR. LENINGRAD, MOSKVA
1954 N3 P155-160 RT-3559 <*>
1954 N3 P198-207 RT-4207 <*>
1954 V3 P219-222 R-127 <*>
1954 V3 P223-231 R-124 <*>
1954 V7 N4 P47-53 61-31053 <=>
1956 N6 P144-149 R-5094 <*>
1956 V6 P2 62-25064 <=>
1956 V6 P148-149 59-15171 <+*>
59-16372 <+*>

TRUDY RADIEVOGO INSTITUTA IM. V. G. KHLOPINA. AKADEMIYA NAUK SSSR. LENINGRAD
1937 N3 P141-159 R-4513 <*>
RT-1874 <*>
1937 N3 P228-238 R-4485 <*>
RT-1872 <*>
1938 V4 P34 I-313 <*>
1938 V4 P253-303 AEC-TR-730 <=*>
60-00279 <*>
1956 V7 ENTIRE ISSUE AEC-TR-4376 <=>
1957 V5 N2 ENTIRE ISSUE AEC-TR-4498 <=>
1957 V5 N2 P89-93 63-23289 <*=>
1957 V6 ENTIRE VOLUME AEC-TR-4208 <=>
1958 V8 P1-267 AEC-TR-4474 <=>
1959 V9 P126-130 UCRL TRANS-1004(L) <=*$>
1959 V9 P131-133 UCRL TRANS-1005(L) <=*$>

TRUDY ROSTOVSKOGO-NA-DONU GOSUDARSTVENNOGO NAUCHNO-ISSLEDOVATELSKOGO PROTIVOCHUMNOGO INSTITUTA
1945 V4 P51-63 61-19895 <=*>
1947 V6 P71-81 61-19902 <=*>
1947 V6 P82-88 61-19894 <=*>

1947 V6 P89-94 61-19896 <=*>
1956 V10 P3-16 62-13995 <*=>
1956 V10 P44-53 62-13996 <=*>
1956 V10 P96-105 62-13997 <=*>
1956 V10 P120-124 62-13998 <*=>
1956 V10 P125-141 62-13999 <*=>
1956 V10 P142-146 62-13993 <=*>
1956 V10 P147-148 62-13994 <=*>
1956 V10 P149-165 62-13992 <=*>
1956 V10 P166 62-13991 <=*>
1956 V10 P167-186 62-15000 <*=>
1956 V10 P197-199 62-15001 <*=>
1956 V10 P200-204 62-15002 <=*>
1956 V10 P205-207 62-15003 <=*>
1956 V10 P229-230 62-15004 <*=>
1956 V10 P317-332 62-15005 <=*>
1956 V10 P333-338 62-15006 <=*>
1956 V10 P413-431 62-15007 <*=>
1956 V10 P475-477 62-15008 <*=>
1956 V11 P3-9 62-15333 <=*>
1956 V11 P11-16 62-15332 <=*>
1956 V11 P17-24 62-15331 <=*>
1956 V11 P49-58 62-15330 <=*>
1956 V11 P65-67 62-15329 <=*>
1956 V11 P69- 62-15328 <=*>
1959 V15 N1 P97-105 AD-638 580 <=$>
1959 V15 N1 P173-181 AD-619 338 <=$>
1961 V18 P3-23 AD-625 630 <=$>

TRUDY SANITARNO-GIGIENICHESKOGO MEDITSINSKOGO INSTITUTA. LENINGRAD
1950 V5 P17-28 RT-2511 <*> O
1950 V5 P30-37 RT-2466 <*>
1950 V5 P38-50 RT-2516 <*>
1950 V5 P51-64 RT-2506 <*>
1950 V5 P65-72 RT-2517 <*>
1950 V5 P73-77 RT-2464 <*>
1950 V5 P78-87 RT-2465 <*>
1950 V5 P88-95 RT-2467 <*>
1950 V5 P96-115 RT-2463 <*>
1950 V5 P116-126 RT-2518 <*>
1950 V5 P127-137 RT-2505 <*>
1950 V5 P138-147 RT-2504 <*>
1950 V5 P149-153 RT-2509 <*>
1950 V5 P154-162 RT-2514 <*>
1950 V5 P163-176 RT-2508 <*>
1950 V5 P177-182 RT-2649 <*>
1950 V5 P183-189 RT-2507 <*>
1950 V5 P190-199 RT-2512 <*>
1950 V5 P200-205 RT-2682 <*>
1950 V5 P206-218 RT-2679 <*>
1950 V5 P219-226 RT-2510 <*>
1950 V5 P227-236 RT-2513 <*>
1950 V5 P237-250 RT-2680 <*>
1950 V5 P251-265 RT-2681 <*>
1950 V5 P266-276 RT-2678 <*>
1955 V25 P5-78 60-11799 <=>
1956 V29 P9-25 62-23533 <=*>
1956 V31 P36-40 60-11522 <=>
1956 V31 P41-60 60-11523 <=> O
1956 V33 P11-70 62-15139 <=*>
1956 V33 P71-81 62-15158 <=*>
1958 V37 P197-213 64-19819 <=$> O
1958 V44 P155-163 62-25404 <=*>
1958 V44 P164-176 62-25405 <=*>
1959 V47 P243-253 62-15334 <*=>
1959 V47 P1693-1705 10281-I <KH>
1960 V58 P158-180 C-5426 <NRC>
1963 V72 P76-83 65-63150 <=$>
1963 V75 P7-56 65-30312 <=$>
1963 V75 P62-68 65-30312 <=$>
1963 V75 P74-90 65-30312 <=$>
1963 V75 P94-118 65-30312 <=$>
1963 V75 P132-143 65-30312 <=$>
1963 V75 P231-251 65-30312 <=$>

TRUDY SEISMOLOGICHESKOGO INSTITUTA SSSR. LENINGRAD
1930 V6 P1-57 UCRL-TRANS-796(L) <=*>
1933 V42 P1-26 UCRL TRANS-799(L) <=*>
1941 N106 P9-11 RT-1277 <*>
1941 V106 P64-73 61-11057 <=>
1945 V116 P1-44 61-11995 <=>
1945 V117 P56-64 61-11051 <=>
1947 P1-42 IGR V2 N5 P418 <AGI>

TRUDY SEKTORA ASTROBOTANIKI. AKADEMIYA NAUK KAZAKHSKOI SSR
1957 V5 P110-125 61-19320 <+*>
1957 V5 P149-161 61-19320 <+*>
1957 V5 P207-211 61-19320 <+*>
1957 V5 P221-227 61-19320 <+*>
1958 V6 P3-54 60-13238 <+*>

TRUDY SEKTORA MATEMATIKI I MEKHANIKI. AKADEMIYA NAUK KAZAKHSKOI SSR. ALMA ATA
1958 V1 P41-45 61-20047 <=*>

TRUDY. SEKTSII METALLOVEDENIYA I TERMICHESKOI OBRABOTKI METALLOV. NAUCHNO TEKHNICHESKOE OBSHCHESTVO MASHINOSTROITELNOI PROMYSHLENNOSTI
1960 V2 P3-11 63-23382 <=*$>
1964 N3 P42-63 M-5677 <NLL> O

TRUDY SELSKOKHOZYAISTVENNOI AKADEMII IMENI K.A.
1947 V37 P3-8 60-23731 <+*>
1947 V37 P88-114 60-23731 <+*>
1947 V37 P119-128 60-23731 <+*>

TRUDY SEMINARA PO FUNKTSIONALNOMU ANALIZU. VORONEZH
1956 N1 P81-105 AMST S2 V47 P1-30 <AMS>
1956 N2 P85-90 AMST S2 V37 P51-57 <AMS>

TRUDY SEMINARA PO VEKTORNOMU I TENZORNOMU ANALIZU S I KH PRILOZHENIYAMI K GEOMETRII, MEKHANIKE I FIZIKE
SEE TRUDY NAUCHNO-ISSLEDOVATELSKOGO INSTITUTA MATEMATIKI I MEKHANIKI. SEMINAR PO VEKTORNOMU I TENZORNOMU ANALIZU

TRUDY SEMINARA PO ZHAROSTOIKIM MATERIALAM
1959 V4 P5-37 63-21046 <=>
1959 V4 P38-64 63-19675 <=*>
1959 V4 P65-71 63-21046 <=>
1959 V4 P96-115 63-19675 <=*>
1961 V6 P38-58 63-19681 <=*>
1961 V6 P64-74 63-19681 <=*>

TRUDY SESII KOMMISII PO OPREDELENIYU ABSOLUTNOGO VOZRASTA GEOLOGICHESKIKH FORMATSII. AKADEMIYA NAUK SSSR
1961 P146-164 IGR V4 N8 P916 <AGI>

TRUDY SEVANSKOI GIDROBIOLOGICHESKOI STANTSII
1951 V12 P29-34 63-15903 <=*$>

TRUDY SEVASTOPOLSKOI BIOLOGICHESKOI STANTSII. AKADEMIYA NAUK SSSR
1936 V5 P297-317 62-23861 <=*> O
1957 V9 P3-13 60-23769 <+*>
1959 V11 P254-261 63-19135 <=*>
1959 V12 P130-152 RTS-3012 <NLL>
1960 V13 P89-91 RTS-2845 <NLL>
1960 V13 P216-244 65-61399 <=>
1960 V13 P269-274 63-11109 <=>
1960 V13 P275-292 63-31523 <=>
1960 V13 P305-308 63-31523 <=>
1960 V13 P309-314 63-15968 <=*>
1960 V13 P309-324 63-31583 <=>
1961 V14 P3-32 63-31612 <=>
1961 V14 P188-201 C-4690 <NRC>
1961 V14 P286-291 63-31613 <=>
1961 V14 P292-302 63-31615 <=>
1961 V14 P303-308 63-31616 <=>
1961 V14 P309-313 63-31614 <=>
1961 V14 P314-328 64-31088 <=>
1961 V14 P329-333 63-31614 <=>
1964 V17 P379-387 NIOT-86 <NLL>

TRUDY SIBIRSKOGO FIZIKO-TEKHNICHESKOGO INSTITUTA.

TOMSK
1958 V36 P153-158 64-14026 <=*$>
64-14650 <=*$>
1958 V36 P159-171 61-20835 <*=>
64-14027 <=*$>
1962 V41 P87-91 64-11889 <=> M

TRUDY SIBIRSKOGO NAUCHNO-ISSLEDOVATELSKOGO INSTITUTA GEOLOGII, GEOFIZIKI I MINERALNOGO SYRYA. NOVOSIBIRSK
1959 N2 P56-77 96P64R <ATS>
1960 N19 P240-241 RTS-3131 <NLL>
1960 V8 P67-71 RTS-3132 <NLL>

TRUDY SOVESHCHANII. IKHTIOLOGICHESKAYA KOMISSIYA. MOSCOW
1953 N1 P19-36 R-5017 <*>
1953 N1 P37-60 59-18359 <+*>
1953 N1 P291-300 C-3230 <NRC>
1953 N1 P354-362 60-51139 <=>
1953 N1 P529-537 59-18358 <+*>
1954 N3 P40-49 C-4458 <NRC>
1954 N4 P10-13 R-181 <*>
1954 N4 P10-21 60-51039 <=>
1954 N4 P22-37 60-51041 <=>
1954 N4 P38-47 60-21123 <=>
1954 N4 P120-128 60-51042 <=>
1954 N4 P144-155 60-51139 <=>
1954 N6 P124-134 C-3201 <NRCC>
1957 N7 P235-242 C-3642 <NRCC>
61-01021 <*>
1958 N8 P82-89 NASA-TT-F-190 <=>
1958 N8 P115-120 59-20952 <+*>
1958 N8 P124-131 59-20951 <+*>
1958 N8 P132-141 59-20950 <+*>
1958 N8 P142-152 C-3410 <NRCC> O
1958 N8 P158-170 C-3835 <NRC>
63-01557 <*>
1958 N8 P372-379 59-20953 <+*>
1958 N8 P380-386 59-20955 <+*>
1958 N8 P387-392 59-20954 <+*>
1959 N9 P192-197 61-15835 <+*>
1960 N10 P66-72 63-11108 <=>
1960 N10 P117-121 63-11112 <=>
1960 N10 P151-164 C-4195 <NRC>
1960 N10 P165-172 C-4198 <NRC>
1960 N10 P188-194 63-11118 <=>
1960 N10 P219-229 63-11108 <=>
1960 N10 P230-234 63-11126 <=>
1960 N10 P235-238 63-11114 <=>
1960 N10 P239-242 63-11108 <=> P
1961 N12 P68-71 65-31268 <=$>
1961 N12 P125-132 M.5145 <NLL>
1961 N12 P138-149 63-19977 <=*$>
1961 N13 P5-6 63-31056 <=>
1961 N13 P7-20 63-23343 <=*$>
1961 N13 P21-23 63-23342 <=*$>
1961 N13 P34-43 64-15991 <=*$> O
1961 N13 P44-46 63-31056 <=>
1961 N13 P82-93 64-19012 <=*$>
1961 N13 P94-107 64-15951 <=*$> C
1961 N13 P108-116 64-19882 <=$>
1961 N13 P117-129 C-4654 <NRC>
1961 N13 P130-131 64-19009 <=*$>
1961 N13 P132-146 C-4447 <NRC>
1961 N13 P147-157 C-4443 <NRC>
63-00883 <*>
1961 N13 P173-179 64-15963 <=*$>
1961 N13 P185-193 63-23524 <=*$>
1961 N13 P228-234 63-23523 <=*$>
1961 N13 P235-237 MAFF-TRANS-NS-55 <NLL>
1961 N13 P238-247 MAFF-TRANS-NS-56 <NLL>
1961 N13 P260-264 63-23522 <=*$> O
1961 N13 P301-306 63-23521 <=*$>
1961 N13 P307-313 C-4466 <NRC>
63-23520 <=*$> C
1961 N13 P344-347 63-23519 <=*$> O
1961 N13 P348-352 64-15962 <=*$>
1961 N13 P364-373 MAFF-TRANS-NS-69 <NLL>
1961 N13 P391-392 63-23526 <=*$>
1961 N13 P402-410 63-23525 <=*$> O
1961 N13 P454-456 63-23527 <=*$>
1961 N13 P487-495 63-31056 <=>

TRUDY SOVESHCHANIYA PO EKSPERIMENTALNOI MINERALOGII I PETROGRAFII. AKADEMIYA NAUK SSSR. MOSKVA
1937 P129-135 R-2581 <*>

TRUDY SOVESHCHANIYA PO GEOKHIMIYA METODY, 1958. MOSKVA
1959 P175-178 99T90R <ATS>
1959 P207-210 00T91R <ATS>

TRUDY. SOVESHCHANIE PO SEREBRISTYM OKLAKAM
1959 V1 P23-24 65-14931 <*>
1961 P7-12 65-14932 <*>
1962 V3 P126-130 65-14933 <*>

TRUDY SOVESHCHANIYA PO VOPROSAM KOSMOGONII. AKADEMIYA NAUK SSSR. MOSCOW
1955 P203-233 R-4148 <*>
1955 P229-241 R-4156 <*>

TRUDY SOVETA PO IZUCHENIYU PROIZVODITELNYKH SIL. SERIYA KOLSKAYA
1933 N6 P5-56 IGR V3 N2 P147 <AGI>
1933 V6 P5-56 62-15603 <=*>

TRUDY SOVETA PO IZUCHENIYU PROIZVODITELNYKH SIL. SERIYA KUZBASSKAYA
1932 N1 P6-8 IGR V3 N1 P69 <AGI>
1932 V1 P1-28 62-25430 <=*>

TRUDY SOVETA PO IZUCHENIYU PROIZVODITELNYKH SIL. SERIYA SIBIRSKAYA
1934 V8 P59-114 62-15997 <=*>

TRUDY SOVETA PO IZUCHENIYU PROIZVODITELNYKH SIL. SERIYA TURKMENSKAYA
1934 V8 P5-7 62-25432 <=*>
1934 V8 P9-23 62-25432 <=*>
1934 V8 P25-30 62-25432 <=*>

TRUDY SOVETA PO IZUCHENIYU PROIZVODITELNYKH SIL. SERIYA YAKUTSKAYA. LENINGRAD
1932 V6 P1-46 62-15604 <=*>
1932 V6 P52-53 62-15604 <=*>

TRUDY SREDNE-AZIATSKOGO NAUCHNO-ISSLEDOVATELSKOGO GIDROMETEOROLOGICHESKOGO INSTITUTA. TASHKENT
1959 N1 P105-110 61-21723 <=>
1959 N2 P139-159 63-19746 <=*> C
1963 N8 P12-16 TRANS-F-122 <NLL>
1963 N9 P3-66 TRANS-F-131 <NLL>
1965 N22 P51-59 65-33574 <=*$>

TRUDY. SUKHUMSKOGO BIOLOGICHESKAYA STATSIYA
1949 V1 P244-257 60-18171 <+*>

TRUDY SVERDLOVSKOGO GORNOGO INSTITUTA IM. V. V. VAKHRUSHEVA. (TITLE VARIES)
1956 N26 P24-44 25K21R <ATS>

TRUDY TADZHIKSKOGO FILIALA, AKADEMIYA NAUK SSSR. MOSCOW
1935 V6 P205-217 62-00156 <*>

TRUDY TALLINSKOGO POLITEKHNICHESKOGO INSTITUTA
SEE TALLINNA POLUTEHNILISE INSTIDUDI TOIMETISED

TRUDY TASHKENTSKOGO GOSUDARSTVENNOGO UNIVERSITETA
1955 V65 N12 P47-53 AEC-TR-4272 <+*>
1957 V91 P5-15 60-13444 <+*>
1957 V91 P57-66 60-13444 <+*>
1958 N134 P43-50 79Q71R <ATS>
1963 N221 P45-52 64-31368 <=>
1963 N221 P61-64 49R79R <ATS>
1963 N221 P71-83 64-31368 <=>
1963 N221 P128-135 64-31368 <=>
1963 N221 P145-148 64-31368 <=>

1963 N221 P180-196 64-31368 <=>

TRUDY TATARSKOI RESPUBLIKANSKOI GOSUDARSTVENNOI SELSKOKHOZYAISTVENNOI OPYTNOI STANTSII
1961 V1 P105-120 64-15708 <=$>

TRUDY TBILISSKOGO MATEMATICHESKOGO INSTITUTA. AKADEMIYA NAUK GRUZINSKOI. (TITLE VARIES)
1949 V17 P1-27 R-494 <*>
1951 V18 P210-219 62-33269 <=*>
1951 V18 P307-313 62-11573 <=>
1953 V19 P1-31 AMST S2 V4 P1-29 <AMS>
1954 V20 P245-278 AMST S2 V5 P305-333 <AMS>
1960 V27 P367-410 63-23437 <*=>

TRUDY TEKHNOLOGICHESKOGO INSTITUTA MIASNOI I MOLOCHNOI PROMYSHLENNOSTI. MOSCOW
1965 N3 P115-120 M.5643 <NLL>

TRUDY TEKHNOLOGICHESKOGO INSTITUTA PISHCHEVOI PROMYSHLENNOSTI. MOSCOW
1956 N13 P92-100 AD-625 431 <=$*>
1958 N14 P139-142 AD-625 431 <=$*>
1958 N14 P143-146 AD-625 431 <=$*>
1958 N14 P147-150 AD-625 431 <=$*>
1958 N14 P151-154 AD-625 431 <=$*>

TRUDY TEKHNOLOGICHESKOGO INSTITUTA TSELLYULOZNO-BUMAZHNOI PROMYSHLENNOSTI. LENINGRAD
1964 N12 P193-198 17S84R <ATS>
1964 N12 P199-205 16S84R <ATS>
1965 N16 P126-138 66-14532 <*>

TRUDY TEKHNOLOGICHESKOGO INSTITUTA. VORONEZH
1960 V16 P52-58 65-61402 <=>

TRUDY TOMSKOGO GOSUDARSTVENNOGO UNIVERSITETA
1924 V75 P76-86 65-61426 <=>
1954 V126 P96-102 67K23R <ATS>
1962 V154 P262-263 66-61535 <=*$>
1963 V157 P261-264 66-11184 <*>

TRUDY TRANSPORTNO-ENERGETICHESKOGO INSTITUTA. SIBIRSKOE OTDELENIE. AKADEMIYA NAUK SSSR. NOVOSIBIRSK
1954 N4 P49-58 C-3903 <NRCC>
1954 V4 P59-69 C-4674 <NRC>
1954 V4 P71-88 C-4675 <NRC>
1954 V4 P89-97 C-4603 <NRC>
1954 V4 P111-118 C-4676 <NRC>
1954 V4 P119-126 C-4604 <NRC>

TRUDY TSENTRALNOGO AEROGIDRODINAMICHESKOGO INSTITUTA
1926 N21 P5-42 RT-646 <*>
1932 V149 P5-37 RT-719 <*>
1935 N191 P2-59 RT-654 <*>
1935 N224 P2-39 RT-652 <*>
1935 N240 P18-22 RT-1182 <*>
1935 V240 P23-27 RT-1183 <*>
1936 N262 P2-39 RT-653 <*>
1936 N279 P3-18 RT-413 <*>
1937 N295 P3-118 RT-651 <*>
1937 N321 P2-20 RT-649 <*>
1937 V284 P1-248 RT-3035 <*>
1937 V311 P3-52 RT-2832 <*>
1938 N346 P3-27 RT-331 <*>
1939 N377 P2-69 RT-648 <*>
1939 N381 P2-20 RT-650 <*>
1939 N395 P2-43 RT-647 <*>
1939 N418 ENTIRE ISSUE RT-1647 <*>
1939 N427 P1-14 63-20550 <=*$>
1939 N437 ENTIRE ISSUE RT-1317 <*>
1939 N439 ENTIRE ISSUE RT-1319 <*>
1939 N440 ENTIRE ISSUE RT-1320 <*>
1939 N449 ENTIRE ISSUE RT-1648 <*>
1939 N452 ENTIRE ISSUE RT-1321 <*>
1939 N453 ENTIRE ISSUE RT-1004 <*>
1939 N454 ENTIRE ISSUE RT-1322 <*>
1939 N455 ENTIRE ISSUE RT-1323 <*>
1939 N462 ENTIRE ISSUE RT-1325 <*>
1939 N490 P3-15 RT-1176 <*>
1940 N478 ENTIRE ISSUE RT-1324 <*>
1957 N702 P22 60-17436 <+*>

TRUDY TSENTRALNOGO AEROLOGICHESKOGO OBSERVATORIYA. LENINGRAD
1959 N32 P46-87 63-23625 <=*>

TRUDY TSENTRALNOGO INSTITUTA PROGNOZOV. LENINGRAD, MOSCOW
1950 N43 P34-48 60-41100 <=>
1957 N55 P57-63 PB 141 230T <=>
1957 N55 P64-71 PB 141 231T <=>
1957 N60 P3-9 60-41114 <=>
1957 N60 P10-16 59-12008 <+*>
1957 N60 P17-21 60-41104 <=>
1957 N60 P22-31 59-12654 <+*>
1958 N71 P3-10 61-13580 <*+>
1958 N71 P11-16 61-13542 <*+> O
1958 N71 P17-26 61-13579 <*+>
1958 N71 P27-39 61-13541 <*+> O
1958 N71 P40-43 61-13578 <*+>
1958 N71 P44-47 61-13314 <+*> O
1958 N71 P48-77 61-13577 <+*> O
1958 N71 P78-113 61-13540 <*+> O
1958 N73 P57-72 60-19979 <+*> O
1958 N73 P73-93 61-13581 <*+>
1958 N78 P5-36 62-24884 <=*> O
1958 N78 P73-82 62-24885 <=*> O
1958 N78 P83-91 62-24883 <=*> O
1958 V77 P3-7 65-60632 <=$>
1958 V78 P37-63 63-19911 <=*>
1958 V78 P64-72 63-19912 <=*>
1958 V78 P105-111 60-41102 <=>
63-19913 <=*>
1959 N74 P3-24 62-24881 <=*>
1959 N74 P47-55 62-24882 <=*> O
1959 N74 P56-68 62-24879 <=*> O
1959 N80 P64-78 63-19802 <=*$> O
1959 N85 P27-39 63-19740 <=*$> O
1959 N85 P40-63 63-19805 <=*$> O
1959 N87 P51-56 63-19739 <=*$>
1959 V83 P28-38 63-19741 <=*> O
1959 V87 P3-25 63-19838 <*=> O
1959 V91 P3-17 63-19808 <*=> O
1960 N92 P13-20 63-19748 <=*$>
1960 N93 P3-24 62-11596 <=>
1960 N93 P25-34 61-19932 <*=>
1960 N93 P35-37 61-19935 <*=>
1960 N93 P38-48 61-19933 <*=>
1960 N93 P49-58 61-19936 <*=>
1960 N93 P59-64 61-19931 <*=>
1960 N93 P65-79 62-11596 <=>
1960 N97 P3-46 63-19745 <=*$> O
1960 N97 P67-81 63-19743 <=*$> O
1960 N98 P48-55 65-11352 <*>
1960 V92 P3-12 63-19747 <*=>
1960 V95 P3-31 63-19753 <=*>
1960 V106 P3-19 63-19906 <=*> O
1960 V106 P20-31 63-19903 <=*> O
1960 V106 P37-44 63-19904 <=*> O
1960 V106 P45-52 65-61242 <=$> O
1960 V106 P61-64 63-19902 <*=> O
1960 V106 P123-131 63-19905 <*=> O
1961 N81 P113-126 M.5783 <NLL>
1961 N110 P32-50 62-25052 <=*>
1961 N110 P70-76 62-25052 <=*>
1961 N111 P3-12 62-01490 <*>
62-11742 <=>
1961 N111 P29-38 62-11683 <=>
1961 N111 P39-43 62-11744 <=>
1961 N111 P44-49 62-01486 <*>
62-11741 <=>
1961 N111 P50-62 63-11503 <=>
1961 N112 P18-31 62-25469 <=>
1961 N112 P32-39 62-23829 <=*>
1961 N112 P40-45 62-25469 <=>
1961 V81 P.48-91 63-19972 <=*>

Citation	Translation
1961 V104 P123-128	64-19003 <=*$> 0
1961 V111 P13-28	63-11611 <=>
1962 N112 P47-52	63-11752 <=>
1962 V102 P3-12	M.5373 <NLL> 0
1962 V102 P13-19	M.5371 <NLL> 0
1962 V102 P20-26	M.5394 <NLL> 0
1962 V102 P27-33	M.5370 <NLL> 0
1962 V102 P34-46	M.5366 <NLL>
1962 V102 P47-52	M.5367 <NLL> 0
1962 V102 P52-59	M.5368 <NLL> 0
1962 V102 P53-59	63-11701 <=>
1962 V102 P60-63	TRANS- 191(M) <NLL> 0
1962 V102 P64-70	M.5372 <NLL> 0
1962 V102 P71-79	M.5395 <NLL> 0
1962 V115 P3-4	65-60967 <=$>
1962 V115 P5-17	65-60958 <=$> 0
1962 V115 P107-125	65-60113 <=$> 0
1962 V116 P3-12	M.5414 <NLL>
1962 V116 P13-23	M.5413 <NLL> 0
1962 V116 P24-33	M.5436 <NLL> 0
1962 V116 P41-64	M.5416 <NLL> 0
1962 V118 P5-12	65-61517 <=$> 0
1962 V118 P19-21	65-61516 <=$> 0
1962 V118 P41-55	65-60970 <=$>
1962 V119 P70-77	65-61287 <=$>
1963 N124 P3-13	64-21565 <=>
1963 N124 P28-32	64-21565 <=>
1963 N125 P3-12	64-21822 <=>
1963 V121 P3-13	65-61361 <=$> 0
1963 V121 P14-17	65-60738 <=> 0
1963 V121 P31-52	65-60764 <=>
1963 V123 P4-17	64-19002 <=*$> 0
1963 V124 P3-13	65-61318 <=$> 0
1963 V124 P14-22	65-60611 <=$> 0
1963 V124 P23-27	65-60606 <=$> 0
1963 V124 P28-32	65-61390 <=$> 0
1963 V124 P33-39	65-61367 <=$> 0
1963 V124 P75-80	TRANS-F.66 <NLL> 0
1963 V126 P3-7	65-60630 <=$>
1963 V126 P8-19	65-60634 <=$> 0
1963 V126 P20-27	65-60633 <=$>
1963 V126 P38-46	64-41897 <=$>
	65-61362 <=$> 0
1963 V126 P47-54	64-51049 <=$>
	65-61388 <=$>
1963 V126 P55-99	65-61389 <=$> 0
1963 V126 P100-103	65-61386 <=$>
1963 V128 P46-55	AD-614 021 <=$>
1963 V128 P155-159	TRANS- F.60 <NLL> 0
1963 V130 P126-140	SOU.HYDROL.1963 P437 <AGU>
1963 V130 P141-151	SOU.HYDROL.1963 P450 <AGU>
1963 V131 P3-12	SOU.HYDROL.1963 P459 <AGU>
1963 V131 P30-41	SOU.HYDROL.1963 P488 <AGU>
1963 V131 P42-52	SOV.HYDROL.1963 P400 <AGU>
1964 N135 P22-44	TRANS-305(M) <NLL>
1964 N135 P45-62	TRANS-306(M) <NLL>
1964 N135 P63-90	TRANS-307(M) <NLL>
1964 N136 P3-11	TRANS-333(M) <NLL>
1964 N136 P12-26	TRANS-334(M) <NLL>
1964 N136 P61-68	TRANS-335(M) <NLL>
1964 N136 P101-115	TRANS-332(M) <NLL> 0
1964 V132 P3-40	65-30114 <=$>
1964 V132 P41-47	TRANS- F.61 <NLL> 0
1964 V132 P48-58	TRANS- F.62 <NLL> 0
1964 V132 P48-74	65-30114 <=$>
1964 V132 P59-63	TRANS-F.63 <NLL> 0
1964 V132 P75-92	TRANS-F.64 <NLL> 0
1964 V135 P3-12	65-60628 <=$> 0
1964 V135 P13-21	TRANS-304(M) <NLL> 0
1964 V137 P11-20	65-30054 <=$>
1964 V137 P27-36	TRANS-323(M) <NLL> 0
1964 V137 P37-43	TRANS-324(M) <NLL> 0
1964 V137 P54-67	TRANS-F.71 <NLL>
1964 V137 P68-82	65-30053 <=$>
1964 V137 P151-159	TRANS-F-72 <NLL>
1965 N139 P22-28	TRANS-353(M) <NLL>
1965 N139 P71-88	TRANS-F-148 <NLL>
1965 N139 P111-116	TRANS-F.123. <NLL>
1966 N148 P34-61	SHSP N6 1966 P578 <AGU>
1966 N151 P64-81	SHSP N3 1966 P262 <AGU>
1966 N151 P93-148	SHSP N6 1966 P602 <AGU>
1966 N151 P149-170	SHSP N3 1966 P276 <AGU>
1966 N156 P31-38	SHSP N2 1966 P199 <AGU>
1966 N156 P99-104	SHSP N2 1966 P206 <AGU>

TRUDY TSENTRALNOGO INSTITUTA USOVERSHENSTVOVANIYA VRACHEI. MOSCOW

Citation	Translation
1963 V62 P14-21	64-51263 <=>
1963 V62 P379-381	64-51263 <=>

TRUDY TSENTRALNOGO NAUCHNO-ISSLEDOVATELSKOGO DIZELNOGO INSTITUTA. LENINGRAD

Citation	Translation
1961 V37 P70-82	63-21210 <=>

TRUDY TSENTRALNOGO NAUCHNO-ISSLEDOVATELSKOGO GEOLOGO-RAZVEDOCHNOGO INSTITUTA. LENINGRAD

Citation	Translation
1941 V139 ENTIRE ISSUE	62-24496 <=*> 0

TRUDY TSENTRALNOGO NAUCHNO-ISSLEDOVATELSKOGO INSTITUTA GEODEZII, AEROSEMSKII I KARTOGRAFII. MOSCOW

SEE TRUDY NAUCHNO-ISSLEDOVATELSKOGO INSTITUTA GEODEZII I KARTOGRAFII. MOSCOW

TRUDY TSENTRALNOGO NAUCHNO-ISSLEDOVATELSKOGO

Citation	Translation
1956 V4 P203-222	C-2957 <NRC>
1957 V7 P185-220	C-2985 <NRCC>
1960 V15 N6 P71-92	C-3959 <NRCC> 0
1960 V18 N6 P3-53	C-3947 <NRCC> 0
1960 V23 N3 P3-32	C-3967 <NRCC> 0

TRUDY TSENTRALNOGO NAUCHNO-ISSLEDOVATELSKOGO INSTITUTA MORSKOGO FLOTA. LENINGRAD

Citation	Translation
1964 N26 P37-42	65-32868 <=*$>
1964 N57 P37-42	AD-627 822 <=*$>
1965 N77 P74-88	66-33835 <=$>

TRUDY TSENTRALNOGO NAUCHNO-ISSLEDOVATELSKOGO INSTITUTA PROTEZIROVANIYA I PROTEZOSTROVENIYA

Citation	Translation
1948 V1/3 P125	60-13339 <+*>
1948 V1/3 P188	60-13341 <+*>
1948 V1/3 P255	60-13340 <+*>
1948 V1/3 P282	60-13450 <+*>

TRUDY TSENTRALNOGO NAUCHNO-ISSLEDOVATELSKOGO INSTITUTA RENTGENOLOGII I RADIOLOGII. MOSCOW

Citation	Translation
1955 V9 P129-136	60-23613 <+*>

TRUDY TSENTRALNOGO NAUCHNO-ISSLEDOVATELSKOGO INSTITUTA SUDOSTROITELNOI PROMYSHLENNOSTI. LENINGRAD

Citation	Translation
1956 N20 P86-162	60-17907 <+*>

TRUDY. TSENTRALNYY NAUCHNO-ISSLEDOVATELSKIY INSTITUT TEKHNOLOGII I MASHINOSTROYENIY. MOSCOW

Citation	Translation
1957 V79 P5-24	58-2285 <*>
1957 V79 P24-35	59-11252 <=>
1961 V101 P205-209	63-13170 <=*>
1961 V102 P45-47	62-32416 <=*>
1961 V103 P1-165	63-23092 <*=>
1962 V105 P5-11	64-23538 <=>

TRUDY TSENTRALNYI NAUCHNO-ISSLEDOVATELSKIY KOTLOTURBINNYI INSTITUT. LENINGRAD

Citation	Translation
1954 V26 P189-230	62-25893 <=*$>

TRUDY TSENTRALNOI AEROLOGICHESKOI OBSERVATORII

Citation	Translation
1952 N7 P3-15	RT-3785 <*>
1952 N7 P16-21	RT-3783 <*>
1952 N7 P22-38	RT-3784 <*>
1952 N7 P39-49	59-15560 <+*>
1952 N7 P50-55	59-15997 <+*>
1952 N7 P56-61	59-17161 <+*> 0
1953 N10 P3-17	RT-4241 <*>
1953 N10 P18-22	RT-4648 <*>
1953 N10 P23-30	RT-4649 <*>
1953 N10 P55-74	RT-4560 <*>
1953 N12 P3-12	59-17286 <+*> 0
1953 N12 P49-55	59-17284 <+*>

1956 N17 P15-35 PB 141 158T <=>
1956 N17 P36-46 32K21R <ATS>
1956 N17 P47-56 36K21R <ATS>
1956 N17 P57-70 35K21R <ATS>
1957 N22 P3-16 61-23171 <*=>
1957 N22 P32-34 61-13316 <*+> 0
1957 N22 P40-50 61-13319 <*+> 0
1957 N22 P51-59 60-19980 <+*>
1957 N22 P100-116 61-13320 <*+> 0
1957 N23 P3-15 59-11728 <=>
1957 N23 P63-77 61-27825 <*=>
1958 N19 P3-32 60-19978 <+*> 0
1958 N19 P33-67 00N58R <ATS>
61-13093 <*+> 0
1958 N19 P68-80 01N58R <ATS>
1958 N20 P3-16 62L34R <ATS>
63-19910 <=*>
1958 N20 P58-66 65L36R <ATS>
1958 N20 P67-72 66L36R <ATS>
1958 N20 P73-80 63L34R <ATS>
1958 N20 P82-87 67L36R <ATS>
1958 N24 P16-31 63-19836 <*=> 0
1959 N25 P5-62 61-21233 <=>
1959 N25 P63-71 61-11889 <=>
1959 N25 P72-76 61-11915 <=>
1959 N25 P77-84 61-11948 <=>
1959 N26 P39-55 62-23273 <=*>
63-24427 <=*$>
1959 N30 P39-52 63-23702 <=*$>
1959 N30 P73-80 63-19771 <=>
1959 N31 P83-92 61-23014 <=*>
1959 N32 P3-36 61M1 <CTT>
1959 N32 P84-87 61M2 <CTT>
1960 N28 P49-57 M.5787 <NLL>
62-11707 <=>
1960 N28 P58-69 00P59R <ATS>
62-11666 <=>
1960 N29 P78-83 65-61359 <=$>
1960 N37 P13-23 66-61838 <=*$>
1960 N37 P24-29 65-61475 <=> 0
1960 N37 P49-54 65-61358 <=$> 0
1960 N37 P55-58 65-61356 <=$> 0
1961 N36 P3-13 65-61289 <=$> 0
1961 N36 P31-36 32Q68R <ATS>
1961 N36 P37-42 33Q68R <ATS>
1961 N36 P53-61 TRANS-F-39 <NLL>
65-10801 <*>
1961 N36 P126-136 65-10800 <*>
1961 V36 P3-12 31Q68R <ATS>
1961 V36 P37-42 63-23714 <=*>
1962 N41 P3-11 65-60638 <=$> 0
1962 N42 ENTIRE ISSUE N64-26961 <=>
1962 N43 P47-56 63-19281 <=*>
63-23719 <=*>
1962 N43 P79-90 63-19282 <=*>
1962 N44 P15-37 63-24247 <=*$>
1963 N46 P63-75 65-13992 <*>
1963 N46 P76-84 N64-24591 <=>
1963 N47 P3-24 M.5786 <NLL>
1963 N47 P55-62 65-60601 <=$> 0
1963 N48 P3-54 AD-621-418 <=$>
1963 N48 P56-97 AD-621 419 <=$>
1963 N48 P56-77 65-30234 <=$>
1963 N48 P56-97 65-60617 <=$> 0
1963 N48 P98-105 AD-621 422 <=$>
65-30233 <=$>
1963 N48 P106-111 64-51933 <=$>
1964 N52 P3-5 65-30165 <=$>
1964 N52 P53-59 65-30166 <=$>
1964 N52 P60-66 65-30164 <=$>
1964 N52 P67-74 65-61263 <=$> 0
1964 N53 P21-34 TRANS-326(M) <NLL> 0
1964 N53 P43-53 AD-637 728 <=$>
TRANS-311(M) <NLL>
1964 N53 P54-78 TRANS-312(M) <NLL>
1964 N54 P4-52 TRANS-302(M) <NLL>
1964 N54 P80-84 TRANS-F.75 <NLL>
1964 N56 P3-8 TRANS-352(M) <NLL>
1964 N56 P121-127 AD-632 315 <=$>
1964 N57 P3-18 AD-631 963 <=$>
1964 N57 P19-23 AD-628 974 <=*$>
1964 N57 P19-23,55-66 65-33018 <=*$>
1964 N57 P24-40 TRANS-F.117. <NLL>
1964 N57 P41-48 AD-630 670 <=$>
1964 N57 P49-54 AD-631 964 <=$>
1964 N57 P67-71 AD-636 628 <=$>
1964 N57 P72-76 TRANS-F.118. <NLL>
65-33018 <=*$>
1964 N57 P77-86 AD-630 556 <=$>
1964 N59 P11-19 TRANS-F.126 <NLL> 0
1964 N59 P20-31 TRANS-F.132. <NLL>
1964 N59 P43-46 TRANS-F-109 <NLL>
1964 N59 P47-52 TRANS-F-110 <NLL>
1964 V53 P91-100 TRANS-327(M) <NLL>
1965 N61 P86-92 66-30123 <=*$>
1965 N63 P31-36 66-31248 <=$>
1965 N63 P77-84 66-31248 <=$>
1965 N64 P11-27 TRANS-358(M) <NLL>
1965 N65 P3-8 66-33979 <=$>
1965 N65 P30-66 66-33979 <=$>

TRUDY UFIMSKOGO NEFTYANOGO NAUCHNO-ISSLEDOVATEL-SKOGO INSTITUTA
1963 N9/10 P194-202 66-13860 <*>

TRUDY UKRAINSKOGC NAUCHNO-ISSLEDOVATELSKOGO GIDROMETEOROLOGICHESKOGC INSTITUTA. KIEV
1958 V12 P31-56 61-23296 <=*> 0
1959 V11 P42-51 62-24880 <=*> 0
1959 V11 P93-142 63-19807 <=*$> 0
1960 V21 P3-15 61-27774 <=*>
1960 V21 P16-22 62-13911 <*=>
1960 V21 P38-49 63-19742 <*=> 0
1962 N34 P3-23 SHSP N2 1962 P194 <AGU>
1962 N34 P39-44 SHSP N2 1962 P211 <AGU>
1962 N34 P45-52 SHSP N2 1962 P217 <AGU>
1963 V35 P84-108 SOV.HYDROL.1963 P568 <AGU>
1963 V35 P109-115 SOV.HYDROL.1963 P592 <AGU>
1965 N47 P17-21 AD-636 629 <=$>
1965 N47 P51-58 AD-630 673 <=$>
1965 N47 P59-64 AD-629 007 <=*$>
1965 N48 P21-38 AD-630 555 <=$>
1966 N60 P25-28 SHSP N2 1966 P161 <AGU>
1966 N60 P54-61 SHSP N2 1966 P164 <AGU>
1966 N60 P62-72 SHSP N2 1966 P170 <AGU>
1966 N60 P73-75 SHSP N2 1966 P179 <AGU>
1966 N60 P97-106 SHSP N2 1966 P182 <AGU>
1966 N63 P96-105 SHSP N2 1966 P190 <AGU>

TRUDY UKRAINSKOGC NAUCHNC-ISSLEDOVATELSKOGO INSTITUTA SOTSIALISTINOGO ZEMLEROBSTVA. KHARKOV
1951 V6 P170-171 R-541 <*>

TRUDY UNIVERSITETA DRUZHBY NARODOV. MOSCOW.
1963 V1 P68-70 AD-639 336 <=$>
1963 V1 N1 P52-55 AD-618 629 <=$>
1963 V1 N1 P56-63 AD-625 093 <=$>
1963 V1 N1 P71-86 AD-622 360 <=*$>
1964 N2 P90-94 AD-636 584 <=$>

TRUDY URALSKOGO NAUCHNO-ISSLEDOVATELSKOGO INSTITUTA GEOLOGII, RAZVEDOK I ISSLEDOVANIYA MINERALNOGO SYRIYA. SVERDLOVSK
1938 N3 P285-293 3298 <HB>

TRUDY URALSKOGO NAUCHNO-ISSLEDOVATELSKOGO KHIMICHESKOGO INSTITUTA. SVERDLOVSK
1957 V5 P236-251 62-24458 <=*>
1957 V5 P252-262 62-24459 <=*>
1957 V5 P263-280 63-24281 <=*$>

TRUDY URALSKOGO POLITEKHNICHESKOGO INSTITUTA IM. S. M. KIROVA: SBORNIK. SVERDLOVSK
1955 N41 P112-116 59-22572 <+*>
1955 N54 P82-102 59-22569 <+*>
1956 V61 P15-24 63-24173 <=>
1957 N58 P128-144 61-28911 <*=>
1957 N67 P78-91 62-24900 <=*>
1957 N72 P21-34 61-11116 <=>
1957 N72 P41-54 61-11117 <=>

1957 N72 P65-75 60-21975 <=>
1958 N68 P54-58 61-19393 <*=>
1960 N96 P82-92 46P64R <ATS>
1962 V121 P91-94 65-10826 <*>
1963 V134 P46-53 AD-614 405 <=$>
1964 V139 P86-95 65-14056 <*>
65-63702 <=$>

TRUDY UZBEKSKOGO NAUCHNO-ISSLEDOVATELSKOGO VETERINARNOGO INSTITUTA. TASHKENT
1959 V13 P173-185 62-25643 <=*>
1959 V13 P186-194 62-25635 <=*>
1959 V13 P195-198 62-25636 <=*>
1961 V14 P3-5 64-15415 <=*$>

TRUDY V. E. I. VSESOYUZNYI ELEKTROTEKHNICHESKII INSTITUT IM. V. I. LENINA.
1958 V62 P5-15 60-19967 <+*>
1958 V62 P16-28 60-19966 <+*>
1958 V62 P29-42 60-19965 <+*>
1958 V62 P43-87 60-19964 <+*>
1958 V62 P88-122 60-19963 <+*>
1958 V62 P123-171 60-19962 <+*>
1958 V62 P172-191 60-19961 <+*>
1958 V62 P192-204 60-19960 <+*>
1958 V62 P205-216 60-19959 <+*>
1958 V62 P217-239 60-19958 <+*>
1958 V62 P240-257 60-19957 <+*>
1958 V62 P258-271 60-19956 <+*>
1958 V62 P272-287 60-19955 <+*>
1958 V62 P288-295 60-19954 <+*>
1958 V63 P5-6 60-19953 <+*>
1958 V63 P7-16 60-19952 <+*>
61-13546 <+*>
1958 V63 P17-37 60-19951 <+*>
1958 V63 P38-47 60-19950 <+*>
1958 V63 P48-53 60-19949 <+*>
1958 V63 P54-87 60-19948 <+*>
1958 V63 P88-106 60-19947 <+*>
61-13547 <+*>
1958 V63 P107-128 60-19946 <+*>
1958 V63 P129-146 60-19945 <+*>
1958 V63 P147-149 60-19944 <+*>
1958 V63 P170-191 60-19943 <+*>
1958 V63 P192-217 60-19942 <+*>
1958 V63 P218-240 60-19941 <+*>

TRUDY VNIIKIMASH. VSESOYUZNYI NAUCHNO-ISSLEDOVATELSKII INSTITUT KISLORODNOGO MASHINOSTROENIYA
1956 V1 P89-101 60-18343 <+*> 0

TRUDY VOENNO-MEDITSINSKOI AKADEMII RKKA. LENINGRAD
1939 V18 P51-57 64-15690 <=$>

TRUDY VOLOGODSKOGO SELSKOKHOZYAISTVENNOGO INSTITUTA
1940 V1 P210-216 50/2972 <NLL>

TRUDY VORONEZHSKOGO UNIVERSITETA
1938 V10 N1 P48-66 62-16407 <=*>
1938 V10 N2 P101-183 61-16961 <=*$>
61-18022 <=*$>
64-18566 <*> PO

TRUDY VOSTOCHNO-SIBIRSKOGO FILIALA AKADEMII NAUK SSSR. MOSKVA
1955 V3 P105-126 59-16545 <+*>
1958 N1 P65-70 59-14288 <+*>
1960 V25 P110-116 UCRL TRANS-787 <=*>
1962 N41 P72-77 UCRL TRANS-958 <=*$>

TRUDY VOSTOCHNO-SIBIRSKOGO GEOLOGICHESKOGO INSTITUTA. SIBIRSKOE OTDELENIE, AKADEMIYA NAUK SSSR
1962 N5 P70-76 <JBS>
1962 N5 P77-108 <JBS>

TRUDY VOSTOCHNO-SIBIRSKOGO GOSUDARSTVENNOGO UNIVERSITETA
1937 N3 P3-22 R-2788 <*>

TRUDY VSESOYUZNOGO AEROGEOLOGICHESKOGO TRESTA
1955 N1 P61-70 IGR V3 N6 P495 <AGI>
1955 N1 P71-81 IGR V3 N6 P501 <AGI>
1955 N1 P82-88 IGR V3 N6 P507 <AGI>
1955 N1 P89-98 IGR V3 N6 P512 <AGI>
1955 N1 P99-117 IGR V3 N7 P598 <AGI>
1955 N1 P118-134 IGR V3 N7 P609 <AGI>
1955 N1 P135-146 IGR V3 N7 P619 <AGI>
1955 N1 P147-151 IGR V3 N7 P626 <AGI>

TRUDY VSESOYUZNOGO ENTOMOLOGICHESKOGO OBSHCHESTVA. AKADEMIYA NAUK SSSR. MOSKVA
1954 V44 P62-68 R-4561 <*>
1954 V44 P62-201 59-15839 <+*>
1954 V44 P105-128 R-4561 <*>
1954 V44 P191-199 R-4561 <*>
1954 V44 P202-239 R-1858 <*>
1956 V45 P128-166 R-4137 <*>
1956 V45 P343-374 61-19187 <+*> 0
1958 V46 P109-161 62-15627 <=*> 0
1962 V51 N3 P672-684 63-31506 <=$>

TRUDY. VSESOYUZNOGO GEOLOGICHESKOGO INSTITUTA. LENINGRAD
1959 V27 P322-340 62-23750 <=*> 0
1960 V35 P276-287 66-14130 <*$>
1960 V35 P304-307 66-14130 <*$>
1960 V35 P325-329 66-14130 <*$>
1960 V35 P345-347 66-14130 <*$>
1960 V35 P376-402 62-20300 <=*>
1960 V35 P423 62-20300 <=*>
1960 V39 P1-130 62-23494 <=*> 0
1961 V50 N7 P34-35 99S85R <ATS>

TRUDY VSESOYUZNOGO GIDROBIOLOGICHESKOGO OBSHCHESTVA. MOSCOW
1949 V1 P194-209 65-62668 <=$>
1950 P231-251 R-1503 <*>
1951 V3 P44-57 CSIR-TR.455 <CSSA>
1951 V3 P227-238 61-31035 <=>
1952 V4 P3-92 C-4387 <NRC>
1953 V5 P258-263 62-23851 <=*>
1955 V6 P46-69 59-10234 <+*>
1955 V6 P70-79 59-16540 <+*>
1955 V6 P217-222 M.5123 <NLL>
1955 V6 P223-226 59-22697 <+*> 0
1956 V7 P52-66 62-00144 <*>
1959 V9 P108-120 61-15712 <+*>
1961 V11 P117-121 63-31402 <=>
1961 V11 P171-188 63-31402 <=>
1961 V11 P265-284 C-4657 <NRC>
1961 V11 P299-308 63-31402 <=>
1961 V11 P323-344 63-31402 <=>
1961 V11 P354-369 63-31402 <=>
1961 V11 P394-410 63-31402 <=>
1962 V12 P235-244 63-21314 <=>
1962 V12 P400-409 63-21314 <=>
1962 V12 P410-415 63-21844 <=>

TRUDY VSESOYUZNOGO INSTITUTA GELMINTOLOGII
1947 V16 N3 P327-352 RT-562 <*>
1947 V16 N4 P403-421 RT-563 <*>
1959 N6 P216-220 61-27442 <*=>
1963 V10 P63-67 64-19960 <=$>
1963 V10 P238-244 65-64391 <=*$>

TRUDY VSESOYUZNOGO INSTITUTA ZASHCHITY RASTENII. LENINGRAD, MOSKVA
1954 N6 P116-132 60-51014 <=>
1954 N6 P133-150 60-51015 <=>
1957 N8 P5-17 66-51076 <=>
1957 V8 P99-113 61-23030 <=*>
1958 N9 P323-340 61-16153 <*=>
1958 N12 P203-219 61-21024 <=>
1958 V13 P62-63 61-15819 <=*>
1960 V14 P201-206 62-32237 <=*>
1960 V14 P207-226 62-23298 <=*>

TRUDY VSESOYUZNOGO MATEMATICHESKOGO SEZDA.

AKADEMIYA NAUK SSSR. MOSKVA
1958 V3 P447-453 UCRL-TRANS 993(L) <=$>

TRUDY VSESOYUZNOGO NAUCHNOGO INZHENERNO-TEKHNICHESKOGO OBSHCHESTVA SUDOSTROENIYA
1955 V6 N33 P66-89 60-23035 <*+> 0
1957 V7 N2 P161-170 59-19815 <+*>
1957 V7 N2 P183-204 59-19816 <+*>

TRUDY VSESOYUZNOGO NAUCHNO-ISSLEDOVATELSKOGO ALYUMINIEVO-MAGNIEVYI INSTITUTA. LENINGRAD
1960 N44 P113-119 63-13588 <=*>
1963 N50 P56-65 66-10254 <*>

TRUDY VSESOYUZNOGO NAUCHNO-ISSLEDOVATELSKOGO GEOLOGO-RAZVEDOCHNOGO NEFTYANOGO INSTITUTA. MOSKVA
1956 N95 P232-265 61-31046 <=>
1957 N4 P66-77 45R78R <ATS>
1957 N105 P131-139 65-11129 <*>
1958 N123 P135-155 62-10442 <*=>
1959 V17 P250-252 85N50R <ATS>
1959 V17 P259-262 86N50R <ATS>
1960 N152 ENTIRE ISSUE 62-13291 <=*> 0
1960 V16 P69-106 RTS-3165 <NLL>
1960 V16 P85-86 61-15079 <+*>
1960 V16 P107-123 61-15894 <=*> 0
1964 N239 P270-274 66-13658 <*>

TRUDY VSESOYUZNOGO NAUCHNO-ISSLEDOVATELSKOGO INSTITUTA AVTOGENNOI OBRABOTKI METALLOV. MOSCOW (TITLE VARIES)
1957 V4 P160-161 61-15742 <+*>
1962 V8 P55-71 AD-614 950 <=$>

TRUDY VSESOYUZNOGO NAUCHNO-ISSLEDOVATELSKOGO INSTITUTA ELEKTROENERGETIKI. MOSCOW. (TITLE VARIES)
1961 V11 P226-236 65-60372 <=$>
1961 V11 P274-288 65-60371 <=$>

TRUDY VSESOYUZNOGO NAUCHNO-ISSLEDOVATELSKOGO INSTITUTA GALURGII. LENINGRAD
1949 N21 P336-358 58Q72R <ATS>
1949 N21 P359-370 59Q72R <ATS>
1949 V21 P160-185 72M40R <ATS>
1959 N36 P136-159 19P64R <ATS>

TRUDY VSESOYUZNOGO NAUCHNO-ISSLEDOVATELSKOGO INSTITUTA GEOFIZICHESKIKH METODOV RAZVEDKI. MOSKVA
1959 N29 P14-35 64-15180 <=*$>

TRUDY VSESOYUZNOGO NAUCHNO-ISSLEDOVATELSKOGO INSTITUTA GIDROMASHINOSTROENIYA
1959 V23 P141-180 61-19078 <*+>

TRUDY VSESOYUZNOGO NAUCHNO-ISSLEDOVATELSKOGO INSTITUTA KHIMICHESKIKH PERERABOTKI GAZOV
1933 V1 P102-111 61-16838 <=*$>
1935 V2 P157-164 61-16968 <=*$>
1936 V3 P1-9 61-16965 <=*$>
1936 V3 P14-26 61-18041 <=*$>
1936 V3 P126-137 61-16967 <=*$>
1936 V3 P137-140 61-16966 <=*$>
1936 V3 P153-167 RT-2835 <*> 0
61-18152 <*>
63-20523 <=*$> 0
1936 V3 P168-173 RT-2838 <*> 0
1936 V3 P197-202 60-18317 <=*$>
79N54R <ATS>
1936 V3 P202-208 63-10981 <=*>
1936 V3 P305-311 61-16834 <=*$>

TRUDY VSESOYUZNOGO NAUCHNO-ISSLEDOVATELSKOGO INSTITUTA KHIMICHESKIKH REAKTIVOV
1958 V22 P60-64 65-13702 <*>

TRUDY VSESOYUZNOGO NAUCHNO-ISSLEDOVATELSKOGO INSTITUTA KONDITERSKOGO PROMYSHLENNOSTOGO. MOSKVA
1954 N10 P43-48 RT-4076 <*>
1958 V12 P20-32 C-3960 <NRCC>
1958 V12 P53-56 C-3951 <NRCC>
1958 V12 P128-133 C-3968 <NRCC>
1958 V12 P134-141 C-3980 <NRCC>

TRUDY VSESOYUZNOGO NAUCHNO-ISSLEDOVATELSKOGO INSTITUTA LESNOGO KHOZYAISTVA. PUTKINO, MOSCOW
1940 V18 P65-112 63-19604 <=*>

TRUDY VSESOYUZNOGO NAUCHNO-ISSLEDOVATELSKOGO INSTITUTA METODIKI I TEKHNIKI RAZVEDKI
1961 V3 P189-217 IGR V6 N6 P985-1002 <AGI>
1962 P100-113 IGR V7 N6 P970 <AGI>
1962 V5 P133-155 IGR V6 N11 P1964 <AGI>

TRUDY VSESOYUZNOGO NAUCHNO-ISSLEDOVATELSKOGO INSTITUTA MEDITSINSKIKH INSTRUMENTOV I OBORUDOVANIYA. MOSCOW
1963 N1 P4-24 64-31107 <=>
1963 N1 P27-35 64-31107 <=>
1963 N1 P41-64 64-31107 <=>
1963 N1 P83-101 64-31107 <=>
1963 N1 P164-171 64-31107 <=>

TRUDY VSESOYUZNOGO NAUCHNO-ISSLEDOVATELSKOGO INSTITUTA METOROLOGII IM. D. I. MENDELEEVA. MOSKVA
1949 V8 N68 P45-58 R-4991 <*>
1954 N23 P49-58 60-19006 <+*>
1955 N25 P35-43 63-24633 <=*$>
1955 N25 P44-53 63-24632 <=*$>
1955 N25 P54-65 AEC-TR-3544 <+*>
1956 N28 P5-19 61-27259 <*=> 0
1957 N31 P5-18 61-11963 <=>
1958 N32 P3-91 60-11731 <=>
1958 N35 P5-10 C-4801 <NRC>
1959 N40 P5-15 62-33099 <=*>
1960 V49 P24-29 63-15975 <=*>
1961 N51 P3-109 63-13396 <=>
1961 N55 P35-41 C-5749 <NRC>
1962 N66 P27-30 65-60132 <=$>
1962 N66 P46-51 64-19383 <=$>
1962 N66 P61-66 64-19384 <=$>
1962 N66 P75-89 64-19385 <=$>
1962 N67 P76-88 64-21660 <=>

TRUDY VSESOYUZNOGO NAUCHNO-ISSLEDOVATELSKOGO INSTITUTA MINERALNOGO SYRYA. MOSKVA
1938 V127 P22-36 63-14554 <=*>
1938 V127 P36-47 63-14555 <=*>
1939 N146 P53-66 3337 <HB>
1939 V142 P18-26 60-14414 <+*>

TRUDY VSESOYUZNOGO NAUCHNO-ISSLEDOVATELSKOGO INSTITUTA MORSKOGO RYBNOGO KHOZYAISTVA I OKEANOGRAFII. MOSCOW, LENINGRAD
1939 V4 P321-338 C-4921 <NRC>
64-00382 <*>
1940 V11 P115-170 R-1021 <*>
RT-3420 <*>
1954 V28 P75-101 64-19156 <=*$> P
1955 V30 P128-145 65-60962 <=$>
1958 V33 P96-100 M.5116 <NLL>
1958 V34 P7-18 C-3237 <NRCC>
1958 V34 P19-29 61-31040 <=>
1958 V34 P30-62 C-3228 <NRCC> 0
1958 V34 P63-86 61-31041 <=>
1958 V34 P102-126 C-3263 <NRCC>
63-01213 <*>
1958 V34 P133-153 C-3254 <NRC>
64-00054 <*>
1958 V34 P178-184 61-31039 <=>
1958 V35 P5-38 60-21865 <=>
1958 V35 P46-83 60-21865 <=>
1958 V35 P102-271 60-21865 <=>
1958 V36 P25-32 59-22363 <+*> 0
1958 V36 P33-51 64-19157 <=*$>
1958 V36 P52-61 59-22354 <+*>
1958 V36 P83-105 60-13264 <+*> 0

1958 V36 P242-249 60-16491 <+*>
1958 V36 P259-269 81M39R <ATS>
1958 V36 P280-294 60-14523 <+*>
1959 V40 P5-20 61-15711 <+*>
1959 V40 P98-107 65-60509 <=$>
1959 V40 P117-120 65-60942 <=$>
1960 V42 P84-98 64-19158 <=*$>
1960 V42 P99-108 64-19159 <=*$>
1960 V43 P76-218 61-23555 <=>
1960 V43 P225-234 61-23555 <=>
1961 V44 P177-186 RTS 2764 <NLL>
1962 V45 P5-14 65-60940 <=$>
1962 V45 P15-25 RTS-2632 <NLL>
1962 V45 P139-142 64-21691 <=>
1962 V46 P38-57 AD-628 195 <=$>
1962 V46 P64-67 65-61529 <=$> P
1962 V46 P74-92 65-61532 <=$> P
1962 V46 P92-102 65-61530 <=$>
1963 V48 P13-76 64-19369 <=$> P
1964 V50 P143-161 C-5446 <NRC>

TRUDY. VSESOYUZNYI NAUCHNO-ISSLEDOVATELSKII INSTITUTA MORSKOGO RYBNOGO KHOZYAISTVA I OKEANOGRAFII. LATVIISKOE OTDELENIE. MOSKVA
1961 V3 P105-138 C-4773 <NRC>

TRUDY VSESOYUZNOGO NAUCHNO-ISSLEDOVATELSKOGO INSTITUTA MYASNOI PROMYSHLENNOSTI
1953 V5 P91-102 60-17441 <+*>
1962 V12 P128-148 RTS-2784 <NLL>
1963 V15 P79-84 M-5569 <NLL>
1963 V15 P84-94 M.5570 <NLL>

TRUDY VSESOYUZNOGO NAUCHNO-ISSLEDOVATELSKOGO INSTITUTA PO PERERABOTKE NEFTI
1963 N9 P81-94 64-19923 <=$>

TRUDY VSESOYUZNOGO NAUCHNO-ISSLEDOVATELSKOGO INSTITUTA PO PERERABOTKE NEFTI I GAZA I POLUCHENIYU ISKUSSTVENNOGO ZHIDKOGO TOPLIVA
1958 V7 P181-202 94L36R <ATS>

TRUDY VSESOYUZNOGO NAUCHNO-ISSLEDOVATELSKOGO INSTITUTA PO PERERABOTKE SLANTSEV
1954 V2 P315-362 62-10427 <*=> O
1954 V3 N2 P207-215 46L36R <ATS>
1958 V6 ENTIRE ISSUE 61-11434 <=>
1958 V7 N6 P120-130 54L30R <ATS>
1959 V7 P107-119 63-19944 <=*$>
1959 V8 N7 P282-293 61-23664 <*=>

TRUDY VSESOYUZNOGO NAUCHNO-ISSLEDOV ATELSKOGO INSTITUTA PEZOOPTICHESKOGO MINERALNOGO SYRYA
1957 V1 N2 P9-29 IGR V3 N7 P575 <AGI>
1957 V1 N2 P41-51 IGR V3 N8 P706 <AGI>
1960 V3 N2 P119-121 64-16378 <=*$>

TRUDY VSESOYUZNOGO NAUCHNO-ISSLEDOVATELSKOGO INSTITUTA PRIRODNYKH GAZOV. MOSCOW
1957 N1 P27-34 27K26R <ATS>
1957 N1 P132-138 22K24R <ATS>
1958 N3 P34-63 69L31R <ATS>
1959 N6 P125-136 00N49R <ATS>
1961 N12 P27-41 64-30145 <*>
1961 N12 P56-60 64-30144 <*>
1961 N12 P187-194 63-18882 <=*>
1962 N17 P75-98 25Q74R <ATS>

TRUDY. VSESOYUZNYI NAUCHNO-ISSLEDOVATELSKII INSTITUT RAZVEDOCHNOI GEOFIZIKI. MOSKVA
1950 V3 P33-38 65-11097 <*>

TRUDY VSESOYUZNOGO NAUCHNO-ISSLEDOVATELSKOGO INSTITUTA SINTETICHESKIKH I NATURALNYKH DUSHISTYKH VESHCHESTV
1958 N4 P25-27 73L34R <ATS>
1958 V4 P125- 61-15096 <*+>
1963 N6 P19-21 RTS-2886 <NLL>

TRUDY VSESOYUZNOGO NAUCHNO-ISSLEDOVATELSKOGO INSTITUTA SPIRTOVOI I LIKERO-VODOCHNOI PROMYSHLENNOSTI. KIEV
1958 V4 P57-73 63-18401 <=*>

TRUDY VSESOYUZNOGO NAUCHNO-ISSLEDOVATELSKOGO INSTITUTA STEKLA
1956 V36 P43-50 62-18205 <=*>

TRUDY VSESOYUZNOGO NAUCHNO-ISSLEDOVATELSKOGO INSTITUTA STROITELNOI KERAMIKI. MOSKVA
1960 N16 P60-69 RTS-3275 <NLL>

TRUDY VSESOYUZNOGO NAUCHNO-ISSLEDOVATELSKOGO INSTITUTA TABACHNOI I MAKHOROCHNOI PROMYSHLENNOSTI. KRASNODAR
1926 V29 P3-55 3893-B <K-H>
1928 V46 P1-49 3881-B <K-H>
1929 V49 P83-98 5454 <K-H>
1931 V81 P77-85 5461-C <K-H>
1931 V81 P87-92 5461-G <K-H>
1931 V81 P93-102 5461-E <K-H>
1931 V81 P103-111 5461-D <K-H>
1931 V81 P113-118 5461-F <K-H>
1936 V129 P43-51 67F2R <ATS>
1937 V133 P3-9 35N51R <ATS>
1958 V150 P123-138 61-28649 <*=>
1958 V150 P191-204 60-14388 <+*>
1958 V150 P205-221 59-17000 <+*>
1958 V150 P222-232 59-15999 <+*>
1960 V151 P23-28 63-13603 <=*>
1960 V151 P83-89 11401-G <K-H>
62-18835 <=*>
1960 V151 P117-126 64-10664 <=*$>
1961 V152 P131-140 64-10490 <=*$>
1963 V153 P193-196 65-63524 <=*$>

TRUDY VSESOYUZNOGO NAUCHNO-ISSLEDOVATELSKOGO INSTITUTA TRANSPORTNOGO STROITELSTVA. BABUSHKIN
1957 V23 ENTIRE ISSUE 61-20440 <=>

TRUDY VSESOYUZNOGO NAUCHNO-ISSLEDOVATELSKOGO INSTITUTA UDOBRENII, AGROTEKHNIKI I AGROPOCHVOVEDENIYA. MOSKVA
1936 V45 P67-84 60-17775 <*=>

TRUDY VSESOYUZNOGO NAUCHNO-ISSLEDOVATELSKOGO INSTITUTA UDOBRENII, AGROTEKHNIKI I AGROPOCHVOVEDENIYA. LENINGRADSKOE OTDELENIE
1938 V49 P108-123 R-2127 <*>

TRUDY VSESOYUZNOGO NAUCHNO-ISSLEDOVATELSKOGO INSTITUTA VETERINARNOI SANITARII I EKTOPARAZITOLOGII. MOSCOW
1957 V12 P196-210 62-23438 <=*>
1958 V13 P107-116 61-27444 <*=>
1959 V14 P34-35 61-27445 <*=>
1959 V15 P204-212 61-23898 <*=>
1959 V15 P213-223 61-23002 <=*>

TRUDY VSESOYUZNOGO NAUCHNO-ISSLEDOVATELSKOGO INSTITUTA VODOSNABZHENIYA, KANALIZATSII, GIDROTEKHNICHESKIKH SOORUZHENII I INZHENERNOI GIDROGEOLOGII, GIDRAVLICHESKAYA LABORATORIYA. MOSCOW
1964 N11 P90-102 RTS-2953 <NLL>
1964 N11 P103-115 RTS-2952 <NLL>

TRUDY VSESOYUZNOGO NAUCHNO-ISSLEDOVATELSKOGO INSTITUTA ZERNA I PRODUKTOV EGO PERERABOTKI
1954 V27 P217-235 59-17116 <+*> O
1958 V35 P28-42 61-15847 <+*>
1958 V35 P62-75 61-15858 <+*>
1961 V41 P17-25 62-32491 <=*>

TRUDY VSESOYUZNOGO NAUCHNO-ISSLEDOVATELSKOGO INSTITUTA ZHELEZNODOROZHNOGO TRANSPORTA. MOSKVA
1961 N207 P127 C-4493 <NRCC>
1961 N214 P52-62 63-19157 <=*>
1964 N264 P30-40 AD-637 428 <=$>
1964 N282 P123-142 AD-638 972 <=$>

TRUDY VSESOYUZNOGO NAUCHNO-ISSLEDOVATELSKOGO INSTITUTA ZHIROV. MOSCOW
1960 V20 P216-222 54S85R <ATS>
64-10596 <*>

TRUDY VSESOYUZNOGC NAUCHNC-ISSLECOVATELSKOGO KINOFOTOINSTITUTA. MOSKVA
1957 N10 P41-49 60-10576 <+*>
1957 N10 P55-67 60-10575 <+*>
1957 N10 P77-93 60-14018 <+*>
1957 N11 P6-16 UCRL TRANS-451 <*+>
1957 N11 P17-29 UCRL TRANS-452 <*+>
1957 N11 P30-35 UCRL TRANS-453 <*+>
1957 N11 P36-42 UCRL TRANS-454 <*+>
1957 N11 P43-48 UCRL TRANS-455 <=*>
1957 N11 P49-57 UCRL-TRANS-456 <*+>
62N56R <ATS>
1957 N11 P58-65 UCRL TRANS-457 <+*>
1957 N11 P66-72 UCRL TRANS-458 <+*>
1957 N11 P73-86 UCRL TRANS-459 <*+>
1957 N11 P87-93 UCRL TRANS-460 <+*>
1957 N11 P94-101 UCRL TRANS-461 <+*>
1957 N11 P102-105 UCRL TRANS-642 <=*>
1957 V21 N11 P102-105 UCRL TRANS-462 <*>
1959 N29 P5-15 11N55R <ATS>
1959 N29 P16-23 10N55R <ATS>
1959 N29 P23-32 61K110 <CTT>
1959 N29 P34-42 08N55R <ATS>
1959 N29 P43-58 09N55R <ATS>
1959 N32 P5-18 39N56R <ATS>
1960 N35 P54-59 61K111 <CTT>
1960 N35 P6C-63 61K112 <CTT>
1960 N35 P64-69 61K113 <CTT>
1960 N37 P5-16 61K120 <CTT>
1960 N37 P17-26 38N56R <ATS>
1960 N37 P45-49 41P66R <ATS>
1960 N37 P5C-57 61K119 <CTT>
1960 N37 P95-106 52N57R <ATS>
1960 N40 P5-11 09N57R <ATS>
1960 N40 P12-20 10N57R <ATS>
1960 N40 P21-25 08N57R <ATS>
1960 N40 P26-33 07N57R <ATS>
1960 N40 P34-49 13N57R <ATS>
1960 N40 P86-94 14N57R <ATS>
1960 N40 P95-105 62-18698 <=*>
1960 N40 P106-118 63-18532 <=*>
1961 N42 P66-123 64-16546 <=*$>
1961 V42 P35-43 64-14777 <=*$>
1961 V43 P71-79 28Q70R <ATS>

TRUDY VSESOYUZNOGO NAUCHNO-ISSLEDOVATELSKOGO I KONSTRUKTORSKOGO INSTITUTA KHIMICHESKOGO MASHINOSTROENIYA. MOSKVA
1960 V34 P12-25 62-15392 <=*>

TRUDY VSESOYUZNOGO NAUCHNO-ISSLEDOVATELSKOGO LESOKULTURNOGO I AGROLESOMELIORATIVNOGO INSTITUTA MOSCOW
1934 N4 P5-143 R-951 <*>

TRUDY VSESOYUZNOGO NAUCHNO-ISSLEDOVATELSKOGO I PROEKTNOGO INSTITUTA MEKHANICHESKOI OBRABOTKI POLEZNYKH ISKOPAEMYKH. (TITLE VARIES)
1959 V122 ENTIRE ISSUE 3021 <BISI>

TRUDY VSESOYUZNOGO NAUCHNO-ISSLEDOVATELSKOGO VITAMINNOGO INSTITUTA
1941 V3 N1 P78-84 87K20R <ATS>
1941 V3 N1 P85-87 88K20R <ATS>
1953 V4 P211-215 RJ-504 <ATS>

TRUDY VSESOYUZNOGO NEFTEGAZOVOGO NAUCHNO-ISSLEDOVATELSKOGO INSTITUTA. MOSCOW
1958 V12 P242-251 63-10223 <=*>
1958 V12 P331-360 63-16988 <=*> 0

TRUDY. VSESOYUZNYI NEFTEGAZOVYI NAUCHNO-ISSLEDOVATELSKII INSTITUT. KRASNODARSKII FILIAL
1962 N10 P235-241 IGR V7 N1 P7 <AGI>
1962 V10 P235-241 IGR V7 N1 P1-10 <AGI>
65-11411 <*>

TRUDY. VSESOYUZNYI NEFTEGAZOVYI NAUCHNO-ISSLEDOVATELSKII INSTITUT. TURKMENSII FILIAL
1962 N2 P212-214 65-11184 <*>

TRUDY VSESOYUZNOGO NEFTYANOGO NAUCHNO-ISSLEDOVATELSKOGO GEOLOGO-RAZVEDOCHNOGO INSTITUTA
1953 N74 P1-21 34K25R <ATS>
1953 N74 P49 34K25R <ATS>
1953 N74 P67-79 34K25R <ATS>
1953 N74 P114-115 34K25R <ATS>
1953 V74 P21-22 63-13868 <=*> 0
1953 V74 P26-31 63-13868 <=*> 0
1953 V74 P32-34 65-61572 <=$> 0
1953 V74 P36-38 63-13868 <=*> 0
1956 N100 P263-266 61-28848 <*=>
1957 N105 P51-57 65-11402 <*>
1957 V105 N4 P11-22 65-14345 <*>
1958 N117 P252-276 82L32R <ATS>
1958 N119 P3-19 61-15960 <*=>
1958 N124 P95-110 86Q73R <ATS>
1959 N132 P242-262 96L36R <ATS>
1959 N133 P112-120 61-23965 <*=>
1959 N133 P233-256 61-23967 <*=>
1959 N133 P257-271 61-23966 <*=> 0
1959 N136 P5-131 61-19566 <=*> 0
1959 N144 P1-52 87P58R <ATS>
1960 N155 P194-212 63-27102 <$>
1960 N163 P576-587 63-24349 <=*$>
1960 N163 P588-597 63-24350 <=*$>
1961 N174 P17-25 63-27409 <$>
1961 N174 P63-74 63-19309 <=*$>
1961 N174 P77-79 63-19261 <=*>
1963 N207 P305-309 65-11095 <*>
1964 N239 P259-266 66-14419 <*$>

TRUDY VSESOYUZNOGO OBSHCHESTVA FIZIOLOGOV, BIOKHIMIKOV I FARMAKOLOGOV. AKADEMIYA NAUK SSSR
1953 V3 P140-146 <CP>

TRUDY VSESOYUZNOI AKADEMIIA SELSKOKHOZYAISTVENNYKH NAUK IM. V. I. LENINA. KAZAKHSKII FILIAL. ALMA ATA. RESPUBLIKASKAYA STANTSIYA ZASHCHITY RASTENII
1953 N21 P12-14 RT-4421 <*>

TRUDY VSESOYUZNOI KONFERENTSII ANALITICHESKOI KHIMII. AKADEMIYA NAUK SSSR. MOSKVA
1943 V2 P603-614 63-18967 <=*$> 0
65-13252 <*>

TRUDY VTOROGO LENINGRADSKOGO MEDITSINSKOGO INSTITUTA. LENINGRAD
1935 N6 P13-16 RT-635 <*>

TRUDY. VYCHISLITELNYI METEOROLOGICHESKII TSENTR. AKADEMIYA NAUK SSSR. MOSCOW
1963 V1 P3-17 65-60600 <=$>
1963 V1 P18-41 65-60610 <=$>
1963 V1 P42-52 65-60605 <=$> 0
1963 V1 P53-61 65-60612 <=$>
1963 V1 P62-71 65-60608 <=$>
1963 V1 P72-87 65-60607 <=$>

TRUDY. VYCHISLITELNYI TSENTR. EREVAN
1963 V1 P1-85 64-41997 <=>
1963 V1 P7-12 64-21585 <=>
1963 V1 P13-29 64-21585 <=>
1963 V1 P30-39 64-21585 <=>
1963 V1 P40-45 64-21585 <=>
1963 V1 P46-58 64-21585 <=>
1963 V1 P59-65 64-21585 <=>
1963 V1 P66-70 64-21585 <=>
1963 V1 P71-85 64-21623 <=>

TRUDY VYCHISLITELNOGO TSENTRA. AKADEMIYA NAUK GRUZINSKOI SSR. TIFLIS
1960 V1 P263-282 63-19159 <=*>

TRUDY VYSOKOGORNOGO GEOFIZICHESKOGO INSTITUTA
ELBRUSSKOGO VYSOKOGORNOGO EKSPEDITSKIYA
1961 N2 P87-92 30P64R <ATS>
1961 N2 P108-126 63-41008 <=>
1961 N2 P146-168 31P64R <ATS>
1961 N2 P169-174 32P64R <ATS>
1961 N2 P180-186 33P64R <ATS>
1961 N2 P187-194 34P64R <ATS>
1961 N2 P195-198 35P64R <ATS>
1961 N2 P199-204 36P64R <ATS>

TRUDY YAKUTSKOGO FILIALA. AKADEMIYA NAUK SSSR.
MOSKVA
1959 N4 P47-74 IGR V1 N12 P21 <AGI>
1960 V3 ENTIRE ISSUE 61-31576 <=>

TRUDY PO ZASHCHITE RASTENII. SER. 1.
ENTOMOLOGICHESKAYA. LENINGRAD
1931 V1 N2 P277-299 RT-166 <*>
1931 V1 N2 P350-355 RT-166 <*>

TRUDY PO ZASHCHITE RASTENII. SERIIA 3. LENINGRAD
1931 S3 N1 P147-153 4069-C <K-H>

TRUDY PO ZASHCHITE RASTENII. SERIIA 4.
POZVONOCHNORIE. LENINGRAD
1932 S4 V2 P65-86 60-13376 <=*$>

TRUDY ZONALNOGO INSTITUTA ZERNOVOGO KHOZYAISTVA
NECHERNOZEMNOI POLOSY SSSR
1946 N8 P32-49 RT-764 <*>

TRUDY ZOOLOGICHESKOGO INSTITUTA. AKADEMIYA NAUK
SSSR. LENINGRAD
1935 V2 N2/3 P509-636 RT-1179 <*>
1941 V6 N4 P68-91 61-11417 <=>
1941 V7 P89-121 C-3607 <NRCC>
1946 V8 N1 P43-88 61-11418 <=>
1948 V10 P5-18 59-15898 <+*>
1949 V8 N4 P870-878 T-2245 <INSD>
1951 V9 N2 P378-404 RT-1587 <*>
1951 V9 N2 P405-455 RT-1560 <*>
1951 V9 N2 P460-461 RT-1561 <*>
1951 V9 N2 P462-475 RT-1122 <*>
1951 V9 N2 P479-507 RT-1562 <*>
1951 V9 N2 P508-511 RT-1588 <*>
1951 V9 N2 P512-553 RT-1548 <*>
RT-3032 <*>
1951 V9 N2 P554-572 RT-1555 <*>
1951 V9 N2 P583-591 RT-1589 <*>
1951 V9 N2 P613-624 RT-1530 <*>
1953 V13 P5-11 RT-2634 <*>
1953 V13 P12-22 60-23544 <+*> O
1953 V13 P390-419 61-15090 <*+>
61-19744 <*=> O
1954 V15 P89-137 64-11081 <=>
1954 V15 P138-145 64-11082 <=>
1954 V16 P427-456 R-512 <*>
1955 V17 P160-199 C-2330 <NRC>
60-15429 <+*>
1955 V17 P346-348 61-15103 <*+>
61-19743 <*=>
1955 V18 P53-62 61-15306 <+*>
1955 V18 P308-313 R-4124 <*>
1955 V18 P343-345 62-23305 <=*>
1955 V18 P349-384 C-3377 <NRCC>
1955 V18 P389-458 61-31032 <=>
1955 V21 P18-35 CSIRO-3346 <CSIR>
R-3636 <*>
1955 V21 P36-43 C-2343 <NRC>
R-1152 <*>
1958 V25 P3-129 59-16366 <+*> O
1959 V14 P72-83 64-15637 <=*$>
1962 V29 N31 P118-136 RTS-2683 <NLL>

TRUDY ZOOTEKHNICHESKO-VETERINARNOGO INSTITUTA.
VYATKA
1961 V10 P109-124 65-60943 <=$>

TRUDY ZOOVETERINARNOGO INSTITUTA. MOSCOW
1958 V22 N1 P230-237 61-13956 <*+>

TRUPPENPRAXIS
1958 N9 P651-654 59-16950 <=*>

TSCHERMAKS MINERALOGISCHE UND PETROGRAPHISCHE
MITTEILUNGEN
1954 S3 V5 N1/23 P7-47 60-14275 <*> O
1965 S3 V10 P120-124 66-12910 <*>

T'SE HUI T'UNG PAO
1958 V4 N2 P91-93 59-13957 <=>
1964 V10 N5 P1-14 65-30212 <=$>

T'SE LIANG CHI SHU HSUEH PAO
1959 V3 N3 P171 60-11124 <=>
1959 V3 N3 P174-176 60-11124 <=>
1964 V7 N4 P292-299 AD-637 667 <=$>
1966 V9 N1 P44-62 66-34559 <=$>
1966 V9 N2 P87-99 66-33521 <=$>

*TSEMENT
1936 V4 N4 P31-36 59-17695 <+*>
1947 V13 N8 P3-7 63-18503 <=*>
1948 V14 N1 P16-19 T.680 <INSD>
1951 V17 N6 P9-13 R-1463 <*>
T-1729 <NLL>
1952 V18 N3 P19-21 RT-1599 <*>
1952 V18 N4 P11-15 R-1497 <*>
1952 V18 N5 P3-5 64-20333 <*> O
1952 V18 N5 P6-7 61-19215 <*=>
64-20382 <*>
1953 V19 N2 P14-16 RT-3474 <*>
1953 V19 N5 P4-8 RT-2638 <*>
1954 V20 N6 P10-13 R-2841 <*>
60-13755 <+*>
1955 V21 N4 P19-22 R-1462 <*>
1955 V21 N4 P24-25 62-25903 <=*>
1957 V23 N3 P11-16 59-19280 <+*>
1957 V23 N6 P11-15 59-15775 <+*> C
1957 V23 N6 P19-21 64-10174 <=*$> O
1958 V24 N2 P15-18 60-15290 <+*>
1959 V25 N3 P1-2 59-13939 <=>
1959 V25 N3 P27-28 65-63522 <=*$>
1960 V26 N3 P1-3 60-41507 <=>
1960 V26 N4 P1-2 61-11722 <=>
1960 V26 N4 P11-15 62-33214 <=*>
1960 V26 N5 P106 61-21388 <=> P
1960 V26 N6 P17-20 61-27055 <=*>
1961 V27 N1 P1-4 61-23063 <=>
1961 V27 N1 P5-9 62-25365 <=*>
1961 V27 N2 P17-22 62-25369 <=*>
65-11700 <*>
1961 V27 N2 P29-31 61-20554 <*=>
1961 V27 N3 P4-8 63-20309 <=*>
1962 V28 N2 P11-12 <CEM>
1962 V28 N2 P13-15 63-23249 <=*>
1962 V28 N3 P3-7 64-13723 <=*$>
1962 V28 N3 P12-14 63-10207 <=*> O
1962 V28 N4 P7-9 63-24215 <=*$>
1962 V28 N4 P21 63-13340 <=> P
1963 V29 N1 P7-9 64-15892 <=*$>
1963 V29 N5 P14-15 65-60991 <=$>
1964 V30 N1 P1-3 CEMBUREAU TR.506 <CEM>
1965 V31 N5 P18-19 66-11942 <*>
1965 V31 N6 P2 AC/66/II-106 <CEMB>
1965 V31 N6 P4-5 <PAR>
1965 V31 N6 P7 AC/66/II-96 <CEMB>
1965 V31 N6 P12-15 <PAR>
1965 V31 N6 P15-17 <PAR>
1965 V31 N6 P19 <PAR>
1966 V32 N1 P2 AC/66/II-87 <CEMB>
1966 V32 N1 P5 AC/66/II-91 <CEMB>
1966 V32 N1 P22 <PAR>
1966 V32 N2 P3 AC/66/II-90. <CEMV>
1966 V32 N2 P6 AC/66/II-85 <CEMB>

TSIFROVAYA TEKHNIKA I VYCHISLITELNYE USTROISTVA. AKADEMIYA NAUK SSSR. INSTITUT ELEKTRONNYKH UPRAVLYAYUSHOLIKH MASHIN
1962 N2 P88-97 AD-608 077 <=> M
1962 N2 P89-97 64-41754 <=$>

TSIRKULYAR ASTRONOMICHESKOI OBSERVATORII, KHARKOVSKOGO GOSUDARSTVENNOGO UNIVERSITETA
1957 V12 P73-165 62-10159 <*=>
1961 N24 P3-13 64-15193 <=*$>
1961 N24 P14-24 63-24233 <=*$>
1961 N24 P36-38 63-19142 <=*>
1963 N26 P14-19 65-30388 <=$>
1963 N26 P43-46 65-30387 <=$>

TSIRKULYAR GLAVNOI ASTRONOMICHESKOI OBSERVATORII V PULKOVE
1941 N32 P74-86 59-19541 <+*>

*TSITOLOGIYA
1959 V1 N2 P141-177 60-11674 <=>
1959 V1 N2 P177-182 65-61606 <=$>
1959 V1 N2 P182-255 60-11674 <=>
1959 V1 N3 P257-354 60-31271 <=>
1959 V1 N4 P422-430 60-18561 <+*>
1959 V1 N4 P431-435 62-14761 <=*>
1959 V1 N5 P527-540 65-60912 <=$>
1960 V2 N1 P29-36 62-14707 <=*>
1960 V2 N3 P379-384 63-10931 <=*> 0
1960 V2 N5 P598-601 62-13732 <*=>
1961 V3 N4 P489-493 61-20749 <*=>
1962 V4 N1 P27- FPTS V22 N2 P.T369 <FASE>
1962 V4 N1 P42- FPTS V22 N1 P.T180 <FASE>
1962 V4 N1 P52-58 64-19974 <=$>
1962 V4 N1 P59- FPTS V22 N1 P.T188 <FASE>
1962 V4 N1 P61- FPTS V22 N1 P.T174 <FASE>
1962 V4 N1 P68- FPTS V22 N1 P.T123 <FASE>
1962 V4 N1 P80- FPTS V22 N1 P.T169 <FASE>
1962 V4 N1 P88- FPTS V22 N1 P.T172 <FASE>
1962 V4 N1 P88-90 RTS-2767 <NLL>
1962 V4 N1 P93-98 63-23257 <=*>
1962 V4 N2 P128- FPTS V22 N2 P.T363 <FASE>
1962 V4 N2 P137-149 66-60974 <=*>
1962 V4 N2 P150-159 63-19007 <=*>
1962 V4 N2 P193- FPTS V22 N2 P.T320 <FASE>
1962 V4 N2 P214- FPTS V22 N2 P.T349 <FASE>
1962 V4 N2 P227- FPTS V22 N2 P.T294 <FASE>
1962 V4 N2 P232-237 63-23259 <=*>
1962 V4 N3 P251- FPTS V22 N3 P.T557 <FASE>
1962 V4 N3 P290- FPTS V22 N3 P.T576 <FASE>
1962 V4 N3 P297- FPTS V22 N2 P.T357 <FASE>
1962 V4 N3 P297-305 64-13202 <=*$>
1962 V4 N3 P306- FPTS V22 N3 P.T507 <FASE>
1962 V4 N3 P318- FPTS V22 N3 P.T571 <FASE>
1962 V4 N3 P323-325 63-15994 <=*>
1962 V4 N3 P339- FPTS V22 N3 P.T433 <FASE>
1962 V4 N3 P347-353 63-15995 <=*>
1962 V4 N3 P359- FPTS V22 N4 P.T632 <FASE>
1962 V4 N3 P365-367 63-15996 <=*>
1962 V4 N3 P370-373 62-33225 <=*>
1962 V4 N4 P391- FPTS V23 N1 P.T202 <FASE>
FPTS,V23,P.T202-T208 <FASE>
1962 V4 N4 P403-408 64-13498 <=*$>
1962 V4 N4 P409-417 64-13497 <=*$>
1962 V4 N4 P427- FPTS V22 N6 P.T1053 <FASE>
1962 V4 N4 P427-433 FPTS,V22,N6,T1053-57 <FASE>
1962 V4 N4 P449-451 63-24460 <=*$>
1962 V4 N4 P457- FPTS V22 N6 P.T1078 <FASE>
1962 V4 N4 P457-459 FPTS,V22,P.T1078-80 <FASE>
1962 V4 N4 P460- FPTS V22 N6 P.T1259 <FASE>
1962 V4 N4 P460-465 FPTS,V22,P.T1258-62 <FASE>
63-13261 <=*>
1962 V4 N4 P465- FPTS V22 N6 P.T1081 <FASE>
1962 V4 N4 P465-467 FPTS,V22,P.T1C81-82 <FASE>
1962 V4 N5 P490- FPTS V22 N6 P.T1017 <FASE>
1962 V4 N5 P490 FPTS,V22,P.T1017-21 <FASE>
1962 V4 N5 P499- FPTS V23 N1 P.T23 <FASE>
FPTS,V23,N1,T23-T27 <FASE>
1962 V4 N5 P519- FPTS V23 N1 P.T31 <FASE>
1962 V4 N5 P519 FPTS,V23,P.T31-T36 <FASE>
1962 V4 N5 P530-537 63-23834 <=*>
1962 V4 N5 P545- FPTS V22 N6 P.T1230 <FASE>
1962 V4 N5 P545-554 FPTS,V22,P.T1230-35 <FASE>
63-23230 <=*$>
1962 V4 N5 P555-558 63-23864 <=*>
1962 V4 N5 P559- FPTS V23 N1 P.T46 <FASE>
1962 V4 N5 P559 FPTS,V23,N1,P.T46-50 <FASE>
1962 V4 N5 P565-570 63-24474 <=*$>
1962 V4 N5 P571-573 64-13505 <=*$> 0
1962 V4 N5 P58C- FPTS V22 N6 P.T1087 <FASE>
FPTS,V22,N6,T1087-89 <FASE>
1962 V4 N6 P652-660 63-23229 <=*$>
1963 V5 N1 P3-23 63-21773 <=>
1963 V5 N1 P61-68 64-23514 <=$>
1963 V5 N2 P194-203 63-31087 <=>
1963 V5 N3 P261-262 63-23546 <=*$>
1963 V5 N3 P323-326 64-15554 <=*$>
1963 V5 N3 P359 NLLTB V5 N12 P1090 <NLL>
1963 V5 N3 P362 NLLTB V5 N12 P1090 <NLL>
1963 V5 N4 P365-378 63-41150 <=>
1963 V5 N4 P391-403 63-41155 <=>
1963 V5 N5 P485-498 64-21450 <=>
1963 V5 N6 P601-614 65-61093 <=$>
1963 V5 N6 P639- FPTS V24 N3 P.T533 <FASE>
1964 V6 N1 P59-66 65-60189 <=$>
1964 V6 N2 P133-151 64-31697 <=>
1964 V6 N2 P152- FPTS V24 N3 P.T411 <FASE>
1964 V6 N2 P162-167 64-31975 <=>
1964 V6 N2 P175-192 64-31697 <=>
1964 V6 N2 P222-225 64-31975 <=>
1964 V6 N2 P27C-271 64-31536 <=>
1964 V6 N3 P346- FPTS V24 N3 P.T375 <FASE>
1964 V6 N4 P432-443 65-62365 <=$>
1964 V6 N4 P443-452 65-63623 <=$>
1964 V6 N4 P452-457 65-62367 <=$>
1964 V6 N4 P531-532 64-51171 <=>
1964 V6 N4 P532-533 64-51235 <=$>
1964 V6 N5 P583- FPTS V24 N5 P.T868 <FASE>
1964 V6 N6 P667- FPTS V25 N1 P.T128 <FASE>
1964 V6 N6 P695- FPTS V24 N6 P.T1135 <FASE>
1964 V6 N6 P754- FPTS V24 N6 P.T1131 <FASE>
1965 V7 N1 P126-128 65-30865 <=$>
1965 V7 N2 P274-276 65-31394 <=$>
1965 V7 N2 P281-284 65-31394 <=$>
1965 V7 N3 P372-377 66-61486 <=$>
1965 V7 N4 P447-466 65-33481 <=*$>
1965 V7 N4 P531-537 65-33481 <=*$>
1965 V7 N5 P664-666 66-61487 <=*>

TSUKEN GEPPO
1952 V5 N12 P6C8-61C 2641 <*>

*TSVETNYE METALLY
1930 N7 P960-977 63-16762 <=*>
1932 N4 P536-542 64-20818 <*>
1935 V10 N8 P102-110 63-16039 <=*> 0
1938 N9 P91-94 RT-3127 <*>
1938 V13 N5 P71-76 RT-780 <*>
63-20760 <=*$>
64-10236 <=*$> 0
1938 V13 N6 P92-98 6048-C <K-H>
64-20202 <*> 0
1938 V13 N7 P66-71 AL-185 <*>
RT-2109 <*>
1938 V13 N8 P84-89 R-2290 <*>
1939 N8 P81-88 65-11633 <*>
1939 N8 P89-93 R-2518 <*>
1939 V14 N7 P84-88 RT-1812 <*>
64-10449 <=*$>
1940 N2 P68-72 65-10132 <*> 0
66-13986 <*>
1940 N7 P79-85 R-2890 <*>
1940 N5/6 P129-134 R-2807 <*>
1946 N5 P45-48 61-11664 <=>
1947 N4 P41-48 R-4555 <*>
1947 V20 N2 P23-25 63-14544 <=*>
1947 V20 N2 P28-31 63-14610 <=*>
1955 N8 P46-47 57-1524 <*>
1955 V28 N4 P35-40 CR-2872 <PS>
1956 N1 P1-10 RT-4404 <*>

1956 N1 P11-19	RT-4403 <*>
1956 N1 P62-65	R-380 <*>
1956 V29 N1 P22-30	RT-4402 <*>
1956 V29 N1 P57-61	62-23056 <=*>
1956 V29 N3 P50-58	62-19854 <=*>
1956 V29 N4 P22-24	59-16375 <+*>
1956 V29 N4 P55-57	62-23031 <=*>
1956 V29 N5 P61-63	61-10718 <+*>
1956 V29 N7 P55-58	CR-2876 <PS>
1956 V29 N9 P1-6	59-11328 <=>
1956 V29 N9 P62-67	20K20K <ATS>
1956 V29 N11 P36-42	CR-2874 <PS>
1957 V30 N1 P49-56	62-11220 <=>
1957 V30 N1 P66-71	29K21R <ATS>
1957 V30 N2 P5-8	CR-2873 <PS>
	61-23283 <*=>
1957 V30 N2 P32-36	4379 <HB>
1957 V30 N2 P61-67	62-13201 <=*>
1957 V30 N4 P7-9	4380 <HB>
1957 V30 N4 P29-32	4381 <HB>
1957 V30 N4 P37-44	60-23582 <+*> 0
1957 V30 N5 P85-90	59-18769 <+*>
1957 V30 N6 P57-61	TRANS-54 <MT>
	62-18184 <=*>
1957 V30 N7 P70-75	59-19437 <+*>
1957 V30 N8 P67-73	71M44R <ATS>
1957 V30 N9 P19-25	4411 <HB>
	60-13574 <+*>
1957 V30 N10 P80-87	60-15824 <+*> 0
1957 V30 N12 P9-11	82K23R <ATS>
1957 V30 N12 P65-70	4415 <HB>
1958 V31 N1 P44-47	60-19180 <+*>
1958 V31 N2 P1-6	62-13752 <*=>
1958 V31 N2 P28-35	60-13686 <+*>
1958 V31 N2 P36-39	60-23621 <+*>
1958 V31 N3 P30-38	60-23565 <+*>
1958 V31 N3 P46-47	61-13494 <*+>
1958 V31 N3 P47-54	60-13614 <+*>
1958 V31 N3 P55-60	59-16371 <+*> 0
1958 V31 N3 P66-69	60-13600 <+*>
1958 V31 N4 P1-5	60-19975 <+*>
1958 V31 N4 P16-28	CR-2826 <PS>
	61-15708 <+*>
1958 V31 N4 P29-34	60-23805 <+*>
1958 V31 N4 P43-49	61-15234 <*+>
1958 V31 N4 P74	TRANS-55 <MT>
1958 V31 N5 P35-38	63-15245 <=*>
1958 V31 N5 P62-70	<PS>
	TRANS-119 <MT>
1958 V31 N6 P26-30	60-11114 <=>
1958 V31 N6 P30-33	61-15204 <*+>
1958 V31 N6 P38-41	59-11343 <=>
1958 V31 N6 P42-52	59-11344 <=>
1958 V31 N6 P57-61	59-11331 <=>
1958 V31 N6 P74-82	59-11345 <=>
	60-15825 <+*> 0
1958 V31 N6 P80-85	59-11348 <=>
1958 V31 N6 P86-93	59-11347 <=>
1958 V31 N7 P3-6	PB 141 232T <=>
	61-10736 <+*>
1958 V31 N7 P7-11	PB 141 232T <=>
1958 V31 N7 P30-34	61-10732 <+*>
1958 V31 N7 P44-50	61-10734 <+*>
1958 V31 N7 P78-84	62-13604 <*=>
1958 V31 N7 P85-87	61-15205 <*+>
1958 V31 N7 P90-91	PB 141 232T <=>
1958 V31 N7 P91-92	59-11337 <=>
1958 V31 N8 P15-20	59-11373 <=> 0
1958 V31 N8 P21-24	59-11366 <=> 0
1958 V31 N8 P34-39	59-11365 <=>
1958 V31 N8 P53-56	CR-2824 <PS>
1958 V31 N8 P61-63	4488 <HB>
	59-18126 <+*>
1958 V31 N8 P72-73	59-11372 <=>
1958 V31 N8 P83-85	59-11358 <=>
1958 V31 N9 P58-61	1328 <BISI>
	62-24212 <=*>
1958 V31 N10 P29-32	59-11646 <=>
1958 V31 N10 P39-43	59-11647 <=>
1958 V31 N10 P61-66	CR-2822 <PS>

1958 V31 N10 P78-80	60-17473 <+*>
1958 V31 N11 P1-8	59-11644 <=> 0
1958 V31 N11 P47-52	<PS>
	60-13575 <+*>
1958 V31 N11 P52-60	59-11645 <=> 0
1958 V31 N12 P18-19	59-11692 <=>
1958 V31 N12 P28-30	59-11692 <=>
1958 V31 N12 P38-44	63-21174 <=>
1958 V31 N12 P67-69	<PS>
1958 V31 N12 P70-77	<PS>
1959 V32 N1 P21-27	59-11553 <=> 0
1959 V32 N1 P27-32	CR-2828 <PS>
1959 V32 N1 P48-53	AEC-TR-3996 <+*>
1959 V32 N1 P73-79	62-25895 <=*> P
1959 V32 N1 P92-93	61-13229 <*+>
1959 V32 N2 P8-9	59-11553 <=> 0
1959 V32 N2 P17-27	59-11553 <=> 0
1959 V32 N2 P24-27	62-18265 <=*>
1959 V32 N2 P49-52	59-19358 <+*>
1959 V32 N2 P58-64	61-23636 <*=>
1959 V32 N3 P19-22	61-27514 <*=>
1959 V32 N4 P53-57	CR-2827 <PS>
1959 V32 N4 P64-68	TRANS-56 <MT>
1959 V32 N4 P64-69	61-15650 <+*>
1959 V32 N4 P90-93	61-15650 <+*>
1959 V32 N5 P10-15	60-11118 <=>
1959 V32 N5 P16-22	61-15221 <*+>
1959 V32 N5 P22-24	60-11118 <=>
1959 V32 N5 P32-37	60-11118 <=>
1959 V32 N5 P38-45	CR-2821 <PS>
1959 V32 N5 P54-58	63-19125 <=*$>
1959 V32 N5 P55-58	CR-2825 <PS>
1959 V32 N5 P67-72	CR-2823 <PS>
1959 V32 N6 P39-41	60-11114 <=>
1959 V32 N6 P57-62	59-13920 <=>
1959 V32 N7 P40-46	61-27511 <*=> 0
1959 V32 N8 P18-24	CR-2877 <PS>
1959 V32 N8 P24-27	60-11463 <=>
1959 V32 N8 P44-50	60-15931 <+*>
1959 V32 N8 P50-54	60-11073 <=>
1959 V32 N8 P57-61	61-15721 <+*>
1959 V32 N8 P69-71	60-15776 <+*> 0
1959 V32 N8 P72-74	CR-2875 <PS>
	61-10232 <*+>
	64-10562 <=*$>
1959 V32 N9 P4-10	61-28974 <*=> 0
1959 V32 N9 P39-44	60-51138 <=>
1959 V32 N9 P53-56	61-15735 <+*>
1959 V32 N9 P56-61	61-19696 <=*>
1959 V32 N9 P63-66	61-13234 <*+>
1959 V32 N10 P22-24	60-11463 <=>
1959 V32 N10 P35-42	61-28164 <*=>
1959 V32 N10 P65-68	61-10213 <*+>
1959 V32 N10 P68-75	TRANS-100 <MT>
	1781 <BISI>
	62-11322 <=>
1959 V32 N10 P88-92	60-41076 <=>
1959 V32 N11 P1-7	61-23510 <*=>
	61-27512 <*=> 0
1959 V32 N11 P31-32	62-24287 <=*>
1959 V32 N11 P43-51	AEC-TR-4030 <+*>
1959 V32 N11 P48-50	61-13338 <*+>
1959 V32 N11 P55-60	61-19789 <*=>
	63-13074 <=*>
1959 V32 N11 P65-77	60-41166 <=> 0
1959 V32 N12 P1-6	60-11693 <=>
	60-11749 <=>
1959 V32 N12 P19-22	60-11693 <=>
1959 V32 N12 P23-28	60-11693 <=>
	62-13914 <*=>
1959 V32 N12 P28-34	60-11749 <=>
1959 V32 N12 P49-54	60-11749 <=>
1959 V32 N12 P78-79	23M46R <ATS>
1959 V32 N12 P78-80	60-11749 <=>
1959 V32 N12 P81-88	60-11801 <=>
1960 V33 N1 P43-48	AEC-TR-5296 <=*>
	TRANS-101 <MT>
	61-13758 <*+> 0
1960 V33 N3 P68-74	60-31138 <=>
1960 V33 N4 P16-18	61-27518 <*=> 0

1960 V33 N4 P28-33 31N58R <ATS>
1960 V33 N4 P84-85 60-31679 <=> 0
1960 V33 N5 P1-4 61-21670 <=>
1960 V33 N6 P43-45 61-16431 <*=>
1960 V33 N7 P72-77 61-31474 <=>
1960 V33 N11 P42-44 62-14593 <=*>
1960 V33 N12 P47-53 61-21701 <=>
1960 V33 N12 P74-80 61-27165 <*=>
1961 V34 N1 P1-4 61-27273 <=>
1961 V34 N1 P75-78 5169 <HB>
1961 V34 N3 P82-86 61-31487 <=>
1961 V34 N5 P77-79 61-28267 <*=>
1961 V34 N6 P81-84 62-23608 <=*>
1961 V34 N8 P56-59 62-15447 <*=>
1961 V34 N9 P78-81 62-25303 <=*> C
1962 N1 P1-5 NLLTB V4 N7 P605 <NLL>
1962 N4 P1-3 NLLTB V4 N10 P906 <NLL>
1962 V35 N2 P60-65 63-13652 <=*>
1962 V35 N3 P48-53 5679 <HB>
1962 V35 N3 P88-91 63-13692 <=*>
1962 V35 N4 P88-92 63-13173 <=*>
1962 V35 N6 P78-80 63-19923 <*=>
1962 V35 N8 P44-47 63-13071 <=*>
1962 V35 N8 P72-76 63-13071 <=*>
1962 V35 N9 P1-7 63-21199 <=>
1962 V35 N9 P91-92 63-13528 <=*>
1962 V35 N10 P63-68 63-21172 <=>
1962 V35 N1C P71-75 63-21172 <=>
63-27824 <$>
1962 V35 N11 P80-95 63-13655 <=>
1962 V35 N11 P85-88 6374 <HB>
1963 V36 N1 P54-57 63-21659 <=>
1963 V36 N2 P68-71 63-41078 <=>
1963 V36 N3 P74-76 63-31840 <=>
1963 V36 N3 P87-89 63-41026 <=>
1963 V36 N3 P89 63-41C26 <=>
1963 V36 N5 P53-58 64-21090 <=>
1963 V36 N6 P64-68 64-13215 <=*$>
1963 V36 N11 P69-74 64-21203 <=>
1963 V36 N11 P85-86 65-61341 <=$>
1963 V36 N12 P54-56 65-61342 <=$>
1964 V37 N1 P58-62 64-31160 <=>
1964 V37 N1 P63-66 65-61320 <=$>
1964 V37 N4 P50-55 40R79R <ATS>
1964 V37 N4 P55-60 65-61169 <=$>
1964 V37 N6 P66-71 AD-618 012 <=$>
65-61178 <=>
1964 V37 N6 P74-81 639 <TC>
1964 V37 N7 P7C-76 64-51046 <=$>
1964 V37 N8 P86-90 64-51751 <=$>
1964 V37 N8 P95-96 65-30128 <=$>
1964 V37 N9 P90-91 M-5616 <NLL>
1964 V37 N10 P67-68 AD-619 333 <=$>
1964 V37 N1C P82-86 65-30172 <=$>
1964 V37 N11 P42-46 M-5635 <NLL>
1964 V37 N11 P88-90 65-30347 <=$>
1964 V37 N12 P25-31 M-5692 <NLL>
1964 V37 N12 P46-52 638 <TC>
1965 N9 P69-71 66-33953 <=$>
1965 N10 P79-83 66-33953 <=$>
1965 N11 P106-111 66-31855 <=$>
1965 N12 P77-82 66-31855 <=$>
1965 N12 P86-89 66-31855 <=$>
1965 V38 N1 P84-89 65-30730 <=$>
1965 V38 N2 P49-52 887 <TC>
1965 V38 N2 P71-72 65-30757 <=$>
1965 V38 N3 P53-60 65S84R <ATS>
1965 V38 N3 P65-68 64S84R <ATS>
1965 V38 N4 P70-72 M.5836 <NLL>
1965 V38 N4 P87 65-13247 <*>
1965 V38 N5 P41-42 M-5835 <NLL>
1965 V38 N5 P53-56 1011 <TC>
95S85R <ATS>
1965 V38 N5 P73-78 65-31576 <=$>
1965 V38 N5 P82-85 65-31576 <=$>
1965 V38 N6 P45-49 39S86R <ATS>
1965 V38 N6 P49-54 40S86R <ATS>
1965 V38 N6 P54-61 1303 <TC>
38S86R <ATS>
1965 V38 N7 P2-8 M.5837 <NLL>
1965 V38 N7 P7C-74 65-33101 <=$>
1965 V38 N7 P76-81 6444 <HB>
1965 V38 N7 P76-86 65-33101 <=$>
1965 V38 N8 P58-60 23S89R <ATS>
1965 V38 N8 P87-89 24S89R <ATS>
1965 V38 N9 P12-17 65-17173 <*>
1965 V38 N9 P43-44 65-17174 <*>
1965 V38 N9 P49-53 84S89R <ATS>
1965 V38 N9 P53-54 1602 <TC>
1965 V38 N10 P59-62 05T92R <ATS>
1965 V38 N10 P62-66 06T92R <ATS>
1965 V38 N11 P90-93 23T93R <ATS>
1966 N2 P80-82 66-31855 <=$>
1966 N6 P72-75 66-33953 <=$>
1966 N6 P77-82 66-33953 <=$>
1966 N7 P77-78 66-33953 <=$>
1966 V39 N2 P39-44 05T95R <ATS>
1966 V39 N2 P44-49 22T95R <ATS>
1966 V39 N2 P5C-53 23T95R <ATS>
1966 V39 N2 P53-59 24T95R <ATS>
1966 V39 N2 P59-63 25T95R <ATS>
1966 V39 N6 P4- 66-34681 <=$>

TU JANG. CHINESE PECPLES REPUBLIC
1961 N6 P15-19 63-21634 <=>
1961 N6 P23-24 63-21634 <=>
1961 N6 P29-3C 63-21634 <=>
1961 N8 P1-5 63-21634 <=>
1961 N8 P55-58 63-15117 <=>
1961 N8 P55-56 63-21634 <=>
1961 V9 N1 P17-19 63-15117 <=>
1961 V9 N1 P22 63-15117 <=>
1961 V9 N3 P32 63-15117 <=>
1961 V9 N4 P30-32 63-15117 <=>

T'U JANG T'UNG PAO
1964 N5 P16-21 65-30431 <=$>

T'U JUNG HSUEH PAO
1956 V4 N2 P99-114 61-28578 <*=>
1956 V4 N2 P117-126 61-28568 <*=>
1956 V4 N2 P129-140 62-13407 <*=>
1956 V4 N2 P143-157 62-13429 <*=>
1956 V4 N2 P159-167 62-15230 <*=>
1956 V4 N2 P169-176 62-13359 <*=>
1956 V4 N2 P179-184 61-28571 <*=>
1956 V4 N2 P185-195 62-15379 <*=>
1958 V6 N1 P1-42 62-13008 <*=>
1958 V6 N1 P44-52 62-13008 <*=>
1958 V6 N1 P54-63 62-13008 <*=>
1958 V6 N1 P65-88 62-13008 <*=>
1962 V10 N1 P62 63-13488 <=>
1962 V1C N1 P65 63-13488 <=>
1962 V10 N2 P62 63-13488 <=>
1962 V10 N2 P65 63-13488 <=>
1963 V11 N1 P1C9-11C 63-31766 <=>

TUBERCULOSIS, JAPAN
SEE KEKKAKU

TUBERKULOSEARZT
1954 V8 N2 P112-113 3178-G <K-H>
1956 V10 P1C5-107 5339-B <KH>
1957 V11 P243-248 63-00556 <*>
1957 V11 N9 P1-6 59-15760 <*> 0

TU MU KUNG CHENG HSUEH PAO
1960 V7 N3 P3-19 60-41167 <=>
1960 V7 N3 P26-40 6C-41167 <=>
1960 V7 N3 P47-48 <CCA>
6C-41167 <=>
1960 V7 N3 P53-54 <CCA>
60-41167 <=>
1960 V7 N5 P1-7 61-21208 <=>
1960 V7 N5 P11-24 61-21232 <=>
1963 V10 N1 P32-36 63-31800 <=>
1963 V10 N1 P95-96 63-31800 <=>
1964 V1C N1 P43-49 AC-636 666 <=$>

TUMORI

1934 V8 P220-230 2041 <*>
1947 V21 N3 P167-172 57-782 <*>
1952 V38 N4 P209-214 2735 <*>
1954 V40 N2 P184-188 57-783 <*>
1955 V41 N1 P108-122 57-684 <*>
1958 V44 N2 P137-151 58-2341 <*>
1961 V47 P326-337 11627-F <K-H>
1963 V49 N3 P203-217 64-00367 <*>
1963 V49 N4 P235-251 64-00366 <*>
1965 V51 N1 P1-12 66-12963 <*>

TUNGSRAM TECHNISCHE MITTEILUNGEN
1935 V4 P1-8 73Q73G <ATS>

TURK IJIYEN VE TECRUBI BIYOLOJII DERGISI. ANKARA
1960 V20 P163-166 62-20033 <*> P
1964 V24 N3 P293-297 66-11581 <*>

TURK VETERINER CEMIYETI BULTENI
1953 V23 N76/7 P459-477 MF-2 <*>

TURK VETERINER HEKIMLERI DERNEGI DERGISI
1953 V26 N76/7 P506-518 MF-2 <*>

TURKMENSKAYA ISKRA
1966 P2 08/14 66-34212 <=$>
1966 P3 05/26 66-33145 <=$>

TURRIALBA
1954 V4 N1 P23-28 58-2096 <*>
1957 V7 N1/2 P8-12 64-00281 <*>

TVORBA
1961 V26 N17 P387-388 62-19362 <*=>

TWORZYWA, GUMA, LAKIERY. (TITLE VARIES)
1960 V5 N1/2 P6-9 61-18679 <*>

TYAGA ELEKTRICHESKAYA TEPLOVOZNAYA
1966 N10 P9-11 67-30300 <=>

TYL I SNABZHENIE SOVETSKOI ARMII. MOSCOW (TITLE VARIES)
1961 V21 N7 P50-54 62-33150 <=*> 0

TZU TUNG HUA
1960 V3 N3 P81-94 AD-636 655 <=$>
1960 V3 N3 P106-110 AD-636 589 <=$>
1960 V3 N3 P113 AD-620 062 <=*>

★TZU TUNG HUA HSUEH PAO
1963 V1 N2 P59-73 AD-623 819 <=*$>
1964 V2 N3 P123-135 AD-640 730 <=$>

USSR STANDARDS
1941 GOST 316-41 61-27293 <*=>
1941 GOST 338-41 61-27294 <*=>
1942 GOST 1875-42 61-27296 <*=>
1943 GOST 2160-43 60-13881 <+*>
1943 GOST 2000-43 61-27783 <*=>
1944 GOST 2507-44 61-27768 <*=>
1946 GOST 2963-45 60-17224 <+*>
1947 GOST 747-47 60-17223 <+*>
1948 GOST 3457-46 62-23906 <=*> 0
1948 GOST 1898-48 62-25107 <=*>
1950 GOST 633-50 60-17217 <+*>
1950 GOST 5473-50 61-27542 <*=>
1951 GOST 5632-51 59-19161 <+*>
1951 GOST 2789-51 61-13302 <*=>
1951 GOST 5954-51 62-24812 <=*>
1951 GOST 2669-51 62-24813 <=*>
1951 GOST 5953-51 62-24814 <=*>
1953 GOST 2940-52 62-23899 <=*> 0
1953 GOST 3465-52 62-23900 <=*> 0
1953 GOST 3462-52 62-23901 <=*> 0
1954 GOST 1903-54 61-27769 <*=>
1956 GOST 3349-56 61-27767 <*=>
1957 GOST 5008-49 60-17673 <+*>
1957 GOST 5944-57 61-27553 <*=>
1957 GOST 4798-57 63-19285 <=*>
1957 GOST 1075-H 63-24317 <=*$>
1958 GOST 5368-58 61-15636 <+*>
1958 GOST 8865-58 61-27554 <*=>
1958 GOST 8608-57 62-24662 <=*>
1958 GOST 8480-57 63-13253 <=*>
1958 GOST 2132-58 63-19878 <=*>
1959 GOST 2256-59 61-13022 <+*>
1959 GOST 9219-59 61-23371 <*=>
1959 GOST 5477-59 61-27539 <*=>
1959 GOST 5476-59 61-27540 <*=>
1959 GOST 5471-59 61-27541 <*=>
1959 GOST 2909-59 61-28669 <*=>
1959 GOST 3450-59 62-23914 <=*> 0
1959 GOST 1845-59 62-24659 <=*>
1960 GOST 2824-60 61-23373 <*=>
1960 GOST 9362-60 61-23531 <=*>
1960 GOST 461-60 61-27295 <*=>
1960 GOST 6570-60 61-27547 <*=>
1960 GOST 9444-60 61-27548 <*=>
1960 GOST 9435-60 61-27550 <*=>
1960 GOST 3461-59 62-23902 <=*> 0
1960 GOST 3460-59 62-23903 <=*> 0
1960 GOST 3459-59 62-23904 <=*> C
1960 GOST 3458-59 62-23905 <=*>
1960 GOST 9171-59 62-23907 <=*> C
1960 GOST 3456-59 62-23908 <=*>
1960 GOST 3455-59 62-23909 <=*> 0
1960 GOST 3454-59 62-23910 <=*>
1960 GOST 3453-59 62-23911 <=*> 0
1960 GOST 3452-59 62-23912 <=*>
1960 GOST 3451-59 62-23913 <=*>
1960 GOST 8038-60 62-23923 <=*>
1960 GOST 8934-58 62-24896 <=*>
1960 GOST 9488-60 62-25457 <=*>
1960 GOST-9378 RTS-2878 <NLL>
1961 GOST 2246-60 AD-607 902 <=>
1961 GOST 9825-61 NS-31 <TTIS>
1961 GOST 1631-52 64-15880 <=*$>
1961 GOST 9783-61 64-26934 <$>
1962 GOST 10178 CEM-TR.584 <CEM>
1962 GOST-9982 RAPRA-1233 <RAP>
1963 GOST 9963-62 62-32501 <=*>
1963 GOST 1707-51 64-13281 <=*$>
1963 GOST-10672 RAPRA-1232 <RAP>
1964 GOST-11012 RAPRA-1235 <RAP>

UCHENI ZAPYSKY KHARKYIVSKOHO DERZHAVNOHO UNIVERSYTETU
1948 V25 P75-104 AEC-TR-3492 <=>
1955 V64 P45-57 63-19169 <=*>
1955 V64 P95-100 63-23791 <=*>

UCHENYE ZAPISKI AZERBAIDZHANSKOGO GOSUDARSTVENNOGO UNIVERSITETA. BAKU
1958 N2 P37-48 IGR V3 N1 P42 <AGI>
1962 N2 P43-48 480 <TC>

UCHENYE ZAPISKI DAGESTANSKOGO GOSUDARSTVENNOGO UNIVERSITETA. MAKHACH-KALA
1960 V5 P103-116 17P64R <ATS>

UCHENYE ZAPISKI GORKOVSKOGO GOSUDARSTVENNOGO
1958 V32 P169-173 65-14721 <*>

UCHENYE ZAPISKI GRODNOGO PEDAGOGICHESKOGO INSTITUTA
1955 V1 P51-58 STMSP V2 P175-181 <AMS>

UCHENYE ZAPISKI IMPERATORSKAGO MOSKOVSKAGO UNIVERSITETA
1904 N21 P1-121 RT-2071 <*>
63-20347 <=*$>

UCHENYE ZAPISKI INSTITUTA FARMAKOLOGII I KHIMIOTERAPII AKADEMII MEDITSINSKIKH NAUK SSSR. MOSKVA
1958 V1 ENTIRE ISSUE 60-21856 <=>
1958 V1 P13-26 65-63170 <=$> 0
1958 V1 P27-49 65-63169 <=$> 0
1958 V1 P59-66 64-19809 <=$> 0

1958 V1 P67-83 65-60578 <=$>
1958 V1 P85-95 65-60579 <=$>
1958 V1 P97-103 65-60577 <=$> 0
1958 V1 P105-111 64-19812 <=$> 0
1958 V1 P113-128 65-60576 <=$>
1958 V1 P129-139 64-19810 <=$> 0
1958 V1 P153-160 64-19811 <=$> 0
1958 V1 P161-165 64-19813 <=$> 0
1958 V1 P167-175 64-19815 <=$>
1958 V1 P177-181 64-19814 <=$>
1958 V1 P183-188 64-19816 <=$> 0

UCHENYE ZAPISKI KAZANSKOGO GOSUDARSTVENNOGO UNIVERSITETA
1953 V113 N8 P23-26 <ATS>
1954 V114 N1 P19-22 62-24621 <=*>
1955 V115 P123-137 R-4360 <*>
1955 V115 P139-150 R-4361 <*>
1955 V115 N10 P50 62-16493 <=*>
1955 V115 N12 P57-71 63-23479 <*=>
1955 V115 N14 P21-28 AMST S2 V10 P311-318 <AMS>
1956 V116 N5 P69-72 AD-615 242 <=$>
1956 V116 N5 P77-81 R-4160 <*>
76P62R <ATS>
1963 V123 N5 P112-122 65-12738 <*>

UCHENYE ZAPISKI KISHINEZ GOSUDARSTVENNOGO UNIVERSITETA
1960 V56 P171-177 62-18699 <=*>
1961 V49 P123-128 63S88R <ATS>

UCHENYE ZAPISKI LENINGRADSKOGO ORDENA LENINA GOSUDARSTVENNOGO UNIVERSITETA
1936 V11 N2 P162-195 61-16681 <*>
1937 V3 N4 P41-52 RT-1234 <*>
1939 V42 N7 P5-60 RT-1123 <*>
1949 N114 P28-71 65-10252 <*>
1949 V16 N3 P142-166 R-5310 <*>
1950 N137 P198-203 63-16992 <=*>
1952 N153 P114- 62-10189 <*=>
1952 V23 P85-101 AMST S2 V18 P1-14 <AMS>
1954 N175 P21-35 31-98 <BISI>
1954 N175 P36-51 3199 <BISI>
1954 N175 P52-70 3200 <BISI>
1954 V175 P7-20 3197 <BISI>
1956 N9 P64-71 R-4440 <*>
1956 N9 P72-82 R-4441 <*>
1956 N210 P47-54 61-31614 <=>
1956 N210 P64-71 59-16160 <+*>
1956 N210 P72-82 59-16159 <+*>
1957 N8 P135-156 62-10534 <=*>
1957 N15 P147-154 R-5111 <*>
1957 N15 P156-162 R-5110 <*>
1957 N211 P92-103 60-15381 <+*>
1957 N215 P84-107 59-13992 <=>
1957 N222 P272-286 R-5373 <*>
1957 N222 P286-297 R-5373 <*>
1957 N228 P3-6 60-15287 <+*> 0
1957 N228 P117-154 C-3212 <NRCC> 0
1957 V6 N6 P1-46 C-4471 <NRC>
1958 N240 P34-51 61-23973 <=*>
65-63156 <=*$>
1958 N240 P52-60 62-15724 <=*>
1958 N240 P88-121 62-13100 <*=>
1958 N246 P167-227 62-24913 <=*>
1959 N269 P24-49 61-27775 <*=>
1960 N286 P125-129 63-16993 <=*>
1960 V18 P3-20 63-20398 <=*$>
1960 V35 N280 P25-30 AIAAJ V1 N1 P271 <AIAA>
1962 N303 P3-15 63-21629 <=>
63-41028 <=>
1962 N303 P49-113 63-21629 <=>
1962 N303 P49-55 63-41028 <=>
1962 N303 P56-66 63-41028 <=>
1962 N303 P118-134 63-21629 <=>
1962 N303 P167-202 63-21629 <=>
1962 N303 P222-225 63-21629 <=>
63-41038 <=>
1962 N303 P241-244 63-21629 <=>
1962 N303 P267-273 09Q73R <ATS>
1962 N303 P278-287 63-21629 <=>
1962 N303 P292-300 63-21629 <=>
1962 N307 P179-186 63-31282 <=>
1962 N307 P234-242 63-31282 <=>
1962 N311 P92-152 63-21656 <=>
1962 V17 N12 P120-125 63-15950 <=*>
1964 N37 P169-182 65-14939 <*>
1964 N324 P17-26 66-13350 <*>
1964 N324 P121-135 66-13352 <=$>
1965 N326 P99-102 AD-631 970 <=$>
1965 N328 P150-159 66-61039 <=*$>

UCHENYE ZAPISKI LVOVSKOGO GOSUDARSTVENNOGO UNIVERSITETA IM. I. FRANKO. SERIYA ASTRONOMICHESKAYA
1946 V2 N3 P22-54 97R78U <ATS>
1954 V32 N2 P66-72 61-19314 <+*>
1955 V35 N8 P162-169 62-18529 <=*>

UCHENYE ZAPISKI LVOVSKOGO GOSUDARSTVENNOGO UNIVERSITETA IM. I. FRANKO. SERIYA FIZIKO-MATEMATICHESKOGO
1949 V15 N4 P45-77 60-19073 <+*>
1954 N6 P5-44 STMSP V2 P183-205 <AMS>

UCHENYE ZAPISKI MINSKOGO PEDAGOGICHESKOGO INSTITUTA
1957 V7 P145-158 AMST S2 V18 P243-255 <AMS>

UCHENYE ZAPISKI MOSKOVSKOGO GOSUDARSTVENNOGO PEDAGOGICHESKOGO INSTITUTA
1953 V71 P199-204 AMST S2 V37 P45-49 <AMS>
1954 V88 P53-56 59-13866 <=>

UCHENYE ZAPISKI MOSKOVSKOGO GOSUDARSTVENNOGO UNIVERSITETA
1936 N6 P207-211 59-17812 <+*>
1936 N6 P213-234 RT-3483 <*>
1937 N9 P5-16 60-13255 <+*> 0
1938 V14 P205-233 RT-2858 <*>
1939 V30 P3-13 AMST S1 V2 233-243 <AMS>
1940 N42 P7-68 PANSDOC-TR.168 <PANS>
1940 N42 P167-177 PANSDOC-TR.365 <PANS>
61-23641 <*=>
1940 V39 P3-81 RT-446 <*>
1944 N74 P167-179 60-16508 <+*>
1946 V1 P104-112 AMST S2 V10 P283-290 <AMS>
1949 N134 P134-145 62-11774 <=>
1949 V134 N5 P3-16 RT-759 <*>
1950 N3 P59-91 STMSP V3 P91-129 <AMS>
1950 V135 P110-133 63-10167 <=*>
1951 V4 P133-143 AMST S2 V32 P139-154 <AMS>
1951 V148 P133-143 AMST,S2,V32,P139-154 <AMS>
1953 N164 P73-86 64-23535 <=$>
1953 N164 P87-114 65-61542 <=$>
1953 V164 P129-143 R-985 <*>
2293 <NRC>
1955 N74 P229-234 59-11744 <=>
1955 N174 P7-16 60-15049 <+*>
1955 N174 P53-74 R-3236 <*>
1956 N8 P3-12 AMST S2 V18 P151-161 <AMS>
1957 V99 P221-226 N65-32968 <=$>

UCHENYE ZAPISKI. NAUCHNO-ISSLEDOVATELSKII INSTITUT PO IZUCHENIIU LEPRY. ASTRAKHOV
1962 N3 P77-80 65-60569 <=$>

UCHENYE ZAPISKI OBLASTNOI PEDAGOGICHESKII INSTITUT MOSCOW
1960 N92 P73-78 63-20822 <=*$>

UCHENYE ZAPISKI PEDAGOGICHESKOGO INSTITUTA. LENINGRAD
1958 V197 P54-112 AMST S2 V57 P207-276 <AMS>
1961 V207 P113-118 64-18539 <*>
1961 V207 P250-251 64-16660 <*>

UCHENYE ZAPISKI. PEDAGOGICHESKII INSTITUT. SMOLENSK
1962 V10 P94-102 AD-607 880 <=>

UCHENYE ZAPISKI PERMSKOGO GOSUDARSTVENNOGO UNIVERSITETA IM. M. GORKOGO. PERM
1959 N169 P3 59-31015 <=> 0
1964 N111 P182-184 66-13952 <*>

UCHENYE ZAPISKI SARATOVSKOGO GOSUDARSTVENNOGO UNIVERSITETA IMENI N. G. CHERNYSHEVSKOGO
1936 V14 N1 P73-82 63-10979 <=*>
1940 V15 N4 P147-150 61-18008 <*=>
1959 V64 P135-138 62-20265 <=*>
1959 V71 P251-259 63-24640 <=*$>

UCHENYE ZAPISKI TARTUSKOGO GOSUDARSTVENNOGO UNIVERSITETA
1957 V52 P184-205 62-20365 <=*>
1958 V62 P169-179 62-18618 <=*>
1958 V62 P180-189 63-18721 <=*>
1960 V95 P38-55 C-4989 <NRC>

UCHENYE ZAPISKI TOMSKOGO GOSUDARSTVENNOGO UNIVERSITETA IM. V. V. KUIBYSHEVA
1959 V29 P37-41 AEC-TR-5529 <=*>
97P61R <ATS>
1961 V7 N8 P35-36 62-13414 <*=>

UCHENYE ZAPISKI YAROSLAVSKOGO TEKHNOLOGICHESKOGO INSTITUTA
1961 V6 P63-65 65-60888 <=$>

UCHET I FINANSY V KOLKHOZAKH I SOVKHOZAKH
1963 V20 N5 P46 63-31203 <=>

UDOBRENIE I UROZHAI. (1929-1931)
1929 P67-71 R-3375 <*>
1929 P259-267 R-3316 <*>
1930 V2 P397-409 R-3315 <*>
1931 V3 N11 P1058-1067 63-10063 <=*>

UDOBRENIE I UROZHAI. (RUSSIA, 1923- USSR). 1956-1959
1956 V1 N1 P7-10 RT-4142 <*>
1956 V1 N1 P11-13 RT-4168 <*>
1956 V1 N8 P13-17 R-1541 <*>
1956 V1 N10 P23-32 R-1569 <*>
1956 V2 P20-24 RT-4385 <*>
1957 N11 P16-25 61-19113 <+*>
1959 N8 P3-7 60-31018 <=>
1959 N9 P61 60-11423 <=>
1959 N10 P63-64 60-11423 <=>

UGESKRIFT FOR LAEGER
1934 V96 N52 P1429-1432 66-12013 <*>
1938 V100 P989-998 64-10836 <*>
1938 V100 N30 P862-863 748E <K-H>
1940 V102 P1325-1326 3118 <*>
1949 V111 P1091-1096 2015 <*>
1950 V112 N41 P1405-1411 65-14324 <*> 0
1950 V112 N51 P1744-1747 2217B <K-H>
1951 V113 P1357-1358 57-1264 <*>
1953 V115 N4 P130-131 2840C <K-H>
1953 V115 N40 P1474-1476 58-1599 <*> P
1954 V116 P1294-1295 3390-A <KH>
1955 V117 P1347-1348 1395 <*>
1955 V117 N44 P43-47 5TM <CTT>
1956 V118 N16 P435-441 1377 <*>
1956 V118 N30 P868-872 62-00295 <*>
1956 V118 N48 P1421-1425 65-14306 <*> 0
1957 V119 P355-357 57-3200 <*>
1957 V119 P458-460 58-1236 <*>
6574-A <KH>
1957 V119 P825-829 0A-51 <RIS>
1957 V119 N31 P1003-1007 48P59D <ATS>
1958 V120 N42 P1407-1408 59-20700 <*>
1958 V120 N47 P1548-1553 59-10647 <*>
1959 V121 N8 P299-300 62-00314 <*>
9462 <K-H>
1959 V121 N14 P532-536 65-14304 <*> 0
1959 V121 N17 P662-664 60-16917 <*> 0
1959 V121 N22 P881-886 62-00329 <*>
1959 V121 N46 P1759-1768 65-14305 <*> 0
1960 V122 P714-715 62-00296 <*>
1960 V122 P890-893 61-00149 <*>
1960 V122 N21 P714-718 10494-B <K-H>
1960 V122 N25 P856-860 65-14303 <*> 0
1960 V122 N43 P1504-1507 61-00845 <*>
1961 V123 P1449-1452 65-17292 <*>
1961 V123 P1452-1455 66-12009 <*>
1961 V123 N11 P384-386 61-00846 <*>
1962 V124 N49 P1797-1804 65-14781 <*> OP
1964 V126 N8 P248 65-13366 <*>
66-12103 <*>
1966 V128 N6 P178-181 66-11435 <*>

UGOL
1939 V14 N7 P48-53 59-22602 <+*>
1947 V22 N12 P9-13 59-22599 <+*>
1949 V24 N12 P29 52/2943 <NLL>
61-13377 <+*>
1950 V25 N5 P29 61-13377 <+*> P
1950 V25 N8 P27-29 59-18675 <+*>
1950 V25 N12 P50- 61-13443 <+*>
1951 V26 N5 P29-30 RT-3823 <*>
1952 V27 N4 P12-14 63-19224 <=>
1952 V27 N6 P1-10 RT-1121 <*>
1953 V28 N2 P50-52 NCB A.2323/JG <NLL>
1953 V28 N3 P32-37 64-19930 <=$>
1953 V28 N10 P19-23 62-25902 <=*>
1953 V28 N10 P27-30 T.833 <INSD>
1953 V28 N10 P34- 60-10052 <*>
1954 V29 N3 P15-20 RT-2501 <*>
1954 V29 N11 P18-24 60-19568 <+*>
1955 N9 P35-37 C-2338 <NRC>
1956 V31 N11 P26-32 60-17233 <+*>
1957 N1 P1-7 C2363 <NRC>
1957 N2 P1-8 C2364 <NRC>
1957 N3 P45-46 59-15282 <+*>
1957 N4 P13-17 C2359 <NRC>
1957 N8 P38-40 59-14717 <+*>
1957 N9 P38-39 C-2551 <NRC>
1957 N10 P35-36 C-2536 <NRC>
1957 N11 P64-69 C-2545 <NRC>
1957 V32 N4 P13-17 59-16369 <+*> 0
1957 V32 N6 P8-13 62-15271 <*=>
1957 V32 N7 P35-37 60-17235 <+*>
1957 V32 N8 P38-40 62-25291 <=*> 0
1958 N4 P39-43 59-12252 <+*>
1958 V33 N1 P34-35 60-17246 <+*>
1958 V33 N3 P22-24 61-13139 <+*>
1958 V33 N3 P41-44 59-18021 <+*>
1958 V33 N5 P16-19 60-13475 <+*>
1958 V33 N5 P35-36 60-17250 <+*>
1958 V33 N6 P11-16 61-13140 <+*>
1958 V33 N7 P20-24 59-18407 <+*>
1958 V33 N7 P24-28 59-19443 <+*>
1958 V33 N7 P28-30 59-19248 <+*>
1958 V33 N7 P30-35 59-19817 <+*> 0
62-24889 <=*>
1959 V34 N2 P12-17 62-24825 <=*>
1959 V34 N3 P23-29 61-27529 <*=>
1959 V34 N6 P10-15 59-19353 <+*>
1959 V34 N8 P27-32 64-13287 <=*$>
1959 V34 N10 P42-44 61-21756 <=>
1960 V35 N4 P51-55 63-19939 <*=>
1960 V35 N5 P27-31 61-19949 <=*>
1960 V35 N6 P1-5 60-31803 <=>
1960 V35 N7 P48-53 63-11092 <=>
1961 V36 N1 P34-36 63-19938 <=*>
1961 V36 N2 P11-16 CE-TRANS-3771 <NLL>
1961 V36 N4 P23-28 63-19930 <*=>
1961 V36 N4 P43-44 63-19854 <=*>
1961 V36 N9 P36-41 63-11092 <=>
1962 N1 P1-6 NLLTB V4 N7 P614 <NLL>
1962 V37 N2 P24-28 64-13028 <=*$>
1962 V37 N2 P49-50 64-13035 <=*$>
1962 V37 N7 P62-64 63-13841 <=*>
1962 V37 N8 P61-63 64-13036 <=*$>
1963 V38 N1 P37-38 64-15267 <=*$>
1963 V38 N2 P45-48 63-24154 <=*$>
1963 V38 N8 P10-12 63-31852 <=>
1963 V38 N9 P37-41 96R77R <ATS>

1964 V39 N5 P42-48 31S81R <ATS>

UGOL UKRAINY
1957 V1 N4 P11-14 60-17248 <+*>
1958 V2 N9 P12-15 61-13236 <+*>
1959 P4-8 60-15298 <+*>
1959 P8-11 60-15296 <+*>
1959 N12 P43 60-31164 <=>
1959 V3 N1 P20-22 62-15269 <*=>
1959 V3 N2 P8-10 61-27527 <*=>
1959 V3 N4 P16-19 62-15262 <*=>
1959 V3 N7 P18-19 62-15258 <*=>
1959 V3 N9 P33-36 63-19936 <=*>
1960 V4 N1 P16-18 62-23884 <=*>
1960 V4 N3 P24-26 62-23892 <=*>
62-24838 <=*>
1960 V4 N3 P41-42 62-23891 <=*>
1960 V4 N5 P28-29 62-15261 <*=>
1960 V4 N7 P16-20 62-24844 <=*>
1960 V4 N9 P9-11 62-24890 <=*>
1960 V4 N10 P35-39 61-28030 <*=>
1960 V4 N12 P13-15 63-19940 <=*>
1960 V4 N12 P19-20 63-19855 <=*>
1961 V4 N5 P41-42 62-15259 <*=>
1961 V5 N4 P23-24 63-11092 <=>
1961 V5 N5 P34 62-24845 <=*>
1961 V5 N7 P16-20 63-15186 <=*>
1961 V5 N8 P27-28 64-13288 <=*$>
1961 V5 N12 P26-27 63-19900 <=*>
1962 V6 N1 P38-39 63-19776 <*=>
1962 V6 N8 P14-15 64-23544 <=$>
1963 V7 N3 P41-42 64-15386 <=*$>
1963 V7 N4 P20-22 SMRE-TRANS-4833 <NLL>
1964 V8 N3 P21-22 NCB-A.243/JG <NLL>
1964 V8 N3 P32-34 NCB-A.2432/JG <NLL>
1964 V8 N6 P26-27 SMRE-TRANS-5052 <NLL>
1964 V8 N9 P42-44 SMRE-TRANS-5054 <NLL>
1964 V8 N10 P41-42 66-61664 <=*$>

UHLI
1957 N2 P52-55 59-00772 <*>
1957 V7 N11 P381-387 CSIRO-3845 <CSIR>

UHR; FACHZEITSCHRIFT DER DEUTSCHEN UHRMACHER
1962 N13 P20- AEC-TR-5556 <*>
1962 N13 P22- AEC-TR-5558 <*>

UIRUSU
1956 V6 P1-7 61-10879 <*>
1957 V7 N1 P1-8 59-12682 <=*> 0
1957 V7 N3 P169-178 59-18115 <=*> 0
65-00395 <*> 0
1958 V8 N1 P59-66 59-14962 <=*> 0
1965 V15 P37-43 66-14091 <*$>

UJ SZOVJET KONYVEK KONYVTARAINKBAN
1961 V14 N1 P6 62-19361 <=>
1962 V28 N3 P287-421 AEC-TR-6193/64 <=>
1962 V28 N4 P427-537 AEC-TR-6193/64 <=>

UJITOK LAPJA
1961 V13 N5 P13 62-19363 <=*>
1961 V13 N5 P27-29 62-19363 <=*>
1964 V16 N8 P4 64-31525 <=>
1964 V16 N8 P29 64-31526 <=>
1965 V17 N13 P28 65-32806 <=*$>

UKRAINA. MOSCOW
1961 N24 P17-18 62-11674 <=>
1963 N4 P2 63-21885 <=>
1963 N4 P12 63-21885 <=>

★ UKRAINSKII KHIMICHESKII ZHURNAL
1925 V1 P97-140 R-2866 <*>
1925 V1 P141-163 R-3354 <*>
1929 V4 P45-63 545 <K-H>
1929 V4 N1 P117-125 4036-B <KH>
1930 V5 P167-184 416 <K-H>
1930 V5 P185-192 5072-A <KH>
1930 V5 P225-239 T-2247 <INSD>
1930 V5 P257-266 R-3048 <*>
1931 V6 P121-134 R-2547 <*>
1931 V6 P123-133 R-12196 <*>
1931 V6 P175-185 63-10362 <=*> 0
1932 V7 P83-93 64-20838 <*>
1932 V7 P141-168 R-2197 <*>
1932 V7 P169-177 R-3016 <*>
1933 V8 P68-71 RT-2329 <*>
59-10727 <+*>
1933 V8 N3/4 P335-365 64-18925 <*>
1935 V10 P242-310 R-2209 <*>
1936 V11 P18-22 R-3302 <*>
1937 V12 P53-60 R-2484 <*>
1937 V12 P190-196 R-2993 <*>
1949 V15 P341-351 3485-B <K-H>
1949 V15 P351-361 3485-C <K-H>
64-14909 <=*$> 0
1949 V15 P460-466 65-14913 <*>
1950 V16 P254-263 3763-E <K-H>
64-16156 <=*$> 0
1950 V16 P414-437 64-20334 <*>
1950 V16 P616-619 62-16232 <=*>
1950 V16 N2 P230-241 64-15360 <=*$>
1950 V16 N3 P414-436 I-128 <*>
1950 V16 N3 P414-437 RT-2055 <*> 0
64-10460 <=*$> 0
1950 V16 N3 P438-446 R-234 <*>
64-14668 <=*$> 0
1950 V16 N5 P569-587 60-21102 <=>
1950 V16 N6 P599-611 63-14097 <=*>
1951 V17 N6 P902-910 RT-2134 <*>
64-10441 <=*$>
64-15361 <=*$>
1952 V18 P339-346 61-13342 <+*>
1952 V18 P673-682 R-701 <*>
1952 V18 P813-818 65-60946 <=$>
1952 V18 N6 P579-588 3397 <HB>
61-17579 <=$>
1952 V18 N6 P667-671 RT-2192 <*>
64-14145 <=*$> 0
1953 V19 P223-229 RJ-136 <ATS>
1953 V19 P255-265 3737-F <K-H>
64-16132 <*> 0
1953 V19 P552-555 AEC-TR-3578 <+*>
1953 V19 N2 P119-127 RT-1731 <*>
1953 V19 N2 P223-229 RJ-136 <ATS>
1953 V19 N4 P372-376 RT-3228 <*>
1953 V19 N6 P610-617 61M38R <ATS>
1953 V19 N6 P687-696 C-2422 <NRC>
R-2059 <*>
1954 V20 P35-38 RJ-351 <ATS>
1954 V20 P194-203 65-61311 <=$>
1954 V20 P343-349 63-16158 <=*> 0
1954 V20 P384-391 R-1504 <*>
1954 V20 P595-601 96N53R <ATS>
1954 V20 P602-614 60-17716 <+*>
1954 V20 P620-625 59-20180 <+*>
1954 V20 N1 P35-38 RJ-351 <ATS>
1954 V20 N2 P169-181 RT-2599 <*>
1954 V20 N4 P399-407 5027-B <K-H>
1954 V20 N4 P456-458 RT-2600 <*>
1954 V20 N5 P502-507 RT-4215 <*>
1954 V20 N5 P555-563 GB60 914 <NLL>
R-2080 <*>
1955 V21 P32-38 R-2846 <*>
1955 V21 P215-217 R-1922 <*>
RJ-647 <ATS>
1955 V21 P291-295 R-4808 <*>
1955 V21 P331-334 13077C <K-H>
65-14875 <*>
1955 V21 P381-383 T-2483 <INSD>
1955 V21 P384-387 T-2482 <INSD>
1955 V21 P484-490 09M41R <ATS>
1955 V21 P561-568 AEC-TR-2538 <*>
1955 V21 P600-613 49K27R <ATS>
1955 V21 P703-709 AEC-TR-2643 <+*> 0
1955 V21 N1 P21-26 5114 B <K-H>
64-16464 <=*$>
1955 V21 N1 P66-70 65-61448 <=>
1955 V21 N1 P76-80 3561 <RT>

Citation	Translation
1955 V21 N2 P143-148	RT-3060 <*>
1955 V21 N2 P176-181	RJ-523 <ATS>
1955 V21 N2 P195-204	RT-4208 <*>
1955 V21 N2 P211-214	R-2024 <*>
1955 V21 N2 P218-221	RJ-582 <ATS>
1955 V21 N3 P314-317	64-16462 <=*$> 0
1955 V21 N4 P472-479	64-10518 <=*$> 0
1955 V21 N5 P554-555	RJ-652 <ATS>
1955 V21 N5 P561-568	RT-4364 <*>
	15H12R <ATS>
1955 V21 N6 P687-693	RT-4353 <*>
1955 V21 N6 P703-709	RT-4659 <*>
1956 V21 P449-456	R-297 <*>
1956 V22 P173-179	RTS-2923 <NLL>
1956 V22 P213-214	65-61501 <=$>
1956 V22 P217-222	61-13345 <=*>
1956 V22 N1 P3-10	60-31201 <=>
1956 V22 N1 P11-18	60-11482 <=>
1956 V22 N1 P24-33	60-31202 <=>
1956 V22 N2 P173-179	82H13R <ATS>
1956 V22 N2 P239-246	61-19913 <=*>
1956 V22 N3 P405-408	54K27R <ATS>
1956 V22 N4 P523-526	TRANS-187 <MT>
1956 V22 N5 P603-607	59-20155 <+*>
1956 V22 N5 P687-690	61-13235 <*+>
1956 V22 N6 P726-730	60-23567 <+*>
	62-10064 <*> 0
1956 V22 N6 P809-812	14J15R <ATS>
	59-11237 <=>
1957 V23 P110-116	60-14312 <+*>
1957 V23 P146-151	RTS-2900 <NLL>
1957 V23 P287-296	R-4969 <*>
1957 V23 P287-297	73TM <CTT>
1957 V23 P689-694	66-26091 <*$>
1957 V23 N1 P54-57	85J18R <ATS>
1957 V23 N2 P110-116	63-26582 <=$>
1957 V23 N2 P180-190	62M38R <ATS>
1957 V23 N2 P191-199	1J19R <ATS>
1957 V23 N4 P443-447	AEC-TR-5803 <=*$>
1957 V23 N4 P493-495	62-10606 <*=> 0
1957 V23 N4 P530-532	62-11614 <=>
1957 V23 N5 P567-572	62-26640 <=$>
1957 V23 N6 P771-776	26L31R <ATS>
1958 V24 P325-335	R-5301 <*>
1958 V24 N1 P7-12	61-27801 <*=>
1958 V24 N1 P37-45	60-15886 <+*>
1958 V24 N1 P46-54	59-22594 <+*>
1958 V24 N2 P177-181	59-22406 <+*>
1958 V24 N2 P182-189	59-22407 <+*>
1958 V24 N2 P190-197	59-22595 <+*>
1958 V24 N3 P305-311	80L29R <ATS>
1958 V24 N3 P312-319	81L29R <ATS>
1958 V24 N4 P453-458	60-16552 <+*>
	65-10086 <*> 0
1958 V24 N4 P462-466	CSIR-TR.186 <CSSA>
1958 V24 N4 P489-490	59-14629 <+*>
1958 V24 N4 P549-554	59-14630 <+*>
1958 V24 N5 P599-601	60-15025 <+*> 0
1958 V24 N5 P629-631	61-10617 <+*>
1958 V24 N6 P749-756	61-14930 <=*>
1959 V25 P217-225	AEC-TR-5291 <=*>
1959 V25 P274-275	61-27879 <*=>
1959 V25 P370-372	UCRL TRANS-801-(L) <=*>
1959 V25 P735-738	AEC-TR-4286 <*+>
1959 V25 N3 P279-284	AEC-TR-4357 <+*>
	66-15702 <=$>
1959 V25 N3 P285-287	AEC-TR-3991 <+*>
1959 V25 N3 P322-325	61-15727 <+*>
1959 V25 N3 P337-343	61-18310 <*=>
1959 V25 N3 P351-353	60-16576 <+*>
1959 V25 N3 P392-394	61-15217 <*+>
1959 V25 N4 P471-476	61-15009 <*+>
1959 V25 N4 P509-511	61-13945 <*+>
1959 V25 N4 P553-556	60-41234 <=>
1959 V25 N5 P644-649	61-15213 <*+>
1959 V25 N5 P668-673	63-15274 <=*>
1959 V25 N5 P680-691	60-41725 <=>
1959 V25 N6 P713-721	61-15067 <+*> 0
1959 V25 N6 P735-738	65-11834 <*> 0
1959 V25 N6 P767-773	60-18990 <+*>
1960 V26 N1 P10-15	61-15066 <+*> 0
1960 V26 N1 P58-65	60-18190 <+*> 0
1960 V26 N1 P138-141	63-20850 <=*$> 0
1960 V26 N2 P227-232	17T90R <ATS>
1960 V26 N2 P273-277	60-31678 <=>
1960 V26 N3 P305-313	73M47R <ATS>
1960 V26 N3 P364-367	61-16568 <*=>
1960 V26 N4 P409-411	62-18729 <=*> 0
1960 V26 N4 P432-439	ICE V1 N1 P26-30 <ICE>
	46N53R <ATS>
1960 V26 N4 P540-542	61-21742 <=>
1960 V26 N5 P588-593	ICE V1 N1 P31-34 <ICE>
	47N53R <ATS>
1960 V26 N5 P594-599	62-24625 <=*>
1960 V26 N5 P669-671	32R78R <ATS>
1961 V27 P550-551	62-33555 <=*>
1961 V27 P739-743	63-27822 <$>
1961 V27 N2 P212-226	65-60496 <=$>
1961 V27 N3 P396-402	63-19144 <=*>
1961 V27 N4 P529-536	CSIR-TR.333 <CSSA>
1961 V27 N6 P732-739	64-10711 <=*$>
1961 V27 N6 P835-836	62-25167 <=>
1962 V28 P682-687	63-10738 <=*>
1962 V28 N1 P3-14	NLLTB V4 N8 P691 <NLL>
1962 V28 N1 P48-52	63-11528 <=>
1962 V28 N2 P260-262	65-60757 <=>
1962 V28 N3 P329-332	65-61295 <=$>
1962 V28 N3 P403-409	63-24614 <=*$>
1962 V28 N3 P412-417	62-11720 <=>
1962 V28 N4 P478-483	63-24397 <=*$>
1962 V28 N5 P565-570	AEC-TR-5601 <=*>
1962 V28 N5 P589-594	63-16533 <=*> C
1962 V28 N6 P675-681	63-14463 <=*>
1962 V28 N9 P1031-	ICE V3 N3 P360-364 <ICE>
1962 V28 N9 P1031-1036	16Q68R <ATS>
1962 V28 N9 P1066-1068	RTS-2747 <NLL>
1962 V28 N9 P1096-1099	64-10485 <=*$>
1963 N7 P667-685	ICE V4 N4 P632-642 <ICE>
1963 V29 N1 P21-24	94Q71R <ATS>
1963 V29 N3 P255-258	AEC-TR-5819 <=*$>
1963 V29 N3 P271-278	64-16589 <=*$>
1963 V29 N3 P287-292	RTS-2742 <NLL>
1963 V29 N4 P365-368	97S80R <ATS>
1963 V29 N4 P372-376	65-12396 <*>
1963 V29 N5 P532-538	64-11856 <=>
1963 V29 N5 P538-547	64-71377 <=>
1963 V29 N7 P667-685	ICE V4 N4 P632-642 <ICE>
	32R76R <ATS>
	64-19379 <=$>
1963 V29 N7 P731-733	98S80R <ATS>
1963 V29 N8 P868-873	RTS-2849 <NLL>
1963 V29 N10 P1088-1092	52R74R <ATS>
1963 V29 N12 P1259-1264	ORNL-TR-31 <=>
1964 V30 P83-85	M-5655 <NLL>
1964 V30 P1318-1321	M-5737 <NLL>
1964 V30 N2 P220-223	M-5752 <NLL>
1964 V30 N5 P457-460	AI-TRANS-62 <=$>
1964 V30 N7 P670-677	65T95R <ATS>
1964 V30 N8 P805-810	65-30121 <=$>
1964 V30 N12 P1241-1246	NLLTB V7 N7 P597 <NLL>
1965 V31 P768-771	NS-556 <TTIS>
1965 V31 N1 P45-48	58S89R <ATS>
1965 V31 N1 P94-100	66T95R <ATS>
1965 V31 N2 P166-171	65-14237 <*>
1965 V31 N5 P449-457	66-14439 <*>
1965 V31 N5 P500-505	65-17214 <*>
1965 V31 N8 P813-817	66-12755 <*>
1965 V31 N9 P918-923	66-11229 <*>
	66-12754 <*>
1965 V31 N11 P1136-1143	66-13910 <*$>

UKRAINSKII MATEMATICHESKII ZHURNAL

Citation	Translation
1949 V1 N3 P68-79	64-23580 <=>
1949 V1 N4 P64-98	AMST S2 V34 P69-108 <AMS>
1950 V2 N1 P10-59	AMST S2 V34 P109-164 <AMS>
1950 V2 N2 P3-24	62-23015 <=*>
1950 V2 N3 P45-69	61-10622 <=*>
1950 V2 N4 P37-63	AMST S2 V1 P111-137 <AMS>
1950 V2 N4 P37-43	61-10317 <=*>
1951 V3 N3 P317-339	AMST S2 V1 P139-161 <AMS>

1952 V4 P230-275 AMST S2 V2 P23-87 <AMS>
1952 V4 P347-372 AMST S2 V2 P23-87 <AMS>
1953 V5 P196-206 63-10305 <=*>
1953 V5 P207-243 STMSP V3 P1-40 <AMS>
1953 V5 P247-290 STMSP V3 P41-90 <AMS>
1953 V5 P291-298 STMSP V1 P135-143 <AMS>
1953 V5 P413-433 STMSP V2 P5-26 <AMS>
1953 V5 N4 P370-374 RT-1487 <*>
1955 V7 P131-133 AMST S2 V27 P31-33 <AMS>
1955 V7 P262-266 AMST S2 V8 P13-17 <AMS>
1955 V7 N5 P5-17 60-15995 <+*>
1956 V8 P273-288 AMST S2 V25 P131-149 <AMS>
1957 V9 N4 P377-388 STMSP V3 P167-180 <AMS>
1958 V10 N1 P3-12 65-13060 <*>
1958 V10 N2 P219 64-71497 <=> M
1958 V10 N3 P333-334 64-71420 <=> M
1958 V10 N4 P375-388 60-15853 <+*>
62-11497 <=>
1958 V10 N4 P405-412 60-13008 <+*>
62-11508 <=>
1959 V11 N1 P105-108 61-10318 <=*>
1959 V11 N1 P111-114 62-11644 <=>
1959 V11 N3 P287-294 60-12508 <=>
1959 V11 N3 P336-338 60-11245 <=>
1959 V11 N4 P413-430 66-12478 <*>
1960 V12 N2 P147-156 61-11899 <=>
62-19859 <=*>
1960 V12 N4 P476-479 61-19807 <*=>
62-11505 <=>
1961 V13 N1 P28-38 62-23188 <=*>
1961 V13 N2 P220-223 62-32685 <=*>
1961 V13 N3 P12-38 AMST S2 V35 P263-295 <AMS>
1961 V13 N3 P97-100 62-32938 <=*>
1961 V13 N4 P3-4 63-13631 <=*>
1961 V13 N4 P104-109 62-23653 <=*>
1962 V14 P119-128 STMSP V5 P259-269 <AMS>
1962 V14 P138-144 STMSP V5 P270-277 <AMS>
1962 V14 P184-190 STMSP V5 P278-284 <AMS>
1962 V14 N1 P109-112 65-13679 <*>
1962 V14 N2 P198-202 65-12640 <*>
1962 V14 N4 P351-361 63-21228 <=>
1962 V14 N4 P407-411 63-21228 <=>
65-13392 <*>
1963 V15 P205-213 AMST S2 V54 P17-28 <AMS>
1963 V15 N3 P305-309 64-10508 <=*$>
1963 V15 N3 P338-343 64-10811 <=*$>
1964 V16 P78-82 AMST S2 V54 P119-124 <AMS>
1964 V16 P319-333 AMST S2 V51 P113-131 <AMS>
1964 V16 P445-462 AMST S2 V50 P295-316 <AMS>
1964 V16 N3 P375-382 AD-635 892 <=$>

UKRAYINSKYI BOTANICHNYI ZHURNAL
1951 V8 N3 P63 61-15929 <+*>
1954 V11 N1 P106-107 61-15912 <+*>
1954 V11 N2 P55-61 61-15706 <+*>
1954 V11 N3 P30-39 61-15707 <+*>
1954 V11 N3 P116-119 61-15913 <+*>
1956 V13 N2 P89-97 61-15408 <*+>
1957 V14 N1 P30-40 61-13242 <*+>
1957 V14 N2 P72-85 61-15395 <*+>
1957 V14 N2 P87-92 61-15914 <+*>
1958 V15 N1 P71-77 63-24631 <=*$>
1958 V15 N2 P74-83 60-15883 <+*>
1958 V15 N2 P84-86 60-15684 <+*>
1959 V16 N1 P74-85 62-13600 <*=>
1959 V16 N1 P87-98 62-13738 <*=>
1959 V16 N4 P73-76 63-24638 <=*$>
1959 V16 N5 P77-79 63-24639 <=*$>
1959 V16 N6 P55-58 63-24636 <=*$>
1960 V17 N1 P12-18 61-28673 <*=>
1960 V17 N1 P19-28 63-24642 <=*$>
1960 V17 N1 P76-83 62-13092 <*=>
1960 V17 N2 P39-41 66-10257 <*>
1960 V17 N2 P113-116 60-41040 <=>
1960 V17 N3 P85-86 61-28720 <=>
1960 V17 N4 P61-65 62-13096 <*=>
1960 V17 N5 P23-31 63-15960 <=*>
1960 V17 N6 P100-102 61-28720 <=>
1961 V18 N1 P96-98 62-13925 <*=>
1961 V18 N2 P3-16 62-25459 <=*>
1961 V18 N2 P99-106 61-28417 <*=>
1961 V18 N3 P78-80 63-19130 <=*>
1961 V18 N3 P111-113 61-28887 <*=>
1961 V18 N3 P113-114 61-28779 <*=>
1961 V18 N4 P100-112 63-21230 <=>
1961 V18 N4 P100-109 65-60342 <=$>
1961 V18 N5 P74-82 63-24637 <=*$>
1962 V19 N1 P31-38 63-21977 <=>
1962 V19 N1 P58-65 65-60343 <=$>
1962 V19 N2 P3-12 65-50025 <=$>
1962 V19 N2 P108-114 62-33127 <=*>
1962 V19 N3 P10-19 63-13593 <=*>
1962 V19 N3 P28-33 RTS-2256 <NLL>
1962 V19 N6 P60-63 63-19331 <=*>
1963 V20 N2 P115-121 63-31753 <=>
1963 V20 N5 P44-47 65-60952 <=$>
1964 V21 N2 P95-103 RTS-2864 <NLL>
1965 V22 N5 P105-106 65-33695 <=*$>

UKRAYINSKYI BIOKHEMICHNYI ZHURNAL
1948 V20 N2 P266-268 R-3234 <*>
64-19408 <=$>
1949 V21 N4 P317-323 R-781 <*>
64-15093 <=*$>
1950 V22 N4 P455-461 RT-4381 <*>
1272 <*>
1952 V24 N2 P137-147 RT-1886 <*>
1952 V24 N2 P149-157 RT-1885 <*>
1952 V24 N4 P401-409 60-18522 <*>
1953 V25 N1 P3-9 RT-2643 <*>
1953 V25 N1 P17-27 RT-1392 <*>
1953 V25 N3 P247-257 RT-1850 <*>
1953 V25 N3 P259-270 R-2700 <*>
RT-1405 <*>
1954 V26 N4 P363-375 R-309 <*>
1954 V26 N4 P397-404 R-331 <*>
1954 V26 N4 P427-434 R-265 <*>
1954 V26 N4 P435-443 R-259 <*>
1954 V26 N4 P444-453 R-76 <*>
1955 V27 N3 P355-363 59-12104 <+*>
1957 V29 P10-19 59-17417 <+*>
1957 V29 P20-24 59-17418 <+*>
1957 V29 P25-31 59-17436 <+*>
1957 V29 P275-284 R-4614 <*>
1957 V29 P285-291 R-4617 <*>
1957 V29 P428-435 R-4871 <*>
1957 V29 N2 P131-141 R-4618 <*>
1957 V29 N3 P375-382 61-23210 <*=>
1957 V29 N4 P530-532 62-11614 <=>
1958 N3 P392-401 PB 141 315T <=>
1958 V30 N1 P3-7 R-5371 <*>
1958 V30 N1 P18-25 62-14702 <=*>
1958 V30 N2 P200-209 59-13226 <=> 0
1958 V30 N2 P212-218 59-13227 <=> 0
1958 V30 N2 P317-318 59-13228 <=>
1958 V30 N3 P323-331 59-17434 <+*>
1958 V30 N4 P506-512 61-23653 <*=>
1958 V30 N4 P513-520 60-18423 <*>
1958 V30 N4 P521-528 60-18417 <*>
1958 V30 N4 P604-639 60-15068 <+*>
1958 V30 N5 P643-651 60-18424 <*>
1958 V30 N6 P803-812 59-13469 <=>
1958 V30 N6 P852-859 60-18421 <*>
1958 V30 N6 P860-864 60-18420 <*>
1958 V30 N6 P880-887 60-17956 <+*> 0
1959 V31 N3 P307-311 64-19543 <=> 0
1959 V31 N3 P322-329 60-18419 <*>
1959 V31 N3 P405-413 60-18422 <*>
1959 V31 N3 P435-443 60-18510 <*> 0
1959 V31 N4 P481-488 60-18416 <*>
62-14744 <=*>
1959 V31 N4 P562-568 61-19461 <+*>
1959 V31 N5 P665-670 60-18415 <*>
1959 V31 N5 P751-758 60-18531 <*>
1959 V31 N6 P799-806 60-11870 <=>
1959 V31 N6 P834-848 62-14689 <=*> 0
1959 V31 N6 P849-858 60-18366 <+*>
1959 V31 N6 P859-867 62-14763 <=*>
1959 V31 N6 P877-882 60-11871 <=>
1959 V31 N6 P906-911 60-18365 <+*>

Citation	Translation
1959 V31 N6 P913-936	60-11872 <=>
1960 V32 N1 P67-76	62-25113 <=*>
1960 V32 N3 P381-391	62-14762 <=*>
1960 V32 N3 P483-484	60-31821 <=>
1960 V32 N4 P491-506	62-33270 <=*>
1960 V32 N4 P614-618	61-11907 <=>
1960 V32 N5 P623-634	63-10173 <=*>
1960 V32 N5 P645-652	61-21627 <=>
1960 V32 N5 P655-666	61-21503 <=>
1960 V32 N5 P678-681	61-21628 <=>
1960 V32 N5 P710-715	63-10170 <=*>
1960 V32 N5 P718-724	63-10171 <=*>
1960 V32 N6 P793-807	62-33087 <=*>
1960 V32 N6 P808-822	62-20315 <=*>
1960 V32 N6 P857-866	63-10174 <=*>
1961 V33 N1 P3-13	63-13033 <=*>
1961 V33 N1 P14-20	29P60U <ATS>
1961 V33 N1 P14-21	63-13034 <=*>
1961 V33 N1 P22-29	61-27060 <*=>
1961 V33 N1 P37-43	61-27061 <*=>
1961 V33 N1 P57-62	61-23608 <*=>
1961 V33 N1 P88-93	61-23941 <*=>
1961 V33 N2 P175-179	61-28276 <*=>
1961 V33 N2 P187-194	61-27286 <*=>
1961 V33 N2 P195-199	61-27288 <*=>
1961 V33 N2 P208-213	61-27283 <*=>
1961 V33 N2 P215-219	61-27231 <*=>
1961 V33 N2 P272-297	61-27281 <*=>
1961 V33 N3 P348-351	61-28926 <*=>
1961 V33 N3 P352-361	61-28882 <*=>
1961 V33 N3 P402-406	RTS 2853 <NLL>
1961 V33 N3 P407-418	61-28734 <*=>
1961 V33 N3 P431-435	61-28904 <*=>
1961 V33 N4 P467-474	61-28907 <*=>
1961 V33 N4 P563-568	63-10172 <=*>
1961 V33 N5 P647-656	62-20320 <=*> 0
1961 V33 N6 P805-809	62-20305 <=*>
1962 V34 N1 P10-22	62-20319 <=*>
1962 V34 N1 P23-31	63-24522 <=*$>
1962 V34 N1 P32-38	62-24383 <=*>
1962 V34 N1 P40-46	62-20318 <=*> P
1962 V34 N1 P47-55	63-23891 <=*$>
1962 V34 N1 P56-	FPTS V22 N5 P.T935 <FASE>
1962 V34 N1 P56-63	1963,V22,N5,PT2 <FASE> P
	62-20317 <=*> P
1962 V34 N1 P81-85	62-20316 <=*> P
1962 V34 N2 P167-175	62-33627 <=*>
1962 V34 N2 P208-216	62-20310 <=*> P
	63-00152 <*>
	63-15577 <=*>
1962 V34 N2 P230-235	63-16455 <=*>
1962 V34 N3 P371-378	AEC-TR-5354 <=*>
1962 V34 N3 P400-405	RTS-2245 <NLL>
1962 V34 N3 P406-418	64-18755 <*>
1962 V34 N6 P846-851	65-18078 <*>
1963 V35 N1 P3-	FPTS V23 N4 P.T821 <FASE>
	FPTS,V23,N4,P.T821 <FASE>
1963 V35 N1 P30-34	64-13904 <=*$>
1963 V35 N1 P84-90	63-21895 <=>
1963 V35 N1 P84-91	64-13837 <=*$>
1963 V35 N1 P104-112	64-19228 <=*$> 0
1963 V35 N2 P202-	FPTS V23 N4 P.T879 <FASE>
	FPTS,V23,N4,P.T879 <FASE>
1963 V35 N2 P220-226	64-15568 <=*$> 0
1963 V35 N2 P256-	FPTS V23 N4 P.T772 <FASE>
1963 V35 N2 P274-279	AD-614 962 <=$>
1963 V35 N3 P367-	FPTS V24 N1 P.T39 <FASE>
1963 V35 N3 P367-376	FPTS V24 P.T39-T42 <FASE>
1963 V35 N3 P377-384	65-60202 <=$>
1963 V35 N3 P428-435	65-60158 <=$>
1963 V35 N3 P448-477	63-31654 <=>
1963 V35 N4 P483-	FPTS V24 N1 P.T68 <FASE>
1963 V35 N4 P483-495	FPTS V24 P.T68-T72 <FASE>
1963 V35 N4 P514-518	63-31948 <=>
1963 V35 N4 P528-534	65-60159 <=$>
1963 V35 N4 P542-547	63-31948 <=>
1963 V35 N4 P580-	FPTS V24 N1 P.T65 <FASE>
1963 V35 N4 P580-587	FPTS V24 P.T65-T67 <FASE>
1963 V35 N4 P593-	FPTS V24 N1 P.T80 <FASE>
1963 V35 N4 P593-605	FPTS V24 P.T80-T84 <FASE>
1963 V35 N5 P656-	FPTS V24 N1 P.T34 <FASE>
1963 V35 N5 P656-667	FPTS V24 P.T34-T38 <FASE>
	N64-27036 <=>
1963 V35 N5 P713-	FPTS V23 N6 P.T1305 <FASE>
	FPTS V23 P.T1305 <FASE>
1963 V35 N6 P816-	FPTS V24 N1 P.T85 <FASE>
1963 V35 N6 P841-851	65-60203 <=$>
1963 V35 N6 P881-887	65-60153 <=$>
1963 V35 N6 P924-929	64-21736 <=>
1963 V35 N6 P924-930	65-60201 <=$>
1964 V36 N1 P14-20	64-31065 <=>
1964 V36 N1 P22-	FPTS V24 N1 P.T73 <FASE>
1964 V36 N1 P22-31	FPTS V24 P.T73-T76 <FASE>
	64-31065 <=>
1964 V36 N1 P46-51	65-60176 <=$>
1964 V36 N1 P52-	FPTS V24 N1 P.T90 <FASE>
1964 V36 N1 P52-58	FPTS V24 P.T90-T92 <FASE>
1964 V36 N1 P72-79	64-31035 <=>
1964 V36 N1 P8C-87	65-60235 <=$>
1964 V36 N1 P132-155	64-31065 <=>
1964 V36 N2 P209-215	64-31323 <=>
1964 V36 N2 P216-	FPTS V24 N2 P.T199 <FASE>
1964 V36 N2 P216-225	
	FPTS V24 P.T199-T202 <FASE>
1964 V36 N2 P226-	FPTS V24 N2 P.T203 <FASE>
1964 V36 N2 P226-233	
	FPTS V24 P.T203-T205 <FASE>
1964 V36 N2 P258-	FPTS V24 N2 P.T283 <FASE>
1964 V36 N2 P258-266	
	FPTS V24 P.T283-T286 <FASE>
1964 V36 N2 P276-281	65-61084 <=$>
1964 V36 N4 P483-491	66-12615 <*>
1964 V36 N4 P492-497	64-51115 <=>
1964 V36 N4 P506-512	65-62320 <=$>
1964 V36 N4 P536-547	65-62319 <=$>
1964 V36 N4 P574-583	65-62321 <=$>
1964 V36 N4 P584-	FPTS V24 N4 P.T641 <FASE>
1964 V36 N4 P607-	FPTS V24 N4 P.T645 <FASE>
1964 V36 N5 P665-	FPTS V24 N6 P.T991 <FASE>
1964 V36 N5 P772-	FPTS V24 N6 P.T983 <FASE>
1964 V36 N6 P814-	FPTS V24 N6 P.T960 <FASE>
1964 V36 N6 P829-835	65-63624 <=$>
1964 V36 N6 P9C7-	FPTS V24 N6 P.T1044 <FASE>
1965 V37 N1 P8-	FPTS V25 N1 P.T181 <FASE>
1965 V37 N1 P43-	FPTS V25 N1 P.T184 <FASE>
1965 V37 N1 P97-104	66-60298 <=*$>
1965 V37 N1 P131-136	66-60299 <=*$>
1965 V37 N2 P2C7-216	66-60849 <=*$>
1965 V37 N2 P251-	FPTS V25 N1 P.T118 <FASE>
1965 V37 N3 P315-323	66-61061 <=$*>
1965 V37 N3 P420-429	66-61060 <=*$>
1965 V37 N4 P483-	FPTS V25 N4 P.T721 <FASE>
1965 V37 N4 P529-537	66-61344 <=*$>
1965 V37 N4 P608-	FPTS V25 N3 P.T473 <FASE>
1965 V37 N6 P920-926	66-61972 <=*$>
1965 V38 N3 P307-	FPTS V25 N3 P.T479 <FASE>
1966 V38 N2 P217-218	67-30625 <=>

UKRAYINSKYI FIZYCHNYI ZHURNAL

Citation	Translation
1956 V1 N1 P5-13	59-10329 <+*>
1956 V1 N2 P158-169	64-71189 <=>
1956 V1 N3 P226-244	64-13300 <=*$>
1956 V1 N3 P308-311	59-18509 <+*>
1956 V1 N4 P366-370	64-13196 <=*$>
1957 V2 N1 P43-52	R-3129 <*>
1957 V2 N1 P95-96	59-18408 <+*>
1957 V2 N2 P43-48	62-13393 <*=>
1957 V2 N2 P149-163	R-3228 <*>
1957 V2 N2 P165-173	AD-610 277 <=$>
1957 V2 N2 P183-188	59-18090 <+*>
1957 V2 N3 P274-291	59-13162 <=> 0
1957 V2 N4 P363-368	59-22440 <+*>
1958 V3 N1 P59-71	61-23854 <*=>
1958 V3 N1 P95-103	63-18439 <=*> 0
1958 V3 N1 P124-134	62-25558 <=*>
1958 V3 N1 P135-138	59-19772 <+*>
1958 V3 N1 P64-67 SUP	62-25558 <=*>
1958 V3 N2 P178-184	59-10239 <+*>
1958 V3 N2 P246-253	59-13498 <=> 0
1958 V3 N3 P351-357	65-10889 <*>

1958 V3 N3 P370-374	61-28172 <*=>
1958 V3 N3 P375-384	61-28171 <*=>
1958 V3 N3 P391-396	60-19116 <+*>
1958 V3 N3 P408-417	60-19116 <+*>
1958 V3 N4 P433-438	69-19849 <=>
1958 V3 N4 P468-473	60-13014 <+*>
1958 V3 N4 P528-535	61-15307 <+*>
1958 V3 N4 P552-559	62-25558 <=*>
1958 V3 N5 P567-570	62-11295 <=>
1958 V3 N5 P571-575	62-19890 <=>
1958 V3 N5 P646-650	61-23687 <=*>
1958 V3 N5 P688-689	61-23687 <=*>
1958 V3 N5 P693-694	61-23687 <=*>
1958 V3 N6 P712-720	62-18642 <=*>
1958 V3 N6 P728-741	64-13327 <=*$>
1958 V3 N6 P788-794	60-19030 <+*>
1958 V3 N6 P802-814	60-15539 <+*>
	62-18848 <=*>
1959 V4 P167-176	UCRL TRANS-670(L) <*>
1959 V4 N1 P5-16	64-21710 <=>
1959 V4 N1 P72-91	64-13195 <=*$>
1959 V4 N1 P125	62-18092 <=*>
	62-24530 <=*>
1959 V4 N1 P127-129	66-12552 <*>
1959 V4 N3 P293-298	61-13955 <+*>
1959 V4 N5 P577-585	63-19976 <=*>
1959 V4 N6 P750-753	62-32475 <=*>
1960 V5 N1 P125-130	60-31869 <=*>
1960 V5 N2 P231-234	2459 <BISI>
1960 V5 N2 P259-269	63-24430 <=*$>
1960 V5 N2 P268-269	61-23711 <=$>
1960 V5 N3 P333-344	63-16550 <=*>
1960 V5 N3 P358-362	62-33554 <=*>
1960 V5 N3 P430-431	64-13197 <=*$>
1960 V5 N4 P576-577	62-32244 <=*>
1960 V5 N5 P252-258	5818 <HB>
1960 V5 N5 P606-614	RTS-2689 <NLL>
1960 V5 N5 P615-619	TIL/T.5523 <NLL>
	65-61141 <=$>
1960 V5 N5 P620-627	62-11319 <=>
	62-13618 <*=>
	62-13618 <=>
1960 V5 N5 P629-633	62-11248 <=>
1960 V5 N5 P677-680	62-28789 <=$>
1960 V5 N5 P687-692	63-19784 <=*>
1960 V5 N6 P781-789	61-23048 <=>
1960 V5 N6 P781-790	61-23048 <=*>
1960 V5 N6 P865-868	61-23472 <=>
1961 V6 N1 P20-23	AEC-TR-5305 <=*>
1961 V6 N1 P25-33	62-11520 <=>
1961 V6 N1 P34-39	62-11236 <=>
1961 V6 N1 P86-92	63-23193 <=*>
1961 V6 N1 P116-120	64-13328 <=*$>
1961 V6 N1 P139-140	63-19680 <=*>
1961 V6 N3 P415-416	62-32989 <=*>
1961 V6 N3 P424-425	5729 <HB>
1961 V6 N4 P461-467	62-23671 <=*>
1961 V6 N4 P486-488	62-33647 <=*>
1961 V6 N4 P525-530	64-13198 <=*$>
1961 V6 N4 P541-546	62-11112 <=>
1961 V6 N4 P549-555	62-18619 <=*>
1961 V6 N5 P618-641	65-10846 <*>
1961 V6 N6 P738-741	63-13683 <=*>
1961 V6 N6 P876-878	04Q70U <ATS>
1962 V7 P602-617	UCRL TRANS-941(L) <=*$>
1962 V7 N1 P73-74	63-19318 <=*>
1962 V7 N1 P75-77	AD-632 312 <=$>
1962 V7 N1 P79-81	63-19318 <=*>
1962 V7 N1 P83-86	63-19318 <=*>
1962 V7 N2 P117-131	64-13644 <=*$>
1962 V7 N2 P132-136	63-13704 <=*> 0
1962 V7 N2 P152-166	63-14466 <=*>
1962 V7 N2 P205-211	AD-621 423 <=*$>
1962 V7 N2 P212-216	63-23963 <=*$>
1962 V7 N4 P448-454	63-21018 <=>
1962 V7 N5 P457-468	AEC-TR-5658 <=*>
1962 V7 N5 P476-481	63-19575 <=*>
1962 V7 N7 P685-691	64-19102 <=*$>
1962 V7 N8 P820-825	65-60929 <=$>
1962 V7 N8 P826-833	64-19934 <=$>
1962 V7 N9 P1003-1012	63-23789 <=*>
1962 V7 N10 P1131-1134	63-21917 <=>
1962 V7 N10 P1152-1157	AD-621 424 <=$>
1962 V7 N11 P1165-1170	N65-21647 <=$>
1962 V7 N12 P1274-1279	63-21080 <=>
1962 V7 N12 P1350-1354	64-13200 <=*$>
1962 V7 N12 P1355-1362	64-19008 <=*$> 0
1962 V7 N12 P1364-1365	63-23636 <=*>
1962 V7 N12 P1367-1368	63-21080 <=>
1963 V8 P1287-1302	ANL-TRANS-86 <*>
1963 V8 N1 P57-60	64-13320 <=*$>
1963 V8 N1 P106-107	64-21710 <=>
1963 V8 N2 P157-162	AD-614 387 <=$>
1963 V8 N2 P183-190	AD-617 665 <=$>
1963 V8 N2 P206-210	AEC-TR-5807 <=$>
1963 V8 N2 P248-251	AEC-TR-5801 <=*$>
1963 V8 N3 P386-390	65-61407 <=>
1963 V8 N4 P405-417	NLLTB V5 N12 P1053 <NLL>
1963 V8 N4 P498-500	64-13721 <=*$>
1963 V8 N5 P537-548	65-10436 <*>
1963 V8 N5 P601-605	63-31970 <=>
1963 V8 N6 P624-627	64-15233 <=*$>
1963 V8 N8 P861-868	RTS-2825 <NLL>
1963 V8 N8 P913-915	AD-630 824 <=$>
1963 V8 N8 P918-921	AD-608 179 <=>
1963 V8 N8 P921-924	AD-615 237 <=$>
1963 V8 N9 P1039	66-11270 <*>
1963 V8 N12 P1328-1334	65-29055 <$>
1964 V9 P1016-1022	726 <TC>
1964 V9 N1 P26-31	64-16032 <=*$>
1964 V9 N2 P226-229	AD-617 159 <=$>
	64-30000 <=$>
1964 V9 N3 P309-317	AD-612 424 <=$>
1964 V9 N6 P593-598	AD-622 451 <=*$>
1964 V9 N6 P659-663	AD-622 451 <=*$>
1964 V9 N7 P733-743	AD-621 005 <=$>
1964 V9 N7 P766-768	AD-621 005 <=$>
1964 V9 N7 P769-791	AD-637 101 <=$>
1964 V9 N7 P777-781	52S84U <ATS>
1964 V9 N9 P948-955	AD-634 426 <=$>
1964 V9 N9 P1023-1025	AD-620 822 <=$>
1964 V9 N9 P1027-1030	AD-639 428 <=$>
1964 V9 N10 P1134-1136	AD-620 100 <=$>
1964 V9 N11 P1240-1247	AD-626 954 <=$>
1964 V9 N12 P1286-1290	AD-625 302 <=*$>
1965 V10 P543-548	66-11114 <*>
1965 V10 N1 P16-20	N65-27725 <=*$>
1965 V10 N2 P172-177	N66-21686 <=$>
1965 V10 N2 P228-229	46S87U <ATS>
1965 V10 N5 P566-568	69T91U <ATS>
1965 V10 N6 P630-635	AD-643 638 <=>
1965 V10 N6 P668-671	03T93U <ATS>
1965 V10 N6 P676-681	AD-643 638 <=>
1965 V10 N8 P912-913	37T92U <ATS>
1965 V10 N8 P915-917	66-11250 <*>
1966 V11 P177-182	AD-636 935 <=$>
1966 V11 N5 P501-506	66-61713 <=$>
1966 V11 N5 P502-506	66-13475 <*>

UKRAYINSKYI ISTORYCHNYI ZHURNAL. KIEV

1966 N6 P27-39	66-34019 <=$>

UKRAYINSKYI KHIMICHNYI ZHURNAL
SEE UKRAINSKII KHIMICHESKII ZHURNAL

UMSCHAU

1923 V27 N12 P179-184	63-16075 <*> 0
1929 V33 N43 P854	62-16153 <*>
1938 V42 N10 P218-219	59-20771 <*>
1938 V42 N10 P221	59-20771 <*>
1951 V51 N19 P598-599	63-20520 <*>
1951 V51 N19 P605	63-20525 <*> 0
1953 N13 P385-388	C-2406 <NRC>
1954 V54 N8 P243-244	64-20309 <*> 0
1955 N5 P140-142	58-2125 <*>
1955 N7 P209-211	2869 <*>
1956 V56 P75-77	SCL-T-246 <+*>
1956 V56 N4 P110-113	60-13044 <=*>
1957 V57 N7 P197-200	58-2530 <*>
1959 V59 N23 P716-717	62-10090 <*> 0

UNDA MARIS
1947 P153-168 AL-226 <*>

UNION MEDICALE DU CANADA
1913 V42 N5 P249- 2818 <*>
1953 V82 P767-774 2856B <K-H>
1955 V84 N6 P672-674 57-1920 <*>
1955 V84 N10 P1163-1166 58-96 <*>
1956 V85 N6 P676-680 65-17418 <*>
1958 V87 N10 P1169-1173 65-00289 <*>
1963 V92 N1 P115-122 64-51439 <=$>
1964 V93 N1 P61-67 65-10961 <*> 0
1964 V93 N6 P748-752 65-14672 <*>

UNSER WERK
1954 V5 P136-138 64-20307 <*>

UNTERNEHMENSFORSCHUNG
1960 V4 N3 P138-142 63-14077 <*>

UPPSALA LAKAREFORENINGS FORHANDLINGAR
1932 V38 N1/2 P1-14 66-12151 <*>

URAL
1963 N2 P145-150 63-31413 <=>

URANIA-WISSEN UND LEBEN
1955 V18 P138-141 57-1925 <*>
1963 V26 N12 P997-1002 AD-614 388 <=$>

URBANISME
1955 V24 P4-11 T1735 <INSD>

UROLOGIA. VENEZIA
1957 V24 N3 P310-313 59-15399 <*>

UROLOGIA INTERNATIONALIS. BASEL, NEW YORK
1961 V12 P307-327 63-00548 <*>

UROLOGIYA
1955 N1 P94-95 RT-3673 <*>
1955 V20 N4 P68-76 62-23544 <=*>
1957 V22 N3 P64-72 62-23502 <=*>
1957 V22 N5 P50-52 C-4112 <NRC>
1957 V22 N6 P46-50 64-19463 <=$>
1959 N3 P94 60-31155 <=>
1961 V26 N3 P34-36 C-4146 <NRC>
1961 V26 N4 P11-17 62-32770 <=*>
1964 V29 P22-25 66-60618 <=*$>

USINE NOUVELLE
1967 P160-161 67-31604 <=$>
1967 P163 67-31604 <=$>
1967 P165 67-31604 <=$>
1967 P167-168 67-31604 <=$>

USPEKHI ASTRONOMICHESKIKH NAUK. MOSKVA, LENINGRAD
1947 V3 P147-154 61-19364 <+*> 0
1954 V6 P281-322 R-4596 <*>
2685 <NLL>

USPEKHI BIOLOGICHESKOI KHIMII
1962 V4 P61-80 64-51222 <=$>
1963 V5 P289-304 64-41329 <=>

★ USPEKHI FIZICHESKIKH NAUK
1937 V18 N2 P153-202 TT.499 <*>
1939 V21 N4 P466-477 NP-TR-993 <*>
65-64299 <=$>
1940 V23 N3 P251-292 RT-1421 <*>
1940 V24 N1 P21-67 61-22436 <=>
63-23308 <=*>
1940 V24 N4 P433-486 RT-1422 <*>
1946 V28 N2/3 P155-201 59-20360 <+*>
1947 V31 N2 P157-173 RT-496 <*>
1947 V33 N2 P157-164 RT-1285 <*>
1947 V33 N3 P294-296 65-60051 <=$>
1947 V33 N3 P353-438 RT-715 <*>
1948 V34 N3 P1-21 RT-1640 <*>
1948 V34 N3 P443-445 RT-880 <*>
1948 V34 N3 P445-450 RT-1116 <*>
1948 V34 N3 P450-452 RT-843 <*>
1948 V34 N3 P452-453 RT-903 <*>
1948 V36 N3 P292-307 TT.128 <NRC>
1949 V37 N4 P414-458 RT-1086 <*>
1949 V38 N1 P133-135 59-20287 <+*>
1949 V38 N3 P413-427 60-19933 <+*>
1949 V38 N3 P456-457 R-4505 <*>
RT-2024 <*>
1949 V38 N3 P490-525 65-12237 <*>
1949 V38 N4 P461-489 RT-938 <*>
1949 V39 N1 P140-142 60-19934 <+*>
1949 V39 N2 P153-200 52/3429 <NLL>
1949 V39 N2 P315-316 59-15567 <+*> 0
1950 V40 P20-27 R-4101 <*>
1950 V40 P120-141 R-4113 <*>
1950 V40 N3 P369-406 TT.509 <NRC>
1950 V42 P76-92 R-3283 <*>
R-3424 <*>
59-19109 <+*>
1950 V42 P505-560 R-4049 <*>
1951 V43 P347-379 AEC-TR-3031 <=>
R-3465 <*>
1951 V43 N1 P11-29 TT.266 <NRC>
1951 V43 N4 P587-599 59-20285 <+*>
1951 V44 N1 P3-6 RT-944 <*>
1951 V44 N1 P7-20 RT-1873 <*>
1951 V44 N1 P33-45 RT-3938 <*>
1951 V44 N1 P46-69 RT-738 <*>
1951 V44 N1 P70-79 RT-913 <*>
1951 V44 N1 P80-88 RT-810 <*>
1951 V44 N1 P89-103 RT-722 <*>
1951 V44 N1 P104-109 RT-669 <*>
1951 V44 N1 P283-287 RT-937 <*>
1951 V44 N2 P291-292 RT-716 <*>
1951 V44 N2 P292-295 34 <FRI>
1951 V44 N3 P331-339 R-517 <*>
1951 V44 N3 P393-436 RT-667 <*>
1951 V44 N3 P437-442 RT-1760 <*>
1951 V44 N3 P463-466 RT-939 <*>
1951 V44 N4 P485-510 RT-933 <*>
1951 V44 N4 P511-526 61-23516 <=*>
1951 V44 N4 P527-557 RT-674 <*>
1951 V44 N4 P618-622 RT-950 <*>
1951 V44 N4 P629-630 RT-943 <*>
1951 V44 N4 P630-633 RT-906 <*>
1951 V45 N3 P313-356 59-15010 <+*> 0
1952 V46 N3 P348-387 61-20730 <*=>
1952 V47 N3 P398-444 60-23228 <+*>
1952 V47 N4 P537-560 61-16143 <=*>
67M38R <ATS>
1952 V48 N1 P25-118 62-23147 <=>
1952 V48 N3 P447-451 62-25865 <=*>
1953 V49 N1 P49-91 61-19344 <+*> 0
1953 V49 N2 P304-305 <INSD>
1953 V49 N3 P449-468 RT-2557 <*>
1953 V49 N4 P601-611 RT-1678 <*>
1953 V50 N1 P123-133 RT-4121 <*>
1953 V50 N2 P161-196 62-24655 <=*>
1953 V50 N2 P197-252 63-23115 <=*>
1953 V50 N2 P309-313 RT-2580 <*>
1953 V50 N4 P481-537 R-637 <*>
RT-2422 <*>
1953 V50 N4 P539-576 59-14096 <+*>
1953 V51 N1 P155-158 RT-2351 <*>
59-17324 <*>
1953 V51 N1 P173-178 R-1394 <*>
1953 V51 N2 P205-251 RT-4517 <*>
1953 V51 N3 P317-341 TT.544 <NRC>
1953 V51 N3 P343-392 RT-3973 <*>
59-20276 <+*>
1954 N4 P551-586 1355TM <CTT>
1954 V52 P561-602 R-4098 <*>
R-419 <*>
1954 V52 N1 P173-178 RJ-600 <ATS>
62-16538 <=*>
1954 V52 N3 P341-376 RT-3943 <*>
1954 V52 N4 P501-559 AEC-TR-5563 <=*>
1954 V52 N4 P561-602 R-2774 <*>
RT-3749 <*>

Citation	Translation
1954 V53 N1 P7-68	TT.543 <NRC>
1954 V54 N2 P185-230	R-4563 <*>
1954 V54 N2 P186-230	R-1440 <*>
1954 V54 N2 P358-360	RT-4496 <*>
1954 V55 N5/6 P67-72	62-18315 <=*> 0
1955 V55 P49-80	R-1747 <*>
1955 V55 P265-287	R-1751 <*>
	765TM <CTT>
1955 V55 N1 P101-110	R-4984 <*>
1955 V55 N1 P126-127	R-4074 <*>
1955 V55 N1 P127-131	R-1418 <*>
	R-1735 <*>
1955 V55 N2 P265-287	R-1751 <*>
1955 V55 N3 P299-314	R-1001 <*>
1955 V55 N3 P315-390	RT-3305 <*>
1955 V55 N4 P595-608	RJ-436 <ATS>
1955 V56 P215-256	R-421 <*>
1955 V56 N1 P111-130	R-1000 <*>
1955 V56 N2 P215-256	RT-3527 <*>
1955 V56 N4 P477-530	UCRL TRANS-556(L) <+*>
1955 V57 N2 P205-278	61-15240 <*+>
1956 V58 P37-68	AEC-TR-3031 <=>
	R-3465 <*>
	60-18992 <+*>
1956 V58 N1 P85-110	UCRL TRANS-642 <*=>
1956 V58 N2 P321-346	60-21966 <=>
1956 V58 N3 P415-432	61-23441 <*=>
1956 V58 N4 P667-684	R-1753 <*>
1956 V59 N2 P229-323	59-19289 <+*>
1956 V59 N2 P325-362	63-14674 <=*>
1956 V59 N3 P543-552	61-19828 <*=>
	62-11564 <=>
	64-14115 <=*$>
1956 V59 N3 P553-582	R-4011 <*>
	59-18518 <+*>
1956 V59 N4 P591-602	63-23061 <=*$>
1956 V60 N1 P69-160	UCRL TRANS-675L <*=>
1956 V60 N2 P179-189	59-22233 <+*>
1956 V60 N2 P191-212	59-18516 <+*>
1956 V60 N2 P225-248	59-19630 <+*>
1956 V60 N2 P249-293	59-18528 <+*>
1956 V60 N4 P617-688	65-63890 <=$>
1957 N3 P485-501	CSIRO-3348 <CSIR>
1957 V61 P137-159	R-2779 <*>
	R-3120 <*>
1957 V61 N1 P3-22	AEC-TR-3971 <=>
1957 V61 N1 P27-43	AEC-TR-3973,PT.2 <=>
	59-14761 <+*> 0
1957 V61 N1 P53-135	AEC-TR-3971 <=>
1957 V61 N2 P137-159	AEC-TR-3973,PT.2 <=>
1957 V61 N2 P161-310	AEC-TR-3971 <=>
1957 V61 N2 P249-276	UCRL TRANS-378 <*+>
1957 V61 N3 P313-330	AEC-TR-3971 <=>
1957 V61 N3 P341-398	AEC-TR-3971 <=>
	R-5138 <*>
1957 V61 N3 P423-450	AEC-TR-3973,PT.2 <=>
	R-2048 <*>
	59-11982 <=>
1957 V61 N3 P451-460	AEC-TR-3971 <=>
1957 V61 N4 P461-490	AEC-TR-3971 <=>
	59-14760 <+*> 0
1957 V61 N4 P535-559	AEC-TR-3971 <=>
1957 V61 N4 P561-612	AEC-TR-3493 <=>
	AEC-TR-3973,PT.2 <=>
	59-20286 <+*>
1957 V61 N4 P673-698	AEC-TR-3971 <=>
1957 V62 P99-128	R-5027 <*>
1957 V62 P381-383	R-2369 <*>
1957 V62 N1 P41-69	SCL-T-217 <+*>
1957 V62 N2 P159-186	AEC-TR-3973,PT.2 <=>
	R-4281 <*>
	61-15104 <+*>
1957 V62 N2 P191-196	AEC-TR-3973,PT.2 <=>
1957 V62 N3 P201-	59-21009 <=>
1957 V62 N3 P247-303	59-20326 <+*>
1957 V63 P51-58	C-2533 <NRC>
1957 V63 P227-238	59-16762 <+*>
1957 V63 N1 P5-16	59-16768 <+*> 0
1957 V63 N1 P33-71	AD-610 964 <=$>
1957 V63 N1 P33-50	59-20807 <+*>
1957 V63 N1 P51-58	R-3057 <*>
1957 V63 N1 P59-71	R-3054 <*>
1957 V63 N1 P123-129	59-10869 <+*>
	59-16769 <+*>
	62-32645 <=*>
1957 V63 N1 P163-180	59-12205 <+*>
	59-16754 <+*> 0
1957 V63 N1 P181-196	59-16757 <+*>
1957 V63 N1 P205-225	R-4874 <*>
	59-16756 <+*>
1957 V63 N1 P253-265	R-4874 <*>
	59-16764 <+*> 0
1957 V63 N1 P267-282	59-16761 <+*>
1957 V63 N3 P461-641	AEC-TR-3973,PT.2 <=>
1957 V63 N3 P461-501	PB 141 303T <=>
1957 V63 N3 P503-525	PB 141 304T <=>
1957 V63 N3 P527-532	60-13905 <+*>
1957 V63 N3 P613-641	60-13085 <+*>
1957 V63 N4 P645-691	AEC-TR-3973,PT.2 <=>
1957 V63 N4 P645-656	C-2667 <NRC>
	R-4022 <*>
	61-17585 <=$>
1957 V63 N4 P801-864	AEC-TR-3973,PT.2 <=>
1957 V63 N1A P33-50	C-2524 <NRC>
1957 V63 N1A P59-71	C-2517 <NRC>
1958 P40-108	R-5195 <*>
1958 V64 N1 P3-154	AEC-TR-5331 <=*>
1958 V64 N1 P3-14	R-3855 <*>
	59-12051 <+*>
	59-20277 <+*>
	61-19928 <=*>
1958 V64 N1 P193-195	AEC-TR-5331 <=*>
1958 V64 N1 P195-196	AEC-TR-3683 <+*>
1958 V64 N2 P197-313	AEC-TR-5331 <=*>
1958 V64 N2 P273-313	62-14846 <=*>
1958 V64 N2 P361-389	AEC-TR-5331 <=*>
1958 V64 N3 P417-434	AEC-TR-5331 <=*>
1958 V64 N3 P417-424	C-2728 <NRC>
	R-4454 <*>
	63-14495 <=*>
	64-71369 <=>
1958 V64 N3 P425-434	R-4609 <*>
	64-71370 <=>
1958 V64 N3 P447-528	AEC-TR-5331 <=*>
1958 V64 N3 P615-623	AEC-TR-5331 <=*>
1958 V64 N4 P641-667	AEC-TR-3594 <+*>
	AEC-TR-5331 <=*>
	59-20279 <+*>
1958 V64 N4 P669-731	AEC-TR-5331 <=*>
1958 V64 N4 P781-798	AEC-TR-5331 <=*>
1958 V65 N1 P2-110	AEC-TR-5332 <=>
1958 V65 N1 P3-38	63-10744 <=*>
1958 V65 N1 P87-110	59-20278 <+*>
1958 V65 N1 P133-140	AEC-TR-3965 <+*>
1958 V65 N1 P133-155	AEC-TR-5332 <=>
1958 V65 N1 P133-140	C-3282 <NRCC>
1958 V65 N1 P141-155	59-11857 <=>
1958 V65 N1 P157-158	AEC-TR-5332 <=>
1958 V65 N1 P733-738	AEC-TR-5332 <=>
1958 V65 N2 P157-161	59-11550 <=>
1958 V65 N2 P161-174	AD-610 974 <=$>
	AEC-TR-5332 <=>
	21K28R <ATS>
	59-21058 <=>
1958 V65 N2 P207-256	AEC-TR-5332 <=>
1958 V65 N2 P313-349	AEC-TR-5332 <=>
1958 V65 N3 P351-385	AEC-TR-3637 <+*>
1958 V65 N3 P387-406	AEC-TR-5332 <=>
1958 V65 N3 P441-488	AEC-TR-5332 <=>
1958 V65 N3 P451-488	TT-809 <NRCC>
1958 V65 N3 P521-543	AEC-TR-5332 <=>
1958 V65 N3 P545-548	AEC-TR-5332 <=>
1958 V65 N4 P551-591	AEC-TR-5332 <=>
	60-19005 <+*>
1958 V65 N4 P593-719	AEC-TR-5332 <=>
1958 V65 N4 P665-688	59-20280 <+*>
1958 V65 N4 P721-726	AEC-TR-5332 <=>
1958 V66 P43-77	AEC-TR-3529 <*>
1958 V66 N1 P99-110	AEC-TR-3778 <+*>
1958 V66 N1 P141-144	59-15724 <+*>

Citation	Translation
1958 V66 N2 P157-161	59-00720 <*>
	59-12735 <+*>
	59-21049 <=>
1959 V67 N4 P743-750	60-13899 <+*>
1959 V68 N3 P513-528	60-11914 <=>
1959 V68 N4 P687-715	61-13331 <=*>
1959 V69 N1 P13-56	61-15500 <+*>
1959 V69 N1 P149-156	60-14079 <+*>
1960 V70 N1 P191-198	60-11809 <=>
1960 V70 N2 P351-360	62-13148 <=*>
1960 V70 N4 P585-619	61-21164 <=>
1960 V70 N4 P679-692	61-31458 <=> 0
1960 V70 N4 P693-714	61-21976 <=> 0
1960 V71 N1 P117-130	61-11981 <=>
1960 V71 N1 P131-160	61-21065 <=>
1960 V71 N2 P339-347	60-41700 <=>
1960 V71 N3/4 P369-410	ARSJ V32 N1 P151 <AIAA>
1960 V72 N2 P3-4	61-21487 <=>
1960 V72 N2 P307-321	61-21487 <=>
1961 V74 N3 P393-434	62-23369 <=*>
1961 V75 N3 P421-458	62-24282 <=*>
1962 V77 N4 P589-596	63-13544 <=*>
1962 V78 N1 P53-64	63-26581 <=$>
1962 V78 N1 P167-179	63-13490 <=*>
1962 V78 N2 P181-265	63-11747 <=>
1963 V79 N3 P377-440	RPQ,V1,N1,P19-66 <IPIX>
	63-20938 <=*$>
1963 V80 N1 P93-124	63-31521 <=>
1963 V80 N4 P553-595	65-11007 <*>
1963 V81 N2 P249-270	64-21044 <=>
1963 V81 N3 P566-573	64-21725 <=>
1963 V81 N4 P774-777	AD-610 409 <=$>
1964 V82 N2 P201-220	AD-618 315 <=$>
1964 V82 N4 P585-647	64-31455 <=>
1964 V83 N1 P35-62	64-31913 <=>
1964 V83 N2 P197-222	AD-637 098 <=$> M
1964 V83 N2 P287-298	65-11296 <*>
1964 V83 N2 P361-369	AD-637 098 <=$> M
1965 V85 N2 P197-258	65-31343 <=$>
1965 V86 N2 P263-302	65-33631 <=*$>
1965 V86 N2 P327-333	66-11186 <*>
1965 V87 N1 P125-142	66-10060 <*>
1965 V87 N3 P471-489	66-30411 <=*$>
1966 V89 N2 P309-319	66-34516 <=$>
1966 V90 P209-236	66-61476 <=*$>

★USPEKHI KHIMII

Citation	Translation
1934 V4 P610-624	R-2546 <*>
1935 V4 N4 P573-609	R-4940 <*>
1937 V6 N6 P777-821	RT-1002 <*>
1938 V7 P1042-1051	R-2584 <*>
1938 V7 N3 P404-435	TT-191 <NRC>
1939 V8 N2 P195-240	60-41263 <=>
1939 V8 N10 P1519-1534	R-766 <*>
	RT-1000 <*>
	66P64R <ATS>
1940 V9 P1333-	R-3244 <*>
1940 V9 N5 P498-521	61-16870 <*=>
1940 V9 N6 P609-628	61-16874 <*=> P
1940 V9 N6 P673-681	RT-1420 <*>
1940 V9 N11/2 P1301-1332	61-16877 <*=>
1941 V10 P262-293	61-16915 <*=>
1941 V10 N2 P121-187	62-10885 <*> P
1941 V10 N2 P224-255	62-10886 <*> P
1941 V10 N4 P373-415	62-10887 <=*> P
1941 V10 N5 P544-558	62-10888 <=*>
1941 V10 N6 P654-661	62-10889 <*=> P
1941 V10 N7 P825-844	62-10890 <*=> P
1943 V12 P189-208	61-16977 <*=>
1943 V12 P209-215	61-16926 <*=>
1943 V12 P472-479	63-24283 <=*$>
1943 V12 N4 P308-317	60-41351 <=>
1944 V13 P203-233	61-16892 <*=>
1944 V13 N3 P234-251	TT.149 <NRC>
1944 V13 N5 P365-374	62-10823 <*=> P
1944 V13 N5 P375-388	62-10824 <*=> P
	64-16120 <=*$>
1945 V14 N1 P3-41	61-20131 <*=>
1945 V14 N1 P56-64	62-10825 <*=> P
1945 V14 N2 P148-153	62-10826 <*=> P
1945 V14 N3 P185-212	62-10827 <*=> P
1945 V14 N3 P255-258	62-10828 <*=> O
1945 V14 N4 P301-329	R-1040 <*>
1945 V14 N5 P367-394	62-10829 <=*> P
1945 V14 N6 P476-500	62-10830 <*=> P
1945 V14 N6 P501-509	174TM <CTT>
	62-10831 <*=> P
1946 V15 N1 P63-80	TT.130 <NRC>
1946 V15 N3 P325-342	64-16475 <=*$>
1946 V15 N3 P326-342	59-17242 <+*>
1946 V15 N3 P343-358	62-10832 <*=> P
1947 V16 N1 P26-68	60-13885 <+*> 0
1947 V16 N4 P461-489	RT-2236 <*>
1947 V16 N6 P744-748	63-20806 <=*$>
1948 V17 P641-662	60-19926 <+*>
1948 V17 N5 P578-584	<ATS>
1948 V17 N5 P585-605	<ATS>
1948 V17 N5 P608-623	<ATS>
1948 V17 N6 P663-691	TT.168 <NRC>
	64-18375 <=*$>
1949 V18 P288-301	R-3999 <*>
1949 V18 P449-461	RCT V30 N2 P531 <RCT>
1949 V18 N2 P237-257	50/3275 <NLL>
1949 V18 N2 P237-260	525 <NRC>
1949 V18 N4 P449-461	59-15938 <+*>
1949 V18 N5 P546-556	6144 <K-H>
	64-20221 <*>
1949 V18 N6 P682-696	TT.253 <NRC>
1950 V19 P52-54	GB43 <NLL>
1950 V19 N1 P59-87	TT.140 <*>
1950 V19 N1 P88-124	TT.304 <NRC>
1950 V19 N1 P128-133	61-27984 <*=>
1950 V19 N4 P419-444	TT.228 <NRC>
1950 V19 N5 P565-575	RT-1460 <*>
1950 V19 N6 P673-696	60-41543 <=>
1950 V19 N6 P697-715	TT-240 <NRC>
1951 V20 P213-230	R-4110 <*>
1951 V20 P410-429	62-10562 <*=>
	65-61304 <=$>
1951 V20 P560-588	64-16123 <*> 0
1951 V20 N1 P54-70	59-16365 <+*>
1951 V20 N2 P194-212	RT-306 <*>
1951 V20 N2 P213-230	RT-564 <*>
1951 V20 N3 P288-308	RT-2300 <*>
1951 V20 N3 P309-335	RJ-71 <ATS>
1951 V20 N4 P473-494	60-41377 <=>
1951 V20 N4 P495-497	82DR <ATS>
1951 V20 N5 P521-532	RT-3837 <*>
1951 V20 N6 P714-733	RJ-84 <ATS>
1951 V20 N6 P734-758	R-102 <*>
	RT-2654 <*>
	64-15054 <=*$>
1952 N5 P513-533	59-13563 <=> 0
1952 V21 P207-236	R-4112 <*>
1952 V21 N3 P280-312	59-12207 <+*>
1952 V21 N4 P369-378	RT-1053 <*>
1952 V21 N5 P534-565	RT-2328 <*>
1952 V21 N5 P615-640	RT-190 <*>
	61-28025 <*=>
1952 V21 N6 P641-713	AEC-TR-3976 <=>
1952 V21 N7 P785-807	RJ-131 <ATS>
1952 V21 N7 P808-823	175TM <CTT>
1952 V21 N8 P881-913	<ATS>
	RJ-214 <ATS>
1952 V21 N9 P1086-1095	AEC-TR-4095 <+*>
1953 V22 N2 P133-178	RT-743 <*>
1953 V22 N2 P179-190	64-18771 <*>
	64-18771 <=*$>
	65-18187 <*>
1953 V22 N3 P253-278	RT-2859 <*>
1953 V22 N7 P838-982	R-1455 <*>
1953 V22 N9 P1138-1156	RT-2287 <*>
1954 V23 P986-1026	T-2008 <INSD>
1954 V23 N2 P199-222	RJ-199 <ATS>
1954 V23 N3 P273-294	RT-2748 <*>
1954 V23 N4 P426-478	RT-2532 <*>
1954 V23 N5 P529-546	RT-2715 <*>
1954 V23 N5 P547-580	RJ-274 <ATS>
	RT-274 <*>
	RT-4388 <*>

Citation	Translation
	1278 <*>
	64-19093 <=*$>
1954 V23 N6 P654-696	59-17353 <+*>
1954 V23 N6 P697-736	R-521 <*>
1954 V23 N8 P901-920	<TCT>
1954 V23 N8 P921-942	<TCT>
1954 V23 N8 P943-966	<TCT>
1954 V23 N8 P967-985	<TCT>
1954 V23 N8 P986-1022	<TCT>
1954 V23 N8 P986-1026	UCRL TRANS-532L <+*>
1955 V24 P14-31	59-18750 <+*>
1955 V24 P527-574	R-409 <*>
1955 V24 P951-984	AEC-TR-4839 <=*>
1955 V24 N1 P3-13	RT-3315 <*>
1955 V24 N1 P14-31	<TCT>
	RT-4201 <*>
1955 V24 N1 P32-51	<TCT>
1955 V24 N1 P52-68	<TCT>
1955 V24 N1 P69-92	<TCT>
1955 V24 N1 P93-119	<TCT>
1955 V24 N2 P220-239	R-4798 <*>
1955 V24 N3 P260-264	RJ-309 <ATS>
1955 V24 N3 P260-274	RJ-309 <ATS>
1955 V24 N4 P377-429	80L28R <ATS>
1955 V24 N4 P430-439	RJ-332 <ATS>
1955 V24 N4 P453-470	R-1948 <*>
	RJ-342 <ATS>
1955 V24 N5 P513-526	<TCT>
1955 V24 N6 P759-778	62-14879 <=*>
	62-16705 <=*>
1955 V24 N7 P842-874	59-18529 <+*>
1956 V25 P57-90	RCT V31 N5 P1105 <RCT>
1956 V25 P632-642	R 772 <RIS>
	R-801 <*>
	176TM <CTT>
1956 V25 N2 P162-189	RJ-619 <ATS>
1956 V25 N2 P190-237	RJ-631 <ATS>
1956 V25 N2 P190-241	59-17699 <+*>
	64-16333 <=*$>
1956 V25 N3 P263-287	C-2513 <NRC>
	R-2759 <*>
1956 V25 N4 P419-485	M-5649 <NLL>
1956 V25 N4 P486-516	64-14436 <=*$>
1956 V25 N8 P921-932	17K23R <ATS>
1956 V25 N9 P1173-1193	62-15132 <=*>
1956 V25 N10 P1223-1240	NP-TR-36 <*>
1956 V25 N1C P1267-1281	60-15565 <+*>
	94K24R <ATS>
1956 V25 N11 P1351-1372	61-15809 <+*>
	65-61484 <=>
	81J18R <ATS>
1956 V25 N12 P1446-1473	63-20013 <=*>
1957 V26 P528-553	R-755 <RIS>
1957 V26 P625-639	R-1986 <*>
1957 V26 P678-688	79K23R <ATS>
1957 V26 P725-	59-11946 <=>
1957 V26 P1007-1035	60 17738 <+*>
1957 V26 N2 P139-163	26K22R <ATS>
1957 V26 N3 P292-344	59-10685 <+*>
1957 V26 N4 P389-398	59-22232 <+*>
1957 V26 N4 P416-458	30J19R <ATS>
	60-19104 <+*>
1957 V26 N4 P494-515	61-15215 <*+>
	63-24311 <=*$>
1957 V26 N6 P678-688	R-4603 <*>
1957 V26 N7 P725-767	59-10247 <+*>
1957 V26 N7 P768-800	68K24R <ATS>
1957 V26 N8 P895-922	47M46R <ATS>
	60-17442 <+*>
1957 V26 N8 P944-964	AEC-TR-3820 <+*>
1957 V26 N8 P965-974	62-19824 <=>
	62-19824 <=*>
1957 V26 N10 P1109-1124	1K22R <ATS>
1957 V26 N10 P1125-1140	60-17443 <+*>
	64-30115 <*>
	6868 <K-H>
1957 V26 N11 P1241-1294	55L30R <ATS>
1957 V26 N11 P1295-1309	56L30R <ATS>
1957 V26 N11 P1310-1319	59L30R <ATS>
1957 V26 N11 P1343-1354	58L30R <ATS>
1957 V26 N12 P1355-1373	59-10565 <+*>
	60-23986 <+*>
1957 V26 N12 P1434-1468	60-41250 <=>
1958 V27 N1 P3-56	63-24293 <=*$>
	65-13223 <*>
1958 V27 N1 P4-56	62-14052 <*>
1958 V27 N1 P57-93	77K23R <ATS>
1958 V27 N1 P94-106	RCT V34 N1 P215 <RCT>
	31K25R <ATS>
1958 V27 N3 P306-315	76N50R <ATS>
1958 V27 N3 P316-352	6K25R <ATS>
1958 V27 N4 P365-402	59-10730 <+*>
1958 V27 N4 P481-487	78K24R <ATS>
1958 V27 N5 P517-550	R 756 <RIS>
1958 V27 N5 P622-642	61-19223 <+*>
1958 V27 N6 PI-IV ADD	61-28456 <*=>
1958 V27 N6 P717-730	56K28R <ATS>
1958 V27 N8 P949-965	61-23379 <=*>
1958 V27 N8 P1010-1024	35L31R <ATS>
1958 V27 N9 P1C56-1083	63-14326 <=*>
	75L31R <ATS>
1958 V27 N9 P1084-1100	59-18410 <+*>
1958 V27 N9 P1101-1114	60-13438 <+*>
	62-13297 <*=> 0
1958 V27 N10 P1177-1197	61-23424 <*=>
1958 V27 N10 P1221-1256	59-15000 <+*>
1958 V27 N11 P1257-1303	59-18847 <+*>
	60-19139 <+*>
1958 V27 N11 P1304-1320	62-16693 <=*>
	62L29R <ATS>
1958 V27 N11 P1354-1360	87L30R <ATS>
1958 V27 N12 P1361-1436	40L32R <ATS>
1958 V27 N12 P1471-1503	60-11253 <=>
1959 V28 N1 P3-32	61-17574 <=$>
1959 V28 N1 P33-43	AEC-TR-3992 <+*>
	59-31026 <=>
1959 V28 N1 P96-116	60-11709 <=>
1959 V28 N2 P189-217	61-18577 <*=>
	61-19787 <=*> 0
1959 V28 N2 P218-234	RTS-2290 <NLL>
1959 V28 N4 P408-435	61-19785 <=*>
1959 V28 N4 P465-483	61L36R <ATS>
1959 V28 N5 P576-604	61-15000 <*+>
1959 V28 N5 P6C5-614	60-23572 <+*>
1959 V28 N5 P615-638	22M45R <ATS>
1959 V28 N6 P741-771	61-15013 <*+>
	61-27159 <*=>
1959 V28 N7 P783-825	60-17183 <+*>
1959 V28 N7 P783-818	60-23867 <+*>
1959 V28 N7 P850-876	PANSDOC-TR.366 <PANS>
1959 V28 N8 P921-947	61-15726 <+*>
1959 V28 N9 P1011-1035	60-31397 <=>
1959 V28 N9 P1086-1113	AEC-TR-4097 <+*>
1959 V28 N9 P1114-1133	61-15641 <+*>
	61-15641 <*+>
1959 V28 N9 P1134	62-11149 <=*>
	62-15090 <*=>
1959 V28 N10 P1201-1215	62-32046 <=*>
1959 V28 N11 P1342-1352	63-20012 <=*>
1959 V28 N12 P1399-1449	60-11835 <=>
1959 V28 N12 P1450-1487	60-11994 <=>
	61-13148 <=*>
1959 V28 N12 P1488-1522	60-11995 <=>
1959 V28 N12 P1523-1543	54M39R <ATS>
1960 V29 N1 P23-54	60-16653 <+*>
1960 V29 N1 P55-73	62-23168 <=*>
1960 V29 N3 P277-297	92M46R <ATS>
1960 V29 N3 P298-363	21M42R <ATS>
1960 V29 N6 P7C9-735	61-13846 <+*>
1960 V29 N6 P736-759	61-10350 <+*>
1960 V29 N6 P774-785	61-23886 <*=> 0
1960 V29 N7 P833-863	10M45R <ATS>
1960 V29 N8 P993-1010	52N49R <ATS>
1960 V29 N9 P1C88-1111	64-13806 <=*$>
	64-14959 <=*$>
1960 V29 N10 P1229-1259	10599 <K-H>
	65-13192 <*>
1960 V29 N12 P1409-1438	99N50R <ATS>
1961 V30 P1013-1049	RCT V35 N1 P1 <RCT>
1961 V30 N1 P60-91	62-19937 <=>

Citation	Translation
1961 V30 N3 P345-385	61N52R <ATS>
1961 V30 N4 P550-560	64-51287 <=>
1961 V30 N5 P563-592	19N54R <ATS>
1961 V30 N5 P679-700	61-27946 <*=>
1961 V30 N7 P846-876	62-13423 <=*>
1961 V30 N8 P982-1012	62-11130 <=>
1961 V30 N9 P1175-1195	62-23754 <=*>
1961 V30 N11 P1391-1409	64-10099 <=*$>
1961 V30 N12 P1429-1432	62-25167 <=>
1961 V30 N12 P1462-1489	19P60R <ATS>
1962 V31 N1 P73-100	68P60R <ATS>
1962 V31 N2 P211-221	64-18711 <*>
1962 V31 N4 P397-416	63-10067 <=*>
1962 V31 N4 P452-473	26P62R <ATS>
1962 V31 N5 P609-655	97P62R <ATS>
1962 V31 N7 P822-837	26P65R <ATS>
	63-23605 <=*$>
1962 V31 N8 P901-939	62P65R <ATS>
1963 V32 N2 P154-194	15Q71R <ATS>
1963 V32 N3 P336-353	63-24265 <=*$>
1963 V32 N4 P501-507	AEC-TR-5920 <=*$>
1963 V32 N7 P882-895	63-31858 <=>
1963 V32 N8 P897-947	24Q74R <ATS>
1963 V32 N9 P1052-1086	64-15366 <=*$>
1964 V33 N2 P205-232	RCT V37 N4 P1153 <RCT>
1964 V33 N3 P261-295	66-10602 <*>
1964 V33 N4 P396-417	66-10618 <*>
1964 V33 N5 P549-579	21R79R <ATS>
1964 V33 N5 P580-601	64-41353 <=>
1964 V33 N10 P1216-1231	65-30301 <=$>
1964 V33 N12 P1409-1464	63-31076 <=$>
1964 V33 N12 P1501-1525	63-31076 <=$>
1965 V34 N1 P132-153	65-30565 <=$>
1965 V34 N4 P585-617	12T90R <ATS>
1965 V34 N4 P586-597	AD-622 476 <=*$>
1965 V34 N4 P630-652	46S85R <ATS>
1965 V34 N9 P1648-1673	66-13394 <*>
1965 V34 N11 P1945-1964	66-13395 <*>
	66-13983 <*>
1966 V35 N1 P93-106	66-31459 <=$>
1966 V35 N5 P779-822	66-33651 <=$>
1966 V35 N8 P1404-1429	66-34959 <=$>
1966 V35 N8 P1477-1494	66-34960 <=$>

USPEKHI KHIMII I TEKHNOLOGII POLIMEROV. MOSKVA

Citation	Translation
1960 V3 P107-129	61-27573 <*=>

★ USPEKHI MATEMATICHESKIKH NAUK

Citation	Translation
1936 N1 P141-174	AMST S2 V34 P1-37 <AMS>
1939 N6 P26-89	RT-932 <*>
1939 N6 P26-89	R-3188 <*>
1941 N8 P171-231	AMST S2 V51 P1-73 <AMS>
1944 N10 P115-165	AMST S1 V11 P1-81 <AMS>
	RT-1390 <*>
1946 V1 N2 P48-146	AMST S2 V5 P115-220 <AMS>
1946 V1 N3/4 P44-70	AMST S1 V4 P373-414 <AMS>
1947 V2 N1 P58-155	AMST S1 V7 P1-148 <AMS>
1947 V2 N3 P18-59	AMST S1 V1 P283-338 <AMS>
	RT-560 <*>
1947 V2 N4 P59-127	AMST S1 V9 P328-469 <AMS>
1947 V2 N6 P159-173	AMST S1 V9 P308-327 <AMS>
1947 V2 N6 P174-214	AMST S1 V10 P326-377 <AMS>
1948 V3 N1 P3-95	AMST S1 V10 P199-325 <AMS>
1948 V3 N2 P47-158	AMST S1 V6 P274-423 <AMS>
	RT-2869 <*>
	RT-452 <*>
1948 V3 N3 P29-112	AMST S1 V10 P84-198 <AMS>
1948 V3 N3 P242	RT-1283 <*>
1948 V3 N5 P52-145	AMST S1 V9 P42-171 <AMS>
1948 V3 N6 P89-185	RT-334 <*>
	60-13157 <+*>
1948 V3 N6 P188-200	AMST S1 V2 P25-37 <AMS>
1949 V4 N1 P3-112	AMST S1 V10 P408-541 <AMS>
	RT-451 <*>
1949 V4 N2 P22-56	AMST S1 V2 P38-80 <AMS>
	RT-561 <*>
1949 V4 N3 P3-68	AMST S1 V3 P107-195 <AMS>
	RT-2871 <*>
1949 V4 N3 P178-179	AMST S2 V39 P165-166 <AMS>
1949 V4 N4 P19-49	AMST S1 V2 P81-124 <AMS>
	RT-449 <*>
1949 V4 N4 P173-178	STMSP V4 P339-344 <AMS>
1949 V4 N4 P187-188	62-10150 <=*>
1949 V4 N5 P14-48	AMST S1 V2 P125-169 <AMS>
	RT-448 <*>
1949 V4 N5 P99-141	AMST S1 V4 P207-267 <AMS>
	RT-450 <*>
1949 V4 N6 P91-153	AMST S1 V5 P414-497 <AMS>
1949 V4 N6 P91-152	RT-1462 <*>
1950 V5 N1 P135-186	AMST S1 V9 P470-534 <AMS>
	RT-660 <*>
1950 V5 N2 P148-154	RT-1157 <*>
1950 V5 N2 P180-190	AMST S2 V1 P27-35 <AMS>
1950 V5 N3 P87-103	52/3063 <NLL>
1950 V5 N3 P152-155	RT-840 <*>
1950 V5 N3 P156-160	RT-849 <*>
1950 V5 N4 P75-120	AMST S1 V2 P367-424 <AMS>
	RT-723 <*>
1950 V5 N4 P144-153	TT-470 <NRC>
1950 V5 N6 P102-135	AMST S1 V4 P331-372 <AMS>
	RT-615 <*>
1951 V6 P100-111	POED-TRANS-2224 <NLL>
1951 V6 N1 P68-90	STMSP V1 P17-40 <AMS>
1951 V6 N1 P91-137	AMST S1 V8 P392-455 <AMS>
	RT-145 <*>
1951 V6 N2 P16-101	AMST S1 V6 P27-146 <AMS>
	RT-2076 <*>
1951 V6 N3 P31-98	AMST S2 V27 P51-124 <AMS>
1951 V6 N4 P3-120	AMST S2 V12 P1-121 <AMS>
1951 V6 N5 P43-68	AMST S2 V1 P1-26 <AMS>
1951 V6 N5 P172-175	RT-783 <*>
1951 V6 N6 P100-111	AMST S1 V8 P62-77 <AMS>
	RT-144 <*>
	59-15132 <+*>
1951 V6 N6 P112-154	AMST S1 V1 P51-107 <AMS>
	RT-886 <*>
1951 V6 N6 P174-181	R-4159 <*>
	RJ-448 <ATS>
1952 V7 N1 P3-117	AMST S2 V2 P207-316 <AMS>
1952 V7 N1 P118-137	AMST S2 V1 P49-65 <AMS>
1952 V7 N2 P3-6	64-11943 <=> M
1952 V7 N2 P31-122	AMST S1 V3 P294-391 <AMS>
	RT-1156 <*>
1952 V7 N2 P197-200	UCRL TRANS-169 <=*>
1952 V7 N3 P139-143	UCRL TRANS-170 <=*>
1952 V7 N6 P3-96	AMST S2 V16 P1-101 <AMS>
1952 V7 N6 P97-178	AMST S2 V3 P163-248 <AMS>
1953 V8 N2 P7-73	AMST S2 V26 P11-86 <AMS>
1953 V8 N3 P3-20	AMST S2 V12 P181-197 <AMS>
	RT-3946 <*>
1953 V8 N3 P135-142	STMSP V1 P87-95 6 <AMS>
1953 V8 N3 P176-200	RT-2908 <*>
	64-15582 <=*$> OP
1953 V8 N4 P3-63	AMST S2 V19 P109-166 <AMS>
1953 V8 N4 P163-167	AMST S2 V27 P19-23 <AMS>
1953 V8 N4 P185-192	RT-1730 <*>
1953 V8 N5 P3-71	AMST S2 V3 P91-162 <AMS>
1953 V8 N5 P73-120	AMST S2 V12 P251-300 <AMS>
1953 V8 N5 P189-191	AMST S2 V39 P167-170 <AMS>
1953 V8 N6 P3-54	AMST S2 V5 P221-274 <AMS>
1954 V9 N2 P3-66	AMST S2 V8 P289-351 <AMS>
1954 V9 N2 P125-128	64-71396 <*>
1954 V9 N2 P163-170	AD-638 626 <=$>
1954 V9 N2 P189-190	STMSP V1 P169-170 <AMS>
1954 V9 N2 P191-198	AMST S2 V12 P141-148 <AMS>
1954 V9 N2 P199-	59-11869 <=>
1954 V9 N2 P233	AMST S2 V12 P251-300 <AMS>
1954 V9 N3 P57-114	AMST S2 V10 P345-409 <AMS>
1954 V9 N3 P61-	R-2111 <*>
1954 V9 N3 P141-148	AMST S2 V5 P275-283 <AMS>
1954 V9 N3 P155-156	AMST S2 V12 P137-139 <AMS>
1954 V9 N3 P163	PB 141 460T <=>
1954 V9 N3 P181-186	AMST S2 V17 P1-7 <AMS>
1954 V9 N3 P226-230	64-11805 <=>
1954 V9 N4 P19-93	AMST S2 V6 P379-458 <AMS>
1954 V9 N4 P95-112	61-16085 <=*>
1954 V9 N4 P195-202	STMSP V6 P1 <AMS>
1955 V10 P167-174	T1626 <INSD>
1955 V10 N1 P3-40	AMST S2 V8 P143-181 <AMS>
1955 V10 N1 P67-78	AMST S2 V12 P123-135 <AMS>

1955 V10 N1 P117-121 AMST S2 V27 P25-29 <AMS>
1955 V10 N1 P243-244 AMST S2 V3 P91-162 <AMS>
1955 V10 N2 P3-110 AMST S2 V6 P1-110 <AMS>
1955 V10 N2 P111-143 AMST S2 V5 P35-65 <AMS>
1955 V10 N2 P161-166 66-10899 <*>
1955 V10 N3 P3-70 AMST S2 V8 P21-85 <AMS>
1955 V10 N4 P3-74 AMST S2 V21 P119-192 <AMS>
1956 V11 P227-231 R-4803 <*>
1956 V11 N1 P17-75 R-2075 <*>
1956 V11 N1 P77- 59-18264 <+*>
1956 V11 N2 P31-66 AMST S2 V19 P47-85 <AMS>
1956 V11 N2 P125-159 AMST S2 V19 P221-252 <AMS>
63-14489 <=*>
1956 V11 N2 P185-190 64-71343 <=> M
1956 V11 N3 P151-158 AMST S2 V12 P155-162 <AMS>
R-3543 <*>
1956 V11 N4 P3-43 AMST S2 V22 P95-137 <AMS>
1956 V11 N5 P26-37 AMST S2 V10 P1-12 <AMS>
1956 V11 N5 P37-44 AMST S2 V13 P177-183 <AMS>
1956 V11 N5 P60-66 AMST S2 V18 P37-43 <AMS>
1956 V11 N5 P107-152 AMST S2 V10 P59-106 <AMS>
1956 V11 N6 P3-12 AMST S2 V16 P315-324 <AMS>
1956 V11 N6 P3- 59-11926 <=>
1956 V11 N6 P13-40 AMST S2 V16 P325-353 <AMS>
1956 V11 N6 P41-97 AMST S2 V10 P223-281 <AMS>
1956 V11 N6 P117-144 AMST S2 V18 P49-79 <AMS>
1956 V11 N6 P183-202 AMST S2 V20 P55-75 <AMS>
1957 V12 P1-51 R-5128 <*>
1957 V12 N1 P3-52 AMST S2 V12 P199-246 <AMS>
1957 V12 N1 P53-98 AMST S2 V14 P59-106 <AMS>
1957 V12 N1 P99-142 AMST S2 V19 P253-297 <AMS>
1957 V12 N1 P143-145 AMST S2 V16 P355-357 <AMS>
1957 V12 N1 P145-146 AMST S2 V16 P357-358 <AMS>
1957 V12 N1 P147-152 AMST S2 V16 P358-364 <AMS>
1957 V12 N1 P152-156 AMST S2 V16 P364-369 <AMS>
1957 V12 N1 P157-160 AMST S2 V16 P369-373 <AMS>
1957 V12 N1 P160-162 AMST S2 V16 P373-374 <AMS>
1957 V12 N1 P162-165 AMST S2 V16 P375-378 <AMS>
1957 V12 N1 P166-169 AMST S2 V16 P378-382 <AMS>
1957 V12 N1 P169-172 AMST S2 V16 P382-385 <AMS>
1957 V12 N1 P173-175 AMST S2 V16 P385-388 <AMS>
1957 V12 N1 P176-177 AMST S2 V16 P389-390 <AMS>
61-11757 <=>
1957 V12 N1 P177-179 AMST S2 V16 P391-392 <AMS>
1957 V12 N1 P179-182 AMST S2 V16 P393-396 <AMS>
1957 V12 N1 P182-186 AMST S2 V16 P396-400 <AMS>
1957 V12 N1 P187-191 AMST S2 V16 P401-406 <AMS>
1957 V12 N1 P192-195 AMST S2 V16 P406-409 <AMS>
1957 V12 N1 P195-199 AMST S2 V16 P410-413 <AMS>
1957 V12 N1 P199-201 AMST S2 V16 P414-416 <AMS>
1957 V12 N1 P201-203 AMST S2 V16 P416-417 <AMS>
1957 V12 N1 P203-208 AMST S2 V16 P418-423 <AMS>
1957 V12 N1 P208-210 AMST S2 V16 P423-426 <AMS>
1957 V12 N1 P211-212 AMST S2 V16 P426-427 <AMS>
1957 V12 N1 P212-218 AMST S2 V16 P427-434 <AMS>
1957 V12 N1 P218-222 AMST S2 V16 P434-437 <AMS>
1957 V12 N1 P222-226 AMST S2 V16 P437-442 <AMS>
1957 V12 N1 P226-229 AMST S2 V16 P442-445 <AMS>
1957 V12 N1 P230-234 AMST S2 V16 P445-450 <AMS>
1957 V12 N1 P234-238 AMST S2 V16 P450-454 <AMS>
1957 V12 N1 P238-240 AMST S2 V16 P454-456 <AMS>
1957 V12 N1 P240-245 AMST S2 V16 P457-462 <AMS>
1957 V12 N1 P246-249 AMST S2 V16 P471-475 <AMS>
1957 V12 N1 P249-251 AMST S2 V16 P477-479 <AMS>
1957 V12 N1 P251-253 AMST S2 V16 P462-464 <AMS>
1957 V12 N1 P254-257 AMST S2 V16 P464-468 <AMS>
1957 V12 N1 P257-258 AMST S2 V16 P468-469 <AMS>
1957 V12 N2 P3-41 AMST S2 V15 P55-93 <AMS>
1957 V12 N2 P43-118 AMST S2 V13 P185-264 <AMS>
1957 V12 N2 P185-192 AMST S2 V26 P87-94 <AMS>
1957 V12 N3 P3-73 AMST S2 V26 P95-172 <AMS>
1957 V12 N3 P205-210 AMS TRANS V44 P109 <AMS>
AMST S2 V44 P109-114 <AMS>
1957 V12 N3 P255-261 61-10189 <*+>
1957 V12 N3 P389-396 AMST S2 V18 P163-171 <AMS>
1957 V12 N4 P41-46 61-21704 <=>
1957 V12 N4 P57-124 AMST S2 V14 P107-172 <AMS>
1957 V12 N4 P335-340 AMST S2 V30 P235-240 <AMS>
1957 V12 N5 P3-122 AMST S2 V20 P239-364 <AMS>
1957 V12 N5 P123-148 AMST S2 V20 P77-104 <AMS>
1957 V12 N5 P191-196 64-71414 <=> M
1957 V12 N5 P270 AMST S2 V28 P210 <AMS>
1958 V13 N1 P3-85 AMST S2 V13 P265-346 <AMS>
1958 V13 N2 P3-72 AMST S2 V14 P217-287 <AMS>
1958 V13 N2 P189-194 AMST S2 V18 289-294 <AMS>
1958 V13 N3 P3-110 AMST S2 V23 P231-335 <AMS>
1958 V13 N3 P111-143 AMST S2 V18 P81-115 <AMS>
1958 V13 N3 P191-196 AMST S2 V24 P13-18 <AMS>
1958 V13 N4 P3-28 AMST S2 V29 P217-245 <AMS>
1958 V13 N4 P29-88 AMST S2 V20 P105-171 <AMS>
1958 V13 N4 P89-172 AMST S2 V17 P29-115 <AMS>
1958 V13 N5 P3-120 AMST S2 V22 P163-288 <AMS>
1958 V13 N5 P167-170 STMSP V2 P237-240 <AMS>
1958 V13 N5 P197-202 AMST S2 V39 P171-176 <AMS>
1959 V14 N1 P3-20 AMST S2 V18 P321-339 <AMS>
03L33R <ATS>
61-27264 <*=>
1959 V14 N1 P21-85 AMST S2 V20 P173-238 <AMS>
1959 V14 N1 P87-130 AMST S2 V37 P291-336 <AMS>
1959 V14 N1 P135-140 62-11701 <=>
1959 V14 N1 P189-196 61-31594 <=>
1959 V14 N2 P3-86 AMST S2 V17 P277-364 <AMS>
1959 V14 N2 P87-158 AMST S2 V29 P295-381 <AMS>
60-11989 <=>
1959 V14 N2 P159-164 AMST S2 V33 P277-283 <AMS>
1959 V14 N2 P165-170 AMST S2 V33 P285-290 <AMS>
1959 V14 N2 P171-194 AMST S2 V26 P173-200 <AMS>
1959 V14 N2 P217-218 AMS-TRANS V42 P37-39 <AMS>
AMST S2 V42 P37-39 <AMS>
1959 V14 N3 P3-19 AMST S2 V26 P201-219 <AMS>
1959 V14 N3 P21-73 AMS-TRANS V42 P71 <AMS>
AMST S2 V42 P71-128 <AMS>
1959 V14 N3 P75-97 AMST S2 V25 P173-197 <AMS>
1959 V14 N3 P99-114 AMST S2 V27 P143-157 <AMS>
1959 V14 N3 P145-152 AMST S2 V26 P221-229 <AMS>
1959 V14 N3 P153-160 AMS-TRANS V43 P1-9 <AMS>
AMST S2 V43 P1-9 <AMS>
1959 V14 N3 P195-202 64-23578 <=$>
1959 V14 N4 P21-56 AMST S2 V38 P301-340 <AMS>
1959 V14 N4 P57-120 61-19685 <=*>
1959 V14 N4 P121-131 AMST S2 V25 P271-282 <AMS>
1959 V14 N4 P237-241 60-15451 <+*>
62-11457 <=>
1959 V14 N5 P3-44 AMS-TRANS V42 P129 <AMS>
AMST S2 V42 P129-173 <AMS>
1959 V14 N5 P97-116 AMST S2 V25 P249-270 <AMS>
1959 V14 N5 P167-180 61-11517 <=>
64-23617 <=$>
1959 V14 N5 P181-195 60-31891 <=>
1959 V14 N6 P3-104 AMST S2 V33 P323-438 <AMS>
1959 V14 N6 P191-195 60-12598 <=>
60-15855 <+*>
62-11272 <=>
1960 V15 P125-132 AMST S2 V28 P211-219 <AMS>
1960 V15 N1 P199-202 AMST S2 V26 P231-234 <AMS>
1960 V15 N2 P165-172 AMST S2 V30 P351-358 <AMS>
1960 V15 N2 P173-180 62-11537 <=>
1960 V15 N2 P189-194 STMSP V4 P189-195 <AMS>
1960 V15 N2 P195-199 AMST S2 V28 P191-196 <AMS>
1960 V15 N3 P133-136 AMST S2 V30 P345-349 <AMS>
1960 V15 N3 P139-146 AMST S2 V28 P197-210 <AMS>
1960 V15 N3 P147-150 AMST S2 V37 P79-83 <AMS>
1960 V15 N3 P157-159 62-11450 <=>
62-11456 <=>
1960 V15 N3 P181-183 AMST S2 V28 P207-209 <AMS>
1960 V15 N4 P129-135 AMST S2 V34 P405-412 <AMS>
61-10797 <+*>
1960 V15 N4 P141-148 AMST S2 V27 P297-304 <AMS>
1960 V15 N4 P149-156 AMST S2 V26 P235-244 <AMS>
1960 V15 N4 P177-184 AMST S2 V26 P245-252 <AMS>
1960 V15 N4 P185-192 AMST S2 V29 P247-254 <AMS>
1960 V15 N5 P143-150 AMST S2 V26 P253-261 <AMS>
1960 V15 N5 P187-190 AMST S2 V30 P291-294 <AMS>
1960 V15 N6 P149-156 AMST S2 V26 P263-272 <AMS>
1960 V15 N6 P169-174 STMSP V4 P295-301 <AMS>
1961 V16 N1 P135-141 AMST S2 V47 P59-65 <AMS>
1961 V16 N1 P179-183 AMST S2 V37 P115-119 <AMS>
1961 V16 N1 P185-188 AMST S2 V28 P187-190 <AMS>
1961 V16 N2 P189-196 61-27722 <*=>
1961 V16 N3 P171-174 63-10299 <=*>

65-12120 <*>
1961 V16 N3 P181-188 AMS TRANS V.46 P57 <AMS>
AMST S2 V46 P48-56 <AMS>
1961 V16 N3 P219-244 62-13421 <*=>
1961 V16 N3 P245-248 62-13547 <*=>
1961 V16 N4 P167-170 AMST S2 V47 P67-71 <AMS>
1962 V17 N1 P3-25 62-24741 <=*>
63-23586 <=*$>
1962 V17 N1 P191-198 62-33550 <=*>
1962 V17 N1 P215-222 AMST S2 V54 P145-152 <AMS>
1962 V17 N4 P141-146 62-20263 <=*>
1962 V17 N6 P3-126 63-20296 <=*>
1962 V17 N6 P145-149 65-12121 <*>
1963 V18 N5 P13-40 64-15691 <=*$>
1963 V18 N6 P91-192 64-19166 <=*$> M
1964 V19 N2 P3-63 AD-615 531 <=$>
1965 V20 N3 P153-174 AD-637 390 <=$>
1965 V20 N4 P3-36 66-11224 <*>

USPEKHI MIKROBIOLOGII
1964 V1 P101-129 RTS-2837 <NLL>

USPEKHI NAUCHNOI FOTOGRAFII. AKADEMIYA NAUK SSSR
1954 V2 P11-27 RT-3543 <*>
1955 V3 P30-34 R-1386 <*>
1955 V3 P59-60 23R79R <ATS>
1955 V3 P66-75 R-1623 <*>
1955 V3 P76-85 R-1832 <*>
1955 V3 P101-109 R-1835 <*>
1955 V3 P110-118 59-15809 <+*>
1955 V3 P119-128 R-4238 <*>
1955 V3 P152-154 R-2321 <*>
1955 V3 P168-173 R-1834 <*>
1955 V3 P174-182 R-1622 <*>
1955 V3 P212-218 R-934 <*>
1955 V4 P7-16 R-930 <*>
1955 V4 P17-22 R-2017 <*>
1955 V4 P23-28 R-2014 <*>
1955 V4 P29-43 R-2013 <*>
1955 V4 P44-53 R-3072 <*>
1955 V4 P67-81 R-2008 <*>
1955 V4 P88-105 R-4236 <*>
1955 V4 P106-110 R-4348 <*>
1955 V4 P111-116 R-4351 <*>
1955 V4 P117-124 R-4353 <*>
1955 V4 P144-149 R-3070 <*>
1955 V4 P150-163 R-3071 <*>
1955 V4 P164-176 R-4916 <*>
1955 V4 P190-201 R-4919 <*>
1955 V4 P210-231 R-4920 <*>
1955 V4 P232-240 R-2320 <*>
1955 V4 P241-262 R-4352 <*>
1957 V5 P81-94 95N50R <ATS>
1957 V5 P182-192 61-15453 <*+>
1959 V6 P35-42 64-10981 <=*$>
1959 V6 P58-61 64-13369 <=*$>
1959 V6 P93-101 61-20875 <=*> 0
1959 V6 P152-154 61-20224 <*=> P
1960 V7 P3-24 79N51R <ATS>
1960 V7 P25-49 62-14465 <=*> 0
62-16521 <=*>
1960 V7 P77-86 62-14464 <=*> 0
1960 V7 P103-108 61K106 <CTT>
1960 V7 P109-114 62-14468 <=*> 0
66N52R <ATS>
1960 V7 P115-119 65N52R <ATS>
1960 V7 P120-133 64N52R <ATS>
1960 V7 P134-136 61K107 <CTT>
1960 V7 P137-149 61K108 <CTT>
1960 V7 P150-160 61K109 <CTT>
1960 V7 P161-169 62-10977 <=*>
62-14466 <=*>
1960 V7 P163-190 88N52R <ATS>
1960 V7 P170-177 61-20226 <*=>
62-14467 <=*>
1960 V7 P191-200 89N52R <ATS>
1960 V7 P210-218 09N58R <ATS>
1960 V7 P219-229 08N58R <ATS>
1962 V8 P13-20 38Q71R <ATS>
1962 V8 P56-60 38Q70R <ATS>
1962 V8 P146-154 39Q71R <ATS>
1962 V8 P161-171 66-13760 <*>
1964 V10 ENTIRE ISSUE AD-460 800 <=*$>
1964 V10 P50-57 1228 <TC>

★USPEKHI SOVREMENNOI BIOLOGII
1942 V15 N3 P283-294 61-15209 <*+>
1944 V17 N3 P273-311 RT-372 <*>
65-2916 <=*$> 0
1944 V18 N1 P105-107 RT-636 <*>
1945 V19 N1 P1-44 61-28026 <*=> 0
1946 V22 N2 P161-180 RT-2435 <*>
1948 V25 N1 P107-122 61-28109 <*=> 0
1948 V25 N2 P231-250 63-13220 <=*>
1948 V26 N1 P515-530 RT-1770 <*>
1949 V27 P185-210 C-2630 <NRC>
1950 V29 N1 P140-144 RT-2368 <*>
1950 V29 N3 P370-378 AD-620 499 <=$>
63-19601 <=*>
1950 V30 P188-221 C-2629 <NRC>
1950 V30 N1 P15-48 RT-2784 <*>
1950 V30 N2 P188-221 64-19083 <=*$> 0
1950 V30 N2 P234-270 63-13358 <=*>
1950 V30 N3 P372-381 59-10991 <+*>
1951 V31 N1 P5-12 RT-2862 <*>
1951 V31 N1 P101-107 R-2754 <*>
R-3511 <*>
RT-3168 <*>
1951 V31 N3 P346-361 RT-4339 <*>
1951 V31 N3 P391-412 TT.303 <NRC>
1951 V32 N3 P309-329 RT-3534 <*>
1952 V33 N1 P47-63 63-13341 <=*>
1952 V33 N1 P81-100 RT-208 <*>
1952 V33 N1 P148-152 RT-2077 <*>
1952 V33 N3 P328-364 RT-3083 <*>
1952 V33 N3 P365-379 RT-2693 <*>
1952 V33 N3 P380-390 RT-2159 <*>
1952 V33 N3 P409-430 R-2751 <*>
R-3514 <*>
1952 V34 N3 P408-422 RT-4245 <*>
1953 V35 N2 P168-188 RT-1949 <*>
1953 V35 N2 P305-310 RT-2694 <*>
1953 V36 N4 P43-63 RT-1370 <*>
1953 V36 N5 P278-280 RT-1200 <*>
64-15578 <=*$>
1954 V37 P291-304 R-4888 <*>
1954 V37 N1 P1-21 RJ-292 <ATS>
1954 V37 N2 P158-176 RT-2304 <*>
1954 V37 N2 P203-208 RJ-367 <ATS>
1954 V37 N3 P255-278 RT-3008 <*>
1954 V37 N3 P291-308 RT-3010 <*>
1954 V37 N3 P341-357 RT-3011 <*>
1954 V37 N3 P358-360 RT-3013 <*>
1954 V37 N3 P361-365 RT-3015 <*>
1954 V37 N3 P366-377 RT-3066 <*>
1954 V37 N3 P378-381 RT-3012 <*>
1954 V37 N3 P382-387 RT-3009 <*>
1954 V38 N1 P18-38 RT-2941 <*>
1954 V38 N1 P39-57 RT-2940 <*>
1954 V38 N1 P58-67 RT-2939 <*>
1954 V38 N1 P68-85 RT-3074 <*>
1954 V38 N1 P86-110 RT-3075 <*>
1954 V38 N2 P183-198 R-2709 <*>
63-11605 <=>
1954 V38 N2 P199-213 RT-3819 <*>
1954 V38 N2 P216-226 TT.549 <NRC>
1954 V38 N2 P227-242 RT-3913 <*>
1954 V38 N3 P280-293 RT-3781 <*>
1955 V39 P328-349 62-15615 <*=> 0
1955 V39 N2 P228-244 RT-3666 <*>
1955 V39 N2 P245-252 64-15297 <=*$>
1955 V39 N3 P257-275 60-23230 <+*> 0
1955 V39 N3 P276-298 R-2714 <*>
62-15130 <*=>
1955 V39 N3 P299-307 62-15152 <*=>
1955 V40 N1 P8-30 62-15135 <=*>
1955 V40 N1 P31-51 62-15103 <*=> 0
1955 V40 N1 P52-67 62-15133 <*=>
1955 V40 N1 P94-107 RT-4269 <*>
1955 V40 N2 P159-178 64-19472 <=$>

1955 V40 N2 P179-191 RT-4250 <*>
86K25R <ATS>
1955 V4C N1/4 P8-30 R-2696 <*>
1955 V40 N2/5 P159-178 R-2677 <*>
1956 V41 P3-25 61-27994 <*=>
1956 V41 N1 P40-54 59-21107 <=> 0
1956 V41 N1 P90-96 61-28007 <*=>
1956 V42 N1 P33-50 R-4781 <*>
1956 V42 N2 P143-159 R-2729 <*>
1957 V44 N1 P3-18 62-15137 <*=>
1957 V44 N1 P37-54 62-15134 <*=>
1957 V44 N1 P103-120 59-12741 <+*>
1957 V44 N1 P121-126 61-13978 <+*> 0
1957 V44 N3 P313-333 60-11402 <=>
62-13295 <*=>
1957 V44 N6 P313-333 R-5352 <*>
1958 V44 N2 P158-172 PB 141 177T <=>
1958 V45 N1 P28-45 59-17376 <+*>
1958 V45 N1 P80-96 59-11690 <=>
1958 V45 N1 P114-129 63-15924 <=*>
1958 V45 N2 P185-199 60-21161 <=>
1958 V45 N2 P218-233 64-15295 <=*$>
1958 V45 N3 P261-271 61-19614 <*=>
1958 V45 N3 P286-312 61-23649 <=*>
1958 V45 N3 P328-348 59-17375 <+*>
1958 V46 N1 P3-18 61-19613 <*=>
1958 V46 N1 P33-47 60-51031 <=>
1958 V46 N1 P48-61 59-11385 <=>
1958 V46 N2 P113-129 61-19615 <=*>
1958 V46 N2 P194-207 59-13200 <=>
1958 V46 N2 P208-216 59-13201 <=>
1958 V46 N2 P230-239 60-23550 <*+>
1958 V46 N3 P337-356 60-19890 <+*>
1959 V47 N2 P133-136 59-11794 <=>
1959 V47 N2 P152-167 59-11795 <=>
1959 V47 N2 P204-219 64-11646 <=>
1959 V47 N2 P235-254 60-15967 <+*>
1959 V47 N3 P362-374 61-15097 <*+>
1959 V47 N3 P390-396 64-15084 <=*$>
1959 V48 N2 P239-244 60-31304 <=>
1960 V49 N2 P156-173 60-41193 <=>
1960 V49 N2 P183-199 60-41194 <=>
1960 V49 N2 P265-271 60-41042 <=>
1960 V49 N3 P305-327 61-21011 <=>
1960 V49 N3 P373-387 61-21011 <=>
1960 V50 N1 P3-28 61-11665 <=>
1960 V50 N1 P44-61 61-11677 <=>
1960 V50 N1 P62-76 61-11832 <=>
1960 V50 N1 P77-100 61-11785 <=>
1960 V50 N2 P121-135 61-21344 <=>
1960 V50 N2 P167-173 61-21457 <=>
1960 V50 N3 P310-321 61-21126 <=*>
1960 V50 N3 P322-336 62-14746 <=*> 0
1961 V51 N1 P3-20 61-27216 <*=>
63-16456 <=*> 0
1961 V51 N1 P62-73 61-27092 <*=>
1961 V51 N1 P74-83 61-28264 <=*>
1961 V51 N1 P84-103 61-31472 <=>
1961 V51 N1 P125-128 61-27387 <*=>
1961 V51 N2 P129-152 61-27931 <*=>
62-25140 <=*>
1961 V51 N2 P170-187 61-27929 <*=>
1961 V51 N2 P188-202 65-23203 <$>
1961 V51 N3 P299-316 61-28863 <*=>
1961 V51 N3 P299-314 62-13596 <*=>
1961 V51 N3 P352-368 63-24380 <=*$>
1961 V52 N1 P19-39 CSIR-TR 276 <CSSA>
1961 V52 N2 P181-207 62-19764 <=*>
1961 V52 N2 P208-223 62-11124 <=>
1961 V52 N3 P347-361 62-23333 <=*>
1961 V52 N14 P19-78 62-13648 <=>
1962 V53 N2 P152-168 63-13368 <=*>
1962 V53 N3 P306-322 62-33694 <=*>
1962 V53 N3 P347-363 63-16453 <=*>
1962 V53 N3 P393-398 62-33694 <=*>
1962 V54 N1 P3-25 63-13372 <=*>
1962 V54 N1 P57-70 63-19556 <=*>
1962 V54 N1 P71-89 63-13372 <=*>
1962 V54 N3 P265-284 63-21200 <=>
1963 V54 N1 P90-97 63-41013 <=>
1963 V55 N1 P9-33 64-13794 <=*$>
1963 V55 N2 P219-238 63-31214 <=>
1963 V55 N3 P339-354 64-19670 <=$>
1963 V55 N3 P428-439 64-19723 <=$>
1963 V56 N1 P90-97 AD-609 141 <=>
1963 V56 N2 P161-179 64-21190 <=>
1963 V56 N3 P341-364 64-21918 <=>
1964 V57 N1 P3-49 64-31925 <=>
1964 V57 N1 P143-158 64-31925 <=>
1964 V57 N2 P211-231 64-31923 <=>
1964 V57 N3 P337-349 64-51319 <=$>
1964 V57 N3 P447-492 64-51098 <=>
1964 V57 N3 P463-476 66-61000 <=*$>
1964 V58 N1 P22-32 65-32074 <=*$>
1964 V58 N1 P74-85 64-51402 <=$>
1964 V58 N2 P177-200 65-30005 <=$>
1964 V58 N2 P242-252 65-30005 <=$>
1964 V58 N2 P262-271 65-30005 <=$>
1964 V58 N2 P307-320 65-30005 <=$>
1964 V58 N3 P346-366 65-30494 <=$>
1964 V58 N3 P367-394 65-30725 <=$>
1964 V58 N3 P395-408 67-30862 <=$>
1964 V58 N3 P423-440 65-30725 <=$>
1964 V58 N3 P453-455 65-30494 <=$>
1965 V59 N2 P246-256 65-31426 <=$>

UZBEKSKII BIOLOGICHESKII ZHURNAL
1959 N5 P55-60 61-19907 <=*>
1960 N1 P67-71 61-11016 <=>
1961 N2 P3-6 62-15486 <*=>
1963 V7 N2 P5-9 63-31247 <=>
1963 V7 N2 P11-15 63-31247 <=>
1963 V7 N2 P21-36 63-31247 <=>
1963 V7 N2 P66-69 63-31247 <=>
1963 V7 N4 P35-36 63-41212 <=>
1963 V7 N4 P55-58 63-41226 <=>
1963 V7 N4 P59-61 63-41211 <=>
1963 V7 N5 P8-14 64-15176 <=*$>
1964 V8 N2 P76-77 64-51276 <=>
1964 V8 N4 P5-9 65-64554 <=*$>
1964 V8 N4 P14-18 65-64553 <=*$>
1964 V8 N5 P23-27 65-64551 <=*$>
1964 V8 N5 P75-78 64-61564 <=*$>
1965 V9 N1 P32-35 66-60297 <=*$>
1965 V9 N2 P9-17 66-60848 <=*$>
1965 V9 N2 P66-68 65-31241 <=$>
1965 V9 N3 P33-36 66-61079 <=$>
1965 V9 N4 P35-39 66-61354 <=*$>
1965 V9 N5 P68-71 66-30313 <=*$>
1965 V9 N6 P10- FPTS V25 N6 P.T1076 <FASE>
1965 V9 N6 P37- FPTS V25 N6 P.T1016 <FASE>

UZBEKSKII GEOLOGICHESKII ZHURNAL
1961 N4 P46-58 IGR V6 N6 P1057 <AGI>
1961 N5 P65-79 IGR V7 N6 P959 <AGI>
1961 N6 P62-84 IGR V6 N12 P2110 <AGI>
1961 V5 N4 P46-58 IGR V6 N6 P1057-1067 <AGI>
1961 V5 N6 P62-84 IGR V6 N12 P2110 <AGI>
1962 N4 P83-87 SHSP N5 1962 P553 <AGU>
1963 V7 N3 P19-25 65-11942 <*>
1964 V8 N2 P89-92 64-41439 <=$>
1964 V8 N4 P19-29 65-30498 <=$>
1966 N3 P58-62 SHSP N1 1966 P86 <AGU>
1966 N3 P64-68 SHSP N1 1966 P90 <AGU>

UZBEKSKII KHIMICHESKII ZHURNAL
1958 N3 P5-17 63-24286 <=*$>
1958 N3 P69-74 AEC-TR-3728 <+*>
1958 N4 P25-32 60-17786 <+> 0
1959 N1 P39-42 60-16958 <*+>
1959 N1 P43-49 61-15063 <=*> 0
1959 N5 P45-49 61-15062 <=*> 0
1959 N5 P81-86 62-24300 <=*>
1960 N4 P67-70 61-21106 <=>
1961 N2 P25-30 62-13400 <*=>
1961 N2 P38-42 62-13430 <*=>
1961 N3 P45-50 62-15780 <*=>
1963 N2 P17-21 ICE V5 N1 P5-8 <ICE>
1963 V7 P25-34 65-64399 <=*$>
1963 V7 N1 P5-14 65-64416 <=*$>

1963 V7 N2 P17-21   ICE V5 N1 P5-8 <ICE>
   74R74R <ATS>
1963 V7 N3 P37-42   65-11024 <*>
1963 V7 N4 P18-22   AD-610 216 <=$>
1963 V7 N5 P13-15   64-30201 <*>
1963 V7 N6 P84-87   65-63675 <=$> 0
1964 V8 N1 P75-81   65-29599 <$>
   65-63399 <=$> 0
1964 V8 N2 P66-72   65-62961 <=$> 0
1964 V8 N3 P54-56   65-63998 <=> 0

V ZASHCHITU MIRA
1959 V8 N103 P18-23   61-21019 <=>

V.D.E. FACHBERICHTE. VEREIN DEUTSCHER ELEKTRO-TECHNIKER. BERLIN
1929 P108-110   57-3251 <*>
1935 V7 P129-132   61-10036 <*> 0
1938 V10 P35-39   II-721 <*>
1951 V15 P2-7   57-2529 <*>
1952 V16 N5 P53-57   62-18151 <*> 0
1956 V19 P91-92   63-14094 <*>
1960 V21 P150-161   62-16304 <*> 0

VDI-BERICHTE
1955 V6 P17-24   12R76G <ATS>
1955 V6 P27-28   12R76G <ATS>
1957 V17 P5-16   57-3296 <*>
1957 V17 P17-22   57-3294 <*>
1957 V17 P23-28   57-3295 <*>
1957 V17 P29-34   57-3293 <*>
1957 V17 P35-64   57-3337 <*>
1957 V20 P117-121   59-10065 <*>
1957 V20 P157-178   C-4260 <NRCC>
1957 V24 P85-90   2175 <BISI>
1957 V25 P45-56   59-17742 <*>
1957 V29 P133-140   58-2368 <*>
1958 V29 P143-148   62-18961 <*>
1961 V49 P43-46   2494 <BISI>
1964 V85 P47-52   4603 <BISI>

VDI FORSCHUNGSARBEITEN
SEE FORSCHUNGSARBEITEN AVF DEM GEBIET DES INGENIEURWESENS

VDI FORSCHUNGSHEFTE
SEE FORSCHUNGSHEFTE. VEREIN DEUTSCHER INGENIEURE

VDI RICHTLINIEN
1958 N-2098   <DAP>
1959 N-2095   <DAP>
1959 N-2099   <DAP>
1960 N-2102   <DAP>
1960 N-2105   <DAP>
1960 N-2106   <DAP>
1960 N-2107   <DAP>
1960 N-2109   <DAP>
1960 N-2110   <DAP>
1961 N-2090   <DAP>
1961 N-2091   <DAP>
1961 N-2092   <DAP>
1961 N-2093   <DAP>
1961 N-2094   <DAP>
1961 N-2101   <DAP>
1961 N-2103   <DAP>
1961 N-2108   <DAP>
1961 N-2115   <DAP>
1961 N-2281   <DAP>
1961 N-2284   <DAP>
1961 N-2285   <DAP>
1961 N-2292   <DAP>
1961 N-2293   <DAP>
1962 N-2104   <DAP>
1962 N-2290   <DAP>
1962 N-2302   <DAP>

VDI ZEITSCHRIFT
V96 N14 P403-409   58-1058 <*>
1954 P585-589   58-898 <*>
1954 V96 P773-777   T-2311 <INSD>
1954 V96 P1085-1090   60-23458 <=*>
1954 V96 N1 P18-21   57-1374 <*>
1954 V96 N2 P43-47   RCT V34 N2 P482 <RCT>
1954 V96 N9 P261-268   58-1154 <*>
1954 V96 N14 P403-409   58-1143 <*>
1954 V96 N24 P802-804   57-1447 <*>
1954 V96 N26 P899-903   62-18792 <*>
1954 V96 N28 P946-950   62-18793 <*>
1954 V96 N29 P973-979   57-819 <*>
1954 V96 N30 P1005-1007   58-1132 <*>
1954 V96 N11/2 P314-346   58-1158 <*>
1954 V96 N15/6 P450-456   58-2300 <*>
1954 V96 N17/8 P506-512   58-1063 <*>
1955 V97 P1111-1121   61-16272 <*>
1955 V97 P1121-1125   61-16271 <*>
1955 V97 N2 P33-41   64-20563 <*> 0
1955 V97 N2 P49-58   1123 <*>
1955 V97 N7 P17-23   2818 <BISI>
1955 V97 N7 P185-232   58-1444 <*>
1955 V97 N7 P226-230   59-20843 <*>
1955 V97 N17 P509-515   58-1157 <*>
1955 V97 N23 P785-788   57-891 <*>
1955 V97 N27 P945-948   57-3297 <*>
1955 V97 N29 P1009-1011   62-18407 <*> 0
1955 V97 N35 P1279-1281   58-1102 <*>
1955 V97 N11/2 P347-350   57-74 <*>
1955 V97 N15/6 P493-496   63-18456 <*> 0
1956 V98 P13-14   8762 <IICH>
1956 V98 P308-310   62-18786 <*>
1956 V98 P425-427   57-2854 <*>
1956 V98 P921-928   57-2853 <*>
1956 V98 N6 P209-214   53K25G <ATS>
1956 V98 N15 P842-843   58-1161 <*>
1956 V98 N16 P869-874   59-17177 <*> C
1956 V98 N22 P1265-1275   59-10902 <*>
1956 V98 N26 P1541-1548   46J15G <ATS>
   57-3005 <*>
1956 V98 N26 P1562   62-18796 <*>
1956 V98 N32 P1789-1794   59-14728 <+*>
1956 V98 N36 P1955-1965   58-783 <*>
1957 V99 P1355-1361   58-1228 <*>
1957 V99 N1 P23-25   60-17399 <*=>
1957 V99 N3 P89-125   57-3298 <*>
1957 V99 N7 P274-279   64-18401 <*> 0
1957 V99 N13 P587-591   59-20845 <*>
1957 V99 N14 P623-   57-1979 <*>
1957 V99 N15 P667-671   59-20149 <*>
   62-32398 <=*>
1957 V99 N24 P1165-1171   TRANS-57 <MT>
1957 V99 N25 P1233-1244   TRANS-57 <MT>
1957 V99 N1/13 P587-591   58-1436 <*>
1958 V100 P216-226   TS-1342 <BISI>
1958 V100 P241-248   TS 1343 <BISI>
1958 V100 N2 P41-46   58-2632 <*>
1958 V100 N5 P165-171   3253 <BISI>
1958 V100 N6 P216-226   62-14178 <*>
1958 V100 N6 P241-248   1343 <BISI>
   62-14196 <*>
1958 V100 N20 P841-850   60-00787 <*>
1958 V100 N22 P1041-1045   63-00532 <*>
1958 V100 N36 P1717-1719   60-16255 <*>
1959 V101 P1-8   TS 1648 <BISI>
1959 V101 P769-773   TS 1720 <BISI>
1959 V101 N1 P1-8   61-20944 <*> P
   62-14524 <*>
   62-18808 <*>
1959 V101 N8 P301-308   60-14551 <*>
1959 V101 N8 P318   2862 <BISI>
1959 V101 N10 P1303-1304   2135 <BISI>
1959 V101 N12 P463-468   61-18881 <*>
   82L35G <ATS>
1959 V101 N12 P1645-1649   1923 <BISI>
1959 V101 N19 P769-773   62-16850 <*>
1959 V101 N20 P832-834   62-32400 <=*>
1959 V101 N22 P1045-1050   61-18836 <*>
1959 V101 N25 P1181-1189   64-10396 <*> 0
1959 V101 N29 P1341-1345   66-11692 <*> 0
1959 V101 N31 P1448-1461   61-20936 <*>
   62-18735 <*>
1959 V101 N32 P1499-1502   60-17396 <*=>

1960 V102 P117-119 2941 <BISI>
1960 V102 P131-132 TS-1790 <BISI>
1960 V102 P262-264 1946 <BISI>
1960 V102 P293-322 AEC-TR-4128 <*>
1960 V102 P347-349 AEC-TR-4126 <*>
1960 V102 P383-384 AEC-TR-4127 <*>
1960 V102 P645-65C 1994 <BISI>
1960 V102 P728-732 AEC-TR-1951 <*>
1960 V102 N2 P193-202 1857 <BISI>
1960 V102 N5 P157-158 60-18132 <*>
1960 V102 N6 P216-224 89M40G <ATS>
1960 V102 N7 P265-267 60-18122 <*>
1961 V103 P1229-1234 3488 <BISI>
1961 V103 P1393-1404 2713 <BISI>
1961 V103 P1429-1433 2678 <BISI>
1961 V103 N20 P869-874 AEC-TR-4807 <*>
1962 V104 N7 P317-323 69P62G <ATS>
1962 V104 N18 P817-819 3131 <BISI>
1962 V104 N25 P1275-1281 3084 <BISI>
1963 V105 N1 P29-39 06Q68G <ATS>
1963 V105 N27 P1268 64-14056 <*>
1964 V106 N2 P57-59 65-13897 <*>
1964 V106 N12 P487-491 65-63402 <=*$>
1964 V106 N19 P829-835 14S82G <ATS>
65-13116 <*>
1964 V106 N30 P1548-1551 25S82G <ATS>
1965 V107 N4 P160-167 39S85G <ATS>
1966 V108 N10 P429-434 38T94G <ATS>

VFDB-ZEITSCHRIFT FUER FORSCHUNG UND TECHNIK IM. BRANDSCHUTZ
1957 V6 N3 P111-117 58-817 <*>
60-15399 <+*>
1957 V6 N3 P118-124 58-1192 <*>
1961 V10 N4 P138-146 62-18007 <*>
1962 V11 P1-4 UCRL TRANS-838(L) <*>
1962 V11 P5-9 UCRL TRANS-839(L) <*>
1962 V11 P9-12 UCRL TRANS-840(L) <*>
1964 V13 N2 P56-59 65-11709 <*>

VVS: TIDSKRIFT FOER VAERME-VENTILATIONS-, SANI-TETS- OCH KYLTECHNIK
1958 V29 P211-213 C-3195 <NRC>
1958 V29 P230 C-3195 <NRC>

VACUUM. LONDON
1957 V7/8 P56-60 UCRL TRANS-719(L) <*>

VACUUM CHEMISTRY. JAPAN
SEE SHINKU KAGAKU

VAG- OCH VATTENBYGGAREN
1961 N1 P9-12 AEC-TR-4964 <*>
69N565 <ATS>

VAG-OCH VATTENBYGGNADSKONST
1943 N10 P129-131 II-774 <*>

VAKBLAD VOOR BIOLOGIEN
1957 N9 P1-10 62-00859 <*>
1957 V37 N9 P1-10 C-3971 <NRCC>

VAKUUM-TECHNIK
1954 V3 N2 P32-36 C-2435 <NRC>
1955 V4 N3 P51-64 59-10214 <*>
1955 V4 N4/5 P82-100 63-18930 <*>
1956 V5 N3 P39-53 65-13338 <*>
1956 V5 N4 P69-82 65-13338 <*>
1958 V7 N1 P13-16 11169 <K-H>
1958 V7 N6 P121-130 62-18538 <*>
1958 V7 N7 P159-171 60-18195 <*>
1958 V7 N2/3 P46-51 11169 <K-H>
1959 V8 N1 P15-19 60-18194 <*>
1959 V8 N2 P39-43 06L34G <ATS>
1959 V8 N2 P44-47 07L34G <ATS>
1959 V8 N3 P59-62 SCL-T-417 <*>
1960 V9 N1 P1-6 62-16307 <*>
1960 V9 N4 P93-100 10798 <K-H>
61-20446 <*>
1960 V9 N7 P199-201 61-20541 <*>

1961 N1 P1-7 NP-TR-908 <*>
1961 N2 P40-45 NP-TR-908 <*>
1961 V10 P113-118 63-10398 <*>
1961 V10 N2 P31-39 AI-TR-3 <*>
1961 V10 N4 P95-99 62-14783 <*>
1961 V10 N4 P100-105 62-16590 <*>
1961 V10 N4 P113-118 64-00025 <*>
1961 V10 N5 P131-138 N65-18339 <=$>
1961 V10 N6 P184-19C N65-18339 <=$>
1962 V11 P17-20 65-00313 <*>
1962 V11 N3 P88-94 16Q71G <ATS>
1962 V11 N6 P161-167 NP-TR-1011 <*>
1964 V13 P80-84 4217 <BISI>
1964 V13 P185-191 4254 <BISI>
1964 V13 P201-209 4218 <BISI>
1965 V14 P157-163 4793 <BISI>
1965 V14 N4 P91-97 POED-TRANS-2248 <NLL>

VALOSAG
1962 V5 N6 P8-16 63-21350 <=>
1964 V7 N9 P23-35 65-30029 <=$>
1965 V8 N5 P19-25 65-32344 <=$>

VALSALVA
1933 V9 P753-770 II-770 <*>
1964 V40 P103-110 65-14708 <*>

VALTION TEKNILLINEN TUTKIMUSLAITOKSEN TIEDOTUS
1951 V95 P66- AL-705 <*>
1953 N126 P21 58-2531 <*>
1961 S2 N9 P5-31 AEC-TR-5085 <*>
1962 S3 N65 P1-8 AEC-TR-6083 <*>

VARME. KOBENHAVN
1951 V16 N5 P89-100 AL-345 <*>
1951 V16 N6 P108-115 AL-345 <*>

VARMLANDSKA BERGSMANNAFORENINGENS ANNALER
1946 P47-94 61-18818 <*>

VASIONA
1960 N4 P120-125 62-19698 <=*>

VASUT
1966 N1 P1-4 66-31051 <=$>

VAXTODLING
1952 V7 P11-24 63-14351 <*>

VAZDUHOPLOVNI GLASNIK
1957 N6 P625-631 59-13229 <=>

VEDA A VYZKUM V PRUMYSLU TEXTILNIM
1958 V4 P73- 61-20679 <*>

VEDOMOSTI VERKHOVNOGO SOVETA SSSR
1960 V23 N41 P923-949 62-13304 <=*>
1961 V24 N39 P920-1075 62-25528 <=>
1963 V26 N24 P655-656 63-31420 <=>
1963 V26 N31 P823 63-31752 <=>
1964 V27 N45 P787-788 65-33390 <=*$>

VEILIGHEID
1952 P121-123 58-1127 <*>

VEREIN ZUR FORDERUNG DER MOORKULTUR IM DEUTSCHEN REICHE LANDESANSTALT FUER MOORWIRTSCHAFT
1918 V2 P14-17 59-17251 <*>

VERFKRONIEK
1953 V26 P61-64 II-235 <*>
1954 V27 N6 P144-147 3035 <*>
1954 V27 N7 P173-177 3035 <*>
1955 V28 N12 P333-335 57-762 <*>
1956 V29 N2 P40- 1619 <*>
1956 V29 N4 P108-110 19Q70DU <ATS>
1958 V31 N2 P61 64-10199 <*>
1960 V33 P56-66 64-10180 <*> 0
1960 V33 N2 P56-66 5014 <TTIS>
1960 V33 N4 P132-140 58M41D4 <ATS>

1960 V33 N12 P486-494 5018 <TTIS>
1962 V35 P284-290 NS-418 <TTIS>

VERHANDELINGEN DER K. AKADEMIE VAN WETENSCHAPPEN
1917 V19 N5 P25-63 57-2553 <*>

VERHANDELINGEN. K. VLAAMSCHE ACADEMIE VOOR GENEESKUNDE VAN BELGIE
1952 V14 N4 P239-272 5092 <K-H>

VERHANDELINGEN VAN DE K. VLAAMSE ACADEMIE VOOR WETENSCHAPPEN. LETTEREN EN SCHONE KUNSTEN VAN BELGIE. KLASSE DER WETENSCHAPPEN. BRUSSELS
1953 N42 P4-76 62-23842 <=*> PO

VERHANDLUNGEN DER ANATOMISCHEN GESELLSCHAFT. JENA
1951 P183-185 57-792 <*>
1954 V52 P197-204 57-80 <*>

VERHANDLUNGEN DER DEUTSCHEN GESELLSCHAFT FUER ARBEITSCHUTZ
1958 V5 P158 61-18736 <*>

VERHANDLUNGEN DER DEUTSCHEN GESELLSCHAFT FUER CHIRURGIE
1908 N37 P204-213 64-16942 <*>
1911 V11 N2 P591-616 57-1598 <*>

VERHANDLUNGEN DER DEUTSCHEN GESELLSCHAFT FUER INNERE MEDIZIN
1951 V57 P142-143 59-17031 <*>
1951 V57 P285-304 63-16688 <*> O
9915-A <K-H>
1954 V60 P316-321 2649 <*>
1954 V60 P630-634 57-3287 <*>
1956 P214-227 57-2955 <*>
1956 V62 P121-125 64-10632 <*>
9149-B <K-H>
1959 V65 P407-410 64-16352 <*>
1959 V65 P591-596 65-13639 <*>

VERHANDLUNGEN DER DEUTSCHEN GESELLSCHAFT FUER KREISLAUFFORSCHUNG
1953 V19 P290-295 I-599 <*>
1956 V22 P113-117 62-10711 <*> O
1957 V23 P375-380 59-10226 <*> O
1960 V26 P269-272 62-16907 <*>
1962 V28 P11-26 63-18860 <*>

VERHANDLUNGEN DER DEUTSCHEN GESELLSCHAFT FUER PATHOLOGIE
1950 V34 P221-226 65-14791 <*> O
1951 P233-244 62-00667 <*>
1951 V34 P232-234 57-3412 <*>
1953 V37 P247-254 I-525 <*>
1954 V38 P314-320 59-18119 <+*> O
1955 V38 P323-327 63-00561 <*>
1957 P361-364 62-00682 <*>
1958 P304-312 62-01043 <*>
1958 V41 P314-320 59-18775 <+*> O
1959 V43 P329-334 62-00132 <*>
1963 V47 P182-186 66-10869 <*> O
1963 V47 P369-372 65-11786 <*>

VERHANDLUNGEN DER DEUTSCHEN PATHOLOGISCHEN GESELLSCHAFT
1904 V6 P247-250 66-10960 <*>
1907 V11 P224-228 66-10953 <*>
1912 V15 P199-207 AEC-TR-4111 <*>
1923 V19 P126-131 3147 <*>
1936 V29 P27-29 61-00374 <*>

VERHANDLUNGEN DER FORTBILDUNGS-LEHRGANG IN BAD NAUHEIM
1960 V25 P34-52 61-10250 <*>

VERHANDLUNGEN DER GESELLSCHAFT DEUTSCHER NATURFORSCHER UND AERZTE
1909 N2 PT.2 P39-43 66-10962 <*>

VERHANDLUNGEN DER JAPANISCHEN PATHOLOGISCHEN GESELLSCHAFT
SEE NIPPON BYORIGAKKAI KAISHI

VERHANDLUNGEN DER NATURFORSCHENDEN GESELLSCHAFT IN BASEL
1956 V67 P447-478 57-734 <*>
1956 V67 P479-499 57-685 <*>
1956 V67 N2 P447-478 57-2055 <*>

VERHANDLUNGEN DES NATURHISTORISCH-MEDIZINISCHEN VEREINS ZU HEIDELBERG
1862 V3 P51-55 61-14086 <*>
1862 V3 P170-171 61-16001 <*>

VERKEHR UND TECHNIK
1954 P282-284 T-1977 <INSD>
1954 P314-316 T-1977 <INSD>
1954 P350-351 T-1977 <INSD>
1956 V9 N4 P37-38 59-20984 <*>

VERMESSUNGS-INFORMATIONEN
1964 P35-46 C-5981 <NRC>

VERMESSUNGSTECHNIK
1959 V7 N6 P147-150 59-13930 <=>
1960 V8 N4 P99-102 61-28934 <=*$>
1960 V8 N6 P153-158 62-24554 <=*>
1960 V8 N6 P179-180 62-24561 <=*>
1960 V8 N9 P263-268 62-24560 <=*>
1961 V9 N2 P33-37 62-24552 <=*>
1961 V9 N2 P47-50 62-24553 <=*> O
1961 V9 N2 P60 62-24552 <=*>
1961 V9 N3 P65-77 62-24558 <=*>
1961 V9 N4 P119-122 62-24551 <=*>
1961 V9 N5 P145-149 62-19737 <=*>
1961 V9 N6 P176-186 62-23453 <=*> P
1963 V11 N7 P241-244 64-31063 <=>
1964 V12 N10 P361-368 65-30105 <=$>
1964 V12 N12 P471-472 AD-628 332 <=*$>

VEROEFFENTLICHUNGEN DER ARBEITSGEMEINSCHAFT FUER FORSCHUNG DES LANDES NORDRHEIN-WESTFALEN
1952 N21A P33-54 64-16016 <*>

VEROEFFENTLICHUNGEN DES ASTRONOMISCHEN RECHEN-INSTITUTS ZU BERLIN-DAHLEM
1901 N14 ENTIRE ISSUE AD-615 367 <=$>

VEROEFFENTLICHUNGEN DER BUNDESANSTALT FUER ALPINE LANDWIRTSCHAFT IN ADMONT
1952 N18 P36 <CBPB>
1953 N1 P14-15 57-1937 <*>

VEROEFFENTLICHUNGEN DER BUNDESANSTALT FUER TABAKFORSCHUNG. KAISERSLAUTERN
1953 P3-4 2161 <*>

VEROFFENTLICHUNGEN. DEUTSCHE GEODAETISCHE KOMMISSION
SEE DEUTSCHE GEODASTISCHE KOMMISSION

VEROEFFENTLICHUNGEN DES DEUTSCHEN AUSCHUSSES FUER STAHLBETON
1952 N107 P1-73 61-18931 <*> P
1952 N108 P1-27 61-18952 <*>
1954 N118 P41-51 65-11705 <*>
1959 N132 P1-17 PT.1 60-14558 <*>
1959 N132 P1-66 PT.2 60-14559 <*> O
1964 N164 P1-18 PT1 66-10177 <*> C
1964 N164 P1-64 PT2 66-10185 <*>

VEROEFFENTLICHUNGEN DES DEUTSCHEN STAHLBAU-VERBANDES
1962 V15 P27-45 5776 <HB>

VEROEFFENTLICHUNGEN AUS DEM GEBIET DER NACH-RICHTENTECHNIK
1931 S1 P57-66 57-1112 <*>
1934 V4 N2 P139-142 66-13311 <*>

1937 V7 P281-284 60-18985 <*> 0
1937 S1 V7 P109-113 57-1479 <*>
1938 V8 N3 P459-462 61-10088 <*> 0
1940 V10 P37-48 57-1449 <*>

VEROEFFENTLICHUNGEN AUS DEM UEBERSEEMUSEUM IN BREMEN. A. NATURWISSENSCHAFTEN
1964 V3 N3 P143-151 65-00400 <*> 0

VEROEFFENTLICHUNGEN DER WISSENSCHAFTLICHEN PHOTO-LABORATORIEN AGFA.
1951 V7 P8-31 66-14135 <*> 0

VERPACKUNG. LEIPZIG
1964 V5 P24-26 66-10544 <*>

VERPACKUNGS-RUNDSCHAU
1955 N10 P67-70 SUP 58-55 <*>
1957 V8 N9 P65-69SUP 64-14535 <*>
8025-B <K-H>
1958 N4 P26-32 SUP 60-17179 <+*> 0
1958 N7 P57-59 SUP 60-17179 <+*> 0
1958 N8 P61-68 SUP 60-17179 <+*> 0
1958 N10 P77-81 SUP 60-17179 <+*> 0
1960 V11 N5 P33-39SUP 68M46G <ATS>
1963 V14 N5 P410,412,415,416
64-14018 <*>
1963 V14 N6 P43-49 SUP 63-20864 <*> 0
1963 V14 N7 P51-58 SUP 63-20863 <*> 0
1964 V15 N1 P1-7 SUP 66-10745 <*>
1964 V15 N10 P73-77SUP 66-10879 <*>
1964 V15 N12 P93-95SUP 66-11992 <*>
1966 V17 N4 P418-426 66-12637 <*>

VERPAKKING. ROTTERDAM
1958 V10 N9 P608-611 62-18182 <*>

VERRE ET SILICATES INDUSTRIELS
1948 V13 N6 P81-86 65-12867 <*>
1948 V13 N7 P166-169 65-12866 <*>

VERRES ET REFRACTAIRES
1947 V1 P19-25 60-16861 <*>
1947 V1 P21-29 63-20384 <*>
1947 V1 N9 P19-25 63-14024 <*>
1948 V2 N3 P158-163 64-14041 <*>
1948 V2 N4 P211-221 59-15331 <*>
1950 V4 N1 P10-19 60-16863 <*>
1951 V5 P15-19 BGIRA-574 <BGIR>
1951 V5 N5 P247-260 59-15330 <*>
1951 V5 N6 P303-311 62-16147 <*>
1952 V6 N4 P209-218 64-10534 <*>
1953 V7 N2 P91-104 57-1027 <*>
1953 V7 N6 P339-345 62-16412 <*>
1955 V9 N1 P3-12 60-14046 <*>
1955 V9 N2 P69-81 62-18105 <*>
1955 V9 N5 P251-263 64-14043 <*>
1960 V14 P113-123 65-17154 <*>
1960 V14 N6 P340-342 64-14031 <*>
1961 N4 P204-208 65-14411 <*>
1961 V15 N2 P63-72 64-14078 <*>
1961 V15 N4 P197-203 64-16595 <*>
1962 V16 N6 P337-343 65-13514 <*>
1962 V16 N6 P344-351 65-17332 <*>
1963 V17 N1 P3-10 575 <BGIR>
1963 V17 N1 P11-15 66-10809 <*>
1963 V17 N3 P151-162 BGIRA-595 <BGIR>
65-17333 <*>
1963 V17 N6 P397-400 65-12753 <*>
1964 V18 N2 P89-97 BGIRA-640 <BGIR>
1964 V18 N2 P98-106 BGIRA-626 <BGIR>
1964 V18 N2 P108-118 BGIRA-637 <BGIR>
1964 V18 N3 P181-190 BGIRA-636 <BGIR>
1964 V18 N3 P191-198 BGIRA-638 <BGIR>
1964 V18 N6 P461-465 BGIRA-641 <BGIR>
1966 N1 P22-32 66-13899 <*>
1966 V20 N2 P113-120 66-14502 <*>

VERSLAGEN VAN HET CENTRAAL INSTITUUT VOOR LANDBOUWKUNDIG ONDERZOEK
1954 P49-55 CSIRO-3370 <CSIR>
1954 P83-90 CSIRO-3376 <CSIR>
1954 P115-120 CSIRO-3372 <CSIR>
1954 P121-126 CSIRO-3373 <CSIR>
1954 P126-132 CSIRO-3375 <CSIR>
1954 P157-163 CSIRO-3434 <CSIR>

VERSLAGEN VAN DE GEWONE VERGADERING DER AFDEELING NATUURKUNDE, K. AKADEMIE VAN WETENSCHAPPEN
1926 V35 P1023- I-100 <*>
1943 V52 N2 P58-68 I-772 <*>

VERSLAGEN VAN DE GEWONE VERGADERING DER WIS- EN NATUURKUNDIGE AFDEELING. K. AKADEMIE VAN WETENSCHAPEN TE AMSTERDAM
1909 V18A P755-768 1179 <*>

VERSLAGEN VAN LANDBOUWKUNDIGE ONDERZOEKINGEN
V15 N62 P1-23 CSIRO-3656 <CSIR>

VERSLAGEN EN MEDEDEELINGEN BETREFFENDE DE VOLKSGEGONDHEID
1958 V8 N9 P807-858 64-10608 <*>
9149-A <K-H>

VESNIK. JUGOSLOVENSKA INVESTICIONA BANKA
1964 V8 N89 P6-9 64-41491 <=>
1964 V8 N90 P10-13 64-51256 <=$>
1965 N98/9 P1-6 65-32219 <=$>
1966 N118 P7-10 67-30503 <=$>

★VESNIK ZAVODA ZA GEOLOSKO I GEOFIZICKO ISTRAZIVANJE N. S. SRBIJE. SERIJA B: INZENJERSKA GEOLOGIJA I HIDROGEOLOGIJA
1962 N2 P5-141 62-11756/2 <=$>
1963 N3 P5-142 63-11454/2 <=$>

★VESNIK ZAVODA ZA GEOLOSKO I GEOFIZICKO ISTRAZIVANJE N. S. SRBIJE. SERIJA C: PRIMENJENA GEOFIZIKA.
1962 V3 P5-99 62-11756/3 <=>

★VESTNIK AKADEMII MEDITSINSKIKH NAUK SSSR
1954 N1 P3-43 RT-1742 <*>
RT-2333 <*>
1954 N1 P75-87 R-522 <*>
1954 N1 P88-91 RT-2049 <*>
1954 N4 P21-29 R-163 <*>
1954 N4 P29-34 62-15799 <*=>
1955 N3 P41-56 RT-4247 <*>
1955 N3 P61-75 RT-4248 <*>
1955 N3 P79-82 R-346 <*>
1955 N4 P57-60 RT-3665 <*>
1956 V11 N2 P13-24 62-15147 <*=>
1956 V11 N2 P35-45 R-2665 <*>
62-15128 <=*>
64-19485 <=$>
1956 V11 N2 P45-56 62-23516 <=*>
1956 V11 N12 P32-36 60-11430 <=>
1957 V12 N2 P56-65 PB 141 226T <=>
1957 V12 N4 P39-45 AEC-TR-3610 <+*>
1957 V12 N5 P9-18 62-15097 <*=>
1958 N2 P8-14 PB 141 205T <=>
1958 N2 P15-26 PB 141 206T <=>
1958 N2 P83-85 PB 141 207T <=>
1958 V13 N1 P12-18 59-17378 <+*>
1958 V13 N3 P3-11 R-4712 <*>
64-19446 <=$> 0
1958 V13 N3 P20-30 59-11515 <=>
1958 V13 N3 P60-62 59-11516 <=>
1958 V13 N3 P86-93 59-11517 <=>
1958 V13 N4 P3-12 59-13310 <=> 0
1958 V13 N4 P12-27 59-13311 <=>
1958 V13 N4 P27-32 59-13312 <=>
1958 V13 N4 P58-61 59-13313 <=>
1958 V13 N4 P62-63 59-13314 <=>
1958 V13 N5 P76-81 59-11448 <=>
1958 V13 N6 P10-22 59-11177 <=> 0
1958 V13 N6 P78-85 59-11157 <=>
1958 V13 N6 P85-88 59-11156 <=>

Citation	Number
1958 V13 N7 P46-61	59-14573 <+*>
1958 V13 N8 P54-57	59-11536 <=>
1958 V13 N8 P79-83	59-11537 <=>
1958 V13 N8 P93-95	59-11538 <=>
1958 V13 N9 P12-21	59-13370 <=> 0
1958 V13 N9 P57-62	59-13371 <=>
1958 V13 N9 P62-64	59-13372 <=>
1958 V13 N9 P79-81	62-11141 <=>
1958 V13 N10 P23-34	59-11683 <=> 0
1958 V13 N10 P74-80	59-11684 <=>
1958 V13 N10 P90-92	59-11685 <=>
1958 V13 N11 P30-41	59-13243 <=> 0
1958 V13 N11 P51-63	59-13244 <=> 0
1958 V13 N11 P63-72	59-13470 <=>
1958 V13 N12 P3-6	59-13436 <=> 0
1958 V13 N12 P7-15	59-13437 <=>
1958 V13 N12 P15-21	59-13438 <=>
1958 V13 N12 P39-46	59-13439 <=>
1958 V13 N12 P55-58	59-13440 <=>
1959 N10 P47-59	NLLTB V2 N2 P73 <NLL>
1959 N12 P3-12	NLLTB V2 N6 P447 <NLL>
1959 V14 N1 P3-7	59-13605 <=>
1959 V14 N1 P14-19	61-27333 <*=>
1959 V14 N1 P54-57	59-13918 <=>
1959 V14 N2 P3-9	60-23175 <+*>
1959 V14 N2 P9-15	60-23171 <+*>
1959 V14 N2 P22-32	60-15081 <+*>
1959 V14 N2 P42-49	60-17977 <+*>
1959 V14 N4 P3-18	59-13772 <=> 0
1959 V14 N4 P36-44	59-13773 <=>
1959 V14 N4 P74-82	59-13774 <=>
1959 V14 N5 P3-10	59-13767 <=>
1959 V14 N5 P27-40	59-13768 <=>
1959 V14 N5 P40-45	59-13769 <=>
1959 V14 N5 P70-76	59-13770 <=>
1959 V14 N5 P91-92	59-13771 <=>
1959 V14 N6 P25-37	61-15120 <+*>
1959 V14 N6 P51-58	59-13938 <=>
1959 V14 N6 P74-78	59-13851 <=>
1959 V14 N7 P57-67	60-11057 <=>
1959 V14 N7 P87-92	60-31385 <=>
1959 V14 N8 P3-6	59-31041 <=>
1959 V14 N8 P7-19	59-31042 <=>
1959 V14 N8 P32-38	59-31043 <=>
1959 V14 N8 P61-67	59-31044 <=>
1959 V14 N8 P86-87	60-12367 <+*>
1959 V14 N9 P64-69	59-31039 <=>
1959 V14 N10 P23-28	65-60244 <=$> 0
1959 V14 N11 P29-37	60-11201 <=>
1959 V14 N11 P46-51	60-11202 <=>
1959 V14 N11 P51-61	60-11203 <=>
1959 V14 N11 P62-72	60-11204 <=>
1959 V14 N12 P13-16	60-11552 <=>
1959 V14 N12 P36-41	60-31076 <=>
1959 V14 N12 P42-53	60-11385 <=>
1959 V14 N12 P54-60	60-31094 <=>
1959 V14 N12 P84-89	60-11254 <=>
1960 V15 N2 P3-14	60-11895 <=>
1960 V15 N2 P20-30	61-28424 <*=>
1960 V15 N2 P46-57	60-11663 <=>
	63-19096 <=*>
1960 V15 N2 P63-72	60-11912 <=>
1960 V15 N2 P84-89	60-41525 <=>
1960 V15 N2 P89-94	60-41526 <=>
1960 V15 N3 P78-82	60-31747 <=>
1960 V15 N4 P17-29	60-41150 <=>
1960 V15 N4 P29-36	60-41151 <=>
1960 V15 N4 P57-62	60-41152 <=>
1960 V15 N4 P82-86	60-41153 <=>
1960 V15 N4 P91-93	61-11032 <=>
1960 V15 N5 P3-7	60-41395 <=>
1960 V15 N5 P8-22	60-41255 <=>
1960 V15 N5 P43-54	60-41396 <=>
1960 V15 N5 P72-75	60-41251 <=>
1960 V15 N5 P75-81	60-41253 <=>
1960 V15 N5 P81-85	60-41254 <=>
1960 V15 N6 P85-90	60-41238 <=>
1960 V15 N7 P3-14	60-41708 <=>
1960 V15 N7 P41-44	60-41661 <=>
1960 V15 N7 P65-72	60-41660 <=>
1960 V15 N7 P72-77	60-41605 <=>
1960 V15 N8 P3-10	61-21625 <=>
	61-27883 <*=>
1960 V15 N8 P41-53	61-21629 <=>
1960 V15 N8 P62-74	61-21464 <=>
1960 V15 N9 P59-65	61-31522 <=>
1960 V15 N9 P68-76	61-31522 <=>
1960 V15 N9 P91	61-31522 <=>
1960 V15 N10 P18-29	61-21317 <=>
1960 V15 N10 P40-44	61-21486 <=>
1960 V15 N10 P55-59	61-21398 <=>
1960 V15 N10 P60-66	61-15818 <+*>
1960 V15 N10 P73-76	61-21455 <=>
1960 V15 N10 P82-85	61-21341 <=>
1960 V15 N11 P10-25	61-21319 <=>
1960 V15 N11 P25-38	RTS 2777 <NLL>
1960 V15 N11 P38-46	61-21297 <=>
1960 V15 N11 P63-66	61-21350 <=>
1960 V15 N11 P67-72	61-21310 <=>
1960 V15 N11 P72-75	61-21309 <=>
1960 V15 N12 P3-14	62-15798 <*=>
1960 V15 N12 P15-21	63-15591 <=*> 0
1961 V16 N1 P3-7	61-21640 <=>
1961 V16 N1 P8-13	61-21642 <=>
1961 V16 N1 P40-49	61-21643 <=>
1961 V16 N2 P52-63	64-19891 <=$>
1961 V16 N2 P79-90	61-21540 <=>
1961 V16 N3 P48-50	63-15533 <=*>
1961 V16 N4 P3-29	61-27690 <*=>
1961 V16 N4 P59-78	61-27690 <*=>
1961 V16 N5 P3-12	61-27394 <*=>
1961 V16 N5 P70-78	62-13503 <=*>
1961 V16 N5 P88-93	63-15593 <=*>
1961 V16 N6 P32-87	61-28292 <*=>
1961 V16 N7 P3-13	61-31621 <=>
1961 V16 N7 P13-20	62-13324 <*=>
1961 V16 N7 P20-30	61-28569 <*=>
1961 V16 N7 P50-55	62-23469 <=*>
1961 V16 N7 P63-71	61-31622 <=>
1961 V16 N7 P72-81	61-31623 <=>
1961 V16 N8 P8-12	62-13637 <*=>
1961 V16 N8 P12-16	62-13614 <*=>
1961 V16 N8 P32-35	62-13636 <*=>
1961 V16 N8 P81-84	62-13638 <*=>
1961 V16 N10 P62-67	62-33461 <=*>
1961 V16 N10 P68-78	62-20368 <=*> 0
1961 V16 N10 P68-86	62-23339 <=*>
1961 V16 N11 P68-74	64-21923 <=>
1961 V16 N11 P75-79	62-19164 <=*>
1961 V16 N11 P88-93	62-19164 <=*>
1961 V16 N12 P87-91	62-23595 <=*>
1962 V17 N1 P77-87	62-23828 <=*>
1962 V17 N1 P93-96	64-19845 <=$>
1962 V17 N2 P3-	FPTS V22 N5 P.T800 <FASE>
1962 V17 N2 P3-9	1963,V22,N5,PT2 <FASE>
1962 V17 N2 P3-62	63-21241 <=>
1962 V17 N2 P9-	FPTS V22 N3 P.T514 <FASE>
1962 V17 N2 P82-89	62-24380 <=*>
1962 V17 N3 P48-58	62-25988 <=*>
1962 V17 N3 P58-59	62-25016 <=*>
1962 V17 N3 P58-74	63-15426 <=*>
1962 V17 N3 P74-82	AD-609 790 <=>
1962 V17 N3 P74-	FPTS V22 N1 P.T3 <FASE>
1962 V17 N4 P5-26	62-33606 <=>
1962 V17 N4 P7-26	62-33092 <=*>
1962 V17 N4 P27-37	62-32897 <=*>
1962 V17 N4 P38-44	62-33092 <=*>
1962 V17 N4 P44-57	62-32897 <=*>
1962 V17 N4 P44-50	63-15530 <=*>
1962 V17 N4 P57-76	62-33606 <=>
1962 V17 N4 P65-70	63-15530 <=*>
1962 V17 N4 P76-81	62-32897 <=*>
	63-15530 <=*>
1962 V17 N4 P82-93	62-33092 <=*>
1962 V17 N4 P82-96	62-33606 <=>
1962 V17 N5 P5-13	63-15164 <=*>
1962 V17 N5 P24-28	63-19310 <=*>
1962 V17 N5 P47-58	63-19310 <=*>
1962 V17 N5 P64-68	63-19310 <=*>
1962 V17 N5 P72-93	63-19964 <=*>

1962 V17 N5 P93-101 63-19310 <=*> C
1962 V17 N6 P3-16 62-33351 <=>
1962 V17 N6 P7-16 63-19024 <=*>
1962 V17 N6 P17-23 63-19027 <=*>
1962 V17 N6 P23-35 63-19026 <=*>
1962 V17 N6 P67- FPTS V22 N2 P.T337 <FASE>
1962 V17 N6 P67-72 62-33351 <=>
1962 V17 N7 P17-22 62-33030 <=*>
66-60564 <=*$>
1962 V17 N7 P74-82 62-33030 <=*>
1962 V17 N8 P3-10 62-33674 <=*>
1962 V17 N8 P16-22 63-15992 <=*>
1962 V17 N8 P25 62-33674 <=*>
1962 V17 N8 P44-50 63-15993 <=*>
1962 V17 N8 P56- FPTS V22 N2 P.T279 <FASE>
1962 V17 N8 P60- FPTS V22 N2 P.T277 <FASE>
1962 V17 N9 P48-58 63-13489 <=*>
1962 V17 N9 P76-84 63-13489 <=*>
1962 V17 N9 P85- FPTS V22 N5 P.T806 <FASE>
1962 V17 N9 P85-93 1963,V22,N5,PT2 <FASE>
1962 V17 N10 P3-14 63-13668 <=*>
1962 V17 N10 P44-48 63-19362 <=*>
1962 V17 N10 P52-59 63-19361 <=*>
1962 V17 N1C P59-67 63-19360 <=*> 0
1962 V17 N10 P76-88 65-63076 <=*$>
1962 V17 N10 P89-92 63-13523 <=*>
1962 V17 N11 P23- FPTS V22 N5 P.T964 <FASE>
1962 V17 N11 P23-31 1963,V22,N5,PT2 <FASE>
1962 V17 N11 P36-4C 63-23842 <=*$>
1962 V17 N11 P51- FPTS V22 N4 P.T671 <FASE>
1962 V17 N11 P82-96 63-21161 <=>
1962 V17 N11 P89-96 63-23841 <=*>
1962 V17 N12 P3-62 63-21143 <=>
1962 V17 N12 P3-13 63-23051 <=*>
1962 V17 N12 P38- FPTS V22 N5 P.T789 <FASE>
1962 V17 N12 P38-49 1963,V22,N5,PT2 <FASE>
1963 N4 P31- FPTS V23 N3 P.T541 <FASE>
1963 N5 P61-67 64-00077 <*>
1963 V18 N1 P1-96 63-21331 <=>
1963 V18 N1 P10-17 63-23831 <*>
1963 V18 N1 P18-23 63-23871 <=*>
1963 V18 N1 P23-27 63-23872 <=*$> 0
1963 V18 N1 P28- FPTS V22 N5 P.T866 <FASE>
1963 V18 N1 P28-32 1963,V22,N5,PT2 <FASE>
1963 V18 N1 P33-35 63-23873 <=*>
1963 V18 N1 P36-41 63-23874 <=*>
1963 V18 N1 P41-47 63-23875 <=*$>
1963 V18 N1 P47-52 64-13518 <=*$>
1963 V18 N1 P52-58 63-23876 <=*$>
1963 V18 N1 P64-68 63-23830 <=*>
1963 V18 N1 P69- FPTS V22 N6 P.T1130 <FASE>
1963 V18 N1 P69-86 FPTS,V22,P.T1130-38 <FASE>
1963 V18 N2 P38-42 63-23838 <=*>
1963 V18 N2 P60-70 63-23549 <=*$>
1963 V18 N2 P77-84 64-13508 <=*$>
1963 V18 N4 P8- FPTS V23 N2 P.T369 <FASE>
1963 V18 N4 P20-28 64-13923 <=*$>
1963 V18 N4 P31- 64-10612 <FASE>
1963 V18 N4 P63-67 64-13442 <=*$> 0
1963 V18 N5 P22-26 64-13550 <=*$>
1963 V18 N5 P27-36 64-13886 <=*$> 0
1963 V18 N5 P49-55 64-13897 <=*$> 0
1963 V18 N5 P55-60 64-15541 <=*$>
1963 V18 N5 P61-68 64-13926 <=*$> 0
64-13956 <=*$>
1963 V18 N5 P81-87 64-13898 <=*$>
1963 V18 N7 P3-101 63-31829 <=>
1963 V18 N7 P27- FPTS V23 N4 P.T667 <FASE>
1963 V18 N7 P37-42 64-15514 <=*$> 0
1963 V18 N7 P50-62 64-15544 <=*$>
1963 V18 N8 P13- FPTS V23 N4 P.T789 <FASE>
1963 V18 N8 P13-20 64-15222 <=*$>
1963 V18 N8 P47-53 FPTS V23 N4 P.T881 <FASE>
1963 V18 N8 P47- FPTS,V23,N4, P.T881 <FASE>
1963 V18 N8 P54- FASE V23 N4 P.T908 <FASE>
FPTS,V23,N4,P.T908 <FASE>
1963 V18 N8 P60- FPTS,V23,N4,P.T837 <FASE>
1963 V18 N8 P68-76 64-13796 <=*$>
1963 V18 N1C P15- FPTS,V23,N5,P.T960 <FASE>
1963 V18 N10 P42-48 63-41294 P71-80 <=>
1963 V18 N11 P29-38 64-19630 <=$> 0
1963 V18 N11 P38-43 64-19700 <=$>
1963 V18 N11 P43-46 64-19623 <=$> 0
1963 V18 N12 P29-34 65-60197 <=$>
1963 V18 N12 P43-51 65-60177 <=$>
1963 V18 N12 P51-62 65-60144 <=$>
1963 V18 N12 P84-89 65-60178 <=$>
1964 V19 N1 P26-33 64-23598 <=$>
1964 V19 N1 P72-84 65-62105 <=>
1964 V19 N2 P33-38 AD-615 216 <=$>
1964 V19 N2 P54-60 65-61080 <=$>
1964 V19 N2 P60- FPTS V23 N6 P.T1374 <FASE>
FPTS V23 P.T1201 <FASE>
1964 V19 N2 P71-78 65-61081 <=$>
1964 V19 N2 P78-81 64-60164 <=$>
1964 V19 N3 P23- FPTS V24 N3 P.T500 <FASE>
1964 V19 N3 P37-42 65-61115 <=$>
1964 V19 N3 P42- FPTS V24 N3 P.T507 <FASE>
1964 V19 N3 P78-84 65-61116 <=$>
1964 V19 N4 P43-51 65-61120 <=$>
1964 V19 N6 P69- FPTS V24 N4 P.T709 <FASE>
1964 V19 N6 P72- FPTS V24 N4 P.T737 <FASE>
1964 V19 N6 P72-75 65-60570 <=$>
1964 V19 N7 P66-76 65-62311 <=$>
1964 V19 N10 P39- FPTS V24 N6 P.T1084 <FASE>
1964 V19 N11 P46- FPTS V24 N6 P.T1067 <FASE>
1964 V19 N11 P51- FPTS V24 N6 P.T1070 <FASE>
1964 V19 N12 P28- FPTS V24 N6 P.T1052 <FASE>
1964 V19 N12 P47- FPTS V25 N2 P.T353 <FASE>
1965 V20 N3 P68-74 66-60854 <=*$>
1965 V20 N4 P1C-22 66-60855 <=*$>
1965 V20 N4 P63- FPTS V25 N1 P.T172 <FASE>
1965 V20 N6 P41-52 66-61070 <=*$>
1965 V20 N7 P47-55 66-61343 <=*$>
1965 V20 N10 P3- FPTS V25 N6 P.T1023 <FASE>
1965 V20 N10 P72- FPTS V25 N5 P.T794 <FASE>
1965 V20 N11 P12- FPTS V25 N5 P.T785 <FASE>
1965 V20 N11 P93- FPTS V25 N5 P.T883 <FASE>
1966 N1 P41-45 NLLTB V8 N6 P499 <NLL>
1966 V21 N2 P34- FPTS V25 N6 P.T1124 <FASE>
1966 V21 N4 P66- FPTS V25 N6 P.T1000 <FASE>

VESTNIK AKADEMII NAUK KAZAKHSKOI SSR. ALMA-ATA
1954 N11 P42-53 R-1581 <*>
1955 V11 N1 P54-59 60-23861 <+*>
1955 V11 N4 P1C8-110 R-4419 <*>
1956 V12 N1 P25-39 6C-11023 <=>
1956 V12 N6 P79-88 59-14289 <=*>
63-16223 <=*> C
1957 V13 N1 P60-69 513TT <CCT>
1957 V13 N2 P63-70 R-4413 <*>
1957 V13 N12 P56-60 AD-626 704 <=*$>
N66-15876 <=$>
1957 V13 N12 P61-70 59-22381 <+*>
63-19037 <=*>
1958 N1 P48-61 R-4434 <*>
1958 N5 P68-74 IGR V1 N9 P28 <AGI>
1958 V14 N5 P68-73 PB 141 211T <=>
1958 V14 N6 P6-9 59-11338 <=>
1958 V14 N6 P73-75 59-11340 <=>
1958 V14 N8 P3-9 59-13258 <=>
1958 V14 N8 P32- 60-13666 <+*>
1958 V14 N9 P48-60 60-11258 <=>
1958 V14 N10 P42-48 60-15830 <+*> 0
1958 V14 N12 P4-10 59-11692 <=>
1958 V14 N12 P71-75 60-13844 <+*>
1959 V15 N1 P38-45 C-139 <RIS>
62-13729 <*=>
1959 V15 N2 P5C-63 83Q68R <ATS>
1959 V15 N4 P3-25 61-21180 <=>
1959 V15 N4 P90-94 UCRL TRANS-743(L) <*>
UCRL TRANS-743(L) <=*>
1959 V15 N5 P3-9 60-11118 <=>
1959 V15 N6 P28-29 60-11311 <=>
1959 V15 N6 P36-46 60-11158 <=>
1959 V15 N6 P96-98 60-31500 <=>
1959 V15 N6 P99-103 59-31055 <=>
1959 V15 N7 P106 60-31064 <=>
1959 V15 N10 P100-109 60-31295 <=>
1959 V15 N11 P90-91 60-11519 <=>

1959 V15 N11 P98-99	60-31200 <=>
1960 V16 N1 P44-49	61-11006 <=>
1960 V16 N2 P85-91	61-11745 <=>
1960 V16 N4 P79-83	60-41117 <=>
1960 V16 N5 P7-26	60-41464 <=>
1960 V16 N6 P82	61-27960 <*=>
1960 V16 N8 P3-19	62-23478 <=*>
1960 V16 N8 P108-109	62-23478 <=*>
1960 V16 N9 P15-24	61-27273 <=>
1960 V16 N11 P101-102	61-21517 <=>
1960 V16 N11 P104-106	61-21516 <=>
1960 V16 N12 P97-101,103	61-11008 <=>
1960 V16 N12 P103	61-11008 <=>
1961 V17 N3 P19-28	65-64413 <=*$>
1961 V17 N5 P10-14	62-13060 <*=>
1961 V17 N5 P31-36	62-13047 <*=>
1961 V17 N5 P99-103	61-31649 <=>
1961 V17 N6 P3-6	61-28618 <*=>
1961 V17 N6 P38-52	61-28888 <=>
1961 V17 N6 P103-104	61-28888 <=>
1961 V17 N7 P2-9	62-15473 <*=>
1961 V17 N8 P15-35	62-13668 <=>
1961 V17 N8 P46-48	62-13668 <=>
1961 V17 N12 P3-80	62-23810 <=*>
1962 V18 N1 P3-14	62-32983 <=*>
1962 V18 N5 P10-19	62-32865 <=*>
1962 V18 N6 P26-29	63-15377 <=*>
1963 N10 P33-42	65-11085 <*>
1963 V19 N2 P60-69	63-31045 <=>
1963 V19 N3 P41-45	RTS 2718 <NLL>
1963 V19 N4 P8-45	63-31409 <=>
1963 V19 N4 P51-57	63-31409 <=>
1963 V19 N4 P65-73	63-31409 <=>
1963 V19 N10 P18-26	66-10605 <*>
1963 V19 N10 P33-42	64-21279 <=$>
1963 V19 N11 P58-67	N65-16592 <=$>
1963 V19 N12 P65-69	32T93R <ATS>
1964 V20 N1 P3-10	65-30127 <=$>
1964 V20 N2 P23-31	65-30089 <=$>
1964 V20 N3 P47-56	AD-615 867 <=$>
1964 V20 N5 P31-39	64-51548 <=$>
1964 V20 N5 P48-53	64-51549 <=$>
1964 V20 N6 P39-45	64-51614 <=$>
1964 V20 N6 P55-60	64-51613 <=$>
1964 V20 N9 P74-76	65-30149 <=$>
1964 V20 N9 P77-81	65-30150 <=$>
1964 V20 N12 P15-24	65-30663 <=$>
1964 V20 N12 P41-50	65-14294 <*>
1965 V21 N2 P54-58	65-31070 <=$>
1965 V21 N2 P70-72	65-31020 <=$>
1965 V21 N3 P3-23	65-32368 <=$>
1965 V21 N3 P28-39	65-31701 <*>
	66-10901 <*>
1965 V21 N10 P31-41	66-12498 <*>

★VESTNIK AKADEMII NAUK SSSR

1934 N10 P9-25	R-414 <*>
1943 V13 N3 P48-66	60-19924 <*+>
1943 V13 N3 P75-	61-13648 <=*>
1943 V13 N6 P75-98	C-2192 <NRC>
1944 V14 N4/5 P57-64	RT-447 <*>
1945 V15 N10/1 P90-93	1-20406 <*>
1947 N5 P24-30	RT-4315 <*>
1947 V17 N7 P28-40	TT.215 <NRC>
1948 N5 P28-31	RT-1404 <*>
1948 V18 N1 P17-31	R-5021 <*>
1948 V18 N7 P39-48	RT-4220 <*>
1950 N5 P94-110	RT-2198 <*>
1950 V20 N2 P22-34	RT-1805 <*>
1950 V20 N3 P118-120	RT-2852 <*>
1950 V20 N3 P120-122	RT-1061 <*>
1950 V20 N5 P95-96	61-13177 <*+>
1950 V20 N5 P112-113	RT-1571 <*>
1950 V20 N12 P109-111	RT-3253 <*>
1951 V20 N10 P112-114	RT-2851 <*>
1951 V21 N4 P68-74	RT-1060 <*>
1951 V21 N4 P77-79	RT-1039 <*>
1951 V21 N9 P87-89	RT-2850 <*>
1951 V21 N10 P23-24	RT-1063 <*>
1951 V21 N10 P107-112	RT-4435 <*>
1952 V22 N4 P15-19	RT-1062 <*>
1952 V22 N4 P113-115	RT-1664 <*>
1952 V22 N6 P91-92	RT-1663 <*>
1953 N2 P38-43	NLLTB V5 N8 P699 <NLL>
1953 N9 P72-75	RT-2601 <*>
1953 N9 P84-85	RT-2242 <*>
1953 N10 P113-114	RT-2244 <*>
1953 N10 P115-123	RT-2232 <*>
1953 N10 P124-126	RT-2243 <*>
1953 N10 P130-133	RT-2257 <*>
1953 V23 N8 P72-79	RT-1288 <*>
1953 V23 N8 P85-86	RT-1139 <*>
1953 V23 N9 P54-55	RT-1172 <*>
1953 V23 N9 P55-56	RT-1170 <*>
1953 V23 N9 P56-57	RT-1171 <*>
1953 V23 N9 P68-71	RT-1173 <*>
1953 V23 N9 P82-84	RT-3868 <*>
1953 V23 N11 P7-11	RT-1351 <*>
1953 V23 N11 P49-64	60-41348 <=>
1953 V23 N11 P104-105	RT-1164 <*>
1953 V23 N12 P75-77	RT-1331 <*>
1954 N1 P42	RT-3218 <*>
1954 N1 P96-100	RT-2202 <*>
1954 N3 P48	RT-2807 <*>
1954 N5 P26-43	IGR V1 N8 P1 <AGI>
	R-4682 <*>
1954 N8 P22-34	RT-2376 <*>
1954 N9 P10-16	RT-4165 <*>
1954 V24 N2 P21-26	CSIRO-3773 <CSIR>
1954 V24 N3 P56-58	RT-2474 <*>
1954 V24 N3 P83-86	RT-4192 <*>
1954 V24 N4 P87-88	RT-1743 <*>
1954 V24 N5 P44-48	RT-1889 <*>
1954 V24 N5 P75-78	RT-2546 <*>
1954 V24 N6 P62-63	R-5151 <*>
1954 V24 N7 P53-56	61-15170 <*+>
1954 V24 N8 P78-80	60-27017 <*+>
1954 V24 N10 P48-56	63-19731 <=*>
1954 V24 N12 P30-34	RT-3299 <*>
1954 V24 N12 P67-71	RT-2916 <*>
1954 V24 N12 P85-89	RT-2952 <*>
1955 N1 P110	RT-2964 <*>
1955 N3 P12-18	RT-3364 <*>
1955 N3 P21-31	R-353 <*>
1955 N9 P19-30	R-984 <*>
1955 N9 P56-60	R-365 <*>
1955 V25 N1 P30-40	59-15909 <+*>
1955 V25 N1 P41-42	RT-2734 <*>
1955 V25 N1 P127-132	RT-2921 <*>
1955 V25 N2 P25-32	61-19365 <+*>
1955 V25 N2 P39-43	RT-3974 <*>
1955 V25 N3 P12-18	C-2256 <NRC>
1955 V25 N3 P19-38	RT-3057 <*>
1955 V25 N3 P39-58	RT-3134 <*>
1955 V25 N4 P75-79	RT-3059 <*>
1955 V25 N6 P21-24	62-25289 <=*>
1955 V25 N7 P87-90	84K25R <ATS>
1955 V25 N8 P34-46	RT-3302 <*>
1955 V25 N9 P19-30	61-15323 <+*>
1955 V25 N9 P31-34	59-15910 <+*>
1955 V25 N10 P90-94	RT-3882 <*>
	59-12892 <+*>
1956 N2 P115-118	C-2257 <NRC>
1956 N3 P3-22	R-2684 <*>
1956 N12 P44-57	R-2537 <*>
1956 V26 N1 P90-93	R-644 <*>
1956 V26 N2 P118-119	60-23859 <+*>
	61-00977 <*>
1956 V26 N3 P8-15	C-2256 <NRC>
1956 V26 N3 P74-78	PANSDOC-TR.45 <PANS>
1956 V26 N4 P23-25	C-2124 <NRC>
	RT-4095 <*>
1956 V26 N5 P29-31	C-2232 <NLL>
	R-2 <*>
	R-294 <*>
	62-18253 <=*>
1956 V26 N6 P78-81	C-2238 <NLL>
	R-25 <*>
	R-290 <*>
1956 V26 N6 P95-96	64-23648 <=$>

Citation	Translation
1956 V26 N6 P1C1-103	85K25R <ATS>
1956 V26 N7 P3-15	TT-791 <NRC>
1956 V26 N8 P1C6-108	60-17373 <+*>
1956 V26 N11 P93-101	64-19401 <=$>
1956 V26 N12 P1-13	R-3484 <*>
1956 V26 N12 P14-23	59-19512 <+*> O
1956 V26 N12 P24-33	R-4957 <*>
	59-12378 <+*>
1957 N8 P998-999	R-2667 <*>
1957 N10 P111-116	R-4199 <*>
1957 V27 N2 P3-42	59-11249 <=>
1957 V27 N2 P50-58	AEC-TR-37C5 <+*>
1957 V27 N2 P59-68	179 <TC>
1957 V27 N2 P114-117	C-2334 <NRC>
	R-1180 <*>
1957 V27 N4 P88-91	64-19488 <=$>
1957 V27 N4 P114-116	PB 141 17CT <=>
1957 V27 N5 P34-36	64-00472 <*>
	64-11834 <=>
1957 V27 N5 P51-55	61-23691 <*=>
1957 V27 N5 P56-59	59-10403 <+*>
1957 V27 N5 P108-110	R-4069 <*>
1957 V27 N6 P25-32	C-2456 <NRC>
	R-2042 <*>
1957 V27 N6 P73-75	59-19009 <+*>
1957 V27 N6 P93	59-19009 <+*>
1957 V27 N6 P1C1-104	59-19009 <+*>
1957 V27 N6 P124-127	59-19009 <+*>
1957 V27 N8 P21-25	64-71160 <=>
1957 V27 N8 P26-36	59-16805 <+*>
1957 V27 N8 P37-48	59-11988 <=>
1957 V27 N9 P31-40	62-15109 <*=>
1957 V27 N9 P77-78	R-4211 <*>
1957 V27 N9 P123-125	R-3163 <*>
1957 V27 N1C P60-65	60-17158 <+*>
1957 V27 N10 P111-116	31L33R <ATS>
1957 V27 N12 P3-7	63-23732 <=*$>
1957 V27 N12 P8-14	59-14908 <+*>
1958 N1 P65-68	R-4267 <*>
1958 N9 P122-123	NLLTB V1 N1 P5 <NLL>
1958 N10 P58-60	NLLTB V1 N2 P24 <NLL>
1958 N10 P64-66	NLLTB V1 N2 P21 <NLL>
1958 N12 P33-35	NLLTB V1 N6 P39 <NLL>
1958 V28 N1 P5-7	R-5304 <*>
1958 V28 N1 P65-68	C-2695 <NRC>
	DS IS-T-289-R <NRC>
	60-21208 <=> O
1958 V28 N1 P69-70	59-15266 <+*>
	59-15784 <+*> O
1958 V28 N1 P91-93	59-15785 <+*>
1958 V28 N1 P131-132	60-13586 <+*>
1958 V28 N2 P13-22	PB 141 251T <=>
1958 V28 N3 P33-39	59-19067 <+*>
1958 V28 N3 P40-47	60-11587 <=>
1958 V28 N3 P106-107	60-11602 <=>
1958 V28 N4 P13-24	C-3650 <NRCC>
1958 V28 N4 P25-29	59-10407 <+*> O
1958 V28 N4 P30-36	59-1C519 <+*>
1958 V28 N5 P4-31	59-13222 <=>
1958 V28 N5 P48-66	59-13222 <=>
1958 V28 N5 P96-103	59-19507 <+*>
1958 V28 N6 P28-34	59-14157 <+*>
1958 V28 N6 P70-73	R-5217 <*>
1958 V28 N6 P77-79	59-19168 <+*>
1958 V28 N7 P7-18	60-13632 <+*>
1958 V28 N7 P19-25	60-13649 <+*>
1958 V28 N7 P75-78	AEC-TR-3632 <+*>
1958 V28 N8 P1C7-109	59-12680 <+>
1958 V28 N8 P113-114	59-12680 <+>
1958 V28 N8 P116-118	59-12680 <+>
1958 V28 N8 P118-119	59-18579 <+*>
1958 V28 N8 P125-126	59-13290 <=>
1958 V28 N8 P155-158	59-13375 <=>
1958 V28 N9 P3-10	59-11551 <=>
	59-15264 <+*>
1958 V28 N9 P22-29	59-11426 <=>
1958 V28 N9 P30-37	59-11427 <=>
1958 V28 N9 P66-71	59-21199 <=>
1958 V28 N9 P109-123	59-22640 <+*>
1958 V28 N10 P11-19	59-12477 <+*>
1958 V28 N10 P20-33	61-19400 <+*>
1958 V28 N1C P58-60	59-22633 <+*>
1958 V28 N10 P64-66	59-22635 <+*>
1958 V28 N11 P3-9	59-15265 <+*>
1958 V28 N11 P10-16	59-22683 <+*>
1958 V28 N11 P73-74	59-16814 <+*>
1958 V28 N11 P120-121	63-13698 <=*>
1958 V28 N12 P97-10C	6C-13661 <+*>
	61-10164 <*+>
1959 N1 P11-23	NLLTB V1 N9 P28 <NLL>
1959 N1 P81-82	NLLTB V1 N7 P30 <NLL>
1959 N2 P43-47	IGR V1 N6 P90 <AGI>
	NLLTB V1 N7 P25 <NLL>
1959 N3 P15-18	NLLTB V1 N11 P14 <NLL>
1959 N3 P19-23	NLLTB V1 N10 P19 <NLL>
1959 N3 P36-44	NLLTB V1 N7 P10 <NLL>
1959 N3 P55-58	NLLTB V1 N9 P42 <NLL>
1959 N6 P49-56	NLLTB V1 N10 P25 <NLL>
1959 N6 P62-65	NLLTB V1 N10 P34 <NLL>
1959 N6 P67-71	NLLTB V1 N12 P37 <NLL>
1959 N7 P27-33	NLLTB V2 N2 P116 <NLL>
1959 N7 P34-38	NLLTB V2 N3 P191 <NLL>
1959 N9 P50-55	NLLTB V2 N1 P14 <NLL>
1959 N11 P47-52	NLLTB V2 N5 P352 <NLL>
1959 V29 N1 P3-10	60-13788 <+*>
1959 V29 N1 P11-23	AEC-TR-3772 <+*>
1959 V29 N1 P52-58	UCRL TRANS-492 <*>
1959 V29 N1 P65-67	60-13791 <+*>
1959 V29 N1 P81-82	59-16813 <+*>
1959 V29 N1 P98-100	59-16815 <+*>
1959 V29 N1 P132-138	59-16816 <+*>
1959 V29 N1 P142-143	59-16817 <+*>
1959 V29 N2 P3C-33	60-17257 <+*>
1959 V29 N2 P64-72	60-13811 <+*>
1959 V29 N3 P3-7	60-17258 <+*>
1959 V29 N3 P8-14	59-16359 <+*>
	59-16482 <+*>
	59-21063 <=>
	6C-13109 <+*>
1959 V29 N3 P15-18	61-19406 <+*>
1959 V29 N3 P8C	59-19669 <+*>
1959 V29 N3 P130-132	60-17261 <+*>
1959 V29 N5 P7-17	60-13787 <+*>
1959 V29 N5 P3C-34	61-13514 <*+>
1959 V29 N6 P26-34	60-11191 <=>
1959 V29 N6 P49-56	61-13035 <*+>
	62-24750 <=*>
1959 V29 N6 P66-97	59-31014 <=>
1959 V29 N6 P111-112	6C-11069 <=>
1959 V29 N8 P3-11	60-15835 <+*>
1959 V29 N8 P87-89	60-11180 <=>
1959 V29 N9 P56-64	60-11285 <=>
1959 V29 N9 P60-64	IGR V2 N4 P357 <AGI>
1959 V29 N9 P65-68	60-11284 <=>
1959 V29 N9 P69-75	60-11247 <=>
1959 V29 N9 P76-78	IGR V2 N4 P354 <AGI>
	6C-11285 <=>
1959 V29 N9 P79-86	60-11293 <=>
1959 V29 N9 P87-89	6C-11292 <=>
1959 V29 N9 P105-106	60-31162 <=>
1959 V29 N10 P32-38	60-31072 <=> O
1959 V29 N10 P60-67	6C-31073 <=>
1959 V29 N10 P68-71	60-31030 <=>
1959 V29 N11 P3-22	60-31109 <=>
	61-28361 <*=>
	61-28363 <*=> P
1959 V29 N11 P58-64	60-31181 <=>
1959 V29 N11 P74-78	60-31182 <=> O
1959 V29 N12 P3-15	60-31484 <=>
1959 V29 N12 P37-41	60-11852 <=>
1959 V29 N12 P57-60	61-13755 <*+>
1959 V29 N12 P94-95	60-31205 <*=>
1959 V29 N12 P106-108	60-11853 <=>
1960 N1 P53	NLLTB V2 N5 P363 <NLL>
1960 N2 P33-37	NLLTB V2 N7 P568 <NLL>
1960 N3 P3-12	NLLTB V2 N9 P764 <NLL>
1960 N3 P13-21	NLLTB V2 N12 P1C29 <NLL>
1960 N3 P22-30	NLLTB V2 N9 P746 <NLL>
1960 N3 P52-54	NLLTB V3 N2 P109 <NLL>
1960 N4 P62	NLLTB V2 N9 P729 <NLL>

Citation	Translation
1960 N4 P63-69	NLLTB V2 N9 P731 <NLL>
1960 N4 P70-86	NLLTB V2 N10 P819 <NLL>
1960 N4 P87-106	NLLTB V2 N10 P855 <NLL>
1960 N10 P97	NLLTB V3 N1 P22 <NLL>
1960 V30 N1 P28-36	60-31123 <=>
	61-15498 <+*>
1960 V30 N1 P108-109	60-17960 <+*>
1960 V30 N2 P12-20	60-11728 <=>
1960 V30 N2 P21-32	60-11730 <=>
	60-17700 <+*>
	61-23528 <=*>
1960 V30 N2 P33-37	60-23158 <+*>
1960 V30 N2 P74-76	64-21286 <=>
1960 V30 N2 P82-85	60-11729 <=>
1960 V30 N3 P13-21	60-11967 <=>
1960 V30 N3 P42-46	63-23339 <=*$>
1960 V30 N3 P101-103	60-31302 <=>
1960 V30 N3 P103-105	60-31303 <=>
1960 V30 N4 P121-124	60-31438 <=>
1960 V30 N5 P10-16	ARSJ V30 N11 P1065 <AIAA>
1960 V30 N5 P32-61	60-31677 <=>
	61-15845 <+*>
1960 V30 N5 P99-102	61-15826 <+*>
1960 V30 N5 P104-106	61-15846 <+*>
1960 V30 N5 P106-107	61-15844 <+*>
1960 V30 N6 P21-37	61-21539 <=>
1960 V30 N6 P38-47	61-21688 <=>
1960 V30 N6 P52-60	61-21624 <=>
1960 V30 N6 P61-69	59M45R <ATS>
1960 V30 N6 P81-88	61-21752 <=>
1960 V30 N6 P116-118	61-21632 <=>
1960 V30 N6 P139-141	61-15843 <+*>
1960 V30 N7 P7-20	61-11197 <=>
	61-15287 <+*>
1960 V30 N7 P57-61	1803 <RIA>
1960 V30 N7 P58-61	61-11074 <=>
1960 V30 N7 P98	61-11737 <=>
1960 V30 N7 P99-100	SCL-T-366 <*=>
	61-27639 <*=>
1960 V30 N7 P108	61-11519 <=>
1960 V30 N8 P3-9	61-11515 <=>
1960 V30 N8 P10-26	ARSJ V31 N3 P399 <AIAA>
	61-11515 <=>
1960 V30 N8 P47-55	61-11912 <=>
1960 V30 N8 P62-68	61-11515 <=>
1960 V30 N8 P115-116	61-15821 <+*>
1960 V30 N8 P120-121	62-32985 <=*>
1960 V30 N8 P125-126	61-15820 <+*>
1960 V30 N9 P3-12	61-11584 <=>
1960 V30 N9 P48-52	61-11848 <=>
1960 V30 N9 P65-72	61-11725 <=>
1960 V30 N9 P103-104	61-11725 <=>
1960 V30 N10 P3-22	61-21518 <=>
1960 V30 N10 P23-31	61-21173 <=>
1960 V30 N10 P32-36	62-15402 <=*>
1960 V30 N10 P37-40	61-21518 <=>
1960 V30 N10 P45-77	61-21518 <=>
1960 V30 N10 P96-98	61-21518 <=>
1960 V30 N10 P99-102	62-15402 <=*>
1960 V30 N10 P102	61-21518 <=>
1960 V30 N10 P106-112	61-21518 <=>
1960 V30 N10 P114-115	61-21518 <=>
1960 V30 N10 P120-126	61-21518 <=>
1960 V30 N11 P3-74	61-21528 <=>
1960 V30 N11 P36-50	39N53R <ATS>
	62-24668 <=*>
1960 V30 N11 P76-132	61-21528 <=>
1960 V30 N11 P78-84	12N53R <ATS>
1960 V30 N11 P135-142	61-21528 <=>
1960 V30 N12 P32-35	61-14986 <=*>
	62-10178 <*=>
1961 N3 P4-11	NLLTB V3 N9 P733 <NLL>
1961 N3 P12-26	NLLTB V3 N10 P823 <NLL>
1961 V31 N1 P10-18	61-28717 <*=>
1961 V31 N1 P19-26	61-28736 <*=>
1961 V31 N1 P39-42	61-31159 <=>
1961 V31 N1 P48-53	61-27417 <*=>
1961 V31 N1 P58-59	61-27492 <*=> 0
1961 V31 N1 P60-64	61-27493 <*=>
1961 V31 N1 P77-79	62-13667 <*=>
1961 V31 N1 P106-119	62-13643 <*=>
1961 V31 N1 P113-114	62-32622 <=*>
1961 V31 N2 P44-47	71Q70R <ATS>
1961 V31 N3 P3-52	61-31517 <=>
1961 V31 N3 P61-66	61-23918 <*=>
1961 V31 N3 P67-77	61-27341 <*=>
1961 V31 N3 P92-130	61-27395 <=>
1961 V31 N3 P111-113	91P62R <ATS>
1961 V31 N4 P19-26	61-28599 <*=>
1961 V31 N4 P27-57	62-13666 <*=>
1961 V31 N4 P68-72	63-19347 <=*>
1961 V31 N4 P73-82	61-28927 <*=>
1961 V31 N4 P83-87	61-28925 <*=>
1961 V31 N4 P99-104	61-28595 <*=>
1961 V31 N4 P105-110	62-13258 <=>
1961 V31 N4 P107-110	62-13665 <*=>
1961 V31 N4 P113-114	62-13258 <=>
1961 V31 N4 P113-115	62-13645 <*=>
1961 V31 N4 P122-123	62-13662 <*=>
1961 V31 N4 P124-127	62-13640 <*=>
1961 V31 N4 P130-131	62-13660 <=*>
1961 V31 N4 P132-133	63-19541 <=*>
1961 V31 N5 P13-22	61-28552 <*=>
	62-19928 <=*>
1961 V31 N5 P23-32	61-28594 <*=>
1961 V31 N5 P33-38	62-23143 <=*>
1961 V31 N5 P39-45	62-16064 <=*>
1961 V31 N5 P55-59	61-28737 <*=>
1961 V31 N5 P98-99	62-25910 <=*>
1961 V31 N6 P3-13	61-28628 <*=>
1961 V31 N6 P19-30	61-28291 <*=>
1961 V31 N6 P41-47	61-28596 <*=>
1961 V31 N6 P81-85	61-28593 <*=>
1961 V31 N6 P86	62-13258 <=>
1961 V31 N6 P90-94	61-28590 <*=>
1961 V31 N6 P95-102	61-28591 <*=>
1961 V31 N8 P24-27	62-13469 <*=>
1961 V31 N8 P53-58	62-23599 <=*>
1961 V31 N8 P59-63	62-15935 <=*>
1961 V31 N9 P25-31	62-23368 <=*>
1961 V31 N9 P40-47	62-25906 <=*>
1961 V31 N9 P48-55	62-15502 <*=>
1961 V31 N9 P56-60	62-15236 <*=>
	62-32546 <=*>
1961 V31 N9 P107	62-19151 <=*>
1961 V31 N9 P113-114	62-19151 <=*>
1961 V31 N9 P137-138	SCL-T-427 <*>
1961 V31 N10 P43-52	62-11133 <=>
1961 V31 N10 P98-101	63-15170 <=*>
1961 V31 N10 P122-123	62-19151 <=*>
1961 V31 N10 P126-128	62-15476 <*=> P
1961 V31 N10 P134-135	63-15929 <=*>
1961 V31 N10 P136-137	62-25857 <=*>
1961 V31 N11 P47-58	62-33651 <=*>
1961 V31 N11 P81-84	62-32668 <=*>
1961 V31 N11 P88-93	62-23587 <=*> 0
1961 V31 N12 P5-93	62-23421 <=*>
1961 V31 N12 P98-100	63-15931 <=*>
	65-13427 <*>
1962 N8 P65-66	RCT V37 N3 P774 <RCT>
1962 V32 N1 P7-15	63-19503 <=*>
1962 V32 N1 P30-34	62-24338 <=*>
1962 V32 N1 P62-67	62-32089 <=*>
1962 V32 N1 P74-76	65-13983 <*> 0
	65-61352 <=$>
1962 V32 N1 P77-79	62-32349 <=*>
	65-17399 <*>
1962 V32 N1 P134-136	AD-609 171 <=>
1962 V32 N2 P23-28	62-32086 <=*>
1962 V32 N2 P51-53	62-32350 <=*>
1962 V32 N2 P90-93	62-11718 <=>
1962 V32 N2 P95-96	62-11718 <=>
1962 V32 N2 P111-112	63-19576 <=*>
1962 V32 N3 P3-36	62-25176 <=>
1962 V32 N3 P9-16	N66-11610 <=$>
1962 V32 N3 P19-23	62-33224 <=*>
1962 V32 N3 P24-34	62-24968 <=*>
1962 V32 N3 P40-50	62-25176 <=>
1962 V32 N3 P51-55	C-4024 <NRCC>
	62-18806 <=*>

Citation	Reference
1962 V32 N3 P75-78	62-25176 <=>
1962 V32 N3 P85-125	62-25176 <=>
1962 V32 N4 P15-20	62-32069 <=*>
1962 V32 N4 P21-24	63-11693 <=>
1962 V32 N4 P25-31	62-25490 <=*>
1962 V32 N4 P56-81	62-32370 <=*>
	62-32592 <=*> P
1962 V32 N4 P86-90	63-13702 <=*>
	65-17392 <*> 0
1962 V32 N4 P100-104	63-21009 <=>
1962 V32 N5 P20-32	62-32690 <=*>
1962 V32 N5 P33-42	63-15399 <=*>
1962 V32 N6 P3-8	62-33327 <=*>
1962 V32 N6 P29-36	62-32540 <=*>
1962 V32 N6 P41-49	62-32543 <=*>
1962 V32 N6 P49-50	63-10839 <=*>
1962 V32 N6 P53-55	62-33327 <=*>
1962 V32 N6 P69-74	63-19138 <=*>
1962 V32 N6 P80-82	62-33026 <=*>
1962 V32 N6 P83-86	62-11775 <=>
1962 V32 N6 P113-115	62-33327 <=*>
1962 V32 N7 P3-9	62-32861 <=>
1962 V32 N7 P10-18	62-33630 <=*>
1962 V32 N7 P19-32	NASA-TT-F-136 <=>
	62-33039 <=*>
1962 V32 N7 P33-37	64-00479 <*>
	64-11829 <=>
1962 V32 N7 P38-43	62-32885 <=*>
1962 V32 N7 P51-53	AD-611 076 <=$>
1962 V32 N7 P6C-63	62-33431 <=*>
1962 V32 N7 P69-72	63-13025 <=*>
1962 V32 N7 P73-76	AD-609 172 <=>
	62-33458 <=*>
1962 V32 N7 P93-95	62-32861 <=>
1962 V32 N7 P95	62-33561 <=*>
1962 V32 N7 P97-99	63-23612 <=*$>
1962 V32 N7 P112-114	62-11795 <=>
1962 V32 N7 P118-120	62-33560 <=*>
1962 V32 N7 P122	62-33559 <=*>
1962 V32 N7 P123-125	AD-610 273 <=$>
1962 V32 N8 P3-10	62-33275 <=>
1962 V32 N8 P25-33	62-33598 <=*>
1962 V32 N8 P53-56	62-33292 <=*>
1962 V32 N8 P57-61	62-33299 <=*>
1962 V32 N8 P72-75	63-13553 <=*>
1962 V32 N8 P82-84	63-13306 <=*>
1962 V32 N8 P110-112	63-31529 <=>
1962 V32 N8 P113-116	62-33275 <=>
1962 V32 N8 P123-125	62-33122 <=>
1962 V32 N8 P125-128	62-33133 <=*>
1962 V32 N9 P9-15	63-13098 <=*>
1962 V32 N9 P16-29	63-11689 <=>
1962 V32 N9 P78-85	64-71599 <=>
1962 V32 N9 P94-95	62-33714 <=*>
1962 V32 N9 P128-131	63-19652 <=*>
1962 V32 N1C P18-25	63-11691 <=>
1962 V32 N10 P36-45	63-21760 <=>
1962 V32 N1C P1C5-108	63-13858 <=*>
1962 V32 N11 P58-63	63-11692 <=>
	63-21289 <=>
1962 V32 N11 P64-69	63-2101 <=>
1962 V32 N11 P76-79	64-19029 <=*$>
1962 V32 N11 P130-131	C-4278 <NRCC>
1962 V32 N11 P135-136	64-00481 <*>
	64-11833 <=>
1962 V32 N12 P9-14	63-21864 <=>
1962 V32 N12 P16-18	63-21864 <=>
1962 V32 N12 P20-21	64-71159 <=>
1962 V32 N12 P92-95	63-21419 <=>
1962 V32 N12 P105-108	63-19648 <=*>
1962 V32 N12 P1C9-110	63-21358 <=>
1963 N1 P32-37	NLLTB V5 N7 P592 <NLL>
1963 N1 P45-50	NLLTB V5 N7 P579 <NLL>
1963 N3 P11-24	NLLTB V5 N9 P781 <NLL>
1963 N3 P21-38	NLLTB V6 N11 P955 <NLL>
1963 N5 P7-19	NLLTB V5 N11 P981 <NLL>
1963 N5 P34-39	NLLTB V5 N10 P891 <NLL>
1963 N7 P21-32	AD-635 076 <=$>
1963 N7 P73-76	NLLTB V6 N1 P44 <NLL>
1963 N8 P70-75	NLLTB V6 N1 P1 <NLL>
1963 N9 P23-29	NLLTB V6 N1 P28 <NLL>
1963 N10 P43-46	AD-618 648 <=$>
	NLLTB V6 N3 P250 <NLL>
1963 N10 P58-62	NLLTB V6 N3 P259 <NLL>
1963 N11 P14-15	SGRT V5 N4 P32 <AGS>
1963 N11 P82-85	NLLTB V6 N4 P324 <NLL>
1963 N12 P12-15	NLLTB V6 N6 P549 <NLL>
1963 N12 P16-25	SGRT V5 N2 P32 <AGS>
1963 V33 N1 P3-10	63-21738 <=>
1963 V33 N1 P11-15	63-21472 <=>
1963 V33 N1 P16-20	63-21849 <=>
1963 V33 N1 P32-37	63-21472 <=>
1963 V33 N1 P38-41	63-21406 <=>
1963 V33 N1 P42-44	63-21472 <=>
1963 V33 N1 P45-50	63-21738 <=>
1963 V33 N1 P121-122	64-15100 <=*$>
1963 V33 N2 P38-43	63-21479 <=>
	64-11864 <=>
	64-15119 <=*$>
1963 V33 N2 P73-76	RTS-2628 <NLL>
1963 V33 N2 P73-79	63-31944 <=>
1963 V33 N2 P76-79	RTS-2629 <NLL>
1963 V33 N2 P88-91	63-21848 <=>
1963 V33 N3 P36-76	63-19460 <=*$> P
1963 V33 N3 P127-128	63-21950 <=>
1963 V33 N4 P54-61	63-21903 <=>
1963 V33 N4 P62-77	63-21954 <=>
1963 V33 N4 P85-88	63-21951 <=>
1963 V33 N4 P98-100	63-31573 <=>
1963 V33 N4 P119-145	63-21952 <=>
1963 V33 N5 P3-6	63-31529 <=>
	64-21797 <=>
1963 V33 N5 P7-19	NLLTB,1963,V5,N11 <HMSO>
1963 V33 N5 P22-26	64-26245 <$>
1963 V33 N5 P34-39	64-71390 <=> M
1963 V33 N5 P58-60	63-31549 <=>
1963 V33 N5 P73-75	64-00468 <*>
	64-11831 <=>
1963 V33 N5 P79-82	63-31978 <=>
1963 V33 N5 P83-86	63-31993 <=>
1963 V33 N5 P101-102	63-31529 <=>
1963 V33 N6 P3-25	63-31408 <=>
1963 V33 N6 P31-38	NLLTB V5 N12 P1101 <NLL>
	64-71317 <=>
1963 V33 N6 P69-71	63-31515 <=>
1963 V33 N6 P73-76	63-31897 <=>
1963 V33 N6 P95-96	63-31529 <=>
1963 V33 N6 P99-102	63-31529 <=>
1963 V33 N6 P113-114	63-31529 <=>
1963 V33 N6 P119-121	63-31514 <=>
1963 V33 N7 P5-9	63-31950 <=>
1963 V33 N7 P33-39	63-31780 <=>
1963 V33 N7 P54-55	63-31950 <=>
1963 V33 N7 P80-83	RTS-2605 <NLL>
1963 V33 N7 P84-89	63-31950 <=>
1963 V33 N7 P113-114	AD-617 147 <=$>
1963 V33 N8 P3-5	63-41056 <=>
1963 V33 N8 P6-7	63-41056 <=>
1963 V33 N8 P8-28	63-41056 <=>
1963 V33 N8 P29-39	63-41056 <=>
1963 V33 N8 P40-42	63-41056 <=>
1963 V33 N8 P43-48	63-41056 <=>
1963 V33 N8 P52-54	63-41056 <=>
1963 V33 N8 P55-61	63-41056 <=>
1963 V33 N8 P62-69	63-41056 <=>
1963 V33 N8 P70-75	63-41056 <=>
1963 V33 N8 P76-77	63-31919 <=>
1963 V33 N8 P77	63-41C56 <=>
1963 V33 N8 P84-88	63-41056 <=>
1963 V33 N8 P88-90	63-41056 <=>
1963 V33 N8 P112-115	63-31919 <=> 0
1963 V33 N9 P3-8	63-41318 <=>
1963 V33 N9 P9-16	63-20913 <*>
1963 V33 N9 P9-22	63-41293 <=>
1963 V33 N9 P9-32	64-11965 <=> M
1963 V33 N9 P9-16	64-71332 <=> M
1963 V33 N9 P17-22	63-20912 <=*$>
1963 V33 N9 P23-29	63-41224 <=>
1963 V33 N9 P38-41	63-41210 <=>
1963 V33 N9 P88-89	64-21442 <=>

Citation	Translation
1963 V33 N9 P92-94	63-41067 <=>
1963 V33 N1C P3-8	64-21050 <=>
1963 V33 N10 P8-10	64-21050 <=>
1963 V33 N10 P43-46	NLLTB,V6,N3,P250-258 <HMSO>
1963 V33 N1C P58-62	64-21628 <=>
1963 V33 N10 P65-66	AD-620 130 <=*$>
1963 V33 N1C P75-79	64-21127 <=> 0
1963 V33 N10 P80-85	64-21050 <=>
1963 V33 N1C P105-106	AD-619 519 <=$>
1963 V33 N1C P1C9-115	64-21050 <=>
1963 V33 N10 P114-115	64-11790 <=>
1963 V33 N11 P82-85	64-21541 <=>
1963 V33 N11 P132-136	64-31210 <=>
1963 V33 N12 P12-15	TRANS.BULL.1963 P549 <NLL>
1963 V33 N12 P37-41	64-21650 <=>
1963 V33 N12 P62-65	64-21833 <=>
1963 V33 N12 P83-85	64-21650 <=>
1964 N2 P13-16	NLLTB V6 N7 P644 <NLL>
1964 N2 P44-46	NLLTB V6 N7 P654 <NLL>
1964 N3 P13-21	NLLTB V6 N10 P869 <NLL>
1964 N3 P38-60	NLLTB V6 N12 P1061 <NLL>
1964 N3 P70-87	NLLTB V7 N1 P1 <NLL>
1964 N7 P74-77	65-14684 <*>
1964 N7 P78-80	GI N1 1964 P112 <AGI>
1964 N9 P58-61	NLLTB V7 N3 P212 <NLL>
1964 N9 P127-129	GI N2 1964 P339 <AGI>
1964 N10 P33-35	NLLTB V7 N5 P447 <NLL>
1964 N10 P36-42	NLLTB V7 N3 P231 <NLL>
1964 N10 P69-78	NLLTB V7 N4 P325 <NLL>
1964 N11 P2C-27	NLLTB V7 N7 P581 <NLL>
1964 N11 P28-34	NLLTB V7 N6 P507 <NLL>
1964 V34 N1 P16-21	64-31233 <=>
1964 V34 N1 P22-25	64-31014 <=>
1964 V34 N1 P36-38	<SIC>
1964 V34 N1 P39-44	<SIC>
1964 V34 N1 P90-91	64-21813 <=>
1964 V34 N1 P100-102	<SIC>
1964 V34 N1 P105-107	<SIC>
1964 V34 N1 P107-108	64-22167 <SIC>
1964 V34 N1 P108-111	64-22128 <SIC>
1964 V34 N1 P114-115	64-21813 <=>
1964 V34 N2 P3-12	64-31178 <=>
1964 V34 N2 P13-16	TRANS.BULL.1964 P644 <NLL>
1964 V34 N2 P29-32	66-11190 <*>
1964 V34 N2 P33-38	65-60328 <=$>
1964 V34 N2 P39-43	66-11201 <*>
1964 V34 N2 P44-46	AD-620 089 <=$>
	TRANS.BULL.1964 P654 <NLL>
1964 V34 N2 P44-49	64-31212 <=>
1964 V34 N2 P105-107	64-41205 <=>
1964 V34 N3 P3-90	64-31690 <=>
1964 V34 N3 P13-21	TRANS.BULL.1964 P869 <NLL>
1964 V34 N3 P21-38	TRANS.BULL.1964 P955 <NLL>
1964 V34 N3 P38-60	TRANS.BULL1964 P1061 <NLL>
1964 V34 N3 P70-87	TRANS.BULL.1965 P1 <NLL>
1964 V34 N3 P122-123	AD-619 470 <=$>
	64-31690 <=>
1964 V34 N3 P127-130	64-31690 <=>
1964 V34 N3 P141-147	64-31690 <=>
1964 V34 N4 P98-105	AD-611 863 <=$>
1964 V34 N5 P3-18	64-51027 <=>
1964 V34 N5 P19-26	64-41688 <=>
	64-51254 <=$>
1964 V34 N5 P27-37	64-51027 <=>
1964 V34 N5 P47-56	<SIC>
1964 V34 N5 P86-92	AD-625 292 <=*$>
1964 V34 N5 P122-124	64-41972 <=$>
1964 V34 N5 P126-127	AD-621000 <=>
1964 V34 N5 P134-135	AD-621000 <=>
1964 V34 N5 P144-148	64-41334 <=>
1964 V34 N5 P156-163	64-51027 <=>
1964 V34 N6 P21-25	AD-616 310 <=$>
1964 V34 N6 P26-30	64-41335 <=>
1964 V34 N6 P51-57	AD-618 631 <=$>
1964 V34 N6 P87-88	64-41805 <=>
1964 V34 N6 P91-92	64-41805 <=>
1964 V34 N6 P105-106	64-41805 <=>
1964 V34 N7 P3-11	64-41805 <=>
1964 V34 N7 P25-3C	64-51838 <=$>
1964 V34 N7 P36-42	64-41640 <=>
1964 V34 N7 P74-77	AD-616 676 <=$>
1964 V34 N7 P88-92	64-51361 <=$>
1964 V34 N7 P99-100	64-41805 <=>
1964 V34 N7 P1C3-106	64-51309 <=$>
1964 V34 N7 P110-111	64-41993 <=$>
1964 V34 N7 P111-113	64-41805 <=>
1964 V34 N7 P118-121	64-41805 <=>
1964 V34 N8 P3-27	65-30377 <=$>
1964 V34 N8 P39-52	64-51926 <=$>
1964 V34 N8 P65-70	AD-619 320 <=$>
1964 V34 N8 P71-76	AD-625 146 <=*$>
1964 V34 N8 P84-88	64-51655 <=$>
1964 V34 N8 P89-94	65-30377 <=$>
1964 V34 N8 P1C0	65-30377 <=$>
1964 V34 N8 P109-117	65-30377 <=$>
1964 V34 N8 P113-115	65-30518 <=$>
1964 V34 N8 P118-119	65-30518 <=$>
1964 V34 N9 P19-33	64-51956 <=$>
1964 V34 N9 P58-61	65-30012 <=$>
1964 V34 N9 P71-74	AD-619 451 <=$>
1964 V34 N9 P114-115	AD-619 451 <=$>
1964 V34 N10 P33-42	65-30C74 <=$>
1964 V34 N10 P43-59	65-30201 <=$>
1964 V34 N1C P69-78	65-30139 <=$>
1964 V34 N10 P79-81	65-30201 <=$>
1964 V34 N10 P108-111	66-62424 <=*$>
1964 V34 N11 P44-46	AD-620 129 <=$>
1964 V34 N11 P66-77	65-30256 <=$>
1964 V34 N11 P110-117	65-30495 <=$>
1964 V34 N12 P105-106	65-30473 <=$>
1965 N1 P37-41	NLLTB V7 N10 P874 <NLL>
1965 N5 P3-1C	NLLTB V7 N11 P930 <NLL>
1965 N6 P48-53	NLLTB V8 N1 P27 <NLL>
1965 N7 P12-13	SGRT V6 N7 P36 <AGS>
1965 N8 P45-50	NLLTB V8 N4 P292 <NLL>
1965 N10 P21-24	NLLTB V8 N4 P308 <NLL>
1965 N10 P55-57	NLLTB V8 N6 P490 <NLL>
1965 V35 P37-41	65-30906 <=$>
1965 V35 N1 P25-29	65-30906 <=$>
1965 V35 N1 P42-50	65-30934 <=$>
1965 V35 N1 P133-135	65-31046 <=$>
1965 V35 N2 P47-50	65-13904 <*>
1965 V35 N2 P51-57	AD-620 208 <=$*>
	65-30867 <=$>
1965 V35 N2 P62-65	65-30677 <=$>
1965 V35 N2 P66-70	65-30933 <=$>
1965 V35 N3 P5-10	65-30872 <=$>
1965 V35 N3 P125-127	65-31024 <=$>
1965 V35 N3 P128-15C	65-31013 <=$>
1965 V35 N3 P151-159	65-30832 <=*>
1965 V35 N4 P3-4	65-31103 <*>
1965 V35 N4 P42-50,52-64	65-31197 <=$>
1965 V35 N4 P65-68	65-31234 <=$>
1965 V35 N4 P1C9-11C	65-31320 <=$>
1965 V35 N4 P121-127	65-31103 <*>
1965 V35 N5 P3-15	65-31818 <=$>
1965 V35 N5 P27-31	65-31818 <=$>
1965 V35 N5 P39-45	65-17110 <*>
1965 V35 N5 P39-59	65-32832 <=*$>
1965 V35 N5 P46-52	65-17109 <*>
1965 V35 N5 P53-59	65-17113 <*>
1965 V35 N5 P6C-66	65-17108 <*>
1965 V35 N5 P60-65	65-32110 <=$>
1965 V35 N5 P72-76	65-31975 <=$>
1965 V35 N5 P81-88	65-31818 <=$>
1965 V35 N5 P94-95	65-63419 <=$>
1965 V35 N5 P1C6-111	65-31365 <=*$>
1965 V35 N6 P3-20	65-31915 <=$>
1965 V35 N6 P27-33	65-32092 <=*$>
1965 V35 N6 P34-47	65-31858 <=*$>
1965 V35 N6 P94-95	65-14432 <*>
	65-32491 <=$>
1965 V35 N6 P103-105	65-32491 <=$>
1965 V35 N6 P114-117	65-32491 <=$>
1965 V35 N6 P122-123	65-32020 <=*$>
1965 V35 N6 P127-133	65-31772 <=$*>
1965 V35 N7 P3-7	65-32297 <=$*>
1965 V35 N7 P14-25	65-32342 <=$>
1965 V35 N7 P39-41	65-32919 <=*$>

1965 V35 N7 P42-58,94-95 65-32510 <=*$>
1965 V35 N7 P99-102 65-32505 <=*$>
1965 V35 N7 P112-115 65-32297 <=$*>
1965 V35 N8 P3-18 65-32873 <=*$>
66-10482 <*>
1965 V35 N8 P27-34 66-10281 <*>
1965 V35 N8 P45-50 65-32578 <=*$>
1965 V35 N8 P45-60 65-32833 <=*$>
1965 V35 N8 P61-62 65-32690 <=$>
1965 V35 N8 P67-69 65-32879 <=*$>
1965 V35 N8 P79-82 65-33363 <=*$>
1965 V35 N8 P91-92 65-32877 <=*>
1965 V35 N8 P116-120 65-32690 <=$>
1965 V35 N9 P3-21 65-33806 <=*$>
1965 V35 N9 P22-23 66-11187 <*>
1965 V35 N9 P24-31 65-33920 <=*$>
1965 V35 N9 P52-56 65-33576 <=*$>
1965 V35 N9 P76-78 65-33364 <=*$>
1965 V35 N10 P3-9 66-30133 <=*$>
1965 V35 N10 P55-57 66-30133 <=*$>
1965 V35 N1C P58-60 66-30645 <=$>
1965 V35 N12 P42-47 66-31095 <=$> 0
1965 V35 N12 P47-50 66-30703 <=$>
1965 V35 N12 P77-78 66-30857 <=$>
1965 V35 N12 P89-91 66-31095 <=$> 0
1965 V35 N12 P98-101 66-30857 <=$>
1965 V35 N12 P120-122 66-31095 <=$> 0
1965 V35 N12 P124-131 66-31095 <=$> 0
1966 N2 P21-29 NLLTB V8 N1C P803 <NLL>
1966 N6 P33-41 NLLTB V8 N12 P1049 <NLL>
1966 V36 N1 P55-60 66-32846 <=$>
1966 V36 N1 P73-86 66-30942 <=$>
1966 V36 N1 P158-161 66-30942 <=$>

VESTNIK CESKOSLOVENSKE AKADEMIE VED
1959 V68 N2 P234-238 59-31038 <*=>
1959 V68 N2 P253-258 60-31004 <=*>
1959 V68 N2 P265-266 60-31004 <=*>
1959 V68 N2 P282-286 60-31729 <*=>
1963 V72 N1 P118-120 64-21086 <=>

VESTNIK CESKOSLOVENSKE SPOLECNOSTI ZOOLOGICKE
1955 V19 N3 P197-214 62-19870 <=*>
1957 V21 P65-84 C-2626 <NRC>

VESTNIK DALNEVOSTOCHNOGO FILIALA. AKADEMIYA NAUK SSSR
1933 N1/3 P137-139 C-3630 <NRCC>
1933 N1/3 P139-141 C-3629 <NRCC>
1934 N9 P137-139 RT-1045 <*>

VESTNIK DERMATOLOGII I VENEROLOGII
1960 V34 N1 P19-21 64-13301 <=*$>
1960 V34 N1 P82-83 30M41R <ATS>
1960 V34 N1 P84-85 63-20905 <=*$>
1960 V34 N4 P3-8 60-41620 <=>
1960 V34 N4 P32-38 60-31799 <=>
1961 V35 N4 P3-8 62-15954 <*=>
1961 V35 N6 P3-9 62-13027 <*=>
1961 V35 N6 P17-20 63-24424 <=*$>
1961 V35 N7 P49-50 62-20172 <=*>
1961 V35 N9 P8C-83 62-15381 <*=>
1962 V36 N3 P14-19 62-25165 <=*>
1962 V36 N3 P74-77 64-11759 <=> 0
1962 V36 N10 P49-53 64-15484 <=*$>
1962 V36 N10 P63-66 63-13283 <=*>
1963 V37 N6 P39-40 63-41068 <=>
1963 V37 N9 P53-56 AD-619 886 <=*$>
1965 V39 N2 P3-10 65-31078 <=$>
1965 V39 N5 P3-9 65-32200 <=$>

VESTNIK ELEKTROPROMYSHLENNOSTI
1940 N9 P32-34 RT-679 <*>
1954 V25 N2 P17-21 76L34R <ATS>
1956 V27 N6 P3-16 59-14109 <+*>
1956 V27 N6 P66-69 RCT V31 N2 P356 <RCT>
1957 V28 N1 P64-68 59-14141 <+*>
1957 V28 N2 P3-10 63-23940 <=$>
1957 V28 N2 P22-24 63-23939 <=$>
1957 V28 N5 P46-51 59-22493 <+*>
1957 V28 N6 P45-49 65-60358 <=$>
1958 V29 N1 P32-34 59-15866 <+*>
1958 V29 N2 P7-12 59-19164 <+*> 0
1958 V29 N4 P47-49 63-24306 <=*$>
1958 V29 N6 P41-45 61-13363 <+*>
1958 V29 N6 P46-48 63-24305 <=*$>
1958 V29 N6 P62-65 6C-13594 <+*>
1958 V29 N6 P77-80 59-15865 <+*>
1958 V29 N8 P25-28 59-11269 <=> 0
1958 V29 N8 P28-32 61-11102 <=>
1958 V29 N9 P29-34 63-24307 <=*$>
1958 V29 N11 P36-39 63-24304 <=*$>
1958 V29 N11 P78-80 59-11467 <=>
1959 N1 P46-47 NLLTB V1 N8 P36 <NLL>
1959 V30 N4 P6C-64 6C-19885 <+*>
1959 V30 N5 P41-43 3388 <BISI>
1959 V30 N6 P13-16 61-15734 <+*>
1959 V30 N7 P32-35 61-15016 <*+>
1959 V30 N9 P5-8 60-11757 <=>
1959 V30 N9 P62-66 6C-11757 <=>
1959 V30 N10 P1-3 60-11305 <=>
1959 V30 N10 P12-13 60-11305 <=>
1959 V30 N1C P54-59 67N56R <ATS>
1959 V30 N10 P79-80 60-11259 <=>
1959 V3C N11 P1-3 6C-11757 <=>
1959 V30 N11 P4-8 60-51135 <=>
1960 N8 P1-3 62-23483 <*=>
1960 N8 P30 62-23483 <*=>
1960 N8 P34 62-23483 <*=>
1960 N8 P47 62-23483 <*=>
1960 V31 N1 P24-30 63-16708 <=*>
1960 V31 N1 P33-36 63-24065 <=*$>
1960 V31 N1 P5C-55 62-13277 <=*> 0
1960 V31 N2 P46-49 62-25828 <=*>
1960 V31 N5 P1-8 61-27620 <*=>
1960 V31 N5 P76-78 61-11033 <=>
1960 V31 N7 P5-17 62-23483 <*=>
1960 V31 N7 P24-28 E-770 <RIS>
1960 V31 N7 P38-43 62-15823 <*=>
1960 V31 N8 P1-3 62-23483 <*=>
1960 V31 N8 P23-26 2365 <BISI>
1960 V31 N8 P30 62-23483 <*=>
1960 V31 N8 P34 62-23483 <*=>
1960 V31 N8 P36-41 E-773 <RIS>
1960 V31 N8 P41-44 61N54R <ATS>
1960 V31 N8 P47 62-23483 <*=>
1960 V31 N8 P66-68 E-774 <RIS>
1960 V31 N9 P5-8 61-19908 <=*>
1960 V31 N9 P26-29 62-25724 <=*>
63-15219 <=*>
1960 V31 N9 P41-47 E-771 <RIS>
1960 V31 N10 P33-36 64-19928 <=$>
1960 V31 N11 P1-4 61-11964 <=>
1960 V31 N11 P78-80 61-11964 <=>
1961 V32 N1 P25-31 62-19651 <=*>
1961 V32 N1 P34-38 61-28818 <*=>
1961 V32 N1 P55-57 61-27829 <*=> 0
1961 V32 N2 P14-17 63-19757 <=*>
1961 V32 N2 P32-41 62-25313 <=*> C
1961 V32 N3 P64-70 65-20012 <$>
1961 V32 N4 P1-5 62-23476 <=>
1961 V32 N4 P74-76 62-23476 <=>
1961 V32 N5 P54-58 65-60729 <=>
1961 V32 N6 P55-60 62-32472 <=*>
1961 V32 N7 P16-25 63-19761 <=*>
1961 V32 N7 P25-28 63-23388 <=*>
1961 V32 N7 P49-51 62-10721 <*=> 0
1961 V32 N8 P31-34 62-24305 <=*>
1961 V32 N10 P7-12 62-25342 <=*>
1961 V32 N11 P33-37 63-19898 <*=>
1961 V32 N12 P5-10 65-23228 <$>
1961 V32 N12 P11-15 65-23231 <$>
1961 V32 N12 P26-32 63-19894 <*=>
1962 V33 N2 P44-49 63-15174 <=*>
1962 V33 N2 P77-80 63-13091 <=*>
1962 V33 N4 P8-11 62-32050 <=>
1962 V33 N5 P24-26 62-32576 <=>
1962 V33 N5 P56-61 62-32576 <=>
1962 V33 N5 P77-80 62-32880 <=*>
1962 V33 N6 P1-3 62-33124 <=*>

1962 V33 N6 P38-40 63-19866 <*=>
1962 V33 N6 P45-46 65-60713 <=>
1962 V33 N8 P5-7 65-61336 <=$>
1962 V33 N8 P69-71 64-15922 <=*$>
1962 V33 N9 P3-7 63-13486 <=>
1962 V33 N10 P79-80 63-13310 <=>
1962 V33 N11 P1-4 63-13486 <=>
1962 V33 N11 P15-22 63-23195 <=*$>
1962 V33 N12 P1-9 63-21381 <=>
1962 V33 N12 P16-19 64-13017 <=*$>
1962 V33 N12 P32-35 3183 <BISI>
1962 V33 N12 P40-42 63-21381 <=>
1963 N3 P1-3 NLLTB V5 N8 P734 <NLL>
1963 V34 N1 P1-5 63-21339 <=>
1963 V34 N1 P36-38 63-21339 <=>
1963 V34 N2 P1-2 63-21979 <=>
1963 V34 N2 P8-13 63-21979 <=>
1963 V34 N2 P28-31 64-15921 <=*$>
1963 V34 N2 P74-75 AD-610 270 <=$>
NS-71 <TTIS>
63-23924 <=*>
1963 V34 N2 P76-77 63-21979 <=>
1963 V34 N3 P1-3 63-31064 <=>
1963 V34 N3 P29 63-31064 <=>
1963 V34 N3 P50-51 63-31064 <=>
1963 V34 N3 P78 63-31064 <=>
1963 V34 N4 P1-3 62-21940 <=>
1963 V34 N4 P65 62-21940 <=>
1963 V34 N4 P76-78 60R77R <ATS>
1963 V34 N5 P1-3 63-31176 <=>
1963 V34 N5 P66 63-31176 <=>
1963 V34 N6 P37-40 63-31399 <=>
1963 V34 N7 P71-77 63-31855 <=>

VESTNIK ELEKTROTEKHNIKI
1931 N8 P247-255 RT-4063 <*>

VESTNIK. GOSUDARSTVENNOGO UNIVERSITETA. KIEV.
1959 N2 P81-84 78R76U <ATS>
1962 N5 P77-80 77R76U <ATS>

VESTNIK. GOSUDARSTVENNOGO UNIVERSITETA. KIEV. SERIIA FIZIKI I KHIMII
1962 N2 P25-32 26S88U <ATS>

VESTNIK INZHENEROV I TEKHNIKOV
1915 N19 P897-908 63-18924 <=*>
1947 N4 P153-157 RT-194 <*>

VESTNIK IRRIGATSII
1925 V3 N11 P45-48 RT-1125 <*>
1926 N2 P81-85 RT-1430 <*>
1926 N3 P67-71 RT-1429 <*>
1926 V4 N11 P43-45 RT-1126 <*>
1927 V5 N9 P7-21 RT-1127 <*>
1928 V6 N1 P65-69 RT.1128 <*>
1928 V6 N7 P37-38 64-18611 <*>
1929 V7 N3 P46-55 RT-1129 <*>
1930 V8 N1 P74-86 59-17388 <+*>
1930 V8 N2 P6-10 59-17394 <+*>
1930 V8 N2 P11-17 59-17393 <+*>
1930 V8 N2 P17-22 59-17392 <+*>
1930 V8 N2 P26-34 59-17391 <+*>
1930 V8 N2 P59-63 59-17390 <+*>
1930 V8 N2 P64-88 59-17395 <+*>
1930 V8 N2 P89-95 59-17389 <+*>
1930 V8 N6 P60-72 60-16738 <=*$>

VESTNIK ISTORII MIROVOI KULTURY
1961 N6 P45-49 62-23759 <=*>

VESTNIK KHIRURGII
1936 V46 P205-208 R-1051 <*>
1939 V58 N2 P176-179 R-1067 <*>
1940 V59 N6 P592-599 R-1053 <*>
1951 V71 N4 P41-43 R-4816 <*>
64-19529 <=$>
1954 V74 N1 P84-85 RT-1741 <*>
1955 V75 N4 P146-158 RT-3098 <*>
1955 V76 N8 P94-103 62-15116 <=*> 0
1955 V76 N10 P84-100 RT-3912 <*>
59-12972 <+*>
1955 V76 N11 P157-166 RT-4249 <*>
1956 V77 N7 P28-37 59-18706 <+*> 0
1956 V77 N7 P124-128 59-12971 <+*>
1956 V77 N9 P90-95 60-23183 <+*>
1956 V77 N9 P95-99 60-23184 <+*>
1956 V77 N9 P100-113 <CP>
62-23508 <=*>
1957 V78 P56-63 R-4850 <*>
1957 V78 N5 P3-12 <CP>
1957 V78 N5 P12-24 62-23518 <=*> 0
1957 V78 N5 P35-45 <CP>
1957 V79 N7 P129-132 62-18264 <=*> 0
1957 V79 N9 P110-120 62-23519 <=*>
1958 V81 N10 P67-71 60-142266 <+*> 0
62-15296 <*=> 0
1959 N12 P129-130 60-31155 <=>
1959 V83 N10 P101-108 64-13632 <=*$> 0
1959 V83 N11 P135-139 60-11756 <=>
1960 V84 N3 P97-102 60-41177 <=>
1960 V84 N3 P135-136 60-41178 <=>
1960 V84 N4 P85-93 60-41249 <=>
1960 V84 N4 P128-134 61-28900 <=*>
1960 V84 N5 P3-6 60-41481 <=>
1960 V84 N5 P60-63 60-41355 <=>
1960 V84 N5 P64-67 60-41359 <=>
1960 V84 N5 P77-81 60-41360 <=>
1960 V84 N5 P82-83 60-41361 <=>
1960 V84 N5 P84-87 60-41362 <=>
1960 V84 N5 P87-92 60-41363 <=>
1960 V84 N5 P109-112 60-41364 <=>
1960 V84 N6 P48-53 60-31892 <=>
1960 V85 N7 P9-17 61-21055 <=>
1960 V85 N7 P18-21 61-21043 <=>
1960 V85 N7 P82-91 60-21081 <=>
1960 V85 N7 P91-98 61-21108 <=>
1960 V85 N7 P98-101 61-21185 <=>
1960 V85 N7 P101-108 61-21184 <=>
1960 V85 N8 P3-7 61-11850 <=>
1960 V85 N10 P43-46 61-21050 <=>
1960 V85 N10 P47-53 61-11886 <=>
1960 V85 N10 P54-59 61-11922 <=>
1960 V85 N10 P97-101 61-11858 <=>
1960 V85 N10 P104-108 61-21170 <=>
1960 V85 N10 P130-133 61-11911 <=>
1960 V85 N11 P96-98 61-21058 <=>
1960 V85 N11 P99-105 61-21057 <=>
1960 V85 N12 P127-136 61-27358 <*=>
1961 V86 N1 P109-115 61-21658 <=>
1961 V86 N1 P115-119 61-21665 <=>
1961 V86 N2 P100-102 61-31129 <=>
1961 V86 N2 P118-119 61-31130 <=>
1961 V86 N4 P65-68 61-27050 <*=>
1961 V86 N4 P70-74 61-23900 <*=>
1961 V86 N6 P8-12 61-28403 <*=>
1961 V86 N6 P13-19 61-28418 <*=>
1961 V86 N6 P23-27 61-31606 <=>
1961 V86 N6 P74-79 61-31607 <=> 0
1961 V86 N6 P80-83 61-28615 <*=>
1961 V86 N7 P10-14 62-13338 <=*>
1961 V86 N7 P90-97 62-13341 <=*>
1961 V86 N7 P110-111 62-13370 <*=>
1961 V86 N7 P112-115 62-13342 <*=>
1961 V87 N7 P15-18 62-13315 <=*>
1961 V87 N7 P30-34 62-13344 <=*>
1961 V87 N8 P18-26 62-13677 <=>
1961 V87 N8 P53-59 62-13677 <=>
1961 V87 N8 P71-75 62-13677 <=>
1961 V87 N8 P79-85 62-13677 <=>
1961 V87 N11 P3-10 62-23640 <=*>
1961 V87 N11 P11-16 62-23389 <=*>
1961 V87 N11 P16-24 62-23398 <=*>
1961 V87 N11 P24-30 62-23394 <=*>
1961 V87 N11 P38-40 62-23407 <=*>
1961 V87 N11 P111-112 62-19768 <=*>
1961 V87 N12 P17-36 62-23343 <=*>
1961 V87 N12 P77-81 62-23343 <=*>
1962 V88 N1 P3-6 62-24006 <=*>
1962 V88 N1 P84-89 62-24013 <=*>

1962 V88 N2 P6-11 62-24402 <=*>
1962 V88 N3 P73-74 62-25539 <=*>
1962 V88 N3 P78-80 62-25539 <=*>
1962 V88 N4 P35-41 62-32553 <=>
1962 V88 N4 P100-102 62-32533 <=>
1962 V88 N4 P137-140 62-32533 <=>
1962 V88 N5 P72-74 62-32077 <=*>
1962 V88 N5 P116-117 62-32077 <=*>
1962 V88 N6 P48-50 62-32542 <=>
1962 V88 N6 P57-61 62-32542 <=>
1962 V89 N7 P79-84 62-33293 <=*>
1962 V89 N8 P24-42 62-33705 <=*>
1962 V89 N8 P74-76 62-33705 <=*>
1962 V89 N8 P86-92 64-13787 <=*$> C
1962 V89 N8 P112-126 62-33698 <=*>
1962 V89 N9 P63-69 63-13634 <=*>
1962 V89 N9 P147-149 63-13634 <=*>
1962 V89 N10 P34-37 63-13499 <=> O
1962 V89 N10 P53-72 63-13499 <=> O
1962 V89 N11 P41-48 63-21004 <=>
1962 V89 N11 P75-80 63-21004 <=>
1962 V89 N11 P131-134 63-21004 <=>
1962 V89 N12 P3-7 63-21683 <=>
1962 V89 N12 P14-22 63-21683 <=>
1963 V90 N3 P81-87 63-31541 <=>
1963 V90 N4 P73-82 64-41064 <=>
1963 V90 N5 P83-92 63-31682 <=>
1963 V90 N6 P14-18 63-31765 <=>
1963 V91 N8 P129-130 63-41230 <=>
1963 V91 N12 P19-32 64-21839 <=>
1963 V91 N12 P51-64 64-21839 <=>
1963 V91 N12 P79-80 64-21839 <=>
1964 V92 N1 P116-118 64-21784 <=>
1964 V92 N2 P132-133 64-31074 <=>
1964 V92 N3 P96-103 64-31566 <=>
1964 V92 N4 P92-96 64-31423 <=>
1964 V92 N4 P107-11 64-31423 <=>
1964 V92 N5 P37-43 64-41199 <=>
1964 V92 N7 P79-85 64-41753 <=$>
1964 V93 N8 P53-83 64-51886 <=*$>
1965 V94 N1 P152-15 65-30789 <=$>
1965 V94 N2 P114-11 65-30797 <=$>
1965 V95 N10 P3-9 66-30400 <=$>
1965 V95 N11 P3-10 66-30413 <=*>
1965 V95 N11 P125-127 66-30414 <=$>

VESTNIK KHIRURGII I POGRANICHNYKH OBLASTEI
1927 V10 P191 RT-640 <*>

VESTNIK LENINGRADSKOGO GOSUDARSTVENNOGO UNIVERSITETA
1946 V1 N1 P13-18 62-16527 <*>
1947 N6 P3-23 R-5048 <*>
RT-4170 <*>
1947 V2 N1 P17-31 RT-1449 <*> O
1947 V2 N4 P5-11 72M47R <ATS>
1947 V2 N6 P3-23 R-784 <*>
1947 V2 N8 P10-29 60-31279 <=>
1947 V2 N12 P3-12 R-5016 <*>
1947 V2 N12 P3-21 RT-3744 <*>
1948 N1 P12-40 1348TM <CTT>
1948 V3 N1 P12-40 RT-4504 <*>
1952 N12 P53-66 RT-2589 <*>
1952 V7 N4 P154-160 RT-2763 <*>
1952 V7 N9 P137-158 10235 <K-H>
61-10553 <+*> O
1953 V8 N5 P73-81 64-71384 <=> M
1954 N11 P67-93 63-00268 <*>
1954 N11 P119-125 R-1984 <*>
1954 V9 N11 P67-93 63-15665 <=*$>
1954 V9 N11 P119-125 772TM <CTT>
1955 N2 P125-134 R-1398 <*>
R-1682 <*>
1395TM <CTT>
1955 N5 P85-100 R-1397 <*>
1955 V10 N4 P75-83 60-21111 <=>
1955 V10 N4 P117-133 59-10459 <+*> O
1955 V10 N8 P29-41 AMST S2 V22 P43-57 <AMS>
1955 V10 N8 P43-65 AMST S2 V8 P183-207 <AMS>
1955 V10 N8 P97-11 R-1991 <*>
1955 V10 N8 P97- 60-16654 <+*>
1955 V10 N11 P3-21 AMST S2 V22 P59-80 <AMS>
1955 V10 N11 P32-42 AMST S2 V22 P81-94 <AMS>
1955 V10 N11 P113-120 64-71577 <=>
1956 N1 P89-100 59-16783 <+*>
1956 V11 N1 P35-48 STMSP V1 P191-206 <AMS>
1956 V11 N1 P49-52 STMSP V1 P207-211 <AMS>
1956 V11 N1 P155-167 64-71577 <=>
1956 V11 N10 P6-11 14L35R <ATS>
1956 V11 N13 P119-123 61-13533 <=*>
1956 V11 N13 P124-138 61-23843 <*=> O
1956 V11 N16 P57-63 59-22511 <+*>
1956 V11 N19 P5-17 AMST S2 V21 P341-353 <AMS>
1956 V11 N19 P100-113 64-16392 <=*$>
1956 V11 N19 P174-185 61-10719 <+*>
1957 N1 P110-129 AD-633 662 <=$>
1957 N4 P145-152 AD-635 924 <=$>
1957 N10 P40-51 66J18R <ATS>
1957 V12 N1 P130-140 64-16393 <=*$>
1957 V12 N1 P152-157 AD-607 959 <=>
1957 V12 N1 P192-196 64-13118 <=*$>
1957 V12 N4 P79-84 61-28766 <*=>
1957 V12 N4 P118-126 72L35R <ATS>
1957 V12 N7 P15-44 AMST S2 V21 P354-388 <AMS>
1957 V12 N10 P93-105 79TM <CTT>
1957 V12 N10 P148-151 59-20790 <+*>
1957 V12 N13 P129-131 63-10308 <=*>
1957 V12 N16 P41-43 85L33R <ATS>
1957 V12 N16 P85-102 61-15278 <+*>
1957 V12 N18 P68-78 61-15890 <*=> O
1957 V12 N19 P87-97 AD-610 292 <=$>
1957 V12 N22 P103-119 59-22405 <+*>
1958 N2 P172-188 62-23082 <=>
1958 N4 P151-156 60-31877 <=>
1958 N13 P12-15 59-11998 <=>
1958 N19 P187-202 59-11766 <=>
1958 V13 N4 P36-44 80Q70R <ATS>
1958 V13 N4 P53-68 AEC-TR-4561 <*=>
1958 V13 N6 P24-35 61-15889 <*=> O
1958 V13 N7 P14-26 AMST S2 V21 P389-403 <AMS>
1958 V13 N13 P27-34 AMST S2 V21 P403-411 <AMS>
1958 V13 N13 P102-111 60-21215 <=> O
1958 V13 N15 P125-132 62-24361 <=*>
1958 V13 N19 P5-8 AMST S2 V21 P412-416 <AMS>
1958 V13 N19 P9-18 AMST S2 V25 P151-160 <AMS>
1958 V13 N19 P19-38 62-23711 <*=>
1958 V13 N21 P110-124 62-24360 <=*>
1958 V13 N24 P118-119 59-16321 <+*>
1959 N1 P14-23 STMSP V2 P159-169 <AMS>
1959 N1 P111-118 59-11639 <=>
1959 N1 P119-129 59-11640 <=> O
1959 N24 P159-160 SGRT V1 N6 P70 <AGS>
1959 V14 N1 P30-33 61-16546 <*=>
1959 V14 N2 P72-77 60-16769 <+*>
1959 V14 N4 P120-131 60-16770 <+*>
1959 V14 N4 P157-158 60-31145 <=>
1959 V14 N6 P132-136 60-51154 <=>
1959 V14 N6 P139-143 60-51146 <*=>
1959 V14 N6 P144-147 62-24310 <+*>
1959 V14 N6 P147-155 59-13907 <=>
1959 V14 N7 P41-49 STMSP V4 P233-244 <AMS>
1959 V14 N7 P50-54 60-11166 <=>
1959 V14 N9 P119-127 60-51145 <=>
1959 V14 N9 P128-136 60-51147 <=>
1959 V14 N9 P137-141 63-13236 <=*>
1959 V14 N10 P25-26 AEC-TR-3869 <+*>
1959 V14 N10 P27-42 63-23710 <=*$>
1959 V14 N10 P66-71 60-16828 <+*>
1959 V14 N10 P83-86 60-16684 <+*>
1959 V14 N13 P5-19 60-21804 <=>
1959 V14 N16 P19-24 60-13186 <+*>
1959 V14 N16 P25-32 62-11244 <=>
1959 V14 N16 P67-70 60-12926 <=>
60-17464 <+*>
1959 V14 N16 P106-118 62-14471 <=*>
62-23096 <=>
1959 V14 N16 P119-127 61-15389 <+*>
1959 V14 N16 P128-133 B-61 <RIS>
1959 V14 N19 P78-105 STMSP V5 P1-31 <AMS>
1959 V14 N19 P106-112 60-11351 <=>

Citation	Translation
1959 V14 N19 P113-120	60-11372 <=>
1959 V14 N19 P121-129	60-11468 <=>
1959 V14 N22 P136-139	01S88R <ATS>
1959 V14 N22 P146-148	61-19915 <=*>
1959 V14 N24 P11-35	61-15382 <+*>
1960 N1 P100-103	ARSJ V31 N1 P134 <AIAA>
1960 N1 P111-122	ARSJ V30 N9 P853 <AIAA>
1960 N1 P152-158	ARSJ V30 N11 P1051 <AIAA>
1960 N2 P162-166	60-31877 <=>
1960 N7 P116-119	ARSJ V31 N5 P686 <AIAA>
1960 N7 P120-153	ARSJ V31 N1 P102 <AIAA>
1960 N7 P154-163	ARSJ V32 N3 P490 <AIAA>
1960 V15 N1 P55-69	STMSP V5 P33-48 <AMS>
1960 V15 N1 P85-99	STMSP V4 P153-170 <AMS>
1960 V15 N3 P90-98	63-19971 <=*$>
1960 V15 N4 P80-87	61-18279 <*=>
	62-23173 <=>
1960 V15 N4 P88-94	61-18621 <*=>
	62-23172 <=>
1960 V15 N4 P106-116	AEC-TR-5176 <=*>
1960 V15 N6 P153-158	61-15228 <=*>
1960 V15 N7 P81-115	STMSP V5 P49-88 <AMS>
1960 V15 N7 P120-153	61-16547 <*=>
1960 V15 N9 P29-36	61-19646 <=*> 0
1960 V15 N9 P147-149	61-11042 <=>
1960 V15 N9 P150-151	61-11004 <=>
1960 V15 N10 P34-44	62-11599 <=>
1960 V15 N10 P104-111	64-13148 <=*$>
1960 V15 N16 P42-66	63-11508 <=>
1960 V15 N16 P102-108	35S86R <ATS>
1960 V15 N16 P113-119	63-13400 <=*>
1960 V15 N17 P76-86	61-21753 <=>
1960 V15 N19 P95-99	61-28708 <*=>
1960 V15 N21 P101-112	63-19039 <=*>
1960 V15 N21 P171-173	61-21141 <=>
1960 V15 N22 P40-48	61-21691 <=>
1960 V15 N22 P126-129	64-19922 <=>
1960 V15 N24 P118-130	61-21606 <=>
1960 V15 N24 P131-138	61-23467 <=*>
1960 V15 N24 P139-143	63-19901 <=*>
1961 N1 P133-143	AIAAJ V1 N6 P1486 <AIAA>
1961 N7 P75-80	AIAAJ V1 N10 P2451 <AIAA>
1961 N7 P81-86	AIAAJ V1 N10 P2454 <AIAA>
1961 N18 P42-55	IGR V5 N2 P157 <AGI>
1961 N22 P135-138	RCT V35 N4 P1063 <RCT>
1961 V16 N4 P86-96	63-18919 <*>
1961 V16 N4 P87-99	63-13032 <=*>
1961 V16 N4 P100-104	63-13401 <=*>
1961 V16 N4 P153-155	62-23858 <=*>
1961 V16 N5 P137-140	62-15770 <=*>
1961 V16 N6 P19-32	61-27039 <*=>
1961 V16 N7 P51-61	STMSP V5 P179-190 <AMS>
1961 V16 N7 P136-141	63-10310 <=*>
1961 V16 N9 P135-136	62-13452 <*=> 0
1961 V16 N10 P13-30	61-20722 <*=>
1961 V16 N10 P70-72	63-13402 <=*>
1961 V16 N10 P112-124	62-19923 <*=>
1961 V16 N10 P156-158	62-19923 <*=>
1961 V16 N16 P66-76	62-15813 <=*>
1961 V16 N16 P143-147	62-10649 <=*>
1961 V16 N19 P122-137	62-32424 <=*>
1961 V16 N21 P156-159	62-23857 <=*>
1961 V16 N22 P86-96	64-13232 <=*$>
1962 N3 P47-65	RTS-3025 <NLL>
1962 N4 P52-55	65-13663 <*>
1962 N4 P83-89	SGRT V7 N7 P40 <AGS>
1962 N6 P36-46	IGR V6 N3 P450 <AGI>
1962 N6 P153-157	SGRT V3 N7 P16 <AGS>
1962 N13 P5-21	64-00374 <*>
1962 V17 N1 P80-88	STMSP V5 P240-250 <AMS>
1962 V17 N1 P116-123	62-33621 <=*>
1962 V17 N3 P21-23	64-11510 <=>
1962 V17 N3 P124-133	62-25482 <=*>
1962 V17 N4 P75-89	65-60368 <=$>
1962 V17 N4 P128-134	62-20164 <=*>
1962 V17 N4 P148-153	4509 <BISI>
1962 V17 N6 P19-34	63-24344 <=*$>
1962 V17 N6 P36-46	IGR,V6,N3,P450-458 <AGI>
1962 V17 N6 P73-79	63-11767 <=>
1962 V17 N7 P5-13	AMST S2 V48 P255-265 <AMS>
1962 V17 N7 P58-61	62-20258 <=*>
1962 V17 N7 P62-74	63-23440 <=*>
1962 V17 N7 P87-116	63-23440 <=*>
1962 V17 N10 P103-106	63-10250 <*=>
1962 V17 N11 P26-40	62-32561 <=*>
1962 V17 N11 P141-144	62-32559 <=*>
	63-15353 <=*>
1962 V17 N11 P152-158	63-18876 <=*>
1962 V17 N12 P116-120	63-15949 <=*>
1962 V17 N13 P5-21	64-11743 <=>
1962 V17 N13 P103-112	64-13074 <=*$>
1962 V17 N18 P109-112	63-18918 <=*>
1962 V17 N18 P120-124	65-61357 <=$> C
1962 V17 N19 P62-78	64-10798 <=*$>
1962 V17 N21 P86-93	65-60928 <=$>
1962 V17 N23 P35-43	63-21247 <=>
1962 V17 N24 P5-16	63-24345 <=*$>
1963 N2 P45-51	66-13764 <*>
1963 N2 P62-71	IGR V7 N2 P225 <AGI>
1963 N3 P88-	FPTS V24 N2 P.T351 <FASE>
1963 N3 P90-	FPTS V23 N3 P.T609 <FASE>
1963 N3 P121-	FPTS V23 N2 P.T337 <FASE>
1963 N3 P136	AD-638 172 <=$>
1963 N10 P91-95	ICE V4 N4 P660-663 <ICE>
1963 N12 P57-61	IGR V7 N2 P202 <AGI>
1963 V4 N21 P155-	FPTS V23 N6 P.T1185 <FASE>
1963 V18 N1 P79-85	63-23731 <=*>
1963 V18 N1 P86-89	65-63894 <=>
1963 V18 N1 P109-115	64-13861 <=*$>
1963 V18 N2 P26-30	64-11532 <=>
1963 V18 N3 P90-	FPTS,V23,N3,P.T609 <FASE>
1963 V18 N3 P121-	FPTS,V23,P.T337-T339 <FASE>
1963 V18 N3 P139-143	64-13672 <=*$>
	64-13818 <=*$>
1963 V18 N4 P93-98	64-27242 <$>
1963 V18 N5 P25-34	63-41272 <=>
1963 V18 N5 P28-33	64-21229 <=>
1963 V18 N9 P73-	FPTS V23 N4 P.T729 <FASE>
1963 V18 N9 P73-80	FPTS V23 N4 P.T729 <FASE>
1963 V18 N9 P150-156	63-31511 <=>
1963 V18 N9 P159-160	63-31511 <=>
1963 V18 N10 P91-95	ICE V4 N4 P660-663 <ICE>
	74R75R <ATS>
1963 V18 N10 P101-107	64-11921 <=> M
	64-16838 <=*$>
1963 V18 N13 P1-152	AD-623 642 <=*$> M
1963 V18 N19 P35-48	AD-609 656 <=>
1963 V18 N19 P85-91	64-14838 <=*$>
1963 V18 N21 P57-64	FPTS V23 P.T1299 <FASE>
1963 V18 N21 P65-71	64-19671 <=$>
1963 V18 N21 P72-85	64-11786 <=>
1963 V18 N21 P148-150	FPTS V23 P.T1303 <FASE>
1963 V18 N21 P155-159	FPTS V23 P.T1185 <FASE>
1963 V18 N22 P135-139	IS-TRANS-3 <=$>
1963 V18 N24 P147-153	64-21579 <=>
1964 N1 P47-55	<JBS>
1964 N1 P116-122	65-64541 <=*$>
1964 N1 P124-129	65-63068 <=$>
1964 N1 P130-141	65-12639 <*>
1964 N3 P103-109	65-13154 <*>
1964 N3 P133-143	65-13153 <*>
1964 N3 P159-162	65-13155 <*>
1964 N3 P162-167	65-13152 <*>
1964 N4 P86-97	65-12645 <*>
1964 N4 P109-114	65-13151 <*>
1964 N4 P114-128	65-12713 <*>
1964 N4 P129-134	N65-15059 <=$>
1964 N4 P153-164	N65-14618 <=$>
1964 N24 P78-88	SGRT V7 N2 P14 <AGS>
1964 N24 P153-157	65-30888 <=$>
1964 V19 N1 P26-35	64-31543 <=>
1964 V19 N1 P76-87	64-31543 <=>
1964 V19 N1 P130-141	64-30076 <=>
1964 V19 N1 P165-167	M.5754 <NLL>
1964 V19 N3 P76-81	65-62327 <=$>
1964 V19 N3 P88-99	FPTS V24 P.T351-T356 <FASE>
1964 V19 N3 P160-170	65-62106 <=>
1964 V19 N3 P162-167	65-63196 <=$>
1964 V19 N4 P34-	FPTS V24 N6 P.T1105 <FASE>
1964 V19 N4 P160-164	65-63992 <=>

1964 V19 N5 P121-124	64-41093 <=>
1964 V19 N6 P155-159	<JBS>
1964 V19 N6 P161-162	65-63409 <=$>
1964 V19 N7 P155-161	N65-32972 <=$>
1964 V19 N9 P73-79	65-62331 <=$>
1964 V19 N13 P103-109	65-63176 <=$>
1964 V19 N13 P133-146	65-62972 <=$>
1964 V19 N13 P159-162	65-63178 <=$>
1964 V19 N22 P94-101	01S85R <ATS>
1964 V19 N24 P69-77	65-30404 <=$>
1964 V19 N24 P158-160	65-63411 <=$>
1965 N1 P73-	FPTS V25 N1 P.T161 <FASE>
1965 N1 P102-109	N65-21004 <=$>
1965 N1 P110-120	65-13150 <*>
1965 N1 P120-130	66-12314 <*>
1965 N1 P137-140	66-11453 <*>
1965 N1 P159-160	65-31036 <=$>
1965 N2 P113-115	65-32581 <=*$>
1965 N2 P130-	FPTS V25 N4 P.T725 <FASE>
1965 N2 P140-150	65-32582 <=*$>
1965 N3 P105-	FPTS V25 N5 P.T874 <FASE>
1965 N3 P125-127	AD-630 867 <=$>
1965 N3 P138-139	AD-631 013 <=$>
1965 N4 P112-119	66-12489 <*>
	66-61037 <*=$>
1965 N4 P140-142	N66-13300 <=$>
1965 N4 P143-146	66-12490 <*>
1965 N12 P95-103	SGRT V6 N9 P48 <AGS>
1965 N18 P112-120	SGRT V7 N2 P27 <AGS>
1965 N19 P147-156	66-61712 <=$>
1965 V20 N1 P77-83	AD-640 487 <=$>
1965 V20 N1 P110-120	65-63679 <=*$>
1965 V20 N2 P118-126	65-14566 <*>
1965 V20 N3 P138-142	66-60296 <=*$>
1965 V20 N4 P115-119	90S85R <ATS>
1965 V20 N4 P120-127	91S85R <ATS>
1965 V20 N6 P56-63	65-31026 <=$>
1966 N1 P154-159	66-12321 <*>
	66-61978 <=*$>
1966 N1 P172-173	AD-639 907 <=$>
1966 N6 P80-87	SGRT V7 N7 P56 <AGS>
1966 V21 N1 P120-130	66-61987 <=*$>
1966 V21 N7 P76-80	66-13422 <*>
	66-61699 <=$>

★VESTNIK MASHINOSTROENIYA

1947 V27 N2 P33-38	2691 <HB>
1947 V27 N8 P35-43	60-00049 <*>
1948 V28 N5 P43-48	62-24202 <=*>
1949 V28 N6 P49-52	TS 3477 <BISI>
1950 V30 N15 P17-18	2430 <BISI>
1952 V32 N1 P44-46	2957 <HB>
1952 V32 N2 P37-40	2953 <HB>
1952 V32 N3 P69-71	3046 <HB>
1952 V32 N3 P76-79	2968 <HB>
1952 V32 N3 P81-92	3047 <HB>
1952 V32 N4 P39-44	63-18961 <=*$>
1952 V32 N5 P53-59	3559 <HB>
1952 V32 N7 P38	4933 <HB>
1952 V32 N7 P48-54	3530 <HB>
1952 V32 N7 P54-57	3450 <HB>
1952 V32 N8 P63-65	3187 <HB>
1952 V32 N9 P60-61	2977 <HB>
1952 V32 N11 P44-49	3284 <HB>
1952 V32 N11 P49-52	3090 <HB>
1952 V32 N11 P59-61	3616 <HB>
1953 V33 N3 P37-39	3307 <HB>
1953 V33 N5 P43-46	3122 <HB>
1953 V33 N5 P59-61	3158 <HB>
1953 V33 N6 P65-67	3134 <HB>
	64-18385 <*>
1953 V33 N7 P67-70	3310 <HB>
1953 V33 N8 P59-66	RT-4096 <*>
1953 V33 N8 P74-79	61-11160 <=>
1953 V33 N9 P59-61	3296 <HB>
1953 V33 N9 P61-65	AEC-TR-4214 <+*>
1953 V33 N11 P54-55	3505 <HB>
1953 V33 N12 P69-70	3316 <HB>
1953 V33 N12 P75-76	3306 <HB>
1954 N8 P16-22	RT-4354 <*>
1954 V34 P18-26	1K25R <ATS>
1954 V34 N1 P12-18	RT-3336 <*>
1954 V34 N1 P79-82	R-1231 <*>
1954 V34 N2 P28-29	R-3141 <*>
1954 V34 N3 P14-20	31TM <CTT>
1954 V34 N3 P60-63	62-20157 <=*>
1954 V34 N4 P79-82	RT-1732 <*>
1954 V34 N5 P18-26	60-15768 <+*>
1954 V34 N5 P64-68	3482 <HB>
1954 V34 N6 P56-60	3468 <HB>
1954 V34 N7 P18-22	3451 <HB>
1954 V34 N7 P45-47	3753 <HB>
1954 V34 N8 P53-55	R-188 <*>
1954 V34 N8 P71-74	R-172 <*>
1954 V34 N8 P85-86	3495 <HB>
1954 V34 N9 P3-9	62-10528 <*=> 0
1954 V34 N11 P56-58	3466 <HB>
1954 V34 N12 P3-6	62-18746 <=*>
1954 V34 N12 P51-55	3496 <HB>
1954 V34 N12 P56-57	3497 <HB>
1955 N2 P80-83	RT-4329 <*>
1955 N5 P67-71	3955 <HB>
1955 V34 N7 P65-70	3640 <HB>
1955 V35 N1 P3-13	RT-2771 <*>
1955 V35 N2 P51-55	3607 <HB>
1955 V35 N2 P55-57	3506 <HB>
1955 V35 N2 P58-67	3560 <HB>
1955 V35 N3 P38-40	3713 <HB>
1955 V35 N3 P49-53	1262 <BISI>
	62-24208 <=*$>
1955 V35 N3 P53-56	RT-3446 <*>
1955 V35 N4 P19-26	R-3682 <*>
1955 V35 N4 P48-51	3630 <HB>
1955 V35 N5 P62-66	4211 <HB>
1955 V35 N6 P30-34	59-22423 <+*> C
1955 V35 N6 P48-53	R-2534 <*>
1955 V35 N7 P47-51	63-41172 <=>
1955 V35 N7 P70-74	3651 <HB>
1955 V35 N7 P75-77	3714 <HB>
1955 V35 N9 P41-45	64-14422 <=*$>
1955 V35 N10 P37-41	4824 <HB>
1956 N11 P47-50	4024 <HB>
1956 V36 N2 P60-64	3863 <HB>
	59-12378 <+*>
1956 V36 N2 P66-68	R-4338 <*>
	59-12378 <+*>
1956 V36 N2 P77-86	59-12378 <+*>
1956 V36 N3 P85-88	60-13295 <+*>
1956 V36 N5 P12-15	60-15748 <+*> C
1956 V36 N5 P43-46	59-11994 <=> 0
1956 V36 N6 P48-51	59-10351 <+*> C
1956 V36 N6 P51-53	59-10304 <+*>
1956 V36 N7 P23-27	3986 <*>
1956 V36 N7 P42-45	59-14179 <+*>
1956 V36 N7 P63-64	3904 <HB>
	4932 <HB>
1956 V36 N7 P89-90	60-13293 <+*>
1956 V36 N8 P58-59	59-12397 <+*>
1956 V36 N8 P71-79	59-12397 <+*>
1956 V36 N9 P65-67	3873 <HB>
1956 V36 N10 P25-26	R-3884 <*>
	4037 <HB>
1956 V36 N10 P56-58	4296 <HB>
1956 V36 N12 P3-13	4238 <HB>
	62-15689 <*=>
1956 V36 N12 P16-18	3962 <HB>
1957 V37 N2 P3-8	61-13618 <*+>
1957 V37 N2 P61-62	3932 <HB>
1957 V37 N4 P20-28	N65-29729 <=*$>
1957 V37 N4 P41-47	59-15781 <+*>
1957 V37 N5 P59-61	4105 <HB>
1957 V37 N5 P62-65	4084 <HB>
1957 V37 N6 P57-62	63-19105 <=*>
1957 V37 N8 P70-72	4809 <HB>
1957 V37 N8 P84-88	60-17439 <+*>
1957 V37 N9 P20-24	59-11989 <=> C
1957 V37 N9 P54-56	4441 <HB>
1957 V37 N9 P72-74	59-16067 <+*>
1957 V37 N11 P47-55	59-14299 <+*> C
1957 V37 N12 P7-14	59-14298 <+*> 0

1957 V37 N12 P14-18 59-12414 <+*>
1957 V37 N12 P28-33 59-12679 <+*>
1957 V37 N12 P71-75 59-15782 <+*>
1958 N1 P42-47 NLLTB V1 N11 P47 <NLL>
1958 N3 P77-81 NLLTB V2 N3 P212 <NLL>
1958 N3 P82-84 NLLTB V1 N12 P44 <NLL>
1958 N5 P42-47 NLLTB V2 N6 P490 <NLL>
1958 N5 P54-57 NLLTB V2 N11 P976 <NLL>
1958 N6 P3-5 NLLTB V2 N7 P590 <NLL>
1958 N6 P33 NLLTB V2 N11 P974 <NLL>
1958 N6 P64-67 NLLTB V2 N11 P985 <NLL>
1958 N8 P3-9 60-23797 <+*>
1958 N11 P25-30 NLLTB V1 N3 P34 <NLL>
1958 N11 P31-35 NLLTB V1 N4 P42 <NLL>
1958 N11 P36-39 NLLTB V1 N3 P41 <NLL>
1958 N11 P40-46 NLLTB V1 N2 P5 <NLL>
1958 N11 P51-58 NLLTB V1 N5 P40 <NLL>
1958 N12 P43-46 NLLTB V2 N7 P596 <NLL>
1958 V38 N1 P27-30 60-23558 <+*>
1958 V38 N2 P6-10 59-19313 <+*>
1958 V38 N2 P10-16 N65-29730 <=*$>
1958 V38 N2 P16-19 65-60671 <=$>
1958 V38 N2 P40-42 61-13239 <*+>
1958 V38 N2 P55-56 4556 <HB>
1958 V38 N2 P56-58 4212 <HB>
1958 V38 N2 P61-63 61-13218 <*+>
1958 V38 N3 P3-8 60-15658 <+*>
1958 V38 N3 P8-11 60-15431 <+*>
1958 V38 N3 P12-13 60-15432 <+*>
1958 V38 N3 P24-27 60-23793 <+*>
1958 V38 N3 P27-28 95K27R <ATS>
1958 V38 N3 P29-31 61-13219 <*+>
1958 V38 N3 P38-39 96K27R <ATS>
1958 V38 N3 P55-56 60-23794 <+*>
1958 V38 N3 P82-84 35K25R <ATS>
1958 V38 N4 P3-11 59-22461 <+*>
1958 V38 N4 P59-62 60-23795 <+*>
1958 V38 N5 P11-19 59-22462 <+*>
1958 V38 N5 P65-69 60-23994 <*+>
1958 V38 N6 P11-16 60-13562 <+*>
1958 V38 N6 P22-25 61-13220 <*+>
1958 V38 N6 P26-28 60-23796 <+*>
1958 V38 N6 P49-52 63-19103 <=*>
1958 V38 N6 P54-55 61-23029 <=*>
1958 V38 N6 P63-64 63-19226 <=*>
1958 V38 N8 P27-29 61-13221 <*+>
1958 V38 N8 P32-38 NLLTB V2 N5 P370 <NLL>
1958 V38 N9 P14-17 64-19926 <=$>
1958 V38 N9 P34-35 60-23798 <+*>
1958 V38 N10 P3-9 60-13563 <+*>
1958 V38 N10 P10-18 61-13222 <*+>
1958 V38 N10 P23-30 2394 <BISI>
60-23799 <+*>
1958 V38 N11 P11-15 60-13834 <+*>
1958 V38 N11 P16-24 63-23383 <=*>
1958 V38 N11 P40-46 59-22634 <+*>
1958 V38 N11 P92-95 60-23756 <+*>
1958 V38 N12 P70-72 60-13651 <+*>
1959 V39 N1 P47-52 60-19976 <+*> 0
1959 V39 N2 P65-69 61-13038 <*=>
1959 V39 N3 P14-19 60-13786 <+*>
1959 V39 N5 P16-19 61-13559 <+*>
1959 V39 N5 P53-56 63-21442 <=>
1959 V39 N5 P61-65 61-31477 <=>
1959 V39 N6 P44-47 NLLTB V2 N4 P286 <NLL>
1959 V39 N8 P10- 61-13557 <+*>
1959 V39 N9 P55-59 60-23607 <+*> 0
1959 V39 N11 P3-9 60-21730 <=> 0
1959 V39 N11 P47-49 TS 1638 <BISI>
1959 V39 N11 P47-50 62-14679 <=*>
1960 N4 P80-82 NLLTB V3 N2 P86 <NLL>
1960 V40 N2 P73-76 61-15065 <+*> 0
1960 V40 N3 P80-81 60-41075 <=>
1960 V40 N4 P33-39 61-27642 <*=>
1960 V40 N5 P35-36 61-14985 <=*> 0
1960 V40 N6 P46-49 61-27638 <*=>
1960 V40 N7 P41-45 62-25947 <=*>
1960 V40 N8 P7-17 61-21343 <=>
1961 V41 N2 P37-45 62-13174 <*=> 0
1961 V41 N9 P39-43 62-33589 <=>
1961 V41 N9 P63-64 5370 <HB>
1962 V42 N1 P61-62 2901 <BISI>
1962 V42 N2 P22-26 64-13226 <=*$>
1962 V42 N2 P26-28 64-13227 <=*$>
1962 V42 N2 P37-40 5566 <HB>
1962 V42 N5 P3-8 62-33018 <=>
62-33018 <=*>
1962 V42 N7 P38-41 63-15354 <=*>
1962 V42 N8 P44-47 63-13099 <=*>
1962 V42 N9 P34-36 62-23360 <=*>
1962 V42 N9 P53-54 62-23360 <=*>
1962 V42 N10 P20-23 64-13011 <=*$>
1962 V42 N12 P65-66 5887 <HB>
1963 V43 N1 P32-34 07R76R <ATS>
1963 V43 N2 P72-74 6147 <HB>
1963 V43 N5 P3-8 63-31419 <=>
1963 V43 N5 P22-25 6099 <HB>
1963 V43 N8 P74-76 6119 <HB>
1963 V43 N10 P84-86 6255 <HB>
1963 V43 N10 P86-88 64-41156 <=>
1963 V43 N11 P3-6 64-21345 <=>
1963 V43 N11 P21-22 AD-630 688 <=*$>
1964 V44 N9 P3-8 64-51855 <=$>
1964 V44 N10 P3-8 65-30140 <=$>
1964 V44 N11 P3-6 65-30588 <=$>
1964 V44 N12 P77-78 65-30615 <=$>
1965 V45 N1 P69-74 65-32609 <=$>
1965 V45 N4 P52-54 M.5838 <NLL>
1965 V45 N8 P3-8 65-33190 <=*$>
1965 V45 N11 P78-79 66-30299 <=$>
1966 V46 N4 P74-78 66-32329 <=$>
1967 N2 P3-6 67-31492 <=$>

VESTNIK METALLOPROMYSHLENNOSTI
1936 V16 P75-82 R-911 <*>
1936 V16 N6 P75-82 RT-3124 <*>
1938 N7 P91-107 60-18956 <=*> 0
1938 N8/9 P59-109 60-18956 <=*> 0
1940 N3 P50-60 61-17202 <=$>
62-23970 <=*>
1940 N4/5 P123-129 61-17183 <=$>
1940 V20 N10 P10-12 64-18374 <*>

VESTNIK MIKROBIOLOGII I EPIDEMIOLOGII I PARAZITOLOGII
1930 V9 N1 P54-59 RT-793 <*>
1935 V14 N2 P183-197 65-62998 <=>
1936 V15 P334-339 60-19707 <=*$>
1936 V15 N3/4 P461-470 61-15653 <+*>
1939 V17 N1/2 P20-27 65-60104 <=$>
1939 V17 N1/2 P43-55 64-23636 <=$>
1940 V19 P450-468 61-19726 <=*>
1940 V19 P538-544 R-1088 <*>

VESTNIK MINISTERSTVA SKOLSTVI A KULTURY. CZECHOSLOVAKIA
1962 V18 N11 P156-171 62-32363 <=*>

VESTNIK MINISTERSTVA ZDRAVOTNICTUI REPUBLIKY CESKOSLAVENSKE
1959 V6 N13 P117-128 59-31013 <=*$>
1961 V9 SECT.1/2,P1-4 62-19368 <*=>
1961 V9 N7 P70-79 62-19534 <=*>
1961 V9 N16 P169-174 62-19551 <=*>
1961 V9 N14/5 P150-158 62-19551 <=*>
1962 V10 N11/3 P101-129 62-32859 <=>
1963 V11 N16/7 P173-195 63-41179 <=>
1965 V13 N11/3 P125-149 65-33002 <=*$>

VESTNIK MOSKOSKOGO GOSUDARSTVENNOGO UNIVERSITETA
1946 V1 P37-55 RT-1147 <*>
1947 V2 N3 P65-83 64-10281 <=*$>
1948 N10 P81-91 RT-279 <*>
1948 V3 N5 P25-37 AMST S1 V11 P356-374 <AMS>
1949 N3 P35-58 R-4402 <*>
1956 V11 N1 P13-16 UCRL-TRANS-875(L) <=*>
1962 N4 P3-12 IGR V6 N12 P2093 <AGI>
1963 N5 P12-16 ICE V4 N2 P329-332 <ICE>
1963 N5 P57-61 ICE V4 N2 P272-274 <ICE>
1964 N1 P20-29 ICE V4 N3 P440-446 <ICE>

1964 N3 P9-13 ICE V4 N4 P685-688 <ICE>
1964 N3 P14-18 ICE V4 N4 P667-669 <ICE>
1964 N4 P28-41 SGRT V5 N8 P50 <AGS>
1964 N6 P17-23 N65-27705 <=*$>
1964 V19 N3 P3-6 AD-626 253 <=*$>
1964 V19 N3 P87-88 RTS-2822 <NLL>
1964 V19 N6 P86-87 65-63568 <=>
1964 V19 N6 P87-88 65-63660 <=$>
1964 V19 N6 P88 65-63661 <=$>
1965 N4 P7-9 ICE V6 N1 P113-114 <ICE>
1965 N4 P34-38 ICE V6 N1 P115-117 <ICE>
1966 N2 P130-135 66-33189 <=$>
1966 N3 P15-24 SGRT V7 N10 P52 <AGS>

VESTNIK MOSKOVSKOGO GOSUDARSTVENNOGO UNIVERSITETA. SERIYA BIOLOGII, POCHVOVEDENIIA, GEOLOGII, GEOGRAFII
1957 N1 P25-35 R-3275 <*>
1957 N1 P39-47 R-3476 <*>
1957 N3 P77-85 R-4203 <*>
1957 V12 N1 P3-17 R-3274 <*>
1957 V12 N1 P17-24 R-3276 <*>
1957 V12 N1 P97-104 R-4437 <*>
1957 V12 N2 P67-73 61-19192 <+*> 0
1959 V14 N1 P45-48 AEC-TR-4004 <+*>
1959 V14 N2 P31-38 C-3386 <NRCC>
1959 V14 N4 P3-12 60-11932 <=>

VESTNIK MOSKOVSKOGO GOSUDARSTVENNOGO UNIVERSITETA. SERIYA FIZIKO-MATEMATICHESIKH I ESTESTVENNYKH NAUK
1950 V5 N8 P41-42 RT-2596 <*>
1951 V6 N8 P61-67 AEC-TR-5642 <=*>
1952 N3 P31-40 R-1399 <*>
1952 N8 P19-39 R-617 <*>
1952 V7 N9 P37-48 61-18817 <*=>
1953 N8 P171-183 RT-4246 <*>
1953 V8 N3 P95-100 RT-1368 <*>
61-27291 <=*>
1953 V8 N3 P119-123 R-780 <*>
64-15092 <=*$>
1953 V8 N8 P79-86 RT-3044 <*>
1953 V8 N9 P109-114 61-10145 <*+>
1953 V8 N12 P142-144 RT-2477 <*>
1954 N1 P17-32 R-3728 <*>
1954 N3 P55-59 19K21R <ATS>
1954 N3 P69-72 RT-4495 <*>
1954 N3 P73-76 RT-4492 <*>
1954 N5 P47-53 RT-4493 <*>
1954 N9 P115-118 R-3893 <*>
1954 N12 P57-61 R-253 <*>
1353TM <CTT>
1954 V9 N2 P69-72 62-18540 <=*>
1954 V9 N3 P85-89 PT.1 NP-TR-879 PT.1 <*>
1954 V9 N3 P85-89 63-15652-1 <=*>
1954 V9 N3 P113-118 R-4131 <*>
1954 V9 N5 P65-69 TT.558 <NRC>
1954 V9 N6 P33-40 62-24771 <=*>
1954 V9 N9 P109-113 PT.2 NP-TR-879 PT.2 <*>
1954 V9 N9 P109-113 63-15652-2 <=*>
1955 N2 P33-40 RT-4494 <*>
1955 V10 N8 P67-69 61-10144 <*+>
1955 V10 N8 P81-86 61-15402 <=*>
1955 V10 N10 P47-57 AMST S2 V29 P255-270 <AMS>
1955 V10 N12 P61-67 63-10742 <=*>
1955 V10 N12 P119-120 59-15900 <+*>
1955 V10 N4/5 P101-113 R-1237 <*>
146TM <CTT>
1955 V10 N4/5 P115-124 AEC-TR-3487 <=>

VESTNIK MOSKOVSKOGO GOSUDARSTVENNOGO UNIVERSITETA. SERIYA MATEMATIKI, MEKHANIKI, ASTRONOMII, FIZIKI I KHIMII
1956 N2 P21-28 R-2324 <*>
1956 N2 P45-48 62-15849 <=*>
1956 V11 N1 P187-189 62-18972 <=*>
1956 V11 N1 P197-199 R-4232 <*>
1956 V11 N2 P3-12 AMST S2 V18 P187-197 <AMS>
1956 V11 N2 P71-76 61-10286 <*+> 0
1956 V11 N2 P185-188 60-16522 <*+>
1956 V11 N5 P55- R-430 <*>
1956 V11 N6 P73-79 62-25936 <=*>
1957 N2 P131-138 R-3227 <*>
1957 N3 P65-73 59-10186 <+*>
1957 N4 P29-40 59-11255 <=>
1957 N4 P125-128 59-11255 <=>
1957 N4 P251-255 PB 141 284T <=>
1957 V12 N2 P45-50 61-11091 <=>
1957 V12 N2 P195-19 66-11452 <*>
1957 V12 N3 P23-29 60-41695 <=>
1957 V12 N3 P145-147 62-18974 <=*>
1957 V12 N4 P9-16 <EEUP>
1957 V12 N4 P17-27 60-21437 <=> 0
1957 V12 N4 P29-40 59-11255 <=>
1957 V12 N4 P49-59 60-21455 <=> C
1957 V12 N4 P125-128 59-11255 <=>
1957 V12 N4 P129-136 60-21437 <=> 0
61-23710 <=*>
1957 V12 N5 P33-39 60-21456 <=>
1957 V12 N5 P181-186 65-14914 <*>
1957 V12 N6 P35-50 61-15099 <=*>
1957 V12 N6 P57-67 65-63678 <=$>
1957 V12 N6 P87-98 59-12457 <+*>
1957 V12 N6 P99-106 61-23185 <*=>
1957 V12 N6 P147-172 RTS-2726 <=$*>
1957 V12 N6 P199-214 60-11725 <=>
1958 V13 N1 P55-68 62-33649 <=*>
1958 V13 N1 P87-95 63-10377 <=*>
1958 V13 N1 P107-114 AD-640 972 <=$>
1958 V13 N1 P133-136 TRANS-136 <FRI>
1958 V13 N1 P215-222 21N52R <ATS>
1958 V13 N2 P3-8 60-15994 <+*>
1958 V13 N2 P67-75 63-10378 <=*>
1958 V13 N2 P93-102 60-10695 <+*>
1958 V13 N2 P137-143 65-12261 <*>
1958 V13 N3 P13-16 62-20220 <=*>
1958 V13 N4 P17-26 59-19794 <=>
1958 V13 N4 P37-44 61-19360 <+*>
1958 V13 N4 P45-46 60-13018 <+*>
1958 V13 N4 P57-64 62-20400 <=*>
1958 V13 N4 P97-108 61-28227 <=*>
1958 V13 N4 P109-118 63-19491 <=*>
1958 V13 N4 P209-213 62-10247 <*=>
81L34R <ATS>
1958 V13 N4 P235-238 82L34R <ATS>
1958 V13 N5 P33-35 62-15804 <=*>
1958 V13 N5 P37-48 60-13426 <+*>
1958 V13 N5 P55-61 63-10376 <=*>
1958 V13 N5 P63-65 63-10375 <=*>
1958 V13 N5 P105-110 60-21765 <=>
1958 V13 N6 P31-37 63-10359 <=*>
1958 V13 N6 P39-44 63-10374 <=*>
1958 V13 N6 P59-64 63-19677 <=*>
1958 V13 N6 P105-109 62-25403 <=*>
1958 V13 N6 P127-138 AEC-TR-4113 <+*>
1958 V13 N6 P139-147 08N53R <ATS>
1958 V13 N6 P181-190 61-15098 <*+>
1958 V13 N6 P223-230 AEC-TR-4096 <+*>
1959 N6 P12-36 ARSJ V32 N3 P472 <AIAA>
1959 N6 P87-98 ARSJ V31 N9 P1323 <AIAA>
1959 N6 P112-119 ARSJ V32 N3 P458 <AIAA>
1959 N6 P131-136 ARSJ V31 N5 P688 <AIAA>
1959 N6 P142-145 ARSJ V31 N1 P136 <AIAA>
1959 V14 N1 P43-50 60-12545 <=>
1959 V14 N1 P105-108 AD-618 091 <=$>
60-12505 <=>
1959 V14 N1 P109-116 60-12577 <=>
62-11498 <=>
62-24872 <=*>
1959 V14 N1 P117-120 60-12575 <=>
62-11340 <=>
1959 V14 N1 P165-169 61-28233 <*=>
1959 V14 N1 P203-210 AWRE/TR/10 <*>
61-27578 <*=>
1959 V14 N2 P15-23 62-19920 <=>
1959 V14 N2 P43-47 62-23919 <=*>
1959 V14 N2 P61-68 63-10143 <=*>
64-15169 <=*$>
1959 V14 N2 P69-78 60-15856 <+*>
62-11448 <=>

1959 V14 N2 P93-99	60-12593 <=>
	62-11338 <=>
1959 V14 N2 P107-111	60-12594 <=>
	62-11339 <=>
1959 V14 N2 P135-141	62-11228 <=>
1959 V14 N2 P149-155	32N58R <ATS>
1959 V14 N2 P189-197	62-15886 <=>
1959 V14 N2 P199-202	61-27989 <*=>
1959 V14 N2 P213-216	AEC-TR-5703 <*=>
1959 V14 N2 P217-224	61-27806 <*=>
1959 V14 N3 P55-62	63-10492 <=*>
1959 V14 N3 P93-96	62-11415 <=>
1959 V14 N3 P97-104	62-11542 <=>
1959 V14 N3 P129-133	62-11253 <=>
1959 V14 N3 P213-220	61-20036 <*=>
1959 V14 N4 P3-18	62-19816 <=>
1959 V14 N4 P19-25	62-23212 <=>
1959 V14 N4 P27-39	60-31870 <=>
1959 V14 N4 P41-53	60-11875 <=>
1959 V14 N4 P55-60	60-41080 <=>
	63-10309 <=*>
1959 V14 N4 P105-115	60-41081 <=>
1959 V14 N4 P141-147	60-41082 <=>
1959 V14 N4 P149-151	62-11548 <=>
1959 V14 N4 P153-158	62-23716 <=*>
1959 V14 N5 P53-63	63-10306 <=*>
1959 V14 N6 P51-57	62-14006 <=*>
1959 V14 N6 P58-63	63-10337 <=*>
1959 V14 N6 P99-105	61-31671 <=>
1959 V14 N6 P112-119	61-23848 <*=>
1959 V14 N6 P137-141	61-21994 <=>
1959 V14 N6 P146-148	62-19130 <=*>

VESTNIK MOSKOVSKOGO GOSUDARSTVENNOGO UNIVERSITETA. SERIYA OBSHCHESTVENNYKH NAUK

1953 V6 N9 P65-77	R-312 <*>
1953 V6 N9 P161-162	R-323 <*>

VESTNIK MOSKOVSKOGO GOSUDARSTVENNOGO UNIVERSITETA. SERIYA 1. MATEMATIKI, MEKHANIKA

1960 N4 P26-36	ARSJ V31 N11 P1614 <AIAA>
1960 V15 N4 P14-25	61-16086 <=*>
1960 V15 N6 P51-59	61-21608 <=>
	63-10307 <=*>
1960 V15 N6 P72-78	61-23055 <=*>
1960 V15 N13 P93-102	62-13058 <*=>
1961 N3 P62-76	AIAAJ V1 N4 P1008 <AIAA>
1961 V16 N1 P22-28	AD-618 015 <=$>
1961 V16 N1 P38-50	62-15826 <*=>
1961 V16 N1 P62-65	63-23424 <=*>
1961 V16 N2 P10-20	62-19847 <*=>
1961 V16 N2 P71-79	63-24234 <=*$>
1961 V16 N4 P48-57	62-25744 <=*>
1961 V16 N6 P19-27	63-13195 <=*>
1961 S1 V16 N2 P28-36	64-71339 <=> M
1962 N3 P15-19	AMST S2 V48 P266-270 <AMS>
1962 V17 N1 P47-52	63-13912 <=*>
	65-13678 <*>
1962 V17 N1 P60-67	63-21841 <=>
1962 V17 N3 P24-33	64-71676 <=$>
1962 V17 N4 P3-12	62-33282 <=*>
1962 V17 N4 P17-27	63-10728 <=*>
1962 V17 N4 P69-74	65-13674 <*>
1962 V17 N4 P75-84	64-10801 <=*$>
1963 N4 P27-48	AMST S2 V52 P50-74 <AMS>
1963 V18 N2 P35-36	64-19968 <=$>
1963 V18 N2 P37-43	64-21466 <=>
1963 V18 N6 P34-44	65-11003 <*>
1963 V18 N6 P55-63	65-10999 <*>
1963 S1 V18 N6 P64-70	64-11860 <=>
1964 V19 N1 P60-64	65-12838 <*>
1964 V19 N3 P75-86	65-30028 <=$>
1964 V19 N5 P39-48	N65-13552 <=$>
1964 S1 V19 N2 P83-96	64-31549 <=>
1965 N1 P76-82	65-14562 <*>
1965 N1 P88-95	65-14537 <*>
1965 V20 N1 P52-60	AD-629 227 <=*$>
1965 V20 N1 P76-82	65-63693 <=$>
1965 V20 N1 P88-95	65-63699 <=$>
1965 V20 N3 P77-82	65-63184 <=$>
1965 V20 N3 P83-87	65-63185 <=$>
1965 S1 N3 P77-82	65-14564 <*>
1965 S1 N3 P83-87	65-14563 <*>
1966 N1 P58-63	AD-639 910 <=$>
1966 N2 P112-120	66-61356 <=*$>

★VESTNIK MOSKOVSKOGO GOSUDARSTVENNOGO UNIVERSITETA. SERIYA 2. KHIMIYA

1960 N4 P11	ARSJ V31 N9 P1350 <AIAA>
1960 V15 N1 P73-78	65-12046 <*>
1960 V15 N4 P19-20	ARSJ V32 N9 P1454 <AIAA>
	61-27137 <*=>
1960 V15 N4 P40-42	63-13577 <=*>
1960 V15 N4 P66-70	65-10440 <*>
1960 V15 N5 P19-24	63-11513 <=>
1960 S2 V15 N1 P55-56	66-11456 <*>
1961 V16 N1 P77-80	61-28630 <=*>
1961 V16 N3 P27-32	RTS-2577 <NLL>
1961 V16 N4 P31-32	62-25835 <=*>
1961 V16 N6 P3-15	63-15276 <=*>
1961 V16 N6 P31-34	62-24283 <=*>
1961 V16 N6 P53-61	ICE V2 N2 P236-241 <ICE>
1962 N6 P41-	ICE V3 N2 P229-230 <ICE>
1962 V17 N1 P3-20	63-13364 <=*>
1962 V17 N2 P29-30	ICE V2 N4 P524-525 <ICE>
1962 V17 N2 P36-39	18P62R <ATS>
1962 V17 N4 P16-18	66-14792 <*$>
1962 V17 N4 P43-47	11P66R <ATS>
1962 V17 N5 P11-13	AEC-TR-6006 <=*$>
1962 V17 N5 P79-82	65-13691 <*>
1962 V17 N5 P83-84	43P66R <ATS>
1962 V17 N6 P48-50	AEC-TR-5806 <=*$>
1962 V17 N6 P58-60	63-11667 <=>
1963 V18 N1 P3-6	AEC-TR-5732 <=*$>
1963 V18 N1 P7-9	AEC-TR-5759 <=*$>
1963 V18 N2 P3-5	64-20944 <*> 0
1963 V18 N2 P58-60	AEC-TR-5855 <=*$>
1963 V18 N3 P37-40	64-15246 <=*$>
1963 V18 N3 P60-63	64-16574 <=*$>
1963 V18 N4 P24-27	66-60918 <=*$>
1963 V18 N5 P12-16	ICE V4 N2 P329-332 <ICE>
1963 V18 N5 P29-31	ORNL-TR-69 <=$>
1963 V18 N5 P57-61	ICE V4 N2 P272-274 <ICE>
1963 V18 N6 P65-69	AD-622 461 <=*$>
1963 S2 V18 N1 P66-68	66-11445 <*>
1963 S2 V18 N2 P46-48	66-14227 <*$>
1963 S2 V18 N3 P10-13	AD-640 473 <=$>
1963 S2 V18 N3 P41-44	64-30143 <*> 0
1963 S2 V18 N3 P48-51	UCRL-TRANS-1099(L) <=*$>
1963 S2 V18 N6 P36-38	64-18126 <*> 0
1964 N3 P24-28	MLM-1248(TR) <*>
1964 N4 P7-12	ICE V5 N1 P68-71 <ICE>
1964 V19 N1 P20-29	ICE V4 N3 P440-446 <ICE>
1964 V19 N1 P53-55	AD-609 783 <=>
1964 V19 N3 P9-13	ICE V4 N4 P685-688 <ICE>
1964 V19 N3 P14-18	ICE V4 N4 P667-669 <ICE>
1964 V19 N3 P60-63	AD-614 955 <=$>
1964 V19 N3 P87-88	RTS-2822 <NLL>
1964 V19 N4 P7-12	ICE V5 N1 P.68-71 <ICE>
1964 V19 N4 P72-74	43R80R <ATS>
1964 S2 V19 N1 P35-38	ORNL-TR-68 <=$>
1965 V20 N4 P92-93	66-13636 <*>
1965 S2 V20 N5 P37-41	RTS-3648 <NLL>

★VESTNIK MOSKOVSKOGO GOSUDARSTVENNOGO UNIVERSITETA. SERIYA 3. FIZIKA, ASTRONOMIA

1960 N5 P47-52	ARSJ V32 N5 P802 <AIAA>
1960 V15 N1 P23-26	61-28706 <=*>
1960 V15 N1 P27-38	61-11064 <=>
1960 V15 N1 P39-49	60-41724 <=>
	62-15696 <*=>
1960 V15 N2 P51-58	60-31839 <=>
1960 V15 N3 P7-10	AEC-TR-4825 <=>
1960 V15 N4 P18-20	62-11250 <=>
	62-15613 <*=>
1960 V15 N4 P32-37	61-27815 <*=>
1960 V15 N4 P43-47	62-11313 <=>
1960 V15 N4 P59-63	UCRL TRANS-731(L) <=*>
1960 V15 N4 P64-70	64-19536 <=$>
1960 V15 N5 P36-46	61-21302 <=>

1960 V15 N5 P47-52 62-13062 <*=>
1960 V15 N5 P74-80 63-13891 <=*>
1960 V15 N6 P64-66 63-14010 <=*>
1960 S3 V15 N5 P81-93 AD-623 448 <=*$>
1961 V16 N1 P13-15 62-33206 <=*>
1961 V16 N1 P49-55 61-27202 <*=>
1961 V16 N1 P67-75 ARSJ V32 N9 P1488 <AIAA>
1961 V16 N1 P79-81 63-11715 <=>
1961 V16 N1 P82-84 62-24873 <=*>
1961 V16 N2 P3-8 61-28536 <*=>
62-24872 <=*>
1961 V16 N2 P19-25 62-24941 <=*>
1961 V16 N2 P46-53 62-25741 <=*>
1961 V16 N2 P60-72 ARSJ V32 N11 P1809 <AIAA>
1961 V16 N5 P71-82 62-15824 <=*>
1962 N4 P83-91 AIAAJ V1 N9 P2218 <AIAA>
1962 N5 P81-89 65-14061 <*>
1962 V17 N1 P63-68 63-13696 <=*>
1962 V17 N2 P51-59 63-11506 <=>
1962 V17 N3 P17-26 63-15978 <=*>
1962 V17 N4 P83-91 64-10838 <=*$>
1962 V17 N5 P81-89 N65-24659 <=*$>
1962 V17 N6 P3-6 65-61619 <=$>
1962 V17 N6 P59-62 63-14445 <=*>
1962 V17 N6 P83-85 63-23205 <=*$>
1962 S3 N5 P81-90 65-14892 <*>
1963 V18 N1 P54-61 63-18165 <=*>
63-31839 <=>
1963 V18 N2 P25-29 63-20068 <=*>
1963 V18 N3 P55-65 RPQ,V1,N1,P114-121 <IPIX>
1963 V18 N3 P75-80 AD-622 354 <=*$>
1963 V18 N4 P14-17 AD-614 766 <=$>
1963 V18 N5 P34-42 AD-614 393 <=$>
1963 V18 N6 P77-84 AD-609 149 <=>
1963 S3 V18 N2 P37-42 AD-611 015 <=$>
1964 N5 P11-14 66-11264 <*>
1964 V19 N2 P52-55 N64-29512 <=>
1964 V19 N4 P83-86 65-10250 <*>
1964 V19 N4 P87 <YLM>
1964 V19 N4 P88-89 <YLM>
1964 V19 N5 P28-30 N65-14427 <=$>
1964 S3 V19 N1 P3-10 N64-26039 <=>
64-18758 <*>
1964 S3 V19 N5 P15-21 N65-13553 <=$>
1965 V20 N1 P3-8 N65-19731 <=*$>
1965 V20 N1 P26-33 65-13932 <*>
1965 V20 N3 P36-45 65-63433 <=>
1965 V20 N4 P3-6 66-11767 <*>
1965 S3 N5 P62-68 18T93R <ATS>
1965 S3 V20 N3 P36-45 65-14429 <*>
1965 S3 V20 N3 P46-50 65-17112 <*>

VESTNIK MOSKOVSKOGO GOSUDARSTVENNOGO UNIVERSITETA. SERIYA 4. GEOLOGIYA
1960 N2 P19-26 IGR V7 N6 P978 <AGI>
1961 V16 N2 P52-60 63-16552 <*=>
1962 N4 P50-54 IGR V6 N11 P1959 <AGI>
1962 V17 N3 P10-30 47P63R <ATS>
1962 V17 N4 P50-54 IGR V6 N11 P1959 <AGI>
1962 V17 N5 P34-35 63-27422 <$>
1962 S4 V17 N4 P3-12 IGR V6 N12 P2093 <AGI>
1962 S4 V17 N5 P55-65 IGR,V6,N4,P573-580 <AGI>
1963 S4 V18 N1 P28-42 SOU.HYDROL.1963 P297 <AGU>
1964 N6 P20-27 02S86R <ATS>
1964 V19 N5 P94-96 59S84R <ATS>
1964 S4 V19 N5 P94-96 66-10319 <*>
1965 V20 N5 P72-78 73T95R <ATS>

VESTNIK MOSKOVSKOGO GOSUDARSTVENNOGO UNIVERSITETA. SERIYA 5. GEOGRAFIYA
1960 N1 P15-22 SGRT V1 N9 P35 <AGS>
1960 N1 P23-30 SGRT V1 N9 P43 <AGS>
1960 N1 P32-42 SGRT V1 N9 P51 <AGS>
1960 N1 P43-48 SGRT V1 N9 P56 <AGS>
1960 N1 P57-62 SGRT V1 N9 P61 <AGS>
1960 N1 P73-74 SGRT V1 N7 P81 <AGS>
1960 N2 P28-32 SGRT V2 N1 P47 <AGS>
1960 N3 P30-36 SGRT V2 N5 P33 <AGS>
1960 N3 P67-69 SGRT V2 N2 P72 <AGS>
1960 N4 P3-12 SGRT V2 N3 P63 <AGS>
1960 N4 P20-25 SGRT V2 N5 P42 <AGS>
1960 N5 P14-19 SGRT V2 N3 P73 <AGS>
1960 N6 P3-9 SGRT V2 N4 P60 <AGS>
1960 N6 P26-32 SGRT V2 N5 P49 <AGS>
1960 N6 P33-39 SGRT V2 N5 P56 <AGS>
1960 N6 P49-54 SGRT V2 N6 P9 <AGS>
1960 N6 P55-60 SGRT V2 N6 P14 <AGS>
1960 N6 P61-63 SGRT V2 N5 P29 <AGS>
1960 V15 N5 P14-19 61-31155 <=>
1961 N1 P3-12 SGRT V2 N7 P29 <AGS>
1961 N1 P13-20 SGRT V2 N7 P34 <AGS>
1961 N2 P3-9 SGRT V2 N8 P39 <AGS>
1961 N2 P19-23 SGRT V2 N8 P46 <AGS>
1961 N2 P27-33 SGRT V2 N8 P51 <AGS>
1961 N2 P56-61 SGRT V2 N8 P75 <AGS>
1961 N6 P25-29 SGRT V3 N3 P69 <AGS>
1961 N6 P66-73 SGRT V3 N2 P64 <AGS>
1961 V16 N1 P40-47 62-24971 <=*>
1961 S5 V16 N2 P3-9 62-25997 <=>
1961 S5 V16 N2 P27-33 62-25997 <=>
1961 S5 V16 N2 P72-75 62-25997 <=>
1962 N1 P37-42 SGRT V3 N4 P58 <AGS>
1962 N1 P71-75 SGRT V3 N3 P74 <AGS>
1962 N3 P18-24 SGRT V4 N5 P30 <AGS>
1962 N4 P22-26 SGRT V3 N7 P39 <AGS>
1962 N5 P3-8 SGRT V4 N1 P47 <AGS>
1962 N5 P12-17 SGRT V4 N3 P24 <AGS>
1962 N5 P65-68 SGRT V4 N1 P53 <AGS>
1962 N6 P83-84 SGRT V4 N3 P21 <AGS>
1963 N4 P3-13 SGRT V5 N1 P3 <AGS>
1963 N4 P16-24 SGRT V4 N8 P14 <AGS>
1963 N6 P3-8 SGRT V5 N2 P3 <AGS>
1963 N6 P24-25 SGRT V5 N2 P11 <AGS>
1963 N6 P77-80 SGRT V5 N2 P17 <AGS>
1963 N6 P81-83 SGRT V5 N2 P21 <AGS>
1964 N1 P37-43 SGRT V5 N3 P52 <AGS>
1964 N3 P7-19 SGRT V5 N10 P45 <AGS>
1964 N4 P18-27 SGRT V6 N9 P66 <AGS>
1964 N6 P7-16 SGRT V6 N9 P56 <AGS>
1964 V19 N3 P31-38 AD-635 540 <=$>
1964 V19 N3 P39-43 65-33167 <=$>
1964 V19 N4 P3-8 65-33168 <=$>
1964 S5 V19 N4 P28-41 SOV.GEOGR. V5 N8 P50 <AGS>
1965 N4 P3-9 SGRT V7 N6 P23 <AGS>
1965 N5 P47-54 SGRT V7 N6 P32 <AGS>
1965 N6 P3-16 SGRT V7 N6 P8 <AGS>
1965 N6 P70-72 SGRT V7 N6 P30 <AGS>
1965 N6 P79-82 SGRT V7 N2 P9 <AGS>
1966 N1 P91-95 SGRT V7 N6 P3 <AGS>
1966 N2 P3-10 SGRT V7 N7 P65 <AGS>
1966 N6 P62-71 SGRT V7 N10 P34 <AGS>

VESTNIK MOSKOVSKOGO GOSUDARSTVENNOGO UNIVERSITETA. SERIYA 6. BIOLOGIYA, POCHVOVEDENIE
1960 V15 N1 P9-19 C-3639 <NRCC>
1961 V16 N1 P26-31 61-28885 <*=>
1961 V16 N3 P27-34 62-15469 <*=>
1962 V17 N2 P7-15 64-31048 <=>
1963 N3 P29- FPTS V23 N6 P.T1337 <FASE>
1963 N6 P8-15 RTS-2988 <NLL>
1963 V18 N1 P17-23 63-31972 <=>
1963 V18 N2 P10-21 64-19245 <=*$> 0
1963 S6 V18 N3 P18-24 64-19693 <=$>
1963 S6 V18 N3 P25-28 64-19694 <=$>
1963 S6 V18 N4 P14-25 64-19718 <=> 0
1964 V19 N1 P3-8 65-30508 <=$>
1964 V19 N1 P39- FPTS V23 N6 P.T1371 <FASE>
FPTS V23 P.T1371 <FASE>
1964 V19 N1 P74-80 65-30508 <=$>
1964 V19 N2 P73-79 64-31747 <=>
1964 V19 N5 P3-10 64-51982 <=$>
1964 V19 N5 P57-62 64-51982 <=$>
1964 V19 N5 P70-79 64-51982 <=$>
1964 V19 N6 P3- FPTS V25 N1 P.T111 <FASE>
1964 V19 N6 P15-25 65-30508 <=$>
1964 V19 N6 P15-19 65-64537 <=*$>
1964 V19 N6 P39-45 65-30508 <=$>
1964 S6 V19 N2 P25-37 64-31746 <=>

1960 V35 N5 P69-70 61-11888 <=>
1960 V35 N5 P85-87 61-11890 <=>
1961 V36 N1 P3-10 61-21557 <=>
1961 V36 N1 P44-49 61-21570 <=>
1961 V36 N1 P49-53 61-21578 <=>
1961 V36 N1 P60-61 62-15679 <=>
1961 V36 N2 P10-14 61-23578 <=*>
1962 V37 N1 P23- FPTS V22 N1 P.T54 <FASE>
1962 V37 N1 P29-32 63-15166 <=>
1962 V37 N1 P50- FPTS V22 N1 P.T98 <FASE>
1962 V37 N1 P62-65 63-15411 <=*> 0
1962 V37 N1 P63- FPTS V22 N2 P.T373 <FASE>
1962 V37 N1 P65-66 63-15986 <=*>
1962 V37 N1 P67-68 63-15415 <=*>
1962 V37 N1 P68-69 63-15202 <=>
1962 V37 N1 P73-77 62-25501 <=>
1962 V37 N1 P77-79 63-13258 <=*>
1962 V37 N1 P84-86 62-25501 <=>
1962 V37 N2 P3-10 63-15240 <=> 0
1962 V37 N2 P37-42 63-15238 <=>
1962 V37 N2 P42-47 63-15158 <=*>
1962 V37 N2 P47-49 63-15239 <=>
1962 V37 N2 P72-81 62-33453 <=*>
1962 V37 N3 P3-13 63-19377 <=*>
1962 V37 N3 P60-62 62-33304 <=*>
1962 V37 N5 P49-63 63-21081 <=>
1962 V37 N6 P41-44 65-60573 <=$>
1962 V37 N6 P45-51 64-13500 <=*$>
1962 V37 N6 P51-55 63-24457 <=*$>
1963 V38 N1 P55-57 64-13845 <=*$>
1963 V38 N2 P16-20 64-15466 <=*$> 0
1963 V38 N4 P22- FPTS V23 N4 P.T758 <FASE>
FPTS V23 N4 P.T759 <FASE>
1963 V38 N4 P52-59 63-31929 <=>
1964 V39 N1 P54-59 64-21932 <=>
1964 V39 N1 P74 64-21932 <=>
1964 V39 N3 P3-6 65-62120 <=$>
1964 V39 N6 P60-66 65-31285 <=$>
1964 V39 N6 P72-74 65-31285 <=$>
1964 V39 N6 P76-77 65-31286 <=$>
1965 V40 N1 P73-78 65-30876 <=$>
1965 V40 N4 P3-12 65-33227 <=*$>

VESTNIK SELSKO-KHOZYAISTVENNOI NAUKI. ALMA-ATA
1961 V4 N6 P44-52 62-24394 <=*>
1963 V6 N3 P39-43 63-24171 <=*$>
1963 V6 N12 P56-73 64-31646 <=>
1965 V8 N3 P59-63 65-30891 <=$>
1965 V8 N7 P73-81 66-30013 <=*$>

VESTNIK SELSKO-KHOZYAISTVENNOI NAUKI. NEW SERIES
1956 N1 P143-149 R-4196 <*>
1956 V1 N1 P69-71 R-4198 <*>
1956 V1 N1 P149- R-4197 <*>
1956 V1 N2 P93-97 88K25R <ATS>
1957 N11 P149-152 R-4202 <*>
1957 N11 P153-155 R-4000 <*>
1957 N12 P93-99 R-4195 <*>
1957 V2 N5 P147-150 C-2717 <NRC>
1957 V2 N7 P145-148 R-3155 <*>
1957 V2 N8 P51-58 59-10085 <+*>
1957 V2 N9 P80-97 R-3088 <*>
1957 V2 N9 P127-131 R-4212 <*>
1957 V2 N11 P17-21 60-15949 <+*>
1958 N10 P94-98 NLLTB V1 N3 P44 <NLL>
1958 N10 P99-103 NLLTB V1 N2 P28 <NLL>
1958 V3 N1 P82-91 61-20707 <*=> 0
1958 V3 N1 P92-117 59-19670 <+*> 0
1958 V3 N2 P33-39 PB 141 214T <=>
1958 V3 N3 P43-55 62-15839 <=*>
1958 V3 N9 P50-58 62-32502 <=*>
1958 V3 N10 P15-22 59-20369 <+*>
1958 V3 N10 P99-103 59-22637 <+*>
1959 N5 P97-102 NLLTB V2 N1 P1 <NLL>
1959 V4 N1 P62-70 59-16812 <+*>
1959 V4 N2 P114-121 59-18271 <+*>
1959 V4 N2 P129-130 59-16828 <+*>
1959 V4 N2 P131-134 59-18272 <+*>
1959 V4 N3 P3-18 59-18273 <+*>
1959 V4 N3 P21-26 59-31004 <=>
1959 V4 N3 P34-45 59-18274 <+*>
1959 V4 N3 P59-65 59-18279 <+*> 0
1959 V4 N3 P137-138 59-18280 <+*>
1959 V4 N3 P139-140 59-18281 <+*>
1959 V4 N3 P143-144 59-18270 <+*>
1959 V4 N3 P144-147 59-22323 <+*>
1959 V4 N4 P23-33 60-15333 <+*>
1959 V4 N4 P58-63 60-15341 <+*> C
1959 V4 N4 P132-134 60-15342 <+*>
1959 V4 N4 P141-142 60-15343 <+*>
1959 V4 N5 P3-10 60-15344 <+*>
1959 V4 N5 P11-20 60-31001 <=>
1959 V4 N5 P97-102 60-15348 <+*> C
1959 V4 N5 P110-113 60-15345 <+*> 0
1959 V4 N5 P133-136 60-15349 <+*>
1959 V4 N5 P140-141 60-15346 <+*>
1959 V4 N6 P25-34 60-15347 <+*>
1959 V4 N6 P35-42 60-15350 <+*>
1959 V4 N6 P43-54 60-15353 <+*>
1959 V4 N6 P55-60 60-15354 <+*>
1959 V4 N6 P88-97 60-15355 <+*> 0
1959 V4 N6 P98-107 60-31001 <=>
1959 V4 N7 P56-66 62-32949 <=*>
1959 V4 N7 P138-140 60-15360 <+*>
1959 V4 N8 P9-19 60-17255 <+*> 0
1959 V4 N8 P47-50 60-15352 <+*> C
1959 V4 N8 P58-67 60-17262 <+*>
1959 V4 N9 P68-78 60-15358 <+*> C
1959 V4 N9 P92-99 60-17259 <+*> C
1959 V4 N11 P15-22 60-31294 <=>
1959 V4 N11 P50-57 60-31294 <=>
1959 V4 N11 P98-102 66-51089 <=>
1959 V4 N11 P133-136 60-31294 <=>
1959 V4 N12 P65-70 66-51087 <=$>
1960 V5 N1 P12-22 60-31408 <=>
1960 V5 N1 P23-27 60-31408 <=>
1960 V5 N1 P36-44 60-31696 <=>
1960 V5 N1 P54-60 61-15851 <+*>
1960 V5 N1 P121-125 61-19642 <=*>
1960 V5 N2 P12-18 60-31412 <=>
1960 V5 N2 P19-23 61-19647 <*=>
1960 V5 N2 P116-119 62-10989 <=*>
1960 V5 N2 P144-146 61-11560 <=>
1960 V5 N2 P148-152 61-11560 <=>
1960 V5 N3 P32-40 61-19751 <=*>
1960 V5 N3 P129-133 61-19749 <=> 0
1960 V5 N4 P31-37 61-15859 <+*>
1960 V5 N4 P58-63 61-15842 <+*>
1960 V5 N4 P92-95 61-15857 <+*>
1960 V5 N4 P110-112 60-23723 <+*>
1960 V5 N4 P126-128 61-13774 <*+> 0
1960 V5 N4 P136-138 60-23724 <+*>
1960 V5 N5 P3-21 61-15827 <+*>
1960 V5 N5 P77-82 61-19652 <=*> C
1960 V5 N6 P22-28 61-11958 <=>
1960 V5 N7 P10-28 61-21300 <=>
1960 V5 N7 P62-65 63-11052 <=>
1960 V5 N7 P81-84 63-11053 <=>
1960 V5 N7 P127-130 63-11054 <=>
1960 V5 N8 P3-45 61-21244 <=>
1960 V5 N9 P72-76 61-27280 <*=>
1960 V5 N10 P3-10 61-23186 <=*>
1960 V5 N10 P110-116 61-23187 <=*> 0
1960 V5 N10 P117-122 61-23188 <=*> 0
1960 V5 N10 P146-148 61-23189 <=*>
1960 V5 N10 P148-150 61-23190 <=*>
1960 V5 N11 P66-79 61-11994 <=>
1960 V5 N11 P124-127 61-19754 <=>
1960 V5 N11 P143-148 61-19750 <=*>
1960 V5 N12 P1-9 61-27077 <=>
1960 V5 N12 P23-34 61-23191 <*=>
1960 V5 N12 P47-53 61-23192 <=*>
1960 V5 N12 P79-83 61-23193 <=*> 0
1961 V6 N1 P3-10 61-23196 <=*>
1961 V6 N1 P45 61-23197 <=*>
1961 V6 N1 P99 61-23198 <=*>
1961 V6 N1 P149-155 61-27481 <*=> C
1961 V6 N3 P45-50 64-15321 <=*$>
1961 V6 N3 P101-106 62-15489 <*=>
1961 V6 N3 P138-139 61-27491 <*=>

1961 V6 N4 P109-115 63-19050 <=*>
1961 V6 N5 P67-74 61-28439 <*=> 0
1961 V6 N5 P129-133 61-28440 <*=> 0
1961 V6 N5 P137-140 62-23293 <=*>
1961 V6 N5 P151-153 62-23294 <=*>
1961 V6 N5 P153-155 62-23295 <=*>
1961 V6 N5 P156 62-13644 <*=>
1961 V6 N7 P122-129 62-23297 <=*>
1961 V6 N7 P143-155 62-23853 <=*> 0
1961 V6 N10 P9-17 62-23332 <=*>
1961 V6 N10 P18-27 64-13538 <=*$>
1961 V6 N10 P147-153 62-24988 <=*>
1961 V6 N12 P47-50 62-23859 <=*>
1961 V6 N12 P57-64 62-23816 <=*>
1961 V6 N12 P151-154 62-24527 <=*>
1962 N3 P59-63 65-29157 <$>
1962 V7 N1 P10-14 62-23852 <=*>
1962 V7 N1 P34-40 62-24755 <=*>
1962 V7 N2 P49-54 62-24528 <=*> 0
1962 V7 N2 P55-63 62-25040 <=*>
1962 V7 N2 P64-68 62-25039 <=*> 0
1962 V7 N2 P104-116 62-25041 <=*> 0
1962 V7 N2 P148-151 62-25038 <=*>
1962 V7 N3 P144-148 64-15590 <=*$>
1962 V7 N6 P3-19 63-21115 <=>
1962 V7 N8 P43-49 64-15177 <=*$>
1962 V7 N8 P140-141 NLLTB V5 N2 P87 <NLL>
1962 V7 N11 P105-110 63-21073 <=>
1962 V7 N11 P135-139 63-21175 <=>
1962 V7 N12 P7-17 63-21495 <=>
1962 V7 N12 P19-24 63-21495 <=>
1963 N2 P21-24 65-64396 <=*$>
1963 V8 N4 P81-84 63-23805 <=*>
1963 V8 N4 P135-138 63-23916 <=*$>
1963 V8 N5 P3-25 63-24063 <=*>
1963 V8 N6 P2-6 63-23547 <=*$>
1963 V8 N9 P47-71 64-31402 <=*$>
1963 V8 N9 P72-80 64-15080 <=>
1963 V8 N10 P1-5 64-13540 <=*$>
1963 V8 N10 P19-22 64-31053 <=>
1963 V8 N10 P141-144 64-13683 <=*$>
1964 V9 N1 P5-15 64-31237 <=>
1964 V9 N1 P19-25 64-31237 <=>
1964 V9 N1 P46-52 64-31237 <=>
1964 V9 N1 P128-130 64-31237 <=>
1964 V9 N1 P139-144 64-31237 <=>
1964 V9 N2 P10-15 64-21945 <=>
1964 V9 N3 P62-65 65-60648 <=$>
1964 V9 N3 P87-91 64-31211 <=>
1964 V9 N3 P110-114 64-19186 <=*$> 0
1964 V9 N4 P12-14 64-19371 <=$>
1964 V9 N4 P52-54 64-19370 <=$>
1964 V9 N5 P148-150 64-51810 <=$>
1964 V9 N7 P89-95 NIAE-TRANS-164 <NLL>
TR-164 <NIAE>
1964 V9 N10 P18-22 65-30567 <=$>
1965 N2 P125-128 RTS-3316 <NLL>
1965 N10 P145-148 NLLTB V8 N5 P390 <NLL>
1965 N11 P62-70 NLLTB V8 N7 P563 <NLL>
1965 V10 N2 P129-134 RTS-3616 <NLL>
1965 V10 N5 P152-155 65-31897 <=$*>
1965 V10 N7 P112-116 66-62012 <=*$>
1965 V10 N10 P28-30 65-34015 <=$>
1966 V11 N6 P149-153 66-34947 <=$>

VESTNIK. SLOVENSKA AKADEMIA VIED. BRATISLAVA
1960 V2 N1 P3-41 62-19521 <=*>
1960 V2 N1 P51-53 62-19529 <=*>
1960 V2 N2/3 P90-121 62-19520 <=*>
1960 V2 N2/3 P170-173 62-19520 <=*>
1960 V2 N2/3 P185-192 62-19519 <=*>

VESTNIK SLOVENSKEGA KEMIJSKEGA DRUSTVA
1959 V6 P1-8 43N55SL <ATS>

VESTNIK SOTSIALISTICHESKOGO RASTENIEVODSTVA
1940 N2 P32-39 RT-1027 <*> 0

VESTNIK STATISTIKI
1955 N2 P17-28 RT-4436 <*>
1955 N5 P31-50 62-16126 <=*>
1958 N8 P3-14 NLLTB V1 N1 P8 <NLL>
1958 N9 P95 NLLTB V1 N1 P40 <NLL>
1959 N4 P51-54 NLLTB V1 N11 P2 <NLL>
1959 N12 P7-19 60-31417 <=>
1960 N6 P6-17 61-11738 <=>
1960 N11 P66-67 61-23932 <=>
1960 N12 P3-21 61-27423 <*=>
1960 N12 P59-64 61-23580 <=>
1961 N2 P58-68 61-23580 <=>
1962 N1 P24-31 NLLTB V6 N6 P519 <NLL>
1962 N1 P24-33 TRANS.BULL.1964 P519 <NLL>
1962 N4 P62-73 62-33149 <=*>
1963 N7 P86-96 63-31843 <=>
1963 N8 P37-40 63-41088 <=>
1964 N1 P44-54 64-31118 <=>
1964 N2 P83-85 64-31047 <=>
1964 N3 P8-15 64-31639 <=>
1964 N3 P63-71 64-316[illegible]9 <=>
1964 N5 P11-23 64-41247 <=>
1964 N6 P16-25 64-51681 <=$>
1965 N2 P36-50 65-32940 <=$>
1965 N3 P16-21 65-32940 <=$>
1965 N5 P54-59 65-32940 <=$>
1965 N6 P3-15 65-32181 <=$>
1965 N8 P22-27 65-33016 <=$*>
1965 N10 P11-17 65-33895 <=*$>
1965 N10 P58-62 65-34012 <=*$>
1965 N12 P85-90 66-30253 <=*$>
1966 N1 P3-13 66-32555 <=$>
1967 N1 P24-29 67-31308 <=$>

★VESTNIK SVYAZI
1945 N7 P11-12 R-976 <*>
1947 N4 P3-4 RT-2750 <*> 0
1947 V7 N10 P19-22 RT-854 <*>
1952 N4 P3-4 R-595 <*>
1952 N4 P7-8 R-596 <*>
1952 N4 P12-15 R-597 <*>
1952 N4 P17-18 R-592 <*>
1952 N6 P6-8 R-458 <*>
1952 N7 P3-6 RT-4574 <*>
1952 N8 P3-5 R-460 <*>
1952 N8 P6-7 R-461 <*>
1952 N8 P11-13 R-605 <*>
1952 N10 P9-10 RT-4575 <*>
1953 N6 P2 RT-3207 <*>
1953 N6 P3-6 R-607 <*>
1953 N6 P22-23 RT-3203 <*>
1953 N6 P24-26 RT-3206 <*>
1953 N6 P27-28 R-608 <*>
1953 N6 P30- R-610 <*>
1953 N8 P3-5 R-593 <*>
1953 N8 P8-11 R-594 <*>
1953 N8 P15-16 R-588 <*>
1954 N2 P3-6 RT-3368 <*>
1954 N2 P9-11 RT-3362 <*>
1954 N4 P3-7 R-611 <*>
1954 V14 N7 P10-12 62-33213 <=*>
1955 N2 P5-8 R-590 <*>
1955 N2 P20-22 R-583 <*>
1955 N5 P3-5 R-632 <*>
1955 N5 P5-7 R-633 <*>
1955 N5 P7-10 R-614 <*>
1955 N5 P10-11 R-615 <*>
1955 N5 P27-28 R-616 <*>
1955 N9 P9-10 R-584 <*>
1955 N9 P12- R-585 <*>
1955 N9 P18-20 R-586 <*>
1955 N9 P25- R-587 <*>
1955 N11 P1-2 RT-4562 <*>
1955 N11 P5-8 RT-4563 <*>
1955 N11 P10-12 RT-4564 <*>
1955 N11 P12-15 RT-4565 <*>
1955 N11 P15 RT-4568 <*>
1955 N11 P20 RT-4566 <*>
1955 N11 P23-24 RT-4567 <*>
1955 V15 N10 P1-2 RT-4472 <*>
1955 V15 N10 P3-4 RT-4475 <*>
1955 V15 N10 P5-6 RT-4477 <*>

1955 V15 N10 P7-8 RT-4474 <*>
1955 V15 N10 P12-13 RT-4473 <*>
1955 V15 N10 P27-28 RT-4476 <*>
1955 V15 N12 P3-5 RT-4461 <*>
1955 V15 N12 P5-7 RT-4462 <*>
1955 V15 N12 P8-10 RT-4464 <*>
1955 V15 N12 P24-25 RT-4463 <*>
1956 V16 N4 P4-6 59-11028 <=>
1957 N2 P3-5 C-2575 <NRC>
R-3672 <*>
1957 N5 P11-14 C-2522 <NRC>
R-3109 <*>
1957 V17 N3 P6-7 R-947 <*>
1957 V17 N4 P11-14 59-16031 <+*>
1958 V18 N1 P7-9 59-11609 <=> 0
1958 V18 N1 P20-21 60-17619 <+*> 0
1958 V18 N1 P31-33 59-11603 <=> 0
1958 V18 N3 P5-7 60-13685 <+*>
1958 V18 N4 P17-19 60-17494 <+*>
1958 V18 N4 P20-22 60-17629 <+*> 0
1958 V18 N5 P4-6 59-11579 <=> 0
1958 V18 N5 P8-11 60-21602 <=> 0
1958 V18 N5 P14-16 60-21721 <=> 0
1958 V18 N6 P3-4 60-21476 <=> 0
1958 V18 N7 P2 60-21735 <=>
1958 V18 N7 P36 60-21478 <=>
1958 V18 N8 P3-5 62-32648 <=*>
1958 V18 N8 P19-21 60-21481 <=> 0
1958 V18 N9 P3-6 60-21451 <=> 0
1958 V18 N9 P11-12 60-21449 <=> 0
1958 V18 N9 P12-14 60-15668 <+*>
1958 V18 N9 P14 60-21452 <=> 0
1958 V18 N10 P3-4 62-32907 <=*>
1958 V18 N12 P10-11 60-21434 <=> 0
1958 V18 N12 P13-14 60-21435 <=> 0
1959 V19 N1 P9-10 62-32643 <=*>
1959 V19 N3 P37-38 62-18088 <=*>
1959 V19 N9 P3-4 62-32786 <*=>
1959 V19 N9 P5-7 62-32787 <=*>
1959 V19 N9 P7-9 62-32788 <=*>
1959 V19 N9 P9-10 62-32789 <=*>
1959 V19 N9 P10-12 62-32790 <=*>
1959 V19 N9 P12 62-32792 <=*>
1960 V20 N1 P1-2 60-31350 <=>
1960 V20 N1 P3-5 60-31351 <=> 0
1960 V20 N1 P6-8 60-31352 <=> 0
1960 V20 N1 P8-10 60-31353 <=> 0
1960 V20 N1 P11-12 60-31354 <=>
1960 V20 N1 P16 60-31355 <=> 0
1960 V20 N1 P17-196 60-31356 <=>
1960 V20 N1 P18 60-31357 <=>
1960 V20 N4 P11-13 61-23277 <*=> 0
1960 V20 N5 P5-8,12 60-41213 <=>
1961 V21 N7 P5-6 64-71525 <=>
1962 N9 P1-2 NLLTB V5 N3 P205 <NLL>
1964 V24 N10 ENTIRE ISSUE 65-30113 <=$>
1965 V25 N4 P1-121 65-31335 <=*$>
1965 V25 N5 P1-96 65-31597 <=*$>
1965 V25 N6 P1-104 65-32064 <=*$>
1965 V25 N7 P1-105 65-32529 <=*$>
1965 V25 N9 P1-34 65-33594 <=$>
1966 V26 N5 ENTIRE ISSUE 66-33904 <=$>
1966 V26 N10 P15-21 67-30102 <=>
1966 V26 N10 P24-27 67-30102 <=>
1966 V26 N10 P29-32 67-30102 <=>

VESTNIK USHNYKH NOSOVYKH I GORLOVYKH BOLEZNEI
1912 V4 P258-266 65-60092 <=$> P

VESTNIK USTREDNIHO USTAVU GEOLOGICKEHO
1955 V30 P197-210 30L37C <ATS>

VESTNIK VENEROLOGII I DERMATOLOGII
1945 N3 P8-16 65-63258 <=>
1945 V2 P23-27 RT-162 <*>
1946 N1 P36-42 65-60067 <=$>
1952 N3 P31-33 64-19515 <=$>
1952 V3 P31-33 R-841 <*>
1954 N1 P58-62 RT-2470 <*>
1954 N6 P3-8 RT-4123 <*>
1954 N6 P26-29 RT-1628 <*>

VESTNIK VOZDUSHNOGO FLOTA
1938 V21 N11 P67-77 RT-1198 <*>
1938 V21 N12 P40-46 RT-1194 <*>
1947 N9 P14-17 RT-1110 <*>
1947 N9 P45-48 RT-1111 <*>
1947 N12 P47-54 RT-495 <*>
1949 V31 N3 P35-39 RT-1106 <*>
1955 N5 P52-57 59-16952 <+*>
1956 V39 N12 P1 61-28433 <=*>
1956 V39 N12 P3-12 61-28433 <=*>
1956 V39 N12 P17-64 61-28433 <=*>
1956 V39 N12 P71 61-28433 <=*>
1956 V39 N12 P76-84 61-28433 <=*>
1956 V39 N12 P90-94 61-28433 <=*>
1957 N12 P80- R-3776 <*>
1957 V39 N11 P56-64 62-18309 <=*>
1959 N2 P69-72 59-18453 <+*> PO

VESTNIK VSESOYUZNOGO NAUCHNO-ISSLEDOVATELSKOGO INSTITUTA ZHELEZNODOROZHNOGO TRANSPORTA
1957 V16 N4 P6-16 59-11571 <=>
1959 N12 P15-20 61-13949 <=*>
1962 V21 N4 P58-62 62-32866 <=*>
1963 V22 N2 P3-9 63-11758 <=>
1963 V22 N2 P9-13 63-11759 <=>
1963 V22 N2 P16-19 63-11760 <=>
1963 V22 N2 P63-64 63-11761 <=>
1963 V22 N5 P61-64 64-31137 <=*$>
1964 V23 N1 P41-45 65-11178 <*>
1964 V23 N1 P45-48 64-20974 <*>
1964 V23 N1 P53-57 64-20973 <*>
1964 V23 N1 P61-64 65-11177 <*>
1964 V23 N2 P39-41 65-13786 <*>
1964 V23 N2 P48-51 65-14797 <*>
1964 V23 N3 P3-8 65-13788 <*>
1964 V23 N3 P9-13 65-13789 <*>
1964 V23 N3 P35-38 65-13792 <*>
1964 V23 N3 P42-46 65-13791 <*>
1964 V23 N6 P37-41 65-13790 <*>
1964 V23 N6 P42-44 65-13787 <*>
1964 V23 N6 P44-46 65-14798 <*>
1965 V24 N3 P34-36 65-17181 <*>
1965 V24 N5 P45-49 66-31040 <=$>
1965 V24 N8 P23-24 AD-641 273 <=$>

VESTNIK VYSSHEI SHKOLY
1952 V10 N2 P19-21 RT-1549 <*>
1955 N9 P8-13 RT-3704 <*>
1955 N11 P55-59 RT-3782 <*>
1958 N7 P9-14 59-11125 <=>
1958 N7 P14-16 59-11126 <=>
1958 N11 P65-72 NLLTB V1 N4 P3 <NLL>
1958 V16 N1 P38-41 64-21793 <=>
1958 V16 N8 P20-22 59-13238 <=>
1958 V16 N8 P22-25 59-13239 <=>
1959 V17 N7 P91-94 60-31153 <=>
1959 V17 N9 P57-69 60-11236 <=>
1960 V18 N8 P48-51 63-15551 <=*>
1961 V19 N1 P69-71 61-21585 <=>
1961 V19 N5 P90-96 61-27968 <=*>
1961 V19 N9 P30-32 63-27968 <$>
1962 V20 N7 P51-55 63-13288 <=>
1963 V21 N1 P12-24 63-31163 <=>
1963 V21 N2 P50-59 63-21804 <=>
1963 V21 N3 P13-18 63-21984 <=>
1963 V21 N3 P18-20 63-21964 <=>
1963 V21 N6 P48-51 64-21797 <=>
1963 V21 N7 P32-34 64-21046 <=>
1963 V21 N8 P51-54 64-21435 <=>
1963 V21 N9 P40-46 65-60506 <=$>
1963 V21 N12 P45-51 64-31006 <=>
1963 V21 N12 P52-59 64-21897 <=>
1964 N4 P45-57 65-30979 <=*$>
1964 V22 N2 P3-6 64-31804 <=>
1964 V22 N2 P7-9 64-41091 <=>
1964 V22 N3 P43-57 65-30676 <=*$>
1964 V22 N3 P67-69 65-30933 <=$>
1964 V22 N5 P20-40 65-30736 <=*$>

1964 V22 N5 P61 65-30587 <=$*>
1964 V22 N5 P85-86 65-30587 <=$*>
1964 V22 N6 P24-28 64-51372 <=> 0
1964 V22 N6 P39-40 64-51372 <=> 0
1964 V22 N6 P57-66 64-51372 <=> 0
1964 V22 N7 P31-35 64-51422 <=$>
1964 V22 N7 P38-40 64-51422 <=$>
1964 V22 N8 P23-34 64-51923 <=*>
1964 V22 N9 P22-31 65-30761 <=$>
1964 V22 N9 P40-42 65-30761 <=$>
1964 V22 N10 P12-34 65-30760 <=$>
1965 V23 N1 P35-45 65-31432 <=$>
1965 V23 N2 P41-47 65-31041 <=$>
1965 V23 N3 P19-24 65-33519 <=*$>
1965 V23 N3 P37-41 65-31798 <=$>
1965 V23 N3 P46-47 65-31798 <=$>
1965 V23 N3 P94-96 65-31799 <=$>
1965 V23 N4 P90-91 65-31814 <=$>
1965 V23 N5 P13-25 65-31735 <=$>
1965 V23 N5 P87-91 65-31735 <=$>
1965 V23 N6 P7-11 65-32330 <=$>
1965 V23 N6 P19-24 65-32330 <=$>
1965 V23 N6 P47-49 65-32330 <=$>
1965 V23 N6 P69-72 65-33488 <=*$>
1965 V23 N6 P81-85 65-32330 <=$>
1965 V23 N9 P17-32 65-33882 <=*$>
1965 V23 N9 P32-40 66-30182 <=*$>
1965 V23 N10 P553-556 66-31681 <=$>
1965 V23 N11 P66-69 66-31537 <=$>
1966 V24 N1 P28-35 66-31070 <=$>
1966 V24 N1 P40-41 66-31070 <=$>
1966 V24 N5 P51-56 66-33185 <=$>
1966 V24 N5 P69-71 66-33185 <=$>
1966 V24 N7 P23-44 66-34492 <=$>
1966 V24 N7 P85-86 66-34360 <=$> 0
1966 V24 N8 P31-34 66-34922 <=$>
1966 V24 N12 P35-40 67-30413 <=>

VESTNIK ZASHCHITY RASTENII
1939 N1 P9-14 RT-3154 <*>
1940 N1/2 P8-14 RT-3160 <*>
1940 N1/2 P276-277 RT-3162 <*>
1941 N1 P100-105 RT-1677 <*>

VESTSI AKADEMII NAVUK BELARUSKAI SSR. MINSK
1953 N1 P71 61-20996 <*=>
1954 N2 P62-72 RCT V29 N3 P1047 <RCT>

VESTSI AKADEMII NAVUK BELARUSKOI SSR. SERIYA BIYALAGICHNYKH NAVUK.
1959 N2 P12-24 60-31641 <=>
1960 N1 P130-139 60-41165 <=>
1960 N1 P143-144 60-41163 <=>
1961 N2 P125-133 62-15459 <=*>
1962 N2 P62-70 66-60561 <=*$>
1963 N1 P21-33 65-61104 <=$>
1963 N1 P34-37 65-62338 <=$>
1963 N1 P87-91 65-61105 <=$>
1964 N1 P5-14 64-41083 <=>
1964 N1 P53-58 65-60196 <=$>
1964 N1 P87-89 64-31637 <=>
1964 N1 P92-96 64-31774 <=>
1964 N1 P101-106 64-31902 <=>
1964 N3 P107-118 65-30541 <=$>
1965 N1 P74-81 65-32325 <=*$>
1965 N1 P116-119 65-31009 <=$>
1965 N2 P128-138 66-30264 <=*$>

VESTSI AKADEMII NAVUK BELARUSKOI SSR. SERIYA FIZIKA-TEKHNICHNYKH NAVUK
1956 N1 P15-24 83J19W <ATS>
1956 N4 P39-50 63-13732 <=*>
1956 N4 P119-131 17L35W <ATS>
60-18514 <+*>
1960 N1 P5-26 60-31713 <=>
1960 N1 P145-147 60-31712 <=>
1960 N2 P137-139 61-21147 <=>
1960 N4 P93-98 64-13121 <=*$>
1961 N1 P19-28 62-23271 <=>
1961 N1 P59-74 62-11134 <=>
1961 N1 P114-119 64-13099 <=*$>
1961 N3 P108-114 62-25185 <=*>
1961 N4 P45-49 63-11671 <=>
1961 N4 P132-133 63-11573 <=>
1962 N1 P30-33 63-19090 <*=>
1962 N2 P122-130 62-15466 <=*>
1963 N1 P33-41 63-21728 <=>
1963 N3 P51-58 64-23515 <=$>
1964 N2 P47-50 AD-627 004 <=$>
1964 N3 P129-131 AD-640 254 <=$>
1964 N4 P128-131 65-14058 <*>
65-62956 <=$>
1965 N1 P133-136 65-32082 <=$>
1965 N2 P84-89 66-11244 <*>
1966 N1 P34-42 ICE V6 N4 P598-603 <ICE>
1966 N2 P42-49 66-61984 <=*$>

VESTSI AKADEMII NAVUK BELARUSKOI SSR. SERIYA KHIMICHNYKH NAVUK
1966 N1 P43-47 ICE V6 N4 P571-574 <ICE>

VESZPREMI VEGYIPARI EGYETEM KOZLEMENYEI
1963 V7 P55-60 65-17101 <*>
66-11109 <*>

VETERINARIA. SARAJEVO
1958 V7 N1 P153-155 60-21632 <=>
1958 V7 N2 P369-372 60-21632 <=>

VETERINARIYA
1946 V23 N4 P33-36 R-2650 <*>
1947 N5 P20-21 RT-1191 <*>
1947 V24 P29-31 RT-234 <*>
1947 V24 N10 P40-43 65-60791 <=>
1948 V25 N2 P42-45 RT-1113 <*>
1949 V26 P26-28 RTS-2752 <NLL>
1949 V26 N9 P15-19 65-61549 <=$>
1950 V27 P34-36 RJ-670 <ATS>
1950 V27 N2 P8-14 RT-1150 <*>
1950 V27 N2 P38-40 716TM <CTT>
1950 V27 N9 P34-35 RT-2274 <*>
1951 N4 P8-15 RT-88 <*>
1951 V28 N5 P36-38 65-61548 <=$>
1951 V28 N6 P20-29 62-13929 <*=>
1952 V29 N3 P19-21 65-64607 <=*$>
1952 V29 N10 P24-27 RTS 2725 <NLL>
1952 V29 N10 P29-32 TT.417 <NRC>
1953 V30 N9 P9-10 60-13496 <+*>
1954 V31 P44 61-00457 <*>
64-15616 <=*$>
1954 V31 N1 P29-33 63-19229 <=*>
1954 V31 N6 P54-57 RT-4172 <*>
1954 V31 N7 P31-33 61-00464 <*>
64-15627 <=*$>
1954 V31 N8 P44-46 RT-2223 <*>
1955 V32 N2 P38-42 61-15385 <*+>
1955 V32 N3 P42-46 61-15973 <+*>
64-00284 <*>
64-15498 <=*$>
1955 V32 N4 P47-48 65-63791 <=*$>
1955 V32 N8 P36-43 64-19853 <=$>
1956 V33 N5 P60-61 64-19526 <=$>
1956 V33 N5 P70-71 60-15379 <+*>
1956 V33 N7 P53-56 61-13849 <+*>
1956 V33 N8 P57-60 59-14227 <+*>
1956 V33 N10 P68-71 R-3615 <*>
1957 V34 N1 P3-9 R-2390 <*>
1957 V34 N1 P80-82 R-5236 <*>
1957 V34 N3 P27-30 64-19378 <=$>
1957 V34 N3 P74-75 R-5226 <*>
1957 V34 N4 P50-54 R-2402 <*>
1957 V34 N4 P83-86 R-5224 <*>
1957 V34 N5 P3-9 R-1829 <*>
1957 V34 N5 P18-22 R-1793 <*>
1957 V34 N5 P25-26 R-1799 <*>
1957 V34 N5 P26-27 59-17292 <+*>
1957 V34 N5 P58-64 R-1795 <*>
1957 V34 N5 P64-66 59-10987 <+*>
1957 V34 N5 P79-81 R-1828 <*>
1957 V34 N5 P84-86 R-1794 <*>

Citation	Translation
1957 V34 N5 P88-	R-1798 <*>
1957 V34 N5 P88-95	R-1800 <*>
1957 V34 N6 P3-10	R-1801 <*>
1957 V34 N6 P59-60	R-1792 <*>
1957 V34 N6 P66-68	R-1789 <*>
1957 V34 N6 P71-75	R-1797 <*>
1957 V34 N6 P77-78	R-1806 <*>
1957 V34 N6 P83-85	R-1796 <*>
1957 V34 N6 P86-89	R-1786 <*>
1957 V34 N7 P41-43	R-1976 <*>
1957 V34 N7 P44-45	R-2032 <*>
1957 V34 N7 P45-47	R-2034 <*>
1957 V34 N7 P48-51	R-2036 <*>
1957 V34 N7 P79-80	R-2011 <*>
1957 V34 N7 P80-	R-2021 <*>
1957 V34 N7 P81-84	R-2041 <*>
1957 V34 N8 P30-35	R-2029 <*>
1957 V34 N8 P50-51	R-5222 <*>
1957 V34 N8 P59-	R-2398 <*>
1957 V34 N8 P59-60	R-4752 <*>
1957 V34 N8 P77-80	59-14229 <+*>
1957 V34 N9 P8-26	R-3093 <*>
1957 V34 N9 P27-31	R-2409 <*>
1957 V34 N9 P38-43	R-3555 <*>
1957 V34 N9 P47-50	R-3091 <*>
1957 V34 N9 P75-78	R-3607 <*>
1957 V34 N9 P81-82	R-2410 <*>
1957 V34 N9 P82-83	R-3610 <*>
1957 V34 N9 P84	R-3602 <*>
1957 V34 N9 P85-94	R-3608 <*>
1957 V34 N9 P95-96	R-3485 <*>
1957 V34 N10 P6-16	R-3867 <*>
1957 V34 N10 P16-24	R-3869 <*>
1957 V34 N10 P33-38	60-15375 <+*>
1957 V34 N10 P39-45	R-3609 <*>
1957 V34 N10 P58-62	R-4315 <*>
1957 V34 N10 P64-67	R-4327 <*>
1957 V34 N10 P67-68	R-4306 <*>
1957 V34 N10 P69-	R-4318 <*>
1957 V34 N10 P70-	R-4329 <*>
1957 V34 N10 P71-	R-4328 <*>
1957 V34 N10 P74-79	R-3871 <*>
1957 V34 N10 P80-83	R-3090 <*>
1957 V34 N10 P89-91	R-2397 <*>
1957 V34 N10 P91-92	R-3502 <*>
1957 V34 N10 P92-96	R-4311 <*>
1957 V34 N11 P25-39	R-3100 <*>
1957 V34 N11 P56-62	R-3750 <*>
1957 V34 N11 P63-65	R-3709 <*>
1957 V34 N11 P66-67	R-3710 <*>
1957 V34 N11 P93-	R-2396 <*>
1957 V34 N11 P94-96	R-3498 <*>
1957 V34 N12 P11-20	R-3611 <*>
1957 V34 N12 P21-26	R-3605 <*>
1957 V34 N12 P37-43	R-3596 <*>
1957 V34 N12 P60-64	R-4755 <*>
1957 V34 N12 P65-70	R-3866 <*>
1957 V34 N12 P74-76	R-4756 <*>
1957 V34 N12 P77-79	R-3489 <*>
1957 V34 N12 P79-80	R-3599 <*>
1957 V34 N12 P80-82	R-3606 <*>
1957 V34 N12 P83-84	R-3601 <*>
1957 V34 N12 P84-86	R-3101 <*>
1958 V35 N1 P5-10	R-3501 <*>
1958 V35 N1 P19-25	R-4346 <*>
1958 V35 N1 P33-37	R-4312 <*>
1958 V35 N1 P37-40	R-4326 <*>
1958 V35 N1 P40-49	R-3500 <*>
1958 V35 N1 P50-51	R-3729 <*>
1958 V35 N1 P87-91	R-4309 <*>
1958 V35 N1 P92-93	R-3598 <*>
1958 V35 N1 P93-96	R-3597 <*>
1958 V35 N2 P5-11	R-4316 <*>
1958 V35 N2 P12-20	R-3495 <*>
1958 V35 N2 P20-25	R-3868 <*>
1958 V35 N2 P43-44	59-10036 <+*>
1958 V35 N2 P94-96	R-3497 <*>
1958 V35 N2 P96	R-3496 <*>
1958 V35 N3 P10-	R-4310 <*>
1958 V35 N3 P10-12	R-4754 <*>
1958 V35 N3 P29-30	R-5175 <*>
1958 V35 N3 P31-34	R-5221 <*>
1958 V35 N3 P38-41	R-5225 <*>
1958 V35 N3 P74-76	59-10035 <+*>
1958 V35 N3 P83	R-4753 <*>
1958 V35 N3 P85-86	R-4761 <*>
1958 V35 N3 P87-88	R-3870 <*>
1958 V35 N3 P89-93	R-3874 <*>
1958 V35 N3 P94	R-4768 <*>
1958 V35 N4 P27-29	59-10034 <+*>
1958 V35 N4 P79-81	R-4345 <*>
1958 V35 N4 P83-88	R-4308 <*>
1958 V35 N4 P91-92	R-4314 <*>
1958 V35 N4 P92-93	R-4317 <*>
1958 V35 N4 P93-96	R-4313 <*>
1958 V35 N5 P10-13	R-4758 <*>
1958 V35 N5 P16-17	R-4762 <*>
1958 V35 N5 P22	R-4760 <*>
1958 V35 N5 P84-89	R-4759 <*>
1958 V35 N5 P100-104	R-4767 <*>
1958 V35 N6 P13-15	59-10041 <+*>
1958 V35 N6 P19-24	R-5174 <*>
1958 V35 N6 P28-31	R-5223 <*>
1958 V35 N6 P34-37	59-18405 <+*>
1958 V35 N6 P37-38	R-5176 <*>
1958 V35 N6 P56-57	R-4764 <*>
1958 V35 N6 P57	R-4765 <*>
1958 V35 N6 P61	59-10040 <+*>
1958 V35 N6 P62-63	59-10091 <+*>
1958 V35 N6 P64-66	59-10032 <+*>
1958 V35 N6 P69-73	R-4766 <*>
1958 V35 N6 P83-91	59-10033 <+*>
1958 V35 N6 P91-96	R-4763 <*>
1958 V35 N7 P15-20	R-5235 <*>
1958 V35 N7 P21-25	59-20320 <+*>
1958 V35 N7 P31-36	R-5177 <*>
1958 V35 N7 P37-46	59-14223 <+*>
1958 V35 N7 P81	59-14225 <+*>
1958 V35 N7 P82-83	59-14224 <+*>
	59-15389 <+*>
1958 V35 N7 P86-87	59-15390 <+*>
1958 V35 N7 P88-91	R-5246 <*>
1958 V35 N7 P92-95	59-10050 <+*>
1958 V35 N8 P11-13	R-5228 <*>
1958 V35 N8 P23-29	59-12720 <+*>
1958 V35 N8 P30-35	59-14673 <+*>
1958 V35 N8 P37-41	59-14230 <+*>
1958 V35 N8 P42-46	59-14226 <+*>
1958 V35 N8 P61-64	R-5227 <*>
1958 V35 N8 P80-81	59-12714 <+*>
1958 V35 N8 P81	59-12716 <+*>
1958 V35 N8 P82	59-12717 <+*>
1958 V35 N8 P94-95	59-10045 <+*>
1958 V35 N8 P95-96	59-10048 <+*>
1958 V35 N9 P19-23	R-5239 <*>
1958 V35 N9 P48-50	59-14666 <+*>
1958 V35 N9 P59-60	59-12718 <+*>
1958 V35 N9 P60-64	R-5238 <*>
1958 V35 N9 P72	59-12715 <+*>
1958 V35 N9 P78	59-15395 <+*>
1958 V35 N9 P92-95	59-10008 <+*>
1958 V35 N10 P28-31	59-12731 <+*>
1958 V35 N10 P31-38	59-14665 <+*>
1958 V35 N10 P38-43	59-12723 <+*>
1958 V35 N10 P44-45	59-12719 <+*>
1958 V35 N10 P49-54	59-12725 <+*> 0
1958 V35 N10 P60-62	59-12730 <+*>
1958 V35 N10 P69-71	59-12721 <+*> 0
1958 V35 N10 P76-80	59-14672 <+*>
1958 V35 N10 P81-83	59-14667 <+*>
1958 V35 N10 P87-88	59-12722 <+*>
1958 V35 N10 P90-92	59-12713 <+*>
1958 V35 N10 P92-94	59-12728 <+*>
1958 V35 N10 P94-95	59-12729 <+*>
1958 V35 N11 P89-92	59-12726 <+*>
1958 V35 N11 P92-94	59-12727 <+*>
1958 V35 N11 P94-96	59-31022 <=>
1958 V35 N12 P59-60	59-16144 <+*>
1958 V35 N12 P77-	59-14228 <+*>
1958 V35 N12 P81-83	59-14232 <+*>

Citation	Reference
1959 V36 N1 P30-32	59-14671 <+*>
1959 V36 N2 P11-13	59-16205 <+*>
1959 V36 N2 P28-34	59-16151 <+*> 0
1959 V36 N2 P41	59-16203 <+*>
1959 V36 N2 P42-44	59-18303 <+*>
1959 V36 N2 P45-48	59-14676 <+*>
1959 V36 N2 P48-51	59-16150 <+*>
1959 V36 N2 P51-55	59-14679 <+*>
1959 V36 N2 P55-56	59-16204 <+*>
1959 V36 N2 P57-62	59-16207 <+*> 0
1959 V36 N2 P92-93	59-14674 <+*>
1959 V36 N2 P93-	59-14675 <+*>
1959 V36 N3 P37-	60-31644 <=>
1959 V36 N3 P68-73	61-19641 <=*>
1959 V36 N3 P76-78	64-11648 <=>
1959 V36 N3 P91	59-16148 <+*>
1959 V36 N3 P94-95	59-16152 <+*>
1959 V36 N4 P8-13	59-19266 <+*>
1959 V36 N4 P71-72	59-19267 <+*>
1959 V36 N4 P79-81	59-18305 <+*>
1959 V36 N4 P84-87	59-18308 <+*>
1959 V36 N4 P88	59-18306 <+*>
1959 V36 N4 P91-94	59-18304 <+*>
1959 V36 N4 P94-96	59-18302 <+*>
1959 V36 N5 P41-45	59-18311 <+*> 0
1959 V36 N5 P62-63	59-19268 <+*>
1959 V36 N5 P70-72	59-19263 <+*>
1959 V36 N5 P85-89	59-18309 <+*>
1959 V36 N5 P90-95	59-18310 <+*>
1959 V36 N6 P87-90	59-19262 <+*>
1959 V36 N6 P91-94	59-19264 <+*>
1959 V36 N6 P94-96	59-19260 <+*>
1959 V36 N7 P30-31	59-22294 <+*>
1959 V36 N7 P69-75	59-19681 <+*>
1959 V36 N7 P74-83	59-22292 <+*> 0
1959 V36 N7 P84-88	59-22298 <+*>
1959 V36 N7 P89-90	59-22293 <+*>
1959 V36 N7 P91	59-2295 <+*>
1959 V36 N7 P92-93	59-19673 <+*>
1959 V36 N8 P79	59-22299 <+*> 0
	60-14127 <+*>
1959 V36 N8 P93	59-22296 <+*>
	59-22300 <+*>
1959 V36 N9 P26-27	60-13483 <+*>
1959 V36 N9 P54-60	60-13485 <+*>
1959 V36 N9 P67-70	60-13484 <+*>
1959 V36 N9 P70-72	60-31644 <=>
1959 V36 N9 P80-82	59-22302 <+*>
1959 V36 N9 P83	59-22303 <+*>
1959 V36 N9 P84-89	60-13482 <+*>
1959 V36 N9 P90-92	59-22291 <+*>,
1959 V36 N10 P58	60-15954 <+*> 0
1959 V36 N10 P70-77	60-13488 <+*>
1959 V36 N10 P91	60-13487 <+*>
1959 V36 N11 P51-55	60-13489 <+*>
1959 V36 N12 P32-33	60-15953 <+*>
1959 V36 N12 P43-50	60-15946 <+*>
1959 V36 N12 P50-54	60-15948 <+*>
1960 V37 N1 P5-7	60-15958 <+*>
1960 V37 N1 P12-17	60-15960 <+*> C
1960 V37 N1 P30-31	60-15964 <+*>
1960 V37 N1 P33-35	60-15961 <+*>
1960 V37 N1 P68-70	60-17967 <+*> 0
1960 V37 N1 P73-74	60-15962 <+*>
1960 V37 N1 P95-96	60-15959 <+*>
1960 V37 N2 P5-9	60-17969 <+*>
1960 V37 N2 P16-19	60-17970 <+*>
1960 V37 N2 P35-38	60-17971 <+*>
1960 V37 N2 P67-69	18M42R <ATS>
1960 V37 N2 P75-76	60-17974 <+*>
1960 V37 N3 P66-69	61-19532 <=*>
1960 V37 N3 P81-84	60-17972 <+*>
1960 V37 N3 P88-90	60-17975 <+*>
1960 V37 N4 P91-93	60-17973 <+*>
1960 V37 N6 P25-29	61-23204 <=*> 0
1960 V37 N6 P31-33	61-27478 <*=>
1960 V37 N6 P73-74	61-19774 <=*>
1960 V37 N6 P94	61-19775 <=*>
1960 V37 N7 P11-13	61-19761 <=*>
1960 V37 N7 P43-44	61-23203 <=*>
1960 V37 N7 P82-83	61-19776 <=*>
1960 V37 N7 P85-86	61-19777 <=*>
1960 V37 N8 P10-16	61-19778 <=*>
1960 V37 N8 P26-27	61-23624 <=*>
1960 V37 N8 P33-46	61-23624 <=*>
1960 V37 N9 P10-23	61-23202 <=*>
1960 V37 N9 P38-40	61-23205 <=*>
1960 V37 N10 P15-18	61-23206 <=*>
1960 V37 N10 P25-32	61-27064 <=*>
1960 V37 N10 P33-35	61-19758 <=*>
1960 V37 N10 P39-43	61-27064 <=*>
1960 V37 N10 P93	61-19759 <=*>
1960 V37 N11 P19-23	61-27483 <*=>
1960 V37 N11 P75-78	61-19651 <=*>
1960 V37 N12 P17-18	61-27484 <*=>
1960 V37 N12 P59-63	61-28148 <*=>
1961 V38 N1 P27-29	61-27485 <*=>
1961 V38 N1 P30-	61-27486 <*=>
1961 V38 N1 P75-77	61-23207 <=*>
1961 V38 N2 P37-38	64-13790 <=*$>
1961 V38 N2 P45-47	61-27487 <*=>
1961 V38 N2 P54-55	RTS-3048 <NLL>
1961 V38 N2 P74-77	61-27488 <*=> C
1961 V38 N3 P15-16	61-28443 <*=>
1961 V38 N3 P69-70	NP-TR-1136 <*>
1961 V38 N5 P36-39	61-28446 <*=>
1961 V38 N5 P43	61-28449 <*=>
1961 V38 N5 P48-51	61-28149 <*=>
1961 V38 N6 P36-42	61-28444 <*=>
1961 V38 N6 P45-46	61-28450 <*=>
1961 V38 N6 P48-52	61-28445 <*=>
1961 V38 N6 P83-84	61-28451 <*=>
1961 V38 N7 P42-44	62-15789 <=>
1961 V38 N7 P78-79	62-15789 <=>
1961 V38 N7 P95-96	62-15789 <=>
1961 V38 N8 P18-20	62-23300 <=*>
1961 V38 N8 P21-22	62-23301 <=*>
1961 V38 N8 P23-24	62-23302 <=*>
1961 V38 N8 P33-36	62-23303 <=*>
1961 V38 N8 P39-40	AD-619 339 <=$>
1961 V38 N8 P70-71	62-23290 <=*>
1961 V38 N9 P12-16	62-23384 <=>
1961 V38 N9 P30-33	AD-619 340 <=$>
1961 V38 N9 P76-81	62-23384 <=>
1961 V38 N9 P84-86	62-23384 <=>
1961 V38 N9 P92-96	62-23384 <=>
1961 V38 N10 P14	62-23629 <=>
1961 V38 N10 P18-19	62-23629 <=>
1961 V38 N10 P38-44	62-23629 <=>
1961 V38 N10 P55-57	62-23629 <=>
1961 V38 N10 P70-71	64-13789 <=*$>
1961 V38 N11 P19-23	62-23291 <=*>
1961 V38 N11 P33-34	62-23288 <=*>
1961 V38 N11 P65-68	62-23289 <=*>
1961 V38 N12 P13-14	62-10654 <=*>
1962 V39 N1 P15-22	62-16086 <=*>
1962 V39 N1 P23-24	62-14852 <=*>
1962 V39 N1 P41-42	62-14851 <=*>
1962 V39 N1 P43-46	62-14850 <=*>
1962 V39 N1 P66-69	62-14848 <=*>
1962 V39 N1 P86-87	62-33556 <=*>
1962 V39 N1 P88	62-33557 <=*>
1962 V39 N1 P92	62-14849 <=*>
1962 V39 N2 P18-23	63-15559 <=*>
1962 V39 N2 P27-30	63-19018 <=*>
1962 V39 N2 P37-41	62-16194 <=*> 0
	64-13889 <=*$>
1962 V39 N2 P76	63-19371 <=*>
1962 V39 N3 P48-50	63-15413 <=*>
1962 V39 N3 P58-60	63-15414 <=*>
1962 V39 N3 P96	62-32130 <=*>
1962 V39 N4 P31-33	63-15412 <=*>
1962 V39 N4 P53-54	64-13892 <=*$>
1962 V39 N4 P86-87	63-24375 <=*$>
1962 V39 N4 P87-91	63-19016 <=*>
1962 V39 N5 P65-72	62-32594 <=*>
1962 V39 N5 P88	62-32131 <=*>
1962 V39 N6 P10-12	63-13232 <=*>
1962 V39 N6 P57	63-23414 <=*>
1962 V39 N8 P67-69	63-15466 <=*>

1962 V39 N8 P69-72 63-15939 <=*>
1962 V39 N8 P77-78 63-15465 <=*>
1962 V39 N10 P54-55 64-13639 <=*$>
1962 V39 N10 P80-82 63-31777 <=>
1962 V39 N1C P92-96 63-21070 <=>
1962 V39 N10 P92 63-31777 <=>
1962 V39 N1C P95-96 63-31777 <=>
1962 V39 N11 P50 66-11296 <*>
1962 V39 N11 P75-79 63-15730 <=*>
1962 V39 N11 P92-94 63-31777 <=>
1962 V39 N12 P47-52 63-23914 <=*$>
1962 V39 N12 P58-61 63-24483 <=*$>
1962 V39 N12 P61-62 63-23915 <=*$>
1962 V39 N12 P69-74 63-24011 <=*>
1962 V39 N12 P76 63-31777 <=>
1963 V40 N1 P5 63-31777 <=>
1963 V40 N1 P30-31 64-13895 <=*$>
1963 V40 N1 P79-81 63-23336 <=*$> 0
1963 V40 N1 P84 63-23694 <=*$>
1963 V40 N1 P89-92 63-23663 <=*$>
1963 V40 N2 P24-27 63-23660 <=*$>
1963 V40 N2 P27-28 63-24008 <=*>
1963 V40 N2 P28-30 63-24009 <=*$>
1963 V40 N2 P73-74 63-24010 <=*$>
1963 V40 N2 P79-83 63-23913 <=*$> 0
1963 V40 N3 P5-9 63-23688 <=*$>
1963 V40 N3 P17-23 63-23689 <=*$>
64-15441 <=*$>
1963 V40 N3 P23-25 63-23690 <=*$>
1963 V40 N3 P26 63-23691 <=*$>
1963 V40 N3 P27 63-23692 <=*$>
1963 V40 N3 P28 63-23806 <=*$>
1963 V40 N3 P64-65 63-23807 <=*$>
64-15442 <=*$>
1963 V40 N3 P73-76 64-15443 <=*$>
1963 V40 N3 P92-94 63-23805 <=*$>
1963 V40 N3 P93 63-31777 <=>
1963 V40 N3 P95-96 63-31777 <=>
1963 V40 N4 P9-92 63-23543 <=*$>
1963 V40 N4 P38-41 64-19258 <=*$>
1963 V40 N4 P56-57 64-19255 <=*$>
1963 V40 N4 P87-90 63-31777 <=>
1963 V40 N5 P30-32 63-23345 <=*>
1963 V40 N5 P58-61 63-23662 <=*$>
1963 V40 N5 P64-67 63-23661 <=*$>
1963 V40 N5 P91-92 63-24112 <=*> 0
1963 V40 N6 P6-10 63-24113 <=*>
1963 V40 N6 P20-26 64-21220 <=>
1963 V40 N7 P19-20 64-19272 <=*$>
1963 V40 N7 P20-21 64-19287 <=*$>
1963 V40 N7 P25-26 63-24391 <=*$>
1963 V40 N7 P26-30 64-13137 <=*$>
1963 V40 N7 P32-34 63-24392 <=*$>
1963 V40 N7 P38-40 64-19628 <=$>
1963 V40 N7 P46 64-19288 <=*$>
1963 V40 N8 P15-16 64-13135 <=*$>
1963 V40 N8 P16-17 64-13134 <=*$>
1963 V40 N8 P18-19 64-13133 <=*$>
1963 V40 N8 P25-26 64-13141 <=*$>
1963 V40 N8 P25 64-13145 <=*$>
1963 V40 N8 P26 64-13143 <=*$>
1963 V40 N8 P41 64-13344 <=*$>
1963 V40 N8 P77-78 63-24440 <=*$>
1963 V40 N10 P29-30 64-15873 <=*$>
1963 V40 N10 P31 64-19363 <=$>
1963 V4C N1C P7C-72 64-13139 <=>
1963 V40 N11 P19-27 64-13539 <=*$>
1963 V40 N11 P28-30 64-13348 <=*$>
1963 V40 N11 P30-31 64-13572 <=*$>
1963 V40 N11 P37 64-13346 <=*$>
1963 V40 N11 P38-43 64-13755 <=*$>
1963 V40 N11 P43-47 64-19035 <=*$>
1963 V40 N11 P47-49 64-19036 <=*$>
1963 V40 N11 P50-51 64-13349 <=*$>
1963 V40 N11 P52-53 64-19037 <=*$>
1963 V40 N11 P63-65 64-13697 <=*$>
1963 V40 N11 P69-70 64-13756 <=*$>
1963 V40 N11 P71-72 64-13696 <=*$>
1963 V40 N11 P77-79 64-13695 <=*$>
1963 V40 N12 P54-55 AD-618 861 <=$>
1963 V40 N12 P55-57 AD-619 008 <=$>
1964 V41 N1 P78-81 65-63263 <=>
1964 V41 N5 P23-25 64-19367 <=$>
1964 V41 N5 P25-27 64-19366 <=$>
1964 V41 N5 P32-37 64-19365 <=$>
1964 V41 N5 P38-39 64-19364 <=$>
1964 V41 N5 P39-42 64-19187 <=*$>
1964 V41 N5 P43-46 64-19840 <=$>
1964 V41 N5 P59-63 CSIR-TR-412 <CSSA>
1964 V41 N5 P73 64-19841 <=$>
1964 V41 N5 P88 65-61505 <=$>
1964 V41 N5 P9C-92 64-19842 <=$>
1964 V41 N5 P94-97 64-19188 <=*$>
1964 V41 N5 P123-126 64-41776 <=$>
1964 V41 N6 P23-24 65-60248 <=$>
1964 V41 N6 P25-26 65-63157 <=$>
1964 V41 N6 P35-40 64-19901 <=$> C
1964 V41 N6 P59 65-60249 <=$>
1964 V41 N8 P1-3 65-60250 <=$>
1964 V41 N8 P6-9 65-60253 <=$>
1964 V41 N8 P13-16 65-60251 <=$> 0
1964 V41 N8 P21-24 64-19908 <=$> C
1964 V41 N8 P37-39 64-19902 <=$>
1964 V41 N8 P43-44 65-64601 <=*$>
1964 V41 N8 P9C-92 66-60978 <=*$>
1964 V41 N9 P7-9 64-19905 <=$>
1964 V41 N9 P11 64-19906 <=$>
1964 V41 N9 P17-20 64-19907 <=$>
1964 V41 N9 P29 65-60240 <=$>
1964 V41 N9 P129 65-60241 <=$>
1964 V41 N10 P26-29 65-60986 <=$>
1964 V41 N10 P29-31 65-63265 <=>
1964 V41 N10 P96-98 65-62965 <=> 0
1964 V41 N10 P109-112 65-60907 <=$>
1964 V41 N11 P7-10 65-63387 <*>
1964 V41 N11 P16-21 65-63388 <=$>
1964 V41 N11 P25-27 65-63390 <=$>
1964 V41 N11 P27-32 65-63389 <=$>
1964 V41 N12 P8-10 65-63384 <*>
1964 V41 N12 P13-15 65-63383 <=$>
1964 V41 N12 P35-36 65-63382 <=$>
1964 V41 N12 P47-49 65-63381 <=$>
1964 V41 N12 P59-61 65-63380 <=$>
1964 V41 N12 P68-70 65-63391 <=$>
1965 V42 N1 P4C-43 65-31416 <=$>
1965 V42 N2 P1-4 65-31132 <=$>
1965 V42 N3 P89-91 65-31416 <=$>
1965 V42 N3 P1C4-108 65-31416 <=$>
1965 V42 N5 P123-124 65-33602 <=*$>
1965 V42 N6 P118-120 65-32504 <=$>
1965 V42 N8 P107-108 65-33568 <=*$>
1965 V42 N9 P15-17 65-33557 <=*$>
1965 V42 N10 P67-69 RTS-3378 <NLL>

VETERINARNA SBIRKA
1962 N10 P3-5 63-15378 <=*>
1963 V40 N7 P9-10 63-24388 <=*$>
1963 V40 N7 P1C-12 63-24389 <=*$>
1963 V40 N7 P12 63-24390 <=*$>
1963 V40 N7 P87-88 63-24393 <=*$>
1963 V40 N8 P11-14 64-13136 <=*$>
1963 V40 N8 P26-27 64-13144 <=*$>
1963 V40 N8 P26 64-13146 <=*$>
1963 V40 N8 P71-74 64-13537 <=*$>
1963 V4C N1C P68-69 63-24387 <=*$>

VETERINARNI MEDICINA
1964 V37 N9 P222 CSIR-TR-468 <CSSA>

VETERINARSKI ARHIV
1952 V22 N3/4 P151-156 60-21682 <=>
1952 V22 N9/10 P375-377 60-21682 <=>
1953 V23 N7/8 P232-242 3932 <K-H>
1956 V26 N3/4 P120-132 60-21627 <=>
1956 V26 N7/8 P192-201 60-21682 <=>
1957 V27 N1/2 P25-32 60-21627 <=>
1958 V28 N7/8 P236-242 60-21682 <=>

VETERINARSKI GLASNIK
1954 V8 N10 P623-631 60-21624 <=>

1960 V14 N5 P323-326 62-19365 <=>
1960 V14 N8 P572-589 62-19364 <*=>
1960 V14 N8 P596-636 62-19364 <*=>
1961 V15 N1 P71-73 62-19367 <=*>
1961 V15 N2 P93-98 62-19710 <=*>
1961 V15 N2 P1C5-108 62-19711 <=*>
1961 V15 N2 P139-144 62-19709 <=*>
1961 V15 N3 P191-200 62-19712 <=*>
1962 V16 N3 P279-285 62-25172 <=*>

VETERINARSTVI. BRNO, PRAHA
1958 P341-344 62-00866 <*>
1958 V8 N3 P93-95 61-27063 <*=>
1959 V9 N11 P402-408 62-19366 <*=> O
1964 V14 N4 P162-166 66-10926 <*> O

VETRO E SILICATI
1960 V4 N19 P9-16 63-18593 <*>
1962 V6 N34 P5-9 BGIRA-564 <BGIR>
1963 V7 N38 P5-9 BGIRA-593 <BGIR>

VIATA ECONOMICA
1964 N33 P8 64-41961 <=$>
1965 N24 P3 65-31603 <=$>
1965 N29 P12-13 65-32190 <=$>
1965 N31 P8-9,14 65-32382 <=*$>
1966 N6 P1 66-30996 <=$> O
1966 N52 P5 67-30410 <=>
1966 N52 P18 67-30410 <=>
1966 V4 N18 P5 66-32281 <=$>
1966 V4 N29 P8-9 06/22 66-34135 <=$>
1966 V4 N37 P5 66-34868 <=$>
66-34899 <$=>
1966 V4 N38 P13 66-34996 <=$>
1967 N9 P5-9 67-31083 <=$>
1967 N9 P6 67-31097 <=$>

VIATA MEDICALA
1961 V8 N15 P833-838 62-19605 <=*>

VIATA STUDENTEASCA
1965 P10 01/09 65-30456 <=*$>

VIDA NUEVA. HAVANA
1952 V69 N5 P103-111 64-16397 <*> O

VIDE. TECHNIQUE, APPLICATIONS
1946 V1 N4/5 P137-142 57-303 <*>
1947 V2 P194-204 I-344 <*>
1947 V2 N7 P194-204 AEC-TR-1945 <*>
1947 V2 N10/1 P304-307 57-370 <*>
1947 V2 N1CC1 P291-304 57-320 <*>
1948 N18 P522-530 AD-440 359 <=$>
1948 V3 P487-495 63-20474 <*> O
1951 N31 P942-950 I-934 <*>
1952 V7 P1262-1266 57-2420 <*>
1954 V9 N52/3 P200-202 65-11472 <*> O
1955 N58/9 P105-118 57-3007 <*>
1955 N58/9 P124-134 57-3006 <*>
1955 V10 N55 P362-365 64-16924 <*>
1956 V11 N61 P3-13 66-12372 <*>
1956 V11 N64 P194-205 64-16960 <*> O
1957 N68 P162-166 58-745 <*>
1957 V12 N68 P162-166 63-16541 <*>
1962 V17 N99 P267-272 NP-TR-101C <*>
1963 V18 N103 P55-68 N65-62760 <=$>
65-11155 <*>

VIE AUTOMOBILE
1931 V27 N977 P322-323 59-17016 <*> O
1955 P29-38 58-2361 <*>
1956 P231-238 58-2365 <*>
1956 P313-318 58-2360 <*>

VIE MEDICALE
1955 P16-26 58-2230 <*>
1960 V41 P1139-1140 11153-B <KH>
62-00372 <*>
63-16652 <*>
1961 V42 N3 P85-92 63-18037 <*> O

VIEH -UND FLEISCHWIRTSCHAFT
1951 V1 N22 P9-11 60-10143 <*>

VIERTELJAHRSSCHRIFT FUER GERICHTLICHE MEDIZIN UND OEFFENTLICHES SANITAETSWESEN
1877 V27 P325-331 58-546 <*>

VIERTELJAHRSSCHRIFT DER NATURFORSCHENDEN GESELLSCHAFT IN ZURICH
1911 V56 P15-41 1093 <*>
64-10443 <*>
1951 V96 N1 P1-21 18J19G <ATS>
1955 V100 P87-113 60-16651 <*>
1960 V105 P1-46 71M43G <ATS>

VIERTELJAHRSSCHRIFT FUER SCHWEIZERISCHE SANITAETS-OFFIZIERE
V34 N3 P262-269 58-2353 <*>

VINODELIE I VINOGRADARSTVO SSSR.
1950 V10 N5 P34-37 62-20148 <=*>
1950 V10 N7 P17-19 63-23407 <=*$>
1951 N1 P36-38 <CBPB>
1951 V11 N3 P16-19 75N54R <ATS>
1951 V11 N3 P19-22 75N54R <ATS>
1951 V11 N4 P43-44 RT-3319 <*>
1951 V11 N5 P42-44 64-15709 <=*$>
1952 V12 N4 P25-28 76N54R <ATS>
1952 V12 N4 P28-29 77N54R <ATS>
1952 V12 N5 P4C RT-3289 <*>
1952 V12 N11 P47-48 RT-3320 <*>
1953 V13 N7 P50 RT-3321 <*>
1953 V13 N8 P36-37 63-13606 <=*>
1953 V13 N8 P38-41 RT-3751 <*>
1953 V13 N9 P22-28 RT-3372 <*>
1954 V14 N2 P23-25 63-15971 <=*>
1954 V14 N6 P26-32 62-20154 <=*>
1955 V15 N6 P41-42 62-20155 <=*>
1956 V16 N7 P42-49 63-19104 <=*>
1956 V16 N8 P27-31 63-19150 <=*>
1960 V20 N2 P8-15 62-32964 <=*>
1961 V21 N7 P32-34 64-19123 <=*$>
1962 V22 N1 P32-34 63-23749 <=*>
1962 V22 N2 P31-36 63-23991 <=*>
1962 V22 N2 P36-42 63-23748 <=*>
1962 V22 N3 P19-21 63-23990 <=*>
1962 V22 N3 P52 63-23989 <=*>
1962 V22 N4 P1-3 63-24072 <=*>
1962 V22 N4 P8-12 63-24070 <=*>
1962 V22 N4 P19-22 63-24069 <=*$>
1962 V22 N4 P26-28 63-24073 <=*$>
1962 V22 N4 P36-38 63-24071 <=*>
1962 V22 N4 P44-50 62-33050 <=*> O
1962 V22 N4 P52-53 62-33050 <=*> O
1962 V22 N5 P30-33 63-23995 <=*>
1962 V22 N6 P3-6 63-23741 <=*$>
1962 V22 N6 P6-7 63-23992 <=*>
1962 V22 N6 P8-11 63-23743 <=*>
1962 V22 N6 P18-21 63-23744 <=*$>
1962 V22 N6 P29-31 63-23745 <=*$>
1962 V22 N6 P47-50 63-23746 <=*$>
1962 V22 N6 P5C-51 63-23747 <=*>
1962 V22 N7 P7-10 63-23993 <=*>
1962 V22 N7 P39-43 63-23994 <=*$>
1962 V22 N8 P21-23 64-13069 <=*$>
1962 V22 N8 P24-26 64-13072 <=*$>
1962 V22 N8 P3C-37 64-13070 <=*$>
1962 V22 N8 P47-49 64-13071 <=*$>
1963 V23 N1 P37-39 64-13077 <=*$>
1963 V23 N1 P46-51 64-13076 <=*$>
1963 V23 N2 P14-16 64-13075 <=*$>
1963 V23 N2 P2C-27 64-19128 <=*$>
1964 V24 N2 P46-52 65-30505 <=$>

VINYL AND POLYMER. JAPAN
SEE ENKA BINIIRU TO PORIMA

VIRCHOWS ARCHIV FUER PATHOLOGISCHE ANATOMIE UND PHYSIOLOGIE UND FUER KLINISCHE MEDIZIN

1861 V20 P381-393 61-14896 <*>
1868 V42 P112-128 57-1542 <*>
1871 V53 N4 P470-484 3344-C <K-H>
1877 V69 P106-143 N65-16304 <=$>
1878 V78 P15-53 59-10348 <*>
1888 N3 P461-474 57-1515 <*>
1889 V118 N3 P391-413 62-00303 <*>
1900 V162 P32-84 63-16589 <*>
9249-D <K-H>
1904 N176 P472-514 57-2710 <*>
1910 V199 P187-192 66-13290 <*>
1910 V199 P210-211 66-13290 <*>
1915 V219 P309-319 66-10952 <*>
1923 V244 P1-7 65-13832 <*> O
1923 V247 P294-361 3175 <*>
1926 V261 P846-857 58-1699 <*>
1928 V266 P697-711 57-804 <*>
1930 V275 P114-133 57-1306 <*>
1932 V286 P702-732 58-1702 <*>
1933 V290 P23-46 63-00314 <*>
1934 V293 P665-673 62-00592 <*>
1937 V300 P113-129 63-20897 <*> O
1940 V305 N1 P171-215 1730 <*>
1947 V314 P534-593 61-20186 <*> P
1950 V319 P116-126 I-576 <*>
1951 V321 P88-100 63-00078 <*>
1953 V323 P133-142 63-20200 <*> O
1953 V324 P489-508 65-00277 <*>
1954 V324 P612-628 57-673 <*>
1954 V324 N5 P612-628 59-17336 <*>
1954 V326 P89-118 61-00592 <*>
1955 V326 P501-562 57-570 <*>
1955 V327 P112-126 57-3207 <*>
1956 V328 P536-551 64-16342 <*> PO
1956 V329 P235-244 58-370 <*>
1958 V331 P249-262 61-00603 <*>
1959 V332 P224-235 62-00895 <*>
1960 V333 N1 P68-80 62-00826 <*>

VIRUS, JAPAN
SEE UIRUSU

VISION RESEARCH
1965 V5 P361-377 AD-621 957 <=*>
AD-626 436 <=*$>

VISITE UND ORDINATION
1963 V9 N2 P17-18 65-13780 <*>
1964 V10 N3/4 P45-46 65-13726 <*>

VISNYK AKADEMIYI NAUK UKRAYINSKOYI RSR. KYYIV
1954 V25 N8 P77-79 RT-2716 <*>
1955 V26 N5 P36-39 5866-H <K-H>
64-18403 <*>
1959 V30 N1 P15-23 59-13963 <=>
1959 V30 N6 P27-37 60-11155 <=>
1959 V30 N7 P64-66 60-31055 <=>
1959 V30 N9 P3-7 60-31021 <=>
1959 V30 N9 P59-62 60-11370 <=>

VISNYK. KIEV UNIVERSITY
SEE VESTNIK. GOSUDARSTVENNOGO UNIVERSITETA. KIEV

VISNYK SILSKOHOSPODARSKOYI NAUKY
1960 V3 N4 P10-11 60-31686 <=>

VISTI AKADEMIYI NAUK UKRAYINSKOYI RSR. KYYIV
1941 V8 N1 P61-66 61-11520 <=>
1941 V14 N1 P61-66 RT-1152 <*>

VISTI INSTYTUTU HIDROLOHYI I HIDROTEKHNIKU
1963 V22 P68-80 AD-617 089 <=$>

VISTI Z UKRAINY
1965 N47 P1 66-30166 <=$>

VITAMINS. JAPAN
SEE BITAMIN

VITAMINY. INSTITUT BIOKHIMII. AKADEMIYA NAUK USSR.
1958 V3 P32-43 66-10256 <*> C
1958 V3 P159-165 61-21482 <=>
1958 V3 P166-173 61-21408 <=>
1959 V4 P56-70 61-21451 <=>
1959 V4 P78-82 61-21465 <=>
1959 V4 P92-100 61-21473 <=>
1959 V4 P123-129 61-21474 <=>
1959 V4 P144-147 61-21466 <=>
1959 V4 P222-233 61-21366 <=>

VIZUGYI ERTESITO
1961 V8 N8 P60-64 62-19376 <=*>
1961 V8 N11/2 P96 62-19546 <=*>

VIZUGYI KOZLEMENYEI
1960 N3 P360-397 62-19377 <=*>
1962 N1 P137-150 65-20C07 <$>

VNADRENIE NOVYKH SPOSOBOV SPARKI V PROMYSHLENNOST
1959 N2 P5-16 2482 <BISI>

VNESHNYAYA TORGOVLYA
1959 N11 P34-38 60-11682 <=>
1959 V29 N9 P30-32 60-11413 <=>
1959 V29 N10 P48-50 60-11413 <=>
1962 V42 N7 P30-38 63-13329 <=> P
1962 V42 N8 P45 62-33664 <=*>
1962 V42 N10 P40-42 63-13511 <=> O
1963 V43 N3 P30-31 63-21783 <=>
1963 V43 N7 P25-26 63-41088 <=>
1964 V44 N1 P49-52 CEMBUREAU-TR-569 <CEM>
1964 V44 N6 P36-39 64-51488 <=>
1966 N6 P20-25 66-34341 <=$>

VNITRNI LEKARSTVI
1956 V2 P598-607 63-00845 <*>
1961 V7 P1242-1246 63-16384 <*> C
1962 V8 P95-100 63-14884 <*> O
1962 V8 N12 P1279-1283 64-00076 <*>
1963 V9 N4 P336-338 63-20902 <*>

VODNI HOSPODARSTVI
1960 V10 N8 P348-349 NP-TR-1073 <*>
1965 V15 N1 P19-20 25T91C <ATS>

VODOSNABZHENIE I SANITARNAYA TEKHNIKA
1955 N4 P38-40 RT-3686 <*>
1958 N1 P1-5 59-18758 <+*> O
1958 N2 P6-9 59-18761 <+*> C
1958 N2 P26-29 59-18760 <+*> C
1958 N2 P35 59-18759 <+*>
1958 N3 P15-21 59-18762 <+*> C
1958 N5 P20-23 59-18763 <+*> O
1958 N9 P10-14 60-15425 <+*>
1958 N10 P24-27 59-18735 <+*> C
1958 N10 P27-30 59-18736 <+*> O
1959 N1 P38-39 59-19177 <+*> C
1959 N12 P1-3 61-23362 <=*>
1959 N12 P4-5 61-23361 <=*>
1959 N12 P5-6 61-23360 <=*>
1959 N12 P6-11 61-23359 <=*>
1959 N12 P11-13 61-23357 <=*>
1959 N12 P11 61-23358 <=*>
1959 N12 P14-15 61-23356 <=*>
1959 N12 P15-18 61-23335 <=*>
1959 N12 P19-22 61-23354 <=*>
1959 N12 P22-25 61-23353 <=*>
1959 N12 P26 61-23352 <=*>
1959 N12 P27-28 61-23351 <=*>
1959 N12 P28-29 61-23350 <*=>
1959 N12 P29-31 61-23349 <=*>
1959 N12 P31-32 61-23348 <=*>
1959 N12 P33-34 61-23346 <*=>
1959 N12 P33 61-23347 <=*>
1959 N12 P34-36 61-23345 <=*>
1959 N12 P36-37 61-23232 <=*>
1959 N12 P37-41 61-23233 <=*>
1960 N6 P37-38 60-31864 <=>
1960 N8 P27-30 61-28929 <=*>

1961 N3 P1-6 61-27098 <*=>
1961 N3 P39-40 61-27072 <*=>
1961 N5 P35-38 61-28906 <*=>
1962 N6 P12-15 63-15954 <=*>
1963 N7 P11-13 64-13628 <=*$> 0
1965 N10 P26-29 65-34011 <=*$>
1965 N11 P9-12 66-30370 <=*$>
1966 N1 P37-39 66-31939 <=$>

VOEDING
1956 V17 N11 P485-498 59-14715 <=*>
1962 V23 N10 P509-510 64-00261 <*>

VOENNAYA MYSL
1955 P19-32 64-71433 <=>
1955 N3 P19-32 R-3590 <*>

VOENNO-ISTORICHESKII ZHURNAL
1963 V5 N2 P37-47 64-71424 <=> M
1963 V5 N8 P38-48 64-41327 <=>
1964 N1 P85-88 AD-626 075 <=*$>
1965 N5 P56-72 AD-626 287 <=*$>
1965 N7 P3-15 65-32424 <=$>
1965 N7 P24-34 65-32424 <=$>
1965 N7 P88-104 65-32424 <=$>
1965 N8 P3-10 65-33739 <=*$>
1965 N8 P40-44 65-33816 <=*$>
1965 N10 P3-12 65-33823 <=*$>
1965 N10 P14-25 65-33826 <=*$>
1965 N10 P26-34 65-33822 <=*$>
1965 N10 P68-74 65-33779 <=*$>
1966 V3 P24-44 66-32015 <=$>

VOENNO-MEDITSINSKII SBORNIK
1944 V1 P109-118 65-60099 <=$>

★ VOENNO-MEDITSINSKII ZHURNAL
1933 V4 N6 P367-370 RT-1467 <*>
1955 N1 P48-53 PB 141 153T <=>
1955 N3 P46-48 59-11082 <=>
1955 N3 P55-59 59-11081 <=>
1955 N3 P75-77 59-11083 <=> 0
1955 N5 P83-84 62-33720 <=*>
1955 N7 P61-69 59-11278 <=>
1955 N8 P52-57 59-11058 <=> 0
1955 N8 P57-65 59-11057 <=>
1955 N8 P65-67 59-11056 <=>
1955 N12 P73-74 RT-4569 <*>
1956 N2 P11-23 62-23546 <=*>
1956 N5 P29-32 62-15227 <=*>
1957 N1 P29-35 PB 141 275T <=>
1957 N1 P35-37 PB 141 276T <=>
1957 N1 P38-50 PB 141 277T <=>
1957 N1 P50-55 PB 141 278T <=>
1957 N1 P61-67 PB 141 461T <=>
1957 N3 P81-82 60-17638 <+*>
1957 N3 P94-96 PB 141 163T <=>
1957 N4 P19-21 AEC-TR-3625 <+*>
1957 N6 P9-20 PB 141 264T <=>
1957 N6 P20-22 59-11041 <=>
1957 N6 P22-25 59-11035 <=>
1957 N6 P25-28 59-11036 <=>
1957 N6 P33-37 59-11040 <=>
1957 N6 P37-40 59-13066 <=>
1957 N6 P65-72 59-11118 <=>
1957 N7 P40-43 59-12128 <+*>
1957 N8 P70-80 61-31586 <=>
1957 N8 P83-84 61-31586 <=>
1957 N9 P31-34 59-11658 <=>
1957 N9 P57-61 59-11659 <=>
1957 N9 P84-85 59-11660 <=>
1957 N9 P87-90 59-11661 <=>
1957 N9 P91-94 R-3604 <*>
1957 N9 P95 59-11662 <=>
1958 N4 P12-15 63-13619 <=*>
1958 N12 P40-44 59-19708 <+*>
1958 N12 P51-59 AD-630 735 <=$>
1959 N5 P23-29 59-22155 <+*>
1959 N5 P29-32 59-19701 <+*>
1959 N5 P73-75 59-13986 <=>
1959 N5 P75-77 59-13987 <=>
1959 N6 P71-75 59-13967 <=>
1959 N6 P76-80 60-11154 <=>
1959 N6 P81-88 60-11154 <=>
1959 N6 P91 59-13968 <=> 0
1959 N7 P3-5 59-22288 <+*>
1959 N7 P6-7 59-22287 <+*>
1959 N7 P83-90 60-11058 <=>
1959 N7 P91-94 60-31007 <=>
1959 N8 P7-10 59-31037 <=>
1959 N8 P27-32 59-22286 <+*>
1959 N8 P42-45 60-23727 <+*>
1960 N4 P51-54 63-19703 <=>
1960 N5 ENTIRE ISSUE 60-41394 <=>
1961 N2 P31-34 61-20551 <*=>
1961 N3 P41-51 61-20552 <*=> 0
1962 N8 P57-62 AD-619 335 <=$>
1966 N5 P91-94 66-32861 <=$>

VOENNO-MEDITSINSKII ZHURNAL. S.-PETERBURG
1877 V130 P1-2 PT.2 RT-2520 <*>
1877 V130 P1-2 SEC. 2 64-19084 <=*$>

VOENNO-MEDITSINSKO DELO
1960 V15 N1 P21-24 62-19378 <=*>
1960 V15 N6 P35-40 62-19618 <=*>
1962 V17 N3 P26-29 63-21149 <=>
1963 V18 N2 P21-28 64-21904 <=>
1965 V20 N2 P37-42 66-31632 <=$>
1966 V21 N2 P29-34 66-33119 <=$>

VOENNO-SANITARNOE DELO. MOSKVA
1935 N2 P25- R-1052 <*>
1936 N9 P13-19 R-1071 <*>
1937 V9 N4 P8-14 AD-613 460 <=$>
1938 N2 P49-57 RT-1438 <*>
1938 V10 N7 P16-24 AD-613 584 <=$>
1940 N10 P57-59 RT-1826 <*>
1941 N11 P57-58 RT-1827 <*>
1942 N11/2 P37-43 RT-1829 <*>

VOENNYI SVIAZIST
1958 V16 N3 P27-33 60-13553 <+*>
1958 V16 N6 P23-27 59-12898 <+*> P
1958 V16 N7 P20-24 59-16329 <+*> P

VOENNYI VESTNIK
1957 V37 N9 P67-72 59-22169 <+*>
1958 V38 N6 P19-29 64-71445 <=>
1959 V39 N6 P1-7 59-13897 <=>
1959 V39 N6 P35-44 59-13898 <=> 0
1959 V39 N8 P61-65 59-13521 <+*>
1959 V39 N8 P77-82 60-17385 <+*> 0
1959 V39 N10 P30-34 60-11510 <=>
1959 V39 N12 P57-61 60-31083 <=>
1960 N11 P26-28 AEC-TR-5790 <=*$>
1961 N3 P40-42 AD-619 466 <=$>
1961 V41 N3 P100-101 62-23320 <=*>
1961 V41 N4 P88-91 65-63570 <=$>
1961 V41 N5 P50-53 61-28413 <*=>
1961 V41 N5 P64-68 61-28408 <*=>
1961 V41 N12 P98-99 62-23622 <=*>
1964 V44 N11 P56-61 AD-617 674 <=$>
AD-625 192 <=*$>

VOENNYE ZNANIYA
1956 V32 N4 P18 60-21800 <=>
1958 V34 N7 P24-25 59-13444 <=> 0
1958 V34 N9 P27-28 61-27020 <*=>
1959 V35 N1 P17-19 62-32782 <=*>
1959 V35 N1 P28-29 60-11148 <=>
1959 V35 N2 P21-23 60-17533 <+*>
1959 V35 N10 P19-20 61-28337 <*=> 0
1959 V35 N11 P14-15 61-28334 <*=>
1961 V37 N2 P17-18 66-13906 <*$>
1961 V37 N6 P24 62-24390 <=*>
1963 V39 N2 P34-36 AD-610 411 <=$>
1963 V39 N4 P23-24 64-13611 <=*$>
1963 V39 N11 P21-22 64-41288 <=>
1964 V40 N4 P1-2 AD-611 747 <=$>

1964 V40 N5 P19-27 64-41235 <=>
1964 V40 N6 P18-27 64-41357B <=>
1964 V40 N6 P33-34 AD-614 392 <=$>
64-41357B <=>
1964 V40 N12 P3 AD-620 806 <=$>
1965 N11 P1-2 66-30045 <=*$>
1965 N11 P14-23 66-30045 <=*$>
1965 N11 P25-27 66-30045 <=*$>
1965 N11 P34-35 66-30045 <=*$>
1965 N12 P16-26 66-30051 <=$>
1965 N12 P28-33 66-30051 <=$>
1965 N12 P36-37 66-30051 <=$>
1965 V41 N1 P36-37 65-30630 <=$>
1965 V41 N7 P15-28 65-32472 <=$>
1966 N12 P16-29 67-30289 <=>
1966 N12 P48 67-30289 <=>

VOGELWARTE. STUTTGART
1950 V15 N3 P194-196 C-5179 <NRC>
1951 V16 N2 P55-59 C-5168 <NRC>
1965 V23 P71-77 66-11083 <*>

VOJNI GLASNIK
1959 V13 N5 P36-45 60-41656 <*=>

VOJNO-DELO
1959 V11 N7/8 P400-416 60-31842 <*=>

VOJNO-SANITETSKI PREGLED
1951 V8 N1/2 P35-38 AL-557 <*>
1957 V14 N9 P279-283 PB 141 162T <=>
1959 V16 N12 P1015-1022 60-11591 <*+>
1960 V17 P550-554 65-17237 <*>
1960 V17 N2 P164-167 62-19373 <=*>
1960 V17 N3 P251-256 62-19704 <=*>
1960 V17 N4 P367-372 62-19372 <=*>
1960 V17 N5 P525-528 62-19703 <=*>
1960 V17 N9 P909-910 62-19705 <=*>
1960 V17 N10 P1008-1011 62-19370 <=*>
1960 V17 N10 P1035-1039 62-19369 <=*>
1960 V17 N7/8 P792-794 62-19464 <=*>
1962 V19 P210-212 12770 <K-H>

VOJNO-TEHNICKI GLASNIK
1960 V7 P510-513 UCRL TRANS-823 <*>

VOKRUG SVETA
1958 N7 P1-8 59-11254 <=>
1958 N8 P1-5 60-31092 <=>
1963 N9 P37-39 AD-627 119 <=$>

VOLKSARMEE
1964 N7 P30-31 64-41516 <=$>

VOLKSSTIMME. EAST GERMANY
1961 N14 P5 62-19371 <*=>

VOM WASSER
1937 V12 P177-180 C-2386 <*>
1949 V17 P164-197 63-14081 <*> 0
1952 V19 P99-121 77J15G <ATS>
1953 V20 P34-71 71J14G <ATS>
1953 V20 P72-126 76J15G <ATS>
1957 V24 P59-70 NP-TR-746 <*>
1957 V24 P160-170 60-14561 <*>
1959 V26 P241-257 63-14093 <*> 0

VOORDRACHTEN VAN DE VACANTIE-LEERGANG VOOR VERWARMINGSTECHNIEK
1942 P89-95 I-770 <*>

VOPROSY ANTROPOLOGII
1960 N3 P12-32 62-13029 <=*>
1960 N3 P106-111 62-13216 <=*>
1960 N3 P155-158 62-13264 <*=>
1963 N13 P52-68 64-31764 <=>

VOPROSY ARKHIVOVEDENIYA
1963 V8 N4 P94-98 64-51068 <=$>

VOPROSY BEZOPASNOSTI V UGOLNYKH SHAKHTAKH
1964 V4 P3-12 SMRE-TRANS-5107 <NLL>
1964 V4 P12-22 SMRE-TRANS-5106 <NLL>
1964 V4 P22-35 SMRE-TRANS-5105 <NLL>
1964 V4 P58-63 SMRE-TRANS-5104 <NLL>
1964 V4 P101-115 SMRE-TRANS-5103 <NLL>
1964 V4 P115-121 SMRE-TRANS-5102 <NLL>
1964 V4 P188-198 SMRE-TRANS-5101 <NLL>
1964 V4 P198-206 SMRE-TRANS-5100 <NLL>
1964 V4 P214-224 SMRE-TRANS-5099 <NLL>
1964 V4 P227-232 SMRE-TRANS-5098 <NLL>
1964 V4 P233-235 SMRE-TRANS-5097 <NLL>
1964 V4 P257-264 SMRE-TRANS-5094 <NLL>

VOPROSY BIOFIZIKI, BIOKHIMI I PATOLOGII ERITROTSITOV
1961 V2 P153-162 63-21798 <=>
1961 V2 P313-314 63-21798 <=>

VOPROSY BIOKHIMII
1963 V3 P5-9 64-51985 <=$>

VOPROSY BIOLOGII I KRAEVOI MEDITSINY
1961 V2 P356-359 64-19879 <=$>
1963 V4 P123-130 AD-611 522 <=$>

VOPROSY BIOLOGII PUSHNYKH ZVERE
1953 N11 P43- C-2473 <NRC>
1953 V13 P53- C-2461 <NRC>
1953 V13 P200- C-2466 <NRC>
1953 V13 P217- C-2468 <NRC>
1955 V14 P197- C-2459 <NRC>

VOPROSY DINAMICHESKOI TEORII RASPROSTRANENIYA SEISMICHESKIKH VOLN
1962 N4 P121-132 64-16407 <=*$>
1962 N4 P133-142 65-11128 <*>
1962 N4 P143-158 65-12706 <*>
1962 N4 P181-193 65-11410 <*>
1962 N4 P230-241 65-12815 <*>
1962 N4 P242-248 65-12708 <*>
1962 N6 P181-184 64-11659 <=>

VOPROSY DINAMIKI I PROCHNOSTI: SBORNIK STATEI
1956 N4 P49-71 61-11159 <=>
1962 V9 P115-126 63-24538 <=*$>
1963 N10 P95-108 65-14702 <*>
1963 V10 P95-108 65-62974 <=$>

VOPROSY DOZIMETRII I ZASHCHITY OT IZLUCHENIYA
1963 V2 P28-39 AD-635 279 <=$> M
1963 V2 P66-77 AD-635 279 <=$> M
1963 V2 P105-108 64-31025 <=>
1964 V3 P77-81 AD-638 871 <=$>

VOPROSY EKOLOGII
1962 N4 P53-55 62-21961 <=>
1962 N4 P66-67 62-21961 <=>
1962 N4 P157-159 62-21961 <=>
1963 N4 P32-34 63-31077 <=>

VOPROSY EKONOMIKI
1956 N2 P101-113 RT-4328 <*>
1957 N3 P52-65 R-3097 <*>
1958 N10 P58-70 59-11460 <=>
1958 N10 P156-157 59-18156 <+*>
1958 N10 P158 59-18404 <+*>
1958 N10 P159-160 59-18403 <+*>
1958 N11 P159-160 59-18123 <+*>
1960 N5 P61-66 61-11863 <=>
1960 N8 P45-59 61-11537 <=> P
1961 N2 P14-24 61-31146 <=>
1961 N9 P102-110 64-31082 <=>
1961 N11 P86-90 62-23762 <=*>
1962 N1 P30-38 62-32010 <=>
1962 N9 P20-30 63-13822 <=>
1962 N9 P127-133 63-21003 <=>
1962 N9 P151-155 63-21003 <=>
1962 N12 P114-117 63-21416 <=>
1963 N1 P10-16 63-21899 <=>

Citation	Reference
1963 N2 P64-70	64-31027 <=>
1963 N2 P143-144	63-21901 <=>
1963 N3 P15-25	63-31112 <=>
1963 N5 P3-14	63-31416 <=>
1963 N7 P59-69	63-31994 <=>
1963 N8 P89-95	63-31962 <=>
1963 N9 P96-102	63-41177 <=>
1964 N9 P63-110	64-51858 <=$>
1965 N5 P3-14	65-31783 <=$*>
1965 N5 P15-21	65-31803 <=$>
1965 N7 P13-22	65-32467 <=$*>
1965 N8 P13-21	65-32857 <=$*>
1965 N8 P126-138	65-32925-B <=*$>
1965 N9 P23-41	65-33621 <=*$>
1965 N9 P62-74	65-33434 <=*$>
1965 N10 P86-97	66-30304 <=*$>
1965 N11 P3-25	66-30578 <=$>
1965 N12 P44-45	66-31686 <=$>
1966 N1 P37-58	66-31827 <=$>
1966 N1 P113-121	66-31688 <=$>
1966 N1 P142-150	66-31590 <=$>
1966 N2 P66-75	66-31936 <=$>
1966 N3 P92-100	66-32208 <=$>
1966 N4 P13-23	66-34699 <=$>
1966 N5 P18-30	66-34914 <=$>
1966 N5 P41-47	66-34914 <=$>
1966 N5 P60-73	66-34914 <=$>
1966 N7 P3-13	66-34891 <=$>
1966 N7 P31-40	66-34905 <=$>
1966 N8 P3-14	66-34895 <=$>
1966 N8 P40-61	66-34895 <=$>
1966 N9 P35-45	67-30082 <=>
1966 N10 P57-67	67-30139 <=>
1966 N10 P115-120	67-30084 <=>
1966 N10 P120-125	66-35699 <=>

VOPROSY FILOSOFII

Citation	Reference
1950 N3 P309-317	60-13735 <+*>
1953 N4 P185-191	RT-1338 <*>
1953 N4 P207-212	RT-1337 <*>
1953 N4 P213-215	RT-1336 <*>
1953 N4 P216-218	RT-1355 <*>
1955 N1 P28-41	62-15048 <=*>
1955 N4 P136-147	59-12218 <+*>
1956 N3 P116-122	PB 141 246T <=>
1957 N4 P158-168	60-11154 <=>
1957 N5 P101-113	PB 141 476T <=>
1958 N2 P35-49	59-13872 <=>
1958 N2 P112-127	59-13873 <=> 0
1958 N4 P78-87	59-13860 <=>
1958 N6 P127-138	59-13833 <=>
1958 N7 P102-112	59-13848 <=>
1958 N8 P3-13	59-13870 <=>
1958 N8 P82-96	59-13871 <=>
1958 N8 P97-107	59-13302 <=>
1958 N9 P58-72	59-11673 <=>
1958 N9 P104-111	59-20499 <+*>
1959 N4 P3-15	60-11051 <=>
1959 N4 P133-138	60-11049 <=>
1959 N4 P156-161	60-11050 <=>
1959 N4 P182-190	60-11048 <=>
1959 N6 P148-154	60-11226 <=>
1959 V13 N9 P3-13	61-15496 <+*>
1959 V13 N11 P79-90	61-11591 <=>
1959 V13 N12 P128-138	60-17263 <+*>
1960 N3 P13-27	60-31684 <=>
1960 N5 P51-62	60-31781 <=>
1960 V14 N1 P82-90	60-31734 <=>
1960 V14 N5 P126-134	61-11591 <=>
1960 V14 N7 P15-23	60-41306 <=>
1960 V14 N7 P24-33	60-41307 <=>
1960 V14 N7 P34-49	60-41308 <=>
1960 V14 N7 P106-110	60-41309 <=>
1960 V14 N7 P113-123	60-41310 <=>
1960 V14 N7 P127-134	61-11591 <=>
1960 V14 N8 P120-131	61-21201 <=>
1960 V14 N8 P126-136	61-15823 <+*>
1960 V14 N9 P93-99	61-11691 <=>
1960 V14 N9 P111-119	61-21125 <=>
1960 V14 N9 P132-142	61-11885 <=>
1960 V14 N10 P100-109	61-28036 <*=>
1961 V15 N1 P150-157	61-21667 <=>
1961 V15 N2 P39-51	61-23003 <=*>
1961 V15 N3 P164-169	61-27661 <*=>
1961 V15 N5 P47-56	61-28601 <*=>
1961 V15 N5 P133-137	63-15784 <=*>
1961 V15 N8 P101-117	63-15786 <=*>
1961 V15 N8 P156-160	63-26959 <=$>
1961 V15 N9 P89-104	63-15785 <=*>
1962 V16 N2 P99-109	62-32886 <=>
1962 V16 N2 P151-153	62-32886 <=>
1962 V16 N3 P62-78	62-24409 <=*>
1962 V16 N4 P32-46	62-25535 <=*>
1962 V16 N6 P25-35	63-15026 <=*>
1962 V16 N7 P163-167	62-33517 <=>
1962 V16 N8 P50-62	63-21013 <=>
1962 V16 N8 P78-87	63-15104 <=*>
1962 V16 N9 P75-87	63-21233 <=*>
1962 V16 N11 P146-153	63-21187 <=*>
1962 V16 N12 P57-62	63-21294 <=>
1963 N11 P66-74	SGRT V5 N2 P24 <AGS>
1963 V17 N1 P36-48	63-21395 <=>
1963 V17 N1 P102-115	63-21605 <=>
1963 V17 N1 P126-142	63-21535 <=>
1963 V17 N2 P26-38	63-21764 <=>
1963 V17 N2 P61-69	63-31080 <=>
1963 V17 N2 P70-81	63-21782 <=>
1963 V17 N4 P76-96	63-31436 <=>
1963 V17 N5 P167-169	64-21348 <=>
1963 V17 N6 P13-21	64-21346 <=>
1963 V17 N6 P155-158	64-21349 <=>
1963 V17 N8 P74-83	63-41108 <=>
1963 V17 N8 P92-103	63-41108 <=>
1963 V17 N8 P104-114	63-41107 <=>
1963 V17 N8 P143-154	63-41107 <=>
1963 V17 N9 P40-50	63-41269 <=>
1963 V17 N9 P78-88	63-41269 <=>
1963 V17 N10 P13-18	64-21163 <=>
1963 V17 N11 P153-161	64-41282 <=>
1964 N2 P35-45	SGRT V5 N4 P34 <AGS>
1964 V18 N2 P14-24	64-31412 <=>
1964 V18 N2 P14-24,52-61	64-31412 <=>
1964 V18 N2 P52-61	64-31412 <=>
1964 V18 N2 P115-137	64-31412 <=>
1964 V18 N2 P178-179	64-31412 <=>
1964 V18 N3 P38-49	64-31433 <=>
1964 V18 N5 P3-12	64-31666 <=>
1964 V18 N5 P65-75	64-41962 <=$>
1964 V18 N5 P173-177	64-41032 <=>
1964 V18 N7 P154-157	65-30027 <=$>
1964 V18 N10 P59-82	65-30010 <=$>
1964 V18 N10 P151-157	65-30010 <=$>
1964 V18 N11 P73-84	65-30361 <=$>
1964 V18 N12 P25-32	65-30425 <=$>
1965 N9 P3-13	66-30038 <=*$>
1965 V19 N1 P67-73	65-31253 <=$>
1965 V19 N1 P84-94	65-31306 <=$>
1965 V19 N1 P135-145	65-31389 <=$>
1965 V19 N1 P154-157	65-31826 <=$>
1965 V19 N2 P57-64	65-31124 <=$>
1965 V19 N3 P3-31	65-31547 <=*$>
1965 V19 N3 P47-57	65-31547 <=*$>
1965 V19 N5 P83-94	65-34090 <=*$>
1965 V19 N5 P141-161	65-34089 <=*$>
1965 V19 N7 P22-30	65-32287 <=*$>
1965 V19 N7 P42-55	65-33556 <=*$>
1965 V19 N7 P56-63	65-33556 <=*$>
1965 V19 N7 P176-177	65-33555 <=*$>
1965 V19 N8 P3-12,68-77	65-33367 <=*$>
1965 V19 N8 P116-125	65-32936 <=*$>
1965 V19 N8 P162-165	65-33367 <=*$>
1965 V19 N9 P57-67	65-33454 <=*$>
1965 V19 N9 P118-131	65-33454 <=*$>
1965 V19 N9 P171-175	66-30090 <=*$>
1965 V19 N10 P3-13	66-30073 <=*$>
1965 V19 N10 P14-21	66-30074 <=*$>
1965 V19 N10 P33-45	66-30073 <=*$>
1965 V19 N10 P65-78	65-34092 <=*$>
1965 V19 N11 P155-159	66-30378 <=$>
1965 V19 N12 P80-90	66-30560 <=*$>

1966 N1 P65-74 66-31628 <=$>
1966 N1 P93-103 66-31628 <=$>
1966 N1 P148-150 66-31628 <=$>
1966 N3 P25-30 66-32758 <=$>
1966 N4 P64-75 66-32719 <=$>
1966 N4 P99-109 66-32874 <=$>
1966 N6 P160-164 66-34551 <=$>
1966 N10 P3-60 67-30393 <=>
1966 N10 P80-102 67-30393 <=>
1966 N10 P125-136 67-30393 <=>
1966 N10 P156-165 67-30393 <=>
1966 V20 N1 P3-11 66-31054 <=$>
1966 V20 N1 P22-31 66-31046 <=$>
1966 V20 N2 P100-109 66-31941 <=$>
1966 V20 N5 P179-180 66-33281 <=$>

VOPROSY FIZIOLOGII INTEROTSEPTSII
1952 P3 RT-1138 <*>
1952 V1 P5-23 RT-1227 <*>
62-15050 <=*> 0
1952 V1 P152-165 RT-1230 <*>
1952 V1 P152-166 62-15056 <=*> 0
1952 V1 P236-254 RT-1226 <*>
1952 V1 P255-264 RT-1225 <*>
1952 V1 P339-352 RT-2194 <*>
1952 V1 P359-368 RT-1962 <*>
1952 V1 P369-381 RT-1963 <*>
1952 V1 P382-389 RT-2926 <*>
1952 V1 P390-395 RT-2844 <*>
1952 V1 P411-428 RT-2927 <*>
1952 V1 P455-468 RT-1228 <*>
1952 V1 P469-483 RT-1229 <*>
1952 V1 P484-499 RT-4058 <*>
1952 V1 P505-514 RT-1168 <*>

VOPROSY GEOGRAFII
1948 N7 P123-129 RT-1363 <*>
1953 V33 P45-64 65-50065 <=$>
1958 N42 P178-188 61-28295 <*=>
1958 N44 P114-134 SGRT V2 N6 P21 <AGS>
1961 N53 P48-62 SGRT V3 N2 P3 <AGS>
1961 N53 P63-84 SGRT V3 N2 P16 <AGS>
1961 N53 P85-100 SGRT V3 N2 P34 <AGS>
1962 N59 P105-131 IGR V6 N2 P238-253 <AGI>
1962 V59 P6-52 IGR,V5,N11,P1403-31 <AGI>
1962 V59 P132-144 IGR,V6,N3,P379-386 <AGI>
1962 V59 P145-157 IGR,V6,N4,P581-588 <AGI>
1962 V59 P158-163 IGR,V6,N5,P763-767 <AGI>
1962 V59 P164-171 IGR V6 N6 P979-984 <AGI>
1963 N61 P5-23 SGRT V4 N7 P3 <AGS>
1963 N61 P24-33 SGRT V4 N7 P19 <AGS>
1963 N61 P34-45 SGRT V4 N7 P26 <AGS>
1963 N61 P122-132 SGRT V5 N4 P46 <AGS>
1963 N61 P153-176 SGRT V4 N7 P36 <AGS>
1965 N66 P6-33 SGRT V7 N1 P3 <AGS>
1965 N66 P118-129 SGRT V7 N1 P60 <AGS>
1965 N66 P130-140 SGRT V7 N1 P52 <AGS>
1965 N68 P14-71 SGRT V7 N3 P3 <AGS>

VOPROSY GEOLOGII KUZBASSA
1956 V1 P99-108 61-28230 <*=>

VOPROSY GEOLOGII I VOSTOCHNOI OKRAINY RUSSKOI PLATFORMY I YUZHNOGO URALA
1960 N7 P67-71 64-13337 <=*$>
1960 N7 P83-92 64-13336 <=*$>

VOPROSY IADERNOI ENERGETIKI
1957 N1 P5-17 59-12217 <+>

VOPROSY IAZYKOZNANIIA
1956 N5 P107-111 NLLTB V1 N8 P12 <NLL>
1956 N5 P111-121 NLLTB V1 N6 P21 <NLL>
1956 N5 P121-124 NLLTB V1 N7 P20 <NLL>
1956 V5 N5 P107-124 PB 141 383T <=>
1956 V5 N5 P107-111 R-4959 <*>
1956 V5 N5 P111-121 61-23286 <*=>
1956 V5 N5 P121-124 R-4959 <*>
1957 N1 P107-113 PB 141 383T <=>
1957 N4 P92-97 PB 141 383T <=>
1957 N5 P117-121 PB 141 383T <=>
1957 N5 P161-162 NLLTB V1 N3 P14 <NLL>
PB 141 383T <=>
1957 N5 P161-163 61-10170 <*+>
1957 V6 N5 P111-116 59-18013 <+*>
60-13855 <+*>
1958 N1 P129-133 59-11300 <=>
1958 N2 P116-121 59-11300 <=>
1958 N2 P166 R-4956 <*>
1958 N3 P136-137 59-11300 <=>
1958 N3 P149 59-11300 <=>
1958 V7 N1 P170-173 61-11737 <=>
1958 V7 N2 P121-125 61-11737 <=>
1958 V7 N2 P166 59-16343 <+*>
1958 V7 N5 P151-152 61-11737 <=>
1959 N5 P102-104 NLLTB V2 N2 P66 <NLL>
1959 V8 N6 P28-35 60-00227 <*>
1959 V8 N6 P128-130 60-00227 <*>
1960 V9 N1 P32-43 61-11745 <=>
1960 V9 N1 P52-59 60-31697 <=>
1960 V9 N1 P100-115 60-31697 <=>
1960 V9 N3 P100-107 60-31848 <=>
1960 V9 N3 P130-131 60-31848 <=>
1960 V9 N3 P142-149 60-31848 <=>
1960 V9 N3 P154 60-31848 <=>
1960 V9 N5 P47-60 61-11800 <=>
1960 V9 N5 P85-88 61-11800 <=>
1960 V9 N5 P96-116 61-11800 <=>
1960 V9 N6 P88-94 61-21399 <=>
1961 V10 N1 P155-164 61-21702 <=>
1961 V10 N4 P105-115 61-28905 <*=>
1961 V10 N5 P122-128 62-11078 <*=>
1962 V11 N1 P91-99 62-25613 <=*>
1962 V11 N3 P147 62-25613 <=*>
1962 V11 N6 P115-130 63-21804 <=>
1963 V12 N1 P113-130 63-21786 <=>
1963 V12 N4 P102-112 64-21096 <=>
1964 V13 N4 P112-119 65-30522 <=$>
1965 N3 P22-43 AD-639 429 <=$>
1965 V14 N1 P56-68 65-30800 <=$>
1965 V14 N2 P92-96 65-31108 <=$>
1966 N1 P37-46 66-31347 <=$>

VOPROSY IKHTIOLOGII
1953 N1 P117-127 66-61848 <=*$> 0
1954 N2 P139-143 C-3627 <NRCC>
61-00661 <*>
1955 N3 P3-8 63-15935 <=*>
1955 N3 P9-20 63-19461 <*=>
1955 N3 P21-31 63-15936 <=*>
1955 N3 P50-53 62-33568 <=*> 0
1955 N4 P3-9 63-15237 <=*>
1955 V4 P57-82 C-2486 <NRC>
1955 V4 P71-81 C-2490 <NRC>
1956 N6 P90-99 60-51139 <=>
1956 N7 P3-20 R-5070 <*>
1956 N7 P21-32 FRBC TRANS SER N354 <NRCC>
1956 N7 P139-148 FRBC TRANS-255 <NRC> 0
61-00864 <*>
1956 N7 P158-173 C-3879 <NRCC>
1956 N7 P209-220 C-3307 <NRCC>
1956 V7 N7 P123-133 C-3354 <NRC> 0
1957 V8 N8 P3-7 C-3407 <NRCC>
60-21095 <=>
1958 N10 P162-171 63-24619 <=*$>
1958 V9 N11 P162-166 61-15435 <*+> 0
1959 N12 P3-7 62-25062 <=*>
1959 N13 P139-155 63-11104 <=>
1959 N13 P156-162 62-00860 <*>
1959 N13 P182-188 64-19925 <=$>
1959 V10 N13 P16-18 C-3378 <NRCC>
1960 N14 P160-165 FRBC TRANS SER N366 <NRCC>
1960 N16 P67-88 C-4646 <NRC>
1960 N16 P89-110 C-3836 <NRCC>
1960 N16 P187-190 C-4757 <NRC>
1961 N3 P481-490 63-21612 <=>
1961 N3 P552-569 63-21612 <=>
1961 N17 P14-23 C-3897 <NRCC>
1961 N17 P159-168 C-4301 <NRCC>
1961 N17 P179-190 C-3901 <NRCC>

1961 V1 N1 P39-45 C-3916 <NRCC>
62-00184 <*>
1961 V1 N1 P46-51 C-3908 <NRCC>
62-00185 <*>
1961 V1 N1 P69-78 C-3904 <NRCC>
1961 V1 N1 P127-135 C-4236 <NRC>
63-00877 <*>
1961 V1 N2 P253-261 RTS-3108 <NLL>
1961 V1 N2 P333-346 C-4287 <NRC> O
1961 V1 N3 P391-394 63-15236 <=*> O
1961 V1 N3 P439-447 62-23279 <=*>
1961 V1 N3 P570-582 62-24363 <=*>
1961 V1 N3 P583-589 64-14302 <=*$>
1961 V1 N4 P599-611 63-31702 <=>
1961 V1 N4 P616-649 63-31702 <=>
1961 V1 N4 P659-665 C-4227 <NRC>
63-00879 <*>
1961 V1 N4 P666-680 63-31702 <=>
1961 V1 N4 P752-761 63-31702 <=>
1962 N1 P3-17 63-31085 <=>
1962 N1 P104-105 63-31085 <=>
1962 N1 P116-126 63-31085 <=>
1962 N1 P169-173 63-31085 <=>
1962 N1 P192-196 63-31085 <=>
1962 N2 P740-742 C-5199 <NRC>
1962 N3 P517-529 63-21480 <=>
1962 N4 P677-686 64-15990 <=*$>
1962 N4 P703-716 64-15960 <=*$>
1962 V2 N1 P140-157 MAFF-TRANS-NS-60 <NLL> O
1962 V2 N2 P299-308 63-21518 <=>
1963 N1 P7-14 63-21781 <=>
1963 N1 P131-143 63-21781 <=>
1963 N1 P163-176 63-21781 <=>
1963 N4 P625-651 65-60322 <=$> O
1963 V3 N2 P209-221 63-41075 <=>
1963 V3 N2 P243-255 63-41095 <=>
1963 V3 N2 P275-287 65-60155 <=$>
1963 V3 N2 P288-303 65-60149 <=$>
1963 V3 N3 P561-562 C-5453 <NRC>
1963 V3 N4 P585-590 64-21525 <=>
1963 V3 N4 P591-609 64-21525 <=>
1963 V3 N4 P610-617 64-21525 <=>
1963 V3 N4 P675-687 64-21525 <=>
1963 V3 N4 P726-733 64-21525 <=>
1963 V3 N4 P734-737 64-21525 <=>
1963 V3 N4 P757-758 64-21525 <=>
1963 V3 N4 P758-761 64-21525 <=>
1963 V3 N4 P761-762 64-21525 <=>
1964 N1 P23-33 64-31188 <=>
1964 N1 P92-96 64-31188 <=>
1964 N1 P136-141 64-31188 <=>
1964 N1 P153-161 64-31188 <=>
1964 N1 P178-183 64-31188 <=>
1964 N1 P196-197 64-31188 <=>
1964 V4 N1 P68-81 MAFF-TRANS-NS-59 <NLL>
1964 V4 N4 P625-631 65-30658 <=$>
1964 V4 N4 P644-657,664-678
65-30658 <=$>
1964 V4 N4 P716-735 65-30658 <=$>
1964 V4 N4 P769-778 65-30658 <=$>
1965 V5 N1 P149- FPTS V25 N1 P.T149 <FASE>
1965 V5 N1 P157- FPTS V25 N2 P.T291 <FASE>
1965 V5 N1 P178-197 65-30833 <=$>
1965 V5 N2 P296-302 66-61062 <=$*>
1965 V5 N2 P331- FPTS V25 N3 P.T531 <FASE>
1965 V5 N3 P532-539 66-61966 <=*$>
1965 V5 N3 P540-547 66-62004 <=*$>
1965 V5 N4 P593-599 66-30294 <=*$>
1965 V5 N4 P639-651 66-62001 <=*$>
1966 V6 N38 P186-188 66-32250 <=$>

VOPROSY ISTORII
1963 N12 P156-158 AD-626 106 <=*$>
1965 N7 P39-52 65-32116 <=$>
1965 N9 P29-39 65-32767 <=$>
1966 N2 P3-13 66-30862 <=$>
1966 N3 P6-17 66-31738 <=$>
1966 N3 P15-30 66-32290 <=$>

VOPROSY ISTORII ESTESTVOZNANIYA I TEKHNIKI
1956 N2 P207-216 R-4982 <*>
1956 V1 P168-178 59-10303 <+*>
1956 V1 N2 P207-216 59-22471 <+*>

VOPROSY KHIMICHESKOI KINETIKI, KATALIZA, I REAKTSIONNOI SPOSOBNOSTI. AKADEMIYA NAUK SSSR. OTDELENIE KHIMICHESKIKH NAUK
1955 P40-53 C-2521 <NRC>
R-3108 <*>
1955 V21 P902-931 C-2718 <NRC>
1955 V50 P59-74 R-389 <*>

★VOPROSY KOSMOGONII
1957 V5 P203-241 59-19525 <+*>
1957 V5 P291-296 59-16075 <+*>
1958 V6 P1-368 64-31190 <=>
1958 V6 P291-329 63-13558 <=*>
1958 V6 P368 63-13558 <=*>
1960 V7 P55-58 64-15165 <=*$>
1960 V7 P333-372 65-63013 <=$>
1963 V9 P200-202 AD-617 657 <=$>
1963 V9 P203-213 N64-29497 <=>
1964 V10 P45-57 65-64389 <=*$>
1964 V10 P70-73 65-64390 <=*$>

VOPROSY KURORTOLGII, FIZIOTERAPII I LECHEBNOI FIZICHESKOI KULTURY
1959 N6 P563-567 60-31075 <=>
1960 N2 P185-188 60-41385 <=>
1960 N2 P191 60-41385 <=>
1960 V25 N3 P281-283 60-41237 <=>
1960 V25 N6 P561-566 61-21515 <=>
1961 V26 N2 P97-99 61-27696 <*=>
1961 V26 N2 P182-184 61-27681 <*=>
1961 V26 N2 P185-190 61-27654 <*=>
1961 V26 N3 P235-238 62-13229 <*=>
1961 V26 N3 P276-285 61-28258 <*=>
1962 V27 N1 P3-11 62-25479 <=>
1962 V27 N1 P16-26 62-25479 <=>
1962 V27 N1 P55-61 62-25479 <=>
1962 V27 N1 P64 62-25479 <=>
1962 V27 N1 P68-69 62-25479 <=>
1962 V27 N1 P74-79 62-25479 <=>
1962 V27 N1 P81 62-25479 <=>
1962 V27 N1 P90-93 62-25479 <=>
1963 V28 N2 P97-104 63-31304 <=>
1963 V28 N2 P135-139 63-31304 <=>
1963 V28 N2 P173-175 63-31304 <=>
1963 V28 N2 P177-178 63-31304 <=>
1963 V28 N2 P184-191 63-31304 <=>
1963 V28 N3 P223-229 63-31755 <=>
1963 V28 N4 P373-377 63-41126 <=>
1963 V28 N4 P377-380 63-41126 <=>
1963 V28 N4 P382 63-41126 <=>
1963 V28 N5 P398-402 64-21016 <=>
1963 V28 N5 P403-404 64-21886 <=>
1963 V28 N5 P404-410 64-21886 <=>
1963 V28 N5 P465-466 64-21016 <=>
1963 V28 N6 P481-487 64-31072 <=>
1963 V28 N6 P510-526 64-31072 <=>
1964 V29 N1 P19-23 64-21979 <=>
1964 V29 N1 P64-70 64-21979 <=>
1964 V29 N2 P104-112 65-31225 <=$>
1964 V29 N2 P144-155 64-31500 <=>
1964 V29 N2 P172 64-31500 <=>
1964 V29 N2 P183-190 64-31500 <=>
1964 V29 N3 P239-242 64-41308 <=>
1964 V29 N4 P337-345 65-30118 <=$>
1964 V29 N5 P433-435 65-31074 <=$>
1964 V29 N6 P513-518 65-30629 <=$>
1965 V30 N1 P40-47 65-30903 <=$>
1965 V30 N1 P90-92 65-31186 <=$>
1965 V30 N2 P124-129 65-31237 <=*$>
1965 V30 N4 P375-376 65-33071 <=*$>
1966 V31 N1 P89-90 66-33833 <=*$>

VOPROSY MAGMATIZMA I METAMORFIZMA
1963 P125-143 IGR V6 N10 P1735 <AGI>

VOPROSY MASHINOVEDENIYA I PROCHNOSTI V

MASHINOSTROENII. AKADEMIYA NAUK UKRAINSKOI SSR. KIEV
1962 V9 N8 P41-46 63-24133 <=*$>

VOPROSY MATEMATICHESKOI FIZIKI I TEORII FUNKTSII
1964 N1 P18-24 65-14590 <*>
1964 V1 P124-144 65-14732 <*>

VOPROSY MEDITSINSKOI KHIMII. AKADEMIYA MEDITSINSKIKH NAUK SSSR.
1949 V1 N1/2 P159-171 R-232 <*>
R-79 <*>
64-19072 <=*$> 0
1956 V2 N1 P47-52 R-5363 <*>
1956 V2 N1 P128-132 R-4835 <*>
1956 V2 N2 P103-108 R-3766 <*>
1956 V2 N2 P128-132 59-17431 <+*>
1956 V2 N2 P140-149 59-17433 <+*>
1956 V2 N3 P203-209 R-406 <*>
64-19462 <=$> 0
1956 V2 N3 P229-233 59-17428 <+*>
1956 V2 N4 P243-251 R-1444 <*>
64-19511 <=$> 0
1956 V2 N4 P309-315 AEC-TR-3726 <+*>
1956 V2 N5 P338-345 59-17427 <+*>
1956 V2 N6 P443-449 59-17432 <+*>
1957 V3 P36-39 65-63172 <=$> 0
1957 V3 N1 P36-38 R-2690 <*>
1957 V3 N2 P121-128 R-4865 <*>
1957 V3 N3 P195-201 R-3233 <*>
64-19410 <=$>
1957 V3 N3 P228-231 59-11613 <=> 0
1957 V3 N4 P243-254 59-15596 <+*> 0
1957 V3 N4 P279-285 59-17374 <+*>
1957 V3 N5 P340-350 62-15136 <*=>
1958 V4 P182-186 PB 141 297T <=>
1958 V4 P274-279 59-17808 <+*>
1958 V4 N1 P15-20 PB 141 154T <=>
1958 V4 N1 P27-32 PB 141 155T <=>
1958 V4 N1 P33-37 PB 141 156T <=>
1958 V4 N2 P120-124 59-13355 <=>
1958 V4 N2 P139-148 5348 <*>
1958 V4 N2 P139-149 59-13356 <=>
1958 V4 N3 P182-186 PB 141 216T <=>
1958 V4 N4 P292-297 64-41257 <=>
1958 V4 N5 P327-338 59-13298 <=> 0
1958 V4 N5 P339-344 59-13299 <=>
1958 V4 N5 P345-349 59-17373 <+*>
1958 V4 N5 P362-364 59-17372 <+*>
1958 V4 N5 P370-372 59-13300 <=>
1958 V4 N5 P373-378 59-13301 <=> 0
1958 V4 N5 P379-384 59-17371 <+*>
1959 V5 N1 P3-5 59-11716 <=>
1959 V5 N1 P32-38 59-11717 <=>
1959 V5 N1 P39-47 61-23211 <=*>
62-15396 <*=>
1959 V5 N1 P48-53 59-11718 <=>
1959 V5 N2 P107-111 C-3385 <NRCC>
1959 V5 N3 P200-205 59-13805 <=>
1959 V5 N5 P323-327 60-18563 <+*>
1959 V5 N6 P403-408 60-31119 <=>
1959 V5 N6 P448-457 60-31120 <=>
1960 V6 P349-353 62-14755 <=*> 0
1960 V6 N1 P49-52 60-11739 <=>
1960 V6 N2 P115-120 60-11848 <=>
1960 V6 N2 P136-145 62-25114 <=*>
1960 V6 N2 P158-165 60-11849 <=>
1960 V6 N2 P188-191 62-14692 <*=>
1960 V6 N3 P254-259 60-31742 <=>
1960 V6 N3 P323-326 60-41582 <=>
1960 V6 N4 P335-344 81M45R <ATS>
1960 V6 N4 P358-364 61-11670 <=>
1960 V6 N4 P365-367 61-11663 <=>
1960 V6 N4 P377-381 61-11654 <=>
1960 V6 N4 P408-411 61-11666 <=>
1960 V6 N5 P464-468 61-11799 <=>
1960 V6 N5 P475-483 61-11799 <=>
1960 V6 N5 P475-479 63-16462 <=*>
1960 V6 N5 P497-505 61-11799 <=>
1960 V6 N6 P641-642 61-19618 <=*>
1960 V6 N6 P645-646 62-20329 <=*>
1961 V7 N1 P3-16 61-23609 <*=>
1961 V7 N1 P28-32 61-23947 <=*>
1961 V7 N1 P52-55 61-23946 <=*>
1961 V7 N1 P55-61 61-23936 <*=>
1961 V7 N1 P61-65 61-23937 <=*>
1961 V7 N1 P65-70 63-16379 <=*> 0
1961 V7 N1 P70-74 61-23935 <=*>
1961 V7 N1 P74-81 61-23940 <*=>
1961 V7 N1 P93-96 63-19690 <=*$>
1961 V7 N2 P120-132 61-27209 <=>
1961 V7 N2 P149-154 61-27209 <=>
1961 V7 N2 P172-178 61-27209 <=>
1961 V7 N2 P186-190 61-27209 <=>
1961 V7 N2 P197-200 61-27209 <=>
1961 V7 N3 P243-245 62-14750 <=*>
1961 V7 N3 P250-258 61-28723 <*=> C
63-21661 <=>
1961 V7 N3 P273-276 62-13221 <=>
1961 V7 N3 P285-291 62-13221 <=>
1961 V7 N3 P309-313 62-20327 <=*> 0
1961 V7 N3 P329-332 62-13221 <=>
1961 V7 N4 P354-357 62-13399 <=*>
1961 V7 N4 P358-362 62-13401 <*=>
1961 V7 N4 P417-424 62-13530 <*=>
1961 V7 N5 P524-527 62-14718 <=*> P
1961 V7 N6 P632-638 62-33364 <=*>
1962 V8 N1 P3-19 62-25000 <=*>
1962 V8 N1 P35- FPTS V22 N4 P.T608 <FASE>
1962 V8 N1 P47- FPTS V22 N3 P.T479 <FASE>
1962 V8 N1 P83-86 62-20328 <=*>
1962 V8 N1 P87-92 63-19370 <=*> 0
1962 V8 N1 P95-99 63-19097 <=*>
1962 V8 N2 P199-204 62-25054 <=*>
1962 V8 N2 P218-221 63-23056 <=*>
1962 V8 N3 P227-236 62-32457 <=*>
1962 V8 N3 P238-241 62-20321 <=*> P
1962 V8 N3 P256- FPTS V22 N2 P.T252 <FASE>
1962 V8 N3 P256-263 62-20322 <=*> P
1962 V8 N3 P270-274 63-19368 <=*> C
1962 V8 N3 P276-279 63-15983 <=*>
1962 V8 N3 P279- FPTS V22 N2 P.T223 <FASE>
1962 V8 N3 P289- FPTS V22 N2 P.T244 <FASE>
1962 V8 N3 P305-306 63-15715 <=*>
1962 V8 N3 P306- FPTS V22 N3 P.T542 <FASE>
1962 V8 N4 P93-94 63-15668 <=*>
1962 V8 N4 P339-353 63-23049 <=*>
1962 V8 N4 P354- FPTS V22 N6 P.T1058 <FASE>
1962 V8 N4 P354-361 FPTS,V22,P.T1058-62 <FASE>
1962 V8 N4 P374- FPTS V22 N5 P.T951 <FASE>
1962 V8 N4 P374-378 1963,V22,N5,PT2 <FASE>
1962 V8 N4 P379-384 63-13314 <=*>
1962 V8 N4 P389- FPTS V22 N4 P.T637 <FASE>
1962 V8 N4 P392-395 64-51287 <=>
1962 V8 N4 P396- FPTS V22 N6 P.T1256 <FASE>
1962 V8 N4 P396-400 FPTS,V22,P.T1256-58 <FASE>
1962 V8 N4 P412- FPTS V22 N6 P.T1045 <FASE>
1962 V8 N4 P412-416 FPTS,V22,P.T1045-48 <FASE>
1962 V8 N4 P420-423 50Q71R <ATS>
63-10209 <=*> 0
63-23907 <=*$>
1962 V8 N4 P424- FPTS V22 N5 P.T915 <FASE>
1963,V22,N5,PT2 <FASE>
1962 V8 N4 P437-439 63-13314 <=*>
1962 V8 N5 P451-453 63-24527 <=*$>
1962 V8 N5 P463- FPTS V22 N6 P.T1063 <FASE>
1962 V8 N5 P463-467 FPTS,V22,P.T1063-66 <FASE>
1962 V8 N5 P468-471 65-64108 <=*$>
1962 V8 N5 P486- FPTS V22 N6 P.T1069 <FASE>
1962 V8 N5 P486-492 FPTS,V22,P.T1069-72 <FASE>
1962 V8 N5 P493-497 AD-607 878 <=>
63-13678 <=>
1962 V8 N5 P514- FPTS V22 N6 P.T1042 <FASE>
1962 V8 N5 P514-517 FPTS,V22,P.T1042-44 <FASE>
1962 V8 N5 P518-525 63-23885 <=*$>
1962 V8 N5 P525- FPTS V22 N6 P.T1248 <FASE>
1962 V8 N5 P525-531 FPTS,V22,T.1248-52 <FASE>
63-13678 <=>
1962 V8 N5 P532- FPTS V22 N6 P.T1049 <FASE>
1962 V8 N5 P532-537 FPTS,V22,P.1049-52 <FASE>

1962 V8 N5 P538-544 63-23884 <=*$> 0
1962 V8 N5 P547- FPTS V22 N5 P.T976 <FASE>
1962 V8 N5 P547-549 1963,V22,N5,PT2 <FASE>
1962 V8 N5 P550-551 63-19098 <=*>
1962 V8 N5 P551-552 63-19099 <=*>
1962 V8 N5 P553-558 63-13678 <=>
1962 V8 N6 P611- FPTS V22 N5 P.T938 <FASE>
1963,V22,N5,PT2 <FASE>
1962 V8 N6 P616-620 63-23845 <=*>
1962 V8 N6 P628- FPTS V22 N5 P.T957 <FASE>
1962 V8 N6 P628-633 1963,V22,N5,PT2 <FASE>
1962 V8 N6 P634- FPTS V22 N5 P.T803 <FASE>
1963,V22,N5,PT2 <FASE>
1962 V8 N6 P638- FPTS V22 N4 P.T619 <FASE>
1963 V9 P303- FPTS V23 N4 P.T673 <FASE>
1963 V9 P373-380 36R77R <ATS>
1963 V9 N1 P19- FPTS V22 N6 P.T1073 <FASE>
1963 V9 N1 P19-26 FPTS,V22,P.T1073-77 <FASE>
1963 V9 N1 P27- FPTS V23 N3 P.T444 <FASE>
FPTS,V23,N3,P.T444 <FASE>
1963 V9 N1 P63- FPTS,V23,N4,P.T803 <FASE>
1963 V9 N1 P84-89 31Q70R <ATS>
1963 V9 N1 P93- FPTS V23 N1 P.T15 <FASE>
FPTS,V23,P.T15-T16 <FASE>
1963 V9 N1 P102- FPTS V23 N1 P.T17 <FASE>
FPTS,V23,P.T17-T18 <FASE>
1963 V9 N2 P142- FPTS V23 N3 P.T562 <FASE>
1963 V9 N2 P142-145 FPTS,V23,N3,P.T587 <FASE>
1963 V9 N2 P146- FPTS V23 N2 P.T326 <FASE>
FPTS,V23,P.T326-T331 <FASE>
1963 V9 N2 P154- FPTS V23 N3 P.T533 <FASE>
FPTS,V23,N3,P.T533 <FASE>
1963 V9 N2 P154-160 63-31185 <=>
1963 V9 N2 P161-166 FPTS V23 N4 P.T832 <FASE>
1963 V9 N2 P161- FPTS,V23,N4,P.T832 <FASE>
1963 V9 N2 P180-188 63-31185 <=>
1963 V9 N2 P184-188 64-23599 <=$>
1963 V9 N2 P218-220 64-15461 <=*$>
1963 V9 N3 P227-232 63-31878 <=>
1963 V9 N3 P232-239 64-15540 <=*$>
1963 V9 N3 P244-250 64-13887 <=*$> 0
1963 V9 N3 P250- FPTS V23 N5 P.T1043 <FASE>
1963 V9 N3 P250-255 FPTS,V23,N5,P.T1043 <FASE>
1963 V9 N3 P256- FPTS V23 N3 P.T536 <FASE>
1963 V9 N3 P256-260 FPTS,V23,N3,P.T536 <FASE>
1963 V9 N3 P261- FPTS V23 N3 P.T503 <FASE>
1963 V9 N3 P261-266 FPTS,V23,N3,PT503 <FASE>
1963 V9 N3 P267- FPTS V23 N3 P.T496 <FASE>
FPTS,V23,N3,P.T495 <FASE>
1963 V9 N3 P273- FPTS V23 N2 P.T427 <FASE>
FPTS,V23,P.T427-T430 <FASE>
1963 V9 N3 P282- FPTS V23 N5 P.T1011 <FASE>
1963 V9 N3 P288- FPTS V23 N3 P.T559 <FASE>
FPTS,V23,N3,P.T559 <FASE>
1963 V9 N3 P301-303 64-15444 <=*$>
1963 V9 N3 P303-308 FPTS V23 N4 P.T673 <FASE>
1963 V9 N3 P317-319 C-5013 <NRC>
1963 V9 N3 P319- FPTS V23 N2 P.T430 <FASE>
1963 V9 N3 P319 FPTS,V23,P.T430-T432 <FASE>
1963 V9 N3 P322- FPTS V23 N3 P.T55T <FASE>
FPTS,V23,N3,P.T503 <FASE>
1963 V9 N3 P324 FPTS V23 N4 P.T677 <FASE>
1963 V9 N3 P324- FPTS V23 N4 P.T677 <FASE>
1963 V9 N4 P365-370 64-15525 <=*$>
1963 V9 N4 P371-373 64-15480 <=*$>
1963 V9 N4 P380- FPTS V23 N3 P.T501 <FASE>
FPTS,V23,N3,P.T501 <FASE>
1963 V9 N4 P393- FPTS V23 N4 P.T876 <FASE>
FPTS,V23,N4,P.T876 <FASE>
1963 V9 N4 P403- FPTS V23 N3 P.T437 <FASE>
FPTS,V23,N3,P.T437 <FASE>
1963 V9 N4 P414- FPTS V23 N4 P.T865 <FASE>
FPTS,V23,N4,P.T865 <FASE>
1963 V9 N4 P421- FPTS,V23,N4,P.T813 <FASE>
1963 V9 N5 P469-475 64-19238 <=*$>
1963 V9 N5 P475- FPTS,V23,N5,P.T957 <FASE>
1963 V9 N5 P495-499 FPTS,V23,N5,P.T964 <FASE>
1963 V9 N5 P495- FPTS,V23,N5,P.T965 <FASE>
1963 V9 N5 P500- FPTS V24 N1 P.T99 <FASE>
1963 V9 N5 P500-506 FPTS V24 P.T99-T102 <FASE>
1963 V9 N5 P507- FPTS V24 N1 P.T51 <FASE>
1963 V9 N5 P507-513 FPTS V24 P.T51-T54 <FASE>
1963 V9 N5 P514-517 65-61103 <=$>
1963 V9 N5 P530-531 C-5014 <NRC>
1963 V9 N6 P575- FPTS V24 N1 P.T103 <FASE>
1963 V9 N6 P575-580 FPTS V24 P.T103-T105 <FASE>
1963 V9 N6 P583-587 64-21756 <=>
1963 V9 N6 P597- FPTS V23 N5 P.T981 <FASE>
1963 V9 N6 P597-600 FPTS,V23,N5,P.T981 <FASE>
1963 V9 N6 P614-621 65-60166 <=$>
1963 V9 N6 P626- FPTS V24 N1 P.T59 <FASE>
1963 V9 N6 P626-632 FPTS V24 P.59-T62 <FASE>
1963 V9 N6 P643-646 65-60200 <=$>
1964 V10 N1 P3-12 CSIR-TR-447 <CSSA>
1964 V10 N1 P12- FPTS V24 N1 P.T93 <FASE>
1964 V10 N1 P12-15 FPTS V24 P.T93-T95 <FASE>
64-31365 <=>
1964 V10 N1 P15-20 65-61C78 <=$>
1964 V10 N1 P20- FPTS V23 N6 P.T1269 <FASE>
FPTS V23 P.T1269 <FASE>
1964 V10 N1 P32-36 65-61079 <=$>
1964 V10 N1 P36- FPTS V24 N2 P.T206 <FASE>
1964 V10 N1 P36-43 FPTS V24 P.T206-T208 <FASE>
1964 V10 N1 P64- FPTS V24 N3 P.T557 <FASE>
1964 V10 N1 P73- FPTS V24 N1 P.T135 <FASE>
1964 V10 N1 P73-76 FPTS V24 P.T135-T136 <FASE>
1964 V10 N1 P77-80 64-31365 <=>
1964 V10 N2 P144-148 64-31551 <=>
1964 V10 N2 P149- FPTS V24 N4 P.T637 <FASE>
1964 V10 N2 P179- FPTS V24 N2 P.T218 <FASE>
1964 V10 N2 P179-184
FPTS V24 P.T218-T220 <FASE>
64-31550 <=>
1964 V10 N2 P185- FPTS.V24 N2 P.T287 <FASE>
1964 V10 N2 P185-189
FPTS V24 P.T287-T290 <FASE>
1964 V10 N2 P197-201 65-60192 <=$>
1964 V10 N3 P227-237 65-62314 <=$>
1964 V10 N3 P252-256 64-31937 <=>
1964 V10 N3 P305- FPTS V24 N4 P.T703 <FASE>
1964 V10 N3 P305-310 64-31939 <=>
1964 V10 N3 P311- FPTS V24 N3 P.T548 <FASE>
1964 V10 N3 P311-315 65-63052 <=$>
1964 V10 N3 P326- FPTS V24 N3 P.T561 <FASE>
1964 V10 N4 P339-351 64-51436 <=$>
1964 V10 N4 P352- FPTS V24 N4 P.T614 <FASE>
1964 V10 N4 P362- FPTS V24 N4 P.T693 <FASE>
1964 V10 N4 P386- FPTS V24 N4 P.T617 <FASE>
1964 V10 N4 P4C7- FPTS V24 N4 P.T589 <FASE>
1964 V10 N5 P469- FPTS V24 N6 P.T977 <FASE>
1964 V10 N5 P520- FPTS V24 N6 P.T1127 <FASE>
1964 V10 N6 P576- FPTS V24 N6 P.T986 <FASE>
1964 V10 N6 P584- FPTS V24 N5 P.T916 <FASE>
1964 V10 N6 P595- FPTS V24 N5 P.T809 <FASE>
1964 V10 N6 P595-600 65-30276 <=$>
1964 V10 N6 P631-633 65-63154 <=$>
1964 V10 N6 P635- FPTS V24 N6 P.T981 <FASE>
1965 V11 N1 P3-17 65-30780 <=$>
1965 V11 N1 P50- FPTS V25 N1 P.T72 <FASE>
1965 V11 N1 P75-81 66-60408 <=*$>
1965 V11 N1 P95- FPTS V25 N1 P.T121 <FASE>
1965 V11 N1 P101-103 65-63153 <=$>
1965 V11 N2 P20- FPTS V25 N2 P.T344 <FASE>
1965 V11 N2 P32-36 66-60889 <=*$>
1965 V11 N3 P3-10 66-60899 <=*$>
1965 V11 N3 P40-45 66-60865 <=*$>
1965 V11 N3 P64- FPTS V25 N2 P.T271 <FASE>
1965 V11 N3 P67- FPTS V25 N3 P.T485 <FASE>
1965 V11 N3 P93-95 66-11708 <*>
1965 V11 N4 P34- FPTS V25 N4 P.T713 <FASE>
1965 V11 N4 P44- FPTS V25 N3 P.T482 <FASE>
1965 V11 N4 P64- FPTS V25 N3 P.T507 <FASE>
1965 V11 N5 P17-22 66-61625 <=*$>
1965 V11 N5 P26- FPTS V25 N5 P.T897 <FASE>
1965 V11 N5 P49- FPTS V25 N6 P.T1009 <FASE>
1965 V11 N5 P91-94 66-12539 <*> 0
1965 V11 N6 P54- FPTS V25 N5 P.T758 <FASE>
1966 V12 N1 P51-54 66-31474 <=$>
1966 V12 N1 P123-124 66-31474 <=$>

VOPROSY MEKHANIKI REALNOGO TVERDOGO TELA
1964 N3 P88-94 AD-622 358 <=*$>

★ VOPROSY MIKRCPALEONTOLOGII
1958 V2 P53-73 04L35R <ATS>
1958 V2 P130-135 03L35R <ATS>
1958 V2 N2 P91-104 69M42R <ATS>
1958 V2 N2 P121-123 10L36R <ATS>
1960 N3 P5-8 63N52R <ATS>
1964 V7 N8 P13-29 66-24684 <$>
1965 N9 P111-128 96T92R <ATS>

VOPRCSY MINERALOGII GEOKHIMII I PETRCGRAFII. AKADEMIYA NAUK SSSR
1946 P199-208 R-1669 <*>
R-4644 <*>

VOPRCSY MINERALOGII CSADCCHNYKH OBRAZCVANII
1956 V3/4 P252-265 62-20297 <=*>

VOPROSY NEUROKHIRURGII
1941 V5 N3 P27-33 RT-642 <*>
1952 V16 N1 P36-40 R-3224 <*>
64-19409 <=$>
1953 V17 N1 P3-8 RT-1840 <*>
1953 V17 N2 P53-55 R-3217 <*>
64-19412 <=$>
1954 V18 N4 P56-57 R-3428 <*>
64-19440 <=$>
1954 V18 N4 P58-59 R-3427 <*>
64-19441 <=$>
1955 V19 N1 P3-8 RT-3628 <*>
1955 V19 N1 P51-53 RT-3780 <*>
1955 V19 N1 P60-62 RT-3876 <*>
1956 V20 N2 P43-48 62-23510 <=*>
1956 V20 N3 P39-40 R-3219 <*>
64-19419 <=$>
1956 V20 N3 P41-42 R-3222 <*>
64-19415 <=$>
1956 V20 N3 P43-44 R-3216 <*>
64-19413 <=$>
1956 V20 N5 P49-55 62-15129 <*=>
1960 V23 N1 P21-26 (IE V24)
60-11680 <=>
1960 V23 N1 P8-10 (IE V24)
60-11681 <=>
1960 V24 N2 P9-12 60-11998 <=>
1960 V24 N3 P3-13 60-41401 <=>
1960 V24 N3 P14-19 60-41402 <=>
1960 V24 N3 P19-23 60-41403 <=>
1960 V24 N3 P24-28 60-41404 <=>
1960 V24 N3 P28-31 60-41405 <=>
1960 V24 N3 P31-37 60-41406 <=>
1960 V24 N3 P37-44 60-41407 <=>
1960 V24 N3 P44-48 60-41408 <=>
1960 V24 N3 P48-52 60-41409 <=>
1960 V24 N3 P52-54 60-41410 <=>
1960 V24 N3 P54-57 60-41411 <=>
1960 V24 N5 P39-43 61-21064 <=>
1960 V24 N5 P44-46 61-11838 <=>
1960 V24 N6 P24-29 61-21431 <=>
1961 V25 N5 P21-23 63-16479 <=*>
1962 V26 N3 P1-6 62-32895 <=*>
1962 V26 N5 P35-40 63-13304 <=*>
1963 V27 N1 P59-63 63-21827 <=>
1963 V27 N3 P37-41 63-31762 <=>
1963 V27 N6 P6-11 64-21724 <=>
1964 V28 N1 P1-5 64-21823 <=>
1964 V28 N3 P23-29 64-31860 <=>
1964 V28 N3 P60-61 64-31860 <=>
1964 V28 N4 P6-10 64-51268 <=>
1964 V28 N4 P41-47 64-51268 <=>
1965 V29 N1 P46-49 65-30803 <=$>
1965 V29 N2 P6-10 65-31782 <*>
1965 V29 N2 P59-60 65-31782 <*>
1966 V30 N3 P2-3 66-33500 <=$>

VOPROSY OKHRANY MATERINSTVA I DETSTVA
1960 V5 N2 P3-13 60-41000 <=>
1962 V7 N3 P47-52 62-24743 <=*>
1962 V7 N9 P67-75 63-21827 <=>
1962 V7 N9 P79-81 63-19595 <=*>
1963 N2 P80- FPTS V23 N4 P.T753 <FASE>
1963 V8 N2 P80-83 FPTS V23 N4 P.T753 <FASE>
1963 V8 N5 P85-86 64-23600 <=$>
1963 V8 N7 P89-92 63-31804 <=>

★VOPROSY ONKOLOGII
1930 V3 N1 P18-21 60-19600 <=*$> 0
1955 N2 P91-95 60-18172 <+*>
1955 N4 P34-38 R-3181 <*>
1955 V1 N3 P63-71 R-447 <*>
64-19509 <=$>
1955 V1 N3 P96-100 5253-B <K-H>
1955 V1 N4 P34-38 R-3768 <*>
63-18465 <=*>
64-19519 <=$>
6769-A <K-H>
1955 V1 N5 P14-20 62-23490 <=*>
1955 V1 N5 P109-116 5253-A <KH>
62-23542 <=*>
1955 V1 N5 P121-126 R-577 <*>
1955 V1 N6 P84-94 R-805 <*>
RT-4327 <*>
1956 N3 P285-294 R-1916 <*>
1956 V2 P205-211 R-945 <*>
1956 V2 P275-284 R-3192 <*>
767TM <CTT>
1956 V2 N1 P10-18 62-15112 <*=>
8535-E <K-H>
1956 V2 N2 P198-201 59Q67R <ATS>
1956 V2 N2 P205-211 64-19480 <=$>
1956 V2 N2 P211-215 R-946 <*>
64-19514 <=$> 0
1956 V2 N3 P285-294 63-19459 <=*>
1956 V2 N3 P323-328 62-13293 <*=>
1956 V2 N5 P528-532 62-11157 <=>
62-15102 <=*>
1957 V3 N2 P179-188 63-19109 <=*>
1959 V5 N7 P34-38 63-18462 <=*>
1959 V5 N10 P498-505 60-11265 <=>
1960 V6 N3 P109-117 60-41475 <=>
1960 V6 N3 P201-213 62-14794 <=*>
1960 V6 N4 P7-22 60-31867 <=>
1960 V6 N8 P25-33 10880-A <KH>
63-16659 <=*>
1960 V6 N11 P3-11 62-13961 <*=>
1960 V6 N11 P12-17 62-13962 <*=> 0
1960 V6 N11 P17-22 62-13963 <=*> 0
1960 V6 N11 P22-26 62-13964 <=*>
1960 V6 N11 P33-38 62-13966 <=*> C
1960 V6 N11 P38-52 62-13967 <*=> 0
1960 V6 N11 P53-60 62-13968 <=*> 0
1960 V6 N11 P6C-66 62-13969 <=*> 0
1960 V6 N11 P66-70 62-13970 <=*> 0
1960 V6 N11 P7C-74 62-13971 <=*> C
1960 V6 N11 P75-83 62-13972 <=*> 0
1960 V6 N11 P84-86 62-13973 <*=> 0
1960 V6 N11 P86-88 62-13974 <*=> C
1960 V6 N11 P89-92 62-13975 <=*> 0
1960 V6 N11 P93-99 62-13976 <=*> C
1960 V6 N11 P99-101 62-13977 <=*> C
1960 V6 N11 P102-103 62-13978 <*=> 0
1960 V6 N11 P1C4-106 62-13979 <=*> C
1960 V6 N11 P107-110 62-13980 <=*>
1960 V6 N11 P110-113 62-13981 <*=>
1960 V6 N11 P114-117 62-13982 <=*>
1960 V6 N11 P118-124 62-13983 <*=>
1960 V6 N11 P125-126 62-13984 <*=>
1961 V7 N1 P108-118 61-31492 <=>
1961 V7 N1 P122-126 61-23593 <=*>
1961 V7 N7 P105-111 11538-A <KH>
64-14602 <=*$>
1961 V7 N9 P46-51 AEC-TR-5684 <=*>
1961 V7 N11 P108-110 63-15163 <=*>
1962 V8 N1 P107-117 AEC-TR-5653 <=*>
1962 V8 N10 P59-64 AEC-TR-6234 <=*$>
1963 V9 N1 P37- FPTS V23 N2 P.T362 <FASE>
1963 V9 N1 P42-44 63-23129 <=*>
1963 V9 N1 P48-54 64-13813 <=*$> 0

1963 V9 N1 P59-63 64-13814 <=*$> 0
1963 V9 N2 P83- FPTS V23 N2 P.T355 <FASE>
FPTS,V23,P.T355-T357 <FASE>
1963 V9 N2 P88- FPTS V23 N2 P.T352 <FASE>
FPTS,V23,P.T352-T354 <FASE>
1963 V9 N2 P92-98 64-13812 <=*$> 0
1963 V9 N2 P98-101 64-13464 <=*$>
1963 V9 N3 P8-12 67-31300 <=$>
1963 V9 N3 P28- FPTS V23 N1 P.T19 <FASE>
FPTS,V23,P.T19-T22 <FASE>
1963 V9 N3 P32-39 64-13476 <=*$>
1963 V9 N3 P40- FPTS V23 N1 P.T59 <FASE>
FPTS,V23,P.T59-T61 <FASE>
1963 V9 N3 P49- FPTS V23 N1 P.T37 <FASE>
FPTS,V23,P.T37-T38 <FASE>
1963 V9 N4 P13-17 64-19200 <=*$>
1963 V9 N4 P18-24 64-13916 <=*$>
1963 V9 N4 P58-61 64-13486 <=*$>
1963 V9 N4 P61- FPTS V23 N2 P.T358 <FASE>
FPTS,V23,P.T358-T361 <FASE>
1963 V9 N4 P61-68 38R77R <ATS>
1963 V9 N4 P68-75 63-31465 <=>
1963 V9 N4 P79- FPTS V23 N2 P.T345 <FASE>
FPTS,V23,P.T345-T348 <FASE>
1963 V9 N4 P93-95 63-31465 <=>
1963 V9 N5 P68- FPTS,V23,N4,P.T809 <FASE>
1963 V9 N6 P34-41 64-19295 <=*$>
1963 V9 N6 P67-70 64-19199 <=*$>
1963 V9 N6 P70-78 64-19202 <=*$>
1963 V9 N7 P15- FPTS V23 N5 P.T1033 <FASE>
FPTS,V23,N5,P.T1033 <FASE>
1963 V9 N7 P33- FPTS V23 N6 P.T1369 <FASE>
FPTS V23 P.T1369 <FASE>
1963 V9 N7 P79-85 64-19203 <=*$>
1963 V9 N8 P103-116 64-15571 <=*$>
1963 V9 N9 P9-14 64-15469 <=*$>
1963 V9 N9 P14-18 64-15470 <=*$>
1963 V9 N9 P18- FPTS,V23,N4,P.T796 <FASE>
1963 V9 N9 P23- FPTS V23 N5 P.T1028 <FASE>
FPTS,V23,N5,P.T1028 <FASE>
1963 V9 N10 P35- FPTS V23 N4 P.T893 <FASE>
FPTS,V23,N4,P.T893 <FASE>
1963 V9 N11 P3-12 64-19204 <=*$>
1963 V9 N11 P18- FPTS V23 N4 P.T890 <FASE>
FPTS,V23,N4,P.T890 <FASE>
1963 V9 N11 P22- FPTS V23 N6 P.T1367 <FASE>
FPTS V23 P.T1367 <FASE>
1963 V9 N11 P22-25 64-15932 <=*$>
1963 V9 N11 P26-30 FPTS,V23,N3,P.T587 <=*$>
1963 V9 N12 P8-17 64-19726 <=$>
1963 V9 N12 P26- FPTS V23 N5 P.T1021 <FASE>
FPTS,V23,N5,P.T1021 <FASE>
1963 V9 N12 P39- FPTS V24 N1 P.T106 <FASE>
1963 V9 N12 P39-43 FPTS V24 P.T106-T108 <FASE>
1963 V9 N12 P44-46 64-21737 <=>
1963 V9 N12 P47- FPTS V23 N5 P.T1018 <FASE>
FPTS,V23,N5,P.T1018 <FASE>
1963 V9 N12 P51-55 64-19638 <=$>
1963 V9 N12 P61- FPTS V23 N5 P.T1037 <FASE>
FPTS,V23,N5,P.T1037 <FASE>
1963 V9 N13 P7- FPTS,V23,P.T362-T364 <FASE>
1964 V10 N1 P7-11 65-60169 <=$>
1964 V10 N1 P17- FPTS V24 N2 P.T113 <FASE>
1964 V10 N1 P17-20 FPTS V24 P.T113-T115 <FASE>
1964 V10 N1 P21- FPTS V24 N1 P.T116 <FASE>
1964 V10 N1 P21-25 FPTS V24 P.T116-T118 <FASE>
1964 V10 N1 P114-118 64-31527 <=>
1964 V10 N2 P73- FPTS V24 N1 P.T121 <FASE>
1964 V10 N2 P73-80 FPTS V24 P.T121-T125 <FASE>
1964 V10 N2 P81-88 64-31070 <=>
1964 V10 N3 P9-16 65-61086 <=$>
1964 V10 N3 P16-26 64-31529 <=>
1964 V10 N3 P22- FPTS V24 N4 P.T706 <FASE>
1964 V10 N3 P27- FPTS V24 N2 P.T277 <FASE>
1964 V10 N4 P41- FPTS V24 N4 P.T699 <FASE>
1964 V10 N4 P47- FPTS V24 N2 P.T221 <FASE>
1964 V10 N4 P47-52 FPTS V24 P.T221-T224 <FASE>
1964 V10 N5 P34-41 65-61087 <=$>
1964 V10 N5 P44-49 64-41217 <=>
1964 V10 N5 P50- FPTS V24 N3 P.T529 <FASE>
1964 V10 N5 P55- FPTS V24 N3 P.T537 <FASE>
1964 V10 N5 P60- FPTS V24 N3 P.T541 <FASE>
1964 V10 N5 P66-72 65-61091 <=$>
1964 V10 N5 P103-107 64-41217 <=>
1964 V10 N6 P3- FPTS V24 N4 P.T711 <FASE>
1964 V10 N6 P7- FPTS V24 N3 P.T545 <FASE>
1964 V10 N6 P65- FPTS V24 N4 P.T717 <FASE>
1964 V10 N7 P38- FPTS V24 N4 P.T714 <FASE>
1964 V10 N7 P61-65 65-62196 <=>
1964 V10 N7 P65-69 65-62197 <=$>
1964 V10 N7 P70-75 65-63626 <=$>
1964 V10 N7 P99-104 67-31300 <=$>
1964 V10 N8 P27- FPTS V24 N4 P.T721 <FASE>
1964 V10 N8 P47-52 65-62199 <=>
1964 V10 N9 P3-7 64-51600 <=$>
1964 V10 N9 P58- FPTS V24 N5 P.T881 <FASE>
1964 V10 N9 P74-76 64-51599 <=$>
1964 V10 N10 P72- FPTS V24 N5 P.T873 <FASE>
1964 V10 N11 P61- FPTS V24 N5 P.T893 <FASE>
1964 V10 N11 P66- FPTS V24 N5 P.T929 <FASE>
1964 V10 N12 P39- FPTS V24 N6 P.T939 <FASE>
1964 V10 N12 P45-63 65-30475 <=$>
1964 V10 N12 P45-48 65-64544 <=*$>
1964 V10 N12 P49-52 65-63627 <*>
1964 V10 N12 P52- FPTS V24 N6 P.T1111 <FASE>
1964 V10 N12 P88-97 66-11697 <*>
1964 V10 N12 P102-104 65-30475 <=$>
1965 V11 N1 P34- FPTS V25 N1 P.T85 <FASE>
1965 V11 N1 P44-45 65-14393 <*>
1965 V11 N2 P30-34 65-30873 <=$>
1965 V11 N2 P52- FPTS V25 N1 P.T88 <FASE>
1965 V11 N2 P52-57 66-60566 <=*$>
1965 V11 N2 P67- FPTS V25 N1 P.T79 <FASE>
1965 V11 N2 P74-77 65-30873 <=$>
1965 V11 N2 P77-85 66-60420 <=*$>
1965 V11 N3 P35-42 66-60858 <=*$>
1965 V11 N3 P48-61 65-31128 <=$>
1965 V11 N3 P57- FPTS V25 N2 P.T283 <FASE>
1965 V11 N3 P73-77 65-31128 <=$>
1965 V11 N3 P104-105 65-31128 <=$>
1965 V11 N4 P31-36 66-60878 <=*$>
1965 V11 N4 P36- FPTS V25 N2 P.T347 <FASE>
1965 V11 N4 P78- FPTS V25 N1 P.T77 <FASE>
1965 V11 N5 P40-43 66-60888 <=*$>
1965 V11 N5 P46-54 66-60867 <=*$>
1965 V11 N5 P72-76 66-60868 <=*$>
1965 V11 N6 P58-63 66-61065 <=*$>
1965 V11 N6 P68-74 66-61083 <=$*>
1965 V11 N6 P115-123 65-32425 <=*$>
1965 V11 N7 P67- FPTS V25 N5 P.T819 <FASE>
1965 V11 N7 P77- FPTS V25 N5 P.T824 <FASE>
1965 V11 N9 P42- FPTS V25 N5 P.T821 <FASE>
1965 V11 N9 P51- FPTS V25 N4 P.T651 <FASE>
1965 V11 N10 P58- FPTS V25 N5 P.T811 <FASE>
1965 V11 N10 P58-63 66-61478 <=*$>
1966 V12 N1 P52- FPTS V25 N6 P.T1073 <FASE>
1966 V12 N2 P86-88 66-31733 <=$>
1966 V12 N3 P114-117 66-32072 <=$>
1966 V12 N5 P115-116 66-33312 <=$>
1966 V12 N7 P17-22 67-31090 <=$>
1966 V12 N7 P40-44 67-31090 <=$>
1966 V12 N8 P61-64 67-31090 <=$>
1966 V12 N10 P123-124 67-31090 <=$>
1966 V12 N11 P44-45 67-31090 <=$>
1966 V12 N11 P126 67-31090 <=$>

VOPROSY PATOLOGII KROVI I KROVOOBRASHCHENIYA
1959 V5 P45-68 62-33604 <=>
1959 V5 P215-216 62-33604 <=>
1961 V6 P27-32 62-33611 <=>
1961 V6 P113-121 62-33611 <=>
1961 V6 P198-204 62-33611 <=>
1961 V6 P236-240 62-33611 <=>

VOPROSY PATLOGII RECHI
1959 V32 N81 P5-7 62-15340 <*=>
1959 V32 N81 P8-9 61-27747 <=*>
62-15317 <*=>
1959 V32 N81 P10-14 62-15321 <*=>
1959 V32 N81 P15-22 62-15324 <*=>

1959 V32 N81 P23-27 62-15323 <*=>
1959 V32 N81 P28-33 61-27790 <=*>
62-15319 <*=>
1959 V32 N81 P34-37 61-27791 <=*>
1959 V32 N81 P59-65 61-27749 <=*>
1959 V32 N81 P66-71 61-27800 <=*>
1959 V32 N81 P72-77 61-27748 <=*>
1959 V32 N81 P78-84 61-27792 <=*>
1959 V32 N81 P85-89 61-27798 <=*>
1959 V32 N81 P94-97 61-27797 <=*>

VOPROSY PATOLOGII SERDECHNO-SOSUDISTOI SISTEMY
1955 V4 N2 P3-13 62-23535 <=*>
1955 V4 N4 P11-19 62-23505 <=*>
1955 V4 N5 P3-13 62-23504 <=*>
1956 V5 N1 P3-16 62-23506 <=*>
1957 V6 N4 P3-13 <CP>
62-24956 <=*>

VOPROSY PEREDACHI INFORMATSII
1964 N3 P118-121 AD-630 996 <=$>

VOPROSY PETROGRAFII I MINERALOGII
1953 V2 P281-305 63-20399 <=*$>

VOPROSY PITANIYA
1938 V7 N3 P132-139 RT-2124 <*>
1954 V13 N2 P41-45 RJ-427 <ATS>
1954 V13 N2 P57-59 RT-2475 <*>
1954 V13 N6 P15-21 R-2710 <*>
1955 V14 N1 P39-44 59-11203 <=>
1955 V14 N2 P3-8 62-15343 <*=>
1955 V14 N3 P43 66-14132 <*$>
1955 V14 N5 P3-12 59-15622 <+*>
1955 V14 N5 P27-35 AD-610 347 <=$>
1956 V15 N5 P32-37 R-2375 <*>
64-19487 <=$> 0
1957 V16 N2 P6-17 62-23532 <=*>
1957 V16 N5 P3-8 62-15096 <=*>
1957 V16 N5 P36-44 64-19443 <=$>
1957 V16 N6 P9-17 63-24367 <=*$>
1957 V16 N6 P52-56 NP-TR-692 <*>
1958 V17 N1 P18-24 62-14714 <=*>
1958 V17 N1 P46-49 62-14773 <=*>
1958 V17 N1 P50-54 C-3496 <NRCC>
1958 V17 N2 P3-14 59-18854 <+*>
1958 V17 N3 P3-9 59-11702 <=>
1958 V17 N3 P9-12 59-11703 <=>
1958 V17 N3 P38-43 59-11704 <=> 0
1958 V17 N4 P24-28 59-19246 <+*>
1958 V17 N4 P58-61 14L32R <ATS>
1958 V17 N5 P3-8 59-11431 <=>
1958 V17 N5 P9-11 59-11432 <=>
1958 V17 N5 P34-39 48L33R <ATS>
59-22254 <+*>
1958 V17 N5 P79-81 59-11433 <=> 0
1959 V18 N1 P3-6 59-13606 <=>
1959 V18 N1 P26-34 61-15816 <+*>
1959 V18 N1 P68-71 61-27793 <=*> 0
1959 V18 N3 P14-17 59-13775 <=>
1959 V18 N3 P36-40 59-13776 <=>
1959 V18 N5 P42-46 62-15320 <=*>
1960 V19 N1 P22-27 60-11736 <=>
1960 V19 N1 P35-38 60-11737 <=>
1960 V19 N1 P54-55 63-19139 <=*>
1960 V19 N1 P63-66 60-11738 <=>
1960 V19 N2 P63-69 60-11761 <=>
1960 V19 N2 P92-95 60-11905 <=>
1960 V19 N4 P54-58 62-25401 <=*>
1960 V19 N4 P83-85 60-31791 <=>
1960 V19 N5 P9-13 62-10255 <=*> 0
1960 V19 N5 P13-18 61-20863 <*=> 0
1960 V19 N5 P18-24 61-20202 <*=> 0
1960 V19 N5 P24-28 61-11562 <=>
1960 V19 N5 P59-61 61-11571 <=>
1960 V19 N5 P79-82 63-11119 <=>
1960 V19 N6 P18-22 61-21757 <=>
1961 V20 N1 P35-38 61-21676 <=>
1961 V20 N1 P69-73 61-21672 <=>
1961 V20 N1 P74-77 63-11113 <=>
1961 V20 N2 P3-10 61-23921 <*=>
1961 V20 N2 P62-64 61-23948 <=*>
1961 V20 N3 P91-94 62-13635 <*=>
1961 V20 N4 P3-8 62-13514 <*=>
1961 V20 N4 P52-56 62-13044 <*=>
1961 V20 N4 P91-92 62-15467 <=*>
1961 V20 N5 P26-31 63-19140 <=*>
1961 V20 N6 P3-11 62-23396 <=*>
1961 V20 N6 P18-22 62-23396 <=*>
1961 V20 N6 P55-58 63-24382 <=*$>
1961 V20 N6 P67-70 62-23396 <=*>
1962 V21 N1 P13-18 62-23589 <=>
1962 V21 N1 P45-48 62-23589 <=>
1962 V21 N1 P54-68 62-23589 <=>
1962 V21 N1 P68-72 96S80R <ATS>
1962 V21 N1 P87-92 62-23589 <=>
1962 V21 N2 P3-10 63-15561 <=*>
1962 V21 N2 P11- FPTS V22 N1 P.T132 <FASE>
1962 V21 N2 P20- FPTS V22 N2 P.T311 <FASE>
1962 V21 N2 P26-32 62-32900 <=*>
1962 V21 N2 P37-39 62-24244 <=*>
1962 V21 N2 P76-80 62-24244 <=*>
1962 V21 N2 P81-84 63-15422 <=*>
1962 V21 N3 P3-8 62-32439 <=>
1962 V21 N3 P22-28 63-19021 <=*>
1962 V21 N3 P28-31 63-19022 <=*>
1962 V21 N3 P37-40 62-32439 <=>
1962 V21 N3 P61-66 62-32439 <=>
1962 V21 N3 P65- FPTS V22 N4 P.T605 <FASE>
1962 V21 N3 P66-71 63-19002 <=*>
1962 V21 N3 P80-81 63-23028 <=*>
1962 V21 N3 P88-90 62-32439 <=>
1962 V21 N3 P89-90 63-23900 <=*$> 0
1962 V21 N4 P3-10 62-33129 <=*>
1962 V21 N4 P25-30 63-15988 <=*>
1962 V21 N4 P42-45 62-33129 <=*>
1962 V21 N4 P71-74 63-15989 <=*>
1962 V21 N5 P13-20 63-24498 <=*$>
1962 V21 N5 P26-31 63-24453 <=*$>
1962 V21 N5 P47-52 64-13501 <=*$>
1962 V21 N5 P58-66 63-24454 <=*$>
1962 V21 N5 P71-81 63-13495 <=*>
1962 V21 N5 P85-86 63-24455 <=*$>
1962 V21 N6 P22-27 63-23055 <=*>
1962 V21 N6 P27-33 63-24451 <=*$>
1962 V21 N6 P33-36 63-23846 <=*>
1962 V21 N6 P49- FPTS V22 N4 P.T611 <FASE>
1962 V21 N6 P64-67 63-23054 <=*>
1963 V22 N1 P17- FPTS V23 N3 P.T462 <FASE>
1963 V22 N1 P17-21 FPTS,V23,N3,P.T462 <FASE>
1963 V22 N1 P32- FPTS V23 N3 P.T450 <FASE>
1963 V22 N1 P32-37 FPTS,V23,N3,P.T583 <FASE>
1963 V22 N1 P50- FPTS V23 N3 P.T456 <FASE>
1963 V22 N1 P50-54 FPTS,V23,N3,P.T456 <FASE>
1963 V22 N1 P66-69 63-21550 <=>
64-15481 <=*$>
1963 V22 N1 P87- FPTS V23 N3 P.T461 <FASE>
1963 V22 N1 P87 FPTS,V23,N3,P.T461 <FASE>
1963 V22 N2 P11-15 64-13912 <=*$>
1963 V22 N2 P39- FPTS V23 N2 P.T423 <FASE>
FPTS,V23,P.T423-T426 <FASE>
1963 V22 N2 P45- FPTS V23 N3 P.T459 <FASE>
FPTS,V23,N3,P.T459 <FASE>
1963 V22 N2 P57- FPTS V23 N3 P.T453 <FASE>
1963 V22 N2 P57-62 FPTS,V23,N3,P.T453 <FASE>
1963 V22 N2 P63- 64-13913 <=*$>
1963 V22 N3 P3-8 63-31358 <=>
1963 V22 N3 P22-28 64-13903 <=*$>
1963 V22 N3 P41- FPTS V23 N4 P.T763 <FASE>
1963 V22 N3 P51-55 63-31358 <=>
1963 V22 N3 P64- FPTS V23 N4 P.T911 <FASE>
FPTS,V23,N4,P.T911 <FASE>
1963 V22 N3 P72- FPTS,V23,N4,P.T799 <FASE>
1963 V22 N3 P90-94 63-31358 <=>
1963 V22 N4 P3-24 63-31803 <=>
1963 V22 N4 P6-12 64-19681 <=$>
1963 V22 N4 P12-19 64-19682 <=$>
1963 V22 N4 P19-25 64-19683 <=$>
1963 V22 N4 P35-38 64-19262 <=*$>
1963 V22 N4 P39- FPTS V23 N5 P.T1069 <FASE>

FPTS,V23,N5,P.T1069 <FASE>
1963 V22 N4 P56-58 64-19263 <=*$>
1963 V22 N4 P59-61 64-19264 <=*$>
1963 V22 N5 P3- FPTS V23 N6 P.T1285 <FASE>
FPTS V23 P.T1285 <FASE>
1963 V22 N5 P22- FPTS V23 N6 P.T1223 <FASE>
FPTS V23 P.T1223 <FASE>
1963 V22 N5 P27- FPTS V23 N6 P.T1277 <FASE>
1963 V22 N5 P27-34 64-13174 <=*$>
1963 V22 N5 P43- FPTS V23 P.T1283 <FASE>
1963 V22 N5 P50-55 65-60213 <=$>
1963 V22 N5 P55-58 64-19669 <=$>
1963 V22 N5 P60- FPTS V23 N6 P.T1281 <FASE>
FPTS V23 P.T1281 <FASE>
1963 V22 N6 P12-15 64-11592 <=*$>
1963 V22 N6 P12-16 65-60214 <=$>
1963 V22 N6 P30-33 64-19719 <=$>
1963 V22 N6 P37- FPTS V23 N6 P.T1288 <FASE>
FPTS V23 P.T1288 <FASE>
1964 V23 N1 P3-16 64-15396 <=*$>
1964 V23 N1 P3-17 65-60199 <=$>
1964 V23 N1 P36-43 64-19635 <=$> O
1964 V23 N1 P43-51 64-19675 <=$>
1964 V23 N2 P13-17 65-60222 <=$>
1964 V23 N2 P17- FPTS V24 N2 P.T313 <FASE>
1964 V23 N2 P17-20 FPTS V24 P.T313-T315 <FASE>
1964 V23 N2 P21-30 64-31694 <=>
65-60223 <=$>
1964 V23 N2 P35-40 65-60224 <=$>
1964 V23 N2 P49- FPTS V24 N1 P.T186 <FASE>
1964 V23 N2 P49-53 FPTS V24 P.T186-T188 <FASE>
1964 V23 N2 P73-76 64-31694 <=>
1964 V23 N2 P79 64-31694 <=>
1964 V23 N2 P82 64-31694 <=>
1964 V23 N2 P90-93 64-31694 <=>
1964 V23 N4 P13- FPTS V24 N4 P.T635 <FASE>
1964 V23 N4 P38-45 65-62313 <=$>
1964 V23 N4 P81-88 65-62317 <=$>
1964 V23 N5 P3-6 64-51497 <=>
64-51497 <=*>
1964 V23 N5 P33-41 65-62837 <*>
1964 V23 N5 P77-80 64-51497 <=>
64-51497 <=*>
1964 V23 N5 P84-90 64-51497 <=>
64-51497 <=*>
1964 V23 N6 P44- FPTS V24 N6 P.T1113 <FASE>
1964 V23 N6 P49- FPTS V24 N6 P.T963 <FASE>
1964 V23 N6 P56-62 65-63628 <=$>
1965 N4 P9-13 AD-624 585 <=*$>
1965 V24 N1 P66-71 65-64529 <=*$>
1965 V24 N1 P71-75 65-64528 <=*$>
1965 V24 N1 P83-87 65-64527 <=*$>
1965 V24 N2 P22-26 66-60421 <=*$>
1965 V24 N2 P37-40 66-60422 <=*$>
1965 V24 N2 P63-68 66-60423 <=*$>
1965 V24 N2 P72-78 66-60850 <=*$>
1965 V24 N3 P83-84,87-91 65-32336 <=*$>
1965 V24 N4 P3-8 65-33220 <=*$>
1965 V24 N4 P25-28 66-61341 <=*$>
1965 V24 N4 P70-76 66-61634 <=*>
1965 V24 N4 P89-94 65-33337 <=*$>
1965 V24 N6 P7- FPTS V25 N6 P.T1055 <FASE>
1966 V25 N6 P3-19 67-30590 <=> O
1966 V25 N6 P72-74 67-30590 <=> O

VOPROSY POROSHKOVOI METALLURGII I PROCHNOSTI MATERIALOV
1955 V2 P150-154 65-11876 <*>
1956 N3 P97-107 63-27825 <$>
1958 P79-80 4706 <HB>
1958 V6 P65-79 61-19670 <*=>
1959 V7 P33-56 62-33622 <=*>
1959 V7 P105-119 62-33622 <=*>

VOPROSY PROIZVODSTVA STALI
1958 V6 P68-76 2732 <BISI>
1958 V6 P87-95 2736 <BISI>
1958 V6 P96-109 2737 <BISI>
1960 V7 P74-81 2733 <BISI>
1960 V7 P117-134 2735 <BISI>
1960 V7 P135-140 2734 <BISI>
1961 N8 P55-69 3641 <BISI>
1961 V8 P88-95 3011 <BISI>
1963 V9 P65-72 3864 <BISI>
1963 V9 P79-95 4153 <BISI>

VOPROSY PSIKHIATRII
1959 V3 P407-436 62-23456 <=*> O
1963 V7 N2 P173-180 63-24241 <=*$>
1965 N3 P165-171 65-31938 <=$>

VOPROSY PSIKHIATRII I NEVROPATOLOGII
1961 V7 P86-94 63-23130 <=*>
1961 V7 P113-120 64-15904 <=*$>

VOPROSY PSIKHOFIZIOLOGII, REFLEKSOLOGII I GIGENY TRUDY
1928 N3 P32-37 64-10044 <*>

VOPROSY PSIKHOLOGII
1955 N2 P101-112 PB 141 239T <=>
1955 N6 P16-38 PB 141 238T <=>
1957 N3 P106-113 PB 141 240T <=>
1957 N4 P25-33 PB 141 259T <=>
1957 V3 N1 P53-60 62-13795 <=*>
1957 V3 N1 P78-87 62-13796 <=*>
1957 V3 N1 P179-182 62-15049 <*=>
1957 V3 N4 P8-24 <CP>
62-23515 <=*>
1957 V3 N4 P34-44 <CP>
1957 V3 N4 P70-79 60-16610 <+*>
1957 V3 N4 P90-101 60-16609 <+*>
1958 V4 N1 P24-36 59-13840 <=>
1958 V4 N1 P50-76 59-13184 <=> O
1958 V4 N1 P184-190 59-13841 <=> O
1958 V4 N2 P23-37 59-13801 <=>
1958 V4 N3 P74-86 59-13842 <=>
1958 V4 N3 P161-167 59-13843 <=>
1958 V4 N4 P105-115 59-13822 <=>
1958 V4 N5 P3-17 59-13854 <=>
1958 V4 N6 P66-78 59-13795 <=>
1958 V4 N6 P79-87 59-13796 <=>
1958 V4 N6 P119-130 59-13797 <=>
1959 V5 N1 P3-22 60-11008 <=>
1959 V5 N1 P64-75 60-11009 <=>
1959 V5 N3 P56-65 60-13520 <+*>
1959 V5 N4 P3-15 59-13999 <=>
1959 V5 N5 P124-133 61-19081 <*=> C
1959 V5 N5 P170-180 60-11276 <=>
1959 V5 N6 P182-186 60-31169 <=>
1960 V6 N1 P165-172 60-11984 <=>
1960 V6 N2 P61-73 61-27311 <*=>
1960 V6 N3 P46-56 64-13149 <=*$>
1961 V7 N2 P133-159 62-23803 <=*>
1961 V7 N3 P47-59 62-23781 <=*> O
1961 V7 N3 P171-176 62-23781 <=*> O
1961 V7 N4 P101-108 63-15944 <=*>
1961 V7 N5 P173-182 63-19265 <=*$>
1962 N6 P26-39 63-01538 <*>
1962 V8 N1 P19-27 64-15200 <=*$>
1962 V8 N1 P28-44 62-25599 <=*>
1962 V8 N1 P56-80 64-13958 <=*$>
1962 V8 N1 P168-173 63-21276 <=>
1962 V8 N2 P19-40 62-32453 <=>
1962 V8 N2 P61-72 62-32453 <=>
1962 V8 N2 P92-100 62-32453 <=>
1962 V8 N2 P151-155 62-32453 <=>
1962 V8 N3 P3-26 63-13333 <=>
1962 V8 N3 P37-44 63-13333 <=>
1962 V8 N3 P45-55 62-11785 <=>
1962 V8 N3 P95-105 63-13333 <=>
1962 V8 N3 P149-155 63-13333 <=>
1962 V8 N3 P173,190 63-13333 <=>
1962 V8 N3 P177-184 63-13333 <=>
1962 V8 N4 P58-73 63-15392 <=*>
1962 V8 N4 P89-98 63-15392 <=*>
1962 V8 N4 P156-160 63-15392 <=*>
1962 V8 N4 P172-182 63-15392 <=*>
1962 V8 N5 P15-30 63-21087 <=>
63-21580 <=>

1962 V8 N5 P70-74 64-21386 <=>
1962 V8 N5 P75-83 63-21580 <=>
1962 V8 N5 P107-112 63-21580 <=>
1962 V8 N5 P184-189 63-21095 <=>
63-21580 <=>
1962 V8 N6 P14-25 65-11930 <=$>
1962 V8 N6 P26-39 64-13064 <=*$>
65-10298 <*>
1963 V9 N1 P3-12 63-21894 <=>
1963 V9 N1 P35-40 63-21845 <=>
1963 V9 N1 P67-79 63-21985 <=>
1963 V9 N1 P104-109 63-31013 <=>
1963 V9 N1 P111-113 63-31013 <=>
1963 V9 N1 P137-141 64-16622 <=*$>
1963 V9 N1 P142-145 63-21833 <=>
1963 V9 N1 P146-150 63-21869 <=>
1963 V9 N2 P17-29 63-31412 <=>
1963 V9 N2 P59-67 63-31481 <=>
1963 V9 N2 P75-90 63-31552 <=>
1963 V9 N3 P3-4 64-41284 <=>
1963 V9 N3 P47-53 64-41284 <=>
1963 V9 N3 P64-72 64-41284 <=>
1963 V9 N3 P125-142 64-41284 <=>
1963 V9 N3 P176-180 63-31478 <=>
1963 V9 N3 P181-183 64-41284 <=>
1963 V9 N4 P3-6 63-41071 <=>
1963 V9 N4 P7-20 63-41253 <=>
1963 V9 N4 P135-142 63-41189 <=>
1963 V9 N4 P143-152 63-41254 <=>
1963 V9 N4 P153-158 63-41191 <=>
1963 V9 N5 P99-117 64-21481 <=>
1963 V9 N5 P176-186 64-21682 <=>
1964 V10 N1 P25-32 64-31173 <=>
1964 V10 N1 P61-70 64-31168 <=>
1964 V10 N1 P145-149 64-31168 <=>
1964 V10 N1 P163-177 64-31168 <=>
1964 V10 N2 P98-102 AD-622 456 <=*$>
1964 V10 N3 P131-139 64-41200 <=>
1964 V10 N3 P177-181 64-41789 <=>
1964 V10 N4 P3-38 64-51096 <=$>
1964 V10 N4 P61-68 AD-615 234 <=$>
1964 V10 N4 P83-93 64-51096 <=$>
1964 V10 N4 P99-114 64-51096 <=$>
1964 V10 N5 P3-10 AD-615 534 <=$>
1964 V10 N5 P20-30 65-30122 <=$>
1964 V10 N6 P3-12 65-30767 <=$>
1965 N1 P149-151 NLLTB V7 N9 P791 <NLL>
1965 N6 P119-125 66-30641 <=*$>
1965 V11 N2 P3-16 65-31560 <=$>
1965 V11 N2 P28-34 65-31560 <=$>
1965 V11 N2 P49-74 65-31560 <=$>
1965 V11 N2 P155-164 65-31331 <=*$>
1965 V11 N2 P173-180 65-31560 <=$>
1965 V11 N2 P181-190 65-31328 <=$>
1965 V11 N3 P62-75 65-31935 <*=$>
1965 V11 N3 P113-123 65-31936 <=*$>
1965 V11 N3 P138-146 65-31937 <=*$>
1966 N2 P131-138 66-32267 <=$>
1966 V12 N3 P3-48 66-33844 <=$>

VOPROSY RADIOBIOLOGII: SBORNIK TRUDOV
1960 V3 ENTIRE ISSUE 61-31017 <=>
1964 V3/4 P53-57 64-51212 <=$>
1964 V3/4 P59-69 64-51212 <=$>
1964 V3/4 P71-109 64-51212 <=$>
1964 V3/4 P197-215 64-51212 <=$>
1964 V3/4 P275-281 64-51212 <=$>
1964 V3/4 P283-287 64-51212 <=$>

VOPROSY RAKETNOI TEKHNIKI
1965 V14 N7 P28-41 66-34326 <=$>

VOPROSY RENTGENOLOGII I ONKOLOGII
1963 V7 P45-51 64-41537 <=>
1963 V7 P145-153 64-41537 <=>
1963 V7 P199-204 64-41537 <=>

VOPROSY REVMATIZMA
1962 V2 N1 P3-12 62-24995 <=*>
1962 V2 N1 P88-91 62-24995 <=*>

1963 V3 N1 P66-70 63-21536 <=>
1964 V4 N1 P7- FPTS V24 N2 P.T248 <FASE>
FPTS V24 P.T248-T252 <FASE>
1964 V4 N1 P16-23 65-60183 <=$>
1964 V4 N2 P53-69 64-51136 <=>
1964 V4 N4 P20-22 65-64536 <=*$>
1964 V4 N4 P33-36 65-64535 <=*$>

VOPROSY RUDNOI GEOFIZIKI: SBORNIK STATEI
1957 V1 P62-68 61-27256 <*=>
1964 N4 P67-73 65-14288 <*>

VOPROSY TEORII MATEMATICHESKIKH MASHIN
1958 V1 P219-230 61-27030 <*=>

VOPROSY TEORII PLAZMY
1963 N1 P98-285 AD-619 987 <=*$> M

VOPROSY VETROENERGETIKI
1959 P5-10 SCL-T-360 <*>

VOPROSY VYCHISLITELNOI MATEMATIKI I TEKHNIKI
1964 N1 P144-179 65-30840 <=*$>

★VOPROSY VIRUSOLOGII
1956 N4 P19-25 R-138 <*>
R-499 <*>
1956 N4 P26-30 R-516 <*>
1442 <*>
1956 V1 N1 P26-30 R-2313 <*>
62-16053 <=*> C
1956 V1 N2 P47-51 R-2339 <*>
1956 V1 N3 P30-32 R-1445 <*>
64-19430 <=$>
1956 V1 N3 P32-35 R-1452 <*>
64-19428 <=$> C
1958 V3 N5 P315-316 59-18300 <+*>
1959 V4 P396-400 59-22289 <+*>
1959 V4 P505-508 59-22290 <+*>
1959 V4 N1 P63-67 59-13836 <=>
1959 V4 N2 P198-204 61-19717 <=*>
1959 V4 N2 P208-213 61-19715 <=*>
1959 V4 N2 P213-218 61-19716 <=*>
1959 V4 N2 P237-238 60-17966 <+*>
1959 V4 N3 P305-310 61-19728 <=*>
1959 V4 N3 P311-314 61-19727 <=*>
1959 V4 N4 P449-456 61-15971 <+*>
1959 V4 N5 P533-537 60-31135 <+*>
1959 V4 N5 P537-544 60-31136 <=>
1959 V4 N5 P580-585 60-31137 <+*>
1960 V5 N5 P515-520 62-13954 <=*>
1960 V5 N5 P521-525 62-13948 <=*>
1960 V5 N5 P525-528 62-13940 <=*>
1960 V5 N5 P528-532 62-13945 <=*>
1960 V5 N5 P532-537 62-13949 <=*>
1960 V5 N5 P538-541 62-13944 <*=>
1960 V5 N5 P542-547 62-13960 <*=>
1960 V5 N5 P560-564 62-13935 <=*>
1960 V5 N5 P564-567 62-13932 <=*> O
1960 V5 N5 P568-572 62-13953 <=*>
1960 V5 N5 P572-577 62-13955 <=*>
1960 V5 N5 P577-582 62-13956 <=*> C
1960 V5 N5 P582-586 62-13946 <=*> O
1960 V5 N5 P586-591 62-13952 <=*>
1960 V5 N5 P591-598 62-13934 <=*> C
1960 V5 N5 P599-602 62-13951 <=*>
1960 V5 N5 P602-611 62-13947 <=*> C
1960 V5 N5 P611-613 62-13943 <=*>
1960 V5 N5 P614-618 62-13938 <=*> O
1960 V5 N5 P618-621 62-13937 <=*>
1960 V5 N5 P622-624 62-13936 <=*>
1960 V5 N5 P624 62-13959 <=*>
1960 V5 N5 P625 62-13957 <=*>
1960 V5 N5 P626-627 62-13933 <=*> O
1960 V5 N5 P627-628 62-13941 <=*> C
1960 V5 N5 P629-632 62-13950 <=*> O
1960 V5 N5 P632-633 62-13958 <=*>
1960 V5 N5 P634-635 62-13942 <*=>
1960 V5 N5 P636-637 62-13939 <=*>
1960 V5 N6 P670-675 61-19654 <=*>

Citation	Translation
1961 V6 N2 P236-237	61-28153 <*=>
1961 V6 N4 P454-469	62-24997 <=*>
1961 V6 N6 P656-668	63-31308 <=>
1961 V6 N6 P743-745	62-23856 <=*>
1962 V7 N1 P18-22	63-21910 <=>
1962 V7 N1 P85-91	63-00288 <*>
1962 V7 N2 P140-144	63-31019 <=>
1962 V7 N2 P162-167	63-31019 <=>
1962 V7 N4 P39-42	66-61753 <=*$>
1962 V7 N4 P110-111	63-00168 <*>
	63-15575 <=*>
1962 V7 N5 P520-528	63-19249 <=*>
1963 V8 N1 P11-	FPTS V22 N6 P.T1013 <FASE>
1963 V8 N1 P11-16	FPTS,V22,P.T1013-16 <FASE>
1963 V8 N1 P17-	FPTS V22 N6 P.T1222 <FASE>
1963 V8 N1 P17-19	FPTS,V22,P.T1222-23 <FASE>
1963 V8 N1 P20-24	63-15938 <=*>
1963 V8 N1 P24-27	64-15572 <=*$>
1963 V8 N1 P27-31	63-24495 <=*$>
1963 V8 N1 P36-40	64-13514 <=*$>
1963 V8 N1 P40-	FPTS V22 N6 P.T1067 <FASE>
1963 V8 N1 P40-43	FPTS,V22,P.T1067-68 <FASE>
1963 V8 N1 P68-72	63-24496 <=*$>
1963 V8 N1 P82-87	64-13513 <=*$>
1963 V8 N1 P87-90	63-15940 <=*>
1963 V8 N1 P98-99	64-13512 <=*$>
1963 V8 N1 P100-	FPTS V23 N1 P.T69 <FASE>
	FPTS,V23,P.T69-T70 <FASE>
1963 V8 N1 P643-649	67-30862 <=$>
1963 V8 N2 P155-159	64-13478 <=*$>
1963 V8 N2 P193-199	64-13477 <=*$>
1963 V8 N2 P199-204	64-13893 <=*$>
1963 V8 N2 P213-216	64-31299 <=>
1963 V8 N2 P217-221	63-23128 <=*>
1963 V8 N2 P232-	FPTS V23 N2 P.T379 <FASE>
1963 V8 N2 P232-233	FPTS,V23,N2,P.T379 <FASE>
	63-23552 <=*$>
1963 V8 N3 P264-	FPTS V23 N5 P.T992 <FASE>
1963 V8 N3 P268-275	64-15471 <=*$>
1963 V8 N3 P297-	FPTS V23 N3 P.T545 <FASE>
	FPTS,V23,N3,P.T545 <FASE>
1963 V8 N3 P301-307	64-15472 <=*$>
1963 V8 N3 P316-321	64-15473 <=*$>
1963 V8 N3 P335-337	65-60252 <=$> 0
1963 V8 N3 P343-346	64-19196 <=*$>
1963 V8 N4 P398-399	64-15474 <=*$>
1963 V8 N4 P400-403	FPTS V23 N4 P.T849 <FASE>
1963 V8 N4 P400-	FPTS,V23,N4,P.T849 <FASE>
1963 V8 N4 P440-444	FPTS V23 N4 P.T852 <FASE>
1963 V8 N4 P440-	FPTS,V23,N4,P.T852 <FASE>
1963 V8 N4 P452-456	64-19344 <=$>
1963 V8 N4 P457-460	64-15908 <=*$>
1963 V8 N4 P466-470	FPTS V23 N4 P.T855 <FASE>
1963 V8 N4 P466-	FPTS,V23,N4,P.T855 <FASE>
1963 V8 N4 P471-	FPTS V23 N3 P.T548 <FASE>
	FPTS,V23,N3,P.T548 <FASE>
1963 V8 N4 P480-	FPTS V23 N5 P.T998 <FASE>
1963 V8 N4 P480-488	FPTS,V23,N5,P.T998 <FASE>
1963 V8 N5 P542-	FPTS V23 N5 P.T1003 <FASE>
1963 V8 N5 P542-546	FPTS,V23,N5,P.T1003 <FASE>
1963 V8 N5 P547-553	64-19296 <=*$>
1963 V8 N5 P564-	FPTS V23 N5 P.T1006 <FASE>
1963 V8 N5 P564-567	FPTS,V23,N5,P.T1006 <FASE>
1963 V8 N5 P568-573	64-19289 <=*$>
1963 V8 N5 P608-	FPTS V23 N6 P.T1349 <FASE>
	FPTS V23 P.T1349 <FASE>
1963 V8 N5 P617-	FPTS V23 N5 P.T1031 <FASE>
1963 V8 N5 P617-618	FPTS,V23,N5,P.T1031 <FASE>
1963 V8 N5 P628-629	AD-619 889 <=$>
1963 V8 N5 P630-631	AD-619 890 <=*$>
1963 V8 N6 P662-	FPTS V24 N2 P.T273 <FASE>
1963 V8 N6 P662-667	FPTS V24 P.T273-T276 <FASE>
	64-30052 <*> 0
1963 V8 N6 P667-	FPTS V24 N1 P.T14 <FASE>
1963 V8 N6 P667-675	FPTS V24 PT14-T18 <FASE>
1963 V8 N6 P679-	FPTS V24 N1 P.T1 <FASE>
1963 V8 N6 P679-687	FPTS V24 P.T1-T5 <FASE>
1963 V8 N6 P688-692	65-62198 <=>
1963 V8 N6 P692-697	39R75R <ATS>
	65-61049 <=$>
1964 V9 N1 P11-	FPTS V24 N1 P.T6 <FASE>
1964 V9 N1 P11-15	FPTS V24 P.T6-T8 <FASE>
1964 V9 N1 P16-	FPTS V24 N1 P.T9 <FASE>
1964 V9 N1 P16-18	FPTS V24 P.T9-T10 <FASE>
1964 V9 N1 P19-24	64-21950 <=>
	65-61050 <=$>
1964 V9 N1 P30-	FPTS V24 N2 P.T266 <FASE>
	FPTS V24 P.T266-T268 <FASE>
1964 V9 N1 P35-39	65-61051 <=$>
1964 V9 N1 P57-	FPTS V24 N2 P.T268 <FASE>
	FPTS V24 P.T268-T270 <FASE>
1964 V9 N1 P61-	FPTS V24 N1 P.T11 <FASE>
1964 V9 N1 P61-65	FPTS V24 P.T11-T13 <FASE>
1964 V9 N1 P66-72	64-21950 <=>
1964 V9 N1 P83-90	64-21950 <=>
1964 V9 N2 P148-	FPTS V24 N3 P.T473 <FASE>
1964 V9 N2 P173-	FPTS V24 N4 P.T743 <FASE>
1964 V9 N2 P184-	FPTS V24 N3 P.T481 <FASE>
1964 V9 N2 P208-	FPTS V24 N3 P.T484 <FASE>
1964 V9 N3 P259-268	64-41152 <=>
1964 V9 N3 P269-271	65-62190 <=>
1964 V9 N3 P280-286	65-61133 <=$>
1964 V9 N3 P287-291	65-61136 <=$>
1964 V9 N3 P291-296	65-63160 <=$>
1964 V9 N3 P296-301	65-62118 <=$>
1964 V9 N3 P315-320	65-62318 <=$>
1964 V9 N4 P417-421	AD-625 626 <=$>
1964 V9 N4 P438-443	65-62368 <=$>
1964 V9 N4 P455-	FPTS V24 N4 P.T739 <FASE>
1964 V9 N5 P550-555	65-11943 <*>
1964 V9 N5 P555-559	65-11878 <*>
1964 V9 N5 P559-564	65-11769 <*>
1964 V9 N5 P564-569	65-63629 <=$>
1964 V9 N5 P597-601	RTS-2909 <NLL>
1964 V9 N6 P3-10	67-30862 <=$>
1964 V9 N6 P652-657	66-60428 <=*$>
1964 V9 N6 P657-661	65-63630 <=$>
1965 V10 N1 P8-	FPTS V25 N2 P.T315 <FASE>
1965 V10 N1 P17-	FPTS V25 N2 P.T325 <FASE>
1965 V10 N1 P30-36	66-60632 <=*$>
1965 V10 N1 P36-41	66-60846 <=*$>
1965 V10 N1 P67-73	65-31037 <=$>
1965 V10 N2 P139-	FPTS V25 N2 P.T318 <FASE>
1965 V10 N2 P142-149	65-63426 <=$>
1965 V10 N2 P165-167	66-60631 <=*$>
1965 V10 N2 P172-	FPTS V25 N3 P.T515 <FASE>
1965 V10 N2 P213-	FPTS V25 N3 P.T509 <FASE>
1965 V10 N2 P221-224	66-61347 <=*$>
1965 V10 N2 P226-228	65-31126 <=$>
1965 V10 N3 P275-	FPTS V25 N6 P.T1105 <FASE>
1965 V10 N3 P280-	FPTS V25 N6 P.T1103 <FASE>
1965 V10 N3 P307-315	65-64614 <=*$>
1965 V10 N3 P349-351	65-33113 <=*$>
1965 V10 N4 P397-402	66-61620 <=*$>
1965 V10 N4 P402-	FPTS V25 N4 P.T692 <FASE>
1965 V10 N4 P414-	FPTS V25 N5 P.T815 <FASE>
1965 V10 N4 P417-420	66-61622 <=*$>
1965 V10 N4 P454-462	65-33569 <=*$>
1965 V10 N4 P483-486	66-61761 <=*$>
1965 V10 N5 P515-519	66-30307 <=$>
1965 V10 N6 P643-648	66-31012 <=$>
1965 V10 N6 P720-	FPTS V25 N6 P.T1119 <FASE>
1966 N2 P197-202	66-32459 <=$>
1966 V11 N1 P80-84	66-32755 <=$>
VOSTOCHNAYA NEFT	
1940 V2 N9 P36-38	62-14106 <=*>
1940 V2 N9 P38-42	62-14107 <=*>
1940 V2 N7/8 P40-42	62-14103 <=*$>
VOTRE ELECTRICITE	
1955 V26 N3 P26-36	1513 <*>
VOX	
1913 V23 P.81-82	57-1656 <*>
VOX SANGUINIS	
1958 V3 N2 P123-142	58-1834 <*>
1961 V6 P435-450	62-00677 <*>
1962 V7 P362-369	63-01137 <*>

Citation	Translation
1963 V8 P627-631	64-00268 <*>

VRACH

Citation	Translation
1885 V6 P213-216	RT-154 <*>

VRACHEBNAYA GAZETA

Citation	Translation
1927 V31 P403-407	60-19747 <=*$>
1928 V32 P1705-1707	R-1083 <*>

VRACHEBNOE DELO

Citation	Translation
1933 N11 P775-782	R-1081 <*>
1935 V18 N8 P695-700	R-753 <*>
	RT-3825 <*>
	64-15066 <=*$>
1939 V21 N2/3 P106-	R-1061 <*>
1939 V21 N2/3 P106-110	RT-637 <*>
1940 V22 N6 P415-423	RT-1819 <*>
1941 V23 N2 P95-100	RT-1820 <*>
1946 V26 N3/4 P105-106	RT-1311 <*>
	65-60087 <=$>
1956 N8 P861-862	63-23118 <=*> O
1956 N11 P1193-1194	61-19640 <=*>
1957 P103-104	62-33365 <=*>
1957 P105-106	59-13123 <=>
1957 P106-107	59-13124 <=>
1957 N1 P5-8	63-31382 <=>
1957 N1 P88	63-31382 <=>
1957 N3 P329-332	59-11396 <=>
1957 N5 P513-514	59-18743 <+*>
1957 N6 P619-622	62-11151 <=>
	62-15073 <*=>
1957 N8 P848-852	60-17637 <+*>
1958 N1 P11-18	59-11386 <=>
1958 N1 P37-39	61-11678 <=>
1958 N2 P203-206	59-13063 <=>
1958 N4 P363-368	R-5345 <*>
1958 N9 P991-994	59-19671 <+*>
1959 N2 P119-124	62-13296 <*=>
1959 N3 P335-	60-31101 <=>
1959 N6 P625-628	63-31382 <=>
1959 N10 P1107-1110	60-11795 <=>
1959 N11 P1215-1218	60-31841 <=>
1959 N11 P1221-1224	60-11287 <=>
1959 N12 P1303-1304	60-31795 <=>
1960 N1 P75-76	60-17976 <+*>
	60-31688 <=>
1960 N2 P173-178	60-31485 <=>
1960 N3 P297-300	60-41504 <=>
1960 N4 P351-354	60-41706 <=>
1960 N6 P663-666	60-41705 <=>
1960 N6 P669-670	60-41705 <=>
1961 N1 P3-9	61-28871 <*=>
1961 N1 P140-142	61-28776 <*=>
1961 N1 P144-149	61-28864 <*=>
1961 N1 P149-157	61-28728 <*=>
1961 N5 P3-9	62-13259 <=*>
1961 N5 P9-15	62-13260 <=*>
1961 N8 P107-110	62-23323 <=*>
1961 N10 P22-26	62-23460 <=*>
1961 N11 P3-7	62-23414 <=*>
1961 N11 P123-127	62-23348 <=*>
1961 N11 P151-154	62-23323 <=*>
1961 N12 P3-11	62-25534 <=>
1961 N12 P126-132	62-25534 <=>
1962 N1 P76-85	62-23643 <=*>
1962 N1 P133-136	62-23643 <=*>
1962 N4 P13-30	62-33296 <=*>
1962 N4 P87-90	62-33296 <=*>
1962 N9 P114-118	63-13285 <=*>
1962 N9 P151-153	63-13285 <=*>
1963 N10 P111-113	AD-612 413 <=$>
1964 N1 P115-119	64-19698 <=$>
1964 N2 P3-8	NLLTB V6 N6 P564 <NLL>
	TRANS.BULL.1964 P564 <NLL>
1964 N2 P23-30	65-60156 <=$>
1964 N3 P22-26	65-61061 <=$>
1964 N3 P34-38	65-60193 <=$>
1964 N3 P110-	FPTS V24 N2 P.T235 <FASE>
1964 N3 P110-113	FPTS V24 P.T235-T236 <FASE>
1964 N3 P117-121	65-61062 <=$>
1964 N4 P85-88	65-62308 <=$>
1964 N5 P89-94	65-61092 <=$>
1964 N6 P20-	FPTS V24 N4 P.T575 <FASE>
1964 N8 P28-35	65-62348 <=$>
1964 N8 P87-92	65-62119 <=$>
1964 N9 P3-5	65-62369 <=$>
1964 N9 P54-58	65-62370 <=$>
1964 N9 P81-84	65-62371 <=$>
1964 N11 P46-49	65-64556 <=*$>
1964 N12 P74-77	65-64533 <=*$>

VRACHEBNYIA ZAPISKI

Citation	Translation
1896 V3 N12/3 P231-238	60-19774 <=$>
1899 V6 N7/8 P135-145	RT-1309 <*> O

VUNSHNA TURGOVIYA

Citation	Translation
1965 V13 N5 P2-5	65-32124 <=$>
1965 V13 N6 P2-4	65-32124 <=$>

VYCHISLITELNAYA MATEMATIKA

Citation	Translation
1957 N2 P20-44	62-16095 <=*>
1957 N2 P146-153	59-12244 <+*>
1957 V1 P3-22	AMST S2 V35 P95-115 <AMS>
1957 V1 P34-61	AMST S2 V24 P279-307 <AMS>
1957 V1 P62-80	AD-618 216 <=$>
	AMST S2 V35 P117-136 <AMS>
1957 V1 P87-115	AMST S2 V35 P137-165 <AMS>
1958 N3 P99-110	62-16391 <=*>
1958 N3 P149-185	64-71620 <=>
1959 N4 P104-149	63-19893 <*=>
1959 N4 P150-158	62-24663 <=*>
1959 N4 P173-183	60-11920 <=>
1961 N7 P151-160	63-19412 <=*>

VYCHISLITELNAYA MATEMATIKA I VYCHISLITELNAYA TEKHNIKA

Citation	Translation
1953 V1 P37-40	R-1150 <*>
1953 V1 P41-45	R-1151 <*>
1955 V2 P116-144	AMST S2 V8 P257-287 <AMS>
1955 V2 P145-150	AMST S2 V12 P149-154 <AMS>
	58-619 <*>
1958 N3 P116-118	59-10533 <*>

VYCHISLITELNAYA TEKHNIKA

Citation	Translation
1961 N2 P3-7	64-11785 <=>
1961 N2 P8-20	AD-634 809 <=$>
1961 N2 P39-50	AD-621-142 <=$> M
1962 V3 P112-122	63-21000 <=>
1962 V4 P98-110	63-24036 <=*>
1963 N3 P74-90	AD-616 269 <=$>
1963 N3 P103-121	AD-616 269 <=$>
1963 N3 P153-164	64-21884 <=$>

VYCHISLITELNAYA TEKHNIKA I VOPROSY PROGRAMMIROVANIYA

Citation	Translation
1962 V1 P48-57	65-31391 <=$>
1964 V3 P63-83	65-31112 <=$>

VYCHISLITELNYE SISTEMY

Citation	Translation
1962 N1/9 ENTIRE ISSUE	AEC-TR-6637/1-9 <=>
1963 N7 P24-37	64-51969 <=$>
1963 N9 P3-29	64-51970 <=$>
1964 N10/4 ENTIRE ISSUE	AEC-TR-6637/10-14 <=>

VYSHKA

Citation	Translation
1965 P2	65-32528 <=$>
	65-32530 <=$>
1966 P3	66-31540 <=$>
	66-33894 <=$>

VYSOKA SKOLA

Citation	Translation
1958 V6 N5 P138-141	59-11425 <=>

★VYSOKOMOLEKULYARNYE SOEDINENIYA. MOSCOW

Citation	Translation
1959 V1 P29-35	RCT V33 N3 P741 <RCT>
1959 V1 P132-137	RCT V33 N2 P469 <RCT>
1959 V1 P230-239	RCT V33 N3 P748 <RCT>
1959 V1 P254-264	RCT V33 N2 P373 <RCT>
1959 V1 P865-868	RCT V33 N4 P942 <RCT>
1959 V1 P878-	RCT V33 N2 P361 <RCT>

Citation	Translation
1959 V1 P978-989	RCT V33 N4 P1166 <RCT>
1959 V1 P1016-1020	RCT V33 N4 P988 <RCT>
1959 V1 P1042-1047	RCT V33 N4 P964 <RCT>
1959 V1 P1200-1206	61-15432 <+*> 0
1959 V1 N1 P17-20	51M38R <ATS>
	61-20638 <*=>
1959 V1 N1 P46-57	RCT V33 N3 P610 <RCT>
1959 V1 N1 P58-67	RCT V33 N2 P401 <RCT>
1959 V1 N1 P94-102	04M39R <ATS>
1959 V1 N1 P114-122	82L36R <ATS>
	83L36R <ATS>
1959 V1 N1 P126-127	44M44R <ATS>
1959 V1 N1 P132-137	50L35R <ATS>
1959 V1 N1 P138-142	60-16649 <+*>
	63-18293 <=*>
1959 V1 N1 P143-148	63-10781 <=*>
1959 V1 N1 P149-151	48N55R <ATS>
	60-14252 <+*>
	60-23810 <+*>
1959 V1 N1 P166-172	60-18524 <+*>
1959 V1 N2 P185-190	52M38R <ATS>
1959 V1 N2 P201-207	61-20637 <*=>
1959 V1 N2 P208-214	37M41R <ATS>
1959 V1 N2 P222-229	61-20644 <*=>
	65-14820 <*>
1959 V1 N2 P230-239	60-16946 <*+>
	61-13484 <*+>
1959 V1 N2 P240-243	61-10209 <*+>
1959 V1 N2 P265-268	26L36R <ATS>
1959 V1 N2 P269-286	61-19243 <+*>
1959 V1 N2 P293-300	38M41R <ATS>
1959 V1 N2 P301-307	61-27261 <=*>
1959 V1 N2 P308-314	61-15643 <*+>
1959 V1 N2 P315-323	RCT V33 N1 P199 <RCT>
1959 V1 N2 P324-329	60-18439 <*+>
1959 V1 N2 P330-331	27L36R <ATS>
1959 V1 N2 P534-538	RCT V33 N2 P457 <RCT>
1959 V1 N3 P349-356	M-5663 <NLL>
1959 V1 N3 P357-361	33L37R <ATS>
	61-18635 <*=>
1959 V1 N3 P362-366	61-19779 <*=>
1959 V1 N3 P367-372	57N56R <ATS>
	61-18631 <*=> 0
1959 V1 N3 P378-386	63-15197 <=*>
1959 V1 N3 P400-403	61-23519 <*=>
1959 V1 N3 P404-409	63-10836 <=*>
1959 V1 N3 P455-459	11R78R <ATS>
1959 V1 N3 P474-481	63-23369 <=*$>
1959 V1 N3 P482-484	56N56R <ATS>
1959 V1 N4 P518-525	863 <TC>
1959 V1 N4 P526-529	864 <TC>
1959 V1 N4 P530-533	865 <TC>
1959 V1 N4 P542-548	48T90R <ATS>
1959 V1 N4 P627-634	05L36R <ATS>
	60-10042 <+*>
1959 V1 N4 P635-640	RAPRA-1255 <RAP>
1959 V1 N5 P670-676	32L37R <ATS>
1959 V1 N5 P677-681	M-5736 <NLL>
	34L37R <ATS>
1959 V1 N5 P730-737	64-13710 <=*$>
1959 V1 N6 P793-798	63-10840 <=*>
1959 V1 N6 P809-818	35M41R <ATS>
1959 V1 N6 P825-828	62-10105 <*=>
1959 V1 N6 P846-851	61-10359 <*+>
1959 V1 N6 P900-906	63-10835 <=*>
1959 V1 N7 P937-945	12M40R <ATS>
1959 V1 N7 P937-944	61-14971 <=*>
1959 V1 N7 P946-950	61-27622 <*=>
1959 V1 N7 P962-965	12R78R <ATS>
1959 V1 N7 P1042-1047	61-10366 <*+>
1959 V1 N7 P1056-1062	60-18435 <*>
1959 V1 N7 P1074-1076	61-27473 <*=>
	65-12451 <*>
1959 V1 N7 P1086-1093	42M40R <ATS>
1959 V1 N7 P1112-1125	61-23544 <*=>
1959 V1 N8 P1143-1147	74N53R <ATS>
1959 V1 N8 P1182-1193	AEC-TR-4409 <*=>
	60-16510 <+*>
	97M37R <ATS>
1959 V1 N8 P1214-1226	64-15263 <=*$>
1959 V1 N8 P1258-2164	C-3481 <NRCC>
1959 V1 N8 P1258-1265	60M41R <ATS>
1959 V1 N8 P1266-1269	60-18428 <*>
1959 V1 N9 P1287-1294	RCT V34 N1 P57 <RCT>
1959 V1 N9 P1295-1304	61-15921 <+*>
1959 V1 N9 P1319-1326	70N48R <ATS>
1959 V1 N9 P1327-1332	00S87R <ATS>
	RCT V34 N3 P953 <RCT>
1959 V1 N9 P1361-1363	44M38R <ATS>
	61-15631 <*+>
1959 V1 N9 P1364-1368	62-11493 <=>
1959 V1 N9 P1390-1395	48M46R <ATS>
1959 V1 N9 P1416-1421	61-19939 <=*>
1959 V1 N9 P1422-1427	59M41R <ATS>
1959 V1 N10 P1453-1455	53M43R <ATS>
1959 V1 N10 P1453-1456	61-13695 <*+>
1959 V1 N10 P1457-1461	61-13696 <*+>
1959 V1 N10 P1463-1468	61-13697 <*+>
1959 V1 N10 P1469-1472	61-13698 <*+>
1959 V1 N10 P1474-1481	61-13699 <*+>
1959 V1 N10 P1482-1486	61-13700 <*+>
1959 V1 N10 P1487-1492	61-13701 <+*>
1959 V1 N10 P1493-1495	61-13702 <+*>
1959 V1 N10 P1496-1501	61-13703 <+*>
1959 V1 N10 P1502-1506	61-13704 <*+>
1959 V1 N10 P1507-1513	61-13705 <*+>
1959 V1 N10 P1514-1518	61-13706 <*+>
1959 V1 N10 P1519-1525	60-16511 <+*>
	61-13707 <*+>
1959 V1 N10 P1526-1530	61-13708 <*+>
1959 V1 N10 P1531-1537	61-13709 <*+>
1959 V1 N10 P1538-1546	61-13710 <*+>
1959 V1 N10 P1547-1551	52M43R <ATS>
	61-13711 <*+>
1959 V1 N10 P1552-1557	61-10983 <+*> 0
	61-13712 <*+>
1959 V1 N10 P1558-1560	61-13713 <+*>
1959 V1 N10 P1561-1565	61-13714 <+*>
1959 V1 N10 P1566-1569	61-13715 <*+>
1959 V1 N10 P1570	61-13716 <+*>
1959 V1 N11 P1573-1579	61-20636 <*=>
1959 V1 N11 P1586-1592	63-18263 <=*>
1959 V1 N11 P1610-1616	60M43R <ATS>
1959 V1 N11 P1634-1642	60-16526 <+*>
1959 V1 N11 P1643-1646	61-15744 <+*>
	83M40R <ATS>
1959 V1 N11 P1647-1651	84M40R <ATS>
1959 V1 N11 P1659-1666	61-10912 <+*> 0
1959 V1 N11 P1691-1695	61M45R <ATS>
1959 V1 N11 P1713-1720	62-18818 <=*> 0
1959 V1 N12 P1754-1757	77T94R <ATS>
1959 V1 N12 P1758-1763	88M42R <ATS>
1959 V1 N12 P1764-1771	18M39R <ATS>
1959 V1 N12 P1788-1794	64-30721 <*> 0
1959 V1 N12 P1817-1820	60-18468 <*+>
1959 V1 N12 P1840-1843	64-10200 <=*$> 0
1959 V1 N12 P1848-1859	89M42R <ATS>
1959 V1 N12 P1862-1867	76T94R <ATS>
1959 V1 N12 P1877-1891	60-31403 <=>
1960 V2 P174-	RCT V34 N1 P318 <RCT>
1960 V2 P514-517	RCT V33 N4 P1068 <RCT>
1960 V2 P541-545	RCT V33 N4 P971 <RCT>
1960 V2 P778-784	RCT V37 N2 PT1 P355 <RCT>
1960 V2 P1148-1153	62-13834 <=*>
1960 V2 P1188-1192	RCT V34 N3 P760 <RCT>
1960 V2 P1608-1612	RCT V34 N3 P959 <RCT>
1960 V2 P1625-1634	RCT V34 N3 P975 <RCT>
1960 V2 P1692-1697	RCT V36 N3 P668 <RCT>
1960 V2 P1746	62-13990 <=*>
1960 V2 N1 P35-37	09M45R <ATS>
1960 V2 N1 P38-45	53M40R <ATS>
	62-10476 <=*> 0
1960 V2 N1 P46-49	61-14972 <=*>
	62-14460 <=*> 0
1960 V2 N1 P61-66	60-18472 <*+>
1960 V2 N1 P82-84	60-18470 <*+>
	62-23898 <=*>
1960 V2 N1 P88-91	62M46R <ATS>
1960 V2 N1 P103-118	60-23831 <+*>
1960 V2 N1 P130-132	60-18426 <*>

Citation	Translation
	63-14323 <=*>
1960 V2 N1 P133-135	60-18469 <*+>
1960 V2 N1 P136-140	44M46R <ATS>
1960 V2 N1 P162-165	65-60396 <=$>
1960 V2 N2 P200-204	63-23754 <=*$>
	69P58R <ATS>
1960 V2 N2 P238-242	RCT V34 N2 P461 <RCT>
1960 V2 N2 P243-246	65-11527 <*>
1960 V2 N2 P303-305	35M47R <ATS>
1960 V2 N2 P310-312	35P65R <ATS>
	62-10439 <=*>
1960 V2 N3 P365-374	61-14970 <=*>
1960 V2 N3 P377-385	61-19740 <=*>
	62-16843 <=*>
1960 V2 N3 P386-389	64-13225 <=*$>
1960 V2 N3 P421-426	60-18427 <*>
	62-18089 <=*>
1960 V2 N3 P427-432	50M41R <ATS>
1960 V2 N3 P451-455	19N52R <ATS>
	61-20642 <*=>
1960 V2 N3 P464-465	91M41R <ATS>
1960 V2 N4 P481-484	24M42R <ATS>
1960 V2 N4 P492-497	45M42R <ATS>
1960 V2 N4 P498-507	44M42R <ATS>
1960 V2 N4 P518-520	60-18108 <+*>
1960 V2 N4 P521-525	62-14281 <=*>
1960 V2 N4 P526-528	32M42R <ATS>
1960 V2 N4 P529-534	61-23380 <*=>
1960 V2 N4 P535-540	61-28143 <*=>
1960 V2 N4 P541-545	61-27557 <*=>
	62-13986 <*=>
1960 V2 N4 P546-548	92N54R <ATS>
1960 V2 N4 P567-571	31M42R <ATS>
1960 V2 N4 P576-580	1381 <TC>
1960 V2 N4 P581-589	63-13733 <=*>
1960 V2 N4 P591-600	63-18380 <=*>
1960 V2 N4 P619-625	15M43R <ATS>
1960 V2 N4 P626	31N48R <ATS>
1960 V2 N5 P633-635	25M43R <ATS>
1960 V2 N5 P662-672	13M43R <ATS>
1960 V2 N5 P673-678	16M43R <ATS>
1960 V2 N5 P685-688	32M43R <ATS>
1960 V2 N5 P689-697	00M43R <ATS>
1960 V2 N5 P698-703	17M43R <ATS>
1960 V2 N5 P704-709	24M43R <ATS>
1960 V2 N5 P710-715	18M43R <ATS>
1960 V2 N5 P716-718	48M44R <ATS>
1960 V2 N5 P728-730	63-23754 <=*$>
	68P58R <ATS>
1960 V2 N5 P756-769	62-13837 <*=>
1960 V2 N5 P770-777	43M46R <ATS>
1960 V2 N5 P791-792	14M43R <ATS>
1960 V2 N5 P793-796	41P60R <ATS>
1960 V2 N5 P802-805	61-10571 <*+> 0
1960 V2 N5 P817-822	60-41245 <=>
1960 V2 N6 P838-844	93N47R <ATS>
1960 V2 N6 P845-850	94N47R <ATS>
1960 V2 N6 P851-856	65-17146 <*>
1960 V2 N6 P857-863	66-10794 <*>
1960 V2 N6 P875-878	40M44R <ATS>
1960 V2 N6 P884-890	61-28699 <*=>
1960 V2 N6 P899-903	41M44R <ATS>
	62-10399 <*=> 0
1960 V2 N6 P926-930	63-16596 <*=>
	63-23754 <=*$>
1960 V2 N6 P931-936	42M44R <ATS>
1960 V2 N7 P977-983	05M45R <ATS>
1960 V2 N7 P1010-1012	61-10703 <+*> 0
1960 V2 N7 P1013-1014	36M46R <ATS>
1960 V2 N7 P1020-1025	07M45R <ATS>
1960 V2 N7 P1031-1038	62-32953 <=*>
1960 V2 N7 P1039-1044	08M45R <ATS>
1960 V2 N7 P1045-1048	62-13836 <=*>
1960 V2 N7 P1049-1055	61-15283 <=> 0
1960 V2 N7 P1056-1062	62-13835 <*=>
1960 V2 N7 P1077-1081	28M47R <ATS>
1960 V2 N7 P1082-1092	10358 <K-H>
	61-10731 <+*>
1960 V2 N7 P1119-1121	06M45R <ATS>
1960 V2 N7 P1122	<ES>
	62-13987 <=*>
1960 V2 N8 P1157-1161	61-14967 <=*>
1960 V2 N8 P1162-1166	86M45R <ATS>
1960 V2 N8 P1171-1175	62-13767 <*=>
1960 V2 N8 P1176-1187	18N48R <ATS>
1960 V2 N8 P1213-1220	62-13833 <*=>
1960 V2 N8 P1239-1245	10610-A <K-H> 0
	65-13191 <*>
1960 V2 N8 P1255-1260	87M45R <ATS>
1960 V2 N8 P1266-1269	<*=> 0
1960 V2 N8 P1279-1282	17N48R <ATS>
1960 V2 N8 P1283-1286	88M45F <ATS>
1960 V2 N8 P1287	49M46R <ATS>
1960 V2 N9 P1297-1300	61-10985 <+*>
1960 V2 N9 P1309-1319	61-23675 <*=>
1960 V2 N9 P1320-1323	61-10984 <=*> C
1960 V2 N9 P1383-1390	42N57R <ATS>
1960 V2 N9 P1391-1397	77M46R <ATS>
1960 V2 N9 P1409-1412	ICE V1 N1 P19-21 <ICE>
	81N52R <ATS>
1960 V2 N9 P1432-1437	40P60R <ATS>
	61-20643 <*=>
1960 V2 N10 P1456-1458	66-10656 <*>
1960 V2 N10 P1459-1462	49N55R <ATS>
1960 V2 N10 P1481-1485	61-10980 <=*>
1960 V2 N10 P1531-1534	62-10596 <*=> C
1960 V2 N10 P1557-1563	62-16626 <=*>
	63N48R <ATS>
1960 V2 N10 P1580-1585	61-10990 <+*> 0
1960 V2 N10 P1586-1587	62-18760 <=*>
1960 V2 N11 P1613-1615	62-13591 <*=>
1960 V2 N11 P1616-1619	62-13739 <*=>
1960 V2 N11 P1635-1638	62-13768 <*=>
1960 V2 N11 P1639-1644	61-20056 <*=>
1960 V2 N11 P1655-1658	62-13838 <*=>
1960 V2 N11 P1659-1664	35N48R <ATS>
	61-16117 <=*> C
1960 V2 N11 P1698-1702	64-13559 <=*$>
1960 V2 N11 P1703-1708	64-13560 <=*$>
1960 V2 N11 P1709-1714	64-13561 <=*$>
1960 V2 N11 P1714-1716	62-13986 <*=>
1960 V2 N11 P1717-1721	13N49R <ATS>
1960 V2 N11 P1722-1727	14N49R <ATS>
1960 V2 N11 P1728-1738	97N51R <ATS>
1960 V2 N12 P1811-1816	63-18412 <=*>
1960 V2 N12 P1824-1827	87N49R <ATS>
1960 V2 N12 P1832-1838	62-13839 <*=>
1960 V2 N12 P1839-1844	61-19205 <+*>
1960 V2 N12 P1870-1874	62-13769 <*=>
1960 V2 N12 P1876-1881	20N52R <ATS>
1961 V3 P46-49	4003 <TTIS>
1961 V3 P134-138	62-13770 <*=>
1961 V3 P139-143	62-13771 <*=>
1961 V3 P164-173	RCT V35 N2 P437 <RCT>
1961 V3 P497-504	RCT V35 N2 P509 <RCT>
1961 V3 P706-710	62-13909 <*=>
1961 V3 P1117-1118	62-10226 <*=>
1961 V3 P1125-1127	RCT V35 N2 P517 <RCT>
1961 V3 P1220-1223	66-10797 <*>
1961 V3 P1441-1445	RCT V35 N3 P700 700 <RCT>
1961 V3 P1572-1579	RCT V35 N2 P501 <RCT>
1961 V3 N1 P21-29	40N53R <ATS>
	62-25360 <=*>
1961 V3 N1 P33-36	41N53R <ATS>
1961 V3 N1 P37-40	ICE V1 N1 P132-134 <ICE>
	63-16595 <=*>
	94P59R <ATS>
1961 V3 N1 P66-71	42N53R <ATS>
1961 V3 N1 P79-80	43N53R <ATS>
1961 V3 N1 P105-112	62-10663 <*=>
1961 V3 N1 P134-138	62-23894 <=*>
1961 V3 N1 P139-143	62-23893 <=*>
1961 V3 N2 P198-204	33N57R <ATS>
1961 V3 N2 P205-207	44N53R <ATS>
1961 V3 N2 P306-309	62-23952 <=*>
	63-18894 <=*>
1961 V3 N2 P328-348	61-13684 <*=>
	65-10027 <*>
1961 V3 N3 P371-375	45N53R <ATS>
1961 V3 N3 P402-407	51N53R <ATS>

Citation	Translation
1961 V3 N3 P441-449	62-16701 <=*> 0
	62-25362 <=*>
1961 V3 N3 P450-455	62-16702 <=*> 0
	62-25363 <=*>
1961 V3 N3 P459-463	89N51R <ATS>
1961 V3 N3 P464-469	63-13241 <=*>
1961 V3 N3 P470-474	63-10811 <=*>
	63-18410 <*=>
1961 V3 N3 P486-487	62-13988 <=*> 0
1961 V3 N4 P505-514	65-10064 <*>
1961 V3 N4 P515-520	62-10980 <=*> 0
1961 V3 N4 P521-529	62-24749 <=*>
1961 V3 N4 P536-540	C-4553 <NRCC>
	65-60778 <=>
1961 V3 N4 P560-561	63-15262 <=*>
1961 V3 N4 P562-569	62-23955 <=*>
1961 V3 N4 P577-581	05N53R <ATS>
1961 V3 N4 P585-593	62-14819 <=*>
1961 V3 N4 P594-601	65-60399 <=$>
1961 V3 N4 P602-606	64-13746 <=*$>
	78S81R <ATS>
1961 V3 N4 P642-649	60N52R <ATS>
1961 V3 N5 P662-664	92N53R <ATS>
1961 V3 N5 P671-678	63-24371 <=*$>
1961 V3 N5 P686-690	G-314A <RIS>
1961 V3 N5 P691-698	G-314B <RIS>
1961 V3 N5 P699-705	28N55R <ATS>
1961 V3 N5 P734-739	ICE V1 N1 P35-38 <ICE>
	67N53R <ATS>
1961 V3 N5 P748-757	20N54R <ATS>
	62-20382 <=*>
1961 V3 N5 P761-767	SCL-T-461 <=*$>
1961 V3 N6 P828-832	90N54R <ATS>
1961 V3 N6 P841-846	21N56R <ATS>
1961 V3 N6 P861-864	89N54R <ATS>
1961 V3 N6 P870-876	64-16534 <=*$>
	65-14262 <*>
1961 V3 N6 P877-881	57N54R <ATS>
1961 V3 N6 P898-900	56N54R <ATS>
	62-13832 <=*>
1961 V3 N6 P925-930	91N54R <ATS>
1961 V3 N6 P943-947	RCT V36 N2 P527 <RCT>
	RUB.CH.TECH.V36 N2 <ACS>
	SCL-T-469 <=*>
1961 V3 N7 P953-955	61-20877 <*=> 0
1961 V3 N7 P953-956	62-14023 <*>
1961 V3 N7 P995-999	20N57R <ATS>
1961 V3 N7 P1027-1030	62-14285 <=*>
1961 V3 N7 P1041-1043	26N56R <ATS>
1961 V3 N7 P1062-1064	52N56R <ATS>
	62-32799 <=*>
1961 V3 N7 P1104-1109	RUB.CH.TECH. V36 N2 <ACS>
1961 V3 N7 P1104-1108	62-26041 <=$>
1961 V3 N7 P1110-1115	62-14283 <=*>
	62-26042 <=$>
1961 V3 N7 P1117	41P63R <ATS>
1961 V3 N8 P1128-1130	03N57R <ATS>
1961 V3 N8 P1131-1134	88N56R <ATS>
1961 V3 N8 P1161-1169	62-18382 <=*>
1961 V3 N8 P1203-1209	63-23362 <=*$>
1961 V3 N8 P1231-1233	62-14422 <=*> 0
1961 V3 N8 P1234-1237	66-10657 <*>
1961 V3 N8 P1238-1242	62-32819 <=*>
1961 V3 N8 P1243-1246	71N57R <ATS>
1961 V3 N9 P1307-1310	13P59R <ATS>
	62-32816 <=*>
1961 V3 N9 P1326-1331	14P59R <ATS>
1961 V3 N9 P1347-1351	UCRL TRANS-906(L) <*=>
	62-10979 <=*> 0
1961 V3 N9 P1383-1388	(CHEM)-261 <CTS>
	15P59R <ATS>
1961 V3 N9 P1410-1414	73P62R <ATS>
1961 V3 N9 P1428	16P59R <ATS>
	63-15206 <=*>
1961 V3 N10 P1482-1490	63-18414 <*=>
1961 V3 N10 P1491-1494	SCL-T-454 <=*$>
1961 V3 N10 P1495-1499	SCL-T-435 <=*>
1961 V3 N10 P1500-1508	64-11643 <=>
1961 V3 N11 P1698-1704	63-13394 <=*>
1961 V3 N11 P1739-1745	63-23755 <=*$>
1961 V3 N12 P1808-1815	63-15215 <=*>
1961 V3 N12 P1827-1832	RCT V35 N4 P1041 <RCT>
1961 V3 N12 P1853-1856	58P61R <ATS>
1962 N11 P1714-1717	65-60817 <=>
1962 V4 P196-200	RCT V35 N3 P813 <RCT>
1962 V4 P935-937	85S85R <ATS>
1962 V4 P961-967	RCT V36 N3 P807 <RCT>
1962 V4 P1060-1063	63-10082 <=*>
1962 V4 P1333-1337	66-12060 <*>
1962 V4 P1682-1687	RCT V36 N2 P541 <RCT>
1962 V4 N1 P20-24	63-23370 <=*$>
1962 V4 N1 P30-36	64-13664 <=*$>
1962 V4 N1 P64	23P62R <ATS>
1962 V4 N1 P66-73	RCT V36 N3 P709 <RCT>
1962 V4 N1 P105-108	64-15367 <=*$>
1962 V4 N1 P109-115	37P60R <ATS>
1962 V4 N1 P140-144	62-25168 <=*> 0
1962 V4 N2 P188-191	65-12880 <*>
1962 V4 N2 P196-200	20P62R <ATS>
1962 V4 N2 P216-220	63-13319 <=*>
1962 V4 N2 P246-249	63-11521 <=>
1962 V4 N2 P256-260	63-13319 <=*>
1962 V4 N2 P276-281	44P65R <ATS>
1962 V4 N2 P285-293	BGIRA-603 <BGIR>
	62-20432 <=*>
	63-14068 <=*>
1962 V4 N2 P285-298	65-12891 <*>
1962 V4 N2 P294-298	BGIRA-604 <BGIR>
	46P63R <ATS>
1962 V4 N2 P304-311	63-10784 <=*>
1962 V4 N3 P366-370	RCT V35 N4 P1047 <RCT>
1962 V4 N3 P376-382	64-11824 <=>
1962 V4 N3 P383-388	29P63R <ATS>
1962 V4 N3 P389-392	90Q72R <ATS>
1962 V4 N3 P414-418	63-15211 <=*>
1962 V4 N3 P448-451	46P61R <ATS>
1962 V4 N4 P481-485	63-19085 <=*>
1962 V4 N4 P528-532	N65-21634 <=$>
1962 V4 N4 P583-590	63-18740 <=*>
1962 V4 N4 P601-604	RCT V36 N1 P111 <RCT>
	63-20065 <=*>
1962 V4 N4 P605-612	64-23640 <=$>
1962 V4 N5 P678-682	63-11554 <=>
1962 V4 N5 P683-689	34P66R <ATS>
1962 V4 N5 P738-742	63-14311 <=*>
1962 V4 N5 P783-784	33P66R <ATS>
1962 V4 N6 P809-814	38P64R <ATS>
1962 V4 N6 P821-827	64-00476 <*>
	64-11830 <=>
1962 V4 N6 P848-850	50P65R <ATS>
1962 V4 N6 P869-875	39P64R <ATS>
1962 V4 N6 P885-888	33P65R <ATS>
1962 V4 N6 P889-893	40P64R <ATS>
1962 V4 N6 P894-900	41P64R <ATS>
1962 V4 N6 P913-916	RCT V36 N3 P799 <RCT>
1962 V4 N6 P917-921	66-60979 <=*$>
1962 V4 N6 P935-939	65-18021 <*> 0
1962 V4 N7 P987-994	50Q67R <ATS>
1962 V4 N7 P995-999	32P66R <ATS>
1962 V4 N7 P1049-1052	58Q70R <ATS>
1962 V4 N8 P1155-1162	63-15210 <=*>
1962 V4 N8 P1178-1185	61P65R <ATS>
	63-23368 <=*$>
1962 V4 N8 P1186-1192	63-18292 <=*>
	63-20478 <=*$> 0
	84Q68R <ATS>
1962 V4 N8 P1214-1218	85Q68R <ATS>
1962 V4 N8 P1219-1222	63-16281 <*=> 0
	86Q68R <ATS>
1962 V4 N9 P1338-1344	64-10106 <=*$>
1962 V4 N9 P1351-1353	41P65I <ATS>
1962 V4 N9 P1380-1384	64-15282 <=*$> 0
1962 V4 N9 P1398-1403	56P66R <ATS>
1962 V4 N9 P1412-1418	63-18927 <=*>
1962 V4 N10 P1528-1536	63-16597 <=*>
1962 V4 N10 P1604-1605	63-18179 <=*>
1962 V4 N11 P1625-1630	65-60884 <=$>
1962 V4 N11 P1631-1638	65-60885 <=$>
1962 V4 N11 P1639-1646	63-23515 <=*$>
1962 V4 N11 P1682-1687	RUB.CH.TECH.V36 N2 <ACS>

Citation	Translation
1962 V4 N11 P1688-1695	64-14361 <=*$>
	64-16973 <=*$>
	65Q67R <ATS>
1962 V4 N11 P1714-1717	63-20272 <=*>
1962 V4 N11 P1732-1738	64-14953 <=*$>
1962 V4 N12 P1761-1769	63-20847 <=*$> 0
1962 V4 N12 P1799-1804	AD-627 005 <=$>
1962 V4 N12 P1822-1828	89Q68R <ATS>
1962 V4 N12 P1829-1832	64-15283 <=*$>
1962 V4 N12 P1833-1838	05Q69R <ATS>
1962 V4 N12 P1858-1862	60Q70R <ATS>
1962 V4 N12 P1867-1871	06Q69R <ATS>
1962 V4 N12 P1887	63-14715 <=*> P
1963 V5 P892-	RCT V38 N3 P666 <RCT>
1963 V5 P1219-1227	65-63002 <=$>
1963 V5 P1725-1728	RCT V37 N2 PT1 P404 <RCT>
1963 V5 N1 P11-17	64-15206 <=*$>
1963 V5 N2 P161-163	13142 <K-H>
	64-10864 <=*$>
	65-14878 <*>
1963 V5 N2 P195-199	63-20280 <=*$>
	64-13088 <=*$>
1963 V5 N2 P212-216	65-11960 <*>
1963 V5 N2 P222-226	63-20060 <=*>
1963 V5 N2 P227-232	63-20061 <*=>
1963 V5 N2 P243-251	64-10867 <=*$>
1963 V5 N2 P269-273	64-16528 <=*$>
1963 V5 N3 P321-324	63-31992 <=>
	92Q70R <ATS>
1963 V5 N3 P412-416	63-18697 <=*$>
1963 V5 N3 P417-423	RCT V37 N1 P291 <RCT> 0
1963 V5 N4 P506-511	64-10868 <=*$>
1963 V5 N4 P593-597	RCT V36 N4 P1019 <RCT>
1963 V5 N4 P629	64-10863 <=*$>
1963 V5 N5 P644-648	13327B <K-H>
	65-11169 <*>
1963 V5 N5 P649-654	76R74R <ATS>
1963 V5 N5 P659-662	65S87R <ATS>
1963 V5 N5 P693-699	64-10865 <=*$>
1963 V5 N5 P706-711	64-10866 <=*$>
1963 V5 N6 P861-867	65-11170 <*>
1963 V5 N6 P873-874	64-10862 <=*$>
	64-14172 <=*$>
1963 V5 N6 P925-931	AD-616 283 <=$>
1963 V5 N7 P1030-1034	65-18151 <*>
1963 V5 N7 P1080-1084	64-19997 <=$>
1963 V5 N7 P1117	65-12168 <*>
1963 V5 N8 P1171-1177	65-12092 <*>
1963 V5 N8 P1196-1200	77Q73R <ATS>
1963 V5 N8 P1263-1267	64-15288 <=*$>
1963 V5 N9 P1292-1296	64-16626 <=*$>
1963 V5 N9 P1297-1302	65-12902 <*>
1963 V5 N9 P1393-1397	RCT V37 N2 PT1 P365 <RCT>
1963 V5 N10 P1441-1445	AD-609 782 <=>
1963 V5 N10 P1558-1561	65-60785 <=>
1963 V5 N11 P1627-1631	66-14142 <*$>
1963 V5 N11 P1638-1640	64-16377 <=*$>
1963 V5 N11 P1684-1690	64-00477 <*>
	64-11823 <=>
1963 V5 N11 P1711-1716	65-11543 <*> 0
1963 V5 N12 P1761-1764	95R75R <ATS>
1963 V5 N12 P1795-1798	69R76R <ATS>
1963 V5 N12 P1805-1808	M-5732 <NLL>
1963 V5 N12 P1870-1874	64-21637 <=>
1963 V5 N12 P1875-1888	RCT V37 N3 P627 <RCT>
1964 P33-36	18S89R <ATS>
1964 P2083-2089	1256 <TC>
1964 N12 P2113-2116	ICE V5 N2 P334-335 <ICE>
1964 V6 P131-136	66-14430 <*>
1964 V6 P635-	RCT V38 N1 P204 <RCT>
1964 V6 P655-658	65-60749 <=>
1964 V6 P896-900	293 <TC>
1964 V6 P1294-	RCT V38 N4 P991 <RCT>
1964 V6 P1330-	RCT V38 N2 P452 <RCT>
1964 V6 P1483-	RCT V38 N3 P657 <RCT>
1964 V6 N1 P31-33	05R76R <ATS>
1964 V6 N1 P37-41	M-5713 <NLL>
1964 V6 N1 P47-51	70R76R <ATS>
1964 V6 N1 P59-63	M-5727 <NLL>
1964 V6 N1 P64-68	64-20946 <*> 0
1964 V6 N1 P137-143	64-18124 <*> 0
1964 V6 N1 P180-181	65-13324 <*> 0
1964 V6 N1 P186-195	M-5728 <NLL>
1964 V6 N1 P192-195	M-5729 <NLL>
	47-S89R <ATS>
1964 V6 N2 P258-264	M-5660 <NLL>
1964 V6 N2 P265-268	AD-618 878 <=$>
1964 V6 N2 P310-313	AD-615 258 <=$>
1964 V6 N2 P329-333	M-5656 <NLL>
1964 V6 N2 P329-334	65-10424 <*>
1964 V6 N3 P448-451	64-18101 <=*$>
1964 V6 N3 P473-479	65-11324 <*>
1964 V6 N3 P545-550	AD-608 051 <=>
1964 V6 N3 P568-570	64-31370 <=>
1964 V6 N4 P605-607	TRC-TRANS-1097 <NLL>
	65-28551 <$>
1964 V6 N4 P659-660	65-10486 <*>
1964 V6 N4 P691-694	AD-620 972 <=$>
1964 V6 N4 P734-736	AD-620 972 <=$>
1964 V6 N5 P777-781	64-23266 <$>
1964 V6 N5 P818-822	66-10752 <*>
1964 V6 N5 P823-826	66-10750 <*>
1964 V6 N5 P832-837	AD-620 999 <=$>
1964 V6 N5 P862-867	64-23265 <$>
1964 V6 N5 P929-933	64-41970 <=$>
	65-10423 <*>
1964 V6 N6 P969-974	900 <TC>
1964 V6 N6 P1001-1005	65-13038 <*>
1964 V6 N6 P1128-1131	65-11880 <*>
1964 V6 N6 P1167-1173	M-5597 <NLL>
1964 V6 N7 P1251-1255	AD-615 865 <=$>
1964 V6 N7 P1261-1266	65-11488 <*>
1964 V6 N7 P1275-1280	65-64142 <=>
1964 V6 N7 P1313-1317	65-11489 <*>
1964 V6 N7 P1330-1334	M-5715 <NLL>
1964 V6 N7 P1346-1349	74T94R <ATS>
1964 V6 N8 P1407-1410	AD-622 362 <=*$>
1964 V6 N8 P1493-1495	AD-615 991 <=$>
1964 V6 N8 P1496-1497	65-12038 <*>
1964 V6 N8 P1515-1521	65-13398 <*>
1964 V6 N9 P1717-1721	C9R80R <ATS>
	65-63744 <=>
1964 V6 N10 P1755-1757	AD-638 975 <=$>
1964 V6 N10 P1795-1797	M-5714 <NLL>
1964 V6 N10 P1802-1805	AD-638 975 <=$>
1964 V6 N10 P1814-1820	AD-638 975 <=$>
1964 V6 N10 P1838-1843	AD-638 975 <=$>
1964 V6 N10 P1848-1852	N66-14624 <=*$>
	65-64379 <=*$>
1964 V6 N10 P1852-1861	AD-638 975 <=$>
1964 V6 N10 P1880-1884	67R80R <ATS>
1964 V6 N10 P1885-1889	M-5717 <NLL>
1964 V6 N11 P1974-1979	65-13418 <*>
	700 <TC>
1964 V6 N11 P2030-2034	65-11522 <*> 0
1964 V6 N11 P2035-2039	65-11521 <*> 0
1964 V6 N11 P2046-2050	54S81R <ATS>
1964 V6 N11 P2078-2082	1255 <TC>
1964 V6 N12 P2117-2121	M-5721 <NLL>
1964 V6 N12 P2145-2148	28S82R <ATS>
1964 V6 N12 P2155-2162	1204 <TC>
	66-10748 <*>
1964 V6 N12 P2189-2192	1146 <TC>
1964 V6 N12 P2193-2196	1145 <TC>
1965 N1 P50-54	RAPRA-1289 <RAP>
1965 N1 P63-64	ICE V5 N3 P518-519 <ICE>
1965 N1 P184	ICE V6 N1 P15-16 <ICE>
1965 N5 P795-801	ICE V6 N1 P16-20 <ICE>
1965 V7 P28-32	1633 <TC>
1965 V7 N1 P19-23	ICE V5 N3 P436-439 <ICE>
1965 V7 N1 P39-44	1024 <TC>
1965 V7 N1 P45-49	66-13372 <*>
1965 V7 N1 P63-64	12S82R <ATS>
1965 V7 N1 P80-87	83S83R <ATS>
1965 V7 N1 P98-100	66-13389 <*>
1965 V7 N1 P129-135	666 <TC>
1965 V7 N1 P135-140	COS83R <ATS>
	1062 <TC>
1965 V7 N1 P141-144	01S83R <ATS>
1965 V7 N1 P143-149	65-13417 <*>

1965 V7 N1 P145-149 65-13417 <*>
1965 V7 N1 P150-155 13S82R <ATS>
1965 V7 N1 P182 65-13123 <*>
96S82R <ATS>
1965 V7 N1 P184 97S82R <ATS>
1965 V7 N1 P188-189 65-14164 <*> 0
1965 V7 N2 P264-268 65-14254 <*>
1965 V7 N2 P308-311 65-14685 <*>
1965 V7 N3 P411-416 65-18052 <*>
1965 V7 N3 P427-431 AD-624 725 <=*$>
1965 V7 N3 P462-467 24S84R <ATS>
1965 V7 N3 P503-508 81S89R <ATS>
1965 V7 N3 P531-535 RAPRA-1288 <RAP>
1965 V7 N4 P650-654 14S85R <ATS>
1965 V7 N4 P741-746 66-10360 <*>
1965 V7 N4 P747-750 66-10361 <*>
1965 V7 N5 P795-801 38S85R <ATS>
65-17215 <*>
1965 V7 N5 P878-883 66-11276 <*>
1965 V7 N5 P884-890 66-11274 <*>
1965 V7 N5 P933-938 31S85R <ATS>
1965 V7 N6 P1010-1015 65-17054 <*> 0
1965 V7 N6 P1051-1055 65-17055 <*> 0
1965 V7 N6 P1070-1074 63S89R <ATS>
1965 V7 N6 P1075-1079 04S87R <ATS>
1965 V7 N6 P1117-1121 64S89R <ATS>
1965 V7 N7 P1134-1139 66-13380 <*>
1965 V7 N7 P1140-1145 66-14210 <*$>
1965 V7 N7 P1154-1158 31T90R <ATS>
1965 V7 N7 P1248-1253 69S89R <ATS>
1965 V7 N7 P1258-1263 66-11484 <*>
1965 V7 N7 P1288-1290 66-12237 <*>
1965 V7 N8 P1306-1309 66-13285 <*>
1965 V7 N9 P1481-1483 35T94R <ATS>
1965 V7 N9 P1592-1596 66-12977 <*> 0
1965 V7 N9 P1604-1609 66-14005 <*>
1965 V7 N9 P1609-1613 42S89R <ATS>
1965 V7 N10 P1743-1745 68S89R <ATS>
1965 V7 N10 P1767-1770 67S89R <ATS>
1965 V7 N11 P1899-1904 66-12976 <*> 0
1965 V7 N11 P1927-1929 01T91R <ATS>
1965 V7 N11 P1997-2000 03T95R <ATS>
1965 V7 N12 P2089-2093 04T95R <ATS>
1965 V7 N12 P2172-2173 88T92R <ATS>
1965 V7 N12 P2177-2178 66-12974 <*> 0
1966 V8 N1 P16-19 02T94R <ATS>
1966 V8 N1 P31-33 97T92R <ATS>
1966 V8 N1 P146-152 66-14434 <*>
1966 V8 N1 P163-168 98T92R <ATS>
1966 V8 N2 P256-260 03T94R <ATS>
1966 V8 N2 P272-277 66-14185 <*$>
1966 V8 N2 P297-301 04T94R <ATS>
1966 V8 N4 P708-712 66-14730 <*$>

VYZIVA LIDU
1959 V14 N3 P35-37 59-21072 <+*>

VZRYVNOE DELO
1960 N45 P20-35 AD-633 490 <=$>
1962 N48 P5-9 64-13615 <=*$>
1962 N48 P10-19 64-13615 <=*$>
1962 N48 P19-21 64-13615 <=*$>
1962 N48 P21-23 64-13615 <=*$>
1962 N48 P23-27 64-13615 <=*$>
1962 N48 P27-33 64-13615 <=*$>
1962 N48 P33-44 64-13615 <=*$>
1962 N48 P44-51 64-13615 <=*$>
1962 N48 P51-59 64-13615 <=*$>
1962 N48 P59-66 64-13615 <=*$>
1962 N48 P66-87 64-13615 <=*$>
1962 N48 P87-92 64-13615 <=*$>
1962 N48 P92-98 64-13615 <=*$>
1962 N48 P98-101 64-13615 <=*$>
1962 N48 P101-104 64-13615 <=*$>
1962 N48 P104-108 64-13615 <=*$>
1962 N48 P108-111 64-13615 <=*$>
1962 N48 P111-112 64-13615 <=*$>
1962 N48 P113-122 64-13615 <=*$>
1962 N48 P123-125 64-13615 <=*$>
1963 N52 P115-129 AD-627 254 <=$>
1963 N51/8 P22-39 UCRL-TRANS-1104(L) <=$>
1963 N52/9 P5-10 AD-630 026 <=*$>
1963 N52/9 P39-56 AD-625 574 <=$>
1963 N52/9 P57-67 AD-610 342 <=$>
1963 N52/9 P108-114 AD-625 605 <=*$>
1963 N52/9 P130-140 AD-634 307 <=$>
1964 N55 P93-97 AD-634 414 <=$>
1964 N55 P132-135 AD-634 414 <=$>
1964 N55 P136-139 AD-634 414 <=$>
1964 V54 N11 P48-52 66-30763 <=$>
1964 V54 N11 P102-113 66-30764 <=$>
1964 V54 N11 P113-125 66-30730 <=$>

WAERME
1935 V58 N4 P54-55 57-1106 <*>
1936 V59 P307-311 57-864 <*>
1938 V61 P823-828 58-151 <*>
1938 V61 P891-898 I-358 <*>
1938 V61 N36 P660-664 2660 <*>
1939 V62 P657 AEC-TR-3999 <*>
1941 V64 N11 P127-129 60-14398 <*>
1961 V69 P121-127 64-14748 <*>

WAERME-UND KAELTETECHNIK
1939 N9 P126-135 AL-130 <*>

WAKAYAMA IGAKU
1955 V6 P203-209 AEC-TR-3355 <+*> 0

WASSER. LEIPZIG
1956 N23 P95-101 NP-TR-736 <*>

WASSER, LUFT UND BETRIEB
1963 N5 P276- 64-10394 <*>
1965 N1 P18-20 8780 <IICH>

WASSERKRAFT UND WASSERWIRTSCHAFT
1944 V39 N1 P5-17 60-10046 <*> 0

WASSERVERSORGUNG UND ABWASSERTECHNIK IN HAUS, HOF UND STRASSE
1962 S2 V13 P4-5 8726 <IICH>

WASSERWIRTSCHAFT
1952 V42 N7 P233-237 I-422 <*>
1954 V44 N3 P60-63 II-195 <*>
1957 V47 N4 P97-100 C-4795 <NRC>

WASSERWIRTSCHAFT UND WASSERTECHNIK
1961 N4 P147-151 62-19695 <*=>
1961 V11 N4 P152-159 62-19748 <=*>
1961 V11 N6 P279-281 63-00534 <*>
1961 V11 N8 P404-405 62-24790 <=*>
1961 V11 N8 P406-408 62-24783 <=*>
1964 V14 N6 P161-163 64-41266 <=>
1966 V16 N1 P1-3 66-31051 <=$>
1966 V16 N10 P321-324 66-35589 <=>

WATER. DEN HAAG
1934 V18 N16 P147-165 AD-617 552 <=$>

WATER CONSERVATION AND HYDROELECTRIC POWER CONSTRUCTION. CHINESE PEOPLES REPUBLIC
SEE SHUI LI SHUI TIEN CHIEN SHE

WATER CONSERVATION AND ELECTRIC POWER. CHINESE PEOPLES REPUBLIC
SEE SHUI LI YU TIEN LI

WATER CONSERVATION IN CHINA
SEE CHUNG KUO SHUI LI

WATER TRANSPORTATION. CHINESE PEOPLES REPUBLIC
SEE SHUI YUN

WEAR-USURE-VERSCHLEISS
1960 V3 N2 P122-141 35M43G <ATS>
1960 V3 N2 P122-143 60-18109 <*>

WEATHER MONTHLY. CHINESE PEOPLES REPUBLIC

SEE TIEN-CHI YUEH-K'AN

WEATHER SERVICE BULLETIN. JAPAN
SEE SOKKO JIHO

WEED RESEARCH
1961 V1 N1 P32-43 65-11673 <*>

WEEKLY MEDICAL TIMES. TOKYO
SEE IGAKU TSUSHIN

WEHR UND WIRTSCHAFT
1957 P40 59-11216 <=*>
1959 N12 P20-21 60-15991 <+*> 0

WEHRMEDIZINISCHE MITTEILUNGEN
1961 N10 P150-152 65-00355 <*>
1961 N10 P152-154 62-01186 <*>

WEHRTECHNISCHE HEFTE
1954 V51 N4 P97-120 57-764 <*>

WEHRTECHNISCHE MONATSHEFTE
1958 V6 N1 P42-46 64-21785 <=>

WEI SHEN WU HSUEH PAO
1959 V7 N3 P145-188 61-27337 <=> 0
1959 V7 N3 P189-195 61-27726 <=*>
1959 V7 N3 P197-204 61-27726 <=*>
1959 V7 N3 P205-208 60-41148 <=>
62-00816 <*>
64-15622 <=*$>
1959 V7 N3 P209-222 61-27726 <=*>
1959 V7 N3 P223-225 61-27337 <=> 0
1959 V7 N3 P226-229 60-41148 <=>
1959 V7 N3 P231-234 60-41148 <=>
1959 V7 N3 P236-239 60-41148 <=>
1959 V7 N3 P241-262 61-27337 <=> 0
1959 V7 N3 P264-272 61-27337 <=> 0
1959 V7 N4 P330-339 60-11961 <=>
1959 V7 N4 P353-359 60-11961 <=>
1959 V7 N4 P370-373 60-11961 <=>
1959 V7 N1/2 P1-14 60-41130 <=>
1959 V7 N1/2 P16-30 62-13440 <=>
1959 V7 N1/2 P32-38 62-13440 <=>
1959 V7 N1/2 P40-41 62-13440 <=>
1959 V7 N1/2 P43-58 62-13440 <=>
1959 V7 N1/2 P60-74 62-13440 <=>
1959 V7 N1/2 P88-92 60-41130 <=>
1959 V7 N1/2 P103-107 60-41130 <=>
1959 V7 N1/2 P128-131 60-41130 <=>
1964 V10 N1 P9-15 AD-646 585 <=>
1964 V10 N1 P24-29 AD-646 585 <=>
1964 V10 N1 P31-38 AD-646 585 <=>

WELTALL
1913 V14 N1/2 P4-5 I-1042 <*>

WELTRAUMFAHRT
1950 N2 P31- 3135 <*>
1952 N2 P53- 3131 <*>
1953 N1 P2- 3218 <*>
1958 V9 N12 P100-106 63-14486 <*>

WEN TZU KAI KO
1963 N2 P1-3 63-31045 <=>

WERK UND WIR
1961 V9 P347-351 2586 <BISI>

WERKEN. GENOOTSCHAP TER BEVORDERING VAN NATURGENEES- EN HEILKUNDE TE AMSTERDAM
1896 S2 V2 P296-335 63-14833 <*>

WERKSTATT
1964 V8 N8 P261-264 64-51226 <=>

★WERKSTATT UND BETRIEB
1935 V68 P257-259 59-15466 <*>
1937 V70 N5 P74-78 59-20914 <*> 0
1937 V70 N6 P135-136 59-20914 <*> 0
1937 V70 N15/6 P203-205 61-10081 <*> 0
1937 V70 N19/2 P270-272 61-10028 <*> 0
1938 V71 P259 60-18598 <*>
1938 V71 P320-322 57-583 <*>
1939 V72 N12 P273-275 57-2191 <*>
1939 V72 N3/4 P43-45 57-1405 <*>
1939 V72 N5/6 P69-73 59-20917 <*> 0
1940 V73 N7 P141-144 60-18596 <*>
1947 V80 N1 P11-14 62-18716 <*> 0
1947 V80 N4 P77-84 63-10045 <*> 0
1947 V80 N7 P169-171 62-18245 <*> 0
1948 V81 N1 P12-13 62-18717 <*> 0
1948 V81 N3 P71-74 62-18343 <*>
1948 V81 N3 P103-104 62-18343 <*>
1948 V81 N5 P128 63-10040 <*> 0
1948 V81 N7 P188-190 62-16980 <*> 0
1949 V82 N5 P165-171 60-18691 <*> 0
1949 V82 N9 P316-317 62-18244 <*> 0
1949 V82 N10 P345-348 39F4G <ATS>
1949 V82 N11 P390-396 61-18426 <*>
1949 V82 N12 P435-440 61-18426 <*>
1951 V84 N5 P225-231 64-18882 <*> 0
1955 V88 N4 P163-164 62-18247 <*> 0
1955 V88 N5 P209-220 59-17200 <*> 0
1955 V88 N6 P267-302 1525 <*>
1955 V88 N10 P649-650 59-17199 <*> 0
1956 V89 N7 P377-380 <CSIR>
1956 V89 N7 P383-390 CSIRO-3387 <CSIR>
1957 V90 N4 P227-232 61-16275 <*>
1958 V91 P649-653 TS 1567 <BISI>
1958 V91 N1 P33-38 3419 <BISI>
1958 V91 N5 P246-248 61-18411 <*>
1958 V91 N9 P565-571 66-10138 <*> 0
1958 V91 N10 P605-608 TS-1467 <BISI>
62-14523 <*>
75L30G <ATS>
1958 V91 N10 P625-629 74L30G <ATS>
1958 V91 N11 P649-653 62-14657 <*>
1959 V92 N4 P195-202 61-16168 <*> 0
61-18914 <*> 0
1961 V94 N8 P529-533 62-16840 <*> 0
1962 V95 N11 P755-758 63-20052 <*>
1963 V96 N6 P333-338 4213 <BISI>

WERKSTATTSTECHNIK
1928 V22 N9 P261-267 59-20536 <*> 0
1929 V23 N1 P3-10 59-20535 <*> 0
1929 V23 N2 P33-42 59-20535 <*> 0
1930 V24 N12 P325-330 59-20547 <*> 0
1930 V24 N13 P354-360 59-20547 <*> 0
1930 V24 N14 P381-384 59-20546 <*> 0
1930 V24 N15 P410-416 59-20546 <*> 0
1930 V24 N19 P525-528 59-20545 <*>
1930 V24 N21 P573-577 59-20544 <*> 0
1930 V24 N23 P629-635 57-3130 <*>
1930 V24 N24 P668-670 59-15696 <*>
1931 V25 N6 P159-161 59-20529 <*>
1931 V25 N7 P187-188 59-20549 <*> 0
1931 V25 N8 P201-205 59-20548 <*> 0
1936 V30 N7 P167-170 59-20543 <*> 0
1936 V30 N8 P186-187 59-20542 <*>
1936 V30 N23 P511-520 57-3022 <*>
1937 V31 N1 P8-11 59-20514 <*> 0
1937 V31 N9 P213-214 59-20513 <*>
1937 V31 N19 P421-427 59-20512 <*> 0
1937 V31 N24 P546-549 59-20511 <*> 0
1938 V32 N18 P401-404 59-20541 <*> 0
1938 V32 N21 P465-467 57-1049 <*>
1938 V32 N21 P473-475 59-20540 <*> 0
1939 V33 N8 P217-220 59-20539 <*>
1939 V33 N9 P240-244 59-20538 <*> 0
1940 V34 N5 P75-77 63-10047 <*> 0
1940 V34 N8 P129-132 59-20537 <*> 0
1942 V36 P228-231 II-747 <*>
1942 V36 N11/2 P228-231 59-17014 <*> 0
1959 V49 N1 P22-26 3132 <BISI>
1959 V49 N4 P223-230 1650 <BISI>
62-16861 <*>
1960 V50 P12-14 TS-1739 <BISI>

1960 V50 N4 P192-195 1787 <BISI>
1961 V51 P23-29 5186 <HB>
1961 V51 P324-326 2434 <BISI>
1961 V51 N2 P102-105 3981 <BISI>
1962 V52 N2 P66-71 2833 <BISI>
1962 V52 N4 P167-172 2915 <BISI>
1963 V53 N11 P561-566 4355 <BISI>
1964 V54 P264-270 4051 <BISI>
1964 V54 P370-373 4224 <BISI>
1964 V54 P384-388 4225 <BISI>
1964 V54 N2 P93-95 6755 <HB>
1964 V54 N4 P189-193 4215 <BISI>
1964 V54 N1C P508-511 4352 <BISI>
1964 V54 N10 P511-515 4226 <BISI>
1964 V54 N1C P530-538 4596 <BISI>
1964 V54 N11 P549-555 4253 <BISI>
1965 V55 P53-57 4288 <BISI>
1965 V55 P301-304 4413 <BISI>
1965 V55 P515-524 4880 <BISI>
1965 V55 N3 P398-403 6754 <HB>

WERKSTATTSTECHNIK UND MASCHINENBAU
1949 V39 N5 P145-147 62-16916 <*> 0
1949 V39 N7 P215-216 62-20277 <*> 0
1949 V39 N8 P233-236 62-18372 <*> 0
1950 V40 N6 P214-220 61-16274 <*>
1952 V42 P140-145 1154 <*>
1953 V43 N3 P90-92 62-18423 <*> 0
1953 V43 N3 P92-97 62-18233 <*> 0
1955 V45 N11 P557-561 59-17202 <*> 0
1956 V46 N5 P218-226 61-18915 <*> 0
63-20470 <*> PO
1956 V46 N6 P274-282 60-10310 <*>
1956 V46 N12 P603-611 61-16273 <*>
1957 V47 N1 P26-35 20K24G <ATS>
1957 V47 N2 P73-78 21K24G <ATS>
1958 V48 N3 P341-343 61-18751 <*>
1958 V48 N12 P631-635 62-10941 <*>

WERKSTOFFE UND KORROSION
1950 V1 N2 P49-51 58-1C99 <*>
1950 V1 N10 P400-404 63-16831 <*>
1950 V1 N11 P433-437 63-16830 <*>
1951 N10 P369-371 65-17328 <*>
1951 V2 P212-221 62-18838 <*>
1951 V2 N5 P173-181 63-14099 <*> 0
1951 V2 N5 P182-185 62-16931 <*> 0
1951 V2 N7 P249-258 63-14350 <*> 0
1951 V2 N8 P289-292 62-18338 <*> 0
1951 V2 N10 P369-371 66-10157 <*>
1951 V2 N11 P409-414 57-730 PT2 <*>
1952 N5/6 P186-188 R-3818 <*>
1952 V3 N9/10 P329-333 63-14620 <*>
1953 V4 P153-156 I-486 <*>
1953 V4 P156-172 62-18248 <*> 0
1953 V4 N1 P4-14 AL-249 <*>
1953 V4 N5 P156-172 64-20505 <*>
1954 V5 N2 P49-54 C-2253 <NRC>
1381 <*>
1954 V5 N11 P433-440 59-17345 <*>
1954 V5 N8/9 P290-295 62-18609 <*> 0
1955 N6 P294-296 57-1195 <*>
1955 V6 N1 P6-8 C-2126 <NRC>
1955 V6 N2 P72-80 C-2416 <NRC>
1955 V6 N3 P117-130 93G7G <ATS>
1955 V6 N5 P237-245 59-17504 <*>
1955 V6 N7 P325-328 57-2C61 <*>
1955 V6 N11 P521-523 62-16984 <*>
1956 P322-330 58-1484 <*>
1956 V7 N1 P448-452 62-18662 <*> P
1956 V7 N5 P254-256 57-2294 <*>
1956 V7 N11 P642-649 AEC-TR-3215 <*> 0
1956 V7 N8/9 P453-460 62-16541 <*> 0
1957 V8 P321 C-2581 <NRC>
1957 V8 N6 P329-344 34M46G <ATS>
1957 V8 N7 P394-401 05N57G <ATS>
1957 V8 N8/9 P483-484 58-253 <*>
1958 V9 P693-698 AEC-TR-5724 <*>
1958 V9 N5 P278-283 3624 <BISI>
1958 V9 N6 P383-391 25L30G <ATS>
1958 V9 N7 P429-434 3160 <BISI>
1958 V9 N7 P44C-443 60-10750 <*>
1958 V9 N10 P623-630 60-16886 <*>
1958 V9 N8/9 P524-526 6C-10749 <*>
1958 V9 N8/9 P533-535 59-15336 <*>
1958 V9 N8/9 P547-549 66-13833 <*>
1958 V9 N8/9 P552-556 66-10156 <*> 0
1959 V10 P239-243 62-01079 <*>
1959 V1C N2 P91-93 60-14502 <*>
1959 V10 N6 P377-383 76M41G <ATS>
1959 V10 N7 P425-429 75M41G <ATS>
1959 V10 N10 P6C5-608 66-11685 <*> 0
1959 V10 N10 P617-621 66-13624 <*>
1959 V10 N11 P668-681 60-18505 <*> C
1959 V10 N12 P739-760 78N51G <ATS>
1960 V11 N3 P147-151 3018 <BISI>
1960 V11 N5 P271-273 62-14775 <*>
1960 V11 N6 P352-356 85M44G <ATS>
1960 V11 N1C P601-616 NP-TR-608 <*>
1961 V12 P148-150 66-14495 <*> 0
1961 V12 N5 P288-295 62-16285 <*>
1961 V12 N8 P486-493 2851 <BISI>
1961 V12 N11 P669-680 3056 <BISI>
1962 V13 N8 P474-479 65-11529 <*> 0
1963 V14 P925-930 128 <TC>
1963 V14 N2 P65-69 13175 <K-H>
65-18146 <*>
1963 V14 N2 P69-80 3339 <BISI>
1963 V14 N4 P257-260 NS 125 <TTIS>
1963 V14 N4 P261-269 3546 <BISI>
1963 V14 N6 P458-464 47Q72G <ATS>
1963 V14 N7 P551 TRANS-187A <FRI>
1963 V14 N8 P646-654 3459 <BISI>
1963 V14 N9 P729-739 3541 <BISI>
1963 V14 N12 P1021-1029 3854 <BISI>
1964 V15 P348-349 4473 <BISI>
1964 V15 N1 P46-51 4032 <BISI>
1964 V15 N6 P457-463 3947 <BISI>
1964 V15 N7 P537-543 4588 <BISI>
1964 V15 N8 P653-66C 4212 <BISI>
1964 V15 N10 P804-808 3946 <BISI>
1965 V16 P1-11 4514 <BISI>
1965 V16 P93-103 4269 <BISI>
1965 V16 P185-189 4554 <BISI>
1965 V16 P453-465 4433 <BISI>
1965 V16 P560-570 4813 <BISI>
1965 V16 P761-768 66-12178 <*> 0
1965 V16 P837-843 4576 <BISI>
1965 V16 N3 P208-212 4234 <BISI>
1965 V16 N9 P750-754 MFA/TS-762 <MFA>
1966 V17 P110-119 4873 <BISI>

WERKZEUGMASCHINE
1929 V33 N3 P41-44 60-10192 <*> 0
1930 V34 N9 P191-193 59-20660 <*> C
1935 P444-447 57-2362 <*>
1940 V44 N22 P465-470 58-124 <*>

WETERNARIA
1957 V3 P169-178 60-21249 <=>

WETTER UND KLIMA
1948 N9/10 P257-27C 66-12126 <*>

WIADOMOSCI CHEMICZNE
1955 V9 N11 P616-634 61-10849 <*> 0
1959 V13 N1 P1-39 62-16085 <*> P
1959 V13 N12 P680-698 62-23171 <=>
1960 V14 P295-309 12075 <K-H>
1960 V14 P759-778 65-12884 <*>
1962 V16 N5 P307-319 27Q68P <ATS>

WIADOMOSCI GORNICZE
1959 V10 N3 P88-90 63-11386 <=>
1965 V16 N10 P290-296 65-33800 <=$>

WIADOMOSCI HUTNICZE
1955 V11 P266-269 62-10455 <*> 0
1956 V12 N4 P111-115 57-3550 <*>
1957 V13 N7/8 P235-242 PB 141 178T <=>

1958 V14 N6 P175-181 PB 141 322T <=>
1958 V14 N6 P182-185 IGR V2 N2 P134 <AGI>
  PB 141 321T <=>
1965 V21 N10 P197-199 66-30070 <=$>
1965 V21 N10 P607-609 66-30070 <=$>

WIADOMOSCI. INSTYTUT MELIORACJI I UZYTKOW ZIELONYCH. WARSAW
1958 N1 P137-148 61-31308 <=>
1958 N1 P158-196 61-31310 <=>

WIADOMOSCI PARAZYTOLOGICZNE
1957 V3 N2/3 P261-272 60-21513 <=>
1959 V5 N1 P53-55 63-11381 <=$>
1959 V5 N6 P577-583 63-11381 <=$>
1959 V5 N6 P616-617 63-11381 <=$>
1959 V5 N2/3 P293-297 63-11381 <=$>
1959 V5 N2/3 P299-306 63-11381 <=$>
1959 V5 N4/5 P453-458 63-11354 <=>
1959 V5 N4/5 P459-465 63-11401 <=>
1959 V5 N4/5 P467-468 63-11401 <=>
1960 V6 N1 P3-10 63-11381 <=$>
1960 V6 N1 P11-19 63-11381 <=$>
1960 V6 N4 P340-341 63-11381 <=$>
1961 V7 N4/6 P828-832 70P61P <ATS>

WIADOMOSCI SLUZBY HYDROLOGICZNEJ I METEOROLOGICZNEJ
1958 V6 N2 P1-29 61-31295 <=>
1958 V6 N2 P53-86 61-31305 <=>
1960 V8 P85-159 61-31302 <=>
1963 V11 N5 P19-24 M.5571 <NLL>

WIADOMOSCI TELEKOMUNIKACYJNE
1956 N4 P90-92 59-21099 <=*>

WIENER ARCHIV FUER INNERE MEDIZIN UND DERN GRENZGEBIETE
1927 N14 P149-162 58-334 <*>

WIENER BEITRAGE GEBURTSHILFOU GYNAKOLOGIE
1955 V2 P76-79 2270 <*>
1955 V2 P107-111 2263 <*>

WIENER KLINISCHE WOCHENSCHRIFT
1893 V6 P211-213 58-1348 <*>
1893 V6 P234-239 58-1578 <*>
1904 V17 N3 P63-64 2562 <*>
1905 V18 N1 P7-10 61-14859 <*>
1920 V33 P333-334 65-00332 <*>
1926 V39 P896-899 64-18986 <*>
1927 V40 P1408-1413 II-604 <*>
1927 V40 N38 P1201-1203 2479 <*>
1933 V46 N52 P1574-1576 59-17045 <*>
1936 V49 N46 P1399-1403 65-00139 <*>
1937 V50 P1180 61-00257 <*>
1937 V50 N41 P1423-1424 63-14583 <*>
1937 V50 N46 P1575-1577 61-00259 <*>
1939 V52 P645-647 58-1378 <*>
1941 V54 N52 P1055-1059 AEC-TR-6137 <*>
1946 V58 N47 P757-758 57-859 <*>
1950 V62 P922-924 57-357 <*>
1951 V63 P364-367 57-372 <*>
1952 V64 P369-374 1374 <*>
1952 V64 N35/6 P707-709 1727 <*>
1953 V65 P88-89 AL-221 <*>
1953 V65 P172-174 AL-192 <*>
1953 V65 P281-283 AL-593 <*>
1953 V65 P601-603 I-897 <*>
1953 V65 N36/7 P1-6 2757 <*>
1953 V65 N36/7 P734-736 61-00859 <*>
1954 V66 P10-15 2052 <*>
1954 V66 P424-427 1841 <*>
1954 V66 N1 P10-15 II-737 <*>
1955 V67 P15-17 3546-D <K-H>
1955 V67 P511-515 2678 <*>
1956 V68 P656-659 3468 <*>
1956 V68 P754-757 57-2536 <*>
1956 V68 P779-781 58-1833 <*>
1957 V69 P538-541 65-00136 <*>
1958 V70 P518-521 59-10832 <*> O
1958 V70 N15 P261-264 95L30G <ATS>
1958 V70 N24 P437-440 59-10842 <*>
1958 V70 N46 P921-924 60-14536 <*>
1958 V70 N49 P978-980 59-15460 <*>
1959 V71 N1 P1-3 59-17667 <*>
1960 V72 P482-483 65-13736 <*>
1961 V73 N7 P131 62-14646 <*>
1961 V73 N14 P247-248 11279-F <KH>
  63-16588 <*>
1961 V73 N31/2 P529-531 11548-C <KH>
  63-16902 <*>
1962 V74 P41-42 63-00281 <*>
1962 V74 P577-580 63-14722 <*>
1962 V74 P700-702 63-00281 <*>
1962 V74 P804-806 63-14718 <*>
1963 V75 P403-405 64-14308 <*>
1963 V75 N46 P820-821 64-00259 <*>
1963 V75 N46 P824-826 64-00259 <*>
1964 V76 P305-309 88S88G <ATS>
1964 V76 P313 88S88G <ATS>
1964 V76 N2 P24-25 66-11616 <*>
1965 V77 P521-525 66-13177 <*>

WIENER MEDIZINISCHE WOCHENSCHRIFT
1869 V19 N66 P1111-1113 58-1206 <*>
1869 V19 N74 P1243-1246 58-1206 <*>
1912 V62 N40 P2632- 1269 <*>
1913 V63 P1986- 57-379 <*>
1930 V80 N4 P133-134 1722 <*>
1933 V83 N19 P541-544 1109F <K-H>
1934 V84 P95-98 881C <K-H>
1949 V99 N25/6 P287-289 C-2412 <NRC>
1949 V99 N25/6 P299-300 65M45G <ATS>
1949 V99 N27/8 P330-331 59-20974 <*>
1951 V101 N38 P734-738 2541 <*>
1952 V102 P317-320 AL-721 <*>
1952 V102 P795- 57-393 <*>
1953 V103 P125- AL-381 <*>
1953 V103 P269- I-529 <*>
1953 V103 N39 P725-727 63-10286 <*>
1953 V103 N33/4 P602-607 63-10287 <*> O
1954 V104 P597-599 II-254 <*>
1954 V104 N18 P357-359 3178-D <KH>
1954 V104 N18 P369 63-10302 <*>
1954 V104 N46 P910-911 63-10288 <*>
1955 V105 P310-311 57-1922 <*>
1955 V105 P480-481 57-2535 <*>
1955 V105 N10 P200 64-20042 <*>
1955 V105 N10 P201 64-20043 <*>
1955 V105 N20/1 P430-432 2748 <*>
1956 P106-930 57-2822 <*>
1956 V106 P425- 57-1856 <*>
1956 V106 N40 P840-842 64-20002 <*>
1957 V107 P307-308 57-3176 <*>
1957 V107 P464-466 OA-90 <RIS>
1959 V109 N50 P1006-1007 75M42G <ATS>
1960 V110 P731-733 G-275 <RIS>
1960 V110 N7 P148-150 61-10823 <*>
1960 V110 N11 P250-255 G-234 <RIS>
1960 V110 N36 P721-722 G-276 <RIS>
1960 V110 N38 P791-792 62-16215 <*> O
1960 V110 N40 P819-823 10897-C <K-H>
  62-00378 <*>
1961 V111 P256-258 64-00114 <*>
1961 V111 P422-424 63-20221 <*> O
1961 V111 P659-664 63-10540 <*>
1961 V111 N1 P30-33 62-00596 <*>
1962 V112 P33-36 62-16909 <*>
1962 V112 P728 63-20702 <*>
1963 V113 N1 P8-10 64-18966 <*> O
1963 V113 N37 P684-686 66-12036 <*>
1963 V113 N38 P704-706 64-16071 <*>
1963 V113 N41 P745-748 66-12703 <*>
1963 V113 N45 P838-841 64-16357 <*>
1964 V114 P825-827 66-13194 <*>
1964 V114 N15 P256-257 64-20146 <*>
1964 V114 N49 P890-891 66-17414 <*>
1965 V115 N3 P33-38 65-14350 <*>
1965 V115 N3 P41-45 66-10370 <*> C

1965 V115 N8 P158-160 65-14240 <*>
1965 V115 N47 P994-995 64T91G <ATS>

WIENER TIERAERZTLICHE MONATSSCHRIFT
1950 V37 N9 P644-649 59-20968 <*> O
1953 V40 N4 P226-229 59-20969 <*>
1953 V40 N9 P513-530 59-20967 <*> P
1961 V48 N6 P348-361 65-13106 <*> O

WIENER ZEITSCHRIFT FUER INNERE MEDIZIN UND IHRE GRENZGEBIETE
1951 V32 N1 P36-41 C-2413 <NRC>
1953 V34 P283-292 I-910 <*>
1956 V37 P306-329 8133-B <K-H>
1956 V37 N4 P137-145 UCRL TRANS-1154 <*>
1958 V39 P334-339 63-16651 <*>
9331-D <K-H>
1959 V40 P135-138 62-18046 <*>
1960 V41 P34-39 62-10334 <*>
1961 V42 P354-361 63-00954 <*>

WIENER ZEITSCHRIFT FUER NERVENHEILKUNDE UND DEREN GRENZGEBIETE
1952 V5 P175-187 63-20266 <*> O

WIES WSPOLCZESNA
1967 N2 P12-20 67-30979 <=$>
1967 N2 P42-52 67-30979 <=$>
1967 N2 P85-91 67-30979 <=$>

WILHELM ROUX ARCHIV FUER ENTWICKLUNGSMECHANIK DER ORGANISMEN
1929 V115 P1-26 2212 <*>
1938 V138 P281-304 C-2484 <NRC>
1954 V147 N2/3 P171-200 57-2550 <*>
1960 V152 P632-654 61-01043 <*>
1961 V153 P158-167 62-00634 <*>

WINTER; ILLUSTRIERTE ZEITSCHRIFT FUER DEN WINTERSPORT
1936 V29 N6 P96- 61-14295 <*>

WINTERBERICHTE DES EIDGENOSSISCHEN INSTITUTES FUER SCHNEE- UND LAWINENFORSCHUNG
1959 N302 ENTIRE ISSUE C-4087 <NRC>

WIRTSCHAFT
1960 V15 N35 P17 62-19374 <=*>
1964 V19 N19 P12-13 64-41029 <=>
1966 P4-5 66-31973 <=$>
1966 P14 66-31973 <=$>
1967 N9 P17 67-31575 <=$>

WIRTSCHAFTGRUEFER
1940 N54 P53-62 61-20181 <*>

WIRTSCHAFTSWISSENSCHAFT
1963 V11 N3 P436-443 63-21620 <=>
1964 V12 N4 P646-648 64-31635 <=>
1966 N11 P1810-1823 67-30270 <=>

WISSENSCHAFT UND FORTSCHRITT
1957 V7 P1-13 59-11208 <=*>
1957 V7 P32-34 59-11208 <=*>
1957 V7 P56-59 59-11208 <=*>
1958 V8 P225-228 60-23948 <*=>

WISSENSCHAFTLICHE ABHANDLUNGEN DER DEUTSCHEN MATERIALPRUEFUNGSANSTALTEN
1950 V2 N7 P76-83 AL-42 <*>
1950 V2 N7 P109-112 64-20400 <*>

WISSENSCHAFTLICHE ABHANDLUNGEN DER PHYSIKALISCH-
1900 V3 P269-424 I-288 <*>
1930 V14 N1 P261-265 60-10436 <*>

WISSENSCHAFTLICHE ABHANDLUNGEN. REICHSAMT FUER WETTERDIENST
1938 V1 P1-34 II-1040 <*>
1938 V5 N1 P1-34 57-543 <*>
1941 SA N13 P21-27 AL-216 <*>

WISSENSCHAFTLICHE UND ANGEWANDTE PHOTOGRAPHIE
1955 V1 P234-257 UCRL TRANS-983 <*>

WISSENSCHAFTLICHE ERGEBNISSE DER DEUTSCHEN ATLANTISCHEN EXPEDITION AUF DEM VERMESSUNGS- UND FORSCHUNGSSCHIFF 'METEOR', 1925-1927
V3 P278-298 PT3 57J15G <ATS>
1934 V3 PT.3 SEC.D P278 57J15G <ATS>

WISSENSCHAFTLICHE FORSCHUNGSBERICHTE. NATURWISSENSCHAFTLICHE REIHE. DRESDEN
1938 V45 P146-167 66-13652 <*> O
1938 V45 P196-216 66-13652 <*> O

WISSENSCHAFTLICHE MEERESUNTERSUCHUNGEN DER KOMMISSION ZUR WISSENSCHAFTLICHEN UNTERSUCHUNG DER DEUTSCHEN MEERE. ABT. HELGOLAND
1927 V16 N10 P1-35 C-4787 <NRC>

WISSENSCHAFTLICHE VEROEFFENTLICHUNGEN AUS DEM SIEMENS-WERKEN
1922 V1 P489-496 56L29G <ATS>
1927 V6 N1 P1-31 66-12257 <*> O
66-13824 <*> P
1930 V9 N2 P323-330 60-10760 <*> O
1931 V10 N2 P55-71 66-13315 <*>
1932 V11 N1 P1-23 58-300 <*>
1934 V13 N3 P63-74 66-13149 <*> O
1935 V14 N3 P1-31 AEC-TR-5179 <*>
1936 V15 P124-128 124-128 <*>
1936 V15 N2 P51-63 57-642 <*>
1936 V15 N2 P64-77 57-2230 <*>
1937 V16 N1 P72-80 61-14907 <*>
1938 V17 N1 P99-106 57-2042 <*>
1938 V17 N1 P107-111 57-1392 <*>
1939 V18 N3 P1-67 65-13282 <*>
1940 P37-44 57-2995 <*>
1940 P78-87 2314 <*>
1940 V19 N1 P28-58 65-10790 <*>
1940 V19 N1 P280-290 59-17507 <*>
1941 V20 P85-103 59-17442 <*>

WISSENSCHAFTLICHE ZEITSCHRIFT DER FRIEDRICH SCHILLER-UNIVERSITET JENA. MATHEMATISCH-NATURWISSENSCHAFTLICHE REIHE
1954 N1 P21-26 5323 <KH>
1957 V7 N2/3 P273-277 C-3595 <NRC>

WISSENSCHAFTLICHE ZEITSCHRIFT DER HOCHSCHULE FUER MASCHINENBAU, KARL-MARX-STADT
1962 V4 P37-42 NP-TR-1247 <*>

WISSENSCHAFTLICHE ZEITSCHRIFT DER HOCHSCHULE FUER SCHWERMASCHINENBAU, MAGDEBURG
1958 V2 P219-223 2506 <BISI>
1959 V3 N2 P215-223 2033 <BISI>
1959 V3 N12 P189-193 2030 <BISI>
1959 V3 N12 P203-206 2031 <BISI>
1959 V3 N12 P207-214 2032 <BISI>

WISSENSCHAFTLICHE ZEITSCHRIFT DER HUMBOLDT-UNIVERSITAET BERLIN. MATHEMATISCH-NATURWISSENSCHAFTLICHE REIHE
1958 V8 N4/5 P737-751 62-01578 <*>

WISSENSCHAFTLICHE ZEITSCHRIFT DER KARL-MARX-UNIVERSITAET, LEIPZIG.
1951 V2 N4 P39-44 65-17364 <*>
1956 V6 N4 P401-404 63-18777 <*>

WISSENSCHAFTLICHE ZEITSCHRIFT DER TECHNISCHEN HOCHSCHULE FUER CHEMIE. LEUNA-MERSEBURG
1963 V5 N3 P263-267 80R80G <ATS>
1965 V7 N1 P9-10 NS-419 <TTIS>

WISSENSCHAFTLICHE ZEITSCHRIFT DER TECHNISCHEN HOCHSCHULE UNIVERSITAT, DRESDEN
1952 V2 N4/5 P719-734 58-1663 <*>

1954 V4 N6 P1003-1019 59-10546 <*>
1958 V8 N2 P349-351 65-11318 <*>

WISSENSCHAFTLICHES ARCHIV FUER LANDWIRTSCHAFT
SEE TIERERNAEHRUNG UND TIERZUCHT

WOCHENBLATT FUER PAPIERFABRIKATION
1948 V76 N3 P51-52 57-2868 <*>
1948 V76 N5 P108-110 63-20632 <*>
1949 V77 P168-169 3160-E <K-H>
1949 V77 N13 P359-363 C2295 <NRC>
1949 V77 N13 P359-362 57-1777 <*>
1950 V78 N19 P567-570 AL-379 <*>
1950 V78 N20 P595-598 AL-379 <*>
1950 V78 N22 P659-661 58-284 <*>
1950 V78 N24 P727-731 AL-379 <*>
1951 V79 N23 P751-756 58-596 <*>
1952 V80 P811-812 58-282 <*>
1952 V80 N21 P779-780 58-283 <*>
1953 V81 N23 P875 57-1422 <*>
1954 V82 N3 P75-84 57-2329 <*>
1955 N20 P820- 57-1437 <*>
1955 V83 N21 P855-856 4056-B <K-H>
1956 N14 P564- 57-1727 <*>
1956 V84 N17 P683- 57-931 <*>
1956 V84 N22 P885-888 57-2863 <*>
57-834 <*>
1956 V84 N23 P960-961 C-2286 <NRC>
1956 V84 N11/2 P427-437 57-2871 <*>
57-831 <*>
1957 V85 P7-9 63-14125 <*>
1957 V85 N6 P204-206 58-1617 <*>
1957 V85 N8 P283- 3434 <*>
1957 V85 N12 P447 58-288 <*>
1957 V85 N12 P452-458 58-2640 <*>
1957 V85 N12 P452 58-294 <*>
1957 V85 N12 P458 58-286 <*>
1957 V85 N18 P695-704 C-2627 <NRC>
58-1566 <*>
1957 V85 N18 P695- 58-758 <*>
1957 V85 N22 P849-850 59-15062 <*>
1958 V86 N8 P290-297 58-1432 <*>
1958 V86 N8 P298-308 59-20100 <*> O
1958 V86 N8 P310-322 58-2357 <*>
1958 V86 N13 P589-594 59-10441 <*>
1958 V86 N17 P753-761 58-2686 <*>
1958 V86 N21 P917-919 60-10448 <*>
1958 V86 N11/2 P489-496 58-2168 <*>
1958 V86 N11/2 P497-511 58-2683 <*>
1958 V86 N11/2 P526-530 59-17704 <*>
1958 V86 N23/4 P1019- 60-14350 <*> O
1958 V86 N23/4 P1037-1043 63-18282 <*> O
1959 V87 N8 P305-321 60-16485 <*> O
1959 V87 N16 P702-705 63-16781 <*> O
1959 V87 N19 P819-823 61-10388 <*> P
1959 V87 N21 P895-900 60-14395 <*> O
1959 V87 N23 P1001-1008 60-14287 <*> O
1959 V87 N24 P1060-1061 60-18929 <*>
1959 V87 N11/2 P467-478 60-14146 <*> O
1960 V88 N1 P7-14 60-16778 <*> O
1960 V88 N8 P299-308 60-18084 <*>
1961 V89 N4 P150-151 61-20896 <*>
1961 V89 N11 P531-546 62-18801 <*> O
1961 V89 N16 P773-777 62-10083 <*> O
1961 V89 N17 P821-831 63-10335 <*> O
1961 V89 N19 P930 62-10593 <*>
1961 V89 N21 P1021-1026 62-16497 <*> O
1961 V89 N22 P1061-1064 62-16531 <*>
1961 V89 N22 P1071-1074 62-16497 <*> O
1961 V89 N11/2 P523-531 61-20738 <*> O
1962 V90 P373-377 62-20453 <*> O
1962 V90 N2 P57-58 62-16532 <*> P
1962 V90 N15 P785-791 64-14343 <*> O
1962 V90 N20 P1045-1052 65-10278 <*> O
1962 V90 N23/4 P1265-1271 63-18283 <*>
1963 V91 N11/2 P554-560 64-14342 <*> O
1963 V91 N15/6 P772-778 64-14340 <*> O
1963 V91 N15/6 P780-788 64-14341 <*> O
1964 V92 N10 P265-267 65-11871 <*> O
1964 V92 N11/2 P287-303 65-13354 <*>
1964 V92 N11/2 P311-314 65-17452 <*>
1964 V92 N23/4 P707-716 96S81G <ATS>
1966 V94 N4 P107-110 66-13131 <*>

WOCHENSCHRIFT FUER THERAPIE UND HYGIENE DES AUGES
1913 V17 N5 P37-38 61-14452 <*>
1916 V19 N25 P121-122 61-14453 <*>

WOJSKOWY PRZEGLAD LOTNICZY
1960 V13 N3 P24-32 AEC-TR-4327 <*>
1963 N2 P72-75 AEC-TR-6384 <*>
1963 N3 P47-55 64-71165 <=>
1963 N8 P22-26 AD-608 451 <=>
1963 N8 P82-84 AD-630 876 <=$>
1965 N1 P86-87 AD-630 877 <=$>
1965 V19 N8 P48-59 AD-630 879 <=$>

WOOD INDUSTRY. JAPAN
SEE MOKUZAI KOGYO

WOODWORKING INDUSTRY. LONDON
1955 V4 P16-17 R-1162 <*>
1955 V4 N3 P24-25 R-169 <*>
1955 V4 N7 P11-14 R-173 <*>

WORLD KNOWLEDGE. CHINESE PEOPLES REPUBLIC
SEE SHIH CHIEH CHIH SHIH

WORLD OF OBSTETRICS AND GYNECOLOGY. JAPAN
SEE SANFUJINKA NO SEKAI

WU HAN TA HSUEH TZO JAN K'O HSUEH HSUEH PAO
1957 N2 P103-118 AD-611 034 <=>
1959 N5 P1-22 63-13223 <=*>
1959 N5 P118-125 60-11818 <=>
1959 N6 P1-28 62-19806 <=>
1959 N6 P38-45 60-11650 <=>

WUHAN UNIVERSITY: JOURNAL OF NATURAL SCIENCES
SEE WU HAN TA HSUEH TZO JAN K'O HSUEH HSUEH PAO

WU-HSIEN-TIEN
1958 N12 P4-6 59-13491 <=>
1958 N12 P18-19 59-13491 <=>
1960 N2 P4 60-31423 <=>
1960 N2 P15 60-31423 <=>
1960 N2 P33 60-31423 <=>
1960 N2 P39 60-31423 <=>
1960 N2 P41 60-31423 <=>
1962 N10 P1-2 63-21527 <=>
1962 N10 P13 63-21527 <=>
1963 N5 P1-2 AD-619 452 <=$>
1963 N11 P1-3 AD-633 684 <=$>
1965 N8 P1-2 AD-636 190 <=$>

WU HSIEN TIEN CHI SHU
1964 N6 P8-11 AD-623 192 <=$>
1964 N6 P21 AD-623 192 <=$>
1965 N12 P9-15 67-31365 <=$>

*WU LI HSUEH PAO
1925 P49-73 I-711 <*>
1956 V12 P177-181 AEC-TR-3521 <*>
AEC-TR-3521 <=>
1956 V12 N4 P281-297 61-28288 <*=>
62-13712 <*=>
1956 V12 N4 P319-337 61-28049 <*=>
1956 V12 N4 P339-359 61-28287 <*=>
1956 V12 N4 P360-372 61-28758 <=*>
1956 V12 N4 P376-378 61-28758 <*=>
1957 V13 N2 P130-141 95K26CH <ATS>
1957 V13 N6 P515-524 60-11819 <=> O
1958 V14 N2 P106-112 60-11774 <=>
1958 V14 N3 P164 60-11600 <=>
1958 V14 N3 P188-189 60-11600 <=>
1958 V14 N3 P191 60-11600 <=>
1958 V14 N3 P202-203 60-11600 <=>
1958 V14 N3 P244 60-11600 <=>
1958 V14 N3 P259-261 60-11600 <=>
1958 V14 N3 P274 60-11600 <=>

Citation	Translation
1958 V14 N3 P283-288	60-11600 <=>
1959 V15 N1 P1-41	61-21460 <=>
1959 V15 N1 P42-53	60-11807 <=>
1959 V15 N2 P55-75	61-28386 <=>
1959 V15 N2 P77-87	61-28386 <=>
1959 V15 N2 P89-111	61-28386 <=>
1959 V15 N2 P113-129	61-28386 <=>
1959 V15 N2 P131-137	61-28386 <=>
1959 V15 N2 P139-165	61-28386 <=>
1959 V15 N3 P113-129	63-24062 <=*$>
1959 V15 N4 P167-177	61-27336 <=>
1959 V15 N4 P178-185	61-27926 <=>
1959 V15 N4 P202-209	61-27336 <=>
1959 V15 N4 P210-216	61-27926 <=>
1959 V15 N5 P219-229	61-27336 <=>
1959 V15 N5 P230-245	61-27926 <=>
1959 V15 N5 P254-261	61-27336 <=>
1959 V15 N5 P262-268	61-27926 <=>
1959 V15 N5 P269-276	61-27336 <=>
1959 V15 N6 P277-340	61-27651 <*=>
1959 V15 N7 P341-352	62-13612 <*=>
1959 V15 N7 P353-367	60-11176 <=>
1959 V15 N7 P368-376	61-27095 <*=>
1959 V15 N7 P377-388	61-27126 <*=>
1959 V15 N7 P389-392	61-27128 <*=>
1959 V15 N7 P393-396	61-27099 <*=>
1959 V15 N9 P449-464	63-23765 <=*$>
1959 V15 N9 P475-506	60-31009 <=>
1959 V15 N10 P507-524	61-23569 <=>
1959 V15 N10 P525-533	60-31323 <=>
1959 V15 N10 P535-549	61-23569 <=>
1959 V15 N10 P550-557	60-31292 <=>
1959 V15 N10 P559-574	61-23569 <=>
1959 V15 N11 P575-623	61-23942 <=>
1959 V15 N11 P588-602	63-24061 <=*$>
1959 V15 N12 P625-650	61-23942 <=>
1959 V15 N12 P664-672	61-23942 <=>
1960 V16 N1 P2-56	61-28521 <=>
1960 V16 N3 P121-185	61-21591 <=>
1960 V16 N3 P155-159	2051 <BISI>
1960 V16 N4 P187-203	61-31144 <=>
1960 V16 N4 P205-212	61-31144 <=>
1960 V16 N4 P214-239	61-31144 <=>
1960 V16 N4 P241-244	61-31144 <=>
1960 V16 N5 P272-279	61-11017 <=>
1960 V16 N5 P281-288	62-32994 <=*>
1960 V16 N5 P289-297	62-15425 <*=>
1960 V16 N6 P305-329	61-31144 <=>
1960 V16 N6 P305-315	63-24058 <=*>
1960 V16 N6 P316-323	63-24060 <=*>
1960 V16 N6 P331-337	61-31144 <=>
1961 V17 N2 P107-110	62-19631 <=*>
1961 V17 N4 P180-190	AD-611 035 <=>
1961 V17 N5 P237-245	AD-619 312 <=$>
1961 V17 N9 P439-444	AD-620 968 <=$>
1961 V17 N9 P445-449	AD-620 968 <=$>
1961 V17 N12 P592-599	AD-631 063 <=$>
1962 V18 P586-593	AEC-TR-6131 <*>
1962 V18 N1 P28-46	64-15400 <=*$>
1962 V18 N4 P188-193	65-10798 <*>
1962 V18 N5 P250-253	AD-630 930 <=$>
1962 V18 N9 P483-485	AD-619 487 <=$>
1962 V18 N11 P594-599	AEC-TR-6130 <*>
1962 V18 N11 P600-604	63-24104 <=*$>
1962 V18 N12 P629-635	AD-621 013 <=$>
1963 V19 P524-537	65-18009 <*>
1963 V19 N1 P1-10	AD-632 403 <=$>
1963 V19 N2 P73-82	AD-608 370 <=>
1963 V19 N3 P139-150	AD-632 404 <=$>
1963 V19 N3 P160-164	64-11977 <=>
1963 V19 N3 P176-190	AD-610 649 <=$>
1963 V19 N3 P191-201	AD-630 930 <=$>
	AD-631 102 <=$>
1963 V19 N5 P273-284	65-10767 <*>
1963 V19 N5 P285-296	AD-609 154 <=>
1963 V19 N5 P306-319	AD-611 028 <=$>
1963 V19 N5 P320-335	AD-611 030 <=$>
	AD-625 038 <=*$>
1963 V19 N6 P384-397	AD-610 647 <=$>
1963 V19 N12 P807-815	AD-619 467 <=$>
1963 V19 N12 P816-823	AD-621 012 <=*$
1964 V20 N1 P41-54	AD-638 976 <=$>
1964 V20 N1 P63-71	AD-619 425 <=$>
1964 V20 N1 P72-82	AD-638 976 <=$>
1964 V20 N2 P147-158	AD-632 410 <=$>
1964 V20 N2 P164-173	66-11207 <*>
1964 V20 N3 P261-269	AD-632 409 <=$>
1964 V20 N3 P285-286	AD-637 067 <=$>
1964 V20 N4 P311-336	65-30223 <=*$>
1964 V20 N4 P368-373	AD-639 161 <=$>
1964 V20 N5 P387-410	65-14582 <*> 0
1964 V20 N6 P528-539	AD-637 441 <=$>
1964 V20 N6 P568-570	AD-636 625 <=$>
1964 V20 N8 P753-759	AD-626 953 <=*$>
1964 V20 N9 P873-889	N66-18449 <=$>
1964 V20 N10 P991-1002	N66-18450 <=$>
1964 V20 N31 P93-206	AD-631 411 <=$>
1965 V21 N5 P889-896	N66-18453 <=$>

WU LI T'UNG PAO

Citation	Translation
1960 N4 P158-162	60-41593 <=>
1960 N4 P184-185	60-41593 <=>
1960 N4 P188-189	60-41593 <=>
1960 N4 P192-193	60-41138 <=>
1960 N5 P217	61-11632 <=*>
1960 N5 P223-225	61-11632 <=*>
1960 N6 P241-246	62-25180 <=>
1960 N6 P254-258	62-25180 <=>
1960 N6 P270-272	62-25180 <=>
1961 N3 P89-110	63-19327 <=*>

WUERZBURGER UNIVERSITATSREDEN

Citation	Translation
1961 N30 P5-19	86S81G <ATS>

Y HOC THUC NANH. HANOI

Citation	Translation
1962 N85 P5-8	63-15098 <=>

YAKKYOKU

Citation	Translation
1958 V9 P176-178	65-14327 <*> 0
1960 V11 P1229-1232	64-14620 <*>

YAKUGAKU KENKYU

Citation	Translation
1957 V29 N6 P588-596	59-18574 <+*>
1962 V34 N2 P39-43	63-10677 <*>

YAKUGAKU ZASSHI

Citation	Translation
1921 V470 P331-335	58-161 <*>
1923 N493 P131-142	65-12131 <*> P
1928 V48 N3 P223-237	2656 <*>
1930 V50 P555-559	62-18805 <*>
1932 V52 N5 P433-458	II-350 <*>
1933 V53 P622-634	63-14153 <*> 0
1933 V53 N8 P807-815	2441 <*>
1937 V57 P985-991	61-16559 <*>
1938 V58 P261-264	3159 <*>
1938 V58 N9 P816-818	67P65J <ATS>
1939 V59 P18-28	59-10726 <*>
1942 V62 N3/4 P95-97	57-2554 <*>
1943 V63 N6 P282-289	62-18191 <*>
1944 V64 N5 P294-298	62-18192 <*>
1947 V67 N1/2 P34-35	57-1972 <*>
1949 V69 N6/10 P159-161	34N56J <ATS>
1949 V69 N6/10 P353-355	61-10841 <*>
1950 V70 P44-45	64-18935 <*>
1950 V70 P741-742	16D1J <ATS>
1950 V70 N3 P164-167	53E2J <ATS>
1951 V71 P1309-1313	61-10331 <*> 0
1951 V71 N1 P20-23	62-18193 <*>
1951 V71 N1 P24-27	62-18194 <*>
1951 V71 N7 P643-645	C-4652 <NRC>
1951 V71 N7 P765-766	16L32J <ATS>
1951 V71 N9 P889-894	AL-971 <*>
1952 V72 P529-532	62-16212 <*>
1952 V72 P533-537	62-16213 <*>
1952 V72 P982-985	25E1J <ATS>
1952 V72 N3 P376-377	55F4J <ATS>
1952 V72 N7 P930-932	1967 <*>
1952 V72 N7 P933-936	1966 <*>
1952 V72 N9 P1128-1131	I-868 <*>
1953 V73 P23-25	I-869 <*>

1953 V73 P290-292 62-16265 <*>
1953 V73 N4 P415 64-18624 <*>
1953 V73 N7 P719-721 71F4J <ATS>
1953 V73 N9 P961-968 C-4652 <NRC>
1953 V73 N11 P1216-1218 1658 <*>
1954 V74 P167-171 59-21026 <=> O
1954 V74 P495-497 59-20329 <*> O
1954 V74 P742-746 59-15380 <*>
1954 V74 P1195-1198 II-883 <*>
1954 V74 P1365-1369 64-16907 <*>
1954 V74 N1 P72-75 II-687 <*>
1954 V74 N5 P498-501 59-15020 <*>
1954 V74 N5 P546-547 1301 <*>
1954 V74 N6 P655-657 C-4652 <NRC>
1954 V74 N6 P658-661 56F4J <ATS>
1954 V74 N6 P694-697 28G5J <ATS>
1954 V74 N7 P721-723 2031 <*>
1954 V74 N7 P742-746 60-10952 <*>
1954 V74 N7 P788- 57-3041 <*>
1954 V74 N8 P797-801 72G8J <ATS>
1954 V74 N8 P801-805 73G8J <ATS>
1954 V74 N8 P806-810 71G8J <ATS>
1954 V74 N8 P810-812 75G8J <ATS>
1954 V74 N8 P827-830 80G5J <ATS>
1954 V74 N9 P984-987 57-184 <*>
1954 V74 N10 P1097-1100 II-1053 <*>
1954 V74 N10 P1127-1129 II-1052 <*>
1954 V74 N11 P1180-1184 II-1051 <*>
1954 V74 N12 P1312-1314 2194 <*>
1954 V74 N12 P1395-1397 59-10271 <*>
1955 V75 P620-622 30G8J <ATS>
1955 V75 P966-969 II-884 <*>
1955 V75 P1287-1288 T-2014 <INSD>
1955 V75 N2 P192-194 2762 <*>
1955 V75 N2 P194-196 2753 <*>
1955 V75 N4 P359-363 95P59J <ATS>
1955 V75 N4 P378-381 59-15021 <*>
1955 V75 N6 P763-764 59-12131 <=*>
1955 V75 N9 P1149-1152 12H14J <ATS>
58-364 <*>
1955 V75 N11 P1318-1321 04M45J <ATS>
1955 V75 N11 P1370-1373 64Q72J <ATS>
1955 V75 N11 P1441-1442 II-544 <*>
1956 V76 P959-960 51H13J <ATS>
58-166 <*>
1956 V76 P1116-1118 58-1429 <*>
1956 V76 P1359-1361 57-2304 <*>
1956 V76 N2 P172-175 58-309 <*>
79H11J <ATS>
1956 V76 N2 P176-180 58-311 <*>
80H11J <ATS>
1956 V76 N3 P325-328 60-17124 <+*> O
1956 V76 N6 P626-628 57-85 <*>
1956 V76 N6 P629-631 22L34J <ATS>
1956 V76 N6 P678-681 58-1074 <*>
1956 V76 N6 P682-686 58-1352 <*>
1956 V76 N6 P747-748 23L34J <ATS>
1956 V76 N8 P883-886 10H14J <ATS>
58-367 <*>
1956 V76 N8 P886-892 11H14J <ATS>
58-366 <*>
1956 V76 N10 P1101-1103 22TM <CTT>
1957 V77 P362-366 65-10060 <*>
1957 V77 P441-442 58-178 <*>
1957 V77 P848-850 59-21027 <=> O
1957 V77 P906-912 66-14147 <*$> O
1957 V77 P912-917 66-14148 <*$> O
1957 V77 N3 P225-237 C-5330 <NRC>
1957 V77 N3 P290-293 57-3445 <*>
1957 V77 N3 P312-316 24K24J <ATS>
1957 V77 N3 P316-318 25K24J <ATS>
1957 V77 N4 P370-375 58-235 <*>
1957 V77 N4 P375-379 58-112 <*>
63-14611 <*> O
1957 V77 N4 P441-442 59-15770 <*>
1957 V77 N7 P743-746 58-628 <*>
1957 V77 N7 P753-758 58-108 <*>
1957 V77 N7 P790-793 C-5230 <NRC>
1957 V77 N7 P815-816 58-1817 <*>
1957 V77 N8 P891-894 58-840 <*>
1957 V77 N9 P955-959 58-548 <*>
1957 V77 N10 P1092-1095 58-1077 <*>
1957 V77 N10 P1095-1100 64-14179 <*> O
1958 V78 P129-131 58-1807 <*>
1958 V78 P671-673 65-13444 <*>
1958 V78 P974-977 66-13016 <*>
1958 V78 P1237-1241 59-15603 <*> O
1958 V78 N4 P346-350 59-12984 <+*> O
1958 V78 N4 P354-361 60-19489 <=*>
1958 V78 N4 P368-375 61-00212 <*>
1958 V78 N5 P486-490 66R80J <ATS>
1958 V78 N6 P632-637 60-10980 <*>
1958 V78 N6 P637-643 60-10969 <*>
1958 V78 N6 P643-646 60-10978 <*>
1958 V78 N6 P647-651 60-10979 <*>
1958 V78 N10 P1166-1170 60-10981 <*>
65S81J <ATS>
1958 V78 N11 P1269-1274 61-00199 <*>
1958 V78 N11 P1274-1279 61-00402 <*>
78L35J <ATS>
1958 V78 N11 P1280-1285 61-00906 <*>
1958 V78 N11 P1286-1290 61-00850 <*>
1958 V78 N11 P1291-1295 62-00423 <*>
1958 V78 N11 P1296-1301 62-00882 <*>
1959 N79 P1305-1309 60-10977 <*>
1959 V79 N1 P32-34 59-18132 <+*>
1959 V79 N2 P131-135 AEC-TR-4378 <*>
1959 V79 N2 P135-139 AEC-TR-4379 <*>
1959 V79 N3 P281-286 60-10970 <*>
1959 V79 N3 P374-377 59-18577 <+*> O
1959 V79 N4 P552-554 60-14094 <*> O
1959 V79 N5 P664-669 60-10975 <*>
1959 V79 N6 P729-737 C-5379 <NRC>
1959 V79 N6 P794-799 60-10972 <*>
62-01057 <*>
1959 V79 N6 P799-802 60-10976 <*>
62-01209 <*>
1959 V79 N6 P814-819 60-10974 <*>
1959 V79 N6 P819-824 60-10973 <*>
1959 V79 N6 P854-855 60-10971 <*>
1959 V79 N7 P949-953 05N55J <ATS>
1959 V79 N8 P1063-1068 61-10341 <*> O
1959 V79 N11 P1381-1385 41M38J <ATS>
1959 V79 N12 P1615-1617 24M40J <ATS>
1960 V80 P298-300 61-10222 <*>
1960 V80 N6 P721-727 27N49J <ATS>
1960 V80 N11 P1530-1533 C-3920 <NRCC>
62-00387 <*>
1960 V80 N11 P1538-1542 27N52J <ATS>
1960 V80 N11 P1567-1573 28N49J <ATS>
1960 V80 N11 P1642-1644 61-20694 <*>
1961 V81 N3 P464-467 61-01049 <*>
1961 V81 N6 P878-882 62-10582 <*>
1961 V81 N6 P931-933 61-20695 <*>
1961 V81 N9 P1229-1233 62-20207 <*>
1961 V81 N9 P1233-1236 62-20056 <*> O
1961 V81 N9 P1245-1248 64-23625 <=$>
1961 V81 N11 P16-55 63-00194 <*>
1962 V82 P1349-1354 63-10906 <*>
1963 N5 P527-531 ICE V4 N3 P541-545 <ICE>
1963 V83 N5 P527-531 ICE V4 N3 P541-545 <ICE>
32Q72J <ATS>
1963 V83 N9 P871-874 65-10983 <*>
1963 V83 N11 P1005-1015 64-31905 <=>
1963 V83 N11 P1052-1055 64-16669 <*>
1963 V83 N12 P1184-1186 C-5346 <NRC>
1964 V84 N7 P621-625 65-11629 <*>
1966 V86 N2 P153-157 66-14398 <*$>
1966 V86 N3 P251-253 66-14092 <*$>

YAO HSUEH HSUEH PAO
1959 V7 N9 P327-334 61-21614 <=>
1959 V7 N9 P336-339 61-21614 <=>
1959 V7 N9 P341-344 61-21614 <=>
1959 V7 N9 P346-383 61-21614 <=>
1960 V8 N1 P1-14 61-19200 <=>
1960 V8 N1 P15-46 61-21467 <=>
1960 V8 N1 P47-52 61-19200 <=>
1960 V8 N1 P53-57 61-21467 <=>
1960 V8 N1 P57-60 61-19200 <=>

1960 V8 N2 P61-64 61-28621 <=>
1960 V8 N2 P66-93 61-28621 <=>
1960 V8 N2 P95-103 61-28621 <=>
1960 V8 N2 P105-108 61-28621 <=>
1960 V8 N5 P197-233 61-21363 <=>

YAO HSUEH T'UNG PAO
1959 V7 N9 P440 60-31322 <=>
1959 V7 N9 P442-444 60-31322 <=>
1959 V7 N9 P446-448 60-31322 <=>
1959 V7 N10 P489-496 60-31322 <=>

YAROVIZATSIYA
1938 N1/2 P198-202 RTS-3113 <NLL>

YAWATA TECHNICAL REPORT. JAPAN
SEE SEITETSU KENKYU

YEH CHIN CHIEN SHE. PEKING
1959 N12 P9-18 60-11345 <=>

YEH CHIN PAO
1959 N31 P3-4 60-11676 <=>
1959 N44 P47 60-11676 <=>
1959 N44 P50 60-11676 <=>
1959 N44 P62-66 60-11676 <=>
1959 N44 P69-71 60-11676 <=>
1959 N47 P19-23 60-11676 <=>
1959 N47 P27 60-11676 <=>
1959 N47 P32-34 60-11676 <=>
1959 N48 P18-19 60-11676 <=>
1959 N48 P26-29 60-11676 <=>
1959 N49 P32-36 60-11676 <=>
1959 N313 P28-30 60-11676 <=>
1959 N313 P37-39 60-11676 <=>

YMER. SVENSKA SALLSKAPET FOR ANTROPOLOGI OCH GEOGRAFI
1885 P147-156 62-00564 <*>

YOGYO KYOKAI-SHI
1943 V51 N607 P381-384 I-431 <*>
1944 V52 P267-269 65-13772 <*>
1948 V56 P7-10 64-18297 <*> 0
1948 V56 P59-62 T-2259 <INSD>
64-18297 <*> 0
1948 V56 P99-101 64-18297 <*> 0
1948 V57 N1/3 P7-10 T-2259 <INSD>
1949 V57 P56-58 T-2259 <INSD>
1951 V59 P323-328 65-11153 <*>
1951 V59 N2 P18-21 T-1465 <INSD>
1951 V59 N658 P146-150 52R79J <ATS>
1951 V59 N664 P482-485 62-14534 <*>
1952 V60 P548-552 61-10753 <*>
1954 V62 P24-27 NP-TR-296 <*>
1954 V62 P666-671 T-1442 <INSD>
1955 V63 P579-582 65-11154 <*>
1956 V64 P43-49 66-12184 <*> 0
1957 V65 P99-101 T-2259 <INSD>
1957 V65 P314-320 66-13386 <*>
1957 V65 N742 P259-268 60-18542 <*>
63-14070 <*>
1958 V66 P11-17 66-12698 <*> 0
1959 V67 P301-311 63-14073 <*>
1959 V67 N1 P20-27 63-18618 <*>
1959 V67 N3 P85-95 63-14071 <*>
1960 V68 N6 P145-153 64-14068 <*>
1960 V68 N10 P223-231 66-10672 <*>
1960 V68 N773 P132-137 62-20171 <*>
1961 V69 N2 P35-43 66-10673 <*>
1961 V69 N2 P52-59 93Q70J <ATS>
1961 V69 N4 P109-118 65-13512 <*>
1961 V69 N5 P137-144 77Q67J <ATS>
1962 V70 N6 P190-194 65-17342 <*>
1962 V70 N12 P330-334 65-12333 <*>
1963 V71 N1 P13-20 63-18933 <*>
1963 V71 N1 P109-116 66-10694 <*>
1964 V72 N5 P71-75 65-13751 <*>
1966 V74 N4 P101-108 66-14606 <*$>

YOKOHAMA MEDICAL JOURNAL
SEE YOKOHAMA IGAKU

YOKUFUEN CHOSA KENKYU KIYO
1962 V35 P35-50 65-17277 <*>
1964 N40 P161-170 66-12006 <*>
1964 N40 P171-180 66-12149 <*>

YONAGO IGAKU ZASSHI
1959 V10 N5 P1268-1277 64-18675 <*>

YOSETSU GAKKAISHI
1950 V19 P6-21 60-21059 <*=> 0
1959 V28 N1 P25-31 04L31J <ATS>
1959 V28 N1 P32-38 69L32J <ATS>
1959 V28 N4 P244-249 87L32J <ATS>
1959 V28 N10 P723-728 70M45J <ATS>

YUKAGAKU
1955 V4 P191-193 T1850 <INSD>
1959 V8 N8 P355-361 42P66J <ATS>
63-20011 <*>
1959 V8 N8 P368-372 60-18125 <+*> 0
1959 V8 N12 P610-615 62-00215 <*>
1961 V10 N1 P2-5 SC-T-515 <*>
1961 V10 N2 P83-88 2657 <BISI>
1961 V10 N9 P540-542 62-20003 <*>
1962 V11 P24-28 63-10323 <*>
1963 V12 P501-505 1133 <TC>
1964 V13 N10 P520-522 66-12571 <*>
1965 V14 N2 P63-66 87S84J <ATS>
1965 V14 N11 P612-614 66-13407 <*>

YUKI GOSEI KAGAKU KYOKAI SHI
1949 V7 P41-43 94S84J <ATS>
1951 V9 N11 P252-253 T-1752 <INSD>
1953 V11 N10 P386-388 64-16444 <*>
64-20326 <*>
1955 V13 P458-470 63-14037 <*> 0
1956 V14 P615-619 65-10005 <*> 0
1956 V14 N3 P109-124 62-20343 <*>
1956 V14 N10 P615-619 50R79J <ATS>
1956 V14 N12 P718-722 58-2483 <*>
1957 V15 N10 P518-527 09N53J <ATS>
1958 V16 N9 P461-467 88S81J <ATS>
1958 V16 N11 P614-620 60S81J <ATS>
1958 V16 N11 P620-624 84S81J <ATS>
1959 V17 N7 P365-371 03N53J <ATS>
1959 V17 N10 P579-594 92P64J <ATS>
1959 V17 N12 P783-788 65-12070 <*>
1960 V18 P97-109 62-18829 <*>
1960 V18 P363-387 1352 <TC>
1960 V18 N1 P15-30 62-16964 <*>
1960 V18 N4 P229-240 45M43J <ATS>
1960 V18 N6 P368-677 66-11563 <*>
1960 V18 N8 P592-601 32N49J <ATS>
1960 V18 N8 P602-608 31N50J <ATS>
1960 V18 N9 P652-655 32N50J <ATS>
1960 V18 N10 P692-700 CODE-182 <SA>
1960 V18 N10 P701-707 CODE-183 <SA>
1960 V18 N10 P743-749 CODE-188 <SA>
1960 V18 N12 P888-894 00P62J <ATS>
1961 V19 P311-317 1110 <TC>
1961 V19 N2 P124-130 63N54J <ATS>
1961 V19 N3 P24-53 M-5651 <NLL>
1961 V19 N4 P332-339 01P63J <ATS>
1961 V19 N12 P853-865 66-10322 <*>
1962 V20 P40-55 62-18865 <*>
1962 V20 N3 P248-251 69P64J <ATS>
1962 V20 N3 P277-283 64-16294 <*>
1962 V20 N4 P326-342 SCL-T-462 <*>
1962 V20 N5 P453-459 54Q68J <ATS>
1962 V20 N5 P501-506 65-10477 <*>
1963 V21 P160-163 64-10595 <*>
1963 V21 P413-424 65-00416 <*>
1963 V21 N1 P37-40 65-13021 <*>
1963 V21 N8 P611-616 65-12892 <*>
1963 V21 N8 P617-620 65-13147 <*>
1963 V21 N8 P621-631 65-12893 <*>
1963 V21 N8 P694-703 65-12894 <*>

1963 V21 N11 P858-869 21R77J <ATS>
66-10005 <*>
1963 V21 N11 P870-879 22R77J <ATS>
1963 V21 N12 P928-935 65-12895 <*>
1964 V22 P138-143 65-13740 <*>
1964 V22 N1 P61-66 65-12896 <*>
1964 V22 N5 P394-400 65-13828 <*>
1964 V22 N12 P1015-1022 1574 <TC>
1965 V23 N2 P99-116 72S85J <ATS>
1965 V23 N5 P447-451 66-10997 <*>
1965 V23 N9 P757-767 66-13979 <*>
1965 V23 N9 P768-777 66-10706 <*>
1965 V23 N11 P1028-1033 66-14380 <*$> O

YU-SE CHEN SHU I SHU
1959 N8 P28 60-31384 <=>
1959 N8 P34-35 60-31384 <=>
1959 N9 P18-26 60-31384 <=>
1959 N10 P18-21 60-31384 <=>
1959 N11 P16-21 60-31384 <=>
1959 N12 P24 60-31384 <=>

ZIS-MITTEILUNGEN
1960 N2 P152-158 62M42G <ATS>

ZA BEZOPASNOST DVIZHENIYA
1958 N6 P14-15 60-15732 <+*>

ZA EKONOMIYU TOPLIVA
1950 V7 N5 P17-18 51-0094 <NLL>
1951 V8 N2 P36 60-13700 <+*>

ZA OBORONY
1946 V22 N10 P19 59-18414 <+*>

ZA RUBEZHOM
1965 N2 P23-24 65-31608 <=$>

ZA RULEM
1958 V16 N11 P2-4 59-21076 <=> O
1959 V17 N1 P3-4 59-16644 <+*> O
1959 V17 N1 P19-20 59-16645 <+*>
1959 V17 N3 P15-16 60-21216 <=>
1960 V18 N1 P14-16 60-23141 <+*> O
1960 V18 N4 P5-9 61-21293 <=>
1960 V18 N5 P12-15 61-21293 <=>
1961 V19 N2 P20 62-33051 <=*>
1961 V19 N3 P17-20 62-33051 <=*>
1961 V19 N9 P4-5 62-33023 <=*>
1961 V19 N9 P19 63-13111 <=>
1961 V19 N10 P8-11 63-13111 <=>
1961 V19 N10 P12-14 62-32002 <=>
1961 V19 N12 P8 62-33051 <=*>
1961 V19 N12 P14-15 63-13111 <=>
1962 V20 N1 P4 62-33023 <=*>
1962 V20 N1 P11-13 62-32002 <=>
1962 V20 N1 P14-15 63-13111 <=>
1962 V20 N2 P15-17 62-11690 <=>
62-32002 <=>
1962 V20 N5 P10 62-33697 <=>
1962 V20 N10 P12-13 63-13850 <=>
1962 V20 N12 P22-23 64-13732 <=*$>
1963 V21 N1 P16-17 63-31300 <=>
1963 V21 N5 P12-14 63-31450 <=>
1963 V21 N7 P10-11 63-41157 <=>
1963 V21 N7 P11 63-41157 <=>
1963 V21 N7 P13 63-41157 <=>
1963 V21 N7 P15 63-31927 <=>
1963 V21 N7 P16-17 63-41157 <=>
1964 V22 N1 P16-17 AD-463 813 <=$>
1964 V22 N8 P4-5 64-51412 <=$>
1964 V22 N11 P1-5 65-30429 <=$>
1964 V22 N11 P12-13 65-30429 <=$>
1965 V23 N1 P27 65-30397 <=$>
1965 V23 N3 P26 65-31092 <=*$>

ZA TEKHNICHESKII PROGRESS. GORKIY
1958 N4 P22-25 C-3925 <NRC>

ZACCHIA
1953 V16 P370-376 62-16415 <*>

ZAHNAERZTLICHE MITTEILUNGEN. BERLIN
1959 V47 N7 P1-5 59-18993 <=*> O

ZAHNAERZTLICHE MITTEILUNGEN. NUERNBERG, KOELN
1940 V39 N21 P888-889 57-680 <*>

ZAHNAERZTLICHE RUNDSCHAU
1960 V69 N11 P402-406 64-00266 <*>
1964 V73 N10 P361-363 66-11730 <*>

ZAHNAERZTLICHE WELT UND ZAHNAERZTLICHE REFORM
1964 V65 P720-721 66-11732 <*> O
1964 V65 N14 P456-460 66-11797 <*>

ZAIRYO
1964 V13 P35-43 66-12833 <*>
1964 V13 P354-357 66-12518 <*>
1965 V14 P101-110 66-12143 <*>

ZAKLAD APARATOW MATEMATYCZNYCH PRACA A. POLSKA AKADEMIA NAUK. WARSAW
1960 SB N4 P1-20 61-11305 <=>

ZAKONOMERNOSTI RAZMESHCHENIIA POLEZNYKH ISKOPAEMYKH
1958 P389-406 IGR V2 N6 P508 <AGI>
1958 V1 P123-141 IGR V2 N6 P461 <AGI>
1958 V1 P252-274 IGR V5 N4 P402 <AGI>
1958 V1 P487-516 IGR V6 N2 P317 <AGI>
IGR V6 N2 P317-343 <AGI>
1959 P136-146 IGR V4 N8 P908 <AGI>
1964 V7 P317-329 IGR V8 N2 P185 <AGI>

ZAKUPKI SELSKOKHOZIAISTVENNYKH PRODUKTOV
1965 N9 P1-3 65-33115 <=*$>

ZAPISKI GEOGRAFICHESKOGO OBSCHCHESTVA SSSR MOSCOW. (TITLE VARIES; PUBLICATION IRREGULAR)
1851 V5 P154-197 62-15980 (=*)
1867 V1 P329-347 62-25442 (=*)
1867 V1 P349-361 62-25442 (=*)
1885 V15 P1-44 62-25443 (=*)
1896 V30 N1 P1-97 62-15961 (=*)
1896 V30 N2 P1-78 62-15973 (=*)
1903 V30 N3 P1-50 62-15977 (=*)
1903 V30 N4 P1-84 62-15972 (=*)

ZAPISKI GEOGRAFICHESKOGO OBSHCHESTVA. ORENBURGSKII (CHKALOVSKII) OTDEL
1870 V1 P259-269 62-25444 (=*)
1871 V2 P259-280 61-28840 (=*)
1875 V3 P424-448 62-25445 (=*)
1881 V4 P26-28 61-28842 (=*)

ZAPISKI PO GIDROGRAFII
1897 V18 P92-148 64-19730 <=$>
1898 V19 P1-45 65-13555 <=*>
1899 V20 P37-52 65-13555 <=*>
1904 V26 P1-36 65-13555 <=*>
1904 V26 P214-258 65-13555 <=*>
1905 V27 P23-46 65-13555 <=*>
1907 V28 P311-359 65-13555 <=*>
1913 V37 P21-26 65-13555 <=*>
1915 V39 P760-773 65-13555 <=*>
1923 V47 P1-28 65-13555 <=*>
1925 V50 P95-114 65-13555 <=*>
1927 V52 P59-95 65-13555 <=*>
1927 V52 P527-532 65-13555 <=*>
1928 V54 P1-28 65-13556 <=*>
1928 V54 P59-85 65-13556 <=*>
1928 V54 P227-229 65-13556 <=*>
1929 V55 P59-70 65-13556 <=*>
1929 V57 P55-66 65-13556 <=*>
1932 N1 P37-58 65-13556 <=*>
1935 N2 P5-18 65-13556 <=*>
1935 N2 P35-59 65-13556 <=*>

1936 N3 P55-73 65-13556 <=*>
1937 N2 P323-326 65-13556 <=*>
1941 N2 P53-71 64-19562 <=$>

ZAPISKI INSTYTUTU KHIMIYI. AKADEMIYA NAUK URSR
1936 V3 N3 P387-400 R-2365 <*>
1937 V4 P99-110 R-2191 <*>
1937 V4 N3 P249- R-4554 <*>
1937 V4 N3 P269-274 61-16515 <=*$>
1938 V4 N4 P405-412 66-10724 <*>
1938 V5 P39-57 64-30092 <*> O
1938 V5 N1 P39-57 66-10836 <*>
1940 V7 N2 P159-172 RT-1507 <*>
1940 V7 N2 P173-175 RT-1911 <*>
1940 V7 N3 P375-382 RT-2016 <*>
64-10459 <=*$>
1946 V8 P23-31 3458-D <K-H>

ZAPISKI LENINGRADSKOGO GORNOGO INSTITUTA
1955 V30 N2 P44-117 60-11640 <=>
1958 V36 N3 P43-64 61-23662 <*=>
1958 V36 N3 P65-73 60-19041 <+*>
1963 V42 N3 P110-120 RTS-2851 <NLL>

ZAPISKI NAUCHNO-ISSLEDOVANIYA INSTITUTA MATEMATIKI I MEKHANIKI I KHARKOVSKOGO MATEMATICHESKOGO OBSHCHESTVA
1933 V4 P38-45 66-11021 <*>
1948 V19 N4 P35-120 RT-1505 <*>
1948 S4 V19 P35-120 AMST S1 V3 P1-78 <AMS>
1964 V30 P116-125 65-12711 <*>
1964 S4 V30 P164-174 65-12710 <*>

ZAPISKI ROSSIISKOGO MINERALOGICHESKOGO OBSHESTVA
1925 S2 V54 P359-376 R-4691 <*>
1925 S2 V54 N2 P377-383 R-4652 <*>

ZAPISKI SELSKO-KHOZYAISTVENNOGO INSTITUTA (IMPERATORA PETRA I) V VORONEZHE
1926 V5 P123-125 64-15159 <=*$>
1926 V5 P145-146 64-15159 <=*$>
1961 V81 P39-43 65-60947 <=$>

ZAPISKI VSEROSSIISKOGO MINERALOGICHESKOGO OBSHCHESTVA
1934 V63 P381-385 60-23738 <+*>
1938 V67 N3 P425-434 R-1160 <*>
1944 V73 N1 P62-73 41P62R <ATS>
1946 V75 N2 P89-93 <INSD>

ZAPISKI. VSESOYUZNOI GEOGRAFICHESKOE OBSHCHESTVO. NOVA SERIYA
1960 V20 P31-37 64-15185 <=*$>

ZAPISKI VSESOYUZNOGO MINERALOGICHESKOGO OBSHCHESTVA. LENINGRAD
1948 V77 N1 P43-54 RT-2042 <*>
1949 N4 P259-264 R-1158 <*>
1950 V79 N1 P15-22 RT-86 <*>
1950 V79 N2 P127-136 62-32960 <=*>
1950 V79 N3 P223-224 51/2155 <NLL>
1950 V79 N3 P224-225 62-13102 <*=>
1952 N2 P120-126 IGR V1 N12 P66 <AGI>
1954 V83 N3 P218-225 RJ-378 <ATS>
1955 V84 N2 P198-208 R-1673 <*>
1955 V84 N2 P253-259 R-26 <*>
1955 V84 N4 P470-492 AD-629 770 <=*$>
1956 V85 P58-74 64-10978 <=*$>
1956 V85 N2 P147-170 77N50R <ATS>
1956 V85 N2 P208-212 52J15R <ATS>
1956 V85 N2 P224-228 64-16046 <=*$>
1956 V85 N2 P228-232 63-23260 <=*$>
1956 V85 N3 P292-296 R-1606 <*>
1956 V85 N4 P491-497 63-10213 <=*>
65-60890 <=$>
1957 V85 N1 P49-57 RJ929 <ATS>
1957 V86 N1 P99-115 62-18525 <=*>
1957 V86 N3 P401-403 61-19909 <=*>
1957 V86 N4 P495-499 R-4656 <*>
1957 V86 N5 P607-621 R-4641 <*>
1957 V86 N5 P622-625 R-4679 <*>
1957 V86 N5 P626-628 R-4687 <*>
1957 V86 N6 P641-644 R-4658 <*>
1957 V86 N6 P657-670 CSIR-TR-243 <CSSA>
62-14576 <=*>
1957 V86 N6 P731-735 60-23627 <+*>
1957 S2 N5 P622-625 60-15702 <+*>
1958 V87 N5 P612-614 60-13458 <+*>
1958 V87 N6 P647-656 61-27603 <*=>
1958 S2 V87 N1 P494-496 59-11374 <=>
1959 N4 P372-373 PT88 IGR V2 N9 P763 <AGI>
1959 V88 N1 P13-20 61-27604 <*=>
1959 V88 N1 P39-47 76M38R <ATS>
1959 V88 N1 P48-59 60-23751 <+*>
1959 V88 N1 P97-99 60-21589 <=>
1959 V88 N3 P234-239 CSIR-TR.199 <CSSA>
1959 V88 N3 P261-274 IGR V3 N6 P463 <AGI>
61-15005 <*+>
1959 V88 N5 P497-511 64-14037 <=*$>
1959 V88 N5 P564-570 61-00782 <*>
1959 V88 N6 P667-671 IGR V3 N1 P1 <AGI>
1959 V88 N6 P693-699 56R80R <ATS>
1959 S2 N5 P497-511PT88 IGR V3 N8 P694 <AGI>
1959 S2 N6 P637-654 PT88 IGR V2 N6 P476 <AGI>
1960 N6 P714-718 65-63991 <=$>
1960 V89 P663-668 RTS-2708 <NLL>
1960 V89 N1 P26-36 61-15206 <*+>
1960 V89 N4 P392-399 61-23000 <=*>
1960 V89 N4 P440-447 61-23016 <=*>
1960 V89 N4 P458-460 63-15977 <=*>
1960 V89 N4 P460-464 64-19832 <=$>
1960 V89 N4 P464-466 61-23017 <=*>
1961 N6 P695-697 PT90 IGR V5 N4 P418 <AGI>
1961 V90 N1 P42-49 61-28898 <*=>
1961 V90 N1 P50-90 62-13330 <*=>
1961 V90 N2 P172-192 62-13741 <*=>
1961 V90 N3 P270-280 63-19778 <=*$> PO
1961 V90 N4 P472-478 63-13870 <=*> O
1961 V90 N5 P500-509 66-10300 <*>
1961 V90 N6 P637-642 IGR,V6,N3,P430-434 <AGI>
1961 S2 N6 P637-642 PT90 IGR V6 N3 P430 <AGI>
1962 V91 N1 P123-124 64-10973 <=*$>
1962 V91 N4 P472-477 64-10974 <=*$>
1962 V91 N4 P487-489 64-10988 <=*$>
1962 V91 N5 P520-536 64-13381 <=*$>
1962 V91 N6 P637-664 65-60485 <=$>
1963 V92 N1 P33-50 RTS-2712 <NLL>
1963 V92 N1 P81-84 63-23600 <=*>
1963 V92 N2 P173-183 64-23526 <=$>
1963 V92 N2 P228-231 63-24622 <=*$>
1963 V92 N4 P499-508 64-21223 <=>
1964 V93 P37-45 GI N2 1964 P365 <AGI>
1964 V93 P357-360 GI N2 1964 P361 <AGI>
1964 V93 N3 P329-334 65-13134 <*>
1965 V94 N1 P10-27 IGR V8 N6 P705 <AGI>
1965 V94 N1 P28-40 IGR V8 N5 P581 <AGI>
1965 V94 N1 P83-104 IGR V8 N5 P608 <AGI>
1965 V94 N3 P253-257 IGR V8 N6 P652 <AGI>
1965 V94 N3 P258-271 IGR V8 N7 P770 <AGI>
1965 V94 N3 P272-287 IGR V8 N6 P689 <AGI>
1965 V94 N5 P571-573 66-14403 <*>
1965 V94 N5 P613-618 66-14124 <*$>
1966 V95 N1 P118-120 66-34786 <=$>

ZAPISKI VSESOYUZNOE MINERALOGICHESKOE OBSHCHESTVO. UKRAINSKOE OTDELENIE
1962 V1 P60-66 64-14016 <*> O

ZASHCHITA RASTENII OT VREDITELEI

Citation	Translation
1935 N1 P5-8	RT-1756 <*>
1935 N2 P5-14	RT-2436 <*>
1938 V17 P119-121	PANSDOC-TR.512 <PANS>

ZASHCHITA RASTENII OT VREDITELEI I BOLEZNEI

Citation	Translation
1956 N4 P6-7	R-1558 <*>
1956 V1 N2 P3-5	R-1533 <*>
1956 V1 N2 P5-7	R-1531 <*>
1956 V1 N2 P11-12	R-1525 <*>
1956 V1 N2 P19-21	R-1526 <*>
1956 V1 N2 P63-	R-1532 <*>
1956 V1 N3 P3-4	R-1538 <*>
1956 V1 N3 P13-15	R-1539 <*>
1956 V1 N3 P29-31	R-1543 <*>
1956 V1 N5 P3-4	R-1577 <*>
1956 V1 N5 P17-20	R-1580 <*>
1956 V1 N5 P27-	R-1579 <*>
1957 N1 P3-6	R-1567 <*>
1957 N1 P47-	R-1563 <*>
1957 N2 P42	R-1568 <*>
1957 N1 P46-	R-1564 <*>
1957 V2 N2 P18-20	R-1814 <*>
1957 V2 N2 P22-	R-1815 <*>
1957 V2 N2 P35-36	R-1816 <*>
1957 V2 N2 P60	R-3173 <*>
1957 V2 N3 P53-55	61-14953 <=*>
1957 V2 N5 P50-51	60-23168 <+*>
1958 N3 P6-7	R-5212 <*>
1958 N3 P20-21	R-5211 <*>
1958 N3 P44-46	R-5220 <*>
1958 N3 P57	R-5200 <*>
1958 N3 P61	R-5219 <*>
1958 V3 N2 P15-17	R-4414 <*>
1958 V3 N2 P28-31	R-4416 <*>
1958 V3 N2 P42-43	R-4417 <*>
1958 V3 N3 P34-35	60-15884 <+*>
1958 V3 N4 P17-18	59-18160 <+*>
1958 V3 N4 P59	60-17981 <+*>
1958 V3 N5 P51-53	59-22587 <+*>
1959 N4 P20-23	60-17269 <+*>
1959 V4 N1 P18-19	59-16807 <+*> 0
1959 V4 N1 P33	59-16865 <+*>
1959 V4 N1 P34-35	59-16808 <+*>
1959 V4 N1 P36-37	59-16809 <+*>
1959 V4 N1 P56-58	59-16810 <+*>
1959 V4 N2 P4-6	59-18275 <+*> 0
1959 V4 N2 P10-12	59-18276 <+*>
1959 V4 N2 P12	59-18277 <+*>
1959 V4 N2 P15-17	59-22306 <+*>
1959 V4 N2 P18	59-18278 <+*>
1959 V4 N2 P21	59-22309 <+*> 0
1959 V4 N2 P27-29	59-22310 <+*>
1959 V4 N2 P29-30	59-22311 <+*>
1959 V4 N2 P33-34	59-22312 <+*>
1959 V4 N2 P34-37	59-22313 <+*>
1959 V4 N2 P46	59-22314 <+*>
1959 V4 N2 P54-55	59-22315 <+*>
1959 V4 N2 P56	59-22316 <+*>
1959 V4 N3 P19-22	60-15340 <+*>
1959 V4 N3 P30-32	60-15334 <+*> 0
1959 V4 N3 P43-44	60-15335 <+*>
1959 V4 N3 P45-47	60-15336 < > 0
1959 V4 N3 P53	60-15337 <+*>
1959 V4 N3 P61	60-15338 <+*>
1959 V4 N3 P63	60-15339 <+*>
1959 V4 N5 P1-3	60-17264 <+*>
1959 V4 N5 P17-18	60-17268 <+*>
1959 V4 N5 P25-26	60-17265 <+*>
1959 V4 N5 P29-32	60-17270 <+*>
1959 V4 N5 P32-33	60-17271 <+*>
	61-11031 <=>
1959 V4 N5 P38-40	61-11031 <=>
1959 V4 N5 P48	61-11031 <=>
1959 V4 N6 P1-3	60-17252 <+*>
1959 V4 N6 P4-5	61-19719 <=*>
1959 V4 N6 P15-17	61-11031 <=>
1959 V4 N6 P17-18	60-15351 <+*> 0
	60-17636 <+*>
1959 V4 N6 P22-23	60-17260 <+*>
	61-11031 <=>
1959 V4 N6 P37-38	61-11031 <=>
1959 V4 N6 P41	60-15359 <+*>
1959 V4 N6 P41-42	60-17253 <+*>
	61-11031 <=>
1959 V4 N6 P42	60-17266 <+*>
1959 V4 N6 P47-48	60-17254 <+*>
1959 V4 N6 P56-57	60-17256 <+*>
1959 V4 N6 P58	60-11441 <=>
1960 V5 N1 P18-22	60-17279 <+*>
1960 V5 N1 P30-33	60-17276 <+*>
1960 V5 N1 P41-42	60-17278 <+*>
1960 V5 N1 P45-46	60-17272 <+*> 0
1960 V5 N1 P53	60-17273 <+*>
1960 V5 N1 P59-60	60-11982 <=>
	60-17274 <+*>
1960 V5 N1 P62	60-11982 <=>
1960 V5 N2 P16-17	60-17280 <+*>
1960 V5 N2 P19	60-17275 <+*>
1960 V5 N2 P29-31	60-17281 <+*> 0
1960 V5 N2 P41-45	60-17283 <+*>
1960 V5 N2 P45-47	60-17284 <+*> 0
1960 V5 N3 P13	60-17940 <+*>
	60-31640 <=>
1960 V5 N3 P15-18	60-31640 <=>
1960 V5 N3 P18-19	60-17941 <+*>
1960 V5 N3 P19-20	60-17939 <+*>
1960 V5 N3 P21-22	60-17944 <+*>
1960 V5 N3 P22-23	60-17945 <+*>
1960 V5 N3 P29-31	60-17946 <+*>
1960 V5 N3 P33-36	60-17942 <+*>
1960 V5 N3 P51-53	60-17947 <+*>
1960 V5 N3 P54	60-17948 <+*>
	60-31640 <=>
1960 V5 N3 P60-61	60-11660 <=>
	60-17949 <+*> 0
1960 V5 N3 P61	60-17943 <+*>
	60-31640 <=>
1960 V5 N3 P62-63	60-17950 <+*> 0
1960 V5 N3 P62	60-17959 <+*>
1960 V5 N4 P1-9	61-11629 <=>
1960 V5 N5 P3-7	60-23712 <+*> C
1960 V5 N5 P28	60-23713 <=*>
1960 V5 N5 P30-31	60-23710 <+*>
1960 V5 N5 P43-45	60-23152 <+*> C
1960 V5 N5 P45-46	60-23159 <+*>
1960 V5 N5 P52	60-23160 <+*> 0
1960 V5 N5 P54	60-23714 <+*>
1960 V5 N5 P57-61	60-23715 <+*> 0
1960 V5 N6 P41-44	61-13781 <*+>
1960 V5 N6 P52	61-13778 <*+>
1960 V5 N6 P54-55	62-15666 <*=>
1960 V5 N6 P55-57	61-13779 <*+>
	62-15475 <*=>
1960 V5 N6 P58	62-15664 <*=>
1960 V5 N6 P58-59	62-15677 <*=>
1960 V5 N6 P61	61-28720 <=>
1960 V5 N6 P62	61-13780 <*+>
1960 V5 N7 P6-8	61-15836 <+*>
1960 V5 N7 P15-16	61-15840 <+*> C
1960 V5 N7 P21-23	61-15841 <+*>
1960 V5 N7 P36-38	61-15839 <+*>
1960 V5 N7 P55-59	61-15848 <+*>
1960 V5 N8 P1-5	61-11551 <=>
1960 V5 N8 P22-23	61-11821 <=>
1960 V5 N8 P38-39	61-11821 <=>
1960 V5 N9 P58-60	61-27364 <*=>
1960 V5 N10 P1-3	61-15831 <+*>
	62-13424 <*=>
1960 V5 N10 P18-21	61-15830 <+*> 0
	62-13434 <*=>
1960 V5 N10 P30-33	61-15832 <+*>
	62-13442 <*=>
1960 V5 N10 P33	62-13439 <*=>
1960 V5 N11 P4-7	61-15833 <+*>
	61-28248 <*=>
1960 V5 N11 P24-27	61-19658 <=*> 0
1960 V5 N11 P34-36	61-19659 <=*>
1960 V5 N11 P36-38	61-19660 <=*> C
1960 V5 N11 P58-59	61-19661 <=*>

Citation	Translation
	61-28284 <*=>
1960 V5 N11 P60-61	61-28249 <*=>
1960 V5 N11 P62-63	61-28720 <=>
1960 V5 N12 P49-55	62-15481 <*=>
1960 V5 N12 P53-54	62-15686 <*=>
1961 V6 N1 P3-6	61-23121 <=>
1961 V6 N1 P10-14	61-23121 <=>
1961 V6 N1 P19-21	61-23121 <=>
1961 V6 N1 P36-40	61-23121 <=>
1961 V6 N1 P41-44	61-19657 <=*>
1961 V6 N1 P50-52	61-23121 <=>
1961 V6 N1 P59-63	61-23121 <=>
1961 V6 N2 P1-2	61-19755 <=*>
1961 V6 N2 P8-11	62-19755 <=*>
1961 V6 N2 P11	61-23194 <=*>
1961 V6 N2 P14-15	61-27480 <*=>
1961 V6 N2 P19	61-23195 <=*> 0
1961 V6 N2 P33-37	61-23199 <=*>
	62-19755 <=*>
1961 V6 N2 P38	61-23200 <*=>
1961 V6 N2 P48-56	62-19755 <=*>
1961 V6 N2 P51-56	61-27490 <*=> 0
1961 V6 N3 P10-11	62-15471 <*=>
1961 V6 N3 P36-38	62-15470 <*=>
1961 V6 N3 P46	62-15668 <*=>
1961 V6 N3 P47	62-15778 <*=>
1961 V6 N3 P57-58	62-15671 <*=>
1961 V6 N3 P57	62-15676 <*=>
1961 V6 N4 P18-19	61-28436 <*=>
1961 V6 N4 P41-47	62-25487 <=>
1961 V6 N4 P57-60	61-28714 <*=> 0
1961 V6 N4 P57-61	62-23604 <=*>
1961 V6 N4 P60-61	61-28438 <*=>
1961 V6 N5 P1-3	62-25981 <=*>
1961 V6 N5 P55-57	62-25487 <=>
1961 V6 N5 P59	62-25487 <=>
1961 V6 N6 P11-13	62-15243 <=>
1961 V6 N6 P34-35	62-15243 <=>
1961 V6 N6 P41-42	62-15243 <=>
1961 V6 N6 P52-54	62-15243 <=>
1961 V6 N7 P1-2	62-25487 <=>
1961 V6 N7 P58-59	62-25487 <=>
1961 V6 N11 P23-24	62-32893 <=*>
1961 V6 N11 P54-55	62-25037 <=*>
1962 N4 P63	NLLTB V4 N10 P936 <NLL>
1962 V7 N1 P40-41	62-23585 <=*>
1962 V7 N1 P49-50	63-19052 <=*> 0
1962 V7 N5 P58-60	63-13279 <=*>
1962 V7 N6 P34-36	63-15711 <=*>
1962 V7 N6 P45	63-19053 <=*> 0
1962 V7 N6 P47-50	64-11079 <=>
1962 V7 N9 P20-23	63-13592 <=*> 0
1963 V8 N2 P34-35	63-11637 <=>
1963 V8 N6 P35-36	AD-638 587 <=$>
1963 V8 N10 P10-13	64-21105 <=>
1964 V9 N3 P29-30	64-19130 <=*$>
1964 V9 N7 P40-41	AD-625 624 <=*$>
1964 V9 N7 P53	AD-619 000 <=$>
1964 V9 N10 P35	65-63215 <=$>
1965 V10 N3 P26-30	65-31216 <=$>

ZASTITA BILJA

Citation	Translation
1954 N23 P44-62	60-21679 <=>
1954 N24 P79-86	60-21679 <=>
1958 N49/0 P113-119	60-21688 <=>
1959 V52 P37-44	60-21662 <=>
1959 V52 P79-87	60-21662 <=>

ZASTITA MATERIJALA

Citation	Translation
1956 V4 N6 P193-197	66-10334 <*>

ZASTOSOWANIA MATEMATYKI

Citation	Translation
1958 V1 P142-168	SCL-T-507 <*>
1958 V3 P307-322	STMSP V3 P225-238 <AMS>

*ZAVODSKAYA LABORATORIYA

Citation	Translation
1934 V3 P29-37	R-2919 <*>
1934 V3 P593-597	R-2210 <*>
1934 V3 P623-625	64-18562 <*>
1934 V3 P1105-1110	R-2548 <*>
1934 V3 N12 P1105-1110	40L30R <ATS>
1936 V5 P947-952	R-4904 <*>
1936 V5 P1071-1072	R-2275 <*>
1936 V5 P1443-1444	48L29R <ATS>
1936 V5 N9 P1071-1072	63-16159 <=*>
1937 V6 P47-51	R-2511 <*>
1937 V6 P432-434	R-5013 <*>
1937 V6 P434-438	R-2879 <*>
1937 V6 P1018	R-2825 <*>
1937 V6 N10 P1218-1220	<INSD>
1937 V6 N11 P1376-1382	RT-860 <*>
1937 V6 N11 P1399-1402	RT-2360 <*>
1938 V7 P24-29	R-2445 <*>
1938 V7 P314-321	59-10458 <+*>
1938 V7 P395-399	63-19697 <=*>
1938 V7 P622-623	63-18120 <=*>
1938 V7 P891-892	R-2214 <*>
1938 V7 P934-936	R-3306 <*>
1938 V7 N1 P29-32	RT-2003 <*>
1938 V7 N2 P1109-1116	RT-3927 <*>
1938 V7 N3 P314-321	RT-461 <*>
1938 V7 N7 P800-801	4637 <HB>
1939 V8 P98-101	R-2298 <*>
1939 V8 P832-836	PANSDOC-TR.450 <PANS>
1939 V8 P916-920	R-2447 <*>
1939 V8 P940-943	R-2924 <*>
1939 V8 N8 P836-837	RT-3241 <*>
1939 V8 N10/1 P1179-1181	R-3311 <*>
1940 V9 P400-407	R-2902 <*>
1940 V9 P1001-1008	R-185 <*>
1940 V9 N2 P229-230	R-3305 <*>
1940 V9 N4 P436-444	62-24183 <=*>
1940 V9 N8 P860-864	61-20362 <*=>
	62-14149 <=*>
1940 V9 N10 P1096-1101	61-20352 <*=>
1941 V10 P462-464	57-1514 <*>
1941 V10 P626-627	R-2063 <*>
1941 V10 N1 P3-4	61-13435 <=*>
1941 V10 N1 P39-42	61-16882 <*=>
1941 V10 N1 P43-46	61-20351 <*=>
1941 V10 N1 P46-50	61-20350 <*=>
1941 V10 N1 P51-61	RT-3242 <*>
1941 V10 N2 P170-176	RT-3243 <*>
1945 V11 P685-694	RT-3240 <*> 0
1945 V11 N6 P504-511	63M45R <ATS>
1945 V11 N6 P567-569	62-00194 <*>
1945 V11 N7 P267-269	T.455 <INSD>
1945 V11 N10 P887-893	2711 <HB>
1946 V12 P203-205	61-20412 <*=>
1946 V12 P286-291	59-10287 <+*>
1946 V12 P908-914	59-19081 <+*>
1946 V12 N2 P161-170	TT.174 <NRC>
1946 V12 N6 P559-568	R-2064 <*>
1946 V12 N6 P569-573	R-2065 <*>
1946 V12 N4/5 P418-421	RT-2451 <*>
1946 V12 N4/5 P500-502	RT-1020 <*>
1946 V12 N7/8 P646-650	3540 <HB>
1946 V12 N11/2 P933-937	RT-1862 <*>
1946 V12 N9/10 P777-793	63-23297 <=*$> P
1947 V13 P680-682	AEC-TR-5250 <=*>
1947 V13 N1 P71-77	RT-1717 <*>
1947 V13 N3 P301-303	3257 <HB>
1947 V13 N4 P404-410	RT-4235 <*>
1947 V13 N5 P521-532	RT-2779 <*>
1947 V13 N5 P532-539	TT.222 <NRC>
1947 V13 N5 P539-547	TT.223 <NRC>
1947 V13 N5 P547-548	TT.224 <NRC>
1947 V13 N5 P565-568	TT.225 <NRC>
1947 V13 N5 P621-623	TT.226 <NRC>
1947 V13 N7 P888-889	R-2078 <*>
1947 V13 N9 P1038-1043	3023 <HB>
1948 V14 P394-396	64-20556 <*> 0
1948 V14 P625-627	65-13254 <*>
1948 V14 P1328-1335	1504TM <CTT>
1948 V14 N2 P131-143	R-4508 <*>
1948 V14 N2 P131-137	TT.206 <NRC>
1948 V14 N2 P138-143	TT-172 <NRC>
1948 V14 N2 P144-148	63-20445 <=*$>
1948 V14 N2 P202-203	2741 <HB>
1948 V14 N3 P326-330	61-19438 <+*> 0

Citation	Reference
1948 V14 N3 P349-350	3141 <HB>
1948 V14 N3 P354	65-60568 <=$>
1948 V14 N4 P503-504	3737 B <K-H> 0
	64-16116 <*> 0
1948 V14 N5 P549-555	3457 <HB>
1948 V14 N5 P555-557	3567 <HB>
1948 V14 N5 P623-624	R.597 <RIS>
1948 V14 N6 P756-758	RT-678 <*>
1948 V14 N7 P783-787	65-61314 <=$>
1948 V14 N7 P787	61-13646 <=*>
1948 V14 N7 P824-825	37 <FRI>
1948 V14 N7 P882-883	61-13436 <+*> 0
1948 V14 N8 P1017-	R-519 <*>
1948 V14 N9 P1070-1079	TT.238 <NRC>
1948 V14 N11 P1328-1335	3029 <HB>
1949 N11 P1365-1366	RJ-47 <ATS>
1949 N15 P338-345	RJ-311 <ATS>
1949 V15 N2 P167-170	2602 <HB>
1949 V15 N2 P191-199	60-13579 <=*>
1949 V15 N2 P248-249	R-184 <*>
1949 V15 N3 P338-345	RJ-311 <ATS>
1949 V15 N5 P523-528	63-18206 <=*>
1949 V15 N5 P542-547	50/1899 <NLL>
1949 V15 N5 P554-557	50/1880 <*>
1949 V15 N5 P570-571	2318 <HB>
1949 V15 N6 P665-669	64-13262 <=*$>
1949 V15 N6 P688-691	3214 <HB>
1949 V15 N7 P797-799	65-11467 <*>
1949 V15 N7 P799-805	3061 <HB>
1949 V15 N7 P870-872	2516 <HB>
1949 V15 N8 P912-914	50/2971 <NLL>
1949 V15 N8 P919-936	3535 <HB>
1949 V15 N8 P957-961	61-23450 <=*> 0
1949 V15 N9 P1027-1030	50/2863 <NLL>
1949 V15 N9 P1031-1034	50/2973 <NLL>
1949 V15 N9 P1034-1038	RCT V25 N1 P152 <RCT>
1949 V15 N9 P1039-1042	RCT V25 N1 P157 <RCT>
1949 V15 N10 P1143-1149	62-18724 <=*>
1949 V15 N10 P1152-1157	758TM <CTT>
1949 V15 N10 P1157-1161	R-1898 <*>
1949 V15 N11 P1247-1248	62-25319 <=*>
1949 V15 N11 P1308-1313	65-60089 <=$>
1949 V15 N11 P1365-1366	R-2061 <*>
	RJ-47 <ATS>
1949 V15 N11 P1367-	R-2060 <*>
1949 V15 N11 P1367	RJ-46 <ATS>
1949 V15 N11 P1367-	RJ-46 <ATS>
1949 V15 N12 P1470	RT-2250 <*>
1949 V15 N12 P1496-1499	50/2306 <NLL>
1950 N2 P168-173	R-3845 <*>
1950 V16 P560-565	PANSDOC-TR.379 <PANS>
	65-17004 <*> 0
1950 V16 P965	AEC-TR-4746 <*>
1950 V16 P1108-1111	59-20569 <+*>
1950 V16 P1136-1139	AEC-TR-4646 <*=>
1950 V16 P1299-1301	64-20329 <*> 0
1950 V16 P1432	59-17845 <+*>
1950 V16 N1 P17-23	3103 <HB>
1950 V16 N1 P24-28	64-20378 <*>
1950 V16 N2 P168-173	2724 <HB>
1950 V16 N3 P283-290	3548 <HB>
1950 V16 N4 P422-429	R-4519 <*>
	RT-1853 <*> 0
1950 V16 N4 P430-438	3203 <*>
1950 V16 N5 P578-579	3193 <HB>
1950 V16 N5 P621-622	64-14287 <=*$>
1950 V16 N5 P632-633	3009 <HB>
1950 V16 N6 P648-650	3164 <HB>
1950 V16 N6 P666-668	2985 <HB>
1950 V16 N7 P777-780	2714 <HB>
1950 V16 N7 P787-792	4036 <HB>
1950 V16 N7 P833-835	RT-2100 <*>
1950 V16 N7 P863-869	61-19397 <=*>
1950 V16 N8 P918-923	2713 <HB>
1950 V16 N8 P955-962	33 <FRI>
1950 V16 N8 P1011-1012	T.958 <INSD>
1950 V16 N9 P1131	51/3367 <NLL>
1950 V16 N10 P1173-1182	3526 <HB>
1950 V16 N10 P1182-1185	3155 <HB>
1950 V16 N11 P1331-1335	R-1227 <*>
1950 V16 N12 P1494-1495	3144 <HB>
1950 V16 N12 P1498-1500	RT-2251 <*>
1952 V18 P764	66-10330 <*>
1952 V18 P891-892	66-10331 <*>
1953 V19 N6 P660-661	62-25565 <=*>
1953 V19 N9 P1033-1034	TS 1721 <BISI>
	62-25701 <=*>
1955 N5 P557	GB70 4088 <NLL>
1955 V21 P152-154	1161 <*>
1955 V21 P841-844	TS 1261 <BISI>
1955 V21 P1482-1483	R-3660 <*>
1955 V21 N1 P17-20	RJ-401 <ATS>
1955 V21 N1 P29-30	3654 <HB>
1955 V21 N1 P62-66	UCRL TRANS-662(L) <*=>
1955 V21 N1 P82-84	3721 <HB>
1955 V21 N2 P152-154	RT-3821 <*>
	62-14041 <=*>
1955 V21 N2 P162-163	R-305 <*>
	R-3628 <*>
	3090 <NLL>
1955 V21 N2 P200-201	4934 <HB>
1955 V21 N2 P212-216	4210 <HB>
1955 V21 N2 P240-241	2214 <BISI>
1955 V21 N3 P259-260	RT-3053 <*>
1955 V21 N3 P281-285	62-11225 <=>
1955 V21 N3 P294-298	RT-3394 <*>
1955 V21 N3 P361-364	R-18 <*>
1955 V21 N3 P371-372	CSIRO-3266 <CSIR>
1955 V21 N4 P396-397	RCT V29 N2 P509 <RCT>
1955 V21 N4 P432-436	3683 <HB>
1955 V21 N4 P443-447	4469 <HB>
1955 V21 N4 P464-466	3608 <HB>
1955 V21 N4 P497-498	CSIRO-3267 <CSIR>
	R-3907 <*>
1955 V21 N5 P579-590	61-19389 <=*>
1955 V21 N5 P612-613	4420 <HB>
1955 V21 N6 P672-675	R-3145 <*>
1955 V21 N6 P678-692	4091 <BISI>
1955 V21 N7 P791-	<ATS>
1955 V21 N7 P808-812	3602 <HB>
1955 V21 N7 P813-820	3568 <HB>
1955 V21 N7 P828-830	3678 <HB>
1955 V21 N7 P837-840	R-3252 <*>
1955 V21 N7 P841-844	62-24206 <=*>
1955 V21 N9 P1109-1110	R-3112 <*>
1955 V21 N9 P1111-1127	3682 <HB>
1955 V21 N10 P1149-1157	R-4983 <*>
1955 V21 N10 P1160-1163	64-16478 <=*$> 0
1955 V21 N10 P1182-1188	09M40R <ATS>
1955 V21 N10 P1224-1229	61-11425 <=>
1955 V21 N10 P1238-1241	59-16732 <+*>
1955 V21 N11 P1285-1288	62-24165 <=*>
1955 V21 N11 P1374-1377	CSIRO-3128 <CSIR>
	R-2332 <*>
1955 V21 N12 P1419-1421	35H12R <ATS>
1955 V21 N12 P1487-1498	61-19385 <+*>
1956 N1 P49-52	R-4466 <*>
1956 N2 P145-154	T2043 <INSD>
1956 N3 P336-340	RT-4480 <*>
1956 N6 P637-639	R-5285 <*>
1956 V22 P316-317	R-3432 <*>
1956 V22 P444-447	62-20100 <=*>
1956 V22 P519-523	766TM <CTT>
1956 V22 P920	60-13569 <+*>
1956 V22 N1 P25-28	64-19112 <=*$>
1956 V22 N1 P30-35	62-32468 <=*>
1956 V22 N1 P46-47	4003 <HB>
1956 V22 N1 P77-80	61-10315 <+*> 0
1956 V22 N1 P80-82	4493 <HB>
1956 V22 N2 P157-160	RJ-529 <ATS>
1956 V22 N2 P161-162	C-3173 <NRCC>
1956 V22 N2 P166-167	3896 <HB>
1956 V22 N2 P169-177	R-5291 <*>
1956 V22 N2 P196-198	3905 <HB>
1956 V22 N2 P218-225	59-10179 <+*> 0
1956 V22 N3 P262-270	R-2081 <*>
1956 V22 N3 P316-317	4629 <HB>
1956 V22 N3 P317	3891 <HB>
1956 V22 N3 P329-331	4515 <HB>
1956 V22 N3 P333-334	60-51037 <=>

Citation	Translation
1956 V22 N3 P345-349	R-4347 <*>
	62-18845 <=*>
1956 V22 N4 P417-418	3919 <HB>
1956 V22 N4 P418	3897 <HB>
1956 V22 N4 P444-447	RJ-632 <ATS>
1956 V22 N4 P472-478	60-13689 <+*>
1956 V22 N4 P482-484	4026 <HB>
1956 V22 N4 P499-501	RJ-642 <ATS>
	3841 <HB>
1956 V22 N5 P525-527	3917 <HB>
1956 V22 N5 P543	3892 <HB>
1956 V22 N5 P556-558	3920 <HB>
1956 V22 N5 P558-560	3936 <HB>
1956 V22 N5 P571-572	3903 <HB>
1956 V22 N5 P573-579	59-18158 <+*>
1956 V22 N5 P584	59-22617 <+*>
1956 V22 N5 P586-588	62-24115 <=*>
1956 V22 N5 P613-614	59-10862 <+*>
1956 V22 N6 P637-639	3990 <HB>
1956 V22 N6 P640-645	2218 <BISI>
	59-11632 <=> 0
	64-18387 <*> 0
1956 V22 N6 P668-673	3899 <HB>
1956 V22 N6 P682-688	3866 <HB>
1956 V22 N6 P688-691	AEC-TR-4276 <*+>
	61-13228 <+*>
1956 V22 N7 P795-801	43J16R <ATS>
1956 V22 N7 P805-806	64-16509 <=*$>
1956 V22 N7 P810-811	4012 <HB>
1956 V22 N7 P840-845	TRANS-58 <MT>
1956 V22 N7 P845-849	UCRL-TRANS- 475L <+*>
	4741 <HB>
1956 V22 N7 P855-856	4694 <HB>
1956 V22 N7 P856-857	60-13046 <+*>
1956 V22 N7 P867-869	4131 <HB>
1956 V22 N7 P869-871	59-19612 <+*>
1956 V22 N8 P907-911	63-23313 <=*>
1956 V22 N8 P967-972	2431 <BISI>
1956 V22 N8 P993-995	59-19619 <+*> 0
1956 V22 N8 P999-1000	4206 <HB>
1956 V22 N9 P1019-1024	C-3168 <NRCC>
1956 V22 N9 P1061-1063	4512 <HB>
1956 V22 N9 P1086-1089	AD-610 960 <=$>
	C-2539 <NRC>
	R-3139 <*>
1956 V22 N10 P1171-1180	AEC-TR-3461 <+*>
1956 V22 N10 P1182-1185	4595 <HB>
1956 V22 N10 P1202-1204	59-22528 <+*>
1956 V22 N10 P1204-1205	3987 <HB>
1956 V22 N10 P1235-1240	AEC-TR-3829 <+*>
1956 V22 N10 P1249-1250	3914 <HB>
1956 V22 N11 P1297-1302	ANL-TRANS-118 <*>
1956 V22 N11 P1303-1306	R-4629 <*>
1956 V22 N11 P1358-1363	4138 <HB>
	62-13208 <=*>
1956 V22 N12 P1406-1407	61-10573 <*+>
1956 V22 N12 P1408-1409	61-10572 <*+>
1956 V22 N12 P1415-1417	61-10377 <*+> 0
1956 V22 N12 P1417-1419	35L30R <ATS>
1956 V22 N12 P1430-1435	4302 <HB>
1956 V22 N12 P1458-1460	4507 <HB>
1956 V22 N12 P1463-1467	64-14852 <=*$>
1956 V22 N12 P1487-1489	60-10309 <+*>
1956 V22 N12 P1491-1492	3915 <HB>
1957 N23 P305-308	R-4357 <*>
1957 V23 P26-	R-1644 <*>
1957 V23 P1066-1067	UCRL TRANS-896(L) <*>
1957 V23 P1153-1161	58-1741 <*>
1957 V23 P1187-1201	R-4927 <*>
1957 V23 N1 P11-14	4039 <HB>
1957 V23 N1 P35-41	N66-22303 <=$>
1957 V23 N1 P64-67	4099 <HB>
1957 V23 N1 P67-69	4382 <HB>
1957 V23 P1292-1294	4290 <HB>
1957 V23 P1327-1328	65-13057 <*>
1957 V23 N2 P140-143	62-24091 <=*>
1957 V23 N2 P153-157	M-5657 <NLL>
1957 V23 N2 P161-162	63-14127 <=*>
1957 V23 N2 P175-184	R-3852 <*>
1957 V23 N3 P263-269	3963 <HB>
1957 V23 N3 P269-272	3959 <HB>
1957 V23 N3 P279-280	3960 <HB>
1957 V23 N3 P305-308	3958 <HB>
1957 V23 N3 P309-311	3968 <HB>
1957 V23 N3 P316-318	61-13226 <*+> 0
1957 V23 N3 P329-332	52K24R <ATS>
1957 V23 N4 P421-422	TS 1642 <BISI>
	62-14655 <=*>
	62-25877 <=*>
1957 V23 N4 P476-484	4232 <HB>
1957 V23 N4 P485-487	4594 <HB>
1957 V23 N5 P533-535	60-19891 <+*>
1957 V23 N5 P537-539	R-3716 <*>
	4239 <HB>
1957 V23 N5 P545-548	4472 <HB>
1957 V23 N5 P564-569	60-11307 <=>
1957 V23 N6 P647-652	61-19668 <=*>
1957 V23 N6 P656-657	5084 <HB>
1957 V23 N6 P679-682	4461 <HB>
1957 V23 N6 P683-686	4466 <HB>
1957 V23 N6 P697-698	4032 <HB>
1957 V23 N6 P699-702	4041 <HB>
1957 V23 N7 P798-800	RCT V35 N2 P498 <RCT>
1957 V23 N7 P803-806	4418 <HB>
1957 V23 N7 P806-808	4419 <HB>
1957 V23 N7 P811-813	4065 <HB>
1957 V23 N7 P821-826	62-18274 <=*>
1957 V23 N7 P858-860	5112 <HB>
1957 V23 N8 P896-900	59-10262 <+*>
1957 V23 N8 P905-907	4133 <HB>
1957 V23 N8 P911-912	4923 <HB>
1957 V23 N8 P925-927	AEC-TR-3613 <+*>
1957 V23 N8 P931	N66-10725 <=*$>
1957 V23 N8 P949-953	61-23157 <=*> 0
1957 V23 N8 P959-961	61-23157 <=*> C
1957 V23 N8 P974-975	4845 <HB>
1957 V23 N9 P1066-1067	UCRL-TRANS-896(L) <=*>
1957 V23 N9 P1088-1091	4825 <HB>
1957 V23 N9 P1093-1094	R-3896 <*>
1957 V23 N9 P1098-1101	4671 <HB>
1957 V23 N9 P1120-1124	64-13256 <=*$> 0
1957 V23 N9 P1126-1127	4292 <HB>
1957 V23 N10 P1162-1167	59-14739 <+*>
1957 V23 N10 P1168-1175	R-4794 <*>
1957 V23 N10 P1202-1211	R-4793 <*>
1957 V23 N10 P1214-1219	R-4791 <*>
1957 V23 N10 P1228-1234	R-4789 <*>
1957 V23 N11 P1278-1283	61-19557 <=*>
1957 V23 N11 P1283-1286	65-61411 <=>
1957 V23 N11 P1362-1365	61-19386 <+*> C
1957 V23 N11 P1391-1392	4264 <HB>
1957 V23 N12 P1410-1412	4130 <HB>
1957 V23 N12 P1412-1413	4132 <HB>
1957 V23 N12 P1420-1424	AEC-TR-3631 <+*>
1957 V23 N12 P1438-1439	R-5005 <*>
1957 V23 N12 P1448-1451	AEC-TR-5280 <=*>
1957 V23 N12 P1462-1466	21L33R <ATS>
1958 V24 N1 P26-29	4204 <HB>
1958 V24 N1 P34-40	60-13618 <+*>
	68N49R <ATS>
1958 V24 N1 P41-43	4265 <HB>
1958 V24 N1 P57-60	4194 <HB>
1958 V24 N1 P62-63	4295 <HB>
1958 V24 N1 P82-83	60-13619 <+*>
1958 V24 N1 P111-112	4844 <HB>
1958 V24 N2 P148-153	59-15030 <+*>
1958 V24 N2 P167-173	60-13621 <+*>
1958 V24 N2 P184-185	59-17510 <+*>
1958 V24 N2 P190-192	59-15029 <+*>
1958 V24 N3 P270-273	59-10775 <+*>
1958 V24 N3 P303-306	61-20699 <*=>
1958 V24 N3 P306-308	2219 <BISI>
1958 V24 N3 P342-346	60-13624 <+*>
1958 V24 N3 P348-352	60-13612 <+*>
1958 V24 N4 P402-403	60-23016 <+*>
1958 V24 N4 P413-415	66-13047 <*>
1958 V24 N4 P434-435	66N57R <ATS>
1958 V24 N4 P443-444	4288 <HB>
1958 V24 N4 P444-446	4326 <HB>
1958 V24 N4 P467-470	<PS>

1958 V24 N5 P549-554 59-15321 <+*>
1958 V24 N5 P579-582 4407 <HB>
1958 V24 N5 P604-613 59-22711 <+*>
1958 V24 N5 P613-617 <PS>
1958 V24 N5 P629-631 59-15802 <+*>
1958 V24 N6 P684-688 39M41R <ATS>
1958 V24 N6 P723-731 <PS>
1958 V24 N6 P736-745 62-25399 <=*>
1958 V24 N7 P837 60-15784 <+*>
1958 V24 N7 P842-843 4345 <HB>
1958 V24 N7 P850-854 TS-1425 <BISI>
62-24235 <=*>
1958 V24 N7 P858-860 4439 <HB>
1958 V24 N7 P874-875 62-15430 <*=>
1958 V24 N7 P889-890 4457 <HB>
1958 V24 N7 P890-892 4383 <HB>
59-22438 <+*>
1958 V24 N7 P902 59-22435 <+*>
1958 V24 N8 P922-925 59-20009 <+*>
1958 V24 N8 P941-947 59-20008 <+*>
1958 V24 N8 P959-968 61-17552 <*=>
1958 V24 N9 P1071- AEC-TR-3643 <+*>
1958 V24 N9 P1097-1106 60-15810 <+*>
62-18901 <=*>
1958 V24 N10 P1171-1178 4496 <HB>
1958 V24 N10 P1214-1217 <PS>
1958 V24 N10 P1258-1261 60-21196 <=> 0
1958 V24 N11 P1308-1314 <PS>
1958 V24 N11 P1356-1358 59-20010 <+*>
1958 V24 N11 P1373-1374 14L30R <ATS>
1958 V24 N11 P1420-1421 61-27495 <*=>
1958 V24 N12 P1458-1459 <PS>
1958 V24 N12 P1464-1467 <PS>
1958 V24 N12 P1514-1516 59-22650 <+*>
1959 V25 N1 P26-27 59-14253 <+*>
1959 V25 N1 P38-41 59-17606 <+*>
1959 V25 N1 P1C1-105 59-17602 <+*>
1959 V25 N2 P161 61-13366 <*+>
1959 V25 N2 P185 61-15562 <*+>
1959 V25 N2 P220-222 60-21453 <=> 0
1959 V25 N3 P285-287 61-27507 <*=>
1959 V25 N3 P288-289 62-18028 <=*>
1959 V25 N3 P302-303 13M39R <ATS>
1959 V25 N3 P375-376 62-18144 <=*>
1959 V25 N4 P413-415 61-27626 <*=>
1959 V25 N4 P484-485 61-10154 <+*>
1959 V25 N4 P485 61-10155 <+*>
1959 V25 N6 P748-749 62-18615 <=*>
1959 V25 N7 P771-778 60-11734 <=>
1959 V25 N7 P807-810 61-13351 <*+>
1959 V25 N7 P816-818 61-13350 <*+>
1959 V25 N7 P850-853 61-15503 <+*>
1959 V25 N9 P1023-1027 60-31156 <=>
1959 V25 N9 P1059-1064 61-23147 <=*> 0
1959 V25 N1C P1205-12C6 61-19110 <+*>
1959 V25 N11 P1283-1287 62M45R <ATS>
1959 V25 N11 P1336-1338 AEC-TR-4107 <*+>

1960 V26 N1 P51-54 99M42R <ATS>

1960 V26 N2 P149-152 58M43R <ATS>
1960 V26 N4 P431-444 64-15254 <=*$>
1960 V26 N4 P509 60-41498 <=>
1960 V26 N5 P523-529 61-21574 <=>
1960 V26 N6 P679-703 61-23120 <=*>
1960 V26 N9 P1043-1046 61-11975 <=>
1960 V26 N9 P1047-1051 61-27629 <*=>
1960 V26 N11 P1317-1318 61-21340 <=>
1961 V27 P1345-1346 65-13603 <*>
1961 V27 N1 P24-28 62-16033 <=*>
1961 V27 N1 P40-43 63-20881 <=*$> 0
1961 V27 N2 P154-157 80N52R <ATS>
1961 V27 N2 P225-226 62-10601 <*=> 0
1961 V27 N4 P420-422 69N53R <ATS>
1961 V27 N4 P490-491 62-23582 <=>
1961 V27 N4 P622-625 62-23582 <=>
1961 V27 N4 P629-630 62-23582 <=>
1961 V27 N6 P730-738 62-23964 <=*>
1961 V27 N7 P839-842 62-15449 <*=>
1961 V27 N9 P1135-1138 63-19782 <=*> 0
1961 V27 N10 P1211-1214 63-23353 <=*$>
1961 V27 N11 P1371-1374 12P63R <ATS>
1961 V27 N11 P1380-1384 63-19789 <=*> 0
1961 V27 N12 P1468-1469 65-23694 <$>
1961 V27 N12 P1486-1490 M-5017 <NLL>
1961 V27 N12 P1498-1501 62-33134 <=*> 0
1962 N1 P49- ICE V2 N4 P486-492 <ICE>
1962 V28 N1 P49-55 73P60R <ATS>
1962 V28 N1 P112 63-19789 <=*> 0
64-13272 <=*$>
1962 V28 N2 P242-245 5557 <HB>
1962 V28 N2 P249 62-18640 <=*>
1962 V28 N3 P316-323 62-32012 <=*>
1962 V28 N4 P416 65-17212 <*>
1962 V28 N4 P447-449 5639 <HB>
1962 V28 N4 P450-462 63-21136 <=>
1962 V28 N4 P462-464 5625 <HB>
1962 V28 N4 P491-493 62-18966 <=*>
1962 V28 N5 P548 57Q68R <ATS>
1962 V28 N5 P590-592 63-19927 <=*>
1962 V28 N7 P865-868 SCL-T-482 <=*$> P
1962 V28 N8 P938-940 5740 <HB>
1962 V28 N8 P940-941 5736 <HB>
1962 V28 N8 P965-966 5737 <HB>
1962 V28 N8 P968-969 5288 <HB>
1962 V28 N9 P1087-1088 63-21329 <=>
1962 V28 N10 P1172-1177 63-23516 <=*$>
1962 V28 N1C P1199-1203 AEC-TR-5983 <=*$>
1962 V28 N11 P1329 5866 <HB>
1962 V28 N11 P1330-1332 5787 <HB>
1962 V28 N11 P1345-1346 5786 <HB>
1962 V28 N11 P1355- RCT V36 N4 P1042 <RCT>
1962 V28 N12 P1443-1446 65-61315 <=$>
1962 V28 N12 P1513-1515 65-61276 <=$>
1963 V29 N2 P189-197 63-21668 <=>
1963 V29 N2 P219-221 63-21668 <=>
1963 V29 N2 P252-253 63-41026 <=>
1963 V29 N3 P175-177 5888 <HB>
1963 V29 N3 P261-271 63-21650 <=>
64-13005 <=*$>
1963 V29 N3 P286-289 64-13092 <=*$>
1963 V29 N3 P356-357 M-5400 <NLL>
64-13090 <=*$>
1963 V29 N3 P364-379 63-21622 <=>
1963 V29 N5 P524-526 65-10756 <*>
1963 V29 N5 P528-53C 65-10897 <*>
1963 V29 N5 P531-536 65-10864 <*>
1963 V29 N5 P540 66-10238 <*>
1963 V29 N5 P557-558 5905 <HB>
1963 V29 N5 P559-560 5966 <HB>
1963 V29 N5 P565-567 62Q72R <ATS>
1963 V29 N5 P579-580 6025 <HB>
1963 V29 N5 P583-584 6022 <HB>
1963 V29 N5 P584-586 6026 <HB>
1963 V29 N5 P618-619 64-13216 <=*$>
1963 V29 N5 P693-695 6024 <HB>
1963 V29 N6 P643-644 63-31606 <=>
1963 V29 N6 P674-675 M-5422 <NLL>
64-15301 <=*$>
1963 V29 N6 P695-696 6023 <HB>
1963 V29 N7 P791-793 M-5420 <NLL>
64-13221 <=*$>
1963 V29 N8 P923-924 64-21752 <=>
1963 V29 N8 P937-938 6087 <HB>
1963 V29 N8 P944-948 64-21132 <=>
1963 V29 N8 P960-962 6121 <HB>
1963 V29 N9 P1076-1077 6086 <HB>
1963 V29 N9 P1148 65-10746 <*>
1963 V29 N10 P1173-1174 66-10000 <*> 0
1963 V29 N10 P1175-1176 65-13689 <*>
1963 V29 N1C P1188-1191 6127 <HB>
1963 V29 N10 P1191-1193 6128 <HB>
1963 V29 N10 P1225-1228 65-61345 <=$>
1963 V29 N10 P1276-1278 64-21112 <=>
1963 V29 N11 P1291-1293 5606 <HB>
1963 V29 N11 P1321-1322 6064 <HB>
1963 V29 N11 P1344-1347 65-60744 <=>
1963 V29 N11 P1352-1354 5723 <HB>
1963 V29 N12 P1451-1452 6189 <HB>

1963 V29 N12 P1452 6190 <HB>
1963 V29 N12 P1463-1464 6241 <HB>
1963 V29 N12 P1503-1504 65-10247 <*>
1964 N30 P364-367 AD-620 969 <=$>
1964 V30 N1 P47-54 64-31199 <=>
1964 V30 N1 P93-94 6351 <HB>
1964 V30 N2 P151-153 65-11218 <*> 0
1964 V30 N2 P213-215 AD-620 063 <=$>
1964 V30 N2 P221-222 6266 <HB>
1964 V30 N3 P305-306 6271 <HB>
1964 V30 N3 P313-314 6267 <HB>
65-61414 <=>
1964 V30 N5 P571-573 M-5583 <NLL>
1964 V30 N5 P598-599 6296 <HB>
1964 V30 N5 P635 IS-TRANS-4 <=$>
1964 V30 N6 P654-655 6354 <HB>
1964 V30 N6 P675 AD-622 453 <=*$>
1964 V30 N6 P725-726 AD-622 453 <=*$>
6355 <HB>
1964 V30 N6 P729-730 6356 <HB>
1964 V30 N6 P733-738 AD-622 453 <=*$>
1964 V30 N6 P743-744 6358 <HB>
1964 V30 N6 P751-752 6366 <HB>
1964 V30 N6 P752-753 6372 <HB>
1964 V30 N6 P753-755 6367 <HB>
1964 V30 N7 P819-821 6359 <HB>
1964 V30 N7 P829-831 6373 <HB>
1964 V30 N7 P862-863 6350 <HB>
1964 V30 N7 P876-879 6353 <HB>
1964 V30 N8 P950-952 6532 <HB>
1964 V30 N8 P997-1005 64-51353 <=$>
1964 V30 N8 P1017-1019 6396 <HB>
1964 V30 N9 P1074 6080 <HB>
1964 V30 N9 P1074-1075 6195 <HB>
1964 V30 N9 P1119-1121 6398 <HB>
1964 V30 N10 P1243-1244 M.5632 <NLL>
1964 V30 N10 P1244 6446 <HB>
1964 V30 N10 P1254-1255 M.5625 <NLL>
1964 V30 N10 P1282-1283 M-5626 <NLL>
1964 V30 N11 P1367-1368 657 <TC>
1964 V30 N11 P1371-1379 M-5633 <NLL>
1965 V31 N1 P77-78 6075 <HB>
1965 V31 N1 P80-83 6567 <HB>
1965 V31 N2 P169-172 4441 <BISI>
1965 V31 N6 P690-692 6618 <HB>
1965 V31 N6 P694 6620 <HB>
1965 V31 N6 P751-752 6617 <HB>
1965 V31 N7 P790-792 6635 <HB>
1965 V31 N7 P806 6644 <HB>
1965 V31 N7 P899-900 6645 <HB>
1965 V31 N8 P943-944 66-11463 <*>
1965 V31 N8 P1009-1010 6619 <HB>
1965 V31 N10 P1205-1207 6758 <HB>
1966 V32 N1 P53-58 66-31758 <=$>
1966 V32 N1 P126-128 66-32617 <=$>
1966 V32 N2 P158-159 6892 <HB>
1966 V32 N2 P170-190 66-32595 <=$>
1966 V32 N2 P196-202 66-32595 <=$>
1966 V32 N2 P212-213 66-32595 <=$>
1966 V32 N2 P251-253 66-32595 <=$>

ZBIRNYK INSTYTUTU ORHANICHNOYI KHEMIYI I TEKHNOLOHIYI. AKADEMIYA NAUK URSR. KIEV
1947 V13 P40-67 61-13522 <=*>

ZBIRNYK PRATS INSTYTUTU MATEMATYKY AKADEMII NAUK URSR
1947 N8 P13-69 AMST S1 V11 P273-321 <AMS>
RT-1486 <*>

ZBIRNYK PRATS INSTYTUTU TEPLOENERHETYKY. AKADEMIYI NAUK URSR
1960 N20 P44-53 64-71483 <=>
65-63969 <=$>
1960 N20 P54-59 65-62549 <=$>

ZBIRNYK PRATS Z OBCHYSLIUVALNOI MATEMATYKY I TEKHNIKY
1961 V3 P42-44 63-19466 <=*>

ZBORNIK ZA KMETIJSTVO IN GOZDARSTVO
1956 V2 P1-258 60-21691 <=>

ZBORNIK RADOVA. MASINSKI INSTITUT, SRPSKA AKADEMIJA NAUKA. BEOGRAD
1952 V24 N4 P125-128 60-21707 <=>

ZBORNIK RADOVA POLJOPRIVREDNOG FAKULTETA, UNIVERSITET U BEOGRADU
1955 V2 P185-194 60-21629 <=>
1959 V7 N277 P1-12 60-21677 <=>

ZDOROVE
1960 N1 P23 NLLTB V2 N4 P316 <NLL>
1960 N11 P2-3 62-24914 <=*>
1960 V6 N5 P11-13 60-31802 <=>
1960 V6 N7 P7-8 60-31820 <=>
1960 V6 N11 P4-5 61-28411 <*=>
1961 V7 N3 P3-5 62-13354 <=*>
1961 V7 N6 P4-5 63-15525 <=*>
1962 V8 N2 P1-3 62-23822 <=*>
1963 V9 N4 P4-7 64-71288 <=>
1963 V9 N5 P4-5 AD-611 053 <=$>
1963 V9 N7 P2-3 AD-611 016 <=$>
1964 V10 N8 P1-5 64-51976 <=$>
1964 V10 N8 P9-11 64-51976 <=$>
1964 V10 N8 P20-21 64-51976 <=$>
1965 V11 P7-8 65-30900 <=$>
1965 V11 N1 P1-6 65-30926 <=*$>
1965 V11 N5 P5-6 65-31322 <=$>
1966 N3 P9 66-32662 <=$>
1966 N4 P3-5 66-32662 <=$>

ZDRAVNO DELO
1957 V10 N2 P1-10 PB 141 262T <=>
1957 V10 N2 P39 PB 141 263T <=>

ZDRAVOOKHRANENIE
1959 N1 P9-12 59-11755 <=>
1959 N1 P24-28 59-11756 <=>
1960 N4 P3-7 61-11894 <=>
1960 N4 P44-48 61-28279 <*=>
1960 N5 P47-54 61-28054 <*=>
1961 N1 P58-59 61-28453 <*=>
1961 N2 P3-8 61-27861 <*=>
1961 N2 P13-16 61-27954 <*=>
1961 N2 P61-64 61-28259 <*=>
1961 N5 P3-7 62-23764 <=*>
1961 N5 P7-10 62-23773 <=*>
1962 V5 N2 P7-11 63-13288 <=>
1963 V6 N2 P38-41 64-31256 <=>

ZDRAVOOKHRANENIE BELORUSSII
1959 V5 N2 P44-45 63-31382 <=>
1959 V5 N3 P3-7 59-13658 <=>
1959 V5 N4 P3-5 59-13863 <=>
1959 V5 N4 P45-49 59-13753 <=>
1960 V6 N1 P15-17 60-31338 <=>
1960 V6 N1 P23-25 60-31339 <=>
1960 V6 N1 P60 60-31340 <=>
1960 V6 N1 P62-68 60-31341 <+*>
1960 V6 N2 P3-10 60-31702 <=>
1960 V6 N2 P11-15 60-31703 <=>
1960 V6 N2 P53-54 60-31704 <+*>
1960 V6 N2 P66 61-28720 <=>
1960 V6 N4 P18-20 60-31745 <=>
1960 V6 N4 P20-23 60-31746 <=>
1960 V6 N4 P75-76 60-41161 <=>
1960 V6 N5 P3-7 60-41386 <=>
1960 V6 N6 P6-26 61-11685 <=>
1960 V6 N7 P48-51 61-27378 <*=>
1960 V6 N7 P67-69 61-11035 <=>
1960 V6 N8 P7-18 61-11516 <=>
1960 V6 N8 P33-36 61-11516 <=>
1960 V6 N9 P24-27 61-11679 <=>
1960 V6 N9 P46-47 61-11634 <=>
1960 V6 N11 P72-76 61-21706 <=>
1961 V7 N2 P65-66 61-28720 <=>
1961 V7 N3 P19-22 61-23951 <=*>
1961 V7 N3 P25-30 61-23950 <*=>

1961 V7 N3 P69-71 63-10175 <=*>
1961 V7 N4 P5-10 61-27199 <=*>
1961 V7 N6 P28-32 61-28286 <*=>
1961 V7 N7 P25-27 62-13226 <*=>
1961 V7 N8 P3-4 62-13382 <*=>
1961 V7 N8 P17-21 62-13378 <=*>
1961 V7 N8 P30-35 62-13372 <*=>
1961 V7 N9 P3-11 62-23818 <=>
1961 V7 N9 P76-77 62-23818 <=>
1961 V7 N11 P31-35 62-23434 <=*>
1962 V8 N3 P38-42 62-25503 <=*>
1962 V8 N11 P34-41 63-21221 <=>
1962 V8 N11 P38-39 65-62986 <=$>
1963 V9 N4 P71-72 63-31248 <=>
1963 V9 N8 P89-91 63-41130 <=>
1963 V9 N9 P28-30 65-60954 <=$>
1963 V9 N9 P38-41 33R8CR <ATS>
1964 V10 N8 P93-94 65-30727 <=$>
1965 V11 N3 P42-46 65-30957 <=$>
1965 V11 N7 P39-43 65-32117 <=$>
1965 V11 N8 P41-43 65-33694 <=$>
1965 V11 N12 P76-77 66-30741 <=$>

ZDRAVOOKHRANENIE KAZAKHSTANA
1957 N8 P11-15 60-11331 <=>
1961 V21 N1 P29-32 61-27069 <*=>
1961 V21 N1 P65-69 61-27062 <*=>
1961 V21 N2 P21-25 61-27399 <*=>
1961 V21 N2 P76-80 61-27400 <*=>
1961 V21 N4 P75-78 61-27284 <*=>
1961 V21 N5 P5-8 61-28877 <=*>
1961 V21 N5 P11-16 61-28877 <=*>
1961 V21 N5 P71-73 61-28877 <=*>
61-28880 <*=>
1961 V21 N5 P78-79 61-28271 <*=>
1961 V21 N6 P3-8 62-15659 <*=>
1961 V21 N10 P3-6 62-23399 <=*>
1961 V21 N10 P16-20 62-15936 <*=>
1961 V21 N12 P29-33 65-60314 <=$>
1961 V21 N12 P59-61 62-23821 <=*>
1962 N7 P63-68 AD-630 734 <=*$>
1962 V22 N3 P67-79 62-32362 <=>
1962 V22 N7 P47-51 64-18752 <*> 0
1962 V22 N7 P56-59 64-18753 <*>
1962 V22 N7 P73-75 63-23119 <=*$>
1962 V22 N9 P47-51 65-60315 <=$>

ZDRAVOOKHRANENIE ROSSIISKOI FEDERATSII
1958 V2 N3 P34-35 AEC-TR-3669 <+*>
1958 V2 N3 P35-36 AEC-TR-3670 <+*>
1960 V4 N3 P40-41 60-11904 <=>
1960 V4 N6 P3-10 60-41651 <=>
1960 V4 N6 P47-49 60-41651 <=>
1960 V4 N8 P16-18 61-21358 <=>
1960 V4 N8 P27-28 61-21358 <=>
1960 V4 N8 P35-41 61-21358 <=>
1961 V5 N3 P41-43 61-27110 <*=>
1961 V5 N3 P43-45 61-27109 <*=>
1961 V5 N3 P46 61-27111 <*=>
1961 V5 N4 P10-13 62-13267 <*=>
1961 V5 N8 P3-7 62-13023 <*=>
1962 V6 N6 P3-8 62-33339 <=>
1962 V6 N6 P40-47 62-33339 <=>
1962 V6 N6 P100-101 62-33339 <=>
1962 V6 N7 P3-7 63-13288 <=>
1962 V6 N9 P44-48 63-13512 <*=>
1963 V7 N9 P37-40 64-21010 <=>
1964 V8 N1 P9-13 64-21866 <=>
1964 V8 N5 P3-4 64-41286 <=>
1964 V8 N5 P4-8 64-41285 <=>
1964 V8 N7 P21-26 64-41932 <=$>
1964 V8 N7 P34-37 64-41932 <=$>
1964 V8 N8 P27-30 64-51321 <=$>
1964 V8 N8 P40-42 64-51359 <=$>
1964 V8 N10 P3-13 64-51888 <=$>
1965 V9 N4 P31-35 65-31304 <=$>
1965 V9 N4 P42-46 65-31300 <=$>
1965 V9 N5 P34-37 65-32228 <=*$>
1965 V9 N8 P25-29 65-32880 <=*$>
1965 V9 N10 P7-9 66-30588 <=$>
1965 V9 N11 P31-34 66-30265 <=*$>
1966 V10 N2 P39-42 66-31353 <=$>
1966 V10 N6 P37-39 66-33458 <=$>

ZDRAVOOKHRANENIE SOVETSKOI ESTONII
1960 N5 P66-68 61-28951 <=>
61-28951 <=*>
1960 N5 P69 61-28918 <*=>
1960 N5 P69-70 62-13080 <*=>
1960 N5 P70-75 61-28951 <=>
61-28951 <=*>
1960 N5 P76-77 61-28790 <*=>
1960 N6 P65-79 62-13049 <*=>
1962 N1 P61-66 62-25003 <=*>
1962 N2 P3-7 62-33052 <=*>

ZDRAVOOKHRANENIE TADZHIKISTANA
1960 V7 N3 P3-7 60-41635 <=>
1960 V7 N3 P11-15 60-41636 <=>
1960 V7 N3 P20-21 60-41637 <=>
1960 V7 N3 P61 60-41638 <=>
1960 V7 N5 P61-62 61-27750 <*=>
1960 V7 N6 P23-28 61-31512 <=>
1961 V8 N1 P3-14 61-31524 <=>
1961 V8 N1 P57-59 61-31524 <=>
1961 V8 N3 P3-6 62-13355 <*=>
1961 V8 N3 P57-58 62-13375 <*=>
1961 V8 N4 P25-31 62-23784 <=>
1961 V8 N4 P55-57 62-23784 <=>
1961 V8 N6 P43-47 62-23441 <=*>
1962 V9 N1 P3-9 62-24719 <=*>
1962 V9 N1 P30-32 62-25164 <=*>
1962 V9 N2 P3-5 62-32558 <=*>
1963 V10 N1 P3-9 63-31646 <=>
1963 V10 N3 P34-37 63-41126 <=>
64-21815 <=>
1963 V10 N3 P37-39 63-41126 <=>
1963 V10 N3 P40-41 63-41126 <=>
64-21815 <=>
1965 V12 N4 P46-47 65-32980 <=*$>

ZDRAVOOKHRANENIE TURKMENISTANA
1957 V1 N3 P41-44 62-11144 <=>
62-15067 <=*>
1960 V4 N3 P20-26 61-11620 <=>
1960 V4 N3 P48-49 61-11570 <=>
1960 V4 N4 P32-36 61-23923 <=>
1960 V4 N4 P39-43 61-23923 <=>
1960 V4 N5 P13-17 61-11900 <=>
1960 V4 N5 P17-19 61-11925 <=>
1960 V4 N6 P38-40 61-21507 <=>
1961 V5 N3 P50-52 62-23436 <=*>
1963 V7 N1 P3-11 63-21520 <=>
1963 V7 N4 P3-6 63-31198 <=>
1964 V8 N2 P36-39 67-51227 <=>
1964 V8 N8 P6-10 64-51523 <=>
1964 V8 N8 P43-45 64-51524 <=>
1964 V8 N9 P3-12 65-30013 <=$>
1965 V9 N4 P11-15 65-31490 <=$>
1965 V9 N5 P5-10 65-32507 <=*$>
1965 V9 N7 P3-8 65-32834 <=*$>

ZDRAVOTNI TECHNIKA A VZDUCHOTECHNIKA
1963 V6 N5 P221-225 LC-1265 <NLL>

ZDRAVSTVENI VESTNIK
1959 V28 N6/7 P193-197 62-00291 <*>
1961 N3/4 P49-52 62-19599 <*=>
1961 V30 N1/2 P1-5 62-19550 <=*>

ZDROWIE
1962 V14 N7 P10 62-33529 <=*>

ZEISS MITTEILUNGEN UEBER FORTSCHRITTE DER TECHNISCHEN OPTIK
1958 V1 N6 P199-211 63N55G <ATS>
1962 V2 N8 P309-334 66-13490 <*>
1963 V3 N1 P32-51 C-4767 <NRC>
1964 V3 N5 P153-192 66-13491 <*>

ZEISS NACHRICHTEN
1942 V4 N5 P126- 3211 <*>
1955 P5-35 T-1865 <INSD>

ZEITSCHRIFT FUER ACKER-UND PFLANZENBAU
1954 V98 P107-118 T2034 <INSD>
1954 V98 N3 P265-278 CSIRO-2860 <CSIR>
1955 V99 N4 P479-484 1505 <*>

ZEITSCHRIFT FUER AEROSOL-FORSCHUNG UND -THERAPIE
1953 V2 P585-595 59-19239 <=*> 0
1954 V3 P500-504 58-1017 <*>
1955 V4 P220-226 58-1030 <*>
1955 V4 N4 P3-11 II-279 <*>
1957 V6 P202-210 60-16601 <*>
1958 V7 N1 P50-61 60-13292 <=*> 0

ZEITSCHRIFT FUER AERZTLICHE FORTBILDUNG
1923 V20 P697-703 3367-F <K-H>
1954 V48 N5 P164-166 1905 <*>
1956 V50 P407-411 T1983 <INSD>
1956 V50 N12 P523-525 59-12342 <+*>
1957 V51 N15 P648-652 60-19481 <*=> 0
1960 V54 N4 P214-215 10177-C <KH>
63-16108 <*>

ZEITSCHRIFT FUER AGRAROKONOMIK
1967 V10 N2 P62-66 67-30952 <=$>

ZEITSCHRIFT FUER ALLGEMEINE MIKROBIOLOGIE
1962 V2 N2 P157-163 63-20816 <*>

ZEITSCHRIFT FUER ALTERSFORSCHUNG
1940 V2 P113-124 57-3429 <*>
1951 V5 P256-277 1357 <*>
1954 V8 N1 P1-19 2201 <*>
1955 V9 N4 P291-309 65-00386 <*> 0
1957 V10 P343-357 58-624 <*>
1958 V13 N1 P7-22 65-00306 <*> 0
1959 V13 N1 P60-67 65-00192 <*> 0

ZEITSCHRIFT FUER ANALYTISCHE CHEMIE
1869 V8 P102-103 65-13967 <*>
1872 V11 P467-469 65-13968 <*>
1890 V29 P332-335 AEC-TR-1718 <*>
II-35 <*>
1902 V41 P710-711 63-16761 <*>
1910 V49 N9/10 P525-596 63-16234 <*>
1923 V62 P284-293 13S85G <ATS>
1923 V62 P321-329 63-10982 <*>
1924 V64 P41-47 UCRL TRANS-1011(L) <*>
1928 V74 P380-386 57-3284 <*>
1930 V79 P452-459 2417F <K-H>
1931 V83 P107-114 4072-C <K-H>
1932 V90 P109-111 59-15302 <*>
1932 V91 P186-187 63-14593 <*>
1933 V91 P81-90 63-10478 <*>
1934 V98 P23-26 63-18672 <*>
1935 V103 P161-166 63-20322 <*> 0
1937 V110 P81-94 II-730 <*>
1938 V113 P419-422 57-941 <*>
1938 V114 P170-181 57-2363 <*>
1939 V116 P312-315 85L30G <ATS>
1939 V117 P392-400 63-18678 <*>
1939 V118 P9-13 23208 <K-H> 0
1940 V119 P161-188 59-20484 <*> 0
1941 V121 N11/2 P385-398 5357 <HB>
1942 V123 P405-407 64-20393 <*>
1942 V124 P216-226 61-18463 <*> 0
1944 V126 P426-452 62-10127 <*> 0
1948 V128 P543-549 59-20718 <*>
1950 V130 N2/3 P105-185 60-10776 <*> 0
1951 V133 P103-109 63-16907 <*>
1951 V133 N4 P241-248 58-1556 <*>
1951 V133 N4 P274-282 RCT V25 N2 P365 <RCT>
1951 V134 P17-22 63-16828 <*>
1951 V134 P183-187 3138-G <K-H>
1951 V134 P188-191 64-10439 <*>
1951 V134 N2 P86-95 60-18062 <*>
1952 V135 P119-122 I-778 <*>
1952 V135 N4 P251-258 64-16839 <*>
1952 V135 N5 P340-349 60-18061 <*>
1952 V136 P185-188 63-18162 <*>
1953 V138 P259-266 29N56G <ATS>
1953 V138 P332-337 1108 <*>
1953 V138 N3 P161-167 64-10336 <*>
1953 V138 N3 P167-179 65-14338 <*> 0
1953 V138 N4 P241-244 64-10334 <*>
1953 V139 P161-181 AEC-TR-4133 <*>
1953 V139 P263-267 30N56G <ATS>
1953 V140 P342-349 58-1060 <*>
1953 V140 P353-355 58-1214 <*>
1953 V140 N5 P330-335 59-15168 <*>
1954 V141 P81-86 II-258 <*>
1954 V142 P1- 2045 <*>
1954 V142 P12-15 57-2604 <*>
1954 V142 P19-27 I-499 <*>
1954 V142 P27-30 I-523 <*>
1954 V143 P182-195 57-766 <*>
1954 V143 P339-348 62-10102 <*>
1954 V143 P401-414 31N56G <ATS>
1955 V144 P8-16 II-218 <*>
1955 V144 N2 P106-108 64-14921 <*>
1955 V144 N3 P165-186 65-13546 <*> 0
1955 V144 N6 P415-420 T2234 <INSD>
1955 V145 P321-322 CSIRO-3014 <CSIR>
1955 V145 P325-334 58-1474 <*>
1955 V145 P338-344 R-3804 <*>
58-833 <*>
1955 V146 P11-17 58-182 <*>
1955 V146 P321-322 II-714 <*>
64-16428 <*> 0
1955 V146 P401-414 CSIRO-2992 <CSIR>
1955 V146 N1 P11-17 63-16044 <*> 0
1955 V146 N3 P174-181 T2233 <INSD>
1955 V148 N1 P6-9 64-16474 <*> 0
1955 V149 N3 P173-184 CSIRO-3220 <CSIR>
1956 P345-355 T-2520 <INSD>
1956 V150 P250-253 T-2272 <INSD>
1956 V150 P339-345 T-2197 <INSD>
1956 V151 N1 P1-15 58-576 <*>
1956 V151 N4 P258-262 57-2732 <*>
1956 V152 P401-411 57-2576 <*>
1956 V152 P421-424 63-16256 <*>
1956 V152 N1/3 P86-96 57-2603 <*>
1957 V154 P182-185 65-10840 <*>
1957 V154 N6 P406-413 CSIRO-3998 <CSIR>
1957 V155 P105-114 57-3406 <*>
1957 V155 P321-327 6922 <K-H>
1957 V156 P241-248 60-19168 <+*>
1957 V156 P248-257 02L31G <ATS>
1957 V156 N3 P169-179 NP-TR-46 <*> 0
1957 V157 P6-13 58-1332 <*>
1957 V157 N3 P161-165 61-10522 <*>
1957 V158 P1-7 C-3926 <NRCC>
1957 V158 P241-251 63-10440 <*>
1957 V159 N2 P81-94 278 <TC>
59-15789 <*>
1958 N6 P403-409 TS-1310 <BISI>
1958 V160 P169-188 63-16136 <*> 0
1958 V160 N3 P161-169 279 <TC>
1958 V160 N4 P277-279 59-15034 <*>
1958 V160 N6 P403-409 61-20587 <*>
1958 V161 N2 P108-114 61-00040 <*>
1958 V162 P423-429 65-18097 <*>
1958 V162 N3 P167-174 63-14128 <*> 0
1958 V162 N4 P266-279 59-17056 <*>
1958 V163 P273-281 280 <TC>
1958 V163 P331-338 281 <TC>
1958 V164 P37-48 62-20451 <*>
1958 V164 P164-181 AEC-TR-4562 <*>
1958 V164 P232-241 42L31G <ATS>
1958 V164 N1 P120-127 20M39G <ATS>
1958 V164 N1 P147-158 59-10569 <*>
1959 V165 N5 P401-415 95L31G <ATS>
1959 V166 P433-439 NP-TR-1271 <*>
1959 V167 N4 P269-271 65-13707 <*>
1959 V167 N4 P271-277 62-16178 <*>
1959 V168 P18-28 64-00314 <*>
1959 V168 N5 P335-339 64-10879 <*>

1959 V169 P401-404 8991 <IICH>
1959 V169 N4 P291 61-10237 <*>
1959 V170 P132-147 RCT V33 N3 P639 <RCT>
63-20243 <*>
1959 V170 P256-263 63-18286 <*>
1959 V170 P333-348 04M43G <ATS>
1959 V170 N1 P29-43 40M41G <ATS>
1959 V170 N1 P132-147 65-10473 <*>
1959 V170 N1 P147-151 65-12445 <*>
1959 V170 N1 P152-154 65-12446 <*>
1959 V170 N1 P278-285 65-13347 <*> 0
1959 V170 N1 P310-317 65-10012 <*> 0
1959 V171 P321-329 66-12176 <*>
1959 V171 P354-359 62-10699 <*>
1960 V173 N1 P84-87 60-18115 <*>
1960 V173 N5 P358-369 62-01482 <*>
1960 V175 N8 P350-355 61-10811 <*> 0
1960 V176 N6 P421-429 65-11280 <*> 0
1960 V177 P14-20 62-20025 <*>
1960 V177 N2 P81-86 62-24559 <=*>
1960 V177 N6 P429-430 65-10788 <*>
1960 V178 P198-201 10985A <K-H>
1960 V178 N2 P81-84 62-10669 <*>
1960 V178 N3 P198-201 65-13218 <*> 0
1961 V179 P175-186 63-00468 <*>
1961 V179 N2 P81-91 63-14092 <*> 0
1961 V179 N5 P355-358 65-13348 <*>
1961 V180 P161-168 66-12175 <*>
1961 V180 N3 P161-168 62-16687 <*>
1961 V181 P54-59 2907 <BISI>
1961 V181 P85-91 62-16302 <*> 0
1961 V181 P303-312 SC-T-64-614 <*>
1961 V181 P514-526 63-00063 <*>
1961 V181 N1 P254-274 64-14960 <*>
1961 V183 P161-172 48P61G <ATS>
1961 V184 P423-427 63-20937 <*>
1961 V184 N5 P322-327 65-14017 <*>
1961 V184 P322-327 11933 <K-H>
1961 V184 N5 P327-341 3281 <BISI>
1962 V185 P99-110 AEC-TR-5201 <*>
1962 V185 P179-201 AEC-TR-5202 <*>
1962 V185 N2 P99-110 <LSA>
1962 V186 P206-212 12315 <K-H>
1962 V186 N1 P63-79 63-18150 <*>
1962 V186 N1 P123-127 63-18151 <*> 0
1962 V186 N1 P127-132 62-16681 <*>
1962 V186 N1 P187-193 63-18166 <*>
1962 V186 N1 P206-212 65-14483 <*> 0
1962 V187 N1 P33-37 12144 <K-H>
63-16945 <*>
1962 V188 P341-344 63-14082 <*>
1962 V188 N4 P266-272 62-20439 <*> 0
1962 V188 N5 P321-334 63-10018 <*>
1962 V189 P111-124 62-18999 <*>
1962 V189 N1 P14-50 BGIRA-565 <BGIR>
64-30627 <*> 0
1962 V189 N1 P66-77 63-20929 <*> 0
1962 V189 N1 P111-124 BGIRA-563 <BGIR>
64-10600 <*>
65-12746 <*>
1962 V189 N1 P124-130 BGIRA-551 <BGIR>
1962 V189 N4 P325-330 64-14962 <*>
1962 V190 P92-97 AEC-TR-5867 <*>
1962 V190 N1 P92-97 3963 <BISI>
1963 V192 P65-81 3495 <BISI>
1963 V192 N1 P202-219 65-10755 <*>
1963 V193 P326-331 66-10289 <*>
1963 V193 N4 P264-270 66-10190 <*>
1963 V194 P81-101 64-15598 <=*$>
1963 V194 P417-422 NS-157 <TTIS>
1963 V196 P251-259 8806 <IICH>
1963 V197 N1 P91-97 SC-T-531 <*>
1963 V197 N1 P98-106 SC-T-537 <*>
1963 V197 N2 P221-228 66-12330 <*>
1963 V198 N1 P40-55 65-10870 <*>
1963 V198 N1 P88-98 4103 <BISI>
1963 V198 N3 P242-253 66-10007 <*> 0
1963 V198 N3 P254-266 66-13570 <*>
1964 V199 N2 P117-121 24S85G <ATS>
1964 V200 P81-134 778 <TC>
1964 V200 P218-225 180 <TC>
1964 V201 N1 P20-29 66-13786 <*> 0
1964 V201 N6 P401-417 4497 <BISI>
1964 V203 P265-272 NS 262 <TTIS>
1964 V204 N2 P97-101 66-10261 <*>
1964 V205 P180-194 530 <TC>
1964 V205 P284-298 671 <TC>
1964 V206 N5 P352-356 66-10998 <*>
1965 V208 N1 P15-26 65-14915 <*>
1965 V208 N5 P338-352 (CHEM)-425 <CTS>
1965 V209 N1 P114-119 66-11861 <*>
1965 V209 N1 P136-150 4548 <BISI>
1965 V211 P429-430 NS-458 <TTIS>
1966 V215 P178-181 66-13520 <*>
1966 V216 N2 P260-272 66-13005 <*>

ZEITSCHRIFT FUER ANATOMIE UND ENTWICKLUNGSGE-SCHICHTE
1923 V68 P29-109 1695 <*>
1924 V73 P247-276 62-00293 <*>
1925 V76 N1 P43-63 62-01414 <*>
1927 V86 P730-775 62-00230 <*>
1950 V115 P32-36 57-1864 <*>
1956 V119 P415-430 62-00896 <*>
1959 V121 P188-240 64-00369 <*>
1960 V121 P388-392 61-00632 <*>

ZEITSCHRIFT FUER ANGEWANDTE BAEDER- UND KLIMAHEIL-KUNDE
1961 V8 P548-557 64-16936 <*>
1962 V9 N5 P481-501 65-10452 <*>

ZEITSCHRIFT FUER ANGEWANDTE CHEMIE
1888 N8 P232-236 63-16361 <*>
1898 N9 P195- 63-14133 <*>
1908 V21 P532-546 2101 <*>
1908 V21 P577-589 2101 <*>
1910 N3 P99-103 58-1901 <*>
1911 V24 P2089-2099 AL-831 <*>
1911 V24 P2459-2469 63-14129 <*> 0
1912 V25 N41 P2110- 57-2089 <*>
1913 V26 P246-250 59-20831 <*>
1918 V31 P45-46 63-10473 <*>
1918 V31 P195-196 64-18310 <*> 0
1919 V32 P21-24 64-18310 <*> 0
1921 V34 P585-586 3249 <*>
1923 V36 P273-276 64-14334 <*>
1923 V36 P575-584 63-10469 <*> 0
1923 V36 N45/6 P293-297 59-20300 <*>
1924 P106-110 2864 <*>
1924 P110-113 2862 <*>
1924 P113-116 2863 <*>
1924 V37 P314-317 66-12097 <*> 0
1925 P439-441 2857 <*>
1925 P545-546 2933 <*>
1926 V39 P88-90 AL-425 <*>
63-20769 <*>
1926 V39 P138-143 3295 <*>
1926 V39 P671-673 65-12100 <*>
1926 V39 P728-729 65-12102 <*>
1926 V39 P883-887 64-18316 <*> P
1926 V39 P1406-1411 II-516 <*>
1926 V39 N6 P194-196 59-17050 <*>
1926 V39 N13 P421-428 59-20721 <*>
1926 V39 N13 P428-431 66-12247 <*> 0
1927 V40 P69-79 58-1549 <*>
1927 V40 P79-85 63-10430 <*>
1927 V40 P166-174 63-10104 <*>
1927 V40 P174-177 62-20054 <*>
1927 V40 P303-314 63-01540 <*>
1928 V41 P130-132 AL-50 <*>
1928 V41 P827-833 58-1239 <*>
63-18124 <*>
1928 V41 P916-924 57-3492 <*>
1928 V41 N38 P1066-1069 66-14667 <*>
1929 V42 P325-331 AL-903 <*>
1929 V42 P541-546 1665 <*>
1929 V42 P981-984 58-2537 <*>
1929 V42 N69 P697-700 63-10498 <*>
1930 V43 P122-124 63-14500 <*>

1930 V43 P286-289 AL-31 <*>
65-14001 <*>
1930 V43 P302-308 63-14130 <*> 0
1930 V43 P459-462 57-3087 <*>
1930 V43 P496-500 64-14322 <*> 0
1930 V43 P6C8-612 I-482 <*>
1930 V43 P877-880 I-864 <*>
1930 V43 P1009-1011 63-10443 <*>
1930 V43 N26 P603-608 57-2232 <*>
1930 V43 N26 P608-612 AEC-TR-3821 <*>
1930 V43 N32 P712-714 59-20659 <*> 0
1931 V44 P273-277 1631 <*>
1931 V44 P391-395 AL-417 <*>
1931 V44 N13 P231-237 59-15640 <*> 0
1931 V44 N21 P383-385 59-15787 <*>
1931 V44 N51 P973-976 58-342 <*>

ZEITSCHRIFT FUER ANGEWANDTE ENTOMOLOGIE
1920 V6 N2 P343-348 63-01775 <*>
1925 V11 N2 P203-215 C-2380 <NRC>
1938 V25 N1 P125-141 63-01684 <*>

ZEITSCHRIFT FUER ANGEWANDTE GEOLOGIE
1955 N3/4 P173-183 IGR V1 N9 P59 <AGI>
1956 N1 P39-44 IGR V2 N2 P159 <AGI>
1961 V7 N3 P122-124 62-24547 <=*>
1962 V8 P4CC-401 63-16551 <*>
1963 V9 P169-176 65-11117 <*>
1963 V9 P225-230 64-16420 <*>
1963 V9 P449-452 65-11096 <*>
1963 V9 P452-455 64-16419 <*>
1963 V9 P507-513 65-11116 <*>
1963 V9 N1 P14-16 64-14012 <*> 0
1964 V10 N4 P172-179 66-10242 <*>
1964 V10 N7 P348-353 31S88G <ATS>
1965 V11 N12 P638-645 18T94G <ATS>

ZEITSCHRIFT FUER ANGEWANDTE MATHEMATIK UND MECHANIK
1923 V3 P279-289 64-14124 <*>
1923 V3 N4 P290-297 63-20149 <*>
1925 V5 N1 P1-16 62-16433 <*>
1925 V5 N1 P36-47 AL-66 <*>
1925 V5 N6 P490-493 58-1454 <*>
1928 V8 P85-91 64-10042 <*>
1928 V8 P161-185 UCRL TRANS-872(L) <*>
1928 V8 P272-283 57-2815 <*>
1928 V8 N5 P341-352 63-10679 <*> 0
1928 V8 N6 P423-424 63-18825 <*>
1929 V9 P49-58 AEC-TR-4649 <*>
1929 V9 N6 P496-498 63-18830 <*>
1929 V9 N6 P507-509 58-1462 <*>
6641-F <K-H>
1930 V10 P346-353 57-2814 <*>
1930 V10 N6 P563-578 57-3219 <*>
1931 V11 P105- 61-15041 <=*>
1931 V11 N1 P15-19 58-480 <*>
1931 V11 N1 P27-40 AEC-TR-4905 <*>
1931 V11 N2 P136-154 59-00722 <*>
66-13009 <*>
1932 V12 P275-279 57-2813 <*>
1932 V12 N1 P1-14 59-17712 <*>
1933 V13 P153-159 AEC-TR-3545 <+*>
1934 V14 N5 P257-276 II-144 <*>
1935 V15 N6 P313-338 AL-661 <*>
1935 V15 N1/2 P56-61 66-14615 <*>
1935 V15 N1/2 P81-87 AEC-TR-4966 <*>
1936 V16 N2 P73-98 BGIRA-555 <BGIR>
1936 V16 N3 P153-164 AL-932 <*>
1936 V16 N5 P275-287 AEC-TR-4967 <*>
1936 V16 N6 P355-358 66-13008 <*>
1937 V17 N5 P249-258 65-12549 <*>
1938 V18 N1 P31-38 2977 <*>
1938 V18 N6 P358-361 AL-579 <*>
1939 P361-371 57-2075 <*>
1940 V20 N6 P297-328 65-13007 <*>
1942 V22 N1 P1-14 58-499 <*>
1942 V22 N4 P206-215 59-17718 <*>
1942 V22 N5 P269-272 59-17717 <*>
1943 V23 N2 P65-71 59-17709 <*>
1943 V23 N2 P81-100 58-477 <*>
1943 V23 N3 P139-149 22M46G <ATS>
59-10290 <*> 0
1943 V23 N3 P169-179 58-476 <*>
1948 V28 N4 P1C4-114 64-00378 <*>
1948 V28 N10 P297-303 64-00377 <*>
1950 V30 N3 P73-83 58-712 <*>
1950 V30 N7 P2C3-215 58-676 <*>
1950 V30 N5/6 P149-153 58-683 <*>
63-18058 <*> 0
1951 V31 P161-169 3116 <*>
1951 V31 P333-338 3116 <*>
1951 V31 N7 P193-228 3193 <*>
1951 V31 N7 P208-219 18G5G <ATS>
1951 V31 N11/2 P349-356 58-1083 <*>
1952 V32 N8/9 P231- 3190 <*>
1952 V32 N8/9 P232- 3216 <*>
1952 V32 N8/9 P281-284 3183 <*>
1953 V33 N7 P221-227 3189 <*>
1953 V33 N1/2 P41- 3124 <*>
1953 V33 N5/6 P200- 1530 <*>
3113 <*>
1954 V34 N6 P2C1-210 58-2451 <*>
1954 V34 N8/9 P344-357 NACA TM1417 <NASA>
1955 V35 N3 P81-88 NACA TM1412 <NASA>
1955 V35 N6/7 P242-263 58-58 <*>
1956 V36 N11/2 P436-443 57-3308 <*>
1956 V36 N9/10 P331-339 91J19G <ATS>
1956 V36 N9/10 P377-395 59-10573 <*> 0
1957 N37 P128-139 58-380 <*>
1957 V37 N1/2 P44-51 3429 <*>
1957 V37 N11/2 P471-485 30N50G <ATS>
1961 V41 N10/1 P408-413 66-11993 <*>
1962 V42 N10/1 P465-476 65-14134 <*>
1963 V43 N4/5 P149-166 N65-23676 <=$>

ZEITSCHRIFT FUER ANGEWANDTE MATHEMATIK UND PHYSIK
1950 V1 P32-42 AL-25 <*>
1951 V2 P421-443 AL-480 <*>
1951 V2 P421- 3182 <*>
1952 V3 P65-74 II-1518 <*>
1952 V3 P271-286 63-20352 <*>
1953 V4 P312-316 C-2434 <NRC>
1953 V4 P376- 3325 <*>
1953 V4 P492-496 2458 <*>
1953 V4 P500-506 2231 <*>
1954 V5 P233-251 62-16586 <*>
1954 V5 P260-263 58-1765 <*>
1954 V5 P496-5C8 63-10154 <*> 0
1955 V6 P387-401 63-10153 <*>
1955 V6 N5 P407-416 CSIRO-3136 <CSIR>
1956 V7 P164-169 63-10155 <*>
1956 V7 N2 P121-129 60-16836 <*>
1957 V8 N1 P64-71 62-18666 <*>
1959 V10 P603-608 63-20126 <*>
1959 V10 N2 P143-159 61-20761 <*> 0P
1959 V10 N3 P256-273 SCL-T-293 <*>
1959 V10 N4 P438-441 NP-TR-422 <*>
1960 V11 N4 P273-306 66-10805 <*>
1960 V11 N6 P508-513 61-20086 <*>
1961 V12 N5 P474-476 63-00588 <*>
63-27138 <$>
1962 V13 N4 P393-401 63-00929 <*>
63-27218 <$>

ZEITSCHRIFT FUER ANGEWANDTE METEOROLOGIE
1940 V57 N7 P214-220 II-1036 <*>
58-2321 <*>

ZEITSCHRIFT FUER ANGEWANDTE PHYSIK
1948 V1 N2 P50-60 RCT V23 N1 P67 <RCT>
1948 V1 N2 P68-75 1537 <*>
1948 V1 N5 P197-211 59-20579 <*>
1948 V1 N5 P228-232 62-16973 <*> 0
1948 V1 N5 P229-231 62-10375 <*> 0
62-10376 <*>
1949 V1 P517-525 AL-551 <*>
1949 V1 N6 P245-252 59-14135 <+*>
1949 V1 N8 P359-366 65-18079 <*>
1949 V1 N10 P462-473 C-2436 <NRC>

1949 V1 N11 P509-516 59-14135 <+*>
1949 V1 N12 P545-549 AL-760 <*>
1949 V1 N12 P569-573 62-16030 <*>
1950 V2 P210-219 II-814 <*>
1950 V2 P261-267 II-12 <*>
1950 V2 N1 P20-24 3314 <*>
1950 V2 N1 P25-33 II-867 <*>
1950 V2 N1 P41-48 62-16974 <*> O
1950 V2 N2 P59-75 3019 <*>
57-777 <*>
1950 V2 N2 P83-88 C-2437 <NRC>
63-00766 <*>
1950 V2 N2 P88-95 59-16793 <+*>
1950 V2 N4 P152-160 AL-552 <*>
1950 V2 N4 P179-188 57-3020 <*>
1950 V2 N5 P204-205 <ATS>
1950 V2 N6 P252-254 C-2439 <NRC>
1950 V2 N11 P437-443 58-1442 <*>
58-701 <*>
1950 V2 N12 P473-478 59-12157 <+*>
1951 V3 P96-103 2883 <*>
1951 V3 P242-246 58-50 <*>
1951 V3 P300-303 3313 <*>
62-20061 <*> O
1951 V3 N5 P161-165 62-18694 <*>
1951 V3 N7 P267-271 63-00901 <*>
63-27382 <$>
1951 V3 N8 P300-303 I-370 <*>
1951 V3 N12 P441-449 65S89G <ATS>
1951 V3 N12 P459-467 62-14282 <*> O
1952 V4 P216-224 57-1193 <*>
1952 V4 P455-458 62-14429 <*> O
62-18964 <*>
1952 V4 N1 P1-12 66S89G <ATS>
1952 V4 N7 P258-267 UCRL TRANS-598(L) <*>
1952 V4 N9 P330-343 AEC-TR-6613 <*> P
1952 V4 N11 P418-424 61-18913 <*> O
1953 V5 P90-94 C-2420 <NRC>
1953 V5 P242-248 65-00359 <*>
1953 V5 P291-292 63-10004 <*> O
1953 V5 N1 P12-14 II-1014 <*>
1953 V5 N3 P95-101 II-23 <*>
1953 V5 N6 P201-206 C-4286 <NRCC>
1953 V5 N6 P207-211 62-10425 <*>
1953 V5 N6 P217-221 C-2450 <NRC>
1953 V5 N6 P221-231 59-10969 <*>
1953 V5 N9 P321-329 62-10572 <*>
1953 V5 N10 P375-376 II-43 <*>
1953 V5 N11 P426-440 1546 <*>
1953 V5 N12 P441-450 SCL-T-228 <*>
1954 V6 P416-417 57-1799 <*>
1954 V6 P481-489 II-414 <*>
1954 V6 P497-499 62-16540 <*> O
1954 V6 N1 P43-48 2163 <*>
1954 V6 N4 P160-163 66-11688 <*>
1954 V6 N5 P198-200 84F4G <ATS>
1954 V6 N5 P213-215 57-1758 <*>
1954 V6 N6 P241-246 57-3009 <*>
69H10G <ATS>
1954 V6 N6 P264-267 62-16631 <*> O
1954 V6 N7 P297-303 65-18121 <*>
1954 V6 N11 P499-507 65-17136 <*>
1955 V7 P50-56 C-2384 <NRC>
3451 <*>
1955 V7 P174-176 61-00669 <*>
1955 V7 P302-311 UCRL TRANS-570(L) <*>
1955 V7 P433-437 II-659 <*>
1955 V7 N1 P1-7 55K26G <ATS>
61Q73G <ATS>
1955 V7 N2 P99-107 58-1582 <*>
1955 V7 N3 P149-163 58-1565 <*>
1955 V7 N4 P195-212 58-1511 <*>
1955 V7 N5 P218-225 NP-TR-238 <*>
1955 V7 N5 P225-229 NP-TR-243 <*>
1955 V7 N5 P229-234 59-19388 <=*$> O
1955 V7 N6 P273-279 62-18277 <*> O
1955 V7 N7 P323-325 66-13909 <*>
1955 V7 N7 P337-344 65-10437 <*> O
1955 V7 N9 P437-443 57-2745 <*>
1955 V7 N9 P453-460 51H9G <ATS>
57-930 <*>
1955 V7 N10 P478-487 64-14204 <*> P
1955 V7 N12 P557-559 62-16582 <*> O
1955 V7 N12 P562-571 08Q70G <ATS>
1956 V8 P186-201 64-15894 <=*$>
1956 V8 P257-263 T-2529 <INSD>
1956 V8 P391-398 58-1629 <*>
1956 V8 P398-405 57-3455 <*>
1956 V8 P581-585 58-2132 <*>
1956 V8 N1 P34-40 57-3057 <*>
1956 V8 N2 P61-69 57-3213 <*>
1956 V8 N2 P69-75 57-2368 <*>
1956 V8 N3 P105-114 58-1619 <*>
1956 V8 N3 P127-135 58-504 <*>
1956 V8 N6 P299-302 63-00762 <*>
1956 V8 N6 P303-311 00L33G <ATS>
1956 V8 N7 P331-337 58-1530 <*>
1956 V8 N8 P372-382 58-33 <*>
1956 V8 N8 P405-406 57-2381 <*>
1956 V8 N9 P8-15 57-1399 <*>
1956 V8 N12 P569-577 57-2810 <*>
1956 V8 N12 P577-580 57-2812 <*>
1957 V9 N2 P66-68 58-2454 <*>
1957 V9 N3 P145-147 66-10702 <*>
1957 V9 N4 P164-171 62-01368 <*>
1957 V9 N5 P213-223 NP-TR-414 <*>
1957 V9 N5 P239-245 62-10238 <*>
1957 V9 N6 P301-305 NP-TR-416 <*>
1957 V9 N6 P305-313 NP-TR-415 <*>
1957 V9 N7 P347-349 59-10741 <*> O
1957 V9 N8 P388-394 62-23024 <=>
1957 V9 N8 P404-407 59-15023 <*>
1957 V9 N9 P429-434 63-10266 <*> O
66-10703 <*> O
1957 V9 N10 P503-513 C-4295 <NRCC>
1957 V9 N10 P520-526 62-10597 <*> O
1957 V9 N10 P526-531 61-18623 <*>
1957 V9 N11 P556-561 62-16409 <*>
1957 V9 N11 P561-566 63-14683 <*>
1957 V9 N12 P617-621 UCRL TRANS-700 <*>
1958 V10 P185-186 UCRL TRANS-589(L) <*>
1958 V10 N1 P1-7 61-16423 <*>
1958 V10 N1 P8-16 61-16098 <*>
1958 V10 N1 P26-31 65-00353 <*>
1958 V10 N1 P43-47 62-18292 <*> O
1958 V10 N3 P127-135 61-00231 <*>
1958 V10 N3 P150-155 97L32G <ATS>
1958 V10 N4 P157-162 59-20472 <*>
1958 V10 N4 P185-186 60-18137 <*>
1958 V10 N4 P195-199 61-00234 <*>
1958 V10 N5 P201-204 63-14006 <*> O
1958 V10 N6 P272-277 ANL-TR-28 <*>
1958 V10 N6 P277-285 60-14075 <*>
1958 V10 N7 P303-309 AEC-TR-3532 <+*>
1958 V10 N7 P309-317 60-14324 <*>
1958 V10 N7 P317-320 60-14490 <*>
1958 V10 N7 P320-327 60-14491 <*>
1958 V10 N7 P337-346 AEC-TR-3686 <+*>
59-10471 <*>
1958 V10 N9 P410-416 59-17741 <*> O
60-16874 <*>
62-14584 <*> O
1958 V10 N9 P424-428 95L28G <ATS>
1958 V10 N9 P428-433 62-18276 <*>
1958 V10 N11 P500-503 65-10077 <*>
1958 V10 N11 P525-531 62-16692 <*> O
1958 V10 N12 P537-546 62-01193 <*>
1958 V10 N12 P553-562 82M41G <ATS>
1959 V11 N2 P51-56 C-3214 <NRCC>
1959 V11 N2 P63-65 59-20471 <*>
1959 V11 N5 P172-174 59-17735 <*>
1959 V11 N5 P184-188 59-20015 <*>
1959 V11 N5 P190-199 65-11640 <*>
1959 V11 N7 P259-264 65-00352 <*>
1959 V11 N7 P264-274 65-00224 <*>
1959 V11 N7 P277-283 61-10319 <*>
1959 V11 N8 P290-296 65-10294 <*>
1959 V11 N10 P399-403 63-16308 <*>
63-20988 <*> O
1959 V11 N11 P441-443 62-10330 <*>

1960 V12 N4 P180-184 63-14014 <*>
1960 V12 N6 P250-253 62-00407 <*>
66-11282 <*> 0
1960 V12 N7 P323-324 62-10095 <*>
1960 V12 N8 P351-353 61-18864 <*>
1961 V13 N1 P11-16 30P65G <ATS>
1961 V13 N2 P59-64 AEC-TR-5695 <*>
1961 V13 N2 P74-76 65-17141 <*>
1961 V13 N3 P161-164 65-17142 <*>
1961 V13 N4 P180-182 63-10682 <*> 0
1961 V13 N4 P187-189 64-10882 <*> 0
1961 V13 N5 P247-250 62-10325 <*> 0
1961 V13 N6 P261-268 62-10761 <*> 0
1961 V13 N6 P283-288 65-17408 <*>
1961 V13 N7 P913-931 65-12794 <*>
1961 V13 N9 P416-419 62-20377 <*> 0
1961 V13 N9 P420-425 63-18739 <*>
1961 V13 N11 P493-500 AEC-TR-5196 <*>
1961 V13 N11 P527-534 3143 <BISI>
1961 V13 N11 P538-541 62-20387 <*>
1961 V13 N12 P560-565 62-20372 <*>
1961 V13 N12 P569-576 3143 <BISI>
1962 V14 N1 P21-22 63-18929 <*>
1962 V14 N3 P53-56 63-18885 <*>
1962 V14 N3 P121-125 62-01525 <*>
1962 V14 N4 P191-193 64-71434 <=>
1962 V14 N5 P297-299 63-16546 <*> 0
1962 V14 N7 P408-412 65-00105 <*>
1962 V14 N7 P412-424 63-18088 <*>
1962 V14 N7 P434-437 63-14745 <*>
1962 V14 N7 P444-448 66-11670 <*>
1962 V14 N8 P458-462 64-16954 <*>
1962 V14 N8 P469-475 63-20580 <*>
1962 V14 N8 P475-481 63-20615 <*>
1962 V14 N11 P681-692 03Q74G <ATS>
1962 V14 N12 P709-713 64-14631 <*>
1962 V14 N12 P734-738 64-10963 <*> 0
1962 V14 N12 P763-765 65-11000 <*>
1963 V15 N2 P61-64 62-10061 <*> 0
1963 V15 N3 P206-213 63-14462 <*>
65-12225 <*>
1963 V15 N3 P220-232 65-10747 <*>
1963 V15 N4 P291-295 65-11868 <*> 0
1963 V15 N6 P491-496 65-11167 <*> 0
1963 V15 N6 P509-518 64-14624 <*>
1963 V15 N6 P547-549 65-13643 <*>
1963 V16 P53-55 66-11100 <*>
1963 V16 N1 P1-5 64-16837 <*> 0
1963 V16 N2 P115-121 63-18177 <*>
1963 V16 N3 P145-150 64-14587 <*>
1963 V16 N3 P190-198 65-11166 <*> 0
1964 V17 N2 P67-68 66-10786 <*>
1964 V17 N2 P108-110 65-12226 <*>
1964 V17 N4 P289-295 65-11616 <*> 0
1964 V17 N7 P504-508 65-17137 <*>
1964 V18 N3 P116-120 28S85G <ATS>
1965 V18 N4 P323-329 09T91G <ATS>
1965 V18 N4 P370-373 66-10782 <*>
1965 V18 N5/6 P414-432 55S86G <ATS>
65-17139 <*>
1965 V18 N5/6 P549-552 66-13800 <*>
1965 V19 N1 P20-23 47S85G <ATS>
1965 V19 N1 P20-33 65-17147 <*>
1965 V19 N5 P401-404 66-12763 <*> 0

ZEITSCHRIFT FUER ANORGANISCHE UND ALLGEMEINE CHEMIE
1899 V19 P67-77 AEC-TR-4370 <*>
1905 V47 P190-202 62-16483 <*> 0
1906 V48 P185-190 57-1565 <*>
1906 V51 P369-390 63-20502 <*> 0
1908 V58 P325-337 62-16482 <*>
1909 V61 P181- 57-1457 <*>
1909 V62 P118-122 63-18173 <*>
1909 V63 N1 P1-48 64-18213 <*>
1909 V64 P245-257 203 <TC>
1910 V67 P113-148 64-10065 <*>
1910 V67 P183-199 AEC-TR-4198 <*>
1910 V67 P234-241 63-10484 <*>
1910 V67 P266-292 AEC-TR-5352 <*>
1910 V67 P293-301 AEC-TR-5385 <*>
1911 V70 P173-177 2084 <*>
1911 V70 P246-251 58-103 <*>
1911 V70 N4 P352-394 57-02250 <*>
1911 V71 P347-355 65-12793 <*>
1911 V71 P356-377 64-18243 <*>
1911 V73 N2 P200-222 64-18218 <*>
1912 V73 N3 P293-303 64-18271 <*> P
1912 V74 P153-156 AL-65 <*>
1912 V75 P30-40 MF-6 <*>
1912 V75 N1 P63-67 64-18249 <*> 0
1912 V76 P357-360 64-18282 <*>
1913 V80 P104-112 AEC-TR-4223 <*>
1913 V81 P243-256 57-2319 <*>
1913 V82 P103-129 63-14291 <*>
65-13298 <*>
1914 V86 P301-304 59-20863 <*>
1914 V87 P209-228 63-14303 <*> OP
1914 V89 P257-278 2292 <*>
1915 V91 P277-298 UCRL TRANS-930(L) <*>
1915 V93 P84-94 UCRL TRANS-990(L) <*>
1915 V93 P271-272 64-18282 <*>
1915 V93 N3 P202-212 66-10206 <*> 0
1916 V97 P312-336 58-1981 <*>
1917 V99 N2 P73-104 AL-621 <*>
1917 V102 P34-40 63-10483 <*>
1920 V112 P233-243 60-18730 <*>
1920 V113 P27-58 2804B <K-H>
63-20463 <*> 0
1920 V113 P179-228 65-18094 <*>
1920 V113 P306-316 59-20131 <*>
1920 V114 P1-23 AEC-TR-5674 <*>
1921 V116 P231-242 I-792 <*>
1921 V118 P193-201 64-18242 <*>
1922 V121 P167-177 I-573 <*>
1922 V123 P196-201 1808 <*>
1922 V123 P208-224 1808 <*>
1923 V127 P273-294 I-849 <*>
1923 V128 P81-95 2104 <*>
5309-J <K-H>
1923 V128 N1 P81-95 64-16741 <*> 0
1923 V128 N1 P96-116 60-16766 <*>
1923 V129 P267-275 63-14304 <*> 0
1923 V133 P321-347 62-20195 <*>
1924 V133 P306-311 AL-58 <*>
63-20768 <*>
1924 V133 P312-324 AL-59 <*>
64-10124 <*>
1924 V135 P201-204 63-16767 <*> 0
1924 V136 P289-294 II-126 <*>
1924 V138 P285-290 AL-423 <*>
63-20780 <*>
1924 V139 P22-36 2841 <*>
1924 V141 P161-227 65-11223 <*>
1925 V146 P225-229 II-732 <*>
1925 V147 P156-170 R-3421 <*>
1925 V148 P345-350 57-2076 <*>
63-20858 <*>
1925 V148 N1 P99- 57-2442 <*>
1925 V149 P181-190 UCRL TRANS-614(L) <*>
1925 V149 P203 58-2021 <*>
1925 V149 P203-216 58-2036 <*>
1925 V149 P217-222 AEC-TR-4804 <*>
1925 V150 P339-342 1143 <K-H>
1926 V150 N4 P355-380 62-18275 <*> 0
1926 V151 P1-20 65-10153 <*>
1926 V152 P52-58 63-16170 <*> 0
1926 V152 P225-234 AL-379 <*>
63-20778 <*> 0
1926 V152 P235-251 2095 <*>
64-10119 <*>
1926 V152 P267-294 AL-60 <*>
64-10125 <*> 0
1926 V152 P295-313 2094 <*>
63-20779 <*> 0
1926 V156 P213-225 63-14131 <*>
1926 V157 N1/3 P290-298 64-18169 <*> 0
1926 V158 P249-263 NP-TR-64 <*>

Citation	Translation
1927 V159 P197-225	63-10049 <*> P
1927 V160 P128-132	63-16150 <*> O
1927 V160 P222-236	65-14264 <*>
1927 V161 P287-303	AEC-TR-5111 <*>
1927 V163 P257-296	I-900 <*>
1927 V164 P287-296	UCRL TRANS-619(L) <*>
1927 V166 P161-	57-2447 <*>
1927 V166 N1/3 P16-26	<AC>
	AEC-TR-3722 <*> O
1927 V167 P230-236	62-16535 <*> O
1927 V168 P46-48	63-16304 <*>
1928 V169 P381-393	66-12936 <*>
1928 V170 N1/2 P25-34	59-20764 <*>
1928 V172 P32-42	63-18145 <*> O
1928 V172 P417-425	II-923 <*>
	59-10243 <*>
	8779 <IICH>
1928 V172 N1/3 P417-425	64-16060 <*>
1928 V173 P164-168	II-610 <*>
1928 V174 P145-160	AL-400 <*>
1928 V176 P209-232	57-2057 <*>
1929 V176 P209-232	63-16764 <*> P
1929 V178 N1/3 P353-365	AEC-TR-3397 <*>
1929 V179 P145	58-1944 <*>
1929 V179 P281-286	63-16040 <*> O
1929 V180 P19-41	60H12G <ATS>
1929 V180 P241	AEC-TR-3513 <*>
1929 V180 P251	AEC-TR-3513 <*>
1929 V181 P193-202	1150E <K-H>
1929 V182 P193-213	II-775 <*>
1929 V183 P127-139	AL-424 <*>
	63-20770 <*>
	8768 <IICH>
1930 V184 P1-38	63-16859 <*>
1930 V185 P225-238	58-1676 <*>
1930 V185 P403-412	63-18127 <*>
1930 V185 P413-416	63-18131 <*>
1930 V188 P72-80	II-344 <*>
1930 V188 P309-324	58-2043 <*>
1930 V190 P178-184	57-01827 <*>
	57-1827 <*>
1930 V190 P321-336	63-10592 <*>
1930 V191 P1-35	I-117 <*>
1930 V191 N3 P246-282	64-20384 <*>
1930 V192 P249-256	64-20559 <*>
1930 V193 P144-160	57-2037 <*>
1930 V193 P176-186	58-1222 <*>
1930 V193 N3 P237-244	59-20436 <*> PO
1930 V194 P1-37	66-10444 <*> O
1931 V195 P366-386	57-1373 <*>
1931 V196 P413-420	90S80G <ATS>
1931 V196 N2 P160-176	64-16415 <*>
1931 V197 P395-398	II-919 <*>
1931 V198 P32-38	II-921 <*>
	58-2647 <*>
	63-10593 <*>
	8778 <IICH>
1931 V198 P88-101	64-16137 <*> O
1931 V198 P233-275	2352 <*>
1931 V198 P243-261	58-1414 <*>
1931 V198 P262-275	63-18878 <*>
1931 V198 P297-309	59-15444 <*> O
1931 V198 N3 P297-309	59-20450 <*> O
1931 V198 N1/2 P32-38	64-16059 <*> O
1931 V199 N1/2 P97-108	SCL-T-289 <*>
1931 V200 P343-384	57-3095 PT1-2 <*>
1931 V201 P245-258	II-335 <*>
	II-519 <*>
1932 V206 P416-424	1208 <TC>
1932 V206 P129-151	I-598 <*>
1932 V207 P21-40	3870 <HB>
1932 V207 N1 P1-20	59-20117 <*> O
1932 V208 P293-303	62-16481 <*> O
1932 V208 P313-316	T-1640 <INSD>
1933 V210 P173-183	8776 <IICH>
1933 V210 N2 P113-124	63-16232 <*> O
1933 V211 P41-48	2343 <*>
1933 V211 P272-276	AL-906 <*>
1933 V211 P321-348	I-170 <*>
1933 V212 P399-400	II-922 <*>
	58-2644 <*>
	65-10122 <*>
1933 V214 P111	58-1984 <*>
1933 V214 N3 P225-238	66-11148 <*>
1933 V215 P321-332	2K21G <ATS>
1933 V215 N3/4 P321-332	NP-TR-288 <*>
1933 V215 N3/4 P388-414	60-10440 <*> O
1934 V216 P209-222	2274 <*>
1934 V217 P1-19	62-16475 <*> P
1934 V217 P93-94	II-920 <*>
1934 V217 P95-104	57-910 <*>
1934 V217 N1 P19-21	62-16484 <*>
1934 V217 N1 P22-26	63-16749 <*>
1934 V219 P143-148	I-574 <*>
1934 V220 P43-48	1640 <*>
1934 V221 P113-123	561 <TC>
1935 V221 N4 P321-333	58-01 <*>
1935 V222 P78-80	AL-353 <*>
	2867P <K-H>
	63-20756 <*>
1935 V222 P126-134	1249 <*>
	66-11951 <*>
1935 V222 P145-160	65K24G <ATS>
1935 V222 P225-239	57-2901 <*>
1935 V222 P289-302	57-2900 <*>
1935 V223 P177-184	AEC-TR-4462 <*>
1935 V223 P213-216	63-18132 <*>
1935 V223 P241-250	63-16752 <*>
1935 V223 P251-252	63-16757 <*>
1935 V226 P23-32	63-20856 <*>
1936 V230 P181-186	1209 <TC>
1936 V227 P81-93	63-10117 <*>
1936 V228 P92-96	59-17271 <*>
1936 V229 P85-86	63-10987 <*>
1936 V230 P123-128	61-10230 <*>
1937 V230 P257-276	1424 <*>
1937 V231 P63-65	63-14811 <*>
1937 V231 P160-171	II-888 <*>
1937 V231 P285-297	I-31 <*>
1937 V234 P313-326	62-18686 <*> O
1938 V235 P257-272	61-20515 <*> O
1938 V235 P337-351	57-2277 <*>
	63-14553 <*>
1938 V236 P45-56	59-17305 <*> O
1938 V236 P339-360	57-1839 <*>
1938 V239 P17-26	58-2115 <*>
1938 V239 P85-89	65-10835 <*>
1938 V239 P145-154	63-20855 <*>
1938 V240 P21-30	57-186 <*>
1939 V240 P167-172	64-20451 <*>
1939 V240 P289-299	60-10084 <*> O
1939 V241 P93-96	63-20267 <*> O
1939 V241 N4 P305-323	57-2498 <*>
1939 V242 P237-248	66-11765 <*>
1939 V243 P69-85	AEC-TR-4197 <*>
	II-17 <*>
1939 V243 N1 P1-13	64-20294 <*>
1940 V244 P133-148	UCRL TRANS-817 <*>
1940 V244 P205-223	57-2372 <*>
1940 V244 N1 P48-56	58-1980 <*>
1940 V245 P8-11	AEC-TR-4355 <*>
1940 V245 P12-15	66-10211 <*>
1941 V246 P149-157	UCRL TRANS-1053(L) <*>
1941 V246 P195-226	57-1150 <*>
1941 V247 P53-64	59-17754 <*>
1941 V247 P65-95	63-10437 <*>
1941 V247 P124-130	1643 <*>
1941 V247 P180-184	T1919 <INSD>
1941 V248 P1-31	57-1798 <*>
1941 V248 P256-268	64-20239 <*> O
1941 V248 P373-382	II-19 <*>
1942 V249 P238-244	58-1683 <*>
1942 V249 P299-307	AL-91 <*>
	I-981 <*>
1942 V249 N1 P113-118	AL-93 <*>
	II-24 <*>
1942 V250 N1 P1-9	63-16751 <*>
1943 V250 P357-372	66-12740 <*>

Citation	Translation
1943 V251 P233-240	I-318 <*>
1943 V251 P424-428	62-10358 <*>
1943 V251 N3 P222-232	C-3317 <NRCC>
1943 V251 N3 P285-294	AL-92 <*>
1943 V252 P144-159	63-10589 <*> 0
1945 V253 P92-94	63-10566 <*> 0
1947 V253 P194-208	65-00328 <*>
1947 V253 P251-262	63-20433 <*>
1947 V253 P281-296	62-16462 <*>
	64-30811 <*>
1947 V254 P1-30	I-77 <*>
1947 V254 P37-51	63-10561 <*>
1947 V254 P107-125	66-11841 <*>
1947 V254 N1/2 P1-30	AL-112 <*>
1947 V255 P79-100	II-312 <*>
1947 V255 P101-124	AEC-TR-5044 <*>
1947 V255 P141-184	59-10240 <*>
1947 V255 P287-293	I-319 <*>
1948 V255 P277-286	II-53 <*>
1948 V256 P125-144	62-10726 <*>
1948 V256 P239-252	58-2426 <*>
1948 V257 P26-40	63-10559 <*>
1948 V257 P340-348	62-16536 <*>
1948 V257 N4 P166-172	65-10289 <*> 0
1948 V258 P58-68	AEC-TR-4657 <*>
1949 V258 P1-14	63-10560 <*>
1949 V258 N3/5 P162-179	65-10128 <*> 0
1949 V259 P1-41	2278 <*>
	57-2852 <*>
1949 V259 P53-74	62-18213 <*>
1949 V259 P201-219	AEC-TR-5125 <*>
1949 V259 P305-308	66-11840 <*>
1949 V259 N1/4 P53-74	65-10129 <*> 0
1949 V259 N1/4 P135-142	65-13054 <*>
1949 V260 P120-126	1351 <*>
1949 V260 P255-266	59-17291 <*> 0
1949 V260 N6 P297-514	I-724 <*>
1950 V261 P26-42	57-248 <*>
1950 V261 P43-51	57-1845 <*>
1950 V261 P106-115	65-10765 <*>
1950 V261 P116-120	39G6G <ATS>
1950 V261 P248-254	65-10050 <*>
1950 V261 N588 P26-42	57-190 <*>
1950 V262 P61-68	57-741 <*>
1950 V262 P229-247	61H12G <ATS>
1950 V262 P258-287	I-943 <*>
1950 V262 P288-296	64-16521 <*>
1950 V262 P309-318	63-18221 <*>
1950 V262 N1/5 P229-247	60-16765 <*> 0
1950 V263 P102-111	65-10766 <*>
1950 V263 P174-184	1383 <TC>
1950 V263 P292-304	63-16110 <*>
1950 V263 P305-309	86T91G <ATS>
1951 V264 P2-16	58-1564 <*>
1951 V264 P226-229	UCRL TRANS-676(L) <*>
1951 V265 P128-138	57-00276 <*>
	57-276 <*>
1951 V265 P186-200	65-10011 <*> 0
1951 V265 P258-	<ES>
1951 V265 P288-302	AEC-TR-5122 <*>
1951 V265 P303-311	C-2663 <NRC>
	58-1423 <*>
1951 V265 N4/6 P186-200	AL-804 <*>
1951 V265 N4/6 P258-272	61-20884 <*>
1951 V265 N4/6 P288-302	65-17132 <*>
1951 V266 N4/5 P225-249	1636 <*>
1951 V267 P1-26	58-2275 <*>
1951 V267 P65-75	I-725 <*>
1951 V267 N1/3 P1-26	62-10384 <*> 0P
1952 V267 P189-197	75G6G <ATS>
1952 V267 P213-237	II-407 <*>
1952 V267 N4/5 P189-197	59-17956 <*>
1952 V268 P1-12	II-600 <*>
1952 V268 P268-278	<ES>
1952 V268 N3 P191-200	57-704 <*>
	59-17957 <*>
1952 V269 P52-66	3736-A <K-H>
1952 V269 P67-75	22L32G <ATS>
1952 V269 P292-307	3288 <*>
1952 V269 N1/2 P1-12	59-17958 <*>

Citation	Translation
1952 V269 N1/2 P52-66	64-16139 <*> 0
1952 V270 P114-144	63-20739 <*>
1952 V270 P240-256	I-1015 <*>
1952 V271 P6-16	77G6G <ATS>
1952 V271 N1/2 P6-16	59-17959 <*>
1953 V271 P150-152	II-277 <*>
1953 V271 P153-170	58-1534 <*>
1953 V271 P257-272	73G6G <ATS>
1953 V272 P40-44	63-16838 <*>
1953 V272 P111-121	64-20076 <*> P
1953 V272 P202-204	63-16112 <*>
1953 V272 P221-232	10464 <K-H>
	65-12506 <*>
1953 V272 P274-278	57-86 <*>
1953 V272 N14 P111-121	59-17916 <*> PO
1953 V272 N5/6 P288-296	62-10225 <*>
	63-14890 <=*>
1953 V274 P48-71	3824 <HB>
1953 V274 P105-113	81G6G <ATS>
1953 V274 P265-270	61-20532 <*>
1953 V274 P297-315	AL-910 <*>
	64-10128 <*>
1953 V274 N6 P297-315	64-20379 <*> 0
1953 V274 N4/5 P250-264	60-18007 <*>
	61-20531 <*>
1954 V275 P331-337	II-985 <*>
1954 V275 N1/3 P32-48	363 <TC>
	64-20312 <*> 0
1954 V276 P193-203	I-388 <*>
1954 V276 P209-	AEC-TR-4653 <*>
1954 V276 P289-315	II-582 <*>
1954 V276 N1/2 P95-112	60-10188 <*> 0
1954 V277 P27-36	61-10163 <*>
1954 V277 P60-72	II-270 <*>
	57-2391 <*>
	64-14377 <*> 0
1954 V277 P113-126	62-10373 <*> 0
1954 V277 P140-145	57-49 <*>
	63-14132 <*> 0
1954 V277 P156-171	59-15293 <*> 0
1954 V277 P249-257	53H10G <ATS>
1954 V277 N3/4 P113-126	65-10781 <*>
1955 P183-186	57-892 <*>
1955 V278 P12-23	58-2272 <*>
1955 V278 P33-41	57-800 <*>
	63-10116 <*>
1955 V278 P184-191	79G6G <ATS>
1955 V279 P38-58	58-2251 <*>
1955 V279 P104-114	58-207 <*>
1955 V279 P289-299	II-783 <*>
	61-10528 <*>
1955 V279 N1/2 P2-17	57-2548 <*>
1955 V280 N5/6 P332-345	302 <TC>
1955 V281 N1/2 P1-17	57-3497 <*>
1955 V282 P141-148	1486 <*>
1955 V282 P293-306	70R80G <ATS>
1955 V282 N1/6 P268-279	66-11008 <*>
1955 V282 N1/6 P345-351	59-15940 <*> P
1956 V283 P58-73	4050-C <K-H>
	64-16443 <*> 0
1956 V283 P88-95	57-750 <*>
1956 V283 P304-313	63-16851 <*>
1956 V283 P372-376	1425 <*>
1956 V283 P401-413	CSIRO-3138 <CSIR>
1956 V283 N1/6 P299-303	CSIRO-3137 <CSIR>
1956 V283 N1/6 P401-413	CSIRO-3138 <CSIR>
	59-10823 <*>
1956 V284 P153-161	13K23G <ATS>
1956 V284 N4/6 P177-190	25P64G <ATS>
1956 V285 P181-190	44L36G <ATS>
1956 V285 P246-261	57-769 <*>
1956 V285 P309-321	57-2596 <*>
1956 V285 N3/6 P181-190	60-16909 <*> 0
	62-10443 <*>
1956 V286 P27-57	63-00764 <*>
1956 V286 P42-55	61-00570 <*>
	63N51G <ATS>
1956 V286 N1 P27-41	61-18252 <*>
1956 V286 N3/4 P146-148	64-18508 <*>
1956 V287 P28-32	61-10948 <*>

1956 V287 P152-168 REPT. 97036 <RIS>
1956 V288 P87-90 AEC-TR-4654 <*>
1956 V288 P291-306 AEC-TR-3566 <+*>
1956 V288 P307-323 AEC-TR-3595 <+*>
1957 V288 P279-287 6094-B <K-H>
1957 V288 N5/6 P279-287 64-20219 <*>
1957 V289 P66-79 65-10009 <*> 0
1957 V289 P81 61-20768 <*>
1957 V289 N1/4 P175-192 62-14449 <*>
1957 V289 N5/6 P279-287 61-10563 <*>
1957 V290 P1-16 AEC-TR-4034 <*>
1957 V290 P184-190 62-14028 <*> 0
1957 V290 P303-319 60-16878 <*>
1957 V290 N5/6 P279-291 60-18004 <*>
1957 V291 P51-66 58-1591 <*>
63-01275 <*>
1957 V291 P221-226 61-18662 <*>
1957 V291 P294-304 61-00571 <*>
64N51G <ATS>
1957 V291 N1/4 P131-133 4143 <HB>
1957 V291 N5/6 P276-293 59-17210 <*>
1957 V292 P293-297 AEC-TR-4869 <*>
1957 V293 P109-131 419 <TC>
1957 V293 N1/2 P33-46 59-15076 <*>
1957 V293 N3/4 P204-213 59-15075 <*> 0
1958 V295 P7-35 65-17180 <*>
1958 V295 N3/4 P156-172 61K28G <ATS>
1958 V295 N3/4 P173-184 62K28G <ATS>
1958 V296 P157-163 60-19160 <=*>
1958 V297 P103-120 60-10702 <*>
1958 V297 P287-295 60-19159 <=*>
61-10227 <*> 0
1959 V298 P12-21 63-18819 <*>
1959 V298 P338-348 62-16447 <*> 0
1959 V298 N5/6 P270-278 65-12203 <*>
1959 V300 P81-83 SCL-T-457 <*>
62-16416 <*>
1959 V300 N1/2 P1-32 63-14134 <*> 0
63-16505 <*> 0
1959 V300 N3/4 P159-174 65-14666 <*>
1959 V301 P316-322 63-10547 <*>
1959 V301 N5/6 P326-335 61-14935 <*>
1960 V303 P133-140 54N48G <ATS>
1960 V303 P227-246 65-11190 <*>
1960 V304 P109-115 66-12995 <*>
1960 V304 N1/2 P12-24 CLAIRA T.18 <CLAI>
1960 V305 P158-168 89Q73G <ATS>
1960 V305 P178-189 63-10767 <*>
1960 V306 P260-265 UCRL TRANS-740(L) <*>
1960 V306 N1/2 P48-62 61-18633 <*> 0
1961 V307 P123-136 AEC-TR-5127 <*>
1961 V307 N3/4 P120-122 65-10857 <*>
1961 V307 N3/4 P187-201 66-13945 <*>
1961 V308 N1/6 P352-368 61-20187 <*>
1961 V309 P189-203 UCRL TRANS-764(L) <*>
1961 V309 N3/4 P43-49 62-24531 <=*>
1961 V309 N5/6 P297-303 61-31608 <=*>
1961 V310 P123-142 63-10585 <*>
1961 V310 P204-216 AEC-TR-5088 <*>
1961 V310 P217-224 AEC-TR-5087 <*>
1961 V310 N1/2 P90-93 62-24556 <=*>
1961 V310 N1/2 P100-109 62-24532 <=*>
1961 V311 N1/2 P13-21 62-14975 <*>
1961 V311 N1/2 P22-31 62-14974 <*>
1961 V311 N1/2 P83-91 62-24542 <=*>
1961 V311 N1/2 P117-120 62-24543 <=*>
1961 V311 N3/4 P151-179 64-16425 <*>
1961 V313 P161-169 AEC-TR-5299 <*>
1962 V314 P206-209 AEC-TR-5208 <*>
1962 V316 N1/2 P1-11 66-12875 <*>
1962 V316 N3/4 P190-196 63-18728 <*>
1962 V317 P21-24 65R79G <ATS>
1962 V318 P212-216 09Q67G <ATS>
1962 V319 N1/2 P1-8 204 <TC>
1963 V319 N5/6 P327-336 65-10872 <*>
1963 V319 N5/6 P404-408 65-10839 <*>
1963 V320 N1/4 P183-204 65-14216 <*>
1963 V320 N3/4 P183-204 65-12808 <*>
1963 V321 P217-223 AD-631 994 <=$>
1963 V321 P246-261 64-14055 <*> 0
1963 V321 N1/2 P41-55 65-10861 <*>
1963 V321 N5/6 P274-276 64-16625 <*>
1963 V322 N3/4 P145-154 63-20785 <*>
1963 V323 N3/4 P114-125 64-16613 <*> 0
1963 V323 N3/4 P207-208 64-16627 <=>
1963 V323 N5/6 P267-274 65-11457 <*>
1963 V326 P78-88 151 <TC>
1964 V327 P28-31 71R77G <ATS>
1964 V327 P32-50 72R77G <ATS>
1964 V328 P34-43 66-12026 <*>
1964 V329 P136-145 66-11084 <*>
1964 V331 N3/4 P169-182 65-12219 <*>
1964 V334 N1/2 P12-14 567 <TC>
1964 V334 N1/2 P50-56 566 <TC>
1964 V334 N3/4 P150-155 565 <TC>
1964 V334 N3/4 P197-201 566 <TC>
1964 V334 N3/4 P202-204 564 <TC>
65-12154 <*>
1965 V335 P225-236 ORNL-TR-649 <=$>
1965 V338 N3/4 P121-133 66-14569 <*>
1965 V340 P253-260 8794 <IICH>
1966 V342 N1/2 P20-24 66-13935 <*>

ZEITSCHRIFT FUER ASTROPHYSIK
1930 V1 P300-313 66-13521 <*>
1935 V10 P285-290 62-16686 <*>
1940 V20 P172-191 59-10112 <*> 0
1941 V20 P246- 59-10106 <*>
1941 V20 P323 59-10108 <*> 0
1941 V20 P332- 59-10121 <*> 0
1944 V22 P319-355 64-15597 <=*$>
1948 V25 P109-134 59-10113 <*>
1948 V25 N3/4 P310-314 AL-117 <*>
1949 V26 P295-304 AL-118 <*>
1950 V27 P42-45 59-10104 <*> OP
1950 V27 P132-152 AL-930 <*>
1950 V27 N1 P58-72 AL-798 <*>
1951 V28 N2 P150-175 66-12471 <*>
1951 V29 P274-286 UCRL TRANS-511(L) <*>
61-10667 <*>
63-14697 <*> 0
1953 N33 P39-73 57-3558 <*>
1953 V33 P39-73 C-2516 <NRC>
1955 V35 P213-232 59-10124 <*>
1955 V36 P181-193 C2223 <NRC>
T-1384 <INSD>
1544 <*>
1955 V37 P217-232 66-11381 <*>
1955 V38 P1-13 59-10125 <*> 0
1955 V38 P14-36 C-5369 <NRC>
1955 V38 P55-68 57-3119 <*>
1955 V38 N3 P166-189 65-13701 <*>
1956 V40 N1 P28-62 CSIRO-3354 <CSIR>
1957 V41 N1 P36-62 63-14888 <*>
1957 V42 N2 P76-86 63-14898 <*> 0
1958 V44 P71-97 N66-11605 <=*$>
1958 V44 N3 P183-202 63-16303 <*> 0
1958 V46 N1 P26-65 65-13703 <*>
1959 V48 P290-297 63-14653 <*>
1959 V48 N2 P140-154 65-13706 <*>
1959 V48 N3 P203-212 65-13705 <*>
1960 V49 N2 P73-94 63-18431 <*> 0
1960 V50 P184-214 UCRL TRANS-1069(L) <*>
1960 V50 N4 P110-120 63-10667 <*> 0
1961 V52 N4 P285-293 65-17161 <*>
1961 V53 N4 P226-236 62-20168 <*>
1962 V54 N2 P67-97 65-17160 <*>
1963 V57 P47-82 66-11855 <*>
1963 V58 P28-56 65-00370 <*>

ZEITSCHRIFT FUER AUGENHEILKUNDE
1905 V13 P220-225 276 <TC>
1920 V44 P132-140 61-14169 <*>
1923 V50 N3/4 P244 61-14264 <*>
1926 V59 N3 P206-207 61-14410 <*>
1927 V62 N6 P405-407 61-14265 <*>
1927 V63 N3 P159-162 61-14249 <*>
1928 V66 N3 P288-289 61-14274 <*>
1932 V77 P333-341 61-14148 <*>
1937 V95 N4 P223-224 61-14137 <*>

ZEITSCHRIFT FUER BIOCHEMIE
1933 V94 P159-170 I-265 <*>

ZEITSCHRIFT FUER BIOLOGIE
1905 V46 P153-166 64-16243 <*>
1905 V46 P441-553 1729 <*>
1918 V68 P163-222 61-14252 <*>
1919 V70 P187-210 N65-23678 <=*$>
1920 V70 P65-94 61-14994 <*>
1923 V78 P119-132 61-14194 <*>
1926 V84 P427-435 61-14423 <*>
1928 V88 P277-285 66-10951 <*>
1930 V90 N3 P260-298 60-17134 <*+> O
1931 V91 N3 P221-230 62-01231 <*>
1935 V96 P146-152 1145E <K-H>
1935 V96 P586-607 62-00502 <*>
1951 V104 P307-314 58-2227 <*>
1952 V105 P49-65 1919 <*>
1953 V105 N6 P454-463 65-12389 <*>
1954 V106 N5 P330-376 63-00731 <*>
63-27747 <= $>
1960 V111 N5 P379-392 10402-B <K-H>
63-16590 <*>
1960 V112 N1 P67-84 63-10588 <*> O
64-12341 <*>
1962 V113 N2 P145-160 C-4759 <NRC>

ZEITSCHRIFT FUER BOTANIK
1937 V31 P200-210 T-1903 <INSD>
1937 V31 P216-248 T-1903 <INSD>
1937 V31 P252-264 T-1903 <INSD>
1940 V35 P385-422 T-1670 <INSD>
1956 V44 P289-296 5526-B <K-H>
1956 V44 P505-514 CSIRO-3578 <CSIR>
1957 V45 P57-76 CSIRO-3577 <CSIR>
1957 V45 N6 P433-473 65-12128 <*> P
1958 V46 N4 P339-354 60-18642 <*>
1959 V47 N5 P421-434 62-00072 <*>
1961 V49 N1 P96-109 62-20471 <*> O

ZEITSCHRIFT FUER CHEMIE
1961 V1 P257-262 66-12061 <*>
1961 V1 N4 P97-105 62-19720 <=*>
1961 V1 N5 P148-152 62-19719 <=*>
1961 V1 N5 P158 62-19722 <=*>
1961 V1 N6 P191 62-19745 <=*>
1961 V1 N8 P247-248 62-19721 <*=>
1962 V2 P158 AEC-TR-5470 <*>
1962 V2 N2 P53 66-10689 <*>
1962 V2 N10 P297-302 65-14593 <*>
1962 V2 N6/7 P208-214 65-14734 <*> O
1963 V3 N5 P196-197 NP-TR-1212 <*>
1963 V3 N9 P329-333 27R76G <ATS>
1964 V4 P1-11 1803 <TC>
1964 V4 N4 P121-124 C-5489 <NRC>
1965 V5 N4 P142 66-12812 <*>

ZEITSCHRIFT FUER CHEMIE UND INDUSTRIE DED KOLLOIDE
1907 V2 P199-208 AEC-TR-4229 <*>
1907 V2 P230-237 AEC-TR-4230 <*>
1907 V2 P275-284 AEC-TR-4234 <*>
1907 V2 P301-307 AEC-TR-4235 <*>
1907 V2 P326-335 AEC-TR-4236 <*>

ZEITSCHRIFT DER DEUTSCHEN GEOLOGISCHEN GESELL-SCHAFT
1941 P213-214 AL-707 <*>
3339 <*>
1948 V100 P158-163 59-17240 <*>
1948 V100 P427-466 C-3206 <NRCC>
1955 V107 P132-139 54T91G <ATS>
1956 V107 P116-119 98K20G <ATS>

ZEITSCHRIFT FUER DIAGNOSTISCHE PSYCHOLOGIE UND PERSOENLICHKEITSFORSCHUNG
1954 V2 N1 P55-64 64-10357 <*>

ZEITSCHRIFT FUER ELEKTROCHEMIE
1954 V54 P478-482 95G7G <ATS>
1954 V58 P9-24 II-317 <*>
1954 V58 P25-39 9034 <IICH>
1954 V58 P49-53 2850 <*>
1954 V58 P66-76 II-308 <*>
64-14160 <*> O
1954 V58 P95-99 I-140 <*>
1954 V58 P334-339 AEC-TR-4220 <*>
1954 V58 P340-345 AEC-TR-4374 <*>
1954 V58 P612-616 61-00104 <*>
1954 V58 P653-655 60-23423 <*+>
1954 V58 P672-673 60-23437 <*+>
1954 V58 P889-899 41M42G <ATS>
1954 V58 N1 P9-24 57-1963 <*>
64-14159 <*> O
1954 V58 N2 P86-95 294 <TC>
65-11423 <*>
1954 V58 N2 P112-118 57-2278 <*>
1954 V58 N3 P156-161 59-17350 <*> O
1954 V58 N4 P230-237 3004 <*>
57-669 <*>
1954 V58 N5 P327-334 57-2832 <*>
1954 V58 N5 P348-354 66-13887 <*>
1954 V58 N7 P467-477 94G7G <ATS>
1954 V58 N9 P686-697 65-10024 <*>
1954 V58 N9 P709-718 68M41G <ATS>
1954 V58 N9 P756-761 23G6G <ATS>
1954 V58 N9 P762-767 12198A <K-H>
65-14468 <*> O
1955 P671-676 58-1435 <*>
1955 V59 P35-45 60-23432 <*+>
1955 V59 P173-184 60-23450 <*+>
1955 V59 P209-211 II-843 <*>
1955 V59 P312-329 II-906 <*>
1955 V59 P338-340 60-23451 <*+>
1955 V59 P494-503 UCRL TRANS-850 <*>
1955 V59 P967-970 57-115 <*>
1955 V59 P1059 UCRL TRANS-850 <*>
1955 V59 N1 P23-31 62-10034 <*> O
1955 V59 N4 P233-245 59-21125 <=>
1955 V59 N4 P290-296 63-14872 <*>
1955 V59 N6 P538-542 64-16559 <*>
1955 V59 N9 P834-839 1676 <*>
57-353 <*>
1955 V59 N9 P891-895 42T92G <ATS>
1955 V59 N10 P942-946 57-116 <*>
1955 V59 N7/8 P596-604 60-18791 <*>
1955 V59 N7/8 P604-612 AEC-TR-2640 <*>
1547 <*>
1955 V59 N7/8 P778-784 5035-B <K-H> O
64-16460 <*> O
1955 V59 N7/8 P784-787 59-10081 <*>
1955 V59 N7/8 P807-818 CSIRO-2976 <CSIR>
59-10825 <*>
1955 V59 N7/8 P807-822 65-13145 <*>
1956 V60 P109-118 58-220 <*>
1956 V60 P229-236 57-67 <*>
9014 <IICH>
1956 V60 P905-911 UCRL TRANS-834(L) <*>
1956 V60 P952-956 57-3401 <*>
1956 V60 P959-964 63-10575 <*>
1956 V60 N1 P67-70 57-1788 <*>
1956 V60 N3 P208-218 44J14G <ATS>
1956 V60 N3 P218-229 59-15080 <*>
1956 V60 N3 P236-245 64-14981 <*> O
1956 V60 N3 P325-333 64-16327 <*>
1956 V60 N4 P400-404 1677 <*>
27H12G <ATS>
1956 V60 N5 P431-454 63-18820 <*> OP
1956 V60 N5 P470-473 5408 <K-H>
64-16764 <*> O
1956 V60 N7 P731-740 57-3453 <*>
1956 V60 N8 P772-773 64-16313 <*>
1956 V60 N8 P774-782 64-16503 <*> O
1956 V60 N8 P789-796 64-16504 <*> O
1956 V60 N8 P804-815 64-16826 <*> O
1956 V60 N8 P816-822 64-16809 <*> O
1956 V60 N8 P823-827 64-16810 <*> O
1956 V60 N8 P828-831 64-16811 <*> O
1956 V60 N8 P831-835 64-16815 <*> O
1956 V60 N8 P835-838 64-16816 <*>
1956 V60 N8 P838-847 59-10572 <*>

Citation	Translation
	64-16892 <*>
	79K26G <ATS>
1956 V60 N8 P848-854	64-16886 <*> 0
1956 V60 N8 P855-859	64-16885 <*> 0
1956 V60 N8 P859-866	64-16898 <*> 0
1956 V60 N8 P866-869	64-18087 <*> 0
1956 V60 N8 P869-874	64-18088 <*> 0
1956 V60 N8 P874-887	64-18089 <*> 0
1956 V60 N8 P887-891	64-18090 <*> 0
1956 V60 N8 P892-898	64-18092 <*> 0
1956 V60 N8 P898-905	64-18091 <*> 0
1956 V60 N8 P905-911	64-18098 <*> 0
1956 V60 N8 P912-921	64-18158 <*> 0
1956 V60 N8 P921-929	64-18157 <*> 0
1956 V60 N8 P929-938	64-18156 <*> 0
1956 V60 N9/10 P978-982	64-16883 <*>
1956 V60 N9/10 P1033-1036	66R78G <ATS>
1956 V60 N9/10 P1053-1056	57-2817 <*>
1956 V60 N9/10 P1056-1063	57-3310 <*>
1956 V60 N9/10 P1080-1088	60-16482 <*> 0
1957 V61 P344-348 PT.3	62-00409 <*>
1957 V61 P315-320	57-1980 <*>
1957 V61 P348-357	60-19165 <+*>
1957 V61 N1 P10-18	66-11499 <*>
1957 V61 N1 P122-123	4036 <BISI>
1957 V61 N1 P163-173	40K28G <ATS>
1957 V61 N1 P222-226	61-18416 <*>
1957 V61 N2 P325-332	4J18G <ATS>
1957 V61 N3 P348-357	27K24G <ATS>
	65-11299 <*>
1957 V61 N3 P380-384	61-18902 <*> 0
1957 V61 N3 P384-393	64-16893 <*>
1957 V61 N4 P463-470	69M41G <ATS>
1957 V61 N4 P484-489	64-30700 <*>
1957 V61 N4 P489-506	65-13437 <*>
1957 V61 N6 P727-733	58-1446 <*>
1957 V61 N7 P827-833	03N50G <ATS>
	64-18517 <*>
1957 V61 N8 P901-907	9K23G <ATS>
1957 V61 N8 P950-956	56K21G <ATS>
1957 V61 N8 P956-961	63-00759 <*>
1957 V61 N8 P993-1000	65-10199 <*>
1957 V61 N8 P1014-1019	64-14264 <*>
1957 V61 N8 P1028-1048	82L33G <ATS>
1957 V61 N8 P1094-1100	04N50G <ATS>
1957 V61 N9 P1127-1137	58-1740 <*>
1957 V61 N9 P1146-1153	2804 <BISI>
1957 V61 N9 P1168-1176	60-16281 <*>
1957 V61 N9 P1184-1196	58-1070 <*>
	59-10962 <*>
1957 V61 N9 P1224-1230	AEC-TR-4749 <*>
1957 V61 N9 P1230-1235	64-30306 <*>
1957 V61 N9 P1241-1246	64-18515 <*>
1957 V61 N10 P1275-1283	61-10523 <*>
1957 V61 N10 P1291-1301	24L33R <ATS>
1958 V62 P335-340	59-17585 <*> 0
1958 V62 P870-873	481 <TC>
1958 V62 P971-977	61-10525 <*>
1958 V62 P977-989	UCRL-TRANS-932 <*=>
1958 V62 N1 P1-28	59-17801 <*>
1958 V62 N1 P33-40	59-20373 <*>
1958 V62 N2 P132-152	59-15695 <*>
1958 V62 N2 P180-188	60-16279 <*> 0
1958 V62 N2 P188-191	64-20227 <*> 0
	88K26G <ATS>
1958 V62 N2 P201-209	31M39G <ATS>
1958 V62 N3 P245-250	66-13891 <*>
1958 V62 N3 P256-264	64-18599 <*>
1958 V62 N3 P296-300	RCT V33 N2 P245 <RCT>
	60-16799 <*>
1958 V62 N3 P373-378	66-10351 <*>
1958 V62 N4 P400-404	59-15031 <*>
1958 V62 N5 P505-519	59-15032 <*>
1958 V62 N5 P544-567	02M39G <ATS>
	61-16144 <*>
1958 V62 N5 P582-587	70M41G <ATS>
1958 V62 N8 P906-914	03M39G <ATS>
	61-16151 <*>
1958 V62 N9 P993-999	62-14093 <*>
1958 V62 N10 P1075-1082	NP-TR-977 <*>
1958 V62 N10 P1124-1130	07L33G <ATS>
1958 V62 N6/7 P638-641	71M41G <ATS>
1958 V62 N6/7 P642-648	72M41G <ATS>
1958 V62 N6/7 P655-660	6490 <HB>
1958 V62 N6/7 P660-663	22L33G <ATS>
1958 V62 N6/7 P739-745	26L33G <ATS>
1958 V62 N6/7 P745-750	25L33G <ATS>
1958 V62 N6/7 P751-759	23L33G <ATS>
1959 V63 P269-274	AEC-TR-4369 <*>
1959 V63 P835-859	3020 <BISI>
1959 V63 P978-987	65-14723 <*>
1959 V63 P1030-1037	9820 <K-H>
1959 V63 N1 P6-12	61-10635 <*>
1959 V63 N1 P34-43	62-10363 <*> 0
1959 V63 N1 P145-152	22T94G <ATS>
1959 V63 N2 P207-213	2766 <BISI>
1959 V63 N2 P249-257	60-16810 <*>
1959 V63 N2 P257-261	73M41G <ATS>
1959 V63 N2 P327-341	TRANS-181 <FRI>
	63-18067 <*>
1959 V63 N3 P360-373	42L33G <ATS>
1959 V63 N4 P484-491	01N49G <ATS>
1959 V63 N4 P537-539	64-30941 <*> 0
1959 V63 N6 P683-688	64-30985 <*> 0
1959 V63 N7 P765-772	2844 <BISI>
1959 V63 N8 P943-950	60-16230 <*>
	62-10119 <*>
1959 V63 N9/10 P1030-1037	65-11244 <*> 0
1959 V63 N9/10 P1096-1110	63-18091 <*>
1959 V63 N9/10 P1120-1133	05N50G <ATS>
1959 V63 N9/10 P1164-1171	65-11941 <*>
1960 V64 P244-251	01P60G <ATS>
1960 V64 P258-269	N65-16306 <=$>
1960 V64 P270-275	N65-16307 <=$>
1960 V64 P650	AEC-TR-5283 <*>
1960 V64 P805-812	64-18554 <*> 0
1960 V64 P1042-1045	AEC-TR-5206 <*>
1960 V64 P1092	AEC-TR-4444 <*>
1960 V64 P1228-1234	AEC-TR-5277 <*>
1960 V64 N1 P43-44	60-18794 <*>
1960 V64 N1 P44-45	65-13810 <*>
1960 V64 N1 P157-165	13T92G <ATS>
1960 V64 N2 P258-269	66-11446 <*>
1960 V64 N3 P391-394	61-18671 <*>
1960 V64 N3 P407-413	02P60G <ATS>
1960 V64 N4 P468-477	16M44G <ATS>
1960 V64 N4 P483-491	27P60G <ATS>
1960 V64 N4 P525-533	65-12953 <*>
1960 V64 N5 P616-631	10486 <K-H>
	65-12523 <*>
1960 V64 N7 P906-919	65-12890 <*>
1960 V64 N7 P930-936	25N48G <ATS>
1960 V64 N7 P945-951	65-13549 <*> 0
1960 V64 N10 P1171-1179	64-14009 <*>
1960 V64 N10 P1215-1219	NP-TR-770 <*>
1960 V64 N8/9 P997-1011	64-10367 <*> 0
1960 V64 N8/9 P1015-1019	NP-TR-679 <*>
1960 V64 N8/9 P1077-1080	63-01693 <*>
1960 V64 N8/9 P1083-1088	63-01696 <*>
1961 V65 P321-327	UCRL TRANS-831(L) <*>
1961 V65 P504-517	63-16404 <*>
1961 V65 P604-615	65-10690 <*> 0
1961 V65 P649-651	8716 <IICH>
1961 V65 N1 P36-50	65-17164 <*>
1961 V65 N1 P50-61	65-10793 <*>
1961 V65 N1 P61-65	NP-TR-768 <*>
1961 V65 N1 P66-70	NP-TR-773 <*>
1961 V65 N5 P421-425	65-14249 <*>
1961 V65 N6 P493-503	65Q72G <ATS>
1961 V65 N9 P776-783	62-16199 <*> 0
1961 V65 N9 P783-786	62-20018 <*>
1961 V65 N7/8 P581-591	66-11479 <*>
1962 V66 P316-325	310 <TC>
1962 V66 P462-466	63-18724 <*>
1962 V66 N1 P29-35	63-16530 <*> 0
1962 V66 N1 P65-73	63-10892 <*> 0
1962 V66 N3 P239-248	64-18606 <*>
1962 V66 N3 P248-255	64-18607 <*>
1962 V66 N3 P255-260	64-18608 <*>
1962 V66 N4 P309-312	66-10350 <*>

1962 V66 N6 P497-502 63-10909 <*> O
1962 V66 N1C P803-813 64-16699 <*> O
1962 V66 N10 P832-838 63-18153 <*>
1962 V66 N8/9 P636-641 65-10432 <*> O
65-13618 <*>

ZEITSCHRIFT FUER ELEKTROCHEMIE UND ANGEWANDTE PHYSIKALISCHE CHEMIE

1905 V11 P809- 57-1453 <*>
1905 V11 P809-813 57-1773 <*>
1906 V12 P463-473 57-2366 <*>
1907 V13 P752-755 57-1750 <*>
1907 V13 N28 P414-434 RAE-LIB-TRANS-601 <NLL>
57-756 <*>
63-14423 <*> O
1908 V14 P768-770 II-18 <*>
1908 V14 P8C9-810 II-106 <*>
1908 V14 P811-813 58-2063 <*>
1908 V14 N2 P17-19 57-756 <*>
19C8 V14 N21 P285-292 57-755 <*>
63-14434 <*> O
1909 V15 P33-34 II-84 <*>
1909 V15 P1C5-106 II-107 <*>
1909 V15 P773-779 I-15 <*>
19C9 V15 P784-790 II-5 <*>
1909 V15 N2C P784-790 64-10447 <*> O
1910 V16 P170-176 MF-6 <*>
1910 V16 P613-620 64-14706 <*> O
1910 V16 N13 P461-498 63-14405 <*> O
1910 V16 N14 P559-566 58-2064 <*>
1911 V17 P348- II-200 <*>
1911 V17 P509-514 64-10110 <*>
1911 V17 P761-764 65-12469 <*>
1911 V17 N2 P45-57 81H11G <ATS>
1912 V18 P718-724 MF-6 <*>
1913 V19 N12 P482-489 1421 <*>
1915 V21 P50-54 58-1860 <*>
1918 V24 P157-162 58-2028 <*>
1920 V26 N7/8 P129-151 65-14752 <*>
1921 V27 P256-268 1153 <*>
1922 V28 P7-16 57-1C42 <*>
57-1056 <*>
1922 V28 P85-89 II-63 <*>
1922 V28 P384-387 I-806 <*>
I-820 <*>
1922 V28 P390-391 1143N <K-H>
1922 V28 P529-542 I-146 <*>
1923 V29 P5-8 66-12371 <*>
1923 V29 P189-192 64-14253 <*>
1923 V29 P354-358 57-1760 <*>
1923 V29 N11/2 P276-285 57-247 <*>
1923 V29 N17/8 P411 66-12898 <*>
1923 V29 N17/8 P411-412 66-12899 <*> O
1924 V30 N21/2 P508-516 68P59G <ATS>
1925 V31 P542-545 AL-340 <*>
2867JJ <K-H>
63-20754 <*>
1925 V31 P655-658 58-2393 <*>
1925 V31 N4 P172-176 63-16746 <*>
1926 V32 P354-362 64-18241 <*> O
1926 V32 P515- 2336 <*>
1926 V32 N8 P399-413 2436 <*>
1927 V33 P477 64-18321 <*>
1927 V33 N7 P290-308 66-13135 <*>
1927 V33 N10 P411-455 64-18837 <*> O
1928 V34 P320-323 58-1240 <*>
1928 V34 P616-625 I-220 <*>
1928 V34 P682-684 1143D <K-H>
1928 V34 N9 P632-637 61-16527 <*>
1929 V35 P165-184 UCRL TRANS-641(L) <*>
1929 V35 N2 P63-65 58-2414 <*>
1929 V35 N2 P65-70 58-2411 <*>
1929 V35 N2 P70-83 59S83G <ATS>
1929 V35 N9 P676-680 57-2129 <*>
1930 V36 N5 P336-346 59-20415 <*>
1930 V36 N9 P807-814 64-18841 <*> O
1931 V37 P674-678 63-10546 <*>
1931 V37 N4 P185-197 59-17626 <*>
1931 V37 N5 P233-237 59-20848 <*>
1931 V37 N8/9 P589-593 65-14245 <*>
1931 V37 N8/9 P740-742 63-16165 <*>
1931 V37 N8/9 P742-745 64-20510 <*>
1932 N38 P774-777 II-1037 <*>
1932 V38 N2 P55-68 58-1890 <*>
1932 V38 N2 P69-70 58-1855 <*>
1932 V38 N3 P137-148 58-2311 <*>
1932 V38 N4 P232-240 64-20319 <*> O
1932 V38 N8A P549-553 59-15816 <*> O
1932 V38 N8A P611-614 C-3928 <NRC>
1933 V39 P283-288 AEC-TR-3635 <+*>
1933 V39 N3 P168-180 61-16552 <*> P
1933 V39 N5 P281-283 45L29G <ATS>
1933 V39 N9 P731-735 II-938 <*>
1933 V39 N9 P756- I-850 <*>
1933 V39 N7B P531-537 AL-626 <*>
1933 V39 N7B P531-538 AL-627 <*>
1933 V39 N7B P531-537 63-20755 <*>
1933 V39 N7B P586-594 61-16497 <*> O
1934 V40 P33-36 AEC-TR-3626 <+*>
1934 V40 P154-158 I-892 <*>
1934 V40 P605-608 I-141 <*>
1934 V40 P702-707 T-2542 <INSD>
1934 V40 N2 P73-78 57-1763 <*>
1934 V40 N3 P154-158 58-1862 <*>
58-1948 <*>
1934 V40 N3 P160-164 AL-300-I <*>
1934 V40 N6 P295-297 58-1431 <*>
1934 V40 N6 P303-306 59-20937 <*> O
1934 V40 N7 P341-344 II-495 <*>
1935 V41 P876-879 R-5343 <*>
1935 V41 N2 P96-101 63-20137 <*> O
1935 V41 N3 P174-183 NP-TR-292 <*>
64-10066 <*>
1935 V41 N8 P596-597 II-496 <*>
1935 V41 N12 P880-883 AEC-TR-3482 <*>
1936 P570-579 I-641 <*>
1936 V42 P499-503 64-18908 <*>
1936 V42 P805-815 T1663 <INSD>
1936 V42 P862-872 3240 <*>
1936 V42 N1 P31-43 66-13310 <*>
1936 V42 N5 P247-258 58-1370 <*>
63-14424 <*> O
1936 V42 N6 P315-319 AEC-TR-3637 <+*>
1936 V42 N6 P319-330 59-20828 <*> PO
1936 V42 N9 P687 57-671 <*>
1936 V42 N12 P881-889 64-10067 <*>
1936 V42 N7B P504-522 59-17341 <*>
1937 V43 P14-28 58-2257 <*>
1937 V43 P390-397 57-868 <*>
1937 V43 P546-557 58-1972 <*>
1937 V43 P891-907 AL-201 <*>
1937 V43 N1 P14-28 63-16257 <*>
1937 V43 N1 P42-52 59-17627 <*>
1937 V43 N2 P104-119 81L32F <ATS>
1937 V43 N2 P127-139 59-20416 <*> O
1937 V43 N12 P891-907 64-10120 <*> O
64-20501 <*> OP
1938 V44 P134-143 1634 <*>
1938 V44 P831-840 57-2572 <*>
1938 V44 P870-877 58-1193 <*>
1938 V44 N2 P134-143 64-20292 <*>
1938 V44 N7 P391-402 58-821 <*>
1938 V44 N8 P513-517 63-16741 <*>
1938 V44 N8 P562-568 27L33G <ATS>
1938 V44 N9 P611-619 63-16768 <*>
1939 V45 P362- 2280 <*>
1939 V45 P541-548 II-587 <*>
1939 V45 P795- 1118 <*>
1939 V45 P881-894 58-2428 <*>
1939 V45 N1 P93-99 63-10701 <*> O
1939 V45 N3 P262-285 57-1022 <*>
1939 V45 N4 P310-313 64-14636 <*> O
1939 V45 N8 P597-608 60-14405 <*>
1939 V45 N8 P608-615 57-639 <*>
1939 V45 N9 P671-674 57-1123 <*>
1939 V45 N11 P820-822 II-447 <*>
1939 V45 N11 P826-829 II-447 <*>
1939 V45 N12 P881-884 2099 <BISI>
1939 V45 N12 P885-888 1635 <*>
1940 V46 P120-129 I-677 <*>

1940 V46 P284-287	58-1574 <*>
1940 V46 P458-480	1820 <*>
1940 V46 P601-626	63-20555 <*> 0
1940 V46 N1 P3-13	62-20341 <*> 0
1940 V46 N3 P170-181	AL-752 <*>
1940 V46 N6 P336-346	63-20773 <*>
	65-10137 <*> 0
1940 V46 N12 P658-661	58-2041 <*>
1941 V47 P155-162	66-12387 <*> 0
1941 V47 N2 P135-143	62-18418 <*> 0
1941 V47 N2 P147-150	57-1005 <*>
1941 V47 N2 P208-211	57-3484 <*>
1941 V47 N6 P441-450	2534 <*>
1941 V47 N7 P519-526	57-3132 <*>
1941 V47 N7 P527-535	58-1951 <*>
1941 V47 N7 P536-545	2533 <*>
1942 V48 P62-82	AL-165 <*>
1942 V48 N1 P9-23	62-18107 <*>
	66-11377 <*>
1942 V48 N10 P513-522	57-171 <*>
	60-00793 <*>
1943 V49 P471-474	8767 <IICH>
1943 V49 N11 P471-474	65-10143 <*>
1943 V49 N4/5 P238-242	2297 <*>
1943 V49 N4/5 P283-287	1143AA <K-H>
1943 V49 N4/5 P292-296	60-14550 <+*>
1944 V50 N11/2 P256-266	61-16779 <*> 0
1946 P336-346	AL-298 <*>
1948 V52 N3 P98-103	07K27G <ATS>
1949 V53 P118-128	57-977 <*>
1949 V53 P151-161	63-18212 <*> P
1949 V53 P241-249	AL-434 <*>
1949 V53 P362-369	8930 <IICH>
1949 V53 N3 P118-125	76G5G <ATS>
1949 V53 N3 P132-142	60-10184 <*> 0
1949 V53 N4 P222-228	2036 <*>
	64-14407 <*> 0
1949 V53 N5 P279-285	65-11308 <*>
1949 V53 N5 P326-331	65-11923 <*>
1949 V53 N6 P339-340	61-20271 <*>
1949 V53 N6 P341-343	61-20272 <*>
1950 V54 P5-12	I-935 <*>
1950 V54 P42-46	3001 <*>
1950 V54 P99-107	AL-852 <*>
1950 V54 P129-132	8942 <IICH>
1950 V54 P353-357	II-188 <*>
1950 V54 P531-535	62-00381 <*>
1950 V54 N2 P120-129	AEC-TR-3170 <*>
	58-975 <*>
1950 V54 N2 P129-132	AEC-TR-3196 <*>
	58-1297 <*>
1950 V54 N3 P184-190	57-2318 <*>
1950 V54 N3 P210-215	66-10625 <*>
1950 V54 N4 P254-263	AL-254 <*>
1950 V54 N4 P318-320	65-14239 <*> 0
1950 V54 N5 P353-357	66-14628 <*> 0
1950 V54 N6 P459-477	63-16742 <*>
1950 V54 N6 P477-485	63-16743 <*>
1950 V54 N6 P485-494	58-703 <*>
1950 V54 N7 P560-566	I-954 <*>
1950 V54 N596 P302-304	57-189 <*>
1951 V55 P229-237	AL-662 <*>
1951 V55 P475-481	63-10292 <*>
1951 V55 N5 P363-366	82G6G <ATS>
1951 V55 N5 P380-384	3018 <*>
	57-677 <*>
1951 V55 N6 P460-469	91K23G <ATS>
1951 V55 N6 P475-481	64-18899 <*> 0
1951 V55 N7 P612-618	64-20506 <*> 0
1951 V55 N7 P633-636	58-2378 <*>
1952 V56 P827-833	59-12031 <+*>
1952 V56 P879-882	63-20246 <*>
1952 V56 N1 P58-61	59-10149 <*>
1952 V56 N1 P61-65	74G6G <ATS>
1952 V56 N3 P223-228	76G6G <ATS>
1952 V56 N4 P324-326	3638 <BISI>
1952 V56 N4 P345-350	29J15G <ATS>
1952 V56 N4 P351-360	57-996 <*>
1952 V56 N4 P366-375	1353 <*>
1952 V56 N4 P366-373	57-2268 <*>
	65-10385 <*> P
1952 V56 N4 P420-428	58-2520 <*>
1952 V56 N5 P465-467	1325 <*>
1952 V56 N5 P490-496	II-77 <*>
1952 V56 N6 P551-561	64-20507 <*>
1952 V56 N7 P602-608	66-11380 <*>
1952 V56 N8 P746-749	64-16555 <*>
1952 V56 N10 P972-979	78G6G <ATS>
1952 V56 N10 P979-985	67K26G <ATS>
1953 V57 P342-350	66-14478 <*>
1953 V57 P445-448	61-18294 <*>
1953 V57 P773-775	58-514 <*>
1953 V57 N3 P172-178	1092 <*>
1953 V57 N3 P189-194	64-18843 <*> 0
1953 V57 N3 P201-207	66-10280 <*>
1953 V57 N3 P216-221	71P58G <ATS>
1953 V57 N4 P282-289	66M44G <ATS>
1953 V57 N4 P305-317	59-10150 <*>
1953 V57 N5 P382-385	67M41G <ATS>
1953 V57 N7 P579-590	I-111 <*>
1953 V57 N8 P632-636	60-15611 <*+>
1953 V57 N9 P801-834	59-17516 <*>
1953 V57 N10 P883-889	2536 <*>

ZEITSCHRIFT FUER ERNAEHRUNGSWISSENSCHAFT

1963 V3 P168-170	13180 <K-H>
	64-10605 <*>

ZEITSCHRIFT FUER ERZBERGBAU UND METALLHUETTENWESEN

1948 V1 P79-83	1825 <*>
1949 N10 P289-296	1826 <*>
1949 V2 P65-69	II-333 <*>
1949 V2 P353-360	63-18557 <*> C
1949 V2 N1 P1-6	II-621 <*>
1950 V3 P65-69	T-1682 <INSD>
1950 V3 P213-220	8937 <IICH>
1951 V4 P14-19	66-10629 <*>
1951 V4 P169-176	2321 <*>
1951 V4 N7 P245-249	2319 <*>
1951 V4 N11 P412-418	2316 <*>
1951 V4 N11 P416-418	1122 <*>
1952 V5 P270-276	1097 <*>
1953 V6 N2 P41-44	1152 <*>
1953 V6 N5 P175-180	1238 <*>
1953 V6 N8 P310-317	I-939 <*>
1954 V7 P1-3	1105 <*>
1954 V7 P18-22	I-593 <*>
1954 V7 P229-234	TS-1192 <BISI>
1954 V7 P389-394	57-2561 <*>
1954 V7 N6 P229-234	61-20594 <*>
1955 P1-	II-663 <*>
1955 V8 P14-18	57-1909 <*>
1955 V8 P117-125	59-10769 <*>
1955 V8 P221-228	NP-TR-60 <*> C
1955 V8 PB109-B114	63-16284 <*> 0
1955 V8 PB117-B126	CSIRO-2989 <CSIR>
1955 V8 PB127-B136	63-16134 <*> 0
1955 V8 PB269-B274	59-20724 <*>
1955 V8 N1 P1-7	62-18601 <*> C
1955 V8 N3 P101-104	64-16210 <*> 0
1955 V8 N3 P104-106	64-16197 <*> 0
1955 V8 N10 P478-485	64-20560 <*>
1956 V9 N1 P11-17	1236 <*>
1956 V9 N1 P17-25	58-1473 <*>
1956 V9 N4 P145-151	64-20952 <*>
	65-17082 <*>
	66-11338 <*> C
1956 V9 N4 P151-158	66-11352 <*> 0
1956 V9 N5 P207-214	CSIRO-3414 <CSIR>
1956 V9 N7 P322-333	58-1478 <*>
1956 V9 N8 P383-388	65-17319 <*> 0
1956 V9 N9 P417-421	59-10043 <*>
1956 V9 N12 P559-563	59-17454 <*>
1957 V10 P158-166	59-18561 <+*> 0
1957 V10 N8 P370-373	59-10001 <*> C
1957 V10 N8 P391-396	4K20G <ATS>
1957 V10 N10 P504-509	66-10139 <*> C
1959 V12 P103-111	1642 <TC>
1959 V12 P164-172	1642 <TC>
1959 V12 P224-226	1642 <TC>

1959 V12 P553-557 4245 <BISI>
1959 V12 N7 P309-321 56L35G <ATS>
1959 V12 N8 P388-393 56L35G <ATS>
1960 V13 P363-373 2475 <BISI>
1960 V13 N4 P151-157 47M45G <ATS>
1960 V13 N5 P205-212 24M46G <ATS>
1960 V13 N5 P216-223 46M45G <ATS>
1961 V14 N1 P32-35 62-14645 <*> 0
1963 V16 N1 P18-21 3286 <BISI>
1963 V16 N12 P611-621 65-17351 <*>

ZEITSCHRIFT FUER ETHNOLOGIE
1963 V87 N2 P203-208 63-01617 <*>

ZEITSCHRIFT FUER EXPERIMENTELLE UND ANGEWANDTE PSYCHOLOGIE
1953 V1 P214-241 59-14574 <+*> 0
1953 V1 P244-264 65-12575 <*>
1953 V1 P420-500 64-10345 <*>
1954 V2 P412-454 60-15645 <=*> 0
1956 V3 P381-412 58-1813 <*>
1958 V5 N1 P108-126 59-20880 <*>
1960 V7 P409-421 65-11667 <*>
1960 V7 N3 P422-426 63-01621 <*>

ZEITSCHRIFT FUER EXPERIMENTELLE PATHOLOGIE UND THERAPIE
1915 V17 P370-382 66-11802 <*>
1916 V18 P139-164 N65-29726 <=*$>

ZEITSCHRIFT FUER FERNMELDETECHNIK,-WERK UND GERAETEBAU
1920 V1 P146- 57-1559 <*>
1920 V1 P168- 57-1559 <*>
1920 V2 P139-141 57-2282 <*>
1923 V4 P27-31 57-1476 <*>
1924 V5 P93- 57-1484 <*>
1925 V6 P105-109 57-1555 <*>
1927 V8 N1 P8- 57-1470 <*>
1928 V9 N8 P113-118 57-1413 <*>
1928 V9 N9 P136-143 57-1413 <*>
1932 V13 N3 P37-42 57-2427 <*>
1936 V17 P129-134 66-13301 <*> 0
1936 V17 N5 P69-73 57-2456 <*>
1937 V18 N12 P185-189 57-2108 <*>
1940 V21 N10 P156-158 57-640 <*>
1942 V23 N12 P177-182 58-317 <*>
1943 V27 N1 P1-10 58-452 <*>
1943 V27 N2 P21-28 58-452 <*>
1943 V27 N2 P35-40 58-452 <*>

ZEITSCHRIFT FUER FLEISCH- U. MILCHHYGIENE
1939 V49 P164-167 57-1601 <*>
1940 V51 P29-32 60-10890 <*>
1941 V51 P267-269 60-10886 <*>
1942 V52 P237-239 60-10887 <*> 0
1942 V52 P247-249 60-10888 <*> 0
1942 V52 P260-263 60-10888 <*> 0

ZEITSCHRIFT FUER FLUGTECHNIK U. MOTORLUFTSCHIFF-FAHRT
1931 V22 N2 P49-52 60-10218 <*> 0
1931 V22 N3 P65-72 65-12961 <*>
1931 V22 N7 P189-195 65-12960 <*>
1931 V24 N4 P99-103 65-12961 <*>

ZEITSCHRIFT FUER FLUGWISSENSCHAFTEN
1953 N7 P175-183 59-16780 <+*>
1953 V1 N5 P109-122 57-2540 <*>
1955 V3 N2 P29-46 3010 <*>
1955 V3 N5 P130-133 C-3281 <NRCC>
1955 V3 N7 P205-216 II-750 <*>
1955 V3 N9 P305-313 61-27356 <=*>
1956 V4 N8 P269-272 65-12671 <*>
1956 V4 N3/4 P95-108 NACA TM1434 <NASA>
1957 V5 N1 P22-26 65-12672 <*>
1957 V5 N6 P181-183 58-352 <*>
1957 V5 N9 P276-280 RE 3-2-59W <NASA>
1957 V6 P161-168 58-1006 <*>
1958 V6 N9 P266-271 60-19145 <=*>

1960 V8 N12 P345-352 62-10050 <*> 0
1961 V9 N7 P222-228 66-13957 <*>
1961 V9 N12 P383-387 64-16031 <*> 0
1961 V9 N4/5 P97-99 62-10921 <*>
1962 V10 N7 P265-277 66-11069 <*>
1965 V13 N5 P158-168 N66-15581 <=$>

ZEITSCHRIFT FUER FORSTGENETIK UND FORSTPFLANZEN-ZUECHTUNG
1953 V2 N5 P96-102 58-2472 <*>
1954 V3 N6 P122-125 62-16623 <*> 0

ZEITSCHRIFT FUER FORST- U. JAGDWESEN
1915 V47 P241-248 57-1235 <*>
1925 V57 P296-303 57-1246 <*>
1925 V57 N3 P140-180 57-1249 <*>

ZEITSCHRIFT FUER GEBURTSHILFE UND GYNAEKOLOGIE
1904 V53 P361-419 5258-A <K-H>
1954 V140 P34-36 II-56 <*>
1954 V140 P77-83 I-922 <*>
1956 V146 N1 P33-42 59-12423 <+*> 0
1961 V157 P5-10 64-14552 <*>
1964 V161 N3 P292-305 65-14947 <*>

ZEITSCHRIFT FUER GEOPHYSIK
1927 V13 P328-348 AL-407 <*>
1939 V15 P304-320 AL-766 <*>
1949 V15 P321-332 AL-111 <*>
1954 V20 P75-83 62-14060 <*>
1955 V21 P57-73 C-2658 <NRC>
1955 V21 N4 P215-228 62-23847 <=*>
1961 V27 P282-299 66-15351 <*>
1961 V27 N6 P282-300 65-12637 <*>
1962 V28 N1 P11-46 63-20714 <*>
1964 V30 N1 P5-12 65-18138 <*>
1964 V30 N1 P43-44 65-18130 <*>
1964 V30 N3 P127-139 65-14821 <*>

ZEITSCHRIFT FUER DIE GESAMTE EXPERIMENTELLE MEDIZIN
1913 V1 P397- 2809 <*>
1923 V31 P215-220 59-15423 <*>
1923 V36 P324-343 62-00425 <*>
1924 P723-730 57-1693 <*>
1924 V42 P267-274 62-11368 <=>
1924 V43 P270-280 59-15422 <*> 0
1926 V51 P48-50 57-1496 <*>
1933 V89 N5/6 P699-764 2396 <*>
1933 V90 P130-137 62-01217 <*>
1933 V90 P651-660 57-386 <*>
1933 V92 P241-250 63-18445 <*> 0
1933 V92 P251-264 63-18444 <*> 0
1936 V98 P428-431 62-00426 <*>
1937 V101 P641-647 57-125 <*>
1939 V106 P457- I-298 <*>
1940 V107 P306-312 I-212 <*>
1942 V110 N4/5 P412-422 59-17883 <*> 0
1943 V113 N2 P256-263 2514 <*>
1951 V117 P227-236 57-3434 <*>
1952 V118 P294-295 II-99 <*>
1952 V119 P266-271 2692B <K-H>
1952 V119 N5 P550-558 65-12605 <*>
1953 V121 P430-441 1845 <*>
1953 V121 P488-496 62-00501 <*>
1953 V121 P574-588 58-1598 <*>
1953 V122 P39-59 58-1703 <*>
1953 V122 N3 P167-169 3324-A <K-H>
1955 V125 P614-625 57-1823 <*>
1956 V127 P90-102 1603 <*>
1957 V128 P340-360 57-3286 <*>
1958 V129 N5 P511-516 61-16138 <*>
1958 V130 N3 P289-300 66-12702 <*> 0
1959 V130 N6 P604-612 61-10259 <*> 0
1960 V134 N3 P310-322 61-20776 <*>
1961 V134 P475-492 62-00640 <*>
1963 V137 N1 P82-84 64-30049 <*>

ZEITSCHRIFT FUER DIE GESAMTE GIESSEREIPRAXIS
1930 P183-184 59-15643 <*>

1930 V51 N5 P41-43 59-20904 <*> 0
1930 V51 N42 P170-172 59-20905 <*>
1934 P75-76 II-443 <*>

ZEITSCHRIFT FUER DIE GESAMTE INNERE MEDIZIN UND IHRE GRENZGEBIETE
1951 V6 P65-68 57-2345 <*>
63-14565 <*>
1953 V8 N24 P1143-114 II-490 <*>
1954 V9 N10 P484-489 I-758 <*>
1955 V10 P961-967 5472-A <K-H>
1956 V11 P640-643 57-3203 <*>
1956 V11 N2 P55-57 57-2598 <*>
1957 V12 N3 P131-134 63-16662 <*>
6713-A <K-H>
1958 V13 N18 P720-723 61-00585 <*>
1958 V13 N22 P881-884 59-16862 <=*>
1959 V14 N4 P209-214 62-00490 <*>
1960 V15 N5 P271-273 64-00199 <*>
1960 V15 N17 P801-804 62-00491 <*>
1961 V16 P386-388 61-20973 <*>
1961 V16 P703-707 62-18099 <*> 0
1961 V16 N22 P961-964 64-16301 <*>

ZEITSCHRIFT FUER DIE GESAMTE KAELTEINDUSTRIE
1929 V36 N10 P189-193 64-20391 <*> 0
1936 V43 P55-58 I-29 <*>
1941 V48 N2 P24-28 63-18119 <*>

ZEITSCHRIFT FUER DIE GESAMTE NEUROLOGIE UND PSYCHIATRIE
1930 V129 P639-646 66-11999 <*>
1931 V135 P657-670 63-00179 <*>
1932 V143 P542-555 59-10703 <*>
1933 V153 N2 P182-192 58-781 <*>
1939 V165 P219-223 61-14310 <*>

ZEITSCHRIFT FUER DAS GESAMTE SCHIESS- UND SPRENGSTOFFWESEN
1919 N22 P361 AL-907 <*>
1925 V20 P146 AL-830 <*>
1925 V20 P165 AL-830 <*>
1928 V23 P44-46 AL-960 <*>
1928 V23 P51-53 AL-739 <*>
1929 V24 P81-88 AL-718 <*>
1929 V24 P129-135 AL-718 <*>
1929 V24 P172-177 AL-718 <*>
1929 V24 P219-223 AL-718 <*>
1929 V24 P259-263 AL-718 <*>
1929 V24 P300-302 AL-718 <*>
1929 V24 P340-346 AL-718 <*>
1929 V24 P383-388 AL-718 <*>
1929 V24 P427-432 AL-718 <*>
1929 V24 P478-484 AL-718 <*>
1931 V26 P145-149 AL-581 <*>
1932 V27 N3 P73-76 73L33G <ATS>
1932 V27 N4 P125-127 73L33G <ATS>
1932 V27 N5 P156-158 73L33G <ATS>
1935 V30 P42-43 AL-88 <*>
1935 V30 P332 AL-5 <*>
1937 P63-65 AL-689 <*>
1937 P88-92 AL-663 <*>
1937 N6 P148- AL-519 <*>
1937 V32 P93-97 AL-759 <*>
1938 P302-305 AL-713 <*>
1938 P310-312 AL-694 <*>
1938 V33 P185-186 AL-758 <*>
1938 V33 P334-338 AL-712 <*>
1939 P126-127 AL-800 <*>
1939 P161-163 AL-801 <*>
1939 P197-201 AL-801 <*>
1940 V35 P193-196 AL-617 <*>
1940 V35 P220-221 AL-617 <*>
1940 V35 P243-245 AL-617 <*>
1942 V37 N8 P148-150 II-576 <*>

ZEITSCHRIFT FUER DIE GESAMTE TEXTILINDUSTRIE
1952 V54 P1053-1056 T-2135 <INSD>
1952 V54 P1059-1066 T-2135 <INSD>
1952 V54 P1069-1075 T-2135 <INSD>
1954 V56 P195- 57-1781 <*>
1954 V56 N4 P195- 1840 <*>
1954 V56 N4 P195 60-10791 <*>
1954 V56 N11 P1265-1269 61-14559 <*> 0
1955 V57 P410-415 61-14561 <*>
1955 V57 P500-502 61-14561 <*>
1955 V57 P1387-1394 T-2143 <INSD>
1955 V57 N1 P107-109 61-14560 <*> 0
1955 V57 N12 P714-724 12K25G <ATS>
1956 V58 N13 P516-518 59-15581 <*> 0
1957 V59 P484-486 TT-786 <NRCC>
1957 V59 P777- 62-10605 <*>
1957 V59 N11 P435-437 62-14049 <*> 0
1957 V59 N19 P808-811 61-14562 <*> 0
1958 V60 N3 P109-112 61-14563 <*> 0
1958 V60 N4 P138-139 61-14563 <*> 0
1958 V60 N11 P450-456 61-14564 <*> 0
1958 V60 N14 P647-648 60-17498 <+*>
1958 V60 N19 P830-832 61-14565 <*> 0
1959 V61 N8 P305-308 C-4524 <NRCC>
1959 V61 N11 P458-461 60-18532 <*>
1960 V62 N1 P23-25 18M46G <ATS>
1960 V62 N2 P43-47 13M46G <ATS>
1960 V62 N21 P921-924 27N54G <ATS>
62-16231 <*>
1961 V63 N1 P47-51 C-3945 <NRCC>
1961 V63 N11 P933-938 64-18130 <*>
1961 V63 N11 P934-938 63-15584 <*>
1963 V65 P108-111 64-18132 <*>
1963 V65 P115-118 66-10556 <*>
1963 V65 N6 P448-453 64-10390 <*> 0
1964 V66 N9 P760-764 751 <TC>
1964 V66 N10 P827-830 752 <TC>
1964 V66 N11 P937-942 66-10744 <*>
1964 V66 N12 P1026-1030 66-10557 <*>
1965 V67 N1 P22-24 1006 <TC>
1965 V67 N8 P602-610 66-10558 <*>

ZEITSCHRIFT DER GESELLSCHAFT FUER ERDKUNDE ZU BERLIN
1928 P3-12 65-12643 <*>

ZEITSCHRIFT FUER GLETSCHERKUNDE UND GLAZIAL-GEOLOGIE
1956 V3 N2 P253-255 57-37 <*>

ZEITSCHRIFT FUER HALS-, NASEN- UND OHRENHEILKUNDE
1926 V16 P521-540 61-14335 <*>
1926 V16 P565-576 61-14793 <*>
1926 V17 P127-135 61-14794 <*>
1926 V17 P259-273 61-14336 <*>
1927 V17 P381-391 61-14773 <*>
1927 V17 P392-402 61-14792 <*>
1927 V18 P411-422 61-16055 <*>
1927 V18 P425-427 61-16032 <*>
1927 V19 P87-103 61-14337 <*>
1927 V19 P139-162 61-14791 <*>
1931 V29 P211-215 61-14660 <*>
1931 V29 P216-218 61-14661 <*>
1932 V30 P595-604 C-2441 <NRC>
1933 V34 P201-211 61-14649 <*>
1935 V39 P194-202 61-14321 <*>
1937 P326-331 57-1680 <*>
1937 V42 P241-255 61-14320 <*>
1939 V45 P22-30 61-14330 <*>
1939 V45 P141-145 61-14300 <*>
1939 V46 P305-320 61-14314 <*>
1941 V48 P478-494 C-2686 <INSD>

ZEITSCHRIFT FUER HAUT- UND GESCHLECHTSKRANKHEITEN UND DEREN GRENZGEBIETE
1951 V11 N8 P327-340 04K21G <ATS>
1953 V15 N8 P265- 66F3G <ATS>
1954 V16 N12 P377-381 3225-C <K-H>
1962 V33 P260-270 64-18638 <*>
1963 V34 N2 P25-30 65-13243 <*> 0
1963 V34 N6 P153-160 65-13242 <*> 0
1963 V35 P252-255 65-13704 <*>
1964 V36 N11 P353-355 66-10112 <*>
1965 V38 N6 P203-208 65-14227 <*>

ZEITSCHRIFT FUER HYGIENE UND INFEKTIONSKRANKHEITEN MEDIZINISCHE MIKROBIOLOGIE, IMMUNOLOGIE UND VIROLOGIE
1903 V42 P255-265 57-1748 <*>
1923 V99 P125-141 57-105 <*>
1923 V99 P142-165 57-104 <*>
1930 V111 P162-179 63-20191 <*>
1931 V112 P559-568 63-20192 <*>
1931 V112 P680-697 5176-A <K-H>
1942 V123 P717-724 57-1520 <*>
1943 V124 P670-682 57-1695 <*>
1953 V138 P94-100 NP-TR-735 <*>
1955 V141 N2 P180-196 II-998 <*>
1956 V143 P1-70 57-3381 <*>
1956 V143 P23-42 58-63 <*>
1957 V143 N4 P416-428 57-1254 <*>
1961 V147 P277-286 62-00532 <*>
1962 V148 P230-240 63-00294 <*>
1964 V149 P414-429 66-13197 <*>
1964 V149 N6 P407-413 65-17033 <*>

ZEITSCHRIFT FUER HYGIENISCHE ZOOLOGIE UND SCHAEDLINGSBEKAEMPFUNG
1951 V39 P129-139 2204 <*>
1952 V40 P257-259 AL-361 <*>

ZEITSCHRIFT FUER IMMUNITAETSFORSCHUNG UND EXPERIMENTELLE THERAPIE
1910 V4 N4 P470-485 62-00600 <*>
1914 V22 P59-78 58-1602 <*>
1922 V34 P75-94 3171 <*>
1928 V58 P434-448 66-13015 <*>
1929 V64 P81-113 57-1887 <*>
1931 V73 P475-493 1171 <*>
1931 V73 P494-504 1172 <*>
1932 V75 P209-216 57-1884 <*>
1932 V75 P513-525 1344 <*>
1933 V78 P86-99 I-729 <*>
1934 V82 P99-123 57-1877 <*>
1935 V84 P177-188 59-12675 <+*>
1938 V94 P436-458 I-1014 <*>
1943 V103 P397-419 64-10824 <*>
1943 V103 P494-508 64-10823 <*>
1950 V107 N1/2 P62-69 3160G <K-H>
1950 V108 P233-244 60-18638 <*>
1955 V112 N4 P34-350 1237 <*>
1957 V114 P378-392 59-14328 <+*> 0
1958 V116 P413-426 66-12965 <*>
1959 V117 N2 P164-176 66-11583 <*>
1961 V121 P239-249 63-20873 <*> 0

ZEITSCHRIFT FUER INDUKTIVE ABSTAMMUNGS- U. VERERBUNGSLEHRE
1942 V80 P210-219 1179A <K-H>
1950 V83 P318-323 64-10016 <*>

ZEITSCHRIFT FUER INFEKTIONSKRANKHEITEN, PARASITAERE KRANKHEITEN U. HYGIENE DER HAUSTIERE
1927 V31 P314-326 66E1G <ATS>
1934 V46 N1/2 P28-47 62-00207 <*>
1935 V48 P138-148 64-10294 <*>
1936 V49 P246-256 2941 <*>
1944 V60 P157-179 II-585 <*>

ZEITSCHRIFT FUER INSTRUMENTENKUNDE
1900 V20 P290-299 57-371 <*>
1902 V22 P14-21 63-14654 <*>
1905 V25 N10 P293-305 61-14379 <*> 0
1905 V25 N11 P329-339 61-14378 <*> 0
1905 V25 N12 P361-371 61-14382 <*> 0
1909 V29 P258 58-2219 <*>
1913 V33 N6 P177-183 16R78G <ATS>
1914 V34 N8 P252-257 66-11073 <*>
1921 V41 N5 P148-152 18R78G <ATS>
1925 V45 P70-84 58-478 <*>
1926 V46 N5 P221-238 2024 <*>
1926 V46 N5 P249-255 60-10307 <*> 0
1929 V49 N12 P595-600 61-14290 <*>
1930 V50 N7 P426-437 59-15635 <*> 0
1931 V51 N6 P318-324 63-16819 <*> 0
1931 V51 N8 P393-396 19R78G <ATS>
1931 V51 N11 P553-559 20R78G <ATS>
1935 V55 N9 P367-373 61-14189 <*>
1938 V58 N8 P316-321 61-14398 <*>
1941 V61 N11 P361-372 15R78G <ATS>
1942 V62 P296-301 57-944 <*>
1942 V62 N1 P7-15 17R78G <ATS>
1943 V63 P46-428 I-860 <*>
1957 V65 P69-72 C-2470 <NRC>
57-3032 <*>
1957 V65 N4 P73-77 59-17779 <*>
1959 V67 P52-59 63-16302 <*> 0
1959 V67 P285-288 66-12366 <*>
1959 V67 N9 P223-231 SCL-T-364 <*>
1960 V68 N3 P64-67 61-18669 <*>
1961 V69 N6 P176-180 AEC-TR-4781 <*>
1962 V70 N1 P9-15 C-5929 <NRC>
1963 V71 N2 P51-54 NP-TR-1061 <*>
1963 V71 N9 P259-262 65-11288 <*>
1964 V72 N9 P259-265 66-12183 <*>
1965 V73 N4 P108-114 65-17194 <*> 0

ZEITSCHRIFT FUER JAGDWISSENSCHAFT
1956 V2 P24-33 59-20163 <*>
1956 V2 N3 P119-123 AEC-TR-4942 <*>

ZEITSCHRIFT FUER KINDERHEILKUNDE
1939 V61 N2 P142- 57-1662 <*>
1949 V67 P85-95 <ATS>
1954 V74 N2 P133-140 II-954 <*>
1956 V77 P577-585 57-3486 <*>
1959 V82 P514-525 62-00107 <*>
1961 V85 P277-285 63-00211 <*>
1962 V86 P525-534 63-00285 <*>
1962 V86 P535-542 63-00200 <*>
63-26700 <=$>
1962 V86 P543-552 63-00286 <*>
1962 V86 P553-559 63-00197 <*>
63-26801 <=$>
1964 V89 N1 P1-17 65-10986 <*>

ZEITSCHRIFT FUER KINDERPSYCHIATRIE
1952 V19 N2 P49-87 62-16469 <*> 0
1952 V19 N2 P88-91 62-16431 <*>
1957 V24 P75-84 58-1180 <*>

ZEITSCHRIFT FUER KLINISCHE CHEMIE
1963 V1 N6 P188-189 65-14333 <*>
1964 V2 N1 P1-6 65-17243 <*>
1964 V2 N2 P44-52 65-17242 <*>

ZEITSCHRIFT FUER KLINISCHE MEDIZIN
1897 V32 P1-28 SUP 66-12969 <*> 0
1899 V38 P127-139 62-00979 <*>
1927 N105 P301-317 58-331 <*>
1931 V116 P88-102 1732 <*>
1949 V145 P61-72 57-3422 <*>
1952 V149 P411-419 AL-918 <*>
1953 V152 P240-243 63-16961 <*>
9827-A <K-H>
1955 V153 N3 P246-255 64-20039 <*>
1957 V154 P439-455 58-232 <*>

ZEITSCHRIFT FUER KREBSFORSCHUNG
1926 V23 N3 P209-228 II-794 <*>
1930 V30 P349-365 64-10631 <*>
9222 <K-H>
1930 V33 P302-319 1354C <K-H>
1932 V37 P49-77 1469 <*>
1933 V40 N2 P189-191 II-989 <*>
1935 V42 P163-177 57-403 <*>
1936 V44 P441-448 57-376 <*>
1937 V45 P471-479 63-16962 <*>
1938 V47 P169-175 57-402 <*>
1938 V47 P198-208 57-377 <*>
1939 V49 P477-495 1114 <*>
1939 V49 N1 P46-56 57-540 <*>
1939 V49 N1 P57-84 10085 <K-H>
1939 V49 N4 P441-442 57-378 <*>
1940 V49 N6 P677-678 57-298 <*>

1940 V50 P196-219 II-991 <*>
1941 V51 P305-321 3510-A <K-H>
1941 V52 P1-16 2381 <*>
1950 V56 P587-595 I-262 <*>
1951 V57 P288-338 58-1334 <*>
1953 V59 N4 P397-401 3199-D <K-H>
1954 V60 P169-195 57-160 <*>
1954 V60 N2 P139-161 4020-B <K-H>
1957 V61 P468-503 57-3238 <*>
1959 V62 P596-610 61-00397 <*>
1960 V63 N3 P215-218 64-16076 <*>
1961 V64 P353-365 63-10711 <*> 0
1962 V64 P499-502 62-00981 <*>
1962 V65 P56-61 63-00154 <*>
1962 V65 P62-68 63-00173 <*>
1962 V65 P69-74 63-00150 <*>
1962 V65 P166-167 63-00551 <*>
1962 V65 P168-170 63-00558 <*>
1963 V65 P385-395 AEC-TR-6161 <*>
1964 V66 P87-108 02R76G <ATS>
66-15343 <*> 0

ZEITSCHRIFT FUER KREISLAUFFORSCHUNG
1948 V37 N9/11 P266-272 C-2414 <NRC>
1954 V43 P217-224 II-463 <*>
1954 V43 P614-624 II-275 <*>
1955 V44 P310-321 2390 <*>
1955 V44 P433-441 II-727 <*>
1955 V44 N1/2 P29-36 3546-F <K-H>
1955 V44 N7/8 P272-278 3769-C <K-H>
1956 V45 N5/6 P167-176 5339-A <K-H>
1957 V46 N1/2 P49-57 62-00629 <*>
1958 V47 N11 P965-971 60-14547 <*>
1958 V47 N5/6 P260-267 59-10651 <*> 0
1959 V48 P230-236 62-18251 <*>
1960 V49 N15/6 P732-744 65-17239 <*>
1960 V49 N19/0 P890-904 61-10751 <*>
1961 V50 P1162-1169 63-01768 <*>
1963 V52 N10 P961-969 64-00196 <*>
1963 V52 N11 P1117-1119 64-18973 <*> 0
1964 V53 N6 P587-593 65-14831 <*>

ZEITSCHRIFT FUER KRISTALLOGRAPHIE, KRISTALLGEOMETRIE, KRISTALLPHYSIK, KRISTALLCHEMIE
1925 V61 P299-314 65-11874 <*>
1925 V62 P113-122 UCRL TRANS-826(L) <*>
1928 V67 N1 P148-162 64-14625 <*> 0
1928 V68 P543-566 2997 <*>
1933 V85 P239-270 298 <TC>
1934 V89 P477-480 II-525 <*>
1935 V90 P82-91 58-1938 <*>
1935 V91 P193-228 57-133 <*>
1935 V92 P89-130 AEC-TR-5934 <*>
66-11356 <*> 0
1935 V92 P190 I-362 <*>
1935 V92 P190- 2102 <*>
1935 V92 P387-391 UCRL TRANS-966(L) <*>
1940 V102 P432-454 18S86G <ATS>
1941 V103 P228-229 61-16803 <*>
1955 V106 P122-144 8663 <IICH>
1955 V106 N6 P430-459 65-14313 <*>
1956 V107 P72-98 AEC-TR-5393 <*>
1956 V107 P265-289 AEC-TR-5391 <*>
1956 V107 P290-317 AEC-TR-5392 <*>
1956 V107 P406-432 ANL-TRANS-46 <*>
1956 V108 N1/2 P52-63 CHEM-110 <CTS>
1957 V109 P338-353 58-2672 <*>
1957 V109 P354-366 58-1531 <*>
1957 V109 N3 P204-225 CHEM-111 <CTS>
1960 V114 N3/4 P232-244 67P66G <ATS>
1961 V116 P126-133 UCRL TRANS-882(L) <*>
1961 V116 N3/6 P173-181 65-11451 <*> 0
1962 V117 P135-145 NP-TR-923 <*>
1964 V120 P19-31 1736 <TC>
1964 V120 N6 P472-475 66-10314 <*>
1965 V122 P185-205 67T91G <ATS>

ZEITSCHRIFT FUER LARYNGOLOGIE, RHINOLCGIE, OTOLOGIE U. IHRE GRENZGEBIETE
1934 V25 N5 P293-304 61-14646 <*>
1935 V26 N1 P29-34 61-14895 <*>
1957 V36 N5 P241-253 58-2310 <*>
1959 V38 P277-294 63-20217 <*> 0
1960 V39 P504-512 63-10709 <*> 0
1960 V39 P512-520 63-10710 <*> 0

ZEITSCHRIFT FUER LEBENSMITTELUNTERSUCHUNG UND-FORSCHUNG
1943 V86 P38-40 63-20167 <*>
1948 V88 P174-190 66-12782 <*>
1948 V88 P499-506 57-1214 <*>
1953 V97 P281-289 58-847 <*>
1953 V97 N4 P253-263 65-10062 <*> 0
1954 V98 P110-118 3046-B <K-H>
1954 V98 P294-301 3161-C <K-H>
1955 V100 P218-222 3614-B <K-H>
1955 V100 P260-266 68G8G <ATS>
1955 V100 P441-449 64-10517 <*> 0
1955 V100 N4 P260-266 59-20746 <*> 0
1955 V100 N4 P275-303 57-1964 <*>
1955 V102 P123-127 3850 <K-H>
1955 V102 P254-268 57-1209 <*>
1955 V102 P330-337 64-10516 <*> 0
1955 V102 N1 P34-37 3799-D <K-H>
1955 V102 N4 P236-245 59-20157 <*>
1955 V102 N5 P321-329 59-20784 <*>
1956 V103 P113-121 5131-A <K-H>
1956 V104 N1 P26-28 59-20148 <*>
1956 V104 N4 P245-260 CSIRO-3424 <CSIR>
1957 V105 N3 P207-210 59-20147 <*>
1957 V105 N5 P361-371 63-16297 <*> 0
1957 V105 N6 P437-445 63-16497 <*>
1957 V106 N3 P187-196 59-20297 <*> P
1957 V106 N3 P196-200 63-16298 <*>
1957 V106 N4 P281-297 CSIRO-3853 <CSIR>
1957 V106 N4 P298-301 59-20298 <*> P
1959 V109 N1 P13-26 60-15977 <*=>
1960 V111 P371-373 11363-B <KH>
63-16885 <*>
1960 V112 N6 P457-463 61-14973 <*>
1961 V114 P13-18 62-14290 <*>
1961 V114 N4 P297-301 64-14771 <*>
1961 V114 N4 P298-301 11071-E <K-H>
1961 V114 N5 P397-406 C-3736 <NRCC>
1961 V115 N1 P20-26 11499-D <K-H>
63-16650 <*>
1961 V115 N3 P244-262 62-18846 <*>
1961 V115 N4 P293-309 62-10903 <*>
1962 V117 P129-131 63-20964 <*>
1962 V118 N1 P51-52 96Q70G <ATS>
1963 V118 N6 P508-522 63-18041 <*>
1963 V119 N3 P201-210 C-4665 <NRC>
1963 V119 N5 P381-389 C-4802 <NRC>
1963 V119 N6 P492-497 C-4666 <NRC>
1963 V120 N4 P277-286 19Q74G <ATS>
1963 V120 N5 P357-363 27Q74G <ATS>
1963 V120 N6 P449-454 28Q74G <ATS>

ZEITSCHRIFT FUER MATHEMATIK UND PHYSIK
1897 V42 N4 P206-213 66-11064 <*>
1909 V57 N1/2 P158-173 AL-826 <*>
1912 V60 P397-398 T-2475 <INSD>
1916 V64 P262-272 65-10787 <*>

ZEITSCHRIFT FUER MATHEMATISCHE LOGIK UND GRUNDLAGEN DER MATHEMATIK
1957 V3 P157-170 AMST S2 V23 P89-101 <AMS>
1958 V4 P66-88 AMS TRANS V.46 P193 <AMS>
AMST S2 V46 P193-212 <AMS>

ZEITSCHRIFT FUER MESSEN, STEUERN, REGELN
1961 V4 N7 P273-276 62-24784 <=*> P
1961 V4 N7 P277-280 62-23386 <=*> P
1961 V4 N7 P281-286 62-19167 <=*> P
1961 V4 N7 P287-294 62-19788 <=*> P
1961 V4 N7 P294-297 62-19765 <=*> P
1961 V4 N7 P298 62-19155 <=*>
1961 V4 N7 P299-303 62-23383 <=*>
1961 V4 N7 P303-304 62-15845 <=*>
1961 V4 N7 P305-310 62-23620 <=*>

Citation	Translation
1961 V4 N8 P333-338	62-23620 <=*>

ZEITSCHRIFT FUER METALLKUNDE

Citation	Translation
1920 V12 P427-430	58-1934 <*>
1921 V13 P259-266	58-99 <*>
1921 V13 P353-367	58-1848 <*>
1921 V13 P5C7-510	57-2242 <*>
1922 V14 P329-334	62-18272 <*> 0
1922 V14 P357-388	57-3217 <*>
1922 V14 P425-449	57-3217 <*>
1925 V17 P310-315	T-1549 <INSD>
1927 V19 P12	58-1947 <*>
1927 V19 P282	58-1847 <*>
1927 V19 P492	58-656 <*>
1927 V19 N2 P41-43	57-3222 <*>
1927 V19 N3 P111-112	57-101 <*>
1927 V19 N3 P111-	57-2176 <*>
1927 V19 N6 P255-	57-2190 <*>
1927 V19 N9 P352-357	58-645 <*>
1928 V20 P323	58-1857 <*>
1928 V20 N4 P145-150	AL-450 <*>
1928 V20 N12 P437-441	57-329 <*>
1929 V21 P46-55	I-249 <*>
1929 V21 P121	58-1882 <*>
1929 V21 N1 P2-6	57-2449 <*>
1929 V21 N5 P145-151	62-16925 <*> 0
1929 V21 N5 P190-196	62-16926 <*> 0
1929 V21 N6 P197-199	57-2185 <*>
1929 V21 N10 P333-337	59-20352 <*> 0
1929 V21 N11 P368-371	59-20352 <*> 0
1930 P273-276	57-3227 <*>
1930 V22 N3 P84-89	57-2448 <*>
1930 V22 N6 P2C7-209	2324 <*>
1930 V22 N6 P210-212	57-3111 <*>
1930 V22 N8 P269-272	59-20336 <*> 0
1930 V22 N8 P277-279	59-20335 <*> 0
1930 V22 N8 P280-283	59-20334 <*> 0
1930 V22 N12 P397-401	59-20333 <*> 0
1931 V23 P105	58-2047 <*>
1931 V23 N1 P18	59-20332 <*>
1931 V23 N2 P47-51	57-3141 <*>
1931 V23 N2 P54-57	58-1987 <*>
1931 V23 N3 P73-76	57-3090 <*>
1931 V23 N3 P95-96	59-20358 <*> 0
1931 V23 N5 P139-142	57-3084 <*>
1931 V23 N5 P151-157	57-2803 <*>
1931 V23 N7 P197-201	63-18090 <*>
1931 V23 N7 P2C5-21C	59-20404 <*> 0
1932 V24 N2 P25-29	59-20587 <*>
1933 V25 P77	58-1989 <*>
1933 V25 P252-259	58-1983 <*>
1933 V25 N4 P84-88	58-1067 <*>
1933 V25 N7 P153-156	58-1878 <*>
1933 V25 N7 P163-165	59-10946 <*>
1933 V25 N9 P197-202	57-340 <*>
	57-3464 <*>
1933 V25 N11 P291	59-17966 <*>
1934 V26 P6-9	ANL-TRANS-87 <*>
1934 V26 P87-90	57-1108 <*>
1934 V26 P220	58-1979 <*>
1934 V26 P227	58-1939 <*>
1934 V26 N3 P49-55	57-3346 <*>
1934 V26 N3 P61	1807 <*>
1934 V26 N3 P61-	1807 <*>
1934 V26 N3 P62-65	59-10946 <*>
1934 V26 N12 P268-271	65-13907 <*>
1934 V26 N12 P274-279	57-1 <*>
1935 V27 P121	58-1880 <*>
1935 V27 P169	58-1992 <*>
1935 V27 N8 P169-174	58-357 <*>
1935 V27 N8 P183-187	57-1489 <*>
1935 V27 N11 P264-266	57-318 <*>
1935 V27 N12 P271-275	II-469 <*>
1936 V28 P133-138	65-13055 <*>
1936 V28 P237-239	59-17280 <*> 0
	63-16854 <*>
1936 V28 P340-343	66-14467 <*> 0
1936 V28 P381-382	57-2107 <*>
1936 V28 N3 P58-63	59-20351 <*> 0
1936 V28 N4 P80-83	62-18346 <*> 0
1936 V28 N6 P159-160	64-16113 <*> 0
1936 V28 N7 P183-188	62-18345 <*> 0
1936 V28 N7 P197-202	62-18604 <*> 0
1936 V28 N8 P230-233	2132 <*>
1936 V28 N9 P248-253	59-20234 <*> 0
1936 V28 N10 P299-308	57-2472 <*>
1936 V28 N10 P312-317	59-20349 <*> 0
1936 V28 N10 P317-319	61-23995 <=*> 0
	63-20832 <*>
1936 V28 N12 P391-393	59-20348 <*> 0
1937 V29 P48-50	57-3518 <*>
1937 V29 P135-137	2277 <*>
1937 V29 P209-	AL-654 <*>
1937 V29 P209-214	AL-655 <*>
1937 V29 P282	58-1881 <*>
1937 V29 P367-	II-79 <*>
1937 V29 P427-433	AL-38 <*>
1937 V29 N1 P16-20	59-20347 <*> 0
1937 V29 N2 P37-39	66-12393 <*>
1937 V29 N2 P50-52	60-14649 <*> 0
1937 V29 N2 P63-66	57-2478 <*>
1937 V29 N3 P84-88	64-20399 <*> 0
1937 V29 N3 P95-101	59-20346 <*> 0
1937 V29 N4 P116-123	I-178 <*>
1937 V29 N6 P173-185	59-20114 <*> 0
	61-10110 <*> 0
1937 V29 N6 P189-192	57-2652 <*>
	63-20857 <*>
1937 V29 N8 P250-251	57-35 <*>
1937 V29 N10 P353-354	59-20344 <*> 0
1937 V29 N11 P361-366	66-12360 <*>
1937 V29 N11 P380-388	57-2031 <*>
1937 V29 N12 P418-420	59-20233 <*> 0
1938 V30 P21-24	57-635 <*>
1938 V30 P299-305	I-322 <*>
1938 V30 P389-	I-972 <*>
1938 V30 P419-422	58-69 <*>
1938 V30 N2 P47-49	59-20342 <*> 0
1938 V30 N2 P50-51	59-20341 <*> 0
1938 V30 N2 P62	58-2012 <*>
1938 V30 N2 P68-70	59-20339 <*> 0
1938 V30 N5 P149-151	58-2662 <*>
1938 V30 N5 P162-173	62-18297 <*> PO
1938 V30 N6 P185-191	59-20338 <*> 0
1938 V30 N9 P294-296	60-14626 <*> 0
1938 V30 N9 P297-298	64-18182 <*>
1938 V30 N9 P350-352	59-20232 <*> 0
1939 V31 P192-203	2131 <*>
1939 V31 N1 P20-23	60-14620 <*> 0
1939 V31 N5 P131-	1817 <*>
1939 V31 N5 P141-143	59-20337 <*>
1939 V31 N5 P143-144	2665 <*>
1939 V31 N6 P161-167	57-721 <*>
1939 V31 N6 P168-170	31L30G <ATS>
	57-722 <*>
1939 V31 N9 P305-310	59-20359 <*> 0
1939 V31 N10 P322-325	59-15205 <*>
1939 V31 N12 P352-358	58-1864 <*>
1940 V32 P39-42	58-1949 <*>
1940 V32 P78-85	II-59 <*>
1940 V32 P120-122	61-00671 <*>
1940 V32 P165-173	UCRL TRANS-750(L) <*>
1940 V32 P376-383	II-52 <*>
1940 V32 P412-414	I-483 <*>
1940 V32 P418-	II-101 <*>
1940 V32 N3 P52-61	I-575 <*>
1940 V32 N6 P165-173	63-20831 <*> 0
1940 V32 N6 P173-177	63-18134 <*>
1940 V32 N6 P184-190	57-2440 <*>
	62-18167 <*> 0
1940 V32 N7 P231-238	2176 <*>
	62-18169 <*>
1940 V32 N8 P184-190	65-00150 <*>
1940 V32 N9 P314-317	1923 <*>
1940 V32 N12 P399-407	I-615 <*>
1941 V33 P98-114	I-179 <*>
1941 V33 P104-111	II-314 <*>
1941 V33 P358-360	AEC-TR-5678 <*>
1941 V33 N4 P168-175	62-14785 <*>
1941 V33 N4 P175-185	62-14786 <*>

Citation	Reference
1941 V33 N10 P337-340	II-7 <*>
1941 V33 N8/9 P289-296	59-17830 <*> 0
1941 V33 N8/9 P297-305	59-17829 <*> 0
1941 V33 N8/9 P323-327	II-85 <*>
1942 V34 P93-96	57-2080 <*>
1942 V34 P121-125	58-1946 <*>
1942 V34 P125-130	58-1413 <*>
1942 V34 P221-	II-15 <*>
1942 V34 P247-253	II-1022 <*>
1942 V34 N7 P145-149	AL-703 <*>
1942 V34 N7 P150-156	AL-72 <*>
1942 V34 N7 P160-165	58-1867 <*>
1942 V34 N8 P193-195	62-18340 <*> 0
1942 V34 N9 P222-224	AL-228 <*>
1943 P181-190	63-16832 <*> 0
1943 V35 P19-21 PT.1	58-648 <*>
1943 V35 P56	58-687 <*>
1943 V35 P117-120	II-385 <*>
1943 V35 P175-181	I-600 <*>
1943 V35 N2 P47-55	60-14360 <*>
1943 V35 N6 P129-136	58-1854 <*>
1943 V35 N10 P185-193	65-12871 <*>
1944 P131-135	57-3080 <*>
1944 V36 P164-169	AEC-TR-4019 <*>
1944 V36 N6 P131-135	59-15196 <*>
1944 V36 N6 P140-141	58-901 <*>
1944 V36 N6 P145-146	II-72 <*>
1944 V36 N8 P197-200	57-3001 <*>
1946 V37 P87-96	II-595 <*>
1946 V37 N6 P161-167	59-20598 <*>
1946 V37 N6 P175-181	62-18497 <*> 0
1946 V37 N1/2 P17-23	64-18123 <*> 0
1946 V37 N1/2 P40-42	63-18483 <*> 0
1946 V37 N1/2 P47-52	62-18498 <*> 0
	63-20972 <*>
1946 V37 N1/2 P53-56	63-20971 <*>
1947 V38 P33-41	AEC-TR-4521 <*>
1947 V38 P69-75	57-705 <*>
1947 V38 P232-235	II-629 <*>
1947 V38 N1 P29-32	62-18377 <*> 0
1947 V38 N3 P65-69	60-18817 <*>
1947 V38 N6 P186-188	60-18819 <*>
1947 V38 N7/8 P207-212	62-18376 <*> 0
1948 V39 P18-22	766 <TC>
1948 V39 P213-216	1064 <*>
1948 V39 P218-220	1256 <*>
	58-1416 <*>
1948 V39 P247-253	58-2039 <*>
1948 V39 P319-320	II-527 <*>
1948 V39 N1 P1-9	62-18402 <*> 0
1948 V39 N1 P9-12	62-18403 <*> 0
1948 V39 N2 P52-56	62-18351 <*> 0
1948 V39 N4 P111-120	62-18404 <*> 0
1948 V39 N8 P233-239	63-20830 <*>
1948 V39 N8 P333-342	2308 <*>
1948 V39 N10 P293-303	2967 <*>
1948 V39 N10 P303-312	62-18350 <*> 0
1949 V40 P201-205	II-394 <*>
1949 V40 P214-217	2296 <*>
1949 V40 N4 P136-140	NP-TR-540 <*>
1949 V40 N5 P187-193	62-16992 <*> 0
1949 V40 N6 P206-214	59-17499 <*>
1949 V40 N8 P281-289	61-18317 <*>
1949 V40 N11 P445-460	637 <TC>
1949 V40 N12 P437-440	66-14134 <*>
1950 V41 P56-59	65-10032 <*> 0
1950 V41 P71-75	2644 <*>
1950 V41 P178-184	1230 <*>
1950 V41 P178-179	66-11862 <*>
1950 V41 P183-184	66-11862 <*>
1950 V41 P228-231	58-1853 <*>
1950 V41 P339-343	63-20450 <*>
1950 V41 P358-367	66-11234 <*>
1950 V41 P370-	2300 <*>
1950 V41 N2 P56-59	61-20752 <*> 0
1950 V41 N4 P97-104	II-390 <*>
1950 V41 N4 P124-127	57-841 <*>
1950 V41 N5 P129-140	65-12382 <*>
1950 V41 N7 P214-218	II-391 <*>
1950 V41 N8 P265-272	AL-607 <*>
1950 V41 N10 P317-321	60-10172 <*>
1950 V41 N11 P381-391	85M45G <ATS>
1951 V42 P17-19	58-2468 <*>
1951 V42 P111-117	UCRL TRANS-1123 <*>
1951 V42 P129-132	AL-436 <*>
1951 V42 P197-201	63-20439 <*>
1951 V42 P238-243	63-20440 <*>
1951 V42 P294-299	62-16918 <*> 0
1951 V42 P336-340	02N49G <ATS>
1951 V42 P371-376	63-10003 <*> 0
1951 V42 P401-409	03N49G <ATS>
1951 V42 N1 P10-13	57-2514 <*>
1951 V42 N2 P34-43	64-18861 <*> 0
1951 V42 N2 P43-54	58-1856 <*>
1951 V42 N10 P294-299	65-10419 <*>
1951 V42 N10 P391-394	60-16481 <*> 0
1952 V43 P1-10	AL-676 <*>
1952 V43 P138-142	AEC-TR-3597 <+*>
1952 V43 P147-150	AEC-TR-6127 <*>
1952 V43 P325-332	2313 <*>
1952 V43 P357-362	AEC-TR-3576 <+*>
1952 V43 P364-369	1158 <*>
1952 V43 N8 P269-284	AEC-TR-4933 <*>
1952 V43 N9 P303-306	I-1021 <*>
1952 V43 N9 P307-310	4111 <BISI>
1952 V43 N11 P373-387	AEC-TR-4949 <*>
1953 P323-324	I-794 <*>
1953 P489-494	I-334 <*>
1953 V44 P27-36	1473 <*>
	62-20071 <*>
1953 V44 P113-114	57-994 <*>
1953 V44 P139-154	II-74 <*>
1953 V44 P198-205	I-337 <*>
1953 V44 P211-216	58-1068 <*>
1953 V44 P264-266	I-1046 <*>
1953 V44 P473-474	4332 <HB>
1953 V44 P503-509	59-15096 <*> 0
1953 V44 P535-550	II-280 <*>
1953 V44 P536-550	62-16913 <*> 0
1953 V44 P570-572	II-272 <*>
1953 V44 N1 P22-26	II-1007 <*>
	57-2537 <*>
1953 V44 N2 P43-50	59-17965 <*> 0
1953 V44 N2 P65-68	19F3G <ATS>
1953 V44 N3 P85-97	1809 <*>
1953 V44 N4 P123-126	II-456 <*>
1953 V44 N4 P154-160	I-193 <*>
1953 V44 N5 P192-197	57-1807 <*>
1953 V44 N8 P333-341	57-3306 <*>
1953 V44 N11 P477-489	57-1797 <*>
1954 V45 P92-93	57-2007 <*>
1954 V45 P233-238	II-584 <*>
1954 V45 P238-242	II-733 <*>
1954 V45 P238-	II-875 <*>
1954 V45 P245-249	II-722 <*>
1954 V45 P436-439	66-11957 <*>
1954 V45 P440-447	57-2403 <*>
1954 V45 N1 P14-22	63-10036 <*> 0
1954 V45 N1 P31-35	3972 <HB>
1954 V45 N2 P86-93	1143 <*>
1954 V45 N4 P154-157	T-2310 <INSD>
1954 V45 N4 P161-165	57-2004 <*>
1954 V45 N4 P180-187	3918 <HB>
1954 V45 N4 P194-196	1132 <*>
1954 V45 N4 P227-230	4347 <HB>
1954 V45 N5 P276-285	57-1386 <*>
1954 V45 N6 P350-356	58-746 <*>
	63-16543 <*>
1954 V45 N6 P371-378	63-18061 <*>
1954 V45 N6 P401-407	3017 <*>
	57-776 <*>
1954 V45 N7 P453-458	59-15048 <*>
1954 V45 N8 P512-513	1244 <*>
1954 V45 N9 P547-550	62-16947 <*> 0
1954 V45 N9 P550-554	57-790 <*>
1954 V45 N10 P584-593	2328 <*>
1954 V45 N11 P623-632	37G5G <ATS>
1954 V45 N11 P651-655	58-1550 <*>
	62-16922 <*> 0
	62-16923 <*> 0

1955 V46 P77-83 AEC-TR-5268 <*>
1955 V46 P582-588 57-3557 <*>
1955 V46 N1 P17-24 64-20847 <*> 0
1955 V46 N1 P52-57 65-12741 <*>
1955 V46 N2 P71-76 59-17546 <*>
1955 V46 N2 P90-94 64-20857 <*>
1955 V46 N2 P94-99 II-969 <*>
1955 V46 N2 P110-118 59-20925 <*> 0
1955 V46 N3 P204-207 57-740 <*>
1955 V46 N3 P210-215 3821 <HB>
1955 V46 N3 P216-224 65-13053 <*>
1955 V46 N3 P225-229 57-1360 <*>
1955 V46 N3 P229-233 57-1356 <*>
57-590 <*>
1955 V46 N4 P242-253 58-1791 <*>
1955 V46 N4 P272-281 57-1382 <*>
6G8G <ATS>
1955 V46 N5 P328-337 AEC-TR-4948 <*>
1955 V46 N6 P442-449 57-3372 <*>
1955 V46 N7 P488-490 2295 <*>
1955 V46 N8 P535-544 II-695 <*>
1955 V46 N8 P545-546 II-740 <*>
1955 V46 N8 P547-551 61-00034 <*>
1955 V46 N8 P560-568 AEC-TR-3873 <*>
1955 V46 N8 P589-592 II-705 <*>
59-12428 <+*>
1955 V46 N9 P623-631 66-11988 <*>
1955 V46 N9 P658-668 57-1280 <*>
1955 V46 N9 P689-692 66R76G <ATS>
1955 V46 N11 P792-800 65-12557 <*>
1956 V47 P1-8 57-25 <*>
1956 V47 P107-117 8775 <IICH>
1956 V47 P418-420 63-18453 <*> 0
1956 V47 P707-713 AEC-TR-4785 <*>
1956 V47 N1 P16-21 59-10367 <*>
66-10432 <*> 0
1956 V47 N1 P24-28 C-2399 <NRC>
1956 V47 N1 P28-36 61-16409 <*>
1956 V47 N2 P57-63 1810 <*>
1956 V47 N2 P64-74 57-2341 <*>
1956 V47 N2 P89-95 AI-TRANS-51 <*>
1956 V47 N2 P97-101 30N54G <ATS>
1956 V47 N2 P102-106 57-2339 <*>
1956 V47 N3 P149-159 62-14463 <*>
1956 V47 N3 P195-198 65-17362 <*>
1956 V47 N3 P199-202 1227 <*>
57-767 <*>
1956 V47 N4 P224-228 61-18421 <*>
1956 V47 N6 P347-356 66-10431 <*> 0
92L32G <ATS>
1956 V47 N6 P357-363 66-10430 <*> 0
93L32G <ATS>
1956 V47 N6 P382-389 66-10429 <*> 0
1956 V47 N7 P499-506 66-10133 <*> 0
1956 V47 N8 P529-534 3909 <HB>
1956 V47 N8 P535-548 66-10415 <*> 0
1956 V47 N8 P564-570 52J18G <ATS>
1956 V47 N9 P614-620 66-10437 <*> 0
1956 V47 N9 P646-649 61-18391 <*>
1956 V47 N9 P649-652 61-18389 <*>
1956 V47 N11 P707-713 66-11054 <*> 0
1956 V47 N12 P743-750 57-3490 <*>
1956 V47 N12 P775-777 68J15G <ATS>
1957 V48 P9-15 57-2935 <*>
1957 V48 P26-32 63-20327 <*>
1957 V48 P126-134 66-11164 <*>
1957 V48 P162-170 AEC-TR-4130 <*>
1957 V48 P341-349 58-1590 <*>
1957 V48 P564-568 61-16410 <*>
1957 V48 P646-649 TS-1177 <BISI>
1957 V48 N2 P57-58 C-2474 <NRC>
1957 V48 N3 P91-100 4111 <HB>
1957 V48 N4 P158-161 61-18253 <*>
63-13450 <=*> 0
1957 V48 N4 P193-200 66-11247 <*>
1957 V48 N5 P241-245 58-1066 <*>
1957 V48 N7 P373-378 58-2220 <*>
58-2516 <*>
1957 V48 N7 P404-409 65-17320 <*> 0
1957 V48 N9 P503-508 61-14909 <*>
1957 V48 N9 P516-522 59-18163 <+*> 0
1957 V48 N11 P592-594 AEC-TR-6344 <*>
1957 V48 N12 P615-624 61-18876 <*>
1957 V48 N12 P624-627 61K26G <ATS>
1957 V48 N12 P646-649 62-10468 <*>
1958 V49 P95-101 65-18103 <*>
1958 V49 P213-220 UCRL TRANS-653(L) <*>
1958 V49 P455-460 TS-1317 <BISI>
1958 V49 N2 P80-86 58-1767 <*>
1958 V49 N4 P165-172 64-14804 <*> 0
1958 V49 N4 P399-403 AEC-TR-5315 <*>
1958 V49 N5 P220-225 TS 1238 <BISI>
61-20623 <*>
1958 V49 N6 P302-311 59-17494 <*>
1958 V49 N6 P312-316 29N54G <ATS>
1958 V49 N7 P361-364 62-01574 <*>
1958 V49 N7 P365-367 63-00184 <*>
1958 V49 N9 P449-455 63-18818 <*> 0
1958 V49 N9 P455-460 62-10515 <*>
1958 V49 N9 P461-463 61-20692 <*> 0
1958 V49 N9 P491-494 59-10732 <*> 0
1958 V49 N10 P527-532 TRANS-146 <FRI>
61-10806 <*>
1958 V49 N11 P549-556 TRANS-137 <FRI>
60-18903 <*>
1958 V49 N11 P557-562 65-10045 <*> 0
1958 V49 N11 P577-583 AEC-TR-3678 <*>
1958 V49 N11 P600-601 59-17513 <*>
1958 V49 N12 P622-626 65-17133 <*> 0
1959 V50 P248-257 AEC-TR-3872 <*>
1959 V50 P299-306 TS 1402 <BISI>
1959 V50 N1 P11-18 61-10566 <*>
1959 V50 N1 P37-41 13L37G <ATS>
1959 V50 N2 P87-93 62-14764 <*>
1959 V50 N2 P112-117 66-10132 <*> 0
1959 V50 N5 P248-257 43M40G <ATS>
1959 V50 N5 P269-274 95L34G <ATS>
1959 V50 N5 P274-277 60-14482 <*>
1959 V50 N5 P299-306 62-14404 <*>
1959 V50 N6 P315-327 66-10442 <*> 0
1959 V50 N6 P315-323 66-15175 <*> 0
1959 V50 N6 P324-327 66-10443 <*> 0
66-15176 <*> 0
1959 V50 N6 P346-350 83L35G <ATS>
1959 V50 N7 P392-395 61-10564 <*>
1959 V50 N8 P442-452 66-10080 <*>
1959 V50 N8 P466-472 TRANS-171 <FRI>
62-16306 <*>
1959 V50 N9 P521-527 44M40G <ATS>
1959 V50 N10 P568-574 62-10590 <*>
1959 V50 N10 P574-581 TRANS-142 <FRI>
61-10787 <*>
1959 V50 N10 P606-610 TRANS-183 <FRI>
64-14473 <*>
1960 V51 P53-58 AEC-TR-4226 <*>
1960 V51 P85-95 AEC-TR-4195 <*>
1960 V51 P101-103 AEC-TR-6284 <*>
1960 V51 P212-218 65-18117 <*>
1960 V51 P535-536 AEC-TR-5603 <*>
1960 V51 N2 P73-80 5184 <HB>
1960 V51 N2 P120-131 1731 <BISI>
61-10549 <*> 0
1960 V51 N3 P186-190 NP-TR-495 <*>
1960 V51 N5 P280-289 79M45G <ATS>
1960 V51 N5 P290-291 AEC-TR-5421 <*>
1960 V51 N6 P303-313 61-18843 <*>
1960 V51 N6 P322-326 63-18592 <*>
1960 V51 N6 P350-353 88M43G <ATS>
1960 V51 N6 P353-365 TRANS-148 <FRI>
61-16094 <*>
1960 V51 N7 P409-413 TRANS-175 <FRI>
62-16599 <*>
1960 V51 N7 P414-420 AEC-TR-4373 <*>
1960 V51 N8 P435-456 C-3677 <NRCC>
1960 V51 N8 P477-481 53M45G <ATS>
1960 V51 N9 P459-501 2059 <BISI>
1960 V51 N11 P663-666 10963 <KH>
2146 <BISI>
61-18299 <*>
62-10319 <*>

1960 V51 N12 P694-699 61-20085 <*>
1960 V51 N12 P722-736 TRANS-182 <FRI>
36N53G <ATS>
61-20206 <*> 0
61-20526 <*>
63-20051 <*>
1961 V52 N1 P47-56 51-20073 <*>
1961 V52 N1 P69-75 2408 <BISI>
1961 V52 N1 P76-85 ANL-TRANS-143 <*>
1961 V52 N2 P95-100 61-20074 <*>
1961 V52 N2 P141-142 5164 <HB>
1961 V52 N3 P152-157 2478 <BISI>
65-17158 <*> 0
1961 V52 N3 P168-179 AEC-TR-6347 <*>
1961 V52 N4 P224-228 61-20754 <*>
1961 V52 N5 P291-309 77N56G <ATS>
1961 V52 N5 P344-352 14N55G <ATS>
62-10079 <*>
1961 V52 N5 P359-364 2593 <BISI>
1961 V52 N5 P368-370 62-14632 <*> 0
1961 V52 N7 P464-476 205 <LS>
1961 V52 N7 P481-488 2867 <BISI>
1961 V52 N8 P529-531 65-11466 <*>
1961 V52 N10 P599-607 TRANS-187 <FRI>
65-11574 <*>
1961 V52 N10 P626-631 62-14837 <*> 0
1961 V52 N10 P671-676 66-14440 <*>
1961 V52 N10 P682-684 62-16175 <*>
1961 V52 N11 P721-727 53P59G <ATS>
1961 V52 N11 P727-735 54P59G <ATS>
1961 V52 N12 P785-794 AEC-TR-5843 <*>
1961 V52 N12 P799-813 65-11468 <*> 0
1962 V53 P122-130 NP-TR-1099 <*>
1962 V53 P541-547 UCRL TRANS-897(L) <*>
1962 V53 N1 P48-51 2905 <BISI>
1962 V53 N2 P78-85 5596 <HB>
1962 V53 N2 P86-89 5635 <HB>
1962 V53 N2 P138-144 3116 <BISI>
1962 V53 N3 P172-177 SCL-T-434 <*>
1962 V53 N9 P593-600 3077 <BISI>
1962 V53 N10 P617-632 64-10368 <*> 0
1962 V53 N11 P729-735 3492 <HB>
1962 V53 N12 P770-774 66-11882 <*>
1963 V54 P206-212 AEC-TR-5960 <*>
1963 V54 P477-483 65-11857 <*>
1963 V54 P493-504 AEC-TR-6255 <*>
1963 V54 P650-655 C-5601 <NRC>
1963 V54 N1 P31-37 64-14127 <*>
1963 V54 N2 P98-111 64-16578 <*>
1963 V54 N3 P142-146 AEC-TR-5964 <*>
1963 V54 N4 P213-226 5632 <HB> P
1963 V54 N5 P257-262 66-11987 <*>
1963 V54 N5 P308-311 ANL-TRANS-67 <*>
1963 V54 N6 P370-380 AEC-TR-5980 <*>
63-20316 <*>
1963 V54 N11 P619-622 C-5348 <NRC>
1963 V54 N12 P724-728 3665 <BISI>
1964 V55 P432-444 365 <TC>
1964 V55 N2 P55-60 ANL-TRANS-88 <*>
1964 V55 N3 P142-145 6263 <HB>
1964 V55 N3 P146-152 66-13229 <*>
1964 V55 N4 P205-216 66-13230 <*>
1964 V55 N5 P261-265 66-11837 <*>
1964 V55 N5 P266-276 66-12947 <*>
1964 V55 N7 P378-382 66-11283 <*> 0
1964 V55 N7 P382-387 66-14181 <*>
1964 V55 N7 P398-405 66-11005 <*>
1964 V55 N8 P453-456 65-13842 <*>
1964 V55 N8 P456-459 TRANS-190 <FRI>
65-13595 <*>
1964 V55 N9 P503-511 66-12660 <*>
1964 V55 N11 P691-698 65-10991 <*> 0
1964 V55 N11 P711-715 65-13142 <*>
1964 V55 N12 P737-740 49S85G <ATS>
65-17151 <*> 0
66-11851 <*>
1965 V56 P176-182 990 <TC>
1965 V56 P403-410 1373 <TC>
1965 V56 P544-554 4782 <BISI>
1965 V56 N3 P155-164 65-17205 <*>
1965 V56 N5 P316-319 65-14115 <*>
1965 V56 N6 P339-345 66-13912 <*>
1965 V56 N7 P450-452 66-12920 <*> 0
1965 V56 N7 P465-469 65-14886 <*>
1965 V56 N7 P484-487 66-12661 <*>
1965 V56 N8 P569-575 66-13401 <*>
1965 V56 N9 P585-598 66-10904 <*>
1965 V56 N9 P619-622 4783 <BISI>
1966 V57 N2 P122-129 66-11779 <*>
1966 V57 N3 P171-175 66-12501 <*>
1966 V57 N3 P181-186 66-12306 <*>
1966 V57 N4 P270-274 66-13610 <*>

ZEITSCHRIFT FUER METEOROLOGIE
1949 V3 N1/2 P32-39 66-11178 <*>
1952 V6 N12 P369-378 1873 <*>
57-623 <*>
1954 V8 N2/3 P76 SCL-T-243 <+*>
1955 V9 N4 P116-120 1584 <*>
1955 V9 N8 P240-248 57-2729 <*>
1956 V10 N3 P76-86 CSIRO-3335 <CSIR>
1956 V10 N7 P217-219 CSIRO-3579 <CSIR>
1956 V10 N10 P308-313 C-2693 <NRC>
1956 V10 N11 P337-341 SCL-T-235 <*>
1957 V11 N2 P43-49 19L31G <ATS>
61-10253 <*>
1957 V11 N10/1 P305-311 59-11483 <=>
1957 V11 N10/1 P339-344 60-00444 <*>
60-17409 <*+>
1965 V18 N3/4 P132-141 65-33678 <=*$>

ZEITSCHRIFT FUER MIKROSKOPISCH-ANATOMISCHE FORSCHUNG
1926 V5 P625-667 63-00554 <*>
1928 V15 N3/4 P599-652 II-537 <*>
1934 V35 P112-117 II-355 <*>
1959 V65 N4 P495-528 61-00269 <*>
1963 V70 P103-117 ANL-TR-7 <*>

ZEITSCHRIFT FUER MORPHOLOGIE UND ANTHROPOLOGIE
1927 V26 P305-332 62-00484 <*>
1950 V42 P169-183 63-01689 <*>
1954 V46 N2 P170-183 64-00045 <*>

ZEITSCHRIFT FUER MORPHOLOGIE UND OEKOLOGIE DER TIERE
1933 V27 P199-261 C-2237 <NRC> P
1933 V27 N2 P199-261 61-00444 <*> 0
1954 V42 P529-549 66-11870 <*>

ZEITSCHRIFT FUER NATURFORSCHUNG
1946 V1 P131-136 II-48 <*>
57-2126 <*>
1946 V1 P137-141 58-1523 <*>
1946 V1 P173-179 II-20 <*>
1946 V1 P214-217 58-82 <*>
1946 V1 P243-252 I-1043 <*>
1946 V1 P305-310 63-23174 <=*>

ZEITSCHRIFT FUER NATURFORSCHUNG. A
1947 V2A P39-46 II-55 <*>
1947 V2A P133-145 19G5G <ATS>
1947 V2A P180-181 58-2466 <*>
1947 V2A P185-201 AEC-TR-5267 <*>
1947 V2A P329-332 I-1037 <*>
1947 V2A P359-362 64-11815 <=>
1947 V2A P505-521 24L29G <ATS>
1947 V2A P534-535 61-16541 <*>
1947 V2A P650-652 29P66G <ATS>
1947 V2A N3 P133-145 59-17447 <*>
1947 V2A N10 P583-584 AL-683 <*>
1948 V3A P78-97 35G5G <ATS>
1948 V3A P211-216 AEC-TR-3842 <*>
1948 V3A P477-481 AL-222 <*>
1948 V3A P500-506 65-10786 <*>
1948 V3A P672-690 58-2286 <*>
1948 V3A N2 P78-97 59-17542 <*>
1948 V3A N7 P392-395 62-20480 <*> 0
1949 V4A P204-217 AL-96 <*>
1949 V4A P337-341 63-14816 <*> 0

Citation	Translation
1949 V4A P378-391	76P60G <ATS>
1949 V4A P424-432	2457 <*>
1949 V4A P449-455	UCRL TRANS-588(L) <*>
1949 V4A P456-463	1548 <*>
1949 V4A P549-559	1327 <*>
1949 V4A N7 P549-559	II-30 <*>
1949 V4A N7 P549-	57-347 <*>
1949 V4A N7 P549-559	64-20564 <*> 0
1950 V5A P1-5	I-158 <*>
1950 V5A P62-63	27P66G <ATS>
1950 V5A P101-108	1055 <*> P
	64-14413 <*> 0
1950 V5A P174-175	I-27 <*>
1950 V5A P192-208	AEC-TR-5269 <*>
1950 V5A P230-	I-926 <*>
1950 V5A P270-275	AL-273 <*>
	58-1084 <*>
1950 V5A P345-346	36Q73G <ATS>
1950 V5A P404-405	I-964 <*>
1950 V5A P467-469	63-14367 <*>
1950 V5A P502-504	I-194 <*>
1950 V5A P569-570	I-172 <*>
1950 V5A P570-571	AL-548 <*>
1950 V5A P612-617	62-00848 <*>
1950 V5A P631-632	63-00765 <*>
1950 V5A P657-660	58-1551 <*>
1950 V5A N1 P35-40	63-18243 <*>
1950 V5A N8 P420-431	64-20133 <*>
1950 V5A N11 P581-589	I-201 <*>
1951 V6A P103-122	I-908 <*>
1951 V6A P184-191	61-16469 <*> 0
1951 V6A P302-304	57-797 <*>
1951 V6A P456-459	UCRL TRANS-892(L) <*>
1951 V6A P459-462	1800 <*>
	1974 <*>
1951 V6A P508-509	SLAC-TRANS-12 <*>
1951 V6A P603-613	C-2708 <NRC>
	58-1426 <*>
1951 V6A P700-705	85F4G <ATS>
1951 V6A N1 P25-31	61-10118 <*> 0
1951 V6A N2 P85-102	MF3 <*>
	66-12327 <*>
1951 V6A N10 P544-550	66-12738 <*>
1952 V7A P202-204	AL-223 <*>
1952 V7A P372-379	59-17355 <*>
1952 V7A P474-480	AL-46 <*>
1952 V7A P511-532	75Q67G <ATS>
1952 V7A P533-541	AL-893 <*>
1952 V7A P542-553	AL-894 <*>
1952 V7A P651-664	AEC-TR-5527 <*>
1952 V7A P664-668	AEC-TR-5526 <*>
1952 V7A P744-749	57-956 <*>
1952 V7A P750-752	77M41G <ATS>
1952 V7A P756-759	59-22718 <+*> 0
1952 V7A P765-771	57-3103 <*>
1952 V7A P781-	I-937 <*>
1952 V7A P822-823	AEC-TR-5394 <*>
1952 V7A N6 P432-439	58-2015 <*>
1952 V7A N11 P744-749	59-17960 <*>
1952 V7A N11 P760-763	64-18931 <*>
1953 V8A P39-46	57-951 <*>
1953 V8A P197-204	57-1749 <*>
1953 V8A P248-251	57-1357 <*>
1953 V8A P448-450	AEC-TR-5670 <*>
1953 V8A P457-459	II-935 <*>
1953 V8A P463-469	57-2522 <*>
1953 V8A P470-479	II-139 <*>
1953 V8A P556-562	58-2080 <*>
1953 V8A P562-564	II-313 <*>
1953 V8A P565-566	1066 <*>
1953 V8A P672-673	SC-T-65-702 <*>
1953 V8A P681-686	61-16456 <*> 0
1953 V8A P686-	58-1260 <*>
1953 V8A P756	60-18144 <*> 0
1953 V8A P796-804	I-909 <*>
1953 V8A N1 P69-73	65-11682 <*>
1953 V8A N1 P74-79	65-12335 <*>
1953 V8A N4 P248-251	59-17912 <*>
1953 V8A N5 P305-307	AL-883 <*>
1953 V8A N5 P333	59-17971 <*>
1953 V8A N6 P361-371	58-1250 <*>
	59-17908 <*>
1953 V8A N8 P457-459	59-17907 <*>
1953 V8A N8 P463-469	59-17906 <*>
1953 V8A N8 P498-499	62-16921 <*> 0
1953 V8A N9 P562-564	64-14399 <*> 0
1953 V8A N11 P681-686	57-988 <*>
1954 V9A P67-	II-976 <*>
1954 V9A P134-147	57-1981 <*>
1954 V9A P167-174	57-1368 <*>
	97H8G <ATS>
1954 V9A P228-235	66-11176 <*>
1954 V9A P368-369	1235 <*>
1954 V9A P508-511	62-10207 <*>
1954 V9A P515-520	57-2647 <*>
1954 V9A P527-534	C-2379 <NRC>
1954 V9A P535-546	2561 <*>
1954 V9A P561-562	1243 <*>
1954 V9A P625-630	58-1719 <*>
1954 V9A P637-653	58-2651 <*>
	63-10613 <*>
1954 V9A P654-663	1495 <*>
1954 V9A P667-674	57-1772 <*>
1954 V9A P716-724	78M41G <ATS>
1954 V9A P735-740	58-2085 <*>
1954 V9A P802-804	AEC-TR-3242 <*>
1954 V9A P836-851	C-2703 <NRC>
	58-1721 <*>
1954 V9A P851-855	C-2694 <NRC>
	58-1427 <*>
1954 V9A P954-958	C-2370 <NRC>
1954 V9A P975-986	II-957 <*>
1954 V9A N3 P185-204	57-28 <*>
1954 V9A N6 P520-526	59-17354 <*> 0
1954 V9A N6 P535-546	59-10574 <*>
1954 V9A N7/8 P611-617	57-1015 <*>
1955 V10A P62-67	58-2079 <*>
1955 V10A P170-171	AEC-TR-4419 <*>
	61-18293 <*>
	64-16316 <*>
1955 V10A P459-	58-743 <*>
1955 V10A P502-503	58-2421 <*>
1955 V10A P565-572	50H9G <ATS>
	57-2253 <*>
1955 V10A P659-667	AEC-TR-5314 <*>
1955 V10A P667-669	1423 <*>
1955 V10A P669-680	UCRL TRANS-630(L) <*>
1955 V10A P877-886	AEC-TR-5279 <*>
	57-2342 <*>
1955 V10A P887-893	AEC-TR-5282 <*>
1955 V10A P898-899	58-2422 <*>
1955 V10A P927-930	62-16588 <*>
1955 V10A N2 P136-145	59-17545 <*>
1955 V10A N3 P219-229	3033 <*>
1955 V10A N3 P230-238	C-3484 <NRCC>
1955 V10A N4 P285-291	57-1383 <*>
	59-17535 <*>
1955 V10A N5 P374-378	59-20500 <*>
1955 V10A N7 P584-586	59-15851 <*>
1955 V10A N11 P857-863	60-18046 <*>
1955 V10A N11 P863-872	60-18096 <*>
1955 V10A N9/10 P736-740	66-11691 <*>
1956 P430-434	57-2384 <*>
1956 V11A P46-54	57-2705 <*>
1956 V11A P80-84	58-2698 <*>
1956 V11A P94-95	C-3590 <NRC>
1956 V11A P95-96	77J17G <ATS>
1956 V11A P131-138	57-1708 <*>
	58-2395 <*>
1956 V11A P167-168	1909 <*>
1956 V11A P238-243	57-1347 <*>
1956 V11A P373-379	74Q67G <*>
1956 V11A P425-429	57-3355 <*>
1956 V11A P459-463	57-3304 <*>
1956 V11A P463-472	57-1857 <*>
	60-16824 <*>
1956 V11A P608-609	58-2423 <*>
1956 V11A P676-677	61-10552 <*>
1956 V11A P679	AEC-TR-2660 <*>
1956 V11A P679-	57-690 <*>

Citation	Translation
1956 V11A P942-945	63-14667 <*>
1956 V11A N1 P1-14	66-11261 <*>
1956 V11A N3 P238-243	58-1371 <*>
1956 V11A N5 P379-382	57-2755 <*>
1956 V11A N5 P420-422	59-10000 <*>
1956 V11A N6 P434-439	58-1623 <*>
1956 V11A N6 P440-456	66-14492 <*>
1956 V11A N6 P472-477	57-3467 <*>
1956 V11A N6 P478-481	58-1624 <*>
1956 V11A N6 P515-517	57-2041 <*>
1956 V11A N8 P649-655	62-16542 <*>
1956 V11A N12 P1023-1030	62-16024 <*>
1957 V12A P80	58-09 <*>
1957 V12A P148-153	AEC-TR-4744 <*>
1957 V12A P167-169	58-1509 <*>
1957 V12A P322-325	58-726 <*>
1957 V12A P451-465	AEC-TR-4743 <*>
1957 V12A P815-821	58-728 <*>
1957 V12A P826-832	58-724 <*>
1957 V12A P833-841	59-2297 <*>
1957 V12A P841-843	58-978 <*>
1957 V12A P844-849	58-730 <*>
1957 V12A P850-854	58-727 <*>
1957 V12A P855-859	58-725 <*>
1957 V12A P974-982	AEC-TR-4078 <*>
1957 V12A P996-1002	1289 <TC>
1957 V12A P1016-	58-1322 <*>
1957 V12A N2 P89-95	62-16031 <*>
1957 V12A N3 P223-225	57-3040 <*>
1957 V12A N3 P247-256	60-00674 <*>
1957 V12A N4 P310-319	66-11246 <*>
1957 V12A N4 P334-337	59-10576 <*> 0
1957 V12A N6 P499-507	59-10577 <*>
1957 V12A N6 P507-513	59-10578 <*>
1957 V12A N6 P528-529	59-10579 <*>
1957 V12A N7 P550-557	59-10944 <*>
1957 V12A N7 P583-598	C-3519 <NRCC>
1957 V12A N8 P609-621	AEC-TR-3839 <*>
1957 V12A N9 P752	61-10530 <*>
1957 V12A N11 P776-881	60-18048 <*>
1957 V12A N11 P926-931	61-10531 <*>
	61-10987 <*>
1957 V12A N11 P943-944	59-10742 <*>
1957 V12A N11 P1020-1021	59-10752 <*>
1958 V13A P105-110	UCRL TRANS-612(L) <*>
	62-00112 <*>
1958 V13A P354-355	UCRL TRANS-830 <*>
1958 V13A P474-483	UCRL TRANS-490(L) <*>
1958 V13A P531-536	63-10880 <*>
1958 V13A P722-740	60-00672 <*>
1958 V13A P757-767	UCRL TRANS-929(L) <*>
1958 V13A P794-796	08L33G <ATS>
1958 V13A P874-885	76Q67G <ATS>
1958 V13A P941-950	AERE-TRANS-825 <*>
1958 V13A P951-961	63-00182 <*>
1958 V13A P988-996	37T91G <ATS>
1958 V13A N1 P16-21	61-10540 <*>
1958 V13A N1 P50-51	61-10539 <*>
	61-10949 <*>
1958 V13A N1 P51-52	61-10538 <*>
1958 V13A N1 P84-89	76M37G <ATS>
1958 V13A N3 P235	21M47G <ATS>
1958 V13A N4 P295-302	59-10724 <*>
	59-17749 <*> 0
1958 V13A N4 P346-347	59-10424 <*>
1958 V13A N4 P355-356	59-20626 <*>
1958 V13A N5 P423-425	60-18797 <*>
1958 V13A N6 P484-489	62-14353 <*>
1958 V13A N7 P510-514	SCL-T-272 <*>
1958 V13A N7 P564-566	66-12994 <*>
1958 V13A N8 P602-608	60-18049 <*>
	60-21012 <=*>
1958 V13A N8 P680-698	62-14475 <*>
1958 V13A N8 P698-699	26P66G <ATS>
1958 V13A N9 P745-753	61-10537 <*>
	65-12242 <*>
1958 V13A N9 P753-757	61-10536 <*>
	61-10988 <*>
1958 V13A N9 P792-794	61-18916 <*>
1958 V13A N10 P856-865	59-10725 <*>
1958 V13A N12 P1039-1043	NP-TR-314 <*>
1959 V14A P74-80	61-14750 <*>
1959 V14A P247-261	63-20989 <*>
1959 V14A P374-379	60-00693 <*>
1959 V14A P866-869	NP-TR-805 <*>
1959 V14A P923-924	63-00531 <*>
1959 V14A P951-958	61-00246 <*>
1959 V14A P1040-1046	63-18831 <*>
1959 V14A P1074-1077	92M38G <ATS>
1959 V14A N2 P106-108	64-30956 <*> 0
1959 V14A N2 P109-112	64-30957 <*> 0
1959 V14A N3 P211-216	65-14402 <*>
1959 V14A N3 P247-261	65-10789 <*> 0
1959 V14A N7 P589-599	64-10202 <*>
1959 V14A N9 P846-847	62-10059 <*>
1959 V14A N9 P848-849	62-16576 <*> 0
1959 V14A N10 P882-889	66-12999 <*>
1959 V14A N10 P926-928	61-10274 <*>
	61-18262 <*>
1959 V14A N12 P1077-1078	60-18070 <*>
1959 V14A N12 P1079-1080	60-16866 <*>
1959 V14A N5/6 P582-584	92L34G <ATS>
1960 V15A P59-65	63-20247 <*> 0
1960 V15A P64-65	AEC-TR-4791 <*>
1960 V15A P108-115	AEC-TR-4363 <*>
1960 V15A P412-417	63-18817 <*> 0
1960 V15A P566-574	NP-TR-814 <*>
1960 V15A P647-648	AEC-TR-5178 <*>
1960 V15A P1004-1007	61-00783 <*>
1960 V15A P1031-1038	UCRL TRANS-1036 <*>
1960 V15A P1082-1086	AEC-TR-5058 <*>
1960 V15A N4 P287-292	66-12873 <*>
1960 V15A N7 P575-584	66-12591 <*>
1960 V15A N9 P839-841	63-18942 <*>
1960 V15A N12 P1056-1067	31N51G <ATS>
	61-16421 <*>
1960 V15A N5/6 P484-489	66-10888 <*>
1961 V16A P210	AEC-TR-4648 <*>
1961 V16A P505-510	UCRL TRANS-710 <*>
1961 V16A P525-527	AEC-TR-5197 <*>
1961 V16A P712-714	C-4308 <NRCC>
1961 V16A P780-790	24P59G <ATS>
1961 V16A P845-849	AEC-TR-5256 <*>
1961 V16A P972-984	NP-TR-886 <*>
1961 V16A P1038-1042	62-01404 <*>
1961 V16A P1096-1097	AEC-TR-5171 <*>
1961 V16A P1124-1130	AEC-TR-5295 <*>
1961 V16A P1130-1135	AEC-TR-5520 <*>
	11P62G <ATS>
1961 V16A P1180-1191	1031 <TC>
1961 V16A P1255-1257	AEC-TR-5192 <*>
1961 V16A P1257-1259	63-10120 <*>
1961 V16A P1259	63-10350 <*>
1961 V16A P1260-1261	63-18031 <*>
1961 V16A P1373-1378	AEC-TR-5511 <*>
	13P6G2 <ATS>
1961 V16A P1401	63-18816 <*>
1961 V16A N1 P25-30	62-18708 <*>
1961 V16A N2 P169-172	61-20826 <*>
1961 V16A N3 P262-267	66-13000 <*>
1961 V16A N3 P268-279	65-11164 <*>
1961 V16A N3 P279-283	66-11032 <*>
1961 V16A N5 P459-466	64-10905 <*> 0
1961 V16A N6 P569-577	AEC-TR-4775 <*>
1961 V16A N7 P676-685	62-14231 <*>
1961 V16A N8 P765-780	62-14440 <*>
	63-10265 <*> 0
1961 V16A N9 P869-872	66-12328 <*>
1961 V16A N12 P1290-1312	NP-TR-940 <*>
1961 V16A N12 P1345-1354	AEC-TR-5063 <*>
1962 V17A P297-305	12P62G <ATS>
1962 V17A P332-334	SLAC-TRANS-1 <=$>
1962 V17A P476-478	AEC-TR-5713 <*>
1962 V17A P550-552	AI-TRANS-45 <*>
1962 V17A P921-924	AEC-TR-5708 <*>
1962 V17A P1081-1085	AEC-TR-5864 <*>
1962 V17A P1085-1088	AEC-TR-5865 <*>
1962 V17A P1103-1111	25Q68G <ATS>
1962 V17A N1 P8-17	63-16559 <*>
1962 V17A N4 P358-360	63-10731 <*>

1962 V17A N5 P397-405 46Q69G <ATS>
1962 V17A N5 P405-410 66-12705 <*>
1962 V17A N5 P414-421 AEC-TR-5767 <*>
1962 V17A N6 P514-521 66-11295 <*> 0
1962 V17A N7 P603-614 AEC-TR-6017 <*>
1962 V17A N9 P799-805 AI-TR-21 <*>
1962 V17A N10 P906-921 AEC-TR-5831 <*>
1962 V17A N11 P967-976 C-5272 <NRC>
1962 V17A N11 P994-1010 AEC-TR-5832 <*>
1962 V17A N11 P1011-1022 65-11126 <*>
1962 V17A N12 P1088-1091 NP-TR-1048 <*>
1963 V18A P95-97 SCL-T-508 <*>
1963 V18A P105-111 AEC-TR-5972 <*>
1963 V18A P228-235 AEC-TR-5973 <*>
1963 V18A P246-250 AEC-TR-5974 <*>
1963 V18A P389-397 63-20401 <*> 0
1963 V18A N1 P20-23 65-11571 <*> 0
1963 V18A N1 P26-31 64-16579 <*>
1963 V18A N1 P95-97 63-18916 <*> 0
1963 V18A N3 P389-397 AEC-TR-6162 <*>
1963 V18A N3 P397-409 64-10166 <*> 0
1963 V18A N3 P410-417 64-10190 <*> 0
1963 V18A N3 P436-438 64-18011 <*>
1963 V18A N6 P724-734 C-5023 <NRC>
1963 V18A N7 P822-828 C-4979 <NRC>
1963 V18A N7 P843-853 66-11290 <*> 0
1963 V18A N11 P1169-1181 64-18792 <*>
1963 V18A N11 P1215-1224 65-12844 <*> 0
1963 V18A N11 P1225-1235 65-12845 <*> 0
1963 V18A N33 P63-367 65-13274 <*> 0
1964 V19 N2 P161-171 65-12220 <*>
1964 V19 N9 P1126-1127 65-11618 <*>
1964 V19A P265-266 65-13600 <*>
1964 V19A N2 P172-177 65-12258 <*>
1964 V19A N2 P177-181 C-5481 <NRC>
1964 V19A N2 P181-190 C-5522 <NRC>
1964 V19A N2 P254-263 39R80G <ATS>
65-10885 <*>
1964 V19A N4 P454-458 65-11545 <*>
1964 V19A N4 P515-516 65-10775 <*>
1964 V19A N11 P1276-1296 66-11245 <*>
1965 V20A N1 P76-81 C-5741 <NRC>
1965 V20A N2 P169-175 65-18068 <*> 0
1965 V20A N2 P249-252 65-17056 <*> 0
1965 V20A N7 P902-917 66-11034 <*>
1965 V20A N7 P971 66-11271 <*>

ZEITSCHRIFT FUER NATURFORSCHUNG. B
1947 V2B P245-246 AL-601 <*>
1948 V3B P376- II-928 <*>
1948 V3B N9/10 P345-348 63-00921 <*>
1948 V3B N9/10 P376- 57-1865 <*>
1949 V4B P80-91 2239 <*>
1949 V4B P133-138 <ATS>
1493 <*>
1949 V4B P226-229 58-506 <*>
1949 V4B P360-367 65-00334 <*>
1950 V5B P307-311 I-538 <*>
64-10151 <*>
1950 V5B N7 P396-397 64-30971 <*>
1950 V5B N7 P397 59-15154 <*>
1950 V5B N7 P397-398 59-15156 <*>
65-10321 <*>
1950 V5B N7 P397 65-10322 <*>
65-11448 <*>
1951 V6B P49-50 63-18253 <*>
1951 V6B P171- AL-825 <*>
2953 <*>
1951 V6B P171-172 2960 <*>
1951 V6B P172-173 2963 <*>
1951 V6B P172- 2964 <*>
1951 V6B P226 63-20797 <*>
1951 V6B P333-334 2962 <*>
1951 V6B P334-335 2961 <*>
1951 V6B P335- 1623 <*>
1951 V6B P338-339 2887 <*>
1951 V6B P339-340 I-541 <*>
64-10135 <*>
1951 V6B P395-397 AL-597 <*>
1951 V6B P410-413 I-540 <*>
64-10136 <*>
1951 V6B P458-459 2955 <*>
1951 V6B P459-460 2954 <*>
1951 V6B P460-461 2950 <*>
1951 V6B P461- 2951 <*>
1951 V6B P461-462 2952 <*>
1951 V6B P463- 2072 <*>
1951 V6B N1 P25-34 59-20795 <*>
1951 V6B N3 P171-172 64-30965 <*>
1951 V6B N3 P172 64-30979 <*>
1951 V6B N7 P392-393 65-11258 <*>
1951 V6B N7 P392 65-11266 <*> 0
1951 V6B N8 P458-459 64-16771 <*> 0
1951 V6B N8 P459-460 64-18552 <*> 0
1951 V6B N8 P460-461 64-18553 <*> 0
1951 S6B P171 58-2631 <*>
1952 V7B P131-132 57-2976 <*>
1952 V7B P148-155 58-171 <*>
1952 V7B P201-207 57-664 <*>
1952 V7B P220-230 57-2764 <*>
1952 V7B P270-277 AL-658 <*>
1952 V7B P280-284 T1901 <INSD>
1952 V7B P341-344 58-119 <*>
1952 V7B P377-379 AL-243 <*>
1952 V7B P377-380 63-20455 <*>
1952 V7B P429-433 58-1743 <*>
1952 V7B P633-634 57-3369 <*>
58-1502 <*>
1952 V7B P635 AL-912 <*>
1952 V7B P676- AL-622 <*>
1952 V7B N2 P102-108 II-538 <*>
1952 V7B N2 P131-132 64-30984 <*>
1952 V7B N5 P270-277 58-708 <*>
1952 V7B N6 P334-337 57-2556 <*>
1952 V7B N6 P338-340 II-457 <*>
1952 V7B N6 P434-443 II-404 <*>
1952 V7B N7 P386-389 58-107 <*>
1952 V7B N12 P676 I-571 <*>
1953 V8B P1-2 AL-244 <*>
1953 V8B P343-355 AEC-TR-4401 <*>
1953 V8B P555-577 NP-TR-913 <*>
1953 V8B P696- I-539 <*>
1953 V8B P777-778 58-2259 <*>
1953 V8B N5 P232-236 59-17904 <*>
1953 V8B N11 P690-691 64-20489 <*> 0
1953 S8B P609-610 58-2487 <*>
1954 V9B P188-203 58-586 <*>
1954 V9B P231-233 63-00577 <*>
1954 V9B P326 2115 <*>
1954 V9B P495-503 60-23045 <*+>
64-10462 <*>
1954 V9B P536-540 63-16486 <*>
1954 V9B P684- II-284 <*>
1954 V9B P713-729 1450 <*>
1954 V9B N1 P1-9 63-00778 <*>
1954 V9B N3 P186-188 64-10916 <*>
1954 V9B N5 P358-371 1168 <*>
1954 V9B N9 P618-619 64-16477 <*>
1955 V10B P140-143 64-14688 <*>
1955 V10B P229-240 3564 <K-H>
64-14925 <*>
1955 V10B P295-296 63-10606 <*>
1955 V10B P420-421 37L30G <ATS>
1955 V10B P463-473 58-955 <*>
1955 V10B P474-478 58-946 <*>
1955 V10B P616-622 8760 <IICH>
1955 V10B N1 P34-37 2213 <*>
1955 V10B N2 P77-80 58-368 <*>
1955 V10B N2 P108-115 59-10353 <*>
1955 V10B N4 P191-198 II-370 <*>
1955 V10B N5 P292-294 64-20122 <*>
1955 V10B N5 P294-295 64-20123 <*>
1955 V10B N5 P295-296 64-20124 <*>
1955 V10B N12 P665-668 62-18067 <*> 0
64-16150 <*> 0
1955 V10B N12 P666-668 62-10385 <*> 0
1955 V10B N12 P698-704 57-2859 <*>
1956 V11B P82-85 58-2089 <*>
1956 V11B P172-174 II-1032 <*>
1956 V11B P363-364 AEC-TR-2542 <*>

1956 V11B P485-491 65-10334 <*>
1956 V11B P485-486 9086 <K-H>
1956 V11B P581-588 59-18077 <+*> 0
1956 V11B N2 P61-64 63-00945 <*>
1956 V11B N5 P278-282 62-16448 <*>
1956 V11B N11 P657-662 58-60 <*>
1956 V11B N11 P677-678 62-10415 <*>
1957 V12B P68-69 57-1821 <*>
1957 V12B P115-118 58-61 <*>
1957 V12B P123-125 63-10605 <*>
1957 V12B P125-126 63-10609 <*>
1957 V12B P172-180 UCRL TRANS-605 <*>
1957 V12B P196-208 53J17G <ATS>
1957 V12B P215-222 58-1646 <*>
1957 V12B P384-394 UCRL TRANS-606 <*>
1957 V12B P393-395 58-1812 <*>
1957 V12B P437-440 NP-TR-523 <*>
1957 V12B P704-715 63-01117 <*>
1957 V12B N1 P23-26 59-10945 <*>
1957 V12B N1 P67-68 64-18805 <*> 0
64-30104 <*>
1957 V12B N4 P232-237 59-10575 <*>
1957 V12B N6 P351-355 59-14943 <=*$>
1957 V12B N7 P478-479 64-20198 <*>
1957 V12B N11 P669-671 59-10370 <*> 0
1957 V12B N11 P697-703 63-00928 <*>
1957 V128 P234-236 63-00515 <*>
1958 V13B P141-147 60-13042 <=*> 0
1958 V13B P756-757 59-22100 <+*> 0
1958 V13B P793-802 AEC-TR-4110 <*>
1958 V13B N4 P203-205 60-15910 <*+> 0
1958 V13B N4 P222-224 60-10718 <*>
1958 V13B N4 P225-234 60-10010 <*> 0
1958 V13B N6 P339-347 64-30116 <*> 0
1958 V13B N6 P349-351 64-30099 <*> 0
1958 V13B N7 P454-456 62-01000 <*>
1958 V13B N7 P459 59-10735 <*>
1958 V13B N7 P460-461 TT-851 <NRCC>
1958 V13B N7 P463-464 59-10631 <*>
1958 V13B N9 P588-590 60-17342 <*+>
1958 V13B N9 P591-596 60-17343 <*+> 0
1958 V13B N10 P633-638 59-20625 <*>
1958 V13B N11 P697-704 61-00602 <*>
1958 V13B N12 P820-821 59-22092 <+*>
1959 N14B P52-56 61-17138 <*+>
1959 V14B P52-56 61-00271 <*>
1959 V14B P188-194 61-00645 <*>
1959 V14B P213-214 65-00421 <*>
1959 V14B P355-358 60-17136 <*+>
1959 V14B P712-724 63-00621 <*>
1959 V14B P770-779 66-10970 <*>
1959 V14B P789-801 62-25211 <=*>
1959 V14B N1 P33-37 61-10565 <*>
1959 V14B N10 P674-689 63-00246 <*>
1959 V14B N11 P703-706 61-00649 <*>
1959 V14B N8/9 P598-599 65-10939 <*>
1959 V14B N8/9 P599-600 65-10941 <*>
1960 V15B P351-364 63-00920 <*>
1960 V15B P465-466 65-11846 <*>
1960 V15B P554-560 64-00066 <*>
1960 V15B P622 65-13588 <*>
1960 V15B P681 63-14749 <*>
1960 V15B P691-693 62-01419 <*>
1960 V15B N5 P332 65-13545 <*>
1960 V15B N6 P372-374 63-00157 <*>
1960 V15B N6 P414 62-00886 <*>
1960 V15B N7 P448-449 62-00499 <*>
1960 V15B N9 P547-551 65-13552 <*> 0
1960 V15B N9 P621-622 65-13551 <*>
1960 V15B N10 P656-661 62-16286 <*>
1961 V16B P283 62-18871 <*>
1961 V16B P469-470 64Q70G <ATS>
1961 V16B P618-620 63-10611 <*>
1961 V16B P730-739 66-12514 <*>
1961 V16B N2 P109-115 61-00843 <*>
1961 V16B N3 P181-185 AEC-TR-4794 <*>
1961 V16B N4 P264-267 63-01056 <*>
1961 V16B N5 P310-321 62-00751 <*>
1961 V16B N11 P730-739 63-00041 <*>
1961 V16B N12 P839-840 62-14644 <*>
1962 V17B P347 AEC-TR-5462 <*>
1962 V17B P639-646 NP-TR-1005 <*>
1962 V17B P751-757 NP-TR-1006 <*>
1962 V17B N3 P150-153 63-10269 <*> 0
1962 V17B N4 P275-276 65-14839 <*> 0
1962 V17B N4 P276-277 65-14840 <*> 0
1962 V17B N5 P346-347 66-11294 <*>
1962 V17B N8 P544-547 63-01064 <*>
1963 V18B P110-114 65-11689 <=$>
1963 V18B P145-153 C-5362 <NRC>
1963 V18B N3 P236-242 63-01390 <*>
1963 V18B N7 P541-543 63-01765 <*>
1963 V18B N11 P871-875 65-11749 <*> 0
1964 V19B P587-593 65-17257 <*>
1964 V19B P655- <ES>
1964 V19B P882-886 65-13265 <*>
1964 V19B N7 P655 65-11870 <*>
1964 V19B N7/8 P947-967 4528 <BISI>
1965 V20B N3 P200-203 66-11035 <*>

ZEITSCHRIFT DES OESTERREICHISCHEN INGENIEUR- U. ARCHITEKTENVEREINS
1904 N11 P165-173 57-3105 <*>
1957 V102 P69-71 57-3413 <*>

ZEITSCHRIFT FUER OHRENHEILKUNDE
1904 V46 P72-83 61-14587 <*>
1912 V64 P358-374 65-13870 <*>

ZEITSCHRIFT FUER OPHTHALMOLOGISCHE OPTIK MIT EINSCHLIESSUNG DER INSTRUMENTENKUNDE
1913 V1 P97-99 61-14293 <*>
1914 V2 P69-77 61-14733 <*>
1916 V3 P1-17 61-14386 <*>
1916 V3 P33-40 61-14383 <*>
1916 V3 P40-41 61-14396 <*> 0
1916 V3 P145-153 61-14388 <*>
1916 V3 P161-163 61-14384 <*>
1916 V3 P161-162 61-14393 <*>
1916 V4 P146-149 61-14376 <*>
1918 V6 P25-35 61-14394 <*>
1919 V7 P10-14 61-14448 <*> 0
1921 V9 P161-170 61-14129 <*>
1923 V11 P41-59 61-16057 <*> C
1924 V12 P97-107 61-14125 <*>
1925 V13 P8-14 61-14397 <*>
1927 V14 P129-132 61-14270 <*>
1927 V15 P1-10 61-14116 <*>
1927 V15 P129-146 61-14207 <*>
1928 V15 P1-10 61-14635 <*>
1931 V19 P156-160 61-10976 <*>

ZEITSCHRIFT DER ORAGANISATION FUER DIE ZUSAMMENARBEIT DER EISENBAHNEN
1963 N1 P15-17 63-21614 <=>
1963 N1 P18-20 63-21590 <=>
1963 N2 P20-22 63-31131 <=>
1965 N4 P1-4 65-33750 <=$>
1965 N4 P5-10 65-34000 <=$>
1965 N4 P17-26 65-33750 <=$>
1965 N4 P29-31 65-33750 <=$>
1965 V8 N2 P4-7 65-32263 <=$>
1965 V8 N2 P10-12 65-32263 <=$>
1965 V8 N2 P23-32 65-32263 <=$>

ZEITSCHRIFT FUER ORTHOPAEDIE UND IHRE GRENZGEBIETE
1960 V3 N5 P72-88 61-11672 <=>
1961 V94 N2 P303-317 63-00292 <*>
1961 V95 N2 P202-212 65-17281 <*>
1962 V96 N1 P7-27 66-11591 <*>
1963 V97 N3 P339-353 66-11598 <*>

ZEITSCHRIFT FUER PARASITENKUNDE
1935 V7 P717-718 66-12104 <*>
1949 V14 N4 P320-363 CSIRO-3840 <CSIR>
1953 V16 P56-83 57-62 <*>
1954 V16 P388-406 I-688 <*>
1955 V17 P217-234 CSIRO-3132 <CSIR>

ZEITSCHRIFT FUER PFLANZENERNAEHRUNG, DUENGUNG UND

BODENKUNDE
1927 V9A P1-16 60-16702 <*>
1942 V30 P137-156 63-18117 <*> 0
1943 V30 P360-370 63-18121 <*>
1948 V40 P40-51 63-16135 <*> 0
1949 V46 P152-169 59-20765 <*>
1950 V50 P25-38 65-11668 <*>
1951 V52 N2 P120-150 C-2389 <NRC>
1951 V54 P36-58 I-632 <*>
1951 V55 N2 P97-105 CSIRO-3835 <CSIR>
1951 SNS V54 N3 P210-240 58-2659 <*>
1954 V66 P193-202 II-997 <*>
1954 V66 P222-226 3296 <*>
1954 V67 P211-219 II-823 <*>
1955 N1/3 P134-137 57-3234 <*>
1955 V68 N2 P158-169 CSIRO-3074 <CSIR>
1955 V69 N3 P71-86 CSIRO-3175 <CSIR>
1955 V69 N1/3 P15-25 CSIRO-3170 <CSIR>
1955 V69 N1/3 P25-31 CSIRO-3171 <CSIR>
1955 V69 N1/3 P38-43 CSIRO-TRANS-3172 <CSIR>
1955 V69 N1/3 P50-57 CSIRO-3173 <CSIR>
1955 V69 N1/3 P65-71 CSIRO-3179 <CSIR>
1955 V69 N1/3 P66-71 CSIRO-3174 <CSIR>
1955 V69 N1/3 P94-97 CSIRO-3176 <CSIR>
1955 V69 N1/3 P155-165 CSIRO-3178 <CSIR>
1955 V70 N2 P141-164 66-10195 <*> 0
1956 V73 N3 P235-245 65-12606 <*>
1957 V77 N1 P37-52 65-12690 <*>
1958 V82 N1 P33-37 59-16579 <+*>
1959 V84 N1/3 P98-102 65-10472 <*>
1960 V91 N2 P122-130 63-10295 <*>
1961 V92 N1 P36-46 66-10282 <*> 0
1961 V94 N1 P39-47 63-10519 <*>

ZEITSCHRIFT FUER PFLANZENKRANKHEITEN, PFLANZENPATHOLOGIE UND PFLANZENSCHUTZ
1954 V35 P112-118 II-206 <*>
1957 V64 N5 P257-271 8214 <K-H>
1957 V149 N5 P594-607 64-10709 <*>

ZEITSCHRIFT FUER PFLANZENZUECHTUNG
1943 V25 P255-283 II-315 <*>
1953 V32 P233-274 T-1495 <INSD>
1954 V33 N4 P367-418 1685 <*>

ZEITSCHRIFT FUER PHYSIK
1920 V3 N4 P224-242 66-13629 <*>
1920 V3 N5 P417-421 NP-TR-998 <*>
1921 V4 P393-409 63-20393 <*> 0
1921 V6 P153-164 2508 <K-H>
63-18686 <*> 0
1921 V6 P405-410 66-13631 <*>
1921 V6 P411-414 66-13630 <*>
1921 V8 P321-362 63-14901 <*> P
63-14902 <*> P
1922 V8 P321-362 AL-43 <*>
1922 V9 P259-266 C-5449 <NRC>
1923 V10 P1-11 T-2243 <INSD>
1923 V14 P63-107 66-12382 <*> 0
1923 V15 P8-23 I-222 <*>
1923 V16 P100-102 57-2216 <*>
1923 V17 P253-265 II-267 <*>
1923 V19 P313-332 I-221 <*>
1924 V23 P1-46 61-16458 <*> 0
1924 V23 P334-336 I-223 <*>
1924 V26 P117-138 59-10372 <*>
1924 V29 P345-355 57-1560 <*>
1925 V33 P335-344 58-149 <*>
1925 V33 N10/1 P760 58-344 <*>
1926 V35 P421-441 II-150 <*>
1926 V35 P652-699 UCRL TRANS-832(L) <*>
1926 V36 P563-580 58-141 <*>
1926 V36 P689-736 66-12401 <*>
1926 V36 P759-774 AEC-TR-5763 <=*$>
1926 V37 N4/5 P243-262 61-18622 <*>
1926 V39 P444-446 T1908 <INSD>
1926 V39 P607-622 C-2451 <NRC>
1926 V39 P636-643 63-10675 <*>
1927 V40 P686-707 I-23 <*>
1927 V41 P103-115 65-11327 <*>
1927 V41 P116-139 65-11328 <*>
1927 V41 P453-492 T1907 <INSD>
1927 V41 P701- 57-2069 <*>
1927 V41 P867-871 3021 <*>
57-791 <*>
1927 V42 P26-42 63-00265 <*>
1927 V42 P283-301 64-30036 <*>
1927 V42 P311-318 62-00145 <*>
1927 V42 P853 AL-268 <*>
1927 V44 P170-189 59-17338 <*> 0
1927 V45 P13-29 SCL-T-351 <*>
1928 V46 P558-567 AL-422 <*>
1928 V46 P747-752 66-12902 <*>
1928 V46 N9 P607-652 58-150 <*>
1928 V47 P184-202 63-10393 <*>
1928 V47 P589-601 1550 <*>
1928 V48 P80-91 61-14291 <*> 0
1928 V48 P323-339 58-148 <*>
1928 V48 P397-425 41H13G <ATS>
1928 V48 N3/4 P192-204 59-17212 <*>
1928 V48 N7/8 P571-582 59-17671 <*>
1928 V50 P537-547 AEC-TR-4486 <*>
1928 V51 P204 AEC-TR-5915 <*>
1928 V51 P545-556 57-2135 <*>
1928 V51 P571-583 AEC-TR-3730 <+*>
1928 V51 P859- 2730 <*>
2834 <*>
1928 V51 N5/6 P355-373 66-12595 <*>
1928 V52 P708-725 57-2402 <*>
1928 V52 N34 P235-248 AEC-TR-3322 <*>
1929 V52 P336-341 58-1673 <*>
1929 V53 P107-129 AEC-TR-3731 <=*>
1929 V53 P255-267 NP-TR-1051 <*>
1929 V53 P747-765 60-10318 <*>
1929 V53 P766-778 I-320 <*>
1929 V53 P779-791 59-16781 <+*>
1929 V53 P840-856 65-12375 <*>
1929 V54 P703-710 UCRL TRANS-917(L) <*>
1929 V54 N112 P53-73 60-18141 <*>
1929 V54 N7/8 P582-587 57-1465 <*>
1929 V56 P147-183 AL-160 <*>
1929 V56 P147-174 63-14906 <*>
1929 V56 P790-801 94L34G <ATS>
1929 V57 P30-70 63-18155 <*>
1929 V57 N5/6 P292-304 57-2134 <*>
1929 V58 P373 58-2014 <*>
1930 V59 P198-207 1570 <*>
58-2178 <*>
1930 V61 P655-659 II-1039 <*>
57-548 <*>
1930 V61 P792-797 65-18084 <*>
1930 V61 N11/2 P792-797 SCL-T-329 <*>
1930 V63 P30-53 57-1566 <*>
1930 V63 P345-369 T1762 <INSD>
1930 V63 P762-770 63-10770 <*>
1930 V63 N1/2 P54-73 11J19G <ATS>
1930 V64 P682-689 66-12146 <*>
1930 V64 N3/4 P237-247 59-20402 <*> 0
1930 V65 P547-563 66-11078 <*>
1930 V66 P471-476 1208 <*>
1931 V67 N11/2 P808-811 57-3136 <*>
1931 V68 N5/6 P289-308 57-1461 <*>
1931 V69 P185-206 AEC-TR-5764 <*>
1931 V69 P298-308 66-11071 <*>
1931 V69 P618-636 I-511 <*>
1931 V70 P204-386 64-20132 <*> P
1931 V70 P353-374 63-00271 <*>
1931 V70 P508-515 66-11975 <*>
1931 V71 P553-592 64-16948 <*>
1931 V71 P689-695 AL-237 <*>
1931 V71 P703-714 58-2019 <*>
1931 V72 P578-586 60-10316 <*>
1931 V73 P312-334 58-209 <*>
1931 V73 P335-347 58-208 <*>
1931 V73 N1/2 P118-120 57-2379 <*>
1932 V75 P145-170 65-18120 <*>
1932 V75 P622-647 63-23724 <=*$>
1932 V75 P763-776 58-1384 <*>
1932 V76 P368-385 C-2492 <NRC>

Citation	Translation
	57-3479 <*>
1932 V76 P608-627	66-13823 <*>
1932 V77 P398-400	58-2470 <*> 0
1932 V77 P459-477	66-11970 <*>
1932 V78 P564-572	AL-630 <*>
1932 V78 N3/4 P230-239	57-633 <*>
1932 V78 N5/6 P318-339	66-13303 <*> 0
1932 V79 P108-121	R-3420 <*>
1933 V80 P755-762	II-738 <*>
1933 V80 N11/2 P755-762	64-16394 <*>
1933 V81 P143-162	AEC-TR-4082 <*>
1933 V81 P363-376	UCRL TRANS-972 <*>
1933 V82 P589-609	AEC-TR-4396 <*>
1933 V84 P1-16	UCRL TRANS-723(L) <*>
1933 V84 P677-685	65-11608 <*>
1933 V86 P583-591	AEC-TR-4989 <*>
1934 N11/2 P826-833	57-588 <*>
1934 V87 P399-408	I-21 <*>
	2238 <*>
1934 V88 P161-177	UCRL TRANS-726 <*>
1934 V89 P210-233	63-10163 <*>
1934 V89 P447-473	AEC-TR-5130 <*>
1934 V89 P634-639	I-462 <*>
1934 V89 P660-664	I-503 <*>
1934 V90 P703-711	AEC-TR-4395 <*>
1934 V90 N3/4 P266-278	57-2312 <*>
1934 V90 N5/6 P384-402	SCL-T-285 <*>
1934 V90 N7/8 P468-479	2182 <*>
1935 V93 P611-619	I-622 <*>
1935 V94 P277-302	63-20120 <*>
1935 V95 N7/8 P440-449	2466 <*>
1935 V95 N7/8 P450-459	2467 <*>
1935 V95 N7/8 P499	AEC-TR-4883 <*>
1935 V95 N11/2 P752-762	57-636 <*>
1935 V96 P251-267	2033 <*>
1935 V97 P355-375	AEC-TR-3984 <*>
1936 V99 P1-23	57-2471 <*>
1936 V99 N1/2 P24-41	57-2174 <*>
1936 V101 P373-377	61-14346 <*>
1936 V101 N9/10 P541-577	79L33G <ATS>
1936 V102 P432-460	AL-138 <*>
1936 V102 P611-625	62-11393 <=>
1936 V102 N11/2 P709-733	AEC-TR-3784 <*>
1936 V103 P170-190	66-12901 <*> 0
1936 V103 P341-349	58-503 <*>
1936 V103 N7/8 P443-453	59-17506 <*>
1936 V104 N3/4 P248-274	66-12277 <*> 0
1937 N1/2 P132-140	66-12598 <*> 0
1937 V105 P399-444	I-752 <*>
1937 V106 P70-92	63-10769 <*>
1937 V106 P474-484	57-2857 <*>
1937 V106 P541-550	I-423 <*>
1937 V106 P675-691	UCRL TRANS-718(L) <*>
1937 V107 P497-	II-764 <*>
1938 N108 P668-680	I-421 <*>
1938 V108 P55-84	64-16993 <*> 0
1938 V108 P244-264	I-323 <*>
1938 V108 P376-390	I-811 <*>
1938 V108 P668-680	58-1527 <*>
1938 V108 N7/8 P459-482	57-2488 <*>
1938 V111 P357-372	64-14983 <*> 0
1938 V111 N1/2 P61-78	60-14077 <*>
1938 V111 N5/6 P357-372	57-3342 <*>
1939 V112 P667-675	65-10878 <*>
1939 V113 P697-	I-502 <*>
1939 V113 P751-768	63-10494 <*> 0
1939 V113 N5/6 P367-414	58-1950 <*>
	58-1968 <*>
1939 V114 P354-367	AL-885 <*>
1939 V114 P579-601	57-979 <*>
1939 V114 P682-704	AL-596 <*>
1940 V115 P530-556	62-16205 <*> 0
1940 V116 P444-453	AERE-TR-11/3/5/1198 <*>
1940 V116 N5/6 P366-369	59-20403 <*> 0
1940 V116 N5/6 P370-378	57-2050 <*>
1940 V116 N7/8 P385-414	66-13147 <*> 0
1940 V117 P565-574	I-814 <*>
1941 V117 P23-40	AL-685 <*>
1941 V117 N3/4 P180-197	57-1414 <*>
1941 V117 N7/8 P437-443	57-882 <*>
1941 V118 P401-	I-904 <*>
1941 V118 P539-592	63-27113 <$>
1941 V118 N3/4 P199-209	57-1376 <*>
1942 V118 P539-592	63-00721 <*>
1942 V118 N9/10 P515-538	57-2155 <*>
1942 V119 P79-99	65-17143 <*>
1942 V119 P366-373	23L30G <ATS>
1942 V119 P493-	I-904 <*>
1942 V120 P212-227	I-938 <*>
1942 V120 P261-269	I-424 <*>
1942 V120 P270-	2488 <*>
1942 V120 P720-740	I-831 <*>
1942 V120 N2 P107-120	57-2120 <*>
1943 V120 P212-227	63-14659 <*>
1943 V120 P437-449	I-903 <*>
1943 V120 P533-536	UCRL TRANS-808 <*>
1943 V121 P604-613	I-812 <*>
1944 V122 N1/4 P137-162	66-13304 <*> 0
1944 V122 N5/8 P510-526	57-341 <*>
1948 V124 P351-	1566 <*>
1948 V124 P628-657	NACA TM1431 <NASA>
1948 V124 P691-699	65-13913 <*> 0
1948 V124 N3/6 P309-314	AL-888 <*>
1949 V125 P394-404	UCRL TRANS-761(L) <*>
1949 V125 N11/2 P739-756	SC-T-519 <*>
1949 V125 N7/10 P497-504	AL-738 <*>
1949 V126 P35-48	62-10630 <*>
1949 V126 P233-260	62-10567 <*> 0
1949 V126 P310-318	UCRL TRANS-1031 <=*$>
1949 V126 P642-665	AEC-TR-4944 <*>
1949 V126 N7/9 P601-618	65-10891 <*>
1949 V126 N7/9 P671-688	60-14198 <*>
1950 V127 P163-167	57-706 <*>
1950 V127 P243-273	59-10429 <*>
1950 V127 P274-288	2979 <*>
1950 V127 P381-390	61-10419 <*>
1950 V127 P468-494	58-2150 <*>
1950 V127 N4 P344-356	65-13592 <=$>
	66-10895 <*>
1950 V127 N4 P416-428	57-997 <*>
1950 V127 N4 P443-454	AL-524 <*>
1950 V128 P166-171	SLAC-TRANS-14 <=$>
1950 V128 P366-367	58-2347 <*>
1950 V128 P500-505	AL-67 <*>
1950 V128 P538-545	SC-T-520 <*>
1950 V128 P546-574	59-10301 <*>
1951 V129 P108-122	58-508 <*>
1951 V129 P123-139	63-01210 <*>
1951 V129 P202-218	AEC-TR-4788 <*>
1951 V129 P219-232	63-20274 <*> 0
1951 V129 P327-342	61-11135 <=>
1951 V129 P343-364	58-509 <*>
1951 V129 P401-415	I-122 <*>
1951 V129 N3 P233-245	66-13632 <*> 0
1951 V130 P174-195	62-01385 <*>
1951 V130 P269-292	57-778 <*>
1951 V130 P356-370	I-123 <*>
1951 V130 P371-384	I-1028 <*>
1951 V130 P385-391	I-124 <*>
1951 V130 P449-456	UCRL TRANS-701(L) <*>
1951 V130 P641-649	AEC-TR-4848 <*>
1951 V130 N2 P227-238	60-14004 <*>
1951 V130 N3 P392-402	59-17502 <*>
1951 V130 N4 P409-414	62-16966 <*> 0
1951 V131 P28-40	58-2208 <*>
1951 V131 N1 P28-40	65-12017 <*>
1952 V131 P395-407	AL-41 <*>
1952 V131 N4 P456-469	59-17940 <*>
1952 V132 P54-71	40F3G <ATS>
1952 V132 P81-106	66-11168 <*>
1952 V132 P261-284	61-18254 <*>
1952 V132 P384	AEC-TR-5783 <*>
1952 V132 P497-507	63-16637 <*>
1952 V132 N4 P468-481	60-18799 <*>
1952 V133 P15-20	C-2427 <NRC>
1952 V133 P101-108	AEC-TR-5061 <*>
1952 V133 P109-117	63-14649 <*> 0
1952 V133 P201-228	AL-708 <*>
1952 V133 P209-228	C-2394 <NRC>
1952 V133 P455-470	I-167 <*>

Citation	Translation
1952 V133 P471-484	I-168 <*>
1952 V133 P499-503	62-10516 <*>
1952 V133 P513-523	I-928 <*>
1952 V133 P589-614	II-1000 <*>
	MF-4 <*>
1952 V133 N3 P340-361	59-17920 <*>
1952 V133 N5 P629-646	SCL-T-270 <*>
1952 V133 N1/2 P2-8	57-1010 <*>
1952 V133 N1/2 P297-308	66-12329 <*>
1952 V133 N3/4 P251-263	NP-TR-453 <*>
1952 V134 N1 P1-8	64-16989 <*>
1952 V134 N1 P21-24	SCL-T-268 <*> P
1953 V134 P435-450	57-2407 <*>
1953 V134 N4 P435-450	59-17941 <*>
1953 V134 N4 P469-482	59-17942 <*>
1953 V135 P177-195	UCRL TRANS-476(L) <*>
1953 V135 P251-254	I-336 <*>
1953 V135 P380-394	49G5G <ATS>
1953 V135 P573-592	57-2928 <*>
1953 V135 P602-614	27T95G <ATS>
1953 V135 P620-638	59-17328 <*>
1953 V135 N2 P177-195	62-18111 <*>
	62-18270 <*> 0
1953 V135 N4 P361-375	C-2432 <NRC>
1953 V135 N5 P573-592	65-12012 <*>
1953 V136 P74-86	2F3G <ATS>
1953 V136 P87-104	3F3G <ATS>
1953 V136 P119-136	7F3G <ATS>
1953 V136 P224-233	24H10G <ATS>
1953 V136 P352-366	NP-TR-321 <*>
1953 V136 P402-410	84J18G <ATS>
1953 V136 N2 P119-136	62-18256 <*>
1954 P212-225	II-182 <*>
1954 V137 P169-174	UCRL TRANS-845(L) <*>
1954 V137 P175-189	57-794 <*>
1954 V137 P228-237	AEC-TR-5702 <*>
1954 V137 P295-308	II-70 <*>
1954 V137 P583-587	1225 <*>
1954 V138 P109-120	62-18837 <*>
	63-10373 <*>
1954 V138 P136-150	62-01195 <*>
1954 V138 P170-182	48G5G <ATS>
1954 V138 P322-329	57-2500 <*>
	64-10371 <*> P
1954 V138 N2 P121-135	61-00037 <*>
1954 V139 P30-34	1576 <*>
1954 V139 P115-146	57-984 <*>
1954 V139 P212-225	91G4G <ATS>
1954 V139 P504-517	62-16480 <*>
1954 V139 P599-618	1134 <*>
	57-3015 <*>
1954 V139 N5 P599-618	66-10161 <*> 0
1955 V140 P47-56	57-1999 <*>
1955 V140 P119-138	G6G <ATS>
1955 V140 P262-273	58-221 <*>
	63-01418 <*>
1955 V140 P409-413	58-1545 <*>
1955 V140 N3 P262-273	62-10202 <*> 0
	65-17395 <*>
1955 V141 P198-216	7G6G <ATS>
1955 V141 P221-236	25G8G <ATS>
	57-2758 <*>
1955 V141 P319-334	57-3192 <*>
1955 V141 N3 P263-276	59-10905 <*>
1955 V141 N3 P335-345	59-10770 <*>
1955 V141 N4 P463-475	AEC-TR-3276 <*> 0
1955 V141 N1/2 P83-86	59-10417 <*>
1955 V142 P182-200	63-14666 <*>
1955 V142 P183-200	UCRL TRANS-504(L) <*>
1955 V142 P463-475	66J17G <ATS>
1955 V142 N1 P21-41	CSIRO-3081 <CSIR>
1955 V143 N1 P77-92	62-16090 <*> 0
1956 V143 P559-590	AI-TRANS-93 <*>
1956 V144 P256-262	62-18263 <*> 0
1956 V144 P449-454	C-3591 <NRCC>
1956 V144 P488-508	UCRL TRANS-681(L) <*>
1956 V144 P586-611	64-71351 <=>
1956 V145 P151-155	57-2338 <*>
1956 V145 P301-318	64-14237 <*> 0
1956 V145 P597-610	61-16459 <*> 0

Citation	Translation
1956 V145 N1 P44-47	62-16308 <*> 0
1956 V145 N4 P486-495	66-13415 <*>
1956 V146 P39-58	58-2207 <*>
1956 V146 P320-332	AEC-TR-4792 <*>
1956 V146 P333-338	AEC-TR-4790 <*>
1956 V146 P505-514	63-14054 <*>
1956 V146 N1 P39-58	66-12739 <*>
1957 V147 P277-287	58-2439 <*>
1957 V147 P323-349	62-10619 <*> 0
1957 V147 P395-418	62-16689 <*>
	62-18864 <*>
1957 V147 P465-480	28N50G <ATS>
1957 V147 N2 P161-183	66-11066 <*>
1957 V147 N5 P544-566	60-14199 <*> 0
1957 V148 P177-191	63-00592 <*>
	63-27144 <$>
1957 V148 N1 P15-27	64-16896 <*>
1957 V149 P111-130	58-2290 <*>
1957 V149 P153-179	61-10706 <*>
1957 V149 P254-266	67K27G <ATS>
1957 V149 P397-411	58-2496 <*>
1957 V149 N1 P40-61	64-14180 <*>
1957 V149 N5 P608-623	64-10907 <*>
1958 V150 P215-230	TRANS-141 <FRI>
	61-10655 <*>
1958 V150 P231-241	AEC-TR-3737 <=*>
1958 V150 P527-550	63-00180 <*>
1958 V150 P632-639	63-14690 <*>
1958 V150 P648-652	C-3593 <NRCC>
1958 V150 N2 P162-171	SCL-T-242 <*>
	59-20621 <*>
1958 V151 P124-143	AEC-TR-3685 <+*>
1958 V151 P252-263	61-20423 <*>
1958 V151 P297-327	SCL-T-368 <*>
1958 V151 P421-430	62-10413 <*>
1958 V151 P487-503	UCRL TRANS-497(L) <*>
1958 V151 N2 P174-186 PT.1	
	61-00036 <*>
1958 V151 N2 P124-143	65-10433 <*>
1958 V151 N2 P159-173	AEC-TR-3745 <+*> 0
	4574 <BISI>
1958 V151 N3 P375-384 PT.2	
	61-00036 <*>
1958 V151 N3 P296-306	59-10734 <*>
1958 V151 N5 P595-612	59-15024 <*>
1958 V152 P143-182	AEC-TR-3484 <*>
1958 V152 P261-271	UCRL TRANS-828(L) <*>
1958 V152 P272-280	62-00196 <*>
1958 V152 P454-469	61-16854 <*> 0
1958 V152 P611-623	UCRL TRANS-639 <*>
1958 V152 N1 P75-86	61-10534 <*>
1958 V153 P1-15	C-3419 <NRCC>
1958 V153 P174-185	62-16476 <*> 0
1958 V153 N3 P257-268	59-17446 <*>
1958 V153 N3 P338-358	AEC-TR-3763 <+*>
1959 V153 N3 P314-316	60-10026 <*>
1959 V154 P319-329	AEC-TR-4129 <*>
1959 V154 P495-511	SCL-T-398 <*>
1959 V154 N3 P330-338	62-16029 <*>
1959 V154 N4 P474-485	66-13908 <*>
1959 V155 P42-58	AEC-TR-3990 <*>
	58L36G <ATS>
1959 V155 P387-394	38L35G <ATS>
1959 V155 N1 P42-58	61-10206 <*>
1959 V155 N2 P206-222	60-18069 <*> 0
	65-17165 <*>
1959 V155 N2 P247-262	TRANS-135 <FRI>
	32L35G <ATS>
	61-18295 <*>
1959 V155 N4 P453-464	60-16807 <*>
1959 V156 P534-554	65-18100 <*>
1959 V156 N1 P1-26	66-12312 <*>
1959 V156 N1 P38-54	66-13416 <*>
1959 V156 N3 P318-347	65-10893 <*>
1959 V156 N3 P360-381	63-00467 <*>
1959 V156 N4 P598-602	NP-TR-1108 <*>
1959 V157 P335-361	61-00225 <*>
1959 V157 N1 P1-29	60-00126 <*>
1960 V158 P444-470	UCRL TRANS-792(L) <*>
1960 V158 N5 P563-571	65-10292 <*> 0

Citation	Translation
1960 V158 N5 P607-622	60-18071 <*>
1960 V159 P125-148	AEC-TR-5593 <*>
1960 V159 P462-477	62-18072 <*>
1960 V159 P584-601	AEC-TR-5921 <*>
1960 V159 N3 P360-368	61-16428 <*>
1960 V160 P1-15	ORNL-TR-71 <*>
1960 V160 P16-32	AEC-TR-5619 <*>
1960 V160 P93-108	653 <TC>
1960 V160 P121-143	63-10274 <*>
1960 V160 P149-161	AEC-TR-5396 <*>
1960 V160 P363-373	62-10535 <*> 0
1960 V160 P406-419	UCRL TRANS-692 <*>
1960 V160 P504-519	64-71127 <=$>
1960 V160 P535-553	62-14150 <*>
1960 V160 N1 P93-108	62-14302 <*>
1960 V161 P339-345	64-71127 <=$>
1961 V161 P205-220	63-10917 <*>
1961 V161 P428-438	11396-E <KH>
1961 V161 N3 P310-329	62-14459 <*>
1961 V162 P203-212	AI-TR-24 <*>
1961 V162 P275-289	AEC-TR-5123 <*>
1961 V162 P290-312	AEC-TR-5694 <*>
1961 V162 P313-328	63-16564 <*>
1961 V162 P557-569	63-20848 <*> 0
1961 V163 P230-239	66-12365 <*>
1961 V164 P563-573 PT1	63-13019 <=*>
1961 V164 P513-521	UCRL TRANS-1161(L) <*>
1961 V164 P581-587	AEC-TR-4987 <*>
1961 V164 P588-594	AEC-TR-4986 <*>
1961 V164 N4 P442-455	65-12628 <*>
1961 V164 N5 P581-587	65-17167 <*>
1961 V164 N5 P588-594	65-17166 <*>
1961 V165 P34-36 PT2	63-13019 <=*>
1961 V165 P202-212	66-13780 <*>
1961 V165 P439-444	C-5593 <NRC>
1961 V165 P502-532	63-01435 <*>
1962 V166 P76-80	76P61G <ATS>
1962 V166 P494-503	NP-TR-1088 <*>
1962 V166 N5 P477-493	NP-TR-1021 <*>
1962 V166 N5 P519-543	64-14010 <*>
1962 V167 P11-19	63-10855 <*>
1962 V167 P501-510	63-20353 <*>
1962 V167 N1 P53-71	65-12218 <*>
1962 V167 N4 P386-402	66-11281 <*> 0
1962 V167 N4 P446-451	65-11699 <*>
1962 V167 N4 P461-467	62-20332 <*>
1962 V168 P42-54	62-20227 <*>
1962 V168 P74-80	AEC-TR-5397 <*>
1962 V168 P148-154	62-20231 <*>
1962 V168 P283-291	62-20230 <*>
1962 V168 P292-297	66-11755 <*>
1962 V168 N3 P333-342	64-14982 <*> 0
1962 V169 P178-197	63-20355 <*>
1962 V169 P273-285	4536 <BISI>
1962 V169 P323-325	65-00212 <*>
1962 V169 P386-408	66-11756 <*>
1962 V169 N4 P564-575	65-10371 <*> 0
1962 V170 P1-21	63-20348 <*> 0
1962 V170 P303-319	UCRL TRANS-1112(L) <*>
1962 V170 P320-335	NP-TR-965 <*>
1962 V170 P526-532	UCRL TRANS-1001 <*>
1962 V170 N3 P274	AEC-TR-5655 <*>
1962 V170 N4 P367-375	64-30649 <*> 0
1963 V172 P512-519	66-14473 <*>
1963 V172 N1 P1-18	66-11074 <*>
1963 V173 P117-134	65-10803 <*>
1963 V173 P413-421	64-14630 <*>
1963 V173 N1 P135-148	65-10797 <*>
1963 V173 N3 P321-346	N66-15920 <=$>
1963 V174 N2 P257-268	64-14126 <*>
1963 V174 N2 P269-279	65-11580 <*>
1963 V174 N2 P280-288	65-11581 <*>
1963 V174 N2 P289-295	65-11579 <*>
1963 V175 P70-104	AEC-TR-6325 <*>
1963 V176 P485-497	65-11534 <*> 0
1963 V176 P498-509	AI-TRANS-36 <*>
1963 V176 N3 P207-220	65-13114 <*>
1963 V176 N4 P510-524	65-10792 <*>
1964 V177 P529-542	66-12829 <*>
1964 V177 N4 P385-423	65-10386 <*> 0
1964 V177 N5 P543-561	65-13126 <*>
	66-12828 <*>
1964 V178 P19-25	30R76G <ATS>
1964 V178 N2 P113-119	65-11542 <*> 0
1964 V178 N3 P319-334	65-11541 <*> 0
1964 V179 P367-378	87R79G <ATS>
1964 V179 N4 P367-378	65-11535 <*> 0
1964 V179 N5 P461-467	C-5344 <NRC>
1964 V180 N2 P184-197	65-11615 <*> 0
1965 V184 N1 P85-91	66-13828 <*>
1965 V184 N3 P293-298	65-14162 <*> 0

ZEITSCHRIFT FUER PHYSIKALISCHE CHEMIE. LEIPZIG

Citation	Translation
V46 P287-292	58-1852 <*>
1887 V1 P631-648	58-2224 <*>
1888 V2 N5 P284-295	64-30299 <*>
1894 V14 P317-	I-326 <*>
1897 V23 P689-692	66-14211 <*> 0
1899 V28 P453-493	62-14517 <*> 0
1899 V29 P527-545	66-13144 <*>
1899 V29 P648-657	64-20456 <*> P
1902 V39 P14-16	58-590 <*>
1906 V54 P55-97	2044 <*>
1906 V56 P43-56	63-10978 <*>
1908 V64 P449-505	58-507 <*>
1909 V67 P212-248	C-5543 <NRC>
1910 V67 P578-597	ANL-TRANS-203 <*>
1910 V72 P723-751	I-773 <*>
1910 V72 N4 P439-450	65-11001 <*> 0
1910 V73 P129-147	64-14278 <*>
1910 V73 P513-546	62-18937 <*>
1910 V73 P667-	AL-889 <*>
1910 V73 P667-684	I-804 <*>
1910 V74 N3 P277-307	NP-TR-276 <*>
1912 N81 P129-139	57-957 <*>
1912 N81 P157-170	57-957 <*>
1912 V78 P564-572	63-20761 <*>
1914 V87 P253-256	AI-TR-29 <*>
1914 V89 P168-178	I-624 <*>
1916 V91 P593-604	8759 <IICH>
1918 V92 P129-168	17J18G <ATS>
1918 V92 P129-147	57-544 <*>
1918 V92 P129-168	58-144 <*>
1919 V93 P596-606	62-18239 <*> 0
1920 V94 P233-253	2978 <*>
1923 V106 P386-398	59-20189 <*>
1923 V107 N5/6 P436-452	59-20325 <*>
1924 V110 P318-342	AI-TR-46 <*>
	63-20776 <*>
1924 V110 P399-416	66-11116 <*>
1924 V110 P559-571	I-338 <*>
1924 V111 P234-250	1143Z <K-H>
	63-16176 <*> 0
1925 V115 P233-238	33K24G <ATS>
1925 V118 P99-106	I-665 <*>
	59-10673 <*>
1925 V118 P107-113	59-10671 <*> 0
1926 V121 P439-455	I-13 <*>
1926 V123 P161-198	63-18135 <*> 0
1926 V124 P427-435	59-17528 <*>
1927 V125 P243-250	59-10686 <*>
1927 V130A P378-389	64-20323 <*>
1928 V132 P161-184	UCRL TRANS-795(L) <*>
1928 V136 P77-92	63-10601 <*> 0
1928 V138 N1/2 P102-122	62-16454 <*>
1928 V139 P169-197	63-20131 <*>
1929 V143 P81-93	AL-821 <*>
1929 V145 N1 P1-26 PT.A	I-517 <*>
1929 V145 N1 P48-56	58-1927 <*>
1929 V140A N5/6 P429-434	59-17552 <*>
1929 V141A N3 P180-194	59-10026 <*>
1929 V144A P340-358	40T91G <ATS>
1929 V144A N5/6 P321-339	39T91G <ATS>
1930 V148 P1-26	57-1140 <*>
1930 V148 P241-260	62H9G <ATS>
1930 V146A P245-280	II-866 <*>
1930 V146A N5 P363-372	59-20388 <*> 0
1930 V146A N5 P373-404	59-20387 <*> 0
1930 V148A P27-35	II-651 <*>
1930 V148A N6 P471-474	59-20386 <*> 0

Citation	Translation
1930 V151A P13-55	C-2477 <NRC>
1931 V152 P110	AEC-TR-14 <*>
1931 V153A N3/4 P161-186	59-10062 <*>
1931 V154A P1-46	ANL-TR-8 <*>
1931 V154A P47-91	ANL-TRANS-34 <*>
1931 V154A P92-96	ANL-TR-37 <*>
1931 V154A N5/6 P385-420	64-20322 <*>
1931 V156A P309-316	59R78G <ATS>
1932 V161 P33-45	AEC-TR-5317 <*>
1932 V160A N3/4 P194-204	57-2479 <*>
1932 V161A P1-32	II-352 <*>
1932 V162A P454-466	I-184 <*>
1933 V163A P409-441	58-1642 <*>
1933 V166A P343-356	64-20324 <*>
1933 V166A N1 P16-22	59-20830 <*>
1933 SA V164 N3/4 P176-200	3022 <*>
1934 V167A P209	58-1935 <*>
1934 V169A P237-240	UCRL TRANS-620(L) <*>
1934 V170A P262-272	62-20080 <*>
1934 V171A P147-180	63-18972 <*> 0
1934 V171A P331-340	66-11849 <*>
1935 V173 P353-364	05G5G <ATS>
1935 V173A P32-34	66-11330 <*> 0
1935 V175A N1/2 P127-139	58-1863 <*>
1936 V176A P65-80	AEC-TR-3507 <*>
1937 V179A P16-22	2371 <*>
1937 V179A P195-209	C-2592 <NRC>
1938 V181A P283-300	64-20404 <*>
1939 V183A P297-317	64-20450 <*> 0
1940 V187A P43-57	63-20650 <*>
1940 V187A P235-249	63-20649 <*>
1941 V189 P126-130	AEC-TR-5017 <*>
1941 V189 P219-316	57-2582 <*>
	57-2861 <*>
1941 V188A P99-108	8918 <IICH>
1941 V188A P272-283	62-18009 <*>
1941 V188A N2 P109-126	59-20188 <*>
1941 V189A P219-316	57-3488 <*>
1942 V191 P95-128	AEC-TR-3911 <*>
1943 V192 P85-100	57-2274 <*>
1943 V192 N1/2 P1-55	43G7G <ATS>
1943 V192 N3/4 P163-210	59-15936 <*> P
1944 V193 P274-286	AEC-TR-3692 <+*>
1944 V193 P386-406	58-2434 <*>
1950 V195 P165-174	57-2352 <*>
1950 V195 P197-212	770 <TC>
1950 V195 N3 P213-224	58-1742 <*>
1950 V195 N1/2 P24-36	57-264 <*>
1950 V196 P127-159	I-55 <*>
1950 V196 P160-180	57-1146 <*>
1950 V196 N1/3 P127-159	57-335 <*>
1951 V197 P256-264	37M44G <ATS>
1951 V197 P265-276	63-14053 <*> 0
1951 V197 N1/2 P1-5	60-10177 <*> 0
1951 V198 N516 P232-247	II-995 <*>
1951 V199 N1/3 P125-141	NP-TR-293 <*>
1952 V199 P142-155	63-20275 <*> 0
1952 V199 P165-169	66-11980 <*>
1952 V199 N1/3 P142-151	83G6G <ATS>
1952 V199 N5/6 P285-299	45N56G <ATS>
1952 V200 P53-69	1886 <*>
1952 V200 P250-292	66-11998 <*>
1952 V200 N1/2 P53-69	64-14381 <*> 0
1952 V200 N3/4 P97-128	II-354 <*>
	64-14397 <*> 0
1952 V201 P168-189	II-73 <*>
1952 V201 N5/6 P268-277	80G6G <ATS>
1953 V202 N5/6 P352-378	61-20273 <*>
	93N52G <ATS>
1954 V202 N5/6 P440-459	19J14G <ATS>
1954 V203 N3/4 P176-200	57-2578 <*>
1954 V203 N5/6 P265-274	79K22G <ATS>
1955 V205 P73-77	25P66G <ATS>
1956 V205 P208-216	57-2461 <*>
1956 V205 P253-260	57-3182 <*>
1957 V207 P138-140	63-10851 <*>
1957 V207 N5/6 P340-349	NP-TR-870 <*>
1958 P143-151	65-13127 <*>
1958 V208 P201-219	61-00226 <*>
1958 V208 P293-308	63-18651 <*> 0
	65-18170 <*>
1958 V209 N1/2 P124-128	61-10203 <*>
1959 V210 P35-49	07M41G <ATS>
1959 V211 P180-193	AEC-TR-4371 <*>
1959 V211 P194-207	AEC-TR-4372 <*>
1960 V215 N3/4 P176-184	2448 <BISI>
1960 V215 N5/6 P269-284	NP-TR-839 <*>
1960 V215 N5/6 P285-292	NP-TR-838 <*>
1961 V218 N1/2 P93-111	65-13989 <*>
1961 V218 N1/2 P108-140	65-12304 <*>
1961 V218 N5/6 P338-353	65-12298 <*>
1961 V219 N1/2 P134-139	AEC-TR-5570 <*>
1962 V220 N4 P240-254	NP-TR-981 <*>
1962 V220 N5/6 P339-354	NP-TR-1058(DRAFT) <*>
1962 V221 P247-258	8743 <IICH>
1962 V221 P301-345	8820 <IICH>
1963 V222 N3/4 P143-160	66-13414 <*>
1963 V223 N5/6 P405-411	64-18527 <*>
1963 V224 N3/4 P177-198	66-14723 <*>
1964 V225 N1/2 P45-56	65-11487 <*>
1964 V225 N3/4 P175-196	66-10614 <*>
1964 V226 N3/4 P146-156	66-12996 <*>

ZEITSCHRIFT FUER PHYSIKALISCHE CHEMIE. SERIES B. LEIPZIG. (1928-1943)

Citation	Translation
1928 V1 P192-204	AEC-TR-4809 <*>
1928 V1B P253-269	63-10536 <*>
1928 V1B N3/4 P275-292	64-18895 <*> 0
1929 V2B P267-281	63-10602 <*>
1929 V5B P467-475	UCRL TRANS-1059(L) <*>
1929 V6B N3 P215-222	59-00796 <*>
1930 V7B P407-422	64-10146 <*>
1930 V8B N5/6 P445-454	62-00673 <*>
1930 V9B P437-475	AEC-TR-5486 <*>
1930 V10B N3 P193-204	60-14078 <*> 0
1930 V11B P163-210	65-10841 <*>
1930 V11B P433-454	AEC-TR-3509 <*>
1930 SB V6 P221-232	2291 <*>
1931 P863-873	65-13912 <*>
1932 V19 N6 P443-450	57-2246 <*>
1932 V15B P77-81	58-2323 <*>
1932 V16B P213-220	64-20831 <*>
1932 V17B N4/5 P310-326	63-18126 <*>
1932 V18B P291-315	63-16745 <*>
1932 V18B N6 P369-379	58-885 <*>
1932 V19B P190-198	63-14813 <*>
1933 V21 N1/2 P25-41	58-1966 <*>
1933 V22 P199-211	57-643 <*>
1933 V22 N3 P212-225	58-1940 <*>
1933 V21B P469-472	UCRL TRANS-754(L) <*>
1933 V23B N2/3 P175-192	62-01072 <*>
1933 SB V21 N25 P25-41	I-808 <*>
1934 V25B P153-156	I-628 <*>
1934 V25B P161-176	UCRL TRANS-545 <*>
1934 V27B N5/6 P359-375	64-20321 <*>
1935 V29 N3 P215-224	57-2644 <*>
1935 V28B P178-188	UCRL TRANS-756(L) <*>
1935 V29B P363-370	59-15314 <*>
1935 V29B N5 P335-355	UCRL TRANS-747(L) <*>
1936 V33B P120-126	I-542 <*>
1936 V33B P179-195	5147-D <K-H>
1937 V35B P363-381	5147-B <K-H>
1937 V37B P75-80	C-5787 <NRC>
1937 V37B P307-314	351 <TC>
1937 V37B P375-391	5147-C <K-H>
1937 V38B P61-71	58-1641 <*>
1937 V38B P221-244	62-20246 <*> 0
1937 V38B P245-269	62-20247 <*> 0
1937 SB V35 P239-255	I-68 <*>
1938 V40B N4 P252-262	64-20509 <*> 0
1938 V41B P98-111	58-1974 <*>
1938 V41B P388-395	65-13413 <*>
1938 V41B P397-426	62-20248 <*> 0
1939 V43 N4 P272-273	RCT V13 N1 P282 <RCT>
1939 V44 P397-450	1854 <*>
1939 V44 P451-473	AL-148 <*>

Citation	Translation
1939 V42B N3/4 P159-172	57-1490 <*>
1939 V42B N3/4 P179-220	42G7G <ATS>
1939 V43B P271-291	64-16053 <*> 0
1939 V44B P58-68	63-18572 <*>
1939 V44B P75-97	62-20458 <*> 0
1939 V44B P397-450	1854 <*>
	2981 <*>
1939 V44B N5/6 P392-396	57-1407 <*>
1940 N46 P242-249	58-2435 <*>
1940 V45B P285-289	58-1524 <*>
1941 V48B P124-130	58-2312 <*>
1941 V50B P13-22	62-14627 <*>
1942 V52B P90-116	UCRL TRANS-1015(L) <*>

ZEITSCHRIFT FUER PHYSIKALISCHE CHEMIE. NEUE FOLGE. FRANKFURT

Citation	Translation
1954 V1 P153-175	II-207 <*>
	1929 <*>
1954 V1 N1/2 P42-62	80K36G <ATS>
1954 V1 N5/6 P275-277	C2350 <NRC>
1954 V2 P142-159	350 <TC>
1954 V2 N3/4 P197-214	16L31G <ATS>
1954 SNS V1 N1/2 P104-128	96G4G <ATS>
1955 V3 P52-64	60-23424 <=*>
1955 V3 P73-102	60-23438 <=*>
1955 V3 P135-182	60-19162 <=*>
1955 V3 N3/4 P183-221	99P59G <ATS>
1955 V3 N3/4 P238-254	62-00214 <*>
1955 V4 P264-285	66-12583 <*>
1955 V4 N3/4 P165-174	57-244 <*>
1955 V5 P32-57	NP-TR-914 <*>
1955 V5 P372-397	64-00041 <*>
1956 V7 P242-264	58-631 <*>
1956 V7 P267-295	57-3340 <*>
1956 V7 P342-359	58-217 <*>
1956 V8 N1 P20-31	62-18396 <*> 0
1956 V9 P285-299	58-2221 <*>
	64-16548 <*>
1956 SNS V8 P192-206	57-2591 <*>
1957 V10 P1-23	AEC-TR-3707 <+*> 0
1957 V10 P98-132	AEC-TR-4991 <*>
1957 V10 P156-160	60-19162 <=*>
1957 V10 P310-322	61-20264 <*>
1957 V10 N3/4 P236-263	65-11852 <*> 0
1957 V11 P16-29	59-10666 <*>
1957 V11 P267-272	60-19164 <=*>
1957 V11 N3/4 P213-233	61-10385 <*> 0
1957 V12 P132-138	44L33G <ATS>
	60-19169 <=*>
1957 V13 P244-247	39R76G <ATS>
1957 V13 P335-352	59-17486 <*>
1957 V13 P353-367	63-10549 <*>
1957 V13 P389-395	58-1661 <*>
1957 V13 N1/2 P95-106	59-10942 <*>
1957 V13 N3/4 P158-183	59-10941 <*>
1958 V14 P32-48	62-20179 <*>
	63-20735 <*>
1958 V14 P56-60	63-10630 <*>
1958 V14 N3/4 P197-207	62-16759 <*> 0
1958 V15 P175-195	AEC-TR-5240 <*>
1958 V15 P223-244	29L37G <ATS>
1958 V15 P409-428	63-01437 <*>
1958 V17 P184-198	61-20269 <*>
1958 V17 P199-219	61-20268 <*>
1958 V17 P368-374	50N48G <ATS>
1958 V17 N3/4 P274-278	64-30118 <*>
1958 V18 P181-200	60-14015 <*>
1958 V18 N1/2 P26-42	61-10801 <*>
1959 N2 P139-180	60-11381 <+*>
1959 V19 P1-20	AEC-TR-4264 <*>
	60-00648 <*>
1959 V19 N1/2 P83-88	C-3219 <NRCC>
1959 V20 N5/6 P332-352	SCL-T-480 <*>
1959 V21 P212-223	308 <TC>
1959 V21 N3/4 P185-191	17L36G <ATS>
1959 V21 N3/4 P213-217	SCL-T-416 <*>
1959 V21 N5/6 P349-357	51N48G <ATS>
1959 V22 P133-138	52N48G <ATS>
1959 V22 N1/2 P81-96	NP-TR-978 <*>
1959 V22 N1/2 P97-114	NP-TR-979 <*>
1959 V22 N1/2 P115-132	NP-TR-980 <*>
1959 V22 N5/6 P336-358	65-10876 <*>
1960 V23 P224-232	61-14915 <*>
1960 V23 P233-256	61-14916 <*>
1960 V23 N1/2 P100-112	61-14914 <*>
1960 V23 N1/2 P113-132	60-18090 <*>
1960 V23 N5/6 P305-312	AI-TR-28 <*>
1960 V24 P183-205	65-00419 <*>
1960 V24 P249-264	AEC-TR-4349 <*>
1960 V24 N1/2 P66-72	65-11452 <*>
1960 V24 N3/4 P217-239	NP-TR-974 <*>
1960 V25 P178-192	UCRL TRANS-762(L) <*>
1960 V25 P267-270	53N48G <ATS>
1961 V27 N1/2 P48-79	55N53G <ATS>
	62-10279 <*>
1961 V27 N3/4 P239-252	NP-TR-1026 <*>
1961 V28 P393-400	04R80G <ATS>
1961 V28 P422-424	66-11981 <*>
1962 V31 P40-70	65-12813 <*>
1962 V31 P309-320	UCRL TRANS-962(L) <*>
1962 V31 P321-327	NP-TR-1040 <*>
1962 V32 P154-167	AEC-TR-5761 <*>
	65P65G <ATS>
1962 V32 N1/2 P27-59	63-01754 <*>
1962 V32 N1/2 P110-126	63-20709 <*>
1962 V32 N5/6 P273-295	63-20708 <*>
1962 V33 P111-128	UCRL TRANS-961(L) <*>
1962 V34 P87-108	AEC-TR-5888 <*>
1962 V34 N1/4 P38-50	65-14212 <*>
1962 V34 N1/4 P87-108	283 <TC>
1962 V35 N1/3 P122-128	64-10959 <*>
1962 V35 N4/6 P343-347	64-10962 <*>
1963 V36 P75-81	65-12475 <*>
1963 V36 N1/2 P103-117	64-16676 <*>
1963 V36 N3/4 P211-220	63-20986 <*>
1963 V36 N5/6 P329-346	65-13136 <*>
1963 V37 N1/2 P99-108	64-16181 <*>
1963 V38 P87-102	SC-T-64-44 <*>
1963 V39 N1/2 P27-44	65-18007 <*>
1963 V39 N3/4 P218-234	77R77G <ATS>
1963 V39 N3/4 P235-240	65-14213 <*>
1963 V39 N3/4 P249-261	64-20087 <*>
1964 V42 N5/6 P329-343	65-13756 <*>
1965 V44 N1/2 P112-121	66-11107 <*>
1966 V48 P35-50	8797 <IICH>

ZEITSCHRIFT FUER PILZKUNDE

Citation	Translation
1953 N14 P15-16	1689 <*>

ZEITSCHRIFT FUER PRAEVENTIVMEDIZIN

Citation	Translation
1957 V2 P397-408	63-20400 <*>

ZEITSCHRIFT FUER PSYCHOLOGIE UND PHYSIOLOGIE DER SINNESORGANE

Citation	Translation
1893 V5 P1-60	61-14234 <*>
1894 V7 P172-187	61-14444 <*>
1896 V12 P331-354	61-14597 <*>
1897 V13 P401-459	61-00302 <*>
1898 V16 P373-398	61-14798 <*>
1902 V27 P264-266	61-14347 <*>
1903 V31 P89-109	61-14650 <*>
1903 V31 P127-150	61-14989 <*>
1906 V43 P268-289	61-14592 <*>
1906 V43 P374-422	61-14601 <*>

ZEITSCHRIFT FUER PSYCHOLOGIE

Citation	Translation
1908 V49 P108-127	61-14100 <*>
1908 V49 P161-205	61-14100 <*>
1909 V50 P209-218	61-14447 <*>
1910 V57 P1-88	61-14077 <*> P
1911 V58 P200-262	61-14372 <*>
1912 V61 P161-265	61-16065 <*>
1917 V78 P273-316	61-14456 <*>
1919 V82 P129-197	61-10971 <*>
1921 V86 N516 P278-367	61-14246 <*>
1924 V94 P134-135	61-14289 <*>
1925 V96 P353-399	61-14411 <*>
1925 V96 N1 P113-170	61-14407 <*>
1934 V133 P134-178	61-14351 <*>
1935 V135 P192-201	61-14350 <*>

1936 V137 P87-108 61-14349 <*>
1938 V141 P283-342 61-14188 <*> 0
1939 V147 P65-132 61-14604 <*>
1940 V149 P31-82 61-14829 <*>
1957 V160 P210-263 66-10942 <*>

ZEITSCHRIFT FUER PSYCHOTHERAPIE UND MEDIZINISCHE PSYCHOLOGIE
1924 V8 P280-303 63-00549 <*>

ZEITSCHRIFT FUER SPYCHOTHERAPIE UND MEDIZINISCHE PSYCHOLOGIE
1957 V7 P67-79 59-14522 <=*>
1963 V13 P57-64 66-10957 <*>

ZEITSCHRIFT FUER RHEUMAFORSCHUNG
1958 V17 N7/8 P259-274 62-00675 <*>
1960 V19 N3/4 P85-92 61-00354 <*>
1961 V20 N1/2 P1-16 62-01219 <*>

ZEITSCHRIFT FUER SAEUGTIERKUNDE
1936 V11 N2 P159-160 62-00817 <*>
1936 V11 N2 P161-236 62-00818 <*>

ZEITSCHRIFT FUER SCHWEISSTECHNIK
1955 V45 N11 P239-245 AEC-TR-3422 <*>
1959 V49 N10 P284-289 66-10458 <*>
66-10827 <*> 0
1959 V49 N11 P330-333 64-10457 <*>
65-10828 <*> 0
1961 V51 N11 P371-384 2697 <BISI>
1963 V53 N2 P42-46 3417 <BISI>

ZEITSCHRIFT FUER SINNESPHYSIOLOGIE
1908 V42 P130-153 61-14627 <*>
1910 V44 P165-181 61-14581 <*>
1910 V44 P428-443 61-14288 <*>
1911 V45 P87-102 61-14868 <*>
1911 V45 P460 61-14867 <*> 0
1912 V46 P362-378 61-14605 <*>
1916 V49 P109-244 61-14712 <*>
1916 V49 P326-364 61-14166 <*> 0
1919 V50 P311-318 57-1625 <*>
1922 V53 P1-35 61-14282 <*>
1930 V61 P87-147 61-14181 <*> 0
1930 V61 P225-247 61-14298 <*> 0
1931 V61 P267-324 61-14340 <*>
1932 V62 P132-136 61-14124 <*>
1933 V63 P38-86 61-14341 <*>
1933 V63 N112 P87-92 61-14267 <*>
1934 V65 N112 P1-26 61-14408 <*>
1940 V69 N1/2 P73-90 61-14345 <*>
1943 V70 P135-183 57-2906 <*>

ZEITSCHRIFT FUER STRAFVOLLZUG
1955 V5 N2 P120-123 57-851 <*>

ZEITSCHRIFT FUER TECHNISCHE PHYSIK
1921 V2 P245-250 61-14197 <*> 0
1923 N3 P93-99 58-42 <*>
1923 V4 N5 P208-213 66-12380 <*>
1923 V4 N12 P468-471 60-10302 <*> 0
1924 V5 P473-475 57-2430 <*>
1925 N1 P11-18 57-694 <*>
1926 V7 P198-200 58-187 <*>
1926 V7 P599-601 57-1350 <*>
57-3462 <*>
1926 V7 P630-635 57-2085 <*>
1926 V7 N9 P417-428 66-13148 <*> 0
1926 V7 N12 P636-640 61-14273 <*> 0
1927 V8 N6 P215-221 57-1036 <*>
1928 V9 P17-19 AL-49 <*>
1928 V9 N1 P26-39 58-1464 <*>
1928 V9 N1 P39-43 58-1457 <*>
1928 V9 N12 P475-478 57-2173 <*>
1929 V10 N1 P25-28 58-318 <*>
1929 V10 N4 P143-146 63-16863 <*>
1930 V11 P182-191 AEC-TR-5598 <*>
1930 V11 N4 P107-111 60-10306 <*>
1930 V11 N6 P182-191 60-10305 <*> 0
1930 V11 N6 P203-206 66-13331 <*>
1930 V11 N6 P221-227 57-2221 <*>
57-3513 <*>
1930 V11 N9 P361-363 60-10304 <*>
1930 V11 N9 P377-379 57-597 <*>
1930 V11 N10 P429-432 57-3221 <*>
1930 V11 N11 P452-455 57-3349 <*>
1930 V11 N12 P506-511 57-2213 <*>
1931 V12 N2 P116-121 58-246 <*>
1931 V12 N3 P142-148 60-10313 <*> 0
1931 V12 N12 P645-647 57-2480 <*>
1931 V12 N12 P647-649 57-1589 <*>
1931 V12 N12 P661-663 58-345 <*>
1932 V13 P501-504 2785 <*>
1932 V13 N1 P17-31 66-13297 <*> 0
1932 V13 N9 P436-441 60-18742 <*> 0
1932 V13 N12 P606-611 66-12274 <*>
1933 V14 N5 P191-197 57-969 <*>
60-10314 <*> 0
1934 V15 P257-262 63-14457 <*>
1934 V15 P469-473 63-10029 <*> 0
1934 V15 N12 P617-622 57-877 <*>
1935 N12 P640-642 I-627 <*>
1935 V16 P80-82 61-16175 <*> 0
1935 V16 P237-242 58-924 <*>
1935 V16 N9 P276-282 57-325 <*>
1935 V16 N12 P516-519 57-36 <*>
1935 V16 N12 P623-627 57-2520 <*>
57-3474 <*>
1936 N11 P407-412 II-145 <*>
1936 V17 P446-452 57-114 <*>
1936 V17 N8 P249-262 57-975 <*>
1936 V17 N11 P441-443 57-316 <*>
1937 P588-593 I-405 <*>
1937 V18 N7 P198-199 1549 <*>
1937 V18 N8 P219-228 66-13342 <*> 0
1937 V18 N11 P426-431 66-12907 <*> 0
1937 V18 N11 P464-466 66-13296 <*>
1938 N4 P113-116 II-146 <*>
1938 N4 P116-118 II-147 <*>
1938 V19 P294-296 57-1771 <*>
1938 V19 N5 P134-146 59-17270 <*> 0
1938 V19 N12 P521-526 58-1369 <*>
1939 V20 N2 P42-46 66-14247 <*> 0
1939 V20 N3 P75-80 60-10303 <*> 0
1939 V20 N11 P316-317 SCL-T-380 <*>
1940 V21 P85-88 65-13000 <*>
1940 V21 N1 P8-15 E-657 <RIS>
1940 V21 N11 P256-261 57-312 <*>
1941 N22 P248-251 58-1489 <*>
1941 V22 N5 P105-110 63-10023 <*> 0
1941 V22 N6 P130-131 60-13123 <=*> 0
1941 V22 N10 P245-248 63-10022 <*> 0
1942 V23 P39-49 60-10074 <*> 0
1942 V23 N10 P245-266 57-916 <*>
1943 N5 P97-104 AL-433 <*>
1943 N10/2 P228-242 AL-862 <*>
1943 V24 P46-53 II-517 <*>
1943 V24 N4 P90-92 62-18320 <*>

ZEITSCHRIFT FUER TIERPSYCHOLOGIE
1961 V18 P84-90 65-11693 <*>

ZEITSCHRIFT FUER TIERZUECHTUNG UND ZUECHTUNGSBIOLOGIE
1951 V59 N3 P309-321 65-12583 <*>
1956 V67 N1 P53-67 C-2296 <NRC>
1956 V67 N1 P68-95 C2383 <NRC>
1963 V78 N2 P179-195 63-01778 <*>

ZEITSCHRIFT FUER TROPENMEDIZIN UND PARASITOLOGIE
1953 V4 N2 P176-186 60-17310 <+*>
1955 V6 P460-470 1598 <*>
1955 V6 N1 P33-52 62-00134 <*>
1955 V6 N3 P257-279 62-00133 <*>
1957 V8 N3 P404-456 C-3284 <NRCC>
1958 V9 P146-156 63-00620 <*>
1959 V10 P375-378 61-00194 <*>
1960 V11 N2 P223-262 62-00376 <*>
1961 V12 N2 P152-160 62-00880 <*>

1962 V13 P175-180 64-10512 <*>
1964 V15 N1 P95-123 65-11281 <*>
1964 V15 N4 P461-463 65-14403 <*>

ZEITSCHRIFT FUER TUBERKULOSE UND HEILSTAETTENWESEN UND ERKRAENKUNGEN DER THORAXORGANE
1939 V82 N2/3 P101-105 2598 <*>
1955 V105 N5/6 P301-306 3998-D <K-H>

ZEITSCHRIFT FUER UNFALLMEDIZIN UND BERUFSKRANKHEITEN
SEE REVUE DES ACCIDENTS DU TRAVAIL ET DE MALADIES PROFESSIONELLES

ZEITSCHRIFT FUER UNTERSUCHUNG DER LEBENSMITTEL
1927 V53 P425-435 62-18932 <*>
1930 V60 N5 P473-484 59-15786 <*> O
1935 V69 P212-220 I-135 <*>
1935 V70 P204-205 60-10144 <*>
1935 V70 N4 P251-255 60-10145 <*>

ZEITSCHRIFT FUER UNTERSUCHUNG DER NAHRUNGS- U. GENUSSMITTEL
1923 V46 P154-159 AL-613 <*>

ZEITSCHRIFT FUER UROLOGIE
1937 V31 P687-708 64-00183 <*>
1955 V48 N9 P584-585 II-701 <*>

ZEITSCHRIFT FUER UROLOGISCHE CHIRURGIE
1925 V18 P73-85 58-120 <*>

ZEITSCHRIFT DES VEREINS DER DEUTSCHEN ZUCKERINDUSTRIE
1929 V79 P1-16 61-10294 <*>

ZEITSCHRIFT DES VEREINS DEUTSCHER INGENIEURE
1871 V15 P1-12 63-20923 <*>
1907 V51 N29 P1133-1141 AEC-TR-4960 <*>
1913 V57 P1174-1182 57-3076 <*>
1919 V63 N29 P677-682 AERE-TRANS-827 <*>
1920 P795-796 66-12601 <*> O
1921 V65 P618-622 58-98 <*>
1922 V66 P938- AL-558 <*>
57-1314 <*>
1923 V67 P692-695 58-933 <*>
1923 V67 P708-711 58-933 <*>
1924 V68 N33 P853-856 64-14333 <*> O
1927 V71 P1395 58-1986 <*>
1927 V71 N6 P199-203 65-14930 <*>
1927 V71 N42 P1471-1474 57-1349 <*>
1928 V72 P734-736 58-2363 <*>
1928 V72 P1155-1157 2281 <*>
1928 V72 N22 P734-736 58-2363 <*>
1929 V73 N11 P1629-1633 57-3088 <*>
1929 V73 N14 P457-460 59-20390 <*> O
1930 V74 P145-148 58-2099 <*>
1930 V74 N39 P1359-1362 59-20389 <*> O
1930 V74 N51 P1735-1738 59-15186 <*>
1931 V75 P481-486 I-879 <*>
1931 V75 P1455-1458 64-16207 <*>
1931 V75 N2 P33-40 59-15349 <*> O
1931 V75 N15 P457-459 57-3106 <*>
1931 V75 N16 P481-486 64-16997 <*>
1931 V75 N49 P1485-1488 58-700 <*>
1932 V76 N18 P433-437 64-20522 <*> O
1932 V76 N48 P1173-1178 40L29G <ATS>
1933 V77 P173-176 64-16198 <*>
1933 V77 N4 P96-97 6874-A <K-H>
1933 V77 N9 P231-238 59-17879 <*> O
1933 V77 N51 P1345-1349 65-13012 <*>
1934 V78 P379-386 2663 <*>
1934 V78 P395-401 58-892 <*>
1934 V78 N2 P53-55 I-755 <*>
58-915 <*>
1934 V78 N3 P81-84 II-590 <*>
1935 V79 P7 58-1963 <*>
1935 V79 N34 P1041-1044 02076 <*>
2076 <*>
1935 V79 N47 P1415-1417 59-20250 <*> O
1936 V80 P687-692 I-932 <*>
1936 V80 N1 P41-44 59-15445 <*>
1936 V80 N2 P49-51 59-20797 <*>
1936 V80 N38 P1163-1165 AL-570 <*>
1936 V80 N39 P1199-1200 59-20391 <*> O
1936 V80 N48 P1433-1439 AL-975 <*>
1936 V80 N52 P1551-1552 39L29G <ATS>
1936 V80 N52 P1570- II-623 <*>
1937 N6 P175-187 64-20554 <*> O
1937 V81 N29 P870 59-20394 <*> O
1937 V81 N30 P889-892 59-17838 <*> O
1937 V81 N32 P935-938 59-20393 <*> O
1938 V82 P167- 57-1473 <*>
1938 V82 N12 P344-346 66-11365 <*>
1938 V82 N20 P566-568 59-16026 <+*>
1938 V82 N22 P684-685 I-368 <*>
1938 V82 N30 P885-889 63-20680 <*>
1938 V82 N39 P1126-1134 63-10702 <*>
1938 V82 N39 P1135-1142 63-10431 <*>
1939 N1 P15-21 64-14161 <*>
1939 N4 P103-115 57-1305 <*>
1939 V83 N3 P69-74 2482 <*>
1939 V83 N6 P141-147 63-14271 <*> O
1939 V83 N9 P269-275 59-20841 <*>
1939 V83 N13 P381-382 63-10470 <*>
1939 V83 N52 P1309-1316 59-15145 <*> O
1940 V84 N11 P181-189 57-1101 <*>
1940 V84 N14 P229-235 61-18523 <*> O
1940 V84 N45 P857-862 59-17688 <*>
1941 V85 N2 P43-44 62-18342 <*> O
1941 V85 N10 P221-228 SCL-T-384 <*>
1941 V85 N11 P265-268 57-2146 <*>
1941 V85 N12 P290-292 AL-876 <*>
1941 V85 N14 P341-344 58-902 <*>
1941 V85 N22 P504-505 61-18879 <*>
1941 V85 N33 P695-701 59-15209 <*> O
1942 N2 P31-34 AEC-TR-3850 <=*>
1942 N2 P31-43 64-10058 <*> O
1942 N2 P49-56 63-18550 <*> O
1942 V86 N39/0 P599-605 61-18555 <*>
1943 N4 P91-98 65-12911 <*>
1943 V87 N1/2 P7-14 58-639 <*>
58-698 <*>
59-17674 <*>
1943 V87 N1/2 P15-19 59-20197 <*> O
1943 V87 N13/4 P177-182 58-639 <*>
58-652 <*>
59-17675 <*>
1943 V87 N15/6 P201-204 59-20392 <*>
1943 V87 N15/6 P215-219 58-639 <*>
58-653 <*>
1943 V87 N21/2 P321-324 58-709 <*>
1943 V87 N23/4 P341-345 II-109 <*>
1943 V87 N23/4 P352-354 60-14402 <*>
1943 V87 N25/6 P406 58-467 <*>
59-17676 <*>
1943 V87 N35/6 P564-571 58-697 <*>
1943 V87 N39/0 P609-615 65-13009 <*>
1943 V87 N41/2 P663-666 58-688 <*>
1943 V87 N47/8 P753-754 58-689 <*>
1943 V87 N49/0 P784 58-690 <*>
1943 V87 N51/2 P808 58-691 <*>
58-694 <*>
1943 V87 N9/10 P123-127 58-647 <*>
1943 V87 N9/10 P137-138 59-20578 <*>
1944 P245- 1793 <*>
1944 V88 P365-371 58-880 <*>
1944 V88 N27 P365-371 62-18236 <*> O
1944 V88 N1/2 P24-25 58-646 <*>
1944 V88 N19/0 P245-256 1793 <*>
1944 V88 N19/0 P256-257 64-10068 <*>
1944 V88 N19/2 P256-257 59-17281 <*>
1944 V88 N37/8 P516-520 AL-214 <*>
63-20835 <*> O
1944 V88 N51/2 P681-686 59-20576 <*> O
1948 V90 N3 P89-90 62-18434 <*> O
1948 V90 N4 P123-124 62-10323 <*>
1948 V90 N5 P135-139 AL-701 <*>
1948 V90 N10 P318-322 I-672 <*>
1948 V90 N11 P331-334 63-10007 <*>

1948 V90 N11 P341-342 63-10050 <*>
1948 V90 N12 P366-368 62-16933 <*> 0
1949 V91 P239-246 UCRL TRANS-503(L) <*>
1949 V91 N2 P25-32 62-18409 <*> 0
1949 V91 N3 P49-56 58-1438 <*>
1949 V91 N18 P476-478 62-16930 <*> 0
1950 V92 P225-230 I-726 <*>
1950 V92 P577-580 2422 <*>
1950 V92 N7 P153-160 63-20459 <*>
1950 V92 N7 P167-168 62-18448 <*> 0
1950 V92 N7 P169-173 62-18158 <*> 0
1950 V92 N8 P189-191 69S82G <ATS>
1950 V92 N10 P225-230 AL-755 <*>
1950 V92 N18 P452-456 <ATS>
1950 V92 N21 P565-570 63-16739 <*>
1951 P219-221 58-1117 <*>
1951 V93 N2 P25-36 63-16230 <*> 0
1951 V93 N9 P215-222 81F4G <ATS>
1951 V93 N9 P236-238 60-10231 <*>
1951 V93 N10 P257-261 59-20567 <*>
1951 V93 N26 P831- 3191 <*>
1951 V93 N29 P932-933 AL-106 <*>
1951 V93 N32 P1012-1014 89L28G <ATS>
1951 V93 N33 P1050-1053 AL-964 <*>
1951 V93 N35 P1098-1100 2715 <BISI>
1951 V93 N19/0 P633-638 70H10G <ATS>
1952 V94 P453-455 62-10062 <*> 0
1952 V94 P729-734 58-911 <*>
1952 V94 P809-815 58-593 <*>
1952 V94 N3 P79-84 59-17933 <*>
1952 V94 N8 P216-218 AL-107 <*>
1952 V94 N19 P595-598 59-17929 <*>
1952 V94 N19 P655-659 62-10369 <*> 0
1952 V94 N25 P829-832 59-20059 <*>
1952 V94 N26 P882-887 58-589 <*>
1952 V94 N35 P1147-1151 AL-105 <*>
1952 V94 N11/2 P305-306 62-18240 <*> 0
1953 V95 P245-247 58-891 <*>
1953 V95 P335-340 62-16938 <*> 0
1953 V95 P933-943 58-890 <*>
1953 V95 N8 P245-247 AL-321 <*>
63-16638 <*>
1953 V95 N21 P733-737 I-321 <*>
1953 V95 N24 P801-831 64-30078 <*> P
1953 V95 N24 P802-809 64-30078 <*> P
1953 V95 N24 P811-817 64-30078 <*> P
1953 V95 N24 P818-822 64-30078 <*>
1953 V95 N24 P822-824 64-30078 <*> 0
1953 V95 N24 P825-826 64-30078 <*> 0
1953 V95 N24 P830-831 64-30078 <*> P0
1953 V95 N30 P1027-1036 59-17852 <*> 0
1953 V95 N35 P1182-1190 58-706 <*>
1953 V95 N38 P1182-1190 AEC-TR-3495 <*> 0

ZEITSCHRIFT DES VEREINS DEUTSCHER INGENIEURE. BEIHEFT FOLGE
1929 V9 N4 P564-612 63-01776 <*>
1930 V11 N4 P702-729 63-01534 <*>
1939 N4 P103-115 I-241 <*>
1958 V41 P300-330 63-00537 <*>

ZEITSCHRIFT FUER VERGLEICHENDE PHYSIOLOGIE
1934 V20 P644-645 63-14950 <*> 0P
1936 V24 P462 63-14951 <*> P
1939 V26 N3 P366-415 61-00152 <*>
1950 V27 P543 63-14952 <*> P
1953 V35 P13-26 1932 <*>
1953 V36 P21-26 1911 <*>
1961 V44 P184-231 62-00800 <*>
1963 V47 P316-338 64-16962 <*>

ZEITSCHRIFT FUER VERKEHRISSICHERHEIT. FRANKFURT
1954 N2 P195-206 T-1610 <INSD>
1954 V2 P21-37 T-1611 <INSD>
1954 V2 N3/4 P2-8 60-18134 <*>

ZEITSCHRIFT FUER VERSUCHSTIER-KUNDE
1962 V1 P173-178 65-17278 <*>
1963 V3 P21-26 66-12003 <*>
1963 V3 P34-40 64-14721 <*> 0

1964 V4 P116-121 65-11767 <*>
1964 V5 P21-37 65-13373 <*>

ZEITSCHRIFT FUER VITAMIN-, HORMON- UND FERMENTFORSCHUNG
1948 V2 N5/6 P403-407 59-17192 <*> 0

ZEITSCHRIFT FUER WELTFORSTWIRTSCHAFT
1934 V1 N7 P439-447 57-1231 <*>

ZEITSCHRIFT. WIRTSCHAFTSGRUPPE ZUCKERINDUSTRIE
1947 V87 P711-715 58-841 <*>
63-16610 <*>

ZEITSCHRIFT FUER WIRTSCHAFTLICHE FERTIGUNG
1956 N2 P53-56 62-16700 <*>
1956 V31 N6 P13-16 63-16861 <*>
1965 P131-136 4839 <BISI>
1965 P269-273 4957 <BISI>

ZEITSCHRIFT FUER WISSENSCHAFTLICHE MIKROSKOPIE UND FUER MIKROSKOPISCHE TECHNIK
1889 V5 N4 P417- 57-2125 <*>
1904 V21 P129-165 66-12675 <*> 0
1904 V21 P273-304 66-12675 <*0>
1924 V41 P356-357 57-1513 <*>
1925 V42 N4 P421-433 64-10927 <*>
1951 V60 N5 P282-289 2251 <*>
1951 V60 N3/4 P181-188 59-14520 <+*> 0
1952 V60 P309-316 2549 <BISI>
1952 V61 N2 P92-94 57-609 <*>
1953 V61 N5 P298-301 57-232 <*>
1955 V62 N3 P147-151 61-10666 <*> 0
1955 V62 N8 P521-524 SCL-T-273 <*>
1956 P110- 57-1730 <*>
1956 V63 N1 P16-21 65-00164 <*>
1956 V63 N2 P110-118 58-1795 <*>
1961 V65 N2 P88-91 15P61G <ATS>
1963 V65 N5 P285-291 ANL-TRANS-108 <*>
1964 V65 N7 P430-433 3763 <BISI>
1964 V65 N7 P434-436 3764 <BISI>
1964 V66 N4 P201-209 77S82I <ATS>

ZEITSCHRIFT FUER WISSENSCHAFTLICHE PHOTOGRAPHIE, PHOTOPHYSIK UND PHOTOCHEMIE
1934 V33 P111-122 66-12502 <*>
1937 V36 N1001 P217-234 59-17333 <*>
1938 V37 P261-274 1999 <*>
1939 V38 N4/6 P137-156 58-1745 <*>
1948 V43 P209-237 59-10543 <*>
1950 V45 N1/3 P32- 62-18814 <*>
1950 V45 N10/2 P226-233 59-10544 <*>
1952 V47 N4/6 P106-118 13N48G <ATS>
1954 V49 N7/12 P137-219 II-755 <*>
1956 V51 N1/6 P17-58 64-10961 <*> 0
1957 V52 N1/3 P1-24 65-13066 <*>
1958 V53 P1-17 62-14048 <*>
1959 V53 N10/2 P226-237 61-18703 <*>
1960 V54 N1/3 P5-24 05N48G <ATS>
1964 V58 N5/8 P109-124 1046 <TC>

ZEITSCHRIFT FUER WISSENSCHAFTLICHE ZOOLOGIE
1887 V45 P227-254 T-1537 <INSD>
1894 V58 P636-660 57-785 <*>
1900 V68 P456-469 57-805 <*>
1905 V79 P325-396 57-2779 <*>
1961 V165 N1/2 P186-220 62-00870 <*>

ZEITSCHRIFT FUER ZELLFORSCHUNG UND MIKROSKOPISCHE ANATOMIE
1937 V25 P559-563 II-334 <*>
1952 V37 P144-183 57-1281 <*>
1953 V38 P475-487 2402 <*>
1954 V39 P392-413 2247 <*>
1955 V42 P1-18 57-1428 <*>
1955 V42 P273-304 3073 <*>
1955 V42 P331-343 II-749 <*>
1955 V43 P121-167 C-2462 <NRC>
1955 V43 P195-205 58-1296 <*>
1956 V44 P496-516 61-00209 <*>

1956 V45 P265-279 57-2346 <*>
1956 V45 P328-338 61-20586 <*> O
1957 V46 P100-107 62-00901 <*>
1958 V48 P130-136 62-01077 <*>
1958 V48 N6 P720-734 61-10251 <*>
1959 V49 P690-693 62-00628 <*>
1959 V50 P297-306 63-00284 <*>
1959 V50 P315-331 61-01046 <*>
1960 V51 P560-570 62-00898 <*>
1961 V55 N2 P148-178 63-18915 <*> O
1962 V56 N2 P203-212 62-14855 <*> O
1965 V65 P199-205 66-11886 <*>

ZEITSCHRIFT FUER DIE ZUCKERINDUSTRIE
1956 V6 P429-432 63-16369 <*> O
1957 V7 N3 P109-115 63-16374 <*>

ZELEZNICAR. SUPPLEMENT
1964 N6 P33-48 SUP 64-41003 <=>

ZELLSTOFF UND PAPIER
1955 V4 N7 P203-215 58-1434 <*>
1957 V6 N8 P234-239 60-14672 <*>
1957 V6 N9 P279-281 61-20905 <*> O
1958 V7 N2 P34-45 59-15967 <*> O
1958 V7 N7 P199-206 59-10796 <*> O
1959 V8 N4 P127-132 60-18932 <*> O
1959 V8 N6 P222- 60-18492 <*>
1959 V8 N6 P222-225 64-14021 <*>
1959 V8 N7 P251-256 63-18271 <*>
1959 V8 N7 P256-261 60-10737 <*> O
1959 V8 N8 P302- 60-18473 <*>
1959 V8 N8 P305-309 61-10025 <*>
1959 V8 N9 P340-345 64-14020 <*>
1959 V8 N10 P375-378 64-14019 <*>
1959 V8 N12 P455-464 61-18570 <*> O
61-20880 <*> O
1960 V9 N3 P99-110 60-18471 <*> O
61-10890 <*>
1960 V9 N3 P110- 60-18447 <*> O
1960 V9 N4 P144-151 61-10709 <*> O
1960 V9 N8 P291-297 61-10631 <*>
1960 V9 N8 P309-312 62-16502 <*>
1960 V9 N11 P422- 62-14492 <*>
1961 V10 P453-461 62-16525 <*> O
1961 V10 N1 P4-15 61-20190 <*>
1961 V10 N1 P18-24 62-10219 <*>
66-10499 <*>
1961 V10 N2 P50-56 61-18267 <*> O
1961 V10 N4 P138-149 61-20043 <*> O
1961 V10 N5 P174-181 63-10910 <*>
1961 V10 N9 P329-334 62-14254 <=*>
1961 V10 N9 P344-348 62-14491 <*>
1961 V10 N10 P370-374 63-10732 <*> O
1961 V10 N10 P375-383 64-14345 <*> O
1961 V10 N11 P410-415 62-18811 <*>
1962 V11 N2 P43-50 63-18920 <*>
1962 V11 N3 P89-94 62-16507 <*> O
1963 V12 N1 P4-10 64-10965 <*>
1963 V12 N2 P40-45 64-14750 <*>
1963 V12 N3 P67-72 63-18012 <*> O
1963 V12 N4 P99-106 66-12246 <*>
1963 V12 N8 P236-239 64-14346 <*> O
1964 V13 N5 P129-136 65-13961 <*>
1964 V13 N5 P147-149 65-11187 <*> O
1964 V13 N6 P161-167 64-30053 <*>
1964 V13 N11 P323-330 65-11610 <*>
1965 V14 N2 P35-41 65-17196 <*>
1965 V14 N11 P327-330 66-14645 <*> O

ZELLWOLLE, KUNSTSEIDE, SEIDE
1942 V47 N1 P2-8 59-17634 <*>

ZEMEDELSKA EKONOMIKA
SEE SBORNIK CESKOSLOVENSKE AKADEMIE ZEM ELSKYCH VED. RADA B: ZEMEDELSKA EKONOMIKA

ZEMENT. CHARLOTTENBURG
1923 V12 P297-298 64-14337 <*>
1923 V12 P307-310 64-14337 <*>
1923 V12 P317-319 64-14337 <*>
1923 V12 P326-329 64-14337 <*>
1924 V13 P235-238 64-14421 <*> O
1924 V13 P247-251 64-14421 <*> C
1924 V13 P262-266 64-14421 <*> O
1924 V13 P277-279 64-14421 <*> O
1924 V13 P289-291 64-14421 <*> C
1930 V19 N35 P818-825 64-18336 <*> O
1930 V19 N36 P847-849 64-18336 <*> O
1931 V20 N2 P26 64-18338 <*>
1931 V20 N3 P48-51 64-18339 <*>
1932 V21 N26 P380-381 64-18342 <*> O
1932 V21 N26 P381-382 II-378 <*>
1932 V21 N38 P536- II-336 <*>
1933 V22 N51 P705-708 64-18224 <*> O
1934 V23 N36 P514-518 64-18252 <*> O
1934 V23 N39 P572-575 64-18317 <*>
1934 V23 N41 P605-609 64-18281 <*>
1934 V23 N45 P665-669 64-18252 <*> O
1934 V23 N46 P677-682 64-18252 <*> O
1935 V24 N7 P94-97 64-18262 <*> O
1935 V24 N13 P183-191 64-18291 <*>
1936 V25 N11 P161-164 64-18260 <*> O
1936 V25 N32 P537-540 64-18250 <*>
1936 V25 N47 P814-816 64-18335 <*>
1940 V29 N15 P193-194 I-692 <*>

ZEMENT. CHARLOTTENBURG
1919 V8 P499-501 64-18265 <*> PO
1919 V8 P511-515 64-18265 <*> PC
1919 V8 P523-527 64-18265 <*> PO
1919 V8 P536-538 64-18265 <*> PO
1919 V8 P548-550 64-18265 <*> PC
1921 V10 P294-295 64-18244 <*> O
1921 V10 P306-308 64-18244 <*> C
1921 V10 P328-332 64-18244 <*> O
1922 V11 P487-489 64-18286 <*>
1922 V11 P499-500 64-18286 <*>
1922 V11 P507-509 64-18286 <*>
1922 V11 P515-516 64-18286 <*>
1923 V12 P279-280 64-18236 <*>
1924 V13 P11-13 64-14324 <*>
1924 V13 P135-136 64-18259 <*>
1924 V13 P147-150 64-18259 <*>
1924 V13 P377 64-14331 <*>
1924 V13 P386-388 64-14330 <*> O
1924 V13 P399-402 64-14330 <*> O
1924 V13 P455-457 64-14329 <*> C
1924 V13 P467-470 64-14329 <*> O
1924 V13 P470-472 64-18255 <*>
1924 V13 P509-511 64-14325 <*> O
1924 V13 P512 64-18232 <*>
1924 V13 P543-544 64-18263 <*>
1924 V13 P589-591 64-18198 <*> P
1924 V13 P605-608 64-18198 <*> P
1924 V13 P625-626 64-18198 <*> P
1924 V13 P626-627 64-14201 <*>
1924 V13 P632-634 64-14418 <*> P
1924 V13 P646-649 64-18198 <*> P
1924 V13 P649-651 64-14201 <*>
1924 V13 P652-654 64-14418 <*> P
1924 V13 P666-669 64-18278 <*>
1924 V13 P679-681 64-18278 <*>
1924 V13 P684-688 64-14418 <*> P
1924 V13 N16 P161-164 64-18275 <*> O
1924 V13 N17 P178-180 64-18275 <*> C
1924 V13 N18 P193-195 64-18275 <*> O
1924 V13 N49 P626-627 64-18273 <*>
1924 V13 N49 P649-651 64-18273 <*>
1925 V14 P379-382 64-14326 <*> O
1925 V14 P397-399 64-14326 <*> C
1925 V14 P859-861 64-18220 <*> O
1925 V14 P880-882 64-18220 <*> O
1925 V14 P898-907 64-18220 <*> O
1925 V14 P916 64-14202 <*>
1925 V14 P917 64-18220 <*> C
1925 V14 P948-950 64-14365 <*>
1925 V14 P963-965 64-14365 <*>
1925 V14 P987-990 64-14365 <*>
1925 V14 N3 P37-39 64-18240 <*> O

1925 V14 N5 P75-78	AL-858 <*>
1925 V14 N9 P177-182	64-18285 <*>
1925 V14 N12 P272-274	64-18292 <*> O
1925 V14 N40 P816-817	64-14428 <*> O
1925 V14 N41 P57-60	AL-858 <*>
1925 V14 N46 P935-936	64-18238 <*>
1925 V14 N19/0 P419-422	64-18298 <*> O
1925 V14 N19/0 P437-439	64-18298 <*> O
1926 V15 N9 P164-168	64-14198 <*> O
1926 V15 N10 P185-187	64-14198 <*> O
1926 V15 N11 P200-203	64-14198 <*> O
1927 V16 P37-40	64-14332 <*>
1927 V16 P55-58	64-14332 <*>
1927 V16 P731-735	64-14328 <*>
1927 V16 P741-746	64-18254 <*>
1927 V16 P1017-1023	64-18323 <*> O
1927 V16 N17 P329-332	64-18318 <*>
1927 V16 N17 P352-356	64-18318 <*>
1928 V17 P296-299	64-14199 <*> O
1928 V17 N7 P256-261	64-18231 <*> O
1928 V17 N7 P299-301	64-18231 <*> O
1928 V17 N26 P1009-1010	64-14200 <*>
1929 V18 P1345-1347	64-18332 <*>
1929 V18 N10/1 P292-296	64-18264 <*> O
1929 V18 N10/1 P338-342	64-18264 <*> O
1930 V19 N34 P796	64-14214 <*> O
1930 V19 N39 P914-915	64-14213 <*>
1930 V19 N40 P936-941	64-14188 <*> O
1930 V19 N46 P1078	64-14223 <*> O
1931 V20 N7 P144-147	64-14187 <*>
1932 V21 P301-305	64-18331 <*>
1932 V21 N27 P392-397	64-18235 <*> O
1932 V21 N28 P405-410	64-18235 <*> O
1933 V22 N51 P705-708	64-14186 <*> O
1934 V23 N1 P658-660	64-14184 <*>
1934 V23 N17 P237-240	64-14185 <*>
1934 V23 N18 P249-256	64-18205 <*> O
1934 V23 N27 P376-388	64-14212 <*> O
1934 V23 N28 P401-411	64-14222 <*> O
1934 V23 N30 P429-435	64-14190 <*> PO
1934 V23 N31 P448-453	64-14221 <*> O
1934 V23 N32 P461-466	64-14219 <*> O
1934 V23 N33 P473-475	64-14220 <*> O
1934 V23 N36 P511-514	64-18196 <*> O
1934 V23 N36 P528-534	64-14211 <*> O
1934 V23 N41 P605-609	64-14189 <*>
1934 V23 N41 P610-613	64-14210 <*> O
1934 V23 N48 P705-708	64-18225 <*> O
1935 V24 N2 P17-22	64-14431 <*> O
1935 V24 N3 P33-37	64-14431 <*> O
1935 V24 N6 P77-82	64-14431 <*> O
1935 V24 N10 P139-146	64-14431 <*> O
1935 V24 N13 P191-197	64-14431 <*> O
1935 V24 N14 P217-219	64-18197 <*>
1935 V24 N28 P429-430	64-18199 <*>
1936 V25 N5 P61-62	64-14191 <*>
1936 V25 N14 P218-221	64-18194 <*> O
1936 V25 N29 P485-490	64-14192 <*> O
1936 V25 N31 P518-523	64-14216 <*> P
1936 V25 N32 P537-540	64-14215 <*>
1936 V25 N38 P633-644	64-14183 <*> OP
1939 V28 N22 P345-350	64-14224 <*> O
1939 V28 N23 P359-362	64-14224 <*> O
1939 V28 N24 P374-379	64-14224 <*> O
1941 V30 N7 P83-85	64-18191 <*> O
1941 V30 N8 P100-103	64-18191 <*> O
1942 V31 P489-494	64-14430 <*> O

ZEMENT UND BETON. WIEN

1959 N16 P10-15	60-10578 <*>
1959 N16 P17-19	60-14014 <*>
1959 N16 P19-20	60-10579 <*>
1959 N16 P21-24	60-10581 <*> O
1959 N16 P32-34	60-10577 <*>
1959 N16 P35-36	60-10580 <*>
1959 N16 P36-37	60-10582 <*>
1961 V21 P1-9	61-20546 <*> O

ZEMENT-KALK-GIPS

1950 V3 P36-38	65-12779 <*>
1950 V3 N1 P5-11	1284 <*>
1951 V4 N2 P41-43	64-14173 <*> O
1951 V4 N6 P152-157	64-10761 <*> P
1951 V4 N7 P185-188 SP IS	I-186 <*>
1951 V4 N7 P192	65-13613 <*>
1952 V5 P142-151	64-14258 <*> O
1952 V5 N6 P175-179	I-403 <*>
1952 V5 N7 P203-205	AL-829 <*>
	61-20260 <*>
1952 V5 N10 P315-319	1127 <*>
1953 V6 P197-210	64-14260 <*> O
1953 V6 N1 P1-8	57-617 <*>
	64-10716 <*> O
1953 V6 N2 P46-53	64-18311 <*> P
1953 V6 N3 P69-73	64-18295 <*> PO
1953 V6 N3 P81-88	AL-869 <*>
1953 V6 N8 P269-270	64-18315 <*> O
1953 V6 N11 P413-415	I-638 <*>
1954 V7 P160-166	I-115 <*>
1954 V7 P337-342	I-847 <*>
1954 V7 P409-415	I-674 <*>
1954 V7 N12 P475-476	65-12745 <*>
1955 V8 N1 P1-6	I-639 <*>
1955 V8 N2 P59-	I-883 <*>
1955 V8 N7 P240-242	II-364 <*>
1955 V8 N8 P268-272	II-405 <*>
1955 V8 N9 P322-325	II-337 <*>
1955 V8 N10 P345-352	II-285 <*>
1957 V10 N1 P61-62	58-658 <*>
1957 V10 N3 P95-99	57-2287 <*>
1957 V10 N5 P187-194	62-18703 <*> O
1957 V10 N9 P354-359	CSIRO-3914 <CSIR>
1957 V10 N10 P394-398	61-18714 <*>
1957 V10 N10 P409-413	59-15068 <*>
1957 V11 N9 P396-408	59-15069 <*>
1958 V11 N2 P41-45	CLAIRA T.5 <CLAI>
1958 V11 N2 P46-49	CLAIRA T.6 <CLAI>
1958 V11 N3 P92-94	59-15071 <*>
1958 V11 N4 P144-154	58-2623 <*>
	63-16130 <*>
1958 V11 N5 P216-218	58-2626 <*>
1958 V11 N6 P264-272	C-3425 <NRCC>
1958 V11 N7 P298-304	59-15070 <*>
1958 V11 N8 P357-364	60-17309 <*+>
1958 V11 N9 P381-383	59-20327 <*>
1958 V11 N12 P529-543	59-15693 <*>
1958 V11 N12 P550-562	CLAIRA T.8 <CLAI>
1959 V12 P392-408	TS-1501 <BISI>
1959 V12 N1 P1-9	61-18968 <*>
1959 V12 N2 P41-55	59-17565 <*> O
1959 V12 N2 P65-67	59-17564 <*> O
1959 V12 N6 P253-257	60-14799 <*> O
1959 V12 N7 P289-294	60-10243 <*>
1959 V12 N7 P294-305	60-10244 <*>
1959 V12 N8 P351-355	C-3425 <NRCC>
1959 V12 N8 P362-369	60-10992 <*> O
1959 V12 N9 P385-392	60-14555 <*> O
1959 V12 N9 P392-408	<MT>
	62-14521 <*> O
1959 V12 N10 P449-456	<MT>
	TS-1500 <BISI>
	62-14182 <*>
1959 V12 N10 P481-484	60-14257 <*>
1959 V12 N11 P516-519	61-10200 <*>
1960 V13 N2 P64-70	TS 1722 <BISI>
	62-16875 <*>
1960 V13 N2 P70-78	TS 1723 <BISI>
	62-16874 <*>
1960 V13 N3 P111-118	TS 1707 <BISI>
	62-16871 <*>
1960 V13 N5 P181-192	62-14300 <*>
1960 V13 N6 P259-264	66-12190 <*>
1960 V13 N6 P270-274	61-10217 <*>
1960 V13 N7 P302-310	CLAIRA T.12 <CLAI>
1960 V13 N7 P310-316	61-10204 <*>
1960 V13 N7 P317-324	61-10205 <*> O
1960 V13 N11 P501-522	<SM>
1960 V13 N11 P522-530	61-18712 <*>
1960 V13 N12 P565-572	61-20557 <*> O
1960 V13 N12 P572-579	61-18709 <*>

1960 V13 N12 P581-583 65-12472 <*>
1961 V14 N1 P1-16 63-18394 <*> 0
1961 V14 N3 P105-108 62-18704 <*> 0
1961 V14 N3 P108-112 61-18708 <*>
1961 V14 N4 P153-158 62-14215 <*> 0
1961 V14 N4 P158-160 66-14438 <*>
1961 V14 N5 P177-179 65-12744 <*>
1961 V14 N7 P297-305 64-30205 <*>
1961 V14 N8 P325-339 63-20307 <*> 0
1961 V14 N8 P339-346 63-18393 <*> 0
1961 V14 N9 P370-384 63-10182 <*>
1961 V14 N11 P507-514 64-14259 <*>
1961 V14 N12 P545-558 64-30026 <*> 0
1961 V14 N12 P558-566 63-10746 <*>
1962 V15 N5 P197-204 64-30024 <*>
1962 V15 N5 P205-207 64-30033 <*>
1962 V15 N10 P437-438 63-18257 <*> 0
1962 V15 N11 P469-478 63-20953 <*>
1962 V15 N11 P479-485 64-30203 <*>
1962 V15 N11 P486-489 64-30204 <*>
1962 V15 N12 P529-534 64-16845 <*> 0
1963 V16 P187-191 65-11710 <*>
1963 V16 P443-445 427F <BISI>
1963 V16 N1 P9-12 63-18258 <*>
1963 V16 N1 P23-27 63-20955 <*>
1963 V16 N2 P41-44 63-20954 <*>
1963 V16 N3 P98-100 65-11338 <*>
1963 V16 N4 P131-134 64-30615 <*>
1963 V16 N5 P170-176 64-16366 <*>
1963 V16 N9 P370-375 64-16365 <*>
66-10175 <*>
1963 V16 N9 P380-389 64-30202 <*>
1963 V16 N11 P483-485 64-16168 <*>
64-18004 <*>
1964 V17 N6 P229-236 65-11356 <*> 0
1964 V17 N9 P392-396 65-11963 <=$>
1964 V17 N12 P543-546 C-6036 <NRC>
1965 V18 N4 P157-163 4342 <BISI>
66-10174 <*>
1965 V18 N5 P233-237 66-10178 <*>
1965 V18 N11 P574-579 66-10927 <*> 0
1966 V19 N2 P45-51 66-12816 <*>
1966 V19 N2 P82-85 66-12817 <*>

ZEMLEDELIE. MOSCOW
1953 N7 P108-109 RT-3953 <*>
1953 V1 N6 P70-73 60-21897 <=>
1954 N2 P114-115 RT-2746 <*>
1954 N5 P62-73 RT-4536 <*>
1954 V2 N2 P99-102 61-13647 <+*>
1954 V2 N6 P45-49 RT-2999 <*>
1954 V2 N8 P39-43 RT-4626 <*>
1954 V2 N8 P60-67 RT-3297 <*>
60-17708 <+*>
1954 V2 N12 P42-49 RT-3298 <*>
1955 N1 P80-89 RT-4359 <*>
1955 V3 N1 P100-103 RT-3903 <*>
1955 V3 N1 P104-105 RT-3904 <*>
1955 V3 N3 P114- R-126 <*>
1955 V3 N6 P115-118 RT-3300 <*>
1955 V3 N7 P92-94 RT-4627 <*>
1955 V3 N8 P65-69 RT-4199 <*>
1955 V3 N11 P78-82 R-366 <*>
1955 V3 N12 P77-81 R-125 <*>
1955 V3 N12 P112-113 R-123 <*>
1956 V4 N1 P80-85 RT-4157 <*>
1956 V4 N2 P87-90 RT-4131 <*>
1956 V4 N3 P107-108 R-1545 <*>
1956 V4 N6 P80-83 RT-4622 <*>
1957 V5 N4 P50-53 R-4421 <*>
1957 V5 N10 P20-22 R-4187 <*>
1957 V5 N10 P23-29 R-4180 <*>
1960 V8 N1 P41-47 60-31412 <=>
1960 V8 N8 P52-60 61-11813 <=>
1960 V8 N9 P3-22 61-21100 <=>
1960 V8 N9 P60-66 61-21100 <=>
1961 V23 N4 P33-39 62-23566 <=*>
1961 V23 N4 P65-67 62-23565 <=*>
1961 V23 N5 P42-49 62-23284 <=*>
1962 V24 N5 P83-86 62-25985 <=>
1963 V25 N9 P42-46 64-15589 <=*$>
1963 V25 N9 P46-47 64-21769 <=>
1963 V25 N10 P30-34 64-31558 <=>
1963 V25 N10 P41-49 64-31558 <=>
1964 V26 N1 P2-8 64-19544 <=$>
1964 V26 N1 P26-33 64-19545 <=$>

ZEMLEDELIE I ZHIVOTNOVODSTVO MOLDAVII
1960 N8 P35-36 61-21258 <=>
1960 N11 P1-4 61-21721 <=>
1960 N12 P24-27 61-23011 <=>
1962 V7 N11 P48 63-15105 <=*>

ZEMEDELSKA TECHNIKA
1964 V10 N4 P151-162 TR.196 <NIAE>
1964 V10 N4 P271-286 TR.159 <NIAE>
1964 V10 N8 P465-474 TR.193 <NIAE>

ZEMLEVEDENIE
1934 V36 N4 P376-395 64-19194 <=*$>

ZEMLIA I VSELENNAYA
1965 N4 P49-53 65-32930-A <=*$>
1965 N4 P58-62 65-32931-A <=*$>
1965 N5 P11-16 65-34052 <=*$>
1965 N5 P46-56 65-34052 <=*$>
1965 V1 N4 P65-69 65-32920 <=*$>
1965 V1 N4 P74-78 66-13920 <*$>
1965 V1 N5 P71-73 AD-641 755 <=$>
1965 V1 N6 P16-23 66-30584 <=$>
1965 V1 N6 P24-30 66-30585 <=$>
1965 V1 N6 P31-35 66-30586 <=$>
1965 V1 N6 P45-52 66-30587 <=$>
1965 V1 N6 P53-55 N66-23536 <=$>
1965 V1 N6 P67-71 N66-23535 <=$>

ZEMLJISTE I BILJKA
1952 V1 N2 P187-232 60-21620 <=>
1953 V2 N2 P193-316 60-21614 <=>
1956 V5 N1/3 P1-49 60-21700 <=>

ZEMNOI MAGNETIZM I ZEMNYE TOKI: SBORNIK STATEI. MEZHDUVEDOMSTVENNYI KOMITET PO PROVEDENIIU MEZHDUNARODNOGO GEOFIZICHESKOGO GODA. AKADEMIYA NAUK SSSR. III. RAZDEL PROGRAMMY MGG.
1959 N1 ENTIRE ISSUE 61-11158 <=>

ZENTRALBLATT FUER ALLGEMEINE PATHOLOGIE UND PATHOLOGISCHE ANATOMIE
1898 V9 P289-292 1989 <*>
1901 V12 N10 P417-421 66-10954 <*>
1916 V27 N9 P193-213 57-571 <*>
1916 V27 N12 P265-272 57-575 <*>
1916 V27 N20 P457-464 57-572 <*>
1949 V85 N6 P192-212 5069-D <K-H>
1953 V90 P394-399 57-243 <*>
1954 V91 N6/8 P317-325 II-327 <*>
1954 V92 P74-80 57-182 <*>
1954 V92 N5/6 P145-153 1434 <*>
1956 V94 P311-312 63-00555 <*>
1957 V96 P51-56 59-14585 <+*> 0
1957 V96 P219-220 59-19894 <+*>
1957 V97 P248-250 59-19053 <+*> 0
1963 V104 N9/11 P578-579 64-00269 <*>

ZENTRALBLATT FUER ARBEITSMEDIZIN UND ARBEITSSCHUTZ
1952 V2 P15-16 63-00307 <*>
1956 V6 P138-140 62-10402 <*>
1959 V9 P185-188 10000 <K-H>
64-14511 <*>
1961 V11 N7 P157-163 65-14522 <*> 0
1964 V15 P75-83 747 <TC>

ZENTRALBLATT FUER BAKTERIOLOGIE, PARASITENKUNDE, UND INFEKTIONSKRANKHEITEN
1892 V12 P859-861 59-10365 <*>
1894 V16 P175-180 59-10349 <*> C

ZENTRALBLATT FUER BAKTERIOLOGIE, PARASITENKUNDE UND INFEKTIONSKRANKHEITEN. ABT. 1. ORIGINALE

1900 V6 N7 P193-206 58-2189 <*>
1913 V29 N15/7 P440-447 58-2190 <*>
61-10891 <*>
1915 V76 N7 P523-535 63-01250 <*>
1930 V117 N7/8 P489-494 2593 <*>
1931 V119 P294-297 57-3432 <*>
1931 V121 P413-420 63-14852 <*>
1932 V124 P361-365 T1840 <INSD>
1932 V126 P517-526 64-10030 <*>
1934 V131 P1-6 63-14832 <*>
1934 V132 N7/8 P382-384 63-00843 <*>
1937 V140 P293-297 57-3427 <*>
1941 V147 P16-25 II-944 <*>
1941 V147 N1 P16-25 59-16655 <+*> 0
1942 V148 P388-395 58-88 <*>
1943 V150 N6 P311-318 65-00329 <*> 0
1953 V159 N5 P335-338 MF-2 <*>
1953 V159 N8 P500-508 MF2 <*>
1953 V159 N6/7 P477-480 1614 <*>
1955 V164 N6/7 P481-492 58-386 <*>
1956 P563-565 58-952 <*>
1956 V165 P122-134 60-15609 <*+>
1956 V166 P54-59 3071 <*>
1956 V166 P546-553 3040 <*>
1956 V167 P55-64 62-16264 <*>
1957 P386-390 58-185 <*>
1957 V169 P461-470 61-01044 <*>
1957 V169 N7/8 P461-469 59-10025 <*> 0
1958 V170 P316-323 65-00187 <*>
1958 V170 N1/5 P288-316 64-00431 <*>
1958 V172 N1/2 P37-43 59-15769 <*> 0
1958 V173 N3/4 P185-195 62-00128 <*>
1959 V176 P259-294 62-01216 <*>
1960 V177 P269-278 63-10532 <*> 0
1960 V177 P504-517 61-00419 <*>
1960 V178 P15-24 65-60299 <=$>
1960 V179 P308-324 63-00204 <*>
1962 V184 P251-258 UCRL TRANS-1024 <*>
UCRL TRANS-1024 <=*$>
1962 V184 N1/3 P60-65 62-01423 <*>
1962 V184 N1/3 P165-168 62-16227 <*> 0
1962 V185 P423-439 64-00111 <*>
1962 V186 N1 P134-138 64-00271 <*>
1962 V187 P2-14 63-00866 <*>
1962 V187 N2 P168-178 65-14879 <*>
1962 V187 N2 P179-194 65-14880 <*>
1962 V187 N2 P243-251 C-4429 <NRC>
1963 V188 P297-304 65-17298 <*>
1963 V188 N3 P403-407 64-00354 <*>
1963 V190 P115-131 66-12968 <*>
1963 V190 N2 P219-224 65-14558 <*>
1963 V190 N3 P322-333 65-17429 <*>
1963 V191 N1/3 P37-50 65-11027 <*>
1963 V191 N1/3 P92-97 65-14948 <*>
1964 V192 N1 P121,125 64-18109 <*>
1964 V193 N2 P140-146 65-17430 <*>

ZENTRALBLATT FUER BAKTERIOLOGIE, PARASITENKUNDE UND INFEKTIONSKRANKHEITEN. ABT 2
1908 V21 P385-386 65-11056 <*>
1910 V25 P178-260 65-11082 <*> PO
1913 V37 P595-609 59-15254 <*> 0
1933 V88 P353-366 C4811 <NRC>
1941 V104 P264-269 65-11046 <*> 0
1956 V109 P516-523 59-10661 <*> 0
1957 V110 P612-617 63-10533 <*> 0
1959 V113 N1/5 P35-41 65-17121 <*> 0

ZENTRALBLATT DER BAUVERWALTUNG
1908 N77 P515- 66-13334 <*>

ZENTRALBLATT FUER CHEMIE UND ANALYSE DER HYDRAULISCHEN ZEMENTE
1910 V1 P39-44 64-18320 <*>

ZENTRALBLATT FUER CHIRURGIE
1898 V25 N16 P425-428 62-00493 <*>
1920 V47 N13 P293-297 2395 <*>
1925 V52 N51 P2877-2878 59-17110 <*>
1937 V64 N47 P2694-2698 60-10656 <*>
1939 V66 P1248-1252 57-1654 <*>
1952 V77 N42A P2119-2123 59-17666 <*> PC
1953 V78 P288-289 AL-125 <*>
1953 V78 P623- AL-497 <*>
1955 V80 P189-194 61-18728 <*>
1955 V80 P1585-1589 1199 <*>
1955 V80 N50 P1987-1991 59-15855 <*>
1956 V81 P1441-1444 61-18726 <*> 0
1956 V81 P2323-2331 64-20032 <*> 0
1960 V85 P1146-1154 63-20220 <*>
1960 V85 N5 P209-219 62-16747 <*> 0
1961 V86 N45 P2352-2353 65-10091 <*> 0

ZENTRALBLATT FUER DAS GESAMTE FORSTWESEN
V30 P387-394 57-1230 <*>

ZENTRALBLATT FUER DIE GESAMTE OPHTHALMOLOGIE UND IHRE GRENZGEBIETE
1914 V1 N1 P430-431 61-14451 <*>
1920 V3 P353-416 61-14164 <*> 0
1922 V8 N4 P177 61-14198 <*>
1923 V8 N9 P405-406 61-14152 <*>
1927 V18 P387-388 61-14403 <*>
1928 V20 P198 61-14399 <*>
1928 V20 P209 61-14400 <*>
1930 V23 N6/7 P366-370 61-14209 <*>
1932 V27 N3 P97-105 61-14119 <*>
1932 V27 N6 P289-298 61-14121 <*>
1934 V30 N10 P568-570 61-14364 <*>
1940 V46 N7 P210 61-14406 <*>
1940 V46 N7 P221-222 61-14406 <*>
1942 V48 N8 P219 61-14146 <*>

ZENTRALBLATT FUER GEWERBEHYGIENE UND UNFALLVERHUETUNG
1927 V4 P312-316 59-10694 <*>
1930 V17 N7 P264-266 64-20538 <*>

ZENTRALBLATT FUER DIE GRENZGEBIETE DER MEDIZIN UND CHIRURGIE
1902 V5 P497-502 58-1774 <*>
1902 V5 P545-554 58-1774 <*>
1902 V5 P601-610 58-1774 <*>
1902 V5 P634-636 58-1774 <*>
1902 V5 P666-682 58-1774 <*>

ZENTRALBLATT FUER GYNAEKOLOGIE
1930 V54 N19 P1164-1172 2400 <*>
1931 V55 P362-369 57-1629 <*>
1937 V61 P516-524 59-15856 <*> 0
1939 V63 P373-375 60-10015 <*> 0
1939 V63 P2057-2063 60-10014 <*>
1940 V64 P1225-1230 60-10016 <*>
1941 V65 N2 P89-92 57-869 <*>
1943 V67 N30 P1160-1164 63-00210 <*>
1950 V72 P694-701 2702 <*>
1953 V75 P21- AL-905 <*>
1958 V80 N7 P275-280 63-26692 <=$>
1958 V80 N45 P1766-1768 63-00780 <*>
1960 V82 N10 P387-395 <LSA>
1960 V82 N46 P1791-1796 64-00052 <*>

ZENTRALBLATT FUER INNERE MEDIZIN
1906 V27 P465-478 61-14588 <*>

ZENTRALBLATT FUER LANDAERZTE
1934 P1-4 SP. NO. 42 61-10436 <*>

ZENTRALBLATT FUER MINERALOGIE, GEOLOGIE UND PALAEONTOLOGIE
1924 P161-166 80L35G <ATS>
1925 N9 P273- 3181 <*>
1930 SA P220-232 AEC-TR-4946 <*>
1930 SA P321-328 AEC-TR-4944 <*>
1933 N12 P300-302 60-14315 <*>

ZENTRALBLATT FUER NEUROCHIRURGIE
1953 V13 P223-243 II-1038 <*>
1953 V13 P290-313 II-1038 <*>
1960 V20 P134-158 63-20223 <*> 0

1960 V20 N5 P257-266 62-16220 <*> 0

ZENTRALBLATT FUER PHYSIOLOGIE
1900 V14 P195-197 58-805 <*>

ZENTRALBLATT FUER PRAKTISCHE AUGENHEILKUNDE
1878 V2 P229-233 61-14102 <*>

ZENTRALBLATT FUER VERKEHRS-MEDEZIN, VERKEHRS-PSYCHOLOGIE UND ANGRENZENDE GEBIETE
1960 V6 N4 P198-199 62-18896 <*>

ZENTRALBLATT FUER VERKEHRS-MEDEZIN, VERKEHRS-PSYCHOLOGIE, LUFT UND RAUMFAHRT-MEDEZIN
1963 V9 N2 P1-5 65-11346 <*>

ZENTRALBLATT FUER VETERINAERMEDIZIN
1955 V2 P451-458 58-77 <*>
1955 V2 N7 P682-692 CSIRO-3269 <CSIR>
1957 V4 P459-484 39K25G <ATS>
1958 V5 N1 P1-16 61-18820 <*>
1959 V6 N4 P441-462 61-14499 <*>
1960 V7 P215-221 64-10805 <*> 0
1960 V7 N2 P222-225 61-10676 <*> 0
1960 V7 N4 P364-387 61-14500 <*> 0
1961 V8 N3 P201-213 65-13355 <*>
1961 V8 N9 P908-922 64-00117 <*>

ZENTRALBLATT FUER VETERINAERMEDIZIN. REIHE A.
1964 V11A N6 P493-501 65-12842 <*>

ZENTRALBLATT FUER VETERINAERMEDIZIN. REIHE B.
1963 V10 N1 P57-66 64-00184 <*>

ZENTRALBLATT FUER DIE ZUCKERINDUSTRIE
1943 V51 N6 P37-38 64-20889 <*>

ZENTRALORGAN FUER DIE GESAMTE CHIRURGIE UND IHRE GRENZGEBIETE
1932 V57 P512- 57-374 <*>

ZENTRALZEITUNG FUER OPTIK U. MECHANIK, ELEKTROTECHNIK
1914 V35 N1 P10 61-14449 <*>
1914 V35 N2 P20 61-14450 <*>
1921 V42 P375-377 AL-443 <*>
1921 V42 N25 P375-377 61-14275 <*>
1922 V43 N12 P190-192 61-14127 <*>
1923 V44 P21-27 61-14208 <*>

ZERNOBOBOVYE KULTURY
1964 N9 P1-3 64-51959 <=$>

ZESZYTY GEOGRAFICZE
1959 V1 P205-215 63-11356 <=>

ZESZYTY NAUKOWE POLITEKHNIKI GDANSK. CHEMIA
1957 N2 P47-57 60-21414 <=>

ZESZYTY NAUKOWE POLITECHNIKI LODZKIEJ. LODZ.
1959 N5 P37-47 60-21548 <=>
1959 N27 P49-90 60-21549 <=>
1963 V10 N53 P25-33 M-5726 <NLL>

ZESZYTY NAUKOWE POLITECHNIKI LODZKIEJ. CHEMIA
1956 V4 N12 P41-65 59-15465 <*>
63-20866 <*> 0

ZESZYTY NAUKOWE POLITECHNIKI LODZKIEJ. LODZ. WLOKIENNICTWO
1961 N7 P3-44 22P60P <ATS>
1962 N9 P5-15 65-14628 <*>

ZESZYTY NAUKOWE POLITECHNIKI SLASKIEJ, GLIWICE CHEMIA
1963 N16 P47-55 29S83P <ATS>

ZESZYTY NAUKOWE POLITECHNIKI SZCZECINSKIEJ. SER. MECHANIKA
1961 N7 P3-10 NEL-TRANS-1631 <NLL>

ZESZYTY NAUKOWE POLITECHNIKI WARSZAWSKIEJ
1963 N8 P3-49 63-11387 <=$>

ZESZYTY NAUKONE POLITECHNIKI WROCLAWSKIEJ. BRESLAU
1964 N85 P43-60 AD-637 065 <=$>
1964 N85 P77-85 AD-637 065 <=$>

ZESZYTY PROBLEMOWE NAUKI POLSKIEJ
1956 N7 P9-40 59-13180 <=> 0
1956 N7 P109-162 59-13180 <=> 0
1956 N7 P199-233 59-13180 <=> 0
1956 N7 P235-279 59-13180 <=> 0

ZESZYTY PROBLEMOWE POSTEPOW NAUK ROLNICZYCH
1957 N8 P175-187 60-21543 <=>
1959 N16 P139-152 63-11081 <=>
1959 N21 P389-397 60-21554 <=>
1960 N20 P63-68 60-21553 <=>

ZHELEZNODOROZHNYI TRANSPORT
1947 N2 P92 RT-273 <*>
1959 V41 N3 P82-84 60-11146 <=>
1959 V41 N8 P55 61-15404 <+*>
1960 V42 N1 P23-27 61-31593 <=>
1960 V42 N1 P68-69 61-31593 <=>
1960 V42 N1 P81 61-31593 <=>
1960 V42 N10 P10-16 62-24782 <=>
1961 V43 N5 P8-14 62-11704 <=>
1961 V43 N7 P22-24 62-11115 <=>
1961 V43 N10 P18-26 62-15496 <*=> 0
1962 V44 N4 P50-52 63-16468 <=*>
1962 V44 N4 P64-66 63-14004 <=*>
1962 V44 N5 P27-32 63-19862 <*=>
1962 V44 N5 P43-47 63-16471 <=*>
1962 V44 N5 P50-54 63-14005 <=*>
1962 V44 N6 P33-37 63-10642 <=*>
1962 V44 N6 P37-41 63-10646 <=*>
1962 V44 N6 P47-50 63-10644 <=*>
1962 V44 N6 P51-56 63-10645 <=*>
1962 V44 N6 P57-59 63-10641 <=*>
1962 V44 N6 P67-72 63-10647 <=*>
1962 V44 N6 P74-76 63-10639 <=*> 0
1962 V44 N6 P76-78 63-10640 <=*>
1962 V44 N6 P80-83 63-10643 <=*>
1962 V44 N7 P32-34 63-10637 <=*>
1962 V44 N7 P45-48 63-10648 <=*> 0
1962 V44 N7 P66-69 63-10638 <=*> 0
1962 V44 N7 P90-94 63-10649 <=*>
1962 V44 N8 P28-32 63-10761 <=*>
1962 V44 N8 P44-49 63-10762 <=*>
1962 V44 N8 P49-52 63-10636 <=*>
1962 V44 N8 P55-59 63-10765 <=*>
1962 V44 N8 P60-62 63-10635 <=*>
1962 V44 N8 P73-78 63-10763 <=*>
1962 V44 N8 P88-91 63-10764 <=*>
1962 V44 N9 P40-44 63-10760 <=*>
1962 V44 N9 P41-43 63-10949 <=*>
1962 V44 N9 P44-47 63-10758 <=*>
1962 V44 N9 P59-63 63-10759 <=*>
1962 V44 N9 P87-89 63-16469 <=*>
1962 V44 N10 P3-7 63-13822 <=>
1962 V44 N10 P18-25 63-13822 <=>
1962 V44 N10 P44-47 63-14727 <=*>
1962 V44 N10 P48-50 63-16470 <=*>
1962 V44 N10 P74-76 63-14725 <=*>
1962 V44 N11 P47-49 63-14726 <=*>
1962 V44 N11 P58-61 63-14003 <=*>
1962 V44 N11 P64-66 63-14002 <=*>
1962 V44 N11 P66-68 63-14724 <=*>
1962 V44 N11 P77-78 63-14747 <=*>
1962 V44 N11 P79-80 63-14001 <=*> 0
1962 V44 N12 P51-53 63-14728 <=*>
1962 V44 N12 P60-62 63-20663 <=*$>
1962 V44 N12 P70-71 63-18794 <=*>
1962 V44 N12 P72-73 63-18081 <=*>
1962 V44 N12 P83-88 63-18082 <=*>
1963 V45 N1 P70-73 63-18062 <=*>
1963 V45 N1 P73-75 63-18080 <=*>
1963 V45 N1 P78-80 63-20664 <=*$>

1963 V45 N2 P44-49 63-20665 <=*$>
1963 V45 N2 P75-76 63-18077 <=*>
1963 V45 N2 P77 63-20378 <=*$>
1963 V45 N2 P83-86 63-18792 <=*>
1963 V45 N3 P31-33 63-20380 <=*$>
1963 V45 N3 P52-54 63-20379 <=*> 0
1963 V45 N3 P78-80 63-18076 <=*>
1963 V45 N4 P31-34 63-20377 <=*>
1963 V45 N4 P48-50 63-18793 <=*>
1963 V45 N4 P51-57 64-11516 <=>
1963 V45 N4 P57-60 64-11540 <=>
1963 V45 N4 P60-64 63-20666 <=*$>
1963 V45 N5 P45-49 64-10941 <=*$>
1963 V45 N5 P51-53 63-20915 <=*$>
1963 V45 N5 P57-61 63-20916 <=*$>
1963 V45 N5 P78-81 63-20667 <=*$>
1963 V45 N6 P5-10 64-11526 <=>
1963 V45 N6 P10-16 64-11527 <=>
1963 V45 N6 P17-21 64-11523 <=>
1963 V45 N6 P22-24 64-11524 <=>
1963 V45 N6 P24-28 64-11531 <=>
1963 V45 N6 P36-39 64-11513 <=>
1963 V45 N6 P49-52 63-20917 <=*$>
64-11529 <=>
1963 V45 N6 P53-56 64-11530 <=>
1963 V45 N6 P62 64-11521 <=>
1963 V45 N6 P69-73 64-11525 <=>
1963 V45 N6 P82-85 63-20668 <=*$>
1963 V45 N7 P26-29 63-20669 <=*$>
1963 V45 N7 P34-37 63-20919 <=*$>
1963 V45 N7 P37-40 63-20914 <=*$>
1963 V45 N7 P51-52 64-10942 <=*$>
1963 V45 N8 P36-40 64-10943 <=*$>
1963 V45 N8 P43-46 64-10944 <=*$>
1963 V45 N8 P50-51 63-20920 <=*$>
1963 V45 N8 P76-77 64-10945 <=*$>
1963 V45 N9 P30-37 64-14464 <=*$>
1963 V45 N9 P37-41 64-14519 <=*$>
1963 V45 N9 P42-45 64-14465 <=*$>
1963 V45 N9 P49-52 64-14466 <=*$>
1963 V45 N9 P53-55 64-18068 <*>
1963 V45 N9 P65-66 64-10946 <=*$>
1963 V45 N9 P67-70 64-14467 <=*$>
1963 V45 N9 P71-75 64-14468 <=*$>
1963 V45 N9 P75-77 64-14469 <=*$>
1963 V45 N10 P24-29 64-16023 <=*$>
1963 V45 N10 P33-36 64-14470 <=*$>
1963 V45 N10 P47-48 64-18069 <*>
1963 V45 N10 P48-51 64-16024 <=*$>
1963 V45 N10 P51-54 64-16025 <=*$>
1963 V45 N10 P73-75 64-14471 <=*$>
1963 V45 N10 P75-76 64-14472 <=*$>
1963 V45 N11 P20-22 64-18070 <*>
1963 V45 N11 P25-29 64-18071 <*>
1963 V45 N11 P29-34 64-18072 <*>
1963 V45 N11 P34-37 64-16026 <=*$>
1963 V45 N11 P43-48 64-16027 <=*$>
1963 V45 N11 P65-67 64-16028 <=*$>
1963 V45 N11 P68-70 64-18073 <*>
1963 V45 N12 P11-14 64-18075 <*>
1963 V45 N12 P18-20 64-18076 <*>
1963 V45 N12 P40-43 64-20960 <*>
1963 V45 N12 P76-77 64-20961 <*>
1963 V45 N12 P80 64-18078 <*>
1963 V45 N12 P82-83 64-18077 <*>
1964 N3 P5-12 NLLTB V6 N8 P707 <NLL>
1964 V46 N1 P37-39 AD-610 328 <=$>
64-20966 <*>
1964 V46 N1 P40-42 64-20963 <*>
1964 V46 N1 P57-60 64-20964 <*>
1964 V46 N1 P71-72 64-20962 <*>
1964 V46 N1 P77 64-20965 <*>
1964 V46 N2 P5-6 64-71162 <=>
1964 V46 N2 P37-41 64-20972 <*>
1964 V46 N2 P54-56 64-20970 <*>
1964 V46 N2 P64-66 65-11107 <*>
1964 V46 N2 P67-69 65-11104 <*>
1964 V46 N2 P69-72 64-20969 <*>
1964 V46 N2 P75-78 64-20971 <*>
1964 V46 N3 P50-55 65-11102 <*>
1964 V46 N3 P67-71 65-11103 <*>
1964 V46 N3 P75-79 65-11105 <*>
1964 V46 N3 P87-89 64-20967 <*>
1964 V46 N3 P89-90 64-20968 <*>
1964 V46 N4 P22-30 65-11978 <*>
1964 V46 N4 P43-48 65-11979 <*>
1964 V46 N4 P57-59 65-11980 <*>
1964 V46 N5 P29-33 65-11977 <*>
1964 V46 N5 P58-64 65-11106 <*>
1964 V46 N5 P78-81 65-11111 <*>
1964 V46 N6 P28-30 65-11100 <*>
1964 V46 N6 P40-42 65-11109 <*>
1964 V46 N6 P42-44 65-11845 <*> 0
1964 V46 N6 P44-49 65-11843 <*>
1964 V46 N6 P70-72 65-11863 <*>
1964 V46 N6 P77-79 65-11108 <*>
1964 V46 N6 P79-81 65-11101 <*>
1964 V46 N7 P40-42 65-11842 <*>
1964 V46 N7 P43-45 65-11841 <*>
1964 V46 N7 P45-48 65-13769 <*>
1964 V46 N7 P49-53 65-11110 <*>
1964 V46 N7 P79-80 65-11840 <*>
1964 V46 N8 P30-34 65-11835 <*>
1964 V46 N8 P34-38 65-11837 <*>
1964 V46 N8 P50-52 65-11836 <*>
1964 V46 N8 P69-72 65-13768 <*>
1964 V46 N8 P79-80 65-11838 <*>
1964 V46 N9 P56-59 65-11827 <*>
1964 V46 N9 P65-68 65-13767 <*>
1964 V46 N9 P73 65-11829 <*>
1964 V46 N9 P79-80 65-11830 <*>
1964 V46 N10 P27-29 65-13766 <*>
1964 V46 N10 P44-47 65-14379 <*>
1964 V46 N10 P51-54 65-11828 <*>
1964 V46 N10 P59-61 65-13764 <*>
1964 V46 N10 P71-72 65-13765 <*>
1964 V46 N11 P31-33 65-13761 <*>
1964 V46 N11 P44-47 65-13763 <*>
1964 V46 N11 P47-49 65-14378 <*>
1964 V46 N11 P82 65-13762 <*>
1964 V46 N12 P17-22 65-14380 <*>
1964 V46 N12 P23-25 65-13759 <*>
1964 V46 N12 P35-38 65-13760 <*>
1965 V47 N1 P12-16 65-17190 <*>
1965 V47 N1 P27-29 65-14388 <*>
1965 V47 N1 P47-50 65-17189 <*>
1965 V47 N1 P61-63 65-14385 <*>
1965 V47 N1 P64 65-14386 <*>
1965 V47 N1 P78-80 65-14387 <*>
1965 V47 N2 P37-38 65-14384 <*>
1965 V47 N2 P68-69 65-14383 <*>
1965 V47 N2 P82 65-14382 <*>
1965 V47 N3 P10-13 65-17188 <*>
1965 V47 N3 P20-23 65-17187 <*>
1965 V47 N3 P42-44 65-17186 <*>
1965 V47 N3 P52-57 65-14381 <*>
1965 V47 N3 P62-64 65-17306 <*>
1965 V47 N4 P8-13 65-17191 <*>
1965 V47 N4 P61-64 65-14389 <*>
1965 V47 N4 P74-77 65-14390 <*>
1965 V47 N5 P36-37 65-17183 <*>
1965 V47 N5 P61-63 65-17185 <*>
1965 V47 N6 P37-38 65-17182 <*>
1965 V47 N6 P89-90 65-17184 <*>

ZHILISHCHNO-KOMMUNALNOE KHOZYAISTVO
1961 V11 N2 P13 61-31550 <=>

ZHILISHCHNOE STROITELSTVO
1959 N10 P20-21 NLLTB V2 N3 P202 <NLL>

ZHIVOTNOVODSTVO
1953 N5 P61-63 44P61R <ATS>
1955 N11 P98-100 63-16096 <=*>
1957 V19 N3 P37-44 65-63206 <=$>
1959 N3 P3-6 59-31004 <=>
1959 N6 P23-25 59-31015 <=> 0
1959 N6 P41-43 59-31015 <=> 0
1959 V21 N11 P83-88 65-20016 <$>
1960 P42-43 60-31630 <=>

1960 N4 P10-15 60-31696 <=>
1960 V22 N12 P43-49 61-27489 <*=> O
1961 N4 P32-34 61-27381 <=>
1961 N4 P43 61-27381 <=>
1961 V23 N1 P40-48 62-25464 <=*>
1962 V24 N1 P56-59 62-32351 <=*> O
1962 V24 N6 P87-88 64-13588 <=*$>
1962 V24 N10 P3-19 63-21346 <=>
1963 V25 N1 P17-19 63-21623 <=>
64-13910 <=*$>
1963 V25 N1 P25-26 64-15702 <=*$>
1964 V26 N4 P65-69 RTS-2812 <NLL>

ZHURNAL AKUSHERSTVA I ZHENSKIKH BOLEZNEI
1914 V29 P73-80 64-10603 <*>
1928 V39 N5 P605-607 66-60571 <=*$>

*ZHURNAL ANALITICHESKOI KHIMII. MOSKVA
1946 V1 N1 P57-63 61-15445 <+*>
1946 V1 N2 P123-128 65-10899 <*>
1946 V1 N3 P198-205 01N55R <ATS>
1946 V1 N5/6 P290-294 TT-827 <NRCC>
1947 V2 N1 P236-238 91R77R <ATS>
1947 V2 N3 P131-134 63-19590 <=*$>
1947 V2 N3 P135-146 C-2508 <NRC>
1947 V2 N4 P210-214 RT-1470 <*>
1947 V2 N5 P265-273 R-1865 <*>
RJ-63 <ATS>
1947 V2 N5 P274-280 60-13722 <+*>
1947 V2 N5 P299-308 RT-1417 <*>
1947 V2 N6 P373-376 64-18514 <*>
1948 V3 N1 P52-62 RT-1423 <*>
1948 V3 N5 P284-289 RT-1695 <*>
1949 V4 P244-247 61-13639 <+*>
1949 V4 N1 P40-45 C-4062 <NRCC>
63-16758 <=*>
1949 V4 N2 P75-79 RT-2935 <*>
66-14754 <*>
1949 V4 N5 P292-297 RJ-166 <ATS>
1950 V5 P58-72 R-4488 <*>
1950 V5 P178-192 61-13640 <+*> P
1950 V5 P319-320 AEC-TR-5468 <=*>
1950 V5 N1 P58-72 RT-2883 <*>
1950 V5 N2 P110-118 RJ-43 <ATS>
1950 V5 N2 P119-122 RJ-44 <ATS>
1950 V5 N5 P262-267 RJ-40 <ATS>
1950 V5 N5 P262-271 RJ-40 <ATS>
1950 V5 N5 P296-299 RJ-41 <ATS>
62-18978 <=*>
1950 V5 N6 P330-338 RJ-48 <ATS>
1950 V5 N11 P28-31 50/2715 <ATS>
1951 V6 P5-14 66-12710 <*>
1951 V6 N1 P59-60 RJ-51 <ATS>
1951 V6 N2 P96-100 RT-798 <*>
1951 V6 N4 P207-210 RT-3828 <*>
1951 V6 N4 P230-233 RJ-68 <ATS>
1951 V6 N4 P230-234 RJ-68 <ATS>
1951 V6 N4 P257-258 R-4010 <*>
RJ-69 <ATS>
1951 V6 N5 P288-296 2930 <HB>
1951 V6 N5 P328 52/2584 <NLL>
1951 V6 N6 P364-370 59-22553 <+*>
1952 V7 P112-115 64-20807 <*>
1952 V7 N1 P52-55 63-20658 <=*$>
1952 V7 N2 P112-115 RT-3798 <*>
1952 V7 N3 P185-187 RT-781 <*>
1952 V7 N5 P281-284 R-928 <*>
RT-2768 <*>
64-15055 <=*$>
1952 V7 N5 P305-311 RT-3266 <*>
1953 V8 N1 P18-21 RT-3041 <*>
1953 V8 N1 P22-32 62-16638 <=*> O
1953 V8 N1 P38-45 RT-1994 <*>
64-10455 <=*$> O
1953 V8 N2 P90-104 RT-2258 <*>
1953 V8 N3 P131-139 60-19921 <*+>
1953 V8 N3 P158-162 62-18021 <=*>
1953 V8 N5 P286-292 RT-1934 <*>
1954 V9 P199-207 R-1523 <*>
1954 V9 N2 P85-95 <INSD>
1954 V9 N3 P155-161 R-4913 <*>
1954 V9 N3 P162-165 RT-3510 <*>
1954 V9 N5 P275-281 R-779 <*>
RT-2755 <*>
64-15095 <=*$>
1955 V10 P94-99 PT2 1686 <NLL>
1955 V10 N1 P63-64 RJ-299 <ATS>
1955 V10 N2 P132-133 RJ-300 <ATS>
1955 V10 N3 P139-157 <CB>
1955 V10 N3 P158-163 <CB>
1955 V10 N3 P164-168 <CB>
1955 V10 N3 P169-174 <CB>
1955 V10 N3 P175-179 <CB>
1955 V10 N3 P180-183 <CB>
1955 V10 N3 P189-190 <CB>
1955 V10 N3 P191 <CB>
1955 V10 N3 P192-198 <CB>
1955 V10 N3 P199-200 <CB>
1955 V10 N5 P315-322 62-25904 <=*>
1955 V10 N6 P349-354 47J14R <ATS>
1955 V10 N6 P355-357 46J14R <ATS>
1955 V10 N6 P358-363 51J14R <ATS>
1955 V10 N6 P364-367 55J14R <*>
1956 N11 P144-148 R-3242 <*>
1956 V11 P697-703 62-00581 <*>
1956 V11 N2 P223-232 50J14R <ATS>
1956 V11 N4 P376-382 52J14R <ATS>
1956 V11 N4 P389-392 10N51R <ATS>
1956 V11 N4 P483-488 55S82R <ATS>
1956 V11 N5 P615-620 62-14070 <=*>
9J18R <ATS>
1956 V11 N5 P638-639 62-18976 <=*>
1957 V12 P338- R-3788 <*>
1957 V12 P1042-1048 UCRL TRANS-530(L) <=*>
1957 V12 N3 P338-341 59-15486 <+*>
1957 V12 N3 P386-389 R-1647 <*>
1957 V12 N4 P495-498 61-13114 <*+>
62-23522 <=*>
1957 V12 N4 P504-508 59-22516 <+*>
1957 V12 N4 P513-515 62-14207 <=*>
1957 V12 N5 P587-592 R-4740 <*>
60-10946 <+*>
1957 V12 N5 P647-664 R-4739 <*>
R-5023 <*>
1958 V13 N1 P88-94 92M45R <ATS>
1958 V13 N2 P230-234 18K25R <ATS>
1958 V13 N3 P274-279 AEC-TR-3485 <+*> O
83L29R <ATS>
1958 V13 N3 P344-348 62-18023 <=*>
1958 V13 N6 P695-701 62-19431 <*=>
1959 V14 P112-117 62-14972 <=*>
1959 V14 N1 P67-70 60-15780 <+*>
1959 V14 N1 P100-103 60-31133 <=>
1959 V14 N1 P141-142 62-14206 <=*>
1959 V14 N3 P360-361 86L33R <ATS>
1959 V14 N3 P375-377 60-13799 <+*>
1959 V14 N3 P378-380 61-10225 <*+>
1959 V14 N5 P595-597 68M43R <ATS>
1959 V14 N6 P658-662 CSIR-TR.139 <CSSA>
1959 V14 N6 P746 61-23376 <*=>
1960 V15 N1 P96-98 18M45R <ATS>
1960 V15 N1 P99-103 11M44R <ATS>
1960 V15 N1 P127-128 60-31277 <=>
1960 V15 N2 P170-174 AEC-TR-4313 <*+> O
1960 V15 N2 P207-210 28M46R <ATS>
1960 V15 N3 P342-346 13M44R <ATS>
61-19572 <*=>
1960 V15 N3 P355-358 46M44R <ATS>
1960 V15 N3 P383-384 60-41738 <=>
1960 V15 N4 P413-418 CSIR-TR.154 <CSSA>
1960 V15 N4 P476-480 CSIR-TR.155 <CSSA>
1960 V15 N4 P487-494 32M45R <ATS>
1960 V15 N5 P601-604 21N54R <ATS>
1960 V15 N6 P686-691 CSIR-TR.319 <CSSA>
1960 V15 N6 P726-730 63-15182 <=*>
1960 V15 N6 P748-749 62-18027 <=*>
1961 V16 P343-347 63-20620 <=*$>
1961 V16 P462-464 62-14976 <=*>
1961 V16 N1 P83-86 62-14977 <=*>
1961 V16 N1 P118-119 61-27873 <*=>

1961 V16 N2 P180-184 62-18389 <=*>
1961 V16 N2 P378-380 61-27873 <*=>
1961 V16 N3 P337-342 62-18032 <=*>
1961 V16 N3 P372-374 62-18975 <=*>
1961 V16 N4 P465-468 63-20674 <=*$>
1961 V16 N6 P657-659 62-24710 <=*>
1961 V16 N6 P747-748 62-18030 <=*>
1962 V17 P142 62-16034 <=*>
1962 V17 N2 P227-230 62-18026 <=*>
1962 V17 N2 P260-262 63-10614 <=*>
1962 V17 N5 P639-641 63-10105 <=*>
1962 V17 N5 P644-646 65-10487 <*>
1962 V17 N6 P748-750 63-16413 <*=>
1962 V17 N6 P759-762 63-23355 <=*$>
1962 V17 N7 P878-881 63-24395 <=*$>
1962 V17 N8 P993-997 66-11451 <*>
1963 V18 P639-643 63-20619 <=*$>
1963 V18 N3 P377-384 65-10866 <*>
1963 V18 N3 P412-413 65-10377 <*>
1963 V18 N4 P450-453 65-10837 <*>
1963 V18 N4 P454-459 65-10845 <*>
1963 V18 N4 P507-509 63-23179 <=*>
1963 V18 N4 P514-519 63-20683 <=*$>
1963 V18 N5 P618-623 04Q71R <ATS>
1963 V18 N7 P840-850 49Q72R <ATS>
1963 V18 N7 P873-879 64-16182 <*>
1963 V18 N9 P1116-1119 64-10670 <=*$>
1964 V19 N3 P366-368 65-10376 <*>
1964 V19 N3 P369-374 65-10396 <*>
1964 V19 N6 P749-752 65-10492 <*>
1964 V19 N6 P754-756 69S81R <ATS>
1964 V19 N7 P821-829 N65-19704 <=$>
1964 V19 N10 P1251-1253 65-12036 <*>
1964 V19 N11 P1417-1418 65-13075 <*>
1965 V20 P612-614 1278 <TC>
1965 V20 P990-993 66-14206 <*$>
1965 V20 P1054-1057 66-14200 <*$>
1965 V20 N2 P145-152 65-63680 <=*>
1965 V20 N2 P153-164 65-30806 <=$>
1965 V20 N2 P183-191 65-30806 <=$>
1965 V20 N2 P196-199 65-30806 <=$>
1965 V20 N3 P276-277 65-63882 <=*$>
1965 V20 N5 P603-607 66-31532 <=$>
1965 V20 N6 P719-726 65-14916 <*>
1965 V20 N7 P832-835 65-14917 <*>
1965 V20 N7 P875-888 65-32525 <=*$>
1965 V20 N8 P860-863 66-11286 <*> 0
1965 V20 N9 P1007-1009 66-13426 <*>
1966 V21 P355-359 66-14208 <*$>
1966 V21 P617-624 66-14207 <*$>
1966 V21 N4 P473-478 66-33252 <=$>

ZHURNAL EKSPERIMENTALNOI BIOLOGII I MEDITSINY
1927 V5 P86-101 61-31005 <=>
1929 V12 P334-337 R-3394 <*>
1929 V13 P73-78 RT-2527 <*>

★ZHURNAL EKSPERIMENTALNOI I TEORETICHESKOI FIZIKI
1932 V2 N1 P77-90 R-480 <*>
1933 V3 N3 P165-172 65-10372 <*>
1933 V3 N3 P165-180 66-11237 <*>
1933 V3 N5 P447-453 818TM <CTT>
1934 V4 N5 P527-530 RT-3871 <*>
1935 V5 P33-44 R-3957 <*>
1935 V5 N7 P563-584 RT-1465 <NRC>
1936 V6 N1 P37-51 71L32R <ATS>
1936 V6 N10 P1060-1061 RT-673 <*>
1936 V6 N10 P1062-1074 RT-727 <*>
1937 V7 P409-423 66-13140 <*>
1937 V7 P424-429 66-13139 <*>
1937 V7 P750-763 65-17168 <*>
1937 V7 N1 P191-197 61-11593 <=>
1937 V7 N1 P198-202 61-11604 <=>
1937 V7 N2 P203-209 62-00405 <*>
63-14673 <=*>
1937 V7 N2 P249-255 61-11626 <=>
1937 V7 N4 P477-482 RT-4018 <*>
1937 V7 N5 P645-649 61-11614 <=>
1937 V7 N6 P696-713 RT-1813 <*>
1937 V7 N6 P818 61-11608 <=>
1937 V7 N12 P1457-1462 C-2220 <NLL>
1937 V7 N12 P1457-1467 R-3184 <*>
1937 V7 N12 P1457-1462 59-11923 <=>
1937 V7 N12 P1466-1468 61-11627 <=>
1938 V8 P1381-1383 63-14605 <=*>
1938 V8 N3 P291-318 R-1390 <*>
RT-3265 <*>
1938 V8 N4 P436-443 N66-15313 <=$>
1938 V8 N6 P716-723 RT-573 <*>
1938 V8 N12 P1249-1254 RT-3592 <*>
1938 V8 N8/9 P894-906 RT-463 <*>
1938 V8 N8/9 P930-944 R-941 <*>
855TM <CTT>
1939 N2 P218-222 59-10504 <+*> 0
1939 V9 P857-863 AEC-TR-3762 <+*>
1939 V9 P1094-1106 RCT V16 N1 P1 <RCT>
61-16972 <=*$>
1939 V9 N2 P167-175 N66-15314 <=$>
1939 V9 N3 P267-279 59-10971 <+*>
1939 V9 N5 P578-585 80K22R <ATS>
1939 V9 N6 P742-759 AEC-TR-3650 <+*>
1939 V9 N8 P963-968 62-18690 <=*>
1939 V9 N9 P1078-1080 62-32586 <=*>
1939 V9 N10 P1169-1181 RT-670 <*>
1939 V9 N10 P1182-1187 RT-531 <*>
1939 V9 N11 P1320-1321 AEC-TR-5615 <=*>
1939 V9 N12 P1530-1534 63-10878 <=*>
1940 V10 P823-827 61-18037 <*=> 0
1940 V10 P1038-1042 RJ-74 <ATS>
1940 V10 P1248-1256 63-14464 <=*>
1940 V10 P1460-1462 65-12723 <*>
1940 V10 N1 P40-57 RT-2173 <*>
65-28890 <$>
1940 V10 N5 P477-482 RT-963 <*>
1940 V10 N5 P542-568 RT-574 <*>
1940 V10 N5 P569-575 61-11093 <=>
1940 V10 N6 P589-600 64-16955 <*>
1940 V10 N6 P601-607 65-11578 <*>
1940 V10 N7 P723-739 AEC-TR-4010 <+*>
1940 V10 N8 P831-834 61-11514 <=>
1940 V10 N11 P1305-1310 R-317 <*>
1940 V10 N12 P1422-1423 61-11511 <=>
1940 V10 N12 P1441-1445 61-11512 <=>
1940 V10 N9/10 P1038-1042 RJ-74 <ATS>
1940 V10 N9/10 P1116-1123 RT-402 <*>
1940 V10 N9/10 P1116-1136 61-23306 <*=>
1940 V10 N9/10 P1137-1145 59-10815 <+*> 0
61-23305 <+*>
1940 V10 N9/10 P1146-1150 RT-3533 <*>
1941 V11 N1 P60-73 59-10972 <+*>
1941 V11 N1 P89-92 RT-480 <*>
1941 V11 N1 P133-140 61-11522 <=>
1941 V11 N2/3 P226-245 UCRL TRANS-565 <+*>
1942 V12 P342-348 NP-2102 <*>
1942 V12 N10 P467-472 60-17220 <+*> 0
1942 V12 N10 P479-480 65-60078 <=$>
1942 V12 N1/2 P79-89 61-11600 <=>
1942 V12 N11/2 P498-524 61-11779 <=>
1942 V12 N11/2 P525-538 61-11777 <=>
1943 V13 P116-120 59-10778 <+*>
1943 V13 P238-242 AEC-TR-3762 <+*>
1943 V13 N5 P181-182 60-41583 <=>
1943 V13 N1/2 P59-64 RT-3130 <*>
1943 V13 N3/4 P65-84 RT-1428 <*> 0
1943 V13 N3/4 P93-100 RT-1014 <*>
1943 V13 N7/8 P306-312 RT-929 <*>
1944 P289-306 R-3975 <*>
1944 V14 N6 P221-223 RT-575 <*>
1944 V14 N9 P330-341 1426TM <CTT>
59-16704 <+*>
1944 V14 N12 P474-480 62-11269 <=>
1944 V14 N1/2 P32-34 60-31836 <=>
1944 V14 N3/4 P65-83 RT-1991 <*>
1944 V14 N10/1 P439-447 TT.400 <NRC>
1945 V15 P81-88 UCRL TRANS-822(L) <=*>
1945 V15 N8 P445-458 RT-953 <*>
1945 V15 N1/2 P25-31 TT.455 <NRC>
1945 V15 N1/2 P57-76 62-19992 <=>
1946 N7 P574-586 1384TM <CTT>
1946 V16 P495-498 R-3958 <*>

1946 V16 P703-716 1315TM <CTT>
1946 V16 N1 P81-85 R-4498 <*>
1946 V16 N1 P81-86 RT-1898 <*>
1946 V16 N2 P119-128 RT-936 <*>
1946 V16 N2 P151-160 R-4496 <*>
RT-1897 <*>
1946 V16 N4 P363-364 61-11693 <=>
1946 V16 N4 P365-368 61-11818 <=>
1946 V16 N7 P614-625 RT-870 <*>
1946 V16 N8 P703-716 RT-4550 <*>
1946 V16 N8 P718-727 RT-4551 <*>
1316TM <CTT>
1946 V16 N8 P744-755 RT-327 <*>
1946 V16 N9 P776-779 62-15999 <*=>
1946 V16 N9 P840-848 N65-32971 <=$>
1946 V16 N11 P990-995 61-11747 <=>
1946 V16 N11 P996-999 RT-724 <*>
1946 V16 N12 P119-127 R-4502 <*>
1947 V17 P577-584 NP-2103 <*> 0
1947 V17 P986-997 60-23115 <*+>
1947 V17 N3 P227-237 TT.305 <NRC>
1947 V17 N3 P251-259 61-10799 <+*>
62-10938 <=*>
1947 V17 N4 P301-314 61-23170 <=*>
1947 V17 N5 P377-385 62-28790 <=$>
1947 V17 N5 P416-426 R-4952 <*>
66-11147 <*>
1947 V17 N7 P601-606 RT-713 <*>
1947 V17 N7 P614-628 60-12509 <=>
61-18702 <*=>
1947 V17 N7 P629-647 TT.330 <NRC>
1947 V17 N7 P655-660 R-571 <*>
1947 V17 N8 P698-707 60-14201 <+*>
1947 V17 N9 P769-782 RT-935 <*>
1947 V17 N9 P802-813 59-16696 <+*>
1947 V17 N9 P833-836 65-28886 <$>
1947 V17 N10 P851-861 R-886 <*>
1947 V17 N10 P889-900 60-41508 <=>
61-10261 <+*>
1947 V17 N11 P949-963 UCRL TRANS-856(L) <=*>
1947 V17 N11 P964-968 61-13393 <*+>
1947 V17 N11 P969-975 61-13392 <*+>
1947 V17 N12 P1106-1113 RT-1806 <*>
1948 V18 P424-428 59-19130 <+*>
1948 V18 P429 59-19131 <+*>
1948 V18 P444-448 R-4491 <*>
1948 V18 P501-509 R-4492 <*>
1948 V18 P519-524 NP-2107 <*>
1948 V18 P750-758 NP-2117 <*>
1948 V18 P867-872 RT-136 <*>
1948 V18 P1012- R-2356 <*>
1948 V18 P1030-1040 T-2296 <INSD>
1948 V18 P1049-1055 R-3455 <*>
1948 V18 N1 P3-18 AEC-TR-5514 <=*>
T.832A <INSD>
1948 V18 N1 P3-28 61-23317 <*=> 0
1948 V18 N1 P3-18 61-27621 <*=>
62-13142 <*=>
85P62R <ATS>
1948 V18 N1 P19-28 AEC-TR-6179 <=$>
T.832B <INSD>
61-13521 <+*>
1948 V18 N1 P29-44 RT-802 <*>
1948 V18 N1 P45-47 RT-853 <*>
1948 V18 N1 P102-103 RT-668 <*>
1948 V18 N2 P105-109 RT-803 <*>
1948 V18 N2 P149-163 RT-671 <*>
1948 V18 N2 P174-186 61-19498 <+*>
1948 V18 N2 P201-209 RT-476 <*>
1948 V18 N2 P210-218 RT-3739 <*>
59-15308 <+*>
1948 V18 N4 P337-345 RT-711 <*>
1948 V18 N4 P392-401 C-4464 <NRC>
1948 V18 N5 P429-433 RJ-549 <ATS>
1948 V18 N5 P438-443 R-4456 <*>
64-71368 <=>
1948 V18 N5 P475-478 RT-1134 <*>
61-17477 <=$>
1948 V18 N6 P501-509 RT-1981 <*>
1948 V18 N6 P525-532 66-11232 <*>
1948 V18 N6 P533-538 RT-672 <*>
1948 V18 N6 P539-547 RT-930 <*>
1948 V18 N7 P603-608 RT-1135 <*>
61-17478 <=$>
1948 V18 N7 P622-630 TT.518 <NRC>
1948 V18 N7 P631-635 RT-4471 <*>
1948 V18 N7 P651-658 RT-1992 <*>
1948 V18 N7 P659-667 61-19690 <=*>
1948 V18 N8 P673-702 TT-101 <NRC>
1948 V18 N8 P703-733 TT.345 <NRC>
1948 V18 N8 P734-736 62-15507 <=*>
1948 V18 N8 P750-758 RT-1896 <*>
1948 V18 N8 P759-766 61-15284 <+*>
1948 V18 N9 P795-798 RT-1899 <*>
1948 V18 N9 P799-806 CSIRO-2960 <CSIR>
R-4572 <*>
1948 V18 N9 P833-836 62-10190 <=*>
62-13706 <*=>
62-13706 <=>
1948 V18 N10 P878-885 RT-710 <*>
1948 V18 N10 P907-912 62-24840 <=*>
1948 V18 N10 P927-936 RT-4021 <*>
1948 V18 N11 P1041-1044 RT-576 <*>
1948 V18 N11 P1049-1055 AEC-TR-3139 <+*>
59-20261 <+*> P
1948 V18 N12 P1057-1069 TT-255 <NRC>
1948 V18 N12 P1070-1080 TT-256 <NRC>
1948 V18 N12 P1145-1146 60-13719 <+*>
1948 V19 N7 P651-663 50/2801 <NLL>
1949 V19 P18-24 R-2219 <*>
1949 V19 P121-125 R-2161 <*>
1949 V19 P181-182 R-2159 <*>
1949 V19 P215-224 R-2160 <*>
1949 V19 P535-542 60-13739 <+*>
1949 V19 P561-564 R-2216 <*>
1949 V19 P584-590 R-2218 <*>
1949 V19 N1 P18-24 62-13138 <=*>
1949 V19 N1 P25-35 66-11253 <*>
1949 V19 N1 P36-41 R-2338 <*>
RT-482 <*>
1949 V19 N1 P42-54 60-13716 <+*>
1949 V19 N1 P54-62 RT-577 <*>
1949 V19 N1 P74-77 RT-701 <*>
1949 V19 N2 P105-120 61-27536 <*=>
1949 V19 N2 P135-145 RT-709 <*>
1949 V19 N2 P146-154 RT-902 <*>
1949 V19 N3 P202-206 RT-852 <*>
1949 V19 N3 P215-224 AEC-TR-3880 <+*>
1949 V19 N3 P251-255 R-1489 <*>
1949 V19 N3 P256-268 R-1487 <*>
1949 V19 N4 P297-299 RT-712 <*>
1949 V19 N4 P300-303 TT.293 <NRC>
1949 V19 N5 P383- 61-13433 <+*>
1949 V19 N5 P413-420 59-10868 <+*>
1949 V19 N5 P451-459 TT-352 <NRC>
1949 V19 N5 P470-472 RT-4240 <*>
60-16585 <+*>
1949 V19 N6 P502-506 RT-1696 <*>
1949 V19 N7 P565-576 63-11690 <=>
64-00397 <*>
1949 V19 N7 P584-590 R-1474 <*>
R-959 <*>
32TM <CTT>
1949 V19 N7 P615-620 RT-1704 <*>
1949 V19 N7 P637-650 RT-1705 <*>
59-19116 <+*>
1949 V19 N7 P664-666 RT-3925 <*>
1949 V19 N7 P667-670 50/2302 <NLL>
60-19935 <+*>
1949 V19 N8 P673-679 61-17465 <=$>
1949 V19 N8 P703-708 59-17155 <+*> 0
59-22605 <+*> 0
1949 V19 N8 P709-726 RT-1701 <*>
1949 V19 N8 P727-730 60-19936 <+*>
1949 V19 N8 P731-738 R-1112 <*>
1949 V19 N8 P739-748 50/2477 <NLL>
1949 V19 N8 P753-755 RT-708 <*>
1949 V19 N8 P756-759 50/2478 <NLL>
1949 V19 N8 P756-757 60-19932 <+*>
1949 V19 N8 P757-759 RT-3748 <*>

Citation	Translation
	59-10418 <+*>
1949 V19 N9 P761-783	TT.161 <NRC>
1949 V19 N9 P784-795	TT.162 <NRC>
1949 V19 N9 P826-850	50/3263 <NLL>
1949 V19 N10 P876-886	<ATS>
	RT-1703 <*>
1949 V19 N1C P887-898	RJ-795 <ATS>
	RT-1702 <*>
1949 V19 N1C P916-929	RT-3108 <*>
1949 V19 N10 P937-939	RT-479 <*>
1949 V19 N10 P940-945	AEC-TR-6059 <=*$>
	RT-1136 <*>
	61-17462 <=$>
1949 V19 N11 P959-964	<ATS>
1949 V19 N11 P965-970	RT-478 <*>
	63-14433 <=*> 0
1949 V19 N11 P971-972	RT-2229 <*>
1949 V19 N12 P1108-1112	62-11382 <=>
1949 V19 N12 P1113-1120	61-13505 <*+>
1949 V19 N12 P1130-1135	61-17466 <=$>
1949 V19 N12 P1136-1140	RT-1700 <*>
1950 N12 P1C64-1082	65-61163 <=>
1950 V20 P385-389	R-4802 <*>
1950 V20 P523-526	R-3398 <*>
	751TM <CTT>
1950 V20 P944-961	R-4743 <*>
1950 V20 P1C64-	60-13723 <+*>
1950 V20 P1145	61-13434 <*+>
1950 V20 N1 P54-61	C-2454 <NRC>
	R-2089 <*>
1950 V20 N1 P91-92	RT-3970 <*>
	61-17449 <=$>
1950 V20 N2 P97-107	RT-1634 <*>
1950 V20 N2 P108-116	RT-1635 <*>
1950 V20 N2 P127-131	RT-1636 <*>
1950 V20 N2 P139-151	RT-1637 <*>
1950 V20 N2 P152-165	RT-1638 <*>
1950 V20 N2 P166-174	RT-491 <*>
1950 V20 N2 P175-182	RT-1649 <*>
	66-11052 <*>
1950 V20 N2 P183-191	RT-979 <*>
1950 V20 N2 P192	RT-1137 <*>
	61-17463 <=$>
1950 V20 N3 P211-217	TT.194 <NRC>
1950 V20 N3 P218-223	TT.314 <NRC>
1950 V20 N3 P243-266	60-13864 <+*>
1950 V20 N4 P330-333	62-11383 <=>
1950 V20 N4 P342-346	RT-707 <*>
1950 V20 N4 P347-350	RT-477 <*>
1950 V20 N4 P356-370	61-13643 <*+>
1950 V20 N5 P385-389	76K25R <ATS>
1950 V20 N5 P395-400	R-170 <RIS>
1950 V20 N5 P427-437	RT-1744 <*>
	62-15920 <=*>
1950 V20 N5 P458-464	R-1389 <*>
1950 V20 N5 P465-47C	RT-1716 <*>
1950 V20 N6 P481-491	60-18509 <+*>
1950 V20 N6 P497-509	64-13147 <=*$>
1950 V20 N6 P511-522	RT-1763 <*>
1950 V20 N6 P566-571	61-19361 <=>
1950 V20 N7 P624-635	TT.284 <NRC>
1950 V20 N8 P705-721	UCRL TRANS-227(L) <*=>
	02J20R <ATS>
1950 V20 N8 P729-733	RT-1977 <*>
1950 V20 N9 P783-786	RT-572 <*>
1950 V2C N9 P811-823	59-10973 <+*>
1950 V20 N9 P834-841	16K27R <ATS>
1950 V20 N9 P860-861	TT-287 <NRC>
1950 V20 N10 P685-870	62-11538 <=>
1950 V20 N10 P865-870	62-11538 <=>
1950 V2C N1C P919-924	58-1002 <*>
1950 V20 N11 P961-978	RT-388 <*>
1950 V20 N11 P979-986	RT-1979 <*>
1950 V20 N11 P987-994	62-11452 <=>
	62-11472 <=>
1950 V20 N11 P1C11-1018	RT-1143 <*>
1950 V20 N11 P1019-1021	RT-773 <*>
1950 V20 N11 P1031-1034	RT-241 <*>
1950 V20 N11 P1035-1038	RT-1866 <*>
1950 V20 N11 P1047-1050	60-21769 <=>

Citation	Translation
1950 V20 N11 P1055-1056	R-4552 <*>
	RT-2008 <*>
	61-13432 <+*>
1950 V20 N12 P1064-1082	R-2341 <*>
	64-18601 <*>
1950 V20 N12 P1098-1108	R-1421 <*>
1950 V20 N12 P1143	77K25R <ATS>
1951 N9 P1001-1008	R-671 <*>
1951 V21 P444-453	R-4111 <*>
1951 V21 P656	61-13645 <+*>
1951 V21 P658	61-13642 <*+>
1951 V21 P659	61-13641 <+*>
1951 V21 P721-730	R-4114 <*>
1951 V21 P770-774	R-755 <*>
1951 V21 P1239-1249	R-1031 <*>
1951 V21 N2 P133-141	TT.433 <NRC>
	63-19232 <=*>
1951 V21 N2 P164-171	C-2190 <NRC>
1951 V21 N2 P172-188	TT.559 <NRC>
1951 V21 N2 P206-220	R-3844 <*>
1951 V21 N2 P220-229	11N49R <ATS>
	61-14755 <=*>
	63-16467 <=*>
1951 V21 N2 P269-274	UCRL TRANS-857(L) <=*>
1951 V21 N2 P305-309	UCRL-TRANS-858(L) <=*>
1951 V21 N2 P338-340	62-18390 <=*$> 0
1951 V21 N2 P341-349	RJ-328 <ATS>
1951 V21 N4 P473-483	<ATS>
	R-4589 <*>
	59-17537 <*>
1951 V21 N4 P484-492	RJ-152 <ATS>
1951 V21 N4 P493-506	<ATS>
1951 V21 N4 P493-505	38M40R <ATS>
1951 V21 N4 P5C7-509	RT-3419 <*>
1951 V21 N4 P510-513	RJ-151 <ATS>
1951 V21 N4 P532-540	RJ-329 <ATS>
1951 V21 N5 P598-609	R-4795 <*>
1951 V21 N5 P627-633	64-15202 <=*$>
1951 V21 N5 P634-641	AD-610 114 <=$>
1951 V21 N6 P717-720	60-13877 <+*>
1951 V21 N6 P748-760	TT.276 <NRC>
1951 V21 N7 P788-794	R-227 <*>
1951 V21 N7 P8C3-808	62-11200 <=>
1951 V21 N7 P814-820	60-21812 <+*>
1951 V21 N8 P389-400	59-10505 <+*>
1951 V21 N8 P917-922	64-10781 <=*$>
1951 V21 N8 P942-944	TT.553 <NRC>
1951 V21 N9 P979-991	62-19925 <=>
1951 V21 N9 P1C01-1008	RT-1768 <*>
1951 V21 N10 P1132-1138	C-3500 <NRCC>
1951 V21 N10 P1139-1145	64-11819 <=>
1951 V21 N10 P1164-1175	61-17305 <=$>
1951 V21 N10 P1172-1175	60-41365 <=>
1951 V21 N10 P1182-1183	RT-2205 <*>
1951 V21 N11 P1201-1208	62-11543 <=>
1951 V21 N11 P1223	IS-TRANS-14 <=$>
1951 V21 N11 P1227-1238	R-899 <*>
	62-15917 <=*>
1951 V21 N11 P1262-1269	AEC-TR-1512 <*>
1951 V21 N11 P1270-1283	R-778 <*>
	RT-1223 <*>
1951 V21 N11 P1284-1302	UCRL TRANS-220L <=*>
1951 V21 N11 P1303-1308	UCRL TRANS-555 <+*>
1951 V21 N12 P1330-1336	RT-1748 <*>
1951 V21 N12 P1354-1363	RJ-545 <ATS>
1951 V21 N12 P1370-1375	RT-4114 <*>
	53/0371 <NLL>
1951 V21 N12 P1384-1388	RT-1769 <*>
1952 V22 P1C2-111	AEC-TR-6088 <=*$>
1952 V22 P562-578	UCRL TRANS-667(L) <*>
1952 V22 N1 P3-10	RT-4552 <*>
1952 V22 N1 P112-119	RT-956 <*>
1952 V22 N2 P172-183	TT.373 <NRC>
1952 V22 N2 P2C0-203	RT-1814 <*> 0
	60-15681 <+*> C
1952 V22 N2 P204-213	RT-2344 <*>
1952 V22 N2 P230-240	59-20357 <+*> P
1952 V22 N2 P249-250	C-4079 <NRCC>
1952 V22 N2 P385-399	60-17633 <+*>

Citation	Translation
1952 V22 N3 P276-283	TT.366 <NRC>
1952 V22 N3 P284-296	RT-512 <*>
1952 V22 N3 P360-366	R-1500 <*>
	61-15324 <+*>
1952 V22 N3 P378-380	RT-2420 <*>
1952 V22 N4 P385-399	AEC-TR-4748 <*=>
1952 V22 N4 P400-405	RT-4553 <*>
1952 V22 N4 P406-413	RT-4554 <*>
1952 V22 N4 P414-420	RT-125 <*>
1952 V22 N4 P471-474	62-18124 <=*>
	63-24251 <=*$>
1952 V22 N4 P475-486	62-18125 <=*>
1952 V22 N4 P507-509	RT-1815 <*>
1952 V22 N5 P513-519	62-15900 <=*>
1952 V22 N5 P539-543	TT.441 <NRC>
1952 V22 N5 P544-561	59-14142 <+*>
1952 V22 N6 P641-657	RT-1442 <*>
1952 V22 N6 P676-686	RT-2847 <*>
1952 V22 N6 P687-704	RT-2673 <*>
1952 V23 P161-168 PT2	R-411 <*>
1952 V23 P49-54	1308TM <CTT>
1952 V23 P151-160	R-1674 <*>
1952 V23 P327-335	T1827 <INSD>
1952 V23 P632-640	65-10862 <*>
1952 V23 N1 P8-20	C.T.S.NO.32 <NLL>
	RT-2672 <*>
1952 V23 N1 P21-34	RT-2905 <*>
1952 V23 N1 P49-54	RT-4555 <*>
1952 V23 N1 P60-67	AEC-TR-3989 <+*>
1952 V23 N1 P126-127	RT-2367 <*>
1952 V23 N2 P129-139	ANL-TRANS-27 <=>
1952 V23 N2 P151-160	1380TM <CTT>
1952 V23 N2 P169-181	RT-3358 <*>
1952 V23 N2 P222-228	R-1947 <*>
1952 V23 N2 P236-238	62-19924 <*=>
1952 V23 N3 P253-264	C.T.S.NO.18 <NLL>
	RT-2671 <*>
1952 V23 N3 P275-289	65-62397 <=$>
1952 V23 N3 P289-305	65-62398 <=$>
1952 V23 N3 P305-314	RT-2210 <*>
1952 V23 N4 P357-370	63-11601 <=>
1952 V23 N4 P371-380	63-11688 <=>
1952 V23 N4 P461-467	RT-1939 <*>
1952 V23 N4 P468-476	RT-4556 <*>
1952 V23 N4 P484	RT-311 <*>
1952 V23 N5 P512-516	61-19034 <+*>
1952 V23 N5 P517-524	ANL-TRANS-25 <=$>
1952 V23 N5 P525-531	R-885 <*>
1952 V23 N5 P552-563	RT-1276 <*>
1952 V23 N5 P564-575	RT-1940 <*>
	1339TM <CTT>
1952 V23 N5 P588-592	04T92R <ATS>
1952 V23 N5 P610	RT-312 <*>
1952 V23 N6 P641-648	63-10298 <=*>
1952 V23 N6 P660-668	RT-4557 <*>
	1347TM <CTT>
1952 V23 N6 P669-677	R-1420 <*>
1952 V23 N6 P678-681	R-1944 <*>
	RT-3087 <*>
	60-12488 <=>
1952 V23 N6 P712-719	RT-1775 <*>
	60-16509 <*+>
1952 V23 N7 P83-90	RT-699 <*>
1952 V23 N7 P91-100	RT-700 <*>
1952 V23 N8 P129-139	61-15943 <+*>
1953 V24 P279-283	66-61671 <=*$>
1953 V24 P445-453	AEC-TR-3710 <+*>
1953 V24 P453-465	R-5066 <*>
1953 V24 P659-672	AEC-TR-3627 <+*>
1953 V24 N1 P69-77	R-738 <*>
1953 V24 N1 P83-89	RT-3670 <*>
1953 V24 N1 P107-113	RJ-422 <ATS>
1953 V24 N2 P167-176	RJ-598 <*>
1953 V24 N3 P269-278	RJ-592 <ATS>
1953 V24 N3 P319-325	RJ-536 <ATS>
1953 V24 N3 P326-338	RJ-537 <ATS>
1953 V24 N3 P339-341	RJ-597 <ATS>
1953 V24 N3 P365-366	62-11559 <=>
1953 V24 N3 P368	56G6R <ATS>
	56G6R(B) <ATS>
1953 V24 N4 P419-428	59-17360 <*>
1953 V24 N4 P501	62-11422 <=>
1953 V24 N4 P501-503	62-15907 <=*>
1953 V24 N4 P503-504	62-11419 <=>
1953 V24 N5 P505-515	UCRL TRANS-655 <*=>
1953 V24 N5 P571-574	59-22446 <+*>
1953 V24 N5 P589-600	60-41658 <=>
1953 V24 N6 P659-672	R-4877 <*>
1953 V24 N6 P681-690	59-10976 <+*> 0
	62-11378 <=>
1953 V25 P429-434	UCRL TRANS-415(L) <=*>
1953 V25 N1 P74-83	2974 <RT>
1953 V25 N1 P74-84	62-15901 <=*>
1953 V25 N1 P84-86	R-996 <*>
	62-15902 <=*>
1953 V25 N1 P107-114	62-11544 <=>
1953 V25 N1 P127-128	64-11666 <=>
1953 V25 N2 P157-168	RT-4171 <*>
1953 V25 N2 P208-214	3546 <HB>
1953 V25 N3 P257-263	62-23017 <=*>
1953 V25 N3 P264-270	62-23159 <=*>
1953 V25 N3 P296-306	AEC-TR-4773 <*>
1953 V25 N3 P327-340	R-417 <*>
1953 V25 N4 P435-440	RT-4643 <*>
1953 V25 N4 P448-454	RJ-596 <ATS>
1953 V25 N4 P471-478	64-13108 <=*$>
1953 V25 N4 P479-484	RT-3373 <*>
1953 V25 N4 P485-490	R-1897 <*>
1953 V25 N4 P498-508	RJ-552 <ATS>
1953 V25 N4 P509-510	6G7R <ATS>
	62-11630 <=>
1953 V25 N5 P560-570	RJ-546 <ATS>
1953 V25 N5 P596-604	RJ-595 <ATS>
1953 V25 N5 P605-613	RT-3137 <*>
1953 V25 N6 P683-687	64-00396 <*>
1953 V25 N6 P718-722	62-11370 <=>
1953 V25 N6 P733-737	RJ-441 <ATS>
1953 V25 N6 P749-750	RJ-438 <*>
	RJ-438 <ATS>
1954 P362-366	R-1755 <*>
1954 V26 P293-299	AEC-TR-4257 <+*>
1954 V26 P300-306	AEC-TR-4258 <*+>
1954 V26 P537-544	T-2218 <INSD>
1954 V26 P723-735	R-4717 <*>
1954 V26 N1 P3-18	R-1941 <*>
	RT-3113 <*>
	60-12611 <=>
1954 V26 N1 P19-34	64-16284 <*>
1954 V26 N1 P35-41	R-2069 <*>
	62-23221 <=*>
1954 V26 N1 P42-56	RJ-601 <ATS>
	RT-4559 <*>
1954 V26 N1 P57-63	R-3468 <*>
	RJ-324 <ATS>
1954 V26 N1 P98-106	R-888 <*>
	RJ-533 <ATS>
	64-00316 <*>
1954 V26 N1 P109-114	RJ-553 <ATS>
	59-20370 <+*>
1954 V26 N1 P120-123	59-20894 <+*> 0
1954 V26 N1 P124-127	64-16568 <=*$>
1954 V26 N1 P576-584	RJ-541 <ATS>
1954 V26 N2 P139-	AEC-TR-4737 <*=>
1954 V26 N2 P139-149	C-4097 <NRCC>
1954 V26 N2 P168-172	RT-3176 <*>
1954 V26 N2 P173-178	RT-3177 <*>
1954 V26 N2 P179-184	RT-3178 <*>
1954 V26 N2 P185-188	RT-3732 <*>
1954 V26 N2 P189-207	RT-3914 <*>
1954 V26 N3 P264-269	62-11635 <=>
1954 V26 N3 P270-274	RT-4221 <*>
1954 V26 N3 P275-280	63-14475 <=*>
1954 V26 N3 P307-316	62-11075 <=>
	64-16621 <*>
1954 V26 N3 P370-376	R-883 <*>
1954 V26 N3 P377-384	64-21710 <=>
1954 V26 N4 P405-416	59-17356 <+*>
1954 V26 N4 P439-446	62-11377 <=>
1954 V26 N4 P446-453	62-23007 <=*>
1954 V26 N4 P459-472	86S89R <ATS>

Citation	Translation
1954 V26 N4 P511-	91G6R <*>
1954 V26 N5 P545-560	62-15899 <=*>
1954 V26 N5 P562	GB43 <NLL>
1954 V26 N5 P562-575	RJ-543 <ATS>
	59-18679 <+*> 0
1954 V26 N5 P585-587	03Q73R <ATS>
1954 V26 N5 P629-639	SCL-T-245 <+*>
1954 V26 N5 P668-679	GB43 <NLL>
1954 V26 N6 P649-667	RT-3682 <*>
1954 V26 N6 P668-697	RJ-544 <ATS>
1954 V26 N6 P668-679	59-18680 <+*> 0
1954 V26 N6 P680-683	63-14449 <=*>
1954 V26 N6 P684-695	GB125 IB 1452 <NLL>
1954 V26 N6 P684-696	R-739 <*>
1954 V26 N6 P696-703	RJ-323 <ATS>
1954 V26 N6 P696-706	RJ-323 <ATS>
1954 V26 N6 P723-735	R-875 <*>
	RT-4558 <*>
1954 V26 N6 P744-750	RT-2817 <*>
1954 V26 N6 P751-754	62-10647 <=*>
1954 V26 N6 P761-763	62-11510 <=>
1954 V27 P195-207	T-1639 <NLL>
1954 V27 P262-264	R-2026 <*>
	66-12348 <*>
1954 V27 P431-438	R-3879 <*>
1954 V27 P458-466	754TM <CTT>
1954 V27 P467-476	775TM <CTT>
1954 V27 P605-614	R-939 <*>
1954 V27 P648-649	R-1790 <*>
1954 V27 P690-698	UCRL TRANS-669(L) <*>
1954 V27 P708-711	UCRL TRANS-674-L <=*>
1954 V27 N1 P19-23	RJ-535 <ATS>
	RT-3099 <*>
	1354TM <CTT>
	59-20268 <+*>
	60-15382 <+*>
1954 V27 N1 P24-28	RJ-325 <ATS>
	59-20267 <+*>
	60-15370 <+*>
1954 V27 N1 P94-96	RJ-327 <ATS>
1954 V27 N1 P94	60-15017 <+*>
1954 V27 N1 P97-102	RJ-237 <ATS>
1954 V27 N1 P115	59-20854 <+*>
1954 V27 N1 P121-123	RT-4530 <*>
1954 V27 N1 P123-124	59-17361 <*>
1954 V27 N1 P595-604	1298TM <CTT>
1954 V27 N2 P142-146	<INSD>
	59-18683 <+*> 0
1954 V27 N2 P147-155	<INSD>
1954 V27 N2 P180-188	RT-4534 <*>
1954 V27 N2 P189-194	GB125 IB 1431 <NLL>
	RT-4535 <*>
1954 V27 N2 P195-207	R-1008 <*>
1954 V27 N2 P208-214	12K23R <ATS>
1954 V27 N2 P224-242	RT-4533 <*>
1954 V27 N2 P262-264	GB125 IB 1461 <NLL>
	RT-4531 <*>
1954 V27 N2 224-242	-2007 <*>
1954 V27 N3 P283-287	RT-3922 <*>
1954 V27 N3 P288-295	RT-4295 <*>
	59-17557 <*>
	63-18396 <=*>
1954 V27 N3 P351-354	RT-2816 <*>
1954 V27 N3 P355-368	RT-2591 <*>
1954 V27 N3 P369-381	R-876 <*>
	62-23074 <=*>
1954 V27 N4 P425-430	ANL-TRANS-16 <=$>
1954 V27 N4 P431-438	3062 <NLL>
	59-15045 <+*> 0
	62-23061 <=*>
1954 V27 N4 P487-500	R-1026 <*>
	SCL-T-257 <+*>
1954 V27 N4 P521-528	64-21710 <=>
1954 V27 N5 P529-541	RT-4529 <*>
1954 V27 N5 P542-548	RT-4527 <*>
	1313TM <CTT>
1954 V27 N5 P549-556	RT-4526 <*>
	1337TM <CTT>
1954 V27 N5 P590-594	RT-4528 <*>
	1299TM <CTT>
1954 V27 N5 P595-604	RT-4532 <*>
1954 V27 N5 P605-614	SCL-T-239 <+*> 0
	60-27011 <*+>
1954 V27 N5 P648-649	RJ-446 <ATS>
1954 V27 N6 P712-718	RT-3363 <*>
1954 V27 N6 P719-727	RT-3408 <*>
1954 V27 N6 P728-734	RT-3908 <*>
	62-28791 <=$>
1954 V27 N6 P761	62-23225 <=*>
1954 V27 N9 P318-332	59-22421 <+*>
1954 V27 N11 P542-548	R-2334 <*>
1954 V27 N1/7 P97-102	RJ-237 <ATS>
1955 V28 P18-27	1286TM <CTT>
1955 V28 P28-37	1285TM <CTT>
1955 V28 P126-127	RJ-548 <ATS>
1955 V28 P191-198	66-10793 <*> 0
1955 V28 P283-286	R-2137 <*>
1955 V28 P378	1305TM <CTT>
1955 V28 P589-602	R-1429 <*>
1955 V28 N1 P13-16	R-4506 <*>
	R-4899 <*>
	RT-4522 <*>
1955 V28 N1 P17-27	RT-4521 <*>
1955 V28 N1 P28-37	RT-4520 <*>
1955 V28 N1 P38-48	R-1027 <*>
1955 V28 N1 P49-60	R-915 <*>
1955 V28 N1 P70-76	RJ-554 <ATS>
	RJ554 <ATS>
1955 V28 N1 P125	RJ-445 <ATS>
1955 V28 N1 P126-127	RJ-548 <ATS>
1955 V28 N2 P223-227	R-1142 <*>
1955 V28 N2 P242	62-11612 <=>
1955 V28 N2 P249-250	62-23061 <=*>
1955 V28 N3 P312	GB43 <NLL>
1955 V28 N3 P312-320	62-14472 <=*>
1955 V28 N3 P330-334	RT-3431 <*>
1955 V28 N3 P343-351	AEC-TR-2616 <=*>
1955 V28 N4 P422-430	1312TM <CTT>
1955 V28 N4 P471-473	RT-4223 <*>
1955 V28 N4 P480-484	59-17555 <*>
1955 V28 N4 P503-505	59-17505 <*>
1955 V28 N4 P505-506	59-17440 <*>
1955 V28 N5 P608-609	RT-3365 <*>
1955 V28 N5 P610-614	RT-3361 <*>
1955 V28 N5 P614-616	RT-3366 <*>
1955 V28 N5 P616-617	RT-2938 <*>
1955 V28 N5 P621-623	59-17488 <*>
1955 V28 N5 P626-627	59-17489 <*>
1955 V28 N5 P628-629	RT-4518 <*>
1955 V28 N5 P636	59-17490 <*>
1955 V28 N6 P639-650	RJ-550 <ATS>
	RT-4119 <*>
1955 V28 N6 P729-732	59-17529 <*>
1955 V28 N6 P732-734	59-17530 <*>
1955 V28 N6 P762	59-17541 <*>
1955 V29 P209-220	R-1708 <*>
1955 V29 P380-381	R-1403 <*>
	1412TM <CTT>
1955 V29 P385	61-16441 <*=>
1955 V29 P406-416	1345TM <CTT>
1955 V29 P651-657	R-1688 <*>
1955 V29 P701-702	R-1886 <*>
	1410TM <CTT>
1955 V29 P874-876	R-1437 <*>
	1409TM <CTT>
1955 V29 P884-886	R-1685 <*>
	1408TM <CTT>
1955 V29 P889-892	R-236 <*>
1955 V29 N1 P94-110	GB43 <NLL>
1955 V29 N2 P140-144	RT-4122 <*>
1955 V29 N2 P140-146	59-17554 <*>
1955 V29 N2 P261-263	RT-4523 <*>
1955 V29 N3 P380-381	62-15869 <=*>
1955 V29 N3 P385	8388-C <K-H>
1955 V29 N4 P422-430	RT-4519 <*>
1955 V29 N4 P479-485	RT-4296 <*>
1955 V29 N5 P559-571	61-13857 <+*>
1955 V29 N5 P658-667	R-3739 <*>
1955 V29 N5 P702-703	62-11621 <=>
1955 V29 N6 P743-747	59-17677 <+*>

	62-11299 <=>
1956 P416-423	65-12620 <*>
1956 V30 P351-361	R-2834 <*>
1956 V30 P414-416	R-2835 <*>
1956 V30 P592-	R-807 <*>
1956 V30 N1 P213-214	62-15847 <=*>
1956 V30 N2 P382-385	3242 <NLL>
1956 V30 N2 P414-416	00H14R <ATS>
1956 V30 N3 P532-536	51K24R <ATS>
1956 V30 N3 P560-564	62-23055 <=*>
1956 V30 N3 P560-563	62-32904 <=*>
1956 V30 N3 P569-570	79J16R <ATS>
1956 V30 N4 P806	59-18681 <+*> 0
1956 V30 N5 P833-839	R-427 <*>
1956 V30 N6 P1047-1051	CSIRO-3474 <CSIR>
1956 V31 N3 P517-518	64-16564 <=*$>
1957 V32 P14-19	R-3062 <*>
1957 V32 P398-399	R-2368 <*>
1957 V32 P1022-1035	R-2367 <*>
1957 V32 P1587-1588	R-2371 <*>
1957 V32 N2 P199-206	12J16R <ATS>
1957 V32 N2 P305-310	RJ-911 <ATS>
	62-14081 <=*>
1957 V32 N3 P510-514	62J16R <ATS>
1957 V32 N3 P559-565	R-1589 <*>
1957 V32 N3 P605-607	R-1631 <*>
1957 V32 N3 P611-612	28K22R <ATS>
1957 V32 N4 P896-914	R-1838 <*>
1957 V32 N5 P1092-1097	60-18500 <+*>
1957 V33 P335-338	R-3104 <*>
1957 V33 P610-613	R-4521 <*>
1957 V33 P730-734	R-3724 <*>
1957 V33 P735-745	R-3725 <*>
1957 V33 P788-790	R-4354 <*>
1957 V33 P873-876	R-3416 <*>
1957 V33 P1221-1226	R-4804 <*>
1957 V33 N1 P3-8	R-2113 <*>
1957 V33 N1 P151-157	63-15684 <=*$>
1957 V33 N4 P851-855	59-17154 <+*>
1957 V33 N4 P1064-1065	58-279 <*>
1957 V33 N5 P1140-1143	59-12046 <+*>
1957 V33 N6 P513-514	R-4718 <*>
1957 V33 N6 P1396-1402	62-11294 <=>
1957 V33 N6 P1417-1427	59-12460 <=>
	60-17376 <+*>
1957 V33 N6 P1454-1457	R-3415 <*>
1957 V33 N6 P1509-1511	59-12453 <=>
	59-20291 <+*>
1957 V33 N6 P1513-1514	59-12047 <+*>
1957 V33 N11 P1140-1143	R-4430 <*>
1958 V34 P501-502	R-5316 <*>
1958 V34 P1349-1350	R-5355 <*>
1958 V34 N1 P58-65	R-3809 <*>
1958 V34 N1 P66-72	R-3812 <*>
1958 V34 N1 P73-79	R-3810 <*>
1958 V34 N1 P97-106	R-4403 <*>
1958 V34 N1 P106-109	R-3853 <*>
1958 V34 N1 P139-150	R-4507 <*>
1958 V34 N1 P237-238	R-4012 <*>
1958 V34 N2 P286-297	AEC-TR-3587 <=*>
1958 V34 N2 P327-330	R-5141 <*>
1958 V34 N2 P483-493	62-16047 <=*>
1958 V34 N2 P512-513	65-13849 <=>
	65-13849 <=$>
1958 V34 N2 P513-514	R-4006 <*>
1958 V34 N2 P528-529	R-3890 <*>
	59-20262 <+*>
1958 V34 N2 P530-531	AEC-TR-3466 <+*>
1958 V34 N3 P670-674	59-11860 <=>
1958 V34 N3 P675-683	59-11867 <=>
1958 V34 N3 P688-693	59-11861 <=>
1958 V34 N3 P772-773	R-4167 <*>
1958 V34 N3 P777	61-23400 <*=>
1958 V34 N4 P1023-1024	61-23400 <*=>
1958 V34 N5 P1142-1147	59-11939 <=>
1958 V34 N5 P1240-1245	R-5120 <*>
1958 V34 N5 P1254-1257	R-4825 <*>
1958 V34 N5 P1325-1327	AEC-TR-3491 <+*>
1958 V34 N6 P1558-1565	62-19830 <=>
1958 V34 N6 P1605-1609	62-19831 <=>
1958 V34 N6 P1641-1642	R-5135 <*>
1958 V34 N6 P1643-1644	R-5187 <*>
1958 V34 N6 P1655-1656	59-10236 <+*>
1958 V34 N6 P1658-1659	62-23193 <=>
1958 V34 N6 P1660-1661	59-12450 <=>
1958 V35 P300	AEC-TR-3547 <+*>
1958 V35 P798	AEC-TR-3546 <+*>
1958 V35 N1 P184-202	UCRL TRANS-446 <+*>
1958 V35 N1 P302-303	59-12451 <=>
1958 V35 N2 P392-400	62-18283 <=*> 0
1958 V35 N2 P408-415	59-14740 <+*>
1958 V35 N2 P535-536	AEC-TR-3861 <+*>
1958 V35 N2 P551	59-20791 <+*>
1958 V35 N3 P808-809	R-5337 <*>
	59-22688 <+*>
1958 V35 N3 P817-	59-11930 <=>
1958 V35 N4 P965-969	R-5376 <*>
1958 V35 N4 P1044-1045	59-16748 <+*>
1958 V35 N4 P1047-1049	R-5336 <*>
1958 V35 N6 P1402-1406	59-20266 <+*>
1958 V35 N6 P1435-1439	59-20265 <+*>
1958 V35 N6 P1466-1470	59-20264 <+*>
1958 V35 N6 P1573-1575	59-20263 <+*>
1959 V36 N2 P629-630	AEC-TR-3864 <+*>
1959 V36 N2 P636-637	AEC-TR-3758 <+*>
1959 V36 N2 P638-639	AEC-TR-3757 <+*>
1959 V36 N3 P954-955	AEC-TR-3862 <+*>
1959 V36 N4 P1319-1321	59-22059 <+*>
1959 V36 N5 P1573-1574	AEC-TR-3915 <+*>
1959 V36 N5 P1608-1609	AEC-TR-3914 <+*>
1959 V37 P318-319	UCRL TRANS-479 <*>
1959 V37 N1 P41-45	AEC-TR-3934 <+*>
1959 V37 N1 P315-	UCRL TRANS-493 <+*>
1959 V37 N2 P399-405	61-13558 <*+>
1959 V37 N2 P587-588	59-20407 <+*>
1959 V37 N3 P880-881	60-14492 <+*>
1959 V37 N6 P1543-1550	60-31149 <=>
1959 V37 N6 P1692-1696	60-21934 <=>
1959 V37 N6 P1770-1774	60-21936 <=>
1959 V37 N6 P1784-1788	60-21937 <=> 0
1960 V38 P1641-1643	AEC-TR-4072 <=>
1960 V38 N1 P292-293	62-10311 <*=>
1960 V38 N4 P1061-1073	60-41488 <=>
1960 V39 P669-679	AEC-TR-4170 <=>
1960 V39 N2 P384-392	61-14756 <=*>
1960 V39 N2 P413-415	61-15725 <+*>
1960 V39 N3 P536-544	AEC-TR-4068 <=>
1960 V39 N4 P997-1000	C-3552 <NRCC>
1961 V40 N6 P1543-1550	62-15239 <=>
1961 V40 N6 P1676-1681	62-15239 <=>
1961 V40 N6 P1775-1787	62-15239 <=>
1961 V40 N6 P1825-1831	62-15239 <=>
1961 V40 N6 P1866-1870	62-15239 <=>
1961 V41 N1 P113-114	62-23686 <=*>
1961 V41 N2 P354-362	62-14384 <=*>
1961 V41 N2 P389-390	63-19770 <=*>
1961 V41 N2 P665-667	62-16591 <=*>
1961 V41 N3 P692-695	62-18711 <=*>
1961 V41 N4 P1015-1022	62-24929 <=*>
1961 V41 N5 P1522-1526	62-14832 <=*>
1961 V41 N5 P1582-1591	62-18002 <=*>
1961 V41 N6 P1809-1810	62-14735 <=*>
	62-18712 <=*>
1962 V42 N3 P833-838	N65-22581 <=$>
1962 V42 N3 P910-912	62-18843 <=*>
1962 V43 N1 P40-50	AEC-TR-5395 <=*>
1962 V43 N1 P347-349	AD-621 803 <=$>
1962 V43 N2 P637-648	63-21055 <=>
1962 V43 N4 P1323-1330	63-10849 <=*>
1962 V43 N5 P1897-1903	63-18589 <=*$>
1962 V43 N6 P2313-2315	63-19349 <=*>
1963 V44 N2 P649-656	63-24046 <=*$>
1963 V44 N3 P1115-1116	63-18330 <=*>
	65-10815 <*>
1963 V44 N4 P1211-1227	RPQ,V1,N1,P1-11 <IPIX>
1963 V44 N4 P1228-1238	RPQ,V1,N1,P12-18 <IPIX>
1963 V44 N5 P1525-1533	RPQ,V1,N1,67-73 <IPIX>
1963 V44 N5 P1567-1576	RPQ,V1,N1,P88-94 <IPIX>
1963 V44 N5 P1644-1649	64-15195 <=*$>
1963 V44 N5 P1650-1660	RPQ,V1,N1,P95-102 <IPIX>

1963 V44 N5 P1703-1718
RPQ,V1,N1,P103-113 <IPIX>
1963 V44 N5 P1723-1741 RPQ,V1,N1,P74-87 <IPIX>
1963 V44 N6 P1818-1822 64-13175 <=*$>
1963 V44 N6 P1884-1888 64-71394 <=> M
1963 V44 N6 P1982-1992
RPQ,V1,N1,P150-157 <IPIX>
1963 V44 N6 P2016-2022 63-23810 <=*$>
1963 V44 N6 P2122-2130 64-16909 <=*$>
1963 V44 N6 P2130-2141
RPQ,V1,N1,P158-165 <IPIX>
1963 V45 N3 P780-782 AEC-TR-6180 <=*$>
1964 V46 N1 P142-147 64-19308 <=$>
1964 V46 N2 P598-611 65-11490 <*>
1964 V46 N3 P1106-1116 65-10759 <*>
1964 V46 N4 P1197-1204 AEC-TR-6285 <=*$>
1964 V46 N4 P1510-1513 AD-610 304 <=$>
1964 V46 N41 P1205-1207 65-10285 <*>
1964 V47 P1905-1918 66-11122 <*>
1964 V47 N2 P518-523 527 <TC>
1964 V47 N2 P778-780 65-10744 <*>
1964 V47 N2 P780-781 92S82R <ATS>
1964 V47 N4 P1588-1590 65-10816 <*>
1964 V47 N5 P2003-2005 65-17013 <*>
1964 V47 N6 P2318-2320 65-14156 <*>
1964 V47 N8 P667-677 N65-15061 <=$>
1965 V48 P777-778 N66-22300 <=$>
1965 V48 N3 P976-999 65-13232 <*>
1965 V49 N1 P159-162 66-31280 <=$>
1965 V49 N2 P447-448 66-12831 <*>
1965 V49 N2 P485-492 CSIR-TR-552 <CSSA>
1965 V49 N2 P515-528 N66-18455 <=$>
1965 V49 N3 P755-759 66-10147 <*> 0
1965 V49 N4 P1083-1090 65-25339 <$>
1966 V50 N1 P131-134 AD-633 412 <=$>

*ZHURNAL EKSPERIMENTALNOI I TEORETICHESKOI FIZIKI. PISMA V REDAKTSIYU
1965 V1 N6 P18-23 AD-634 161 <=$>
1965 V2 N1 P48-50 65-22513 <$>
1965 V2 N2 P83-87 65-32992 <=*$>
1965 V2 N3 P125-129 65-23423 <$>
1966 V3 N11 P436-439 66-62588 <=$>

ZHURNAL EVOLIUTSIONNOI BIOKHIMII I FIZIOLOGII
1965 V1 N1 P3-6 65-30935 <*>
1965 V1 N1 P7- FPTS V25 N2 P.T303 <FASE>
1965 V1 N1 P16- FPTS V25 N2 P.T277 <FASE>
1965 V1 N1 P26-31 65-31000 <=*$>
1965 V1 N1 P32- FPTS V25 N2 P.T273 <FASE>
1965 V1 N1 P59- FPTS V25 N2 P.T191 <FASE>
1965 V1 N1 P59-66 65-32229 <=$>
1965 V1 N1 P67- FPTS V25 N1 P.T34 <FASE>
1965 V1 N1 P84- FPTS V25 N2 P.T221 <FASE>
1965 V1 N2 P117-126 66-60429 <=*$>
1965 V1 N2 P126-133 66-60430 <=*$>
1965 V1 N2 P145-151 66-60847 <=*$>
1965 V1 N2 P151-156 RTS-3280 <NLL>
1965 V1 N3 P227- FPTS V25 N3 P.T489 <FASE>
1965 V1 N4 P311-319 66-61349 <=*$>
1965 V1 N4 P320-324 66-61348 <=*$>
1965 V1 N5 P385-390 66-62000 <=*$>
1965 V1 N5 P391-397 66-61350 <=*$>
1965 V1 N5 P413- FPTS V25 N5 P.T879 <FASE>
1965 V1 N6 P550- FPTS V25 N5 P.T863 <FASE>

*ZHURNAL FIZICHESKOI KHIMII
1931 V2 N6 P799-805 65-63514 <=$>
1932 V3 P406-418 64-16066 <=*$> C
1933 V4 N1 P92-102 64-10770 <=*$> 0
1933 V4 N3 P380-382 AEC-TR-5271 <=*>
1933 V4 N4 P461-464 RT-2078 <*>
1934 V4 N7 P969-974 63-14558 <=*>
1934 V5 P1452-1458 65-11875 <*>
1934 V5 N1 P32-37 AEC-TR-5153 <=*>
1934 V5 N5 P557-573 RT-2846 <*>
1934 V5 N6 P793-801 RT-1799 <*>
63-20742 <=*$> 0
1935 N6 P189-205 AEC-TR-5185 <=*>
1935 V6 P130- 61-18074 <=*>
1935 V6 P478-495 R-2520 <*>
1935 V6 P496-510 R-2521 <*>
1935 V6 P649-663 64-10728 <*>
1935 V6 P870-881 R-4001 <*>
1935 V6 N4 P496-510 CSIRO-2966 <CSIR>
R-4566 <*>
1935 V6 N5 P632-639 R-2109 <*>
1935 V6 N8 P1079-1085 RT-1811 <*>
63-20743 <=*$> OP
64-10111 <=*$> OP
1936 V7 P546-548 R-2610 <*>
1936 V7 P587-591 R-2594 <*>
1936 V7 P899-908 R-2512 <*>
1936 V7 N3 P405-417 65-60889 <=$>
1936 V7 N3 P438-443 R-3260 <*>
1936 V8 P504-513 60-18716 <+*> C
1936 V8 N1 P159-164 59-17525 <*>
1936 V8 N2 P290-294 TT.148 <NRC>
1936 V8 N3 P364-371 RT-578 <*>
1936 V8 N3 P438-457 60-41586 <=>
1937 V9 N1 P143-146 61-11048 <=>
1937 V9 N2 P239-251 RT-954 <*>
63-18298 <*>
63-20767 <=*$>
1937 V9 N3 P387-392 RJ-458 <ATS>
1937 V9 N4 P542-552 61-11053 <=>
1937 V9 N6 P854-866 RT-1941 <*>
1937 V10 P627-635 R-3318 <*>
1937 V10 P788-792 R-2300 <*>
1937 V10 N2 P279-282 64-10570 <=*$>
1937 V10 N3 P479-483 61-11054 <=>
1938 V11 P390-399 R-2452 <*>
1938 V11 P569-577 R-2453 <*>
1938 V11 P664-669 R-2454 <*>
1938 V11 N1 P6-67 RT-2845 <*>
1938 V11 N6 P852-857 RT-1798 <*>
2867Z <K-H>
63-20744 <=*$>
1938 V12 N1 P100-105 62-16704 <=*>
1939 V13 P56-70 R-4107 <*>
1939 V13 P106-118 R-2451 <*>
1939 V13 P248-258 R-2867 <*>
1939 V13 P281-285 R-2184 <*>
1939 V13 P541-546 R-3355 <*>
1939 V13 P547-550 R-2627 <*>
1939 V13 P561-571 R-2458 <*>
1939 V13 P586-592 AEC-TR-5186 <=*>
1939 V13 P690-692 1267 <*>
1939 V13 P767-770 R-2178 <*>
1939 V13 P851-867 R-2932 <*>
1939 V13 P889-895 R-2840 <*>
1939 V13 P934-941 R-3663 <*>
1939 V13 P1472-1482 R-2450 <*>
1939 V13 P1512-1522 63-16025 <=*>
1939 V13 P1559-1562 R-2936 <*>
64-18578 <*>
1939 V13 N1 P147 RT-2018 <*>
64-10478 <=*$>
1939 V13 N1 P148 RT-2017 <*>
64-10452 <=*$>
1939 V13 N2 P169-173 61-11759 <=>
1939 V13 N2 P302 RT-2261 <*>
1939 V13 N3 P397-405 R-5062 <*>
1939 V13 N5 P561-571 AEC-TR-2869 <*>
1939 V13 N5 P621-630 RT-2288 <*>
64-14374 <=*$> 0
1939 V13 N6 P738-755 R-2346 <*>
RT-1068 <*>
40L35R <ATS>
40TM <CTT>
1939 V13 N7 P851-867 84J15R <ATS>
1939 V13 N7 P965-982 TT.473 <NRC>
1939 V13 N8 P1087-1091 RT-2403 <*>
64-14391 <=*$> 0
1939 V13 N8 P1105-1116 RT-3975 <*>
1939 V13 N8 P1124-1130 RT-1208 <*>
1939 V13 N9 P1260-1270 61-21009 <=>
1939 V13 N9 P1315-1321 RT-2182 <*>
64-14147 <=*$> 0
1939 V13 N10 P1472-1482 RT-2183 <*>

Citation	Translation
	64-14148 <=*$> O
1939 V13 N11 P1559-1562	RT-2184 <*>
	64-14146 <=*$> O
1939 V13 N11 P1635-1641	63-24114 <=*$>
	64-19151 <=*$>
1939 V13 N12 P1695-1727	61-21124 <=>
1940 V14 P220-224	R-2444 <*>
1940 V14 P349-356	59-15408 <+*>
1940 V14 P413-417	R-2446 <*>
1940 V14 P505-512	R-2477 <*>
1940 V14 P571-581	R-2939 <*>
1940 V14 P582-588	59-10285 <+*>
1940 V14 P650-662	R-2448 <*>
1940 V14 P774-781	R-2449 <*>
1940 V14 P985-988	R-1014 <*>
1940 V14 P1118-1127	R-2801 <*>
1940 V14 P1229-1240	R-2961 <*>
1940 V14 P1241-1245	R-2944 <*>
1940 V14 P1246-1249	R-2960 <*>
1940 V14 P1250-1259	R-2975 <*>
1940 V14 P1260-1270	R-2966 <*>
1940 V14 P1347-1352	R-2962 <*>
1940 V14 P1495-1497	R-5011 <*>
1940 V14 N1 P6-9	RJ-661 <ATS>
	59-12204 <+*>
1940 V14 N1 P10-15	65-62977 <*>
1940 V14 N1 P936-940	RT-138 <*>
1940 V14 N2 P175-179	1502TM <CTT>
1940 V14 N2 P220-224	62-11369 <=>
1940 V14 N2 P235-243	62-18318 <=*> O
1940 V14 N2 P253-256	AEC-TR-4347 <=*>
1940 V14 N3 P287-290	61-21172 <=>
1940 V14 N3 P418-421	32K25R <ATS>
1940 V14 N3 P427-428	RT-2668 <*>
	64-14667 <=*$> O
1940 V14 N4 P461-473	RT-405 <*>
1940 V14 N4 P474-487	RT-404 <*>
1940 V14 N4 P488-493	RT-409 <*>
1940 V14 N4 P494-504	61-11052 <=>
	63-20979 <=*$> O
1940 V14 N4 P582-588	61-15355 <*+>
1940 V14 N7 P853-858	61-11953 <=>
1940 V14 N7 P859-862	61-21167 <=>
1940 V14 N7 P953-963	AEC-TR-5228 <=*>
1940 V14 N7 P1004-1006	15L37R-C <ATS>
1940 V14 N8 P1009-1025	RT-1023 <*>
1940 V14 N8 P1026-1042	61-11058 <=>
1940 V14 N9 P877-885	TT.571 <NRC>
1940 V14 N11 P1381-1395	RT-410 <*>
1940 V14 N11 P1517-1519	61-11049 <=>
1940 V14 N12 P1521-1527	61-11979 <=>
1940 V14 N12 P1528-1534	61-11956 <=>
1940 V14 N12 P1566-1568	61-27795 <=*>
1940 V14 N5/6 P605-614	TT.561 <NRC>
1940 V14 N5/6 P759-767	15L37R-A <ATS>
1940 V14 N5/6 P768-773	15L37R-B <ATS>
1940 V14 N9/10 P1153-1158	61-18110 <*=>
1940 V14 N9/10 P1229-1240	64-10737 <=*$> O
1940 V14 N9/10 P1241-1245	64-10743 <=*$> O
1940 V14 N9/10 P1246-1249	64-10742 <=*$> O
1940 V14 N9/10 P1250-1259	64-10736 <=*$>
1940 V14 N9/10 P1260-1270	64-10741 <=*$>
1940 V14 N9/10 P1301-1307	61-11960 <=>
1940 V14 N9/10 P1308-1311	61-11959 <=>
1940 V14 N9/10 P1362-1369	AEC-TR-5227 <*=>
1941 V15 P1-30	R-2952 <*>
1941 V15 P40-49	R-2947 <*>
1941 V15 P145-155	AEC-TR-5231 <=*>
1941 V15 P164-173	R-2951 <*>
1941 V15 P333-345	61-18012 <*=>
1941 V15 P470-474	61-18009 <*=>
1941 V15 P525-531	60-23516 <+*>
1941 V15 P652-658	62-14614 <=*>
1941 V15 P708-730	R-2941 <*>
1941 V15 P768-775	65-13636 <*>
1941 V15 P776-792	65-12792 <*> O
1941 V15 P927-933	R-2967 <*>
1941 V15 P966-973	R-2963 <*>
	61-14955 <=*>
1941 V15 P1055-1056	R-2974 <*>
1941 V15 P1061-1071	R-2972 <*>
1941 V15 P1121-1128	R-2954 <*>
1941 V15 N1 P78-90	RT-411 <*>
1941 V15 N1 P101-108	TT.254 <NRC>
1941 V15 N2 P184-192	R-2950 <*>
1941 V15 N2 P193-197	R-2948 <*>
1941 V15 N2 P228-233	59-17524 <*>
1941 V15 N4 P525-531	64-13555 <=*$>
1941 V15 N6 P731-738	60-41510 <=>
1941 V15 N9 P1059	R-2973 <*>
1941 V15 N10 P1072-1081	61-16694 <*=>
1941 V15 N7/8 P831-846	RT-1140 <*>
1941 V15 N7/8 P847-864	RT-1141 <*>
1941 V15 N7/8 P865-881	RT-1221 <*>
1941 V15 N7/8 P934-941	RT-1443 <*>
1941 V15 N7/8 P966-973	80J15R <ATS>
1942 V16 P13-17	R-2925 <*>
1942 V16 P18-22	R-2242 <*>
1942 V16 P48-58	R-2645 <*>
1942 V16 P59-71	R-2626 <*>
1942 V16 N3/4 P106-114	62-14640 <=*>
1942 V16 N3/4 P148-151	62-14638 <=*>
1943 V17 P115-125	61-16947 <*=>
1943 V17 P172-186	R-2282 <*>
1943 V17 P187-214	R-3343 <*>
1943 V17 P215-225	R-2281 <*>
1943 V17 P236-	<ATS>
1943 V17 P264-268	R-2174 <*>
1943 V17 P269-270	R-2186 <*>
1943 V17 P295-310	R-2283 <*>
1943 V17 P336-346	R-4796 <*>
1943 V17 P381-390	61-16918 <*=>
1943 V17 P391-407	61-16902 <*=>
1943 V17 P414-423	R-2297 <*>
1943 V17 N1 P24-31	RT-3486 <*>
1943 V17 N1 P32-44	RT-1006 <NRC>
1943 V17 N2 P87-96	RT-1209 <*>
	61-16953 <*=>
1943 V17 N3 P115-125	R-2288 <*>
1943 V17 N3 P134-144	RT-973 <*>
	63-20561 <=*$>
1943 V17 N3 P145-158	RT-412 <*>
1943 V17 N4 P226-235	TT.138 <NRC>
1943 V17 N4 P269-270	62-16811 <=*>
1943 V17 N7 P347-380	TT.142 <NRC>
1943 V17 N5/6 P300-312	R-2269 <*>
1943 V17 N5/6 P313-317	R-2295 <*>
1943 V17 N5/6 P408-413	R-2296 <*>
	RT-1617 <*>
1943 V17 N5/6 P424-445	R-2268 <*>
1943 V18 P356-363	R-2429 <*>
1944 V18 P121-130	R-2432 <*>
1944 V18 P187-193	R-2433 <*>
1944 V18 P258-259	61-20130 <*=>
1944 V18 P315-320	61-18195 <*=>
1944 V18 P321-328	61-18194 <*=>
1944 V18 P335-355	R-2235 <*>
1944 V18 P380-382	61-18226 <*=>
1944 V18 P386-388	R-2430 <*>
1944 V18 P389-394	R-2431 <*>
1944 V18 N3 P174-182	TT-794 <NRCC>
1944 V18 N9 P389-394	RT-1022 <*>
1944 V18 N9 P1550-1554	R-1918 <*>
1944 V18 N10 P453-465	RT-1021 <*>
	61-18227 <*=>
1944 V18 N11 P537-546	TT.248 <NRC>
1944 V18 N3/4 P102-109	RJ-211 <*>
	RJ-211 <ATS>
1944 V18 N3/4 P110-114	60-41344 <=>
	61-28107 <*=> O
1944 V18 N5/6 P253-257	33K25R <ATS>
1944 V18 N7/8 P268-282	RTS-3065 <NLL>
1945 V19 P15-47	R-3370 <*>
1945 V19 P48-71	R-3369 <*>
1945 V19 P72-82	R-3332 <*>
1945 V19 P92-95	R-3368 <*>
1945 V19 P142-151	R-2811 <*>
1945 V19 P160-166	R-4045 <*>
1945 V19 P250-252	R-3382 <*>
1945 V19 P314-322	R-2799 <*>

Citation	Translation
1945 V19 P497-506	R-2635 <*>
1945 V19 P676	R-2638 <*>
1945 V19 N3 P167-168	RT-3485 <*>
1945 V19 N9 P417-428	63-19831 <=*$>
1945 V19 N10 P497-506	RT-1167 <*>
1945 V19 N12 P657-664	63-19832 <=*$>
1945 V19 N4/5 P250-252	R-1236 <*>
	756TM <CTT>
	88K22R <ATS>
1945 V19 N7/8 P392-404	<ATS>
	64-10746 <*> 0
1946 V20 P165-174	R-2640 <*>
1946 V20 P399-402	63-19563 <=*>
1946 V20 P403-408	R-3976 <*>
1946 V20 P869-880	R-2791 <*>
1946 V20 P881-894	R-3308 <*>
1946 V20 P941-946	R-3307 <*>
1946 V20 P1025-1028	R-2153 <*>
1946 V20 P1085-1093	59-20896 <+*> P
1946 V20 P1209-1211	R-2154 <*>
1946 V20 P1459-1470	R-3328 <*>
1946 V20 N1 P105-110	2357 <HB>
1946 V20 N2 P189-193	5141 A <K-H>
	64-16467 <=*$>
1946 V20 N2 P221-222	65-10142 <*>
1946 V20 N3 P225-238	RT-2635 <*>
	59-17322 <*>
	83H13R <ATS>
1946 V20 N3 P297-302	T.623 <INSD>
1946 V20 N6 P575-	60-23017 <+*> 0
1946 V20 N7 P613-622	RT-1007 <*>
1946 V20 N7 P667-677	RTS-3081 <NLL>
1946 V20 N7 P679-692	63-23295 <=*>
1946 V20 N8 P769-772	RT-4186 <*>
1946 V20 N8 P773-778	RT-4214 <*>
1946 V20 N8 P789-801	RT-2046 <*>
1946 V20 N8 P881-894	TT-286 <NRC>
1946 V20 N9 P1043-1047	63H13R <ATS>
1946 V20 N10 P1065-1080	RT-127 <*>
	62-20460 <=*>
1946 V20 N10 P1187-1198	TT.229 <NRC>
1946 V20 N11 P1259-1272	RT-1905 <*>
1946 V20 N11 P1273-1282	RT-1902 <*>
1946 V20 N11 P1283-	61-23276 <=*>
1946 V20 N11 P1301-1317	R-1228 <*>
1946 V20 N11 P1319-1323	09J16R <ATS>
	62-18170 <=*>
1946 V20 N11 P1325-	61-23275 <=*>
1946 V20 N11 P1347-1358	63-18238 <=*> 0
1946 V20 N11 P1359-1362	61-11502 <=>
1946 V20 N11 P1363-1366	61-11769 <=>
1946 V20 N11 P1367-1371	AEC-TR-3781 <+*>
1946 V20 N11 P1371-1375	61-11778 <=>
1946 V20 N11 P1381-1389	61-11991 <=>
1946 V20 N11 P1391-1397	61-21123 <=>
1946 V20 N4/5 P345-354	61-21056 <=>
1946 V20 N4/5 P355-362	3582 <HB>
1946 V20 N4/5 P365-368	RT.565 <*>
1946 V20 N4/5 P379-386	52/2944 <NLL>
	61-13431 <*+> 0
1947 P389-390	R-2620 <*>
1947 N3 P331-336	RJ-37 <ATS>
1947 N8 P41-53	R-4786 <*>
1947 V21 P11-13	R-2633 <*>
1947 V21 P119-123	R-2631 <*>
1947 V21 P197-200	R-3309 <*>
1947 V21 P391-396	60-17723 <+*>
1947 V21 P397	R-2632 <*>
1947 V21 P449-457	R-2621 <*>
1947 V21 P459-462	R-2619 <*>
1947 V21 P569-574	R-2639 <*>
1947 V21 P629-630	R-2622 <*>
1947 V21 P659-674	R-3005 <*>
1947 V21 P749-758	R-3006 <*>
1947 V21 P759-760	R-2613 <*>
1947 V21 P927-952	R-762 <*>
1947 V21 P983-985	277TM <CTT>
1947 V21 P1057-1068	R-2612 <*>
1947 V21 P1135-1142	R-2988 <*>
1947 V21 P1159-1162	R-3004 <*>
1947 V21 P1163-1176	R-3003 <*>
1947 V21 P1293-1298	R-3325 <*>
1947 V21 P1303-1316	R-3002 <*>
1947 V21 P1413-1433	R-3324 <*>
1947 V21 P1449-1461	R-2872 <*>
1947 V21 N1 P3-10	64-13987 <=*$>
1947 V21 N1 P11-13	63-16807 <*=> 0
1947 V21 N1 P51-63	RT-3757 <*>
1947 V21 N1 P85-96	RT-2452 <*>
1947 V21 N1 P97-107	RT-665 <*>
1947 V21 N2 P163-178	TT.120 <*>
1947 V21 N2 P231-232	89P62R <ATS>
1947 V21 N3 P245-255	AEC-TR-4758 <*=> 0
	RT-2000 <*>
1947 V21 N3 P257-260	R-5276 <*>
1947 V21 N3 P331-336	RT-799 <*>
1947 V21 N3 P355-359	61-15356 <*=>
1947 V21 N4 P405-410	AEC-TR-5902 <=*$>
1947 V21 N5 P613-622	RT-2890 <*>
1947 V21 N6 P707-714	RT-2891 <*>
1947 V21 N6 P745-748	RT-1913 <*>
1947 V21 N7 P761-768	60-41500 <=>
1947 V21 N7 P879-880	AEC-TR-5918 <=$>
1947 V21 N8 P919-926	TT.471 <NRC>
1947 V21 N8 P927-952	83J15R <ATS>
1947 V21 N8 P983-985	62-16107 <=*> P
1947 V21 N8 P987-988	66-12748 <*>
1947 V21 N9 P1027-1032	TT.556 <NRC>
1947 V21 N10 P1111-1124	TT.442 <NRC>
1947 V21 N10 P1135-1142	TT.152 <NRC>
1947 V21 N10 P1183-1204	60-15831 <*+>
	91L34R <ATS>
1947 V21 N10 P1213-1222	TT.185 <NRC>
1947 V21 N11 P1263-1268	59-19074 <+*>
1947 V21 N11 P1289-1292	TT.154 <NRC>
1947 V21 N11 P1303-1316	RT-1956 <*>
1947 V21 N11 P1317-1334	TT.167 <NRC>
	60-13726 <+*>
1947 V21 N12 P1373-1386	AD-610 898 <=$>
	RT-1608 <*>
	TT.186 <NRC>
1947 V21 N12 P1387-1401	RT-2887 <*>
1947 V21 N12 P1461-1469	TT.108 <NRC>
1948 V22 P3-10	R-2990 <*>
1948 V22 P21-26	R-3000 <*>
1948 V22 P27-49	63-20808 <*>
1948 V22 P103-106	R-3001 <*>
1948 V22 P179-194	R-2989 <*>
1948 V22 P195-208	R-2989 <*>
1948 V22 P209-216	R-2611 <*>
1948 V22 P233-242	R-2610 <*>
1948 V22 P243-254	R-3023 <*>
1948 V22 P261-264	R-2809 <*>
1948 V22 P289-301	R-2608 <*>
1948 V22 P339-348	R-2607 <*>
1948 V22 P439-444	R-2863 <*>
1948 V22 P575-586	R-2874 <*>
1948 V22 P575-585	R-4359 <*>
1948 V22 P587-590	R-3020 <*>
1948 V22 P633-636	R-2875 <*>
1948 V22 P637-639	R-2877 <*>
1948 V22 P711-720	R-2220 <*>
1948 V22 P711-719	R-4448 <*>
1948 V22 P847-858	R-3013 <*>
1948 V22 P859-869	R-2189 <*>
1948 V22 P925-	R-3760 <*>
1948 V22 P1002-1015	R-2166 <*>
1948 V22 P1081-1089	R-2168 <*>
1948 V22 P1081	60-15851 <+*> 0
1948 V22 P1146-1148	R-3327 <*>
1948 V22 P1271-1279	R-2188 <*>
1948 V22 P1302-1311	280TT <CCT>
1948 V22 P1322-1330	R-2130 <*>
1948 V22 P1374-1380	R-2170 <*>
1948 V22 P1385-1389	R-2165 <*>
1948 V22 P1397-1404	R-2167 <*>
1948 V22 P1405-1407	R-5293 <*>
1948 V22 P1443-1453	R-2190 <*>
1948 V22 P1478-1485	R-2129 <*>
1948 V22 P1486-1495	R-2129 <*>

Citation	Translation
1948 V22 N1 P3-10	RT-406 <*>
1948 V22 N1 P11-19	RT-408 <*>
1948 V22 N1 P21-25	RT-407 <*>
1948 V22 N1 P27-48	RT-900 <*>
	63-20808 <=*$>
1948 V22 N1 P49-51	64-19353 <=$>
1948 V22 N1 P53-57	RT-3747 <*>
1948 V22 N1 P59-68	TT.300 <NRC>
1948 V22 N1 P103-106	61-10300 <*+> 0
1948 V22 N1 P1374-1380	RT-1534 <*>
1948 V22 N2 P129-131	RT-1904 <*>
1948 V22 N2 P145-155	RT-3924 <*>
	63-10977 <=*>
1948 V22 N2 P157-160	63-16041 <=*> 0
1948 V22 N2 P161-167	63-16042 <=*>
1948 V22 N2 P221-231	<ATS>
1948 V22 N3 P311-330	R-1876 <*>
	RJ-61 <ATS>
	RJ61 <ATS>
1948 V22 N3 P351-361	RT-878 <*>
1948 V22 N3 P395-408	R-4475 <*>
1948 V22 N4 P445-455	TT.153 <*>
	59-10729 <+*>
1948 V22 N4 P483-486	RT-1131 <*>
1948 V22 N4 P487-494	RT-1266 <*>
1948 V22 N4 P513-520	RT-1346 <*>
1948 V22 N5 P541-548	RT-1901 <*>
1948 V22 N5 P565-574	TT.569 <NRC>
1948 V22 N5 P625-632	60-15832 <+*>
1948 V22 N6 P641-646	60-13717 <+*>
1948 V22 N6 P647-653	60-13718 <+*>
1948 V22 N6 P655-667	RT-1219 <*>
1948 V22 N6 P683-	59-19105 <+*>
1948 V22 N6 P697-710	AEC-TR-5814 <=*$>
	82Q71R <ATS>
1948 V22 N6 P711-720	53/0097 <NLL>
	62-13129 <*=>
1948 V22 N6 P721-729	61-13506 <*+>
	62-13133 <*=>
1948 V22 N7 P769-782	RT-2029 <*>
1948 V22 N7 P801-803	21N58R <ATS>
1948 V22 N8 P925-929	RT-78 <*>
1948 V22 N8 P930-940	TT.159 <NRC>
	60-13883 <+*>
1948 V22 N8 P953-955	TT.343 <NRC>
1948 V22 N8 P956-960	RT-2222 <*>
1948 V22 N8 P961-967	59-13783 <=> 0
1948 V22 N9 P1027-1033	TT.592 <NRC>
1948 V22 N9 P1097-1099	236-TT <CCT>
1948 V22 N9 P1108-1115	R-410 <*>
	63-18988 <=*>
1948 V22 N10 P1214-1218	RJ-35 <ATS>
1948 V22 N10 P1246-1250	RT-362 <*>
1948 V22 N10 P1251-1255	61-19104 <+*>
	64-13257 <=*$>
1948 V22 N11 P1302-1311	RT-2122 <*>
1948 V22 N11 P1331-1343	RJ-644 <ATS>
1948 V22 N11 P1344-1359	RJ-141 <ATS>
1948 V22 N11 P1360-1373	RT-1612 <*>
1948 V22 N11 P1397-1404	RT-94 <*>
1948 V22 N12 P1427-1442	AEC-TR-5230 <=*>
	62-16109 <=*> P
1948 V22 N12 P1443-1453	TT.114 <NRC>
	63-20807 <=*$>
1948 V22 N17 P1478-1485	59-19094 <+*>
1949 V23 P71-77	R-2171 <*>
1949 V23 P101-103	R-2155 <*>
1949 V23 P105-114	R-2157 <*>
1949 V23 P379-382	R-2221 <*>
1949 V23 P388-405	R-2234 <*>
1949 V23 P484-496	R-2236 <*>
1949 V23 P647-655	R-2225 <*>
1949 V23 P714-718	60-14006 <+*>
1949 V23 P723-728	R-2223 <*>
1949 V23 P729-735	R-3019 <*>
1949 V23 P800	AEC-TR-4123 <+*>
1949 V23 P858-860	R-2232 <*>
1949 V23 P863-870	R-2845 <*>
1949 V23 P942-952	63-18994 <=*>
1949 V23 P1177-1179	R-2231 <*>
1949 V23 P1266-1274	60-17709 <+*>
1949 V23 N2 P115-123	TT.116 <NRC>
1949 V23 N2 P124-130	TT.111 <NRC>
1949 V23 N2 P156-179	61-11787 <=>
1949 V23 N2 P180-191	61-11741 <=>
1949 V23 N2 P931-935	61-11742 <=>
1949 V23 N3 P281-288	TT.213 <NRC>
1949 V23 N3 P304-314	62-13581 <=*>
1949 V23 N4 P383-387	RT-3921 <*>
1949 V23 N4 P422-427	2759E <K-H>
	63-18997 <=*>
1949 V23 N4 P441-444	60-14005 <+*>
1949 V23 N4 P452-468	TT.237 <NRC>
1949 V23 N5 P505-515	TT.177 <NRC>
1949 V23 N5 P572-576	61-23285 <=*>
1949 V23 N5 P616-624	234TT <CCT>
1949 V23 N6 P676-688	07Q72R <ATS>
	62-14025 <=*> 0
1949 V23 N6 P695-713	RT-3484 <*>
	60-13697 <+*>
1949 V23 N6 P695-714	81J15R <ATS>
1949 V23 N7 P790-799	TT.123 <NRC>
1949 V23 N7 P800-808	AEC-TR-4122 <+*>
1949 V23 N7 P839-848	RT-3436 <*>
	63-18991 <=*> 0
1949 V23 N7 P861-862	63-14847 <=*>
1949 V23 N8 P873-884	TT.200 <NRC>
1949 V23 N8 P897-916	TT.124 <NRC>
1949 V23 N8 P917-930	TT.126 <*>
	61-17461 <=$>
1949 V23 N8 P953-958	R-1791 <*>
1949 V23 N9 P1031-1035	RT-1914 <*>
1949 V23 N9 P1095-1105	13J19G <ATS>
	13J19R <ATS>
	62-16087 <=*> P
1949 V23 N10 P1129-1140	52/2936 <NLL>
1949 V23 N10 P1161-1171	61-17260 <=$>
	61-28120 <*=>
1949 V23 N10 P1172-1176	RT-1903 <*>
1949 V23 N10 P1187-1196	TT.151 <NRC>
1949 V23 N11 P1361-1374	RJ-32 <ATS>
1949 V23 N12 P1410-1420	TT.193 <NRC>
1949 V23 N12 P1421-1425	59-20725 <*>
1949 V23 N12 P1464-1476	2759B <K-H>
	63-20437 <=*$>
1949 V23 N12 P1477-1482	RT-77 <*>
1949 V23 N12 P1495-1501	64-19851 <=$>
1950 V24 P3-9	62-13135 <=*>
1950 V24 P43-57	62-13143 <*=>
1950 V24 P244-253	R-3762 <*>
1950 V24 P337-344	R-3761 <*>
1950 V24 P366-378	AEC-TR-5220 <=*>
1950 V24 P445-454	R-106 <*>
1950 V24 P981-987	62-13134 <*=>
1950 V24 P1312-1325	62-16039 <=*>
1950 V24 P1371-1382	60-19912 <+*>
	64-18062 <=*$>
1950 V24 P1471-	R-3895 <*>
1950 V24 N2 P161-176	LA-TR-65-4 <=$>
1950 V24 N2 P207-213	TIL/T-5548 <NLL>
1950 V24 N2 P214-223	TT.219 <NRC>
1950 V24 N2 P244-253	RT-80 <*>
1950 V24 N3 P272-278	TT-195 <NRC>
1950 V24 N3 P279-291	LA-TR-65-7 <=$>
1950 V24 N3 P311-320	R.707 <RIS>
1950 V24 N3 P321-336	60-13179 <+*>
1950 V24 N3 P337-344	RT-79 <*>
1950 V24 N4 P385-393	3074 <HB>
1950 V24 N4 P403-408	RJ-645 <ATS>
1950 V24 N4 P445-454	60-13754 <+*>
	63-20508 <=*$>
1950 V24 N5 P513-518	TT.207 <NRC>
1950 V24 N5 P567-573	RT-1609 <*>
1950 V24 N6 P641-649	AEC-TR-4523 <*=>
	64-14661 <=*$> 0
1950 V24 N6 P670-682	04P59R <ATS>
1950 V24 N6 P706-713	RT.203 <*>
	TT.203 <NRC>
1950 V24 N6 P718-720	TT.201 <NRC>
1950 V24 N7 P778-785	60-17378 <+*>

Citation	Translation
	61-11661 <=>
1950 V24 N7 P856-859	R-1499 <*>
1950 V24 N7 P875-880	RT-2830 <*>
	63-18996 <=*$>
1950 V24 N8 P994-1003	RCT V26 N1 P81 <RCT>
1950 V24 N9 P1068-1082	RT-3482 <*>
	TT.230 <NRC>
1950 V24 N9 P1090-1093	TT.227 <NRC>
1950 V24 N10 P1210-1218	RT-730 <*>
1950 V24 N10 P1219-1234	TT.563 <NRC>
1950 V24 N10 P1235-1251	39TM <CTT>
1950 V24 N11 P1299-1301	3980-B <K-H> 0
1950 V24 N11 P1312-1325	82J15R <ATS>
1950 V24 N11 P1361-1370	64-00464 <*>
	64-11828 <=>
1950 V24 N11 P1371-1382	RJ-654 <ATS>
	TT.272 <NRC>
1950 V24 N11 P1407-1408	RT-2060 <*>
1950 V24 N12 P1416-1419	R.704 <RIS>
1950 V24 N12 P1437-1441	61-17233 <=$>
1950 V24 N12 P1502-1510	TT.288 <NRC>
1951 V25 P24-28	60-17724 <+*>
1951 V25 P127-128	60-13725 <+*>
1951 V25 P203-211	60-13871 <+*>
1951 V25 P612-623	60-15848 <+*>
1951 V25 P1181-1185	R-1762 <*>
1951 V25 P1347-1354	R-720 <*>
1951 V25 P1374-1383	R-2830 <*>
1951 V25 P1503-1511	64-16131 <*> 0
1951 V25 N1 P61-70	63-23238 <=*$>
1951 V25 N1 P121-126	TT.242 <NRC>
1951 V25 N2 P143-146	TT.236 <NRC>
1951 V25 N2 P153-160	64-19986 <=$>
1951 V25 N2 P161-169	63-24284 <=*$>
1951 V25 N2 P203-211	TT.233 <NRC>
1951 V25 N2 P221-223	R-1670 <*>
1951 V25 N2 P241-254	TT.241 <NRC>
1951 V25 N3 P261-273	RT-2565 <*>
1951 V25 N3 P274-282	RT-3187 <*>
1951 V25 N3 P312-322	60-41494 <=>
1951 V25 N3 P362-368	TT.243 <NRC>
1951 V25 N4 P398-403	RT-2188 <*>
	64-14138 <=*$> 0
1951 V25 N4 P404-408	R-561 <*>
1951 V25 N5 P523-537	60-41495 <=>
	64-14674 <=*$> 0
1951 V25 N5 P538-541	69K22R <ATS>
1951 V25 N5 P565-576	TT.283 <NRC>
1951 V25 N5 P594-606	TT.292 <NRC>
1951 V25 N5 P612-623	TT.274 <NRC>
1951 V25 N6 P682-687	61-17304 <=$>
1951 V25 N6 P710-718	TT.277 <NRC>
1951 V25 N7 P779-790	62-15160 <*=$>
1951 V25 N7 P814-822	TT.302 <NRC>
1951 V25 N7 P841-853	60-17470 <+*>
	62-23888 <=*>
1951 V25 N7 P863-868	RCT V25 N4 P872 <RCT>
1951 V25 N8 P960-970	3313 <HB>
1951 V25 N8 P981-987	<INSD>
1951 V25 N8 P995-1000	66-61757 <=$>
1951 V25 N9 P1033-1042	61-13638 <*+>
1951 V25 N9 P1103-1110	R-1992 <*>
1951 V25 N10 P1214-1227	RT-3658 <*>
1951 V25 N10 P1228-1238	62-23148 <=*>
1951 V25 N10 P1262-1270	RT-1052 <*>
1951 V25 N11 P1300-1305	62-11358 <=>
1951 V25 N11 P1365-1373	RT-2181 <*>
	64-14143 <=*$> 0
1951 V25 N12 P1460-1468	60-15753 <+*>
1951 V25 N12 P1503-1511	3763-F <K-H>
1952 V26 P39-47	64-16130 <*> P
1952 V26 P106-111	1681 <NLL>
1952 V26 P156-164	R-3770 <*>
1952 V26 P253-269	RJ-206 <ATS>
1952 V26 P509-519	61-18286 <*=>
1952 V26 P560-580	R-3713 <*>
1952 V26 P647-658	60-13638 <+*> 0
1952 V26 P1061-1083	R-4547 <*>
1952 V26 P1184-	R-3764 <*>
1952 V26 P1244-1258	R-570 <*>
1952 V26 P1785-1790	R-2278 <*>
	3763 G <K-H>
	64-16432 <=*$>
1952 V26 P1814-1823	R-4857 <*>
	R-909 <*>
1952 V26 N1 P27-30	62-11362 <=>
	63-18736 <=*>
1952 V26 N1 P60-71	88L28R <ATS>
1952 V26 N1 P106-111	RJ-167 <ATS>
1952 V26 N2 P145-155	TT.322 <NRC>
	61-18701 <*=>
1952 V26 N2 P165-172	62-11361 <=>
1952 V26 N2 P193-200	RT-2578 <*>
1952 V26 N2 P201-210	RT-2579 <*>
1952 V26 N2 P253-269	RJ-206 <ATS>
1952 V26 N2 P310-314	62-14559 <=*>
1952 V26 N3 P358-363	RT-4216 <*>
1952 V26 N3 P377-388	61-17286 <=$>
1952 V26 N3 P425-437	RT-2665 <*>
1952 V26 N4 P492-499	R-1868 <*>
	RJ-117 <ATS>
	RT-2353 <*>
	59-19145 <+*>
1952 V26 N5 P647-658	RJ-163 <ATS>
1952 V26 N5 P672-679	R-4283 <*>
1952 V26 N6 P773-786	63-14677 <=*>
1952 V26 N6 P810-812	52/2938 <NLL>
	61-13430 <+*>
1952 V26 N6 P834-841	R-5085 <*>
1952 V26 N6 P898-902	65-60925 <=$>
1952 V26 N7 P1024-1035	RT-2350 <*>
1952 V26 N7 P1036-1039	TT.421 <NRC>
1952 V26 N8 P1090-1096	59-11767 <=>
1952 V26 N8 P1198-1211	R-299 <*>
1952 V26 N9 P1337-1348	60-15063 <+*>
1952 V26 N9 P1349-1373	RTS-3495 <NLL>
1952 V26 N10 P1463-1471	AEC-TR-4043 <+*> 0
1952 V26 N12 P1804-1807	60-23566 <+*>
1952 V26 N12 P1814-1823	RJ-515 <ATS>
1953 V27 P57-62	RJ-119 <ATS>
1953 V27 P588-595	R-2831 <*>
1953 V27 P1157-1162	61-20989 <=*> 0
1953 V27 P1437-1442	AEC-TR-6003 <*>
1953 V27 N1 P6-25	59-13511 <=> 0
1953 V27 N1 P26-40	64-27256 <$>
1953 V27 N1 P57-62	RJ-119 <ATS>
1953 V27 N1 P100-105	60-23623 <+*>
1953 V27 N1 P106-110	60-23674 <+*>
1953 V27 N1 P125-129	61-13145 <*+>
1953 V27 N2 P157-158	RT-3742 <*>
1953 V27 N2 P233-236	RT-4054 <*>
1953 V27 N3 P321-329	RT-2286 <*>
	64-14144 <=*$> 0
1953 V27 N3 P330-340	TT.486 <NRC>
1953 V27 N3 P341-354	RT-2409 <*>
1953 V27 N3 P355-361	05Q72R <ATS>
1953 V27 N3 P389-393	AD-634 298 <=$>
1953 V27 N3 P394-398	RT-2370 <*>
1953 V27 N4 P491-494	RT-1986 <*>
	775TM <CTT>
1953 V27 N4 P502-504	3172 <HB>
1953 V27 N4 P534-541	TT.527 <NRC>
1953 V27 N4 P554-564	R-4993 <*>
1953 V27 N5 P631-639	RT-2285 <*>
1953 V27 N5 P657-661	61-19948 <=*>
1953 V27 N5 P703-712	R-4716 <*>
	RT-3303 <*>
1953 V27 N6 P830-839	64-14698 <=*$> 0
1953 V27 N6 P889-918	RT-1044 <*>
1953 V27 N7 P1013-1024	TT.520 <NRC>
1953 V27 N8 P1133-1136	RT-3487 <*>
1953 V27 N8 P1163-1171	62-16539 <=*> 0
1953 V27 N8 P1176-1180	RT-3657 <*>
1953 V27 N8 P1208-1212	R-2775 <*>
1953 V27 N9 P1362-1369	RT-2751 <*>
1953 V27 N9 P1430-1436	64-13230 <=*$> 0
1953 V27 N10 P1437-1442	RT-4335 <*>
1953 V27 N10 P1443-1445	RT-4336 <*>
	4486 <HB>
1953 V27 N10 P1545-1555	AEC-TR-4289 <*+>

Citation	Translation
1953 V27 N11 P1617-1621	RT-3141 <*>
1953 V27 N11 P1622-1630	RT-3140 <*>
1953 V27 N11 P1636-1641	RT-2131 <*> 0
1953 V27 N11 P1676-1681	TT.566 <NRC>
1953 V27 N11 P1725-1730	3549 <HB>
1953 V27 N12 P1737-1747	62-20340 <=*> 0
1953 V27 N12 P1808-1815	3300 <HB>
1953 V27 N12 P1827-1836	RJ-223 <ATS>
1953 V27 N12 P1837-1847	63-23310 <*=>
1953 V27 N12 P1887-1890	74S84R <ATS>
1953 V27 N12 P2195-2198	827TM <CTT>
1954 V28 P81-86	R-4221 <*>
1954 V28 P186-190	R-2849 <*>
1954 V28 P459-472	R-4527 <*>
1954 V28 P883-901	R-1432 <*>
1954 V28 P1076-1082	R-1117 <*>
	R-2661 <*>
	R-5147 <*>
1954 V28 P1272-1279	62-10436 <*=>
1954 V28 P1280-1285	62-16634 <=*>
1954 V28 P1523-1527	AEC-TR-4116 <+*>
	64-10690 <=*$>
1954 V28 P2042-2046	R-2147 <*>
1954 V28 P2107-2115	R-2848 <*>
1954 V28 P2195-2198	RJ-256 <ATS>
1954 V28 N1 P30-35	61-17269 <=$>
1954 V28 N1 P51-59	65-60347 <=$>
1954 V28 N1 P60-66	RT-3392 <*>
1954 V28 N1 P127-134	64-10910 <=*$>
1954 V28 N2 P240-253	RT-3314 <*>
1954 V28 N2 P299-304	RT-2571 <*>
	64-14400 <=*$> 0
1954 V28 N2 P317-327	20K27R <ATS>
1954 V28 N3 P414-421	RT-3181 <*>
1954 V28 N3 P422-432	59-13535 <=> 0
1954 V28 N3 P433-434	63-19883 <*=>
1954 V28 N3 P435-439	R-2000 <*>
1954 V28 N3 P440-452	3789 <BISI>
1954 V28 N3 P473-489	61-19[illegible]2 <=*>
1954 V28 N3 P490-497	60-23017 <+*>
1954 V28 N3 P507-518	RT-2248 <*>
	64-15562 <=*$> P
1954 V28 N3 P552-553	28M41R <ATS>
1954 V28 N4 P645-649	C-2447 <NRC>
	R-2057 <*>
1954 V28 N4 P688-691	RT-4602 <*>
	489 <NLL>
1954 V28 N4 P697-699	60-16834 <+*>
1954 V28 N5 P772-779	59-15232 <+*>
1954 V28 N5 P785-796	R-916 <*>
	RJ-312 <ATS>
	RT-2331 <*>
1954 V28 N5 P797-800	61-13561 <*+>
1954 V28 N5 P824-836	61-27187 <*=>
1954 V28 N5 P856-864	3121 <NRCC>
1954 V28 N6 P1017-1024	59-13509 <=>
1954 V28 N6 P1095-1097	65-13125 <*>
1954 V28 N7 P1169-1173	AEC-TR-2236 <*>
	RT-3263 <*>
1954 V28 N7 P1174-1185	TT.586 <NRC>
	63-14333 <=*>
1954 V28 N7 P1235-1242	61-10529 <*+>
1954 V28 N7 P1243-1256	60-11628 <=>
1954 V28 N7 P1272-1279	R-1802 <*>
1954 V28 N7 P1280-1285	R-1788 <*>
1954 V28 N7 P1286-1291	AD-634 598 <=$>
1954 V28 N8 P1361-1370	TT.587 <NRC>
1954 V28 N8 P1386-1394	R-3931 <*>
	2954 <NLL>
	59-22593 <+*>
1954 V28 N8 P1395-1398	R-4598 <*>
	2953 <NLL>
	59-22166 <+*>
1954 V28 N8 P1439-1450	R-4579 <*>
	2913 <NLL>
1954 V28 N8 P1458-1464	97N54R <ATS>
1954 V28 N8 P1489-1496	59-10064 <+*>
1954 V28 N8 P1497-1506	59-10066 <+*>
1954 V28 N9 P1555-1561	RT-3940 <*>
1954 V28 N9 P1599-1614	59-10737 <+*>
1954 V28 N9 P1615-1622	59-17551 <*> 0
1954 V28 N9 P1623-1627	RT-4545 <*>
1954 V28 N9 P1676-1677	59-15035 <+*> C
1954 V28 N9 P1678-1679	RT-4546 <*>
	68S84R <ATS>
1954 V28 N10 P1713-1719	39M42R <ATS>
	65-10057 <*> 0
1954 V28 N10 P1755-1764	59-13536 <=> C
1954 V28 N10 P1770-1773	RT-4236 <*>
1954 V28 N10 P1804-1811	16N54R <ATS>
1954 V28 N10 P1825-1830	63-19040 <=*>
1954 V28 N10 P1837-1844	59-13835 <=> 0
1954 V28 N10 P1845-1853	RT-4491 <*>
	491TT <CCT>
1954 V28 N10 P1869-1871	64-20806 <*>
1954 V28 N11 P1878-1881	R-2098 <*>
	RJ-267 <ATS>
	63-14332 <=*>
1954 V28 N11 P1882-1888	C-2265 <NRC>
	R-434 <*>
1954 V28 N11 P1935-1949	64-14935 <=*$>
1954 V28 N11 P1987-1998	RT-3727 <*>
	5114-A <K-H>
	64-16463 <=*$> 0
1954 V28 N11 P2047-2066	63-23087 <=*>
1954 V28 N11 P2067-2080	R-3237 <*>
1954 V28 N12 P2088-2094	<TCT>
1954 V28 N12 P2095-2106	<TCT>
	RT-3490 <*>
1954 V28 N12 P2116-2128	R-3247 <*>
	RT-3445 <*>
1954 V28 N12 P2137-2141	<TCT>
1954 V28 N12 P2170-2177	59-10069 <+*>
1954 V28 N12 P2195-2198	RT-2931 <*>
1954 V28 N12 P2222-2231	56H12R <ATS>
1954 V28 N12 P2232-2233	NP-TR-552 <*>
	63-23302 <=*>
1954 V28 N12 P2234-2252	R-3195 <*>
1954 V28 N12 P2255-2257	27M41R <ATS>
1955 N9 P1684-1695	R-5252 <*>
1955 V29 P11-19	R-1556 <*>
1955 V29 P39-47	R-2148 <*>
1955 V29 P111-119	R-2136 <*>
1955 V29 P463-468	R-2135 <*>
1955 V29 P629-634	R-971 <*>
1955 V29 P745-749	R-2142 <*>
1955 V29 P959-974	R-4408 <*>
1955 V29 P1278-	RCT V30 N1 P61 <RCT>
1955 V29 P1486-1498	R-2529 <*>
1955 V29 P1549-1554	AEC-TR-3856 <+*>
1955 V29 P1582-1590	T-1620 <INSD>
1955 V29 P1668-1677	242TM <CTT>
1955 V29 P1725-1727	T-1619 <INSD>
1955 V29 P1746-	R-736 <*>
1955 V29 N1 P3-14	65-60919 <=$>
1955 V29 N1 P28-38	AEC-TR-2217 <*>
	RT-3121 <*>
	3642 <K-H>
1955 V29 N1 P51-61	RT-3122 <*>
	64-20562 <*>
1955 V29 N1 P76-83	64-18407 <*>
1955 V29 N2 P244-249	59-20834 <+*>
1955 V29 N2 P272-279	3570 <HB>
1955 V29 N2 P291-304	64-16387 <=*$>
1955 V29 N2 P327-344	RT-3905 <*>
1955 V29 N2 P380-389	66-11067 <*>
1955 V29 N2 P390-391	RT-3907 <*>
1955 V29 N3 P424-427	65K20R <ATS>
1955 V29 N3 P435-449	RJ-449 <ATS>
1955 V29 N3 P469-475	27J14R <ATS>
	61-20706 <*=> 0
	64-16299 <=*$>
1955 V29 N3 P485-495	R-5108 <*>
1955 V29 N3 P496-501	RT-4406 <*>
1955 V29 N3 P508-512	64-14120 <=*$> C
1955 V29 N3 P518-522	63-15912 <=*>
1955 V29 N3 P523-532	C-2137 <NLL>
	RT-4104 <*>
	61-17384 <=$>
1955 V29 N3 P539-541	R-4990 <*>

Citation	Translation
1955 V29 N3 P576-583	R-1642 <*>
1955 V29 N4 P597-601	3642 <HB>
1955 V29 N4 P668-670	R-3815 <*>
1955 V29 N4 P677-681	<INSD>
	62-11232 <=>
1955 V29 N4 P692-698	RT-3385 <*>
1955 V29 N4 P718-722	70K22R <ATS>
1955 V29 N4 P750-751	60-31111 <=>
1955 V29 N5 P791-796	64-14936 <=*$> 0
1955 V29 N5 P839-845	3997 <HB>
1955 V29 N5 P855-859	65-61404 <=>
1955 V29 N5 P892-897	60-11626 <=>
1955 V29 N5 P898-903	RT-3428 <*>
	69J18R <ATS>
1955 V29 N6 P959-974	SCL-T-269 <+*>
1955 V29 N6 P980-982	348 <TC>
1955 V29 N6 P996-1000	AD-611 332 <=$>
1955 V29 N6 P1010-1019	18N49R <ATS>
1955 V29 N6 P1073-1079	RT-3400 <*>
1955 V29 N6 P1080-1086	64-16561 <=*$>
1955 V29 N7 P1152-1161	72L36R <ATS>
1955 V29 N7 P1173-1180	61-15386 <*+>
1955 V29 N7 P1201-1203	SCL-T-437 <=*>
	62-18212 <=*>
	65-12295 <*>
1955 V29 N7 P1240-1247	59-10005 <+*>
1955 V29 N7 P1311-1317	64-18579 <*> 0
1955 V29 N8 P1372-1382	AEC-TR-4359 <=*>
1955 V29 N8 P1383-1395	J19R <ATS>
1955 V29 N8 P1508-1512	RT-4024 <*>
1955 V29 N9 P1537-1548	AEC-TR-4361 <+*>
1955 V29 N9 P1569-1581	R-3179 <*>
	64-19518 <=$> 0
1955 V29 N10 P1746-1750	RT-4034 <*>
1955 V29 N10 P1755-1770	RT-4279 <*>
1955 V29 N10 P1784-1803	59-15060 <+*>
1955 V29 N1C P1815-1821	R-3834 <*>
	RT-4660 <*>
1955 V29 N10 P1827-1846	61-19700 <=*>
1955 V29 N1C P1854-1865	48R77R <ATS>
1955 V29 N10 P1921-1923	RJ-393 <ATS>
1955 V29 N11 P1937-1941	25K26R <ATS>
1955 V29 N11 P1958-1973	30K24R <ATS>
1955 V29 N11 P1974-1983	C-2588 <NRC>
	R-3671 <*>
1955 V29 N11 P2048-2053	RJ-570 <ATS>
1955 V29 N11 P2054-2073	59-15705 <+*>
1955 V29 N11 P2076-2080	R-3639 <*>
1955 V29 N11 P2086-2089	RJ-514 <ATS>
1955 V29 N12 P2173-2184	61-13346 <+*>
1956 N1 P50-55	RJ-648 <ATS>
1956 N5 P1007-1018	R-4237 <*>
1956 V30 P28-33	R-2150 <*>
1956 V30 P61-68	R-1022 <*>
1956 V30 P241-	AEC-TR-4119 <+*>
1956 V30 P277-285	AEC-TR-2531 <*>
1956 V30 P500-507	R-2145 <*>
1956 V30 P593-6C7	R-2143 <*>
1956 V30 P896-900	R-2239 <*>
1956 V30 P901-911	R-2833 <*>
1956 V30 P928-933	C-2731 <NRC>
1956 V30 P968-985	R-1370 <*>
1956 V30 P1C19	C-2727 <NRC>
1956 V30 P1104-1117	R-2144 <*>
1956 V30 P1228-1237	R-2258 <*>
1956 V30 P1455	59-12883 <+*>
1956 V30 P1537-1539	R-960 <*>
1956 V30 P1626-1635	RCT V32 N1 P77 <RCT>
1956 V30 P1652-1661	R-2025 <*>
1956 V30 P1672-1675	R-3215 <*>
	R-498 <*>
1956 V30 P1727-1728	61-20736 <*=>
1956 V30 P1830-1839	R-3664 <*>
	59-19070 <+*>
1956 V30 P2051-2056	R-3838 <*>
1956 V30 P2C57-2060	62-11652 <=>
1956 V30 P2373-2383	R-2140 <*>
1956 V30 P2387-2398	R-3306 <*>
1956 V30 P2441-2443	R-2139 <*>
1956 V30 P2568-2579	R-2138 <*>
1956 V30 P2580-2581	274TT <CCT>
1956 V30 N1 P1-19	27M38R <ATS>
1956 V30 N1 P23-27	16K25R <ATS>
1956 V30 N1 P50-55	R-394 <*>
1956 V30 N1 P69-75	62-25562 <=*>
1956 V30 N1 P76-78	RJ-522 <ATS>
1956 V30 N1 P161-171	61-13408 <+*>
1956 V30 N1 P172-176	C-2527 <NRC>
	R-3107 <*>
1956 V30 N1 P190-195	RJ-617 <ATS>
1956 V30 N1 P223-227	59-15230 <+*>
1956 V30 N2 P251-260	PB 141 456T <=>
1956 V30 N2 P277-285	RT-4351 <*>
1956 V30 N2 P321-328	62-18866 <=*>
1956 V30 N2 P379-381	62-20149 <=*>
1956 V30 N2 P383-390	RJ-510 <ATS>
1956 V30 N2 P417-423	RJ-509 <ATS>
1956 V30 N2 P470-473	TT-808 <NRCC>
1956 V30 N3 P492-499	60-11687 <=>
1956 V30 N3 P515-521	61-20966 <*=>
	76M44R <ATS>
1956 V30 N3 P556-565	59-10072 <+*>
1956 V30 N3 P581-588	R-3676 <*>
1956 V30 N3 P589-592	62-20153 <=*>
1956 V30 N3 P639-645	R-110 <*>
1956 V30 N3 P676-684	R-2680 <*>
1956 V30 N3 P700-703	56J19R <ATS>
1956 V30 N3 P704-705	57J19R <ATS>
1956 V30 N4 P763-768	64-16306 <=*$>
1956 V30 N4 P769-783	62-25390 <=*>
1956 V30 N4 P784-788	CSIRO-3395 <CSIR>
1956 V30 N4 P794-797	22J14R <ATS>
	66-12769 <*>
1956 V30 N4 P798-810	23J14R <ATS>
	61-15637 <+*>
1956 V30 N4 P811-820	R-403 <*>
	63-14321 <=*>
1956 V30 N4 P847-855	69J17R <ATS>
1956 V30 N4 P856-861	75J16R <ATS>
1956 V30 N4 P862-870	64-30039 <*>
1956 V30 N4 P871-877	61-20967 <*=>
	77M44R <ATS>
1956 V30 N4 P878-881	61-20968 <*=>
	78M44R <ATS>
1956 V30 N4 P882-888	R-1975 <*>
	4089 <HB>
1956 V30 N4 P912-921	64-16324 <*>
1956 V30 N4 P928-933	R-4748 <*>
1956 V30 N4 P934-936	64-16300 <=*$>
1956 V30 N5 P986-994	C-2562 <NRC>
1956 V30 N5 P995-1006	59-16983 <+*>
1956 V30 N5 P1C19-1027	62-20383 <=*>
1956 V30 N5 P1044-1047	C-2568 <NRC>
1956 V30 N5 P1C75-1081	13H13R <ATS>
1956 V30 N5 P1118-1125	R-1759 <*>
1956 V30 N5 P1140-1144	R-1758 <*>
1956 V30 N5 P1179-1181	68K22R <ATS>
1956 V30 N5 P1189-1190	61L31R <ATS>
1956 V30 N6 P1202-1206	14H13R <ATS>
1956 V30 N6 P1207-1222	69J19R <ATS>
1956 V30 N6 P1258-1266	61-28160 <*=>
1956 V30 N6 P1291-1296	C-2532 <NRC>
	R-3106 <*>
1956 V30 N6 P1327-1342	R-940 <*>
1956 V30 N6 P1349-1355	R-1116 <*>
	63-20484 <*>
1956 V30 N6 P1356-1366	60-31107 <=>
1956 V30 N6 P1376-1379	6C-31106 <=>
1956 V30 N6 P1416-1418	62-10191 <*=> 0
1956 V30 N6 P1425-1426	284TM <CTT>
1956 V30 N7 P1433-1437	R-712 <*>
1956 V30 N7 P1469-1472	61-28914 <*=>
	62-11628 <=>
1956 V30 N7 P1529-1536	RCT V31 N1 P49 <RCT>
1956 V30 N7 P1615-1622	65-12789 <*>
1956 V30 N7 P1662-1667	60-11625 <=>
1956 V30 N8 P1696-1701	R-983 <*>
1956 V30 N8 P1746-1751	RCT V31 N2 P244 <RCT>
1956 V30 N8 P1801-1806	25H13R <ATS>
1956 V30 N8 P1840-1851	62-16188 <=*>

Citation	Translation
	98N50R <ATS>
1956 V30 N8 P1860-1866	41J18R <ATS>
	64-16890 <=*$>
1956 V30 N8 P1871-1876	60-11457 <=>
1956 V30 N8 P1885-1888	R-2016 <*>
1956 V30 N8 P1896-1899	59-17598 <+*>
1956 V30 N8 P1901-1902	59-12149 <+*>
1956 V30 N8 P1904-1905	CSIRO-3441 <CSIR>
1956 V30 N9 P1921-1931	59-18522 <+*>
	62-10592 <*=>
1956 V30 N9 P1941-1947	RCT V31 N2 P250 <RCT>
1956 V30 N9 P1948-1958	R-4741 <*>
	20P60R <ATS>
	565TM <CTT>
	60-17712 <+*>
1956 V30 N9 P1983-1989	233TT <CCT>
1956 V30 N9 P1990-2002	AEC-TR-4350 <*+>
1956 V30 N9 P2C57-2C60	R-1434 <*>
1956 V30 N9 P2070-2076	60-11453 <=>
1956 V30 N9 P2085-2089	60-11497 <=>
1956 V30 N9 P2106-2121	61-28193 <*=>
1956 V30 N9 P2126	54T94R <ATS>
1956 V30 N1C P2144-2148	RCT V31 N2 P257 <RCT>
1956 V30 N10 P2251-	AEC-TR-4862 <*=>
1956 V30 N10 P2327-2336	SNSRC-152 <SNSR>
	18K23R <ATS>
1956 V30 N10 P2337-2339	50J17R <ATS>
1956 V30 N1C P2340-2347	R-1632 <*>
1956 V30 N1C P2348-2352	60-11520 <=>
1956 V30 N10 P2353-2355	42J8R <ATS>
1956 V30 N11 P2373-2383	61-13367 <*+>
1956 V30 N11 P2387-2398	C-3325 <NRCC>
	R-4750 <*>
1956 V30 N11 P2435-2440	59-19614 <+*>
1956 V30 N11 P2489-2498	63-19257 <=*>
	65-60918 <=$>
1956 V30 N11 P2499-2510	SNSRC-153 <SNSR>
1956 V30 N11 P2499-2509	31J19R <ATS>
1956 V30 N11 P2519-2524	60K20R <ATS>
1956 V30 N11 P2539-2546	79J14R <ATS>
1956 V30 N11 P2555-2559	80J14R <ATS>
1956 V30 N11 P2560-2567	28K20R <ATS>
1956 V30 N11 P2580-2581	81J14R <ATS>
1956 V30 N12 P2621-2639	R-4737 <*>
	R-5295 <*>
	612TM <CTT>
1956 V30 N12 P2649-2655	62-16633 <=*> 0
1956 V30 N12 P2656-2661	59-19615 <+*>
1956 V30 N12 P2691-2704	64-14989 <=*$>
1956 V30 N12 P2713-2723	63-19257 <=*>
	64-19998 <=$>
1956 V30 N12 P2724-2733	59-21020 <=>
1956 V30 N12 P2812	62-25298 <=*>
1956 V30 N12 P2813-2819	RTS-3009 <NLL>
1957 N1 P61-71	36K20R <ATS>
1957 N1 P109-116	93J16R <ATS>
1957 V31 P232-237	240TM <CTT>
	8024-A <K-H>
1957 V31 P350-353	R-4222 <*>
1957 V31 P354-361	ORNL-TR-541 <=*$>
1957 V31 P515	R-4805 <*>
1957 V31 P519-521	C-2571 <NRC>
1957 V31 P704-707	R-2674 <*>
1957 V31 P1C13-1018	R-4215 <*>
1957 V31 P1377-1386	R-4218 <*>
1957 V31 P1445-1447	AEC-TR-4442 <=*>
1957 V31 P1593-1599	R-3125 <*>
1957 V31 P1650-1653	AEC-TR-3310 <*>
1957 V31 P1721-1731	R-5081 <*>
1957 V31 P2381-2390	R-5109 <*>
1957 V31 N1 P3-26	C-2448 <NRC>
1957 V31 N1 P42-48	70L35R <ATS>
1957 V31 N1 P72-82	64-19936 <=$>
1957 V31 N1 P83-92	63-11097 <=>
1957 V31 N1 P93-99	59-15304 <+*>
	68J17R <ATS>
1957 V31 N1 P117-127	AEC-TR-5820 <=*$>
1957 V31 N1 P157-164	C-2457 <NRC>
1957 V31 N1 P232-237	64-30096 <*> 0
1957 V31 N1 P250-262	61-28189 <*=>
1957 V31 N2 P283-290	R-4549 <*>
1957 V31 N2 P283-291	59-10870 <+*>
1957 V31 N2 P321-327	60-13309 <+*>
	63J16R <ATS>
1957 V31 N2 P328-339	65-60361 <=$>
1957 V31 N2 P340-349	SNSRC-154 <SNSR>
	96J17R <ATS>
1957 V31 N2 P387-394	65-10326 <*> 0
1957 V31 N2 P395-402	71L35R <ATS>
1957 V31 N2 P444-451	R-5039 <*>
1957 V31 N2 P467-473	64-30093 <*> 0
	8024-B <K-H> C
1957 V31 N2 P481-484	61-13323 <*+>
1957 V31 N2 P493-497	64-30126 <*> C
1957 V31 N2 P514	59-11234 <=>
1957 V31 N2 P516-518	C-2458 <NRC>
1957 V31 N2 P519-521	37K20R <ATS>
1957 V31 N3 P537-553	94K21R <ATS>
1957 V31 N3 P554-562	SNSRC-155 <SNSR>
	64-18410 <*>
1957 V31 N3 P573-581	59-10411 <+*>
1957 V31 N3 P599-611	06K21R <ATS>
	59-20150 <+*> PO
1957 V31 N3 P612-625	5J17R <ATS>
1957 V31 N3 P626-64C	61-15737 <+*>
1957 V31 N3 P641-647	R-4896 <*>
1957 V31 N3 P692-698	60-17703 <+*>
1957 V31 N3 P7C8-711	6J17R <ATS>
	60-17902 <+*>
1957 V31 N3 P724-725	7J17R <*>
1957 V31 N3 P726-731	43K20R <ATS>
1957 V31 N3 P732-733	C-2725 <NRC>
	R-4747 <*>
1957 V31 N4 P745-769	R-4749 <*>
	47K24R <ATS>
1957 V31 N4 P780-791	59-13830 <=> 0
1957 V31 N4 P802-807	64-30101 <*> 0
	8024-C <K-H> 0
1957 V31 N4 P836-841	33J17R <ATS>
1957 V31 N4 P851-864	61-27302 <=*>
1957 V31 N4 P865-873	35J17R <ATS>
1957 V31 N4 P874-881	60-15579 <+*>
1957 V31 N4 P9C4-910	36J17R <ATS>
1957 V31 N4 P911-914	60-11410 <=>
1957 V31 N5 P937-951	09L30R <ATS>
1957 V31 N5 P952-959	21P60R <ATS>
1957 V31 N5 P952-958	60-17712 <+*>
1957 V31 N5 P1C19-1026	59-22585 <+*>
1957 V31 N5 P1027-1032	62-25563 <=*>
1957 V31 N5 P1033-1041	10J19R <ATS>
1957 V31 N5 P1C42-1C55	56J18R <ATS>
1957 V31 N5 P1063-1071	R-5317 <*>
	64-30103 <*>
1957 V31 N5 P1072-1090	18K21R <ATS>
1957 V31 N6 P1227-1234	60-23668 <*+>
1957 V31 N6 P1248-1255	6C-17156 <+*>
1957 V31 N6 P1266-1275	41K20R <ATS>
	59-22412 <+*>
1957 V31 N6 P1305-1310	62-25381 <=*> P
1957 V31 N6 P1337-1343	28K23R <ATS>
1957 V31 N6 P1352-1359	6C-11608 <=>
1957 V31 N6 P1377-1386	61-17633 <=$>
1957 V31 N6 P1426-1428	59-17267 <+*>
1957 V31 N7 P1459-1467	65-11847 <*>
1957 V31 N7 P1473-1484	53K20R <ATS>
1957 V31 N7 P1492-1500	59-15484 <+*>
1957 V31 N7 P1532-1541	AEC-TR-3572 <+*> 0
1957 V31 N7 P16CC-1605	62-25564 <=*>
1957 V31 N7 P1619-1626	60K23R <ATS>
1957 V31 N7 P1650-1653	60-15699 <+*>
1957 V31 N7 P1667-1668	85K20R <ATS>
1957 V31 N8 P1687-1692	M-5709 <NLL>
1957 V31 N8 P1721-1731	R-5061 <*>
	61-19204 <+*>
1957 V31 N8 P1756-1761	62-18531 <=*>
1957 V31 N8 P1757-1762	R-3126 <*>
1957 V31 N8 P1812-1819	59 <MT>
1957 V31 N8 P1820-1824	TRANS-60 <MT>
1957 V31 N8 P1839-1842	6C-11529 <=>
1957 V31 N8 P1843-1850	TT-788 <NRCC>

Citation	Translation
1957 V31 N8 P1843-1849	59-11873 <=>
1957 V31 N8 P1843-1850	60-17161 <+*>
1957 V31 N8 P1851-1860	60-11512 <=>
1957 V31 N9 P1917-1925	65-11137 <*>
1957 V31 N9 P2005-2011	R-5302 <*>
1957 V31 N9 P2032-2035	64-30129 <*> 0
	8024-D <K-H>
1957 V31 N9 P2036-2041	59-15019 <+*> 0
1957 V31 N9 P2074-2077	60-27004 <+*>
1957 V31 N9 P2127-2137	60-10584 <+*>
1957 V31 N9 P2138-2139	88L33R <ATS>
1957 V31 N10 P2162-2183	19K25R <ATS>
	59-10305 <+*>
1957 V31 N10 P2213-2223	22K21R <ATS>
1957 V31 N10 P2224-2228	60-11514 <=>
1957 V31 N10 P2269-2277	61-10271 <*+> 0
1957 V31 N10 P2281-2286	59-19566 <+*>
1957 V31 N10 P2281-2287	61-20969 <*=>
	79M44R <ATS>
1957 V31 N10 P2295-2301	AEC-TR-5663 <*=>
1957 V31 N10 P2306-2309	UCRL TRANS-708 <=*>
1957 V31 N10 P2322-2327	60-17711 <+*>
1957 V31 N10 P2328-2335	60-23081 <+*>
1957 V31 N10 P2375-2376	60-21590 <=>
	66-12298 <*>
1957 V31 N11 P2391-2399	20K25R <ATS>
1957 V31 N11 P2428-2434	58M47R <ATS>
1957 V31 N11 P2438-2444	61-13498 <*+>
1957 V31 N11 P2522-2525	56K24R <ATS>
1957 V31 N11 P2562-2570	14K25R <ATS>
1957 V31 N11 P2575-2576	R-3705 <*>
1957 V31 N11 P2580-2583	AEC-TR-3539 <+*>
1957 V31 N12 P2649-2655	59-19320 <+*>
1957 V31 N12 P2682-2689	61-20970 <*=>
	80M44R <ATS>
1957 V31 N12 P2690-2696	59-17093 <+*> 0
1957 V31 N12 P2720-2724	61-28161 <*=>
1958 V32 P209-218	R-5250 <*>
1958 V32 P460-464	UCRL TRANS-1022(L) <=*$>
1958 V32 P819-823	AEC-TR-3680 <*>
1958 V32 P1016-1021	65-10087 <*> 0
1958 V32 P1149-1152	R-5251 <*>
1958 V32 N1 P62-65	R-4478 <*>
1958 V32 N1 P79-85	R-5312 <*>
1958 V32 N1 P87-92	65-14277 <*>
1958 V32 N1 P164-169	59-22255 <+*>
1958 V32 N1 P178-187	59L36R <ATS>
1958 V32 N1 P194-195	R-4429 <*>
1958 V32 N2 P209-218	59-15305 <+*>
	59-18411 <+*>
	93K26R <ATS>
1958 V32 N2 P219-223	59-18949 <+*>
1958 V32 N2 P262-269	62-23790 <=*>
1958 V32 N2 P274-281	60L36R <ATS>
1958 V32 N2 P291-297	59-18950 <+*>
1958 V32 N2 P332-339	3171 <NRCC>
1958 V32 N2 P413-417	R-5349 <*>
1958 V32 N2 P433-441	60-16898 <*+>
1958 V32 N2 P454-459	AEC-TR-3679 <+*>
1958 V32 N3 P506-511	60-13459 <+*>
1958 V32 N3 P620-627	61-15640 <*+>
1958 V32 N3 P640-643	62-32125 <=*>
1958 V32 N3 P670-677	60-13780 <+*>
1958 V32 N3 P670-678	60-17379 <+*>
1958 V32 N3 P697-698	61-13115 <+*>
1958 V32 N3 P717	59-10261 <+*>
1958 V32 N4 P737-745	60-14116 <+*>
1958 V32 N4 P762-768	TRANS-61 <MT>
	61-20971 <*=>
1958 V32 N4 P762-767	81M44R <ATS>
1958 V32 N4 P785-790	TRANS-62 <MT>
	34L33R <ATS>
1958 V32 N4 P797-805	TT-814 <NRCC>
1958 V32 N4 P806-810	59-22164 <+*>
1958 V32 N4 P819-824	60-13448 <+*>
1958 V32 N4 P831-833	29L33R <ATS>
1958 V32 N4 P841-847	62-14033 <=*>
	77N48R <ATS>
1958 V32 N4 P856-863	60-14115 <+*> 0
1958 V32 N4 P864-868	60-31171 <=>
1958 V32 N4 P869-874	60-11515 <=>
1958 V32 N4 P882-893	87L33R <ATS>
1958 V32 N4 P930-931	60-18123 <+*>
1958 V32 N5 P969-980	59-17412 <+*>
1958 V32 N5 P981-985	24K26R <ATS>
1958 V32 N5 P1008-1015	14M42R <ATS>
	59-17256 <+*>
	61-10122 <*+>
1958 V32 N5 P1043-1048	72K26R <ATS>
1958 V32 N5 P1049-1054	59-20290 <+*> 0
	64L31R <ATS>
1958 V32 N5 P1055-1058	29L35R <ATS>
1958 V32 N5 P1077-1080	61-19179 <+*> 0
1958 V32 N5 P1095-1102	60-17730 <+*>
	63-10235 <=*>
1958 V32 N5 P1103-1106	59-14306 <+*>
1958 V32 N5 P1107-1115	61-19179 <+*> 0
1958 V32 N5 P1122-1130	61-19179 <+*> 0
1958 V32 N5 P1153-1154	05L31R <ATS>
1958 V32 N5 P1160-1162	59-13213 <=>
1958 V32 N5 P1175-1176	TRANS-525(L) <=*>
1958 V32 N5 P1182-1183	60-41627 <=>
1958 V32 N6 P1211-1213	60-17789 <+*>
1958 V32 N6 P1214-1217	SCL-TR-234 <+*> 0
1958 V32 N6 P1218-1225	SNSRC-156 <SNSR>
	28S83R <ATS>
1958 V32 N6 P1269-1276	SNSRC-157 <SNSR>
1958 V32 N6 P1313-1318	61-23271 <*=>
1958 V32 N6 P1393-1403	AEC-TR-3667 <+*>
1958 V32 N6 P1420	96L31R <ATS>
1958 V32 N6 P1423-1424	60-13838 <+*>
1958 V32 N7 P1476-1480	59-10410 <+*>
	60-14060 <+*>
1958 V32 N7 P1532-1535	AEC-TR-3601 <+*>
1958 V32 N7 P1565-1575	60-18784 <+*>
	65-11632 <*>
1958 V32 N7 P1591-1596	60-10369 <+*>
1958 V32 N7 P1618-1621	63-19591 <=*$>
1958 V32 N7 P1622-1631	AEC-TR-4856 <=*>
1958 V32 N7 P1648-1657	60-17725 <+*>
1958 V32 N7 P1671-1672	AEC-TR-3955 <+*>
	61-28168 <*=>
1958 V32 N8 P1715-1725	60-23042 <*+>
1958 V32 N8 P1763-1773	60-17719 <+*>
1958 V32 N8 P1779-1783	57K28R <ATS>
1958 V32 N8 P1785-1795	AEC-TR-3658 <+*>
1958 V32 N8 P1831-1841	45L33R <ATS>
	60-10583 <+*>
1958 V32 N8 P1842-1846	63M41R <ATS>
1958 V32 N8 P1851-1858	60-11768 <=>
1958 V32 N8 P1859-1862	59-15788 <+*>
1958 V32 N8 P1869-	59-11928 <=>
1958 V32 N9 P224-	59-11925 <=>
1958 V32 N9 P1967-1970	59-15004 <+*>
1958 V32 N9 P2035-2041	00P65R <ATS>
1958 V32 N9 P2068-2072	59-21207 <=>
1958 V32 N9 P2080-2086	60-14057 <+*>
1958 V32 N9 P2155-2164	62-23069 <=>
1958 V32 N9 P2174-2182	AEC-TR-4416 <=*>
1958 V32 N9 P2187-2191	88M37R <ATS>
1958 V32 N9 P2221-2223	59-15003 <+*>
1958 V32 N10 P2250-2255	59-10445 <+*>
1958 V32 N10 P2266-2273	62-11551 <=>
1958 V32 N10 P2282-2286	59-21208 <=>
1958 V32 N10 P2308-2314	03L29R <ATS>
1958 V32 N10 P2315-2323	18L33R <ATS>
	60-21928 <=>
1958 V32 N10 P2324-2332	63-23246 <=*>
1958 V32 N10 P2339-2346	19L33R <ATS>
	59-10444 <+*>
1958 V32 N10 P2362-2373	63-10216 <=*>
1958 V32 N10 P2383-2391	62-23152 <=>
1958 V32 N10 P2415-2417	60-21788 <=>
1958 V32 N10 P2418-2423	60-21784 <=>
1958 V32 N10 P2424-2429	2191 <BISI>
	55L31R <ATS>
1958 V32 N11 P2483-2486	61-23288 <*=>
1958 V32 N11 P2492-2499	61-20705 <*=>
1958 V32 N11 P2543-2549	52N51R <ATS>
	63-18418 <=*>

Citation	Translation
1958 V32 N11 P2565-2570	59-19457 <+*>
1958 V32 N11 P2580-2585	60-13082 <+*> 0
1958 V32 N11 P2641-2643	60-11521 <=>
1958 V32 N12 P2663-2673	1221 <TC>
1958 V32 N12 P2679-2685	61-13171 <*+> 0
1958 V32 N12 P2686-2689	59-21010 <=>
1958 V32 N12 P2711-2716	63-15923 <=*>
1958 V32 N12 P2717-2724	59-18543 <+*>
1958 V32 N12 P2731-2738	08M42R <ATS>
1958 V32 N12 P2739-2747	63-14330 <=*>
1958 V32 N12 P2767-2771	26M38R <ATS>
1958 V32 N12 P2780-2786	60-13435 <+*>
1958 V32 N12 P2792-2796	11L31R <ATS>
	60-15733 <+*>
1958 V32 N12 P2797-2802	63-20684 <=*$>
1958 V32 N12 P2810-2816	61-19731 <*=> 0
1958 V32 N12 P2817-2823	61-15173 <*+>
1958 V32 N12 P2847	09M42R <ATS>
1959 V33 P793-798	RCT V33 N4 P953 <RCT>
1959 V33 P1401-	RCT V34 N1 P334 <RCT>
1959 V33 N1 P8-14	C-3399 <NRCC>
1959 V33 N1 P50-56	60-16594 <+*> 0
1959 V33 N1 P50-57	60-21777 <=>
1959 V33 N1 P58-64	60-11985 <=>
1959 V33 N1 P74-77	4-126 <RIS>
1959 V33 N1 P96-99	11L36R <ATS>
	62-10257 <*=>
1959 V33 N1 P129-136	61-19693 <=*>
1959 V33 N1 P156-160	AEC-TR-4298 <+*>
	79N49R <ATS>
1959 V33 N1 P165-173	60-16595 <+*> 0
1959 V33 N1 P197-203	30L33R <ATS>
1959 V33 N2 P262-270	53L33R <ATS>
	61-13487 <*+>
1959 V33 N2 P310-317	60-23033 <+*>
1959 V33 N2 P326-327	76L33R <ATS>
1959 V33 N2 P367-373	61-19672 <=*>
1959 V33 N2 P405-410	60-16906 <*+>
1959 V33 N2 P411-415	61-19544 <=*>
1959 V33 N2 P437-440	60-21780 <=>
1959 V33 N2 P441-446	59-16138 <+*>
1959 V33 N2 P452-455	60-15782 <+*> 0
1959 V33 N2 P480-486	RCT V36 N2 P473 <RCT>
	RUB.CH.TECH.V36 N2 <ACS>
1959 V33 N2 P503	65-10881 <*>
1959 V33 N3 P550-553	60-17726 <+*>
1959 V33 N3 P572-580	AD-613 450 <=$>
1959 V33 N3 P610-616	60-23812 <+*>
1959 V33 N3 P656-661	RCT V33 N4 P946 <RCT>
	RCT V36 N2 P480 <RCT>
	RUB.CH.TECH.V36 N2 <ACS>
1959 V33 N3 P662-664	60-41626 <=>
1959 V33 N3 P700-709	60-11839 <=>
	62-16023 <=*>
1959 V33 N3 P728-731	00L32R <ATS>
1959 V33 N3 P734-735	60-11599 <=>
1959 V33 N3 P736-737	62-14525 <=*>
	62-25896 <=*>
1959 V33 N4 P771-779	61-13493 <*+>
1959 V33 N4 P799-807	62-33409 <=*>
1959 V33 N4 P822-827	65L33R <ATS>
1959 V33 N4 P835-839	12N49R <ATS>
1959 V33 N4 P844-851	91M42R <ATS>
1959 V33 N4 P863-868	61-10408 <*+>
1959 V33 N4 P869-876	50L33R <ATS>
	61-10544 <*+>
1959 V33 N4 P913-919	62-11516 <=>
1959 V33 N4 P922-929	51L33R <ATS>
1959 V33 N4 P936-942	61-13398 <*+>
	62-10229 <*=>
1959 V33 N4 P943-947	60-21736 <=>
1959 V33 N5 P983-987	69L36R <ATS>
1959 V33 N5 P988-991	RCT V33 N4 P959 <RCT>
1959 V33 N5 P1002-1015	65-18065 <*>
1959 V33 N5 P1016-1022	61-19671 <=*>
1959 V33 N5 P1023-1029	06L37R <ATS>
1959 V33 N5 P1030-1034	63-19134 <=*>
1959 V33 N5 P1065-1070	60-17727 <+*>
1959 V33 N5 P1071-1079	61-13281 <*+>
1959 V33 N5 P1080-1086	05L37R <ATS>
	60-17620 <+*>
1959 V33 N5 P1100-1110	61-15732 <+*>
1959 V33 N5 P1129-1133	AEC-TR-4299 <*+>
1959 V33 N5 P1140-1146	62-10307 <*=>
1959 V33 N6 P1198-1208	64-18412 <*>
1959 V33 N6 P1209-1213	60-11595 <=>
1959 V33 N6 P1246-1252	23M44R <ATS>
1959 V33 N6 P1256-1262	60-13094 <+*> 0
1959 V33 N6 P1324-1327	61-15219 <*+>
1959 V33 N6 P1328-1335	32L34R <ATS>
1959 V33 N6 P1353-1359	60-17729 <+*>
1959 V33 N6 P1365-1373	61-13488 <*+>
	61-15634 <*+>
1959 V33 N6 P1409-1414	C-4770 <NRC>
1959 V33 N6 P1409-1413	97P64R <ATS>
1959 V33 N6 P1440-1445	AEC-TR-4271 <+*>
1959 V33 N7 P1467-1472	18M38R <ATS>
1959 V33 N7 P1492-1494	60-18930 <+*>
1959 V33 N7 P1625-1631	60-21038 <=>
1959 V33 N8 P1681-1686	49L35R <ATS>
	62-15708 <*=>
1959 V33 N8 P1709-1714	80N49R <ATS>
1959 V33 N8 P1769-1773	61-20416 <*=>
	81M41R <ATS>
1959 V33 N9 P1928-1932	11M46R <ATS>
1959 V33 N9 P1945-1950	06M44R <ATS>
1959 V33 N9 P2030-2035	RCT V34 N2 P600 <RCT>
1959 V33 N9 P2081-2089	60-16900 <*+>
1959 V33 N10 P2128-2134	62M41R <ATS>
1959 V33 N10 P2259-2263	60-16590 <+*>
1959 V33 N11 P2424-2428	60-17732 <+*>
1959 V33 N11 P2435-2441	60-17733 <+*>
1959 V33 N11 P2513-2516	19M42R <ATS>
1959 V33 N11 P2516-2520	85M43R <ATS>
1959 V33 N11 P2517-2520	60-18081 <+*>
1959 V33 N11 P2631-2632	62-14551 <=*>
1959 V33 N12 P2755-2766	60-11690 <=>
1959 V33 N12 P2778-2785	60-18765 <+*>
1960 N5 P960-	ICE V2 N3 P415-418 <ICE>
1960 V34 P192-195	49M40R <ATS>
1960 V34 P1380-1381	62-24448 <=*>
1960 V34 P1382-1383	62-24449 <=*>
1960 V34 P1763-1767	62-24453 <=*>
1960 V34 P1883-1884	62-24454 <=*>
1960 V34 P2431-2447	62-24456 <=*>
1960 V34 P2489-2494	AEC-TR-4634 <*=>
1960 V34 N1 P32-38	62-16116 <=*> 0
1960 V34 N1 P57-62	33P59R <ATS>
1960 V34 N1 P135-137	65-64411 <=*$>
1960 V34 N1 P150-158	11M41R <ATS>
1960 V34 N1 P177-181	60-31785 <=$> 0
1960 V34 N1 P229-230	61-27513 <*=> 0
1960 V34 N2 P477	60-23179 <+*>
1960 V34 N3 P603-610	60-41562 <=>
1960 V34 N3 P611-617	61-11066 <=>
1960 V34 N3 P623-626	60-18442 <*+>
1960 V34 N3 P679-683	61-27517 <*=> 0
1960 V34 N4 P789-794	62-13605 <*=>
1960 V34 N4 P860-865	64M42R <ATS>
1960 V34 N5 P1004-1008	60-41162 <=>
1960 V34 N6 P1176-1185	10M47R <ATS>
	63-18886 <=*>
1960 V34 N6 P1299-1306	AEC-TR-4301 <*+>
1960 V34 N6 P1357-1363	65-61573 <=$>
1960 V34 N7 P1420-1424	61-14978 <=*>
1960 V34 N7 P1517-1523	48M47R <ATS>
1960 V34 N7 P1634-1637	AEC-TR-4348 <*>
1960 V34 N8 P1728-1733	61-14976 <=*>
1960 V34 N8 P1745-1752	30N49R <ATS>
1960 V34 N9 P1952-1959	62-19872 <=*>
1960 V34 N9 P2030-2040	95N47R <ATS>
1960 V34 N9 P2130-2133	61-21751 <=>
1960 V34 N11 P2619-2620	61-21417 <=>
1960 V34 N12 P2682-2686	62-11563 <=>
1960 V34 N12 P2694-2703	76N52R <ATS>
1960 V34 N12 P2709-2718	63-19729 <*=>
1960 V34 N12 P2804-2807	62-11266 <=>
1960 V34 N12 P2844	01N52R <ATS>
1961 N10 P2328-	ICE V2 N2 P274-278 <ICE>
1961 V35 P488-491	61-18706 <*=>

1961 V35 P1907-1910 62-10705 <*=>
1961 V35 P2389-2391 62-24463 <=*>
1961 V35 N1 P9-14 52N52R <ATS>
1961 V35 N1 P118-128 53N52R <ATS>
1961 V35 N1 P181-188 02N52R <ATS>
1961 V35 N2 P298-305 D-773 <RIS>
1961 V35 N2 P384-388 62-11265 <=>
1961 V35 N2 P460- ICE V2 N2 P207-208 <ICE>
1961 V35 N2 P460 82N57R <ATS>
1961 V35 N3 P509-512 66-12283 <*>
1961 V35 N3 P543-547 62-11256 <=>
1961 V35 N3 P600-604 62-14856 <=*>
1961 V35 N4 P776-781 62-18387 <=*>
1961 V35 N4 P821-824 ICE V1 N1 P22-24 <ICE>
62-10228 <*=>
1961 V35 N4 P842-847 36N56R <ATS>
1961 V35 N5 P1102-1104 62-13501 <*=>
1961 V35 N6 P1177-1185 62-16573 <=*>
1961 V35 N6 P1186-1191 11P59R <ATS>
62-25642 <=*>
1961 V35 N6 P1212-1214 61-28791 <*=>
1961 V35 N6 P1215-1218 62-10350 <*=> O
1961 V35 N6 P1235-1239 65-29153 <$>
1961 V35 N6 P1331-1336 62-16574 <=*>
1961 V35 N6 P1343-1350 62-25641 <=*>
1961 V35 N6 P1359-1361 65-12885 <*>
1961 V35 N7 P1425-1429 23N56R <ATS>
1961 V35 N7 P1478-1480 64-16598 <=*$>
1961 V35 N7 P1622-1628 62-14212 <=*> O
1961 V35 N8 P1853-1859 62-18386 <=*>
1961 V35 N8 P1896-1900 62-13680 <*=>
1961 V35 N9 P2010-2015 23P59R <ATS>
1961 V35 N9 P2039-2046 62-18385 <=*>
1961 V35 N9 P2083-2089 62-23520 <=*>
1961 V35 N9 P2166-2167 62-10643 <=*>
1961 V35 N1C P2341-2347 62-24349 <=*>
1961 V35 N11 P2448- ICE V2 N2 P261-266 <ICE>
1961 V35 N11 P2475-2479 97P58R <ATS>
1961 V35 N11 P2514-2523 63-19791 <=*> O
1961 V35 N11 P2540-2545 62-14261 <=*>
1961 V35 N11 P2553-2556 62-19636 <=*>
1961 V35 N11 P2576-2581 62-14255 <=*>
1961 V35 N11 P2582-2588 26P59R <ATS>
1961 V35 N11 P2652-2654 65-17337 <*>
1961 V35 N12 P2736-2750 63-24282 <=*$>
1962 V36 P1219-1221 63-10766 <=*>
1962 V36 P1246-1254 63-10672 <=*>
1962 V36 P1369-1371 AEC-TR-15 <*>
1962 V36 P1578-1579 63-10673 <=*>
1962 V36 N1 P3-14 62-23788 <=*>
1962 V36 N1 P3-52 62-32211 <=*>
1962 V36 N1 P72-80 62-33073 <=*>
1962 V36 N1 P93-97 62-11710 <=>
1962 V36 N1 P98-102 62-16691 <=*>
1962 V36 N1 P218-219 14P62R <ATS>
1962 V36 N1 P219-222 62-25840 <=*>
1962 V36 N2 P274-281 62-25855 <=*>
1962 V36 N2 P336-344 63-13194 <=*>
1962 V36 N2 P362-364 62-18908 <=*>
1962 V36 N3 P443-448 63-10748 <=*>
1962 V36 N4 P765-769 62-18226 <=*>
1962 V36 N4 P895-897 63-15527 <=*>
1962 V36 N4 P897-899 63-16560 <=*>
1962 V36 N5 P989-992 63-15554 <=*>
1962 V36 N5 P1001-1004 64-14008 <=*$>
1962 V36 N5 P1005-1009 62-20326 <=*>
1962 V36 N5 P1010-1015 63-18413 <=*>
1962 V36 N5 P1086-1088 62-18968 <=*>
63-10772 <=*>
1962 V36 N5 P1089-1094 63-15554 <=*>
1962 V36 N6 P1205-1209 24P64R <ATS>
1962 V36 N7 P1401-1407 63-18749 <=*$>
1962 V36 N7 P1420-1425 63-19259 <=*>
1962 V36 N7 P1554-1555 63-15705 <=*>
1962 V36 N7 P1560-1562 73P65R <ATS>
1962 V36 N7 P1574-1577 62-33712 <=*>
1962 V36 N8 P1785-1787 98Q67R <ATS>
1962 V36 N8 P1806-1809 97Q67R <ATS>
1962 V36 N9 P2066-2071 63-14735 <=*>
1962 V36 N10 P2153-2161 28Q69R <ATS>
1962 V36 N10 P2222-2224 64-10678 <=*$>
1962 V36 N10 P2244-2245 M-5201 <NLL>
64-23570 <=>
1962 V36 N11 P2393-2399 26S87R <ATS>
1962 V36 N12 P2616-2620 64-10679 <=*$>
1962 V36 N12 P2804-2805 63-20635 <=*$>
1963 V37 P698-701 66-12240 <*>
1963 V37 N1 P179-181 64-14439 <=*$>
1963 V37 N1 P226-227 64-14892 <=*$>
1963 V37 N1 P237-241 64-10686 <=*$>
1963 V37 N2 P413-419 64-14449 <=*$>
1963 V37 N3 P517-525 64-14437 <=*$>
1963 V37 N4 P887-888 63-20469 <=*>
1963 V37 N4 P927-928 98Q72R <ATS>
1963 V37 N5 P1119-1123 64-71413 <=> M
1963 V37 N5 P1137-1138 19Q73R <ATS>
1963 V37 N5 P1169-1172 AEC-TR-6074 <=$>
1963 V37 N6 P1251-1257 64-16967 <=*$>
1963 V37 N6 P1297-1303 99Q72R <ATS>
1963 V37 N6 P1304-1310 64-11938 <=> M
1963 V37 N6 P1377-1381 65-18161 <*>
1963 V37 N8 P1851-1854 66-14175 <*$>
1963 V37 N9 P1949-1957 63-41299 <=>
1963 V37 N9 P2007-2011 64-10714 <=*$>
1963 V37 N9 P2041-2047 64-10713 <=*$>
1963 V37 N12 P2689-2705 65-11777 <*>
1963 V37 N12 P2721-2727 RTS-3037 <NLL>
1963 V37 N12 P2728-2733 06R76R <ATS>
1963 V37 N12 P2763 65-12408 <*>
1963 V37 N12 P2771-2773 AD-611 087 <=$>
1964 V38 P514-517 65-10259 <*>
1964 V38 N1 P52-58 65-10204 <*>
1964 V38 N1 P2662-2663 M-5697 <NLL>
1964 V38 N3 P562-564 LA-TR-64-9 <=$>
1964 V38 N3 P798-801 65-10415 <*>
1964 V38 N4 P850-857 34R77R <ATS>
1964 V38 N4 P1059-1060 65-18048 <*>
1964 V38 N5 P1276-1280 81T94R <ATS>
1964 V38 N6 P1442-1449 65-12318 <*>
1964 V38 N7 P1734-1742 N65-11312 <=$>
1964 V38 N8 P1956-1962 M-5615 <NLL>
1964 V38 N8 P2023-2029 65-13425 <*>
1964 V38 N9 P2176-2180 M-5605 <NLL>
1964 V38 N10 P2408-2414 07S82R <ATS>
1964 V38 N11 P2633-2639 65-13423 <*>
1965 V39 P313-320 1733 <TC>
1965 V39 P762-764 51S85R <ATS>
1965 V39 P2273-2275 66-12235 <*>
1965 V39 N1 P84-91 65-14740 <*>
1965 V39 N1 P230-231 32S83R <ATS>
1965 V39 N2 P328-334 65-11634 <*>
65-17085 <*>
784 <TC>
1965 V39 N2 P428-429 65-14775 <*>
1965 V39 N2 P516-519 65-63663 <=>
1965 V39 N4 P855-858 M-5746 <NLL>
1965 V39 N7 P1572-1576 65-64503 <=*$>
1965 V39 N7 P1590-1594 66-10642 <*>
1965 V39 N7 P1640-1646 65-64612 <=*$>
1965 V39 N8 P1809-1916 66-10379 <*> O
1965 V39 N8 P1817-1822 66-14203 <*$>
1965 V39 N8 P1890-1894 66-12225 <*>
1965 V39 N8 P1938-1943 66-12224 <*>
1965 V39 N12 P3081-3083 66-12234 <*>
1966 V40 N3 P746-748 66-14202 <*$>

ZHURNAL GEOFIZIKI
1931 V1 N20 P207-222 62-20298 <=*>
1932 V2 N3/4 P350-375 RT-947 <*>
1933 V3 N4 P440-450 RT-664 <*>
1935 V5 N2 P204-221 RT-2001 <*>

ZHURNAL GEOFIZIKI I METEORCLOGII
1927 V4 N1 P24-37 RT-1120 <*>

ZHURNAL INSTYTUTU BOTANIKY. AKADEMIYA NAUK URSR
1936 N9 P91-122 61-19137 <*=>

ZHURNAL KHIMICHESKOI PROMYSHLENNOSTI
1927 N4 P840-842 RJ-34 <ATS>

Citation	Reference
1927 V4 P49-51	R-2815 <*>
1927 V4 P49	R-2822 <*>
1927 V4 P122-129	R-3049 <*>
1927 V4 P485-489	63-14608 <=*>
1927 V4 P552-553	30T92R <ATS>
1927 V4 P642-650	R-1125 <*>
1927 V4 N10 P840-842	RJ-34 <ATS>
1928 V5 P682-689	R-3364 <*>
1928 V5 P941-942	R-2795 <*>
1928 V5 P1273-1276	60-15009 <+*>
1928 V5 N11/2 P528-529	61-28111 <=*$>
1929 V6 P8-12	R-2907 <*>
1929 V6 P538-540	T-2302 <INSD>
1929 V6 N1 P3-7	R-3345 <*>
1929 V6 N7/9 P538-540	61-28095 <=*$>
1929 V6 N23/4 P1721-1731	R-2192 <*>
1930 V7 P34-38	R-2206 <*>
1930 V7 P332-335	64-10726 <=*$> 0
1930 V7 P962-964	64-10239 <=*$>
1930 V7 P1224-1233	53J18R <ATS>
1930 V7 P1414-1419	T2156 <INSD>
1931 V8 P248-252	R-302 <*>
1931 V8 P351-359	59-10738 <+*>
1931 V8 P861-863	T2157 <INSD>
1931 V8 N13 P1-11	R-2819 <*>
1931 V8 N13 P11-15	R-2818 <*>
1931 V8 N14 P35-38	R-3367 <*>
1931 V8 N17 P17-25	63-18107 <=*> 0
1931 V8 N18 P23-27	R-2797 <*>
1931 V8 N15/6 P36-40	R-2428 <*>
1932 N8 P38-39	R-3290 <*>
1932 N12 P21-24	R-2997 <*>
1932 V9 N3 P17-26	07T92R <ATS>
1932 V9 N12 P16-17	63-18110 <=*>
1933 N2 P37-41	R-2598 <*>
1933 N3 P14-18	R-2812 <*>
1933 N3 P22-24	R-2597 <*>
1933 N5 P48-51	R-4365 <*>
1933 N6 P613-620	R-3351 <*>
1933 N7 P32-35	R-3018 <*>
1933 N7 P41-47	R-2469 <*>
1933 N7 P47-56	R-2307 <*>
1933 N8 P41-49	R-2091 <*>
	R-2475 <*>
1933 N8 P58-60	R-2208 <*>
1933 N10 P21-27	R-3291 <*>
1933 N10 P37-45	R-2641 <*>
1933 V10 N8 P31-40	61-18322 <=*$>
1933 V10 N8 P41-49	RJ-77 <ATS>
1933 V10 N8 P50-58	08T92R <ATS>
1934 N6 P46-48	54J18R <ATS>
1934 N9 P32-38	R-2284 <*>
	07L30R <ATS>
1934 N9 P38-43	R-2285 <*>
1934 N10 P59-62	R-2943 <*>
1934 V11 N1 P66-72	R-3352 <*>
1934 V11 N2 P33-37	56T92R <ATS>
1934 V11 N4 P60-65	R-3338 <*>
1934 V11 N5 P45-53	61-18338 <=*$>
1935 N1 P28-37	R-2201 <*>
1935 N1 P37-44	R-3289 <*>
1935 V12 P134-139	R-2800 <*>
1935 V12 P265-268	R-2938 <*>
1935 V12 P597-603	R-2933 <*>
1935 V12 P1050-1054	R-2937 <*>
1935 V12 P1054-1057	R-2816 <*>
1935 V12 P1057-1061	R-2480 <*>
1935 V12 P1167-1171	64-18936 <*>
1935 V12 P1240-1249	R-2806 <*>
1935 V12 N3 P259-265	RT-1433 <*>
1935 V12 N3 P279-281	64-20073 <*>
1935 V12 N7 P687-690	64-10740 <=*$> 0
1935 V12 N8 P795-802	RT-2449 <*>
	65-64401 <=*$>
1936 N18 P1109-1114	05L30R <ATS>
1936 N19 P1170-1173	06L30R <ATS>
1936 V13 P92-102	R-2864 <*>
1936 V13 P102-104	R-2478 <*>
1936 V13 P215-220	R-2794 <*>
1936 V13 P408-411	R-2910 <*>
1936 V13 P466-469	R-2413 <*>
1936 V13 P718-722	R-2906 <*>
1936 V13 P847-850	R-2226 <*>
1936 V13 P866-867	R-2498 <*>
1936 V13 P912-915	R-3313 <*>
1936 V13 P922-924	61-16698 <=*$>
1936 V13 P987-991	R2588 <*>
1936 V13 P1109-1114	R-2180 <*>
1936 V13 P1221-1225	R-2128 <*>
1936 V13 P1225-1228	R-2524 <*>
1936 V13 P1345-1348	R-2971 <*>
1936 V13 P1413-1417	R-2306 <*>
1936 V13 P1462-1467	R-2905 <*>
1936 V13 N2 P92-102	RJ-265 <ATS>
1936 V13 N5 P262-265	<INSD>
1936 V13 N7 P408-411	RT-3606 <*>
1936 V13 N14 P866-867	51/1899 <NLL>
1936 V13 N21 P1273-1283	RJ-183 <ATS>
1936 V13 N21 P1292-1296	RT-1629 <*>
1936 V13 N24 P1486-1489	RT-1630 <*>
1937 P917-924	R-2823 <*>
1937 V14 P86-92	R-3043 <*>
1937 V14 P99-104	R-2970 <*>
1937 V14 P244-249	R-2419 <*>
1937 V14 P337-341	R-2904 <*>
1937 V14 P342-346	R-2934 <*>
1937 V14 P365-369	R-2552 <*>
1937 V14 P429-435	R-3347 <*>
1937 V14 P496-504	R-2416 <*>
1937 V14 P504-507	R-2803 <*>
1937 V14 P567-576	R-3015 <*>
1937 V14 P581-584	R-2810 <*>
1937 V14 P631-636	R-2301 <*>
1937 V14 P646-654	R-4495 <*>
1937 V14 P707-710	R-2916 <*>
1937 V14 P719-725	R-2847 <*>
1937 V14 P738-744	R-2599 <*>
1937 V14 P799-804	R-2182 <*>
1937 V14 P807-817	R-2514 <*>
1937 V14 P818-824	R-2847 <*>
1937 V14 P831-836	R-2793 <*>
1937 V14 P991-993	R-2181 <*>
1937 V14 P993-996	61-16948 <=*$>
1937 V14 P1065-1067	R-3377 <*>
1937 V14 P1136-1150	R-2805 <*>
1937 V14 P1225-1229	R-2213 <*>
1937 V14 P1303-1307	R-2870 <*>
1937 V14 P1323-1329	R-2304 <*>
1937 V14 P1381-1386	R-2870 <*>
1937 V14 P1404-1407	R-2922 <*>
1937 V14 P1408-1410	R-2998 <*>
1937 V14 P1526-1532	R-2870 <*>
1937 V14 P1618-1622	R-2870 <*>
1937 V14 P1688-1693	61-32015 <=$>
1937 V14 P1693-1699	R-100 <*>
	R-2294 <*>
1937 V14 N1 P1704-1707	R-2870 <*>
1937 V14 N13 P907-917	<INSD>
1937 V14 N14 P991-993	RT-1435 <*>
1937 V14 N11/2 P799-804	RT-1434 <*>
1937 V14 N17/8 P1225-1229	64-10739 <=*$> 0
1938 N10 P31	R-3333 <*>
1938 N11 P33-40	R-3319 <*>
1938 V15 P28-38	R-3366 <*>
1938 V15 N3 P31-35	R-3373 <*>
1938 V15 N4 P25-27	R-2207 <*>
1938 V15 N4 P34-39	R-3358 <*>
1938 V15 N6 P24-28	R-2257 <*>
1938 V15 N8 P15-18	R-2999 <*>
1938 V15 N9 P13-17	R-2911 <*>
1938 V15 N9 P19-22	R-2802 <*>
1938 V15 N9 P23-24	R-2422 <*>
1938 V15 N11 P41-46	R-2519 <*>
1938 V15 N11 P46-51	R-2211 <*>
1938 V15 N12 P23-26	R-2979 <*>
	R-3146 <*>
1938 V15 N12 P27-	R-2302 <*>
1939 V16 P31-35	R-2093 <*>
	RJ-79 <ATS>
1939 V16 N1 P10-14	61-23319 <*=>

1939 V16 N1 P19-22	RJ-185 <ATS>
1939 V16 N2 P26-	R-2183 <*>
1939 V16 N2 P26	RT-2057 <*>
1939 V16 N2 P31-37	64-10738 <=*$> 0
1939 V16 N2 P37-40	62-24462 <=*>
	64-13554 <=*$>
1939 V16 N3 P11-14	R-2892 <*>
1939 V16 N4 P54-61	R-2592 <*>
1939 V16 N6 P25-27	R-2229 <*>
1939 V16 N6 P30-32	R-2909 <*>
1939 V16 N6 P32-36	R-2415 <*>
1939 V16 N6 P43-44	RT-2061 <*>
1939 V16 N7 P48-50	R-3292 <*>
1939 V16 N8 P24-28	RJ-184 <ATS>
	62-18765 <=*>
1939 V16 N9 P12-15	RT-2167 <*>
	63-18265 <=*>
1939 V16 N9 P26-30	R-3295 <*>
1939 V16 N10 P31-35	R-2472 <*>
	RJ-79 <ATS>
	61-13637 <=*>
1939 V16 N10 P36-37	R-2481 <*>
1939 V16 N12 P28-32	RT-1582 <*>
1939 V16 N4/5 P29-31	RT-2058 <*>
1939 V16 N4/5 P33-38	R-2580 <*>
1939 V17 N2 P25-29	59-17422 <+*>
1940 V17 P25-29	R-2092 <*>
1940 V17 N1 P28-31	R-2590 <*>
	RJ-81 <ATS>
1940 V17 N2 P25-29	RJ-78 <ATS>
	59-17422 <+*>
1940 V17 N3 P5-8	R-2455 <*>
1940 V17 N3 P8-13	PANSDOC-TR.285 <PANS>
1940 V17 N6 P3-9	R-2461 <*>
1940 V17 N6 P35-39	R-2946 <*>
1940 V17 N6 P39-41	R-3297 <*>
1940 V17 N6 P55-56	R-2423 <*>
1940 V17 N7 P25-28	R-2185 <*>
	RT-1019 <*>
1940 V17 N8 P48-	R-2427 <*>
1940 V17 N9 P8-13	R-2945 <*>
1940 V17 N9 P14-19	R-2476 <*>
1940 V17 N9 P20-24	R-2965 <*>
1940 V17 N12 P11-13	R-2424 <*>
1940 V17 N12 P14-16	62-13140 <*=>
1940 V17 N12 P16-18	R-2489 <*>
1940 V17 N12 P25-28	R-2953 <*>
1940 V17 N4/5 P28-32	R-2920 <*>
	RT-2731 <*>
1940 V17 N4/5 P44-45	63-24285 <=*$>
1941 N22 P30-31	<ATS>
1941 V18 N1 P24-29	61-16921 <*=>
1941 V18 N2 P6-7	60-17722 <+*>
1941 V18 N2 P18-23	RT-3605 <*>
1941 V18 N2 P28-31	63-18140 <=*>
1941 V18 N3 P6-11	61-16823 <*=>
1941 V18 N4 P6-8	61-16901 <*=>
1941 V18 N4 P10-15	R-2652 <*>
	63-18139 <=*>
1941 V18 N5 P7-13	61-16922 <*=>
1941 V18 N5 P21-26	61-16946 <*=>
1941 V18 N6 P25-27	61-18134 <*=>
1941 V18 N7 P26-27	RT-1366 <*>
1941 V18 N11 P8-16	R-2842 <*>
1941 V18 N12 P8-13	63-14548 <=*>
1941 V18 N14 P40-43	61-16929 <*=>
1941 V18 N17 P7-12	61-16931 <*=>
1941 V18 N17 P12-18	61-16932 <*=>
1941 V18 N18 P9-15	61-16933 <*=>
1941 V18 N19 P4-7	RT-4004 <*>
1941 V18 N21 P3-10	R-2842 <*>
1941 V18 N21 P15-23	61-16924 <*=>
1941 V18 N22 P3-7	R-2420 <*>
1941 V18 N22 P8-10	61-18157 <*=>

*ZHURNAL MIKROBIOLOGII, EPIDEMIOLOGII I IMMUNOBIOLOGII

1940 N7 P3-11	65-62993 <=>
1941 N5/6 P125-131	RT-1304 <*>
1942 N3 P59-62	65-60103 <=$>
	65-62967 <=>
1942 N11/2 P82-86	65-63270 <=>
1943 N1 P36-43	64-23628 <=$>
1943 N6 P76-80	65-63242 <=>
1943 N1/2 P33-36	64-10818 <=*$>
	65-60068 <=$>
1943 N1/2 P43-48	64-10819 <=*$>
	65-60063 <=$>
1943 N1/2 P48-51	65-60050 <=$>
1943 N1/2 P59-64	R-222 <*>
	R-782 <*>
	64-15091 <=*$>
1943 N1/2 P65-68	65-60053 <=$>
1943 N1/2 P68-73	RT-1305 <*>
	65-60047 <=$> 0
1943 N7/8 P59-66	59-10070 <+*>
1944 N3 P56-61	59-13212 <=>
1944 N3 P67	R-5153 <*>
1944 N6 P71	64-10839 <=*$>
1944 N1/2 P26-29	R-1105 <*>
1944 N1/2 P29-34	R-1066 <*>
	RT-638 <*>
1944 V15 N6 P71	64-23632 <=$>
1945 N6 P62-66	R-1084 <*>
	RT-1372 <*>
1945 V16 N7/8 P32-38	65-60065 <=$>
1945 V16 N10/1 P16-20	60-17640 <+*>
1946 N11 P85-86	R-2405 <*>
1947 N1 P33-35	RT-153 <*>
1947 N5 P21-28	65-60027 <=$>
1947 N5 P28-33	64-23639 <=$>
1947 N5 P33-37	65-60026 <=$>
1947 N6 P3-10	65-62995 <=>
1947 N6 P11-17	65-63254 <=$>
1947 N6 P35-38	65-62996 <=$>
1947 N7 P3-9	59-10009 <+*>
1947 N11 P52-56	R-2404 <*>
1948 N10 P10-25	R-2033 <*>
	R-2035 <*>
1948 N10 P81-84	R-2031 <*>
1948 N10 P84-85	R-2030 <*>
1949 N3 P73-	R-2407 <*>
1949 N4 P69-76	59-14231 <+*>
1950 N6 P37-44	59-14233 <+*>
1950 V21 N10 P66-73	61-19713 <=*>
1951 N5 P5-12	59-14234 <+*>
1951 N5 P12-15	59-14235 <+*>
1951 N8 P71-72	59-14236 <+*>
1951 N8 P72-74	59-14237 <+*>
1951 N8 P91-93	59-14664 <+*>
1951 V22 N11 P34-37	59-13396 <=> 0
1953 N4 P6-12	R-90 <*>
1953 N6 P3-6	RT-2609 <*>
1953 N6 P14	RT-2677 <*>
1953 N6 P23-24	RT-2684 <*>
1953 N6 P24-25	RT-2685 <*>
1953 N6 P25-31	RT-2686 <*>
1953 N6 P34-40	RT-2690 <*>
1953 N6 P40-44	RT-2689 <*>
1953 N6 P47-50	RT-2688 <*>
1953 N6 P53-56	RT-2607 <*>
1953 N11 P24-27	RT-2562 <*>
1953 N12 P19-21	RT-3021 <*>
1953 N12 P21-27	RT-3019 <*>
1953 N12 P28-32	RT-3007 <*>
1953 N12 P33	RT-3022 <*>
1953 V24 N12 P14-18	RT-3020 <*>
1954 N1 P7-13	RT-4057 <*>
1954 N3 P43-46	RJ-667 <ATS>
1954 N3 P53-56	RT-2639 <*>
1954 N3 P61-63	RT-3091 <*>
1954 N5 P91-92	RT-1839 <*>
1954 N6 P97-98	RT-2932 <*>
1954 N7 P3-8	RT-2388 <*>
1954 N7 P44	59-10029 <+*>
1954 N7 P55-57	RT-4442 <*>
1954 N7 P113-114	RT-3095 <*>
1954 N8 P52-	R-207 <*>
1954 N8 P106-110	R-625 <*>
1954 N9 P13-15	R-209 <*>

Citation	Report
1954 N9 P16-20	R-624 <*>
1954 N9 P20-22	R-626 <*>
	65-63437 <=$>
1954 N9 P33-37	R-621 <*>
1954 N9 P49-56	R-620 <*>
1954 N9 P61-63	R-283 <*>
1954 N9 P124-126	RT-2676 <*>
1954 N10 P3-11	RT-2787 <*>
1954 N11 P11-15	R-2095 <*>
1954 V25 N3 P61-63	64-15661 <=*$>
1954 V25 N5 P24-32	59-13396 <=>
1954 V25 N6 P55-61	64-15647 <=*$>
1954 V25 N8 P52	64-19498 <=$>
1954 V25 N8 P106-110	64-19497 <=$>
1954 V25 N9 P13-15	64-19499 <=$>
1954 V25 N9 P16-20	64-19500 <=$>
1954 V25 N9 P33-37	64-19495 <=$>
1954 V25 N9 P49-56	64-19501 <=$>
1954 V25 N10 P38-44	62-24756 <=*>
1955 N1 P48-56	RT-3649 <*>
1955 N1 P56-60	RT-3650 <*>
1955 N1 P61-65	RT-3651 <*>
1955 N1 P70-76	RT-3652 <*>
1955 N1 P76-81	RT-3653 <*>
1955 N1 P82-87	RT-3654 <*>
1955 N1 P87-92	RT-3655 <*>
1955 N1 P92-97	RT-3656 <*>
1955 N1 P97-98	RT-3636 <*>
1955 N1 P98-102	RT-3637 <*>
1955 N1 P103-107	RT-3638 <*>
1955 N1 P114-116	RT-3648 <*>
1955 N1 P122-126	RT-3639 <*>
1955 N2 P82-90	RT-3096 <*>
1955 N2 P108-109	R-66 <*>
1955 N4 P20-26	RT-4481 <*>
1955 N5 P76-82	59-12754 <+*>
1955 N6 P28-31	R-4757 <*>
1955 N6 P75-78	R-433 <*>
	R-456 <*>
1955 N6 P120-124	RT-3351 <*>
1955 N7 P3-7	RT-3472 <*>
1955 N7 P38-40	R-327 <*>
1955 N7 P40-42	R-307 <*>
1955 N7 P42-47	R-369 <*>
1955 N7 P64-69	R-364 <*>
1955 N7 P80-84	R-280 <*>
1955 N8 P7-13	R-315 <*>
1955 N8 P42-	R-316 <*>
	R-87 <*>
1955 N8 P44-50	<INSD>
	R-574 <*>
1955 N8 P53-	R-88 <*>
1955 N8 P60-64	R-332 <*>
1955 N8 P127-128	R-333 <*>
1955 N10 P40-42	R-86 <*>
1955 N10 P55-56	R-113 <*>
1955 N10 P62-66	R-149 <*>
1955 N10 P83-89	RT-3635 <*>
1955 N11 P15-21	R-313 <*>
1955 N11 P36-48	R-349 <*>
1955 N11 P71-72	R-350 <*>
1955 N11 P91-94	AD-630 726 <=*$>
1955 N11 P96-99	R-351 <*>
1955 N11 P100-106	R-378 <*>
1955 N11 P107-109	R-374 <*>
1955 N11 P110-114	R-373 <*>
1955 N12 P29-34	R-1457 <*>
1955 N12 P54-61	R-2740 <*>
	R-3509 <*>
1955 N12 P97-105	RT-4270 <*>
1955 N12 P121-126	AD-633 208 <=$>
1955 V26 N1 P82-87	64-15619 <=*$>
1955 V26 N4 P20-26	64-19052 <=*$>
	66-60975 <=$>
1955 V26 N4 P27-31	63-41203 <=>
1955 V26 N6 P28-31	63-23091 <=*$>
1955 V26 N6 P75-78	64-19496 <$>
1955 V26 N11 P72-77	63-10006 <=*>
1955 V26 N11 P77-83	59-20944 <+*>
1955 V26 N12 P29-34	64-19429 <=$>
1956 N5 P6-13	RT-4244 <*>
1956 N5 P30	R-2749 <*>
	R-3442 <*>
1956 N5 P48-52	R-2746 <*>
	R-3444 <*>
1956 N5 P52-53	R-2741 <*>
	R-3510 <*>
1956 N5 P53-54	R-2742 <*>
	R-3508 <*>
1956 N5 P54-59	R-2743 <*>
	R-3507 <*>
1956 N5 P97-101	R-2745 <*>
	R-3561 <*>
1956 N6 P58-61	AD-630 724 <=*$> 0
	59-10042 <+*>
1956 N9 P20-22	C-2254 <NRC>
1956 N11 P27-30	AD-630 727 <=*$>
1956 V27 N1 P3-5	R-370 <*>
1956 V27 N1 P19-22	AD-625 629 <=*$>
	R-371 <*>
1956 V27 N1 P32-36	R-83 <*>
1956 V27 N1 P36-42	R-2744 <*>
	R-754 <*>
	R-89 <*>
1956 V27 N1 P43-49	R-282 <*>
1956 V27 N1 P49-50	R-244 <*>
1956 V27 N1 P62-63	R-155 <*>
1956 V27 N1 P63-65	R-287 <*>
1956 V27 N1 P69-76	R-368 <*>
1956 V27 N1 P77-83	R-156 <*>
1956 V27 N1 P101-105	R-263 <*>
1956 V27 N1 P115-118	R-279 <*>
1956 V27 N2 P3-7	59-12748 <+*>
1956 V27 N2 P8-13	65-60243 <=>
1956 V27 N3 P3-7	R-391 <*>
1956 V27 N3 P12-15	R-392 <*>
1956 V27 N3 P21-22	R-393 <*>
1956 V27 N3 P42-49	R-214 <*>
1956 V27 N3 P56-	R-213 <*>
1956 V27 N3 P69-73	R-212 <*>
1956 V27 N3 P73-79	R-211 <*>
1956 V27 N3 P79-83	R-210 <*>
1956 V27 N3 P84-90	R-208 <*>
1956 V27 N3 P90-94	R-206 <*>
1956 V27 N3 P94-97	R-205 <*>
1956 V27 N5 P22-29	R-2085 <*>
1956 V27 N7 P47-51	R-3067 <*>
1956 V27 N7 P96-100	59-19269 <+*>
1956 V27 N7 P126-127	R-2389 <*>
1957 N9 P101-107	R-3550 <*>
1957 V28 N4 P37-38	59-15751 <+*>
1957 V28 N9 P35-41	59-15734 <+*>
1957 V28 N9 P82-86	64-19459 <=$>
1957 V28 N11 P40-46	59-13018 <=>
1957 V28 N11 P65-70	59-13019 <=>
1957 V28 N11 P84-91	59-13009 <=>
1958 N2 P130-131	R-4925 <*>
1958 N5 P33-37	59-10037 <+*>
1958 N5 P83-87	59-10038 <+*>
1958 N10 P102-109	59-13091 <=>
1958 V29 N2 P74-78	64-19404 <=$>
1958 V29 N6 P63-68	65-31964 <*>
1958 V29 N7 P56-59	65-31964 <*>
1958 V29 N10 P129-132	59-13090 <=> 0
1958 V29 N11 P151-153	59-12732 <+*>
1958 V29 N12 P43-46	59-19253 <+*>
1959 V30 N1 P24-28	59-19675 <+*>
1959 V30 N1 P36-37	59-14668 <+*>
1959 V30 N2 P74-78	59-13823 <=>
1959 V30 N2 P148-151	65-61132 <=$>
1959 V30 N3 P13-16	59-19259 <+*> 0
1959 V30 N3 P22-26	59-19255 <+*>
1959 V30 N3 P26-28	59-19672 <+*>
1959 V30 N3 P33-35	59-19256 <+*>
1959 V30 N3 P41-46	59-19258 <+*>
1959 V30 N4 P3-9	59-18299 <+*>
1959 V30 N4 P19-23	59-11768 <=>
1959 V30 N4 P79-92	59-16206 <+*>
1959 V30 N8 P3-9	59-31009 <=>
1959 V30 N8 P9-12	59-31010 <=>

Citation	Report
1959 V30 N8 P12-17	59-31008 <=>
1959 V30 N8 P18-21	59-31011 <=>
1959 V30 N8 P21-26	59-31012 <=>
1959 V30 N8 P67-72	60-23172 <+*>
1959 V30 N9 P126	60-15965 <+*>
1959 V30 N10 P3-8	60-13490 <+*>
1959 V30 N10 P8-16	60-15951 <+*>
1959 V30 N10 P16-20	60-13491 <+*>
1959 V30 N10 P20-24	60-13493 <+*>
1959 V30 N10 P72-75	60-13492 <+*>
1959 V30 N10 P130-136	60-15955 <+*>
1959 V30 N10 P155-156	60-31056 <=>
1959 V30 N11 P15-18	60-19898 <+>
	61-13792 <*+>
1959 V30 N11 P45-47	60-13494 <+*> 0
1959 V30 N11 P119	60-13495 <+*>
1959 V30 N12 P6-11	60-11505 <=>
1959 V30 N12 P48-54	60-11506 <=>
1960 V31 N1 P16-22	60-15963 <+*>
1960 V31 N1 P106-111	60-17963 <+*>
1960 V31 N1 P156-158	60-31705 <=>
1960 V31 N2 P133-139	60-17965 <+*> 0
1960 V31 N3 P18-21	60-23161 <+*>
1960 V31 N3 P22-27	60-23166 <+*>
1960 V31 N4 P10-15	60-23164 <+*>
1960 V31 N4 P60-64	60-23170 <+*>
1960 V31 N4 P64-66	60-23708 <+*>
1960 V31 N4 P84-87	60-23169 <+*>
1960 V31 N4 P87-93	60-23162 <+*>
1960 V31 N4 P94-99	60-17980 <+*>
1960 V31 N4 P102-107	60-23163 <+*>
1960 V31 N5 P22-25	60-23726 <+*>
1960 V31 N5 P62-64	60-23717 <+*>
1960 V31 N5 P117	60-23718 <+*>
1960 V31 N5 P119	60-23719 <+*>
	60-23720 <+*>
1960 V31 N6 P7-11	61-13808 <*+> 0
1960 V31 N6 P111-112	61-13807 <*+>
1960 V31 N6 P150-152	61-13810 <*+>
1960 V31 N7 P26-30	61-13811 <*+> 0
1960 V31 N7 P92-97	61-13806 <*+>
1960 V31 N7 P144-148	61-13809 <*+>
1960 V31 N8 P47-50	61-15834 <+*>
1960 V31 N10 P44-50	61-15860 <=*>
1960 V31 N10 P50-55	61-15822 <=*> 0
1960 V31 N10 P56-62	61-19648 <=*>
1960 V31 N10 P89-94	61-19649 <=*>
1960 V31 N11 P10-15	61-19756 <=*>
1960 V31 N11 P35-39	61-19760 <=>
1960 V31 N11 P43-49	61-19653 <=*> 0
1960 V31 N11 P166-167	61-23954 <*=>
1960 V31 N12 P38-44	61-19638 <=*>
1960 V31 N12 P90-94	61-19637 <=*> 0
1961 N5 P42-46	AD-630 736 <=*$>
1961 V32 N1 P40-46	61-19636 <=*> 0
1961 V32 N3 P129-135	61-27649 <*=>
1961 V32 N4 P6-10	61-28452 <*=>
1961 V32 N5 P9-15	61-27501 <*=>
1961 V32 N5 P115	61-27502 <*=>
1961 V32 N6 P104-110	61-27504 <*=>
1961 V32 N6 P147-151	61-28442 <*=>
1961 V32 N7 P12-19	62-13478 <=*>
1961 V32 N7 P56-62	62-13455 <=*>
1961 V32 N7 P74-77	62-13454 <=*>
1961 V32 N7 P114-117	62-13468 <=*>
1961 V32 N7 P117-123	62-23299 <=*> 0
1961 V32 N7 P137-139	62-13453 <=*>
1961 V32 N8 P74-78	63-21861 <=>
1961 V32 N9 P3-7	62-24754 <=*>
1961 V32 N11 P26-29	63-31323 <=>
1961 V32 N11 P52-56	63-31323 <=>
1961 V32 N11 P138-139	62-23292 <=*>
1961 V32 N12 P96-99	62-23287 <=*>
1962 N1 P35-	FPTS V22 N4 P.T639 <FASE>
1962 N2 P20-	FPTS V22 N4 P.T644 <FASE>
1962 N2 P52-	FPTS V22 N4 P.T649 <FASE>
1962 N2 P56-	FPTS V22 N2 P.T226 <FASE>
1962 N2 P116-	FPTS V22 N4 P.T647 <FASE>
1962 V33 N1 P57-62	63-19005 <=*>
1962 V33 N2 P8-14	62-25036 <=*>

Citation	Report
1962 V33 N2 P25-28	AD-638 588 <=$>
1962 V33 N2 P37-41	63-19006 <=*>
1962 V33 N2 P41-45	63-19385 <=*>
1962 V33 N2 P45-48	63-19009 <=*>
1962 V33 N2 P68-70	63-19386 <=*> C
1962 V33 N2 P70-73	63-19387 <=*>
1962 V33 N3 P9-14	63-31319 <=>
1962 V33 N3 P18-27	63-31319 <=>
1962 V33 N3 P103-108	63-19107 <=*> 0
1962 V33 N3 P123-127	63-19108 <=*>
1962 V33 N4 P9-14	62-32593 <=*>
	63-24528 <=*$>
1962 V33 N4 P23-26	63-24529 <=*$>
1962 V33 N4 P27-30	63-24530 <=*$>
1962 V33 N4 P110-115	63-24531 <=*$>
1962 V33 N5 P19-23	62-32087 <=*>
1962 V33 N5 P23-26	62-32088 <=*>
1962 V33 N5 P126-129	62-32093 <=*>
1962 V33 N6 P103-107	63-21928 <=>
1962 V33 N6 P126-130	63-19236 <=*>
1962 V33 N7 P32-35	63-19240 <=*>
1962 V33 N7 P46-50	62-32595 <=*>
	63-19234 <=*>
1962 V33 N7 P92-95	62-32596 <=*>
1962 V33 N7 P101-107	63-21913 <=>
1962 V33 N7 P114-116	63-11552 <=>
1962 V33 N7 P125-130	63-11552 <=>
1962 V33 N7 P134	63-11552 <=>
1962 V33 N8 P101-104	63-13027 <=*>
1962 V33 N9 P77-82	63-21911 <=>
1962 V33 N9 P105-111	63-19239 <=*>
1962 V33 N9 P111-115	63-19238 <=*>
1962 V33 N10 P3-7	63-31159 <=>
1962 V33 N10 P12-15	64-15203 <=*$>
1962 V33 N10 P26-30	63-31159 <=>
1962 V33 N10 P121-126	63-19106 <=*> C
	63-19552 <=*>
1962 V33 N10 P128-131	63-13031 <=*>
1962 V33 N11 P31-37	63-23686 <=*$>
1962 V33 N11 P47-52	63-24169 <=*$>
1962 V33 N11 P63-67	63-15464 <=*>
	63-19554 <=*>
1962 V33 N11 P76-82	63-23685 <=*$>
1962 V33 N11 P106-110	63-15463 <=*>
1962 V33 N11 P123-126	64-13345 <=*$>
1962 V33 N12 P44-49	63-24168 <=>
1962 V33 N12 P49-54	63-19463 <*=>
1962 V33 N12 P95-102	63-19609 <=*>
1962 V33 N12 P102-107	63-24170 <=*$>
1962 V33 N12 P112-114	63-31062 <=>
1963 N1 P52-57	63-31072 <=>
1963 N1 P73-79	AD-617 073 <=$> 0
1963 N1 P152	63-23090 <=*$>
1963 N2 P93-101	63-31044 <=>
1963 N3 P32-34	AD-630 738 <=$>
1963 N3 P55-58	63-31279 <=>
1963 N4 P46-49	63-31357 <=>
1963 N4 P57-71	63-31357 <=>
1963 N5 P3-7	64-21000 (P.1-11) <=>
1963 N5 P14-17	63-41306 <=>
	64-15298 <=*$>
1963 N5 P17-23	63-41306 <=>
1963 N5 P23-26	64-11754 <=>
1963 N5 P31-34	63-41306 <=>
	64-15205 <=*$> 0
1963 N5 P35-41	64-21000 <=>
	64-21000 (P.12-25) <=>
1963 N5 P41-47	64-21000 (P.26-36) <=>
1963 N5 P48-51	63-24386 <=*$>
1963 N5 P56-60	64-21000 (P.37-45) <=>
1963 N5 P60-64	64-21000 (P.46-54) <=>
1963 N5 P65-68	63-41306 <=>
1963 N5 P107-112	64-11756 <=>
1963 N5 P148-150	64-21000 (P.55-62) <=>
1963 N5 P152-153	63-41168 <=>
1963 N6 P79-84	64-13900 <=*$> 0
1963 N6 P99-101	63-41306 <=>
1963 N6 P101-105	63-41306 <=>
1963 N6 P108-114	64-13901 <=*$>
1963 N6 P129	63-41168 <=>

Citation	Reference
1963 N7 P40-45	63-24381 <=*$>
1963 N7 P77-82	63-41306 <=>
1963 N7 P94-97	63-31588 <=>
1963 N9 P120-125	AD-630 737 <=*$>
1963 V34 N6 P54-58	63-31518 <=>
1963 V34 N6 P74-78,131	63-31517 <=>
1963 V40 N1 P5-	FPTS V22 N6 P.T1028 <FASE>
1963 V40 N1 P5-13	FPTS,V22,P.T1028-32 <FASE>
1963 V40 N1 P20-	FPTS V22 N6 P.T1033 <FASE>
1963 V40 N1 P28-32	63-23120 <=*>
1963 V40 N1 P69-	FPTS V23 N1 P.T85 <FASE>
	FPTS,V23,P.T85-T87 <FASE>
1963 V40 N1 P73-	FPTS V22 N6 P.T1227 <FASE>
1963 V40 N1 P73-78	FPTS,V22,P.T1227-29 <FASE>
1963 V40 N1 P88-91	AD-618 652 <=$>
1963 V40 N1 P92-98	AD-620 497 <=$>
1963 V40 N1 P103-107	AD-619 002 <=$>
	63-23122 <=*>
1963 V40 N1 P127-132	AD-618 868 <=$>
	63-23121 <=*$>
1963 V40 N1 P132-	FPTS V22 N6 P.T1207 <FASE>
1963 V40 N1 P132-136	FPTS,V22,P.T1207-09 <FASE>
1963 V40 N2 P13-	FPTS V23 N1 P.T79 <FASE>
	FPTS,1964,V23,N1,T79 <FASE>
1963 V40 N2 P28-33	64-13524 <=*$>
1963 V40 N2 P37-42	64-13523 <=*$>
1963 V40 N2 P42-48	64-13522 <=*$>
1963 V40 N2 P51-57	64-13521 <=*$>
1963 V40 N2 P62-66	64-13520 <=*$>
1963 V40 N2 P82-88	64-13427 <=*$>
1963 V40 N2 P88-93	64-13347 <=*$>
	64-13891 <=*$> 0
1963 V40 N2 P103-	FPTS V23 N3 P.T547 <FASE>
1963 V40 N2 P103	FPTS,V23,N3,P.T547 <FASE>
1963 V40 N3 P3-9	64-13439 <=*$>
1963 V40 N3 P15-	64-13880 <=*$>
1963 V40 N3 P21-26	64-13438 <=*$> 0
1963 V40 N3 P26-	FPTS V23 N3 P.T551 <FASE>
	FPTS,V23,N3,P.T551 <FASE>
1963 V40 N3 P27-31	AD-625 611 <=*$>
1963 V40 N3 P50-55	64-13881 <=*$>
1963 V40 N3 P64-68	64-13838 <=*$>
1963 V40 N3 P91-95	65R75R <ATS>
1963 V40 N3 P96-100	64-13437 <=*$>
1963 V40 N3 P120	63-23123 <=*$>
1963 V40 N4 P117-122	63-31774 <=>
1963 V40 N4 P128-131	63-31774 <=>
1963 V40 N5 P17-23	AD-616 976 <=$>
	64-19246 <=*$>
1963 V40 N5 P27-	FPTS,V23,N4,P.T867 <FASE>
1963 V40 N5 P65-68	AD-616 971 <=$>
1963 V40 N5 P68-72	AD-616 972 <=$>
1963 V40 N5 P72-77	AD-616 768 <=$>
1963 V40 N5 P78-83	AD-616 769 <=$>
1963 V40 N5 P113-117	AD-625 610 <=*$>
1963 V40 N5 P128-	FPTS,V23,N4,P.T860 <FASE>
1963 V40 N5 P133-137	64-15413 <=*$>
1963 V40 N5 P152-153	AD-616 694 <=$>
1963 V40 N6 P45-48	64-15482 <=*$>
1963 V40 N6 P84-91	64-15412 <=*$>
1963 V40 N7 P51-55	64-19248 <=*$>
1963 V40 N7 P55-60	64-15477 <=*$>
1963 V40 N7 P61-64	64-19249 <=*$>
1963 V40 N7 P68-72	64-13694 <=*$>
1963 V40 N7 P72-77	64-13693 <=*$>
1963 V40 N7 P77-82	64-19284 <=*$>
1963 V40 N7 P98-102	AD-625 642 <=$>
1963 V40 N7 P98-	FPTS,V23,N4,P.T845 <FASE>
1963 V40 N7 P116-120	64-19210 <=*$>
1963 V40 N8 P19-24	64-19285 <=*$>
1963 V40 N8 P24-29	64-19343 <=$>
1963 V40 N8 P39-44	AD-616 975 <=$>
1963 V40 N8 P49-53	64-13691 <=*$>
1963 V40 N8 P76-82	64-30051 <*>
1963 V40 N8 P101-104	64-19250 <=*$>
1963 V40 N8 P141-142	AD-617 375 <=$>
	64-13691 <=*$>
1963 V40 N9 P49-52	64-15478 <=*$> 0
1963 V40 N9 P73-78	64-15479 <=*$>
1963 V40 N9 P87-92	64-15077 <=*$>
1963 V40 N9 P93-96	64-15078 <=*$>
1963 V40 N9 P134-135	65-13466 <*>
1963 V40 N10 P5-12	64-19679 <=$>
1963 V40 N10 P17-21	64-19251 <=*$>
1963 V40 N10 P26-33	64-19252 <=*$> 0
1963 V40 N10 P33-38	65-60172 <=$>
1963 V40 N10 P43-47	64-19211 <=*$>
1963 V40 N10 P51-58	AD-619 001 <=$>
1963 V40 N10 P70-74	64-15667 <=*$>
1963 V40 N10 P85-92	64-15668 <=$>
1963 V40 N10 P92-96	AD-619 933 <=$>
	64-15666 <=*$>
1963 V40 N10 P125-130	AD-619 936 <=$>
1963 V40 N11 P3-6	AD-619 937 <=$> 0
1963 V40 N11 P3-7	64-15671 <=*$>
1963 V40 N11 P7-12	AD-618 864 <=$>
1963 V40 N11 P16-21	64-19631 <=$>
1963 V40 N11 P29-32	AD-619 938 <=$>
1963 V40 N11 P33-39	64-19632 <=$>
1963 V40 N11 P65-69	AD-618 866 <=$>
	64-19633 <=$> 0
1963 V40 N11 P111-114	64-15665 <=*$>
1963 V40 N11 P127-132	64-15664 <=*$>
1963 V40 N11 P142	AD-618 867 <=$>
1963 V40 N11 P146	AD-617 389 <=$>
	64-15663 <=*$>
1963 V40 N12 P9-13	64R75R <ATS>
1963 V40 N12 P22-28	64-15861 <=*$>
1963 V40 N12 P46-51	65-62629 <=$>
1963 V40 N12 P56-60	66-60973 <=$*>
1963 V40 N12 P84-88	64-15870 <=*$>
1963 V40 N12 P117-118	64-15871 <=*$>
1963 V40 N40 P85-92	64-15668 <=*$>
1964 V41 P51-53	65-63430 <=*$>
1964 V41 N1 P3-5	AD-619 336 <=$>
1964 V41 N1 P96-101	64-15670 <=*$>
1964 V41 N1 P101-108	64-15416 <=*$> 0
1964 V41 N1 P112-	FPTS V24 N3 P.T489 <FASE>
1964 V41 N1 P119-125	AD-618 651 <=$>
	64-15420 <=*$>
1964 V41 N1 P126-129	64-15406 <=*$>
1964 V41 N1 P135-140	AD-617 393 <=$>
1964 V41 N1 P135-141	64-15421 <=*$>
1964 V41 N1 P141-142	64-15874 <=*$>
	65-63162 <=*$>
1964 V41 N2 P10-	FPTS V23 N6 P.T1340 <FASE>
	FPTS V23 P.T1340 <FASE>
1964 V41 N2 P16-21	AD-618 863 <=$>
1964 V41 N2 P38-43	64-21927 <=>
1964 V41 N2 P43-48	64-15422 <=*$>
1964 V41 N2 P60-64	64-19031 <=*$>
1964 V41 N2 P70-	FPTS V24 N1 P.T27 <FASE>
1964 V41 N2 P70-75	FPTS V24 P.T27-T30 <FASE>
1964 V41 N2 P85-88	64-19032 <=*$>
1964 V41 N2 P102-108	65-60180 <=$>
1964 V41 N2 P122-128	66-61481 <=*$>
1964 V41 N2 P143	64-19033 <=*$>
1964 V41 N2 P149	64-19034 <=*$>
1964 V41 N3 P3-	FPTS V24 N1 P.T31 <FASE>
1964 V41 N3 P3-6	FPTS V24 P.T31-T33 <FASE>
1964 V41 N3 P7-11	64-31086 <=>
1964 V41 N3 P24-27	66-61482 <=*$>
1964 V41 N3 P27-33	64-31086 <=>
1964 V41 N3 P65-69	64-19899 <=$>
1964 V41 N3 P109-114	64-31086 <=>
1964 V41 N3 P126-129	65-61046 <=$>
1964 V41 N4 P22-	FPTS V24 N4 P.T735 <FASE>
1964 V41 N4 P93-	FPTS V24 N3 P.T512 <FASE>
1964 V41 N4 P100-105	FPTS V24 N3 P.T470 <FASE>
	64-31389 <=>
1964 V41 N4 P115-118	AD-619 006 <=$>
1964 V41 N5 P3-14	64-31586 <=>
1964 V41 N5 P14-19	64-19189 <=*$>
1964 V41 N5 P19-24	AD-620 501 <=*$>
1964 V41 N5 P24-33	64-31586 <=>
1964 V41 N5 P36-60	64-31586 <=>
1964 V41 N5 P100-101	64-31586 <=>
1964 V41 N5 P101-104	66-61484 <=*$>
1964 V41 N5 P105-109	65-62209 <=$>

Citation	Translation
1964 V41 N5 P138-146	64-31586 <=>
1964 V41 N5 P149-152	64-19190 <=*$>
1964 V41 N5 P152-154	64-19191 <=*$>
1964 V41 N6 P3-7	AD-638 596 <=$>
	64-19838 <=$>
1964 V41 N6 P12-17	AD-619 423 <=*$>
1964 V41 N6 P29-34	64-19839 <=$> 0
1964 V41 N6 P60-67	64-41870 <=$>
1964 V41 N6 P70-	FPTS V24 N3 P.T497 <FASE>
1964 V41 N6 P80-85	65-62165 <=>
1964 V41 N6 P125-130	65-60904 <=$>
1964 V41 N7 P3-5	AD-619 337 <=$>
1964 V41 N7 P5-10	64-19846 <=$>
1964 V41 N7 P15-19	65-11750 <*>
	65-60901 <=$> 0
1964 V41 N7 P15-20	65-63635 <=$>
1964 V41 N7 P39-42	64-19847 <=$>
	64-41968 <=$>
1964 V41 N7 P54-59	65-63070 <=$>
1964 V41 N7 P79-84	64-19848 <=>
1964 V41 N7 P113-	FPTS V24 N4 P.T745 <FASE>
1964 V41 N7 P117-123	64-41968 <=$>
1964 V41 N7 P123-	FPTS V24 N5 P.T921 <FASE>
1964 V41 N7 P129-134	65-62360 <=*>
	65-63149 <=$>
1964 V41 N7 P134-140	64-19903 <=$>
1964 V41 N7 P147-148	64-19904 <=$>
1964 V41 N7 P152-153	65-63071 <=$>
1964 V41 N8 P3-7	AD-619 401 <=*$>
	64-51183 <=>
1964 V41 N8 P8-12	AD-619 403 <=$>
1964 V41 N8 P16-18	65-63072 <=$>
1964 V41 N8 P26-29	AD-638 594 <=$>
	64-51183 <=>
	65-62361 <=$>
1964 V41 N8 P41-50	64-51183 <=>
1964 V41 N8 P45-50	AD-619 341 <=$>
1964 V41 N8 P93-120	64-51183 <=>
1964 V41 N8 P140-141	AD-638 584 <=$>
1964 V41 N9 P55-59	AD-619 405 <=*$>
1964 V41 N9 P60-64	AD-619 009 <=$>
1964 V41 N9 P65-	FPTS V24 N6 P.T1081 <FASE>
1964 V41 N9 P104-107	AD-619 010 <=$>
1964 V41 N9 P129-135	65-63636 <=$>
1964 V41 N9 P147-148	AD-619 007 <=$>
1964 V41 N10 P3-7	AD-625 651 <=*$>
1964 V41 N10 P3-16	65-51891 <=*$>
1964 V41 N10 P7-12	65-63634 <=$>
1964 V41 N10 P17-22	AD-618 865 <=$>
1964 V41 N10 P27-31	65-51891 <=*$>
1964 V41 N10 P32-37	AD-619 402 <=*$>
1964 V41 N10 P47-50	65-51891 <=*$>
	65-63633 <=$>
1964 V41 N10 P61-66	AD-619 406 <=*$>
1964 V41 N10 P61-70	65-51891 <=*$>
1964 V41 N10 P66-70	AD-625 620 <=*$>
1964 V41 N10 P85-88	AD-625 645 <=$>
1964 V41 N10 P85-89	AD-628 492 <=$>
	65-51891 <=*$>
1964 V41 N10 P90-93	65-11751 <*>
1964 V41 N10 P94-98	AD-625 647 <=*$>
1964 V41 N10 P98-102	65-51891 <=*$>
1964 V41 N10 P107-	FPTS V24 N6 P.T1079 <FASE>
1964 V41 N10 P116-120	AD-620 134 <=*$>
	AD-625 623 <=*$>
1964 V41 N10 P125-130	65-51891 <=*$>
1964 V41 N10 P135-139	AD-619 893 <=$>
	65-51891 <=*$>
1964 V41 N11 P43-	FPTS V24 N6 P.T1073 <FASE>
1964 V41 N11 P48-54	AD-619 404 <=$>
1964 V41 N11 P48-53	65-63220 <=>
1964 V41 N11 P54-58,65-68	65-30346 <=*$>
1964 V41 N11 P99-103	65-63221 <=> 0
1964 V41 N11 P128-133	AD-625 631 <=*$>
1964 V41 N11 P128-	FPTS V24 N6 P.T1076 <FASE>
1964 V41 N11 P134-136	AD-625 657 <=$>
1964 V41 N11 P138-139	65-31121 <=$>
1964 V41 N11 P139-140	AD-625 628 <=*$>
1964 V41 N11 P142-143	AD-625 655 <=*$>
	65-31121 <=$>
1964 V41 N11 P145	AD-625 608 <=*$>
1964 V41 N12 P3-8	65-63067 <=*$>
1964 V41 N12 P32-35	AD-638 586 <=$>
1964 V41 N12 P43-48	65-63066 <=*$>
1964 V41 N12 P48-53	65-63379 <=*$>
1964 V41 N12 P71-74	65-63386 <=$*>
1964 V41 N12 P78-83	65-12091 <*>
1964 V41 N12 P83-	FPTS V24 N6 P.T1087 <FASE>
1964 V41 N12 P104-109	65-11993 <*>
1965 N3 P148	AD-638 592 <=$>
1965 V42 N1 P3-5	AD-638 576 <=$>
1965 V42 N1 P5-10	AD-619 939 <=$>
1965 V42 N1 P14-21	AD-620 500 <=*$>
1965 V42 N1 P14-20	65-30807 <=$>
1965 V42 N1 P37-	FPTS V25 N1 P.T143 <FASE>
1965 V42 N1 P37-41	65-13667 <*>
1965 V42 N1 P47-52	AD-618 885 <=$>
1965 V42 N1 P57-60	AD-619 940 <=$>
1965 V42 N1 P61-69	65-30807 <=$>
1965 V42 N1 P70-71	65-13460 <*>
1965 V42 N1 P70-72	66-60396 <=*$>
1965 V42 N1 P82-85	AD-619 884 <=$>
	65-30808 <=$>
1965 V42 N1 P91-96	AD-638 579 <=$>
1965 V42 N1 P107-111	65-13749 <*> 0
1965 V42 N1 P120-129	65-30807 <=$>
1965 V42 N1 P123-129	AD-620 498 <=*$>
1965 V42 N1 P130-135	66-60397 <=*$>
1965 V42 N1 P150-152	65-31022 <=$>
1965 V42 N2 P66-73	65-31214 <=$>
1965 V42 N2 P68-73	65-30863 <=$>
1965 V42 N2 P90-95	66-60863 <=*$>
1965 V42 N2 P105-110	65-14234 <*> 0
1965 V42 N2 P150-151	AD-638 585 <=$>
1965 V42 N2 P156-157	AD-638 595 <=$>
1965 V42 N3 P3-7	AD-638 581 <=$>
1965 V42 N3 P14-	FPTS V25 N1 P.T141 <FASE>
1965 V42 N3 P58-65	65-63560 <=$>
1965 V42 N3 P81-87	66-60859 <=*$>
1965 V42 N3 P111-122	65-31017 <=$>
1965 V42 N3 P130-	FPTS V25 N1 P.T139 <FASE>
1965 V42 N3 P154-157	65-31017 <=$>
1965 V42 N4 P27-32	65-31016 <=$>
1965 V42 N4 P41-47	66-61063 <=*$>
1965 V42 N4 P69-	FPTS V25 N2 P.T341 <FASE>
1965 V42 N4 P78-80	AD-630 857 <=*$>
1965 V42 N4 P105-	FPTS V25 N2 P.T333 <FASE>
1965 V42 N5 P33-	FPTS V25 N2 P.T300 <FASE>
1965 V42 N5 P114-119	66-60898 <=*$>
1965 V42 N5 P130-133	66-60874 <=*$>
1965 V42 N5 P143-	FPTS V25 N2 P.T339 <FASE>
1965 V42 N6 P56-60	66-61074 <=$>
1965 V42 N6 P77-80	66-61043 <=*$>
1965 V42 N7 P16-20	65-32322 <=$*>
1965 V42 N7 P16-21	65-64510 <=*>
1965 V42 N7 P58-	FPTS V25 N4 P.T687 <FASE>
1965 V42 N7 P118-	FPTS V25 N4 P.T701 <FASE>
1965 V42 N7 P139-144	66-61632 <=*$>
1965 V42 N8 P27-30	65-32729 <=*$>
1965 V42 N9 P3-	FPTS V25 N5 P.T913 <FASE>
1965 V42 N9 P6-	FPTS V25 N5 P.T907 <FASE>
1965 V42 N9 P66-70	66-11769 <*>
1965 V42 N9 P129-134	65-33691 <=*$>
1965 V42 N10 P7-	FPTS V25 N5 P.T839 <FASE>
1965 V42 N10 P22-	FPTS V25 N5 P.T827 <FASE>
1965 V42 N10 P54-	FPTS V25 N5 P.T829 <FASE>
1965 V42 N10 P54-61	66-30436 <=$>
1965 V42 N10 P105-111	66-62003 <=*$>
1965 V42 N10 P111-	FPTS V25 N5 P.T833 <FASE>
1965 V42 N10 P125-131	66-30251 <=$>
1965 V42 N11 P108-111	66-61495 <=*>
1966 N2 P3-10	66-31387 <=$>
1966 N4 P17-21	66-32368 <=$>
1966 N4 P84-86	66-32365 <=$>
1966 N4 P98-103	66-32367 <=$>
1966 N4 P139-140	66-32366 <=$>
1966 V43 N2 P94-	FPTS V25 N6 P.T1099 <FASE>
1966 V43 N3 P3-8	66-13017 <*>
1966 V43 N3 P22-27	66-13151 <*>

ZHURNAL MOSKOVSKOI PATRIARKII

1965 N7 P25-29	65-32697 <=*$>
1966 N4 P35-36	66-33798 <=$>
1966 V2 N1/2 P40-43	66-32314 <=$>

ZHURNAL NAUCHNOI I PRIKLADNOI FOTOGRAFII I KINEMATOGRAFII

1956 V1 P266-271	61-13328 <*+>
1956 V1 P354-358	61-13343 <*+>
1956 V1 P419-424	61-13348 <*+>
1956 V1 N1 P39-41	R-925 <*>
1956 V1 N2 P89-97	61K114 <CTT>
1956 V1 N3 P174-182	R-926 <*>
1956 V1 N5 P359-361	02Q73R <ATS>
1956 V1 N6 P413-418	R-3074 <*>
1956 V2 N1 P15-21	R-3074 <*>
1956 V2 N4 P293-303	AEC-TR-3502 <+*>
1957 N2 P106-109	R-4350 <*>
1957 V2 P7-12	R-1624 <*>
1957 V2 N2 P81-90	R-4930 <*>
	61-15620 <*+>
1957 V2 N3 P172-175	R-4918 <*>
1957 V2 N3 P191-197	R-3277 <*>
1957 V2 N3 P198-201	R-3076 <*>
1957 V2 N3 P228-233	R-4917 <*>
1957 V2 N4 P253-259	R-4915 <*>
	61-15617 <+*>
1957 V2 N4 P260-269	R-3075 <*>
1957 V2 N4 P270-276	R-4929 <*>
	61-15619 <*+>
1957 V2 N5 P325-329	59-17281 <+*>
1957 V2 N5 P330-339	R-3280 <*>
1957 V2 N5 P340-343	R-3279 <*>
1957 V2 N5 P344-348	R-3284 <*>
1957 V2 N5 P389-399	59-17636 <+*>
1957 V2 N6 P437-444	R-4926 <*>
1957 V2 N6 P459-468	60-14001 <+*>
1958 N1 P51-52	59-10246 <+*>
1958 V3 N1 P19-24	59-15027 <+*>
1958 V3 N1 P34-38	60-18484 <*+>
1958 V3 N1 P52-53	59-10237 <+*>
1958 V3 N1 P53-54	59-10245 <+*>
1958 V3 N1 P55-56	59-11048 <=>
1958 V3 N1 P55-65	60-10872 <+*>
1958 V3 N1 P68-72	59-10238 <+*>
1958 V3 N2 P81-87	59-15028 <+*>
1958 V3 N2 P88-95	59-10915 <+*>
1958 V3 N2 P96-100	59-10773 <+*>
1958 V3 N2 P101-111	94N50R <ATS>
1958 V3 N2 P112-116	59-10774 <+*>
1958 V3 N2 P117-119	59-10776 <+*>
1958 V3 N2 P136-137	59-10771 <+*>
1958 V3 N2 P139-140	59-10772 <+*>
1958 V3 N3 P161-165	59-15591 <+*>
1958 V3 N3 P166-172	59-15590 <+*>
1958 V3 N3 P173-181	59-15589 <+*>
1958 V3 N3 P182-190	59-15704 <+*>
1958 V3 N3 P191-196	59-15703 <+*> 0
	59-19814 <+*>
1958 V3 N3 P197-206	59-15793 <+*>
1958 V3 N3 P207-209	59-15588 <+*>
1958 V3 N3 P224-225	59-15181 <+*>
1958 V3 N3 P225-226	59-15702 <+*>
1958 V3 N4 P246-250	59-17696 <+*>
1958 V3 N4 P251-255	59-17695 <+*>
1958 V3 N4 P256-266	59-17603 <+*>
1958 V3 N4 P281-282	59-17744 <+*>
1958 V3 N4 P282-284	59-17748 <+*>
1958 V3 N4 P284	59-17745 <+*>
1958 V3 N4 P285	59-17746 <+*>
1958 V3 N4 P286	59-17604 <+*>
1958 V3 N4 P287-288	59-17747 <+*>
1958 V3 N4 P289-293	59-20838 <+*>
1958 V3 N4 P299-305	59-17605 <+*>
1958 V3 N4 P306-310	59-17601 <+*>
1958 V3 N5 P335-344	54L35R <ATS>
	61-15553 <*=>
1958 V3 N5 P363-367	TS-1839 <BISI>
1958 V3 N5 P379-380	59-20837 <+*>
1958 V3 N6 P416-418	60-18484 <*+>
	63-16710 <=*>
1958 V3 N6 P419-426	59-15797 <+*>
1958 V3 N6 P427-429	60-14157 <+*>
1958 V3 N6 P430-439	60-14156 <+*>
	61-11178 <=>
1958 V3 N6 P440-442	60-14155 <+*>
1958 V3 N6 P452-468	59-20839 <+*>
1959 V4 P64-66	61-20983 <=*>
1959 V4 N1 P3-11	60-10591 <+*>
1959 V4 N1 P32-34	35P63R <ATS>
	60-14000 <+*>
	61-20978 <*=> C
1959 V4 N1 P49-55	60-10593 <+*>
1959 V4 N1 P62-64	60-10870 <+*>
1959 V4 N2 P94-99	53L35R <ATS>
	60-10273 <+*>
1959 V4 N2 P100-105	33L34R <ATS>
1959 V4 N2 P116-120	60-10595 <+*>
1959 V4 N2 P121-126	60-10875 <+*>
1959 V4 N2 P133-135	60-10876 <+*>
1959 V4 N2 P135-136	60-10874 <+*>
1959 V4 N2 P136-138	63-16635 <=*>
1959 V4 N2 P144-148	60-10873 <+*>
1959 V4 N2 P149-152	60-10594 <+*>
1959 V4 N3 P161-171	60-10275 <+*>
1959 V4 N3 P172-174	59-20007 <+*>
1959 V4 N3 P175-182	60-14278 <+*>
1959 V4 N3 P193-197	60-10274 <+*>
1959 V4 N3 P202-209	61-20225 <*=>
1959 V4 N3 P235-236	60-14154 <+*>
1959 V4 N3 P237-238	59-17600 <+*>
1959 V4 N4 P241-252	60-14276 <+*>
1959 V4 N4 P253-258	60-14153 <+*>
1959 V4 N4 P259-268	63-18782 <=*>
1959 V4 N4 P276-284	60-16232 <+*>
1959 V4 N4 P292-295	60-16558 <+*>
1959 V4 N4 P300-301	63-16711 <*=>
1959 V4 N4 P311-312	60-14152 <+*>
1959 V4 N4 P313-316	60-14151 <+*>
1959 V4 N4 P317-319	60-14150 <+*>
1959 V4 N4 P319-320	60-14149 <+*>
1959 V4 N5 P321-328	60-16301 <+*>
1959 V4 N5 P329-333	60-16300 <+*>
1959 V4 N5 P345-355	60-16232 <+*>
1959 V4 N5 P356-362	60-16532 <+*>
1959 V4 N5 P380-381	60-14669 <+*>
1959 V4 N5 P381-383	60-14673 <+*>
1959 V4 N5 P385-389	60-16533 <+*>
1959 V4 N5 P393-399	60-16299 <+*>
1959 V4 N6 P401-410	60-16541 <+*>
1959 V4 N6 P411-416	60-16532 <+*>
1959 V4 N6 P417-422	60-16746 <+*>
1959 V4 N6 P423-426	60-16540 <+*>
	63-16712 <=*>
1959 V4 N6 P427-429	60-16539 <+*>
1959 V4 N6 P430-432	60-16536 <+*>
1959 V4 N6 P433-437	60-16535 <+*>
1959 V4 N6 P446-457	60-16537 <+*>
1959 V4 N6 P458-459	60-16531 <+*>
1959 V4 N6 P460	60-16297 <+*>
1959 V4 N6 P461-462	60-16530 <+*>
1959 V4 N6 P462-463	60-16298 <+*>
1959 V4 N6 P463-464	60-16542 <+*>
1959 V4 N6 P465-476	60-16538 <+*>
1960 V5 P367-368	36P63R <ATS>
1960 V5 N1 P7-9	60-18862 <+*>
1960 V5 N1 P10-14	60-18863 <+*>
1960 V5 N1 P15-19	60-18864 <+*>
1960 V5 N1 P20-27	60-18865 <+*>
1960 V5 N1 P28-33	61-10280 <*+>
1960 V5 N1 P34-38	61-10276 <*+>
1960 V5 N1 P54-55	61-10278 <*+>
1960 V5 N1 P56-57	61-10277 <*+>
1960 V5 N1 P57-58	60-18866 <+*>
1960 V5 N1 P58-60	60-18872 <+*>
1960 V5 N1 P60-61	60-18873 <+*>
1960 V5 N1 P62-63	60-18874 <+*>
1960 V5 N1 P63-64	60-18875 <+*>
1960 V5 N1 P65-74	60-18876 <+*>
1960 V5 N1 P74-76	61-10279 <*+>

Citation	Translation
1960 V5 N2 P81-83	60-18870 <+*>
1960 V5 N2 P84-89	60-18871 <+*>
1960 V5 N2 P90-97	61-18283 <*=>
1960 V5 N2 P98-100	61-18282 <*=>
1960 V5 N2 P101-107	61-20229 <*=>
1960 V5 N2 P108-113	60-18869 <+*>
1960 V5 N2 P114-122	60-18868 <+*>
1960 V5 N2 P133-140	61-18284 <*=>
1960 V5 N2 P141-142	61-18763 <*=>
1960 V5 N2 P142-143	61-18764 <*=>
1960 V5 N2 P145-146	61-10325 <*+>
1960 V5 N2 P146-147	61-10324 <=*>
1960 V5 N2 P158-159	60-18867 <+*>
1960 V5 N3 P161-167	61-20064 <*=>
1960 V5 N3 P168-172	61-20065 <*=>
1960 V5 N3 P173-182	61-20071 <*=>
1960 V5 N3 P183-186	61-20070 <*=>
	63-16713 <=*>
1960 V5 N3 P187-194	61-20062 <*>
1960 V5 N3 P207-208	61-18672 <*=>
1960 V5 N3 P218-219	61-20227 <*=>
1960 V5 N3 P219-220	61-20058 <*=>
1960 V5 N3 P221-223	61-18673 <*=>
1960 V5 N3 P221	61-20063 <*=>
1960 V5 N3 P225-230	61-20072 <*=>
1960 V5 N3 P234-238	61-18309 <*=>
1960 V5 N4 P241-246	62-10984 <=*>
1960 V5 N4 P247-254	61-20069 <*=>
1960 V5 N4 P255-261	62-10978 <=*>
1960 V5 N4 P274-279	61-31689 <=>
1960 V5 N4 P293-294	62-10985 <=*>
1960 V5 N4 P294-295	61-20061 <*=>
	62-10026 <*=>
1960 V5 N4 P295-296	61-20757 <*=>
1960 V5 N4 P297-298	62-10973 <=*>
1960 V5 N4 P299-301	63-16714 <=*>
1960 V5 N5 P327-330	62-18580 <=*>
1960 V5 N5 P331-333	61-18311 <*=>
1960 V5 N5 P334-342	97N52R <ATS>
1960 V5 N5 P352-355	62-18579 <=*>
1960 V5 N5 P361-363	62-18581 <=*>
1960 V5 N5 P363-364	62-18879 <=*>
1960 V5 N5 P380-395	62-18680 <=*>
1960 V5 N6 P401-402	63-16715 <*=>
1960 V5 N6 P413-418	15N56R <ATS>
1960 V5 N6 P433-438	61-18313 <*=>
1960 V5 N6 P452-453	61-20066 <*=>
1960 V5 N6 P463-472	62-18770 <=*>
1961 V6 N1 P6-13	61K116 <CTT>
1961 V6 N1 P34-38	61K115 <CTT>
1961 V6 N1 P64-66	61-20758 <*=> 0
1961 V6 N2 P92-96	87N53R <ATS>
1961 V6 N2 P154-159	61-28508 <=>
	66-33956 <=$>
1961 V6 N3 P186-192	61K121 <CTT>
1961 V6 N3 P193-202	63-24238 <=*$>
1961 V6 N3 P220-224	46N57R <ATS>
1961 V6 N3 P225-226	47N57R <ATS>
1961 V6 N3 P227-229	48N57R <ATS>
1961 V6 N4 P247-250	63-14477 <=*>
1961 V6 N4 P264-273	62P59R <ATS>
1961 V6 N4 P274-285	61P59R <ATS>
1961 V6 N4 P289-293	60P59R <ATS>
1961 V6 N4 P294-296	63-16633 <*=>
1961 V6 N4 P300-301	64-10901 <=*$>
1961 V6 N5 P323-333	63-20132 <=*>
1961 V6 N5 P349-352	65-60689 <=$>
1961 V6 N5 P358-362	13P61R <ATS>
	63-16709 <*=>
1961 V6 N5 P363-366	62-18582 <=*>
1961 V6 N5 P382-385	14P61R <ATS>
1961 V6 N6 P397-403	63-19181 <=*>
1961 V6 N6 P418-420	84P61R <ATS>
1961 V6 N6 P440-447	62-18597 <=*>
	63-10726 <=*>
1961 V6 N6 P462-475	62-18598 <=*>
1962 V7 N1 P5-19	62-18589 <=*>
1962 V7 N1 P20-25	63-18727 <=*>
1962 V7 N1 P26-29	62-18585 <=*>
1962 V7 N1 P30-35	62-18592 <=*>
	62-20374 <=*>
	63-10727 <=*>
	63-18767 <=*>
1962 V7 N1 P57-58	62-18586 <=*>
1962 V7 N1 P59-60	62-18591 <=*>
	62-20373 <=*>
	65-14817 <*>
1962 V7 N1 P61-62	62-18587 <=*>
1962 V7 N1 P62-64	62-18588 <=*>
1962 V7 N1 P64-66	62-18583 <=*>
1962 V7 N1 P66-67	62-18590 <=*>
1962 V7 N1 P68-69	62-18584 <=*>
1962 V7 N1 P76-80	62-24717 <=*>
1962 V7 N2 P86-90	63-16634 <*=>
1962 V7 N2 P91-95	62-18878 <=*>
1962 V7 N2 P96-102	62-18599 <=*>
1962 V7 N2 P103-111	63-18073 <=*$>
	63-18183 <=*>
1962 V7 N2 P112-120	62-18600 <=*>
	64-13167 <=*$>
1962 V7 N2 P121-127	62-18594 <=*>
1962 V7 N2 P128-132	62-18595 <=*>
1962 V7 N2 P133-140	63-14474 <=*>
1962 V7 N2 P141-142	63-18074 <*=>
1962 V7 N2 P142-145	63-18075 <=*> C
1962 V7 N2 P145-146	62-18596 <=*>
1962 V7 N2 P146-148	63-18072 <=*>
1962 V7 N2 P151-157	63-16640 <=*>
1962 V7 N2 P157-160	62-18593 <=*>
1962 V7 N3 P161-168	63-18725 <=*>
1962 V7 N3 P176-181	03Q68R <ATS>
1962 V7 N3 P182-186	63-18184 <=*>
1962 V7 N3 P187-194	97Q73R <ATS>
1962 V7 N3 P209-218	63-20073 <*>
	64-14779 <=*$>
1962 V7 N3 P225-226	63-14476 <=*>
1962 V7 N3 P229-239	28Q68R <ATS>
1962 V7 N4 P245-251	77P65R <ATS>
1962 V7 N4 P252-256	79P65R <ATS>
1962 V7 N4 P262-267	98Q73R <ATS>
1962 V7 N4 P268-271	52P66R <ATS>
1962 V7 N4 P297-298	95Q73R <ATS>
1962 V7 N4 P300-304	78P65R <ATS>
1962 V7 N4 P304-306	00Q74R <ATS>
1962 V7 N4 P306-308	38P66R <ATS>
1962 V7 N5 P333-340	39Q68R <ATS>
1962 V7 N5 P341-347	63-20019 <=*>
1962 V7 N5 P369-379	65Q68R <ATS>
1962 V7 N5 P380-381	74Q71R <ATS>
1962 V7 N5 P381-383	62Q68R <ATS>
1962 V7 N5 P384-385	63Q68R <ATS>
1962 V7 N5 P386-388	64Q68R <ATS>
1962 V7 N5 P394-402	75Q71R <ATS>
1962 V7 N6 P409-417	66-13724 <*>
1962 V7 N6 P447-453	63-20528 <=*$>
1962 V7 N6 P461-462	66Q68R <ATS>
1962 V7 N6 P463-464	67Q68R <ATS>
1962 V7 N6 P465-467	68Q68R <ATS>
1962 V7 N6 P470-471	69Q68R <ATS>
1963 V8 N1 P3-10	63-18779 <=*>
1963 V8 N1 P10-21	63-18786 <*>
1963 V8 N1 P29-36	63-18787 <=*$>
1963 V8 N1 P36-41	70Q72R <ATS>
1963 V8 N1 P47-50	63-18726 <=*>
1963 V8 N1 P57-58	71Q72R <ATS>
1963 V8 N1 P59-61	63-18783 <=*>
1963 V8 N1 P67-69	63-18784 <=*$>
1963 V8 N1 P69-70	63-18785 <=*$>
1963 V8 N1 P71-74	63-18778 <=*>
1963 V8 N2 P81-86	64-10902 <=*$>
1963 V8 N2 P87-91	63-18533 <=*>
	64-10903 <=*$>
1963 V8 N2 P92-97	64-10887 <=*$>
1963 V8 N2 P101-105	40Q71R <ATS>
	63-20261 <=*>
1963 V8 N2 P105-109	63-18599 <=*>
	81Q72R <ATS>
1963 V8 N2 P135-136	63-18597 <=*>
1963 V8 N2 P135-137	63-18730 <=*>
	82Q72R <ATS>

Citation	Translation
1963 V8 N2 P137-139	63-18729 <=*>
1963 V8 N2 P140-142	98Q71R <ATS>
1963 V8 N2 P142-144	64-10888 <=*$>
1963 V8 N2 P144-146	64-10889 <=*$>
1963 V8 N2 P148-149	64-10890 <=*$>
1963 V8 N2 P149-151	64-10891 <=*$>
1963 V8 N2 P151-152	64-10892 <=*$>
1963 V8 N2 P153-156	64-10893 <=*$>
1963 V8 N3 P161-164	64-10904 <=*$>
1963 V8 N3 P165-166	64-10885 <=*$>
1963 V8 N3 P167-174	41Q71R <ATS>
	63-20020 <=*>
1963 V8 N3 P174-184	64-14588 <=*$>
1963 V8 N3 P185-189	64-14589 <=*$>
1963 V8 N3 P189-193	88R74R <ATS>
1963 V8 N3 P193-199	64-20085 <*>
1963 V8 N3 P203-204	64-14445 <=*$>
1963 V8 N3 P205-206	64-14447 <=*$> 0
1963 V8 N3 P206-209	64-10886 <=*$>
1963 V8 N3 P209-210	64-10884 <=*$>
1963 V8 N3 P211-212	63-20262 <=*>
	64-10883 <=*$>
1963 V8 N3 P212-213	63-20260 <=*>
1963 V8 N3 P212-214	64-14441 <=*$>
1963 V8 N3 P214-215	83Q72R <ATS>
1963 V8 N3 P216-228	64-18529 <*>
1963 V8 N3 P230-244	64-18542 <*>
1963 V8 N4 P249-252	64-10894 <=*$>
1963 V8 N4 P253-261	64-10895 <=*$>
1963 V8 N4 P261-267	64-10896 <=*$>
1963 V8 N4 P267-270	72Q72R <ATS>
1963 V8 N4 P270-275	AD-614 398 <=$>
	64-10897 <=*$>
1963 V8 N4 P284-292	64-14444 <=*$>
1963 V8 N4 P293-302	64-20084 <*>
1963 V8 N4 P303-304	64-10898 <=*$>
1963 V8 N4 P304-305	64-10899 <=*$>
1963 V8 N4 P309-310	64-10900 <=*$>
1963 V8 N4 P311-312	64-14442 <=*$>
1963 V8 N5 P321-326	64-10410 <=*$>
	64-16667 <=*$>
1963 V8 N5 P327-334	64-14590 <=*$>
1963 V8 N5 P334-337	64-14448 <=*$>
1963 V8 N5 P337-342	AD-614 389 <=$>
	RTS-2968 <NLL>
1963 V8 N5 P342-348	64-16666 <=*$>
1963 V8 N5 P348-352	64-14591 <=*$>
1963 V8 N5 P359-361	64-14592 <=*$>
1963 V8 N5 P375-376	64-14593 <=*$>
1963 V8 N5 P376-378	64-14443 <=*$>
1963 V8 N5 P379-380	64-14594 <=*$>
1963 V8 N5 P381-384	64-14446 <=*$>
1963 V8 N6 P401-404	64-18534 <*>
1963 V8 N6 P405-409	66-13723 <*>
1963 V8 N6 P410-415	64-16663 <=*$>
1963 V8 N6 P419-426	64-18543 <*>
1963 V8 N6 P434-437	64-16542 <=*$>
	64-18538 <*>
1963 V8 N6 P437-446	64-16541 <=*$>
	64-16661 <=*$>
1963 V8 N6 P449-454	02S85R <ATS>
	64-16698 <=*$>
1963 V8 N6 P454-459	64-16665 <=*$>
1963 V8 N6 P460-461	64-16544 <=*$>
	64-18536 <*>
1963 V8 N6 P461-463	64-16662 <=*$>
1963 V8 N6 P463-465	64-18535 <*>
1963 V8 N6 P468-470	64-18528 <*>
1963 V8 N6 P471-474	64-20086 <*>
1964 V9 N1 P11-14	64-18530 <*>
1964 V9 N1 P14-18	64-20089 <*>
1964 V9 N1 P18-21	64-20090 <*>
1964 V9 N1 P21-27	64-20093 <*>
1964 V9 N1 P27-31	64-20091 <*>
1964 V9 N1 P38-46	64-20101 <*>
1964 V9 N1 P46-50	64-20097 <*>
1964 V9 N1 P51-53	64-20095 <*>
1964 V9 N1 P53-55	64-20094 <*>
1964 V9 N1 P57-59	AD-614 767 <=$>
	64-18531 <*>
1964 V9 N2 P83-90	64-20102 <*>
1964 V9 N2 P90-92	64-20100 <*>
1964 V9 N2 P92-95	64-20103 <*>
1964 V9 N2 P96-102	64-20104 <*>
1964 V9 N2 P102-108	64-20105 <*>
1964 V9 N2 P108-111	64-20083 <*>
1964 V9 N2 P111-114	64-20106 <*>
1964 V9 N2 P116-121	N65-31648 <*>
1964 V9 N2 P122-124	64-20107 <*>
1964 V9 N2 P124-126	64-20108 <*>
1964 V9 N2 P126-127	64-20109 <*>
1964 V9 N2 P128-129	65-11197 <*>
1964 V9 N2 P129-131	65-11198 <*>
1964 V9 N2 P132-140	65-11151 <*>
1964 V9 N2 P142-151	65-11199 <*>
1964 V9 N2 P155-156	AD-614 768 <=$>
1964 V9 N3 P161-167	65-11201 <*>
1964 V9 N3 P171-174	65-11200 <*>
1964 V9 N3 P184-189	65-11196 <*>
1964 V9 N3 P189-191	07R79R <ATS>
1964 V9 N3 P197-201	65-11193 <*>
1964 V9 N3 P202-203	65-11203 <*>
1964 V9 N3 P203-205	65-11202 <*>
1964 V9 N3 P205-206	65-11204 <*>
1964 V9 N3 P207-208	65-11205 <*>
1964 V9 N3 P209-210	65-11206 <*>
1964 V9 N3 P210-211	65-11195 <*>
1964 V9 N3 P231-236	65-11194 <*>
1964 V9 N4 P241-248	65-11159 <*>
1964 V9 N4 P248-254	65-11160 <*>
1964 V9 N4 P254-259	65-11945 <*>
1964 V9 N4 P259-261	65-11174 <*>
1964 V9 N4 P262-263	65-11162 <*>
1964 V9 N4 P263-266	65-11163 <*>
1964 V9 N4 P266-276	65-11209 <*>
1964 V9 N4 P276-282	65-13721 <*>
1964 V9 N4 P283-285	65-14532 <*>
1964 V9 N4 P286-288	65-11208 <*>
1964 V9 N4 P289-296	65-11176 <*>
1964 V9 N4 P298-300	17S82R <ATS>
	589 <TC>
	65-11175 <*>
1964 V9 N4 P300-302	65-11161 <*>
1964 V9 N4 P302-303	65-11207 <*>
1964 V9 N4 P401-404	65-14578 <*>
1964 V9 N5 P321-327	65-13722 <*>
1964 V9 N5 P333-336	AD-627 916 <=*$>
	65-13725 <*>
1964 V9 N5 P336-341	65-13724 <*>
1964 V9 N5 P341-351	65-13918 <*>
1964 V9 N5 P352-357	65-13992 <*>
1964 V9 N5 P357-359	65-13921 <*>
1964 V9 N5 P360-363	65-13919 <*>
1964 V9 N5 P363-368	65-13920 <*>
1964 V9 N5 P368-378	65-13923 <*>
	65-14361 <*> 0
1964 V9 N5 P396-398	65-13924 <*>
1964 V9 N6 P405-410	65-14617 <*>
1964 V9 N6 P411-413	65-14624 <*>
1964 V9 N6 P414-419	65-14623 <*>
1964 V9 N6 P422-425	65-14622 <*>
1964 V9 N6 P425-435	65-14606 <*>
1964 V9 N6 P436-440	65-14610 <*>
1964 V9 N6 P440-447	66-11291 <*>
1964 V9 N6 P448-451	65-14618 <*>
1964 V9 N6 P451-457	46S84R <ATS>
	65-13925 <*>
	65-17074 <*> 0
1964 V9 N6 P458-459	65-14816 <*>
1964 V9 N6 P460-461	65-14619 <*>
1964 V9 N6 P461-463	65-14621 <*>
1964 V9 N6 P464-466	65-14620 <*>
1964 V9 N6 P466-469	65-14625 <*>
1965 V10 N1 P3-7	65-14823 <*>
1965 V10 N1 P8-10	65-14824 <*>
1965 V10 N1 P10-15	65-14811 <*>
1965 V10 N1 P16-22	66-10397 <*>
1965 V10 N1 P22-27	AD-619 455 <ATS>
	66-10396 <*>
1965 V10 N1 P28-34	65-14626 <*>

1965 V10 N1 P34-38 07S83R <ATS>
65-14822 <*>
1965 V10 N1 P38-46 66-10395 <*>
1965 V10 N1 P46-61 66-10393 <*>
1965 V1C N1 P56-61 08S83R <ATS>
1965 V10 N1 P62-71 AD-621 421 <=$>
N65-27700 <=*$>
66-10392 <*>
1965 V10 N2 P81-83 66-10407 <*>
1965 V1C N2 P84-9C 66-10406 <*>
1965 V1C N2 P91-93 66-10400 <*>
1965 V10 N2 P93-103 66-10399 <*>
1965 V1C N2 P1C3-107 66-10398 <*>
1965 V10 N2 P107-112 47S84R <ATS>
65-17073 <*> 0
66-12467 <*>
1965 V10 N2 P112-116 66-10402 <*>
1965 V1C N2 P116-118 65-14812 <*>
1965 V10 N2 P118-123 48S84R <ATS>
66-10401 <*>
1965 V10 N2 P131-143 66-10403 <*>
1965 V10 N2 P144-146 66-10404 <*>
1965 V1C N2 P146-147 66-10405 <*>
1965 V10 N3 P161-169 66-10387 <*>
1965 V10 N3 P169-173 66-10391 <*>
1965 V10 N3 P174-178 66-10388 <*>
83S85R <ATS>
1965 V1C N3 P178-181 66-10384 <*>
1965 V10 N3 P182-185 66-10386 <*>
1965 V10 N3 P186-193 66-12456 <*>
1965 V1C N3 P193-20C 66-12455 <*>
1965 V10 N3 P200-206 84S85R <ATS>
1965 V10 N3 P2C6-216 66-12453 <*>
1965 V10 N3 P217-218 66-12457 <*>
1965 V10 N3 P219-220 66-10389 <*>
1965 V1C N3 P220-221 66-10390 <*>
1965 V10 N3 P222-234 66-10385 <*>
1965 V10 N4 P241-247 66-12444 <*>
1965 V10 N4 P247-250 66-12445 <*>
1965 V10 N4 P250-258 66-12446 <*>
1965 V1C N4 P258-261 66-12447 <*>
1965 V10 N4 P273-276 66-13748 <*>
1965 V10 N4 P278-286 66-13727 <*>
1965 V10 N4 P286-291 66-13726 <*>
1965 V10 N4 P292-294 66-13725 <*>
1965 V1C N4 P294-296 66-13750 <*>
1965 V10 N4 P296-297 66-13749 <*>
1965 V10 N4 P298-312 66-13728 <*>
1965 V1C N5 P321-329 66-13729 <*>
1965 V10 N5 P330-343 66-12462 <*>
1965 V10 N5 P344-347 66-12461 <*>
1965 V10 N5 P347-351 66-13731 <*>
1965 V10 N5 P351-353 66-13730 <*>
1965 V1C N5 P354-36C 66-13732 <*>
1965 V10 N5 P360-365 66-13733 <*>
1965 V10 N5 P365-369 66-10143 <*> 0
66-13734 <*>
1965 V10 N5 P369-377 66-13735 <*>
1965 V1C N5 P378-390 66-12463 <*>
1965 V10 N5 P391-392 66-13736 <*>
1965 V10 N5 P394-395 66-13708 <*>
1965 V10 N5 P395-397 66-13709 <*>
1965 V10 N5 P397-399 66-13710 <*>
1965 V1C N6 P401-410 66-13711 <*>
1965 V10 N6 P411-418 66-12460 <*>
1965 V10 N6 P418-422 66-13712 <*>
1965 V1C N6 P422-424 66-12459 <*>
1965 V10 N6 P424-429 66-12464 <*>
1965 V1C N6 P429-433 66-12465 <*>
1965 V10 N6 P433-438 66-13713 <*>
1965 V10 N6 P438-445 66-12466 <*>
1965 V10 N6 P446-448 66-13714 <*>
1965 V10 N6 P448-449 66-13715 <*>
1965 V10 N6 P450-451 66-13716 <*>
1965 V10 N6 P451-452 66-13718 <*>
1965 V10 N6 P453-455 66-13717 <*>
1966 V11 N1 P3-11 66-13766 <*>
1966 V11 N1 P12-13 66-13769 <*>
1966 V11 N1 P14-18 66-13767 <*>
1966 V11 N1 P23-27 66-13771 <*>
1966 V11 N1 P27-30 66-13770 <*>
1966 V11 N1 P39-46 66-13721 <*>
1966 V11 N1 P46-56 66-13768 <*>
1966 V11 N1 P57-58 66-13722 <*>

*ZHURNAL NEORGANICHESKCI KHIMII
1956 N2 P193-211 R-1978 <*>
1956 V1 P125 R-5042 <*>
1956 V1 P17C-178 60-19982 <*+>
1956 V1 P274-281 R-5105 <*>
1956 V1 P1285-1291 61-19574 <*=>
1956 V1 P1523-1532 6C-17705 <+*>
1956 V1 P1613-1618 59-19242 <+*>
1956 V1 P1722-1730 60-13171 <+*>
1956 V1 P1843-1849 R-1896 <*>
1956 V1 P1850-1856 59-19243 <+*>
1956 V1 P1937-1942 298TM <CTT>
1956 V1 P2210-2215 R-3678 <*>
1956 V1 P2272-2277 AEC-TR-3870 <+*>
1956 V1 P2294-2299 UCRL TRANS-537(L) <*=>
1956 V1 P2316-2326 R-4214 <*>
1956 V1 P2708-2711 299TM <CTT>
1956 V1 N1 P24-26 59-17228 <+*>
1956 V1 N2 P193-211 35TM <CTT>
1956 V1 N2 P274-281 79J18R <ATS>
1956 V1 N2 P282-297 R-2820 <*>
R-963 <*>
178TM <CTT>
1956 V1 N3 P357-361 63-10844 <=*> 0
1956 V1 N3 P362-365 AEC-TR-5477 <=*>
92F13R <ATS>
1956 V1 N3 P478-484 R-2310 <*>
3CCTM <CTT>
1956 V1 N3 P506-515 R-2860 <*>
R-735 <*>
179TM <CTT>
1956 V1 N4 P615-618 61-20894 <*=> 0
1956 V1 N4 P618-631 3C1TM <CTT>
1956 V1 N4 P628-631 R-1995 <*>
1956 V1 N4 P777-782 47J16R <ATS>
1956 V1 N5 P894-902 59-10230 <+*>
1956 V1 N5 P1013-1018 SCL-T-338 <+*>
1956 V1 N5 P1042-1046 302TM <CTT>
1956 V1 N5 P1074-1090 R-4933 <*>
R-5249 <*>
63-19225 <=*>
1956 V1 N6 P1131-1149 73K22R <ATS>
1956 V1 N6 P1271-1278 UCRL TRANS-553L <=*>
1956 V1 N6 P1383-1386 531TM <CTT>
1956 V1 N7 P1473-1478 19J16R <ATS>
1956 V1 N7 P149C-15C0 3C3TM <CTT>
1956 V1 N8 P1703-1712 59-10228 <+*>
1956 V1 N8 P1722-1730 AEC-TR-4810 <=*> 0
1956 V1 N8 P1843-1849 R-3618 <*>
1956 V1 N8 P1867-1876 64-13233 <=*$>
64-26098 <$>
1956 V1 N9 P2071-2082 R-4355 <*>
1956 V1 N9 P21C6-2109 64-18303 <=*$>
1956 V1 N9 P2118-2130 R 770 <RIS>
59-18525 <+*>
1956 V1 N1C P2390-24C2 59-13533 <=>
1956 V1 N11 P2556-256C 64-13601 <=*$>
1956 V1 N11 P2577-2587 15P60R <ATS>
1956 V1 N11 P2601-2605 59-18734 <+*>
1956 V1 N12 P2749-2752 59-10228 <+*>
1956 V1 N12 P2772-2777 62-25566 <=*>
1956 V1 N12 P2778-2791 47K25R <ATS>
1957 P1761-1762 296TM <CTT>
1957 N2 P259-262 3C4TM <CTT>
1957 N4 P897-902 87TM <CTT>
1957 N9 P2259-2263 71TM <CTT>
1957 V2 P13-22 R-3863 <*>
1957 V2 P68-72 R-4217 <*>
1957 V2 P206-2C8 R-1450 <*>
1957 V2 P233-237 R-3187 <*>
1957 V2 P263-267 305TM <CTT>
1957 V2 P510-514 295TM <CTT>
1957 V2 P1175-1182 59-19176 <+*>
1957 V2 P1281-1288 59-14759 <+*>
1957 V2 P1522-1527 AEC-TR-3600 <+*>

Citation	Translation
1957 V2 P1883-1887	8653-C <K-H>
1957 V2 P2109-2114	UCRL TRANS-5166 <+*>
1957 V2 P2154-2158	8653-G <K-H>
1957 V2 P2168-2173	8653-H <K-H>
1957 V2 N1 P13-22	82J18R <ATS>
1957 V2 N1 P34-41	60-17904 <+*>
	61-20755 <*=>
1957 V2 N1 P80-91	R-3191 <*>
	297TM <CTT>
1957 V2 N1 P160-166	60-14508 <+*>
1957 V2 N1 P206-208	4788 <HB>
1957 V2 N1 P212-221	C-3312 <NRCC>
1957 V2 N2 P233-237	61-10327 <*+>
	63-14694 <=*>
1957 V2 N2 P238-243	C-2531 <NRC>
	R-3063 <*>
1957 V2 N2 P417-421	64-30981 <*> 0
	8653 <K-H> 0
1957 V2 N3 P475-490	61-15728 <+*>
1957 V2 N3 P523-531	59-14762 <+*> 0
1957 V2 N3 P604-605	60-13226 <+*>
1957 V2 N3 P611-622	R-4931 <*>
	R-5012 <*>
1957 V2 N4 P890-896	60-23629 <+*>
1957 V2 N4 P928-932	60-15588 <+*>
1957 V2 N4 P970-974	60-11494 <=>
1957 V2 N4 P975-979	60-11496 <=>
1957 V2 N5 P985-994	60-11486 <=>
1957 V2 N5 P1015-1024	59-13832 <=>
1957 V2 N5 P1145-1148	64-30117 <*> 0
	8050 <K-H>
1957 V2 N5 P1164-1166	62-25415 <=*>
1957 V2 N6 P1223-1231	60-15696 <+*>
1957 V2 N6 P1232-1241	60-13318 <+*>
1957 V2 N6 P1242-1247	60-15572 <+*>
1957 V2 N6 P1248-1253	60-13319 <+*>
1957 V2 N6 P1441-1447	NP-TR-500 <*>
	61-23280 <*=>
1957 V2 N7 P1087-1692	R-5059 <*>
1957 V2 N7 P1471-1473	<ES> 0
1957 V2 N7 P1494-1496	60-17285 <+*> 0
1957 V2 N7 P1535-1537	R-5148 <*>
1957 V2 N7 P1587-1590	R-5103 <*>
1957 V2 N7 P1649-1654	19K27P <ATS>
1957 V2 N7 P1655-1657	62K23R <ATS>
1957 V2 N7 P1677-1681	60-11492 <=>
1957 V2 N7 P1682-1686	60-11493 <=>
1957 V2 N8 P1757-1760	60-17900 <+*>
1957 V2 N8 P1768-1774	59-22533 <+*>
1957 V2 N8 P1807-1811	SCL-T-255 <+*>
1957 V2 N8 P1848-1854	60-23584 <+*> 0
1957 V2 N8 P1864-1882	65-12049 <*>
	8653 <K-H>
1957 V2 N8 P1883-1887	64-30987 <*> 0
1957 V2 N8 P1888-1894	64-30986 <*> 0
	8653-D <K-H>
1957 V2 N8 P1895-1906	65-10301 <*> 0
	8653 <K-H> 0
1957 V2 N8 P1907-1914	64-30980 <*> 0
	8653 F <K-H>
1957 V2 N8 P1915-1921	AEC-TR-3741 <+*>
	60-23083 <+*>
1957 V2 N8 P1956-1969	59-10080 <+*>
1957 V2 N8 P1972-1974	59-10095 <+*>
1957 V2 N9 P2129-2135	AEC-TR-3748 <+*>
	59-18733 <+*>
1957 V2 N9 P2145-2153	63-24294 <=*$>
1957 V2 N9 P2154-2158	64-30988 <*> 0
1957 V2 N9 P2168-2173	64-30974 <*> 0
1957 V2 N9 P2259-2263	20M45R <ATS>
	62-10313 <*=>
1957 V2 N9 P2274-2276	59-10603 <+*>
	61-15362 <+*>
1957 V2 N9 P2276-2278	59-10602 <+*>
	61-15361 <+*>
1957 V2 N10 P2281-2303	59-18506 <+*>
1957 V2 N10 P2334-2345	59-15727 <+*>
1957 V2 N10 P2416-2422	61-15360 <+*>
1957 V2 N10 P2423-2425	60-23589 <+*>
1957 V2 N11 P2530-2538	60-13852 <+*>
	60-23987 <*+>
1957 V2 N11 P2539-2542	60-23988 <+*>
1957 V2 N12 P2716-2722	61-15359 <+*>
1958 V2 N7 P1649-1654	R-4693 <*>
1958 V3 P121-128	R-4807 <*>
1958 V3 P195-203	NP-TR-519 <*>
1958 V3 P986-995	AEC-TR-3533 <+*>
1958 V3 P1799-1803	UCRL TRANS-623 <*>
1958 V3 P2433-2436	65-10331 <*>
1958 V3 N1 P40-45	AEC-TR-3678 <+*>
1958 V3 N1 P82-87	UCRL TRANS-523(L) <+*>
1958 V3 N1 P88-94	UCRL TRANS-535(L) <=*>
1958 V3 N1 P136-138	AEC-TR-3499 <+*>
1958 V3 N1 P155-159	AEC-TR-3598 <+*>
1958 V3 N1 P165-166	AEC-TR-3490 <+*>
1958 V3 N1 P181-183	60-11641 <=>
1958 V3 N1 P195-203	59-16383 <+*>
1958 V3 N1 P212-214	AEC-TR-3617 <+*>
1958 V3 N1 P215-221	AEC-TR-3618 <+*>
1958 V3 N1 P225-230	AEC-TR-3655 <+*>
1958 V3 N2 P508-516	60-13480 <+*>
1958 V3 N2 P542-545	60-11549 <=>
1958 V3 N3 P607-610	14P60R <ATS>
1958 V3 N3 P659-667	60-10938 <+*> 0
1958 V3 N3 P699-707	TRANS-63 <MT>
1958 V3 N4 P841-846	R-739 <RIS>
	R-739 <RIS> 0
	61-13084 <*+>
1958 V3 N4 P853-859	R-740 <RIS> 0
1958 V3 N4 P889-897	R-741 <RIS>
1958 V3 N4 P898-903	R-742 <RIS>
1958 V3 N4 P924-933	R-743 <RIS>
1958 V3 N4 P936-938	R-744 <RIS>
1958 V3 N4 P939-944	R 745 <RIS>
1958 V3 N4 P951-955	AEC-TR-3505 <+*>
	R-746 <RIS>
1958 V3 N4 P956-961	R-747 <RIS>
1958 V3 N4 P962-974	R-748 <RIS>
1958 V3 N4 P986-995	R-749 <RIS>
1958 V3 N4 P996-1001	R-750 <RIS>
1958 V3 N4 P1002-1007	R-751 <RIS>
1958 V3 N4 P1008-1027	R-752 <RIS>
1958 V3 N4 P1028-1036	R-753 <RIS>
	R-753 <RIS> 0
1958 V3 N4 P1054-1059	R 754 <RIS>
	R-754 <RIS>
1958 V3 N5 P1071-1074	R 723 <RIS>
	59-15381 <+*>
	59-22542 <*+>
1958 V3 N5 P1075-1078	R 724 <RIS> 0
1958 V3 N5 P1079-1088	R-725 <RIS> 0
1958 V3 N5 P1089-1094	R-726 <RIS>
1958 V3 N5 P1100-1104	R-727 <RIS>
1958 V3 N5 P1131-1138	R-728 <RIS>
1958 V3 N5 P1177-1180	R-729 <RIS>
1958 V3 N5 P1192-1199	R 730 <RIS> 0
1958 V3 N5 P1200-1204	R-731 <RIS>
1958 V3 N5 P1205-1209	R-732 <RIS>
1958 V3 N5 P1210-1223	11Q73R <ATS>
1958 V3 N5 P1214-1219	R-733 <RIS>
1958 V3 N5 P1227-1231	R-734 <RIS>
1958 V3 N5 P1232-1236	R 735 <RIS> 0
1958 V3 N5 P1237-1240	R-736 <RIS>
1958 V3 N5 P1241-1244	R-737 <RIS>
1958 V3 N5 P1245-1253	64-14064 <=*$>
1958 V3 N5 P1254-1260	R-738 <RIS>
1958 V3 N5 P1267	62-10644 <=*>
1958 V3 N6 P1281-1285	62-18091 <=*>
1958 V3 N6 P1286-1294	64-16594 <=*$>
1958 V3 N6 P1403-1409	60-11507 <=>
1958 V3 N7 P1487-1489	61-19330 <+*>
1958 V3 N7 P1605-1607	64-15359 <=*$>
1958 V3 N7 P1632-1636	59-15397 <+*>
1958 V3 N7 P1637-1643	60-23992 <+*>
1958 V3 N7 P1694-1702	UCRL TRANS-607(L) <*+>
1958 V3 N7 P1714-1715	59-15485 <+*>
1958 V3 N7 P1715-1716	61-19330 <+*>
1958 V3 N8 P1727-1730	REPT. R-765 <RIS>
	65-14370 <*>
1958 V3 N8 P1731-1733	61-15259 <*+> 0

1958 V3 N8 P1772-1780 18P64R <ATS>
1958 V3 N8 P1791-1798 UCRL TRANS-658(L) <*=>
1958 V3 N8 P1799-1803 UCRL TRANS-628L <=*>
1958 V3 N8 P1891-1895 RTS-3314 <NLL>
1958 V3 N8 P1914-1924 AEC-TR-4423 <=*>
1958 V3 N9 P2039-2044 62-18886 <=*>
1958 V3 N9 P2143-2149 62-11550 <=>
1958 V3 N9 P2213-2214 59-19154 <+*>
1958 V3 N10 P2225-2230 R 760 <RIS>
1958 V3 N10 P2236-2239 59-15729 <+*>
1958 V3 N10 P2244-2252 59-15728 <+*>
1958 V3 N10 P2366-2374 R-762 <RIS>
60-13436 <+*>
61-15258 <*+>
1958 V3 N10 P2382-2389 62-18749 <=*> 0
1958 V3 N10 P2390-2394 61-18576 <*=>
1958 V3 N11 P2433-2436 60-13248 <+*>
9077 <K-H>
1958 V3 N11 P2487-2490 59-13979 <=>
1958 V3 N11 P2512-2522 65-60483 <=$>
1958 V3 N11 P2553-2561 62-18196 <=*>
1958 V3 N12 P2597-2598 61-19730 <=*>
1959 V4 P985-988 62-24450 <=*>
1959 V4 N1 P28-32 60-11011 <=>
1959 V4 N2 P253-256 63-18369 <=*>
1959 V4 N2 P257-259 63-18474 <=*>
1959 V4 N2 P352-355 65-10090 <*>
1959 V4 N2 P488-491 60-13792 <+*>
1959 V4 N4 P956-957 60-10939 <+*>
1959 V4 N5 P961-962 60-41086 <=> 0
1959 V4 N5 P985-988 61-18718 <*=> 0
1959 V4 N6 P1461-1463 60-31029 <=>
1959 V4 N8 P1710-1714 54M41R <ATS>
1959 V4 N9 P1958-1960 65-11237 <*>
9766 <K-H>
1959 V4 N9 P1961-1966 57M45R <ATS>
1959 V4 N9 P1967-1969 53M41R <ATS>
61-23417 <=*>
1959 V4 N9 P2052-2057 56M45R <ATS>
60-16833 <+*>
61-23417 <=*>
1959 V4 N9 P2169-2171 61-23417 <=*>
1959 V4 N10 P2384-2389 60-12503 <=>
62-11347 <=>
1959 V4 N11 P2443-2448 61-27508 <*=> 0
1959 V4 N11 P2544-2550 62-10386 <*=> 0
1959 V4 N11 P2647-2654 60-31172 <=>
1959 V4 N12 P2670-2681 61-18592 <*=>
1959 V4 N12 P2688-2696 60-41087 <=>
1960 V5 P477-480 62-24465 <=*>
1960 V5 P2844-2847 K-53 <RIS>
1960 V5 N2 P297-305 63-23606 <=*$>
1960 V5 N4 P775-777 62-18285 <=*>
1960 V5 N6 P1241-1247 61-23440 <=*>
1960 V5 N6 P1257-1282 61-23440 <=*>
1960 V5 N6 P1391-1392 61-11025 <=>
1960 V5 N8 P1688-1695 65-13101 <*>
1960 V5 N9 P1938-1942 62-10463 <*=> 0
1960 V5 N11 P2462-2465 64-10491 <=*$>
1960 V5 N11 P2466-2470 63-18501 <=*>
1961 V6 P212-215 63-20935 <=*$> 0
1961 V6 P1443-1452 65-20000 <$>
1961 V6 N1 P18-20 62-16174 <=*>
1961 V6 N1 P182-185 61-27832 <*=> 0
62-23135 <=*>
1961 V6 N1 P208-211 AEC-TR-4835 <=*>
1961 V6 N2 P261-264 62-11371 <=>
1961 V6 N4 P985-993 61-20858 <*=>
1961 V6 N5 P1189-1197 62-23089 <*=>
1961 V6 N5 P1198-1203 62-15312 <*=>
1961 V6 N5 P1242-1244 36P60R <ATS>
1961 V6 N5 P1256-1258 61-27692 <*=>
61-27692 <=*>
1961 V6 N6 P1466-1470 63-24287 <=*$>
1961 V6 N6 P1490-1491 62-23580 <=*>
1961 V6 N8 P1891-1901 62-24821 <=>
1961 V6 N9 P2014-2018 62-14069 <=*>
1961 V6 N9 P2086-2090 AEC-TR-4936 <=*>
1961 V6 N9 P2139-2141 62-23371 <=*>
1961 V6 N10 P2225-2236 62-15412 <*=>
62-24336 <=*>
1961 V6 N11 P2470-2480 62-15412 <*=>
62-24335 <=*>
1961 V6 N12 P2724-2726 76P65R <ATS>
1962 V7 N3 P628-632 20P61R <ATS>
1962 V7 N3 P701-703 63-19303 <=*> 0
1962 V7 N3 P703-705 62-18022 <=*>
1962 V7 N4 P844-849 62-25979 <=*>
63-19874 <*=>
1962 V7 N5 P975-979 63-21005 <=>
1962 V7 N6 P1336-1342 63-24082 <=*$>
1962 V7 N7 P1636-1639 AEC-TR-5691 <=*>
1962 V7 N8 P1779-1782 63-24288 <=*$>
1962 V7 N8 P2025-2026 63-13351 <=*>
1962 V7 N9 P2115-2121 12464 <K-H>
65-14506 <*>
1962 V7 N9 P2200-2205 63-21005 <=>
1962 V7 N10 P2383-2387 64-23559 <=>
1963 V8 N1 P18-23 UCRL-TRANS-977 <=*$>
1963 V8 N1 P89-93 AEC-TR-6073 <=*$>
1963 V8 N1 P186-191 63-23782 <=*$>
1963 V8 N2 P278-284 63-20473 <=*$> 0
1963 V8 N2 P403-406 63-24618 <=*$>
1963 V8 N3 P567-572 64-13168 <=*$>
1963 V8 N3 P772-774 64-11880 <=> M
1963 V8 N4 P954-958 63-24617 <=*$>
1963 V8 N6 P1520-1522 64-11734 <=>
1963 V8 N7 P1567-1573 AEC-TR-6157 <=*$>
1963 V8 N8 P1806-1808 64-11671 <=>
1963 V8 N8 P1915-1920 64-13222 <=*$>
1963 V8 N9 P2136-3139 64-11973 <=>
1963 V8 N10 P2254-2257 64-11784 <=>
1963 V8 N11 P2417-2419 64-21931 <=>
1963 V8 N12 P2819 IS-TRANS-2 <=>
1964 V9 N2 P330-334 54R76R <ATS>
1964 V9 N2 P372-377 65-61437 <=>
1964 V9 N5 P1305-1306 65-13139 <*>
1964 V9 N6 P1519-1520 AD-619 331 <=$>
1964 V9 N7 P1676-1680 ANL-TRANS-104 <=$>
1964 V9 N7 P1681-1683 ANL-TRANS-105 <=$>
1964 V9 N7 P1776-1779 64-41971 <=$>
1964 V9 N9 P2085-2090 65-11788 <*>
1964 V9 N10 P2424-2432 AD-637 375 <=$>
1964 V9 N12 P2701-2704 65-17363 <*>
1965 V10 N2 P336-343 65-14776 <*>
1965 V10 N2 P539-541 90S83R <ATS>
1965 V10 N6 P1338-1343 37S86R <ATS>
1965 V10 N7 P1716-1722 2109 <TC>
1965 V10 N9 P1966-1970 66-11194 <*>
1966 V11 N1 P20-24 66-14213 <*$>
1966 V11 N1 P33-38 66-13412 <*>
1966 V11 N1 P120-127 66-13397 <*>
1966 V11 N1 P151-155 66-14340 <*$>
1966 V11 N1 P195-197 66-13399 <*>
1966 V11 N1 P216-219 66-13402 <*>
1966 V11 N8 P1992-1998 66-34171 <=$>

ZHURNAL NEVROPATOLOGII I PSIKHIATRII
1927 V20 P403-419 65-63165 <=$>
1927 V20 P627-632 R-1103 <*>
1943 V12 P1537-1542 R-5366 <*>
1945 V14 N2 P34-41 60-19396 <=*$>
1946 V15 N6 P3-13 60-13720 <+*>
1950 V19 N1 P85-86 64-15707 <=*$>
1951 V20 N1 P18-22 60-13724 <+*>
1951 V20 N1 P32- 60-13710 <+*>
1951 V20 N5 P70-72 63-15165 <=*$>
1952 V52 N5 P26-30 64-10361 <=*$>
1952 V52 N9 P49-56 RT-2012 <*>
1953 V53 N7 P483-485 RT-2610 <*>
1953 V53 N7 P486-490 RT-2551 <*>
1953 V53 N7 P491-494 RT-2552 <*>
1953 V53 N7 P552-557 RT-2553 <*>
1953 V53 N9 P735-739 RT-1987 <*>
1953 V53 N10 P753-758 RT-2483 <*>
1953 V53 N10 P790-795 RT-1658 <*>
1953 V53 N11 P854-860 RT-2592 <*>
1953 V53 N11 P906-909 RT-2907 <*>
1954 V54 N2 P203-204 RT-2555 <*>
1954 V54 N4 P328-335 R-2724 <*>

Citation	Translation
	R-3526 <*>
	RT-1887 <*>
1954 V54 N7 P583-589	RT-2948 <*>
1954 V54 N11 P934-940	R-278 <*>
1954 V54 N12 P979-986	RT-3779 <*>
1954 V54 N12 P996-1005	RT-3504 <*>
1955 P663-668	R-2733 <*>
	R-3521 <*>
1955 V55 P856-861	65-63055 <=$>
1955 V55 N1 P6-16	RT-3894 <*>
1955 V55 N1 P29-30	RT-3471 <*>
1955 V55 N1 P41-42	RT-3773 <*>
	62-15065 <*=>
1955 V55 N1 P43-47	RT-3560 <*>
1955 V55 N1 P50	RT-3509 <*>
1955 V55 N1 P51-53	RT-3774 <*>
1955 V55 N7 P535-539	R-147 <*>
1955 V55 N8 P637-639	R-159 <*>
1955 V55 N1C P755-758	61-23829 <*=>
	62-00102 <*>
1955 V55 N11 P825-830	R-2726 <*>
	R-3523 <*>
	62-23511 <=*>
1955 V55 N12 P881-889	PB 141 267T <=>
1956 V56 P81-93	65-63C48 <=$> 0
1956 V56 P94-103	65-63239 <=*$> 0
1956 V56 P146-154	65-63043 <=$> 0
1956 V56 P155-161	65-63056 <=$>
1956 V56 P162-165	65-63045 <=$>
1956 V56 P166-171	65-63C46 <=$>
1956 V56 P294-295	65-63044 <=$>
1956 V56 P3C0-3C6	65-63225 <=$> 0
1956 V56 P575-576	65-63145 <=$> 0
1956 V56 P605-611	65-63042 <=$> 0
1956 V56 P645-653	61-19836 <=*> 0
1956 V56 N1 P49-55	R-2719 <*>
	R-3518 <*>
1956 V56 N2 P81-93	RT-4276 <*>
1956 V56 N2 P103	R-690 <*>
1956 V56 N2 P136-138	R-277 <*>
1956 V56 N2 P139-145	RT-4309 <*>
	62-24333 <=*>
	85K22R <ATS>
1956 V56 N2 P146-154	R-285 <*>
1956 V56 N2 P171-	R-288 <*>
1956 V56 N2 P192-198	R-284 <*>
1956 V56 N3 P201-210	62-15315 <=*>
1956 V56 N3 P248-252	62-15344 <=*>
1956 V56 N4 P281-287	R-2699 <*>
1956 V56 N4 P307-310	R-2693 <*>
1956 V56 N5 P361-369	62-23509 <=*>
1956 V56 N5 P385-388	62-15347 <=*>
1956 V56 N5 P389-394	62-11147 <=>
	62-15070 <=*>
1956 V56 N5 P395-400	62-23543 <=*>
1956 V56 N7 P567-570	62-24334 <=*>
1956 V56 N7 P591-599	62-23500 <=*>
1956 V56 N8 P631-633	R-3180 <*>
	64-19406 <=$>
1956 V56 N8 P668-675	<CP>
1956 V56 N9 P757-763	62-23537 <=*>
1956 V56 N10 P765-777	R-2685 <*>
1956 V56 N1C P765	62-15303 <*=>
1956 V56 N12 P925-936	61-00088 <*>
1956 V56 N12 P949-962	R-2708 <*>
	62-23536 <=*>
1956 V56 N12 P963-966	62-15060 <=*>
1956 V56 N12 P967-968	62-15061 <=*>
1957 V57 P203-207	65-63787 <=$>
1957 V57 P208-213	62-24331 <=*>
1957 V57 P220-224	65-63561 <=$> 0
1957 V57 P389-392	61-28467 <*=> 0
	62-15336 <*=> 0
1957 V57 N1 P95-103	60-41634 <=>
1957 V57 N1 P104-113	62-24620 <=*>
1957 V57 N2 P214-219	62-24336 <=*> 0
1957 V57 N5 P545-555	62-23512 <=*>
1957 V57 N6 P681-693	62-15062 <=*>
1957 V57 N6 P706-711	62-15059 <=*>
1957 V57 N6 P761-767	63-24332 <=*>
1957 V57 N8 P1C38-1044	<CP>
	62-23513 <=*>
1957 V57 N8 P1C44-1050	<CP>
1957 V57 N9 P1101-1105	R-3232 <*>
	61-19180 <+*> 0
	64-19407 <=$>
1957 V57 N9 P1171-1178	R-4885 <*>
	61-19184 <+*>
	62-15051 <*=>
1957 V57 N10 P1205-1209	R-4613 <*>
1957 V57 N11 P1409-1417	61-23831 <=*>
1958 V58 P208-211	62-24329 <=*>
1958 V58 P585-591	64-19868 <=$>
1958 V58 P1248-1252	61-19835 <=*> C
1958 V58 N1 P46-54	PB 141 203T <=>
1958 V58 N1 P55-56	PB 141 204T <=>
1958 V58 N2 P137-149	59-17732 <+*>
	62-24327 <=*>
1958 V58 N2 P150-157	62-24328 <=*>
1958 V58 N2 P158-162	64-13750 <=*$> 0
1958 V58 N2 P215-217	64-19867 <=$> C
1958 V58 N2 P255-256	59-11507 <=>
1958 V58 N3 P282-287	62-13928 <*=>
	64-19522 <=$>
1958 V58 N5 P560-566	59-17730 <+*>
1958 V58 N5 P567-572	59-17731 <+*>
1958 V58 N5 P592-599	64-19863 <=$> 0
1958 V58 N5 P600-615	64-19864 <=$>
1958 V58 N5 P616-624	64-19866 <=$> 0
1958 V58 N5 P625-627	64-19865 <=$>
1958 V58 N5 P638-640	59-11507 <=>
1958 V58 N6 P641-649	59-13784 <=> 0
	59-17733 <+*>
1958 V58 N6 P703-704	6C-16471 <+*>
	62-24326 <=*>
1958 V58 N6 P7C5-709	64-19821 <=$> C
1958 V58 N7 P861-866	59-13791 <=>
1958 V58 N8 P883-886	64-19861 <=$> 0
1958 V58 N8 P897-906	60-16608 <+*>
1958 V58 N9 P1079-1089	65-60580 <=$> 0
1958 V58 N9 P1C96-1105	64-19862 <=$> C
1958 V58 N9 P1106-1110	64-19823 <=$> 0
1958 V58 N10 P1153-1163	64-19825 <=$> 0
1958 V58 N1C P1176-1182	64-19828 <=$>
1958 V58 N10 P1183-1186	64-19824 <=$>
1958 V58 N10 P1187-1189	64-19860 <=$>
1958 V58 N10 P1190-1195	64-19826 <=$> 0
1958 V58 N10 P1248-1252	64-15312 <=*$>
1958 V58 N10 P1259-1264	59-13793 <=>
1958 V58 N10 P1264-1268	59-13794 <=>
1958 V58 N1C P1269-1274	64-19827 <=$> 0
1958 V58 N11 P1326-1332	59-13839 <=>
1958 V58 N11 P1354-1358	60-17639 <+>
1958 V58 N11 P1399-1406	59-13901 <=>
1959 V59 P172-176	62-24322 <=*>
1959 V59 P402-409	62-24321 <=*>
1959 V59 P981-985	64-13044 <=*$> C
1959 V59 P1001-1004	62-15341 <=*>
1959 V59 P1198-1200	64-13043 <=*$> 0
1959 V59 P1222-1223	64-13042 <=*$> 0
1959 V59 P1444-1446	61-19834 <*=>
1959 V59 N2 P129-134	63-24256 <=*$> C
1959 V59 N2 P135-142	62-24324 <=*> 0
1959 V59 N2 P143-150	63-24258 <=*$> 0
1959 V59 N2 P151-155	63-24257 <=*$>
1959 V59 N2 P156-159	62-24323 <=*> 0
1959 V59 N2 P160-166	64-13046 <=*$> C
1959 V59 N2 P167-171	63-24259 <=*$>
1959 V59 N2 P177-181	64-13047 <=*$> 0
1959 V59 N2 P222-232	64-13048 <=*$> 0
1959 V59 N2 P232-244	60-15647 <+*>
1959 V59 N3 P313-320	61-19723 <=*>
1959 V59 N4 P385-395	64-13193 <=*$> C
1959 V59 N4 P396-401	64-13045 <=*$>
1959 V59 N5 P575-58C	64-13189 <=*$> 0
1959 V59 N5 P581-585	64-13188 <=*$>
1959 V59 N5 P586-589	64-13187 <=*$> C
1959 V59 N5 P590-592	64-13185 <=*$> C
1959 V59 N5 P609-620	60-31257 <=>
1959 V59 N5 P636-639	60-31257 <=>

1959 V59 N8 P972-980 64-13194 <=*$> 0
1959 V59 N9 P1151 60-31155 <=>
1959 V59 N10 P1172-1178 64-13178 <=*$>
1959 V59 N10 P1215-1221 64-13192 <=*$> 0
1959 V59 N10 P1253-1255 60-11263 <=>
1959 V59 N12 P1436-1443 61-23827 <=*> 0
63-23176 <=*$>
1959 V59 N12 P1444-1446 62-15346 <*=>
1959 V59 N12 P1462-1469 64-13183 <=*$> 0
1959 V59 N12 P1470-1479 64-13181 <=*$>
1959 V59 N12 P1480-1484 64-13186 <=*$> 0
1959 V59 N12 P1485-1488 64-13184 <=*$> 0
1959 V59 N12 P1489-1493 64-13190 <=*$>
1959 V59 N12 P1494-1498 64-13182 <=*$> 0
1959 V59 N12 P1499-1500 64-15244 <=*$>
1960 V60 P220-223 62-24312 <=*>
1960 V60 P234-237 62-24313 <=*>
1960 V60 P242-247 62-24314 <=*>
1960 V60 P585-592 62-24319 <=> 0
1960 V60 N1 P77-84 62-24311 <=*>
1960 V60 N2 P135-139 60-11671 <=>
1960 V60 N2 P182-193 63-19722 <*=> 0
1960 V60 N2 P194-201 63-19116 <=*$> 0
1960 V60 N2 P202-209 63-19718 <=*> 0
1960 V60 N2 P210-219 63-19115 <=*$>
1960 V60 N2 P224-233 63-19117 <=*>
1960 V60 N2 P238-241 63-19720 <*=> 0
1960 V60 N4 P458-461 61-11717 <=>
1960 V60 N5 P522-528 60-41393 <=>
1960 V60 N5 P529-534 60-41414 <=>
1960 V60 N5 P568-576 62-24318 <=*> 0
1960 V60 N5 P577-584 62-24316 <=*> 0
1960 V60 N5 P593-594 62-24315 <=*> 0
1960 V60 N5 P602-611 62-24317 <=*> 0
1960 V60 N5 P632-636 60-41415 <=>
64-13041 <=*$>
1960 V60 N6 P758-762 63-19711 <*=>
1960 V60 N8 P994-1001 63-19120 <=*>
1960 V60 N8 P1002-1008 63-19714 <=*> 0
1960 V60 N8 P1015-1018 63-19704 <*=> 0
1960 V60 N8 P1019-1023 63-19715 <=*> 0
1960 V60 N8 P1024-1026 63-19716 <=*> 0
1960 V60 N8 P1027-1032 63-19119 <=*$>
1960 V60 N8 P1049-1053 59-12869 <*+>
1960 V60 N9 P1096-1100 63-23134 <=*>
64-31651 <=>
1960 V60 N9 P1101-1105 63-19250 <*>
1960 V60 N9 P1198-1203 63-19709 <=*> 0
1960 V60 N10 P1342-1351 61-31514 <=>
1960 V60 N10 P1391-1396 61-31514 <=>
1960 V60 N11 P1510-1514 61-11955 <=>
1960 V60 N11 P1515-1517 61-21196 <=>
1960 V60 N11 P1518-1521 61-21199 <=>
1960 V60 N11 P1523-1527 61-11990 <=>
1961 V61 N2 P161-164 62-14751 <=*>
1961 V61 N2 P166-175 61-31497 <=>
62-20323 <=*>
1961 V61 N2 P201-207 61-27096 <*=>
1961 V61 N3 P387-395 62-15949 <=*>
1961 V61 N8 P1122-1128 62-13322 <=*>
1961 V61 N10 P1574-1578 62-15932 <=*>
1961 V61 N11 P1744-1746 62-19124 <=*>
1961 V61 N12 P1785-1788 62-20366 <=*>
1961 V61 N12 P1828-1834 62-23618 <=*>
1961 V61 N12 P1864-1870 62-20370 <=*> P
1962 V62 P1143-1148 65-11158 <*>
1962 V62 N1 P3-14 62-25010 <=*>
1962 V62 N1 P59- FPTS V22 N2 P.T240 <FASE>
1962 V62 N2 P178-182 62-23798 <=*>
1962 V62 N2 P183-189 63-19388 <=*>
1962 V62 N2 P216-218 65-17036 <*>
1962 V62 N3 P321-332 63-23030 <=*>
1962 V62 N3 P422-427 65-63200 <=$>
1962 V62 N4 P537-542 63-24524 <=*$>
1962 V62 N5 P641-647 62-32574 <=*>
1962 V62 N5 P657-660 62-32574 <=*>
1962 V62 N5 P688- FPTS V23 N2 P.T340 <FASE>
FPTS,V23,P.T340-T342 <FASE>
1962 V62 N8 P1211- FPTS V23 N2 P.T375 <FASE>
FPTS,V23,P.T375-T378 <FASE>
1962 V62 N9 P1293-1299 64-21977 <=*$>
1962 V62 N9 P1350-1355 64-13468 <=*$>
1962 V62 N9 P1356-1358 64-13467 <=*$> 0
1962 V62 N9 P1359- FPTS V23 N2 P.T268 <FASE>
1962 V62 N9 P1359-1363
FPTS,V23,P.T268-T271 <FASE>
62-33706 <=*>
1962 V62 N9 P1396- FPTS V23 N4 P.T741 <FASE>
1962 V62 N10 P1508- FPTS,V23,N4,P.T885 <FASE>
1962 V62 N12 P1777- FPTS V23 N3 P.T630 <FASE>
1962 V62 N12 P1777-1783
FPTS,V23,N3,P.T630 <FASE>
1962 V62 N12 P1821-1831 64-41837 <=$>
1962 V62 N114 P1143-1148 65-64357 <=$>
1963 V63 N1 P3-18 63-21771 <=>
1963 V63 N1 P92-95 63-21771 <=>
1963 V63 N4 P582- FPTS V23 N3 P.T479 <FASE>
FPTS,V23,N3,P.T479 <FASE>
1963 V63 N4 P610-613 63-20622 <=*$>
1963 V63 N5 P690-692 64-15502 <=*$> 0
1963 V63 N5 P693-695 64-15547 <=*$>
1963 V63 N6 P804-813 63-31389 <=>
1963 V63 N6 P814- FPTS V23 N3 P.T621 <FASE>
FPTS,V23,N3,P.T621 <FASE>
1963 V63 N7 P958-960 64-19280 <=*$>
1963 V63 N7 P1052- FPTS V23 N4 P.T873 <FASE>
FPTS,V23,N4,P.T873 <FASE>
1963 V63 N8 P1189- FPTS V23 N4 P.T719 <FASE>
1963 V63 N9 P1281-1296 63-41326 <=>
1963 V63 N9 P1322-1328 63-41164 <=>
1963 V63 N9 P1368- FPTS V23 N4 P.T715 <FASE>
1963 V63 N10 P457-465 66-61755 <*=$>
1963 V63 N11 P1679-1687 64-21144 <=>
1963 V63 N12 P1835- FPTS V23 N5 P.T1145 <FASE>
FPTS,V23,N5,P.T1145 <FASE>
1964 V64 N1 P116-124 AD-614 403 <=$>
1964 V64 N1 P137-141 64-21680 <=>
1964 V64 N1 P142-147 64-21681 <=>
1964 V64 N1 P148-155 64-21685 <=>
1964 V64 N2 P295-302 64-31026 <=>
1964 V64 N3 P326-330 65-62210 <=>
1964 V64 N3 P453-464 64-31328 <=>
1964 V64 N3 P464-467 64-31230 <=>
1964 V64 N4 P515-519 64-31265 <=>
1964 V64 N4 P515-526 64-31602 <=>
1964 V64 N4 P544-549 64-31264 <=>
1964 V64 N4 P550-554 64-31602 <=>
1964 V64 N4 P559- FPTS V24 N3 P.T450 <FASE>
1964 V64 N4 P559-586 64-31602 <=>
1964 V64 N8 P1172-1176 66-61745 <=*$>
1964 V64 N9 P1345-1347 65-62310 <=$>
1964 V64 N11 P1618- FPTS V24 N5 P.T799 <FASE>
1964 V64 N11 P1661-1666 65-62346 <=$>
1964 V64 N12 P1759-1765 65-30810 <=$>
1965 N3 P361-366 65-31028 <=$>
1965 V65 N3 P386-393 65-31029 <=$>
1965 V65 N4 P547- FPTS V25 N2 P.T239 <FASE>
1965 V65 N5 P696- FPTS V25 N2 P.T243 <FASE>
1965 V65 N11 P1753-1757 66-30114 <=*$>
1965 V65 N12 P1789- FPTS V25 N6 P.T915 <FASE>

ZHURNAL OBSHCHEI BIOLOGII
1942 N5 P96-122 RT-4379 <*> 0
1943 V4 N1 P15-27 RT-2778 <*>
1943 V4 N2 P65-72 61-31008 <=>
1944 V5 N6 P325-356 61-11476 <=>
1945 V6 N4 P259-277 RT-2975 <*>
1947 V8 N1 P3-35 RT-811 <*>
64-23634 <=$>
1948 V9 N3 P181-201 <ANSP>
1950 V11 P91-103 61-28224 <*=> 0
1951 V12 N3 P161-232 R-3878 <*>
1951 V12 N5 P363-367 RT-1145 <*>
1952 V13 N5 P346-362 RT-2978 <*>
1953 V14 N6 P461-469 63-31090 <=>
1953 V14 N6 P469-473 RT-2045 <*>
1953 V14 N6 P475-476 63-31090 <=>
1954 V15 N1 P18-30 RT-2245 <*>
60-21118 <=>
1954 V15 N6 P460-467 PANSDOC-TR.373 <PANS>

1955 V16 N2 P106-118 63-13421 <=*>
1955 V16 N3 P222-237 61-23573 <=*>
1956 V17 N3 P169-184 RT-4623 <*>
1956 V17 N4 P317-320 R-3171 <*>
1956 V17 N5 P396-399 62-23718 <=*>
1956 V17 N6 P413-435 59-12742 <+*>
1957 V18 N3 P194-207 AEC-TR-3611 <+*> 0
1957 V18 N5 P381-394 61-00437 <*>
1958 V19 N3 P234-238 59-13360 <=> 0
1958 V19 N5 P329-337 63-11035 <=>
1958 V19 N5 P376-386 64-19873 <=$>
1958 V19 N6 P457-466 59-13330 <=> 0
1959 V20 N2 P81-84 60-11034 <=>
1959 V20 N2 P94-103 61-27525 <*=>
1959 V20 N2 P104-114 60-11035 <=>
1959 V20 N3 P161-173 61-15908 <+*>
1959 V20 N3 P230-238 61-13077 <*+>
1959 V20 N4 P299-307 62-23282 <=*>
1959 V20 N6 P428-438 61-00438 <*>
1960 V21 N2 P81-88 65-20424 <$>
1960 V21 N5 P387-389 61-28739 <*=>
1961 V22 N2 P120-127 UCRL TRANS-1029 <=*$>
1961 V22 N3 P233-240 63-15467 <=*> 0
1961 V22 N4 P281-291 63-21389 <=> 0
1961 V22 N4 P305-310 63-13023 <=*>
1961 V22 N5 P321-324 62-15483 <*=>
1961 V22 N5 P325-332 62-11107 <=>
1962 V23 N2 P81-89 63-13183 <=*>
1962 V23 N2 P90-97 63-31624 <=>
1962 V23 N5 P396-400 63-21897 <=>
1963 V24 N1 P3-22 63-21688 <=>
1963 V24 N2 P81-97 63-31184 <=>
1963 V24 N3 P182-193 63-31698 <=>
1963 V24 N3 P215-220 64-19226 <=*$>
1963 V24 N3 P221-225 63-31697 <=>
1963 V24 N4 P241-260 64-21189 <=>
1963 V24 N4 P261-275 63-41298 <=>
1963 V24 N5 P313-323 63-41341 <=>
1963 V24 N5 P324-333 63-41339 <=>
1963 V24 N5 P352-359 63-41340 <=>
1963 V24 N5 P380-381 63-41340 <=>
1963 V24 N6 P435-444 64-21316 <=>
1964 V25 N1 P3- FPTS V24 N3 P.T463 <FASE>
1964 V25 N1 P3-16 64-21708 <=>
1964 V25 N1 P18-21 64-21712 <=>
1964 V25 N1 P22- FPTS V24 N2 P.T239 <FASE>
1964 V25 N1 P22-38 FPTS V24 P.T239-T247 <FASE>
1964 V25 N1 P62-68 65-62366 <=$>
1964 V25 N1 P75- FPTS V24 N2 P.T271 <FASE>
1964 V25 N1 P75-77 FPTS V24 P.T271-T272 <FASE>
1964 V25 N2 P141-144 64-31349 <=>
1964 V25 N3 P177-187 64-41504 <=>
1964 V25 N3 P210-229 64-41505 <=>
1964 V25 N3 P224- FPTS V24 N3 P.T431 <FASE>
1964 V25 N4 P267-276 64-51365 <=$>
1964 V25 N4 P277- FPTS V24 N5 P.T886 <FASE>
1964 V25 N4 P298-310 64-51365 <=$>
1964 V25 N5 P321- FPTS V24 N5 P.T907 <FASE>
1964 V25 N5 P321-327 64-51921 <=$>
1964 V25 N5 P347- FPTS V24 N5 P.T836 <FASE>
1964 V25 N5 P371- FPTS V24 N5 P.T896 <FASE>
1964 V25 N6 P401-411 65-64542 <=*$>
1964 V25 N6 P417-432 65-30226 <=$>
1964 V25 N6 P434-442 65-30265 <=$>
1965 V26 N1 P83- FPTS V25 N1 P.T45 <FASE>
1965 V26 N1 P96-101 RTS-3103 <NLL>
66-60398 <=*$>
1965 V26 N2 P161- FPTS V25 N1 P.T146 <FASE>
1965 V26 N2 P176-190 66-60399 <=*$>
1965 V26 N3 P372- FPTS V25 N2 P.T337 <FASE>
1965 V26 N3 P372-374 65-31856 <=*$>
1965 V26 N4 P451-457 66-61630 <=*$>
1965 V26 N4 P494-501 66-61307 <=*$>
1965 V26 N4 P677-684 66-61995 <=*$>
1965 V26 N5 P563- FPTS V25 N6 P.T1087 <FASE>
1965 V26 N5 P569-576 66-61969 <=*$>
1965 V26 N6 P711- FPTS V25 N6 P.T1094 <FASE>

*ZHURNAL OBSHCHEI KHIMII
1931 N10 P1249-1257 RJ-180 <ATS>
1931 V1 P91-104 63-16803 <*=>
1931 V1 P826-844 R-2628 <*>
1931 V1 P917-925 R-2817 <*>
1931 V1 P1108-1113 R-2587 <*>
1931 V1 N2 P330-339 64-30152 <*> 0
1931 V1 N6 P717-728 63-10420 <=*>
1931 V1 N6 P736-739 AEC-TR-4307 <*+>
1931 V1 N1C P1171-1176 25Q70R <ATS>
1931 V1 N10 P1249-1257 RJ-180 <ATS>
1931 V1 N3/4 P411- 61-13183 <+*>
1932 V2 P415-420 R-2287 <*>
1932 V2 N6 P506-514 RJ-646 <ATS>
1932 V2 N6 P524-528 63-23242 <=*>
1932 V2 N6 P542-552 R-4987 <*>
1932 V2 N4/5 P371-375 RT-3595 <*>
1933 V3 P351-359 AEC-TR-6097 <=*$>
1933 V3 P534-539 R-2968 <*>
1933 V3 P670-678 64-10763 <=*$> C
1933 V3 P747-758 61-16675 <=*$>
1933 V3 N1 P21-27 RT-3598 <*>
1933 V3 N5 P59C-595 RT-2104 <*>
1933 V3 N7 P825-830 R-4961 <*>
1934 V4 P23-30 R-3964 <*>
1934 V4 P856-865 61-16969 <=*$>
1934 V4 P948-950 61-18088 <=*$>
1934 V4 P969-974 65-14991 <*>
1934 V4 P988-994 R-4965 <*>
1934 V4 P1027-1033 61-16817 <=*$>
1934 V4 P1088-1090 61-18154 <=*>
1934 V4 P1206-1210 60-13003 <*+>
1934 V4 P1274-1278 61-16830 <=*$>
1934 V4 N1 P13-22 61-16963 <=*$>
61-17840 <=$>
1934 V4 N1 P23-30 61-16615 <=*$>
1934 V4 N7 P975-978 61-16900 <=*$>
1934 V4 N9 P1250-1257 63-10421 <*> 0
1934 V4 N10 P1347-1352 62-20150 <=*>
64-18026 <=*$>
1934 V4 N1C P1385-1389 61-16632 <=*$> P
63-10370 <=*>
1934 V4 N10 P1428-1433 PANSDOC-TR.385 <PANS>
1935 V5 P117-120 61-18116 <=*>
1935 V5 P143-148 <ATS>
1338F <K-H>
1935 V5 P149-154 R-2539 <*>
1935 V5 P182-184 R-2265 <*>
1935 V5 P444-449 61-16833 <=*$>
1935 V5 P555-561 R-3971 <*>
1935 V5 P764-766 61-18115 <=*>
1935 V5 P818-829 61-16818 <=*$>
1935 V5 P924-933 63-14305 <=*> 0
1935 V5 P1415-1420 61-13428 <*+>
61-18046 <=*$>
1935 V5 P1517-1526 65-12336 <*>
1935 V5 P1772-1780 61-16610 <=*$>
1935 V5 P1839-1858 61-16602 <=*$> P
1935 V5 P1859-1865 61-18045 <=*$>
1935 V5 P2128-2134 RT-1528 <*>
1935 V5 N2 P211-215 65-12608 <*>
1935 V5 N4 P555-561 RT-1012 <*>
1935 V5 N10 P1460-1467 88L34R <ATS>
1935 V5 N11 P1517-1526 RT-775 <*>
2867I <K-H>
63-20746 <=*$>
1935 V5 N12 P1639-1641 59-10783 <+*>
61-16607 <=*$>
1936 P227-231 R-2929 <*>
1936 V6 P197-202 61-16812 <=*$>
1936 V6 P549-554 R-2522 <*>
1936 V6 P732-747 R-2523 <*>
1936 V6 P748-756 62-14608 <=*>
1936 V6 P988-998 R-3973 <*>
1936 V6 P1096-1100 R-2913 <*>
65-63244 <=>
1936 V6 P1112-1115 R-2928 <*>
1936 V6 P1729-1736 61-16626 <=*$>
1936 V6 P1806-1814 61-16836 <=*$>
1936 V6 P1892-1896 61-16811 <=*$>
1936 V6 N1 P68-74 61-16679 <=*$>
1936 V6 N1 P75-77 63-16036 <=*>

Citation	Translation
1936 V6 N1 P129-136	61-16807 <=*$>
1936 V6 N1 P144-160	RT-371 <*>
1936 V6 N2 P283-288	AD-610 884 <=$>
	RT-1529 <*>
1936 V6 N2 P1823-1827	RT-1653 <*>
1936 V6 N3 P470-477	R-826 <*>
1936 V6 N4 P576-583	RT-2352 <*>
1936 V6 N5 P774-779	RT-1436 <*>
1936 V6 N6 P821-838	16K28R <ATS>
1936 V6 N10 P1418-1429	63-18044 <=*>
1936 V6 N10 P1506-1509	61-16829 <=*$>
1936 V6 N10 P1553-1558	36L37R <ATS>
1936 V6 N11 P1694-1697	59-17526 <*>
1936 V6 N11 P1701-1714	RT-1968 <*>
1936 V6 N11 P1736-1743	65-10782 <*>
1937 P1026-1032	<ATS>
1937 N7 P20-21	TT.474 <NRC>
1937 V7 P56-64	RJ-140 <ATS>
1937 V7 P350-352	61-18076 <=*$>
1937 V7 P353-356	59-19117 <+*>
1937 V7 P363-368	R-2470 <*>
1937 V7 P538-544	1092 <K-H>
1937 V7 P591-594	61-18087 <=*$>
1937 V7 P632-636	61-18114 <=*>
1937 V7 P637-640	61-16974 <=*$>
1937 V7 P641-645	61-16973 <=*$>
1937 V7 P667-672	61-18086 <=*$>
1937 V7 P722-728	61-18082 <=*$> P
1937 V7 P747-749	61-18085 <=*$>
1937 V7 P839-841	R-2901 <*>
1937 V7 P923-927	61-16601 <=*$>
1937 V7 P956-958	61-18025 <=*$>
1937 V7 P1413-1416	R-2900 <*>
1937 V7 P1868-1873	64-20502 <*>
1937 V7 P1900-1903	65-14992 <*>
1937 V7 P2175-2182	RJ-178 <ATS>
1937 V7 N1 P14-17	61-17847 <=$>
1937 V7 N1 P56-64	RJ-140 <ATS>
1937 V7 N1 P84-92	RT-800 <*>
1937 V7 N1 P90-92	AD-610 942 <=$>
	RT-1569 <*>
1937 V7 N1 P122-130	63-10407 <*> 0
1937 V7 N1 P258-261	RT-871 <*>
1937 V7 N1 P270-272	AD-610 887 <=$>
	RT-1570 <*>
1937 V7 N1 P273-276	RT-3426 <*>
1937 V7 N2 P545-549	TT.307 <NRC>
	61-18160 <=*>
1937 V7 N5 P879-881	RT-3864 <*>
1937 V7 N9 P1366-1377	62-14611 <=*>
1937 V7 N12 P1721-1728	63-14306 <*> P
1937 V7 N14 P2007-2008	64-18584 <*>
1937 V7 N16 P2175-2182	RJ-178 <ATS>
	RT-1633 <*>
1937 V7 N22 P2715-2718	65-62903 <=*$>
1937 V7 N22 P2760-2766	RT-872 <*>
1937 V7 N22 P2774-2778	RT-976 <*>
1937 V7 N24 P2941-2944	64-20445 <*>
1937 V7 N3/4 P623-631	65-13852 <*>
1938 V8 P291-295	64-10021 <=*$>
1938 V8 P341-345	R-2930 <*>
1938 V8 P685-713	63-10413 <=*>
1938 V8 P746-750	R-2438 <*>
1938 V8 P751-758	63-20795 <=*$> 0
1938 V8 P981-985	65-12573 <*>
1938 V8 P986-990	65-12574 <*>
1938 V8 P1385-1389	R-2333 <*>
1938 V8 N1 P22-34	(CHEM)-108 <CTS>
1938 V8 N2 P981-985	65-62905 <=>
1938 V8 N3 P266-272	R-908 <*>
	RT-2738 <*>
	64-15122 <=*$>
1938 V8 N4 P319-329	64-11078 <=>
1938 V8 N6 P524-528	59-10004 <+*>
1938 V8 N6 P524-529	64-10045 <=*$>
1938 V8 N6 P538-543	54L36R <ATS>
1938 V8 N6 P552-557	59-20729 <+*>
1938 V8 N6 P578-580	41R79R <ATS>
1938 V8 N7 P583-587	61-18102 <=*> P
1938 V8 N7 P642-650	61-17237 <=$>
1938 V8 N7 P656-661	RT-387 <*>
1938 V8 N9 P830-833	4521 <BISI>
1938 V8 N11 P981-985	RJ626 <ATS>
1938 V8 N12 P1120-1124	RT-1590 <*>
	62-24444 <=*>
1938 V8 N13 P1255-1263	63-18610 <=*>
1938 V8 N17 P1784-1785	RT-1508 <*>
	60-19512 <=*$> 0
1938 V8 N14/5 P1298-1301	<LSA>
1938 V8 N14/5 P1314-1324	61-32017 <=$>
1939 V9 P65-68	R-2829 <*>
1939 V9 P119-125	R-3974 <*>
1939 V9 P460-466	61-16671 <=*$>
1939 V9 P577-586	R-3353 <*>
1939 V9 P753-758	R-2889 <*>
	62-23942 <=*>
	66-10244 <*> 0
1939 V9 P792-793	R-2486 <*>
	61-13633 <+*>
1939 V9 P845-854	R-3969 <*>
1939 V9 P996-1006	64-14699 <=*$>
1939 V9 P1819-1824	63-20672 <=*$>
1939 V9 N2 P182-192	RT-2739 <*>
1939 V9 N4 P304-313	RT-3936 <*>
1939 V9 N4 P314-324	RT-3937 <*>
1939 V9 N5 P436-441	55J16R <ATS>
1939 V9 N7 P625-627	57-1576 <*>
1939 V9 N8 P735-758	RJ-653 <ATS>
1939 V9 N9 P814-818	63-18045 <=*>
1939 V9 N9 P845-854	RT-3761 <*>
1939 V9 N12 P1077-1085	RT-2861 <*>
1939 V9 N13 P1155-1157	RT-1600 <*>
	62-24445 <=*>
1939 V9 N14 P1332-1341	R-4909 <*>
1939 V9 N15 P1435-1440	RT-1381 <*>
	65-62951 <=>
1939 V9 N16 P1479-1488	R-4797 <*>
1939 V9 N18 P1666-1673	R-4516 <*>
	RT-1142 <*>
1939 V9 N19 P1739-1741	RT-874 <*>
1939 V9 N19 P1755-1758	R-2005 <*>
	RT-774 <*>
	2867AA <K-H>
	63-20747 <=*$>
1939 V9 N19 P1759-1763	R-2053 <*>
	RT-1041 <*>
	2867BB <K-H>
	64-10126 <=*$>
1939 V9 N19 P1777-1782	RJ-72 <ATS>
	64-10764 <=*$>
1939 V9 N22 P2061-2066	R-2004 <*>
	RT-714 <*>
	2867DD <K-H>
	63-20749 <=*$>
1939 V9 N24 P2291-2301	61-16857 <=*$>
1940 V10 P21-30	65-12390 <*>
1940 V10 P165-171	R-2187 <*>
1940 V10 P233-246	R-2471 <*>
1940 V10 P305-318	R-4089 <*>
1940 V10 P655-666	66-11317 <*>
1940 V10 P1065-1068	1058B <K-H>
1940 V10 P1457-1461	65-10131 <*> 0
1940 V10 P1843-1854	65-62978 <=>
1940 V10 N1 P1-10	58M40R <ATS>
	61-16860 <*=>
1940 V10 N2 P154-157	RT-875 <*>
1940 V10 N8 P673-676	RT-3972 <*>
1940 V10 N8 P677-682	RT-3033 <*>
1940 V10 N8 P730-732	62-10014 <*=>
1940 V10 N8 P736-744	RT-876 <*>
1940 V10 N11 P981-996	60-18491 <+*>
1940 V10 N13 P1231-1240	66-12472 <*>
1940 V10 N14 P1333-1342	64-18573 <*> P
1940 V10 N16 P1457-1461	62-18218 <=*>
1940 V10 N17 P1543-1546	73K23R <ATS>
1940 V10 N17 P1600-1604	64-30151 <*>
1940 V10 N21 P1813-1818	65-63241 <=>
1940 V10 N21 P1843-1854	73L36R <ATS>
1940 V10 N5/6 P497-506	RT-1382 <*>
1940 V10 N19/0 P1751-1756	RT-1378 <*>

1940 V10 N19/0 P1793 65-60079 <=$>
1940 V10 N19/2 P1751-1756 60-13138 <+*> O
1940 V10 N23/4 P2041-2046 RT-1169 <*>
1941 V11 P358-362 PT4 RCT V15 N1 P693 <RCT>
1941 V11 P267-275 R-2299 <*>
1941 V11 P309-313 61-18039 <*=>
1941 V11 P402-404 66-13500 <*>
1941 V11 P683-685 61-18071 <*=>
1941 V11 P887-890 R-2855 <*>
1941 V11 P1092-1095 61-16934 <*=>
1941 V11 P1096-1099 61-16935 <*=>
1941 V11 P1100-1103 61-16936 <*=>
1941 V11 P1104-1106 61-16937 <*=>
1941 V11 P1107-1110 61-16938 <*=>
1941 V11 N1 P16-22 RT-3759 <*>
1941 V11 N1 P23-40 TT.115 <NRC>
1941 V11 N2 P117-126 61-15817 <+*>
1941 V11 N2 P149-156 RJ-457 <ATS>
1941 V11 N3 P190-196 RJ-650 <ATS>
1941 V11 N3 P198-202 RT-1380 <*>
1941 V11 N3 P200-202 R-4501 <*>
1941 V11 N3 P209-212 RT-2254 <*>
1941 V11 N3 P225-231 746TM <CTT>
1941 V11 N4 P305-308 64-18583 <*> O
64-20852 <*>
1941 V11 N4 P349-352 C-5622 <NRC>
1941 V11 N4 P353-357 RCT V15 N1 P698 <RCT>
64-18585 <*>
1941 V11 N4 P358-362 64-18586 <*>
1941 V11 N4 P471-482 61-16876 <*=>
1941 V11 N7 P483-492 61-16878 <*=>
1941 V11 N7 P527-532 61-16880 <*=>
1941 V11 N9 P707-712 50/1889 <NLL>
1941 V11 N9 P713-721 RT-2497 <*>
61-18080 <*=>
1941 V11 N9 P722-728 61-18084 <*=>
1941 V11 N10 P765-767 61-13634 <*+>
1941 V11 N10 P817-820 61-18017 <*=>
1941 V11 N10 P821-823 61-18016 <*=>
1941 V11 N10 P824-828 61-18015 <*=>
1941 V11 N10 P835-838 RT-2255 <*>
1941 V11 N10 P851-858 RT-1379 <*>
1941 V11 N12 P1019-1022 65-60045 <=$>
1941 V11 N13 P1081-1091 5866-G <K-H>
1941 V11 N5/6 P405-410 59-10696 <+*>
1941 V11 N5/6 P423-424 65-62980 <=>
1941 V11 N5/6 P461-466 61-18105 <*=>
1941 V11 N5/6 P467-470 61-28099 <*=>
1941 V11 N13/4 P1081-1091 64-18404 <*> O
1941 V11 N13/4 P1111-1120 60-15846 <+*>
1942 V12 P73-86 61-16617 <*=>
1942 V12 P87-94 61-16637 <*=>
1942 V12 P112-123 61-16657 <*=>
1942 V12 P112-122 63-16037 <=*>
1942 V12 P220-226 61-16638 <*=>
1942 V12 P229-239 61-16666 <*=>
1942 V12 P240-245 61-16639 <*=>
1942 V12 P246-254 61-16640 <*=>
1942 V12 P296-305 61-16916 <*=>
1942 V12 P306-320 61-16893 <*=>
1942 V12 P321-328 TT.347 <NRC>
1942 V12 P329-336 TT.348 <NRC>
1942 V12 P409-414 64-20220 <*>
1942 V12 P422-432 61-16664 <*=>
1942 V12 P468-473 R-2646 <*>
1942 V12 P533-567 61-16649 <*=>
1942 V12 P646-648 61-16650 <*=>
1942 V12 N1/2 P99-103 63-16801 <=*>
1942 V12 N3/4 P153-159 62-20386 <=*>
1942 V12 N3/4 P186-192 RT-328 <*>
1942 V12 N3/4 P220-226 RCT V18 N1 P24 <RCT>
1942 V12 N7/8 P398-402 RJ-628 <ATS>
1942 V12 N7/8 P409-414 6108 <K-H>
1942 V12 N9/0 P510-517 65-60100 <=$>
1943 V13 P272-278 65-14351 <*>
1943 V13 P353-357 61-16660 <*=>
1943 V13 P510-515 65-13979 <*>
1943 V13 P540-551 61-16913 <=*>
1943 V13 P579-583 61-16897 <*=>
1943 V13 P609-615 61-16896 <*=>
1943 V13 P717-721 61-16899 <*=>
1943 V13 P733-735 61-16898 <*=>
1943 V13 N3 P125-130 RT-1010 <*>
61-16618 <*=>
1943 V13 N3 P136-144 RT-3841 <*>
1943 V13 N3 P184-188 RT-3852 <*>
1943 V13 N3 P189-194 RT-3853 <*>
61-16619 <*=>
1943 V13 N3 P195-201 RT-3854 <*>
61-16631 <*=>
1943 V13 N3 P202-212 RT-3858 <*>
61-16980 <*=>
1943 V13 N6 P391-397 RJ-259 <ATS>
57-1515 <*>
1943 V13 N6 P428-435 63-20181 <=*>
1943 V13 N6 P477-480 RT-3830 <*>
1943 V13 N1/2 P1-20 RT-3760 <*>
61-16955 <*=>
1943 V13 N1/2 P21-35 61-16982 <*=>
1943 V13 N1/2 P36-40 61-16981 <*=>
1943 V13 N1/2 P62-67 RT-3813 <*>
1943 V13 N1/2 P102-107 RT-3817 <*>
1943 V13 N1/2 P108-112 RT-3860 <*>
1943 V13 N4/5 P304-308 RT-3808 <*>
61-16659 <*=>
1943 V13 N4/5 P319-321 TT.278 <NRC>
1943 V13 N11/2 P830-833 TT.279 <NRC>
1944 V14 P148-160 61-18193 <*=>
1944 V14 P236-244 61-18168 <*=>
1944 V14 P274-279 R-3960 <*>
1944 V14 P337-342 61-18163 <*=>
1944 V14 P343-349 61-18162 <*=>
1944 V14 P359-364 61-18161 <*=>
1944 V14 P420-427 R-3956 <*>
1944 V14 P492-494 61-18204 <*=>
1944 V14 P495-497 61-18203 <*=>
1944 V14 P498-500 61-18202 <*=>
1944 V14 P501-504 65-63376 <=*$>
1944 V14 P914-919 R-2559 <*>
1944 V14 P955-959 61-18214 <*=>
1944 V14 P1025-1029 R-3028 <*>
1944 V14 N6 P385-402 RT-2187 <*>
1944 V14 N6 P420-427 RCT V19 N1 P385 <RCT>
59-15848 <*+> OP
62-10848 <*=> P
1944 V14 N6 P428-434 62-10849 <*=> P
1944 V14 N6 P435-437 62-10850 <*=> P
1944 V14 N1/2 P57-69 RT-2656 <*>
61-16903 <*=>
1944 V14 N1/2 P77-80 61-16906 <*=>
1944 V14 N1/2 P113-115 RT-1009 <*>
1944 V14 N1/2 P120-127 RT-3833 <*>
61-16905 <*=>
1944 V14 N7/8 P833-841 59-17827 <+*> O
1944 V14 N11/2 P1030-1037 RT-2398 <*>
63-20255 <=*> O
1944 V14 N11/2 P1138-1141 61-28119 <*=>
1945 V15 P120-130 61-18251 <*=>
1945 V15 P273-276 61-20126 <*=>
1945 V15 P358-361 61-20129 <*=>
1945 V15 P461-468 65-12544 <*>
1945 V15 P587-590 61-20143 <*=>
1945 V15 P635-638 61-20141 <*=>
1945 V15 P699-703 61-20142 <*=>
1945 V15 P724-728 R-2647 <*>
63-16069 <=*>
1945 V15 P796-798 61-20146 <*=>
1945 V15 P799-801 61-20147 <*=>
1945 V15 P820-824 61-20148 <*=>
1945 V15 P895-901 61-20149 <*=>
1945 V15 P947-951 61-20112 <*=> P
1945 V15 P981-987 61-20150 <*=>
1945 V15 P1001-1006 61-20152 <*=>
1945 V15 N3 P131-134 AEC-TR-3756 <+*>
1945 V15 N3 P135-141 61-18248 <*=>
1945 V15 N3 P156-164 61-18246 <*=>
1945 V15 N3 P165-168 RT-1444 <*>
1945 V15 N3 P173-176 RT-1008 <*>
1945 V15 N3 P177-186 RCT V21 N1 P48 <RCT>
1945 V15 N3 P225-236 RT-2502 <*>

Citation	Translation
1945 V15 N3 P237-244	65-63377 <=*$>
1945 V15 N6 P412-420	64-20810 <*>
1945 V15 N6 P565-573	RT-3901 <*>
1945 V15 N1/2 P37-41	61-18249 <*=>
1945 V15 N1/2 P42-59	TT.415 <NRC>
1945 V15 N1/2 P70-74	RT-1418 <*>
1945 V15 N1/2 P86-89	TT.368 <NRC>
1945 V15 N4/5 P328-331	RT-2503 <*>
1945 V15 N4/5 P341-352	TT.133 <*>
	61-20127 <*=>
1945 V15 N4/5 P353-357	59-15838 <+*>
1945 V15 N4/5 P353-356	61-20128 <*=>
1945 V15 N4/5 P353-357	62-14057 <=*>
1945 V15 N7/8 P603-607	RT-2172 <*>
1945 V15 N7/8 P608-618	RT-2588 <*>
1945 V15 N7/8 P628-634	65-60069 <=$> P
1945 V15 N7/8 P678-683	62-10864 <*=> P
1945 V15 N7/8 P690-698	62-10865 <*=> P
1945 V15 N11/2 P903-	AEC-TR-4870 <=*>
1946 V16 P17-32	R-2648 <*>
1946 V16 P65-70	61-20177 <*=>
1946 V16 P71-76	61-20161 <=*>
1946 V16 P77-93	61-20164 <*=>
1946 V16 P171-178	R-2649 <*>
1946 V16 P193-198	R-2625 <*>
1946 V16 P261-276	65-12547 <*>
1946 V16 P283-294	61-20178 <*=>
1946 V16 P295-299	65-12697 <*>
1946 V16 P415-426	R-5158 <*>
1946 V16 P427-434	61-20167 <*=>
1946 V16 P445-446	61-20168 <*=>
1946 V16 P531-536	65-12696 <*>
1946 V16 P657-664	61-20118 <*=> P
1946 V16 P821-824	61-20121 <*=>
1946 V16 P825-828	61-20122 <*=>
1946 V16 P829-834	61-20123 <*=>
1946 V16 P933-936	61-20124 <*=>
1946 V16 P937-950	61-20125 <*=>
1946 V16 P1003-1019	R-1176 <*>
1946 V16 P1025-1028	R-3995 <*>
1946 V16 P1317-1358	R-3985 <*>
1946 V16 P1421-1430	66-12334 <*>
1946 V16 P1495-1504	61-13429 <*+>
1946 V16 P1561-1566	63-18359 <=*>
1946 V16 P1643-1653	63-20804 <=*$>
1946 V16 P1835-1844	65-61485 <=>
1946 V16 P1871-1872	60-17213 <+*>
1946 V16 P1914-1922	170TM <CTT>
1946 V16 P2126-2131	59-22480 <+*>
1946 V16 N1 P27-32	RT-1603 <*>
1946 V16 N1 P141-144	85S80R <ATS>
1946 V16 N2 P171-178	60-18044 <+*>
1946 V16 N2 P243-250	59-17826 <+*> 0
1946 V16 N3 P471-476	PB 141 454T <=>
1946 V16 N6 P805-810	61-13632 <*+>
	61-20120 <*=>
1946 V16 N6 P851-854	63-20805 <=*$>
1946 V16 N6 P891-896	RT-106 <*>
1946 V16 N6 P897-900	RT-105 <*>
1946 V16 N6 P937-951	64-20464 <*>
1946 V16 N8 P1269-1278	RT-3596 <*>
1946 V16 N8 P1311-1316	RT-3116 <*>
1946 V16 N9 P1421-1430	3730 <K-H>
	63-13910 <=*>
	63-18683 <=*>
1946 V16 N9 P1481-1484	24S82R <ATS>
1946 V16 N9 P1505-1520	59-15923 <+*>
1946 V16 N10 P1729-1736	63-14826 <=*>
1946 V16 N11 P1896-1906	99L35R <ATS>
1946 V16 N12 P2065-2071	63-14292 <=*>
1946 V16 N12 P2141-2144	65-10138 <*>
1946 V16 N4/5 P709-719	65-12580 <*>
1946 V16 N4/5 P767-770	RT-3597 <*>
1947 V17 P122-129	R-1451 <*>
1947 V17 P269-272	65-14993 <*>
1947 V17 P347-354	LSA-ZH-1-47 <LSA>
1947 V17 P347-353	R-5198 <*>
1947 V17 P783-807	748TM <CTT>
1947 V17 P1024-1028	R-3977 <*>
1947 V17 P1203-1207	65-60070 <=$> P
1947 V17 P1385-1400	AEC-TR-1139 <*>
1947 V17 P2149-2157	61-10741 <+*>
1947 V17 P2268-2276	60-13721 <+*>
1947 V17 N1 P100-104	TT.377 <NRC>
1947 V17 N1 P122-129	64-18576 <*> 0
1947 V17 N2 P175-180	RT-470 <*>
1947 V17 N2 P185-192	R-1388 <*>
	RT-1697 <*>
	63-14408 <=*> 0
1947 V17 N2 P199-207	50S83R <ATS>
1947 V17 N2 P299-301	RT-3811 <*>
1947 V17 N2 P343-346	RT-2814 <*>
1947 V17 N3 P436-442	59-15924 <+*>
1947 V17 N3 P443-449	61-16159 <*=> 0
1947 V17 N3 P457-459	64-20342 <*>
1947 V17 N3 P485-488	86K20R <ATS>
1947 V17 N3 P502-506	65-12594 <*>
1947 V17 N3 P522-527	65-12595 <*>
1947 V17 N3 P560-564	RJ-253 <ATS>
1947 V17 N4 P669-672	RT-435 <*>
	63-20750 <=*$>
1947 V17 N4 P681-685	RT-3992 <*>
1947 V17 N5 P923-928	RCT V22 N2 P287 <RCT>
1947 V17 N5 P941-956	RCT V21 N3 P605 <RCT>
1947 V17 N6 P1043-1046	63-16991 <=*>
1947 V17 N6 P1089-1098	R-906 <*>
	RT-3186 <*>
	64-19090 <=*$> 0
1947 V17 N6 P1203-1207	RT-1376 <*>
1947 V17 N8 P1453-1467	<INSD>
1947 V17 N8 P1522-1527	65-12607 <*>
1947 V17 N9 P1710-1713	R-4013 <*>
1947 V17 N9 P1714-1717	RT-2363 <*>
1947 V17 N9 P1718-1727	18N48R <ATS>
1947 V17 N10 P1752-1757	59-12136 <+*>
1947 V17 N10 P1758-1767	59-12135 <+*>
1947 V17 N10 P1768-1773	RT-3818 <*>
1947 V17 N10 P1774-1787	TT.106 <*>
1947 V17 N10 P1834-1842	RT-3801 <*>
1947 V17 N10 P1876-1887	RT-3930 <*>
	TT.324 <*>
	TT.324 <NRC>
1947 V17 N11 P2028-2047	64-16329 <*> P
1947 V17 N11 P2058-2065	62-25894 <=*>
1947 V17 N12 P2158-2165	27L29R <ATS>
	61-10745 <+*>
1947 V17 N12 P2166-2177	28L29R <ATS>
	61-10746 <+*>
1947 V17 N12 P2200-2207	RJ-353 <ATS>
1947 V17 N12 P2201-2207	RJ-353 <ATS>
1947 V17 N12 P2222-2227	RCT V22 N2 P310 <RCT>
1947 V17 N12 P2253-2255	RT-3444 <*>
1948 V18 P98-99	R-2769 <*>
	R-4851 <*>
1948 V18 P117-123	1296B <K-H>
1948 V18 N1 P48-	61-23323 <*=> 0
1948 V18 N1 P87-97	65-12386 <*>
1948 V18 N2 P299-305	65-12383 <*>
1948 V18 N3 P388-397	RT-1900 <*>
1948 V18 N4 P594-600	RT-1163 <*>
1948 V18 N4 P601-604	RT-1076 <*>
1948 V18 N4 P612-628	RT-4072 <*>
1948 V18 N4 P775-784	64-19951 <=$>
1948 V18 N5 P855-858	RT-1213 <*>
1948 V18 N5 P917-920	RT-2498 <*>
1948 V18 N6 P1027-1032	RT-3884 <*>
1948 V18 N6 P1198-1202	RT-1674 <*>
1948 V18 N7 P1361-1369	RT-982 <*>
1948 V18 N8 P1427-1439	53/0367 <NLL>
1948 V18 N8 P1440-1451	TT.281 <NRC>
	61-17210 <=$>
	63-20791 <=*$>
1948 V18 N8 P1470-1474	72K23R <ATS>
1948 V18 N8 P1517-1524	65-12531 <*>
1948 V18 N8 P1525-1536	RT-3885 <*>
1948 V18 N8 P1545-1549	60-13715 <+*>
1948 V18 N9 P1626-1638	RT-3874 <*>
	60-18073 <+*>
	64-20524 <*> 0
1948 V18 N9 P1674-1680	87N56R <ATS>

Citation	Reference
1948 V18 N9 P1696-1698	63-20996 <=*$>
1948 V18 N9 P1703-1709	RT-2404 <*>
1948 V18 N9 P1724-1732	RT-1510 <*>
	65-61595 <=$>
1948 V18 N9 P1736-1740	RT-2215 <*>
1948 V18 N10 P1766-1774	RJ-181 <ATS>
1948 V18 N10 P1836-1842	RT-2268 <*>
1948 V18 N11 P1989-1999	64-20462 <*>
1948 V18 N11 P2023-2032	RT-1414 <ATS>
1949 V19 P981-982	RT-93 <*>
1949 V19 P1C14-1C27	R-2141 <*>
1949 V19 N4 P649-659	63-20303 <=*>
1949 V19 N4 P720-723	61-17389 <=$>
1949 V19 N4 P724-729	61-17387 <=$>
1949 V19 N4 P736-739	TT.132 <NRC>
1949 V19 N5 P826-842	RT-2896 <*>
	61-16787 <*=>
1949 V19 N5 P967-979	60-13699 <+*>
1949 V19 N7 P1382-1386	RT-2120 <*>
1949 V19 N9 P1596-1603	64L32R <ATS>
1949 V19 N9 P1627-1634	60-19919 <+*>
1949 V19 N12 P2175-2181	RJ-27 <ATS>
1949 V19 N12 P2201-2207	RJ-28 <ATS>
1950 P1404-1407	R-2241 <*>
1950 V20 P31-37	R-2132 <*>
1950 V20 P38-44	R-2133 <*>
1950 V20 P92-99	R-2899 <*>
1950 V20 P100-106	R-2654 <*>
1950 V20 P1C7-115	R-2134 <*>
	61-10744 <+*>
1950 V20 P201-207	R-2931 <*>
1950 V20 P221-227	R-3044 <*>
1950 V20 P284-288	61-17000 <=$>
1950 V20 P338-345	R-3045 <*>
1950 V20 P755-761	R-194 <*>
1950 V20 P774-782	R-3031 <*>
1950 V20 P1336-1346	R-2571 <*>
1950 V20 P1347-1357	R-3040 <*>
	63-16122 <=*>
1950 V20 P1384-1387	R-2227 <*>
1950 V20 P1392-1393	R-3037 <*>
1950 V20 P1394-1398	R-3036 <*>
1950 V20 P1420-1424	R-2570 <*>
1950 V20 P1543-1548	R-2571 <*>
1950 V20 P1560-1567	R-3981 <*>
1950 V20 P1747-1759	R-3035 <*>
1950 V20 P178C-1785	R-2569 <*>
1950 V20 P1929-1936	R-2567 <*>
1950 V20 N2 P3C4-309	RT-782 <*>
1950 V20 N3 P469-471	RT-1965 <*>
1950 V20 N3 P531-538	RT-655 <*>
	61-28014 <*=>
1950 V20 N4 P648-660	61-28016 <*=> 0
1950 V20 N4 P661-670	61-28017 <*=> 0
1950 V20 N4 P744-748	68K21R <ATS>
1950 V20 N5 P910-916	RT-1413 <*>
1950 V20 N6 P986-998	61-13014 <*+>
1950 V20 N6 P1068-1072	RT-1403 <*>
1950 V20 N7 P1199-1208	RT-1808 <*>
1950 V20 N8 P1347-1357	RT-3594 <*>
1950 V20 N8 P1468-1477	RJ-70 <ATS>
1950 V20 N8 P1514-1523	RT-3532 <*>
	64-19062 <=*$>
1950 V20 N10 P1826-1828	61-10305 <*+>
1950 V20 N1C 1904-1913	-1928 <*>
1950 V20 N11 P1955-1961	<INSD>
1950 V20 N11 P1981-1990	RT-1011 <*>
1950 V20 N12 P2121-2123	3161-A <K-H>
1951 V21 P28-46	R-2556 <*>
1951 V21 P456-467	10K20R <ATS>
1951 V21 P1132-1139	RJ-142 <ATS>
1951 V21 P1267-1273	R-3886 <*>
1951 V21 P1625-1632	R-4043 <*>
1951 V21 P1788-1794	R-2538 <*>
1951 V21 P1837-1841	61-18825 <*=>
1951 V21 P2217-2220	R-2562 <*>
1951 V21 P2220-2235	R-2555 <*>
1951 V21 P2235-2245	R-2561 <*>
1951 V21 N1 P3-10	R-737 <*>
1951 V21 N1 P86-9C	RT-2499 <*>
1951 V21 N2 P382-388	RT-3630 <*>
1951 V21 N5 P967-973	TT.269 <NRC>
1951 V21 N6 P1132-1139	RJ-142 <ATS>
1951 V21 N9 P1588-1602	66-12221 <*>
1951 V21 N9 P1659-1660	R-3997 <*>
1951 V21 N10 P1867-1869	61-18831 <*=>
1951 V21 N10 P1897-1902	59-20129 <+*>
1951 V21 N11 P1931-1940	T.794 <INSD>
1951 V21 N11 P2046-2050	RT-3631 <*>
1951 V21 N11 P2055-2064	RT-3632 <*>
1951 V21 N11 P2064-2068	RT-3679 <*>
1951 V21 N11 P2068-2074	<INSD>
1951 V21 N12 P2199-2205	RT-3564 <*>
1951 V21 N12 P2205-2210	RT-3593 <*>
1952 V22 P41-43	R-2079 <*>
1952 V22 P76-81	R-2558 <*>
1952 V22 P489-491	R-4052 <*>
1952 V22 P935-945	RCT V26 N4 P787 <RCT>
1952 V22 P935-	61-13636 <*+>
1952 V22 P1282-1284	R-2554 <*>
1952 V22 P1506-1516	61-18840 <*=>
1952 V22 P1546-1550	78N54R <ATS>
1952 V22 P1594-1598	11Q72R <ATS>
1952 V22 N1 P66-76	52/2933 <NLL>
1952 V22 N2 P274-278	RT-1415 <*>
1952 V22 N3 P361-368	64-10301 <=*$>
1952 V22 N3 P408-414	61-17287 <=$>
1952 V22 N3 P462-467	6C-16875 <+*>
1952 V22 N3 P467-473	60-16876 <+*>
1952 V22 N3 P473-477	60-16877 <+*>
1952 V22 N3 P5C9-513	6C-16879 <+*>
1952 V22 N4 P632-637	R-859 <*>
1952 V22 N5 P781-784	R-4833 <*>
	R-910 <*>
	61-17236 <=$>
1952 V22 N5 P784-789	61-17234 <=$>
1952 V22 N5 P792-797	RT-126 <*>
1952 V22 N6 P1C47-1055	TT.362 <NRC>
1952 V22 N6 P1055-1061	RT-363 <*>
1952 V22 N7 P1173-1175	R-900 <*>
1952 V22 N7 P1266-1270	64-15281 <=*$>
	65-61427 <=>
1952 V22 N8 P1457-1461	RT-2174 <*>
1952 V22 N8 P1465-1467	RT-2121 <*>
1952 V22 N9 P1528-1534	RT-440 <*>
1952 V22 N9 P1534-1542	RJ-484 <ATS>
	RT-425 <*>
1952 V22 N10 P1711-1715	T.644 <INSD>
1952 V22 N10 P1726-1731	61-17203 <=$>
1952 V22 N10 P1783-1787	RT-3467 <*>
1952 V22 N11 P2021-2035	RT-3695 <*>
1952 V22 N12 P2083-2090	60-10585 <+*>
1953 V23 P161-163	3189-A <K-H>
1953 V23 P450-463	63-14634 <=*>
1953 V23 P1075-1085	R-2792 <*>
1953 V23 P1278-1283	R-2256 <*>
1953 V23 P1399-1401	R-2768 <*>
	R-4852 <*>
1953 V23 N1 P1C0-106	RT-3659 <*>
1953 V23 N1 P107-116	69Q70R <ATS>
1953 V23 N1 P157-158	RT-2066 <*>
1953 V23 N1 P158-161	RT-3313 <*>
1953 V23 N2 P223-229	M-6270 <NLL>
1953 V23 N3 P392-396	59-15375 <+*>
1953 V23 N3 P450-463	64-18577 <*> P
1953 V23 N5 P771-776	RT-2065 <*>
	64-15561 <=*$>
1953 V23 N7 P1199-1203	59-10232 <+*>
1953 V23 N7 P1203-1205	59-10231 <+*>
1953 V23 N8 P1241-1245	65-10823 <*>
1953 V23 N8 P1278-1283	RT-4092 <*>
1953 V23 N8 P1323-1330	RT-1643 <*>
1953 V23 N8 P1352-1357	RT-3326 <*>
1953 V23 N8 P1357-1364	RT-3453 <*>
1953 V23 N9 P1504-1509	RT-1644 <*>
	65-63175 <=*$>
1953 V23 N9 P1522-1525	R-2067 <*>
	RJ-284 <ATS>
1953 V23 N9 P1522-1555	RJ-284 <ATS>
1953 V23 N9 P1539-1542	RT-3325 <*>

1953 V23 N9 P1547-1552 64-15560 <=*$>
1953 V23 N11 P1785-1794 <CB>
1953 V23 N11 P1794-1799 <CB>
1953 V23 N11 P1799-1802 <CB>
1953 V23 N11 P1802-1808 <CB>
1953 V23 N11 P1808-1813 <CB>
1953 V23 N11 P1813-1822 <CB>
1953 V23 N11 P1822-1825 <CB>
1953 V23 N11 P1826-1828 <CB>
1953 V23 N11 P1828-1830 <CB>
1953 V23 N11 P1830-1835 <CB>
1953 V23 N11 P1835-1842 <CB>
1953 V23 N11 P1842-1845 <CB>
1953 V23 N11 P1845-1848 <CB>
1953 V23 N11 P1849-1854 <CB>
1953 V23 N11 P1854-1867 <CB>
1953 V23 N11 P1867-1873 <CB>
1953 V23 N11 P1873-1877 <CB>
1953 V23 N11 P1878-1884 <CB>
1953 V23 N11 P1884-1886 <CB>
1953 V23 N11 P1886-1889 <CB>
1953 V23 N11 P1890-1892 <CB>
1953 V23 N11 P1893-1896 <CB>
1953 V23 N11 P1896-1899 <CB>
1953 V23 N11 P1900-1904 RJ-218 <ATS>
1953 V23 N11 P1905-1911 <CB>
1953 V23 N11 P1912-1920 <CB>
1953 V23 N11 P1920-1922 <CB>
1953 V23 N11 P1922-1927 <CB>
1953 V23 N11 P1927-1930 <CB>
1953 V23 N11 P1931-1934 <CB>
1953 V23 N11 P1934-1936 <CB>
1953 V23 N11 P1936-1939 <CB>
1953 V23 N11 P1940-1944 <CB>
1953 V23 N11 P1944-1947 <CB>
1953 V23 N11 P1947-1952 <CB>
1953 V23 N11 P1953-1960 <CB>
1953 V23 N11 P1960-1964 <CB>
1953 V23 N12 P1965-1981 <CB>
1953 V23 N12 P1981-1986 <CB>
1953 V23 N12 P1986-1990 <CB>
1953 V23 N12 P1990-1994 <CB>
1953 V23 N12 P1994-1997 <CB>
1953 V23 N12 P1998-2001 <CB>
1953 V23 N12 P2002-2006 <CB>
1953 V23 N12 P2006-2009 <CB>
1953 V23 N12 P2009-2014 <CB>
1953 V23 N12 P2014-2020 <CB>
1953 V23 N12 P2020-2023 <CB>
1953 V23 N12 P2023-2027 <CB>
65-60310 <=$>
1953 V23 N12 P2027-2034 <CB>
1953 V23 N12 P2035-2036 <CB>
1953 V23 N12 P2037-2043 <CB>
1953 V23 N12 P2043-2046 <CB>
1953 V23 N12 P2047-2052 <CB>
1953 V23 N12 P2053-2058 <CB>
1953 V23 N12 P2056-2059 <CB>
1953 V23 N12 P2060-2063 <CB>
1953 V23 N12 P2063-2066 <CB>
1953 V23 N12 P2067-2074 <CB>
1953 V23 N12 P2074-2076 <CB>
1954 V24 P581-592 T-1278 <INSD>
1954 V24 P664-670 06L36R <ATS>
1954 V24 P1101-1108 R-1961 <*>
1954 V24 P1422-1427 R-3789 <*>
1954 V24 P1782-1787 07L36R <ATS>
1954 V24 N3 P423-427 61-17271 <=$>
1954 V24 N3 P447-450 64-15266 <=*$>
1954 V24 N4 P733-738 TT.466 <NRC>
1954 V24 N5 P745-751 <CB>
1954 V24 N5 P751-759 <CB>
1954 V24 N5 P755-780 <CB>
1954 V24 N5 P759-762 <CB>
1954 V24 N5 P762-766 <CB>
1954 V24 N5 P766-775 <CB>
1954 V24 N5 P780-790 <CB>
1954 V24 N5 P790-795 <CB>
1954 V24 N5 P795-798 <CB>
1954 V24 N5 P799-802 <CB>
1954 V24 N5 P803-807 <CB>
1954 V24 N5 P807-814 <CB>
1954 V24 N5 P814-816 <CB>
1954 V24 N5 P816-820 <CB>
1954 V24 N5 P821-824 <CB>
1954 V24 N5 P825-833 <CB>
1954 V24 N5 P833-839 <CB>
1954 V24 N5 P840-848 <CB>
1954 V24 N5 P849-852 <CB>
1954 V24 N5 P852-856 <CB>
1954 V24 N5 P855-858 59-10014 <+*>
1954 V24 N5 P856-860 <CB>
1954 V24 N5 P861-868 <CB>
1954 V24 N5 P868-870 <CB>
1954 V24 N5 P870-876 <CB>
1954 V24 N5 P877-881 <CB>
1954 V24 N5 P882-887 <CB>
1954 V24 N5 P887-894 <CB>
1954 V24 N5 P895-898 <CB>
1954 V24 N5 P898-905 <CB>
1954 V24 N5 P905-909 <CB>
1954 V24 N5 P909-916 <CB>
1954 V24 N5 P917-919 <CB>
1954 V24 N5 P919-922 <CB>
1954 V24 N5 P922-924 <CB>
1954 V24 N6 P1033-1038 25P61R <ATS>
1954 V24 N7 P1142-1150 61-18574 <=*>
1954 V24 N7 P1271-1276 59-15044 <+*>
1954 V24 N8 P1277-1278 62-11384 <=>
1954 V24 N8 P1315-1321 64-16007 <=*$>
1954 V24 N8 P1465-1473 R-1529 <*>
RT-3276 <*>
64-19063 <=*$>
1954 V24 N11 P2040-2044 1672 <*>
1955 V25 P286-293 RJ-283 <ATS>
1955 V25 P675-683 62-10475 <=*>
1955 V25 P793-802 GB2 <NLL>
1955 V25 P1041-1043 T-2324 <INSD>
T-2335 <INSD>
1955 V25 P1680-1685 62-16067 <=*>
1955 V25 P1844-1859 T-2392 <INSD>
1955 V25 P2173-2182 RJ-347 <ATS>
1955 V25 P2182-2188 RJ-348 <ATS>
1955 V25 P2445-2448 GB2 <NLL>
1955 V25 P2457-2464 RJ491 <ATS>
1955 V25 P2469-2474 RJ-486 <ATS>
1955 V25 P2522-2527 GB2 <NLL>
1955 V25 N2 P286-293 RJ-283 <ATS>
1779 <TC>
1955 V25 N3 P425-433 <CB>
1955 V25 N3 P433-444 <CB>
59-10039 <+*>
1955 V25 N3 P444-451 <CB>
1955 V25 N3 P451-458 <CB>
1955 V25 N3 P458-463 <CB>
1955 V25 N3 P463-468 <CB>
1955 V25 N3 P469-470 <CB>
1955 V25 N3 P471-477 <CB>
1955 V25 N3 P477-479 <CB>
1955 V25 N3 P479-485 <CB>
1955 V25 N3 P485-488 <CB>
1955 V25 N3 P489-492 <CB>
1955 V25 N3 P493-495 <CB>
1955 V25 N3 P495-499 <CB>
1955 V25 N3 P499-503 <CB>
1955 V25 N3 P504-508 <CB>
1955 V25 N3 P508-517 <CB>
1955 V25 N3 P517-523 <CB>
1955 V25 N3 P523-526 <CB>
1955 V25 N3 P526-529 <CB>
1955 V25 N3 P530-536 <CB>
1955 V25 N3 P536-539 <CB>
1955 V25 N3 P539-544 <CB>
1955 V25 N3 P545-550 <CB>
1955 V25 N3 P551-558 <CB>
1955 V25 N3 P558-562 <CB>
1955 V25 N3 P563-565 <CB>
1955 V25 N3 P565-571 <CB>
1955 V25 N3 P571-574 <CB>
1955 V25 N3 P574-576 <CB>

1955 V25 N3 P576-580 <CB>
1955 V25 N3 P580-583 <CB>
1955 V25 N3 P583-589 <CB>
1955 V25 N3 P589-594 <CB>
1955 V25 N3 P594-598 <CB>
1955 V25 N3 P599-603 <CB>
1955 V25 N3 P603-609 <CB>
1955 V25 N3 P610-611 <CB>
1955 V25 N3 P611-617 <CB>
1955 V25 N3 P617-622 <CB>
1955 V25 N3 P622-626 <CB>
1955 V25 N3 P627-631 RJ-282 <ATS>
1955 V25 N4 P749-751 59-15376 <+*>
1955 V25 N5 P919-921 RT-3270 <*>
1955 V25 N5 P1035-1039 R-752 <*>
RT-3126 <*>
64-15072 <=*$>
1955 V25 N7 P1413-1417 RT-3274 <*>
1955 V25 N8 P1469-1473 RT-3275 <*>
1955 V25 N8 P1473-1477 RT-3273 <*>
1955 V25 N9 P1680-1685 65-12305 <*>
1955 V25 N10 P2009-2016 CSIRO-3227 <CSIR>
1955 V25 N11 P2028-2038 66-10190 <*> O
1955 V25 N11 P2138-2143 RJ-346 <ATS>
1955 V25 N11 P2173-2182 RJ-347 <ATS>
1955 V25 N11 P2182-2188 RJ-348 <ATS>
1955 V25 N12 P1707-1717 RT-3434 <*>
1955 V25 N12 P2204-2208 RJ-572 <ATS>
1955 V25 N12 P2294-2299 RT-3945 <*>
1955 V25 N13 P2464-2469 RJ-493 <ATS>
1955 V25 N13 P2474-2480 RT-3949 <*>
1955 V25 N13 P2520-2522 RT-3950 <*>
1955 V25 N13 P2529-2531 RT-3939 <*>
1956 V26 P131 C2346 <NRC>
1956 V26 P138-142 RJ-14 <ATS>
1956 V26 P656-660 R-2839 <*>
1956 V26 P735-739 RJ-623 <ATS>
1956 V26 P808-810 RJ-621 <ATS>
1956 V26 P819-832 PANSDOC-TR.163 <PANS>
1956 V26 P1212-1216 R-1134 <*>
1956 V26 P1269-1310 R-2833 <*>
1956 V26 P1368-1371 RJ-574 <ATS>
1956 V26 P1450-1456 R-4164 <*>
1956 V26 P1452-1453 62-26000 <=$>
1956 V26 P1626-1628 RCT V31 N3 P588 <RCT>
1956 V26 P1656-1659 R-1224 <*>
1956 V26 P2014-2019 71H12R <ATS>
1956 V26 P2201-2209 RCT V31 N2 P278 <RCT>
1956 V26 P2376-2380 R-2603 <*>
1956 V26 P2385-2393 R-3024 <*>
1956 V26 P2476-2485 RCT V32 N1 P220 <RCT>
1956 V26 P2607-2611 70K27R <ATS>
1956 V26 P2679-2680 R-2602 <*>
1956 V26 P2980-2984 R-1135 <*>
1956 V26 P3198-3201 59-15735 <+*>
1956 V26 N3 P699-706 RJ-615 <ATS>
1956 V26 N3 P712-714 65-11012 <*>
1956 V26 N3 P739-745 59-20227 <+*>
1956 V26 N3 P773-782 R-793 <*>
64-19517 <=$> O
1956 V26 N4 P971-980 23J16R <ATS>
1956 V26 N4 P992-996 13H12R <ATS>
1956 V26 N5 P1326-1335 RJ-629 <ATS>
1956 V26 N5 P1335-1340 RJ-630 <ATS>
1956 V26 N5 P1368-1371 64-16760 <=*$>
1956 V26 N5 P1466-1471 <ATS>
1956 V26 N6 P1381-1390 RCT V31 N1 P156 <RCT>
1956 V26 N6 P1651-1653 RJ-616 <ATS>
1956 V26 N6 P1666-1672 74H13R <ATS>
1956 V26 N6 P1677-1678 75H13R <ATS>
1956 V26 N6 P1679-1682 62-20156 <=*>
1956 V26 N7 P1907-1909 67J15R <ATS>
1956 V26 N7 P2014-2019 60-00713 <*>
1956 V26 N7 P2042-2046 59-20727 <+*>
1956 V26 N8 P2151-2155 65H13R <ATS>
1956 V26 N8 P2201-2209 64-16506 <=*$>
1956 V26 N8 P2228-2233 66H13R <ATS>
1956 V26 N8 P2233-2237 67H13R <ATS>
1956 V26 N8 P2238-2243 68H13R <ATS>
1956 V26 N8 P2285-2289 59-20726 <+*> O
70H13R <ATS>
1956 V26 N8 P2290-2294 64H13R <ATS>
1956 V26 N9 P2418-2420 40H13R <ATS>
1956 V26 N9 P2426-2433 53H13R <ATS>
1956 V26 N9 P2493-2498 52H13R <ATS>
1956 V26 N9 P2531-2537 95K21R <ATS>
1956 V26 N9 P2569-2575 54H13R <ATS>
1956 V26 N9 P2574-2577 <ATS>
1956 V26 N9 P2577-2579 55H13R <ATS>
1956 V26 N9 P2579-2581 60H13R <ATS>
1956 V26 N9 P2581-2583 61H13R <ATS>
1956 V26 N9 P2583-2588 56H13R <ATS>
1956 V26 N9 P2588-2592 62H13R <ATS>
1956 V26 N9 P2602-2604 57H13R <*>
1956 V26 N9 P2642-2648 58H13R <ATS>
59-20731 <+*> P
1956 V26 N10 P2697-2704 58J16R <ATS>
1956 V26 N10 P2704-2710 54J16R <ATS>
1956 V26 N10 P2891-2896 49Q71R <ATS>
1956 V26 N11 P2980-2986 64-16311 <*> O
1956 V26 N11 P3030-3036 49J14R <ATS>
1956 V26 N11 P3201-3206 59-15736 <+*>
1956 V26 N12 P3426-3430 63-15969 <=*>
1957 V27 P295- AEC-TR-4118 <+*>
1957 V27 P295-299 58-2366 <*>
1957 V27 P411-414 89J16R <ATS>
1957 V27 P775-780 66-12050 <*>
1957 V27 P1195-1201 RCT V32 N1 P288 <RCT>
1957 V27 P1276-1279 47J18R <ATS>
1957 V27 P1389-1391 79J19R <ATS>
1957 V27 P1487-1489 RCT V32 N1 P284 <RCT>
1957 V27 P1570-1575 R-4162 <*>
1957 V27 P1575-1581 R-4163 <*>
1957 V27 P1591-1593 38J19R <ATS>
1957 V27 P2546-2553 64-18162 <*>
1957 V27 P3075-3078 37K22R <ATS>
1957 V27 P3160-3162 <CP>
1957 V27 P3386-3390 R-5143 <*>
1957 V27 N1 P72-77 64-19425 <=$>
1957 V27 N4 P1002-1006 59-20732 <+*>
1957 V27 N5 P1167-1174 7317R <ATS>
1957 V27 N5 P1215-1217 64-16895 <=*$>
1957 V27 N5 P1330-1339 34J18R <ATS>
1957 V27 N6 P1454-1459 09K21R <ATS>
1957 V27 N6 P1727-1730 R-4141 <*>
64-19476 <=$>
1957 V27 N7 P1905-1908 63J18R <ATS>
1957 V27 N7 P1908-1910 62J18R <ATS>
1957 V27 N8 P2039-2049 65-10215 <*>
1957 V27 N8 P2138-2142 97J18R <ATS>
1957 V27 N8 P2142-2145 98J18R <ATS>
1957 V27 N8 P2152-2160 39J19R <ATS>
1957 V27 N8 P2226-2232 64-18096 <=*$>
1957 V27 N10 P2644-2648 R-4225 <*>
1957 V27 N10 P2697-2699 64-30100 <*>
1957 V27 N11 P2972-2974 64-18484 <*>
1957 V27 N11 P2977-2983 42K27R <ATS>
65-14256 <*> O
1957 V27 N11 P3064-3067 R-4224 <*>
1957 V27 N11 P3148-3151 R-4230 <*>
1957 V27 N11 P3152-3154 R-4233 <*>
1957 V27 N12 P3357-3361 R-4444 <*>
64-19474 <=$>
1957 V27 N12 P3365-3367 R-4445 <*>
64-19475 <=$>
1957 V27 N89 P72-77 R-3849 <*>
1958 V28 P47-51 59-10077 <+*>
1958 V28 P617-619 R-5170 <*>
1958 V28 P1999-2000 13K27R <ATS>
1958 V28 P2303-2304 R-5149 <*>
1958 V28 P2305-2307 R-5150 <*>
1958 V28 N1 P267-272 59-15749 <+*>
1958 V28 N2 P330-333 67K22R <ATS>
1958 V28 N2 P383-388 R-5184 <*>
1958 V28 N3 P657-661 74K26R <ATS>
1958 V28 N3 P1568-1573 21K25R <ATS>
1958 V28 N4 P1056-1059 59-22547 <+*>
1958 V28 N5 P1125-1127 68K28R <ATS>
1958 V28 N5 P1144-1150 69K28R <ATS>
1958 V28 N5 P1279-1284 70K26R <ATS>

1958 V28 N6 P1501-1503 49K28R <ATS>
1958 V28 N6 P1524-1528 63-23518 <=*$>
1958 V28 N6 P1563-1568 22K25R <ATS>
1958 V28 N6 P1628-1631 17K25R <ATS>
1958 V28 N7 P1901-1904 59-15907 <+*>
1958 V28 N7 P1904-1907 59-20130 <+*>
1958 V28 N7 P1959-1962 61-17628 <=$>
1958 V28 N7 P1984-1987 59-15382 <+*>
1958 V28 N7 P1997-1998 60-18151 <+*>
1958 V28 N8 P2089-2094 59-15383 <+*>
1958 V28 N8 P2C94-2096 59-15373 <+*>
1958 V28 N8 P2174-2178 59-17150 <+*>
1958 V28 N8 P2296-2302 89K26R <ATS>
1958 V28 N9 P2360-2371 70K28R <ATS>
1958 V28 N9 P2407-2412 59-10469 <+*>
1958 V28 N9 P2500-2502 59-10468 <+*>
1958 V28 N9 P2502-2504 76L29R <ATS>
1958 V28 N10 P2638-2644 60-31241 <=>
1958 V28 N10 P2644-2652 60-31242 <=>
1958 V28 N10 P2656-2662 78L29R <ATS>
1958 V28 N1C P2706-2710 61-19407 <+*>
1958 V28 N11 P2909-2915 70L31R <ATS>
1958 V28 N11 P3020-3023 87L29R <ATS>
1958 V28 N11 P3046-3051 88L29R <ATS>
1958 V28 N11 P3051-3055 89L29R <ATS>
1958 V28 N11 P3056-3C58 59-10470 <*>
1958 V28 N11 P3059-3061 98L28R <ATS>
1958 V28 N12 P3192-3202 47L32R <ATS>
1958 V28 N12 P3214-3216 48L32R <ATS>
1958 V28 N12 P3220-3224 82L30R <ATS>
1958 V28 N12 P3317-3319 05L35R <ATS>
1959 V29 N1 P158-165 62L32R <ATS>
1959 V29 N1 P174-179 45L31R <ATS>
1959 V29 N1 P283-285 60-10387 <+*>
1959 V29 N1 P329-332 59-18481 <+*>
1959 V29 N1 P338-339 60-10924 <+*>
1959 V29 N2 P565-575 74L31R <ATS>
1959 V29 N3 P735-744 41L33R <ATS>
1959 V29 N3 P986-989 15L32R <ATS>
1959 V29 N4 P1064-1072 60-10127 <+*> P
1959 V29 N4 P1262-1269 75M45R <ATS>
1959 V29 N4 P1285-1289 60-15805 <+*>
1959 V29 N5 P1413-1423 62-24447 <=*>
1959 V29 N5 P1522-1527 61-13517 <*+>
1959 V29 N5 P1541-1545 94L35R <ATS>
1959 V29 N6 P1819-1821 48L34R <ATS>
1959 V29 N6 P1853-1856 60-16563 <+*>
1959 V29 N6 P1998-2006 26L34R <ATS>
1959 V29 N6 P2042-2045 47L34R <ATS>
1959 V29 N7 P2125-2129 15M45R <ATS>
1959 V29 N7 P2310-2314 15L35R <ATS>
1959 V29 N8 P2591-2594 60-16818 <*>
1959 V29 N8 P2643-2646 81L35R <ATS>
1959 V29 N8 P2750-2760 61-10541 <=*>
1959 V29 N9 P2811-2816 60-31155 <=>
1959 V29 N9 P2969-2973 60-16803 <*>
1959 V29 N9 P2973-2975 60-16805 <*>
1959 V29 N9 P3010-3019 81L36R <ATS>
1959 V29 N10 P3276-3278 72M38R <ATS>
1959 V29 N10 P3322-3329 99M40R <ATS>
1959 V29 N10 P3411-3416 31L37R <ATS>
1959 V29 N1C P3504-3510 68M38R <ATS>
1959 V29 N11 P3598-3601 61-19336 <+*>
1959 V29 N11 P3627-3631 60-16804 <*>
1959 V29 N11 P3652-3657 61-19336 <+*>
1959 V29 N11 P3752-3757 61-19336 <+*>
1959 V29 N12 P3947-3953 73M38R <ATS>
1959 V29 N12 P3957-3959 60-16800 <+*>
1960 V30 P794-798 RCT V33 N4 P1062 <RCT>
1960 V30 N1 P186-190 02M40R <ATS>
1960 V30 N1 P191-192 88M40R <ATS>
1960 V30 N1 P227-230 60-16548 <+*>
1960 V30 N2 P542-545 45M4CR <ATS>
1960 V30 N2 P545-556 46M40R <ATS>
1960 V30 N2 P596-600 61-18632 <*=>
1960 V30 N2 P697-698 99N48R <ATS>
1960 V30 N3 P717-723 61-16139 <=*>
1960 V30 N3 P875-884 01M41R <ATS>
1960 V30 N3 P907-912 00M41R <ATS>
1960 V30 N3 P967-972 27M47R <ATS>
1960 V30 N3 P1048-1050 61-16140 <=*> 0
1960 V30 N4 P1208-1216 26M42R <ATS>
1960 V30 N4 P1250-1255 93M43R <ATS>
1960 V30 N4 P1256-1258 94M43R <ATS>
1960 V30 N4 P1321-1325 35M42R <ATS>
1960 V30 N4 P1342-1346 25M42R <ATS>
62-14031 <*>
1960 V30 N5 P1441-1444 56M43R <ATS>
1960 V30 N5 P1684-1685 61-10671 <+*> 0
1960 V30 N6 P1753-1755 23M45R <ATS>
1960 V30 N6 P1837-1841 03M44R <ATS>
1960 V30 N6 P1946-1950 99M43R <ATS>
1960 V30 N6 P1950-1954 61-27005 <+*> 0
1960 V30 N6 P1964-1968 30M44R <ATS>
1960 V30 N6 P2014-2016 96M43R <ATS>
1960 V30 N6 P2062-2067 60-23904 <+*>
1960 V30 N7 P2134-2138 06M47R <ATS>
65-13203 <*>
1960 V30 N7 P2138-2141 07M47R <ATS>
63-10841 <=*>
1960 V30 N7 P2172-2176 RCT V34 N1 P200 <RCT>
1960 V30 N7 P2319-2322 02N48R <ATS>
1960 V30 N7 P2371-2374 61-10016 <+*>
1960 V30 N7 P2374-2377 61-10017 <+*>
1960 V30 N8 P2498-2506 10863 <K-H>
65-13216 <*> 0
1960 V30 N8 P2556-2562 61-16141 <=*>
1960 V30 N8 P2586-2590 21N48R <ATS>
1960 V30 N8 P2624-2630 68M45R <ATS>
1960 V30 N8 P2763-2767 19N48R <ATS>
1960 V30 N9 P2854-2863 20N48R <ATS>
1960 V30 N9 P2967-2970 61-16549 <*=>
1960 V30 N9 P3C00-3004 61-16550 <*=>
1960 V30 N10 P3412-3414 61-16554 <*=>
1960 V30 N10 P3504-3505 61-18586 <*=>
1960 V30 N11 P3755-3759 61-18628 <*=>
1960 V30 N12 P3845-3846 61-18584 <*=>
1960 V30 N12 P4044-4048 72N51R <ATS>
1960 V30 N12 P4066-4069 61-18808 <*=>
1960 V30 N12 P4107-4108 61-18686 <*=>
1961 V31 P516-522 <LSA>
1961 V31 P739-742 62-23875 <=*>
1961 V31 P1132-1135 62-10227 <*=>
1961 V31 P2667-2674 62-10250 <*=>
1961 V31 P3027-3030 62-10683 <=*>
1961 V31 N1 P245-249 54N51R <ATS>
1961 V31 N1 P259-263 53N51R <ATS>
1961 V31 N3 P880-883 61-18704 <*=>
1961 V31 N5 P1517-1518 61-18692 <*=>
1961 V31 N5 P1689-1692 85N57R <ATS>
1961 V31 N6 PI-VI ADD 61-28604 <*=>
1961 V31 N6 P1763-1773 62-15364 <=>
1961 V31 N6 P1770-1773 00N55R <ATS>
1961 V31 N6 P2012-2013 62-15364 <=>
1961 V31 N7 P2085-2089 62-13438 <*=>
1961 V31 N7 P2134-2138 62-18388 <=*>
1961 V31 N7 P2159-2161 65-10092 <*> 0
1961 V31 N7 P2161-2166 62-10347 <*=>
1961 V31 N7 P2192-2198 62-24386 <=>
1961 V31 N7 P2263-2270 56P60R <ATS>
1961 V31 N7 P2270-2274 92P60R <ATS>
1961 V31 N7 P2316-2321 62-24386 <=>
64-13685 <=*$>
1961 V31 N7 P2367-2380 62-13546 <*=>
1961 V31 N7 P2438-2439 42P60R <ATS>
1961 V31 N8 P2675-2682 62-18893 <=*>
1961 V31 N9 P2851-2853 62-24386 <=>
1961 V31 N9 P2983-2984 62-24386 <=>
1961 V31 N9 P3C27-3030 67P60R <ATS>
1961 V31 N10 P3202-3205 62-24386 <=>
1961 V31 N10 P3251-3255 62-24386 <=>
1961 V31 N10 P3281-3287 62-24386 <=>
1961 V31 N10 P3316-3319 62-24386 <=>
1961 V31 N1C P3411-3414 62-16005 <=*>
1961 V31 N10 P3417-3420 62-16004 <=*>
1961 V31 N10 P3421-3424 62-16003 <=*>
1961 V31 N10 P3469-3477 62-24386 <=>
1961 V31 N11 P3571-3585 62-24386 <=>
1961 V31 N11 P3605-3610 62-24386 <=>
1961 V31 N11 P3750-3756 62-24386 <=>

1961 V31 N12 P4028-4033 62-16737 <=*>
1962 V32 P1006-1007 65-14491 <*>
1962 V32 P1993-1997 63-10081 <=*>
1962 V32 N1 P9-12 64-16318 <=*$>
1962 V32 N1 P144-150 62-18979 <=*>
1962 V32 N1 P291-293 91R79R <ATS>
1962 V32 N2 P367-369 55P62R <ATS>
1962 V32 N2 P553-557 62-18025 <=*>
1962 V32 N2 P562-567 64-14961 <=*$>
1962 V32 N2 P588-590 62-18024 <=*>
1962 V32 N3 P689-692 12337-A <K-H>
65-14488 <*>
1962 V32 N3 P718-721 62-18997 <=*>
1962 V32 N3 P823-827 93P62R <ATS>
1962 V32 N3 P829-833 63-10749 <=*>
1962 V32 N3 P1006-1007 12337-B <K-H>
1962 V32 N4 P1293-1296 80Q66R <ATS>
1962 V32 N5 P707-708 60P62R <ATS>
1962 V32 N5 P1423-1427 63-18430 <=*>
1962 V32 N5 P1567-1572 62-18920 <=*>
1962 V32 N7 P2273-2278 07Q70R <ATS>
1962 V32 N8 P2664-2670 63-20700 <=*$>
63-23763 <=*>
1962 V32 N9 P2812-2816 63-24172 <=*$>
1962 V32 N9 P2954-2957 63-20204 <=*>
1962 V32 N10 P3340-335 73P66R <ATS>
1962 V32 N1C P3351-336 63-16415 <*=>
1962 V32 N11 P3595-3599 65-18067 <*> 0
1962 V32 N11 P3611-3613 63-18381 <=*>
1962 V32 N11 P3676-3681 UCRL-TRANS-939 <*=>
1963 V33 N1 P234-237 68Q69R <ATS>
1963 V33 N2 P549-551 AD-619 566 <=$*>
1963 V33 N3 P851-853 64-16688 <=*$>
1963 V33 N4 P1294-1299 63-18893 <=*>
1963 V33 N4 P1335-1342 AD-619 565 <=*$>
63-31430 <=>
1963 V33 N6 P2027-2030 64-16692 <=*$>
1963 V33 N9 P2803-2804 65-10873 <*>
1963 V33 N10 P3239-3242 73R74R <ATS>
1963 V33 N1C P3426-3433 17R75R <ATS>
1963 V33 N11 P3455-3460 64-21329 <=>
1963 V33 N11 P3683-3684 47R76R <ATS>
65-11779 <*>
1963 V33 N11 P3731-3733 65-11780 <*>
1963 V33 N12 P3833-3838 02R75R <ATS>
1963 V33 N12 P3852-3857 64-21451 <=>
1964 V34 N1 P359-361 65-60574 <=$>
1964 V34 N2 P340-343 65-64398 <=*$>
1964 V34 N2 P700-701 96R76R <ATS>
1964 V34 N3 P748-751 65-12489 <*>
1964 V34 N3 P914-916 64-71452 <=>
1964 V34 N3 P977-980 92S80R <ATS>
1964 V34 N4 P1174-1179 65-63745 <=$>
1964 V34 N5 P1484-1487 01S81R <ATS>
1964 V34 N5 P1693-1695 AD-610341 <=$>
1964 V34 N6 P1948-1951 65-13403 <*>
1964 V34 N6 P2081-2084 65-12327 <*>
1964 V34 N7 P235-240 65-11611 <*>
1964 V34 N8 P2798-2800 65-63511 <=*$>
1964 V34 N10 P3296-3300 45S83R <ATS>
1964 V34 N1C P3444-3449 56S81R <ATS>
1964 V34 N10 P3509-3510 65-10494 <*>
1964 V34 N11 P3621-3624 54S82R <ATS>
1964 V34 N11 P3697-3703 52S82R <ATS>
65-14736 <*>
1964 V34 N11 P3703-3709 53S82R <ATS>
65-14735 <*>
1964 V34 N12 P3979-3982 65-63158 <*>
1964 V34 N12 P4120 590 <TC>
1965 V35 P190-193 66-60914 <=*$>
1965 V35 N2 P396-397 AD-625 054 <=*$>
1965 V35 N3 P485-489 65-14163 <*> 0
1965 V35 N3 P584-587 66-60610 <=*$>
1965 V35 N3 P587-588 65-14739 <*>
1965 V35 N3 P588 65-14738 <*>
1965 V35 N4 P621-628 65-31050 <*$=>
1965 V35 N4 P723-728 65-31050 <*$=>
1965 V35 N4 P737-738 65-31050 <*$=>
1965 V35 N6 P1001-1005 66-10923 <*>
1965 V35 N8 P1412-1415 66-11287 <*>
1965 V35 N8 P1416-1418 66-11288 <*>
1965 V35 N10 P889-1781 1874 <TC>
1965 V35 N1C P1774-1778 1873 <TC>

*ZHURNAL ORGANICHESKOI KHIMII
1965 V1 P1848-1853 66-12761 <*>
1965 V1 N1 P167-171 65-33416 <=*$>
1965 V1 N3 P448-450 CSIR-TR-531 <CSSA>
1965 V1 N10 P1734-1736 66-11486 <*>
1965 V1 N10 P1747-1748 66-11485 <*>
1965 V1 N11 P1984-1987 66-13790 <*>
1966 V2 N3 P942-946 66-34091 <=$>

*ZHURNAL PRIKLADNOI KHIMII
1928 V1 P70-77 R-2921 <*>
1928 V1 P96-109 R-2292 <*>
1929 V2 N5 P621-627 RT-656 <*>
1929 V2 N6 P663-673 RT-2881 <*>
53/0370 <NLL>
1930 V3 P797-805 R-2813 <*>
1930 V3 N7 P981-998 RT-3604 <*>
1930 V3 N7 P1041-1054 PANSDOC-TR.451 <PANS>
1931 V4 P994-1001 R-3342 <*>
1931 V4 N4 P532-534 R-2491 <*>
1931 V4 N7/8 P994-1001 63-18144 <=*>
1932 V5 P557-566 R-2980 <*>
1932 V5 P705-714 R-2982 <*>
1932 V5 P715-721 R-2871 <*>
1932 V5 N1 P31-33 RJ-264 <ATS>
1932 V5 N3/4 P332-343 R-3330 <*>
RT-2788 <*>
1933 V6 P266-273 61-16809 <=*$>
1933 V6 P470-478 R-3314 <*>
1933 V6 P479-493 R-3344 <*>
1933 V6 P577-587 R-2914 <*>
1933 V6 P588-606 R-2303 <*>
1933 V6 P607-612 R-3348 <*>
1933 V6 P769-771 R-2506 <*>
1933 V6 P1133-1139 RT-1577 <*>
64-18565 <*>
1933 V6 N1 P36-44 61-16813 <=*$> P
1933 V6 N2 P224-227 R-2814 <*>
1933 V6 N2 P240-244 RT-797 <*>
1933 V6 N4 P607-612 RT-2749 <*>
1934 V7 P39-46 R-2515 <*>
1934 V7 P123-126 61-16959 <=*$>
1934 V7 P497-502 R-2212 <*>
1934 V7 P760-763 61-17928 <=$>
1934 V7 P825-829 61-18023 <=*$>
1934 V7 P831-836 61-16822 <=*$>
1934 V7 P881-886 R-2935 <*>
1934 V7 P906-915 R-2903 <*>
1934 V7 P1121-1124 R-3027 <*>
1934 V7 P1125-1127 R-2986 <*>
1934 V7 P1127-1129 R-2985 <*>
1934 V7 P1144-1146 R-2983 <*>
1934 V7 P1166-1180 61-16970 <=*$>
1934 V7 N3 P455-468 63-16231 <=*> 0
1934 V7 N4 P489-494 TT.398 <NRC>
1934 V7 N5 P669-686 <INSD>
1934 V7 N5 P687-724 <INSD>
1934 V7 N5 P764-769 63-18540 <=*>
1934 V7 N6 P1029-1035 TT.588 <NRC>
1934 V7 N6 P1099-1102 RT-2301 <*>
1934 V7 N8 P1398-1422 61-13138 <+*>
1935 V8 P64-91 R-820 <*>
1935 V8 P177-183 R-2926 <*>
1935 V8 P429-438 R-2509 <*>
1935 V8 P587-597 R-2493 <*>
1935 V8 P1380-1387 R-3359 <*>
1935 V8 P1476-1477 64-10261 <=*$>
1935 V8 N1 P35-43 61-16612 <=*$>
1935 V8 N1 P64-91 61-16613 <=*$>
1935 V8 N2 P193-194 <INSD>
1935 V8 N3 P439-445 R-3374 <*>
1935 V8 N3 P501-504 59-20895 <+*>
1935 V8 N4 P589-597 T.760 <INSD>
1935 V8 N4 P685-694 63-18541 <=*>
1935 V8 N7 P1237-1247 RT-978 <*>
1936 V9 P583-588 R-2437 <*>

Citation	Report
1936 V9 P779-788	R-2496 <*>
1936 V9 P986-993	R-2243 <*>
1936 V9 P1082-1089	AL-6 <*>
1936 V9 P1183-1190	63-14618 <=*>
1936 V9 P1278-1288	61-18054 <=*$>
1936 V9 P1605-	61-16633 <=*$>
1936 V9 P1613-1625	61-16551 <=*$>
1936 V9 P2155-2174	R-3008 <*>
1936 V9 N2 P217-224	R-438 <*>
1936 V9 N3 P439-445	51/1897 <NLL>
1936 V9 N4 P589-590	R-2865 <*>
1936 V9 N4 P737-741	PANSDOC-TR.387 <PANS>
1936 V9 N5 P885-888	RJ-893 <ATS>
	63-20324 <=*>
	64-18567 <*>
1936 V9 N6 P1033-1037	RT-2189 <*>
1936 V9 N6 P1038-1048	RT-2004 <*>
1936 V9 N6 P1082-1089	RT-680 <*>
1936 V9 N6 P1090-1095	RT-681 <*>
1936 V9 N10 P1770-1780	RT-1384 <*>
1937 V10 P435-455	R-2468 <*>
1937 V10 P853-860	63-14587 <=*>
1937 V10 P1165-1172	R-2443 <*>
1937 V10 P1178-1182	R-2158 <*>
1937 V10 P1183-1193	R-2175 <*>
	63-14619 <=*> 0
1937 V10 P1211-1215	63-10352 <=*> 0
1937 V10 P1523-1529	R-3017 <*>
1937 V10 P1531-1536	R-3294 <*>
1937 V10 P1847-1866	R-2600 <*>
1937 V10 N1 P6-18	RT-891 <*>
1937 V10 N1 P19-34	64-10744 <=*$> 0
1937 V10 N1 P208-211	50/3283 <NLL>
	61-13425 <+*>
1937 V10 N3 P403-413	R-2868 <*> P
1937 V10 N3 P435-455	64-10747 <=*$> 0
1937 V10 N6 P1011-1019	RT-1527 <*>
1937 V10 N7 P1165-1172	RT-2888 <*>
1937 V10 N7 P1211-1215	R-4960 <*>
1937 V10 N8 P1500-1503	RT-1580 <*>
1937 V10 N9 P1548-1557	RT-682 <*>
1937 V10 N9 P1696-1699	59-18753 <+*>
1937 V10 N12 P1957-1967	R-861 <*>
1937 V10 N10/1 P1847-1866	64-10745 <=*$> 0
1938 V11 P4-11	65-11927 <*>
1938 V11 P214-222	R-3339 <*>
1938 V11 P523-527	R-2545 <*>
1938 V11 P734-739	R-2898 <*>
1938 V11 P734-753	RT-2211 <*>
1938 V11 P750-753	R-2898 <*>
1938 V11 P823-852	65-63233 <=>
1938 V11 P1103-1107	64-18571 <*>
1938 V11 P1161-1171	R-2956 <*>
1938 V11 P1398-1422	T-2126 <INSD>
1938 V11 P1543-1547	R-2577 <*>
1938 V11 P1548-1555	R-2579 <*>
1938 V11 N1 P4-11	374 <TC>
1938 V11 N1 P105-108	914E <K-H>
1938 V11 N3 P515-522	9098 <K-H>
1938 V11 N5 P818-822	61-17364 <=$>
1938 V11 N6 P967-974	59-20733 <+*>
1938 V11 N9 P1344-1347	65-62950 <=>
1938 V11 N9 P1348-1360	60-18704 <*>
1938 V11 N7/8 P1217-1221	30J17R <ATS>
1938 V11 N10/1 P1471-1474	61-27469 <=*>
1939 V12 P167-180	R-2957 <*>
1939 V12 P635-641	57-1513 <*>
1939 V12 P704-718	62-24849 <=*>
1939 V12 P1330-	R-765 <*>
1939 V12 P1644-1652	R-2177 <*>
1939 V12 P1900-1906	R-3310 <*>
1939 V12 N1 P3-12	46Q73R <ATS>
1939 V12 N2 P226-233	RT-873 <*>
1939 V12 N3 P346-351	55J18R <ATS>
1939 V12 N6 P826-830	R-2459 <*>
	63-18130 <=*>
1939 V12 N6 P923-933	59-22351 <+*>
1939 V12 N6 P949-950	1244 <K-H>
1939 V12 N6 P951-952	R-4727 <*>
	64-19392 <=$> 0
1939 V12 N7 P976-980	63-18301 <=*>
1939 V12 N7 P1013-1021	63-20995 <=*$>
1939 V12 N7 P1092-1096	RJ-38 <ATS>
1939 V12 N9 P1286-1294	R-2474 <*>
1939 V12 N9 P1387-1392	TT.257 <NRC>
1939 V12 N9 P1430-1432	63-14876 <=*> 0
1939 V12 N12 P1816-1825	61-16858 <=*$>
1939 V12 N12 P1826-1834	61-16859 <=*$>
1940 V13 P29-36	R-2595 <*>
1940 V13 P170-180	AEC-TR-710 <*>
	64-18572 <*>
1940 V13 P643-651	R-3349 <*>
1940 V13 P769	168TM <CTT>
1940 V13 P823-830	R-2488 <*>
1940 V13 P1204-1207	43J14R <ATS>
1940 V13 P1417-1426	R-3029 <*>
1940 V13 P1833-1838	59-10007 <+*>
1940 V13 N1 P29-36	61-17854 <=$>
	63-14609 <=*>
1940 V13 N1 P51-55	RT-1583 <*>
1940 V13 N1 P102-117	61-16864 <*=>
1940 V13 N2 P198-209	61-16865 <*=>
1940 V13 N2 P227-234	61-16866 <*=>
1940 V13 N2 P235-243	61-16867 <*=>
1940 V13 N4 P483-498	TT.399 <NRC>
1940 V13 N7 P978-1002	59-10542 <+*>
	60-18399 <+*>
1940 V13 N7 P1022-1027	RT-3846 <*>
	61-18337 <*=>
1940 V13 N7 P1053-1059	RT-2213 <*>
1940 V13 N9 P1310-1314	RT-877 <*>
1940 V13 N10 P1512-1516	UCRL TRANS-572(L) <*+>
1940 V13 N10 P1517-1522	<ATS>
1940 V13 N11 P1545-1555	RT-1800 <*>
1940 V13 N11 P1612-1619	61-18104 <*=>
1941 V14 P11-13	R-2955 <*>
1941 V14 P19-29	R-2517 <*>
1941 V14 P30-38	R-2513 <*>
1941 V14 P317-334	59-10133 <+*>
	8215 <K-H>
1941 V14 P435-445	61-18011 <*=>
1941 V14 P500-506	R-2959 <*>
1941 V14 P555-558	61-16971 <*=>
1941 V14 P652-661	61-18038 <*=>
1941 V14 P800-804	61-18001 <*=>
1941 V14 P805-808	61-18000 <*=>
1941 V14 P827-833	61-18033 <*=>
1941 V14 P849-855	61-16939 <*=>
1941 V14 P928-938	R-2244 <*>
1941 V14 N1 P3-10	RT-371 <*>
1941 V14 N1 P11-13	RJ-526 <ATS>
1941 V14 N1 P110-119	62-18189 <=*> 0
1941 V14 N2 P161-172	65-10144 <*>
1941 V14 N3 P302-304	61-13294 <+*>
	62-24461 <=*>
1941 V14 N3 P405-409	61-16881 <*=>
1941 V14 N6 P716-733	61-13427 <=*>
1941 V14 N6 P816-826	RT-2234 <*>
1941 V14 N4/5 P449-465	61-16917 <*=>
	62-15246 <*=>
1941 V14 N4/5 P528-538	61-18077 <*=>
1941 V14 N4/5 P542-550	61-16887 <*=>
1941 V14 N4/5 P579-586	59-10728 <+*>
1941 V14 N4/5 P600-604	R-1453 <*>
	64-19427 <=$>
1941 V14 N4/5 P618-631	61-16923 <*=>
	61-17231 <=$>
1942 V15 P128-138	61-16636 <*=>
1942 V15 P139-150	61-16919 <*=>
1942 V15 P173-178	749TM <CTT>
1942 V15 P275-289	R-2884 <*>
1942 V15 P319-330	61-16648 <*=>
1942 V15 P437-446	R-2286 <*>
1942 V15 P447-452	R-2094 <*>
	R-2267 <*>
1942 V15 N4 P249-259	63-20173 <=*> 0
1942 V15 N4 P260-266	RT-3601 <*>
1942 V15 N5 P290-301	T.654 <INSD>
1942 V15 N6 P380-395	60-19970 <+*>
1942 V15 N6 P396-404	60-18057 <+*>

Citation	Translation
1942 V15 N6 P447-452	RJ-80 <ATS>
1942 V15 N1/2 P25-46	62-20273 <=*> 0
1942 V15 N1/2 P71-78	59-10605 <+*>
1943 V16 P72-77	61-16957 <*=>
1943 V16 P87-94	61-16958 <*=>
1943 V16 P129-133	61-16985 <=*> P
1943 V16 P143-150	61-16984 <*=>
1943 V16 P206-210	68F2R <ATS>
1943 V16 P219-226	61-18169 <*=>
1943 V16 P283-294	61-16911 <*=>
1943 V16 P345-350	61-18178 <*=>
1943 V16 P356-364	61-16894 <*=>
1943 V16 N11 P397-416	UCRL TRANS-821(L) <=*>
1943 V16 N1/2 P15-19	RT-3603 <*>
1943 V16 N5/6 P206-210	68F2R <ATS>
1943 V16 N7/8 P253-257	63-16137 <=*>
1943 V16 N11/2 P351-355	74Q72R <ATS>
1943 V16 N9/10 P426-432	RT-3602 <*>
1944 P183-187	R-2435 <*>
1944 V17 P16-21	R-2434 <*>
1944 V17 P52-59	R-3967 <*>
1944 V17 P83-93	R-3099 <*>
1944 V17 P165-169	R-2541 <*>
1944 V17 P259-265	61-13423 <+*>
1944 V17 P266-273	R-2436 <*>
1944 V17 P381-393	R-2149 <*>
	63-14640 <=*>
1944 V17 P409-416	R-2152 <*>
	63-14616 <=*>
1944 V17 P435-444	R-3968 <*>
1944 V17 P458-462	61-18218 <*=>
1944 V17 P463-470	61-20929 <*=> 0
1944 V17 P520-526	R-3356 <*>
1944 V17 P538-545	R-4004 <*>
1944 V17 P552-556	61-18215 <*=>
1944 V17 P570-575	399 <TC>
1944 V17 P576-579	61-18232 <*=>
1944 V17 P613-618	R-2781 <*>
	R-4037 <*>
1944 V17 N3 P107-113	RCT V18 N1 P780 <RCT>
1944 V17 N3 P156-158	63-15134 <=*>
1944 V17 N6 P319-325	62-28788 <=$>
1944 V17 N6 P346-353	452 <TC>
1944 V17 N1/2 P52-59	61-16904 <*=>
1944 V17 N1/2 P60-64	61-16907 <*=>
1944 V17 N1/2 P107-113	61-16908 <*=>
1944 V17 N1/2 P125-136	62-14612 <=*>
1944 V17 N4/5 P259-265	64-10771 <=*$>
1944 V17 N7/8 P435-444	RT-1017 <*>
	61-18201 <*=>
1944 V17 N7/8 P463-470	62-14156 <=*> P
1944 V17 N11/2 P588-593	RT-3406 <*>
1944 V17 N11/2 P623-627	61-18347 <*=>
1944 V17 N11/2 P629-633	61-18346 <*=>
1944 V17 N11/2 P650-655	61-18345 <*=>
1944 V17 N9/10 P520-526	60-18075 <+*> 0
1944 V17 N9/10 P529-532	61-18217 <*=> P
1944 V17 N9/10 P533-537	59-22467 <+*>
1944 V17 N9/10 P538-545	RT-1018 <*>
	61-18216 <*=>
1945 V18 P90-96	61-18348 <*=>
1945 V18 P158-171	61-18239 <*=>
1945 V18 P193-206	62-23951 <=*>
1945 V18 P221-229	R-2634 <*>
1945 V18 P294-300	R-2644 <*>
1945 V18 P301-308	61-18344 <*=>
1945 V18 P367-371	61-18341 <*=>
1945 V18 P425-429	61-20132 <*=>
1945 V18 P439-449	R-2516 <*>
1945 V18 P450-458	61-20117 <*=>
1945 V18 P459-468	61-20133 <*=>
1945 V18 P509-517	R-2636 <*>
1945 V18 P509-516	61-13415 <+*>
1945 V18 P518-520	R-2643 <*>
1945 V18 P521-528	R-2637 <*>
1945 V18 P548-555	61-20140 <*=>
1945 V18 P609-611	61-20144 <*=>
1945 V18 P686-689	61-20145 <=*>
1945 V18 N6 P289-293	2584 <HB>
1945 V18 N6 P313-321	RT-3405 <*>
1945 V18 N11 P658-665	62-25322 <=*>
1945 V18 N45 P251-258	RT-363 <*>
1945 V18 N1/2 P3-9	RT-3438 <*>
	61-18222 <*=>
	62-10048 <*=> 0
1945 V18 N1/2 P10-14	61-18223 <*=>
1945 V18 N1/2 P35-42	61-18224 <*=>
1945 V18 N1/2 P43-49	55L36R <ATS>
	61-18250 <*=>
1945 V18 N1/2 P50-54	62-14159 <=*>
1945 V18 N1/2 P62-68	61-18242 <*=>
1945 V18 N1/2 P81-85	RT-2358 <*>
1945 V18 N7/8 P381-392	1807 <K-H>
	60-18411 <+*>
	63-14848 <=*>
1945 V18 N7/8 P420-424	RJ-54 <ATS>
	63-18535 <=*>
1945 V18 N7/8 P439-449	62-14163 <=*> P
1945 V18 N7/8 P459-468	96S86R <ATS>
1946 V19 P30-34	61-20160 <*=>
1946 V19 P172-175	T-1774 <NLL>
1946 V19 P200-206	R-5356 <*>
	60-17198 <+*>
1946 V19 P207-216	R-4055 <*>
	61-20159 <*=>
1946 V19 P322-328	61-20119 <*=>
1946 V19 P416-419	61-20113 <*=>
1946 V19 P428-434	61-20114 <*=>
1946 V19 P492-503	61-20116 <*=>
1946 V19 P504-510	63-18534 <=*>
1946 V19 P511-516	279TM <CTT>
1946 V19 P684-692	R-2501 <*>
1946 V19 P833-840	R-2508 <*>
1946 V19 P973-988	RJ-64 <ATS>
1946 V19 P1007-1017	R-2502 <*>
1946 V19 P1156-1168	R-3321 <*>
1946 V19 P1181-1196	R-2574 <*>
1946 V19 P1181-1188	RT-2166 <*>
1946 V19 P1201-1213	R-2573 <*>
1946 V19 P1214-1224	R-2629 <*>
1946 V19 P1350-1357	R-2623 <*>
1946 V19 P1953-1957	R-2507 <*>
1946 V19 N1 P63-70	66-10247 <*>
1946 V19 N1 P90-96	63-18766 <=*$> 0
1946 V19 N2 P180-186	59-19085 <+*>
1946 V19 N4 P363-370	6065-A <K-H>
	64-20201 <*> 0
1946 V19 N4 P371-378	65-10139 <*>
1946 V19 N4 P435-439	RT-1395 <*>
1946 V19 N4 P440-444	RT-1650 <*>
1946 V19 N7 P632-650	60-41321 <=>
1946 V19 N7 P678-683	3186 <HB>
1946 V19 N8 P826-832	R-4484 <*>
	RT-1762 <*>
1946 V19 N9 P869-879	R-1873 <*>
	RJ-14 <ATS>
1946 V19 N9 P869-880	RT-2445 <*>
	61-17837 <=$>
1946 V19 N9 P869-879	63-23235 <=*>
1946 V19 N9 P973-988	R-1869 <*>
	RJ-64 <ATS>
1946 V19 N9 P1007-1017	62-16465 <=*>
1946 V19 N12 P1259-1264	RT-1573 <*>
1946 V19 N12 P1265-1270	RT-2056 <*>
1946 V19 N12 P1277-1280	64-20366 <*>
1946 V19 N12 P1350-1357	63-16429 <=*> C
	64-20853 <*> 0
1946 V19 N5/6 P542-552	RJ-473 <ATS>
	61-18627 <*=>
1946 V19 N10/1 P1093-1106	RT-1118 <*>
1946 V19 N10/1 P1143-1148	RCT V21 N4 P830 <RCT>
1946 V19 N10/1 P1181-1188	R-2574 <*>
1947 V20 P23-35	R-3323 <*>
1947 V20 P69-76	59-19077 <+*>
1947 V20 P239-244	R-1643 <*>
1947 V20 P327-330	R-3007 <*>
1947 V20 P502-514	R-2615 <*>
1947 V20 P630-634	R-2653 <*>
1947 V20 P685-692	R-3326 <*>
1947 V20 P714-720	R-2614 <*>

Citation	Translation
1947 V20 P721-738	R-3322 <*>
1947 V20 P800-808	<ATS>
	1338A <K-H> 0
1947 V20 P911-913	R-3022 <*>
1947 V20 P997-1004	R-3293 <*>
1947 V20 P1053-1082	R-3021 <*>
1947 V20 P1248-1254	R-2873 <*>
1947 V20 P1270-1282	60-13868 <+*>
1947 V20 N3 P163-169	RT-3403 <*>
1947 V20 N3 P163-170	61-14751 <=*> 0
1947 V20 N3 P225-238	RT-96 <*>
1947 V20 N4 P331-344	RT-1594 <*>
1947 V20 N4 P353-359	64-18955 <*>
1947 V20 N5 P385-390	RT-1936 <*>
1947 V20 N6 P523-531	63-20162 <=*$>
1947 V20 N7 P577-583	RT-1591 <*>
1947 V20 N7 P597-604	63-20431 <=*$>
1947 V20 N7 P630-634	RT-2446 <*>
1947 V20 N7 P660-668	RT-1801 <*>
	2876B <K-H>
	63-20751 <=*$>
1947 V20 N8 P721-738	61-17806 <=$>
1947 V20 N9 P813-817	RT-1863 <*>
1947 V20 N9 P875-882	59-17259 <+*>
	64-20156 <*>
1947 V20 N9 P883-886	59-17260 <+*>
	63-20611 <=*$>
1947 V20 N10 P963-975	63-18566 <=*$> 0
1947 V20 N10 P1013-1018	RT-3404 <*>
1947 V20 N11 P1053-1082	RT-4060 <*>
1947 V20 N11 P1115-1159	59-19129 <+*>
1947 V20 N11 P1133-1144	RT-1592 <*>
1947 V20 N11 P1145-1154	RT-1574 <*>
1947 V20 N12 P1223-1234	R-2101 <*>
	RJ-88 <ATS>
1947 V20 N12 P1235-1241	RT-1575 <*>
1947 V20 N12 P1242-1247	59-19124 <+*>
	63-14831 <=*>
1947 V20 N1/2 P69-76	RJ-112 <ATS>
1947 V20 N1/2 P81-88	RT-2854 <*>
1947 V20 N1/2 P89-96	RT-2855 <*>
1948 V21 P10-17	R-2604 <*>
1948 V21 P18-25	R-2606 <*>
1948 V21 P35-41	R-2609 <*>
1948 V21 P249-250	R-2876 <*>
1948 V21 P362-371	R-2492 <*>
1948 V21 P389-393	T-1403 <NLL>
1948 V21 P441-447	R-759 <*>
1948 V21 P486-495	R-3011 <*>
1948 V21 P496-501	R-3010 <*>
1948 V21 P543-547	R-3012 <*>
1948 V21 P717-731	R-2169 <*>
1948 V21 P802-806	R-3009 <*>
1948 V21 P807-809	R-2173 <*>
1948 V21 P820-823	R-2164 <*>
1948 V21 P917-926	R-2557 <*>
1948 V21 P927-936	R-2193 <*>
1948 V21 P937-947	R-2194 <*>
1948 V21 P1037-1044	R-2195 <*>
1948 V21 P1073-1082	R-2467 <*>
1948 V21 P1249-1260	60-15928 <+*>
1948 V21 P1261-1271	60-15940 <+*>
1948 V21 P1278-1281	R-2917 <*>
1948 V21 N1 P82-100	RT-1593 <*>
1948 V21 N2 P134-138	CSIR-TR.231 <CSSA>
1948 V21 N2 P151-155	RT-2567 <*>
1948 V21 N3 P195-203	RT.110 <*>
1948 V21 N3 P218-226	63-20164 <=*$>
1948 V21 N3 P251-259	61-19704 <=*>
1948 V21 N4 P362-371	61-23303 <*=>
1948 V21 N5 P429-436	R-4707 <*>
1948 V21 N5 P473-485	64-20119 <*>
1948 V21 N7 P717-731	50/3262 <NLL>
	61-13374 <+*>
1948 V21 N7 P775-780	62-23085 <=*>
1948 V21 N8 P816-819	60-17182 <*+>
1948 V21 N9 P887-902	61-13630 <+*>
1948 V21 N9 P966-975	TT.381 <*>
1948 V21 N10 P1037-1044	61-13424 <=*>
1948 V21 N10 P1065-1072	RT-1066 <*>
1948 V21 N11 P1137-1146	RT-1983 <*>
1948 V21 N12 P1199-1209	RT-1982 <*>
1948 V21 N12 P1210-1227	TT.595 <NRC>
1949 V22 P13-23	R-2156 <*>
1949 V22 P24-32	R-2856 <*>
1949 V22 P41-44	R-2630 <*>
1949 V22 P342-353	R-2224 <*>
1949 V22 P448-454	R-2233 <*>
1949 V22 P499-503	R-2215 <*>
1949 V22 P611-624	64-20120 <*>
1949 V22 P778-779	R-2217 <*>
	RT-95 <*>
1949 V22 P1063-1067	RT-76 <*>
1949 V22 P1078-1082	R-2222 <*>
1949 V22 P1263-1272	R-4855 <*>
1949 V22 P1263-1273	R-902 <*>
1949 V22 P1279-1283	R-3026 <*>
1949 V22 N2 P157-171	RT-469 <*>
1949 V22 N2 P995-1001	RJ-6 <ATS>
1949 V22 N3 P223-234	61-13368 <+*>
1949 V22 N3 P235-239	RJ-29 <ATS>
1949 V22 N3 P278-289	RJ-30 <ATS>
1949 V22 N5 P491-498	RJ-5 <ATS>
	RJ05 <ATS>
1949 V22 N5 P508-517	67Q70R <ATS>
1949 V22 N5 P518-523	RCT V24 N4 P916 <RCT>
1949 V22 N6 P545-552	59-19148 <+*>
1949 V22 N6 P553-557	<ATS>
	R-1872 <*>
	RJ-2 <ATS>
1949 V22 N6 P558-567	R-1871 <*>
	RJ-3 <ATS>
	RT-2455 <*>
	61-17290 <=$>
	63-10414 <=*>
1949 V22 N6 P611-624	25J19R <ATS>
1949 V22 N7 P683-688	RJ-93 <ATS>
1949 V22 N7 P734-746	59-22479 <+*>
1949 V22 N7 P778-779	R-02217 <*>
1949 V22 N8 P848-852	RJ-4 <ATS>
1949 V22 N8 P853-856	RJ-7 <ATS>
1949 V22 N8 P874-881	68Q70R <ATS>
1949 V22 N9 P943-946	RJ-11 <ATS>
1949 V22 N9 P947-951	50/3116 <NLL>
1949 V22 N9 P952-964	RJ-10 <ATS>
1949 V22 N9 P970-977	RJ-12 <ATS>
1949 V22 N9 P995-1001	RJ-6 <ATS>
1949 V22 N10 P1068-1077	RJ-419 <ATS>
	53/0372 <NLL>
1949 V22 N10 P1124-1132	25L34R <ATS>
	61-17879 <=$>
1949 V22 N12 P1225-1230	RJ-420 <ATS>
1949 V22 N12 P1310-1320	TT.555 <NRC>
1950 V23 P8-20	R-3046 <*>
1950 V23 P32-50	R-2995 <*>
1950 V23 P66-80	R-2651 <*>
1950 V23 P113-117	R-2893 <*>
1950 V23 P127-135	R-2882 <*>
1950 V23 P136-139	R-2786 <*>
1950 V23 P230-232	R-2893 <*>
1950 V23 P396-408	R-2787 <*>
1950 V23 P444-445	R-2895 <*>
1950 V23 P470-471	R-2894 <*>
1950 V23 P545-551	R-828 <*>
1950 V23 P575-579	R-2896 <*>
1950 V23 P823-836	R-3039 <*>
1950 V23 P1040-1046	R-3038 <*>
	T-1397 <NLL>
1950 V23 P1133-1141	R-2568 <*>
1950 V23 P1249-1259	R-3032 <*>
1950 V23 N1 P8-20	64-10755 <=*$> 0
1950 V23 N1 P60-65	RJ-25 <ATS>
1950 V23 N1 P99-111	R-1870 <*>
	RJ-26 <ATS>
1950 V23 N2 P145-152	R-1848 <*>
	RJ-31 <ATS>
1950 V23 N2 P145	59-22565 <+*> 0
1950 V23 N2 P145-152	61-17291 <=$>
1950 V23 N3 P125-129	60-13872 <+*>
1950 V23 N6 P580-598	RT-1975 <*>

1950 V23 N8 P853-863 RT-2447 <*>
1950 V23 N11 P1176-1186 47L36R <ATS>
61-17248 <=$>
1950 V23 N11 P1187-1190 RT-1526 <*>
1950 V23 N12 P1271-1276 2707 <HB>
1950 V23 N12 P1277-1279 R-822 <*>
1950 V23 N12 P1290-1298 RT-3092 <*>
1950 V23 N12 P1326-1330 R-821 <*>
1951 V24 P32-36 R-2565 <*>
1951 V24 P252-263 R-2556 <*>
1951 V24 P353-360 R-2861 <*>
1951 V24 P404-412 R-2564 <*>
1951 V24 P433-438 R-2563 <*>
1951 V24 P444-452 R-3033 <*>
1951 V24 P495-501 R-3980 <*>
1951 V24 P567-582 R-2542 <*>
1951 V24 P690-697 R-3034 <*>
1951 V24 P698-709 62-24446 <=*>
1951 V24 P893-903 R-2237 <*>
1951 V24 P904-914 R-2543 <*>
1951 V24 P940-944 R-2255 <*>
1951 V24 P1210-1212 R-2544 <*>
1951 V24 N1 P28-31 10Q73R <ATS>
1951 V24 N1 P56-60 52/2519 <NLL>
1951 V24 N1 P85-92 29K22R <ATS>
1951 V24 N1 P93-94 RT-2448 <*>
1951 V24 N1 P95-101 61-13426 <+*>
1951 V24 N2 P197-202 RJ-55 <ATS>
1951 V24 N4 P353-360 61-15403 <*+>
1951 V24 N4 P373-382 AD-630 848 <=$>
1951 V24 N4 P421-432 R-818 <*>
61-13422 <*=>
1951 V24 N5 P460-470 AEC-TR-1596 <+*>
1951 V24 N5 P477-484 63-10037 <=*> 0
1951 V24 N5 P502-508 R-816 <*>
1951 V24 N5 P542-545 RT-2880 <*>
53/0096 <NLL>
1951 V24 N5 P545-547 62-18443 <=*> 0
1951 V24 N6 P674-677 RCT V25 N1 P36 <RCT>
1951 V24 N8 P851-857 53/0825 <NLL>
1951 V24 N8 P870-875 2692A <K-H>
1951 V24 N9 P984-988 66-14378 <*$> 0
1951 V24 N10 P1084-1089 R-815 <*>
1951 V24 N11 P1151-1155 59-17475 <*> 0
1951 V24 N12 P1257-1262 61-13631 <=*>
1952 V25 P57-63 64-16902 <=*$>
1952 V25 P313-321 64-18164 <*> 0
1952 V25 P345-349 64-20350 <*>
1952 V25 P662-668 64-18163 <*> 0
1952 V25 P748-751 RJ-270 <*>
1952 V25 P943-954 64-18167 <*> 0
1952 V25 N2 P134-147 3121 <HB>
1952 V25 N2 P154-158 RT-2195 <*>
1952 V25 N2 P207-212 RJ-176 <ATS>
1952 V25 N2 P224-228 61-13422 <*=>
1952 V25 N7 P696-717 48L36R <ATS>
1952 V25 N7 P748-751 RJ-270 <ATS>
1953 V26 P277-285 R-2260 <*>
1953 V26 P549-553 T1975 <INSD>
1953 V26 P1114-1121 R-2262 <*>
1953 V26 N2 P117-123 RT-2277 <*>
1953 V26 N2 P148-154 RJ-135 <ATS>
1953 V26 N2 P225-227 RJ-143 <ATS>
1953 V26 N3 P328-331 TT.493 <NRC>
1953 V26 N3 P332-335 TT.494 <NRC>
1953 V26 N4 P364-367 3301 <HB>
1953 V26 N4 P368-381 RJ-128 <ATS>
1953 V26 N5 P495-499 RJ-263 <ATS>
1953 V26 N5 P505-511 61-17204 <=$>
1953 V26 N5 P558-560 RJ-254 <ATS>
1953 V26 N6 P605-611 RJ-134 <ATS>
1953 V26 N6 P640-644 RJ-398 <ATS>
1953 V26 N6 P652-655 3262 <HB>
1953 V26 N7 P673-6800 RT-2349 <*>
1953 V26 N7 P752-756 64K24R <ATS>
1953 V26 N8 P854-859 RJ-153 <ATS>
1953 V26 N9 P859-906 40M42R <ATS>
1953 V26 N9 P907-911 RJ-213 <ATS>
1953 V26 N9 P931-940 3391 <HB>
1953 V26 N9 P976-981 TT.501 <NRC>
1953 V26 N9 P982-990 R-5043 <*>
RT-4197 <*>
62-16443 <=*>
1953 V26 N10 P1005-1013 RJ-164 <ATS>
1953 V26 N10 P1089-1091 69H12R <ATS>
1953 V26 N11 P1176-1185 RT-2281 <*>
1953 V26 N11 P1200-1204 TT.502 <NRC>
1953 V26 N12 P1233-1237 <CB>
1953 V26 N12 P1238-1244 <CB>
1953 V26 N12 P1245-1251 <CB>
1953 V26 N12 P1252-1257 <CB>
1953 V26 N12 P1258-1267 <CB>
1953 V26 N12 P1268-1275 <CB>
1953 V26 N12 P1276-1283 <CB>
1953 V26 N12 P1284-1289 <CB>
1953 V26 N12 P1290-1298 <CB>
1953 V26 N12 P1299-1303 <CB>
1953 V26 N12 P1304-1309 <CB>
1953 V26 N12 P1310-1313 <CB>
1953 V26 N12 P1314-1317 <CB>
1953 V26 N12 P1317-1320 <CB>
1953 V26 N12 P1320-1322 <CB>
1954 V27 P843-850 RJ-307 <ATS>
1954 V27 P1227-1230 RCT V29 N1 P95 <RCT>
1954 V27 N1 P33-42 T2119 <INSD>
1954 V27 N1 P55-63 3377 <HB>
1954 V27 N1 P97-104 RT-2800 <*>
1954 V27 N1 P118-120 RJ-201 <ATS>
1954 V27 N2 P171-181 R-1044 <*>
RT-3696 <*>
1954 V27 N2 P189-197 RCT V28 N3 P793 <RCT>
1954 V27 N2 P201-206 RJ-233 <ATS>
1954 V27 N2 P206-209 3379 <HB>
1954 V27 N2 P214-217 RCT V28 N1 P186 <RCT>
1954 V27 N3 P229-236 <CB>
1954 V27 N3 P237-247 <CB>
1954 V27 N3 P248-257 <CB>
1954 V27 N3 P258-264 <CB>
1954 V27 N3 P265-272 <CB>
1954 V27 N3 P273-279 <CB>
1954 V27 N3 P280-297 <CB>
1954 V27 N3 P298-309 <CB>
1954 V27 N3 P310-313 <CB>
1954 V27 N3 P314-318 <CB>
1954 V27 N3 P319-326 <CB>
1954 V27 N3 P327-331 <CB>
1954 V27 N3 P332-334 <CB>
1954 V27 N3 P334-340 <CB>
1954 V27 N3 P341-346 <CB>
1954 V27 N3 P346-348 <CB>
1954 V27 N3 P349-351 <CB>
1954 V27 N4 P382-390 60-19916 <+*>
1954 V27 N4 P413-420 3389 <HB>
1954 V27 N4 P450-454 60-19913 <+*>
1954 V27 N5 P246-248 <CB>
1954 V27 N5 P469-472 <CB>
1954 V27 N5 P473-479 <CB>
1954 V27 N5 P480-484 <CB>
1954 V27 N5 P485-492 <CB>
1954 V27 N5 P493-500 <CB>
1954 V27 N5 P501-505 <CB>
1954 V27 N5 P506-513 <CB>
1954 V27 N5 P514-526 <CB>
1954 V27 N5 P527-532 <CB>
1954 V27 N5 P533-538 <CB>
1954 V27 N5 P539-544 RT-3180 <*>
1954 V27 N5 P548-551 <CB>
1954 V27 N5 P552-557 <CB>
1954 V27 N5 P557-560 <CB>
1954 V27 N5 P560-562 <CB>
1954 V27 N5 P562-565 <CB>
1954 V27 N5 P566-567 <CB>
1954 V27 N5 P568-570 <CB>
1954 V27 N5 P570-572 <CB>
1954 V27 N5 P573-575 <CB>
1954 V27 N5 P575-576 <CB>
1954 V27 N6 P577-593 RT-3501 <*>
1954 V27 N6 P600-612 60-15754 <+*>
1954 V27 N6 P639-644 RJ-492 <ATS>
1954 V27 N6 P681-684 RJ-459 <ATS>

1954 V27 N6 P687-688 R-306 <*>
R-3627 <*>
3097 <NLL>
1954 V27 N7 P693-698 <CB>
1954 V27 N7 P699-711 <CB>
1954 V27 N7 P722-734 <CB>
1954 V27 N7 P735-756 RJ-609 <ATS>
1954 V27 N7 P757-768 <CB>
1954 V27 N7 P769-773 <CB>
1954 V27 N7 P774-781 <CB>
1954 V27 N7 P789-791 <CB>
1954 V27 N8 P801-803 RJ-293 <ATS>
1954 V27 N8 P831-842 RJ-268 <ATS>
1954 V27 N8 P843-850 RJ-307 <ATS>
1954 V27 N8 P901-906 RJ-277 <ATS>
1954 V27 N9 P921-929 <INSD>
1954 V27 N9 P930-938 RJ-317 <ATS>
1954 V27 N9 P945-952 RJ-423 <ATS>
1954 V27 N9 P975-982 RJ-275 <ATS>
1954 V27 N9 P1012-1016 RJ-243 <ATS>
61-17266 <=$>
1954 V27 N1C P1056-1066 R-1225 <*>
1954 V27 N10 P1067-1081 RJ-320 <ATS>
1954 V27 N1C P1094-1100 61-17275 <=$>
1954 V27 N10 P1115-1120 R-987 <*>
RT-3503 <*>
RT-3724 <*>
1954 V27 N10 P1128-1131 RT-3502 <*>
1954 V27 N1C P1133-1136 RJ-279 <ATS>
1954 V27 N11 P1170-1179 63-10577 <=*> 0
1954 V27 N11 P1184-1193 RJ-319 <*>
RJ-319 <ATS>
64-16531 <=*$>
1954 V27 N11 P1207-1212 RJ-269 <ATS>
1954 V27 N12 P1263-1268 RJ-314 <ATS>
66-60259 <=$*>
1954 V27 N12 P1302-1306 RJ-278 <ATS>
1954 V27 N12 P1307-1312 RJ-618 <ATS>
1954 V27 N12 P1327-1330 64-16140 <=*$> 0
1955 V28 P222-226 59-14284 <+*> 0
1955 V28 N1 P30-39 61-23331 <*=>
65-63C12 <=$>
1955 V28 N1 P81-86 RJ-302 <ATS>
1955 V28 N1 P111-114 RJ-389 <ATS>
1955 V28 N2 P156-165 60-C0692 <*>
1955 V28 N3 P245-253 R-3644 <*>
1955 V28 N3 P311-316 RCT V29 N2 P607 <RCT>
1955 V28 N4 P428-431 RT-3384 <*>
1955 V28 N5 P533-542 RCT V29 N3 P933 <RCT>
1955 V28 N6 P565-572 RT-3383 <*>
1955 V28 N6 P573-578 RT-3520 <*>
1955 V28 N6 P608-615 RT-3961 <*>
1955 V28 N6 P651-655 RJ-608 <ATS>
1955 V28 N7 P748-750 09H13R <ATS>
1955 V28 N11 P1156-1160 08K21R <ATS>
1955 V28 N12 P1275-1284 36L30R <ATS>
1955 V28 N12 P1308-1313 3820 <HB>
1956 V29 P27- 61-23336 <*=>
1956 V29 P40-46 R-1754 <*>
1956 V29 P40- 61-23316 <*=>
1956 V29 P46-52 62-15879 <=*>
1956 V29 P166-170 RJ-475 <ATS>
1956 V29 P170- 61-23313 <*=>
1956 V29 P463-467 T-2219 <INSD>
1956 V29 P655-659 R-2393 <*>
1956 V29 P988-996 59-19283 <+*>
1956 V29 P1070-1080 T-2503 <INSD>
1956 V29 P1479-1483 39J15R <ATS>
1956 V29 P1512-1521 62-23140 <=*>
1956 V29 N1 P27-32 76H13R <ATS>
1956 V29 N1 P73-82 R-4842 <*>
1956 V29 N1 P90-97 RCT V29 N4 P1276 <RCT>
1956 V29 N1 P1C5-110 RJ-490 <ATS>
1956 V29 N1 P130-131 RJ-474 <ATS>
1956 V29 N1 P136-138 46K24R <ATS>
1956 V29 N2 P229-236 10J16R <ATS>
1956 V29 N3 P321-330 RJ-478 <ATS>
1956 V29 N3 P463-467 RJ-477 <ATS>
1956 V29 N4 P577-583 62-11417 <=>
1956 V29 N4 P589-595 65-13876 <*> 0
1956 V29 N4 P621-627 64-16305 <*>
1956 V29 N6 P914-918 31J14R <ATS>
1956 V29 N6 P935-940 08H13R <ATS>
1956 V29 N6 P955- 67H12R <ATS>
1956 V29 N6 P955-957 68H12R <ATS>
1956 V29 N7 P996-1006 32J14R <ATS>
1956 V29 N7 P1C29-1035 17H12R <ATS>
1956 V29 N8 P1142-1147 79H12R <ATS>
1956 V29 N8 P1152-1159 30J14R <ATS>
1956 V29 N8 P1283-1287 80H12R <ATS>
1956 V29 N8 P1292-1295 <ATS>
1956 V29 N9 P1401-1404 70H12R <ATS>
1956 V29 N9 P1405-1411 66H12R <ATS>
1956 V29 N9 P1456-1459 96J15R <ATS>
1956 V29 N11 P1664-1673 26J14R <ATS>
1956 V29 N29 P908-914 62-11356 <=>
1957 V30 P233-239 65J17R <ATS>
1957 V30 P486-489 59-12175 <+*>
1957 V30 P917-926 RCT V32 N1 P164 <RCT>
1957 V30 P1750-1755 R-4449 <*>
1957 V30 P1760-1763 R-4541 <*>
1957 V30 N1 P38-44 61-23318 <*=>
1957 V30 N1 P1C4-114 63-18402 <=*>
1957 V30 N1 P120-124 11J16R <ATS>
64-18083 <*>
1957 V30 N2 P293-297 60-13172 <+*>
1957 V30 N3 P439-446 64-20243 <*> 0
6457 <K-H>
1957 V30 N4 P546-552 63-18849 <=*>
1957 V30 N4 P567-575 61-23326 <*=>
1957 V30 N4 P576-581 65K25R <ATS>
1957 V30 N4 P586-598 77M39R <ATS>
1957 V30 N4 P643-645 28L33R <ATS>
1957 V30 N4 P646-649 R-4537 <*>
98J16R <ATS>
1957 V30 N4 P650-652 62-14046 <=*>
99J16R <ATS>
1957 V3C N4 P653-654 J17R <ATS>
1957 V30 N4 P660-663 01J17R <ATS>
1957 V30 N5 P710-716 55J19R <ATS>
1957 V30 N5 P716-723 61-13137 <*=> P
1957 V30 N5 P744-750 62-10211 <*=>
81J17R <ATS>
1957 V30 N5 P763-767 64-16888 <=*$>
1957 V30 N5 P768-774 36J19R <ATS>
1957 V30 N5 P781-786 00M40R <ATS>
1957 V30 N5 P796-799 61-13306 <*=>
1957 V30 N5 P811-813 99J19R <ATS>
1957 V30 N6 P862-875 81K22R <ATS>
1957 V30 N6 P893-903 13J18R <ATS>
1957 V30 N6 P927-932 14J18R <ATS>
1957 V30 N6 P960-962 80K23R <ATS>
1957 V30 N6 P963-964 84M37R <ATS>
1957 V30 N7 P1C12-1016 74K22R <ATS>
1957 V30 N8 P1135-1141 61-23524 <=>
1957 V30 N8 P1141-1148 67K21R <ATS>
1957 V30 N8 P1160-1169 89J18R <ATS>
1957 V30 N8 P1277-1279 55K24R <ATS>
1957 V30 N9 P1356-1361 12K20R <ATS>
1957 V30 N9 P1368-1374 13K20R <ATS>
60-13324 <+*>
1957 V30 N9 P1378-1382 66-10217 <*> 0
1957 V30 N10 P1547-1552 R-4858 <*>
1957 V3C N10 P1553-1558 48K23R <ATS>
1957 V30 N11 P1716-1719 68K27R <ATS>
1957 V30 N12 P1741-1746 08L30R <ATS>
1957 V30 N12 P1750-1755 78K26R <ATS>
1957 V30 N12 P1866-1868 11K23R <ATS>
1958 V31 P1876-1879 60-15774 <+*>
1958 V31 N1 P5-13 20K22R <ATS>
1958 V31 N1 P19-25 83K23R <ATS>
1958 V31 N1 P25-33 60-23585 <+*>
1958 V31 N1 P45-54 21K22R <ATS>
1958 V31 N1 P68-81 59-19889 <+*>
1958 V31 N1 P77-83 62-23959 <=*>
1958 V31 N1 P84-89 79K25R <ATS>
1958 V31 N1 P116-124 59K22R <ATS>
1958 V31 N1 P140-142 58K23R <ATS>
1958 V31 N1 P146-148 45K22R <ATS>
1958 V31 N2 P186-191 36K24R <ATS>

Citation	Translation
1958 V31 N2 P191-197	59K23R <ATS>
1958 V31 N2 P265-273	26K23R <ATS>
1958 V31 N3 P402-408	62-23960 <=*>
1958 V31 N3 P452-458	50K23R <ATS>
1958 V31 N3 P476-484	77K24R <ATS>
1958 V31 N4 P533-542	61-23329 <*=>
1958 V31 N4 P542-549	33K26R <ATS>
1958 V31 N4 P550-558	61-23321 <*=>
1958 V31 N4 P595-601	59-10158 <+*>
1958 V31 N4 P625-630	13K24R <ATS>
1958 V31 N5 P706-710	61-23314 <*=>
1958 V31 N5 P711-718	61-23312 <*=>
1958 V31 N5 P754-765	79K24R <ATS>
1958 V31 N5 P79C-799	<ATS>
1958 V31 N5 P810-813	91L29R <ATS>
1958 V31 N5 P813-816	K26R <ATS>
1958 V31 N6 P946-948	31L34R <ATS>
1958 V31 N7 P1007-1012	59-10267 <+*>
1958 V31 N7 P1C13-1118	57K25R <ATS>
1958 V31 N7 P1013-1018	60-10919 <+*>
	61-23333 <*=>
1958 V31 N7 P1019-1025	60-10918 <+*>
1958 V31 N7 P1101-1102	51K26R <ATS>
	59-22554 <+*>
1958 V31 N7 P1118-1122	61K27R <ATS>
1958 V31 N8 P1215-1220	59-17257 <+*>
1958 V31 N8 P1241-1245	59-15758 <+*>
1958 V31 N8 P1245-1251	61-15577 <+*>
1958 V31 N8 P1252-1258	69K26R <ATS>
1958 V31 N8 P1259-1265	73K26R <ATS>
1958 V31 N9 P1281-1284	55K27R <ATS>
1958 V31 N9 P1323-1332	61-23328 <*=>
1958 V31 N9 P1348-1354	87K27R <ATS>
1958 V31 N9 P1397-1402	60L30R <ATS>
1958 V31 N9 P1415-1419	59-1C177 <+*>
1958 V31 N9 P1419-1426	59-10176 <+*> 0
1958 V31 N9 P1437-1440	08K28R <ATS>
1958 V31 N10 P1453-1460	22K28R <ATS>
1958 V31 N10 P1472-1477	75M38R <ATS>
1958 V31 N10 P1478-1483	74M38R <ATS>
1958 V31 N10 P1560-1565	59K28R <ATS>
1958 V31 N11 P1655-1661	60-16550 <+*>
1958 V31 N11 P1759-1761	21L3CR <ATS>
1958 V31 N12 P1792-1799	61L30R <ATS>
1958 V31 N12 P1805-1809	57L29R <ATS>
1958 V31 N12 P1809-1816	58L29R <ATS>
1958 V31 N12 P1817-1823	59-17226 <+*> 0
1958 V31 N12 P1898-1900	59L29R <ATS>
	60-13656 <+*>
1959 V32 P170-173	RCT V33 N2 P587 <RCT>
1959 V32 P1126-1132	25S89R <ATS>
1959 V32 N1 P6-21	86L30R <ATS>
1959 V32 N1 P50-54	03N51R <ATS>
1959 V32 N1 P99-1044	63L32R <ATS>
1959 V32 N1 P128-132	59-20317 <+*>
1959 V32 N1 P138-141	97L31R <ATS>
1959 V32 N1 P141-150	98L31R <ATS>
1959 V32 N2 P326-336	63-24319 <=*$>
1959 V32 N2 P342-346	61-27045 <*=>
	67L31R <ATS>
1959 V32 N2 P363-369	52L33R <ATS>
1959 V32 N2 P418-423	60-16571 <+*>
	64-10875 <=*$>
1959 V32 N2 P452-453	72L31R <ATS>
1959 V32 N3 P536-542	20L36R <ATS>
1959 V32 N3 P548-556	60-16570 <+*>
1959 V32 N3 P603-608	60-16546 <+*>
1959 V32 N3 P649-655	02L33R <ATS>
	60-13813 <+*>
1959 V32 N3 P68C-686	61-20635 <*=>
1959 V32 N4 P742-749	62-18109 <=*>
1959 V32 N4 P801-807	31L36R <ATS>
1959 V32 N4 P913-919	52L32R <ATS>
1959 V32 N5 P948-952	89L32R <ATS>
1959 V32 N5 P1050-1055	60-14668 <+*>
1959 V32 N5 P1060-1065	26P61R <ATS>
1959 V32 N5 P1066-1071	14L37R <ATS>
1959 V32 N5 P1112-1121	18L36R <ATS>
1959 V32 N5 P1112-11125	59-17208 <+*>
1959 V32 N5 P1177-1178	60-15804 <+*>
1959 V32 N6 P1186-1192	35L33R <ATS>
1959 V32 N6 P1238-1244	50M38R <ATS>
1959 V32 N6 P1348-1353	53P61R <ATS>
1959 V32 N6 P1354-1359	60-18773 <+*>
	84L33R <ATS>
1959 V32 N6 P1399-1401	60-15778 <+*> 0
1959 V32 N6 P1404-1407	60-13789 <+*>
	61-10543 <+*>
1959 V32 N7 P1467-1470	AEC-TR-3919 <+*>
	59-22140 <+*>
	65-10216 <*>
1959 V32 N7 P1496-1502	98L34R <ATS>
1959 V32 N7 P1620-1623	97L36R <ATS>
1959 V32 N8 P1695-1707	61-13029 <+*> 0
1959 V32 N8 P1707-1711	61-13028 <+*> 0
1959 V32 N8 P1738-1743	19M38R <ATS>
	60-14391 <+*>
1959 V32 N8 P1744-1750	20M38R <ATS>
1959 V32 N8 P1751-1755	60-16547 <+*>
1959 V32 N8 P1763-1766	61-23523 <=*>
1959 V32 N8 P1774-1781	61-23894 <*=>
1959 V32 N8 P1787-1793	61-28154 <*=> 0
1959 V32 N8 P1795-1797	52L35R <ATS>
1959 V32 N8 P1811-1818	21M38R <ATS>
1959 V32 N8 P1819-1824	60-14390 <+*>
1959 V32 N8 P1824-1830	60-16902 <*+>
1959 V32 N8 P1855-1857	52M41R <ATS>
1959 V32 N8 P1880-1883	60-10708 <+*>
1959 V32 N9 P1992-2000	97M39R <ATS>
1959 V32 N9 P2001-2005	98M39R <ATS>
1959 V32 N9 P2107-2115	32M40R <ATS>
1959 V32 N10 P2129-2138	60-11954 <=>
1959 V32 N10 P2294-2299	61-19335 <+*> 0
1959 V32 N10 P2332-2335	61-19335 <+*> 0
1959 V32 N10 P2336-2339	64-10573 <=*$>
1959 V32 N10 P2366-2367	61-19335 <+*> 0
1959 V32 N11 P2600-2601	30N52R <ATS>
1959 V32 N12 P2782-2787	60-18066 <+*>
1960 V33 P1008-1012	UCRL TRANS-628(L) <*>
1960 V33 N1 P49-55	03N52R <ATS>
1960 V33 N1 P1C2-110	AEC-TR-4132 <+*>
1960 V33 N1 P186-190	64M40R <ATS>
1960 V33 N1 P209-212	76M47R <ATS>
1960 V33 N1 P251-253	60-16809 <+*>
1960 V33 N2 P323-332	61-23325 <*=>
1960 V33 N2 P448-454	60-18749 <+*>
1960 V33 N2 P468-469	62-24523 <=*>
1960 V33 N3 P533-535	61-10394 <*+>
1960 V33 N3 P628-632	61-27515 <*=> 0
1960 V33 N3 P690-694	61-10395 <*+>
1960 V33 N4 P753-759	60-41523 <=>
1960 V33 N4 P774-779	62-24455 <=*>
1960 V33 N4 P835-841	AEC-TR-4405 <=>
1960 V33 N4 P896-902	63-18379 <=*>
1960 V33 N4 P962-968	33M42R <ATS>
	61-10568 <*+>
1960 V33 N5 P1C08-1012	AEC-TR-4595 <*=>
	UCRL TRANS-628(L) <=*>
1960 V33 N5 P1C52-1059	61-10569 <*+> 0
1960 V33 N5 P1068-1075	85M42R <ATS>
1960 V33 N5 P1154-1157	78M43R <ATS>
1960 V33 N5 P1158-1167	59M44R <ATS>
1960 V33 N5 P1158-1159	65-12066 <*>
1960 V33 N5 P1224-1226	62-15762 <*=>
1960 V33 N6 P1281-1285	TRANS-149 <FRI>
	61-28177 <*=>
1960 V33 N7 P1449-1465	45M44R <ATS>
1960 V33 N7 P1477-1482	39M44R <ATS>
1960 V33 N7 P1495-1499	62-15702 <*=>
1960 V33 N7 P1500-1506	62-15706 <*=>
1960 V33 N8 P1641-1653	61-10754 <+*>
1960 V33 N8 P1883-1893	61-16127 <=*>
1960 V33 N8 P1897-1899	67M45R <ATS>
1960 V33 N8 P1912-1913	61-10793 <+*>
1960 V33 N8 P1913-1915	96N57R <ATS>
1960 V33 N9 P2009-2014	98N55R <ATS>
1960 V33 N9 P2C71-2078	61-27612 <*=>
1960 V33 N9 P2128-2132	61-16128 <=*>
1960 V33 N10 P2153-2165	61-18591 <*=>
	61-21193 <=>

Citation	Translation
1960 V33 N10 P2290-2300	12N48R <ATS>
	62-23882 <=*>
1960 V33 N1C P2329-2335	05N49R <ATS>
1960 V33 N10 P2371-2373	09N50R <ATS>
1960 V33 N11 P2521-2526	62-14791 <=*>
1960 V33 N11 P2593-2597	61-18589 <*=>
1960 V33 N11 P2600-2603	61-18590 <*=>
1960 V33 N12 P2633-2637	76P64R <ATS>
1960 V33 N12 P2657-2664	96N50R <ATS>
1960 V33 N12 P2664-2671	97N50R <ATS>
1960 V33 N12 P2671-2675	61-31675 <=>
1960 V33 N12 P2685-2692	61-18588 <*=>
1961 N12 P2669-	ICE V2 N2 P247-250 <ICE>
1961 V34 P113-130	62-10431 <*=>
1961 V34 P943-946	63-14327 <=*>
1961 V34 N1 P107-112	61-19632 <*=>
1961 V34 N1 P220-223	62-23880 <=*>
1961 V34 N1 P227-229	63-15173 <=*>
1961 V34 N2 P323-331	62-14823 <=*>
1961 V34 N2 P350-356	62-14242 <=*>
1961 V34 N2 P370-375	15N55R <ATS>
	61-18587 <*=>
1961 V34 N2 P448-451	62-25885 <=*>
1961 V34 N2 P468-469	61-18684 <*=>
	78N52R <ATS>
1961 V34 N3 P505-512	62-23361 <=*>
1961 V34 N3 P541-547	62-14824 <=*>
1961 V34 N3 P558-564	16N55R <ATS>
1961 V34 N3 P581-584	62-14825 <=*>
1961 V34 N3 P613-620	62-10429 <*=>
1961 V34 N3 P632-64C	62-16118 <=*>
1961 V34 N3 P703-705	61-18594 <*=>
1961 V34 N3 P711-712	40N57R <ATS>
1961 V34 N4 P821-825	35P6CR <ATS>
1961 V34 N4 P821-826	61-27834 <*=>
1961 V34 N4 P836-841	62-14821 <=*>
1961 V34 N4 P841-847	62-16572 <=*>
1961 V34 N5 P1061-1065	62-16127 <=*>
1961 V34 N5 P1143-1146	19N55R <ATS>
1961 V34 N6 P1281-1285	63-20911 <=*$> O
1961 V34 N6 P1369-1371	62-23962 <=*>
1961 V34 N7 P163-1632	81N55R <ATS>
1961 V34 N7 P1444-1448	AEC-TR-4997 <=*>
1961 V34 N7 P1461-1469	62-25638 <=*>
1961 V34 N7 P1507-	ICE V2 N1 P50-53 <ICE>
1961 V34 N7 P1537-1542	65-10082 <*> O
1961 V34 N7 P1543-1547	65-10083 <*> O
1961 V34 N7 P1583-1587	31N55R <ATS>
1961 V34 N7 P1591-1597	RCT V35 N3 P644 <RCT>
1961 V34 N7 P1630-1632	81N55R <ATS>
1961 V34 N8 P1714-1722	18N57R <ATS>
1961 V34 N8 P1744-1747	16P64R <ATS>
1961 V34 N8 P1760-1775	62-14676 <=*> O
1961 V34 N8 P1834-1841	62-18892 <=*>
1961 V34 N8 P1879-1880	62-25296 <=*> O
1961 V34 N9 P1920-	ICE V2 N1 P120-123 <ICE>
1961 V34 N9 P1936-1939	AEC-TR-5040 <=*>
1961 V34 N9 P2073-2079	RCT V35 N3 P652 <RCT>
	19N57R <ATS>
1961 V34 N9 P2124-2125	31N57R <ATS>
1961 V34 N10 P2133-2146	62-20409 <=*>
1961 V34 N10 P2146-2153	62-18715 <=*>
1961 V34 N10 P2175-2183	63-13391 <=*>
1961 V34 N10 P2206-2216	62-20410 <=*>
1961 V34 N10 P2248-2253	62-25306 <=*> O
1961 V34 N10 P2282-2288	62-32417 <=*>
1961 V34 N1C P2295-	ICE V2 N2 P177-181 <ICE>
1961 V34 N10 P2295-2300	63-10233 <=*>
1961 V34 N10 P2295-2302	70P58R <ATS>
1961 V34 N11 P2441-2446	63-10234 <=*>
1961 V34 N12 P2609-2613	63-10237 <=*>
1961 V34 N12 P2613-2622	91P59R <ATS>
1961 V34 N12 P2692-2699	62-25320 <=*> O
1961 V34 N12 P2792-2796	63-14324 <=*>
1962 N1 P90-	ICE V2 N3 P421-425 <ICE>
1962 V35 P132-138	62-33074 <=*>
1962 V35 N1 P37-42	63-14322 <=*>
1962 V35 N1 P43-47	63-14331 <=*>
1962 V35 N1 P56-60	CSIR-TR.318 <CSSA>
1962 V35 N1 P196-197	63-19880 <=*$>
1962 V35 N2 P250-256	62-20462 <=*> O
1962 V35 N2 P334-337	62-20225 <=*>
1962 V35 N2 P385-389	63-15127 <=*>
1962 V35 N2 P394-397	AD-621 818 <=$>
1962 V35 N2 P398-404	AD-621 818 <=$>
1962 V35 N2 P452-455	63-19785 <=*>
1962 V35 N3 P5C3-511	63-20257 <=*> C
1962 V35 N3 P511-516	63-15560 <=*>
1962 V35 N3 P638-647	62-18374 <=*>
1962 V35 N4 P756-759	62-18705 <=*>
1962 V35 N4 P777-781	63-19876 <=*>
1962 V35 N4 P781-786	63-19877 <*=>
1962 V35 N4 P9C9-910	64-13280 <=*$>
1962 V35 N5 P984-989	63-19076 <=*>
1962 V35 N5 P1CC0-1008	62-33320 <=*>
1962 V35 N5 P1136-1138	29P64R <ATS>
1962 V35 N5 P1148-1150	63-14313 <=*>
1962 V35 N6 P1206-1209	63-18403 <=*>
1962 V35 N6 P1293-1302	64-13282 <=*$> O
1962 V35 N6 P1379-1382	27P65R <ATS>
1962 V35 N7 P1597-1601	63-18417 <=*>
1962 V35 N7 P1629-1633	63-21203 <=>
1962 V35 N8 P1715-1722	20Q67R <ATS>
1962 V35 N9 P1956-1960	AEC-TR-5654 <*=>
1962 V35 N9 P1976-1980	AEC-TR-5645 <=*>
1962 V35 N9 P1996-2007	63-23199 <=*>
1962 V35 N9 P2093-2099	63-19036 <=*>
1962 V35 N10 P2213-2219	64-10681 <=*$>
1962 V35 N11 P2359-2362	63-21689 <=>
1962 V35 N11 P2363-2368	63-23676 <=*>
1962 V35 N11 P2386-2393	15Q67R <ATS>
1962 V35 N11 P2520-2526	64-10682 <=*$>
1962 V35 N11 P2566-2567	65-61428 <=>
1962 V35 N11 P2574-2577	16Q67R <ATS>
1962 V35 N12 P2588-2591	64-10488 <=*$>
1962 V35 N12 P2653-2661	33Q71R <ATS>
1962 V35 N12 P2688-2693	64-10683 <=*$>
1962 V35 N12 P2693-2700	RAPRA-1265 <RAP>
1962 V35 N12 P2715-2719	65-61429 <=>
1962 V35 N12 P2729-2734	47Q68R <ATS>
1963 N5 P151-153	RTS-2872 <NLL>
1963 N6 P166-169	RTS-2835 <NLL>
1963 V36 P415-423	65-12090 <*>
1963 V36 P2217-2224	65-11778 <*>
1963 V36 N1 P95-101	63-18300 <=*> O
1963 V36 N1 P2527-2533	65-11775 <*>
1963 V36 N2 P362-372	63-23922 <=> O
1963 V36 N2 P424-428	64-13228 <=*$>
1963 V36 N2 P428-430	65-17442 <*> O
	66-13617 <*>
1963 V36 N3 P594-604	64-14888 <=*$>
1963 V36 N4 P730-736	64-16976 <=*$>
1963 V36 N4 P914-917	64-14889 <=*$>
1963 V36 N5 P977-981	85R78R <ATS>
1963 V36 N5 P1040-1045	64-13004 <=*$>
1963 V36 N5 P1101-1106	64-16694 <=*$>
1963 V36 N5 P1116-1123	64-14887 <=*$>
1963 V36 N5 P1146-1149	64-10685 <=*$>
1963 V36 N6 P1224-1231	64-16968 <=*$>
1963 V36 N7 P1604-1607	64-16974 <=*$>
1963 V36 N7 P1614-1615	10R75R <ATS>
1963 V36 N8 P1637-1641	65-10463 <*> O
1963 V36 N8 P1784-1793	64-16982 <=*$>
1963 V36 N9 P1955-1958	65-61340 <=$>
1963 V36 N9 P1958-1965	313 <TC>
1963 V36 N10 P2184-2192	314 <TC>
1963 V36 N11 P2437-2445	65-63519 <=$>
1963 V36 N11 P2460-2464	65-11776 <*>
1963 V36 N11 P2502-2506	66-10024 <*>
1963 V36 N11 P2527-2533	11R75R <ATS>
1963 V36 N11 P2552-2554	66-10025 <*>
	97R75R <ATS>
1963 V36 N12 P2587-2595	65-11708 <*>
1963 V36 N12 P2620-2625	64-19361 <=$>
	64-26434 <$>
	65-13027 <*>
1963 V36 N12 P2631-2636	64-26433 <$>
1963 V36 N12 P2725-2728	93R75R <ATS>
1963 V36 N12 P2781-2782	312 <TC>
	65-11418 <*>

1964 N1 P1-137 AD-615 211 <=$>
1964 V37 P568-574 64-31398 <=>
1964 V37 P2008-2016 TC-1594 <TC>
1964 V37 N1 P36-41 64-20874 <*>
65-13037 <*>
1964 V37 N1 P41-45 65-64410 <=*$>
1964 V37 N1 P58-64 64-30691 <*>
1964 V37 N1 P165-170 42R76R <ATS>
1964 V37 N1 P194-196 65-60019 <=$>
1964 V37 N2 P233-239 65-13036 <*>
1964 V37 N2 P433-441 64-18477 <*>
1964 V37 N4 P796-800 65-12202 <*>
1964 V37 N4 P918-920 18S81R <ATS>
66-10883 <*>
1964 V37 N5 P946-951 665 <TC>
1964 V37 N6 P1256-1260 AD-612 497 <=$>
1964 V37 N6 P1371-1372 65-13035 <*>
1964 V37 N7 P1414-1419 627 <TC>
1964 V37 N7 P1597-1601 65-10464 <*> 0
1964 V37 N7 P1606-1609 65-13034 <*>
1964 V37 N7 P1609-1611 65-12761 <*>
1964 V37 N7 P1627-1628 65-12132 <*>
1964 V37 N7 P1638-1640 65-12821 <*>
1964 V37 N8 P1695-1699 626 <TC>
1964 V37 N8 P1790-1797 65-11494 <*>
1964 V37 N8 P1851-1854 65-13033 <*>
1964 V37 N9 P1872-1878 AD-620 813 <=*$>
1964 V37 N9 P1951-1958 65-12133 <*>
1964 V37 N9 P1984-1988 65-11623 <*>
1964 V37 N9 P2008-2016 65-11622 <*>
1964 V37 N9 P2052-2055 526 <TC>
1964 V37 N9 P2074-2077 65-13032 <*>
1964 V37 N10 P2089-2093 65-13300 <*>
65-64408 <=*$>
1964 V37 N10 P2140-2145 65-13301 <*>
1964 V37 N10 P2223-2228 65-13299 <*>
1964 V37 N10 P2263-2268 09S83R <ATS>
1964 V37 N11 P2375-2382 N65-23683 <=*$>
1964 V37 N11 P2382-2387 574 <TC>
1964 V37 N11 P2528-2530 M-5684 <NLL>
1964 V37 N12 P2596-2600 570 <TC>
1964 V37 N12 P2640-2643 572 <TC>
1964 V37 N12 P2662-2668 65-14678 <*>
1964 V37 N12 P2692-2695 573 <TC>
1964 V37 N12 P2692-2696 70S83R <ATS>
1964 V37 N12 P2706-2708 66-12781 <*>
1964 V37 N12 P2748 46S82R <ATS>
1964 V37 N12 P2750-2752 568 <TC>
1964 V37 N12 P2758-2761 571 <TC>
1964 V37 N12 P2761-2763 569 <TC>
1964 V37 N12 P2771-2772 65-63676 <$=> 0
1965 V38 P437-439 1206 <TC>
1965 V38 P1068-1072 65-64613 <=*$> 0
1965 V38 N1 P47-51 65-14694 <*>
1965 V38 N1 P51-54 65-14683 <*>
1965 V38 N1 P78-83 1397 <TC>
1965 V38 N1 P87-92 1395 <TC>
1965 V38 N1 P121-128 65-14682 <*>
1965 V38 N1 P128-134 65-14692 <*>
1965 V38 N1 P143-148 65-14693 <*>
1965 V38 N1 P170-173 65-14691 <*>
1965 V38 N1 P188-192 AD-632 291 <=$>
1965 V38 N1 P197-201 AD-627 006 <=$>
1965 V38 N1 P201-205 1396 <TC>
1965 V38 N2 P365-369 M-5400 <NLL>
1965 V38 N2 P442-443 65-14695 <*>
1965 V38 N2 P449-451 57S89R <ATS>
1965 V38 N3 P579-589 66-11036 <*> 0
1965 V38 N3 P627-632 66-13256 <*>
1965 V38 N3 P674-676 65-14696 <*>
1965 V38 N4 P791-800 46S86R <ATS>
1965 V38 N4 P919-925 2129 <TC>
1965 V38 N5 P1091-1097 66-10641 <*>
1965 V38 N5 P1097-1101 66-10640 <*>
1965 V38 N5 P1102-1105 45S88R <ATS>
1965 V38 N5 P1109-1113 26S89R <ATS>
1965 V38 N5 P1178-1181 66-12971 <*>
1965 V38 N5 P1183-1185 66-12724 <*>
1965 V38 N6 P1327-1331 66-11667 <*>
1965 V38 N6 P1378-1380 80S87R <ATS>
1965 V38 N6 P1404-1407 M.5840 <NLL>
1965 V38 N7 P1556-1562 M.5842 <NLL>
1965 V38 N8 P1732-1735 66-12203 <*>
1965 V38 N8 P1894-1896 1712 <TC>
66-12223 <*>
1965 V38 N9 P2017-2026 N66-18451 <=$>
1965 V38 N11 P2401-2406 66-12919 <*>
1965 V38 N11 P2533-2537 73T90R <ATS>
1965 V38 N12 P2728-2734 66-12975 <*> 0
1966 V39 P1024-1028 66-14204 <*$>

ZHURNAL PRIKLADNCI KHIMI. ZASHCHITNYE METALLICHESKOE I OKSIDNIE POKRYTIYA, KORROZIYA METALLOV I ISSLEDOVANIYA V OBLASTI ELEKTROKHIMII. SBORNIK STATEI
SEE ZASHCHITNYE METALLICHESKIE I OKSIDNIE POKRYTIYA KORROZIYA METALLCV I ISSLEDOVANIYA V OBLASTI ELEKTROKHIMII. SBORNIK STATEI. AKADEMIYA NAVK SSSR. OTDELENIE OBSHCHEI I TEKHNICHESKOI KHIMII

*ZHURNAL PRIKLADNOI MEKHANIKI I TEKHNICHESKOI FIZIKI
1960 N1 P18-20 AEC-TR-5036 <=*>
1960 N1 P21-35 62-14552 <=*>
1960 N1 P103-109 62-18101 <=*>
1960 N1 P133-134 ARSJ V32 N5 P807 <AIAA>
1960 N2 P100-111 63-19642 <=*>
1960 N2 P116-119 AD-633 317 <=$>
1960 N2 P120-124 62-14010 <=*>
1960 N2 P124-127 ARSJ V32 N11 P1791 <AIAA>
1960 N2 P147-149 62-18852 <=*>
1960 N3 P152-156 UCRL TRANS-968(L) <*>
UCRL TRANS-968(L) <=*$>
1960 N4 P3-12 AIAAJ V1 N7 P1735 <AIAA>
1960 N4 P43-48 63-13623 <=*>
1960 N4 P51-54 UCRL TRANS-806(L) <*>
UCRL TRANS-806(L) <=*>
1960 N4 P54-55 AD-640 525 <=$>
1960 N5 P103-109 62-32993 <=*>
1960 N5 P135- ICE V2 N2 P208-209 <ICE>
1960 V1 N1 P138-141 62-11460 <=>
1960 V1 N3 P75-82 62-18802 <=*> 0
1960 V1 N3 P111-118 65-63189 <=$>
1961 N1 P3- ICE V2 N1 P90-94 <ICE>
1961 N1 P3-9 62-19279 <=*>
1961 N1 P56-65 65-13378 <*>
1961 N1 P121- ICE V2 N1 P54-56 <ICE>
1961 N2 P29-35 M.5774 <NLL>
1961 N2 P40-46 AD-612 883 <=$>
1961 N3 P36-42 UCRL TRANS-859(L) <=*>
63-15199 <=*>
1961 N3 P85-92 62-18823 <=*>
1961 N3 P101-102 NEL-TRANS-1082 <NLL>
1961 N3 P103-104 AIAAJ V1 N7 P1744 <AIAA>
1961 N4 P3-56 62-32991 <=*> 0
1961 N4 P82-93 62-23663 <=*>
1961 N4 P94-101 62-23663 <=*>
63-15169 <=*>
1961 N4 P102-108 63-18487 <=*>
1961 N4 P118-127 63-15175 <=*>
1961 N4 P128-132 <AIAA>
AIAAJ V1 N8 P2006 <AIAA>
1961 N4 P133-135 AEC-TR-5100 <=*>
1961 N5 P3-9 62-25554 <=*>
1961 N5 P10-15 63-23668 <=*$>
1961 N5 P16- ICE V2 N2 P200-207 <ICE>
1961 N5 P16-25 62-33407 <=*>
1961 N5 P26-38 AD-637 444 <=$>
1961 N5 P57-60 AD-615 985 <=$>
1961 N5 P71-75 62-25554 <=*>
1961 N5 P119-123 64-18012 <=*$>
1961 N6 P3-7 62-24344 <=*>
1961 N6 P17-23 63-13386 <=*>
1961 N6 P24-35 62-24344 <=*>
1961 N6 P36-43 AIAAJ V1 N12 P2911 <AIAA>
1961 N6 P54-68 62-24344 <=*>
1961 N6 P73-77 62-20004 <=*>
63-19517 <=*>
1961 N6 P78-87 UCRL-TRANS-934(L) <=*>
1961 N6 P88-92 AD-624 924 <=*$>

1961 N6 P118-122	64-19113 <=*$>
1961 N6 P135-144	64-15228 <=*$>
1961 N6 P163-164	63-15300 <=*>
1962 N1 P12-14	63-13215 <=*>
1962 N1 P15-19	AEC-TR-5982 <=$>
	63-19511 <=*>
1962 N1 P20-24	AD-620 814 <=$>
1962 N1 P39-43	63-19511 <=*>
1962 N1 P51-60	63-13896 <=*>
1962 N1 P68-	ICE V3 N1 P99-104 <ICE>
1962 N1 P76-81	63-13690 <=*>
1962 N1 P153	63-19511 <=*>
1962 N2 P3-13	63-19636 <=*>
1962 N2 P7-13	63-23938 <=$>
1962 N2 P20-31	63-19636 <=*>
1962 N2 P32-36	62-33665 <=*>
1962 N2 P37-41	65-62410 <=>
1962 N2 P42-	ICE V2 N4 P514-519 <ICE>
1962 N2 P50-58	ICE V2 N4 P554-559 <ICE>
	63-19636 <=*>
1962 N2 P59-71	AD-637 477 <=$>
1962 N2 P95-103	63-21099 <=>
1962 N2 P110-112	63-19636 <=*>
1962 N3 P10-14	63-11625 <=>
1962 N3 P44-52	AD-621 772 <=*$> M
1962 N3 P92-	ICE V3 N1 P52-54 <ICE>
1962 N3 P96-	ICE V3 N1 P33-35 <ICE>
1962 N3 P117-	ICE V3 N1 P64-67 <ICE>
1962 N4 P10-20	63-19178 <=*>
1962 N4 P121-128	AD-630 875 <=$>
	RTS-3085 <NLL>
1962 N5 P39-41	64-11972 <=> M
1962 N5 P83-88	63-19196 <=*>
1962 N5 P89-95	63-19653 <=*>
1962 N5 P96-106	AEC-TR-5762 <=*$>
1962 N5 P107-116	UCRL-TRANS-1087(7) <=*$>
1962 N5 P150-153	UCRL TRANS-995(L) <=*$>
1962 N5 P175-176	63-18944 <*> 0
	64-13315 <=*$>
1962 N6 P3-7	AD-630 769 <=$>
	64-71125 <=> M
1962 N6 P22-28	63-24000 <=*> 0
1962 N6 P29-	ICE V3 N2 P230-235 <ICE>
1962 N6 P45-49	AD-620 763 <=$*>
1962 N6 P50-59	64-71304 <=> M
1962 N6 P68-80	AD-630 286 <=*$>
1962 N6 P121-124	63-23693 <=*$>
1963 N1 P32-37	ICE V3 N P379-382 <ICE>
1963 N1 P67-76	63-24250 <=*$>
1963 N1 P77-	ICE V3 N4 P491-495 <ICE>
1963 N1 P130-132	65-13660 <*>
1963 N2 P1-176	AD-637 118 <=$>
1963 N2 P3-20	64-15051 <=*$>
	65-13661 <*>
1963 N2 P21-30	M.5775 <NLL>
1963 N2 P93-	ICE V3 N4 P478-482 <ICE>
1963 N2 P1CC-107	68R80R <ATS>
1963 N2 P113-123	RTS-2899 <NLL>
1963 N2 P130-131	AD-630 650 <=$>
1963 N2 P173-174	AD-619 347 <=$>
	64-19355 <=$>
1963 N3 P36-40	64-13313 <=*$>
1963 N3 P84-90	UCRL-TRANS-1158(L) <=*$>
1963 N3 P111-116	64-14271 <=*$>
1963 N3 P155-159	RTS-3347 <NLL>
1963 N4 P21-26	63-24132 <=*$> 0
1963 N4 P40-47	ICE V4 N1 P55-60 <ICE>
1963 N4 P68-73	64-11892 <=> M
1963 N4 P78-82	N65-16594 <=$>
	64-15238 <=*$>
1963 N4 P88-93	ICE V4 N1 P32-35 <ICE>
	64-11892 <=> M
1963 N4 P94-99	63-24132 <=*$> 0
1963 N4 P99-101	LA-TR-64-3 <=>
1963 N4 P103-105	63-20677 <=*$>
1963 N4 P114-115	63-20676 <=*$>
1963 N4 P116-118	63-20675 <*>
1963 N4 P117-119	63-24132 <=*$> 0
1963 N4 P128-131	RTS-2736 <NLL>
1963 N5 P117-120	LA-TR-64-15 <*>
1963 N6 P1-192	AD-618 314 <=*$>
1963 N6 P3-6	N64-29498 <=>
1963 N6 P7-12	N64-29499 <=>
1963 N6 P19-28	N64-26837 <=>
1963 N6 P74-79	64-19352 <=$>
1963 N6 P104-107	64-21926 <=>
1963 N6 P108-112	N64-29493 <=>
1963 N6 P124-127	N65-34508 <=$*>
1963 N6 P131-134	LA-TR-64-10 <=$*>
1963 N6 P141-143	N64-29492 <=>
1963 N6 P148-	M.5778 <NLL>
1963 N6 P148-156	64-16017 <=*$>
1964 N1 P134-136	RTS-2832 <NLL>
1964 N2 P8-21	AD-634 275 <=$>
	AD-637 997 <=$>
1964 N2 P99-105	RTS-2972 <NLL>
1964 N2 P160-163	M.5777 <NLL>
1964 N3 P1-185	AD-636 992 <=$> M
1964 N3 P43-51	ICE V4 N4 P705-711 <ICE>
1964 N3 P145-149	66-61882 <=*$>
1964 N4 P10-15	65-12733 <*>
1964 N4 P40-48	AD-640 609 <=$>
1964 N4 P49-56	AD-616 733 <=$>
1964 N4 P57-65	AD-629 670 <=*$>
1964 N4 P124-126	AD-629 732 <=*$>
1964 N4 P126-129	AD-611 820 <=$>
1964 N4 P137-138	RTS-3267 <NLL>
1964 N5 P38-43	66-13585 <*>
1964 N6 P69-76	AD-629 472 <=$>
1964 N6 P104-1C9	65-13928 <*>
1964 V3 P118-125	RTS-3285 <NLL>
1964 V4 N6 P63-68	65-13929 <*>
	65-63701 <=$>
1964 V4 N6 P99-104	65-13930 <*>
	65-63722 <=$>
1964 V4 N6 P104-109	65-63718 <=$>
1964 V4 N6 P115-119	65-13931 <*>
	65-63197 <=$>
1965 N1 P45-56	AD-633 118 <=$>
	65-14536 <*>
1965 N1 P57-61	AD-625 059 <=*$>
1965 N1 P62-67	AD-625 151 <=*$>
1965 N1 P88-92	AD-633 118 <=$>
1965 N1 P98-102	65-14565 <*>
1965 N1 P106-108	AD-626 965 <=$>
1965 N1 P115-119	RTS-3426 <NLL>
1965 N2 P50-53	RTS-3268 <NLL>
1965 N2 P97-105	M.5776 <NLL>
1965 N2 P119-130	AD-625 298 <=*$>
1965 N3 P68-70	AD-626 972 <=*$>
1965 N3 P85-88	AD-627 126 <=$>
1965 N4 P10-20	66-10485 <*>
1965 N4 P168-169	66-11227 <*>
1965 N5 P3-8	N66-23731 <=$>
1965 V1 P103-105	AD-626 702 <=$>
1965 V4 P10-20	65-64507 <=*$>
1965 V5 N1 P45-56	65-63420 <=$>
1965 V5 N1 P98-102	65-63721 <=$>
1965 V5 N1 P125-127	65-63183 <=$>
1966 N1 P15-20	ICE V6 N4 P594-598 <ICE>
1966 N1 P83-92	66-34133 <=$>
1966 N2 P122-125	66-13421 <*>
	66-61711 <=*$>
1966 N3 P58-62	66-62302 <=*$>

*ZHURNAL PRIKLADNCI SPEKTROSKOPII

1965 V2 N1 P37-44	AD-647 738 <=> M
1965 V2 N1 P87-89	AD-647 738 <=> M
1965 V2 N4 P341-345	66-11514 <*>
1965 V2 N2/3 P269-272	15T91R <ATS>
1965 V3 N1 P9-13	65-64506 <=*$>
	66-10475 <*>
1965 V3 N1 P32-37	65-64508 <=*$>
	66-10486 <*>
1965 V3 N1 P96-98	66-11230 <*>
1965 V3 N3 P225-229	AD-638 973 <=$>
1965 V3 N3 P279-282	AD-638 973 <=$>
1965 V3 N4 P311-319	AD-634 229 <=$>
1965 V3 N5 P449-455	AD-647 720 <=>
1965 V3 N5 P463-467	AD-647 720 <=>

Citation	Translation
1966 V4 N5 P410-414	AD-641 876 <=$>

ZHURNAL REZINCVOI PROMYSHLENNOSTI

Citation	Translation
1933 V9 P34-37	59-15692 <+*>
1935 V12 N7 P665-670	RT-3519 <*>

ZHURNAL RUSSKAGO FIZIKO-KHIMICHESKAGO OBSHCHESTVA PRI IMPERATORSKAGO ST-PETERBURGSKAGO UNIVERSITETE

Citation	Translation
1876 V8 P212-214	63-14307 <=*>
1881 V13 P149-	63-10346 <=*>
1881 V13 N4 P153-175	RT-507 <*>
	RT-927 <*>
1884 V16 P14-20	R-3828 <*>
1884 V16 P242-246	64-10466 <=*$>
1884 V16 P478-511	64-10278 <=*$>
1891 V23 P507-511	61-18767 <=*$>
1893 V25 P439-456	65-10912 <*>
1896 V28 N1 P1-15	60-12144 <=>
1897 V29 N2 P87-90	61-17968 <=$>
1898 V30 P826-842	R-5193 <*>
1898 V30 P1036-1040	R-825 <*>
1899 V31 N8 P1251-1257	57-1517 <*>
1901 V33 P496-501	64-20473 <*>
1903 V35 P448-452	61-16991 <=*$>
1906 V38 P656-659	RJ-627 <ATS>
1908 V39 P777-787	R-2540 <*>
1909 V41 P1763-1778	63-10404 <*> 0
1910 V42 P55	58-494 <*>
1910 V42 P364-365	61-18032 <*>
1910 V42 P581-585	63-20809 <=*$>
1910 V42 P691-701	63-20810 <=*$>
1910 V42 P1180-1194	63-10404 <*> 0
1911 V42 N2 P195-205	R-4907 <*>
1913 V45 N5 P995-1003	RT-1966 <*>
1914 V46 N5 P975-994	RT-2216 <*>
1915 V47 P1374-1401	59-10697 <+*>
1915 V47 P1401-1441	59-10675 <+*>
	64-30206 <*>
1915 V47 P1441-1453	59-10676 <+*>
	64-30207 <*>
1915 V47 P1453-1461	64-30208 <*>
1915 V47 P1462-1467	64-30209 <*>
1915 V47 P1467-1471	64-30210 <*>
1915 V47 P1472-1494	59-10677 <+*>
	61-16962 <=*$>
1915 V47 P1494-1506	59-10678 <+*>
	61-16964 <=*$>
1915 V47 P1507-1509	59-10679 <+*>
1915 V47 P1509-1529	61-18020 <=*$>
1915 V47 P1590-1606	63-20812 <=*$>
1915 V47 P1611-1615	63-20811 <=*$>
1915 V47 P1807-1808	63-10401 <=*>
1915 V47 P1910-1915	R-1123 <*>
1915 V47 P1928-1931	R-1124 <*>
1915 V47 P1932-1936	R-1019 <*>
1915 V47 P1941-1947	R-1018 <*>
1915 V47 P1947-1978	R-1122 <*>
1915 V47 N8 P1947-1978	RT-1162 <*>
1916 V48 P1071-1114	59-10759 <+*>
1920 V50 P512-519	14H14R <ATS>
1920 V52 N5 P451-455	1945 <*>
1921 V53 P20-65	R-1020 <*>
1921 V53 N1 P376-377	R-3378 <*>
1924 V54 N2 P40-58	RT-2791 <*>
1925 V56 P11-14	59-10068 <+*>
1925 V57 P301-304	15H14R <ATS>
1928 V60 P1045-1051	R-2593 <*>
1928 V60 N6 P929-932	59188 <*>
1929 V61 P997-1009	R-2912 <*>
1929 V61 P1035-1044	R-3050 <*>
1929 V61 P1323-1328	R-2596 <*>
1929 V61 P1353-1368	R-2978 <*>
1929 V61 N1 P109-118	RT-1101 <*>
1929 V61 N10 P2175-2187	61-17841 <=$>
1930 N4 P1317-1321	34M42R <ATS>
1930 N7 P2039-2042	R-325 <*>
1930 N11 P3817-3822	61-18585 <*=>
1930 V62 P133-149	R-3303 <*>
1930 V62 P975-988	61-23514 <=*>
1930 V62 P1343-1354	64-10735 <=*$>
1930 V62 P1385-1393	61-16999 <*>
1930 V62 P1395-1406	61-16687 <*>
1930 V62 P2235-2241	R-2503 <*>
1930 V62 N1 P1-29	61-18091 <=*$>
1930 V62 N3 P683-692	RT-2460 <*>
1930 V62 N5 P451-455	RT-2879 <*>
1930 V62 N8/9 P1947-1951	RT-2712 <*>

*ZHURNAL STRUKTURNOI KHIMII

Citation	Translation
1960 V1 N1 P36-38	62-14447 <=*>
1960 V1 N1 P64-65	62-11258 <=>
1960 V1 N1 P109-121	66-12228 <*>
1960 V1 N4 P458-463	62-11471 <=>
1961 V2 N1 P49-58	66-12213 <*>
1961 V2 N4 P424-433	62-19632 <=*>
1961 V2 N5 P640-641	62-24718 <=*>
1961 V2 N6 P734-739	62-18914 <=*>
1962 V3 N3 P339-342	63-13022 <=*>
1962 V3 N3 P345-346	63-26202 <=$>
1963 V4 P629-631	65-64415 <=*$>
1963 V4 N2 P235-244	65-10784 <*>
1964 V5 N5 P697-701	TIL/T-5596 <NLL>
1964 V5 N6 P829-833	1400 <TC>
1965 V6 N3 P465-467	66-61860 <=$>

*ZHURNAL TEKHNICHESKOI FIZIKI

Citation	Translation
1931 V1 N7 P718-724	RT-4056 <*>
1934 V4 N10 P1867-1876	64-15285 <=*$>
1935 V5 N6 P1047-1056	RT-2493 <*>
1936 V6 N3 P553-560	TT-824 <NRCC>
1937 V7 N4 P335-342	64-71206 <=>
1937 V7 N6 P559-578	RT-1357 <*>
1937 V7 N6 P579-613	RT-1358 <*>
1937 V7 N6 P627-641	RT-535 <*>
1938 V8 N3 P226-231	63-23942 <=*>
1938 V8 N11 P1023-1033	63-24016 <=*>
1938 V8 N24 P2103-2106	RT-1754 <*>
1938 V8 N13/4 P1229-1234	63-23944 <=*$>
1939 V9 P537-544	61-16810 <=*$>
1939 V9 P1916-1922	61-16826 <=*$>
1939 V9 N1 P63-70	RT-1448 <*>
1939 V9 N3 P171-187	RT-1447 <*>
1939 V9 N4 P275-282	RCT V13 N1 P316 <RCT>
1939 V9 N4 P287-294	RT-4176 <*>
	53/0200 <NLL>
1939 V9 N8 P680-686	SCL-T-317 <*>
1939 V9 N9 P808-818	RT-532 <*>
1939 V9 N14 P1249-1260	RCT V13 N1 P886 <RCT>
1939 V9 N14 P1261-1266	RCT V13 N1 P886 <RCT>
1939 V9 N14 P1267-1279	RCT V13 N1 P886 <RCT>
1939 V9 N21 P1923-1931	RT-1670 <*>
1939 V9 N22 P2032-2041	AD-633 411 <=$>
1940 V10 P574-577	61-23307 <*=>
1940 V10 P725-732	61-18004 <*=>
1940 V10 P1173-1182	R-2414 <*>
1940 V10 P1201-1206	61-16825 <*=>
1940 V10 P1533-1540	61-18035 <*=>
1940 V10 P1843-1848	R-3962 <*>
1940 V10 P1849-1856	R-3961 <*>
1940 V10 P1887-1896	R-4040 <*>
	R-4109 <*>
1940 V10 N1 P20-23	RT-1514 <*>
1940 V10 N4 P342-343	66-14229 <*$>
1940 V10 N5 P369-375	RT-1222 <*>
1940 V10 N8 P633-639	63-14699 <=*> 0
1940 V10 N9 P757-760	61-17186 <=$>
1940 V10 N12 P988-998	61-18111 <*=>
1940 V10 N12 P1021-1026	RT-2877 <*>
1940 V10 N13 P1063-1073	RJ-585 <ATS>
1940 V10 N15 P1251-1255	<ATS>
1940 V10 N15 P1256-1267	59-10781 <+*> 0
1940 V10 N15 P1297-1300	RT-3538 <*>
1940 V10 N16 P1311-1323	50/3342 <NLL>
1940 V10 N16 P1324-1330	RT-2930 <*>
1940 V10 N16 P1331-1339	AEC-TR-3868 <+*>
	AEC-TR-3868 <*>
1940 V10 N18 P1502-	61-20715 <*=>
1940 V10 N18 P1519-1532	65-12584 <*>
1940 V10 N22 P1913-1918	R-717 <*>
1940 V10 N22 P1919-1923	64-71451 <=>

Citation	Translation
1941 P509-518	R-2976 <*>
1941 V11 P367-377	61-18072 <*=>
1941 V11 P607-612	R-767 <*>
1941 V11 P613-616	61-18003 <*=>
1941 V11 P689-699	R-768 <*>
1941 V11 P775-776	R-3986 <*>
1941 V11 P801-808	61-18002 <*=>
1941 V11 P1133-1139	61-18034 <*=>
1941 V11 N3 P197-204	RJ-306 <ATS>
	61-18036 <*=>
	62-18807 <=*>
1941 V11 N4 P317-322	E-642 <RIS>
1941 V11 N5 P431-442	65-13253 <*>
1941 V11 N1/2 P37-43	RT-1345 <*>
1941 V11 N1/2 P160-169	1503TM <CTT>
	3392 <HB>
1942 V12 N1 P14-30	62-18273 <=*> 0
1942 V12 N10 P632-639	64-00467 <*>
	64-11821 <=>
1942 V12 N2/3 P65-75	<ATS>
	RT-1698 <*>
1942 V12 N11/2 P715-721	40K22R <ATS>
1943 V13 P637-640	R-4102 <*>
1943 V13 N3 P109-115	62-32641 <=*>
1943 V13 N3 P116-122	R-1177 <*>
1943 V13 N1/2 P3-34	63-19569 <*=>
1943 V13 N11/2 P627-636	RT-1978 <*>
1943 V13 N9/10 P520-530	RT-541 <*>
1944 P18-23	R-3970 <*>
1944 V14 P24-28	R-3966 <*>
1944 V14 P427-	61-23334 <*=>
1944 V14 N3 P163-171	RT-1948 <*>
1944 V14 N3 P181-192	N65-22620 <=$>
1944 V14 N6 P289-313	RCT V19 N1 P360 <RCT>
1944 V14 N9 P481-506	R-1951 <*>
1944 V14 N9 P561-567	RT-527 <*>
1944 V14 N12 P730-748	RT-1124 <*>
1944 V14 N10/1 P605-610	RT-1715 <*>
1945 V15 P63-70	R-425 <*>
1945 V15 P193-199	R-4056 <*>
1945 V15 P200-205	R-4033 <*>
1945 V15 P206-211	R-3987 <*>
1945 V15 P718-724	R-4053 <*>
1945 V15 P725-731	R-4054 <*>
1945 V15 N3 P157-172	R-3781 <*>
1945 V15 N12 P916-923	62-25576 <=*>
1946 V16 P199-206	R-3963 <*>
1946 V16 P277-282	R-5298 <*>
	61-28423 <*=>
1946 V16 P437-442	R-2666 <*>
	290TM <CTT>
1946 V16 P1135-1140	R-4058 <*>
1946 V16 P1141-1144	R-3989 <*>
1946 V16 P1383-1388	RCT V21 N3 P621 <RCT>
1946 V16 N2 P231-242	AEC-TR-4772 <=*>
1946 V16 N2 P243-248	AEC-TR-4778 <=*>
1946 V16 N2 P249-254	R-865 <*>
1946 V16 N3 P361-364	50/3075 <NLL>
	62-26680 <=*$>
1946 V16 N3 P371-373	RT-1604 <*>
1946 V16 N4 P381-394	RT-291 <*>
1946 V16 N4 P413-416	RT-894 <*>
1946 V16 N4 P417-426	RT-702 <*>
1946 V16 N4 P437-442	RT-1807 <*>
1946 V16 N4 P443-454	RT-904 <*>
1946 V16 N4 P463-468	R-00560 <*>
	RT-280 <*>
1946 V16 N5 P527-538	R-1847 <*>
1946 V16 N5 P551-554	63-14490 <=*>
1946 V16 N6 P669-680	59-10788 <+*> 0
1946 V16 N6 P733-736	62-11681 <=>
1946 V16 N6 P737-743	RT-2369 <*>
1946 V16 N7 P745-770	RT-3755 <*>
	64-15290 <=*$> P
	64-15291 <=*$> P
	64-15292 <=*$> P
	64-15293 <=*$> P
1946 V16 N8 P857-878	64-15585 <=*$>
	66-13628 <*>
1946 V16 N10 P1101-1104	R-1808 <*>
1946 V16 N11 P1307-1318	RT-3109 <*>
1946 V16 N11 P1325-1336	RT-3110 <*>
1946 V16 N12 P1409-1426	64-19977 <=$>
1946 V16 N12 P1475-1482	3404 <HB>
1947 V17 P435-438	AEC-TR-448 <*>
1947 V17 P543-548	R-4014 <*>
	292TM <CTT>
1947 V17 P1143-1148	RT-137 <*>
1947 V17 N1 P3-26	RT-1178 <*>
1947 V17 N1 P37-42	62-20345 <=*> 0
1947 V17 N1 P43-52	3008 <HB>
1947 V17 N1 P111-114	RT-1694 <*>
1947 V17 N2 P129-142	RT-2889 <*>
1947 V17 N2 P161-176	59-17040 <+*> 0
1947 V17 N2 P215-230	TT.232 <NRC>
1947 V17 N4 P435-438	TT.204 <NRC>
1947 V17 N4 P461-468	RT-2292 <*>
1947 V17 N5 P537-542	RT-703 <*>
1947 V17 N5 P613-618	60-17148 <+*>
1947 V17 N7 P765-	AEC-TR-6071 <=*$>
1947 V17 N7 P809-818	RT-4502 <*>
1947 V17 N7 P835-838	3213 <HB>
1947 V17 N8 P881-890	61-13011 <=>
1947 V17 N9 P965-982	RT-1186 <*>
1947 V17 N9 P983-986	RT-462 <*>
1947 V17 N9 P1027-1034	RT-2228 <*>
1947 V17 N9 P1051-1065	RT-1242 <*>
1947 V17 N10 P1097	RT-4062 <*>
1947 V17 N10 P1171-1180	TT.218 <NRC>
1947 V17 N12 P1513-1530	3037 <HB>
1947 V17 N12 P1521-1526	3010 <HB>
1948 V18 P93-97	R-2660 <*>
1948 V18 P1073-1083	60-13737 <+*>
1948 V18 N1 P71-74	2134 <HB>
1948 V18 N2 P153-160	C-5389 <NRC>
1948 V18 N2 P197-206	RT-2267 <*>
1948 V18 N2 P239-246	3493 <HB>
1948 V18 N3 P329-332	RT-1238 <*>
1948 V18 N3 P345-354	RT-3500 <*>
	50/2803 <NLL>
1948 V18 N3 P355-376	60-13734 <+*>
	63-20980 <=*$>
1948 V18 N3 P377-382	35 <FRI>
1948 V18 N3 P389-394	2735 <HB>
1948 V18 N4 P509-516	RT-728 <*>
1948 V18 N6 P827-830	12 <FRI>
1948 V18 N6 P831-842	60-16953 <+*>
1948 V18 N7 P949-954	RT-1445 <*>
1948 V18 N7 P955-958	RT-534 <*>
1948 V18 N8 P999-1025	RT-3553 <*>
	2300 <HB>
	2412 <HB>
	2423 <HB>
1948 V18 N9 P1136-1148	60-13729 <+*>
1948 V18 N9 P1149-1155	63-14488 <=*> 0
1948 V18 N9 P1156-1158	R-3619 <*>
	3109 <NLL>
1948 V18 N10 P1290-1305	2987 <HB>
1948 V18 N11 P1347-1355	R-1005 <*>
1948 V18 N11 P1369-1377	RT-1865 <*>
1948 V18 N12 P1494-1497	R-4518 <*>
	RT-704 <*>
1948 V18 N12 P1498-1510	62-16162 <=*>
1948 V18 N12 P1543-1564	RT-1482 <*>
1949 V19 P225-230	R-4035 <*>
1949 V19 P578-595	R-3993 <*>
1949 V19 N1 P1-29	RT-1964 <*>
1949 V19 N1 P30-33	3208 <HB>
1949 V19 N1 P76-83	RCT V24 N3 P541 <RCT>
1949 V19 N1 P88-94	RT-1713 <*>
1949 V19 N1 P95-99	RT-4055 <*>
1949 V19 N1 P100-110	R-997 <*>
1949 V19 N2 P153-183	RT-492 <*>
1949 V19 N2 P205-217	51/2154 <NLL>
1949 V19 N2 P225-230	14R78R <ATS>
1949 V19 N2 P271-273	RT-962 <*>
1949 V19 N3 P408-411	64-14755 <=*$>
1949 V19 N3 P421-430	50/1892 <NLL>
1949 V19 N4 P441-447	50/1878 <NLL>
1949 V19 N4 P465-470	2588 <HB>

Citation	Translation
1949 V19 N4 P492-506	50/1896 <NLL>
1949 V19 N5 P525-531	3041 <ATS>
1949 V19 N5 P560-566	R-1477 <*>
	RT-3441 <*>
1949 V19 N5 P611-615	60-10028 <+*> O
1949 V19 N6 P653-660	RT-1714 <*>
1949 V19 N7 P773-781	50/3021 <NLL>
1949 V19 N8 P871-881	3056 <HB>
1949 V19 N8 P911-930	RT-1483 <*>
1949 V19 N8 P931-942	TT.309 <NRC>
1949 V19 N8 P943-951	RT-483 <*>
1949 V19 N9 P1032-1040	R-474 <*>
1949 V19 N10 P1126-1135	R-1235 <*>
1949 V19 N10 P1146-1153	50/3175 <NLL>
1949 V19 N10 P1199-1210	RT-549 <*>
1949 V19 N10 P1211-1214	61-13629 <+*>
1949 V19 N11 P1231-2159	R-1126 <*>
1949 V19 N11 P1231-1259	62-32700 <=*>
1949 V19 N11 P1301-1311	RT-4046 <*>
1949 V19 N12 P1445-1447	<ATS>
	R-1483 <*>
1950 V20 P1389-1392	AEC-TR-1868 <*>
1950 V20 N1 P3-26	R-881 <*>
	RT-2754 <*>
1950 V20 N1 P4-15	GB43 <NLL>
1950 V20 N1 P38-44	69J16R <ATS>
1950 V20 N2 P149-159	60-15389 <+*>
1950 V20 N2 P180-192	RT-1953 <*>
1950 V20 N2 P230-231	RT-445 <*>
1950 V20 N3 P258-281	81L33R <ATS>
1950 V20 N3 P295-307	RT-940 <*>
1950 V20 N3 P308-326	RT-970 <*>
1950 V20 N3 P345-352	RT-1747 <*>
1950 V20 N3 P353-362	RT-1187 <*>
1950 V20 N4 P424-427	11 <FRI>
1950 V20 N4 P428-430	RT-2657 <*>
1950 V20 N4 P480-482	62-18165 <=*> O
1950 V20 N5 P501-515	RT-4222 <*>
1950 V20 N5 P516-525	RT-3437 <*>
1950 V20 N5 P564-570	TT.312 <NRC>
1950 V20 N5 P629-632	2858 <HB>
1950 V20 N6 P666-669	12L35R <ATS>
1950 V20 N7 P802-804	13R78R <ATS>
1950 V20 N7 P805-808	RT-2402 <*>
1950 V20 N7 P822-824	R-783 <*>
	RT-3163 <*>
1950 V20 N7 P880-883	21 <FRI>
1950 V20 N7 P884-887	28 <FRI>
1950 V20 N8 P923-927	RT-2180 <*>
1950 V20 N8 P928-936	E-641 <RIS>
1950 V20 N8 P944-953	RT-3932 <*>
1950 V20 N8 P970-974	R-995 <*>
1950 V20 N8 P1001-1004	2754 <HB>
1950 V20 N8 P1005-1014	3013 <HB>
1950 V20 N9 P1141-1144	R-4487 <*>
	RT-3129 <*>
1950 V20 N10 P1210-1212	R-1395 <*>
	R-4255 <*>
1950 V20 N10 P1273-1274	RT-2079 <*>
1950 V20 N11 P1382-1388	RT-2293 <*>
1950 V20 N12 P1413-1421	R-1393 <*>
1951 V21 P129-134	R-3991 <*>
1951 V21 P316-327	R-3984 <*>
	850TM <CTT>
1951 V21 P448-452	AEC-TR-1781 <+*>
1951 V21 P1544-1554	R-1914 <*>
1951 V21 N1 P12-17	R-1428 <*>
	R-962 <*>
	63-14664 <=*>
1951 V21 N1 P26-31	61-19662 <=*>
1951 V21 N1 P104-111	RT-2302 <*>
	64-14369 <=*$>
1951 V21 N2 P160-166	RT-2238 <*> P
1951 V21 N2 P206-220	RT-3451 <*>
1951 V21 N3 P267-286	65-11286 <*>
1951 V21 N3 P287-303	RT-801 <*>
1951 V21 N3 P316-327	GB43 <NLL>
	R-1480 <*>
1951 V21 N4 P427-437	R-1392 <*>
	289TM <CTT>
1951 V21 N4 P448-452	RT-2149 <*>
1951 V21 N4 P453-457	AD-616 338 <=$>
	RT-705 <*>
1951 V21 N5 P532-546	62-23128 <=*>
1951 V21 N5 P564-572	ANL-TRANS-24 <=$>
1951 V21 N6 P663-666	60-16584 <+*>
1951 V21 N6 P667-685	60-15842 <+*>
1951 V21 N7 P733-745	R-1391 <*>
	48J16R <ATS>
1951 V21 N7 P746-752	C-3220 <NRC>
	60-14134 <+*>
1951 V21 N7 P766-771	R-1953 <*>
	809TM <CTT>
1951 V21 N7 P772-786	1421TM <CTT>
	59-16699 <+*>
1951 V21 N7 P787-794	R-1950 <*>
	1422TM <CTT>
	59-16700 <+*>
1951 V21 N7 P795-801	1423TM <CTT>
	59-16701 <+*>
1951 V21 N8 P994-1003	63-14426 <=*>
1951 V21 N9 P1021-1028	63-13904 <=*>
1951 V21 N9 P1066-1075	R-1946 <*>
	RT-3086 <*>
	60-12614 <=>
1951 V21 N9 P1121-1136	AEC-TR-3838 <+*> C
1951 V21 N10 P1157-1163	3020 <HB>
1951 V21 N10 P1164-1169	3021 <HB>
1951 V21 N12 P1492-1496	RT-2119 <*>
1951 V21 N12 P1497-1503	45 <FRI>
1952 V22 P177-178	64-15286 <=*$>
1952 V22 P193-202	R-4092 <*>
1952 V22 P2038-2060	66-11026 <*>
1952 V22 N1 P33-39	R-4787 <*>
1952 V22 N1 P63-66	48 <FRI>
1952 V22 N1 P101-104	R-661 <*>
1952 V22 N1 P129-142	R-887 <*>
1952 V22 N2 P325-331	62-15898 <=*>
1952 V22 N3 P381-393	RT-2418 <*>
1952 V22 N3 P394-407	62-10477 <*=> C
1952 V22 N3 P436-445	R-5014 <*>
1952 V22 N4 P585-592	62-18288 <=*>
1952 V22 N4 P593-601	62-18287 <=*>
1952 V22 N4 P656-669	61-17205 <=$>
1952 V22 N5 P729-735	RT-2339 <*>
1952 V22 N5 P736-743	GB43 <NLL>
1952 V22 N5 P744-746	RT-2410 <*>
1952 V22 N5 P759-764	RT.2343 <*>
1952 V22 N5 P765-772	RT-2741 <*>
1952 V22 N5 P888-889	RT-1446 <*>
1952 V22 N6 P921-931	41K22R <ATS>
1952 V22 N6 P968-972	62-16110 <=*>
1952 V22 N6 P973-980	61-17211 <=$>
1952 V22 N6 P1029-1042	RT-3536 <*>
1952 V22 N6 P1043-1049	R-998 <*>
1952 V22 N6 P1050-	60-15841 <+*> O
1952 V22 N6 P1062-1075	62-11519 <=>
1952 V22 N7 P1195-1200	62-11622 <=>
1952 V22 N8 P1281-1289	R-1381 <*>
1952 V22 N9 P1409-1427	RT-2416 <*>
1952 V22 N10 P1552-1555	66-61519 <=*$>
1952 V22 N10 P1592-1598	61-10856 <+*>
1952 V22 N11 P1730-1740	62-24129 <=*>
1952 V22 N11 P1741-1750	60-21195 <=> O
1952 V22 N11 P1773-1782	TT.475 <NRC>
1952 V22 N11 P1794-1802	RT-2362 <*>
	57-2251 <*>
1952 V22 N11 P1834-1837	R-4278 <*>
1952 V22 N11 P1857-1866	RT-2617 <*>
1952 V22 N11 P1881-1884	R-3932 <*>
	3130 <NLL>
1952 V22 N12 P2005-2013	R-863 <*>
1952 V22 N12 P2014-2025	RT-3452 <*>
1953 V23 P32-34	GB57 460 <NLL>
1953 V23 P1502-1520	RCT V30 N2 P487 <RCT>
1953 V23 P1521-1540	RCT V30 N2 P507 <RCT>
1953 V23 N2 P341-351	RT-3455 <*>
1953 V23 N3 P904-912	95J14R <ATS>
1953 V23 N5 P796-801	34K22R <ATS>
1953 V23 N5 P802-805	38K22R <ATS>

1953 V23 N5 P894-903 94J14R <ATS>
1953 V23 N5 P904-912 62-14062 <=*>
1953 V23 N6 P1064-1114 59-18101 <+*>
1953 V23 N7 P1125-1134 61-15720 <+*>
1953 V23 N7 P1192-1201 RT-3989 <*>
59-18346 <+*>
1953 V23 N8 P1343-1349 RCT V29 N3 P718 <RCT>
1953 V23 N9 P1558-1568 R-3263 <*>
1953 V23 N10 P1677-1689 RTS-2925 <NLL>
1953 V23 N11 P1915-1919 R-2020 <*>
1953 V23 N11 P2025-2037 61-23842 <*=>
1954 V24 P292-298 61-17382 <=$>
1954 V24 P854-858 RJ-447 <ATS>
1954 V24 P1179-1186 GB4 T/1037 <NLL>
1954 V24 P1724-1728 59-18005 <+*>
1954 V24 P1773-1785 GB57/458 <NLL>
1954 V24 N1 P50-59 RT-3664 <*>
1954 V24 N1 P67-95 R-1945 <*>
RT-3085 <*>
60-12486 <=>
1954 V24 N1 P103-112 62-15921 <=*>
1954 V24 N2 P184-192 294-TT <CCT>
1954 V24 N2 P250-257 R-17 <*>
61-20529 <*=>
78N53R <ATS>
1954 V24 N2 P308-319 59-18644 <+*> 0
1954 V24 N2 P329-336 61-19405 <+*>
1954 V24 N2 P337-347 RT-4490 <*>
1334TM <CTT>
1954 V24 N3 P369-374 RT-2608 <*>
1954 V24 N3 P527-536 61-19395 <+*>
1954 V24 N4 P577-598 RCT V29 N3 P946 <RCT>
1954 V24 N4 P622-636 T-2022 <INSD>
1954 V24 N5 P769-783 RJ-540 <ATS>
1954 V24 N5 P784-796 RJ-547 <ATS>
1954 V24 N5 P797-810 50K24R <ATS>
1954 V24 N5 P811-817 59-18656 <+*>
1954 V24 N5 P818-825 GB43 <NLL>
1954 V24 N5 P826-832 GB43 <NLL>
1954 V24 N5 P854-858 RJ-447 <ATS>
1954 V24 N5 P871-874 R-1426 <*>
RT-4083 <*>
62-14040 <=*>
1954 V24 N5 P879-883 SCL-T-305 <+*>
1954 V24 N5 P889-894 RJ-534 <ATS>
59-20253 <+*>
1954 V24 N6 P961-977 62-19927 <=*>
1954 V24 N6 P1049-1054 CSIR-TR.331 <CSSA>
60-11089 <=>
1954 V24 N6 P1111-1124 GB125 IB 1408 <NLL>
1954 V24 N6 P1136-1138 R-3490 <*>
1954 V24 N6 P1139-1145 R-3491 <*>
1954 V24 N7 P1169-1178 RJ-326 <ATS>
1954 V24 N7 P1179-1186 RJ-593 <ATS>
1954 V24 N7 P1187-1193 RJ-532 <ATS>
1954 V24 N7 P1205-1208 GB125 IB 1432 <NLL>
1954 V24 N7 P1228-1243 62-11518 <=>
1954 V24 N7 P1229-1243 63-14692 <=*>
1954 V24 N7 P1291-1297 R-5324 <*>
1954 V24 N7 P1329-1332 RJ-542 <ATS>
1954 V24 N7 P1354-1357 RT-4202 <*>
60-10036 <+*> 0
1954 V24 N7 P1359-1360 R-256 <*>
1954 V24 N8 P1367-1370 3650 <HB>
59-10863 <+*>
1954 V24 N8 P1371-1374 RT-2699 <*>
RT-3045 <*>
1954 V24 N8 P1375-1386 R-719 <*>
R-978 <*>
RT-4489 <*>
1336TM <CTT>
1954 V24 N8 P1392-1400 RT-4488 <*>
57-2252 <*>
1954 V24 N8 P1428-1440 R-4443 <*>
1954 V24 N8 P1474-1482 R-4204 <*>
1954 V24 N9 P1553-1560 RT-2826 <*>
1954 V24 N9 P1579-1583 62-23206 <=*>
1954 V24 N9 P1584-1593 RT-4487 <*>
1317TM <CTT>
1954 V24 N9 P1687-1696 62-23116 <=*>
1954 V24 N9 P1697-1706 R-2777 <*>
1954 V24 N10 P1745-1750 R-3250 <*>
1954 V24 N10 P1751-1760 RT-4080 <*>
1954 V24 N10 P1912-1914 R-1400 <*>
R-1678 <*>
1398TM <CTT>
1954 V24 N10 P1915-1919 RT-4486 <*>
63-10658 <=*>
1954 V24 N11 P1974-1982 GB125 IB 1462 <NLL>
1954 V24 N11 P2036-2044 GB125 IB 1442 <NLL>
1954 V24 N11 P2046-2053 RJ-421 <ATS>
59-17501 <*>
1954 V24 N11 P2054-2063 R-2384 <*>
1954 V24 N11 P2090-2096 3459 <HB>
1954 V24 N12 P2136-2149 RT-3928 <*>
1954 V24 N12 P2169-2182 RAPRA-1261 <RAP>
1954 V24 N12 P2190-2201 R-4158 <*>
1954 V24 N12 P2217-2220 62-11375 <=>
62-15612 <*=>
1954 V24 N12 P2285-2288 RJ-418 <ATS>
1955 V25 P541-543 AMST S2 V8 P353-356 <AMS>
1955 V25 P569-576 62-18857 <=*>
1955 V25 P984-994 R-1446 <*>
R-4099 <*>
R-420 <*>
1955 V25 P1157-1159 R-1002 <*>
1955 V25 P1209-1216 R-1410 <*>
1955 V25 P1248-1256 R-1715 <*>
1955 V25 P1352-1356 R-4071 <*>
1955 V25 P1386-1398 61-15123 <+*>
1955 V25 P1825-1842 UCRL-TRANS-734 <=*>
1955 V25 P2179-2182 R-1479 <*>
R-4100 <*>
1955 V25 P2203- R-1989 <*>
1955 V25 P2329-2335 R-1478 <*>
R-4083 <*>
1955 V25 P2365-2368 UCRL TRANS-1149(L) <*>
1955 V25 P2451-2457 R-248 <*>
1955 V25 P2571-2573 R-3613 <*>
1955 V25 N1 P3-10 62-23137 <=*>
1955 V25 N1 P11-17 62-23049 <=*>
1955 V25 N1 P18-20 62-23117 <=*>
1955 V25 N1 P21-28 62-19929 <=*>
1955 V25 N1 P66-73 62-23179 <=*>
1955 V25 N1 P81-84 62-15871 <=*>
1955 V25 N1 P108-111 62-23060 <=*>
1955 V25 N1 P117-124 60-10029 <+*> 0
62-11218 <=>
1955 V25 N1 P125-134 R-4081 <*>
1955 V25 N1 P144-148 RT-2818 <*>
62-15915 <=*>
1955 V25 N1 P158-166 62-23091 <=*>
1955 V25 N1 P167-176 62-23090 <=*>
1955 V25 N2 P177-181 60-16971 <*> 0
1955 V25 N2 P182-187 GB70 4099 <NLL>
3575 <HB>
1955 V25 N2 P221-235 R-1120 <*>
1955 V25 N2 P323-332 1440TM <CTT>
59-16717 <+*>
1955 V25 N2 P358-363 62-15846 <=*>
1955 V25 N2 P364 62-15853 <=*>
1955 V25 N3 P397-401 62-18858 <=*>
1955 V25 N3 P402-409 GB43 <NLL>
1955 V25 N3 P421-429 62-19866 <=>
1955 V25 N3 P430-435 62-23016 <*=>
1955 V25 N3 P478-496 62-23201 <=*>
1955 V25 N3 P534-540 62-15903 <=*>
1955 V25 N3 P546-557 RT-3037 <*>
60-12600 <=>
1955 V25 N3 P558-561 RJ-385 <ATS>
1955 V25 N4 P590-594 R-860 <*>
1955 V25 N4 P595-600 R-862 <*>
1955 V25 N4 P595-6005 59-10354 <+*>
1955 V25 N4 P601-609 62-16724 <=*> 0
1955 V25 N4 P649-670 R-5159 <*>
1955 V25 N4 P707-710 59-16691 <+*>
1955 V25 N4 P711-719 61-13285 <+*>
1955 V25 N4 P733-741 62-23059 <*=>
1955 V25 N4 P747-762 R-3695 <*>
1955 V25 N4 P767 GB43 <NLL>

Citation	Translation
1955 V25 N5 P805-811	63-13731 <=*>
1955 V25 N5 P812-816	RT-4485 <*>
	1325TM <CTT>
1955 V25 N5 P841-846	1453TM <CTT>
	59 16693 <+*>
1955 V25 N5 P847-852	1454TM <CTT>
	59-16694 <+*>
1955 V25 N5 P864-876	59-17492 <*>
1955 V25 N5 P877-880	61-23158 <*=>
1955 V25 N5 P945-946	R-3690 <*>
1955 V25 N5 P952-954	62-11588 <=>
1955 V25 N6 P955-977	R-5097 <*>
1955 V25 N6 P984-994	GB43 <NLL>
	47T90R <ATS>
1955 V25 N6 P1C19-1025	GB43 <NLL>
1955 V25 N6 P1038-1044	GB4 T/1038 <NLL>
	R-1436 <*>
1955 V25 N6 P1111-1123	61-15068 <*+>
	61-20003 <*=>
1955 V25 N6 P1140-1154	TT.590 <NRC>
	1437TM <CTT>
	59-16714 <+*>
1955 V25 N7 P1190-1197	GB43 <NLL>
	62-18078 <=*>
1955 V25 N7 P1204-1208	RTS-2612 <NLL>
1955 V25 N7 P1248-1256	CSIRO-3349 <CSIR>
	R-405 <*>
1955 V25 N7 P1257-1264	AEC-TR-5356 <=*>
1955 V25 N7 P1307-1315	R-3251 <*>
	RT-4645 <*>
1955 V25 N7 P1316-1322	R-1409 <*>
	1365TM <CTT>
	62-23121 <=*> 0
1955 V25 N7 P1323-1325	R-1408 <*>
1955 V25 N8 P1347-1351	66-12295 <*>
1955 V25 N8 P1364-1375	63-23768 <=> P
1955 V25 N8 P1399-1404	61-23862 <*=>
1955 V25 N8 P1426-1435	62-23006 <=*>
1955 V25 N8 P1449-1457	GB125 IB 1496 <NLL>
	R-1435 <*>
	1406TM <CTT>
1955 V25 N8 P1458-1461	R-1406 <*>
1955 V25 N8 P1508-1517	R-1407 <*>
	R-1675 <*>
	R-2528 <*>
1955 V25 N8 P1520	GB43 <NLL>
1955 V25 N9 P1581-1596	R-1724 <*>
	1404TM <CTT>
1955 V25 N9 P1603-1638	62-23710 <=*> P
	62-25375 <=*> P
1955 V25 N9 P1604-1621	R-2391 <*>
1955 V25 N9 P1622-1638	R-2392 <*>
1955 V25 N9 P1649-1652	RT-4308 <*>
1955 V25 N9 P1667-1670	R-1405 <*>
	R-1722 <*>
1955 V25 N9 P1667-	R-3693 <*>
1955 V25 N10 P1675-1682	R-3692 <*>
1955 V25 N10 P1754-1767	RT-4230 <*>
1955 V25 N1C P1788-1799	1447TM <CTT>
	59-16687 <+*>
1955 V25 N10 P1804-1816	AEC-TR-5355 <=*>
1955 V25 N10 P1819-1824	1429TM <CTT>
	59-16707 <+*>
1955 V25 N11 P1878-1882	59-14123 <+*> 0
1955 V25 N11 P1883-1892	59-1C140 <+*>
1955 V25 N11 P1944-1953	59-20330 <*>
1955 V25 N11 P1957-1964	60-17353 <+*>
1955 V25 N11 P1965-1971	R-1387 <*>
	1428TM <CTT>
	59-16706 <+*>
1955 V25 N11 P1972-1982	1445TM <CTT>
	59-16698 <+*>
1955 V25 N12 P2050-2060	59-18645 <+*> 0
1955 V25 N12 P2093-2096	65-13898 <*>
1955 V25 N12 P2097-2103	59-18646 <+*>
1955 V25 N12 P2143-2149	RT-4016 <*>
1955 V25 N12 P2157-2162	RCT V29 N4 P1209 <RCT>
	59-18532 <+*>
1955 V25 N13 P2239-2255	R-1404 <*>
	RJ-594 <ATS>
1955 V25 N13 P2254-2276	RJ-588 <ATS>
1955 V25 N13 P2264-2276	62-10130 <*=>
1955 V25 N13 P2280-2281	62-11209 <=>
1955 V25 N13 P2282-2285	62-11207 <=>
1955 V25 N13 P2286-2295	60-17348 <+*>
1955 V25 N13 P2313-2318	1441TM <CTT>
	59-16718 <+*>
1955 V25 N13 P2329-2335	16J14R <ATS>
	59-18667 <+*>
1955 V25 N13 P2339-2354	UCRL TRANS-473L <=*>
1955 V25 N13 P2371-2380	59-16692 <+*>
1955 V25 N13 P2381-2394	62-19853 <=*>
1955 V25 N13 P2395	62-11301 <=>
1955 V25 N14 P2411-2418	RT-3995 <*>
1955 V25 N14 P2419-2421	05Q71R <ATS>
1955 V25 N14 P2451-2457	C-2264 <NRC>
1955 V25 N14 P2477-2485	64-13557 <=*$>
1956 V26 P3-5	T-1633 <INSD>
1956 V26 P310-	60-15016 <+*>
1956 V26 P945-957	R-1903 <*>
	R-445 <*>
1956 V26 P1004-1014	R-4395 <*>
1956 V26 P1730-	R-3894 <*>
1956 V26 P2448-	R-2040 <*>
1956 V26 N1 P157-174	RT-4271 <*>
1956 V26 N1 P220-231	R-435 <*>
1956 V26 N2 P310-322	RJ-657 <ATS>
1956 V26 N2 P375-378	RJ-584 <ATS>
	62-11620 <=>
1956 V26 N2 P385-397	59-20282 <+*>
	62-19895 <=*>
1956 V26 N2 P430-435	62-11423 <=>
1956 V26 N2 P439-441	R-4339 <*>
	RJ-587 <ATS>
1956 V26 N2 P457-474	62-19844 <=*>
1956 V26 N2 P478-482	R-2657 <*>
	R-809 <*>
1956 V26 N2 P482-483	59-17315 <+*>
1956 V26 N3 P683-685	R-441 <*>
1956 V26 N3 P686-692	GB4 T/1046 <NLL>
	R-870 <*>
1956 V26 N3 P693-694	R-451 <*>
	RT-4386 <*>
	59-22693 <+*>
1956 V26 N4 P772-778	R-3462 <*>
1956 V26 N4 P772-794	R-923 <*>
1956 V26 N4 P853-856	R-5288 <*>
1956 V26 N4 P887-894	59-16985 <+*>
1956 V26 N4 P913-916	R-924 <*>
1956 V26 N5 P945-957	60-17681 <+*>
	61-17484 <=$>
	61-17591 <=$>
1956 V26 N5 P1C15-1020	59-16554 <+*>
1956 V26 N5 P1064-1069	CSIRO-3422 <CSIR>
1956 V26 N6 P1141-1149	59-16555 <+*>
1956 V26 N6 P1150-1162	59-18517 <+*>
1956 V26 N6 P1174-1176	R-716 <*>
1956 V26 N6 P1233-1250	CSIRO-3423 <CSIR>
1956 V26 N7 P1373-1378	59-10079 <+*>
1956 V26 N7 P1428-1432	R-931 <*>
1956 V26 N7 P1456-1460	86J15R <ATS>
1956 V26 N7 P1611-1621	R-933 <*>
1956 V26 N8 P1646-1650	R-558 <*>
1956 V26 N8 P1669-1670	62-11264 <=>
1956 V26 N8 P1723-1729	59-14174 <+*>
1956 V26 N8 P1802-1814	62-15914 <=*>
1956 V26 N9 P2021-2031	59-18730 <+*>
1956 V26 N9 P2C32-2036	61-10156 <*+>
	62-18332 <=*>
1956 V26 N10 P2248-2253	AEC-TR-2883 <+*> 0
1956 V26 N10 P2398-2399	R-790 <*>
	61-19849 <=*>
1956 V26 N11 P2570-2577	61-23393 <*=>
1956 V26 N12 P2628-2632	59-22567 <+*>
1956 V26 N12 P2661-2666	61-13125 <+*>
1956 V26 N12 P2667-2683	62-16625 <=*> C
1956 V26 N12 P2755-2765	59-20284 <+*>
1957 V27 P262-268	R-1833 <*>
1957 V27 P516-521	R-4530 <*>
1957 V27 P1019-1028	RCT V32 N3 P651 <RCT>

1957 V27 P1387-1391 4057 <HB>
1957 V27 P2229- R-4066 <*>
1957 V27 P2303-2306 RCT V32 N4 P1016 <RCT>
1957 V27 P2461-2470 R-4016 <*>
1957 V27 P2472- R-4017 <*>
1957 V27 N1 P119-122 R-3131 <*>
1957 V27 N1 P123-126 R-3132 <*>
1957 V27 N1 P212-213 R-1223 <*>
1957 V27 N2 P360-363 59-22464 <+*>
1957 V27 N3 P478-483 R-3433 <*>
1957 V27 N5 P1051-1055 R-1971 <*>
1957 V27 N6 P1174-1181 R-3123 <*>
1957 V27 N6 P1347-1356 R-3240 <*>
1957 V27 N8 P1676-1685 R-3092 <*>
1957 V27 N9 P1940-1944 R-3266 <*>
1957 V27 N9 P1944-1949 59-18728 <+*>
1957 V27 N9 P2109-2128 59-20283 <+*>
1957 V27 N9 P2177-2180 R-3268 <*>
1957 V27 N9 P2181-2182 R-3267 <*>
1957 V27 N10 P2285-2290 60-15010 <+*>
1957 V27 N10 P2291-2302 63-10889 <=*> 0
1957 V27 N11 P2669-2671 62-11211 <=>
1957 V27 N12 P2702-2703 59-22501 <+*>
1957 V27 N12 P2791-2792 61-13155 <=*>
1958 P3-17 R-4019 <*>
1958 V28 P114-131 RCT V33 N3 P669 <RCT>
1958 V28 P345-350 R-4897 <*>
1958 V28 P2487-2492 RCT V33 N3 P689 <RCT>
1958 V28 N2 P254-255 R-4168 <*>
1958 V28 N2 P256-258 R-4166 <*>
1958 V28 N4 P779-781 R-4468 <*>
1958 V28 N4 P786-788 R-5122 <*>
1958 V28 N4 P908-909 R-5339 <*>
61-15612 <*+>
1958 V28 N5 P925-931 59-17467 <+*>
1958 V28 N5 P932-935 59-12882 <+*>
1958 V28 N5 P962-964 R-5119 <*>
1958 V28 N5 P1029-1031 R-5116 <*>
1958 V28 N6 P1174-1176 59-22680 <+*>
1958 V28 N6 P1177-1186 59-10306 <+*>
1958 V28 N6 P1201-1226 66-12292 <*>
1958 V28 N8 P1642-1645 59-10130 <+*>
1958 V28 N8 P1662-1669 59-16751 <+*>
1958 V28 N8 P1670-1671 R-5244 <*>
1958 V28 N8 P1672-1675 R-5245 <*>
1958 V28 N8 P1727-1733 59-10766 <+*>
1958 V28 N8 P1768-1774 59-10765 <+*> 0
59-16750 <+*>
60-13835 <+*>
1958 V28 N10 P2302-2309 59-16742 <+*>
1958 V28 N11 P2548-2566 59-16752 <+*>
1959 V29 P189- AEC-TR-3639 <+*>
1959 V29 N1 P27-44 59-19690 <+*> P0
1959 V29 N3 P344-345 62-33408 <=*>
1959 V29 N3 P358-364 RCT V33 N2 P462 <RCT>
1959 V29 N3 P381-383 63-19945 <=*$> 0
1959 V29 N4 P63-71 61-13539 <*=>
1959 V29 N4 P506-513 59-19690 <+*> P0
59-19713 <+*> P
1959 V29 N8 P933-938 60-21803 <=>
60-21932 <=>
1959 V29 N8 P962-966 59-22146 <+*> 0
60-15475 <+*>
1959 V29 N10 P1273-1276 62-13132 <*=>
1959 V29 N11 P1354-1359 60-31187 <=> 0
1959 V29 N11 P1360-1367 60-31188 <=> 0
1959 V29 N11 P1368-1372 60-31189 <=>
1959 V29 N11 P1381-1387 60-21795 <=> 0
1960 V30 P907-912 AEC-TR-4280 <=>
1960 V30 P1152-1164 AEC-TR-4189 <*>
1960 V30 N1 P57-62 60-16591 <+*>
1960 V30 N1 P63-73 60-21808 <=> 0
1960 V30 N2 P242-247 60-41623 <=>
1960 V30 N4 P450-459 61-11643 <=>
1960 V30 N5 P504-507 60-41023 <=>
1960 V30 N5 P512-521 60-41023 <=>
1960 V30 N8 P893-898 AEC-TR-4183 <=>
1960 V30 N8 P899-906 AEC-TR-4177 <=>
1961 V31 N4 P407-427 61-27808 <*=>
1961 V31 N9 P1049-1056 63-23386 <=*>
1961 V31 N9 P1057-1060 63-23391 <=*>
1961 V31 N9 P1112-1118 62-15792 <=*>
1961 V31 N11 P1294-1297 N65-22587 <=$>
62-23373 <=*>
1961 V31 N11 P1317-1323 62-11662 <=>
1961 V31 N11 P1324-1328 62-25067 <=*>
1961 V31 N11 P1329-1336 62-11685 <=>
1961 V31 N12 P1431-1438 62-24030 <=*>
1961 V31 N12 P1477-1484 62-23445 <=*>
1962 V32 N1 P30-33 65-29149 <$>
1962 V32 N1 P129-132 62-33396 <=*>
1962 V32 N2 P133-138 62-11709 <=>
1962 V32 N2 P161-167 65-20002 <$>
1962 V32 N2 P197-201 62-32314 <=*>
1962 V32 N2 P214-219 62-11709 <=>
1962 V32 N4 P457-460 63-23510 <=*$>
1962 V32 N10 P1259-1265 64-14002 <=*$>
1962 V32 N12 P1418-1427 63-23626 <=*>
1963 N33 P1098-1103 64-13606 <=*$>
1963 V33 N1 P90-99 63-21747 <=>
1963 V33 N3 P257-262 AEC-TR-5971 <=*$>
1963 V33 N6 P719-723 84Q72R <ATS>
1963 V33 N7 P835-838 64-15003 <=$>
1963 V33 N9 P1137-1138 64-71293 <=> M
1963 V33 N12 P1444-1448 64-11881 <=>
1964 V34 N1 P142-145 AD-609 138 <=>
1964 V34 N2 P354-368 64-11789 <=>
1964 V34 N3 P519-522 64-41757 <=$>
1964 V34 N4 P624-629 TRC-TRANS-1111 <NLL>
1964 V34 N6 P1031-1036 N65-14608 <=$>
1964 V34 N6 P1050-1056 N65-14565 <=*$>
1964 V34 N8 P1417-1423 N65-19506 <=$>
1964 V34 N8 P1444-1461 64-41980 <=$>
1964 V34 N8 P1521-1525 64-41980 <=$>
1964 V34 N10 P1721-1731 N65-12273 <=$>
1964 V34 N11 P81-82 AD-610 416 <=$>
1964 V34 N12 P2089-2111 N65-22619 <=*$>
1964 V34 N12 P2194-2196 AD-612 421 <=$>
1965 V35 N8 P1461-1470 66-12486 <*>

ZHURNAL TEORETICHESKOI I PRAKTICHESKOI MEDITSINY
1929 V3 P317-322 6K22R <ATS>

ZHURNAL USHNYKH, NOSOVYKH I GORLOVYKH BOLEZNEI
1938 V15 N3 P247-260 RT-3831 <*>
1962 N4 P51- FPTS V23 N5 P.T1133 <FASE>
1962 V22 N1 P49-55 64-11970 <=> M
1962 V22 N4 P9-12 64-13825 <=*$>
1962 V22 N4 P51- FPTS,V23,N5,P.T1133 <FASE>
1963 V23 N1 P88-96 63-41002 <=>
1964 V24 N3 P56-62 65-62208 <=$>

*ZHURNAL VSESOYUZNOGO KHIMICHESKOGO OBSHCHESTVA
1960 V5 N1 P2-9 61-11797 <=>
1960 V5 N1 P10-17 61-21072 <=>
1960 V5 N1 P97-101 61-21059 <=>
1960 V5 N1 P102 61-21084 <=>
1960 V5 N1 P103-104 61-21082 <=>
1960 V5 N1 P103- 61-21085 <=>
1960 V5 N1 P112-113 96N55R <ATS>
1960 V5 N1 P114-116 97N55R <ATS>
1960 V5 N1 P119-120 63-13713 <=*>
1960 V5 N2 P122-125 61-11876 <=>
1960 V5 N2 P156-168 61-11568 <=>
61-27043 <=*>
1960 V5 N2 P168-172 61-11877 <=>
1960 V5 N2 P173-180 61-11645 <=>
61-19728 <=*>
1960 V5 N2 P186-191 63-23513 <=*>
65-17336 <*>
1960 V5 N2 P226-228 61-11583 <=>
1960 V5 N2 P228-229 61-11607 <=>
1960 V5 N3 P242-249 61-21005 <=>
1960 V5 N3 P260-267 11M45R <ATS>
1960 V5 N3 P275-285 61-21005 <=>
1960 V5 N3 P292-297 61-21169 <=>
1960 V5 N3 P298-306 61-21005 <=>
1960 V5 N3 P346-347 61-27190 <*=>
1960 V5 N3 P356-357 64-23528 <=$>
1960 V5 N4 P371-376 80M46R <ATS>

Citation	Translation
1960 V5 N4 P438-442	72M46R <ATS>
1960 V5 N4 P460-465	61-21075 <=>
1960 V5 N5 P482-492	61-23009 <*=>
1960 V5 N5 P498-543	61-23009 <*=>
1960 V5 N5 P498-506	62-14316 <=*>
1960 V5 N5 P507-514	62-13115 <*=>
1960 V5 N5 P515-521	61-23090 <=*>
	62-11473 <=>
1960 V5 N5 P522-534	61-28167 <*=>
	62-14648 <=*> 0
	62-23136 <=*>
1960 V5 N5 P535-543	62-23039 <=*>
	62-23358 <=*>
1960 V5 N5 P544-552	62-11261 <=>
1960 V5 N5 P553-557	61-23009 <*=>
	62-11324 <=>
1960 V5 N5 P562-582	61-23009 <*=>
1960 V5 N5 P562-568	62-11489 <=>
1960 V5 N5 P591-592	ICE V1 N1 P129-131 <ICE>
1960 V5 N5 P594-595	24N49R <ATS>
1960 V5 N6 P686-688	62-13230 <*=>
1960 V5 N6 P689-690	62-13235 <*=>
1960 V5 N6 P697-699	62-13239 <*=>
1960 V5 N6 P699-701	62-13240 <*=>
1960 V5 N6 P701-703	62-13242 <*=>
1960 V5 N6 P712-713	61P66R <ATS>
1961 V6 P236-237	63-19132 <=*>
1961 V6 P635-642	66-10688 <*>
1961 V6 N1 P2-16	61-27771 <=*>
1961 V6 N1 P81-87	2303 <BISI>
1961 V6 N1 P94	62-13519 <*=>
1961 V6 N1 P102-107	62-15484 <*=>
1961 V6 N1 P114-116	67P59R <ATS>
1961 V6 N1 P118-119	18N53R <ATS>
1961 V6 N2 P130-141	59N54R <ATS>
1961 V6 N2 P193-199	61-28921 <*=>
1961 V6 N2 P224-226	61-28788 <*=>
1961 V6 N3 P244-267	62-23840 <=>
1961 V6 N3 P268-275	AEC-TR-4903 <=*> 0
1961 V6 N3 P268-274	62-13533 <*=> 0
1961 V6 N3 P285-292	62-23840 <=>
1961 V6 N3 P305-313	62-23840 <=>
1961 V6 N3 P319-324	62-13533 <*=> 0
1961 V6 N3 P325-342	62-23840 <=>
1961 V6 N3 P357-358	62-15499 <*=>
1961 V6 N4 P389-393	62-33068 <=*>
1961 V6 N4 P394-403	69P60R <ATS>
1961 V6 N4 P428-434	RCT V36 N2 P337 <RCT>
	RUB.CH.TECH. V36 N2 <ACS>
1961 V6 N5 P482-498	62-24014 <=*>
1961 V6 N5 P596-597	62-32243 <=*$>
1961 V6 N6 P612-618	62-25848 <=*>
1961 V6 N6 P629-635	63-19699 <=>
1962 V7 N1 P87-91	50P61R <ATS>
1962 V7 N2 P122-131	63-19346 <=*>
1962 V7 N2 P131-140	13P63R <ATS>
1962 V7 N2 P194-200	12337-C <K-H>
	65-14492 <*>
1962 V7 N2 P229-230	SCL-T-442 <=*>
1962 V7 N3 P343-347	65-13072 <*>
1962 V7 N4 P362-366	C-5475 <NRC>
1962 V7 N4 P401-410	SCL-T-499 <=*$>
1962 V7 N5 P482-488	63-21116 <=>
1962 V7 N5 P489-494	65-64394 <=*$>
1962 V7 N5 P520-523	64-14364 <=*$>
1962 V7 N5 P520-	66-10187 <*> 0
1962 V7 N5 P524-529	63-19137 <=*>
1962 V7 N5 P583-584	SCL-T-492 <=*$>
1962 V7 N5 P590-600	72R75R <ATS>
1963 V8 P155-162	AD-615 218 <=$>
1963 V8 N1 P2-94	63-31747 <=>
1963 V8 N2 P148-154	AD-615 218 <=$>
1963 V8 N2 P192-197	63-24041 <=*>
1963 V8 N3 P242-244	63-31553 <=>
1963 V8 N3 P354-356	800Q3R <ATS>
1963 V8 N4 P424-448	64-21678 <=>
1963 V8 N5 P586-587	64-11865 <=>
	64-16488 <=*$> 0
1963 V8 N5 P592	AD-614 963 <=$>
	65-60245 <=$>
1963 V8 N6 P691-692	66-10033 <*>
1964 N3 P280-289	ICE V4 N4 P670-679 <ICE>
1964 V9 N1 P12-18	64-31116 <=>
1964 V9 N1 P41-46	64-31116 <=>
1964 V9 N1 P65-86	64-31116 <=>
1964 V9 N1 P117-119	63S85R <ATS>
1964 V9 N3 P280-289	ICE V4 N4 P670-679 <ICE>
1964 V9 N3 P348-350	65-63684 <=$>
1964 V9 N4 P374-380	65-30226 <=$>
1964 V9 N4 P395-432	65-30226 <=$>
1964 V9 N4 P462-466	65-30869 <=$>
1964 V9 N4 P472-473	GI N3 1964 P563 <AGI>
1964 V9 N4 P479-480	GI N3 1964 P565 <AGI>
1964 V9 N5 P496-503	65-30380 <=*$>
1964 V9 N5 P518-523	65-30770 <=$>
1964 V9 N5 P524-536	65-33453 <=$>
1964 V9 N5 P561-566	61R80R <ATS>
1965 N1 P2-11	ICE V5 N4 P623-632 <ICE>
1965 V10 N2 P234-235	65-14174 <*> 0
1965 V10 N2 P235-236	65-14176 <*> 0
1965 V10 N3 P287-294	65-33202 <=$>
1965 V10 N3 P312-319	65-33202 <=$>
1965 V10 N5 P594-595	66-13286 <*>
1966 N11 P334-340	66-33913 <=$>

*ZHURNAL VYCHISLITELNOI MATEMATIKI I MATEMATICHESKOI FIZIKI

Citation	Translation
1961 V1 N1 P129-143	63-10823 <=*>
	63-10900 <=*>
1961 V1 N1 P151-168	62-15436 <*=>
1961 V1 N2 P246-254	62-13842 <*=>
1961 V1 N2 P267-279	C-4022 <NRCC>
1961 V1 N2 P361-364	62-15216 <*=>
1961 V1 N2 P545-548	62-15215 <*=>
1961 V1 N3 P371-411	62-15506 <*=>
1961 V1 N3 P499-512	62-11109 <=>
1961 V1 N3 P513-522	62-15221 <*=>
1961 V1 N4 P622-637	AIAAJ V2 N1 P208 <AIAA>
1961 V1 N5 P856-868	62-24038 <=*> 0
1961 V1 N5 P869-883	AD-624 761 <=*$> M
1961 V1 N5 P884-902	63-23632 <=*$>
1961 V1 N6 P1020-1050	AD-614 916 <=$>
	65-13278 <*>
1961 V1 N6 P1051-1060	N65-15060 <=$>
1961 V1 N6 P1136-1139	62-33642 <=*>
1962 V2 P972-982	UCRL TRANS-980(L) <=*$>
1962 V2 N1 P3-14	UCRL-TRANS-101 (L) <=$>
1962 V2 N1 P97-106	62-32104 <=*>
	63-13212 <=*>
1962 V2 N1 P125-132	62-33628 <=*>
1962 V2 N2 P195-205	63-23099 <=*>
1962 V2 N2 P217-240	N65-15157 <=$>
1962 V2 N2 P255-289	63-23099 <=*>
1962 V2 N2 P303-316	62-25508 <=*>
1962 V2 N2 P317-335	62-24740 <=*>
1962 V2 N2 P343-348	63-13892 <=*>
1962 V2 N3 P398-410	64-11879 <=> M
1962 V2 N3 P459-466	62-11716 <=>
	63-10623 <=*>
	65-17169 <*>
1962 V2 N3 P488-491	62-32073 <=*>
1962 V2 N3 P499-502	AIAAJ V1 N11 P2708 <AIAA>
	63-18370 <=*>
	63-23425 <*=>
1962 V2 N4 P684-694	63-21024 <=>
	65-13394 <*>
1962 V2 N4 P695-702	63-13259 <=*>
1962 V2 N5 P902-911	63-11502 <=>
1962 V2 N6 P1044-1053	65-13882 <*>
1962 V2 N6 P1062-1085	63-01218 <*>
1962 V2 N6 P1086-1101	AIAAJ V1 N9 P2224 <AIAA>
1962 V2 N6 P1102-1110	64-11853 <=>
1962 V2 N6 P1132-1144	63-21110 <=>
1962 V2 N6 P1132-1139	63-24025 <=*>
1963 V3 N3 P431-466	64-13156 <=*$>
1963 V3 N3 P574-580	64-13156 <=*$>
1963 V3 N4 P730-741	64-11737 <=>
1963 V3 N4 P742-754	AD-619 564 <=$>
1963 V3 N5 P887-904	64-31009 <=>
1963 V3 N5 P942-949	64-31010 <=>

1963 V3 N6 P1006-1013	EECL-LIB-TRANS-1294 <NLL>
1963 V3 N6 P1067-1076	AD-612 372 <=$>
1963 V3 N6 P1089-1102	64-21401 <=>
1963 V3 N6 P1103-1111	64-21401 <=>
1963 V3 N6 P1126-1130	AD-610 208 <=$>
	64-14319 <=*$>
1963 V3 N6 P1137-1139	64-21401 <=>
1964 N4 P139-143	AD-638 615 <=$>
1964 V4 N1 P52-60	64-21999 <=>
1964 V4 N1 P61-77	64-16371 <=*$>
1964 V4 N1 P78-95	64-21999 <=>
1964 V4 N1 P96-112	64-21949 <=>
1964 V4 N1 P113-119	64-21999 <=>
1964 V4 N1 P146-149	N65-32727 <=$>
1964 V4 N1 P149-155	N65-32970 <=$>
1964 V4 N1 P171-177	AD-625 236 <=*$>
1964 V4 N1 P178-183	64-16372 <=*$>
1964 V4 N2 P292-305	64-31908 <=>
1964 V4 N2 P306-316	AD-614 630 <=$> M
	65-10998 <*>
1964 V4 N2 P376-379	AD-614 630 <=$>
	64-31909 <=>
1964 V4 N3 P495-511	AD-640 611 <=$>
1964 V4 N3 P512-524	AD-634 832 <=$> M
1964 V4 N4 P660-670	AD-640 474 <=$>
1964 V4 N5 P791-802	65-30083 <=$>
1964 V4 N5 P896-904	65-12649 <*>
1964 V4 N5 P905-911	65-30084 <=$>
1964 V4 N5 P920-926	65-12648 <*>
1964 V4 N5 P963-967	65-30085 <=$>
1964 V4 N5 P970-973	65-30086 <=$>
1964 V4 N6 P983-994	AD-637 996 <=$>
1964 V4 N6 P1065-1077	AD-621 771 <=*$> M
1965 V5 P157-160	66-10722 <*>
1965 V5 N1 P21-33	AD-616 611 <=$>
1965 V5 N1 P67-76	AD-624 727 <=*$>
1965 V5 N3 P395-453	66-10916 <*>
1965 V5 N3 P503-518	N66-13482 <=$>
1965 V5 N4 P587-607	66-61038 <=*$>
1965 V5 N5 P903-907	66-12480 <*>
1965 V5 N5 P938-944	66-12479 <*>
1965 V5 N5 P949-951	66-12488 <*>
1965 V5 N5 P960-966	AD-631 971 <=$>
1965 V5 N5 P960-965	66-10481 <*>
1965 V5 N6 P1107-1115	66-13423 <*>
	66-61358 <=$>
1965 V5 N6 P1138-1141	66-12487 <*>
1966 V6 N4 P714-726	66-62306 <=*$>

*ZHURNAL VYSSHEI NERVNOI DEYATELNOSTI IMENI I. P. PAVLOVA

1951 V1 N2 P160-164	RT-1281 <*>
	64-15368 <=*$>
1951 V1 N3 P28-42	62-13777 <*=>
1951 V1 N4 P579-602	TT.537 <NRC>
1952 V2 N2 P169-181	R-2704 <*>
1952 V2 N2 P258-282	R-2689 <*>
1952 V2 N3 P437-440	RT-3199 <*>
1952 V2 N4 P481-500	RT-4542 <*>
1952 V2 N4 P501-508	RT-2311 <*>
1952 V2 N4 P518-534	RT-2762 <*>
1952 V2 N4 P572-581	RT-2323 <*>
	62-13788 <=*>
1952 V2 N5 P723-733	66-11136 <*>
1953 P495-509	RT-2623 <*>
1953 V3 N1 P55-70	60-10105 <+*>
1953 V3 N2 P192-202	62-13787 <=*>
	62-19262 <*=>
1953 V3 N2 P247-259	62-13773 <*=>
1953 V3 N3 P353-368	RT-1758 <*>
1953 V3 N3 P381-392	RT-2667 <*>
1953 V3 N3 P416-427	RT-2966 <*>
1953 V3 N5 P755-764	RT-4219 <*>
1953 V3 N6 P955-959	RT-2554 <*>
1954 V4 N2 P153-158	RT-3018 <*>
1954 V4 N2 P267-281	RT-2965 <*>
1954 V4 N2 P289-304	RT-2970 <*>
1954 V4 N3 P406-414	62-13783 <*=>
1954 V4 N4 P581-585	R-158 <*>
1954 V4 N4 P591-598	R-154 <*>
1954 V4 N6 P791-798	65-63202 <=>
1954 V4 N14 P568-573	59-16163 <+*>
1955 V5 N2 P145-156	62-15058 <*=>
1955 V5 N2 P173-186	59-11361 <=> 0
1955 V5 N3 P305-317	PB 141 241T <=>
1955 V5 N4 P449-462	R-2712 <*>
	RT-4273 <*>
	62-15052 <*=>
1955 V5 N4 P463-473	R-2706 <*>
	62-15055 <=*> 0
1955 V5 N4 P480-491	R-2722 <*>
	R-318 <*>
	R-3516 <*>
1955 V5 N4 P511-519	R-2725 <*>
	R-3525 <*>
1955 V5 N4 P574-581	R-319 <*>
1955 V5 N5 P636-643	59-11050 <=>
1955 V5 N5 P665-676	59-11016 <=>
1956 V6 P881-890	65-62127 <=$>
1956 V6 N1 P76-86	62-15338 <=*>
1956 V6 N1 P87-92	64-19869 <=$> 0
1956 V6 N1 P157-163	59-11049 <=> 0
	62-15046 <0>
1956 V6 N3 P394-398	59-11388 <=> 0
1956 V6 N3 P415-425	59-11389 <=> 0
1956 V6 N3 P443-450	59-11390 <=> 0
1956 V6 N3 P479-481	59-11391 <=> 0
1956 V6 N3 P482-493	59-11392 <=> 0
1956 V6 N5 P776-785	59-11008 <=>
1956 V6 N6 P863-871	59-11038 <=> 0
1957 N5 P673-682	59-11150 <=>
1957 N5 P706-710	59-11149 <=>
1957 N5 P741-753	59-11072 <=> 0
1957 V7 P254-262	65-61202 <=$>
1957 V7 P442-446	65-63144 <=$>
1957 V7 P722-726	65-61201 <=> 0
1957 V7 N1 P75-82	R-4616 <*>
	62-15063 <*=>
1957 V7 N2 P291-298	PB 141 258T <=>
1957 V7 N3 P402-409	61-28692 <*=>
1957 V7 N4 P554-559	R-5372 <*>
1957 V7 N4 P626-628	PB 141 250T <=>
1957 V7 N6 P805-818	60-15608 <+*> 0
1958 V8 N1 P3-16	59-13060 <=> 0
1958 V8 N1 P56-63	62-13782 <*=>
1958 V8 N1 P192-202	62-13776 <=*>
1958 V8 N3 P358-365	62-13785 <=*> 0
1958 V8 N3 P431-436	62-13778 <=*>
1958 V8 N5 P766-773	60-18359 <+*>
1959 V9 N3 P398-408	60-11796 <=>
1959 V9 N6 P941-947	60-11406 <=>
1960 V10 N2 P274-279	63-19706 <=*> 0
1960 V10 N2 P280-284	63-19710 <=*> 0
1960 V10 N3 P305-312	60-41681 <=>
1960 V10 N3 P330-336	60-41682 <=>
1960 V10 N3 P443-448	63-19705 <=*> 0
1960 V10 N3 P449-458	60-41674 <=>
1960 V10 N4 P630-633	62-14719 <=*>
1960 V10 N6 P809-813	62-13904 <=*> 0
1960 V10 N6 P814-820	62-13900 <=*> 0
1960 V10 N6 P821-826	62-13906 <=*>
1960 V10 N6 P827-831	62-13903 <=*> 0
1960 V10 N6 P832-841	62-13901 <*=>
1960 V10 N6 P842-850	62-13886 <=*> 0
1960 V10 N6 P851-859	62-13888 <=*> 0
1960 V10 N6 P860-863	62-13899 <*=> 0
1960 V10 N6 P864-868	62-13894 <=*> 0
1960 V10 N6 P869-873	62-13897 <=*>
1960 V10 N6 P874-879	62-13891 <=*>
1960 V10 N6 P880-883	62-13890 <=*> 0
1960 V10 N6 P884-891	62-13902 <=*> 0
1960 V10 N6 P892-895	62-13887 <=*>
1960 V10 N6 P896-902	62-13892 <=*> 0
1960 V10 N6 P903-907	62-13905 <=*>
1960 V10 N6 P908-912	62-13893 <=*>
1960 V10 N6 P913-917	62-13896 <*=> 0
1960 V10 N6 P918-921	61-13907 <=*>
1960 V10 N6 P922-932	62-13898 <=*>
1960 V10 N6 P933-938	62-13895 <=*>
	62-20330 <*=>

1960 V10 N6 P939-940 62-13889 <=*>
1961 V11 N1 P22-28 61-27393 <*=>
1961 V11 N1 P165-168 61-23903 <=*>
1961 V11 N1 P169-170 61-23902 <=*>
1961 V11 N3 P566-575 62-13681 <=>
1961 V11 N4 P586-592 62-13859 <=>
1961 V11 N4 P609-622 62-13859 <=>
1961 V11 N4 P674-679 62-13859 <=>
1961 V11 N4 P697-702 62-13859 <=>
1961 V11 N4 P730-737 62-13859 <=>
1961 V11 N4 P759-762 62-13859 <=>
1961 V11 N4 P766-768 62-15498 <*=>
1961 V11 N5 P795-805 62-20308 <=*>
1961 V11 N5 P806-813 62-20314 <=*>
1961 V11 N5 P814-822 62-20361 <=*> P
1961 V11 N6 P1134-1136 62-23459 <=*>
1962 V12 N1 P37- FPTS V22 N1 P.T26 <FASE>
1962 V12 N1 P74-81 63-15424 <=*>
1962 V12 N1 P145- FPTS V22 N1 P.T28 <FASE>
1962 V12 N1 P161-168 63-15425 <=*>
1962 V12 N1 P169- FPTS V22 N1 P.T34 <FASE>
1962 V12 N2 P202-207 62-11732 <=>
1962 V12 N2 P223-228 63-19092 <=*>
1962 V12 N2 P318- FPTS V23 N4 P.T896 <FASE>
FPTS,V23,N4,P.T896 <FASE>
1962 V12 N2 P332-337 62-11732 <=>
1962 V12 N2 P358-360 62-33049 <=*>
1962 V12 N3 P563-566 63-13465 <=*>
1962 V12 N4 P569-577 63-13527 <=*>
1962 V12 N4 P623-629 63-13517 <=*>
1962 V12 N4 P727-732 63-13527 <=*>
1962 V12 N4 P762-768 63-13517 <=*>
1962 V12 N5 P797-807 64-23644 <=$>
1962 V12 N5 P969-974 63-13837 <=*>
1962 V12 N6 P985-1010 63-21399 <=>
1962 V12 N6 P1001-1010 63-21107 <=>
1962 V12 N6 P1021-1028 63-21107 <=>
1962 V12 N6 P1055-1064 63-21107 <=>
1962 V12 N6 P1085-1096 63-21399 <=>
1963 V13 N1 P62-72 64-15909 <=*$>
1963 V13 N1 P112-120 FPTS,V22,P.T1115-19 <FASE>
1963 V13 N1 P131-139 64-13528 <=*$> O
1963 V13 N2 P28-285 64-13397 <=*$>
1963 V13 N2 P193- FPTS V23 N1 P.T105 <FASE>
FPTS,V23,P.T105-T112 <FASE>
1963 V13 N2 P207- FPTS V23 N2 P.T241 <FASE>
FPTS,V23,P.T241-T246 <FASE>
1963 V13 N2 P228- FPTS V23 N2 P.T264 <FASE>
1963 V13 N2 P228 FPTS,V23,N1,T264-267 <FASE>
1963 V13 N2 P235- FPTS V23 N2 P.T260 <FASE>
FPTS,V23,P.T260-T263 <FASE>
1963 V13 N2 P291-300 64-13396 <=*$> O
1963 V13 N2 P309- FPTS V23 N4 P.T733 <FASE>
1963 V13 N2 P309-315 FPTS V23 N4 P.T733 <FASE>
1963 V13 N2 P316-325 63-31270 <=>
1963 V13 N2 P330- FPTS V23 N1 P.T96 <FASE>
1963 V13 N3 P482- FPTS V23 N5 P.T1117 <FASE>
1963 V13 N3 P543- FPTS V23 N5 P.T1080 <FASE>
1963 V13 N4 P585-592 64-15248 <=*$>
64-31422 <=>
1963 V13 N4 P638- FPTS V23 N4 P.T737 <FASE>
1963 V13 N4 P638-645 FPTS V23 N4 P.T737 <FASE>
1963 V13 N4 P680-685 64-13939 <=*$>
1963 V13 N4 P699- FPTS V23 N3 P.T633 <FASE>
FPTS,V23,N3,P.T633 <FASE>
1963 V13 N4 P715- FPTS V23 N3 P.T636 <FASE>
FPTS,V23,N3,P.T636 <FASE>
1963 V13 N5 P391- FPTS V24 N3 P.T443 <FASE>
1963 V13 N5 P798-815 65-60141 <=$>
1963 V13 N5 P859-869 65-60233 <=$>
1963 V13 N5 P870- FPTS V24 N1 P.T150 <FASE>
1963 V13 N5 P882-891 65-60142 <=$>
1963 V13 N5 P904- FPTS V23 N6 P.T1189 <FASE>
FPTS V23 P.T1189 <FASE>
1963 V13 N5 P939-945 65-60236 <=$>
1963 V13 N5 P946-950 64-21751 <=>
1963 V13 N6 P1,108-110 64-21879 <=>
1963 V13 N6 P953-962 AD-609 152 <=>
1963 V13 N6 P953-961 64-21874 <=>
1963 V13 N6 P1039-1045 64-21874 <=>
1963 V13 N6 P1059-1071 64-19710 <=$>
1963 V13 N6 P1077-1086 64-21874 <=>
1963 V13 N6 P1108-1110 AD-610 418 <=$>
1964 V14 P326-336 64-41685 <=>
1964 V14 N1 P3-31 64-31098 <=>
1964 V14 N1 P33-38 64-31098 <=>
1964 V14 N1 P55-67 64-31098 <=>
1964 V14 N1 P95-101 64-31098 <=>
1964 V14 N1 P122-146 64-31098 <=>
1964 V14 N1 P161-171 64-31098 <=>
1964 V14 N1 P181-183 64-31098 <=>
1964 V14 N2 P195-222 64-41685 <=>
1964 V14 N2 P239-245 64-41685 <=>
1964 V14 N2 P277- FPTS V24 N3 P.T437 <FASE>
1964 V14 N2 P277-300 64-41685 <=>
1964 V14 N2 P346-349 64-31424 <=>
1964 V14 N2 P358-362 64-41132 <=>
1964 V14 N2 P364-368 AD-620 784 <=*$>
64-31425 <=>
1964 V14 N3 P377-386 64-41330 <=>
1964 V14 N3 P417-426 64-41330 <=>
1964 V14 N3 P459-467 64-41330 <=>
1964 V14 N3 P475-479 64-41330 <=>
1964 V14 N3 P489-497 64-41330 <=>
1964 V14 N3 P489-498 65-62496 <=$>
1964 V14 N3 P498- FPTS V24 N3 P.T434 <FASE>
1964 V14 N3 P520- FPTS V24 N4 P.T655 <FASE>
1964 V14 N3 P520-526 64-41330 <=>
1964 V14 N3 P544-561 64-41330 <=>
1964 V14 N4 P608- FPTS V24 N5 P.T758 <FASE>
1964 V14 N4 P667-677 64-51533 <=$>
1964 V14 N4 P687-694 65-30667 <=$>
1964 V14 N4 P695-700 65-30666 <=$>
1964 V14 N4 P701-706 64-51541 <=*$>
1964 V14 N4 P707- FPTS V24 N5 P.T781 <FASE>
1964 V14 N4 P707-713 65-30709 <=$>
1964 V14 N4 P726-731 64-51580 <=$>
1964 V14 N5 P745- FPTS V24 N5 P.T753 <FASE>
1964 V14 N5 P745-762 64-51862 <=$>
1964 V14 N5 P755-762 AD-615 995 <=$>
1964 V14 N5 P876- FPTS V24 N5 P.T773 <FASE>
1964 V14 N5 P885-891 AD-627 853 <=$>
1964 V14 N6 P1 65-31194 <=$>
1964 V14 N6 P104-109 65-31194 <=$>
1964 V14 N6 P953-957 65-63631 <=$>
1964 V14 N6 P957- FPTS V24 N6 P.T1023 <FASE>
1964 V14 N6 P1079- FPTS V24 N6 P.T1028 <FASE>
1964 V14 N6 P1090-1100 65-63632 <=$>
1964 V14 N6 P1104-1107 65-30474 <=$>
1965 V15 N1 P3-10 65-30955 <=$>
1965 V15 N1 P45-51 65-30955 <=$>
1965 V15 N1 P53-61 65-30955 <=$>
1965 V15 N1 P86- FPTS V24 N6 P.T1003 <FASE>
1965 V15 N1 P120- FPTS V24 N6 P.T995 <FASE>
1965 V15 N1 P156- FPTS V24 N6 P.T1011 <FASE>
1965 V15 N1 P188-189 65-30955 <=$>
1965 V15 N2 P311- FPTS V25 N2 P.T263 <FASE>
1965 V15 N2 P364- FPTS V25 N2 P.T253 <FASE>
1965 V15 N2 P405- FPTS V25 N2 P.T248 <FASE>
1965 V15 N3 P481- FPTS V25 N4 P.T546 <FASE>
1965 V15 N3 P491- FPTS V25 N3 P.T384 <FASE>
1965 V15 N3 P500- FPTS V25 N3 P.T397 <FASE>
1965 V15 N3 P513- FPTS V25 N3 P.T373 <FASE>
1965 V15 N3 P521- FPTS V25 N3 P.T377 <FASE>
1965 V15 N3 P529- FPTS V25 N3 P.T391 <FASE>
1965 V15 N4 P585-593 65-33598 <=*$>
1965 V15 N4 P603- FPTS V25 N4 P.T565 <FASE>
1965 V15 N4 P677-682 65-33598 <=*$>
1965 V15 N4 P695- FPTS V25 N4 P.T569 <FASE>
1965 V15 N4 P705- FPTS V25 N4 P.T577 <FASE>
1965 V15 N4 P739- FPTS V25 N3 P.T389 <FASE>
1965 V15 N5 P903- FPTS V25 N5 P.T753 <FASE>
1965 V15 N5 P927- FPTS V25 N5 P.T750 <FASE>
1965 V15 N5 P934- FPTS V25 N5 P.T765 <FASE>
1965 V15 N6 P977- FPTS V25 N6 P.T924 <FASE>
1965 V15 N6 P1072- FPTS V25 N6 P.T919 <FASE>
1965 V15 N6 P1088- FPTS V25 N6 P.T943 <FASE>
1966 V16 N1 P34-51 66-31751 <=$>
1966 V16 N1 P62-66 66-31751 <=$>
1966 V16 N1 P82-87 66-31751 <=$>

1966 V16 N1 P96-101 66-31751 <=$>
1966 V16 N1 P103-111 66-31751 <=$>
1966 V16 N1 P117-123 66-31751 <=$>
1966 V16 N1 P125-130 66-31751 <=$>
1966 V16 N1 P134-135 66-31751 <=$>

ZIEGELINDUSTRIE
1959 V12 N24 P726-730 C-3702 <NRCC>

ZINATNISKIE RAKSTI. LATVIJAS VALSTS UNIVERSITATE. RIGA
1960 N14 P159-171 IGR V7 N6 P951 <AGI>

ZINC, CADMIUM ET ALLIAGES
1961 V7 P34-45 3061 <BISI>
1963 V33 P1-10 TRANS-532 <MT>

ZISIN
1956 V9 N4 P276-280 SCL-T-337 <*>
1957 V9 N2 P200-208 SCL-T-340 <*>

ZIVA. PRAHA
1958 V6 P58-59 62-19903 <=>

ZIVOT STRANY
1964 N15 P953-958 64-51030 <=$>

ZNAMYA. MOSCOW
1965 V35 N4 P155-173 65-33157 <=$>

ZNANIE-SILA
1955 N10 P10-14 R-117 <*>
1957 V32 N11 P27-41 61-15257 <+*> 0
1958 N3 P9 59-16343 <+*>
1958 N6 P5-6 59-19709 <+*>
1958 N8 P17-18 60-13447 <+*>
1958 N9 P20-23 59-11884 <=>
1958 N12 P4-6 59-19498 <+*>
1958 V33 N12 P19 62-33538 <=*>
1959 N4 P8 60-11874 <=>
1959 N4 P10-11 60-11874 <=>
1959 N4 P42-43 60-11896 <=>
1959 V34 N4 P23-26 61-23685 <*=>
1959 V34 N4 P44-45 61-23685 <*=>
1959 V34 N5 P16-17 61-15272 <*+>
1959 V34 N6 P32-36 61-23416 <*=>
1959 V34 N10 P29- 61-19447 <+*>
1959 V34 N11 P15-16 60-31717 <=>
1960 V35 N4 P34-36 61-23176 <*=>
1960 V35 N5 P38-39 60-41203 <=>
1960 V35 N7 P31-33 AD-611 872 <=$>
1960 V35 N10 P1-3 61-28703 <*=>
1960 V35 N10 P6-8 61-28535 <*=>
1960 V35 N11 P40-42 61-31509 <=>,
1960 V35 N12 P18-23 61-31493 <=>
1961 V36 N1 P10-13 62-13683 <*=>
1961 V36 N1 P19-21 62-32423 <=*>
1961 V36 N3 P28-31 61-27106 <=>
1961 V36 N3 P33-34 61-27106 <=>
1961 V36 N3 P42-43 61-27106 <=>
1961 V36 N4 P23-24 63-15707 <=*>
1961 V36 N4 P26-29 61-27352 <*=>
1961 V36 N5 P23-25 63-13185 <=*>
1961 V36 N6 P34 61-28410 <*=>
1961 V36 N7 P8-9 62-23665 <=*>
1961 V36 N7 P10-13 62-19257 <*=>
1961 V36 N9 P43 62-23660 <=*>
1961 V36 N10 P46-50 62-15665 <*=>
1961 V36 N11 P11-14 63-15326 <=*>
1961 V36 N12 P6-10 62-33076 <=*>
1961 V36 N12 P11-12 62-11697 <=>
1961 V36 N12 P24-27 62-11648 <=> 0
64-71602 <=> 0
1961 V36 N12 P54 62-33324 <=>
1962 V37 N1 P14-17 62-23391 <=*>
62-25601 <=*>
1962 V37 N1 P23-26 62-23613 <=*>
1962 V37 N2 P10-12 AD-637 385 <=$>
1962 V37 N3 P12-15 63-13900 <=*>
1962 V37 N3 P17-19 63-13900 <=*>
1962 V37 N4 P26-29 62-32842 <=>
1962 V37 N4 P36-38 62-25993 <=*>
1962 V37 N5 P17-20 62-33432 <=*>
1962 V37 N6 P16-19 62-33607 <=*>
1962 V37 N8 P20-23 62-33693 <=*>
1962 V37 N9 P4-6 63-23725 <=*$>
1962 V37 N10 P13-16 63-13506 <=*>
1962 V37 N10 P29-32 AD-609 784 <=>
1962 V37 N10 P29-35 63-11534 <=>
1962 V37 N10 P29-32 63-13492 <=*>
1962 V37 N10 P33-35 AD-609 156 <=>
1962 V37 N11 P28-30 63-21278 <=>
1962 V37 N12 P34-37 63-21146 <=>
1963 V38 N1 P19-20 63-21718 <=>
1963 V38 N3 P10-13 AD-618 312 <=$>
1963 V38 N3 P46-49 63-31018 <=>
1963 V38 N5 P16-23 63-31286 <=>
1963 V38 N8 P19-21 63-41062 <=>
1963 V38 N12 P16 AD-611 017 <=$>
1964 V39 N2 P47-48 64-11863 <=>
1964 V39 N3 P1-5 64-41243 <=>
1964 V39 N3 P42-46 64-41166 <=>
1964 V39 N4 P38-41 AD-609 139 <=>
1964 V39 N5 P6-8 64-41797 <=>
1964 V39 N6 P20-24 64-41116 <=>
1964 V39 N7 P19-21 64-51887 <=$>
1964 V39 N11 P25-27,34 65-30710 <=$>
1965 V40 N1 P8-10 65-31528 <=*$>
1965 V40 N3 ENTIRE ISSUE 65-31616 <=$> 0
1965 V40 N6 P6-10 65-32915 <=*$>
1965 V40 N6 P18-19 65-32914 <=*$>
1965 V40 N8 P34-37 AD-627 236 <=$>
1965 V40 N9 P12-14 66-30116 <=*$>
1965 V40 N9 P48-49 66-30116 <=*$>
1965 V40 N10 P6-9 65-34078 <=*$>
1965 V40 N11 P7-10 66-30259 <=*$>

ZNANNYA TA PRATSYA
1961 N2 P18-19 61-28851 <*=>
1961 N3 P8-9 61-27956 <*=>
1961 N5 P12-13 61-27969 <*=> 0
1961 N6 P6-7 62-13385 <=*>

ZOLNIERZ POLSKI. WARSAW
1963 N8 P16-17 63-21464 <=> 0

ZOOLOGICAL MAGAZINE. TOKYO
SEE DOBUTSUGAKU ZASSHI

ZOOLOGICHESKII ZHURNAL
1934 V13 N3 P540-582 MAFF-TRANS-NS-42 <NLL>
1936 N2 P191-198 62-32832 <=*>
1937 V16 N2 P247-278 61-31006 <=>
1937 V16 N3 P511-546 PANSDOC-TR.406 <PANS>
1938 V17 N1 P175-179 59-20957 <+*>
1938 V17 N3 P540-548 TT.183 <NRC>
1940 V19 N2 P333-335 63-19242 <=*>
1940 V19 N3 P445-455 C-2622 <NRC>
1942 V21 N6 P282-283 RTS-3150 <NLL>
1942 V21 N1/2 P10-18 65-63251 <=> 0
1943 V22 N2 P214-221 63-23175 <*>
1943 V22 N3 P178-192 TT.184 <*>
1943 V22 N4 P214-221 RT-2123 <*>
63-23175 <=*$>
1944 V23 N4 P146-155 61-13042 <=>
1944 V23 N2/3 P98-101 RT-182 <*>
1946 V25 N5 P451-457 C-2421 <NRC>
R-5072 <*>
1946 V25 N55 P451-457 R-1844 <*>
1949 V28 N1 P7-38 62-25671 <=*>
1949 V28 N1 P71-78 RTS-2995 <NLL>
1949 V28 N5 P419-420 61-00460 <*>
1949 V28 N5 P479-481 64-15652 <=*$>
1949 V28 N6 P523-526 63-23406 <=*$>
1950 V29 N6 P489-500 RT-1882 <*>
1950 V29 N6 P523-529 R-3861 <*>
1950 V29 N6 P566-568 62-14901 <=*>
1951 V30 N2 P143-148 63-19148 <=*>
1951 V30 N3 P211-216 61-28117 <*=>
1951 V30 N6 P523-539 60-19929 <+*>

1952 N2 P169-174	R-1058 <*>
1952 V31 N2 P292-296	65-60936 <=$>
1952 V31 N2 P297-304	RT-529 <*>
1952 V31 N4 P523-527	RT-2401 <*>
1953 V32 N1 P105-111	C-2619 <NRC>
1953 V32 N3 P495-505	59-16984 <+*>
1953 V32 N5 P955-963	65-50046 <=$>
1953 V32 N6 P1052-1063	RT-3397 <*>
1953 V32 N6 P1211-1216	RT-3622 <*>
1954 V33 N1 P14-25	R-5071 <*>
1954 V33 N1 P65-68	R-92 <*>
	RT-2490 <*>
	64-15064 <=*$>
1954 V33 N1 P69-76	R-91 <*>
	RT-2573 <*>
	64-15063 <=*$> 0
1954 V33 N2 P243-268	64-15623 <=*$>
1954 V33 N2 P356	AD-625 644 <=$>
1954 V33 N3 P677-692	R-5077 <*>
1954 V33 N4 P755-768	R-4154 <*>
1954 V33 N4 P841-847	63-41316 <=>
1954 V33 N5 P1053-1057	62-01508 <*>
	64-15629 <=*$> 0
1954 V33 N5 P1116-1125	62-00213 <*>
1954 V33 N6 P1201-1205	60-15535 <+*>
1954 V33 N6 P1312-1324	60-21099 <=>
1954 V33 N6 P1325-1335	RT-3386 <*>
	60-14294 <+*>
1954 V33 N6 P1403-1409	R-4876 <*>
1955 V34 P291-294	T2215 <INSD>
1955 V34 P965-974	GB146 <NLL>
1955 V34 N2 P371-379	R-1016 <*>
	RT-3304 <*>
1955 V34 N2 P399-407	64-00291 <*>
	64-15714 <=*$>
	65-63990 <=$>
1955 V34 N3 P631-639	R-4879 <*>
	59-16551 <+*>
1955 V34 N3 P640-651	R-4147 <*>
1955 V34 N4 P800-805	RT-3888 <*>
1955 V34 N4 P948-950	CSIRO-3294 <CSIR>
	R-3951 <*>
	64-15499 <=*$> 0
1955 V34 N5 P1037-1051	CSIRO-3162 <CSIR>
	R-4557 <*>
	60-14037 <+*> 0
1955 V34 N6 P1189-1202	61-28222 <*=> 0
1955 V34 N6 P1202-1223	C-2502 <NRC>
1955 V34 N6 P1210-1223	60-17108 <+*>
1956 V35 P709-718	R-2119 <*>
1956 V35 P1450-1453	59-14865 <+*> 0
1956 V35 N1 P29-31	CSIRO-3313 <CSIR>
	R-3926 <*>
1956 V35 N1 P32-44	62-15633 <=*>
1956 V35 N1 77-88	-3793 <*>
1956 V35 N2 P266-274	R-3630 <*>
	3073 <NLL>
1956 V35 N2 P290-299	59-22702 <+*>
	59-22706 <+*> 0
1956 V35 N3 P379-383	CSIRO-3379 <CSIR>
	R-3904 <*>
	60-14036 <+*>
1956 V35 N3 P384-391	CSIR-TR-436 <CSSA>
1956 V35 N4 P511-528	R-3905 <*>
1956 V35 N4 P614-615	R-3912 <*>
1956 V35 N5 P781-783	60-23597 <+*>
1956 V35 N6 P868-875	62-15626 <=*> 0
1956 V35 N9 P1281-1298	60-16679 <*>
1956 V35 N9 P1415-1417	64-15497 <=*$> 0
1956 V35 N10 P1441-1449	R-5273 <*>
1956 V35 N11 P1601-1607	60-51152 <=>
1956 V35 N11 P1617-1625	59-19599 <+*>
1956 V35 N11 P1681-1684	R-5089 <*>
1956 V35 N11 P1728-1730	60-51139 <=>
1956 V35 N11 P1747-1764	63-31207 <=> 0
1956 V35 N12 P1780-1790	R-4130 <*>
1956 V35 N12 P1849-1852	60-16683 <=*>
1956 V35 N12 P1902	62-23315 <=*>
1957 V36 P548-560	61-23975 <=*>
1957 V36 P938-945	61-23976 <=*>
1957 V36 N2 P204-213	R-4132 <*>
1957 V36 N4 P617-620	CSIR-TR.293 <CSSA>
	59-15307 <+*> 0
1957 V36 N4 P620-622	CSIR-TR.294 <CSSA>
1957 V36 N6 P801-819	60-17792 <+*>
1957 V36 N6 P864-869	R-4728 <*>
	62-01506 <*>
	64-15630 <=*$> 0
	64-19469 <=$> 0
1957 V36 N6 P870-873	65-63999 <=$> 0
1957 V36 N9 P1420-1421	61-19193 <+*>
1958 V37 P542-547	61-23715 <*=>
1958 V37 P939-942	61-15037 <+*> 0
1958 V37 P1012-1023	62-15426 <=*>
1958 V37 N1 P41-56	C-3826 <NRCC>
	62-00173 <*>
1958 V37 N3 P474-475	63-19228 <=*$>
1958 V37 N4 P584-593	65-62669 <=$>
1958 V37 N6 P820-831	RTS-2998 <NLL>
1958 V37 N6 P838-845	62-32833 <=*>
1958 V37 N8 P1252-1253	59-13443 <=>
1958 V37 N9 P1345-1351	62-15624 <=*> 0
1958 V37 N9 P1409-1411	62-33567 <=*> 0
1958 V37 N9 P1412-1416	63-23141 <=*$>
1958 V37 N10 P1433-1448	63-19073 <=*>
1958 V37 N10 P1521-1530	59-18307 <+*>
1958 V37 N10 P1563-1568	63-41317 <=>
	65-63151 <=*$>
1958 V37 N10 P1568-1571	59-15830 <+*>
1959 V38 N1 P3-14	RTS-2997 <NLL>
1959 V38 N1 P123-126	63-23338 <=*$>
1959 V38 N2 P182-187	59-20947 <+*>
1959 V38 N2 P208-220	62-15638 <=*> 0
1959 V38 N2 P228-242	59-20948 <+*>
1959 V38 N3 P362-377	65-60507 <=$>
1959 V38 N3 P465-468	63-15159 <=*> 0
1959 V38 N5 P734-744	C-3895 <NRCC>
1959 V38 N5 P781-782	60-17591 <+*>
1959 V38 N7 P970-986	60-23660 <+*>
1959 V38 N7 P1060-1068	61-28223 <*=> 0
1959 V38 N8 P1214-1229	62-15642 <=*> 0
1959 V38 N8 P1268-1269	61-31043 <=>
1959 V38 N9 P1282-1291	63-15937 <=*>
1959 V38 N9 P1383-1387	64-15642 <=*$>
1959 V38 N9 P1426-1428	M.5319 <NLL>
1959 V38 N10 P1537-1552	61-15301 <+*>
1959 V38 N10 P1537-1553	61-15468 <+*>
1959 V38 N11 P1634-1648	63-15149 <=*>
1959 V38 N11 P1643-1662	63-15145 <=*>
1959 V38 N11 P1663-1673	61-19534 <=*>
1959 V38 N11 P1683-1688	61-27451 <=*> 0
1959 V38 N11 P1689-1701	C-3768 <NRCC>
1959 V38 N11 P1741-1744	65-64365 <=*$>
1959 V38 N12 P1797-1805	61-28648 <*=>
1959 V38 N12 P1886-1887	60-31098 <=>
1959 V38 N12 P1889-1891	64-15986 <=*$>
1959 V38 N12 P1894-1896	63-21729 <=>
1959 V38 N12 P1907-1911	60-31099 <=>
1960 V39 P768-769	61-28651 <*=>
1960 V39 N1 P45-52	62-32240 <=*>
1960 V39 N1 P53-62	61-19535 <=*>
1960 V39 N2 P176-187	61-21163 <=>
1960 V39 N2 P244-249	61-28763 <*=>
1960 V39 N2 P263-276	64-21255 <=>
1960 V39 N2 P318-320	60-41397 <=>
1960 V39 N3 P424-428	61-21137 <=>
1960 V39 N3 P429-432	61-31050 <=>
1960 V39 N3 P478-479	60-41233 <=>
1960 V39 N4 P481-489	C-3772 <NRCC>
1960 V39 N4 P494-499	FRBC TRANS SER N364 <NRCC>
	61-27576 <*=> 0
1960 V39 N4 P509-513	62-32484 <=*>
1960 V39 N5 P777-779	C-4385 <NRC>
1960 V39 N5 P794-796	60-41709 <=>
1960 V39 N7 P1056-1061	63-23140 <=*>
1960 V39 N7 P1062-1068	61-15967 <+*>
1960 V39 N7 P1080-1087	64-13371 <=*$>
1960 V39 N7 P1096-1099	64-15650 <=*$>
1960 V39 N8 P1169-1173	64-13204 <=*$>
1960 V39 N8 P1174-1178	61-28771 <*=>

1960 V39 N8 P1218-1231 RTS-2685 <NLL>
1960 V39 N9 P1446-1448 61-27366 <*=>
1960 V39 N10 P1525-1530 63-11110 <=>
1960 V39 N11 P1747-1750 63-11117 <=>
1960 V39 N12 P1774-1778 63-15376 <=*>
1960 V39 N12 P1779-1782 62-13709 <*=>
1961 V40 N1 P60 61-27059 <=*>
1961 V40 N1 P122-128 63-11106 <=>
64-13268 <=*$>
1961 V40 N2 P159-163 61-28435 <*=>
1961 V40 N2 P258-263 61-28434 <*=> 0
1961 V40 N3 P351-357 62-24297 <=*>
1961 V40 N3 P434-446 62-13535 <*=>
1961 V40 N4 P616-619 63-23132 <=*>
1961 V40 N5 P776-777 CSIR-TR 244 <CSSA>
1961 V40 N6 P818-825 M.5316 <NLL>
1961 V40 N6 P953-955 62-24401 <=>
1961 V40 N7 P1058-1069 64-13306 <=*$>
1961 V40 N7 P1070-1078 64-15644 <=*$>
1961 V40 N9 P1299-1303 62-32829 <=*>
63-21304 <=>
1961 V40 N9 P1314-1334 62-32830 <=*>
1961 V40 N9 P1354-1363 66-61174 <=*$>
1961 V40 N9 P1384-1394 63-21304 <=>
1961 V40 N11 P1619-1623 C-4942 <NRC>
1961 V40 N11 P1656-1664 65-50052 <=>
1961 V40 N12 P1815-1826 64-15621 <=*$> 0
1961 V40 N12 P1868-1873 62-32488 <=*>
1961 V40 N12 P1883-1891 64-13140 <=*$>
1962 V41 N1 P24-57 63-31138 <=>
1962 V41 N1 P58-62 CSIR-TR 309 <CSSA>
1962 V41 N1 P85- FPTS V22 N5 P.T825 <FASE>
1962 V41 N1 P85-91 1963,V22,N5,PT2 <FASE>
63-31138 <=>
1962 V41 N1 P155-160 63-31138 <=>
1962 V41 N3 P358-363 64-00287 <*>
64-15711 <=*$>
65-63980 <=*$> 0
1962 V41 N5 P666-674 64-15640 <=*$>
1962 V41 N5 P797-800 63-13468 <=*>
1962 V41 N6 P905-912 64-15653 <=*$>
1962 V41 N7 P969-985 63-31455 <=>
1962 V41 N7 P1004-1012 63-31457 <=>
1962 V41 N7 P1056-1060 63-31454 <=>
1962 V41 N7 P1061-1066 62-33422 <=*>
1962 V41 N8 P1132-1141 63-31510 <=>
1962 V41 N8 P1162-1165 64-15649 <=*$>
1962 V41 N8 P1203-1209 AD-623 256 <=*$>
1962 V41 N8 P1210-1219 63-31510 <=>
1962 V41 N8 P1220-1225 62-01504 <*>
64-15632 <=*$>
1962 V41 N10 P1477-1487 65-60348 <=$>
1962 V41 N10 P1516-1528 63-21396 <=>
1962 V41 N10 P1575-1578 66-61172 <=$*>
1962 V41 N11 P1618-1630 63-21238 <=>
1962 V41 N11 P1749-1751 63-21238 <=>
1962 V41 N12 P1875-1882 C-4699 <NRC>
1963 V42 N1 P3-10 63-21502 <=>
1963 V42 N1 P20-31 63-21502 <=>
1963 V42 N1 P78-91 63-21502 <=>
1963 V42 N1 P138-139 63-21502 <=>
1963 V42 N2 P161-196 63-21609 <=>
1963 V42 N2 P256-267 63-21609 <=>
1963 V42 N2 P274-281 64-19239 <=*$>
1963 V42 N3 P338-344 63-21955 <=>
1963 V42 N3 P345-358 64-19883 <=$>
65-63521 <=*>
1963 V42 N3 P400-407 64-13131 <=*$> 0
1963 V42 N3 P473-475 63-21955 <=>
1963 V42 N3 P477-478 63-21955 <=>
1963 V42 N4 P500-505 66-61538 <=*$>
1963 V42 N4 P513-517 64-15643 <=*$>
1963 V42 N4 P546-550 63-31306 <=>
1963 V42 N4 P609-617 64-13724 <=*$>
1963 V42 N4 P618-620 63-31306 <=>
1963 V42 N4 P637-638 63-31306 <=>
1963 V42 N5 P641-651 63-31447 <=>
1963 V42 N5 P704-715 63-31519 <=>
1963 V42 N5 P752-762 63-31520 <=>
1963 V42 N6 P801-808 63-31649 <=>
1963 V42 N6 P947-949 63-31650 <=>
1963 V42 N6 P967 63-31661 <=>
1963 V42 N8 P1161-1173 66-61543 <=*$>
1963 V42 N8 P1187-1199 64-21131 <=>
1963 V42 N8 P1232-1244 64-21135 <=>
1963 V42 N8 P1257-1260 AD-618 653 <=$>
1963 V42 N8 P1260-1264 AD-618 654 <=$>
1963 V42 N8 P1283-1285 64-21134 <=>
1963 V42 N9 P1329-1337 63-41104 <=>
1963 V42 N10 P1465- FPTS,V23,N5,P.T945 <FASE>
1963 V42 N10 P1472-1483 64-19707 <=$> 0
1963 V42 N10 P1546-1553 65-63526 <=*$>
1963 V42 N11 P1631-1636 CSIR-TR-409 <CSSA>
1963 V42 N12 P1849-1855 64-21834 <=>
1963 V42 N12 P1866-1869 64-21834 <=>
1963 V42 N12 P1892-1901 64-21834 <=>
1964 V43 N1 P3-15 64-21789 <=>
1964 V43 N1 P138-139 65-63985 <=$>
1964 V43 N1 P149-150 65-60662 <=$>
1964 V43 N1 P154-158 64-21789 <=>
1964 V43 N2 P161-171 64-31081 <=>
1964 V43 N2 P172-177 64-31463 <=>
1964 V43 N2 P214-219 65-60179 <=$>
1964 V43 N2 P272-273 64-31081 <=>
1964 V43 N2 P282-284 65-63988 <=$>
1964 V43 N2 P292-293 64-31081 <=>
1964 V43 N2 P310-317 64-31081 <=>
1964 V43 N3 P346-354 64-41631 <=>
1964 V43 N3 P385-397 64-41631 <=>
1964 V43 N3 P418-423 65-63767 <=>
1964 V43 N4 P481-482 64-31701 <=>
1964 V43 N4 P502 64-31701 <=>
1964 V43 N4 P518-524 64-31701 <=>
1964 V43 N4 P549-559 64-31701 <=>
1964 V43 N4 P581-589 CSIR-TR.448 <CSSA>
1964 V43 N5 P647-651 64-41436 <=$>
1964 V43 N5 P647-650 65-63422 <=$>
1964 V43 N5 P720- FPTS V24 N3 P.T390 <FASE>
1964 V43 N5 P720-734 64-41436 <=$>
1964 V43 N5 P788-793 64-41436 <=$>
1964 V43 N6 P801-807 64-41086 <=>
1964 V43 N6 P809-814 64-41086 <=>
1964 V43 N7 P953-965 65-62312 <=$>
1964 V43 N7 P1077- FPTS V24 N4 P.T683 <FASE>
1964 V43 N7 P1084- FPTS V24 N5 P.T936 <FASE>
1964 V43 N8 P1105-1119 64-51184 <=>
1964 V43 N8 P1173-1181 64-51182 <=>
1964 V43 N8 P1243-1246 64-51598 <=$>
1964 V43 N9 P1355-1360 65-63155 <=$>
1964 V43 N9 P1391-1394 65-63897 <=*$>
1964 V43 N11 P1638-1647 65-30176 <=$>
1964 V43 N12 P1826-1837 65-64539 <=*$>
1965 V44 N2 P192- FPTS V25 N2 P.T217 <FASE>
1965 V44 N2 P192-197 65-30943 <=$>
1965 V44 N4 P626-630 65-31243 <=$>
1965 V44 N8 P1131-1138 66-30816 <=$>
1965 V44 N9 P1291-1307 66-30814 <=$>
1965 V44 N9 P1423-1424 66-61602 <=*$>
1965 V44 N9 P1437-1438 66-30814 <=$>
1965 V44 N10 P1585-1588 65-34071 <=*$>
1966 V45 N2 P245-260 66-31652 <=$>

ZOOLOGICKE A ENTOMOLOGICKE LISTY
PHYSIOLOGIE DER TIERE
1954 V3 N1 P37-46 60-14300 <*>

ZOOLOGISCHE JAHRBUECHER. ABTEILUNG FUER ANATOMIE UND ONTOGENIE DER TIERE
1908 V25 P637-674 C-2302 <NRC>
1921 V42 N2 259-282A 57-3416 <*>
1955 V74 N3 P339-490 60-15643 <+*> 0

ZOOLOGISCHE JAHRBUECHER. ABTEILUNG FUER SYSTEMATIK, OEKOLOGIE UND GEOGRAPHIE DER TIERE
1904 V20 P131-138 59-20876 <*>
1950 V78 P531-546 65-12360 <*> 0

ZOOLOGISCHER ANZEIGER
1938 V124 N7 P182-190 C-2567 <NRC>

ZOOLOGISCHES ZENTRALBLATT
1899 V6 P65-67 T-1724 <INSD>

ZOOPROFILASSI:RIVISTI MENSILE DI SCIENZA E TECNICA VETERNIARIA
1949 V4 P127-135 57-2013 <*>
1949 V28 N5 P419-420 64-15618 <=*$>

ZOSEN KYCKAI RONBUNSHU
1951 V54 P779-781 779 <TC>
1954 V57 P379-382 780 <TC>
1960 V106 P275-284 62-26010 <=$>
1961 N109 P185- 2159 <BSRA>
1961 N110 P43- 2159 <BSRA>
1961 V110 P31- 2165 <BSRA>
1963 N113 P201- 2092 <BSRA>
1963 N114 P169- 2032 <BSRA>
1963 V114 P218- TRANS-1885 <NLL>
1963 V114 P230 1894 <BSRA>
1963 V114 P242 1861 <BSRA>
1964 N115 P49- 2068 <BSRA>
1964 N115 P56- 2032 <BSRA>
1964 N115 P175-194 2054 <BSRA>
1964 N115 P195 2073 <BSRA>
1964 N115 P618- 2056 <BSRA>
1964 N116 P154 2076 <BSRA>
1964 V115 P218- 2091 <BSRA>
1965 N116 P154- 2139 <BSRA>
1965 N117 P20- 2265 <BSRA>
1965 N117 P85- 2267 <BSRA>
1965 N117 P93- 2259 <BSRA>
1965 N117 P256- 2272 <BSRA>
1965 N117 P292- 2271 <BSRA>

ZPRAVODAJ VYZKUMNEHO A ZKUSEBNIHO LETECKEHO USTAVU. LETNANY, CZECHOSLOVAKIA
1963 N6 P9-15 AD-630 284 <=*$>

ZUCKER
1950 V3 P240-241 63-16358 <*>
1953 V6 P126-133 63-14861 <*> O
1953 V6 N5 P92-94 66-10215 <*> O
1954 V7 P309-311 63-14860 <*> O
1954 V7 P407-410 63-16365 <*>
1955 V8 P129-134 63-16363 <*> O
1956 V8 N7 P129-134 1621 <*>
59-17638 <*>
1956 V9 P575-578 63-16360 <*> O
1957 V10 P78-85 63-16362 <*> O
1958 V11 N8 P182-187 63-16366 <*>

ZUCKER-BEIHEFTE. SUPPLEMENT
1950 N1 P1-3 63-16367 <*>
1962 V4 N3 P72-76 620 <TC>

ZUCKER-UND SUESSWARENWIRTSCHAFT
1954 V7 P796-800 63-14859 <*>
1957 V10 P787-788 59L32G <ATS>
1960 V13 N10 P487-488 64M43G <ATS>

ZUECHTER
1931 V3 N2 P33-38 5389-B <K-H>
1934 V6 P25-30 58-851 <*>
1942 V14 P112-117 AL-666 <*>
1947 N17/8 P161-223 AL-667 <*>
1949 V13 N8/9 P247-254 59-15464 <*>
1952 V23 N10/1 P327-334 2069 <*>
1954 V24 P252-256 57-707 <*>
1954 V24 P305-306 II-784 <*>
1954 V24 P313-316 T-1897 <INSD>
1954 V24 P346-352 57-2569 <*>
1954 V24 N11/2 P313-316 59-10540 <*>
1957 V27 N2 P69-76 58-2247 <ATS>
1957 V27 N2 P85-89 59-20152 <*>
1957 V27 N5 P223-231 NP-TR-848 <*>
1958 V28 P37-40 61-20506 <*> O
1958 V28 P161-166 61-20428 <*>
1959 V29 N4 P187-191 61-20427 <*>
1960 V30 N6 P247-250 65-12701 <*>
1961 V31 N6 P263-265 12900 A <K-H>
63-14392 <*>

ZUECHTUNGSKUNDE
1934 V9 P444-452 II-230 <*>
1948 V20 P29-39 60-10815 <*>
1951 V22 P204-215 64-10004 <*>
1951 V23 P93-101 60-18024 <*>
1952 V23 N4 P141-150 64-10029 <*>
1952 V23 N6 P263-273 64-10040 <*>
1955 V26 P257-264 65-12527 <*>
1956 V28 P333-336 CSIRO-3707 <CSIR>
1956 V28 P423-425 CSIRO-3706 <CSIR>
1962 V34 N7 P296-306 65-11698 <*>

ZVARACKY SBORNIK
1958 V7 N3 P312-327 TRANS-64 <MT>
1960 V9 N1 P51-68 TRANS-196 <MT>
1961 V10 N4 P343-352 3878 <BISI>

ZVARANIE
1957 V6 N3 P327-346 426 <MT>
1957 V6 N4 P105-109 427 <MT>
1957 V6 N11 P327-329 428 <MT>
1958 V7 N1 P8-15 TRANS-66 <MT>
1958 V7 N6 P161-168 TRANS-67 <MT>
1958 V7 N12 P363-368 TRANS-70 <MT>
1958 V7 N9/10 P284-286 TRANS-68 <MT>
1958 V7 N9/10 P288-289 TRANS-69 <MT>
1958 V7 N9/10 P307-314 429 <MT>
1959 V8 N9 P258-261 TRANS-112 <MT>
1960 V9 P177-178 2299 <BISI>
1960 V9 P242-245 2227 <BISI>
1960 V9 N3 P78-80 2487 <BISI>
1960 V9 N4 P103-107 TRANS-113 <MT>
1961 V10 N4 P104-106 2943 <BISI>

ZVEZDA
1958 N10 P154-162 61-19315 <+*>
1964 N2 P130-138 65-60397 <=$>
1966 N1 P167-177 66-31297 <=$>

ZWANGLOSE ABHANDLUNGEN AUS DEM GEBIET DER NORMALEN UND PATHOLOGISCHEN ANATOMIE
1959 N5 P1-7 63-10528 <*> O

ZYCIE GOSPODARCZE
1964 V19 N25 P1-6 64-31971 <=>
1964 V19 N40 P4 64-51641 <=$>
1965 V20 N22 P1 65-31404 <=$>
1965 V20 N22 P3 65-31404 <=$>
1965 V20 N22 P6-8 65-31404 <=$>

ZYCIE SZKOLY WYZSZEJ
1959 V7 N4 P8-14 63-21334 <=>
1961 V9 N6 P3-15 62-19530 <*=>

ARGENTINA

123,307 63-14336 <*>
154,513 60-16675 <+*>

AUSTRIA

1,732 I-198 <*>
2,709 64-19910 <=$>
5,196 63-14825 <=*>
13,116 59-17886 <+*>
28,586 65-60271 <=$>
31,354 ADD 63-13739 <=*> 0
34,197 60-15540 <+*> 0
34,243 63-16306 <=*>
35,642 59-16829 <+*> 0
40,410 64-30160 <*>
45,502 59-12491 <+*> 0
48,984 65-12025 <*>
55,275 59-12069 <+*> 0
63,959 63-13739 <=*> 0
70,348 64-30161 <*>
75,862 65J18G <ATS>
83,892 64-30159 <*> 0
88,667 63-14337 <=*>
89,575 65-64755 <=*$>
89,789 63-14338 <=*>
96,185 732 <TC>
99,179 59-12070 <+*>
99,583 63-23559 <=*$> 0
102,250 60-18899 <=*$>
112,237 62-16403 <=*>
119,959 AL-308 <*>
63-20738 <=*$>
122,486 I-686 <*>
124,286 63-16699 <=*>
125,471 1285 <TC>
125,931 63-15437 <=*>
125,943 59-12071 <+*>
128,301 62-18294 <=*>
129,236 62-18293 <=*>
129,462 59-12492 <+*> 0
136,544 62-32926 <=*> 0
138,020 59-14201 <+*> 0
138,779 64-19791 <=$> 0
143,021 59-12072 <+*>
143,482 1284 <TC>
149,470 <LSA>
150,822 64-20724 <*>
151,541 59-22271 <+*> 0
159,405 61-27457 <*=> 0
162,142 62-33240 <=*> 0
162,445 8640 <IICH>
163,215 59-14202 <+*>
164,263 63-14934 <=*>
164,714 64-19169 <=*$> 0
165,319 61-15676 <+*>
167,432 63-19267 <=*>
167,455 63-19071 <=*> 0
167,974 3437 <*>
168,059 62-16239 <=*>
168,357 63-15778 <=*> 0
168,640 63-19268 <=*>
170,254 1603 <TC>
170,626 AL-24 <*>
170,634 59-15737 <+*>
171,023 63-24401 <=*$> 0
171,168 64-20057 <*>
171,420 98N47G <ATS>
173,467 548 <TC>
174,207 64-20067 <*>
176,205 64-20428 <*>
176,414 62-16402 <=*>
176,609 62-16420 <=*>
176,931 1211 <*>
177,661 59-12073 <+*>
179,905 61-15247 <+*>
179,953 61-15043 <*+>
180,122 60-16464 <+*> 0
180,680 61-23986 <*=> 0
181,405 59-12915 <+*> 0
181,431 63-13740 <=*>
181,478 237 <TC>
183,532 61-19477 <+*> 0
183,758 63-18899 <=*$>
63-18900 <=*>
63-23769 <=*$>
185,832 59-22270 <+*>
189,475 4TK26G <ATS>
190,535 64-30158 <*> 0
192,824 60-17598 <+*> 0
65-10526 <*>
193,015 61-23214 <=*> 0
193,363 64-20725 <*>
193,396 62-14273 <=*>
195,073 63-24402 <=*$> 0
197,585 58-2456 <*>
198,061 8744 <IICH>
199,193 63-23770 <=*$>
199,376 64-10636 <*>
199,793 65-13709 <*>
199,794 65-13708 <*>
200,578 65-60274 <=$>
200,712 65-11921 <*>
202,685 63-24403 <=*$> 0
203,934 63-15777 <=*> 0
205,985 65-10597 <*>
207,301 60-23133 <+*>
207,827 65-64486 <=*$>
208,223 64-14966 <*>
208,575 61-23987 <*=> 0
210,132 1089 <TC>
212,023 NS-9 <TTIS>
212,564 NS-11 <TTIS>
214,632 842 <TC>
215,609 63-23162 <=*$> 0
217,008 66-13976 <*>
217,009 66-13977 <*>
217,585 60-15548 <+*> 0
223,668 65 11901 <*>
242,673 66-14004 <*> 0
282,609 I-478 <*>

BELGIUM

404,537 63-14512 <=*>
433,112 65-10583 <*> 0
440,360 64-30167 <*>
442,575 65-10580 <*>
445,724 64-20717 <*>
445,907 64-20529 <*>
445,946 64-20827 <*>
446,643 61-16509 <*=>
447,044 65-10582 <*>
448,187 63-14339 <=*>
449,653 65-10528 <*>
454,113 63-14364 <=*> 0
454,540 65-10581 <*>
485,904 13G5F <ATS>
487,688 63-14630 <=*>
491,132 63-18963 <=*>
493,580 I-887 <*>
494,237 61-28953 <*=> 0
496,600 63-18959 <=*>
496,841 63-18018 <*=>
496,995 59-12916 <+*>
497,236 31S83F <ATS>
497,868 59-12074 <+*> 0
498,380 77G8F <ATS>
498,777 6H10F <ATS>
499,917 63-14341 <=*>
500,222 60-17599 <+*>
500,535 59-12310 <+*>
64-14673 <=*$>

Patent	Citation
500,644	I-886 <*>
501,638	64-30165 <*>
502,159	07H10F <ATS>
503,546	194 <TC>
	62-16319 <=*>
503,549	65-10553 <*>
504,274	63-18990 <=*>
504,423	63-18017 <*=>
504,852	64-30164 <*>
505,421	64-30163 <*>
506,664	61-23215 <=*> C
506,991	8754 <IICH>
508,016	60-17600 <+*>
508,450	64-30170 <*> O
508,544	AL-917 <*>
	64-10112 <=*$>
509,732	64-20514 <*>
510,151	60-17601 <+*> O
510,282	63-14342 <=*>
510,283	63-14343 <=*>
510,284	63-14345 <*>
510,371	I-508 <*>
	64-10132 <=*$>
510,816	I-415 <*>
	64-10152 <=*$>
512,208	1661 <*>
	64-10470 <=*$>
512,267	63-16813 <=*>
513,216	64-16439 <=*$>
513,838	64-16442 <=*$>
516,623	64-20718 <*>
517,042	62-16334 <=*> O
517,560	2727 <*>
518,215	59-14687 <+*> O
518,938	64-30174 <*>
519,547	8745 <IICH>
520,196	59-18282 <+*>
520,338	60-14823 <+*>
520,445	60-14824 <+*>
520,677	65-60273 <=$>
520,694	64-13782 <=*$> C
521,159	63-16081 <*=>
521,244	63-20395 <=*>
523,230	65-62858 <=$> O
523,438	65-62601 <=$> O
524,931	63-13741 <=*> O
525,162	65-30840 <*>
525,519	98H8F <ATS>
525,535	64-30841 <*>
525,585	43P61F <ATS>
527,135	60-15857 <+*> O
527,736	64-20206 <*> P
	96372 <RIS>
530,617	62-16336 <=*>
532,316	62-25676 <=*> O
532,474	63-24615 <=*$> C
532,543	64-20474 <*>
	65-63401 <=$>
533,252	60-14825 <+*>
533,362	62-16335 <=*>
	63-10871 <=*>
	64-30828 <*>
	65-11854 <*>
	96369 <RIS>
533,428	64-30173 <*>
534,212	60-14674 <+*>
534,792	65-11855 <*>
	96370 <RIS>
534,888	64-20992 <*>
	96371 <RIS>
534,888 SUP	96866 <RIS>
535,043	59-16364 <+*> O
535,082	1476 <*>
535,235	64-30886 <*>
	96373 <RIS>
535,727	60-14826 <+*>
536,061	90N57F <ATS>
536,855	62-16337 <=*>
536,877	62-16338 <=*> O
537,091	62-16339 <=*$> O
537,792	65-11856 <*>
538,087	60-14827 <+*>
538,241	62-16340 <=*>
538,768	62-16341 <=*>
538,782	64-16146 <=*$>
	96480 <RIS>
538,815	62-16342 <=*>
538,911	60-14675 <+*>
539,197	47H10F <ATS>
539,338	60-16449 <+*> P
539,548	<AC>
540,449	61-18356 <*=>
540,459	96542 <RIS>
540,575	60-14828 <+*>
540,863	65-10584 <*> C
541,504	63-14346 <=*>
542,094	63-10886 <=*> P
542,113	60-14829 <+*>
542,366	89J15DU <ATS>
542,658	64-16459 <=*$>
	96587 <RIS>
542,914	65-13556 <*>
542,915	65-11824 <*>
543,082	64-16452 <=*$>
543,083	64-16458 <=*$>
543,259	96588 <RIS>
543,292	64-18363 <*>
	65-62859 <=$> O
543,452	65-10569 <*> C
543,576	64-20490 <*>
	96593 <RIS>
543,888	96371 <RIS>
543,913	96866 <RIS>
544,234	60-14830 <+*>
544,418	64-30862 <*>
	96592 <RIS>
544,419	64-16457 <=*$>
	96590 <RIS>
544,471	64-16450 <=*$>
	64-30863 <*>
	96591 <RIS>
544,518	54021 <RIS>
544,576	96593 <RIS>
544,599 ADD	45R75F <ATS>
544,829	96589 <RIS>
545,087	96638 <RIS>
545,326	64-30937 <*>
545,376	64-30938 <*>
	96639 <RIS>
545,952	96634 <RIS>
546,150	96635 <RIS>
546,151	64-20469 <*>
	96633 <RIS>
546,216	96637 <RIS>
546,376	64-20472 <*>
546,377	64-20476 <*>
546,432	96644 <RIS>
546,474	96642 <RIS>
546,742	64-30939 <*>
	96641 <RIS>
546,808	96643 <RIS>
546,843	96645 <RIS>
546,848	60-14831 <+*>
546,893	65-60261 <=$>
546,893 ADD	64-13781 <=*$> C
547,019	96636 <RIS>
547,184	96640 <RIS>
547,259	60-17983 <+*> C
547,618	65-11858 <*>
547,874	65-10538 <*>
547,992	63-13742 <=*> O
548,025	65-10537 <*>
548,145	60-14676 <+*>
548,250	64-18365 <*>
548,485	63-15776 <=*> O
548,547	60-14832 <+*>
548,895	96756 <RIS>
548,927	96758 <RIS>
549,055	65-60836 <=$>
549,412	65-11261 <*>

549,489	65-10656 <*> 0
549,553	96755 <RIS>
549,554	96757 <RIS>
549,629	96761 <RIS>
549,638	96753 <RIS>
549,767	57-3373 <*>
549,891	96751 <RIS>
549,950	96759 <RIS>
549,951	96760 <RIS>
550,093	96752 <RIS>
550,396	66-10627 <*>
550,412	96762 <RIS>
550,413	96754 <RIS>
550,429	65-11559 <*>
550,495	62-10135 <*=>
550,506	22K20F <ATS>
550,573	62-14377 <=*>
	65-13555 <*>
550,605	72J17F <ATS>
551,276	<RIS>
	64-30823 <*>
	96779 <RIS>
551,295	60-14175 <+*>
551,308	61-10619 <+*>
551,319	<RIS>
551,488	64-18381 <*>
551,494	65-13262 <*>
551,526	64-13781 <=*$> 0
551,633	65-62600 <=$> 0
551,788	65-11378 <*>
551,789	<RIS>
551,790	<RIS>
551,905	<RIS>
	64-13780 <=*$> 0
551,993	<RIS>
551,997	<RIS>
552,060	65-10568 <*> 0
552,256	65-61499 <=$>
552,351	<LSA>
552,401	<RIS>
552,462	<RIS>
552,644	57-3331 <*>
553,174	96844 <RIS>
553,277	96839 <RIS>
553,655	96848 <RIS>
553,694	64-19792 <=$>
553,792	64-20719 <*> P
553,815	76J19F <ATS>
554,242	49N51F <ATS>
	96851 <RIS>
554,409	60-14834 <+*>
554,435	63-19070 <=> 0
554,506	63-16334 <=*>
554,685	96843 <RIS>
554,919	60-14835 <+*>
555,097	64-30110 <*>
555,180	65-10316 <*>
555,218	64-30168 <*>
555,219	64-30169 <*>
555,252	96850 <RIS>
555,277	62-10529 <*=>
555,380	63-24404 <=*$>
555,618	62-16248 <=*>
555,759	23K20F <ATS>
556,105	10L37F <ATS>
	61-10827 <*>
556,902	96849 <RIS>
556,974	618 <TC>
557,123	60-14836 <+*>
557,766	64-30838 <*>
557,830	65-10655 <*>
557,831	65-10654 <*>
557,832	64-30839 <*>
557,833	65-10658 <*>
557,834	65-10657 <*>
557,853	96900 <RIS>
558,502	60-14837 <+*>
558,563	96902 <RIS>
558,616	63-15775 <=*> 0
559,228	REPT. 96956 <RIS>
559,298	59-15219 <+*>
559,497	96905 <RIS>
559,602	63-24614 <=*$>
559,808	65-64756 <=*$>
559,842	96960 <RIS>
559,884	563 <TC>
560,234	REPT. 96983 <RIS>
560,446	63-14347 <=*>
560,450	60-14838 <+*>
560,503	60N57F <ATS>
560,522	REPT. 96985 <RIS>
560,523	96909 <RIS>
560,648	63-10882 <=*>
560,782	63-14348 <=*>
561,346	REPT. 96984 <RIS>
561,458	REPT. 96978 <RIS>
561,459	96980 <RIS>
561,515	REPT. 96977 <RIS>
561,812	REPT. 96982 <RIS>
561,915	REPT. 96976 <RIS>
562,022	64-10182 <=*$>
562,893	96972 <RIS>
562,899	65-10304 <*>
562,900	65-10303 <*>
563,173	60-14207 <+*>
563,247	65-10313 <*>
563,251	97013 <RIS>
563,262	60-14839 <+*>
563,347	12Q72FL <ATS>
563,350	REPT. 96955 <RIS>
563,585	62-16343 <=*>
563,591	REPT. 96961 <RIS>
563,683	63-13206 <*>
563,834	REPT. 96954 <RIS>
	59-17988 <+*> 0
563,859	REPT. 96969 <RIS>
564,188	63-24613 <=*$> 0
564,206	62-16344 <=*> 0
564,218	60-18997 <+*>
564,316	REPT. 96974 <RIS>
564,359	62-16333 <=*>
564,375	64-30955 <*>
565,083	65-10314 <*>
565,270	63-10881 <=*> 0
565,407	933 <TC>
565,604	04L29FL <ATS>
565,739	62-18141 <*>
566,124	REPT. 96966 <RIS>
566,144	34L29F <ATS>
566,265	60-18996 <+*>
566,385	65-60275 <=$>
566,436	65-10318 <*>
566,460	61-10613 <=*>
566,656	06M41F <ATS>
566,713	61-10425 <=*>
566,797	REPT. 96975 <RIS>
566,910	REPT. 96965 <RIS>
567,135	AEC-TR-6129 <*>
567,633	62-16345 <=*>
567,744	64-30970 <*>
567,815	60-14840 <+*>
568,002	REPT.97056 <RIS>
568,017	65-60262 <=$>
568,054	65-10352 <*>
568,231	65-10341 <*>
568,481	<ES>
	60-14242 <+*>
569,094	64-13933 <=*$>
569,162	64-10543 <=*$>
569,355	65-10355 <*>
569,481	92L30F <ATS>
569,816	97137 <RIS>
569,921	65-10942 <*>
569,953	64-30172 <*> 0
570,168	REPT. 97057 <RIS>
571,009	65-11224 <*>
571,010	65-11231 <*> 0
571,104	65-10346 <*>
571,475	65-11225 <*> 0
571,718	43L35F <ATS>

Patent	Citation
571,731	65-10917 <*>
571,890	REPT. 97068 <RIS>
571,921	65-13570 <*>
572,125	REPT. 97055 <RIS>
572,237	60-14206 <+*>
572,649	61-20710 <*=>
572,685	65-64485 <=*$>
572,726	97136 <RIS>
572,737	65-10926 <*>
572,902	65-10928 <*>
572,952	65-10347 <*>
573,134	60-14841 <+*>
573,233	1925 <TC>
	65-10929 <*> O
573,406	62-20418 <=*> O
	63-10790 <=*>
573,458	65-12766 <*>
573,629	61-20285 <*=> O
573,680	60-14236 <+*>
573,694	62-14768 <=*>
573,872	50N51F <ATS>
574,102	60-14842 <+*>
574,275	1927 <TC>
574,524	65-10350 <*>
575,326	60-14024 <+*>
575,475	62-18518 <=*>
575,507	60-16202 <+*>
575,595	REPT. 97072 <RIS>
	65-10924 <*>
575,641	REPT. 97071 <RIS>
	65-10925 <*>
575,671	60-14255 <+*>
576,280	62-16348 <=*>
576,945	60-16865 <+*>
577,573	227 <TC>
577,609	60-14677 <+*>
577,627	65-11831 <*>
577,781	65-13207 <*>
578,143	65-11832 <*>
578,587	60-14469 <+*>
578,670	65-10531 <*> O
578,671	65-10571 <*>
579,074	G-283 <RIS>
579,882	65-11816 <*>
579,883	65-11817 <*>
579,972	07R75F <ATS>
580,436	61-20284 <*=>
580,479	60-14470 <+*>
580,553	60-18104 <+*>
580,855	60-16502 <+*>
581,218	G-258 <RIS>
581,466	65-11257 <*>
582,447	62-18535 <=*>
582,597	64-20928 <*> O
582,793	65-12515 <*>
583,440	66-10667 <*>
584,755	65-12503 <*>
584,756	65-11857 <*>
584,941	65-11851 <*>
585,045	60-16864 <+*>
585,563	REPT. 97135 <RIS>
585,705	65-12504 <*>
586,481	N.S.6 <TTIS>
586,808	63-18579 <=*>
587,662	62-10703 <*=>
588,113	64-30102 <*>
588,134	04N49FL <ATS>
	65-12509 <*>
588,211	74N56F <ATS>
588,493	63-20372 <=*>
588,539	65-13195 <*>
588,577	61-10825 <+*>
588,662	64-30171 <*>
588,939	83N53F <ATS>
588,970	N.S.5 <TTIS>
589,336	61-10826 <+*>
590,037	65-13202 <*>
590,159	64-30179 <*>
590,314	65-12522 <*>

Patent	Citation
590,453	61-10685 <+*>
590,494	63-20367 <=*>
590,495	63-20369 <=*>
590,573	63-13710 <*>
	65-13204 <*>
590,574	65-11398 <*>
590,575	65-13717 <*>
590,766	62-18504 <=*>
590,819	64-14067 <=*$>
590,908	63-18581 <=*>
591,336	06N50F <ATS>
591,481	64-20927 <*> O
591,541	25N58F <ATS>
592,488	64-10535 <=*$>
592,605	63-16403 <=*>
592,803	62-14573 <=*>
592,986	58N52FL <ATS>
592,987	59N52FL <ATS>
593,063	61-18616 <*=>
593,155	62-10667 <*=>
593,172	64-30178 <*>
593,558	63-14349 <=*>
593,672	62-14308 <=*>
593,946	64-14038 <=*$>
594,218	62-10702 <*=>
594,591	61-20872 <*=>
594,694	62-16350 <=*>
594,963	62-16352 <=*>
595,093	90N53FL <ATS>
596,036	61-20864 <*=>
596,094	62-20402 <=*>
596,418	C-179 <RIS>
596,419	65-13624 <*>
596,572	65-10619 <*>
596,574	62-10328 <*=>
596,904	64-20720 <*>
	65-13584 <*>
597,116	62-10201 <*=>
597,316	65-13540 <*>
597,319	65-13574 <*>
597,383	62-10674 <*=>
	65-12762 <*>
597,826	62-14889 <=*>
597,827	62-14891 <=*>
597,934	65-10610 <*>
598,678	63-14353 <=*>
598,860	62-16356 <=*>
599,968	65-13995 <*>
600,026	65-10612 <*>
600,143	63-10814 <=*>
600,255	62-16357 <=*>
600,256	64-30177 <*>
600,262	AEC-TR-5496 <*>
600,263	AEC-TR-5497 <*>
600,554	65-14472 <*>
600,665	62-14890 <=*>
600,949	62-10658 <*=>
601,082	62-16359 <=*>
601,091	62-16349 <=*>
601,132	62-14407 <=*>
601,347	63-10142 <=*> O
601,484	62-10657 <*=>
601,608	65-14003 <*>
601,857	<LSA>
	874 <TC>
601,907	62N57F <ATS>
602,617	66-14010 <*>
602,841	64-30176 <*>
602,903	98R76F <ATS>
602,975	62-18685 <=*>
603,067	62-18980 <=*>
603,224	62-16362 <=*$>
603,254	65-14002 <*>
603,270	64-30175 <*>
603,329	66-10655 <*>
604,445	88P59F <ATS>
604,448	89P59F <ATS>
604,611	65-14009 <*>
604,741	8722 <IICH>

Patent	Citation
604,881	65-14475 <*>
605,315	62-20014 <=*> 0
605,502	285 <CTS>
605,765	63-14354 <=*>
605,810	62-16197 <=*> 0
606,003	62-20013 <=*>
606,222	65-14499 <*>
606,312	62-20007 <=*>
606,313	62-20008 <=*>
606,447	63-14355 <=*>
606,488	62-18503 <=*>
606,644	290 <CTS>
606,698	62-18507 <=*>
606,699	62-18571 <=*> 0
606,700	62-18508 <=*>
606,701	62-18505 <=*>
606,702	62-18506 <=*>
606,703	62-18574 <=*>
606,704	62-18575 <=*>
606,705	62-18543 <=*>
606,706	62-18682 <=*>
606,707	62-18572 <=*>
606,708	62-18573 <=*>
606,709	62-18683 <=*>
606,711	65-14496 <*>
606,712	65-14497 <*>
607,024	62-18656 <=*>
	65-10523 <*>
607,198	62-20447 <=*>
607,355	62-18500 <=*> 0
607,673	906 <TC>
607,821	62-20444 <=*>
607,823	65-14860 <*>
608,146	62-20417 <=*>
608,646	62-18509 <=*> 0
608,678	982 <TC>
608,718	64-20721 <*>
608,775	60P61FL <ATS>
608,803	301 <CTS>
609,408	62-20012 <=*>
609,409	62-20011 <=*>
609,489	62-20009 <=*>
609,540	97Q68F <ATS>
609,541	16R77F <ATS>
609,550	63-18313 <*>
609,841	65-12858 <*>
610,018	64-30996 <*>
610,019	64-30995 <*>
610,112	64-30994 <*>
610,118	13S83F <ATS>
610,223	65-14495 <*>
610,268	62-20404 <=*>
610,289	65-14884 <*>
610,607	64-30997 <*>
611,101	62-18660 <=*>
611,102	62-18661 <=*>
611,224	64-20929 <*> 0
611,273	65-12059 <*>
611,549	66-11414 <*>
611,565	64-30217 <*>
611,792	NS-65 <TTIS>
611,880	52P64F <ATS>
612,832	65-11986 <*>
612,901	65-14466 <*>
613,039	63-16452 <=*>
613,078	66-11418 <*> 0
613,292	63-16324 <=*>
613,471	63-16450 <=*>
613,495	63-20010 <=*>
613,661	65-14857 <*>
613,802	64-30180 <*>
614,149	65-11558 <*>
614,387	63-18320 <=*>
614,581	64-30181 <*>
614,946	66-10031 <*>
615,204	1630 <TC>
615,230	63-16439 <=*>
615,269	66-11415 <*>
615,435	63-10875 <=*> 0
615,499	66-10636 <*>

Patent	Citation
615,601	NP-TR-1218 <*>
615,683	63-18060 <=*>
615,684	63-18261 <=*>
615,733	458 <TC>
615,849	66-10593 <*>
616,311	63-18299 <=*>
616,613	63-18319 <*=>
616,636	2438 <TTIS>
616,737	03Q67F <ATS>
616,756	66-12052 <*>
616,762	1924 <TC>
616,815	64-20930 <*> 0
616,933	66-10630 <*>
617,026	63-10873 <=*>
617,223	1828 <TC>
	66-10009 <*>
617,390	230 <TC>
	65-11394 <*>
617,519	63-24405 <=*$> 0
617,581	63-16322 <=*>
617,901	64-20931 <*> 0
617,913	65-14516 <*>
618,327	66-10761 <*>
618,668	66-10760 <*>
619,716	87Q69F <ATS>
619,879	63-18093 <=*>
620,345	66-12805 <*>
620,363	43Q72F <ATS>
620,364	1102 <TC>
	65-12178 <*>
620,447	27Q72F <ATS>
620,574	74R76F <ATS>
621,091	247 <TC>
	65-11392 <*>
621,574	64-10846 <=*$>
621,846	14R80F <ATS>
622,220	65-11991 <*>
622,473	251 <TC>
622,949	1359 <TC>
622,976	235 <TC>
623,422	66-13263 <*>
623,450	63-18889 <=*>
623,478	64-20932 <*>
623,552	63-20058 <=*>
623,869	686 <TC>
623,923	65-11815 <*>
623,980	256 <TC>
624,020	64-18821 <*>
624,215	66-10126 <*>
624,340	405 <TC>
624,690	178 <TC>
624,705	685 <TC>
624,841	680 <TC>
625,002	369 <TC>
625,056	66-11419 <*>
625,138	65-11804 <*>
625,440	65-11790 <*>
625,698	246 <TC>
625,905	1360 <TC>
625,982	64-20800 <*>
626,042	64-10847 <=*$>
626,076	64-10845 <=*$>
626,150	64-30182 <*>
626,227	27S83F <ATS>
626,273	65-12769 <*>
626,275	234 <TC>
626,362	36R75F <ATS>
626,453	64-30183 <*> 0
626,491	76S84F <ATS>
626,759	216 <TC>
626,760	215 <TC>
626,761	214 <TC>
626,796	681 <TC>
626,901	64-18547 <*>
626,919	63-20276 <=*$>
627,602	66-10508 <*> 0
627,623	64-16409 <=*$> 0
627,625	64-16480 <=*$> 0
627,626	64-16481 <=*$> 0
628,004	65-11271 <*>

Patent	Citation
628,518	242 <TC>
	65-11393 <*>
628,650	64-18110 <*>
628,933	65-11990 <*>
629,240	65-12050 <*>
629,537	66-10615 <*>
630,049	684 <TC>
630,161	66-12513 <*>
630,224	1605 <TC>
630,363	64-16169 <*>
630,633	847 <TC>
630,699	66-11427 <*>
630,738	907 <TC>
630,749	64-20933 <*> 0
630,963	981 <TC>
631,155	908 <TC>
631,451	909 <TC>
631,471	23R76F <ATS>
631,539	66-10125 <*>
631,557	65-17066 <*> 0
632,044	983 <TC>
632,439	65-11916 <*>
632,807	65-11531 <*>
632,894	683 <TC>
633,383	64-20934 <*> 0
633,610	66-11420 <*>
633,621	1349 <TC>
634,034	181 <TC>
634,174	891 <TC>
634,605	65-11917 <*>
634,798	1620 <TC>
634,890	65-12179 <*>
635,059	66-11476 <*>
635,094	679 <TC>
635,097	677 <TC>
635,132	47R80FL <ATS>
635,349	492 <TC>
635,350	493 <TC>
635,352	1044 <TC>
635,549	64-20935 <*> 0
635,557	1021 <TC>
635,738	682 <TC>
635,848	689 <TC>
635,891	688 <TC>
636,147	66-13974 <*>
636,217	65-11989 <*>
636,234	66-10293 <*>
636,283	66-13264 <*>
636,286	65-11985 <*>
636,335	647 <TC>
636,350	691 <TC>
636,357	8650 <IICH>
636,443	65-12177 <*>
	66-11315 <*>
636,782	66-13975 <*>
	706 <TC>
636,856	980 <TC>
637,005	65-17454 <*>
637,397	645 <TC>
637,399	644 <TC>
637,444	66-11911 <*>
637,537	66-11307 <*>
	984 <TC>
637,575	848 <TC>
637,692	690 <TC>
638,252	66-11308 <*>
638,396	65-11719 <*>
638,453	687 <TC>
638,958	1335 <TC>
639,089	1336 <TC>
639,090	1338 <TC>
639,091	1337 <TC>
639,092	1339 <TC>
639,094	1340 <TC>
639,174	66-12794 <*>
639,261	362 <TC>
640,243	65-13878 <*>
640,396	1996 <TC>

Patent	Citation
640,397	1997 <TC>
640,658	1016 <TC>
641,141	8678 <IICH>
641,383	851 <TC>
641,442	66-14009 <*>
641,455	66-11366 <*>
641,567	65-13087 <*>
642,092	1203 <TC>
642,156	849 <TC>
642,304	1999 <TC>
643,057	813 <TC>
643,085	2000 <TC>
643,262	986 <TC>
643,263	985 <TC>
643,291	66-10124 <*>
643,501	737 <TC>
643,575	65-14107 <*> 0
643,649	1149 <TC>
	65-14160 <*>
643,650	1334 <TC>
	65-13944 <*>
643,700	1812 <TC>
643,758	86R79F <ATS>
643,909	65-13720 <*>
644,090	65-14236 <*>
644,182	1147 <TC>
	65-17467 <*> C
644,189	66-13249 <*>
644,588	65-14108 <*>
644,614	65-11995 <*>
644,615	65-11782 <*>
645,039	1813 <TC>
645,459	931 <TC>
645,617	66-14524 <*>
645,618	65-13741 <*>
	920 <TC>
645,836	625 <TC>
645,841	66-10742 <*>
645,989	649 <TC>
	65-12782 <*>
645,990	65-13080 <*>
	650 <TC>
646,120	1150 <TC>
646,151	65-14112 <*>
646,152	1148 <TC>
646,342	65-12034 <*>
646,518	65-12181 <*>
646,636	65-13695 <*>
	915 <TC>
646,736	65-13738 <*>
	919 <TC>
646,834	65-11645 <*>
646,845	1350 <TC>
646,975	65-14318 <*>
646,995	66-10123 <*>
646,996	66-10122 <*>
647,004	66-10524 <*>
647,140	65-12060 <*>
647,926	96S84F <ATS>
647,927	647-926 <ATS>
647,947	669 <TC>
648,299	431 <TC>
648,476	65-14918 <*>
649,326	2002 <TC>
649,333	66-10121 <*>
649,334	66-10120 <*>
649,390	65-14910 <*>
649,489	2003 <TC>
649,539	66-10292 <*>
650,476	66-12053 <*>
650,678	1745 <TC>
650,856	65-14911 <*>
651,264	2095 <TC>
651,747	2001 <TC>
651,761	66-11082 <*>
652,105	97S85FL <ATS>
652,106	96S85FL <ATS>
652,718	91T90F <ATS>
652,825	996 <TC>
653,231	66-14001 <*>

Patent	Citation
653,590	66-10588 <*>
653,869	1928 <TC>
654,281	1929 <TC>
654,539	2137 <TC>
654,556	98S85F <ATS>
654,751	1414 <TC>
656,373	66-11447 <*>
656,469	1612 <TC>
656,475	1836 <TC>
656,583	1763 <TC>
	66-10421 <*>
657,689	827 <TC>
657,721	66-12553 <*>
657,838	66-10589 <*>
658,374	66-11512 <*>
658,452	8700 <IICH>
658,972	66-10590 <*>
659,382	64-14008 <*>
659,399	65-11805 <*>
659,956	2036 <TC>
660,574	1217 <TC>
660,734	66-11367 <*>
661,907	66-14527 <*>
663,194	66-11450 <*>
664,294	1935 <TC>
666,636	66-12718 <*>

BRAZIL

Patent	Citation
40,657	64-20213 <*> P
48,152	64-20212 <*> P

CANADA

Patent	Citation
449,687	65-10506 <*>
496,583	59-17295 <+*> 0
544,675	61-14917 <=*>

CUBA

Patent	Citation
27,217	60-14678 <+*>
27,642	60-14679 <+*>
27,667	60-14680 <+*>
28,341	60-14681 <+*>
28,415	60-14682 <+*>
28,476	60-14683 <+*>
28,555	60-14684 <+*>
28,627	60-14685 <+*>

CUBA APPLICATION

Patent	Citation
147,578	64-18398 <=*$>

CZECHOSLOVAKIA

Patent	Citation
42,340	64-15356 <=*$> 0
69,690	65-60276 <=$>
72,671	65-62860 <=$> 0
76,678	61-28954 <*=> 0
79,895	61-28955 <*=> 0
83,795	72K25C <ATS>
83,911	24M45C <ATS>
85,465	65-12033 <*>
85,505	59M40C <ATS>
86,889	<MT>
88,063	66-11648 <*>
89,170	27R75C <ATS>
89,694	66-11651 <*>
89,919	62-10327 <*=>
89,975	62-33230 <=*>
90,657	80R77C <ATS>
91,880	31N49C <ATS>
92,270	62-13488 <*=> 0
92,456	11P65C <ATS>
92,697	66-11650 <*>
93,066	63-13362 <=*>
93,557	61-18612 <*=>
	98R79C <ATS>
93,617	28N52C <ATS>
93,618	63-24045 <=*>
93,741	63-23716 <=*>
93,834	61-18611 <*=>
94,033	61-18609 <*=>
95,034	63-23659 <=*>
95,115	62-16486 <=*>
95,316	AD-620 461 <=$>
95,337	61-18610 <*=>
95,459	15Q69C <ATS>
96,154	62-16485 <=*>
96,758	66-11649 <*>
97,206	63-19631 <=*>
97,291	64-14787 <=*$>
97,292	64-16605 <*>
97,559	79P62C <ATS>
97,594	66-14537 <*>
97,901	74P62C <ATS>
98,093	40R77C <ATS>
98,413	63-20359 <=*$>
98,797	43P64C <ATS>
99,196	64-13636 <=*$>
100,322	48Q68C <ATS>
100,329	63-20194 <=*>
100,870	27R77C <ATS>
101,393	64-20716 <*>
102,859	05Q73C <ATS>
103,012	77Q70C <ATS>
103,680	65-13711 <*>
106,302	66-11647 <*>
106,904	83T90C <ATS>
107,101	66-12668 <*>
107,321	66-12521 <*>
107,891	66-12525 <*>
107,966	99R79C <ATS>
108,762	1780 <TC>
108,845	65-14768 <*>
109,049	65-11336 <*>
109,653	79R77C <ATS>
109,957	66-12523 <*>
110,397	NS 242 <TTIS>
111,226	66-12524 <*>
112,040	NS-399 <TTIS>

DENMARK

Patent	Citation
7,710	59-17203 <+*> C
14,102	59-17840 <+*>
21,091	63-18657 <=*>
22,259	65-12882 <*>
27,102	59-17193 <+*> 0
33,316	64-20715 <*>
35,339	60-10225 <+*> 0
37,551	65-60277 <=$>
38,435	63-14356 <*>
40,609	3483 <K-H>
44,490	64-15128 <=*$> C
45,776	64-30184 <*>
56,607	6600-B <K-H>
56,692	66CC-C <K-H>
56,828	63-14935 <=*>
57,551	63-14357 <=*>
57,609	64-15127 <=*$> 0
59,092	63-14358 <=*>
60,881	AL-19 <*>
61,453	62-18655 <*>
62,400	63-14359 <*=>
62,791	59-17822 <+*>
63,061	7061 <K-H>
63,553	35Q70G <ATS>
64,957	AL-20 <*>
66,246	60-17286 <+*> 0
66,435	59-12075 <+*>
66,573	10041-B <K-H>
67,279	60-16450 <+*>
67,806	62-23723 <=*> C
68,698	60-17287 <+*>
69,594	60-17288 <+*> 0
70,844	60-14851 <+*>
71,630	62-25678 <=*> C
72,134	62-16249 <=*>
74,302	59-16164 <+*> 0
75,814	59-14203 <+*> C
76,954	63-15438 <=*$> 0

78,856	65-60835 <=$>
79,469	62-11173 <=>
80,088	60-13071 <+*> 0
80,977 ADD	60-14690 <+*>
81,282	12S88D <ATS>
81,494	T.3 <CLAI>
82,686	61-15045 <+*>
84,196	60-14693 <+*>
84,963	60-14694 <+*>
86,158	60-14695 <+*>
86,314	65-62861 <=$> 0
86,639	1750 <TC>
86,643	8688 <IICH>
87,556	60-16452 <+*> 0
89,161	1751 <TC>

EAST GERMANY (SEE ALSO GERMANY, EAST)

11,379	66M37G <ATS>
13,586	63-10796 <=*>
13,661	59-22337 <+*>
21,562	NS 14 <TTIS>
22,030	NS-65 <TTIS>
	63-16421 <=*>
22,305	63-19059 <=*> 0
23,172	WS-38 <TTIS>
23,207	WS-41 <TTIS>
23,467	NS-64 <TTIS>
23,488	WS-56 <TTIS>
23,498	NS-63 <TTIS>
23,509	63-16420 <=*>
23,595	NS-67 <TTIS>
23,685	NS-68 <TTIS>
23,856	NS-75 <TTIS>
23,987	NS-66 <TTIS>
24,059	NS 123 <TTIS>
24,144	NS-76 <TTIS>
24,170	NS-83 <TTIS>
24,196	NS-74 <TTIS>
24,239	NS-79 <TTIS>
24,279	NS-85 <TTIS>
24,302	NS-86 <TTIS>
24,305	NS-84 <TTIS>
24,547	NS-99 <TTIS>
24,579	NS-73 <TTIS>
24,671	NS-88 <TTIS>
24,672	NS-90 <TTIS>
24,674	NS 114 <TTIS>
24,726	NS 135 <TTIS>
24,757	NS-89 <TTIS>
24,843	NS 136 <TTIS>
24,885	NS 124 <TTIS>
24,896	NS-108 <TTIS>
24,904	NS119 <TTIS>
24,958	NS 126 <TTIS>
25,494	NS 137 <TTIS>

FINLAND

16,597	63-20510 <=*$>
18,630	60-13272 <+*>
23,169	61-19479 <+*> 0
25,785	64-15842 <=*$> 0
26,122	10819 <K-H>
	60-13273 <+*>

FRANCE

11,061 ADD	64-15145 <=*$>
86,481	63-15439 <=*> 0
94,642	63-18631 <=*>
107,512	63-14501 <=*>
115,164	62-18559 <=*> 0
165,283	65-61498 <=$> 0
253,363	63-20331 <=*>
285,164	64-15353 <=*$> 0
300,751	59-16165 <+*> 0
319,237	64-14967 <=*$>
319,436	63-16000 <=*>
323,499	62-18549 <=*> 0
330,083	63-15773 <=*> 0
334,369	I-667 <*>
342,112	60-16137 <+*>
353,370	59-20063 <+*> 0
353,858	64-30218 <*>
363,039	2027 <*>
368,393	65-64757 <=*$>
372,370	60-15541 <+*>
373,884	64-30220 <*>
382,188	63-20694 <=*$>
384,751	66-60257 <=*$>
386,627	64-30221 <*>
390,637	64-20668 <*> 0
392,766	61-10093 <=*$>
393,655 ADD	64-13241 <=*$> 0
399,313	59-17832 <+*> 0
412,304	64-19911 <=$>
412,377	I-308 <*>
415,352	IICH-8635 <IICH>
415,888	64-13583 <=*$> 0
417,925	62-18202 <=*>
423,494	63-23561 <=*$> 0
426,343	60-16144 <+*>
430,463	63-19269 <=*> 0
433,999	59-17856 <+*>
	62-16545 <*>
436,297	I-477 <*>
436,529	59-12077 <+*>
439,753	63-16001 <*=>
440,516	62-18555 <=*>
444,011	64-13779 <=*$> 0
444,809	59-17026 <+*> 0
448,774	62-18552 <=*> 0
453,323	62-25677 <=*> 0
453,393	64-30222 <*>
455,539	10L31F <ATS>
456,056	64-20670 <*>
461,676	I-400 <*>
464,999	62-18653 <=*>
466,112	64-13582 <=*$> 0
466,721	64-30223 <*>
466,824	63-19069 <*=> 0
470,392	63-14936 <=*>
475,594	59-12493 <+*> 0
476,913	64-15843 <=*$> 0
485,904	II-132 <*>
489,155	8632 <IICH>
494,246	60-16147 <+*>
495,638	01G5F <ATS>
498,304 ADD	63-24612 <=*$> 0
499,406	61-10100 <=*$>
504,001	<ATS>
505,741	61-10096 <=*$>

Patent	Citation
506,479	59-18283 <+*> 0
509,875 ADD	63-23560 <=*$> 0
513,495	65-64568 <=*$>
517,084	62-18362 <=*> 0
517,328	66-60256 <=*$>
518,736	59-12494 <+*> 0
521,526	02G5F <ATS>
522,572	65-63474 <=$>
523,896	63-24115 <=*>
527,832	63-19270 <=*> 0
528,456	65-10546 <*> 0
534,249	60-15552 <+*> 0
537,207	65-10585 <*>
543,252	62-25679 <=*>
543,409	63-16002 <=*>
548,342	60-15097 <+*> 0
553,049	61-10097 <=*$>
553,448	66-60255 <=*$>
554,122	64-30130 <*>
554,425	63-13743 <=*> 0
558,265	I-659 <*>
560,035	59-22273 <+*> 0
561,787	59-19214 <+*> 0
563,031	63-23562 <=*$> 0
563,666	63-14781 <=*>
575,633	64-20669 <*>
575,683	61-10056 <=*$> 0
578,105	59-17118 <+*> 0
578,445	64-20671 <*>
581,194	64-19793 <=$> 0
584,091	58-1415 <*>
584,842	62-18560 <=*> 0
586,375	59-14204 <+*> 0
588,617	59-19741 <+*> 0
588,649	218 <TC>
590,562	59-17850 <+*> 0
592,722	59-12495 <+*> 0
593,863	61-10112 <=*$>
595,221	64-15144 <=*$> 0
598,330	59-22274 <+*> 0
599,711	63-20360 <*>
605,026	64-30224 <*>
605,097	I-1000 <*>
607,136	64-13242 <=*$> 0
607,283	62-18553 <=*>
608,302	64-30225 <*>
609,970	57-3135 <*>
610,585	64-13581 <=*$> 0
611,929	63-18267 <=*>
612,699	65-64569 <=*$>
616,804	59-12155 <+*>
618,847	63-19568 <=*>
619,612	64-19912 <=$>
621,077	61-10059 <=*$> 0
621,277	63-10033 <=*> 0
621,565	61-10066 <=*$> 0
621,728	59-16830 <+*>
624,631	64-30226 <*>
625,238	<INFO>
628,934	62-32229 <=*> 0
630,718	62-33241 <=*> 0
631,596	59-18776 <+*> 0
634,184	61-10094 <=*$>
636,466	61-10055 <=*$>
636,586	63-16006 <=*>
638,103	63-18652 <=*>
638,186	61-10113 <=*$> 0
640,535	63-16007 <=*>
641,587	60-23917 <=*> 0
642,140	2412 <*>
642,392	63-16307 <*=>
643,302	61-10041 <=*$>
644,054	54Q69F <ATS>
645,395	<ATS>
	63-16008 <=*>
647,090	63-16009 <=*>

Patent	Citation
649,126	24G6F <ATS>
655,737	64-30227 <*>
656,482	62-18611 <=*> 0
656,582	59-17857 <+*>
	62-16543 <=*>
656,647	64-30228 <*>
657,799	63-16010 <=*>
657,880	59-12496 <+*> 0
658,094	63-16011 <=*>
659,043	65-63475 <=$> 0
659,676	62-18364 <=*> 0
660,529	63-16161 <*=>
660,732	8667 <IICH>
662,375	62-18411 <=*> C
664,447	63-14780 <=*>
664,455	64-30229 <*>
669,100	63-14516 <=*>
669,815	63-14517 <=*>
670,099	5009-C <K-H>
670,881	62-16388 <=*>
670,889	63-14779 <=*>
670,938	65-60280 <=$>
674,507	11917A <K-H>
	63-16581 <=*>
679,998	65-10650 <*>
682,785	62-16387 <=*> 0
682,877	59-20062 <+*> C
683,202	62-18546 <=*> 0
683,946	63-16012 <=*>
685,567	64-20672 <*>
686,445	62-18410 <=*> 0
690,008	65-60268 <=$>
693,723	64-13243 <=*$>
694,257	63-16013 <=*>
697,313	57-1129 <*>
697,435	60-14769 <+*>
697,521	60-18897 <=*$>
698,355	61-10051 <=*$> 0
699,975	61-10092 <=*$>
701,032	64-20673 <*>
701,187	63-16305 <=*>
703,555	63-18138 <=*>
704,471	59-19215 <+*> 0
705,664	64-14694 <=*$>
707,526	58-1555 <*>
709,044	63-24116 <=*>
710,496	64-14690 <=*$>
710,975	43F4F <ATS>
711,095	63-18872 <=*>
	64-20674 <*>
711,595	63-20186 <=*>
712,037	64-30230 <*>
712,416	60-16168 <+*> 0
712,775	63-18185 <=*>
713,382	63-16014 <=*>
714,791	65-60269 <=$>
714,968	59-16166 <+*> C
716,299	62-23724 <=*> 0
717,067	63-16015 <*=>
717,232	63-14778 <=*>
718,140	62-18147 <=*>
719,067	62-16544 <=*>
721,675	59-17104 <+*>
721,986	59-17027 <+*> C
722,752	63-16016 <=*>
723,566	61-10040 <=*$>
723,704	63-16018 <*=>
725,245	60-16172 <=*>
725,864	63-14777 <=*> 0
726,339	II-509 <*>
728,737	61-10119 <=*$> 0
729,730	64-20532 <*>
730,874	63-20157 <=*>
732,512	62-18363 <=*> 0
732,569	62-33243 <=*>
733,825	63-16019 <*=>
736,231	63-23563 <=*>
736,379	09L35F <ATS>
736,717	63-16083 <=*>

Patent	Citation
736,736	59-12918 <+*> O
738,200 ADD	63-24406 <=*$> O
738,285	65-63476 <=$> O
741,858	59-16831 <+*> O
742,153	63-18186 <=*>
743,302	64-10122 <=*$>
744,156	61-28957 <=*> O
744,397	58L33F <ATS>
745,293	64-20859 <*>
745,316	64-30309 <*>
745,578	59-16832 <+*> O
745,987	63-18187 <=*>
747,246	62-16386 <=*>
748,019	59-18777 <+*> O
750,154	59-18284 <+*> O
750,873	64-30231 <*>
752,150	63-18188 <=*>
753,189	63-18357 <=*>
753,729	18F3F <ATS>
754,361	63-18189 <*>
755,273	62-16385 <=*> O
755,701	57-177 <*>
757,871	64-20675 <*>
758,454	64-30232 <*>
758,829	59-22275 <+*> O
759,171	I-526 <*>
759,442	I-276 <*>
760,355	63-19566 <=*>
762,064	62-16384 <=*>
763,379	59L33F <ATS>
764,240	64-30233 <*>
766,051	63-10976 <=*>
	64-30234 <*>
766,369	64-30235 <*>
767,191	63-18190 <=*> P
767,816	59-16167 <+*> O
768,057	59-16168 <+*> O
770,326	59-16169 <+*> O
770,391	60-16189 <+*>
772,724	64-30236 <*>
772,805	64-30237 <*>
773,656	63-18191 <=*>
773,849	59-12497 <+*> O
773,885	63-20343 <=*>
775,380	AEC-TR-4881 <*>
776,346	63-18192 <=*>
777,899	64-30238 <*>
778,634	59-18285 <+*> O
779,493	3366-A <K-H>
779,655	59-16170 <+*> O
780,057	1655 <*>
	63-20273 <=*>
	64-10471 <=*$>
783,484	63-20321 <=*$>
784,348	63-14518 <=*>
785,075	1230 <TC>
785,881	62-16383 <=*> O
786,533	63-18193 <=*>
786,909	60L33F <ATS>
787,477	61-28495 <*=>
787,529	63-20174 <=*>
787,655	19J17F <ATS>
788,074	<AC>
788,981	63-16091 <=*>
789,304	750 <TC>
789,522	63-10487 <=*>
790,193	62-18551 <=*> O
790,945	64-20677 <*>
791,296	63-14519 <=*>
791,479	63-19068 <=*> O
792,611	1669 <*>
792,880	63-18194 <=*>
793,173	63-18195 <=*>
793,226	64-20678 <*>
793,250	64-20679 <*>
795,285	59-22276 <+*> O
795,429	64-10680 <*>
796,588	I-524 <*>
796,917	63-18196 <=*>
797,415	62-10900 <*=> O
798,727	64-20858 <*>
	64-30239 <*>
799,345	63-18197 <=*>
800,330	64-30249 <*>
800,385	61-10109 <=*$>
801,106	63-18198 <=*>
801,277	63-14520 <=*> O
802,021	64-30240 <*>
802,571	18E2F <ATS>
802,736	1604 <TC>
802,840	63-18517 <=*>
804,595	61-10067 <=*$>
804,958	64-15844 <=*$> O
806,279	64-20567 <*>
806,796	62-16381 <=*>
807,447	64-10681 <*>
808,319	64-20682 <*>
808,612	61-10068 <=*$>
808,655	64-20683 <*>
811,104	59-19216 <+*> O
811,600	59-16833 <+*> O
811,628	62-10049 <*=>
812,183	59-18778 <+*> O
812,348	66-10014 <*>
813,333	<INFO>
813,375	64-20684 <*>
813,427	63-18199 <=*>
814,759	62-18558 <=*> O
815,247	64-23594 <=$> C
817,010	64-30241 <*>
817,537	63-24408 <=*$> O
818,399	62-16380 <=*> C
818,737	64-20685 <*>
820,698	64-30242 <*>
821,791	63-18200 <=*> C
822,572	59-12919 <+*> O
824,022	58-1116 <*>
825,782	65-60281 <=$>
827,380	62-18548 <=*>
827,388	59-14688 <+*> C
827,469	63-18201 <=*>
827,867	64-30243 <*>
828,573	64-20686 <*> O
831,194	61-10069 <=*$> O
831,282	58-1113 <*>
832,725	64-30244 <*>
832,821	II-706 <*>
833,113	64-20687 <*>
833,633	59-18779 <+*> O
834,018	65-11022 <*>
834,613	63-15774 <=*> C
835,445	64-30245 <*>
835,592	64-14695 <=*$>
835,834	65-10649 <*>
836,457	62J15F <ATS>
836,705	62-16379 <=*> C
836,776	59-19742 <+*> O
837,280	59J14F <ATS>
837,828	64-30246 <*>
837,965	65-62863 <=$>
839,037	64-20688 <*>
840,422	66-10525 <*>
841,403	II-709 <*>
842,001	61-10038 <=*$> O
842,186	1125 <*>
842,651	60-15542 <+*>
843,182	64-20689 <*>
843,227	64-20690 <*>
843,891	8752 <IICH>
843,940	63-13744 <=*> O
844,322	64-30248 <*>
844,945	64-20691 <*>
845,116	63-18202 <=*>
845,658	<INFO>
846,628	65-10643 <*>
846,704	63-13745 <=*> C

Patent	Citation
846,890	64-30247 <*>
847,001	63-18203 <=*>
847,456	64-30250 <*>
847,587	62-14268 <=*>
848,421	64-30251 <*>
849,151	63-18204 <=*>
849,804	59-22277 <+*>
849,920	63-13746 <=*> 0
852,763	8681 <IICH>
853,322	62-16377 <=*> 0
855,694	57-716 <*>
856,250	64-30252 <*>
856,487	60-15543 <+*>
857,564	94N57F <ATS>
859,473	63-18225 <=*>
860,963	60-16266 <+*>
861,306	62-10296 <*=>
861,390	I-99 <*>
863,089	62-16376 <=*>
863,445	64-10381 <=*$>
863,630	63-18654 <=*>
	65-11818 <*>
864,361	59-12311 <+*>
864,561	19E2F <ATS>
866,780	64-20692 <*> 0
867,323	62-18550 <=*>
867,571	3454 <*>
868,233	54F3F <ATS>
87C,080	64-30253 <*>
871,031	59-17891 <+*>
872,207	60-15544 <+*>
872,801	62-13490 <=*>
873,288	61-18682 <*=>
873,983	64-14924 <=*$>
874,248	1012 <TC>
874,681	22E2F <ATS>
875,570	61-23217 <=*> 0
876,032	3455 <*>
876,558	64-18486 <*>
876,573	61-19482 <+*> 0
876,601	59-19743 <+*> 0
877,077	62-18557 <=*> 0
877,873	62-25032 <=*>
878,782	<ATS>
879,189	59-17887 <+*>
879,343	8730 <IICH>
879,689	262 <TC>
	64-13244 <=*$>
880,418	59-16171 <+*> 0
881,578	62-33776 <=*>
881,643	65-63473 <=$>
881,646	59-17890 <+*>
881,858	64-20693 <*> 0
881,969	59-17889 <+*>
882,048	59-17888 <+*> 0
882,127	62-18554 <=*> 0
883,348	60-17605 <+*> 0
883,701	64-30254 <*>
884,017	64-20695 <*>
884,271	64-30255 <*>
884,329	60-13274 <+*> 0
884,391	17F3F <ATS>
885,521	64-30256 <*>
885,615	77H9F <ATS>
885,694	64-14671 <=*$>
885,886	59-22278 <+*>
886,597	65-10554 <*>
886,716	61-15678 <+*> 0
886,890	63-23771 <=*$> 0
887,056	60-13072 <+*> 0
889,727	64-20483 <*>
890,906	60-17987 <+*> 0
891,884	63-15772 <=*> 0
892,851	63-24117 <=*>
894,056	65-64758 <=*$>
894,546	63-18229 <=*>
895,100	60-15860 <+*> 0
896,049	65-62854 <=$> 0
896,092	63-18386 <*=>
897,634	63-14644 <=*> 0

Patent	Citation
897,853	64-20060 <*>
898,269	64-20482 <*>
898,577	1124 <*>
899,215	63-18625 <=*>
899,955	62-16375 <=*>
900,188	60-23087 <+*> 0
901,551	63-20046 <=*$>
902,127	62-13491 <=*> 0
902,385	61-28078 <*=> 0
902,403	64-30257 <*>
903,040	62-18547 <=*> 0
903,462	61-13823 <+*>
904,201	63-18626 <=*>
904,255	62-16374 <=*>
904,509	59-14689 <+*> 0
904,831	64-30258 <*>
905,038	3460 <*>
	63-18627 <=*>
905,039	64-20068 <*>
905,927	62-14313 <=*>
906,502	60-16648 <+*>
	63-14776 <=*>
906,701	62-23725 <=*>
908,076	62-25031 <=*> 0
911,443	64-15354 <=*$> 0
913,966	64-30259 <*>
913,993	63-18628 <=*>
916,194	61-15251 <+*>
917,287	59-18286 <+*> 0
918,167	61-11126 <=>
919,464	63-23163 <=*>
919,583	62-25681 <=*>
919,818	59-18312 <+*> 0
920,437	65-63472 <=$>
922,554	61-16191 <=*>
924,327	62-33244 <=*> 0
926,470	58-1109 <*>
927,841	1109 <TC>
928,399	59-22185 <=>
928,732	63-18629 <=*>
929,234	61-23879 <*=>
929,874	63-14775 <=*> 0
930,609	63-18630 <=*>
931,881	65-60909 <=$> 0
932,001	64-10148 <=*$>
932,560	9372-A <K-H>
933,424	62-10036 <*=>
934,220	58-1302 <*>
934,336	64-23585 <=$>
935,298	59-16834 <+*>
940,731	AL-188 <*>
941,139	66-13247 <*> 0
943,184	64-30260 <*>
944,320	64-16777 <=*$>
945,164	60-16620 <+*>
946,650	60-13275 <+*> 0
946,944	65-64759 <=*$>
947,875	61-19839 <=*> 0
949,844	AL-4 <*>
	63-20454 <=*$>
950,626	AL-545 <*>
951,159	65-10570 <*> 0
952,380	64-15352 <=*$>
952,571	63-19067 <=*> 0
953,800	63-15792 <=*>
955,816	63-18653 <=*>
956,267	59-12078 <+*>
956,459	62-16378 <=*>
956,583	63-18632 <=*>
957,082	60-18047 <+*>
957,737	60-16619 <+*>
958,306	64-15143 <=*$> 0
958,437	59-12920 <+*> 0
959,731	64-13778 <=*$> 0
961,310	65-60282 <=$>
962,951	63-18964 <=*>
963,465	58-2419 <*>
963,540	59-19217 <+*> 0
963,621	60-13073 <+*> 0

964,056 61-23218 <=*> O
965,036 64-30261 <*>
966,208 59-17997 <+*> O
966,988 62-13882 <*=>
967,362 II-908 <*>
968,596 64-16776 <=*$>
970,519 62-33245 <=*> O
971,647 63-18960 <=*>
972,735 65-10695 <*>
973,692 62-25683 <=*> O
973,774 63-18957 <=*>
974,374 64-18359 <*>
974,767 62-25030 <=*> O
974,938 61-27459 <*=> O
975,099 62-16260 <=*>
975,099 ADD 62-16244 <=*>
975,729 64-30262 <*>
976,263 59-16172 <+*> O
976,384 64-30263 <*>
976,665 AL-189 <*>
976,815 63-23564 <=*$> O
977,304 62-18545 <=*> O
977,460 64-20077 <*> O
978,686 60-23918 <+*> O
978,997 64-19171 <=*$> O
979,406 63-24410 <=*$>
979,652 64-14810 <=*$>
979,658 61-13824 <*+> O
980,709 AL-669 <*>
63-20453 <=*>
982,463 64-20842 <*> O
982,837 59-12498 <+*> O
983,423 61-10802 <=*>
984,088 61-13825 <+*>
984,099 60-16267 <+*> O
984,360 63-18633 <=*>
984,809 63-18634 <*>
985,016 51K27F <ATS>
985,444 59-12312 <+*> O
985,905 61-15984 <+*> O
986,473 59-14690 <+*> O
986,637 62L35F <ATS>
986,834 61-13254 <+*>
987,118 62-23307 <=*> O
987,173 60-14307 <+*>
987,343 60-17606 <+*> O
989,286 61-15985 <+*> O
989,396 64-15142 <=*$>
989,548 59-16173 <+*> O
990,161 65-10525 <*>
991,512 98H9F <ATS>
992,111 60-16268 <+*>
992,188 62-23726 <=*> O
992,215 64-30264 <*>
992,360 61-18683 <=*>
993,097 63-18665 <=*>
993,520 59-18780 <+*> O
995,188 59-17142 <+*>
995,280 96H9F <ATS>
995,367 63-15771 <=*> O
997,262 61-13255 <+*> O
997,728 60-16269 <+*>
998,079 63-13435 <=*> O
998,475 64-15350 <=*$> O
998,481 59-17306 <+*>
64-30265 <*>
998,652 5412-C <K-H>
998,673 58-1544 <*>
1,000,158 60-16270 <+*>
1,000,310 63-15770 <=*> O
1,000,746 63-18635 <=*>
1,000,961 60-17607 <+*> O
1,001,514 64-20696 <*>
1,001,936 64-20358 <*>
1,001,957 AL-763 <*>
1,002,017 64-30266 <*> O
1,002,422 60-16283 <+*>
1,002,423 60-16284 <+*>
1,002,441 554 <TC>
1,002,732 63-18260 <=*>
1,002,800 61-13826 <+*> O
1,005,037 60-16285 <+*> P
1,005,231 64-30267 <*>
1,005,681 60-14785 <+*>
1,005,700 59-12499 <+*> O
1,005,747 60-17608 <+*> O
1,006,151 64-30268 <*>
1,006,342 60-14786 <+*>
1,006,393 60-16286 <+*>
1,006,458 61-19483 <+*> O
1,008,583 60-16287 <+*>
1,008,615 60-16288 <+*>
1,008,895 60-16289 <+*>
1,008,957 60-16290 <+*> P
1,009,208 60-16291 <+*>
1,009,260 60-16292 <+*>
1,009,482 63-15440 <=*>
1,009,955 64-20211 <*>
1,010,729 60-16293 <+*>
1,010,818 59-19218 <+*> O
1,011,528 60-16294 <+*> P
1,011,905 59-17303 <+*> O
1,012,968 64-15351 <=*$> O
1,013,428 60-14788 <+*>
1,013,692 60-16295 <+*>
1,013,951 64-20410 <*>
1,015,018 61-19840 <=*> O
1,017,110 59-17810 <+*>
1,017,513 64-20697 <*> O
1,017,761 60-14789 <+*>
1,018,241 63-20045 <=*>
1,018,504 64-30269 <*>
1,018,846 97H9F <ATS>
1,019,413 62-13881 <*=> O
1,019,596 63-18636 <=*>
1,022,411 64-30270 <*>
1,022,419 59-18287 <+*> O
1,024,013 60-16304 <+*>
1,024,565 64-20698 <*>
1,025,484 63-19066 <=*> O
1,026,768 60-16305 <+*>
1,027,404 <AC>
63-15441 <=*> O
1,031,424 61-13827 <*+> O
1,031,467 61-19484 <+*>
1,032,840 64-15845 <=*$> O
1,032,945 60-16307 <+*>
1,033,433 63-13747 <=*> O
1,033,447 64-19170 <=*$> O
1,034,734 63-18637 <*>
1,034,776 59-14691 <+*> O
1,034,998 59-12313 <+*>
1,035,259 62-16401 <=*>
1,035,523 64-20699 <*> O
1,035,691 61-19451 <=*>
1,035,946 64-10380 <=*$>
1,037,521 63-13433 <=*> O
1,037,792 62-33777 <=*> O
1,038,606 3457 <*>
1,039,055 9H10F <ATS>
1,039,264 64-20700 <*>
1,040,795 60-23088 <+*> O
1,040,978 63-24610 <=*$> O
1,041,535 64-15860 <=*$> O
1,042,011 64-20315 <*>
1,042,084 63-18638 <*>
1,043,380 59-17299 <+*> O
1,043,435 64-20407 <*>
1,043,456 64-30271 <*>
1,044,393 60-15091 <+*> O
1,045,031 61-15046 <*+>
1,046,537 57-178 <*>
1,046,698 II-850 <*>
1,046,910 59-19219 <+*> O
1,047,146 62-25684 <=*> O
1,047,261 63-20047 <=*>
1,047,573 II-827 <*>
1,049,818 62-16419 <=*>
1,050,249 62-24502 <=*> O
1,050,286 96543 <RIS>

1,053,147 II-717 <*>
1,054,009 57-1954 <*>
1,054,555 62-23727 <=*>
1,055,831 63-18639 <=*>
1,056,2C3 63-10259 <=*> O
1,056,327 59-19220 <+*> O
1,058,495 63-15442 <=*> O
1,058,697 63-14937 <=*>
1,058,934 63-10815 <=*>
1,059,53C 60-14790 <+*>
1,059,988 60-15092 <+*> O
1,060,511 64-20701 <*> O
1,061,133 62-25685 <=*>
1,061,184 61-23219 <=*> O
1,061,960 61-23660 <=*> O
1,062,041 65-10591 <*> O
1,062,587 64G1F <ATS>
1,062,632 60-16308 <+*>
1,062,803 57-21 <*>
1,064,849 63-15443 <=*> O
1,065,026 60-16309 <+*>
1,065,292 61-20712 <*=>
1,068,273 63-15444 <=*> O
1,068,677 61-19841 <=*> O
1,069,192 <ES>
1,069,431 64-16440 <=*$>
1,070,846 65-61497 <=$>
1,071,233 61-20868 <*=>
1,071,289 59-12079 <+*>
1,071,783 64-19913 <=$>
1,071,910 186 <TC>
1,073,038 62-33242 <=*> O
1,073,302 64-30272 <*>
1,073,382 60-14791 <+*>
1,074,014 60-16310 <+*>
1,074,C96 304 <LS> C
1,075,964 200 <TC>
1,076,501 59-16174 <+*>
1,076,533 63-18640 <=*>
1,076,558 59-17117 <+*> O
59-19744 <+*> O
1,076,924 99H9F <ATS>
1,078,506 63-18641 <*>
1,078,580 64-23586 <=$>
1,078,586 62-25029 <=*> O
1,078,733 61-15252 <+*> O
1,078,856 64-30273 <*>
1,079,860 ADD 61-19837 <*=>
1,079,997 60-23089 <+*> O
1,081,322 60-15862 <+*> O
1,081,584 64-30274 <*> O
1,081,691 62-33246 <=*>
1,082,151 60-13074 <+*> O
1,082,154 64-30275 <*>
1,082,852 58-1263 <*>
1,085,458 62-25686 <=*> O
1,085,983 64-30276 <*>
1,087,019 2846 <*>
1,092,276 61-28952 <*=> O
1,092,302 64-30277 <*> O
1,092,393 61-15679 <+*>
1,C92,604 59-18781 <+*> O
1,092,797 60-23090 <+*> O
1,C93,096 62-16400 <=*>
1,094,06C 64-30278 <*>
1,094,578 60-16313 <+*>
1,094,887 60-16312 <+*>
1,094,978 62-16395 <=*> O
1,094,988 66-60254 <=*$>
1,095,251 62-32928 <=*> O
1,095,320 60-16311 <+*>
1,095,395 65-62604 <=$> O
1,096,252 60-16616 <+*>
1,096,593 60-16314 <+*>
1,098,766 64-18393 <*>
1,099,062 60-13276 <+*> O
1,099,953 64-20210 <*>
1,100,943 62-16242 <=*>
1,101,229 59-16175 <+*> O
1,101,469 61-23221 <=*> O
1,101,776 65-60283 <=$>
1,102,216 64-16048 <=*$>
1,102,771 62-25028 <=*> O
1,103,314 59-18782 <+*> O
1,103,549 63-15445 <=*> O
1,103,895 64-16158 <=*$>
1,104,210 60-17289 <+*>
1,105,254 8631 <IICH>
1,105,990 62-24508 <=*> O
1,106,011 63-14875 <*=>
1,106,055 62-13880 <*=> O
1,106,060 ADD 63-19612 <=*$> O
1,106,384 63-23164 <=*$> O
1,106,435 60-23919 <+*> O
1,106,527 59-18288 <+*>
1,106,759 57-22 <*>
1,108,424 57-1996 <*>
1,108,534 <AC>
1,108,787 62-23308 <=*> O
1,109,415 <ATS>
1,109,536 66-10669 <*> O
1,109,975 62-23309 <=*> O
1,110,081 II-904 <*>
1,110,334 62-14380 <=*> O
1,112,068 60-16315 <+*>
1,112,823 64-13777 <=*$> O
1,113,385 63-24611 <=*$> O
1,113,883 64-30279 <*>
1,114,501 64-19914 <=$> O
1,115,645 64-15141 <=*$> O
1,115,647 59-14692 <+*>
1,115,820 60-14792 <+*>
1,116,299 64-30280 <*> O
1,116,341 59-12500 <+*> PO
1,118,129 59-10361 <+*>
1,118,311 63-15446 <=*>
1,120,344 65-12514 <*>
1,120,910 <INFO>
1,122,406 62-16243 <=*>
1,122,568 61-23989 <*=> O
1,122,602 60-16772 <+*>
1,123,598 64-15349 <=*$> O
1,124,515 315 <TC>
1,125,876 61-15047 <+*>
1,126,110 64-30310 <*> O
1,126,481 63-16330 <*>
1,127,274 60-16316 <+*>
1,127,687 64-30311 <*>
1,127,936 59-12616 <+*> O
1,128,423 65-62605 <=$>
1,128,613 62-24507 <=*> O
1,128,746 64-20034 <*>
1,129,111 65-62867 <=$>
1,129,123 64-19794 <=$> O
1,131,423 10929A <K-H>
63-16888 <=*>
1,132,051 64-30312 <*>
1,132,156 59-12921 <+*> O
1,132,401 62-23728 <=*>
1,132,506 59-16176 <+*>
1,132,791 64-15348 <=*$>
1,133,052 61-28079 <*=> O
1,133,654 ADD 61-15677 <*=>
1,133,803 NP-TR-866 <*>
1,133,854 59-17307 <+*>
1,135,575 64-15347 <=*$> O
1,136,039 60-15545 <+*> O
1,136,866 59-19221 <+*> O
1,137,005 64-15140 <=*$> O
1,137,175 64-30313 <*>
1,137,918 64-30314 <*>
1,137,949 65-60837 <=$> O
1,138,694 63-14938 <=*>
1,138,800 60-16317 <+*>
1,139,009 57-3167 <*>
1,139,070 57-3168 <*>
1,139,081 60-16318 <+*>
1,139,359 64-30315 <*> P
1,139,717 63-18642 <*>
1,140,534 64-15139 <=*$> O

Patent	Translation
1,140,683	60-16319 <+*>
1,140,801	64-15859 <=*$> 0
1,141,001	61-28958 <=*> 0
1,141,071	65-63471 <=$>
1,141,446	60-16320 <+*>
1,141,985	59-19222 <+*> C
1,142,158	65-10626 <*>
1,142,370	60-15093 <+*> 0
1,142,720	63-18273 <=*>
1,143,692	60-15546 <+*> 0
1,143,793	<ES>
1,144,881	58-569 <*>
1,146,192	64-30316 <*>
1,146,275	63-23565 <=*>
1,146,808	63-14981 <=*>
1,147,248	62-13879 <*=> C
1,147,642	64-13245 <=*$> C
1,147,707	64-20702 <*>
1,147,728	63-24411 <=*$> C
1,147,756	64-20703 <*>
1,149,169	63-18314 <*=>
1,149,758	64-30317 <*>
1,150,025	60-17609 <+*> 0
1,151,554	64-20704 <*>
1,151,724	64-30318 <*>
1,153,593	60-13277 <+*> 0
1,153,810	60-16321 <+*>
1,156,356	65-10522 <*>
1,156,395	65-10540 <*>
1,159,491	64-13776 <=*$> 0
1,159,894	65-62868 <=$> 0
1,159,950	60-16322 <+*>
1,160,056	59-19223 <+*> 0
1,160,152	59-14205 <+*> C
1,160,925	64-30319 <*>
1,161,079	AEC-TR-4184 <*>
1,161,239	63-18643 <=*>
1,161,299	60-18107 <+*>
1,161,445	59-19745 <+*> 0
1,161,498	65-60284 <=$>
1,161,824	65-10590 <*>
1,162,232	62-18127 <=*> 0
1,162,270	<INFO>
1,162,473	66-10512 <*> 0
1,162,672	60-14793 <+*>
1,162,747	60-23091 <+*> 0
1,162,773	60-16323 <+*>
1,163,284	59-14206 <+*> 0
1,163,932	60-23920 <+*> 0
1,164,393	59-16356 <+*>
1,165,103	65-17116 <*>
1,165,410	199 <TC>
1,165,757	60-16324 <+*>
1,165,781	60-16325 <+*>
1,165,801	64-15846 <=*$>
1,165,852	61-27460 <*=> 0
1,166,173	65-10604 <*>
1,166,607	65-10603 <*>
1,166,608	65-10602 <*>
1,166,652	65-10589 <*>
1,166,653	65-10605 <*>
1,166,679	65-11799 <*>
1,167,007	63-18644 <*>
1,167,489	65-11386 <*>
1,167,683	62-16241 <=*>
1,167,826	64-13775 <=*$> 0
1,167,926	60-16326 <+*> 0
1,168,262	60-16327 <+*>
1,169,029 ADD	64-15355 <=*$> 0
1,169,121	63-19613 <*=>
1,169,208	NP-TR-605 <*>
1,169,524	62-13878 <*=>
1,169,802	65-11374 <*>
1,170,103	60-16328 <+*>
1,170,299	63-18645 <=*>
1,170,688	60-16329 <+*>
1,170,821	59-15008 <+*>
1,171,134	65-13197 <*>
1,171,718	65-63506 <=$> 0
1,172,527	AEC-TR-3773 <+*>

Patent	Translation
1,172,578	63-13434 <=*> 0
1,172,832	62-16394 <=*> 0
1,173,573	64-30320 <*>
1,174,009	64-20705 <*>
1,174,118	65-60285 <=$>
1,174,252	65-63505 <=$> 0
1,175,309	64-30321 <*>
1,175,661	65-14251 <*>
1,175,883	60-15094 <+*> 0
1,176,932	60-16330 <+*>
1,178,079	63-24118 <=*>
1,178,661	AEC-TR-4528 <*>
1,178,718	60-16331 <+*>
1,180,088	65-63470 <=$>
1,180,320	10704 <K-H>
	64-10427 <=*$>
1,180,921	64-30322 <*>
1,181,312	66-60253 <=*$>
1,182,656	60-16332 <+*>
1,182,945	64-30323 <*>
1,183,516	64-30324 <*>
1,183,538	717 <TC>
1,183,644	60-10730 <+*>
1,183,854	62-32228 <=*> 0
1,184,272	65-10527 <*>
1,184,303	62-15746 <*=>
1,184,613	<ES>
1,185,741	61-28080 <*=> C
1,185,889	LJ-ILKUM <=*$>
1,186,512	60-16333 <+*>
1,186,750	66-11947 <*>
1,186,897	63-23772 <=*$> 0
1,187,247	65-10536 <*>
1,187,432	65-62869 <=$> 0
1,188,967	62-25026 <=*> 0
1,189,947	61-13828 <*+>
1,191,114	62-32227 <=*> 0
1,191,408	65-64483 <=*$> 0
1,191,434	62-33780 <=*> 0
1,191,694	60-16334 <+*>
1,192,840	80M37F <ATS>
1,193,579	60-17988 <+*> 0
1,193,700	64-30325 <*> 0
1,193,847	63-23566 <=*>
1,195,037	64-14070 <=*$>
1,195,425	65-11884 <*>
1,195,572	64-30326 <*> 0
1,197,675	61-28959 <*=> 0
1,198,054	61-19486 <+*> 0
1,198,235	AEC-TR-4824 <*>
1,198,432	63-19065 <=*> C
1,198,870	NP-TR-1097 <*>
1,199,460	C-129 <RIS>
1,199,461	C-130 <RIS>
1,201,550	62-11696 <=>
1,202,165	64-15847 <=*$> 0
1,204,627	<AC>
1,205,864	65-62870 <=$> 0
1,206,191	U-141 <RIS>
1,206,359	63-16323 <=*>
1,206,368	64-19795 <*> 0
1,211,577	<AC>
1,212,060	61-19842 <=*> 0
1,212,252	63-18646 <*>
1,213,459	AEC-TR-6238 <*>
1,213,624	AEC-TR-4764 <*>
1,213,809	<INFO>
1,214,398	62-14682 <=*>
1,215,629	61-20026 <=*>
1,215,655	C-131 <RIS>
1,215,972	8680 <IICH>
1,217,175	64-16482 <=*$>
1,217,202	64-15138 <=*$> 0
1,218,434	D-771 <RIS>
1,218,717	63-14982 <=*> 0
1,220,932	65-12103 <*>
1,221,242	672 <TC>
1,224,158	62-18919 <=*> 0
1,225,481	61-23222 <=*> 0
1,225,704	8689 <IICH>

Patent	Citation
1,226,260	64-18481 <*>
1,226,310	63-10833 <=*>
1,226,661	65-60286 <=$>
1,226,679	61-28081 <*=> 0
1,226,883	62-13492 <*=> 0
	8634 <IICH>
1,227,482	61-18578 <*=>
1,228,433	65-62871 <=$> 0
1,228,463	8704 <IICH>
1,228,861	62-10698 <*=>
1,229,113	62-33778 <=*> 0
1,229,702	64-18591 <*>
1,230,077	64-30650 <*>
1,231,192	62-25027 <=*> 0
1,231,593	AEC-TR-5199 <*>
1,232,050	62-24509 <=*> 0
1,232,067	64-10522 <=*$>
1,232,068	64-10521 <=*$>
1,232,431	61-28082 <=*> 0
1,233,010	63-16325 <=*>
1,233,145	266 <TC>
1,233,189	62-13877 <*=> 0
1,234,835	8668 <IICH>
1,236,895	64-13580 <=*$> 0
1,237,042	61-23990 <*=> 0
1,238,411	65-63469 <=$> 0
1,238,450	1372 <TC>
	65-63504 <=$> 0
1,240,638	62-32929 <=*>
1,240,803	65-62872 <=$> 0
1,241,725	64-30327 <*> 0
1,242,380	65-11887 <*>
1,242,907	AEC-TR-5193 <*>
1,243,346	64-13246 <=*$> 0
1,244,286	63-15447 <=*> 0
1,244,428	65-62873 <=$>
1,244,449	65-11883 <*>
1,245,215	62-10661 <*=>
1,245,940	65-62606 <=$> 0
1,246,644	64-15137 <=*$> 0
1,247,044	282 <TC>
1,247,168	27N57F <ATS>
1,248,223	64-30328 <*>
1,249,410	62-32930 <=*>
1,249,913	62-16390 <=*>
1,250,825	64-10664 <*>
1,250,965	65-63503 <=$> 0
1,251,127	65-13560 <*>
1,252,047	62-33015 <=*>
1,252,417	65-17120 <*>
1,252,722	8637 <IICH>
1,252,835	65-62874 <=$> 0
1,253,028	66-10515 <*>
1,254,251	64-20706 <*>
1,254,348	62-18511 <=*>
1,254,587	62-14309 <=*>
1,255,349	64-15345 <=*$> 0
1,256,764	65-11897 <*>
1,256,765	65-11895 <*>
1,256,766	65-11894 <*>
1,256,848	C-178 <RIS>
1,257,893	62-148-93 <=*> 0
1,258,197	790 <TC>
1,258,476	64-15848 <=*$> 0
1,260,410	65-13510 <*>
1,260,411	65-13509 <*>
1,260,510	66-12786 <*>
1,261,465	62-14807 <=*>
1,261,704	65-11889 <*>
1,261,831	65-14464 <*>
1,264,029	1081 <TTIS>
1,264,034	65-10534 <*>
1,264,494	62-16372 <=*>
1,264,672	64-15344 <=*$> 0
1,265,086	64-15849 <=*$>
1,265,242	66-14471 <*> 0
1,266,368	331 <TC>
1,266,417	64-20707 <*>
1,266,511	63-18647 <=*>

Patent	Citation
1,266,663	62-14892 <=*>
1,266,782	62-10664 <*=>
1,267,377	65-10613 <*>
1,267,389	64-14354 <=*$>
1,267,474	64-30329 <*>
1,267,623	62-10659 <*=>
1,267,686	64-16700 <*>
1,268,172	66-10665 <*>
1,268,696	62-18569 <=*>
1,269,042	64-30330 <*>
1,269,057	64-30331 <*>
1,269,065	62-10991 <=*>
1,269,780	65-60287 <=$>
1,270,372	65-60288 <=$>
1,271,356	62-14408 <=*> 0
1,271,532	62-14409 <=*>
1,272,059	62-14683 <*=>
1,273,106	62-18570 <=*>
1,273,165	63-20342 <=*$>
1,273,725	65-61496 <=$> 0
1,273,982	8674 <IICH>
1,274,247	62-18512 <=*>
1,274,325	64-30332 <*> 0
1,274,384	66-11896 <*>
1,274,529	267 <CTS>
1,274,561	64-15343 <=*$> 0
1,274,718	62-18576 <=*>
1,274,802	62-16209 <=*>
1,275,411	64-15850 <=*$> 0
1,276,531	63-18648 <=*>
1,276,804	64-30333 <*>
1,277,755	64-23587 <=$>
1,277,886	8672 <IICH>
1,278,007	65-62607 <=$> 0
1,278,642	63-20025 <=*>
1,278,818	289 <CTS>
1,279,316	62-16196 <=*>
1,281,034	62-18514 <=*> 0
1,281,123	65-14008 <*>
1,281,304	8682 <IICH>
1,281,444	63-18344 <=*>
1,281,774	62-20114 <=*>
1,282,141	65-12763 <*>
1,282,219	63-10874 <=*>
1,282,253	62-18513 <=*>
1,282,837	64-19173 <=*$> 0
1,283,378	64-16483 <=*$> 0
1,283,432	62-20115 <=*>
1,283,616	66-11898 <*>
1,284,511	64-20027 <*>
1,284,537	64-18111 <*>
1,284,995	62-20006 <=*>
1,285,286	65-12764 <*>
1,286,109	62-20421 <=*> 0
1,286,118	66-11406 <*>
1,286,934	66-13248 <*>
1,288,059	62-20403 <=*>
1,289,540	66-11899 <*>
1,290,872	63-10278 <=*>
1,291,655	64-19915 <=$>
1,291,965	63-20569 <=*>
1,293,072	1090 <TTIS>
1,293,373	65-14489 <*>
1,294,386	63-10270 <=*>
1,294,662	65-14851 <*>
1,294,745	63-20026 <=*>
1,294,872	2096 <TC>
1,295,189	66-11407 <*>
1,295,575	63-18556 <=*>
1,295,737	63 <INT>
1,295,757	66-10591 <*>
1,296,819	524 <TC>
1,296,879	63-18339 <=*>
1,297,619	63-18855 <=*>
1,297,834	63-10884 <=*>
1,297,928	66-11408 <*>
1,298,157	63-16430 <=*>
1,298,761	66-10518 <*>
1,299,089	65-14111 <*>
1,299,557	63-20027 <=*>

1,299,604	63-16424 <=*>
1,299,731	63-20054 <=*>
1,299,859	66-10531 <*> 0
1,300,404	1259 <TC>
	1260 <TC>
1,300,500	63-10887 <=*>
1,300,501	63-20028 <=*>
1,300,521	63-20029 <=*>
1,300,795	64-18747 <*>
1,300,819	66-11900 <*>
1,301,035	63-18391 <=*>
	63-20066 <=*>
	65-12857 <*>
1,302,340	63-16319 <=*>
1,302,610	63-16442 <*=> 0
1,303,136	65-61486 <=$> 0
1,303,137	65-60911 <=$>
1,303,313	65-60838 <=$> 0
1,304,072	831 <TC>
1,304,333	65-11644 <*>
1,304,891	64-10953 <=*$>
1,304,911	221 <TC>
	270 <TC>
1,305,343	63-18332 <=*>
1,306,946	65-14091 <*>
1,307,280	63-18898 <=*>
1,307,421	63-20030 <=*>
1,308,181	65-18163 <*>
1,308,567	744 <TC>
1,309,443	63-18350 <=*>
1,309,809	64-30334 <*>
	65-10509 <*>
1,310,122	64-20708 <*>
1,310,547	<TTIS>
1,312,627	66-10505 <*>
1,312,820	2154 <TC>
1,313,247	65-13086 <*>
1,313,290	8673 <IICH>
1,313,723	66-14513 <*>
1,314,407	64-30335 <*>
1,314,557	64-30336 <*> 0
1,315,167	65-11375 <*> 0
1,315,168	65-11376 <*> 0
1,315,191	1844 <TC>
1,315,413	65-62876 <=$> 0
1,315,773	326 <TC>
1,316,044	66-10592 <*>
1,316,401	AD-619,051 <=$>
1,317,031	NS-142 <TTIS>
1,317,185	327 <TC>
1,317,335	65-62877 <=$>
1,317,641	66-10550 <*>
1,318,147	65-14853 <*>
1,318,496	631 <TC>
1,318,863	1759 <TC>
1,319,216	NS 141 <TTIS>
1,319,454	558 <TC>
1,319,732	65-14876 <*>
1,319,733	65-14877 <*>
1,320,310	66-10504 <*> 0
1,321,708	829 <TC>
1,321,751	65-11553 <*>
1,321,757	8664 <IICH>
1,322,358	63-20341 <=*>
1,322,379	63-20340 <=*$> 0
1,322,470	63-20339 <=*>
1,322,589	64-20709 <*>
1,322,626	64-10849 <=*$>
1,323,071	66-11946 <*>
1,323,377	63-20841 <=*$>
1,323,379	63-20842 <=*$>
1,323,589	63-20840 <=*$>
1,323,725	NP-TR-1111 <*>
1,323,780	63-20843 <=*$>
1,323,787	231 <TC>
1,323,831	63-20338 <=*>
1,323,849	64-10850 <=*$>
1,324,070	64-10848 <=*$>
1,324,451	64-10851 <=*$>
1,325,888	65-11886 <*>
1,326,053	64-30337 <*>
1,326,717	216 <TC>
1,327,119	152 <TC>
1,327,120	153 <TC>
1,327,309	153 <TC>
1,327,505	64-14968 <=*$> 0
1,328,692	64-20710 <*> 0
1,328,936	64-18112 <*>
1,329,117	66-11844 <*> 0
1,329,249	64-10691 <=*$>
1,329,505	176 <TC>
1,329,982	64-18492 <*>
	64-30710 <*>
1,330,175	64-10954 <=*$> 0
1,330,367	364 <TC>
	64-20711 <*> 0
1,330,516	64-30338 <*> 0
1,331,135	8636 <IICH>
1,331,152	64-14743 <=*$>
1,331,246	64-30339 <*> 0
1,331,260	430 <TC>
1,331,287	171 <TC>
1,331,412	64-30340 <*>
1,331,450	65-18155 <*> 0
1,332,000	64-18113 <*>
1,332,186	632 <TC>
1,332,815	65-11914 <*>
1,332,816	65-11915 <*>
1,332,817	66-10548 <*>
1,332,824	65-18165 <*>
1,333,111	65-63502 <=$> 0
1,333,247	198 <TC>
1,334,263	65-11885 <*>
1,334,279	172 <TC>
1,335,152	173 <TC>
1,335,166	65-14533 <*>
1,335,203	174 <TC>
1,335,665	64-30341 <*>
1,335,692	341 <TC>
1,335,737	66-12405 <*>
1,336,195	2127 <TC>
1,336,354	66-14470 <*> 0
1,336,379	156 <TC>
	271 <TC>
1,336,411	64-30342 <*> 0
1,336,637	175 <TC>
1,336,641	217 <TC>
1,336,706	NS-197 <TTIS>
1,337,144	177 <TC>
1,337,187	265 <TC>
1,337,309	185 <TC>
1,337,324	8628 <IICH>
1,337,325	8629 <IICH>
1,337,379	8653 <IICH>
1,337,536	65-18191 <*>
1,337,544	65-18188 <*>
1,337,592	342 <TC>
1,337,681	318 <TC>
1,338,380	65-11905 <*>
1,338,591	66-13277 <*>
1,338,597	66-10566 <*>
1,338,618	65-11910 <*>
1,338,846	65-13883 <*>
1,339,082	64-14969 <*>
1,339,149	64-14970 <=*$>
1,339,383	154 <TC>
1,339,384	155 <TC>
1,340,155	206 <TC>
1,340,385	64-14238 <=*$>
1,340,488	209 <TC>
1,341,068	65-11890 <*>
	65-18190 <*>
1,341,268	210 <TC>
1,341,535	64-30343 <*> 0
1,341,711	219 <TC>
1,341,818	208 <TC>
1,342,312	65-63501 <=$> 0
1,342,529	65-13884 <*>
1,342,620	205 <TC>
1,342,813	65-11813 <*>

Patent	Citation
1,342,919	64-16295 <=*$>
1,343,120	241 <TC>
1,343,753	64-14786 <=*$>
1,343,836	66-10503 <*>
1,343,907	264 <TC>
1,343,990	245 <TC>
1,344,042	65-11913 <*>
1,344,055	65-13341 <*>
1,344,859	207 <TC>
1,345,138	64-18114 <*>
1,345,217	1045 <TC>
1,345,836	66-10010 <*>
1,345,933	1261 <TC>
1,346,111	65-11912 <*>
1,346,484	64-30628 <*>
1,346,504	243 <TC>
1,346,533	64-30629 <*>
1,346,548	64-16484 <=*$>
1,346,550	782 <TC>
1,346,650	240 <TC>
1,346,878	244 <TC>
1,347,124	64-18115 <*>
1,347,318	66-10042 <*>
1,347,398	482 <TC>
1,347,399	483 <TC>
1,347,473	65-11911 <*>
1,348,596	232 <TC>
	65-12771 <*>
1,348,656	263 <TC>
1,348,904	66-11431 <*>
1,349,076	599 <TC>
1,349,484	233 <TC>
1,349,897	65-60913 <=$>
1,349,923	474 <TC>
1,349,990	471 <TC>
1,350,056	66-10637 <*> O
1,350,167	472 <TC>
1,350,458	473 <TC>
1,350,624	65-64570 <=*$>
1,350,772	66-11088 <*>
1,350,880	229 <TC>
1,351,093	1415 <TC>
1,351,484	66-11475 <*>
1,352,034	606 <TC>
1,352,298	977 <TC>
1,353,162	64-20936 <*> O
1,353,187	65-11791 <*>
1,353,230	65-11892 <*>
1,353,543	608 <TC>
1,353,669	550 <TC>
1,354,548	65-11560 <*>
1,355,208	885 <TC>
1,355,279	64-30630 <*>
1,355,698	66-10526 <*>
1,355,951	66-11124 <*>
1,355,976	65-11803 <*>
1,356,617	398 <TC>
1,356,799	8642 <IICH>
1,356,939	605 <TC>
1,357,010	604 <TC>
1,357,214	64-30631 <*>
1,357,634	65-11812 <*>
1,357,736	66-13283 <*>
1,357,952	476 <TC>
1,357,990	66-10552 <*>
1,358,062	607 <TC>
1,358,473	477 <TC>
1,359,016	361 <TC>
1,359,090	353 <TC>
	65-11368 <*>
1,360,163	162 <WFK>
1,360,251	66-12795 <*>
1,360,473	623 <TC>
1,360,488	65-11802 <*>
1,360,683	66-11432 <*>
1,360,715	66-11915 <*>
1,360,905	604 <TC>
1,361,089	66-11889 <*>
1,361,220	601 <TC>
1,361,247	8755 <IICH>
1,361,264	66-11919 <*>
1,361,296	600 <TC>
1,362,100	65-12774 <*>
1,362,612	65-62608 <=$> O
1,362,622	65-14137 <*>
1,362,623	8683 <IICH>
1,362,624	65-62609 <=$> O
1,362,954	66-10553 <*>
1,363,062	66-11417 <*>
1,363,069	65-11801 <*>
1,363,312	65-11552 <*>
1,363,499	444 <TC>
1,363,673	66-10509 <*>
1,363,746	65-11021 <*>
1,364,519	66-11409 <*>
1,365,023	65-11811 <*>
1,365,633	161 <WFK>
1,365,885	66-11416 <*>
1,366,686	643 <TC>
1,367,023	65-11551 <*>
1,367,074	66-12771 <*>
1,368,374	652 <TC>
1,368,589	66-11405 <*>
1,368,809	66-12772 <*>
1,369,911	65-11714 <*>
1,370,058	8656 <IICH>
1,370,354	1842 <TC>
1,370,473	519 <TC>
	65-11646 <*>
1,370,485	1127 <TC>
1,370,722	65-13085 <*> O
1,370,810	1084 <TC>
1,370,905	8657 <IICH>
1,371,067	NS-434 <TTIS>
1,371,163	578 <TC>
1,371,169	8659 <IICH>
1,371,338	66-11127 <*>
1,371,720	577 <TC>
	65-11648 <*>
1,372,987	65-13264 <*>
1,373,020	551 <TC>
	65-11647 <*>
1,373,189	1774 <TC>
1,373,236	66-12777 <*>
1,373,614	1746 <TC>
1,373,709	902 <TC>
1,374,460	1319 <TC>
1,374,463	1320 <TC>
1,374,998	1749 <TC>
1,375,795	8658 <IICH>
1,376,129	138 <WFK>
1,376,200	1748 <TC>
1,376,351	130 <WFK>
1,376,599	901 <TC>
1,376,809	903 <TC>
1,377,244	65-12874 <*>
	660 <TC>
1,377,276	65-14235 <*>
1,377,343	1026 <TC>
1,377,435	850 <TC>
1,377,943	66-12793 <*>
1,377,964	65-14837 <*>
1,378,027	463 <TC>
1,378,130	66-10519 <*>
1,378,365	66-60252 <=*$>
1,378,812	774 <TC>
1,379,684	2005 <TC>
1,380,043	166 <WFK>
1,380,154	66-12773 <*> O
1,380,360	2006 <TC>
1,380,458	830 <TC>
1,380,970	65-13949 <*>
	832 <TC>
1,381,029	65-13948 <*>
	833 <TC>
1,381,215	789 <TC>
1,381,442	1315 <TC>
1,381,786	788 <TC>
1,382,072	66-13284 <*>

1,382,454	65-13084 <*> 0
1,383,076	882 <TC>
1,383,362	1644 <TC>
1,383,381	65-13716 <*>
	834 <TC>
1,383,951	2007 <TC>
1,385,124	65-14377 <*>
1,385,764	1357 <TC>
1,386,234	587 <TC>
1,386,285	8661 <IICH>
1,386,289	8648 <IICH>
1,386,295	911 <TC>
1,386,901	1318 <TC>
1,387,492	1747 <TC>
1,388,082	1986 <TC>
1,388,083	1985 <TC>
1,388,307	164 <WFK>
1,388,525	66-14539 <*>
1,388,580	66-14540 <*>
1,388,593	1348 <TC>
1,389,360	8646 <IICH>
1,389,500	913 <TC>
1,389,817	1077 <TC>
1,390,143	66-12804 <*>
1,390,168	8645 <IICH>
1,390,398	8643 <IICH>
1,390,470	8644 <IICH>
1,390,487	1055 <TC>
	65-14159 <*>
1,390,874	65-14750 <*>
1,391,580	2024 <TC>
1,391,984	1742 <TC>
1,392,729	1202 <TC>
1,394,382	8652 <IICH>
1,395,399	2008 <TC>
1,396,152	2010 <TC>
1,397,818	1368 <TC>
	65-17313 <*>
1,397,820	1408 <TC>
1,397,859	2011 <TC>
1,397,897	1814 <TC>
1,398,153	66-13262 <*>
1,398,551	1754 <TC>
1,398,775	66-10493 <*>
1,398,826	1387 <TC>
1,398,827	1388 <TC>
1,399,077	66-10118 <*>
1,399,078	66-10116 <*>
1,399,518	2013 <TC>
1,399,949	2014 <TC>
1,400,431	2015 <TC>
1,403,325	66-11945 <*>
1,405,884	66-10583 <*>
1,406,132	66-13993 <*>
1,406,804	1902 <TC>
1,407,624	2104 <TC>
1,408,589	66-11944 <*>
1,409,564	1871 <TC>
1,412,512	66-10766 <*> 0
1,413,372	66-12570 <*>
1,413,893	8751 <IICH>
1,414,509	66-12507 <*>
1,429,509	66-13868 <*>

FRANCE APPLICATION

609,407	91Q70F <ATS>
778,117	60-14205 <+*>
818,761	65-11859 <*>
821,306	65-10530 <*>
	8625 <IICH>
839,994	65-62875 <=*$>

FRANCE, ADDITION

10,231	64-13241 <=*$> C
21,878	63-24612 <=*$> 0
22,597	63-23560 <=*$> 0
22,956	62-18731 <=*>
43,543	63-24406 <=*$> 0
61,249	62-33775 <=*> C
65,273	61-19837 <*=> 0
67,911	63-19612 <=*$> C
68,796	61-15677 <*=> 0
70,956	61-28956 <=*>
72,237	63-10300 <=*>
73,430	64-15355 <=*$> 0
84,722	1815 <TC>

FRANCE, MEDICAL

1,156	63-20031 <=*>

GERMANY

1,303	63-23567 <=*>
2,173	63-16060 <*=>
4,933	63-16066 <*=>
5,786	64-30344 <*> 0
7,235	63-13438 <=*> 0
7,890	63-13439 <=*> C
9,376	64-30347 <*>
11,210	63-13440 <=*> 0
12,537	SCL-T-432 <*>
13,155	59-12635 <=>
13,397	64-30346 <*>
13,661	59-00831 <*>
15,540	63-13441 <=*> 0
15,909	<LSA>
16,273	65-10513 <*>
19,731	63-13443 <=*> 0
21,589	64-20594 <*>
23,266	REPT. 97048 <RIS>
23,580	62-18412 <=*> C
25,150	65-61495 <=$>
27,923	63-19271 <=*> 0
30,909	63-13444 <=*> C
38,332	63-13445 <=*> 0
40,354	63-18248 <=*>
43,213	64-30345 <*>
43,235	63-16155 <*=> 0
44,640	63-13446 <=*> C
45,105	63-16154 <*=> 0
45,378	63-16153 <=*> 0
49,222	65-64482 <=*$>
55,868	I-33 <*>
60,460	64-30348 <*> C
61,724	59-17028 <+*> C
63,180	63-20206 <=*>
64,928	64-20722 <*>
70,674	63-14295 <=*> 0
71,384	62-18481 <=*>
78,601	65-62878 <=$>
82,326	59-12080 <+*>
86,607	II-743 <*>
88,999	64-20300 <*>
89,060	62-18208 <=*> 0
90,460	64-15851 <=*$> 0
91,203	63-14050 <=*>
91,591	62-25682 <=*> 0
93,042	60-16439 <+*>
96,160	57-3564 <*>
102,241	65-12051 <*>
102,315	64-20595 <*>
103,706	62-18301 <=*>
105,823	60-16434 <+*>
108,053	[illegible]-14947 <=*>
108,342	[illegible]-20599 <*>
108,400	[illegible]-30349 <*>
110,953	[illegible]-10574 <*>
112,449	63-20229 <=*>
114,210	61-10115 <+*>
115,722	65-10516 <*>
116,706	63-23773 <=*$> 0
117,748	64-20596 <*>
121,530	60-16431 <+*>
124,286	119178 <K-H>
127,156	[illegible]-459 <*>

Patent	Citation
127,273	62-18456 <=*> 0
129,181	65-12027 <*>
133,345	65-64572 <=*$> 0
133,533	62-32912 <0>
134,186	60-14735 <+*>
135,722	63-14521 <=*>
135,875	64-20597 <*>
137,994	60-16044 <+*>
138,531	65-10607 <*>
138,562	63-14522 <=*>
143,042	60-14737 <+*>
145,880	64-30350 <*> 0
147,193	60-10222 <+*> 0
151,541	59-22279 <+*> C
152,724	62-18427 <=*> 0
154,657	64-20598 <*>
156,269	60-16465 <=*$>
158,817	59-17109 <+*> 0
160,172	59-15090 <+*>
160,540	65-10306 <*>
160,851	62-18451 <=*> 0
160,906	60F2G <ATS>
161,023	63-24609 <=*$> 0
161,526	60-10223 <+*> 0
162,395	63-20208 <=*>
165,330	<INFO>
166,898	63-20207 <=*$>
166,899	63-20041 <=*>
170,133	63-20231 <=*>
179,304	64-30351 <*>
182,042	<ATS>
	64-20600 <*> 0
184,269	64-20621 <*>
186,469	62-18458 <=*>
188,571	63-20230 <=*$>
188,712	63-20042 <=*>
189,482	63-20043 <=*$>
189,843	57-1991 <*>
194,753	65-10308 <*>
195,523	63-14523 <=*>
197,800	64-20602 <*> 0
198,715	64-30352 <*>
199,753	II-400 <*>
214,229	59-18783 <+*> 0
215,480	I-425 <*>
223,909	60-15549 <+*> 0
229,634	59-14207 <+*> 0
231,800	65-10548 <*> 0
	8715 <IICH>
234,016	64-15852 <=*$> 0
234,306	64-20603 <*> 0
235,185	65-64481 <=*$>
240,960	58-702 <*>
241,705	61-10060 <+*>
243,892	66-15189 <*>
244,779	64-30353 <*>
245,150	I-658 <*>
245,195	63-23568 <=*$> 0
249,116	64-13774 <=*$> 0
252,322	63-20583 <=*$>
256,856	64-20604 <*>
258,364	62-33235 <=*> 0
259,929	59-22280 <+*> 0
260,295	59-12922 <+*> C
260,379	63-20584 <=*$>
261,689	64-20605 <*>
264,992	65-63479 <=$> 0
265,210	63-20646 <*=>
265,260	63-20813 <=*$>
268,816	64-30354 <*> 0
268,858	63-14646 <=*>
269,826	59-16835 <+*>
271,381	64-30355 <*>
271,522	15H9G <ATS>
272,428	56K25G <ATS>
275,166	64-20606 <*>

Patent	Citation
276,400	5044-B <K-H>
277,128	62-32913 <=*> 0
280,907	64-30356 <*> C
281,175 SUP	64-14923 <=*$>
281,388	64-20607 <*>
281,688	64-30357 <*>
282,694	57-3559 <*>
284,367	59-17194 <+*> C
285,037	61-13258 <=*> 0
285,110	61-10075 <=*>
287,932	741 <TC>
	742 <TC>
289,103	59-17886 <+*>
290,054	59-17886 <+*>
293,921	3964-H <K-H>
294,267 ADD	57-1801 <*>
295,547	57-1801 <*>
296,867	64-20608 <*>
301,446	64-13773 <=*$> 0
305,192	63-20585 <=*$>
305,620	62-18300 <=*> C
306,092	59-14693 <+*> 0
307,153	59-18289 <+*> C
307,791	63-20586 <=*$>
308,011	64-13772 <=*$> 0
308,386	63-20587 <=*$>
308,992	59-14694 <+*> 0
310,510	62-25689 <=*> 0
311,048	62-25687 <=*> C
311,712	62-18450 <=*> 0
312,959	64-30358 <*>
313,053	63-16064 <*=>
316,292	63-13748 <=*> 0
318,223	64-10071 <=*$>
318,900	64-10072 <=*$>
318,901	64-10073 <=*$>
319,023	63-20588 <=*$>
323,665	64-10074 <=*$>
323,665 ADD	64-10076 <=*$>
325,511	65-63500 <=$> 0
325,647	64-10076 <=*$>
329,366	64-20611 <*> 0
329,422	62-18260 <=*>
333,985	62-18460 <=*> C
335,603	64-10075 <=*$>
336,029	59-14695 <+*> 0
336,558	63-20589 <=*$>
336,583	64-30359 <*> 0
337,672	65-11909 <*>
339,495	61-16746 <=*$>
340,591	62-18459 <=*> 0
340,872	64-10758 <=*$>
342,310	64-30360 <*>
342,564	65-64574 <=*$> C
342,610	II-28 <*>
344,450	63-14985 <=*>
346,949	63-20590 <*>
349,328	62-18449 <=*> 0
351,104	61-16747 <=*$>
355,851	64-20612 <*>
357,757	61-16748 <=*$>
359,784	63-23569 <=*$> C
360,202	64-20613 <*>
360,414	64-30361 <*>
360,767	65-64573 <=*$> 0
360,890	64-30362 <*>
362,990	66-10494 <*>
363,192	63-20647 <=*$>
363,553	59-12081 <+*>
363,666	63-20591 <=*$>
364,519	65-10606 <*>
365,160	63-20592 <=*$>
367,768	64-20614 <*>
368,721	63-24119 <=*>
370,422	64-19174 <=*$> C
372,648	62-33779 <=*>
372,925	59-17858 <+*>
373,369	64-30363 <*>
375,638	61-16750 <=*$>
378,119	62-18475 <=*> C

Patent	Citation
379,337	64-30364 <*>
379,410	64-20615 <*>
380,103	33L29G <ATS>
382,626	59-17836 <+*> 0
383,270	64-14971 <=*$> 0
384,147	61-16749 <=*$>
384,351	64-10077 <=*$>
385,049	64-30365 <*>
386,848	62-25690 <=*> 0
388,219	63-20593 <=*$>
388,285	1025 <TC>
389,024	I-861 <*>
392,831	II-710 <*>
394,499	64-30366 <*>
394,520	63-24120 <=*>
404,170	64-30367 <*>
404,214	63-23165 <=*$> 0
405,043	59-17119 <+*>
406,113	61-19487 <+*> 0
407,895	64-10078 <=*$>
409,277	63-23570 <=*>
411,052	63-18680 <=*>
	64-14923 <=*$>
411,107	64-10123 <=*$>
411,950	65-12026 <*>
412,169	65-11860 <*>
414,593	63-13000 <=*>
415,191	64-30368 <*> 0
415,482	934 <TC>
416,647	59-17149 <+*> 0
419,717	64-20616 <*>
420,035	65-12052 <*>
420,232	59-16177 <+*> C
420,894	62-18454 <=*> 0
422,807	65-62879 <=$> 0
423,826	59-18784 <+*> 0
426,479	64-13579 <=*$> 0
426,957	59-21209 <=>
427,858	64-10145 <=*$>
429,272	63-20594 <=*$>
430,539	64-30369 <*>
	65-13585 <*>
430,741	65-62610 <=$> 0
431,537	62-18464 <=*> 0
434,924	63-20595 <=*$>
438,191	59-18785 <+*> 0
439,651	59-18786 <+*> 0
439,971	57-3081 <*>
443,020	65-13189 <*>
443,960	63-20597 <=*$>
446,174	62-33236 <=*> 0
446,268	59-10360 <+*>
446,290	63-20598 <=*$>
447,392	63-16331 <=*>
448,526	64-30370 <*>
452,922	64-30371 <*> 0
453,473	63-10032 <=*>
454,142 ADD	62-18472 <=*> C
454,458 SUP	64-10225 <=*$>
455,504	65-64480 <=*$>
456,751	65-62880 <=$> 0
456,841	I-480 <*>
457,179	64-30372 <*>
457,992	59-19746 <+*> 0
464,142	63-20645 <=*$>
464,510	60-15864 <+*> 0
465,381	59-17120 <+*>
466,844	62-18472 <=*> 0
467,289	63-23571 <=*$> 0
469,994	62-18466 <=*> 0
471,737	61-15683 <+*> C
472,421	64-30373 <*> 0
472,680	63-20599 <=*$>
475,229	62-18448 <=*> 0
475,789	64-20617 <*>
476,074	64-30374 <*> 0
478,666	63-10031 <=*>
479,119	62-32225 <=*> 0
479,349	64-30375 <*>
479,629	64-30376 <*> 0
481,795	62-18479 <=*> C
482,857	16F3G <ATS>
484,167	64-30377 <*>
485,055	64-10692 <=*$>
485,124	62-18612 <=*> 0
485,139	63-20600 <=*$>
485,362	62-18463 <=*> 0
485,944	63-20601 <=*$>
488,298	65-62881 <=$> C
491,309	62-18474 <=*>
492,064	61J15G <ATS>
492,612	65-60289 <=$>
492,768	62-18465 <=*> C
493,319	II-898 <*>
494,575	64-14368 <=*$>
494,691	64-10092 <=*$>
494,950	64-10093 <=*$>
496,062	63-18539 <=*>
496,380	25G6G <ATS>
496,436	63-13426 <=*> 0
497,630	3964-C <K-H>
498,848	59-12617 <+*> 0
498,974	64-30378 <*>
499,550	65-10358 <*>
500,202	253 <TC>
500,294	3964-D <K-H>
500,998	3964-E <K-H>
501,124	64-10204 <=*$>
501,135	64-30379 <*>
501,852	64-20486 <*>
502,833	60-10224 <+*>
503,594	64-30380 <*> 0
503,766	59-12153 <+*>
505,265	64-30381 <*>
505,278	64-23588 <=$>
505,318	64-20618 <*>
507,194	3869-D <K-H>
508,667	1126 <*>
	64-14372 <=*$>
508,795	64-10091 <=*$>
508,895	62-18462 <=*> 0
509,016	3964-B <K-H>
509,467	59-16178 <+*> 0
510,376	5513-F <K-H>
510,376 SUP	5513-E <K-H>
510,377	5513-D <K-H>
511,983	5513-E <K-H>
513,027	59J15G <ATS>
513,942	64-20619 <*>
514,593	64-30382 <*>
515,315	6E2G <ATS>
516,283	3964-F <K-H>
516,595	64-30383 <*>
516,793	65-62882 <=$> 0
	65-64470 <=*$> 0
520,502	63-19064 <=*> 0
523,052	64-30384 <*> C
523,436	65-10930 <*>
523,601	64-14691 <=*$>
524,943	62-00403 <*>
525,332	65-10302 <*>
526,405	63-15769 <=*> C
526,478	63-20320 <=*>
526,491	64-10090 <=*$>
526,626	65-12032 <*>
527,739	61-10042 <+*>
529,817	62-18490 <=*> 0
530,179	61-13831 <=*$> 0
530,687	63-13001 <=*> C
530,732	64-10205 <=*$>
531,936	65-62684 <=$> 0
532,106	5550-C <K-H>
533,395	64-23589 <=$>
535,235	96373 <RIS>
537,696	65-10608 <*>
537,716	65-10653 <*>
538,422	5550-B <K-H>
539,104	64-20485 <*>

Patent	Citation
539,228	64-30385 <*>
539,945	63-16087 <*=>
540,701	64-10089 <=*$>
541,739	64-15341 <=*$> O
541,961	59-16179 <+*> O
542,640	59-18290 <+*> O
543,171	<INFO>
544,387	63-20176 <=*>
545,264	64-10208 <=*$>
545,763	64-10209 <=*$>
546,335	64-10693 <=*$>
548,731	62-18457 <=*>
548,939	62-18455 <=*> O
549,413	10860B <K-H>
	64-14480 <=*$>
550,243	64-10087 <=*$>
551,169	65-64575 <=*$>
551,424	64-10211 <=*$>
551,869	60-16024 <+*>
552,251	64-10086 <=*$>
552,328 SUP	64-14489 <*>
552,466	64-30386 <*>
552,584	64-30387 <*> O
552,987	60-16025 <+*>
553,071	64-30388 <*>
553,123	6J19G <ATS>
553,648	62-11555 <=>
553,811	64-10085 <=*$>
554,695	64-20628 <*>
554,700	64-10214 <=*$>
554,906	59-18787 <+*> O
556,753	II-826 <*>
557,151	5434B <K-H>
557,203	64-10084 <=*$>
557,220	64-10083 <*>
557,285	5434C <K-H>
558,028	59-17121 <+*>
558,581	64-10080 <=*$>
559,201	61-13832 <=*> O
559,328	64-10081 <=*$>
559,833	RJ42F4G <ATS>
560,146	64-10082 <=*$>
561,082	65-63499 <=$*>
562,010	64-14394 <=*$>
562,738	261 <TC>
562,820	64-30389 <*>
564,793	64-20626 <*>
564,899	64-14696 <=*$>
565,160	64-20625 <*>
565,318	64-20484 <*>
566,823	66-60251 <=*$>
567,099	64-10220 <=*$>
567,648	64-10221 <=*$>
567,683	64-10222 <=*$>
567,705	64-30390 <*>
567,844	63-14524 <=*>
568,375	59-17105 <+*>
569,699	59-14208 <+*> O
570,148	62-18519 <=*>
571,124	64-10223 <=*$> O
572,062	I-604 <*>
573,048	64-10224 <=*$>
573,105	64-30391 <*>
575,593	64-20452 <*>
577,051	63-16067 <*=>
577,428 ADD	64-10653 <=*$>
578,351	64-23590 <=$>
579,655	64-10225 <=*$>
579,948	64-20629 <*>
580,452	64-10226 <=*$>
580,517	60-14731 <+*>
580,533	62-15751 <=*>
582,384	64-15340 <=*$>
582,670	62-18467 <=*> O
583,259	64-13247 <=*$>
583,343	64-10227 <*>
584,844	II-209 <*>
585,063	II-27 <*>
585,185	64-14087 <=*$>
585,657	64-20624 <*>
585,793	64-20623 <*>
586,162	64-30392 <*>
587,653	64-30393 <*>
589,970	64-30394 <*>
590,818	64-10228 <*>
591,196	64-10640 <=*$>
591,592	59-12314 <+*>
591,715	62-20466 <*=>
591,962	06N49G <ATS>
592,053	64-10641 <=*$>
592,777	64-10642 <=*$>
593,384	64-30395 <*>
593,416	64-30396 <*>
593,425	2353 <*>
593,923	63-16068 <*=>
594,219	64-10643 <=*$>
595,464	64-20568 <*>
595,604	64-10644 <=*$>
596,576	64-10645 <=*$>
597,305	64-10646 <=*$>
599,383	64-20621 <*>
599,565	76T90G <ATS>
601,323	59-17885 <+*>
601,508	60-17610 <+*> O
601,700 ADD	64-10647 <=*$>
602,064	65-62686 <=$>
603,257	<INFO>
603,302	64-10647 <=*$>
604,278	65-64479 <=*$>
605,016	I-263 <*>
605,121	64-30397 <*> O
605,746	63-14525 <=*>
605,973	64-10648 <=*$>
605,984	64-10649 <=*$>
606,439	64-10650 <=*$>
606,776	64-10651 <=*$>
607,018	64-10652 <=*$>
608,362	64-10653 <=*$>
608,413	64-10654 <=*$>
608,692	64-14088 <=*$>
608,693	64-14089 <*>
608,973	64-10655 <=*$>
609,164	64-10656 <=*$>
609,300	64-10657 <=*$>
609,336	64-10658 <=*$>
609,456	64-10659 <=*$>
609,634	64-10660 <=*$>
609,636	64-10661 <=*$>
609,697	64-10662 <=*$>
609,731	60-14961 <+*> O
610,409	64-14090 <*>
610,409 SUP	64-14504 <*>
610,600	64-14091 <*>
611,043	64-14092 <*>
611,283	60-14959 <+*>
611,373	64-14093 <*>
611,443	64-14094 <*>
611,924	60-14958 <+*>
612,073	64-14697 <=*$>
612,417	64-14095 <*>
612,489	59-17977 <+*> O
612,509	59-19747 <+*> C
612,843	63-24412 <=*$> O
613,124	60-16378 <+*> PO
613,308	64-14096 <*>
614,567	64-30398 <*>
614,702	64-14097 <=*$>
615,087	62-18453 <=*> O
615,470	64-14098 <=*$>
616,084	64-15339 <=*$> O
616,443 SUP	64-14506 <*>
618,182	64-30399 <*>
618,216	61-19843 <=*> O
618,592	64-14099 <=*$>
618,727	64-30400 <*> O
618,836	64-14100 <*>
619,019	64-14101 <*>
619,233	64-14102 <=*$>
619,241	15G6G <ATS>
619,386	64-14103 <=*$>

Patent	Citation
621,396	64-14104 <=*$>
621,962	64-14105 <*>
622,100	64-14106 <*>
622,262	64-14107 <*>
622,268	64-14108 <*>
622,296	64-14109 <*>
622,298	64-14110 <=*$>
622,581	62-32914 <=*> O
622,648	63-14526 <=*>
622,773	62-18461 <=*> O
623,108	64-14111 <*>
623,542	60-16370 <+*>
623,657	64-14112 <=*$>
623,876	64-30401 <*>
624,091	63-19063 <=*> O
624,843	64-14113 <=*$>
624,871	65-60290 <=$>
625,148	64-20622 <*>
625,477	64-20620 <*>
625,976	61-15050 <+*> O
626,880	64-14484 <*>
626,970	61-10072 <+*>
627,043	63-23572 <=*$> O
627,636	64-14485 <=*$>
627,925	65-64478 <=*$>
627,927	65-64576 <=*$> O
628,273	64-14486 <=*$>
628,715	64-14487 <=*$>
629,220	8725 <IICH>
629,723	64-14488 <=*$>
629,798	I-681 <*>
630,679	64-14489 <*>
630,790	64-14490 <*>
632,015	64-14491 <*>
633,056	64-14492 <=*$>
633,083	63-18518 <=*>
633,517	64-14493 <*>
633,871	64-14494 <=*$>
633,895	65-13104 <*>
633,921	1082 <*>
634,119	59-16836 <+*> O
634,278	64-30402 <*> O
634,305	62-20065 <=*>
634,306	63-10079 <=*>
635,184	64-14495 <*>
635,396	64-30403 <*>
635,397	65-10552 <*>
635,903	64-14496 <*>
635,926	64-14497 <*>
635,989	59-16837 <+*>
636,092	64-14498 <*>
636,193	64-14499 <*>
636,259	64-14500 <=*$>
636,560	63-14986 <=*>
637,942	64-14502 <=*$>
638,117	63-20318 <=*>
638,302	64-14503 <*>
638,400	64-14504 <*>
638,594	64-30404 <*>
638,963	63-19567 <=*>
639,281	64-14505 <*>
639,308	64-14506 <*>
639,454	64-14507 <*>
639,527 SUP	64-14864 <*>
639,583	64-14508 <*>
639,625	64-14509 <*>
639,733 SUP	64-14510 <*>
640,018	64-14856 <*>
640,019	64-14857 <*>
640,405	64-14510 <*>
640,681	64-14858 <*>
640,997 SUP	64-14870 <*>
641,004	64-14859 <*>
641,086	64-14860 <*>
642,307	64-16248 <=*$>
642,344	64-14861 <*>
642,361	64-30131 <*>
642,414	64-14862 <*>
642,454	64-14863 <*>
642,518	64-14864 <*>
642,531	64-30405 <*>
642,533	64-14865 <*>
642,574	64-14866 <*>
642,590	64-14867 <*>
642,829	64-14868 <*>
642,950	64-14869 <*>
643,052	64-14870 <*>
643,062	64-14871 <*>
643,157	64-20630 <*>
643,301	62-18480 <=*>
643,668	64-14872 <*>
643,993	64-14873 <*>
644,077	63-10832 <=*>
644,131	64-14874 <*>
644,486	63-20278 <=*>
644,686	64-14875 <*>
644,759	59-12082 <+*>
645,255	64-14876 <*>
645,545	66-10501 <*> O
645,608	64-14877 <*>
646,109	64-18489 <*>
646,290	64-14878 <*>
646,478	2647 <*>
	64-14682 <=*$>
646,503	64-14879 <*>
646,575	59-16838 <+*>
647,324	64-14880 <*>
647,429	59-14696 <+*> C
647,451	64-14881 <*>
647,523	64-14882 <*>
647,649	64-14883 <*>
647,988	64-14884 <=*$>
648,381	64-10822 <=*$> O
648,558	64-14885 <*>
649,118	64-20631 <*>
649,659	63-16326 <=*>
649,831	63-14987 <=*>
650,137	65-10319 <*>
651,109	1147 <*>
651,111	63-20184 <*>
651,797	1071 <TC>
653,175	63-14984 <=*>
653,883	60-14932 <+*>
654,113	59-17708 <+*> O
654,250	64-30406 <*>
654,714	1157 <*>
654,757	62-18190 <=*>
655,934	64-16255 <=*$>
656,567	64-16256 <=*$> O
656,642	<LSA>
656,956	62-20066 <=*>
657,010	64-30407 <*> O
659,434	64-20632 <*>
659,766	I-243 <*>
659,882	64-16257 <=*$>
660,003	60-14926 <+*>
660,164	NP-TR-802 <*>
660,268	64-19805 <=>
661,173	63-16065 <=*>
661,883	64-16258 <=*$>
662,027	64-13248 <=*$> O
662,032	59-12923 <+*> O
663,685	64-14681 <=*$>
663,762	64-20633 <*>
664,021	64-16259 <=*>
664,086	64-16260 <=*$>
	64-30408 <*>
664,088	64-16261 <=*$>
664,320	59-12083 <+*>
664,745	64-16262 <=*$>
665,209	59-12084 <+*>
665,790	65-10616 <*>
667,279	MF-5 <*>
	2824 <*>
667,344	59-19748 <+*> O
667,744	64-16264 <=*$>
668,142	260 <TC>
668,227	65-63489 <=$> O
668,443	64-30409 <*>

668,602	59-12618 <+*> 0
668,688	64-30410 <*>
668,952	65-10551 <*>
669,605	59-14697 <+*> 0
669,857	59-17971 <+*> 0
670,106	62-18452 <=*> 0
671,098	63-20277 <=*>
672,022	64-20431 <*>
672,110	64-30411 <*>
672,432	64-16455 <=*$>
673,016	64-30412 <*> 0
673,139	63-16049 <=*>
673,277	64-16628 <=*$>
673,682	64-16267 <=*$>
673,796	62-16164 <=*>
673,828	65-10550 <*>
674,250	64-20634 <*>
675,039	64-20635 <*>
675,278	62-32915 <=*> 0
675,657	63-18027 <=*>
676,117	64-20548 <*>
677,071	64-30413 <*>
677,405	59-14698 <+*> 0
677,436	I-156 <*>
677,977	63-10813 <=*>
678,623	64-20636 <*>
679,592	64-20637 <*C>
681,916	64-16269 <=*$>
682,048	63-18984 <=*>
684,332	63-15767 <=*> 0
684,466	63-23166 <=*$> 0
684,975	64-16270 <=*$>
685,048	60-16104 <=*>
685,125	64-30414 <*>
685,367	14R76G <ATS>
685,659	59-12924 <+*> 0
685,941	59-18788 <+*> 0
687,409	65-62883 <=$> 0
689,151	64-16271 <=*$>
692,642	20N58G <ATS>
692,711	64-16272 <=*$>
692,954	65-10549 <*>
693,276	63-23167 <=*$> 0
693,302	63-13749 <=*> 0
	64-30415 <*>
693,373	64-16629 <=*$>
693,416	64-16630 <=*$>
694,043	65-10647 <*>
694,408	62-16255 <=*>
694,823	64-30416 <*>
695,097	64-16631 <=*$>
695,619	62-18473 <=*> 0
695,995	AL-531 <*>
696,182	58J15G <ATS>
696,317	64-13771 <=*$>
696,393	64-15853 <=*$>
698,108	63-14983 <=*>
698,485	59-16180 <+*> 0
698,653	64-16632 <=*$>
698,742	64-16633 <=*$>
698,852	99 <TC>
699,448	64-30417 <*> 0
699,664	62-18483 <=*>
700,620	64-20638 <*>
700,918	8677 <IICH>
701,006	59-12085 <+*> 0
701,318	64-30418 <*> 0
701,423	59-14699 <+*> 0
703,976	60-16107 <+*>
704,494	60-16108 <+*>
704,530	65-13105 <*>
704,757	62-24515 <=*> 0
705,419	59-19749 <+*> 0
705,450	64-16634 <=*$>
705,568	62-18477 <=*>
706,197	64-30419 <*> 0
706,273	59-14700 <+*> 0
706,782	60-16109 <+*>
707,557	65-63481 <=$>
708,261	64-16476 <=*$>
708,900	62-18478 <=*>
710,048	63-14527 <=*>
710,365	62-18468 <=*>
710,439	I-683 <*>
710,637	I-885 <*>
710,694	64-30420 <*>
712,002	I-884 <*>
712,564	60-16113 <+*>
712,648	62-18469 <=*> 0
712,901	62-18470 <=*> 0
713,514	63-19272 <=*> 0
713,910	54K21G <ATS>
714,033	65-11134 <*>
714,246	I-684 <*>
714,365	62-18471 <=*> 0
714,456	63-14819 <=*>
714,819	61-16759 <*=>
714,870	62-13873 <=*>
715,337	<ATS>
715,581	I-828 <*>
715,748	<INFO>
715,780	62-32224 <=*> 0
715,931	<INFO>
716,917	64-19796 <=$> 0
717,078	64-18039 <*>
717,188	62-18482 <=*> 0
717,273	I-257 <*>
717,456	62-32223 <=*> 0
717,547	64-14972 <=*$>
717,766	II-4 <*>
718,013	64-20374 <*>
718,071	60-16115 <+*>
718,510	63-14528 <=*>
718,606	I-685 <*>
719,027	60-16116 <+*>
719,051	62-23732 <=*>
719,394	I-275 <*>
720,102	60-16443 <+*>
720,178	63-16332 <=*>
720,870	62-23733 <=*> 0
721,885	60-16117 <+*>
721,991	64-18040 <*>
722,282	64-30421 <*>
722,566	63-24413 <=*$> 0
724,028	59-18291 <+*> 0
724,204	62-32916 <=*> 0
724,946	63-14647 <=*>
725,082	1360 <*>
725,135	60-23929 <+*> 0
725,276	64-20639 <*>
725,967	I-682 <*>
726,290	II-1045 <*>
726,665	64-16635 <=*$>
726,949	63-14529 <=*>
727,352	61-13834 <*+>
727,800	59-14209 <+*> 0
727,907	59-19521 <+*>
728,275	549 <TC>
728,834	64-20558 <*>
728,982	I-163 <*>
729,059	I-76 <*>
729,274	60-14751 <+*>
729,429	62-18476 <=*> 0
729,523	62-23734 <=*> 0
729,751	64-30422 <*>
729,904	65-10533 <*>
730,720 ADD	59-14210 <+*>
731,204	60-14746 <+*>
733,574	<LSA>
733,750	64-20640 <*>
734,723	63-18146 <=*>
735,764	59-14210 <+*> 0
735,864	63-20506 <=*$>
736,315	65-10630 <*>
736,486	59-12620 <+*> 0
737,933	60J15G <ATS>
738,073	64-16461 <=*$>
738,131	3461 <*>
738,994	60-14691 <+*>

Patent	Citation
739,247	6C-17992 <+*> O
740,335	63-14530 <=*>
74C,637	1913 <*>
740,804	64-30423 <*>
741,411	63-24414 <=*$> O
741,478	64-20488 <*>
741,687	65-10646 <*>
741,982	I-444 <*>
	64-10134 <=*$>
742,360	<INFO>
742,395	64-18887 <*>
742,561	63-15410 <=*> O
742,644	64-20641 <*>
742,797	64-20642 <*>
742,817	61-16512 <*=>
743,411	50M46G <ATS>
743,510	59-17973 <+*> O
744,279	64-10694 <=*$>
744,729	300 <TC>
744,862	63-14944 <=*>
744,994	64-30424 <*>
745,425	<LSA>
745,587	59-17979 <+*>
746,708	65-11904 <*>
747,209	64-30425 <*>
747,257	63-19062 <=*> C
747,917	62-00402 <*>
	62-23354 <=*>
747,961	64-30426 <*>
748,729	66-60250 <=*$>
749,016	64-18041 <*>
749,053	65-12C31 <*>
749,184	II-824 <*>
749,607	67G5G <ATS>
749,639	59-16839 <+*> O
750,915 ADD	62-24516 <=*> O
751,103	59-12086 <+*>
751,621	64-19797 <=$>
753,618	57-23 <*>
753,831	61-23227 <=*> O
755,177	59-17108 <+*> O
756,117	63-19273 <=*> O
757,142	1454 <TC>
757,355	8724 <IICH>
758,570	65-13718 <*>
759,126	65-63497 <=$>
76C,163	T62-1 <MVL>
763,024	1455 <TC>
763,141	1456 <TC>
763,557	64-14052 <=*$>
763,609	I-940 <*>
765,193	1457 <TC>
766,203	1458 <TC>
767,075	61-16745 <*=>
767,842	2H9G <ATS>
768,914	63-14531 <=*>
800,407	60-16091 <+*>
800,651	65-64760 <=*$>
800,662	62-16275 <=*>
800,663	3459 <*>
80C,86C	64-30427 <*>
801,233	64-30428 <*>
801,634	64-30429 <*> O
802,364	6C-23930 <+*> O
802,613	62-32222 <=*>
803,856	2J15G <ATS>
803,956	64-18042 <*>
804,536	64-20387 <*>
804,62C	59-18292 <+*> C
804,724	64-18873 <*> O
805,005	II-573 <*>
806,019	59-14701 <+*> O
806,456	1459 <TC>
806,467	60-16092 <+*>
807,262	63-14532 <=*>
808,586	64-10750 <=*$>
808,877	60-14747 <+*>
808,991	64-30430 <*>
809,561	16E2G <ATS>

Patent	Citation
809,812	1460 <TC>
809,897	8747 <IICH>
810,814	57-1992 <*>
811,060	65-62884 <=$> C
811,649	59-12621 <+*> C
812,251	63-20462 <=*$>
812,253	63-14567 <=*>
812,314	65-10618 <*>
812,786	64-10748 <=*$>
812,805	59-1C754 <+*>
812,806	59-1C756 <+*>
813,169	62-33237 <=*> C
813,707	64-1C751 <=*$>
814,231	64-30431 <*> O
814,432	64-20069 <*> C
814,442	1461 <TC>
815,107	58-2418 <*>
815,548	64-13770 <=*$> O
816,506	63-19614 <=*>
817,C77	62-33238 <=*> C
817,255	59-16181 <+*> O
817,300	64-30432 <*> O
817,408	6C-15868 <+*> C
817,626	64-10752 <=*$>
818,347	1462 <TC>
820,419	62-14902 <=*> C
821,207	65-10645 <*>
821,335	I-609 <*>
821,375	284 <TC>
822,030	64-30107 <*>
822,094	64-13769 <=*$> O
822,551	1463 <TC>
823,148	1464 <TC>
823,518	II-599 <*>
823,970	1200 <TC>
824,004	63-19615 <=*>
824,052	64-10147 <=*$>
824,237	64-16636 <=*$>
825,549	63-16113 <INT>
825,738	63-16212 <=*>
826,640	63-19061 <=*> C
826,883	63-14638 <=*>
826,884	63-14639 <=*>
827,391	61-28084 <*=> C
827,798	64-30433 <*>
827,909	62-32918 <=*> O
830,431	64-30434 <*>
830,585	61-15051 <+*> C
831,131	6C-16093 <+*>
831,134	60-16094 <+*>
831,490	6C-16095 <+*>
831,494	60-16096 <+*>
831,723	64-2C643 <*>
832,152	65-10617 <*>
832,205	64-20811 <*>
832,333	65-13083 <*>
832,499	65-10943 <*>
832,548	62-33239 <=*> O
832,882	259 <TC>
832,886	8728 <IICH>
833,034	258 <TC>
833,731	41G5G <ATS>
834,095	58-2430 <*>
834,152	64-30435 <*>
834,296	61-15988 <=*>
834,317	61-14487 <=*> C
834,799	59-18789 <+*> O
834,991	64-30436 <*>
835,117	60-13278 <+*> O
835,142	1465 <TC>
836,040	1466 <TC>
836,354	58-1481 <*>
836,416	61-15989 <+*> C
837,088	59-12087 <+*>
837,424	62P66G <ATS>
837,537	64-10149 <=*$>
837,645	64-10872 <=*$>
838,375	59-17809 <+*>
838,512	64-19798 <=$> C

838,604	59-15863 <+*>
838,933	65-60291 <=$>
839,122	60-23096 <+*> 0
840,685	64-30437 <*>
841,359	59-15405 <+*>
841,589	64-30438 <*>
841,613	62-18310 <=*>
	63-13427 <=*> 0
841,794	60-17993 <+*> 0
842,119	64-30439 <*>
842,241	60-17994 <+*> 0
842,267	63-24121 <=*$> 0
	8627 <IICH>
842,420	<ES>
842,545	64-30440 <*>
843,038	64-30108 <*>
843,843	1469 <TC>
844,644	64-30441 <*>
845,267	64-20652 <*>
845,292	60-18105 <+*>
845,438	65-61494 <=$> 0
845,518	60-16097 <+*>
845,586	58-2431 <*>
845,800	64-16637 <=*$>
845,841	65-61493 <=$> 0
846,163	65-64577 <=*$>
846,651	64-18043 <*>
846,847	1470 <TC>
847,003	1471 <TC>
847,047	31J15G <ATS>
847,178	65-20426 <*>
847,286	60-16098 <+*>
847,900	268 <TC>
848,044	95K25G <ATS>
848,145	60-16099 <+*>
848,188	1472 <TC>
848,189	1473 <TC>
848,277	63-23574 <=*>
848,352	1474 <TC>
848,680	60-16100 <+*>
849,534	70G8G <ATS>
849,883	64-30442 <*> P
849,987	60-16101 <+*>
850,003	64-30443 <*>
850,328	59-19750 <+*>
850,967	42G5G <ATS>
851,237	64-10475 <=*$>
851,612	59-19224 <+*> 0
852,622	43G5G <ATS>
852,762	62-13872 <*=>
852,913	61-16445 <*=> 0
853,927	57-3565 <*>
854,304	60-16102 <+*>
854,497	63-14617 <=*>
855,453	64-20651 <*>
855,480	65-11819 <*>
855,845	64-30444 <*>
856,145	2155 <*>
	64-14910 <=*$>
856,439	64-20650 <*>
857,604	64-30445 <*>
858,694	64-30446 <*>
859,002	64-20649 <*>
859,464	<ES>
859,597	60-13078 <+*> 0
859,887	64-13586 <=*$>
	64-20648 <*>
860,109	61-28085 <*=>
860,639	64-30447 <*>
861,155	57-1330 <*>
861,373	63-15448 <=*> 0
861,485	32J15G <ATS>
861,718	63-24415 <=*$> 0
861,794	57-1833 <*>
861,838	64-16638 <=*$>
861,926	64-16639 <=*$>
862,043	3464 <*>
862,954	64-30448 <*>
863,035	21J17G <ATS>
863,047	3458 <*>
864,865	AL-657 <*>
	AL-70 <*>
	63-20781 <=*$>
865,100	60-16103 <+*>
865,774	64-14922 <=*$>
865,885	63-15768 <=*> 0
866,441	59-12622 <+*> 0
867,474	64-20481 <*> 0
867,848	64-20647 <*>
867,888	65-60292 <=$>
868,031	65-61487 <=$>
868,296	64-20646 <*>
868,669	II-693 <*>
	57-175 <*>
869,129	19J19G <ATS>
869,193	64-20645 <*>
870,090	<ATS>
870,242	64-10852 <=*$>
870,400	63-13003 <=*>
870,666	59-14211 <+*> 0
870,672	59-12315 <+*>
871,488	21J91G <ATS>
871,514	64-20409 <*>
872,939	64-30449 <*>
873,937	59-18293 <+*> 0
874,215	64-30450 <*>
874,874	60-17995 <+*>
874,901	1484 <TC>
875,101	44G5G <ATS>
875,975	60-15550 <+*> 0
875,992	60-14990 <+*>
876,024	64-15854 <=*$> 0
876,094	64-30451 <*> 0
876,334	64-30452 <*>
876,475	64-13578 <=*$> 0
876,503	64-30453 <*>
876,691	64-10473 <=*$>
876,692	64-10474 <=*$>
876,854	62-16274 <=*>
877,139	64-16640 <*>
877,230	46G5G <ATS>
877,449	1485 <TC>
877,598	64-30454 <*>
877,868	59-19751 <+*> 0
877,894	64-30455 <*>
878,530	62-24516 <=*> 0
878,544	64-16641 <=*$>
878,560	96262 <RIS>
878,561	II-198 <*>
878,797	22J19G <ATS>
878,863	64-30456 <*>
878,935	64-10163 <=*$>
879,138	64-10476 <=*$>
879,315	64-30457 <*>
879,439	60-17293 <+*> 0
879,466	65-61491 <=$> 0
879,546	64-30458 <*> 0
880,140	1487 <TC>
880,746	64-20644 <*>
881,345	63-10070 <=*>
881,792	66-13260 <*>
881,861	63-24416 <=*$> 0
882,153	61-13835 <+*>
883,589	64-16642 <=*$>
883,647	60-10729 <+*>
883,653	64-16643 <=*$>
883,697	64-20654 <*>
884,257	45G7G <ATS>
884,354	65-17353 <*> 0
885,174	64-30459 <*>
886,528	64-30460 <*>
887,042	1488 <TC>
887,099	59-12316 <+*>
887,200	64-10140 <=*$>
887,270	II-318 <*>
887,473	65-62611 <=$> 0
887,500	66-13259 <*>
887,642	65-10592 <*> 0
887,809	1489 <TC>

Patent	Reference
887,944	60-14989 <+*>
888,036	47G5G <ATS>
888,172	64-30461 <*>
888,173	66-10638 <*>
888,180	AL-624 <*>
888,614	64-30462 <*>
888,926	64-30463 <*>
888,949	AL-484 <*>
889,229	96263 <RIS>
889,349	64-16644 <=*$>
889,366	64-16645 <=*$>
889,739	65-62685 <=$>
889,981	64-30464 <*>
890,013	63-24417 <=*$> O
891,450	64-30465 <*>
891,842	64-30466 <*> O
891,863	SCL-T-227 <+*>
892,382	I-797 <*>
892,470	61-13836 <+*>
892,855	1850 <TC>
893,197	59-10753 <+*>
893,241	65-10594 <*>
893,639	63-13750 <=*> O
893,946	1492 <TC>
894,633	61-15254 <+*>
895,063	78N57G <ATS>
895,591	63-16397 <=*> O
895,723	63-16336 <=*> O
896,086	62-25023 <=*> O
896,106	63-20692 <=*$>
896,114	AL-625 <*>
896,229	63-23575 <=*>
896,556	64-20403 <*>
896,641	65-10637 <*>
897,014	64-18044 <*>
897,196	1849 <TC>
897,552	1493 <TC>
897,568	61-15052 <+*>
898,024	1878 <TC>
898,308	64-30467 <*>
899,191	64-10431 <=*$>
899,216	59-14212 <+*>
899,435	64-30468 <*>
899,487	59-12623 <+*> O
900,182	61-23993 <=*> O
900,201	59-18294 <+*> O
900,228	61-19844 <=*>
901,413	63-11755 <=*$>
901,471	64-18045 <*>
901,650	63-13002 <=*>
901,774	1494 <TC>
901,776	II-16 <*>
902,518	60-17611 <+*> O
902,556	65-61492 <=$> O
903,034	64-30469 <*>
903,639	59-16840 <+*> O
903,890	61-13264 <+*>
903,929	64-20655 <*>
904,389	1135 <TC>
904,593	64-16646 <=*$>
904,832	61-28962 <*=> O
905,368	64-18368 <*>
905,544	64-30470 <*>
905,610	II-412 <*>
	64-14392 <=*$>
906,444	64-13249 <=*$>
907,753	63-15449 <=*> O
907,891	64-16647 <=*$>
907,996	59-18723 <+*> O
908,729	64-20656 <*>
908,775	64-30471 <*>
908,978	62-20420 <=*>
909,266	45G5G <ATS>
909,343	<INFO>
909,358	36S82G <ATS>
910,142	60-15869 <+*> O
910,164	1496 <TC>
910,408	8676 <IICH>
910,502	8738 <IICH>
910,601	65-61488 <=$> O
911,211	59-12088 <+*>
911,731	96260 <RIS>
911,952	59-19225 <+*> O
912,022	64-30472 <*>
	66-10639 <*>
912,061	61-28540 <*=> O
913,033	63-24418 <=*$> O
913,474	202 <TC>
913,537	1498 <TC>
913,892	64-16928 <=*$>
914,438	64-19177 <=*$>
915,072	63-14799 <=*>
917,006	64-30821 <*>
	96259A <RIS>
917,603	506 <TC>
918,586	64-20657 <*>
918,928	63-16817 <=*>
919,290	64-30473 <*>
919,727	60-10088 <+*>
920,109	61-13265 <+*> O
921,450	96261 <RIS>
921,986	64-30815 <*>
922,158	59-14702 <+*> O
922,378	8718 <IICH>
922,386	62-15754 <*=> O
922,796	66-11848 <*>
922,922	59-12925 <+*> O
923,185	43K22G <ATS>
923,422	62-23735 <=*>
924,057	64-30474 <*>
924,922	64-30475 <*>
925,348	96258 <RIS>
925,348 SUP	64-14704 <=*$>
925,464	64-30476 <*>
925,612	65-10639 <*>
925,947	64-20658 <*>
	64-30477 <*>
925,949	1499 <TC>
926,022	2439 <TTIS>
926,487	1500 <TC>
927,292	66-10117 <*>
927,925	58-2420 <*>
927,929	60-14988 <+*>
928,185	65-63482 <=$> O
928,645	1651 <*>
928,896	61-10614 <+*>
929,185	64-30478 <*>
929,643	64-30479 <*>
929,788	622 <TC>
929,968	II-697 <*>
930,749	61-15685 <+*>
930,749 ADD	61-15686 <+*>
931,017	59-14703 <+*> O
931,080	<ATS>
931,084	59-17172 <+*>
932,152	60-14987 <+*>
932,323	305 <TC>
932,425	32L29G <ATS>
932,556	59-18790 <+*> O
932,633	65-10547 <*>
	8740 <IICH>
933,231	64-16524 <=*$>
933,689	65-10113 <*>
934,362	60-16343 <+*>
934,701	65-11250 <*>
935,426	60-14986 <+*> O
935,644	63-18908 <=*>
935,867	1264 <TC>
	65-10638 <*>
	65-14898 <*>
935,965	1502 <TC>
936,294	59-16182 <+*> O
936,632	64-30480 <*>
936,685	1503 <TC>
936,922	1616 <TC>
937,326	61-19845 <=*>
937,645	85S84G <ATS>
937,919	1265 <TC>
	65-14887 <*>
938,062	64-30481 <*>

938,801	60-23098 <+*> 0
939,148	65-10678 <*>
939,266	64-18819 <*>
939,293	63-14798 <*>
939,324	1504 <TC>
939,529	306 <TC>
940,110	65-10709 <*>
940,781	61-18271 <*=>
941,460	60-14985 <+*> 0
941566	59-12448 <=>
941,566	62-11530 <=>
943,207	63-14797 <=*>
943,404	65-10710 <*> 0
943,472	64-13768 <=*$> 0
943,760	60-15712 <+*>
944,032	61-20022 <*=>
944,311	64-30482 <*>
944,549	1505 <TC>
944,789	63-23576 <=*$>
944,831	60-14984 <+*>
945,685	64-30483 <*>
946,315	61-19846 <=*>
946,435	65-10502 <*> 0
946,681	60-14983 <+*>
946,682	64-16648 <=*$>
946,891	1506 <TC>
946,891 ADD	1509 <TC>
947,550	65-10508 <*>
947,571	65-10578 <*> 0
949,049	65-10503 <*>
949,314	64-15338 <=*$> 0
949,469	63-10077 <=*>
949,616	60-14981 <+*>
949,963	62-32221 <=*> 0
949,970	57-1215 <*>
950,050	59-16841 <+*> 0
950,064	65-18171 <*>
950,104	60-14980 <+*>
950,836	59-19226 <+*> 0
951,003	65-10677 <*>
951,158	64-30484 <*>
951,751	61-15686 <+*>
951,953	60-14978 <+*>
952,086	64-18046 <*>
952,089	64-30485 <*>
952,168	1507 <TC>
953,010	64-16649 <=*$>
953,068	63-16398 <=*>
953,078	65-10599 <*>
953,342	64-30486 <*>
953,835	8729 <IICH>
954,056	64-30487 <*>
954,149	64-30488 <*> 0
954,414	64-30489 <*> 0
955,234	62-16273 <=*>
955,377	59-19752 <+*> 0
955,412	64-30495 <*> 0
955,765	1786 <TC>
955,943	65-62885 <=$> 0
956,612	60-16840 <+*>
956,993	64-20659 <*> 0
957,179	60-14721 <+*>
957,474	39L34G <ATS>
957,630	60-14720 <+*>
958,736	65-10507 <*> 0
959,449	64-30632 <*> 0
959,450	64-30633 <*> 0
959,884	60-23097 <+*> 0
959,999	61-15053 <*> 0
960,206	274 <TC>
960,214	63-23577 <=*>
960,272	40L34C <ATS>
960,324	27M44G <ATS>
960,331	63-14943 <=*>
960,707	10281L <K-H>
	64-14478 <=*$>
960,774	1775 <TC>
960,788	10281K <K-H>
	64-14479 <=*$>

960,812	1509 <TC>
961,250	63-14795 <=*>
961,576	64-30490 <*> 0
961,813	65-10596 <*>
962,062	60-15551 <+*> 0
962,828	303 <TC>
963,425	8623 <IICH>
963,822	48M42G <ATS>
964,316	65-10598 <*>
964,675	65-10627 <*>
965,065	304 <TC>
965,125	65-10566 <*> 0
965,226	65-10560 <*> 0
965,231	66-10502 <*>
965,236	62-10237 <*=>
965,768	8742 <IICH>
965,921	61-18272 <*=>
966,085	64-30491 <*> 0
966,955	58-1395 <*>
967,011	1510 <TC>
967,144	64-14481 <=*$>
967,444	11071-D <K-H>
967,823	65-10563 <*> 0
968,056	60-10727 <+*>
969,068	1877 <TC>
969,191	1512 <TC>
969,774	61-18273 <*=>
970,084	64-14045 <=*$>
970,196	63-24419 <=*$> 0
970,423	61-18274 <*=>
970,433	64-30492 <*> 0
970,942	1513 <TC>
970,965	64-30493 <*>
971,098	714 <TC>
971,835	8701 <IICH>
973,004	65-64761 <=*$>
973,270	65-13375 <*>
974,149	66-10521 <*>
1,000,119	65-62886 <=$>
1,000,225	60-16409 <+*>
1,000,300	63-14794 <=*>
1,000,379	65-13554 <*>
1,000,388	60-16414 <+*>
1,000,747	65-10539 <*>
1,000,801	20J17G <ATS>
1,001,367	59-19753 <+*> 0
1,001,429	AEC-TR-5809 <=*$>
1,001,559	64-30496 <*>
1,001,866	60-17612 <+*> 0
1,002,298	65-10561 <*> 0
1,002,381	61-28541 <*=> 0
1,002,754	64-30497 <*>
1,003,578	84N55G <ATS>
1,003,639	63-16592 <=*>
	9730 <K-H>
1,003,701	64-20661 <*>
1,003,726	64-18047 <*>
1,003,985	64-18397 <*>
1,004,179	65-11379 <*>
1,004,380	64-30498 <*>
1,004,763	44J17G <ATS>
1,004,810	64-30499 <*>
1,004,986	11349 <K-H>
	63-16569 <=*>
1,005,726	63-13004 <=*>
1,005,924	80J160 <ATS>
1,005,947	64-18048 <*>
1,006,162	64-30500 <*>
1,006,184	57-2301 <*>
1,006,553	60-18000 <+*>
1,006,618	8748 <IICH>
1,006,722	61-23994 <=*> 0
1,006,733	65-62887 <=$> 0
1,007,305	64-20662 <*>
1,007,342	59-17294 <+*>
1,007,479	63-18582 <=*>
1,007,725	64-18049 <*>
1,007,744	65-10559 <*> 0

1,008,728 64-20663 <*>
1,008,733 65-11380 <*>
96871 <RIS>
1,008,959 64-20225 <*> O
1,009,172 64-20665 <*>
65-10535 <*>
1,009,357 57-3504 <*>
1,009,630 65-11377 <*>
1,009,743 63-10076 <*=>
1,010,202 58-297 <*>
61-13773 <+*>
1,010,504 269 <TC>
1,010,516 65-10576 <*>
1,010,707 63-18591 <=*>
1,010,951 65-11906 <*>
1,010,955 65-10557 <*>
1,010,971 8713 <IICH>
1,011,575 60-14995 <+*>
1,011,622 58-518 <*>
1,011,623 66-11388 <*>
1,011,853 61-16510 <*=>
1,011,855 65-10567 <*> O
1,012,018 <ES>
1,012,137 64-20801 <*>
1,012,580 59-22281 <+*> O
1,012,941 44J18G <ATS>
1,012,999 AEC-TR-4950 <*>
1,013,494 63-14627 <=*>
1,013,501 64-18050 <*>
1,013,555 59-18074 <+*>
1,013,564 65-10565 <*>
1,013,657 59-17290 <+*> P
1,013,841 59-16842 <+*>
1,013,861 61-28964 <*=> O
1,013,951 64-18051 <*>
1,013,953 64-18052 <*>
1,014,080 64-30501 <*> O
1,014,093 64-30128 <*>
1,014,595 62-24517 <=*> O
1,014,744 1266 <TC>
65-14888 <*>
1,015,775 63-14,942 <=*>
1,016,654 63-14941 <=*>
1,017,308 64-20668 <*>
1,017,369 64-30502 <*>
1,017,600 65-10564 <*> O
1,017,807 62-18302 <*>
1,018,299 59-10221 <+*>
1,018,409 64-18053 <*>
1,018,770 10359 <K-H>
1,018,777 NP-TR-924 <*>
1,019,285 64-30952 <*>
1,019,361 65-10555 <*>
1,019,387 62-11481 <=>
1,019,416 65-62888 <=$>
1,019,459 64-30503 <*>
1,019,467 64-16519 <=*$>
1,019,468 64-30953 <*>
1,019,504 65-11822 <*>
1,019,805 59-17091 <+*> O
60-16854 <+*> O
1,019,817 58-1628 <*>
1,020,024 64-30504 <*>
1,020,075 62-15755 <*=>
1,020,355 65-63496 <=$> O
1,020,762 62-23736 <*> O
1,020,962 65-10562 <*>
1,020,966 65-10573 <*>
1,021,167 8733 <IICH>
1,021,962 AEC-TR-3786 <*>
1,022,005 64-19133 <=*$>
1,022,167 62-15756 <*=> O
1,022,383 65-10635 <*>
1,022,561 64-30505 <*>
1,022,669 63-13428 <=*> O
1,022,796 64-30506 <*>
1,023,017 63-20691 <=*$>
64-15337 <=*$>
1,023,195 61-18275 <*=> O
1,023,257 AEC-TR-3804 <*>

1,024,061 65-10558 <*>
1,024,062 65-10310 <*>
65-11373 <*>
1,024,854 61-28965 <*=> O
1,024,900 63-14793 <=*>
1,025,145 8734 <IICH>
1,025,198 65-10919 <*>
1,025,274 731 <TC>
1,025,302 <AC>
1,025,478 SC-T-536 <*>
1,025,533 AEC-TR-5542 <*>
1,025,886 60-14977 <+*>
1,026,038 60-14976 <+*>
1,026,522 62-16272 <=*>
1,026,524 64-30507 <*>
1,026,805 65-64477 <=*$>
1,026,961 64-30508 <*>
1,026,962 66-11389 <*>
1,026,963 64-30509 <*>
1,026,964 64-30510 <*>
1,027,113 61-10501 <+*> C
1,027,403 8750 <IICH>
1,027,410 8617 <IICH>
1,027,418 AEC-TR-5756 <=*$>
1,027,649 65-10593 <*>
1,027,728 63-19060 <=*> C
1,027,896 62-20017 <=*>
1,028,050 59-16843 <+*> O
1,028,265 64-30511 <*>
1,028,328 60-17613 <+*>
1,028,337 64-30512 <*>
1,028,339 64-30513 <*>
1,028,741 63-14792 <=*>
1,029,351 64-30514 <*>
1,029,941 61-19847 <=*> O
1,029,988 63-18619 <=*>
1,029,989 63-18620 <=*>
1,030,232 60-23071 <+*>
1,030,718 60-14975 <+*>
1,030,821 65-10504 <*>
1,030,953 64-30944 <*>
1,031,619 60-14974 <+*>
1,031,620 60-14973 <+*>
1,031,687 60-23047 <+*>
1,031,766 65-60839 <=$> O
1,032,025 65-11267 <*>
1,032,240 64-30945 <*>
1,032,530 60-10731 <+*>
1,032,542 66-11385 <*>
1,033,179 65-62612 <=$> O
1,033,406 65-11903 <*>
1,033,409 62-16271 <=*>
1,033,808 63-24420 <=*$> O
1,033,852 60-14972 <+*>
1,034,356 62-16270 <=*> O
1,034,596 64-14511 <*>
1,034,818 61-19488 <+*>
1,034,858 64-18054 <*>
1,034,861 64-30515 <*>
1,035,635 8732 <IICH>
1,036,191 64-15336 <=*$> O
1,036,426 65-14098 <*>
1,036,884 60-17294 <+*>
1,037,245 61-20901 <*=>
1,037,319 60-17295 <+*>
1,037,652 60-16411 <+*>
1,038,035 65-10317 <*>
1,038,913 255 <TC>
1,039,041 65-10311 <*>
1,039,053 62-16269 <=*>
1,039,070 61-10021 <+*>
1,039,380 60-13079 <+*> O
1,039,894 61-19489 <+*> C
1,039,904 10573-F <K-H>
1,040,009 62-10995 <=*>
1,040,174 60-14971 <+*>
1,040,192 61-16758 <*=> P
1,040,196 64-10529 <=*$>
1,040,198 64-14061 <=*$>
1,040,312 65-11821 <*>

Patent	Citation
1,040,625	60-15096 <+*> O
1,040,792	64-30516 <*>
1,040,900	65-11520 <*>
1,040,907	63-13005 <=*> O
1,041,023	64-30969 <*>
1,041,486	64-30517 <*> O
1,041,727	65-11820 <*>
1,041,844	65-62889 <=$>
1,042,183	61-16762 <*=> P
1,042,361	60-14970 <+*>
1,042,439	61-28087 <*=>
1,042,550	63-16242 <*=>
1,043,338	60-14969 <+*>
1,043,776	60-14968 <+*>
1,044,070	62-16279 <=*> O
1,044,323	61-10772 <+*>
1,044,505	65-10325 <*>
1,044,805	65-10307 <*>
1,045,050	61-16761 <*=> P
1,045,089	65-14863 <*>
1,045,103	64-30518 <*>
1,045,649	62-16278 <=*>
1,045,775	60-14967 <+*>
1,046,033	60-14966 <+*>
1,046,319	64-30519 <*>
1,046,331	65-10623 <*>
1,046,842	63-10262 <=*> O
1,046,881	64-30520 <*>
1,046,885	64-30521 <*>
1,047,349	64-30522 <*>
1,047,447	65-10356 <*>
1,047,545	62-24518 <=*> O
1,047,701	61-20962 <*=>
1,048,275	65-10357 <*>
1,048,276	65-10342 <*>
1,048,583	65-10615 <*>
1,048,586	65-10344 <*>
1,048,790	60-14759 <+*>
1,049,095	65-10927 <*>
1,049,694	11033-A <K-H>
	61-20015 <*=>
1,049,707	62-16168 <=*>
1,049,845	8693 <IICH>
1,050,023	60-14965 <+*>
1,050,062	64-30523 <*>
1,050,064	8756 <IICH>
1,050,342	60-16342 <+*>
1,050,538	65-14099 <*>
1,050,697	65-60997 <=$> O
1,050,740	65-10348 <*> O
1,050,752	60-14308 <+*>
1,050,768	63-19274 <=*$>
1,050,769	65-10312 <*>
1,051,175	65-62613 <=$>
1,051,265	62-16277 <=*>
1,051,267	62-16280 <=*>
1,051,439	61-14963 <*=>
	826 <TC>
1,051,647	63-13006 <=*> O
1,051,895	62-25691 <=*> O
1,052,400	65-10324 <*>
1,052,458	64-13250 <=*$> O
1,052,673	8737 <IICH>
1,052,972	62-16281 <=*O>
1,052,986	65-10349 <*>
1,053,011	60-17296 <+*>
1,053,121	66-10528 <*>
1,053,145	66-10668 <*>
1,053,304	64-30524 <*>
1,053,310	64-14973 <=*$>
1,053,476	65-10351 <*>
1,053,904	60-14964 <+*>
1,054,079	65-10338 <*>
1,054,080	65-10339 <*>
1,054,081	65-10340 <*>
1,054,185	60-23931 <+*> O
1,054,455	65-10354 <*>
1,054,456	65-10353 <*>
1,054,976	61-20025 <*=>
1,055,320	61-31457 <=>
1,055,508	AEC-TR-4751 <*>
1,055,534	59-17236 <+*>
	65-10359 <*>
1,056,105	64-20667 <*>
1,056,138	65-10921 <*>
1,056,155	64-30525 <*>
1,056,369	61070 <RIS>
1,056,830	66-11411 <*>
1,056,831	66-11428 <*>
1,056,834	64-30526 <*>
1,056,953	62-32220 <=*> O
1,056,990	10920-B <KH>
	63-16671 <=*>
1,057,067	65-10922 <*>
1,057,099	64-30527 <*>
	65-10915 <*>
1,057,434	64-18474 <*>
1,057,600	65-10936 <*>
1,057,613	65-10948 <*> O
1,057,973	61-16190 <=*>
1,058,024	AEC-TR-4752 <*>
1,058,024 ADD	AEC-TR-6479 <*>
1,058,302	65-10911 <*>
1,058,643	8707 <IICH>
1,058,671	11006A <K-H>
	63-16886 <*=>
1,058,968	63-10802 <=*>
1,059,120	60-23932 <+*> O
1,059,310	61-15990 <+*> O
1,059,413	60-17996 <+*> O
1,059,439	1323 <TC>
	66-13250 <*>
1,059,688	715 <TC>
1,059,743	61-23228 <=*> O
1,059,883	61-20873 <*=>
1,059,903	61-10399 <+*>
1,060,139	60-14237 <+*>
1,060,187	60-18103 <+*>
1,060,246	63-18295 <=*>
1,060,360	AEC-TR-4990 <*>
1,060,398	65-10947 <*>
1,060,626	62-33247 <=*> O
1,060,846	AEC-TR-4754 <*>
1,060,857	61-15245 <+*>
1,061,308	65-10595 <*>
1,061,351	62-32219 <=*> O
1,061,755	65-10505 <*> O
1,061,971	61-16757 <*=> P
1,062,009	64-30528 <*>
	66-11412 <*>
1,062,225	65-10501 <*>
1,062,524	61-27463 <*=> O
1,062,705	97075 <RIS>
1,062,836	63-24421 <=*$>
1,063,151	61-10396 <+*>
1,063,190	65-17355 <*>
1,063,293	63-19616 <=*>
1,063,389	8618 <IICH>
1,063,390	8619 <IICH>
1,063,832	63-10911 <=*> O
1,064,053	66-13251 <*>
	976 <TC>
1,064,343	61-14463 <*=> O
1,064,496	65-10579 <*>
1,064,503	65-10600 <*>
1,064,809	63-23578 <=*>
1,064,861	T.9 <CLAI>
1,064,934	65-10620 <*>
1,064,952	65-10679 <*>
1,065,101	63-19617 <=*> O
1,065,118	2247 <TTIS>
1,065,170	63-10838 <=*>
1,065,178	61069 <RIS>
1,065,399	49N53G <ATS>
1,065,401	63-24608 <=*$>
1,065,607	61068 <RIS>
1,065,610	66-11413 <*>
1,065,845	64-30529 <*> O
1,065,962	61-18617 <*=>
1,066,743	60-14208 <+*>

Patent	Citation
1,067,009	65-11391 <*>
1,067,207	64-18055 <*>
1,067,598	64-30530 <*> O
1,067,814	65-11255 <*>
1,067,986	61-16754 <*=> P
1,068,251	11033-B <K-H>
	63-16568 <=*>
1,068,440	65-14463 <*>
1,068,682	62-10758 <*=>
1,068,841	2248 <TTIS>
1,068,971	63-10837 <=*>
1,069,112	1791 <TC>
1,069,387	66-11901 <*>
1,069,689	63-24122 <=*>
1,069,840	61-16753 <*=>
1,070,027	64-18056 <*>
1,070,373	62-16282 <=*>
1,070,376	65-63494 <=$>
1,071,593	AEC-TR-4406 <*>
1,071,945	REPT.61099 <RIS>
1,072,244	64-30531 <*>
1,072,766	2246 <TTIS>
1,072,812	66-11390 <*>
1,072,820	1126 <RIS>
1,072,918	730 <TC>
1,073,050	65-62890 <=$> O
1,073,743	66-11386 <*>
1,074,024	61-20857 <*=>
1,074,052	62-16276 <=*>
1,074,079	61-15810 <*=> O
1,074,080	61-15333 <=*> O
1,074,188	64-13767 <=*$>
1,074,262	62-16238 <=*>
1,074,265	REPT.61105 <RIS>
1,074,281	64-13577 <=*$> O
1,074,824	61-10020 <+*>
1,074,854	62-18186 <=*>
1,075,495	63-10258 <=*>
1,075,612	65-11248 <*>
1,075,888	64-23591 <=$>
1,076,101	61-20281 <*=>
1,076,103	61-19358 <+*>
1,076,105	64-30532 <*>
1,076,111	65-11251 <*>
1,076,125	08P60G <ATS>
1,076,130	65-13559 <*>
1,076,374	66-11392 <*>
1,076,545	64-20137 <*>
1,076,636	61-20280 <=*>
1,076,651	62-10681 <*=>
1,076,810	65-63493 <=$>
1,076,861	64-30533 <*>
1,077,212	65-13569 <*>
1,077,217	65-10622 <*>
	65-17227 <*>
1,077,325	65-63492 <=$> O
1,077,358	65-11825 <*>
1,077,503	64-30534 <*>
1,077,785	65-63491 <=$> O
1,077,858	63-15766 <=*> O
1,078,037	T.13 <CLAI>
1,078,106	65-12507 <*>
1,078,488	T.16 <CLAI>
1,079,390	867 <TC>
1,079,621	65-11853 <*>
1,080,289	64-15335 <=*$> O
1,080,303	65-11862 <*>
1,080,527	63-20009 <=*>
1,080,647	64-15334 <=*$> O
1,080,854	85N55G <ATS>
1,080,931	AEC-TR-4984 <*>
1,081,436	63-16201 <=*>
1,081,482	63-16253 <=*>
1,081,606	62-14808 <=*>
1,082,004	63-10797 <=*>
1,082,008	61-20563 <*=> O
1,082,010	61-16752 <*=>
1,082,016	62-23737 <=*>
1,082,034	63-15765 <=*> O
1,082,359	2229 <TTIS>
1,082,541	11572 <K-H>
1,082,618	62-23314 <=*> O
1,082,740	64-30999 <*>
1,083,055	64-30535 <*> O
1,083,082	61-10022 <+*>
1,083,230	61-10023 <+*>
1,083,251	61-16774 <*=>
1,083,550	64-30536 <*>
1,083,790	62-10708 <=*>
1,084,570	63-10277 <=*>
1,084,714	61-10019 <+*>
1,084,799	63-15450 <=*> O
1,085,146	92N48G <ATS>
1,085,514	775 <TC>
1,085,806	62-14809 <=*>
1,086,433	64-30537 <*>
1,086,896	2235 <TTIS>
1,086,897	2236 <TTIS>
1,087,810	2237 <TTIS>
1,087,902	65-12511 <*>
	66-10610 <*>
1,088,061	63-10069 <=*>
1,089,208	65-12512 <*>
1,089,211	66-14007 <*>
1,089,239	816 <TC>
1,090,241	65-10512 <*>
1,090,399	65-12859 <*>
1,090,659	65-13564 <*>
	67N50G <ATS>
1,090,851	62-16259 <=*>
1,091,161	841 <TC>
1,091,757	66-11928 <*>
1,092,448	AEC-TR-6479 <*>
1,093,507	843 <TC>
1,094,170	65-62891 <=$> O
1,094,249	65-10611 <*>
1,094,723	65-10514 <*>
1,094,746	NS-7 <TTIS>
1,094,830	1070 <TC>
1,094,981	201 <TC>
1,095,186	64-14030 <=*$>
1,095,268	65-10624 <*> O
1,095,290	65-13210 <*>
1,095,441	65-64578 <=*$>
1,095,814	NS-8 <TTIS>
1,095,829	NS-10 <TTIS>
1,096,323	8670 <IICH>
1,096,529	1410 <TC>
1,096,850	935 <TC>
1,097,121	63-10943 <=*>
1,097,963	63-14940 <=*>
1,097,994	64-18473 <*>
1,098,467	64-13766 <=*$> O
1,098,716	66-11393 <*>
1,098,799	63-14939 <=*>
1,098,921	65-17356 <*>
1,099,519	65-10683 <*>
1,100,001	62-14766 <*>
	62-18732 <=*>
1,100,002	62-14756 <=*>
1,100,020	65-13565 <*> O
1,100,024	65-13566 <*>
1,100,025	65-13567 <*>
1,100,214	65-60999 <=$> O
1,100,287	62-16246 <=*>
1,100,616	1324 <TC>
1,100,657	61-20900 <*=>
1,100,888	62-14539 <=*>
1,101,115	64-15855 <=*$> O
1,101,243	65-62856 <=$> O
1,101,753	62-14266 <=*>
1,102,151	N.S. 12 <TTIS>
1,102,299	64-16590 <=*$>
1,102,519	64-14352 <=*$>
1,102,736	N.S.13 <TTIS>
1,103,571	403 <TC>
	63-10895 <=*>
1,104,693	62-16247 <=*>
1,104,757	65-13222 <*>

1,105,395	65-10515 <*>
1,105,414	NS-12 <TTIS>
1,105,616	66-11394 <*>
1,106,311	1325 <TC>
1,108,205	65-60293 <=$>
1,108,779	64-10957 <=*$> O
1,109,156	65-10500 <*>
1,109,278	65-63483 <=$> O
1,109,365	62-14538 <=*>
1,110,207	65-62614 <=$> O
1,110,420	65-14858 <*>
1,110,862	64-18057 <*>
1,110,866	66-11925 <*>
1,110,873	65-14501 <*>
1,111,583	62-18869 <=*>
1,112,807	62-25021 <=*$> O
1,112,832	66-11926 <*>
1,113,088	65-14502 <*>
1,113,444	64-30538 <*>
1,113,531	64-14956 <*>
1,113,565	62-10660 <*=>
1,113,711	64-19178 <=*$> O
1,114,008	65-14462 <*>
1,114,279	64-30539 <*>
1,114,330	65-11900 <*>
1,114,472	66-10510 <*> O
1,115,236	62-10960 <*>
1,115,891	65-11896 <*>
1,116,171	39P60G <ATS>
1,116,647	62-16245 <=*>
1,117,092	65-10517 <*>
1,117,298	63-13007 <=*>
1,117,549	63-20050 <=*>
1,117,766	62-14894 <=*> O
1,117,869	NS-30 <TTIS>
1,118,112	228 <TC>
1,118,924	62-18502 <=*> O
1,119,513	66-11429 <*>
1,120,137	546 <TC>
1,120,690	62-18501 <=*>
1,121,258	249 <TC>
1,121,615	62-18657 <=*>
1,121,812	300 <CTS>
1,122,289	62-18510 <=*> O
1,122,933	62-18658 <=*> O
1,123,305	64-30991 <*>
1,123,312	8694 <IICH>
1,123,661	1326 <TC>
1,123,837	65-10544 <*>
1,123,986	<LSA>
1,125,176	66-11430 <*>
1,125,877	8671 <IICH>
1,125,891	64-30540 <*>
1,126,053	65-14004 <*>
1,128,653	292 <TC>
1,128,657	62-18515 <=*> O
1,128,667	62-18516 <=*>
1,128,832	65-10518 <*>
1,128,870	62-18816 <=*>
1,128,871	62-18815 <=*>
1,130,250	8714 <IICH>
1,130,419	62-20413 <=*> O
1,130,992	62-20414 <=*>
1,132,609	64-19179 <=*$> O
1,132,716	66-10041 <*>
1,132,786	64-18058 <*>
1,133,229	62-20401 <*> O
1,133,542	63-18337 <=*>
1,133,547	66-10631 <*>
1,133,721	64-18472 <*>
1,133,740	62-20405 <*> O
1,134,093	63-10257 <=*>
1,134,587	63-10263 <=*> O
1,134,974	64-30541 <*>
1,134,982	64-18059 <*>
1,135,451	1327 <TC>
1,135,599	4014 <TTIS>
1,135,755	63-10870 <=*>
1,135,897	633 <TC>
1,136,098	63-10885 <=*>
1,136,109	66-10632 <*>
1,136,818	932 <TC>
1,137,214	66-10633 <*>
1,137,216	63-20055 <=*>
1,137,370	8666 <IICH>
1,137,731	63-20059 <=*>
1,137,858	64-20937 <*>
1,138,760	65-18164 <*>
1,138,895	1068 <TC>
1,138,937	66-11387 <*>
1,139,329	868 <TC>
1,139,610	8705 <IICH>
1,139,626	257 <TC>
1,139,654	64-16485 <=*$>
1,139,741	63-16451 <*>
1,140,564	1328 <TC>
1,140,706	63-23579 <=*>
1,140,721	NP-TR-1101 <*>
1,141,453	66-10551 <*>
1,141,531	63-20844 <=*$>
1,141,973	63-18383 <=*>
1,142,161	1329 <TC>
1,144,580	65-11907 <*>
1,144,702	8699 <IICH>
1,145,142	65-14852 <*>
1,145,418	8698 <IICH>
1,146,257	<TTIS>
1,146,447	63-20337 <=*>
1,146,473	63-20336 <=*$>
1,146,617	63-20335 <=*>
1,147,931	8706 <IICH>
1,148,448	65-63495 <=$>
1,149,006	773 <TC>
1,149,048	64-14974 <=*$>
1,149,488	64-10853 <=*$>
1,149,987	64-18116 <*>
1,150,692	64-14744 <=*$>
1,151,160	1363 <TC>
1,151,498	886 <TC>
1,151,556	64-19180 <=*$> O
1,151,789	64-30542 <*>
1,152,391	275 <TC>
1,152,537	64-14175 <=*$>
1,152,543	64-14975 <*>
1,152,825	291 <TC>
1,153,056	64-10956 <=*$> O
1,153,891	160 <TC>
1,154,081	65-14261 <*>
1,154,097	195 <TC>
1,154,289	64-10955 <=*$> O
1,154,455	8633 <IICH>
1,154,793	66-11120 <*>
1,155,097	65-12783 <*>
1,155,763	64-30543 <*>
1,155,906	161 <TC>
1,156,061	64-10164 <=*$>
1,156,220	<ICS>
1,156,308	10T90G <ATS>
1,156,922	NS-320 <TTIS>
1,157,128	66-10716 <*>
1,157,601	162 <TC>
1,158,290	285 <TC>
1,158,366	64-14976 <=*$>
1,158,367	1364 <TC>
1,159,413	250 <TC>
1,159,422	636 <TC>
1,159,597	197 <TC>
1,159,630	65-11908 <*>
1,159,669	66-14521 <*>
1,159,936	64-18820 <*>
1,160,172	196 <TC>
1,160,845	1447 <TC>
1,161,248	1448 <TC>
1,161,479	57S83G <ATS>
1,161,547	528 <TC>
1,161,587	64-20938 <*>
1,161,876	1449 <TC>
1,162,365	AD-620 736 <=*$>
1,162,433	65-11888 <*>
1,162,576	NS217 <TTIS>

Patent	Citation
1,163,672	64-20939 <*>
1,163,785	220 <TC>
1,163,827	518 <TC>
1,164,487	64-30634 <*> O
1,164,675	65-11893 <*>
1,165,054	596 <TC>
1,165,408	65-11794 <*>
1,165,666	64-30635 <*> O
1,165,839	65-13342 <*>
1,165,845	239 <TC>
1,165,850	65-11395 <*>
1,165,856	66-14000 <*>
1,166,473	238 <TC>
1,166,474	66-10606 <*>
1,166,475	66-10611 <*>
1,166,476	66-10612 <*>
1,166,522	64-30636 <*> O
1,166,619	65-11479 <*>
1,167,077	64-30637 <*> O
1,167,149	65-11933 <*>
1,167,319	65-11806 <*>
1,167,992	334 <TC>
1,168,058	<ICS>
1,168,161	64-30544 <*>
1,168,237	64-30638 <*> O
1,168,767	64-20940 <*>
1,168,887	66-10506 <*>
1,169,133	64-30639 <*>
1,169,898	597 <TC>
1,170,384	598 <TC>
1,170,780	64-30640 <*>
1,171,145	64-30641 <*>
	65-11798 <*>
1,171,878	485 <TC>
1,171,959	65-64762 <=*$>
1,172,239	461 <TC>
1,173,085	1451 <TC>
1,173,244	391 <TC>
1,173,246	393 <TC>
1,173,641	392 <TC>
1,173,646	390 <TC>
1,173,649	389 <TC>
1,174,069	1067 <TC>
1,174,075	491 <TC>
1,174,679	1816 <TC>
1,175,213	701 <TC>
1,175,383	395 <TC>
1,175,524	65-11792 <*>
1,175,916	66-14522 <*> O
1,176,116	65-11020 <*>
1,176,419	65-11724 <*>
1,177,100	65-12770 <*>
1,177,852	65-12030 <*>
1,178,705	65-17059 <*>
1,178,838	65-12028 <*>
1,179,718	651 <TC>
1,180,124	65-13094 <*>
1,180,622	NS-321 <TTIS>
1,180,730	65-13092 <*>
1,181,031	512 <TC>
1,181,198	66-10511 <*> O
1,181,402	595 <TC>
1,181,407	66-13244 <*>
1,181,537	66-10549 <*>
1,181,912	66-10516 <*>
1,182,649	702 <TC>
1,182,660	624 <TC>
1,182,826	1365 <TC>
1,182,827	65-13103 <*>
1,182,954	821 <TC>
1,182,955	822 <TC>
1,182,956	823 <TC>
1,183,074	65-13082 <*>
1,183,344	NS-435 <TTIS>
1,184,337	845 <TC>
1,184,745	66-14476 <*>
1,185,037	997 <TC>
1,185,604	1382 <TC>
1,186,035	794 <TC>
1,186,038	1082 <TC>
1,186,455	787 <TC>
1,186,540	65-14100 <*> O
1,186,736	65-14101 <*>
1,186,857	66-10527 <*>
1,187,512	1788 <TC>
1,187,513	1789 <TC>
1,188,048	1139 <TC>
1,188,049	1818 <TC>
1,188,463	1056 <TC>
	65-14320 <*>
1,188,625	65-14102 <*> O
1,188,804	1905 <TC>
1,188,805	8791 <IICH>
1,189,032	1819 <TC>
1,189,066	1019 <TC>
1,189,219	65-14103 <*>
1,190,005	65-17051 <*>
1,191,299	1151 <TC>
	65-14627 <*>
1,192,359	1117 <TC>
1,192,410	1821 <TC>
1,193,015	1820 <TC>
1,194,404	65-17052 <*>
1,194,879	65-17071 <*>
1,195,428	66-11943 <*>
1,195,592	66-13691 <*>
1,195,600	65-17070 <*>
1,195,633	1790 <TC>
1,197,143	65-17069 <*>
1,197,440	66-11498 <*>
1,199,763	1906 <TC>
1,200,134	66-10115 <*>
1,200,531	66-10114 <*>
1,200,848	66-14368 <*>
1,201,053	1852 <TC>
1,202,135	66-10767 <*>
1,202,137	66-10768 <*>
1,202,292	66-10769 <*>
1,202,293	66-10770 <*>
1,202,501	1988 <TC>
1,202,785	1770 <TC>
1,203,135	66-10773 <*>
1,203,608	66-10772 <*>
1,204,111	66-10771 <*>
1,209,741	66-11362 <*>
1,221,050	66-14550 <*>

GERMANY APPLICATION

Application	Citation
A-14,598	AL-303 <*>
A-29,865	4755 <RIS>
A-33,960	5E2G <ATS>
A-44,278	64-14670 <=*$>
B-199	64-20995 <*>
B-245	64-30902 <*>
B-3,040	65-10673 <*>
B-3,117	64-30071 <*>
B-3,537	64-20993 <*>
B-6,266	64-20448 <*>
B-7,371	64-20331 <*>
B-7,541	64-10749 <=*$>
B-7,567	65-10669 <*>
B-7,944	64-20989 <*>
B-11,626	3462 <*>
B-12,275	64-30017 <*>
B-13,286	64-30077 <*>
B-13,985	64-20978 <*>
B-14,051	64-20503 <*>
B-14,517	64-20576 <*>
B-15,286	64-30860 <*>
B-16,708	REPT. 25658 <RIS>
B-18,223	63-20039 <=*>
B-18,232	64-30545 <*>

Patent	Citation
B-18,458	64-20976 <*>
B-18,532	65-30893 <*>
B-18,680	64-30920 <*>
B-19,500	64-30935 <*>
B-19,595	64-20443 <*>
B-19,844	65-10949 <*>
B-19,899	65-10931 <*>
B-20,028	64-20991 <*>
B-20,132	60-14753 <+*>
B-20,158	65-10706 <*>
B-20,371	65-10666 <*>
B-21,108	65-10667 <*>
B-21,133	65-10670 <*>
B-21,813	96294 <RIS>
B-22,104	64-18408 <*>
B-22,576	65-10671 <*>
B-22,876	64-16765 <*>
B-22,899	64-30917 <*>
B-22,986	64-20494 <*>
B-23,003	64-20575 <*>
B-23,004	64-30878 <*>
B-23,131	<RIS>
B-23,526	64-30877 <*>
B-24,041	64-30881 <*>
B-24,110	64-30861 <*>
B-24,574	65-10668 <*>
B-25,064	64-30915 <*>
B-25,145	64-30901 <*>
B-26,022	64-30818 <*>
B-26,023	<RIS>
B-26,024	64-30817 <*>
B-26,205	REPT. 61021 <RIS>
B-26,206	64-30124 <*>
B-26,959	64-30546 <*>
B-27,506	<RIS>
B-28,386	64-30883 <*>
B-28,832	64-30547 <*>
B-29,038	64-20573 <*>
B-29,521	54103 <RIS>
B-29,806	64-18367 <*>
B-29,904	96730 <RIS>
B-29,967	96731 <RIS>
B-30,045	96746 <RIS>
B-30,527	64-30880 <*>
	96548 <RIS>
B-30,615	<RIS>
B-30,824	<RIS>
B-31,069	96608 <RIS>
B-31,215	62-13871 <*=> 0
B-31,584	63-14372 <*=>
B-31,758	96702 <RIS>
B-31,855	64-30882 <*>
B-32,159	96748 <RIS>
B-32,292	59-17841 <+*> 0
B-32,614	64-30548 <*> 0
	64-30906 <*>
B-32,974	65-10609 <*>
B-33,146	<RIS>
B-34,331	64-20572 <*>
B-34,505	REPT. 61063 <RIS>
B-35,499	REPT. 96698 <RIS>
B-35,597	96737 <RIS>
B-35,750	<RIS>
B-36,753	<RIS>
B-36,767	63-10078 <=*>
B-36,895	96604 <RIS>
B-37,100	<RIS>
B-38,451	<RIS>
B-40,025	REPT.61022 <RIS>
B-40,284	REPT. 96889 <RIS>
B-40,490	REPT. 54080 <RIS>
B-41,178	REPT. 61059 <RIS>
B-41,403	REPT. 54081 <RIS>
B-41,485	REPT. 6497 <RIS>
B-41,537	REPT.61079 <RIS>
B-41,839	REPT. 97103 <RIS>
B-42,483	REPT. 6496 <RIS>
B-43,298	61005 <RIS>
B-44,023	REPT. 97011 <RIS>
B-44,268	61044 <RIS>
B-44,384	REPT.97102 <RIS>
B-44,548	REPT. 97106 <RIS>
B-44,575	REPT. 97105 <RIS>
B-45,396	REPT. 61071 <RIS>
B-46,704	REPT. 97045 <RIS>
B-46,887	97125 <RIS>
B-47,818	REPT. 97044 <RIS>
B-48,330	REPT. 97098 <RIS>
B-50,167	K-44 <RIS>
B-50,308	61108 <RIS>
B-50,871	C-133 <RIS>
C-1,115	57-2642 <*>
C-4,171	65-10660 <*>
C-4,548	65-10675 <*>
C-4,791	63-20033 <=*>
C-4,871	64-30843 <*>
C-5,238	64-30004 <*>
C-5,324	64-20980 <*>
C-5,534	64-20332 <*>
C-5,643	64-30002 <*>
C-5,718	64-20571 <*>
C-5,737	64-20981 <*>
C-5,826	57-3123 <*>
C-6,425	REPT. 6300 <RIS>
C-6,576	60-14993 <+*>
C-6,748	64-30021 <*>
C-7,074	REPT. 54091 <RIS>
C-7,269	<RIS>
C-7,610	<RIS>
C-7,762	64-30003 <*>
C-8,049	64-30549 <*> P
C-8,058	96606 <RIS>
C-8,437	64-30550 <*>
C-8,497	64-30009 <*>
C-8,751	<RIS>
C-8,859	<RIS>
C-10,060	64-30900 <*>
	96743 <RIS>
C-10,125	63-24123 <=*$>
C-10,292	64-30551 <*> 0
C-10,293	64-30552 <*> 0
C-10,439	REPT. 6298 <RIS>
C-10,493	96603 <RIS>
C-10,649	64-30015 <*> 0
C-11,630	REPT. 6412 <RIS>
C-11,654	96949 <RIS>
C-11,896	REPT. 96973 <RIS>
C-12,208	REPT. 61048 <RIS>
C-13,031	REPT. 97035 <RIS>
C-13,162	61087 <RIS>
C-13,338	61015 <RIS>
C-14,116	REPT. 96950 <RIS>
C-14,489	REPT. 97046 <RIS>
C-14,531	REPT. 97116 <RIS>
C-14,676	REPT. 53012 <RIS>
C-14,967	REPT. 97095 <RIS>
C-15,296	REPT. 97117 <RIS>
C-15,325	61076 <RIS>
C-15,524	REPT. 97054 <RIS>
C-16,481	REPT. 61061 <RIS>
C-16,515	REPT. 97101 <RIS>
C-16,523	REPT. 97094 <RIS>
C-20,836	64-30936 <*>
C-25,628	62-20016 <=*>
C-26,226	62-20015 <=*>
D-5,906	64-30856 <*>
D-5,931	63-20259 <=*>
D-9,614	64-30926 <*>
D-9,926	64-20990 <*>
D-10,161	64-30934 <*>
D-10,769	AL-9II <*>
D-10,903	64-30000 <*>
D-11,711	65-10664 <*>
D-11,797	64-30933 <*>
D-13,001	64-30012 <*>
D-14,763	54065 <RIS>
D-15,494	<RIS>
D-15,882	REPT. 6294 <RIS>
D-16,645	62-32919 <=*>

Patent	Citation
D-17,461	63-13436 <=*> 0
D-17,955	64-30020 <*>
D-18,164	96707 <RIS>
D-18,523	96733 <RIS>
D-18,895	REPT. 96926 <RIS>
D-19,282	REPT. 97008 <RIS>
D-19,629	64-30011 <*>
D-20,736	REPT. 61016 <RIS>
D-21,060	REPT.61100 <RIS>
D-21,278	60-17300 <+*> 0
D-23,273	54066 <RIS>
D-24,036	REPT. 1101 <RIS>
D-24,600	96937 <RIS>
D-24,933	G-242 <RIS>
D-25,653	REPT. 96986 <RIS>
D-26,858	REPT. 61060 <RIS>
D-27,374	REPT. 97049 <RIS>
D-29,361	REPT. 61072 <RIS>
D-34,918	6055 <RIS>
D-51,291	97140 <RIS>
D-93,847	REPT. 5948 <RIS>
E-35,899	59-17855 <+*>
F-2,853	63-20032 <=*>
F-2,916	REPT. 95957 <RIS>
	64-30929 <*>
F-2,961	6277 <RIS>
F-3,974	64-30830 <*>
F-4,826	64-30827 <*>
F-5,384	REPT. 6114 <RIS>
F-5,436	64-20996 <*>
F-5,731	64-30924 <*>
F-6,417	64-30928 <*>
F-6,672	64-30075 <*>
F-6,794	64-30834 <*>
F-6,821	6241 <RIS>
	63-20034 <=*>
	64-30820 <*>
F-7,154	65-10661 <*>
F-8,063	REPT. 25732 <RIS>
F-8,103	64-20330 <*>
F-8,170	1498 <*>
F-8,233	6207 <RIS>
F-8,356	1511 <*>
	64-10463 <=*$>
F-8,417	60-16420 <+*>
F-8,783	63-20036 <=*>
F-8,882	64-30859 <*>
	96605 <RIS>
F-9,115	63-15451 <=*> 0
F-9,159	64-30904 <*>
F-9,205	63-20037 <=*>
F-9,388	64-30824 <*>
F-9,430	REPT. 54027 <RIS>
F-9,447	63-20038 <=*>
F-9,556	REPT. 96346 <RIS>
F-9,624	64-20999 <*>
	64-30852 <*>
F-9,957	64-30826 <*>
F-9,973	64-30921 <*>
F-9,989	2025 <*>
	57-176 <*>
F-10,094	64-30016 <*>
F-10,185	64-30554 <*>
F-10,231	64-30842 <*>
F-10,412	REPT. 6457 <RIS>
F-10,554	64-30553 <*> 0
F-10,712	<RIS>
F-10,759	1507 <*>
F-10,974	64-30907 <*>
F-11,407	64-30555 <*> 0
	64-30845 <*>
F-11,430	<RIS>
	64-30896 <*>
F-11,560	64-30867 <*>
F-11,596	54070 <RIS>
F-11,912	64-30825 <*>
	64-30829 <*>
F-11,991	REPT.61030 <RIS>
F-12,116	6317 <RIS>
F-12,145	60J14G <ATS>
F-12,298	6293 <RIS>
	64-30846 <*>
F-12,574	<RIS>
F-12,585	64-30844 <*>
F-12,647	6363 <RIS>
	65-10707 <*>
F-12,669	6364 <RIS>
	64-30912 <*>
F-12,976	64-30914 <*>
F-12,977	64-30911 <*>
F-12,992	64-30876 <*>
	96552 <RIS>
F-13,053	64-30908 <*>
F-13,119	<RIS>
F-13,543	96740 <RIS>
F-13,562	64-16465 <=*$>
F-13,716	6383 <RIS>
F-14,286	64-30870 <*>
F-14,442	6276 <RIS>
F-14,487	57-131 <*>
F-14,557	64-30858 <*>
F-14,755	<RIS>
F-14,832	REPT. 61004 <RIS>
F-15,019	64-30556 <*>
F-15,172	64-30894 <*>
F-15,173	64-30890 <*>
F-15,175	64-30889 <*>
F-15,266	64-16154 <=*$>
	64-30819 <*>
F-15,300	REPT. 6324 <RIS>
F-15,431	64-30910 <*>
	65-13374 <*>
F-15,521	64-30905 <*>
F-15,583	96747 <RIS>
F-15,622	REPT. 61009 <RIS>
F-15,858	57-29 <*>
F-15,911	6382 <RIS>
F-15,992	96741 <RIS>
F-16,038	REPT. 6427 <RIS>
F-16,278	REPT. 61020 <RIS>
F-16,297	64-10979 <=*$> 0
F-16,407	65-10652 <*>
F-16,448	64-30875 <*>
F-16,764	57-817 <*>
F-16,782	6381 <RIS>
F-16,912	REPT. 61019 <RIS>
F-17,150	54072 <RIS>
F-17,152	96739 <RIS>
F-17,373	64-30869 <*>
F-17,591	<RIS>
F-17,672	REPT. 54104 <RIS>
F-17,717	6362 <RIS>
	64-30913 <*>
F-17,783	96507 <RIS>
F-17,878	REPT. 96720 <RIS>
F-17,901	57-1858 <*>
F-18,054	61043 <RIS>
F-18,241	REPT. 61029 <RIS>
F-18,557	REPT. 61031 <RIS>
F-18,643	<RIS>
F-18,678	REPT. 96919 <RIS>
F-18,712	REPT. 61027 <RIS>
F-18,865	REPT. 97064 <RIS>
F-19,012	REPT. 61041 <RIS>
F-19,062	REPT. 96971 <RIS>
F-19,123	REPT. 61018 <RIS>
F-19,357	REPT. 61064 <RIS>
F-19,439	REPT. 6709 <RIS>
F-20,019	<RIS>
F-20,038	REPT. 97000 <RIS>
F-20,372	REPT. 6498 <RIS>
F-20,468	REPT. 61038 <RIS>
F-20,480	REPT. 61036 <RIS>
F-20,569	4752 <RIS>
F-20,797	REPT.61028 <RIS>
F-20,857	REPT. 61025 <RIS>
F-20,858	REPT. 61039 <RIS>
F-21,259	REPT. 61007 <RIS>
F-21,339	U-143 <RIS>
F-21,470	REPT.61024 <RIS>

F-21,552	REPT. 61013 <RIS>
F-21,625	REPT. 96988 <RIS>
F-21,694	REPT. 61037 <RIS>
F-21,864	REPT. 61026 <RIS>
F-21,892	REPT. 96931 <RIS>
F-22,016	REPT. 97020 <RIS>
F-22,145	REPT. 97034 <RIS>
F-22,163	REPT. 97003 <RIS>
F-22,199	96947 <RIS>
F-22,200	REPT.61032 <RIS>
F-22,271	REPT. 53013 <RIS>
F-22,796	59-17089 <+*>
F-22,797	59-17099 <+*>
F-23,041	U-142 <RIS>
F-23,715	REPT.97129 <RIS>
F-23,994	REPT. 61103 <RIS>
F-24,245	REPT. 61065 <RIS>
F-24,387	65-10945 <*>
F-24,472	REPT. 97051 <RIS>
F-24,738	REPT. 61104 <RIS>
F-25,093	97121 <RIS>
F-25,446	65-13571 <*>
F-26,081	REPT.61082 <RIS>
F-26,192	REPT. 97114 <RIS>
F-26,284	REPT.61098 <RIS>
G-174	64-10754 <=*$>
G-2,109	64-30832 <*>
G-2,571	64-30833 <*>
G-7,529	AL-270 <*>
G-12,802	REPT. 61066 <RIS>
G-14,427	REPT. 97043 <RIS>
G-16,399	65-10640 <*>
G-17,273	60-16050 <+*>
G-18,381	REPT. 96918 <RIS>
G-20,237	REPT. 97012 <RIS>
G-28,610	65-11241 <*>
H-7,525	64-30930 <*>
H-10,650	61-28086 <*=> O
H-11,399	64-20977 <*>
H-11,780	REPT. 61049 <RIS>
H-12,561	64-30018 <*>
H-13,896	64-30835 <*>
H-14,548	5498 <RIS>
H-14,666	65-10672 <*>
	96551 <RIS>
H-14,688	REPT. 6634 <RIS>
H-15,185	63-13437 <=*> O
H-15,579	64-30871 <*>
H-15,776	64-30892 <*>
H-17,425	96734 <RIS>
H-18,578	96735 <RIS>
H-19,793	62-10294 <*=>
	62-10295 <=*>
H-19,921	60-17297 <+*> O
H-20,721	60-17299 <+*> O
H-20,850	64-30013 <*>
H-21,244	96607 <RIS>
H-21,805	REPT. 61012 <RIS>
H-21,834	REPT. 96943 <RIS>
H-22,699	96738 <RIS>
H-23,523	61-27461 <*=> O
H-23,534	60-14997 <+*>
H-24,402	61-10146 <+*>
H-25,016	REPT. 61014 <RIS>
H-26,638	96736 <RIS>
H-26,922	REPT. 96990 <RIS>
H-27,158	REPT. 96942 <RIS>
H-28,196	96944 <RIS>
H-28,232	REPT. 96940 <RIS>
H-28,335	REPT. 96941 <RIS>
H-29,044	61-20282 <=*>
H-29,393	REPT. 96939 <RIS>
H-32,951	REPT. 97099 <RIS>
I-2,175	64-18848 <*>
	95940 <RIS>
I-2,361	63-18553 <*>
I-4,186	64-20495 <*>
I-14,609	S-2111 <RIS>
I-15,377	S-2112 <RIS>

I-31,568	63-18240 <=*>
I-40,266	63-10972 <=*>
I-55,196 ADD	64-18920 <*>
I-66,695	63-16021 <=*>
I-67,946	63-10974 <=*> O
I-70,852	95840 <RIS>
I-70,879	64-18902 <*>
I-71,427	64-30925 <*>
I-71,744	64-30891 <*>
I-73,726	64-30854 <*>
I-75,976	64-30855 <*>
I-76,488	6209 <RIS>
I-76,828	5829 <RIS>
I-77,296	63-14809 <=*>
I-78,329	65-10659 <*>
I-78,590	64-30829 <*>
I-125,947	65-10529 <*>
J-70,592 ADD	63-10957 <=*>
K-11,714	62-23738 <=*>
K-12,285	64-30073 <*>
K-14,320	59-18295 <+*> O
K-16,568	64-30557 <*> O
K-20,071	59-20183 <+*>
	65-11381 <*>
	65-14867 <*>
K-20,355	60-17298 <+*> O
K-21,845	64-20854 <*>
K-22,047	64-30008 <*>
K-22,148	REPT. 6479 <RIS>
K-22,285	65-14870 <*>
K-23,364	64-30007 <*>
K-24,881	64-30558 <*> O
K-29,359	REPT. 97042 <RIS>
K-32,451	97014 <RIS>
K-33,295	REPT. 97110 <RIS>
K-34,784	REPT. 61073 <RIS>
K-35,580	97107 <RIS>
K-35,703	REPT. 97109 <RIS>
K-154,822	51G5G <ATS>
L-1,936	60-14757 <+*>
L-10,506	1918 <*>
L-17,095	REPT. 4741 <RIS>
L-21,111	II-894 <*>
L-23,358	REPT.61101 <RIS>
L-26,600	REPT. 97108 <RIS>
L-27,083	REPT. 97039 <RIS>
L-27,119	REPT. 97041 <RIS>
L-31,149	97076 <RIS>
L-90,419	62-20269 <=*>
L-115,667	58-930 <*>
M-5,815	64-30848 <*>
M-20,249	59-20184 <+*> O
M-22,306	64-30864 <*>
M-22,307	64-30868 <*>
M-26,352	64-30918 <*>
M-26,353	65-10543 <*>
M-27,624	64-30014 <*>
M-27,682	<RIS>
M-27,743	REPT. 97115 <RIS>
M-29,976	61107 <RIS>
M-32,768	REPT.61106 <RIS>
M-33,717	REPT. 54105 <RIS>
M-34,587	REPT.61080 <RIS>
M-64,515	801 <TC>
M-64,712	802 <TC>
M-142,440	65-10665 <*>
M-151,848	64-20570 <*>
N-2,973	64-20975 <*>
N-3,020	64-30932 <*>
N-5,600	64-30559 <*> O
N-6,225	64-19916 <=$>
N-9,898	64-30560 <*> O
N-10,785	96932 <RIS>
N-12,844	REPT. 97033 <RIS>
	96945 <RIS>
N-13,184	REPT.61075 <RIS>
N-15,201	K-47 <RIS>
O-3,865	65-63478 <=$> O

Patent	Citation
P-1,866	64-20979 <*>
P-2,644	64-18849 <*>
P-5,700	64-18840 <*>
P-6,343	64-30919 <*>
P-7,009	REPT. 54079 <RIS>
P-7,224	4714 <RIS>
P-9,993	6292 <RIS>
P-10,386	54067 <RIS>
P-10,432	64-16143 <=*$>
P-10,489	REPT. 6466 <RIS>
P-10,879	54068 <RIS>
P-11,968	60-13279 <+*> C
P-12,516	6367 <RIS>
P-13,155	62-11631 <=>
P-13,874	64-30561 <*>
P-13,916	01J15G <ATS>
P-17,1C5	REPT. 61035 <RIS>
P-17,938	REPT. 97050 <RIS>
P-19,591	97077 <RIS>
P-21,228	4762 <RIS>
P-33,878	64-10753 <=*$>
P-36,021	64-30070 <*>
P-54,857	65-10674 <*>
	95854 <RIS>
P-55,651	REPT. 96951 <RIS>
P-82,274	64-20589 <*>
R-1,516	60-16437 <+*> P
R-2,366	57-3108 <*>
R-7,986	64-14905 <=*$>
R-8,137	1671 <*>
	64-14382 <=*$>
R-8,517	63-20035 <=*>
R-9,422	REPT. 6335 <RIS>
R-9,513	57-3107 <*>
R-9,564	REPT. 96109 <RIS>
R-9,639	64-30001 <*>
R-9,640	<RIS>
	64-30865 <*>
R-9,713	64-30873 <*>
R-10,277	64-30899 <*>
R-11,293	REPT. 61C11 <RIS>
R-11,422	64-30006 <*>
R-12,583	<RIS>
R-12,677	64-30562 <*>
R-12,684	57-816 <*>
R-13,195	57-815 <*>
R-13,473	64-30897 <*>
R-14,794	64-30563 <*>
R-14,900	64-30879 <*>
R-15,616	57-814 <*>
R-15,622	64-30564 <*>
R-16,980	57-813 <*>
R-17,969	REPT.61081 <RIS>
R-18,813	61008 <RIS>
R-19,146	REPT. 97096 <RIS>
R-20,212	REPT. 96987 <RIS>
R-25,549	A-121 <RIS>
R-93,021	6965G <ATS>
R-96,765	95839 <RIS>
R-102,012	65-10165 <*>
R-103,24C	65-10164 <*>
R-103,379	65-10163 <*>
R-104,54C	65-10161 <*>
R-104,875	65-10162 <*>
S-15,359	64-20988 <*>
S-18,721	64-20987 <*>
S-31,176	65-10663 <*>
S-32,050	64-20590 <*>
S-36,512	64-14376 <=*$> 0
S-36,927	REPT. 97016 <RIS>
S-40,030	64-30565 <*> 0
S-41,845	T-22 <RIS>
S-41,936	60-14998 <+*>
S-42,054	64-30909 <*>
S-42,168	REPT.61077 <RIS>
S-42,687	REPT. 61050 <RIS>
S-43,547	64-30010 <*>
S-44,397	64-30566 <*> 0
S-47,979	REPT. 96995 <RIS>
S-54,542	REPT.97139 <RIS>
S-55,096	97113 <RIS>
S-55,390	REPT. 97112 <RIS>
S-55,671	REPT. 97018 <RIS>
S-56,480	REPT. 97122 <RIS>
S-57,408	REPT.1106 <RIS>
S-61,231	REPT.97138 <RIS>
SCH-4,308	64-20591 <*>
SCH-8,022	64-20592 <*>
SCH-9,697	64-30927 <*>
SCH-11,546	REPT. 25733 <RIS>
SCH-11,832	64-30872 <*>
SCH-13,993	64-30567 <*>
SCH-18,852	64-30568 <*>
SCH-25,854	63-16239 <*=>
ST-4,654	64-20593 <*>
	64-30569 <*>
ST-4,892	60-14992 <+*> C
ST-5,783	64-30005 <*>
ST-6,061	64-30837 <*>
ST-8,190	64-30874 <*>
ST-12,180	97015 <RIS>
ST-13,064	63-24129 <=*>
ST-14,492	A-123 <RIS>
T-13,006	REPT. 97117 <RIS>
T-14,282	K-46 <RIS>
T-16,172	37N57G <ATS>
T-16,186	38N57G <ATS>
T-54,845	61-10030 <*+>
T-55,995	64-20982 <*>
U-1,413	64-30923 <*>
U-1,851	60-14991 <+*> C
U-5,393	REPT. 97097 <RIS>
V-2,298	AL-587 <*>
V-4,275	64-30895 <*>
	96727 <RIS>
V-4,499	96728 <RIS>
V-5,430	96729 <RIS>
V-7,016	REPT. 54028 <RIS>
V-10,355	REPT. 61051 <RIS>
V-10,579	REPT. 6499 <RIS>
V-13,709	97078 <RIS>
V-13,72C	REPT. 61062 <RIS>
V-14,673	C-173 <RIS>
W-1,016	64-30898 <*>
W-7,062	63-16816 <*=>
W-7,258	65-10662 <*>
W-7,659	64-30857 <*>
W-8,238	REPT. 6185 <RIS>
W-10,357	REPT. 6262 <RIS>
W-12,605	REPT. 6286 <RIS>
W-13,988	<RIS>
W-14,358	63-20040 <=*>
W-15,627	REPT. 61001 <RIS>
W-16,319	4751 <RIS>
W-16,971	64-30570 <*>
W-17,573	REPT.61102 <RIS>
W-18,266	61023 <RIS>
W-18,842	REPT. 96883 <RIS>
W-19,784	REPT. 61002 <RIS>
W-21,848	REPT. 97017 <RIS>
W-23,120	S-2097 <RIS>
W-23,713	REPT. 61067 <RIS>
Y-38	61-16189 <=*>
Z-1,118	6106 <RIS>
Z-2,635	<RIS>
	96252 <RIS>
Z-2,636	57-139 <*>
	64-18388 <*>
	96253 <RIS>
Z-2,647	96256 <RIS>
Z-2,648	96693 <RIS>
Z-2,679	64-14904 <=*$>
	64-30887 <*>
	96254 <RIS>
Z-2,681	96255 <RIS>
Z-2,838	63-16814 <=*>
Z-2,857	96264 <RIS>
Z-2,971	64-20530 <*>
	63-16815 <*=>
Z-3,172	64-30571 <*>

Z-3,509	96505 <RIS>
Z-3,584	64-14704 <=*$>
	64-30821 <*>
	96259 <RIS>
Z-3,614	<RIS>
	96573 <RIS>
Z-3,672	64-30572 <*>
Z-3,803	64-30019 <*>
	96257 <RIS>
Z-3,915	REPT. 96948 <RIS>
Z-4,269	<RIS>
	96483 <RIS>
Z-4,357	96853 <RIS>
Z-4,603	REPT. 97019 <RIS>
Z-4,846	64-30866 <*>
	96595 <RIS>
Z-6,162	65-11256 <*>

GERMANY, EAST (SEE ALSO EAST GERMANY)

9,657	64-30998 <*>
11,318	64-30199 <*>
12,916	NS-207 <TTIS>
16,242	NS 243 <TTIS>
21,328	NS-177 <TTIS>
21,542	NS-16 <TTIS>
21,731	N.S. 29 <TTIS>
22,489	64-13610 <=*$>
22,629	NS-32 <TTIS>
22,937	NS-301 <TTIS>
23,246	64-13584 <=*$> C
23,467	NS-52 <TTIS>
24,307	NS-87 <TTIS>
25,043	NS-143 <TTIS>
25,127	65-63477 <=$>
25,177	NS-144 <TTIS>
25,329	NS-145 <TTIS>
25,391	NS-158 <TTIS>
25,535	NS-146 <TTIS>
	8SR76G <ATS>
25,584	NS-185 <TTIS>
25,585	NS-159 <TTIS>
25,736	NS-147 <TTIS>
26,C21	NS 273 <TTIS>
26,022	NS-302 <TTIS>
26,239	NS-362 <TTIS>
26,361	NS-229 <TTIS>
26,402	NS 252 <TTIS>
26,403	NS-303 <TTIS>
26,514	NS 259 <TTIS>
26,541	NS-184 <TTIS>
26,571	NS-275 <TTIS>
26,640	NS 218 <TTIS>
26,707	NS 274 <TTIS>
26,710	NS 260 <TTIS>
26,722	NS-276 <TTIS>
26,973	NS-328 <TTIS>
27,013	NS-230 <TTIS>
27,020	NS-204 <TTIS>
27,022	NS-253 <TTIS>
27,025	NS-304 <TTIS>
27,142	NS-305 <TTIS>
27,145	NS 244 <TTIS>
27,161	NS-306 <TTIS>
27,627	NS-307 <TTIS>
27,894	NS-364 <TTIS>
28,C31	NS 245 <TTIS>
28,057	NS-460 <TTIS>
28,095	NS-363 <TTIS>
28,098	NS-365 <TTIS>
28,142	NS-420 <TTIS>
28,154	NS 254 <TTIS>
28,217	NS-366 <TTIS>
28,279	NS-329 <TTIS>
28,295	NS-330 <TTIS>
28,320	NS-331 <TTIS>
28,346	NS 256 <TTIS>
28,357	NS-367 <TTIS>
28,480	NS-421 <TTIS>
28,566	NS-368 <TTIS>
28,569	NS-369 <TTIS>
28,595	NS-370 <TTIS>
28,600	NS-332 <TTIS>
28,740	NS-371 <TTIS>
28,752	NS-372 <TTIS>
28,788	NS-373 <TTIS>
28,822	NS-375 <TTIS>
28,829	NS-376 <TTIS>
28,940	NS-377 <TTIS>
29,032	NS-378 <TTIS>
29,076	NS-379 <TTIS>
29,143	NS 246 <TTIS>
29,304	NS 277 <TTIS>
29,408	NS-380 <TTIS>
29,409	NS-422 <TTIS>
29,459	NS-381 <TTIS>
29,513	NS-382 <TTIS>
29,606	NS 278 <TTIS>
29,707	NS-383 <TTIS>
29,811	NS-423 <TTIS>
29,812	NS-424 <TTIS>
29,895	NS-384 <TTIS>
29,900	NS-308 <TTIS>
29,902	NS-425 <TTIS>
30,095	NS-385 <TTIS>
30,311	NS-426 <TTIS>
30,316	NS-427 <TTIS>
30,406	NS-279 <TTIS>
31,061	NS-459 <TTIS>
31,929	NS-428 <TTIS>
32,580	NS-430 <TTIS>
33,848	NS-464 <TTIS>
33,890	NS-431 <TTIS>
	66-12803 <*>
34,074	NS-341 <TTIS>
	65-12785 <*>
34,615	NS-432 <TTIS>
35,119	NS-463 <TTIS>
36,632	NS-462 <TTIS>
36,725	NS-461 <TTIS>
38,258	1404 <TC>
38,316	NS-433 <TTIS>

HUNGARY

37,938	62-32920 <=*>
48,513	63-23168 <=*$> 0
112,277	64-16466 <=*$>
113,819	60-16368 <=*$>
130,209	60-14866 <+*>
130,804	60-14711 <+*>
131,128	AL-967 <*>
133,573	63-19275 <=*> C
142,508	63-24607 <=*$>
150,991	65-18074 <*>

ITALY

244,484	59-16844 <+*> 0
268,259	64-13251 <=*$> 0
269,130	65-62854 <=$>
271,056	63-19276 <=*> C
271,263	34S86I <ATS>
276,911	59-12624 <+*> 0
278,648	64-15856 <=*$> 0
281,639	59-12105 <+*> 0
299,706	65-13261 <*>
310,672	63-19058 <=*> C
312,837	64-30573 <*>
317,767	59-17185 <+*> 0
321,470	64-13576 <=*$> 0
325,890	64-30574 <*>
328,002	64-30576 <*> 0
329,884	64-13252 <=*$> 0
331,686	63-19618 <=*> C
331,754	64-10620 <=*$>
333,444	61-20865 <*=>
341,126	63-13429 <=*> 0
353,296	62-33248 <=*> 0
353,773	64-19799 <=$> C
354,131	59-12625 <+*> 0

Patent	Citation
354,510	63-18776 <=*>
355,426	64-30577 <*>
355,502	63-14533 <=*>
361,149	59-19754 <+*> 0
363,163	60-16404 <=*$>
363,380	62-32921 <=*> 0
366,926	59-12626 <+*>
367,704	3253 <K-H>
369,098	59-16183 <+*> 0
374,500	59-16184 <+*> 0
379,301	60-17998 <+*> 0
382,799	63L33I <ATS>
389,980	5185 <K-H>
402,961	65-10641 <*> 0
403,697	60-17615 <+*> 0
404,396	59-12627 <+*>
405,301	62-15757 <*=> 0
405,361	4047 <K-H>
410,106	4022 <K-H>
410,281	63-13751 <=*> 0
410,573	61-27464 <*=> 0
412,709	64-15136 <=*$> 0
412,709 ADD	64-15135 <=*$> 0
414,395	63-19619 <=*> 0
415,64C	64-15135 <=*$> 0
415,982	60-14864 <+*>
416,280	60-14863 <+*>
417,939	63-14945 <=*>
418,088	61-13839 <+*> 0
418,823	59-18791 <+*> 0
419,810	2413 <*>
421,443	59-19755 <+*> 0
422,782	59-16185 <+*> 0
423,864	61-15687 <+*> 0
424,371	60-17616 <+*> 0
424,653	63-23169 <=*$> 0
428,289	64-19181 <=*$> 0
429,377	59-17094 <+*>
431,313	61-19491 <+*> 0
431,407	64-18892 <*>
431,484	64-18881 <*>
432,082	63-18561 <=*>
433,260	62-24520 <=*> 0
433,333	64-30575 <*>
433,554	42K211 <ATS>
433,855	64-30578 <*>
439,665	61-13840 <+*>
440,655	63-18827 <=*>
441,817	64-30579 <*>
442,987	60-13284 <+*> 0
443,057	63-14534 <*=>
444,275	AL-974 <*>
448,583	59-12628 <+*>
450,020	62-15758 <*=> 0
450,973	66-11047 <*>
453,963	63-14535 <*=>
454,358	62-10233 <*=>
455,785	64-15333 <=*$> 0
457,227	59-19756 <+*> 0
458,677	59-12926 <+*> 0
459,171	64-23593 <=$>
459,422	64-10794 <=*$> 0
460,196	63-18807 <=*>
	63-23184 <=*$>
460,863	63-19277 <=*> 0
462,021	62-20010 <=*>
463,918	63-18808 <=*>
	63-23738 <=*$>
465,376	59-16186 <+*> 0
465,991	63-15452 <=*> 0
466,456	63-19057 <=*> 0
466,792	64-15332 <=*$> 0
467,300	60-15870 <+*> 0
467,508	66-10658 <*>
468,238	60-15871 <+*> 0
468,568	41J17I <ATS>
	58-1399 <*>
469,458	64-30580 <*>
469,574	64-16161 <=*$>
	64-20578 <*>

Patent	Citation
	65-10681 <*> 0
470,886	63-19278 <=*> 0
470,904	<AC>
473,891	1914 <*>
474,516	64-30581 <*> 0
478,826	63-23774 <=*$> 0
480,005	63-18770 <=*>
480,353	61-13266 <+*> 0
480,404	61-15688 <+*> 0
480,407	64-30582 <*>
481,887	64-20851 <*>
482,794	61-13267 <+*> 0
482,999	64-19182 <=*$> 0
483,244	60-17999 <+*> 0
483,639	65-11722 <*>
486,564	61-15055 <+*>
487,433	63-10273 <=*> 0
489,348	59-16187 <+*> 0
491,777	96R79I <ATS>
491,942	65-10588 <*>
492,286	63-13008 <=*> 0
492,469	64-30583 <*>
492,729	30J151 <ATS>
494,489	59-22283 <+*> 0
495,082	553 <TC>
495,814	64-19800 <=$>
496,366	63-16812 <=*>
497,244	59-12089 <+*>
498,899	64-20215 <*> P
502,414	65-62852 <=$> 0
504,291	62-33234 <=*> 0
504,930	65-60294 <=$>
506,548	63-24126 <=*>
508,233	65-10587 <*>
508,340	59-14213 <+*> 0
509,694	65-14866 <*>
510,368	62-15759 <*=> 0
510,939	65-10300 <*>
511,194	61-19492 <+*> 0
511,764	66-60249 <=*$>
512,162	59-16188 <+*> 0
512,428	63-15764 <=*> 0
514,583	64-14482 <=*$>
	8003-E <K-H>
514,965	63-24125 <=*>
	64-18107 <*>
515,093	<ES>
515,228	64-16164 <*>
515,240	64-20577 <*>
516,436	96482 <RIS>
517,229	61-15689 <+*> 0
517,363	60-15872 <+*> 0
518,844	63-19279 <=*> 0
519,469	61-23984 <*=> 0
521,507	65-62651 <=$> 0
521,521	64-30847 <*>
	96481 <RIS>
522,838	61-23230 <=*> 0
	62-13869 <*=> 0
522,868	65-64476 <=*$> 0
525,664	61-13268 <+*> 0
526,101	<LSA>
	62-25019 <=*>
528,221	65-61489 <=$> 0
528,697	64-15134 <=*$> 0
529,370	64-15133 <=*$> 0
529,640	61-27465 <*=> 0
530,211	63-16890 <*>
530,753	04N51I <ATS>
	64-16468 <=*$>
531,509	64-14474 <=*$>
531,511	10934-C <K-H>
	64-14475 <=*$>
533,751	63-19620 <=*> 0
534,115	64-20214 <*> P
535,712	1271 <TC>
	60-10091 <+*>
536,353	61-27466 <*=> 0
537,164	CHEM-330 <CTS>
537,203	60-15873 <+*> 0

537,229	63-14791 <=*>
540,614	57-3509 <*>
543,096	64-15132 <=*$> 0
543,706	63-15453 <=*>
544,089	65-62853 <=$> 0
544,461	60-15874 <+*> 0
545,307	65-10572 <*>
545,576	65-10676 <*> 0
546,143	65-62850 <=$> 0
546,379	65-10642 <*>
546,380	60-13081 <+*> 0
546,404	63-14790 <=*>
546,891	59-19227 <+*> 0
547,128	64-19801 <=$> 0
547,583	58K26I <ATS>
549,517	61-27467 <*=> 0
550,379	AEC-TR-4821 <*>
551,361	61-15995 <+*> 0
552,738	60-10090 <+*>
553,976	60-15875 <+*> 0
554,015	<LSA>
554,198	65-60840 <=$> 0
554,803	62-10759 <*=> 0
556,500	65-10327 <*>
558,434	61-13269 <+*> 0
560,271	63P59I <ATS>
560,287	61-19848 <=*> 0
563,676	64-30584 <*> 0
564,394	63-23775 <=*$> 0
565,200	63-13752 <=*> 0
566,252	60-16753 <+*>
566,294	65-60295 <=$>
566,623	60-16682 <+*>
566,944	60-14305 <+*>
566,945	60-14304 <+*>
567,214	791 <TC>
568,630	64-19183 <=*$> 0
571,970	63-24606 <=*$> 0
577,545	1272 <TC>
577,685	62-13496 <*=> 0
580,541	253 <CTS>
580,979	47M37I <ATS>
581,135	66-60248 <=*$>
582,349	64-15131 <=*$> 0
582,493	65-60841 <=$> 0
583,501	CHEM-158 <CTS>
584,862	869 <TC>
586,834	65-63485 <=$>
587,349	64-13765 <=*$> 0
588,211	1586 <TC>
589,326	66-60247 <=*$>
590,250	212 <TC>
591,082	64-19917 <=$>
591,906	65-11268 <*>
592,308	65-11235 <*>
593,152	64-23594 <=$>
594,018	24P61I <ATS>
594,084	64-15331 <=*$> 0
594,612	65-11234 <*>
594,971	65-63486 <=$>
597,165	64-13575 <=*$> 0
598,220	CHEM-197 <CTS>
600,498	65-11265 <*>
600,817	CHEM-331 <CTS>
601,506	65-63490 <=$>
603,859	44P63I <ATS>
609,886	65-60996 <=$>
612,116	04Q73I <ATS>
615,048	1270 <TC>
	65-63489 <=$>
866,288	61-19838 <=*> 0

JAPAN

2,963	63-14652 <=*>
3,097	63-16058 <=*>
3,623	AL-465 <*>
4,642	63-16063 <=*> 0
5,020	77H12J <ATS>
32,311	62-16330 <=*>
32,783	64-20770 <*>
86,574	63-16793 <=*>
91,248	64-20779 <*>
93,152	63-14537 <=*>
100,833	64-20780 <*>
100,964	CODE-P-157 <SA>
102,330	64-20781 <*>
104,003	64-20782 <*>
109,412	PANSDOC-TR.649 <PANS>
109,884	64-16454 <=*$>
110,287	64-20783 <*>
111,612	64-20850 <*>
112,088	64-20784 <*>
112,278	64-20785 <*>
116,453	64-20792 <*>
134,221	63J15J <ATS>
145,417	64-20786 <*>
154,815	97G6J <ATS>
155,238	64-20787 <*>
156,256	58-291 <*>
	64-16055 <=*$>
158,610	63-16071 <=*>
158,669	63-20158 <=*>
162,228	75P60J <ATS>
172,313	4904 <HB>
172,596	63-20159 <=*>
172,740	36S89J <ATS>
173,575	37S89J <ATS>
173,717	<INT>
174,320	59-17364 <+*>
174,363	64-16881 <=*$>
175,044	64-30150 <*>
175,175	35E2J <ATS>
175,771	65-11997 <*>
176,029	64-20726 <*>
176,130	CODE P-382 <SA>
176,514	61-10400 <+*>
176,576	AL-325 <*>
176,793	64-20728 <*> 0
176,795	64-20727 <*>
176,810	59-14401 <ATS>
176,907	63-15763 <=$> 0
177,371	CCDE-P-54 <SA>
177,553	11F4J <ATS>
178,340	64-20799 <*>
179,838	59-17878 <+*>
180,004	59-17876 <+*>
180,077	64-20797 <*>
180,137	64-20729 <*>
188,695	63-19853 <=*>
189,015	CODE-P-225 <SA>
190,201	64-20741 <*>
194,591	62-16208 <=*> 0
194,648	54P61J <ATS>
201,869	64-20233 <*> P
206,133	64-10514 <*>
206,418	CODE-P-224 <SA>
206,998	CODE-P-118 <SA>
211,024	64-20234 <*> P
214,074	64-14477 <=*$>
218,829	64-14483 <=*$>
228,850	CODE-P-220 <SA>
230,110	CODE P-383 <SA>
233,138	65-11246 <*>
233,609	CODE-P-146 <SA>
244,168	
	CODE-P-223 <SA>
245,707	CODE-P-185 <SA>
251,846	251,846 JAPAN <INT>
253,283	CODE-P-221 <SA>
253,649	CODE-P-265 <SA>
255,539	CODE-P-153 <SA>
261,634	<INT>
266,501	CODE P-385 <SA>
272,926	CODE P-386 <SA>
281,165	65-62992 <=>

285,265	CODE P-388 <SA>
324,117	NP-TR-1030 <*>
337,627	NP-TR-1029 <*>
384,561	<INT>

JAPAN PUBLICATION

10-103	64-20783 <*>
10-2,208	64-20785 <*>
11-1,124	64-20792 <*>
11-37,834	<ATS>
16-3,049	03R78J <ATS>
17-4,531	64-20786 <*>
17-4,853	CODE-P-410 <SA>
17-4,944	64-20787 <*>
18-1,535	<INT>
18-2,028	<INT>
22-1,423	<INT>
	64-20726 <*>
22-2,078	CODE-P-336 <SA>
23-768	64-20728 <*>
23-1,799	63F3J <ATS>
23-2,729	64-20799 <*>
24-326	65-11814 <*>
24-1,218	64-20798 <*>
24-1,717	64-20797 <*>
24-4,627	63-18965 <=*$> C
25-580	<ATS>
25-883	60-14802 <+*>
25-1,302	63-18684 <=*>
	64-20730 <*>
25-1,807	63-14536 <=*>
25-2,266	59-15607 <+*>
25-2,363	MF-5 <*>
25-2,378	64-20731 <*> O
25-2,579	01M45J <ATS>
25-3,114	64-20732 <*>
25-3,406	59-15367 <+*>
25-3,485	64-20733 <*>
25-3,639	36F3J <ATS>
25-3,705	64-20734 <*>
25-3,718	18H9J <ATS>
25-4,351	<AC>
25-4,373	63-18975 <=*$>
25-4,375	97E1J <ATS>
25-4,484	60-14809 <+*>
25-4,549	64-20735 <*>
25-7,400	62-19911 <=>
25-13,827	70F4J <ATS>
26-16	95E1J <ATS>
26-263	<INT>
26-278	64-20737 <*>
26-279	CODE-P48 <SA>
26-286	94E1J <ATS>
26-294	63-18974 <=*$> O
	64-20738 <*> O
26-296	63-18962 <=*>
26-892	64-20739 <*> O
26-1,301	63-19853 <=*>
26-2,229	<INT>
26-2,238	25S81J <ATS>
26-2,420	<INT>
26-2,690	62-11278 <=>
26-2,709	64-20740 <*>
26-2,720	CODE-P-278 <SA>
26-2,723	<INT>
26-2,781	<SA>
26-3,019	63-20581 <=*$>
	64-20061 <*>
26-4,082	41S89J <ATS>
26-4,147	64-20741 <*>
26-4,540	CODE-P4 <SA>
26-5,249	64-20742 <*>
26-5,587	62-19911 <=>
26-5,979	64-18490 <*>
26-6,169	64-20743 <*>
26-6,433	60-14806 <+*>
26-6,747	64-20752 <*>
26-6,865	<INT>
26-6,915	64-20349 <*>
26-7,130	<INT>
26-7,288	<INT>
27-148	7E2J <ATS>
27-323	64-20789 <*>
27-349	<INT>
27-399	<INT>
27-814	80Q69J <ATS>
27-1,139	63-18993 <=*>
27-1,275	64-20744 <*> C
27-1,925	<INT>
27-2,572	64-20745 <*>
27-2,959	<INT>
27-3,211	CODE-P-155 <SA> C
27-3,229	64-20336 <*>
	64-30149 <*>
27-3,426	<INT>
27-3,759	64-20747 <*>
27-4,156	59-20102 <+*>
27-4,567	<INT>
27-4,738	62-11626 <=>
27-4,876	1414 <*>
	57-1334 <*>
27-5,050	CODE-P-192 <SA>
27-5,120	64-20748 <*> O
27-5,254	64-20749 <*>
27-5,267	64-20750 <*>
27-5,361	64-20751 <*>
28-74	64-20796 <*>
28-609	79Q69J <ATS>
28-975	CODE-P-129 <SA> O
28-1,061	64-20753 <*>
28-1,209	69F4J <ATS>
28-1,570	CODE-P-131 <SA> O
28-1,575	64-20754 <*> O
28-1,694	64-20316 <*>
28-1,719	<INT>
28-1,863	<INT>
28-1,974	48G6J <ATS>
28-1,994	17F4J <ATS>
28-2,154	3138-D <K-H>
28-2,222	53K22J <ATS>
28-2,286	40S89J <ATS>
28-3,180	34S89J <ATS>
28-3,181	38S89J <ATS>
28-3,182	39S89J <ATS>
28-3,669	64-30148 <*>
28-3,844	65-18159 <*>
28-4,226	64-20755 <*>
28-4,472	60-14810 <+*>
28-4,770	09J15J <ATS>
28-4,919	20J16J <ATS>
28-5,016	<INT>
	65-13219 <*>
28-5,188	CODE-P-172 <SA>
28-5,529	<INT>
28-5,765	<INT>
28-6,008	PANSDOC-TR.169 <PANS>
28-6,179	<INT>
28-6,325	11296-B <K-H>
28-6,639	64-20756 <*>
28-8,109	60-15625 <+*>

Patent	Citation
29-115	10J15J <ATS>
29-116	64-14676 <=*$> 0
29-117	64-14675 <=*$> 0
29-144	CODE-P-297 <SA>
29-830	<INT>
29-1,283	58-1815 <*>
29-1,481	CODE-P-224 <SA>
29-1,493	62-11310 <=>
	64-10514 <*>
29-2,068	<INT>
29-2,426	CODE-P44 <SA>
29-3,766	64-16429 <=*$>
29-5,219	60-10928 <+*>
29-5,376	60-10927 <+*>
29-5,398	<INT>
29-5,614	<INT>
29-5,617	<INT>
29-6,183	35S89J <ATS>
29-6,200	63-00733 <*>
29-6,428	<INT>
29-6,627	<INT>
29-7,018	63-14622 <=*>
29-7,042	64-20757 <*>
29-7,245	<INT>
29-7,330	63-23170 <=*$> 0
29-7,463	<INT>
29-7,716	64-20795 <*>
29-7,718	58-1090 <*>
29-7,719	58-1212 <*>
	64-20794 <*>
29-8,008	<INT>
29-8,010	64-20758 <*>
29-8,017	64-20759 <*>
29-8,449	54G5J <ATS>
29-8,655	<INT>
30-317	65-13847 <*>
30-330	CODE-P-130 <SA> 0
30-422	<INT>
30-439	CCL-65/3 <STC>
30-967	CODE-P-124 <SA>
30-968	63-14624 <=*>
30-1,071	<INT>
	CODE-P-264 <SA>
30-1,871	64-20760 <*>
30-2,151	11252 <K-H>
	64-14476 <=*$>
30-2,299	<INT>
30-2,619	61-10165 <+*>
30-2,750	CODE-P-241 <SA>
30-2,953	64-10549 <=*$>
30-3,217	<INT>
30-3,218	63-14631 <=*>
30-3,221	82S83J <ATS>
30-3,672	66-12806 <*>
30-3,809	<INT>
30-3,810	63-14633 <=*>
30-3,812	CODE-P-220 <SA>
30-3,924	60-10930 <+*>
30-3,931	64-20761 <*>
30-4,015	<INT>
30-5,758	<INT>
30-6,122	64-20791 <*>
30-6,469	64-30157 <*>
30-6,661	24Q72J <ATS>
30-6,851	<INT>
30-7,026	CODE-P-254 <SA>
30-7,089	62-16328 <=*>
30-7,199	<WBI>
30-8,C87	62-16327 <=*>
30-8,098	62-16254 <=*>
30-8,732	62-16326 <=*>
30-9,072	60-14805 <+*>
30-9,073	60-14804 <+*>
31-66	<INT>
31-295	79L30J <ATS>
31-422	63-16290 <=*> 0
31-2,152	<INT>
31-2,688	64-20762 <*>
31-2,900	<INT>
31-2,972	63-16097 <=*>
31-3,108	<INT>
31-3,217	65-13848 <*>
31-3,653	<INT>
31-3,721	62-18038 <*>
31-3,722	<INT>
31-3,768	63-16098 <=*>
31-4,117	63-16074 <*=>
31-4,118	63-16073 <*=>
31-4,544	CODE-P-273 <SA>
31-5,221	<INT>
31-5,223	64-20763 <*>
31-5,224	CODE-P45 <SA>
31-5,352	66-10834 <*>
31-5,520	<INT>
31-5,713	63-16079 <*=>
31-5,740	64-20764 <*>
31-5,894	<INT>
31-6,414	<INT>
31-6,720	<INT>
31-7,002	65-10510 <*>
31-7,021	65-10511 <*>
31-8,298	64-20765 <*>
31-8,566	CODE-P-234 <SA>
31-8,581	CODE-P-305 <SA>
31-8,797	CODE P-383 <SA>
31-8,863	93K25J <ATS>
31-8,896	<INT>
31-9,459	63-16694 <=*>
31-9,492	64-13574 <=*$>
31-9,997	<INT>
31-10,218	63-16086 <*=>
32-13	63-16089 <*=>
32-328	60-10929 <+*>
32-420	63-16090 <=*>
32-530	65-11246 <*>
32-1,119	<INT>
32-1,418	60-15622 <+*>
32-1,422	59P60J <ATS>
32-1,546	CODE-P-67 <SA>
32-2,045	<INT>
32-2,699	64-20293 <*>
32-3,215	<INT>
32-3,251	<INT>
32-3,374	<INT>
	CODE-P-283 <SA>
32-3,482	CODE-P-128 <SA> 0
32-3,589	CODE-P-248 <SA>
32-3,710	<INT>
32-3,859	65-10125 <*>
32-3,971	CODE-P-282 <SA>
32-4,306	<INT>
32-4,307	<INT>
32-4,599	60-14808 <+*>
32-4,851	CODE-P-269 <SA>
32-4,852	<INT>
32-5,595	64-30156 <*>
32-6,087	65-10820 <*>
32-6,104	64-10539 <=*$>
32-6,105	64-10538 <=*$>
32-6,613	<INT>
32-6,624	97K25J <ATS>
32-6,720	<INT>
32-6,848	63-23171 <=*$> 0
32-7,338	<INT>
32-8,061	66-12808 <*>
32-8,177	<INT>
32-8,492	<INT>
32-8,498	<INT>
32-8,611	<INT>
32-8,867	AEC-TR-4823 <*>
32-8,959	<INT>
32-9,068	CODE-P-223 <SA>
32-9,089	<INT>
32-9,858	<INT>

Patent	Citation
32-9,864	<INT>
32-10,308	<INT>
32-10,495	<INT>
32-28,641	76Q73J <ATS>
33-781	60-14813 <+*>
33-1,205	<INT>
33-1,473	CODE-P-222 <SA>
33-1,571	<INT>
33-2,274	CODE-P-243 <SA>
33-2,374	<INT>
33-2,473	60-14812 <+*>
33-2,787	60-14705 <+*>
33-3,189	CODE-P-185 <SA>
33-3,215	<INT>
33-3,525	63-16123 <*=>
33-3,769	<LSA>
33-4,569	66-12807 <*> 0
33-4,7C8	65-10601 <*>
33-5,434	53S88J <ATS>
33-5,480	<INT>
33-5,848	65-62855 <=$*>
33-7,676	33S89J <ATS>
33-8,133	<INT>
33-8,261	<INT>
33-8,959	<INT>
33-8,965	98R77J <ATS>
33-9,455	978 <TC>
33-10,579	60-14702 <+*>
33-10,765	23M42J <ATS>
33-17,790	<SA>
33-20,212	CODE-P-353 <SA>
33-31,101	<SA>
34-55	CODE-P-190 <SA>
34-89	CODE-P-76 <SA>
34-344	1078 <TC>
34-416	979 <TC>
34-893	CODE-P-85 <SA>
34-1,072	CODE-P-221 <SA>
34-1,748	<INT>
34-2,142	64-20766 <*>
34-2,271	CODE-P-378 <SA>
34-2,275	65-10684 <*>
34-2,310	65-10631 <*>
34-2,311	65-10633 <*>
34-2,594	64-30155 <*>
34-2,619	65-11800 <*>
34-2,661	65-10632 <*>
34-2,662	CODE P-384 <SA>
34-2,969	42L37J <ATS>
34-2,989	60-10726 <+*> 0
34-3,768	<INT>
	65-10519 <*>
34-4,457	<INT>
34-4,686	65-10628 <*> 0
34-4,794	60-14704 <+*>
34-5,204	<INT>
34-5,285	64-20653 <*>
34-5,286	64-30154 <*>
34-5,444	66-11313 <*>
34-5,581	65-10577 <*>
34-5,659	65-10556 <*>
34-5,668	64-30153 <*>
34-5,669	CODE-P-151 <SA> 0
34-5,899	63-23550 <=*$>
34-6,619	65-10520 <*>
34-7,750	CODE-P-235 <SA>
34-7,920	CODE-P-92 <SA>
34-8,842	<INT>
34-9,858	<INT>
34-10,183	CODE-P-8C <SA>
34-10,309	22N52J <ATS>
34-10,344	65-10542 <*>
34-10,461	D 664 <RIS>
34-10,619	65-10685 <*>
34-10,753	D-66C <RIS>
35-3	<INT>
35-41	CODE-P-214 <SA>
35-42	CODE-P-211 <SA>
35-641	19R77J <ATS>
35-643	1079 <TC>
35-743	<INT>
35-912	77P59J <ATS>
35-1,831	62-16329 <=*>
35-1,839	CCDE-P-215 <SA>
35-2,751	<INT>
35-2,752	<INT>
35-2,757	<INT>
35-3,514	65-10532 <*>
35-3,781	10S82J <ATS>
35-4,988	<INT>
35-5,292	CODE-P-150 <SA> 0
35-5,608	<INT>
35-5,609	<INT>
35-5,768	11S82J <ATS>
35-6,888	CCDE P-385 <SA>
35-7,C58	CCDE-P-91 <SA>
35-7,097	64-30162 <*>
35-7,373	<INT>
35-7,807	CODE-P41 <SA>
35-7,837	CODE-P42 <SA>
35-8,202	<INT>
35-8,466	<INT>
35-8,479	AEC-TR-4993 <*>
35-8,946	64-30993 <*> 0
35-9,436	CODE-P-249 <SA>
35-9,889	<INT>
35-10,143	66-11448 <*>
35-10,505	66-12505 <*>
35-10,614	CCDE-P50 <SA>
35-12,374	CODE-P-180 <SA>
35-12,375	CODE-P-191 <SA>
35-12,793	CODE-P-154 <SA>
35-13,221	CCDE-P-113 <SA>
35-13,674	32N55J <ATS>
35-13,844	CODE P-386 <SA>
35-13,904	CCDE-P-93 <SA>
35-14,781	66N50J <ATS>
35-14,782	67N50J <ATS>
35-14,792	CODE P-358 <SA>
35-15,247	CODE-P-158 <SA>
35-15,394	CCDE-P-275 <SA>
35-15,67C	<INT>
35-15,682	65-10521 <*>
35-16,598	<INT>
35-16,966	63-27260 <INT>
35-17,492	<INT>
35-17,594	66-12797 <*>
35-17,658	CODE-P-355 <SA>
35-18,418	CCDE-P-272 <SA>
36-37	CODE-P-411 <SA>
36-291	CODE-P1 <SA>
36-787	<SA>
36-892	<INT>
36-893	63-20334 <=*>
36-947	32S82J <ATS>
36-1,129	63-24605 <=*$>
36-1,487	CODE-P47 <SA>
36-1,792	<INT>
36-1,794	CCDE-P-75 <SA>
36-3,093	<INT>
36-3,390	CODE-P-83 <SA>
	66-11361 <*>
	66-12798 <*>
36-3,394	CODE-P2 <SA>
	64-30992 <*> 0
36-3,495	62-16325 <=*>
36-3,565	30R75J <ATS>
36-3,772	CODE-P-63 <SA>
36-3,903	<INT>
36-3,977	62-14802 <=*>
36-4,492	<INT>
36-5,395	CCDE P-387 <SA>
36-6,499	65-11561 <*>
36-7,351	<INT>
36-7,360	31P59J <ATS>
36-7,390	CODE P-359 <SA>
36-7,728	<INT>
36-7,832	<INT>
36-8,190	CODE-P-306 <SA>

Patent	Citation
36-8,340	<INT>
36-8,425	<INT>
36-8,558	63-13753 <=*> 0
36-8,581	<INT>
36-8,595	CODE-P-55 <SA>
36-8,938	CODE-P-53 <SA>
36-9,120	63-27286 <INT>
36-9,150	63-23180 <=*>
36-9,156	<INT>
36-9,170	63-23776 <=*$>
36-9,323	<WBI>
36-9,326	<WBI>
36-9,518	<WBI>
36-9,519	<WBI>
36-9,529	<WBI>
36-9,811	<WBI>
36-9,828	<WBI>
36-9,893	CODE P-388 <SA>
36-9,990	CODE-P49 <SA>
36-10,016	<WBI>
36-10,065	62-16324 <=*>
36-10,266	<WBI>
36-10,491	CODE-P-216 <SA>
36-10,728	<WBI>
36-10,729	<WBI>
36-10,732	<WBI>
36-10,733	<WBI>
36-10,734	<WBI>
36-11,073	<WBI>
36-11,077	<WBI>
36-11,087	<WBI>
36-11,088	
	62-16323 <=*>
36-11,511	<WBI>
36-11,522	<WBI>
36-12,182	<INT>
36-12,553	<INT>
36-12,666	<WBI>
36-12,913	<WBI>
36-12,920	<WBI>
36-13,280	<WBI>
36-13,440	<INT>
	30P59J <ATS>
36-13,446	CODE-P3 <SA>
36-13,461	<WBI>
36-13,470	<WBI>
36-13,474	<WBI>
36-13,872	<WBI>
36-13,873	<WBI>
	62-16322 <=*>
	62-18828 <=*> 0
36-14,019	<INT>
36-14,204	<WBI>
36-14,411	<WBI>
36-14,412	<WBI>
36-14,428	<WBI>
36-14,608	62-18763 <=*>
36-14,643	CODE-P-78 <SA>
36-14,826	<WBI>
36-14,827	<WBI>
36-14,828	<WBI>
36-14,830	<WBI>
36-15,107	<INT>
36-15,204	<WBI>
36-15,207	<WBI>
	62-16321 <=*> 0
36-15,208	<WBI>
36-15,209	<WBI>
	62-16320 <=*> 0
36-15,542	CODE-P-61 <SA>
36-15,683	<WBI>
36-15,796	CODE P-389 <SA>
36-15,878	<WBI>
36-15,940	CODE-P-73 <SA>
36-16,177	<WBI>
36-16,374	<WBI>
36-16,742	29P59J <ATS>
36-16,956	<WBI>
36-16,969	<WBI>
36-16,973	<WBI>

Patent	Citation
36-17,060	CODE-P-163 <SA>
36-17,725	<INT>
36-18,930	<INT>
36-19,296	<INT>
36-19,401	<INT>
36-19,495	CODE-P-212 <SA>
36-20,016	<INT>
36-20,242	CODE-P-197 <SA>
36-20,376	CODE-P-156 <SA>
36-20,805	62-10994 <=*> 0
36-21,044	65-62892 <=$*>
36-21,376	<INT>
36-22,149	65-11716 <*>
36-22,848	65-13945 <*> 0
	835 <TC>
36-23,020	CODE-P-187 <SA>
36-23,318	<INT>
36-24,040	<INT>
36-24,047	<INT>
36-24,062	CODE-P46 <SA>
36-24,069	CODE-P46 <SA>
37-78	CODE-P-114 <SA>
37-188	CODE-P-267 <SA>
37-217	62-18659 <=*>
37-694	64-20768 <*>
37-695	64-20767 <*>
37-696	64-20769 <*>
37-761	<INT>
37-824	63-16441 <=*>
37-852	<INT>
37-1,087	<INT>
37-1,471	<INT>
37-1,589	CODE-P-147 <SA> 0
37-1,592	CODE-P-51 <SA>
37-1,593	CODE-P-57 <SA>
37-1,893	CODE-P-56 <SA>
37-2,015	CODE-P-58 <SA>
37-2,030	CODE-P-250 <SA>
37-2,247	62-20481 <=*> 0
37-2,248	62-20407 <=*> 0
37-2,597	<INT>
37-2,987	14S86J <ATS>
	63-18906 <=*>
37-2,996	<INT>
37-2,999	65-11370 <*>
37-3,140	CODE-P-59 <SA>
37-3,192	62-20483 <=*> 0
37-3,194	62-20411 <=*> 0
37-3,289	CODE-P-171 <SA>
37-3,295	<INT>
37-3,300	<INT>
37-3,353	CODE-P-178 <SA>
37-3,464	63-18907 <=*>
37-3,471	CODE-P-169 <SA>
37-3,484	<INT>
37-3,701	CODE-P-166 <SA> 0
37-4,109	CODE-P-368 <SA>
37-4,122	58R76J <ATS>
37-4,212	<INT>
37-4,311	<INT>
37-4,642	CODE-P-123 <SA>
37-4,644	CODE-P-276 <SA>
37-4,646	CODE-P-285 <SA>
37-4,791	<INT>
37-4,851	CODE-P-103 <SA>
37-4,854	CODE-P-162 <SA>
37-4,857	CODE-P-161 <SA> 0
37-5,561	CODE-P-167 <SA>
37-5,749	CODE-P-168 <SA>
37-5,823	63-16402 <=*>
37-6,106	66-11934 <*> 0
37-6,108	<INT>
37-6,132	CODE-P-170 <SA>
37-6,506	CODE-P-219 <SA>
37-6,521	<INT>
37-6,556	<WBI>
37-6,557	<WBI>
37-6,642	CODE-P-198 <SA>
37-6,703	<INT>

37-6,739	CODE-P-72 <SA>
37-6,747	CODE-P-71 <SA>
37-6,811	<INT>
37-6,846	CODE-P-111 <SA>
37-6,855	<WBI>
37-6,857	<WBI>
37-6,858	<WBI>
37-6,859	<WBI>
37-6,860	<WBI>
37-6,861	<WBI>
37-6,862	<WBI>
37-6,864	<WBI>
37-6,865	<WBI>
37-6,866	<WBI>
37-6,867	<WBI>
37-6,953	<INT>
37-7,108	<INT>
37-7,141	63-20332 <=*>
37-7,145	CODE-P-90 <SA>
37-7,221	CODE-P-145 <SA> O
37-7,457	<INT>
37-7,564	CODE-P-174 <SA>
37-7,643	<INT>
37-7,721	66-10507 <*> O
37-7,738	<INT>
37-7,750	64-15251 <=*$>
37-7,805	<WBI>
37-7,827	<WBI>
37-7,854	CODE-P-247 <SA>
37-7,880	CODE-P-160 <SA>
37-7,896	<INT>
37-7,953	<INT>
37-7,954	<INT>
37-8,000	CODE-P-81 <SA>
37-8,178	CODE-P-84 <SA>
37-8,179	CODE-P-165 <SA>
37-8,188	CODE-P-133 <SA> O
37-8,191	CODE-P-202 <SA>
37-8,192	CODE-P-203 <SA>
37-8,262	CODE-P-137 <SA> O
37-8,276	<INT>
37-8,412	<WBI>
37-8,413	<WBI>
37-8,414	<WBI>
37-8,415	<WBI>
37-8,424	<WBI>
37-8,425	<WBI>
37-8,478	CODE-P-82 <SA>
37-8,479	CODE-P-52 <SA>
37-8,723	CODE-P-95 <SA>
37-8,724	CODE-P-94 <SA>
37-8,758	<WBI>
37-8,759	<WBI>
37-8,760	<WBI>
37-8,762	<WBI>
37-8,763	<WBI>
37-8,764	<WBI>
37-8,765	<WBI>
37-8,801	<INT>
37-8,803	CODE-P-159 <SA> O
37-8,830	CODE-P-74 <SA>
37-8,831	CODE-P-69 <SA>
37-8,832	CODE-P-68 <SA>
37-8,993	CODE-P-173 <SA>
37-9,059	CODE-P-189 <SA>
37-9,060	CODE-P-140 <SA> O
37-9,084	<INT>
37-9,157	<WBI>
37-9,158	<WBI>
37-9,159	<WBI>
37-9,162	<WBI>
37-9,163	<WBI>
37-9,209	<INT>
37-9,293	<INT>
37-9,337	CODE-P-112 <SA>
37-9,504	<WBI>
37-9,505	<WBI>
37-9,506	<WBI>
37-9,507	<WBI>
37-9,508	<WBI>
37-9,509	<WBI>
37-9,510	<WBI>
37-9,516	<WBI>
37-9,517	<WBI>
37-9,580	<INT>
37-9,687	CODE P-360 <SA>
37-9,763	CODE-P-183 <SA>
37-9,766	CODE-P-102 <SA>
37-9,836	<LSA>
	CODE-P-193 <SA>
37-9,838	<INT>
37-9,839	<INT>
37-9,840	<INT>
37-9,841	<INT>
37-9,957	CODE-P-280 <SA>
37-9,969	<INT>
37-10,069	CODE-P-201 <SA>
37-10,091	CODE-P-184 <SA>
37-10,095	<INT>
37-10,318	CODE-P-64 <SA>
37-10,322	09P65J <ATS>
37-10,567	CODE-P-122 <SA>
	71Q71J <ATS>
37-10,593	CODE-P-87 <SA>
37-10,595	CODE-P-86 <SA>
37-10,668	CODE-P-177 <SA>
37-10,778	<INT>
37-10,865	CODE-P-105 <SA>
37-10,873	<INT>
37-10,895	CODE-P-182 <SA>
37-10,941	<INT>
	CODE-P-196 <SA>
37-10,944	CODE-P-206 <SA>
37-11,011	<INT>
37-11,033	CODE P-412 <SA>
37-11,089	<WBI>
37-11,121	<INT>
	CODE-P-99 <SA>
37-11,141	CODE-P-104 <SA>
37-11,248	CODE-P-79 <SA>
37-11,402	<INT>
37-11,436	CODE-P-62 <SA>
37-11,523	CODE-P-117 <SA>
37-11,524	CODE-P-175 <SA>
37-11,536	CODE-P-204 <SA>
37-11,537	CODE-P-205 <SA>
37-11,538	CODE-P-70 <SA>
37-11,540	CODE-P-66 <SA>
37-11,542	CODE-P-65 <SA>
37-11,550	<INT>
37-11,614	<WBI>
37-11,658	<INT>
	CODE-P-218 <SA>
37-11,698	<INT>
37-11,746	<INT>
37-11,802	CODE-P-152 <SA>
37-11,805	CODE-P-143 <SA> O
37-11,836	CODE-P-138 <SA> O
37-11,906	<WBI>
37-11,907	<WBI>
37-11,910	<WBI>
37-11,916	<WBI>
37-12,016	81P65J <ATS>
37-12,141	CODE-P-200 <SA>
37-12,202	<WBI>
37-12,253	CODE-P-101 <SA>
37-12,254	CODE-P-176 <SA>
37-12,255	CODE-P-141 <SA> O
37-12,256	CODE-P-88 <SA>
37-12,537	CODE-P-213 <SA>
37-12,543	CODE-P-251 <SA>
	14Q69J <ATS>
37-12,544	49P65J <ATS>
37-12,606	<WBI>
37-12,607	<WBI>
37-12,608	<WBI>
37-12,609	<WBI>
37-12,610	<WBI>
37-12,611	<WBI>

Patent	Citation
37-12,659	CODE-P-109 <SA>
37-12,666	CODE-P-149 <SA> O
37-12,901	CODE-P-181 <SA>
37-12,932	CODE-P-77 <SA>
37-12,941	87S88J <ATS>
37-13,010	<WBI>
37-13,011	<WBI>
37-13,013	<WBI>
37-13,017	<WBI>
37-13,018	<WBI>
37-13,019	<WBI>
37-13,020	<WBI>
37-13,071	<INT>
37-13,346	CODE-P-188 <SA>
37-13,348	CODE-P-144 <SA> O
37-13,357	<WBI>
37-13,361	<WBI>
37-13,368	<WBI>
37-13,455	CODE-P-142 <SA> O
37-13,791	CODE-P-97 <SA>
37-13,795	CODE-P-227 <SA>
	22Q70J <ATS>
37-13,798	CODE-P-96 <SA>
37-13,860	CODE-P-107 <SA>
37-13,864	CODE-P-207 <SA>
37-13,877	CODE-P-139 <SA>
37-13,882	CODE-P-208 <SA>
37-13,885	CODE-P-110 <SA>
	81Q67J <ATS>
37-13,893	<INT>
	CODE-P-116 <SA>
37-13,896	CODE-P-98 <SA>
37-13,897	<INT>
37-13,898	<INT>
37-13,899	<INT>
37-13,911	<WBI>
37-13,912	<WBI>
37-13,914	<WBI>
37-13,915	<WBI>
37-13,916	<WBI>
37-13,918	<WBI>
37-13,919	<WBI>
37-13,920	<WBI>
37-13,959	CODE-P-136 <SA> O
37-14,063	CODE-P-135 <SA> O
37-14,220	CODE-P-100 <SA>
37-14,246	13Q69J <ATS>
37-14,255	<WBI>
37-14,256	<WBI>
37-14,258	<WBI>
37-14,259	<WBI>
37-14,260	<WBI>
37-14,420	CODE-P-341 <SA>
37-14,421	CODE-P-119 <SA>
37-14,507	<WBI>
37-14,633	CODE-P-194 <SA>
37-14,634	CODE-P-120 <SA>
37-14,641	CODE-P-121 <SA>
37-14,718	<INT>
37-14,758	<WBI>
37-14,759	<WBI>
37-14,760	<WBI>
37-14,761	<WBI>
37-14,762	<WBI>
37-14,763	<WBI>
37-14,764	<WBI>
37-14,765	<WBI>
37-14,766	<WBI>
37-14,767	<WBI>
37-14,768	<WBI>
37-14,769	<WBI>
37-14,770	<WBI>
37-14,797	<INT>
37-14,798	63-16446 <=*> O
37-14,822	63-16327 <=*>
37-15,037	CODE-P-217 <SA>
37-15,039	CODE-P-60 <SA>
37-15,041	CODE-P-164 <SA>
37-15,042	CODE-P-115 <SA>
37-15,049	63-10877 <=*>
37-15,050	CODE-P-134 <SA> O
37-15,159	<WBI>
37-15,160	<WBI>
37-15,164	<WBI>
37-15,237	63-16328 <=*> O
37-15,246	51P66J <ATS>
37-15,466	CODE-P-210 <SA>
37-15,469	CODE-P-125 <SA>
37-15,484	CODE-P-132 <SA>
37-15,489	<INT>
	CODE-P-108 <SA>
37-15,833	CODE-P-209 <SA> O
37-15,866	<WBI>
37-15,867	<WBI>
37-15,868	<WBI>
37-15,877	63-16449 <=*> O
37-15,878	65-13093 <*>
37-15,893	<WBI>
37-15,930	<LSA>
37-15,932	<INT>
37-15,937	CODE-P-127 <SA>
37-15,946	63-16643 <=*>
37-16,019	<INT>
37-16,027	<INT>
37-16,046	35Q73J <ATS>
37-16,058	<WBI>
	63-16445 <=*> O
37-16,059	<WBI>
37-16,060	<WBI>
37-16,061	<WBI>
37-16,062	<WBI>
37-16,063	<WBI>
37-16,064	<WBI>
37-16,065	<WBI> P
37-16,066	<WBI>
37-16,067	<WBI>
37-16,068	<WBI>
37-16,069	<WBI>
37-16,070	<WBI>
37-16,071	<WBI>
37-16,097	<INT>
37-16,098	63-16444 <=*>
37-16,195	63-16321 <*=>
37-16,277	CODE-P-266 <SA>
37-16,309	<WBI>
37-16,310	<WBI>
	63-17190 <WBI>
37-16,313	<WBI>
37-16,438	63-16320 <*=>
	89Q71J <ATS>
37-16,602	<WBI>
37-16,640	<WBI>
37-16,648	<WBI>
37-16,848	63-16317 <=*>
37-16,906	<WBI>
37-16,907	<WBI>
37-16,909	<WBI>
37-16,910	<WBI>
37-16,911	<WBI>
37-16,912	<WBI>
37-16,913	<WBI>
37-16,914	<WBI>
37-16,915	<WBI>
37-16,916	<WBI>
37-16,917	<WBI>
37-16,921	63-17189 <WBI>
37-17,141	<LSA>
37-17,207	CODE-P-126 <SA>
37-17,229	<INT>
37-17,386	CODE-P-260 <SA>
37-17,393	<INT>
	CODE-P-256 <SA>
37-17,394	<INT>
	CODE-P-257 <SA>
37-17,398	<INT>
37-17,693	72P66J <ATS>
37-17,697	CODE-P-186 <SA>
37-17,698	CODE-P-148 <SA> O
37-17,719	<WBI>

	63-17188 <WBI>
37-17,722	<WBI>
37-17,723	<WBI>
37-17,724	<WBI>
37-17,725	<WBI>
37-17,726	<WBI>
37-17,727	<WBI>
37-17,732	63-17187 <WBI>
37-17,739	<INT>
	13R80J <ATS>
37-17,995	<INT>
37-18,156	CODE-P-263 <SA>
37-18,166	CODE-P-262 <SA>
37-18,313	<INT>
37-18,400	64-14893 <=*$>
37-18,466	CODE-P-268 <SA>
37-18,517	<WBI>
37-18,521	<WBI>
37-18,522	<WBI>
37-18,644	CODE-P-271 <SA>
37-18,954	<WBI>
37-18,987	<INT>
37-18,990	65-11550 <*>
37-18,998	<INT>
37-20,009	<WBI>
37-20,667	<WBI>
37-23,315	<WBI>
37-23,316	<WBI>
37-23,317	<WBI>
37-23,318	<WBI>
37-23,866	<WBI>
37-23,868	<WBI>
37-24,661	<WBI>
37-25,710	<WBI>
37-25,711	<WBI>
37-25,788	<WBI>
37-27,339	<WBI>
37-27,341	<WBI>
37-27,356	<WBI>
37-27,358	<WBI>
37-28,610	<WBI>
37-28,612	<WBI>
37-28,615	<WBI>
37-29,674	<WBI>
37-31,217	<WBI>
37-31,218	<WBI>
37-31,219	<WBI>
37-32,105	<WBI>
37-32,106	<WBI>
37-32,107	<WBI>
37-40,522	CODE P-390 <SA>
38-14	<INT>
38-143	63-18315 <=*>
38-222	CODE-P-391 <SA>
38-446	17R79J <ATS>
8-641	CODE-P-255 <SA>
8-642	CODE-P-259 <SA>
38-643	CODE-P-244 <SA>
38-756	<INT>
38-757	<INT>
38-774	<INT>
	CODE-P-258 <SA>
38-831	63-16316 <=*>
38-1,216	CODE-P-229 <SA>
38-1,495	<INT>
38-1,499	<INT>
38-1,500	<INT>
38-1,610	CODE-P-238 <SA>
38-1,785	85Q72J <ATS>
38-1,996	CODE-P-253 <SA>
38-2,254	CODE-P-239 <SA>
38-2,388	CODE-P-236 <SA>
38-2,505	CODE-P-252 <SA>
38-2,549	<INT>
38-2,665	<INT>
38-2,734	CODE-P-231 <SA>
38-2,735	CODE-P-232 <SA>
38-2,736	CODE-P-233 <SA>
38-2,737	CODE-P-[illegible]7 <SA>
38-2,738	CODE-P-342 <SA>
38-2,986	CODE-P-274 <SA>
38-3,111	<INT>
38-3,112	<INT>
38-3,116	<INT>
	CODE-P-228 <SA>
38-3,117	<INT>
	CODE-P-230 <SA>
38-3,143	<INT>
38-3,342	<INT>
38-3,745	1234 <TC>
38-3,746	18R79J <ATS>
	64-20941 <*> 0
38-4,496	CODE-P-261 <SA>
38-4,498	CODE-P-307 <SA>
38-4,551	CODE-P-279 <SA>
38-4,582	CODE-P-287 <SA>
38-4,689	63-20565 <=*$>
38-4,690	63-20606 <=*$>
38-4,692	64-14356 <=*$>
38-4,693	CODE P-308 <SA>
38-4,694	CODE-P-309 <SA>
38-4,867	<INT>
38-4,868	<INT>
	CODE-P-288 <SA>
38-4,939	<INT>
38-5,269	<INT>
	CODE-P-310 <SA>
38-5,579	CODE-P-291 <SA>
38-5,661	CODE-P-245 <SA>
38-5,804	CODE-P-284 <SA>
38-5,805	CODE-P-281 <SA>
38-5,814	CODE-P-292 <SA>
38-5,820	CODE-P-321 <SA>
38-5,833	65-11397 <*>
38-5,848	CODE-P-413 <SA>
38-5,897	<INT>
38-5,976	CODE-P-293 <SA>
38-6,154	CODE-P-289 <SA>
38-6,156	CODE-P-294 <SA>
38-6,157	CODE-P-300 <SA>
38-6,158	CODE-P-301 <SA>
38-6,199	CODE-P-312 <SA>
38-6,459	CODE-P-246 <SA>
38-6,488	CODE-P-296 <SA>
38-6,494	13Q73J <ATS>
38-6,764	66-11927 <*>
38-6,780	CODE-P-299 <SA>
38-6,781	CODE-P-302 <SA>
38-7,038	CODE-P-313 <SA>
38-7,039	CODE-P-314 <SA>
38-7,042	CODE-P-315 <SA>
38-7,043	CODE-P-326 <SA>
38-7,044	CODE-P-327 <SA>
38-7,045	CODE-P-328 <SA>
38-7,133	CODE-P-333 <SA>
38-7,142	66-10607 <*>
38-7,318	CODE P-402 <SA>
38-7,546	<INT>
38-7,620	CODE-P-298 <SA>
38-7,887	CODE-P-343 <SA>
38-7,890	63-20693 <=*$>
38-7,892	CODE-P-295 <SA>
38-7,895	63-20786 <=*$>
38-7,992	CODE-P-270 <SA>
38-8,118	<INT>
38-8,147	CODE-P-303 <SA>
38-8,746	64-10695 <=*$>
38-8,996	CODE-P-316 <SA>
38-9,148	CODE-P-325 <SA>
38-9,149	CODE-P-330 <SA>
38-9,217	<INT>
38-9,239	64-14977 <=*$>
38-9,436	43R77J <ATS>
38-9,437	46S83J <ATS>
38-9,439	CODE-P-344 <SA>
38-9,440	CODE-P-345 <SA>
38-9,441	CODE-P-346 <SA>
38-10,035	CODE-P-304 <SA>
38-10,244	63-20845 <=*$>

38-10,733 CODE-P-392 <SA>
38-10,739 CODE P-361 <SA>
38-10,741 CODE P-362 <SA>
38-10,836 <INT>
38-11,077 <INT>
38-11,27C 66-11888 <*>
38-11,276 CODE-P-347 <SA>
38-11,442 CODE-P-331 <SA>
38-11,867 CODE-P-332 <SA>
38-11,937 <INT>
38-12,096 <INT>
38-12,522 <INT>
38-12,805 CODE-P-339 <SA>
38-12,995 72R78J <ATS>
38-12,996 <INT>
38-13,249 66-11895 <*>
38-13,354 CODE-P-348 <SA>
38-13,534 CODE-P-381 <SA>
38-13,609 CODE-P-317 <SA>
38-13,791 CODE-P-349 <SA>
38-13,853 CODE-P-318 <SA>
38-13,856 CODE-P-379 <SA>

38-14,348 77R78J <ATS>
38-14,356 <INT>
38-14,357 <INT>
38-14,491 64-10696 <=*$>
38-14,495 <INT>
38-14,561 <INT>
38-14,564 <INT>
38-14,711 CODE-P-242 <SA>
38-14,833 CODE-P-350 <SA>
38-14,895 69R78J <ATS>
38-14,993 65-18136 <*>
38-15,25C 65-11996 <*>
38-15,422 1235 <TC>
38-15,495 CODE-P-334 <SA>
38-16,010 <INT>
38-16,016 <INT>
38-16,02C <INT>
38-16,287 65S82J <ATS>
38-16,288 89R76J <ATS>
38-16,335 CODE-P-351 <SA>
38-16,342 CODE-P-335 <SA>
38-16,345 <INT>
38-16,457 CODE-P-352 <SA>
38-16,496 <INT>
38-16,499 CODE-P-323 <SA>
38-16,780 CODE P-405 <SA>
38-16,854 CODE-P-319 <SA>
38-16,873 CODE-P-324 <SA>
38-17,083 CODE-P-369 <SA>
38-17,105 71R78J <ATS>
38-17,245 90R76J <ATS>
38-17,491 CODE-P-337 <SA>
38-17,656 CODE-P-354 <SA>
38-17,969 CODE-P-277 <SA>
38-18,148 74R78J <ATS>
38-18,244 CODE-P-370 <*>
38-18,247 CODE-P-371 <SA>
38-18,385 CODE-P-356 <SA>
38-18,511 CODE-P-340 <SA>
38-18,542 CODE-P-320 <SA>
38-18,662 CODE P-363 <SA>
38-19,103 ORNL-TR-334 <*>
38-19,646 C-5292 <NRC>
38-19,708 CODE-P-372 <SA>
38-19,791 75R78J <ATS>
38-19,792 76R78J <ATS>
38-20,367 65-18177 <*>
38-20,643 CODE-P-240 <SA>
38-20,733 CODE P-364 <SA>
38-20,735 CODE-P-322 <SA>
38-20,951 77R74J <ATS>
38-20,987 CODE P-365 <SA>
38-20,990 66-10608 <*>
38-21,636 CODE P-393 <SA>
38-22,139 CODE-P-366 <SA>
38-22,343 73R78J <ATS>
38-22,489 64-14894 <=*$>

38-24,346 CODE-P-373 <SA>
38-24,364 CODE P-394 <SA>
38-24,577 CODE P-406 <SA>
38-24,596 CODE-P-375 <SA>
38-24,597 24R78J <ATS>
38-24,599 66-13973 <*>
38-24,897 31R77J <ATS>
38-25,048 64-16540 <=*$>
38-25,358 <INT>
38-25,937 584 <TC>
38-26,085 CODE-P-367 <SA>
38-26,086 CODE-P-338 <SA>
38-26,128 25S83J <ATS>
38-26,136 <INT>
38-26,140 65-11807 <*>
38-26,279 17R76J <ATS>
38-26,282 64-18104 <=*$>
38-26,564 66-11897 <*>
38-26,585 25R78J <ATS>
38-26,766 CODE-P-377 <SA>
38-26,767 CODE P-407 <SA>
38-26,943 <INT>
CODE-P-414 <SA>

38-57,748 66-11403 <*>
39-2 65-11810 <*>
39-70 CODE P-395 <SA>
39-950 <INT>
39-1,042 CODE-P-396 <SA>
39-1,207 <INT>
39-1,420 66-10500 <*>
39-1,515 CODE P-408 <SA>
39-1,668 64-30642 <*>
39-1,691 <INT>
39-1,844 29R77J <ATS>
39-1,846 28R77J <ATS>
39-2,600 65-12767 <*>
39-2,759 65-11389 <*>
39-3,324 CODE-P-397 <SA>
39-3,414 65-11371 <*>
39-3,569 08R77J <ATS>
39-3,574 45R77J <ATS>
39-3,746 <INT>
39-3,828 65-11387 <*>
39-3,953 65-11388 <*>
39-3,965 CODE-P-380 <SA>
39-3,994 <INT>
39-4,273 <INT>
39-4,452 CODE P-398 <SA>
39-4,747 92R77J <ATS>
39-5,758 64-30643 <*> 0
39-5,771 66-10635 <*> 0
39-5,790 CODE P-415 <SA>
39-6,246 64-20957 <*>
39-6,248 64-20959 <*>
39-6,620 <INT>
39-6,808 65-11891 <*>
39-6,850 65-11015 <*>
39-7,265 64-30644 <*>
39-7,266 95S80J <ATS>
39-7,753 42S81J <ATS>
39-7,765 CODE-P-416 <SA>
39-7,929 CODE-P-399 <SA>

39-8,806 65-11532 <*>
39-9,698 65-11549 <*> 0
39-10,517 66-11924 <*>
39-10,520 66-11931 <*>
39-10,521 66-11933 <*>
39-11,247 66-11917 <*>
39-12,119 66-11425 <*> 0
39-12,123 66-11887 <*>
39-12,139 <INT>
39-12,141 38S81J <ATS>
39-13,803 65-14105 <*> 0
39-14,064 66-11932 <*>
39-14,205 65-11808 <*>
39-14,450 69S85J <ATS>

39-14,503	CODE P-417 <SA>
39-14,654	65-14836 <*>
39-15,659	66-11312 <*> 0
39-16,606	65-11369 <*>
39-16,681	65-14104 <*>
39-16,919	66-12799 <*>
39-17,067	66-12802 <*>
39-17,262	65-11385 <*>
39-18,105	65-13946 <*>
	837 <TC>
39-18,108	838 <TC>
39-18,547	CODE P-418 <SA>
39-18,746	33S87J <ATS>
39-18,955	66-12800 <*>
39-19,027	33S85J <ATS>
39-19,282	65-11564 <*> 0
39-19,431	CODE P-400 <SA>
39-19,433	66-11433 <*>
39-19,535	81S82J <ATS>
39-19,544	CODE P-419 <SA>
39-19,633	74S88J <ATS>
39-20,078	32S89J <ATS>
39-20,152	65-11533 <*>
39-20,201	05S82J <ATS>
39-20,295	65-11563 <*> 0
39-20,440	65-11562 <*> 0
39-20,914	792 <TC>
39-22,067	768 <TC>
39-22,068	769 <TC>
39-22,946	65-13947 <*>
	836 <TC>
39-22,960	48S88J <ATS>
39-23,396	839 <TC>
39-24,305	765 <TC>
39-24,957	CODE P-403 <SA>
39-25,113	CODE P-404 <SA>
39-25,114	66-11891 <*> 0
39-25,637	62S86J-A,B <ATS>
39-26,000	65-12768 <*>
39-26,005	65-12055 <*>
39-26,399	66-10514 <*>
39-26,831	65-13693 <*>
39-27,629	74S85J <ATS>
39-27,636	73S85J <ATS>
39-28,305	66-11311 <*> 0
39-28,634	66-11357 <*>
39-29,357	65-14919 <*>
39-29,471	65-13714 <*>
39-30,147	65-13096 <*>
39-36,240	65-14106 <*> 0
39-39,352	1086 <TC>
40-84	19S84J <ATS>
40-539	CODE P-401 <SA>
40-1,572	CODE P-409 <SA>
40-1,660	18S84J <ATS>
40-1,711	1083 <TC>
40-1,867	65-14109 <*> 0
40-4,130	1347 <TC>
40-5,309	66-12801 <*>
40-5,609	66-10352 <*>
40-5,610	66-10353 <*>
40-6,019	1568 <TC>
40-8,026	65-17068 <*> 0
40-8,027	65-17067 <*> 0
40-8,262	13S86J <ATS>
40-9,230	40S88J <ATS>
40-9,965	12S86J <ATS>
40-10,586	66-10728 <*>
40-11,305	66-10354 <*>
40-13,932	66-14483 <*>
40-14,586	66-13276 <*>
40-16,341	1896 <TC>
40-16,522	66-12051 <*>
40-19,008	66-11500 <*>
40-19,177	66-11497 <*>
40-19,189	66-11496 <*>
40-19,190	66-11495 <*>
40-19,191	66-11494 <*>
40-20,193	66-11493 <*>
40-20,250	66-11511 <*>
40-20,389	66-11510 <*>
40-20,487	66-11509 <*>
40-20,876	66-14011 <*>
40-20,986	66-11508 <*>
40-21,145	66-11507 <*>
40-21,263	66-11506 <*>
40-21,275	66-11505 <*>
40-23,354	66-12055 <*>
40-25,329	66-11482 <*>
40-26,208	66-12302 <*>
40-27,371	66-11949 <*>
40-27,544	66-12340 <*>
40-27,548	66-12339 <*>
41-571	66-10839 <*>
41-575	66-14016 <*>
41-599	66-14510 <*>
41-989	66-14509 <*>
41-990	66-14514 <*>
41-7,466	66-12717 <*>

NETHERLANDS

1,586	63-18658 <=*>
3,014	I-702 <*>
4,270	59-18792 <+*> 0
6,711	63-15762 <=*> C
8,490	65-62602 <=$>
10,555	6600-A <K-H>
13,521	5360 <K-H>
15,419	63-18222 <=*>
19,339	60-16341 <=*$>
23,759	I-479 <*>
29,525	10317 <K-H>
30,503	63-18668 <=*>
31,007	64-30185 <*> 0
32,029	64-20546 <*>
33,796	64-30186 <*> 0
33,866	63-14362 <=*>
35,089	673 <TC>
37,649	59-14214 <+*> 0
37,919	63-14363 <=*>
40,273	64-30187 <*>
40,627	64-30188 <*>
42,514	674 <TC>
43,005	59-17188 <+*> 0
43,220	65-63468 <=$> 0
43,899	63-18659 <=*>
44,150	59-12927 <+*> 0
44,160	63-18660 <=*>
46,908	83Q70DU <ATS>
47,568	64-30189 <*>
47,761	66-10522 <*>
47,835	65-10644 <*>
47,836	63-18661 <=*>
48,664	64-30190 <*> 0
50,106	64-30191 <*>
50,563	8288 <K-H>
52,896	63-10276 <*>
53,694	64-20723 <*>
54,641	65-11720 <*>
55,399	63-14360 <=*>
55,728	62-16251 <=*>
57,814	3105 <K-H>
57,939	60-17133 <+*> 0
58,091	62-23740 <=*$> 0
58,511	64-20712 <*>
59,343	T-2041 <INSD>
59,870	63-14980 <=*>
62,435	59-14215 <+*>
63,302	64-30192 <*>
63,604	64-18870 <*>
64,136	64-30193 <*>

64,204	48J18DU <ATS>
65,014	65-62862 <=$> 0
65,772	63-14513 <=*>
67,091	1565 <TC>
67,932	489 <TC>
68,351	64-30194 <*> 0
68,392	59-20077 <+*>
68,567	64-30195 <*>
68,706	64-10772 <=*$>
	64-30196 <*>
69,011	60-14696 <*>
69,290	63-23580 <=*>
69,883	64-20048 <*>
70,152	97R77DU <ATS>
71,057	64-20836 <*>
71,073	AL-403 <*>
71,791	60-14697 <+*>
71,829	64-20713 <*>
72,448	63-14514 <=*>
72,449	63-14515 <=*>
74,261	63-13009 <=*> 0
74,766	65-64484 <=*$>
75,058	59-17197 <+*>
75,733	63-14636 <=*>
76,326	58-305 <*>
76,886	62-16250 <=*>
76,888	62-16252 <=*>
77,286	61-19478 <+*> 0
77,300	60-14858 <+*>
77,845	64-30197 <*>
80,063	61-18615 <=*>
80,669	63-10264 <=*>
81,350	63-14365 <=*>
83,379	63-18935 <=*>
83,819	64-30198 <*>
84,445	63-10268 <=*>
84,526	88L32D4 <ATS>
84,715	59-20078 <+*>
87,310	59-14216 <+*>
87,740	AEC-TR-4822 <*>
88,425	62-13885 <*=> 0
89,614	65-62603 <=$>
91,017	63-16348 <*=> 0
92,382	62-15760 <=*> 0
92,780	97R79DU <ATS>
96,760	66-13265 <*>
98,129	AEC-TR-4992 <*>
98,545	63-10267 <=*>
99,202	1684 <TC>
102,468	65-60278 <=$>
103,069	65-63467 <=$> 0
103,433	ORNL-TR-265 <*>
103,918	87S85DU <ATS>
106,100	65-17218 <*>
109,239	66-11652 <*>
110,370	145 <WFK>
110,656	65-14912 <*>
110,975	107 <TC>
111,349	30S85DU <ATS>
123,152	I-430 <*>
123,072	63-14361 <=*>
126,948	I-70 <*>
142,456	59-20177 <+*> 0
145,445	63-18235 <*>
176,750	64-16772 <=*$>
177,984	II-909 <*>
185,496	60-14698 <+*>
192,519	60-14862 <+*>
198,776	57-03482 <*>
	57-3483 <*>
204,433	65-13626 <*>
204,923	49N48DU <ATS>
205,548	90N48DU <ATS>
205,759	62-14806 <=*>
216,199	64-20714 <*>
218,714	65-13625 <*>
220,064	517 <TC>
220,612	63-20119 <=*>
225,922	61R74DU <ATS>

229,618	91N48DU <ATS>
232,547	11N48DU <ATS>
237,385	62-10329 <*=>
243,343	62-16332 <=*>
244,010	NS-400 <TTIS>
247,433	2128 <TC>
247,771	65-12180 <*>
252,969	852 <TC>
257,551	NS-322 <TTIS>
257,826	2126 <TC>
261,086	65-13097 <*>
262,819	66-10517 <*> 0
263,104	57S87DU <ATS>
278,345	66-12796 <*>
278,722	20S84DU <ATS>
279,550	853 <TC>
281,000	66-12767 <*>
281,001	66-12775 <*>
283,779	66-11918 <*>
283,936	66-12792 <*>
285,136	66-10757 <*>
286,115	66-10759 <*>
286,116	66-10758 <*>
286,744	58S86DU <ATS>
288,599	186 <WFK>
292,133	66-10816 <*> 0
294,032	66-10530 <*>
294,912	1073 <TC>
639,186	66-10520 <*>
64-00,368	1212 <TC>
64-01,117	65-13712 <*>
	870 <TC>
64-01,475	66-10762 <*> 0
64-02,175	552 <TC>
	65-11994 <*>
64-02,557	89S86DU <ATS>
64-02,711	66-13257 <*>
64-03,352	1433 <TC>
64-03,395	20S83DU <ATS>
64-03,759	66-10523 <*>
64-04,217	646 <TC>
64-04,833	1434 <TC>
64-05,008	66-13254 <*>
64-06,435	65-12773 <*>
	735 <TC>
64-06,944	66-14528 <*>
64-07,772	65-17117 <*>
64-08,222	1037 <TC>
64-08,245	871 <TC>
64-08,666	65-13410 <*>
	795 <TC>
64-08,800	169 <WFK>
64-08,884	57S86DU <ATS>
64-09,043	783 <TC>
64-09,145	65-13608 <*> 0
	881 <TC>
64-09,146	854 <TC>
64-09,198	855 <TC>
64-09,211	1782 <TC>
64-09,318	65-17118 <*>
64-09,341	66-10763 <*>
64-09,546	65-17119 <*>
64-09,737	1412 <TC>
64-09,959	66-11314 <*> 0
64-10,073	856 <TC>
64-10,224	66-10764 <*>
64-10,260	1213 <TC>
64-10,490	817 <TC>
64-10,504	65-13696 <*>
	923 <TC>
64-10,530	65-13851 <*>
	926 <TC>
64-10,561	66-14012 <*>
64-10,654	66-14013 <*>
64-10,655	65-13697 <*>
	925 <TC>

Patent	Citation
64-10,656	65-13694 <*>
	924 <TC>
64-10,702	65-13950 <*>
	927 <TC>
64-11,003	65-14881 <*>
64-12,869	65-13713 <*>
	896 <TC>
64-13,286	22T90DU <ATS>
64-13,334	66-10529 <*>
64-13,480	21T90DU <ATS>
64-13,549	65-17064 <*> O
64-13,550	65-17063 <*> O
64-13,551	65-17062 <*> O
64-13,552	65-17060 <*> O
64-14,424	65-17061 <*> O
64-14,673	66-10119 <*>
64-14,963	65-17058 <*> O
64-14,983	66-11442 <*>
64-15,097	65-17455 <*>
64-15,185	65-17456 <*>
65-00,190	66-14515 <*>
65-00,459	1860 <TC>
	66-10587 <*>
65-00,955	65-17471 <*>
65-00,998	66-10765 <*>
65-02,102	1614 <TC>
65-05,021	1835 <TC>
65-05,404	66-14534 <*>
65-08,467	66-12499 <*>
65-08,477	66-12048 <*>
65-08,496	66-12503 <*>
65-08,672	66-12049 <*>
65-08,812	66-12500 <*>
65-11,117	66-12497 <*>
NORWAY	
12,660	59-17034 <+*> O
20,993	63-18656 <=*>
25,744	59-16845 <+*>
29,615	59-16846 <+*>
32,626	59-16847 <+*>
40,988	64-20579 <*>
44,909	<AC>
55,103	MF-5 <*>
55,479	MF-5 <*>
56,447	8863 <K-H>
57,331	63-20527 <=*$>
60,026	MF-5 <*>
60,162	MF-5-2C <*>
60,988	MF-5 <*>
63,335	59-12317 <+*>
64,134	65-60908 <=$>
64,934	63-13754 <=*> O
65,728	MF-5 <*>
66,258	59-12090 <+*>
67,595	NS-406 <TTIS>
67,696	NS-407 <TTIS>
67,853	65-62692 <*>
70,507	61-27468 <*=> O
71,236	65-64475 <=*$>
71,454	AL-18 <*>
72,162	59-17107 <+*>
74,890	65-10648 <*> O
80,452	<AC>
84,182	58-1487 <*>
85,376	61-13270 <+*>
86,281	60-13285 <+*> O
87,650	59-12629 <+*> O
89,488	<AC>
89,517	64-30585 <*> O
89,518	64-30586 <*>
90,904	64-30587 <*>
90,936	64-19918 <=$>
91,156	59-16848 <+*> O
91,382	59-16849 <+*> O
91,441	59-16850 <+*> O
91,517	65-62849 <=$> O
91,959	59-16189 <+*> O
92,290	59-16851 <+*> O
92,820	64-30588 <*>
93,100	62-10271 <*=>
93,343	62-32922 <=*> O
93,770	64-30047 <*> P
94,094	62-23714 <=*> O
94,554	64-18470 <*>
94,965	63-23581 <=*>
95,115	93R74N <ATS>
95,498	64-18469 <*>
95,499	64-18468 <*>
95,500	64-18467 <*>
98,287	62-24521 <=*>
98,880	63-23172 <=*$> O
99,423	65-60296 <=$>
POLAND	
40,188	63-19649 <=*>
40,975	61-10999 <=*>
	63-18587 <=*>
42,060	64-30589 <*>
42,081	63-19406 <=*>
42,867	62-10688 <*=>
43,443	62-25120 <=*>
43,444	63-24004 <=*>
44,034	63-20394 <=*>
44,375	63-19634 <=*>
44,523	63-23793 <=*$>
44,923	63-24229 <=*$>
45,546	NS-436 <TTIS>
45,789	NS-343 <TTIS>
45,791	64-10797 <=*$>
46,051	NS-437 <TTIS>
46,167	NS-148 <TTIS>
46,204	NS-187 <TTIS>
46,265	NS-344 <TTIS>
46,567	NS-171 <TTIS>
46,639	NS 138 <TTIS>
46,645	NS-160 <TTIS>
46,689	NS 117 <TTIS>
46,747	NS-172 <TTIS>
46,767	NS-161 <TTIS>
46,813	NS-150 <TTIS>
46,821	<TTIS>
46,822	NS 139 <TTIS>
46,829	NS-149 <TTIS>
46,842	NS-152 <TTIS>
46,843	NS-153 <TTIS>
46,844	NS 140 <TTIS>
46,925	NS-154 <TTIS>
46,936	NS-173 <TTIS>
46,937	NS-194 <TTIS>
47,009	NS-162 <TTIS>
47,050	NS-175 <TTIS>
47,054	NS-174 <TTIS>
47,067	NS-163 <TTIS>
47,140	NS-263 <TTIS>
47,191	NS-309 <TTIS>
47,240	NS 231 <TTIS>
47,245	NS-176 <TTIS>
47,340	NS 220 <TTIS>
47,344	NS 221 <TTIS>
47,364	NS219 <TTIS>
47,424	NS-211 <TTIS>
47,473	NS-186 <TTIS>
47,486	NS-196 <TTIS>
47,514	NS-311 <TTIS>
47,522	NS-264 <TTIS>
47,626	NS 257 <TTIS>
47,631	NS-195 <TTIS>
47,647	NS-312 <TTIS>
47,648	NS-313 <TTIS>
47,664	NS 222 <TTIS>
47,680	NS-266 <TTIS>
47,694	NS-333 <TTIS>
47,695	NS-267 <TTIS>
47,811	NS-268 <TTIS>
47,821	NS-258 <TTIS>
47,823	NS 269 <TTIS>
47,840	NS-438 <TTIS>

Patent	Citation
	64-23595 <=$>
47,892	NS-314 <TTIS>
47,895	66-11310 <*>
47,919	NS 270 <TTIS>
47,929	NS 223 <TTIS>
47,940	NS-315 <TTIS>
47,941	NS 224 <TTIS>
47,942	NS-316 <TTIS>
47,958	NS 238 <TTIS>
47,965	NS-317 <TTIS>
47,979	NS 232 <TTIS>
47,984	NS-318 <TTIS>
48,000	NS 239 <TTIS>
48,015	NS-334 <TTIS>
48,023	NS-271 <TTIS>
48,057	NS-233 <TTIS>
48,065	NS-465 <TTI>
48,139	NS 240 <TTIS>
48,180	NS 241 <TTIS>
48,252	35S85P <ATS>
48,280	NS 272 <TTIS>
48,292	NS-439 <TTIS>
48,293	36S85P <ATS>
48,304	34S85P <ATS>
48,326	NS-310 <TTIS>
48,328	NS-440 <TTIS>
48,333	NS-319 <TTIS>
48,587	NS-441 <TTIS>
48,710	NS-442 <TTIS>
48,711	NS-443 <TTIS>
48,910	NS-444 <TTIS>

RUMANIA

Patent	Citation
41,642	63-18083 <*=>

SPAIN

Patent	Citation
120,892	63-18050 <*=>
121,650	63-16798 <*=>
121,651	63-18047 <*>
123,631	63-18048 <*>
124,703	63-16795 <*=>
128,591	63-18051 <*>
128,592	63-16796 <*=>
128,593	63-18049 <*>
128,594	63-16797 <*=>
168,254	64-30595 <*>
196,171	I-645 <*>
	MF-5 <*>
198,328	MF-5 <*>
224,700	06J16SP <ATS>
231,107	1776 <TC>
255,065	62-14897 <=*>
255,221	09S82SP <ATS>
257,550	65-62888 <=$*> 0
277,800	63-16448 <*=>
288,912	64-18118 <=*$>
296,420	66-11930 <*>

SWEDEN

Patent	Citation
148	57-2902 <*>
2,758	63-14538 <*=>
3,761	<ATS>
4,048	64-30596 <*>
17,695	64-30613 <*>
26,015	62-18654 <=*>
34,133	59-12091 <+*>
34,187	3470 <K-H>
35,571	<ATS>
41,295	<ATS>
41,723	64-20582 <*>
41,758	<ATS>
42,253	64-20581 <*> 0
45,871	3096 <K-H>
45,904	5034 <K-H>
47,702	9864 <K-H>
53,028	03H11S <ATS>
53,077	59-17195 <+*> 0
56,092	61-10058 <=*$> 0
57,812	64-16451 <=*$>
69,349	12462 <K-H>
70,794	60-10728 <+*>
71,289	8828 <K-H>
72,858	<ATS>
74,272	61-10034 <=*$>
76,374	61-10035 <=*$>
76,801	63-18142 <=*>
77,606	10516 <K-H>
79,261	62-18486 <=*> 0
80,941	<ATS>
82,279	62-18360 <=*> 0
83,018	63-19621 <=*> 0
89,261	64-18466 <*>
91,902	63-19056 <=*> 0
93,501	11530 <K-H>
94,212	10483 <K-H>
97,854	90R74S <ATS>
98,668	62-15761 <=*> 0
98,674	<ATS>
98,938	64-20429 <*>
100,468	2843-A <HB>
104,202	9981 <K-H>
104,721	65-63488 <=$> 0
105,422	10565-A <K-H>
105,422 ADD	10565-B <K-H>
106,116	64-30614 <*>
106,325	60-14794 <+*> 0
106,671	64-20580 <*>
108,293	59-17313 <+*> 0
108,525	61-13271 <+*>
109,090	59-20331 <+*> 0
109,198	61-19850 <=*> 0
109,975	59-18793 <+*> 0
110,303	59-20068 <+*> 0
111,729	60-13286 <+*> 0
112,292	92R74S <ATS>
113,423	63-14539 <=*>
113,509	63-14540 <*=>
113,778	62-25693 <=*> 0
115,681	<ATS>
115,846	10565-B <K-H>
116,111	61-19851 <=*> 0
117,339	59-12092 <+*>
117,418	59-16852 <+*>
119,359	<ATS>
120,428	63-23173 <=*$>
120,780	<ATS>
121,301	91R74S <ATS>
121,639	62-33249 <=*> 0
121,716	64-30612 <*>
121,737	43E2S <ATS>
121,751	61-15690 <+*>
122,608	62-13497 <*=> 0
123,137	56E2S <ATS>
123,920	AEC-TR-4180 <*>
123,994	65-60297 <=$>
124,202	<ATS>
124,374	<ATS>
124,436	<ATS>
124,849	65-10629 <*>
125,001	1814 <*>
125,951	63-13755 <=*> 0
128,174	64-20802 <*>
128,815	59-12630 <+*> 0P
128,823	64-30611 <*>
129,143	60-19001 <+*> 0
130,374	59-17293 <+*>
131,062	60-17617 <+*> 0
131,381	84Q70S <ATS>
133,324	61-15691 <+*> 0
133,866	64-30610 <*>
133,997	95H9S <ATS>
135,148	65-10586 <*> 0
135,421	64-20396 <*>
135,534	59-12318 <+*> 0

Patent	Citation
135,902	61-15056 <+*>
137,341	63-13756 <=*> 0
137,515	64-30609 <*>
138,011	64-30608 <*>
138,532	10294 <K-H>
140,574	I-103 <*>
140,860	61-15057 <+*>
143,999	59-12631 <+*> 0
144,520	62-23742 <=*> 0
145,047	64-30 07 <*> 0
146,526	63-13010 <=*> 0
147,473	63-19622 <=*> 0
148,488	78J15S <ATS>
148,559	64-20460 <*>
148,719	NP-TR-806 <*>
150,459	61-28074 <*=> 0
151,019	63-13430 <=*> 0
152,025	65-10682 <*>
152,491	1302 <*>
153,085	64-30606 <*>
153,137	59-10248 <+*>
153,597	61-19494 <+*> 0
153,961	8396 <K-H>
154,158	59-10249 <+*>
154,491	63-24127 <=*>
154,758	60-15099 <+*> 0
155,276	64-30605 <*>
155,354	63-20333 <=*>
156,878	60-15877 <+*> 0
157,612	AEC-TR-3645 <+*>
157,654	AEC-TR-3787 <*>
158,621	63-15713 <=*> 0
158,930	59-12093 <+*>
159,478	61-15058 <+*>
161,343	61-13841 <*+> 0
161,659	60-14907 <+*>
162,694	60-14795 <+*>
162,759	60-14908 <+*>
163,484	59-22284 <+*> 0
163,584	60-14909 <+*>
166,235	62-15763 <*=> 0
166,418	60-10079 <+*>
166,561	60-16507 <+*>
168,354	10573A <K-H>
	63-16883 <*=>
168,939	60-18530 <+*>
169,993	1765 <TC>
176,586	892 <TC>
176,588	65-14409 <*>
186,610	1758 <TC>
186,781	64-30604 <*> P

SWEDEN APPLICATION	57-66 <*>

SWITZERLAND

Patent	Citation
6,458	57-3093 <*>
9,308	63-19623 <=*> 0
11,546	63-13011 <=*> 0
71,324	63-13431 <=*> 0
71,801	3964-G <K-H>
75,568	62-15764 <=*> 0
75,722	59-19757 <+*> 0
77,143	64-20583 <*>
80,584	59-12094 <+*> 0
85,550	62-18652 <=*>
86,263	<AC>
88,299	64-14978 <=*$> 0
89,285	62-13866 <=*> 0
93,102	64-13253 <=*$> 0
128,512	8639 <IICH>
128,684	62-25694 <=*$> 0
130,503	65-62691 <=$>
130,506	59-16190 <+*> C
134,370	64-18465 <*>
134,819	64-15130 <=*$> 0
139,170	60-14900 <+*>
139,758	62-18361 <=*> 0
144,670	25H9G <ATS>
149,346	59-12928 <+*> 0
149,980	I-244 <*>
150,020	1287 <TC>
154,531	63-23777 <=*$> 0
157,953	64-18464 <*>
159,988	62-32923 <=*>
161,679	64-15129 <=*$> 0
166,486	57R75G <ATS>
167,162	16R76G <ATS>
167,667	63-24604 <=*$> 0
169,179	64-18463 <*>
174,257	59-16853 <+*> 0
176,912	63-14789 <=*>
178,151	63-13012 <=*> 0
183,570	64-13254 <=*$> 0
184,282	64-13764 <=*$> C
190,640	64-18462 <*>
191,188	64-20584 <*>
191,561	64-30603 <*>
194,163	63-23778 <=*$> 0
194,451	64-18461 <*>
194,807	62-18568 <=*>
196,850	59-19758 <+*> 0
200,972	59-17029 <+*> 0
201,169	64-18460 <*>
201,834	64-18459 <*>
201,940 ADD	59-12930 <+*>
203,739	8638 <IICH>
206,380	65-64474 <=*$>
206,995	64-18458 <*>
208,957	1286 <TC>
210,719	60-14884 <+*>
211,222	59-12319 <+*> C
213,959	59-14217 <+*> 0
214,237	59-14704 <+*> 0
214,515	AL-743 <*>
	62-13294 <=*>
217,126	59-12930 <+*>
217,716	57-3110 <*>
218,086	64-30602 <*>
218,907	62-13499 <*=> 0
221,938	63-13432 <=*> 0
222,945	59-14218 <+*>
224,731	63-18228 <=*>
225,802	57S81G <ATS>
225,886	60-14879 <+*>
226,459	99N47G <ATS>
226,920	63-14788 <=*>
227,120	64-18457 <*>
227,212	59-19228 <+*> 0
228,931	65-11902 <*>
230,686	60-14878 <+*>
231,533	65-64764 <=*$>
232,973	61-13842 <=*> 0
233,011	I-469 <*>
	62-33250 <=*> C
233,313	59-14705 <+*> 0
233,433	59-14219 <+*> 0
233,459	65-13808 <*> 0
236,383	64-18456 <*>
236,562	62-14272 <=*>
239,522	61-16511 <*=>
239,765	8736 <IICH>
240,251	64-13783 <=*$> C
240,836	64-19802 <=$> 0
241,515	63-15454 <=*>
242,300	64-30601 <*>
245,926	63G7G <ATS>
246,302	61-27470 <*=> C
246,989	64-18455 <*>
247,514	63-19055 <=*> 0
247,699	63-14787 <=*>
247,708	64-20585 <*>

248,421	64-30600 <*> 0
249,369	62-10256 <*=>
250,118	14G8G <ATS>
251,111	59-17139 <+*> 0
251,635	I-464 <*>
251,637	I-634 <*>
251,824	65-60258 <=$>
252,012	59-18794 <+*> 0
252,302	64-18454 <*>
256,479	64-18453 <*>
257,968	8641 <IICH>
258,134	59-16191 <+*> 0
258,304	64-18452 <*>
259,369	59-16192 <+*> 0
259,878	13G8G <ATS>
260,075	63-14946 <=*>
261,981	64-20586 <*>
263,791	62-18413 <=*>
264,760	60-14754 <+*>
266,379	63-14541 <=*>
267,550	61-15061 <+*> 0
267,746	59-19229 <+*> 0
271,589	33J15G <ATS>
272,061	63-13013 <=*> 0
272,263	64-14373 <=*$>
273,020	59-18795 <+*> 0
273,309	59-18296 <+*> 0
274,225	60-16889 <+*>
	63-16329 <=*>
275,161	64-20588 <*>
275,527	66-60246 <=*$>
275,904	59-12320 <+*>
277,667	64-30599 <*>
279,553	65-10713 <*>
280,178	65-10545 <*> 0
280,261	59-12321 <+*>
280,485	66-14551 <*>
281,053	60-15100 <+*> 0
281,501	64-30598 <*>
282,704	64-18451 <*>
284,177	59-16854 <+*>
284,967	63-23582 <=*>
285,235	60-14877 <+*>
285,248	65-64579 <=*$>
285,458	64-18450 <*>
286,574	1867 <*>
286,658	62-26596 <=*> 0
287,262	59-16855 <+*> 0
287,279	64N57F <ATS>
288,158	64-30597 <*>
289,581	64-18449 <*>
289,582	64-18448 <*>
295,071	8735 <IICH>
295,078	64-20587 <*>
297,087	57-3546 <*>
297,792	59-12095 <+*>
299,720	59-16856 <+*>
300,321	59-16193 <+*>
300,354	64-15857 <=*$> 0
300,581	5877-A <K-H>
300,900	59-14220 <+*> 0
300,927	5877-B <K-H>
301,835	09K28G <ATS>
301,939	59-19230 <+*> 0
303,515	60-14876 <+*>
303,700	60-19002 <+*> 0
304,727	1084 <*>
305,732	1069 <*>
306,069	1452 <*>
306,584	61-28542 <=*> 0
306,614	59-16194 <+*> 0
307,005	59-16864 <+*>
308,281	63-10260 <=*>
308,295	75T90G <ATS>
311,215	753 <TC>
312,591	59-12385 <+*> 0
314,011	58-1189 <*>
314,340	63-19280 <=*> 0
314,391	58-1069 <*>
314,725	64-13255 <=*$> 0
317,984	59-12931 <+*> 0
318,524	48M38G <ATS>
320,352	10K28G <ATS>
321,359	65-62690 <=$> 0
322,133	59-12932 <+*> 0
322,980	60-18413 <+*>
328,951	61-15692 <+*> C
330,282	65-14110 <*>
331,120	825 <TC>
332,919	464 <TC>
333,868	61-10612 <+*> 0
337,846	64-15329 <=*$>
339,463	62-13500 <*=> 0
341,743	64-13573 <=*$> 0
347,641	2238 <TTIS>
354,307	63-10883 <=*>
354,902	64-10523 <=*$>
372,955	65-63487 <=$>
375,191	NS-208 <TTIS>
376,845	65-62689 <=$> 0
379,771	NS-280 <TTIS>
382,587	65-14763 <*>
382,894	NS-445 <TTIS>
397,237	66-13866 <*>
403,708	66-13867 <*>

USSR (AUTHOR CERTIFICATE)

24,874	62L33R <ATS>
30,268	41L35R <ATS>
32,489	01J16R <ATS>
36,392	42L35R <ATS>
39,979	65-17220 <*>
40,348	22J18R <ATS>
40,972	66-12522 <*>
44,934	<ES>
49,921	1600 <TC>
54,391	21J15R <ATS>
68,469	62-10927 <*=>
69,481	PANSDOC-TR.438 <PANS>
91,076	M.5682 <NLL>
96,261	61-28983 <=*>
100,527	41R77R <ATS>
100,875	4455 <HB>
102,033	14L29R <ATS>
102,692	65-62848 <=$>
103,570	64-20230 <*> 0
104,551	5049 <HB>
104,713	8K25R <ATS>
105,243	19L32R <ATS>
105,415	9K25R <ATS>
105,456	12L29R <ATS>
105,505	13L29R <ATS>
105,697	9CL31R <ATS>
105,932	83P61R <ATS>
106,258	4837 <HB>
106,689	60-23021 <+*>
106,837	4K24R <ATS>
106,914	82P61R <ATS>
107,200	63-16411 <=*>
107,658	64-30590 <*>
	72Q69R <ATS>
108,514	35M38R <ATS>
108,817	36M38R <ATS>
109,017	61K24R <ATS>
109,161	34M38R <ATS>
109,422	62-18973 <=*>
110,143	65-10343 <*>
110,387	106 <LSB>
110,579	65-10345 <*>
110,587	5456 <HB>
110,592	59-12881 <+*>
110,736	79M47R <ATS>
110,887	63-19468 <=*>
111,490	59-15332 <+*>
112,274	62-14767 <=*>
112,498	65-17354 <*>
112,633	27P62R <ATS>
	62-18971 <=*>
112,923	63-14750 <=*>

Patent	Citation
113,015	65-17352 <*>
113,474	63-13916 <=*>
113,527	63-16188 <=*>
113,549	63-13920 <*=>
113,592	64-30591 <*>
114,172	63-13924 <=*>
114,646	63-13918 <=*>
115,652	<INT>
115,796	61-31641 <=>
116,082	62-18210 <=*>
116,114	63-13014 <=*>
116,152	64-14040 <=*$>
116,384	64-15858 <=*$> O
116,643	65-64580 <=*$>
116,653	64-19184 <=*$>
116,715	61-19448 <=*>
116,809	62-10204 <*=>
117,148	S-2117 <RIS>
117,167	61-27577 <=*>
117,379	61-28185 <*=>
117,460	65-11226 <*>
117,465	63-24124 <=*>
117,506	61-18906 <*=>
117,521	64-15328 <=*$> O
117,553	630 <TC>
117,657	61-13482 <+*>
118,042	59-17088 <+*>
	61-19496 <+*>
118,405	64-15111 <=*$>
118,470	64-11793 <=>
118,563	62-11336 <=>
118,653	629 <TC>
118,806	118,806 USSR <INT>
118,818	64-16176 <*>
119,225	02R78R <ATS>
119,364	AD-613 173 <=$>
119,403	63-23735 <=*$>
119,876	63-13757 <=*>
119,900	61-11130 <=>
120,265	60N54R <ATS>
120,269	60-21939 <=>
120,273	60-21949 <=>
120,342	60-21950 <=>
120,343	60-21946 <=>
120,344	60-21943 <=> O
120,675	S-2118 <RIS>
120,991	65-60272 <=$>
121,268	1580 <TC>
122,021	64-11603 <=>
122,166	62-13566 <*=>
122,481	23Q72R <ATS>
122,742	65-11882 <*>
122,747	63-13758 <=*>
122,780	G-286 <RIS>
123,096	61-27560 <*=>
123,209	63-15292 <=*>
123,311	72M37R <ATS>
123,513	62-13868 <*=>
123,523	63-23779 <=*$>
123,529	1366 <TC>
123,648	34P61R <ATS>
123,752	63-19072 <=*>
123,957	P-75 <RIS>
124,402	62-13867 <*=>
124,435	A-110 <RIS>
124,510	62-24638 <=*>
124,572	64-30592 <*>
124,628	65-10575 <*>
124,719	AD-622 348 <=*$>
124,805	63-13699 <=*>
125,134	64-11604 <=>
125,207	61-27770 <*=>
125,239	A-122 <RIS>
125,937	63-10834 <=*>
126,259	64-30593 <*>
126,263	C-124 <RIS>
126,265	C-125 <RIS>
126,507	63-18613 <=*>
126,612	63-16318 <*=>

Patent	Citation
	94S81R <ATS>
126,861	<INT>
126,872	M-82 <RIS>
127,047	61-11129 <=>
127,054	65-60843 <=$> C
127,055	65-60842 <=$> O
127,392	87M41R <ATS>
127,474	63-26961 <=*>
	64-23572 <=$>
127,654	63-11665 <=>
127,940	65-11523 <*>
127,962	62-33232 <=*>
128,030	62-23677 <=*>
128,193	63-20193 <=*>
	63-23780 <=*$>
128,460	40P63R <ATS>
128,462	68P64R <ATS>
128,853	65-10541 <*>
128,907	62-13564 <=>
128,950	61-23839 <=*>
	62-24473 <=*>
129,192	65-10634 <*>
129,486	64-71563 <=>
129,613	62-32289 <=*>
129,647	63-19679 <=*>
129,652	64-30594 <*>
129,677	63-15332 <=*>
129,690	62-23370 <=*>
129,739	62-15169 <*=>
129,765	62-32924 <=*> O
129,837	62-15159 <*=>
129,943	63-23583 <=*>
129,945	AD-611 420 <=$>
129,950	63-23642 <=*>
130,071	63-19179 <=*>
130,073	62-15168 <*=>
130,242	62-15185 <*=>
130,323	63-19079 <*=>
130,512	63-16412 <=*>
130,547	62-15201 <*=>
130,688	62-19250 <=*>
130,691	62-13558 <*=>
130,929	62-15167 <*=>
130,931	62-24036 <=*>
131,089	64-19803 <=$> C
131,101	62-24926 <=*>
131,125	AD-620 074 <=$>
131,141	63-15548 <=*>
131,162	62-33055 <=*>
131,350	E-805 <RIS>
	65-13568 <*>
131,433	AD-619 011 <=$>
131,777	63-15552 <=*>
131,793	62-19249 <=*>
131,800	62-15166 <*=>
131,943	63-19183 <=*>
131,977	62-15165 <*=>
132,586	99N54R <ATS>
132,616	62-10293 <*=>
132,617	62-10259 <*=>
132,663	62-32287 <=*>
132,680	62-32288 <=*>
132,716	63-13202 <=*>
132,724	AD-612 782 <=$>
132,813	65-10621 <*>
132,951	64-11652 <=>
132,954	62-25748 <=*>
132,957	62-25836 <=*>
133,133	61-23840 <*=>
133,503	62-15279 <*=>
133,538	62-15187 <*=>
133,540	62-15280 <*=>
133,611	62-15188 <*=>
133,616	62-32312 <=*>
133,684	63-15339 <=*>
134,037	62-24934 <=*>
134,039	62-23661 <=*>
134,086	62-15200 <*=>
134,453	63-13199 <=*>

Patent	Citation
134,495	62-15281 <*=>
134,499	62-25726 <=*>
134,557	63-10334 <=*>
134,641	62-32513 <=*>
134,804	AD-614 961 <=$>
134,872	62-14986 <=*>
134,902	62-19252 <=*>
134,920	63-15329 <=*>
134,930	64-15401 <=*$>
134,986	AD-621 011 <=*$>
135,078	90P59R <ATS>
135,157	4781 <HB>
135,258	62-24474 <=*>
135,271	63-23108 <=*>
135,309	62-24475 <=*>
135,344	62-32282 <=*>
135,345	63-23110 <=*>
135,471	1832 <TC>
135,476	65-13553 <*>
135,500	62-32925 <=*>
135,516	62-25540 <=*>
135,519	62-19253 <=*>
135,696	62-24346 <=*>
135,741	62-25733 <=*>
136,048	61-28088 <*=>
136,070	62-33644 <=*>
136,099	63-15542 <=*>
136,313	89N56R <ATS>
136,327	66-11243 <*>
136,362	A-174 <RIS>
136,481	5106 <HB>
136,556	AD-618 009 <=$>
136,707	64-15327 <=*$> C
136,798	62-32515 <=*>
136,823	64-15326 <=*$>
136,917	62-24345 <=*>
137,036	63-15436 <=*>
137,262	65-10680 <*>
137,277	62-13203 <=*>
137,386	62-23680 <=*>
137,507	NS-205 <TTIS>
137,511	38Q73R <ATS>
	66-60245 <=*$>
137,513	59P59R <ATS>
137,545	62-25407 <=*>
137,767	62-24280 <=*>
137,895	65-10614 <*>
138,029	65-13850 <*>
138,259	63-15331 <=*>
138,278	62-32300 <=*>
138,279	62-25548 <=*>
138,283	AD-620 123 <=*$>
138,373	62-11713 <=>
138,401	AD-620 128 <=$>
138,583	138,583 USSR <INT>
138,930	12P61R <ATS>
138,931	NS-206 <TTIS>
138,974	62-32309 <=*>
139,001	62-33056 <=*>
139,048	63-10831 <=*>
139,087	62-32679 <=*>
139,098	62-32978 <=*>
139,150	62-33548 <=*>
139,309	74P61R <ATS>
139,375	62-25550 <=*>
139,450	62-32399 <=*>
139,470	62-25742 <=*>
139,478	62-25549 <=*>
139,942	62-32310 <=*>
140,059	65-13882 <*>
140,060	65-13880 <*>
140,067	62-25746 <=*>
140,104	62-32537 <=*>
140,122	62-33544 <=*>
140,123	62-25750 <=*>
140,226	63-15540 <=*>
140,262	62-32521 <=*>
140,267	63-13887 <=*>
140,501	62-24269 <=*>
140,579	SCL-T-446 <=*>
140,918	63-11591 <=>
140,920	63-15328 <=*>
140,962	66-10666 <*>
141,018	62-33070 <=*>
141,127	65-11524 <*>
141,219	63-13907 <=*>
141,298	65-11899 <*>
141,301	18P66R <ATS>
141,339	62-32536 <=*>
141,356	62-33057 <=*>
141,381	62-33393 <=*>
141,407	4018 <TTIS>
141,472	63-13917 <=*>
141,502	63-15334 <=*>
141,560	62-25729 <=*>
141,583	63-10843 <=*>
141,592	63-23730 <=*$>
141,652	62-32942 <=*>
141,654	63-15337 <=*>
141,655	63-15336 <=*>
141,658	63-15335 <=*>
141,679	62-33058 <=*>
141,682	62-32311 <=*>
141,758	63-24245 <=*$>
141,861	64-19185 <=*$> O
141,961	63-13695 <=*>
141,962	63-13701 <=*>
142,157	63-24135 <=*$>
142,502	63-19488 <=*>
142,702	62-33546 <=*>
142,706	63-23102 <=*$>
142,818	63-19344 <=*>
142,847	63-19155 <=*>
143,099	63-13193 <=*>
143,172	63-19472 <=*>
143,195	63-18365 <=*>
143,399	63-10194 <=*>
143,554	62-18867 <=*>
143,658	63-19485 <=*>
143,800	65-13879 <*>
143,808	AD-615 289 <=$>
143,982	63-23615 <=*>
143,983	63-23449 <=*>
144,140	64-11605 <=>
144,240	62-33628 <=*>
144,522	63-23623 <=*>
144,546	63-19480 <=*>
144,553	63-23447 <=*$>
144,559	63-15550 <=*>
144,743	AD-619 550 <=$>
144,880	AD-619 549 <=$*>
145,011	63-23434 <=*>
145,019	63-23448 <=*>
145,020	63-23433 <=*>
145,021	63-23422 <=*>
145,064	63-23105 <=*$>
145,067	63-19409 <=*>
145,069	63-19411 <=*>
145,070	63-19410 <=*>
145,073	63-23435 <=*>
145,238	74P66R <ATS>
145,469	65-62693 <=*$>
145,539	145,539 USSR <INT>
145,549	18Q73R <ATS>
145,569	60P64R <ATS>
145,748	63-19252 <=*>
145,797	85Q67R <ATS>
145,919	63-15779 <=*>
145,921	AD-620 092 <=$>
146,057	63-19184 <=*>
146,058	63-19185 <=*>
146,136	63-11644 <=>
146,179	30Q72R <ATS>
146,387	AD-620 095 <=$>
146,432	63-18342 <*=>
146,503	AD-619 454 <=$>
146,733	65-17433 <*>
146,743	65-64473 <=*$>
146,803	AD-620 761 <=$>
146,891	63-19204 <=*>

147,024 63-15780 <=*>
147,044 63-23669 <=*$>
147,162 63-19426 <=*>
65-13148 <*>
147,183 AD-615 290 <=$>
147,236 63-11641 <=>
147,240 63-11643 <=>
147,337 AD-627,016 <=*$>
147,369 64-13159 <=*$>
147,576 63-19418 <=*>
147,608 63-23815 <=*>
147,672 63-23656 <=*$>
147,709 4026 <TTIS>
147,730 65-11919 <*>
147,832 63-15909 <=*>
148,117 AD-621 822 <=$>
148,181 64-71548 <=>
148,190 63-19419 <=*>
148,232 63-18851 <=*>
148,261 AD-617 680 <=$>
148,318 AD-611 421 <=$>
148,398 65-12765 <*>
148,448 63-19156 <=*>
148,517 63-11623 <=>
148,551 64-71547 <=>
148,663 34P65R <ATS>
148,724 64-11854 <=>
148,875 64-71331 <=> M
148,970 64-16486 <=*$> O
148,971 64-16487 <=*$> O
149,356 65-11918 <*>
149,426 64-10854 <=*$>
149,520 64-10855 <=*$>
149,560 64-71148 <=>
149,764 71Q69R <ATS>
149,934 63-11669 <=>
150,085 63-23618 <=*>
150,131 63-11771 <=>
150,262 63-11707 <=>
150,329 65-13846 <*>
150,353 64-16603 <=*$>
150,625 22S84R <ATS>
150,723 AD-610 279 <=$>
150,727 AD-610 652 <=$>
151,025 55T9CR <ATS>
151,346 39S82R <ATS>
151,699 63-24022 <=*>
151,739 AD-615 248 <=$>
151,744 65-13881 <*>
151,755 63-23822 <=*$>
151,885 64-71168 <=>
152,107 63-24143 <=*$>
152,108 63-24141 <=*$>
152,115 AD-618 646 <=$>
152,117 64-71192 <=>
152,133 63-23783 <=*>
152,149 63-23101 <=*$>
152,151 63-24253 <=*$>
152,526 65-18147 <*>
152,725 99S80R <ATS>
152,738 65-12736 <*>
152,741 64-71473 <=>
153,041 64-15325 <=*$> O
153,138 AD-611 051 <=$>
153,139 AD-615 992 <=$>
153,145 AD-617 668 <=$>
153,153 AD-611 050 <=$>
153,286 AD-614 771 <=$>
153,323 64-71185 <=>
153,324 64-71174 <=>
153,492 AD-613 172 <=$>
153,585 64-11792 <=>
153,586 64-11769 <=>
153,588 64-11771 <=>
153,760 AD-614 772 <=$>
153,762 AD-620 080 <=$>
153,826 AD-614 942 <=$>
153,943 AD-611 044 <=$>
153,985 AD-628 299 <=*$>
154,001 65-11920 <*>
154,007 64-71322 <=>
154,042 64-11788 <=>
154,066 AD-620 996 <=$>
154,070 64-11738 <=>
154,096 AD-614 763 <=$>
154,109 64-71287 <=>
154,275 AD-626 052 <=*$>
154,301 64-71468 <=>
154,420 AD-613 162 <=$>
154,447 AD-615 291 <=$>
154,567 64-71152 <=>
154,587 AD-610 375 <=$>
154,623 AD-627 992 <=*$>
154,670 AD-615 028 <=$>
154,674 AD-615 981 <=$>
154,690 AD-614 949 <=$>
154,699 AD-615 292 <=$>
154,743 AD-609 137 <=>
154,895 64-71477 <=>
155,047 64-15876 <=*$>
155,102 64-71151 <=>
155,161 AD-620 078 <=$>
155,181 64-11978 <=>
155,188 AD-611 550 <=$>
155,190 AD-615 293 <=$>
155,291 AD-614 964 <=$>
155,317 AD-614 397 <=$>
155,318 AD-625 748 <=*$>
155,359 AD-625 246 <=*$>
155,541 AD-620 155 <=$>
155,554 AD-620 068 <=$>
155,641 AD-612 414 <=$>
155,670 AD-610 340 <=$>
155,813 AD-620 079 <=$>
155,881 AD-614 965 <=$>
155,949 AD-609-789 <=>
155,953 AD-620 069 <=$>
156,038 64-71177 <=>
156,146 64-71193 <=>
156,185 AD-620 122 <=*$>
156,193 64-11979 <=>
156,367 AD-615 529 <=$>
156,550 64-18479 <*>
156,553 64-71552 <=>
156,621 64-71312 <=>
156,693 AD-609 134 <=>
156,702 64-71318 <=>
156,713 AD-617 907 <=$>
156,735 AD-614 966 <=$>
156,752 AD-620 127 <=$>
156,822 AD-614 022 <=$>
156,975 AD-620 762 <=$*>
156,977 AD-613 457 <=$>
156,978 AD-611 524 <=$>
157,149 AD-619,483 <=$>
157,176 AD-622 384 <=*$>
157,420 497 <TC>
65-11992 <*>
157,519 AD-610 370 <=$>
157,557 AD-620 995 <=$>
157,566 AD-620 093 <=$>
157,694 40S82R <ATS>
157,749 AD-611 858
NS-188 <TTIS>
157,785 64-71316 <=>
157,798 64-71285 <=>
157,845 AD-610 414 <=$>
157,887 64-11844 <=>
157,958 AD-615 249 <=$>
158,063 AD-620 792 <=*$>
158,085 AD-611 049 <=$>
158,153 AD-622 346 <=$*>
158,427 64-71147 <=>
158,600 AD-625 197 <=*$>
158,699 AD-610 653 <=$>
158,702 AD-615 294 <=$>
158,718 AD-610 369 <=$>
158,738 64-71146 <=>
158,965 AD-620 070 <=$>
159,034 AD-625 248 <=*$>

159,068 AD-622 455 <=*$>
159,193 AD-620 132 <=*$>
159,197 AD-620 091 <=$>
159,206 AD-618 092 <=$>
159,407 AD-610 296 <=$>
159,492 620-768 <=$>
159,577 AD-615 295 <=$>
159,591 AD-625 247 <=*$>
159,650 AD-620 071 <=$>
159,667 AD-611 547 <=$>
159,668 AD-611 074 <=$>
159,808 AD-615 296 <=$>
159,831 66-11309 <*>
159,847 418 <TC>
159,920 87R76R <ATS>
159,922 AD-615 297 <=$>
159,938 AD-611 048 <=$>
159,977 AD-620 124 <=*$>
160,038 AD-610 339 <=$>
160,225 AD-615 251 <=$>
160,251 AD-613 175 <=$>
160,339 AD-628 300 <=*$>
160,727 AD-625 196 <=*$>
160,778 AD-617 422 <=$>
160,780 AD-615 866 <=$>
160,785 AD-620 131 <=$>
160,838 AD-611 530 <=$>
160,844 AD-615 987 <=$>
160,916 AD-613 174 <=$>
160,944 AD-620 097 <=$>
160,956 AD-611 532 <=$>
160,975 AD-613 454 <=$>
161,067 AD-619 428 <=$>
161,074 AD-617 684 <=$>
161,087 AD-615 252 <=$>
161,506 AD-617 667 <*>
161,520 AD-614 958 <=$>
161,527 AD-615 253 <=$>
161,559 AD-620 789 <=$>
161,574 AD-620 787 <=$>
161,576 AD-621 798 <=*$>
161,577 AD-625 240 <=*$>
161,601 AD-622 448 <=$*>
161,701 AD-613 455 <=$>
161,702 AD-615 284 <=$>
161,832 AD-620 125 <=$>
161,840 AD-620 066 <=$>
161,844 66-14512 <*>
161,847 AD-614 959 <=$>
161,856 AD-614 960 <=$>
161,857 AD-613 159 <=$>
161,920 AD-620 126 <=$>
161,932 AD-620 090 <=$>
162,246 AD-620 786 <=$>
162,268 AD-622 349 <*$=>
162,269 AD-621 778 <=*$>
162,310 AD-615 254 <=$>
162,554 AD-620 072 <=$>
162,574 AD-625 063 <=*$>
162,671 AD-620 776 <=*$>
162,739 AD-621,040 <=$>
162,748 AD-616 307 <=$>
162,896 AD-620 997 <=*$>
163,047 AD-625 184 <=*$>
163,079 AD-621 777 <=*$>
163,224 AD-619 316 <=$>
163,228 AD-621 004 <=*$>
163,299 AD-626 046 <=*$>
163,326 AD-620 857 <=*$>
163,351 AD-620 067 <=$>
163,444 AD-620 971 <=*>
163,678 AD-620 783 <=$>
163,874 AD-620 774 <=$>
164,036 AD-620 094 <=$>
164,038 AD-624 849 <=*$>
164,289 65-14886 <*>
164,325 AD-625 796 <=*$>
164,510 AD-621 782 <=*$>
164,512 AD-620 073 <=$>
AD-620 791 <=$>
164,968 AD-625 792 <=*$>
165,036 AD-621 776 <=*$>
165,038 AD-621 781 <=*$>
165,042 AD-619 468 <=$>
165,317 AD-620 818 <=$>
165,319 AD-620 778 <=*$>
165,321 AD-617 654 <=$>
165,328 AD-625 064 <=*$>
165,550 92S88R <ATS>
165,562 AD-620 860 <=*$>
165,586 AD-620 861 <=*$>
165,620 AD-619 469 <=*$>
165,624 AD-623 210 <=*$>
165,629 AD-625 285 <=*$>
165,718 AD-619 426 <=$>
165,721 AD-622 447 <=*$>
165,780 AD-625 072 <=*$>
165,781 AD-624 859 <=*$>
165,782 AD-620 760 <=*$>
165,786 AD-625 780 <=*$>
165,787 AD-622 376 <=*$>
165,799 AD-625 287 <=*$>
165,861 AD-625 000 <=*$>
166,025 AD-620 815 <=*$>
166,081 AD-625 160 <=*$>
166,132 AD-620 764 <=$>
166,153 AD-618 904 <=$>
166,183 AD-624 906 <=*$>
166,202 AD-625 001 <=*$>
166,244 AD-622 375 <=*$>
166,247 AD-625 141 <=*$>
166,283 65-13719 <*>
166,336 94S85R <ATS>
166,384 AD-619 317 <=$>
166,390 AD-626 047 <=*$>
166,436 AD-625 053 <=*$>
166,532 AD-625 185 <=*$>
166,555 AD-618 647 <=$>
166,556 AD-625 144 <=*$>
166,660 AD-622 450 <=*$>
166,683 AD-622 449 <=*$>
166,745 AD-626 048 <=*$>
166,808 AD-620 819 <=$>
166,821 AD-621 779 <=*$>
166,824 AD-619 453 <=$>
166,829 AD-622 344 <=$*>
166,832 AD-625 798 <=*$>
166,839 AD-620 820 <=*$>
166,974 AD-625 789 <=*$>
166,977 AD-620 790 <=$>
166,981 AD-625 068 <=*$>
166,985 AD-620 809 <=*$>
167,033 AD-625 033 <=*$>
167,045 83S84R <ATS>
167,092 AD-619 427 <=$>
167,097 AD-622 373 <=*$>
167,103 AD-623 190 <=*$>
167,135 AD-620 855 <=*$>
167,223 AD-625 797 <=*$>
167,260 AD-625 070 <=*$>
167,263 AD-625 186 <=*$>
167,323 AD-623 188 <=*$>
167,367 AD-625 288 <=*$>
167,582 AD-620 856 <=$>
167,589 AD-625 069 <=*$>
167,625 AD-625 228 <=*$>
168,334 AD-625 071 <=*$>
168,347 AD-625 009 <=*$>
168,362 AD-620 824 <=$>
168,754 AD-625 795 <=*$>
170,961 49S88R <ATS>
171,262 20T90R <ATS>
176,270 66-12331 <*>

YUGOSLAVIA
11,067 64L33CR <ATS>